National Association of Broadcasters

ENGINEERING HANDBOOK

The *NAB Engineering Handbook* is the definitive resource for broadcast engineers. From audio and video contribution, to broadcast facilities and antennas, this valuable resource provides in-depth information about each aspect of the broadcast chain. New topics include ultra-high definition television, internet radio interfacing and streaming, ATSC 3.0, digital audio compression techniques, video format and standards conversion. You will also find relevant updates to topics such as AM, shortwave, FM and television transmitting systems, studio lighting, microphones, cameras and principles of acoustics.

The big-picture, comprehensive nature of the *NAB Engineering Handbook* offers useful information to all broadcast engineers—from broadcast chief engineers, who need expanded knowledge of specialized areas they encounter in the field, to technologists in specific areas, such as IT and RF. Also covered are a wide range of topics, including broadcast documentation, FCC practices, technical standards, security, safety, disaster planning, facility planning, project management and engineering management.

National Association of Broadcasters

ENGINEERING HANDBOOK

11th Edition

Editor-in-Chief

GARRISON C. CAVELL

Associate Editors

THOMAS G. OSENKOWSKY
DAVID H. LAYER
SKIP PIZZI
WILLIAM T. HAYES

NEW YORK AND LONDON

Eleventh edition published 2018
by Routledge
711 Third Avenue, New York, NY 10017

and by Routledge
2 Park Square, Milton Park, Abingdon, Oxon, OX14 4RN

Routledge is an imprint of the Taylor & Francis Group, an informa business

First edition published by Focal Press 1935
Tenth edition published by Focal Press 2014

Library of Congress Cataloging-in-Publication Data
Names: Cavell, Garrison C. (Garrison Clayton) editor. | National Association of Broadcasters.
Title: National Association of Broadcasters engineering handbook / editor-in-chief, Garrison C. Cavell ; associate editors, Tom Osenkowsky, David Layer, Skip Pizzi, Bill Hayes.
Other titles: Engineering handbook of the National Association of Broadcasters.
Description: 11th edition. | New York, NY : Routledge, 2017. | Earlier editions cataloged under titles: Engineering handbook of the National Association of Broadcasters; NAB engineering handbook; National Association of Broadcasters engineering handbook, and Engineering handbook. | Includes bibliographical references and index.
Identifiers: LCCN 2016051257 | ISBN 9781138930513 (hbk) | ISBN 9781315680149 (ebk)
Subjects: LCSH: Radio—Transmitters and transmission—Handbooks, manuals, etc. | Radio stations—Design and construction—Handbooks, manuals, etc. | Television—Transmitters and transmission—Handbooks, manuals, etc. | Television stations—Design and construction—Handbooks, manuals, etc.
Classification: LCC TK6561 .N3 2017 | DDC 621.384—dc23
LC record available at https://lccn.loc.gov/2016051257

ISBN: 978-1-138-93051-3 (hbk)
ISBN: 978-1-315-68014-9 (ebk)

Typeset in Palatino
by Apex CoVantage, LLC

Visit the companion website: www.routledge.com/cw/cavell

Contents

SECTION 4: AUDIO PRODUCTION AND STUDIO TECHNOLOGY

SECTION 5: VIDEO PRODUCTION AND TV STUDIO SYSTEMS

SECTION 6: FACILITY INTERCONNECTION

SECTION 7: RADIO TRANSMISSION

SECTION 8: TELEVISION TRANSMISSION

SECTION 9: BROADCAST TOWERS AND TRANSMISSION LINES

SECTION 10: SIGNAL MEASUREMENT AND ANALYSIS

SECTION 11: ANCILLARY BROADCAST SYSTEMS

Foreword

"Good results without good planning is good luck, not good management." That was the sage advice of Ben Waters, Vice President of Administration at Capitol Broadcasting Company in the mid-nineties. Ben was one of my mentors in the broadcast business and decades later his words remain true: a good plan must be informed by sound information.

The curation of such sound information is what makes this 11th Edition of the *NAB Engineering Handbook* so important. This handbook is the definitive reference on the subject of broadcast engineering, including the increased role of information technology (IT) in broadcast operations. In it, you will find detailed information on everything from the fundamentals of the electromagnetic spectrum and weather radar systems to the operations of electronic newsgathering systems and tower design and maintenance.

It has been 10 years since the last version of the handbook was written. In that time, we have seen tremendous and accelerating change. This trend is sure to continue. Increased connectivity permeates our industry and demands a new understanding of Internet Protocol (IP) technologies. Broadcasters have become multi-platform digital distributors. We combine broadcasting and IP connectivity to serve our communities in new and powerful ways. Looking ahead, we are on the precipice of a next-generation television standard that has IP at its core. As engineers, we are tasked with building, deploying, and supporting these efforts, and the handbook is a vital reference.

Authoring and editing such a definitive resource is a massive multi-year undertaking, and doing so in a rapidly evolving environment makes it all the more challenging. I offer my hearty thanks and congratulations to all of the experts and engineers that contributed their time and talent to produce this book.

We are in a time when people want instant results and the pressure to deliver them is great. Working to deliver positive results under such pressures reminds me of the words of one of my favorite football coaches, Steve Logan. When Coach Logan was at East Carolina University he said, "You can't coach the scoreboard." For Coach Logan that meant he had to focus on his team—the blocking, the tackling, the offensive and defensive schemes and developing the plan and people that would yield success. For the broadcast engineer, the *NAB Engineering Handbook* is a valuable tool in producing the desired results and the "score" you want and need.

We are proud of this 11th Edition of the *NAB Engineering Handbook*, as we know it provides broadcast engineers the information needed to excel in their jobs. Please contact us with comments and feedback on the handbook as an applicable and authoritative collection of information on broadcast engineering topics.

Sam Matheny
Executive Vice President and Chief Technology Officer
National Association of Broadcasters

Notes and Acknowledgments

In this completely revised and substantially updated 11th Edition of the *NAB Engineering Handbook*, we have urged the authors to make their manuscripts both readable and "forward looking" when writing or revising a particular chapter. The intent was to make this an approachable reference, not only for seasoned engineers, but also for those who are just starting in the industry who want to learn about any of these topics. Authors were also encouraged to provide links and references to allow the reader to pursue further investigation. While it is true that the Internet can offer many wonderful resources for independent learning, this handbook offers authoritative, vetted material written by some of the leading experts in the industry.

In this handbook of over 90 chapters, many new topics are offered. I am particularly pleased that several topics that haven't been addressed in several decades but that are still relevant have been once more addressed in this book.

Of necessity, all the images in the printed and e-book version of this handbook are rendered in black and white; however, many are reproduced in color in a companion website that has been made available by the publisher.

The editors of the NAB Engineering Handbook thank the authors, co-authors, and contributors for the time and expertise that they have given to make this book an important part of a broadcast engineer's library of reference materials.

Assisting the Editor-in-Chief, and working one-on-one with more than 100 authors, were four Associate Editors who commissioned, assembled, edited, reviewed, formatted, and verified the material:

Thomas G. Osenkowsky
Radio Engineering Consultant
Brookfield, Connecticut

David H. Layer
Senior Director, Advanced Engineering
National Association of Broadcasters
Washington, DC

Skip Pizzi
Vice President, Technology Education and Outreach
National Association of Broadcasters
Washington, DC

William T. Hayes
Director of Engineering and Technology
Iowa Public Television
Johnston, Iowa

While this undertaking could not have been accomplished without the help of these colleagues, I want to give particular thanks to Tom Osenkowsky, who consistently went above and beyond what was expected throughout this project. His considerable expertise, patient perspective with a touch of New York humor, and willingness to take on some of the more difficult challenges was the "special ingredient" that made this book possible. For that reason, I would like to dedicate this book to him, with deep appreciation and thanks.

I would also like to give special thanks to David H. Layer, who undertook some extra heavy lifting to get this book over the finish line, and who crafted the organizational structure with which we put this book together. His endless enthusiasm helped us stay focused so that we could meet our goals.

Of course, such an undertaking could not have been accomplished without the considerable support of many others, including NAB Staffers Lynn Claudy, Senior Vice President, Technology, who volunteered his advice, perspective, and considerable editorial skills, and Sharon Devine, Senior Administrative Assistant, who helped us keep order, stay on task, and process the reams of necessary paperwork. I also am grateful for the expert help and guidance given to us by the professional staff at our publisher, and in particular, Ross Wagenhofer and Nicole Salazar. This was their first foray into working with broadcasters and I am happy to report that they handled it quite well and with grace and humor.

I must especially thank my wife and colleague, Cindy Hutter Cavell, whose moral support, perspective, assistance with editing, and idea generation proved invaluable. Last, but certainly not least, I would like to thank Renee Bijou Gonzalez, Project Coordinator, at Cavell Mertz & Associates, Inc., who provided hours of assistance to me and all of the Associate Editors, helping to maintain organization, backstopping all of us with proof reading and grammar, formatting errant documents, and ensuring that all of the graphics were of the highest quality.

We hope you find this newest edition of the *NAB Engineering Handbook* useful in your work in broadcast engineering and operations, and that it remains a valuable reference throughout your career.

Garrison C. Cavell
Editor-in-Chief

The Editors

Garrison C. Cavell
Editor-in-Chief

Over his 40-year career, Garrison C. Cavell has been on the air, worked in radio and television station engineering management, was a radio station General Manager, a radio station owner, and a serial entrepreneur. His latest and most enduring venture was to found the firm that is now Cavell Mertz & Associates, Inc., where he remains a principal. He has a love for working with RF systems, and AM antenna arrays in particular. He has designed, built, and certified countless station antenna systems over the years.

Gary has served as a lecturer and faculty member for several different NAB Educational Foundation courses. He has presented at numerous NAB Spring conventions and NAB Radio conventions. For several years, he was a lecturer for the Voice of America's/International Broadcasting Bureau's US Technical Training Institute (USTTI). Gary was also an invited panelist at the 2015 North American Broadcasters Association "Future of Radio Audio Symposium" at CBC's Glenn Gould Studio in Toronto.

Gary is a member of several professional organizations including the IEEE, the Advanced Television Systems Committee (ATSC), the Society of Broadcast Engineers (SBE), and is a Life Member of the Society of Motion Picture and Television Engineers (SMPTE). He is a Past President of the IEEE Broadcast Technology Society. His other professional organization memberships or associate memberships include the Antenna Measurement Techniques Association, the Royal Television Society, and the Radio Club of America. He holds an amateur radio Extra Class operator's license (call sign K4MRL) and an FCC General Radiotelephone (formerly first class) Operator's License.

Thomas G. Osenkowsky
Associate Editor

Thomas G. Osenkowsky began his career broadcasting as a news reporter while attending high school. He attended Central Connecticut State College and worked as a Final Quality Assurance Technician for Automation Industries in Danbury, CT. Tom is a Senior Member of the IEEE, iNARTE, and SBE. He holds a Certified Professional Broadcast Engineer life certification from the SBE and a Master Engineer with RF Radiating Endorsement from iNARTE.

Tom has held positions at various radio stations as announcer, Chief Engineer, Operations Manager, and General Manager. He has served as an engineering consultant in the United States and Caribbean Islands, written engineering-themed computer programs, and made presentations at the IEEE Broadcast Technology Society (BTS) annual meeting, Broadcast Engineering Conference at the NAB Annual Convention, and SBE Annual Convention. Tom is a contributor to *Radio World* and has served as Associate Editor for the Tenth Edition of the *NAB Engineering Handbook.*

David H. Layer
Associate Editor

David H. Layer is Senior Director, Advanced Engineering in NAB's Technology department. David has been with NAB since 1995 and has been very active in the radio technology and standards setting area. He is the primary NAB staff person on NAB's radio technology committee and the National Radio Systems Committee (NRSC), a technical standards setting body co-sponsored by NAB and the Consumer Technology Association (CTA).

David's duties within NAB are varied and broad. His current principal responsibilities include serving as a project manager for technology projects being conducted by PILOT (formerly NAB Labs) and as principal administrator of the NAB Labs radio technology committee, a group of technical executives from NAB member companies that advises NAB on technology development and technical regulatory matters. Additionally, David provides technical expertise to NAB's Government Relations and Legal departments on regulatory and legislative matters that pertain to the broadcasting industry.

David is actively involved in NAB's technical conference planning and technical publication activities, and has been an author and contributing author for numerous technical publications, including *IEEE Spectrum* magazine (a leading journal of the electrical engineering profession), the *McGraw-Hill Yearbook of Science and Technology*, and the 9th and 10th editions of the *NAB Engineering Handbook*, as well as the 11th Edition which is now under development. David is an occasional author of NAB's Broadcast Blog covering timely radio and TV broadcasting-related technical topics.

Outside of NAB, David is a Senior Member of the IEEE and is active in the IEEE Broadcast Technology Society (BTS), where he serves on the administrative committee, is a distinguished lecturer, and has been the chair or co-chair of numerous annual IEEE Broadcast Symposia (most recently in 2016). David is an associate member of the Association of Federal Communications Consulting Engineers (AFCCE) where he is currently a board member and is past chair of the AFCCE Scholarship Committee. David was the recipient of Radio World's 2015 Excellence in Engineering Award and was recognized by *Radio Ink Magazine* in November 2010 as among the top ten best engineers in radio. In 2014, David received the Consumer Electronics Association Technology Leadership Award. The IEEE BTS awarded David the Matti Siukola Memorial Award for the Best Paper of the 2014 IEEE Broadcast Symposium.

Skip Pizzi
Associate Editor

Skip Pizzi is Vice President, Technology Education and Outreach, at the National Association of Broadcasters in Washington, DC. His career includes earlier work as an engineer and technical trainer at NPR, as an Editor at *Broadcast Engineering* magazine, and as an Audio Engineering Manager, Director of Media Standards, and Senior Technical Policy Analyst at Microsoft.

He has been a regular columnist for several technical trade publications, and has authored or contributed to several books—the latest of which is *A Broadcast Engineering Tutorial for Non-Engineers (4th edition)*, published by Focal Press, which Skip co-authored with Graham Jones.

Skip currently serves as Vice Chair of ATSC's *Technology Group 3* (TG3), which is developing standards for next-generation digital television. He is a graduate of Georgetown University, where he studied international economics, electrical engineering, and fine arts.

William T. Hayes
Associate Editor

William T. Hayes received his Bachelor's degree in communications in 1977. He has worked in broadcasting since 1973 in both radio and television with extensive experience planning, designing, and constructing television stations. He is currently the Director of Engineering and Technology for Iowa Public Television and is responsible for the planning and development of all technology projects, including the studio origination facility and 17 RF transmission facilities throughout Iowa.

Bill is active in future broadcast planning nationally and internationally and serves on a number of committees developing standards for emerging media technologies and the next generation of terrestrial digital television. He is a Fellow of the Society of Motion Picture and Television Engineers (SMPTE) and the former Central Regional Governor. Bill is a senior member of the Institute of Electrical and Electronics Engineers (IEEE) and currently serves as the president of the IEEE Broadcast Technology Society. He is a member of the Society of Broadcast Engineers (SBE) and was honored as their Educator of the Year in 2007 for his work educating the industry and the public regarding digital television. He is also a member of the Society of Cable Telecommunications Engineers.

Contributors to the 11th Edition

W. Cris Alexander (Chapter 7.13) started his broadcast career in the mid-1970s, working in both radio and television stations before settling into a radio career and coming to work for Crawford Broadcasting Company as Director of Engineering in 1984. In that capacity, he has overseen the growth of the company from six stations to the nationwide entity it is today. While his job deals with every aspect of AM and FM technical facilities and operations, he has a particular love for AM directional antennas. Cris was an active participant in the ad hoc committee from which came the moment method directional antenna proof rules and the new AM antenna protection rules. He is the managing partner of Au Contraire Software, which produces moment-method modeling and AM/FM design software. He is active in the SBE and holds CPBE, AMD, and DRB certifications.

Stephen Alhart (Chapter 5.17), as a junior in high school, was hired as "the kid" for odd jobs and summer relief at KTRK-TV Houston. This progressed to full-time employment as part of the floor crew, operating camera for local news broadcasts and other local programming and commercial productions. Moving up through the ranks at KTRK, he programmed and operated graphics on a CBS Labs VidiFont, edited 2" video tape, and hung film on telecine systems, eventually moving into maintenance and ENG operations with the birth of the "Live Truck." He operated, designed systems, and later managed the ENG operation at KTRK. In conjunction with progressive and supporting news management, he accomplished several firsts for a local TV market: first for live aerial broadcasts from a helicopter and blimp; first ever news live remote satellite transmission from Edwards Air Force Base for the initial "drop test" of the space shuttle Enterprise. From this date in 1977, ENG/SNG has never been the same.

In 1982, Stephen was recruited by ABC Broadcast Operations & Engineering in NYC to join as a Field Technical Manager, where he was responsible for remote broadcasts of *Good Morning America*, *World News Tonight with Peter Jennings*, and *Nightline* and participated in several Olympic Games. In this capacity, he was witness to many broadcast historic moments. Notable were the *Nightline* broadcasts from South Africa in March 1985 and for Mr. Nelson Mandela's release from prison February1990. Papal, presidential and the occasional ABC Sports assignments filled out the world travel schedule.

In 1995, Stephen was recruited by George Hoover of NEP Pittsburgh to join the growing NEP as a mobile unit engineer-in-charge, primarily working events for NBC Sports (including NFL on NBC, Super Bowl, NBA on NBC, NBA Finals, and the Atlanta Olympic games, plus a fair share of other clients, including NFL Films as Technical Manager (International broadcast) for the past six Super Bowls. Through a progression of mobile units and job tasks, he remains with NEP to this day working on and designing a variety of projects.

Wendy Allen (Chapter 5.9) has been in the broadcast industry since 1994. She has worked for a number of companies including PBS, CNN, Turner Studios, Concept House, and New Ideas. Since 2013, Wendy has worked at Fox Networks Engineering and Operations in Los Angeles. She was most recently a part of the FOX, Media Service division of Networking Engineering and Operations (NEO) that underpins broadcast and cable networks businesses. NEO support 35 networks, which includes FOX Network groups such as Sports, FX, FBC, etc., helping to move the vision forward, with cloud supply chain implementations, metadata management, automated processes, and efficiencies across the board. Wendy's education includes a Master's degree in architecture: electronic visualization and an undergraduate degree in business.

Tim Anderson (Chapter 7.14) has 40 years in the broadcast engineering field. Tim joined GatesAir in 2007 and was most recently the Manager of Radio Product and Market Development. In this role, he led the conceptual product development efforts for the current generation of GatesAir's digital radio transmitters.

Prior to joining GatesAir, he served as a broadcast systems engineer with iBiquity Digital Corporation, where he assisted in the development, testing, and deployment of the HD Radio transmission system.

Previously he was a Broadcast Systems Engineer for Avid Technology. Over the years, he has served as a consultant and chief engineer of several major market radio stations, groups, networks, and production facilities including Clear Channel, NBC, and Walt Disney.

Tim has published numerous technical articles on the subject of digital radio broadcasting and radio transmitter systems. He holds The Society of Broadcast Engineers' Professional Broadcast Engineer/Digital Radio Broadcast and Broadcast Network Engineer certifications. He serves on the SBE National Board of directors and Digital Radio Mondial Steering Committee.

David M. Baden (Chapter 3.2) began his career as a musician in 1974 and continued to perform and do session work until 1980. During the same period, David also worked as an engineer for various theatrical lighting, recording studios, and sound reinforcement companies in the Greater Washington DC area.

From 1980 to 1984, he was one of the founding principles at db Tech Inc., a company that designed, installed, and maintained professional recording studios, commercial sound reinforcement systems, and manufactured specialized audio devices.

In 1984, David began his career in professional broadcasting with Radio Free Europe/Radio Liberty, Inc. (RFE/RL). As Director of Broadcast Operations, USA, he provided oversight for the technical and production aspects of RFE/RL's U.S.-based operations. In 1995, David relocated to Prague, Czech Republic, to become the Manager of Technical Construction for the Prague Task Force. In this capacity, he was responsible for all technical systems design, integration, logistics, and physical facility construction for the RFE/RL Munich to Prague headquarters relocation.

In 1996, he moved to Radio Free Asia (RFA) where he was an organizational founding member of senior management as Chief Technical Officer. In this capacity, he was responsible for the initial build out and subsequent expansions of RFA's technical facility.

Over the course of his career, David has been intimately involved with all aspects of technical documentation. He has created and maintained technical documentation from the preliminary proposal to as-built stage for multiple projects worldwide.

Baron Services, Inc., (Chapter 5.15) provides the broadcast industry with critical weather intelligence, helping meteorologists deliver actionable information to their audiences when it matters most. The company has a heritage of weather radar and data processing innovation, leveraging Doppler and dual-pol radar information to deliver value-added weather information to aid in saving life and property.

Baron worked in partnership with L-3 STRATIS to upgrade the nationwide fleet of NEXRAD radars to dual-polarization. This boost in capability allows NWS forecasters to more accurately detect and forecast significant weather, and issue more timely alerts. Through this project, Baron technology now reaches every U.S. resident with a need for weather information. Numerous innovative value-added data products have resulted from this pioneering work. The company has long been a leader in dual-pol technology, and in 2004 delivered the first dual-pol commercial weather radar for broadcast television.

The company's new line of weather radar, the Baron Gen3 Radar series, will provide innovations in weather radar, from enhanced remote monitoring and continuous automated calibration to superior clutter suppression, delivered through an exclusive agreement with the University of Oklahoma.

The information provided by Baron Gen3 Radar systems can be used to feed on-air display systems and mobile and web apps, from the powerful Baron Lynx display to a series of white-label mobile apps, website population, and web widgets that broadcasters can use to provide improved information through every available avenue.

Daniel M. Barton (Chapter 10.9) guides a team of engineers and designers in supporting the ever-changing and growing telecommunications marketplace. He has presented papers at key engineering symposiums and authored papers that were published in technical journals. Dan gives ALIVE the opportunity to approach any communications project with unique solutions including general consulting, system design, or component design for specialized RF projects.

ALIVE has developed antennas for broadcast, government/municipal two-way and wireless applications, including 4G Long Term Evolution (LTE) MIMO antennas.

Under Dan's direction, ALIVE has designed and implemented Distributed Antenna Systems (DAS) for in-building repeater coverage—some of these systems include those for adult correctional facilities and mass transit terminals.

Before ALIVE, Dan worked at Andrew Corporation, now Commscope, where he was the Lead Antenna System Engineer for the Broadcast products group. He advanced the product line that enabled Andrew to differentiate themselves from the competition. Dan left Andrew after assisting their start-up In-Building Antenna products group; where he designed and assisted in the marketing of the antenna product line.

Dan studied electrical engineering for a BSEE, specializing in electromagnetic theory, at the University of Illinois, Urbana, and is also a member of the IEEE.

Fred Baumgartner (Chapters 8.2, 8.3), is a fellow of the Society of Broadcast Engineers, a trustee of the Ennes Foundation, Fellow of the Radio Club of America, and Nautel's TV product manager. Fred was Director of Broadcast Engineering for Qualcomm's MediaFLO project. Previously, he directed Leitch/Harris's Systems

Engineering group. Up to that time, he served as Director of Engineering for the Comcast Media Center in Denver, Director of New Product Development through the center's AT&T ownership, and Director of Broadcast Satellite Operations during its TCI ownership era. Before joining the satellite and cable origination world, he held the position of Chief Engineer at KDVR-TV and KFCT-TV, Denver; WTTV-TV, WTTK-TV, Indianapolis; KHOW AM & FM, Denver; and WIBA AM & FM, Madison, Wisconsin; Operations Manager at KWGN-TV, Denver; and others beginning with the overnight gig and a newly minted FCC 1st Class license at WBIZ AM & FM, Eau Claire, Wisconsin in 1972. Fred was also heavily involved with the development of EAS and has authored several hundred articles and a few books on radio and TV engineering. He also operates amateur radio station KØFMB, sometimes from a sailboat, the skies, or on a hike or a bicycle.

Gregory L. Best, P.E. (Chapter 1.1) is President of Greg Best Consulting, Inc. His firm performs broadcast consulting services for a large variety of customers and serves the RF communications industry in general. Greg earned his BSEE from the University of Missouri-Rolla and MSEE from Illinois Institute of Technology. He has over 35 years' experience in the design and product management of RF communications equipment. His corporate experience includes 16 years with the Broadcast Division of Harris, where he was responsible for TV transmitter design and management. While there, he was responsible for coordinating the development of the Platinum Series VHF TV transmitter, as well as many other VHF and UHF TV transmitters. He has also worked for Motorola on the original 800 MHz AMPS cellular phone system transmitter development and for IFR Systems Test and Measurement division developing 3 GHz spectrum analyzers. He has published several papers on TV Transmitter Design Architecture. Greg is a registered Professional Engineer and a member of the Association of Federal Communications Consulting Engineers, IEEE, and SBE. Greg currently heads the IEEE DTV RF Measurement Standards activities and serves as an associate editor for the IEEE Transactions on Broadcasting.

Clarence M. Beverage (Chapter 10.5) has been involved in broadcasting and amateur radio from his teen years. In 1975, he and two partners purchased the assets of The Paul Godley Company, Consulting Engineers located in Montclair, New Jersey. In 1985, he formed Communications Technologies, Inc., Broadcast Engineering Consultants, Marlton, New Jersey, and continues with that firm today.

His duties include providing broadcast engineering and telecommunications design services to satisfy the specific requirements of AM, FM, and TV broadcast stations and other Radio Frequency transmission entities. These services include allocation studies; development of site selection criteria; design of AM, FM, and DTV antenna systems; preparation of the engineering portion of FCC applications for construction permit and license, and aural STL FCC 601 broadcast systems. Extensive in-house computer software capabilities are used wherever possible to insure accuracy, improve efficiency, and aid innovative design efforts.

Specialized work includes preparation of RF system specifications; design of AM medium wave directional phasing and branching equipment and diplex equipment; numerical modeling of MW, VHF, and UHF antenna systems; field strength measurements including design of temporary test transmitter facilities; station inspection and equipment analysis; non-ionizing radiation analysis; expert testimony; VHF and UHF antenna pattern modeling and antenna range study analysis and Radio Frequency propagation studies using FCC and NBS methods.

Mark H. Bricker (Chapter 10.6) received his ASET degree from Spring Garden College in 1977. From graduation until 2009, he worked with Acrodyne Industries as a lab technician, field service technician, customer service engineer, and systems engineer. From 2009 to the present, he has been employed by Acrodyne Services as a systems engineer.

Jeff Brugger (Chapter 4.4) attended Ohio University and graduated with a major in audio engineering. Since then, he has built his career in Atlanta, starting at Doppler Studios where he recorded and mixed radio commercials, jingles, and music. Then he moved into consultant work focusing on the developing digital audio technologies. His consulting clientele covered broadcast facilities, advertising agencies, musicians, and recording studios, as well as medical technology companies. His consultant work took him to Asia to help locate facilities and train personnel in audio post production. There were stringent delivery requirements and language barriers to overcome in order to produce work that would be suitable for international networks that were beginning to move into the region.

In 2001, Jeff joined Turner Studios as Manager of the Sound Design department. Since then, the group has doubled in size and now employs a staff of 22, with 18 Sound Design rooms. This department produces the majority of the short form audio needs for all of the Turner Entertainment networks, original music, and corporate productions.

In addition to his work in Sound Design, Jeff has been involved in the Audio Council at Turner, a group that is focused on improving audio quality and standards throughout the company. Jeff is active with the local AES chapter and is an active mountain biker, dad, and sports car racing enthusiast and instructor.

Tim Carroll (Chapters 2.8, 5.18) is Senior Director Technology, Sound Group, in the Office of the CTO at Dolby Laboratories. Before this, he was CTO of the Telos Alliance, comprising Telos, Omnia, 25-Seven Systems, Axia, Minnetonka Audio, and Linear Acoustic, a company he founded. Previously, Tim worked for Dolby Laboratories in New York mastering films and DVDs, moving to San Francisco to manage the development of professional broadcast products for Dolby Digital (AC-3), Dolby E, and metadata. He has been honored with several Prime Time, Technology and

Engineering, and Sports Emmy Awards for his work with Dolby E, television Audio Processing, and the Beijing, Vancouver, and London Olympic Games. He is a member of AES, ASA, IEEE, SBE, and SMPTE, and is an active participant in the work of the ATSC. He holds multiple patents in the field of television audio and data and currently resides with his patient wife and family in California.

Ron Castro (Chapter 7.19) is the Chief Technical Officer and part owner of Results Radio, LLC, a chain of small-market FM stations in northern California. He started as an air personality and engineer in Pennsylvania in 1966 and later joined the US Navy as a Communications Technician. After working in Honolulu and San Francisco, he became the owner of a small FM station in Santa Rosa, California. Ron is a resident of Petaluma, CA, an active ham radio operator (callsign N6IE), holder of an FCC Lifetime General Class Radiotelephone license, and a member of the Society of Broadcast Engineers. Ron's most recent published work is "FM Translators and Boosters" from the *NAB Engineering Handbook, 10th Edition.*

Cindy Hutter Cavell (Chapters 5.3, 6.2) joined Cavell Mertz & Associates in 2010, after spending 15 years at NBC and ABC News (traveling with Presidents Reagan and G.H.W. Bush, for three Olympic Games for three technical Emmy Awards, and on weird remotes to strange countries), a few years as a Chief Engineer at major market TV stations for Hearst, Gannett, and Fox, a stint in Houston as GM of the Fox Sports Net technical playout center, and at Sprint as Director of Engineering for the 2GHz BAS Relocation Project. Cindy works primarily with TV clients on TV Spectrum Re-Pack issues, as well as doing pre-purchase due diligence (and post-purchase cleanup), workflow analysis, and TV station technical design/build. Additionally, she designed and managed the construction of a 670-mile, bi-directional, 37-hop microwave system from Aurora, IL, to Carteret, NJ, which was built for a financial trading client.

Richard Chernock, ScD (Chapter 5.10) is currently Chief Science Officer at Triveni Digital. Previously, he was a Research Staff Member at IBM Research, investigating digital broadcast technologies. He is chair of the ATSC Technology Group on ATSC 3.0 (TG3) and chairs the AHG on service delivery and synchronization for ATSC 3.0. He was previously chair of the ATSC Technology and Standards Group (TG1). He is also the Distinguished Lecturer Chair for IEEE BTS and a member of FOBTV.

In another life, Richard used transmission electron microscopy to study materials characteristics for advanced ceramics packaging and semiconductor technology at IBM. His ScD is from MIT in the field of nuclear materials engineering.

Kenneth E. Colwell (Chapter 7.7) holds a PhD from the University of Iowa in instructional design and technology. He is also a Certified Senior Radio Engineer, Digital Radio Broadcast Specialist, and a Certified Broadcast Networking Technologist by the Society of Broadcast Engineers. He is a Professor of Communication at St. Ambrose University, Davenport, IA, and is the Director of the Ambrose Communication Center and General Manager of KALA(FM).

He has worked in broadcasting since 1969, getting his First Class Radiotelephone license in 1971.

He has worked in both commercial and noncommercial broadcasting and served as chief engineer of WQUA, Moline, IL, KRVR(FM), Davenport, IA, and KALA(FM), Davenport, IA.

Ken is a member of the Society of Broadcast Engineers, the Audio Engineering Society, and several academic organizations. He also is an Amateur Extra operator and ARRL VE, with call sign N9CI.

Bobby Cox, PhDEE (Chapter 7.9), is a Senior Staff Engineer with Kintronic Laboratories, where he has been employed since 1994. He holds a BSEE 1989, MEE 1991, and PhDEE 1994 from Mississippi State University. His graduate studies specialized in classical electromagnetics, computational electromagnetics, signal processing, and communications. His doctoral dissertation involved NEC modeling of an electrically large antenna structure and design of an exponentially tapered transmission line antenna. He has been a Member of IEEE since 1985. At Kintronic Laboratories, he is responsible for the design engineering, project management, production details, and testing of hundreds of medium wave (AM) antenna systems, both domestic and international. His work includes low, medium, and high-power directional arrays; multiplexed stations; LF and VLF matching networks; custom high-power RF components; co-location of broadcast facilities; and military mobile broadcast antenna systems. He specializes in numerical modeling of unusual antenna configurations.

Aldo G. Cugnini (Chapter 2.7) is a technology and business consultant who has held various technical and management positions at Philips Electronics and elsewhere. While at Philips, Aldo had a leadership role in the development of the ATSC digital television system (and its progenitor, the "Grand Alliance" digital HDTV system), and was a key member of the Advanced Television Research Consortium (ATRC) HDTV development team. He was also project manager for the MSTV Terrestrial Digital Converter Box Project, which made possible the highly successful NTIA TV Converter Box Coupon Program.

Aldo received his BS and MS from Columbia University, is an inventor on 13 patents in the fields of digital television and broadcasting, with two more pending, and has served on the Board of Directors of the Advanced Television Technology Center. Prior to his DTV work, Aldo developed various audio, content delivery, and RF technologies at Broadcast Technology Partners and CBS Laboratories, and provided product support and field engineering at RCA Broadcast

Systems. He was also Chief Engineer at WKCR-FM and a broadcast engineer at WABC and WPLJ.

A speaker at several industry conferences, he is the author of numerous technical papers and industry reports, including chapters in the 10th edition of the *NAB Engineering Handbook*, and has been a regular contributor to several trade publications including *Broadcast Engineering*. Aldo holds FCC Commercial General (formerly First Class) Radiotelephone and Amateur General Radiotelephone (W2AGC) Operator's licenses. His other interests include the timpani, astronomy, radio-controlled helicopters, writing for Wikipedia, and the occasional blog.

Jeff Detweiler (Chapter 7.6) directs broadcast product development and the introduction and launch of its HD Radio™ brand of In-Band On-Channel (IBOC) technology to radio stations worldwide. In his role, he manages the technical relationships with broadcast equipment manufacturers and coordinates the transfer of technology to DTS licensees. He is a frequent presenter at NAB, IEEE, and other industry events and a regular contributor to broadcast publications. Presently, Jeff chairs the Digital Radio Working Group with the North American Broadcasters Association.

He has more than 37 years of experience in the radio industry and over 16 years focused on HD Radio Technology. Prior to joining DTS, he spent 12 years in management at QEI Corporation, last serving as Worldwide Sales & Marketing Manager. Prior to joining QEI, he was the Northeast Sales Engineer for Allied Broadcast Equipment. He served as Chief Engineer at Lake Erie Radio (WWWE and WDOK) in Cleveland, OH, and in the same capacity at Nassau Broadcasting (WHWH and WPST) Princeton, NJ.

Ronald Eligator (Chapter 4.1) is a Principal in the acoustic consulting firm of Acoustic Distinctions. He has over 30 years' experience in noise control and acoustic design for broadcast, recording and production facilities, performing arts, cultural buildings, worship spaces, and commercial office space. An expert in architectural acoustics, sound isolation, mechanical noise and vibration control, and environmental acoustics, he has served broadcast clients nationwide, including all the major television and cable networks, as well as numerous local stations, film studios, and music recording and production studios.

Among other accomplishments, Ronald has developed sophisticated computer modeling approaches for speech intelligibility analysis for large, reverberant spaces; designed cost-effective sound barrier construction for broadcast and performing arts facilities subject to aircraft overflights by applying standard building materials and system in novel ways; and mentored early career acoustic and architectural professionals in his firm and through presentations at other consulting firms and universities. He is also the author of numerous technical papers on acoustics and has lectured widely at industry conferences.

Ronald is a member of the Technical Committee on Architectural Acoustics of the Acoustical Society of America and the ASHRAE Technical Committee on Sound and Vibration. He holds a Master of Science in mechanical engineering from Yale University and a Bachelor of Arts in music from Williams College, and is Adjunct Professor in Sustainable Interior Environments at the State University of New York Fashion Institute of Technology.

Dane E. Ericksen, P.E. (Chapters 1.3, 2.3) is now a semi-retired consultant to Hammett & Edison, Inc., Consulting Engineers, San Francisco, California. For the prior 31 years, from 1982 to 2014, he was a senior engineer with H&E. He is a registered Professional Engineer (Electrical) in the state of California, and holds SBE CSRTE, 8-VSB, and CBNT certifications. He has served eight terms as a national SBE director, multiple terms as an SBE Chapter 40, San Francisco chapter officer, is an SBE Fellow, and was the 1998 SBE Engineer of the Year. Dane was chair of the SBE FCC Liaison Committee from 1987 to 2007, and has served as a member of the SBE Certification Committee from 1987 to present. He served on the NAB Engineering Conference Committee from 1993 to 2000, and on Subcommittee 4 (SC4) of IEEE Standards Coordinating Committee 28 (SCC28) from 1985 to 2000; this was the IEEE committee responsible for the ANSI C95.1–1992 and C95.1–1999 radio frequency exposure standards. He served as chair of ATSC Technology and Standards Group S3 Specialist Group on Digital ENG from 2004 to 2009, which created the ATSC A/82 Data Return Link (DRL) standard. Dane is founder and co-chair of Engineers for the Integrity of Broadcast Auxiliary Services Spectrum (EIBASS).

Prior to joining H&E, from 1970 to 1982, he was a field engineer (radio inspector) for the FCC, based out of the Commission's San Francisco office. He was an FM/TV/CATV Specialist, operating the Western FM/TV/CATV Enforcement Unit. He received his BSEE from California State University Chico in 1970.

Daniel S. Fallon (Chapter 8.8) received a BSEE from Clarkson University, Potsdam, NY, and an MSEE from the University of Massachusetts, Amherst. He then served in the U.S. Air Force as a radar and communication system engineer. After his service, he worked at Sage Laboratories, Natick, MA, designing microwave components for satellites. In 1999, he moved to Maine and joined the Passive Power Products division of Andrew Corporation as engineering manager. In this capacity, Daniel designed filters and RF systems for broadcast use. In 2009, he joined Dielectric as Sr. RF Engineer and has designed a variety of RF systems and components for wireless, broadcast, and scientific applications.

Walter Fischer (Chapters 8.4, 8.5), born in Freising, near Munich, studied electrical engineering at the Munich University of Applied Sciences.

Since 1985, working at Rohde & Schwarz in Munich, he spent 14 years in the development department for broadcast test and measurement, and since 1999 he has been a trainer at the Rohde & Schwarz training center, responsible for all kinds of video, audio, and broadcast trainings.

He is the author of *Digital Video and Audio Broadcasting—A Practical Engineering Guide* (Springer) and *Digitale Fernseh- und Hörfunktechnik in Theorie und Praxis* (Springer Verlag) and a lecturer at the Munich University of Applied Sciences and at the Deggendorf Institute of Technology.

Brad Gilmer (Chapter 5.8) is President of Gilmer & Associates, Inc. He is a founding member of the Joint Task Force on Networked Media, the Executive Director of the Video Services Forum, and the Executive Director of the Advanced Media Workflow Association. Brad is a member of the Society of Motion Picture and Television Engineers and has been an active participant within the SMPTE since 1984. Brad is the first recipient of the SMPTE Workflow Systems Medal.

Brad was previously employed at Turner Broadcasting System in Atlanta, where he and his staff were responsible for Engineering and Operations for the Entertainment Division Worldwide.

Matthew S. Goldman (Chapters 5.10, 5.11, 5.20) is Senior Vice President of Technology, TV & Media, for Ericsson, where he is focused on video processing and media delivery solutions.

Matthew has been actively involved in the development of digital television (DTV) systems since 1992. He was a prominent participant in the Moving Picture Experts Group where he helped create the MPEG-2 Systems standard, the baseline transport technology used in direct broadcast satellite, digital cable, terrestrial broadcasting, and DVD/Blu ray® video. He also served as project editor for the MPEG-2 DSM-CC standard, the control signaling used in video-on-demand systems, switched digital video, and for DTV data downloads and carousels. Matthew was a co-developer of the Program and System Information Protocol (PSIP) standard, which is used by ATSC 1.0 DTV receivers to navigate and tune broadcast DTV services. He also has been influential in the Society of Motion Picture and Television Engineers (SMPTE), the Digital Video Broadcasting project, the Society of Cable Telecommunications Engineers, and the Consumer Technology Association.

Until 1996, he was a Consulting Engineer at Digital Equipment Corporation, where his last role was as the systems architect for a first-generation digital video server system. From 1996 to 2000, he was Director of Engineering, Advanced Systems Development, at DiviCom, where he specified MPEG-2 based systems solutions. From 2001 to 2003, he was a technology consultant specializing in DTV system solutions and expert witness research; one of his projects included the definition of the broadcasting industry's first compressed-domain high definition program splicer for the FOX Broadcasting Company. In 2004, he joined TANDBERG Television as Director of Technology for the Americas region. He was later promoted to Vice President of Technology in a global role. Ericsson acquired TANDBERG Television in 2007.

Matthew has presented numerous papers at industry trade events and has been a contributing writer to various trade magazines. A technical paper on high dynamic range that he co-wrote received the Best Paper Award of the 2015 NAB Broadcast Engineering Conference. He is an author of the 2007 *NAB Engineering Handbook*, has been published in the IEEE Proceedings, and is an IEEE Broadcast Technology Society Distinguished Lecturer.

He received Bachelor (high honors) and Master of Science degrees in electrical engineering from Worcester Polytechnic Institute. He holds six patents related to digital video transport. Four projects on which he has worked have been recognized by Technology & Engineering Emmy® Awards. A SMPTE Fellow, he is also a Senior Member of the IEEE and an inductee of the Academy of Digital Television Pioneers. He is an active participant in the National Academy of Television Arts and Sciences Technology & Engineering Emmy Awards Committee.

Martyn Gregory (Chapter 10.4) graduated from the University of Durham (United Kingdom) in 1981, with honors in applied physics and electronics. He spent 20 years with the British Broadcasting Corporation's transmission engineering division, working first in AM and HF transmitter operations; moving into FM and TV operations; finally specializing in all aspects of antenna and RF systems. He was heavily involved with the initial DTV antenna program in the United Kingdom as well as FM national radio expansion followed by introduction of digital radio around 1992. In 1999, he moved to the United States as Broadcast Engineering manager for Crown Castle Corporation, subsequently joining Richland Towers, a nationwide broadcast facility owner and integrator as Director of Broadcast and RF Systems Engineering. During this time, he specialized in broadband system development and broadcast facility analysis for both TV and FM services and station co-location. In 2004, he returned to Crown Castle as VP of Service Delivery for their mobile TV initiative using DVB-H technology at 1.7GHz, and was responsible for the implementation, operation, and maintenance of their DVB-H based mobile media networks focused initially throughout New York City. In 2007, Martyn joined Howell Laboratories' subsidiary, Shively Labs, with responsibility for broadcast products, including the introduction of several new products during this period.

He has extensive experience from design and specification of RF systems to testing and commissioning. Martyn has been certified to work on tall structures, has held a professional license as a Chartered Engineer (CEng) in Europe since 1993, and is a member of the IET.

Dave Guerrero (Chapter 5.2) is currently Chief Operating Officer for WLVT-TV. Most recently he was the Vice President and General Manager of the Videotek product line for Harris, Inc. (NYSE: HRS), where he managed Harris Broadcast's business unit that designed, marketed, and manufactured world-class (ISO) broadcast Test and Measurement products. David attended the Darden School for Executive Education at the University of VA.

A member of SMPTE, AES, SBE, and IEEE, he has also spent his lifetime engineering live and studio, audio, and video productions utilizing his degree in electronics technology. He has received two Emmy Awards for contributing engineering expertise to regional and national television productions, and is also named as co-inventor to four U.S. and multiple international patents.

David's past experience as chief engineer at broadcast radio and television stations as well as engineering supervisor for network level remote productions, including all U.S. broadcast networks of Olympics, NFL Super Bowl, and Triple Crown racing, has given him experience in all aspects of the "real world." David brings over 40 years of broadcast experience as the leader of the team at WLVT-TV, PBS39.

Lawrie Hallett, PhD (Chapter 2.5) is a Senior Lecturer in Radio at the University of Bedfordshire in the United Kingdom. A writer and broadcaster, with a background in broadcast radio engineering, he has been researching digital radio technologies since the 1990s. A former Senior Associate at the U.K. broadcast regulator, Ofcom, with particular responsibility for Community Radio and digital radio policy development, he is also the Technical Director of Future Digital Norfolk Limited, a company which provides digital audio broadcasting facilities to the Greater Norwich area of Norfolk in the East of England.

Bill Harland (Chapter 9.4) is Vice President of Marketing for Electronics Research, Inc., and has responsibility for all corporate communications, product management, and applications engineering since 2003. Prior to joining ERI, Bill worked for a number of broadcast equipment manufacturers including Harris Corporation, Broadcast Electronics, and Andrew Corporation. He is a graduate of the University of Idaho and holds a degree in radio and television.

Kirk Harnack (Chapter 4.5) brings nearly 40 years of hands-on experience in broadcast engineering and education to his position at The Telos Alliance. His expertise in putting technology to work in broadcast facilities has driven notable expansion in IP-Audio, VoIP for broadcast, and other new technology adoption by audio content creators.

Kirk maintains an active, hands-on role in broadcast engineering through his positions as a partner and VP-Engineering of South Seas Broadcasting, Inc., licensee of KKHJ-FM and WVUV-FM in American Samoa. He is also partner and Director of Engineering for Delta Radio, LLC, licensee of six radio stations in Cleveland, Indianola, and Greenville, Mississippi. Kirk is a certified Meteorologist, Private Pilot, and licensed General Class Amateur Radio operator (KD5FYD). He has served as a member of the Board of Directors of the Society of Broadcast Engineers and as Chair of the SBE Chapter in Memphis, Tennessee, and is currently Program Chair of SBE Chapter 103 in Nashville, Tennessee.

Kirk founded the Internet video netcast, "This Week in Radio Tech" or "TWiRT." This one-hour weekly video netcast features regular contributors and guests from the world of radio engineering.

Stephen N. Heazlewood (Chapter 10.7) has enjoyed a career in the broadcasting and telecommunications industries spanning more than 20 years and multiple continents including Asia, the United Kingdom, Australia, and the United States. During this time, he has built a strong reputation as a talented engineer through his work on many unique and innovative antenna systems.

Stephen played a key role as Radio Frequency Systems' lead antenna designer for the U.K. Digital Switch-Over project where he developed industry-leading high-power panel antenna solutions for large cross structures. His solutions met all the challenging coverage requirements in addition to maintaining a high bandwidth not previously thought possible in such situations.

Currently fulfilling a systems automation role at BAI Communications (based in Australia), Stephen combines his extensive RF design and project delivery experience with his software development skills to provide automation solutions for all areas of the business. He is well known in the Australian broadcast industry for his development of Vector Analysis software that allows engineers to analyze Vector Network Analyser data in the time domain and model changes. This experience as a software developer led to him playing a critical role in the award-winning Australian Spectrum Restack Project (700 MHz clearance), where he was responsible for the automation of many of the tools used by the project team, including the documentation system, the automation of the test equipment, and testing protocols and the commissioning test data reports.

Stephen's work on automation of these common processes was instrumental in Australia's achieving the release of the Digital Dividend spectrum ahead of the required schedule.

Richard G. Hickey (Chapter 9.3) was introduced to FAA/FCC regulatory compliance in 1996 at Flash Technology as FAA Determination Development Representative. Later at Flash, he served as High Intensity/Airport/International Sales, High Intensity and Service Sales Manager, Inside Sales and Spare Parts Manager, Product Manager, Broadcast Sales Manager, and Director of Business Development-Broadcast, Telecom, and Utility Infrastructure. Richard then joined First Light Resources as Chief Compliance Officer in 2013, as Business Development Manager at Hilights, Inc., a premier service company specializing in aviation obstruction lighting in the broadcast industry, and currently serves as Senior Manager of National Regulatory Compliance at Vertical Bridge.

Richard has designed lighting systems within the applicable guidelines of the FAA, FCC, ICAO, TP 312/382, DGAC, NEC, NFPA, and IEEE standards for many global applications. His expertise has been solicited for many high-profile projects, including a nationwide communication project for the Egyptian Ministry of Aviation, the Suez and Panama Canals, wind farms in Austria and Germany, catenary spans in the United States and Belgium, a U.S. Observatory in the Antarctic, the former World Trade Center Twin Towers in New York, the Trump Spire in Chicago, Sutro Tower, and many of the tallest bridges, buildings, and broadcast towers around the world.

Richard is an active member of SBE Chapter 39, served on the Board of Directors for the UtiliSite Council, received the Utilities Telecom Council Chairman's Award in 2013 and the Silver Scribe Award from ENTELEC, and was a contributing author in the 10th edition of the *NAB Engineering Handbook*.

Paola Hobson, PhD (Chapter 5.13) has extensive experience in communications and media industries, including public safety and consumer mobile products, and professional broadcast systems. She has a track record in delivery of innovative new products and services, strategic partnership development, and business growth. Paola is currently Managing Director for InSync Technology Ltd, an employee-owned company, which develops signal processing hardware and software for the broadcast industry. Paola holds a BSc, a PhD, and an MBA, as well as numerous patents in image processing, content personalization, and mobile video.

Chris Homer (Chapter 5.7) is Vice President of Engineering for Public Broadcast Service (PBS). Chris has spent over 35 years in television engineering and operations for terrestrial and satellite broadcast, media operations, post-production, and studio operations.

His past experiences cross a diverse range of industry experience: 2013–present, Vice President Engineering & Operations PBS; 2006–2013, Vice President of Broadcast Operations at DIRECTV managing broadcast centers throughout the United States; 1999–2006, Chief Engineer for two Los Angeles–based television stations (KWHY and KMEX) during the digital transition; and 1990–1999, Vice President of Production & Post Production for Fox Broadcasting Company. Prior to working at Fox, he worked as an engineering support engineer for Harris Broadcast, Grass Valley Group, and Digital F/X.

Chris has a degree in business management and served in the United States Armed Forces as a Naval Avionics Technician. He is currently serving on Board of Directors for the Advanced Television Standards Committee (ATSC) as chair of S34–5 Ad Hoc Group on Accessibility and on the Board of Directors for the Advanced Media Workflow Association (AMWA). He is an active member of SMPTE and SBE, and is the former Secretary and President of Los Angeles–based Society Television Engineers (STE).

Doug Irwin (Chapter 6.4) has more than 30 years of experience in broadcast engineering, specializing in radio transmission. He currently serves as the Vice-President of Engineering for iHeartMedia's radio stations in Los Angeles, California. In addition, he has been contributing articles on technical subjects to *Radio* magazine for over 15 years. Doug became the Technical Editor of *Radio* magazine in March of 2015.

Tom Jennings (Chapter 5.17) is President and Co-Founder of Accelerated Media Technologies in Auburn, MA, the leading supplier of news vehicles in the United States. Prior to founding Accelerated Media in 2010, he served for over 18 years as Director of Sales for Wolf Coach in Auburn, MA, a leader in the production of mobile platforms for both broadcast and government clients worldwide. A native of Massachusetts and graduate of Emerson College in Boston, MA, Tom has been working with local and national broadcasters for well over 30 years, educating and equipping field crews. His mission at Accelerated Media has been to evolve the art of mobile news acquisition beyond the legacy solutions. He holds several U.S. patents for transmission and other associated technologies relating to news acquisition.

Jeff Juniet (Chapter 6.6) started in television in the production department at WINK-TV and overnight master control at WRXY in Ft. Myers, FL. When WRXY installed a new transmitter, he dove full time into the engineering side of television and hasn't looked back. He later went to work for ACME Television as an Engineering Manager and later as the Director of Technical Operations, making stops in Albuquerque, NM, Madison, WI, and back to Florida, this time in Orlando. While at ACME, he helped guide seven stations through the digital transition. After that, he spent three and a half years in the NBA with the Orlando Magic as the Sr. Video Systems Engineer, opening a state-of-the-art arena and hosting the 2012 NBA All-Star Game. He moved to Cox Media Group in October of 2013 and immediately started working on projects for upgrading emergency power for the studio, creating a new news set, and designing how they grow their workflows for the future.

Alan W. Jurison (Chapter 7.16) started at age 15 with Pilot Communications in Syracuse, NY. During this time, he developed the technical aspects of the industry's first remote voice tracking for stations in distant cities, which became commonplace in the industry. He graduated with a BS in information management and technology from Syracuse University and was promoted to Regional Information Systems Manager and Broadcast Engineer for Citadel Broadcasting. For 11 years, he led many IT and engineering projects companywide. In 2012, Alan joined iHeartMedia as a Senior Operations Engineer for the corporate Engineering and Systems Integration group. Much of his focus has been on deploying and advancing iHeartMedia's digital HD

Radio data services. Alan is a member of the Society of Broadcast Engineers (SBE) and holds several certifications, including Certified Professional Broadcast Engineer (CPBE), AM Directional Specialist (AMD), Digital Radio Broadcast (DRB), and Broadcast Networking Engineer (CBNE). Alan has been in the radio industry for over 20 years and offers a unique prospective on the convergence of IT and broadcast engineering.

John Kean (Chapters 6.1, 10.1) joined National Public Radio as a Senior Engineer in 1980, working on technology projects and FCC regulatory activities for the network. In 1986 he joined Jules Cohen and Associates, followed by Moffet, Larson and Johnson in 1988, where he consulted in a range of broadcast and communications projects. He returned to NPR in 2004 to help establish NPR Labs, which develops and evaluates new technologies, procedures and standards on behalf of public radio. His work ranged from audio metrics and loudness to digital audio broadcasting, RF signal propagation analysis and signal coverage mapping.

He retired from NPR in 2015 and now works with Cavell Mertz and Associates on broadcast consulting projects and performs audio consulting in private practice as Kean Consultants LLC. He contributed to the Audio Engineering Society's recent *Recommendation for Loudness of Audio Streaming and Network File Playback* and is currently active in an AES Technical Committee to develop audio metrics for online video services.

John was named the winner of the 2017 NAB Radio Engineering Achievement Award for Radio Engineering Achievement and was a recipient of the Association of Public Radio Engineers' Engineering Achievement Award. He was also a past president of the IEEE Broadcast Symposium and the Washington DC Section of the AES, and is a contributing author to the *NAB Engineering Handbook*, editions 7, 8 and 9. He is a frequent author and presenter for the NAB's Broadcast Engineering Conference. He holds a U.S. patent for an algorithm to determine digital broadcast radio coverage.

Christopher Kelly (Chapter 5.16) is a Technical Sales Specialist for Ross Video Ltd. Before moving into the sales role, he was a trainer and demonstration artist specializing in switchers, graphics, and production automation. Before joining Ross Video, Christopher worked in various production and promotions roles at television stations in Florida, Illinois, Indiana, and Oklahoma. Christopher is a 2000 graduate of the University of Illinois in Urbana/Champaign with a degree in speech communication and resides in Carmel, IN, with his wife and son.

Tom King (Chapter 7.9) is currently the President of Kintronic Labs, Inc., and the recipient of the 2015 NAB Radio Engineering Achievement Award. His BSEE and MSEE are from the University of Tennessee.

He did two years' study as research assistant toward a PhDEE at the University of Arizona. From 1973 to 1977, he was the Staff Engineer in Electro-Optical System Design at the Naval Weapons Center in China Lake, California. He was also the Project Engineer in Infrared Countermeasures Systems for Navy and Marine Corps aircraft at the Special Warfare Laboratory Division of General Research Corporation in McLean, VA. He is the author or co-author of numerous technical papers on the subject of AM broadcast antenna systems and related topics and is the co-holder of one patent while employed by the U.S. Navy.

Tom is an Associate Member of the Association of Federal Communications Consulting Engineers, a Member of IEEE Broadcast Technology Society, a Member of the National Radio Standards Committee and the Analog and Digital Radio Broadcast Subcommittees. Lastly, he is married to Leigh King and has two sons, Dr. Benjamin King and Joshua King, and two daughters, Katherine Kreuser and Geneva King.

Karl J. Kuhn (Chapter 2.9) is the Senior Video Systems Application Engineer, Tektronix, where he supports Digital Audio and Video, Compression, IP, File Based, and RF test and measurement. He has been at Tektronix since 2000. Prior to Tektronix, he was the lead Video Test Engineer for IBM in their Digital Video Development Laboratory in Bethesda, MD. Karl holds three U.S. patents and one Japanese patent that cover In-service Testing of Digital Broadcast Video. Karl served on the NATAS Technical and Engineering Emmy Committee from 2006 to 2010. He is the Past SMPTE Eastern Region Governor and now SMPTE North American Sections Director. In 2015, Karl was raised to SMPTE Fellow. He is also a Certified Project Management Professional thru PMI and the George Washington University.

Paul Lacombe (Chapter 5.16) is a graduate of SUNY NY where he earned a BTEE. In 1985, Paul started with General Motors working on Finite Element Modeling and Analysis. After a few years in Detroit, he moved to California to work for Silicon Graphics, where he became involved in launching Silicon Studio. He spent one year in Mallorca building a virtual production studio and returned to Los Angeles to launch a virtual set studio and a production with Disney, *One Saturday Morning*.

In 1998, Paul returned to the East Coast to work with Doug Trumbull at Entertainment Design Workshop (EDW), landing another Disney project, *Book of Pooh*, and then set off to start UNREEL, focused on real-time 3D graphics and virtual-set solutions for feature film and broadcast markets. Customers include AccuWeather, CNBC, DVDonTV, ESPN, NASDAQ, NBCUniversal, Pentagon, PGATour, and Reuters. Feature Films include *I'Robot*, *Stealth*, *Shark Boy & Lava Girl*, *Speed Racer*, and *Alice in Wonderland*.

In April 2014, UNREEL was purchased by Ross Video. Paul's role at Ross Video is Creative Director for AR & VS, focused on strategic projects and emerging technologies.

Chris Lennon (Chapter 5.10) is President and CEO of MediAnswers, specialists in advertising and media workflow. He has over 30 years of experience in the media business, leading large projects on every continent except for Antarctica. He is a past Standards Director for the Society of Motion Picture and Television Engineers, and is a SMPTE Fellow and award winner. In his spare time, he is an accomplished racecar driver and high-performance driving coach.

Todd R. Loney (Chapter 10.7) has specialized in RF engineering for over 30 years. Todd has worked in product development and field service for several transmitter and passive RF equipment manufacturers. For the past 19 years, he has been a consulting engineer, providing service for both broadcast transmission and high-power RF for physics research. Todd has designed, managed, and commissioned broadcast transmission system installations on six continents. Todd has co-authored papers for the NAB Engineering Conference and IEEE Broadcast Technology Symposium. He is a member of the IEEE and SBE.

Vinnie Macri (Chapter 5.19) is presently the Product Outreach Manager at Clear-Com. His responsibilities include working with R&D/product development engineering teams, as well as acting as educator and liaison to consultants and sales-channel customers.

Vinnie has over 40 years of experience in communications/intercom and audio distribution technologies. He has recently written the intercom chapter for the 2016 *SBE Broadcast Engineering Handbook*. Industry memberships include AES, Infocomm, NAB, and SVG.

Martin Marietta (Chapters 5.5, 5.6) is a fashion and entertainment entrepreneur turned Producer, Technologist, and Consultant. He regularly works in New York City, and internationally, on some of the highest-level video productions, including long-running television shows, commercials, feature films, music tours, operas, and award shows, as both a Robotic Camera Movement Specialist and a Camera Operator. His show awards include two Emmys and a Peabody Award.

Martin is the creator of the LEDA System: an Operator's real-time see-in-the-dark visual location prompter for remotely operated camera movement devices. Proud to be a member of the Society of Motion Picture and Television Engineers, he is a long-time student of Advanced Motion Imaging, including Stereography and Virtual applications.

Whenever possible, he spends his time outdoors. Often, he can be found exploring nature, regularly participating in sports, such as mountain biking, skiing, kayaking, adventure swimming, and sailing. He lives with his loving wife Katitia, and is soon to be a father.

David P. Maxson (Chapters 3.4, 10.2) has been involved in broadcast engineering for four decades. For two of those decades he was Director of Engineering for Boston's classical music broadcaster, overseeing a classical broadcasting network and contributing to innovations in radio broadcasting. During and after that time, David has been providing consulting services to the industry. He has been a participant on the National Radio Systems Committee since 1998, supporting the development of digital and analog broadcasting standards. With particular experience in spectrum analysis, he has presented technical papers to the NAB and PREC conferences and other white papers on techniques in measuring AM and FM analog and digital broadcast signals. His experience spans managing regulatory compliance in the workplace, including RF safety and other workplace hazards. David is the author of *The IBOC Handbook—Understanding HD Radio(TM) Technology*, published by NAB/Focal Press.

William R. Meintel (Chapter 2.4) holds a degree in electrical engineering and has more than 40 years' experience in the communications field. After graduation, Bill was employed by the Federal Communications Commission (FCC), first as a field engineer and then in the Mass Media Bureau's Policy and Rules Division. While in Policy and Rules, Bill served as the division's computer expert and directed the development of several major computer-modeling projects related to spectrum utilization and planning.

He entered private practice in 1989, and has been heavily involved in technical consulting, computer modeling, and spectrum planning for the broadcast industry. In April 2005, Bill merged his consulting practice (Techware) into the firm Meintel, Sgrignoli, & Wallace, where he is a partner and principal engineer.

Bill co-authored an initial report for the NAB on spectrum requirements for Digital Audio Broadcasting (DAB) and has been extensively involved in spectrum planning for digital television (DTV) in both the United States and internationally, most recently creating an analog-to-DTV transition plan for Haiti.

His DTV work included developing the software analysis tools used by both the broadcast industry and the FCC in developing the DTV allotment tables, while at the same time being deeply involved in the actual allotment planning on behalf of the industry. Subsequently, Bill incorporated many of the DTV planning tools into the OET-69 interference analysis software that he created for FCC.

Bill is a member of IEEE, being the immediate past president of the IEEE Broadcast Technology Society, and a member of the engineering honor society Tau Beta Pi.

Geoff N. Mendenhall, P.E. (Chapter 7.14) has spent most of his 50-year career developing broadcast equipment technology. He has always had a passion for radio broadcasting, where his many contributions have made him one of the industry's best-known experts on FM transmission.

Geoff has authored over 50 technical papers, including the chapter on FM and Digital Radio Broadcast Transmitters in the 8th, 9th, and 10th editions of the *NAB Engineering Handbook*. His technical writing was recognized by the 2010 NAB Best Paper Award and the 2011 Matti S.Siukola memorial award for best IEEE Broadcast Symposium technical paper.

Geoff received the 1999 National Association of Broadcasters, Radio Engineering Achievement Award recognizing his many innovations and contributions to the broadcast industry, and he was named an Engineering Fellow by Harris Corporation in 2008. He retired from Harris in early 2013 to form Mendenhall Engineering, LLC, an RF Engineering Consulting firm, and he is a Technology Advisor to GatesAir (formerly Harris Broadcast). He earned his electrical engineering degree from the Georgia Institute of Technology and is the inventor of nine U.S. patents related to broadcast equipment.

Geoff is a registered Professional Engineer and is a member of the Association of Federal Communications Consulting Engineers (AFCCE), a member of the FCC Technological Advisory Council (FCC-TAC), a contributor to the National Radio Systems Committee (NRSC), and a Senior Member of the IEEE.

Peggy Miles (Chapter 8.9) is president and founder of InterVox Communications in Washington, DC, a recognized expert and futurist in digital media and communications. She founded InterVox Communications in 1995 and provides digital media, strategy, advisement and business development services to leading companies around the world.

A best-selling author, her books include the *NAB's Internet Age Broadcaster I and II, the Radio Advertising Bureau's Selling the Net Video/Book* and others. Peggy was the founder of the International Webcasting Association bringing together Microsoft, Apple, Real Networks, CBS and others to work on issues facing the industry.

James Moser (Chapter 7.9) received his MSEE from the University of Michigan in 1982, with the focus of his graduate studies in electromagnetics, microwave devices, and network analysis. His formal education includes a BSEE (1980) from the University of Michigan and an AS (1978) from Grand Rapids Community College. He has continued his education with course work in C/C++ programming and in computational scientific modeling. After 12 years of experience as a member of the technical staff at MIT Lincoln Laboratory, he joined Kintronic Laboratories in 1995, where he is presently a Senior Staff Engineer. He has been a member of the IEEE since 1978 and is in the Antenna and Propagation, Microwave Theory and Techniques, and Signal Processing societies.

At Kintronic Laboratories, he is responsible for the engineering design, production, testing, and project management of medium wave antenna systems. His experience includes both domestic and international installations, those involving directional arrays and multiplexed sites, and those with unique requirements ranging in scope from low power nondirectional stations to a 1-megawatt directional skirted array. He specializes in the development of custom in-house software for the design and engineering of the antenna systems.

Eugene Novacek, P.E. (Chapter 4.7) is the CEO of ENCO Systems, Inc., established in 1983. He has engineering degrees from the Massachusetts Institute of Technology (MIT). ENCO is headquartered in Southfield, Michigan, the creators of DAD (Digital Audio Delivery) and MOM (Media Operations Manager) with installations all over the world.

Shane O'Donoghue (Chapter 8.9) is Director of Broadcasting for New York's landmark 1,454-foot Empire State Building, responsible for the largest over-the-air U.S. broadcast transmission site. It is the over-the-air transmission home for major NYC radio and TV stations, telecommunications providers, and city and federal essential communication services.

Shane works with every broadcaster and is responsible for all planning, construction, and maintenance requests for critical 24/7 operations. He is managing, planning, and implementing required transmission/antenna system changes for the FCC's DTV spectrum repack affecting TV and all radio, broadband, and telecommunications providers who must remain on air 24/7 during construction and reconfiguration.

The site is home to the largest FM broadcast combiner in the Western Hemisphere and the planned site of a larger integrated system of provisioned/concurrent broadcasting transmissions complying with FCC, OSHA, regulatory code, and safety requirements.

Prior to the Empire State Building, Shane was ODA Consulting President, providing technical solutions for broadcast, cable, and satellite networks, including ABC-TV, CBS-TV/Radio, Showtime, Sirius XM, and Sony. He was CBS Television Network VP (Technical Services and Quality Assurance), rising in the ranks from an engineering position. He automated and standardized the technical irregularities reporting system improving signal performance and maintained studio operations and technical broadcast infrastructure at the CBS Broadcast Center (NYC) for news, sports, and entertainment. He gained experience in engineering and management, as Manufacturing Engineer at Lambda Electronics, Chief Engineer and Quality Assurance Director at Vero Electronics, and Manager Quality Assurance at PCK Technology (Kollmorgan Corporation).

Greg Ogonowski (Chapter 7.3) relocated studio facilities and constructed a new efficient alternative use AM transmission facility as Technical Director for KTNQ/KLVE, Heftel Broadcasting, Los Angeles, from 1985 to 1991. Later becoming Technical Director for KBIG/KLAC, Los Angeles, from 1998 to 2000, he installed a new computer network and digital audio delivery system throughout the business and technical facility. KBIG/KLAC was one of the first radio stations to stream audio on the Internet with high fidelity sound, all from internal encoders and servers.

Greg joined Orban in 2000 as VP Product Development and led the team that created Optimod-PC, a digital audio signal processed PCI sound card for any digital audio or streaming application. This was the first economical professional audio processor to become available.

Destined to change the way Internet streaming audio is perceived and consumed, he was the architect of the first commercial high-quality file and streaming audio encoder using standards-based MPEG-4 AAC/HE-AAC, Orban Opticodec-PC. Continuing with the same goal, he also created a high-quality HE-AAC/xHE-AAC streaming player, supporting standards-based protocols, StreamS HiFi Radio App.

In 2015, Greg left Orban to concentrate on new streaming technologies, providing professional high-quality streaming encoder software and systems under the StreamS brand and Modulation Index, LLC. This includes HLSdirect(tm) and DASHdirect(tm), whereby streaming servers are no longer needed to deliver live streams. A standard web server or cloud storage may be used for streaming delivery, dramatically improving performance and economics.

Clients include many high-profile streaming content providers all over the world.

James E. O'Neal (Chapter 2.6) has been involved in broadcasting for more than half a century, beginning with part-time employment at a hometown radio station during his high school years, and continuing on that basis in both radio and television operations throughout his high school and college years. Following graduation from the University of Arkansas in 1969, he embarked on a full-time broadcast engineering career that spanned nearly 37 years, with more than three decades at U.S. Government television operations, including the United States Information Agency's Television and Film service, WORLDNET Television, and VOA-TV.

Following his retirement from broadcast engineering in 2005, James began a second career in writing and publishing, which continues to the present. He served for some 10 years as Technology Editor at *TV Technology* magazine, and is now a technical consultant to that publication. He is Editor-in-Chief of the IEEE's *Broadcast Technology* publication and serves as an associate editor of the SMPTE *Motion Imaging Journal*. He is a member of the SMPTE Board of Editors and the IEEE's Panel of Editors, and has been a presenter at SMPTE, IEEE, and NAB engineering conferences.

James enjoys researching and writing about broadcasting history and persons making important contributions in this field. He also enjoys collecting and restoring both consumer radio and television items and commercial broadcasting equipment. He holds both commercial and amateur radio licenses and has earned SBE Professional Broadcast Engineer Life Certification. James serves as a member of the board of directors of the National Capital Museum of Radio and Television and the Early Television Foundation.

Robert Orban (Chapter 7.2) received his BSEE from Princeton University in 1967 and MSEE from Stanford University in 1968. In 1970, he founded Orban Associates, which introduced the original Optimod-FM 8000 in 1975. This was the first of many broadcast audio processors for AM, FM, TV, and digital broadcasting.

Robert has been involved in professional recording for many years and has mixed several records released on the Warner Bros. label. A composer, his music has been heard on the air in New York, Chicago, and San Francisco. He has designed studio reverberators, stereo synthesizers, compressors, parametric equalizers, enhancers, and de-essers under both the Orban and dbx brand names.

Robert has been widely published in the trade and refereed press, including *Journal of the Audio Engineering Society*, *Journal of the SMPTE*, and *SAE Transactions*. He holds over 20 U.S. patents. In 1973, he was elected a Fellow of the Audio Engineering Society. In 1993, he shared with Dolby Laboratories a Scientific and Engineering Award from the AMPAS. In 1995, he received the NAB Radio Engineering Achievement Award. In 2002, he received the Innovator award from *Radio* magazine.

David Oxenford (Chapter 2.2) is a partner at the Washington DC–based law firm of Wilkinson Barker Knauer LLP. Wilkinson has a broad practice, representing broadcast and other communications companies on all of their regulatory needs.

David has represented broadcasters for over 35 years on a wide array of matters. His regulatory expertise includes all areas of broadcast law, including the FCC's multiple ownership limitations, the political broadcasting rules, EEO policy, advertising issues, and other programming matters and FCC technical rules.

His clients range from family-owned broadcasters to active mid-market venture-backed companies to one of the largest noncommercial radio licensees in the country. In addition to representing station groups, David represents a number of state broadcast associations, the trade association for media brokers, program and service providers to the broadcast and digital media industry, and banks and others providing financing to media companies.

David also represents digital music companies on copyright matters, particularly dealing with music rights issues.

David is a regular speaker at broadcasting and digital media conferences, conventions, and trade shows across the country, and regularly conducts webinars for many trade associations. He has also served as an expert witness on broadcasting and music licensing issues.

Karl Paulsen (Chapter 5.12) is Chief Technology Officer for Diversified, where he provides company-wide emerging technology advising and strategic planning focused on advanced broadcast and media systems. He directs Diversified's consulting and technology-driven broadcast engineering services with a focus on long-term planning, workflow analysis, media asset management, and storage technologies.

Karl is a Fellow of the Society of Motion Picture and Television Engineers (SMPTE), an officer of the SMPTE Pittsburgh Section, and participates in the SMPTE Standards Development processes. He is an SBE Life

Member, SBE Certified Professional Broadcast Engineer, and a member of the AES and IEEE.

Karl held positions of CTO, VP, and Director of Engineering for leading systems integration firms in the United States and Canada. As its Chief Engineer, he engineered the facilities and launched KTZZ-TV Channel 22, Seattle, in 1985, becoming VP Engineering for CGI/VXF firm Digital Post & Graphics in 1989. His projects and consulting activities as a systems integrator include the major networks, CNN, HBO, Turner, Discovery, IMF, and Al Jazeera and broadcast group television stations including Sinclair Broadcast, ACME Television, and Nexstar Broadcasting.

A past contributor to the *NAB Engineering Handbook*, Karl has published industry-related technical articles for SMPTE's *Motion Imaging Journal*, and for over 25 years, his continuing TV Technology series *Media Servers and Storage Technologies* has yielded over 250 articles on the state and future of media technologies for the digital age. Karl has authored two books: *Moving Media Storage Technologies* and *Video and Media Servers: Applications and Technology*.

Nick Piggott (Chapter 7.21) is the Project Director of the RadioDNS Project and an independent consultant. Nick has spent the majority of his career working in the radio industry, initially in programming and then moving to digital radio and online innovation. He created a much-awarded multi-skilled team that launched the world's first commercial digital radio stations, and then went on to consistently deliver technology-driven innovation for over a decade. Nick is one of the founders of RadioDNS Hybrid Radio, the open technology project that lets broadcast radio and IP work together, combining the power of broadcasting to reach many people and the power of the Internet to deliver enhanced or personalized content. In addition to leading the RadioDNS project, Nick works with broadcasters globally on developing strategies to benefit from change created by technology.

Joseph Pollet (Chapter 3.6) has been involved in virtually all aspects of radio broadcasting, including on-air operations, engineering, and station management for over 50 years. He founded The Traffic Channel, the nation's first video-based traffic information network, in 1985 and subsequently served as National Engineering Director for Shadow Traffic NYC from 1990 through 1995.

Joe is currently employed as Director of Technical Operations for the Entercom New Orleans stations (WWL-AM/FM, WLMG, WEZB, WKBU, WWWL and The New Orleans Saints Radio network). He is also Regional Technical Operations Director for Entercom's South Central markets (Austin, Memphis, New Orleans, and Wichita). At night and on weekends, Joe can frequently be found on the 20-meter ham band operating under Amateur Extra class call sign W5ABC.

Stephen Poole, CBRE-AMD (Chapter 4.2) is market chief engineer, Alabama, for Crawford Broadcasting Company (Kimtron, Inc.). He has more than 40 years of experience in radio broadcasting and has overseen the construction of many new studio facilities during that career, including the current facility for Kimtron's five primary stations in Alabama: WDJC-FM, WYDE AM and FM, and WXJC AM and FM. Stephen lives with his wife, Sandy, in Hayden, AL.

Stephen F. Pumple, M Eng, MBA, PMP (Chapter 3.3). Stephen is President at AZTECH-ENGINTL, a consulting and project management company working on the 2017 television spectrum repack and the implementation of ATSC 3.0 He is also a Distinguished Visiting Professor at Ryerson University on its Faculty of Engineering and Architectural Science, and CEIE Director in their Industrial Partnership Program. Stephan was the Founder of AZCAR Technologies, which provided engineering, design, and systems integration of large-scale, leading-edge television media projects for television, the Olympics, satellite, new media, and Internet facilities, both domestically and internationally.

Schuyler Quackenbush (Chapter 4.8) is an expert in digital audio technology, from capture and compression to transmission and presentation. He is active in the area of standardization of audio coding algorithms, is currently the chair of the International Standards Organization Motion Picture Experts Group (ISO/MPEG) Audio subgroup, and was one of the authors of the ISO/IEC MPEG Advanced Audio Coding standard. He is the author of more than 30 publications in these areas, including one book, *Objective Measures of Speech Quality*.

Schuyler received his BS from Princeton University in 1975. After four years in industry as a design engineer, he entered Georgia Institute of Technology, where he received an MS in electrical engineering in 1980, and a PhD in electrical engineering in 1985. For the latter half of 1985, he was a staff research associate at Georgia Tech. He joined AT&T Bell Laboratories in 1986 as Member of Technical Staff, where he was in the Digital Signal Processing Research department, and in 1996 he joined AT&T Laboratories as Principal Technical Staff Member, where he was in the Speech and Audio Research Department. In 2002, he founded Audio Research Labs, an audio technology consulting company. He is a Fellow of the Audio Engineering Society (AES) and a Senior Member of the Institute of Electrical and Electronics Engineers (IEEE).

Ronald D. Rackley, P.E. (Chapters 7.8, 7.10) is an electrical engineering graduate of Clemson University, where his studies were specialized in the area of electromagnetic fields. He worked as a radio station chief engineer and as a consultant during college. Following his formal education, he worked for two different engineering firms—Palmer A. Greer and Associates of Greenville, South Carolina, and Jules Cohen and Associates of Washington, DC—as well as a major manufacturer of antenna system equipment—Kintronic Laboratories of Bristol, Tennessee—before co-founding du-Treil-Rackley Consulting Engineers with Bob du

Treil in 1983. His present firm, du Treil, Lundin & Rackley, Inc., was formed in 1988 when his firm was combined with A.D. Ring & Associates, a firm that had played a pioneering role in the development of radio and television technology since its founding in 1941.

Ron has worked in AM radio engineering for over 46 years, and as a consultant specialized in the design, troubleshooting, and testing of broadcast antennas for over 43 of those years. He has been among the pioneers in developing modern computer modeling techniques for both antenna and RF network analysis, as used in system design and proofing of antenna radiation patterns.

Ron has been a member of the IEEE for over 40 years and has served as a board member and Vice President of the Broadcast Technology Society in the past. He has served as President of the Association of Federal Communications Consulting Engineers. He is a registered Professional Engineer.

Steven Ramer (Chapter 10.6) is President of SJ Ramer Associates in Warminster, PA, and provides television RF transmission equipment and installation services for clients nationwide. He served as Field Service Engineer and the System Test/Service Manager for Acrodyne Industries, was a television broadcast transmission equipment manufacturer between 1990 and 1997, and served as Field Service Engineer and Service Manager for Wang Laboratories Inc. between 1980 and 1990. Steven served in the U.S. Navy in the field of aircraft electrical and electronic systems from 1970 to 1980.

Rich Rarey (Chapter 4.4) is founder of Rareworks LLC, a broadcast and technology consultancy whose portfolio includes aural accessibility; broadcast facility vulnerability studies; private national emergency alerting systems; mobile high-resolution audio development; and public safety search and rescue projects involving small Unmanned Aircraft Systems, among other projects.

In his past work as Director of NPR Labs (at NPR), he was the architect and lead in the Accessible CAP RDS Demonstration for Gulf Coast States project, which received a National Association of Broadcasters 2011 Technology Innovation Award, and was a 2014 CES Innovations Design and Engineering Awards Honoree. He shares a patent for a novel method of real-time captioning and rapid transcription intended for broadcast radio, developed while at NPR Labs.

Prior to joining NPR Labs, Richard was a Technical Director of a number of NPR programs, created innovative technology for NPR Master Control, and created NPR's first Internet-based audio intake service.

His interest in broadcasting began when he spent his allowance on a used Webcor tape recorder at age 13, and earned his 3rd Class License with Broadcast Endorsement at age 14. Richard graduated cum laude from Ohio University with a BSC and, while at OU, was honored to receive the Edmund A. Williams Student Engineering Award for his work with the university's public radio and TV stations.

He holds the CEA and CBNT certifications from the Society of Broadcast Engineers, and is an FAA-certified private pilot and remote pilot.

Thomas R. Ray III (Chapter 7.1) started his career as a board operator for the Buckley Broadcasting–owned WDRC AM/FM in Hartford, CT, in 1977. He became Assistant Chief Engineer while in college. He has worked at several legendary stations as Chief Engineer or Director of Engineering, among them WTIC AM/FM and WOR, New York City.

He was VP/Corporate Director of Engineering for Buckley/WOR in New York City from 1997 to 2012. While at WOR, he built a new transmitter facility, moved the studios and created the first large-scale Axia studio facility in the United States, made WOR the first full-time high-power AM HD Radio station in the world, and made WOR one of the first high-power MDCL stations in the United States. He participated in much of the AM HD Radio testing with iBiquity and experimented with Harris with the MDCL technology.

He is author of *HD Radio Implementation*, published by Focal Press for SBE. He holds a degree in Communications Electronics from the University of Hartford, is a Certified Professional Broadcast Engineer through the Society of Broadcast Engineers, holding sub-certifications in AM Directionals and Digital Radio Broadcast, has served on the Board of Directors of SBE, and has served on the NAB Engineering Conference Committee, in addition to presenting during numerous NAB Engineering Conferences. He also holds an FCC General Class license and an Amateur Extra license. He is President of Tom Ray Broadcast Consulting and resides in New Windsor, NY.

Jeffrey Riedmiller (Chapter 5.18) is currently Vice President of the Sound Group in the Office of the CTO at Dolby Laboratories in San Francisco, where he leads a creative group of forward-thinking researchers and technology experts responsible for defining and driving the technical vision for Dolby's sound-related technologies and ecosystems. Jeffrey's professional career in audio, broadcast, and electronics engineering spans 27 years. Joining Dolby in 1998, he worked extensively on the design, development, and launch of multichannel audio for over 80 digital cable, terrestrial, and satellite television services throughout North America. He is also the creator of several technology and product innovations that have become the foundation of how television loudness is estimated and controlled worldwide today. Two well-known examples are the Dolby LM100 Broadcast Loudness Meter with Dialogue Intelligence and the DP600 Program Optimizer, which received multiple Emmy Awards for Outstanding Achievement in Engineering Development in 2004, 2009, and in 2011 respectively. Over the past several years, Jeffrey and his team, along with their close colleagues, drove the evolution of the Dolby Digital Plus (E-AC-3) format to fully support immersive audio as well as Dolby's latest audio codec and delivery system, Dolby AC-4. He is an active member of IEEE, ATSC, and SCTE and its Standards Committees. He holds several patents and has authored and presented several technical papers for the Audio Engineering Society (AES), National Association of Broadcasters (NAB), Institute of Electrical and Electronics Engineers (IEEE), National Cable Telecommunications Association (NCTA), and the Society of Motion Picture and Television Engineers (SMPTE).

J. Fred Riley, JD, MBA (Chapter 7.20) holds degrees in physics and mathematics, a master of business administration, and a juris doctor. A lifetime member of IEEE, he is also a member of the West Virginia Bar. He served as Chief Engineer of WKNA, WJLS, WBLW, and WNOE and has worked for both Gates Radio and Continental Electronics. During 1966–1969, he was a field radio operator in the U.S. Army, serving with 1st Special Forces Group (Special Action Force, Asia), and returned to broadcast engineering after finishing his undergraduate degrees after being discharged. While at Gates, he specialized on the high-power PDM transmitters, i.e., MW-50, VP-100, and SW-50 Transmitters. He has published articles on DRM, antenna-bandwidth optimization, and variable carrier power for trade magazines, amateur radio magazines, and manufacturer-sponsored white papers. He started to specialize in high-power High-Frequency Broadcasting at CEC in 1993 after designing two new-design 600 kW MW transmitters for Kuwait and two 300 kW MW transmitters for Taiwan. At CEC, he was responsible for major HF projects in Saipan, Tinian, Malaysia, Australia, Romania, Kuwait, Botswana, Spain, Costa Rica, Saudi Arabia, and Lebanon as either Program Manager or Program Engineer or both. He last commissioned four 300 kW HF transmitters, 6 X 7 Matrix, and four HF curtain antennas—all integrated into a computer-controlled system.

A ham radio operator since the age of 14, he received his Amateur Extra Class License in 1965 and has been an active experimenter with software-defined radios. And, like all HF operators, dreams of bigger and better antennas.

Idin Roshan (Chapter 5.5) has been refining the craft of motion and still imaging since 1987. As an Imaging Scientist, Idin has innovated industry processes that have reshaped electronic capture and colorimetric process. Referred to by imaging technicians across the United States and manufacturers of camera, lighting, and test instruments, the design input based on scientific analysis of image capture has been consulted by Tektronix, Grass Valley, Sony Corporation, Leader, Canon, Fujinon, and Arri, to name a few. Idin serves as Production consultant to a series of multinational media companies, including NBC Universal, Google, CBS Paramount, Nike, Sony Broadcast and Professional Group, AT&T, and Disney. The work, both technical and creative, has garnered numerous industry awards.

Stanley Salek, P.E. (Chapter 2.5) is a Senior Engineer in the consulting engineering firm of Hammett & Edison, Inc., Sonoma, California. Since joining the firm in 1991, he has managed numerous projects related to broadcast radio and television technology, including analog and digital radio and television transmission analysis, signal coverage evaluation, and RF safety compliance analysis and mitigation.

Prior to his present position, Stan was Director of Radio Engineering at the National Association of Broadcasters, a design engineer with RF and audio broadcast equipment manufacturers, and managed technical projects at radio broadcast stations. He is a graduate of Florida Institute of Technology, earning a BS in electrical engineering, and is a Registered Professional Engineer in the State of California. He holds a U.S. patent for an RF modulator design and FCC commercial and amateur radio licenses; is a member of the IEEE, AES, SMPTE, and AFCCE; and has authored articles and book chapters on various topics related to broadcast engineering.

In his capacity as co-chair of the AM and FM Analog Broadcasting Subcommittee of the National Radio Systems Committee, Stan works on the development and enhancement of industry-wide technical standards.

David Sanderford (Chapter 8.1) is the Co-owner/Vice President of Marsand, Inc. He is a graduate of the Georgia Institute of Technology (Georgia Tech '93 Suma Cum Laude) with a Bachelor of Science in electrical engineering. He is a Microsoft Certified Systems Engineer (MCSE) and a member of the Society of Broadcast Engineers (SBE).

David leads research and development at Marsand, where he is involved in system design and specification. He oversees the mechanical and electrical installations, acting as project manager, but also performs "hands-on" work to deliver a quality product based upon years of experience. Over the years, David has worked with Matthew Sanderford to establish internal procedures for engineering solutions to broadcast projects, allowing a small firm to handle multiple, concurrent digital rollouts for broadcast clients.

Matthew A. Sanderford, P.E., CPBE (Chapter 8.1) graduated from the University of Texas El Paso with a Bachelor of Science in electrical engineering. He is a Licensed Professional Engineer in the State of Texas and the President of Marsand, Inc., a Registered Professional Corporation in the State of Texas. He is certified by the SBE as CPBE (Certified Professional Broadcast Engineer). Matthew served in the TARNG as Staff Sergeant E6 with Honorable Discharge, and worked as electronic technician at White Sands Missile Range and Globe Universal Sciences in the R & D department until 1972.

He began his broadcasting career in radio in 1967 and in television in 1974. As Director of Engineering for multiple TV station groups from 1978 to 1987, he designed and built, hands on, studio and transmitter facilities for 35 full-powered television stations, primarily UHF. Matthew is proficient in AutoCAD and VidCAD, and has rebuilt, replaced, and proofed over 100 full power transmitters, both in analog and in the conversion to digital. He has also installed LPTVs, DTLPTVs, FMs, and DFMs.

The consulting firm of Marsand was established in 1981 and is heavily involved in the continuing digital transition for television, replacing, constructing new, or adding on to TV transmitter facilities. Matthew is schooled and authorized for installation and troubleshooting of all the major TV transmitter manufacturers, both IOT and Solid State. He has undertaken extensive manufacturer's training for troubleshooting and optimizing of antennas, transmission line and filter systems (fixed and tunable). He is the author of three technical papers presented at NAB, is an active member of AFCCE, NAB, IEEE, NSPE, SBE, and TAB, and currently resides in Alvarado, Texas.

John L. Schadler (Chapter 10.8) has been with Dielectric for 30 years and is currently the Vice President of Engineering. After receiving his BS in electrical engineering from Penn State University, he began working for Dielectric Communications in New Jersey, formerly the RCA Antenna Division. He then obtained his Master's degree in electrical engineering from Drexel University. His research and development achievements have been focused on unique TV, MobileMedia, FM, cellular, wireless, and many other special bi-directional and broadcast antenna designs.

John has 44 issued and multiple pending patents. In 2005, he was awarded the distinction of joining the ranks as one of SPX's Master Inventors.

Jack S. Sellmeyer, P.E. (Chapter 10.3) worked for seven years as the Senior Engineer in the Broadcast Department at Collins Radio Company in Dallas, TX, where his responsibilities included managing cost reduction projects for the existing one kilowatt AM Transmitter and two kilowatt FM transmitter, and providing engineering support for the rest of the broadcast product line. He headed a small group that included a mechanical engineer, mechanical draftsman, and a P.C. board designer. They reviewed the reasons for unreasonably high costs throughout the line. In most cases, significant cost reductions were obtained by reduction of labor-intensive assembly operations and replacement of certain high-cost Mil Spec parts, where commercial specs would not affect reliability or product quality. During this period, he designed and new FM Exciter, which reduced the cost by more than 50 percent, and went on to establish all-time sales records for the company. He also oversaw the design of three new audio consoles.

For the past 35 years, Jack has been a Consulting Engineer in private practice and has performed AM & FM allocation studies for clients to allow improvement of AM and FM facilities. He has provided optimization services for existing and proposed AM and FM facilities, including significant power increases in many cases, and detailed facility design for Studio-Transmitter Plant and Directional Antenna Systems, including supervision of plant construction for many facilities.

Paul Shulins (Chapters 6.5, 7.1) has been the Director of Technical Operations for Greater Media's Boston, Massachusetts, five FM radio stations for the past 28 years. Paul has been working as a Radio Chief Engineer for 38 years, and has been responsible for the planning and day-to-day management of one of the first major market studio consolidations in 1998 for Greater Media's Boston radio stations. He has designed and constructed many custom, on-air program playback systems, technical facility monitoring, and control systems. He is a member of the Society of Broadcast Engineers, and has held an SBE Radio Engineer Certification since 1982. Paul served on the National Association of Broadcasters' Broadcast Engineering Conference Committee in 2006 and 2007, and is also a member if the IEEE Broadcast Technology Society and an elected member of the BTS AdCom. He is currently serving as a member of NAB's Digital Radio Committee, and served two years on the NAB's Broadcast Engineering Conference Planning Committee. Paul has a long history of contributing papers to the NAB Broadcast Engineering Conference at NAB, as well as the NAB Radio Show, and is authoring several chapters in the 11th Edition of the *NAB Engineering Handbook*. In 2015, he became an associate member of AFCCE.

He earned a Bachelor of Arts from the University System of New Hampshire in 1979 with majors in physics, chemistry, and natural sciences. He is a private pilot, enjoys astronomy, skiing, and photography, and is a newly licensed ham radio operator.

Thomas B. Silliman, P.E. (Chapter 7.17) attended Cornell University, graduating in 1969 with a Bachelor's degree in electrical engineering, and received his Master's degree in electrical engineering from Cornell in 1970. After graduating, Tom went to work at Electronics Research as Director of Engineering. He is currently the President of Electronics Research. He is a registered Professional Engineer in the field of electrical engineering, and is currently registered in three states. He is a two-term past president of the Association of Federal Communications Consulting Engineers (AFCCE), the current secretary of the IEEE Broadcast Technology Society, and a board member for Tri-State Public Radio and TV in Evansville, Indiana. Tom is also the 2008 NAB Engineering Achievement Award recipient for Radio. He is a Senior Fellow of the IEEE (Institute of Electrical and Electronic Engineers) and was awarded the 2015 Broadcast Technology Society Jules Cohen award for Outstanding Broadcast Engineering.

Johannes Sinnhuber (Chapters 8.4) has a diploma degree in communications engineering at TU Munich. He also has a graduate degree in business administration at TU Munich and an Exchange Fellowship Award at National University of Singapore. He worked on a project on optical communication systems at Nokia Siemens Networks. Since 2013, he has been the Product Manager at R&S for liquid-cooled high-power transmitters and for ATSC 3.0 solutions.

Andre J. Skalina (Chapter 8.7) is one of the experts in TV and FM antenna design, specializing in the development of complex broadcast antenna systems, which include many stations on multiple antennas at a single location. He has more than 39 years of experience in the field.

Andre obtained a Bachelor's degree in electrical engineering and a Master's degree in electrical engineering at McGill University in Montreal. After graduation, he started working in the R&D group at RCA Broadcast Antenna Division in Gibbsboro, NJ, where he participated in the design of circularly polarized TV broadcast antennas and developed many new antenna products for U.S. and international markets. When he moved to Maine, he managed an engineering group at Dielectric, where he was instrumental in designing new antennas and improving engineering processes to increase the production from several antennas a year to about 250 a year. He was promoted to the Vice President of R&D and later to the Vice President of Technology, a position he has held for five years. Currently, he is concentrating on complex antenna system designs required for New York City.

He holds 17 patents, has appeared on many technical panels, and has some publications. He is an associate member of AFCCE and a member of AFCEE Scholarship Committee. Andre is a Life Member of IEEE and a member of the IEEE Broadcast Technology Society and the IEEE Antenna and Propagation Society.

Sidney M. Skjei (Chapter 6.3) is President of Skjei Telecom, a mid-sized independent technical consulting company providing engineering and operational support for telecommunications, satellite, and broadcasting applications. Since its founding in 1994, Skjei Telecom has provided consulting and support services to over 100 companies or organizations in a broad spectrum of industries related to telecommunications, broadcasting, or satellite communications. Skjei Telecom is a completely independent consulting services provider and does not sell or market hardware or software products. Consequently, Skjei Telecom frequently finds itself working for end users such as broadcasters, and serves the commercial, federal, and state government markets.

Prior to founding Skjei Telecom, Sidney held executive- or senior-level engineering and management positions with GTE, Southern Pacific Satellite Company (SPRINT), and COMSAT World Systems Division. He has over 40 years' experience in engineering and developing a wide range of hardware and software telecommunications products, systems, and services and in providing engineering support to the design, procurement, operation, and marketing of a large fleet of communications satellites.

Author of numerous publications on telecommunications and broadcasting, Sidney holds a BS from the U.S. Naval Academy with Merit and an MSEE with Distinction from the Naval Postgraduate School. He is a registered professional engineer in the Commonwealth of Virginia.

Stephen F. Smith (Chapter 7.11) was a Senior Development Staff Member and Director of the Position, Navigation, and Timing (PNT) Program at the Oak Ridge National Laboratory in Oak Ridge, Tennessee, until his retirement on January 1, 2014. He is currently serving as a consultant to the communications and broadcast industries. His areas of interest include: navigation and timing; instrumentation and control systems; wireless communications devices and networks; electrical systems diagnostics; analog and digital signal processing; broadcasting systems; MEMS devices; and advanced sensor systems. He holds a BS, an MS, and a PhD in electrical engineering from the University of Tennessee and has served as an adjunct faculty member in the Electrical & Computer Engineering Department there. He has over 60 publications and 30 years' experience in AM/FM/TV broadcast engineering, and holds 40 U.S. and 29 international patents in the areas of communications and signal processing. He has also received two NASA Technology Utilization Awards for developments in the satellite communications field. He has been a member of IEEE, ISA, ION, and SBE and has served on the IEEE 1451.3 (Vice Chair), 1451.5, and 1588 Standards Committees.

James Snyder (Chapter 11.3) is an Emmy Award–winning digital media engineering, data and media archiving, preservation, production, and project management specialist. His 37 years' experience includes television, film, radio, Internet, and data technologies from traditional analog to cutting-edge digital audio, video, and data technologies. His career spans the commercial, non-commercial, and government sectors. He was awarded a 2007 Technology Emmy for his work on the ATSC digital television standard.

James serves as the Senior Systems Administrator for the Library of Congress' National Audio-Visual Conservation Center (NAVCC) on the Packard Campus for Audio Visual Conservation in Culpeper, Virginia (www.loc.gov/avconservation/packard/). His responsibilities include all audio, video, and film preservation and digitization technologies, including long-term planning and implementation, long-term data preservation planning and implementation, technology services to the United States Congress and organizations on Capitol Hill, as well as standards participation and technology liaison with media content producers worldwide.

He has worked for many organizations in media, entertainment, education, engineering, and communications, including MCI, Verizon, Intelsat, PBS, the ABC Radio and Television networks, Harris Corporation, the Advanced Television Test Center, Fox News, Communications Engineering Inc., Reuters, and Discovery Communications. He has consulted on many types of projects for organizations, including Sarnoff Corporation, Turner Engineering, CBS, NBC, ABC, Fox, the News Corporation, FedNet, and multiple agencies of the U.S. federal government. He has worked on key projects in the creation and standardization of the ATSC digital television standard, the HD Radio digital radio standard, the AXF Archive eXchange Format (SMPTE 2034) standard, as well as the first consumer commercial HDTV satellite service Unity Motion. He has taught courses on analog and digital audio, video, television transmission, and engineering basics for the SMPTE, the Harris/PBS DTV Express, PBS, and the U.S. Technical Training Institute (USTTI), as well as many public and private organizations.

He is a member of the Audio Engineering Society (AES), the Institute of Electrical and Electronic Engineers (IEEE), the International Association of Audiovisual and Sound Archives (IASA), the Association of Motion Imaging Archivists (AMIA), and the National Academy of Television Arts and Sciences (NATAS). He is a member and serves as an officer and on standards committees of the Society of Motion Picture and Television Engineers (SMPTE). He serves on technical committees of the AMWA (Advanced Media Workflow Association) and the Academy of Motion Picture Arts and Sciences. He lives and works in the Civil War battlefield country of central Virginia between Washington, DC, and Richmond, VA.

John O. Stanley (Chapters 7.5, 7.20) graduated from MIT in 1962 in electrical engineering before taking seven years of graduate studies in theology, languages, and cross-cultural studies while working at General Radio Company and Texas Instruments. John has taught at the Vanguard Institute of Electronics, Nave Technical Institute in Shahjahanpur, India, ESPOL in Guayaquil Ecuador, John Brown University, and LeTourneau University.

Mainly specializing in short wave broadcasting, he worked for HCJB in Quito, Ecuador, as Chief Engineer at the Pifo transmitter site and as Engineering Director. He has trained staff, installed and repaired equipment, and consulted with FEBC, TWR, Christian Voice, Voice of Hope, and many smaller Christian Broadcasters. He has consulted with Kintronic labs on propagation, transmitters, and antennas for short wave.

John has been an active radio amateur for over 60 years with the call sign of K4ERO. He has operated under many other call signs from various foreign countries. John has been a technical adviser for the ARRL for 25 years and his writings have appeared in many of the league publications.

John's wife Ruth (WB4LUA) got her amateur license in a class which John taught in 1968 and has been his editor, technical assistant, and companion ever since. They have one son, Andrew, who trains dogs in the Atlanta area.

Jim Starzynski (Chapter 5.14) is Director and Principal Audio Engineer for NBCUniversal Advanced Engineering, overseeing audio technologies and practices for all NBCUniversal television properties. This includes the company's two broadcast networks, two national multicast networks, 28 owned stations, and 16 cable channels. Jim is responsible for establishing NBC's audio strategy for Digital Television.

Jim is chairperson of Advanced Television Systems Committee's technical subgroup S34–2 on Next Generation Audio and chairs S6–3, the group that developed ATSC-A/85, the Emmy Award–winning Recommended Practice on Loudness for Digital Television. He is chair of the Sports Video Group's DTV Audio Group, chairs the AES Group for On-Line Loudness Guidelines, and is on the board of directors for the Audio Division of the Consumer Technology Association. He is Technical Committee Co-Chair for the North American Broadcasters Association, is an SMPTE Fellow and a member of AES. Jim gave expert testimony to the 2009 U.S. Congress on the Commercial Advertising Loudness Mitigation Act (C.A.L.M. Act). He holds four NATAS Emmy Awards for technical achievement for multiple Olympic broadcasts and received the 2006 NATAS George Wensel Innovative Technical Achievement Award. Jim is the 2011 recipient of the ATSC's highest technical honor, the Bernard J. Lechner Outstanding Contributor Award.

Robert A. Surette (Chapter 7.18) was Director of Sales Engineering with Shively Labs, a Division of Howell Laboratories, from 1981 to July 2015. He graduated from Lowell Technological Institute, Lowell, Massachusetts, in 1973 with a Bachelor of Science in electrical engineering. He has been directly involved with the design and development of broadcast antennas, filter systems, and RF transmission components since 1974, as an RF Engineer for six years with the original Shively Labs in Raymond, ME, and for a short period with Dielectric Communications.

He is currently a Life Senior Member of IEEE. He has authored a chapter on filters and combining systems for the latest edition of the *CRC Electronics Handbook* and for the 9th and 10th editions of the *NAB Engineering Handbook*.

He took early retirement in July 2015 from Shively Labs and took a one-year position in September 2015 at Bridgton Academy teaching algebra II and trigonometry, a bucket list item.

Norman R. Swan, PhD (Chapter 3.1) began his broadcasting career in 1967 as a radio announcer at KFVS-AM in Cape Girardeau, Missouri. After work in radio production in Sri Lanka and at the University of Missouri, Norman completed a PhD at the latter in 1978. He then led a group of radio and television writers and producers at the University of Minnesota, and in 1981 was named Head of the Department of Radio-Television at Southern Illinois University in Carbondale. In 1985, he became Head of the Department of Broadcasting at the University of Tennessee in Knoxville, also serving as General Manager of WUTK-FM and WUTK-AM. He continues to produce a weekly news magazine program for WBIR-TV in Knoxville, and teaches television news courses at the University of Tennessee, where he also directs the college's International and Outreach Programs.

Since 1996, Norman has also served as a media trainer and consultant to hundreds of radio and television stations around the world. He has conducted workshops in over 50 countries on broadcast journalism and management for the Voice of America, Radio Free Europe/Radio Liberty, IREX, and the U.S. State Department and other agencies. He writes a monthly newsletter for Voice of America and RFE/RL, and has published a book on broadcast management distributed by the International Broadcasting Bureau. He has received Fulbright Senior Specialist assignments to teach at universities in Croatia and Ethiopia, and received the Ed Bliss Award for Distinguished Broadcast Journalism Education from the Association of Education in Journalism and Mass Communication in 2009.

Joe Talbot (Chapter 4.5) is the Telephony Product Manager for the Telos Alliance in Cleveland, Ohio.

He has been maintaining, building, and managing radio stations since 1978. Joe started out at KWIZ in Santa Ana, California, and provided contract technical services through the late 80s and into the 90s. In 2001, he was recalled to Naval military service for three years in response to the 9/11 emergency. During his active duty service period, he was in charge of underwater search and communications and became involved in several Navy R&D programs. After his service ended, he was hired by ABC Radio in San Francisco as Director of Engineering. Joe is a specialist in the design and implementation of telephone systems for talk radio, news, and music-oriented radio stations. He resides in Southern Nevada near Las Vegas.

Gary E. Timm (Chapter 2.10) has served as the Broadcast Chair of the Wisconsin EAS Committee (FCC State Emergency Communications Committee) for the past 24 years. He retired from broadcast engineering in 2010 after 37 years of service at the Journal Broadcast Group in Milwaukee. For the next four-and-a-half years, he worked part time as an alerts and warnings consultant with Washington DC–based SRA International, where his work supported the Department of Homeland Security. He is presently retired, while continuing to devote time to EAS volunteer efforts such as membership on the FCC's Communications Security, Reliability and Interoperability Council (CSRIC) and emergency-alerting advocacy with other EAS experts through the Broadcast Warning Working Group (BWWG).

Doug Vernier (Chapter 1.2) is president of V-Soft Communications and senior consulting engineer at Doug Vernier Telecommunications Consultants. V-Soft is a broadcast engineering software development company supporting radio and TV consulting professionals, the FCC, and station engineers. He served for seven years as the principal consulting engineer for the Corporation for Public Broadcasting's digital radio conversion program, and has provided his services as a broadcast consulting engineering for clients all across the United States while authoring numerous papers, including the chapter on radio propagation in the 10th edition of the *NAB Engineering Handbook*. Doug served two terms on the Board of Directors of National Public Radio and is a past president of Public Radio in Mid America. He is a founding member of the Board of Directors of the Association of Public Radio Engineers (APRE). In 2011, he was awarded the APRE Engineering Achievement award of the year. He is recognized by the Society of Broadcast engineers as a "lifetime" Professional Broadcast Engineer. Doug retired from his long-time position as Director of Broadcasting Services at the University of Northern Iowa. He has degrees from the University of Michigan, where he studied engineering.

John Wahba, PhD, P Eng, P.E. (Chapter 9.1) is Turris's CEO and principal engineer and is recognized internationally for his tower engineering—a reputation earned through 25 years of experience in structural engineering at the design and research level. John has designed some the world's most complex towers—the most recent being the "hybrid" in Jakarta, Indonesia. He has co-authored several publications in the field of dynamic and static analyses of guyed towers. John holds engineering licenses in over 40 jurisdictions across North America and is/has been active on many technical committees (CSA, TIA, ASCE dynamics of latticed structures), and the CSA Technical committee on Communication Towers and Wind Energy.

Eric R. Wandel, P.E. (Chapter 7.17) is the principal engineer and President of Wavepoint Research, Inc. where he provides RF and communications engineering services to the broadcast industry as well as to other communications providers. In addition, he conducts research and provides engineering services for military customers in efforts related to antenna development, radar system optimization and electronic warfare signal processing. In addition, he conducts studies related to antenna design and development, RF radiation hazard assessment, and a variety of applications requiring computational electromagnetic (CEM) modeling, including antenna pattern and EM optimization studies. Eric received dual Bachelors' Degrees in electrical engineering and optics from Rose-Hulman Institute of Technology and a Masters' Degree in electrical engineering from Rensselaer Polytechnic Institute. He is a Registered Professional Engineer in Indiana. His prior corporate experience includes systems engineering and product development positions at GE Aerospace, Lockheed Martin and Electronics Research, Inc. (ERI). He has used a variety of computational electromagnetic tools such as custom Matlab and C code, Method of Moments software (NEC-3, NEC-4), NEC-BSC and finite element analysis tools such as HFSS, CST and FEKO. Eric is a Member and Past-President of the Association of Federal Communications Consulting Engineers (AFCCE). He is a Senior Member of the IEEE and currently serves on the IEEE Broadcast Technology Society (BTS) Administrative Committee representing BTS as a liaison to the International Committee on Electromagnetic Safety (ICES) and the IEEE Committee on Man and Radiation (COMAR).

William P. Weeks (Chapter 10.5) has been engaged full time in broadcast engineering for about 45 years. The most fun part of that, of course, is trying to keep up with the advances in technology—or at least trying to keep close enough to see them in the distance. Most recently, he has been working on various freelance projects as the proprietor of Hungry Wolf Electronics.

S. Merrill Weiss (Chapter 8.6) is a consultant in electronic media technology, technology management, and management. He has spent his five-decade career designing and building systems for broadcast and related-industry entities. During most of that time, he also has worked on developing many of the technologies that underlie the digital television systems currently being implemented. He has led several development efforts and tests and has chaired many committees that have prepared resulting standards for the Society of Motion Picture and Television Engineers (SMPTE), the FCC Advisory Committee on Advanced Television Service (ACATS), and the Advanced Television Systems Committee (ATSC).

One technology that Merrill has pursued for over 25 years is the use of multiple transmitters by television broadcasters to cover their service areas. Such service can be provided by distributed transmitters, distributed translators, and digital on-channel repeaters, or combinations of them. The use of multiple transmitters requires that the transmitters be synchronized with one another, but it was at one time believed that it was not possible with the ATSC 8-VSB system (now known as ATSC 1.0) to synchronize transmitters for distributed transmitter and distributed translator applications. To make such uses practicable, Merrill invented technology that allows synchronization of 8-VSB modulation. That technology was adopted and standardized by ATSC for both ATSC 1.0 and ATSC 3.0, preparation of the documentation for both of which he has chaired.

He is a Fellow of SMPTE and received its 1995 Sarnoff Gold Medal Award and its 2005 Progress Medal. He was the 2006 recipient of the NAB Television Engineering Achievement Award and the 2012 recipient of the ATSC Bernard J. Lechner Outstanding Contributor Award. He won the IEEE Broadcast Technology Society Matti S. Siukola award in both 2012 and 2013. He is certified by the Society of Broadcast Engineers at the level of Certified Professional Broadcast Engineer (CPBE), having been certified at SBE's highest level since 1980. He was nominated for an Emmy Award by the National Academy of Television Arts and Sciences (NATAS) in 1993 for his work on the concept for the Serial Digital Interface (SDI) for television. He is a graduate of the Wharton School of the University of Pennsylvania and holds four U.S. and two international patents on terrestrial broadcast transmission technology.

Robert D. Weller, P.E. (Chapter 3.5) is Vice President for Spectrum Policy at the National Association of Broadcasters, where he is responsible for developing strategies to protect broadcast spectrum. Prior to joining NAB, he was Chief of Technical Analysis at the Federal Communications Commission and was responsible for developing mathematical models relating to propagation, RF Safety, frequency allotments, and other areas. Robert is a graduate of the University of California, Berkeley, and of The George Washington University, where he majored in electromagnetics.

He has split his 30-year career between the FCC and private sector, holding a number of technical and management positions at the FCC and working as consulting engineer to the telecommunications industry. Robert is a Registered Professional Engineer, a Senior Member of the IEEE, a Member of Subcommittee 4 (RF) of the International Committee on Electromagnetic Safety, the IEEE Committee on Man and Radiation, the Bioelectromagnetics Society, and a past president of the AFCCE. He has written extensively on signals analysis, interference mitigation, RF exposure mitigation, and various regulatory matters.

Jeff Welton (Chapters 7.5, 7.15, 11.2) took his training in the Radio College of Canada (RCC) Electronics Engineering Technologist program, finishing in 1984. Jeff has performed component level repair, field installation and service, technical support and quality assurance with various companies and has been with Nautel for over 26 years, the first 17 of which were spent in field service and technical support positions, as well as assisting Engineering with design review of new products and improvement of existing systems. Since moving to Sales in 2007, Jeff keeps finding ways to get his hands dirty and can frequently be found assisting in the install of a transmitter he's sold, as well as performing site inspections and the occasional repair. Jeff performs several presentations every year on the topics of lightning protection, grounding, transmitter site safety, and various other subjects of interest in the broadcast engineering field.

John N. West (Chapter 11.1) is recognized internationally as a consultant, speaker, and trainer. He has conducted hundreds of seminars nationally and internationally with thousands of attendees. His military clients include USAF Intel & Recon Command, U.S. Military all services and commands, Guam USA. He has trained technicians and engineers from the Bahamas Electric Corp., Bahamas Telephone Co., Bahamian Ministry of Labor, and Ministry of Insurance in Nassau, Bahamas, and he has held training seminars for Siemens, Simplex Grinnell, Diebold, AT&T, ADT, and hundreds of electricians. John has spoken to and trained members of the National Association of Broadcasters and attendees at Electric East, NFPA, National Fuel Gas Code, Plumbing Code, Americas Security Expo, Annual Copper Industry Meeting, and Florida Alarm Association. He is a member of the ATIS-Protection Engineers Group.

John's consulting clients represent industry, government agencies ranging from agricultural to the Secret Service. His commercial consulting clients include companies in the copper industry, gaming industry, and computer, financial, and broadcast facilities.

He is a proponent of high performance vs. code/standard minimum electrical systems including their layout and design, grounding, and bonding and lightning protection installations. He employs a proven approach to power quality and electrical system design and has applied this approach at hundreds of client sites. The ROI of his designs is significant, forward looking, and sustainable over the long term.

Peter Wharton (Chapter 5.4) is BroadStream's Vice President for Technology and Business Development.

In his role at BroadStream, he is responsible for the company's technical and strategic direction.

Prior to joining BroadStream, Peter was at Grass Valley, formerly Miranda Technologies, where he held a wide range of roles in sales, business development, strategic account management, product management, and engineering. While at Miranda, he played an instrumental role in the centralization initiatives of several station groups, as well as the design and implementation of numerous specialty channel and TV network playout facilities.

Prior to Grass Valley, Peter held engineering and management roles with PanAmSat, Fox News Channel, and ABC Radio and Television Networks. He currently serves as the Secretary-Treasurer of SMPTE and has previously served as the Eastern Regional Governor and as a past manager of the Washington DC SMPTE section, where he has produced for almost 20 years the annual "Bits By the Bay" two-day SMPTE technology conference in the Washington DC area.

He has presented at numerous SMPTE, SBE, and NAB conferences.

Jerry Whitaker (Chapter 5.1) is Vice President of Standards Development at the Advanced Television Systems Committee (ATSC). He supports the work of the various ATSC technology and specialist groups and assists in the development of ATSC Standards and related documents. He currently serves as Secretary of the Technology and Standards Group and Secretary of the Technology Group on Next Generation Broadcast Television, and is closely involved in work relating to educational programs.

Jerry is a Fellow the Society of Broadcast Engineers and a Fellow of the Society of Motion Picture and Television Engineers. He has served as a Board member and Vice President of the Society of Broadcast Engineers.

He is the author and editor of more than 35 books on technical topics, including *The Standard Handbook of Video and Television Engineering*, 4th ed., *NAB Engineering Handbook*, 9th ed., *DTV Handbook*, 3rd ed., and *The Electronics Handbook*, 2nd ed. Prior to joining the ATSC, Jerry headed the publishing company Technical Press, based in Morgan Hill, CA.

He was previously Editor, Editorial Director, and Associate Publisher of *Broadcast Engineering* magazine and *Video Systems* magazine.

Loren White (Chapter 2.1) is the Director of Research at the National Association of Broadcasters (NAB). Prior to joining NAB, he was employed in academia as a lecturer at San Jose State University and later as a research interviewer for the University of Chicago's National Opinion Research Center. Loren also has a research and policy background, working as a research consultant for New America Foundation and a policy analyst for multiple Middle East news networks. He has authored op-eds and blogs featured on Huffington Post, Al Jazeera English and Foreign Policy's Mid-East Channel. Loren co-authored and collaborated on multiple policy briefs for policy institutes like the Atlantic Council, New America Foundation, and the Institute for Social Policy and Understanding. Loren holds a BA and MA in philosophy from San Jose State University and a Master Certification in international affairs from Texas A&M University.

Lorna White (Chapter 4.6) is the Technical Director for NPR's "Wait, Wait. . . Don't Tell Me!" She has been involved with the program since it began in 1998, and is part of the team that won a 2007 George Foster Peabody award. She is responsible for the logistics and overall sound of the show, which tours around the United States. Previously Lorna was the NPR Chicago Bureau Engineer from 1996 to 2008. She has traveled throughout the Midwest, recording news and feature stories for NPR. She came to NPR in 1984 because of her love of radio and provided technical support to "Morning Edition," "All Things Considered," "Weekend Edition," and "Performance Today." During her time in Washington, DC, she also operated NPR's Master Control. A graduate of Illinois State University, Lorna began her radio career at station WGLT while a student, where she announced classical music programs and recorded music events.

Phillip Young (Chapter 9.4) is currently the Corporate Director of Quality Assurance for Electronics Research, Inc. He has over 10 years of experience in total quality management systems, manufacturing management, engineering and engineering management, materials management, and technical management positions. He joined ERI in 2009 as RF Engineer and assumed his current position in 2012. He is a graduate of the University of Southern Indiana with a degree in electrical engineering. He is a member of the American Society for Quality (ASQ), the American Welding Society (AWS), and the American Institute of Steel Construction (AISC).

Authors whose material was used from the 10th edition and not listed above are:

Michael Bergeron (Chapter 5.5)

Ralph S. Blackman (Chapter 5.2)

Kevin Entrekin (Chapter 5.16)

Kari Eythorsson (Chapter 5.19)

Ty Ford (Chapter 4.3)

Ernest M. Hickin (Chapter 6.4)

Randall Hoffner (Chapter 2.8)

Thomas Kite (Chapter 10.1)

Edward A. Lobnitz (Chapter 9.2)

Peter Long (Chapter 10.4)

Steve Mahrer (Chapter 5.5)

Bill Marshall (Chapter 5.3)

David Mathew (Chapter 10.1)

Chip Morgan (Chapter 2.8)

John Norgard (Chapter 1.1)

Nicholas Paulin (Chapter 9.4)

James H. Rooney III (Chapter 6.4)

Richard Rudman (Chapter 3.6)

Shively Labs (Chapter 7.18)

Stanley Salek, P.E. (Chapter 10.1)

Greg Silsby (Chapter 4.3)

Peter Symes (Chapter 5.11)

John F. Warner (Chapter 7.12)

Scott A. Wright (Chapter 7.16)

Fred Wylie (Chapter 4.8)

NAB Engineering Achievement Award Recipients

NAB ENGINEERING ACHIEVEMENT AWARD WINNERS 1959–2017

2017

RADIO
John Kean, Cavell, Mertz & Associates, Inc., Manassas, VA.

TELEVISION
John Lyons, The Durst Organization, New York, N.Y.

2016

RADIO
Andrew Laird, Journal Broadcast Group, Milwaukee, Wis.

TELEVISION
Richard Chernock, Triveni Digital, Princeton, N.J.

2015

RADIO
Thomas F. King, Kintronic Labs, Bristol, Tenn.

TELEVISION
Richard Friedel, Fox Networks Engineering and Operations, Los Angeles, Calif.

SERVICE TO BROADCAST ENGINEERING
Ray Conover, Hubbard Broadcasting, St. Paul, Minn.

2014

RADIO
Jeff Littlejohn, Clear Channel Media and Entertainment, Cincinnati, Ohio

TELEVISION
Robert Seidel, CBS, New York, N.Y.

2013

RADIO
Frank Foti, Telos Alliance, Cleveland, Ohio

TELEVISION
Jay Adrick, GatesAir, Mason, Ohio

SERVICE TO BROADCAST ENGINEERING
Leonard J. Charles, Morgan Murphy Media, Madison, Wisc.

2012

RADIO
Paul Brenner, Emmis Communications, Indianapolis, Ind.

TELEVISION
Glenn Reitmeier, NBCUniversal, New York, N.Y.

2011

RADIO
L. Robert du Treil, du Treil, Lundin & Rackley, Inc., Sarasota, Fla.

TELEVISION
Thomas B. Keller, Renowned Television Technology Innovator

2010

RADIO
Steve Church, Telos Systems, Cleveland, Ohio

TELEVISION
Mark Richer, ATSC, Washington, D.C.

2009

RADIO
Jack Sellmeyer, Sellmeyer Engineering, Lucas, Texas

TELEVISION
Sterling Davis, Cox Broadcasting, Atlanta, Ga.

2008

RADIO
Thomas B. Silliman, Electronics Research, Inc., Chandler, Ind.

TELEVISION
Antoon Uyttendaele, ABC Inc., New York, N.Y.

2007

RADIO
Louis A. King, Kintronic Laboratories, Inc., Bristol, Tenn.

TELEVISION
Victor Tawil, Association for Maximum Service Television, Washington, D.C.

2006

RADIO
Benjamin Dawson, Hatfield and Dawson, Seattle, Wash.;
Ronald Rackley, du Treil, Lundin and Rackley, Inc., Sarasota Fla.

TELEVISION
S. Merrill Weiss, Merrill Weiss Group LLC, Metuchen, N.J.

2005

RADIO
Milford Smith, Greater Media, Inc., East Brunswick, N.J.

TELEVISION
Oded Bendov, TV Transmission Antenna Group, Inc., Cherry Hill, N.J.

2004

RADIO
Glynn Walden, CBS Radio, Philadelphia, Pa.

TELEVISION
Ira Goldstone, Tribune Broadcasting, Los Angeles, Calif.

2003

RADIO
John W. Reiser, Federal Communications Commission (Retired), Mt. Vernon, Va.

TELEVISION
Robert P. Eckert, Federal Communications Commission, Washington, D.C.

2002

RADIO
Paul Schafer, Schafer International, Bonita, Calif.

TELEVISION
Bernard Lechner, Consultant, Princeton, N.J.

2001

RADIO
Arno Meyer, Belar Electronics Laboratory, Devon, Pa.

TELEVISION
Larry Thorpe, Sony Electronics, Inc., Park Ridge, N.J.

2000

RADIO
Michael Dorrough, Dorrough Electronics, Woodland Hills, Calif.

TELEVISION
Max Berry, Capital Cities/ABC (retired) Elkins Park, Pa.

1999

RADIO
Geoffrey Mendenhall, P.E. Harris Corporation, Quincy, Ill.

TELEVISION
John Turner, Turner Engineering, Mountain Lakes, N.J.

1998

RADIO
John Battison, P.E., John Battison, Consultant, Loudonville, Ohio

TELEVISION
Robert Hopkins, Sony Pictures High Definition Center, Culver City, Calif.

1997

RADIO
George Jacobs, George Jacobs & Associates, Silver Spring, Md.

TELEVISION
Michael Sherlock, NBC, New York, N.Y.

1996

RADIO
Ogden Prestholdt, A.D. Ring & Associates, Nakomis, Fla.

TELEVISION
Charles Rhodes, Advanced Television Test Center, Alexandria, Va.

SERVICE TO BROADCAST ENGINEERING
Gerald Robinson, Hearst Broadcasting, Milwaukee, Wisc.

1995

RADIO
Robert Orban, Chief Engineer, AKG Acoustics, Inc. San Leandro, Calif.

TELEVISION
Carl G. Eilers, Manager of Electronic Systems R&D, Zenith Electronics, Glenview, Ill.

1994

RADIO
Charles T. Morgan, Susquehanna Radio Corporation, York, Pa.

TELEVISION
Thomas J. Vaughan, PESA Micro Communications, Inc., Manchester, N.H.

1993

RADIO
Robert M. Silliman, Silliman and Silliman, Silver Spring, Md.

TELEVISION
Stanley N. Baron, Technical Development, NBC, New York, N.Y.

SERVICE TO BROADCAST ENGINEERING
Herb H. Schubarth, Gannett Broadcasting, Denver, Colo.

1992

RADIO
Edward Edison and Robert L. Hammett, Hammett & Edison, San Francisco, Calif.

TELEVISION
James C. McKinney, Chairman, Advanced Television Systems Committee, Washington, D.C.

1991

RADIO
George Marti, Marti Electronics, Cleburne, Texas

TELEVISION
Kerns Powers, David Sarnoff and NBC Consultant, Princeton, N.J.

1990

Hilmer Swanson, Harris Corporation, Broadcast Division, Quincy, Ill.

1989

William Connolly, Sony Advanced Systems, Montvale, N.J.

1988

Jules Cohen, Jules Cohen and Associates, Washington, D.C.

1987

Renville H. McMann, CBS Technology Center, Stamford, Conn.

1986

Dr. George Brown, RCA Laboratories, Princeton, N.J.

1985

Carl E. Smith, Smith Electronics, Cleveland, Ohio

1984

Otis S. Freeman, WPIX, Inc., Tribune Broadcasting, New York, N.Y.

1983

Joseph Flaherty, Technology, CBS Inc., New York, N.Y.

1982

Julius Barnathan, Broadcast Operations and Engineering, American Broadcasting Companies, Inc., New York, N.Y.

1981

Wallace E. Johnson, Association for Broadcast Engineering Standards, Washington, D.C.

1980

James D. Parker, CBS Television Network, New York, N.Y.

1979

Robert W. Flanders, McGraw Hill Broadcasting Co., Inc., Indianapolis, Ind.

1978

John A. Moseley, President, Moseley Associates, Inc., Goleta, Calif.

1977

Daniel H. Smith, Capital Cities Communications, Inc., Philadelphia, Pa.

1976

Dr. Frank G. Kear, Consulting Engineer, Washington, D.C.

1975

John D. Silva, Golden West Broadcasters, Los Angeles, Calif.

1974

Joseph B. Epperson, Scripps Howard Broadcasting Co., Cleveland, Ohio

1973

A. James Ebel, KOLN TV, Lincoln, Neb.

1972

John M. Sherman, WCCO, Minneapolis, Minn.

1971

Benjamin Wolfe, Post Newsweek Stations, Washington, D.C.

1970

Philip Whitney, General Manager, WINC, Winchester, Va.

1969

Jarrett L. Hathaway, NBC Television Network, New York, N.Y.

1968

Howard A. Chinn, CBS Television Network, New York, N.Y.

1967

Robert M. Morris, American Broadcasting Company, New York, N.Y.

1966

Carl J. Meyers, WGN Continental Broadcasting Co., Chicago, Ill.

1965

Edward W. Allen, Jr., Federal Communications Commission, Washington, D.C.

1964

John H. DeWitt, Jr., WSM, Inc., Nashville, Tenn.

1963

Dr. George R. Town, Iowa State University, Ames, Iowa

1962

Ralph N. Harmon, Westinghouse Broadcasting Co., New York, N.Y.

1961

Raymond F. Guy, Consultant

1960

Commissioner T. A. M. Craven, FCC, Washington, D.C.

1959

John T. Wilner, Hearst Corporation, Baltimore, Md.

SECTION

1

FUNDAMENTALS

CHAPTER

1.1

The Electromagnetic Spectrum

JOHN NORGARD

Air Force Research Laboratory
Rome, New York

Updated for the 11th Edition by

GREGORY L. BEST, P.E.

Greg Best Consulting, Inc.
Stilwell, KS

INTRODUCTION

The electromagnetic (EM) spectrum consists of all forms of EM radiation, from DC to light to gamma rays. A chart of the EM spectrum can be arranged in order of frequency or wavelength into a number of regions,[1] usually wide in extent, within which the EM waves have some specified common characteristics (e.g., those characteristics relating to the production or detection of the radiation). A common example is the spectrum of the radiant energy in the region referred to as *white light*, which when dispersed by a prism will produce a rainbow of its constituent colors.

The EM spectrum is typically displayed as a function of frequency (or wavelength), as shown schematically in Figure 1.1-1. In air, frequency (f) and wavelength (λ) are inversely proportional ($f = c/\lambda$, where $c = 2.998 \times 10^8 \approx 3 \times 10^8$ m/s is the speed of light in a vacuum). The meter-kilogram-second (MKS) unit of frequency is the hertz (Hz, where 1 Hz = 1 cycle per second); the MKS unit of wavelength is the meter.

Frequency is also measured in the following subunits:

- Kilohertz (1 kHz = 10^3 Hz)
- Megahertz (1 MHz = 10^6 Hz)
- Gigahertz (1 GHz = 10^9 Hz)
- Terahertz (1 THz = 10^{12} Hz)
- Petahertz (1 PHz = 10^{15} Hz)
- Exahertz (1 EHz = 10^{18} Hz)

Electromagnetic energy at a particular frequency f has a photon energy (E) associated with it as follows:

$$E = hf \text{ (units are electron volts, eV)}$$

where h is Planck's constant ≈ 4.13567 μeV/GHz, and f is the frequency (in Hz).

Wavelength is also measured in the following subunits:

- Centimeters (1 cm = 10^{-2} m)
- Millimeters (1 mm = 10^{-3} m)
- Micrometers (microns) (1 μm = 10^{-6} m)
- Nanometers (1 nm = 10^{-9} m)
- Angstroms (1 Å = 10^{-10} m)
- Picometers (1 pm = 10^{-12} m)
- Femtometers (1 fm = 10^{-15} m)
- Attometers (1 am = 10^{-18} m)

Another property of electromagnetic energy is time duration. Most EM waves are envisioned as continuous, but this is not always the case. For example, a radar pulse is a type of EM wave and it can be characterized by its amplitude, its time duration, and the spectrum used to make the pulse.

SPECTRAL SUBREGIONS

For convenience, in this chapter the overall EM spectrum is divided into three main subregions:

- DC to light spectrum
- Optical spectrum (this spectrum is treated first)
- Light to gamma ray spectrum

Note that the boundaries between some of the spectral regions are somewhat arbitrary. Some spectral bands

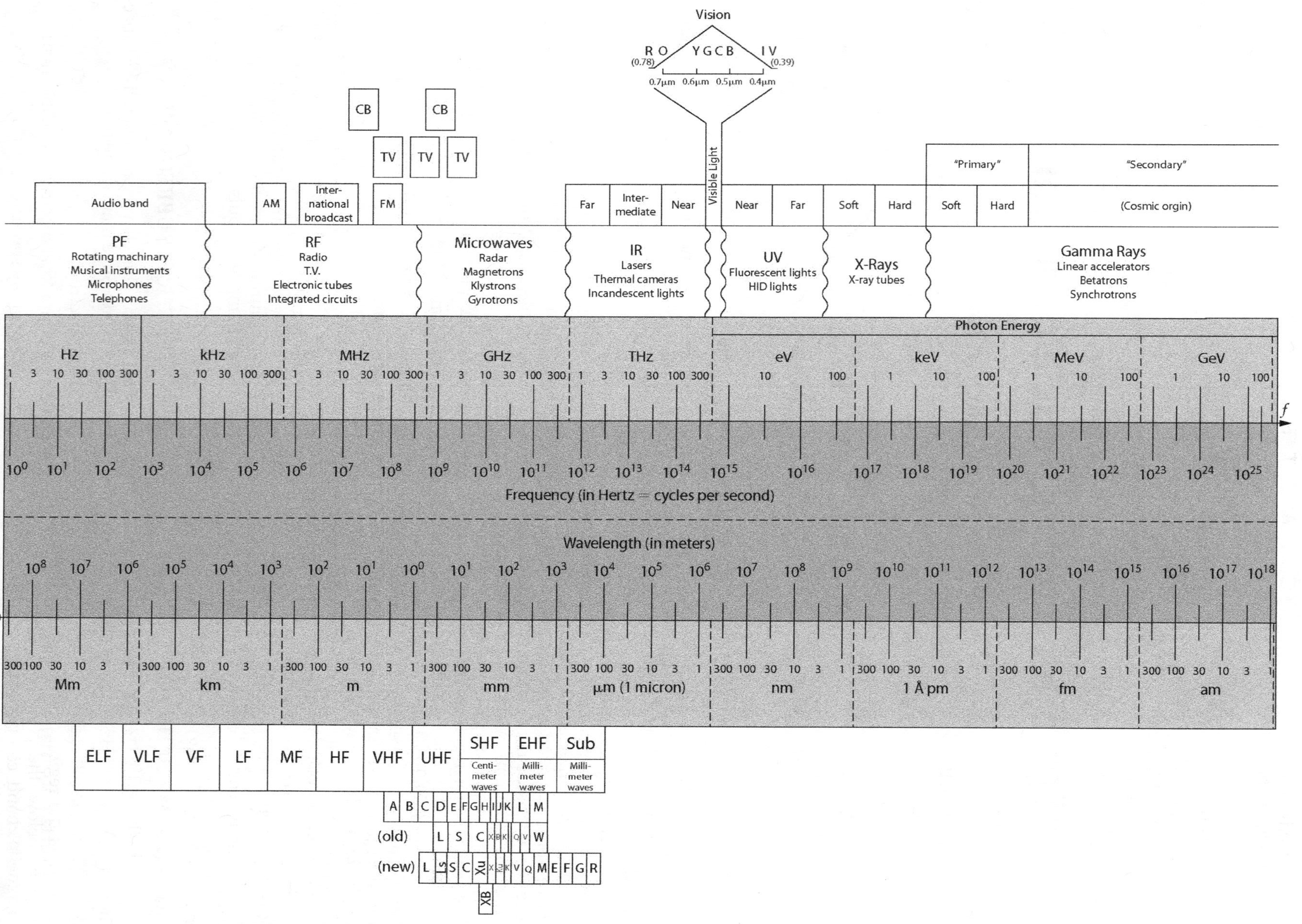

FIGURE 1.1-1 Simplified Chart of the Electromagnetic Spectrum

Reprinted with permission from Whitaker, J.C., Ed., *The Electronics Handbook*, CRC Press, Boca Raton, FL, 1996

have no sharp edges and merge into each other, and some spectral segments overlap each other slightly.

Optical Spectrum

The optical spectrum is the middle frequency/wavelength region of the EM spectrum. It is defined here as the visible and near-visible regions of the EM spectrum and includes the infrared, visible, and ultraviolet bands (listed by wavelength/frequency):

- Infrared (IR) band, ≈300–0.7 μm/≈1–429 THz
- Visible light band, ≈0.7–0.4 μm/≈429–750 THz
- Ultraviolet (UV) band, ≈0.4 μm–≈10 nm/≈750 THz–≈30 PHz

Because frequencies in the optical spectrum are so high, these regions of the EM spectrum are usually described in terms of their wavelengths. Atomic and molecular radiation produce radiant light energy. Atomic radiation (outer shell electrons) and radiation from arcs and sparks produce EM waves in the UV band. Molecular radiation and radiation from hot bodies produce EM waves in the IR band.

Visible Light Band

The visible light band, in the middle of the optical spectrum, extends in wavelength from approximately 0.4 μm (violet) to 0.7 μm (red) and in frequency from approximately 750 THz (violet) to 429 THz (red). EM radiation in this region of the EM spectrum, when entering the eye, gives rise to visual sensations (colors), according to the spectral response of the eye, which responds only to radiant energy in the visible light band extending from the extreme long wavelength edge of red to the extreme short wavelength edge of violet.[2] This visible light band is further subdivided into the various colors of the rainbow (listed in decreasing wavelength/increasing frequency):

- Red—a primary color;[3] peak intensity at 700.0 nm (429 THz)
- Orange
- Yellow
- Green—a primary color; peak intensity at 546.1 nm (549 THz)
- Cyan
- Blue—a primary color; peak intensity at 435.8 nm (688 THz)
- Indigo
- Violet

IR Band

The IR band is the region of the EM spectrum lying immediately below the visible light band in frequency. The IR band consists of EM radiation with wavelengths extending between the longest visible red, approximately 0.7 μm (429 THz), and the shortest microwaves, 300 μm–1 mm (1 THz–300 GHz). The IR band is further subdivided in wavelengths into the near (shortwave), intermediate (midwave), and far (longwave) IR segments (listed by wavelength/frequency):[4]

- Near IR segment, 0.7–3 μm/429–100 THz
- Intermediate IR segment, 3–7 μm/100–42.9 THz
- Far IR segment, 7–300 μm/42.9–1 THz

Note that the submillimeter region of wavelengths is sometimes included in the very far region of the IR band:

- Submillimeter band, 100 μm–1 mm/3 THz–300 GHz

In addition to emanating from electronic devices specifically designed for EM radiation purposes, EM radiation is produced in all matter by the oscillation and rotation of the molecules and atoms of which that matter is comprised; therefore, all objects at temperatures above absolute zero emit EM radiation by virtue of their thermal motion (warmth) alone. Objects near room temperature emit most of their radiation in the IR band; however, even relatively cool objects emit some IR radiation, and hot objects, such as incandescent filaments, emit strong IR radiation.

IR radiation is sometimes incorrectly referred to as radiant heat, because warm bodies emit IR radiation and bodies that absorb IR radiation are warmed; however, IR radiation is not itself heat. This radiant energy is more properly referred to as black body radiation. Such waves are emitted by all material objects; for example, background cosmic radiation (≈2.7 K) emits microwaves, room temperature objects (≈293 K) emit IR rays, the sun (≈6000 K) emits yellow light, and the solar corona (≈1 million K) emits X-rays.

IR astronomy uses the 1 μm to 1 mm portion of the IR band to study celestial objects by their IR emissions. IR detectors are used in night vision systems, intruder alarm systems, weather forecasting, and missile guidance systems. IR photography uses multilayered color film, with an IR-sensitive emulsion in the wavelengths between 700 and 900 nm, for medical and forensic applications and for aerial surveying.

UV Band

The UV band is the region of the EM spectrum lying immediately above the visible light band in frequency. The UV band consists of EM radiation with wavelengths extending between the shortest visible violet (≈0.4 μm) and the longest X-rays (≈10 nm);[5] that is, from 750 THz (≈3 eV) to ≈30 PHz (≈100 eV). The UV band is further subdivided in frequency into the near and the far UV segments (listed by wavelength/frequency/photon energy):[6]

- Near UV segment, ≈0.4 μm–100 nm/≈750 THz–3 PHz/≈3–10 eV
- Far UV segment, 100–≈10 nm/3–≈30 PHz/≈10–100 eV

UV radiation is produced by electron transitions in atoms and molecules, as in a mercury discharge lamp. Radiation in the UV range is easily detected, can cause fluorescence in some substances, and can produce photographic and ionizing effects. In UV astronomy, the emissions of celestial bodies in the wavelength band between 50 and 320 nm are detected and analyzed to study the heavens. The hottest stars emit most of their radiation in the UV band.

DC to Light

Below the IR band are the lower frequency (longer wavelength) regions of the EM spectrum, subdivided generally into the microwave, radiofrequency, and power spectral bands (listed by frequency/wavelength):

- Microwave band,[7] 300 GHz–300 MHz/1 mm–1 m
- Radiofrequency (RF) band, 300 MHz–10 kHz/1 m–30 km
- Power frequency (PF)/telephony band, 10 kHz–DC/30 km–∞

These regions of the EM spectrum are usually described in terms of their frequencies. EM radiation for which the wavelengths are of the order of millimeters and/or centimeters are called microwaves, and those still longer are radiofrequency (RF) or hertzian waves. Radiation from electronic devices produces EM waves in both the microwave and the RF bands. Power frequency energy is generated by rotating machinery. Direct current (DC) is produced by batteries or rectified alternating current (AC).

Microwave Band

The microwave band is the region of wavelengths between the far IR/submillimeter region and the conventional RF region. The boundaries of the microwave band have not been definitively fixed but are commonly regarded as the region of the EM spectrum extending from about 1 mm to 1 m (300 GHz–300 MHz). The microwave band is further subdivided into centimeter and millimeter segments (listed by frequency/wavelength):

- Millimeter waves,[8] 300–30 GHz/1 mm–1 cm (EHF band)
- Centimeter waves, 30–3 GHz/1–10 cm (SHF band)

The microwave band usually includes the UHF band from 3 GHz to 300 MHz (10 cm–1 m). Microwaves are used in satellite and radar applications, in communication links spanning moderate distances as radio carrier waves in radio broadcasting, and for mechanical heating and cooking (e.g., in microwave ovens).

Radiofrequency (RF) Band

The RF range of the EM spectrum is the wavelength band suitable for utilization in radio communications from 10 kHz to 300 MHz (30 km–1 m).[9] Some radio waves serve as the carriers of low-frequency audio signals; other radio waves are modulated by video and digital information.

The amplitude modulated (AM) broadcasting band uses waves with frequencies between 550 and 1705 kHz; the frequency modulated (FM) broadcasting band uses waves with frequencies between 88 and 108 MHz. In the United States, there are two organizations that are responsible for the assignment and regulation of RF spectrum. The National Telecommunications and Information Administration (NTIA, a department of the U.S. Department of Commerce) is responsible for all government agency spectrum utilization and the Federal Communications Commission (FCC) is responsible for all other spectrum uses. Some frequency segments are shared. The International Telecommunication Union Radiocommunication Sector (ITU-R) coordinates frequency band allocation and coordination on a worldwide basis.

Radio astronomy uses radio telescopes to receive and study radio waves naturally emitted by objects in space. Radio waves are emitted from hot gases (thermal radiation), from charged particles spiraling in magnetic fields (synchrotron radiation), and from excited atoms and molecules in space (spectral lines), such as the 21 cm line emitted by hydrogen gas.

Power Frequency (PF) Band

The PF range of the EM spectrum is the wavelength band suitable for generating, transmitting, and consuming low-frequency prime power (e.g., electrical power provided by electric utility companies), extending from 10 kHz to DC (zero frequency; 30 km to ∞ wavelength). In the United States, most prime power is generated at 60 Hz (some military and computer applications use 400 Hz); in Europe, prime power is generated at 50 Hz.

FREQUENCY BAND DESIGNATIONS

The combined microwave, radiofrequency, and power/telephone spectra are subdivided into the specific bands shown in Table 1.1-1. An alternative and more detailed subdivision of the UHF (9), SHF (10), and EHF (11) bands is shown in Table 1.1-2.

The U.S. military tri-service designations for radio communication bands are shown in Table 1.1-3. Another set of designations that predate the U.S. tri-service designations are given in Table 1.1-4. The European Community (EC) radar band designations in prior use are listed in Table 1.1-5. Several other frequency bands of interest (not exclusive) are shown in Tables 1.1-6 and 1.1-7.

A comprehensive and informative chart of the U.S. frequency allocations of the radio spectrum, most recently updated in 2016, has been prepared

TABLE 1.1-1
International Radiofrequency Band Designations and Numerical Designations

Band Name	Band Designation	Frequency Range	Wavelength Range
Extremely low frequency	ELF (1)	3 Hz–30 Hz	100,000 km–10,000 km
Super low frequency	SLF (2)	30 Hz–300 Hz	10,000 km–1000 km
Ultra low frequency	ULF (3)	300 Hz–3 kHz	1000 km–100 km
Very low frequency	VLF (4)	3 kHz–30 kHz	100 km–10 km
Low frequency	LF (5)	30 kHz–300 kHz	10 km–1 km
Medium frequency	MF (6)	300 kHz–3 MHz	1 km–100 m
High frequency	HF (7)	3 MHz–30 MHz	100 m–10 m
Very high frequency	VHF (8)	30 MHz–300 MHz	10 m–1 m
Ultra high frequency	UHF (9)	300 MHz–3 GHz	1 m–10 cm
Super high frequency	SHF (10)	3 GHz–30 GHz	10 cm–1 cm
Extremely high frequency	EHF (11)	30 GHz–300 GHz	1 cm–1 mm
Tremendously high frequency	THF* (12)	300 GHz–3 THz	1 mm–100 μm

*THF is not a universally accepted designation for this band.

TABLE 1.1-2
Alternate and More Detailed Subdivision of the UHF (9), SHF (10), and EHF (11) Bands

Band Designation	Frequency Range	Wavelength Range
L	1.12–1.7 GHz	26.8 cm–17.6 cm
LS	1.7–2.6 GHz	17.6 cm–11.5 cm
S	2.6–3.95 GHz	11.5 cm–7.59 cm
C(G)	3.95–5.85 GHz	7.59 cm–5.13 cm
XN(J, XC)	5.85–8.2 GHz	5.13 cm–3.66 cm
XB(H, BL)	7.05–10 GHz	4.26 cm–3 cm
X	8.2–12.4 GHz	3.66 cm–2.42 cm
Ku(P)	12.4–18 GHz	2.42 cm–1.67 cm
K	18–26.5 GHz	1.67 cm–1.13 cm
V(R, Ka)	26.5–40 GHz	1.13 cm–7.5 mm
Q(V)	33–50 GHz	9.09 mm–6 mm
M(W)	50–75 GHz	6 mm–4 mm
E(Y)	60–90 GHz	5 mm–3.33 mm
F(N)	90–140 GHz	3.33 mm–2.14 mm
G(A)	140–220 GHz	2.14 mm–1.36 mm
R	220–325 GHz	1.36 mm–0.923 mm

TABLE 1.1-3
Current U.S. tri-service radar band designations

Band Designation	Frequency Range	Wavelength Range
A	0 Hz–250 MHz	∞–1.2 m
B	250 MHz–500 MHz	1.2 m–60 cm
C	500 MHz–1 GHz	60 cm–30 cm
D	1 GHz–2 GHz	30 cm–15 cm
E	2 GHz–3 GHz	15 cm–10 cm
F	3 GHz–4 GHz	10 cm–7.5 cm
G	4 GHz–6 GHz	7.5 cm–5 cm
H	6 GHz–8 GHz	5 cm–3.75 cm
I	8 GHz–10 GHz	3.75 cm–3 cm
J	10 GHz–20 GHz	3 cm–1.5 cm
K	20 GHz–40 GHz	1.5 cm–7.5 mm
L	40 GHz–60 GHz	7.5 mm–5 mm
M	60 GHz–100 GHz	5 mm–3 mm
N	100 GHz–200 GHz	3 mm–1.5 mm
O	200 GHz–300 GHz	1.5 mm–1 mm

by the NTIA's Office of Spectrum Management (OSM). The chart graphically partitions the radiofrequency spectrum (3 kHz–300 GHz) into over 450 frequency bands and uses distinct colors to distinguish the allocations for 30 different radio services and for 3 different radio activities. The chart presents a graphical summary of the detailed allocations contained in the U.S. Table of Frequency Allocations found in:

- The NTIA Manual (Manual of Regulations and Procedures for Federal Radio Frequency Management, Chapter 4)
- The FCC Rules (47 CFR, Part 2)

TABLE 1.1-4
Designations Predating the U.S. Tri-Service Designations

Band Designation	Frequency Range	Wavelength Range
I	100 MHz–150 MHz	3 m–2 m
G	150 MHz–225 MHz	2 m–1.33 m
P	225 MHz–390 MHz	1.33 m–76.9 cm
L	390 MHz–1.5 GHz	76.9 cm–19.4 cm
S*	1.5 GHz–3.9 GHz	19.4 cm–7.69 cm
C*	3.9 GHz–6.2 GHz	7.69 cm–48.4 mm
X*	6.2 GHz–10.9 GHz	48.4 mm–27.5 mm
K	10.9 GHz–36 GHz	27.5 m–8.33 mm
Q	36 GHz–46 GHz	8.33 mm–6.52 mm
V	46 GHz–56 GHz	6.52 mm–5.36 mm
W	56 GHz–	5.36 mm–

*An alternative prior S band designation extended from 1.5 GHz to 5.85 GHz, and similarly a prior X band extended from 5.85 GHz to 10.9 GHz, eliminating the C band in that designation.

TABLE 1.1-5
European Community Radar Band Designations in Prior Use

Band Designation	Frequency Range	Wavelength Range
L	1–2 GHz	30 cm–15 cm
S	2–4 GHz	15 cm–7.5 cm
C	4–8.2 GHz	7.5 cm–3.66 cm
X	8.2–12.4 GHz	3.66 cm–2.42 cm
Ku	12.4–18 GHz	2.42 cm–1.67 cm
K*	18–26.5 GHz	1.67 cm–11.3 mm
Ka*	26.5–40 GHz	11.3 cm–7.5 mm
V	40–75 GHz	7.5 mm–4 mm
W	75–110 GHz	4 mm–2.73 mm
mm	110–300 GHz	2.73 mm–1 mm

*The prior K band sometimes included the Ka band and extended from 18 to 40 GHz.

TABLE 1.1-6
Power Band

Band Designation	Frequency Range
Subsonic band	0 Hz–10 Hz
Audio band	10 Hz–10 kHz
Ultrasonic band	10 kHz and up

TABLE 1.1-7
RF Band

Band Designation	Frequency Range
Longwave broadcasting	150–290 kHz
AM broadcasting	550–1705 kHz
International broadcasting	3–30 MHz
Shortwave broadcasting (8 bands)	5.95–26.1 MHz
VHF TV (channels 2–4)	54–72 MHz
VHF TV (channels 5–6)	76–88 MHz
FM broadcasting	88–108 MHz
VHF TV (channels 7–13)	174–216 MHz
UHF TV (channels 14–51)*	512–698 MHz

*As part of the U.S. transition to digital television, TV channels 52 through 69 (698–806 MHz) were removed from the TV band, leaving a core TV band that contains channels 2 through 51.

This radiofrequency allocation chart is available online at www.ntia.doc.gov/page/2011/united-states-frequency-allocation-chart. In the chart, the radio spectrum is divided into the following three subareas according to radio activity:

- Federal exclusive (NTIA)
- Federal/nonfederal shared (NTIA/FCC)
- Nonfederal exclusive (FCC)

The chart is also subdivided into the following 30 subareas according to radio service:

- Aeronautical mobile
- Aeronautical mobile satellite
- Aeronautical radio navigation
- Amateur
- Amateur satellite
- Broadcasting
- Broadcasting satellite
- Earth exploration satellite
- Fixed
- Fixed satellite
- Inter-satellite
- Land mobile
- Land mobile satellite
- Maritime mobile
- Maritime mobile satellite
- Maritime radio navigation
- Meteorological
- Meteorological satellite
- Mobile
- Mobile satellite

- Radio astronomy
- Radiodetermination satellite
- Radiolocation
- Radiolocation satellite
- Radio navigation
- Radio navigation satellite
- Space operations
- Space research
- Standard frequency and time signal
- Standard frequency and time signal satellite

Light to Gamma Rays

Above the UV spectrum are the higher frequency (shorter wavelength) regions of the EM spectrum, subdivided generally into the X-ray and gamma ray spectral bands (listed by electron voltage/wavelength/frequency):[10]

- X-ray band, ≈10 eV – 1 MeV/≈10 nm – ≈1 pm/≈3 PHz – ≈300 EHz
- Gamma ray band, ≈1 keV – ∞/≈300 pm – 0 m/≈1 EHz – ∞

These regions of the EM spectrum are usually described in terms of their photon energies in electron volts. Note that the bottom of the gamma ray band overlaps the top of the X-ray band. Radiation from atomic inner shell excitations produces EM waves in the X-ray band. Radiation from naturally radioactive nuclei produces EM waves in the gamma ray band.

X-Ray Band

The X-ray band is further subdivided into soft and hard X-rays (listed by photon energy/wavelength/frequency):

- Soft X-rays, ≈10 eV – 10 keV/≈ 10 nm – 100 pm/≈3 PHz – 3 EHz
- Hard X-rays, ≈10 keV – 1MeV/100 pm – ≈1 pm/3 EHz – ≈300 EHz

Because the physical nature of these rays was at first unknown, this radiation was named X-rays by Wilhelm Roentgen, the German scientist who discovered them (with X being used as the symbol for an unknown quantity). The more powerful X-rays are the hard X-rays, which are of high frequencies and, therefore, more energetic; less powerful X-rays are the soft X-rays, which have lower energies.

X-rays are produced by transitions of electrons in the inner levels of excited atoms or by rapid deceleration of charged particles (*brehmsstrahlung*, or "breaking radiation"). An important source of X-rays is synchrotron radiation. X-rays can also be produced when high-energy electrons from a heated filament cathode strike the surface of a target anode (usually tungsten) between which a high alternating voltage (approximately 100 kV) is applied.

X-rays are a highly penetrating form of EM radiation, and applications of X-rays are based on their short wavelengths and their ability to easily pass through matter. X-rays are very useful in crystallography for determining crystalline structure and in medicine for photographing the body. Because different parts of the body absorb X-rays to a different extent, X-rays passing through the body provide a visual image (negative) of its interior structure when striking a photographic plate. X-rays are dangerous and can destroy living tissue. They can also cause severe skin burns. X-rays are useful in the diagnosis and nondestructive testing of products for defects.

Gamma Ray Band

The gamma ray band is subdivided into primary and secondary gamma ray segments (listed by photon energy/wavelength/frequency):

- Primary gamma rays, ≈1 keV – 1 MeV/≈ 300 pm – 300 fm/≈1 EHz – 1000 EHz
- Secondary gamma rays, ≈1 MeV – ∞/≈300 fm – 0 m/1000 EHz – ∞

Secondary gamma rays are created from collisions of high-energy cosmic rays with particles in the Earth's upper atmosphere. The primary gamma rays are further subdivided into soft and hard gamma ray segments (listed by photon energy/wavelength/frequency):

- Soft gamma rays, ≈1 keV – ≈300 keV/≈300 pm – ≈3 pm/≈1 EHz – ≈100 EHz
- Hard gamma rays, ≈300 keV – 1 MeV/≈3 pm – 300 fm/≈100 EHz – 1000 EHz

Gamma rays are essentially very energetic X-rays. The distinction between the two is based on their origin. X-rays are emitted during atomic processes involving energetic electrons; gamma rays are emitted by excited nuclei or other processes involving subatomic particles.

Gamma rays are emitted by the nucleus of radioactive material during the process of natural radioactive decay as a result of transitions from high-energy excited states to low-energy states in atomic nuclei. Cobalt 90 is a common gamma ray source (with a half-life of 5.26 years). Gamma rays are also produced by the interaction of high-energy electrons with matter. Cosmic gamma rays cannot penetrate the Earth's atmosphere.

Applications of gamma rays are found both in medicine and in industry. In medicine, gamma rays are used for cancer treatment, diagnosis, and prevention. Gamma ray emitting radioisotopes are used as tracers. In industry, gamma rays are used in the inspection of castings, seams, and welds.

Defining Key Terms

- Cosmic rays—Highly penetrating particle rays from outer space. Primary cosmic rays (particles) that enter the Earth's upper atmosphere consist mainly of protons. Cosmic rays of low energy have their origin in the sun, those of high energy in galactic or extragalactic space, possibly as a result of supernova explosions. Collisions with atmospheric particles result in secondary cosmic rays (particles) and secondary gamma rays (EM waves).
- Electromagnetic spectrum—EM radiant energy arranged in order of frequency or wavelength and divided into regions within which the waves have some common specified characteristics (e.g., the waves are generated, received, detected, or recorded in a similar way).
- Gamma rays—Electromagnetic radiation of very high energy (<30 keV) emitted after nuclear reactions or by a radioactive atom when its nucleus is left in an excited state after emission of alpha or beta particles.
- Infrared (IR) radiation—Electromagnetic radiation with wavelengths in the range of 0.7 nm (the long wavelength limit of visible red light) to 1 mm (the shortest microwaves). A convenient subdivision is as follows: near, 0.7 µm to 2–5 µm; intermediate, 2–5 µm to 10 µm; far, 10 µm to 1 mm.
- Light—White light, when split into a spectrum of colors, is composed of a continuous range of merging colors: red, orange, yellow, green, cyan, blue, indigo, and violet.
- Microwaves—Electromagnetic waves that have wavelengths between approximately 0.3 cm (or 1 mm) and 30 (or 10) cm, corresponding to frequencies between 1 GHz and 100 GHz. Note that there are no well-defined boundaries distinguishing microwaves from infrared or radio waves.
- Radio waves—Electromagnetic radiation suitable for radio transmission in the range of frequencies from about 10 kHz to about 300 MHz.
- Ultraviolet (UV) radiation—Electromagnetic radiation with wavelengths in the range of 0.4 nm (the shortest wavelength limit of visible violet light) to 3 nm (the longest X-rays). A convenient subdivision is as follows: near, 0.4 µm to 100 nm; far, 100 nm to 3 nm.
- X-rays—Electromagnetic radiation of short wavelengths (3 nm to 30 pm) produced when cathode rays impinge on matter.

Notes

[1] Note that specific frequency ranges are often called *bands;* several contiguous frequency bands are usually called *spectrums;* and subfrequency ranges within a band are sometimes called *segments.*

[2] The spectral response of the eye is sometimes described as extending from 0.38 µm (violet) to 0.75 or 0.78 µm (red) (i.e., from 789 THz to 400 or 385 THz).

[3] *Primary colors* are those that cannot be created by mixing other colors. The three *additive* primary colors are red, green, and blue, and the three *subtractive* primary colors are magenta, yellow, and cyan.

[4] Some reference texts use 2.5 µm (120 THz) as the breakpoint between the near and the intermediate IR bands and 10 µm (30 THz) as the breakpoint between the intermediate and far IR bands. Also, 15 µm (20 THz) is sometimes considered as the long wavelength end of the far IR band.

[5] Some reference texts use 4, 5, or 6 nm as the upper edge of the UV band.

[6] The far UV band is also referred to as the *vacuum* UV band, because air is opaque to all UV radiation in this region.

[7] Some reference works define the lower edge of the microwave spectrum at 1 GHz.

[8] Some reference articles consider the top edge of the millimeter region to stop at 100 GHz.

[9] Some authors consider the RF band to extend from 10 kHz to 300 GHz, with the microwave band as a subset of the RF band from 300 MHz to 300 GHz.

[10] Note that *cosmic rays* (from astronomical sources) are not EM waves (rays) and, therefore, are not part of the EM spectrum. Cosmic rays are high-energy charged particles (electrons, protons, and ions) of extraterrestrial origin moving through space, which may have energies as high as 10^{20} eV. Cosmic rays have been traced to cataclysmic astrophysical or cosmological events, such as exploding stars and black holes. Cosmic rays are emitted by supernova remnants, pulsars, quasars, and radio galaxies. Cosmic rays that collide with molecules in the Earth's upper atmosphere produce secondary cosmic rays and gamma rays of high energy, which also contribute to the natural background radiation. These gamma rays are sometimes called *cosmic* or *secondary* gamma rays. Cosmic rays are a useful source of high-energy particles for experiments.

Bibliography

The Cambridge Encyclopedia, Cambridge University Press, London, 1990.

Collocott, T. C. and Dobson, A. B., Eds., *Chambers Dictionary of Science and Technology,* rev. ed., HarperCollins, New York, 1983.

Condon, E. U. and Odishaw, H., Eds., *Handbook of Physics,* McGraw-Hill, New York, 1958.

The Concise Columbia Encyclopedia, Columbia University Press, New York, 1993.

Judd, D. B. and Wyszecki, G., *Color in Business, Science and Industry,* 3rd ed., John Wiley & Sons, New York, 1975.

Kaufman, J. E. and Christensen, J. F., Eds., *IES Illumination Handbook,* Illumination Engineering Society, New York, 1984.

Lapedes, D. N., Ed., *The McGraw-Hill Encyclopedia of Science and Technology,* 2nd ed., McGraw-Hill, New York.

Stimson, A., *Photometry and Radiometry for Engineers,* John Wiley & Sons, New York, 1974.

Webster's New World Encyclopedia, Prentice Hall, Englewood Cliffs, NJ, 1992.

Wyszecki, G. and Stiles, W. S., *Color Science: Concepts and Methods, Quantitative Data, and Formulae,* 2nd ed., John Wiley & Sons, New York, 1982.

CHAPTER

1.2

The Fundamentals of Broadcast Coverage

DOUG VERNIER

President, Doug Vernier Telecommunications Consultants
Cedar Falls, IA

In today's crowded media environment, having a large share of listeners and viewers is a must for the survival of a radio or television station. The station that has noise on its signal or is weak will not be in a position to build audience, no matter how captivating its programming. The television station's picture that blacks out or pixelates due to weak reception will not be watched. In this chapter, we hope to give the reader a better understanding of what good coverage is all about, how to predict it, and how to achieve it. In addition to the process of estimating where a station's coverage should be, age-old questions will be covered, such as antenna height versus power (which is more important?), dealing with multipath fading, extending coverage through boosters and translators, in-band/on-channel (IBOC) digital radio coverage and interference caused by IBOC, optimizing coverage (and the new danger it presents), using directional antennas, short spacing, upgrading coverage by upgrading class, moving the city of license, and moving the transmission tower.

PREDICTING COVERAGE

As a start, to improve coverage it is necessary to know how to analyze what coverage a station should have. Today, computers are used for this purpose, with various propagation models that have different qualities and benefits. The two primary models, accepted by the U.S. Federal Communications Commission (FCC), that are in wide use today are the standard FCC model and the Longley-Rice model. While being able to generate results from these models is the first order of business, knowing how to interpret the results is the key.

FCC Method

For FM radio, the FCC method of predicting coverage requires calculating the effective antenna height of a station antenna along paths of a minimum of eight evenly spaced radials, starting at 3 kilometers from the transmitter and ending at 16 kilometers. These eight radial paths, often called the "cardinal radials," are therefore spaced every 45 degrees. For FM, at least 50 terrain elevation points (heights above mean sea level) are collected and averaged together on each radial to form the "average elevation." The radial elevation averages are then averaged together to determine the "average elevation" of the land at the base of the tower or supporting structure. Finally, the antenna height above ground is added to this average elevation figure to determine the "height above average terrain" (HAAT, see Figure 1.2-1)

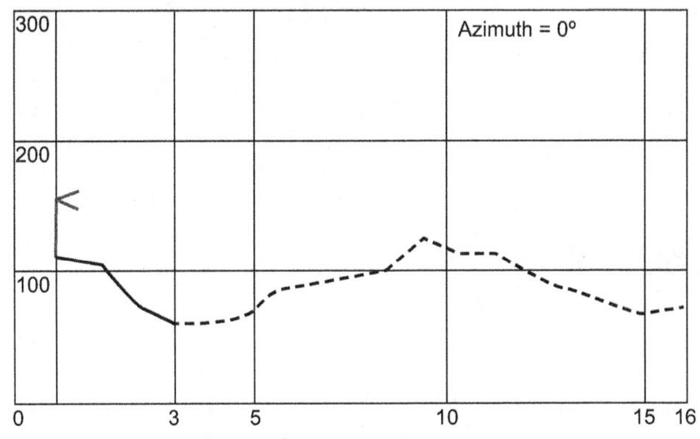

FIGURE 1.2-1 Illustration of height above average terrain

TV uses the distance from 3.2 kilometers to 16.1 kilometers. For many years, topographic maps were used to derive the elevation points, which is why the FCC still requires just a minimum of 50 points, since reading those elevations off a topographic map was difficult and time consuming for the applicant and the FCC staff. However, today nearly all such work is performed by computers using either the National Geophysical Data Center's 30 arc-second digital terrain database or another of the now-available databases offering higher resolution. Modern computer programs can be designed to use the minimum of elevation points or more, such as 131 points, spaced 0.1 kilometers apart.[1] For FM, the distance to a signal contour is calculated using the curves found under 47 CFR §73.333, where the FCC has published two sets, the F(50,50) for coverage and the F(50,10) for interference calculations. These curves were based on actual measurements with the receiving antenna at 9.14 meters or 30 feet above ground. The F(50,50) curves (see Figure 1.2-2) show signal strength as a measure of distance under the statistical probability that the predicted signal will be at 50% of the locations for 50% of the time.

The interfering signal curves are based on a transmitter's signal strength for 50% of the locations and for only 10% of the time. Therefore, since the interfering signal only needs to be at the predicted level for a smaller amount of the time, it travels greater distances. For digital TV, coverage is predicted using F(50,90) calculations. In other words, the desired signal from a digital TV transmitter is at 50% of the locations for 90% of the time. The propagation charts in 47 CFR §73.699 provide a set of special DTV curves for calculating F(50,90) coverage.

The FCC curves were digitized and are now contained in the FCC's TVFMINT FORTRAN program. To do this digitization, the Commission placed a large number of engineers in a room with enlarged graphs of the curves and had them read the graphs and insert the values into a table. It is interesting to note that the engineers could not all agree, so the values they produced were averaged and the results are still used in the FCC's curves tables today. Consequently, although not widely known, the computer program has a standard deviation from the "real" curves of approximately 0.25 dB.

Minimum Distance Separations

Commercial and noncommercial stations have different rules that impact how much coverage these stations may have. The primary allocation system for commercial band stations is based on the minimum distance separations found in 47 CFR §73.207 of the rules (see Table 1.2-1).

Short Spacing

When a commercial band station has a frequency relationship with another station and its distance from the station does not meet the minimum required distance under 47 CFR §73.207, the station may qualify to use the contour-to-contour "short-space" rules under 47 CFR §73.215. First, before the contour-to-contour protections can be applied, the two stations must meet another set of minimum separations found in §73.215 (e) (see Table 1.2-2).

If the stations meet this first test, the contour protection procedure can be applied. The protected service contour of an FM station is based on the class of the station. Class A, C3, C2, C1, C0, and C stations are protected by other stations to the FCC F(50,50), 60 dBµ contour at their maximum class facilities (see Table 1.2-3 and Figure 1.2-3). Class B1 stations are protected to the 57 dBµ and Class B stations are protected to the 54 dBµ contour.

The "short-space" rules are very similar to the rules for noncommercial educational (NCE) FM stations operating in the educational reserved portion of the FM band (88.1 to 91.9 MHz), where the allocation protections are based on preventing contour overlap of the interfering contour with the protected contour. If the station qualifies to use contour protection, and the station being protected has already filed as a §73.215 station or if the protected station is NCE, the process does not require that the station must be protected to its maximum class power and antenna HAAT. It need be protected only to the contours produced at its current HAAT and operating power.

AM Method

AM broadcast band signals propagate using the ground as one portion of the radiofrequency conductor, as opposed to higher frequency waves, such as FM and TV, which travel primarily through the air. The better the ground conducts, the greater distance the AM waves will travel. The M3 conductivity map, shown in Figure 1.2-4, appears in §73.190 of the Commission's rules. It shows the ground conductivity regions in various areas of the United States. A similar map for ITU-R Region 2 defines the conductivity, in less detail, for the entire Western Hemisphere. 47 CFR §73.183 and §73.184 refer to field strength curves that, when used with the M3 or R2 charts, will predict the distance to contours for AM stations during daylight hours.

Nighttime propagation for the AM band is calculated using a combination of the FCC's skywave calculations and ground conductivity calculations. Class C stations use only groundwave for both day and night. Class A stations, sometimes called clear channels, are protected in the daytime to the 0.1 millivolt per meter groundwave contour. The nighttime protection to a Class A station is the 0.5 V/m 50% skywave contour. Stations operating at night also have to protect the root-sum-square (RSS) limit, sometimes called the "interference free" signal contours of other stations.[2] Calculation of nighttime interference-free service is accomplished by evaluating the signals on co- and first-adjacent frequencies in order of decreasing magnitude by adding the squares of the values and extracting the square root of the sum, excluding those

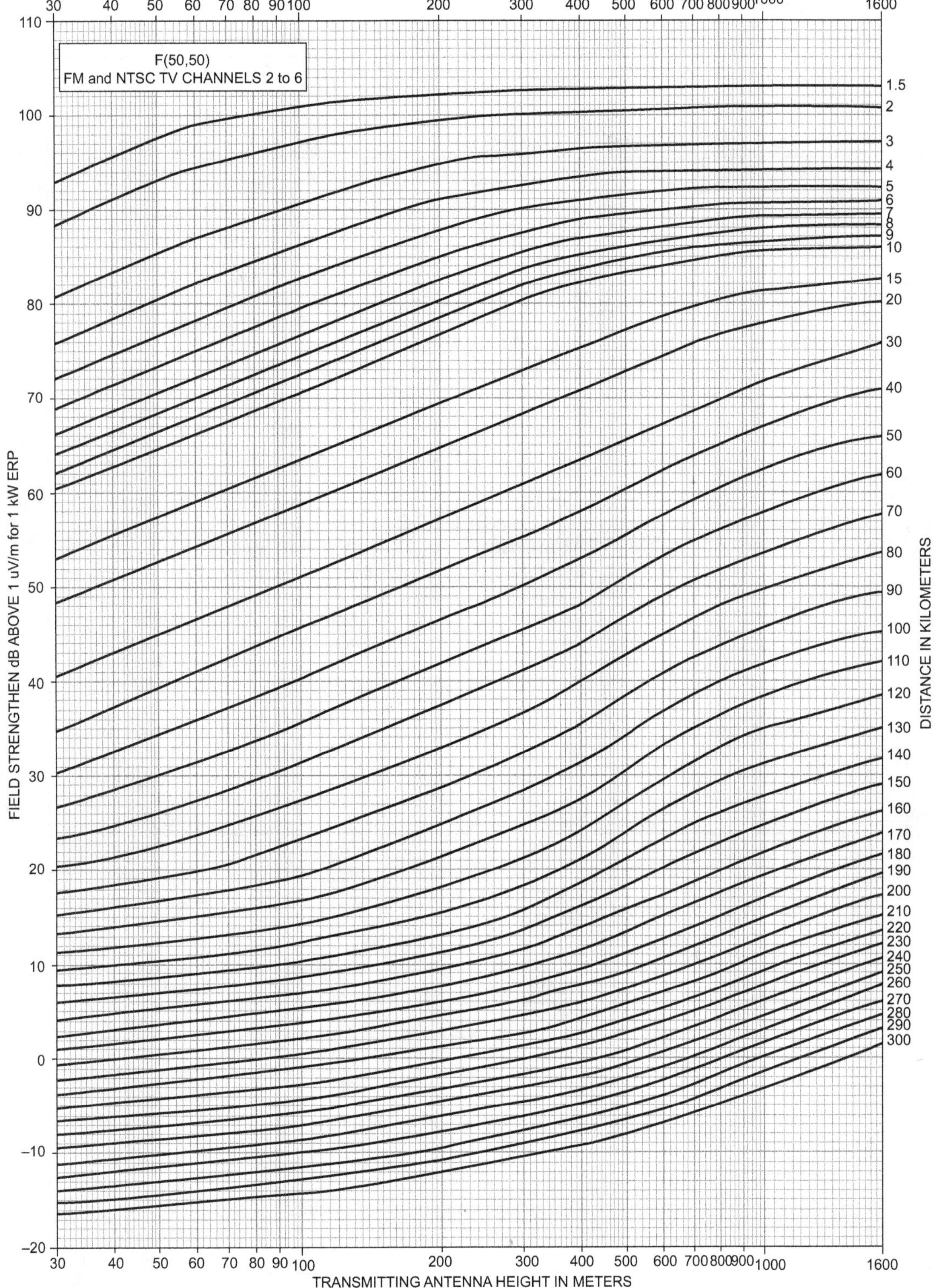

FIGURE 1.2-2 Example of FCC F(50,50) curves

TABLE 1.2-1
Excerpt from FCC table of FM minimum separation distances (47 CFR §73.207)

Minimum Distance Separation Requirements in Kilometers (miles)				
Relation	**Co-channel**	**200 kHz**	**400/600 kHz**	**10.6/10.8 MHz**
A to A	115 (71)	72 (45)	31 (19)	10 (6)
A to B1	143 (89)	96 (60)	48 (30)	12 (7)
A to B	178 (111)	113 (70)	69 (43)	15 (9)
A to C3	142 (88)	89 (55)	42 (26)	12 (7)
A to C2	166 (103)	106 (66)	55 (34)	15 (9)
A to Cl	200 (124)	133 (83)	75 (47)	22 (14)
A to CO	215 (134)	152 (94)	86 (53)	25 (16)
A to C	226 (140)	165 (103)	95 (59)	29 (18)
B1 to B1	175 (109)	114 (71)	50 (31)	14 (9)
B1 to B	211 (131)	145 (90)	71 (44)	17 (11)
B1 to C3	175 (109)	114 (71)	50 (31)	14 (9)
B1 to C2	200 (124)	134 (83)	56 (35)	17 (11)
B1 to C1	233 (145)	161 (100)	77 (48)	24 (15)

TABLE 1.2-2
Excerpt from FCC table of FM minimum separation distances for "short-spaced" stations (47 CFR §73.215)

Relation	Co-Channel	200 kHz	400/600kHz
A to A	92 (57)	49 (30)	25 (16)
A to B1	119 (74)	72 (45)	42 (26)
A to B	143 (89)	96 (60)	63 (39)
A to C3	119 (74)	72 (45)	36 (22)
A to C2	143 (89)	89 (55)	49 (30)
A to C1	178 (111)	111 (69)	69 (43)
A to C0	193 (120)	130 (81)	80 (50)
A to C	203 (126)	142 (88)	89 (55)
B1 to Bl	143 (89)	96 (60)	44 (27)
B1 to B	178 (111)	114 (71)	65 (40)
B1 to C3	143 (89)	96 (60)	44 (27)
B1 to C2	175 (109)	114 (71)	50 (31)
B1 to C1	200 (124)	134 (83)	71 (44)
B1 to C0	0215 (134)	153 (95)	81 (50)
B1 to C	233 (145)	165 (103)	99 (61)
B to B	211 (131)	145 (90)	68 (42)
B to C3	178 (111)	114 (70)	65 (40)
B to C2	211 (131)	145 (90)	68 (42)
B to C1	241 (150)	169 (105)	73 (45)
B to C0	266 (165)	195 (121)	83 (52)
B to C	268 (163)	195 (121)	99 (61)
C3 to C3	142 (88)	89 (55)	37 (23)
C3 to C2	166 (103)	106 (66)	50 (31)

TABLE 1.2-2 *(continued)*
Excerpt from FCC table of FM minimum separation distances for "short-spaced" stations (47 CFR §73.215)

C3 to C1	200 (124)	133 (83)	70 (43)
C3 to C0	215 (134)	152 (94)	81 (50)
C3 to C	226 (140)	165 (103)	90 (56)
C2 to C2	177 (110)	117 (73)	52 (32)
C2 to C1	211 (131)	144 (90)	73 (45)
C2 to C0	227 (141)	163 (101)	83 (52)
C2 to C	237 (147)	176 (109)	96 (61)
C1 to C1	224 (139)	158 (98)	76 (47)
C1 to C0	239 (148)	176 (109)	88 (55)
C1 to C	249 (155)	188 (117)	99 (61)
CO to C0	259 (161)	196 (122)	90 (56)
CO to C	270 (168)	207 (129)	99 (61)
C to C	270 (168)	209 (130)	99 (61)

TABLE 1.2-3
FM station classes

FM Station Classes
Zone I and Zone 1A
East U.S., Southern California, & Puerto Rico
• A (6 kW at 100 m HAAT)
• B1 (25 kW at 100 m HAAT)
• B (50 kW at 150 m HAAT
Zone II—Most of the rest of the U.S.
• A (6 kW at 100 m HAAT)
• C3 (25 kW at 100 m HAAT)
• C2 (50 kW at 150 m HAAT)
• Cl (100 kW at 299 m HAAT)
• C0 (100 kW at 450 m HAAT)
• C (100 kW at 600 m HAAT)

signals that are less than 50% of the RSS values of the higher signals already included. Computer programs now carry out all these functions and will map the AM station's coverage and interference contours.[3]

AM Stations

At the time of this printing, the FCC had published various "AM Revitalization" proposals for comment. These proposals could enable many AM stations to increase power in both day and nighttime operations and otherwise make technical changes that could be highly beneficial to some AM station owners, but could also result in increased interference in the band.

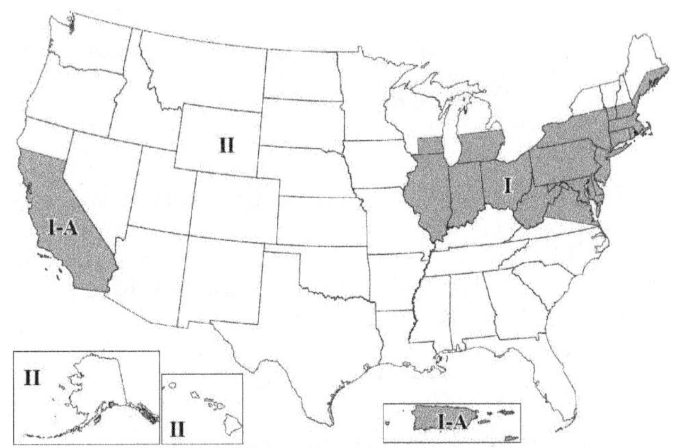

FIGURE 1.2-3 FM broadcast zones

LONGLEY-RICE

The Longley-Rice propagation model is used by the FCC in DTV coverage and interference analysis. Also, the method can be used for alternative showings of FM city grade coverage. In the mid-1960s, the National Bureau of Standards (now the National Institute of Standards and Technology, NIST) published Technical Note 101. Anita Longley, Phil Rice, A. Norton, and A. P. Barsis authored this two-volume propagation treatise in the course of their work at the Institute for Telecommunications Sciences (ITS) at Boulder, Colorado. The concepts expressed in these documents were incorporated into a series of computer routines that came to be known as the "Longley-Rice Model." Going well beyond the propagation factors taken into account by the FCC curves, the Longley-Rice method considers atmospheric absorption, including absorption by water vapor and oxygen, loss due to sky-noise temperature, and attenuation caused by rain and clouds. It considers terrain roughness, knife-edge refraction (with and

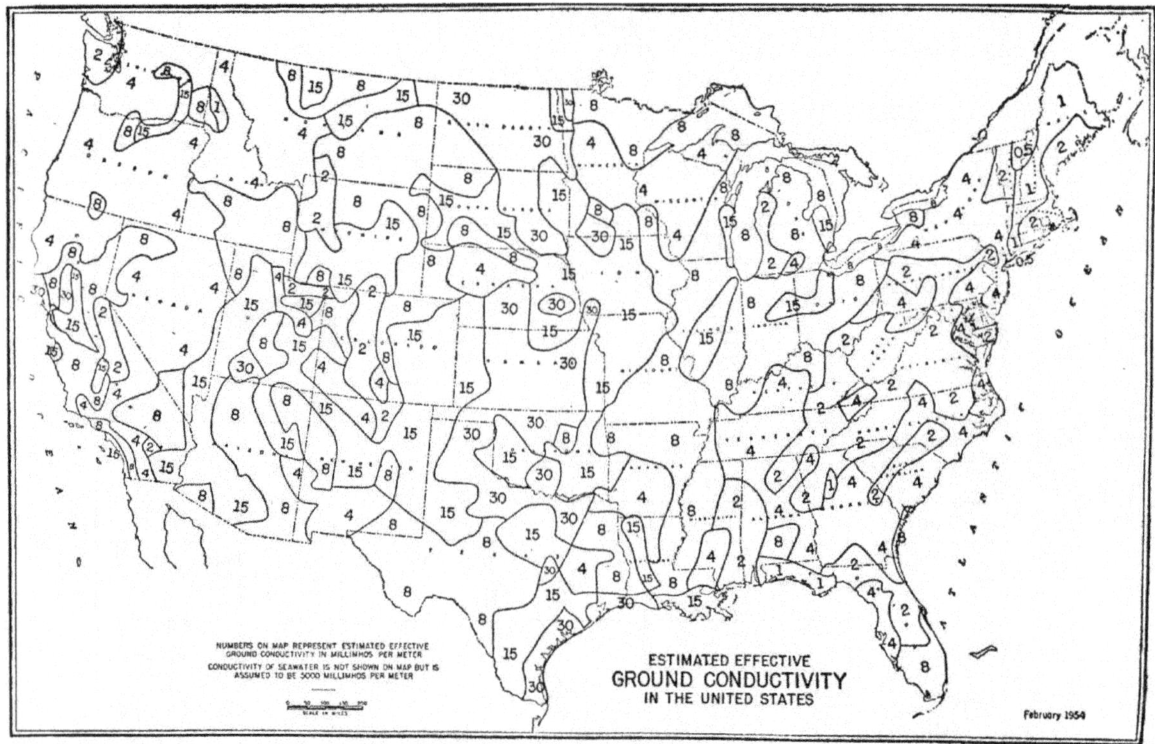

FIGURE 1.2-4 Estimated effective ground conductivity in the United States (1954)

without ground reflections), loss due to isolated obstacles, diffraction, forward scatter, and long-term power fading. Listed below are some of the input parameters required of the user when doing a Longley-Rice propagation analysis:

- Frequency:
  - 20–20,000 MHz
- Transmitter antenna parameters:
  - Transmitter radiated power
  - Transmitter antenna height (above mean sea level—meters)
  - Transmitter antenna height (above ground—meters)
  - Transmitter antenna pattern
  - Receiver antenna height (above ground—meters)
- System antenna polarization (vertical or horizontal)
- System Ground Conductivity (mhoS/m):
  - 0.001 = Poor ground
  - 0.005 = Average ground
  - 0.020 = Good ground
  - 5.000 = Sea water
  - 0.010 = Fresh water
- System dielectric constant (Permittivity):
  - 4.0 = Poor ground
  - 15.0 = Average ground
  - 25.0 = Good ground
  - 81.0 = Sea and fresh water
- System minimum monthly mean surface refractivity (adjusted to sea level):
  - 200 to 450 (available from refractivity map, 301 N-units is default)
- Climate Code:
  - 1 = Equatorial
  - 2 = Continental subtropical
  - 3 = Maritime subtropical
  - 4 = Desert
  - 5 = Continental temperate (default for U.S. continent)
  - 6 = Maritime temperate
  - 7 = Maritime temperate overseas
- Probability Factors:
  - Qt = Time variability—the percentage of time the actual path loss is equal or less than the predicted path loss (standard broadcast coverage = 50%)
  - Ql = Location variability—the percentage of paths (all with similar characteristics) whose actual path loss is less than or equal to the predicted path loss (used with area mode only)
  - Qc = Prediction confidence or "quality"—the percentage of the measured data values the model is based on that are within the predicted path loss (standard broadcast = 50%, ATSC 1.0 DTV = 90%)

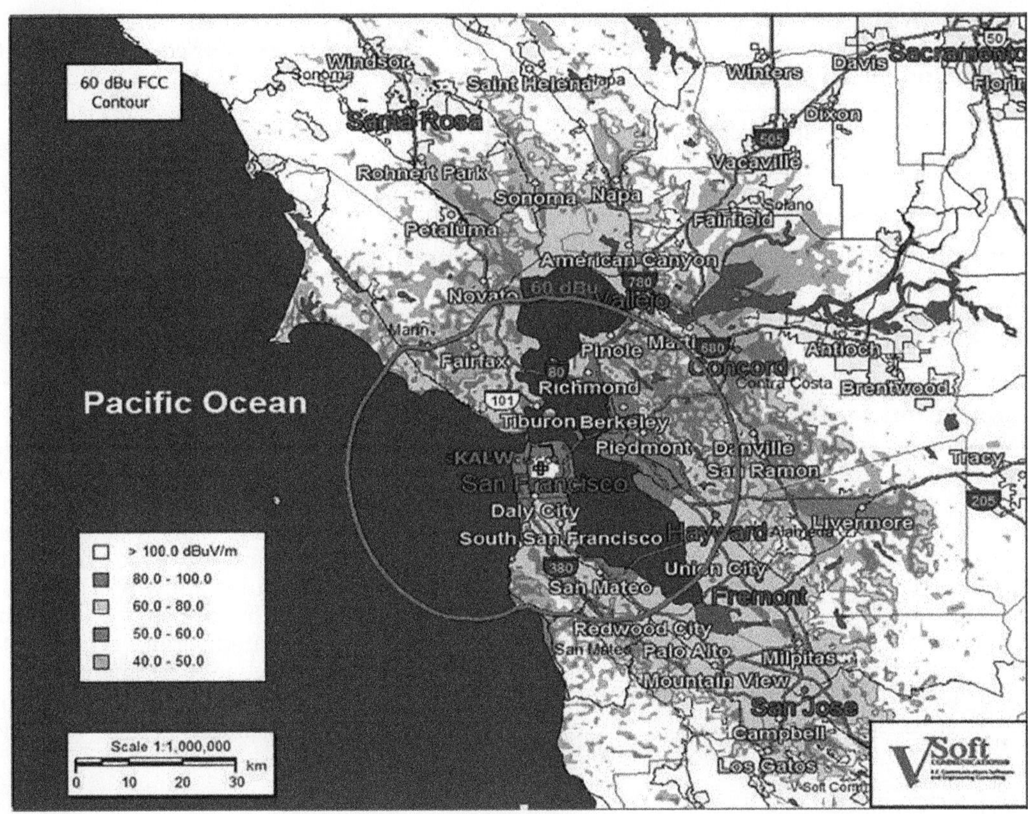

FIGURE 1.2-5 Comparison of FCC contour and Longley-Rice propagation prediction methods

Longley-Rice Computer Implementation

Since Longley-Rice evaluates the terrain along the entire path between the transmitter and receiver, a computer is essential to keep up with the high number of calculations required. The current Longley-Rice computer program used by the FCC is version 1.2.2.[4] In practice, Longley-Rice calculations are often used to evaluate signal strength within grid points. The grid point spacing can usually be set by the user. For display purposes, the signal strength value calculated for each grid point can then be coded either by color or by black-and-white line patterns. This procedure produces identifiable pools of coverage at certain signal values that make it easy for a reader to understand how geography and terrain interact to affect signals.

Further, population centroids within the grid points can be interrogated to produce population totals and other demographics at the locations of user-selected signal values. The map in Figure 1.2-5 contrasts displays of the Longley-Rice method and the FCC contour method on the same map. The Longley-Rice portion of the map was produced using a receive antenna of 9.1 meters, which is the same receiver height used in the FCC curves, 47 CFR §73.313. The receive antenna height for a Longley-Rice study can be set to 1.5 meters above ground, and the polarization can be set for vertical, which would replicate mobile listening coverage.

Longley-Rice, OET-69, and Digital TV "Noise-limited" Coverage Analysis

The FCC's Office of Engineering Technology (OET) Bulletin #69 provides detailed information on using Longley-Rice to determine digital television (DTV) coverage and interference. The bulletin is divided into three parts: service calculation of the coverage limited by terrain and existing interference, interference calculations, and the use of Longley-Rice in the methodology.

In the early 2000s, the Commission used OET-69 to analyze the service contours of existing analog TV stations in its original allocation proceedings. Since the modulation of analog TV and digital TV signals was entirely different, to facilitate the transition to DTV, the FCC used OET-69 to assign each qualifying analog TV station a digital counterpart channel that had comparable coverage. A decade and a half later, along came repacking of the TV band when the FCC used a slightly modified version of the original OET-69 procedure. Today, Longley-Rice has become the staple of the Commission's updated "TVStudy" program (see Tables 1.2-4, 1.2-5, and 1.2-6).

DTV coverage calculations were based on the DTV planning factors in Table 1.2-5. The planning factors assumed for home DTV receiving equipment including antenna systems. The values from Table 1.2-5 are calculated from the equation:

$$C/N = \text{Field} + Kd + Ka + G - L - N1 - Ns.$$

TABLE 1.2-4
Field strengths defining the area subject to calculation for DTV stations

Channels	Defining Field Strength, dBμ, to be predicted for 50% of locations, 90% of time
2–6	28
7–13	36
14–69	41–20 log[615/(channel mid-frequency in MHz)]

TABLE 1.2-5
Planning factors for DTV reception

Planning Factor	Symbol	Low VHF	High VHF	UHF
Geometric mean frequency (MHz)	F	69	194	615
Dipole factor (dBm-dBμ)	Kj	−111.8	−120.8	−130.8
Dipole factor adjustment	K_a	none	none	see text
Thermal noise (dBm)	N_t	−106.2	−106.2	−106.2
Antenna Gain (dBd)	G	4	6	10
Download line loss (dB)	L	1	2	4
System noise figure (dB)	N_s	10	10	7
Required Carrier to Noise ratio (dB)	C/N	15	15	15

OET-69 Interference Calculations

The calculated service area is divided into square cells (typically 2 km on a side), and the Longley-Rice point-to-point propagation model is applied to a point in each cell to determine whether the predicted field strength is above the "threshold for reception," which is the value calculated from the planning factors. If the observed signal is above the threshold, the 10% interfering signal strengths of the undesired stations are calculated at each point. For co- and adjacent-channel relationships, if the Desired-to-Undesired (D/U) relationship does not meet the minimum expressed in Table 1.2-6, the point is said to have interference. If the interference is masked at the point by another station's interference, the interference is not counted. Once the area of interference is determined, the population within the interference area is calculated using 2010 U.S. census population centroids.

TVStudy

With the advent of TVStudy, the Commission has upgraded the older OET-69 procedure to use a new, more accurate, 30-meter terrain elevation database.

TABLE 1.2-6
Interference criteria for co- and adjacent channels

Channel Offset	D/U Ratio. dB			
	Analog into Analog	DTV into Analog	Analog into DTV	DTV into DTV
−1 (lower adjacent)	−3	−14	−48	−28
0 (co-channel)	+28	+34	+2	+15
+1 (upper adjacent)	−13	−17	−49	−26

The program code is available online at the FCC.[5] Another difference in TVStudy compared to OET-69 is the use of a global grid rather than a station-based grid and the introduction of the ability to properly consider mechanical beam tilt. At the time of this writing, the Commission's rules allow a DTV station to cause interference up to 0.5% of the population in a station's calculated noise limited service contour. It is doubtful that this figure will change for the immediate future.

Other popular propagation prediction methods used in the United States include the following:

TIREM

TIREM stands for Terrain Integrated Rough Earth Model. The model is licensed by Alion Science and Technology Corporation, Annapolis, Maryland. This model started with a Tech Note 101 base, but it has been modified over the years to make up for believed inaccuracies in the Longley-Rice model. TIREM is used in numerous modeling and simulation (M&S) tools at the Department of Defense, including specialized routines for evaluating radio paths over sea water.

Since TIREM is a proprietary model, it is not possible to tell exactly what the code is doing, which makes the model less attractive to the FCC and other users. The use of the TIREM code must be licensed by the company.

Okumura Propagation Model

The basic Okumura propagation model uses the height above average terrain to calculate path loss and does not consider specific terrain obstacles. Developed initially for populated areas of Japan for frequencies between 150 and 1920 MHz, the current Okumura model uses the Okumura/Hata/Davidson implementation. Hata developed a set of equations that provide model predictions for computer use. The Davidson correction factors extend the frequency and base antenna height range. The Okumura model is one of the most widely used models for signal prediction in urban areas.

Hata/Davidson Model

Hata/Davidson does not use some of the adjustment factors included in Okumura, such as the slope of the terrain, street orientation, and correction for location on hills. Hata/Davidson uses the area type (urban, suburban, quasi-open, open) and includes corrections for the receiver antenna height. The model also includes frequency and distance corrections to extend the limitations on Hata, particularly the distance range to 300 km. The model computes the basic median field strength and applies adjustments based on the 3 to 16 kilometer path segment HAAT of the transmit antenna, rather than the topography of the entire path.

Cost-231 Propagation Model

The original Cost-231 model was modified by Hata to create the Cost-231/Hata version. The Cost-231/Hata model uses the HAAT along each radial to determine the attenuation based the following equation:

$$\text{Path Loss (dB)} = 46.3 + 33.9*\log(F) - 13.82*\log(H) + [44.9 - 6.55*\log(H)]*\log(D) + C \text{ where}$$

F = Frequency (MHz)

D = Distance between base station and receiver (km)

H = HAAT in the direction of the receiver (m)

C = Environmental-correction factor (dB)

The Hata corrections for receiver height and frequency are applied to calculate the final attenuation.

International Telecommunications Union—ITU-R P.1546

The ITU-R P.1546 propagation model was developed for field strength predictions for terrestrial services in the 30 MHz to 3000 MHz frequency range. It uses a set of propagation curves that are based on measurement data mainly relating to mean climatic conditions in temperate climates. The model considers the transmitter height above average terrain and the receive antenna height, and incorporates a correction for terrain clearance angle when making field strength predictions. The ITU model is used widely in Central America and in Europe.

INFLUENCES ON COVERAGE—REAL-WORLD FACTORS

Antenna Height vs. Power

Generally, the more power a transmitter has, the greater its coverage will be. However, antenna height and antenna patterns also have large influences.

Propagation at FM and TV frequencies is basically line-of-sight, so the antenna height should, at least, be above the tree lines and other obstacles so the radio path to the desired coverage area is unobstructed. §73.315 (b) of the FCC rules says:

> The location of the antenna should be so chosen that line-of-sight can be obtained from the antenna over the principle city or cities to be served; in no event should there be a major obstruction in this path.

While this provision of the rules is seldom enforced by the Commission, it is a good yardstick to be followed by all stations in locating transmitter sites. While FM and TV radio waves will go beyond the radio horizon, the area served there will be impacted by multipath interference (also called multipath fading). Since over-horizon radio waves must bounce around terrain and objects, the waves often take different paths to the radio, causing phase distortion. For FM radio, distortion of the audio is the result. Multipath causes what has been called "picket fencing," where the audio flutters with a vehicle's movement. If the multipath is bad enough, the FM stations will suffer audio distortion, where the voices become muffled and specific notes in music become lost. For DTV, using 8VSB modulation, modern DTV tuners have been programmed to ignore the echoes of multipath as long as the delayed multipath signal fits within the receiver's timing window specifications. However, multipath can still be a problem when the bounced signals exceed the timing window, resulting in reception with pixilated images or lost reception altogether. FM radio broadcast receivers cannot correct for multipath in the way DTV receivers do.

To calculate the radio horizon without correction for Earth bulge, the following formula can be used:

$$\text{Radio Horizon (km)} = 3.57 * \text{square root (antenna height)}$$

To many in the broadcast business, the question that arises is, "Which is better, antenna height or transmitter power?" While this is a very good question, the answer may not be as straightforward . . . "It depends."

As an example, a Class A FM station can have up to 6 kW at an HAAT of 100 meters; however, if the antenna is mounted on a tower at a location where the HAAT is greater than 100 meters, the FCC requires an "equivalent" reduction in power to keep the station within the class maximum, 60 dBμ, service distance of 28 kilometers. If the antenna of this Class A station is mounted on a tower where it has an HAAT of 305 meters (1,000 ft.), the equivalent power would have to be reduced to 0.64 kW. So, to answer the question, we need to ask, "Where is the tower?" Is it city bound or at a more rural location outside the city? If the tower is within the city, 6 kW would provide nearly 10 dB more power at nearby population areas than an antenna located outside the city limits. Higher power will support better building penetration in an electrically noisy environment; however, for whatever reason, if the station's antenna cannot be moved closer to the city, a higher antenna height will provide a better line of sight to the city, even though it may be deficient in its ability to penetrate buildings.

With FCC approval, a number of station owners have moved station antennas closer to metropolitan areas to increase the population within the service area but, due to the allocation requirement, the stations are unable to be moved close enough to competitively serve the metropolitan area. This results in what has been called "rim-shot" stations, where the important part of the station's signal must travel beyond the radio horizon. In this case, power trumps antenna height as the key element in pushing the radiofrequency signal over the horizon to serve areas past a station's "rim." Yet, as said previously, stations transmitting beyond the radio horizon will have more problems with multipath distortion.

Using the Class A FM station example, a higher antenna sometimes provides an unusual benefit to stations that are short spaced, since beyond a certain height, at the equivalent power,[6] the F(50,10) interference contour is actually reduced, so the station would cause less interference. For example, the 0.64 kW station at 305 meters HAAT produces a co-channel 40 dBμ interference contour that travels 62.2 km, while the Class A station at 6 kW, from a maximum antenna height of 100 meters, produces the same 40 dBμ interference, but this time it travels 67.9 km. So, in a contour-to-contour relationship, the high antenna will allow more power to be used than if the antenna were at a lower height.

Antenna Pattern

Directional antennas are used by AM, FM, and TV stations to protect other stations from interference and to present a good signal over a desirable area. Some directional antennas are used to direct the signal away from unpopulated areas, such as ocean, mountains, or large bodies of water. This, in turn, increases the antenna gain and results in the need for less power at the transmitter, consequently reducing the station's electric bills.

In mountainous areas, a directional antenna can be used to direct the signal away from nearby hills to prevent multipath reflections from bouncing off the hills and reflecting back into the desired coverage area. FM radio and TV services have their own rules regarding the use of directional antennas.[7] The FM regulations specify that for an antenna providing contour protection to other stations, the relative field pattern of the antenna must not exceed a change of more than 2 dB for each 10 degrees of azimuth change. The rules also say that the ratio of the maximum to minimum horizontal radiation cannot exceed 15 dB. The TV rules have no such requirements.

When an Omnidirectional Antenna Becomes Directional

The way an omnidirectional antenna is mounted to a tower can make a big difference in coverage, resulting in the actual pattern becoming directional in nature. Any side-mounted antenna will have a significantly distorted pattern due to the interaction between the antenna and the tower. The vertical plane radiation is particularly vulnerable in the direction opposite to the side of the tower on which the antenna is installed. The azimuth patterns shown in Figures 1.2-6 and 1.2-7 display what can happen. Note that the vertical pattern is sucked in due to the influence on the pattern of the vertical elements in the tower. The horizontal plane radiation on the opposite side of the tower is severely limited when the antenna is mounted on a tower face. The optimal rotation and bracketing of the antenna on the tower will keep the vertical and horizontal pattern together as much as possible. This helps to offer a true circularly polarized wave.

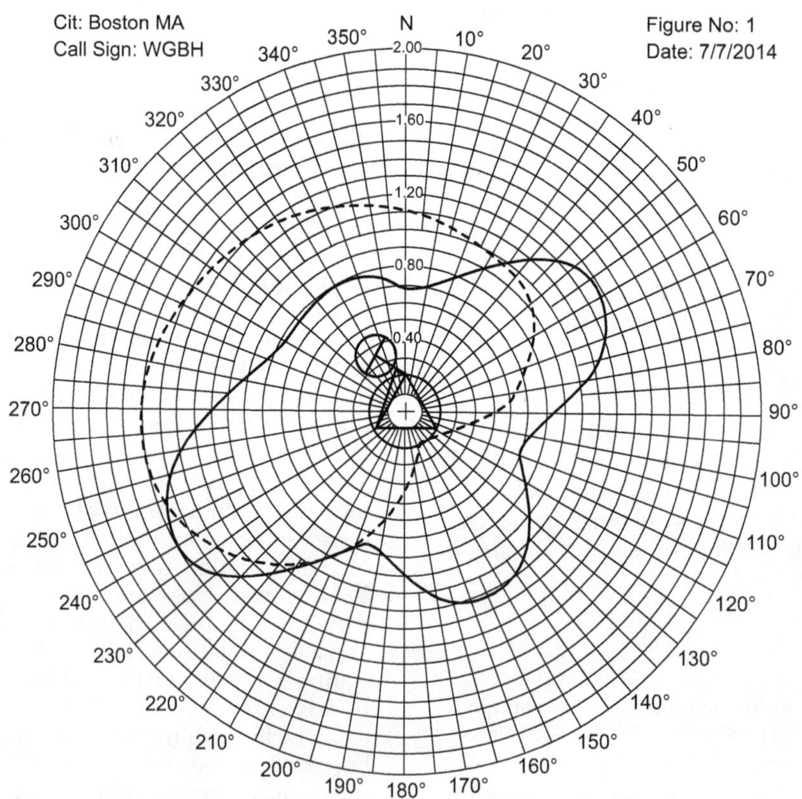

FIGURE 1.2-6 Example antenna pattern for a skewed, leg mount antenna. Dashed black = vertical plane radiation envelope; red solid = horizontal plane.

Courtesy of ERI Inc.

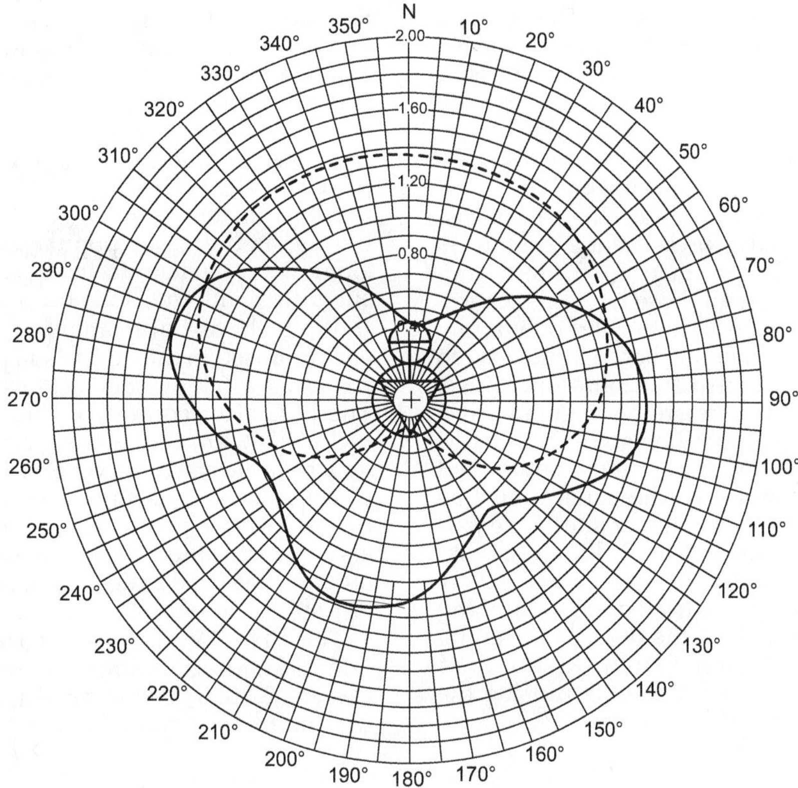

FIGURE 1.2-7 Example antenna pattern for a face-mount antenna. Dashed black = vertical plane radiation envelope; red solid = horizontal plane.

Courtesy of ERI Inc.

Nearly all antenna manufacturers offer an "optimization" package, where the omnidirectional antenna is mounted on a scaled tower at the testing range and rotated while the pattern is being recorded. Numerous mounting arrangements are tried until the best pattern to fit the needs of the client is established. This is done for the purpose of filling holes and making the omnidirectional coverage more uniform. The process should not be undertaken for the purpose of solely transmitting a higher than licensed power in a given direction. In a recent case, the FCC chided a North Texas FM station for going well beyond reasonable omnidirectional optimization by using a testing range to adjust antenna mountings and using parasitic elements to shape the pattern that resulted in a radiated power in one direction of nearly 2.75 times its licensed power and that caused serious interference to another station.[8] After the other station complained to the FCC, the Commission took its own measurements and required the offending station to permanently reduce power to nearly 1/3 of its licensed power. This was the first time the Commission took this kind of interest in the optimization of an "omnidirectional" antenna, so the issue has become somewhat murky. How does one tell the difference between purposely adjusting the mounting of the antenna to do a better job covering a licensed coverage area or clandestinely trying to get more coverage than granted under the station's license? As related by an FCC engineer, an antenna becomes directional as soon as a parasitic element is used. Nevertheless, optimization of an omnidirectional antenna is good engineering practice and can help a station do a better job in serving its assigned coverage area.

For FM, one way to achieve an omnidirectional antenna pattern is to mount the antenna on a pole at the top of the tower (although there is still a limited effect on directionality) or to use a panel antenna, where reflector-backed radiators are mounted on all sides of the tower. Depending on the size of the pole, the pattern may still be affected, particularly in the vertical plane; however, the resulting pattern will be much more uniform than a side-mounted antenna. Panel antennas that wrap around a tower can produce a waffling pattern due to the phase cancellation between the panels; however, the end product will be a pattern that has much better uniformity than a side-mounted antenna. Optimization for TV antennas is typically not required, since such antennas are usually predesigned to achieve the desired pattern even when side-mounted.

AM Interference Caused by Re-radiation

The directional patterns of an AM station are also impacted by nearby vertical structures such as cell towers.[9] These structures can become part of the station's directional pattern by re-radiating RF energy, so that more energy may be directed toward a protected station, thus causing prohibited interference. The FCC has developed rules for protecting the AM patterns of existing stations. These rules impact new towers, where FCC licensed facilities are assigned, that were built within the smaller of either 10 wavelengths (at the AM frequency) or 3 kilometers from a directional array and within one wavelength from an omnidirectional tower.[10]

Exclusions for short towers within this critical distance affecting directional arrays include all towers having electrical heights less than 36 degrees and, for omnidirectional towers, all those having electrical heights less than 60 degrees.[11] These rules only apply to towers, not to other structures that, if built close to AM towers, may cause pattern distortion, so buildings, power lines, and water towers are not included. If a new tower falls under the protection guidelines of this section of the rules, the AM station must be notified and an analysis of the impact on the pattern must be made. If it is found that the new tower affects the pattern in any direction by more than 2 dB, a remediation by the tower owner must take place. This usually amounts to detuning the new tower, often by running skirt wires along the length of the tower and tuning the skirts with an L/C circuit so that the RF from AM frequency is routed to ground, therefore causing the tower to be invisible to the AM frequency.

Short Spacing with Directional Antennas

FM station owners who wish to locate their antennas at convenient locations that do not meet the minimum distance separations required under 47 CFR §73.207 of the FCC rules may choose to use §73.215 short-spacing rules. In making application to the FCC for permission to construct at these short-spaced locations, the consulting engineer will often develop a directional antenna pattern that prevents the station from causing contour overlap interference to other stations and, at the same time, protect the station from receiving contour overlap interference. Assuming that it is an existing commercial band FM station, the engineer must design a pattern that will cover the principal city with a 70 dBμ signal or better over 80% of the principal city's area or population.[12] New FM stations require 100% coverage within the 5 mV/m contour, while new AM stations require 80% coverage to the principal city. AM stations already licensed require 50% coverage with the 5 mV/m contour.

A proposed directional FM pattern must meet the FCC's antenna pattern requirements with an azimuth field that cannot change more than two dB for every 10 degrees of azimuth change in the directions of contour-protected stations, and the minimum-to-maximum ratio, sometimes called the "front-to-back" ratio, of the proposed pattern may not exceed 15 dB. Once this pattern is developed, it becomes the "theoretical pattern" and is submitted to the FCC with the application for construction permit.

Assuming the FCC approves the application and issues the permit to build the station, the owner and engineer will work with an antenna manufacturer to design a real antenna meeting the pattern limits. The resulting product is called the "as-built" antenna. For the license application, the engineer will study the antenna "proof" from the manufacturer to determine if the as-built pattern envelope will fill up to 85% of the proposed area of the theoretical pattern envelope. If this can be done, the engineer will take the as-built pattern and plot the city-grade signal contour to show that the antenna will provide the principal city with the required area or population coverage. This data along with the manufacturer's antenna proof must be submitted with the FCC license application.

While the as-built pattern may not fill all the coverage promised in the theoretical pattern, the theoretical pattern is what must be protected by other stations.[13] The FCC also requires two additional exhibits to be included with the license application, a certification by a licensed surveyor that the azimuth of the directional antenna's pattern main lobe is within one degree of that promised in the construction permit and a certification by a qualified engineer that the antenna has been installed according the manufacturer's specifications.

For VHF TV stations channels 2 through 13, the minimum-to-maximum azimuth field ratio cannot exceed 10 dB, while for the UHF region of the TV band the minimum-to-maximum ratio cannot exceed 15 dB.

Polarization

In the early days of FM, all antennas were horizontally polarized, but eventually stations began adding a vertical component to the antenna patterns. Initially, this was done by adding a set of vertical antenna bays under the horizontal bays and splitting the transmitter power into each plane. This resulted in a 3 dB reduction of power to each plane and required a larger transmitter to make up for the power loss.

Today, only a few installations still use separate horizontal and vertical antennas. Modern FM broadcast transmitting antennas are typically cross- or circularly polarized, where both the vertical and horizontal components transmit from the same antenna. This type of polarization improves the signal reception with listeners' randomly polarized antennas. With the exception of translators, boosters, LPFM stations and noncommercial FM stations protecting channel-six TV stations, the FCC will not permit FM stations to have vertical-only polarization. The vertical component of the azimuth pattern must be contained entirely within the horizontal component azimuth pattern. Through propagation studies, it has been determined that the vertical component in areas of highly divergent terrain can lead to multipath reception problems, therefore many stations in mountainous areas use "elliptical" polarization, where the vertical component is present but at a significantly lower power ratio than the horizontal component. Some television stations, particularly those in the lower VHF channels, have introduced a vertical component to their antennas with improved coverage results.

Vertical Elevation Field and Beam Tilt

There are tales of stations that found high mountaintops for their transmission sites, but when the transmitter was turned on, the city just below the elevated tower had a very poor signal. In such cases, the station's RF signal was going over the heads of the listeners below. The graph shown in Figure 1.2-8 shows a 10-bay, full-wave spaced antenna (without null fill).[14]

Looking at Figure 1.2-8, one can see that after a downward angle from the antenna center of 8.5 degrees, there is a big null and a significant reduction in emissions at larger angles toward the ground after that. For illustration, the graph in Figure 1.2-8, displays 2.5 degrees of electrical beam tilt. This is a rather large amount of beam tilt. Usually, beam tilt depression angles are kept at angles at or below one degree. Beam tilt is often introduced for tall antennas where there is a sizeable population center at a distance between 1 to 10 miles from the antenna.[15] To provide good all-around coverage with fewer nulls, an antenna with fewer bays could be used, as is shown in Figure 1.2-9. This antenna has two bays and much higher downward emissions closer in to the antenna.

Note that this antenna produces one large null at 30 degrees. If there is significant population at this angle from the antenna, most manufacturers can add "null-fill" to the vertical elevation pattern to partially fill the null.

Another way to solve the problem of reaching the city would be to use mechanical beam tilt, where the entire antenna is angled downward to serve the city below. A combination of both techniques can also be used.

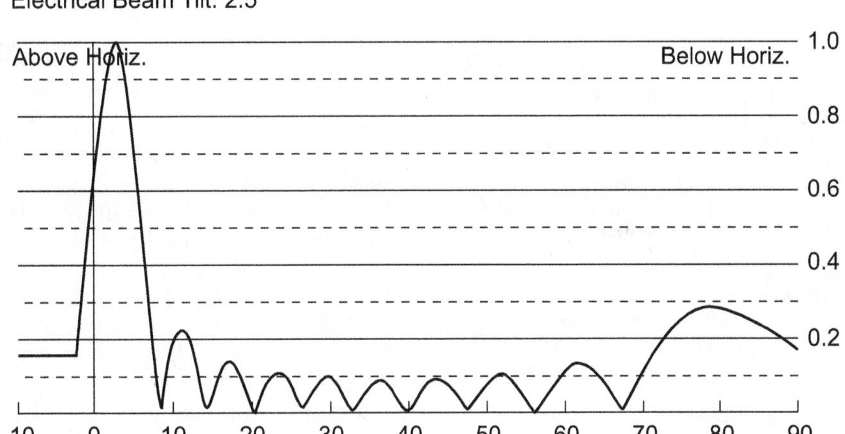

FIGURE 1.2-8 Vertical elevation example of a 10-bay, full-wave spaced antenna (without null fill). X-axis (angle in degrees) is the angle with respect to the horizon.

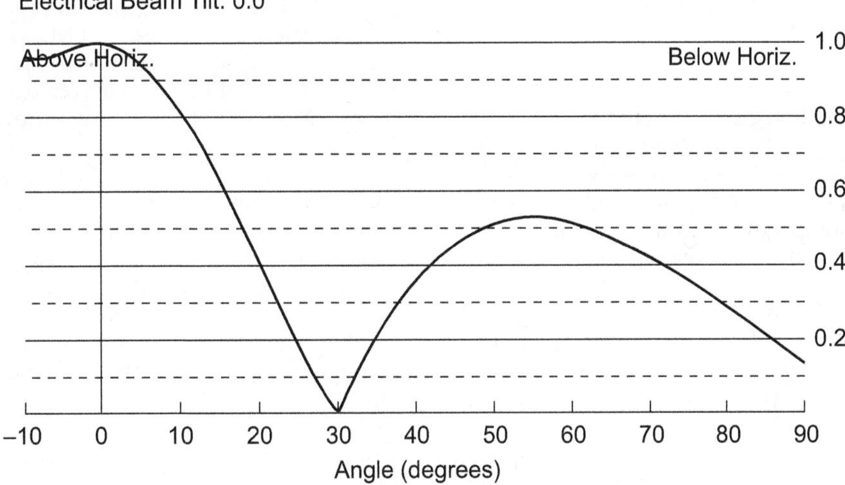

FIGURE 1.2-9 Vertical elevation pattern of antenna example showing good all-around coverage. X-axis (angle in degrees) is the angle with respect to the horizon. This antenna has two bays and much higher downward emissions closer to the antenna compared to what is shown in Figure 1.2-8.

A simple formula can used to calculate the beam tilt angle of a place in respect to the transmitting point:

$$\text{angle} = \tan^{-1}\left(\frac{(h_1 - h_2)}{d}\right)$$

Where:

h_1 = Elevation of base antenna (ft)

h_2 = Elevation of mobile antenna (ft)

d = Distance (ft)

Depression Angles

FCC rules for TV require that an applicant for a TV station submit a table of the depression angles calculated from the antenna's center of radiation. 47 CFR §73.625 (b) "Determining Coverage" defines the calculation procedure of the radiated power producing the coverage contour while considering the depression angle as follows:

> The effective radiated power to be used is that radiated at the vertical angle corresponding to the depression angle between the transmitting antenna center of radiation and the radio horizon as determined individually for each azimuthal direction concerned. In cases where the relative field strength at this depression angle is 90% or more of the maximum field strength developed in the vertical plane containing the pertaining radial, the maximum radiation shall be used. The depression angle is based on the difference in elevation of the antenna center of radiation above the average terrain and the radio horizon, assuming a smooth spherical earth with a radius of 8,495.5 kilometers (5,280 miles) and shall be determined by the equation A = 0.0277 square root of H where:
>
> - A is the depression angle in degrees.
> - H is the height in meters of the transmitting antenna radiation center above average terrain of the 3.2–16.1 kilometers (2–10 miles) sector of the pertinent radial.
>
> This formula is empirically derived for the limited purpose specified here. Its use for any other purpose may be inappropriate.

In-Band/On-Channel (IBOC) Digital Radio Coverage

On October 10, 2002, the FCC approved IBOC for FM stations at an injection level of 1%, which is –20 dBc (dB below carrier). IBOC system designers had indicated that this level should approximately match the analog coverage, but this did not prove to be the case. After receiving various petitions to increase the IBOC injection level, the Commission gave broadcasters a blanket authorization to increase to –14 dBc.[16] This permission was granted "across the board" to all FM stations and no further permission was required. The FCC also agreed to allow the injection level to reach as high as –10 dBc with proof of noninterference to first-adjacent stations.

A comprehensive study of FM IBOC coverage was the 2008 Digital Radio Coverage and Interference Analysis study that National Public Radio (NPR) Labs carried out with funding from the Corporation for Public Broadcasting.[17] NPR studied 850 public radio stations, 75 of which were studied in great detail using actual field measurements. The results of the study were summarized in the final report to CPB.

NPR Labs determined that 85% of the mobile analog coverage area was matched at the 1% level. Indoor coverage for IBOC matched only 38% of the analog coverage. When the IBOC power was increased to 10%, indoor and portable receiver coverage was found to be reduced from analog by 22% and 6%, respectively. However, mobile IBOC coverage averaged an increase of 117% over the analog coverage.

NPR Labs concluded that the percentage coverage of IBOC relative to analog FM varied widely from station to station. The wide variance is tied to the variance among stations of first-adjacent signals, which act as interferers to the digital sidebands of the IBOC transmissions. The IBOC coverage was worse when there were two first-adjacent channel stations on both sides of the station transmitting IBOC.

USING TRANSLATORS AND BOOSTERS TO EXTEND COVERAGE

Originally, FM translators were approved by the FCC to provide coverage in mountainous or hilly areas where terrain blocks the path from the transmitter to the listener's location. Often these translators provided the only signal to a location in the foothills that were blocked on all sides from receiving the FM broadcast band. Translators were not allowed to originate programming and they could only re-transmit the primary station's audio from an off-the-air pickup.

Slowly, over the years, the original purpose of translators has morphed into something else. Now the FCC allows any translator in the noncommercial educational reserved band to extend its coverage beyond the primary station's 60-dBμ protected contour. In fact, a number of noncommercial groups have translators located in several U.S states, far from their primary stations, that are allowed, under the rules, to receive programming from the primary station via satellite.

All translators and boosters are secondary in the eyes of the FCC (as are LPFM stations which are governed by similar rules, although LPFM stations have protections that FM translators do not). In other words, if a full-service station files an application for a new frequency or an upgraded class that conflicts with the frequency of a translator, the translator will have to move to a new channel or go off the air. New translator licenses are not issued except for when the FCC opens an official translator "window." The last translator window opened in 2003 and it resulted in applications for more than 13,000 translators.[18] Since the FCC put no limit on the number of translators for which any single applicant could apply, the number of translator applications from single owners or groups was often well over the hundreds and in some case thousands. The Commission has since suggested that if they would open another translator application window, they would not do it the same way.

47 CFR §74.1235 of the Commission's rules sets the maximum power that a translator can run at a given antenna radial HAAT.[19] For stations in zone 2 (see

Figure 1.2-3), on the west side of the Mississippi River, the maximum power is 250 watts at a radial HAAT of 107 meters or less. This produces a 60-dBμ coverage area contour of 13.3 km from the antenna. For antenna radial HAATs above 107 meters, the power is scaled down until it reaches a low of 10 watts for a radial HAAT of 481 meters. As a noteworthy example, when a translator is on a mountaintop where the highest radial HAAT reaches 1,500 meters, the power would be scaled to 10 watts, but this would create a 60-dBμ contour that travels approximately 20 km, which produces a coverage area equal to some full-service, Class A stations.

Translators on the east side of the Mississippi River or in California's zone 1-A are limited to 250 watts at a radial HAAT that is less than or equal to 32 meters. This produces a 60-dBμ coverage area contour that travels, on the average, 7.1 km. These Zone 1 and 1-A translators have to scale power down to 10 watts when the radial HAAT is equal to or greater than 141 meters. The Commission also has some specific policies regarding moving a translator's location. A translator cannot be moved to a new site unless the 60-dBμ service contour at the new site overlaps, even minimally, the 60 dBμ at the existing site.[20]

There are three kinds of translators:

Standard Range Extending Translator

Translators on noncommercial education-reserved frequencies from 88.1 to 91.9 MHz may be owned by the primary FM station to provide coverage to a distant community that may lie past the station's 60 dBμ. On the other hand, translators in the commercial band that extend the station's licensed protected contour in any direction are not allowed. However, a translator in the commercial band can be owned by a third party, and that translator can be located outside the station's protected service contour. Under the rules, the commercial station cannot pay a third party, in any manner, for this service; however, the commercial station can provide technical assistance to the owners of the translator as long as no money changes hands. Translators for commercial stations may run short 30-second commercial announcements, not more than once per hour, to help pay for the translator's operations. Commercial station translators must receive the translator from a direct off-air broadcast and not through alternative delivery means such as satellite, microwave, or Internet. Translators in the reserved band may use any means of incoming alternate program delivery including satellite, microwave, and Internet.

Fill-in Translators

Fill-in translators, whether for noncommercial educational stations or commercial stations, may not extend the protected service contour of the primary station. Fill-in translators do not have many of the power restrictions found in 47 CFR §74.1235 of the FCC rules. While the maximum power is still restricted to 250 watts, there is no requirement to reduce the power from the maximum for higher antenna heights. Fill-in translators may use any means of "terrestrial" delivery, so delivery of programming by satellite is not allowed. This prohibition does not apply to commonly owned translators operating in the reserved band.

While not specifically approved by the Commission in its rules, primary stations that are operating with IBOC have been allowed to use any of the primary station's IBOC audio channels for input. In this manner, numerous stations have been able to place their IBOC multicast channels on the FM band as analog services.

FM Translators for AM Stations

The FCC now approves FM translator stations for use by AM stations. These translators act as fill-in translators, having basically the same rules as FM translators, with the exception that the translators must remain within the 2 V/m service contour of the AM station and, as of this writing, within a 25-mile radius of the AM station, whichever is the lesser.[21] Translators licensed as fill-in stations for AM stations may also originate programming in the evening, when, due to skywave interference, the power of the primary station must be reduced or turned off. These rules have been made a part of the Commission's AM Revitalization initiative to help struggling AM stations gain a better foothold in their markets.

Boosters

All boosters operate on the same frequency as the primary station. A commercial or noncommercial station can apply for a booster at any time without waiting for an application window. The rules for boosters are similar to fill-in translators in that the primary station's service contour cannot be extended by the booster's service contour. The maximum effective radiated power for a booster station is 20% of the main station's maximum class power.

Co-channel boosters can be trouble for the primary station if they have not been properly planned. Many boosters have been torn down (perhaps as many as those that remain on the air) due to the interference they have caused to the primary station. For a booster, it is always important to have a significant terrain blockage between the primary station and the booster. If not designed properly, the booster can interfere with the primary at places where the signals of both transmitters become closely equivalent. While limited synchronization can be introduced, it is costly and difficult to maintain. The synchronization works most efficiently in areas when the listener is located on a straight line connecting both the primary station and the booster, and where the booster signal can lock in to the timing of the primary station. When the booster is not time synchronized, the listener will hear distortion and experience the fluttering of picket fencing when listening in a moving automobile. This makes widespread optimization difficult, because there will always be some areas where the booster and primary station are not in sync. However, where significant terrain blockage comes into play, boosters can be a handy way of improving coverage to isolated areas.

Single Frequency Networks

The single frequency network (SFN) concept uses several transmitters to serve a market within the protected contour of the original primary station. The transmitters are actually boosters that occupy the same frequency as the FCC-assigned primary frequency. The idea is to synchronize the boosters to cause the least amount of interference to the reception of the primary station over populated areas. SFNs can be designed to work as a single station, thus using several smaller towers rather than one large tower. SFNs are found more often in the DTV service, where the digital characteristics of the transmitter signal make for better synchronization.

Thus far, SFNs for analog FM are rarely successful. One company coined a word for its newly advertised SFN product, calling it "ZoneCasting," and they have patented a technique that combines FM radio, cell towers, and wireless technologies to separate the primary station's signal into various geographic zones.[22] Using the ZoneCasting system, within each zone the FM transmitters can transmit simulcast programming (i.e., the programming on the primary signal) or discrete content, determined by the licensed station on a minute-by-minute basis (however, at present, the FCC rules prohibit origination of content by boosters). In this manner, the zone boosters can theoretically break off and deliver localized commercials to different service areas of the primary station.

SFN transmitters are usually placed in populated areas of high density and synchronized with other SFN transmitters. The capture ratio of modern receivers will favor the stronger signal of the SFN transmitters in areas close to the booster antennas. However, problems arise between the SFN transmitters in areas where the signal strength from each transmitter is close or equal and the time of arrival of the SFN signals do not match. During system design, the engineering planner should try to locate these areas of interference in the least populated areas. The FCC has approved SFN operation on an experimental basis.

INTERFERENCE

Coverage can often be marred by interference. The FCC protects all (primary) licensed stations to their protected contour based on the FCC curves. As stated under the section "FCC Method" above, the FCC method uses the terrain elevation evaluation to calculate the radial HAATs and to determine the power along the azimuths where protection is needed. It has been argued that for FM, looking only at the terrain from 3 to 16 kilometers does not reflect the real world, and whole mountains that block interference may be beyond the 16-kilometer distance and be missed entirely. However, while some may think this method is unfair, it is a more conservative approach, since if there is terrain beyond the 16-km point that blocks the interfering station's path, the FCC treats that signal without the attenuation. Consequently, the method tends to overprotect some stations in hilly areas, thus the FCC is handling protection from interference in a very conservative manner.

As stated previously, commercial FM stations use the minimum separation rules under 47 CFR §73.207 first and, if they need to, they can apply the short-space rules under §73.215. Commercial channel stations that short space use the contour-to-contour method. Full-service stations are not allowed to receive contour overlap interference or cause it. FM translators may receive contour-overlap interference from another station; however, they may not cause it. The F(50,10) curves are used to determine the location of a station's interference contour and the undesired to desired U/D ratios are applied at the F(50,50) contour of the protected station to determine if overlap of these contours exist.[23] The undesired-to-desired (U/D) ratio, established by the FCC for co-channel stations, is −20 dB, −6dB for first-adjacent stations, and 40 dB for second- and third-adjacent channels. As an example, if the class of station is B and the interfering station is co-channel, the interference contour will be the 48-dBμ F(50–,10) contour and the protected contour will be the Class B station's F(50,50) 54-dBμ contour.

The map in Figure 1.2-10 shows the proper FCC contour-to-contour relationship between two first-adjacent FM stations having a U to D ratio of −6 dB, in the noncommercial educational reserved band. The black contour lines are the F(50,50), 60-dBμ protected contours, and the red lines are the F(50,10), first-adjacent, 54-dBμ interfering contours. As the reader can see, the contour lines do not cross.[24]

While the FCC's allocation procedures are used to protect station-to-station interference, other types of interference also need to be addressed here. Environmental interference from electric and electronic equipment can be very harmful to good coverage. This problem is particularly acute for AM stations, but it can be a problem for FM and TV reception as well. Here is a short list of possible causes of environmental interference:

- Dimmer switches
- Overhead power lines
- Cable TV leakage
- Florescent lights—ballasts
- LED lights
- Switching power supplies
- Computers—big and small
- Televisions, particularly with plasma screens
- Motors
- Switches
- Car ignitions
- Lightning
- Electronic HVAC filters
- Internet routers and modems
- Improperly installed TV antenna pre-amplifiers
- Cell phones
- Microwave ovens

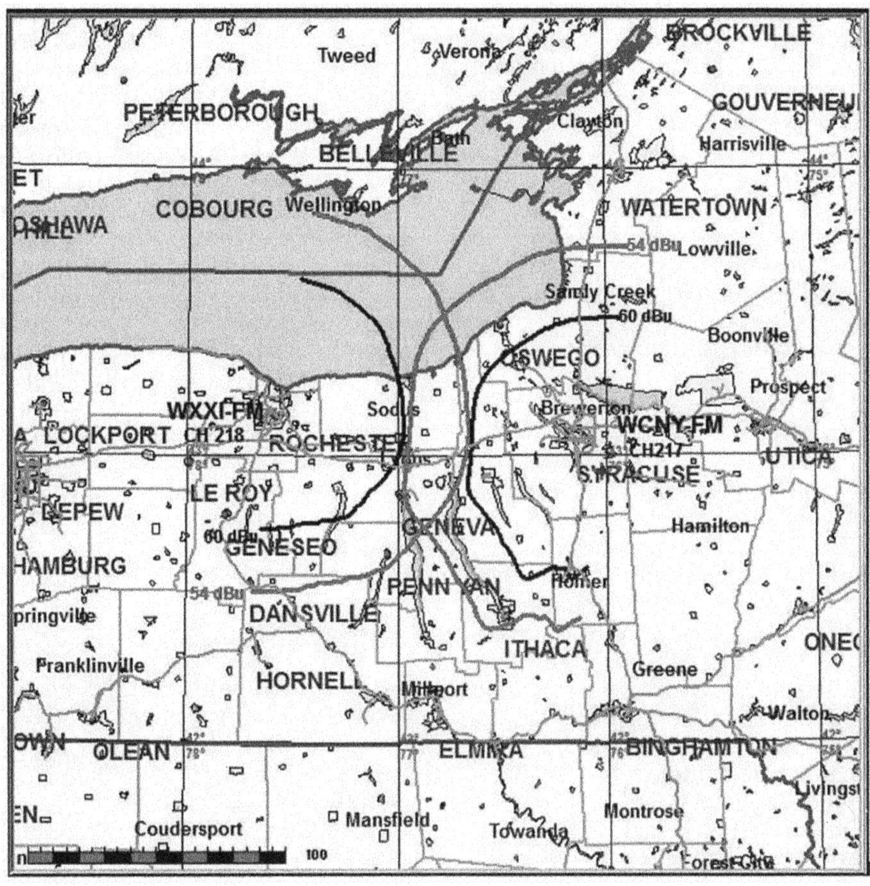

FIGURE 1.2-10 Example of proper FCC contour-to-contour relationship between two first-adjacent FM stations

Increasing the signal level above the noise floor is one way of coping with environmental notice. This can be done by increasing the transmission power. However, that is limited by FCC licensed power requirements as well as by station-to-station interference. Modern digital receiver processing can use digital signal processor (DSP) filtering to provide a noteworthy assist in lowering the noise level.

Grandfathered short-spaced stations are primarily older FM stations in the eastern half of the country that do not meet the minimum separations for non-interference. Much of the history is lost as to why there are so many of these stations in the East. Contour overlaps are commonplace, and all of these stations cause some degree of interference to each other. This situation has been exacerbated by the addition of IBOC to these already short-spaced stations. Add in environmental noise and what results is a difficult interference situation to correct.

Translators and LPFM Interference

Full-service FM station owners are concerned about how co- and adjacent channel LPFM or FM translators may affect their listenership, some of which may be beyond the protected contours of these full-service stations. 47 CFR §74.1203 (a) of the Commission's rules reads as follows:

> (a) An authorized FM translator or booster station will not be permitted to continue to operate if it causes any actual interference to:
>
> . . . (3) The direct reception by the public of the off-the-air signals of any authorized broadcast station including TV Channel 6 stations, Class D (secondary) noncommercial educational FM stations, and previously authorized and operating FM translators and FM booster stations. Interference will be considered to occur whenever reception of a regularly used signal is impaired by the signals radiated by the FM translator or booster station, regardless of the quality of such reception, the strength of the signal so used, or the channel on which the protected signal is transmitted.

Similar rules apply to LPFM stations.[25] Clearly, such rules were written to protect the full-service stations from interference from secondary stations. Translator and LPFM operators should be aware of these provisions in the rules so that the channels they choose do not cause interference to the reception of any FM station having listeners in the area of the translator. This may be difficult to accomplish in many markets due to the lack of channel choices, particularly in medium and large population areas.

The map shown in Figure 1.2-11 illustrates the density of translators and low-power FM stations at various cities in an area in the United States centered on New York City. Nearly all of these translators and LPFM stations fit in between the existing primary station channels.

The FCC typically protects licensed stations from interference caused by co- and up to the six adjacent channels, three on each side. However, to make more frequencies available for LPFM service, they now require LPFM channels to protect only to second-adjacent channels.[26] Translator stations must protect adjacent channels to the third adjacent.

In-Band/On-Channel Digital Radio Interference

Self-interference from IBOC FM (also called "host interference") is another problem for stations that use separate analog and digital transmitting antennas. This is when the IBOC digital sidebands cause interference to the analog host, which is located between them.

The simplest, but not the best, solution for adding IBOC to a station plant is to mount a second antenna with fewer bays than the main antenna on the same tower as the existing multi-bay antenna. This allows a low-power IBOC-only transmitter, having between 1% and 10% of the analog transmitter power, to operate separately without having to be combined to a common antenna. Other methods include using a transmitter that combines the IBOC with the main FM transmissions at the exciter level, or a high-level combiner that combines analog and digital signals coming out of separate analog and digital transmitters. High-level combining requires much more powerful transmitters, because the combiner at these power levels is not typically efficient, resulting in significant wasted power (usually dissipated into a dummy load).

While adding a second antenna for IBOC to an existing tower may be a cost saver, it can cause interference to the main analog transmitter due to inequalities in the vertical elevation fields of the two antennas. This can cause locations, at some fairly close distances to the tower, to receive interference to the analog from the station's own IBOC digital sidebands. These locations are places where the IBOC digital signal level rises well above the threshold for noninterference to the adjacent analog signal. Many modern transmission plants combine the IBOC and analog at the transmitter using the same antenna. Since both the analog and the IBOC transmit from a common antenna, the vertical elevation fields are matched and no host interference results.

An interesting case of interference between IBOC stations was observed by NPR Labs and will be described in some detail below. Since IBOC transmits on one-half of the upper and lower first-adjacent channels, interference sometimes occurs between the originating station and its first-adjacent partners, as shown in Figure 1.2-12.

There can also be second-adjacent interference caused by the addition of IBOC to one or both of the

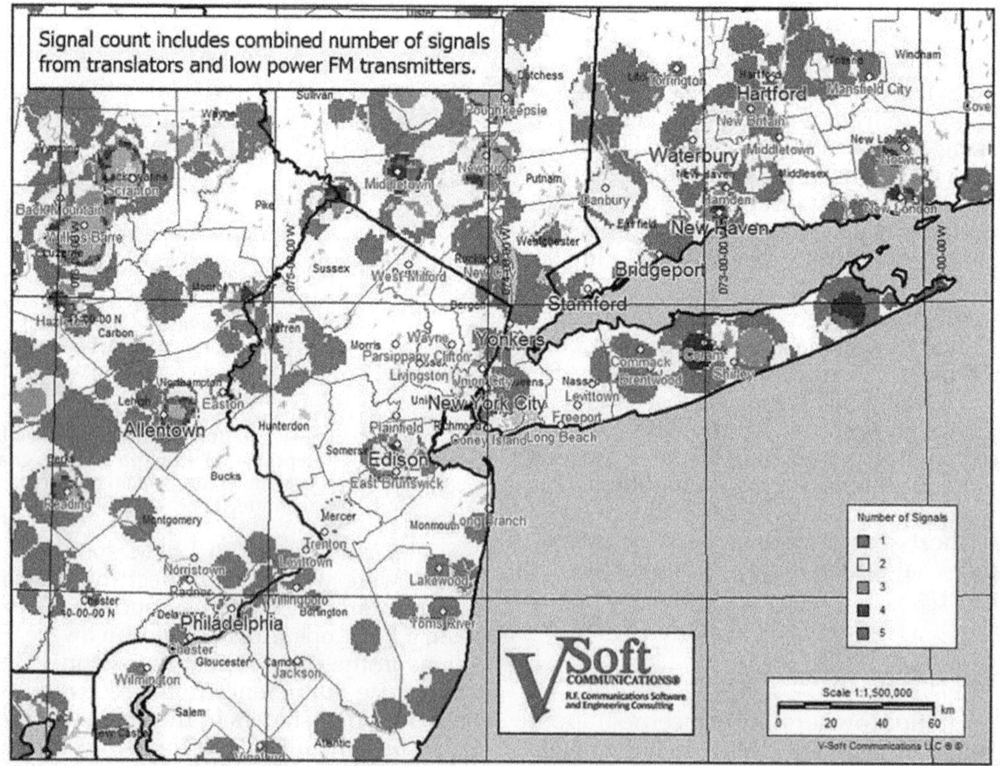

FIGURE 1.2-11 Map showing FM translator and LPFM stations near New York City

stations involved. This is because the IBOC OFDM carriers fall in the second-adjacent pass band between the stations.

Through reports, it was learned that first-adjacent stations WRUR-FM and WBFO were causing mutual interference to their analog transmissions after adding their IBOC carriers. Some of the IBOC interference for both stations falls within their normally protected 60-dBμ service contours. The stations are located 97 kilometers apart. Listeners traveling by car reported an area of interference between the two stations where analog reception was affected by the opposite station's IBOC carriers. Mapping the interference using the U/D ratios established by NPR Labs and V-Soft Communications' Probe 4 propagation prediction program, a prediction of where the interference falls was generated (see Figure 1.2-13). This map shows pockets of predicted IBOC interference to the 60-dBμ contour of WRUR-FM when WBFO operates at an IBOC injection level of – 20 dBc (1% IBOC injection). Figure 1.2-14 shows the level of interference at −10 dBc (10% IBOC injection). Figure 1.2-15 shows the interference caused to WBFO, with WRUR-FM operating at −20 dBc. Figure 1.2-16 is a similar map shown with WRUR-FM at −10 dBc.

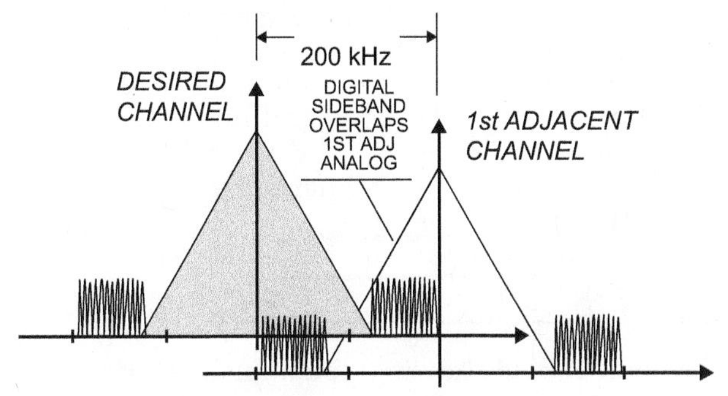

FIGURE 1.2-12 Illustration of first-adjacent channel interference between FM IBOC stations

As previously discussed, tests done by NPR Labs have indicated that analog FM reception is most affected by IBOC when there are two IBOC stations transmitting from the upper *and* lower first-adjacent channels. For digital reception, since the IBOC digital data is being transmitted redundantly on both the upper and lower digital sidebands, interference to one of the sidebands is usually not enough to avoid serious digital reception problems; however, when both upper and lower sidebands are compromised, IBOC reception falls off.

There have been occasions when stations have experienced a loss of coverage within their analog or digital service due to powerful first-adjacent stations that have increased their IBOC injection levels. IBOC presents a unique opportunity for FM stations to run digital main and multicast channels. Its transmissions appear to be significantly less affected by multipath so, in most cases, station owners decide that the good it provides to stations and their listeners outweighs the negative issues of increased interference.

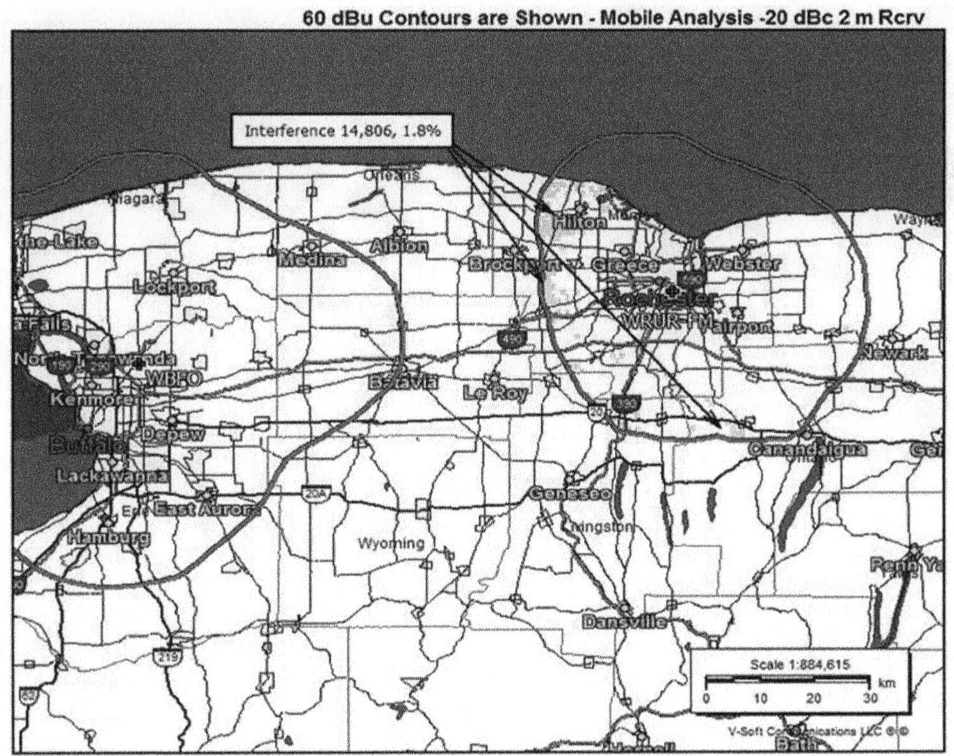

FIGURE 1.2-13 Predicted simulation between stations WRUR-FM and WBFO showing potential areas of interference (shown in light blue) with WBFO digital signals at −20 dBc

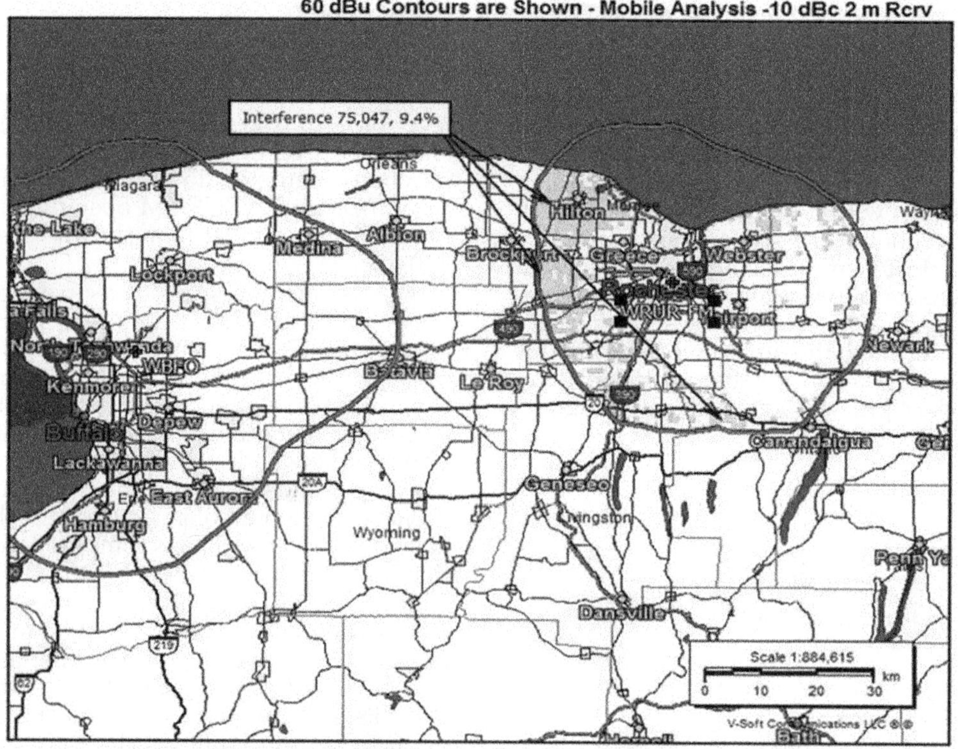

FIGURE 1.2-14 Predicted simulation between stations WRUR-FM and WBFO showing potential areas of interference (shown in light blue) with WBFO digital signals at −10 dBc

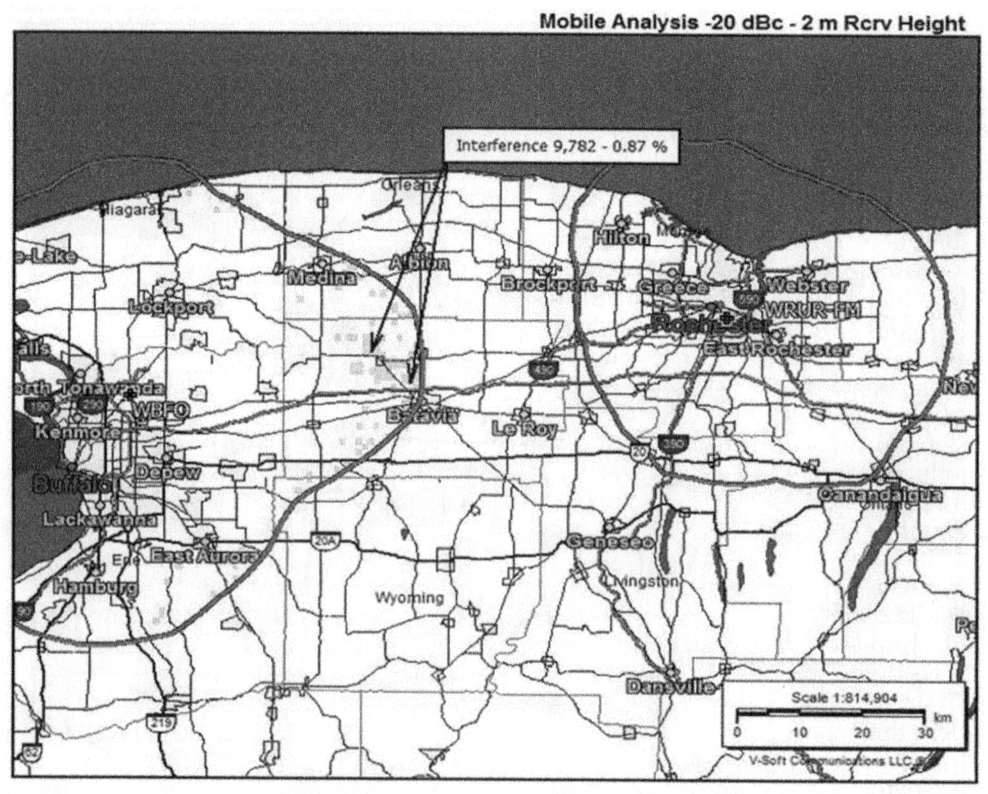

FIGURE 1.2-15 Predicted simulation between stations WRUR-FM and WBFO showing potential areas of interference (shown in light blue) with WRUR-FM digital signals at −20 dBc

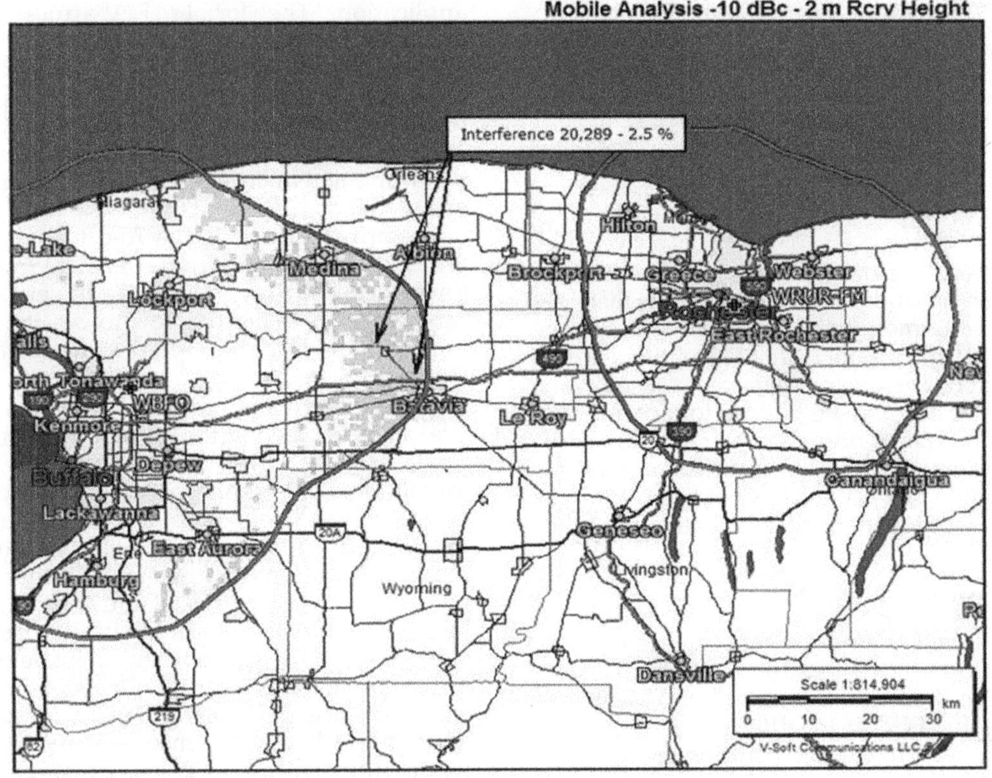

FIGURE 1.2-16 Predicted simulation between stations WRUR-FM and WBFO showing potential areas of interference (shown in light blue) with WRUR-FM digital signals at −10 dBc

Intermodulation

A harmonic is a periodic wave and is of concern to all broadcast services, because harmonics are usually out of band and they can easily cause interference. Harmonic waves are generated from the fundamental wave; for example, harmonics of 95.1 MHz will exist on double and triple of the frequency of the originating wave. Another example in the AM broadcast service is the station that transmits on 640 kHz and has a harmonic on 1280 kHz, which falls within the AM band.

Interference due to intermodulation is caused by a mix of the fundamental frequencies coming into the front end of a receiver that has one or more nonlinear stages. These strong signals can mix with the original and harmonic frequencies to produce new sum and difference products. A newly created frequency is called an "intermodulation product." The type of mix is called the "order." For example, if one transmitter is on 95.1 MHz, its second-order harmonic will be two times the fundamental (2 * 95.1 = 190.2 MHz), the third-order harmonic is 285.3 MHz (3 * 95.1) and so on. The "intermodulation product order" refers to the sum of the absolute values of the coefficients on the terms being considered in the intermodulation product.

Take for example, a three-station mix, the intermodulation frequency (IM) = A + B − C, that is (1 * A) + (1 * B) − (1 * C), so the product order would be, 1 + 1 + 1 = 3, and this becomes a "third-order intermodulation product." For a mix of a harmonic of one station with the fundamental frequency of another station, you have an IM Frequency = 2 * A − B, which is (2 * A) − (1 * B), the intermodulation product order would be 2 + 1 = 3. The result is a type of interference called "intermodulation distortion."[27]

The FCC is concerned with interference that may be generated by a harmonic that may mix with another station's fundamental frequency to produce a product on or near 455 kHz, the intermediate frequency (IF) used in most AM radios.[28] For the FM band, the IF taboo frequency is generally 10.7 MHz or in terms of FM channel numbers, 53 and 54 FM channels removed. The FCC minimum separations FM tables require Class A stations located 53 or 54 channels apart to be separated by at least 10 km. The Class C to Class A IF separation is 29 km (see Table 1.2-1). Some modern receivers use digital processing and direct conversion and, by design, can avoid the problem of IF interference.

It is also possible for an FM transmitting antenna to pick up the signal of a nearby transmitter. This signal can travel down the transmission line causing a product mix to occur at the transmitter that can be amplified and then transmitted by the station. These are called spurs and are prohibited from transmission. Some spurs can be generated in the transmitter itself without mixing with an external signal. 47 CFR §73.317 requires that:

> (b) Any emission appearing on a frequency removed from the carrier by between 120 kHz and 240 kHz inclusive must be attenuated at least 25 dB below the level of the unmodulated carrier. Compliance with this requirement

will be deemed to show the occupied bandwidth to be 240 kHz or less.

(c) Any emission appearing on a frequency removed from the carrier by more than 240 kHz and up to and including 600 kHz must be attenuated at least 35 dB below the level of the unmodulated carrier.

(d) Any emission appearing on a frequency removed from the carrier by more than 600 kHz must be attenuated at least 43 + 10 Log10 (Power, in watts) dB below the level of the unmodulated carrier, or 80 dB, whichever is the lesser attenuation.

Before a new FM station transmitter is given permission to use a common transmitter tower, a good site manager will perform an intermodulation study so the frequency mix of all the external products can be identified and, if possible, filtered out before interference is caused.[29] The FCC requires AM, FM, and TV stations to carry out performance measurements on a regular repeating basis, so adherence to good engineering practices is highly important.[30] Station engineers use a spectrum analyzer to show compliance to the spurious radiation rules.

Blanketing Contours

AM and FM stations may cause interference to "consumer protected devices" within certain distances from towers. The "blanketing contour" for FM stations has been set by the FCC at 115 dBμ, while for AM the contour is the 1-mV/m (60-dBμ) contour. Broadcasters are required to resolve all complaints of interference from their station that occur within a station's "blanketing contour" during the first year of a station's operation from a particular transmitter site. Protected devices include radios, TVs, and VCRs with tuners in them. The broadcaster is not required to resolve complaints made about reception from mobile receivers. Telephones, digital audio or tape recorders, or devices using high-gain antennas with preamplifiers also are not covered. After the first year of operation, stations have the responsibility to provide information and assistance about how to resolve the interference to persons suffering blanketing interference. According to the FCC, a station has to respond to all complaints filed and provide to the FCC the details of what they have done to address interference complaints.

UPGRADING A STATION

Upgrading the class of a station to one that results in greater coverage is much more difficult now that so many stations have taken to the air. It is a fact that nearly all the available broadcast channels in our larger communities have been used up. Chances of finding new broadcast channels for a full-service FM station are rare indeed, unless the location is in one of the unpopulated areas of the western states where the opportunities of generating income through noncommercial donations or commercial sales is severely limited.

Upgrading the Station Class

There are still some occasions where an FM station can upgrade class in what the FCC calls a "one-step" application. The former FCC rules allowed class upgrades only through the somewhat complicated allocation process where full public disclosure was required. This often brought counter proposals and responses from the station proposing the new class. Today, the FCC accepts one-step applications where the proponent makes a showing that a fully spaced site exists that will serve 100% of the area or population of the principal city with a 70-dBμ signal (see 47 CFR §73.207). The applicant must have permission from the tower owner to use an existing tower, or if the applicant proposes to build a new tower, the site proposed must be at a place where new towers can actually be located.

Occasionally, the proposed, one-step, fully spaced site can be located at the current location of the applicant's existing tower. If the applicant's current transmission tower is short spaced at the higher class, the applicant can propose to use the tower by applying the §73.215 short-spacing rules, assuming the site meets maximum short-space distance, no contour overlap interference exists, and the principal city can be served with the required city service contour. The site proposed by the applicant must be one in which a tower can actually be built. One-step applications are commonplace, particularly for Class A stations that are limited to 6 kW at 100 meters. A move to a Class C3 would give the station 25 kilowatts, which would significantly expand its coverage.[31]

There are times when the owners of a station may find that they cannot upgrade their station on the current frequency, yet the upgrade can take place on a different frequency. The FCC no longer considers a move of a commercial FM station's channel by more than three channels to be a "major change" that would open up the proposal for competing applications. Once the station changes to the better frequency and receives a license, the station owner can then apply for a one-step to increase the class (perhaps this should really be called a "two-step" process!). The map in Figure 1.2-17 provides an example of §73.207 minimum separation circle distances of all stations that have a class, frequency, and distance relationship with a proposed C3 station on the channel being studied. Any location inside the station circles would not meet the minimum separations and would not be allowed. The "area-to-locate" is shown by the arrow, and it is not overlapped by any minimum separation circles. The violet-colored complete circle marks the 70-dBμ service area that must include 100% of the principal community.[32]

Upgrading Through an Allocation Proposal

In this case, the station owner's engineer finds that an upgrade of the station can be made if certain other stations are also moved in frequency. An allocation proposal is then made to move up to two other stations' frequency to make room for the applicant's upgraded class. These allocation proposals often draw objections or counterproposals. If the allocation proposal is approved, a "show cause" order will be issued to the stations being moved. Generally, all costs to move

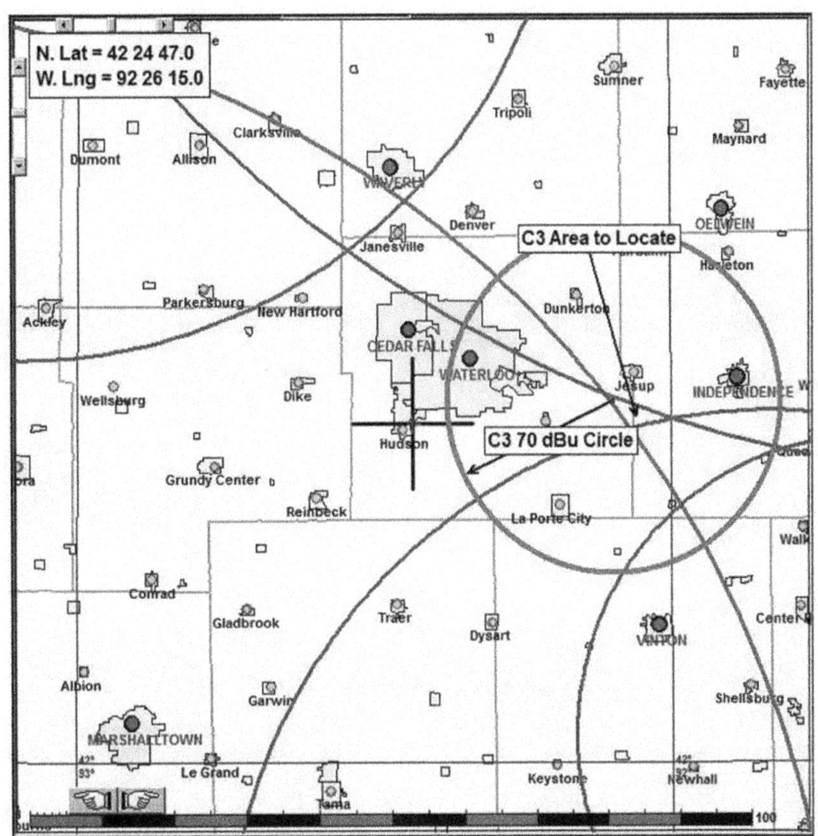

FIGURE 1.2-17 Example of §73.207 minimum separation circle distances of all stations that have a class, frequency, and distance relationship with a proposed C3 station on the channel being studied

the channel of an existing station will be borne by the one making the proposal. Due to the typical number of proposals, counterproposals, and responses, this process can be long and drawn out. Once the allocation proposal is approved by the Commission and the stations in the way are moved and licensed to new channels, the applicant can complete the upgrade to a higher class.[33]

For TV stations, the FCC has been in the process of making fewer frequencies available to broadcasters in the TV band through "re-packing" of the TV frequencies to provide cellular services with additional bandwidth for their mobile services. It is expected that no new TV stations will be added to the band in the short- or long-range future. Low-power, secondary TV stations and TV translators owners complain that the FCC is not protecting their licenses and they may lose their channels, even though these stations have always been considered secondary.

Changing the Principal City

There are cases where a commercial-band FM station owner wants to upgrade its class; however, although a fully spaced area-to-locate can be found that meets all the minimum separations of 47 CFR §73.207, the site cannot provide the 70-dBμ allocation circle to 100% of the principal city.[34] Under such a circumstance, the licensee can file with the FCC to change the principal city to one where the 70 dBμ will provide 100% coverage. The FCC reviews these proposals to make sure that the change will result in a favorable arrangement of allotments, making sure that the distribution of radio channels is in the public interest. This can be accomplished if the proposal meets the following FCC priorities:

1. First priority is proposals that provide coverage of "white areas"—areas currently receiving no other primary signals from any radio station;
2. Second priority is given to coverage of "gray areas" that currently receive only one other service; therefore, the new service provides a first competitive "reception" service;
3. The third criterion is equal in importance to the second. The first "transmission service" to a community, such that the first station licensed to that community will have as its primary responsibility to cover the community's needs and interests;
4. Forth in priorities is composed of "other public interest factors," which traditionally consist of the coverage population provided to the communities.

Under the Commission's policy, there is a presumption that when a station proposes to serve a new community and that community is a part of a whole urbanized area around a major city and where the station would cover more than 50% of the whole urbanized area, either from the proposed site or another possible site, the proposal would be removing service from a rural area and is actually deemed to serve the larger city and, therefore, it would not meet the Commission's priorities and public interest factors. Consequently, principal city moves where rural coverage is lost in favor of new major city coverage are prohibited. This follows that to move a station from a city that has only one full-service radio station is also prohibited.

However, if a licensee is moving from one rural community to another and it is not the only station to serve the original community, the odds are greater for FCC approval.

RECEIVERS

While this chapter has focused on coverage topics, the other half of the equation is the receiver's ability to pick up a station. The NPR IBOC study, referenced herein, and previous receiver studies have found considerable differences between the sensitivity and selectivity of AM and FM radios in buildings, at portable locations, and in mobile vehicles. Automobile radios historically have better receiving capabilities than portables. Unfortunately, the station engineer has no choice in the selection of the universe of radios that are being used to pick up the station's signal. The best a station can do is to provide sound engineering practices to produce the greatest coverage available to it and to let the programmers provide competitive content.

Notes

[1] See 47 CFR §73.313.
[2] See 47 CFR §73.182 (k),(1–8).
[3] An example of such a program is V-Soft Communications AM-Pro 2™.
[4] FORTRAN and C computer code for the Longley-Rice implementation can be found in an appendix of Report 82–100, A Guide to the use of the I.T.S. Irregular Terrain Model in the Area Prediction Mode, G.A. Hufford, A.G. Longley, and W.A. Kissack, U.S. Dept. of Commerce, April 1982.
[5] See FCC TVStudy Interference Analysis Software/www.fcc.gov/engineering-technology/electromagnetic-compatibility-division/general/tvstudy-interference-analysis.
[6] Based on a common distance of the standard Class A, 60-dBμ service contour of 28 kilometers.
[7] See 47 CFR §73.316 and §73.625.
[8] *In the Matter of LKCM Radio Licenses, L.P., Involuntary Modification of License of Station KFWR(FM), Jacksboro, Texas*, Order, DA 15–886, August 5, 2015.
[9] See 47 CFR §1.30002(d).
[10] Formula for wavelength, Lambda (meters) = 300 / Frequency (MHz).
[11] Three hundred sixty electrical degrees equals one wavelength at the frequency of the AM transmitter (540 kHz has an electrical length of 553.6 meters).
[12] For noncommercial FM stations operating in the reserved band, the percentage is 50%.
[13] The FCC does not require the "as-built" pattern to meet the 15 dB front to back rule.
[14] You can tell the number of bays that compose an FM antenna by counting the bumps in the vertical elevation pattern. The nulls between the bumps are where the phasing between the bays produces cancelation.
[15] See *ERI—Analysis of Antenna Beam-Tilt and Broadcast Coverage*, Myron D. Fanton, PE.
[16] The −14 dBc level was not studied by NPR Labs.
[17] National Public Radio Final Report to the Corporation for Public Broadcasting Digital Radio Coverage & Interference Analysis (DRCIA) Research Project, DRCIA Final Report, National Public Radio, Inc., July 9, 2008.
[18] Construction permits for translators expire after 36 months.
[19] Translators use 12 cardinal evenly spaced radials. A translator may be limited in one radial direction but not in another, since the HAAT on each radial will vary.
[20] There are exceptions to this rule for FM translators being moved to serve AM stations.
[21] As part of the FCC's "AM Revitalization" proceeding, the Commission is considering changing this to the greater of a 25-mile radius or the AM 2 V/m contour.
[22] See http://geobroadcastsolutions.com/zonecasting.
[23] See CFR 47 §73.333 for the FCC (50,50) and FCC (50,10) curves.
[24] Figure 1.2-10 map is produced by V-Soft Communications' FMCommander™ computer program.
[25] See 47 CFR §73.809.
[26] Except in the case of a station having a radio reading service subcarrier. These stations must be protected to the third-adjacent channel.
[27] See Internet link "http://users.tpg.com.au/users/ldBmtler/Intermodulation.htm" by Lloyd Butler, published in "Amateur Radio," August 1997.
[28] 47 CFR §74.128 (s) states: "Two stations, one with a frequency twice of the other, should not be assigned in the same groundwave service area unless special precautions are taken to avoid interference from the second harmonic of the station operating on the lower frequency. Additionally, in selecting a frequency, consideration should be given to the fact that occasionally the frequency assignment of two stations in the same area may bear such a relation to the intermediate frequency of some broadcast receivers as to cause interference."
[29] A low-pass filter is in common use at all broadcast transmitters.
[30] See 47 CFR §73.1590 (b) "Equipment Performance Measurements" of the Federal Communications Rules and Regulations.
[31] See Table 1.2-3.
[32] Since this is a one-step C3 allocation map, the coverage to the principal city must be shown at the maximum 70-dBμ circle distance, rather than as a 70-dBμ FCC contour. The coordinates shown on the map are for the current Class A facility, located at the large plus mark.
[33] Figure 1.2-17 map was produced by FMCommander, a computer program.
[34] Policies to Promote Rural Radio Service and to Streamline Allotment and Assignment Procedures, FCC MB Docket 09–52, RM-11528, October 12, 2012, Second Order on Reconsideration.

CHAPTER

1.3

Distance and Bearing Calculations

DANE E. ERICKSEN, P.E.

Hammett & Edison, Inc., Consulting Engineers
San Francisco, CA

INTRODUCTION

The current FCC Rules for distance calculations use two methods: "flat-earth" and "spherical-earth." The flat-earth method assumes the distance between two points to be the hypotenuse of a right triangle whose sides are determined by the difference in latitude and longitude of the starting and ending points multiplied by the length per degree of latitude and longitude at the mid-latitude of the two points, as shown in Figure 1.3-1.

The "flat-earth" term is not meant to be disparaging, and merely refers to the use of a right triangle to calculate the distance. Because the lengths of a degree of longitude and a degree of latitude used in the flat-earth method are derived from an ellipsoid rather than a spheroid model of the earth, the flat-earth method is actually more accurate than the spherical-earth method for short to moderate distances.

The spherical-earth method uses conventional spherical trigonometry to determine the distance. Section 73.208 of the FCC Rules requires that the flat-earth method be used for distances up to and including 475 km. Distances greater than 475 km must be calculated using the spherical-earth method, which becomes more accurate than the flat-earth method for large distances. Section 73.208 is silent on how azimuths are to be calculated, nor does it specify the earth radius to be used for spherical-earth calculations.

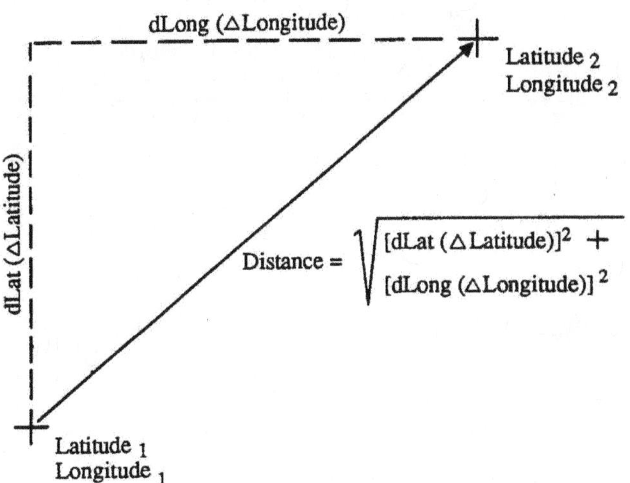

FIGURE 1.3-1 Flat-earth method

FCC FLAT-EARTH METHOD

In FCC Docket 80–90,[1] formulas were substituted for the tables previously used for determining the length of a degree of latitude or of longitude as a function of latitude. However, the coefficients adopted in Docket 80–90 truncated to only two terms the trigonometric series used to generate the tables and "fudged" the coefficients by a factor of (1.609/1.609347) because of the Docket 80–90 decision to define the conversion factor from U.S. statute miles to kilometers as 1.609,[2] rather than the value of (5,280 ft/mile) × (1200/3937 m/ft) × (1/1000 km/m), or 1.609347219 km/mile (approximately).[3]

In the Second Report and Order to Docket 86–144,[4] the FCC corrected these problems by adopting the full-precision, non-truncated trigonometric series for the arc length formulas given in the 1966 edition of U.S. Naval Hydrographic Office Publication Number 9, also known as "H.O. 9," or the "American Practical Navigator," or simply "Bowditch," after Nathaniel Bowditch (1773–1838), its original author. These trigonometric series are based upon a binomial theorem expansion[5] of an ellipsoid model of the earth corresponding to the Clarke spheroid of 1866, upon which topographic maps in the United States are currently based.[6]

The trigonometric series defining the length of one degree of latitude and one degree of longitude for the Clarke spheroid of 1866 are:

$$dLat = 111.13209 - 0.56605\cos(2L) + 0.00120\cos(4L)\ldots$$

$$dLong = 111.31513\cos(L) - 0.09455\cos(3L) + 0.00012\cos(5L)\ldots$$

where dLat is the length in kilometers of one degree of latitude at latitude L and dLong is the length in kilometers of one degree of longitude, again at latitude L. The latitude, L, is taken as the mid-latitude of the two points between which the distance is to be calculated, as follows:

$$L = (Latitude_1 + Latitude_2) / 2$$

where $Latitude_1$ and $Latitude_2$ are the latitudes of the starting and ending points. Similarly, $Longitude_1$ and $Longitude_2$ are the longitudes of the starting and ending points. In all cases, north latitudes are treated as positive and south latitudes as negative, and west longitudes are treated as positive and east longitudes as negative. The distance between two points is then given by the Pythagorean Theorem:

$$D = \sqrt{[(dLat)(Lat_1 - Lat_2)]^2 + [(dLong)(Long_1 - Long_2)]^2}$$

Plots showing how the lengths of one degree of latitude and longitude vary with latitude are given in Figures 1.3-2 and 1.3-3.

The FCC has a calculator that uses this method on the Media Bureau website, at www.fcc.gov/media/radio/distance-and-azimuths.

Canadian Method

In August 1987,[7] the Canadian government adopted the truncated and fudged arc-length formulas, which had been implemented by the FCC in Docket 80–90. Namely,

$$dLat = 111.108 - 0.566\cos(2L)$$

$$dLong = 111.391\cos(L) - 0.095\cos(3L)$$

Unfortunately, the Canadian Government has not yet adopted the corrected, more accurate formulas that were implemented by the FCC in Docket 86–144. Nor was the flat-earth versus spherical-earth break point changed from 350 km to 475 km. This, then, is the source of current discrepancies between the U.S. and Canadian distance calculation methods. Although the differences between the two methods will usually not be significant when the calculated distance is rounded to the nearest kilometer (for FM) or to the nearest one-tenth kilometer (for TV), one should always check to see whether there is a difference between roundings. For calculations involving Canadian stations, the Canadian (Docket 80–90) version of the Clarke spheroid formulas is controlling.

Mexican Method

The August 11, 1992, U.S.-Mexican FM Agreement specifies the spherical-earth method exclusively; there is no provision for using the FCC flat-earth method for short to moderate distances. However, the Mexican Agreement specifies a spherical earth arc of 111.18 kilometers per degree, corresponding to an earth radius of 6,370.14 kilometers. The Mexican Agreement further specifies that azimuths are to be calculated on a spherical-earth basis. Finally, the Mexican Agreement specifies the rounding of distances to the nearest kilometer, so in that respect it matches the FCC rounding practice for FM station distances.

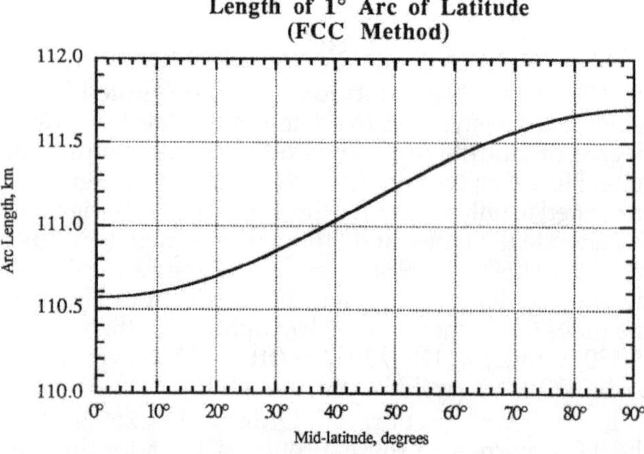

FIGURE 1.3-2 Length of 1° arc of latitude (FCC method)

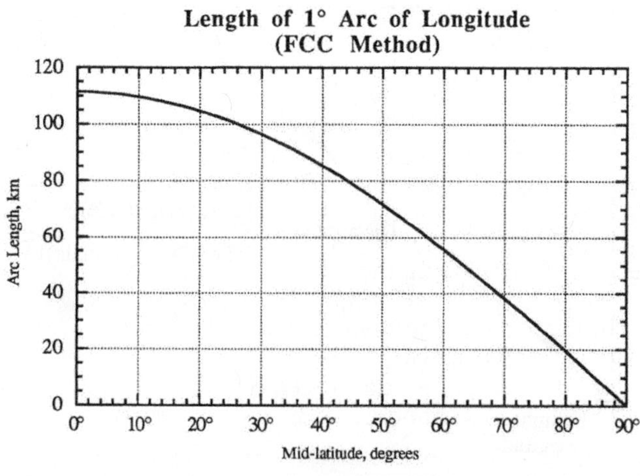

FIGURE 1.3-3 Length of 1° arc of longitude (FCC method)

SPHERICAL-EARTH METHOD

The formula for the spherical-earth distance, or great-circle distance, is:

$$D = K \cos^{-1}[(\sin Lat_1)(\sin Lat_2) + (\cos Lat_1)(\cos Lat_2)\cos(Long_2 - Long_1)]$$

The constant K is in km/degree and is determined by the radius of the sphere being modeled. The FCC has never defined the earth radius to be used for spherical-earth calculations. The example given in Section 73.185(d) of the FCC Rules suggests an earth radius of 6,365 km (K = 111.090 km/degree). A 6,373 km radius (K = 111.230 km/degree) is implied by the 5,280-mile 4/3-earth radius given in Section 73.684(c)(1) of the FCC Rules. This 4/3-earth radius was also used in FCC Report No. R-6410, "Elevation and Depression Angle Tables," September 15, 1964. An earth radius of 6,367 km (K = 111.125 km/degree) can be deduced from the 1,852-meter definition of a nautical mile.[8] Finally, an earth radius of 6,371 km (K = 111.195 km/degree) corresponds to the mean radius of the Clarke spheroid of 1866.

In Docket 86–144 comments prepared by the author, the FCC was asked to adopt an official earth radius to be used when performing spherical-earth distance computations, but unfortunately the FCC did not do so, apparently by oversight (that is, there was simply no discussion of this issue in the Report and Order). Until such time as the FCC so specifies, the author suggests using the mean Clarke spheroid value of 6,371 km (K = 111.195 km/degree).[9]

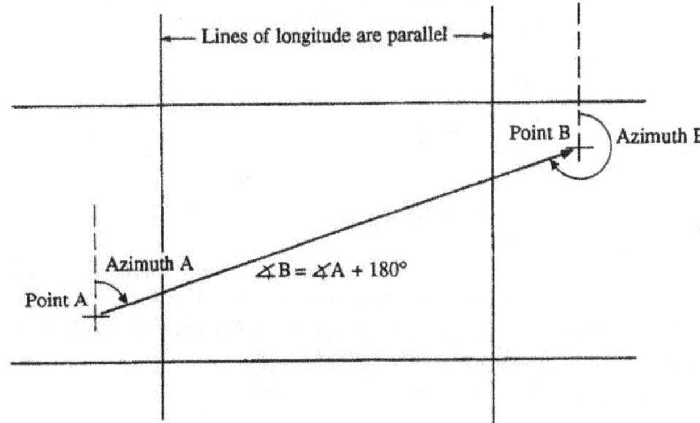

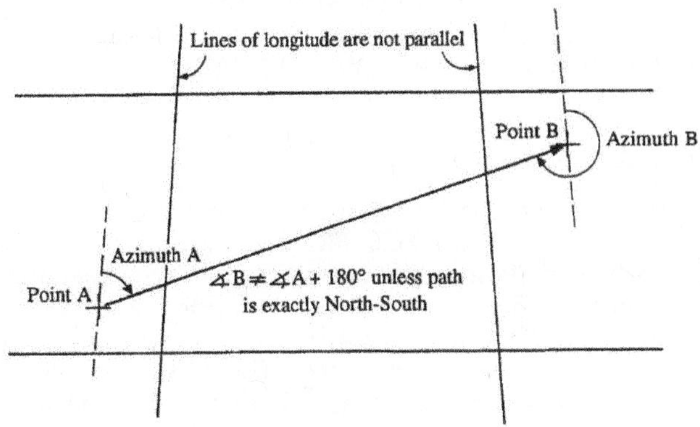

FIGURE 1.3-4 Azimuth calculations

AZIMUTH CALCULATIONS

Because the FCC Rules are silent on how azimuth, or bearing, calculations are to be performed, both the flat-earth and spherical-earth methods are commonly in use. The flat-earth method determines azimuth using the arctangent of the right triangle defined in the FCC method. The spherical-earth method uses standard spherical trigonometry. It is the author's recommendation that azimuth always be determined using the spherical-earth method, even when the FCC flat-earth method is used to determine distance. The flat-earth method will be in error by up to two degrees at distances approaching 500 km, whereas the spherical-earth azimuth will be correct within about 0.1 degrees at such distances.[10] Figure 1.3-4 shows why this is so.

Azimuths determined using the arctangent of the right triangle defined by the FCC flat-earth method assume that lines of longitude are parallel, whereas they are not. This is why the forward and back (reciprocal) azimuths using the spherical-earth method will generally not be exactly 180° apart, whereas the forward and back azimuths using the flat-earth method are always 180° apart.

The formula for determining azimuth by the spherical-earth method is:

$$C = \cos^{-1}\frac{\sin Lat_2 - \sin Lat_1 \cos(D/K)}{\sin(D/K)\cos Lat_1}$$

if $\sin(Long_2 - Long_1) < 0$, Azimuth = C

if $\sin(Long_2 - Long_1) \geq 0$,

Azimuth = 360° − C

The ratio (D/K) is the great circle arc length *in degrees,* and is obtained from equation by rearranging the formula for spherical earth distance given above, namely:

$$D/K = \cos^{-1}[(\sin Lat_1)(\sin Lat_2) + (\cos Lat_1)(\cos Lat_2)\cos(Long_2 - Long_1)]$$

It should be noted that the Industry Canada rules applying to FM Broadcasting (BPR-3), the U.S.-Canada FM Agreement, and the U.S.-Mexico FM Agreement *do* specify that azimuth is to be calculated using the spherical-earth method and, further, that for Canadian azimuth calculations, the bearing is to be rounded to the nearest degree.[11] It should also be noted that the radius of the sphere is irrelevant for azimuth calculations.

Clarke Spheroid vs WGS Ellipsoid

The original FCC distance tables were based upon the Clarke spheroid (or ellipsoid) of 1866, with a major-axis radius of 6,378.2064 km and ellipticity[12] of 1/294.98. The current edition[13] of the *American Practical*

Navigator now bases Table 7, "Length of a Degree of Latitude and Longitude," on the World Geodetic System (WGS) ellipsoid of 1972. The coefficients for dLat and dLong for this ellipsoid are:

$$dLat = 111.13292 - 0.55982\cos(2L) + 0.001175\cos(4L) \ldots$$

$$dLong = 111.31282\cos(L) - 0.0935\cos(3L) + 0.000118\cos(5L) \ldots$$

The difference between the Clarke 1866 and WGS 1972 trigonometric series is inconsequential when coordinate databases are maintained only to the nearest second and distances are rounded to the nearest kilometer or one-tenth kilometer. Over a zero- to ninety-degree arc of latitude, at one-degree increments, the RMS difference between the two arc length formulas is only 0.0041% for latitude and 0.0033% for longitude. For this reason, the FCC decided not to amend the formulas given in Section 73.208(c) of its Rules from the Clarke 1866 coefficients to WGS 1972 coefficients.

Rounding Practices

There continues to be an inconsistency in rounding practices between the FCC FM and TV Rules. For FM distance calculations, Section 73.208(c)(8) specifies rounding to the nearest kilometer. For TV distance calculations, Section 73.611(d) specifies rounding to the nearest one-tenth kilometer. The FCC was asked to eliminate this inconsistency between the FM and TV Rules in the author's comments to Docket 86–144, as there appears to be no need for such inconsistency. Unfortunately, no action was taken. The Docket 86–144 Report and Order again did not discuss this point, so it is not known whether the failure to adopt a uniform rounding policy was inadvertent or deliberate. Persons performing distance calculations must, therefore, be mindful to the different rounding criteria between the FM and TV Rules. The AM Rules and the Broadcast Auxiliary Rules are silent on distance rounding practices. The Part 74 Broadcast Auxiliary Rules also do not prescribe how distances are to be calculated, even though portions of those Rules include minimum distance requirements (for example, Subpart G, *Low Power TV, TV Translator, and TV Booster Stations*, specifies certain minimum distances between LPTV/TV Translator stations and land mobile stations). The FCC was urged to make its distance calculation rules applicable to all Part 73 and Part 74 services in the Docket 86–144 comments, but again did not do so. However, in WT Docket 98–20, the Commission created Section 1.958 addressing distance calculations for wireless services. Since Section 1.901 defines wireless services as including Part 74, that meant the flat earth method of distance calculation should be used where minimum separation distances apply. Unfortunately, when Section 1.958 was created, the first term of the dLong series was given as 111.31513cos(5ML) rather than the correct 111.31513cos(ML). Since this was the first and largest term of the series, it was a huge error, resulting in negative distances for some mid-latitude values. However, since the $D_{longitude}$ term is then squared, this created a complex error plot where for some mid-latitudes a distance *shorter* than actual results, and for other mid-latitudes a distance *longer* than actual results. Further, the errors are not so large as to be obviously incorrect; for example, between Station KGO-TV in San Francisco and Station KCRA-TV in Sacramento the separation is 101.65 km, whereas the incorrect Section 1.958 formula gives 119.68 km, a value that is incorrect but not obviously so, such as 1,196.8 km or 11.97 km. For an explanation of the error plot, see the article *FCC Invents Negative Distance*, in the October 4, 2011, issue of *TV Technology*; the link is www.tvtechnology.com/miscellaneous/0008/fcc-invents-negative-distance/210440.

When this error was discovered by Hammett & Edison, the FCC was contacted and asked to correct this clearly typographic error in Section 1.958. However, the attorneys at the FCC said that couldn't be done without a rulemaking, since the WT Docket 98–20 R&O had long passed its Petition for Reconsideration period. So, on March 21, 2011, H&E filed a formal Petition for Rulemaking to have the $D_{longitude}$ formula in Section 1.958 corrected. The FCC used a 2012 rulemaking involving changes to the Part 22 and Part 27 wireless rules to implement the fix, at paragraph 66 of the 87-page February 15, 2012, WT Docket 12–40 Notice of Proposed Rulemaking (NPRM). The correction was, as expected, formally adopted in the resulting November 10, 2014, Report & Order (R&O). That R&O was published in the Federal Register on December 22, 2014, and therefore became effective thirty days later, on January 9, 2015. So, after 17 years, the error in the Commission's distance calculation formula for wireless services, including Part 74 stations (but not for Part 73 broadcast stations, which are governed by the Section 73.208 version of the formulas rather than the Section 1.958 version of the formulas), has finally been corrected. But, be careful to not inadvertently grab an old code of federal regulations (CFR) copy of Section 1.958, with the incorrect formula. Fortunately, the FCC website version of the distance calculation formula never suffered from this error, and so FCC webpage distance calculations have always been correct.

NAD27/NAD83 Datums

Most USGS maps are still based upon the Clarke spheroid of 1866, as opposed to the 1983 North American Datum (NAD83). NAD83 differs from the 1927 North American Datum (NAD27) in that it is referenced to the earth's center of mass, making it fully compatible with satellite systems for position determination. NAD83 is based upon the Geodetic Reference System ellipsoid of 1980 (GRS 1980), with a major axis of 6,378.135 km and an ellipticity of 1/298.26. These are the same parameters as for the WGS 1972 ellipsoid, and the trigonometric series for the one-degree arc length formulas are identical to those for WGS 1972.[14] The FCC has converted all of its databases to NAD83 except for the Consolidated Data Base System (CDBS), which still uses NAD27 coordinates. The CDBS is used for Part 73 radio broadcast stations and Part 74 FM translator, FM booster, and LPFM stations. Part 73 TV stations, Class A TV stations, and Part 74 TV Translator, TV Booster, and LPTV stations have been converted to the new Licensing & Management System (LMS), which uses NAD83 coordinates.

7.5-minute topographic quadrangle maps published by the Geological Survey specify the datum in the lower left map corner, and most USGS maps still use NAD27 as the reference datum. The Geological Survey has begun issuing maps based on the NAD83 datum. Initial attempts at printing dual-sided (NAD27/NAD83) maps proved impractical.[15]

In order to prevent intermixing of numerical information using two different map datums, the FCC has stated[16] that the following procedures will be in effect until further notice:

1. Broadcast station applicants still using the CDBS must continue to furnish coordinates based on NAD27. As of the time of this writing (March 2016), FCC Audio Division stations still use the CDBS.
2. Broadcast station applicants using the new LMS must furnish coordinates based on NAD83. As of the time of this writing, only FCC Video Division stations have been converted to the LMS.
3. All Form 601 Universal Licensing System (ULS) applications must use only NAD83 coordinates.
4. All Form 854 Antenna Structure Registration (ASR) filings must use only NAD83 coordinates.
5. All International Bureau Form 312 satellite applications must use only NAD83 coordinates.

Conversion Between Datums

The National Geodetic Survey (NGS) has developed an algorithm known as the North American Datum Conversion program, or NADCON. The program is bi-directional, meaning that it will convert NAD83 coordinates to NAD27 coordinates, as well as NAD27 coordinates to NAD83 coordinates. It also applies in Alaska, Hawaii, and Puerto Rico. But, most importantly, it is the conversion algorithm recommended by statute. The Federal Register dated August 10, 1990, Volume 55, Number 155, Page 32681 stated:

> The intent of this notice is to standardize a horizontal datum transformation method when a mathematical transformation is desired. FGCC [Federal Geodetic Control Committee] selected the method incorporated in the software identified as NADCON. It is not the intent of the notice to declare when to use a datum transformation or by what method but only to declare that when a mathematical transformation is appropriate, NADCON is recommended.

Thus, while use of NADCON is not mandatory, use of conversion algorithms other than NADCON may result in a rounding error that will cause the results to be inconsistent with those obtained by the FCC or by the Federal Aviation Administration (FAA), which do use NADCON.

NADCON uses a minimum-curvature approach to transforming between the two datums, and comes with four reference files: a large file for the contiguous United States and smaller files for Alaska, Hawaii, and Puerto Rico. By having a separate set of polynomial coefficients for each 7.5-minute topographic map, NADCON achieves an improved conversion accuracy of approximately ±0.0003 seconds in latitude and ±0.0005 seconds in longitude. Because of the size of these data files, programmable calculators are unable to implement the program, but today's personal computers are certainly capable of loading the NADCON data files. Information regarding the NADCON software is available at www.ngs.noaa.gov/TOOLS/Nadcon/Nadcon.shtml.

For a more detailed discussion of the NAD27/NAD83 datums, see *NAD83: What It Is and Why You Should Care*, by the author of this chapter, published in the Proceedings of the 1994 SBE Engineering Conference. A copy of this paper is also available on the H&E website, at www.h-e.com; the link is:

http://h-e.com/sites/default/files/sites/default/files/tech_docs/de_sbe94.pdf.

USE OF GPS TO DETERMINE COORDINATES

With the widespread availability of low-cost Global Positioning System (GPS) receivers and the termination of "selective availability," a technique formerly used to deny the full accuracy of GPS to parties not friendly to the United States, a typical consumer-grade GPS receiver that can simultaneously receive the signals from multiple GPS satellites will generally return geographic coordinates accurate to the nearest second of latitude and longitude, which is generally acceptable accuracy for FCC filings. A further potential source of error exists because most GPS receivers have user-driven menus allowing display of coordinates in either the NAD27 or NAD83 datum. So, there is always the possibility that the user does not realize which datum the GPS receiver has been set to display. On more than one occasion, the author has discovered parties reporting GPS-derived coordinates as NAD27 when in fact the GPS receiver is in its factory default mode to display NAD83 coordinates. If NADCON is then used to convert the coordinates from NAD27 to NAD83, in the mistaken belief that the reported coordinates were NAD27, a bogus set of coordinates, corresponding to neither datum, are then created. Therefore, persons using GPS receivers to determine coordinates must ensure they are aware of which datum their receiver is displaying.

Notes

[1] Docket 80–90 Report and Order, May 26, 1983.
[2] Ibid., Page 29, Footnote 35.
[3] ANSI/IEEE Standard 268–1982, "Metric Practice," Page 31, Note 14.
[4] Docket 86–144 Second Report and Order, September 10, 1987.
[5] Personal correspondence between the author and Mr. Adam W. Mink, Chief, Hydrography and Navigation Department, Defense Mapping Agency, Washington, D.C. 20315-0030. May 24, 1985.
[6] *Maps for America*, Second Edition, 1981, Page 238. Published by the U.S. Department of the Interior, Geological Survey National Center, Reston, Virginia 22092.

[7] Broadcast Procedure No. 13 (BP-13), Issue 2, Broadcasting Regulation Branch, Department of Communications, Government of Canada. Effective August 6, 1987. Now Industry Canada BPR-3 (see footnote 11).

[8] H.O. No. 9, 1981 Edition, Volume 2, Page 862, defines a nautical mile as one minute of any great circle of the earth. In 1929, the International Hydrographic Bureau proposed a standard length of 1,852 meters (exactly), which as known as the International Nautical Mile. A nautical mile of 1,852 meters implies an earth radius of 6,366.707 km [(1852 meters/minute × 60 minutes/degree × 360 degrees/circumference) / 2ϖ].

[9] H.O. No. 9, Volume II, 1981 Edition, Appendix D, Page 648. The mean radius of an ellipsoid is defined as (2a + b)/3, where a is the major or equatorial radius and b is the minor or polar radius. The full precision value is 6,370.9989 km (111.1949075 km/degree), but use of the 111.195 km/degree rounded value is suggested.

[10] Azimuth errors are referenced to values obtained from Andoyer-Lambert formulas. Andoyer-Lambert formulas model the earth as a true ellipsoid and are used extensively in Loran computations. The complexity of Andoyer-Lambert formulas do not warrant their routine use for FCC calculations.

[11] See Industry Canada *Broadcasting Procedures and Rules*, Part 3 (BPR-3), August 2002, Section C-4, for Canadian calculations; see Appendix 2, Item 2, of the U.S.-Mexico FM Agreement of 1992, for Mexican calculations.

[12] Ellipticity, or flattening, f, is defined as f = (1 – b / a), where a is the equatorial radius and b is the polar radius.

[13] The American Practical Navigator is available online at https://en.wikisource.org/wiki/The_American_Practical_Navigator/.

[14] The major axis and ellipticity for WGS1972 and GRS1980 are identical, according to NOAA Technical Memorandum NOS NGS-16, "Determination of North American Datum 1983 Coordinates of Map Corners (Second Prediction)," by T. Vincenty, National Geodetic Survey, January 1987. Because the binomial theorem expansion of an ellipsoid model starts with only two constants, the major axis dimension and the ellipsoid models of the earth with the same major axis and ellipticity values must also be identical.

[15] "Implementing North American Datum 1983 for the National Mapping Program (Ashaway Quadrangle)," U.S. Department of the Interior, Geological Survey, Reston, Virginia 22092. Undated.

[16] FCC Public Notice "FCC Interim Procedure for the Specification of Geographic Coordinates," March 14, 1988.

SECTION

2

BROADCAST ADMINISTRATION, STANDARDS, AND TECHNOLOGIES

CHAPTER

2.1

Broadcast-Related Organizations and Information

LOREN WHITE

National Association of Broadcasters
Washington, DC

INTRODUCTION

This chapter provides a general overview of entities whose activities affect the work of broadcast engineers. It is divided into five categories:

- Federal government
- State and local government
- Trade associations
- Professional associations
- Related broadcast-oriented organizations

Website references are included to direct broadcast engineers to more information on the listed entity, including publications, conferences, committees, and other resources. Information contained in this chapter was current as of May 2016. Much of the information contained in this chapter has been obtained from the websites of the entities described herein. The chapter concludes with a list of broadcast engineering and related periodicals.

FEDERAL GOVERNMENT

U.S. Department of Agriculture (USDA), Forest Service

1400 Independence Avenue, S.W.
Washington, DC 20250
Phone: (800) 832-1355
Website: www.fs.fed.us

The U.S. Department of Agriculture (USDA) Forest Service covers over 190 million acres of public land under the domain of the National Forest System. Often, the use of a mountaintop transmitter site is administered by the Forest Service. Leases authorizing communications facilities are issued by the Forest Supervisor for the particular forest of interest. Specific information can be found on the Service's website under "Working with Us."

Broadcasting Board of Governors (BBG)

330 Independence Avenue, S.W.
Washington, DC 20237
Phone: (202) 203-4400; fax: (202) 203-4585
Website: www.bbg.gov/

The Broadcasting Board of Governors (BBG) is an independent federal agency that oversees all U.S. civilian international media. The BBG's stated mission is to inform, engage, and connect people around the world in support of freedom and democracy. The Board oversees Radio Free Europe, Radio Free Asia, and the Middle East Broadcasting Networks and Voice of America. The International Broadcasting Bureau (IBB) provides the transmission services for BBG and acts as their administrative and marketing arm.

U.S. Department of Commerce, National Institute of Standards and Technology (NIST)

100 Bureau Drive
Gaithersburg, Maryland 20899
Phone: (301) 975-2000
Website: www.nist.gov

The National Institute of Standards and Technology (NIST) is a nonregulatory agency within the U.S.

Department of Commerce Technology Administration. NIST's mission is to promote U.S. innovation and industrial competitiveness by advancing measurement science, standards, and technology in ways that enhance economic security and improve our quality of life.

U.S. Department of Commerce, Time and Frequency Division

325 Broadway
Boulder, Colorado 80305
Phone: (303) 497-3295
Website: http://tf.nist.gov

The functions of the Time and Frequency Division, part of the Physics Laboratory of NIST, include:

- Maintaining the primary frequency standard for the United States
- Developing and operating standards of time and frequency
- Coordinating U.S. time and frequency standards with other world standards
- Providing time and frequency services for U.S. clientele
- Performing basic and applied research in support of improved standards and services

Precise time and frequency information is required by radio and television stations, electric power companies, telephone companies, air traffic control systems, participants in space exploration, computer networks, scientists monitoring data of all kinds, and navigators of ships and planes. These users need to compare their own timing equipment to a reliable, legally traceable internationally recognized standard. NIST provides this standard for civilian users in the United States. Broadcast services include shortwave radio signals from NIST radio stations WWV, WWVH, WWVB, and Loran-C. Time synchronization can also be accomplished via global positions system (GPS) satellites. The division offers a fee-supported Frequency Measurement and Analysis Service (FMAS) that allows users to make accurate, NIST-certified frequency calibrations at their site, rather than sending their oscillators to NIST or elsewhere for calibration.

National Telecommunications and Information Administration (NTIA)

1401 Constitution Avenue, N.W.
Washington, DC 20230
Phone: (202) 482-2000
Website: www.ntia.doc.gov

The National Telecommunications and Information Administration (NTIA) is the President's principal adviser on telecommunications and information policy issues, and in this role frequently works with other Executive Branch agencies to develop and present the Administration's position on these issues.

In addition to representing the Executive Branch in both domestic and international telecommunications and information policy activities, NTIA also manages the federal use of spectrum; performs telecommunications research and engineering, including resolving technical telecommunications issues for the federal government and private sector; and administers infrastructure and public telecommunications facilities grants. NTIA's mission is to promote the development of an advanced telecommunications and information infrastructure that efficiently serves the needs of all Americans, creates job opportunities for American workers, and enhances the competitiveness of U.S. industry in the global marketplace. NTIA addresses a broad range of telecommunications issues and concerns. Many efforts cross-cut across the agency, drawing on telecommunications policy expertise, radiowave propagation knowledge, spectrum engineering, and lessons learned from actual applications.

Of particular interest to broadcast engineers is NTIA's Office of Spectrum Management (OSM), which is responsible for managing the federal government's use of the radiofrequency spectrum. To achieve this, OSM receives assistance and advice from the Interdepartment Radio Advisory Committee.

U.S. Department of the Interior, Bureau of Land Management (BLM)

1849 C Street, N.W., Room 5665
Washington, DC 20240
Phone: (202) 208-3269
Website: www.blm.gov

The Bureau of Land Management (BLM), an agency within the U.S. Department of the Interior, works to manage, maintain, and preserve public lands. They administer 245 million surface acres of America's public lands, located primarily in 12 western states.

U.S. Department of the Interior, U.S. Geological Survey (USGS)

12201 Sunrise Valley Drive
Reston, Virginia 20192
Phone: (703) 648-5953
Website: www.usgs.gov/

The U.S. Geological Survey (USGS) serves the nation by providing reliable scientific information to describe and understand the Earth; minimize loss of life and property from natural disasters; manage water, biological, energy, and mineral resources; and enhance and protect our quality of life. In support of the U.S. Geological Survey's mission to provide information about the earth and its physical resources, the National Map provides geographic, cartographic, and remote sensing information, maps, and technical assistance. The USGS also makes available maps, images, spatial data, remote sensing data, and related information; provides assistance in selecting, acquiring, and using geographic and cartographic products; and designs, prints, and distributes maps of the National Atlas. It coordinates federal topographic mapping and digital cartographic activities and provides leadership in the development and advancement of surveying and mapping technology. Topographic maps usually used in the prediction of coverage and other engineering studies that require accurate information about the position and elevation of terrain features may be obtained from the U.S. Geological Survey website.

Department of Labor (DOL), Occupational Safety and Health Administration (OSHA)

200 Constitution Avenue, N.W.
Washington, DC 20210
Phone: (800) 321-6742
Website: www.osha.gov

Broadcast engineers should be aware that two federal laws—the Federal Labor Standards Act and the Occupational Safety and Health Act—empower the Department of Labor (DOL) to regulate workplace safety standards and the wages and hours of employment for broadcast employees. Occupational safety standards are enforced by the Occupational Safety and Health Administration (OSHA) of the DOL, and the minimum wage and hours of employment are enforced by the DOL's Wage and Hour Division (WHD). The applicability of OSHA regulations depends on hazards, but WHD regulations encompass many factors, such as the station's geographic location, the number of station employees, and the type of work they perform. Broadcast engineers should consult their station's attorney to determine what regulations apply to them and how to abide by them.

Department of Transportation, Federal Aviation Administration (FAA)

800 Independence Avenue, S.W.
Washington, DC 20590
Phone: (866) 835-5322
Website: www.faa.gov

When construction or alteration of a broadcast tower is proposed, the Federal Aviation Administration (FAA) conducts an aeronautical study to determine the potential impact that the proposal may have on the navigable airspace. Broadcasters are required to notify the FAA of new proposals or alterations to existing towers. The obstructions standards in Title 14 of the Code of Federal Aviation Regulations (Part 77) are applied to determine the effect the proposal would have on aeronautical operations. It may be necessary for broadcasters to amend their proposal by modifying the tower height or location in order to eliminate a determination of hazard to air navigation. When an aeronautical study results in a determination of no hazard to air navigation, the FAA will recommend the marking and/or lighting for that structure, if appropriate. The Federal Communication Commission's license will contain the marking and/or lighting recommended by the FAA. Conspicuity is achieved only when all recommended lights are working; therefore, it is important that any outage be corrected as soon as possible. The FAA must receive notice immediately of any failure or malfunction that lasts more than 30 minutes and affects a top light or flashing obstruction light, regardless of its position.

Environmental Protection Agency (EPA)

1200 Pennsylvania Ave., N.W.
Washington, DC 20460
Phone: (202) 564-4700
Website: www.epa.gov

The Environmental Protection Agency (EPA) protects public health and safeguards and improves the natural environment—air, water, and land—upon which human life depends. The EPA's purpose is to ensure that:

- Federal environmental laws are implemented and enforced fairly and effectively.
- Environmental protection is an integral consideration in U.S. policies concerning economic growth, energy, transportation, agriculture, industry, international trade, and natural resources.
- National efforts to reduce environmental risk are based on the best available scientific information.
- All parts of society—business, state and local governments, communities, and citizens—have full access to information so they can become full participants in preventing pollution and protecting human health and the environment.

When broadcasters construct station facilities, such as antenna towers or large satellite earth stations, they should be aware that these activities may fall under the scope of the National Environmental Protection Act (NEPA), which grants the EPA authority to regulate activities that may affect the "quality of the human environment." The Federal Communications Commission cooperates with the EPA in enforcing provisions of NEPA that relate to telecommunication licensees

Federal Communications Commission (FCC)

445 12th Street, S.W.
Washington, DC 20554
Phone: (202) 225-5322
Website: www.fcc.gov

The Federal Communications Commission regulates interstate and international communications by radio, television, wire, satellite, and cable in all 50 states, the District of Columbia, and U.S. territories. An independent U.S. government agency overseen by Congress, the commission is the United States' primary authority for communications laws, regulation, and technological innovation.

National Labor Relations Board (NLRB)

1099 14th Street, N.W.
Washington, DC 20570
Phone: (866) 667-6572
Website: www.nlrb.gov

Broadcast engineers should be aware that the National Labor Relations Act (NRLA) protects the right of station employees to bargain collectively with management over the "terms and condition of employment." The National Labor Relations Board (NLRB), an independent federal agency, was established to enforce the right of workers to organize and engage in "concerted activity." As with the Department of Labor regulations, the extent of these rights depends upon many factors, such as the type and number of employees involved.

STATE AND LOCAL GOVERNMENT

Several aspects of a broadcast engineer's job are affected or controlled by state or local government agencies. It is important for engineers to have some familiarity with the laws, codes, and zoning ordinances governing such matters as building construction, electrical wiring, and fire safety. Regulations may vary from one community to another, even within the same state or county. Although there are model national codes, these codes may or may not be adopted by a state or local government. If adopted, there may be some changes from the national model. The only way to determine this is to check with the local agency or agencies having jurisdiction over the matter in question. For more information on state or local regulations, see the following:

- The broadcast station's local lawyer
- The county or city business licensing office
- The county or city building inspector or fire marshal
- A licensed local contractor who performs the type of work in question

TRADE ASSOCIATIONS

Asia-Pacific Broadcasting Union (ABU)

P.O. Box 12287
50772 Kuala Lumpur, Malaysia
Phone: (603) 2282-3592; fax: (603) 2282-5292
Website: www.abu.org.my

The Asia-Pacific Broadcasting Union (ABU) is a nonprofit, nongovernment, professional association of broadcasting organizations, formed in 1964 to facilitate the development of broadcasting in the Asia-Pacific region and to organize cooperative activities among its members. It has over 279 members in 68 countries, which spans from Turkey to Samoa. The ABU provides a forum for promoting the collective interests of television and radio broadcasters and engages in activities to encourage regional and international cooperation among broadcasters.

Association of Public Television Stations (APTS)

2100 Crystal Drive, Suite 700
Arlington, Virginia 22202
Phone: (202) 654-4200; fax: (202) 654-4236
Website: www.apts.org

The Association of Public Television Stations (APTS) is a nonprofit membership organization established in 1980 to support the continued growth and development of a strong and financially sound noncommercial television service for the American public. APTS provides advocacy for public television interests at the national level, as well as consistent leadership and information in marshaling support for its members: the nation's public television stations. APTS's affiliated organization, APTS Action, Inc., provides legislative advocacy and seeks grassroots and congressional support.

Association of Radio Industries and Businesses (ARIB)

Website: www.arib.or.jp/english

The objectives of the Association of Radio Industries and Businesses (ARIB) are to investigate, conduct research and development, provide consultation regarding the utilization of radiowaves from the view of developing radio industries, and promote the realization and popularization of new radio systems in the field of telecommunications and broadcasting.

Cable Television Laboratories, Inc. (CableLabs)

858 Coal Creek Circle
Louisville, Colorado 80027
Phone: (303) 661-9100
Website: www.cablelabs.com

Cable Television Laboratories, Inc. (CableLabs), a nonprofit research and development consortium, is dedicated to helping its cable operator members integrate new cable telecommunications technologies into their business objectives. CableLabs serves the cable television industry by:

- Researching and identifying new broadband technologies
- Authoring specifications
- Certifying products
- Disseminating information

Canadian Association of Broadcasters (CAB)

770-45 O'Connor Street
Ottawa, Ontario K1P 1A4
Phone: (613) 233-4035; fax: (613) 233-6961
Website: www.cab-acr.ca

The Canadian Association of Broadcasters (CAB) represents Canada's private broadcasters, which include radio and television stations, networks, and pay and pay-per-view services. Their goal is to both represent and advance the interests of Canada's private broadcasters in the social, cultural, and economic fabric of the country. The CAB's stated mission is to advocate and lobby on their members' behalf and to act as a central point of action on matters of joint interest.

Consumer Technology Association (CTA)

1919 S. Eads Street
Arlington, Virginia 22202
Phone: (703) 907-7600
Website: www.cta.tech

The Consumer Technology Association (CTA), formerly known as the Consumer Electronics Association (CEA), promotes growth in the consumer technology industry through technology policy, events, research, promotion, and the fostering of business and strategic relationships. The CTA represents more than 2,200 corporate members involved in the design, development, manufacturing, distribution, and integration of audio, video, mobile electronics, wireless and landline communications, information technology, home networking, multimedia, and accessory products, as well as related services that are sold through consumer channels. The CTA also sponsors and manages the annual International Consumer Electronics Show, the world's largest consumer technology tradeshow. The CTA develops standards in such areas as product safety, audio and video systems, antennas, mobile electronics, and cable compatibility. In addition, the CTA is the cosponsor, along with the National Association of Broadcasters, of the National Radio Systems Committee (described below).

European Broadcasting Union (EBU)

Ancienne Route 17A
Postal Box 45
1218 Grand-Saconnex/Geneva, Switzerland
Phone: + 41 (0) 22-717-2111; fax: +41 (0) 22-747-4000
Website: www.ebu.ch

The European Broadcasting Union (EBU) is one of the world's leading professional associations of national broadcasters in the world. They have 73 full members and an additional 34 associate members, which operate in 56 countries and broadcast in 123 different languages. Working on behalf of its members in the European area, the EBU negotiates broadcasting rights for major sports events, operates the Eurovision and Euroradio networks, organizes program exchanges, stimulates and coordinates coproductions, and provides a full range of other operational, commercial, technical, legal, and strategic services. In the technical area, the EBU is involved in the research and development of new broadcast media, such as Radio Data System (RDS), digital audio broadcasting (DAB), digital video broadcasting (DVB), and high-definition television (HDTV).

Joint Electron Device Engineering Council (JEDEC)

3103 N. 10th Street, Suite 240-S
Arlington, Virginia 22201
Website: www.jedec.org

The Joint Electron Device Engineering Council (JEDEC) is an international group with more than 300 member companies that develops open standards for the microelectronics industry. JEDEC brings together manufacturers and supplies for collaboration in over 50 committees and subcommittees.

International Association of Broadcasters (IAB)

Carlos Quijano 1264. 11100
Montevideo, Uruguay
Phone: (598) 2901-1319; fax: (598) 2908-0458
Website: www.airiab.com/en

The International Association of Broadcasters (IAB) is the oldest international organization of private broadcasters. The IAB was founded in 1946 and includes radio and television broadcast groups throughout the Americas. They represent over 17,000 stations and are a founding member of the World Broadcasting Unions (WBU).

Japan Electronics and Information Technology Industries Association (JEITA)

Washington Office
1025 Connecticut Avenue, N.W., Suite 1200
Washington, DC 20036
Phone: (202) 974-2476
Website: www.jeita.or.jp/english

The Japan Electronics and Information Technology Industries Association (JEITA) aims to promote healthy manufacturing, international trade, and consumption of electronics products and components in order to contribute to the development of the electronics and information technology (IT) industries, thereby furthering Japan's economic development and cultural prosperity. JEITA's mission is to foster a digital network society for the 21st century, in which IT advancement improves quality of life for everyone. The association is also engaged in promoting environmental preservation, especially on the issue of climate change.

National Association of Broadcasters (NAB)

1771 N Street, N.W.
Washington, DC 20036
Phone: (202) 429-5300
Website: www.nab.org

The National Association of Broadcasters (NAB) represents the broadcasting industry before Congress, the courts, regulatory agencies, the White House, and the general public. The NAB's Technology team works with other NAB departments to represent the industry before the Federal Communications Commission and other agencies on issues affecting spectrum management and technical regulations. The Technology Department's mission is to provide leadership as technology innovators, educators, and advocates, as well as to strengthen current broadcast services and foster new media opportunities. They conduct a series of technical conferences, including the NAB Broadcast Engineering Conference (BEC) at the annual NAB Show, as well as the technical program portion of the annual NAB Radio Show. The department also provides timely, useful, and accurate technical information in the Broadcast Blog and through numerous other NAB technical publications, including the annual Broadcast Engineering Conference Proceedings. The Technology Department also oversees PILOT, formerly the NAB Labs, which works with industry stakeholders to help facilitate technological innovation in the broadcast industry.

North American Broadcasters Association (NABA)

P.O. Box 500, Station A
Toronto, Ontario M5W 1E6
Canada
Phone: (416) 598-9877; fax: (416) 598-9774
Website: www.nabanet.com

The North American Broadcasters Association (NABA) is a nonprofit group of broadcasting organizations throughout North America focused on furthering the interests of broadcasters at home and abroad. As a member of the World Broadcasting Unions, NABA creates the opportunity for North American broadcasters to share information, identify common interests, and reach consensus on issues of an international nature. NABA provides representation for North American broadcasters in global forums.

National Association of Broadcast Employees and Technicians-Communications Workers of America (NABET-CWA)

501 3rd Street, N.W.
Washington, DC 20001
Phone: (202) 434-1254; fax: (202) 434-1426
Website: www.nabetcwa.org

The National Association of Broadcast Employees and Technicians (NABET) is a union representing more than 10,000 broadcast and related industry workers. Through collective bargaining, contract enforcement, job training, political activity, and organizing, NABET works to improve wages, working conditions, and job security. NABET is a sector of the larger Communication Workers of America (CWA) parent union, which has 650,000 members in the communication field.

National Cable & Television Association (NCTA)

25 Massachusetts Avenue, N.W., Suite 100
Washington, DC 20001
Phone: (202) 222-2300
Website: www.ncta.com

The National Cable and Telecommunications Association (NCTA), formerly the National Cable Television Association, is the principal trade association of the cable television industry in the United States. Founded in 1952, NCTA's primary mission is to provide its members with a strong national presence by providing a single, unified voice on issues affecting the cable and telecommunications industry. The NCTA represents more than 200 cable program networks, as well as equipment suppliers and providers of other services to the cable industry. In addition to offering traditional video services, NCTA's members also provide high-speed Internet access and telecommunications services to customers across the United States. The NCTA also hosts the industry's annual trade show, the Internet and Television Expo (INTX), which serves as a national showcase for the cable industry's services.

National Cable Television Cooperative (NCTC)

11200 Corporate Avenue
Lenexa, Kansas 66219
Phone: (913) 599-5900; fax: (913) 599-5903
Website: www.nctconline.org

The National Cable Television Cooperative (NCTC) is group of more than 850 small cable companies that come together to improve their ability negotiate as a collective whole rather than as single companies. NCTC also provides their members with marketing support and the opportunity to be a part of membership conferences.

Pearl TV

One City Center
850 10th Street, N.W., Suite 312N
Washington, DC 20001
Website: www.pearltv.com

Pearl TV is a partnership of U.S. broadcast companies coming together to explore innovative ways to promote local broadcast television and find ways to further develop digital media and wireless platforms for the broadcast industry. Pearl's member companies reach nearly two-thirds of the U.S. population through their 173 television stations in 98 markets. Pearl's members include Cox Media Group, the E.W. Scripps Company, Graham Media Group, Hearst Television Inc., Media General Inc., Meredith Local Media Group, Raycom Media, Schurz Communications, and TEGNA Inc.

World Broadcasting Unions (WBU)

P.O. Box 500, Station A
Toronto, Ontario M5W 1E6
Canada
Phone: (416) 598-9877; fax: (416) 598-9774
Website: www.worldbroadcastingunions.org

The World Broadcasting Unions (WBU) is the coordinating body for broadcasting unions that represent broadcaster networks across the globe. It was established in 1992 as a coordinating body at the international broadcasting level. The North American Broadcasters Association (NABA) acts as secretariat for the WBU. The broadcasting unions that belong to the WBU include the Asia-Pacific Broadcasting Union (ABU), the Arab States Broadcasting Union (ASBU), the International Association of Broadcasting (IAB/AIR), the Caribbean Broadcasting Union (CBU), the European Broadcasting Union (EBU), the North American Broadcasters Association (NABA), and the Organización de la Televisión Iberoamericana (OTI). The WBU has four working committees: the International Satellite Operations Group (WBU-ISOG), the Technical Committee (WBU-TC), the Sports Committee, and the WBU-WIPO Broadcaster Treaty Group (WBU-WIPO).

PROFESSIONAL ASSOCIATIONS

Association of Federal Communications Consulting Engineers (AFCCE)

P.O. Box 19333
Washington, DC 20036
Website: www.afcce.org

The Association of Federal Communications Consulting Engineers (AFCCE), founded in 1948, is a professional association of communications engineers practicing before the Federal Communications Commission (FCC). AFCCE members offer their professional engineering services for broadcast stations in the AM, FM, and television services; for microwave, cellular radio, PCS, paging, and cable systems; and for satellite facilities. In addition, associate membership is offered to technical personnel and professionals sharing an interest in the technical aspects of communications. The primary purpose behind the AFCCE is to assist and promote the proper administration and regulation of the engineering and technical phases of communications regulated by the FCC.

Broadcast Education Association (BEA)

1771 N Street, N.W.
Washington, DC 20036
Phone: (202) 602-0587; fax: (202) 609-9940
Website: www.beaweb.org

The Broadcast Education Association (BEA) is the premiere international academic media organization, driving insights, excellence in media production, and career advancement for educators, students, and professionals, which has served the industry for more than 60 years. The association's publications, annual convention, web-based programs, and regional district activities provide opportunities for juried production competition and presentation of current scholarly research related to aspects of the electronic media. These areas include media audiences, economics, law and policy, regulation, news, management, aesthetics, social effects, history, and criticism, among others. BEA publishes the *Journal of Broadcasting & Electronic Media*, *Journal of Radio & Audio Media*, and the *Journal of Media Education*. BEA administers a variety of scholarships annually to honor broadcasters and the broadcast industry.

Digital Media Association (DiMA)

1050 17th Street, N.W., Suite 220
Washington, DC 20036
Phone: (202) 639-9509
Website: www.digmedia.org

Digital Media Association (DiMA) is an advocate for the digital media industry, which includes webcasters, online media, digital services, and technology innovators. They represent its members in Congress and lobbies actively in support of favorable legislation or in opposition to unfavorable opposition. DiMA is actively working on copyright and royalty issues, consumer protection, competition, and network neutrality, as well as digital taxation.

Institute of Electrical and Electronics Engineers (IEEE) Broadcast Technology Society

3 Park Avenue, 17th Floor
New York, New York 10016
Phone: (732) 562-6061
Website: http://bts.ieee.org

The Institute of Electrical and Electronics Engineers (IEEE) has 425,000 members in more than 160 countries, including students, engineers, and scientists in electrical engineering, electronics, computers, and allied fields. The IEEE Broadcast Technology Society is one of the Institute's 39 societies created to addressing member interests. The IEEE holds numerous meetings and special technical conferences (for example, the Broadcast Technology Society holds an annual Broadcast Symposium), conducts lecture courses at the local level on topics of current engineering and scientific interest, assists student groups, and awards medals, prizes, and scholarships for outstanding technical achievement. Publications of the IEEE include their *Proceedings of the IEEE* (monthly), *IEEE Spectrum* (monthly), and more than 700 industry standards. The societies and councils publish journals, magazines, and conference proceedings.

Satellite Broadcasting and Communications Association (SBCA)

1100 17th Street, N.W., Suite 1150
Washington, DC 20036
Phone: (800) 541-5981; fax: (202) 318-2618
Website: www.sbca.org

Satellite Broadcasting and Communications Association (SBCA), founded in 1986, is a national trade organization that represents the consumer satellite industry. Its focus is to help create value-added programing in order to promote stability and growth of their members through member benefits and training. SBCA also provides leadership to the industry by advocating for consumer access to the satellite services and assuring its availability at a fair price.

Society of Broadcast Engineers (SBE)

9102 North Meridian Street, Suite 150
Indianapolis, Indiana 46260
Phone: (317) 846-9000
Website: www.sbe.org

The Society of Broadcast Engineers (SBE), with its more than 5,100 members, has been devoted to the advancement of all levels and types of engineers in the field of broadcast for over 50 years. The SBE promotes communication among and offers educational opportunities for its members, as well as provides national representation for those members before federal and state regulatory agencies, manufacturers, and the general public. The SBE, through its 114 local chapters, holds regular meetings and technical conferences. It administers a certification program that recognizes various levels of engineering experience: entry level, five years, ten years, and twenty-plus years. Its three specialist certifications (8-VSB Specialist[tm], AM Diectional Specialist[tm], and Digital Radio Broadcast Specialist[tm]) establish a benchmark of individual strengths for those already certified at specific certification levels. The certification designations are as follows:

- Certified Radio Operator (CRO)
- Certified Television Operator (CTO)
- Certified Broadcast Technologist (CBT)
- Certified Broadcast Networking Technologist (CBNT)
- Certified Audio Engineer (CEA)
- Certified Video Engineer (CEV)
- Certified Broadcast networking Engineer (CBNE)
- Certified Broadcast Radio Engineer (CBRE)
- Certified Broadcast Television Engineer (CBTE)
- Certified Senior Broadcast Radio Engineer (CSRE)
- Certified Senior Broadcast Television Engineer (CSTE)
- Certified Professional Broadcast Engineer (CPBE)
- Certified 8-VSB Specialist (8-VSB)
- Certified AM Directional Specialist (AMD)
- Certified Digital Radio Broadcast Specialist (DRB)

The SBE National Frequency Coordinating Committee was established in 1982 when the Federal Communications Commission (FCC) asked the broadcast industry to identify local contacts for Part 74 frequency coordination. The SBE formed a national network of volunteers to develop local databases of frequencies and users to assist the FCC and Part 74 users. Voluntary frequency coordination is handled on a local basis by SBE coordinators. The Ennes Educational Foundation Trust was founded in 1981. Each year, this trust awards scholarships to qualified individuals pursuing an education in broadcast engineering or to those pursuing continuing education in broadcast technology. The trust also presents workshops around the country, providing in-depth instruction to members, and underwrites educational publications of a technical broadcast nature.

Society of Cable Telecommunications Engineers (SCTE) and International Society of Broadband Experts (ISBE)

140 Phillips Road
Exton, Pennsylvania 19341
Phone: (800) 542-5040; fax: (610) 884-7237
Website: www.scte.org

The Society of Cable Telecommunications Engineers (SCTE), along with its international brand known as the International Society of Broadband Experts (ISBE), is a professional organization, was formed in 1969 to provide meaningful resources for their members, while promoting the sharing of operational and technical knowledge in the field of telecommunications. Through the efforts of both the national and local chapter levels of the organization, the SCTE provides training opportunities, standards development, and certification at the technician and engineer levels. Additional information and exchange opportunities are provided through the SCTE's national conferences, board of directors, committees, newsletter, and associated trade journals. The SCTE has 15,000 members and over 70 national and international chapters and meeting groups.

Society of Motion Picture and Television Engineers (SMPTE)

3 Barker Avenue, 5th Floor
White Plains, New York 10601
Phone: (914) 761-1100; fax: (914) 761-3115
Website: www.smpte.org

The more than 6,000-person membership of the Society of Motion Picture and Television Engineers (SMPTE) is comprised of professional engineers, motion-imaging executive, creative and technology professionals, researchers, scientists, educators, and students. SMPTE advances engineering technology, disseminates scientific information, and sponsors lectures, exhibitions, and conferences to advance the theory and practice of motion picture and television engineering. Additionally, it publishes the *SMPTE Motion Imaging Journal* and, as an accredited standards developer under the American National Standards Institute, has developed more than 800 national standards, guidelines, and recommended practices for motion pictures, television, and sound associated with motion picture and television images. (For copies of SMPTE engineering documents, contact their Standards Department.) SMPTE also makes available picture and sound test films and videotapes for use as standardized measuring tools and serves as administrator of the Secretariat of International Organization for Standardization (ISO) Technical Committee 36 on Cinematography and of the U.S. Technical Advisory Groups for ISO/TC 36. The Society presents nine annual awards for outstanding contributions to motion picture and television engineering. Some of the notable SMPTE engineering committees include:

- Cinema Sound Systems Committee
- Digital Cinema Technology
- Technology Committee
- Television and Broadband Media Committee

RELATED BROADCAST-ORIENTED ORGANIZATIONS

Advanced Media Workflow Association (AMWA)

436 North Westfield Road
Madison, Wisconsin 53717
Phone: (608) 513-5992
Website: www.amwa.tv

The Advanced Media Workflow Association, Inc. (AMWA), formerly known as the Advanced Authoring Format Association, is a broad-based trade association that works closely with industry stakeholders to promote the development and adoption of AMWA technology, which enables content creators to easily exchange digital media and metadata across platforms and between systems and applications throughout the media industry. The Advanced Authoring Format simplifies project management, saves time, and preserves valuable metadata that was often lost when transferring media between applications in the past. With representatives from several major players in the industry, AMWA intends to help deliver the full benefits of digital media to content creators, including film, television, Internet, and postproduction professionals.

Advanced Television Systems Committee (ATSC)

1776 K Street, N.W., 8th Floor
Washington, DC 20006
Phone: (202) 872-9160; fax: (202) 872-9161
Website: www.atsc.org

The Advanced Television Systems Committee (ATSC) is an international, nonprofit organization developing voluntary standards for digital television. ATSC has nearly 140 members representing the spectrum of broadcast, broadcast equipment, motion picture, consumer electronics, computer, cable, satellite, and semiconductor industries. ATSC creates and fosters implementation of voluntary standards and recommended practices to advance terrestrial digital television broadcasting and to facilitate interoperability with other media. ATSC was formed in 1982 by the member organizations of the Joint Committee on Intersociety Coordination: the Electronic Industries Alliance, the Institute of Electrical and Electronics Engineers, the National Association of Broadcasters, the National Cable and Telecommunications Association, and the Society of Motion Picture and Television Engineers. ATSC digital television standards include digital high-definition television (HDTV), standard-definition television (SDTV), data broadcasting, multichannel surround-sound audio, and interactive television. In 1996, the Federal Communications Commission adopted the major elements of the ATSC Digital Television (DTV) Standard (A/53). Today they are working on developing the next generation of broadcast ATSC 3.0 standard.

American National Standards Institute (ANSI)

25 W. 43rd Street, 4th Floor
New York, New York 10036
Phone: (212) 642-4900; fax: (212) 398-0023
Website: www.ansi.org

The American National Standards Institute (ANSI) is a private, nonprofit organization that administers and coordinates the U.S. voluntary standardization and conformity assessment system. The mission of ANSI is to improve global competitiveness of U.S. business and quality of life through facilitating the creation of voluntary consensus standards and conformity assessment systems and by safeguarding their integrity. ANSI serves as the clearinghouse for nationally coordinated voluntary safety, engineering, and industrial standards. One of ANSI's important functions is accreditation. ANSI accredits standards developers, certification bodies, and technical advisory groups (TAGs) to both the International Organization for Standardization and the International Electrotechnical Commission. The Society of Motion Picture and Television Engineers, Society of Cable Telecommunications Engineers, and Consumer Electronics Association are ANSI-accredited standards development organizations (SDOs).

American Radio Relay League (ARRL)

225 Main Street
Newington, Connecticut 06111
Phone: (860) 594-0200; fax: (860) 594-0259
Website: www.arrl.org

Founded in 1914, the American Radio Relay League (ARRL) is a not-for-profit organization working to promote interest in amateur radio communications and experimentation, represent U.S. radio amateurs in legislative matters, and maintain the high standard of conduct among amateur radio operators. ARRL is also international secretariat for the International Amateur Radio Union, which is made up of similar societies in more than 160 countries around the world. ARRL publishes the monthly journal *QST*, as well as newsletters and many publications covering all aspects of amateur radio. Its headquarters station, W1AW, transmits bulletins of interest to radio amateurs and Morse code practice sessions. The ARRL also coordinates an extensive field organization, which includes volunteers who provide technical information for radio amateurs and public-service activities. In addition, ARRL represents U.S. amateurs with the Federal Communications Commission and other government agencies in the United States and abroad.

Association for Computing Machinery (ACM)

2 Penn Plaza, Suite 701
New York, New York 10121
Phone: (800) 342-6626; fax: (212) 944-1318
Website: www.acm.org

The Association for Computing Machinery (ACM) works with educators, researchers, and professionals to deliver resources that advance computing as a science and a profession. ACM provides the computing field's premier digital library and serves its members and the computing profession with leading-edge publications, conferences, and career resources.

Association of Public Radio Engineers (APRE)

8801 JM Keynes Drive
Charlotte, North Carolina 28262
Website: www.apre.us

The Association of Public Radio Engineers (APRE) works to advance the art and science of public radio engineering through research, education, and public service. The association consists of both public radio engineering professionals and students. APRE hosts a number of events throughout the year, including the Annual Public Radio Engineering Conference.

Audio Engineering Society (AES)

551 Firth Ave., Suite 1225
New York, New York 10176
Phone: (212) 661-8528
Website: www.aes.org

The Audio Engineering Society (AES) is devoted exclusively to audio technology. Its more than 14,000 membership includes engineers, scientists, students, and other authorities throughout the world. The AES serves its members, the industry, and the public by stimulating and facilitating advances in the constantly changing field of audio. It encourages and disseminates new developments through annual technical meetings, exhibitions of professional equipment, and the *Journal of the Audio Engineering Society*, the professional archival publication in the audio industry.

British Kinematograph Sound and Television Society (BKSTS)

BKSTS Pinewood Studios
Pinewood Road
Iver Heath, Buckinghamshire, SL0 0NH
United Kingdom
Phone: +44 (0) 175-365-6656
Website: www.bksts.com

Originally called the British Kinematograph Society and founded in London in 1931, it was established to help serve the growing film industry in the country. The name has since been changed to the British Kinematograph Sound and Television Society (BKSTS) to reflect the wide range of interests of the membership. In addition to meetings, presentations, seminars, international exhibitions, and conferences, BKSTS also organizes an extensive program of training courses, lectures, workshops, and special events and publishes its quarterly journal *Cinema Technology*.

Computer and Communications Industry Association (CCIA)

900 17th Street, N.W., Suite 1100
Washington, DC 20006
Phone: (202) 783-0070; fax: (202) 783-0534
Website: www.ccianet.org

The Computer and Communications Industry Association (CCIA) is an international nonprofit organization whose members come from the computer, Internet, information technology, and telecommunications industries. CCIA works to ensure and enhance society's access to information and communications. They advocate for open markets, open systems, open networks, and fair and open completion in their member's industries. CCIA lobbies both domestically and internationally on high-tech-related policy and legislative issues.

Digital Cinema Society

P.O. Box 1973
Studio City, California 91614
Phone: (818) 762-2214
Website: www.digitalcinemasociety.org

The Digital Cinema Society is a nonprofit corporation dedicated to educating and informing the entertainment industry about digital motion picture production, postproduction, delivery, and exhibition. Rather than serve as an advocate for digital technology, the Society's purpose is to provide an objective examination of all media, solutions, services, and technologies without showing favor to any one brand or service over another.

Digital Television Group (DTG)

89 Albert Embankment
Vauxhall, London, SE1 7TP
United Kingdom
Phone: +44 (0) 207-840-6500
Website: www.dtg.org.uk

The Digital Television Group (DTG) is the industry association for digital television in the United Kingdom. It is independent, platform neutral, and technology agnostic. The group is currently focused on supporting the next generation of video delivery technology, which includes work on mobile device video delivery, spectrum coexistence management, home networking, connected TV, accessibility, and Ultra HD.

DIGITALEUROPE

14 Rue de la Science
1040 Brussels, Belgium
Phone: +32 (2) 609-53-10; fax: +32 (2) 431-04-89
Website: www.digitaleurope.org

DIGITALEUROPE represents the digital technology industry in Europe and their members include some of the world's largest IT, telecom, and consumer electronics companies and associations. They are dedicated to improving the business environment where their members operate and to promoting the industry's contribution to economic growth and social progress in the European Union. DIGITALEUROPE promotes the collective European interests in the ICT and CE sectors, seeks to participate in the development and implementation of EU policies, and facilitates long-term business generation for the digital technology industry in Europe by supporting the diffusion and usage of ICT and CE technologies.

European Telecommunications Standards Institute (ETSI)

650, Route des Lucioles
06921 Sophia-Antipolis Cedex, France
Phone: +33 (0) 4-92-94-42-00; fax: +33 (0) 4-93-65-47-16
Website: www.etsi.org

The European Telecommunications Standards Institute (ETSI) is an independent, nonprofit organization responsible for producing internationally applicable standards relating to information and communication technologies within Europe. These technologies include fixed, mobile, radio, converged, broadcast, and Internet technologies. ETSI unites more than 800 members from 66 countries inside and outside Europe, including manufacturers, network operators, administrations, service providers, research bodies, and users. ETSI's prime objective is to support global harmonization by providing a forum in which all the key players can contribute actively.

Guild of Television Cameramen (GTC)

Briar Cottage
Holyhead Road
Llanfairpwll, Gwynedd LL61 5YX
United Kingdom
Phone: +44 (0) 300-111-4123
Website: www.gtc.org.uk

The Guild of Television Cameramen (GTC) is an international nonprofit organization offering a way for camera manufacturers, when designing new equipment, to consult with working camera operators to produce outline specifications for such things as cameras, lenses, and mountings. The GTC is not a trade union and avoids any political involvement, but it is an authoritative source of advice and information on all matters concerning television camera operators. Its aim is to preserve the professional status of the television camera operator and to establish, uphold, and advance the standards of qualification and competence of the television camera operator.

Hollywood Professional Association (HPA)

846 S. Broadway, Suite 601
Los Angeles, California 90014
Phone: (213) 614-0860; fax: (213) 614-0890
Website: www.hpaonline.com

The Hollywood Professional Association (HPA), formerly known as the Hollywood Post Alliance, is a trade association representing the businesses and individuals who provide expertise, support, tools, and infrastructure for the creation and finishing of motion pictures, television, commercials, digital media, and other dynamic media content. The HPA's stated mission is to provide a networking forum for colleagues and peers who have dedicated their careers to postproduction; represent their members' large and diverse interests and experience; facilitate the exchanging of information on issues relating to business, technology, skills training, and industry education; act as an industry advocate on common issues and topics relating to government affairs, local community and business issues, and technology; and provide a platform for industry events, seminars, e-mail exchange, web-based information outlets, and professional special interest groups.

Institution of Engineering and Technology (IET)

Michael Faraday House
Six Hills Way
Stevenage, Hertfordshire, SG1 2AY
United Kingdom
Phone: +44 (0) 143-831-3311; fax: +44 (0) 143-831-3311
Website: www.theiet.org

The Institution of Engineering and Technology (IET) was formed by the Institution of Electrical Engineers (IEE) and the Institution of Incorporated Engineers

(IIE) and is one of the world's largest professional engineering societies with more than 167,000 members in 127 countries. IET's mission is to inspire, inform, and influence the global engineering community, supporting technical innovations to meet the needs of society.

International Association of Broadcasting Manufacturers (IABM)

3 Bredon Court
Brockeridge Park
Twyning, Gloucestershire GL20 6FF
United Kingdom
Phone: +44 (0) 168-445-0030; fax: +44 (0) 168-445-0024
Website: www.theiabm.org

The International Association of Broadcasting Manufacturers (IABM) is an international trade association that represents manufacturers and suppliers of products and services to the broadcasting and electronic media industries. The Association's aim is to identify and promote the interests of its members and to provide benefits and services that enhance their business performance.

International Broadcasting Convention (IBC)

10 Fetter Lane, 34d Floor
London, EC4A 1BR
United Kingdom
Phone: +44 (0) 207-832-4100; fax: +44 (0) 207-832-4130
Website: www.ibc.org

The International Broadcasting Convention (IBC) is a European trade show established in 1967 as a showcase for broadcasting technology. The IBC consists of conferences and exhibits; exhibitors on the show floor include every major supplier of technology for the creation, management, and delivery of entertainment content. The IBC is designed to serve everyone involved in the creation, management, and delivery of content for the entertainment industry. It is owned by a partnership of six sponsoring organizations: the International Association of Broadcasting Manufacturers, the Institution of Engineering and Technology, the IEEE Broadcast Technology Society, the Royal Television Society, the Society of Cable TV Engineers, and the Society of Motion Picture and Television Engineers.

International Electrotechnical Commission (IEC)

3, rue de Varembé
P.O. Box 131
CH-1211 Geneva, Switzerland
Phone: +41 (22) 919-0211; fax: +41 (22) 919-0300
Website: www.iec.ch

Founded in 1906 with British scientist Lord Kelvin as its first president, the International Electrotechnical Commission (IEC) is a global organization for the preparation and dissemination of international standards for all electrical, electronic, and related technologies. The IEC charter embraces all electrotechnologies, including electronics, magnetics and electromagnetics, electroacoustics, multimedia, telecommunications, and energy production and distribution, as well as associated general disciplines such as terminology and symbols, electromagnetic compatibility, measurement and performance, dependability, design and development, safety, and the environment.

International Standard Audiovisual Number (ISAN) International Agency

5 rue Cornavin
CH-1201 Geneva, Switzerland
Phone: +41 (22) 545-1000; fax: +41 (22) 545-1040
Website: www.isan.org

The International Standard Audiovisual Number (ISAN) system is administered by the ISAN International Agency, who has responsibility for the overall ISAN system maintenance and administration. The ISAN is a voluntary numbering system for the identification of audiovisual works. It provides a unique, internationally recognized, and permanent reference number that distinguishes one audiovisual work from all other audiovisual works registered in the ISAN system. As mandated by International Standards Organization (ISO), the ISAN International Agency creates a network of appointed Registration Agencies; develops and maintains the ISAN registry; stores ISAN metadata, develops, monitors and implements policies and procedures; and implements generic marketing strategies.

International Organization for Standardization (ISO)

8, Chemin de Blandonnet
Case Postale 401
CH-1214 Vernier, Geneva, Switzerland
Phone: +41 (22) 749-0111
Website: www.iso.org

The International Organization for Standardization (ISO) is a nongovernmental organization comprised of a network of the national standards institutes of 161 countries, with a central secretariat in Geneva, Switzerland, that coordinates the system. The ISO occupies a special position between the public and private sectors by bridging organizations to meet both the requirements of business and the broader needs of society. ISO standards support innovation and help provide solutions to global problems. They also make trade between countries easier and more equitable, as well as provide governments with a technical base for health, safety, and environmental legislation and aid in transferring technology to developing countries.

International Telecommunications Union (ITU)

Place des Nations
CH-1211 Geneva, Switzerland
Phone: +41 (22) 730-5111; fax: +41 (22) 733-7256
Website: www.itu.int

The International Telecommunications Union (ITU) is an international organization within the United Nations system where governments and the private sector coordinate global telecommunications and radiocommunications networks and services. The Union's standardization activities, which have already helped foster the growth of new technologies such as mobile telephony and the Internet, are now being put to use in defining the building blocks of the emerging global information infrastructure and designing advanced multimedia systems that deftly handle a mix of voice, data, audio, and video signals. The ITU is organized into three sectors:

- The Radiocommunication Sector (ITU-R) is charged with determining the technical characteristics and operational procedures for wireless services and plays a role in the international management and use coordination of the radiofrequency spectrum.
- The Telecommunication Standardization Sector (ITU-T) embodies ITU's oldest activity—developing internationally agreed-upon technical and operating standards and defining tariff and accounting principles for international telecommunication services. The work of the ITU-T aims to foster seamless interconnection of the world's communication network and systems.
- The Telecommunication Development Sector (ITUD) seeks to promote investment and foster the expansion of telecommunications infrastructures in developing nations throughout the world.

Internet Society (ISOC)

1775 Wiehle Avenue, Suite 201
Reston, Virginia 20190
Phone: (703) 439-2120
Website: www.internetsociety.org

The Internet Society (ISOC) is a professional membership society composed of more than 143 organizations and over 80,000 individual members. It provides leadership in addressing issues that confront the future of the Internet and is the organizational home for the groups responsible for Internet infrastructure standards, including the Internet Engineering Task Force and the Internet Architecture Board. Through its International Networking (INET) conferences and other sponsored events, developing-country training workshops, tutorials, statistical and market research, publications, public policy and trade activities, regional and local chapters, standardization activities, committees, and an international secretariat, the Internet Society serves the needs of the growing global Internet community. From commerce to education to social issues, the Internet Society's goal is to enhance the availability and utility of the Internet on the widest possible scale.

Internet Engineering Task Force (IETF)

Website: www.ietf.org

The Internet Engineering Task Force (IETF), an organized activity of the Internet Society, is a large open international community of network designers, operators, vendors, and researchers concerned with the evolution of the Internet architecture and the smooth operation of the Internet. It is open to any interested individual. The actual technical work of the IETF is done in its working groups, which are organized by topic into several areas (e.g., routing, transport, security). Much of the work is handled via mailing lists. The IETF holds meetings three times per year and publishes its tri-annual *IETF Journal*.

Illuminating Engineering Society of North America (IES)

120 Wall Street, 17th Floor
New York, New York 10005
Phone: (212) 248-5000; fax: (212) 248-5018
Website: www.ies.org

The Illuminating Engineering Society of North America (IESNA) is the recognized technical authority for the illumination field. For over 100 years, its objective has been to communicate information on all aspects of good lighting practice to its members, to the lighting community, and to consumers through a variety of programs, publications, and services. IES is a forum for the exchange of ideas and information and a vehicle for its members' professional development and recognition. Through its technical committees, with hundreds of members from the lighting community, the IES correlates vast amounts of research, investigations, and discussions to guide lighting experts and nonprofessionals on research- and consensus-based lighting recommendations. Complete lists of current and available recommendations may be obtained by writing to the IES Publication Office. In addition to the *IES Lighting Handbook*, IES also publishes *Lighting Design + Application* (*LD+A*) and the *Journal of the Illuminating Engineering Society* (*LEUKOS*).

Media Communications Association International (MCA-I)

P.O. Box 5135
Madison, Wisconsin 53705
Phone: (888) 899-6224
Website: www.mca-i.org

The Media Communications Association International (MCA-I) is a not-for-profit, member-driven, global community that provides opportunities for networking, members-only benefits, forums for education, and the resources for information to media communications professionals. Through facilitating effective communication, MCA-I offers media professionals the connections needed to succeed in a highly competitive environment.

National Captioning Institute (NCI)

3725 Concorde Parkway
Chantilly, Virginia 20151
Phone: (703) 917-7600; fax: (703) 917-9853
Website: www.ncicap.org

The National Captioning Institute (NCI) was established in 1979 as a nonprofit corporation with the mission of ensuring that deaf and hard-of-hearing people, as well as others who can benefit from the service, have access to television's entertainment and news through the technology of closed-captioning. With a highly skilled captioning staff and state-of-the-art facilities, NCI provides the highest quality captioning services for broadcast and cable television, home video and DVD, and government and corporate video programming. NCI also provides subtitling and language translation services in over 40 languages and dialects.

National Center for Accessible Media (NCAM)

One Guest Street
Boston, Massachusetts 02135
Phone: (617) 300-3400; fax: (617) 300-1035
Website: http://ncam.wgbh.org

The Carl and Ruth Shapiro Family National Center for Accessible Media (NCAM) is a research and development facility dedicated to the issues of media and information technology for people with disabilities in their homes, schools, workplaces, and communities. NCAM acts as the research and development arm of WGBH's Media Access Group; its mission is to expand access to present and future media for people with disabilities; to explore how existing access technologies may benefit other populations; to represent its constituents in industry, policy, and legislative circles; and to provide access to educational and media technologies for special needs students.

National Radio Systems Committee (NRSC)

1771 N Street, N.W.
Washington, DC 20036
Phone: (202) 429-5339
Website: www.nrscstandards.org

The National Radio Systems Committee (NRSC) is jointly sponsored by the National Association of Broadcasters (NAB) and the Consumer Technology Association (CTA). Its purpose is to study and make recommendations for technical standards that relate to radio broadcasting and the reception of radio broadcast signals. The NRSC is a vehicle by which broadcasters and receiver manufacturers can work together toward developing solutions to common problems in radio broadcast systems. Anyone who has a business interest in the technology being investigated by the NRSC is welcome to join and participate in its activities. Members of the NRSC are generally engineers, scientists, or technicians with in-depth knowledge of the subject being studied. To promote the free exchange of ideas during committee work, members of the press are not allowed to attend NRSC meetings; however, members of the press are free to contact committee chairpersons, the NAB, or the CTA with general questions about meetings. NRSC meetings are held on an as-needed basis; however, meetings are regularly held at the annual International CES in January, The NAB Show in April, and the Radio Show in September. NRSC members participate at their own expense.

National Society of Professional Engineers (NSPE)

1420 King Street
Alexandria, Virginia 22314
Phone: (888) 285-6773; fax: (703) 836-4875
Website: www.nspe.org

The National Society of Professional Engineers (NSPE) was formed in 1934 to meet the needs and professional concerns of professional engineers. With more than 31,000 members, NSPE works to create an inclusive, nontechnical organization dedicated to furthering the interests of licensed professional engineers, regardless of practice area, that protects them from unqualified practitioners, builds public recognition for the professional, and stands against unethical practices and inadequate compensation.

Radio Television Digital News Association (RTNDA)

529 14th Street, N.W., Suite 1240
Washington, DC 20045
Fax: (202) 223-4007
Website: www.rtdna.org

The Radio Television Digital News Association (RTDNA) is the world's largest professional organization devoted exclusively to electronic journalism. Its membership includes local and network news executives, news directors, producers, reporters, photographers, editors, multimedia journalists, and digital news professionals in broadcasting and cable and digital media, as well as journalism educators and students. RTDNA's purpose is to protect the rights of electronic journalists in the courts and legislatures, to promote ethical standards in the industry, and to provide members with training and education. The RTDNA honors outstanding work in the industry through its annual Edward R. Murrow Awards. Additionally, RTDNA runs the Radio Television Digital News Foundation (RTDNF), which provides scholarships, fellowships, and professional development assistance to aspiring and early-career electronic journalists.

Royal Television Society (RTS)

3 Dorset Rise
London EC4Y 8EN
United Kingdom
Phone: +44 (0) 207-822-2810
Website: www.rts.org.uk

The Royal Television Society (RTS) is a British-based educational charity dedicated to promoting the art and science of television. Formed in 1927 and originally a meeting place for television engineers—both amateur and professional—the Society's earliest publications chart the birth of television and document the pioneering work of, among others, John Logie Baird. In the late 1960s and early 1970s, the balance of the industry shifted from engineering to program production and broadcasting. The Society broadened to become an independent forum, hosting regular symposia on topical subjects at which members from all the different television companies could meet together. These symposia have developed into a wide range of different events that continue to be an essential focus for the television industry.

Special Interest Group on Computer Graphics and Interactive Techniques (SIGGRAPH)

Website: www.siggraph.org

The Special Interest Group on Computer Graphics and Interactive Techniques (SIGGRAPH) is dedicated to promoting the generation and dissemination of information on computer graphics and interactive techniques by fostering a membership community, whose core values help them to catalyze the innovation and application of computer graphics and interactive techniques. Also known as ACM SIGGRAPH (because the Association for Computing Machinery is SIGGRAPH's parent organization), SIGGRAPH sponsors not only the annual SIGGRAPH conferences but also focused symposia, chapters in cities throughout the world, awards, grants, and educational and online resources.

Society for Information Display (SID)

1475 South Bascom Avenue, Suite 114
Campbell, California 95008
Phone: (408) 879-3901; fax: (408) 879-3833
Website: www.sid.org

The Society for Information Display (SID) is more than 50 years old and is composed of professionals in the technical and business disciplines that relate to display research, design, manufacturing, applications, marketing, and sales. Each member belongs to the SID chapter of his or her choice. Chapters hold periodic meetings, conferences, and trade shows that have national as well as international appeal. SID's largest international gathering is the annual Display Week Symposium and Exhibition, which attracts thousands of attendees, speakers, and exhibitors from around the world. They also publish a monthly *Journal of SID, Information Display* magazine, and the annual *Digest of Technical Papers.*

Society of Television Lighting Design (STLD)

Website: www.stld.org.uk

The Society of Television Lighting Directors (STLD), formerly the Society of Television Lighting Directors, provides a forum that stimulates a free exchange of ideas in all aspects of the television profession, including discussion of techniques and the use and design of equipment. Professional meetings are organized throughout the United Kingdom and abroad; also, technical information and news of members and their activities are published in the Society's magazine. The STLD has no union affiliations and is therefore recognized as a valuable platform for open discussion and demonstration by production, management, and industry.

Underwriters Laboratories (UL)

333 Pfingsten Road
Northbrook, Illinois 60062
Phone: (847) 854-3577
Website: www.ul.com

Underwriters Laboratories (UL) is an independent, not-for-profit, product-safety testing and certification organization. UL has been evaluating products in the interest of public safety since 1894. UL seeks by scientific investigation, study, experiments, and tests to determine the relation of various materials, devices, products, equipment, constructions, methods, and systems to identified hazards and how their use affects life and property. It also seeks to ascertain, define, and publish standards, classifications, and specifications for materials, devices, products, equipment, construction, methods, and systems affecting such hazards, and other information tending to reduce loss of life and property from such hazards.

Video Services Forum (VSF)

Phone: (609) 410-6767
Website: www.videoservicesforum.org

Video Services Forum (VSF) is an international not-for-profit association open to businesses, public sector organizations, and individuals worldwide and dedicated to video transport technologies, interoperability, quality metrics, and education. VSF is composed of service providers, users, and manufacturers. The organization's activities include providing forums to identify issues involving the development, engineering, installation, testing, and maintenance of audio and video services; exchanging nonproprietary information to promote the development of video transport service technology; fostering resolution of issues common to the video services industry; identifying video services applications and educational services utilizing video transport services; promoting interoperability; and encouraging technical standards for national and international standards bodies.

World Wide Web Consortium (W3C)

32 Vassar Street, Room 32-G515
Cambridge, Massachusetts 02139
Phone: (617) 253-2613; fax: (617) 258-5999
Website: www.w3.org

The World Wide Web Consortium (W3C) is an international consortium where member organizations, staff, and the public work together to create web standards. W3C's mission is to lead the World Wide Web to its full potential by developing standards, protocols, and guidelines that ensure long-term growth for the Web. Since 1994, W3C has published more than 265 such standards, referred to as W3C Recommendations. W3C also engages in education and outreach, develops software, and serves as an open forum for discussion about the Web.

BROADCAST ENGINEERING AND RELATED PRINT AND DIGITAL PERIODICALS

ABU Technical Review

Asian Pacific Broadcast Union
P.O. Box 12287, 50772
Kuala Lumpur, Malaysia
Phone: (603) 2282-3592; fax: (603) 2282-5292
Website: www.abu.org.my

Above Ground Level (AGL)

P.O. Box 2090
Ashburn, Virginia 20146
Phone: (540) 882-4290
Website: www.aglmediagroup.com

Broadcasting and Cable

NewBay Media
28 East 28th Street, 12th Floor
New York, New York 10016
Phone: (212) 378-0400; fax: (212) 378-0470
Website: www.broadcastingcable.com

Digital Video

NewBay Media
28 East 28th Street, 12th Floor
New York, New York 10016
Phone: (212) 378-0400; fax: (212) 378-0470
Website: www.creativeplanetnetwork.com

EBU Technical Review

European Broadcasting Union
L'Ancienne-Route 17A
CH-1218 Grand-Saconnex, Geneva, Switzerland
Phone: +41 (22) 717-2745
Website: https://tech.ebu.ch/techreview

ECN Electronic Component News

100 Enterprise Drive, Suite 600
Rockaway, New Jersey 07866
Phone: (646) 746-6400
Website: www.ecnmag.com

Electronic Design

Penton Media, Inc.
1166 Avenue of the Americas, 10th Floor
New York, New York 10036
Phone: (212) 204-4200
Website: http://electronicdesign.com

Electronic Engineering Times

UBM Canon
2901 28th Street, Suite 100
Santa Monica, California 90405
Phone: (310) 445-4200
Website: www.eetimes.com

Engineering and Technology Magazine (E&T)

Savoy Place
London, WC2F 0BL
United Kingdom
Phone: +44 (0) 207-240-1871
Website: http://eandt.theiet.org

Government Video

NewBay Media
28 East 28th Street, 12th Floor
New York, New York 10016
Phone: (212) 378-0400; fax: (212) 378-0470
Website: www.governmentvideo.com

GPS World

North Coast Media
1360 E. 9th Street, Suite 1070
Cleveland, Ohio 44114
Website: www.gpsworld.com

IEEE Transactions on Broadcasting

3 Park Avenue, 17th Floor
New York, New York 10016
Website: http://ieeexplore.ieee.org/xpl/RecentIssue.jsp?punumber=11

IEEE Transactions on Consumer Electronics

3 Park Avenue, 17th Floor
New York, New York 10016
Website: http://ieeexplore.ieee.org/servlet/opac?punumber=30

Journal of the Audio Engineering Society

551 Fifth Avenue, Suite 1225
New York, New York 10176
Phone: (212) 661-8528
Website: www.aes.org/journal

Mix

NewBay Media
28 East 28th Street, 12th Floor
New York, New York 10016
Phone: (212) 378-0400; fax: (212) 378-0470
Website: www.mixonline.com

Pro Sound News

NewBay Media
28 East 28th Street, 12th Floor
New York, New York 10016
Phone: (212) 378-0400; fax: (212) 378-0470
Website: www.prosoundnews.com

QST

American Radio Relay League
225 Main Street
Newington, Connecticut 06111
Phone: (860) 594-0200; fax: (860) 594-0259
Website: www.arrl.org/qst

Radio Guide

P.O. Box 20975
Sedona, Arizona 86341
Phone: (928) 284-3700; fax: (866) 728-5764
Website: www.radio-guide.com

Radio Ink

1901 S. Congress Ave., Suite 118
Boynton Beach, Florida 33426
Phone: (561) 655-8778
Website: http://radioink.com/

Radio Magazine

NewBay Media
28 East 28th Street, 12th Floor
New York, New York 10016
Phone: (212) 378-0400; fax: (212) 378-0470
Website: www.radiomagonline.com

Radio World

NewBay Media
28 East 28th Street, 12th Floor
New York, New York 10016
Phone: (212) 378-0400; fax: (212) 378-0470
Website: www.radioworld.com

SMPTE Motion Imaging Journal

3 Barker Avenue, 5th Floor
White Plains, New York 10601
Phone: (914) 761-1100; fax: (914) 761-3115
Website: www.smpte.org/publications/journal

Streaming Media Magazine

Phone: (920) 342-6263
Website: www.streamingmedia.com/Previous_Issues

Television Week

Website: www.tvweek.com

TV Technology

NewBay Media
28 East 28th Street, 12th Floor
New York, New York 10016
Phone: (212) 378-0400; fax: (212) 378-0470
Website: www.tvtechnology.com

TWICE (This Week in Consumer Electronics)

NewBay Media
28 East 28th Street, 12th Floor
New York, New York 10016
Phone: (212) 378-0400; fax: (212) 378-0470
Website: www.twice.com

CHAPTER

2.2

Broadcast-Related Regulatory Considerations

Environmental, Aeronautical, Safety, Online

DAVID OXENFORD

Partner, Wilkinson Barker Knauer, LLP
Washington, DC

INTRODUCTION

Broadcast engineers are often called on to do much more than simply engineering. In all phases of broadcast operations, from construction of new broadcast facilities to the routine selection and maintenance of equipment, legal issues and regulatory agencies play a significant role. To avoid delays and costs that can be incurred from ignoring necessary regulatory requirements, the engineer needs to be aware of the many regulatory issues that can arise. In some cases, an entire project can be scuttled if the proper approvals have not been timely obtained. And in today's complex world of digital communications, where the over-the-air operations of radio and television stations more and more merge with the online world, broadcast engineers have new responsibilities, implicating issues and agencies that just a few years ago would never have been on a broadcaster's radar.

In a publication of this type, it is impossible to provide a detailed regulatory review of all of the rules that govern any project in which an engineer may be involved. Instead, this chapter will highlight some of the agencies and legal authorities whose rules need to be considered in connection with any project in which a broadcast station may be involved. As the rules change constantly, and as the requirements may change from state to state and even city by city, in the initial stages of planning any project the broadcast engineer will need to check the rules that apply in the station's area, as well as the federal regulations that apply to the planned project wherever it may be located. And even in day-to-day operations, broadcast engineers need to be sure that they are observing all legal requirements and undertaking the kinds of review and diligence necessary to ensure continued compliance with all regulatory obligations.

The review will start with the agency with which engineers will have the most contact—the Federal Communications Commission. The FCC is the agency that has the primary responsibility for setting the rules for over-the-air operations. Everything from getting approval for new construction of a station to the modification of existing stations (including broadcast auxiliaries) needs FCC approval. But that is not all the FCC does. The FCC has many rules governing day-to-day operations of stations and, in some cases, these rules even extend to the digital media properties of broadcasters.

From there, we will look at some of the other governmental authorities that may become involved in broadcast operations. We will look at the issues involved in any sort of construction, whether it be of a tower or of a building housing broadcast operations. We will look at some operational considerations for broadcasters. Finally, we will review the many privacy, security, and intellectual property issues that can impact broadcasters' digital operations, as well as the more traditional means of reaching audiences.

As will be discussed, there are plenty of legal issues to think about. As we urge later in this chapter, an engineer needs to get to know the lawyer working with the station, because only by developing a close working relationship between a station's legal and engineering advisors can broadcasters avoid the costs and delays that legal missteps can bring about.

FCC OVERVIEW

Broadcast engineers are likely to encounter a wide array of regulatory agencies in the course of performing their duties. Of course, for broadcasters, none is as crucial as the FCC. The FCC grants broadcasters the authority to operate on over-the-air frequencies. Without that FCC license, the station is essentially out of business. As the FCC's reach is so pervasive in the broadcast industry, engineers should have a basic understanding of the structure of the agency.

Structure

As the name implies, the FCC is made up of five Commissioners. All of the Commissioners are nominated by the President, and must be confirmed by the Senate. They have staggered five-year terms, meaning that each year, one seat needs to be filled (or the person holding that seat needs to be renominated). Of the five seats, a majority (including the Chairman) can be held by the political party that holds the White House. The remaining two Commissioners must come from other political parties. Normally, Senators will make suggestions to the President of potential nominees.

The FCC has a staff of hundreds of employees, and thus the Commissioners themselves are usually involved only in decisions involving broad policy matters. Other employees, organized into Bureaus (with smaller "Divisions" forming part of each Bureau) and Offices, tend to do most of the real work of the FCC.

Most of a broadcaster's interactions on day-to-day matters will be at the Bureau level, principally with the Media Bureau and two of its Divisions, the Audio and Video Divisions, as explained in more detail below. Each of the Bureaus and Offices are focused on a particular mode of communication or a defined segment of the communications industry (e.g., there is one Bureau, for instance, that deals with media issues, another for issues for the wireless industry, and a third for wireline communications) or a unique regulatory function (e.g., there is an Office of the General Counsel that deals with legal issues, a Bureau dealing with the enforcement of FCC rules, and Offices dealing with technical and economic issues). From time to time, these Offices and Bureaus change name, and functions may be moved from one to another.

Rulemaking Proceedings

The Commissioners, for the most part, consider issues of broad policy and the adoption of new rules (or the modification of existing rules). Virtually all rule changes, and most policy changes, are done through a "rulemaking" proceeding. In a rulemaking proceeding, the FCC will release a document called a "Notice of Proposed Rulemaking" suggesting changes to its rules and policies. The public will have the opportunity to comment on the proposed changes before the FCC acts in the rulemaking proceeding. Interested parties can also make more informal presentations to the Commissioners and their staffs (essentially, they can lobby), provided that written descriptions of such communications are filed in the written record of the proceeding. The comments and the lobbying help to shape the final decision of the Commissioners. Because of the amount of business that the FCC does, the complexity of the issues, and the number of interested parties who comment in these proceedings (often with radically different views as to what the outcome should be), changes in the FCC rules can sometimes take years.

The record developed in these rulemaking proceedings (a record that, in complicated proceedings, can comprise thousands or even millions of pages) is considered by members of the FCC staff in the appropriate office or division. These staff members will usually compile a draft decision, often along parameters directed by the FCC Chairman, which will be reviewed by the other Commissioners and the Commissioners' personal office staffs. Negotiations over the specific provisions contained in the decisions, and revised drafts, are common. Decisions in rulemaking proceedings can either be adopted by the FCC Commissioners at the monthly meeting, which is open to the public or "on circulation," where the written decision is simply released to the public when all of the Commissioners have reviewed it and registered their votes.

Rulemakings can be very important proceedings, often setting the direction for an entire segment of the communications industry. As such, they can have important ramifications for the broadcast engineering community. Thus, part of the job of the broadcast engineer needs to be staying on top of the currently pending proceedings, as the matters that are now under consideration could profoundly affect decisions about broadcast operations, including potentially imposing new equipment and operational obligations on stations.

Examples of recent proceedings that have imposed such obligations are many. Some may be as profound as the adoption of the digital television standard in the 1990s and its implementation in the first decade of this century, or the broadcast incentive auction repurposing part of the television spectrum for wireless uses, a process that has unfolded since the DTV transition in 2009. At the time that this chapter is being written, the FCC is considering a new rulemaking, looking at a brand new digital television transmission standard (ATSC 3.0), for which broadcast engineers should already be planning.

Rulemakings can also be less sweeping for the industry, but still impose burdens that engineers will need to implement. Over the past decade, the FCC has imposed numerous captioning and accessibility obligations on broadcast television to assist audience members who have visual or auditory limitations. Many of these compliance measures even involve station Internet operations—for instance, requiring the captioning of broadcast programming that is later distributed through Internet channels. The accessibility measures almost all require technical solutions, including equipment changes, to ensure compliance. The FCC, on an ongoing basis, also reviews emergency communications, in the last decade adopting many technical changes in the way that alerts are transmitted to the public and requiring more reporting by the broadcaster about emergency test results.

Other changes to technical rules and operational requirements are also considered on a routine basis. In fact, Congress has imposed on the FCC the obligation to review each of its rules periodically to ensure that they are still necessary and operating in the public interest. While these reviews are often perfunctory, from time to time the FCC does engage in a more in-depth review of some aspect of broadcast operations. Over the past quarter century, as the AM band has been eclipsed by FM, the FCC has been continually reviewing how rules can be changed to assist AM station operators. The FCC has been constantly reviewing the efficient and effective use of broadcast auxiliary spectrum. Alternative uses of broadcast spectrum beyond just the incentive auction (e.g., expanding digital operations, providing for the use of broadcast spectrum to transmit nonbroadcast data and other information services) have also been routinely considered by the FCC. Rules for new services, and changes in existing operations of existing stations, may affect interference protections afforded to broadcast stations, or may make changes that can affect whether technical changes in a broadcast facility are easier or harder to make. Even procedural rules for getting approval for FCC applications and the handling of routine FCC requirements are considered by the FCC in rulemaking proceedings on a regular basis.

Broadcast engineers need to be on top of these changes and participate in proceedings that the Commission initiates to consider such changes to make sure that technical and operational realities are considered as new obligations are adopted. First, it is important for engineers to assess any proposed rule changes to determine if the changes will affect current operations or, perhaps, provide future opportunities that could enhance the business operations of their (or their client's) stations. Missing new rules that impose new obligations can end up costing stations in fines and other penalties should the FCC discover that a station has not complied with new requirements. But as changes can also present opportunities, not following new FCC regulatory developments may result in a station missing potentially lucrative business opportunities.

In addition, engineers can help shape new FCC rules before they are adopted. While the FCC staff includes a substantial number of engineers and others with a technical background, there is no requirement that the Commissioners themselves have any engineering background, or even that their immediate staff include an engineer. The engineers and other technical employees who work in the Bureaus and Offices may or may not have real-world business experience. Many will likely have never worked at a broadcast station—in fact, some may have never even visited a station, especially a station outside of a major market. Thus, it is important that the viewpoint of the broadcaster (and broadcast engineers, in particular) be provided to the FCC when it considers rule changes, to give the FCC the perspective on which changes would be most beneficial to broadcasters generally and to ensure that such changes can be implemented in the real world. The opportunity for public comment on new proposals after the adoption of a Notice of Proposed Rulemaking allows engineers and other broadcasters to become part of the process and to help shape its outcome.

Review of Staff Decisions

In addition to the consideration of broad policy issues through the rulemaking process, the Commissioners also consider appeals of decisions made by its Bureaus. So, for broadcasters, the denial of an application by the Media Bureau, which initially deals with most broadcast applications, can be appealed to the Commissioners. Proposed fines on broadcasters for violations of the rules can also be reviewed by the Commissioners. A broadcast company seeking review of an FCC staff decision will file a request with the Commissioners, following a process outlined in the FCC rules, setting out the reasons why the staff decision was incorrect and why the Commissioners should overturn that decision.

The review of staff decisions can be important to more than just the participants in the particular case being considered by the Commissioners. These decisions can have an impact on the entire industry. Decisions in individual cases will set precedent that the Commission and its staff will use in the future for interpreting a rule, regulation, or other requirement. A decision in any single case may set the obligations for all broadcasters when dealing with the same rule or the same set of factual circumstances. Thus, while it is important for engineers to stay on top of rulemaking decisions, and they must also be aware of significant decisions by the FCC and its staff in particular cases, as these cases may give the best, most detailed interpretation of what the FCC believes to be required to stay in compliance with a particular rule or regulatory obligation.

Keeping Informed About FCC Decisions

As stated above, it is important that broadcast engineers stay on top of new developments at the FCC—both in rulemaking proceedings and in specific cases that interpret FCC rules and policies. But with so much happening at a station, how does a broadcast engineer do that?

There are many broadcast trade publications, some available for free. These publications routinely carry news about new proposals being considered by the FCC and about decisions in particular cases. The National Association of Broadcasters (NAB), state broadcast associations, the Society of Broadcast Engineers (SBE), and many other associations and societies also send emails, publish newsletters, host webinars and list servers, and provide other information that can help to keep a broadcast engineer on top of developments at the FCC. Also, many law and engineering firms publish newsletters or blogs, many available to the general public, that cover many of these FCC developments. Finally, don't ignore the FCC's own website, www.fcc.gov, which provides a wealth of information about the activities in which the FCC is engaged.

FCC Bureaus

As set forth above, the FCC's staff is organized into several Bureaus that deal with specific areas of communications regulation and Offices that perform certain functions at the FCC. Each of the Bureaus has primary responsibility for processing applications relating to the services that they regulate, as well as providing the manpower to formulate the initial take on policy issues that are being considered by the Commissioners dealing with the services that the Bureau regulate. The Bureaus and Offices that an engineer may deal with are described below (as they exist at the time of the drafting of this chapter, Fall 2016).

The FCC's Media Bureau is the portion of the FCC with which broadcasters will have the most contact. The Media Bureau is responsible for the licensing and regulation of broadcast radio and television stations. It also regulates cable television and assists in the regulation of direct broadcast satellite systems. The Media Bureau, like each FCC Bureau, is composed of various divisions, among them the Audio Division, which regulates radio, the Video Division, which regulates television, and the Policy Division, which (among other things) regulates cable systems.

The Wireless Bureau licenses and regulates non-broadcast wireless communications and oversees the FCC's auctions, including auctions of broadcast licenses. While the Media Bureau issues the primary broadcast licenses of radio and television stations, most broadcast stations also hold auxiliary licenses issued by the Wireless Bureau that are crucial to day-to-day operations, such as studio-transmitter links, remote pickups, or point-to-point microwave relays. The Wireless Bureau also oversees the registration of towers and other antenna structures.

The International Bureau licenses and regulates satellite and international communications, including undersea telecommunications cables. Broadcast engineers are most likely to work with the International Bureau in connection with the licensing of satellite uplink or downlink dishes for television stations, including mobile satellite trucks.

The Enforcement Bureau, as the name implies, is tasked with investigating potential violations of the Communications Act and the FCC's rules and, when a violation is found, imposing a fine or other sanction on the responsible party. The Enforcement Bureau not only has employees at the FCC's main office in Washington, but its employees also staff the FCC's field offices and perform random inspections of broadcast stations. Its staff will also usually be the one to conduct any physical inspection of a broadcast station resulting from a complaint or other Commission action. Fines and other sanctions from the Enforcement Bureau typically are initiated in response to a complaint filed by the public or a competitor or in connection with a visit to a station by an FCC field agent.

The Public Safety and Homeland Security Bureau is most likely to be of interest to broadcast engineers in connection with its consideration of issues relating to the Emergency Alert System (EAS) and other related emergency communications issues. The Consumer and Government Affairs Bureau deals with accessibility issues, so broadcast engineers may encounter the work of this Bureau in connection with rules dealing with captioning and similar requirements, as well as consumer complaints. Broadcast engineers are less likely to deal with the FCC's Wireline Competition Bureau.

Getting FCC Approvals

One of the primary areas where broadcast engineers will need to deal with the FCC will be in connection with the filing of applications for FCC approval—particularly approvals for technical changes in broadcast facilities. The FCC has detailed rules about when prior approval is necessary for actions taken by a broadcaster in connection with its station and when changes can be made without prior FCC approval (and, in such cases, whether or not notice to the FCC is required). The broadcast engineer must become very familiar with circumstances where FCC prior approval is necessary as, if such approval is not obtained before any new station is constructed or any change in the transmission facilities of an existing station is made, penalties can be imposed on the broadcaster. In an extreme case, it is possible that the construction must be undone—and it can even jeopardize the station's FCC license.

In most cases, FCC applications are filed by completing electronic forms available on websites maintained by the FCC. For commercial operators, most applications will require the payment of a processing fee. Processing times vary based on many factors, including the type of application, the workload of the FCC at the time that the application is filed, and the complexity of the actions being requested. If any deviation or waiver of FCC rules or policies is required for the application's approval, it is likely that the application will take even longer to process. In the planning process for any project, the broadcast engineer should always inquire in advance as to the expected time for processing of the FCC application. Even if an estimate is obtained, plans should be made in the event of unanticipated delays, as unexpected delays can occur in almost any case.

While FCC rules change from time to time, and there may be certain circumstances where the FCC approvals listed below may not apply or may be simplified in some way or another, there are certain technical situations where prior FCC approval generally is necessary. Generally speaking, broadcasters must secure FCC approval before implementing certain changes to station operations, including the following:

- Construction of a new facility: This includes primary broadcast facilities, backup (auxiliary) broadcast facilities, and satellite and wireless facilities that support broadcast operations;
- Technical modification of an existing facility: This requirement covers not just physical facilities changes but also most modifications of existing licenses to correct errors in coordinates or other authorized parameters, even if no facilities change is contemplated;

- Licensing of constructed stations: In most cases, once authorized construction or modification of broadcast facilities is complete, broadcasters must return to the FCC to request a license to operate the facilities as constructed or modified, or at least provide notice of how the new facilities were constructed and whether there was any deviation in that construction from what was authorized by the FCC;
- Temporary operations at variance from license: If it becomes necessary to operate at reduced power, or with facilities that are in some way different from those authorized in the station's FCC license, or to discontinue operations (e.g., because of equipment failure or damage), FCC notice or approval may be required. In certain cases, the FCC may also allow temporary operations of broadcast facilities that will not be permanently authorized by the FCC (e.g., special authorizations for the use of broadcast auxiliaries near big events like a Super Bowl or Presidential Inauguration);
- Experimental operations: In certain circumstances, the FCC will authorize stations to operate using nonstandard fundamental parameters for testing or experimental purposes (for example, a TV station testing a new DTV standard prior to its adoption by the FCC).

In connection with these applications, in most cases the FCC will require a technical showing that the proposed facilities comply with the FCC rules regarding the interference protection accorded to other stations, the proposed coverage area will provide an adequate level of service to the station's city of license or designated service area, and the power levels set out in the application comply with limitations imposed on the type of service being proposed. The FCC will also routinely look at legal issues in connection with the application, some of which may involve some degree of technical analysis, including whether any new or modified station facilities will comply with the FCC's multiple ownership rules that limit the number of stations that one party can own in a particular market. Other issues, including whether the applicant is current on paying FCC fees, also come into play in processing and evaluating these applications.

Other changes in a station's operations, not involving its transmission plant, may still trigger the need for an engineer to assess the compliance of the proposal with applicable FCC rules. For instance, at the time this chapter was written, the FCC required that a station maintain a main studio in close proximity to its city of license. Any change in that main studio needs to be reviewed to assure the licensee that the location of the studio will meet FCC requirements. Other FCC approvals for such a change may also be required, e.g., any changes in the locations of studio-transmitter links and other broadcast auxiliaries.

The sale of a station or a licensee company also must be approved by the FCC before it can proceed. While, for the most part, engineers are not likely to be deeply involved in the preparation of the related applications, the sale process will nevertheless likely be a busy time for broadcast engineers. Potential buyers will normally conduct what is referred to as "due diligence." This will involve reviewing the physical facilities of a broadcast station both to assure its compliance with the FCC rules, checking to see if the operational facilities match those specified in the FCC licenses, and to assess the general operating condition of the technical equipment at the station. Assessment of whether any real estate involved in the transaction has any environmental toxins, or could otherwise involve any environmental issues, may also occur. Discrepancies in documentation or other problems identified in the presale review process often must be addressed prior to the closing of the sale, frequently by the seller's engineer. Buyers may even demand that certain equipment be repaired or upgraded as a condition for the sale.

Every eight years, a broadcaster must seek the renewal of its operating license. All stations in the same state file in the same window in the same year. Engineers are routinely called on in connection with the completion of such applications, as the applications require certifications as to compliance with technical rules, including those dealing with RF radiation (as described below). The renewal application must be submitted timely and all information must be accurate. Stations sometimes get in more trouble for submitting false information about issues at a station than they would have if they had provided the accurate information in the first instance—even when the information admitted issues with station operations. Stations can be fined for issues identified in connection with the renewal, and in extreme cases, the renewal can be denied, stripping a station of its right to broadcast. Thus, whenever an engineer is asked to help with the license renewal, extreme care should be exercised.

The construction and most modifications of towers used by FCC licensees require an environmental review process. Virtually all antenna support structures over 200 feet tall, including the construction of buildings on which auxiliary antennas will be located, require registration with the FCC. Before the FCC will issue an antenna structure registration (ASR), it must ensure that the construction will not constitute a significant environmental action. Thus, applicants must conduct a review to ensure that their proposals would not constitute an adverse action under rules dealing with environmental preservation (including preservation of wetlands), historic preservation, wildlife protection, and the protection of certain Native American sites. This often requires that an engineer coordinate with an expert in such reviews, as notices need to be sent to numerous interested government agencies and other parties, including Indian tribes, wildlife officials, state historical preservation offices, and others. This can be a time-consuming process, so in planning any project involving new tower construction, this process must be taken into account. The FCC will not issue an ASR without this process having been completed. Plus, if the FCC approval for a new station or the modification of an existing station is involved, those approvals will not be forthcoming until the ASR has been issued.

Related to the environmental review is a review of RF radiation that will be created in connection with any application for new broadcast facilities. The Occupational Safety and Health Administration (OSHA) has adopted certain regulations concerning the exposure of workers and others to radiation created by different sources, including broadcast transmitters. Broadcasters must certify in their FCC applications, and in some cases provide details, as to how their proposed facilities will comply with OSHA guidelines to ensure that workers and other visitors to a broadcast site will not suffer exposure to radiation in excess of OSHA limits.

The government agencies that are involved in the environmental review are not the only government agencies that may become involved in the approval of FCC applications for new broadcast stations or changes in existing stations. The FCC will routinely require Federal Aviation Administration (FAA) approval of any proposal involving the construction of towers of certain heights to ensure that these towers don't impinge on flightpaths or otherwise interfere with airspace use. From time to time, broadcast operations, particularly on higher FM frequencies (near 108 MHz, the upper end of the FM frequency band), can also be predicted to cause interference to certain frequencies used in electronic equipment in certain aircraft. If FAA approval is not obtained, FCC applications will not be granted.

International issues can also arise in connection with broadcast station applications. Applications for broadcast operations near to the Canadian or Mexican borders will require "coordination" with the agencies in these countries that deal with broadcast matters. The coordination process involves the FCC forwarding the broadcaster's proposal to the appropriate foreign government for approval if that proposal could cause interference or otherwise limit broadcast uses in these countries. Sometimes, the approval process from these foreign agencies can move quite slowly. As the area of potential interference can encompass stations several hundred miles from the border with these countries, when planning construction in any state near to an international border, check with experts as to whether any applications will require international coordination, as this can, in some cases, materially delay the FCC approval process.

Routine FCC Compliance

While the broadcast engineer may have entered the business thinking that his or her primary responsibility would be the maintenance of equipment to ensure its optimal technical operation, day-to-day broadcast operations require constant reference to FCC regulatory obligations, because broadcasting is a regulated industry. Of particular importance is ensuring that station operations are conducted in accordance with the terms of its FCC authorizations. The FCC authorizations dictate frequencies, power levels, and a host of other technical parameters that must all be regularly monitored to ensure that the station is not substantially deviating from these specified parameters. In some cases, the FCC rules specify how the monitoring must be done, and even when it must be done. But in all cases, the station's engineers must be sure that the station is operating in line with the facilities specified in its FCC authorizations. Where there is substantial deviation from licensed parameters, repairs must be made and, if the repairs cannot be made quickly, the FCC may need to be notified and in some cases authority must be requested for the station to operate with facilities at variance from those authorized.

The FCC also has specific rules about interference to other stations and even to the operations of electronic equipment by local residents. Sometimes, such interference can be caused by damaged or malfunctioning equipment. At other times, new or modified facilities of a station can cause interference to consumer radios, TVs, or other electronic devices in the immediate vicinity of a new transmitter site. The FCC has specific rules about how quickly instances of interference must be resolved. Obviously, it is not only a regulatory requirement, but also a good business practice, to keep good relations with other broadcasters and potential audience members. Engineers need to be particularly attentive when interference issues are raised.

In addition to observing the station's transmission facilities, many other station functions must be monitored on a regular basis. The FCC requires that stations conduct periodic tests of EAS systems, both initiating periodic tests and monitoring other stations and other sources to detect tests initiated by those sources. Stations that have broadcast towers that have lighting obligations must routinely observe those lights to make sure that they are functioning properly. Along with routine monitoring of the station's transmission, the observations of tower lighting and the conducting of EAS tests must be documented in a station log required by the FCC, and that log must routinely be reviewed by a chief operator designated by the station. Care needs to be maintained to ensure that these required records are maintained by stations and available for FCC inspection should issues arise.

Monitoring other station operations often falls within the responsibility of the broadcast engineer. Many tower sites, particularly those of AM stations and other stations where there are high levels of RF radiation near ground levels, require controlled access to the transmitter sites—usually accomplished by secured tower fences. Those fences need to be monitored to ensure that they are in good repair and that locks and other security features are maintained. Required information about the ownership of the tower and, in some cases, about which stations are located at the site, may also need to be posted on a sign at points where there is public access to the transmitter site. Engineers need to ensure that such information is posted and the signs are maintained and updated as necessary. In some cases, warning signs about high RF radiation levels are also required—and the status of those signs need to be checked routinely to ensure that they are still in place and readable by site visitors.

Engineers at television stations are usually tasked to ensure that the station's over-the-air programming (and, now, its online video) contains all required closed-captioning and other accessibility information. Certain emergency information must be conveyed on secondary audio channels of TV stations, so engineers need to ensure that such channels are operational and systems are in place to transmit the required information on these channels. Loudness of commercials on television stations must be monitored to ensure that the audio levels don't exceed that of adjacent programming. And many other routine record-keeping obligations (e.g., maintenance of the station's public file) are often delegated to the broadcast engineer. Station management also appreciates the engineer keeping on top of other regulatory deadlines, such as those for filing license renewal applications, ownership and EEO reports, children's television reports for television stations, and annual regulatory fees. There are plenty of regulatory obligations to keep an engineer very busy.

The FCC can always visit a station—sometimes doing an unannounced inspection by an FCC Enforcement Bureau field inspector. The station's engineer should be prepared to take the inspector through the station and transmitter site to demonstrate that its facilities are rule compliant. Inspections can occur randomly, or they can be initiated because the FCC has received a complaint about station operations. Management should be alerted when such inspections occur, and a call to the station's attorney is also advisable.

Most state broadcast associations run alternative broadcast inspection programs (ABIP), in which the station will contract with the association for a technical inspection provided by a private engineer. If a station passes this inspection, or corrects identified deficiencies in a timely manner, the station will get a pass on routine unscheduled FCC inspections (though the FCC can still inspect the station if there is a complaint or for other targeted inspections in connection with specific FCC investigations). Broadcast engineers should look into ABIP, as this not only helps ensure that the station complies with FCC rules, but also acts as insurance against a random FCC inspection.

Even hiring employees for the engineering department at a station needs to be done in accordance with FCC's equal employment opportunity (EEO) rules. The FCC has rules and policies that have been adopted to try to bring about a diverse workforce at broadcast stations. These rules require that most hiring at broadcast stations be done only after there is outreach to all significant groups in a station's service area, with information about job openings to encourage these groups to refer qualified individuals to the station.

The outreach is designed to provide stations with a broad pool of applicants for any position from which they can choose the most qualified and to cut down on word-of-mouth hiring, where only those already in the know about broadcast opportunities get the chance at open positions. The EEO rules also require that stations use a number of outreach services to inform the public about job openings. The rules also require that broadcasters take actions, even when there are no job openings, to educate the public in their area (e.g., through participation in school and community group events) about the kinds of occupational opportunities available at broadcast stations, the training necessary for those jobs, and the career path to enter and advance in the broadcast industry. Documentation of all outreach, both that which is job specific and that which does not relate to any particular job opening, must be maintained, and is subject to FCC inspection. Engineering personnel are likely to be involved in these EEO outreach efforts at broadcast stations.

Penalties

Compliance with FCC rules is an imperative, not just as a matter of course, but also because there can be substantial penalties for noncompliance. The FCC can issue monetary fines for violations of its rules and policies that can reach over $30,000 per violation, and each day that a technical violation continues can be considered by the FCC to be a new violation—meaning that fines can add up fast. Moreover, if a pattern of violations is established, the FCC can even revoke a station's license, essentially putting it out of business. So staying compliant with FCC rules is essential to the broadcast station and thus central to the role of a broadcast engineer.

OTHER GOVERNMENTAL AND REGULATORY AGENCIES

Broadcast stations and their business operations are subject to the oversight and regulations of many entities other than the FCC. Some of these governmental obligations are those that are imposed on any business. Others arise due to the unique nature of the operations of broadcast stations. Engineers will encounter these regulators in the course of performing their day-to-day responsibilities, as well as in connection with the development of new facilities and the implementation of new ways of reaching the audience of the broadcaster. In fact, digital communications channels, including station websites, streaming services, mobile apps, text-based services, and other means of communicating with the audience, have required many broadcasters to deal with a whole new host of regulatory agencies and a whole new set of legal requirements.

We cannot hope to provide a comprehensive list of all of these obligations in this one chapter. Instead, we will highlight some of the governmental and regulatory entities that are most likely to be encountered, directly or indirectly, by any broadcast engineer. Because of the web of regulation and legal obligations that exist in running any business, we urge all engineers to get to know the legal counsel for their stations, both counsel representing the station in connection with federal matters like FCC regulation and counsel who helps the station navigate more local issues like zoning and land-use regulation. Engineers should be comfortable in seeking legal guidance when questions arise and in seeking that advice in advance when considering new projects. Only by timely seeking legal advice when appropriate can a broadcaster avoid the penalties and delays that can arise if legal obligations are overlooked.

Construction Projects—General

The construction or modification of any kind of station facilities will typically require approvals from, or at least a review of compliance with the regulations of, a broad group of local, state, and national regulators other than the FCC. A consideration of legal and regulatory issues will be needed for everything from the construction of a new studio building, to the modification of an existing office structure to house a station's main studio, to erecting a new broadcast tower (to be discussed in more detail below).

Even the most basic construction project will routinely require local building permits. More significant building projects may require the approval of local zoning or land use authorities. With broadcast facilities, in particular, there may well be questions about and concerns with all of the broadcast equipment typically maintained at any facility—from satellite dishes to antennas and transmission equipment for stations and auxiliary facilities.

Routinely, a plethora of local building and occupancy codes must be observed to ensure that broadcast facilities comply with the law. These may regulate everything from access to the facility (including mandating emergency fire exits), the maintenance of fire extinguishers or sprinkler systems, electrical codes and structural requirements to host broadcast equipment, rules on allowable types of paint and insulation, and even requirements for adequate rest rooms for employees. Some of the electronic equipment used by stations may heighten the need to carefully review these local codes to make sure that full compliance is achieved.

Federal laws also come into play. The Americans with Disabilities Act (ADA) requires that most businesses make reasonable accommodations in their facilities and practices for those with disabilities. This may include everything from ensuring wheelchair access to the main studio to determining whether station equipment must be modified to facilitate its use by, for instance, individuals who are visually impaired. OSHA also has a variety of workplace safety laws that must be observed, including, as discussed above in the section on FCC rules, rules about RF radiation and the protection of workers from such radiation in areas where equipment may cause hazards.

Federal and state environmental laws may also become involved in any construction. Before the acquisition of any land for the construction of a building or tower, the broadcaster may want to consider contracting for an environmental assessment of the property. Spending a few thousand dollars for a "Phase 1" environmental assessment to review the grounds and the history of the site can uncover whether any potential pollution issues exist at the site. If possible issues are discovered, but the property is still attractive to the broadcaster, a more extensive Phase 2 study can be done, usually involving soil samples and a more in-depth on-site review. The purchase of a site with significant pollution issues can become an incredibly costly issue, especially if such pollution is not adequately contained. It can also make the site difficult to sell in the future and hard to insure, as parties do not want to assume the potential liability for a pollution cleanup. Here, too, engineers can be of assistance in thinking about this kind of issue in advance of any construction project.

Tower Construction

Tower construction involves its own unique and sometimes very challenging set of regulatory issues that must be considered. As mentioned in the FCC section above, most towers that are 200 feet tall or more (and shorter towers that fall within a flight path of a local airport or, under a law that was just enacted as this chapter was being written, possibly shorter towers in rural areas) require review and approval of the Federal Aviation Administration. In most cases, this review is required before the location of any FCC-licensed facility on the tower will be approved by the FCC. Such approval is also usually a condition to obtaining insurance for the tower. The FAA will review the proposal to determine whether it could pose a hazard to air travel. It will also determine the appropriate marking and lighting for the structure. The FAA has an evolving set of standards for the lighting of broadcast towers, and engineers need to stay on top of changes in these standards. While changes in standards usually provide some grandfathering for existing towers, that grandfathering period may be limited, and any modification of the tower may trigger the new obligations.

Local zoning and land-use issues can be particularly complicated, and sometimes may be very contentious and time-consuming for broadcast stations. For towers planned anywhere near populated areas, beware, as there have been numerous well-publicized cases of community opposition, both on aesthetic and health-related grounds. Such opposition can sometimes result in adversarial proceedings before local land-use authorities. Flashing and strobe lights, as sometimes required by the FAA, can be of particular concern to local residents. There are often concerns about the radiation emitted by broadcast antennas, with there being some widely circulated but not particularly well-documented concerns about the health effects of exposure to broadcast radiation. From time to time, engineers should be prepared to employ experts familiar with academic studies on these issues to rebut such unfounded fears.

Local construction codes often can present their own challenges to tower construction. Industry standards have been adopted for the "wind loading" of towers to assure that they can withstand significant storms. These standards, revisions to standards adopted by the Telecommunications Industry Association (TIA), are often referred to a "Rev" (for revision), with an alphabetical letter following to denote the current version. These are often incorporated into local building codes, with local codes sometimes reflecting different standards based on when they were last revised. Engineers need to be aware of the status of these revisions, as the adoption of a new standard in the middle of the planning of a new tower could increase costs to comply with that new "Rev."

Other common issues that arise in building codes involve the "fall radius" for a tower, requiring cleared land in the area in which a tower might fall if it were to collapse. Requirements for ice abatement in areas with colder climates (e.g., requiring some heating of upper tower elements to limit the formation of ice that could fall on people below the tower) are sometimes faced by broadcasters. We have seen some local jurisdictions hire "experts" on towers to evaluate proposals for new construction, and sometimes even for the evaluation of any change to a tower, even adding a small tenant to an existing tower. These costs are often passed on to the broadcaster asking for approval for a tower. Any review will add to the time and expense of a project. Engineers need to review local codes to see what may be in store for them in connection with any tower construction.

As set out in the section above on FCC regulation, the FCC obligates companies that it regulates to do an environmental assessment in connection with tower construction. This assessment will require the broadcaster to coordinate with numerous groups and government agencies. In some cases where concerns are expressed by such agencies, further communications and perhaps even formal processes at those agencies may be required. The environmental review process, for instance, requires that broadcasters assess the impact on historical landmarks in the area of tower construction, which may involve the broadcaster with the local State Historical Preservation Office (SHPO) in doing studies and reviews as to whether the proposed construction would have any visual or other impact on these local sites. An environmental review may also implicate the U.S. Fish and Wildlife Service to determine if the construction could impact wildlife, most often an assessment of the tower's potential impact on birds. If the proposed construction is in or near a flood plain, the Army Corps of Engineers may be required to opine on the project. Native American organizations need to be notified, so they can opine as to whether the project will have any impact on their cultural sites. Clearly, these environmental assessments require significant coordination and can be very time-consuming, so they must be planned in advance and often conducted with the assistance of an expert.

Many towers, particularly in western states, may be constructed on federal lands. In such cases, the Federal Bureau of Land Management or the Forest Service may need to approve the underlying ground lease for the site. Sometimes, they will also be involved in approving new tenants at the site. In addition, additional regulations can be imposed on access to and the use of sites on these federally owned lands, sometimes making repairs and replacements of equipment at the sites very difficult.

Operational Issues

While the principal regulator of a broadcast station's day-to-day operations is the FCC, engineers should be aware of other regulations. We have already mentioned the FAA standards for tower marking and lighting guidelines that will be relevant to most broadcasters. Engineers should be mindful of evolving FAA standards. For example, if the FAA relaxed the standards applicable to particular circumstances, an astute engineer could request that a stations' lighting and marking obligations specified in their FAA and FCC authorizations be modified to reflect these changes. Ensuring that the required lighting is maintained properly is also crucial. In the event of any tower lighting failure, the broadcaster must give notice to the FAA of the problem, so that notice can be given to aviators in the affected area. The failure to provide such notice could impact insurance coverage if there were some tragedy involving the tower while the lights were malfunctioning.

The FAA also regulates the commercial use of "drone" aircraft, so if a station wants to use drones as part of newsgathering (or otherwise), it should be sure to secure the necessary licenses and authorizations. These regulations limit who can pilot drones and where the drones can be used, so all relevant regulations in this fast-changing regulatory environment need to be observed.

Other regulatory obligations extend to areas not wholly within an engineer's purview, but of which engineers nevertheless should be aware. A station or station group's fleet vehicles, as commercial vehicles, likely are subject to broad obligations with respect to maintenance, registration, insurance, and inspection. In the event of an accident, if these obligations have not been met, potential liability could extend to the station.

Equal employment opportunity and antidiscrimination laws beyond just those that are enforced by the FCC apply to the hiring and firing of all station personnel, including members of the engineering staff. Thus, the station needs to ensure that hiring, firing, and the payment and advancement of employees is done in a nondiscriminatory fashion in compliance with all applicable rules.

Engineers should not lose sight of the basic obligation to ensure that all station facilities—from steps, sidewalks, and fences to towers hundreds of feet tall—are structurally sound and otherwise safe. It doesn't matter if an individual is injured by a defect in a 4-foot fence or a 400-foot tower—at the end of the day a broadcaster could have liability if it does not take all reasonable steps to ensure that all its facilities are safe.

And this obligation extends below ground as well as above it: if a station works with dangerous chemicals, it should be sure to store and dispose of those chemicals appropriately. This can be particularly important in connection with the storage and disposal of old equipment, especially some very old AM transmitters that contained polychlorinated biphenyls (PCBs). These and newer electronics that contain other potentially hazardous chemicals cannot simply be disposed of by putting them out for the trash the day before the collection is done in the station's neighborhood. Instead, careful disposal should be undertaken to avoid potential liabilities. Where PCBs and other chemicals still exist at a station, notice may also need to be given to local emergency responders, especially the fire department, so that hazards can be avoided in the event of a fire or other emergency.

Pirate Radio

While "pirate" broadcasters can be encountered anywhere, certain locations across the country have proven to be particularly fertile ground for these operations of unlicensed facilities operating in the radio broadcasting bands. Wherever pirates are encountered, licensed stations should not hesitate to pursue all avenues for relief. In any instance, the first, or one of the first, calls should be to the FCC's local field office. Not only will the field office have authority under the Communications Act to address the issue, but also the local FCC field officer likely will be familiar with the practical issues and legal tools a broadcaster can bring to bear against pirates.

However, the FCC does not have the resources to pursue all pirate radio stations. Some states have also passed laws making the operation of pirate radio stations a criminal offense. In these areas, local prosecutors and police may have the authority to shut down pirate operators independent of the FCC and potentially quicker than the FCC can act. In such situations, local law enforcement may provide the most efficient and effective method to resolve interference issues posed by pirate operators. Check locally, for instance with a state broadcast association, to see if there are state or local laws that can be used to combat pirate radio operators.

Online Operations

A broadcast station's operations are increasingly focused online (i.e., over the Internet) as well as over the air, and stations that do not even operate a website likely have some elements of their business in contact with the Internet. As a result, even if a station has separate IT staff, broadcast engineers need to be familiar with the risks and obligations that relate to a station's interactions with the online world.

Online Operations—Security

Above all else, stations should ensure that they take appropriate security precautions with respect to all of their facilities. The FCC has said that it expects all the entities it regulates, including broadcasters, to operate their businesses in a manner that mitigates cybersecurity risks. This includes not only a station's online presence but also network-connected elements of its operations, including its EAS system, any networked content delivery mechanisms, and its internal computer network. Unless properly secured, any of these elements could be breached by a cyberattack and result in embarrassment, or worse, for a broadcast station.

Cybersecurity attacks and potential liabilities can come in many forms. Broadcast stations have seen hackers successfully attack and compromise on-air program streams (particularly through hacks of IP-connected program delivery systems, including studio-to-transmitter links), EAS systems, the over-the-air "metadata" systems that allow show or song titles to be displayed on compatible receivers, and even station social media accounts. Much of the potential vulnerability can be addressed with simple fixes, such as changing default passwords on network-connected devices, but any vulnerability left unprotected could result in a security breach. Station technical staff should stay on top of industry-recommended security best practices and implement those practices into their operations.

Online Operations—Privacy

Stations that collect personal information from website listeners, viewers, customers, and others with whom they interact, whether online or otherwise (for example, in connection with station contests or listener clubs), should take great care to ensure that this information is secure, as additional data security obligations attach to personal information, and additional liability could result from a breach of such data. While any data breach is serious, protection of personal information is an area of particular regulatory concern. Thus, even the sale of station T-shirts can provide a station with confidential information (e.g., credit card numbers) that needs to be protected. Civil liability to those affected by breaches in the security of personal information, plus government fines from agencies including the Federal Trade Commission, are possible, so technical staff must carefully assess the best way to keep all such information secure.

A station's privacy obligations extend beyond just data security, although protecting personal information from breach is a core online obligation. Privacy rules also provide strict limitations on the information that can be collected from children younger than 13 and how such information can be used. Thus, stations need to think twice before trying to collect data from their youngest listeners and viewers.

Federal law (both through the FCC and through civil law suits) can give rise to liability for the use of automated systems to call or text messages to individuals without their explicit consent to do so. These rules require that very specifically worded permissions be obtained from individuals before they can be contacted by any autodialing system. Thus, contacting a station's audience and promoting a station's programs and the services of a station's advertisers can be costly if it is done through calls or text messages that are delivered without permission. Observe all obligations under the Telephone Consumer Privacy Act (TCPA) in detail to avoid liability in this area.

Some privacy obligations are not limited to the digital world. Federal and state privacy protections can limit the manner in which stations gather and report news, by limiting a reporter's ability to record phone calls without the other party's knowledge or to record in areas where a person could reasonably expect privacy. FAA rules limit the ability to use drones to gather news or other footage on private property. Civil law suits, and even criminal liability, can be imposed for trespass or certain invasions of privacy. Be aware of these rules, and tread carefully.

Online Operations—Intellectual Property

Finally, engineers should be aware of the complex layers of intellectual property rights that limit how a station may use software and hardware, music, and other content created by the station or found online.

The various technical systems that stations use—from automation and traffic systems to digital broadcasting facilities—each contain countless technologies that may be covered by one or more federal patents. Each patented technology must be properly licensed, and the patent holders can sue individual stations if they use these technologies without a license. Patent law allows users of patented technologies to be sued, even if they had no knowledge that the vendor from whom they acquired the hardware or software involved infringed on the patent. Thus, stations should be sure that when they purchase or lease software or hardware, their contract with the vendor contains a promise that all necessary licenses have been secured and that the vendor agrees to indemnify the station if the rights to use these products are ever challenged.

Stations should also be mindful of the copyrights relating to content made available to the public. For example, stations that stream music online are obligated to pay royalties on this music and must provide to royalty collection companies detailed data regarding not just what songs were played when, but also how many people were listening to each of those songs. This data can be voluminous, and it requires the station's online audience tracking software to interact with its music scheduling software. Systems need to be developed or purchased to provide this information.

There are also requirements that, to qualify for certain royalties, the station identify on its website or mobile app, in text, the name of the song that is playing, the artist, and the album from which the song came. The requirements prohibit giving out this information before the song is played, so it is again a tricky technical job to ensure that the information lines up with the music that is playing, and does not appear on the site too early.

The rebroadcast of the station's content by bars, restaurants, and retail outlets in the station's service area can also cause problems. Broadcast engineers may be approached by some local business to facilitate the business' use of the station programming to entertain its customers. In many cases, those businesses may need to obtain copyright licenses before turning on the radio or TV programming in their facility. While the Copyright Act does provide certain small businesses with exemptions from copyright liability for playing a broadcast station on their premises, these exceptions are very precisely limited under the law. Engineers should be aware of this potential liability, so that they don't inadvertently push a local business (who is possibly a station advertiser) into a situation where they are sued by a copyright holder.

Taking any content from a broadcast environment to an online one, or vice versa, can create copyright issues. For instance, TV stations don't usually have the rights to stream network and syndicated content on their websites. If a station licenses a music cue for its news product, it should be sure that the license provides online streaming rights before making available online any clips including the cue. And program producers should understand that just because content—like a photograph or a video clip—is posted online, that does not mean that it can be incorporated into a broadcast program. For instance, not all YouTube clips can be broadcast in a TV program unless rights from their producers have been obtained. Indeed, the explicit permission of the person that created any online media production may be necessary before it can be used as part of any broadcast product. Engineers need to know that while it is technically possible to repurpose content from one medium to another, this may violate copyright laws. Engineers may need to remind programmers that, just because content can be gathered from the Internet and supplied on a station website or broadcast on the air, caution needs to first be taken to ensure that proper rights are secured before this online content is exploited in other places.

SUMMARY

This chapter has provided a broad overview intended merely to identify areas where a broadcast engineer should be mindful of the regulatory landscape. Engineers should not hesitate to contact internal or external counsel for guidance in any area where an issue may arise. In most cases, asking the right question in advance is much cheaper and easier that failing to do so and winding up with a costly problem!

CHAPTER

2.3

Frequency Coordination for Broadcast Auxiliary Services

DANE E. ERICKSEN, P.E.

Hammett & Edison, Inc., Consulting Engineers
San Francisco, CA

INTRODUCTION

This chapter discusses techniques and rules used for United States frequency coordination of the Broadcast Auxiliary Service (BAS). Not only does the electromagnetic spectrum provide broadcasters the means for reaching listeners and viewers via the AM, FM, and TV bands, but under the BAS, electromagnetic spectrum is used by broadcasters for point-to-point links, remote-pickup links, voice channels for cueing and coordination, remote control of cameras and transmitters, wireless microphones, and other applications. Yet, as varied as these applications appear to be, their treatment in the FCC rules is contained in only one of four subparts in Part 74.[1] Further, these applications all have two things in common: all require some form of frequency coordination to minimize the likelihood of interference since they operate in shared frequency bands, i.e., they use frequencies that are also used by other services, and all applications are impacted by the ever-increasing demand for radio spectrum. This chapter provides an overview of the different ways frequency coordination is employed in an attempt to mitigate the effects of spectrum crowding on these essential auxiliary services.

The purpose of this chapter is not to provide an encyclopedic guide to all things encompassed by the phrase "frequency coordination." Rather, it is to provide information about frequency coordination services, whether those services are provided by a local frequency coordination committee, a volunteer appointed by a local chapter of the Society of Broadcast Engineers, Inc. (SBE), or a commercial firm in the business of providing frequency search services.

WHAT IS FREQUENCY COORDINATION?

A typical place to begin such a discussion would be with a definition of terms. Unfortunately, the phrase "frequency coordination" does not lend itself to a single, tidy definition. This simple term may mean different things in different radio services. Even within the range of BAS applications, frequency coordination will mean different things in different situations. It may mean working out how scarce TV pickup channels are to be shared on an hour-by-hour basis in the most crowded markets. Alternatively, it may mean knowing how to find an available channel for a fixed point-to-point link and being able to demonstrate quantitatively why that channel can in fact be used on that path. These are quite different tasks requiring different skills.

The FCC rules provide some help. In the relevant subparts of Part 74, one finds the statements, "licensees shall endeavor to select frequencies or schedule operation in such manner as to avoid mutual interference" (74.403); "they shall take such steps as may be necessary to avoid mutual interference, including consultation with the local coordination committee, if one exists" (74.604); and "licensees shall endeavor to select frequencies or schedule operation in such manner as to avoid mutual interference" (74.803). The theme is clear: licensees in the shared-frequency bands assume the obligation of choosing their frequencies and patterns of use so as to avoid interfering with one another. Frequency coordination in this context can be defined as the act of assisting licensees to meet that obligation, by means of knowledge of the uses being made by other affected licensees.

Local Coordination

Historically, such assistance has come through a local frequency coordination committee or from a local volunteer frequency coordinator. Either entity would most likely be equipped with some sort of database or directory of local BAS users, along with the locations of the various BAS facilities and the frequencies and other technical parameters of these BAS channels. The local licensees most likely know one another, share a common interest in avoiding interference, and achieve interference avoidance or mitigation primarily through the particular skill of knowing whom to call when interference is encountered or suspected. The process is informal, local, flexible, and generally effective.

While what constitutes "local coordination" is not defined in the FCC's Rules, it continues to be the applicable requirement for most BAS services, other than for fixed links in the 950 MHz and 18 GHz Aural BAS band and 2.5, 7, 13, and 18 GHz TV BAS bands (i.e., all TV BAS bands except 2 GHz) in the television auxiliary services.

Local coordination, however, is not the whole story. FCC rules make a distinction between those services that may rely on informal, local coordination and those for which more formal protocols must be followed. As a result of a rule change that took effect in October 2003, broadcasters must know which of three different frequency coordination protocols to observe, based on frequency range and whether operation is fixed or temporary/mobile. The rest of this chapter is devoted to describing and comparing the three different protocols, which break down as follows:

- For operations in the 6425–6525-MHz band, and for fixed-link, point-to-point stations in the 950-MHz and 18-GHz Aural BAS bands and the 2.5-, 7-, 13-, and 18-GHz TV BAS bands (but not the 2-GHz band), the prior coordination notice (PCN) protocol of Part 101 applies. Part 101 PCNs are a *negative* option, meaning that if no response to a PCN is received other licensees are deemed to not object.
- For fixed links in the 2-GHz band—now something of a rarity in the major TV markets owing to this band's widespread use for TV pickup (colloquially referred to as electronic news gathering, or ENG)—the applicable standard is still local coordination, but with a *positive* option procedure, described in more detail in a subsequent section.
- For all other BAS services, including all mobile or temporary fixed operations (except 6425–6525 MHz, as noted above), informal local coordination still applies.

FIXED-LINK COORDINATION

Frequency coordination of fixed links is fundamentally different from frequency coordination involving remote pickup (RPU) stations, Part 74 Subpart H Low-Power Auxiliary (LPA) stations (e.g., wireless microphones), TV pickup, or electronic news gathering (ENG) stations, where generally the available frequencies must be shared. For fixed links, frequency reuse is appropriate. For the purposes of this discussion, "fixed links" refers to point-to-point BAS microwave stations. Frequency reuse rather than frequency sharing is practical for these fixed-link stations, because such stations use highly directive parabolic dish transmitting and receiving antennas, as opposed to often omnidirectional transmitting and receiving antennas used by RPU, wireless microphone, and some ENG systems.

Fixed-link frequency coordination involves first identifying all co-channel and adjacent-channel links sufficiently close to require consideration, and then selecting a frequency that will neither cause interference to these existing links nor receive interference from those links. While the concept is simple, the implementation can be complex. It is the intent of this portion of the chapter to give a sufficiently detailed overview of the fixed-link frequency coordination process to at least ensure that a station engineer knows the steps involved and when the services of a commercial microwave frequency coordinator (CMFC) should be retained.[2]

Preparatory Steps

Before commencing a fixed-link frequency coordination study, it should be established that the applicant is eligible to use the band in question and that an unobstructed microwave path exists. There is little point in going to all the work involved in coordinating a fixed link, only to discover that the proposed use is not a permitted one, or that the proposed path is blocked by a nearby structure or by intervening terrain. The applicant should determine that a reasonable path length for the microwave band in question is involved:

- For 13 GHz, path lengths of more than about 20 miles (32 km) are generally not a good idea.
- For 18 GHz, path lengths of more than about 5 miles (8 km) are generally to be avoided.
- For 2, 2.5, and 7 GHz, path lengths up to 62 miles (100 km) are entirely practical from a noise-limited standpoint. However, in most major metros, microwave paths become interference-limited long before they become noise-limited.

New BAS Fixed-Link Frequency Coordination Paradigm

As a result of the ET Docket 01-75 rulemaking, which undertook a general updating and harmonization of the Part 74 BAS Rules with other services, as of October 16, 2003, formal Fixed Service (FS) Section 101.103 frequency coordination protocols became applicable to 950 MHz Aural BAS stations, and to 7- and 13-GHz TV BAS fixed-link stations (Part 101 frequency coordination protocols, often referred to as the PCN process, already applied to 2.5 GHz TV BAS fixed-link stations, to 6.5 GHz TV BAS stations, and to 18 GHz Aural and TV BAS stations). Frequency coordination for 2 GHz TV BAS fixed-link stations remain subject to "local" coordination, although now a frequency coordination certification must be included with such applications; see Section 74.638(c) of the FCC rules. This "certification" is a less formalized process than the Part 101 PCN process, but is still a significant change from the prior requirement, which did not require any frequency coordination showing at all; previously, a simple "yes" answer to Form 601, Schedule I, Question 7 ("Has frequency coordination been completed for this application?") was sufficient.

The PCN Process

Under the PCN process, one first identifies all existing co-channel and adjacent-channel links sufficiently close to the proposed link to require study. "Sufficiently close" is defined by Telecommunications Industry Association (EIA) Bulletin TSB-10F, "Interference Criteria for Microwave Systems."[3] As shown by Figure 2.3-1, this bulletin defines a coordination "keyhole" for determining whether a co-channel or adjacent-channel link is sufficiently close to require study. There are two "keyholes"; one for stations below 15 GHz and a smaller keyhole for stations above 15 GHz. Further, Section 101.105(c) of the FCC rules does not *require* use of TSB-10F methodologies: "Other procedures that follow generally accepted good engineering practice are acceptable to the Commission." But the TSB-10F procedures can be thought of as a "safe harbor," in that if those procedures are followed, then the PCN would presumably not be second-guessed by the FCC.

Each of the stations identified as being sufficiently close to require study must then be checked for interference from the proposed newcomer link. Although not an FCC requirement, the PCN study should also check to ensure that each of these existing links will not cause interference *to* the proposed new link. Making these calculations requires detailed technical data for each of the studied stations, although in some cases conservative assumptions, such as assuming only the use of Category B receiving antennas, parallel polarization, and a high receiving antenna height (i.e., forgoing any terrain blockage that might exist), can be employed when station technical data is missing or suspect. If, based on these conservative assumptions, the newcomer link is calculated to not be an interference threat to an existing link, one can probably forgo the time and effort to obtain sometimes hard-to-find *accurate* data for the to-be-protected path.[4] However, if the interference calculations are "close," or if moderate

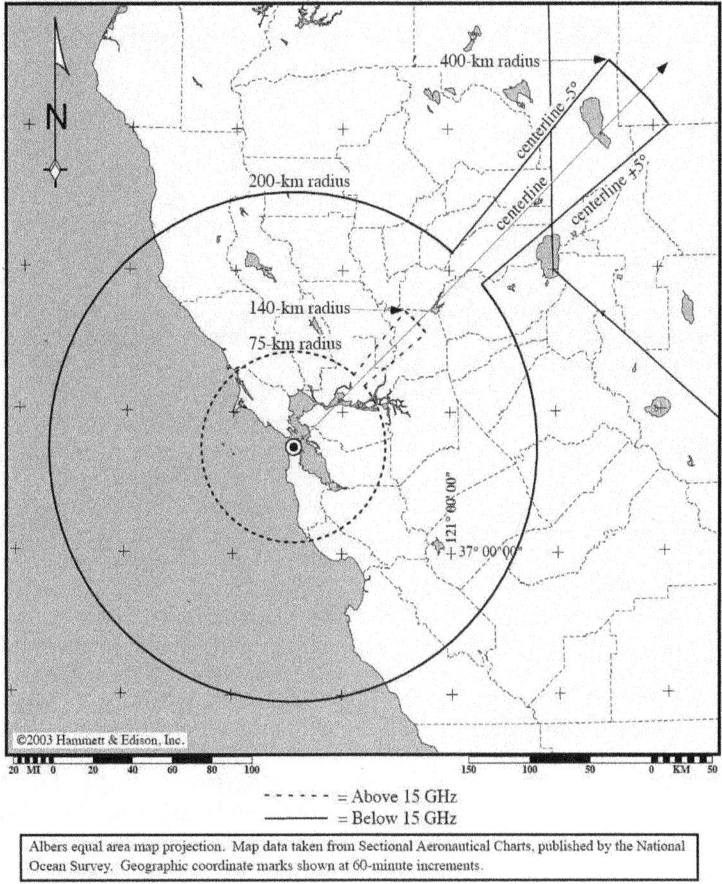

FIGURE 2.3-1 Example of TIA/EIA TSB10-F coordination keyholes (From Section 3.4.4 of TIA/EAI TSB10-F standard. Albers equal area map projection; map data taken from Sectional Aeronautical Charts, published by the National Ocean Survey. Geographic coordinate marks shown at 60-minute increments.)

interference is predicted, then the actual facilities of the to-be-protected link must be ascertained.

If the first-cut interference calculations predict massive interference, an alternative frequency should be sought. The one exception would be if one suspects that an FCC Universal Licensing System (ULS) record is for a station that no longer exists, but the licensee has not informed the FCC of that fact (either due to oversight, or perhaps because the licensee thinks it might want the path at a future date). Such frequency "warehousing" is, of course, prohibited by the FCC rules (Section 74.632(g) for TV BAS, Section 1.955(a)(3) for Wireless Telecommunications Services[5] in general), but this doesn't mean that the practice still doesn't exist. So, if only one ULS record is causing a preclusion, and if there are no other available channels, checking to see if the link is actually in use might prove fruitful.

Note that assumptions that are conservative as to the amount of interference, if any, a newcomer link would cause to an existing and must-be-protected link are generally also conservative insofar as predicting whether an existing link will cause interference to a proposed new link. For example, the assumption of only a Category B transmitting antenna at great height for an existing link would tend to overpredict any interference that existing link might cause to a proposed new link. Again, if based on such conservative assumptions no interference is predicted to the new link, it is not necessary to ferret out missing data for the existing, and potentially interfering, link.

Special Cases

When looking for potential to-be-interfered with paths, beware of special cases, such as split-path links. Split-path links may use separate transmitting antennas, or a single transmitting antenna, splitting the difference in the azimuths and elevation (or depression) angles to two closely spaced receiving locations. Another situation to watch out for is an existing path with an unusually long receive-end waveguide run, which could cause the to-be-protected signal to be significantly (10 to 20 dB) weaker than one might otherwise predict.

Finally, some microwave paths employ a passive reflector, meaning that the transmit-dish azimuth is not aimed toward the receive-end site, but rather towards the passive reflector that the receive-end dish can see (and, similarly, this means that the receive-end dish is aimed at the passive reflector, and not towards the microwave transmit site). While these situations are rare, the consequences of failing to properly study a split-path link or a link with a passive reflector could potentially be severe to the newcomer station, which could then be viewed as the at-fault party.

INTERFERENCE CRITERIA

One of most difficult aspects of a PCN study, at least one involving BAS links, is deciding the criteria to use to define whether interference will exist. For analog FM video-into-analog FM video, an interference criteria of 60 dB or better desired-to-undesired (D/U) signal ratio for co-channel, and a 0 dB or better D/U ratio for adjacent channel, has been used for many years by this author with acceptable results. However, the interference criteria becomes more complex when the proposed or protected links are using digital modulation. Often a digital signal more fully occupies the channel bandwidth, resulting in greater signal strength near the channel edge than would exist for the FM video analog case; to allow for this, a 10-dB desired-to-undesired (D/U) adjacent-channel interference ratio, rather than just a 0-dB adjacent-channel interference ratio, is suggested.

When the protected link is one that uses digital modulation, the interference criteria becomes a two-step process. First, the required receive carrier level (RCL) to give an agreed-upon path reliability must be calculated. Path reliability calculations are outside the scope of this chapter, but generally use a multipath fading model such as Vigants's model[6] and, above 7 GHz, a rain attenuation model, such as Crane's.[7]

Let's assume that the reliability criteria is a path availability of 99.95% or better, as used in the WT Docket 10-153 rulemaking, which as previously mentioned opened the 7- and 13-GHz TV BAS bands to shared use by Part 101 FS stations. Let us further assume that an RCL of −55 dBm is needed to give this path reliability. Finally, assume that the required signal ratio between the desired signal and the undesired signal is 30 dB (i.e., much lower co-channel D/U ratios than required for analog-into-analog situations are generally possible for digital-into-digital situations). Then, the to-be-protected RCL for the desired signal becomes −55 dBm minus 30 dB, or −85 dBm. TSB10-F allows up to a 1-dB degradation of the protected digital link's noise floor.[8] Therefore, the undesired digital signal could be no stronger than −90.9 dBm, as −85 dBm plus −90.9 dBm = −84 dBm, or a 1-dB degradation of the noise floor.[9] Finally, it should be noted that if the protected receiver is an ENG receive-only site (ENG-RO), then one simply protects the noise threshold of the receiver, since it is the operational range of the ENG-RO that is being protected. For example, if the receiver's noise threshold is −95 dBm, then the maximum RCL of an undesired signal having the same bandwidth would be −100.9 dBm, since −100.9 dBm plus −95 dBm = −94 dBm, a 1-dB degradation of the noise floor. If the interfering signal has a different bandwidth than the protected signal, a $10\log(BW_{interfering} / BW_{protected})$ factor can be applied. Again for example, if the interfering signal has a bandwidth of 2.5 MHz and the protected signal has a bandwidth of 25 MHz, the level of the interfering signal could be increased by 10 dB, since log(2.5 MHz / 25 MHz) = −10 dB; that is, for digital-into-digital, a narrower interfering signal has less interference potential than a broader interfering signal.

Another issue is what constitutes "co-channel" and "adjacent-channel" operation, especially when stations with different channel widths are involved. For example, consider a 25-MHz-wide 13-GHz TV STL channel versus 6-MHz-, 12.5-MHz-, or 25-MHz-wide cable television relay service (CARS[10]) microwave channels. One approach is to use the channel width of the protected station as the criteria: if any portion of the proposed channel overlaps any portion of the protected channel, no matter how small, the proposed station is treated as "co-channel." And if any portion of the proposed channel overlaps any portion of one channel bandwidth from the lower edge of the protected channel, or any portion of one channel bandwidth from the upper edge of the protected channel, then the proposed channel would be treated as "adjacent channel."

Under these criteria, a 6-MHz-wide CARS channel would have adjacent-channel windows extending 6 MHz below the lower channel edge, and 6 MHz above the upper channel edge, whereas a 25-MHz-wide CARS channel would have 25-MHz adjacency windows. Finally, if a link is neither co-channel nor adjacent-channel, it need not be studied, on the assumption that the receiver's selectivity should be able to handle second and higher adjacencies.

ANTENNA CONSIDERATIONS

The FCC rules require all fixed-link BAS stations to use directional antennas. Except for 950-MHz aural BAS stations, the FCC rules specify the minimum acceptable performance of these antennas as either Category B or Category A. All fixed-link stations except 950-MHz aural BAS stations must normally use at least Category B transmitting antennas, and in "frequency-congested areas" must normally use at least Category A transmitting antennas. However, in the most frequency-congested areas, even a Category A antenna often won't suffice, and "super" Category A antennas, with off-axis suppression ratios significantly better than those required to achieve Category A status, must be used.

There are presently no minimum antenna performance requirements for BAS fixed-link receiving antennas, except that the receiving antenna must be directional. However, Section 74.641(a)(3) of the FCC rules provides a powerful incentive for also using highly directive receiving antennas, as follows:

> The choice of receiving antennas is left to the discretion of the licensee. However, licensees will not be protected from interference that results from the use of antennas with poorer performance than identified in the table of this section.

This means that an existing link in a non-frequency-congested area using a sub-Category B receiving antenna may be studied (and protected) as if the receiving antenna met Category B performance. Or, in a frequency-congested area, an existing link using just a Category B receiving antenna (or even a sub-Category B antenna) may be studied (and protected) as if the receiving antenna met Category A performance. Thus, it is beneficial to always use at least Category B receiving antennas in non-frequency-congested areas, and to always use at least Category A receiving antennas in frequency-congested areas.

The FCC rules define antenna performance using two of the following three parameters: the on-axis minimum gain or the maximum half power beam width (HPBW), and the minimum required suppression at larger and larger off-axis angles (i.e., up to ±180° from the main beam). This off-axis suppression is referred to as the antenna's radiation pattern envelope (RPE). As an example, the RPE of an Andrew (now CommScope) Model UHX10-59J 10-foot diameter, ultra-high-performance, cross-polarized microwave antenna is shown in Figure 2.3-2. Note that the actual antenna performance may be even better than the suppression shown by the RPE, but this additional suppression is not guaranteed. Therefore, one should always base interference calculations on the guaranteed performance RPE.

It is important to note that for the same diameter dish, standard-performance (SP), high-performance (HP), and ultra-high-performance (UHP) antennas all have essentially the same gain and half-power beam width; it is only on the off-axis rejection do HP and UHP dishes differ from a SP dish. Because a UHP dish can cost up to four times more than an SP dish of the same diameter, and because HP and UHP dishes represent a greater structural load than an SP dish of the same size (diameter), HP or UHP antennas should only be specified when required for frequency coordination purposes. That is, use of HP or UHP dishes adds nothing to the path's link budget compared to SP dishes of the same size.

HP and UHP parabolic dish antennas are created by adding a "cake-pan" shroud around the periphery of the dish, and then lining that shroud with microwave-absorbing material. Figure 2.3-3 shows a comparison of the off-axis rejection capabilities of SP,

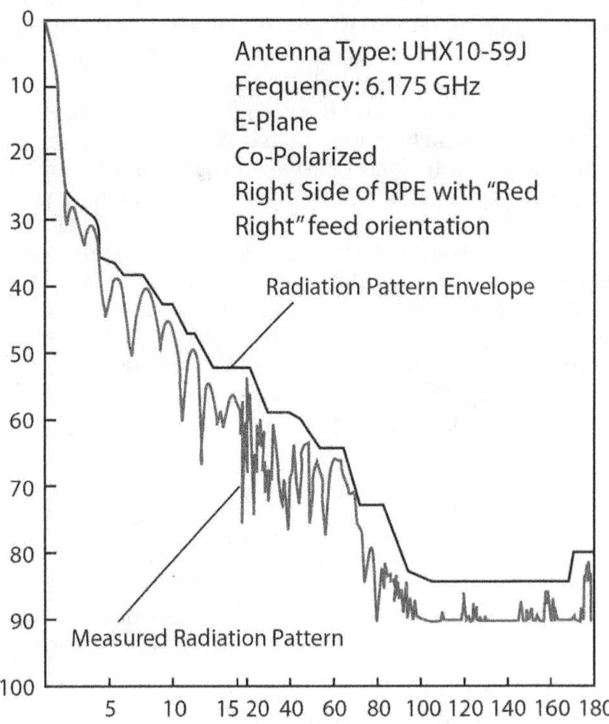

FIGURE 2.3-2 Radiation pattern envelope for a CommScope Model UHX10-59J antenna

Courtesy of CommScope

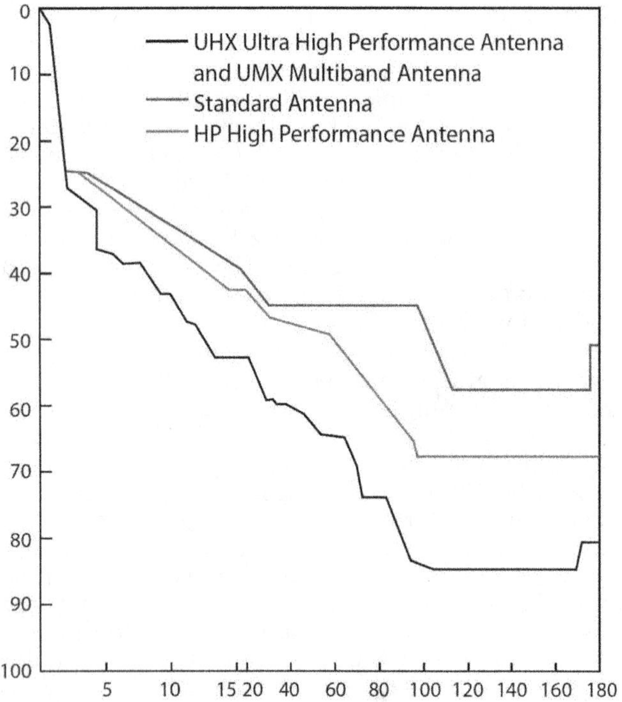

FIGURE 2.3-3 Comparison of the off-axis rejection capabilities of SP, HP, and UHP antennas having the same diameter

Courtesy of CommScope

HP, and UHP antennas all having the same diameter. A UHP dish can have up to 40 dB better off axis rejection than that of a SP dish of the same size, and so is a powerful, albeit expensive, frequency coordination tool. The benefits of an HP or UHP antenna can apply at either the transmit end or receive end. If the problem is one of interference to an existing link, using an upgraded antenna at the transmit end of the new path can sometimes be a solution. If the problem is one of interference from an existing link to a newcomer link, using an upgraded antenna at the receive end of the new path can sometimes be a solution. In the most congested TV markets, HP or UHP antennas for both ends of the new path are often the only way to make the new path frequency coordinate.

RH and LH Feeds

When a UHP antenna is used, especially a cross-polarized version with two feed horns, the RPE is generally no longer symmetrical around the main beam. This non-symmetry can be used to the newcomer's advantage, by selecting a right-hand (RH) or left-hand (LH) feed orientation, so that the non-symmetrical portion of the RPE with the greater suppression is aimed toward a problematic co-channel or adjacent-channel link. Thus, for UHP antennas, there are two degrees of freedom for minimizing interference: horizontal vs. vertical polarization, and RH vs. LH feeds.

Frequency-Congested Areas

Section 74.641(b) of the FCC rules defines "frequency-congested area" for the 13-GHz TV BAS band as one where a newcomer station shows that the use of a Category B transmitting antenna is precluding the new path, and that upgrading the existing station to a transmitting antenna meeting Category A criteria would eliminate the preclusion. However, this "frequency-congested" criteria only applies to the 13-GHz TV BAS band. For 2-, 2.5-, 7-, and 18-GHz band fixed links, the term "frequency-congested area" is undefined.

Section 74.641(c) provides exceptions to the use of Category A or B microwave antennas, where a showing can be made that structural, zoning, or other restrictions preclude use of a microwave antenna with a diameter sufficient to meet Category A or B specifications. However, the showing threshold is high, so in most cases the prudent choice will be to simply install the normally required Category A or B antenna.

Parallel-Polarized and Cross-Polarized Signal Interference Calculations

Figure 2.3-4 shows how parallel-polarized (PPOL) and cross-polarized (XPOL) signals between two fixed-link paths combine (for a given antenna, the RPEs for the PPOL and XPOL signal will always be different, as demonstrated in Figure 2.3-5). The drawings and tables in this figure reflect the fact that an HPOL or VPOL TX antenna will always have off-axis leakage for both polarizations, as will an RX antenna always be susceptible to off-axis interference for both polarizations. Thus, when the protected and interfering paths have different polarizations, both the PPOL portion of the interfering signal, as well as the XPOL portion of the interfering signal, must be taken into account. When the protected and interfering paths have the same polarizations, the PPOL leakage is always predominant.

Referring to Figure 2.3-4, there are four possible combinations between the desired and undesired signal paths. For the PPOL case, use (A + C) when calculating the level of the undesired signal. For the XPOL case, use either (A + D) or (B + C), whichever gives the smaller dB number (i.e., the worst case off-axis rejection). If the (A + D) and (B + C) dB values are within 10 dB of each other, then a power combining should be done. For example, if (A + D) is 45 dB, and (B + C) is 41 dB, the "sum" is 39.5 dB of isolation. But if (A + D) is 45 dB and (B + C) is 55 dB, the "sum" is 44.6 dB. Thus, when (A + D) differs from (C + D) by more than about 10 dB, the smaller dB number is controlling and the larger dB number can be ignored, with little loss in accuracy.

DESIGN EXAMPLE

Figure 2.3-6 shows an example worksheet used for manually performing a microwave path interference calculation. The goal of this exercise is to determine the amount of interference created by a proposed,

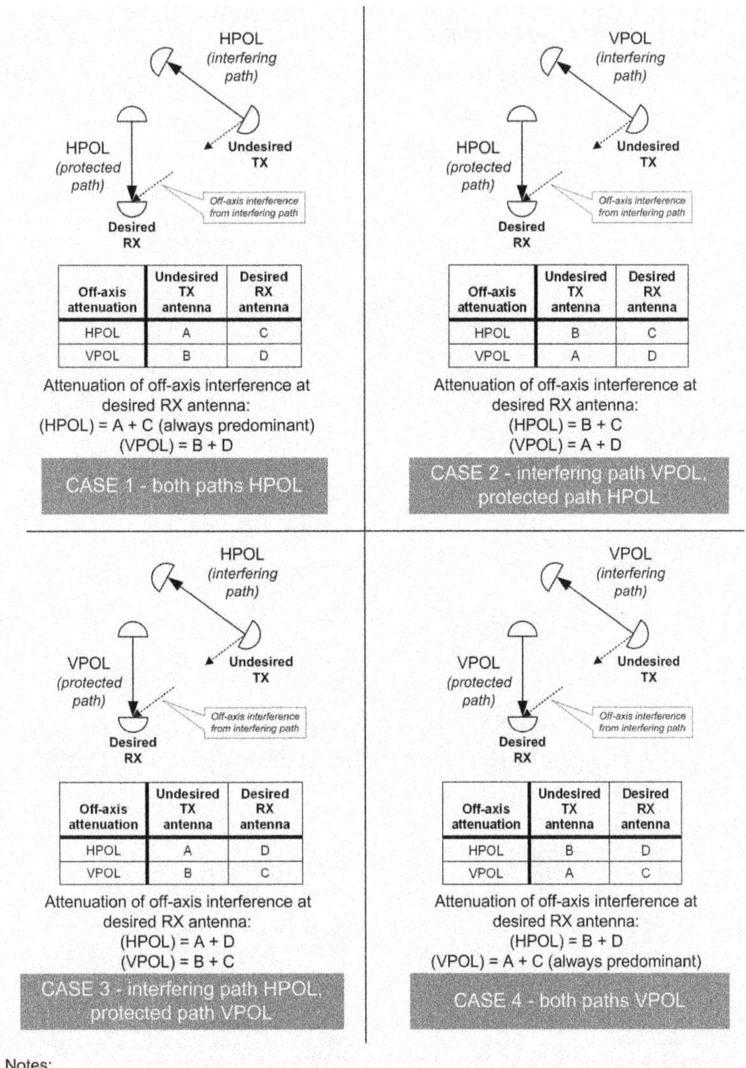

Notes:
1) For undesired TX antenna, A=PPOL off-axis suppression (dB), B=XPOL off-axis suppression (dB)
2) For desired RX antenna, C=PPOL off-axis suppression (dB), D=XPOL off-axis suppression (dB)
3) For PPOL case, (A+C) will always give lowest attenuation (worst interference)
4) For XPOL case, the smaller of (A+D) or (B+C) will give lowest attenuation (worst interference)

FIGURE 2.3-4 Examples of parallel-polarized (PPOL) and cross-polarized (XPOL) interference

new co-channel path with respect to an existing path, WAA-1234, which is horizontally polarized. For this example, both paths are using Channel A07 (12,850–12,875 MHz) in the 13-GHz TV BAS band.

First, the unfaded RCL of the existing signal, WAA-1234, is calculated in the right-hand column, by completing a "link budget" that consists of the following values:

- TX power output: output power of the existing path transmitter (located at site 3)—for this example, equals 0.4 W (26 dBm);
- Line loss (TX): this is the loss (in dB) due to the waveguide between the transmitter and the antenna;
- TX antenna gain, in dBi (at site 3, on-axis): this is specific to the transmit antenna used;
- FSPL: free space path loss for 9.5-mile distance between site 3 and site 4 where
  - $\text{FSPL} = 92.5 + 20\log(D_{\text{miles}}) + 20\log(F_{\text{GHz}})$
- RX antenna gain in dBi: a function of the receive antenna used;
- Line loss (RX): loss (in dB) due to waveguide between antenna and receiver;
- RX level: sum of all of the above values—this is the unfaded signal level received at site 4, considered the "desired" signal level for the purposes of this exercise.

Next, the signal level of the new path (the "undesired" signal) at the input to the existing WAA-1234 receiver is calculated. For this part of the calculation, the RPEs of the newcomer TX dish (at site 1), and the

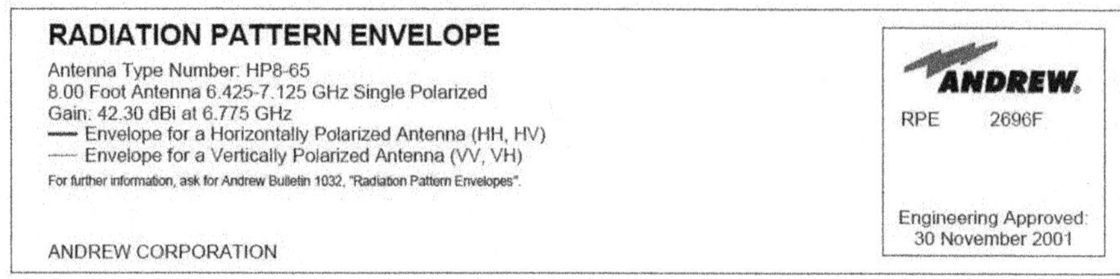

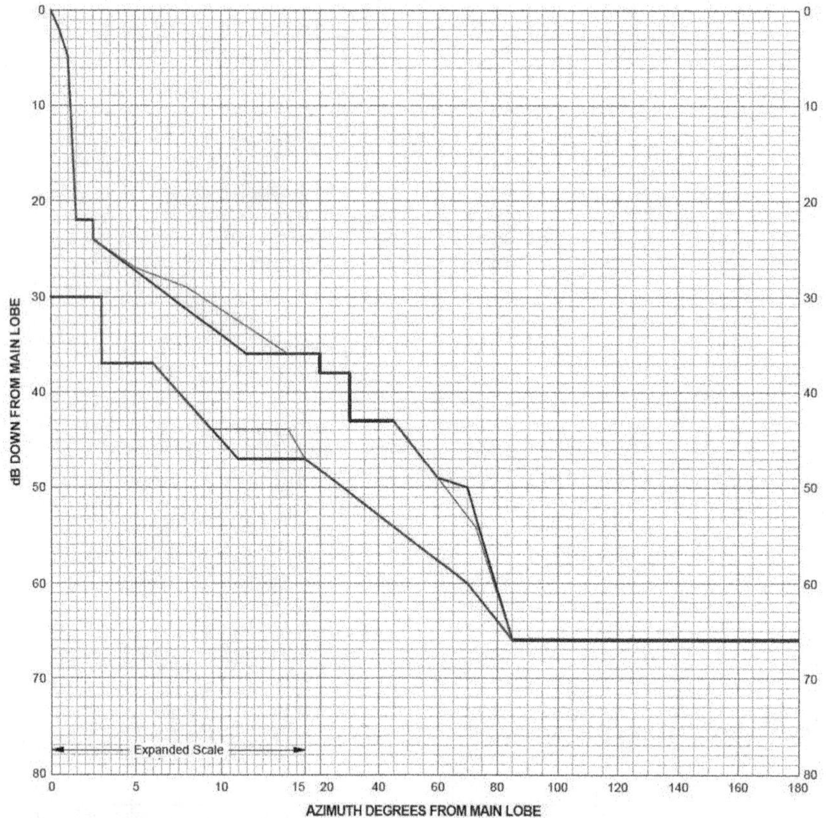

FIGURE 2.3-5 Illustration of differences in radiation pattern envelope (RPE) as a function of polarization

Courtesy of CommScope

to-be-protected RX dish (at site 4), must be known. This link calculation is performed in the left-hand column of Figure 2.3-4F as follows:

- TX power output: output power of the new path transmitter (located at site 1) for this example, equals 1.0W (30 dBm);
- Line loss (TX): this is the loss (in dB) due to the waveguide between the transmitter and the antenna;
- Loss due to off-axis interference path (TX): knowledge of the RPE of the TX antenna at site 1 is required here. First, the *off-axis angle* of the interference path is calculated—for this calculation, the baseline is the site 1-to-site 2 path (since the TX antenna main beam is aligned with that path); in this example, the angle is 15.3°. Next, both the HPOL and VPOL responses of the site 1 TX antenna at this angle must be determined since ultimately the potential interference for both cases needs to be determined (that is, case 1, where the new path is the same polarization as the existing path, and case 2 where the new path is cross-polarized to the existing path). For the antenna in this example, for 15.3° off-axis the HPOL response is found to be 43 dB below the main beam gain, and the VPOL response is 54 dB below the main beam gain. Both values are written into the link budget, and both will be used later as discussed below;
- FSPL: free space path loss for 12.3-mile distance between site 1 and site 4;
- RX antenna gain in dBi (on-axis, site 4): a function of the receive antenna used;
- Loss due to off-axis interference path (RX): knowledge of the RPE of the RX antenna (site 4) is required here. First, the *off-axis angle* of the

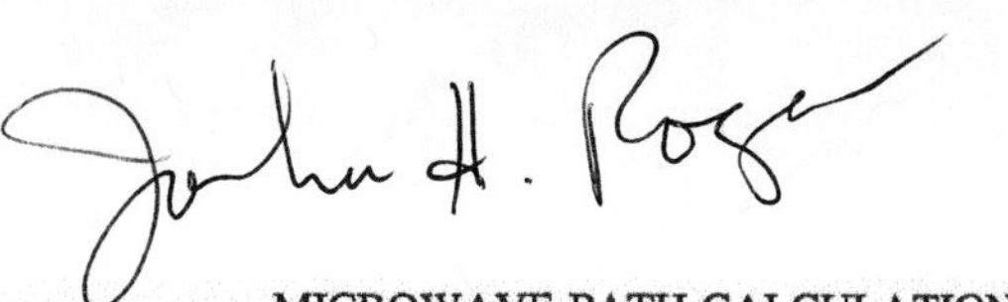

MICROWAVE PATH CALCULATIONS

PROPOSED NEW PATH: Site 1 to Site 2, A07, polarization to be determined

EXISTING PATH TO BE CHECKED: WAA-1234, A07, Site 3 to Site 4

(√) Check for IX to existing path ( ) Check for IX from existing path

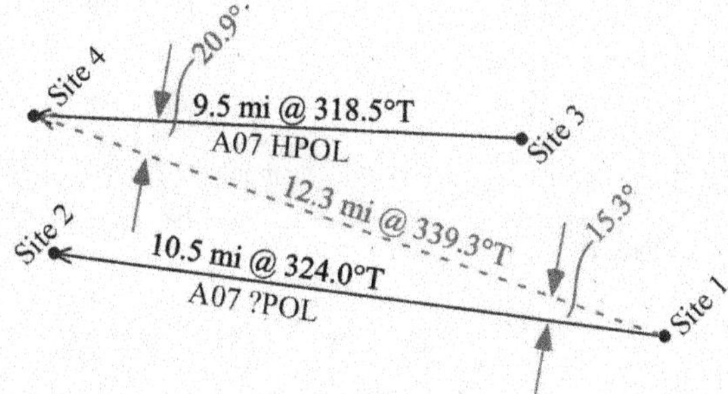

		WAA-1234 DIRECT PATH	
TX power output: 1W	30 dBm	TX power output: 0.4 w	26.0 dBm
Line loss:	-2.2 dB	Line loss:	-1.8 dB
TX antenna gain: Andrew P6-122F	45.1 dBi 72.9 dBm EIRP	TX antenna gain: 6' Prodelin 213-740	45.1 dBi 69.3 dBm EIRP
Loss 15.3° off axis:	-43.0 / -54.0 dB		
FSPL, 12.3 miles:	-140.6 dB	FSPL, 9.6 miles:	-138.4 dB
RX antenna gain: 8' standard	47.6 dBi	RX antenna gain: 8' standard	47.6 dBi
Loss 20.9° off axis:	-45.0 / -58.0 dB		
Line loss:	-3.5 dB	100' EW127 Line loss:	-3.5 dB
RX level, PPOL:	-111.6 dBm	RX level:	-25.0 dBm
RX level, XPOL:	-122.6 dBm		
D/U, PPOL:	84.4 dB	D/U, Xpol:	97.6 dB

FIGURE 2.3-6 Example of microwave path interference calculation worksheet

interference path is calculated—for this calculation, the baseline is the site 3 to site 4 path (since the RX antenna main beam is aligned with that path); in this example, the angle is 20.9°. Next, both the HPOL and VPOL responses of the site 4 RX antenna at this angle must be determined—for the antenna in this example, the HPOL response is found to be 45 dB below the main beam gain, and the VPOL response is 58 dB below the main beam gain. As before, both values are written into the link budget;

- Line loss (RX): loss (in dB) due to the waveguide between antenna and receiver;
- RX level, PPOL: if the newcomer link is HPOL (which means it is parallel polarized to the existing path) then the HPOL RPE values are used for this calculation, that is, 30 dBm – 2.2 dB + 45.1 dBi – 43.0 dB (HPOL value for TX antenna) – 140.6 dB + 47.6 dBi – 45.0 dB (HPOL value for RX antenna) – 3.5 dB = –111.6 dBm (equals undesired signal level at site 4 RX antenna, PPOL);
- RX level, XPOL: if the newcomer link is VPOL, and thus cross-polarized to the existing protected path, then the smaller of the two XPOL sums (in dB) must be determined. The two possibilities are 43 dB (TX HPOL) + 58 dB (RX VPOL) and 54 dB (TX VPOL) + 45 dB (RX HPOL). Since the first combination totals 101 dB of isolation, and the second combination totals 99 dB of isolation, the smaller 99 dB figure is used. Note that the exact value is actually the power summation of these two values; that is, (–99 dB) + (–101 dB) = (–96.9 dB); for manual calculations, though, this refinement can usually be skipped without significantly affecting the end result; i.e., the worst case error for this simplification is 3 dB.

Now the level of the undesired signal, as XPOL, is 30 dBm – 2.2 dB + 45.1 dBi – 54.0 dB (VPOL value for TX antenna) – 140.6 dB + 47.6 dBi – 45.0 dB (HPOL value for RX antenna) – 3.5 dB = –122.6 dBm (equals undesired signal level at site 4 RX antenna, XPOL).

Finally, the desired-to-undesired (D/U) ratios for the PPOL and XPOL cases can be calculated. For PPOL, the D/U ratio is –25.0 dBm (desired signal into RX antenna at site 4) – 111.6 dBm (undesired signal into RX antenna at site 4 in the PPOL case), which equals 84.4 dB. For XPOL, the D/U ratio is –25.0 dBm minus –122.6 dBm (undesired signal into RX antenna at site 4 in XPOL case), which equals 97.6 dB. For co-channel analog links, a D/U ratio of 60 dB or better is generally required. In this case, that target ratio is met even if the newcomer link is PPOL to the existing link, so the newcomer would be free to choose either HPOL or VPOL. Additionally, in this case the undesired signal is calculated to be well below the –85 dBm typical 13-GHz receiver threshold, and so for this example case the new link is no threat to the existing link.

Custom computer programs now typically do the interference calculations, using RPEs that have been tabulated from manufacturer's data so the program can "look up" the PPOL and XPOL suppression values at the pertinent off-axis angle for a specified microwave antenna. For such automated calculations, the software can routinely mathematically "add" the two suppression values by converting each dB suppression value to power, adding the two powers, and converting back to a combined dB suppression number, even when one XPOL dB value is 10 or more dB different from the lower, or controlling, XPOL dB value.

FIXED-LINK COORDINATION—2 GHZ

As noted previously, for the 2025–2110-MHz band, no Part 101 PCN process has been established for new or major-modification fixed, point-to-point links. Rather, a procedure has been established that combines local coordination with quasi-PCN requirements. Here is the procedure for frequency coordination is specified in Section 74.638(c) of the FCC's rules:

- General requirements: Applicants are responsible for selecting the frequency assignments that are least likely to result in mutual interference with other licensees in the same area. Applicants may consult local frequency coordination committees, where they exist, for information on frequencies available in the area. Proposed frequency usage must be coordinated with existing licensees and applicants in the area whose facilities could affect or be affected by the new proposal in terms of frequency interference on active channels, applied-for channels, or channels coordinated for future growth. Coordination must be completed prior to filing an application for regular authorization, for major amendment to a pending application, or for major modification to a license.
- To be acceptable for filing, all applications for regular authorization, or major amendment to a pending application, or major modification to a license, must include a certification attesting that all co-channel and adjacent-channel licensees and applicants potentially affected by the proposed fixed use of the frequency(ies) have been notified and are in agreement that the proposed facilities can be installed without causing harmful interference to those other licensees and applicants.

LOCAL COORDINATION

The third paradigm—informal local coordination—is now addressed. As stated previously, coordination of shared frequencies is the act of facilitating frequency sharing, not performing path engineering to establish a basis for frequency reuse. This is the proper coordination paradigm for non-point-to-point services, notably remote pickup (FCC service code RP, more commonly referred to as RPU for remote pickup unit), television pickup (FCC service code TP, commonly referred to as TVPU or ENG), and low-power auxiliary devices (LPA) such as wireless microphones and similar short-range devices typically used in a remote production environment.

The remainder of this chapter is intended as a primer on current issues affecting the practice of frequency sharing and coordination. Because the issues described here are in flux, the reader is encouraged to seek up-to-date information on the status of those matters of immediate interest. The two most useful resources for current information on spectrum management issues are the Federal Communications Commission website (www.fcc.gov) and the SBE website (www.sbe.org).

Event-Specific Coordination

Local frequency coordination is seen in many locales as taking on ever-greater importance due to the prevalence of radio-frequency devices around major events. These may include sporting events, political conventions, or major news stories. The devices in use may comprise a broad range of RPU, ENG, and low-power auxiliary devices. The nature and location of the event often determines the mixture of local (or in-market) users and itinerant (or out-of-market) users. Quite a number of spectrum-sharing issues typically arise that require the attention (sometimes on-site and in real time) of a local frequency coordinator, for example:

- Itinerant users may travel with LP equipment that is incompatible with local TV "white spaces" (i.e., unused TV channels) where most such devices operate;
- The sheer number of frequencies desired may overwhelm the sharing plan; this has been known to shift the coordination paradigm from one of "facilitating sharing" to one of "gatekeeping";
- Irrespective of the fact that all the users competing for spectrum availability may be equally eligible, an unofficial priority-of-use paradigm sometimes insinuates itself, based on which entities have broadcast rights to a commercial sporting event. This can lead to conflict, underscoring the importance of the local coordinator as a disinterested third party.

To meet these challenges, there has been a growing acceptance of the role of a local, event-specific frequency coordinator. Perhaps the best-known program of event coordination is the game day coordinator (GDC), a cooperative venture between the National Football League (NFL) and the SBE. The GDC program has been cited favorably by the FCC—in an October 2005 ruling, in response to the NFL's request for a waiver of station-identification rules on its wireless headset intercom system, the Commission explicitly noted the GDC's contribution to a more controlled RF environment:

> The NFL indicates that the game day frequency coordinators monitor the RF environment in real time during the game, and that they are careful to prevent any interference both inside and outside the stadium. In addition, the NFL indicates that by using prior coordination the potential for interference is greatly reduced if not eliminated.[11]

The specific case of NFL game coordination points to a more general trend that poses an ongoing challenge for local coordination. This is actually the convergence of two trends—the ever-greater use of wireless devices, especially in the LP class, and the shrinking availability of spectrum in which they can operate.

Section 74.802 of the FCC rules states that the majority of the channels available for LP devices consist of television broadcast channels that are unused in a particular location (also known as "white spaces"). However, the Incentive Auctions scheduled for 2016 will undoubtedly reduce the number of television channels available for BAS users. Upon completion of the Incentive Auctions process, broadcasters will lose all or a portion of the 600-MHz band, which has historically been a popular portion of spectrum for wireless microphones and similar LPA devices. As of this writing (March 2016), no replacement spectrum is on the horizon. It remains to be seen what combination of procedural and technological changes may be able to mitigate this spectrum management issue.

One technological fix that some vendors are trying is an intelligent, scanning receiver capable of communicating with the LPA transmitter to set its operating frequency. On power-up, the receiver quickly scans the RF environment and programs the best choice of frequencies to the transmitter. While such a feature may be useful and even admirable in an isolated system, concerns have been raised about this type of automatic operation in a complex, but coordinated, event. Vendors, users, and coordinators are advised to proceed with caution when introducing such "smart" systems into an environment in which a human coordinator is tracking frequency usage and making recommendations to multiple users on site.

REGULATORY CHANGES

Other issues facing local event coordinators arise from decisions in the regulatory realm. For example, one effect of the ET Docket 01-75 rulemaking, mentioned previously, was to replace the existing RPU channel plan at 450 MHz and 455 MHz with a new band plan, compatible with similar Part 90 land mobile radio services and based on a narrower fundamental unit of bandwidth (6.25 kHz, down from 12.5 kHz). Ironically, over a decade later, the RPU channel splits are still problematic, resulting in WT Docket 15-36 addressing fixes proposed by EIBASS in an October 7, 2011, Petition for Rulemaking (RM-11649), and a similar petition by SBE, filed on November 2, 2011 (RM-11648).

This change has resulted in placing a mixture of voice-grade, program-quality, and telemetry channels in the same allocation. As of this writing, the FCC is still deciding whether to adopt the RPU rebanding proposed by EIBASS or SBE. The reader is encouraged to check the latest status at either the FCC, EIBASS, or SBE websites.

A final frequency coordination issue is the Department of Defense (DoD) entry into the 2-GHz TV BAS band, as a result of the AWS-3 auction. The AWS-3 auction, General Docket 13-185, opened the 1.8-GHz L-band to commercial mobile radio service (CMRS), or cellular stations, by relocating a significant portion of DoD L-band operations to the S-band, namely the 2025–2110-MHz TV BAS band. While the rulemaking requires DoD to not only "protect" TV BAS, but further to "not constrain" TV BAS, details of how such band sharing can be successfully accomplished is still a work in progress.[12]

SUMMARY

An understanding of the concepts presented in this chapter will help the broadcast engineer know what type of frequency coordination services are required for projects making use of BAS spectrum. For events where heavy BAS usage is anticipated, especially involving use of LP auxiliary devices, the broadcaster is encouraged to seek out the coordinator responsible, either the local frequency coordinator or an event-specific coordinator such as a GDC. To ensure favorable outcomes, it is essential for both user and coordinator to be versed in the technological and regulatory challenges related to the BAS bands.

With the advent of mandatory PCN protocols for 7- and 13-GHz TV BAS fixed-link stations and 950-MHz aural BAS stations, most stations planning new or modified facilities in these bands will likely find that it makes sense to retain the services of a CMFC. But even when this task is so delegated, the user should be familiar with the methodology to know what questions to ask when selecting a CMFC.

The author wishes to express his appreciation of permission granted by Mr. David P. Otey to use still-pertinent portions of text from the prior edition of the *NAB Engineering Handbook* regarding frequency coordination for RPU and LPA stations.

Notes

[1] Specifically, they are as follows: Subpart D—Remote Pickup Broadcast Stations; Subpart E—Aural Broadcast Auxiliary Stations; Subpart F—Television Broadcast Auxiliary Stations; and Subpart H—Low Power Auxiliary Stations.

[2] See the FCC Wireless Telecommunications Bureau (WTB) website, at http://wireless.fcc.gov/microwave/coordinators.html, for a list of CMFCs.

[3] TSB-10F was published in June 1994. Unfortunately, it is unlikely that a TSB-10G version will be released anytime soon. TSB-10F was written primarily for FS and common carrier microwave links, although the standard does address FM video microwave links. However, TSB-10F is silent regarding interference criteria for 950 MHz Aural BAS microwave links.

[4] In the April 4, 2003, Society of Broadcast Engineers, Inc. (SBE) filed a Request for Stay of the PCN requirement for 950-MHz Aural BAS, and 2.5-, 7-, and 13-GHz TV BAS, fixed-link stations, scheduled to go into effect on April 16, 2003, asking for a one-year delay. The reason was the amount of missing or incomplete data for Part 74 fixed-link stations in the FCC's ULS. SBE pointed out that an analysis of 21,033 fixed-link BAS records showed that 6,163 of those records, or 29.3%, were either lacking receive-end geographic coordinates or had corrupted data for the receive-end coordinates (e.g., null coordinates, or coordinates that were identical to the transmit-end coordinates). In response, the FCC granted a sixth-month stay, until October 16, 2003, for the new PCN requirement.

As a result of the 7- and 13-GHz TV BAS bands being opened to Part 101 FS stations by the WT Docket 10–153 rulemaking, Working Group 3 (WG3) of the National Spectrum Management Association (NSMA) has been working with EIBASS, NAB, and SBE to further improve the accuracy of TV BAS records in the ULS. NSMA has also drafted a recommend practices document, *Fixed Service Frequency Coordination in the Broadcast Auxiliary Service and Cable Television Relay Service Bands of 6875–7125 MHz and 12700–13,150 MHz.*

On October 24, 2014, NSMA made a detailed, 41-page ex parte filing to WT Docket 10–153, documenting continued data accuracy problems for BAS records in the ULS. The filing showed that the primary cause of the inaccuracies was the manual conversion (by the FCC or its contractor) of BAS records for existing stations into the ULS (identifiable by the "license converted" annotation in the "history" portion of the station's ULS record). The NSMA filing asked that BAS licenses with missing or inaccurate data be allowed relaxed filing protocols to correct the data. Examples of incorrect data include missing receive-end coordinates or receive-end coordinates identical to the transmit-end coordinates; very short or very long paths; and paths where the transmitting antenna azimuth of record does not match the calculated azimuth for records with both transmitting and receiving coordinates. EIBASS and SBE both filed in support of the NSMA ex parte filing.

On February 24, 2016, representatives of NSMA, EIBASS, and a Washington DC engineering consulting firm met with FCC WTB staff to discuss implementing the mitigation steps suggested by the 2014 NSMA ex parte filing. However, as of this time of this writing, March 2016, the matter remains pending at the FCC.

Therefore, conducting a valid PCN study sometimes requires "detective work" by a CMFC to obtain missing critical data for a to-be-protected, earlier-in-time, fixed-link station.

[5] Wireless telecommunications services (WTSs) include Parts 13, 20, 22, 24, 26, 27, 74, 80, 87, 90, 95, and 101 of the FCC Rules. Thus, all BAS stations are WTS stations, and subject to the Part 1, Subpart F, WTS procedural provisions, in addition to any Part 74 provisions.

[6] "Space Diversity Engineering," A. Vigants, *Bell System Technical Journal*, Volume 54, Number 1, January 1975, pp. 103–142.

[7] "Prediction of Attenuation by Rain," R. K. Crane, *IEEE Transactions on Communications*, Volume COM-28, Number 9, September 1980, pp. 1717–1732.

[8] See TSB-10F, at Section 2.5.5.

[9] To add powers in dBm, each dBm value must first be converted to mW, added, and then converted back to dBm.

[10] Formerly "community antenna relay service"; the "CARS" abbreviation was retained.

[11] See Order, DA 05–2870, in the matter of National Football League request for waiver, October 31, 2005.

[12] See the November 23, 2015, issue of *TV Technology Broadcast Engineering Extra*, DoD Increases Presence in the 2 GHz TV BAS Band.

References

www.sbe.org

www.nsma.org

Federal Government Spectrum Compendium www.ntia.doc.gov/other-publication/2015/federal-government-spectrum-compendium

www.fcc.gov

March 31, 2014, General Docket 13-185 R&O (1.8 GHz AWS3 Auction and DoD Entry to 2 GHz)

August 9, 2014, WT 10-153 R&O (Opening 7- and 13-GHz TV BAS to Part 101 FS)

November 13, 2002, ET 01-75 R&O (Updating and Harmonizing of the Part 74 BAS Rules)

February 18, 2015, WT 15-36 NPRM (Proposing Fixes to the Part 74 RPU Band Plan)

FCC-Recognized Commercial Microwave Frequency Coordinators (CMFCs)

CHAPTER

2.4

Frequency Allocation for Broadcasting and the Broadcast Auxiliary Services

WILLIAM R. MEINTEL
Meintel, Sgrignoli & Wallace
Warrenton, VA

INTRODUCTION

This chapter provides an overview of terrestrial frequency allocations[1] and a listing of the frequencies available for AM, FM, and TV broadcast stations, as well as for the "auxiliary" broadcast services that support broadcasting operations. Necessarily, such an overview must include a description of the decision-making process that is involved in allocating a frequency band for a specific purpose. This complex subject of allocations involves more than just the location of the service in the frequency spectrum; it includes decisions as to the number and width of channels and power and antenna limitations, as well as decisions concerning the technical standards that define how the spectrum will be shared with other users. Because each broadcast service presents unique service and interference objectives, the allocation process has been and will continue to be different for each of these services. Here it is possible to provide only a brief description of the allocation process before turning to the current situation for each of the services. Where appropriate, the discussion will touch on changes expected in the foreseeable future.

HOW SPECTRUM IS ALLOCATED

It is a fundamental characteristic of radiowave propagation that these waves follow the laws of physics and thus ignore political or geographic boundaries. As a result, decisions concerning radiofrequency allocations cannot be made solely at the local level but must take into account their anticipated impact outside the station's coverage area. Recognizing that coordination in the allocation and use of spectrum is essential, an international mechanism has been established to perform this function, and in the United States, a parallel coordination system has been established at the federal level.

INTERNATIONAL ALLOCATION PROCESS

At the international level, frequency allocation decisions are made by the International Telecommunication Union (ITU), a specialized agency of the United Nations headquartered in Geneva, Switzerland. Like the United Nations, the ITU is a consortium of more than 190 governments, whose purpose is to propose, develop, revise, and administer worldwide frequency allocation plans. In the terminology of the ITU, participating governments are referred to as "states," "members," or "administrations." Such international cooperation serves to minimize interference and maximize use of the spectrum.

Although the ITU maintains a permanent staff, the power of the organization, as indicated by the structure shown in Figure 2.4-1, resides with the member nations who direct its activities through periodic meetings. At the highest level of this structure is the Plenipotentiary Conference, which adopts the fundamental policies of the organization and decides on the organization and activities of the Union through a treaty known as the International Telecommunication Constitution and Convention. Plenipotentiary Conferences are held approximately every 4 years and are open to ITU member countries, the United Nations and its specialized agencies, the International Atomic Energy Agency, regional telecommunications organizations, and intergovernmental satellites operators.

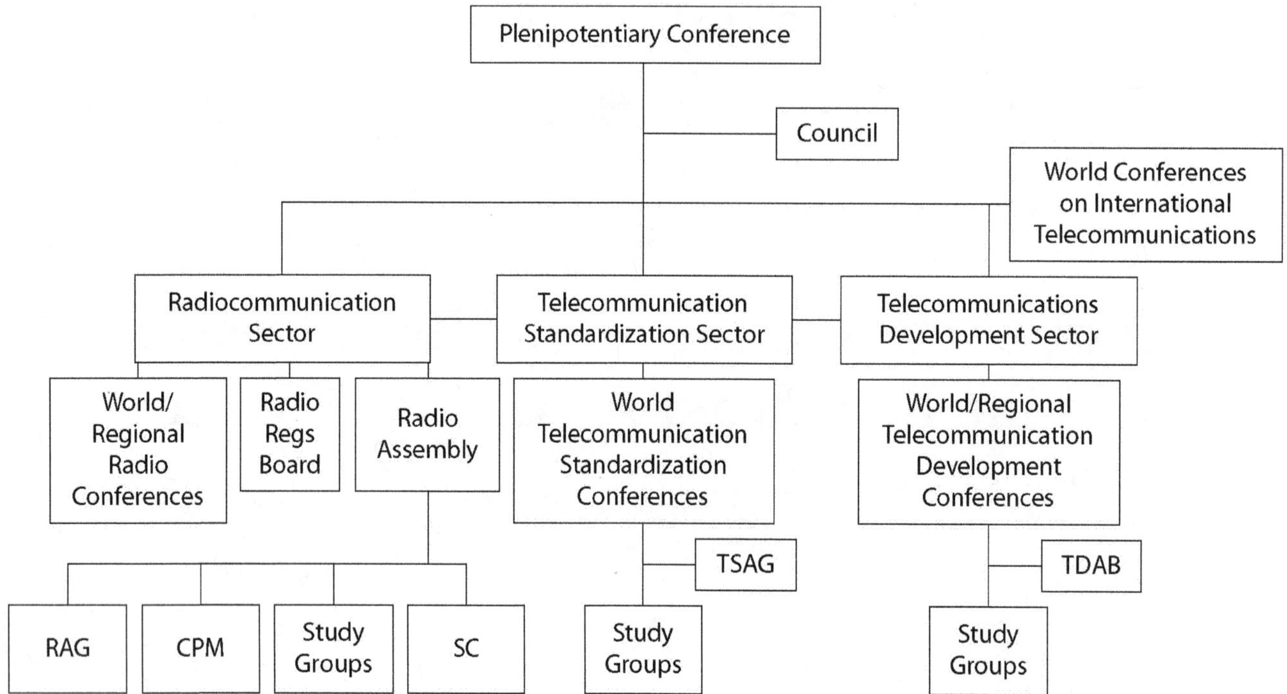

FIGURE 2.4-1 International Telecommunication Union (ITU) structure

Proceeding down the structure, the ITU Council consists of a representative group of members elected by the Plenipotentiary Conference. The task of the council is to ensure that in the period between Plenipotentiary Conferences the ITU's policies and strategy are adequately responding to the constantly changing telecommunication environment. The council is also responsible for ensuring efficient coordination of work, effective financial management, and assisting members in the implementation of the provisions and regulations of the Union. The day-to-day work of the ITU is performed in three sectors, of which the Radiocommunication Sector is relevant to this discussion. The other sectors, as their names imply, deal with standardization and development of telecommunications.

Worldwide allocations for radio services are made through decisions made at international conferences of the Radiocommunication Sector called World Radio Conferences, or WRCs (pronounced "warcs"). These WRCs deal with frequency allocations and related matters of concern on a worldwide basis through review and revision of the international Radio Regulations. Through a process of give and take, decisions are made for the allocation of spectrum. WRCs are held every 2 to 3 years and deal with a specific agenda that is generally established 4 to 6 years in advance, with a final agenda established by the ITU Council 2 years prior to the WRC. The early establishment of the agenda is intended to allow for proper preparation so the work of the conference can be concluded in a timely manner; however, due to the rapid changes occurring in radiocommunications, last-minute requests for agenda changes are not uncommon.

Because of the geographic separation of certain parts of the world, it was determined to be practical to divide the world into three regions (Figure 2.4-2). Region 1 consists of Europe and Africa; Region 2 consists of North and South America, Greenland, and the Caribbean; and Region 3 consists of Asia and Oceania. In view of this, the ITU sometimes convenes Regional Radio Conferences (RRCs) to consider questions that are unique to a specific ITU region. Often these RRCs consider implementation of decisions made at an earlier WRC.

Technical Submissions

Radiocommunication Assemblies are held in conjunction with and prior to the WRCs and provide the technical basis for the work of the WRC.[2] The Assemblies also approve and set priorities for the work of the study groups that operate under the Assemblies' supervision. In addition, the Assemblies also set up or abolish study groups as deemed appropriate. This work is currently carried out by six study groups, listed in Table 2.4-1a, each of which is devoted to a particular radio communication service or specific technical issue. The work most relevant to broadcasting is carried out by Study Group 6. This study group is further divided into subgroups or working parties that deal with the specific areas of interest listed in Table 2.4-1b. There also are sometimes joint study groups that deal with questions that are of relevance to more than one study group. Each study group has an international chair and one or more vice chairs who are provided by interested participating administrations.

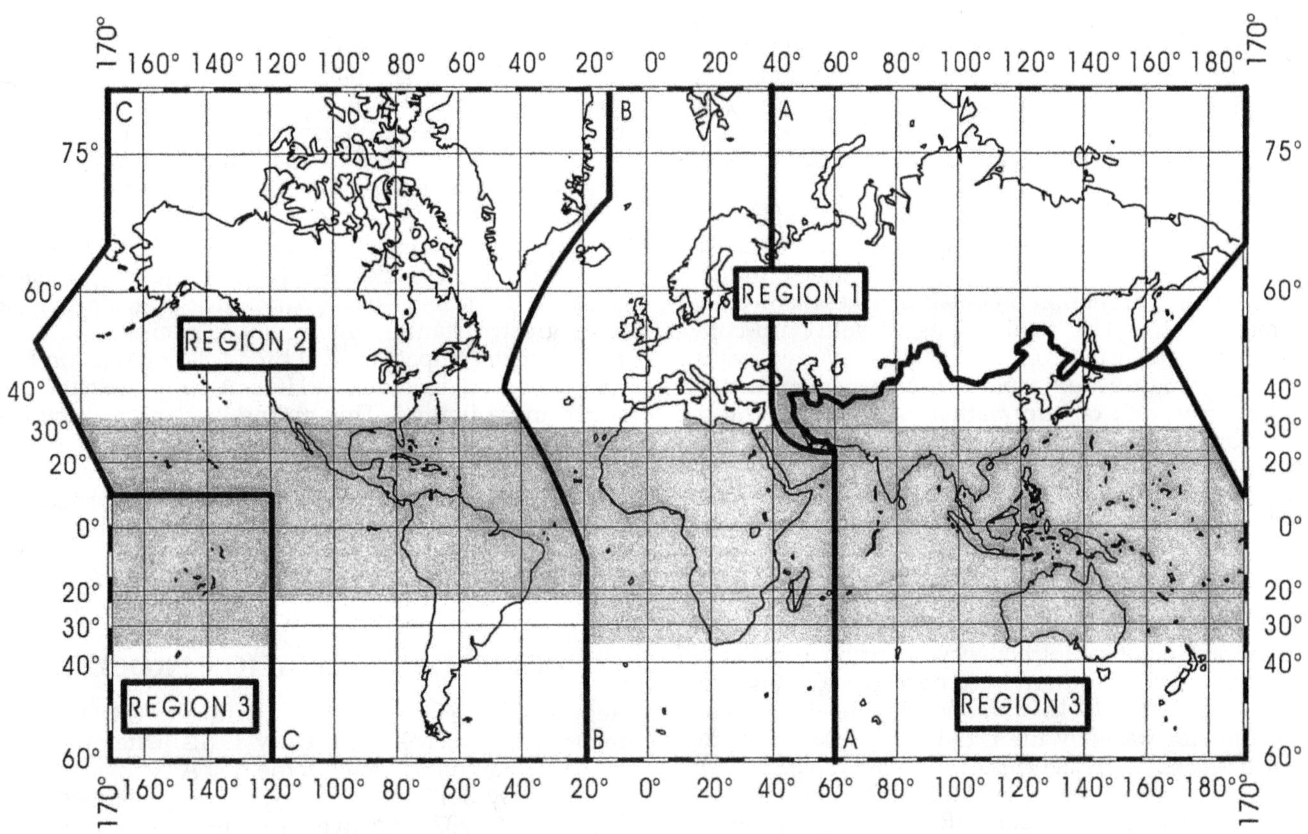

FIGURE 2.4-2 ITU regions. (From 47 CFR §2.104; the shaded area indicates the Tropical Zone as defined in 47 CFR §2.104(c)(4).)

TABLE 2.4-1A
ITU-R Study Groups*

Study Group	Subject
1	Spectrum management
3	Radiowave propagation
4	Satellite services
5	Terrestrial services
6	Broadcasting service
7	Science services

*Currently there is no Study Group 2

TABLE 2.4-1B
Subgroups of ITU Study Group 6

Subgroup	Subject
Working Party 6A	Terrestrial broadcasting delivery
Working Party 6B	Broadcast service assembly and access
Working Party 6C	Program production and quality assessment

ITU-R study group work is carried out by the membership of the ITU (member states, sector members, associates, and academia) that includes recognized users, standard setting groups, the European and other broadcasting unions and U.S. broadcasting networks. Within individual administrations, organized structures often are created to provide input for the administration to submit to the ITU meetings (World Radio Conferences, Radio Assemblies, and study groups). In the United States, this activity is performed by the National Committee, which is chartered by and operates under the Department of State (DOS). A U.S. national chair is appointed by the DOS and oversees the National Committee's approval of U.S. submissions to the ITU, relying on support from representatives of the Federal Communications Commission (FCC) and the National Telecommunications and Information Administration (NTIA). The DOS also appoints a U.S. national chair for each of the individual study groups. Each is responsible for the work of the group and heads the U.S. delegation to international meetings of the study group. In the United States, study group activities are open to participation by the public, and contributions to study group work come mainly from the private sector.

In general, the work of the study groups is timed to coincide with scheduled WRCs and RRCs, with an international meeting of each study group usually occurring about once a year. In the interim, work is carried out by correspondence. Administrations may suggest any matter of interest for study that is within the purview of the Radiocommunication Sector, but priority is given to issues that are relevant to a scheduled WRC or RRC.

The cumulative recommendations and reports of the various study groups are updated and published periodically in electronic format. Interested parties may obtain information on obtaining these as well as numerous other ITU publications, in both hard copy and electronic form, from the ITU Web site (www.itu.int). The study group reports and recommendations contain a wealth of information not only relating to spectrum allocation but also concerning measurement procedures and standards for audio and video recording equipment used for the exchange of broadcast programming.

The Allocation Process

The allocation of radiofrequency spectrum occurs as a result of a series of interrelated decisions. On the first, most basic level, blocks of frequencies are allocated on a worldwide basis by the ITU WRC process. Exactly which blocks of frequencies are allocated to particular services is determined by evaluating the many specific proposals submitted to the WRC for each frequency band. Technical input is obtained from propagation studies and other engineering analyses undertaken as part of the study group process and from the submissions of individual administrations. Frequently, blocks of spectrum are allocated for the same purpose on a worldwide basis but may also be allocated for different purposes on a regional basis.

Member nations theoretically retain the sovereign right to domestic use of the spectrum so long as such use is not in contravention of international radio regulations or the international agreements to which that administration is a party. However, as a practical matter, the flexibility of administrations to use the spectrum is limited by the worldwide allocation system and the need to avoid harmful interference.

Unlike the FCC, the ITU does not license users of the spectrum. Instead, it operates only as a coordinator, maintaining a Master International Frequency Register (MIFR, or Master Register) of radio stations worldwide that is maintained by the Radiocommunication Bureau within the Radiocommunication Sector. Member administrations have agreed to provide notifications of new stations or modifications in existing stations operating within their respective countries. The Radiocommunication Bureau studies these notifications for compliance with the existing world or regional agreements and provides the results of its studies to the member nations. Only those notifications that comply with the existing agreements are placed in the Master Register of stations. Once a station has been placed in the Master Register, the member nations are obliged to provide it with the internationally agreed level of interference protection.

Because the WRC and RRC agreement texts provide only a general framework, many specific matters are left to individual nations to resolve and implement. In the United States, the FCC and NTIA share responsibility for implementing agreements to which the country has assented, including the bilateral or multilateral agreements negotiated with our neighbors to deal with concerns that are unique to the countries involved.

There are differences in the treatment of the various broadcast services. Because lower frequency signals, such as those used by AM broadcasting, propagate over great distances, international decisions have a much greater impact on AM broadcasting than on FM and television, for which VHF or UHF propagation is much more limited. This means that the restrictions on FM and TV allocations imposed by international agreements are usually applied only to areas near the borders; however, international agreement on technical transmission standards often is desired in order to foster the absence of interference and the worldwide free flow of communication.

DOMESTIC ALLOCATION PROCESS

Regulation of spectrum began with the U.S. Department of Commerce in the early 1920s, when the Secretary of Commerce granted the first AM broadcasting licenses. By 1927, the number of AM stations had increased to 733, and over 6 million radio receivers had been manufactured; however, because of an unfortunate court decision that precluded the Secretary of Commerce from dealing with the specific choice of location, power, and operating frequency, these matters were left largely to the discretion of the broadcaster. This led to chaotic use of the spectrum and widespread interference, a situation that led to creation of the Federal Radio Commission in 1927. Seven years later, the Federal Radio Commission was replaced by the Federal Communications Commission (FCC), formed pursuant to the Communications Act of 1934. Ever since, anyone desiring to operate a broadcast station, or almost any kind of radio transmitting device, must apply to the FCC and be granted a license before commencing operation. Today, a broadcast license sets forth all essential technical parameters of station operation. The NTIA performs a similar function and coordinates the spectrum used by government agencies. The FCC works with the NTIA where there is a need for coordination between government and private uses of the spectrum.

Spectrum for use domestically must be allocated by FCC rulemaking proceedings. Domestic allocation rulemaking proceedings can be initiated by the FCC on its own motion or in response to requests from the public ("petitions"), but in so doing it must not act in contravention of international agreements to which the United States is a party. The FCC rulemaking process is a complex subject in its own right, but for our purposes here only a brief description is required. "Rulemaking" proceedings are based on a public record developed through responses to the issuance of a notice of proposed rulemaking (NPRM) that is filed by interested parties. In addition to filing comments on the FCC's proposal, the public may reply to the comments of other parties. After the Commission evaluates the responses to its notice, it may decide to adopt the proposal as originally set forth, modify it based on the comments received, or possibly reject the proposal.

Many FCC proceedings are controversial in nature. Allocation proceedings may be especially controversial, as a particular communications industry's livelihood may depend in part on how much spectrum is allocated. Thus, FCC allocation decision-making is not simply a matter of technical evaluation but must be seen as part of the political process as well, as happens when entire industries may compete for a limited amount of spectrum. Where the number of users in a particular frequency band is expected to be relatively small or their use is sporadic, the FCC may propose sharing of this spectrum with other users. Such proposals may also be controversial, because sharing spectrum with a dissimilar service invites the possibility of interference and difficulties in coordinating the use of the frequencies. Because of these and other factors, FCC allocation proceedings consume a great deal of its time and energy and can impose burdens on the organizations that participate in them.

From time to time, alternative methods have been proposed for allocating spectrum. These ideas usually envision the removal of the FCC as the arbiter of mutually exclusive requests for spectrum and instead substituting marketplace forces. Under a market allocation system, frequencies would be used by entities that would pay for them; noneconomic, social, or public policy aspects would not be considered. In recent years, such a system has been employed for some frequency allocations in the form of spectrum auctions. In these cases, the Commission's role is reduced to that of a technical "traffic cop" of the airwaves.

At the time this chapter is being written, the FCC is conducting an unusual auction of television broadcast spectrum that is discussed in more detail in the section dealing with television spectrum allocation.

U.S. AM Broadcasting Frequency Allocation

In the United States, amplitude-modulated (AM) stations operate with carrier frequencies in the center of channels assigned every 10 kHz. For many years, the AM broadcast band in the United States included 107 channels in the band from 535 to 1605 kHz; however, the frequency band allocated to AM broadcasting now totals 117 channels in the band from 535 to 1705 kHz. Currently, nearly 4,700 commercial and noncommercial AM stations are operating in the United States. These stations operate with various power levels, up to a maximum of 50 kilowatts (kW). About half of these stations use multi-tower directional antennas to restrict radiation in certain directions for the purpose of controlling interference or maximizing radio service in particular directions.

Allocation decisions for the AM broadcast band are probably the most complex of the broadcast services. Because propagation varies with time of day, geographic latitude, soil conductivity, and frequency, the engineering analyses necessary to establish interference protection for other stations can be quite complicated. Engineers, the FCC, and the ITU have sophisticated computer programs that analyze the impact of a new or modified AM station proposal. Before going into the details of the AM broadcast allotment system currently in place in the United States, it is useful to provide a brief history of AM broadcasting allocations.

Over the years, there have been many changes in the nature of AM broadcasting in the United States. Initially, *clear channel* stations (high-powered, omnidirectional stations with a large coverage area) provided the only service available in many areas of the United States, but with the end of World War II, demand increased greatly and many AM stations were established in all areas of the country. Because of its early development of AM radio service, the U.S. experience has been used as a model for regional and bilateral agreements.

Early in the history of AM broadcasting, the countries in the North American area recognized the need to cooperate in the use of AM frequencies, and in 1937, they reached agreement on how to proceed. Soon, however, this agreement was found to be inadequate and negotiations began on a new agreement. Although the North American Regional Broadcasting Agreement (NARBA) was signed on November 15, 1950, it did not go into effect until 10 years later, on April 16, 1960. Signatories to NARBA include the United States, Canada, Cuba, the Dominican Republic, and the United Kingdom on behalf of Jamaica and the Bahamas. Mexico, an earlier participant, removed itself from these negotiations, and a bilateral agreement between the United States and Mexico was reached in 1957. These international agreements became necessary principally because nighttime AM propagation has the potential for causing widespread interference to neighboring countries unless mutual allocation criteria and related technical standards could be agreed upon and implemented by the parties to the agreement. To this end, NARBA provided for a partitioning of AM broadcast channels into three basic classes.

Clear Channel Stations

The first of these classes includes the so-called *clear channels*, whose high-powered stations would have primary access to the frequency and other stations could use the channel subject to providing full protection to the dominant station(s). Clear channel dominant stations were designed to provide service over extensive areas by means of skywave as well as groundwave signals. NARBA set aside 60 of the 107 channels then available for clear channel use. Each NARBA country, except Jamaica, received a priority on one or more clear channels, with the United States receiving a major portion of available priorities. NARBA countries without a priority on a given clear channel could still assign stations on that channel, provided these stations protected the wide-area service of the dominant station in the country with the NARBA priority.

Regional Channels

The second class of channels, *regional channels*, occupies an additional 41 channels. Unlike clear channels, these channels were shared on an equal basis by all the NARBA countries. Stations operating on these channels were intended to provide service to a considerable area, but unlike the clear channel stations that received protection for their skywave as well as groundwave service, only the groundwave service provided by the regional stations was protected.

Local Channels

The remaining six frequencies were the *local channels*, which provide an even more limited type of groundwave service. Only limited interference protection was provided to these relatively low-power operations.

1981 Rio Agreement

Recognizing the need for updating these agreements and for developing more efficient coordination throughout Region 2 (North and South America), the 1979 WRC called for a conference to be held in Region 2 to address AM broadcasting and sharing criteria. That conference was held in two sessions in 1980 and 1981 and resulted in the adoption of an agreement among most of the countries of the hemisphere. Included as part of that agreement, referred to as the 1981 Rio Agreement (for Rio de Janeiro, the location of the second conference session), was a list of all of the operating stations in the hemisphere along with information indicating whether or not the stations were receiving or causing harmful interference according to the technical criteria set forth in the agreement. Stations not causing interference were placed in the ITU Master Register and accorded protection from interference as defined by the agreement. In situations where interference already existed, the countries involved were asked to meet and work out mutually satisfactory solutions. Because the general framework of the Rio Agreement did not deal with the particular needs and desires of the United States, Canada, and Mexico, separate new bilateral agreements have been negotiated that incorporate the required additional items concerning coordination and technical parameters.

Although the Rio Agreement applies throughout most of Region 2, as of this writing, relations with the Bahamas and the Dominican Republic continue to be governed by NARBA, as neither country has taken the necessary steps to replace the NARBA provisions with the Rio 1981 Agreement. Relations with Cuba regarding the 535- to 1605-kHz band are governed solely by the international Radio Regulations rather than by agreement.

Daytime-Only Stations

As the AM broadcasting system in the United States continued to evolve, the demand for more stations also grew. One result of the demand for facilities was the increased use of directional antennas to provide required interference protection while enhancing coverage in other directions. With the spectrum becoming crowded, another development was a large increase in the number of daytime-only stations. These are stations that were authorized to operate only during daylight hours. Because propagation conditions during these daylight hours do not normally support significant skywave transmission, there are many locations where a station can be operated during the daytime without causing harmful interference to other stations. Based on this concept, the FCC over the years licensed approximately 2,500 stations for daytime-only operation.

Recognizing that daytime-only stations (and even some full-time stations that operate with restrictive directional antenna patterns during the night) are unable to provide effective service during early morning hours, the FCC originally allowed these stations to operate during this early morning period so long as no interference complaint had been received. Ultimately, this proved to be unworkable, and a more formalized approach was adopted. The FCC began granting pre-sunrise authorizations (PSRAs), permitting many of these stations to operate their daytime facilities with powers up to 500 W during the pre-sunrise period between 6:00 a.m. and local sunrise. Although some interference occurred, the FCC believed the interference was balanced by the public's need for local informational services during this important morning period.

While pre-sunrise operation did provide some relief, it did not end the economic problems many stations faced in effectively competing with full-time stations. The FCC was pressed to provide relief in the form of post-sunset operation for daytime-only stations. With the removal of international impediments, the FCC did provide such relief. Once again, taking into consideration the need for more service of a local nature and recognizing the changing nature of propagation conditions in which full nighttime conditions do not exist until several hours after sunset, the FCC granted post-sunset authorizations (PSSA) for most of the daytime-only stations. These authorizations permit operation for periods of up to 2 hours past sunset with power reduced to prevent interference. The FCC also changed its rules concerning the minimum power at which a station is permitted to operate. This, in turn, led to a subsequent decision allowing many PSSA stations to operate throughout the night, albeit with reduced power.[3]

Class I-A Clear Channel Stations

Still another change pertained to the use of the Class I-A clear channels. At one time, only a single station was permitted to operate on these channels at night, but in two FCC decisions, first some and now all of these channels have been broken down to permit the authorization of additional nighttime operations. Clear channel stations that at one time provided service for a major portion of the country during nighttime hours now are protected only out to a distance averaging 750 miles.

Use of the AM Expanded Band

At an ITU Regional Radio Conference that concluded in 1988, participating administrations reached agreement on the criteria for expanding the AM band in Region 2 by adding 10 new channels between 1605 and 1705 kHz. Although the regional agreement established specific technical criteria for the implementation of the new channels (including the granting of priority usage of certain channels), the United States still retains considerable latitude in its domestic implementation. In a large country like the United States, the use of the channels is unrestricted except in the relatively few areas near the borders with its neighbors.

The basic criteria set forth in the regional agreement are as follows: Stations may operate with 1 kW of power with a nondirectional antenna height of 90°, or stations may operate with a power not in excess of 10 kW by employing a directional antenna to provide equivalent protection to stations in other countries. The channels allotted to the United States in the border areas vary from location to location, but, as noted, over a large portion of the United States all 10 of the channels may be used. Non-allotted channels are not precluded from use, but the allotted channels in the other countries must remain fully protected. After a lengthy proceeding, the FCC developed a system whereby the channels in the expanded band would be used to improve the current interference situation by giving selected stations the opportunity to move to the new band. In view of this, applications for new stations are not currently being accepted for operation in the 1605- to 1705-kHz band.

Persons seeking an authorization for a new AM broadcast station on frequencies below 1605 kHz may do so by filing an application with the FCC. Applications must provide documentation that the proposed operation will comply with all applicable FCC rules as well as the appropriate international regulations. The details and methodology for allocating AM radio stations along with basic design specifications for AM directional antennas can be found in the FCC rules, §73.14-73.190.

AM Stations Reclassified

The 1981 Rio Agreement changed the NARBA station classifications. No longer are the channels themselves classified. Stations are now classified without regard to the channel on which they operate. Stations providing wide area service, both groundwave and skywave, are now designated Class A stations, and stations providing the equivalent of regional and local services are designated as Classes B and C, respectively. The 1981 Rio Agreement permits any class of station to operate on any channel so long as it provides protection to other stations based on their classification. Tables 2.4-2 through 2.4-4 show the relationships between these international classifications and those that existed in the United States prior to 1990. In 1990, the FCC proposed to align the U.S. domestic classification system for AM stations with the 1981 Rio Agreement system. In addition to the three international classes of stations, the FCC has added a fourth: Class D. Class D stations are those stations that operate at nighttime with a power of less than 250 W and an equivalent RMS antenna field of less than 141 V/m at 1 km. Such stations are usually former daytime-only stations that have been granted some limited amount of nighttime operation. The current status of AM station classification is contained in the FCC rules, §73.21. In that the allocation system in the United States had been long established, the station classifications still are primarily related to frequency. The current scheme is shown in Tables 2.4-5 and 2.4-6.

TABLE 2.4-2
International and Domestic Classifications of AM Stations and Channels Prior to 1990

International Classes of AM Stations	Corresponding U.S. Classes Prior to 1990	Classes of Channels Available in the U.S. for Each Class
Class A	I-A	Clear channels
	I-B	
	I-N	
Class B	II	Clear channels
	II-A	
	II-B	
	II-C	
	II-D	
	II-S	
	III	Regional channels
	III-S	
Class C	IV	Local channels

TABLE 2.4-3
Channel Utilization in the United States Prior to 1990

Frequency (kHz)	Class of Station*	Frequency (kHz)	Class of Station*	Frequency (kHz)	Class of Station*
540	II	900	II	1260	III, III-S
550	III, III-S	910	I-A, III, III-S	1270	III, III-S
560	III, III-S	920	I-A, III, III-S	1280	III, III-S
570	III, III-S	930	I-A, III, III-S	1290	III, III-S
580	III, III-S	940	II	1300	III, III-S
590	III, III-S	950	III, III-S	1310	III, III-S
600	III, III-S	960	III, III-S	1320	III, III-S
610	III, III-S	970	III, III-S	1330	III, III-S
620	III, III-S	980	III, III-S	1340	IV
630	III, III-S	990	II	1350	III, III-S
640	I-A	1000	I-B	1360	III, III-S
650	I-A	1010	II	1370	III, III-S
660	I-A	1020	I-A	1380	III, III-S
670	I-A	1030	I-A	1390	III, III-S
680	I-B	1040	I-A	1400	IV
690	II	1050	II	1410	III, III-S
700	I-A	1060	I-B	1420	III, III-S
710	I-B	1070	I-B	1430	III, III-S
720	I-A	1080	I-B	1440	III, III-S
730	II	1090	I-B	1450	IV
740	II	1100	I-A	1460	III, III-S
750	I-A	1110	I-B	1470	III, III-S
760	I-A	1120	I-A	1480	III, III-S
770	I-A	1130	I-B	1490	IV
780	I-A	1140	I-B	1500	1-B
790	III, III-S	1150	III, III-S	1510	I-B
800	II	1160	I-A	1520	I-B
810	I-B	1170	I-B	1530	I-B
820	I-A	1180	1-A	1540	I-B
830	I-A	1190	I-B	1550	II
840	I-A	1200	I-A	1560	I-B
850	I-B	1210	I-A	1570	II
860	II	1220	II	1580	II
870	I-A	1230	IV	1590	III, III-S
880	I-A	1240	IV	1600	III, III-S
890	I-A	1250	III, III-S		

*Refer to Table 2.4-2; in addition to the Class I-A or Class I-B stations that could be assigned to the above channels, various Class II stations could also be assigned.

TABLE 2.4-4
Power Limitation by Class of Station

	Power (kW)			
	Daytime		Nighttime	
Class of Station	Minimum	Maximum	Minimum	Maximum
I-A	10.0	50.0	10.0	50.0
I-B	10.0	50.0	10.0	50.0
I-N	10.0	50.0	10.0	50.0
II	10.0	50.0	10.0	50.0
II-A	10.0	50.0	10.0	50.0
II-B	0.25	50.0	0.25	50.0
II-C	0.25	50.0	0.25	1.0
II-D	0.25	50.0	N/A	N/A
II-S	0.25	50.0	N/A	<0.25
III	0.25	5.0	0.25	5.0
III-S	0.25	5.0	N/A	<0.25
IV	0.10	1.0	0.10	1.0

TABLE 2.4-5
Current Channel Utilization in the United States

Frequency (kHz)	Class of Station	Frequency (kHz)	Class of Station	Frequency (kHz)	Class of Station
540	A, B, D	930	B, D	1320	B, D
550	B, D	940	A, B, D	1330	B, D
560	B, D	950	B, D	1340	C
570	B, D	960	B, D	1350	B, D
580	B, D	970	B, D	1360	B, D
590	B, D	980	B, D	1370	B, D
600	B, D	990	A, B, D	1380	B, D
610	B, D	1000	A, B, D	1390	B, D
620	B, D	1010	A, B, D	1400	C
630	B, D	1020	A, B, D*	1410	B, D
640	A, B, D*	1030	A, B, D*	1420	B, D
650	A, B, D*	1040	A, B, D*	1430	B, D
660	A, B, D*	1050	A, B, D	1440	B, D
670	A, B, D*	1060	A, B, D	1450	C
680	A, B, D	1070	A, B, D	1460	B, D
690	A, B, D	1080	A, B, D	1470	B, D
700	A, B, D*	1090	A, B, D	1480	B, D
710	A, B, D	1100	A, B, D*	1490	C
720	A, B, D*	1110	A, B, D	1500	A, B, D

TABLE 2.4-5 *(continued)*
Current Channel Utilization in the United States

Frequency (kHz)	Class of Station	Frequency (kHz)	Class of Station	Frequency (kHz)	Class of Station
730	A, B, D	1120	A, B, D*	1510	A, B, D
740	A, B, D	1130	A, B, D	1520	A, B, D
750	A, B, D*	1140	A, B, D	1530	A, B, D
760	A, B, D*	1150	B, D	1540	A, B, D
770	A, B, D*	1160	A, B, D*	1550	A, B, D
780	A, B, D*	1170	A, B, D	1560	A, B, D
790	B, D	1180	A, B, D*	1570	A, B, D
800	A, B, D	1190	A, B, D	1580	A, B, D
810	A, B, D	1200	A, B, D*	1590	B, D
820	A, B, D*	1210	A, B, D*	1600	B, D
830	A, B, D*	1220	A, B, D	1610	B, D
840	A, B, D*	1230	C	1620	B, D
850	A, B, D	1240	C	1630	B, D
860	A, B, D	1250	B, D	1640	B, D
870	A, B, D*	1260	B, D	1650	B, D
880	A, B, D*	1270	B, D	1660	B, D
890	A, B, D*	1280	B, D	1670	B, D
900	A, B, D	1290	B, D	1680	B, D
910	B, D	1300	B, D	1690	B, D
920	B, D	1310	B, D	1700	B, D

*Only one Class A station will be assigned on these channels.

TABLE 2.4-6
Power Limitation by Class of Station

	Power (kW)			
	Daytime		Nighttime	
Class of Station	Minimum	Maximum	Minimum	Maximum
A	10.0	50.0	10.0	50.0
B	0.25	50.0	0.25	50.0 (below 1605 kHz)
B	0.25	10.0	0.25	10.0 (1605–1705 kHz)
C	0.25	1.0	0.25	1.0
D	0.25	50.0	N/A	<0.25

FM Broadcasting Frequency Allocation

The 88- to 108-MHz frequency band is allocated for FM broadcasting in Region 2 and, with some exceptions, for Regions 1 and 3, as well; however, unlike AM, FM broadcast allotments are largely a domestic matter (especially in large countries such as the United States) due to the limited nature of signal propagation at these frequencies. Although there are some international regulations regarding FM broadcasting, there is no region-wide FM agreement in Region 2. Instead, there are bilateral agreements between the United States and Canada and between the United States and Mexico. Both regulate the use of FM channels in the border areas and specify technical standards in order to ensure system compatibility. The FM broadcast band is divided into 100 channels, each 200 kHz wide. In the United States, the lower 20 channels, located between 88 and 92 MHz, have been reserved for noncommercial broadcasting; however, such stations are not restricted solely to these channels. In addition, although it is part of television channel 6 (82–88 MHz), the frequency 87.9 MHz is used for low-power noncommercial FM stations, but its use is severely restricted.

FM Station Classes and Broadcast Zones

As with AM broadcasting, different classes of FM stations are designed to provide different types of service. As the demand for more stations has increased so has the number of different classes of stations. In response to this demand, the FCC has significantly modified the criteria concerning the use of the frequencies. In June 1983, the FCC concluded a lengthy rule-making proceeding and modified the domestic allotment criteria for FM broadcasting. Prior to this action, there were three classes of stations in the 80-channel commercial band. Twenty of these channels were used for lower power Class A stations having a maximum effective radiated power (ERP) of 3 kW and a maximum antenna height above average terrain (HAAT) of 91.4 m (300 ft). Class A stations had a 1 V/m service radius of about 15 miles. Higher power Class B or C stations operated on the remaining 60 channels. Whether a station was designated Class B or C depended on where it was located (Figure 1.5-3).[4] Class B stations were located in Zone I or I-A. Zone I is the northeast United States, extending south to the Virginia–North Carolina border and west to the Mississippi River. Zone I-A is all but the northernmost portion of California, plus Puerto Rico and the Virgin Islands. Class C stations operate elsewhere in the country in what is referred to as Zone II.

Class B stations operate with a maximum ERP of 50 kW at 150 m (492 ft.) HAAT and have a service radius of about 52 km (32 miles). Class C stations operate with a maximum power of 100 kW at 600 m (1968 ft.) HAAT, for a service radius of approximately 92 km (57 miles). FM stations in each of these classes may elect to operate at a HAAT above the maximum, but in such cases they are required to make a compensatory reduction in ERP as noted in FCC rules, §73.211.

This system was changed substantially by the above-noted 1983 proceeding, when the FCC:

- Permitted Class A stations to operate on channels previously reserved for Class B or C stations;
- Created three new classes of FM stations. Class B1 stations are permitted to operate in Zones I and I-A with a maximum ERP of 25 kW at 100 m (328 ft.) HAAT; Class C1 and Class C2 stations are permitted to operate in Zone II. Class C1 stations are permitted a maximum ERP of 100 kW at 299 m (981 ft.) HAAT and Class C2 stations are permitted a maximum ERP of 50 kW at 150 m (492 ft.) HAAT;
- Required stations that were previously licensed as Class B or C and were not operating at the minimum level specified for their class under the new rules to upgrade their facilities within 3 years;

FIGURE 2.4-3 FM broadcast zones; I and I-A indicated by shading

otherwise, the under-minimum facilities would be reclassified to the appropriate lower class based on the facilities they used;

- Increased the maximum antenna HAAT for Class A stations to 100 m (328 ft.).

In 1989, the FCC further modified the rules to permit Class A stations to operate with an ERP of 6 kW and a HAAT of 100 m and added an additional classification, C3, that permits operation in Zone II with a maximum of 25 kW at 100 m HAAT. In 2000, the FCC again modified the rules to create an additional station classification, C0. Stations in this class operate with a maximum ERP of 100 kW with a HAAT of between 300 m (980 ft.) and 450 m (1476 ft.).

In addition to the above, there also are Class D stations that operate as noncommercial educational stations with power not in excess of 10 W; however, applications for this class of station are no longer being accepted (see 47 CFR §73.512(c)). A complete list of the station classes and a summary of FM allotment standards can be found in Tables 2.4-7 and 2.4-8.

FM Table of Allotments

Unlike AM broadcasting, where a new station may be applied for at any location where it can meet applicable criteria, the use of commercial FM channels (channels 221 to 300) is governed by the FM Table of Allotments found in the FCC rules, §73.202. This table lists all FM channel allotments that have been made available for use. Most already are in use. If the table does not list a vacant channel in the desired community, the prospective applicant must file a rulemaking petition with the FCC seeking to add such a channel for the community. The rulemaking petition proposing such addition must provide a showing that the proposal meets the separation requirements that are applicable to the class of station being proposed. A complete list of the spacing requirements, including those that pertain to stations located near Canada and Mexico, is provided in Tables 2.4-9 through 2.4-12. Alternatively, a petitioner can propose to modify the table by deleting a vacant existing allotment or by changing the frequencies of an existing station and thereby achieve compliance with these spacing requirements.

Once a location has been added to the Table of Allotments, the FCC will announce a period of time (called a *window*) when it will accept applications for the location. If, as often is the case, there are multiple applicants, the winning applicant will be determined by an auction. It also should be noted that in cases where it is not possible to locate a transmitter site that meets the mileage separation requirements, the FCC does permit the use of reduced power or antenna height, as well as directional antennas, in order to provide equivalent protection to other stations. However, this only applies to the filing of an application for a location that is already in the Table of Allotments. The FCC will not accept proposals to modify the table unless it is shown to meet applicable separation requirements.

Assignment of stations on the noncommercial educational channels (200–220) is accomplished more in the manner that is followed in AM, where an application includes a showing that interference will not be caused to other stations. In addition, for proposals to

TABLE 2.4-7
Standards for FM allotments in Puerto Rico and the Virgin Islands

Station Class	Maximum ERP kW	Maximum HAAT m (ft.)	Expected Service Radius km (mi)
A	6	240 (787)	42 (26.1)
B1	25	150 (492)	46 (28.6)
B	50	472 (1549)	78 (48.5)

TABLE 2.4-8
Standards for FM Allotments for Locations Other Than Puerto Rico and the Virgin Islands

Station Class	Maximum ERP kW	Maximum HAAT m (ft.)	Expected Service Radius km (mi)
A	6	100 (328)	28 (17.4)
B1	25	100 (328)	39 (24.2)
B	50	150 (492)	52 (32.3)
C3	25	100 (328)	39 (24.2)
C2	50	150 (492)	52 (32.3)
C1	100	299 (981)	72 (44.7)
C0	100	450 (1476)	83 (51.6)
C	100	600 (1968)	92 (57.2)

TABLE 2.4-9
Minimum Distance Separation Requirements

Relation	Minimum Separation km (mi)			
	Co-Channel	200 kHz	400/600 kHz	10.6/10.8 MHz
A to A	115 (71)	72 (45)	31 (19)	10 (6)
A to B1	143 (89)	96 (60)	48 (30)	12 (7)
A to B	178 (111)	113 (70)	69 (43)	15 (9)
A to C3	142 (88)	89 (55)	42 (26)	12 (7)
A to C2	166 (103)	106 (66)	55 (34)	15 (9)
A to C1	200 (124)	133 (83)	75 (47)	22 (14)
A to C0	215 (134)	152 (94)	86 (53)	25 (16)
A to C	226 (140)	165 (103)	95 (59)	29 (18)
B1 to B1	175 (109)	114 (71)	50 (31)	14 (9)
B1 to B	211 (131)	145 (90)	71 (44)	17 (11)
B1 to C3	175 (109)	114 (71)	50 (31)	14 (9)
B1 to C2	200 (124)	134 (83)	56 (35)	17 (11)
B1 to C1	233 (145)	161 (100)	77 (48)	24 (15)
B1 to C0	248 (154)	180 (112)	87 (54)	27 (17)
B1 to C	259 (161)	193 (120)	105 (65)	31 (19)
B to B	241 (150)	169 (105)	74 (46)	20 (12)
B to C3	211 (131)	145 (90)	71 (44)	17 (11)
B to C2	241 (150)	169 (105)	74 (46)	20 (12)
B to C1	270 (168)	195 (121)	79 (49)	27 (17)
B to C0	272 (169)	214 (133)	89 (55)	31 (19)
B to C	274 (170)	217 (135)	105 (65)	35 (22)
C3 to C3	153 (95)	99 (62)	43 (27)	14 (9)
C3 to C2	177 (110)	117 (73)	56 (35)	17 (11)
C3 to C1	211 (131)	144 (90)	76 (47)	24 (15)
C3 to C0	226 (140)	163 (101)	87 (54)	27 (17)
C3 to C	237 (147)	176 (109)	96 (60)	31 (19)
C2 to C2	190 (118)	130 (81)	58 (36)	20 (12)
C2 to C1	224 (139)	158 (98)	79 (49)	27 (17)
C2 to C0	239 (148)	176 (109)	89 (55)	31 (19)
C2 to C	249 (155)	188 (117)	105 (65)	35 (22)
C1 to C1	245 (152)	177 (110)	82 (51)	34 (21)
C1 to C0	259 (161)	196 (122)	94 (58)	37 (23)
C1 to C	270 (168)	209 (130)	105 (65)	41 (25)
C0 to C0	270 (168)	207(129)	96 (60)	41 (25)
C0 to C	281 (175)	220 (137)	105 (65)	45 (28)
C to C	290 (180)	241 (150)	105 (65)	48 (30)

TABLE 2.4-10
Minimum Distance Separation Requirements: Canadian Agreement

Relation	Minimum Separation km (mi)				
	Co-Channel	200 kHz	400 kHz	600 kHz	10.6/10.8 MHz
A1 to A1	78 (49)	45 (28)	24 (15)	20 (12)	4 (2)
A1 to A	131 (82)	78 (49)	44 (27)	40 (25)	7 (4)
A1 to B1	164 (102)	98 (61)	57 (35)	53 (33)	9 (6)
A1 to B	190 (118)	117 (73)	71 (44)	67 (42)	12 (7)
A1 to C1	223 (139)	148 (92)	92 (57)	88 (55)	19 (12)
A1 to C	227 (141)	162 (101)	103 (64)	99 (62)	26 (16)
A to A	151 (94)	98 (61)	51 (32)	42 (26)	10 (6)
A to B1	184 (115)	119 (74)	64 (40)	55 (34)	12 (7)
A to B	210 (131)	137 (85)	78 (49)	69 (43)	15 (9)
A to C1	243 (151)	168 (105)	99 (62)	90 (56)	22 (14)
A to C	247 (154)	182 (113)	110 (68)	101 (63)	29 (18)
B1 to B1	197 (123)	131 (82)	70 (44)	57 (35)	24 (15)
B1 to B	223 (139)	149 (93)	84 (52)	71 (44)	24 (15)
B1 to C1	256 (159)	181 (113)	108 (67)	92 (57)	40 (25)

TABLE 2.4-10 *(continued)*
Minimum Distance Separation Requirements: Canadian Agreement

Relation	Minimum Separation km (mi)				
	Co-Channel	200 kHz	400 kHz	600 kHz	10.6/10.8 MHz
B1 to C	259 (161)	195 (121)	116 (72)	103 (64)	40 (25)
B to B	237 (148)	164 (102)	94 (59)	74 (46)	24 (15)
B to C1	271 (169)	195 (121)	115 (72)	95 (59)	40 (25)
B to C	274 (171)	209 (130)	125 (78)	106 (66)	40 (25)
C1 to C1	292 (182)	217 (135)	134 (83)	101 (63)	48 (30)
C1 to C	302 (188)	230 (143)	144 (90)	111 (69)	48 (30)
C to C	306 (190)	241 (150)	153 (95)	113 (70)	48 (30)

Note: Class A1 stations are defined as having an ERP of 0.25 kW and an antenna HAAT of 100 m; U.S. Class C2 stations are considered as Class B. Class A1 stations are defined as having an ERP of 0.25 kW and an antenna HAAT of 100 m; U.S. Class C2 stations are considered as Class B.

TABLE 2.4-11
Minimum Separation Requirements: Mexican Agreement

Class Relation	Minimum Separation km			
	Co-Channel	200 kHz	400 kHz or 600 kHz	10.6 or 10.8 MHz (I.F.)
A to A	100	61	25	8
A to AA	111	68	31	9
A to B1	138	88	48	11
A to B	163	105	65	14
A to C1	196	129	74	21
A to C	210	161	94	28
AA to AA	115	72	31	10
AA to B1	143	96	48	12
AA to B	178	125	69	15
AA to C1	200	133	75	22
AA to C	226	165	95	29
B1 to B1	175	114	50	14
B1 to B	211	145	71	17
B1 to C1	233	161	77	24
B1 to C	259	193	96	31
B to B	237	164	65	20
B to C1	270	195	79	27
B to C	270	215	98	35
C1 to C1	245	177	82	34
C1 to C	270	209	102	41
C to C	290	228	105	48

Notes:
1. U.S. or Mexican assignments or allotments which have been notified internationally as Class A are limited to a maximum of 3.0 kW ERP at 100 meters HAAT, or the equivalent;
2. U.S. or Mexican assignments or allotments which have been notified internationally as Class AA are limited to a maximum of 6.0 kW ERP at 100 meters HAAT, or the equivalent;
3. U.S. Class C3 assignments or allotments are considered Class B1;
4. U.S. Class C2 assignments or allotments are considered Class B; and
5. Class C1 assignments or allotments assume maximum facilities of 100 kW ERP at 300 meters HAAT. However, U.S. Class C1 stations may not, in any event, exceed the domestic U.S. limit of 100 kW ERP at 299 meters HAAT, or the equivalent.

TABLE 2.4-12
Minimum Distance Separation Requirements to TV Channel 6 from FM Stations on Channel 253 (98.5 MHz)

FM Class	Minimum Separation km (mi)	
	TV Zone I	TV Zones II and III
A	17 (11)	22 (14)
B1	19 (12)	23 (14)
B	22 (14)	26 (16)
C3	19 (12)	23 (14)
C2	22 (14)	26 (16)
C1	29 (18)	33 (21)
C	36 (22)	41 (26)

use channels 218, 219, and 220, compliance with applicable separation requirements to any allotments on higher, adjacent commercial channels is required.

Low-Power FM Stations

In addition to the provisions for noncommercial FM stations discussed above, the FCC also licenses low-power FM (LPFM) radio stations. This service was created by the Commission in January 2000 for noncommercial educational broadcasting only. LPFM stations will be authorized to operate with maximum facilities of 100 watts ERP at 30 meters HAAT. An LPFM station with a HAAT that exceeds 30 meters will not be permitted to operate with an ERP greater than that which would result in a 60-dBμ contour of 5.6 kilometers. In no event will an ERP less than one watt be authorized, and no facility will be authorized in excess of one watt ERP with a HAAT of 450 meters or greater. In addition, LPFM stations may not operate with facilities less than 50 watts ERP at 30 meters HAAT or the equivalent necessary to produce a 60-dBμ contour that extends at least 4.7 kilometers.

LPFM stations are not protected from interference that may be received from other classes of FM stations; however, LPFM stations must protect other authorized radio broadcast stations on the same channel (co-channel), as well as stations on first-, second-, or third-adjacent channels above or below the LPFM station. In addition, LPFM stations must protect other LPFM stations.

The protection is accomplished using minimum distance separation requirements that are detailed in the FCC rules, §73.807. However, the third-adjacent channel spacing requirement does not necessarily prevent the grant of an LPFM authorization but does subject the station to the provisions of FCC rules, §73.810, should interference occur. The required separation distances are detailed in Tables 2.4-13 to 2.4-17.

An LPFM station may be licensed only to (1) nonprofit educational organizations and upon a showing that the proposed station will be used for the advancement of an educational program; (2) state and local governments and nongovernment entities that will provide noncommercial public safety radio services, and (3) tribal applicants, where the tribal lands are within the service area of the proposed LPFM station. LPFM stations are not available to individuals or for commercial operations. Also, current broadcast licensees with interests in other media (broadcast or newspapers) are not eligible to obtain LPFM stations.

TABLE 2.4-13
Minimum Separation Requirements from LPFM to Other Stations

Class Relation	Minimum Separation km		
	Co-Channel	200 kHz	400 kHz or 600 kHz
LPFM	24	14	None
D	24	13	6
A	67	56	29
B1	87	74	46
B	112	97	67
C3	78	67	40
C2	91	80	53
C1	111	100	73
C0	122	111	84
C	130	120	93

TABLE 2.4-14
Additional Minimum Separation Requirements from LPFM to Other Stations in Puerto Rico and the U.S. Virgin Islands

Class Relation	Minimum Separation km		
	Co-Channel	200 kHz	400 kHz or 600 kHz
A	80	70	42
B1	95	82	53
B	138	123	92

TABLE 2.4-15
Minimum Separation Requirements from LPFM to FM Translators

Distance to FM translator 60-dBμ contour	Minimum Separation km		
	Co-Channel	200 kHz	400 kHz or 600 kHz
13.3 km or greater	39	28	21
Greater than 7.3 km, but less than 13.3 km	32	21	14
7.3 km or less	26	15	8

FM Translator and Booster Stations

In addition to the regular FM broadcast stations discussed above, two other types of stations are permitted

TABLE 2.4-16
Minimum Separation Requirements from LPFM to Other Stations in Canada

Canadian station class	Minimum Separation km				
	Co-channel	First-adjacent channel	Second-adjacent channel	Third-adjacent channel	Intermediate frequency (IF) channel
A1 & Low Power	45	30	21	20	4
A	66	50	41	40	7
B1	78	62	53	52	9
B	92	76	68	66	12
C1	113	98	89	88	19
C	124	108	99	98	28

TABLE 2.4-17
Minimum Separation Requirements from LPFM to Other Stations in Mexico

Mexican station class	Minimum Separation km				
	Co-channel	First-adjacent channel	Second-adjacent channel	Third-adjacent channel	Intermediate frequency (IF) channel
Low Power	27	17	9	9	3
A	43	32	25	25	5
AA	47	36	29	29	6
B1	67	54	45	45	8
B	91	76	66	66	11
C1	91	80	73	73	19
C	110	100	92	92	27

to operate in the FM band on a secondary basis: FM translator stations and FM booster stations. An FM booster station retransmits the signal of a primary station on the primary station's channel in order to serve areas where the primary station's signal is inadequate. An FM translator station is similar to an FM booster station, except that the signal is not retransmitted on the same channel but instead is translated to a different channel. These are authorized in accordance with Part 74, Subpart L, of the FCC rules, which, among other things, requires that such stations provide protection from interference to all regular FM broadcast stations. For more information on FM translators and boosters, see Chapter 7.19.

TELEVISION BROADCASTING: FREQUENCY ALLOCATION

Three different frequency bands are used for television broadcasting within the United States. The plan, as shown in Table 2.4-18, includes the low-VHF band TV channels 2 to 4 (54–72 MHz) as well as channels 5 and 6 (76–88 MHz), the high-VHF band channels 7 through 13 (174–216 MHz), and the UHF band channels 14 through 51 (470–698 MHz). The greater portion of all of these bands is allocated for broadcasting throughout the world, but this allocation is not uniform and, in many areas, other uses such as land mobile are permitted on a secondary basis. This is also the case in the United States, where certain UHF television channels are now used for land mobile in some major cities. In general, due to the limited extent of radiowave propagation in the television band, TV allocations, like FM, are basically a domestic matter, with few international regulations. Although no regional agreement exists, the United States does have agreements with both Canada and Mexico concerning television allocation.

TABLE 2.4-18
TV Channel Allotments

Frequency Band (MHz)	TV Channels
54–72	2–4
76–88	5–6
174–216	7–13
470–698	14–51

The amount of spectrum allocated to television broadcasting in the United States has been shrinking over the past 60-plus years. In 1953, 486 MHz of spectrum was allocated for television broadcasting.

This consisted of the VHF channels that are still in use today, along with 414 MHz of UHF spectrum (channels 14–83). Some years later, channels 70 to 83 were reallocated to other radio services and, when the television service was transitioned from analog to digital, channels 52 to 69 were also reallocated, reducing the allocation by 39.5% to 294 MHz. In addition to these reductions and as noted above, certain television channels in the range of 14–20 are shared with land mobile in a number of major cities. This reduction, coupled with the high noise levels that are present on low-VHF channels making them undesirable for digital transmission, further reduces the amount of spectrum available for television broadcasting.

As the demand or the perceived demand for spectrum for other uses has grown, the FCC has devised a plan to satisfy some of that requirement by conducting an unusual auction of broadcast spectrum. The auction, that was in progress at the time this chapter was originally written (Fall 2016), had both reverse and forward components. In the reverse auction, the FCC will purchase spectrum from existing broadcast license holders, and then offer the purchased spectrum in a forward auction where the intended use is wireless broadband services. Participation was voluntary on the part of broadcasters and was open to full-service and Class A stations only. Low-power television (LPTV) stations and translator stations were neither eligible to participate in the auction nor protected and therefore could be displaced without compensation. Those eligible stations had three participation options: (1) give up their channel, (2) if currently on a UHF channel, move to the VHF bands, or (3) agree to share a channel with another eligible station.

Stations that did not wish to participate or were not needed to satisfy the spectrum-clearing target were subject to being "repacked" to a new channel. The repacked stations cannot be forced to change bands, and the FCC is required to make a reasonable effort to not reduce their current service. The result of this auction, completed in 2017, modified the U.S. television band, such that the highest available television channel would be channel 36. (Channel 2 through 36 are now the only channels available for U.S. television broadcasting; channel 37 remains reserved for radio astronomy.)

Like FM broadcasting, TV allocations in the United States are governed by a Table of Frequency Allotments. The current allotment table can be found in §73.622(i) and is titled Post-Transition Table of DTV Allotments. This current DTV allotment table was mainly developed during the transition from analog to digital through a complex analysis that took into consideration actual computations of coverage and interference. The table contains all commercial as well as noncommercial allotments (the latter are identified by an asterisk). Please note that the FCC rules also currently contain the original DTV allotment table in §73.622(a). The table in §73.622(a) was the table established when analog stations were still in operation. It is should further be noted that although the use of analog by full-service television stations is no longer permitted, the FCC rules also still contain the analog table of allotments in §73.606.

The DTV allotment table can be amended to change an allotted channel or add a new allotment using the following procedure. A petition for rulemaking may be filed with the FCC seeking to modify the table to change or add a new allotment. Such petitions must provide a showing that the proposal complies with the applicable separation criteria, shown in Table 2.4-19, that serve to prevent mutual interference. You will note that there are different separation distances listed in Table 2.4-19 depending on the zone location of the station. The different zones are shown in Figure 2.4-4.

Zone I is the northeast United States extending south to the Virginia–North Carolina border and west to the Mississippi River. Zone II consists of that portion of the United States that is not in Zone I or Zone III, including Alaska, Hawaii, Puerto Rico, and the Virgin Islands. Zone III is that portion of the southeast United States extending from the east coast of Georgia westward to the Mexican border. An exact description of the zones is contained in §73.609 of the FCC rules. As with FM, the TV zones reflect the differing population densities in various parts of the country. In addition, differences in propagation conditions were also considered. The closer spacings in Zone I recognize the fact that, in the northeastern portion of the United States, many large population centers require stations of their own and are close enough to one another to lessen the need for wide area service. Zone II is characterized by fewer population centers, usually smaller and farther apart; for them, wide area service is a necessity. Finally, because the area of the country along the Gulf Coast (Zone III) is susceptible to high levels of tropospheric propagation, stations in that area must be spaced farther apart to

TABLE 2.4-19
Separation Requirements for New DTV Allotments

		Required Separation Distance (km)	
Band	**Channel Relationship**	**Zone I**	**Zones II & III**
VHF (Channels 2–13)	Co-channel	244.6	273.6
	Adjacent Channel	< 20.0 or > 110.0	< 23.0 or > 110.0
UHF	Co-channel	196.3	223.7
	Adjacent Channel	< 24.0 or > 110.0	< 24.0 or > 110.0

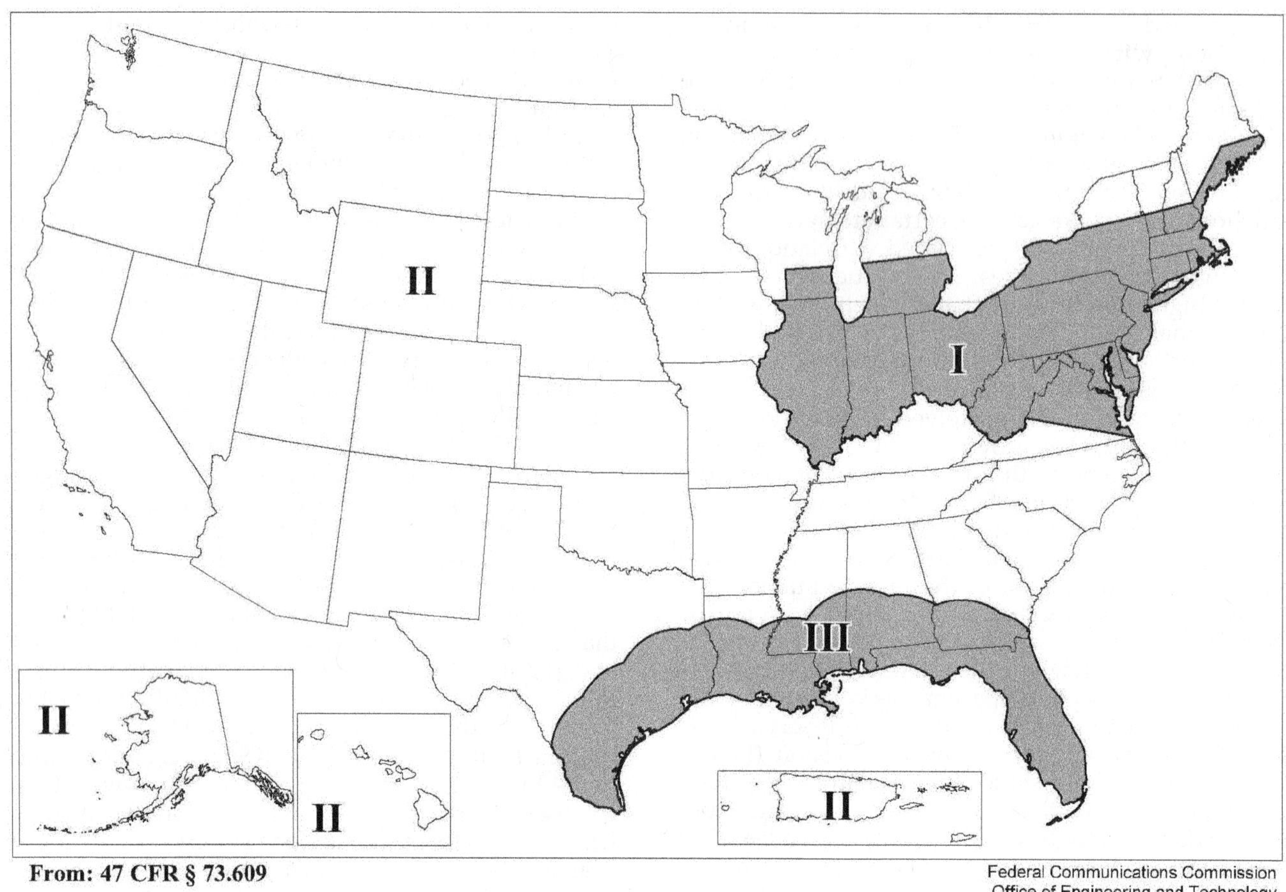

FIGURE 2.4-4 TV broadcast zones; Zone II is all of the United States not in Zones I or III.

minimize interference. Although there are differences in the powers authorized for low-VHF, high-VHF, and UHF stations, unlike AM and FM broadcasting there are no class designations as such in television. The differing power limitations reflect the differences in signal propagation for low-VHF, high-VHF, and UHF.

In addition to complying with the spacing requirements of Table 2.4-19, applications to construct a station on a new allotment must demonstrate that the proposed facility will provide applicable coverage to the city of license and that no new interference would be caused to existing stations, either analog or digital. The required coverage and interference analyses are carried out using the methodology described in the FCC Office of Engineering and Technology Bulletin No. 69 (OET-69) entitled "Longley-Rice Methodology for Evaluating TV Coverage and Interference" and dated February 6, 2004. This methodology is also used to study proposed changes to existing stations that are not proposing a channel change.

Low-Power Television Stations, Translators, and Boosters

In addition to the regular full-service television broadcast stations, three other types of analog and digital stations are permitted to operate in the television bands with low power on a secondary basis: *low-power television (LPTV) stations*, *TV translator stations*, and *TV booster stations*. A TV booster station retransmits the signal of a primary station on the primary station's channel in order to serve areas where the primary station's signal is inadequate. A TV translator station is similar to a TV booster station, except that the signal is not retransmitted on the same channel but instead is translated to a different channel. LPTV stations may retransmit the signals of another station or they may originate programming. All of these stations are authorized in accordance with Part 74, Subpart G, of the FCC rules that requires full protection from interference to all existing television broadcast stations, including full-service and other LPTV stations.

These stations are not accorded any protection from the operation of full-service television broadcast stations. As part of the transition to an all-digital television system, the FCC in September 2004 established rules to allow LPTV stations, TV translator stations, and TV booster stations to make the digital transition. These rules allow existing analog LPTV stations to file an application at any time to "flash cut" to digital. The rules also established interference protection criteria that will be used to assess these applications, which are much like those used for regular DTV stations. In addition, these rules also set up criteria for allowing digital companion channels for LPTV stations similar to those established for full-service television stations. Applications for the companion channels were accepted during a filing window that closed on June 30, 2006. Like the current low-power stations, the low-power digital stations are authorized on a secondary basis and are not afforded protection from full-service stations. It should be noted that all LPTV stations will eventually be required to transition to digital or cease operation; however, because of the ongoing spectrum auction and their unknown future status, the deadline for transitioning to digital has been delayed until the auction has been completed.

Class A TV Stations

In March 2000, the FCC established a new class of television stations, designated as Class A stations. These are former LPTV stations that have been granted primary status based on meeting specific programming and operational requirements. To maintain their Class A status, these stations must broadcast a minimum of 18 hours per day and broadcast an average of 3 hours per week of locally produced programming each quarter. Due to their primary status, these stations are afforded protection from increased interference from all other television broadcast stations. The rules that established Class A stations also allow these stations to "flash cut" to digital at any time if they provide interference protection to all other primary and secondary stations. These stations also will need to transition to digital at the conclusion of the auction.

AUXILIARY BROADCAST SERVICES: FREQUENCY ALLOCATION

Although all auxiliary broadcast services share a common role in support of AM, FM, or TV broadcast operations, there are important differences between them. The nature of the service they provide varies, as does the frequency band in which they operate. For these reasons, the allotments available for each auxiliary service are discussed separately; however, before turning to a specific discussion of each service, some general comments are necessary. The steady growth in the number of AM, FM, and TV stations, including DTV, as well as their desire to use more advanced technology in every aspect of their operation, has greatly increased demand for spectrum in all auxiliary services. Because of the continuing demand for spectrum from a multitude of other broadcast and nonbroadcast activities, it has been difficult for the FCC to allocate more spectrum to alleviate the congestion faced by the auxiliary services.

At the same time, some relief has come through changes in the FCC rules to facilitate the use of newer, more spectrum-efficient technology. When reading the material in this section, the reader should be aware that additional changes may take place and should check the current regulations governing that particular radio service. Recognizing that available spectrum is limited, broadcasters need to use it efficiently (see Chapter 2.3 in this handbook, which deals with frequency coordination for broadcast auxiliary services). To do this, it is first necessary to understand what spectrum is allocated, how it may be used, and what advantages or disadvantages may be involved in the use of a particular band for each of the auxiliary services discussed below. It should also be noted that due to the overall demand for spectrum, it is becoming necessary for some of the broadcast auxiliary services to share spectrum with other services, including those used by the U.S. government agencies.

Remote Pickup Broadcast Stations

Remote pickup (RPU) stations are mobile or portable facilities used to transmit live on-the-air programming from a temporary remote location, such as a shopping center or football game, to the station's studio facilities. This material can be taped for later rebroadcast, or it can be incorporated into actual ongoing, live broadcasts. Radio stations typically have several RPUs that may be licensed to one or more frequencies.

In November 1984, the FCC significantly revised its radio broadcast auxiliary frequency allocations to permit operational use of narrowband technologies in the Broadcast Remote Pickup Service. The goal was to foster spectrum efficiency in a flexible manner. Broadcasters and equipment manufacturers who wanted to operate narrowband equipment, the Commission believed, should not be precluded from doing so by rigid FCC rules; however, the FCC never formally implemented these changes due to delays in setting up the mechanism to handle the expected applications. In the meantime, the FCC accepted applications based on this revised allocation. Then, in November of 2002, the FCC once again revised the broadcast remote pickup rules, which are now in effect and are detailed below.

Frequency Allocations for Radio Broadcast Remote Pickup Stations

- 25.87–26.47 MHz—There are 26 frequencies available in this band for use by remote pickup broadcast stations. Bandwidth is limited to 20 kHz except between 25.87 and 26.03 MHz, where 40 kHz is permitted. Note that the use of the frequencies between 25.87 and 26.09 MHz are subject to the condition that no harmful interference is caused to broadcast stations sharing this band. Note also that the frequencies between 26.100 and 26.175 MHz have been allocated on a worldwide basis to the Maritime Mobile Services and are shared with the Maritime Service in the United States. Note further that the frequencies between 26.175 and 26.47 MHz are allocated on a worldwide basis for use by various types of fixed and mobile operations. Hence, when selecting a frequency in this band for remote pickup use, the implications of the above should be taken into consideration.

- 152.8625–153.3575 MHz—There are 67 frequencies available in this band for use by remote pickup broadcast stations. Each channel is 7.5 kHz wide, and channels may be stacked to form a single channel with a maximum bandwidth of 30 kHz. This band is shared with the Private Land Mobile Radio Service, and operation of remote pickup stations is subject to the condition that no harmful interference is caused to these other services. Note that these frequencies are not available to network entities or for use on board aircraft.
- 160.860–161.400 MHz—There are 73 frequencies available in this band for use by remote pickup broadcast stations in Puerto Rico or the Virgin Islands, where they are shared with the Public Safety and Industrial/Business Pool. Each channel is 7.5 kHz wide, and channels may be stacked to form a single channel with a maximum bandwidth of 30 kHz.
- 161.625–161.775 MHz—There are 21 frequencies available in this band for use by remote pickup broadcast stations. Each channel is 7.5 kHz wide, and channels may be stacked to form a single channel with a maximum bandwidth of 30 kHz. These frequencies are not available to network entities and are not available for use in Puerto Rico or the Virgin Islands. Also, Public Safety and Industrial/ Business Pool stations may continue to operate on these frequencies on a noninterference basis.
- 166.25 and 170.15 MHz—These frequencies may be used by remote pickup stations with a maximum bandwidth of 12.5 kHz; however, the area in which they may be used is restricted. A description of the area in which they may be used is found in the FCC rules, §74.402(e)(8).
- 450.00625, 450.0125, 450.01875, 450.025, 450.98125, 450.9875, 450.99375, 455.00625, 455.0125, 455.01875, 455.025, 455.98125, 455.9875, and 455.99375 MHz—These frequencies may be used by remote pickup stations only for the transmission of operational communications, including tones for signaling and for remote control and automatic transmission system control and telemetry. They may be stacked for a maximum bandwidth of 12.5 kHz and comply with the emission and bandwidth restrictions detailed in FCC rules §74.462. In addition to the noted restrictions, the above groups of VHF and UHF frequencies are also subject to the following restrictions; licensed stations or those with applications filed prior to April 16, 2003, must have complied with the above channel plan by March 17, 2006, or, alternatively, may continue to operate on a secondary, noninterference basis.
- 450.03125–455.61875 MHz—There are 190 frequencies available in this band for use by remote pickup broadcast stations. Each channel is 6.25 kHz wide, and channels may be stacked to form a single channel with a maximum bandwidth of 50 kHz.
- 450.6375–455.8625 MHz—There are 20 frequencies available in this band for use by remote pickup broadcast stations. Each channel is 25 kHz wide, and channels may be stacked to form a single channel with a maximum bandwidth of 50 kHz. Users committed to 50 kHz bandwidths to transmit program material will have primary use of these channels.
- 450.900, 450.950, 455.900, and 455.950 MHz—These frequencies are available for use by remote pickup broadcast stations. Each channel is 50 kHz wide, and channels may be stacked to form a single channel with a maximum bandwidth of 100 kHz. Users committed to 100 kHz bandwidths to transmit program material will have primary use of these channels.

Aural Broadcast Auxiliary Stations

Aural broadcast auxiliary stations include studio-to-transmitter link (STL), intercity relay (ICR), and microwave booster stations used by radio broadcast stations. STL stations are fixed stations used for transmitting program material between the studio and the transmitter of a broadcasting station. ICR stations are fixed stations used for the transmission of program material between broadcasting stations, except international, for simultaneous or delayed broadcast. ICRs may also be used on a secondary basis to transmit program material from FM stations to co-owned FM translator or booster stations. Microwave booster stations are used to relay the signals of an STL or ICR station over a path that cannot be covered with a single station. They receive and transmit on the same frequency. One or more microwave booster stations may be authorized to licensees of STLs or ICRs. Note that stations in the aural broadcast auxiliary service may be authorized on a secondary noninterference basis to licensees of TV broadcast stations to transmit aural material. The following frequencies are available for assignment to STL, ICR, and microwave booster stations:

- 942.5, 943.0, 943.5, and 944 MHz—These frequencies are available for use in Puerto Rico and the Virgin Islands. Also, stations licensed in other parts of the United States prior to November 21, 1984, may continue to operate on a co-equal primary basis in this band.
- 944–952 MHz—There are 320 channels (25-kHz wide) available in this band. The channels may be stacked to form a single channel up to 300-kHz wide. Separately, stations also may be authorized additional 25-kHz wide channels up to a grand total of 20 channels. The use of these frequencies by ICR stations is subject to the condition that no harmful interference is caused to other classes of stations.
- 18,760–18,820 and 19,100–19,160 MHz—There were 24 channels (5-MHz wide) available in these bands. These frequencies were shared on a co-primary basis with other fixed services, and their use is subject to the rigorous coordination requirements of the FCC rules, §21.100(d). No new applications are being accepted for this band, and

it was scheduled to be relocated by June 8, 2010. After that date, stations may continue to use this band but will lost their co-primary status, will not be protected from interference from other users with primary status, and must not cause harmful interference to services with primary status.

Television Broadcast Auxiliary Stations

The demand for the spectrum allocated for television auxiliary services is significantly greater than for spectrum allocated for radio broadcasting services. In addition to the extensive local demand, network remote units travel extensively and compete with local broadcasters for available frequencies, resulting in increased spectrum congestion. It is important to recognize that the variety of activities undertaken by television broadcasters usually requires more complex auxiliary systems than is the case in the radio industry. The following are the types of television broadcast auxiliary stations:

- TV pickup stations are land mobile stations used for the transmission of TV program material and related communications from scenes of events occurring at points removed from TV station studios to a TV broadcast, Class A TV, or low-power TV station.
- TV STL station (studio-transmitter link) are fixed stations used for the transmission of TV program material and related communications from the studio to the transmitter of a TV broadcast, Class A TV, or low-power TV station.
- TV relay stations are fixed stations used for transmission of TV program material and related communications for use by TV broadcast, Class A TV, and low-power TV stations.
- TV translator relay stations are fixed stations used for relaying programs and signals of TV broadcast stations or Class A TV stations to Class A, LPTV, or TV translator stations.
- TV microwave booster stations are fixed stations used to receive and amplify signals of TV pickup, TV STL, TV relay, or TV translator relay stations and retransmit them on the same frequency. These stations are used to transmit signals over a path that cannot be covered by a single transmitter.

In addition to the uses described above, stations in this service may also be used on a temporary or in some cases on a regular basis for other purposes. Please refer to §74.631 for more details on the other permitted uses and the allowed duration.

The following bands are available for assignment to TV pickup stations:

- 2025–2110 MHz—Seven 12-MHz wide channels are available in this band, as well as 40 0.25-kHz wide channels to be used as data return links (DRLs). The DRL frequencies are equally divided at either end of the band (2025.0 to 2025.5 and 2109.5 to 2100.0). These frequencies are also available for assignment to all the other types of television broadcast auxiliary stations (see below).
- 2450–2483.5 MHz—Two channels are available in this band, which is also available for assignment to all the other types of television broadcast auxiliary stations. This band is shared with industrial, scientific, and medical (ISM) devices and is not afforded any protection from these ISM devices.
- 6425–6525 MHz—The channels available for mobile use only in this band are co-equally shared with mobile stations licensed under Parts 78 and 101 of the FCC rules. The available channel bandwidth varies from 1 to 25 MHz. Section 74.602(i) of the FCC rules contains further explanation concerning the usage of this band.
- 6875–7125 MHz—Ten channels, each 25-MHz wide, are available in this band, which is also available for assignment to all the other types of television broadcast auxiliary stations.
- 12700–13250 MHz—This band contains 43 channels; however, the channels overlap. If use of this band is contemplated, then §74.602 of the FCC rules should be consulted for a more complete understanding of their usage. This band is also available for assignment to all the other types of television broadcast auxiliary station.

The following are available for assignment to TV STL, TV relay, or TV translator relay stations:

- 2025–2110 MHz—Seven channels are available in this band, which is also available for assignment to TV pickup.
- 2450–2483.5 MHz—Two channels are available in this band, which is also available for assignment to TV pickup stations.
- 6875–7125 MHz—Ten channels, each 25-MHz wide, are available in this band, which is also available for assignment to TV pickup stations.
- 12,700–13,250 MHz—This band contains 43 channels; however, the channels overlap. If use of this band is contemplated, then §74.602 of the FCC rules should be consulted for a more complete understanding of their usage. This band is also available for assignment to TV pickup stations. In addition, the channels between 13,150 and 13,200 MHz are not available within 50 km of the top 100 markets.
- 17,700–19,700 MHz—Stations operating on frequencies in the sub-bands 18.3–18.58 GHz and 19.26–19.3 GHz that were licensed or had applications pending before the Commission as of September 18, 1998, may continue those operations on a shared co-primary basis with other services under parts 21, 25, 78, and 101 of this chapter. Such stations, however, are subject to relocation by licensees in the fixed-satellite service. Such relocation is subject to the provisions of §101.85 through 101.97. No new applications for new licenses will be accepted in the 19.26–19.3-GHz band after June 8, 2000, and no new applications for new licenses will be accepted in the 18.3–18.58-GHz band after November 19, 2002.

In addition to the above frequencies, TV STL and TV relay stations also may be authorized to use UHF-TV channels 14 to 51 on a secondary basis provided no interference is caused to TV, Class A, and LPTV stations operating in this band. Furthermore, the aural portion of television broadcast program material may be transmitted over an aural broadcast STL or ICR station on a secondary, noninterference basis. Likewise, remote pickup stations may be used to transmit the aural portion of television program material.

SUMMARY

Frequency allocation is a complex matter and is subject to frequent changes. The reader is advised to consult a current version of the FCC rules and regulations for a complete description of the current allocation situation, including the procedures and policies being applied by the FCC concerning a particular band.

Notes

[1] Technically the term "allocation" refers to the process by which a frequency band is made available for a specific purpose; however, it is used here in a broader sense that includes the *allotment* of frequencies within a band and the *assignment* of individual stations.

[2] In a reorganization of the ITU, the work of developing recommendations and providing reports dealing with technical issues formerly performed by the International Radio Consultative Committee (CCIR), with which some readers may be familiar, is now performed by the Radiocommunications Assemblies.

[3] The rules pertaining to PSSAs and PSRAs are contained in §73.99 of the FCC Rules and Regulations (47 CFR §73.99).

[4] The concept of allowing different classes of stations in different areas or zones is based on the population density of the areas. The Commission has assumed that there is less need for wide areas of service in areas of dense population.

CHAPTER

2.5

Analog and Digital Terrestrial Radio Standards

STANLEY SALEK, P.E.

Hammett & Edison, Inc. Consulting Engineers
Sonoma, CA

LAWRIE HALLETT, PhD

Senior Lecturer in Radio and Technical Director,
Future Digital Norfolk Limited
Norwich, Norfolk, UK

PART 1: U.S. DOMESTIC NRSC STANDARDS AND GUIDELINES

Introduction

The National Radio Systems Committee (NRSC)[1] is jointly sponsored by the National Association of Broadcasters (NAB)[2] and the Consumer Technology Association (CTA).[3] Its purpose is to study and make recommendations for technical standards that relate to radio broadcasting and the reception of radio broadcast signals. The NRSC is a vehicle by which broadcasters and receiver manufacturers work together toward solutions to common issues and problems in radio broadcast systems.

Since beginning its active standards setting activities in the mid-1980s, the NRSC has produced a number of voluntary industry standards and related guidelines:

- Two standards are related to the transmission and reception of analog AM radio signals in the medium-wave broadcast band (535 to 1705 kHz), with two companion guideline documents;
- One documents the U.S. implementation of the Radio Data System (RDS) used in the FM broadcast band (88 to 108 MHz), and includes two companion guideline documents;
- One covers in-band, on-channel (IBOC) digital radio broadcasting, used in both bands, and includes three companion guideline documents.

These standards and guidelines, along with supporting reports and other related documents, are available at no charge on the NRSC website (www.nrscstandards.org).

AM Radio Broadcasting Standards

The two NRSC AM radio broadcasting standards, first released in 1987 and 1988, cover recommended broadcast audio parameters and transmission systems. In the late 1980s, the Federal Communications Commission (FCC) incorporated portions of these standards into its Part 73 broadcast rules,[4] effectively making them mandatory for U.S. broadcasters. A third standard covering receiver performance was released in 1990, but was subsequently retired in 2007. Two related NRSC guidelines cover analog RF bandwidth considerations and implementation of power-saving modulation-dependent carrier level (MDCL) technology.

NRSC-1

The "NRSC AM Preemphasis/Deemphasis and Broadcast Audio Transmission Bandwidth Specifications" (NRSC-1), first adopted on January 10, 1987, specifies the preemphasis audio characteristic of AM broadcasts, the complementary deemphasis characteristic contained in AM broadcast receivers, and a 10-kHz audio bandwidth limitation prior to modulation. The Standard applies to both monophonic and the (L+R) component of AM stereophonic transmissions, as well as to both single- and multiple-bandwidth receivers. The purpose of the Standard is to create a transmission and reception system that allows AM broadcast stations to know, with a degree of certainty, the audio response characteristics of receivers, and receiver manufacturers to know, with a similar degree of certainty, the likely audio response characteristics of AM broadcasts. The Standard was revised and reaffirmed in 2007 (NRSC-1-A) and 2012 (NRSC-1-B); related changes incorporated compatibility provisions of hybrid IBOC digital broadcasting and portions of receiver evaluation techniques of the retired NRSC-3 Standard.

Audio preemphasis is incorporated at the transmission end to boost high audio frequencies. The characteristic curve shown in Figure 2.5-1 is known as "modified 75-μsec" preemphasis, because its mathematical description (or transfer function) consists of a single zero at 2122 Hz, constituting 75-μsec preemphasis (as used in analog FM broadcasting), followed by a single pole at 8700 Hz, which limits the maximum boost to 10 dB at 10 kHz.

In the receiver, a complementary deemphasis curve is specified, employing a maximum of 10 dB attenuation at 10 kHz, theoretically resulting in a flat end-to-end response; however, most AM receivers recover much less than 10-kHz audio bandwidth, so the overall receiver response must be considered to determine the degree of deemphasis, if any, to employ that will best complement the preemphasis characteristic. For receivers that do have usable response out to 10 kHz, NRSC-1 recommends that notch filters be employed to attenuate the 10-kHz beat ("whistle") caused by first-adjacent-channel AM stations. Figure 2.5-2 illustrates the 10-kHz audio input spectrum limitation to the AM transmitter. The slopes and step in the response are included to accommodate the capabilities of practical analog low-pass filters and limiters available at the time of standard adoption. The 2007 revision (NRSC-1-A) added analog audio bandwidth options of 5, 8, and 9.4 kHz, for implementation in conjunction with hybrid IBOC digital transmission systems.

NRSC-1 also includes sections on test signals and methods for determining compliance with the Standard. Although normal program material can be used in conjunction with an audio spectrum analyzer or fast Fourier transform (FFT) analyzer operating in peak-hold mode, a standard test signal, known as pulsed-USASI noise, is also defined. Use of the standard test signal allows for easier comparison of compliant equipment from different manufacturers.

NRSC-2

The "NRSC Emission Limitation for AM Broadcast Transmission" (NRSC-2) became effective on June 1, 1988. It builds upon the NRSC-1 transmitter audio input specifications, taking into account the AM transmission process, to provide for a maximum occupied bandwidth limit, as shown in Figure 2.5-3. The steps and slopes in the emission mask were designed to accommodate the capabilities of transmitters in existence at the time of Standard adoption. The solid-line border represents the maximum limits of the Standard, and the dashed-line border represents test limits, which are intended to serve as a guide for transmitter manufacturers using the NRSC-1 standard test signal. While the ultimate attenuation is 80-dB or greater below carrier reference level beyond ±75 kHz from the carrier frequency, a relaxation allowing lesser attenuation is incorporated for stations employing less than 5-kW carrier power.

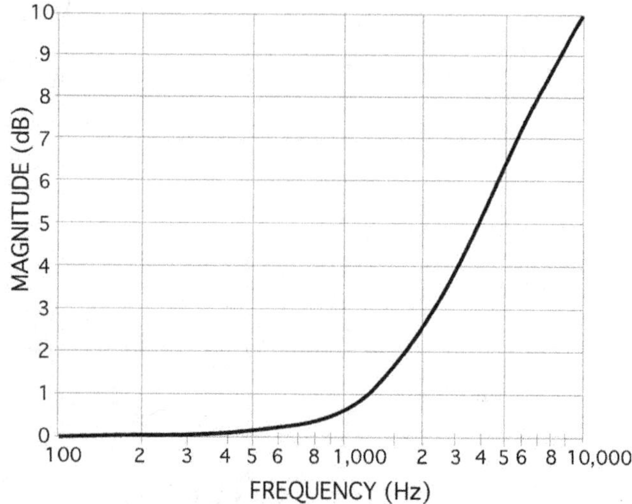

FIGURE 2.5-1 Modified 75 μsec AM standard preemphasis curve of NRSC-1

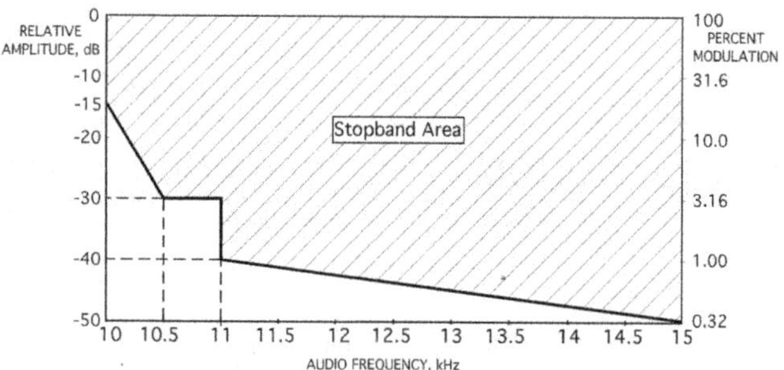

FIGURE 2.5-2 Audio stopband specification of NRSC-1

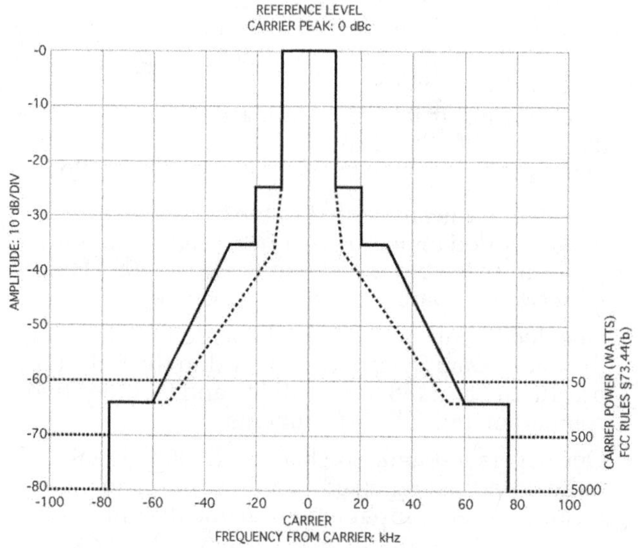

FIGURE 2.5-3 AM broadcast radiofrequency emission limits of NRSC-2

Field measurement of mask compliance is made using a radiofrequency (RF) spectrum analyzer or other device that can effectively monitor emissions exceeding the RF mask. To accomplish this task, the RF spectrum analyzer should be set for 300-Hz resolution bandwidth in peak-hold mode for 10 minutes or more, referenced to the peak carrier level. Using normal program modulation, the system is in compliance if all stored spectrum products are contained within the mask area.

Only minor revisions have been made to the NRSC-2 Standard since its initial adoption. The 2007 revision (NRSC-2-A) harmonized the passband cutoff frequency (from 10 to 10.2 kHz) with the FCC occupied bandwidth requirements of FCC rules §73.44(b); this Standard was also reaffirmed in 2012 (NRSC-2-B).

NRSC-3 (Retired)

While this Standard has been retired, a summary is provided for historical purposes.

Known by the title "NRSC Audio Bandwidth and Distortion Recommendations for AM Broadcast Receivers," NRSC-3 became effective on October 15, 1990. The purpose of the Standard was to provide specific performance goals for AM broadcast receivers, building upon the receiver elements of NRSC-1. Specified parameters included limits for frequency response and total harmonic distortion plus noise, along with incorporation of sufficient attenuation of the 10-kHz "beat" frequency in wideband receivers. The Standard also specified methods for coupling of test signals into receivers under evaluation, connection of audio signals from receivers to test equipment, AM modulation levels used during testing, and RF test frequencies.

In 1992, the NRSC sponsoring organizations introduced AMAX, a certification program for AM radio receivers that met the technical specifications of NRSC-3 and that also exhibited other desirable characteristics, including adjustable reception bandwidth and the availability of an accessible external antenna connection on nonautomotive receivers. While some receiver manufacturers initially implemented elements of the program, by 2007, when NRSC-3 was being considered for reaffirmation, continued interest in the Standard was no longer perceived to be present, and NRSC-3 was retired at that time. Some elements of test procedures employed in NRSC-3 were transferred to NRSC-1-A, which was reaffirmed during that same year.

NRSC-G100

In September 2007, the NRSC published its first guideline, NRSC-G100, "Bandwidth Options for Analog AM Broadcasters," which was reaffirmed in 2012 as NRSC-G100-A. This rather extensive document describes the outcome of studies, both objective and subjective, designed to evaluate practical perceptions of transmitted RF bandwidths of AM analog signals, while considering the performance of receivers typically available to the public.

At the time preceding release of the Guideline, some broadcasters were implementing or considering implementation of transmitted RF bandwidths less than the ±10 kHz specified in the NRSC-2 standard. The core concern was that since AM broadcast stations in the United States are allocated on frequencies in 10-kHz increments, the implementation of a ±5-kHz transmitted RF bandwidth would significantly reduce first-adjacent-channel interference between stations, especially during nighttime hours when skywave propagation is active. Further, it was widely believed that such a reduction in transmitted bandwidth would not be noticed on the vast majority of AM receivers in regular use, as their recovered audio bandwidths were believed to be well under 5 kHz.

In 2005, the NRSC, in conjunction with NPR Labs, undertook studies to characterize a sampling of AM receivers available at the time, and also to conduct subjective studies to determine listener perception of varying transmitted RF bandwidths between ±5 and ±10 kHz, both without and with inclusion of the effects of first-adjacent-channel interference. The graph of Figure 2.5-4 illustrates the results of objective laboratory receiver bandwidth evaluation measurements, presented as a composite average and standard deviation for all tested receivers. An inset table also presents −3-dB and −10-dB bandwidths of receivers in various categories. The average −3-dB bandwidth of 2.45 kHz and −10-dB bandwidth of 4.1 kHz were surprising, albeit not unexpected.

Laboratory tests of first-adjacent-channel interference effects under varying desired-to-undesired (D/U) ratios were also performed with respect to transmission bandwidth conditions between ±5 and ±10 kHz, and the degrading effects were measured on sample receivers. Not unexpectedly, it was found that the narrower transmission bandwidths could improve measured interference conditions by up to 12 dB.

Finally, subjective listening tests were performed with a panel of volunteer participants to determine perceptions of varying first-adjacent-channel interference levels and receiver bandwidth on reception quality. In general, listeners preferred lower receiver bandwidths

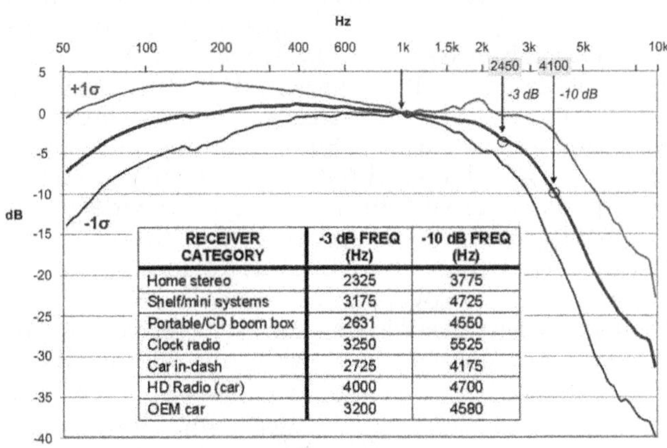

RECEIVER CATEGORY	-3 dB FREQ (Hz)	-10 dB FREQ (Hz)
Home stereo	2325	3775
Shelf/mini systems	3175	4725
Portable/CD boom box	2631	4550
Clock radio	3250	5525
Car in-dash	2725	4175
HD Radio (car)	4000	4700
OEM car	3200	4580

FIGURE 2.5-4 AM receiver frequency response mean and standard deviation for all measured receivers

when evaluating talk programming, as the effects of interference were more noticeable than for music programming. However, even for music programming along with minimal or no adjacent channel interference added, listeners often preferred 5- or 7-kHz audio bandwidth over the full 10 kHz.

The Guideline includes a discussion of analog audio bandwidth in conjunction with the use of hybrid IBOC digital radio broadcasting. While implementation of AM hybrid IBOC broadcasting offers analog audio bandwidths of 5, 8, or 9.4 kHz, depending on selected IBOC transmission mode, the lower analog audio bandwidths allow a higher fidelity and more robust digital signal to be transmitted.

The Guideline concludes that transmitted audio bandwidths of less than 10 kHz could be desirable for analog AM broadcasters due to (1) listener preference of lower audio bandwidth, (2) reduction of adjacent channel interference, and (3) potential of improved signal-to-noise ratio of the transmitted audio if the channel energy is concentrated in a bandwidth comparable to the capabilities of available receivers.

NRSC-G101

Originally released in April 2013 and updated three years later (NRSC-G101-A), the NRSC published NRSC-G101, "AM Modulation Dependent Carrier Level (MDCL) Usage Guideline." While not directly related to an NRSC standard, this Guideline presents information for AM broadcasters interested in implementing electrical power-saving technology, for which authorization procedures were first established for U.S. AM broadcasters by the FCC in September 2011.

For standard double-sideband AM amplitude modulation, when using MDCL the carrier level remains constant while the sideband energy varies in conjunction with the modulating signal. Incorporation of MDCL generally involves use of an algorithm that reduces the carrier and/or sideband power as a function of the modulating signal. Since most of the power of an AM broadcast signal is contained in the carrier, incorporation of MDCL can lead to significant electrical cost savings; while these techniques can theoretically lead to recovered audio degradation in some receivers, the history of implementation by international high-power medium-wave and shortwave broadcasters in the 1980s indicated that the tradeoffs were workable when the technology was employed with modern transmitters. Since that time, implementation of MDCL has essentially become standard operating practice for such broadcasters, with widespread availability of necessary conversion hardware, with some transmitters having MDCL as a standard feature.

The Guideline covers an overview of MDCL algorithms that have been in general use, expected power savings, and compatibility with AM hybrid IBOC digital radio broadcasting. Two MDCL algorithms are covered in some detail, called amplitude modulation companding (AMC) and dynamic carrier system (DCS). Expected power savings is covered in a table, broken down by MDCL system/settings and transmitted audio source material. To address IBOC compatibility, testing was performed in 2015 by NPR Labs to evaluate differences between static MDCL, where the IBOC carriers remain constant, and dynamic MDCL, where the amplitude of the IBOC carriers is varied by the MDCL system in the same way it affects the AM carrier level. The testing results showed no appreciable effect on static carrier IBOC performance, but some degradation of received digital radio performance for IBOC carriers that are dynamically varied by the MDCL algorithm. The full laboratory study report is included as an appendix to the Guideline.

The Guideline also includes discussions on transmitter linearity considerations and the lack of observed effects of MDCL implementation on audio "watermark" encoders used for audience measurement.

NRSC-4 (U.S. RBDS Standard)

The NRSC "United States Radio Broadcast Data System (RBDS) Standard" (NRSC-4) was released by the NRSC's RBDS Subcommittee in January 1993. It was updated in 2005 (NRSC-4-A, redesignated from "RBDS Standard" using the updated NRSC standard naming convention) and most recently in 2011 (NRSC-4-B). The Standard describes usage of the Radio Data System (RDS) in the United States and is originally based on and essentially compatible with the European Broadcast Union (EBU)/Cenelec Standard EN50067:1998 (Specifications of the Radio Data System).[5] Note that although the U.S. version is the RBDS Standard, all hardware implementations of this technology are referred to as RDS, and the logos that appear on consumer equipment are RDS logos. See Chapter 7.16 of this handbook for a more detailed description of RDS.

A primary objective of the RDS system is to improve the functionality and user-friendliness of FM broadcast receivers through an added transmitted data stream. That data stream allows a compatible receiver to provide a visual display of the program service (station call sign or identifying name), potentially including a real-time updated display of the current program, song title and artist, or other information being broadcast. RDS also includes a switching capability to allow a receiver to tune to the strongest signal offering the same program; this function is most commonly used in conjunction with nationalized networks outside the United States. A goal of the most recent release of the Standard, NRSC-4-B, was to streamline its content and to highlight differences between the U.S. and European implementations.

Another objective of the NRSC-4-B standard is to provide for harmonization with similar program data streams in IBOC digital radio, such that digital–analog and analog–digital blends will maintain the appropriate receiver display and functionality. This and several other aspects of the Standard are covered in guideline document NRSC-G200, "Harmonization of RDS and IBOC Program Service Data (PSD)," described below.

RDS information is transmitted on a 57-kHz subcarrier contained within the transmitted FM baseband that also carries stereophonic and possibly other

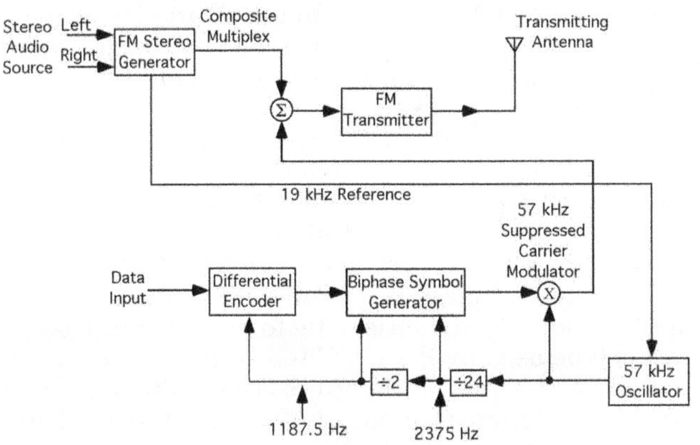

FIGURE 2.5-5 Simplified block diagram of RDS encoder subsystem

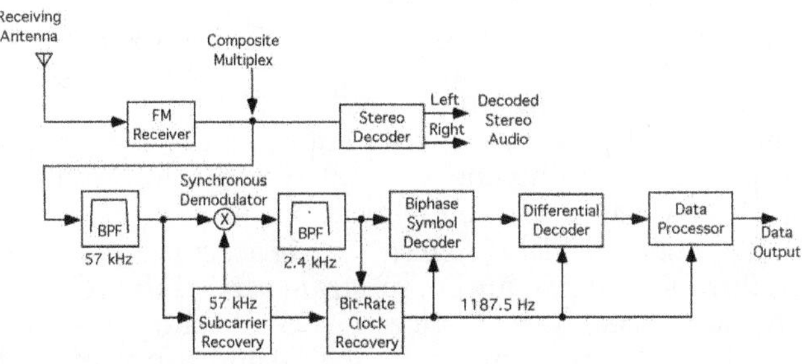

FIGURE 2.5-6 Simplified block diagram of RDS decoder subsystem

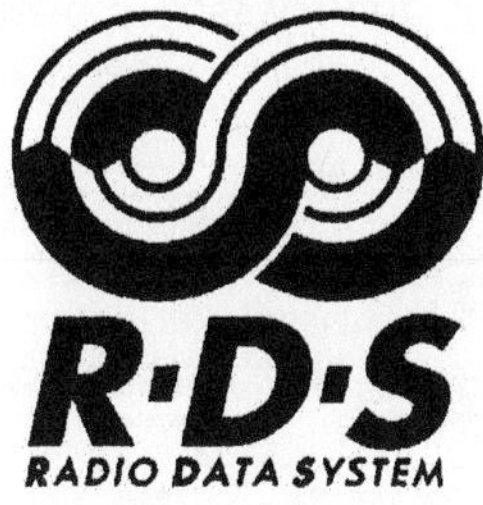

FIGURE 2.5-7 RDS logos for use on RDS- and RBDS-compatible equipment

subcarriers. The RDS subcarrier is locked to the 19-kHz stereo pilot tone and uses biphase suppressed-carrier modulation at 1187.5 bits per second. Figure 2.5-5 is a simplified block diagram of an RDS encoder, and Figure 2.5-6 provides the complementary block diagram of an RDS decoder.

Prior to modulation, the RDS data stream is formatted in accordance with a specified baseband coding structure. The largest element in the structure is called a *group*, which contains 104 bits of information. Each group is further broken down into *blocks* of 26 bits each, consisting of a 16-bit data word and a 10-bit "check word." Data packets such as program information (PI) and program type (PTY) codes, as well as real-time clock updates and other information such as potential emergency information, are formatted into the data stream. Encoders and receivers built to the NRSC-4 standard may display the RDS logo, two variations of which are shown in Figure 2.5-7.

NRSC-G300

First published in September 2012, NRSC-G300, "Radio Data System Usage Guideline," emphasizes practical implementation of RDS technology in typical environments. Subsequent revisions to the guideline, in April 2014 (NRSC-G300-A) and September 2014 (NRSC-G300-B), have expanded covered topics, such as usage with Emergency Alerting Systems (EAS) and providing recommendations for RDS encoder security.

Topics on practical implementation of RDS include hardware installation, subcarrier injection calibration, and system programming using static and live data, where the latter can update the RDS display as a function of the program material being broadcast at any given time. While RDS encoders are designed to be relatively easy to implement, with Internet connectivity, this feature also leaves encoders vulnerable to compromise by outside attacks, sometimes resulting in a station transmitting false or profane RDS information. The Guideline includes a comprehensive section covering recommended security practices to minimize the potential for such breaches.

NRSC-G300-B also includes an expanded discussion of the most widely implemented features of RDS. These features include program identification (PI), program type (PT), program type name (PTN), traffic programming (TA), music/speech (MS), decoder identification, e.g., monaural/stereo audio (DI), alternative frequency (AF), and clock time (CT) codes. Usage of program-associated data (PAD) is also addressed, covering practical applications for display of program information, such as song title, with consideration of typical receiver display capabilities. Another section details many features and recommendations for implementing radio text (RT) transmissions.

The role of RDS in emergency alerting is covered in an extensive section of the Guideline. Included is a method for emergency responders to take advantage of the "one to many" aspect of broadcasting to create an ability to send nonpublic messages across great distances over networked RDS-equipped stations to specialized receivers. Such transmissions can be structured so that they do not disrupt normal uses of RDS, and can provide an important alternative to wireless telephone-based systems that may be adversely affected during emergency conditions.

Finally, the Guideline discusses differences between display character sets commonly used in the computer world in comparison to the RDS character set, which was developed independently. The most common RDS character issue in the United States involves the dollar sign ($), which is not always uniformly displayed on receivers. Other less common display characters with potential translation issues with computer character sets are also discussed, as are the differences between RDS and the character set used to display program-related data for the IBOC digital radio broadcasting system.

NRSC-G301

The NRSC-G301 Guideline, "Creation and Distribution Procedures for Audio Program Metadata" was published in April 2013. It covers the complex issue of correlating audio program material being broadcast with its associated metadata that feeds RDS and other systems used to relay real-time program data to listeners. Metadata can arrive in real-time with the program to be broadcast, as an associated data file or as a completely separate data file or files. Additionally, this metadata from network sources must be successfully merged with other program data sources, including locally generated metadata.

While radio automation systems in use are most often relied upon to parse and assemble metadata needed for broadcast, related unresolved issues remain that the guideline addresses. These include the structure of data files, usually XML, that are not always consistent. Detailed descriptions and recommendations for metadata structuring are provided.

NRSC-5

The complex NRSC "IBOC Digital Radio Broadcasting Standard," first released in 2005 and most recently updated in 2011 (NRSC-5-C),[6] describes an IBOC digital radio broadcasting system developed by DTS, Inc., formerly iBiquity Digital Corporation. The system is designed to permit a smooth evolution from current analog radio broadcasting to fully digital radio broadcasting, within the existing AM and FM terrestrial radiofrequency bands and utilizing the current FCC allocation system and channel assignments.[7] The system delivers digital audio and data services to mobile, portable, and fixed receivers from terrestrial transmitters on existing AM and FM broadcast channels. In the hybrid mode, broadcasters may continue to broadcast analog signals simultaneously with the IBOC digital signals, providing a means for listeners to make the transition from analog to digital radio. A more spectrally efficient all-digital mode, which is fully supported in hybrid-capable IBOC receivers, can be used in a future post-transition environment, after cessation of analog broadcasting. See Chapter 7.6 of this handbook for a more detailed description of radio systems built to this standard.

The system accepts as input compressed digital audio bit streams and ancillary data information (which may or may not be related to the audio) and utilizes baseband signal processing techniques, such as interleaving and forward error correction, to increase the robustness of the signal in the transmission channel. These methods allow audio and ancillary data signals to be transmitted using power levels and spectral band segments designed to minimize interference to existing analog signals.

The block diagram of Figure 2.5-8 illustrates the three major subsystems specified by NRSC-5 and how they relate to one another. The *RF/transmission subsystem* is an AM or FM broadcast facility that is compliant with references specified in the NRSC-5 Standard. This subsystem takes the multiplexed bit stream and applies coding and interleaving to the transmitted data that can be used by the IBOC receiver to reduce transmission and propagation errors. The multiplexed and coded bit stream is modulated onto orthogonal frequency-division multiplexed (OFDM) subcarriers and up-converted to the AM or FM broadcast channel.

The *transport and service multiplex subsystem* feeds the information to be transmitted to the RF/transmission subsystem. It takes the audio and data information, organizes it into packets, and multiplexes the packets into a single data stream. Each packet is uniquely identified as containing audio or data. Data packets containing program service data are added to the stream of packets carrying the associated audio information before they are fed into the multiplexer.

The *audio and data input subsystem* accepts coded and compressed digital audio, program service data (PSD), and advanced data services (ADS). Audio data consists of main program service (MPS) and supplemental program service (SPS) streams (also referred to as *multicast* streams), all of which have their own source coding, compression, and transport subsystems. NRSC-5 does not include specifications for audio source coding and compression. MPS analog audio is delayed and modulated directly onto the AM or FM radiofrequency carrier portion of the IBOC signal to facilitate reception by analog-only receivers. The delay is set such that digital and analog information is decoded by IBOC receivers in a time-aligned fashion, enabling seamless transitions from digital to analog reception when the received signal quality is not sufficient for digital

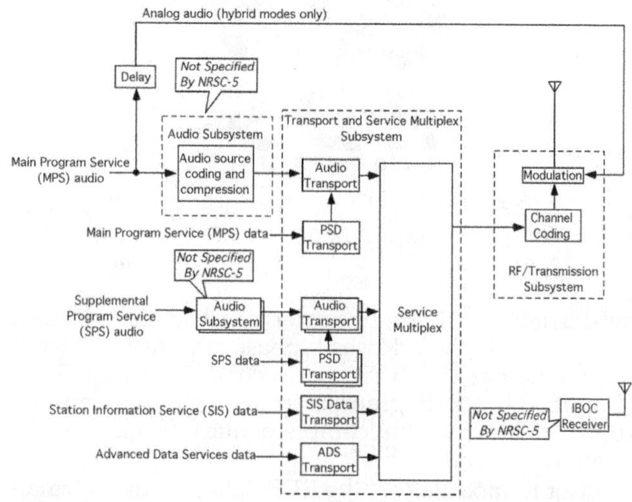

FIGURE 2.5-8 Overview of IBOC digital radio system

audio reception. This blend capability is also used for fast channel changes, allowing the IBOC receiver to demodulate the analog audio first and then blend into the digital audio stream. SPS streams do not have an analog backup.

Program service data is intended to describe or complement the audio program heard by the radio listener. Station information service (SIS) data provides more general information about the station's programming, as well as some technical information that is useful for non-program-related applications. ADS gives broadcasters the ability to transmit information that may be unrelated to the other audio or data streams and may be any content that can be expressed as a data file or data stream, including audio services.

The NRSC-5-C version of the Standard was a comprehensive update that adds the following additional features: (1) operation with asymmetrical digital sidebands, (2) increased digital power in the FM band, and (3) options for decreased digital bandwidth in the AM band, as well as enhancements to data broadcasting, including provisions for conditional access and emergency alerting.

The NRSC-5-C Standard, along with 12 incorporated reference documents, more fully describes each of these subsystems and features. Included are graphs and specifications for AM and FM spectral emission limits, which are fully compliant with current FCC requirements for AM and FM analog stations.

NRSC-G200

The NRSC-G200 Guideline document, "Harmonization of RDS and IBOC Program Service Data (PSD)," was first published in September 2007 and last updated in March 2010 (NRSC-G200-A). It was developed to provide improved correlation between the usage of RDS and the program data system used in IBOC digital radio broadcasting, such that transitions between RDS analog and IBOC digital service areas would result in consistent program information to a listener's radio display.

The Guideline discusses specific text message features of the IBOC (NRSC-5) and RDS (NRSC-4) standards, and defines an intended usage that would allow the broadcaster to send, and the receiver manufacturer to display, common information independent of whether the data signal is being received via IBOC or RDS decoders. The Guideline is designed to be used by broadcasters, automation software vendors, transmission equipment manufacturers, and consumer electronics manufacturers.

Information discussed in the Guideline includes preferred character sets, character translation, and character string length limitations of RDS, of IBOC, and in receivers implementing data display. Also covered is extraction of desired payload text from the transport streams of the two standards and how specific fields should be handled. These include the fields used for title (e.g., song title), artist, album, program type, genre, and comment (i.e., information not covered by the other fields). Additional text fields cover station name (in both short and long formats) and less commonly used information that nevertheless requires correlation between IBOC and RDS.

NRSC-G201

NRSC-G201, "NRSC-5 RF Mask Compliance: Measurement Methods and Practice," is a guideline document that was first adopted in April 2009 and updated in April 2010 (NRSC-G201-A) and April 2016 (NRSC-G201-B). It covers, in detail, with included reference documents, recommended methods to evaluate occupied RF bandwidth of AM and FM radio broadcast transmitting facilities that include hybrid IBOC subcarriers. The measured results are then compared to an RF mask, i.e., graphical template, depicting recommended performance limits. The mask is necessary to ensure compliance with regulations established to avoid interference with other stations, and to ensure that performance is sufficiently robust to avoid potential degraded broadcast quality or self-interference.

The architecture of hybrid IBOC transmitting facilities can vary widely, especially for FM systems. Combining of the FM analog RF signal and IBOC digital subcarriers can be done at low level (e.g., in the FM exciter) for modern transmitters utilizing linear amplification. However, many FM stations have used a high-level combining technique, where the output of a standard FM transmitter is combined with the output of an IBOC-only transmitter, or so-called "space combining," where the FM and IBOC transmitters feed a dual-/multiple-input or separate transmitting antennas. While AM IBOC transmission is implemented more uniformly at the transmitter, site-specific issues, primarily related to antenna bandwidth and load matching, can create significant implementation challenges. For either FM or AM hybrid IBOC transmission, the most common test methods utilize swept-spectrum analyzers. Recent advancements in these devices have made measurements easier to take and results easier to evaluate with a simultaneous overlay of the mask on the measurement display.

The Guideline includes methods of evaluating transmitting facilities to determine a measurement location, as well as connection and configuration of the measuring equipment. Guidance on interpretation of results is also provided, such as evaluation of intermodulation effects between the analog and digital RF signals and potential ingress of unwanted signals. For FM hybrid IBOC signal evaluation, a method of measuring the qualitative modulation error ratio (MER) is discussed. (As of 2016, a similar MER measurement method was under development by the NRSC.)

NRSC-G202

The guideline document NRSC-G202, "FM IBOC Total Digital Sideband Power for Various Configurations," was first released in September 2010 and updated in April 2016 (NRSC-G202-A). It is a companion to NRSC-G101, covering specific information related to increased power of FM IBOC subcarriers authorized by the FCC in 2007 and 2010.

When first authorized in 2002, FM hybrid IBOC transmission was implemented with the total power of the IBOC subcarriers set to 20 dB below the amplitude of the FM carrier (−20 dBc). Addition of the MP2, MP3, and MP11 "extended hybrid" operating modes in 2007 added additional digital carriers, and hence theoretically higher power for stations operating in these modes, but power calculations remained referenced to the original MP1 IBOC operating mode. The most recent authorization in 2010 permits this injection level to be raised to −14 dBc or even higher to −10 dBc in some cases, depending on the outcome of interference evaluations. Further, it is also possible to increase IBOC carrier power asymmetrically, i.e., unequally on the lower and upper sidebands, to meet custom interference reduction criteria.

NRSC-G202 provides a series of tables that can be used to determine IBOC carrier levels under different operating modes and injection levels for both symmetrical and asymmetrical operations. The figures presented in these tables are to be used in conjunction with the measurement procedures contained in the NRSC-G201 Guideline.

Conclusion

The NRSC provides a forum for the broadcasting and receiver manufacturing industries to work together toward solutions to common issues and problems in radio broadcast systems. As the radio industry further transitions to digital broadcasting, the NRSC stands ready to develop standards and guidelines that will benefit the industry and the listening public.

PART 2: EUROPEAN ANALOG AND DIGITAL BROADCASTING STANDARDS

Introduction

In Europe there is no direct equivalent to the NRSC, and radio broadcasting policies are developed at a national level, under International Telecommunication Union (ITU)[8] rules, and in some instances promoted and coordinated by the European Broadcasting Union (EBU),[9] which represents the interests of public service broadcasters and is based in Geneva. International band planning is agreed to through the World Radiocommunication Conferences (WRC),[10] (formerly the World Administrative Radio Conference, WARC). Technical transmission parameters are defined in standards published by the European Technical Standards Institute (ETSI).[11]

In terms of detailed spectrum planning, because of the crowded nature of the continent and the relatively small geographical size of many nation-states, bilateral negotiations are commonplace, adding detail to broad agreements overseen by the International Telecommunications Union, Radiocommunication Sector (ITU-R) regional framework (ITU region one). Although the European Union is involved in the development of television broadcasting policy, it has no direct involvement in radio broadcasting policy, which remains the responsibility of individual nation-states.

Analog Broadcasting

Analog AM and FM broadcasting standards are both long established and well developed across Europe, with a broad degree of similarity across the continent. At a national level, regulations take into account the various international regulatory frameworks as outlined above, implementing these through national policies. A useful example of how regulations are localized can be found in the British "Site Engineering Code for Analogue Radio Broadcast Transmission Systems,"[12] published by the UK's broadcast radio regulator, Ofcom, (The Office of Communications).[13]

In broad terms, for historical reasons to do with the early dominance of public service broadcasting in many European states, analog broadcast radio spectrum in the region continues to see considerable use of networked national and wide-area regional services on both AM and FM frequencies. Today the dominance of public service broadcasting has somewhat diminished and commercial broadcasting now forms a large component of the wider radio industry. Commercial broadcasters tend to dominate at the local level, with an increasing presence at the national level, and are, to some extent, supplemented by not-for-profit community radio services, predominantly on FM, at the local and hyper-local level.

AM Broadcasting Standards

In a European context, domestic analog AM radio is broadcast on what are known locally as the medium-wave (526.5–1606.5 kHz) and long-wave (148.5–283.5 kHz) bands. International AM shortwave broadcasting, while in gradual decline, continues on various frequencies above medium wave and up to 30 MHz. Because Europe falls within ITU region one, the bandwidth of transmissions on both long-wave and medium-wave frequencies is narrower than in North America, with channels allocated on a 9 kHz spacing grid. Moreover, the audio bandwidth allowed is only within the operational channel itself, (i.e., ±4.5 kHz).

The core technical specifications for European AM broadcasting are contained in the 2005 ETSI standard covering transmission equipment for AM (EN 302 017) parts one and two.[14] An example of information about AM service planning can be found in the Ofcom document "Coverage & Planning Policy for Analogue [SIC] Radio Broadcasting Services," published in 2011.[15]

FM Broadcasting Standards

In terms of frequency allocations, FM broadcasting in Europe has standardized on the use of 87.5–108 MHz (also known as Band II). In many jurisdictions, this band is used for a combination of networked national, regional, and local services (both public service and commercial). More recently, a third tier of not-for-profit community radio services has also been introduced in various jurisdictions, in many countries and, particularly, in major conurbations—the FM band is effectively full to capacity.

Although European FM transmissions are in many ways similar to their American FM equivalents, there are some fundamental differences. For example, the service-planning grid employed uses both odd and even decimals, such that stations, with sufficient geographical separation of service areas, can be spaced 100 kHz apart. In addition, audio preemphasis is applied to transmissions using a 50-µsec time constant, instead of the 75-µsec one applied in the Americas. In various countries, subsidiary communications authorization (SCA) subcarrier deviation is not permitted to increase total signal deviation beyond ±75 kHz. In other words, the capacity to add subcarriers to stereo FM transmissions is extremely limited, and adding any subcarrier service requires main service deviation to be commensurately reduced to compensate.

Technical specifications for European FM broadcasting are contained in the 2006 ETSI standard covering transmission equipment for FM (EN 302 018) parts one and two.[16] The planning standards used for FM broadcasting in Europe are based on ITU-R Recommendation BS.412-9 "Planning Standards for Terrestrial FM Sound Broadcasting at VHF."[17]

Radio Data System (RDS)

The Radio Data System (RDS) is commonplace in European FM broadcasting. The Standard is older and slightly different from RBDS, discussed in more detail earlier in this chapter, but the technical implementation is identical. In many countries, it is rare to find any broadcasters not transmitting at least a bare minimum of RDS data. The RDS standard used in Europe was originally defined in 1998 and is contained within the European Committee for Electrotechnical Standardisation (CENELEC) European Standard, EN50067.[18]

Digital Radio Broadcasting

Although approaches to analog AM and FM broadcasting are reasonably consistent across Europe, the same cannot be said in relation to digital radio broadcasting. Some countries (for example, Denmark, Norway, and the United Kingdom) have extensively developed networks using the original Eureka-147 DAB system, which, more recently in some locations have begun to be supplemented and in some cases replaced by DAB+ services. However, digital radio technology has not been pursued to anything like the same extent in many other countries and, in some, it remains altogether absent.

Some regular scheduled international use of the Digital Radio Mondiale (DRM) system takes place, and there have been a variety of successful trials of both modes/systems (DRM30 and DRM+) for domestic consumption. However, both modes of DRM still suffer from a lack of available receivers, and neither can yet be considered viable in terms of availability to members of the general public in Europe.

DAB and DAB+

Certainly the most developed digital radio broadcasting standards in Europe, DAB and DAB+, share a common transmission platform, differing only in terms of audio coding scheme applied. The original DAB standard was first developed by the EBU from the late 1980s onwards, and was funded under the European Union's Eureka technology program as Eureka Project Number 147 in 1987.[19] It was designed as a potential replacement technology for wide-area networked FM broadcast services, in particular those operated by the various multi-network, national public service broadcasters operating in many European countries.

The Standard is based on the principle of transmitting a number of stations through a single transmitter or network, delivering a multiplex of services occupying 1.5 MHz of spectrum and each covering an identical service area. Although it was originally envisaged that DAB would gradually supplant FM within VHF Band II (87.5–108 MHz), in fact this has not happened, and it has generally been implemented within VHF Band III (174–240 MHz), on frequencies that were previously used for analog monochrome terrestrial television services.

The DAB standard can also be used on microwave frequencies and was designed with satellite delivery in mind.[20] In Europe, part of the 1.5-GHz L-Band (1452–1492 MHz) is allocated for use by terrestrial or satellite DAB transmissions. However, although various tests have been successfully completed using terrestrial networks, the allocation has not proved to be popular, primarily due to the added costs associated with receiver production and the less forgiving propagation characteristics when compared to those of the alternative Band III allocation.

The DAB family of standards runs to approaching 40 separate documents, which are helpfully summarized in the 2012 ETSI Guide to DAB technical standards.[21] However, the core technical specifications for European DAB broadcasting are contained in the 2006 ETSI standard covering transmission equipment for DAB (EN 300 401).[22] A later addition to the Standard, for DAB+, which modifies the original system to use AAC audio compression for enhanced service carrying capacity and introduces additional error correction for improved robustness of reception, is further defined in the relevant 2010 ETSI technical standard (TS 102 563).[23] Key differences between the original DAB standard and the enhanced DAB+ version are helpfully summarized by the EBU in its "Report on Frequency and Network Planning Parameters Related to DAB+" (EBU, 2013).[24]

The key distinctive characteristic of DAB is its use of a multiplex, encoding multiple broadcast services onto a single-coded orthogonal frequency division multiplex (COFDM) bearer, occupying approximately 1.5 MHz of spectrum. Operationally, this multiplex can carry a number of separate services at differing bitrates (see Table 2.5-1). These can be dynamically altered, for example, to add temporary services or "opt-outs" where a station may wish to broadcast additional services at a particular time or to cover a local special

TABLE 2.5-1
DAB Multiplex Program Carrying Capacity

Type	Mode	Data Rate (kbps)	Capacity Units Per Service	Number of Services	Capacity Units Remaining
DAB	Stereo	192	140	6	24
DAB	Stereo	160	116	7	52
DAB	Joint Stereo	128	96	10	0
DAB	Joint Stereo	112	84	10	24
DAB	Mono	96	70	12	24
DAB	Mono	80	58	14	52
DAB	Mono	64	48	18	0
DAB	Mono/Speech	56	42	20	24
DAB	Mono/Speech	48	35	24	24

Notes:
1. The above table assumes the use of Error Protection level 3 (UEP)
2. The total number of capacity units available in a DAB multiplex is 864
3. Capacity units can be allocated to services of various data rates
4. For a given data rate, DAB+ services require a broadly similar number of capacity units for delivery
5. DAB+ services are typically operated at data rates of between 64 and 32 kbps (stereo/mono). For more information, see the "Tech" section of the Future Digital website at www.futuredigital.info.

event. Provided all the program data carried is identical, the same block of frequencies can be reused multiple times in a single frequency network (SFN) to provide very wide area coverage. Unlike in the case of FM, there is no need to use multiple frequencies to create a network, as DAB receivers are effectively able to sum or ignore transmissions from multiple transmitters to enhance reception. However, it should be made clear that if any element of the program materials or other data carried is not identical, then the SFN will not function correctly.

Using the early MPEG-1 Audio Layer 2 (MP2) audio encoding standard, the original version of DAB requires stereo services to be broadcast at 192 kbps in order for them to be considered as being of relatively "transparent" broadcast quality. However, in practice, lower bitrates are often used with encoding artifacts that are relatively obvious to the trained ear. Leaving aside its classical music service, the BBC regularly broadcasts in stereo at 128 kbps and in mono at 80 kbps, with some less popular services using as little as 64 kbps, a bitrate which, when using MP2 audio encoding, delivers very obviously degraded audio quality. Although DAB+ offers much improved audio quality, delivering stereo audio of similar quality to 128 kbps (original DAB) in just 48 kbps, it is currently avoided by the majority of mainstream broadcasters because of the number of early DAB receivers still in use that are unable to decode it.

Perhaps the most interesting development within DAB over the past few years has been the emergence of open source approaches for implementing the transmissions standard. Based on work originally started by the Canadian Government's Communications Research Centre, Geneva-based Open Digital Radio (ODR)[25] has created open source stream encoders and multiplexers that can be used in conjunction with software-defined radios to deliver low-cost DAB services. Ofcom, the UK equivalent of the FCC, has been running long-term, low-power, DAB transmissions[26] based primarily on this software since mid-2015, and further multiplexes designed to test it have subsequently been launched in other European countries. Because of their reduced spectrum impact when compared to higher power wide-area services, there is the potential for lower-power DAB transmissions to be introduced without the need for international coordination. Such developments may allow additional, smaller DAB transmissions to be introduced in some European countries over the coming years.

Digital Radio Mondiale (DRM30 and DRM+)

Just as in the case of DAB, the original DRM project was a product of a European Eureka research project (scheme number 1559), launched in 1996.[27] Whereas the DAB project was designed to replace analog FM transmissions within Band II, DRM was initially intended to replace analog AM transmissions on frequencies below 30 MHz. Later, from 2005 onwards, the Standard was expanded to include a version known as "Mode E" for frequencies above 30 MHz, up to and also including VHF Band III frequencies (below 230 MHz), thus including frequencies currently used in some European countries for DAB transmissions. For clarity, DRM is the name of the overarching standard. When this is used on frequencies below 30 MHz, it is referred to as DRM30. On higher frequencies (Bands I, II, and III), it is known as DRM+. A list of the various DRM transmission modes is provided in Table 2.5-2.

TABLE 2.5-2
DRM Transmission Modes

DRM Mode	Application
A	Designed for domestic local and regional broadcasting, where multipath propagation and Doppler effects are minimal.
B	Designed for medium-range transmissions, where multipath propagation is expected.
C	Designed for long-range transmissions, similar to Mode B, but with increased carrier spacing for enhanced robustness against Doppler effects.
D	Designed for very long-range transmissions, similar to Mode B, but with reduced carrier numbers to improve resistance against large delay spread and Doppler spread.
DRM+ Mode	
E	Designed for use in the VHF spectrum with wider channel bandwidth, it can transmit FM quality stereo within a 100-kHz channel and offers reduced co-channel and adjacent-channel protection requirements when compared to analog FM transmissions (also known as ITU Digital System G) (ITU-R, 2011).

As with DAB, the DRM system uses a COFDM bearer with multiple carriers, each one being modulated with quadrature amplitude modulation (QAM). However, unlike DAB, DRM takes a narrow-band approach. Although capable of delivering more than one audio service (up to three audio services plus one data channel depending upon configuration), typically DRM is used to deliver independent stand-alone services. When using the advanced x-HE-AAC codec, DRM is particularly efficient at low bit rates.

Because DRM can operate using the same channel spacing as employed by traditional AM and FM transmissions, bands can be planned to include both analog and digital services operating alongside each other. DRM30 is suitable for European AM transmissions; offering improved audio quality within a 9-kHz channel mask. On frequencies above 30 MHz, DRM+ provides an ideal complimentary system to DAB, useful in situations where a DAB multiplex might not be available or viable. DRM is an open global digital audio broadcasting standard, compatible with the various other open ITU recommended digital audio broadcasting standards (DAB/DAB+, etc.), with which it shares coding as well as various operational functionalities.

The full specifications for DRM on all frequencies, both shortwave and VHF, are contained within the 2014 ETSI document "Digital Radio Mondiale (DRM); System Specification."[28] The DRM Consortium has also published a detailed "Introduction and Implementation Guide," which provides various practical information about the system.[29] A detailed comparison of DRM30 against other digital systems on frequencies below 30 MHz is contained within the relevant ITU-R Recommendation, BS.1514-2.[30] Comparisons between DRM+ and other digital systems on frequencies above 30 MHz can be found in another ITU-R Recommendation, BS.1114-7.[31]

Since 2001, various European broadcasters have used shortwave frequencies to deliver international shortwave services[32] and various tests using DRM have been carried out on domestic AM frequencies (both medium-wave and long-wave). The BBC completed a one-year public trial of DRM30, which looked at nighttime reception and SFN operations in particular.[33] Some higher frequency shortwave trials of services intended for local reception have also been completed, as have tests of DRM+ on Band I, II, and III frequencies. For example, in Scotland, during 2011 and working in conjunction with the DRM Consortium and Ofcom, the BBC completed relatively high-power DRM+ tests on 107.0 MHz at 1 kW equivalent monopole radiated power (e.m.r.p.).[34]

Although the various DRM tests have generally proven successful in terms of technical performance and coverage, etc., the difficulty for broadcasters is the current lack of domestic DRM30 and DRM+ capable receivers. Recently, minimum specifications have been defined in the UK for digital receivers,[35] but these do not currently include a requirement to build in DRM reception capability. Some car radio manufacturers use multi-standard software-defined radio (SDR) technology, which brings multiple standards together to meet European market requirements, and these could accommodate DRM relatively easily. In spite of its operational flexibility and robustness, until receivers are manufactured and used on a wide scale, the take-up of the DRM standard in Europe is likely to remain minimal. Were individual jurisdictions to mandate the provision of more DRM30 and DRM+ services, that situation could begin to change, as indeed it did in the case of DAB.

Conclusion

The European broadcast radio sector is remarkably diverse, in terms of the types of services broadcast, as well as the transmission systems and frequencies used to deliver them. In terms of digitization, the various country-by-country approaches taken, along with the emergence of nonbroadcast content delivery systems, mean that it remains uncertain which path or paths the industry will have taken by the time the next edition of this handbook is published.

Notes

[1] See www.nrscstandards.org for committee rules, membership information, a list of committees, and access to the standards and related guidelines.
[2] See http://www.nab.org.
[3] See www.cta.tech. CTA was formerly known as the Consumer Electronics Association (CEA), the Consumer Electronics Manufacturers Association (CEMA), and the Electronic Industries Association (EIA).
[4] Title 47 of the Code of Federal Regulations, Part 73, §73.44.

[5] In 2000, the Cenelec Standard was retired and superseded by an International Electrotechnical Commission (IEC) standard, IEC 62106 (Specification of the Radio Data System [RDS] for VHF/FM Sound Broadcasting in the Frequency Range from 87.5 to 108.0 MHz), which is essentially an updated version (with only minor changes) of EN50067:1998.

[6] Two versions of the Standard were released in 2005; the NRSC-5-A Standard added information relating to ADS that was not in the first version. The third and fourth versions added coverage of operation with asymmetrical digital sidebands, increased digital power for the FM system, digital bandwidth options for AM broadcasters, and enhancements to data broadcasting. The next update to the Standard is scheduled to be completed in 2017.

[7] DTS markets its technology under the trade name of "HD Radio." Where the term "HD Radio" is used within the NRSC-5 standard, it is interpreted to reflect the generic term "IBOC" for an NRSC-5-compliant system.

[8] See www.itu.int/rec/R-REC-BS/en for a full list of ITU broadcasting standards.

[9] See http://ebu.ch/home.

[10] See www.itu.int/en/ITU-R/conferences/wrc/Pages/default.aspx.

[11] See www.etsi.org/standards.

[12] See "Site Engineering Code for Analogue Radio Broadcast Transmission Systems" at http://stakeholders.ofcom.org.uk/binaries/broadcast/guidance/tech-guidance/code2013.pdf

[13] See www.ofcom.org.uk.

[14] See https://portal.etsi.org/TBSiteMap/broadcast/ARB.aspx for a list of ETSI Analogue (SIC) Radio Broadcasting Standards.

[15] See "Coverage & Planning Policy for Analogue Radio Broadcasting Services" at http://stakeholders.ofcom.org.uk/binaries/broadcast/radio-ops/coverage/analogue-coverage-policy.pdf.

[16] See https://portal.etsi.org/TBSiteMap/broadcast/ARB.aspx for a list of ETSI Analogue Radio Broadcasting Standards.

[17] See "Planning Standards for Terrestrial FM Sound Broadcasting at VHF" at www.itu.int/dms_pubrec/itu-r/rec/bs/R-REC-BS.412-9-199812-I!!PDF-E.pdf.

[18] See "Specification of the Radio Data System (RDS) for VHF/FM Sound Broadcasting in the Frequency Range From 87,5 to 108,0 MHz" at www.interactive-radio-system.com/docs/EN50067_RDS_Standard.pdf.

[19] See "Digital Audio Broadcasting System" at www.eurekanetwork.org/project/id/147.

[20] See Brugger, Roland (1993). "Single-Frequency Networks at 1.5 GHz for Digital Audio Broadcasting." (*EBU Technical Review*, Winter 1993 (3–13)) EBU, Geneva, Switzerland.

[21] See "Digital Audio Broadcasting (DAB); Guide to DAB Standards; Guidelines and Bibliography" at www.etsi.org/deliver/etsi_tr/101400_101499/101495/01.04.01_60/tr_101495v010401p.pdf.

[22] See "Radio Broadcasting Systems; Digital Audio Broadcasting (DAB) to Mobile, Portable and Fixed Receivers" at www.etsi.org/deliver/etsi_en/300400_300499/300401/01.04.01_40/en_300401v010401o.pdf.

[23] See "Digital Audio Broadcasting (DAB); Transport of Advanced Audio Coding (AAC) Audio" at www.etsi.org/deliver/etsi_ts/102500_102599/102563/01.02.01_60/ts_102563v010201p.pdf.

[24] See "TR 025 Report on Frequency and Network Planning Parameters Related to DAB+" at https://tech.ebu.ch/docs/techreports/tr025.pdf.

[25] See www.opendigitalradio.org.

[26] See "Small Scale DAB Trial" at http://stakeholders.ofcom.org.uk/consultations/small-scale-dab.

[27] See "Digital Broadcasting in a.M. Frequency Bands" at www.eurekanetwork.org/project/id/1559.

[28] See "Digital Radio Mondiale (DRM); System Specification" at www.drm.org/wp-content/uploads/2012/10/DRM-System-Specification.pdf.

[29] See Laflin, Nigel (Editor) (2012), "The DRM Digital Broadcasting System, Introduction and Implementation Guide," DRM Consortium, Geneva, Switzerland, available online at: www.drmplus.com/i/DRM-introduction-and-implementation-guide.pdf.

[30] See "System for Digital Sound Broadcasting in the Broadcasting Bands Below 30 MHz" at www.itu.int/dms_pubrec/itu-r/rec/bs/R-REC-BS.1514-2-201103-I!!PDF-E.pdf.

[31] See "Systems for Terrestrial Digital Sound Broadcasting to Vehicular, Portable and Fixed Receivers in the Frequency Range 30–3000 MHz" at www.drm.org/wp-content/uploads/2012/10/Recommendation-ITU-R-BS.1114-7-201112-IPDF-E.pdf.

[32] See Giefer, Andy (2002), "Digital Radio Mondiale Long-Term Test Results," (White Paper WHP174, February 2002), BBC Research & Development, Kingswood Warren, Surrey. See also Lauterbach, Thomas & Hofmann, Frank (2008), "Local Radio Using Digital Radio Mondiale (DRM) in the 26 MHz Band—a Résumé at the End of the Nuremberg/Dillberg Field Trial," Fraunhofer IRB Verlag, Stuttgart, Germany, available online at: www.researchgate.net/publication/228557081_Local_Radio_using_Digital_Radio_Mondiale_DRM_in_the_26_MHz_Band-A_Resume_at_the_end_of_the_Nuremberg Dillberg_field_trial.

[33] See Murphy, Andrew (2009), "The Plymouth Digital Radio Mondiale (DRM) Trial Long-Term Reception Results," (White Paper WHP032, January 2002), BBC Research & Development, Kingswood Warren, Surrey.

[34] See ITU Radiocommunication Study Groups (2011), "Digital Radio Mondiale and British Broadcasting Corporation (BBC), Results of the DRM High Power Field Trial in the United Kingdom," ITU, Geneva. See also Cornell, Lidsay (2011), "DRM+ Tests & Trials, 16th February 2011," (Presentation), available online at: https://tech.ebu.ch/docs/events/radiosummit11/presentations/ebu_radiosummit11_cornell.pdf.

[35] See "What Is the Digital Tick Mark?" at www.getdigitalradio.com/industry/what-is-the-tick-mark.

CHAPTER

2.6

HF Shortwave Radio

Allocation, Design Methods, and Regulation

JAMES E. O'NEAL

Technology Consultant, TV Technology magazine
Alexandria, VA

Of all of the radiofrequency spectrum available or in use terrestrially, the portion that exists between approximately 2 and 30 MHz is unique, as it is here that global communications without the use of intermediate relay facilities is possible on a fairly reliable and predictable basis. Interestingly, this range of frequencies, with their potential for long-distance communication with modest amounts of transmitter power, was ignored by early adopters of wireless communication in the early 20th century, as the prevailing belief was that only the long wavelengths (greater than 200 meters or so) were of any commercial value. Consequently, little serious study of properties of the shorter wavelengths took place. (Efforts to explore or utilize higher frequencies may have been thwarted too due to lack of equipment for effectively generating higher frequencies.)

This situation changed following the cessation of hostilities in the "Great War" or First World War (1914–1918). The value of radio as a communication tool was underscored by the war, and not long after its end, efforts were initiated by a number of countries to "nationalize" RF spectrum for both military and commercial use. To this end, amateur radio hobbyists or "hams," who, prior to the enactment of the Radio Act of 1912 had enjoyed virtually unlimited access to all of the electromagnetic spectrum then in use, found themselves explicitly forbidden from transmitting with longer wavelengths (in the United States, below 200 meters or 1.5 MHz), as this region was considered "prime real estate," and amateur operations had no place there.[1]

This prohibition forced an investigation of the shorter wavelengths (less than 200 meters for amateurs) by individual amateurs, as well as by amateur groups and associations. Such efforts were greatly aided by the wartime refinement of the vacuum tube, as well as the substantial improvements in transmitter and receiver technologies. These individuals and groups were soon able to demonstrate some rather remarkable achievements within the "worthless" spectrum. By the early 1920s, amateurs were communicating across the Atlantic Ocean on a fairly regular basis with relatively modest amounts of power, and equipment that was completely dwarfed by the multi-ton high-frequency alternators being used then by commercial entities for generating the hundreds of kilowatts necessary for transatlantic communications via longer wavelengths (as low as 20 kHz) in use for this purpose.

Such lessons were not wasted, and before long, others were exploiting the long-distance capability afforded by the shorter wavelengths. By 1924, many commercial radio broadcasters had established "shortwave" service, and serious studies of propagation were being conducted.[23]

THE IONOSPHERE

In order to better understand both the early and continuing interest in shortwave broadcasting, a review of the physical principles making it possible is in order. Propagation of electromagnetic radiation within the HF frequency band over very long distances is only possible due to a region of electrically charged atoms and molecules (plasma) that exists some 60 km to 1,000 km (37 miles to 620 miles) above the surface of the earth. This layer exists due to the action of ultraviolet, x-ray, and higher frequency photons emanating from the sun (and to some extent from cosmic radiation sources). Such radiations possess sufficient energy to dislodge electrons from the small amount of atmospheric gases that exist at such altitudes. The relative sparsity of such gases in this region forms the basis for the ionosphere, as it allows atoms and molecules to exist in an ionized state for a significant length of time before they encounter and recombine with electrons. Studies have shown that the ionosphere exists in more or less discrete layers, with the portion closest to the earth being designated the "D" layer and that farthest away termed the "F" layer (150 km to 400 km above the earth), which is further subdivided into the "F1 and "F2" regions, with the "F2" being that most distant from the earth. In the nighttime absence of sunshine, ionization is only present to any extent in these "F" layers. All layers of the ionosphere affect HF radio propagation to some degree. However, the "F2" layer is of most interest in enabling over-the-horizon radio communication. During daylight hours the "D" layer is a region of intense ionization due to its relative proximity to the earth and the proportionally larger number of gas atoms and molecules available. (For this reason, it is also more stable, as the increased density of both ions and free electrons is conducive to frequent recombination.)

Reflection (or, more technically correct, refraction) of HF signals by the ionosphere involves excitation of free electrons there by the electric field of the radio signal striking them. The excited electrons oscillate at the same frequency as the signal reaching them. While some of this energy is lost due to recombination, a certain amount is reradiated, which makes the ionosphere useful in "relaying" HF signals over great distances.[4]

Certain naturally occurring phenomena greatly affect the usefulness of the ionosphere in propagating terrestrial radio signals, with extensive investigations having been conducted to ascertain its availability for propagating HF signals. While an in-depth examination of each factor is beyond the scope of this work, some of the more significant will be examined.

As previously noted, the existence of the ionosphere is largely due to solar radiation, with variations following daytime/nighttime conditions on earth. This diurnal effect is conditioned by seasonal changes affecting the number of hours of daylight.

During daylight hours, the lower "D" layer effectively blocks (absorbs) medium-wave and lower frequencies, preventing them from reaching the higher ionospheric layers. This is also true to some degree for the "E" layer. Solar flare activity (coronal mass ejection) also greatly affects propagation due to the intense ionization created in all layers. Such conditions can largely override the normal absorption of signals by the "D" and "E" layers, with propagation throughout the entire HF spectrum available during peaks of solar flare activity. (Such activity occurs on a predictable 11-year cycle.)

Aurora ionization (also influenced by solar activity) can be a factor in HF propagation at higher latitudes. Auroral propagation (sometimes referred to as auroral backscattering) is generally in an east-west direction from the transmitting station, with best signal delivery to receiving sites at the same latitude. (It is for this reason that locating shortwave broadcast stations at higher latitudes is usually avoided, as this propagation modality may limit their effectiveness.)

It should be noted a limiting or "critical" upper frequency exists at which ionospheric propagation becomes impossible. This is due to an insufficient number of free electrons available to support reradiation. In such situations, terrestrially generated RF energy passes through the ionosphere and is lost. This critical frequency, also known as the maximum usable frequency (MUF), varies with time of day, season, and certain atmospheric events. (As mentioned, solar flare activity also has a great influence.) The angle at which the RF energy strikes the ionosphere is also a factor, with a nearly vertical (normal to the ionosphere) signal largely passing through the excited layers. (It should be noted that this is not always the case, with near vertical incidence skywave (NVIS) propagation occurring during periods of intense ionization (daytime); however, this mode of propagation ceases to function with the arrival of darkness.)

In selecting an operating frequency, HF broadcasters generally operate some 10% to 20% below the MUF to ensure reliability of signal delivery.

Conversely, there is also a lowest usable frequency (LUF) due to the absorption of RF energy primarily in the "D" layer of the ionosphere.

Due to the constantly changing nature of the ionosphere, shortwave broadcasts inevitably suffer from fading—which can be slow or rapid. This typically results from the creation of multiple signal paths between the transmission point and the listener's receiver. These can add constructively or destructively and may result in a drop in the demodulated audio output and/or distortion due to attenuation of the signal's carrier component. Commercial users of HF radio utilized frequency space and/or frequency diversity technology to help overcome such problems, with multiple communication paths established at different frequencies within a particular band, or sometimes in separate frequency bands. When partial or complete loss of signal was detected either by a human operator or by electronic means, audio from a receiver tuned to a different frequency was selected. In most cases, propagation conditions between the two or more paths being used were sufficiently different as to allow continued good-quality reception.

HF BAND SPECTRUM ASSIGNMENT

By international agreement, the radio spectrum is broken into various blocks: low frequency (LF) at 30–300 kHz; medium wave (MW) at 300 kHz–3 MHz; very high frequency (VHF) at 30–300 MHz, etc. The high-frequency (HF or shortwave) international broadcasting portion of this spectrum encompasses frequencies between 5.9 MHz and 26.1 MHz, and has been further divided into the following bands:

- Band 6: 5,900–6,200 kHz (49 meters)
- Band 7: 7,200– 7,300 kHz (41 meters) (ITU Regions 1 [includes Europe, Africa, the Middle East west of the Persian Gulf, the former Soviet Union,

and Mongolia] and 3 [the majority of non-former Soviet Union Asia east of and including Iran, and Oceania] only)

- Band 7: 7,300–7,450 kHz (41 meters) (ITU Region 2; includes North and South America, Greenland, and a portion of the eastern Pacific Islands)
- Band 9: 9,400–9,900 kHz (31 meters)
- Band 11: 11,600–12,100 kHz (25 meters)
- Band 13:13,570–13,870 kHz (22 meters)
- Band 15: 15,100–15,800 kHz (19 meters)
- Band 17: 17,480–17,900 kHz (16 meters)
- Band 18: 18,900–19,020 kHz (15 meters)
- Band 21: 21,450–21,850 kHz (13 meters)
- Band 25: 25,670–26,100 kHz (11 meters)

Although not recognized by the ITU as international HF broadcasting spectrum, readers should be aware of the so-called "tropical" bands (120, 90, 75, and 60 meters) in the 2,300 to 5,060 kHz region. Typically, these are used for domestic shortwave broadcasting but are capable of providing communications at great distances. The FCC allows U.S. shortwave broadcasters to use this spectrum on a noninterfering basis, but as it does not sanction domestic HF broadcasting, the understanding is that transmissions within these frequency bands are intended for international audiences.

SHORTWAVE BAND PROPAGATION AND UTILIZATION CHARACTERISTICS

Since nature does not always respect human-made definitions and boundaries, there is no sharp "cutoff" or shift of characteristics at either end of the HF band. Frequencies near 3 MHz share some of the propagation characteristics of the MW band, and those near 30 MHz tend to share some of the behavior of the low-band VHF spectrum. Generalized propagation characteristics are listed below by band; however, these are strictly generalizations, as there are no "absolutes" in ionospheric propagation.

- 49 meters—generally good for nighttime broadcasting; little long distance reception during daylight hours
- 41 meters—generally good both night and day
- 31 meters—good for nighttime use year-round; daytime openings vary seasonally
- 25 meters—useful night and day, with best conditions existing in summer months
- 22 meters—generally good for daytime broadcasting; nighttime conditions vary seasonally, with summer months being best
- 19 meters—generally good for daytime broadcasting; nighttime conditions vary seasonally
- 16 meters—generally good for daytime broadcasting; nighttime conditions vary seasonally
- 5 meters—generally good for daytime broadcasting; nighttime conditions vary seasonally, with best conditions occurring in summer months
- 13 meters—useful in daylight hours; little activity at night
- 11 meters—little nighttime propagation; daylight availability varies with sunspot activity
- tropical bands mentioned above (120, 90, 75, and 60 meters)—The 120-meter band is actually medium-wave spectrum and shares many of the characteristics of the that region. Typically, propagation in all of these bands is best in afternoon and early evening hours, with communications limited to a few hundred kilometers or miles; however, this is a generalization, with much greater distances possible at times.

PREDICTION OF HF OPERATING CONDITIONS

As explained, many variables and vagaries are associated with ionospheric propagation. However, due to the very large amount of study that has been performed over the course of several decades and the use of modern computer modeling techniques, it is possible to predict HF propagation with a fairly high degree of accuracy. Such information is invaluable to international shortwave broadcasters and is easily available in the 21st century. A prediction tool employed by a number of shortwave broadcasters is VOACAP (Voice of America Coverage Analysis Program). The program has its roots in work performed by the Naval Research Laboratory and the National Telecommunications and Information Administration Institute's NTIS (National Institute for Telecommunication Sciences). This was released to the public domain as IONCAP by the NTIS in 1985, and adopted by the VOA the same year as the basis for its propagation modeling software, VOACAP. This in turn became part of the public domain in 1993. The software provides a user-friendly graphical interface and delivers graphs and maps with detailed information on such things as maximum usable frequency, performance as a function of receiver distance and time of day, human-made noise projections for a given area, and both hourly and 24-hour projections for coverage over the entire HF spectrum region.

REGULATION OF INTERNATIONAL HF BROADCASTING

Early in the history of HF broadcasting, there was little in the way of formal agreement between nations to prevent interference or to resolve potential conflicts between these special stations that could be heard around the globe. A first step was taken in 1927 at the International Radiotelegraph Conference in Washington. Some 79 nations were represented at the meetings, as well as a number of nonvoting entities, and agreement was reached in the establishment of specific bands delegated for HF broadcasting purposes. However, international broadcasting was not specifically precluded in other portions of the HF spectrum as long as it was done on a noninterfering basis.[5] Early shortwave stations largely policed themselves to prevent interference, as all players were aware that simultaneous operation on a particular frequency benefitted no one. Later, through the auspices of the International Telecommunication Union (ITU), a registry was established to identify shortwave operations and assist in coordinating broadcasting efforts. Such coordination is now accomplished by the High Frequency Coordinating Conference (HFCC) a nonprofit, nongovernmental association sector member of the ITU that was created in 1990 to manage and coordinate international shortwave broadcaster databases and activity in accord with ITU regulations governing the allocation and use of HF spectrum. HFCC members are provided with tools and services necessary for minimizing radiofrequency interference among these members. The organization arranges two meetings of its members per year for purposes of coordinating summer and winter broadcasting schedules. The organization also provides an Internet-based system to help identify and resolve interference issues between members.

THE RISE OF INTERNATIONAL HF BROADCASTING

It was not long after reports of success by amateur radio operators in spanning great distances with HF radio that several early commercial broadcasters became interested in the technology and launched shortwave services.[6,7,8] These were usually simulcasts of programming being aired on the stations' medium-wave outlets. The number of shortwave broadcasters grew considerably during the late 1920s in both the United States and other nations, with more than 40 broadcast services above the AM broadcast band being operated on a regular basis by countries including Australia, Austria, Africa (North, East, and South), Denmark, Finland, France, Germany, Great Britain, Holland, Italy, Spain, Japan, Java, Norway, Sweden, Switzerland, and the Union of Soviet Socialist Republics.[9] By late 1928, more than 25 commercial shortwave licenses had been issued by the U.S. Federal Radio Commission. (Not all of these can be assumed to be for broadcast services, but such established broadcasting groups as CBS, Crosley, General Electric, Westinghouse, and individual radio stations (KFWB, WCFL, WOWO, and WRNY) seem certain to have been using the short waves for long distance program dissemination.[10,11])

Despite the arrival of the Great Depression and the economic hardships that came with it, HF broadcasting continued into the 1930s and prospered, with more and more receiver manufacturers offering models with shortwave tuning capability. Reception of international stations became something of a novelty and provided set owners with bragging rights about who had received the most distant station. (Most international broadcasters provided confirmation of reception if a listener wrote the station and reported what he or she had heard.) Public awareness of the shortwave bands was stoked with reports of this portion of the spectrum being used for relay of news about public figures in distant locations (Admiral Richard Byrd's polar explorations and Amelia Earhart's flying feats were prominent examples).

As the decade ended and hostilities began to erupt in Europe and the Far East, shortwave broadcasting also proved useful in providing U.S. citizens with war news and information about international political situations, as broadcast journalism was still in its infancy and U.S. radio networks had yet to develop foreign news bureaus and the frequent live reporting from distant locales that would evolve later.

WARTIME HF BROADCASTING

With the spread of war to a global basis in the early 1940s, HF radio found itself being used as a propaganda tool by leaders of "belligerent" nations for espousing their beliefs internationally. (At the time of the entry of the United States into WWII in December 1941, Germany was operating some 68 shortwave transmitters and Japan had at least 42.) Operating powers were reported to be as high as 200 kW.[12] However, up until that point the United States government had shown little interest in HF broadcasting, leaving this enterprise in the hands of the private sector (CBS, Crosley, General Electric, RCA, Westinghouse, and the Worldwide Broadcasting Corp.).

This changed shortly after the Japanese attack on Pearl Harbor, with the United States launching its own international service under the auspices of a wartime agency (the Office of War Information, or OWI). The Voice of America took to the air in early 1942 through leased facilities of the commercial shortwave broadcasters, and a crash program was launched to construct three government-owned HF stations with 200-kW capabilities in California and Ohio. (These stations, located in Delano and Dixon, California, and Bethany, Ohio, continued in operation for several decades after the end of the war.) HF radio was also used by the United States to provide news and entertainment to U.S. troops via the Armed Forces Radio Service, with transmissions handled by both private sector and government-owned shortwave facilities. This practice continued well into the communications satellite era.

> As an historical note, during the war years the U.S. government mandated that shortwave-capable radios in the hands of non-U.S. citizens living in the United States had to have the shortwave tuning capability disabled, as it was feared that shortwave broadcasts might be used to issue instructions to "fifth columnists" or saboteurs. Radio servicemen helped ensure compliance by removing coils or otherwise disabling HF tuning components.

The postwar years and the outbreak of the Cold War between the Soviet Union and the United States spurred a great rise in international HF broadcasting as a global propaganda tool. Both nations committed many millions of dollars (and rubles) to expand shortwave broadcasting activities. The Soviet Union also funded a number of high-power jamming operations in an attempt to block the availability of foreign broadcasts from Western nations to its citizens.

These high-power propaganda broadcasts and jamming operations continued until the end of the Cold War period in the 1980s.

DIFFERENCES BETWEEN HIGH-FREQUENCY AND MEDIUM-WAVE BROADCASTING OPERATIONS

As pointed out in the above discussion, the operation of a shortwave broadcast facility is quite different from operating an AM broadcast station, with the necessity to continuously follow frequency band openings and closings that occur within a 24-hour period and which vary from season to season. Medium-wave transmitters are generally designed to operate at one specific frequency, with RF tank circuit components factory-selected for operation within a fairly narrow frequency range (high-band, mid-band, and low-band AM broadcast band). Transmitters engineered for HF broadcasting must provide a much greater tank circuit tuning range and be designed to facilitate quick changes over a wide range of frequencies. (The author is familiar with one high-power commercial HF transmitter model that featured interchangeable tank circuits. The coil/capacitor unit was housed in a metal enclosure about the size of two 4-drawer filing cabinets. The assembly was on wheels and connected with the transmitter's RF amplifier via a set of large contacts. When operations on one frequency ceased, technicians stood by to physically wheel the tank unit away from the rear of the transmitter and roll in another unit, which had been pre-tuned to the new frequency band. After some minor fine-tuning, the transmitter was ready for service. Band changing operations could typically be accomplished in 30 seconds or less with this arrangement.) Some more modern transmitters utilize tank circuit inductors with sufficient range to allow operation within all international shortwave bands. These are wound as a spiral with adjacent turns spaced some distance apart, thus allowing a motor-driven contactor to navigate the entire length inductor very rapidly and achieve resonance at the desired operating frequency.

Another difference is methodology for setting transmitter operating frequency. This is governed at MW stations by a crystal-controlled master oscillator that is part of the transmitter and allows the station to operate on a single frequency. As HF stations are changing bands and operating frequencies more or less constantly to follow propagation conditions, crystal control is not practical and most international broadcasters rely on frequency synthesizers to set transmitter operating frequency.

Further, transmitters typically used for international broadcasting in the HF bands are designed to develop considerably more power than is used in MW broadcasting. Units capable of developing output power levels of 250 kW or more are not uncommon in this service, with RF amplifier, modulator, and power supply components being proportionally larger than those encountered in MW transmitters designed for 50 kW or less. RF inductors and associated capacitors are also proportionally much larger, with the high RF peak voltages encountered generally mandating the use of vacuum capacitors.

Methodologies for removing waste heat from modern MW transmitter power devices differ considerably from that employed at HF international broadcasting operations. The use of solid-state radiofrequency power-producing devices is almost universal in MW stations today and in most cases fairly simple air-cooling systems suffice for this purpose. However, many of the large shortwave transmitters in use today still rely on large vacuum tubes for RF amplification, with proportionally large amounts of waste heat energy being generated. These transmitters typically rely on conventional closed-loop pressurized water-cooling systems, with large heat exchangers used for removing heat from the circulating coolant and delivering the waste heat to the atmosphere. Another methodology sometimes employed for removing heat from large power tubes is "vapor phase" cooling. In such systems, water contacting high-power tube anode structures is allowed to boil. The resulting vapor or steam generated has a higher specific heat than liquid water and is able to remove waste heat more efficiently. The steam is allowed to condense and give off the heat to air within a closed chamber, with the condensate returned to the tube anode by gravity flow to continue the cycle. A great advantage with this cooling methodology is the elimination of the pumps and blowers found in conventional air- and water-cooling systems; this results in a savings in operational energy as well as a reduction in maintenance requirements. Additionally, it allows for a somewhat more compact transmitter installation.

A note about water purity is in order in connection with liquid- and vapor phase-cooling systems. As the cooling water is in direct contact with vacuum tube anodes that are operated at thousands (in some instances, tens of thousands) of volts above ground potential, great care must be exercised to ensure that there is little leakage of potential through the liquid-cooling system. Water from which all mineral and other impurities have been removed is a very poor conductor of electricity (few free electrons or ions) and is used to advantage in this situation. Fairly long channels made of nonconductive material (borosilicate glass or ceramics) are used to electrically isolate the cooling system from the tube, and only highly purified water is allowed to flow through these channels and contact the anode structure. With pumped liquid-cooling systems, the purified water flows through a heat exchanger, with fans or blowers removing the excess heat and transferring it outside the transmitter building. While such cooling systems are "closed loops," inevitably replacement or "makeup" water is needed due to losses incurred during tube replacement or other repairs. The quantities can be significant, and very large HF transmitting plants are equipped with distillation systems or ion-exchange column apparatus for purifying well or tap water to the required level.

As noted earlier in this discussion, another difference between MW and high-power shortwave transmitters are the high anode operating potentials and currents associated with large power tubes. When such vacuum devices are operated at elevated potentials, x-rays are inevitably generated by the action of accelerated electrons striking and being stopped by the anode structure (*bremsstrahlung* radiation). With the higher operating voltages utilized in such high-power transmitters, the x-rays generated have sufficient energy to pass through the tube and associated cooling water jackets and escape the transmitter cabinet, thus posing an ionizing radiation hazard to personnel working nearby. The discovery of this radiation hazard has required the installation of shielding material within some transmitters and the posting of radiation hazard warnings.[13] Such ionizing radiation hazards normally do not exist in AM broadcasting stations, even those still using legacy vacuum transmitters, as at operating power levels commonly employed in medium-wave broadcasting, anode potentials are not great enough to generate x-radiation with sufficient penetrating power to be harmful. (Such x-ray generation potential does exist, however, in high-power television transmitters employing klystrons or inductive output tubes.)

Common to transmitter plants in all broadcast services are the requirements for vigilance and the observance of safety measures when working around RF sources and both high-voltage transmitter DC sources and associated AC power distribution systems. The need for such cautions is proportionally increased at large HF transmitter plants due to the level of RF power generated and the sometimes significantly higher AC primary voltage levels entering the building.

In the interest of efficiency and energy savings, newer transmitters designed for HF service employ solid-state modulation technology, and a number of older units have been successfully converted from conventional high-level Class B plate (Barton) and other vacuum tube modulation technologies to some form of solid-state modulation. (One popular technology is the so-called "pulse step" modulator, which employs a large number of solid-state power supplies of varying voltages connected in series and turned on and off at an audio rate by digital bit combinations derived from digitization of the audio signal being transmitted. The output voltage levels of the supplies are weighted, with that producing the highest voltage corresponding to the most significant bit in the digital word created and those with lesser outputs corresponding to less significant bits. The summed output of these supplies provides plate potential to the RF amplifier tube or tubes. Amplitude modulation of the carrier is thus achieved. The use of such technology eliminates the large audio power tubes, iron-core inductors, and transformers found in conventional plate modulation schemes.)

The transfer of such RF power levels to transmitting antennas also requires physically larger transmission lines than those typically employed in medium-wave broadcasting. At international broadcasting operations, these typically take the form of balanced feeders, with two large conductors separated at regular intervals by insulated spacers; however, "open" coax unbalanced lines may also be employed. These are constructed with five or more individual wires supported and spaced at regular intervals, with the insulators designed to place one of the conductors in the center and surround this conductor to form a pseudo coaxial "shield" arrangement. While not in common use, very large Heliax-type corrugated shield coaxial transmission line is available for HF transmission line applications also.

As the typical shortwave transmitting station has a multiplicity of antennas directed at specific geographic locations, it is necessary during normal operating cycles to switch these antennas between one or more transmitters. This is done via contacting devices sufficiently large to handle the very large RF currents in use. Many schemes have been devised for such switching applications, with some of the earliest designs based on manually operated "cross-bar" switching architecture, located outside the transmitter building, and operated manually.[14] More modern plants rely on remotely controlled switching, housed either in the main transmitter building or in a nearby shelter.

ANTENNAS USED FOR HF BROADCASTING

Although virtually any established type of transmitting antenna could be employed for use in the HF band (one early station is known to have employed an insulated-base, series-fed radiator[15]), international broadcasters are generally interested in aiming their signals at a particular region of the world and employ highly directional antennas to do this. Curtain arrays, log periodics, and rhombics are typically utilized for this purpose.

"Curtains" with multiple wire radiators are especially attractive, as they can provide gains of approximately 20 dB region and be designed so that their patterns can be electronically "slewed" by as much as 45 to 50 degrees in azimuth and some 25 degrees in elevation. Despite these advantages, curtain radiators are complex, costly to erect, difficult to maintain, and are not efficient over the whole range of shortwave frequency bands, necessitating the use of separate arrays to cover the upper, medium, and lower portions of the HF broadcasting band. Log periodic antennas constructed with wire elements are less complex than curtain arrays, can provide substantial gain and be designed to be highly directional, making these the antennas of choice for some services. The simplest HF antenna to construct and maintain is the rhombic, which consists of little more than four supporting masts arranged in a "diamond" pattern, with one end aimed at the great circle heading of the desired region of coverage. Wire conductors are suspended from the masts, with power fed to the conductors where they converge at the "back end" of the antenna and a suitable terminating resistor used to join the conductors at the other end of the array. Such antennas do provide moderate gain

FIGURE 2.6-1 Example of a curtain array

and are fairly directional. A big disadvantage is the power-wasting termination resistor. It is possible to eliminate this resistor by replacing it with a properly designed transmission line to feed the otherwise wasted RF energy in-phase back into the feeder end of the antenna. The disadvantage of this "re-entrant" rhombic approach is that the bandwidth of such an antenna is of necessity quite narrow.

It should be noted that despite the overwhelming popularity of curtains, log periodics, and rhombics, a number of other of antenna designs may also be used at HF broadcasting plants, including simple dipoles and dipole/reflector arrays.

MODULATION MODES EMPLOYED IN INTERNATIONAL HF BROADCASTING

Historically, the overwhelming majority of shortwave broadcasting has been done with conventional double sideband amplitude modulation, due to simplicity of both transmitter and receiver design and the listener's ease of tuning-in broadcasts. However, a proposal within the ITU a number of years ago mandated a conversion to single sideband operation at the end of 2015 in the interest of spectrum conservation. This requirement was later dropped, with the ITU suggesting that international HF broadcasters should consider moving to digital modulation formats such as DRM (Digital Radio Mondiale).

OTHER HF SPECTRUM USES

In addition to the transmission of news and entertainment to distant audiences, the HF spectrum has been employed for a number of other uses. Prior to the successful laying and inauguration of service in 1956 of a transatlantic submarine cable with audio capability, all telephone calls between North America and Europe were conducted via shortwave radio, a service that was initiated in 1928. Initially, overseas telephone service users were afforded no privacy, as anyone who possessed a receiver with HF tuning capability could "listen in" to calls being made. However, as full duplex telephony was achieved via the use of separate "outbound" and "inbound" frequencies, two receivers would have been necessary to overhear both sides of the conversation. Overseas telephone service providers later adopted rudimentary "scrambling" technology to make such eavesdropping more difficult. Due to the fading nature of HF communications, dual and triple frequency diversity operation was used to maintain the clarity and intelligibility of audio. Radio networks frequently used this overseas HF telephone service to receive news reports from foreign correspondents and exchange live program material with other domestic broadcasters. Surviving recordings from the 1940s and later attest to the quality of service that could be provided in most cases. Occasionally, however, unusual geophysical or meteorological conditions could cause interruptions of service for varying periods.

FIGURE 2.6-2 VOA Greenville, NC, Site B antennas

FIGURE 2.6-3 Open wire transmission lines feeding antennas

FIGURE 2.6-4 250 kW shortwave transmitter at former VOA Delano, CA, site

HF radio was also used to provide passengers on ocean-going vessels with telephone service and for other maritime communications. This continued well into the 1980s, until improvements in communication satellite technology augmented such service and eventually replaced it entirely. HF was also used for many years to provide communication between aircraft and ground facilities. Even after the post-WWII opening of VHF spectrum for aircraft communication purposes, it was many years before the use of HF radio in aircraft ended, as prior to the implementation of communications satellites this was the only means of communication for aircraft flying long overwater routes.

The shortwave bands are still used to provide precision time and frequency information by government-operated stations such as the U.S.'s WWV,

FIGURE 2.6-5 Electrical distribution system at VOA Greenville, NC, Site B

FIGURE 2.6-6 Audio routing and monitoring of multiple shortwave transmitters

FIGURE 2.6-7 Control console for a 500-kW shortwave transmitter

which is operated under the auspices of the National Institute of Standards and Technology (NIST) and transmits continuously on 2.5, 5, 10, 15, and 20 MHz, with carrier frequencies and time signals derived from "atomic" clocks. In addition, to the very precise one-second interval "ticks," WWV also transmits a digital time code and voice announcements of the exact time and other information, including maritime storm warnings, geophysical information such as solar activity, and shortwave propagation information. The stations operate from Boulder, Colorado, and are receivable over most of North America. (NIST operates a similar facility, WWVH, in Kauai, Hawaii, to provide additional coverage.) Similar HF time and frequency information is provided by Canada's CHU, which is operated by that country's National Research Council's Institute for National Measurement Standards, and operates on 3.3, 7.85, and 14.67 MHz. CHU's signals reach much of eastern Canada and can be heard in the many United States locations.

(As an historical note, for a number of years in the 1930s, the FCC authorized what came to be termed "Apex" wide-band or increased fidelity AM audio broadcasting in both upper U.S. MF broadcast spectrum, the upper portion of the HF band [approximately 25 MHz], and VHF spectrum [termed then "UHF"]. While a number of broadcasters experimented with Apex operation, it was largely a failure due to the inability of most consumer shortwave receivers to tune above 20 MHz and the fact that propagation in this region did not always allows signals to reach the intended local audiences. The service was universally discontinued by its adopters, as frequency modulation technology was developed within the same period and ultimately proved much more attractive for high fidelity delivery of programming.)

Another use of the HF spectrum was developed in the latter half of the 20th century for military purposes, and became somewhat controversial, due to interference caused to other users of shortwave communications. Several countries, including the United States and the former Soviet Union, developed radar surveillance systems that could "see" over the horizon, a limitation that existed with the initial VHF (and later higher) radio frequencies used in radar technology. By using lower frequencies, the ionosphere could serve as a tool for sending RF pulses over great distances and for relaying returned signals from the ground. There are inherent disadvantages in using the substantially longer wavelengths for imaging of distant objects, including the requirement for very large amounts of transmitted power (while there is no published literature available to support this, it is generally believed that the Soviet installations in Ukraine operated with ERPs of 10 megawatts or greater) and the necessity of constructing very large antenna arrays to provide sufficient resolution of distance ground-based and aerial objects.

While such over-the-horizon (OTH) radar is capable of imaging objects at great distances (2,000 km or more), to provide the sort of resolution to be militarily useful required a transmitted signal with a very narrow beamwidth. To achieve such a beamwidth at

HF frequencies, a very large antenna has to employed (a one-degree beamwidth necessitates an antenna stretching one kilometer or more). Further, to be of use in detecting objects at great distances, the "takeoff" angle of the transmitted signal has to be low, relative to the surrounding terrain, requiring an antenna that is ponderously large in the other dimension also. When surveillance photographs of the Soviet Ukraine OTH antenna were made available, it was dubbed "the steel-yard" by Western interests due to the massive amount of supporting structures employed.

It is believed that the Soviet government ordered the construction of several such OTH radars; however, the Ukraine installation was first discovered through triangulation techniques when it became operational in 1976. Its operation was no secret, as the very powerful pulses in the HF band were heard worldwide and interfered with legitimate users of the shortwave bands. This included international broadcasters, amateur radio operators, and aeronautical and maritime communicators (even Radio Moscow broadcasts were interfered with). As the RF pulses were typically transmitted at a 10-Hz repetition rate, the resultant tapping sort of interference was quite annoying and was referred to by those being interfered with as "the Russian Woodpecker." Operating frequencies of this (and other OTH radars) changed regularly to follow ionospheric conditions, with the Ukraine installation generally operating in the 8- to 20-MHz region.

With the increased availability of satellite imaging technology and the easing of Cold War tensions, OTH radar deployment has fallen somewhat out of favor, although some installations still exist and could be made available for use. The Ukraine site was last observed on the air in 1989, with several theories existing as to the reasons for its shutdown. The most likely scenario seems to be that as it was located in the "exclusion zone" established after the 1986 Chernobyl nuclear reactor disaster, radiation levels and general radioactive contamination of the area have made it difficult or dangerous for operating and maintenance personnel. Also, the ending of the Cold War within that period may have also contributed to its demise.

IONOSPHERIC HEATING EXPERIMENTS

While obviously not "broadcasting," the reader should be aware of experimental work engaged in by the United States and others, which involves directing very high intensity radiofrequency energy in the 3- to 30-MHz range towards the ionosphere. Such operations are referred to as "ionospheric heating," and employ transmitters with very high output powers and large arrays of high-gain antennas. The objective of such research is to study the effects of radiowaves on the ionosphere itself, and is generally conducted during periods of low activity from natural phenomena, such as geomagnetic storms, which can affect ionospheric conditions. Perhaps the best known of the ionospheric heating facilities is the U.S. HAARP (High-frequency Active Auroral Research Program) facility located in Alaska. It is said to be capable of producing HF-band RF energy at levels of up to 10 gigawatts.

INTERNATIONAL BROADCASTING AT LOWER FREQUENCIES

A number of countries still engage in another form of international broadcasting outside of the HF spectrum, and while this is not "shortwave," mention is made for the sake of completeness. These broadcasters operate in the LF region (30 kHz to 300 kHz) of the spectrum and are termed "longwave" broadcasters. As radiowaves in this region propagate almost exclusively along the surface of the earth with little impediment from terrain, they are able to span vast distances (1,000 miles or more) across land and water, with none of the signal strength variation or selective fading that characterizes shortwave propagation. While longwave entertainment broadcasting never became popular in the North America, it was adopted early in radio's history in certain portions of Europe, Asia, and the Middle East. As of early 2016, longwave broadcasting continues on some 30 stations in such countries as France, Germany, Morocco, Iceland, Algeria, Norway, Denmark, Mongolia, Luxembourg, Turkmenistan, Romania, Poland, and the United Kingdom.[16]

Although longwave broadcasting is able to deliver programming to very large audiences with a signal level little affected by night/day or seasonal changes, there are a number of disadvantages. The greatest of these is the design and construction of efficient transmitting antenna structures. (At approximate band center of 150 kHz, a half-wave vertical radiator would have a height of 1,000 meters [3,280 feet or approximately 0.62 mile]). A corresponding ground radial system would be equally ponderous. To overcome the inefficiency of reasonably sized antennas, longwave broadcasters have to resort to very large input power levels (in some cases, one megawatt or greater), with proportionally high operating expenses. Another disadvantage of longwave broadcasting is the limitation on transmission of higher audio frequencies due to the long periodicity of the carrier wave.

Frequencies within the longwave spectrum have been assigned in North America for aircraft navigation purposes (nondirectional beacons), and NIST provides time signals at 60 kHz via its WWVB operation located in Colorado. (A big advantage of using longwave spectrum in this application is the very accurate estimation of the time required for such signals to reach the user's receiver. Precise path length prediction is not possible at HF frequencies.)

SHORTWAVE BROADCASTING IN THE 21ST CENTURY

With the rise of the Internet availability of an almost ever-increasing number of undersea fiber optic paths and communication satellite transponders, international shortwave broadcasting activity has fallen considerably from its "golden" period in the mid-1900s. Several of the larger international shortwave broadcasting entities, including the Voice of America, Radio Canada International, BBC World Service, Deutsche Welle, and others, have sharply reduced or altogether eliminated some shortwave broadcasting activities, moving instead to newer modalities for reaching global audiences. However, there is still considerable interest in using shortwave radio for reaching distant audiences. This is evidenced by the listing in the most recent edition of the *World Radio TV Handbook* of known governmental or private shortwave operations in no fewer than 82 countries.[17] A number of long-established manufacturers continue to offer radio receivers with shortwave tuning capability and enhanced feature sets, attesting to the fact that there is still a market for such devices. While international shortwave broadcasting may never again experience the popularity it once enjoyed, it still remains viable and will likely continue well into this century.

Notes

[1] Winters, S.R., "100 Meter C.W. Transmitter," *Radio News*, Nov. 1923, pp. 530–531. [The article describes the successes achieved in long-distance communication at approximately 3 MHz with only a small amount of power.]

[2] Austin, L.W., "Receiving Measurements and Atmospheric Disturbances at the United States Naval Radio Research Laboratory, Bureau of Standards, Washington, November and December, 1922," *Proceedings of the Institute of Radio Engineers*, Apr., 1923, p. 83.

[3] "WRNY's New 31-Meter Transmitter," *Radio News*, Aug., 1927, p. 127.

[4] Terman, Frederick E., *Radio Engineering*, 2nd edition, McGraw Hill Book Co. Inc., New York, 1937, pp. 602–605.

[5] Berg, Jerome S., *The Early Shortwave Stations*, McFarland & Company, Inc. Publishers, Jefferson, NC, 2013, pp. 59, 60.

[6] "KDKA's Powerful Short Wave Station," *Radio News*, Sept. 1924, pp. 292, 293, 422.

[7] Conrad, Frank, "Short Wave Radio Broadcasting," *Proceedings of the Institute of Radio Engineers*, Dec. 1924, p. 723.

[8] Breckel, Lieut H.F., "In the Future—Intermediate- or Short-Wave Broadcasting?" *Radio News*, Nov. 1927, pp. 460–461.

[9] "Foreign Broadcast Stations," *Citizens Radio Call Book Magazine and Scientific Digest*, Spring Edition, Jan. 1929, pp. 34, 35.

[10] Breckel, Lieut H.F., "In the Future—Intermediate- or Short-Wave Broadcasting?" *Radio News*, Nov. 1927, pp. 460–461.

[11] "The List of Broadcast Stations in the United States," *Radio News*, Nov. 1927, pp. 401, 505.

[12] Rockwell, R.J., "OWI 200-kW Transmitters at Bethany, Ohio," *Communications Magazine*, Nov. and Dec., 1944.

[13] Hunter, E. Bruce, "X-Ray Emission from Broadcast Transmitters," *IEEE Transactions on Broadcasting*, vol. 36, issue 1, March 1990, pp. 1–10.

[14] Ibid.—Rockwell.

[15] "KDKA's Powerful Short Wave Station," *Radio News*, Sept. 1924, pp. 292–293.

[16] *World Radio TV Handbook*; 2015 edition, WRTH Publications Ltd., Oxford, vol. 69.

[17] Ibid.—*World Radio TV Handbook.*

CHAPTER

2.7

Worldwide Standards for Analog and Digital Television

ALDO G. CUGNINI

AGC Systems LLC
Long Valley, NJ

INTRODUCTION

Television systems throughout the world incorporate, as their basis, a method for transmission of the synchronized combination of sound and moving pictures, first realized by analog means, and later, by digital technologies. Fundamentally, the distribution of audio-visual content relies on the multiplexing of three different signals: audio, video, and synchronizing information. Although many regions of the world have now adopted digital television broadcasting, it is useful to consider legacy analog transmission systems, not only for historical reasons, but also because the fundamentals of color image processing in use by digital systems have evolved from principles developed for analog television systems. In addition, some pockets of analog broadcasting still remain, albeit usually in a transition scenario; likewise, some professional and consumer television equipment continues to use analog signals for device interfacing, and archives of analog audio-video recordings are still abundant.

ANALOG TELEVISION

The Eleventh Plenary Assembly of the International Radio Consultative Committee (CCIR), held in Oslo in 1966, was an attempt at standardization of analog color television systems by the participating countries of the world. Preceding that date, various systems, incompatible with each other, had been developed around the globe, the difference having emerged as a consequence of the differing region-based communications channel allocations, power source characteristics, and international telecommunications agreements, as well as for political considerations. Although CCIR issued a report describing the characteristics of a variety of proposed systems, it was ultimately left to the controlling organizations of the individual countries to make their own choice as to which standard to adopt.

MONOCHROME-COMPATIBLE ANALOG COLOR TV SYSTEMS

In order to achieve success in the deployment of an analog color television system, it was recognized as essential that the color system be compatible with the existing black-and-white system, in both a forward and backward sense. The first such analog color television system to be placed into commercial broadcast service was developed in the United States. On December 17, 1953, the Federal Communications Commission (FCC) approved transmission standards and authorized broadcasters, as of January 23, 1954, to provide regular service to the public using these standards. This decision was the result of the work of the National Television System Committee (NTSC), upon whose recommendation the FCC action was based [1]; subsequently, this system, commonly referred to as the NTSC system, was adopted by Canada, Japan, Mexico, and a handful of other countries.

The countries of Europe delayed the adoption of an analog color television system, and between 1953 and 1967 a number of alternative systems were devised that were compatible with the existing 625-line, 50-field monochrome systems. Most of the basic techniques of NTSC were incorporated into the other analog television systems. For example, the use of wideband luminance and relatively narrowband chrominance, utilizing the principle of "mixed highs" (discussed later), can be found in all analog systems. Similarly, the concept of using horizontal interlace to reduce the visibility of the color subcarrier is incorporated in all approaches; this feature reduces the visibility of signals carrying color information that are contained within the same frequency range as the coexisting monochrome signal, thus maintaining a high degree of visual compatibility.

An early analog system that received approval was one proposed by Henri de France of the Compagnie Française de Télévision of Paris (later acquired by Thomson). It was argued that if color could be relatively band-limited in the horizontal direction, it could also be band-limited in the vertical direction; thus, the two components of color information (hue and saturation) that must be added to the one component of monochrome information (brightness) could be transmitted using subcarrier modulation that is sequentially transmitted on alternate lines—thereby avoiding the possibility of unwanted crosstalk between color signal components.

At the receiver, a one-line memory, (i.e., a *1-H* delay element), must be employed to store one line to then be concurrent with the following line, and then a linear matrix of the red and blue signal components (*R* and *B*) is used to produce the third, green, component (*G*). Of course, this necessitates the addition of a line-switching identification technique. This approach, designated as *sequential couleur avec memoire* (SECAM; the phrase translates as "sequential color with memory"), was developed and officially adopted by France and the former U.S.S.R., and broadcast service using the technique began in France in 1967.

The need for a *1-H* delay element led to the development of the phase alternation by line (PAL) system, largely through the efforts of Walter Bruch of Telefunken GmbH. This approach was aimed at overcoming an implementation challenge of NTSC that requires a high degree of phase and amplitude integrity (skew symmetry) of the transmission path characteristics about the color subcarrier to prevent color quadrature distortion. The line-by-line alternation of the phase of one of the color signal components averages out any colorimetric distortions to the observer's eye.

The system in its simplest form (simple PAL), however, results in line flicker (sometimes called Hanover bars[1]). The use of a *1-H* delay device in the receiver greatly alleviates this problem (standard PAL), but also results in the requirement of a line-identification technique.

The standard PAL system was adopted by numerous countries in continental Europe, as well as in the United Kingdom. Public broadcasting began in 1967 in Germany and the United Kingdom using two slightly different variants of the PAL system.

NTSC, PAL, AND SECAM SYSTEMS OVERVIEW

Because many basic techniques of NTSC are involved in PAL and SECAM, a thorough knowledge of NTSC can aid the understanding of PAL and SECAM. The same R, G, and B primary color imaging and display devices are used in all systems. The basic camera function is to capture the spectral distribution of the light from the scene in terms of its red, green, and blue components on a point-by-point basis as determined by the scanning rates. The three resulting electrical signals must then be transmitted over a band-limited communications channel to control the three-color display device to make the perceived object colors at the receiver appear essentially the same as the perceived object colors at the scene.

Color can be described as a psycho-physical property of light—specifically, as the combination of those characteristics of light that produce the sensations of brightness, hue, and saturation. Brightness refers to the relative intensity; hue refers to that attribute of color that allows separation into spectral groups perceived as red, green, yellow, and so on (in scientific terms, the *dominant wavelength*); and saturation is the degree to which a color deviates from a neutral gray of the same brightness—the degree to which it is "pure," "pastel," or "vivid." These three characteristics represent the total information necessary to define or recreate a specific color stimulus. This concept is useful to communication engineers in the development of encoding and decoding techniques to efficiently compress the required information within a given channel bandwidth.

A preferred signal arrangement was developed that resulted in reciprocal compatibility with monochrome pictures and was transmitted within the existing monochrome channel, as shown in Figure 2.7-1. One signal (luminance) is chosen in all approaches to occupy the wideband portion of the channel and to convey the brightness as well as the detail information content. A second signal (chrominance), representative of the chromatic attributes of hue and saturation, is assigned less channel bandwidth in accordance with the mixed-highs principle: in human vision, full three-color reproduction is not required over the entire range of resolution.

In order to fit the new color-encoded signal within the same transmission bandwidth as the legacy monochrome signal, another fundamental principle employed in all systems involves arranging the chrominance and luminance signals within the same frequency band without excessive mutual interference. The vertical scanning process, being equivalent to a sampled-data process, produces signal components largely concentrated in uniformly spaced groups across the channel width. This realization led to the introduction of the concept of horizontal frequency interlace (dot interlace): the color subcarrier frequency is chosen so as to be an odd multiple of one-half the line rate (in the case of NTSC), such that the phase of the subcarrier is exactly opposite on successive scanning lines. This substantially reduces the subjective visibility of the luminance dot-pattern components generated by the color signal.

The major differences among the three main analog systems of NTSC, PAL, and SECAM are in the specific modulating processes used for encoding and

- **COMPATIBILITY WITH CO-EXISTING MONOCHROME SYSTEM.**
- **ENCODE WIDEBAND R, G, B COLOR PRIMARY SIGNALS.**
  - **WIDEBAND LUMINANCE (BRIGHTNESS)**
  - **NARROW-BAND MODULATION OF A COLOR SUBCARRIER (Hue and Saturation)**

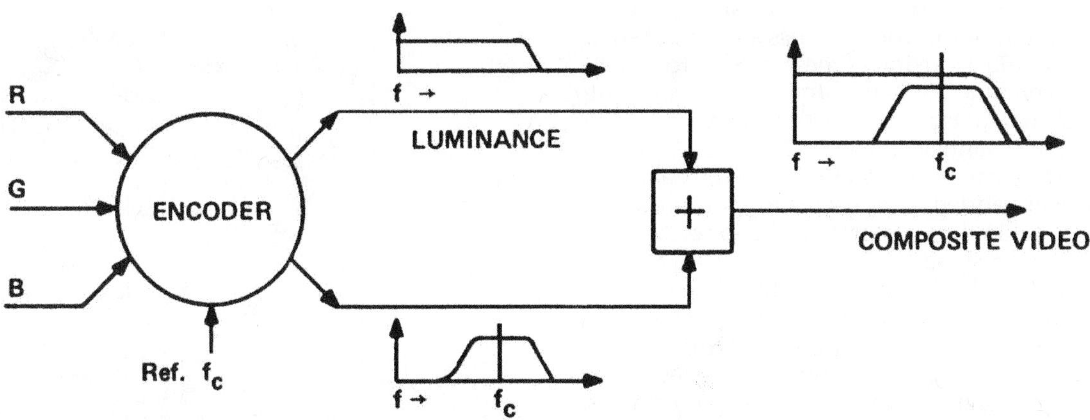

- **SUBCARRIER FREQUENCY INTERLACE**
  - **ODD MULTIPLE OF ½ H TO REDUCE VISIBILITY OF CHROMINANCE INFORMATION SUBCARRIER.**

FIGURE 2.7-1 Original preferred approach to compatible analog color TV systems

transmitting the chrominance information. The similarities and differences are summarized here:

- All systems:
  - Three primary additive colorimetric principles
  - Similar camera pick-up and receiver display technology
  - Wideband luminance and narrowband chrominance
- Compatibility with coexisting monochrome system:
  - First-order differences:
    - Line number
    - Field/frame rates
    - Bandwidth
    - Frequency allocation
- Major differences in color encoding techniques:
  - NTSC—Phase-and-amplitude quadrature modulation of interlaced subcarrier
  - PAL—Similar to NTSC but with line alternation of one chrominance component
  - SECAM—Frequency modulation of line-sequential color subcarriers

NTSC Color System

The colorimetric concepts of brightness, hue, and saturation comprise the three pieces of information necessary to analyze or recreate a specific color value; this information directly maps into the components defining the composite color television NTSC format. The luminance, or monochrome, signal is formed by the addition of specific proportions of the red, green, and blue signals and occupies the total available video bandwidth of 0 to 4.2 MHz. The NTSC, PAL, and SECAM systems all use the same luminance (Y) signal formation, differing only in the available bandwidths.

The Y signal components have relative voltage values representative of the brightness sensation in the human eye; therefore, the red, green, and blue voltage components are tailored in proportion to the standard luminosity curve at the particular values of the dominant wavelengths of the three color primaries chosen for color television. Thus, the signal makeup for all systems, as normalized to white, is described by Eq. 1 [2]:

$$E'_Y = 0.299E'_R + 0.587\,E'_G + 0.114\,E'_B \qquad (1)$$

The signal of Eq. 1 would be exactly equal to the output of a linear monochrome sensor with ideal spectral sensitivity if the red, green, and blue elements were also linear devices with theoretically correct spectral sensitivity curves. In actual practice, the red, blue, and green primary signals are deliberately made nonlinear to accomplish gamma correction (i.e., adjustment of the input/output transfer characteristic to take into account the nonlinear behavior of the CRT—the cathode-ray display tube) [3]. The prime mark (′) is used to denote a gamma-corrected signal. In order to differentiate between the linear and nonlinear signals, it has been recommended that the linear Y (or E_Y) signal be defined as the *luminance* signal, and that the gamma-corrected Y' (or E'_Y) signal be described as the *luma* signal [4].

In order to produce a backwards-compatible correspondence between the monochrome and color reproduction, signals representative of the chromaticity information (hue and saturation) that relate to the differences between the luminance signal and the basic red, green, and blue signals are generated. These new signals are termed *color-difference* signals and are designated as $R - Y$ and $B - Y$. These signals are generated in a linear matrix, and a version of these modulates a subcarrier that is combined with the luminance component; the final combination is then passed through a common communications channel. At the receiver, the $R - Y$ and $B - Y$ color difference signals are detected, separated, and individually added in the right proportion to the luminance signal in three separate paths to recreate the original R, G, and B signals. (The possible $G - Y$ signal is not transmitted, as the G signal can be resynthesized at the receiver from the Y, $R - Y$, and $B - Y$ signals.)

In the specific case of NTSC, two other color-difference signals, designated as I and Q (which are related to the $R - Y$ and $B - Y$ signals), are formed at the encoder, and these are used to quadrature-modulate the color subcarrier. The complete "composite" NTSC signal, as transmitted in North America and specified by the FCC [5], is as shown in Eq. 2:

$$E_M = E'_Y + E'_I \cos(\omega_{sc}t + 33°) + E'_Q \sin(\omega_{sc}t + 33°) \quad (2)$$

where:

$$E'_Y = 0.30E'_R + 0.59\,E'_G \;+\; 0.11\,E'_B \quad (2a)$$

$$E'_I = -0.27(E'_B - E'_Y) + 0.74(E'_R - E'_Y) \quad (2b)$$

$$E'_Q = 0.41(E'_B - E'_Y) + 0.48(E'_R - E'_Y) \quad (2c)$$

Another reason for the choice of signal values in the NTSC system is that the eye is more responsive to spatial and temporal variations in luminance than it is to variations in chrominance; therefore, the visibility of luminosity changes resulting from random noise and interference effects may be reduced by properly proportioning the relative chrominance gain and encoding angle values with respect to the luminance values. For this reason, the principle of constant luminance is incorporated into the system standard [1], [6].

The signal levels from the three camera sensors are adjusted to be equal when a scene reference white or neutral gray object is being scanned for the color temperature of the scene ambient. Under this condition, the color subcarrier also becomes zero. The colorimetric values have been formulated by assuming that the reproducer will be adjusted for Illuminant C, representing the color of average daylight.

Figure 2.7-2 is a CIE chromaticity diagram indicating the primary color coordinates for NTSC, PAL, and SECAM. It is interesting to compare the available color gamut relative to that of all color paint, pigment, film, and dye processes.

In the NTSC color standard, the chrominance information is carried as simultaneous amplitude and phase modulation of a subcarrier located in the high-frequency portion of the 0- to 4.2-MHz video band. The hue information is assigned to the instantaneous phase of the subcarrier, and saturation is determined by the ratio of the instantaneous amplitude of the subcarrier to that of the corresponding luminance signal amplitude. The choice of the I and Q color modulation components relates to the variation of color acuity characteristics of human color vision as a function of the field of view and spatial dimensions of objects in the scene.

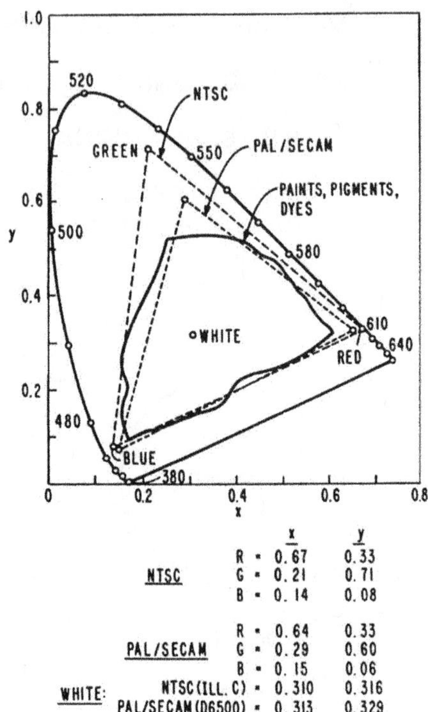

FIGURE 2.7-2 CIE chromaticity diagram

The color acuity of the eye decreases as the apparent size of the viewed object is decreased and thereby occupies a small part of the field of view. Small objects, represented by frequencies above about 1.5 to 2.0 MHz, produce little or no color sensation (mixed highs). Intermediate spatial dimensions (approximately in the 0.5- to 1.5-MHz range) are viewed satisfactorily if reproduced along a preferred orange–cyan axis. Large objects (0 to 0.5 MHz) require full three-color reproduction for subjectively pleasing results. The I and Q bandwidths are chosen accordingly, and the preferred colorimetric reproduction axis is obtained when only the I signal exists by rotating the subcarrier modulation vectors by 33 degrees, as shown in Figure 2.7-3. In this way, the principles of mixed highs and I, Q color-acuity axis operation are exploited.

At the encoder, the Q signal component is band limited to about 0.6 MHz and is representative of the green–purple color-axis information. The I signal component has a bandwidth of about 1.5 MHz and contains the orange–cyan color axis information. These two signals are then used to individually modulate the color subcarrier in two balanced modulators operated in phase quadrature. The sum products are selected and added to form the composite chromaticity subcarrier. This signal is then added to the luminance signal along with the appropriate horizontal and vertical synchronizing and blanking waveforms to include the color-synchronization burst. The result is the total composite color video signal.

Quadrature synchronous detection is used at the receiver to retrieve the individual color signal components. When individually recombined with the luminance signal, the desired *R*, *G*, and *B* signals are recreated. The receiver designer is free to demodulate at either *I* or *Q* and matrix to form, as in many typical receivers, *B* – *Y* and *R* – *Y*, and maintain 500-kHz equiband color signals.

The chrominance information can be recovered without loss of accuracy, given maintenance of the proper phase relationship between the encoding and decoding processes. This is accomplished by transmitting a reference burst signal consisting of eight or nine cycles of the subcarrier frequency at a specific phase [–(*B* – *Y*)] following each horizontal synchronizing pulse.

The specific choice of color subcarrier frequency in NTSC was dictated by at least two major factors. First, the necessity of providing horizontal interlace to reduce the visibility of the subcarrier requires that the frequency of the subcarrier be precisely an odd multiple of one-half the horizontal line rate. Figure 2.7-4 shows the energy spectrum of the composite NTSC signal for a typical stationary scene. This interlace provides line-to-line phase reversal of the color subcarrier, thereby reducing its visibility (and thus improving compatibility with monochrome reception). Second, it is advantageous to also provide interlace of the beat frequency (about 920 kHz) occurring between the color subcarrier and the average value of the FM sound carrier. For compatibility reasons, the legacy sound carrier was left unchanged at 4.5 MHz and the picture-scanning line number remained at 525; thus, the resulting line scanning rate and field rate varied slightly from those for the monochrome values but stayed within the previously existing tolerances. The difference is exactly 1 part in 1000; specifically, the line rate is 15,750 × 1000 / 1001 = 15,734 Hz, the field rate is 60 × 1000 / 1001 = 59.94 Hz, and the color subcarrier is 3.578545 MHz.

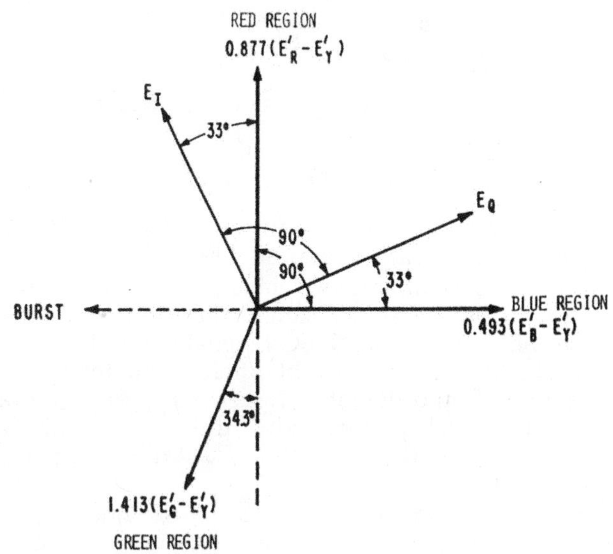

FIGURE 2.7-3 NTSC color-modulation phase diagram

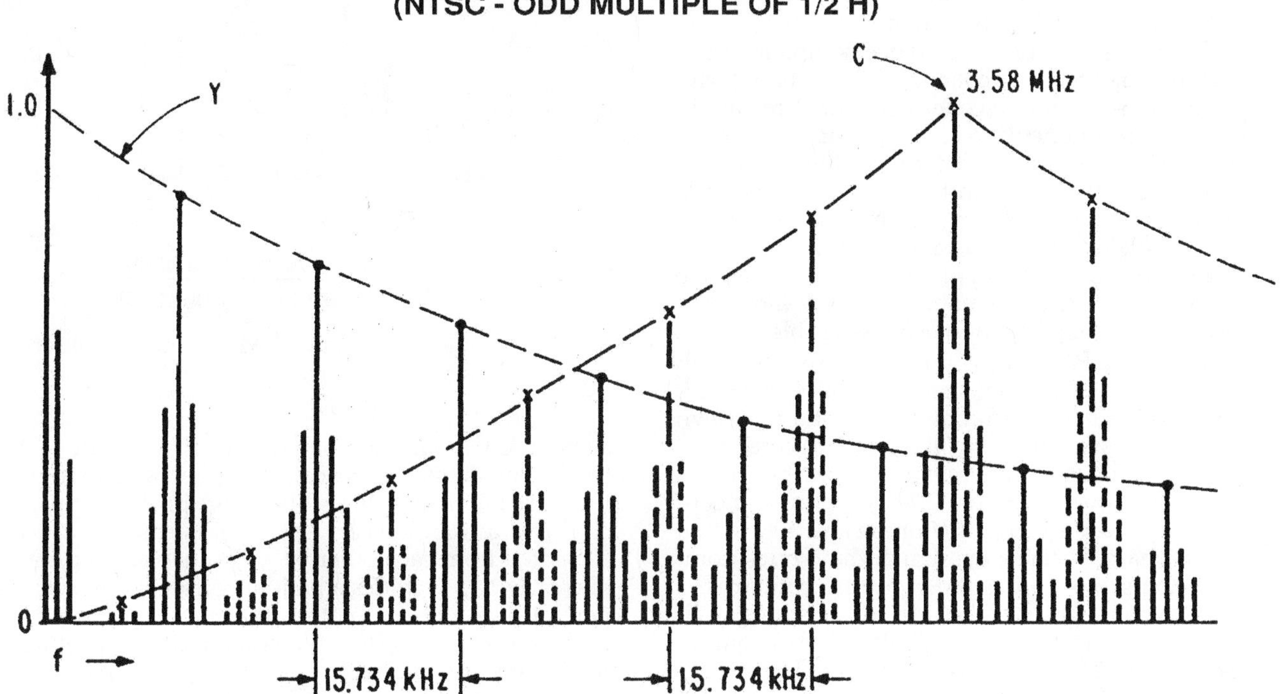

FIGURE 2.7-4 Luminance/chrominance horizontal frequency interlace principle (energy spectrum of the composite NTSC signal for a typical stationary scene)

PAL Color System

Except for some minor details, the color encoding principles for PAL are the same as those for NTSC; however, the phase of the color signal, $E_V = R - Y$, is reversed by 180° from line to line. This is done for the purpose of averaging, or canceling, certain color errors resulting from amplitude and phase distortion of the color modulation sidebands; such distortions might occur as a result of equipment or transmission path problems.

The NTSC chroma signal expression within the frequency band common to both I and Q is given by:

$$C_{NTSC} = \frac{B-Y}{2.03}\sin\omega_{sc}t + \frac{R-Y}{1.14}\cos\omega_{sc}t \qquad (3)$$

The PAL chroma signal expression is given by:

$$C_{NTSC} = \frac{U}{2.03}\sin\omega_{sc}t \pm \frac{V}{1.14}\cos\omega_{sc}t \qquad (4)$$

where U and $\pm V$ have been substituted for the $B - Y$ and $R - Y$ signal values, respectively.

The PAL format employs equal bandwidths for the U and V color-difference signal components that are about the same as the NTSC I signal bandwidth (1.3 MHz at 3 dB). There are slight differences in the U and V bandwidth in different PAL systems because of the differences in luminance bandwidth and sound carrier frequencies. (See the applicable ITU-R documents for specific details.)

The V component was chosen for the line-by-line reversal process because it has a lower gain factor than U and, therefore, is less susceptible to switching rate ($1/2\, f_H$) imbalance. Figure 2.7-5 provides a vector diagram for the PAL quadrature-modulated and line-alternating color modulation approach.

The result of switching the V signal phase at the line rate is that any phase errors produce complementary errors from V into the U channel. Likewise, a corresponding switch of the decoder V channel results in a constant V component with complementary errors from the U channel. Any line-to-line averaging process at the decoder, such as retentivity of the eye (simple PAL) or an electronic averaging technique, such as the use of a *1-H* delay element (standard PAL), produces cancellation of the phase (hue) error and provides the correct hue but with somewhat reduced saturation (this error being subjectively much less visible).

Obviously, the PAL receiver must be provided with some means by which the V signal switching sequence may be identified. The technique employed is known as *AB sync*, *PAL sync*, or *swinging burst*, and consists of alternating the phase of the reference burst by ±45° at a line rate as shown in Figure 2.7-5. The burst is constituted from a fixed value of U phase and a switched value of V phase. Because the sign of the V burst component is the same sign as the V picture content, the necessary switching "sense" or identification information is available. At the same time, the fixed-U component is used for reference carrier synchronization.

Figure 2.7-6 illustrates the manner in which horizontal frequency (dot) interlace of the color subcarrier components with the luminance components is achieved in PAL, summarized as follows. In NTSC, the Y components are spaced at f_H intervals as a result of the horizontal sampling (blanking) process; thus, the choice of a color subcarrier whose harmonics are also separated from each other by f_H (as they are odd multiples of $1/2\, f_H$) provides a half-line offset and results in a perfect dot interlace pattern that moves upward (the so-called "crawling dot" phenomenon). Four complete field scans are required to repeat a specific picture element dot position.

In PAL, the luminance components are also spaced at f_H intervals. Because the V components are switched

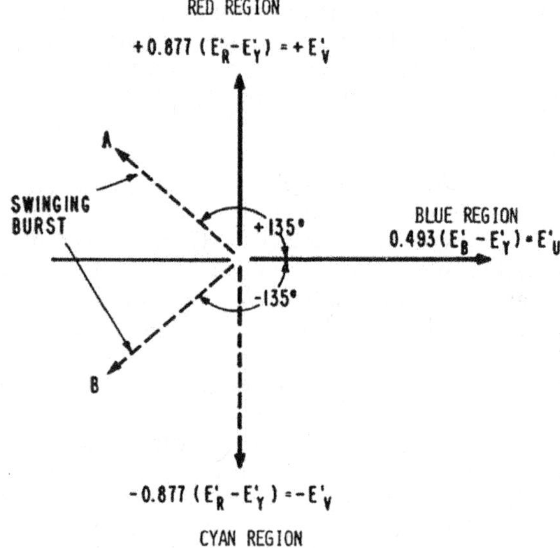

FIGURE 2.7-5 PAL color-modulation phase diagram

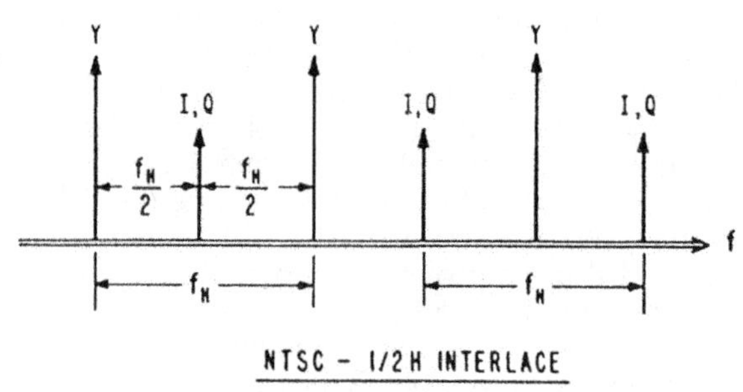

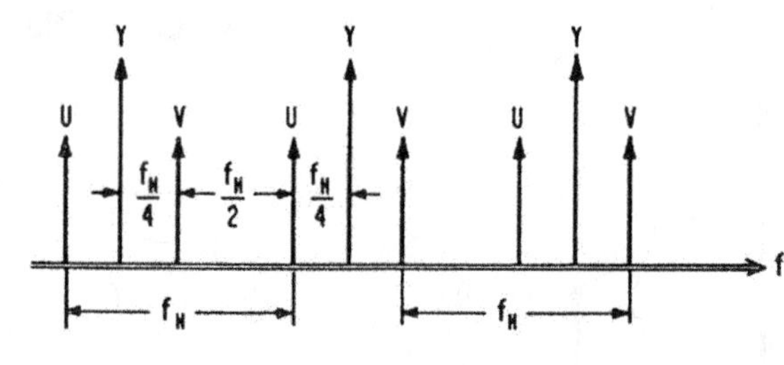

FIGURE 2.7-6 NTSC and PAL frequency interface relationships

symmetrically at half the line rate, only odd harmonics exist, with the result that the V components are spaced at intervals of f_H. They are spaced at half-line intervals from the U components which, in turn, have f_H spacing intervals due to blanking. If half-line offset were used, the U components would be perfectly interlaced, but the V components would coincide with Y and thus not be interlaced, creating vertical stationary dot patterns. For this reason, in PAL a quarter-line offset for the subcarrier frequency is used, as shown in Figure 2.7-6. The expression for determining the PAL subcarrier specific frequency for 625-line/50-field systems is given by:

$$F_{SC} = \frac{1135}{4} f_H + \frac{1}{2} f_V \tag{5}$$

The additional factor $1/2\, f_V = 25$ Hz is introduced to provide motion to the color dot pattern, thereby reducing its visibility. The degree to which interlace is achieved is, therefore, not perfect but is acceptable, and eight complete field scans must occur before a specific picture element dot position is repeated.

One additional function must be accomplished in relation to PAL color synchronization. In all systems, the burst signal is eliminated during the vertical synchronization pulse period. Because, in the case of PAL, the swinging burst phase is alternating line by line, some means must be provided for ensuring that the phase is the same for the first burst following vertical sync on a field-by-field basis; therefore, the burst reinsertion time is shifted by one line at the vertical field rate by a pulse referred to as the *meander gate*.

Simple PAL relies on the eye to average the line-by-line color-switching process and can be plagued with line beats known as Hanover bars, caused by system nonlinearities introducing visible luminance changes at the line rate. Standard PAL employs a *1-H* delay line element to separate U color signal components from V color signal components in an averaging technique, coupled with summation and subtraction functions. Hanover bars can also occur in this approach if an imbalance of amplitude or phase occurs between the delayed and direct signal paths.

In a PAL system, vertical resolution in chrominance is reduced because of the line-averaging processes. The visibility of the reduced vertical color resolution as well as the vertical time coincidence of luminance and chrominance transitions differs depending upon whether the overall system, from transmitter to receiver, includes one or more averaging (comb filter) processes. PAL provides a system similar to NTSC and has gained favor in many areas of the world, particularly for 625-line/50-field systems.

SECAM Color System

The optimized SECAM system, known as SECAM III, is the analog color system adopted by France and the former U.S.S.R. in 1967. The SECAM method has several features in common with NTSC, such as the same E'_Y signal and the same $E'_B - E'_Y$ and $E'_R - E'_Y$ color-difference signals; however, this approach differs considerably from NTSC and PAL in the manner in which the color information is modulated onto the subcarriers. First, the $R - Y$ and $B - Y$ color difference signals are transmitted alternately in time sequence from one successive line to the next, the luminance signal being common to every line. Because there is an odd number of lines, any given line carries $R - Y$ information on one field and $B - Y$ information on the next field. Second, the $R - Y$ and $B - Y$ color information is conveyed by frequency modulation of different subcarriers. Thus, at the decoder, a *1-H* delay element, switched in time synchronization with the line switching process at the encoder, is needed to obtain the simultaneous existence of the $B - Y$ and $R - Y$ signals in a linear matrix, which are then combined with the Y signal to produce the G component.

The $R - Y$ signal is designated as D'_R and the $B - Y$ signal as D'_B. The undeviated frequency for the two subcarriers, respectively, is determined by:

$$F_B = 272\, f_H = 4.250000 \text{ MHz}; \tag{6a}$$

$$F_R = 282\, f_H = 4.406250 \text{ MHz} \tag{6b}$$

These frequencies represent zero color-difference information (zero output from the FM discriminator), or a neutral gray object in the televised scene.

The accepted convention for the direction of frequency change with respect to the polarity of the color difference signal is opposite for the D_B and D_R signals. This choice relates to the idea of keeping the frequencies representative of the most critical color away from the upper edge of the available bandwidth to minimize the instrumentation distortions.

The deviation for D'_R is 280 kHz; for D'_B, 230 kHz. The maximum allowable deviation, including pre-emphasis, for D'_R is –506 kHz and +350 kHz; the values for D'_B are –350 kHz and +506 kHz.

Two types of pre-emphasis are employed simultaneously in SECAM, as shown in Figure 2.7-7. First, a pre-emphasis of the low-frequency color difference signals is introduced. The characteristic is specified to have a reference-level break point at 85 kHz (f_1) and a maximum emphasis of 2.56 dB. The expression for the characteristic is given as:

$$A = \frac{1 + j \cdot (f / f_1)}{1 + j \cdot (f / 3f_1)} \tag{7}$$

A second form of pre-emphasis is introduced at the subcarrier level also, where the amplitude of the subcarrier is changed as a function of the frequency deviation. The expression for this inverted bell shaped characteristic is given in Eq. 8:

$$G = M_0 \frac{1 + 16j \cdot \left(\dfrac{f}{f_C} - \dfrac{f_C}{f} \right)}{1 + 1.26j \cdot \left(\dfrac{f}{f_C} - \dfrac{f_C}{f} \right)}, \tag{8}$$

where f_C = 4.286 MHz and $2M_0$ = 23% of the luminance amplitude (100 IRE). This type of pre-emphasis is intended to further reduce the visibility of the frequency-modulated subcarriers in

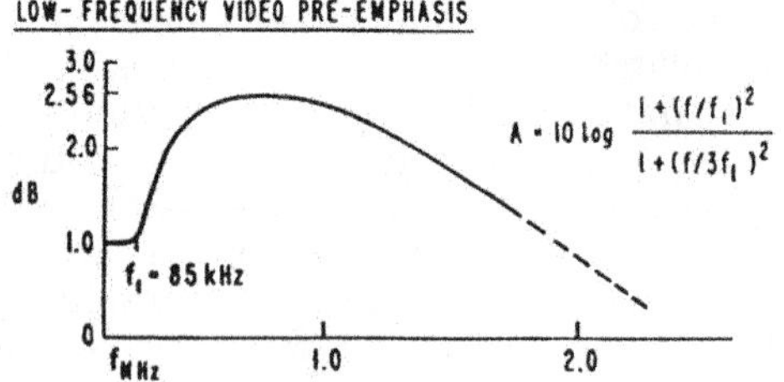

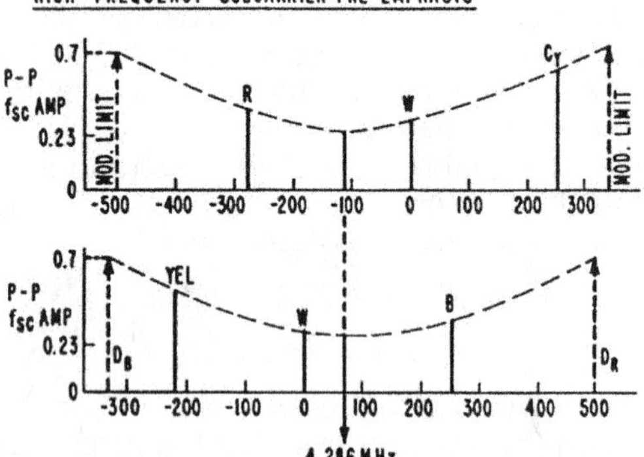

FIGURE 2.7-7 SECAM color signal pre-emphasis

low-luminance-level color values and to improve the signal-to-noise ratio (SNR) in high-luminance and highly saturated colors; thus, monochrome compatibility is better for pastel average picture level objects but sacrificed somewhat in favor of SNR in saturated color areas.

Because the precise interlace of frequency modulated subcarriers for all values of color modulation cannot occur, the visibility of the interference represented by the existence of the subcarriers may be reduced somewhat by the use of two separate carriers, as is done in SECAM. At the undeviated "resting" frequency situation, the two-to-one vertical interlace in relation to the continuous color difference line-switching sequence produces adjacent line pairs of f_B and f_R signals.

To further reduce the subcarrier dot visibility, the phase of the subcarriers (phase carries no picture information in this case) is reversed 180 degrees on every third line and between each field. This, coupled with the bell-shaped pre-emphasis, produces a degree of monochrome compatibility considered subjectively adequate. As in PAL, the SECAM system must provide some means for identifying the line-switching sequence between the encoding and decoding processes. This is accomplished by introducing alternate D_R and D_B color identifying signals for nine lines during the vertical blanking interval following the equalizing pulses after vertical sync. These bottle-shaped signals occupy a full line each and represent the frequency deviation in each time sequence of D_B and D_R at zero luminance value; the signals (which would produce a "fictitious" green color) are used at the decoder to determine the line-switching sequence. During horizontal blanking, the subcarriers are blanked and a burst of f_B / f_R is inserted and used as a gray-level reference for the FM discriminators to establish their proper operation at the beginning of each line; thus, the SECAM system is a line-sequential color approach using frequency-modulated subcarriers. A special identification signal is provided to identify the line-switch sequence and is especially adapted to the 625-line/50-field wideband systems available in France and the former U.S.S.R. It also should be noted that SECAM, as practiced, employs amplitude modulation of the sound carrier as opposed to the FM sound modulation used in other systems.

OTHER SYSTEMS

In the early 1990s, a number of enhanced-definition television (EDTV) systems were proposed for analog terrestrial service, offering a combination of widescreen picture and increased resolution. These systems include PALplus and Enhanced SECAM [7]. Although various EDTV systems based on NTSC were also proposed, that work eventually gave way to the development of incompatible terrestrial digital television systems. Of the various EDTV systems proposed, PALplus saw the most widespread use. The system delivered a 16:9 picture with 574 active lines to a PALplus receiver. Appearing on a conventional television as a 16:9 letterboxed image with 430 active lines, the additional vertical luminance information was carried by a "helper" signal within the letterbox black bars. A transmission in this mode is indicated by a special wide screen signaling (WSS) data carried on line 23. The specification for PALplus was standardized by the European Broadcasting Union (EBU) in 1997 [8], and the signal was transmitted by various broadcasters in the 1990s.

SUMMARY AND COMPARISON OF ANALOG TELEVISION SYSTEMS

In 1998, the successor to the CCIR, the International Telecommunications Union, Radiocommunication Sector (ITU-R), formalized the definition of all worldwide analog television systems then in use in a Recommendation [9], which was subsequently amended [10]. The ITU-R documents define recommended standards for worldwide analog color television systems in terms of the three basic color approaches—NTSC, PAL, and SECAM [4]. The variations—at least 13 of them—are given alphabetical letter designations; some represent major differences, while others signify only very minor frequency-allocation differences in channel spacings or differences between the VHF and UHF bands.

Some general statements can be made comparing the underlying monochrome systems and analog color standards:

- There are three different scanning standards: 525 lines/60 fields, 625 lines/50 fields, and 819 lines/50 fields.
- There are six different spacings of video-to-sound carriers: 3.5, 4.5, 5.5, 6.0, 6.5, and 11.15 MHz.

- Some systems use FM and others use AM for the sound modulation.
- A number of countries using PAL transmission approved the use of a two-sound-carrier FM system [11]. The second carrier is placed at a frequency $f_H \times 15.5 = 242.1875$ kHz above the first (traditional) carrier, at a power level 20 dB below peak visual power, and employing the same modulation as the first sound carrier. The second carrier can carry a second audio program or the right channel of a stereo program (wherein the first carrier carries the $(L + R) / 2$ signal).
- A number of European countries using PAL or SECAM transmission approved the use of an additional digital carrier for stereophonic or multichannel sound transmission [12]. Based on the NICAM-728 system introduced in 1987, a high-frequency subcarrier is digitally modulated with a 728-kbps data stream consisting of 728-bit packets. Using a sampling rate of 32 kHz, the 14-bit samples are digitally companded to 10-bit words.
- Some systems use positive polarity (luminance proportional to voltage) modulation of the video carrier, but others, such as the U.S. NTSC-M system, use negative modulation.
- There are differences in the techniques of color subcarrier encoding represented by NTSC, PAL, and SECAM, and in each case many differences can be found in the details of various pulse widths, timing, and tolerance standards.
- Various countries have their own schemes for using the vertical interval for some form of data. Although each analog standard can accommodate data in various ways, it is up to the individual permitting authority as to what data will be allowed. Examples of such data include closed-captioning, teletext, and the ghost-cancelation reference signal.

DIGITAL TELEVISION

When the possibility of digital transmission of television signals became apparent in the late 1980s, various efforts were started in an attempt to define a new, worldwide set of standards for digital image capture, manipulation, storage, and distribution. The most well-known of these were assembled under the auspices of the Moving Picture Experts Group, MPEG (a working group of the International Organization for Standardization, ISO, and the International Electrotechnical Commission, IEC). At about the same time, work was underway on the practical aspects of high-definition television (HDTV) production and distribution, leading to an effort to develop an analog HDTV system. In the early 1990s, ongoing work based on the encryption of satellite transmissions made it viable to encode and transmit HDTV using a 6-MHz broadcast channel. These interrelated developments spawned the first serious efforts to define digital television systems for satellite, cable, and terrestrial media around the world.

Earlier, in 1982, the Advanced Television Systems Committee (ATSC) was founded in the United States, and was chartered with developing voluntary standards for the emerging technologies of advanced television systems—at first analog—including high-definition television. Ultimately, this led to the development of a digital terrestrial television standard, which became the U.S. digital television (DTV) standard in 1996, and replaced analog broadcasting over full-powered stations in the United States in 2009. (This system is now retrospectively referred to as "ATSC 1.0"; see Chapter 5.10 for more information.) Canada and Mexico also adopted the ATSC system, as did South Korea.

In Europe, work begun in 1991 by the so-called European Launching Group (ELG) culminated in the formation of the Digital Video Broadcasting (DVB) Project, which led in 1993 to a digital satellite standard, DVB-S, followed later by cable and terrestrial standards (DVB-C and DVB-T, respectively). There, the transition to digital (i.e., analog shutoff) started in 2006, and at the time of this writing, only a few countries (mostly in Western Europe) had not yet transitioned. Parallel work in Japan also resulted in another family of digital broadcast standards, called Integrated Services Digital Broadcasting (ISDB), for satellite, cable, and terrestrial broadcasting for that country (later adapted for use elsewhere, as well). Japan's digital broadcast transition started in 2010 and was completed in 2012.

In 2006, the Chinese government ratified its own digital television standard, called Digital Television Terrestrial Multimedia Broadcasting (DTMB), and by 2008, high-definition television programs were broadcast using the DTMB standards in the cities hosting the Beijing Olympic Games. The Chinese digital transition started in 2014, with a final date set for mid-2016.

The difference in initial (analog TV) transmission mediums was probably responsible as much as any other factor for the divergence of the successor digital systems from a worldwide perspective. The standards used in the fixed-service systems for terrestrial digital television broadcasting, and their principal zones of deployment worldwide, are shown in Table 2.7-1. Each of the systems is also described in more detail in the sections below.

Comparison of Systems

The various characteristics of the different transmission systems are due to the varying technologies, objectives, politics, industry structure, and channel usage at the time of the particular system development. To some extent, each region or grouping drew somewhat on what had been developed beforehand when it developed its own system. In addition, each implementation has planned on a degree of sophistication for the receiver, based on the ever-increasing power of semiconductor chips then available. Table 2.7-2 shows the main transmission characteristics of the systems.

The first system to be developed in the early 1990s was the ATSC system. It uses eight-level vestigial sideband (8-VSB) modulation with a fixed level of error

TABLE 2.7-1
Worldwide Terrestrial Digital Television Systems (as currently deployed)

Region	Transmission	Transport	Video	Audio
Africa[2]	DVB-T	MPEG-2	MPEG-2	MPEG-2
Asia				
China	DTMB	MPEG-2	MPEG-2	MPEG-2
India	DVB-T	MPEG-2	MPEG-2	MPEG-2
Japan	ISDB-T	MPEG-2	MPEG-2	MPEG-2/AAC
Russia	DVB-T	MPEG-2	MPEG-2	MPEG-2
S. Korea	ATSC[3]	MPEG-2	MPEG-2	AC-3
Australia	DVB-T	MPEG-2	MPEG-2	MPEG-2, AC-3
Europe	DVB-T	MPEG-2	MPEG-2, MPEG-4 AVC	MPEG-1, -2
North America				
Canada	ATSC	MPEG-2	MPEG-2	AC-3
Cuba	DTMB	MPEG-2	MPEG-2	MPEG-2
Mexico	ATSC	MPEG-2	MPEG-2	AC-3
United States	ATSC[4]	MPEG-2	MPEG-2 MPEG-4 AVC	AC-3
South America				
Argentina	ISDB-T	MPEG-2	MPEG-2	MPEG-2/AAC
Brazil	ISDB-T	MPEG-2	MPEG-2	MPEG-2/AAC

TABLE 2.7-2
Comparison of ATSC (1.0), DVB-T, ISDB-T, and DTMB Systems

Systems	ATSC 1.0	DVB-T	ISDB-T	DTMB
Transmission	Single carrier	Multiple carrier (OFDM)		TDS-OFDM
Bandwidth	6 / 7 / 8 MHz			
Modulation	8VSB	QPSK / 16 QAM /64 QAM	DQPSK/QPSK /16 QAM / 64 QAM	4 QAM / 16 QAM / 32 QAM / 64 QAM
Error correction	Trellis code + RS	Convolutional code + RS		BCH outer code LDPC inner code
Characteristic	Distributed Transmission capability	SFN capability	SFN capability Segmented OFDM Time interleaving	Time interleaving Frequency interleaving
Proposing Region	**USA**	**Europe**	**Japan**	**China**
Standard	**A/52, A/53, A/54**	**EN 300 744**	**ARIB STD-B31**	**GB 20600-2006**

correction and was designed to replicate the coverage of analog NTSC terrestrial television using rooftop antennas, for new digital HDTV broadcasting services.

When the DVB-T system was developed, it was possible to draw on the ATSC system and add new features. These included 2k and 8k Coded Orthogonal Frequency Division Multiplexing (COFDM), which addressed reception in difficult multipath environments, and an agile forward error correction (FEC) level for more flexibility.

Several years later, when the ISDB-T system was developed in Japan, it was possible to draw on the DVB-T system and further add new features. These include a 4k OFDM mode and a multiple frequency-segment structure to bring more flexibility in spectrum management.

Since ISDB-T was developed, the DVB-H system emerged for broadcasting to handheld devices. Although DVB-H does not take advantage of the modular approach of ISDB-T, and was designed only for the concurrent analog TV channeling plans, it does include a 4k OFDM mode and new features including "time slicing," which helps reduce receiver power consumption. It also took advantage of technical progress and the evolution of compression tools in ISO/IEC JTC MPEG, including the addition of high-efficiency MPEG-4 AVC video compression. At this writing, however, the DVB-H system has not been commercially deployed.

A revision of DVB-T, called DVB-T2, added several tools to the standard, including the concept of physical layer pipes (PLPs). More advanced than the FDM-based multiple-segment physical layer previously established in ISDB-T, PLPs were introduced in DVB-S2 and then carried into DVB-T2, and are logical channels within the multiplex carrying one or more services. PLPs allow various elements of the broadcast content to be assigned to different modulation schemes and robustness levels specific to the individual "pipe," all within a single broadcast channel. This can allow, for example, the provision of one service robust enough to be received by mobile devices in HD quality, while another service on the same broadcast channel might be configured for fixed reception in UHD quality, via different PLPs. A number of countries that had already deployed DVB-T services have added DVB-T2 service since it became available. Some concepts utilized in DVB-T2 (such as PLPs) have also been adapted into the ATSC 3.0 standard (see Chapter 8.4).

Subsequently, the Chinese government introduced a different standard for terrestrial digital television, called DTMB (Digital Terrestrial Multimedia Broadcast). It has been deployed for mobile and fixed terminals in the People's Republic of China, Cuba, Hong Kong, and Macau. DTMB resulted from a merger of several standards: ADTB-T (developed by the Shanghai Jiao Tong University), DMB-T (developed by Tsinghua University), and TiMi (Terrestrial Interactive Multiservice Infrastructure), which was proposed by China's Academy of Broadcasting Science.

More detail on each of the DTV standards development organizations (SDOs) mentioned in the historical summary above, and their respective outputs, follows.

ADVANCED TELEVISION SYSTEMS COMMITTEE (ATSC)

The history and technology of ATSC are covered elsewhere in this handbook (see Chapters 5.10 and 8.3). At this writing, ATSC is in the process of developing a next-generation successor standard for terrestrial television, called "ATSC 3.0" (see Chapter 8.4).

DIGITAL VIDEO BROADCASTING (DVB) PROJECT

The DVB Project is a European-originated worldwide alliance of about 250–300 companies, developing specifications for digital media delivery systems, including broadcasting [13]. Its formation in the early 1990s was the result of coinciding ideas about the need for such an initiative from European industry and the European Broadcasting Union (EBU), the collective organization of Europe's national broadcasters. The project is managed by the DVB Project Office, whose staff are employees of the EBU in Geneva, Switzerland, but who work exclusively in the interests of the members of the DVB Project. The members of the DVB Project develop and agree on specifications, which are then passed on to the European standards body for media systems, the EBU/CENELEC/ETSI Joint Technical Committee (JTC), for approval. They appear as European Telecommunications Standards Institute (ETSI) standards or reports [14].

History

The DVB Project began the first phase of its work in 1993. The project's basic philosophy was as follows:

- The initial task was to develop a complete suite of digital satellite, cable, and terrestrial broadcasting technologies in one "pre-standardization" body.
- The systems would be "containers" which carry any combination of image, audio, or multimedia; there is not a one-to-one correspondence between a delivery channel and a program channel.
- The work should result in ETSI standards for the physical layers, error correction, and transport for each delivery medium.

The DVB Project draws extensively on standards from ISO/IEC JTC MPEG. The transport for all systems is the MPEG-2 transport stream [15].

Overview

DVB-S is the specification for the first generation version of the DVB digital satellite system. Other areas include DVB-S2 (the second-generation version of the digital satellite system), DVB-C (the digital cable system), and DVB-T (the digital terrestrial broadcasting system). DVB-H (the digital terrestrial broadcasting system serving handheld receivers), DVB-DATA (the data delivery system), DVB-SI (the service information system), and DVB-MHP (the multimedia delivery system or home platform) were added later. The various systems are summarized in Table 2.7-3.

While some of the features of DVB systems are outlined here, readers may wish to consult the references for more implementation information, including the DVB, ETSI, and DTG websites [13], [14], [16].

Satellite, Cable, and Terrestrial: DVB-S, DVB-C, and DVB-T

The DVB-S system [17] for digital satellite broadcasting was developed in 1993. It comprises a relatively straightforward transmission system using QPSK. The specification describes different tools for channel coding and error protection that are also used for other delivery media systems.

The DVB-C system [18] for digital cable networks was developed in 1994. It is centered on the use of 64 QAM and can convey a complete satellite channel

TABLE 2.7-3
DVB Transmission Systems

	DVB-H	DVB-T	DVB-C	DVB-S	DVB-S2
Channel Bandwidth	5, 6, 7, 8 MHz	6, 7, 8 MHz	6, 7, 8 MHz	Any sat. transp. (e.g., 26–72 MHz) & MCPC/SCPC	Any sat. transp. (e.g., 26–72 MHz) & MCPC/SCPC
Modulation Scheme	QPSK, 16 QAM, 64 QAM on COFDM	QPSK, 16 QAM, 64 QAM on COFDM	16–256 QAM	QPSK	QPSK, 8PSK, 16APSK, 32APSK
Modulation Modes [1]	2K, 4K, 8K	2K, 8K	Constant Coding & Modulation (CCM)	CCM	Adaptive Coding & Modulation (ACM)
FEC	RS(204,188), Conv. Code & MPE-FEC	RS(204,188) & Conv. Code	RS(204,188) & byte interleaver (I=12)	RS(204,188) & Conv. Code	BCH outer code LDPC inner code
Roll-off	N/A	N/A	0.15	0.35	0.35, 0.25, 0.20
C/N range	4.5 dB to 18.4 dB [2]	3.1 dB to 20.1 dB [3]	23.1 dB to 25.1 dB	+4 dB to +8 dB	–2.4 dB to +16 dB
Data Rate in 8MHz channel	4.98~15 Mb/s [4]	4.98–31.67 Mb/s	6.4 (16 QAM)-48.1 (256 QAM) Mb/s	28.1–45.4 Mb/s (in 36 MHz transp.)	35% improvement on DVB-S [5]

[1] The Modulation "Mode" of single carrier systems like DVB-S, DVB-S2, and DVB-C cannot be equivalently defined when describing multicarrier COFDM systems like DVB-T and DVB-H.

[2] The values are calculated using the theoretical C/N figures given in EN 300 744, plus an implementation margin of 2.5 dB, using the UK DTG D-Book noise model with a receiver excess noise source power P_x of –33 dBc and simulated MPE-FEC. An ideal transmitter is assumed, and the quality criterion is Quasi Error Free IP-stream (QEF-IP), meaning an IP-packet error ratio of 10^{-3} calculated over the whole multiplex. (The D-Book is available only to members of the DTG.)

[3] C/N calculated for a Gaussian channel and a BER of 2×10^{-4} after Viterbi decoding, corresponding to Quasi-Error-Free operation. For DVB-T the figures correspond to DVB-T in nonhierarchical mode, C/N figures are usually quoted are for a Rayleigh channel and range from 5.4 dB to 27.9 dB.

[4] Data rate is same as for DVB-T (minus some overhead if optional MPE-FEC is used). However, the practical upper limit is approximately 15 Mb/s in an 8MHz channel.

[5] Data-rates for DVB-S2 do not directly compare with other DVB systems given the wide range of options.

multiplex on a cable channel. The DVB-CS [19] specification describes a version that can be used for satellite master antenna television installations.

The digital terrestrial television system DVB-T [20] is more complex, because it is intended to cope with a more severe noise and bandwidth environment, as well as with multipath. The system has several dimensions of receiver "agility," where the receiver is required to adapt its decoding according to signaling. The key transmission element is the use of OFDM. There are two modes: 2K and 8K carriers, each using QAM. The input is demultiplexed into multiple substreams: two for QPSK, four for 16 QAM, and six for 64 QAM. The 8K mode can allow more multipath protection, but the 2K mode can offer Doppler advantages where the receiver is moving [21].

There are two systems for MMDS (multi-channel microwave distribution systems), one for systems that operate at radio frequencies below 10 GHz, DVB-MC [22], which is like the DVB-C system, and one for systems that operate at radio frequencies above 10 GHz, DVB-MS [23], which is like the DVB-S system. An MMDS system, like DVB-T, DVB-MT [24], is also available.

DVB-S2 and DVB-H

A higher-efficiency digital satellite broadcasting system, DVB-S2 [25], was later developed, providing both DVB-S backwards-compatible and non-backwards-compatible versions. The non-compatible version uses 8-PSK and LPDC (low density parity check) concatenated with BCH coding to achieve an efficiency increase of about 30% more data capacity for the same receiving dish size compared to DVB-S. DVB-S2 is likely to be used for all new European digital satellite multiplexes, and satellite receivers will be equipped to decode both DVB-S and DVB-S2.

A more flexible and robust digital terrestrial system, DVB-H [26], was also developed. The system was designed to be receivable on handheld receivers and thus includes features that reduce battery consumption (time slicing) and a 4K OFDM mode, together with other measures. Published by ETSI as EN 302 304, DVB-H services supported the use of an IP-based mechanism to deliver content, and included metadata delivery and access systems. The introduction of DVB-T2 and DVB-NGH essentially replaced the need for DVB-H.

DVB Conditional Access

Broadcasters need a system to ensure that only authorized or paying viewers can view pay-TV or other pay services. The DVB Conditional Access system is a single private-key system.

The simplified operation is as follows. At the broadcaster, the output from a complex pseudo-random binary sequence (PRBS) generator is added (modulo-2) to the digital stream to be transmitted. This process is termed "scrambling." In the receiver, there is a matching PRBS generator that can "descramble" the signal. A signal or key is delivered over the broadcast path at intervals that resynchronize the two PRBS generators; this is the "control word" or scrambling key.

The version of the control word that is transmitted is itself scrambled, though the term "encryption" is used to avoid confusion with the scrambling of the program. The encrypted version is called the "ECM" or entitlement control message. To decrypt the ECM requires another transmitted signal called the entitlement management message or EMM. Versions of this EMM are only transmitted to viewers who have paid their subscription or pay-per-view fee. They are encrypted with a key contained in the specific paid-up receiver itself. This is how overall control of the signal path is achieved.

The DVB Project has specified the PRBS generator [27] (the common scrambling algorithm DVB-CSA). However, it is not available "over the counter" from ETSI; applicants have to prove they have a genuine reason for needing it. Though currently secure, there are moves to agree on an even more sophisticated CSA. The ECM and EMM systems are not standardized, but are left to the individual broadcaster.

The DVB Project described two strategies to make it possible to use the same receivers for several pay-TV services that have different ECM and EMM systems.

The first strategy is termed Multicrypt. In this case, the receiver is equipped with a PC-Card (PCMCIA) slot. This is the DVB Common Interface (DVB-CI) [28]. The ECM and EMM systems can be included in secure PCMCIA cards, which can be changed depending on the pay services desired. This card interface interrupts the main signal path and so the slot can also have other uses [29].

The second strategy is termed Simulcrypt. In this case, the receiver is only able to cope with one of the ECM/EMM systems, but other pay-TV operators agree to send their signals in a form that can be understood by the native system of the receivers. Several sets of ECM and EMM keys are allowed to be transmitted at the same time. This requires quite complex rules [30] between the pay-TV operators.

Adoption and Use of the DVB Systems

The DVB-S system was adopted in 1994, and the first DVB broadcast services in Europe started in spring 1995, by pay-TV operator Canal Plus in France. The DVB-T system was adopted later, in 1997. The first DVB-T broadcasts began in Sweden and the United Kingdom in 1998. DVB-T services started in parts of Germany in 2002, and in 2003, Europe's first analog switch-off took place in Berlin.

The DVB-S system is used across the world, though in some countries such as Japan and the United States other digital satellite systems are used as well as DVB-S. The DVB-C system is also widely used throughout the world. In the United States, the cable system is similar in some respects to DVB-C, but was standardized by SCTE. The DVB-T system is the least widely used, though the roll out of digital terrestrial television throughout the world has been slower than digital satellite and cable.

DVB in Australia

Because of some interesting regional variations, it is worth noting the implementation of DVB-T in Australia. The variations include, but are not limited to, the following [31]:

- 7-MHz channels are used both at VHF and UHF (i.e., covering Australian channels 6–12 and 28–69).
- Transmissions may be in either 2K or 8K COFDM carrier modes.
- Transport streams include MPEG encoded video at 25 frames, 50 fields or 50 frames per second. The video formats broadcast are up to 1920 pixels by 1080 lines.
- Television services transmitted in SDTV will include at a minimum an MPEG-1 Layer II audio stream. This may be in mono, stereo, or stereo with surround components. In addition, any SDTV service may also contain a Dolby Digital™ (AC-3) audio stream with up to 5.1 discrete channels. The preferred audio stream for a television service transmitted in HDTV is a Dolby Digital™ (AC-3) audio stream with up to 5.1 discrete channels. Alternatively, an HDTV service may contain an MPEG-1 Layer II audio stream.
- Transport streams will include closed-caption subtitles based on ETSI EN 300 472 or ETSI EN 300 743, or both.
- Transmissions may include data broadcasting as specified in ETSI EN 301 192.
- Transmission filter masks for Australia's 7-MHz channel spacing take into account the relative proximity of lower adjacent-channel dual analogue sound carriers.
- Australian Broadcasters shall implement Single Frequency Networks in accordance with ETSI TS 101 191.
- Australian Broadcasters may transmit DVB-T modes suitable for reception by mobile and portable receivers.
- Transmissions may include DVB-T hierarchical modes. The mode should be determined or

confirmed from TPS signaling and identification of the High Priority and Low Priority streams.

- Some Australian broadcasters' transmissions have a +125-kHz or −125-kHz frequency offset.
- Australian broadcasters have implemented a scheme for network identification coding through registration with the DVB Project Office and the Australian Broadcasting Authority.

As specified by DVB, modulation is by means of COFDM, using QPSK, 16 QAM or 64 QAM. When HDTV is broadcast, 64 QAM is used with a 2/3-code rate and 19.4 Mb/s net data rate. Also, the channel allocation plan has attempted to assign digital service on a channel adjacent to or within the same band as the analog service, and analog and digital transmitters are recommended to be co-located.

The digital television conversion plan made by the Australian Broadcasting Authority required that digital and analog broadcasts be simulcast for at least eight years from the start of digital transmission in a given area, that all programs be transmitted in digital SDTV mode, and that at least 1040 hours per year of programs be transmitted in HDTV mode. This requirement resulted in a so-called "triplecast" scenario when HDTV is transmitted. Australian broadcasters have implemented a scheme for network identification coding through registration with the DVB project office and the Australian Broadcasting Authority.

Because of the different parameters in use, receivers must (dynamically) identify and receive all COFDM modes, including hierarchical modulation, and should allow for manual selection of modulation parameters. In order to deploy this as seamlessly as possible, Digital Broadcasting Australia (DBA) advises that receivers should check the DVB-T Transmission Parameter Signaling (TPS); in addition, the performance requirements of receivers are mandated, as defined in a Digital Television Receiver Standard [32].

INTEGRATED SERVICES DIGITAL BROADCASTING (ISDB)

The digital broadcasting system adopted in Japan is named ISDB, which embodies the concept of expanding flexibility across the physical layer. A common format is used overall, while local modulation systems are defined according to transmission and reception characteristics.

The history of ISDB goes back to the idea of a "Digital broadcasting system for the 21st century," which was conceived at the NHK Science and Technical Research Laboratories in the 1980s. The ISDB of the time was aimed at a flexible system that would be open to new services, such as high-quality audio-video and data broadcasting, through the digitization of valuable broadcasting transmission channels, and based on the following fundamental concepts:

- All information related to broadcasting, such as audio-video and multimedia data, is converted to a digital format in an integrated manner for processing.
- All digital signals are available, regardless of the type of transmission media, including satellite, terrestrial, and cable TV networks.
- By using the features of the respective transmission media, users can easily view high-quality services, or obtain desired information while either at home or on the road.

ISDB has several features that are not available with either DVB or ATSC.

One of the initial issues facing multiplexing technology for digital data was how to construct for various services effective, integrated handling methods for audio graphics, which combines HD still pictures and high-quality audio, as well as facsimile broadcasting, high-function teletext, closed-captioning, and HDTV service. Thorough research was conducted in areas such as transmission packet length and the control functions needed for a multimedia broadcasting service. ARIB (Association of Radio Industries and Businesses), the standardizing organization in Japan, examined and evaluated MPEG-2 systems based on these research results, resulting in the ISDB multiplexing scheme adoption.

ISDB is a broadcasting system that realizes HDTV-based multimedia services. It features a transmission scheme designed to make use of respective transmission media, through the incorporation of MPEG-2 video/audio for its audio-video coding and MPEG-2 systems for its digital signal multiplexing scheme. The four variants of ISDB are summarized in Table 2.7-4.

Satellite: ISDB-S

Satellite broadcasting using frequencies in the 12-GHz band faces the problem of rain attenuation, with precipitation generated at a considerable hourly rate in Japan, as the country is located in a rainy region. The ISDB-S digital satellite television broadcasting system incorporates a hierarchical transmission scheme that modifies the transmission rate as a rain attenuation countermeasure. It defines a 204-byte slot in the bit stream, capable of allotting signals combined with BPSK/QPSK/Trellis coded 8PSK modulation and inner coding, by slot unit, based on programming data. This hierarchical transmission aims to provide continuous reception of comprehensible program at a low bit rate when severe rain attenuation prevents the transmission of the program in HDTV.

Terrestrial: ISDB-T

While satellite broadcasting provides nationwide programming, terrestrial broadcasting offers services tailored to each region. The transmission bandwidth for ISDB-T digital terrestrial television broadcasting was determined based on both the required condition that "HDTV is essential even in terrestrial service," and on the requirement that it coexist with current analog service. Using this transmission bandwidth, the BST-OFDM (band-segmented transmission-OFDM) scheme was constructed with the purpose of providing

TABLE 2.7-4
Comparison of ISDB Systems

Systems		ISDB-T	$ISDB\text{-}T_{SB}$	ISDB-S	ISDB-C
Video		MPEG-2 Video		MPEG-2 Video	
Audio		MPEG-2 Audio AAC			
Multiplex		MPEG-2 Systems			
Transmission		Multiple carrier (OFDM)		Single carrier	
Bandwidth		6 MHz	429 kHz	34.5 MHz	6 MHz
Modulation		DQPSK/QPSK/16 QAM/64 QAM (OFDM)		BPSK/QPSK/TC8PSK	64 QAM
Error correction	Inner	8PSK: Trellis others: Convolutional code			—
	Outer	RS			

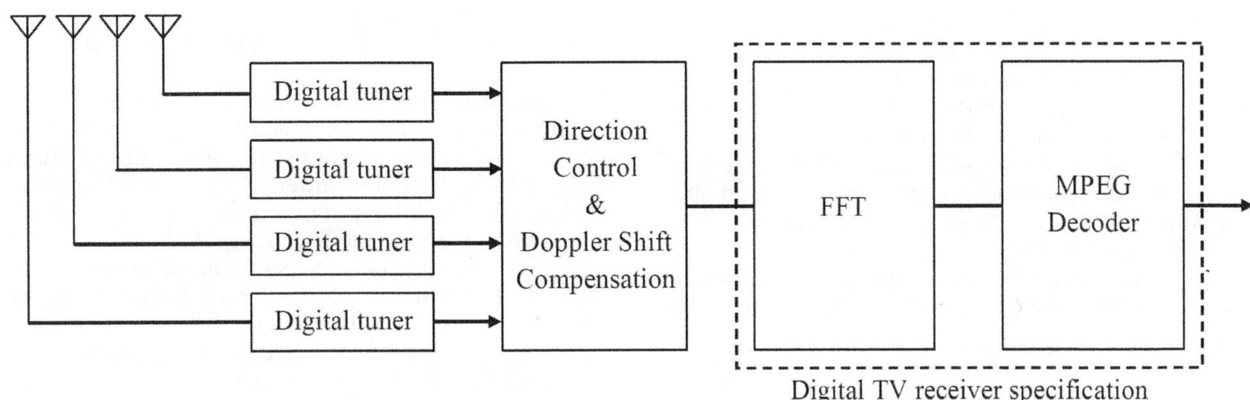

FIGURE 2.7-8 Band-segmented structure of ISDB-T

services for mobile and portable reception, together with HDTV service. This scheme defines an OFDM segment by dividing a 6 MHz channel into 13 segments for signal transmission, as shown in Figure 2.7-8.

Data carriers and reference signal carriers are allotted in OFDM subcarrier segments at regular intervals, with the necessary transmission band consisting of a group of segments. Using this segment group unit, QPSK/16-QAM/64-QAM modulation is used, together with error correction inner coding and time interleaving. Each of these parameters can be changed according to the service purpose, such as fixed reception or mobile reception. It is also feasible to realize mobile reception service that employs one segment that can be partially received (disregarding the others), simultaneously with HDTV broadcasting.

Terrestrial Sound: $ISDB\text{-}T_{SB}$

Digital terrestrial sound broadcasting employs OFDM segments with an identical structure to that of ISDB-T, sharing the same transmission scheme except for the number of segments. There are two types of transmission, a one-segment scheme and a three-segment scheme. The central segment for the three-segment scheme can also be received partially, similar to ISDB-T. This allows a common one-segment receiver to be used with both ISDB-T and $ISDB\text{-}T_{SB}$. In the future, when $ISDB\text{-}T_{SB}$ is established nationwide, integrated mobile terminals for the three media of cellular phone, ISDB-T, and $ISDB\text{-}T_{SB}$ are expected to appear on the market.

Cable: ISDB-C

Full-scale cable-TV system digitization using ISDB-C started at the same time as digital satellite broadcasting. To transmit digital broadcasting, including HDTV programs, together with the retransmission of conventional analog broadcasting, various retransmission schemes were constructed. A "pass-through scheme" is utilized to transmit a digital program over a cable TV network without modulation scheme conversion, in order to employ direct-reception digital broadcasting receivers. The "trans-modulation scheme" demodulates received signals first, then remodulates them to 64 QAM. In this case, signal processing differs by the number of MPEG-2 transport streams (TS) included in the received signals, categorized as either a single-TS scheme or a multiple-TS scheme. A "re-mux scheme," suitable for independent broadcasting service, has also been standardized.

ARIB Standards

All the ISDB systems have been standardized and published as ARIB standards, as shown in Table 2.7-5. ARIB standards are voluntary technical standards, which supplement governmental regulations for telecommunications and broadcasting radio systems, and are set for the purpose of guaranteeing compatibility of radio facilities and transmission quality as well as offering greater convenience to radio equipment manufacturers and users.

Service Features of ISDB-T for Digital Terrestrial Television

ISDB-T utilizes the OFDM transmission system, which is effective for SFN (single frequency networks), and is robust for multipath interference. Because of heavy usage of TV frequencies, and ghost image interference by high-rise buildings, OFDM was chosen for Japan.

ISDB-T separates the 6-MHz channel into 14 segments, using 13 segments for signal transmission and the remaining segment for a guard band between channels. Each segment occupies 429 kHz of bandwidth. By using 13 segments flexibly, multi-programming is possible. Various combinations are possible, including:

- 1 HDTV (12 segments) + mobile service (1 segment)
- 3 SDTV (3 × 4 segments) + mobile service (1 segment)

ISDB-T has some commonality with DVB-T such as the OFDM transmission system. However, the segment structure, time interleaving, and TMCC (transmission and multiplexing configuration control) are unique features.

One-Segment Mobile Reception

In the ISDB-T system, mobile reception service for handheld terminals is also possible by using one segment located in the center of 13 segments. In this manner, the "One Seg" (or "1seg") service has the capacity for 312-kbps data transmission using QPSK modulation. See Table 2.7-6.

The strength of the ISDB-T system is that a 6-MHz channel can simultaneously carry an HDTV program using twelve segments and a one-segment video signal for cellular phone service. Stable TV reception has been demonstrated over a cellular phone by means of one-segment TV broadcasting using MPEG-4 AVC/H.264 video compression coding. At the time of this writing, several prototypes of cellular phone with a TV receiver are under development, and several broadcasting experiments are being conducted using the existing facilities for digital terrestrial television broadcasting. A commercial service for simultaneous broadcasting of twelve-segment HDTV and one-segment QVGA video programs was implemented in 2006.

Cellular phones are a convenient tool for providing information, as they can display TV and data broadcasting within the same display. Recognizing that the data capacity for one-segment TV broadcasting is small and that all cellular phones in Japan have a wireless Internet connection facility, integrated services of data broadcasting and the wireless Internet are expected to be introduced to supplement data capacity and provide detailed information via the Internet. This kind of application is expected as a converged business model of broadcasting and communication.

HDTV Mobile Reception

Though one-segment TV broadcasting is appropriate in size for cellular phones and PDAs, it is not feasible for a large-screen display that needs higher resolution—and the demand for a large-screen display

TABLE 2.7-5
ARIB STANDARDS

Systems	ISDB-T	ISDB-S	ISDB-T_{SB}
Transmission	STD-B31	STD-B20	STD-B29
Source coding & Multiplex	STD-B32 Image encoding, sound encoding and multiplexing		
Service information	STD-B10 Program lineup information		
Data broadcasting	STD-B24 (Presentation engine; BML) STD-B23 (Execution engine; GEM-based)		
Server-type broadcasting	STD-B38 System based on home servers		
Access control	STD-B25 Conditional access		
Receiver	STD-B21		STD-B30
Operational Guideline	TR-B14	TR-B15	TR-B13

TABLE 2.7-6
One-Segment Parameters

ISDB-T One-Segment Parameters		
Video	Bit rate	Max 384 kbps (Operation example: 128 kbps)
	Encoding method	MPEG-4 AVC/ITU-T H.264
	Format	QVGA 4:3 or 16:9
	Frame rate	15 fps
Audio	Bit rate	Max 256 kbps (Operation example: 48 kbps)
	Encoding method	MPEG-2 AAC
	Quantization	16 bit @ 24 kHz or 48 kHz sample rate
Still image	Encoding method	JPEG, GIF, aGIF
Caption	Number of Characters/ Number of line	12 words × 4 lines (Vertical display) 16 words × 3 lines (Horizontal display)

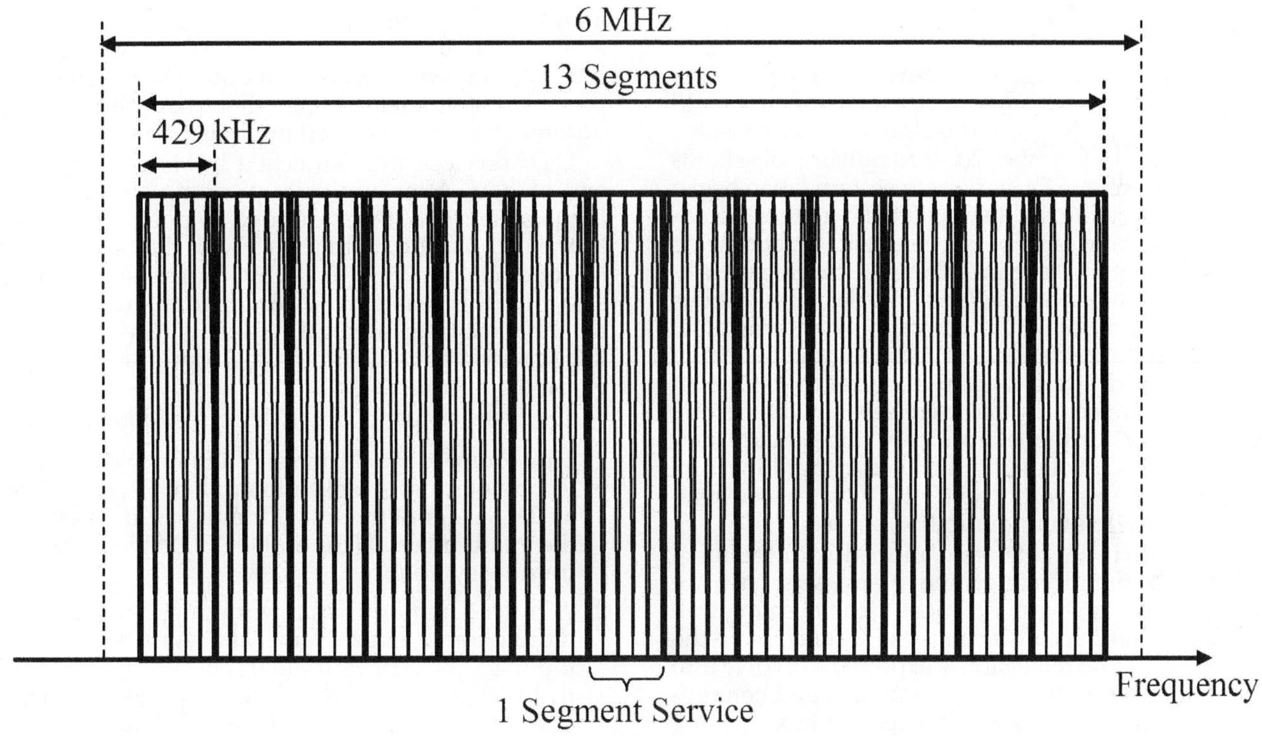

FIGURE 2.7-9 HDTV mobile reception system

for mobile reception is expected to be high. Mobile reception of full HDTV pictures was considered difficult because the receiver direction changes rapidly and the signal frequency is often affected because of Doppler shift. To overcome these difficulties, an HDTV mobile reception system was developed using reception direction-control techniques and Doppler-shift compensation techniques, which enable the transmission parameters for fixed HDTV reception to be used for mobile reception. See Figure 2.7-9.

The stability and feasibility of these techniques have been proved viable in a variety of applications. Installing prototype receivers in a large bus, public demonstrations were conducted in December 2004, showing stable HDTV reception at speeds of 100 km/h. HDTV reception in sightseeing buses, trains, and ferryboats is expected to be realized, together with car navigation systems and traffic information services within a few years.

Status of Digital Terrestrial Television Service

Terrestrial television stations are part of the fundamental media in Japan. With over 50 million households and 100 million TV sets, terrestrial broadcasters have established many relay stations to provide maximum coverage throughout the mountainous archipelago; there are more than 3,000 transmitter sites. Due to the heavy usage of UHF channels by existing analog relay stations, it is impossible to assign digital channels without migrating analog stations, and many analog TV channels are forced to shift to other UHF channels. Therefore, digitization of terrestrial television broadcasting is a project of high importance to Japan.

The government issued digital TV licenses to TV broadcasters in November 2003 with the following conditions and requirements:

- Assign 6-MHz channels for incumbent terrestrial broadcasters.
- At least two-thirds digital/analog simultaneous broadcasting each day.
- Half of all broadcasting each week should be in HDTV.
- Expand digital broadcasting coverage in the same areas as analog broadcasting, as soon as possible, until the analog service is discontinued.

Since the start of digital terrestrial television broadcasting in December 2003 in the three major metropolitan areas of Tokyo, Nagoya, and Osaka, service coverage expanded; as of December 2004, 18 million households (38%) were covered. TV stations in eastern Japan began digital broadcasting in 2005, with 27 million households (57%) planned to be covered, and a nationwide service was put in place in 2006, with coverage reaching 37 million households (79%). Smaller relay stations were set up until full digital migration was completed. As mandated by the Japanese Radio Law, the switch-off of analog broadcasting in 44 prefectures took place in 2011, with 3 prefectures that were severely affected by the 2011 Tohoku earthquake and resulting tsunami transitioning in 2012. Following the analog shutdown, the VHF and the upper portion of UHF TV bands were reallocated for wireless communication service.

All digital HDTV sets are equipped with the function of data broadcasting, and many sets have an Internet web browsing function. People enjoy web browsing without facing a complex PC function. In the near future, digital television is expected to evolve into a home gateway to the ICT (information and communications technology) society, converging broadcasting and communication function.

DTMB SYSTEM

After initially considering then-existing DTV systems, the Chinese government decided to develop their own system, owing to a number of technological and economic factors [33]. In 1999, a new terrestrial DTV transmission system was planned by the State Council of China. Five proposals were presented for evaluation, together with lab and field tests. From 2004 to 2005, a "merging scheme" was developed based on three proposals:

1. A single carrier offset-QAM system, Advanced Digital Television Broadcast-Terrestrial (ADTB-T), developed by a group at Shanghai Jiao Tong University in Shanghai;
2. A non-single-carrier system using "Time-Domain Synchronous-OFDM" (TDS-OFDM), Digital Multimedia Broadcast-Terrestrial (DMB-T), developed by a group at Tsinghua University in Beijing; and
3. A multi-carrier COFDM system, Terrestrial Interactive Multi-service Infrastructure (TiMi), developed by a group at the Academy of Broadcasting Science (ABS) in Beijing.

At the beginning of 2006, the draft standard was in its final stage, representing a combination of the above systems. Following comments and the results of lab and field testing, the formal publication of the draft standard took place later that year. Transmission of the signal allows several different options to be employed: a single-carrier mode, or a multi-carrier mode with 3,780 carriers with carrier spacing of 2 kHz. By some accounts, the latter mode is a hybrid of single-carrier and multi-carrier modulation (i.e., not a "pure" multi-carrier system). As with other DTV systems, the provision of different options should in principle allow different modes to be used in different areas.

DTV services were launched in Hong Kong at the end of 2007. After originally planning on a 2015 analogue switch-off, it was decided to postpone the date to the end of 2020, with a review of the target date to be conducted in 2017–18. A government spokesperson said that this was done in order to allow more time for the remaining analog TV viewers to switch to digital services, as well as to allow optimization of the vacated spectrum [34].

The basic elements of DTMB are as follows:

1. Time-domain synchronous (TDS)-OFDM technology, using a pseudo random PN sequence as the header of the signal frame, serving as both the guard interval (GI) of the OFDM block and the training symbol;
2. A layered frame structure with calendar day frame, minute frame, super-frame, and signal frame, from top to bottom. The minute frame synchronizes with real time, a function developed for power saving. The super-frames are composed of data frames and system information;
3. Two subcarrier options:
   a) "SC = 1" (direct pass) for single carrier operation, and
   b) "SC = 3780" (with IDFT module) for multi-carrier operation;
4. Five different constellation modes: 64 QAM, 32 QAM, 16 QAM, 4 QAM, and 4-QAM-NR (rate ½ Nordstrom-Robinson code);
5. FEC using a BCH (752, 762) outer code and LDPC inner codes. The equivalent FEC code rates are 0.4 (3008/7488), 0.6 (4512/7488), and 0.8 (6016/7488).

An occupied bandwidth of 7.56 MHz in an 8-MHz channel offers a data throughput of up to 32 Mbps. The

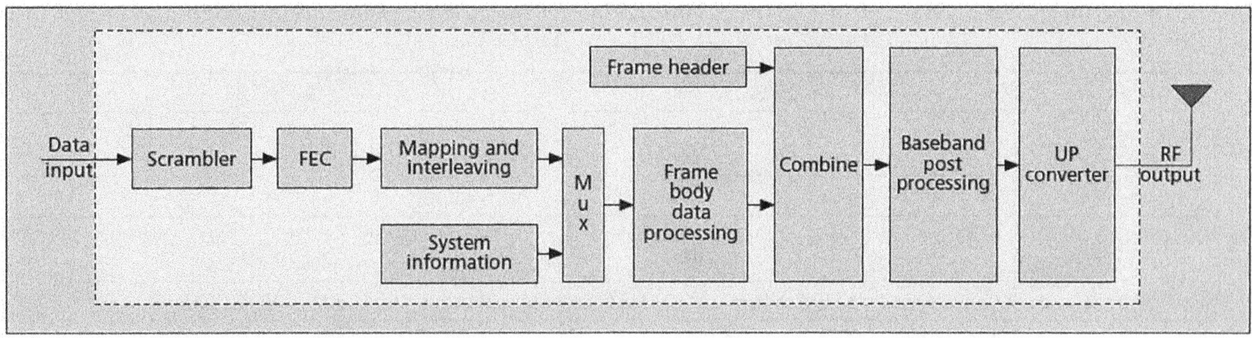

FIGURE 2.7-10 Basic block diagram of the DTMB system

system payload data rates are calculated from the formula shown in Eq. n:

$$R_U = R_S \times b \times R_{FEC} \times (L_{data} / L_{Frame}) \qquad \text{(n)}$$

where:

R_U = System payload data rate

R_S = Symbol rate (7.56 Msymbols/s)

B = Number of bits per symbol

R_{FEC} = FEC code rate

L_{data} = Number of effective data symbols in one signal frame (3744 symbols after excluding system information symbols)

L_{Frame} = Length of the signal frame (4200, 4375, or 4725 symbols)

The system layer used is MPEG-2, with video and audio coding using MPEG-2 compression provisionally, until a new standard called "AVS" (audio-video standard) was finalized. The AVS Video Coding draft standard was published in March 2006, and the AVS Audio Coding draft standard was published soon afterward. A basic block diagram of the DTMB system is shown in Figure 2.7-10.

NEXT GENERATION BROADCAST TELEVISION

Following the deployment of the first generation of digital television systems, the advent of a new generation of cellular networks known as Long-Term Evolution (LTE) demonstrated a strong consumer demand for improved mobile coverage and increased content delivery data rates. Consequently, various standards development organizations, including DVB and ATSC, embarked on a path for development of what has been collectively called Next Generation Broadcasting. This work, which incorporates the latest advances in wireless communications, is aimed at providing significant performance improvements compared to the first-generation mobile broadcast systems. At the time of this writing, this work has resulted in two significant developments, DVB-T2 [35] and ATSC 3.0 [36]. Work on a successor to DVB-H has produced a hybrid profile, applying to both terrestrial and/or satellite. The new profile is defined as a subset of DVB-T2, called DVB-NGH (Next Generation Handheld), and is codified in ETSI EN 303 105. Table 2.7-7 shows the high-level features of DVB-T2 and ATSC 3.0.

DVB-T2

In June 2006, a study group called TM-T2 reported to the joint DVB TM/CM meeting that the group set forth to define the physical layer for a next generation of terrestrial DVB. The goal was to specify a system that improved on the performance of the legacy DVB-T standard according to requirements specified by the CM-AMT (Advanced Modulation for Terrestrial DTV) group. The DVB-T2 draft standard was ratified by the DVB Steering Board in June 2008, published as the DVB-T2 standard BlueBook, and adopted as an ETSI standard in September 2009. The key features of the standard include:

- Channel bandwidth is 1.7, 5, 6, 7, 8, or 10 MHz.
- Multiple PLPs enable service-specific robustness/bitrate tradeoff.
- COFDM modulation with QPSK, 16-QAM, 64-QAM, or 256-QAM constellations.
- OFDM modes are 1k, 2k, 4k, 8k, 16k, and 32k. In 32k mode, the symbol length is about 4 ms, and a larger part of the standard 8-MHz channel can be used, adding about 2% extra capacity.
- Guard intervals are 1/128, 1/32, 1/16, 19/256, 1/8, 19/128, and 1/4. In 32k mode, the maximum is 1/8.
- FEC uses concatenated LDPC and BCH codes (as in DVB-S2 and DVB-C2), with rates 1/2, 3/5, 2/3, 3/4, 4/5, and 5/6.
- Eight different pilot patterns are supported.
- Equalization can be based on the RAI CD3 system.
- MISO (multiple-input single-output) may be used (Alamouti scheme).

DVB-T2 uses OFDM (orthogonal frequency division multiplexing) with a large number of subcarriers delivering a robust signal, and offers a range of different modes. DVB-T2 uses the same error correction

TABLE 2.7-7
Comparison of DVB-T2 and ATSC 3.0 Systems

Systems	DVB-T2	ATSC 3.0
Transmission	Multiple carrier (OFDM)	
Bandwidth	6 / 7 / 8 MHz	
Modulation	QPSK, 16 QAM, 64 QAM, 256 QAM	QPSK, 16 QAM, 64 QAM, 256 QAM, 1024 QAM, 4096 QAM
Error Correction	LDPC + BCH	LDPC + BCH, CRC
FEC Code Rates	1/2, 3/5, 2/3, 3/4, 4/5, 5/6	n / 15, n = {2, 3, . . . 15}
Characteristic	SFN capability	SFN capability
Stream Encapsulation	• Transport Stream (TS) • Generic Stream Encapsulation (GSE) • Generic Continuous Stream (GCS) • Generic Fixed-length Packetized Stream (GFPS)	Internet Protocol (IP)
Proposing Region	Europe	USA
Standard	ETSI EN 302 755	ATSC A/322[1]

[1] ATSC 3.0 Physical Layer standard.

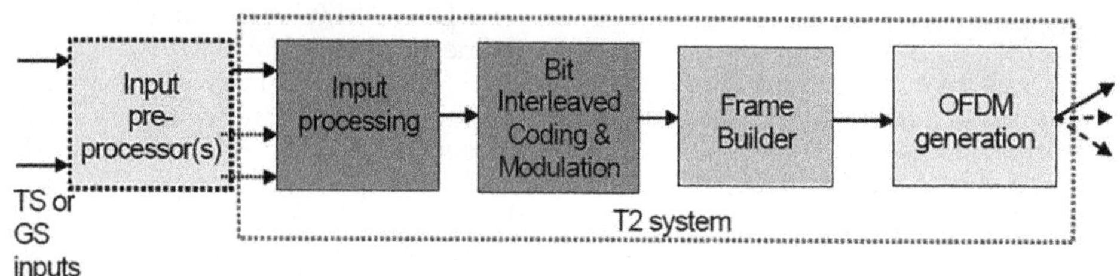

FIGURE 2.7-11 Basic block diagram of the DVB-T2 system

coding as used in DVB-S2 and DVB-C2: LDPC (low density parity check) coding combined with BCH (Bose-Chaudhuri-Hocquengham) coding. The number of carriers, guard interval sizes, and pilot signals can be adjusted, so that the overhead can be optimized for any target transmission channel.

Compared with DVB-T, several new technologies are used in DVB-T2. Multiple physical layer pipes (PLPs) allow separate adjustment of the robustness of each service delivered within a channel, to meet specific target reception conditions. PLPs also allow receivers to save power by decoding only a single service rather than the entire multiplex.

Each PLP may accept one of the following input formats:

- Transport stream (TS) [37].
- Generic stream encapsulation (GSE) [38].
- Generic continuous stream (GCS) (a variable length packet stream where the modulator is not aware of the packet boundaries).
- Generic fixed-length packetized stream (GFPS), a form retained for compatibility with DVB-S2 [39].

Other options within DVB-T2 include:

- Alamouti coding is a transmitter-and-antenna diversity method using space–time block coding that improves coverage in small-scale single-frequency networks.
- Constellation rotation provides additional robustness for low-order constellations.
- Extended interleaving is provided, including bit, cell, time, and frequency interleaving.
- Future extension frames (FEF) allow the standard to be enhanced in the future, with backward compatibility.

A high-level block diagram of DVB-T2 is shown in Fig. 2.7-11.

DVB-NGH

In 2012, at the 72nd Meeting of its Steering Board, the DVB Project announced approval for a new specification, DVB-NGH (Next Generation Handheld). DVB-NGH supports using multiple-input multiple-output MIMO

TABLE 2.7-8
Modulations and Code Rates supported in DVB-NGH

Modulation	Code rate	
	1/3, 2/5, 7/15	8/15, 3/5, 2/3, 11/15
QPSK	2D rotated	4D rotated
16 QAM	2D rotated	
64 QAM or NU-64-QAM	2D rotated	
256 QAM or NU-256-QAM	No rotation	

antenna schemes to overcome the Shannon limit of single-antenna wireless communications. For this reason, NGH has been considered the first "3G broadcasting system," [40] although DVB-T2 also incorporated a limited form of antenna diversity. The 3GPP cellular telephony specifications include advanced transmission techniques such as spatial beam-forming and spatial coding.

Based on DVB-T2, DVB-NGH has several improvements and extensions to aid mobile and portable reception [41]. In addition to MIMO, these new techniques include Time Frequency Slicing (TFS) with a single tuner, 2D and 4D rotated constellations, non-uniform constellations, improved and extended LDPC codes for lower code rates, more efficient time interleaving, and ultra-robust layer-1 signaling. There is also a hybrid profile where terrestrial and satellite transmission schemes can be combined. Hierarchical modulation supports local service insertion.

As codified in the ETSI EN 303 105 standard, DVB-NGH supports four profiles:

- Base profile: Terrestrial transmission with single receiver antenna and tuner.
- MIMO profile: Terrestrial transmission with multiple antennas on the transmitters and receiver sides; two tuners are needed.
- Hybrid profile: Supports a combination of terrestrial and satellite transmissions with a single tuner.
- Hybrid MIMO profile: A combination of the MIMO and Hybrid profiles.

ATSC 3.0

A brief overview of the ATSC 3.0 system is presented below. More details are available in Chapter 8.4. ATSC 3.0 specifies a hybrid content delivery system, where programming can be delivered using a combination of broadcast and broadband media, and which is defined in three functional layers: a physical layer, a management and protocols layer, and an application and presentation layer, as shown in Figure 2.7-12.

The ATSC 3.0 physical layer supports a range of operating points in the channel coding subsystem that can provide performance very close to the Shannon limit (the theoretical limit for the amount of information that can be carried in a noisy channel). The physical layer provides the capability to operate in a robust/low bitrate mode for mobile and indoor services and/or a less robust/high bitrate mode for services to fixed receivers such as those potentially using a higher-performance receiving antenna. Broadcasters can also operate with a simultaneous mixture of different types of services using either time division multiplexing (TDM) or layer division multiplexing (LDM), or both.

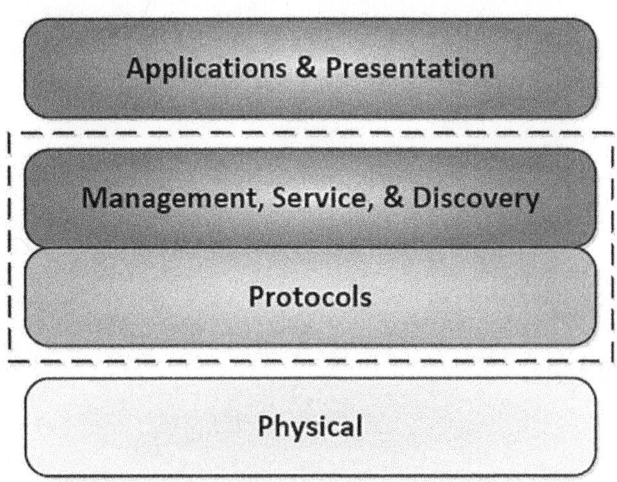

FIGURE 2.7-12 Layered structure of the ATSC 3.0 system

The ATSC 3.0 transport layer uses Internet Protocol (IP) data encapsulation for both streaming and file delivery, instead of the MPEG-2 transport stream (TS) used in earlier systems, allowing for a wider range of business models for content delivery, including hybrid services using both wireless and wired means.

The ATSC 3.0 applications and presentation layer includes video and audio coding, interactivity, accessibility, and other services. For video, HEVC/H.265 coding is specified, and supports 4k ultra-high-definition (UHD) video, high dynamic range (HDR), wider color gamut approaching Rec. ITU-BT.2020, 10 bits per pixel, and frame rates up to 120 Hz. Two audio codecs were specified, Dolby AC-4 and MPEG-H. Among the features universally supported are personalized audio rendering, immersive audio, and a large number of presentation channels. Interactivity is provided by an application runtime environment supporting HTML5.

ACKNOWLEDGMENTS

The section on Analog Television above is based on material from the *NAB Engineering Handbook, 10th Edition* by Dalton H. Pritchard, J. James Gibson, and the author, which was in turn based on an article first published in the *SMPTE Journal*. The section on Digital Television above is based on material from the *NAB Engineering Handbook, 10th Edition* by David Wood, Hiroshi Asami, and the author. Readers are directed to that edition for additional details regarding legacy television systems.

Notes

[1] German spelling *Hannover*.

[2] The DVB Project has reported the support of all African nations for DVB-T.

[3] South Korea plans to add ATSC 3.0 services in newly authorized channels in 2018, and eventually shut down ATSC 1.0 services in subsequent years.

[4] It is expected that the United States will conduct a gradual transition to the ATSC 3.0 system after the standard is ratified.

References

[1] Herbstreit, J.W. and Pouliquen, J., International Standards for Color Television, *IEEE Spectrum*, March 1967, pp. 104–111.

[2] CCIR Characteristics of Systems for Monochrome and Colour Television: Recommendations and Reports, Recommendation 470-1 (1974–1978) of the Fourteenth Plenary Assembly of CCIR, Kyoto, Japan, 1978.

[3] SMPTE Standard for Television, Composite Analog Video Signal: NTSC for Studio Applications, SMPTE 170M-1999, Revision of ANSI/SMPTE 170M-1994, Society of Motion Picture and Television Engineers, White Plains, NY, 1999.

[4] Engineering Guideline EG 28, "Annotated Glossary of Essential Terms for Electronic Production," SMPTE, 1993.

[5] Federal Communications Commission, Radio and Television Broadcast Rules, 47 CFR Part 73, Section 73.682.

[6] Fink, D.G. (Ed.), *Color Television Standards*, McGraw-Hill, New York, 1955.

[7] Recommendation ITU-R BT.1117-2, Studio Format Parameters for Enhanced 16:9 Aspect Ratio 625-Line Television Systems (D- and D2-MAC, PALplus, Enhanced SECAM), International Telecommunication Union, Geneva, Switzerland, 1994, 1995, 1997.

[8] European Telecommunication Standard ETS 300 731, Television Systems; Enhanced 625-Line Phased Alternate Line (PAL) Television; PALplus, European Telecommunications Standards Institute (ETSI), Sophia Antipolis, France, 1997.

[9] Recommendation ITU-R BT.470-6, Conventional Television Systems, International Telecommunication Union, Geneva, Switzerland, 1970, 1974, 1986, 1994, 1995, 1997, 1998.

[10] Recommendation ITU-R BT.1700, Characteristics of Composite Video Signals for Conventional Analogue Television Systems, International Telecommunication Union, Geneva, Switzerland, 2005.

[11] Recommendation ITU-R BS.707-5, Transmission of Multi-Sound in Terrestrial Television Systems PAL B, B1, D1, G, H, and I, and SECAM D, K, K1, and L, International Telecommunication Union, Geneva, Switzerland, 1990, 1994, 1995, 1998, 2005.

[12] Report ITU-R BT.2043, Analogue Television Systems Currently in Use Throughout the World, International Telecommunication Union, Geneva, Switzerland, 2004.

[13] Digital Video Broadcasting Project, www.dvb.org.

[14] European Telecommunications Standards Institute, www.etsi.org.

[15] ISO/IEC Standard 13818-1, 2, and 3 "Information Technology—Generic Coding of Moving Pictures and Associated Audio."

[16] D-Book Technical Specification for UK Digital Terrestrial Television, Digital TV Group.

[17] ETSI EN 300 421 "Digital Video Broadcasting (DVB)—Framing Structure, Channel Coding and Modulation for 11/12 GHz Satellite Services."

[18] ETSI EN 300 429 "Digital Video Broadcasting (DVB)—Framing Structure, Channel Coding and Modulation for Cable Systems."

[19] ETSI EN 300 473 "Digital Video Broadcasting (DVB)—Satellite Master Antenna Television (SMATV) Distribution Systems."

[20] ETSI EN 300 744 "Digital Video Broadcasting (DVB)—Framing Structure, Channel Coding and Modulation for Digital Terrestrial Television (DVB-T)."

[21] ETSI TR 101 190 "Implementation Guidelines for DVB-T."

[22] ETSI EN 300 749 "Digital Video Broadcasting (DVB)—Microwave Multipoint Distribution Systems (MMDS) Below 10 GHz."

[23] ETSI EN 300 748 "Digital Video Broadcasting (DVB)—Multipoint Video Distribution Systems (MVDS) at 10 GHz and Above."

[24] ETSI EN 301 701 "Digital Video Broadcasting (DVB)—OFDM Modulation for Microwave Digital Terrestrial Television."

[25] ETSI EN 302 307 "Digital Video Broadcasting (DVB)—Second Generation Framing Structure, Channel Coding and Modulation Systems for Broadcasting, Interactive Services, News Gathering and Other Broadband Satellite Applications."

[26] ETSI EN 302 304 "Digital Video Broadcasting (DVB)—Transmission System for Handheld Terminals (DVB-H)."

[27] ETSI ETR 289 "DVB Common Scrambling Distribution Agreements."

[28] ETSI EN 50221 "Common Interface Specification for Conditional Access and Other Digital Video Broadcasting Decoder Applications."

[29] ETSI TS 101 699 "Digital Video Broadcasting (DVB)—Extensions to the Common Interface Specification."

[30] ETSI TS 101 197-1 "Digital Video Broadcasting SimulCrypt; Part 1: Head-End Architecture and Synchronization."

[31] Australian Standard AS 4599.1-2005, Digital Television—Terrestrial Broadcasting—Characteristics of Digital Terrestrial Television Transmissions. Reprinted by Permission of SAI Global Ltd.

[32] Australian Standard AS 4933.1-2005, Digital Television—Requirements for Receivers—VHF/UHF DVB-T Television Broadcasts.

[33] Chung-yen Ong, "Technology and Standards of Digital Television Terrestrial Multimedia Broadcasting", IEEE Trans. Broadcasting, 2011.

[34] "Working Target for Analogue Switch-Off Changed to 2020," Press Release, Government Information Centre, Hong Kong Special Administrative Region Government, December 9, 2014.

[35] ETSI EN 302 755, Digital Video Broadcasting (DVB)—Frame Structure Channel Coding and Modulation for a Second Generation Digital Terrestrial Television Broadcasting System (DVB-T2).

[36] ATSC Standard A/322: Physical Layer Protocol.

[37] ISO/IEC 13818-1, Information Technology—Generic Coding of Moving Pictures and Associated Audio Information: Systems.

[38] ETSI TS 102 606, Digital Video Broadcasting (DVB)—Generic Stream Encapsulation (GSE) Protocol.

[39] ETSI EN 302 307, Digital Video Broadcasting (DVB)—Second Generation Framing Structure, Channel Coding and Modulation Systems for Broadcasting, Interactive Services, News Gathering and Other Broadband Satellite Applications (DVB-S2).

[40] Gómez-Barquero, David, *Next Generation Mobile Broadcasting*, Taylor and Francis, New York, 2010.

[41] Physical Layer Specification (DVB-NGH), Next Generation Broadcasting System to Handheld, DVB Document A160, Digital Video Broadcasting (DVB), November 2012.

Bibliography

"China Embraces Digital TV," Masaharu Tanaka, *Nikkei Electronics Asia*, February 2004.

"DTV Standards Submitted To NDRC," Eric Wan, *Pacific Epoch*, July 11, 2005.

"DVB—The Family of International Standards for Digital Video Broadcasting," second edition, Ulrich Reimers, Springer, 2005.

DVB Standards and Specifications, CD issued by the DVB Project Office, Version 7.0 August 2004.

"Interactive TV Standards," first edition, S. Morris and A. Smith-Chaigneau, Focal Press, New York, 2005.

"A Review into High Definition Television Quota Arrangements I Issues Paper," Australian Government, Department of Communications, Information Technology, and the Arts, May 2005.

"Satellites, Science, and Success—The DVB Story," D. Wood, EBU Technical Review, winter 1995.

"Technical Review of the Chinese Digital Terrestrial Television Broadcasting Standard," Jian Song, Zhixing Yang, Lin Yang, Ke Gong, Changyong Pan, Jun Wang, and Youshou Wu, Tsinghua University, Beijing, China, presented at the 56th Annual IEEE Broadcast Symposium, Washington, DC, September 2006.

"Technology and Standards of Digital Television Terrestrial Multimedia Broadcasting," Jian Song and Changyong Pan, Tsinghua University, Yangang Li, Hong Kong Applied Science and Technology Research Institute, *IEEE Communications Magazine*, May 2010.

CHAPTER

2.8

Digital Audio Standards and Practices

CHIP MORGAN
CMBE
El Dorado Hills, California

RANDALL HOFFNER
ABC, Inc.
New York, New York

Updated for the 11th Edition by

TIM CARROLL
Dolby Laboratories
San Francisco, CA

INTRODUCTION

Since the last edition of this book, existing digital audio standards and practices have not seen a dramatic change in their basic definitions, but additional standards have appeared. Digital audio is no longer a format used only by the largest broadcasters and postproduction facilities; it has come to be relied upon as the only way to handle modern broadcast audio requirements. Digital audio has proven to be as robust and flexible as it was originally designed to be. Routing, distribution, storage, and signal processing have advanced to the point of making it difficult or impossible to find analog versions of these processes with the same features and capabilities.

Some useful new formats and standards have emerged along with necessary revisions to existing standards. For example, new transport methods based on TCP/IP computer networks are now standardized via AES67. This is just in time for the need to handle next generation audio systems, which include channels, now objects, and non-optional tightly coupled metadata as well. A new timing signal has also been introduced that finally enables proper synchronization of audio and video in the IP domain.

DIGITAL AUDIO STANDARDS

Below are standards that are relevant to digital audio in broadcast facilities. Many of these have been presented before and remain important parts of standardized digital audio systems. In recent years, there have been significant advances made by the Society of Motion Picture and Television Engineers (SMPTE), and it is necessary to consider these standards along with their Audio Engineering Society (AES) counterparts if they exist.

AES3—AES Recommended Practice for Digital Audio Engineering—Serial Transmission for Two-Channel Linearly Represented Digital Audio Data

This is the baseline standard for digital audio developed by the AES and the European Broadcasting Union (EBU) and is commonly referred to as the AES/EBU standard.

AES3 defines a digital protocol and physical and electrical interfaces for the carriage of two discrete audio channels, accompanied by various housekeeping, status, and user information in a single serial digital bitstream. As its title indicates, AES3 was designed to carry linearly quantized (uncompressed PCM) digital audio. Compressed digital audio may be carried on the IEC 958 digital audio interface. IEC 958 is identical to AES3 in protocol, but can have slightly different electrical characteristics for support of consumer electronics. It addresses a professional implementation (AES/EBU) and a consumer implementation (S/PDIF). The AES3 interface has the capacity to carry linearly sampled digital audio at bit depths from 16 to 24, data descriptive of such factors as channel status and sample validity, along with parity checking data and user data. Total bit count per sample, including audio and housekeeping, is 32 bits. An ancillary standard, AES5, recommends use of the professional audio sample rate of 48 kHz on AES3, while recognizing the use of sample rates of 44.1 kHz, 32 kHz, and 96 kHz.

AES3 carries audio samples using time-division multiplexing, in which samples from each of the two represented audio channels alternate.

Data Structure

The data carried on the AES/EBU interface is divided into blocks, frames, and subframes. An AES block is constructed of 192 frames, each frame being composed of two subframes, each containing a single audio sample. A subframe begins with a preamble that provides sync information and describes what type of subframe it is and ends with a validity bit, a user bit, a channel status bit, and a parity bit.

The subframe is divided into 32 *time slots*, each time slot being one sample bit in duration. The first four time slots are filled with a 4-bit preamble. The 24 time slots following the preamble may be filled in one of two ways. As shown in Figure 2.8-1(a), an audio sample word of up to 24 bits may fill all the time slots. Figure 2.8-1(b) illustrates that the first four time slots of the audio sample word space may be filled with auxiliary bits, which can represent user data or low-quality audio for informational or cueing purposes, for example. In all cases, the audio word is represented least significant bit (LSB) first, most significant bit (MSB) last. If digital audio words of bit depth less than the maximum are represented, the unused bits are set to logic 0. Time slots 28, 29, 30, and 31 are filled with a validity bit (V), a user bit (U), a channel status bit (C), and a parity bit (P), respectively. The subframes are assembled into frames and blocks as shown in Figure 2.8-2.

Each subframe begins with one of three preambles. The first subframe in the 192-frame block, a Channel 1 subframe, starts with Preamble Z. All other Channel 1 subframes in the block start with Preamble X. All Channel 2 subframes start with Preamble Y. Figure 2.8-2 represents the last frame of a block, and the first two frames of the following block. Subframe 1 of Frame 0, the first subframe of the block, begins with Preamble Z, uniquely identifying the beginning of the block. After the first subframe, the successive subframes are marked by Preamble Y and Preamble X, to identify Channel 2 and Channel 1 subframes respectively.

A frame, consisting of two 32-bit subframes, is made up of 64 bits, and the data rate of the interface signal may be readily calculated by multiplying the sampling rate times 64. In the case of the 48-kHz sample rate, the

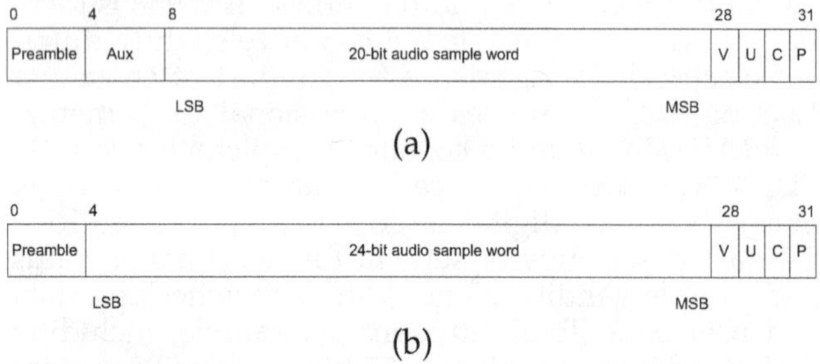

FIGURE 2.8-1 Subframe formats: (a) 16–20-bit audio word; (b) 16–24-bit audio word

total data rate of the signal is 64 times 48,000 or 3.072 Mbps. As will be explained later, the interface employs an embedded clock signal that is twice the sample rate, making the actual frequency of this signal about 6.1 MHz.

Encoding

All time slots except the preambles are encoded using biphase-mark coding to prevent the transmission of long strings of logic 0's or logic 1's on the interface, and thereby minimize the DC component on the transmission line; to facilitate rapid clock recovery from the serial data stream; and to make the interface insensitive to the polarity of connections. The preambles intentionally violate the rules of biphase-mark coding by differing in at least two states from any valid biphase code to avoid the possibility of other data being mistaken for a preamble. Biphase-mark coding requires a clock that runs at twice the sample rate of the data being transmitted, and each bit that is transmitted is represented by a symbol that is composed of two binary states. Figure 2.8-3 illustrates these relationships.

The top sequence of Figure 2.8-3 illustrates the interface clock pulses, running at a speed twice the source coded sample rate. The middle sequence shows the source coding, which is the series of pulse code modulated (PCM) digital audio samples. The bottom sequence shows how the source coded data is represented in biphase-mark coding.

In biphase-mark coding, each source coded bit is represented by a symbol that is composed of two consecutive binary states. The first binary state of a biphase-mark symbol is always different from the second state of the symbol preceding it. A logic 0 is represented in biphase-mark coding by a symbol containing two identical binary states. A logical 1 is represented in biphase-mark coding by a symbol containing two different binary states.

This relationship may be seen by examining in the first full source coded bit at the left in Figure 2.8-3, which is a logic 1. Note that the duration of this bit is two clock pulses. Because the symbol immediately before it ended with a logic 0, the biphase-mark symbol representing it begins with a logic 1. As the bit to be transmitted is a logic 1, the second state of the biphase-mark symbol representing it is different from the first, a logic 0. The second source coded bit to be transmitted is a logic 0. Its first biphase-mark binary state is a logic 1, because the immediately previous state was a logic 0, and the second state is also a logic 1.

The fact that the first binary state of a biphase-mark signal is always different from the last binary state of the previous symbol ensures that the signal on the interface does not dwell at either logic 0 or logic 1 for a period longer than two clock pulses. Because biphase-mark coding does not depend on the absolute logic state of the symbols representing the source coded data, but rather on their relative states, the absolute polarity of a biphase-mark coded signal has no effect on the information transmitted, and the interface is insensitive to the polarity of connections.

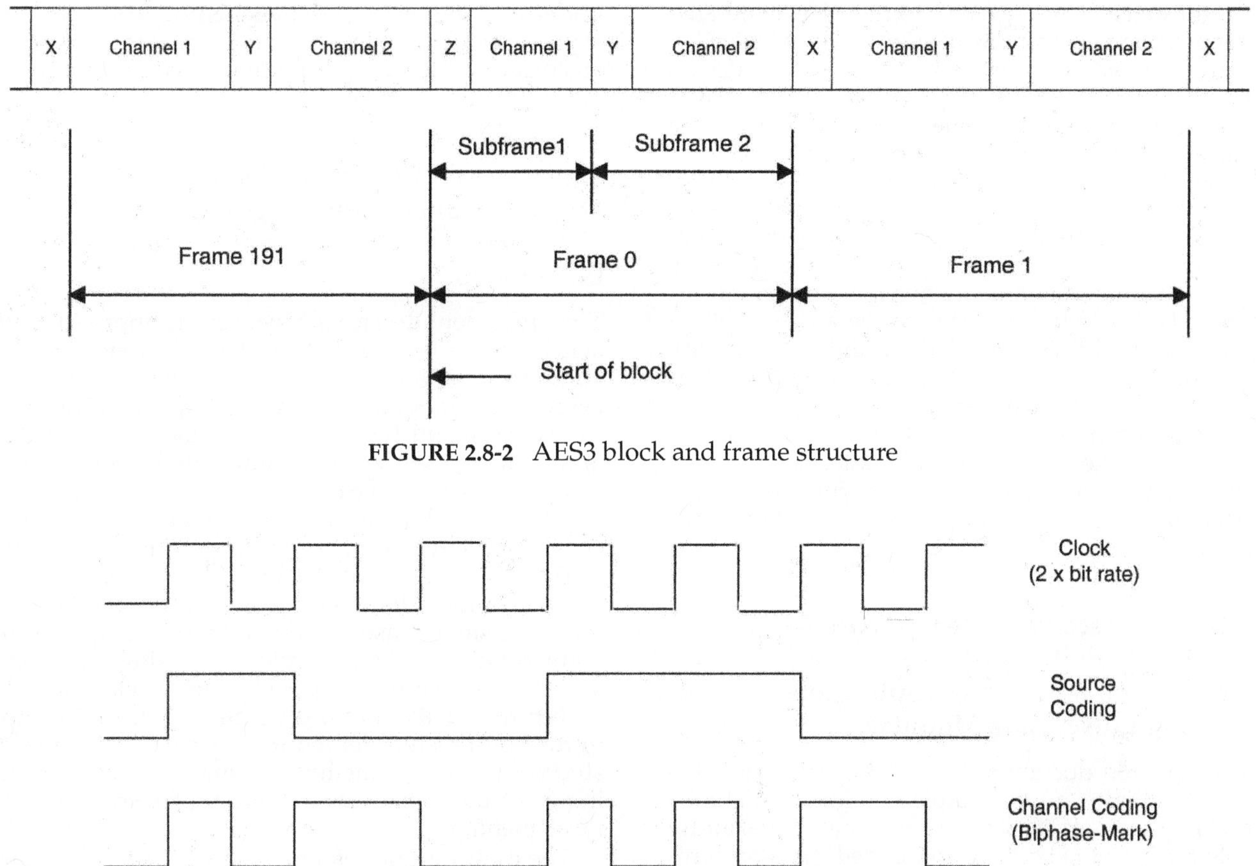

FIGURE 2.8-2 AES3 block and frame structure

FIGURE 2.8-3 AES3 channel coding

Ancillary Data

The last four time slots in a subframe are occupied by various housekeeping and user data. The validity bit (V) indicates whether the audio sample word is suitable for conversion to an analog audio signal. The channel status bit (C) from each subframe is assembled into a sequence spanning the duration of an entire AES3 block, and these 192-bit blocks of channel status data describe a number of aspects of the signal. Examples of channel status data include:

- The length of audio sample words;
- Sampling frequency;
- Number of audio channels in use;
- Emphasis information;
- Consumer or professional interface implemented;
- Audio or data being transmitted on the interface;
- And a host of other possible information.

The 192-bit channel status bits (per block) are subdivided into 24-byte units. There is a separate channel status block for each audio channel, so channel status may be different for each of the audio channels. User data, or U-bits, may be used in any way desired. The parity bit (P) facilitates the detection of data errors in the subframe by applying even parity, ensuring that time slots 4–31 carry an even number of logic 1's and logic 0's.

Electrical Interface

The electrical interface specified by AES3 is a two-wire transformer balanced signal. The AES interface was devised by audio engineers, with the intent of creating a digital audio signal that could be carried on the same balanced, shielded, twisted pair cables and XLR-3 type connectors that are used for analog audio signals. The specified source impedance for AES3 line drivers and the specified input impedance for AES3 line receivers is 110 Ω, which is the approximate characteristic impedance of shielded twisted pair cable as used for analog audio. The permitted signal level on the interface ranges from 2 to 7 V peak-to-peak.

The balanced, twisted pair electrical interface can give rise to some problems in implementation. XLR type connectors and audio patch panels, for example, are not impedance matched devices. This is not critical when the highest frequency of interest is 20 kHz, but it can cause serious problems when a 6-MHz signal must be passed. These considerations, plus the familiarity of television engineers with unbalanced coaxial transmission of analog video and the need for higher connector density for a given product size, generated the requirement for standardization of an unbalanced, coaxial electrical interface for the AES3 signal. Such an electrical interface is standardized in SMPTE 276M, which describes carriage of the AES/EBU interface on standard 75 Ω video cable using BNC connectors, at a signal level of 1 V peak-to-peak. The fact that the 110 Ω balanced and 75 Ω unbalanced signal formats coexist in many systems frequently presents the requirement to translate between these two signals. Devices to perform such translations are available for purchase, and SMPTE 276M has an informative annex explaining how to build them. For density and compatibility issues, most modern multichannel audio equipment is being designed to support SMPTE 276M.

AES2-id is an information document containing guidelines for the use of the AES3 interface. *AES3-id* is an information document containing descriptive information about the unbalanced coaxial interface for AES3 audio.

AES5—AES Recommended Practice for Professional Digital Audio—Preferred Sampling Frequencies for Applications Employing Pulse-Code Modulation

This companion document to AES-3 contains the recommended digital audio sample rate for signals to be carried on the interface. The professional digital audio sample rate of 48 kHz is recommended, with recognition given to the use of the compact disc sample rate of 44.1 kHz, a low bandwidth sample rate of 32 kHz, and new for the 1998 review, the use of a 96-kHz sampling frequency for applications requiring a higher bandwidth or more relaxed anti-alias filtering.

SMPTE EG 32, engineering guideline on AES/EBU audio emphasis and sample rates for use in television systems, also recommends that the 48-kHz sample rate be used. Variations on these sample rates are encountered. Varispeed operation requires the ability to adjust these sample rates by about ±12%, and of course accommodation to 59.94-Hz video requires operation at 48 kHz/1.001.

AES10—AES Recommended Practice for Digital Audio Engineering—Serial Multichannel Audio Digital Interface

This standard describes the data organization for a multichannel audio digital interface. It includes a bit-level description, features in common with the AES3 two-channel format, and the data rates required for its utilization. The specification provides for the serial digital transmission of 32, 56, or 64 channels of linearly represented digital audio data at a common sampling frequency within the range 32 kHz to 96 kHz, having a resolution of up to 24 bits per channel. The format makes possible the transmission and reception of the complete 28-bit channel word (excluding preamble) as specified in AES3, providing for the validity, user, channel status, and parity information allowable under that standard. The transmission format is of the asynchronous simplex type and is specified for a single 75-coaxial cable point-to-point interconnection or the use of fiber optic cables.

AES11—AES Recommended Practice for Digital Audio Engineering—Synchronization of Digital Audio Equipment in Studio Operations

This document describes a systematic approach to the synchronization of AES3 digital audio signals. Synchronism between two digital audio signals is defined as that state in which the signals have identical frame frequencies, and the timing difference between them is maintained within a recommended tolerance on a sample-by-sample basis.

AES11 recommends that each piece of digital audio equipment have an input connector that is dedicated to the reference signal. Two methods of synchronization are proposed: (a) the use of a master digital audio reference signal, insuring that all input/output equipment sample clocks are locked to a single reference; and (b) the use of the sample rate clock embedded within the digital audio program signal that is input to the equipment. Method (a) is preferred for normal studio practice, as method (b) may increase the timing error between pieces of equipment in a cascaded implementation.

The digital audio reference signal is to have the format and electrical configuration of the two-channel AES3 interface, but implementation of only the basic structure of the interface format, where only the preamble is active, is acceptable as a reference signal. A digital audio reference signal may be categorized in one of two grades. A grade 1 reference signal must maintain a long-term frequency accuracy within 61 ppm, while a grade 2 reference signal has a tolerance of less than 610 ppm.

AES17—AES Standard Method for Digital Audio Engineering—Measurement of Digital Audio Equipment

This standard defines a number of tests and test conditions for specifying digital audio equipment. Many of these tests are substantially the same as those used for testing analog audio equipment, but the unique nature of digital audio dictates that additional tests are necessary beyond those used for analog audio equipment.

AES18—AES Recommended Practice for Digital Audio Engineering—Format for the User Data Channel of the AES Digital Audio Interface

This standard describes a method of formatting the user data channels within the AES3 digital audio interface using a packet-based transmission format. This method has gained popularity in some broadcast facilities for carrying non-audio ancillary data such as song titles and other information. It is critical to note, however, that user and other channel status bits are notoriously unreliable. In an effort to save data space, most storage equipment does not preserve this data and instead generates static values prior to output. If a facility design relies on using this data space, it is imperative to verify that all equipment in the chain supports it.

AES42—AES Standard for Acoustics—Digital Interface for Microphones

This standard describes an extension of the existing digital audio interface AES3 to provide a digital interface for microphones. Now in its third edition, AES42 supports synchronous and asynchronous modes of operation, an expansion of the control structure to enable processing internal to the microphone to be controlled by external devices such as mixing consoles, and a further update to the command and control structure to enable higher firmware update rates.

AES67—AES Standard for Audio Applications of Networks—High-Performance Streaming Audio-Over-IP Interoperability

This standard describes how high-performance media networks can support professional quality audio (16 bit, 44.1 kHz and higher) with low latencies (less than 10 milliseconds) compatible with live sound reinforcement. The level of network performance required to meet these requirements is available on local-area networks and is achievable on enterprise-scale networks. A number of networked audio systems have been developed to support high-performance media networking, but until now there were no recommendations for operating these systems in an interoperable manner. This standard provides comprehensive interoperability recommendations in the areas of synchronization, media clock identification, network transport, encoding and streaming, session description, and connection management. Interoperable synchronization is based on IEEE 1588 Precision Time Protocol (PTP), which is also the basis for SMPTE ST 2059, thus finally providing a hard link between audio and video timing.

ATSC A/52:2015—Digital Audio Compression (AC-3) Standard

Digital television broadcasting necessitates the carriage of AC-3 compressed digital audio streams between pieces of DTV equipment. An example is the interface between an AC-3 encoder and the program data stream multiplexer of a DTV transmission system. The former Annex B of the ATSC AC-3 Digital Audio Standard for digital television broadcast that describes the carriage of compressed AC-3 elementary streams on the IEC 958 digital audio interface has been replaced by IEC 61937.

IEC 60958 Digital Audio Interface

IEC 60958 (IEC 958) is logically identical to the AES3 digital audio interface. Electrically, it provides for both the 110 Ω balanced and the 75 Ω unbalanced interfaces. Two versions are described, a consumer version, the S/PDIF interface, in which bit 0 of the channel status word is set at logic 0, and a professional version, the AES/EBU interface, in which bit 0 of the channel status word is set at logic 1. Provision is made in the location of time slots 12–27, which are normally used to carry linear 16-bit PCM audio words, to permit some recording equipment to record and play back either linear 16-bit PCM audio or encoded data streams (compressed digital audio). The consumer implementation permits only the 32-bit mode, in which channel 1 and channel 2 subframes are simultaneously employed to carry 32-bit words. The professional implementation permits either the 32-bit mode or the 16-bit mode, in which each subframe carries a 16-bit digital audio word.

The consumer implementation may carry either two channels of linear PCM digital audio, or one or more compressed audio bitstreams accompanied by time stamps. The professional implementation may carry two channels of linear PCM digital audio, two sets of compressed audio bitstreams with time stamps, or one channel of linear PCM digital audio and one set of compressed audio bitstreams with time stamps. Note that the consumer implementation may also present output levels that are lower than the specified 1 V peak-to-peak of the professional version, and care is advised when connecting consumer and professional devices.

SMPTE STANDARDS AND RECOMMENDED PRACTICES CONCERNING THE USE OF AES DIGITAL AUDIO IN TELEVISION SYSTEMS

ST 272:2004—Formatting AES/EBU Audio and Auxiliary Data Into Digital Video Ancillary Data Space

This standard defines the embedding of AES/EBU digital audio into the standard definition serial digital interface specified in SMPTE 259M, *10-Bit 4:2:2 Component and 4fsc NTSC Composite Digital Signals—Serial Digital Interface*. With such embedding, up to eight channels of digital audio in the AES3 format may be carried on the serial digital video interface signal that travels on a single coaxial cable.

ST 276:1995 (Archived 2004)—Transmission of AES/EBU Digital Audio Signals over Coaxial Cable

This SMPTE standard defines the unbalanced 75-Ω coaxial cable electrical interface for the AES3 bitstream.

ST 299-1:2009—24-Bit Digital Audio Format for HDTV Bit-Serial Interface

This standard defines the embedding of AES/EBU digital audio data into the high-definition serial digital video interface specified in SMPTE 292M, *Bit Serial Digital Interface for High-Definition Television Systems*. This is the high-definition counterpart to SMPTE 272M.

ST 302:2007—Mapping of AES3 Data Into MPEG-2 Transport Stream

This SMPTE standard describes how the 20-bit audio payload of an AES/EBU signal is mapped into an MPEG-2 transport stream in a bit-for-bit accurate manner. This format can be found in most modern MPEG-2 encoders and integrated receiver/decoders (IRDs), and is a method used to carry uncompressed 20-bit PCM audio as well as compressed audio such as Dolby E, apt-X, and high-density multiplexed AC-3. Although it can be used to carry a single AC-3 stream, it is very inefficient and is incompatible with consumer equipment. IEC 13818 describes the proper manner for multiplexing an AC-3 stream into an MPEG-2 transport stream.

ST 337–ST 340—Formatting of Non-PCM Audio and Data in AES3 Serial Digital Audio Interface

These standards describe a standardized method for carrying compressed audio and other data types within AES3 signals and have become the universal method for supporting the carriage of compressed audio formats and data in place of PCM audio. A compliant device will format data per the appropriate standard below and map it into an AES3 signal, which can then be carried over a physical AES connection or be further mapped into a digital video (SDI) signal per SMPTE ST 272:2004 or 299-1:2010.

- ST 337:2015—Format for non-PCM Audio and Data in AES3 Serial Digital Audio Interface
- ST 338:2016—Data types
- ST 339:2015—Generic data types
- ST 340:2015—ATSC A/52 (AC-3) data type
- ST 341:2000 (Archived 2006)—Captioning data type

ST 2035:2009 (Replaces SMPTE 320M) Audio Channel Assignments for Digital Television Recorders (DTRs)

This standard replaces and dramatically expands ST 320M and specifies the allocation of audio channels to the audio tracks of a digital television recorder for purposes of identification, program distribution, and interchange. This standard has proven useful beyond just mapping to a digital television recorder.

IMPLEMENTATION ISSUES

The key to realizing the benefits of digital audio on a system-wide scale is a thorough understanding of the principles underlying digital signal distribution, routing, and switching. There are, as explained, two electrical interfaces available for AES3 signals, and both require good engineering practices for successful implementation. Digital audio's data rate dictates that uncompressed digital audio signals occupy a bandwidth similar to that of analog video.

Regardless of the electrical interface, a well-engineered interconnect requires proper match of source, destination, and characteristic cable impedances. Prior to the 1992 revision of AES3, any equipment manufactured to AES3-1985 violated this principle, as that standard specified a 250-Ω load impedance for receivers and a 110-Ω source impedance for transmitters. Beginning in 1992, AES3 specifies impedance matching among transmitter, receiver, and cable.

Choice of Cable

The use of the unbalanced coaxial cable interface for AES3 data transmission is often preferred by video engineers. SMPTE 276M and *AES3-id* provide guidance for using the 75-Ω unbalanced AES3 interface. Any high-quality video cable will be found quite acceptable for unbalanced AES3 signals. Those engineers designing facilities dealing only with audio may prefer the use of balanced, shielded, twisted pair cables with XLR-type connectors to carry AES3 signals, but should be aware of the cable length restrictions of this implementation, and of the possibility that problems will arise from impedance mismatches at connectors and patch panels. For balanced transmission of AES3 signals, special low-capacitance twisted pair cables intended especially for digital audio use are recommended over the standard twisted pair cables used for analog audio, as the higher capacitance of analog audio cable tends to distort square wave signals by rolling off the higher frequency components.

Digital Audio Distribution

The use of analog video distribution and routing equipment is generally not recommended for AES3 signals, as such equipment may distort AES signal shapes and rise times, adversely affecting the decoding of the signal at receiving equipment. The spurious high-frequency signal energy that may be generated by such distortions of signal shape can cause crosstalk-related bit errors that are difficult to dissect and analyze. Distribution of the AES3 signal using high-quality digital audio distribution amplifiers will maintain the proper frequency and phase relationships, as well as signal shapes and rise times.

System Synchronization

When possible, all digital audio signals should be synchronous in order to avoid objectionable digital artifacts. In a large plant, it is necessary to provide a single master reference signal to which all interconnected systems are synchronized. The master reference, fed to all pieces of equipment, allows audio data to be retimed and synchronized within specified tolerances.

Larger facilities, in particular, will benefit from the conversion of digital audio signals from sources without external sync capability to a standard, synchronized audio sample rate. Broadcast digital audio plants typically contain consumer and other non-synchronizable equipment that require sample rate conversion. Audio sample rate converters perform a function similar to video standards converters, in that a dynamic low-pass filter continually adjusts the offending signal's phase at the output of the converter. In some cases, the output and input sample rates can be locked together via an integer relationship in a process known as *synchronous sample rate conversion.* For example, 48 kHz and 44.1 kHz are related by the integer ratio of 160 to 147. Modern sample rate conversion can be accomplished with full 24-bit resolution and THD+N below −140dB and as such has become an audibly lossless process. However, it is extremely important to note that in systems utilizing compressed audio such as AC-3, Dolby E, DTS, and apt-X, bit-for-bit accuracy of the AES3 audio payload is imperative and will be corrupted by sample rate conversion—even when used in 1:1 modes for retiming (i.e., 48 kHz is reclocked to local 48 kHz reference).

AES3 Synchronization

AES3 is inherently synchronous, the clock signal being readily recovered from the AES3 bitstream. However, the use of a master digital audio reference ensures that all digital audio equipment in a system will be frequency and phase locked and free of cascaded timing errors, and is highly recommended by AES11. The master reference signal may come from the digital audio console in a facility on the scale of a single room, or from an external reference generator in larger facilities. The master sync signal should be sent to all equipment capable of accepting external sync signals.

Digital audio phase integrity must remain intact during the conversion of multiple audio channels between the digital and analog domains. Perfect phase synchronization requires use of an SDIF-2 word clock or an AES3 signal as the common master clock. Digital audio recording and processing equipment forces any AES3 input signal into a common AES3 frame phase. When such an AES3 frame alignment is performed, a phase error will result if there are any deviations in the frame phase of A/D converters.

When digital audio signals are transferred to a piece of equipment that is not synchronized using a master sync signal, sample rate converters must be used at the inputs to the receiving equipment to prevent clicks and pops.

Word Clock Synchronization

SDIF-2 word clock, commonly referred to as simply *word clock,* is a square wave signal at the digital audio sample rate. Word clock is commonly used as a reference signal in small, audio-only facilities. In facilities that handle both video and audio, black burst is commonly used as the reference for both video and AES audio signal synchronization. It should be noted that most professional audio equipment intended for use in video facilities does not accept word clock as a reference signal, but instead relies upon the AES11 standard, whereby an AES/EBU signal with its embedded clock reference is used to derive proper synchronization. This eliminates the difficulties of distributing a high-frequency word clock square wave throughout a facility.

Signal Routing

Asynchronous routing is the simplest and most cost-efficient method of routing digital audio. It passes digital audio signals at any sample rate, a degree of flexibility that is ideal in situations where a number of different audio sample rates are encountered. However, the lack of synchronization to a master reference makes it a poor choice for on-air applications or any other situation where frame accurate switching or editing is required.

An asynchronous router may be thought of as an electronic patching system, functioning as though simple wires were used to connect inputs to outputs. In an asynchronous system, it is imperative that the destination equipment be capable of locking to the sample rate of the signal routed to it, otherwise muting usually takes place.

The disadvantage of asynchronous routers is that their output signal is almost always corrupted when a switch is made between input signals. A switch typically results in one or more AES frames being damaged, and this may cause destination equipment to momentarily lose lock, causing muting or the generation of pops and clicks.

Synchronous routing ensures precise timing and no corruption of the data stream during switches. It is considerably more complex and costly than asynchronous routing, as it requires that a transition between two inputs be made at an AES frame boundary. All inputs to a synchronous router must be locked to a common digital audio reference. A digital audio console is essentially a synchronous router with many controls. Note that when routing compressed audio such as Dolby E, switching must occur not only at an AES frame boundary, but also at an AES frame boundary located near the video vertical interval switch point to prevent corruption of the compressed audio packets. Systems based on the apt-X format require switching only on the AES frame boundary. Routing and switching AC-3 encoded signals is of greater difficulty as the encoded packets from one stream to the next are not phase aligned.

Jitter

Jitter is short-term frequency variation in the input data stream to a digital audio device. It can result from a number of causes, including such things as the coupling of excessive noise into a transmission link. Some jitter build-up is inevitable in a system, as certain components of the system inherently generate some amount of jitter. For example, noise in the phase-locked loops that control clock frequencies in the components of the system unavoidably generates some jitter. The presence of out-of-specification jitter on a digital audio signal or clock can result in bit errors that generate clicking and popping sounds. High levels of jitter may cause a receiving device to lose lock, while a relatively small amount may have no apparent negative effect unless present in devices performing A/D or D/A conversions. Excessive jitter is seldom a problem when only two pieces of equipment are involved, but typically builds up when larger numbers of equipment are interconnected. Jitter may be eliminated by using synchronizing D/As or a common synchronization signal. Jitter on the synchronization signal itself can cause degradation of all digital audio in devices locked to it.

Levels and Metering

(See also Chapter 10.1, Audio Signal Analysis, which includes a detailed discussion of audio level and loudness metering.) When an analog audio signal is converted to digital, the greatest analog voltage level that may be represented digitally is called full-scale digital (FSD). When quantized, this voltage level causes all digital audio bits to be set to logic 1, and this level is called 0 dBFS (full scale). This is an inflexible limit, and any excursion of the analog signal above this level will be clipped off, as the digital audio word does not have the capacity to faithfully represent it. In practice, the FSD level is often set about 1 dB above the analog clip level in an effort to assure that digital clipping never occurs.

When signals are converted between the analog and digital domains, the analog reference levels of A/D and D/A converters may be set to any number of values. If the analog reference level is improperly calibrated in any of the converters in the path, A/D and D/A conversions may result in an increase or a decrease in the level of the recovered analog signal.

Consistency in the type of digital audio metering device used, good operator training, and the establishment of strict house standard reference levels and alignment practices are the best defenses when it comes to accurate audio level control.

There is no U.S. standard for a specific digital audio level meter. Digital audiometers are often of the instantaneous response type, with no integration time, permitting them to respond with full excursion to a peak as brief as a single digital audio sample. Contrast this with the standard volume indicator (VU meter), which is an average-responding device, and the typical peak-program meter, which does not respond with full excursion to peaks with durations less than 10 ms. Typically, digital audio metering devices display a maximum value of 0 dBFS, and reference level lineup tone is set to a designated point below 0 dBFS to accommodate peaks without digital clipping.

Figure 2.8-4 shows a representative digital audio meter, the display device of which is usually an array of light emitting diodes or other such devices. This representative meter displays a range of 140 dB to 0 dBFS, with lineup tone being calibrated at −20 dBFS.

Loudness metering is best accomplished with meters designed to measure loudness. VU and PPM-type meters are not truly appropriate for accurately judging loudness, as the results are often a mixture of meter readout and user interpretation and are thus unreliable for producing consistent results. It is recommended that loudness metering be accomplished per ITU-R BS.1770, and loudness management per ATSC A/85:2013 as described in Chapter 5-14 of this handbook.

SUMMARY

Digital audio, with its many advantages, is not inherently susceptible to many of the problems that are encountered in analog audio systems. It does harbor some potential hazards of its own, however. With care and attention to good engineering practices in the design and maintenance of digital audio facilities and

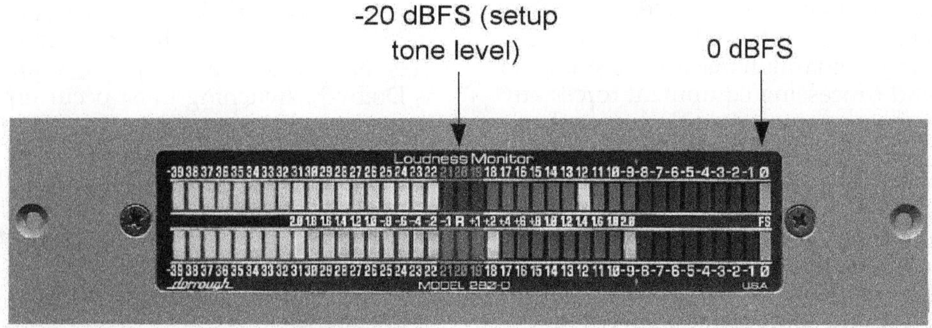

FIGURE 2.8-4 Representative digital audio level meter

Courtesy Dorrough Electronics
For television applications, SMPTE RP 155 recommends adjusting the level of lineup tone to read −20 dBFS on digital audio meters used on digital videotape recorders. Other industry segments have variously used lineup tone levels of −15, −18 and −20 dBFS. These varying reference levels may cause inconsistent results when digital audio recordings are interchanged. It is therefore important to establish common digital audio reference and operating levels when exchanging digital audio recordings.

observance of the recommendations described in AES/EBU, ATSC, IEC, and SMPTE standards, outstanding results will be realized.

References

Standards and Recommended Practices

AES3-2009 (r2014)—*AES Recommended Practice for Digital Audio Engineering—Serial Transmission Format for Two-Channel Linearly Represented Digital audio Data,* New York, Audio Engineering Society, 2014.

AES5-2008 (r2013)—*AES Recommended Practice for Professional Digital Audio—Preferred Sampling Frequencies for Applications Employing Pulse-Code Modulation,* New York, Audio Engineering Society, 2013.

AES10-2008 (r2014)—*AES Recommended Practice for Digital Audio Engineering—serial Multichannel Audio Digital Interface (MADI),* New York, Audio Engineering Society, 2014.

AES11-2009 (r2014)—*AES Recommended Practice for Digital Audio Engineering—Synchronization of Digital Audio Equipment in Studio Operations,* New York, Audio Engineering Society, 2014.

AES17-1998—*AES Standard Method for Audio Engineering—Measurement of Digital Audio Equipment,* New York, Audio Engineering Society, 1998.

AES18-1996—*AES Recommended Practice for Digital Audio Engineering—Format for the User Data Channel of the AES Digital Audio Interface,* New York, Audio engineering Society, 1996.

AES42-2010 (r2015)—*AES Standard for Acoustics—Digital Interface for Microphones,* New York, Audio engineering Society, 2015.

AES67-2015—*AES Standard for Audio Applications of Networks—High-Performance Streaming Audio-Over-IP Interoperability,* New York, Audio engineering Society, 2015.

ATSC A/52-2015—*Digital Audio Compression (AC-3) Standard,* Washington, Advanced Television Systems Committee, 2015.

IEC 60958-1999—*Digital Audio Interface,* Geneva, International Electrotechnical Commission, 1999.

ST 259-2008—*10-Bit 4:2:2 Component and 4fsc NTSC Composite Digital Signals—Serial Digital Interface,* White Plains, Society of Motion Picture and Television Engineers, 2008.

ST 272-2004—*Formatting AES/EBU Audio and Auxiliary Data into Digital Video Ancillary Data Space,* White Plains, Society of Motion Picture and Television Engineers, 2004.

ST 276-1995—*Transmission of AES/EBU Digital Audio Signals Over Coaxial Cable,* White Plains, Society of Motion Picture and Television Engineers, 1995 (Archived 2003).

ST 292-1-2012—*Bit-Serial Digital Interface for High-Definition Television Systems,* White Plains, Society of Motion Picture and Television Engineers, 2016.

ST 299-1-2009—*24-Bit Digital Audio Format for HDTV Bit-Serial Interface,* White Plains, Society of Motion Picture and Television Engineers, 2009.

ST 302-2007—*Mapping of AES3 Data Into MPEG-2 Transport Stream,* White Plains, Society of Motion Picture and Television Engineers, 2007.

ST 2035-2009 (Replaces SMPTE 320M)—*Audio Channel Assignments for Digital Television Recorders (DTRs),* White Plains, Society of Motion Picture and Television Engineers, 2009.

IEC 61937-1-2007—*Digital Audio—Interface for Non-Linear PCM Encoded Bitstreams Applying IEC 60958, Part 1—General,* Geneva, International Electrotechnical Commission, 2007.

IEC 61937-3-2007—*Digital Audio—Interface for Non-Linear PCM Encoded Bitstreams Applying IEC 60958, Part 3Non-linear PCM Bitstreams According to the AC-3 Format,* Geneva, International Electrotechnical Commission, 2007.

SMPTE Engineering Guideline EG 32-1996—*Emphasis of AES/EBU Audio in Television Systems and Preferred Audio Sampling Rate,* White Plains, Society of Motion Picture and Television Engineers, 1996.

ATSC A/85-2013—Techniques for Establishing and Maintaining Audio Loudness for Digital Television, 2013.

CHAPTER

2.9

Digital Video Standards and Practices

KARL J. KUHN
Tektronix, Inc.
Beaverton, OR

INTRODUCTION

The progress made in the industry so far would have been wholly impossible without the considerable efforts of organizations such as the AIMS, ARIB, ATSC, CableLabs, CTA, DVB, IEEE, IETF, ISO, NAB, NABA, SCTE, SMPTE, and others. The work done by these organizations is the foundation for bringing order to the benefit of all in the industry. Standards, specifications, and practices should be an integral part of all aspects of project and product life cycles as well as the end-to-end workflow of content. Standards, whether for a new OTT production facilities or for legacy VTR connector pin assignments, are vital for the continued growth of the communications industry.

Standards should be of interest to anyone involved with purchasing and integrating equipment or products into a greenfield or existing facility. The use of equipment that is proprietary may limit its current and future utility because of its ability (or inability) to integrate with recognized interface standards for connectors, input/output levels, control, timing, and test specifications. Manufacturers may find a limited market for a viable solution, and they could miss out on the economies of scale to optimize their return on investment.

Embracing standards reduces the interoperability issues of the integration of equipment from different manufacturers. The record pace of technology advancement in this industry attests to the need for and benefits of standards. Standards should be referenced and be part of the requirement documents for all phases of project management. This reduces ambiguity and helps reduce project risks.

Quality protects the brand of those involved with production and postproduction. Standards should be agreed upon and adopted throughout the workflow to maintain quality and reduce the need to rework deliverables.

Operations personnel should be concerned with standards that apply to their infrastructure for any system updates as well as for business as usual. The more dynamic the workflow, the more routinely standards play a major role in system QoS/QoE and availability metrics. Audio loudness, closed-captioning, and other mandates require careful monitoring, logging, and trending to avoid violations.

Content delivery mechanism pathways to end users have proliferated and diversified in recent years, creating new challenges. Standards are important to maintain a high quality of experience levels by reducing the likelihood of annoyances for blockiness, freeze frame, and loss of video and audio. Loudness and closed-captioning are important for QoE, as well as government mandates. It is paramount to adhere to required standards in administering mandates for public safety.

Standards in consumer electronics are critical in assuring interoperability of the interconnection and usability of devices. The manufacturers agreeing upon and implementing standards increases the chance of successful product launches and higher rankings in user reviews, facilitating a greater return on investment. Consumers have very high expectations regarding the ease of interconnecting and performance of their mobile and home devices.

Some standards become obsolete by the time they are adopted due to rapid improvements in the industry.

Our technology-based society accepts obsolescence as a fact of life, but this should not deter involvement and implementation of standards development and acceptance. Standards help reduce the unknowns and provide stable platforms for developers, manufacturers, and integrators to assure a high degree of compatibility, user satisfaction, and business success. Of the many standards-setting organizations in the professional video field, the most familiar are:

- Advanced Television Systems Committee (ATSC)
- The Association of Radio Industries and Businesses (ARIB)
- Consumer Technology Association (CTA)
- Digital Video Broadcasting (DVB)
- Institute of Electrical and Electronics Engineers (IEEE)
- The Internet Engineering Task Force (IETF)
- International Standards Organization (ISO)
- Society of Cable and Telecommunications Engineers (SCTE)
- Society of Motion Picture and Television Engineers (SMPTE)

Web Resources

The data contained in this chapter was drawn heavily from material made available by leading standards organizations. Website addresses are given where applicable, and readers are encouraged to explore these valuable resources. Standards documents can be downloaded or ordered online from most of the sites. Because of the rapidly changing nature of digital audio and video implementation, readers are encouraged to check in regularly.

Another valuable resource is the SMPTE Digital Library containing more than 23,000 documents, including all *SMPTE Journal* articles back to 1916, conference papers dating back to 1969, and more than 800 standards documents covering all aspects of motion image technology, including the iconic color bar test patterns, time code, timed text, and digital cinema.

Video Standards and Related Documents

It is impractical within the constraints of this chapter to list all standards, specifications, and recommended practices with related documents of interest to the industry. The following sections, however, contain a representative sample of key documents in alphabetical order by organization relating to the evolving technologies of digital television.

Advanced Television Systems Committee (ATSC)

The Advanced Television Systems Committee, Inc. (ATSC) is an international, nonprofit organization developing voluntary standards for digital television. ATSC was formed in 1982 by the member organizations of the Joint Committee on InterSociety Coordination (JCIC): the Electronic Industries Association (EIA), the Institute of Electrical and Electronic Engineers (IEEE), the National Association of Broadcasters (NAB), the National Cable Telecommunications Association (NCTA), and the Society of Motion Picture and Television Engineers (SMPTE).The ATSC membership has grown and now represents the television broadcast, broadcast equipment, motion picture, consumer electronics, computer, cable, satellite, and semiconductor industries.

Specifically, ATSC is working to coordinate television standards among different communications media focusing on digital television, interactive systems, and broadband multimedia communications. ATSC is also developing digital television implementation strategies and presenting educational seminars on the ATSC standards.

International adopters of the ATSC standard include Canada, Dominican Republic, El Salvador, Guatemala, Honduras, Mexico, and South Korea.

ATSC A/52:2015: Digital Audio Compression (AC-3) (E-AC-3) Standard

This document specifies coded representation of audio information and the decoding process, as well as information on the encoding process. The coded representation specified is suitable for use in digital audio transmission and storage applications, and may convey from one to five full bandwidth audio channels, along with a low-frequency enhancement channel. A wide range of encoded bit rates is supported by this specification. Typical applications of digital audio compression are in satellite or terrestrial audio broadcasting, delivery of audio over metallic or optical cables, or storage of audio on magnetic, optical, semiconductor, or other storage media. Revision B added a new annex, "Enhanced AC-3 Bit Stream Syntax" that specifies an additional syntax that offers additional coding tools and features.

ATSC A/53: ATSC Digital Television Standard

The Digital Television Standard describes the system characteristics of the advanced television (ATV) system. The document and its normative parts provide detailed specification of the parameters of the system, including the video encoder input scanning formats and the preprocessing and compression parameters of the video encoder, the audio encoder input signal format and the preprocessing and compression parameters of the audio encoder, the service multiplex and transport layer characteristics and normative specifications, and the VSB RF/transmission subsystem.

ATSC A/64B: Transmission Measurement and Compliance for Digital Television

This ATSC-recommended practice describes methods for testing, monitoring, and measurement of the transmission subsystem intended for use in the digital television (DTV) system adopted as the U.S. standard by the Federal Communications Commission (FCC). Both 8-VSB and 16-VSB are described in this document, which was reaffirmed December 11, 2014.

ATSC A/74: Receiver Performance Guidelines

This recommended practice addresses the front-end portion of a receiver of digital terrestrial television broadcasts. The recommended performance guidelines enumerated in this document are intended to assure that reliable reception will be achieved. Guidelines for interference rejection are based on the FCC planning factors that were used to analyze coverage and interference for the initial DTV channel allotments. Guidelines for sensitivity and multipath handling reflect field experience accumulated by testing undertaken by the Advanced Television Test Center (ATTC), Maximum Service Television (MSTV), the National Association of Broadcasters (NAB), and receiver manufacturers.

ATSC A/76B: Programming Metadata Communication Protocol Standard, Revision B

This standard defines a method for communicating metadata related to PSIP (program and system information protocol), including duplicate data that needs to be entered in other locations in the transport stream. Communication is based on a protocol utilizing XML message documents generated in accordance with a programming metadata communication protocol (PMCP) XML schema, which is defined within the above-referenced document.

ATSC A/79: Conversion of ATSC Signals for Distribution to NTSC Viewers

This recommended practice provides guidance to broadcasters and other creators of ATSC high-definition (HD) or standard-definition (SD) content and to the operators of multichannel video programming distributor (MVPD) systems (such as cable or direct-to-home satellite). It recommends the equipment capabilities needed to provide the highest quality programming to viewers who only receive NTSC.

ATSC A/85: Techniques for Establishing and Maintaining Audio Loudness for Digital Television

This recommended practice (RP) provides guidance to broadcasters and creators of audio for high-definition (HD) or standard-definition (SD) television content. It recommends production, distribution, and transmission practices needed to provide the highest quality audio soundtracks to the digital television audience. This RP focuses on audio measurement, production, and postproduction monitoring techniques and on methods to effectively control loudness for content delivery or exchange. It recommends methods to effectively control program-to-interstitial loudness, discusses metadata systems and use, and describes modern dynamic range control. It should be noted that the FCC rules and regulations requires adherence to this RP for the broadcast of commercial advertisements by television broadcast stations, cable operators, and other multichannel video programming distributors.

ATSC A/91: Implementation Guidelines for the Data Broadcast Standard

This document provides a set of guidelines for the use and implementation of the ATSC Data Broadcast Standard. As such, the guidelines facilitate the efficient and reliable implementation of data broadcast services. The information contained herein applies to data service providers as the primary entity that assembles the elements of each data channel. It also applies to broadcasters, network operators, and infrastructure manufacturers. The rules are specified in the form of constraints on the data broadcast implementation.

Note that A/91 has not been updated to reflect the latest revision of the Data Broadcast Standard, A/90:2013.

ATSC A/107:2015, ATSC 2.0 Standard

This document is a top-level specification of the ATSC 2.0 fixed-broadcast digital television services, which augment the digital television services defined in ATSC A/53. As a top-level document, this standard references other ATSC standards, as well as standards developed by other organizations.

A/110:2011: ATSC Standard for Transmitter Synchronization

This document defines a standard for synchronization of multiple transmitters emitting trellis-coded 8-VSB signals in accordance with ATSC A/53 Part 2 and of both single and multiple transmitters emitting mobile DTV signals in accordance with ATSC A/153 Part 2. The emitted signals from transmitters operated according to this standard comply fully with the requirements of both ATSC A/53 and A/153. This document specifies mechanisms necessary to transmit synchronization signals to the one or several transmitters using a dedicated PID value, including the formatting of packets associated with that PID value and without altering the signal format emitted from the transmitters. It also provides for adjustment of transmitter timing and other characteristics through additional information carried in the specified packet structure. Techniques are provided for cascading transmitters in networks of synchronous translators. In addition, it specifies an alternative method for transmitting synchronization signals to single transmitters operating according to A/153.

TSC A/111: Design of Synchronized Multiple Transmitter Networks

Many of the challenges of radiofrequency transmission are the same regardless of whether the information carried is in analog or digital form. Because of the signal processing applied when the information carried is digital, however, there are techniques to overcome some of those challenges that are more applicable to digital signals than to analog signals. Among such techniques is the use of multiple transmitters in single frequency networks (SFNs) and multiple frequency networks (MFNs). In the past, SFNs have been considered mostly for applications in multi-carrier systems, such as those using COFDM modulation. This recommended practice applies SFNs to the single-carrier 8-VSB system adopted by the ATSC and the FCC, but has not been updated yet to reflect the additional techniques needed to support mobile DTV (as documented in A/153).

SFNs can be implemented with digital on-channel repeaters (DOCRs), distributed transmitters (DTxTs), distributed translators (DTxRs), or with a combination of them. MFNs generally involve the use of translators. This recommended practice examines all three types of transmitters used in SFNs and MFNs and then concentrates on the design aspects of SFNs in the absence of A/153 enhancements.

ATSC A/321:2016: System Discovery and Signaling

This document describes the system discovery and signaling architecture for the ATSC 3.0 physical layer. Broadcasters anticipate providing multiple wireless-based services, in addition to conventional broadcast television, in the future. Such services may be time-multiplexed together within a single RF channel. The bootstrap provides a universal entry point into a broadcast waveform. The bootstrap employs a fixed configuration (e.g., sampling rate, signal bandwidth, subcarrier spacing, time-domain structure) known to all receiver devices and carries information to enable processing and decoding of the wireless service associated with a detected bootstrap. This capability ensures that broadcast spectrum can be adapted to carry new services and/or waveforms for public interest to continue to be served in the future.

Alliance for IP Media Solutions (AIMS)

The Alliance for IP Media Solutions (AIMS), is a nonprofit trade alliance that promotes the open standards that broadcast and media companies use to move from legacy SDI systems to a virtualized, IP-based future—quickly and profitably.

SMPTE 2022-6

This widely implemented standard in the industry is recommended for continued use and adoption as a baseline for interoperability. Because the SMPTE 2022-6 payload is identical to SDI, it is seen as an effective way to create hybrid IP/SDI systems and will continue to serve that purpose for the foreseeable future.

VSF TR-04

VSF TR-04 is a technical recommendation to use two existing standards: SMPTE 2022-6 for video with embedded audio and AES67 for separate IP addressable audio streams. In VSF TR-04, AES67 effectively operates as the "discrete audio" equivalent in IP. This makes TR-04 an ideal solution for systems that require both the compatibly benefits of SMPTE 2022-6 and the flexibility of discrete audio.

VSF TR-03

This is the third stage in the standards-based AIMS roadmap. TR-03 recommends the transport of uncompressed elementary stream media over IP. TR-03 is very much about essence and separating out the media type, whether video, audio, metadata, or timing events. With TR-03, video, audio, and metadata are individually packetized into separate IP streams.

Association of Radio Industries and Businesses (ARIB)

The Association of Radio Industries and Businesses (ARIB) was established to promote research and development of new radio systems and to serve as a standards development organization to advance the unification of international standards and related activities in the telecommunications and broadcasting fields. The ARIB was also aimed at providing quick and accurate responses to such trends as growing internationalization, the convergence of telecommunications and broadcasting, and the need to promote radiowave usage in business.

ARIB was chartered by Japan's minister of posts and telecommunications as a public service corporation on May 15, 1995. Its activities include those previously performed by the Research and Development Center for Radio System (RCR) and the Broadcasting Technology Association (BTA). On April 1, 2011, ARIB renewed its organization under the provisions of the General Incorporated Association and General Incorporated Foundations Act.

The objectives of ARIB are to carry out studies, research and development, and consultation work related to radiowave use. These activities are conducted to promote the practical application and dissemination of radio systems in the telecommunications and broadcasting fields, as well as to achieve robust advancement and development of radio industries, thereby contributing to public welfare.

ARIB STD-B56 HDTV System Parameters for Program Production Date, March 19, 2013

This standard specifies the system parameters of the UHDTV for program production system. It is a subset of ITU-R BT.2020 and its parameter values are selected to meet program production for a broadcasting service in Japan.

ARIB STD-B58 Interface for UHDTV Production Systems, March 18, 2014

This standard specifies the interface for UHDTV equipment that meets UHDTV system parameters for program production (ARIB STD-B56).

ARIB STD-B66 UHDTV Multi-Format Color Bar, July 3, 2015

This standard establishes a color bar signal for the UHDTV video format specified in ARIB standard STD-B56 to facilitate video level control and monitor adjustment.

This standard applies to the color bar signal for UHDTV program production, especially in multi-format environment.

ARIB STD-B67 Essential Parameter Values for the Extended Image Dynamic Range Television (EIDRTV) System for Program Productions, July 3, 2015

This standard specifies the system parameters essential for the extended image dynamic range television (EIDRTV), including system colorimetry, signal format, and digital representation. The signal format is characterized by opto-electronic transfer function (OETF). For the other system parameters, other television system standards may be referred to.

This standard applies to the video equipment for program production of the extended image dynamic range television.

ARIB STD-B68 Time Code Format in the Interface for UHDTV Production Systems

This standard regulates the data structure and multiplexing method in the case of transmitting the time code in the studio equipment interface of the ultra high definition television (UHDTV) signal.

ARIB STD-B69 Exchange Format of the Digital Closed Caption File for Digital Television Broadcasting System (second generation)

This standard regulates the subtitle file exchange format compatible with the subtitle broadcasting system stipulated in STD-B 62 "Multimedia Coding System (2nd Generation in Digital Broadcasting)."

CableLabs

CableLabs is a nonprofit innovation and R&D lab founded in 1988 by members of the cable television industry. With a strong focus on innovation, CableLabs develops technologies and specifications for the secure delivery of high-speed data, video, voice, and next generation services. It also provides testing, certification facilities, and technical leadership for the industry.

CableLabs' mission is to enable cable operators to be the providers of choice to their customers. Cable operators from around the world are members.

OpenCable Specifications, Adaptive Transport Stream Specification, OC-SP-ATS-I01-140214

There is a variety of adaptive streaming wire formats. Some are based on an MPEG-2 transport stream container such as HLS (HTTP live streaming: Apple) and others on a fragmented MP4 container such as HSS (HTTP smooth streaming: Microsoft) and HDS (HTTP dynamic streaming: Adobe). DASH (Dynamic Adaptive Streaming over HTTP: MPEG) supports both containers. While different, they utilize common video and audio compression formats, namely ISO/IEC 14496-10 (AVC) and ISO/IEC 14496-3 (AAC). Additional audio formats, such as Dolby Digital Plus, may also be supported by these or a subset of these adaptive bit rate (ABR) formats.

In a unified ABR encoding and encapsulating system, video and audio data are encoded and conditioned for adaptive streaming purposes, and the resultant elementary compressed access units are fed to one or more ABR encapsulations or packagers to be formatted into ABR-specific wire formats.

OpenCable Specifications, Encoder Boundary Point Specification, OC-SP-EBP-I01-130118

The Encoder Boundary Point (EBP) structure is a signaling mechanism that can provide a hinting mechanism for taking continuous streams conditioned for adaptive streaming and creating discrete chunks of decodable content with boundaries in one component stream in the multiplex (fragment) or across the multiplex (segment). The EBP is carried as private data of the adaptation private field of an MPEG-2 TS packet for video or audio and can be applied to each video packetized elementary stream (PES) and audio PES packet, resolved down in many cases to a single video access unit (AU) or a group of audio AUs.

Additionally, it can provide a timing field to indicate the acquisition time of the stream that can be used for synchronization purposes. Lastly, a labeling mechanism is introduced for a single access unit or region of access units in the content stream.

Consumer Technology Association (CTA)

The Consumer Technology Association's mission is to grow the consumer technology industry.

The CTA unites 2,000 companies within the consumer technology industry. Members tap into valuable and innovative members-only resources: unparalleled market research, networking opportunities with business advocates and leaders, up-to-date educational programs and technical training, exposure in extensive promotional programs, and representation from the voice of the industry, promoting and advancing member needs and interests. The CTA is also the producer of the Consumer Electronics Show (CES). The following is a partial list of standards and related technical documents.

CTA CEB11-C: NTSC/ATSC Loudness Matching (Formerly CEA-CEB11-C)

This bulletin provides guidance on maintaining uniform audio program loudness between various combinations of analog sources, NTSC services, and digitally delivered terrestrial DTV, satellite, and/or cable services while preserving the dynamic range delivery capability of the DTV service.

It also addresses the source levels, gain structure, and output levels of any consumer digital television broadcast receiver, set-top box, integrated AV receiver, or similar device that is capable of receiving any or all of the above services.

CTA CEB12-B (R2015): DTV Recommended Practice for Locating and Navigating Among ATSC Television Channels (Including PSIP Recommendations) (Formerly CEA-CEB12-B (R2015)

This bulletin provides guidance for the development and implementation of ATSC Program and System Information Protocol (PSIP)-compliant receivers, cable-ready DTV receivers, personal recorders, and other consumer electronics devices. It includes recommendations and suggestions for device functionality.

CTA 608-E (R2014): Line 21 Data Services (Formerly ANSI/CEA-608-E)

This is a technical standard and guide for using or providing closed-captioning services or other data services embedded in line 21 of the vertical blanking interval of the NTSC video signal. It includes provisions for encoding equipment and/or decoding equipment to produce such material, as well as for manufacturers of television receivers that are required to include such decoders in their equipment as a matter of regulation. It is also a usage guide for producing material using such equipment and for distributing such material.

This standard describes the specifications for creation, transmission, reception, and display of caption data, plus the relationship of caption mode data to other line 21 data. A comparison of decoders meeting FCC rules to all decoders designed prior to the drafting of those rules and a timetable for the implementation of features that are unique to the different generations of decoders are retained from the prior version.

CTA CEB16-A: Active Format Description (AFD) and Bar Data Recommended Practice (Formerly CEA-CEB16-A)

The CEA-CEB16-A bulletin provides guidance for the development and implementation of consumer devices that process information related to aspect ratio signaling, active format description (AFD), and bar data. Such devices may respond to this signaling by creating an optimum (or more desirable) display, by passing this signaling through to an analog or digital output, or in preparing a digital recording. CEA-CEB16-A reviews the standardized methods for representing AFD, aspect ratio information, and bar data on various analog and digital SD and HD interfaces. For each of various transmitted display formats, CEA-CEB16-A provides recommendations for output signaling and display appearance, taking into account possible decoder capabilities, scaling options, and user preferences regarding display format.

CTA CEB20 (R2013): A/V Synchronization Processing Recommended Practice (Formerly CEA-CEB20)

This recommended practice outlines the steps that an MPEG decoder should take to ensure and maintain audio/video synchronization. Such synchronization is necessary for end-viewer satisfaction.

CTA 679-C (R2013): National Renewable Security Standard (NRSS) (Formerly CEA-679-C (R2013))

This describes the physical and electrical parameters of the interface for conditional access (CA) devices for consumer electronics. The purpose of this standard is to specify the interface to a CA device that performs the signal access and, optionally, descrambling and key generation.

CTA 708-E: Digital Television (DTV) Closed Captioning (Formerly ANSI/CEA-708-E)

This standard defines DTV closed-captioning (DTVCC) and provides specifications and guidelines for caption service providers, distributors of television signals, decoder and encoder manufacturers, DTV receiver manufacturers, and DTV signal processing equipment manufacturers. The use of the term DTV throughout is intended to include, and apply to, high-definition television (HDTV) and standard-definition television (SDTV).

CTA 766-D: U.S. and Canadian Rating Region Tables (RRT) and Content Advisory Descriptors for Transport of Content Advisory Information Using ATSC Program and System Information Protocol (PSIP) (Formerly ANSI/CEA-766-D)

This standard augments ATSC A/65 [A65] and designates the rating region tables (RRT) that provides the receiver with the definition of the rating system and the content advisory descriptors that provide the receiver with the specific program rating for each program. Specifically, this standard specifies the exact syntax to be used to define the U.S. and Canadian RRTs in accordance with ATSC A/65 [A65] Section 6.4, as well as the exact syntax to be used in the content advisory descriptors that convey the rating information for each program in accordance with ATSC A/65 [A65] Section 6.9.3. Thus, DTV receivers are able to block unwanted programs as determined by the user.

This standard does not indicate and is not meant to indicate techniques or procedures for blocking at the DTV receiver (for example, remote control, other human interface devices, on-screen displays, and/or graphical user interfaces).

CTA 774-C: TV Receiving Antenna Performance Presentation and Measurement (Formerly ANSI/CEA 774-C)

This standard defines test and measurement procedures for use by manufacturers of television receive antennas to categorize antennas in accordance with CTA-2028-B, color codes for outdoor TV receiving antennas, for use with the CEA/NAB TV antenna selector program (see www.Antennaweb.org). Essential elements include procedures to determine antenna gain, front-to-back ratio, average gain to null ratio, directivity, and distortion performance of active antennas with integrated amplifiers.

CTA 796-A (R2012): NRSS Copy Protection Systems (Formerly CEA-796-A (R2012))

The copy protection systems that have been included in CTA-796 are itemized for the purpose of identification. The systems outlined in CTA-796 all support the copy protection frameworks described in CTA-679-B, Parts A and B.

CTA 799-A (R2012): On-Screen Display Specification (Formerly ANSI/CEA-799-A)

This specifies syntax and semantics for bitmapped graphics data typically used for on-screen display (OSD). CTA-799-A is applicable whenever it is necessary to specify a standard method for delivery of bitmapped graphics data. The pixel formats include optional alpha-blend and transparency attributes to support composition of graphics over analog or digitally decoded video within the display.

CTA 863-B: Connection Color Codes for Home Theater Systems (Formerly CEA 863-B)

This standard defines the colors for marking connections commonly used for electronic devices in a home theater system. This standard adds continuity to installation information and ensures consistency of information to installers.

CTA 2028-B: Color Codes for Outdoor TV Receiving Antennas (Formerly ANSI/CEA 2028-B)

This standard defines color codes to be associated with minimum performance parameters of outdoor television receiving antennas. When used in conjunction with the CTA/NAB TV antenna selector program at www.AntennaWeb.org, these color codes can help both consumers and professional installers select appropriate outdoor TV antennas for their particular reception environments.

CTA 2032-B: Indoor TV Receiving Antenna Performance Standard (Formerly ANSI/CEA 2031-B)

This standard defines test and measurement procedures for determining the performance of indoor TV receiving antennas. Indoor TV antennas are defined as devices that enable TV reception in indoor areas where there is adequate signal level above noise and interference to enable production of a viewable picture and sound.

CTA 849-B (R2013: Application Profiles for CTA-775 Compliant DTVs (Formerly ANSI/CEA 849-B)

This standard defines transport and content coding formats a compliant DTV must support in order to interoperate with various digital audio and video sources. A DTV compliant with this standard should also comply with the requirements of CTA-775-C.

CTA 909-B (R2016): Antenna Control Interface (Formerly ANSI/CEA 909-B)

CTA-909-B describes an antenna control subsystem for receiving terrestrial transmissions. The primary use is to facilitate television reception. The receiver controls the antenna apparatus to optimize the signal automatically for best reception by adjusting its configuration. CTA-909-B allows any receiver to operate with any antenna, regardless of manufacturer. CTA-909-B defines the data algorithms used, connection standards, and other requirements. The antenna configuration is neither specified nor implied, leaving certain antenna design considerations to the manufacturer.

Digital Video Broadcasting (DVB)

Digital Video Broadcasting (DVB) is an industry-led consortium of the world's leading digital TV and technology companies, such as manufacturers, software developers, network operators, broadcasters, and regulators, committed to designing open technical standards for the delivery of digital TV and other broadcast services.

ETSI TR 101 290 V1.2.1 (2001-05) Technical Report Digital Video Broadcasting (DVB): Measurement Guidelines for DVB Systems

The present document provides guidelines for measurement in DVB satellite, cable, and terrestrial and related digital television systems. The present document defines a number of measurement techniques, such that the results obtained are comparable when the measurement is carried out in compliance with the appropriate definition.

ETSI TS 102 034 V2.1.1 (2016-04) Digital Video Broadcasting (DVB): Transport of MPEG-2 TS Based DVB Services Over IP-Based Networks

DVB-IPTV is the collective name for a set of open, interoperable technical specifications, developed by DVB, that facilitate the delivery of digital TV using IP over bi-directional fixed broadband networks. Key specifications already published include:

- TS 102 034: Transport of MPEG-2 TS-Based DVB Services over IP Based Networks
- TS 102 539: Carriage of Broadband Content Guide (BCG) Information over Internet Protocol
- TS 102 824: Remote Management and Firmware Update System for DVB IP Services
- TS 102 905: Technical Specification for DVB Services in the Home Network Phase 1
- TS 102 542: Guidelines for the Implementation of DVB-IPTV Phase 1 Specifications
- DVB's interactive middleware specifications, DVB-MHP and GEM, also include IPTV profiles

ETSI TS 103 285 V1.1.1 (2015-05) Digital Video Broadcasting (DVB): MPEG-DASH Profile for Transport of ISO BMFF Based DVB Services Over IP-Based Networks

The present document defines the delivery of TV content via HTTP adaptive streaming. This includes the following:

- A profile of the features defined in MPEG DASH specification (referred to by MPEG as an "interoperability point"), largely based on the "ISOBMFF live" profile defined by MPEG
- Constraints on the sizes or complexity of various parameters defined in the MPEG DASH specification
- A selection of the video and audio codecs from the DVB toolbox that are technically appropriate with MPEG DASH constraints and/or requirements for the use of these, without mandating any particular codec
- Using MPEG Common Encryption for content delivered according to the present document
- Use of TTML subtitles with MPEG DASH
- Requirements on player behavior needed to give interoperable presentation of services
- Guidelines for content providers on how to use MPEG DASH
- Among others, the following subjects are not covered in the present document:
    - Mandatory codecs
    - Specific content protection schemes

ETSI EN 300 744 V1.6.2 (2015-10) Digital Video Broadcasting (DVB): Framing Structure, Channel Coding and Modulation for Digital Terrestrial Television

The present document describes a baseline transmission system for digital terrestrial television (TV) broadcasting. It specifies the channel coding/modulation system intended for digital multi-program LDTV/SDTV/EDTV/HDTV terrestrial services.

ETSI EN 302 307-1 V1.4.1 (2014-11) Digital Video Broadcasting (DVB): Second Generation Framing Structure, Channel Coding and Modulation Systems for Broadcasting, Interactive Services, News Gathering and Other Broadband Satellite Applications; Part 1: DVB-S2

DVB-S (EN 300 421) was introduced as a standard in 1994 and DVB-DSNG (EN 301 210) in 1997. The DVB-S standard specifies QPSK modulation and concatenated convolutional and Reed-Solomon channel coding, and is now used by most satellite operators worldwide for television and data broadcasting services. DVB-DSNG specifies, in addition to DVB-S format, the use of 8PSK and 16QAM modulation for satellite newsgathering and contribution services.

ETSI EN 302 307-2 V1.1.1 (2014-10) Digital Video Broadcasting (DVB): Implementation Guidelines for the Second Generation System for Broadcasting, Interactive Services, News Gathering and Other Broadband Satellite Applications; Part 2—S2 Extensions (DVB-S2X)

DVB-S2X is an extension of the DVB-S2 specification that provides additional technologies and features. DVB-S2X has been published as ETSI EN 302 307 part 2, with DVB-S2 being part 1. DVB-S2X offers improved performance and features for the core applications of DVB-S2, including direct to home (DTH), contribution, VSAT, and DSNG. The specification also provides an extended operational range to cover emerging markets such as mobile applications.

ETSI EN 302 755 V1.4.1 (2015-07) Digital Video Broadcasting (DVB): Frame Structure Channel Coding and Modulation for a Second Generation Digital Terrestrial Television Broadcasting System (DVB-T2)

DVB-T2 like its predecessor DVB-T is an orthogonal frequency division multiplex (OFDM) base modulation with a large number of subcarriers for robustness while offering a range of different modes, which provides flexibility. DVB-T2 uses the same error correction coding as the DVB-S2 and DVB-C2 standards. The number of carriers, guard interval size, and pilots are adjustable to allow optimization for any target transmission channel.

Additional technologies that have been incorporated into DVB-T2 to improve its performance compared to the DVB-T standard include: multiple physical layer pipes (PLPs) that allow for separate adjustment of the each delivered service to meet various reception conditions—these PLPs also allow receivers to save power by only decoding a single service rather than the whole multiplex; Alamouti coding provides a methodology that improves coverage in small-scale single frequency networks; constellation rotation that provides additional robustness; extended interleaving; and future extension frames that allow for compatible enhancements. The DVB-T2 offers higher data rates and more robustness compared to DVB-T.

ETSI EN 302 769 V1.3.1 (2015-10) Digital Video Broadcasting (DVB): Frame Structure Channel Coding and Modulation for a Second Generation Digital Transmission System for Cable Systems (DVB-C2)

DVB-C2 is a digital cable transmission system developed by DVB. It uses the latest modulation and coding techniques to enable highly efficient use of cable networks where, up to now, in many cases downstream transmission capacity is already being used to its limit. DVB-C2 will initially be used for the delivery of innovative new services, such as video-on-demand (VOD) and high-definition television (HDTV), helping digital operators to remain competitive and also to meet retransmission requirements; in the longer term the migration of current DVB-C services to DVB-C2 is also foreseen.

Institute of Electrical and Electronics Engineers (IEEE)

IEEE is the world's largest technical professional organization dedicated to advancing technology for the benefit of humanity. IEEE and its members inspire a global community to innovate for a better tomorrow through its highly cited publications, conferences, technology standards, and professional and educational activities. IEEE is the trusted "voice" for engineering, computing, and technology information around the globe.

There are more than 420,000 IEEE members in over 160 countries. IEEE publishes a third of the world's technical literature in electrical engineering, computer science, and electronics and is a leading developer of international standards that underpin many of today's telecommunications, information technology, and power-generation products and services.

IEEE 1588-2008: IEEE Standard for a Precision Clock Synchronization Protocol for Networked Measurement and Control Systems

This standard defines a protocol enabling precise synchronization of clocks in measurement and control systems implemented with technologies such as network communication, local computing, and distributed objects. The protocol is applicable to systems communicating by local area networks supporting multicast messaging including, but not limited to, Ethernet. The protocol enables heterogeneous systems that include clocks of various inherent precision, resolution, and stability to synchronize to a grandmaster clock. The protocol supports system-wide synchronization accuracy in the sub-microsecond range with minimal network and local clock computing resources. The default behavior of the protocol allows simple systems to be installed and operated without requiring the administrative attention of users. The standard includes mappings to User Datagram Protocol (UDP)/Internet Protocol (IP), DeviceNet, and a layer-2 Ethernet implementation. It includes formal mechanisms for message extensions, higher sampling rates, correction for asymmetry, a clock type to reduce error accumulation in large topologies, and specifications on how to incorporate the resulting additional data into the synchronization protocol. The standard permits synchronization accuracies better than 1 ns. The protocol has features to address applications where redundancy and security are requirements. The standard defines conformance and management capabilities. There are provisions to support unicast and multicast messaging. The standard includes an annex on recommended practices. Annexes defining communication medium-specific implementation details for additional network implementations are expected in future versions of this standard.

1680.3-2012: IEEE Standard for Environmental Assessment of Televisions

The standard establishes a clear and consistent set of environmental performance criteria for the design of televisions, providing an opportunity for manufacturers to secure market recognition for efforts to reduce the environmental impact of electronic products. This standard is also intended to provide a tool for government, institutional, corporate, and consumer purchasers to identify products that demonstrate environmental leadership. This standard is intended to be updated and revised on a periodic basis to continue to set a higher performance standard for leadership products.

Internet Engineering Task Force (IETF)

The mission of the IETF is to make the Internet work better by producing high-quality, relevant technical documents that influence the way people design, use, and manage the Internet.

The IETF pursues this mission in adherence to the following cardinal principles:

Open process—Any interested person can participate in the work, know what is being decided, and make his or her voice heard on the issue. Part of this principle is a commitment to making documents, mailing lists, attendance lists, and meeting minutes publicly available on the Internet.

Technical competence—The issues on which the IETF produces its documents are issues where the IETF has the competence needed to speak to them and is willing to listen to technically competent input from any source. Technical competence also means that IETF output is to be designed to sound network engineering principles, also often referred to as "engineering quality."

Volunteer Core—Participants and leadership are people who come to the IETF because they want to do work that furthers the IETF's mission of "making the Internet work better."

Rough consensus and running code—Standards are based on the combined engineering judgement of participants and real-world experience in implementing and deploying specifications.

Protocol ownership—When the IETF takes ownership of a protocol or function, it accepts the responsibility for all aspects of the protocol, even though some aspects may rarely or never be seen on the Internet. Conversely, when the IETF is not responsible for a protocol or function, it does not attempt to exert control over it, even though it may at times touch or affect the Internet.

RFC 768: User Datagram Protocol, August 1980

UDP is defined to make available a datagram mode of packet-switched computer communication in the environment of an interconnected set of computer networks. This protocol assumes that the Internet Protocol (IP) is used as the underlying protocol. This protocol provides a procedure for application programs to send messages to other programs with a minimum of protocol mechanism. The protocol is transaction oriented, and delivery and duplicate protection are not guaranteed. Applications requiring ordered reliable delivery of streams of data should use the Transmission Control Protocol (TCP).

RFC 791: Internet Protocol, September 1981

The Internet Protocol (IP) is designed for use in interconnected systems of packet-switched computer communication networks. Such a system has been called a "catenet." IP provides for transmitting blocks of data called datagrams from sources to destinations, where sources and destinations are hosts identified by fixed length addresses. IP also provides for fragmentation and reassembly of long datagrams, if necessary, for transmission through "small packet" networks.

RFC 1112: Host Extensions for IP Multicasting

IP multicasting is the transmission of an IP datagram to a "host group," a set of zero or more hosts identified by a single IP destination address. A multicast datagram is delivered to all members of its destination host group with the same "best-efforts" reliability as regular unicast IP datagrams, i.e., the datagram is not guaranteed to arrive intact at all members of the destination group or in the same order relative to other datagrams. The membership of a host group is dynamic; that is, hosts may join and leave groups at any time. There is no restriction on the location or number of members in a host group. A host may be a member of more than one group at a time. A host need not be a member of a group to send datagrams to it.

RFC 3550 RTP: A Transport Protocol for Real-Time Applications, July 2003

This memorandum describes RTP, the real-time transport protocol. RTP provides end-to-end network transport functions suitable for applications transmitting real-time data, such as audio, video, or simulation data, over multicast or unicast network services. RTP does not address resource reservation and does not guarantee quality-of-service for real-time services. The data transport is augmented by a control protocol (RTCP) to allow monitoring of the data delivery in a manner scalable to large multicast networks and to provide minimal control and identification functionality. RTP and RTCP are designed to be independent of the underlying transport and network layers. The protocol supports the use of RTP-level translators and mixers. Most of the text in this memorandum is identical to RFC 1889, which it obsoletes. There are no changes in the packet formats on the wire, only changes to the rules and algorithms governing how the protocol is used. The biggest change is an enhancement to the scalable timer algorithm for calculating when to send RTCP packets in order to minimize transmission in excess of the intended rate when many participants join a session simultaneously.

RFC 4175: RTP Payload Format for Uncompressed Video, September 2005

The representation of uncompressed HDTV is specified in SMPTE standards 274M [274] and 296M [296]. SMPTE 274M defines a family of scanning systems with an image format of 1920x1080 pixels with progressive and interlaced scanning, while SMPTE 296M defines systems with an image size of 1280x720 pixels and progressive scanning. In progressive scanning, scan lines are displayed in sequence from top to bottom of a full frame. In interlaced scanning, a frame is divided into its odd and even scan lines (called fields) and the two fields are displayed in succession. SMPTE 274M and 296M define images with aspect ratios of 16:9, and define the digital representation for RGB and YCbCr components. In the case of YCbCr components, the Cb and Cr components are horizontally sub-sampled by a factor of two (4:2:2 color encoding). Although these formats differ in their details, they are structurally very similar. This memo specifies a payload format to encapsulate these and other similar video formats for transport within RTP.

RFC 6416: RTP Payload Format for MPEG-4 Audio/Visual Streams, October 2011

This document describes Real-time Transport Protocol (RTP) payload formats for carrying MPEG-4 audio and MPEG-4 visual bitstreams without using MPEG-4 systems. This document obsoletes RFC 3016. It contains a summary of changes from RFC 3016 and discusses backward compatibility to RFC 3016. It is a necessary revision of RFC 3016 in order to correct misalignments with the 3GPP Packet-switched Streaming Service (PSS) specification regarding the RTP payload format for MPEG-4 audio. For the purpose of directly mapping MPEG-4 audio/visual bitstreams onto RTP packets, this document provides specifications for the use of RTP header fields and also specifies fragmentation rules. It also provides specifications for media type registration and the use of the Session Description Protocol (SDP). The audio payload format described in this document has some limitations related to the signaling of audio codec parameters for the required multiplexing format. Therefore, new system designs should utilize RFC 3640, which does not have these restrictions. Nevertheless, this revision of RFC 3016 is provided to update and complete the specification and to enable interoperable implementations.

RFC 7798: RTP Payload Format for High Efficiency Video Coding (HEVC), March 2016

This memo describes an RTP payload format for the video coding standard ITU-T Recommendation H.265 and ISO/IEC International Standard 23008-2, both also known as High Efficiency Video Coding (HEVC) and developed by the Joint Collaborative Team on Video Coding (JCT-VC). The RTP payload format allows for packetization of one or more Network Abstraction Layer (NAL) units in each RTP packet payload, as well as fragmentation of a NAL unit into multiple RTP packets. Furthermore, it supports transmission of an HEVC bitstream over a single stream as well as multiple RTP streams. When multiple RTP streams are used, a single transport or multiple transports may be used. The payload format has wide applicability in videoconferencing, Internet video streaming, and high-bitrate entertainment-quality video, among others.

RFC 7933: Adaptive Video Streaming Over Information-Centric Networking (ICN), August 2016

This document considers the consequences of moving the underlying network architecture from the current Internet to an Information-Centric Networking (ICN) architecture on video distribution. As most of the traffic in future networks is expected to be video, it considers how to modify the existing video streaming mechanisms. Several important topics related to video distribution over ICN are presented. The wide range of scenarios covered include the following: evolving Dynamic Adaptive Streaming over HTTP (DASH) to work over ICN and leverage the recent ISO/IEC MPEG standard, layering encoding over ICN, introducing distinct requirements for video using Peer-to-Peer (P2P) mechanisms, adapting the Peer-to-Peer Streaming Protocol (PPSP) for ICN, creating more stringent requirements over ICN because of delay constraints added by Internet Protocol Television (IPTV), and managing digital rights in ICN. Finally, in addition to considering how existing mechanisms would be impacted by ICN, this document lists some research issues in the design of an ICN-specific video streaming mechanism.

International Organization for Standardization (ISO)

ISO is an independent, nongovernmental international organization with a membership of 163 national standards bodies. Through its members, it brings together experts to share knowledge and develop voluntary, consensus-based, market-relevant international standards that support innovation and provide solutions to global challenges.

ISO/IEC 13818-1: 2015 Information Technology, Generic Coding of Moving Pictures and Associated Audio Information—Part 1: Systems

ISO/IEC 13818-1:2015 specifies the system layer of the coding. It was developed principally to support the combination of the video and audio coding methods defined in Parts 2 and 3 of ISO/IEC 13818. The system layer supports six basic functions:

1. The synchronization of multiple compressed streams on decoding
2. The interleaving of multiple compressed streams into a single stream
3. The initialization of buffering for decoding start up
4. Continuous buffer management
5. Time identification
6. Multiplexing and signaling of various components in a system stream

SO/IEC 27001: Information Security Management

The ISO 27000 family of standards helps organizations keep information assets secure. Using this family of standards will help an organization manage the security of assets such as financial information, intellectual property, employee details, or information entrusted by third parties. ISO/IEC 27001 is the best-known standard in the family providing requirements for an information security management system (ISMS).

ISO 50001: Energy Management

Using energy efficiently helps organizations save money as well as helping to conserve resources and tackle climate change. ISO 50001 supports organizations in all sectors to use energy more efficiently, through the development of an energy management system (EnMS).

ISO 50001 is based on the management system model of continual improvement also used for other well-known standards such as ISO 9001 or ISO 14001. This makes it easier for organizations to integrate energy management into their overall efforts to improve quality and environmental management.

ISO 50001:2011 provides a framework of requirements for organizations to:

- Develop a policy for more efficient use of energy
- Fix targets and objectives to meet the policy
- Use data to better understand and make decisions about energy use
- Measure the results
- Review how well the policy works
- Continually improve energy management

International Telecommunication Union (ITU)

The International Telecommunication Union (ITU; French: Union Internationale des Télécommunications), originally the International Telegraph Union (French: Union Télégraphique Internationale), is a specialized agency of the United Nations (UN) that is responsible for issues that concern information and communication technologies.

The ITU coordinates the shared global use of the radio spectrum, promotes international cooperation in assigning satellite orbits, works to improve telecommunication infrastructure in the developing world, and assists in the development and coordination of worldwide technical standards. The ITU is active in areas including broadband Internet, latest-generation wireless technologies, aeronautical and maritime navigation, radio astronomy, satellite-based meteorology, convergence in fixed-mobile phone, Internet access, data, voice, TV broadcasting, and next-generation networks.

Its membership includes 193 Member States and around 700 public and private sector companies, as well as international and regional telecommunication entities, known as Sector Members and Associates, which undertake most of the work of each Sector.

Recommendation ITU-R BS.1864-0: Operational Practices for Loudness in the International Exchange of Digital Television Programs (2010)

This recommendation specifies operational practices intended to improve the consistency of loudness of digital television programs that are intended for international exchange. These practices apply to the production of those programs. They do not apply to the distribution or emission of broadcast programs.

Recommendation ITU-R BT.500-13: Methodology for the Subjective Assessment of the Quality of Television Pictures (2012)

This recommendation provides methodologies for the assessment of picture quality including general methods of test, the grading scales, and the viewing conditions. It recommends the double-stimulus impairment scale (DSIS) method and the double-stimulus continuous quality-scale (DSCQS) method, as well as alternative assessment methods such as single-stimulus (SS) methods, stimulus-comparison methods, single-stimulus continuous quality evaluation (SSCQE), and the simultaneous double stimulus for continuous evaluation (SDSCE) method.

Recommendation ITU-R BT.709-6: Parameter Values for the HDTV Standards for Production and International Program Exchange (2015)

This recommendation defines the image format parameters and values for HDTV.

Recommendation ITU-R BT.814-2: Specifications and Alignment Procedures for Setting of Brightness and Contrast of Displays (1992-1994-2007)

This recommendation defines a test signal for the adjustment of cathode ray tubes (CRTs) and flat panel display (FPD) devices. The signal may be used to set displayed image values for both SD and HD television.

Recommendation ITU-R BT.1120-8:
Digital Interfaces for HDTV Studio Signals
(1994-1998-2000-2003-2004-2005-2007-2012)

This HDTV interface operates at two nominal clock frequencies, 1.485 GHz and 2.97 GHz, and conveys the uncompressed payload defined in Part 2 of Recommendation ITU R BT.709. The interface may also be used for carrying packetized data.

Recommendation ITU-R BT.2020-2: Parameter Values for Ultra High Definition Television Systems for Production and International Program Exchange (2012-2014-2015)

Ultra-high-definition television (UHDTV) provides viewers with an enhanced visual experience primarily by having a wide field of view both horizontally and vertically with appropriate screen sizes relevant to usage at home and in public places. UHDTV applications require system parameters that go beyond the levels of HDTV. This recommendation specifies UHDTV image system parameters for production and international program exchange.

Recommendation ITU-RBT.2022: General Viewing Conditions for Subjective Assessment of Quality of SDTV and HDTV Television Pictures on Flat Panel Displays (2012)

This recommendation specifies general viewing conditions for subjective assessment of quality of SDTV and HDTV television pictures on flat panel displays.

Recommendation ITU-R BT.2100: Image Parameter Values for High Dynamic Range Television for Use in Production and International Programme Exchange (2016)

High Dynamic Range Television (HDR-TV) provides viewers with an enhanced visual experience by providing images that have been produced to look correct on brighter displays, that provide much brighter highlights, and that provide improved detail in dark areas. This recommendation specifies HDR-TV image parameters for use in production and international program exchange using the Perceptual Quantization (PQ) and Hybrid Log-Gamma (HLG) methods.

North American Broadcasters Association (NABA)

The North American Broadcasters Association (NABA) is member of the World Broadcasting Union (WBU) and serves as the WBU Secretariat. NABA is an association of broadcasting and distribution organizations in Canada, Mexico, and the United States. NABA represents member North American broadcasters in national and international entities: The three North American regulators—the FCC in the United States, Industry Canada/Canadian Radio-television and Telecommunications Commission (CRTC) in Canada, and Comisión Federal de Telecomunicaciones (CoFeTel) in Mexico—and, internationally, the World Broadcasting Unions (WBU), International Telecommunications Union-Radiocommunication Sector (ITU-R), and the World Intellectual Property Organization (WIPO). NABA also provides a forum for its members to share and develop information, specifications, and best practices, discuss common challenges, and reach consensus on key issues. The following is a partial list of key published documents that are freely downloadable from the NABA website.

NABA—DPP Technical Specifications for the Delivery of Air Ready Master Television Programs

These documents form a standards-based approach to specifying the technical requirements for the delivery of programming as files, delivered for on-air operations. The standards-based fulfillment of these technical requirements for file-based program interoperability is provided by citing a series of standards and then addressing the issues that frequently cause interoperability problems, by specifically citing the appropriate modes and options supported by these standards, which are necessary to ensure interoperability for broadcasters. These documents were created in conjunction with the Digital Production Partnership in response to requests from major North American Broadcasters to address interoperability issues in program delivery as files. There are two Air Ready Master technical specifications, and while they both utilize MXF, one was created for MPEG 2 codec-based files and the other for AVC codec-based files.

NABA—DPP Technical Specifications for the Delivery of Library Master Television Programs

These documents form a standards-based approach to specifying the technical requirements for the delivery of programming as files, delivered for processing to create files for VOD, online, OTT, and on-air operations. The standards-based fulfillment of these technical requirements for file-based program interoperability is provided by citing a series of standards and then addressing the issues that frequently cause interoperability problems, by specifically citing the appropriate modes and options supported by these standards, which are necessary to ensure interoperability for broadcasters. There are two Library Master technical specifications. Both of these specifications are based upon the SMPTE IMF standards. One specification is for the delivery of HD-based programs and the other is for the delivery of UHD-based programs.

Society of Cable and Telecommunications Engineers (SCTE)

SCTE/ISBE is proud to serve as the technical and applied science leader for the cable telecommunications industry. Since its beginning in 1969, SCTE has been dedicated to providing meaningful resources and programs for its members and the industry.

SCTE's "Boots to Suits" philosophy is designed to not only enhance the careers of technical employees throughout the network, but also provide operators with a well-trained workforce that ultimately means a more efficient system, economic benefits, and an improved customer experience.

SCTE 07 2013: Digital Transmission Standard for Cable Television

This standard describes the framing structure, channel coding, and channel modulation for a digital multi-service television distribution system that is specific to a cable channel. The system can be used transparently with the distribution from a satellite channel, as many cable systems are fed directly from satellite links.

SCTE 18 2007: Emergency Alert Message for Cable, Approved as a Joint Standard with CEA as ANSI-J-STD-042-2002 (formerly DVS 208)

Emergency Alert Message for Cable, approved as a joint standard with CEA as ANSI-J-STD-042-2002, (formerly DVS 208).

SCTE 20 2012: Method for Carriage of CEA-608 Closed Captions and Non-Real Time Sampled Video

This document defines a standard for the carriage of CEA-608 closed captions and certain other Vertical Blanking Interval (VBI) services in MPEG-2 compliant bitstreams constructed in accordance with ISO/IEC 13818-2.

SCTE 21 2012: Standard for Carriage of VBI Data in Cable Digital Transport Stream

This document defines a standard for the carriage of Vertical Blanking Interval (VBI) services in MPEG-2 compliant bitstreams constructed in accordance with ISO/IEC 13818-2. The approach builds upon a data structure defined in ATSC A/53 Part 4 (Digital Television Standard: Part 4 MPEG-2 Video System Characteristics), and is designed to be backwards compatible with that method.

SCTE 35 2014: Digital Program Insertion Cueing Message for Cable

This standard supports frame accurate signaling of events in MPEG-2 transport streams along with associated descriptive data. This standard supports the splicing of MPEG-2 transport streams for the purpose of digital program insertion, which includes advertisement insertion and insertion of other content types. An in-stream messaging mechanism is defined to signal splicing and insertion opportunities and is not intended to ensure seamless splicing. As such, this recommendation does not specify the splicing method used or constraints applied to the streams being spliced, nor does it address constraints placed on splicing devices.

SCTE 43 2015: Digital Video Systems Characteristics Standard for Cable Television

This document describes the characteristics and normative specifications for the Video Subsystem Standard for Cable Television.

SCTE 65 2008: Service Information Delivered Out-of-Band for Digital Cable Television

This document defines a standard for Service Information (SI) delivered out-of-band on cable. This standard is designed to support navigation devices on cable. The current specification defines the syntax and semantics for a standard set of tables providing the data necessary for such a device to discover and access digital and analog services offered on cable.

SCTE 128 2010: AVC Video Systems and Transport Constraints for Cable Television

This document defines the video coding and transport constraints on ITU-T Rec. H.264 / ISO/IEC 14496-10 [4] video compression (AVC) for cable television. In particular, this document describes the transmission of AVC-coded video elementary streams in an MPEG-2 service multiplex (single- or multi-program transport stream).

SCTE 138 2013: Stream Conditioning for Switching of Addressable Content in Digital Television Receivers

This document describes the stream conditioning required to enable client-DPI receivers to implement switching in a both non-seamless fashion ("Level 0" or "L0"), and in a seamless fashion ("Level 1" or "L1").

Society of Motion Picture and Television Engineers (SMPTE)

For more than 100 years, the people of the Society of Motion Pictures and Television Engineers (SMPTE) have sorted out the details of many significant advances in entertainment technology, from the introduction of "talkies" and color television to HD and UHD (4K, 8K) TV. Since its founding in 1916, SMPTE has earned an Oscar and multiple Emmy awards for its work in advancing moving-imagery education and engineering across the communications, technology, media, and entertainment industries. SMPTE has developed thousands of standards, recommended practices, and engineering guidelines, more than 800 of which are currently in force. The following is a partial list of SMPTE standards and related technical documents:

SMPTE ST 12-1:2014 - Time and Control Code

This standard specifies a time and control code for use in television and accompanying audio systems operating at nominal rates of 60, 59.94, 50, 48, 47.95, 30, 29.97, 25, 24, and 23.98 frames per second. This standard defines a time address, binary groups, and flag bit structure. The standard also defines a binary group flag assignment, a linear time code transport, and a vertical interval time code transport.

This standard defines primary data transport structures for Linear Time Code (LTC) and Vertical Interval Time Code (VITC). It defines the LTC modulation and timing for all video formats. This standard also defines the VITC modulation and location for 525/59.94 and 625/50 analog composite and component systems only.

SMPTE ST 12-2:2014 - Transmission of Time Code in the Ancillary Data Space

This standard defines a transmission format for conveyance of linear (LTC) or vertical interval (VITC) time code data formatted according to SMPTE ST 12-1 in 8-, 10-, or 12-bit digital television data interfaces. Time code information is transmitted in the ancillary data space as defined in SMPTE ST 291-1. Multiple time codes can be transmitted within a single digital video data stream. Other information (such as real time clock, film transfer, DTTR tape timer, and user-defined information) may also be carried in the ancillary time code packet. This standard is applicable to digital television data interfaces, whether high definition or standard definition. This 2014 document is a revision of ST 12-2:2008, incorporating changes published as Amendment ST 12-2:2008 AM1:2013. No further changes have been made.

SMPTE ST 12-3:2016 - Time Code for High Frame Rate Signals and Formatting in the Ancillary Data Space

This standard specifies time code formats with the frame counts 72, 96, 100, and 120 and the frame count 120 with drop-frame compensation. This standard also specifies a transmission format for conveyance of the time code and frame count in the ancillary data space of serial digital interface.

SMPTE ST 170-2004 - Television—Composite Analog Video Signal—NTSC for Studio Application

This standard describes the composite analog color video signal for studio applications: NTSC, 525 lines, 59.94 fields, 2:1 interlace with an aspect ratio of 4:3. This standard specifies the interface for analog interconnection and serves as the basis for the digital coding necessary for digital interconnection of NTSC equipment. Note: Parts of the NTSC signal defined in this standard differ from the final report of the Second National Television System Committee (NTSC 1953) due to changes in technology and studio operating practices.

SMPTE 259M-2008 - Television—SDTI Digital Signal/ Data—Serial Digital Interface (Revision of SMPTE 259M-2006)

This standard describes a 10-bit serial digital interface operating at 143/270/360 Mb/s. The serial interface may carry uncompressed SDTV signals, or data. This standard has application in the television studio over lengths of coaxial cable where the signal loss does not exceed an amount specified by the receiver manufacturer. Typical loss amounts may be in the range of 20 dB to 30 dB at one-half clock frequency with appropriate receiver equalization. Receivers designed to work with lesser signal attenuation are acceptable, but are not recommend for new designs.

SMPTE 266:2012 - Television—4:2:2 Digital Component Systems—Digital Vertical Interval Time Code

This standard defines the digital data representation of the digitized values of the Digital Vertical Interval (VITC) code words defined in SMPTE ST 12-1. The intended use is with the standard definition (SD) digital coding given in SMPTE ST 125 (for 525-line, 59.94-Hz field rate, SD digital component signals) or Recommendation ITU-R BT.601-7 and Recommendation ITU-R BT.656 (for 625-line, 50-Hz field rate, SD digital component signals).

SMPTE 272M-2004 - Television—Formatting AES/EBU Audio and Auxiliary Data Into Digital Video Ancillary Data Space

This standard defines the mapping of AES digital audio data, AES auxiliary data, and associated control information into the ancillary data space of serial digital video conforming to ANSI/SMPTE 259M. The audio data and auxiliary data are derived from ANSI S4.40, hereafter referred to as AES audio. This standard provides a minimum of two audio channels and a maximum of 16 audio channels based on available ancillary data space in a given format (four channels maximum for composite digital). Audio channels are transmitted in pairs combined, where appropriate, into groups of four. Each group is identified by a unique ancillary data ID.

SMPTE ST 274:2008 - For Television—1920 x 1080 Image Sample Structure, Digital Representation and Digital Timing Reference Sequences for Multiple Picture Rates (Revision of SMPTE 274M-2005

This standard defines a family of image sample structures for the representation of stationary or moving two-dimensional images sampled temporally at a constant frame rate and having an image format (sample structure) of 1920 x 1080 and an aspect ratio of 16:9.

This standard specifies multiple frame and field rate formats (Table 1) and 8-bit, 10-bit, and 12-bit systems. It is not necessary for an implementation to support all formats to be compliant with this standard. However, an implementation must state which of the formats are supported. Interfaces for 12-bit systems require more than a single link.

SMPTE 276M-1995 - Television—Transmission of AES-EBU Digital Audio Signals Over Coaxial Cable

This standard describes a point-to-point coaxial cable interface for the transmission of AES/EBU digital audio signals throughout television production and broadcast facilities. The purpose of this standard is to ensure that a level of compatibility exists between signals generated to this standard and analog video equipment, such as non-clamping distribution amplifiers, switchers, cables, and connectors, as normally used in television applications.

SMPTE ST 292-1:2012 - 1.5 Gb/s Signal Data Serial Interface

This standard defines a bit-serial data structure and the coaxial cable interface specifications for 1.5 Gb/s [nominal] Signal/Data Serial Interface to carry either 1280 x 720, 1920 x 1080, or 2048 x 1080 active pixel formats mapped into the 1.5 Gb/s payload. This standard defines the method of serializing the parallel source format data to a serial bitstream. This interface may also carry packetized data mapped into the 1.5 Gb/s transport payload providing the constraints imposed by the reference source image formats are observed.

This standard specifies a coaxial cable interface suitable for application where the signal loss does not exceed an amount specified by the receiver manufacturer, typical loss amounts would be in the range of up to 20 dB at one-half the clock frequency.

SMPTE ST 296:2012 - Television—1280 x 720 Progressive Image Sample Structure—Analog and Digital Representation and Analog Interface (Revision of SMPTE 296M-2001

This standard defines a family of progressive image sampling systems for the representation of stationary or moving two-dimensional images sampled temporally at a constant frame rate and having an image format of 1280 x 720 px lines and an aspect ratio of 16:9 as given in a supporting table. This standard specifies RGB color encoding; RGB analog and digital representation; YPBR color encoding, analog representation, and analog interface; and YCBCR color encoding and digital representation. A bit-parallel digital interface is incorporated. This standard specifies multiple frame rates in 8-bit and 10-bit systems.

SMPTE 299-0:2010 - 24-Bit Digital Audio Format for SMPTE 292 Bit-Serial Interface—Roadmap for the 299 Document Suit

The SMPTE 299 suite of documents specifies 24-bit digital audio formats for SMPTE 292 bit-serial interfaces. This informative roadmap describes the documents in the 299 suite.

SMPTE 299-1:2009 - 24-Bit Digital Audio Format for SMPTE 292 Bit-Serial Interface

This standard defines the mapping of 24-bit AES digital audio data and associated control information into the ancillary data space of a serial digital video conforming to SMPTE 292. The audio data are derived from AES3 (AES audio). The AES audio data may contain linear PCM audio or non-PCM data formatted according to SMPTE 337. Not all implementations compliant with this standard may support all sampling frequencies. Manufacturers are encouraged to indicate which sampling frequencies are supported.

Audio sampled at a clock frequency of either 48 kHz or 96 kHz locked (synchronous) to video is the preferred implementation for intra-studio applications. As an option, this standard supports AES audio at synchronous or asynchronous sampling rates from 32 kHz to 48 kHz.

Audio channels are transmitted in groups of four, up to a maximum of 16 audio channels in the case of 32 kHz, 44.1 kHz, or 48 kHz sampling, and up to a maximum of 8 audio channels in case of 96-kHz sampling. Each group is identified by a unique ancillary data ID.

Audio data packets are multiplexed (embedded) into the horizontal ancillary data space of the Cb/Cr data stream, and audio control packets are multiplexed into the horizontal ancillary data space of the Y data stream.

SMPTE 299-2-2010 - Extension of the 24-Bit Digital Audio Format to 32 Channels for 3 Gb/s Bit-Serial Interface

SMPTE ST 299-2 defines an SMPTE 291M Type 1 packet structure for identifying audio channels numbered from 17 to 32, beyond the 16 channels defined in SMPTE ST 299-1. Four extended audio data packets and four extended audio control packets are identified. One of the extended audio control packets and one of the extended audio data packets is assigned to transport each of the four extended audio groups. Each extended audio group has four channels that carry up to four 24-bit audio channels with 32, 44.1, or 48-kHz sample rates, or up to two 24-bit audio channels with a 96-kHz sample rate. The purpose of defining four extended audio groups is to allow for multiplexing up to 32 audio channels into the horizontal ancillary (HANC) data space of the video source image formats mapped to a 3 Gb/s Level A serial data interface. The 3 Gb/s Level B serial data interface is capable of carrying 32 audio channels via application of SMPTE ST 299-1 to each of two 1.5 Gb/s SDI channels.

SMPTE 302M-2007 - Television—Mapping of AES3 Data Into MPEG-2 Transport Stream

This standard specifies the method of transporting AES3 data in an MPEG-2 transport stream for television applications. For these television applications, the AES3 frame rate is 48 kHz.

Some applications may require linear pulse code modulated (PCM) digital audio in conjunction with compressed video specified in the MPEG-2 4:2:2 profile. The MPEG audio standard defines compressed audio, but does not define uncompressed audio for carriage in an MPEG-2 transport system. This standard augments the MPEG standards to address the requirement to carry AES3 streams, which may consist of linear PCM audio (at 48k sample/s), or data carried within the AES3 format.

SMPTE 308M-2004 - Television—MPEG-2 4:2:2 Profile at High Level

ISO/IEC 13818-2, commonly known as MPEG-2 video, includes specification of the MPEG-2 4:2:2 profile. Based on ISO/IEC 13818-2, this standard provides additional specification for the MPEG-2 4:2:2 profile at high level. It is intended for use in high-definition television production, contribution, and distribution applications. As in ISO/IEC 13818-2, this standard defines bitstreams, including their syntax and semantics, together with the requirements for a compliant decoder for 4:2:2 profile at high level, but does not specify particular encoder operating parameters.

SMPTE 325M-1999 - Digital Television—Opportunistic Data Broadcast Flow Control (R2004)

This standard defines the flow control protocol to be used between an emission multiplexer and data server for opportunistic data broadcast. Opportunistic data broadcast inserts data packets into the output multiplex to fill any available free bandwidth. The emission multiplexer maintains a buffer from which it draws data to be inserted. The multiplexer will request additional MPEG-2 transport packets from the data server as its buffer becomes depleted. The number of packets requested depends upon the implementation, with the most stringent requirement being requesting a single MPEG-2 transport packet where the request and delivery can occur in less than the emission time of an MPEG-2 transport packet from the multiplexer. This protocol is designed to be extensible and provide a basis for low-latency, real-time backchannel communications from the emission multiplexer.

SMPTE 333-2008 - DTV Closed-Caption Server to Encoder Interface (Revision of SMPTE 333M-1999)

This standard defines a serial interface for transmission of CEA-708 digital television closed-caption data from a caption server to video encoder to be used for populating the closed-caption structure in the compressed video stream.

SMPTE ST 334-1:2015 - Vertical Ancillary Data Mapping of Caption Data and Other Related Data

This standard defines a method of mapping closed-captioning and certain other related data services into the 10-bit vertical ancillary (VANC) data space. It specifies the structure of packets carrying CEA-708, as well as CEA-608 closed-captioning data, time code data, and ATSC A/65 closed-caption descriptor information, which may be carried in vertical ancillary data space, via serial data communications interfaces or other means.

SMPTE ST 334-2:2015 - Caption Distribution Packet (CDP) Definition

This standard defines the structure of the Caption Distribution Packet (CDP). A CDP may contain time code, CEA-608 data, CEA-708 closed-captioning data, and ATSC A/65 closed-caption service information.

SMPTE ST 337:2015 - Format for Non-PCM Audio and Data in an AES3 Serial Digital Audio Interface

This standard specifies an interface format for the transport of non-PCM audio and data in professional applications using the AES3 serial digital audio interface. This standard includes both physical and logical specifications, based on the existing AES3 format, to allow exchange of non-PCM data between different devices. The standard accommodates multiple non-PCM audio and data formats and allows carriage of multiple data streams within a single interface. This standard provides means for carrying time code or time alignment information so that the information conveyed over this interface may be synchronized with information content delivered over other interfaces.

SMPTE 348M-2000 - Television—High Data Rate Serial Data Transport Interface (HD-SDTI)

This standard provides the mechanisms necessary to facilitate the transport of packetized data over a synchronous data carrier. The HD-SDTI data packets and synchronizing signals provide a data transport interface which is compatible with SMPTE 292M (HD-SDI) such that it can be readily used by the infrastructure provided by this standard.

SMPTE ST 352:2013 - Payload Identification for Serial Digital Interface

This standard defines the specification of a 4-byte payload identifier that describes aspects of the payload carried on the SMPTE serial digital interface (SDI), such as the digital interface standard, picture rate, sampling structure, aspect ratio, colorimetry, bit depth, and channel or link assignment. Other application-specific aspects of the payload, such as audio channel usage, may optionally be described in application-specific documents that reference SMPTE ST 352.

The payload identifier is applicable to all serial digital interfaces and all payloads that produce a directly viewable image on a display device. Furthermore, the viewable image must be a recognizable representation of the source image content.

The payload identifier is mandatory for some reference standards and optional in other reference standards. Refer to the applicable reference standards for further information on mandatory or optional usage of the SMPTE ST 352 payload identifier.

The standard defines how the payload identifier is placed into a 10-bit ancillary data packet according to SMPTE ST 291-1.

It also specifies the repetition rate of the ancillary data packet and provides the preferred horizontal and vertical location for the placement of the packet in different digital interfaces.

The standard defines values for some of the bit fields of each byte of the 4-byte payload identifier, although the values for other bit fields can have custom definitions.

The standard also defines the payload identification code assignment process and associated on-line register for byte 1 values.

SMPTE ST 372:2011 - Dual Link 1.5 Gb/s Digital Interface for 1920 x 1080 and 2048 x 1080 Picture Formats

This standard specifies the mapping of source image data onto a dual-link 1.5 Gb/s HD-SDI (Link A and Link B) in conformance with SMPTE ST 292-1. The total data rate of the dual link connection is 2.970 Gb/s or 2.970/1.001 Gb/s. This dual link also specifies carriage of the embedded audio, ancillary data, and the payload ID of the stream in conformance with SMPTE ST 291, SMPTE ST 299-1, and SMPTE ST 352.

The image source digital representations are defined in SMPTE ST 274, SMPTE ST 428-9, SMPTE ST 428-19, and SMPTE ST 2048.

SMPTE ST 390:2011 - Material Exchange Format (MXF)—Specialized Operational Pattern "OP-Atom" (Simplified Representation of a Single Item)

This standard defines the specialized Operational Pattern "OP-Atom" or OP-Atom for the storage and exchange of an MXF file with a tightly defined structure for a single item of essence described by a single Essence Track. It defines the operating restrictions, Structural Metadata objects, and individual attributes applied to the MXF file format specification to achieve interoperability when exchanging a single item of audio-visual material.

Operational Pattern "OP-Atom" is intended to satisfy the demands of applications where there is a requirement for exchange of material and associated metadata together in a single file with an open and standardized format, yet also a demand for a simple, predictable layout with minimum scope for variation. OP-Atom is specifically intended for applications where each Essence Track is held separately. Operational Pattern "OP-Atom" provides for a subset of MXF features.

SMPTE 392:2013 - Material Exchange Format (MXF)—Operational Pattern 2a (Play-List Items, Single Package)

This standard defines operational pattern 2a for the exchange of an MXF file with a play-list of playable essence containers, each comprising either a single essence element or interleaved essence elements. It defines the operating restrictions, structural metadata objects, and individual attributes that must be applied to the MXF file format specification to achieve interoperability when exchanging an MXF file as a list of continuously playable items of audio-visual material.

Operational pattern 2a is intended to meet the requirements of acquisition, storage, and interchange applications that are satisfied by a play-list of individual items of content packaged in two or more essence containers each of the same type. Operational pattern 2a requires the use of body partitions for internal essence containers. Subdivision of each essence container by using body partitions is optional.

SMPTE 424:2012 - 3 Gb/s Signal/Data Serial Interface

This standard defines a bit-serial data structure and coaxial cable interface for the transport of signals with a total payload of 2.970 Gb/s or 2.970/1.001 Gb/s.

This standard specifies the electrical and physical characteristics of a coaxial cable interface suitable for applications where the signal loss does not exceed an amount specified by the receiver manufacturer. Typical loss amounts would be in the range of up to 30 dB at one-half the clock frequency.

SMPTE 425-0:2014 - SMPTE Bit Serial Interfaces at 3 Gb/s—Roadmap for the 425 Document Suite

The SMPTE 425 suite of documents defines the mapping of various source image formats onto a single link, dual link, and quad link serial digital interfaces operating at a nominal rate of 3 Gb/s. This informative "roadmap" describes the documents in the SMPTE 425 suite. The SMPTE 425 series documents specify a virtual interface that is carried on a physical link specified by SMPTE ST 424.

SMPTE ST 425-1:2014 - Source Image Format and Ancillary Data Mapping for the 3 Gb/s Serial Interface

This standard defines three mapping formats: Level A, Level B Dual-Link mapping, and Level B Dual-Stream mapping as described below.

Level A specifies:

- The direct mapping of various uncompressed video image formats
- The direct mapping of packetized data
- The carriage of ancillary data such as the audio data, the audio control packets, the payload ID, the time code, etc.
- Into a serial digital interface operating at a nominal rate of 3 Gb/s

Level B Dual-Link mapping specifies the mapping of the SMPTE ST 372 Dual-Link interface (Dual-Link mapping) into a serial digital interface operating at a nominal rate of 3 Gb/s.

Level B Dual-Stream mapping specifies the mapping of 2 X SMPTE ST 292-1 (HD-SDI) interfaces (Dual-Stream mapping) into a serial digital interface operating at a nominal rate of 3 Gb/s.

Uncompressed video image formats or packetized data, and all applicable ancillary data, such as the audio data, the audio control packets, the payload ID, and the time code, are mapped into SMPTE ST 372 Dual Link and 2 X SMPTE ST 292-1 interfaces prior to mapping into the virtual interfaces.

SMPTE ST 425-2:2012 - Source Image Format and Ancillary Data Mapping for Stereoscopic Image Formats on a Single-Link **3 Gb/s Serial Interface**

This standard defines a means of transporting a stereoscopic image pair consisting of a Left eye and Right eye image (Le and Re) using an interface consisting of a single 3 Gb/s (nominal) link.

SMPTE ST 425-3:2015 - Image Format and Ancillary Data Mapping for the Dual-Link 3 Gb/s Serial Interface

This standard defines 1080-line and 2160-linemapping formats as described below.

1080-line mapping specifies the mapping of the image formats of 1920 x 1080 and 2048 x 1080 pixels into four parallel 10-bit data streams (known as data stream one, data stream two, data stream three, and data stream four) of a 40-bit virtual interface, followed by the mapping of these four 10-bit data streams onto the Dual-Link 3 Gb/s Serial Interface (3G-SDI Link 1 and Link 2) via two 20-bit virtual interfaces.

2160-line mapping specifies the mapping of the image formats of 3840 x 2160 and 4096 x 2160 pixels listed into four parallel 10-bit data streams (known as data stream one, data stream two, data stream three, and data stream four) of a 40-bit virtual interface, followed by the mapping of these four 10-bit data streams on to the Dual-Link 3 Gb/s Serial Interface (3G-SDI Link 1 and Link 2) via two 20-bit virtual interfaces.

This standard also defines the carriage of ancillary data such as the audio data, the audio control packets, the payload identifier, and the time code.

It is not necessary for implementations to include support for all formats that are included in this standard. Implementers should indicate supported formats in commercial publications.

SMPTE ST 425-4:2012 - Dual 3 Gb/s Serial Digital Interface for Stereoscopic Image Transport

This standard defines a means of transporting stereoscopic images (Left eye and Right eye images) using an interface consisting of two streams based on the SMPTE ST 425-1 data structures. The Le images are carried on one stream of the interface and the Re images are carried on the other stream.

The stereoscopic image formats to be transported using this standard are the 4:2:2 and 4:4:4 image formats enumerated in Table 1, where each image format is defined by SMPTE ST 274, SMPTE ST 296, SMPTE ST 2048-2, SMPTE ST 428-9, or SMPTE ST 428-19 has a payload capacity of nominally 3 Gb/s and can be transported by a single SMPTE ST 424 serial interface.

Audio and other associated ancillary data may also be transported. This standard also defines a payload identifier.

Mapping structures for the video essence and ancillary data should be as defined in SMPTE ST 425-1 Level A or Level B-DL.

It is not necessary for implementations to include support for all formats defined the supporting table, nor is it a requirement to support both mapping modes to conform to this standard. Implementers should indicate supported formats and supported mapping modes in commercial publications.

SMPTE ST 425-5:2015 - Image Format and Ancillary Data Mapping for the Quad Link 3 Gb/s Serial Interface

This standard defines the mapping of the image formats of 3840 x 2160 and 4096 x 2160 pixels into eight parallel 10-bit data streams (known as data stream one, data stream two, data stream three, data stream four, data stream five, data stream six, data stream seven, and data stream eight) of an 80-bit virtual interface, followed by the mapping of these eight 10-bit data streams onto the Quad Link 3 Gb/s Serial Interface (3G-SDI Link 1, Link 2, Link 3, and Link 4) via four 20-bit virtual interfaces.

This standard also defines the carriage of ancillary data such as the audio data, the audio control packets, the payload identifier, and the time code.

It is not necessary for implementations to include support for all formats that are included in this standard. Implementers should indicate supported formats in commercial publications.

SMPTE ST 425-6:2014 - Quad 3 Gb/s Serial Digital Interface for Stereoscopic Image Transport

This standard defines a means of transporting stereoscopic images (Left eye and Right eye images) using an interface consisting of pairs of four data streams based on the SMPTE ST 425-3 data structures. The Le images are carried on four data streams of the interface and the Re images are carried on the other four data streams. The Le and Re images are each mapped onto the respective four data streams in accordance with the mapping rules of SMPTE ST 425-3.

The stereoscopic image formats to be transported using this standard are the 4:2:2 and 4:4:4 image formats enumerated in SMPTE ST 425-3. Mapping structures for the video essence and ancillary data are as defined in SMPTE ST 425-3.

This standard also defines the carriage of ancillary data such as the audio data, the audio control packets, the payload identifier, and the time code.

It is not necessary for implementations to include support for all image formats, nor is it a requirement to support both mapping modes to conform to this standard. Implementers are encouraged to indicate supported formats and supported mapping modes in commercial publications.

SMPTE 435-0: 2012 - 10 Gb/s Serial Signal/Data Interface—Document Roadmap

SMPTE ST 435 10 Gb/s Serial Signal/Data Interface is a multi-part document.

Part 1 defines the source data of the 10 Gb/s data stream. It is composed from multiple 1.5 Gb/s basic streams. The standard classifies applicable source image formats carried on 10 Gb/s interfaces and defines their system numbers. The standard also defines how the source image formats are subdivided into one or more basic streams. The basic streams comply with the interleaved data stream structure defined in SMPTE ST 292-1.

Part 2 defines the multiplexing schemes to map up to eight basic streams defined by part 1 into the 10 Gb/s serial interface. The mapping also supports transmission of the embedded audio, payload ID, and other ancillary data defined in SMPTE ST 291 in the source stream.

Part 3 defines the optical fiber interface for the serial stream operating at data rates of 10.692 Gb/s and 10.692/1.001 Gb/s. The interface specification defined in this part applies to implementations covering a distance up to 2 km using single-mode fiber.

SMPTE ST 2010:2008 - Vertical Ancillary Data Mapping of ANSI/SCTE 104 Messages

This standard specifies a mechanism for mapping ANSI/SCTE 104 messages (with a maximum size of 2000 bytes) into 10-bit vertical ancillary data space. It is equally applicable to SD and HD systems.

SMPTE 2016 - Ancillary Data Handling

This standard includes AFD and bar data generation and entry, and handling of that data throughout production and postproduction, on into the broadcast chain up through editing/switching. SMPTE 2016 has five parts:

SMPTE 2016-1: "Format for AFD & Bar Data" specifies the raster format and the active image area

SMPTE 2016-2: "Format for Pan-Scan Information"

SMPTE 2016-3: "VANC Data Mapping of AFD & Bar Data" defines carriage in SMPTE 259M and SMPTE 292M serial bitstreams

SMPTE 2016-4: "VANC Data Mapping of Pan-Scan Information"

SMPTE 2016-5: "KLV Coding for AFD, Bar Data & Pan-Scan Information"

SMPTE ST 2022-1:2007 - Forward Error Correction for Real-Time Video/Audio Transport Over IP Networks

This standard defines a forward error correction technique for the carriage of real-time video/audio content over IP networks.

SMPTE ST 2022-2:2007 - Unidirectional Transport of Constant Bit Rate MPEG-2 Transport Streams on IP Networks

This standard defines a unidirectional transport protocol for the carriage of real-time Constant Bitrate (CBR) MPEG-2 transport streams over IP networks. For professional applications, MPEG-2 using the 4:2:2P@ML profile is currently the normal practice. However, transport streams containing other forms of MPEG-2 and newer MPEG standards encapsulated as MPEG-2 transport streams are also supported by this standard.

SMPTE ST 2022-3:2010 - Unidirectional Transport of Variable Bit Rate MPEG-2 Transport Streams on IP Networks

This standard defines a transport protocol for the carriage of real-time piecewise constant variable bit rate (VBR) MPEG-2 transport streams over IP networks, either with or without Forward Error Correction for recovery from network transmission errors. A piecewise constant VBR transport stream can only vary at the PCRs of the program under consideration per ISO/IEC 13818-1:2007 § 2.4.2.2. This standard covers the encapsulation and transmission of MPEG-2 transport streams but does not cover other processes such as MPEG-2 encoding or multiplexing.

SMPTE ST 2022-4:2011 - Unidirectional Transport of Non-Piecewise Constant Variable Bit Rate MPEG-2 Streams on IP Networks

This standard defines a transport protocol for the carriage of real-time non-piecewise constant variable bit rate (VBR) MPEG-2 transport streams over IP networks, either with or without Forward Error Correction for recovery from network transmission errors. A non-piecewise constant VBR transport stream has no predictable time base to reconstruct the signal if the inter-packet timing is altered in transit through the network. This standard defines two methods to maintain the inter-packet timing through the non-synchronous IP network transport. This standard covers the encapsulation and transmission of MPEG-2 transport streams but does not cover other processes such as MPEG-2 encoding or multiplexing.

SMPTE ST 2022-5:2013 - Forward Error Correction for High Bit Rate Media Transport Over IP Networks (HBRMT)

This standard defines a forward error correction technique for the carriage of high bit rate real-time media over IP networks.

The term high bit rate is used to distinguish from other media-over-IP applications in which primarily compressed signals are transported. The FEC system described is differentiated by an ability to accommodate media payloads at rates of 270 Mb/s and higher.

SMPTE ST 2022-6:2012 - Transport of High Bit Rate Media Signals Over IP Networks (HBRMT)

This standard defines a unidirectional IP-based protocol for the transport of real-time video, audio, and ancillary signals. In particular, this standard defines a method for the encapsulation of the payloads of a variety of existing SMPTE serial digital video standards. The term high bit rate is used to distinguish from other media-over-IP applications in which compressed signals are transported. The uncompressed signals in this document are at rates of 270 Mb/s and higher.

SMPTE ST 2022-7:2013 - Seamless Protection Switching of SMPTE ST 2022 IP Datagrams

This standard defines requirements for redundant streams of SMPTE 2022 packets to allow for creation of a single reconstructed output stream through seamless protection switching at the RTP datagram level.

SMPTE ST 2034-1:2014 - Archive eXchange Format (AXF)—Part 1: Structure & Semantics

This standard is Part 1 of a series of documents that specify a general-purpose format for the storage and/or communication of information in bulk form. The format is named the Archive eXchange Format (AXF). The format described is intended both for interchange between systems and to serve as a native format within systems.

This standard identifies two major categories of data storage media and specifies the basic structures of data stored on those storage media types. It specifies a number of structural elements for use in constructing the appropriate structures for use on each of the storage media types. It defines the semantics of data contained within fields specified for use in the structural elements. The structural elements themselves are documents coded in the eXtensible Markup Language (XML), and this document defines an XML Schema Description (XSD) file for use in formulating the XML documents to be used for the structural elements of AXF objects.

SMPTE OV 2036-0:2015 - Ultra High Definition Television—Overview for the SMPTE ST 2036 Document Suite

SMPTE ST 2036 Ultra High Definition Television suite of documents is in multiple parts.

SMPTE ST 2036-1 defines a family of progressive image sample structures for the representation of stationary or moving two-dimensional images sampled temporally at a constant frame rate and having an image format (sample structure) of 3840 x 2160 (UHDTV1) or 7680 x 4320 (UHDTV2) which has a hierarchical relationship with 1920 x 1080 and an aspect ratio of 16:9.

SMPTE ST 2036-2 specifies the characteristics of digital audio for UHDTV program production and distribution. This document also defines the mapping and labeling of 22.2 multichannel audio for UHDTV program production.

SMPTE ST 2036-3 defines UHDTV1 and UDHTV2 image mapping into a single-link or a multi-link 10 Gb/s Serial Signal/Data Interface (10G-SDI) Mode D defined in SMPTE ST 435-2. This interface has a maximum frame rate capability of 60 Hz.

Work is underway to define an interface(s) to permit 120-Hz frame rate operation.

SMPTE ST 2031:2015 - Carriage of DVB/SCTE VBI Data in VANC

This standard specifies how to carry the digital values resulting from defined standardized data streams carried in MPEG-2 transport streams as specified by either DVB/ETSI EN 301 775 or ANSI/SCTE 127 (or both) as ancillary data packets in the vertical ancillary data space (VANC).

SMPTE ST 2042-1:2012 - VC-2 Video Compression

This standard defines the VC-2 video compression system through the stream syntax, entropy coding, coefficient unpacking process, and picture decoding process. The decoder operations are defined by means of a mixture of pseudo-code and mathematical operations.

VC-2 is an intra-frame video codec that uses wavelet transforms together with entropy coding that can be readily implemented in hardware or software at very high bit rates. Additional standards and recommended practices may define specific constraints on the encoding for particular applications.

SMPTE 2042-2:2009 - VC-2 Level Definitions

VC-2 is a video codec that includes parameter values to identify the complexity of the coded VC-2 bit-stream through the use of profiles and levels. These values provide a fast assessment for a decoder to establish whether it has the capability to decode the VC-2 bit-stream.

Profiles are defined in the VC-2 standard (SMPTE ST 2042-1). This standard defines the coding constraints for generalized level values lying in the range 0 to 63. This standard also informatively registers the values of specialized levels with values exceeding six.

SMPTE 2046-1:2009 - Specifications for Safe Action and Safe Title Areas for Television

This standard defines and specifies Safe Action and Safe Title Areas for 1920 x 1080, 1280 x 720, 720 x 576, and 720 x 480 television formats. This document is intended for application in program production where the image aspect ratio of the acquired essence is the same as that of the display.

The safe areas defined in this standard differ from past practice; they take into account changes in image acquisition, transmission, and display technology. They do not offer guidance in situations where material generated in one aspect ratio may need to be displayed in a different aspect ratio.

SMPTE RP 2047-3:2016 - VC-2 Level 65 Compression of High Definition Video Sources for Use With a Standard Definition Infrastructure

This recommended practice defines constraints on the VC-2 video compression specification such that 10-bit high definition video signals can be compressed to only occupy the bit rate of 8-bit standard definition signals. Specifically, it defines constraints for compressing interlaced source video having 1920 or 1440 pixels/line by 1080 lines/frame at 25 or 30/1.001 frames/s, or progressive source video having 1280 pixels/line by 720 lines/frame at 50 or 60/1.001 frames/s, using 4:2:2 sampling in all cases. This document defines a VC-2 bitstream for video compressed according to this recommended practice. It also defines an application specific level for VC-2 compression.

SMPTE 2048-0:2012 - 2048 x 1080 and 4096 x 2160 Digital Cinematography Production Image Formats FS/709—Roadmap for the 2048 Document Suite

The SMPTE 2048 suite of documents (ST 2048-1, ST 2048-2, ST-2048-3) defines a family of progressive sample structures of 2048 x 1080 and 4096 x 2160 images intended for digital cinematography content creation.

SMPTE ST 2059-1:2015 - Generation and Alignment of Interface Signals to the SMPTE Epoch

This standard defines:

1. A point in time, the SMPTE Epoch, which is used for alignment of all real-time signals referenced in this standard
2. The alignment of these signals to the SMPTE Epoch
3. Formulae that specify the ongoing alignment of these signals to time since the SMPTE Epoch
4. Formulae that specify the calculation of SMPTE ST 12-1 Time Address values and SMPTE ST 309 date values from SMPTE Profile IEEE 1588-2008 PTP data

Note: Implementers are encouraged to read both SMPTE ST 2059-1 and SMPTE ST 2059-2 as a pair.

SMPTE ST 2059-2:2015 - SMPTE Profile for Use of IEEE-1588 Precision Time Protocol in Professional Broadcast Application

This standard specifies a Precision Time Protocol profile specifically for the synchronization of audio/video equipment in a professional broadcast environment.

The profile is based on IEEE Std 1588-2008 and includes a self-contained description of parameters, their default values, and permitted range.

SMPTE ST 2065-1:2012 - Academy Color Encoding Specification (ACES)

The specification defines the Academy Color Encoding Specification (ACES).

SMPTE ST 2065-2:2012 - Academy Printing Density (APD)—Spectral Responsivities, Reference Measurement Device and Spectral Calculation

This standard defines Academy Printing Density (APD) by specifying the spectral responsivities and the APD reference measurement device.

SMPTE ST 2065-3:2012 - Academy Density Exchange Encoding (ADX)—Encoding Academy Printing Density (APD) Values

This standard defines the Academy Density Exchange Encoding (ADX) including the encoding method, the 16-bit and 10-bit component value encoding metrics, and the method for converting between the 16-bit and 10-bit component value encoding metrics.

SMPTE ST 2065-4:2013 - ACES Image Container File Layout

This standard specifies the layout of, and metadata in, files containing images that conform to SMPTE ST 2065-1, Academy Color Encoding Specification (ACES).

SMPTE ST 2081-10:2015 - 2160-Line and 1080-Line Source Image and Ancillary Data Mapping for Single-Link 6G-SD

This standard defines the mapping of:

Mode 1: 2160-line source image formats identified in a supporting table into a single-link 6-Gb/s [nominal] SDI bit-serial interface

Mode 2: 1080-line source image formats identified in a supporting table and into a single-link 6-Gb/s [nominal] SDI bit-serial interface

This standard also defines the carriage of ancillary data and the SMPTE ST 352 payload IDs for the single-link 6-Gb/s SDI interface.

It is not necessary for implementations to include support for all operating modes or image formats that are included in this standard. Implementers should indicate supported operating modes and image formats in commercial publications.

SMPTE ST 2081-1:2015 - 6-Gb/s Signal/Data Serial Interface—Electrical

This standard describes the electrical and physical characteristics of a 6-Gb/s coaxial cable interface suitable for applications where the signal loss does not exceed an amount specified by the receiver manufacturer. Typical loss amounts would be in the range of up to 40 dB at one-half the clock frequency.

SMPTE ST 2081-10:2015 - 2160-Line and 1080-Line Source Image and Ancillary Data Mapping for Single-Link 6G-SDI

This standard defines the mapping of:

Mode 1: 2160-line source image formats identified in a supporting table into a single-link 6-Gb/s [nominal] SDI bit-serial interface

Mode 2: 1080-line source image formats identified in a supporting table and into a single-link 6-Gb/s [nominal] SDI bit-serial interface

This standard also defines the carriage of ancillary data and the SMPTE ST 352 payload IDs for the single-link 6-Gb/s SDI interface.

It is not necessary for implementations to include support for all operating modes or image formats that are included in this standard. Implementers should indicate supported operating modes and image formats in commercial publications.

SMPTE 2082-0:2015 - 12G-SDI Bit-Serial Interfaces—Overview for the SMPTE ST 2082 Document Suite

The SMPTE ST 2082 suite of documents defines the mapping of various source image formats onto a single-link, dual-link, and quad-link serial digital interface operating at a nominal rate of 12 Gb/s. This informative overview describes the documents in the SMPTE ST 2082 suite.

The SMPTE ST 2082 series specify a common virtual interface that is carried on both electrical and optical physical interfaces, which are also defined in the document suite.

A diagram is provided that illustrates a simplified "image mapping data flow" overview for the 12G-SDI profile as defined by the ST 2082 document suite. The individual roadmaps of the SMPTE ST 2082 standards include full details of reference standards.

SMPTE ST 2082-1:2015 - 12-Gb/s Signal/Data Serial Interface—Electrical

This standard describes the electrical and physical characteristics of a 12G-SDI coaxial cable interface suitable for applications where the signal loss does not exceed an amount specified by the receiver manufacturer. Typical loss amounts would be in the range of up to 40 dB at one-half the clock frequency.

SMPTE ST 2082-10:2015 - 2160-Line Source Image and Ancillary Data Mapping for 12G-SD

This standard defines the mapping of Mode 1: 2160-line Source image formats and ancillary data into a 12-Gb/s [nominal] SDI bit-serial interface. This standard also defines the carriage of the SMPTE ST 352 payload IDs for the 12-Gb/s SDI interface. It is not necessary for implementations to include support for all formats that are included in this standard.

Implementers should indicate supported formats in commercial publications.

SMPTE ST 2084:2014 - Dynamic Range Electro-Optical Transfer Function of Mastering Reference Displays

This standard specifies an Electro-Optical Transfer Function (EOTF) characterizing high-dynamic-range reference displays used primarily for mastering non-broadcast content. This standard also specifies an inverse-EOTF derived from the EOTF.

SMPTE ST 2086:2014 - Mastering Display Color Volume Metadata Supporting High Luminance and Wide Color Gamut Images

This standard specifies the metadata items to specify the color volume (the color primaries, white point, and luminance range) of the display that was used in mastering video content. The metadata is specified as a set of values independent of any specific digital representation.

This standard is applicable to three-color additive display systems, such as RGB displays.

This standard does not specify the measurement methodologies and procedures for capturing the parameters of the metadata as well as any description of the mastering environment.

Additionally, this standard does not fully specify all the information that would be necessary to preserve the creative intent on displays with color volumes different from the mastering display color volume. As a specific example, this standard does not specify the Electro-Optical Transfer Function (EOTF) of the mastering display since it does not affect the color volume.

This standard only defines the mastering display color volume metadata items. System-specific solutions for storing, packaging, transmission, and end-use of this metadata are allowed; however, the specification of these solutions is outside the scope of this standard.

SMPTE ST 2094-1:2016 - Dynamic Metadata for Color Volume Transform—Core Components

The SMPTE ST 2094 suite of documents define metadata for use in color volume transforms of content. The metadata are content-dependent and can vary scene by scene or image by image. The metadata are intended to transform High Dynamic Range and Wide Color Gamut (HDR/WCG) image essence for presentation on a display having a smaller color volume than that of the mastering display used for mastering the image essence. Multiple applications provide particular color volume transforms. This standard, a part of the 2094 suite, specifies common definitions, including a transform metadata set, and metadata parameters, for use by other parts of the 2094 suite. Other parts of the 2094 suite provide representation for carriage and additional application-specific metadata parameters for color volume transforms.

SMPTE ST 2094-10:2016 - Dynamic Metadata for Color Volume Transform—Application #1

This standard specifies the content-dependent color volume transform metadata items for Application #1, a specialized model of the generalized color volume transform defined by the core components document SMPTE ST 2094-1. This color volume transform is based on a parametrically defined tone-mapping curve, the shape of which is defined both by the image essence characteristics (algorithmically computed from the input image essence) as well as by manually set adjustments. The metadata set supporting the application is generated as a part of the mastering process. The adjustment parameters are decided on as a creative adjustment.

SMPTE ST 2094-20:2016 - Dynamic Metadata for Color Volume Transform—Application #2

This standard specifies the content-dependent Color Volume Transform metadata set for Application #2, a specialized model of the color volume transform defined by the core components document SMPTE ST 2094-1. The Color Volume Transform is based on a tone-mapping curve that operates on image data in a perceptually uniform domain and a luminance-dependent saturation gain curve. Both the tone-mapping curve and the saturation-gain curve are fully characterized by a set of parameters.

SMPTE ST 2094-30:2016 - Dynamic Metadata for Color Volume Transform—Application #3

This standard specifies Dynamic Metadata for Color Volume Transform Application #3, Reference-based Color Volume Remapping. It is a specialization of the content-dependent transform metadata entries and processing blocks of the generalized color volume transform model defined in the SMPTE ST 2094-1 Dynamic Metadata for Color Volume Transform—Core Components standard.

SMPTE ST 2094-40:2016 - Dynamic Metadata for Color Volume Transform—Application #4

This standard specifies the metadata for Color Volume Transform Application #4, Scene-based Color Volume Mapping. It is a specialization of the content-dependent transform metadata entries and processing blocks of the generalized color volume transform model defined in the SMPTE ST 2094-1 Core Components standard. Scene-based Color Volume Mapping consists of scene-based tone mapping and scene-based color saturation mapping processing blocks. These processing blocks make use of the peak luminance behaviors of the display used for mastering the image essence (i.e., the "mastering display") and the targeted system display as their ability to achieve peak luminance varies with the average brightness level of the pixels and the number of bright pixels within the scene. In addition to these adjustments that can be algorithmically determined from known display and content characteristics, Scene-based Color Volume Mapping also allows creatively approved adjustments.

Video Services Forum (VSF)

Video Services Forum, Inc. (VSF) was founded on October 7, 1998, to support uniform delivery of video services across the separate regional Bell operating companies after the Bellcore-sponsored organization, the Video Services Industry Forum (VSIF), was disbanded. Today the VSF is an international association comprised of service providers, users, and manufacturers dedicated to interoperability, quality metrics, and education for video networking technologies. The organization's activities include:

- Providing forums to identify issues involving the development, engineering, installation, testing, and maintenance of video networking technologies
- Exchanging non-proprietary information to promote the development of video networking technology and foster resolution of issues common to the video services industry
- Promoting interoperability by contributing to and supporting development of standards by national and international standards bodies

TR-01—Transport of JPEG 2000 Broadcast Profile video in MPEG-2 TS over IP

TR-02—Using RTCP for In Band Signaling of Media Flow Status

TR-03—Transport of Uncompressed Elementary Stream Media over IP

TR-04—Utilization of ST-2022-6 Media Flows within a VSF TR-03 Environment

Acquiring Reference Documents

Contact information is given below for a selection of standards developing organizations (SDOs) working in the digital video realm:

AIMS—Alliance for IP Media Solutions
23117 39th AVE SE
Bothell, WA 98021
USA
+1 (425) 870-6574
www.aimsalliance.org

ARIB—Association of Radio Industries and Businesses
Nittochi Bldg., 1-4-1
Kasumigaseki, Chiyoda-ku, Tokyo 100-0013
Japan
+81-3-5510-8590
www.arib.or.jp/english/

ATSC—Advanced Television Systems Committee
1776 K Street, N.W., 8th Floor
Washington, DC 20006-2340
USA
+1 (202) 872-9160
www.atsc.com

CableLabs
858 Coal Creek Cir.
Louisville, CO 80027
USA
+1 (303) 661-9100
http://cablelabs.com

CTA—Consumer Technology Association
1919 S. Eads Street
Arlington, VA 22202
USA
+1 (703) 907-7600
www.cta.tech

DVB—Digital Video Broadcasting
c/o EBU
17a Ancienne Route
CH-1218 Grand Saconnex, Geneva
Switzerland
+ 41 22 717 27 14
http://dvb.org

IEEE—Institute of Electrical and Electronics Engineers
3 Park Avenue, 17th Floor

New York, NY 10016-5997
USA
+1 (212) 419-7900
www.ieee.org

IETF—The Internet Engineering Task Force
Secretariat
c/o Association Management Solutions, LLC (AMS)
5177 Brandin Court
Fremont, CA 94538
USA
Phone: +1 (510) 492-4080
www.ietf.org

ISO—International Organization for Standardization
ISO Central Secretariat
Chemin de Blandonnet 8
CP 401
1214 Vernier, Geneva Switzerland
+41 22 749 01 11
http://iso.org

NAB—National Association of Broadcasters
1771 N Street, NW
Washington DC 20036
+1 (202) 429-530
http://nab.org

NABA—North America Broadcaster Association
P.O. Box 500, Station A,
Toronto, ON M5V 3G7
Canada
+1 (416) 587-9877
http://nabanet.com

SCTE—Society of Cable Telecommunications Engineers, Inc.
140 Philips Road
Exton, PA 19341-1318
+1 (610) 363-6888
http://scte.org

SMPTE—Society of Motion Picture and Television Engineers
3 Barker Ave., 5th Floor
White Plains, NY 10601
+1 (914) 761-1100
www.smpte.org

CHAPTER

2.10

Broadcasting Emergency Information

GARY E. TIMM

Broadcast Chair, Wisconsin State Emergency Communications Committee
Milwaukee, WI

OVERVIEW

Most broadcasters carry emergency information, both as a public service to their audience and because of regulatory requirements. Along with this practice come certain responsibilities, primarily described in the Federal Communications Commission (FCC) rules.[1] The bulk of these requirements are found in FCC rules Part 11 addressing broadcaster participation in the Emergency Alert System (EAS), but the Commission also has certain other requirements found in FCC rules Part 79 regarding the broadcast of both EAS and non-EAS emergency information by TV stations. Both of these FCC rules sections will be covered in this chapter. It should be noted that although these rules apply to both broadcast stations and cable TV systems, which the FCC refers to collectively as EAS Participants, this chapter will focus solely on the aspects of the rules applicable to broadcasters. The specific details outlined in this chapter are current with the FCC rules up through the FCC EAS Report and Order, released July 11, 2016. Broadcasters need to heed advisories from their communications counsel regarding any additional FCC rules adopted subsequent to that date; for example, an updating of the *EAS Operating Handbook* and basic EAS security guidelines.

Not only must broadcast engineers and management stay up to date on EAS rules, but they must keep station staff informed as well. Although FCC fines (called "forfeitures") for EAS infractions have tended to range up to $10,000, already a severe amount, in 2015 the Commission levied a $1,000,000 forfeiture against a large broadcast entity for the false use of EAS tones during a comedy bit on a nationally syndicated radio show. In particular, engineers who are designated as the station chief operator may wish to subscribe to one or both of the nationwide EAS email lists to stay current on new FCC EAS rules and developments. Those email lists, along with other useful links, are shown in the Resources section at the end of this chapter. Topics covered in this chapter include:

- EAS and IPAWS
- Stations exempt from needing some or all EAS equipment
- Minimum FCC requirements for EAS
- Accessibility rules affecting TV stations
- Multilingual EAS alerting
- Optional state and local EAS alerting
- Emergency alerting in Canada
- Emergency alerting resources

EAS AND IPAWS

The original and still primary purpose of EAS is to enable the president of the United States to speak directly to the public over radio and TV broadcast and cable TV channels during a national emergency. A secondary purpose is to enable state and local government officials and the National Weather Service (NWS) to speak directly to the public during state and local emergencies. Broadcasters are required by FCC rules to carry the president's message,[2] but relaying state and local EAS alerts, including all weather alerts, is voluntary.[3]

The original method of distributing alerts via EAS, now often referred to as "legacy EAS," is via broadcasters receiving alerts over the air either from other broadcasters or from background communications channels, such as FM Radio Data System (RDS) subcarriers or two-way radio channels. In addition to legacy EAS, the Federal Emergency Management Agency (FEMA) has created the Integrated Public Alert and Warning System (IPAWS), which distributes EAS messages to broadcasters via the Internet. Broadcasters are required by FCC rules to monitor both the legacy EAS and the IPAWS EAS feeds for alert messages.[4]

Legacy EAS Message Dissemination

The legacy EAS network is a blending of "daisy-chain" and web-like signal distribution. It is like a daisy chain in the sense that there is a well-defined hierarchy of radio and TV broadcasters and cable TV systems in which national EAS alerts enter the system at national primary (NP) sources and flow from there to state primary (SP) sources, state relay (SR) sources, local primary (LP) sources, and finally to all other local broadcast stations and cable TV systems which the FCC designates as participating national (PN) sources.[5] It is web-like in the sense that each station is required to monitor at least two sources for EAS alerts, thus ensuring that a failure at a single source does not prevent a message from getting through to all stations below it. See Figure 2.10-1 for an example of a typical legacy EAS monitoring architecture, demonstrating how each source receives alerts from two other sources upstream back to the originating NP sources. Although the primary method of distributing the president's message is a nationwide group of primary entry point (PEP) radio stations, the federal government also has agreements with two nongovernmental broadcasting organizations—Premiere Networks and National Public Radio (NPR)—to act as NP sources; the diagram shows a PEP station and NPR as NP sources in this example.

States utilize the daisy chain described above to varying degrees. Not all states exactly follow the PEP/NP to SP to SR to LP to PN architecture, and these sources will not always be broadcast stations. In addition, some states advise monitoring more than two sources. Some states overcome the single-point liability of daisy-chain distribution with wireless and wired local relay networks (LRNs). An LRN is a one-to-many distribution system in which each EAS participant has the capability of receiving EAS messages at the same time. Consult your State EAS Plan for specific details on legacy EAS monitoring and the alert path in your state for your local market.

The decisions regarding what type of source an individual station will be and what two sources it is required to monitor are found in the State EAS Plan, which is typically developed by the State Emergency Communications Committee (SECC). The SECC chair is appointed by the FCC; the chair and all members of the SECC are volunteers from the broadcast and cable TV industry, state emergency management agency, NWS, and other concerned organizations. Some localities may also have a Local Area Emergency Communications Committee (LECC) that develops procedures for local emergencies. A list of all SECC chairs and links to State EAS Plans and the FCC EAS rules are found on the FCC EAS webpage listed in the Resources section at the end of this chapter. Legacy EAS message protocol is discussed later in this section.

IPAWS EAS Message Dissemination

In addition to the two legacy EAS monitoring assignments stipulated in each State EAS Plan, broadcasters are also required by the FCC rules to monitor the FEMA IPAWS EAS feed. Stations must connect their EAS unit to the Internet and ensure that the IPAWS alert gateway address is programmed in to allow the EAS unit to poll IPAWS. Note that IPAWS EAS alerts are "pull" not "push," thus a station does not need to open an Internet port in order to receive IPAWS EAS alerts. The station EAS unit will poll the IPAWS alert gateway for pertinent messages at the polling interval set in the EAS unit, usually about every 30 seconds. Stations may receive the same message via both legacy EAS and by IPAWS EAS. The alert received first will be the one to air, with the other alert detected as a duplicate and logged but not aired.

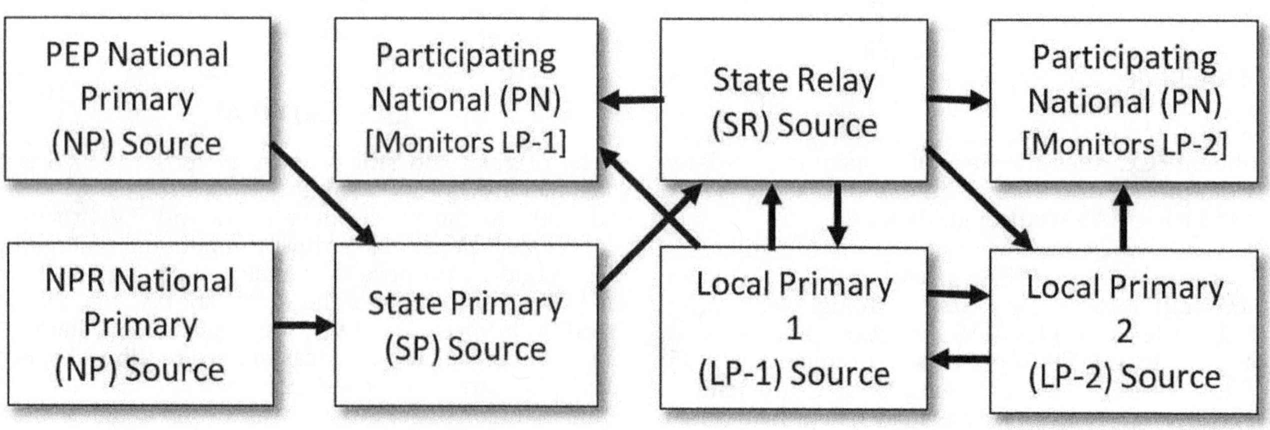

FIGURE 2.10-1 Typical Legacy EAS Monitoring Diagram

In addition to the EAS feed, IPAWS also distributes alerts via Wireless Emergency Alerts (WEA) over cellular telephone networks, NOAA Weather Radio (NWR) using the weather radiofrequency band, public websites like the popular search engine sites, and other dissemination paths. WEA alerts are received on cell phones and other wireless devices, and those messages often end with the phrase "Check local media." It is therefore important for broadcasters to relay IPAWS EAS alerts as soon as possible, knowing that the companion WEA message may be drawing audience to the station. The public is also likely to hear alerts on NWR, as well as see alerts returned in web searches for the affected area, further driving audience to the broadcast station for long-form and real-time information.

FEMA offers IPAWS access free of charge to state and local authorities, which can be authorized to issue IPAWS alerts after completing FEMA's application process. However, those authorities need to first contract with a FEMA-approved third-party IPAWS alert origination service provider (AOSP), which typically will have an associated annual cost paid from the subscribing state or local authority's emergency alerting agency to the AOSP vendor. Once that alert origination capability is acquired, FEMA online training for warning center personnel is completed, and the state or local agency's IPAWS application is approved, those authorities can originate alerts into IPAWS. Figure 2.10-2 demonstrates how a single originated alert from state or local authorities reaches all the dissemination methods described above. For more information on IPAWS, including an email address for IPAWS questions, see the links in the Resources section at the end of this chapter. Details of the IPAWS EAS message protocol are discussed later in this section.

Legacy EAS Message Protocol

A legacy EAS message consists of four elements in the following order: digital header code (repeated three times); two-tone attention signal; audio, video, or text message describing the actual alert; and digital end-of-message (EOM) code (repeated three times).[6] Note that EAS messages transmitted by radio broadcast facilities will include only audio components of an EAS message. EAS message elements are illustrated in Figure 2.10-3.

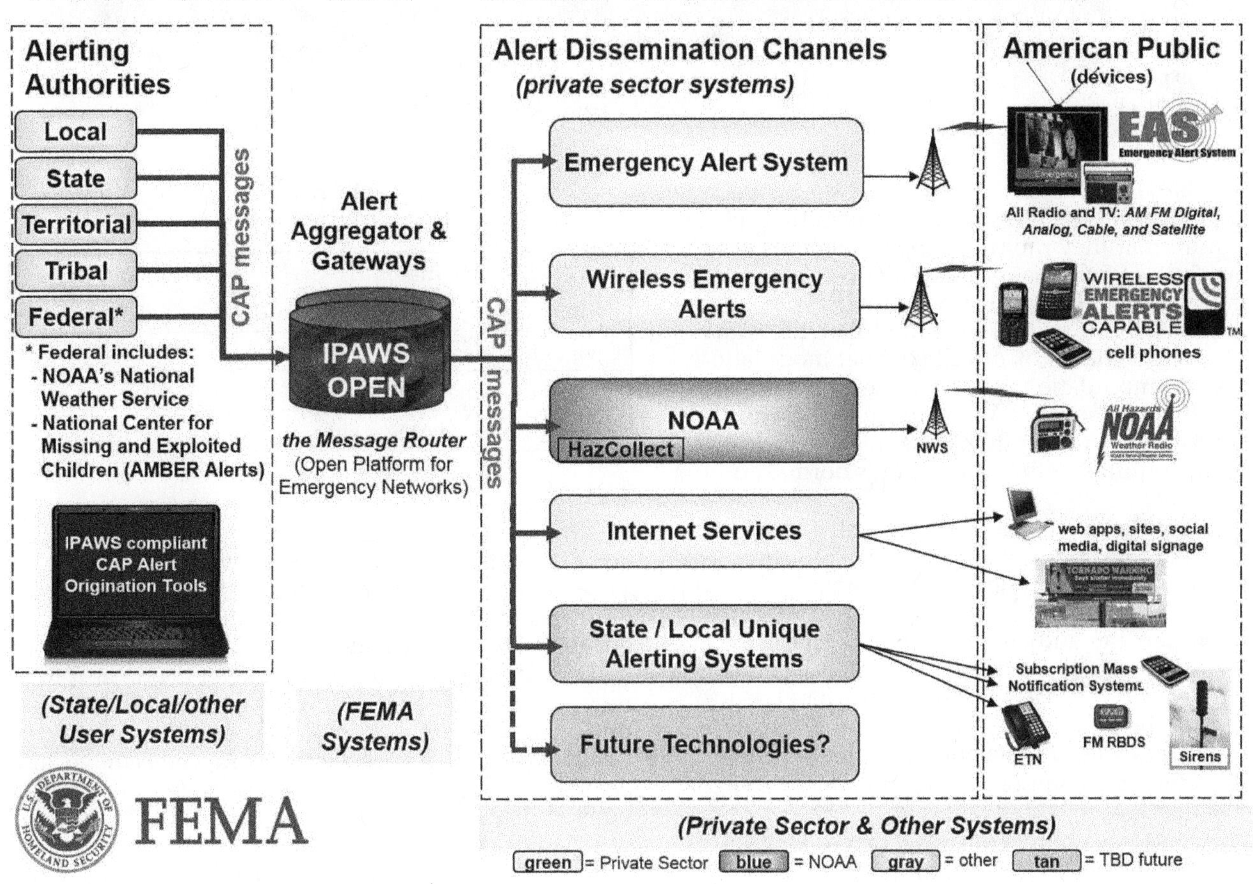

FIGURE 2.10-2 FEMA IPAWS Architecture Drawing

FEMA Graphic

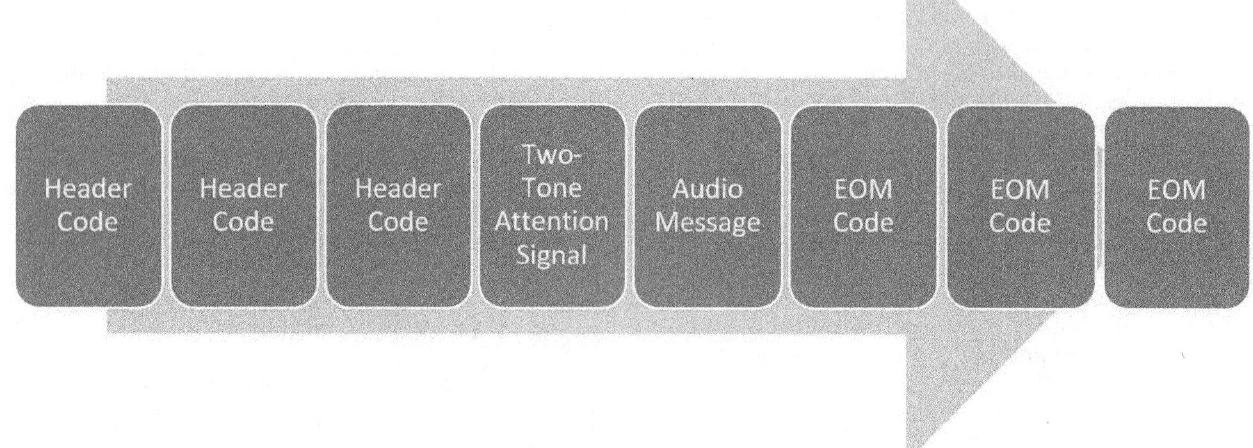

FIGURE 2.10-3 EAS Message Elements

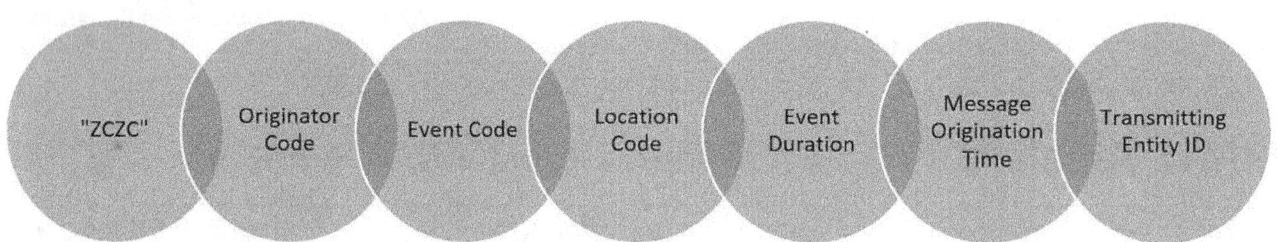

FIGURE 2.10-4 EAS Header Code Components

When the digital components (i.e., header code, EOM code) of an EAS message are transmitted, FCC rules require that they modulate the transmitter at no less than 50% of full channel modulation. When the two-tone attention signal is transmitted, FCC rules require that each individual tone modulate the transmitter at no less than 40% of full channel modulation and that the modulation levels of the two tones be within 1 dB of one another. When the audio portion of an EAS message is transmitted, it is subject to the same modulation requirements as the station's normal audio programming.[7]

Except for the actual audio message, the digital header code is the portion of the EAS message that contains the most information. The two-tone attention signal is not intended to serve any purpose other than to audibly alert the audience that an EAS message is about to be broadcast, and the end-of-message code is simply a series of four ASCII "N" characters used to indicate that the alert is over and that the EAS equipment should reset itself to its normal, non-alert state.

The header code components are illustrated in Figure 2.10-4 and are explained below.

The first code segment in the header code consists of the characters "ZCZC," which indicates that an ASCII code sequence is about to begin. The next code segment contains the originator code, which indicates the type of entity that originated that EAS alert. A list of acceptable originator codes[8] is provided in Table 2.10-1. The event code, which follows the originator code, indicates the type of emergency that is the subject of the alert. Event codes[9] are listed in Table 2.10-2.

TABLE 2.10-1
EAS Originator Codes

Originator	Originator Code
Broadcast station or cable system	EAS
Civil authorities	CIV
National Weather Service	WXR
Primary Entry Point System	PEP

The location code follows the event code and indicates the state (or U.S. territory), county, and sometimes the subdivision (northwest, south, etc.) affected by the EAS alert. Special location codes have also been assigned for specific offshore marine areas. Stations should consult their State EAS Plan for a list of location codes used in their state, or see the link to the location code standard listed in the FCC rules shown in the Resources section at the end of this chapter. Location subdivision codes, which are universal across states, are shown in Table 2.10-3. Use of these

TABLE 2.10-2
EAS Event Codes (covers only one column width, likely two columns long)

Nature of activation	Event Code
National Codes (Required):	
Emergency Action Notification (National only)	EAN
National Information Center	NIC
National Periodic Test	NPT
Required Monthly Test	RMT
Required Weekly Test	RWT
State and Local Codes (Optional):	
Administrative Message	ADR
Avalanche Warning	AVW
Avalanche Watch	AVA
Blizzard Warning	BZW
Child Abduction Emergency	CAE
Civil Danger Warning	CDW
Civil Emergency Message	CEM
Coastal Flood Warning	CFW
Coastal Flood Watch	CFA
Dust Storm Warning	DSW.
Earthquake Warning	EQW
Evacuation Immediate	EVI
Extreme Wind Warning	EWW
Fire Warning	FRW
Flash Flood Warning	FFW
Flash Flood Watch	FFA
Flash Flood Statement	FFS
Flood Warning	FLW
Flood Watch	FLA
Flood Statement	FLS
Hazardous Materials Warning	HMW
High Wind Warning	HWW
High Wind Watch	HWA
Hurricane Warning	HUW
Hurricane Watch	HUA
Hurricane Statement	HLS
Law Enforcement Warning	LEW
Local Area Emergency	LAE
Network Message Notification	NMN
911 Telephone Outage Emergency	TOE
Nuclear Power Plant Warning	NUW
Practice/Demo Warning	DMO
Radiological Hazard Warning	RHW

Nature of activation	Event Code
State and Local Codes (Optional):	
Severe Thunderstorm Warning	SVR
Severe Thunderstorm Watch	SVA
Severe Weather Statement	SVS
Shelter in Place Warning	SPW
Special Marine Warning	SMW
Special Weather Statement	SPS
Storm Surge Watch	SSA
Storm Surge Warning	SSW
Tornado Warning	TOR
Tornado Watch	TOA
Tropical Storm Warning	TRW
Tropical Storm Watch	TRA
Tsunami Warning	TSW
Tsunami Watch	TSA
Volcano Warning	VOW
Winter Storm Warning	WSW
Winter Storm Watch	WSA

TABLE 2.10-3
EAS Location Subdivision Codes (covers only one column width)

Subdivision	Code
All	0
Northwest	1
North	2
Northeast	3
West	4
Central	5
East	6
Southwest	7
South	8
Southeast	9

location subdivision codes is not common, but when implemented, the subdivision code is inserted as the first digit in the state or county location code. If subdivision is not used, the first digit of the location code is defaulted to "0," representing all of that location.[10]

The duration period of the alert, in HHMM format, follows the location code. It indicates the length of time after the origination of the alert during which it is effective. This valid period of an EAS alert must be in 15-minute increments for periods that are no longer than 1 hour and in 30-minute increments for periods that are longer than an hour.

The next element is the origination time of the alert indicating the day and time the message was issued by the originator. It is in JJJHHMM format, where the JJJ represents the day of the year in Julian calendar days (e.g., July 4 would be 185 in a non-leap year), and HHMM is the 24-hour Universal Coordinated Time (UTC).

Following the origination time is the transmitting station identifier. This is the identification of the station transmitting the alert. It is not the station that originated the alert but instead the station that is actually transmitting or retransmitting it. When EAS is activated and an alert is making its way through the network of stations and sources, the transmitting station identifier will be the only part of the header code that is changed as the message is relayed from one station or source to the next. All other components of the header code will remain unchanged.

The ASCII dash/hyphen character is used at the end of all but one code segment within the header code as a separator. The separator following the location code, thus preceding the alert duration element, uses the ASCII plus character ("+").

IPAWS EAS Message Protocol

IPAWS messages, also called "CAP alerts," use the Common Alerting Protocol (CAP), an international non-proprietary open source standard for originating emergency alerts. A CAP alert is an eXtensible Markup Language (XML) message, a text document with each aspect of the alert on a separate line of text. In the United States, the information conveyed in the basic CAP standard is supplemented by the USA IPAWS Profile, which adds elements found in U.S. EAS alerts. In this way, EAS/CAP units at broadcast stations are able to detect duplicates of the same alert received in both legacy EAS and IPAWS CAP formats, preventing the duplicative alert from airing.

A CAP alert, in addition to carrying the same information as a legacy EAS alert, can include capabilities such as alert messages in languages other than English and alert details in the CAP Description and Instruction elements. CAP alerts can also carry audio (MP3) or visual resources (video or graphics) either by direct encoding into the XML message itself or through a reference URL that the EAS/CAP unit then uses to retrieve the audio/video resource via the Internet. If no audio file is provided, the EAS/CAP unit uses an internal text-to-speech function to create audio from the text of the CAP message.

The FCC EAS rules require that the audio/video presentation of a CAP alert contain the same basic information as a legacy EAS alert, followed by any additional information found in the CAP Description and Instruction elements. The ability to deliver this additional information is one of the advantages of CAP alerts over legacy EAS messages. The text portion of the CAP alert (legacy EAS header code + CAP Description + CAP Instruction) is limited to 1,800 characters. Figure 2.10-5 illustrates the elements of the CAP alert text. The presentation order, as well as the algorithm to pare down lengthy CAP elements to stay within the character limit, are described in a FEMA/FCC-adopted Implementation Guide; for more information see the link in the Resources section at the end of this chapter.

Note that although an EAS alert may be issued by an alerting authority in CAP format, stations will not necessarily air the alert from the CAP message. Due to the FCC-mandated continuation of legacy over-the-air EAS monitoring, it is possible that some stations will receive a legacy EAS alert converted from the original CAP message before their EAS/CAP unit polls IPAWS for the original CAP version of the alert. Thus for any given CAP-originated alert, some stations will air the CAP version and other stations will air a legacy EAS version. There is no way to predict which will air for any given alert, and the result will always vary; this possibility of some stations not airing the enhanced CAP text will persist as long as the FCC continues to require legacy EAS monitoring. Station engineers will thus note on their EAS/CAP unit log that the first of the two alerts to be received will be the one to air, with the second alert logged as a duplicate.

IPAWS CAP alerts are utilized by more than just broadcasters for EAS. For instance, some popular

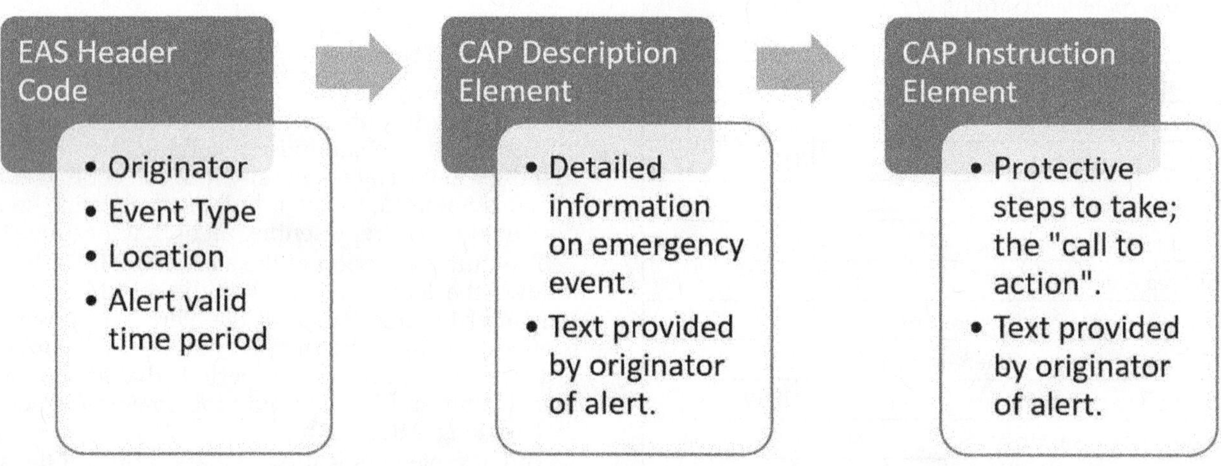

FIGURE 2.10-5 Elements of Enhanced CAP Text Presentation

Internet search engines receive these CAP alerts and display them in search results for any searches in the affected area of the alert. Some government agencies are using CAP to activate their sirens as well as other computer-controlled alerting functions. All of this is possible due to the standardized format of the CAP message; receiving computers know just where to look in the XML of the CAP message to extract the information needed. In the future, more and more devices and applications are expected to leverage the standard CAP format to receive and relay these IPAWS CAP alerts. There is a link to more information on CAP in the Resources section at the end of this chapter.

STATIONS EXEMPT FROM NEEDING SOME OR ALL EAS EQUIPMENT

Some broadcast stations are exempt from being required to purchase any EAS equipment, while some others are exempt from needing an EAS encoder. To meet the needs of those latter stations, some manufacturers offer EAS decoder-only units at a somewhat reduced price from the full EAS encoder/decoder units. The types of stations that are exempt from needing some or all EAS equipment are described below.

EAS Encoder/Decoder General Exceptions

Low-Power TV (LPTV) stations that operate as television broadcast translator stations are not required to have EAS equipment. FM broadcast booster stations and FM translator stations that entirely rebroadcast the programming of other local FM broadcast stations are not required to have EAS equipment. Broadcast stations that operate as satellites or repeaters of a hub station (or common studio or control point if there is no hub station) and rebroadcast 100% of the programming of the hub station (or common studio or control point) are not required to have their own EAS equipment.[11]

EAS Decoder Rules

Other than the exceptions above, all broadcast stations are required to have an EAS decoder. Stations that are co-owned and co-located with a combined studio or control facility can share one EAS decoder.[12]

EAS Encoder Rules

All full-power AM, FM, and TV stations and Class A TV stations are required to have EAS encoders, except those stations fitting the general exceptions section above. Stations that are co-owned and co-located with a combined studio or control facility can share one EAS encoder.[13] Digital radio and television stations must broadcast EAS alerts on all program streams, including when sending the Required Weekly Test (RWT) and Required Monthly Test (RMT).[14] Class D noncommercial educational FM stations, low-power FM (LPFM) stations and low-power TV (LPTV) stations are not required to have an EAS encoder.[15] Thus, these stations are not required to transmit the RWT;[16] however, they are required to transmit the script of the received RMT, even though they cannot generate the EAS tones.[17] Although these stations are not required to have an EAS encoder, they can at their discretion purchase a full EAS encoder/decoder unit in order to fully participate in all EAS alerts and tests as other broadcasters do.

MINIMUM FCC REQUIREMENTS FOR EAS

For broadcasters required to have EAS equipment, the following are the minimum requirements to comply with the FCC rules.

1. *Install and keep in good working order an EAS encoder/decoder.* See exceptions to this requirement in the text above. Be aware that digital radio and television stations must broadcast EAS alerts on all program streams (i.e., both main and multicast channels), including when sending the RWT and RMT, so be certain that the EAS unit interrupts all program streams. Stations with defective EAS equipment removed for repair may operate without the equipment for a period of 60 days without further FCC authority, but entries shall be made in their broadcast station EAS log showing the date and time the equipment was removed and restored to service. If receiving an RMT during this period, the RMT test script must still be transmitted even though the equipment for generating the EAS codes is not functioning.[18] If repair or replacement of defective equipment is not completed within 60 days, an informal request shall be submitted to the district director of the FCC field office serving the area in which the station is located.[19]
2. *Program your EAS unit to auto-forward all EAN and NPT messages with National Location Code 000000 and your State Location Code.* Stations are required to immediately broadcast all received alerts containing national EAS event codes, which are the Emergency Action Notification (EAN) and National Periodic Test (NPT) codes.[20] EAS unit programming of these alerts should be paired with National EAS Location Code 000000, as well as your State Location Code. FEMA has stated that it may employ state-targeted national alerts. Digital radio and television stations must transmit national alerts on all program streams.
3. *Monitor your two legacy EAS monitoring assignments.* Stations are required to monitor the two legacy EAS monitoring assignments specified for that station in the State EAS Plan.[21] Contact your SECC if you have issues monitoring your assigned sources. A list of SECC Chairs and State EAS Plans is found on the FCC EAS webpage listed in the Resources section at the end of this chapter.
4. *Monitor the FEMA IPAWS Internet-based EAS feed.* Stations are required to monitor the FEMA

IPAWS EAS feed via the Internet for appropriate CAP-based EAS messages.[22]

5. *Transmit and log a weekly RWT.* Unless your station is exempt from having an EAS encoder as outlined in the section above, your station must send a Required Weekly Test (RWT) once every week on random days within the week and at random times throughout the day or night and must log the fact that the test was sent. Digital radio and television stations must transmit the RWT on all program streams. A video message is not required to be part of the weekly test. No RWT is required during any week that either an RMT or a national test is sent, nor is an RWT required during any week when a state or local EAS message containing the EAS header and EOM codes is transmitted.[23] When transmitting the RWT, the required location codes are the state and county for the broadcast station city of license; other location codes may be included upon approval of station management.[24]
6. *Log weekly RWTs received from your two legacy EAS monitoring assignments.* Stations are required to log the date and time that the weekly test messages are received from the two EAS monitoring assignments. If any RWT is not received in a given week, stations must investigate and document the cause and any remedies in their broadcast station EAS log. The EAS log is to be retained for a period of two years.[25] Post-legacy EAS equipment has the ability to send emails to those responsible for EAS, greatly simplifying compliance issues.
7. *Log weekly RWT received from the IPAWS EAS feed.* Every Monday morning, FEMA sends a sequence of RWT tests grouped by states on the IPAWS EAS feed. The state codes sent in each test are grouped by time zone, and timed to arrive at approximately 11:00 a.m. local time. These tests should be logged in the same manner as other logged RWTs. If the IPAWS RWT is not received in a given week, check the EAS email lists referenced in the Resources section at the end of this chapter for discussion on the missed test. IPAWS RWTs that FEMA fails to send are documented in a monthly report sent by FEMA to these EAS email lists at the end of each month.[26]
8. *Forward within 60 minutes the received RMT every month.* Stations are required to retransmit the Required Monthly Test (RMT) within 60 minutes of receiving it every month, and log both the receipt and transmission of this message. Digital radio and television stations must transmit the RMT on all program streams. The scheduled times of the RMTs are typically documented in the State EAS Plan and must be conducted between 8:30 a.m. and sunset, local time, in odd-numbered months, and between sunset and 8:30 a.m., local time, in even-numbered months.[27] During the week that the RMT occurs, no RWT is required to be sent by stations.[28] If during a given month a station airs a national test, that station is not required to transmit the RMT that month.[29] If during a given month a station transmits an actual EAS emergency message containing all the normal elements of an EAS message, the EAS header codes, attention signal, emergency message, and end-of-message code, that station is not required to transmit the RMT that month.[30] Stations exempt from having an EAS encoder must still rebroadcast the RMT script, even though they can't produce the EAS tones.[31] If the RMT is not received in a given month, stations must investigate and document the cause and any remedies in their broadcast station EAS log. The EAS log is to be retained for a period of two years.[32]
9. *Post copies of the FCC* EAS Operating Handbook *and your State EAS Plan.* Stations are required to post a copy of the appropriate FCC *EAS Operating Handbook* at normal duty positions or EAS equipment locations (there are currently different versions of the handbook for AM/FM radio and for TV stations). It is advisable to post a copy of the State EAS Plan as well. Links to these documents are found on the FCC EAS webpage listed in the Resources section at the end of this chapter. NOTE: At the time of publication, the FCC *EAS Operating Handbooks* still contained outdated information, and have been scheduled for an update by the FCC. Stations should remain aware of developments on this matter, and ensure that they have the most recent version of the appropriate FCC *EAS Operating Handbook* posted at all times, as required.[33]
10. *Enter all required information into the EAS Test Reporting System (ETRS).* The FCC launched the ETRS webpage on June 27, 2016, and gave EAS participants 60 days to enter certain station information on ETRS Form One, including station transmitter location, EAS contact, EAS unit make, model and software version, and the station's two EAS monitoring assignments. Stations need to be mindful that this ETRS Form One information is required to be renewed on a yearly basis. In addition, following any nationwide EAS test, stations must enter into ETRS Form Two their day-of-test data within 24 hours of the test, and detailed post-test data on ETRS Form Three within 45 days of the test.[34] See the Resources section at the end of this chapter for a link to the FCC's ETRS webpage and an ETRS email contact address.

ACCESSIBILITY RULES AFFECTING TV STATIONS

While these are the current rules at the time of publication of this chapter, TV stations in particular need to maintain their awareness of possible new FCC accessibility rules. This is a priority topic for the Commission. While Part 79 detailed below has always required EAS alerts to be presented on TV both aurally and visually, the FCC has now extended this requirement to non-EAS emergency information, as well.

Non-EAS Emergency Information Is Not Exempt From the Rules

In addition to the EAS rules in Part 11, the Commission also imposes requirements in Part 79 regarding the visual and aural broadcast of non-EAS emergency information by TV stations. These rules deal largely with making sure that emergency information is accessible to those with sight and hearing impairment. The basic rule is that all emergency information must be presented both aurally and visually, and this applies whether it is presented within a newscast or during normal programming. The visual presentation can be via closed-captioning or an open visual display. The aural message must be preceded by a three-beep audio tone on both the main and secondary audio stream, with the aural message itself only required to be presented on the secondary audio stream, and it must be conveyed there at least twice. Emergency information provided using text-to-speech technologies must be intelligible and must use the correct pronunciation of relevant information to allow consumers to learn about and respond to the emergency, including, but not limited to, the names of shelters, school districts, streets, districts, and proper names noted in the visual information. Broadcasters sometimes refer to this aural presentation requirement as the "Audible Crawl Rule."[35]

Concerning this rule, the FCC defines "emergency information" as information about a current emergency that is intended to further the protection of life, health, safety, and property, i.e., critical details regarding the emergency and how to respond to the emergency. Examples of the types of emergencies covered include tornadoes, hurricanes, floods, tidal waves, earthquakes, icing conditions, heavy snows, widespread fires, discharge of toxic gases, widespread power failures, industrial explosions, civil disorders, and warnings and watches of impending changes in weather.[36]

Regarding the term "critical details" used in the paragraph above, the Commission explains that critical details include, but are not limited to, specific details regarding the areas that will be affected by the emergency, evacuation orders, detailed descriptions of areas to be evacuated, specific evacuation routes, approved shelters or the way to take shelter in one's home, instructions on how to secure personal property, road closures, and how to obtain relief assistance.[37]

The FCC has further clarified that as part of this rule it expects TV stations to present an aural description of visual but non-textual emergency information, such as maps or other graphic displays.[38] At the time of publication, this part of the rule was set to take effect on November 30, 2016,[39] but as yet there appears to be no mechanism to perform this function and the description would need to be entered manually by station staff. Check with your communications counsel to determine how revisions to this part of the rule and their effective dates will affect your facility.

Visual Display Rules

In the 2015 EAS Sixth Report and Order, the Commission clarified that EAS visual messages must be readable and accessible to all members of the public, including people with disabilities.[40] Adding to the existing requirement that the EAS visual display be at the top of the television screen or where it will not interfere with other visual messages, the FCC addressed other aspects of the display. The EAS rules now require that the EAS visual message must be displayed in a font size, color, contrast, screen location, and at a speed that is readily readable and understandable, does not contain overlapping lines of text, fits onto the viewable display, and that the text of a message must be displayed in full at least once during any EAS message. The audio message must also be presented in full at least once during any EAS message.[41]

In stating these rules, the Commission does not define "readily readable and understandable" nor does it mandate a specific crawl speed or font size, but instead it affords stations the flexibility to implement this requirement in accordance with the station's particular best practices and equipment capabilities.[42]

MULTILINGUAL EAS ALERTING

The EAS community and the FCC are just beginning to make advances in multilingual EAS alerting. Some EAS units are capable of accessing non-English text when sent in a CAP message, although this capability typically requires additional software be installed in the EAS unit. On November 17, 2015, FEMA originated a CAP message that contained both English and Spanish as part of a regional NPT test. Demonstrations in at least three states presented both the English and Spanish text crawls and audio messages. Although proven possible, multilingual EAS message decoding is not yet commonplace in the typical EAS unit. Check with your EAS unit manufacturer if you are interested in this option.

On March 30, 2016, the FCC released an Order mandating the collection of data on the use of multilingual EAS alerting in the United States. The Order requires all broadcasters to furnish information to their SECC regarding their station's current and planned non-English alerting, which is then summarized by the SECC in its State EAS Plan. The results are expected to be due by the fall of 2017. While the Order does not mandate multilingual alerting, it does establish a baseline for assessing the current use of non-English alerts. For more details, see the link to this FCC Order in the Resources section at the end of this chapter.

Canada has been doing alerts in English, French, and First Nation languages for a number of years already. In the United States, current FCC rules allow stations which carry programming in a foreign language to broadcast state and local EAS alerts "in the primary language of the EAS Participant" for legacy EAS alerting[43] and CAP alerting,[44] and for the RMT.[45] Foreign language stations in particular may want to pay close attention to future FCC developments in multilingual alerting and to EAS equipment manufacturer advances in making non-English alerting available to the average station.

OPTIONAL STATE AND LOCAL EAS ALERTING

Rebroadcast of state and local EAS alerts, including all weather alerts, is voluntary.[46] However, if stations choose to carry these alerts, then the procedures in the State EAS Plan and any Local Area EAS Plan are required to be followed. In addition, FCC rules require that any state or local EAS alert broadcasted be logged.[47]

In the past, under legacy EAS, state and local authorities often injected their EAS messages into the system by manually contacting a state primary or local primary station to originate the message. Now with IPAWS, all state and local authorities can make arrangements to easily originate an EAS message themselves and deliver it directly to every broadcaster in the area via the Internet. Broadcasters should encourage local and regional emergency management authorities to use this system and help ensure these authorities are aware of IPAWS.

Stations should have a written plan describing how they will stay on the air during an emergency in order to get ongoing news and information to their audience. This plan should take into account the impact of many types of emergencies and ways to mitigate those risks, and should include storing emergency supplies to keep both staff and equipment operational for periods that could last two weeks or more.

In addition to individual station broadcast continuity plans, a Local Area EAS Plan can establish the communications means to be used by local county and municipal emergency management agencies to get information to broadcasters during a disaster. This ongoing information is sometimes thought of as "beyond EAS" emergency public information.

Encourage your state to establish broadcaster ID cards to allow engineers to cross police lines to maintain broadcast transmitter sites. Broadcasters can take online National Incident Management System (NIMS) basic courses (IS-100 and IS-700) to learn the procedures their emergency manager is required to follow and what goes on behind those police lines. Many of these principles of establishing a path from emergency management to broadcasters, maintaining broadcast transmitters, and broadcasters supporting each other to stay on the air in disasters are detailed in guidelines developed by the Media Security and Reliability Council (MSRC), an FCC advisory committee. Links to the MSRC website, to a current state Broadcaster ID Card program, and to FEMA's NIMS online training website are listed in the Resources section at the end of this chapter. For more information on other disaster preparation, see Chapter 3.6 of this *NAB Engineering Handbook*, Broadcast Facility Security, Safety, Disaster Planning, and Recovery.

EMERGENCY ALERTING IN CANADA

Canada has a national emergency alert system called the National Public Alerting System (NPAS, similar to EAS in the United States), also known by the more consumer-friendly term "Alert Ready." The Canadian Radio-Television and Telecommunications Commission (CRTC, similar to the FCC in the United States) required that all commercial broadcasters begin fully participating in NPAS as of March 31, 2015, with campus, community, and Native radio and television broadcasters, as well as radio communication distribution undertakings, given until March 31, 2016, to comply. A link to the CRTC NPAS rules is provided in the Resources section at the end of this chapter.

NPAS is completely Internet-delivered via Canada's National Alert Aggregation and Dissemination (NAAD) System (similar to IPAWS in the United States), and does not use any over-the-air relay or embedded emergency data in the broadcasted alert, as does the United States. Although NPAS utilizes a somewhat similar two-tone attention signal to EAS alerts, it differs in that the two-tone signal is alternated with another tone. There is a sample of the Canadian attention signal on the Alert Ready website found in the Resources section at the end of this chapter.

NAAD uses CAP messaging, as does IPAWS in the United States, but the two systems are not directly compatible. The version of CAP used north of the border is known as CAP-CP, for Canadian Profile, and it uses a number of different parameters than the U.S. version of CAP, which also has its own IPAWS profile with unique parameters. In addition, Canada has over 150 event types, about triple the number of EAS event codes, and Canada uses a location code called the Standard Geographical Classification (SGC) code, which identifies areas down to the city level, much more granular than U.S. county location codes.

With both of these North American systems routed in CAP, might U.S. broadcasters on the Canadian border consider airing Canadian alerts in addition to U.S. EAS alerts? At the present time, this would not be advisable. While it sounds like a great idea, and at least one EAS unit manufacturer feels that their unit could most likely handle receiving alerts from both systems, numerous complications remain. Guidance from both the FCC and CRTC would be a presumed precursor to any station desiring to pursue this practice. Then there is the issue that at U.S. stations, EAS alerts would need to take precedence over NPAS alerts; how would it sound on the air to abort an in-progress NPAS alert and immediately launch into an EAS alert? So while it might sound virtuous for stations to try to serve their audience on both sides of the border, this is not yet feasible.

Notes

[1] FCC rules are contained in the Code of Federal Regulations (CFR) 47, Part 11 and Part 79. Links to these rules are provided in the Resources section at the end of this chapter.

[2] CFR 47, Part 11, §11.52 (e) and §11.54.

[3] CFR 47, Part 11, §11.52 (d) (5) and §11.55.

[4] CFR 47, Part 11, §11.52 (d) and §11.56.
[5] CFR 47, Part 11, §11.2.
[6] CFR 47, Part 11, §11.31.
[7] CFR 47, Part 11, §11.51 (f).
[8] CFR 47, Part 11, §11.31 (d).
[9] CFR 47, Part 11, §11.31 (e).
[10] CFR 47, Part 11, §11.31 (c).
[11] CFR 47, Part 11, §11.11 (b).
[12] CFR 47, Part 11, §11.52 (c).
[13] CFR 47, Part 11, §11.51 (l).
[14] CFR 47, Part 11, §11.51 (c) and §11.61 (a) (2) (i) (A).
[15] CFR 47, Part 11, §11.11 (b) and §11.51 (e).
[16] CFR 47, Part 11, §11.61 (a) (2) (ii).
[17] CFR 47, Part 11, §11.61 (a) (1) (i).
[18] CFR 47, Part 11, §11.35 (b).
[19] CFR 47, Part 11, §11.35 (c).
[20] CFR 47, Part 11, §11.52 (e) and §11.54 (a).
[21] CFR 47, Part 11, §11.52 (d) (1).
[22] CFR 47, Part 11, §11.52 (d) (2).
[23] CFR 47, Part 11, §11.61 (a) (2), (3) and (4).
[24] CFR 47, Part 11, §11.51 (m).
[25] CFR 47, Part 11, §11.35 (a).
[26] CFR 47, Part 11, §11.56.
[27] CFR 47, Part 11, §11.61 (a) (1) (i).
[28] CFR 47, Part 11, §11.61 (a) (2) (iii).
[29] CFR 47, Part 11, §11.61 (a) (3) (ii).
[30] CFR 47, Part 11, §11.61 (a) (4).
[31] CFR 47, Part 11, §11.61 (a) (1) (i).
[32] CFR 47, Part 11, §11.35 (a).
[33] CFR 47, Part 11, §11.15 and §11.41.
[34] CFR 47, Part 11, §11.61 (a) (3) (iv).
[35] CFR 47, Part 79, §79.2.
[36] CFR 47, Part 79, §79.2 (a) (2).
[37] CFR 47, Part 79, §79.2, Note to paragraph (a) (2).
[38] Detailed in Report and Order FCC 13–45, Paragraph 24. See link in Resources section.
[39] Waiver from original implementation date detailed in FCC Memorandum Opinion and Order DA 15–632, Paragraph 17. See link in Resources section.
[40] FCC EAS Sixth Report and Order, FCC 15–60, Paragraph 38. See link in Resources section.
[41] CFR 47, Part 11, §11.51 (d).
[42] FCC EAS Sixth Report and Order, FCC 15–60, Paragraph 40. See link in Resources section.
[43] CFR 47, Part 11, §11.55 (c) (4).
[44] CFR 47, Part 11, §11.55 (d) (2).
[45] CFR 47, Part 11, §11.61 (a) (1) (i).
[46] CFR 47, Part 11, §11.52 (d) (5).
[47] CFR 47, Part 11, §11.55 (c) (7) and (d) (1).

Emergency Alerting Resources

Resources are listed in their order of appearance in this chapter.

FCC Rules Regarding Emergency Alerting

FCC rules Part 11: www.ecfr.gov/cgi-bin/text-idx?c=ecfr&sid=75127c72007aa6a3f1ce8fda8cb814e2&rgn=div5&view=text&node=47:1.0.1.1.12&idno=47

FCC rules Part 79: www.ecfr.gov/cgi-bin/retrieveECFR?gp=&SID=224cc2e8f467905b578f6422dfe0c292&mc=true&r=PART&n=pt47.4.79

Blogs Tracking Current EAS Developments

EAS issues blog hosted by the Society of Broadcast Engineers (SBE) EAS Education Committee: www.sbe.org/sections/gov_eas.php

The Broadcasters' Desktop Resource EAS blog hosted by the Broadcast Warning Working Group (BWWG): www.thebdr.net/articles/fcc/eas/eas.html

EAS Email Lists

SBE EAS Exchange List: https://pairlist7.pair.net/mailman/listinfo/sbe-eas

BWWG EAS Forum Discussion List: http://lists.radiolists.net/mailman/listinfo/eas

FCC EAS Information

FCC EAS webpage: www.fcc.gov/general/emergency-alert-system-eas

With questions on EAS, call: FCC Public Safety and Homeland Security Bureau: 202-418-1300

Alternate FCC Contact: FCC Public Call Center: 1-888-CALL-FCC (225-5322)

FEMA IPAWS Information

FEMA IPAWS webpage: www.fema.gov/ipaws.

With questions on IPAWS, write to ipaws@fema.gov

Legacy EAS County Location Codes

County Location Codes are derived from a standard listed in the FCC rules: American National Standards Institute (ANSI) International Committee for Information Technology Standards (INCITS) standard 31-2009, Information Technology—Codes for the Identification of Counties and Equivalent Areas of the United States, Puerto Rico, and the Insular Areas. Purchase the standard at http://webstore.ansi.org/RecordDetail.aspx?sku=INCITS+31-2009%5bR2014%5d

Common Alerting Protocol (CAP) Details:

CAP is a standard of the Organization for the Advancement of Structured Information Standards (OASIS): http://docs.oasis-open.org/emergency/cap/v1.2/CAP-v1.2-os.html

CAP Details Unique to the United States

FEMA and FCC have both adopted the following two documents, which further customize the U.S. implementation of the basic CAP standard.

USA IPAWS Profile: http://docs.oasis-open.org/emergency/cap/v1.2/ipaws-profile/v1.0/cap-v1.2-ipaws-profile-v1.0.pdf

EAS-CAP Industry Group (ECIG) Implementation Guide: www.eas-cap.org/ECIG-CAP-to-EAS_Implementation_Guide-V1-0.pdf

EAS Test Reporting System (ETRS) Information

FCC Public Notice DA 16-721 launching ETRS webpage on June 27, 2016: https://apps.fcc.gov/edocs_public/attachmatch/DA-16-721A1.pdf

ETRS home page: www.fcc.gov/general/eas-test-reporting-system

FCC ETRS email contact address: ETRS@fcc.gov

Accessibility Rules

Report and Order FCC 13-45, Paragraph 24, addressing requirement for an aural description of visual but non-textual emergency information: https://apps.fcc.gov/edocs_public/attachmatch/FCC-13-45A1.pdf

FCC Memorandum Opinion and Order DA 15-632, Paragraph 17, detailing waiver from original implementation date for aural description of visual but non-textual emergency information: https://apps.fcc.gov/edocs_public/attachmatch/DA-15-632A1.pdf

EAS Sixth Report and Order FCC 15-60, Paragraphs 38–40, regarding visual display rules: https://apps.fcc.gov/edocs_public/attachmatch/FCC-15-60A1.pdf

Multilingual EAS Order FCC 16-32: https://apps.fcc.gov/edocs_public/attachmatch/FCC-16-32A1.pdf

Improving Local Alerting

Example of state Broadcaster ID Card program: www.wi-broadcasters.org/Broadcaster-ID

FEMA's NIMS online training website: www.training.fema.gov/IS

MSRC website: www.mediasecurity.org

Example of a state MSRC implementation: www.wi-broadcasters.org/msrc

Canada Emergency Alerting Information

Alert Ready website: www.AlertReady.ca

CRTC NPAS Rules: www.crtc.gc.ca/eng/archive/2014/2014-444.htm

SECTION

3

TECHNICAL MANAGEMENT AND SAFETY

CHAPTER

3.1

Broadcast Engineering Management

NORMAN R. SWAN, PhD

School of Journalism and Electronic Media
University of Tennessee
Knoxville, TN

INTRODUCTION

Most broadcast station engineering staff plan a career in which they will work in studio or remote operations, maintenance, or transmitters. Few think about or plan for a career in management. What happens in many stations is that engineers are promoted from within to become managers, in positions such as chief engineer or director of engineering. However, while engineers may be excellent at the technical aspects of their work, they may not be at all prepared to handle the challenges of managing an engineering department. Very few broadcast station engineers have had specialized courses or training in management. The result is that many chief engineers feel unprepared for dealing with personnel and financial aspects of their new positions. Some fail, and either resign or are replaced.

This need not be the case if engineers who aspire to lead engineering teams fully understand the many challenges they will face as managers. This chapter covers the concepts of engineering management for radio and television broadcast stations, and discusses the relationship the broadcast engineer has with other members of the station management team.

The term chief engineer (sometimes known simply as "the chief") is used throughout the chapter as a general term for the head of a station's engineering department. Different organizations may have different names for specific positions. The term engineer is also used here as a general term; staff working in an engineering department may include persons who are not actually qualified engineers, such as technicians and others. Use of the word "he" applies to both male and female staff.

THE BUSINESS OF BROADCASTING

Mission

One of the first principles chief engineers must understand is the *mission* of the broadcast station. The Federal Communications Commission (FCC) licenses all broadcast stations in the United States. The FCC requires that stations operate in the "public interest, convenience, and necessity." This is often referred to as the *public service* mission of broadcasters. All stations are required to provide programming that serves the public.

A second mission for all commercial stations, though unwritten in FCC code, is that stations must earn a profit. As a result, station general managers seek ways to serve their audience and make money. A chief engineer, and all engineers for that matter, must understand how they fit into these two missions. The station cannot achieve either mission without a quality product to put on the air. Engineers must provide the technical expertise to make that happen. They must also do it in a manner that takes the station's profit margin into consideration.

Businesses

Engineers need to understand that commercial broadcast stations are unlike most other businesses, in that they operate in at least five different business areas.

The first business is that of news. Engineers must assist the news department in providing equipment and resources to cover breaking news.

The public depends on broadcast stations to provide the latest news and weather information.

The second business relates to *information*. Stations produce a wide variety of programs designed to inform their audiences. The engineering department assists with the production of these programs to ensure quality.

A third business is one of *education*. This is especially true for public stations but also applies to commercial stations. Programming for children, documentaries, and other programs are designed to educate the audience. Engineers must provide the technical expertise to develop these educational programs.

The fourth business is that of *entertainment*. Many people listen to radio for the music and many people watch television for the entertainment programs. Engineers must provide the best signal quality possible so that the audience will want to continue to listen or watch the station.

The last business broadcast stations are in is that of *advertising*. Advertisers pay the bills. Stations produce and broadcast commercials in exchange for revenue. Advertisers are more interested in placing their spots in programming from technically superior facilities that reach the largest possible audience.

The result is that engineers must work on a daily basis with *all* departments of a broadcast station. Chief engineers work alongside the general manager, program director, news director, promotions manager, and sales manager to meet the missions of the station.

Customers

Broadcast stations serve four distinct constituent groups, or "customers." Engineers need to understand who these customers are and how they influence decisions that are made.

The first group of customers is the *audience*. All decisions made by the station management team should be made with the audience in mind. The simple fact is that broadcast stations strive to attract and keep the largest possible audience. The station with the "biggest stick" or largest transmitter usually has the potential to reach the largest audience. Quite apart from any programming decisions, a station needs to have a strong, quality signal to make it possible to compete for audience ratings.

The second group of customers the broadcast engineer serves is *advertisers*. Advertisers want to associate with the "best" station, as they perceive it. They can see or hear the station and they will notice obvious technical problems.

The third customer group is *employees*. Employees of the station are a vital resource. Employees who are proud of their station will usually perform at a higher level than those who are indifferent.

The fourth customer group is the *investors* or *stockholders*. They are primarily interested in making money.

These four groups largely determine the priorities for the station. The management team should consider the impact of major decisions on the audience, advertisers, employees, and ownership.

MANAGERS AND LEADERS

There is a difference between management and leadership, and many engineers tend to be better managers than leaders. A manager may manage things—money, equipment, schedules, buildings, vehicles, and so forth. A manager can move things around as he sees fit. In contrast, a leader leads people. Without special knowledge and abilities in leadership, it is not always possible to get people to do what you want them to do.

It is possible for a person to be a manager without being a leader. For instance, a single engineer for a small radio station will most likely function as a manager. He manages the equipment of the station but may have no personnel to lead. On the other hand, a chief engineer for a larger radio station or for a television facility must function as both a manager and leader, because he will have equipment to manage and people to lead. As a result, chief engineers must develop an appropriate leadership style.

Leadership Styles

There are four major leadership styles often found in broadcast stations, depending on the department and the situation. Considerable research has been done to examine the effectiveness and appropriateness of each style to each position.

Autocratic Style

The first style some chief engineers may adopt is what social scientists call the X or autocratic style of leadership. This style of leadership is often compared with the military style of leadership. The leader is in command of his troops. He tells them what to do and how to do it. The X leader directs his employees to perform their tasks according to his wishes. Getting the job done is the most important goal for the X manager. He is socially distant from his employees. He barks out commands and demands performance. If he does not get it, he fires the employees.

This style of leadership may be found in broadcast stations. Directors in the control room are clearly "X-style" leaders—they bark out commands to all crew members in the studio—but is this style effective? The research has examined that question on three measures—employee morale, job turnover, and job productivity. The results are that morale is generally low, job turnover is generally high, but productivity is also high. The job gets done because the manager demands that it is done. The X style seems to be most effective for employees who are not self-directed. Engineering units may have technicians who need to have managers tell them what to do and how to do it. In that case, the X style may be most appropriate.

Democratic Style

The second style of leadership some chief engineers may adopt is the opposite of the X style. It is often called the Y style or democratic style of leadership. Y leaders are people-oriented. The style is follower-driven. Managers will likely have many meetings and take votes on every course of action. The overarching philosophy of the Y manager is that "happy workers will be productive workers." The Y manager is socially close to his employees. He sees all of his employees as members of his extended family and treats them that way.

In broadcast stations, Y managers may include sales managers, creative services directors, and perhaps some others. The research shows that employee morale under a Y manager is generally high and job turnover is very low. Employees like their managers and they enjoy going to work every day, but productivity is low. The reason seems to be that employees have no fear and spend much of their time socializing with coworkers. The kinds of employees who would thrive with Y managers are creative and self-directed employees. That is why creative services, programming, or sales managers are Y managers. Very few chief engineers find this style appropriate for engineers.

Goal-Oriented Style

The third style of leadership found in broadcast stations is the Z style. Z leaders are goal-oriented. They understand the importance of people to the organization and the need to get the job done. Z managers manage by objective. They ask all employees to set measurable and realistic goals for themselves. From individual goals, Z managers develop plans for their units.

Many general managers and sales managers function as Z managers. They are used to setting revenue goals for the station. The findings for this style of leadership are that employee morale is relatively high, job turnover is relatively low, and productivity is relatively high. The main problem with this style is that it does not work in all departments and with all people. Some people simply cannot set goals for themselves. Some units cannot set goals. Very few chief engineers seem to function exclusively in this style.

Contingency Style

The fourth style of leadership found in broadcast stations is the most common and is a combination of the other three styles. It is often called the contingency or C style of leadership. This style can change according to different situations and different people. For instance, in times of crisis, a type-X management style would be best. When weather emergencies occur, chief engineers must go into a type-X mode. When a general manager wants the chief engineer to create plans and budgets for the coming year, the "chief" must function with type-Z style in communicating with others on the management team. On other days, the chief engineer may use the type-Y style to get the most from some of his employees.

Which Style to Use

The research indicates that the contingency type-C style of leadership is most commonly found among news directors and may be a good choice for chief engineers as well. There are some engineers who are professional and self-directed individuals. A type-Y approach may be the best for them. Other staff may need more of a type-X style to motivate them. What is important for the chief engineer to understand is that the management team may be comprised of very different personalities and leadership styles. For instance, the general manager and sales manager are most likely Z or goal-oriented leaders. That is why they seem to see eye-to-eye on most issues. Program directors, production managers, and creative services or promotions directors may tend to be more Y or people-oriented personalities. The news director and chief engineer may be contingency-style leaders and be more likely to see eye-to-eye on many issues.

FUNCTIONS, DUTIES, AND QUALITIES

All chief engineers must perform four main functions as part of their responsibilities. This section discusses the functions of planning, leading, organizing, and controlling, the duties that go with them, and some of the qualities that lead to success.

Planning

Broadcast stations engage in three kinds of planning—short-term, long-term, and strategic.

Short-term planning is done each year by most stations, principally in preparing the annual budget. From three to six months ahead of the fiscal year, stations begin the planning and budgeting process. Each department will be asked to submit plans for the coming year along with a budget. The chief engineer must submit an operating budget including personnel, operating, and equipment expenditures. The short-term plan may also call for special equipment purchases needed in the coming year.

The long-term plan is typically a five-year plan. The chief engineer will be asked to put together a plan related to equipment maintenance and replacement over a five-year period so that the station's general manager can set aside money to meet the needs. This plan will normally include major system upgrades and capital purchases.

Strategic planning is also done periodically by many stations. Chief engineers may be asked to put together what are called "what if" plans. For example: "What if the FCC changes the ownership rules to allow television stations to own and operate more than one station in a market?" How will the station respond from an engineering perspective? Many chief engineers will spend 25% or more of their time engaged in planning. The process requires time away from the day-to-day operation of the station.

Leading

Leadership of the engineering department involves all personnel decisions. Many engineers are more prepared to deal with the management of equipment and other resources but may not be adequately prepared to handle the myriad problems that leading people can bring. The larger the engineering unit, the more personnel problems the manager will usually face. Many of them can be avoided with careful hiring, training, motivating, and evaluating strategies. These will be addressed later in this chapter. In addition, communicating with and coordinating all employees are other functions of a leader. Many managers will spend more than 25% of their time on personnel issues.

Organizing

The chief engineer will be expected to create systems to ensure the proper functioning of the unit. These systems include an organizational structure for the unit and ways to monitor productivity. Time and resource management is an important organizational skill.

Controls

The chief engineer must establish quality-control procedures and systems for his department and for the station. This includes regular maintenance checks on all equipment. It would also include monitoring the quality of the signal, the sound, and/or look of the station, and monitoring the quality of studio productions from a technical standpoint. These tasks may be delegated to other staff or to some extent be automated, but the responsibility rests with the chief engineer.

Another responsibility is cost control. The goal is to be both efficient and effective; this is covered later on.

Duties

In addition to these four basic functions that all chief engineers must perform, several other management duties should be mentioned. Many chief engineers will say that their most important duty is to protect the license. Broadcast stations are granted licenses by the FCC for a period of time. If the station operates according to FCC rules and regulations, the station will most likely be relicensed. Many of the FCC rules and regulations that result in fines are technical in nature. Station general managers rely on their chief engineers to operate the station on the right frequency or channel and at the licensed power. If the station loses its license, it is out of business.

An additional key duty is financial management. This includes the budgeting process (mentioned previously), and monitoring the budget and expenditures. This area of management is important and more attention will be given to it later in this chapter.

A further duty is to manage the station's inventory. This usually includes all technical equipment—not only the electronics, but also the vehicles, buildings, electrical, and mechanical systems—in other words, most of the tangible assets of the station.

The chief engineer must also manage the time and resources of his unit in relation to all other departments of the station. Engineers are usually required to help with all aspects of the station's technical operation. Allocating engineering resources for these activities is a duty of the chief engineer.

Other duties include liaison to all departments in the broadcast station. The chief engineer is part of the management team of the station and must work together with the programming, production, news, and sales departments. He may also be asked to work with engineering unions and perhaps be involved in community relations.

Qualities of the Successful Engineering Manager

It should be clear by now that broadcast engineering managers play a very important role in broadcast stations. Not everyone can do this well, but there are some qualities that seem to be characteristic of those who are successful.

First, effective chief engineers must be excellent communicators. This is a problem for some engineers who find themselves in leadership roles, but it is very important for chief engineers to communicate to the general manager, sales manager, and program director about the needs and issues facing the engineering unit.

It is equally important for the chief engineer to be able to communicate from the management team to engineering employees. Chief engineers must become excellent listeners and team players if they are going to be successful. They must also be able to motivate employees to work together as a team to achieve the goals of the station.

Finally, they must be good decision makers. Many decisions regarding equipment are made each year, and the general manager depends on the chief engineer to make good decisions about all equipment purchases. It is vital that the chief engineer should stay abreast of all the technological changes occurring in broadcasting.

ORGANIZATION

An engineering department at a broadcast station can range in size from one person at a small radio station to as many as 10–20 engineers and other staff at a medium-market television station, to a much larger staff for a large market or network operation. One of the most important functions of a chief engineer is to help the station general manager determine the right number of employees needed to operate and maintain the station properly. Too much staff results in waste and inefficiency. Inadequate numbers of staff results in burnout and excessive overtime wages. The structure and organization of the unit should be periodically evaluated to determine if each area of the unit has the right number of people. The needs of broadcast stations have changed over the years as stations have moved from analog to digital operations, to an increasing use of automation, and to information technology (IT)–based systems. The station may perhaps need fewer transmitter and studio engineers and more IT specialists, or perhaps staff with a mix of different skills.

Where there are lower-level managers below the chief engineer, they are generally responsible for scheduling employees and day-to-day management in their respective areas. It is important for structures to be functional. If the organizational structure is too tall with too many levels between the workers of the station and the top management, there could be problems of coordination and communication. Generally, the flatter the organizational structure, the better the coordination and communication within the unit and across the organization. Stations should strive to minimize the number of managers while ensuring that the responsibilities and workload do not result in overload for any particular position.

Organization Structure

A well-defined organization structure is important for units to function effectively. The purpose of an organizational chart is to indicate chains of command and relationships among staff members and the various areas of the station or unit. However, in small- and medium-sized markets the positions perhaps tend to be filled by ability and willingness rather than going strictly by organizational flowcharts.

The organizational structure of the engineering and operations departments may vary widely from station to station. Three typical examples follow. Other structures may, of course, be selected for particular requirements.

Figure 3.1-1 represents the simplest station engineering department structure, which may apply to either radio or television. In this example, the chief engineer heads the department that is responsible for equipment installation, maintenance, and FCC technical compliance. He is usually responsible for preparing engineering operating and capital budgets and is the designated chief operator for the station. Directly under his supervision is the assistant chief engineer, who is more responsible for the day-to-day maintenance and scheduling maintenance personnel. The number of technical maintenance personnel will vary depending on how much equipment the station has. The hardware includes that in the studio, master control, news department, transmitters, translators,

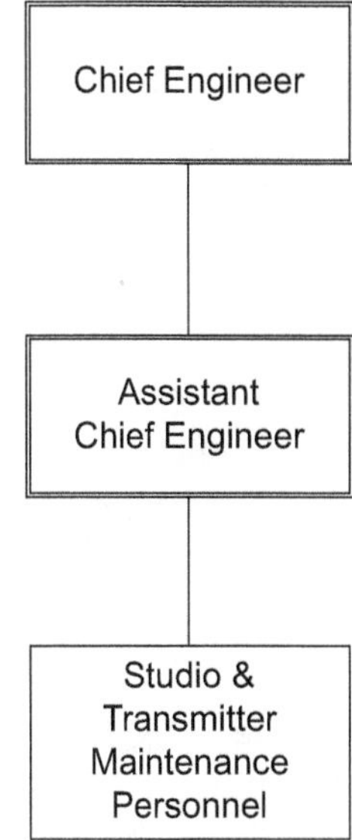

FIGURE 3.1-1 Simple engineering department organization

low-power/satellite stations, and sometimes all IT computer systems.

In this arrangement, staff working in technical operations for studios, news, and master control would report through a separate chain of command to the program director or news director for the station. Responsibility for buildings and grounds may come under the chief engineer or with staff reporting separately to the general manager.

Figure 3.1-2 outlines a very common structure for a television station, in which the chief engineer and assistant chief have responsibilities for the technical aspects of the "on-air" station operations as well as the maintenance responsibilities as outlined above. This includes master control, production operations technicians, as well as IT. Again, sometimes buildings and grounds are added. A radio station equivalent structure would be similar.

In this arrangement for television, staff in "creative" technical positions, such as camera and studio floor crew, would normally report to the production manager and not to engineering.

Figure 3.1-3 illustrates a staffing structure where engineering and operations are combined into a single department with four (or more) sub-departments. Each sub-department is responsible for a different aspect of station operations and maintenance and all report to a broadcast operations and engineering (BO&E) manager. This arrangement may improve coordination of the operational and engineering functions of the station and is often the most efficient and cost-effective arrangement. Clearly, the BO&E manager should have a good understanding of the technical aspects of all the areas and how they tie together.

Broadcast facilities today consist of many sophisticated systems, including heating and air conditioning,

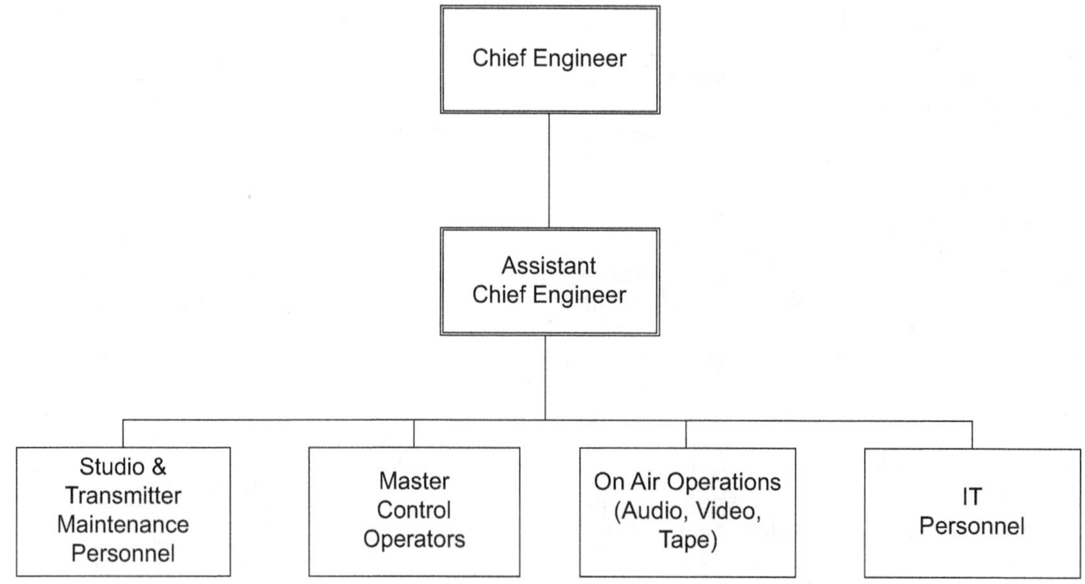

FIGURE 3.1-2 Typical television station engineering organization

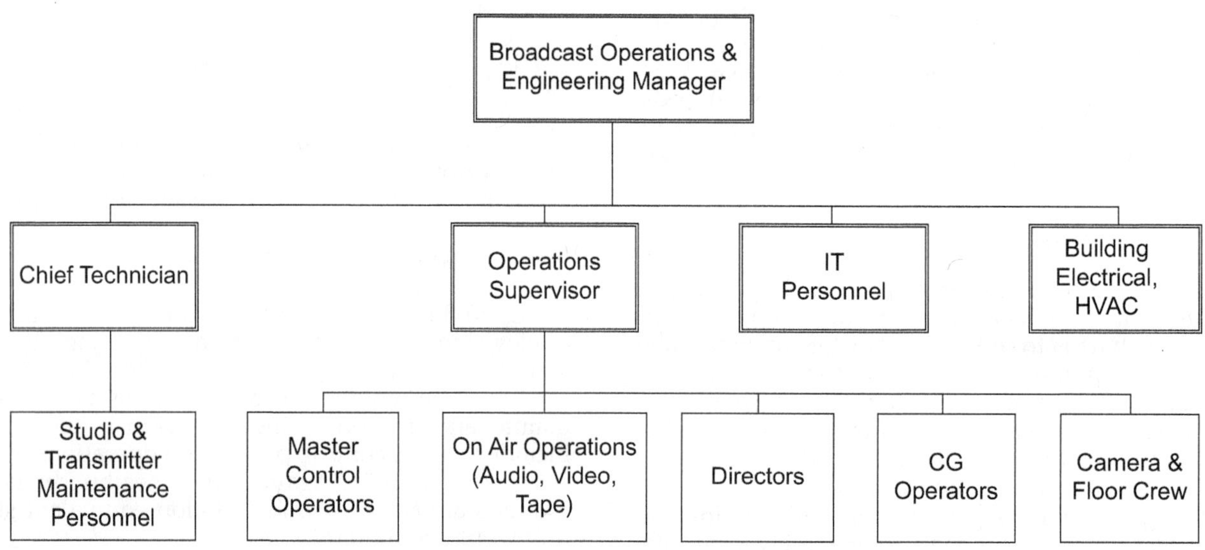

FIGURE 3.1-3 Joint operations and engineering organization

telephone, and security, as well as the broadcast systems for newsroom, newscast automation, master control, remote control, transmitters, traffic, and many more. In modern plants, most such systems rely on computer technology and IT networks, and many of them have to communicate with each other. A structure that combines responsibility for all these systems under one department has many advantages.

MANAGING STATION CLUSTERS

Over the past several years, many stations have been consolidated into what have become known as station clusters. A cluster is a group of stations all managed by the same management team with shared responsibilities and tasks. Large radio group owners such as Cumulus and Citadel have found that forming station clusters can save the station money by consolidating some functions of the station. One general manager may oversee all stations in the group. One general sales manager may oversee all sales team members. One news director may oversee all news activities. Similarly, one chief engineer may be responsible for providing all engineering functions for all stations.

Some clusters have reduced the total number of employees, including engineers, as a way of reducing costs. What has happened in some cases, though, is that station clusters have been left with an inadequate number of engineers to support all stations in the cluster. This can lead to more overtime and increased costs. One strategy some chief engineers are using is to use freelance engineers in the community.

Freelance engineers are independent contractors and former station engineers. They are available on an as-needed basis by stations and clusters, and typically charge by the hour. The advantage to stations and clusters is that independent contractors and engineers are not on the payroll of the station and do not receive benefits. Stations only pay for the time they actually use.

Job Descriptions

Whatever the particular organizational structure, it is important to prepare and maintain detailed job descriptions for each position. A job description should include the job title, specific duties, and desired qualifications. This is useful both for staff recruitment and as a guide for the employee on what is expected from him.

MANAGING PEOPLE

Recruitment

One of the biggest problems facing general managers and chief engineers is finding and hiring qualified engineers for vacant positions. Few technical or vocational schools today are adequately preparing people for broadcast engineering positions. Nevertheless, vacant positions must be filled. What is important is that chief engineers approach the process of filling positions in a positive manner, not just by finding someone who is readily available. The process should be approached as an opportunity to improve the overall quality of the unit and the station.

There are three main goals in filling key positions in the engineering department:

- Try to find someone who is better at the specific job than the person who is leaving. If you replace the person with someone who is not as good, your department's performance will likely decline.
- Aim to fill the vacant position with someone who is better than other employees on your staff performing similar duties. By bringing in someone from the outside that is better, you will cause others to work harder and, therefore, improve the quality of the unit and the station.
- If possible, hire someone better than the chief engineer is at the specific task needing to be done. Recognize that the chief engineer is a manager and cannot be an expert at every kind of engineering, particularly for the more specialized technologies in a modern broadcast facility.

The first place to look for qualified applicants should be at other stations within the same market. Broadcast engineering is a relatively small fraternity and most chief engineers will likely know who the good engineers are at local radio and television stations. By hiring someone away from your competition, you are helping your station and hurting your competition. A second strategy is to look for engineers who are employed at radio or television stations in smaller stations in adjacent markets. Most stations in larger cities can pay larger salaries than stations in smaller markets. It may be possible to recruit engineers from smaller cities to larger cities. The third strategy is to seek engineers from related fields. There may be engineers working for other companies who might want to learn broadcast engineering.

Job advertisements can be placed in the trade press or on the Internet, including the station's website. NAB [1] maintains a "Broadcaster Career Link" web page that has job postings for NAB member stations and also lists a large number of other broadcasting "Job Bank" websites that carry broadcast job advertisements. The most appropriate places to advertise should be selected based on the type of job and geographical area. Another possibility, particularly for larger organizations, is participation in job fairs organized by state broadcasting associations and at the annual NAB convention in Las Vegas, NV.

A more recent recruitment strategy is for stations to use social media as an inexpensive way to reach qualified applicants. Stations may use their own Facebook and Twitter accounts to post job openings. Many stations now have hundreds of thousands of Facebook Likes or Twitter followers. A more professional approach that stations are using is to use LinkedIn. Stations may list job openings, job descriptions, and applications. Job applicants are using LinkedIn as a way to network with companies seeking new employees. LinkedIn also provides a way to search for individuals with broadcast engineering credentials.

Training

Once a person is hired, many managers believe the task of filling the job position is complete—but it is actually often just beginning. Chief engineers must take responsibility for training the new hire to ensure that he has the knowledge and skills to do the job properly.

When new employees need to be trained, the chief should either do it himself or, where appropriate, delegate the responsibility to others with appropriate knowledge and experience.

There are three main levels of training required: orientation, on-the-job, and continuing education.

Orientation: For any new hire, orientation should take place over the first week of employment. Employees should not be just left to their own devices and "thrown to the wolves." They should be introduced to all employees in the engineering unit and taken on a tour of the station. The new engineer should meet key members of staff and managers, especially the general manager. Orientation should also include a detailed explanation of the specific job responsibilities and a review of the station policies and producers manual or station handbook.

On-the-Job Training: Many chief engineers will assign the on-the-job training function to an experienced engineer in the specific area in which the new employee will be working. This person will serve as a mentor for the new employee and show him how to do his job properly. Depending on the work and skills involved, this kind of training can take from one to two weeks to many months.

Continuing Education: The third type of training is continuing education. Technology is constantly changing and nearly all engineering staff can benefit from ongoing training. Stations should take advantage of specialized training offered by equipment manufacturers. Sometimes companies will send trainers to the station to do on-site training. Alternatively, engineers may be sent to the company for specialized training at the factory, with courses typically from one to three weeks in duration.

Chief engineers should encourage staff to attend relevant meetings, seminars, and training events arranged by professional organizations such as the NAB, the numerous state associations of broadcasters, the Society of Broadcast Engineers (SBE) [2], and the Society of Motion Picture and Television Engineers (SMPTE) [3]. Some of these take place at the local level in larger cities where SBE and SMPTE have local sections or chapters. Other events take place at a regional or national level and require a greater commitment of time and money for staff to attend but may be justified for key personnel.

Engineering staff should be encouraged to become SBE Certified Broadcast Engineers. Full details of qualification levels, training materials, and examinations may be found on the SBE website.

Chief engineers should also make a commitment to ongoing education for themselves to maintain their awareness of new technologies.

Motivation

Hiring and training are extremely important, but motivating engineering staff is equally important. Motivation can be accomplished essentially in two ways—intrinsic and extrinsic. Intrinsic motivation involves recognition and praise. Many managers fail to adequately recognize the contributions of their employees, but sometimes simple words of praise can go a long way in keeping people motivated. Managers should praise publicly and criticize privately.

Staff can also be motivated with extrinsic means, money in particular. Stations can use merit salary increases tied directly to performance evaluations as one effective way to motivate individuals. Those who exceed expectations should receive larger salary increases than those who only meet expectations. Those who do not meet expectations should not receive a salary increase.

For professional development, it is highly desirable for engineering staff to join and take part in the activities of organizations, such as the SBE and SMPTE, the Audio Engineering Society [4], and the IEEE Broadcast Technology Society [5]. Additional motivation for staff can be provided if stations agree to pay the annual subscriptions and cost of attending special events for one or more of these organizations.

Evaluation

Evaluating employees is the fourth personnel responsibility of managers. All employees should be evaluated regularly. Most stations do this at least once a year. The evaluation should be based on the employee's job description or goals statement. The evaluation should be a personal one-to-one meeting, and the manager should meet with each member of staff individually to go over the employee's performance. A summary should be sent following the meeting and placed in the employee's file.

COMMUNICATIONS

Communicating with employees is a major responsibility for chief engineers. There are three kinds of employee communication: downward, upward, and across the organization.

Downward Communication

Chief engineers are members of the station management team and must communicate policies and procedures from company management to employees in clear and understandable ways. There can be many misunderstandings if downward communication is not handled properly. Some stations have tried to communicate policies and procedures by having meetings or sending out memos or emails. What happens all too often is that some employees fail to understand what they are told or do not read what they are sent. What has worked in many organizations is a more formalized way of communicating to engineers from the management. The technique is the signed memorandum. Any policy or procedure change is sent individually to all employees asking for a signature from the employee that he has read the communication.

Upward Communication

Another form of communication within the unit is upward communication. Engineering employees may want to share their opinions with the station's top management. The chief engineer's job is to take their concerns and communicate them accurately to the management. Some stations will place suggestion boxes for employees, but few employees ever submit suggestions. Another approach sometimes used is the so-called open-door policy. Employees are encouraged to take their concerns directly to their manager. The problem with this strategy is that, again, few employees take advantage of the opportunity because they do not want to be perceived as whistle-blowers by their colleagues. An effective technique is known as MBWA or "management by walking around." The chief engineer should spend some time each day in each area of the operation just to see how things are going and to make himself available to employees who may want to communicate directly with him. The chief should develop effective listening skills and develop the ability to serve as an advocate for his staff.

Interdepartmental Communication

The third form of communication is communication across the organization. What happens in many broadcast stations is that each department functions as a separate entity. Programming does its own thing. Sales departments operate independently, and so forth. One of the most effective ways to break down walls separating departments is to institute a system of project teams. Project teams are made up of employees of all relevant departments involved in a specific project. For instance, the station may decide to do a live broadcast from an advertising client's place of business, and station management will appoint a team to see the project through to completion. The sales person who works with the client will be on the team, a promotions person will also be added, a programming/on-air person will be needed, a production person, and an engineering person. The project team will work together to plan and execute the event. The result will be that engineers will feel a more integral part of the overall station effort. The chief engineer does not need to be part of every special project team.

STATION OPERATIONS AND MAINTENANCE

The broadcast station relies on technical equipment, for which the engineering department is always responsible for installation, alignment, and maintenance. Responsibility for technical operations is less clear-cut, and operation of equipment may be a responsibility split between engineering and production or programming departments. For critical technical systems, staff operating and monitoring the equipment usually report to the engineering department. However, creative technical operators, such as video editors, camera operators, lighting directors, and others, may report to the production or news department. As discussed in the Organization Structure section, the structure with a BO&E manager avoids this split.

Operations

Many broadcast engineers see their role as a narrow one of installing and maintaining equipment. However, all engineering staff should be encouraged to see their role in the overall operation of the station. Chief engineers serve on the management team with a program director, news director, and sales manager, and other engineering staff may serve on project teams with employees from other station units. It is extremely important for the chief engineer to communicate the goals of the station clearly to his staff. Similarly, the chief must communicate the concerns of engineers to the management team.

Conflicts can occur when communication is not clear. One of the areas of most frequent conflict is between engineering and programming. Program departments may come up with ideas for productions or live remote broadcasts. It often seems to programming personnel that engineers most often throw up red flags or reasons why the station cannot or should not do something programming wants to do. However, the chief engineer and his staff should view their role as a service unit for the rest of the station. They must be perceived as problem solvers rather than obstructionists to progressive ideas presented by other units in the station. The engineering department should offer help and guidance to help the station achieve its goals.

Maintenance

Maintenance is one of the most important responsibilities for the chief engineer and his staff. Some stations rotate engineers across the studio, transmitter, and maintenance work. This approach may keep staff members from getting bored. However, other stations find that assigning engineers to a specific area is more productive in the long run, as maintenance skills become increasingly specialized. Assignments should generally be made by the chief engineer.

Techniques for maintaining broadcast equipment have changed as equipment has become more sophisticated, but there are still two basic categories: scheduled maintenance and troubleshooting.

Scheduled Maintenance

Scheduled maintenance can be predicted and planned for. It includes tasks as varied as cleaning filters for cooling systems, lubricating mechanical assemblies, cleaning and adjusting tape transports, and aligning signal levels. With computer-based systems, it also now includes functions such as regular data backup. Much of the regular maintenance has to be performed in-house, but in some cases it may be more efficient to send equipment back to the manufacturer for routine maintenance. Getting the maximum life out of existing equipment should be the goal of good engineering departments, and the chief engineer should establish schedules and routines for maintenance and ensure that they are carried out. Discrepancy reports should be filed when problems are discovered.

Troubleshooting

Modern broadcast equipment has greatly improved reliability compared to older technologies. While this reduces the chances of failure, it means that, when faults do occur, it is very likely that the maintenance engineer will not have seen the fault before. This potentially increases the time for faultfinding and repair. Good system design, with fault-tolerant systems, built-in redundancy, and automatic fail-over systems, can greatly reduce the likelihood that equipment failures will take the station off the air.

By their nature, most equipment failures are unpredictable. The chief engineer should analyze potential failure scenarios, establish routines to be used in each circumstance, and make sure that maintenance staff know what is expected of them. Traditionally, maintenance staff often did faultfinding to component level. That is usually not possible with modern equipment with high levels of circuit integration and multilayer printed circuit boards. These days, the best approach is often to replace modules or complete items of equipment with a spare, allowing the defective unit to be returned to the manufacturer for service.

Techniques for troubleshooting IT-based systems are fundamentally different from that for traditional equipment. This highlights the need for specialist staff with training in computer and network technologies and software-based systems.

Spares

Comprehensive sets of equipment spares are vital to avoid disruption to station operations from equipment failures and to allow maintenance staff to get systems back to full functionality as soon as possible. However, spare parts held in inventory and not used are a drain on station capital. The chief engineer should establish spares levels needed for equipment and systems based on the quantities of equipment used, manufacturers' recommendations, and his own experience.

Test Equipment and Tools

Considerable quantities of expensive test equipment and specialized tools are required for modern equipment and systems maintenance and alignment. The chief engineer should establish what is essential and what is desirable, again balancing the need to maintain and align equipment to the highest standards with the need to control costs.

Service Contracts

An increasing amount of broadcast equipment is sold with the expectation that the user will purchase a service contract. Service may be provided without such contracts, but typically with slower response and with greater cost for parts and labor. Particularly for software-based systems, service contracts can be a significant proportion of the original purchase price and may effectively be a mandatory purchase if prompt support and upgrades for critical systems are required. It may be possible to obtain more favorable terms for long-term support contracts, and the time to do this is usually when the initial equipment purchase contracts are being negotiated.

Support Systems

Software-based systems may be used to provide support for many aspects of asset management, including equipment management and maintenance. These may range from simple customized spreadsheets and databases using standard office software to more sophisticated purpose-built systems. An example of a system suitable for large organizations is the Maximo® Enterprise Suite from MRO Software (now part of IBM) [6] and, for mid-sized operations, the Datastream 7i® from Infor [7]. These provide a multitude of recordkeeping and automated support functions for tasks including:

- Equipment inventory
- Parts inventory and ordering management
- Maintenance schedules
- Trouble tickets
- Work orders
- Repair and maintenance logs
- Purchase orders
- Financial records

Such systems may include capabilities for staff scheduling and similar functions.

A smaller-scale integrated support package that was designed especially for broadcast stations was known as WinBCAM based on a relational database system called MainTrack from Integrated Tracking Systems. This is still used by many stations but is no longer a supported product.

INVENTORY MANAGEMENT

Part of the chief engineer's duties is to keep track of all broadcast equipment inventory at the station. As well as being useful records for the engineering department, most stations require there to be an annual inventory of all equipment for the annual financial report.

The inventory should identify the equipment and indicate where it is located, when it was purchased, and its acquisition cost. A simple inventory form might include categories as shown in Table 3.1-1.

TABLE 3.1-1
Sample Equipment Inventory Form

Inventory No.	Asset Type	Manufacturer/ Model	Serial Number	Location	Building/ Room	Date of Purchase	Cost

Preparing and maintaining this inventory can be a major task, and the software-based support systems mentioned in the previous section all include modules to help chief engineers keep track of their equipment. Such systems require staff to enter data into the system when equipment arrives and when changes are made. They may use techniques such as barcode technology to track all equipment including parts.

An alternative approach used by some stations is to outsource the inventory process to an inventory company that comes into the station and for a fee does a complete inventory of all equipment.

SAFETY AND SECURITY

One of the most important duties of chief engineers involves safety and security. The safety of all employees at a broadcast station is of paramount importance, and it is the responsibility of the chief engineer to ensure the physical plant is free of safety hazards for employees and visitors. Broadcast stations must comply with Occupational Safety and Health Administration (OSHA) standards for workplace safety and with FCC rules. Because the nature of the broadcasting business can easily result in dangerous situations arising from electrical and radio frequency (RF) sources, safety procedures should be clearly communicated, and training should be provided as required so that accidents can be prevented. Tower safety procedures are vital, and only trained engineers should be allowed in the transmitter and tower area. Drivers of electronic newsgathering (ENG) microwave or satellite newsgathering (SNG) vehicles should be thoroughly trained in safety procedures to prevent electrocutions and other accidents involving microwave masts and satellite antennas.

The security of the physical plant is equally important. Efforts must be made to provide secure facilities for all employees and for station operations. Many stations institute procedures allowing only employees and registered guests of the station into secure areas, including the studio, master control, and other critical areas. The transmitter and tower area must be secured so that no one can interfere with the broadcasts of the station.

For more on managing workplace and environmental hazards, RF hazards, electrical shock, and facility security and disaster planning, see Section 3—Technical Management and Safety.

LIAISON WITH THE FCC

The chief engineer should be the authority on all FCC matters pertaining to technical matters. He should be knowledgeable about all FCC rules and regulations affecting the technical operation of the station and make sure the station abides by the rules. He may utilize consultants to assist in the interpretation of rules and regulations. The chief engineer must keep the general manager and the station management team apprised of any possible violations or potential issues and should bring all new policy changes to their attention.

WORKING WITH UNIONS

Many broadcast stations have unionized engineering departments. Chief engineers are usually asked to represent the station in dealing with the engineering union. Chief engineers will be required to work with the shop steward and local union officials in negotiating the union contract with the station. For many chief engineers, this aspect of their job may be the most difficult. They may have been members of the union before becoming part of management. The advantage of having a union is that virtually all policies and procedures can be negotiated with the union rather than having to be agreed with individual employees. The disadvantage is that all engineers are often treated the same—not allowing for individual motivation and evaluation. The key is to maintain constant communication between the station and the union.

FINANCIAL MANAGEMENT

A key area of responsibility for the chief engineers is financial management. This involves planning, budgeting, and controlling costs.

Budgets

As mentioned earlier, planning and budgeting work hand-in-hand. Budgeting for broadcast stations may be done in a variety of ways. Some stations follow a prior-year basis. What this means is that stations look at expenses and revenues for the previous year in making projections for the coming year. Changes may be based on the rate of inflation or an expectation of increased expenses in one category or another. One drawback with budgeting in this manner is that the process does not take into account unexpected expenses or expansion plans.

Another option for budgeting is the zero-based budget. This is based on the premise that all departments within broadcast stations begin the budgeting process with zero allocated for expenses. Each department head is expected to propose a plan for expenditures for the coming year based in part on goals and objectives as well as prior experience. Many zero-based budgets identify fixed- and variable-cost projections in their budgets. Fixed costs refer to recurring expenses including salaries, operating expenses, supplies, and other expenses. Variable expenses include possible nonrecurring expenses or one-time expenses to buy new equipment or to allow for a contingency budget in anticipation of unforeseen expenses.

The budget process begins approximately six months before the beginning of the fiscal year. Each unit is asked to set goals for the coming year. Programming and production will submit goals for both purchasing programs and producing local programs. Engineering will be asked to determine approximate costs for supporting these plans and will also develop its own goals for replacing equipment.

Once a station budget is developed, the management team will be asked to monitor its execution to make sure the station is operating profitably. It is, therefore, vital for the chief engineer to establish and maintain good financial records. The software-based support systems mentioned previously include modules to help keep track of engineering department expenditure.

Financial Statements

While chief engineers are not expected to become experts in financial analysis, it is important for them to understand the two basic documents that relate to accounting: the income statement and balance sheet. They should know how operations within the engineering department affect the line items on these documents. They can then contribute to and understand the significance of financial decisions being discussed in management team meetings.

Typical examples of these documents are included in Tables 3.1-2 and 3.1-3.

Income Statement

The income statement (also known as the profit-and-loss statement or statement of operations; see Table 3.1-2) is a summary of a company's profit or loss during any one given period of time, such as a month or a year. It lists all revenues during this given period, as well as the operating expenses.

The statement enables the management team to monitor the operating performance of the business to be determined over a period of time, compared with the previous year and goals for the current year. It should be able to show what areas of the business are over budget or under budget and identify specific items that are causing unexpected expenditures.

Balance Sheet

The balance sheet (see Table 3.1-3) is a snapshot of the business' financial condition at a specific moment in time, usually at the close of an accounting period. It is often used by stations when borrowing money or buying new equipment. It comprises a simple list of assets, liabilities, and owners' or stockholders' equity. At any given time, assets must equal liabilities plus owners' equity. An asset is anything the business owns that has monetary value. Liabilities are the claims of creditors against the assets of the business. The difference between assets and liabilities is known as net worth.

For input to this statement, the general manager will ask the chief engineer to make a list of all tangible assets, including equipment, vehicles, and buildings, to help determine the station's value. This inventory, and software-based systems that make it easier to prepare, is mentioned in a previous section.

Controlling Costs

General managers are constantly looking for ways to control all costs for operating the station. This is because they are usually evaluated on their ability to return a profit to the owners of the station. One approach to this strategy is to control costs so that the total costs for operating the station do not increase more than 4–5% per year. Stations also set goals to increase revenues from 8–10% per year. The goal of using this strategy is to increase the station's annual profit margin while building the value of the station over the long term.

Chief engineers will be asked to help with controlling costs. There are four main areas of the annual budget that they need to monitor carefully—personnel, operating expenses, equipment (or "capital") expenditures, and program production expenses.

Personnel

Personnel costs often amount to as much as 80% of the total budget for the station. It is extremely important, therefore, to have the right number of engineers on staff. It is equally important for those engineers to be fully occupied without being overworked. One personnel cost that can sometimes lead to conflicts with the general manager is overtime. Chief engineers must manage their units carefully to reduce or eliminate overtime wherever possible.

Another arrangement sometimes used for managing personnel costs—more often in radio than in television—is employment of freelance staff for some activities. By employing freelance engineering staff on an "as-needed" basis rather than full-time permanent members of staff, the station may save some of the costs of a fully staffed engineering department.

Operating Expenses

There are many operating expenses under the control of the chief engineer, such as replacement parts and supplies. Even utility costs incurred by the station may be included in the engineering budget. Chief engineers must monitor all procedures to make sure that resources are used efficiently and waste does not occur.

Equipment

This major area of expense is the most obvious. Equipment replacement may be necessary. The general manager may want to spend the least possible for replacement equipment, but the chief engineer should consider the life of the equipment before making a recommendation. It may be more cost effective to buy better equipment designed to last three to five years or more rather than to buy something that will likely wear out in one year. If it is at all possible to repair equipment to keep it operating, engineers should strive to do so.

Chief engineers need to become familiar with the concept of depreciation for equipment and plant.

TABLE 3.1-2
Profit-and-Loss Statement Form

INCOME STATEMENT						
As of __________ (Date)						
MONTH: __________				YEAR TO DATE:		
Actual	Budget	Last Year		Actual	Budget	Last Year
			OPERATING REVENUE			
			Local			
			National			
			Network			
			Other Broadcast			
______	______	______	Misc. Revenue	______	______	______
			Gross Revenues			
______	______	______	Less: Commissions	______	______	______
______	______	______	Net Revenue	______	______	______
			OPERATING EXPENSES			
			Technical			
			Programming			
			News			
			Sales and Traffic			
			Research			
			Advertising			
			General and Admin.			
______	______	______	Depreciation and Amortization	______	______	______
______	______	______	Total Operating Expenses	______	______	______
______	______	______	Operating Profit (Loss) before Taxes	______	______	______
______	______	______	Provision for Taxes	______	______	______
______	______	______	Net Income	______	______	______

Source: Reprinted courtesy of the Broadcast Cable Financial Management Association.

Accounting rules for tax purposes allow the cost of capital purchases to be written off as expenses against income over different numbers of years, depending on the type of asset. Periods may range from as short as 3 years, to 5 years, to 10 years or more.

One strategy some stations employ as a means of controlling equipment costs is through leasing equipment rather than purchasing it. This technique avoids large upfront expenditures while allowing the station to account for actual equipment costs as recurrent Balance Sheet Form expenses each year.

Program Production Expenses

The last main area of operating expense is in local production and news. Programming and news departments often want to produce live programs, newscasts, or cover major events. Each program or event must be budgeted within the budgeting guidelines of the station. If specific projects are going to be very costly from an engineering perspective, the management team must be made aware of it. It is often necessary for stations to cover hurricanes, tornadoes, or other breaking news. It should be made clear whether these costs are news or engineering expenses.

CAPITAL PROJECTS

The chief engineer will regularly be asked to present proposals to the management team for new equipment and other capital projects. Sometimes other departments request new equipment, such as camcorders, editing systems, and weather graphics for news. The engineering department will be asked for expert advice on the selection and purchasing of this equipment. On other occasions, engineers will request major system upgrades and items of equipment.

TABLE 3.1-3
Balance Sheet Form

BALANCE SHEET	
As of ___________ (Date)	
ASSETS	
Current Assets	
Cash	
Marketable Securities	
Accounts Receivable	
Program Rights—Current	
Prepaid Expenses and Deferred Charges	
Other Current Assets	
Total Current Assets	_____
Property, Plant, and Equipment at Cost	
Less: Accumulated Depreciation	
Program Rights—Long Term	
Other Noncurrent Assets	
Intangible Assets	
Total Assets	_____
LIABILITIES AND STOCKHOLDERS' EQUITY	
Current Liabilities	
Accounts Payable	
Notes Payable	
Accrued Expenses	
Income Taxes Payable	
Program Rights Payable	
Other Current Liabilities	
Total Current Liabilities	_____
Long-term Debt	
Deferred Income Taxes	
Program Rights—Payable Long Term	
Other Noncurrent Liabilities	
Total Liabilities	_____
Stockholders' Equity	
Capital Stock	
Paid-in Capital	
Retained Earnings	
Treasury Stock	
Total Stockholders' Equity	_____
Total Liabilities and Stockholders' Equity	_____

Source: Reprinted courtesy of the Broadcast Cable Financial Management Association.

It may be difficult for engineers to explain the rationale for a new transmitter, tower, amplifiers, and other technical equipment that non-engineers do not understand. Therefore, it is important for engineers to learn the "language" of the other managers when presenting proposals to purchase new equipment. Engineers should avoid using jargon in presentations. Simple, easy-to-understand language should be used for all non-engineers, without talking down to them. Present proposals with a rationale detailing long-term cost savings due to greater efficiency or effectiveness. Describe the proposal in terms of making the programming more appealing to more people. Program directors and sales managers understand the benefits of improving signal quality as a way to attract and keep more viewers or listeners.

Once authorization is given to buy new equipment, agreed procedures should be followed for procurement. Specifications for the desired new equipment should be drawn up and made available to all potential vendors. Bid requests may include quotations for installation, on-site and/or factory training for station engineers, and sometimes the cost of spare parts. Bids should be sought from three to five companies so that cost comparisons can be made. Bids should be carefully reviewed to ensure that all specifications are being met and, in general, chief engineers should go with the lowest compliant bid.

MANAGING THE MANAGER

Working with station general managers (GMs) can be a daunting task for many chief engineers, especially if they do not fully understand the missions and business of the station and its owners or the motivations of the GM. The more the chief engineer knows about the GM and what "drives" him, the more successful he is likely to be. GMs are interested in meeting the two missions of all licensed broadcast stations—to serve the public and to make a profit. The GM entrusts the service component to the programming, production, and news departments, but an important part of being able to serve the public is the engineering support provided to these other units.

Ratings

The GM is interested in achieving the highest ratings possible for all programs. Whatever you can do as the chief engineer to assist him in achieving higher ratings will be received in a positive light. Ways to assist in achieving higher ratings is to make sure that all equipment in the news department, including microwave and satellite equipment, is functioning at the highest level. Local television stations know that as the news department goes, so goes the station. The news director will need your help in providing him with the latest information and guidance for camcorders, editing systems, and ENG/SNG systems. The weather radar and graphics systems should also be maintained at the highest possible level.

Revenue and Costs

The second major goal of general managers is to return a profit to owners and investors. Chief engineers should do everything they can to assist the GM in controlling unnecessary costs. Engineers also need to assist sales departments in generating revenues for the station. Sales departments often want to produce live remote broadcasts for advertising clients or produce quality commercials. In both cases, station engineers need to be as cooperative as possible in assisting sales personnel in meeting the needs of advertising clients. There are various technologies that stations can use to increase revenue, in which engineers may provide leadership. Whether it is services using SCA channels for FM radio stations or opportunities such as data broadcasting for television, engineers can help with the technical requirements.

When chief engineers present budgets or proposals for capital expenditures, they should present their proposals within the framework of achieving higher ratings and generating greater revenues. Engineers need to speak in the language that general managers understand—ratings and revenues. A new transmitter or tower will be seen by management as a major expense unless it can see the benefits as measured in increased ratings and revenues. New and improved technology should improve the quality of the station's signal or extend the reach. In both cases, the station should increase its chances of attracting a larger audience. The more people who listen or watch the station, the higher the ratings; the higher the ratings, the greater the revenues.

SUMMARY

Chief engineers have a challenging position in broadcast stations, requiring careful preparation and training. Leading a unit of engineering staff requires skills above and beyond solving engineering problems. The chief engineer must see himself as a leader responsible for leading a staff and managing the resources of the station. How well he is able to balance the leadership and management functions of his position will determine not only his success but also the success of the station. Chief engineers should seek continuing education in management and leadership, available through professional organizations such as the NAB, the state broadcasting associations, the SBE, and elsewhere.

ACKNOWLEDGMENTS

Thanks are extended to Bob Williams, chief engineer at WATE-TV, Knoxville, TN, for providing the three organization charts and associated text on alternative engineering staffing structures. Thanks to Clyde Smith of Turner Broadcasting System, Atlanta, GA, and Ron Peters of WUSA-TV, Washington, DC, for information on software-based asset management support systems.

References

[1] National Association of Broadcasters, see "Broadcaster Career Link" at www.nab.org/careers/default.asp.

[2] Society of Broadcast Engineers, see information at www.sbe.org/.

[3] Society of Motion Picture and Television Engineers, see information at www.smpte.org/.

[4] Audio Engineering Society, see information at www.aes.org/.

[5] IEEE BTS, see information at http://bts.ieee.org/.

[6] IBM MRO Software, see information at www.ibm.com/software/tivoli.

[7] Infor Datastream, see information at http://go.infor.com/eam-demo2/.

CHAPTER

3.2

Broadcast Engineering Documentation Management

DAVID M. BADEN

Chief Technology Officer, Radio Free Asia
Washington, DC

INTRODUCTION

Documentation is a statement of pride and ownership. It is a tangible public declaration of competence and ability. This chapter covers various document creation, maintenance, and information management processes. It will outline basic "best practices" and hopefully inspire a more collaborative and comprehensive approach toward the broadcast facility's technical information management.

Even small broadcast facilities are a compilation of multiple complex components and systems. By utilizing the wide variety of design, CAD, and documentation tools available and by following best practices as outlined in this chapter and instituting a proactive document management process, all broadcast engineers should be capable of maintaining comprehensive facility documentation.

The necessity for documentation has increased with the complexity of broadcast facilities. With the new generations of digital components and systems, not only do hardware updates need to be constantly monitored and applied, but also firmware and software updates. Furthermore, there is an ever-expanding interconnectivity between systems and components. This interconnectivity is not only through the normal audio, video, and control interfaces but also through local area networks (LAN) and wide area networks (WAN) such as the Internet. These multiple connections are dynamic in nature and require constant upgrading. No longer are a few technical schematics of static systems adequate to document the complex and dynamic systems used in the modern broadcast facility.

Fortunately, over recent years, document creation and document management software have gone through a technical revolution. With a new generation of computer devices from smart phones to mobile computer pads, laptops, and workstations, all equipped with new applications supported by powerful processing power and Internet connectivity access, review and editing of multiple forms of document files can now be done from virtually any location. Real-time collaborative review and editing with strict version control is now the norm rather than the exception.

The proper management for the creation and constant updating of critical information in technical documentation is essential for preventing downtime or degradation of service in a broadcast facility. Regardless of the complexity of a broadcast facility, proper documentation will help to ensure a successful and efficient daily operation.

The technical documentation process starts in the design phase of broadcast facility construction (or upgrade) and is part of an unending continuous process. From construction to normal online operations to the daily routine maintenance of any facility, a comprehensive approach toward documentation is beneficial. The relevance of the information conveyed in a facility's technical documentation can only be maintained if it is:

- Comprehensive. In modern broadcast facilities, independent stand-alone systems have rapidly become a thing of the past. Even in the simplest radio facility with one air studio and a single production studio, all components are in some way a modular part of larger systems as a whole. This interconnectivity is needed to support the transfer of digital content over LAN-to-network or LAN-to-external network, for cloud-based services, from multiple creation points to file storage servers to management and control applications to traffic control for playout to air across multiple distribution platforms. In an environment where any small change can impact the facility as a whole, all information pertaining to these changes must be tracked and incorporated into the facility's documentation.

- Accurate. With the wide variety of components, computers, software, and systems that comprise the modern broadcast facility, the accuracy of all information is critical. Consider, for example, a networked broadcast component that is only compatible with a limited number of manufacturer-certified network switches. Failure to accurately note this compatibility could cause a system failure at startup or in the future during what should be a routine network upgrade. Inaccurate information can cause facility downtime; hours of troubleshooting can be avoided with proper documentation.
- Dynamic. In a broadcast facility, there will never be a point in time where the technology will be static. Computer and network equipment in particular is advancing at a rapid pace. With a short component life cycle comes a never-ending need to upgrade and patch operating systems, firmware, and application software. With rapid obsolescence of equipment and systems, there is an associated rapid obsolescence of documentation—documentation needs constant upgrading and revision if it is to be useful.
- Accessible. The most concise and well-structured documentation is worthless if it is not readily available to the engineers who need it, when they need it. Technical documentation should be online, searchable, and available to staff needing the information at all times.

CAD, design applications, word processor, spreadsheet, and database software once provided all the tools needed to enable broadcast engineers to effectively create and manage technical information. In the modern broadcast facility, the larger challenge facing today's engineer is the long-term management of a wide variety of electronic information resources, which is often dynamic and needs to be accessed on a frequent and collaborative basis.

ENDORSEMENT DISCLAIMER

Specific software applications, operating systems, and cloud-based services mentioned in this section should not be considered either an independent review or a product endorsement. For the sake of brevity, all documentation creation and management applications and platforms that are currently available could not be covered. Likewise, the omission of any applications or platforms should not be misconstrued to represent a non-endorsement.

INITIAL DOCUMENT CREATION FROM PLANNING TO THE FINISHED BUILD

The creation and maintenance of technical documentation is a cyclical process. The depth and complexity of the documentation will mature with the broadcast facility. The initial process for technical documentation mirrors the same life-cycle phases of the facility creation or facility-wide upgrade. These phases are conceptual, build, and as-built (Figure 3.2-1).

While the documentation process overall is simpler if it starts and grows with a broadcast facility, the process can be started in any phase, but it is important to keep in mind there is no such thing as too much documentation or starting the process too soon. Consider taking advantage of the new generation of collaborative software. Most applications now offer some

FIGURE 3.2-1 The three phases of documentation: conceptual, build, and as-built

variation of group review, comment, and mark-up features in either a secure closed in-house server-based system or a more open "cloud-hosted" platform that can be access from practically any computer or mobile device with Internet access.

Cloud services offer a third party hosted Internet-based solution for a range of ubiquitous on-demand access to a shared pool of high-availability configurable computing resources, such as servers, storage, applications, and services. Various cloud services and options offered by multiple vendors will be covered in more detail later in this chapter.

While this section covers open-source, proprietary, cloud-based, and subscription software, no judgment or opinion should be inferred over the superiority of one distribution model or product over another. This section discusses a small cross-section of the numerous tools that can be used for document creation and management. As with all tools, you should choose the best tool with which you are most familiar that helps you accomplish the task.

Conceptual Phase

The conceptual phase is defined as the point where a new broadcast facility, or a facility-wide upgrade, is a proposed project. This is the "what we want to do" stage of documentation. The conceptual phase is where an overall vision for the facility is conceived and work processes are defined. The vision for the facility should include how the facility will function as a whole, the production workflow is planned, and how the various systems will be integrated. The conceptual phase is the point of the project where the operational parameters of the new facility are stated.

Most importantly, the conceptual phase of any facility design is where budgets begin to finalize. As a draft facility workflow is thought out, consideration can be given to the type of equipment used in the various workflow points. Additionally, the TOC (total cost of ownership), which includes purchase cost, installation cost, lifetime maintenance cost, and expected obsolescence (replacement cost) of equipment, should be considered so that the most economic equipment choices can be made.

Documentation tools used for conceptual design can be as simple as a basic design or CAD program, a spreadsheet, and/or a word processor. Documentation at this phase generally consists of workflow diagrams, basic system block diagrams, manufacturers' supplied product data, and discussion papers geared toward sharing the vision of the new facility with the user community. It is important that flexible documentation tools are used, as the documentation created during the conceptual phase will serve as the starting platform for all the documentation that will follow.

Workflow diagrams are needed when designing an upgrade, as it is important to document how a facility currently operates and how any facility changes will impact these workflow processes. Only by documenting and comparing workflow processes can possible negative impacts on normal daily operation be discovered and designed out. For any change to be implemented, it must have an overall beneficial impact, because if not, then the need for change is negated.

Basic, initial system block diagrams are useful for primary overall systems planning. These can show what new systems are to be used, what existing systems and resources will be incorporated, and how these will all interface. In creating these initial block diagrams, design tools should be used that allow for systems to be easily interchanged as the designs are fleshed out and more suitable alternatives are discovered.

Not to be overlooked are the user "white papers," which can be invaluable. These are high-level descriptions of how users will ultimately work in the facility, for example, where each task will be performed, the tools for accomplishing these tasks, etc. The measure of success for any new system or upgrade is ultimately based on the end user's acceptance. This is usually subjective and based on perceptions of the benefits of any new system versus the old. Bringing the users in at the early stages of any project creates realistic expectations and promotes total ownership within an organization.

Whether the facility is a new physical build or a rebuild of an existing facility, it is during the conceptual phase that initial space use is decided. Where will the studios, satellite support room, master control, electric closets, and related facilities be located? During this phase, the broadcast engineers will find themselves working closely with architects or general contractors, or will become the ad-hoc design architect. This close working relationship increases the likelihood that these technical rooms are in the optimum location. If it can be avoided, technical space usage decisions should never be left solely to an architect who may want to favor aesthetics over practicality.

Regardless, if ultimately the facility will be installed by an outside integrator or by the in-house engineering staff, as much of the conceptual design should be done in-house as possible, since it is the in-house engineering staff that will ultimately have to live with and maintain all new systems. Additionally, internal control of the design process will ensure control over future expenditures.

For the conceptual phase of documentation, it is best to use either collaborative software or a platform that is easily accessible to multiple users either in-house or in a cloud solution. This will be the phase where the documentation will be changed rapidly by multiple engineers making comments, suggestions, and corrections.

It is also strongly recommended during the conceptual phase of the project to bring in as wide a participatory group as possible, which should include not only the engineering staff but system operators, production staff, and talent, especially in regards to any change in the facilities overall workflow. This has two advantages: (1) it allows a sense of ownership for the entire staff working at the broadcast facility, and (2) input from the actual users can sometime streamline and simplify the workflow process, with a real-life, rather than a written spec, perspective.

Build Phase

The build phase documentation starts with the maturation of the conceptual documentation. This documentation is now either a "ready to bid" package, ready to go out for competitive bid to system integration company/contractors, or the "how it is going to be done" documentation, if the work will be done by the in-house engineering staff. It is at this stage that all specifics have been laid out and documented as an overall blueprint for moving forward and starting the installation. Specifics include:

- Physical
  - New room construction
  - HVAC systems
  - Electrical upgrades
- Equipment and Systems
  - What equipment and systems will be used
  - Where it will reside
  - How it will be interfaced
  - Where it will be purchased
- Systems Integration
  - How systems and equipment integrate to function as a whole
  - Numbers and types of cables needed to connect the systems and cable routes

Regardless of whether the facility installation or upgrade is being done by in-house engineering staff or is outsourced partially or in its entirety, this documentation is critical; it must exist or it would be impossible to order the equipment, order various systems and all the basic hardware needed to complete the job, and accomplish the installation in an efficient manner.

The document tools that are used for the build phase should be as varied as necessary for the required project documentation. Consideration should be given to the use of an all-encompassing project management software; several offerings of which will be covered later in this chapter.

The use of databases, either stand-alone or embedded in a design program, should be considered or, at minimum, a master spreadsheet used for tracking multiple items, connections, cable numbers, and other facility specifications. Databases that are integrated with or linked to design software whenever available are recommended. The use of such capabilities within a design program eliminates a redundancy of effort. With many design programs, the drag-and-drop of a graphical representation of a piece of equipment will also update the associated database regarding how many extra connectors need to be ordered, how much additional rack space is required, and other relevant information.

Facility Standard

Early in the build phase of a broadcast facility, a facility standard should be determined, finalized, and documented. The facility standard is a statement of what standards and conventions, on multiple levels, will be used within the broadcast facility. This is a listing of the multiple choice "givens" in a facility that need to be decided on and, once decided on, will not change. It is important to clearly define, document, and adhere to these standards before any actual work starts in order to ensure facility-wide compatibility. Facility standards should include (but are not limited to):

- Grounding architecture
- Cable conventions (shielded cable, plenum cable, manufacturer, etc.)
- Wiring conventions (shields grounded, hot pin, etc.)
- Wiring details (terminations, cross connects, location of raceways, etc.)
- Facility signal-level standards
- Digital media format (sampling rates, bit rate, clocking, etc.)
- Electrical power (backup, location of breaker boxes, breaker types, etc.)
- Network type and topology
- Shutdown procedure
- Evacuation procedure
- Location of exits, alarms, and fire extinguishers
- Location and use of lockouts
- Firewall rules
- Virus protection rules
- Intrusion protection rules
- Allowed system access levels

Having a facility standard, especially at this stage of the documentation process, will help to ensure that any and all work (in progress or in the future) will be accomplished in a consistent manner. Further, understanding and having this information readily available for the engineering staff will make routine maintenance and troubleshooting more effective and efficient.

Facility Operation and Training

Not all documentation is created for the engineering staff. Documentation for the facility content creators and operators should also be included as a section of the build phase facility documentation. This documentation includes operational instructions on how to use all equipment, rooms, systems, and software and is the basis of training documentation for the production and broadcast staff.

Operational documentation designed for the actual facility users should include simplified tutorials on how the facility is operated as a whole and the proper workflow processes that the facility was designed to work under. User white papers/SOPs (standard operating procedures) should be created during the build phase consisting of manufacturer's product manuals that have been either augmented or simplified for actual operations and used as study material for staff training.

User manuals are generally already available from the manufacturers in their product manuals. These manufacturers' manuals should be incorporated into the facility's overall operations documentation whenever possible. Additionally, the engineering staff should create a corresponding quick reference guide for equipment and systems if it is not available from the manufacturer. A quick reference guide is usually a one-page sheet that is a brief step A to step B single-line narrative flowchart that describes minimal operations. Consider too that these quick reference guides could also be a series of "how to" videos. These can be produced simply using any one of a number of screen capture applications with a voice-over walk-through tailored to your facility's operation.

Optimally, operational and training documentation should be made accessible on a facility intranet. An intranet is a private internal IP network that is only accessible inside of the broadcast facility. The intranet serves as a platform for internal communication, collaboration, and other multiple internal organizational resources. Applications and services that can be hosted on a facilities intranet are covered in later sections of this chapter.

Training for both the engineering and production staff can never start too soon or be such a thing as too much. Training material should become a dynamic part of a broadcast facility's entire documentation package and should begin at the latest during the build phase.

As-built

As-built documentation is the updated and final build documentation that reflects the reality of what actually happened during the construction or upgrade of a facility and how it now really works. This is the "how it was done" stage of documentation. As-built documentation should consist of three main sections: physical, operations, and maintenance.

As the build phase documentation started with the maturation of the conceptual phase documentation, the as-built documentation is the final version of the build phase. During the actual build of the facility, the existing documentation requires constant modification, as the reality of how it was actually done is often at odds with how it was envisioned. Real-world details, such as the actual location of cable runs, cable numbers, and numerous converters included that may have been overlooked, have now also been included in the documentation.

The as-built documentation is the basis for the dynamic documentation that will be maintained and updated throughout the life of the broadcast facility, but it is important for the historical record to maintain a hard copy and/or an electronic snapshot of the facility in its original completed configuration. The ability to reference back to the facility's starting point and compare it to how the facility evolved through the years is helpful for any future facility redesign or upgrade planning. This will be critical especially in looking at the workflow, as how a broadcast facility is designed to work often mutates over the years. A master set of the as-built documentation should be preserved in access-controlled secure storage platform in the original file formats in which it was created.

With the as-built documentation becoming the starting point for a broadcast facility's current documentation, special attention should now be paid to setting up a documentation access system to preserve the integrity of the information. The level of access for all the sections of the as-built documentation should be set to "write" or "admin" privileges for a small trusted number of the staff who have primary daily responsibility for the systems/processes covered in the documentation. Read-only privileges should be set for the wider user population requiring access to help them perform their daily job function.

As the as-built set of documentation becomes the foundation for the broadcast facility's current dynamic documentation, careful consideration should be given to how the information will be parsed out and what sections of documentation will reside on which platform. The platforms must be specific to the target admin/user group and how they will be expected to access the documentation. The platform must also maintain some sort of trusted version control and be compatible with a facility document standard. Multiple platform choices are available, including content management systems (CMS), help desk systems, wikis, and project management systems. These platforms can be simple or complex and selected as dictated by the technical ability of the staff to set up and maintain them. Various platforms are available for a wide range of prices, from paid proprietary commercial software to shareware to free open-source software.

Facility Standards Revisited

An update of the facility standard should be included in the physical as-built section of the documentation. The revised facility standard should be updated to include all previously unknown critical information. This facility standard update should include such information as:

- How to disconnect utility power to service-line voltage wiring within equipment racks
- Where breakers are located, and how they are marked
- What equipment is on UPS power, generator, or utility power
- A list of telephone numbers for important contacts (e.g., power company, building systems maintenance, etc.)
- Where keys are located and how they are marked
- Telecommunications circuit designations, associated vendor contact information, and "D-mark" locations
- A facility shutdown plan
- An occupant emergency plan
- A facility recovery plan
- An organizational continuance of operation plan

All additional essential information should be updated into the facility standard. This information will greatly assist the engineering staff with routine operations, emergency situation response, basic maintenance, and facility troubleshooting. Having this information in one location that is easily accessible, preferably on the intranet, will accelerate troubleshooting and problem solving in the future and keep the technical staff all on the same page.

Inventory and Maintenance

A critical section of the as-built documentation will be the maintenance and inventory section. Entering and tagging new equipment and systems with property IDs as they arrive on-site for installation during the build phase of a broadcast facility will save an enormous amount of time, compared to starting the process later after the build. Even if your only option to start an equipment inventory is a simple spreadsheet program, start doing it. In most cases, data can easily be exported and imported between file formats. A spreadsheet can be saved to a simple format, such as CSV (comma-separated values), which is commonly used to transfer data between one database and spreadsheet format to another.

Each piece of equipment within a broadcast facility can potentially be modified, reconfigured, or removed from service. This is especially true for the digital broadcast facility, where routine equipment maintenance regularly includes software and firmware upgrades. System-wide records are especially relevant in a digital broadcast facility, where systems are often based on computers as a core platform. There are several overall common denominators in these systems that need upgrading or maintenance as a whole. Examples include the version of operating system (OS) installed on all servers and workstations and the patch level installed for the OS. System-wide records ensure overall compatibility and uniformity between the common pieces of individual equipment and components that comprise any given system as a whole and set the expected upgrade schedule based on known obsolescence. Whenever possible, maintenance schedules and records should be kept in the same document, database, or application in which the equipment inventory is maintained.

Along with maintenance records, a broadcast facility's inventory should include as much information as possible on a piece-by-piece basis, including purchase date, manufacturer, model, and vendor. This information is especially relevant when tracking vendor and manufacturer warranties and patch and upgrade information.

Facility Operations and Training

Operational and training documentation should now consist of white papers, SOPs, tutorials, quick reference guides, how-to videos, and manufacturer manuals. This documentation will exist in multiple file and media formats. Operational and training documentation should be widely available to all staff on the broadcast facility's intranet. The documentation can be served on-demand to the user base in multiple ways, by a simple public access directory structure on a file server, as a wiki, or as a web page managed by a CMS, to name a few hosting options.

What is important is to have operational and training documentation available to the staff. It will be accessed often for refresher courses, quick how-tos, and new staff training.

DOCUMENT MANAGEMENT

A well-implemented document management process is important to realizing the true potential of a broadcast facility's documentation in regards to long-term information access and retention (Figure 3.2-2). Document management entails more than organizing and archiving large numbers of documents. Document management facilitates collaboration, ensures standard compliance, and improves work processes overall. Access to critical information when needed empowers all staff members in a broadcast facility.

Document management can be implemented either by enforcing a well-defined document standard and work processes within an organization or by employing one of several document management applications available. Document management by adhering to strict facility document standards and work processes is proactive and requires the cooperation of all staff that creates, edits, and/or accesses the information. Automated document management applications, once set up, are typically designed to allow much of the document management activities to be passive and hidden from the staff.

Any document management process used, regardless of whether it is a fully automated program or the manual adherence to specific guidelines and work processes, should, at minimum:

- Be a process for the collaborative creation and maintenance of documents
- Prevent the accidental replacement of the latest version of a document
- Prevent the accidental (or purposeful) deletion of the latest version of a document by casual users

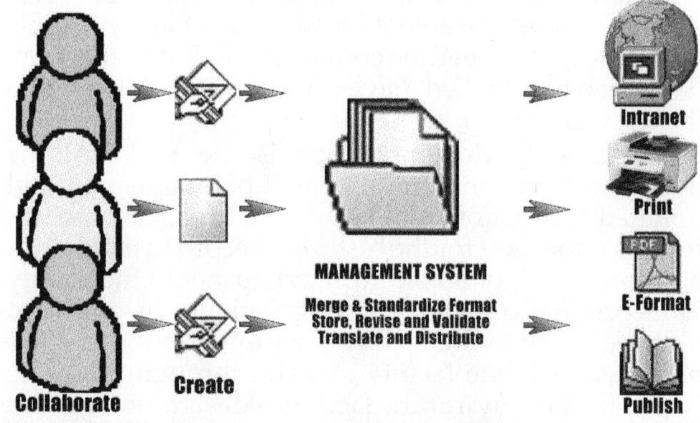

FIGURE 3.2-2 Technical documentation requires the intelligent management of the information, from the creation to the multiple-format distribution of documentation

- Provide a method to track changes to documents by date, time, and by who did what
- Provide a logical structure in which documents can easily be located

Document management, whether automated or manual, is simply the adherence to and enforcement of a set of rules and guidelines regarding how documents are created, what style they follow, what format they are saved in, where they are stored, who has access rights, and the privilege level of access rights. The failure of any document management system is usually not tied to a specific system methodology but is the failure of acceptance by the end users which results in improper or reluctant usage. Successful document management must create an environment of positive improvement in the work process that all users buy into in order to have long-term viability.

Document Standard

Defining the document standard is the first step required for any document management process. The document standard defines how documents are to be created, stored, and accessed, as well as what level of access is granted to what user groups.

As any platform that a document repository resides on always starts with individual files created in a variety of applications, it is important to make sure that files of the same type are created to the same standard. For example, in merging two text documents, if the settings such as font, font size, and numbering conventions are different, the merge will be sloppy at best and require an effort to clean up. The same is true with more complicated documents, such as CAD files, where conflicting layer names and properties can cause problems with a multiple-file merge.

A well-defined document standard that is adhered to by all users is in itself a document management process. Before a single file is created, a document standard should be established for a broadcast facility. A well-defined document standard, at minimum, should include the following.

File Type

Most software document and CAD tools have the capability of saving files in multiple file types or formats (e.g., dwg, dxf, doc, html, rtf, etc.). File types should be selected for each type of document being created based on the maximum compatibility of all documentation tools and CAD programs being used. For example, the standard for all CAD drawings could be to save as dxf (drawing exchange format) files. This would ensure that multiple CAD programs could access the CAD file. Proprietary file types that can only be opened by a single program should be avoided whenever possible.

File-Naming Conventions

The purpose of a file should be reflected in the file title. Therefore, a naming system needs to be standardized. A file name such as "drawing110.dxf" conveys nothing to a user browsing a list of files. In this case, the file would need to be opened in order to determine exactly what it was. On the other hand, a file name such as "STU1_RCK1.dxf," even if the user did not know the defined naming schema, could easily be read as a drawing of Rack 1 in Studio 1. In this naming schema, the first four characters in the file name are reserved to denote location, the fifth character is a "_" used as a space separator, and the next four characters designate the subject of the drawing. File-naming schemas can be as simple or complicated as needed but should be structured in a way that most users can, at a glance, be able to determine what information the file contains.

Layer-Naming Conventions

CAD and other design applications allow multiple layers in a single drawing; layers are provided to separate information within a drawing. Layers can then be controlled individually (e.g., turned on or off, plot or not, locked). Therefore, a schema for layer naming should also be established. The layer-naming schema, like the file-naming schema, should be structured to allow the average user to identify what information is contained in the layer at a glance.

Metadata

This is "informational data about other data." Metadata for specific files can be found in the document properties. Document properties are details about a file that help to identify it; typically this is metadata or information about a file that can be accessed without opening the file. These properties include a descriptive title, author, subject, creation date, modified date, and keywords to identify important information in the file. Often, document properties are overlooked and left blank. The document standard should define what is required to be entered into the document property information box. This information should include the original author, as well as later authors who edited the file and a notation of what these edits were.

Metadata is an often-overlooked option when creating or editing a file, but its usefulness cannot be overstated. This is especially the case if an individual file becomes orphaned from the main collection of documentation. Viewing the properties or metadata of a rogue file (if the information was entered correctly) can reveal if the file was separated from the repository for updating by a careless user, if it is now the current authoritative document compared to the repository, or if it is just a straggler that can be safely deleted.

Document Format and Styles

Format and styles for a document are internal definitions that describe the presentation of a document. These include the font type, font size, page size, page layout, margins, tabs, outline numbering conventions, paragraph justification, spacing, line weights (in CAD), line types, and other parameters. While this may seem trivial, the need for standardization becomes apparent when merging or linking documents into a consolidated documentation package. Without format and style standardization, multiple merged documents will appear to be disjointed rather than one cohesive document. In most programs, styles and format can be defined in templates that can be used by all users when creating a document.

File Storage Organization

File storage organization involves considering where document files are to be stored and how they are to be organized. It is critical, as files that become lost when stored in the wrong place might have well have never been created. Directories and folders on file servers are today's filing cabinets. Just as with a filing cabinet, a filing system needs to be established. A well-defined file organization structure will alleviate any confusion on where to store and subsequently retrieve files. File storage organization should also include how the original creation documents are organized and archived. Since multiple files are often used to create one single final published file, it is important to have ready access to all files that went into the creation processes to ensure that the document can be properly revised as needed.

Access Rights

Access rights determine who has the rights to create, edit, and view a document. Access needs to be defined in order to minimize potential data corruption or accidental deletion of information. Access rights can be set in various ways, either at the file level or at the server level. For the best practices, access rights should be set at all levels possible.

Common Master Index

The common master index is the overall index of a facility's total documentation and indicates where each file resides in hierarchical importance. A master index can be used when publishing the documentation, as it can be converted to a common-use file, such as html, and then linked to the various relevant associated files.

Database Links

Whether external or internal databases are used to synchronize data, a definition of these links and their relationship to each other needs to be defined. Defining these links early helps to establish the authoritative source database that all other database applications and linked applications use as a master and synchronize to.

Retention Policy

Will documents be deleted when no longer relevant, or are they retained as historical reference? How are transient documents that were used to create final documents treated? A retention policy defines how long documents are kept, how a document is determined to no longer be of use, and a deletion policy.

Version Control

Version control is the management of changes to documents, files, and other collections of information. A base master file is committed to a version-control application, and any change made to that file becomes a revised new file in the repository that is typically identified with a revision number, while the original file is preserved.

Version control systems can run as stand-alone applications, be cloud based, or be a component of various software applications. Version control allows for the ability to update a file while keeping the original, track changes, and importantly revert the file back to a previous version when needed. Several version-control systems also allow for the branching of a file into separate editing paths, comparing changes between the branches, and merging these branches back to a single file.

In a sense, the document management and standards policy, when properly adhered to, acts as a manual version-control system for a documentation repository. The document standard defines who is allowed write access and where files are stored and metadata change notations are to be made (i.e., in the property box). A version-control program also tracks similar changes to files (Figure 3.2-3). Version-control programs typically offer advanced automated features, such as allowing for file rollback to a previous version, the automated processing of file and directory comparison, and document merge. Version-control programs provide not only a recorded history of a file but also the ability to restore a file back to a point in that history.

Additionally, version-control programs assist collaboration within groups working on the same files by preventing the accidental overwriting of document changes. This is accomplished by preventing files from being modified by multiple users at the same time. Version control usually prevents overwrite conflicts, insulating the different editors from each other by requiring that individual editors work in a privately owned directory. The version-control program then coordinates the merging of all work back into the common repository directory.

Version control is not a substitute for a proper review and approval process. Version control ensures that the most currently available file version is foremost in the depository. Version control cannot assume responsibility for data integrity regarding accuracy or correctness. When groups of individuals are responsible for making multiple changes in documents, the areas of responsibilities for these changes still need to be communicated. Work, when completed, still needs to be reviewed whenever possible and files still need final approval before they can be moved to the next level of access, such as publishing for online access.

As well as managed version control for facility-wide documentation, in a modern broadcast facility version-control systems can be especially helpful in managing configuration files. Theses configuration files on which multiple broadcast systems now rely are often subject to change with hardware/software/firmware upgrades or changes in routine facility operational parameters.

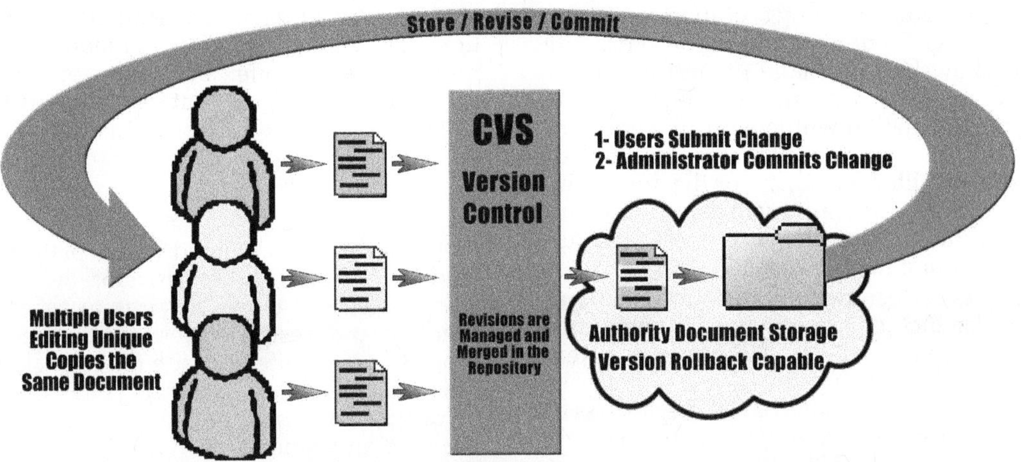

FIGURE 3.2-3 Version control for documents that are under constant revision by multiple users. Changes submitted, accepted, and published. Repeat or roll back.

There are multiple version-control systems in the open source, commercial, and hybrid realms. Open-source offerings include:

- Concurrent Versions System (CVS)—a client-server revision control system
- Subversion (SVN)—versioning control system inspired by CVS
- Vesta—versioning file system and support for distributed repositories
- Git—a distributed revision control system

Commercial proprietary version-control offerings include:

- CADES—Designer productivity and version-control system by International Computers Limited
- Synergy—version control by and for IBM Rational
- Team Project Visual and SourceSafe—version-control offerings from Microsoft

Hybrid offerings include systems such as GitHub, which is a web-based Git repository hosting service that offers free accounts and advanced features and expanded storage subscription accounts.

Trustworthiness

A well-defined document standard that is adhered to (regardless if it is an enforced process or a document management platform) and a reliable system for version control will provide trustworthiness for the broadcast facility's documentation. Information trustworthiness refers to the assumed reliability of the capture, maintenance, preservation, and presentation processes for information. Trustworthiness is important to ensure management accountability, operational continuity, and historical legacy for information.

It is also important, as with a facility standard, to make sure that the document standard is widely distributed, can be understood, and is easily accessed to be trusted. The user community should be informed that technical information within the facility is to be well managed and the maintenance of data integrity is foremost.

Collaboration

Special consideration for collaboration should also be factored into the document creation and management processes that a facility establishes. Collaboration makes critical information widely accessible, improves communication, shortens work cycles, and increases overall ownership and accountability.

If a document management program or platform is not being used, then collaboration is controlled by access rights and the file storage structure defined by the document standard. Various work folders can be created, and access controlled, by allowing write privileges to any particular work folder to only a limited number of people. On a more micro scale, write access in a folder can be defined on a file-by-file basis.

The trustworthiness of the file that was created by a joint collaborative effort can be maintained by defining who has final authority to approve a file or file set as complete and accurate. Once a file has gone through the approval process, it can then be moved into the public folder and be accessible with read-only privileges.

In the cases where collaborative work involves engineers from multiple organizations, online editing access via the Internet, usually through a cloud-based solution, should be considered. Online collaboration allows for people not physically located in the same facility to create, edit, or provide input to facility documentation. Several document tools and services are available that can provide online collaboration.

Online access options are included in many CAD programs, document creation tools, and document management programs as direct editing, markup, and/or commenting. Comments and markup provide a vehicle that allows for a change request and comments to be made on documents in progress while not altering the information within a document. Comments and markups can then either be accepted or rejected by the controlling authority for the document. Most of these applications are available in a cloud option, allowing the collaborative team access on multiple devices, including mobile.

Some cloud-based document and project collaboration platforms also include extended group communication features, such as web-based conferencing, text messaging, and document upload for collective editing that usually includes version control. Regardless of what type or platform of collaboration methodology is used, it should support the following goals:

- Support the sharing and flow of information and allow for real-time collaboration and communications.
- Manage a document creation through all the stages of its processing.
- Capture, organize, and distribute knowledge.

COLLABORATIVE APPLICATION AND PLATFORM OPTIONS

While several of the offerings listed below could be used as a single platform for all document management in a broadcast facility, consideration should be given to using a combination of application and/or platforms. The use of multiple applications, systems, or platforms allows collaborative work groups to manage and control their own document environment and then share their final work with a larger collaborative work group or the general staff.

Consider three editing workgroups that are tasked with keeping current a facility's online intranet training documentation. One group of engineers are tasked with updating and maintaining CAD drawings and are using A360 to coordinate their updates within their workgroup. A second group of engineers are tasked with updating the text and are using Dropbox to coordinate their editing efforts in their workgroup. The third group are tasked with updating the facility's video tutorials using Adobe Creative Cloud. All three workgroups can coordinate their activities with each other using Slack to communicate and meet deadlines. Using Slack, the tree groups in unison update the pages on the facility's intranet CMS, such as Drupal. The administrator in Drupal, seeing the pages being created or updated, can review the changes for accuracy and approve the publishing of the updates for the staff to access.

The following is a brief descriptive overview of a few options that are currently available to create a collaborative work environment. The listings are available in multiple distribution models—commercial proprietary, open source, web-based, cloud, or as local stand-alone installs. Listed are only a few of the available offerings, since all of them cannot possibly be covered, especially with new products and services constantly coming into existence. The brief descriptions also cannot list all of the features these collaborative platforms offer, but they serve to demonstrate the variety of choices that can be used to establish a collaborative work environment at a broadcast facility.

A360

Autodesk A360 is a cloud-based platform that offers paid subscription access to storage, a collaborative workspace, and cloud services to assist in the visualization and design process for a range of Autodesk products. A360 offers a wide range of cloud-based services, such as collaborative editing, web-based review, mark-up, cloud storage, and more. A360 can also be accessed and synchronized across mobile devices, and review and editing can be done with device-specific or web-based apps.

Creative Cloud

Adobe Creative Cloud is a paid software and cloud subscription service offered by Adobe. The Creative Cloud, depending on the subscription level, allows for Adobe applications to be downloaded, installed on a local computer, and used with updates as long as the subscription remains valid. The Creative Cloud also offers file storage and the ability to share files in a virtual workspace for collaboration. The Creative Cloud also offers mobile device access, as well as cross-device synchronization with device apps and on a web access basis.

SharePoint

Microsoft SharePoint is a commercial offering, used as either a stand-alone or a web application platform, that provides integration between several services that were previously run as separate applications. These services include intranet, extranet, document creation, document management, personal cloud, and workflow management. SharePoint can be deployed on internal servers or on a cloud platform using Office 365 and OneDrive. The Microsoft Office Suite is the primary document-processing application used with SharePoint and includes Word, Outlook, Excel, PowerPoint, Access, and Visio for word processing, e-mail, spreadsheet, presentation, database, and design, respectively. The Office applications can be installed locally or be cloud based with access by multiple devices, including mobile.

Google Drive and Apps

Google Drive and Google Apps are cloud-based file creation, storage, and synchronization collaborative workspace platforms provided by Google. Google Drive allows subscribers to store, share, and edit files in a collaborative workspace. Google Apps encompasses Google Docs, Google Sheets, Google Slides, Google Forms, and Google Drawings, respectively word processor, spreadsheet, presentation, survey, and diagramming programs. All Google Apps, with the exception of Google Forms, are available as Chrome browser apps for offline work. Mobile device versions are also available for Google Apps. The Google Apps platform integrates with the Google communication platform Hangouts, which provides voice, text, and video messaging and conferencing. The Google offerings are available as both a free service or a paid subscription with expanded storage and services.

Dropbox

Dropbox is cloud-based file storage and sharing platform that provides collaborative access to documents across multiple devices, including mobile devices. Dropbox is available as both a free and paid expanded storage subscription service. Dropbox is capable of interfacing with multiple third-party applications and platforms, such as Slack and Basecamp.

Slack

Slack is a cloud-based team collaboration platform. Slack features include persistent chat rooms organized by topic, file storage, private groups, and direct messaging. All files, conversations, and Slack members on the Slack platform are searchable. Slack integrates third-party services such as Google Drive, Dropbox, Box, and GitHub. It is available both as a free service and as a paid subscription with expanded storage and service features for large multiple teams or an organization-wide enterprises platform.

Basecamp

Basecamp is a web-based paid subscription collaborative project management platform. It offers to-do lists, wiki-style web-based text documents, milestone management, file sharing, time tracking, messaging, message boards, and comment threads. Basecamp's document and files storage provides a visual management organization of multiple file formats for team access.

Wiki

A wiki is web-server software that allows users to freely create and edit web page content using any web browser. A wiki supports hyperlinks and has simple text syntax for creating new pages and crosslinks between internal pages on the fly. It provides the organization of contributions to be edited with editor and version tracking. Wiki content, including all current and previous revisions, is stored in either a file system or a database. Wiki editing access levels can be assigned by rights and roles. Users can be denied access to view and/or edit given pages, depending upon their role in the content creation process. There are several wiki software offerings, both open source (such as MediaWiki, DokuWiki, and Tiki Wiki) and commercial (such as DrupalWiki, SharePoint Wiki Plus, and ThoughtFarmer). Several Internet wiki comparison websites, such as Wiki Matrix (www.wikimatrix.org/), can be referenced when considering the use of a wiki.

CMS

A CMS (content management system) is web-server software for the creation and modification of digital content for a website using a common user interface. A CMS supports multiple users working in a collaborative environment. CMS features include web-based publishing, format, or presentation management with editing history and version control, indexing, search, and retrieval. CMSs are designed to support the separation of content and presentation.

Generally consisting of two parts, the content management application (CMA) component of a CMS is the front-end user interface that allows editors to create, modify, and delete content from a website without the intervention of a webmaster. Content created and maintained by the CMA is typically stored in a database.

The content delivery application (CDA) component of a CMS compiles that information into a presentation format and updates the site. The CDA uses stylesheets or templates that interpret how the presentation should render and serve the site pages. The presentation or appearance of the content in a CMS can be redesigned by updating or creating new stylesheets or templates.

There are several open-source CMS programs available—Drupal, Joomla, and WordPress are three of the most popular, as well as multiple commercial offerings. A number of CMS comparison websites, such as CMS Matrix (http://cmsmatrix.org/), can be referenced when considering the use of a CMS in your facility.

PROJECT MANAGEMENT SOFTWARE

Project management applications assist with planning, organization, resource management, and schedule management. While not specifically document creation tools, when linked to communications and other collaborative platforms, the applications can assist with tracking tasks, resources, and schedules throughout the life of a broadcast facility. Project management software simplifies complex work processes by breaking the work down into smaller, more manageable specific tasks. Once these multiple tasks are defined, resources can then be assigned, schedules can be set, and dependencies to other tasks can be linked. Time is typically set by start dates, milestones, and completion dates that are also linked with tasks and resources.

It is this linking of tasks, resources, time, and scheduling that makes the use of project management software especially useful. Any change in one of these variables, such as a late equipment delivery, ripples through the master timeline, indicating how the one variable change affects all other dependencies. There are numerous stand-alone, network, and cloud-based project management applications available, and a comparison of project management software can be found on multiple websites, including Wikipedia (https://en.wikipedia.org/wiki/Comparison_of_project_management_software). The following is a brief descriptive overview of only a few project management options currently available.

Microsoft Project

Project by Microsoft is part of the MS Office family but not included in the Office suites. It is available currently in two editions. The software is designed to facilitate project management by assisting in the development of plans, defining resources, and assigning them to task, progress tracking, budget management, and analyzing workloads. Project offers the standard features of Gantt charts, network diagram, project diagram, resource histograms, and project schedule calendars.

Microsoft Project releases include Project Pro for Office 365, Project Professional, and Project Standard. Enterprise collaborative support is provided with Project Online and Project Server. For mobile access to collaborative project review and management, Project Lite is available. The online options for Project allow for a team collaborative workflow and are part of the Microsoft 365 cloud infrastructure. Project team communication and notifications are integrated with other Microsoft offerings, such as Outlook Mail, One Note, and Skype for real-time voice, video calling, conferencing, and text communications.

ProjectLibre

ProjectLibre is an open-source alternative to Microsoft Project. The application runs on Linux, Mac OS, and Windows workstations. ProjectLibre features include Microsoft Project compatibility, Gantt charts, network diagram, resource assignment, task assignment, task-resource-schedule linking, and resource histograms.

While ProjectLibre closely emulates Microsoft Project as far as interface and basic features offered, at the time of this writing, ProjectLibre is not available as a web-based application. Further, ProjectLibre does not have the robust collaborative real-time communications and scheduling interface options that Microsoft Project provides. As a result, ProjectLibre is more viable in smaller, less collaborative deployments and is not suited for larger enterprises' project management use.

Hosted Cloud-Based Project Management

Subscription cloud-based project management platforms generally include all the standard project management software features—planning development, resources assignment, task assignment, progress tracking, budget management, Gantt charts, and project schedule calendars. Being cloud based, these online project management platforms are multi-device accessible, including mobile, and typically have integrated real-time communications and notification features. Subscription cloud-based project management platforms include:

- Smartsheet—online project management and task collaboration that uses a spreadsheet-like interface, file sharing, Gantt charts, and work automation features.
- Wrike—project management and collaboration software that supports multiple collaborative teams and projects to enhance holistic workflow visibility and accelerates project delivery.
- Clarizen—project management that leverages a social collaboration and real-time visibility for projects, resources, and tasks.

DOCUMENT CREATION TOOLS

Document creation tools in the past often consisted of a drafting table, T-square, triangle, typewriter, and templates. Today's broadcast engineer can choose from a wide variety of computer platforms and documentation, database, graphic, and CAD software applications. These tools are available in stand-alone, server-based, and cloud-based platforms. Choosing the right document-creation tools can help to save an enormous amount of time and effort when creating and maintaining technical documentation.

COMPUTER-AIDED DESIGN TOOLS (CAD)

The benefit of computer-aided design (CAD) relies on its ability to organize, share, and reuse information. Most CAD programs offer connectivity to databases and are capable of creating robust database applications.

In the past, the selection of CAD options was somewhat limited. CAD software applications were both platform limited and costly. This situation has since changed with the increase in commonly available high-powered processor workstations. There are now dozens of CAD programs on the market, available for all workstation platforms and in all price ranges, including several free open-source solutions.

Current generation CAD programs now offer collaborative features, such as cloud-based file storage and synchronization of files, as well as version control, cross-device editing, review, and drawing comment and markup.

In a broadcast facility, CAD programs are used for a variety of purposes, including general architectural facility design, studio design, studio set and furniture design, cabling layout, and facility system layout, to name a few.

2D or 3D

When a facility is documented using CAD, consideration should be given to the use of 3D. While the initial drawing process will take longer using 3D, on the backend more accurate and flexible drawings will be realized.

For example, to draw a studio desk table in 2D, multiple views have to be drawn, including top, front, and isometric. With a 3D CAD drawing, the object only has to be drawn once. Once drawn, the 3D object can be represented from any angle and rendered into a photorealistic image. 3D CAD drawings also allow for ergonomic studies and a virtual walkthrough of facilities prior to actual construction.

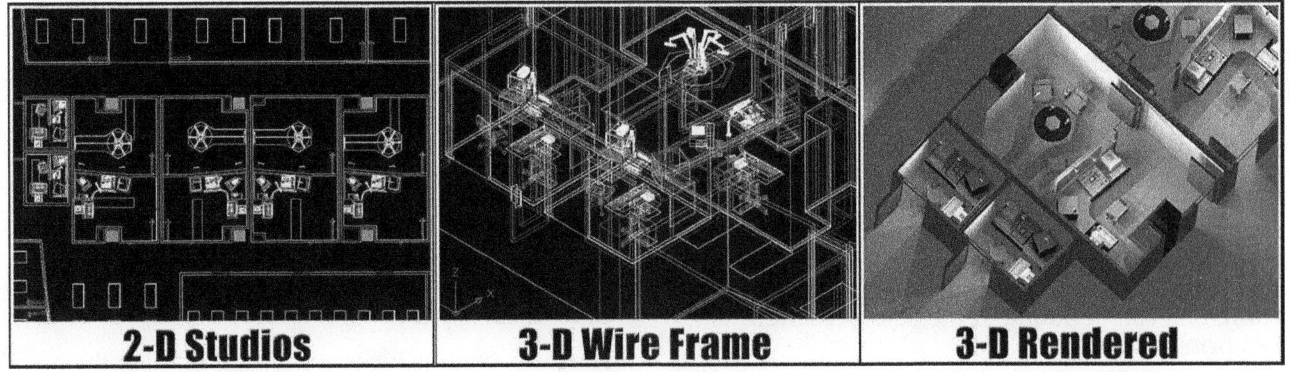

FIGURE 3.2-4 A studio shown as a 2D rendering, a 3D wireframe, and a 3D photo-realistic image

The 2D drawing in Figure 3.2-4 resembles any standard architectural plan and shows two small offices at left and four studio/control room combinations to the right of the offices. The studios can be recognized by the hexagon-shaped interview tables and the control rooms by the workstation tables with computers located in the corners. The same multi-studio facility drawing begins to take on depth and perspective as an isometric wireframe. Again, obvious in this drawing is the hexagon-shaped table (with boom microphones now more apparent) and the workstation areas in the control rooms. Finally, a photo-realistic 3D rendering of the same drawing is shown, completing the transition from flat plan conceptualization to virtual realization.

The most time-consuming task in migrating documentation to a 3D platform is creating the numerous drawings for objects that make up a facility. These drawings are the individual objects (e.g., broadcast equipment items and furnishings) that are inserted into multiple final drawings. The availability of preexisting 3D object drawings can save a broadcast engineer's time when migrating to a 3D platform. Therefore, broadcast equipment manufacturers should be asked if 3D CAD drawings of their products exist.

Once component 3D drawings are created or procured, the drawing creation process goes quicker, as a library is created that can be reused. In most CAD applications, multiple component or building base drawings can be inserted into new master drawings. The inclusion of existing drawings is typically done by the use of external references, drawing blocks, or layer overlays or simply cut and pasted from one drawing to another.

The decision to use 3D drawings does not negate the use of the 2D format, and most facilities will end up with a mix of 2D and 3D CAD drawings.

CAD programs

The following is a brief descriptive overview of only a few CAD programs that are available. The programs are available in multiple distribution models—commercial proprietary, open source, web-based, cloud-based, or local stand-alone installs. The brief descriptions also cannot list all of the features that these CAD programs offer, but they serve to demonstrate the variety of choices that can be considered when selecting a CAD program. A comparison of CAD software can be found on Wikipedia at https://en.wikipedia.org/wiki/Comparison_of_computer-aided_design_editors.

AutoCAD

AutoCAD has had the longest tenure in the CAD software market and is the most widely used CAD program. Most other CAD programs can either convert or import directly AutoCAD DWG-format drawings. AutoCAD has internal and external database support, as well as several additional tools to manage data objects and multiple drawing sets, such as sheet manager, block manager, and reference manager. Collaborative cloud-based A360 offers version control and coordinates workgroups on the Autodesk cloud platform, which was covered more extensively in a previous section of this chapter.

AutoCAD also provides compatibility to several programming scripts that allow third-party vendors and users to create custom programs and macro routines inside of AutoCAD. Scripting languages supported include Microsoft. NET, Visual Basic, Visual C++, Delphi, Java, and Visual LISP.

Included since the AutoCAD 2002 release are drawing standards checking tools. These tools ensure that established CAD drawing standards are being applied during the creation of a CAD file. Additional batch standard checking tools allow for comparing the established standard against the standard used in existing documents. This function runs externally to AutoCAD and allows multiple files to be checked against a standards file and logs any discrepancies into an XML-based report. More information on AutoCAD and all Autodesk products can be found at the Autodesk website at www.autodesk.com.

BricsCAD

Founded in 2002, BricsCAD is a founding member of the Open Design Alliance (ODA). The Open Design Alliance is a member-driven, nonprofit organization that promotes open industry-standard formats for CAD. First established in 1998, the ODA has over 1,000 members in 40 countries. The ODA's mission is to facilitate interoperability between CAD programs for the various drawing file formats and promote cooperation between CAD software providers.

CAD offerings include BricsCAD Pro and BricsCAD Platinum. BricsCAD Pro contains 3D direct modeling, 2D and 3D hardware libraries, and high-definition rendering. BricsCAD Platinum contains the additional features of 3D constraint system creation, surface entity creation and lofting, deformable modeling, imported geometry healing, automatic design intent recognition, and assemblies creation and editing.

Both BricsCAD Pro and Platinum have an AutoCAD ObjectARX-compatible development system that supports third-party application programs. Product information can be found at the BricsCAD website at www.bricsys.com/en_INTL.

SketchUp

A 3D modeling application rather than a true CAD program, SketchUp Pro has the ability to import and export many standard CAD format files, including dwg and dxf. SketchUp Pro is used for multiple-drawing applications, such as architectural, interior design, and civil and mechanical engineering. SketchUp includes drawing layout functionality, allows surface rendering in variable styles, supports third-party plug-ins, and enables placement of its models within Google Earth.

Between 2005 and 2012, Google-owned SketchUp released a free version named Google SketchUp, which had less functionality than SketchUp Pro. At this time, integrated tools for uploading content to Google Earth and to the Google 3D Warehouse were added to the program. Now owned by Trimble Navigation, SketchUp features an online open-source library of free models, such as windows, doors, and automobiles. In addition, the application includes an online community open-source library, the 3D Warehouse, where SketchUp users contribute and share 3D models. These models can be downloaded and imported directly into a SketchUp drawing.

For additional functionality when using SketchUp, hundreds of add-on tools are offered in the Extension Warehouse for download and local install. These extensions add extra functionality to SketchUp, such as drawing, editing, and rendering tools, as well as a range of automatic content generation for drawings.

A free version of SketchUp is still available for download and use. The paid version, SketchUp Pro, is more advanced than the free version and has more functionality. There are also SketchUp viewers available for both computer and mobile platforms. Information on SketchUp can be found at www.sketchup.com/.

IntelliCAD

While not specifically a CAD program, IntelliCAD is a CAD engine used as the core in numerous commercially available CAD programs. IntelliCAD is developed and maintained by the IntelliCAD Technology Consortium (ITC), a nonprofit organization owned and governed by its members. First released in 1998, IntelliCAD is not available as a stand-alone CAD application for purchase directly by the public. The IntelliCAD engine is only available for purchase in products marketed by ITC members, to whom it is exclusively licensed.

IntelliCAD's native file format is DWG and, therefore, when opening an AutoCAD-created DWG file, IntelliCAD does not perform a file conversion. The IntelliCAD engine is compatible with the AutoCAD command set, as well as programming APIs such as LISP.

At the time this was written, IntelliCAD 8.0 is the latest engine release available in both 64-bit and 32-bit versions. More information can be found about the consortium and IntelliCAD technology, including a listing of ITC members and their product offerings, at www.intellicad.org.

CAD File Types

The following file types are typical of those the broadcast engineer encounters when working with software documentation programs.

DWG

DWG is the native binary file type used for saving CAD vector graphics and is the default file format for the Autodesk AutoCAD program. Autodesk has controlled and updated the DWG format since its founding and has made programming libraries available for other applications to access the DWG format since 1994; as a result, the DWG format can be exported or imported into most CAD programs.

DXF

DXF, the drawing interchange format, is an ASCII or binary file format for CAD vector graphics. The DXF format is used for exporting and importing CAD drawings between CAD applications. ASCII-format DXF files can be read with a text editor and are compatible with a wide range of CAD programs and applications. Binary-format DXF files contain all of the information of an ASCII DXF file but are more compact. DXF files, while more portable between applications, are typically 2.5 to 3 times larger than DWG files. The DXF format was created by Autodesk as a human-readable format that could be used by programs needing to access CAD data.

DWF

Design web format (DWF) is an Autodesk format that provides a highly compressed file created from a DWG file. DWF is designed to be viewed on the Internet. DWF files can be displayed using Microsoft Internet Explorer 5.01 or later. DWF files support real-time panning and zooming and the display of layers and named views. Additionally, DWF files can be used for comment and markup distribution by using DWF specific tools.

ANSI

ANSI-standard format for digital representation and exchange of information between CAD/CAM systems includes the STEP and IGES formats. STEP (standard for the exchange of product data) is an ISO standard data representation and exchange format. The STEP file structure is a "class" modular structure that is easy to adapt for specific CAD uses. IGES (initial graphics exchange specification) is the most popular neutral file format for the transfer of 3D files. A neutral file format is nonproprietary and can usually allow for basic data to be imported by a wide variety of applications (i.e., the TXT text format). A disadvantage of neutral format translation is that it lacks interoperability. Native CAD files include a wide range of data types, not just CAD vector data. STEP is more extensible and as such has the ability to deal with more varied data types and applications than IGES. However, the ever-expanding scope of data availability in native CAD formats requires a high cost for these formats to be continually developed and supported.

DGN

DGN is a CAD file format used by Bentley Systems' MicroStation and Intergraph's interactive graphics design system (IGDS) CAD programs. The DGN format is used for large projects such as buildings, bridges, and manufacturing plants. In 2008, Autodesk and Bentley exchanged software libraries to improve their applications' ability to exchange data and read/write the DWG and DGN formats in mixed environments.

DATA VISUALIZATION TOOLS

Data visualization programs are capable of synchronizing directly with data sources to provide up-to-date diagrams, usually in the form of flowcharts or network diagrams. While appearing to be 2D stencil-based programs, they often contain powerful database interfaces capable of updating or, in some cases, generating complex drawings from information contained in the database. 2D stencils generally represent various manufacturers' equipment, and specifications on the product are integrated into a database.

The following is a brief descriptive overview of only a few available data visualization programs. The brief descriptions cannot list all of the features that these data visualization programs offer, but they serve to demonstrate the variety of choices that can be considered when selecting a program.

VISIO

While primarily thought of as a 2D stencil-based drag-and-drop program, Visio is a diagramming program capable of organizing complex technical diagrams. Visio can automate data visualization by synchronizing directly with data sources to provide up-to-date diagrams. As a Microsoft product, Visio is part of the Office Suite and can integrate business processes and systems by extracting data from Visio diagrams and importing or exporting them into Microsoft Office Access, Excel, and Word, and Microsoft SQL Server. Visio is also capable of opening and exporting to native AutoCAD DWG files. Additional information on Visio can be found at www.microsoft.com/office/visio.

LibreOffice Draw

LibreOffice Draw is an open-source versatile drawing, flowchart, and diagram program that is part of the LibreOffice Suite, with releases available for the Microsoft Windows, OS X, and Linux operating systems. LibreOffice Draw, along with the rest of the LibreOffice Suite, is maintained and developed by the nonprofit organization The Document Foundation (TDF).

Draw is a vector graphics editor and diagramming tool similar to Microsoft Visio and provides publishing features similar to Microsoft Publisher. LibreOffice Draw is used for creating drawings of building floorplan layouts, flowcharts, organization charts, network diagrams, and various technical drawings. Draw can directly open and edit multiple file formats, such as Microsoft's Visio vsd, CAD file format dxf, Adobe Photoshop psd, and portable document format pdf files. Additional information about LibreOffice Draw can be found at www.libreoffice.org/discover/draw/.

PROJECT DESIGN APPLICATIONS

Project design applications are software tools that interface with one or more programs, such as AutoCAD, and provide common database access control to manage document file sets that together comprise a project. The following is a brief descriptive overview of only a couple of project design applications available. The brief descriptions cannot list all of the features that these project design applications offer, but they serve to demonstrate some of the functionality of these applications.

VidCAD

VidCAD is specific to the broadcast industry. VidCAD tools were designed to allow for more efficiency concerning time and resources for the design and documentation aspects of broadcast facility planning and installation through automation. VidCAD products natively interface with various CAD engines such as AutoCAD and BricsCAD. Future compatibility with Microsoft Visio is planned in the near future.

With an integrated equipment product database, diagram generation, report generation, and wire-routing capabilities, VidCAD helps streamline the technical documentation process. VidCAD also allows for collaboration with global editing access and synchronous automatic updating of data and drawing files.

The newest release of the VidCAD product, named "Phoenix," is a rewrite of the original code base. A few of the VidCAD Phoenix release features include drawing automation through customizable templates, the ability to automatically import equipment and cable information into drawings, reports, and labels, and more efficient team collaboration. More information on VidCAD can be found on the company's website at http://vidcad.com/.

D-Tools System Integrator

D-Tools System Integrator, another extension and augmentation tool, is a design solution capable of interfacing with both AutoCAD and Visio. D-Tools System Integrator ties together the processes of design, engineering, estimating, documentation, and installation into a single portable project file. Using both Microsoft Visio and AutoCAD for the graphic interface allows for the use of precise drawing tools to automatically create elevation, flow diagrams, and line views from database information. D-Tools System Integrator dynamically links product information stored in a D-Tools-maintained manufacturer database to documents and drawings to create a dynamically linked project file. Additional information about D-Tools can be found at http://d-tools.com/.

GRAPHIC IMAGE EDITING

Graphic images can be separated into two basic types: vector or raster. Understanding the difference between vector and raster images and when it is best to use one or the other can greatly enhance the appearance, portability, and ultimately the readability of facility documentation. A raster image inserted improperly into a document file can at times be illegible if resized to fit using the wrong resolution.

All CAD programs utilize vector graphics. Vector graphics store information about a drawing as a data set of described shapes that make up an image, including their attributes and their coordinates in a drawing. Vector drawings do not have a set resolution and will display and/or print at the highest resolution the output device supports.

Raster images, also referred to as bitmap images, have a fixed resolution, as they are a stored collection of pixels. A computer screen generally displays 72 pixels per inch; printers operate on average anywhere from 300 to 2,400 pixels per inch. Raster images describe a resolution grid and how the grid is laid out.

Vector-based images, being resolution independent, have an advantage when being inserted into a document—the resolution can be resized to fit without the loss of image quality. The disadvantage of vector images is that most photographs cannot be described in vector format. Continuous-tone images, such as photographs, are more efficiently described as raster images. Further, for web use, there is limited support for vector images.

Drawing programs such as Adobe Illustrator and Freehand produce vector-based images. As these are more creative drawing programs, they will not be covered in the chapter. Vector images that are used predominately in a broadcast facility's documentation process are CAD and data visualization programs.

Raster image file types include tif, jpg, gif, bmp, and png. The difference between raster file types is typically in the amount of compression that they are capable of supporting.

Compression is used to make the file size smaller (trading off image quality for file size) for use in limited bandwidth applications, such as displaying multiple images on the Internet. Bmp and tif files use very little compression and are typically large files compared to compressed bitmap formats such a jpg, gif, or png. Some differences in quality as a function of file type are shown in Figure 3.2-5. Note in particular the definition and sharpness of edged surfaces and the smoothness of color and shading transitions.

Raster image editing programs are used in a broadcast facility's document process primarily for editing photo images for inclusion into documents. Raster images are generally imported into other programs for inclusion in the documentation.

The following is a brief descriptive overview of only a few available graphic editing programs. The brief descriptions cannot list all of the features that these graphic editing programs offer, but they serve to demonstrate the variety of choices that can be considered when selecting a graphic editing program.

Photoshop

Adobe Photoshop is a raster graphics editor for the Windows and OS X operating systems. Photoshop creates and edits raster images in multiple layers and supports masks, alpha compositing, and several color models including RGB, CMYK, LAB color space, spot color, and duotone. Photoshop imports and exports multiple graphic file formats. PSD is the native file format for Photoshop and preserves layer, effects, and all other application-specific custom settings. Photoshop has multiple standard editing and text tools, and

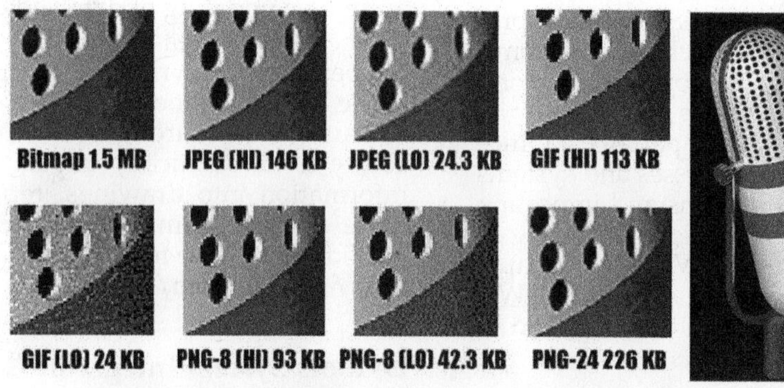

FIGURE 3.2-5 Graphics file type compression examples

functionality can be expanded by installing third-party plug-ins.

The Adobe Photoshop family of applications includes Photoshop Elements (a reduced feature version), Photoshop Lightroom, Photoshop Express, and Photoshop Touch. Photoshop is available on a subscription basis through the Adobe Creative Cloud. The Creative Cloud allows the download and use of the application, cloud storage, and a collaborative shared work environment.

GIMP

The GNU Image Manipulation Program, GIMP, is an open-source raster graphics editor available for free use for Linux, OS X, and Microsoft Windows operating systems. GIMP is a Photoshop-like application with many similar standard editing and text tool features. GIMP also can manipulate raster images in multiple layers and supports masks and alpha compositing. GIMP's basic functionality can be expanded on with numerous additional third-party editing, text, and effect tools that can be installed as plug-ins or add-ons. GIMP imports and exports multiple graphic file formats, including Photoshop psd files and portable document format (pdf) files. The native file format for the GIMP application is xcf.

PaintShop

Corel's PaintShop Pro is a raster and vector graphics editor for the Microsoft Windows operating system. The program is available in two versions—PaintShop Pro, the basic editing program, and PaintShop Pro Ultimate, which includes additional stand-alone programs and extended plugins. PaintShop, like Photoshop and GIMP, supports multiple layers, CMYK, and HSL color modes and similar editing and text tools. PaintShop imports and exports multiple graphic file formats, and cpp is the native file format for PaintShop.

OPERATIONAL SUPPORT TOOLS

Operational support tools, while not specifically document creation tools, often contain relevant data that can be harvested for the overall documentation package. Operational support tools can also drive the need to create additional documentation. Consider that in a facility's workflow, several independent systems interact as a whole. Each system performs a separate function in the production workflow, from ingest to production to scheduling to actual air and back to archive, including online accessibility of broadcast. As most of these systems are now software based, a version upgrade to one system often requires additional training and therefore drives the need to revise SOPs, tutorials, and other training resources.

Because of this ever-changing workflow environment, it is imperative that all operational documentation be constantly updated. Large facilities often employ a dedicated training staff tasked with keeping the training material up to date and ensuring that the broadcast and production staff receives additional training when needed. In smaller facilities, updated tutorials should be made available on the organization's intranet, available for staff to access at their convenience.

The following subsections briefly cover a few of the operational support tools that should be considered for use in a broadcast facility.

Performance Monitoring

Performance-monitoring tools (stand-alone or as part of a facility's overall quality assurance program) should not be overlooked as an essential part of a broadcast facility's operation and documentation. With performance monitoring, small problems within a facility can be identified in systems or equipment before they become major problems. Major problems generally tend to be noticeable, as they can be system stoppers or adversely affect other systems (and perhaps become harder to trace down). If performance can be shown to have measurably degraded, there must be a reason; it is difficult to establish performance degradation if past performance is not documented.

Several performance measurement tools help document baseline technical systems and equipment. These tools are available from many manufacturers, such as HP, Cisco, and Dell. There is also a large offering of performance measurement tools, both open source and commercially available, from manufacturer-independent sources. Performance measurement tools can be applied to track workstations, servers, storage, network performance, and printers. These are all components and systems that comprise the backbone of a broadcast facility.

Where applicable, these tools should be dynamic monitors that establish a documented baseline and then constantly monitor the systems or equipment. Dynamic monitoring is the real-time comparison of current performance against initial baselines. Real-time alert notifications can then be received when a system's performance starts to degrade. As a result, these problem systems can be identified and corrective measures taken with near real-time reaction.

Documented baselines can also be used for planning replacements and upgrades, since these baselines can be compared to the performance levels of new systems and equipment when they become available. For example, a parameter can be established such that when an online system performs at 50% below the newest available systems version, the old system will be upgraded or replaced.

Finally, proper and objective performance documentation can help to eliminate user's complaints in regards to subjective perceived performance issues. For example, a common complaint for a digital editing station could be that "my workstation is running slower." This complaint can be objectively verified and corrected or dismissed if past performance measurements for that workstation have been documented and no performance degradation exists.

Help Desk Software

The help desk is a staff resource with the function to provide end user technical support. Help desk software is an application that addresses the end users' problems and queries and enables the automation of help desk support. At its most simple level, a help desk system provides a way for end users to log an issue or problem and track their request to its final resolution. Typically, a help request is made by e-mail to the system. The help request is logged into the system, where the problem is triaged and given a priority, generating a ticket and placing the ticket in a queue for action. Technicians pick up the ticket from the queue based on priority and resolve the issue. Throughout the whole process, from request to resolution, the progress is tracked and documented. The initial requestor either receives an e-mail or can check the real-time status on the progress being made to resolve the issue.

With a more sophisticated help desk program, the resolutions are tracked, and any commonalties are identified and relationships are linked. With this functionality, vital information compiled for repetitive problems is linked back to the cause and stored in a knowledge database. When the same problem occurs again, the solution would already be documented in the system.

Other features of a sophisticated help desk system would be the inclusion of a bug tracker and equipment inventories with maintenance databases that track updates and repair to equipment, all of which can be displayed in analytics reports.

A robust help desk system could also include a dynamic FAQ (frequently asked questions) or wiki-style information function that allows technicians to log, search, and review past resolutions for the problems being reported. Help desk software with this knowledge database functionality can consolidate a broadcast facility's online documentation by eliminating the need for separate technical wikis, inventory databases, and maintenance tracking programs.

Multiple offerings of help desk software are found under both commercial and open-source distribution. Many cloud-based, web-hosted offerings of help desk software are also now available that allow collaboration of the help desk function across multiple facilities. Comparison of help desk software can be found on multiple websites, including Wikipedia at https://en.wikipedia.org/wiki/Comparison_of_help_desk_issue_tracking_software.

The Use of Custom Databases

Several of the programs and applications covered previously have application-specific internal databases, as well as the ability to link to external multi-user database engines. The use of either internal or external databases provides expanded search capability for documents and information, as well as facilitates multiple document linking and updating.

Figure 3.2-6 shows an example of such a linked database as it appears on a computer screen (top) along with the data display (bottom left) and data entry (bottom right) screens. A data record, such as for a piece of equipment, can include several separate items of information about the equipment stored in data fields. Records can then be linked by either all or selected fields and then linked into tables for data display.

For example, a table could be defined that shows all equipment housed in a rack, to show the total power consumption for the rack or the total cost of the equipment. Data entry can also be allowed in individual records by individual fields or updated from defined tables that would simultaneously update several records.

A database is a collection of logically related information, usually presented in a table format similar to a spreadsheet. Databases are extremely flexible and can be modified by adding new fields or removing existing ones from the data tables. Likewise, one can add or delete records from database tables and edit existing records, as well as define relationships between multiple database tables. For example, if there are two database tables sharing a common field, such as cable numbers, a relational database can join these two database tables using their common field. This ability is what eliminates the need to maintain and input duplicate data across multiple database tables or linked files.

As a facility expands in size and complexity, the complexity of the documentation will also expand. The use of a common linked database can allow for the synchronization of data across multiple documents. For example, AutoCAD can exchange data with Microsoft Access, which in turn can synchronize with Microsoft Excel and Word. Therefore, when using Access database synchronization, a change in a CAD drawing can provide updates to the facility master database, which in turn can provide updates into an Excel spreadsheet or Word document. This eliminates the redundancy of effort, whereas in the past it was required to update the same information in different document files using separate applications.

Allowing applications to efficiently access data where it resides without replication, transformation, or conversion is referred to as universal data access. Universal data access allows open database interface connectivity among multiple data sources. Independent database services, such as Access or Oracle, provide for distributed queries, caching, updating, data marshaling, distributed transactions, and content indexing among these sources.

Data contained in an external database can be associated with graphical objects through the process of linking. These links are pointers to database tables that reference data from one or more records in that table. Linked database records to graphical objects in a drawing can provide powerful database queries to filter and sort information and represent the new results graphically.

The use of databases is especially important in facility documentation for tracking facility-common systems and interfaces, such as cabling origination points, destination points, and cable numbering or LAN network topography, which also includes cable

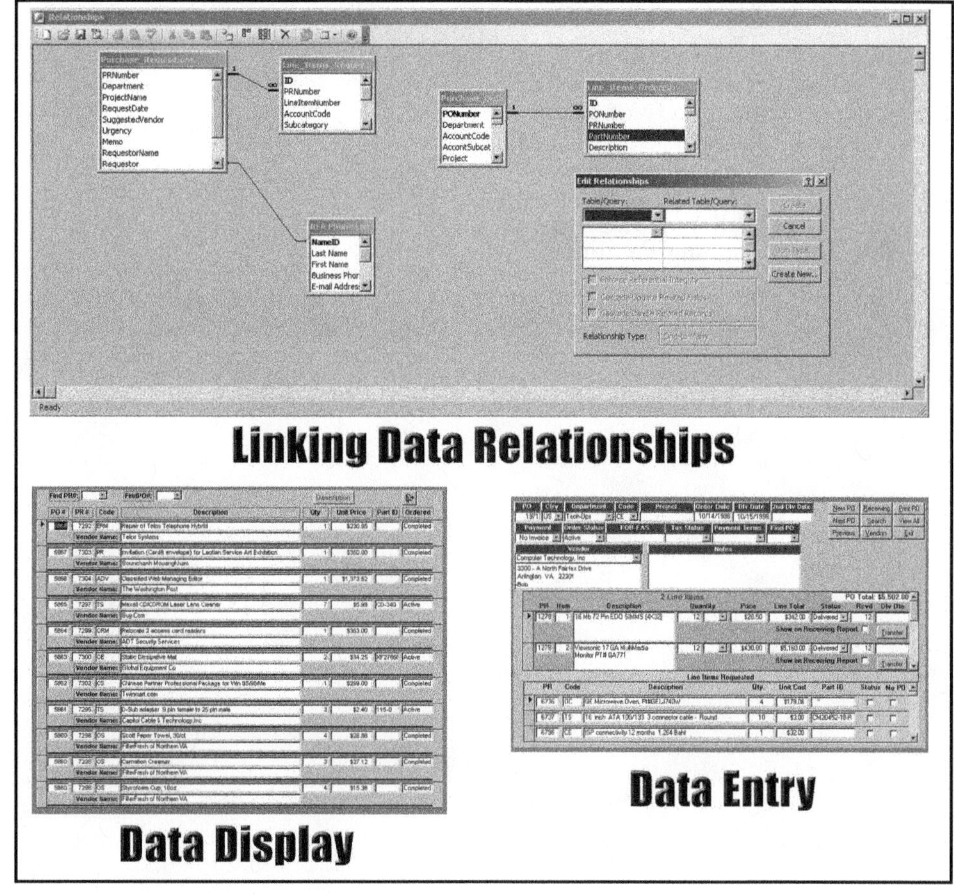

FIGURE 3.2-6 Database Functionality

numbering and routes. Database use can also be expanded to include equipment maintenance records, such as firmware and software updates.

There are multiple offerings of database applications under both commercial and open-source distribution. Comparison of database software can be found on multiple websites, including Wikipedia at https://en.wikipedia.org/wiki/Comparison_of_relational_database_management_systems.

Publishing

With online and cloud-based technology, actual paper publishing is rapidly vanishing. The primary function of many of the applications covered previously, such as a CMS, is to organize and display, or publish, information online..

When an actual paper hard copy is needed, the most popular publishing format is the pdf, or portable document format, for both printed and online presentation. Most programs used to create any type of document file should be capable of exporting to pdf.

The pdf file format preserves the fonts, images, and layout of source documents, regardless of which application or platform they were created on. They are relatively compact documents that are viewable with a wide range of free reader applications, including web browser plug-ins, making pdf ideal for electronic document distribution, such as e-mail.

For the broadcast engineer, pdf is one of the most common formats used by manufacturers to supply systems and equipment documentation. Fortunately, tools are available, such as Adobe Acrobat Pro, that allow the broadcast engineer to easily manipulate these manuals in order to incorporate them into a facility's documentation package. Multiple-source pdf files can be consolidated and reindexed into one common pdf file. Or, multiple manufacturers' source pdf files can have their indexes exported into a common new index file that is actively linked to a collection of document files.

No matter how pdf documents are consolidated, the reindexing of multiple manuals will enhance searching speed and efficiency. These electronic manuals can be organized in the same manner that a file cabinet would be (e.g., alphabetically, one pdf file: manuals A through E; or by location, one pdf file: Studio #1).

The pdf is simple to print when hard copies are needed, as are several of the applications covered previously in this chapter. A CMS, for example, always has special templates to print a page or section or export to pdf.

SUMMARY

In the modern broadcast facility, even small facilities are a compilation of multiple complex components and systems. By utilizing the wide variety of design, CAD, documentation creation tools, and cloud-based options available and by following best practices as outlined in this chapter and instituting a proactive document management process, all broadcast engineers should be capable of maintaining comprehensive facility documentation.

CHAPTER

3.3

System Integration and Project Management

STEPHEN F. PUMPLE, M Eng, MBA, PMP

AZTECH-ENGINTL

Toronto, Ontario, Canada

INTRODUCTION

The technologies employed in the generation and distribution of programs and content are in a constant state of flux and development. These developments can be incremental or disruptive. Although the technology continues to advance, major capital projects are expensive and hence major system upgrades do not happen on a regular basis.

When the decision is made to move forward with a major systems upgrade, many factors come into play. The most obvious requirement is the need for a team of skilled design engineers knowledgeable about the current systems, but who are also actively engaged with and designing systems around the latest technologies. That design, when finalized, is then implemented by a team of skilled installers and commissioning engineers. The important role of a strong and disciplined project management process is somewhat invisible, but it is essential for the project to be successful. The design, installation and integration of systems within the broadcast/new media environments require a unique combination of business, technical and people management skills.

This chapter captures, identifies and discusses the phases, which collectively comprise a structured project management approach to the processes and are integral to the overall design and systems integration processes. While the discussion and techniques center on larger projects and include individual processes which may be combined on smaller projects, the fundamentals of identifying the goals and objectives, timelines, assignment of responsibilities, acceptance criterion and budgets are applicable to the successful implementation of any project. The structure of the project management processes groups and associated knowledge areas, as defined in PMBOK (see References), are identified along with their role in the different phases of the systems integration process. This chapter does not attempt to delve into or address the details, considerations and complexities or the technologies the engineering/design team is required to wrestle with in developing or implementing a final system design. However, it does identify and discuss the various processes.

Also discussed is the role and contribution that external professionals and consultants such as architects, nonbroadcast engineers, subject matter experts and system integrators can have in the process.

BACKGROUND

When a project is first talked about or considered, the focus is often on the end goals or objectives, and insufficient time is taken to consider what is required to execute it successfully when/if the decision is made to move forward with the project. In the broadcast environment, the responsibility to implement technology-based projects has historically fallen to the chief engineer. Over the years, the role and responsibilities of the chief engineer has evolved along with the technologies, from that of a video or RF engineer to include the management of the IT infrastructure. These individuals, who in most instances are already fully occupied with keeping the existing systems up and running, are often tasked with and accept the responsibility of implementing the project without fully appreciating the implications attached to that commitment.

The broadcast business, or more aptly, the business of providing a platform for the delivery of content, whether it is program playback, a live newscast, the production of a program or documentary, or the delivery of the content via satellite, the Internet or conventional over-the-air (OTA) transmission, is a dynamic environment heavily dependent on an ever-changing technology landscape. Some of the changes are incremental and easily incorporated by a business without significantly impacting the current operations, while others are not. Examples of incremental changes can include updating the firmware or software of an individual piece of equipment. Examples of changes that have a broader impact are the transition from an analog to a digital infrastructure, operating system upgrades, and the more recent (2009) and disruptive transition from analog NTSC TV to ATSC 1.0/digital high-definition TV (HDTV).

There are constant technology changes on the horizon for the broadcasters, whether they are traditional OTA service providers, specialty channel providers such as HBO or other content delivery services such as Netflix. Some changes are incremental; others are far more significant and disruptive. Among those are the FCC's TV broadcast spectrum repack program, the upcoming adoption of ATSC 3.0 and ultra-high definition TV (UHDTV, also commonly referred to as "4K"). Other changes that need to be considered include the constantly evolving landscape presented by the advancement in delivering content, including over the top (OTT) and the Internet of Things (IoT).

The transition of the broadcast plant to an IP infrastructure is now underway. This infrastructure will serve as the backbone for the major technology changes on the horizon. As of this writing, the FCC-mandated "re-pack" program is underway. This program ultimately will have a dramatic impact on the operation of all OTA terrestrial broadcast operations in North America. This program is in its early stages and involves multiple stakeholders with different goals and objectives. However the FCC repack initiative unfolds, system designers/engineers will face substantial challenges.

Other pressing technologies that will be adopted in the near future include 4K/UHD high dynamic range (HDR) and 8K UHD. Experienced engineers with strong skills and disciplined project managers will be required to make the integration of these technologies and programs a success.

Broadcasters face many challenges other than changes in the delivery and production technologies. One major consideration is the rapidly evolving viewing habits of consumers. The transition from traditional linear broadcasting into on-demand, targeted channels is well underway, as is the transition to OTT delivery of audio, video and other media over the Internet without the involvement of a multiple system operator in the control or distribution of the content. The ways and the devices used to watch content is changing from the traditional living room TV to personal computers, tablets and smartphones, and the pace of change is accelerating. These changes in the technologies, regulatory rules and viewing habits have many traditional broadcasters reviewing their strategic plans. The outcome of these discussions will be the genesis of many projects.

Most broadcasters do not embark on major capital projects on a regular basis. As mentioned previously, the design and subsequent installation and integration of systems within the broadcast/new media environment requires a unique combination of business, technical and people management skills. The chief engineer/IT manager is first and foremost a highly skilled individual knowledgeable in the particular technologies currently deployed in the broadcast operation(s) of their employer. The focus of the chief engineer's staff generally consists of maintaining the equipment in an operating condition and the occasional replacement or upgrading of existing equipment. It does not normally include the wholesale re-engineering of the facilities.

While most chief engineers, as well as senior staff people, are very aware of the evolving technologies and may even have some experience with them, these people are primarily engaged with managing, maintaining and supporting the systems and technologies currently in place. The assumption that the chief engineer can handle the additional workload of a major project is fraught with problems for several reasons, the first being that the chief engineer is often fully engaged in keeping the current systems operational, i.e., "keeping the lights on," and has little or no spare time to manage a large-scale project. Because of this focus, the chief engineer often lacks a big-picture view of where the technologies are going and how to best integrate and deploy one or more of them to accommodate the company's strategic plan. This broader knowledge is critical to the thinking behind the engineering and design of any new infrastructure.

Second, there is often a lack of appreciation for what is actually involved in successfully and properly managing a project. Third and equally important, while the chief engineer may have the abilities, he/she lacks the requisite training, experience and discipline that a professional project manager brings to the table.

Project management is a profession and like engineers, professional project managers (PMs) have a specific set of core skills that are not generally found in the engineer's quiver. These encompass organizational, business, people and communications skills. They are required in order to orchestrate and manage the various parties who are collectively responsible for the many disciplines and processes involved in a systems integration project. A professional project manager, with a Project Management Professional designation, brings proven project management techniques and methodologies, which will orchestrate and provide overall management and discipline to the project and in doing so help bring it to a successful conclusion.

PROJECT MANAGEMENT

Before we discuss a structured project management process, we should first define what a project is. The Project Management Institute, PMI, defines a project as

> a temporary endeavor undertaken to create a unique product, service or result. It has a specific beginning and end. The end is reached when the objectives have been achieved or when the project is terminated because the objectives cannot be met.[1]

There are five project process groups defined by the PMI and 47 knowledge areas associated with them. The process groups, including some of the knowledge areas, are:

- Initiating—project charter, stakeholder identification
- Planning—scope, schedule, budget, quality, HR, communications, risk, procurement, stakeholder
- Executing—direct and manage, quality assurance, staffing, communications, purchasing, stakeholder
- Monitoring and Control—monitor work, change control, scope, costs, quality, communications, risks, purchasing stakeholders
- Closing—project, purchasing

SYSTEMS INTEGRATION

What is meant by "systems integration"? The Wikipedia engineering definition is "the process of bringing together the component subsystems into one system and ensuring that the subsystems function together as a system." For the purpose of this chapter, component subsystems include hardware, software and workflow processes. The system integration process encompasses many different activities, which can be summarized by the following phases:

- High-level planning
- Conceptual design
- Detailed design and documentation
- Equipment procurement
- Construction and installation
- Configuration and testing
- Training and support

Integrating the Two Groups—Project Management and Systems Integration

Initiating

A project comes about because there is a need to "create a unique product, service or result." There are multiple reasons that may justify and lead up to the genesis of a project. These include one or more of the following:

- A business case
- An agreement
- Political considerations
- Government regulations
- Perceived risks

These and other factors are reviewed on a regular basis by the management team. If the outcome of those high-level discussions results in a decision that identifies one or more reasons to initiate a project, it will ultimately lead to the development of a project charter (PC).

The PC document, prepared by management, will identify the guidelines and objectives of what is to be achieved, a timeline and a budget. Figure 3.3-1 is an example of a project guidelines and objectives statement that may be found in a PC. The PM and the lead engineer are often identified at this time and may participate in the development of the PC. An approved PC requires the signature of a project sponsor. The sponsor represents the company and stakeholders and is connected to the source of funding. The PC is the document that formally authorizes the project and provides the PM with the authority to proceed with the project.

Functions/Goals/Schedules and Milestones

Even though the technology continues to advance, major capital projects are expensive and hence system upgrades do not happen on a regular basis. However, when the decision is made to move forward, the primary responsibility of the lead engineer and his/her team is to design a system that meets the objectives as outlined in the PC. The project manager's role is to oversee and ensure the project adheres to the objectives, timeline and budget as set forth in the PC.

Project Manager

By definition, the PM is the individual assigned through the authority of the PC to lead the team responsible for meeting the project's objectives. A professional

PROJECT CHARTER
Goal: Repack station with WXXX-TV
Objectives
Participate in the FCC auction and release spectrum
Consolidate MCR operations with WXXX-TV
Maintain channel ID
Share WXXX-T's transmitter
Develop a comprehensive facilities sharing agreement with WXXX
The design must allow for the future adoption of ATSC 3.0, IP and 8K UHD technologies
Financial
Re-use existing equipment whenever possible
Identify and sell off any redundant equipment
Develop an RFP for a systems integrator
Engage an independent Project Manager
Develop a project budget
Timeline - assumes a 9/16 start
Sr. management develops the Project charter - 10/16
Identify the internal project team members - 10/16
Meet with staff and explain the plan and objectives - 11/16
Develop RFP - 1/17
Distribute RFP to pre-qualified System Integrators - 1/17
Review System Integrator responses - 2/17
Review Budgets - 3/17
Select System Integrator - 3/17
Project kick off meeting - 4/17
Complete Engineering design work - 6/17
Complete equipment orders - 7/17
Begin installation work - 8/17
Complete installation work - 11/17
Test and commissioning - 12/17
Operator Training - 1/18
Cut over to new MCR 2/18

FIGURE 3.3-1 Example of a project guidelines and objectives statement that may be found in a project charter

PM brings proven project management techniques and methodologies and will orchestrate and provide overall management and discipline to the project and in doing so help bring it to a successful conclusion. It is good practice to assign a PM who is knowledgeable and familiar with the business and the technologies in use. It is also important to recognize the PM's responsibility is not to lead the design work but to oversee, manage, monitor, control the implementation and close out the project as set out in the PC.

Lead Engineer

A mentioned earlier, it is important that the lead engineer on the project has a big-picture view of the various technologies in use and where they are going. If the chief engineer does not assume the role of the lead engineer, the chief engineer will be very much involved and consulted on the decision on who is engaged in the lead engineer role. The chief engineer will also be engaged in the overall design process, provide input to the PM on the status of the design as it progresses and participate in the design review meetings.

Develop PMP and High-Level Planning

Using the guidelines set out in the PC, the PM, with assistance from the other team members, will develop the project management plan (PMP). The PMP document incorporates information from several of the other planning processes. Examples of these are scope, schedule, budgets, change management, staffing plan (human resources), risk, purchasing and quality.

Depending on the size and complexity of the project, a PM experienced in the art of delegation may utilize a linear responsibility chart, which is also referred to as a responsibility, accountability, consult and inform (RACI) matrix. This is a helpful tool in the planning stage to ensure the appropriate people are assigned to the right jobs. The planning process also includes the development of the plans for human resources (HR), project communications, review of risks and development of a risk strategy, equipment purchasing and stakeholder management. An example of a RACI matrix is shown in Figure 3.3-2.

As the project progresses, the PM will receive reports that assist in monitoring the schedule, budget and objectives, but there is a great deal more involved in properly managing a project. The PM is also responsible for managing/overseeing all the plans/processes developed and incorporated into the PMP. These include:

- Managing the scope
- Managing the implementation
- Managing the change order process
- Managing the quality assurance
- Managing and controlling quality
- Managing the human resources
- Managing communication with the stakeholders
- Identifying, managing and mitigating risks
- Managing the purchasing process
- Communicating with the stakeholders

Scope Management and Conceptual Design

Before the conceptual design work begins, the PM, the lead engineer, chief engineer and, depending on the complexity of the project, any other key individual(s) with specialized skills required for the project will review the PMP and develop a scope management plan. The scope management plan establishes the "goal posts" for the project and therefore the activities that ultimately impact the budget and schedule. This process will identify a plan and processes for collecting the requirements information.

The conceptual design team begins their work by undertaking a hierarchical decomposition of the total scope of work for the project. This information is used to develop a work breakdown structure (WBS). As with the determination of the scope, the individuals involved in developing the WBS must understand the impact and interrelationships the tasks have on one another.

A poorly defined scope is a recipe for a myriad of future problems with the schedule, budget, objectives and ultimately the end users. Developing a WBS is

	Individual				
ACTIVITY	**Alice**	**Dave**	**Tom**	**Ralph**	**Zoey**
Create Charter	R	R	A	C	I
Collect Requirements	C	I	I	R	A
Submit Change Request	A	I	R	I	C
Risk Management	I	R	I	A	R
Develop Test Plan	R	A	I	I	C
	R=Responsible A=Accountable C=Consult I=Inform				

FIGURE 3.3-2 A sample RACI matrix

one of the key ways to prevent "scope creep." Work not included in the scope is only undertaken after it has been reviewed and approved through the "change control" process. The approved scope and any subsequently approved changes are then authorized by a change order and incorporated into the PMP.

The PC provides the PM and lead engineer with the parameters for the project, i.e., objectives, timeline and budget. The project work begins with an information gathering process. It is critical that all the stakeholders are consulted and have input as the requirements are being collected, reviewed and analyzed. The work continues until the overall functionality, processes and relationships are identified. The conceptual design process is complex and requires inputs from many people and departments within the company.

As the information is collected and analyzed, the lead engineer and project team can begin work on the high-level conceptual design. This process generally begins with a series of block diagrams. The various relationships are then identified by drawing single lines between the blocks. The conceptual design is an iterative process. This process leads to many additional discussions with the stakeholders as the design and dependencies are identified, tweaked and refined and the scope baseline is established. This high-level conceptual design will identify the relationships between the various systems impacted by the project objectives, will involve a review of the budget and time lines, and will identify those points where the design interfaces with the existing infrastructure. It is important that the team keeps the scope and project objectives in mind as the conceptual design are refined. As the work progresses, a particular core technology may be identified, and this will subsequently flow into the detailed design work.

All the inputs need to be reviewed and the impact on the project assessed, understood and factored into the conceptual design. The project "tentacles" may include technologies or other factors that will impact existing business practices and processes. Examples are the need to make changes to the building, such as providing temporary office space in another part of the building or another building during the project, or changes to the electrical infrastructure, mechanical systems, and fire and security systems.

Schedule and Milestones

As the work on the scope continues and the interrelationships and dependencies are identified in the work packages, a preliminary project schedule will begin to take shape. As the schedule develops, the information will be used to determine the optimum sequence in which the various activities should be undertaken. At the point where the schedule begins to firm up, the conceptual design team will communicate with the installation supervisor assigned to the project. The installation supervisor will review the preliminary time line and activity list in order to identify and determine the specific resources required to complete each identified activity, as well as the time required to do so. This information will be compiled and ultimately reflected in the project schedule.

As with the detailed determination of the scope and the development of the design from its conceptual state, the project schedule is an iterative process. The initial schedule will show the logical sequence of executing a series of tasks but when overlaid on the existing operations, it may highlight conflicts with one or more existing activities. For example, if the work involves major renovations in a part of the building currently in use, then the current operation and supporting infrastructure may have to be relocated temporarily or permanently while remaining fully functional. Once all the dependencies in the schedule have been identified, the team will modify the schedule to reflect the changes.

As the schedule comes together, the PM, lead engineer and chief engineer will review it and compare it with the schedule and milestones identified in the PC. Variations from the approved plan will be reviewed and adjustments made to amend the schedule so it is compliant with the PC. As the detailed design work progresses and the activity timelines are refined, the PM will be able to identify the "critical path." The critical path is defined as the longest path through the schedule and conversely represents the shortest time in which the project can be completed. A careful review of the critical path will identify any opportunities to shorten the schedule.

The schedule is typically shown in one of two formats, either Gantt or PERT, with the Gantt being the more commonly used. The Gantt chart is the more common way of highlighting the project schedule. It uses straight lines and arrows and presents a linear representation of a project schedule. It shows the start and finish dates and dependencies of the different elements of a project. It is named after Henry Gantt, the engineer who developed it. See Figure 3.3-3 for an example Gantt chart.

The PERT or network chart is used to schedule, organize and coordinate tasks within a project. PERT is short for "Program Evaluation Review Technique." It might be of interest to know that the PERT chart was developed by the US Navy in the 1950s as a method to manage the Polaris submarine missile program. An example PERT chart is shown in Figure 3.3-4. The project team, including the installation supervisor, will use this schedule information to determine the specific resources required to implement the activities and identify where there might be an overlap in skill sets.

There are several computer-based program management software packages that support and produce both PERT and Gantt charts (see list of some examples at the end of this chapter).

The design and implementation of a new system within the broadcast plant requires many people with specific skills. These include engineers, drafters, purchasers, installation supervisors, wiring technicians (wiremen), commissioning engineers and trainers. In addition to the aforementioned resources, the sequencing of the activities may have identified that certain skills are needed due to the inclusion of some equipment/infrastructure that requires specialized skills/equipment to install/commission. The installation

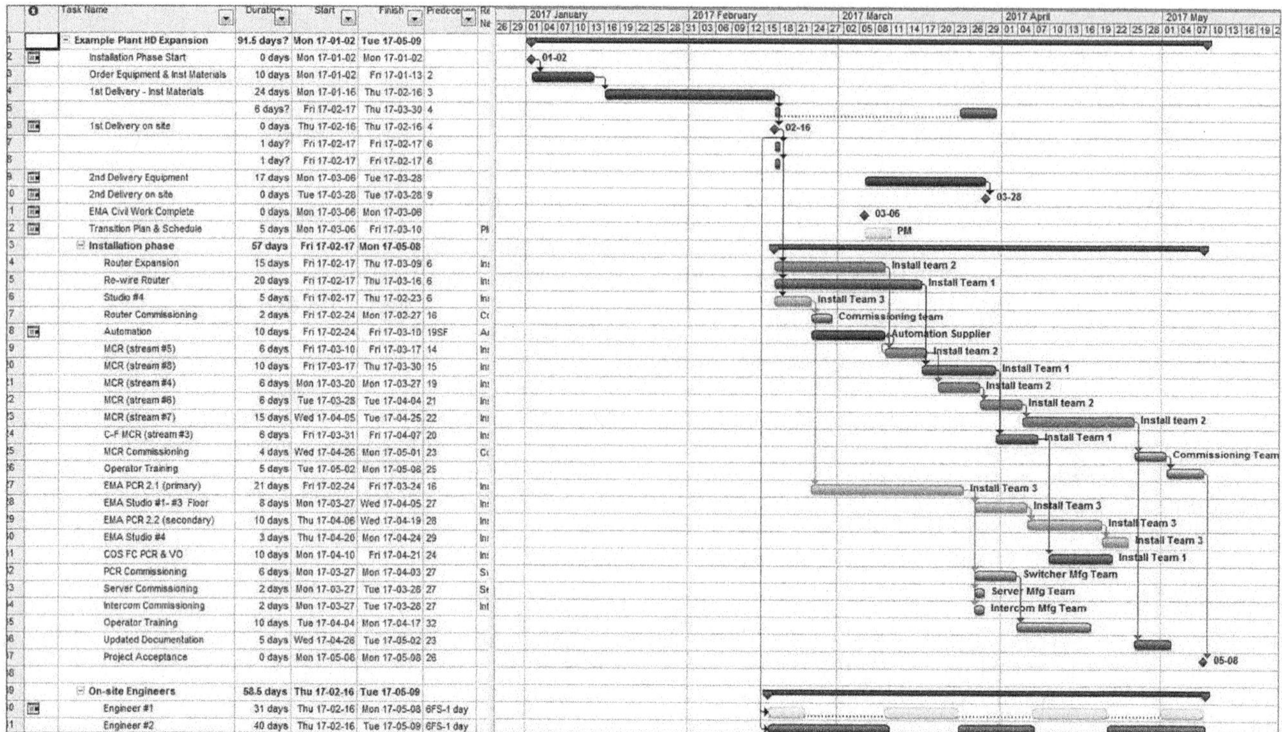

FIGURE 3.3-3 A Gantt chart uses straight lines and arrows to present a linear representation of a project schedule

supervisor will identify those areas that might require specific accommodations to facilitate the installation, as well as installers with specialized skills. Examples of this might be fiber terminations, testing or the setup of specialized terminal equipment or adequately sized and accessible cable raceways. The installation supervisor will also identify any specialized equipment needed for the installation phase of the project. This might include specialized test equipment, scaffolding, lifts or tools not normally available on site. For those items not readily available, the installation supervisor will also identify where these items may be purchased or rented.

There are two techniques for speeding up a project—"crashing" and "fast tracking." Each technique is different and each has different consequences. Crashing involves adding additional resources to shorten the schedule. The downside of this technique is that it adds additional costs. Fast tracking is a way to compress the schedule by undertaking activities normally done in sequence by executing all or a portion of the activity in parallel with other activities for all or a portion of their scheduled duration. The risk that arises is the possibility of rework being required because a subsequent activity is worked on before information from the previous activity is available. It may also require the use of additional resources and the costs associated with them. Once the activities, their durations and relationships are known and understood, the project team can determine the critical path and review ways to optimize it and therefore reduce the project durations by reducing the critical path. An example of a critical path and its calculation is shown in Figure 3.3-5.

Detailed Design

The finalization and formal approval of the conceptual design heralds the start of the major component of the system engineering, that being the detailed design process. Often the conceptual design process has identified a preferred core technology that the detailed design will incorporate. At this point, the engineering team begins converting the approved conceptual design into a workable design. This process involves multiple steps and depending on the complexity of the design and prescribed timeline may require additional resources being added to the team.

It is important to remember that the PM is responsible and accountable for the management of the project. The PM's role is to orchestrate and provide overall management and discipline to the project, but the PM is not the person responsible for the project design. It is the engineering team, led by the lead engineer, which is responsible for developing the design. The requirements, as outlined in the PC, the PMP, the broadcaster's short- and long-term goals and current best operating practices, must be reflected in the completed design.

The lead engineer and design team begin the detailed design process by undertaking a hierarchical decomposition of the total scope of work for the project and generating a WBS. As with the determination of the scope, the individuals involved in developing the WBS must understand the impact and interrelationships the tasks have on one another. For example, if the project involved the introduction of a new IT infrastructure, how does it impact the existing electrical and HVAC systems capacity? This kind of information is critical,

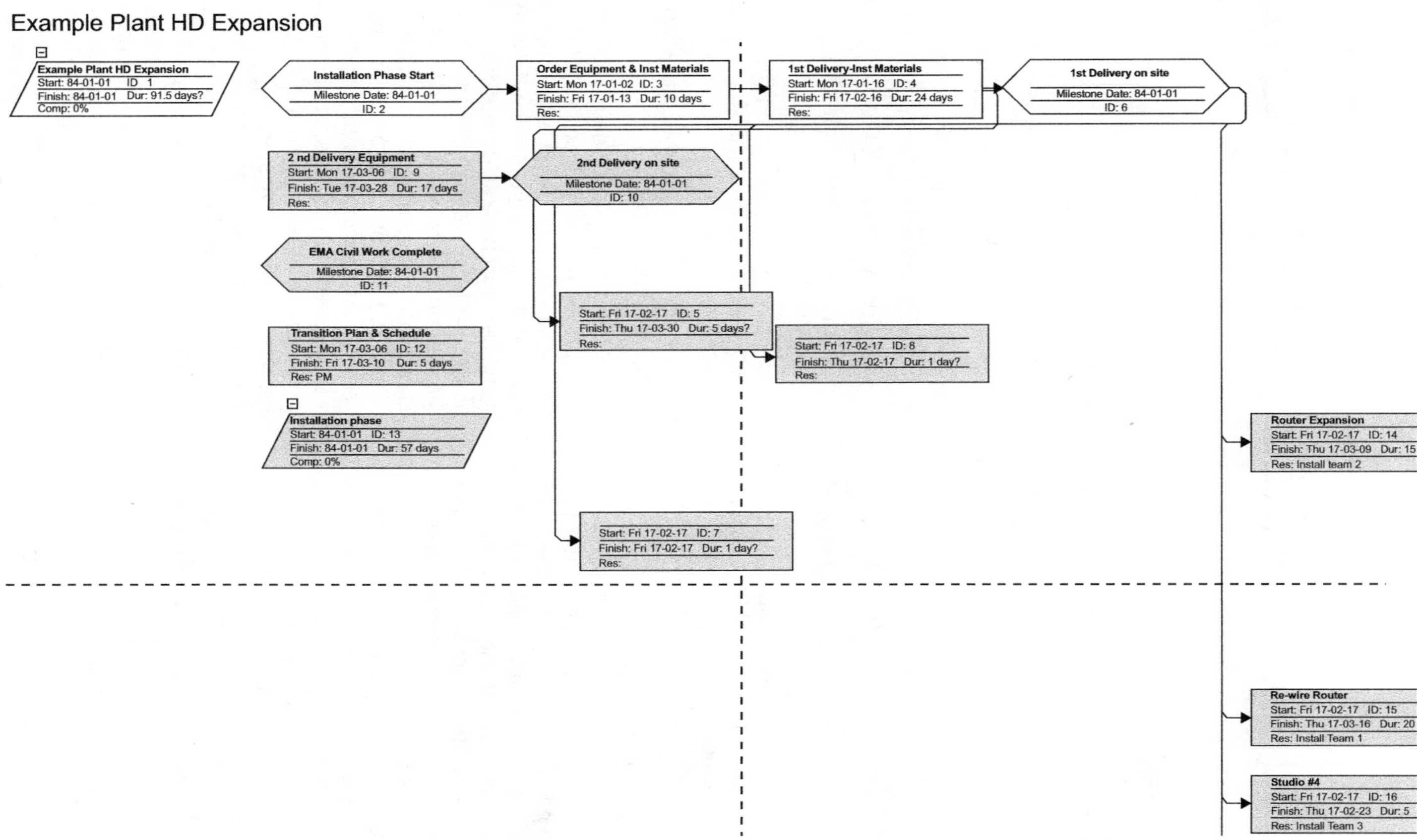

FIGURE 3.3-4 A PERT chart is used to schedule, organize and coordinate tasks within a project

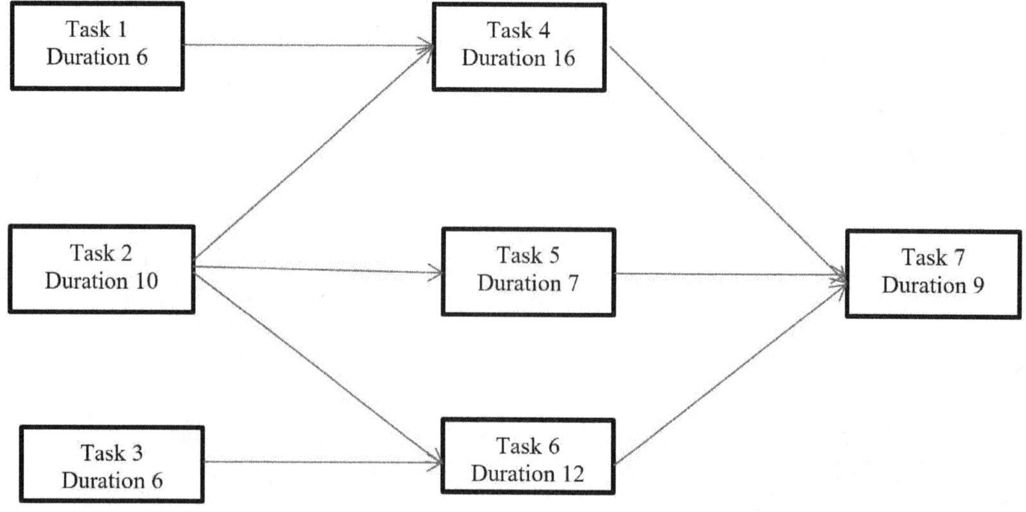

Critical path = Task 2 (10) + Task 4 (16) + Task 7 (9) = 35

FIGURE 3.3-5 Example critical path diagram and critical path calculation

as it is taken into consideration as the WBS is further broken down into work packages. The work packages are the basis on which the schedule, resources and project costs will ultimately be calculated, monitored and measured.

The detailed design, like the conceptual design, is an iterative process. It is guided by the objectives set out in the PC, the scope management plan and the conceptual design. The lead engineer may augment the design team with specialists in operational workflows, HVAC, IT, and electrical engineers and architects, depending on the extent of any renovations or civil work required. As the design progresses, the PM will work with the lead engineer and review the goals and objectives outlined in the PC and PMP to confirm the project is on target. One of the PM's roles is to encourage the design team to extract the maximum benefits for the money spent without compromising the longer-term objectives of the project. The process of simultaneously balancing the budget and meeting the project objectives and time line established in the PC is commonly referred to as "value engineering."

As the design begins to solidify and the engineering team begins putting functional drawings together, the installation supervisor will have an opportunity to review the drawings to gain an overview of the equipment involved and information on any building modifications. The installation supervisor will review the schedule and allowances that have been made for cable access and where trays, conduits or ducts are provided, and determine that they are adequately sized. The installation supervisor will have an opportunity to identify those areas of the design that are potentially difficult to implement and may be able to offer suggestions on ways to mitigate any installation issues. The PM, lead engineer and installation supervisor will jointly:

- Review and validate the schedule time line
- Identify the critical path
- Identify those items that have long lead times
- Identify and confirm the cable paths and lengths
- Confirm the proper sizing for cable trays, conduits and ducts
- Identify and confirm that the appropriate resources required to accomplish the project are identified and available
- Identify any specialized installation tools or other equipment required

The installation supervisor will review the integration phase of the project with the PM and validate the resources and skill sets needed for the project. The installation supervisor will also review the report from the purchasing department showing the status of the project-related equipment on order, especially the longer lead-time items required at the front end of the installation. An example of this would be a routing switcher, custom bulkheads or special cables or connectors. After the installation supervisor has undertaken these reviews, it is an opportune time for the PM to have a discussion with the installation supervisor to see if he/she has any suggestions on ways to minimize the cost of the installation work without impacting the objectives, schedule or budget.

Throughout the project, the PM will hold regular meetings with the members of the project team, and have minutes taken and distributed to those in attendance. The minutes will include action items and assign responsibilities and time lines. The project team includes the design team as well as others, such as finance/purchasing and management. In addition to reviewing the engineering work, the meetings will follow up on the status of any zoning, building or other permits and the people managing them to confirm they are on track and if not, to initiate the appropriate action to get them on track. Also, this is an opportunity to monitor the work of any external consultants or subcontractors on the project. Any significant delays in one or more permitting process may have serious implications on the schedule, budget and potentially the viability of the project. The PM will provide regular communications to the stakeholders in accordance with the stakeholders' communication plan.

Documentation

During the design process, the drafting team will have been engaged in the preparation of various drawings, which are updated to accommodate any changes following design review meetings. Once the design is agreed to and signed off, the drafting team will begin preparing the construction drawing set. These construction drawings and supporting documents will incorporate the additional details required by the installation team. This includes connector pin-out details, rack locations and elevations, wire numbers, specific nomenclature labelling and any other details required by the installation team. The drawings will undergo a final review by the PM, chief engineer and lead engineer, at which point the lead engineer will sign and stamp the drawings "Approved for Installation." The design is now considered final.

Any changes will only be incorporated after being reviewed and approved by the change control process set out in the PMP. The drafting team will prepare two identical sets of drawings, one for use by the installation team on-site, the other as the reference set. During the installation, any approved changes made during the installation will be marked up in red on the installation supervisor's set of drawings. At the conclusion of the installation, the installation drawing set will be returned and used to update the reference set. Once the drawings have been updated, they will be reviewed and confirmed by the chief engineer, lead engineer and installation supervisor as being complete. Following that review, a set of "as-built" drawings will be prepared.

Equipment Procurement

As the detailed design work progresses, the design team will develop a set of detailed equipment lists that incorporates all the equipment called out in the design, including details on any options, long lead items and when the equipment is required on-site. One list will identify the equipment now on hand that will be incorporated in the new design. The second list will identify the equipment that needs to be purchased and is passed onto the purchasing group. A third list may also be prepared listing the equipment that will be retired and available for sale or disposal following the implementation of the new system. The list will identify:

- Manufacturer
- Model
- Options
- Vendor, contact name and information (phone, email)
- Quantity
- Estimated cost
- Required delivery date

The purchasing department will use this list and add the following additional information:

- Order date
- PO number
- Scheduled delivery date
- Order status
- List price
- Discount
- Budgeted cost
- Variance

Communications are an important ingredient for a successful project. Collaborative computer programs such as Microsoft SharePoint can be employed to improve communications within the project team. The administrator of the collaboration software site can establish viewing privileges so that only the information required by an individual can be seen, and editing privileges so documents can only be edited/updated by the responsible individuals. For example, having an equipment list posted on SharePoint or an equivalent site will allow the project team to communicate and share information while simultaneously maintaining the integrity of the information. This will facilitate identifying and tracking those items that have a long lead-time or critical delivery dates. The list can be updated by the purchasing department and allow members of the project team to track the status of all the equipment orders, their delivery dates against the schedule, the funds spent and costs remaining against the budget. The installation supervisor needs to know what equipment has been ordered and delivered and when other items are scheduled for delivery, but does not necessarily need to have access to the related financial information. A list of other collaborative tools is included at the end of this chapter. An example of a typical equipment list is shown in Figure 3.3-6.

Construction and Installation

When the job site is deemed ready for the installation phase, but prior to the installation work beginning, the installation supervisor should visit and survey the site to:

- Identify those areas that may present difficulties
- Identify a secure space to store inventory and tools
- Identify and review access points, cable paths, conduits and cable trays
- Identify a source of electrical power for the installation team
- Determine what local telephone and Internet connections are available
- Determine what, if any, special equipment will be required

Qty	Part Number	Description	List $	Total
		ROUTING SYSTEM		
1	XE8-3232HX+2PS	Xenon 128X128 HD/SD 8RU Router frame W/32X32		
3	XE-IP32HX	32 HD inputs		
3	XE-OP32HX	32 HD/SD outputs		
1	Mafnum-SE-R32P	3RU Controller hardware & software		
1	CP-3100E-RMB	Rack mount bracket		
3	CP-2116E	2 RU Control Panel with 16 programmable buttons		
5	CP-2048E	2RU Control Panel with 48 programmable buttons		
10	CP-2032E	2RU Control Panel with 32 programmable buttons		
7	CP-1024E	2RU Control Panel with 24 programmable buttons		
1	Training	One day on site training		
1	Commissioning	One day on site commissioning		
				-
		SPG & DISTRIBUTION		-
2	SRG-4400	Master Reference and Test Signal Generator		-
1	ACO-4400	Automatic Changeover System		-
1	PS-4400	Spare Power Supply for SRG-4400 and ACO-4400		-
2	NTP-4400	NTP Time Synchronizing option		-
2	GPS-4400-35M	GPS receiver/antenna option with 35m cable		-
10	VEA-8707A-R2L	Analog Video Equalizing DA		-
60	SRA-8802-R2	High Performance 3G Reclocking DA		-
10	SFS-8622-AIC- R2AIC	3G/HD/SD-SDI Frame Synchronizer		-
10	ADA-8405-C-R2C	Analog Audio & Timecode DA w/Remote Gain		-
10	OG3-FR-CN	OpenGear 3.0 Frame W/Advanced Networking		-
10	PS-OG3	450 Watt Universal Power Supply for OG3 Frame		-
				-
		PRODUCTION SWITCHER		-
2	A2M-PANEL	Acuity 2M Control Panel		-
2	A2MP-REDPSU	Redundant Power (Control Panel only)		-
2	AP-AUX2RU24	Auxiliary Control Panel 2RU W/ Source Mnemonics		-
2	AP-SIDESLIDE	SideSlide - Audio Fader Module in a SideBox		-
2	AP-SIDESHOT	SideShot - ShotBox mounted in a SideBox		-
2	ACU4-FRAME	Acuity 4RU Base Frame (NO PANEL INCLUDED)		-
2	ACU4-REDPSU	Redundant Power (Rack Frame only)		-
4	ACU4-IN20	3G Inputs, 20 Additional		-
2	ACU4-ME1	Adds ME 1		-
2	ACU4-ME2	Adds ME 2		-
2	ACU4-ME3	Adds ME 3		-
2	ACU4-3DDVE-ME3	3D DVE with Warp for ME 3		-
2	ACU4-AUDMIXLG	Large Audio Mixer Interface (More than 16 Inputs)		-
2	ACU4-ROBOCAM	Robotic Camera System Interface		-
2	ACU4-ROUTER	Routing Switcher Interface		-
1	ACU4-MV1	20 Source Display MultiViewer 1		-
1	ACU4-MV2	20 Source Display MultiViewer 2		-

FIGURE 3.3-6 Sample equipment list

- Determine the location of the nearest medical facility/hospital, hours of operation and contact information
- Identify and obtain the contact details for the security and maintenance supervisors
- Determine the building procedures for workers to access the building at irregular hours and on weekends and what security procedures/passes/badges are required for access
- Review how materials get to the job site from the shipping/receiving dock to the location(s) where the work is being undertaken and if there are any restrictions on delivery times, special handling procedures or limits on the types/sizes of the delivery vehicle(s)

Following the site survey and prior to bringing the installation team on-site, the installation supervisor and lead installer will review the following:

- The signed "approved for installation" construction drawing set

- A copy of the equipment list for all the equipment the team is responsible for installing, showing what has been received and when the remaining items are scheduled to arrive
- The termination details for any nonstandard interfaces
- The approved cable-numbering schedule

When the installation crew arrives on-site, but before the actual work begins, the installation supervisor will meet with the crew and:

- Review workplace safety practices
- Review the location(s) of the fire extinguishers
- Review the location of the work place first aid kit
- Post the phone number(s) of the local hospital/ emergency services and review their location(s)
- Review any special conditions relating to site security or building access procedures
- Identify the meeting point for all staff in the event the building needs to be evacuated for any reason

The installation supervisor will confer with the lead engineer to arrange the installation schedule so that the various subsystems are completed in a logical order. This will allow the engineers to turn on, test and check the operability of the various subsystems. Managing the installation in this manner will optimize the schedule and reduce the overall project time. It will also allow the engineers to identify and rectify any problems that might show up within the subsystems.

The installation supervisor has a number of key responsibilities to keep the installation schedule on track. These include:

- Ensuring the delivery schedule for the installation materials and equipment required for the project is maintained
- Checking the quality of the workmanship to confirm the proper installation procedures are being followed. This includes ensuring that the cables are properly terminated, the minimum bending radius for the different cables are adhered to, adequate service loops are provided for equipment and the proper cable numbering protocol is followed
- Keeping the worksite drawings current and "red lined" to reflect any changes as approved by the change control committee
- Ensuring safe working practices are being followed
- Keeping the project manager apprised of the status of the installation and identifying anything that will negatively impact the installation schedule

The systems in the modern broadcast plant are no longer based primarily on point-to-point bundles of coax, shielded and multi-conductor cables. Increasingly, there is less of the former, as they are being replaced with a network of cat 5/6 cables and fiber. Current designs specify 3-GHz speed, but 10-GHz speeds and perhaps higher will be required as 4K UHD is implemented, and 8K is on the horizon. Proper terminations of cat 5/6 cables are critical for the higher bandwidth and high data throughput necessary for the modern broadcast plant to function. The same applies to the fiber terminations. In addition to the traditional installation skills of managing large bundles of cables and routing them safely and leaving sufficient loops for equipment servicing as may be required, the installation team needs to understand that DC continuity testing is no longer sufficient to confirm the integrity of the wiring. In many instances, it is necessary to undertake some basic equipment set-ups/configurations to confirm that the actual connections/paths are valid and functioning properly.

Very few projects do not incorporate some level of changes during the installation phase. Not all changes are bad, but it is important that all of them are reviewed by the change review process. Only after they are approved and a change order issued can they be implemented. The later changes are made during the installation process, the progressively more expensive they are and pose a greater risk to the project schedule and project budget.

Configuration and Testing

Before beginning this phase of the project, the PM will review the protocols set out in the project management plan for testing, configuration and commissioning with the senior engineer, chief engineer and installation supervisor.

As the installation progresses, various subsystems may be completed ahead of others. In those instances, the engineering team has the opportunity to power-on the equipment and confirm it is functioning properly. Very little equipment today is not to some degree managed through software. Manufacturers are constantly upgrading the operating systems and firmware in their products, and it is not uncommon for an equipment manufacturer to have released an update to the operating system or firmware or both after the product has been shipped and installed on the customer's premises.

The first part of the configuration process is to check the manufacturer's website for the most current version of the software and, if there is an update, to install the most current version. After any updates to the software and firmware have been installed and the equipment has been shown to be operating correctly, the process of configuring the equipment so it works as per the system design begins. Among other things, this work includes configuring ports, obtaining IP addresses and confirming the correct interoperability with other equipment.

Although there are many commonalities in most systems, the configuration of each individual piece of equipment may vary depending on its application and the role it plays in the overall system. As with the testing procedures, the engineer should record any changes to the default settings of each piece of equipment after they are made. Fortunately, many of the equipment configurations can likely be printed out or stored in a file for future reference, which will come in handy should changes to the configurations be required during troubleshooting or when new equipment is incorporated into the system.

Prior to checking the functionality of the various systems, the commissioning crew needs to verify that the interconnecting cables and patch field paths are correct and properly installed. Depending on the nature of the signal path connections, different signals and protocols are used to test and confirm the signal paths. For example, testing for proper and reliable operation in the ATSC environment requires test signals, such as SMPTE 259M and 292 and test equipment for local area networks (Cat 5/6), and testing for MPEG and ATSC formats with PSIP capability. Typical examples of the test equipment include the following:

- Tektronix WFM/WVR 7200
- MTS4SAV Transport Stream analyzer
- AES, Dolby ACS, Dolby E audio test equipment
- Fluke DSX-5000 cable analyzer
- Fluke FL-7000 FiberInspector

Commissioning

Now that the signal paths have been verified and tested and the equipment has been configured, it is time to begin the commissioning process. This process is an overall checkout of the interoperability of the systems. Today's equipment and workflows are heavily dependent on software, actually many different pieces of software, which need to communicate properly with one another. Fine-tuning the system can be a complex, challenging and tedious task. It is in this phase that everything must be configured so the system operates as designed and intended.

The commissioning plan guides the process and helps keep it on track. During the testing and checkout phase, the various signal paths and controls will be tested to confirm their functionality. At the conclusion of the commissioning process, the system should be checked and the measurements recorded as part of the configuration process. If a third-party integrator is involved, they will often have their own test equipment that can be used when the initial testing and configurations are done, but when it comes to the final testing and formal commissioning tests, the better practice is to use the station's test equipment. The rationale for using the station's test equipment is that it establishes a reference point, which the station engineers can refer to in the event of any future troubleshooting or follow-on testing. When the commissioning work is completed and the chief engineer has signed off on the test results, the system is deemed accepted and ready for use.

One process remains following the completion of the system commissioning. In order to realize the full potential of this new system, the end users must be properly trained on the new technologies and new workflows.

Training

If the investment in the new systems is to be realized, a well-thought-out and thorough training plan must be part of the PMP. Continuous advances and software upgrades introduce ever-increasing complexity in the system components. The increasing use and dependency on automation is a double-edged sword. The complex operation of producing a show, or putting multiple channels through a common master control, is increasingly reliant on software tools. The tools produce consistency while reducing operational errors, but unless the people using them are properly trained, the full potential of these new tools will not be realized.

It must also be kept in mind that automated systems fail on occasion, and when they do, the master control operators need to understand how to bypass the various failed system(s) to maintain operations while the repairs are underway. The various equipment manufacturers offer training on the operation and maintenance of their products. While the engineers get the technical training directly from the manufacturers, the operational training is often undertaken by freelance individuals, hired by the manufacturers. The rational for this is that those individuals, who have years of experience using the equipment, are more familiar with the operating environment and challenges that come with it. Their experiential knowledge and ability to speak the language reinforce and enhance the training program.

The new systems, in particular those employing a higher degree of automation, are sometimes initially seen as overly complex. The master control operators who sit on the front line must understand and be confident in the operation of the automation system(s). Ideally, the training will be done running the "new" system in real time and in parallel with the regular operations but with the new system off-line. Training is an ideal time to simulate potential failures, thereby allowing the operators an opportunity to practice their skills at identifying and working around a problem. Having practiced what to do in the event of a minor or major failure will instill a sense of confidence on the part of the operations staff.

Cutover

After the training is completed and the staff is up to speed on system operation, the various systems, such as the master control/automation, can be cut-over and go live to air. The same applies to the off-line systems, which can be interconnected into the plant once the training on them has been completed.

Documentation

At the commencement of the project, the design team will have prepared documents outlining the subsystems and the role of the various components in the design. Following the commissioning and acceptance of the system, the red-lined installation set of drawings will be returned to engineering and updated by the drafting group to reflect any changes introduced during the installation of the original design. This would include a complete drawing set identifying the key components, signal flows and paths for the various signals such as audio, video, intercom, data, automation, control, etc., and wiring diagrams for any special connectors and custom equipment interfaces. There should also be sections describing the system and one with instructions on how to identify and bypass or work around a failed piece of equipment or the entire system in the event of a catastrophic failure.

External Resources—Consultants and System Integrators

The PM's responsibilities include managing, anticipating and avoiding risks to the project. Risks materialize in a variety of flavors. Some are directly under the control of the station, some partially and others not at all.

The station has direct control of the budget and the allocation of personnel and internal schedules. Projects entirely contained within a building generally may require a number of local/municipal permits such as building, plumbing and electrical. If the building is older, it may require a report confirming there is no exposure to asbestos or, if there is so, an asbestos abatement plan from a qualified contractor. If the project involves substantial civil work, the historical weather patterns should be investigated to see what, if any, potential impact they might have. If the civil work requires zoning amendments/easements or an environmental impact study or has a political impact, these factors must be considered when the project is first contemplated and addressed before being approved.

These are all risks that need to be addressed and managed. Another risk is the availability of the specialized resources and skills required to implement the project. Earlier, in this chapter, we highlighted the pressure often put on the chief engineer to manage and see the project through while continuing to run his/her own department. Too often the chief engineer is assumed to have intimate knowledge of all the technologies available in the market that should be considered in the design and the skills, not to mention the extra time and energy, to manage the project while simultaneously keeping his/her department in order.

As discussed earlier, a PM, just as a chief engineer, has a variety of skills in his/her quiver. Several of their skills are similar but many are not. A professional PM brings proven project management techniques and methodologies which will orchestrate and provide overall management and discipline to the project and, in doing so, help bring it to a successful conclusion. The PM is focused on the project and what is required to properly execute the project. The PM is not required to be a technical person, although experience in similar projects is a valuable asset. The PM requires focus, strong communication, and analytical and, most importantly, leadership skills.

In addition to the PM, the chief engineer, whose staff is already taxed, may choose to outsource some of the other skills required to successfully implement the project. These include the lead engineer, design engineers, installation supervisors, installation labor, fiber termination specialists, drafting and professional consultants. It is very difficult for one person to stay current on the multitude of rapidly evolving technologies, no less than when they have a full-time job keeping the station up and running. This is where hiring a consultant or systems integration company for assistance in a specific area can be beneficial and play a role in mitigating the risk to the schedule and budget.

Consultants

Technology-based projects can be very complex. A consultant with specialized knowledge and experience deploying one or more of the technologies being considered or decided upon for the project can add a great deal and contribute to the success of the project. When considering the use of a consultant, it is important to identify in writing or through a request for proposal (RFP) exactly what specific services are required, where they are required, the timeline and what the budget is. There may be several different areas and phases of the project where an outside consultant can provide assistance; for example, acoustics, civil, mechanical, electrical, architectural or city planners can help navigate through the various permitting processes and bureaucracies.

With a defined scope of work in hand, the next thing is to search out a consultant(s). There are many ways to do this, beginning with referrals from other engineers, professional organizations, trade magazines, trade organizations, industry trade shows and "Dr. Google." When interviewing a consultant, it is important to consider the context of the consultant's experience and match it to the tasks and nature of the project at hand. For example, if you are hiring a PM for a two million dollar project in Ames, Iowa, you probably don't need the services of a person responsible for building out a thirty million dollar studio in New York City. Once you have narrowed your search, it is important to check the consultant's references. Unfortunately, many do not do take this step, which is a potential recipe for disaster. Assuming the references check out, prepare and execute an agreement that clearly identifies the reporting structure, work, deliverables, timeline and budget.

System Integrator

Most stations today operate on very lean budgets and are more often short of personnel rather than having personnel with time on their hands. As with the chief engineer, adding an additional workload to a department already fully burdened is not a recipe for success. The system integrator offers a variety of services and experience implementing projects. They have specific skills, an abundance of experience, the specialized resources and many of the specialized tools required to undertake the implementation of a project. Using a system integrator is another way of controlling the costs and schedule of the project and capitalizing on their experience to mitigate the risk to the project.

While several ways are available for selecting a system integration company, there are two general approaches. The first is to generate an RFP; the second is through a referral:

- An RFP will identify the scope of work, the time line and the specific qualifications and experiences required in order to bid on the project. As part of the pre-qualification process, the RFP will identify any specific project experience history and legal requirements, financial requirements, licensing and bonding capabilities, environmental practices, workplace policies such as sexual harassment, or the ability to comply with other local regulations as may be required. Incorporating this information at the outset will prequalify the potential bidders and save everyone time.

- The referral process is two-fold. It can be based on feedback from trusted associates on the competence and skills of a particular system integrator, because either they have direct knowledge/experience with the capabilities of a specific system integrator or they are familiar with a similar project the system integrator has recently completed. Alternatively, the chief engineer can identify a similar project recently completed and speak with their counterpart to find out what company did the work and the level of satisfaction with the way the project was handled.

System integrators come in many sizes and with a variety of experience in some or all of project management, design engineering, system integration and equipment supply. System integrators can tailor their services to supply some or all of the services required for the project. The larger and more experienced the system integrator, the more likely they are current on the latest and evolving technologies and have direct experience of knowing which technologies work and interface well with other technologies and which do not. In spite of the claims manufacturers make regarding interoperability with other systems and standards compliance, the knowledge and experience the system integrator has is very valuable and has the potential to save considerable time and money on the project.

Figure 3.3-7 shows a typical reference check list, offered as an example of how one might go about prequalifying/scoring the various consultants or system integrators (C/SI) responding to an RFP.

Working with a system integrator brings other advantages that are not necessarily apparent. System integrators are constantly purchasing equipment from multiple vendors. Unlike your project, the likes of which might happen every five to ten years, manufacturers see the system integrator as a regular customer and a source of ongoing sales. The system integrator often enjoys a volume discount which, depending on the terms of the agreement the end user negotiates with the system integrator, may be passed along to the end user. Another advantage is access to the technical people with the manufacturer who have an in-depth knowledge of their product. This direct access to the right people can be very helpful when initially setting up the system. It is during this phase of the configuration work that the subtle conflicts between operating platforms become apparent and are often difficult to resolve. This is when access to the right technical people within the various manufacturers can be the critical key to solving a tough technical problem.

The customer benefits from the many relationships the system integrator has with multiple manufacturers and the skills and talents of the people the system integrator employs. The manufacturer benefits because the system integrator provides direct feedback from the client's experience on those pieces of equipment that are problematic because of design issues or manufacturing problems. The system integrator is also a source of new product ideas and improvements to existing products.

The judicious use of consultants in conjunction with a system integrator can often reduce the real cost of undertaking the project by completing the project in less time. It will also help mitigate the risks. If the company is contemplating additional projects in the future, they might consider establishing a longer term relationship with a system integrator or consultant. There are several reasons for doing so:

- It reduces the need to bring on highly skilled and specialized additional staff
- Provides a ready source of skilled engineers current with the latest technologies
- The system integrator/consultant will have inside knowledge on the facilities, thereby ensuring continuity on future projects
- The system integrator's purchasing power may reduce the capital cost of new equipment

Vendor	Proven Experience	Adequate Resources	Legal Status	References	Bondable	Liability Insurance	Workers Comp Clearance	Banking Reference	Credit Check
C/SI #1									
C/SI #2									
C/SI #3									
C/SI #4									
Totals									

FIGURE 3.3-7 Typical reference check list

SUMMARY

This chapter has outlined and identified the many important factors and processes that play important roles in the successful design and implementation of a system or systems in a broadcast plant. There are two critical components on which all others depend—strong project management accompanied by a comprehensive planning process. Without these two components in place, the success of the project is at risk.

Note

[1] Page 5—Project Management Institute. *Guide to the Project Management Body of Knowledge 5th ed.: PMBOK® Guide*, Newtown Square, PA: Project Management Institute, 2013.

Bibliography

Ensworth, Patricia. *The Accidental Project Manager*, New York, NY: Wiley, 2001.

Frame, Davidson J. *Managing Projects in Organizations: How to Make the Best Use of Time, Techniques*, 3rd edition, New York, NY: Jossey-Bass/Wiley, 2003.

Heisler, Sanford I. *The Wiley Project Engineers Desk Reference—Project Engineering, Operations and Management*, 2nd edition, New York, NY: Wiley, 1998.

Kendrick, Tom. *Identifying and Managing Project Risk, American Management Association*, New York, NY: AMACOM, 2003.

Kendrick, Tom. *The Project Management Tool Kit: 100 Tips and Techniques for Getting the Job Done Right*, New York, NY: AMACOM, 2013.

Larson, Erik; Gray, Cliford. *Project Management: The Managerial Process With MS Project*, New York, NY: McGraw-Hill/Irwin, 2013.

Project Management Institute. *Guide to the Project Management Body of Knowledge 5th ed.: PMBOK® Guide*, Newtown Square, PA: Project Management Institute, 2013.

Schwaber, Ken. *Agile Project Management With Scrum*, Redmond, WA: Microsoft Press, 2004.

Software—Project Management

LeanKit: http://leankit.com

Microsoft Project (versions—standard, Professional & enterprise): http://office.microsoft.com/project

ProWorkflow: http://proworkflow.com

Software—Collaborative Tools

Asana: http://asana.com

Microsoft SharePoint: https://products.office.com/en-us/sharepoint/collaboration

OneHub—Office and Mobile Collaboration: http://onehub.com

Wrike: http://wrike.com

CHAPTER

3.4

Managing Workplace and Environmental Hazards

DAVID P. MAXSON

Isotrope, LLC
Medfield, MA

INTRODUCTION

Broadcast engineers are perhaps most familiar with workplace safety issues relating to human exposure to radio frequency (RF) energy, which can be found in Chapter 2.4 of this handbook. However, as places of employment, broadcast facilities also must comply with a broad range of workplace safety requirements. In addition, materials are present at broadcast facilities that could be hazardous to humans or the environment that must be handled in accordance with environmental requirements.

This chapter provides a review of key aspects of federal regulations regarding workplace safety and environmental protection. When preparing safety programs or addressing environmental protection issues, broadcasters are encouraged to read further and seek the advice of specialists and attorneys in these fields.

Two federal agencies administer workplace safety and environmental protection—the Occupational Safety and Health Administration (OSHA) and the Environmental Protection Agency (EPA). In addition to these, it is important to be aware of any local or state laws or regulations that may differ from the federal ones.

This chapter covers four main areas:

- OSHA background and the General Duty Clause
- OSHA standards of particular interest to the broadcasting workplace
- Toxic and hazardous substances
- Environmental impacts of new facilities

OSHA AND THE GENERAL DUTY CLAUSE

OSHA was formed by the U.S. Department of Labor under the 1970 Occupational Safety and Health Act (OSH Act). Its mission is to "assure so far as possible every working man and woman in the Nation safe and healthful working conditions" [1]. The act contains a section commonly called the "General Duty Clause" [2], which is the umbrella that covers all workplace safety.

Section 5. Duties

(a) Each employer—
  (1) shall furnish to each of its employees employment and a place of employment which are free from recognized hazards that are causing or are likely to cause death or serious physical harm to his employees;
  (2) shall comply with occupational safety and health standards promulgated under this act.

(b) Each employee shall comply with occupational safety and health standards and all rules, regulations, and orders issued pursuant to this act which are applicable to his own actions and conduct.

The General Duty Clause makes it clear that both the employer and the employee have duties to operate safely. At one extreme, if the employer has a sound record of informing, training, and enforcing compliance among its employees, it has reduced its legal exposure if an employee is injured because he or she failed to comply. Here the employee has failed to meet the General Duty Clause. At the other extreme, in the absence of the employer providing suitable workplace safety practices, the employee has no guidance on what compliance is. Here, the employer has failed to meet the General Duty Clause.

Of course, the causes of workplace accidents and injuries are often not so clear-cut. Much workplace litigation revolves around who was at fault for an accident. If not from the obvious legal and ethical responsibilities, then at least from the perspective of self-preservation, broadcasters should implement a thorough and consistent workplace safety program to prevent injury and provide a strong defense in the event of an injury.

The employer's responsibility under the General Duty Clause is two-pronged. While the second part, (a)(2), requires employer compliance with OSHA regulations (called standards) on certain workplace practices, standards compliance is not enough. It is the first part of the General Duty Clause, (a)(1), that unambiguously makes a safe workplace the responsibility of the employer. If an OSHA standard does not exist on a topic, the employer must rely on other expert sources for guidance on safe practices. If an OSHA standard does exist, and there is a more restrictive national consensus standard, OSHA, via the General Duty Clause, may rely on the more restrictive national consensus standard [3].

Under the General Duty Clause, four conditions must be met to result in a violation [4]:

1. The employer failed to keep the workplace free of a hazard to which employees of that employer were exposed.
2. The hazard was recognized.
3. The hazard was causing or was likely to cause death or serious physical harm.
4. There was a feasible and useful method to correct the hazard.

Section 6 of the OSH Act addresses the creation of OSHA Occupational Safety and Health Standards. New OSHA standards must pass a test comparing them against national consensus standards. If a new OSHA standard "differs substantially from a national consensus standard" [5], OSHA must publish the reasons why OSHA's "will better effectuate the purposes of this act than the national consensus standard."

Why is the consensus standard principle important? OSHA cannot be expected to maintain up-to-date standards for all working environments, including broadcasting. Industry standards are regularly reviewed and revised as needed. Therefore, it is incumbent on broadcasters to identify or create national consensus standards that apply to their unique workplace practices, and to rely on other standards that may be applicable in broadcasting. For example, national consensus electrical safety standards and RF safety standards certainly apply to broadcast facilities.

In addition to OSHA requirements, states and municipalities may have different requirements or more restrictive ones that apply to broadcasting facilities. Broadcast engineering managers must identify such requirements as they prepare workplace safety programs.

Broadcast facilities also must be compliant with state workplace standards if they are more restrictive than federal standards. Streamlining workplace safety regulation in 26 states, OSHA has approved these states' OSHA programs. OSHA monitors and provides funding to support these state programs. So long as they implement standards that are at least as restrictive as OSHA's, these states take on OSHA responsibilities for monitoring workplace safety, providing safety management consultation, and enforcing standards.

States operating OSHA-approved plans:

Alaska	*New Jersey**
Arizona	*New York**
California	North Carolina
*Connecticut**	Oregon
Hawaii	Puerto Rico
Indiana	South Carolina
Iowa	Tennessee
Kentucky	Utah
Maryland	Vermont
Michigan	*Virgin Islands**
Minnesota	Virginia
Nevada	Washington
New Mexico	Wyoming

*These four jurisdictions have implemented state plans that apply only to public sector employees' workplaces. State-owned broadcast facilities in these places may be subject to these plans, but private workplaces are not.

Specific OSHA standards that may be relevant to the broadcasting workplace are discussed below. While this chapter provides an overview of several OSHA requirements, it is not a substitute for a professional evaluation of the broadcasting workplace.

OSHA STANDARDS OF PARTICULAR INTEREST TO THE BROADCASTING WORKPLACE

Reviewing the OSHA regulations in §1910 of part 29 of the Code of Federal Regulations, several topics are applicable to broadcasters. What follows is a brief treatment of some of the more germane regulations.

§1910.151, First Aid

OSHA requires that:

- First aid kits are required at the workplace. Consider kits that are compliant with ANSI Z308.12003, "Minimum Requirements for Workplace First Aid Kits."
- If medical help is not available "in proximity" to the workplace, first aid–trained employees should be available.
- Provide suitable personal protective equipment (PPE) to protect employees from blood or other potentially infectious materials (§1910.1030(d)(3)). At a minimum, latex gloves and pocket CPR masks should be in first aid kits.

- Add suitable items and equipment to the kits for industry-specific hazards.

In addition to these OSHA requirements, it is a good idea to maintain an oxygen bottle for first response to certain emergencies. Be sure first responders are familiar with the reasons for administering oxygen and seek advice from emergency medical services in its application. As well, even if not OSHA-required, employees benefit by employer-sponsored first aid training programs. To keep first aid kits up to date, consider subscribing to a service that periodically checks and updates first aid kits. These companies can also provide advice on regional or site-specific first aid supply needs.

§1910.132-148, Personal Protective Equipment

Personal protective equipment (PPE) is anything that is used to provide employees (or guests) protection from specific hazards. These sections of the regulations address requirements for protecting eyes and face, head, hands, feet, and lungs from workplace hazards. PPE for electrical hazards is also addressed here. In general, these regulations are nonspecific and reference various ANSI standards for protective footwear, eyewear, head protection, and the like.

Eye protection should be readily available in broadcast operations that involve power tools or flying or blowing debris. Employees involved in construction work or operations that require moving or lifting materials or equipment should have protective footwear and "handwear" to prevent crush or pinch accidents (e.g., steel-toe shoes, leather gloves). In addition to requiring employees who work with power tools to wear eye protection, operations managers might look at other potential eye hazards that may warrant requiring protection. For example, an operations manager might consider questions like:

- Should field crews and reporters be required to wear eye protection outdoors in very windy conditions or at sites where particles might be airborne?
- Does the facility perform certain construction or cleanup work that may generate airborne particles in addition to those that involve power tools?
- Do any employees handle materials, such as glass lamps or lenses, that may shatter accidentally?
- Do any lamps have removable UV filters that are necessary for the protection of individuals under the lights?

In addition to the OSHA PPE standards cited here, there are many kinds of PPE addressing many hazards. It is important to ensure employees know when and how to use their PPE. Also, PPE wears out, or is subject to retirement after use or a specified number of hours or uses. Gloves or shoes may wear out, while a safety harness may either wear out or need to be replaced after experiencing "impact loading" when its wearer falls. Personnel should have the ability to determine whether their PPE is in good order and have the authority to cease hazardous work until PPE issues are resolved.

Fall Protection

Broadcast operations may include work that involves climbing or working near high openings. Roof hatches, rooftops, towers, stages, staging, sets, stage lighting, microphone placement, scaffolding, and moveable lifts are among the special conditions in which broadcast facilities may place workers at risk of falling. OSHA regulates fall hazards in myriad standards. Subpart D of the Occupational Safety and Health Standards (29 CFR 1910.21–30) contains OSHA standards for general industrial fall risks. These sections address industrial stairs, fixed and portable ladders and scaffolding, and floor and wall openings. Subpart F, Powered Platforms, Manlifts, and Vehicle-Mounted Work (29 CFR 1910.66–68), addresses these mechanized systems for elevating workers. Generally, these sections address design characteristics and appropriate uses of the various safety measures. Railings, toe-kicks, gates, and the like are required as applicable. Fall-arrest systems are required in some circumstances, such as on lifts and high ladders.

OSHA references several American National Standards Institute (ANSI) standards, which can be obtained from www.ansi.org:

- American National Safety Standard for Manlifts, ANSI A90.1-1969
- Safety Code for Mechanical Power Transmission Apparatus, ANSI B15.1-1953 (R 1958)
- Safety Code for Fixed Ladders, ANSI A14.3-1956
- Safety Requirements for Floor and Wall Openings, Railings and Toe boards, ANSI A12.1-1967
- Safety Requirements for Personal Fall Arrest Systems, ANSI Z359.1-1992

The OSHA construction regulations provide more detailed standards for fall protection. 29 CFR §1926.1053 addresses ladders used in construction work. Ladders that are part of a rise of more than 24 feet must be equipped with a ladder safety device (such as a climbing lifeline with rope-grab attachment), a self-retracting lifeline with rest platforms at regular intervals, or a cage or well wrapped around the ladder climbing space. Other locations with exposed heights that are not satisfactorily fall-protected by railing, fencing, or netting are also subject to fall-arrest requirements.

Workers involved in activities that require fall protection must be trained periodically in the safe setup and use of fall protection systems. 29 CFR §1926.502, Fall Protection Systems Criteria and Practices, addresses the various fall-protection methods, including railings, fencing, safety nets, fall-arrest systems, and positioning devices.

Full-body harnesses are necessary for fall-arrest systems, with the attachment point being at the center of the upper back of the worker. A deceleration device, such as a tearing or deforming lanyard or a retractable lifeline, is employed to absorb the shock of a fall. Any equipment that has borne the impact of a fall must be retired from service. Climbing belts are not suitable as fall-arrest equipment, but may be used with a proper lanyard as a positioning device, if the worker so positioned can fall no further than 2 feet.

Generally, it is easily assumed that work on a broadcast tower will require fall-arrest and fall-positioning devices. Tower ladders should be equipped with fall-arrest lines or rails that should be used by all climbers. It is also important, however, to apply fall-protection and fall-arrest criteria to other locations frequented by broadcast personnel. Work on rooftops, whether it is to install or orient antennas, examine rooftop mechanical systems, or shoot pictures, requires fall protection. If suitable railings and such are not available, a fall-arrest system may be required. When exiting a roof hatch, it may be necessary to be wearing a harness and attach to a fall-arrest point.

Overhead work, for instance, in theatrical environments placing microphones, cameras, or lights, may require the use of fall-arrest systems, particularly if one leaves the protected confines of a properly railed or fenced catwalk.

§1910.101, Compressed Gases (General Requirements)

OSHA references Compressed Gas Association standards and publications for dealing with compressed gas cylinders. Broadcasters who have air compressors with storage tanks, who feed dry nitrogen from tanks to charge their transmission lines, or who maintain propane or natural gas storage tanks should review their cylinder-handling practices.

In general, compressed gas cylinders should be kept at least 20 feet away from highly flammable materials and secured from falling. When moving a cylinder, a cover or shield should be in place to protect the valve assembly, and should only be moved with a suitable wheeled hand truck. Cylinders held for long periods may require periodic inspection and pressure testing. Check with a compressed gas supplier for advice on how to determine what a cylinder's maintenance requirements are.

Compressed gases, other than air, have the potential to displace oxygen in a workspace. Depending on the space, the gas, and the normal ventilation of the place where the cylinder is used or stored, precautions should be taken to inform employees about on-site gas storage hazards and, if warranted, set up sensors to trigger an alarm in the event of a gas leak or displaced room air. Compressed air for cleaning should have an outlet pressure of less than 30 PSI (§1910.242).

§1910.242, Hand and Portable Powered Tools and Equipment (General)

While an employer may permit employees to supply their own tools or equipment, the employer is responsible for assuring that all tools used on the job are in safe condition.

§1910.147, The Control of Hazardous Energy (Lockout/Tagout)

Broadcasting facilities contain equipment that is permanently connected to a power source and that requires maintenance and repair work with the equipment turned off. Lockout/tagout regulations require each employee to have positive control over the power source(s) to the equipment upon which he or she is working.

The most obvious equipment that fits this category is the broadcast transmitter. When powered, it is capable of generating lethal or injurious voltages. In addition to electrical or RF energy hazards, kinetic and potential energy hazards are a concern as well. The transmitter is likely to contain moving parts in its air handling system that must also be made safe prior to and during maintenance.

Examples of other equipment at broadcast facilities that may require energy control are:

- Chillers, heaters, and air handlers
- Robotic studio equipment (lighting and camera supports)
- Cranes, booms, lifts, and the like
- Elevators (building or tower)
- Steerable antennas (satellite, radar, ENG, etc.)
- Motor and solenoid actuated devices (e.g., coaxial switches, AM RF relays, ventilation doors, and louvers)
- Motorized vehicles (e.g., remote trucks)
- Transmitters and other equipment hard-wired to power sources

Equipment that can simply be unplugged, with the plug remaining under the control of the affected employee(s), does not have to be locked out.

To manage these hazards, OSHA §1910.147 states:

> (c) General—(1) Energy control program. The employer shall establish a program consisting of energy control procedures, employee training, and periodic inspections to ensure that before any employee performs any servicing or maintenance on a machine or equipment where the unexpected energizing, startup, or release of stored energy could occur and cause injury, the machine or equipment shall be isolated from the energy source and rendered inoperative.

To accomplish this important task, OSHA requires that equipment be "locked out." Each employee working on a device must be able to physically secure the power source(s) in the de-energized position while working on it. Broadcasters should design their facilities with lockout capabilities on their circuit breaker and shutoff switch systems. Lockout fixtures are available from electrical and safety supply houses that permit multiple employees to each apply their own locks to the selected shutoff switch.

Tagging out a power source is an alternative that is permitted if locking out is not possible. Tagging out is also permitted if it demonstrably provides the same degree of safety as locking out. Tagging out a power source involves the placement of a clearly understood tag on the control to the power source with a nonreusable fastener that requires at least 50 lbs. of force to break, to ensure that inadvertent loss of the tag does not occur. Fortunately, the common cable tie is explicitly mentioned by OSHA as a suitable tagout device. The tag must warn of the hazard and give a clear instruction, such as Equipment Being Serviced—Do Not Energize. Locks and tags must contain the employee name.

In addition to the actual locking/tagging requirements, OSHA requires that personnel be trained and their use of the procedures be inspected at least annually. The inspector must individually review each affected employee's knowledge of the procedures and understanding of the hazards. Lockout/tagout programs have employer recordkeeping requirements as well.

The employer or an authorized employee must inform affected employees of the application and removal of lockout/tagout devices. This is to ensure there is no confusion among a crew about systems going off- or on-line. Locks and tags may only be applied and removed individually by each employee working on the system. An exception can be made for a single group lockout instead of individual lockouts, with a number of additional procedural requirements to ensure worker safety.

In preparing a written lockout/tagout program, the whole of §1910.147 should be reviewed by the program's author to absorb the details of program creation and management. Section 1910.147 also contains a model lockout/tagout procedure document.

Subpart S, OSHA Electrical Standards

For electrical systems, Subpart S of the OSHA regulations apply additional criteria to the lockout/tagout procedures. Once a system is de-energized and locked/tagged out, it is not considered safe to work upon until it has been tested. Section 1910.333(b)(2)(iv)(B) states:

> A qualified person shall use test equipment to test the circuit elements and electrical parts of equipment to which employees will be exposed and shall verify that the circuit elements and equipment parts are de-energized. The test shall also determine if any energized condition exists as a result of inadvertently induced voltage or unrelated voltage backfeed even though specific parts of the circuit have been de-energized and presumed to be safe. If the circuit to be tested is over 600 volts, nominal, the test equipment shall be checked for proper operation immediately after this test.

High-capacity capacitors should be discharged, shorted, and grounded if they might endanger personnel. Interlocks, selector switches, and pushbutton switches are not sufficient to meet lockout/tagout requirements. OSHA does anticipate that under some circumstances it may be necessary for a qualified person to temporarily defeat an interlock to service a device. When a qualified person is working near energized components, he or she must maintain a safe distance from energized components or employ appropriate PPE. While OSHA provides Table S-5 (shown here in Table 3.4-1) for qualified persons to maintain safe distances from overhead power lines, the table may be informative for broadcast engineers working near high-voltage power supplies.

The "qualified person" or "qualified employee" in Subpart S is one who "has training in avoiding the electrical hazards of working on or near exposed energized parts." Broadcast engineers familiar with the hazards of high-voltage power supplies may be likely to employ safe work practices with electrical systems; however, it is the responsibility of the employer to provide the necessary safety program and training to ensure that qualified persons are truly qualified and have the procedures and equipment necessary to work safely with electrical power.

Antennas Near Power Lines

There are instances when broadcast operations may come near overhead power lines, such as remote trucks with extendable antenna masts or the installation of an antenna and mast on a roof or an antenna and tower on the ground. OSHA requires a minimum distance of 10 feet from any part of the antenna or mast to the power line. The distance is increased for certain high-voltage lines. In §1910.333(c)(3)(iii)(A), OSHA states:

> Any vehicle or mechanical equipment capable of having parts of its structure elevated near energized overhead lines shall be operated so that a clearance of 10 ft. (305 cm) is maintained. If the voltage is higher than 50 kV, the clearance shall be increased 4 in. (10 cm) for every 10 kV over that voltage.

This regulation concerning operations near energized overhead power lines is contained within Subpart S and is subject to the subpart's personnel protection requirements. Personnel must be trained to follow corporate safety procedures, including identifying and observing safety distances from overhead power lines when erecting antennas and masts. It may be prudent to require greater distances than the OSHA minimums. Safety is improved with the use of alarm/interlock systems that detect the proximity of high-voltage lines, but such systems cannot be a substitute for human observation. Companies such as Sigalarm and Will-Burt offer proximity warning devices for mobile mast systems.

Title 8 of the California Code of Regulations, Chapter 4, Article 40 contains ENG antenna safety requirements for operators in California that operators in other states may find instructive. In addition to antenna safety, operators of ENG news trucks have numerous other safety management concerns. The website at

TABLE 3.4-1
29 CFR 1910.333(c)(3)(ii), Table S-5, Approach Distances for Qualified Employees—Alternating Current

Voltage Range (Phase to Phase)	Minimum Approach Distance
300 V and Less	Avoid Contact
Over 300 V, not over 750 V	1 ft. 0 in. (30.5 cm)
Over 750 V, not over 2 kV	1 ft. 6 in. (46 cm)
Over 2 kV, not over 15 kV	2 ft. 0 in. (61 cm)
Over 15 kV, not over 37 kV	3 ft. 0 in. (91 cm)
Over 37 kV, not over 87.5 kV	3 ft. 6 in. (107 cm)
Over 87.5 kV, not over 121 kV	4 ft. 0 in. (122 cm)
Over 121 kV, not over 140 kV	4 ft. 6 in. (137 cm)

www.engsafety.com, provided by Mark Bell, is a useful resource for all ENG safety matters. ENG operators must be aware of vehicle maintenance, safe driving practices, weather-related safety, generator power and carbon monoxide safety, and protection of/from bystanders, among others.

29 CFR §1910.95, Occupational Noise Exposure

Occupational noise exposure is limited by a combination of duration and intensity. Noise exposure is evaluated by considering the amount of time in the workday an employee is exposed to certain levels of noise. Noise level is expressed in dBA ("A-weighted" sound pressure level) as measured by a "slow response" instrument that integrates sound pressure level in one-second-long samples. Employees may not be exposed to peak sound levels in excess of 140 dBA. Exposure to lesser levels is regulated by duration.

As shown in OSHA's Table G-16 (shown here in Table 3.4-2), it is permissible to be exposed to sound pressure levels of 90 dBA for an 8-hour shift. However, when the average daily exposure is greater than 85 dBA, a "hearing conservation program" must be implemented. OSHA describes the components of such a program in §1910.95(c). Whenever practicable, it would be less cumbersome simply to ensure some workers not exceed the 85-dBA daily average.

In an environment where producers, talent, and technical people have enclosed spaces with speakers (studios and production rooms) and have headphones available, it may be difficult to be certain that average daily exposures are below the 85-dBA threshold. For comparison, normal level conversations tend to fall in the 60–65 dBA range. A pair of headphones turned up to a level that blocks out conversational level and ambient sounds could be in the 75–85 dBA range. It is not uncommon to find some users running headphones at higher levels.

The assessment of daily average noise exposure is complicated by the level-dependent time limits. For instance, Table 3.4-2 shows that there is a 2-hour daily limit on exposure to 100-dBA noise. If an exposure occurred for 1 hour at 100 dBA, that would be an exposure of one-half of the daily limit. Meanwhile, the exposure limit to 90 dBA is 8 hours. If an exposure of 4 hours at 90 dBA were to occur that day, that would also be half of the daily limit. Thus, the employee who is exposed to 100 dBA for 1 hour and 90 dBA for 4 hours has reached the daily limit, and the remaining 3 hours of the 8-hour workday must be below 85 dBA.

Total daily exposure is presented as a sum of the weighted ratios of the various levels of exposure during the day. The calculation is performed with the variable C_n representing an exposure duration to a certain noise level, n, and T_n representing the allowed time limit of exposure to that level, n. The duration of exposure to each sound level is divided by the exposure limit for that sound level, and the ratios are summed:

$$C_1/T_1 + C_2/T_2 + \ldots + C_n/T_n = \text{daily noise exposure} \quad (1).$$

TABLE 3.4-2
29 CFR §1910.95, Table G-16, Permissible Noise Exposures

Duration Hr./Day	Sound Level dBA, Slow Response
8	90
6	92
4	95
3	97
2	100
1.5	102
1	105
0.5	110
0.25 or less	115

The daily noise exposure is not permitted to exceed the value of 1, which represents 100% of the exposure limit.

Facilities managers should determine whether workers are persistently exposed to sound levels in excess of 85 dBA. If so, then a more detailed evaluation of occupational exposure in the facility should be undertaken. It may be necessary to establish a hearing conservation program, provide training, monitoring, and control of exposure levels, and perform a medical surveillance program in which affected employee hearing is routinely tested.

Radio Frequency Energy Exposure (§1910.97, Nonionizing Radiation)

The OSHA standard on RF energy exposure maintains several antiquated elements, including the use of the term "nonionizing radiation," which has generally fallen out of favor; the use of a fixed power density level of 10 mW/cm^2 from 10 MHz to 100 GHz; and an old radiation sign design that has been superseded by several ANSI standards.

Consequently, in radio-transmitting facilities, OSHA relies on the exposure standard adopted by the FCC in 47 CFR §1.1310. OSHA applies the General Duty Clause not only in the application of the FCC standard, but also in the expectation that employers will develop a published safety program and implement employee training.

Broadcast facilities should have an RF energy exposure program developed under the same principles that OSHA applies to other workplace safety programs. The following elements are excerpted from OSHA's *Elements of a Comprehensive RF Protection Program: Role of RF Measurements* [6]:

Element 1: Utilization of RF source equipment that meets applicable RF and other safety standards when new and during the time of use, including after any modifications.

Element 2: RF hazard identification and periodic surveillance by a competent person who can effectively assess RF exposures.

Element 3: Identification and control of RF hazard areas.

Element 4: Implementation of controls to reduce RF exposures to levels in compliance with applicable guidelines (e.g., ANSI, ICNIRP), including the establishment of safe work practice procedures.

Element 5: RF safety and health training to ensure that all employees understand the RF hazards to which they may be exposed and the means by which the hazards are controlled.

Element 6: Employee involvement in the structure and operation of the program and in decisions that affect their safety and health, to make full use of their insight and to encourage their understanding and commitment to the safe work practices established.

Element 7: Implementation of an appropriate medical surveillance program.

Element 8: Periodic (e.g., annual) reviews of the effectiveness of the program so that deficiencies can be identified and resolved.

Element 9: Assignment of responsibilities, including the necessary authority and resources to implement and enforce all aspects of the RF protection program.

For additional information on human exposure to radio frequency fields, see Chapter 3.5.

Perhaps the only element that requires elaboration in this context is medical surveillance (element 7). Medical surveillance involves collecting and acting upon medical information relevant to maintaining workplace safety. OSHA explains that events, such as RF burns or heating sensations, should be recorded in a log to enhance evaluation and improvement of protection programs. Maintaining exposure records collected from personal RF exposure monitors is a form of medical surveillance. As a means of accident or injury prevention, there may be circumstances where it is appropriate to have employees report, for instance, any medical implant devices that might be affected by or cause injury from exposure to high RF energy levels. An employer may have a policy that requires a medical examination in the event of an injury or overexposure. Whatever the elements of the medical surveillance program are, they should be part of a written RF safety program.

In summary, RF energy exposure should be managed under an appropriate workplace safety program that should:

- Be written and available
- Be constructed with worker input
- Be reviewed periodically
- Include periodic training
- Provide for reporting and recordkeeping appropriate to the conditions

TOXIC AND HAZARDOUS SUBSTANCES

This section addresses toxic and hazardous substances. Since this is a regulatory matter of equal interest in protecting workers and the environment, OSHA regulations and EPA regulations are discussed in tandem. The topics discussed below include:

- Asbestos
- Hazard communication
- Material safety data sheets
- Hazardous substance release into the environment
- Underground storage tanks
- PCBs (polychlorinated biphenyls)
- Environmental impacts of new broadcast facilities

29 CFR Subpart Z (§§1910.1000-1450) contain OSHA standards for managing toxic and hazardous substances in the workplace. More than 30 specific substances have their own standards in these regulations. In addition, scores of substances are listed under the classification of air contaminants. Broadcast professionals who have responsibilities for administering worker safety are encouraged to review the OSHA substances and air contaminants to identify any that might be in the workplace and to implement the OSHA standards relating to them.

Asbestos

One substance, asbestos, which is also a potential air contaminant, might be encountered by broadcasters in their facilities. Its OSHA standard is found at §1910.1001. Asbestos is a fire-resistant fiber that is used to reinforce materials that may be exposed to high heat. When airborne, the fibers can be breathed into the lungs where they can provoke disease. Asbestos-containing materials (ACM) that are "friable" are considered hazardous. When a material can be readily crushed or broken in a fashion that releases fibers into the atmosphere, it is friable.

Asbestos may be found in an older broadcasting facility in a variety of places. Boilers and heating pipes could be wrapped or coated in an asbestos material. Some vinyl-composition tile may contain asbestos. Older buildings may have asbestos-containing shingles. Poured-in insulation, such as that consisting of the mineral vermiculite, may be contaminated with asbestos. Roofing felt may be an ACM. Some types of mineral board used as a fireproof backing or liner to fire-prone spaces (boiler rooms, workshops, and the like) may be ACMs.

So long as the ACM remains undisturbed, it may be considered safe. Rather than removing such materials, it may be sufficient to *encapsulate* or *envelop* them to prevent them from being disturbed. Encapsulation is the act of sealing the ACM in a material that keeps it stable and prevents the release of fibers into the air.

Enveloping an ACM is the act of containing the ACM within a confined space to prevent the disturbance of the material and the release of the material outside the envelope.

> **CAUTION:** If any removal or repair work involves asbestos-containing materials, it should be evaluated by a licensed asbestos removal professional.

Nonfriable materials, such as vinyl-asbestos floor tile, mineral board, or roof felt, should not be cut, broken, or sanded. Before removing such materials, consult a professional to be certain the removal process will not create friable conditions.

The EPA publishes several guides on dealing with asbestos in buildings, including the *Green Book* on managing asbestos in place, the *Blue Book* on controlling ACM in buildings, and the *Custodial Brochure*, a guide for maintenance personnel, and guidelines on maintaining asbestos flooring. Visit www.epa.gov/asbestos/buildings.html for more information. The National Institute of Building Sciences also publishes a book on the subject, available at www.nibs.org.

§1910.1200, Hazard Communication Standard

In addition to the substances that have specific OSHA standards, all potentially hazardous substances in the workplace are subject to the OSHA Hazard Communication Standard (HCS), with limited explicit exceptions. From §1910.1200(b)(1):

> This section requires . . . all employers to provide information to their employees about the hazardous chemicals to which they are exposed, by means of a hazard communication program, labels and other forms of warning, material safety data sheets, and information and training.

Potentially hazardous materials should be evaluated and include "any chemical which is known to be present in the workplace in such a manner that employees may be exposed under normal conditions of use or in a foreseeable emergency." The most applicable exception to this is

> where the employer can show that it is used in the workplace for the purpose intended by the chemical manufacturer or importer of the product, and the use results in a duration and frequency of exposure that is not greater than the range of exposures that could reasonably be experienced by consumers when used for the purpose intended.

In short, if a chemical or mixture of chemicals is used in the workplace to a degree greater than a consumer would use it and if it presents a potential hazard in regular use or in a foreseeable emergency (fire, spill, etc.), then it is subject to the HCS.

Nonionizing radiation is not considered a covered "substance" in this context. OSHA has separate requirements for it.

Material Safety Data Sheet

The focal point of the HCS is the material safety data sheet (MSDS). This document contains specific information about the chemical or material, including its name, composition, hazard mechanisms, safe exposure levels, target organs, safety procedures, and the like. All producers and distributors of chemicals and substances are required to make an MSDS available to the purchaser. A corporate hazard communication program will review the MSDS of all chemicals at the workplace and determine which ones, if any, should be included in the program.

Once hazardous materials are identified, several steps must be taken:

- A written program should be prepared.
- An individual should be assigned responsibility for maintaining the program and executing periodic training. A procedure for periodically reviewing the program's effectiveness should also be implemented.
- A list of hazardous materials and their respective MSDS must be posted in a location where employees have access to it. A binder containing the written program and MSDS is suitable. Access via computer is also acceptable.
- Employees should receive regular training on the use, handling, and emergency response to each hazardous material with which they work.
- Safety equipment, and if necessary, medical monitoring, should be provided as appropriate.
- All containers should be labeled in a manner that conforms to the standard, including the identity of the material as given on the MSDS and a description of the hazard(s).

Broadcast facilities may contain, for example, cleaners, solvents, paints, compressed gases, fuels, industrial fire extinguishing material, and solders that could be used in a manner that causes greater than consumer-level exposure. Some materials may be in solid, granular, or particulate form instead of liquid or gas, such as wood cuttings, sawdust, dust from cleaning operations, grinding dust, theatrical smoke, real smoke, or breakable lamps containing heavy metals. These materials may present *physical hazards* and/or *health hazards*. For example, paint may present the risk of fire or explosion (physical hazard) and the risk of toxicity or disease (health hazard). Hazard communication programs are supplemented with the provision of suitable PPE, such as respirators, glasses, gloves and such, and the provision of suitable protection systems, such as ventilated and filtered work booths, spill-containing curbs, fire-suppression systems, and the like.

Hazardous Substance Release

Once such hazardous materials are identified, catalogued, and labeled at the broadcast facility, another aspect of these materials must be considered—environmental effects in use or accidental release. This is the purview of the EPA. The enactment of the Comprehensive Environmental Response, Compensation, and Liability Act of 1980 (CERCLA) resulted in certain EPA regulations relating to environmental release of hazardous substances.

Those substances are listed in 40 CFR §302.4 along with their "reportable quantities." If an accidental release exceeds the reportable quantity of the substance, the EPA must be notified.

Underground Storage Tanks

CERCLA also controls the use of underground storage tanks (USTs) that may contain petroleum products or other hazardous substances. EPA regulations in 40 CFR §280 apply to the care and use of underground storage tanks. A tank is considered an underground storage tank if at least 10% of its capacity is below grade. Tanks storing heating oil for on-premises use are not considered USTs and are exempt.

Many broadcast facilities have standby power systems fueled by petroleum products (e.g., diesel or gasoline). USTs of 110 gallons or less capacity are exempt from these regulations. Larger USTs are subject to certain design requirements intended to prevent accidental release (leaks), such as corrosion protection, double-wall construction, corrosion-free materials, and/or overfill protection. They are also subject to complex leak-monitoring requirements in 40 CFR §280 Subpart D. Fortunately, USTs for standby power systems are exempt from the leak-monitoring requirements.

Old USTs must meet current requirements through upgrade or replacement if necessary.

State and local regulations may be more stringent than federal regulation of USTs, so it is important to be familiar with local requirements. EPA has permitted 26 states and jurisdictions to administer the EPA UST program.

List of States with EPA-Approved UST Programs:

Alabama	North Carolina
Arkansas	North Dakota
Connecticut	Oklahoma
Georgia	Rhode Island
Iowa	South Dakota
Kansas	Tennessee
Louisiana	Texas
Maine	Utah
Massachusetts	Vermont
Mississippi	Virginia
Nevada	West Virginia
New Hampshire	Puerto Rico
New Mexico	

Polychlorinated Biphenyls: Occupational Exposure

Except in its standards for workplace air contaminants, OSHA does not specifically address polychlorinated biphenyls (PCBs) in the workplace. PCBs are a persistent environmental pollutant. Their manufacture in the United States ended in 1977. PCBs were used as a heat-stable material in various applications.

With respect to broadcast facilities, PCBs were often used in capacitors and transformers associated with high-power equipment such as transmitters and power supplies. Because of the manner in which they are used at broadcast facilities, PCBs are not likely to become airborne. However, because they are not soluble in water, PCBs were often dissolved in solvents to perform their function. Thus, the carrier fluid that contains PCBs may also contain other hazardous materials such as benzenes [7].

A PCB MSDS provides the necessary information about PCBs in the event of "accidental release" (a leak or spill) of the chemical, and should be kept in the company MSDS file if there are PCB devices in the facility. Generally stable at room temperature, hot PCB fluids may produce irritation in eyes and lungs. PCBs can be absorbed through the skin by direct contact with the fluid. According to the GE Industrial MSDS, the evidence of possible PCB connection to chronic human disease is not strong, but a significant exposure could be injurious to the subject.

PCBs: Environmental Exposure

The EPA regulates the use, storage, and disposal of PCBs as potential environmental pollutants in 40 CFR §761. Table 3.4-3 contains a matrix of the different classes of electrical equipment that may contain PCBs, the degree of contamination in parts per million of PCBs in the fluid, and the disposal requirements.

PCB items that are not in use and are not stored for a clearly identified reuse are considered PCB waste. If a broadcaster has equipment removed from service that contains PCB items, or has PCB items, such as capacitors, sitting separately on a shelf, it is advisable to determine whether each item can be reused or is intended to be reused. If not reusable, the items should be properly disposed of within certain time limits. Generally, there is a 1-year grace period from removal from service to disposal (40 CFR §761.65), with some exceptions.

Disposal of PCB items must conform to EPA requirements. Only those PCB items listed in Table 3.4-3 as being eligible for disposal in municipal waste require no special treatment. All other PCB items require licensed services to handle, store, transport, and dispose of them. When disposing of such items, contact a qualified waste handler or environmental services provider. For the remainder of its existence, the PCB item that your waste handler picks up from your facility is tracked with transportation manifests, storage records, and ultimately, a disposal document.

If a PCB item has leaked at the broadcast facility, an environmental cleanup may also be necessary. Tools, flooring, and such may need to be decontaminated according to EPA criteria. The cleanup materials will also have to be disposed by the waste carrier.

PCB items must be marked with the EPA PCB mark (40 CFR §761.50; see Figure 3.4-1). The equipment or container within which the PCB item resides should also be marked.

TABLE 3.4-3
PCB Equipment Disposal Matrix

Equipment	Disposal	<50 ppm (Non-PCB)	<500 ppm (PCB Contaminated)	>500 ppm (PCB Transformer or PCB Capacitor)
Transformers with <3 lb. fluid	Municipal solid waste	Assumed		
Circuit breakers, reclosers, oil-filled cable, rectifiers	Municipal solid waste	Assumed		
Mineral oil-filled equipment manufactured before July 2, 1979	EPA incineration		Assumed	
Pole and pad-mount transformers with >3 lb. fluid manufactured before July 2, 1979, or if date is uncertain	(1) EPA incineration, or (2) clean and dispose in chemical waste landfill, then dispose of fluid separately via EPA incineration		Assumed	
Capacitor stamped at time of manufacture "No PCBs"	Normal disposal	Assumed		
Capacitor or electrical equipment made after July 2, 1979	Normal disposal	Assumed		
Capacitor made before July 2, 1979, or with uncertain date of manufacture	See large and small capacitors below			Assumed
Small capacitor: Less than 3 lb. fluid, or assume less than 3 lb. fluid if less than 100 cubic inches overall size, or less than 200 cubic inches overall size and less than 9 lb.	Municipal solid waste			Assumed
Large high voltage capacitor 2000-volt or greater rating and greater than 3 lb. fluid	EPA incineration			Assumed
Large low-voltage capacitor: <2,000 volt rating and greater than 3 lb. fluid	EPA incineration			Assumed

In lieu of assumptions indicated above, PCB concentration may be determined by fluid test, permanent label at time of manufacture, or full documentation of fluid servicing history. Documented communication with manufacturer to obtain concentration is also permissible.

From 40 CFR §§761.2, 761.60

FIGURE 3.4-1 The EPA PCB mark

ENVIRONMENTAL IMPACTS OF NEW FACILITIES

The construction of broadcast facilities is subject to the National Environmental Policy Act (NEPA) when they involve FCC-licensed facilities such as transmitters. In addition to any local regulations for permitting and constructing such facilities, 47 CFR §1.1306 et seq., implements NEPA for FCC licensees.

Of all the environmental impacts with which broadcasters could be concerned, the most familiar may be RF exposure of the general public. Radio-frequency emissions of broadcast facilities are among the criteria to be assessed under the FCC NEPA regulations. Also covered by these regulations are more common national environmental issues, listed in 47 CFR §1.1307 and summarized below.

Environmental Assessment

If there is a potential environmental impact from a new or changed facility, such as a broadcast tower, it is subject to "environmental processing." Environmental processing begins with the preparation of an environmental assessment (EA). If the EA finds that there is an environmental impact, the FCC must respond by preparing an environmental impact statement (EIS).

Some facilities are assumed to have no environmental impact under NEPA and are "categorically excluded" from making an EA. According to 47 CFR §1.1306 and §1.1307, if a proposed licensed facility does not involve any of the following, it is categorically excluded from environmental processing and need not prepare an EA:

1. Officially designated wilderness area
2. Officially designated wildlife preserve
3. Existing or proposed endangered species and/or their habitats
4. Historical, cultural, archeological, architectural, or engineering places or structures that are listed, or are eligible for listing, in the National Register of Historic Places
5. Indian religious sites
6. Flood plain
7. Significant change in surface features (filling wetlands, diverting rivers, clearcutting, etc.)
8. High-intensity white FAA lights in residential neighborhoods
9. Human exposure to radio-frequency energy in excess of the applicable safety standards

With a little legwork, a civil engineering or real estate development firm employing a common NEPA checklist can address the first seven points. It will need the broadcaster's input on the last two—tower lighting and RF energy exposure.

If one or more of these conditions apply, the broadcaster must prepare and file an EA with the FCC. The FCC must then determine if the EA demonstrates there will be no significant environmental impact, in which case the construction of the new or modified facility may proceed. However, if there is a significant environmental impact, the FCC must prepare an EIS. The EIS process is cumbersome, time-consuming, and costly. It may result in an unfavorable decision. Consequently, licensees work intently first to avoid the nine conditions above by carefully designing their proposed facilities. Only if necessary, a broadcaster may have to propose a facility that requires the preparation of an EA. With a well-crafted facility and a well-crafted EA, it may be possible to demonstrate that there is no significant environmental impact, despite an impact on one or more of the items in the checklist. However, local opposition may organize and challenge any favorable conclusions of the EA.

The National Historic Preservation Act

In addition to the specialized nature of items 8 and 9 above, tower lights and RF energy exposure, items 4 and 5 are subject to special treatment at the federal level. The National Historic Preservation Act (NHPA) outlines requirements that the federal government must observe relating to impacts on historical sites and native religious sites. To streamline its handling of its responsibilities, the FCC has entered into a "nationwide programmatic agreement for review of effects on historic properties for certain undertakings approved by the Federal Communications Commission" (NPA) with historic preservation authorities.

The FCC explains that the NPA is a "win-win" agreement because it succeeds in [8]:

- Refining the process for identifying "eligible properties," by requiring the use of records in the SHPO (state historic preservation offices) or in THPO (tribal historic preservation offices)
- Excluding certain categories of undertakings from review that, as determined by the working group, are not likely to adversely affect historic properties. These undertakings include:
  - Enhancements to towers
  - Temporary towers
  - Replacement towers
  - Certain towers constructed in industrial and commercial areas
  - Certain towers constructed in utility corridor rights-of-way
  - Towers constructed in SHPO/THPO designated areas
- Establishing specific procedures for contacting SHPOs, including a provision authorizing tower constructors to proceed with construction if a SHPO does not respond within thirty days
- Creating standard forms, known as the submission packet, establishing a uniform, nationwide standard for filings to SHPOs. On January 18, 2005, OMB approved FCC forms 620 and 621 (see OMB No. 3060-1039).
- Establishing guidelines for consulting with federally recognized Native American tribes and NPOs (national preservation offices)

SUMMARY

This chapter has provided a review of key aspects of federal regulations regarding workplace safety and environmental protection. When preparing safety programs or addressing environmental protection issues, broadcasters are encouraged to read further and seek the advice of specialists and attorneys in these fields.

References

[1] OSH Act of 1970, 29 USC 651 (b).
[2] OSHA, 29 USC 654.
[3] Curtis, Robert, OSHA.
[4] OSHA Standard Interpretation, Elements Necessary for a Violation of the General Duty Clause, Dec. 18, 2003.
[5] OSHA, 29 USC 655 (b)(8).
[6] Curtis, Robert A., director, U.S. DOL/OSHA Health Response Team, Presentation at the National Association of Broadcasters Broadcast Engineering Conference, Las Vegas, NV, April 12, 1995.
[7] GE Industrial, PCB MSDS.
[8] FCC, Learning Interactive Unit: Nationwide Programmatic Agreement, see www.fcc.gov/general/tower-and-antenna-siting.

CHAPTER

3.5

Human Exposure to Radio Frequency Energy

ROBERT D. WELLER, P.E.

National Association of Broadcasters
Washington, DC

INTRODUCTION

The awareness and use of electromagnetic energy for communications continues to increase with consumers routinely having WiFi access points, cell phones, and a host of other wirelessly connected devices in their homes. This chapter updates previous editions of the *NAB Engineering Handbook* and is focused on the requirements of broadcasters, including new and revised standards in force, proposed regulatory actions, and updated methods for measuring radio frequency energy. Since the last version of the handbook was published, a large number of new studies have evaluated the potential for acute and chronic RF field exposures to elicit possible effects on a wide range of biological endpoints, including human cancers; tumor initiation, promotion and co-promotion; mutagenicity and DNA damage; EEG activity; memory, behavior, and cognitive functions; gene and protein expression; and perceived electromagnetic hypersensitivity, among others. Numerous authoritative reviews have summarized the current literature.[1,2] While this chapter will not cover biological research in detail, two regulatory issues important to broadcasters remain fluid:[3]

- The Federal Communications Commission (FCC) is considering the comments it received on a 2013 Further Notice of Proposed Rulemaking, ET Docket 03-137, to broadly revise and harmonize its criteria for assessing single or multiple fixed, mobile, or portable RF sources and generally adopting IEEE Standard C95.7 recommendations for exposure categories, RF safety programs, and signs.
- The FCC is considering comments it received on a related 2013 Notice of Inquiry, ET Docket 13-84, which among other things sought comment on changing its underlying exposure limits.

These ongoing activities mean that the current regulatory state continues to be in flux, so the reader is urged to check, perhaps using the web resources given at the end of this chapter, for the latest information.

STANDARDS

The first standards limiting human exposure to RF energy were promulgated in 1953 by Bell Telephone Laboratories to protect against reduction in visual acuity.[4] At that time, this adverse effect was observed at a power density of 1000 W/m^2 (100 mW/cm^2),[5] and the exposure limit was set 30 dB below that level. A succession of private and public standards appeared between 1954 and 1966, when a uniform exposure limit of 100 W/m^2 (10 mW/cm^2) was established as a standard by the United States of America Standards Institute. Exposure limits were uniform with frequency until 1982, when research demonstrated that the typical adult human body exhibits a strong resonance at frequencies near 70 MHz owing to its physical dimensions. In 1985, the FCC adopted ANSI Standard C95.1-1982, one of the first standards that exhibited frequency dependence, to be used for evaluating human exposure to RF energy.

Since 1986, there has been a trend toward two-tiered safety standards limiting human exposure to RF energy. Today, nearly all of the prevailing standards are two-tiered, typically setting more restrictive limits on public exposure than on worker exposure. The reasons for adopting different limits, however, are largely nonscientific. That is, the general public should not be regarded as being more susceptible to injury from RF exposure than workers. Rather, workers are allowed higher levels of exposure, because they are aware of their exposure, are educated about its risks, and are able to control the level and extent of their exposure. The more restrictive public limits, however, do afford some additional measure of protection against nonbiological hazards, such as interference with implanted pacemakers.

Except at low frequencies (AM broadcast frequencies and below), all of the prevailing standards have identical or similar exposure limits. This makes sense because all science-based standards are based on the same scientific data. The exposure limits of all of the prevailing standards are based upon the lowest exposure level at which any adverse health effect has been scientifically established. Safety margins have been incorporated into the exposure limits to ensure that even worst-case exposures within those limits remain well below the threshold for harm.

NCRP Report No. 86

In 1996, the FCC adopted the recommended power density limits specified in a 1986 report published by the congressionally chartered National Council on Radiation Protection and Measurements (NCRP).[6] The NCRP Report covers 0.3 MHz–100 GHz, which includes all of the broadcast and broadcast-auxiliary bands, except for longwave (frequencies below 300 kHz). Although other standards, such as C95.1, were more recent, the Environmental Protection Agency (EPA) and other U.S. health agencies endorsed the scientific basis of the NCRP guidelines as being superior to those of other standards. Despite being over 30 years old, this standard remains in place and enforced by the FCC, and with respect to its field-strength limits at most broadcast frequencies, is consistent with most standards in the world.

ANSI/IEEE Standard C95.1

The C95.1 standard is part of a family of standards (including techniques of measurement, terminology, electro-explosive devices, and RF safety programs), developed by the International Committee on Electromagnetic Safety (ICES), under sponsorship of the Institute of Electrical and Electronics Engineers (IEEE).[7] Standard C95.1 covers 3 kHz to 300 GHz. The current edition was published in 2005 with an amendment (C95.1a) published in 2010. The companion extremely low frequency standard, C95.3, is presently being merged with the C95.1 RF standard. One outcome of that merger is C95.1-2345, published in 2014, which covers exposure in NATO military workplaces over the entire frequency range, 0 Hz to 300 GHz. The 2005 edition of C95.1 (with its 2010 amendment) expires at the end of 2018.

Safety Code 6

In Canada, Health Canada[8] adopted Safety Code 6 (SC-6, currently published as the 2015 edition). SC-6 is a standard similar to the other standards, but with more conservative (lower) public exposure limits at FM and TV broadcast frequencies. The current edition is a relatively short document (24 pages).

ICNIRP

In most countries other than the United States and Canada, the 1998 recommendations of the International Commission on Non-Ionizing Radiation Protection (ICNIRP)[9] are used to assess human exposure to RF energy. These recommendations were reconfirmed by ICNIRP in August 2009.

Summaries of the NCRP-86 power density guidelines, SC-6 exposure limits, C95.1 exposure limits, and ICNIRP limits are shown in Figures 3.5-1 and 3.5-2 for public and occupational exposures, respectively. Contact and induced body current limits are given in Table 3.5-4.

Unit Conversion

The NCRP power density exposure limits are specified in terms of equivalent plane-wave power densities, while most other standards express the limits in terms of both E- and H-fields (typically below about 300 MHz), and/or power density (typically above 300 MHz). Since most practical equipment measures either the E- or H-field only, it is helpful to define the relationship between these various exposure metrics. The conventional relationship is limited only to free space locations (where the intrinsic impedance is approximately 120π ohms), which are in the far field of the antenna (so that the E- and H-field vectors are orthogonal). In SI units, the relationship is:

$$|\vec{S}| = \frac{|\vec{E}^2|}{120\pi} = 120\pi|\vec{H}^2|$$

where $\vec{S}$ is the power density (Poynting) vector in W/m^2, $\vec{E}$ is the electric field vector in V/m, and $\vec{H}$ is the magnetic field vector in A/m. In CGS units (used in the NCRP report), the relationship is

$$|\vec{S}| = \frac{|\vec{E}^2|}{1200\pi} = 12\pi|\vec{H}^2|$$

where $\vec{S}$ is in units of mW/cm^2. Most broadband measurement used at broadcast frequencies appropriately sums the field strength contributions in all planes of polarization so the vector notation in these relationships can often be ignored.

Guidelines Used by the FCC

As discussed above, the FCC presently uses the public and occupational power density guidelines recommended by the NCRP. At typical broadcast sites, compliance with the power density limits is determined on a whole-body spatial average basis. It is important to note that, although most standards include protection against shock and burn through established limits on induced body current and contact current, the FCC has not adopted these current limits.

Establishment of limits on induced body current and contact current is one topic of ET Docket 13-84, so this policy may be subject to change. Additionally, OSHA *does* enforce those limits in the workplace. Thus, it is important to ensure that significant RF potentials and currents are not created on guy wires, fall-arrest cables, crane cables, metal fences, or other locations that may be subject to human contact.

Guidelines Used by OSHA

The U.S. federal OSHA requires employers to "furnish to each employee a place of employment which is free from recognized hazards that are causing or are likely to cause death or serious physical harm to his employees."[10] Additionally, some 28 states have OSHA-approved plans and have adopted their own standards and enforcement policies. For the most part, these states have adopted standards that are identical to federal OSHA. However, some states have adopted different standards applicable to this industry or may have different enforcement policies, so review of state OSHA regulations is recommended.

Some of the federal OSHA regulations are considered to be out of date, and the single-exposure limit specified in its regulations (10 mW/cm^2 or 100 W/sq. m) has been ruled unenforceable. Additionally, the regulation specifying the design of an RF warning sign is also out of date. Instead of providing enforceable, up-to-date regulations, OSHA applies a complex patchwork of policies and interpretations that are available from its website.[11]

Lockout/Tagout

In addition to exposure guidelines, OSHA applies a "lockout/tagout" requirement in many situations involving maintenance and servicing of high-power RF transmitting equipment. For example, climbing a TV transmitting antenna to replace a top-mounted aviation warning beacon would require that the associated transmitter be shut down. To prevent unexpected activation of the transmitter while the climber is aloft, the circuit breaker feeding the transmitter should be locked (using a padlock) into the off position, and a warning tag placed to indicate that the transmitter may not be operated until the lock and tag are removed by the person who installed them.[12]

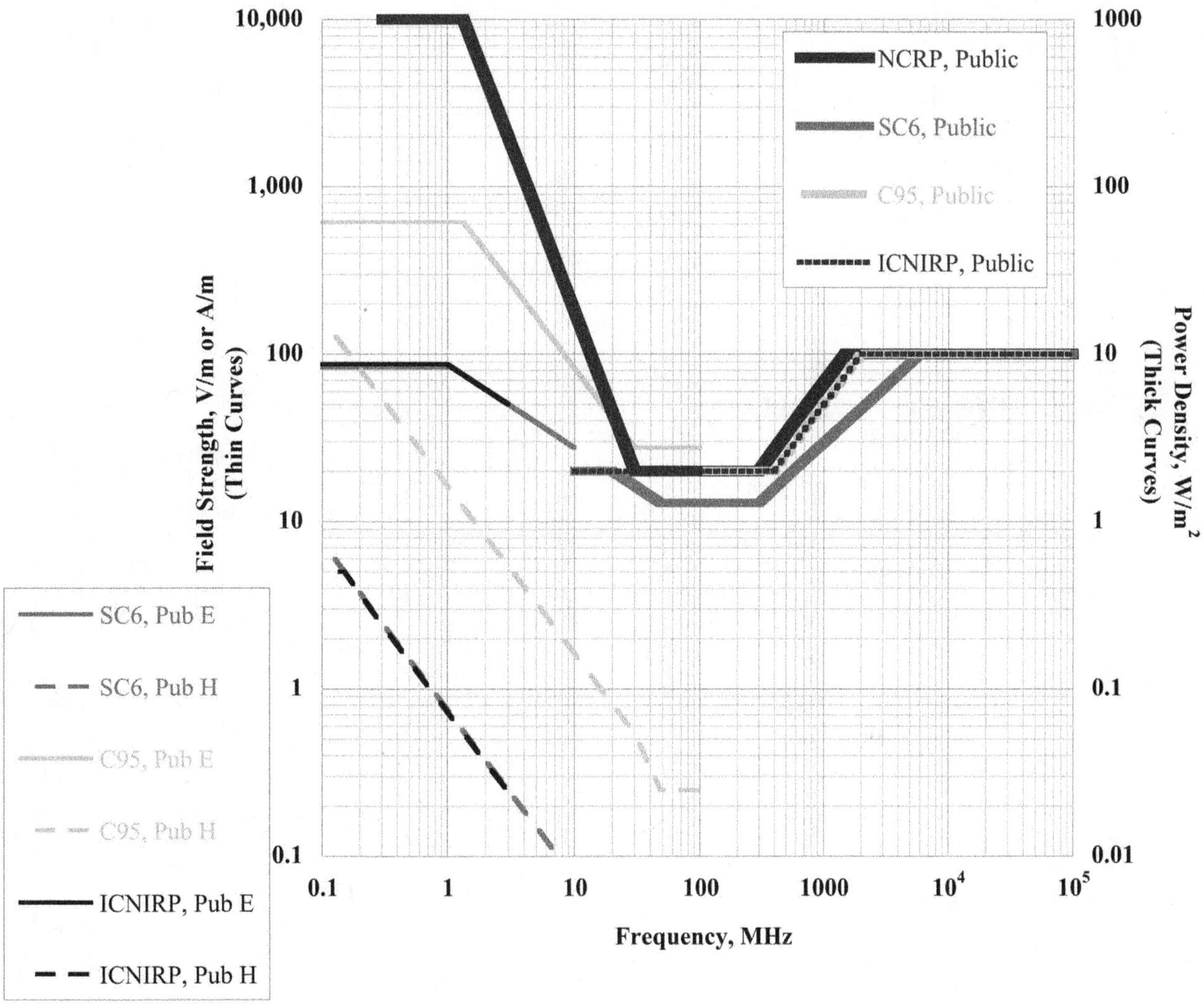

FIGURE 3.5-1 Public exposure limits

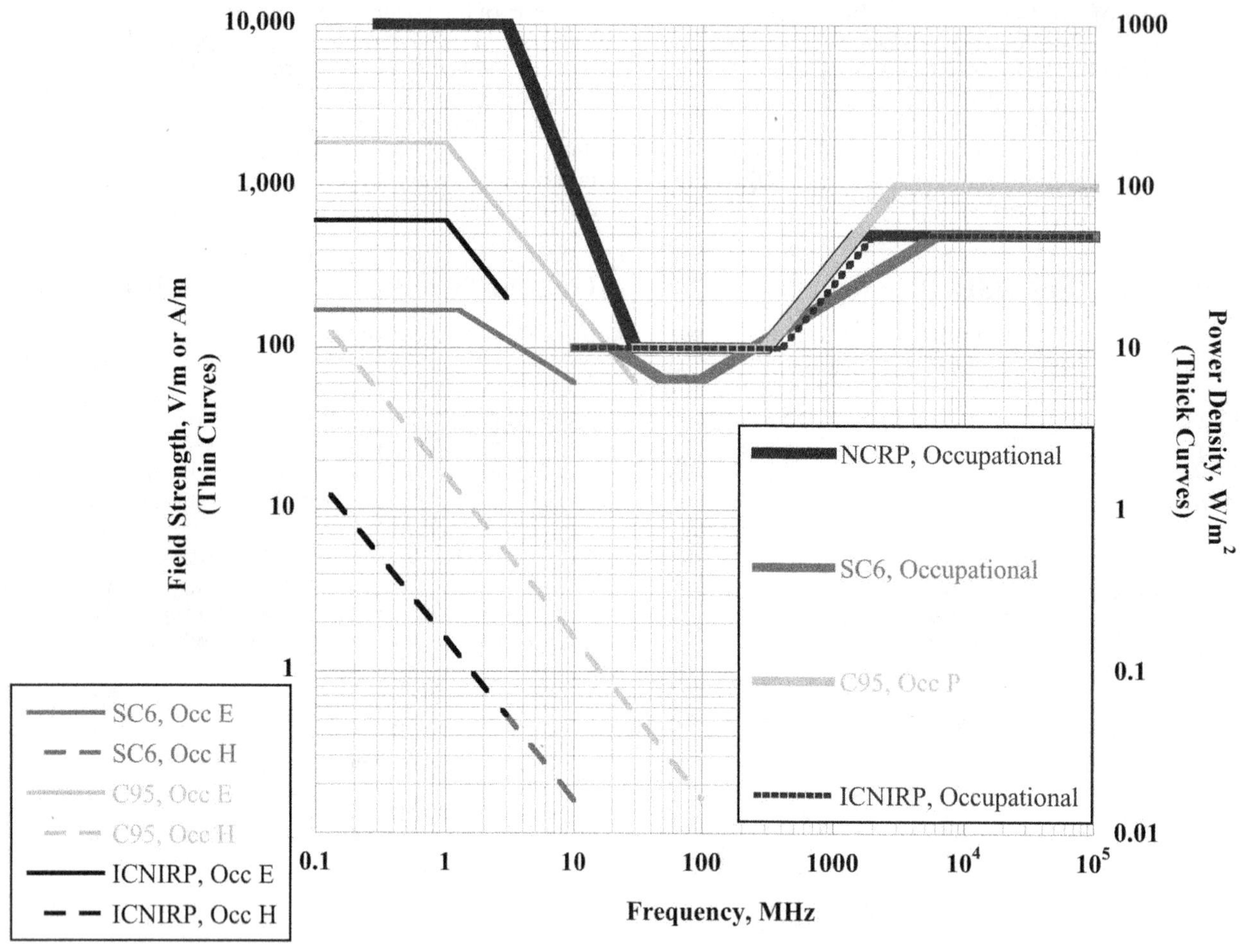

FIGURE 3.5-2 Occupational exposure limits

ESTABLISHING A CONTROLLED ENVIRONMENT

The first step in considering whether an area complies with FCC or other exposure guidelines is to determine whether it is a controlled or uncontrolled environment. Controlled environments are those where persons within exposed under known conditions are "fully aware" of and can "exercise control" over their exposure to RF energy. However, the less restrictive occupational exposure limits cannot be applied *unless* access to the area in question is controlled by fencing, signs, or other measures and individuals with access to the area have received written and/or verbal (orally communicated) information. With the exception of individuals transient through such areas, an educational program or worker training is also required to be provided. The mere posting of warning signs is not sufficient, since all persons with access to areas where the public limits may be exceeded must understand the levels of potential RF exposure and the nature of the environment.

Fencing

Fencing is *required* under FCC rules around series-fed AM towers to prevent people from being injured through direct contact with energized components. For liability reasons, it is preferred that *all* tower bases be fenced to prevent climbing into areas having high RF fields and to protect unauthorized climbers from falling. The fence should provide an effective barrier and be difficult to climb. For example, a few strands of barbed wire that can be easily climbed through probably is not adequate. A 7-foot chain-link fence with a locked gate probably is adequate.

Near AM towers and some other towers, use of nonconductive fencing materials may be preferred to avoid potential shock and contact current hazards, and to reduce the possibility of intermodulation distortion created by "rusty joints" as the fence ages. However, nonmetallic fences are often less effective as physical barriers and may be subject to rapid degradation by sunlight (e.g., ultraviolet rays), so there is often no choice but to construct a fence of metal. A metal fence should be adequately grounded and electrically continuous. This may mean that jumper cables be used to bond the fence to the gates and posts.

Signs

RF exposure warning signs should include at least the following components:

- RF Energy Advisory Symbol (see IEEE C95.2)
- Appropriate signal word (Danger, Warning, Caution, or Notice)
- Explanation of the RF source (e.g., transmitting antennas)
- Avoidance statement (e.g., "Do not climb tower while antennas are energized")
- Contact information

Commercial warning signs are available for a variety of circumstances,[13] but custom signs will often be required. Signs should be posted at locations where they can be read without entering a high-RF area, and there should be a sufficient number of signs that they will not be missed because the area is being accessed by other than the main entrance. If non-English-speaking persons (including subcontractors and trespassers) are likely to be in the area, signs should include the other languages.

The choice of the appropriate sign depends upon the severity of the hazard. ANSI Standard Z535.2-2011 specifies four "signal words" that designate an appropriate degree of alerting, as follows:

- **DANGER**: Indicates a hazardous situation that, if not avoided, will result in death or serious injury.
- **WARNING**: Indicates a hazardous situation that, if not avoided, could result in death or serious injury.
- **CAUTION**: Indicates a hazardous situation that, if not avoided, may result in minor or moderate injury.
- **NOTICE**: Is used to address practices not related to physical injury.

IEEE Standard C95.7 (2014 edition) offers recommendations concerning RF safety programs, including signage. This new standard divides RF environments into four categories, as described in Table 3.5-1.

The minimum letter height recommended in ANSI Z535.2 is one unit for every 150 units of viewing distance for the signal word (e.g., about 5/8-inch tall at a viewing distance of 8 feet), and one-half that height for the rest of the sign text. When visibility conditions may not be favorable (e.g., poor lighting), the letter height should be at least doubled. The FCC has proposed requiring signs meeting these requirements. Examples of some suitable signs are shown in Figures 3.5-3, 3.5-4, 3.5-5, and 3.5-6.

FIGURE 3.5-3 AM RF Danger sign

TABLE 3.5-1
Recommended Categories for RF Safety Programs and Example Actions Required

Category	Characteristic	Example Actions Required
1	No areas above *public* MPE limit	None
2	Areas above public limit, but no accessible areas above *occupational* MPE limit	Signs plus access controls; Designate RF Safety Program Administrator
3	Accessible areas above *occupational* limit if mitigating controls are not applied	As in Category 2, plus training program, personal RF monitors
4	Accessible areas more than 10 times the *occupational* limit	As in Category 3, plus lockout/tagout procedures, physical barriers. Output must be reduced to achieve Category 1, 2, or 3.

FIGURE 3.5-4 RF Warning sign

FIGURE 3.5-5 RF Caution sign

FIGURE 3.5-6 RF Notice sign

Other Factors

In 1986, the FCC published guidance to broadcasters in the form of a list of situations that are likely to arise where the public may be exposed above the public limit, and an explanation of the mitigation required. This guidance document remains Commission policy, but as discussed above, additional measures may be specified in IEEE C95.7. Those situations are paraphrased as:

- On-tower exposure (above ground). If the tower is marked by appropriate warning signs, it may be assumed that there is adequate protection of the general public.
- Ground-level exposure in a remote area. If the area of concern is remote and/or there are natural barriers to access *and* marked by appropriate warning signs, it may be assumed that there is adequate protection of the general public. It is recommended that fences also be used where feasible.
- Ground-level exposure in accessible area (not remote). Fencing and signs are required.
- Exposure in accessible area (not remote), where access cannot be restricted. An environmental assessment must be filed with the FCC to determine what actions are required.

Significant Contributors

The FCC recognizes all stations contributing more than 5% of their exposure limit in an area that exceeds the standard as being equally responsible for bringing that area into compliance. Calculations demonstrating that a station should not contribute more than 5% of its exposure limit do not provide a "safe harbor," if field measurements demonstrate a contribution above 5%. There is also some day-to-day variation in relative contribution levels. It should *not* be assumed that other stations will take care of problem areas. All stations may need to be involved in ensuring compliance.

CALCULATING COMPLIANCE

In many practical situations, a determination of presumptive compliance with the prevailing standards can be done by calculation. Basic to the use of far-field calculation methods is an assumption that the source can be modeled as a point source with known azimuth and elevation plane patterns. This assumption is reasonable in the case of VHF and UHF antennas installed on tall towers. Far-field calculations may not be a good choice near the antennas, however, as in the case of rooftop installations, on-tower situations, or near AM towers.

The Near- and Far-Field Conditions

The exposure from a transmitting antenna's radiated field is characterized by its power density (power radiated per unit area), which is a complex quantity called the Poynting vector, usually abbreviated $\vec{S}$. Mathematically, $\vec{S}$ is defined as the vector cross product of the electric and magnetic field vectors (times a constant). Close to the antenna (called the "near-field zone"), $\vec{S}$ is imaginary, meaning that all of the power is reactive, while farther away (called the "far-field zone"), $\vec{S}$ is real, meaning that all of the power radiates. A transition occurs between the reactive and radiating regions, where the power is composed of both reactive and radiating components.

A determination of whether the area in question is in the near-field, transition, or far-field region of the antenna is critical to determining whether calculations are appropriate and what measurements must be taken. As a rule-of-thumb, one is in the far-field zone of the antenna at a distance, *r*, when:

$$r > \frac{2D^2}{\lambda} \text{ and/or } r > 5D$$

where D is the largest dimension of the antenna and λ is the wavelength of operation; both D and λ must be in the same units. The former inequality is generally applicable to VHF frequencies and above, while the latter one is most applicable at MF (AM broadcast) frequencies. The manufacturer's published directional antenna patterns are applicable *only* in the far field.

Far-Field Calculations

In the far-field zone, the antenna directional patterns, both azimuth and elevation, are assumed to be fully formed. For typical VHF and UHF transmitting antennas, the power density magnitude, $|\vec{S}|$, at distance, *r*, can be calculated from:

$$|\vec{S}| = \frac{2.56 \times 1.64 \times 1000 \times REF_{AZ}^2 \times REF_{EL}^2 \times ERP}{4 \times \pi \times r^2}$$

where the factor 2.56 accounts for typical ground reflection, the factor 1.64 is the gain of half-wave dipole over an isotropic radiator, the factor of 1000 converts to units of W/m² (use 100 instead to convert to mW/cm²), the *RFF* factors are the relative field values from the published azimuth and elevation antenna patterns in the direction of the evaluation point, *ERP* is effective radiated power in kilowatts and *r* is in meters. This is the formula recommended by FCC Office of Engineering and Technology (OET) Bulletin No. 65 (1997).

Some caveats must be applied to the application of this equation; namely, the azimuth pattern is typically defined only in or near the plane of the main beam of the antenna. At steep elevation or depression angles (i.e., angles above or below the antenna), the azimuth pattern degrades and should not be included in the calculation (that is, assume a relative field of 1 for RFF).

Near-Field Calculations

Because many VHF and UHF broadcast antennas consist of a vertical array of radiating elements, they can be treated to a first-order approximation as a line source having the input power distributed uniformly over the antenna aperture. Within the far-field boundary of the antenna, the power density magnitude at distance, *r*, can be calculated from

$$|\vec{S}| = \frac{180 \times P_{IN}}{\theta_{HPBW} \times \pi \times D \times r}$$

where P_{IN} is the average input power to the antenna, θ_{HPBW} is the half-power azimuth beamwidth of the antenna (an omnidirectional antenna has a HPBW of 360°), and *D* is the antenna aperture. The units of power density in this equation are W/m²; multiply the result by 0.1 to convert to mW/cm². This near-field calculation is applicable *only* in the aperture of the antenna and may very well overstate levels above and below the antenna. For broadcast antennas, there are several ways to refine these calculations, such as moment-method analysis, although none is specified in OET Bulletin No. 65.

Intrinsic Compliance

In some situations, the radiating elements of the antennas are insulated or otherwise separated from direct contact, and the power levels are so low that the public limit cannot be exceeded, even at the point of closest approach. Such sources are called intrinsically compliant. In earlier versions of some standards, power levels of, say, 6 watts and less were considered intrinsically compliant, but that exemption is no longer recognized.

The FCC has identified certain circumstances where routine evaluation is not required. Two of the more common include LPTV and FM translators of less than 100 watts ERP. It is important to understand that this *exclusion* from routine evaluation is not the same as an *exemption* from compliance, and such stations may exceed the exposure limits in the vicinity of the antennas.

In the FCC's Further Notice of Proposed Rulemaking to ET Docket No. 03-137, an intrinsic compliance determination is proposed based on a combination of power, frequency, and separation distance. While this approach has not yet been adopted as a rule by the FCC, it may be useful and is reproduced as Table 3.5-2 below.

Antenna Farms

At sites with several antennas on several towers, power reductions can often be used to enable work on a particular tower without the need for all stations to shut down. Computer modeling or, less practically, measurements can be performed to determine, for any section of any tower, which antennas have the largest impact on localized power density levels and which of those sources should be cut back, and by how much, in order to bring total power density levels within the occupational limit. The results of these calculations typically are organized as tables and can form the basis for the cooperative agreements anticipated by the FCC:

> The licensee also certifies that it, in coordination with other users of the site, will reduce power or cease operation as necessary to protect persons having access to the site, tower, or antenna from radiofrequency electromagnetic exposure in excess of FCC guidelines.[14]

The Commission encourages all site occupants, owners, leasers, and managers to cooperate in sharing information and developing accurate site modes, and notes that "site user agreements are particularly useful and desirable to achieve this end."[15]

TABLE 3.5-2
Single RF Sources Subject to Routine Environmental Evaluation

Transmitter Frequency, MHz	Threshold ERP, Watts
Regardless of ERP, evaluation is required if the separation distance *R* is less than $\lambda/2\pi$ from the radiating structure, where λ is the free-space operating wavelength, unless the available maximum time-averaged power is less than one milliwatt. In addition, evaluation is required if the ERP in watts is greater than the value given by the formula below for the appropriate frequency *f* in MHz at the separation distance *R* in meters.	
0.3–1.34	$ERP \geq 1{,}920\ R^2$
1.34–30	$ERP \geq 3{,}450\ R^2/f^2$
30–300	$ERP \geq 3.83\ R^2$
300–1,500	$ERP \geq 0.0128\ R^2 f$
1,500–100,000	$ERP \geq 19.2\ R^2$

Source: FCC FNPRM.

AM Sites

Most types of AM towers present a shock and burn hazard at or near the base, and the FCC requires that the area surrounding the base be fenced. Using a variant of the Numerical Electromagnetic Code (NEC), the FCC developed tables of the predicted distances necessary for compliance with its limits for public exposure. These limits are functions of frequency, tower height, and power, so only the most common case—that of one-quarter wavelength tall towers—is presented in Table 3.5-3. Additional guidance is found in Supplement A to FCC/OET Bulletin No. 65 (August 1997 edition).

MEASURING COMPLIANCE

Broadband Measurements

Surveys of broadcast transmitting sites, especially multi-user sites, are generally taken using broadband instruments. Manufacturers of calibrated, survey-grade instruments include:

- Narda Safety Test Solutions[16]
- ETS-Lindgren[17]

Broadband instruments typically include an integrated readout and probe, which covers a substantial frequency range, such as 0.3–3,000 MHz. While broadband measurements can suffer from reduced accuracy due to interaction of several frequencies within the instrument,[18] proper techniques and an awareness of its limitations can result in the rapid survey of a large area.

Note that there are several manufacturers of handheld EMF meters that include RF capabilities. Most of these meters do not have traceable calibration and may have wildly different responses with frequency, modulation, or orientation. While these units are priced at a few hundred dollars, you should expect to pay at least $3,000 for a survey-quality instrument and substantially more if special features are included.

Narrowband Measurements

Narrowband measurements, typically of a single station in isolation, are sometimes useful in assessing the contribution of that station without taking other

TABLE 3.5-3
Recommended Distances From Tower Base to Fence for Quarter-Wavelength AM Towers at Various Power Levels

	Transmitter Power, kilowatts			
Frequency, kHz	50	10	5	1
535–1540	4 m	2 m	2 m	1 m
1540–1705	5	2	2	1

Source: From FCC/OET Bulletin No. 65

stations out of service. Typically, narrowband measurements are conducted using a field strength meter or a spectrum analyzer with a calibrated antenna. Care must be taken to account for all planes of polarization, and this can become burdensome at frequencies below UHF. Note that some narrowband field strength meters, such as the Potomac Instruments FIM-41, use a magnetic field (loop) antenna, but report results in equivalent electric field units. This situation makes the determination of compliance complex.

Narda offers a combination broadband/narrowband instrument, the SRM, which is small enough to be used for surveys. Essentially, it is a spectrum analyzer with near real-time digital signal processing capabilities. The digital signal processor (DSP) engine allows the same meter to display in different units, account for frequency dependence of various safety standards, integrate power over a particular frequency range, and has other useful features. This instrument can measure the contribution of individual stations directly as a percentage of any standard. One drawback is that the available triaxial probe does not cover frequencies below 27 MHz, making a comprehensive survey difficult if there are nearby AM or HF stations. In addition, care must be used to ensure that the resolution bandwidth (RBW) and amplitude range are appropriate to the emission(s) being measured. Because they have different bandwidths, simultaneous measurement of, for example, FM and DTV emissions may not be possible using a single RBW setting.

E-Field Measurements

Measurements of the electric field are most commonly conducted with electrically short dipole or monopole elements. Care must be taken to account for all planes of polarization, and many manufacturers make triaxial probes, which are designed to provide an isotropic response, so that repetitive measurements in each plane are not necessary. At typical installations above 30 MHz, only measurements of the electric field are necessary at ground level, since the observer will be in the far field. Below 30 MHz, and especially at AM frequencies and below, both E- and H-field measurements are required, because the observer is typically not in the far field. E-field measurements are typically reported in units of volts per meter (V/m), although some instruments report in units of $(V/m)^2$.

So-called hot spots may be encountered near conductive objects, such as tower bracing. These hot spots are generally created by reradiation and hence are often a near-field phenomenon. Because much of the energy in the near field is reactive, there is little real power present that can result in a hazardous condition. To account for this, the C95.3 standard recommends that measurements near reradiating objects be conducted at a distance of at least 20 cm (about 8 inches).

H-Field Measurements

Except in the far field, measurements of both the electric and magnetic fields are recommended. As a practical matter, there are few calibrated magnetic field antennas available for use above 300 MHz, so magnetic field measurements at higher frequencies may not be possible using conventional instruments. H-field measurements are typically most important below 30 MHz, where conventional (coil) inductors are often used for impedance matching and can generate substantial magnetic fields. High magnetic fields are often encountered near AM antenna tuning units, tower bases, and phasors, and it is often the magnetic field that predominates.

Some H-field probes have out-of-band (VHF or UHF) resonances, so care should be used when surveying a site having sources in several frequency bands.

Shaped Probes

Some manufacturers offer electric field probes that have a frequency dependence matching the standard, so that exposure at multi-user sites could be assessed directly. These probes offer the ability for rapid assessment, but typically are less accurate than their "flat response" counterparts, because the abrupt frequency transitions in the standards cannot be precisely realized. The cost of a shaped probe is often twice that of a flat one, and it becomes obsolete if the standard changes. While "flat response" probes often require calibration only at a single frequency, the nature of shaped probes requires calibration at multiple frequencies, which not only increases the cost of the calibration service but also requires that different calibration factors be used, depending upon the measurement frequency.

Spatial Averaging

Although most standards contain both spatial and time-averaging provisions, reliance on these factors at marginally compliant sites is not recommended. There is simply too much uncertainty associated with RF exposure assessment to state unambiguously that a location having a peak power density of, say 110% of the standard, but an average power density of just 90%, will always be in compliance. The FCC uses spatial averaging in its measurements, and this can work to a station's advantage, since spatially averaged measurements will always be lower than peak measurements. Typically, spatially averaged results at FM/TV sites are about 60% of peak readings.

For a number of reasons, the use of spatial averaging is generally not recommended as a means to achieve compliance. Specifically, there is no single approved method of obtaining a spatial average, so different observers may obtain different results. Also, the use of spatial averaging is subject to administrative review and could be eliminated at any time.

If spatial averaging is to be used, a common method (though not necessarily the most accurate) is to take measurements over a vertical line, from 0.2 to 1.8 meters above ground with the probe oriented in each of the four compass directions. The results of the four linear averages are themselves averaged together. A plastic pipe can be used to ensure that each measurement is taken at the same location and over the same vertical segment.

Time Averaging

Like spatial averaging, the use of time averaging to achieve compliance in marginal situations is generally not recommended. Certainly, in most uncontrolled environments, it is not possible to control the actions of the public, including, for example, how long they might remain at some location admiring the view. In controlled environments, it may be possible to control how long a worker remains at a particular location, but such restrictions are often inconvenient or difficult to enforce. The use of time averaging in situations typically encountered by broadcasters should be limited to situations where the source is highly variable, such as a weather radar system, which employs a low-duty cycle pulsed emission and typically rotates. In that situation, involving source-based time averaging, the worker is unlikely to be exposed to the peak power of the radar for any length of time because of its intrinsic properties, and time averaging of the emission would be appropriate.

Uncertainties and Variability

The power density at broadcast sites is subject to variation over time, due to changes in atmospheric conditions, soil conductivity, foliage, and physical changes to the site environment. The day-to-day variation at FM/TV sites can be on the order of 20–30%, so a site having measured power densities exceeding 70–80% of the applicable limit may actually be out of compliance at times. The FCC has *imposed fines* on stations that were determined to exceed the 5% threshold of "significant contribution" in areas exceeding the public limit at one time but were measured to be under that threshold at another time.[19]

The worst-case uncertainty of broadband measurement equipment comes about if all the possible sources of error were at their extreme values and in such directions as to add together constructively, and is on the order of 3–6 dB. Most of the various contributing errors are not systematic, so random errors of perhaps 1 dB can be expected in a well-conducted survey.[20]

Detectors using metal barrier or Schottky diodes are perhaps the most common type used in broadband instruments. Diodes have a *major limitation* when used at typical broadcast sites: often, they do not respond properly in a multiple-station environment.[21] Multiple source and frequency (MSF) errors can result in errors up to 3 dB, with an average error of about 1.2 dB. Modern meters include circuitry to compensate for this effect and appear to be less susceptible to MSF errors.

Contact and Induced Currents

Contact (or touch) currents flow when the human body comes into contact with a metallic object at a different electric potential. Typically, this occurs when a person touches a tower or wire at RF potential, but can also occur when fields are coupled to implanted medical devices. The exposure limits *are not* intended specifically to protect against such situations. Induced currents are those induced in the human body immersed in an RF field, which flow to ground through a foot or both feet.

Although the FCC does not presently enforce contact or induced current limits, they are specified in the C95.1, Safety Code 6, and ICNIRP standards, and are *enforced* by OSHA. Table 3.5-4 summarizes the RMS current limits over the range 0.1–110 MHz (there are no limits above 110 MHz).

Contact and induced currents can be measured using a toroid current probe, which surrounds the limb being measured. Induced currents can also be measured using a parallel-plate current meter.

Except in the case of direct contact with an energized radiator, the contact and induced currents are a direct function of the electric field strength in which the object or body is immersed.

Contact Voltage

In addition to the limits specified for contact and induced currents, the C95.1 standard also suggests a limit for the open circuit voltage that exists on objects in the range 0.1–100 MHz with which an individual may come into contact. The limit is generally 140 volts (rms) between any two points of contact with the body.

TABLE 3.5-4
Contact and Induced Body Current Limits

Standard and Exposure Condition	Current Limit, Milliamperes
C95.1-2005—Public	
Induced through one foot	45
Induced through both feet	90
Contact through one finger	16.7
C95.1-2005—Occupational	
Induced through one foot	100
Induced through both feet	200
Contact through one finger	50
SC-6—Public	
Induced through one foot (0.4–110 MHz)	40
Induced through both feet (0.4–110 MHz)	80
Contact through one finger (0.1–110 MHz)	20
SC-6—Occupational	
Induced through one foot (0.4–110 MHz)	90
Induced through both feet (0.4–110 MHz)	180
Contact through one finger (0.1–110 MHz)	40

Source: From C95.1-2005 and Safety Code 6.

TERMINOLOGY AND UNITS

Electric field—The effect of an electric force exerted on an electrically charged object, divided by the charge. Typically, volts per meter (V/m).

Magnetic field—The effect of moving electric charges that exert a force on other moving charges. Typically, amperes per meter (A/m).

Power density—The magnitude of the cross-product of the electric and magnetic field vectors. Also, equivalent far-field power density, derived from individual electric or magnetic field. Typically, watts per square meter (W/m^2). Some standards use microwatts or milliwatts per square centimeter ($\mu W/cm^2$ or mW/cm^2).

Contact current—The electric current between a metallic object and a biological medium as a result of contact with a source of RF current. Normally, finger contact is assumed, although other limits may apply for so-called "grasping" contact by trained individuals. Typically, milliamperes (mA).

Induced body current—The electric current in a biological medium that occurs when the medium is in an electric or magnetic field, but not in direct contact with a source of RF current. Typically, milliamperes (mA).

Radiation—Streams of photons or waves from a source. Radiofrequency radiation is nonionizing, that is, of insufficient energy to break atomic bonds. Not to be confused with radioactivity. Typically, for radio frequencies, expressed in terms of power density.

Radiofrequency—That portion of the electromagnetic spectrum in which waves or photons can be generated by an alternating current fed to an antenna. Typically taken as the frequency range from 3 kHz to 300 GHz.

Maximum permissible exposure—Those limits on exposure allowed for persons in particular environments, defined as either controlled (occupational) or uncontrolled (public). Sometimes called reference levels.

Specific absorption rate—The measure of energy deposition. Typically, watts per kilogram (W/kg). The SAR forms the basic restrictions from which the maximum permissible exposure (MPE) limits are derived.

Spatial average—At frequencies above 100 kHz, the average of the power densities, computed over the planar area of a human. Typically expressed in power density units.

INTERACTION OF RADIOFREQUENCY ENERGY WITH ORGANIC MATTER

The prevailing standards in the U.S. are intended to protect against established adverse health effects identified in the reviewed studies. An important distinction is that not all reported effects are established or adverse. The standards generally define *basic restrictions*, which are based directly on the established effect. Depending upon the frequency of exposure, the physical quantities of the basic restrictions are current density, specific energy absorption rate (SAR), and power density. Only power density outside the body can be readily measured for exposure of individuals, so the maximum permissible exposure limits established in the standards are derived from the basic restrictions (plus a safety factor) but are specified in indirect, equivalent units of electric or magnetic field strength, power density, and current flow.

RF energy affects the human body differently depending upon the frequency of the source. Below about 5 MHz, the standards minimize effects associated with electro-stimulation to prevent adverse effects on nervous system functions. In the frequency range 100 kHz to about 5 GHz, the standards protect against adverse health effects associated with whole-body heating and excessive localized heating. Above about 5 GHz, the standards protect against excessive heating at or near the body surface.

Although the standards protect primarily against the effects of nerve stimulation and tissue heating, the biological bases for limiting exposure considered a wide variety of studies on potential or reported effects, including epidemiological studies of cancer risk, reproductive outcome (teratology), cellular effects, behavioral effects, and central nervous system (CNS) effects. For exposure to radiofrequency fields (as opposed to contact and induced currents), the lowest threshold of adverse effect was found to be behavioral change in animals due to thermal stress.

ANSWERS TO COMMONLY ASKED QUESTIONS

I work at a radio station. Why am I allowed to be exposed to greater RF levels than a member of the public is?

The basic restrictions on exposure are the same for both workers and the public. It is the underlying philosophy of the safety factor that differs. Since workers are aware of their exposure, the safety factor is smaller, but there is no credible evidence that anyone has been harmed by RF exposure at or below the occupational MPE level.

Can a local jurisdiction establish its own arbitrary exposure limits?

In the case of "Personal Wireless Facilities," the U.S. Congress prohibits local government entities from applying exposure limits more restrictive than the FCC limits. However, neither Congress nor the FCC has established a regulatory preemption for RF exposure from broadcast facilities.

What is the ambient exposure of a typical person in the United States?

The EPA conducted several surveys in the 1970s and found that exposure levels varied considerably depending upon whether a location was urban or rural. Typical exposures in urban areas today are around 1 μW/cm^2.

Do the new digital modes (ATSC television and IBOC radio) present different and unknown biological effects?

Some research has suggested that RF signals containing substantial modulation at extremely low frequencies (below about 200 Hz) may lead to so-called nonthermal effects. To date, the literature review and evaluation process has not shown modulation effects to be adverse except for very high peak power and short duration pulses (e.g., some radars), which is far different from the modulations used in broadcasting.[22,23]

I've heard that some countries, Italy, Russia, and China, for example, have exposure limits many times more restrictive than in the United States. Is that true?

Yes. In the case of Italy, there is a more restrictive limit that applies near hospitals and schools, but it has no scientific basis. In the case of Russia and China, the more restrictive limits are largely historical (dating from the 1960s) and are due to a different philosophical basis for standard setting. Specifically, the more restrictive standards are believed to have been set to preclude all known effects, regardless of whether the effects were adverse or even whether they occurred in whole beings (as opposed to cell cultures).

I worked on a tower on which one of the antennas that was supposed to be off was in fact energized. What adverse effects should I expect?

Likely, none. The safety standards are not set at the boundary of adverse health effects. Rather, they include a significant safety margin below that threshold. So, even at exposure at levels of several times the standard, there would be little likelihood of health problems. If the exposure was sufficiently intense to cause health problems, it is likely that a heating or tingling sensation would have been felt while on the tower. See a doctor if any unusual symptoms such as blurred vision or headaches occur.

What are the cumulative effects of all of these broadcast towers along with all of the cell sites and other RF equipment that is around?

Cumulative usually means that the effects accumulate with time, which is not the case with exposure to RF energy at levels below the standards. A better term might be *additive*, because the exposure levels from all of the sources mentioned add together. A properly conducted broadband survey would include the contributions of all of these sources.

I read about some research that suggests new adverse effects. What do you know about that?

Largely because of the widespread use of personal communications devices, the public has become sensitized to the issue of RF exposure and has demanded further assurances that exposure is safe. Research continues, and a positive result from one study is often offset by negative results from other studies. Information on many specific studies and how those results might be interpreted can be found on the Internet.[24]

Notes

1 Health Canada, "Limits of Human Exposure to Radiofrequency Electromagnetic Energy in the Frequency Range From 3 kHz to 300 GHz," *Safety Code 6* (2015), p. 2. www.hc-sc.gc.ca/ewh-semt/pubs/radiation/radio_guide-lignes_direct/index-eng.php, Accessed February 29, 2016.

2 ICNIRP, "Exposure to High Frequency Electromagnetic Fields, Biological Effects and Health Consequences (100 kHz—300 GHz)," *ICNIRP 16/2009* (2009). www.icnirp.org/cms/upload/publications/ICNIRPStatementEMF.pdf, Accessed February 29, 2016.

3 FCC, First Report and Order, Further Notice of Proposed Rule Making in ET Docket No. 03-137, and Notice of Inquiry in ET Docket 13-84, *FCC 13-39*, Adopted March 27, 2013. https://apps.fcc.gov/edocs_public/attachmatch/FCC-13-39A1.pdf, Accessed February 29, 2016.

4 W.F. Hammett, *Radio Frequency Radiation* (New York: McGraw-Hill, 1997), p. 45.

5 SI (MKS) units are used in this chapter in keeping with the latest standards and FCC publications. For clarity, CGS units may also appear parenthetically.

6 See Scientific Committee 53 on Biological Effects and Exposure Criteria for Radiofrequency Electromagnetic Radiation, "Report No. 086—Biological Effects and Exposure Criteria for Radiofrequency Electromagnetic Fields," Bethesda, MD (1986), www.ncrponline.org/, Accessed February 29, 2016.

7 See IEEE ICES, "IEEE Standard for Safety Levels with Respect to Human Exposure to Radio Frequency Electromagnetic Fields, 3 kHz to 300 GHz," www.ices-emfsafety.org, Accessed February 29, 2016.

8 Health Canada, *Safety Code 6*, *op. cit.*

9 See International Commission on Non-Ionizing Radiation Protection, "ICNIRP Guidelines for Limiting Exposure to Time—Varying Electric, Magnetic and Electromagnetic Fields (Up to 300 GHz)," www.icnirp.de/, Accessed February 29, 2016.

10 See United States Department of Labor—OSHA, "Safety and Health Topics, Radiofrequency and Microwave Radiation," www.osha.gov/SLTC/radiofrequencyradiation/index.html, Accessed February 29, 2016.

11 *Ibid.*

12 OSHA Regulations, Title 29 Code of Federal Regulations §1910.147.

13 See: www.radhaz.com, www.rfsigns.com, Accessed February 29, 2016.

14 Environmental Effects section of FCC licensing application forms. See www.fcc.gov.

15 *FCC 13-39*, *op. cit.*, para. 34.

16 Narda Safety Test Solutions, 435 Moreland Road, Hauppauge, NY 11788 Tel. 631/231-1700. www.narda-sts.com/, Accessed February 29, 2016.

17 ETS-Lindgren, 1301 Arrow Point Drive, Cedar Park, TX 78613 Tel. 512/531-6400. http://ets-lindgren.com, Accessed February 29, 2016.

18 J. Randa and Motohisa Kanda, "Multiple-Source, Multiple-Frequency Error of an Electric Field Meter," *IEEE Transactions on Antennas and Propagation*, Vol. AP-33, No. 1 (Jan. 1985), pp. 2–9.

19 See *Forfeiture Order* FCC 04-281, Released December 10, 2004.

20 Robert Weller and Stanley Salek, "FCC RFR Guidelines: Is Your Facility Really in Compliance?" *Proceedings of the NAB Broadcast Engineering Conference*, 1999.

21 Randa and Kanda, *op. cit.*

22 See *NCRP Commentary No. 18*, "Biological Effects of Modulated Radiofrequency Fields" (2003).

23 R. Peterson, W. Kainz, C-K. Chou, A. Thansandote and M. Ziskin, "Follow-Up to 17 October 2011 Teleconference With Members of GAO and Officers of the IEEE ICES: Response to Questions," October 21, 2011 (unpublished).

24 An excellent resource is at the Medical College of Wisconsin. www.mcw.edu/display/router.aspx?DocID=196547, Accessed February 29, 2016.

CHAPTER

3.6

Broadcast Facility Security, Safety, Disaster Planning, and Recovery

RICHARD RUDMAN
Remote Possibilities
Studio City, California

THOMAS G. OSENKOWSKY
Radio Engineering Consultant
Brookfield, CT

Updated for the 11th Edition by

JOSEPH POLLET
Director of Engineering, Entercom
New Orleans, LA

INTRODUCTION

Risk assessment and disaster preparedness and recovery plans may look good on paper, but the ultimate test is always the next real emergency. This chapter is for designers and managers of critical communications facilities that may be at risk from natural and human-made disasters and terrorist attacks. The premises are the same here as in any other area of disaster and contingency planning:

- Disaster readiness should be an integral part of the design process.
- It is not possible to protect against every possible risk.
- Periodic testing is the only way to know if a plan has a chance of working, short of an actual event.

Adapting to major emergencies that take a station from normal on-the-air operations into the uncertain world of survival in a split second can be daunting. Offering specialized emergency training courses to key personnel can be beneficial to maintaining continuity in the face of emergencies. Personnel in a broadcast organization can receive training so that they know how to shift their work style to emergency-response mode at a moment's notice.

LEGAL OBLIGATIONS AND ISSUES

Before embarking on a disaster planning process, a station should carefully research all legal obligations for preparedness under federal, state, and local obligations as they apply to location, ownership, and investment makeup. This is best accomplished by working in concert with legal counsel.

If outside workers set foot in the facility, some local and state statutes require preparation of a special safety manual, which should include installation standards, compliance with lockout/tagout[1] procedures, and emergency contact names and phone numbers. Ensure outside contractors carry proper insurance and are qualified, licensed, or certified to do the contracted work. If audiences are hosted at the facility, there may be an obligation to plan for their well-being for some specified period should a disaster occur while they are on the premises.

WORKPLACE SAFETY BEFORE THE EMERGENCY

Employers are expected to assure safety in the workplace at all times. Some states, such as California, have passed legislation that mandates that most employers identify hazards and protect workers from them. At the federal level, rules issued by the Occupational Health and Safety Agency (OSHA) are relevant to broadcasters. A strong foundation of built-in day-to-day safety practices can lessen the impact of major emergencies. Some safety procedures are not obvious, for instance:

- Plate glass in doors should have a high safety rating to avoid an accidental workplace injury.
- An earthquake, hurricane, or a tornado can hurl heavy objects through the air at lethal velocities.

- Equipment racks and storage shelves are often not properly secured to floors or load-bearing walls that would prevent them from tipping over during an earthquake.
- Equipment racks should be tethered, rather than firmly bolted.
- While securing heavy objects is mostly common sense, consult experts for special cases.
- Ensure that safety chains and wires are attached to heavy objects in the studio, such as TV studio lights, video monitors, or large speakers.
- Special anchors and fasteners should be installed that can restrain heavy desktop peripherals such as computer monitors if the work surface should fall over.
- For mission-critical areas, consider bolting workstations to the floor and securing heavy equipment to wall studs with properly rated fasteners.
- Cables can snap, monitors can implode, and delicate electronics can be smashed.

While some regions have had no recent seismic activity, note that some seismically quiet parts of the earth have been given a long overdue rating by respected seismologists.

Maintaining safety standards is difficult in any size organization. A regularly updated written safety manual with specific practices and procedures for normal workplace hazards as well as emergency-related identified hazards may be required under law, and could help lower insurance rates.

EMERGENCY PLANNING PROCESS

Emergency planning is integral to a functional and reliable facility. The designer, consultant, or person responsible for maintenance of a facility that will be able to survive and operate under emergency conditions must first obtain operational and financial commitment from management. The planning process must support the main mission as well as the people who must implement the plan under very challenging circumstances.

The technical support group of a critical communications facility may have responsibility over the total physical structure. This group must work closely with top management and legal counsel. Without competent oversight, critical electronic systems may be subject to damage or destruction from environmental hazards. Internal emergencies can be triggered by failures in air-supply systems, roof leaks, or uncoordinated telephone, computer, or AC wiring changes. While most of these emergencies are foreseeable and preventable, it requires the right people to have oversight and access to management in order to plan, test, and refine emergency response.

Successful practitioners of emergency systems' design and support must realize the potential for daily internal emergencies as well as major external emergencies in the overall planning process. Seemingly innocent acts such as plugging an electric space heater into the wrong AC outlet to which an automation or environmental computer controller is connected, for example, can take down an entire facility for hours.

FACILITIES OPERATIONS AND MANAGEMENT CONSIDERATIONS

Government and business often attempt to shift to a different management mode when disaster strikes. Doing this requires the staff to shift gears at the worst time possible—in the middle of chaos and disaster. Since the bookshelf plan is rarely exercised, these shifts often add to confusion and delayed response and recovery. The most successful facilities response and recovery plans have their roots in day-to-day operations.

It may be appropriate to review on a regular basis, such as quarterly or annually, the operation's mission, goals, and objectives to determine if there could be a benefit from a change in day-to-day operations and management. The day-to-day challenges and minor crises (e.g., sewer backup or temporary loss of water, power, and telephone due to local construction) of the typical broadcast facility lend themselves to lessons learned from emergency management practice. Compounding the problem would be a cluster of multiple studio facilities under one roof.

Incident Commander

Programming, editorial, or engineering management personnel at single or multiple studio locations may be already doing many of the functions of an emergency incident commander.[2] A person holding such a job should have good judgment and the ability to stay cool and calm—traits that are vital to incident management during real emergency conditions.

RISK ASSESSMENT FOR COMMUNICATIONS FACILITIES

Stations should perform a realistic assessment of the risks and all the steps needed to avert or mitigate them. Consider that the ultimate expert threat-risk advice might be to move the facility before a disaster strikes. In addition to larger scene events, such as severe weather and earthquakes, do not overlook the obvious but often smaller risks. If computers, transmitters, or telephone equipment depend on cool air, it is unlikely they can continue to operate during a heat wave when the air conditioner has failed. If a water main or wastewater pipe runs through the central power and telecommunications room, a leak in that pipe could disable nearby equipment.

Reliability calculation is an educated prediction based on a number of factors. Emergencies can be counted on to introduce variables into the reliability equation. A system with a calculated reliability of 99.999% seems as if it should be quite reliable. However, even this high percentage may represent several minutes of outage over a 1-year period that may occur in multiple short instances or one prolonged period. Strive to design beyond that level of reliability to have a greater chance to stay on-line during major emergencies. Double, triple, and even quadruple redundancy might be realistic options when 100% uptime is mandated by facility management. However, facility designers must balance realistic redundancy with realistic uptime expectations.

Today's facility designer must have a deep professional commitment to ensure continued operation during and after a major catastrophe. Disaster planning, for example, for a force 3 hurricane, may simply entail boarding up the windows, installing a satellite receiver to provide programming continuity, and leaving town for the duration. Some facilities plan for uninterrupted operation and even staffing in the face of severe weather. In some communities, the communication facility's mission may be an essential tool of local government emergency management.

Realistic Risk Assessment

A *realistic risk* list must contain specific hazards based on local conditions, such as:

- Finding the high-water marks for the 100- and 150-year storms and other flood potentials and whether they will affect the facility.
- Assessing social, political, and governmental conditions.
- Determining commercial electrical power reliability.
- Tracking weather conditions, including icing conditions, lightning, and tornado potential.
- Noting the geography and the related risks from adjacent structures or trees.
- Researching the local geology for earth movement.
- Assessing the potential for terrorism and security risks.

Assess the specific local hazards (e.g., towers, power systems) that could be triggered by deliberate vandalism by present and former employees or external parties.

Other factors that can make the facility an easy target include:

- Nearby human-made hazards or risks (e.g., dams).
- Special on-site hazards (e.g., towers or antennas).
- Disgruntled neighbors.
- Construction of the facility with respect to earth movement or other local geologic conditions.[3]
- Hazardous materials on the premises (e.g., propane tanks).
- Communications links to the outside world (e.g., satellite antennas and Internet e-mail.
- Electrical power (e.g., transformers).
- Other utilities (e.g., exposed gas meters).
- Buried pipelines (e.g., gas).
- Computer virus and security breaches.

Examine the effect that specific threats to critical points in the facility's infrastructure would have on the ability to maintain service and provide safety to personnel:

- Damage to towers, antenna tuning units (ATUs), transmission lines, buildings, generators, and fuel tanks.
- Fire.
- Component failure due to neglect or wear.
- Tower failure due to overloading.
- Equipment failure due to vehicular contact.
- Power surges, improper mains voltage, or phase loss.
- Flooding due to pipe breakage.
- Prolonged outage due to lack of spares.
- Tower failure due to improper tensioning, out of plumb, or galvanic action eroding guy anchors.

FIRST RESPONDER COMMUNICATIONS BACKUP SYSTEM

Broadcasting operations vary by market from single-story licensee-owned buildings, to rentals in a high-rise office building, to locations in office parks. Transmitter sites in some of the larger cities are often located in high-rise buildings, such as the Empire State Building, 4 Times Square, Sears Tower, or Prudential Center. All need to have plans to cover varying types of incidents whether natural or human-made and the involvement of the first responders whose task it is to safely deal with the occupants of the property.

Disaster planning is no longer a back-burner issue, but an everyday business continuity necessity. Included in that planning are issues concerning first responders and how they will deal with the incident at hand and how they will safely evacuate station personnel.

Starting with the simple stand-alone station, first responders, whether police, firefighters, EMS, or any other municipal agency, have to assess the incident, establish a working perimeter, and be able to respond to the needs of the occupants and deliver them to safety. The occupants, hundreds or even thousands, may be on additional floors, above and below the incident.

A backup communications system for these first responders within the structure is as important as their own communications outside the building and, in some cases, more so. Most radio communications used by municipal law enforcement, fire departments, and emergency services work well in an open environment (outside), but may fail quickly inside steel and concrete structures. Emergency stairwells, elevator lobbies, below-grade parking, building support equipment, and storage areas are of primary concern, and these are the areas most lacking in communications reliability.

Whether the structure is owned by the station or the station is a tenant/licensee in a multi-user building, as part of a station's disaster planning, a resilient and redundant communications system that covers the entire structure must be in place for use by the building's emergency team as well as first responders. Both primary and secondary communications infrastructures are necessary so that a failure of one system, whether due to electronic, natural, or human-made events, does not negate the communications for the first responders and even the station's communications system within the building/facility.

By using different building risers and having the primary and secondary systems in upper and lower building areas, the chance of having at least one system operable during an incident is substantially increased. External high-power repeaters can also be fed into the system should there be an internal catastrophic system failure disabling both primary and secondary systems. With such a system, a wireless interoperability system can be installed in the buildings, thus enabling the fire department, police, EMS, OEM, or any other agency to communicate when outside systems fail or are inadequate. The incident commander could, for example, issue an evacuation order to all first responders instantly, without having to worry about whether the information was relayed between services.

Voice Over Internet Protocol (VOIP) control of the communications systems can also be provided that will allow a remote site (e.g., fire headquarters) to communicate directly to the firefighters at the incident, query the equipment remotely to check operating parameters, and send data from the incident to the surrounding hospitals about incoming patients' vital signs, electrocardiograms, and other information.

Data networks, fiber-building backbones, distributed antenna systems, and other innovative techniques are among the present-day technologies being discussed and utilized. When preparing disaster plans, make certain to remember the first responder's communications. Lives may depend on it and stations can help implement this kind of planning.

INSURANCE

An insurance policy can cover a broadcaster for a variety of emergency conditions. In many cases, law or corporate mandate may require insurance coverage. Some special areas of coverage worth considering are:

- Flood insurance to cover rising water damage.
- Loss of income or business interruption coverage.
- General liability coverage.
- Inland property coverage.
- Marine policy to cover items specific to broadcasters.

Federal flood insurance coverage is dependent on geographic eligibility. Mountaintop transmitter sites and studios located in high-rise office buildings, for example, are not good candidates for flood insurance, since it is unlikely these facilities have or will fall victim to rising waters. Flood insurance is determined by location in or near a 100-year flood zone and nearest body of water.

Flooding caused by ruptured pipes, leaking roofs, or accidental spills, for example, is covered by a property coverage policy and not by flood insurance. Loss of income or business interruption insurance provides compensation when it is no longer possible for a facility to broadcast normally. When a station is unable to broadcast, listeners or viewers will select another station. Insurance cannot predict the amount of loss the station will suffer due to the listeners who may never return to the station once they have switched away. In addition, it is not possible to accurately gauge the degree of lost ratings points and subsequent revenue loss during the outage period.

Business interruption insurance may also compensate the insured for expenses incurred for temporary measures taken to resume broadcasting; for example, the rental of a box truck used to house an AM and FM transmitter for a period while a new transmitter building is installed following an arson fire that destroyed the transmitter building. The coverage could also provide payments for security guards on duty 24 hours a day during the construction project.

General liability insurance is important for broadcasters whether they be locally or remotely programmed. It is possible that a listener, advertiser, company, or individual may sue the station for comments aired during a broadcast. Of specific concern are talk-formatted stations. Liability insurance may also cover the station for the words and actions of its employees while on duty.

Inland property coverage will insure the broadcaster for general items in the facility, such as furniture, fixtures, office equipment, and computers.

So-called "marine" policies generally underwrite equipment that is specific to broadcasters. Examples include towers, ATUs, antennas, transmission lines, isocouplers, power generators, ground systems, transmitters, STLs, audio processors, remote controls, and nonstationary equipment such as that employed for remote broadcasts.

Most insurance policies provide for "replacement cost" or "new for old without depreciation." This coverage will provide for replacement of an item with its closest available match. For example, an aged tube transmitter may be replaced with its modern-day solid-state equivalent.

Most insurance policies will compensate for the loss of an item and the labor to properly install and test it. Not all policies will cover services 100%. For example, a fallen tower may be replaced with one of the exact same height and physical construction. However, if a road must be constructed to allow access to the tower site, insurance may not cover this cost. The coverage may not compensate for soil boring tests or other such activities even though required by local or state regulation. Insurance may not cover other regulatory items such as removal of underground storage tanks, etc.

Underwriters will often send inspectors to a station to verify that good engineering practices have been followed and that the insured has taken reasonable care of assets. Frequently, the station engineer will be asked specific questions pertaining to tower structures. The engineering data on towers should be kept readily available for this purpose. Towers with elevators may be subject to periodic city, county, or state inspection. The inspector may also check the main electrical service panel and other utilities.

Consider seeking help from experienced emergency communications contingency planning professionals to help devise a well-written and comprehensive emergency plan. This is especially useful if personnel tasked with day-to-day work to keep the facility operating do not have time to devote to the planning. Professionals can also help with research on factors such as geology and dealing with hazardous materials.

COMMONSENSE PRECAUTIONS

There are commonsense precautions a broadcaster can take to ensure reliable and safe operation. For facilities in locations prone to hurricanes or high winds, the tower should be evaluated to ensure it is able to withstand expected wind forces given its present loads (antennas, dishes, etc.). Many tower companies now offer such evaluation services, as do independent structural consultants. Tornados produce extreme wind forces. While few towers or building structures can survive direct hits, recent revisions to tower standards (TIA-222-G) have addressed tower design and structural integrity in high wind–prone areas of the country. See Chapter 9.1 for additional information.

Lightning protection for towers and equipment buildings is an important aspect of maintaining continuity of transmission. See Chapter 9.2 for a detailed discussion.

Falling ice is responsible for a great deal of damage to broadcast equipment and buildings. An ice bridge can protect horizontal runs of transmission line between tower and building. Small antennas on the tower, such as those used by STL or RPU systems, can be protected by an ice shield above them. Transmitter buildings and carports should have icebreakers installed above them. Guy wire ends should have icebreakers installed to prevent damage to anchor plates from ice sliding down the cables. Hard hats should be provided at ice-prone transmitter sites. Safety procedures should be established, including prudent measures to protect personnel from slip-and-fall injuries on icy catwalks or access ladders.

Flooding caused by rising waters can be anticipated. Transmitter buildings in low-lying areas should be raised above the level of the worst expected water level. In some instances, transmitters and racks can be mounted on 2 × 4 pressure-treated lumber so that floor-mounted components are reasonably protected in case of either rising or flooding water.

Earth movement may be defined as earthquakes, mudslides, avalanches, or unstable earth. Facilities located in earthquake-prone areas can be designed to withstand these forces. Equipment inside these buildings can and should be anchored to prevent damage from shaking or horizontal movement. Unstable earth can be buttressed with suitable retaining walls. A licensed structural engineer specializing in geologic structural mitigation should be consulted on these matters.

Some earth movements may be attributed to human-made causes, such as nearby blasting. Blasting can be dangerous in high radio frequency (RF) environments due to the possibility of blasting cap detonation by radio waves. Nearby blasting may loosen earth and cause guy wire anchor movement and possible tower failure.

Large nearby trees also pose a potential hazard. High winds, either alone or combined with ice accumulation, can cause otherwise healthy trees to topple. Heavy rain can saturate earth and destroy the ability to adequately hold tree roots in place, causing what appears to be a healthy mature tree to fall onto guy wires and buildings. Trees or limbs that could fall on buildings, guy wires, generators, or equipment should be kept properly trimmed by a licensed tree expert.

FACILITY SECURITY

FCC Rule §73.127 and others state, "the licensee or permittee must retain control over all material transmitted in a broadcast mode via the station's facilities with the right to reject any material that it deems inappropriate or undesirable."

While this rule covers programming on subcarrier services on AM, FM, TV, LPFM, LPTV, as well as all digital broadcast services, it could be extrapolated to mean transmitting inappropriate or undesirable programming on the main broadcast services, as well as preventing unauthorized people from entering a broadcast facility to go on camera or on mike.

Aside from protection of employees while in the workplace and protection of assets critical to staying on the air, every broadcast facility has that legal duty under its license to prevent unauthorized access to the airwaves. Stations that exercise due diligence to assess risks to life and property should extend security measures aimed at controlling on-the-air programming.

No one in a broadcast facility is better trained and equipped than the technical staff to carry out the safety and security assessment process or act as key advisors for outside experts brought in to help. The technical staff will not only know all the facility's vulnerable points, but also are in the best position to advise on and oversee implementation of recommendations of the experts.

While design of a broadcast facility reception area is beyond the scope of this chapter, security concerns require a review of new and existing designs to limit entrance of unauthorized persons. Some facilities are placing receptionists behind bullet-resistant glass with intercom communication to visitors. This approach may not be practical in high-rise buildings with multiple tenants or at stations with limited space, but simply adding better locks, security cameras, and peepholes can reduce the potential for casual walk-ins. Further security measures may be needed at the lobby and garage levels as well as parking lots.

Hiring trained security personnel to guard facilities on a 24-hour basis may represent a significant ongoing expense, but it may be prudent in clustered facilities or in cases when physical security cannot be improved. Before proceeding, station management should consult its legal department or outside counsel. Making insurance underwriters aware of risks and obtaining their guidance is advisable as well.

High-quality video surveillance systems with motion-sensing digital recording capability and off-site Internet monitoring can be a reasonable and wise investment. Reports of transmitter site break-ins, thefts, and vandalism have increased since remote control came into practice. While the best option for transmitter site security is live guards, few stations have taken this step. A transmitter site security package that includes video surveillance is now an important and necessary requirement for many facilities. Many sophisticated security features are now economical choices for broadcasters. Most modern electronic security systems can easily be set up for motion sensing. Alarm closures built into the security system can be wired to existing remote control systems even if there is no means for Internet visual supervision. A studio operator monitoring the remote control can make a call to the police, or the remote control can dial a series of numbers including security, police, and the fire department. Many systems can be programmed to do this in addition to calling for transmitter technical failures.

Other types of detectors are readily available that can be wired in to remote control channels that can give a measure of advanced warning to many conditions:

- Heat
- Smoke
- Water
- Motion
- Outdoor and indoor temperatures
- Door and window position switches.

Perimeter Security

Perimeter security starts with assessing the condition of all doors, windows, and access hatches for a structure. A careful inspection of door frames, window frames, windows, hinges and existing locks, dead bolts, and hasps and other fasteners for signs of rot or termites may reveal a general lack of integrity. The assessment may suggest installing metal doors in place of wooden doors, and solid doors in place of hollow-core doors.

Eliminating some windows or installing iron bars or sturdy metal grilles over vulnerable windows may reduce security risk. Roof hatches along with all doors and windows that can be opened should have security switches installed that are tied into a master alarm panel. At-risk windows could be equipped with glass shatter sensors tied into the alarm system. Interior motion sensors can be installed in areas secured for nights and weekends. These need to be set so they do not falsely trigger on rodents and other wildlife that frequent many remote transmitter sites. A reputable and experienced security company can assess risks and do installation, but it usually needs help to cover all the site's vulnerabilities known to the technical staff.

There is a wide choice of secure keys, swipe cards, and biosensors for entrances to studio and technical facilities. The best key systems have features that make it almost impossible for anyone, including employees, to make duplicates.

Card-swipe and biosensor systems not only make it easy to give everyone unique access, but many systems also have the capability of printing out entry and exit logs and allowing selective lockout of employees in off hours. Coupled with a video security system, they make for a powerful security combination. Readily available in the biosensor category are reasonably priced fingerprint recognition sensor pads and even retinal scanners.

High-intensity outdoor lighting should be considered for parking lots and transmitter sites. Some utility companies offer a special fixed rate for such lighting systems. Consider installing current monitors on outdoor lighting circuits. These monitors, similar to the same ones used to monitor tower lights, can be tied in to remote control systems to warn if exterior lights are not working or to expose tampering.

A functional alarm system is essential and must report to live humans to be truly effective. The alarm should monitor for intrusion, heat, smoke, and fire. It is wise to have a wireless alarm in the event of intentional telephone line failure. Under some circumstances, the transmitter building can be surrounded by a fence at least 12-ft. high with razor wire installed inside the fence perimeter at the upper level. Each tower should have its own fence. Another protection approach would be to install an anti-climbing gate on the tower itself. In some cases, it may be necessary to locate external video cameras to monitor the most vulnerable points within the transmitter site.

Locks

Padlock quality is sometimes overlooked in security reviews. Some locks in common use can be opened with keys purchased with new locks, or a medium-sized bolt cutter. If padlocks must be used, remove the key codes from the lock bodies.

For critical locking situations, consult a competent and trustworthy locksmith. There are locks used by contractors, railroads, and the motor freight industry that are close to tamper proof. These padlocks are often referred to as contractor's locks. Taking an example from the storage container industry, it may be possible to add a welded cage around the hasp, making it practically impossible to cut off the padlock. Replace inexpensive padlocks with heavy-duty barn door–type hardware. Door locks may have to be a specific type depending on code.

Assess how facility gates are constructed and hung, and make sure they cannot be unbolted, disconnected, or easily climbed over or under to bypass locks and chains. Trip-wire sensors woven into chain-link fences or buried on or near ground level give a facility added warning if there is a breach.

Special Security Precautions

It is a fact of modern life that human-caused disasters must now enter into the planning and risk-assessment process. The FCC's Media Security and Reliability Council (MSRC) published a comprehensive list of recommendations to harden facilities to survive.[4] Often overlooked, basic security precautions have led to serious incidents at a number of places throughout the country. Here are the basics:

- Approve visits from former employees through their former supervisors.
- Always have non-employees escorted in critical areas.
- Assure that outside doors are never propped open.
- Secure roof hatches from the inside, and have alarm contacts on the hatch.
- Use photo ID badges when employees might not know each other by sight.
- Check for laws that require a written safety and security plan.
- Use video security and card key systems where warranted.
- Install entry alarms at unattended sites. Test weekly.
- Consider remote video surveillance using the Internet for monitoring.
- Maintain fences, especially at unstaffed sites.
- Redesign to limit places bombs could be planted.
- Redesign to prevent unauthorized entry.
- Redesign to limit danger from outside windows.
- Plan for fire, bomb threats, hostage situations, and terrorist takeovers.
- Plan a safe way to shut the facility down in case of invasion.
- Plan guard patrol schedules to be random, not predictable.
- Plan for off-site relocation and restoration of services on short notice.

FIRE PREVENTION

Fires can start through intentional or unintentional means. Local codes will likely require fire extinguishers at certain locations throughout the facility. Be certain that the extinguishers are the appropriate type for a particular area or equipment. Local ordinances and good practice require monthly checks to make sure the pins in fire extinguishers are in place, are holding pressure, are mounted at the correct height, and are not obstructed. Yearly checks by a licensed extinguisher inspection company is not just a good idea, it is usually the law. Most jurisdictions and insurance underwriters also mandate entries on the tag on each extinguisher to be initialed and dated by the person making the check each month. Automatic sprinkler systems may be a requirement depending on state and local regulations.

OTHER DISASTER PREVENTION MEASURES

Component failure can occur due to neglect, such as overlooking cleaning air filters, lubricating blower motors, and control systems. Preparing and adhering to a maintenance schedule can prevent failures before they occur. A site-specific inspection schedule form and maintenance shifts that are long enough to allow personnel time to do careful inspections can, in many cases, catch small problems in time to prevent much larger ones. For example, a transmission line dehydrator that is not working at rated capacity can be identified easily as needing maintenance with simple periodic inspections that include log entries for status and readings. Not catching such a fault can lead to the inability of the dehydrator to keep up with a transmission line leak. Excessive moisture in the transmission line can cause arcing that can take a station off the air with spectacular and expensive results.

Towers sometimes fail due to improper or excessive loading in an attempt to obtain extra rental income. Prior to the addition of any appurtenance to a tower, a structural analysis should be performed. Such an analysis can be contracted to many tower firms or independent structural consultants. A few extra income dollars may not be worth the risk of a total tower loss.

Vehicular contact with tower guy anchor points, power generators, or propane storage tanks can occur if these critical components are located near roadways or parking lots or side roads. Most local regulations require concrete or metal pilings to protect natural gas tanks. The same precaution can be taken for guy anchors and generators. Where guy wires cross low over a parking lot or driveway, the lowest guy wire should be decorated with red fluorescent flagging to help prevent contact with ground vehicles. Aircraft contact with towers falls under FAA jurisdiction. While this is less common, it does occur so it is important that the tower lights are fully functional at all times.

Power surges may be caused by other users on the facility's utility mains feed as well as new equipment installed on an old service at the site. Surge suppressors should be installed on all incoming AC power mains. Uninterruptible power supply (UPS) devices should be installed on each computer. Occasionally a utility will accidentally feed improper mains voltage or phasing. See Chapter 11.1 for a discussion on AC power-conditioning issues.

Strategic location of water feed pipes can prevent accidental flooding. Runs along ceilings in general should be avoided, but avoid routing pipes near or over electrical equipment. Overhead pipe leaks and bursts can cause thousands of dollars in equipment and time loss. If overhead pipes cannot be avoided, then a water shield above sensitive equipment may be worth consideration.

Many prolonged outages are caused by not having spare parts on hand. While managers consider spare parts to be an interruption in cash flow, to the engineer they are insurance. Adequate spares should be maintained for each critical piece of equipment. The station manager and chief engineer may differ on which equipment is critical. Both could list mission-critical equipment. The chief engineer could then compile a list of reasonable spare parts that might include fuses, IC chips, tubes, pots, switches, power supplies, and functional modules. The component budget could then be used to determine the reasonable cost for what amounts to an insurance policy against prolonged outage or failure.

OUTSIDE PLANT LINKS

Interfacility links such as wire, microwave, or fiber communications circuits are single-point failure opportunities that could cause significant outages. Outside plant links discussed below presuppose proper installation by people and organizations not under station control. Despite this lack of control, a visual inspection of the local utility and telecommunication infrastructure could be revealing. For wire and fiber, this means an inspection to check for adequate service loops so quake and wind stresses will not snap lines. It means that telephone and telecommunications service providers have installed terminal equipment so it will not be vandalized, fall over in a quake, or easily be flooded out. No matter what the results of the survey, a range of backup options are available and should be considered.

Outside Plant Wire

Local telephone companies still use wire, even for T1 and DSL connections. If the facility is served only by wire on telephone poles or underground in flood-prone areas, investigate alternate telecommunications routing. Alternate routing from the facility to another central office (CO) may be very costly, since the next nearest CO is rarely close. But, if an alternate route is only to the next block, or duplicating existing telephone pole or underground risk, the advantage gained may be minimal.

Many telephone companies can designate as an essential service a limited block of telephone numbers at a given location for lifeline communications. Lines so designated are usually found at critical facilities like hospitals and public safety headquarters. Contact the local phone company to ascertain if the broadcast facility can apply for such service. Applicants with close ties to local government emergency management and who are vital links to the public during emergencies (such as broadcasters) may easily qualify.

Microwave Links

Wind, explosions, and seismic activity likely will cause microwave dishes to go out of alignment. Quake-resistant towers and mounts can help prevent alignment failure, even in the face of high wind or explosions. Redundant systems should be considered part of the solution, but an exact duplicate microwave system may lead to a false sense of security. Consider a non-microwave backup such as a fiber link.

Fiber Optics Links

If a broadcast facility is not a fiber customer today, it probably will be tomorrow. The facility may be fortunate enough to be served by separate fiber vendors with separate fiber systems and routing to enhance reliability and uptime. Special installation techniques are essential to make sure fiber links will not be bothered by earth movement or subject to vandalism or other single-point failure causes.

Fiber should be installed underground in a sturdy plastic sheath with a warning ribbon above the cable. The sheath offers protection from sharp rocks or other forces that might cause a nick or break in the armor of the cable or actually sever one or more of the bundled fibers. The ribbons are usually colored bright orange to make them stand out in trenches, manholes, and other places where careless digging and prodding could spell disaster. From a risk-assessment standpoint, their color easily identifies what they hold as targets to technically aware terrorists or vandals.

Cable systems that use aerial rights-of-way on utility poles for their fiber may not prove as reliable in some areas as fiber installed underground. Terminal equipment for fiber should be installed in quake-secure (where applicable) equipment racks and away from flood hazards. Fiber terminal electronics should have at least two DC power supplies in parallel that are, in turn, on line with rechargeable battery backup.

Satellite Links

Ku or C Band satellite is a costly but effective way to link critical communications elements. C Band has an added advantage over Ku during heavy rain or snow conditions. A significant liability of satellite transmission for ultra-reliable facilities is the possibility that a storm could cause a deep fade, even for C Band links. Another liability is short but deep semiannual periods of "sun outage," when a link is lost while the sun is focused directly into a receive dish. While these periods are predictable and last for short periods, service will be lost unless there is alternate service on another satellite with a different sun outage time, or terrestrial backup.

SINGLE-POINT FAILURES

Single-point failure can occur in any system without backups or bypasses. Single-point failure analysis and prevention are based on a simple concept: A chain is as strong as its weakest link, but two chains, equally strong, should never share the same weak link. The lesson may be that each chain should be constructed with totally independent links, or, even better, where the risk and the consequences are both high, use three chains.

STANDBY POWER SYSTEMS

A variety of standby power systems is available in the event of commercial power interruption. Chapter 11.1 addresses AC power conditioning. Clean, reliable, and stable AC power is essential in modern broadcast facilities due to the use of computer- and microprocessor-based technology currently employed in nearly all stations.

AIR-HANDLING SYSTEMS

People and equipment both do not operate well when their environment becomes too hot. Adequate, clean, cool, dry, and pollutant-free air from the facility's heating, ventilating, and air conditioning (HVAC) system is important for modern communications equipment and operating staff. If the facility occupies leased space in a high-rise building, total control of the facility air system may not be possible. Many large buildings have no backup systems and do not provide night and weekend supervision or emergency service. The best protection is to establish precise conditions for air heating and cooling in the lease. Adding a separate backup system in such cases is prudent but may be costly. A short-term emergency option might be portable industrial-strength air-conditioning systems that are available from many rental companies. A contract for emergency equipment that can be obtained with a phone call could save hours or days of downtime. Consider buying a portable emergency HVAC unit if the facility and its operations provide essential services, or negotiate a contract for emergency rapid delivery of a portable HVAC with a vendor providing such a service.

HVAC Full Recirculation

Wherever cooling air comes from, there may be occasions when there is a need to set the system to recirculate air within the building. This temporarily makes it a closed system. Smoke or toxic fumes from a fire in the neighborhood can enter an open system and incapacitate the staff and even cause some equipment to malfunction. Also consider the effect of smoke detectors accidentally setting off a sprinkler system from outside events. With advanced warning, having the ability to configure the air-handling system to full recirculation might avoid or forestall personnel discomfort and increase their safety. It could provide enough time to arrange an orderly evacuation and transition to an alternate site, or time for outside toxic conditions to dissipate.

WATER HAZARDS

Water in the wrong place at the wrong time can be part of a larger emergency or be its own emergency. A simple mistake like locating a water heater where it can flood out electrical equipment can cause fatal electrical problems when it finally wears out and begins to leak. Unsecured water heaters can tear away from gas lines, possibly causing an explosion or fire in an earthquake. At best, the water in that heater could be lost, depriving employees of a source of emergency drinking water.

ELECTROMAGNETIC PULSE PROTECTION

The electromagnetic pulse (EMP) protection phenomenon associated with nuclear explosions can disable almost any electronic component in a communications system. EMP energy can enter any component or system coupled to a wire or metal surface directly, capacitively, or inductively. Some chemical weapons can produce EMP, albeit on a smaller scale.[5] The Federal Emergency Management Agency (FEMA), now part of the Department of Homeland Security, has been involved in EMP protection since 1970 and is charged at the federal level with the overall direction of the EMP program. FEMA provides detailed guidance and, in some cases, direct assistance on EMP protection to critical communications facilities in the private sector. People responsible for a site that needs EMP protection should discuss the issue with a knowledgeable consultant before installing such protection devices for RF circuitry.

PREPAREDNESS TIPS BASED ON EXPERIENCE

Experience is often the best teacher when considering security and preparation of a site to reduce safety and potential risks. Examples include:

- Evidence of previous high-water marks on walls and in cable and transformer vaults means it could happen again. Such evidence could present itself as calcium or mineral salt deposits that stand out against normal wall or soil color.
- Evidence of roof and wall leaks that may have happened years ago could cause new problems when major rain or wind storms strike.
- Review old maintenance logs looking for disasters (not just water related) that occurred in the past and could happen again if no remedial measures are taken.
- In shared facilities, such as in high-rise buildings, get to know the neighbors. Find out if others near the station are using hazardous chemicals or otherwise conducting their business in a way that could affect station operations.

ALTERNATE SITES

Prudent disaster planning should include the potential of abandoning the facility for a substantial period. Government emergency planners usually arrange for an alternate site for their emergency operations centers (EOCs). Communications facilities can sign mutual aid agreements with each other. This may be a way to access telephone lines, satellite uplink equipment, microwave, or fiber on short notice.

PRESERVING VITAL SITE-SPECIFIC INFORMATION

Management should have an information access document containing:

- The supervisor username password for the network.
- Telephone system control and voicemail usernames and passwords.
- Engineering area keys and lock combinations.
- Transmitter site alarm codes, keys, and lock combinations.
- Passwords to digital audio processors.
- Names and telephone numbers for tower, telephone system repair, and computer service personnel.
- Other contractors, such as snow-plowing contractors, consulting engineers, electricians, plumbers, HVAC contractors, and others.
- Circuit numbers for telephone, ISDN, and communications systems for use when contacting repair services.

A security and risk-assessment checklist should be compiled and updated quarterly or annually. Check the facility maintenance calendar annually for critical items that apply to each predicted risk or after some serious event. For example, guy points and tower plumb need to be checked periodically, as well as after every seismic event.

SELF-DISPATCHING PLANNING FOR EMERGENCIES

A key part of each station's disaster planning should be a section that clearly outlines where and when key personnel should self-dispatch when emergency conditions exist. This is especially important when major disasters take out phone service, including cell phones. Management should know in advance where engineering resources will be when their expertise is required.

HURRICANE KATRINA

In the post-9/11 world, most businesses have realized the importance of having an emergency plan in place that will help them survive potentially disastrous events, be they either natural or human-made. Broadcasters, for the most part, realized the importance of emergency planning long before it became a front-burner issue for their counterparts in the general business environ. In general, WWL Radio and many other broadcast facilities have had detailed emergency plans in place many years prior to the devastations of 9/11. For the minority of broadcast facilities who had not yet done so, the unprecedented events of 9/11 served as an undeniable wakeup call to get a viable emergency plan in place as soon as possible.

Today, with the vast majority of broadcast facilities having some form of emergency contingency plan in place, just how protected are we? In order to answer that question accurately, we first need to identify some of the issues that we all have in common, regardless of size and geographic region. Power interruptions, equipment failure, and staffing problems, just to name a few, all fit into the common issues area. Additionally, we also need to identify geographically specific issues that can either enhance or minimize certain exposures depending upon location. Snow and ice storms, tsunamis, mud slides, tornados, and hurricanes all fit into the geographically specific issues list.

The preceding issues are actually easy to identify and enumerate, especially by someone who has been involved in broadcasting and based at the same physical location for an extended period. Over time, the pertinent issues tend to become repetitious in nature and thereby obvious to the observant. What about the less obvious, seldom, or never before experienced issues?

On August 29, 2005, Hurricane Katrina came ashore just east of New Orleans and confirmed that the old adage "experience is the best teacher" is indeed still true. As director of engineering for the Entercom New Orleans stations during that time frame, and as a lifelong New Orleans area resident, I would like to share some of the more salient, yet almost totally unexpected issues that revealed themselves during and in the extended aftermath of Hurricane Katrina.

Prior to Hurricane Katrina, those of us who had been involved in broadcasting for the better part of our lives tended to believe that, in general, we knew what to expect from a hurricane. Many of us had already dealt with hurricanes on multiple occasions and had always somehow managed to come out none the worse for wear. The first indication that we might not be as knowledgeable about hurricane behavior as we had originally thought came on August 27. At that time, Katrina was still a minimal Category 3 storm moving north across the Gulf of Mexico but, unknown to many, it was rapidly approaching an unusually warm section of water known as a Loop Current ring. This large ring, or eddy vortex, had broken off from the main loop current, which is the source that eventually feeds the Gulf Stream. It was now perfectly positioned to "super-charge" the approaching hurricane. The National Weather Service was now forecasting that Katrina, a Category 3 storm at that time, might possibly increase to Category 5 status while over the unusually warm Loop ring area. Even the most seasoned amongst us had never before experienced a direct hit from a Category 5 hurricane and concern was growing.

By this time, our then existing five-level emergency plan had already been implemented and was already at level three. Per the plan, essential core staff members would now remain onsite 24/7 to keep the station on the air throughout the event. With Katrina now forecast to possibly increase to a maximum Category 5 status, and on an ever-shifting path that was bringing it closer and closer to a direct hit on New Orleans, government officials decided that it was now time to pull the trigger and call for a first-ever mandatory evacuation of the entire area. Those of us who were going to stay behind packed our families, friends, and pets into pre-loaded vehicles, wished them well, and saw them off on their way to what we hoped would be safer distant inland locations. What we did not know at that time and would later prove to be a major concern was that an unusually large percentage of nonessential station employees were not planning on evacuating as they were supposed to.

In an effort to avoid the now-crowded and slow-moving evacuation routes, a significant number of nonessential station employees had unilaterally decided not to evacuate and would be staying in the city. Many who had chosen not to evacuate were apparently having second thoughts about riding out a Category 5 hurricane at home. These people unexpectedly started showing up at station seeking shelter, some arriving just prior to the storm. Everyone was seeking what they foolishly thought would be a safe place of refuge. In reality, they would have been better off had they evacuated while they still had the opportunity. Many who had parked at street level eventually lost their vehicle to the rising flood waters.

Based on our prior collective hurricane experiences, no one ever expected Katrina to be more than a one- or two-day event at the most. In previous typical hurricane experiences, the storm blows in, usually an overnight experience, and by the end of the following day, the skies are clearing, winds are subsiding, and everyone returns home to clean up the debris and await the restoration of commercial power. Typically, within one or two weeks, the debris is gone, commercial power is restored, and life returns to some semblance of normalcy. Unfortunately for all concerned, Katrina was not by any stretch of the imagination a typical storm.

We started to realize that we might have a serious supply-related problem on the afternoon just before the storm's arrival. At that time, we noticed that in addition to the 20 or so staffers that were expected to stay at the station for the storm's duration, we now had many more unexpected guest on hand. An additional 40 to 50 unexpected employees were now on site, in many cases with additional family members and even pets too. It became readily apparent that the supplies that were intended to feed 20 people for two to three days were going to be woefully inadequate for the 60 to 70 people now on site! The matter of supplies became even more critical early on the following day when the levees broke and the city began filling with water, cutting off most vehicular evacuation options.

As a direct result of these lessons, we now also maintain a very large stockpile of military style MREs (meals ready to eat) on hand at all times. The MRE supplies are now estimated to be sufficient to feed the expected staff members for a full month or more. Additionally, required staff members now sign a voluntary emergency event participation form, agreeing in advance that they will be available for the duration of an emergency event and that they will make advance preparations for the safety and evacuation of family members. Nonessential employees are also made aware that, in the event of a mandatory evacuation, shelter space will not be available at the station. They must follow government directives and be personally responsible for the safety of themselves and their families. The unexpectedly large number of employees that showed up at the last moment before Katrina's arrival, along with building security concerns, factored heavily into a post-storm corporate decision to evacuate everyone from our downtown studio. That evacuation occurred via helicopter two days after Katrina, while floodwaters were still rising in the streets around the station. In retrospect, flying in additional supplies and having a security presence might have been a better option than totally evacuating what was, at that time, a still fully functional studio site.

Our large natural gas–powered generator continued to operate flawlessly, along with a fully functional HVAC system, for almost a full month following Katrina. Additionally, although the POTS-based business phone lines were down, our ISDN-based on-air phone system and several ISDN-based CODECs remained 100% operational throughout the storm and its aftermath. Our primary corporate Internet connection went down at the height of the storm; however, a smaller DSL-based Internet connection remained fully functional at all times. It would have been sufficient to meet the needs of our news department had we not abandoned that site.

On the positive side, a strategic pre-storm alliance with neighboring Jefferson Parish (County) Emergency Operations officials provided us with a relatively safe temporary studio site inside the Jefferson EOC. Three of our news people and one engineer had been preassigned to that location. They kept WWL on the air from the EOC when we were required to evacuate from our downtown studios. Nonstop on-air operation continued from that EOC studio for several days. Beyond that initial term, daily broadcast and updates continued from the EOC for several months after Katrina. Following the downtown studio evacuation, the Entercom New Orleans Engineering staff was also invited in and lived in the basement of the Jefferson EOC for several weeks. Fortunately, the Jefferson EOC was located near the WWL transmitter site, which allowed us to setup a temporary STL link.

After the storm ended, some of the Entercom New Orleans programming staff evacuated to Baton Rouge where they met with Clear Channel Baton Rouge management. Following a quick impromptu agreement, improvised studio facilities were quickly set up and operating from what had previously been a storage closet at the Clear Channel Baton Rouge site. Information originating from that site was originally intended to only be simulcast on the Entercom and Clear Channel New Orleans stations. However, audio from that improvised shared facility would soon be simulcast live on virtually every functional radio outlet in the region. This was accomplished quickly, utilizing the satellite distribution facilities of The Louisiana Radio Network.

With nationwide satellite distribution now available, United Radio Broadcasters audio could now be carried by many other stations throughout the country, especially in areas with high concentrations of evacuees from the New Orleans area. It was even simulcast on an international shortwave radio station. This all became a shining example of cooperation between broadcasters for the common good. Obstacles were quickly overcome and goals rapidly accomplished.

All considered, the entire post-Katrina episode is an excellent example of broadcasters working together to accomplish something that harkens back to our prime directive as originally expressed in the Communications Acts of 1927 and 1934, which contained the directive "to serve the public interest, convenience and necessity." In my 51 years in broadcasting, I've never encountered a better example!

SUMMARY

Preparing for natural and human-made disasters, ensuring personnel safety, and securing critical facilities are all key elements in keeping broadcast stations on the air when they are needed most. Assessing risks within all aspects of station operations, studio, transmitter, and towers, and planning how to deal with disruptions, can reduce downtime, improve the safety and welfare of station personnel, and instill confidence that the facility can continue to operate when disasters of all kinds occur.

ACKNOWLEDGMENTS

The authors thank John Lyons, Assistant Vice President and Director of Broadcast Communications for the Durst Organization in New York City, for material for the First Responder Communications Backup System section.

KEY TERMS

Business impact analysis (BIA): A formal study of the impact of a risk or multiple risks on a specific business. A properly conducted BIA becomes critical to the business recovery plan.

Business recovery plan (BRP): A blueprint to maintain or resume core business activities following a disaster. Three major goals of a BRP are the resumption of products and services, customer service and cash flow, and recovery of the business to normal operations.

Central office (CO): The building where local switching is accomplished.

Emergency operations centers (EOCs): Locations where emergency managers receive damage assessments, allocate resources, and begin recovery.

Electromagnetic pulse (EMP): A high burst of energy associated most commonly with nuclear explosions. EMP can instantly destroy the functionality of many electronic systems and components.

HVAC: Architectural acronym for heating, ventilation, and air conditioning.

Incident commander (IC): The title of the person in charge at an emergency scene from a street corner incident to an emergency involving an entire state or region.

Incident command system (ICS): An effective emergency management model invented by firefighters in California.

USAR: Emergency management acronym for urban search and rescue.

Notes

[1] *Lockout/tagout* is one of a set of standard safety procedures that assures that energy is removed from equipment during installation and maintenance. It assures that every member of a work detail is clear before power is reapplied.

[2] For more on disaster management organization, please refer to the FEMA website at http://training.fema.gov/EMIWeb/IS/IS100CM/ICS0103summary.htm.

[3] Information stemming from the Northridge earthquake in 1994 has influenced seismic building codes in California and beyond for many types of structures. The Northridge earthquake showed that some high-rise structures thought to be quake safe really are not. Designers should also be aware that seismic building codes usually only allow for safe evacuation. They do not embody design criteria that prevent major structural damage. Quake safe may not be quake proof, just as water safe may not be waterproof.

[4] See www.fcc.gov for more information on the MSRC, or www.mediasecurity.org.

[5] FEMA publishes a three-volume set of documents on EMP referenced in the Bibliography. They cover the theoretical basis for EMP protection, protection applications, and protection installation.

Bibliography

Baylus, Ellen. *Disaster Recovery Handbook.* Chantico Publishing Company, Carrolton, TX, 1991.

Federal Emergency Management Agency (FEMA). *Electromagnetic Pulse Protection Guidance*, vols. 1–3. Federal Emergency Management Agency, Washington, DC, 1991.

Federal Emergency Management Agency (FEMA).. *Federal Response Plan.* Federal Emergency Management Agency, Washington, DC, 1998.

Hadden, Josh. "Disaster Preparedness and Business Continuity: Ideas, Financing, Implementation," *Broadcast Engineering Conference Proceedings*, National Association of Broadcasters, 2006.

Handmer, John, and Parker, Dennis. *Hazard Management and Emergency Planning.* James and James Science Publishers, Abingdon, Oxfordshire, 1992.

Janitschek, Andrew. "Common Sense Rules to Uncommon Events," *Broadcast Engineering Conference Proceedings*, National Association of Broadcasters, 2006.

Osenkowsky, Thomas G. "Disaster Planning . . . Before and After the Fact," *Broadcast Engineering Conference Proceedings*, National Association of Broadcasters, 1997.

Osenkowsky, Thomas G. "Emergency Planning for Radio and Television," *Broadcast Engineering Conference Proceedings*, National Association of Broadcasters, 2000.

Rothstein Associates. *The Rothstein Catalog on Disaster Recovery and Business Resumption Planning.* Rothstein Associates, Brookfield, CT, 1993.

Further Information

The firms listed below are shown as representative only and are not specifically endorsed by the authors.

Associations/Groups

Association of Contingency Planners (www.acp-international.com).

Business and Industry Council for Emergency Planning and Preparedness (BICEPP), P.O. Box 1020, Northridge, CA 91328.

Disaster Recovery Institute (DRI), 1810 Craig Road, Suite 125, St. Louis, MO 63146. DRI holds national conferences and publishes the *Disaster Recovery Journal.*

Earthquake Engineering Research Institute (EERI), 6431 Fairmont Avenue, Suite 7, El Cerritos, CA 94530.

National American Red Cross, 2025 E Street, N.W., Washington, DC 20006.

National Center for Earthquake Engineering Research, State University of New York at Buffalo, Science and Engineering Library-304, Capen Hall, Buffalo, NY 14260.

National Coordination Council on Emergency Management (NCCEM), 7297 Lee Highway, Falls Church, VA 22042.

National Hazards Research & Applications Information Center, Campus Box 482, University of Colorado, Boulder, CO 80309.

Business Recovery Planning

Harris Devlin Associates, 1285 Drummers Lane, Wayne, PA 19087.

Industrial Risk Insurers (IRI), 85 Woodland Street, Hartford, CT 06102.

MLC & Associates, Mary Carrido, President, 15398 Eiffel Circle, Irvine, CA 92714.

Price Waterhouse, Dispute Analysis and Corporate Recovery Dept., 555 California Street, Suite 3130, San Francisco, CA 94104.

Resource Referral Service, P.O. Box 2208, Arlington, VA 22202.

The Workman Group, Janet Gorman, President, P.O. Box 94236, Pasadena, CA 91109.

Life Safety/Disaster Response

Caroline Pratt & Associates, 24104 Village #14, Camarillo, CA 93013.

Industry Training Associates, 3363 Wrightwood Drive, Suite 100, Studio City, CA 91604.

Emergency Supplies

BEST Power Technology, P.O. Box 280, Necedah, WI 54646.

Exide Electronics Group, Inc., 8521 Six Forks Road, Raleigh, NC 27615.

Extend-A-Life, Inc., 1010 South Arroyo Parkway, #7, Pasadena, CA 91105.

Velcro® USA, P.O. Box 2422, Capistrano Beach, CA 92624.

Worksafe Technologies, 25133 Avenue Tibbets, Building F, Valencia, CA 91355.

Construction/Design/Seismic Bracing

American Institute of Architects, 1735 New York Avenue, N.W., Washington, DC 20006.

DATA Clean Corporation (800-328-2256), Geotechnical/Environmental Consultants.

H. J. Degenkolb Associates, Engineers, 350 Sansome Street, San Francisco, CA 94104.

Leighton and Associates, Inc., 17781 Cowan, Irvine, CA 92714.

Miscellaneous

Data Processing Security, Inc., 200 East Loop 820, Fort Worth, TX 76112.

EDP Security, 7 Beaver Brook Road, Littleton, MA 01460.

ENDUR-ALL Glass Coatings, Inc., 23018 Ventura Blvd., Suite 101, Woodland Hills, CA 91464.

Mobile Home Safety Products, 28165 B Front Street, Suite 121, Temecula, CA 92390.

SECTION

4

AUDIO PRODUCTION AND STUDIO TECHNOLOGY

CHAPTER

4.1

Principles of Acoustics and Noise Control for Broadcast Applications

RONALD ELIGATOR

Acoustic Distinctions
New York, NY

Note: The mentioning of any commercial products, processes or services or the use of names, logos or brands is for informational purposes and the convenience of readers. It is not to be construed as an endorsement or recommendation. All product names, logos and brands are property of their respective owners.

INTRODUCTION

Acoustics refers to the way sound and vibration are heard and felt. In broadcast and related applications, it relates to the sources, paths and receivers of audible vibrations in air, and physically perceptible vibrations in structures, that affect enclosed rooms and exterior spaces used for creation, production, recording, listening and viewing of audio and video material.

A successful broadcast facility is characterized by room acoustic environments having proper control of direct, reflected and reverberant sound energy; appropriately low background sound and vibration levels; and sound isolation between spaces that prevents undesirable intrusive noise from interfering with the production activities. This chapter defines these attributes, relates them to the range of spaces normally found in broadcast facilities, and presents general guidelines for design and construction to best ensure that broadcast spaces can support the scheduled functions taking place.

As a chapter on acoustics in a handbook whose primary readership is expected to be radio and TV broadcast engineers, the focus here is on basic guidelines of acoustics for facility planning, design and construction in both new construction and renovation. Fundamentals of acoustic theory, which have been far more comprehensively addressed in acoustic and engineering textbooks, is addressed to only a limited extent. The bibliography at the end of this chapter includes a list of additional references for those interested in learning more about first principles of acoustics and related fields.

Acoustics is a discipline that integrates all aspects of facility design and construction, including architecture, structural, mechanical and electrical engineering, plumbing, lighting, etc. Proper integration of these design and construction disciplines is required from the earliest stage in project design to ensure achievement of appropriate acoustical environments. Successful projects require collaboration between experienced studio designers, architects, engineers and consultants from the start. This is just as true in renovation projects, or where corrective acoustic constructions are required, as it is in new construction.

Acoustic design also needs to be considered in relation to electrical and electronic component and systems integration in virtually every aspect of the broadcast chain, from newsgathering and program origination, program production and post-production, through master control and distribution links, to transmission, antennas, RF propagation, cable and satellite.

While acoustics is a key component characterizing the quality of a broadcast facility, it is far from the only attribute that impacts a facility's successful use. It is vital that from the start of any construction project, an accurate and clear-headed determination be made regarding the functional and performance requirements of each space in a broadcast facility, and that acoustic quality be prioritized, along with other attributes, such as workflow, lighting, indoor air quality, sustainability, schedule and budget. At the end, success is predicated on ensuring that all aspects of a project are designed and constructed to support and enhance the way in which the facility will be used.

STUDIO TYPES

With respect to acoustics, broadcast facilities can be characterized by their use relative to recording, production and listening.

Studios

Open-microphone recording spaces are the most acoustically sensitive spaces in a broadcast facility, requiring the lowest levels of background noise, maximum limitations on intrusive noise and highest control of reflected and reverberant sound energy. The goal of the construction should be to maximize the signal-to-noise ratio at the microphone location, where noise sources may include mechanical equipment and the air distribution system, external noise sources or reverberant sound within the space.

Open-microphone recording spaces in which optimum control of the acoustic environment can generally be obtained include radio studios, television studios, film studios, Foley stages, automatic dialog replacement (ADR) studios, scoring studios, and others. In many spaces used for open microphone recording, such as newsrooms, or in field recording, control of room acoustics may be more difficult to achieve.

The quality of the required acoustic environment may also depend on the anticipated use of the recorded material and how it will eventually be produced or distributed. While it is hard to generalize given the rapidly changing landscape of the industry, higher quality acoustic environments are generally required, for example, when recording for cable, gaming or film distribution when compared with material being recorded for distribution over channels with higher compression rates, such as characterized currently by podcasts or terrestrial radio.

The optimum acoustic design of a recording space is also dependent on the audio production requirements, such as mono, stereo, multi-channel surround (e.g. 5.1, 7.1 or proprietary, such as Dolby Atmos or Meyer Constellation systems). In general, the greater the number of discrete audio channels being recorded, the greater the benefit of increased direct-to-reverberant sound energy ratio. Achieving the required environment can be accomplished through proper room shaping, sizing and appropriate design and location of sound absorptive and diffusive finishes. Adjustability of these attributes through moveable finish systems can also help improve the flexibility of a single space for different recording applications.

Control Rooms

Control rooms serve many purposes, such as master control, production control, video control and audio control. The acoustic quality requirements of control rooms varies according to their use, with audio control rooms generally having the most stringent requirements for listening. In these spaces, every effort should be made to ensure that noise-producing equipment (i.e. equipment with cooling fans or noisy transformers) are located outside the room or in a sound-attenuated enclosure. In other control rooms, a critical listening environment is generally not required, and the noise of cooling fans and transformers is often not an impediment to use. In all control rooms, significant areas of sound absorptive materials are normally required.

In large control rooms, speech communication between personnel is enhanced by controlling equipment noise, as well as controlling noise from mechanical equipment, air distribution systems and external noise sources. This is beneficial even though headset-type intercom systems are frequently used for communication.

Audio and Video Production and Post-Production

Audio and video production spaces include linear and nonlinear edit rooms and mix rooms. Spaces used for audio production have a greater need for quiet and controlled acoustic environments, but a high-quality sound reproduction and a controlled listening environment is required to ensure excellence in the finished product.

Recently, there has been significant interest in moving these production functions to editing stations located in open plan work areas, in an effort to foster collaboration and reduce real estate costs. In an open work environment, desktop monitor loudspeakers may cause distraction to nearby workstations. The use of headphones for editing in open work environments is a common work-around for this, although listener fatigue with headphone editing, especially when in-ear headphones are used, can become a concern.[1]

Open Plan Newsrooms and Assignment Desks

Originating newscasts for radio, television or cable news from an open plan newsroom or assignment desk can provide a sense of energy and immediacy. These environments, however, can be challenging locations for controlling background and intrusive noise and reflected and reverberant sound. Some of these challenges can be mitigated by the use of directional microphones, but difficulties still exist. For example, the use of directional lavalier microphones in noisy environments can improve signal-to-noise when the talker's mouth is facing the microphone capsule (which may be secured to a lapel clip), but when the talker's head is turned, significant dropout may occur. An excellent (and humorous) example of this effect can be seen in the movie *Singing in the Rain*, during the scene where Lena Lamont (Jean Hagen) and Don Lockwood (Gene Kelley) are filming a scene from *Dueling Cavalier*![2]

Another set of acoustic design challenges in newsrooms, whether or not used for audio program origination, involve sound propagation around the assignment desk. Reporters and producers working in open plan workstations surrounding an assignment desk need to concentrate on their work tasks, which include writing, reading and talking on the phone.

Distractions may occur from people and loudspeakers around them. At the same time, the assignment desk editors need to be able to communicate easily to those in the newsroom, with good unamplified verbal communication among people. Further, the assignment desk editors and producers are often listening with half an ear to audio from police and fire scanners located on the assignment desk. Accommodating these multiple and contradictory acoustic performance requirements requires careful design effort regarding noise control and integration of sound reflective and absorptive surfaces into the newsroom.

PLANNING, DESIGN, AND CONSTRUCTION PROCESS

Studio projects may vary in size and complexity, from a small all-in-one-room project studio to a large broadcast studio complex with many large stages, control rooms, and production and support spaces. Budgets may range from a few thousand dollars to the hundreds of millions of dollars. Regardless of the size and scope of a project, making the project a success involves implementing the same basic steps. Of course, the duration and complexity of the issues to be investigated and resolved in each phase will vary according to the project, but the basics remain the same.

The major phases of a studio project are:

a. Establish criteria. This may include design aesthetic, structural requirements, mechanical cooling requirements, electrical requirements and of course acoustic performance goals. The acoustic performance requirements should be quantified and reflect the intended functions of the spaces and facility. These may be based on industry standards, prior experience and the unique requirements of the project at hand. For a small renovation project, the criteria may be easily stated, such as "eliminate the intrusive noise in the on-air studio when people walk through the adjacent space." Whatever the complexity of the project, prioritizing the goals related to all the disciplines of design and construction, including acoustic goals, is necessary to ensure the appropriate allocation of the client's budget.

b. Develop a conceptual design to achieve these goals. A collaborative design approach, involving all disciplines of design, is the most productive way to ensure achievement of acoustic performance criteria through cost-effective and practical integration of acoustic design elements. This means, for example, addressing structural isolation issues as part of the initial project design, rather than trying to overlay an ineffective finish on a partition that will radiate structure-borne sound due to plumbing embedded in the wall. It can also mean, for example, developing a conceptual approach to space planning, which maintains noisy equipment as far as practical from sound-sensitive spaces and achieves cost-effective sound-lock construction at the entrance to open microphone recording spaces in lieu of the use of expensive sound-rated doors.

c. Develop a realistic budget required to implement the concept design. Determining the cost to implement the design is a reality check on the design team's vision. The most successful projects (i.e. the ones that get built and go on to fulfill their mission to support the needs of the client) have design concepts which cost-effectively integrate the requirements of all the disciplines, including acoustics, into a unified whole. In a small renovation project, where budgets are severely limited, it is no less important that the available funds be appropriately allocated to those remedial constructions that will mitigate the situation at hand.

d. Prepare construction documents with details and specifications coordinated among all disciplines. Once it has been established that a project meeting the client's goals can be realized within the budget restraints, the design team needs to produce construction documents that unambiguously describe the project, so that a contractor can accurately bid on and build the facility. Regardless of the size of the problem or project, clear direction to the contractor regarding the intent of the work and the expected outcome, effective designs and practical and achievable construction means and methods will result in successful outcomes.

e. Construction supervision is possibly the most important phase. Drawing a project on paper (or in 3D model space) is one thing; realizing the project in steel, concrete, wood, gypsum board, etc., is quite another. The architect and engineers have important roles to play, working with the contractor to ensure the project is built in strict accordance with the construction documents. In addition, the acoustician, whose discipline bridges across all the design professions, is in a unique position to contribute to the resolution of any questions that arise during construction, insofar as the solutions to field conditions may affect the achievement of the acoustic performance goals required by the client.

f. Perform proof-of-performance testing to verify achievement of design goals. Once the project is complete, it is important that objective measurements be made to verify that the project meets the design criteria established at the start of the project. If corrective actions need to be taken, the acoustician should be part of the team that determines what steps should be made.

ACOUSTIC DESIGN CRITERIA

As noted above, the design of a broadcast facility starts with the establishment of objective acoustic design criteria required to support the scheduled functions; the project ends with successful use, and perhaps with field measurements confirming that the acoustic goals have been achieved. Of what do these criteria consist, and how are they established?

Among the primary sources for determining appropriate acoustic performance criteria for a broadcast facility are established industry standards and measurements in existing facilities. Both of these have an important roles to play in determining design goals for a new facility.

For example, a studio facility wants to expand by adding several new recording spaces. They have a long-standing operation, the success of which is at least partially a function of the high esteem in which their clients hold the acoustic quality of their existing facility. A survey of the successful spaces in their existing facility, to measure background and intrusive noise, sound isolation and impulse response and/or reverberation time can be used to establish the appropriate acoustic performance goals for the expansion.

Alternatively, a successful startup studio facility has decided to move to a larger space, with a goal of increasing revenue by offering more sophisticated technology to its clients. While benchmarking the quality of their existing acoustic environment may be important, the owners understand that the design of the new facility needs to result in a marked improvement in room acoustic quality and noise control. In this case, understanding both what exists and what industry standards for excellent acoustics are can help set realistic expectations for the new facility, especially in light of budget limitations.

Another example: a small radio studio, after many years in use, needs a new console, with new electronics and larger desk that can accommodate several guests for talk radio programs. The opportunity, and need, is there to modify the room finishes—add more absorptive panels—and increase the air supply to address the resulting increased cooling and ventilation requirements when there are more people in the room for a longer period of time. Determining the minimal changes needed in ductwork and wall treatments is important to bringing the project in at the limited budget available.

The acoustic performance goals for broadcast facilities are defined below.

Background and Intrusive Noise

It is helpful in facility design, and in troubleshooting noise issues in completed facilities, to differentiate between *background* (or ambient) noise and *intrusive* noise.

Background noise refers to sound in a space generated by the air distribution system serving that space. It is characterized by a broadband noise spectrum and can be defined in terms of octave- or third-octave-band sound pressure levels in dB from 63 Hz to 8000 Hz. Background noise can be expressed using a single figure of merit, such as noise criteria (NC) or room criteria (RC) ratings, both of which are defined in the *ASHRAE Fundamentals Handbook*.[3]

Figures 4.1-1 and 4.1-2 show the relationship between NC and RC values and the octave-band sound pressure levels in each octave band.

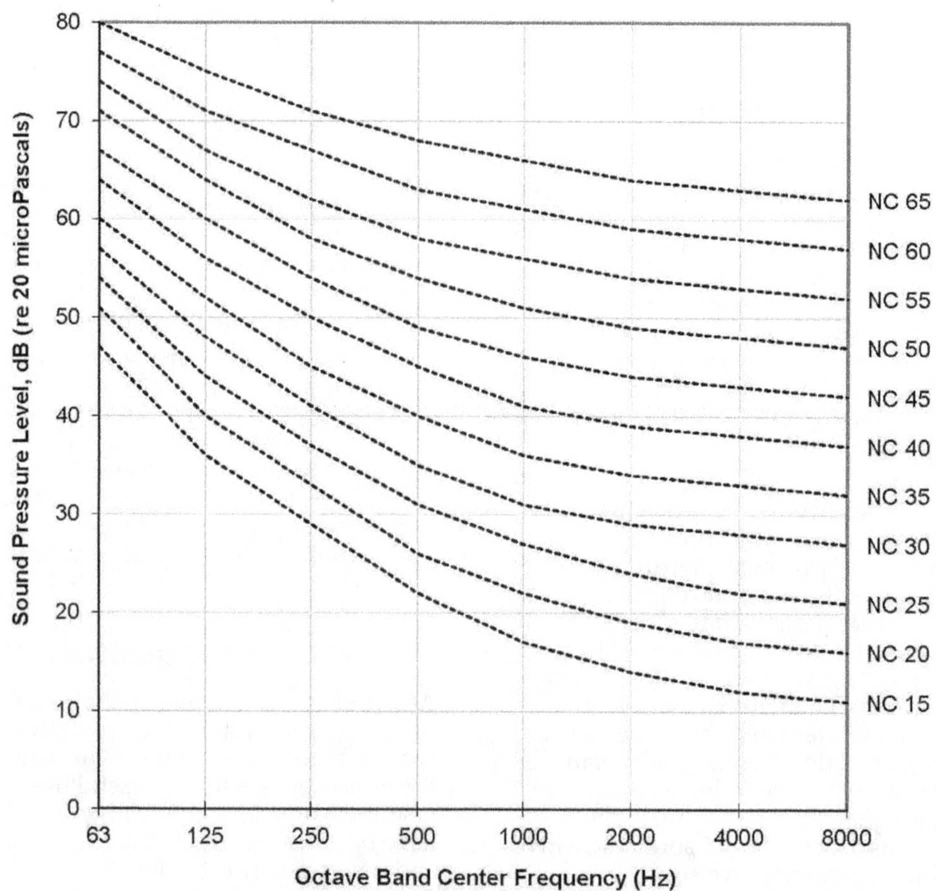

FIGURE 4.1-1 Noise Criteria (NC) Levels

©ASHRAE, www.ashrae.org, *ASHRAE Fundamentals Handbook*, 2015 Edition, Chapter 48, Table 5

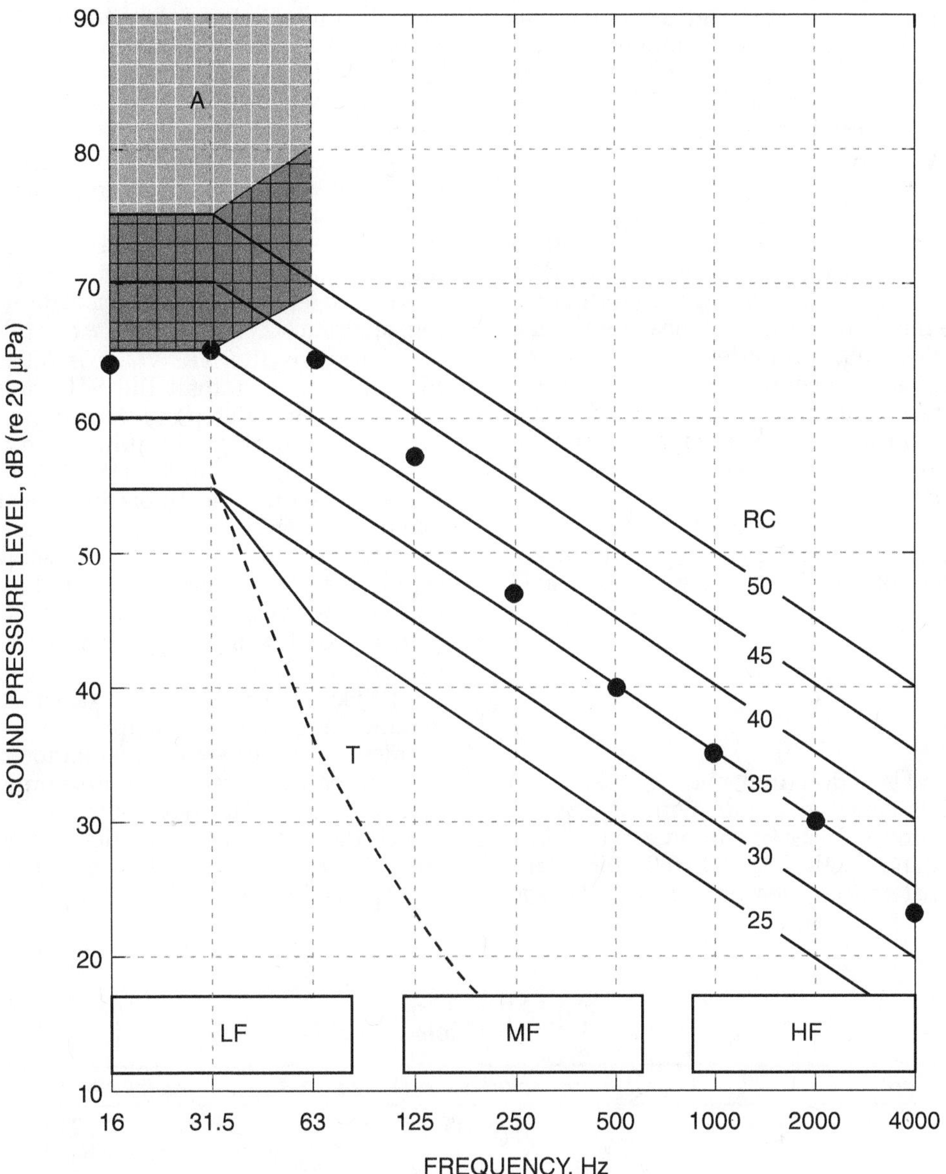

Note:

- Noise levels for lightweight wall and ceiling constructions:
  - In shaded region B are likely to generate vibration that may be perceptible. There is a slight possibility of rattles in light fixtures, doors, windows, etc.
  - In shaded region A have a high probability of generating easily perceptible noise-included vibration. Audible rattling in light fixtures, doors, windows, etc. may be anticipated.
- Regions LF, MF, and HF are explained in the reference.
- Solid dots are sound pressure levels for the example discussed in the reference.

FIGURE 4.1-2 Room Criteria (RC) Levels

While recent work has shown that sounds that fit the RC curves are considered more "neutral" in sound quality than those that fit the NC curves, the use of NC values as descriptors of sound levels in studio design and construction remains the current industry standard.

The term "intrusive noise" is used to refer to sound that intrudes into a space, the source of which may be inside or outside the space. It may be caused by mechanical equipment (e.g. fans, compressors, elevators, etc.), activity noise in adjacent spaces, outdoor noise sources, traffic, etc. Often, intrusive noise is characterized by discrete frequency components, (e.g. music, fan blade-pass noise, truck rumble, etc.). It is also often intermittent in nature. To avoid problems, intrusive noise levels should be limited to sound levels at least 5 dB, and ideally 10 dB or more, below the sound pressure levels in each octave band that defines the ambient noise level (NC level) in each space.

The NC level of a space is determined by plotting measured octave-band sound levels due to air distribution system noise on an NC chart, and finding the lowest NC curve that is tangent to the measured sound levels. While this does not preclude a space that has a background sound level meeting a specific NC design goal from having a sound that is characterized by rumble or hiss, it does ensure that a space meeting the NC design goal can be effectively used for the broadcast function for which it is designed.

The table below summarizes the background and intrusive sound level design goals typically appropriate for a variety of broadcast facilities and related spaces.

Sound Isolation

Airborne

The airborne sound isolation (transmission loss, or TL) provided by an assembly (e.g. partition) or product (e.g. door or window), measured in an acoustic laboratory in accordance with ASTM E90-09 (Standard Test Method for Laboratory Measurement of Airborne Sound Transmission Loss of Building Partitions and Elements), is used to determine the sound transmission class (STC) rating of the assembly or product in accordance with ASTM E413-10 (Classification for Rating Sound Insulation).

When tested in the field, sound isolation measurements between rooms made in accordance with ASTM E336-14 (Standard Test Method for Measurement of Airborne Sound Attenuation Between Rooms in Buildings) result in noise reduction (NR) values that can be used to determine the noise isolation class (NIC) rating, again in accordance with ASTM E413-10.

In practice, the NIC value is often approximately five points lower than the STC rating of a similar assembly. This five-point reduction is commonly a consequence of (1) field conditions (e.g. sound leakage, sound flanking, etc.) and (2) the difference between the degree to which the laboratory construction is built in strict accordance with the detailed specification requirements of the laboratory assembly construction standard and the degree to which the contractor in the field follows these same requirements.

The use of a single figure of merit rating, whether STC or NIC, as a criterion in characterizing the sound isolation performance of an assembly is problematic, since these descriptors are not sensitive to frequency-specific sound attenuation characteristics of a sound barrier assembly. For example, two partitions each with an STC (or NIC) rating of 45 may have transmission loss (or noise reduction) values in the 125-Hz one-third octave band that can differ by as much as eight points. Furthermore, neither STC nor NIC ratings

TABLE 4.1-1
Background and Intensive Noise Level Design Goals

Space	*Background Sound Level Design Goal*	*Intrusive Noise Level Design Goal*
TV Studio	NC-25	5 dB below NC-25 in each octave band
Voice/Announce Booth	NC-20	10 dB below NC-20 in each octave band
Control Room	NC-30	5 dB below NC-30 in each octave band
Audio Control Room	NC-25	5 dB below NC-25 in each octave band
Newsroom	NC-35	NC-35
Podcast Studio (recording)	NC-25	5 dB below NC-25 in each octave band
Streaming (live distribution)	NC-25	5 dB below NC-25 in each octave band
Edit Room	NC-30	5 dB below NC-30 in each octave band
Open Plan Office	NC-40	NC-40
Enclosed Offices	NC-35	NC-35
Conference Rooms, Training Rooms	NC-30	NC-30
Other occupied tenant space	NC-40	NC-40

address sound in the frequency range below 100 Hz. In rating the sound isolation performance of an assembly, more useful information is obtained by augmenting a single number rating with one-third-octave band or one-octave band sound attenuation values.

Structure-Borne

Similarly to airborne sound isolation, there are both laboratory and field rating standards for characterizing structure-borne sound isolation. The laboratory measurement method is ASTM E492-09 (Standard Test Method for Laboratory Measurement of Impact Sound Transmission Through Floor-Ceiling Assemblies Using the Tapping Machine). The single figure of merit impact isolation class (IIC) value is determined from the impact sound transmission value using ASTM E989-06(2012) (Standard Classification for Determination of Impact Insulation Class (IIC)).

The associated field-measured structure-borne sound isolation standard is ASTM E1007-14 (Standard Test Method for Field Measurement of Tapping Machine Impact Sound Transmission Through Floor-Ceiling Assemblies and Associated Support Structures). Using ASTM E989, the field IIC can be determined.

As with airborne sound attenuation measurements, the use of a single figure of merit rating system can be problematic with respect to characterizing the structure-borne sound isolation performance of an assembly or product. In addition to the inability of the single number rating to accurately characterize the difference in impact sound isolation between two different assemblies having similar IIC (or FIIC) ratings but varying frequency responses, it is important to understand that E492 and E1007 (and hence IIC and FIIC ratings) do not address low-frequency impact isolation below 100 Hz: the "thud," for example, of footfall or kick drum that carries significant low-frequency energy into building structures. Standardized test methods for measuring low-frequency impact sound isolation are in development, and a Japanese standard (JIS A1418) is occasionally applied to this situation.

Airborne and Structure-Borne Sound Isolation Design Goals

It is not possible in a handbook of this type to provide numerical values of airborne and structure-borne sound isolation design goals applicable to all spaces in broadcast facilities, since these goals depend on the space plans, adjacencies building structural design and construction, and project-specific functional requirements of each space. However, the table below provides what should be considered at best a starting point for discussion for some common room adjacencies. The use of both an STC and 125-Hz TL octave-band transmission loss (TL) goal is recommended to address the lack of frequency sensitivity in the STC value alone. While the use of an NIC and NR, rather than STC and TL, value can be considered—since the former pair of descriptors addresses the entirety of the assembly separating two spaces (which may include, for example, both a partition and a window), including the effect of sound flanking—the use of STC and TL values for specifying sound attenuation requirements makes assignment of the required sound barrier construction more straightforward.

TABLE 4.1-2
Airborne and Impact Sound Isolation Design Goals

Adjacent Spaces		*Design Goal Guideline for Sound Isolation*		
Space	*Adjacency* (Source)	*Airborne* (vertical or horizontal adjacency)		*Structure-borne* (vertical adjacency)
		STC	NR @ 125 Hz	IIC
TV Studio	Occupied Space or Corridor	65	50	65
TV Studio	Door (single door—personnel or large loading door)	55	40	
TV Studio	Doors (each of two personnel-sized doors separated by sound lock vestibule	40	25	
On-Air Studio	Production or Control Room	55	40	80
On-Air Studio	Other occupied space	60	45	80
On-Air Studio	Corridor	55	40	80
Audio Control	Other occupied space	55	40	70
Audio Control	Corridor	50	35	70
Master Control	Other occupied space	50	35	60
Master Control	Corridor	45	30	60
Audio Post Production	Other occupied space	55	40	70
Audio Post Production	Corridor	50	35	70

Sound Decay Characteristics

In all spaces used for speech, music, recording, listening, production or other activities (which means, in all spaces), the decay characteristics of sound are a critical aspect of the quality of the space, and significantly affect the ability of a space to be used for its intended purpose.

The sound-decay characteristics of a room are a function of the size, shape and finishes of the room. The simplest descriptor of the way sound decays in a room is the reverberation time, which was defined circa 1895 by Wallace Sabine as the time it takes for the sound of a loud organ to decay to inaudibility, which is approximately a 60-dB reduction in sound level. Reverberation time, therefore, is given the symbol RT_{60}. Sabine also developed an equation relating the size and finishes of a room to the reverberation time that remains in common use.

$$RT_{60} = (.049 * V) / (S * \alpha)$$

where:
RT_{60} = Reverberation Time (in seconds)
V = Room Volume (in cubic feet)
S = Total Surface Area of Room (in square feet)
α = Average Absorption Coefficient of All Surfaces

While the Sabine equation represented a significant step forward in the science of acoustics in the late 19th century, it has significant deficiencies. Foremost among these is its assumption that the sound field in a room is diffuse, a characteristic that does not apply to the sound field in real rooms, especially small rooms. Improvement in the prediction of reverberation time have been made over the last century by addressing this issue, through terms in equations that address room shape, the distribution of finishes and more sophisticated measurements of the sound absorptive properties of materials.[4]

This work continues, with the latest research focused on the use of computer models to address the effect of specular reflections, diffusion and diffraction from surfaces and edges. This work has led to greater appreciation that reverberation time (RT_{60}) alone is inadequate in describing the acoustic quality of a room, and nowhere is this more true than in spaces for broadcast and audio production. Other important factors include:

- Early Decay Time (RT_{30})
- Reverberant-to-Direct Energy Ratio
- Impulse Response and Shape of Decay Curve
- Comb Filtering Effects
- Bass Energy Ratio
- Modal Density

The way in which these factors interact in a room are difficult to analyze and quantify in design and challenging to measure in practice. Further complicating the issue, there is little consensus in the industry regarding the relationship between the values of the various room acoustic parameters and the subjective assessment of sound quality in a room.

Despite (or maybe because of) this challenge, the basic reverberation time parameter has remained the go-to descriptor of room acoustics. In most broadcast spaces in which audio recording, production and listening/playback is important, minimizing reverberation time is important for maximum signal-to-noise and direct-to-reverberant energy. A workable standard goal for reverberation time is defined by ITU-R BS.1116-3 (02/2015) (Methods for the subjective assessment of small impairments in audio systems, Paragraph 8.2.3.1, Figure 1):

In which

$$T_m = 0.25\ (V / V_0)^{1/3}$$

where:
V = Volume of room
V_0 = Reference volume of 100 m^3.

The tolerances to be applied to T_m over the frequency range 63 Hz to 8 kHz are given in Figure 4.1-3. Note there are difficulties in measuring small values of reverberation time at low frequencies.

Speech Intelligibility

Many broadcast rooms are used primarily for speech and require excellent intelligibility for live, unamplified speech as well as for amplified speech; excellent signal to noise for speech signals at microphone locations; and excellent intelligibility for post-production or listening to recorded speech played back through loudspeakers that may be near to or far from the listener. These goals encompass recording of and listening to program material, as well as communication

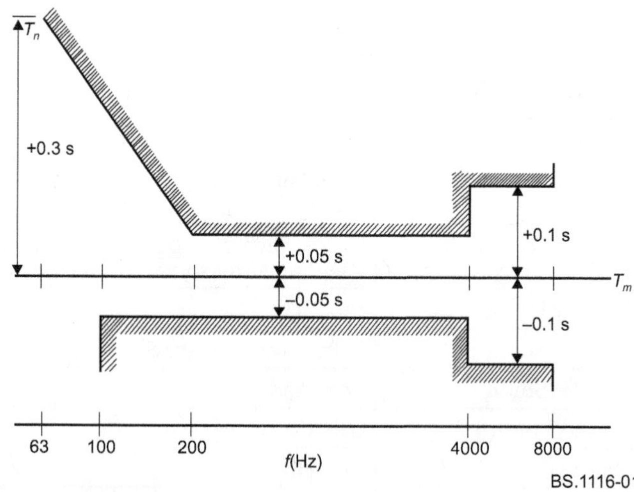

FIGURE 4.1-3 Tolerance Limits for the Reverberation Time, Relative to the Average Value, Tm

ITU-R BS.1116-3 (02/2015) (Methods for the subjective assessment of small impairments in audio systems, Paragraph 8.2.3.1, Figure 1)

among broadcast personnel in control rooms, studios, newsrooms and other technical spaces.

There are several descriptors used to characterize speech intelligibility. Three of the most common are:

Speech Transmission Index (STI)

STI is a rating method for speech intelligibility that considers signal-to-noise and reverberant sound energy effects on intelligibility. It is defined in IEC 60268-16:2011 (Objective rating of speech intelligibility). This standard addresses speech intelligibility with and without a sound reinforcement system.

Speech Intelligibility Index (SII)

Formerly called articulation index (AI), SII is a rating method for speech intelligibility that considers speech, noise, and hearing threshold levels. It is defined in ANSI S3.5-1997 (R2012) (Methods for Calculation of the Speech Intelligibility Index). The standard cautions that "the value of SII required for a given communication system is . . . dependent upon factors whose importance can be evaluated only by users of the system." Nevertheless, the following relationship between SII and speech intelligibility has become widely acceptable:

Achieving STI and SII values greater than 0.75 generally requires signal-to-noise ratio at the listener (or microphone) location greater than 6 dB.

Reverberant-to-Direct Sound Energy Ratio

In a free field, sound from a point source (approximated by a talker or loudspeaker) falls off at a rate of 6 dB per doubling of distance. In a real room (not anechoic), there will be distance from the source at which the reverberant sound energy developed by the source is equal to that of the direct sound. The point at which this occurs (where the ratio of reverberant (*Er*) to direct (*Ed*) energy equals 1) is the critical distance (D_{crit}).

$$D_{crit}\text{(in feet)} = (Q * R / 50) \wedge 0.5$$

Where:
Q = Directivity of source
R = Room Constant = $[A / (1 - \alpha)]$
A = Total Absorption (in Sabines) of All Surfaces
α = Average Absorption Coefficient of All Surfaces

Where the reverberant sound energy level exceeds the background sound, level optimum speech intelligibility occurs where the direct sound energy is at least 6 dB above the reverberant sound level. An omnidirectional microphone ($Q = 1$) should be placed at least this close to the sound source, which is generally about 1/3 of the critical distance. Directional microphones may be placed farther from the sound source, up to 1/2 to 2/3 of the critical distance from the source.

TABLE 4.1-3
STI: Subject Evaluation of Speech Intelligibility (STI values may range from 0.00 to 1.00)

STI Value	*Quality according to IEC 60268-16*
0.00–0.30	Bad
0.30–0.45	Poor
0.45–0.60	Fair
0.60–0.75	Good
0.75–1.00	Excellent

TABLE 4.1-4
SII: Subject Evaluation of Speech Intelligibility

SII Value	*Speech Intelligibility*
0.00–0.45	Poor
0.45–0.75	Fair
0.75–1.00	Good

CONTEXT

Choosing the location of a broadcast studio facility can have a significant effect on acoustics, noise control and facility cost. A studio on a noisy site, such as in proximity to an airport or major highway, can require substantial sound barrier construction to be incorporated into the building envelope to mitigate intrusive environmental noise. Space planning within the building to locate sound-sensitive spaces in interior locations can reduce these requirements, but at the expense of loss of daylight and exterior views, unless larger budgets can accommodate sound isolating exterior walls, windows and roof construction. An acoustic survey of environmental noise is recommended at the earliest stage of planning to ensure realistic construction budgets are developed which reflect a facility construction that achieves the required noise control for exterior noise sources.

In any building, sound-sensitive spaces should be located as far from mechanical equipment and noisy activities as possible. Mechanical equipment, such as cooling towers, chillers, air conditioning and air handling units, fans, pumps, compressors, elevators, etc., all represent significant sources of intrusive noise that can significantly compromise audio recording or production. In addition to the benefits of distance with respect to noise reduction, the building structure itself has an impact on the efficacy of noise and vibration control for mechanical equipment. Lightweight structures (e.g. wood-framed buildings or buildings with long-span metal decks and open web joists) provide significantly less sound attenuation than reinforced concrete structures or buildings having shorter-span steel beams with metal decks and thicker concrete slabs. A lightweight structure will also tend to have a lower natural frequency of vibration, making achievement of adequate vibration isolation of mechanical equipment, especially on upper floors or the roof, more challenging and costly.

Within the facility, as well, space planning is an important tool to achieve good sound isolation between spaces. Space planning should reflect workflow, aesthetics and limitations of the building layout and structure, and should also reflect the impact of acoustic control on the facility. For example, spaces within a multi-room suite consisting of control rooms and a recording or broadcast studio clearly need to be immediately adjacent to each other, and significant sound barrier construction, perhaps including isolated floors, walls and ceiling, may be required. However, interposing nonacoustically sensitive spaces, such as storage or break rooms, between such suites or adjacent to stand-alone spaces, such as post-production audio spaces, can help reduce construction budgets by reducing the magnitude of constructions required for sound isolation.

Recent emphasis on sustainable construction goals also affects studio design and construction. Among the most important advances in video, film and television studios in this regard is the changeover from incandescent to LED lighting instruments. The significant power savings afforded by LED lighting results in much lower cooling loads in studio facilities, reducing air quantities, fan power requirements, and as a result, noise levels. Reduced air quantities also allow reduction in duct sizes, further reducing costs.

Concurrent with the advancements in lighting power reduction comes improvements in video display technology, which has resulted in a significant increase in the use of LED and LCD displays in broadcast studios and production spaces. This has had a tendency to increase heat loads, offsetting part, or occasionally all, of the savings that are obtained by using LED lighting.

While changes in lighting and equipment heat output affect acoustics only indirectly, the increased interest in the use of sustainable materials has led to the development of new acoustic products, primarily in the area of sound absorptive materials. Sound absorptive panels made from recycled fabrics top the list of product types in this category, but sound diffusive and diffractive products are also being fabricated from sustainable materials using lower-energy fabrication techniques.[5]

In space planning and site selection, the addition of design goals relating to energy efficiency, improved workplace environments and sustainable construction, materials and waste reduction can have an effect on achievement of acoustic design goals. For example, the laudable goal of bringing more daylight into workspaces can significantly increase the cost of construction of edit rooms and other production spaces where good sound isolation is required; multi-layer glass walls or windows are needed to simultaneously achieve the acoustic performance and daylighting requirements. Careful definition and prioritization of goals, including those linked to rating systems such as LEED (Leadership in Energy Efficiency in Design) and GBI (Green Building Initiative) is needed to ensure that the functional needs related to acoustic performance in studio facilities are achieved sustainably and cost effectively.

RECOMMENDATIONS FOR DESIGN AND CONSTRUCTION

Room Design

The acoustic quality of any space is primarily a function of its size (volume) and shape, and secondarily a function of the room finishes. In studio facilities, as in any building, room size and shape should be driven by the functions that will occur therein. The table below summarizes some typical desirable room dimensions to support the indicated functions. Many spaces not meeting these goals are pressed into service for the functions shown with concomitant acoustic challenges:

In spaces where recording, critical listening or audio production occurs, the shape of the room (walls and ceiling) is important to creating the best recording

TABLE 4.1-5
Broadcast Facility Spaces—Typical Sizes

Studio Type	*Typical Floor Area* (sq ft)	*Typical Room Volume* (cu ft)
Radio		
Announce Booth	64	500–600
News Studio	120	1100–1200
On-Air Combo Operator Studio	120	1100–1200
Panel Discussion Studio	260	2400–2600
Dedicated Control Room	200	1600–1800
Edit Bay	80	650–750
Music Studio	300	3000–4000
Television		
Announce Booth	100	1000
Flash Studio	210	2300–2700
Small Insert Studio	1000	13000–16000
Midsize Production Studio	2000	30000–36000
Large Production Studio	5000	100000–130000
Large Sound Stage	10000+	250000–350000+
Video Production Control Room	750	8000
Audio Control Room	200	2000
Edit Bay	120	1000
Audio Post Production Room	200	2000
Small Screening Room	300	3600

and listening environment. The goals of shaping the walls and ceiling include creating sound reflections that enhance the reproduced sound from the loudspeakers and avoiding echoes and other undesirable acoustic anomalies. A fan-shaped room (where the rear of the room is wider than the front), for example, results in side wall reflections being pushed to the rear of the room, which can result in a reflection-free zone. A reverse-fan-shaped room can result in the buildup of early energy from the front-of-room loudspeakers at the mix position, which can reduce the localizability of the sound sources. Multi-speaker reproduction systems (such as 5.1, 7.1, etc.) require consideration of the speaker locations and the goals and expectations of the mix engineer with respect to source localization and signal smoothing.

The judicious implementation of sound absorptive and diffusive materials, in conjunction with room shaping, is also used to enhance the acoustic environment in recording and listening spaces. The sound at a listener location, such as the mix position, is a combination of direct sound from the loudspeakers or live source, reflected sound energy from walls, ceiling, floor or equipment, and reverberant sound that makes up the diffuse field. Appropriately balancing the energy in these components of the sound field is required to achieve the optimum recording or listening conditions.

Sound absorptive materials are rated in terms of the percentage, expressed as a decimal, of sound they absorb in a reverberant or diffuse field. This percentage, called the absorption coefficient, varies with frequency. Measurements of the absorption coefficient of a material are made in accordance with ASTM C423-09a (Sound Absorption and Sound Absorption Coefficients by the Reverberation Room Method). For example, a 1-inch thick fabric-wrapped fiberglass panel directly applied to a solid surface ("Type A Mounting") absorbs the fractions of reverberant sound energy incident on the panel in each of the indicated octave bands shown in Table 4.1-6.

The average of the absorption coefficients in the 250-, 500-, 1000- and 2000-Hz octave bands results in a value called the noise reduction coefficient, or NRC. This is the value most often seen in acoustic specification of materials. It provides some useful information about the acoustic performance of materials, but insofar as it is a single figure of merit, it has limits regarding its usefulness in design. Absorption coefficients and NRC values for specific products can often be obtained from their manufacturers.

It is important to recognize that the NRC (and octave-band absorption coefficients) are related to the *diffuse field* sound absorptive properties of a material, and do not provide information about whether the materials will reflect *specular* sound energy incident on the panel from a specific direction. This is especially important in small-room design (such as control rooms, edit rooms and post-production audio rooms), since a significant amount of the sound energy from front-of-room loudspeakers, for example, is often incident on panels on the side walls of these rooms at very oblique angles. The behavior of this specularly-incident sound energy—whether it is absorbed, reflected or diffused—is not characterized by the NRC or sound absorption coefficient, which is measured in a reverberant field.

Sound scattering and diffusive materials cause incident sound to be reflected in a random direction, rather than to be absorbed or specularly reflected. These materials can be useful in creating more diffuse sound fields in small rooms, improving direct-to-reverberant sound energy ratios and increasing the spatial impression provided by reproduced or live sound. The degree to which a material scatters and diffuses sound can be characterized by the scattering and diffusion coefficients: the scattering coefficient indicates how much of the incident energy is scattered in nonspecular directions, whereas the diffusion coefficient indicates how uniformly the incident energy is scattered.

The efficacy of a surface to scatter or diffuse incident sound is a function of the relationship of the dimensions of surface articulations to the wavelength of incident sound. Maximum scattering occurs when the surface articulations are of the same order of magnitude as the wavelength of sound; small perturbations (less than 1/4") in a surface (such as rough-surfaced concrete or plaster wall) will have little or no sound scattering or diffusive effect at audible frequencies. Sound diffusive panels with areas of randomly varying depth, size and shape can obtain sound scattering over relatively wide frequency ranges.

Mechanical Equipment and Air Distribution System Noise and Vibration Control

Providing proper noise and vibration control for mechanical equipment and air distribution systems is mandatory to ensure broadcast facilities can operate successfully. Achieving the design goal background sound levels suggested earlier in this chapter requires careful design and construction of vibration isolation, air distribution and attenuation, air terminal devices and sound barriers, as well as appropriate specification and construction of the equipment itself to minimize noise and vibration, and balancing of air and water systems to minimize turbulent flow.

TABLE 4.1-6
Typical Acoustic Panel Sound Absorption Coefficients

Octave-Band Center Frequency, Hz	125	250	500	1000	2000	4000	NRC
Absorption Coefficient, 1-inch thick fabric-wrapped fiberglass panel	0.09	0.27	0.80	1.01	1.03	1.01	0.80

Air Distribution System

There are two aspects of noise control relative to air distribution systems: (a) control of fan noise that is carried through ductwork and (b) control of air turbulence noise generated by fittings, air valves and terminal boxes (such as VAV boxes) balancing devices such as dampers, and grilles and diffusers.

Reducing fan noise to acceptable levels starts by correct specification of fan type and operating characteristics. Backward-curved centrifugal fans with airfoil blades tend to be the quietest type for a given static pressure, BHP (brake horsepower) and cfm, although manufacturers of fan equipment have criteria besides acoustics (e.g. size, electrical power consumption, ease of manufacture, price) for incorporating one type of fan rather than another into their equipment, especially for smaller air handling units (AHUs) or package rooftop equipment. Equipment having low-noise fans, better maintenance records and greater enclosure attenuation are normally manufactured from higher quality components and command a premium price; these systems may or may not be cost effective for a small studio facility.

Regardless of the fan type, it is important to obtain the manufacturer's sound power data for the fan operating point that is expected in the field. The sound power varies exponentially with fan size, RPM, BHP and static pressure, and small changes in these variables can have a significant effect on noise levels.

All of which is to emphasize the importance of proper air balancing to ensure that the fan is operating at the minimum required power output to obtain the design goal air deliver to each space. Air balancing in studio facilities should be performed in accordance with National Environmental Balancing Bureau (NEBB) *Procedural Standards for Testing Adjusting and Balancing of Environmental Systems* (www.nebb.org/assets/1/7/PST_TAB_2005.pdf).

Even when proper air balancing procedures are followed, fan noise and aerodynamic (turbulence) noise may be a problem in a broadcast facility. Among the techniques that should be considered:

1. Use sheet metal ductwork throughout the air distribution system. Flex duct can become misshapen or deformed, resulting in significant increase in air turbulence noise. Fiberglass duct has very low transmission loss, resulting in break-in noise (noise entering ductwork through the duct wall) and break-out noise (noise leaving through the duct wall) that can cause excessive noise problems in studios and other occupied spaces through which the duct passes.
2. Provide appropriate in-duct sound attenuation. The three primary options are internally lined ducts, duct silencers and acoustic plenums.
    a. Unlined rectangular sheet metal duct, whether externally wrapped or not, provides sound attenuation of about 0.05 to 0.20 dB per foot, depending on duct cross-section and frequency. Rectangular sheet metal duct with 1-inch fiberglass internal duct lining provides attenuation of about 0.5 to 4.0 dB per foot (again, depending on duct size and frequency), or 10 to 20 times the sound attenuation of unlined duct. Round duct, whether internally lined or not, provides less attenuation per foot than rectangular duct.
    b. Duct silencers provide more attention per foot, but at a higher cost, and greater pressure drop, than internally lined ducts. Frequently, however, a duct silencer is required to provide the required attenuation especially in short duct runs.
    c. An acoustical plenum is a (usually) large sheet metal box connected to ductwork that serves as an expansion chamber, resulting in a sudden change in the duct cross-section as air enters and leaves the plenum. This change in cross-section, in addition to the sound absorptive fiberglass duct lining with which the plenum is lined, results in high values of sound attention, especially at low frequency, in the duct path. Plenums can be more desirable than silencers in some situations due to their low-frequency attenuation, cost and ability to be fabricated to fit in oddly shaped spaces. They can also be used to good effect when mounted directly on top of a top-discharge fan unit. Typically, acoustic plenums should have 4-inch fiberglass duct lining (for maximum sound attenuation at low frequency) with perforated metal facing (to protect the lining from degradation due to turbulence in the plenum).
3. Size ductwork to avoid excessive turbulence noise. Guidelines for duct sizing are based on the background sound level design goal in the space the ductwork is serving, or the space through which the duct passes. Industry-standard guidelines are given in the *ASHRAE Fundamentals Handbook*, Applications volume (latest edition at the time of this writing: 2015). The recommended velocity limits are reprinted in Tables 4.1-7 and 4.1-8.
4. In spaces having background sound level design goal of NC-30 or less, do not install dampers closer than seven duct diameters to the air outlet. This includes not installing face dampers (e.g. butterfly dampers) in diffusers. Air balancing should be accomplished well upstream of the air outlets.
5. Where manufactured air outlets (grilles, diffusers, etc.) are used, select air outlets from the manufacturer's catalog at least five NC points (preferably 10) below the space design goal. This is because the manufacturer's catalog data is based on laminar airflow conditions that can only be found in the test laboratories; in the real world, duct conditions always generate more turbulence, and the sound power generated by the air outlets are higher than cataloged.
6. Where variable air volume (VAV) boxes are installed above ceilings, the five to ten NC point rule also applies to the selection of VAV boxes for both radiated and discharge noise. Where VAV boxes are installed without a ceiling below them, keep in mind that the NC values shown in the box manufacturer's catalog for radiated noise are based on the attention of mineral tile acoustical ceilings;

TABLE 4.1-7
Air Velocity Limits in Distribution Ductwork[3]

Location	*Design Goal (NC)*	*Velocity (FPM)*		
		Main	*Branch*	*Final Runout*
In shaft or above drywall ceiling	45	3500	2800	1750
	40	3000	2400	1500
	35	2500	2000	1250
	30	2100	1680	1050
Above suspended ACT	45	2500	2000	1250
	40	2125	1700	1063
	35	1750	1400	875
	30	1475	1180	738
Duct located within occupied space	45	2000	1600	1000
	40	1725	1380	863
	35	1450	1160	725
	30	1200	960	600

Note: Elbows and other fittings can increase airflow noise substantially, depending on type; airflow velocities should be reduced accordingly.

TABLE 4.1-8
Air Velocities Limits at Neck of Grilles, Diffusers or Open-Ended Ducts[3]

Location	*Design Goal (NC)*	*"Free" Opening Airflow Velocity (FPM)*
Supply	45	625
	40	560
	35	500
	30	425
	25	350
Return	45	750
	40	675
	35	600
	30	500
	25	425

without this ceiling in place, the radiated sound in the occupied space generated by VAV boxes will be 10 to 15 dB greater than shown in the catalog.

Mechanical Equipment—Airborne and Structure-Borne Noise and Vibration Control

In addition to the air distribution noise control required for fan equipment, all rotating and reciprocating equipment generates noise and vibration that must be appropriately attenuated when located near (and sometimes not so near) sound-sensitive spaces in broadcast facilities. This equipment includes fans, package indoor units, rooftop air handling and package units, cooling towers, chillers, computer room air conditioning, emergency generators, elevators, escalators, etc. It can also include kitchen, server and pantry equipment such as refrigerators, freezers, icemakers, coffee and espresso machines, etc., as well hand dryers, chilled-water drinking fountains, soda machines, etc.

Airborne noise control requirements for mechanical equipment can result in sound barrier construction attenuation requirements greater than those shown in Table 4.1-2, Airborne and Impact Sound Isolation Design Goals. Large mechanical equipment can generate high levels of sound energy at low frequencies (below 100 Hz), requiring substantial mass to separate these spaces from production areas. Multi-wythe partitions, including mortar-filled block walls, can be required to achieve the appropriate degree of acoustic separation. Isolated room-in-room construction for mechanical equipment rooms or production spaces, or both, may be required, including the incorporation of acoustic joints between structural slabs, or even separate structures supporting mechanical equipment and highly sound-sensitive spaces adjacent or nearby.

At a minimum, each piece of rotating or reciprocating equipment should be supported by resilient isolation mounts or hangers. Duct and electrical connections to this equipment, and often piping connections, should be made using flexible connectors.

A detailed list of standard approaches to vibration isolation can be found in the *ASHRAE Fundamentals Handbook*, Applications edition, Noise and Vibration Control Chapter (Table 47, Chapter 48, *ASHRAE Fundamentals Handbook* 2015 Edition). Some general guidelines:

1. Floor-mounted air handling and air conditioning equipment may be provided by the manufacturer with spring isolation mounting of fans internal to the equipment. In this case, the unit itself should be mounted on rubber or neoprene mounts with 0.3-inch deflection; in some cases, rubber or neoprene pads may be used. If the fans are internally isolated on springs, do not use external springs—springs in series can resonate and cause large problems.
2. Floor-mounted air handling and air conditioning equipment in which the fans are not internally isolated on springs should be externally mounted on bare-stable spring isolators. Housed or sleeved isolators can be problematic in that they can transmit vibration through the vertical walls of the isolators (designed to limit lateral motion).
3. Rooftop package air conditioning units maybe be mounted (in increasing order of vibration isolation effectiveness) on nonisolated curbs, isolated curbs, or on external isolators on steel dunnage. If placed on isolated curbs or external isolators on steel dunnage, the same caveat as noted for indoor fans and air conditioning units, precluding the use of external springs when fans are internally isolated, applies. If placed on external isolators on steel dunnage, the isolators should be unhoused, but have limit stops.

FIGURE 4.1-4 Unhoused (left) and Housed (right) Spring Isolators

Mason Industries

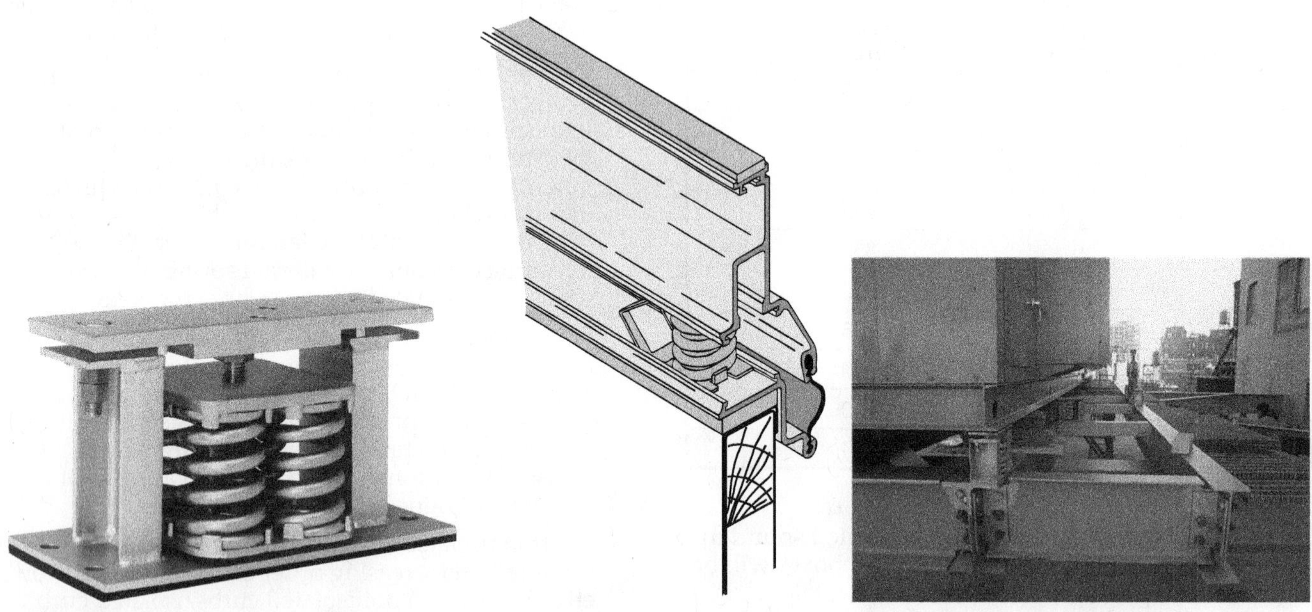

FIGURE 4.1-5 Restrained Isolator (left); Isolated Curb (center); Bare Spring on Steel Dunnage (right)

Mason Industries (left and center); Shmerykowsky Consulting Engineers, Modern Steel Construction Magazine, January 2010 (right).

Electrical Equipment Noise and Vibration Control

The potential sources of noise from electrical and electronic equipment include:

i. **Small cooling fans** can be found in lighting instruments, computers, television cameras, video projectors, transformers and all manner of electrical and electronic equipment. In addition, motors in moving lights, robotic cameras and other such gear utilize motors that make noise. While these sources are technically mechanical in nature (noise is generated by fans), the solutions to addressing them require significant coordination with the electrical properties of this equipment. Given the ubiquitous nature of equipment with fans, it is inevitable that fans will show up in spaces used for open microphone recording. Whether this becomes an acoustic problem or not depends on a number of factors, including the background sound level of the room, the expectations of the production in terms of the quality of the recording, the sound source being recorded, the number and type of microphones being used, the skill of the recording, mix and post-production engineers, and the channel over which the recorded (or live) material will be distributed. When noise from these mechanical sources becomes problematic, the equipment needs to be enclosed; have a barrier built; have sound absorptive material placed nearby; or be removed entirely from the recording space.

ii. **Dimming** of tungsten and other lighting filaments can cause vibration in lamps when using waveform dimmers; these vibrations can be radiated as audible noise by lighting instruments or the lamps themselves. Sinewave dimmers result in less filament and power line noise when dimmed. LED lamps don't use filaments, and therefore that potential source of audible noise doesn't exist, but the LED drivers contain components that can vibrate when pulse wave modulation (PWM) dimmers are used. LED dimmer and driver design and fabrication can reduce or eliminate audible noise from LEDs, but care must be taken in specification.

iii. **Lighting ballast noise** is rated on a scale of A (quietest) to F (noisiest). Although there is no standardized measurement approach, Type A ballasts create noise levels that are generally in the 20- to 25-dBA range. This is generally suitable for all broadcast facility spaces except those used for recording and critical listening. In those spaces, remote mounting of ballasts is generally desirable. Magnetic ballasts generate audible hum due to vibration of the magnetic cores. Electronic ballasts have no magnetic core and are generally quieter, although a variety attributes of the fixture, such as mounting method and location of the ballast in the fixture, may exacerbate noise issues.

iv. **Transformers** in a wide variety of capacities may be found in broadcast facilities. In 60-cycle electrical systems, they generate their highest levels of vibration at 120 Hz. They should never be located in sound-sensitive spaces, but should be mounted in enclosed electrical closets with appropriate sound barrier constructions. They should be mounted or suspended using neoprene mounts or hangers. Large transformers require spring and neoprene mounts, especially if located near studio and audio production areas.

FIGURE 4.1-6 ND Mount (left) and HD Hanger (right)

Mason Industries

Sound Barrier Construction

Achieving adequate levels of sound isolation between spaces in broadcast facilities is mandatory to ensure high-quality production standards are achieved.

The degree of attenuation provided by a construction element (e.g. partition, door, window, etc.) is a function of its mass, stiffness and damping. Figure 4.1-7 summarizes the relationship between these factors and sound attenuation.

As noted in the chart, this curve of transmission loss versus frequency is based on isotropic (homogeneous) barriers, such as poured-in-place concrete, gypsum panels or monolithic glass panels. It does not accurately describe the performance of cavity constructions, such as gypsum board installed on both sides of a metal stud, or insulated glass (IGU) panels, but it does describe the performance of the components of these systems.

The most straightforward portion this curve is the central portion, in which the transmission loss of the panel is controlled by its mass. The mass law states that the transmission loss of a panel increases at a rate of 6 dB per octave: if the TL of a panel is 15 dB at 500 Hz, its TL will be 21 dB at 1000 Hz. It also states that doubling the weight of a panel will increase its TL by 6 dB.

Due to the mass law, it quickly becomes cost-, space- and weight-prohibitive to create partitions between spaces with high sound isolation using monolithic panels. For example, the transmission loss of a 6-inch thick concrete wall is about 30 dB at 125 Hz. As can be seen in the chart earlier in this chapter, the partition enclosing a TV studio should have a TL of about 50 dB at 125 Hz. To achieve this through a monolithic concrete wall, the mass laws tell us that it would require a concrete wall over 48 inches thick, which is impractical, to say the least.

The frequency and sharpness of the first panel resonance depends on the material properties of the panel; a monolithic concrete wall, for example, will not exhibit as much of a "dip" between the stiffness and mass controlled ranges as will a single glazed window. The frequency of the coincidence dip is a function of the thickness of the material; the depth of this dip is a function of the damping: a 1/2-inch-thick glass panel will exhibit a reduction in transmission loss at a lower frequency than a 1/4-inch-thick glass panel; a 1/2-inch-thick laminated glass panel will exhibit less reduction in transmission loss at the coincidence frequency than will a 1/2-inch-thick monolithic glass panel.

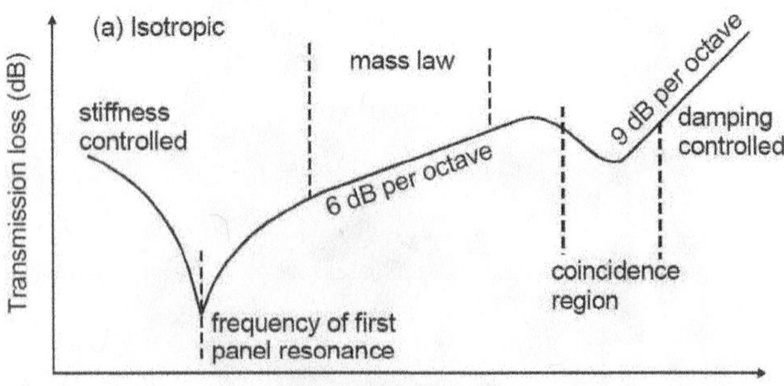

FIGURE 4.1-7 Typical Single Panel Sound Transmission Loss Curve

Bies, D.A. and C.H. Hansen, *Engineering Noise Control: Theory and Practice*, 4th edition, Spon Press, 2009

While the sound isolation performance of isotropic panels is interesting, such panels do not represent the majority of sound barriers found in construction: most sound barriers are made of multiple layers of disparate materials, and many are cavity constructions, with multiple layers separated by air spaces or insulation materials. These attributes significantly impact the acoustic performance of sound barriers, as well as making partitions thinner, lighter, more cost effective and easily able to accommodate within them electrical infrastructure necessary for studio operations.

Partitions

Cavity wall partition constructions using gypsum board on metal or wood studs with batt insulation in the stud cavity are the most common partition assemblies in interior construction. These can be combined with block or concrete walls for maximum broad-frequency-range sound isolation performance.

The use of metal studs generally results in better sound barrier performance, due to the lower stiffness of the stud web compared with wood studs. Using the lightest gauge metal stud at the largest spacing appropriate for the partition height and loading is also beneficial for maximizing sound barrier performance.

The use of resilient clips to secure layers of gypsum board to studs can be an effective method of increasing transmission loss; careful attention to fastening details and edge conditions is needed to ensure the isolation provided by the resilient attachment method is not shorted out.

Gypsum board products have been developed with a resilient interlayer (septum) designed to provide damping and increase mass with the goal of improving sound barrier performance using thinner assemblies. The septum does reduce the ability of vibration to transmit through the panel, although it makes the product more difficult to work with than standard gypsum board. Achieving superior performance with these products requires careful attention to the attachment method, spacing of fasteners and penetrations for junction boxes, conduit, piping, etc., which are more difficult to address due to the septum. It also comes at a significant cost premium over the standard product. Many practitioners have found that the same acoustical benefit can be obtained by using two layers of standard gypsum board that would be obtained through the use of one layer of damped gypsum board. Where space is of an extreme premium, the use of these products can be worthwhile when carefully installed.

Sound leakage at the perimeter of partitions can have a significant detrimental effect on the overall sound barrier performance. For example, an 8″ × 1/4″ slot opening through a 15′ × 10′ wall (less than 0.01% of the wall area) that otherwise provides 50 dB of noise reduction will reduce the noise reduction by

10 dB. To avoid this degradation of performance, partitions should be sealed with nonhardening acoustical caulk at all terminations. Penetrations of partitions by ductwork, piping, conduit or other services should also be sealed.

Typical partition details:

Legend:
a - 2 layers 5/8" gypsum board
b - 3-5/8" 25 ga. metal stud
c - 3" thick 2.5 pcf mineral wool batt insulation
d - non-hardening acoustic caulk
e - 1" air space

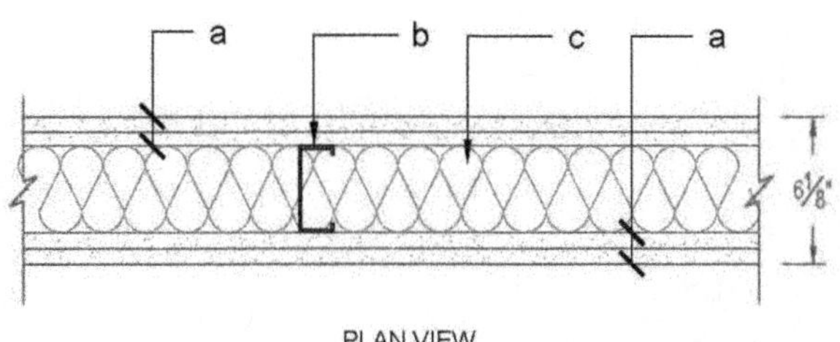

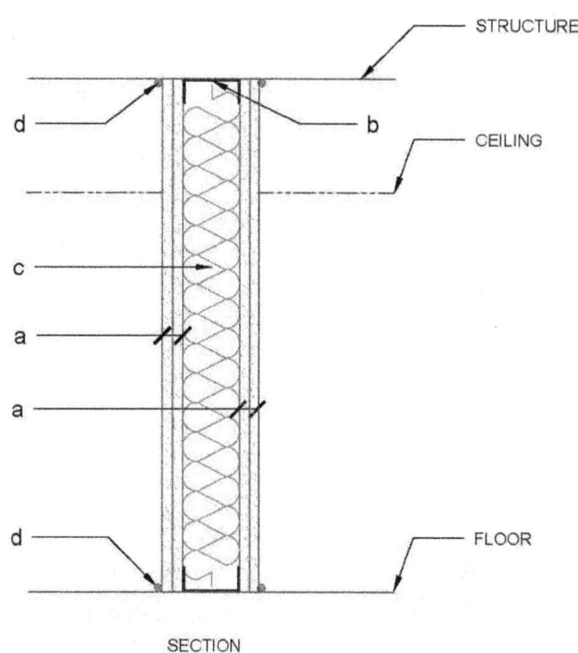

FIGURE 4.1-8 STC-50 Partition

Acoustic Distinctions

Legend:
a - 2 layers 5/8" gypsum board
b - 3-5/8" 25 ga. metal stud
c - 3" thick 2.5 pcf mineral wool batt insulation
d - non-hardening acoustic caulk
e - 1" air space

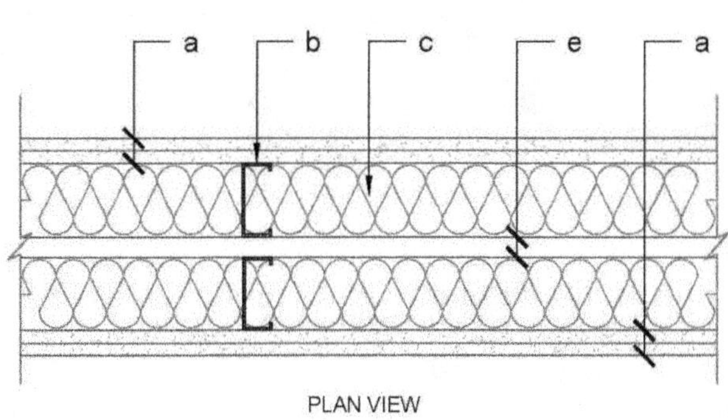

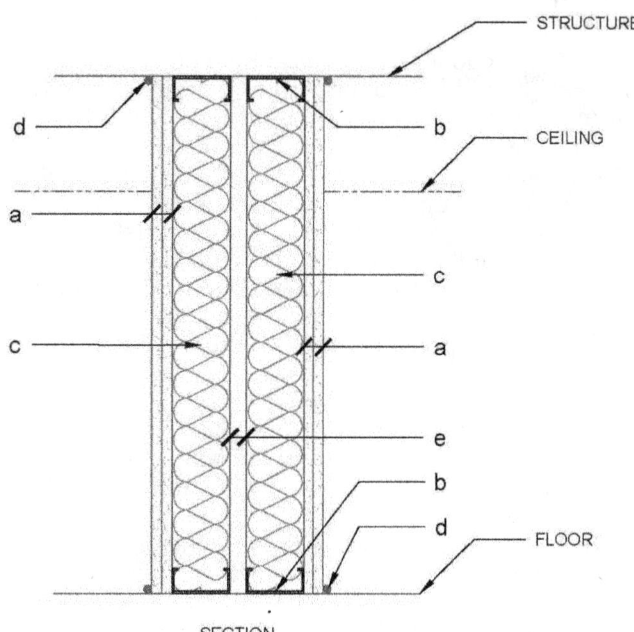

FIGURE 4.1-9 STC-60 Partition

Acoustic Distinctions

Penetration detail:

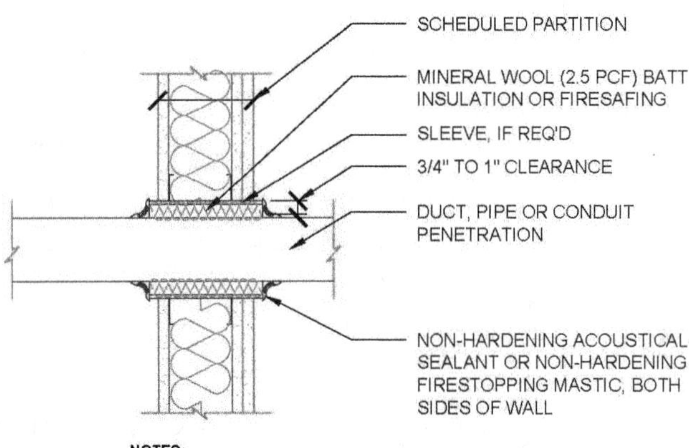

FIGURE 4.1-10 Penetration Detail

Acoustic Distinctions

Doors

There are two type of "acoustic" doors:

1. Sound-rated (STC-rated) assemblies that are tested in an acoustic laboratory. The frames, door panels and seals, as well as other hardware, such as saddles or thresholds, are provided by the door manufacturer and certified to achieve specific transmission loss values in one-third octave bands.
2. Non-sound-rated doors consisting of a door panel and frame to which acoustic seals are applied.

Sound-rated doors are significantly more expensive than non-sound-rated doors, and provide certified performance in terms of transmission loss is third-octave bands, typically from 100 Hz to 5000 Hz.

For personnel-size doors (typically 3 feet wide and 7 or 8 feet high), STC ratings of sound-rated doors range from STC-40 for 1-3/4-inch-thick solid core wood or hollow metal doors with a single set of compression or magnetic seals up to STC-60 to 65 for doors that may be 3-1/2 inches to 5 inches thick. A number of acoustic door manufacturers also make acoustic sliding doors.

A non-sound-rated insulated panel hollow metal door, or solid-core wood door, in a steel frame with butt hinges, compression seals at the head and jambs and an automatic drop seal will obtain an STC rating of approximately 35.

Large loading doors, such as those found on TV studios, may be obtained in almost any size desired, and may be 6 inches to 12 inches thick in the largest sizes. STC ratings in the low 60s are possible. Large openings are most easily accommodated with electrified sliding doors; these ensure ease of use with reliable acoustic performance.

Regardless of whether sound-rated or non-sound-rated doors with applied acoustic seals are used, the door frame should be installed plumb, square and true to ensure good contact between the door panel and acoustical seals. Fully welded frames are also more reliable than knock-down assemblies. Some manufacturers ship sound-rated door frames, panels and seals fully assembled, which improves the reliability of field installation when compared with doors shipped as a kit of parts to be assembled in the field.

A number of spaces in broadcast facilities, such as edit rooms, may require doors with relatively low acoustic performance, although for these spaces, some degree of sound separation from adjacent spaces is required. With an emphasis on daylighting for comfort and energy efficiency, many facilities consider glass sliding or swing doors and glass-front walls for such spaces. Care should be taken, however, with the design and specification of these systems where even a moderate degree of sound isolation is required, since with glass-front office system with glazed swinging doors, it is difficult to achieve ratings higher than STC-35 (NIC-30 in the field), or higher than STC-25 (NIC-20 in the field) with sliding doors.

Details:

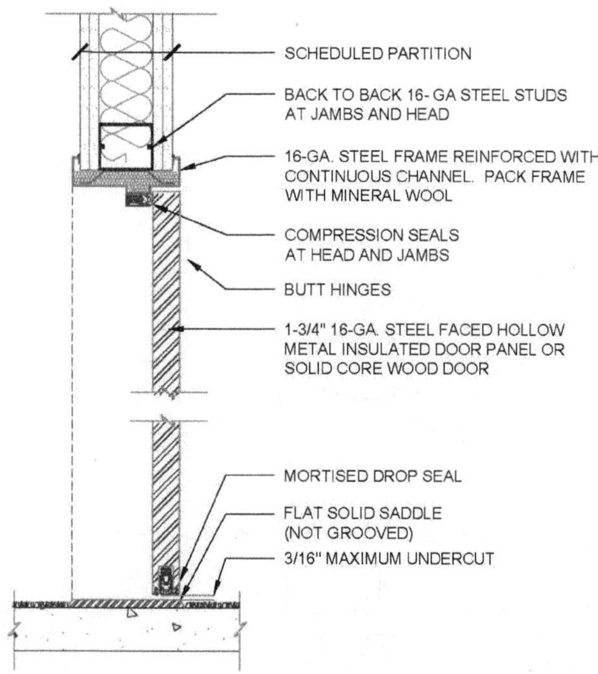

FIGURE 4.1-11 Non-Sound-Rated Door with Acoustic Seals

Acoustic Distinctions

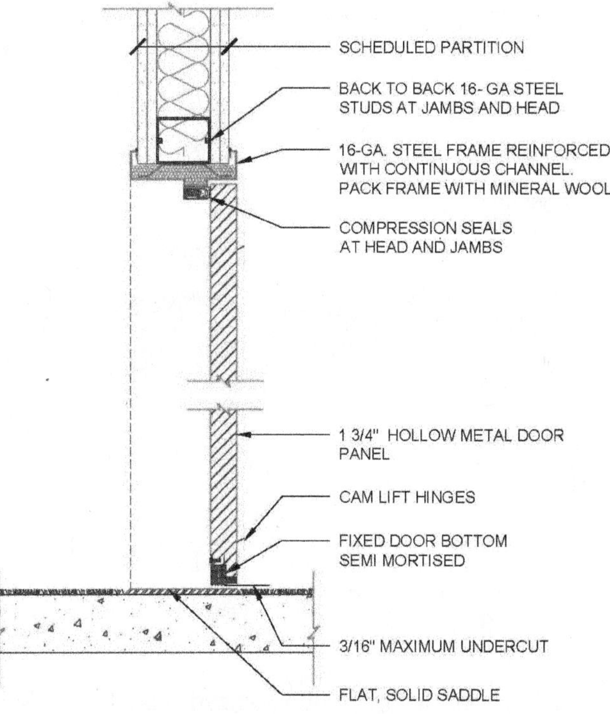

FIGURE 4.1-12 STC-51 Door

Acoustic Distinctions

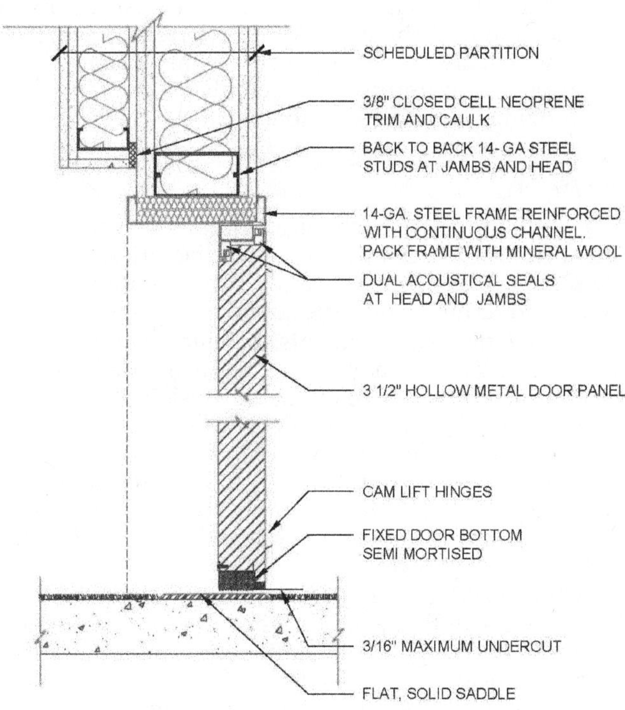

FIGURE 4.1-13 STC- 55 Door

Acoustic Distinctions

Windows

The acoustic performance of fixed and operable windows is a function of the number of layers of glazing, the glazing thickness and whether the glass is laminated or monolithic. However, while the glazing is an important contributor to the sound barrier performance of the window, the frame and the way the glass terminates to it, as well as how the frame terminates to the rough opening, also contribute significantly to the overall acoustic performance of the assembly.

A single lite of 1/4 inch-thick monolith glass can achieve an STC rating of STC-31. A laminated lite of the same thickness (made up of two pieces of 1/8-inch-thick glass with a .030-inch plastic interlayer) is rated STC-35. The weight of these two glass lites is similar, but the plastic interlayer provides damping, reducing the magnitude of sound reduction at the critical frequency, as well as reducing the critical frequency. For example, the critical frequency of 1/4-inch-thick monolithic glass is approximately 2000 Hz; the transmission loss at 2000 Hz of monolithic 1/4-inch-thick glass is 30 dB. For laminated glass of the same thickness, the TL at 2000 Hz is 35 dB.

In addition to improving the sound barrier performance of a window by using laminated glass, further improvement can obtained by using multiple panes of glass separated by an airspace. Large airspaces provide significant improvement; a window consisting of two lites of 1/4-inch monolithic glass separated by a 4-inch airspace could have a STC rating up to 50, assuming there is no flanking of leakage through the frame. In typical assemblies, one would expect an approximate 5 dB reduction in field performance (NIC rating, and in each one-third octave band) when compared to laboratory STC and TL ratings.

Window details are shown in Figure 4.1-14.

Ceiling

Like any sound barrier, monolithic sound barrier ceilings (e.g. reinforced concrete slabs that comprise both the floor of one space and the ceiling of the space below) provide good, but not necessarily the most efficient performance in terms of weight-to-transmission loss, and hence weight-to-cost performance. Cavity-type ceiling construction that incorporates a variety of materials, including sound absorptive materials in the cavity between the ceiling and the underside of the floor or roof above, is generally the most effective approach.

Where high levels of sound isolation are needed between vertically aligned spaces, for best performance, ceilings should be suspended using a hanger system of neoprene or fiberglass (good), spring (better) or a combination (best). The sound barrier ceiling itself should be comprised of multiple layers of drywall. A substantial thickness of sound absorptive materials in the cavity above the ceiling is desirable to ensure optimum attenuation of low-frequency sound in the ceiling cavity is obtained.

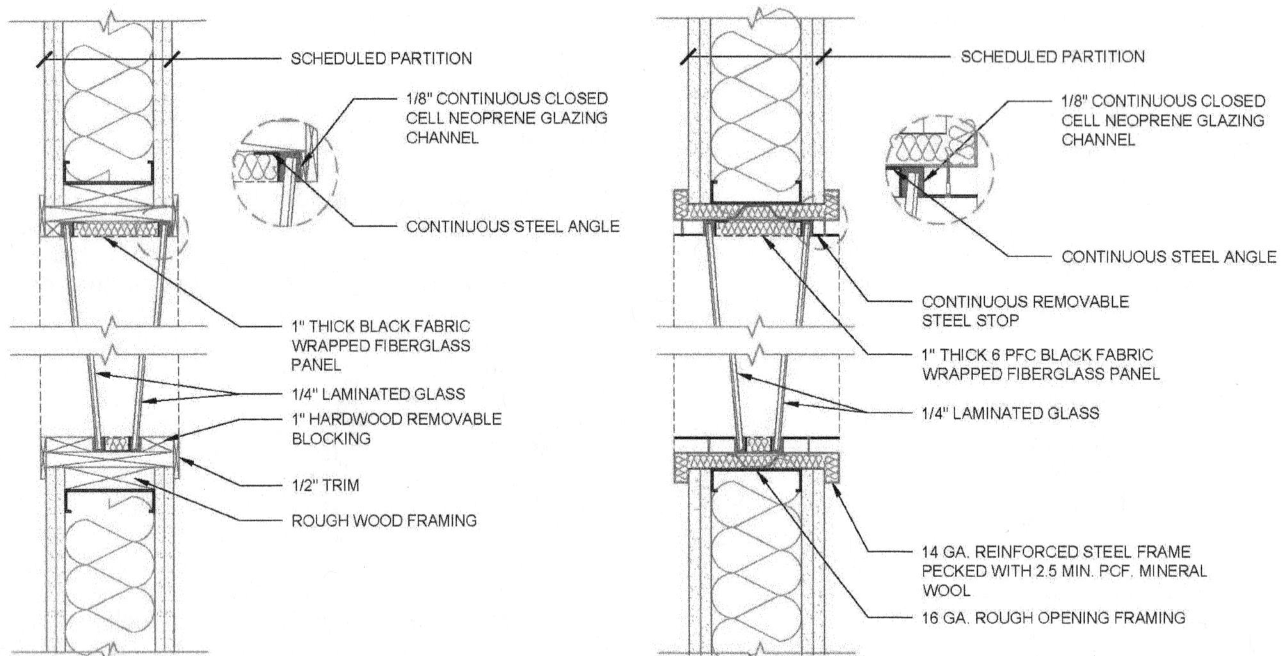

FIGURE 4.1-14 Double-Glazed Fixed Window, Wood Frame, STC-50 (left); Double-Glazed Fixed Window, Steel Frame, STC-50 (right)

Acoustic Distinctions

Penetrations of an isolated ceiling by ductwork, conduit, sprinklers, lighting, suspended ceiling panel or acoustic tile ceiling hangers, etc., can reduce the sound barrier performance of a ceiling and must be well sealed. It is often desirable to suspend these services, or a finished acoustic tile or other sound absorptive ceiling, from the underside of the isolated ceiling to minimize the number of penetrations required. Some services, such as sprinkler pipes, must be supported by the structure above the ceiling; in these cases, hanger penetrations are inevitable. One must rely on the contractor to carefully seal around all hanger penetrations. The convex round hanger rod and round sprinkler piping are not efficient radiators of structure-borne sound, and therefore penetrations do not generally result in substantial degradation of the overall sound isolation performance of the ceiling.

In wood-frame buildings, the reduction of impact sound from footfall or other activity is often as much or more of a concern as airborne sound attenuation. Carpeting the floor above, adding additional layers of plywood underlayment, stiffening the structure with additional joists or through the sistering of additional structure, and the use of insulated isolated ceilings with high-deflection springs, can all help improve the vertical isolation efficiency.

As an alternative to suspended sound barrier ceilings, secondary lightweight steel joists may be used below a structural slab or wood-framed floor system to independently span across a room. A gypsum board ceiling may be secured to the underside of these joists; plywood may be laid on the joists to provide a walking surface if appropriate. Insulation in the joists cavity is also desirable. If the joists are supported by isolated walls, which are in turn supported by an isolated floor, a complete room-in-room assembly is created, thus maximizing sound attenuation.

Roofs

Noise from rooftop mechanical equipment, road or air traffic can all be a concern for studio facilities. Many of the same issues noted above concerning ceilings apply to roofs in terms of airborne sound attenuation.

In low-noise environments, adequate sound attenuation can be assured if the roof structure includes several inches of concrete. The underside of this roof, whether on metal deck or as part of a reinforced concrete structure, can be exposed in a studio space, although adding sound absorptive material to the underside (e.g. duct liner as described earlier in this chapter) should still be implemented to ensure appropriate control of reverberant sound is achieved.

Where outdoor sound levels are higher, such as would occur in the area of rooftop HVAC equipment, on an airport flight path or where thunder is common, significant sound barrier performance of the roof may be required. A heavy suspended spring isolated ceiling or interior "roof" construction supported on joists spanning between interior isolated walls may be required.

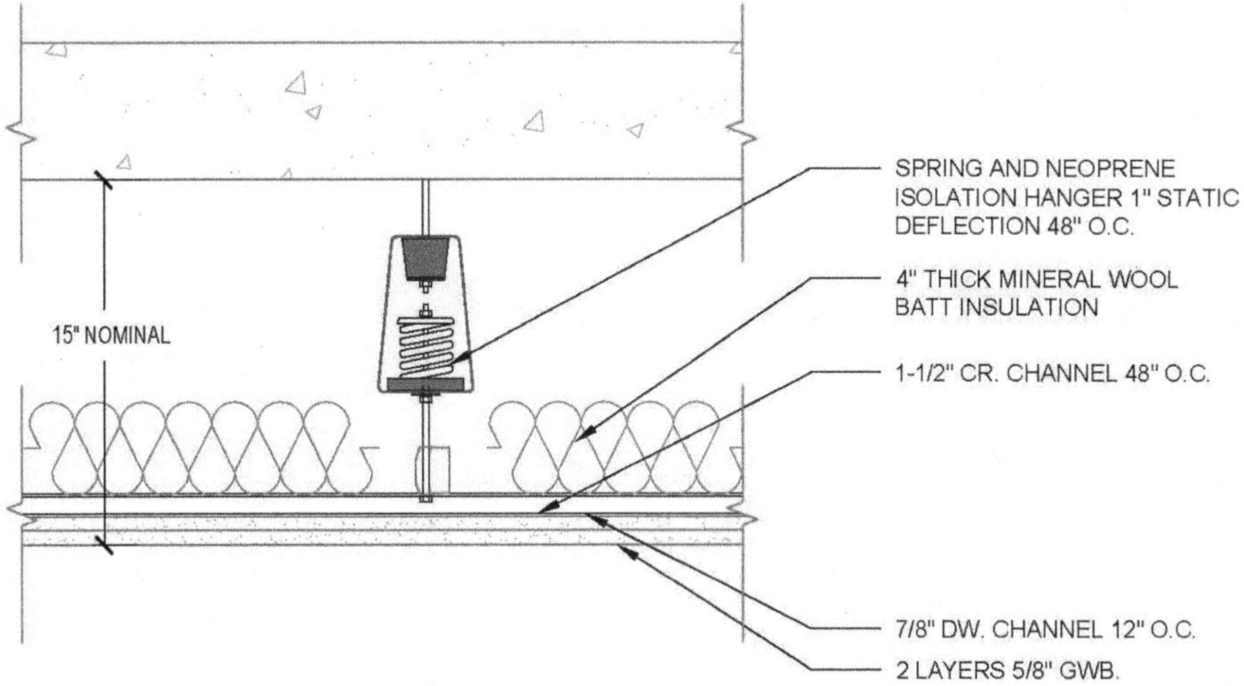

FIGURE 4.1-15 Spring Isolated Gypsum Board Ceiling

Acoustic Distinctions

Roofs can also be a source of intrusive rain noise, especially lightweight roofs without concrete. Multiple layers of heavy-duty gypsum based roof boards can be used below roofing systems to mitigate rain noise on lightweight roofs.

Avoiding Acoustic Deficiencies in Design

Good practice should result in an integrated approach to design that avoids acoustic deficiencies in the three broad categories of acoustic performance addressed above: room acoustics sound isolation, and mechanical and electrical systems noise and vibration control. Common issues to be addressed include:

i. Sound Leakage: The ability of a sound barrier construction to isolate sound between two spaces can be significantly degraded by small openings in the barrier or poor termination conditions at the edges of the barrier. The best ways to avoid these problems include shiplapping of joints between construction elements and sealant applied at penetrations and terminations. Construction documents should detail and specify these features.

ii. Sound Flanking: Sound may be transmitted past even well-sealed penetrations or terminations due to flanking through abutting construction. Consider, for example, two adjacent edit rooms along the exterior curtain wall of a building. The partition between the edit rooms may be a substantial, insulated multi-layer gypsum board assembly that terminates to the curtainwall mullion. Sound in one edit room is incident on the curtainwall glazing; vibrational energy is transmitted through the curtain wall to the glazing at the exterior of the adjacent edit room, where it is radiated as audible sound, thus bypassing (flanking) the demising partition. Depending on the severity of the flanking condition, mitigating constructions may be required, such as interior glazing along the curtain wall in one or both of the adjacent spaces.

iii. Sound Focusing: For sound incident on a plane surface, the angle of reflection is equal to the angle of incidence. If the surface off of which the sound is reflected is concave, the sound will be focused in such a way as to create a sound level increase at the center of curvature of the concave surface. In recording and listening rooms, concave curved surfaces should generally be avoided. If they are required, surfaces having a center of curvature that falls near the possible locations of listeners or microphones, or at the location of loudspeakers or talent, should be avoided.

On the other hand, convex surfaces can add significant and desirable sound diffusion to broadcast spaces, resulting in improved recording, production and listening environments. The integration into audio production spaces of convex surfaces having a variety of scales relative to the wavelength of sound is often used as a technique of increasing sound diffusion and signal to noise. For example, convex shapes made from sound reflective materials ranging from 3-inch to 36-inch radius used as wall treatments can help mitigate against undesirable sound reflections, reducing reverberant sound levels without adding excessive absorption.

SUMMARY

This chapter presents a brief overview of the primary issues to be addressed in the acoustic design and construction of broadcast facilities. It is hoped that the reader has been left with the understanding that proper acoustic design is not a field unto itself, but requires careful integration of all the disciplines of design and construction, including architecture as well as mechanical, electrical, structural, plumbing, etc.

As a chapter in a handbook with wide-ranging subject matter and wide-ranging readership, the goal is to provide an exposure to the variety of concerns one must consider in designing and building a successful studio. For those interested in more detailed explanations of the first principles on which the information presented here is based, or for examples of how to build a studio space, the bibliography presents a variety of reference materials.

ACKNOWLEDGEMENTS

The author would like to thank Lynn Claudy of NAB, who wrote the original chapter in 9th edition of the handbook, and to Howard K. Pelton, Ted N. Carnes and Gary L. McAuliffe, who wrote the acoustics chapter in the 10th (2007) edition. Also, thanks to Ken Andria for insightful review notes and to John Gebbie for helpful background regarding lighting noise.

Notes

[1] See NSF Press Release 11–098, *Sound Safety*, www.nsf.gov/news/news_summ.jsp?cntn_id=119406.

[2] www.youtube.com/watch?v=m6jsXQm5IrM, starting at 2:55

[3] ASHRAE. *ASHRAE Fundamentals Handbook*, 2015 Edition, Chapter 48, Tables 5 and 6, Atlanta, GA: ASHRAE.

[4] Neubauer, R.O. *Prediction of Reverberation Time in Rectangular Rooms with Non-Uniformly Distributed Absorption Using a New Formula.* Acustica 2000 / Tecni Acustica Madrid 2000. Print.

[5] Asdrubali, F., Schiavoni, S., Horoshenkov, K.V., A Review of Sustainable Materials for Acoustic Applications. *Building Acoustics*, Vol. 19 No. 4, 2002, Pages 283–312.

Bibliography

American Society for Testing Materials. *Thermal Insulation, Building and Environmental Acoustics*, Vol. 04.06. West Conshohocken, PA: American Society for Testing and Materials, 2014. Print. Annual Book of ASTM Standards.

Backus, J. *The Acoustical Foundations of Music.* New York, NY: W. W. Norton, 1977.

Ballou, G. *Handbook for Sound Engineers: The New Audio Cyclopedia*, 2nd ed. Boston, MA: Focal Press, 1991.

BBC Engineering. *Guide to Acoustic Practice*, 2nd ed. London: British Broadcasting Corporation, 1990. See www.bbc.co.uk/rd/bbc-guideacousticpractive.pdf

Beranek, L. L. *Acoustics*. Eds. Frederick Emmons Terman, W. W. Harman, and J. G. Truxal. New York, NY: McGraw-Hill Book Company, 1954. Print. Ser. McGraw-Hill Electrical and Electronic Engineering.

Beraneck, L. L. *Acoustics*. Melville, NY: Acoustical Society of America, 1986.

Beraneck, L. L. *Acoustic Measurements*. Melville, NY: Acoustical Society of America, 1988.

Cavanaugh, W. J., G. C. Tocci, and J. A. Wilkes, eds. *Architectural Acoustics: Principles and Practice*, 2nd ed. Hoboken, NJ: John Wiley & Sons, 2010. Print.

DuPree, R. B., and California Department of Health Services, Office of Noise Control. *Catalog of STC and IIC Ratings for Wall and Floor/Ceiling Assemblies*. Berkeley, CA: California Office of Noise Control, 1980. Print.

Egan, M. D. *Architectural Acoustics*. New York, NY: McGraw-Hill, 1988.

Ermann, M. *Architectural Acoustics Illustrated*. Hoboken, NJ: John Wiley & Sons, 2015. Print.

Everest, F. A., and K. C. Pohlmann. *Master Handbook of Acoustics*, 5th ed. New York, NY: McGraw-Hill Companies, 2009. Print.

Halliwell, R. E., T. R. T. Nightingale, A. C. C. Warnock, and J. A. Birta. *Gypsum Board Walls: Transmission Loss Data*. Internal Report no. 761. Ottawa, ON: National Research Council of Canada, 1998. Print.

Harris, C. M., ed. *Handbook of Noise Control*, 2nd ed. New York, NY: McGraw-Hill Book, 1979. Print.

Huber, D. M., and R. E. Runstein. *Modern Recording Techniques*, 5th ed. Boston, MA: Elsevier Science, 1997. Print.

Kinsler, L. E., and A. R. Frey. *Fundamentals of Acoustics*, 2nd ed. New York, NY: John Wiley & Sons, 1962. Print.

Knudsen, V. O., and C. M. Harris. *Acoustical Designing in Architecture*. Melville, NY: Acoustical Society of America, 1978. Print.

Neubauer, R. O., and B. Kostek. *Prediction of Reverberation Time in Rectangular Rooms with Non-Uniformly Distributed Sound Absorption*. ResearchGate, January 2000. Web.

Rettinger, M. *Handbook of Architectural Acoustics and Noise Control*. Blue Ridge Summit, PA: Tab Books, 1998.

RPG. *Scattering versus Diffusion Coefficients*. Tech Topic 08092013, 2013. Web.

Walker, R. *Acoustic Criteria for Listening Rooms and Control Rooms*. BBC RD 1994/6.

Other Sources of Information

The National Council of Acoustical Consultants (NCAC, see www.ncac.com/), which can provide information and guidance on working with qualified acousticians.

The American Society of Heating, Refrigerating, and Air Conditioning Engineers (ASHRAE, see www.ashrea.org/) publishes the indispensable *ASHRAE Fundamentals Handbook*.

The Acoustical Society of American (ASA, see http://asa.aip.org/) research papers.

The Sheet Metal and Air Conditioning Contractor's National Association (SMACNA, see www.smacna.org/xs) publishes guidelines on ductwork design.

CHAPTER

4.2

Planning a Radio Studio or Audio Production Facility

STEPHEN POOLE, CBRE-AMD

Market Chief Engineer, Crawford Broadcasting Company
Birmingham, AL

INTRODUCTION

The material in this chapter has been substantially revised for this edition of the *NAB Engineering Handbook*. While the technology embedded in a radio studio has undergone a radical change, other fundamental aspects to audio studio design remain the same. No matter the technology utilized, a room designed for audio production will need sound absorption to tame unwanted reflections. The walls, ceiling and flooring must attenuate outside noises. There will be audio consoles, microphones and loudspeakers (and engineers will doubtless continue to argue the merits of different brands).

Signal processing and transport have changed dramatically. Digital audio workstations (DAWs) are currently mature technology, and are found in even the smallest studios. Other systems have been used over the years to transport audio digitally, but audio over IP (AoIP) is rapidly becoming the most popular. The AES67 and AES70 standards (and industry standards such as the Telos Alliance's *Livewire*+ and others) have as their goal(s) the possibility for AoIP equipment from many different vendors to "plug and play" together. For budget-conscious planners, the prices for AoIP continue to drop. Many devices, from audio processors to telephone interfaces, now offer AoIP connectivity as a standard option, at no additional charge. Therefore, the examples and illustrations in this chapter will feature AoIP interconnections, taken directly from the author's own experience in designing and constructing practical, cost-effective studios for radio broadcasting and audio production.

Regardless of what the reader might be planning or what equipment is chosen, the basic approach is essentially the same. It is hoped that anyone who is contemplating a facility for audio production and distribution will find this chapter useful.

The key component of an audio production facility is the *studio*. Generically, the term refers to any room in which artists are likely to work, and it fits here as well. Whether the room is used specifically for voice recording, editing, post-production or the control of audio distribution, there will be certain standard features included. A typical live broadcast studio (for a talk radio format) is shown in Figure 4.2-1.

Item (1) in the diagram is optional: It is a separate, sound-isolated room specifically for recording talent. There will be one or more microphones in the room, carefully selected and positioned, to capture a particular "sound" or effect. See Chapter 4.3, "Microphones for Broadcast Applications" for more on microphone selection, usage and placement.

Item (2) is usually referred to as a mixer (or audio console) because its primary task is to blend different audio sources to the appropriate levels, then to distribute them as required. The actual mixing equipment chosen will vary widely depending on the application.

Item (3) is a computer monitor for the digital audio workstation (DAW), which is the computer-based system used for audio studio recording and editing; other computer monitors (or a KVM switch to select devices) may be included for other functions. These monitors may be mounted to one side of the mixer, on a small shelf, or on movable arm assemblies that allow them to be positioned as desired.

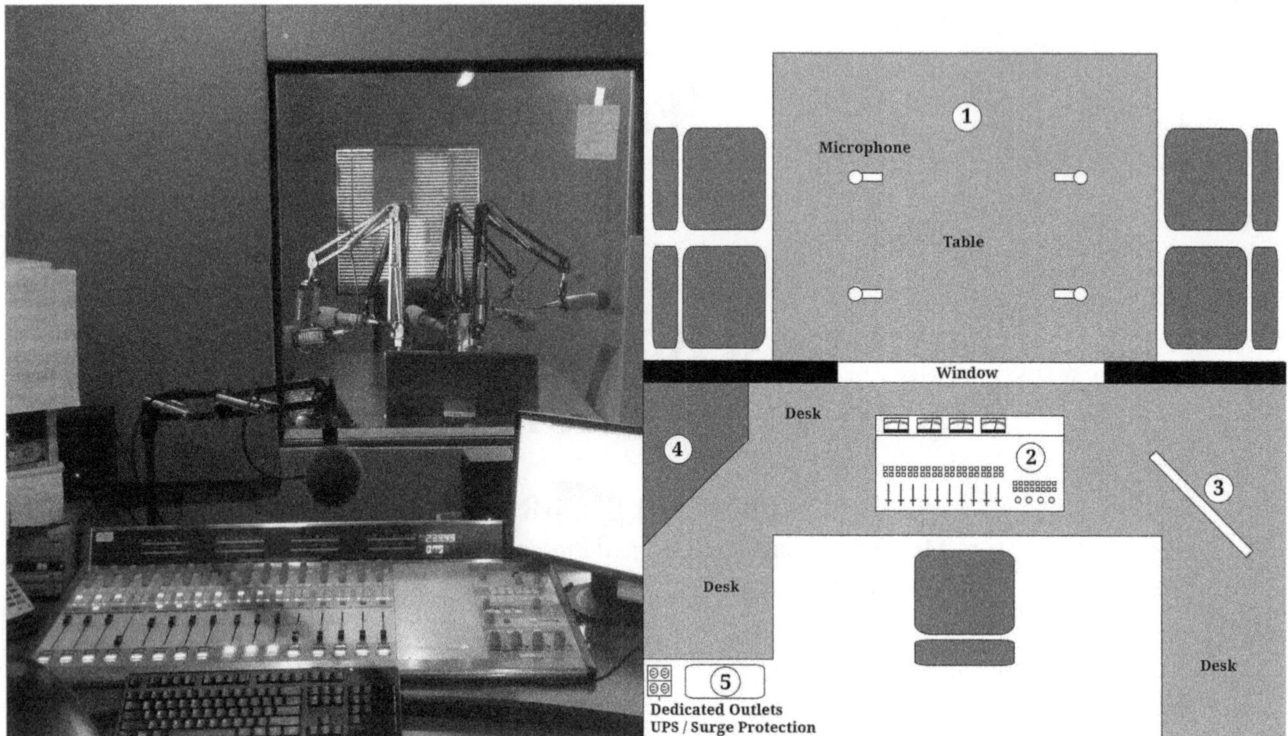

FIGURE 4.2-1 A typical studio for live talk radio

Item (4) is a rack containing other equipment, including special effects, inputs and outputs for the AoIP system (or other types of audio signals), and other ancillary systems. Storage and/or additional rack space may be required and can be located elsewhere in the room; for example, under the desk.

Item (5) includes AC power outlets and an uninterruptible power supply (UPS). The arrangement will vary from one studio to the next, but it underscores an important point: power to the studio will be supplied directly from the facility's main AC distribution point (via so-called dedicated outlets). A UPS system and surge suppression equipment are essential to protect against power outages, surges and electrical power spikes.

In years past, it was common to see things like reel-to-reel tape machines in roll-around enclosures or carts with thick umbilical cables for interconnections. These have largely been superseded by newer technologies. One is only likely to see these in older installations or in a "retro" studio that specifically advertises this as a feature to clients, or where retrieval of archival audio stored in legacy formats is frequently required.

AUDIO OVER IP (AOIP)

A Brief Review

This topic is covered in more depth in Chapter 6.1, but a brief mention is appropriate here.

In a traditional professional analog audio facility, there is typically one twisted-pair cable per signal path, and all signal paths are unidirectional. To transport a number of different audio channels, multiple pairs are needed. This tradition was retained in large part even in the early days of digital audio, as the AES3 interface replaced analog interconnections. AES3 did generally reduce cable count by a factor of two (it carried a stereo *pair* of signals on a single cable), but this still meant that many separate cables and lots of other wiring were still required within and between studios. Eventually, however, digital audio facilities began to transition away from the tradition of discrete, unidirectional audio signal paths, and the first digital audio networks used vendor-specific (and usually proprietary) approaches to reduce the need for multiple pairs.

AoIP systems are the latest implementation of this approach, and as serial network-based systems, they allow many discrete signals to be multiplexed onto the same network, thereby reducing what previously required dozens of individual cable pairs to a single Ethernet cable. By specifically using IP-based networking, cost reductions and economies of scale can be realized with the utilization of standard network hardware. In the simplest case, the end user simply dedicates a standard local area network (LAN) to the AoIP, thereby eliminating the punch blocks, patch bays and other fixtures of the analog/discrete digital eras. Cable count is further reduced because network-based systems are inherently bidirectional.

"Connect once, use anywhere" is the ability to feed a source to any input and then share it throughout the facility, through use of the "multicast" or "broadcast" modes of digital networking. This eliminates the need

FIGURE 4.2-2 A production/recording room using AoIP equipment

for splitters and distribution amplifiers. An AoIP system is very easily expanded, as well, via the intrinsic scalability of IP networking. A facility on a budget can choose to first convert a single studio to AoIP, and then continue expanding over the years, spreading the cost over time.

The features of systems from different AoIP system manufacturers will vary, of course, but a sampling might include "virtual mixers" and audio processors, implemented in software. Sound cards may be eliminated by using software drivers that present AoIP to production DAWs as selectable inputs and outputs. Software may be available to support a direct interface with popular computer-based radio station automation systems as well. See Chapter 4.7 for more information.

Note that different AoIP systems may not be fully interoperable with one another. Some manufacturers have cooperated to bilaterally minimize this issue between their respective systems, and more broadly, a common AoIP interface has been developed, called AES67, that allows at least basic interoperability among different AoIP systems.

AoIP Components

Various AoIP system manufacturers may use slightly different terminology, but what follows is a list of a few common terms that may be encountered when working with AoIP systems.

Node: An interface point from traditional audio sources and destinations (see below) with the AoIP network. A node will typically include audio inputs and/or outputs, analog and/or digital. Some nodes may be task specific, such as a processor that presents a microphone as a source on the network. Other nodes might contain additional switching, routing or processing. Some nodes are combinations of part or all of the above. Some task-specific nodes may be offered by a number of individual AoIP system vendors.

Sources and *Destinations:* These are interchangeable with the terms "input" and "output" and are applied from the viewpoint of the node. If a signal enters a device or a node, it is a *source;* a signal leaving the node is a *destination.* GPIO (general purpose I/O) switch closures may also be transported via the network as control signals and may be part of the AoIP system's source/destination matrix as well.

Control Surface: This may look like a typical audio mixer, but it is simply a *controller*. It sends commands to *other* nodes on the network, which perform the requested tasks, in much the same way that a keyboard or mouse sends commands to a PC. More powerful functions may be supported, such as the ability to change several sources and destinations (with associated GPIO) via a single programmed button.

Mix Engine or Mix Node: These are usually utilized with one or more control surfaces (sometimes in the same physical cabinet, but more often as separate physical nodes). This node does the actual source mixing. It applies algorithms to the audio data to adjust levels, add effects, create mixes and then send those mixes to different destinations.

The AoIP Studio

Figure 4.2-2 illustrates a node in a production room for a radio station. Typical of such studios, the production mix node is used to record, edit and assemble creative content, from commercials to music to news elements.

The different nodes in the figure are numbered: (1) is the control surface and (2) is the mix engine. Node (3) provides additional audio inputs and outputs for the room. The DAW (4), which in this case, but not always, doubles as a workstation for the radio automation system, uses a virtual driver that directly interfaces to the AoIP network, and therefore a hardware sound card isn't needed.

Nodes are connected with "CAT6" network cabling to a central network switch (not shown). Other nodes, or even complete studios, may be added as needed by simply adding the new devices into the network switch. The "connect once, use anywhere" capability means that anything from a single microphone to *an entire studio as a unit* connected to the central switch can be used as a source of audio anywhere else in the facility.

FIGURE 4.2-3 A Trango tower-mounted, high-speed microwave data link

Latency

A typical AoIP facility will be built on an Ethernet platform, which itself uses serial protocols to stream *packets* of data across the network. While today's networks can operate at very high speeds, it still takes a finite amount of time for the packets to travel through a network. *Latency* is a measure of the delay from input to output on the network. *Packet-switched* networks like the Internet suffer from congestion, routing delays and other slow-downs that will increase latency. On the other hand, Ethernet with a modest-sized LAN, if designed carefully, will add very little latency. Some AoIP systems also include proprietary latency mitigation schemes to further alleviate the issue.

Further, in an AoIP system, additional latency can come from repeated analog to digital conversions and vice versa, signal processing and other data manipulation. In general, *overall latency* for any given route through the AoIP network should be kept to no more than a few milliseconds for real-time audio. In some situations, enough delay may be accrued within the network such that the resulting audio throughput may no longer be classified as "real time." This must be discussed in detail with the equipment vendor when planning a studio based on AoIP technologies.

Audio data compression (MP3 for example) can introduce significant latency and loss of fidelity. As a result, AoIP systems generally use uncompressed, linear PCM audio. There may be links in the station's internal architecture or distribution chain, especially those that must exit the AoIP network and get distributed by some other technology, where audio data compression will be required. If so, a fast, high-quality codec, such as FLAC or AAC+, should be chosen with the highest feasible bit rate.

Ethernet

By far, Ethernet is the most common networking protocol. Even if other technologies are used, the end user will normally be presented with Ethernet-compatible RJ45 or fiber connection interface that support the Internet Protocol (IP). Figure 4.2-3 is a typical example, showing an "Ethernet over microwave" radio. The red arrows point to the Ethernet and GBIC (gigabit interface converter) ports. From the standpoint of the user, an IP microwave link like this one can simply be treated as an especially long run of network cable.

AoIP Networking Architecture

The AES67 AoIP interface specifies RTP/IP, the Real-Time Transfer Protocol over Internet Protocol, which provides increased robustness and reduced latency. In terms of basic network design, the same methodology for creating a reliable and fast local area network is also good for an AoIP network. Nevertheless, 802.11x ("Wi-Fi") wireless networking technology is not up to the task of transporting a multitude of RTP signals; AoIP networks within the facility should run exclusively on *wired* Ethernet.

Further, each manufacturer of AoIP-connected system components may recommend specific models of "smart" network switch(es) for the LAN backbone, CAT6 cable or fiber lines, and may require Gigabit Ethernet connectivity as opposed to the older 100BaseT. Many vendors provide detailed information for switch configuration and other setup as it relates to their particular hardware and software. These guidelines should be followed explicitly, and they will also form a critical part of the planning process.

The AoIP network should be on a completely isolated subnet with its own exclusive range of IP addresses for two reasons. First, isolating it will enhance security. Second, the network should not be loaded with additional traffic. Dedicate this subnet solely to AoIP. The office desktop network, the audio automation and traffic systems and any other Ethernet-based hardware which isn't strictly part of the AoIP system should *not* be on this network. If a workstation for the audio automation needs direct access to the AoIP network, use two network interface cards (NICs) in that workstation with appropriate IP addresses: one for the automation system subnet and another for the AoIP subnet.

Finally, the network should be firewalled to guard against unauthorized outside access. If remote access is required, it should be via a virtual private network (VPN) or the equivalent. Inside the facility, physical and virtual access to the AoIP system by nonauthorized personnel should be restricted and controlled. For instance, if station personnel can easily access the physical jacks and wiring, it is possible that someone would attempt to insert a small network switch in order to access the network architecture to insert a personal recording or playback device. Not only will this degrade packet flow through that part of the network, but also if the personal item is infected with malware, the entire AoIP network could be compromised. For more in-depth material on networking in broadcast facilities, refer to Chapter 11.1.

DESIGNING A LARGE, MULTI-STUDIO FACILITY

With the groundwork laid, we can now start planning a complete facility, with several studios, for audio production and distribution. A typical radio broadcast station will be considered in this example, but the approach will be essentially the same for either a small recording studio or a large international complex for an audio network headquarters. The differences are mostly of scale.

The Planning Team

Planning a facility that will create and distribute audio is a collaborative effort. Even a single, small studio will be constructed by using the ideas, recommendations and experience of others. This is especially the case for a larger facility that will contain many interconnected studios.

Each member of the design team will typically be assigned different aspects for oversight. Engineering will primarily be concerned with the studios and associated equipment, while others might work on office space and other facilities. It is essential that this team should meet regularly for updates and discussion. The design team can be made up of all, or only a few, of the stakeholders: the engineer in charge, who is often also the project manager and/or general contractor, the rest of the engineering team, radio station management, the station attorney, the end users of the studio and equipment, outside tradespeople, and possibly a structural engineer, an architect, an acoustic expert, a consulting engineer and the equipment vendors.

The engineer who is tasked with the design/build of a large, multi-studio facility will need to wear many hats and must be prepared to understand the nonengineering issues surrounding the design. Thus, we need to mention the following, at least briefly.

Regulations and Building Codes

The planning group, including the engineer(s), should be familiar with the regulations that govern the project. Common sense dictates that the team would prefer an architect and/or building contractor with experience in studio design and construction, but *it is important that they be familiar with the applicable laws, rules and building and occupancy codes.* (The catch-all term "codes" will be used here for brevity.) In many cases, the station and engineering team will end up working with an architect that has never designed a radio, production or sound studio before. This will require a commitment to communication, diplomatic information sharing and professional mutual respect in order for the project to succeed.

For a small project, the design team might deal with local governing boards directly. To aid with this, these agencies sometimes provide useful checklists and handy guides. But if this course is taken, *proceed with care.*

Here are two examples taken from actual projects. First, studio personnel often prefer dimly lit rooms, but new regulations could require brighter illumination for safety reasons. This may be solved with *two* lighting systems, on separate wall switches, in each studio—a few track lights for talent and brighter overhead fixtures to satisfy code. Second, many localities now have rules mandating the use of riser and/or plenum-rated network cable if it is run in the ceiling space or under the floor; the jurisdiction may also require that *all* wiring (including audio and network lines) be inside of conduit in these locations. This is most important to know when designing raised-floor rack rooms and studios with the intent to run wiring in the floor.

Issues like this must be considered during the initial planning, and frequent checks with experts on building code should be made throughout. A failure to meet code could result in fines or an expensive rework before a permit to occupy the new premises will be granted.

PLANNING STEP ONE: WHAT IS NEEDED?

The first, and most obvious, part of the planning phase is determining, What is the intended purpose? What is the goal? What is needed to achieve this purpose? As stated above, the framework example for this chapter will be a multi-studio radio facility. Regardless of the scope of the project, this first step is the initial "sanity check."

A rough estimate of the project's overall cost will help station management decide whether to expand or limit the project scope, or if it is worthwhile to proceed at all. What may be a dream facility in initial discussions will often have to be "value engineered" once the final cost of the project is known.

Estimating Square Footage

The engineering team will be working on estimates for the needed equipment. Others might price office furniture and supplies. But for the facility itself, a rough idea of the total square footage will be needed. There are no fixed rules for the size of a studio (save that some localities require a minimum square footage per occupant), but each should be as roomy as the budget will permit. While it's possible to build a control room for a single radio station into a 6′ × 9′ closet, the staff will be cramped and unhappy. Smaller rooms, generally speaking, can be more difficult to treat acoustically. More sound absorption may be required on the walls, making the room feel even more crowded. One hundred to two hundred square feet should be considered a minimum for any control room, and other studios may need to be larger. Local building codes and affording access to persons with disabilities will also be key to room size and layout factor.

Consult the staff and listen carefully to the people who will have to use each room. Paper, pencil and rulers (or the digital equivalent, such as CAD software) should be used to sketch ideas and to derive rough measurements, to check sight paths, and so on. *Overestimate and be generous with the square footage.*

Don't forget the thickness of the walls: a good rule of thumb is 5″ to 6″ for office walls and 8″ to 12″ for those with sound treatment. Space for wiring troughs and thruways, conduit and/or wire trays will need to be reserved as well.

Space Planning Needs for Each Studio

Questions like these should be considered when space planning for each studio:

1. How many people will be in the room on a regular basis?
2. How many microphones will be needed? How many headphone stations?
3. For mixers, how many inputs will be needed?
4. Will the mixer be located in one big room with the talent, or in a separate control room?
5. Will talent need access to telephones?
6. How many computer workstations will be required?
7. Does talent prefer to sit or to stand? Should you allow for both possibilities?
8. If this is to be an interview studio, how many guests should be planned for?
9. Will large musical instruments or vocal groups need to be accommodated?
10. How many outputs will be needed, and will any be outside of the studio itself? With AoIP, "remote" destinations can be accommodated with network routing, but for discrete wiring, thought must be given to the physical routing and marking of cables.

Armed with this information for each desired studio, the project manager should now obtain dimensions for typical equipment and appropriate studio furniture. These specifications can normally be provided by the manufacturers and should allow the design team to calculate the space required for each studio.

Project managers should have discussions with station management and other staff to determine what other space is needed for support staff, such as restrooms and perhaps a break/kitchen area. If the facility will receive guests, a nice reception area will likely be included, with a special sitting area (a "green room") for guests. Space should be reserved for these as well.

If several studios are planned, some form of master control or technical operations center (TOC) will be needed. This can take up considerable space, and the engineer should be prepared to justify the request.

Space for wire trays, plenums and/or conduit will need to be considered. At this point, a rough drawing should begin to take shape, with lines to indicate the interconnections between rooms.

Estimate 24″ for the width of each equipment rack. The depth should be generous as well; in the absence of other guidelines, choose at least 36″. Modern equipment is often designed to fit into a one-rack-unit height profile. As a result, the depth of these appliances is increasing, so that sometimes 42″-deep racks are needed in some locations for server equipment. It is particularly necessary to allow for wire lacing and power distribution at the rear of each rack, and rack depths can become a large issue in a hurry when it comes to space planning. Nothing will mar the appearance of a TOC more than equipment that protrudes from the back of an equipment rack (not to mention that this protrusion can be a personnel hazard), and this can easily be prevented before any orders have been placed.

Space should be left between each rack-mounted unit to allow for cooling. Keep that as a standard practice—racks tend to fill up very quickly. To prevent tampering and provide a nice, clean appearance, plan to purchase plenty of single- and double-rack space "filler" panels, preferably vented to improve cooling.

Be generous when estimating the number of racks and the needed floor space. In all rooms that have racks, enough space must be included for engineers to be able to work comfortably in between rows of racks, or between racks and furniture or walls. There should be enough space behind each rack for the rear door (if used) to swing completely open, and for an engineer to be seated behind the rack to work on the equipment.

Finally, square footage will also be needed for electrical and mechanical systems. Some of this may be in the TOC; some may be in a separate "telco demarc" or "utility" closet. If the latter, it should be properly lit and large enough to allow comfortable access for maintenance. A separate closet with equipment in it should be ventilated, implying that it will need to be included in the heating and air conditioning plan.

PLANNING STEP TWO: THE ROUGH ESTIMATE

Now, a rough estimate of the needed square footage can be calculated. Again, remember to be generous when allocating space. Watch for doors that obstruct one another. Watch for walkways that are too narrow and/or that don't permit normal traffic flow. Create enough hallways so that technical areas don't become high-traffic areas. Think through workflow and the workflow of other departments that will have space in the building. Make sure that the building plans enhance the workflow, so that areas that need to be adjacent can fit next to one another in the plan. For example, if the TOC is going to double as master control, make sure that it can be designed to be adjacent to all of the studios that it is controlling, and that good sight lines are available between master control and all of the studios if necessary. If line of sight isn't possible to all studios, a good intercom and video system might be used.

At this point in the planning process, the design team could have an initial meeting with an architect or a building contractor to discuss feasibility and to review for obvious problems and overlooked items. The contractor may be able to estimate a per square foot cost for construction, which will provide a general guideline dollar figure for the required amount of space.

There are handy calculators online that can be used to obtain an estimate, but be warned that these will invariably underestimate the cost of studio construction. Studios need thicker walls, doors and sound treatment, additional wiring, unique heating and air ducting, etc. If forced to rely on such a calculator, the provided figure(s) may need to be at least doubled.

A firm price won't be available until formal drawings have been finalized and the contractors return hard bids. But armed with these rough estimates, the owners of the facility can decide whether the project is worth pursuing. For the purpose of this chapter, we will assume that the answer is "yes."

PLANNING STEP THREE: CHOOSING THE PHYSICAL LOCATION

A radio or television facility will usually decide to remodel or to add to current premises, if at all possible. However, we should briefly discuss the alternative of choosing a new location as opposed to renovating or making additions to an existing facility.

The Relocation Option

If the decision is made to relocate, a knowledgeable broker who specializes in commercial properties should be consulted. This person will be able to provide valuable insight into which areas might be the most desirable for the facility and, conversely, which areas to avoid.

Local laws will impact these choices: the real estate agent should be familiar with local zoning. Finding a bargain on a building that is not zoned to accommodate a radio station is no bargain at all. A broker who is experienced and knowledgeable, in fact, might even recommend that certain localities be avoided entirely because of known problems with zoning and permitting. This advice should be heeded.

The broker may provide you with several choices, ranging from unimproved land to an existing building. Obtain opinions from the architect, as well as from other businesses in the proposed location(s). Local law enforcement should be able to provide figures on crime.

Finally, armed with this information, discuss the choices with management and the staff. Point out that sometimes there is no such thing as a "perfect" location or an "ideal" facility; planning like this usually involves creative compromises.

Things Often Overlooked

When new locations are being vetted, don't forget reliable utilities, including *good high-speed Internet access.* A gorgeous location out in the countryside may be ideal from a creative and aesthetic standpoint, but could lack high-speed Internet service (both download *and* upload), which is a necessity these days and is required not only for content acquisition, but often also needed for the station's studio-to-transmitter link (STL) or for feeds to and from a separate, distant network.

If there is a need for a satellite downlink or uplink, does line of sight exist to the orbital arc? Would an interference study show that reception would be problematic at the new location due to terrestrial interference? Could an uplink be prohibited because of potential interference to others? Is there room available for a satellite dish or several dishes? Would they be secure? Might zoning be an issue?

Similarly, if a microwave-based studio-to-transmitter or remote pickup (RPU) radio system will be needed, does line-of-sight exist between the proposed studio and the other end of the link(s)? Is the studio building tall enough for the antenna? Would a tower be necessary, and if so, would there be enough room to build it? If there is, would the zoning board allow the tower construction?

Additionally, routine utilities such as water supply and trash pickup may also be problematic at the proposed new location. Is there adequate secure parking for station personnel?

Is the proposed building near heavy industry that would vibrate the building (train tracks) or create unwanted noise, even in soundproofed studios? Similarly, is there an airport nearby? Runway approach paths can change over time, so even if there are no audible aircraft during site selection, any location within a mile or two of an airport should raise a red flag. For space in a multi-tenant building, are any neighboring tenants or businesses likely to be sources of excessive noise (e.g., the disco upstairs)? Often, structure-borne noise (noise that is carried from another floor or section of the building by direct coupling to the building's superstructure) is often the most difficult to isolate. While many of these problems can be solved by proper acoustical treatment, they can add considerably to the expense of the project. For further detail on acoustics and noise control, see Chapter 4.1.

There may be a need for extra ceiling height to accommodate the oversized ductwork required for quiet HVAC operation (more on this later). Commercial space intended for retail use, including ground-level space in a multi-floor building, is often well suited in this regard.

If this is a broadcast facility, in the United States, FCC rules (Section 73.1125) come into play as well. Generally speaking, the FCC will require that the main studios be located

> (1) Within the station's community of license; (2) At any location within the principal community contour of any AM, FM or TV broadcast station licensed to the station's community of license; or (3) Within twenty-five miles from the reference coordinates of the center of its community of license.

This must be kept in mind as well, especially for a multi-station facility for which it's possible that one of the stations will fall just outside of this requirement.

As you can see, before spending money on a new location, a complete due diligence, which could include a thorough physical inspection and environmental study, should be done. A surveyor should be hired to determine property boundaries and to check for easements that could affect construction. For the title work and sales contract, a local attorney should be retained.

The Remodeling Option

If an existing building is to be renovated, the architect will come on site and make a complete set of measurements. He or she will also examine the structure and determine, for example, which walls are load bearing and which can safely be demolished. If a load-bearing wall must be removed, some other means of supporting the upper floors or roof will be needed. This may require changes in the layout.

The planning team can take the architect's numbers and recommendations to draft final rough sketches for the desired facility. Better yet, the architect will normally draw the "shell" of the structure with no walls, but with obstructions marked, for his or her own later use. If a copy of this can be obtained, it will be ideal for planning.

PLANNING STEP THREE: WORKING WITH THE ARCHITECT

Unique Features of Studio Design and Construction

Ideally, the station should work with an architect and/or contractor that have experience designing and constructing radio and television stations. If none is available in the area in which the building is to be built or renovated, the station engineer or outside engineering consultant will need to be diligent in discussions with the architect that radio studio requirements are different from ordinary commercial building requirements.

Floor to Ceiling Height: The slab-to-slab height of each floor in the building should be taken into consideration. No matter if the audio, signal and Ethernet cabling and the AC distribution is run overhead or under the floor, additional space is needed to accommodate it. Even more vertical space may be required for the large cross-section of ductwork used for low-velocity, acoustically isolated (i.e., internally lined with absorbent material) HVAC systems preferred for sound-sensitive facilities. Spaces with a short ceiling height therefore should be avoided if possible.

Heating and air conditioning: Due to the amount of heat-generating equipment installed, (and sometimes the number of people working in a studio), studios tend to be warmer than a typical office. The equipment in the technical operations center will also generate a large amount of waste heat. These areas will need special cooling, and may require a separate HVAC (heating, ventilation and air conditioning) unit from the office and utility areas of the building. Using a typical "per wattage, per person" heat load calculator will invariably result in studios that are too warm, so any heat load calculations provided to the builder should include the AC power load of all of the equipment in each technical area, converted to BTUH (British thermal units of heat). All HVAC ductwork should be carefully planned and isolated to avoid unwanted sound transmission between offices and studios.

Electrical: Typical residential and commercial wiring may string a number of outlets together in parallel to reduce electrical costs. Studios need numerous dedicated outlets, each with a primary run to central power distribution. Each control room will also need many more outlets than is typical, and perhaps additional wiring to power distribution strips in an equipment rack. When renovating an existing building or constructing a new building, breaker panels and incoming utility power may need to be replaced or upgraded. The radio engineers will need to understand the entire complement of equipment that they plan on using in each technical area and determine (often with the help of the equipment manufacturer) how much current each piece of equipment draws to be able to understand how much power is needed for the entirety of the new construction. Be generous with the estimate; you can always cut back later if the budget requires it.

It is possible that additional utility power in the form of a new power company transformer will be needed, and the cost for the utility company to provide that must be included in the budget, as well as provisions to run the entire facility from a rented generator if necessary while the utility cutover is completed.

A standby generator to maintain critical operations during a power failure may be included, and UPS and power surge suppression equipment should be used as well. Having these items may seem like an unnecessary investment to station management, but they can be extremely important for keeping the facility on the air, especially if commercial power is unreliable or the area is subject to electrical storm activity.

Proper facility grounding, bonding and lightning protection is covered in Chapter 11.2, but it is important to note that proper electrical and signal grounding is critical in any radio technical facility. It becomes even more important if the facility is tied to appliances on a nearby tower, such as STL or satellite antennas.

Lighting: Ensure that the architect understands that studio personnel may want a dimly lit room. Emergency and brighter lighting options for equipment servicing will likely be required as well. Separate rack rooms should be properly lit so that engineers working behind the racks can see the equipment they are working on. The engineering staff should be a part of the design team for their own workspaces, and plenty of properly placed lighting should be provided in the aisles between racks.

Additional Conduit, Plenum and Wire Trays: These will be used for the wiring between studios. As an alternative, raised flooring (or "computer floor") can be used with all wiring run under the floor. Either case would need a good bit of space and could conflict with HVAC ducts and plumbing. Conduits and thruways should be as straight and direct as possible; however, unwanted sound transmission paths should be eliminated. Care should be taken so that if both electrical service and signal cabling have to share the same path, either under the floor or overhead, that they are separated from one another in conduit.

It is possible for the telephone and noncritical network services to be run by the building contractor and would thus need to be specified by the architect, but there are arguments for and against. As a cost-effective alternative, consider having the electrical contractor install any needed conduit, particularly since they would likely have the necessary "low-voltage wiring" electrical license. The engineering staff would then pull the studio wiring using an inexpensive wire "snake" and twine.

HVAC Ductwork: In most commercial buildings, a typical air/heat installation uses a central unit with "tapped" ductwork for each room. A plenum return is also typical. Neither is suitable for studios, because sound will leak between rooms. In fact, the priorities for ductwork design in general construction (such as minimum number of main duct feeds and minimum cross-section) are often just the opposite of those optimized for studios. Individual "home-run" supply and return ducts to each studio are preferred, with large cross-sections allowing requisite air exchange in cubic feet per minute (CFM) to be achieved with minimum air velocity, for quieter operation. HVAC equipment (compressors, fans, etc.) should be physically isolated from the building's structural frame via spring mounting or other techniques to minimize structure-borne vibration and rumble. Special consideration will have to be given to avoid unwanted sound transmission between studios, offices and outside spaces. See also Chapter 4.1.

Walls, Doors and Windows: Use staggered 2×4 studs in the walls, double-layer wallboard, insulation and angled windows mounted with cushioned dual panes of glass. Sound absorption material should be used on the inner walls of studios.

Finally, but not least importantly, the building contractor should endeavor to complete the studios and TOC first. That way, the engineering staff may install equipment while the other work is being finished.

By emphasizing these things to the architect, the final drawings will correctly detail what must be done. This will be invaluable if questions or disputes arise during the construction.

It is very important that the architect and those who will be doing the actual construction work according to the building plans understand the peculiarities of studio construction. When construction starts, someone from the planning team should remain on site to ensure that these guidelines are followed as well.

PLANNING STEP FOUR: THE STUDIOS

At this point, the engineering team should immediately begin to finalize the furniture and physical equipment layout for the studios and the inter-studio wiring plans. Equipment should be chosen and interconnections mapped out. As mentioned before, the number and routing of conduit or wire trays will need to be provided for the architect. If the building contractor is to run conduit, or at least do some of the phone and network wiring, this will be sketched as well. The architect and/or contractor will likely produce a set of drawings for the project based upon your sketches. Double check these up front, as any subsequent changes may require repermitting, change orders and added costs.

A sketch of the TOC should show the locations and sizes of the racks. If electrical is to be included in the same room, remember that codes typically require a 3-feet (or greater) clearance around each breaker panel or disconnect, so space will need to be provided for that easement.

Larger facilities may have 480V (or higher) primary AC service, so space planning is needed for a large transformer to step this down to 240V and 120V for the rest of the facility. These can be quite bulky. If a generator is used for backup power, the transfer switch will need space (and clearance to meet code) as well. Lightning protection and/or power conditioning equipment will also need adequate space.

Our Working Example

The example that will be used here is for an AM/FM combo, two radio stations in one facility. The requirements are:

1. Two control rooms, each with an on-air phone system.
2. One large live performance studio, shared by both stations.
3. One small live studio, shared by both stations.
4. Two production rooms for creative content. One should be able to use the small live studio to record guests.
5. A TOC sufficient for managing signals into and out of the above studios, and two STLs, one for each station.
6. All critical equipment will be on a UPS, and a generator will be used to back up commercial power.

A suggested layout is shown in Figure 4.2-4. This is essentially the rough sketch as it might be presented to the architect. We assume that electrical distribution will be from the TOC to everything else, so plenty of clearance is needed on that side of the room.

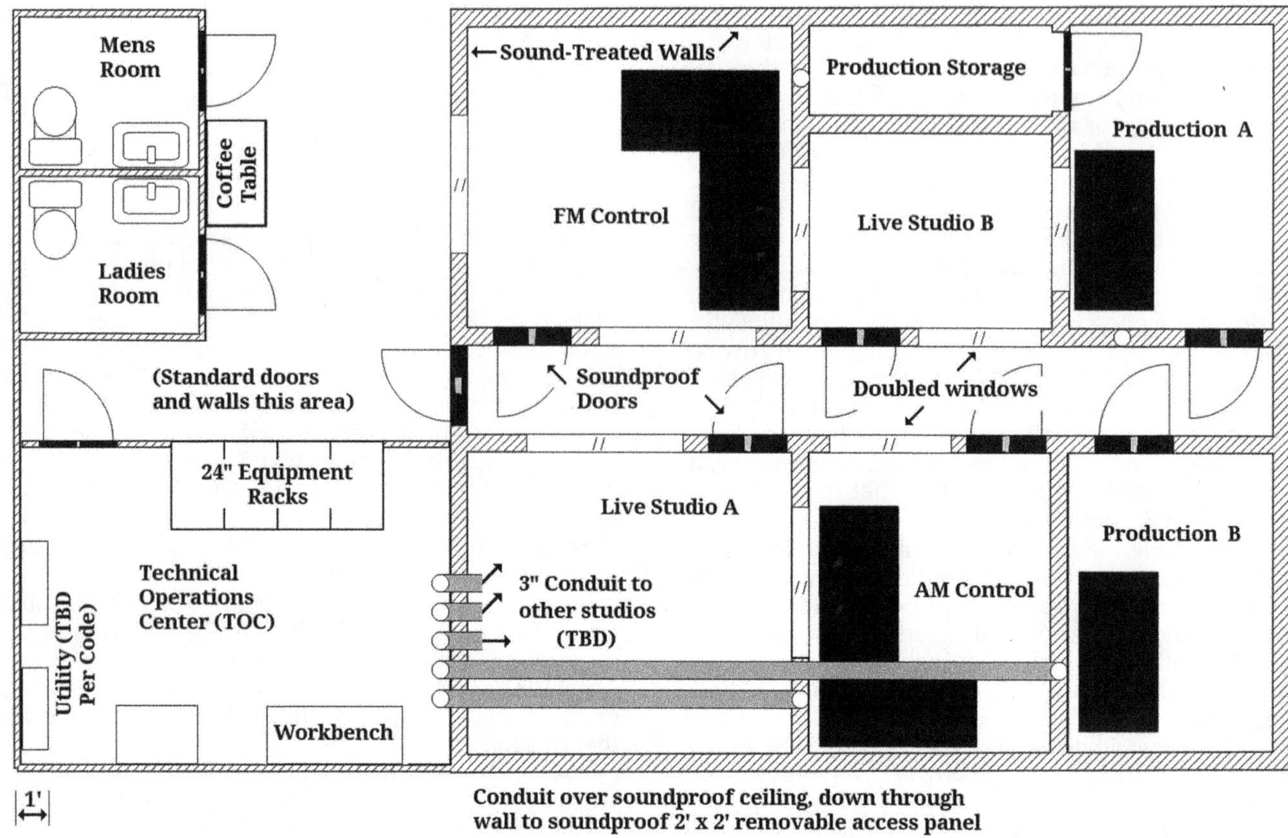

FIGURE 4.2-4 A rough sketch for the architect

PLANNING STEP FIVE: EQUIPMENT AND INTERCONNECTS

A copy of the rough sketch should be given to the architect to aid in creating the formal drawings (the actual blueprints). Engineering staff can use this same sketch for final planning of the equipment layout and interconnections, understanding that things may change between the initial sketch and the final drawings. Especially for an AoIP facility, this plan should be a detailed as possible, and should include:

1. Determination of the number of needed AoIP nodes.
2. Routing of the underlying Ethernet (or equivalent) that will tie the AoIP together. (Remember, this should be a separate, isolated subnet.)
3. Routing of other network systems between studios, and to/from the TOC.
4. Assignment of unique IP addresses and names (if applicable) to each AoIP node.
5. Determination of where and how wire trays and/or conduit will be placed for inter-studio wiring.
6. Any other wiring, especially in the TOC, such as coax to satellite equipment and STLs.

Figure 4.2-5 is part of one spreadsheet (of over a dozen) from a project recently done at the author's studios. This is an example of the level of planning required in advance. The ID of this node (called a "Blade" by Wheatstone, the manufacturer), the IP address and the inputs and outputs are filled in as the planning proceeds.

Equipment Orders

At this time, the equipment orders should be finalized. If it's a large order, it could take several weeks (or even months) for everything to arrive. Obtain copies of the installation manuals before the equipment arrives. The engineering team should complete as much prep work as possible prior to installation.

Typically, the equipment will arrive as a number of separate shipments over many days. The project lead should plan for a secure area in which to store everything while construction on the facility is being completed. It is essential to inspect each box as it arrives, and note the apparent condition of the equipment and check off its arrival against your master equipment checklist. If shipping damage is found, it is important to notify the shipper, the manufacturer and the insurance company at once, not several weeks later when the box is unearthed out of storage for the install.

Preconfigured Equipment

Especially for larger orders, the vendor may be willing to preconfigure each node before shipment, but it

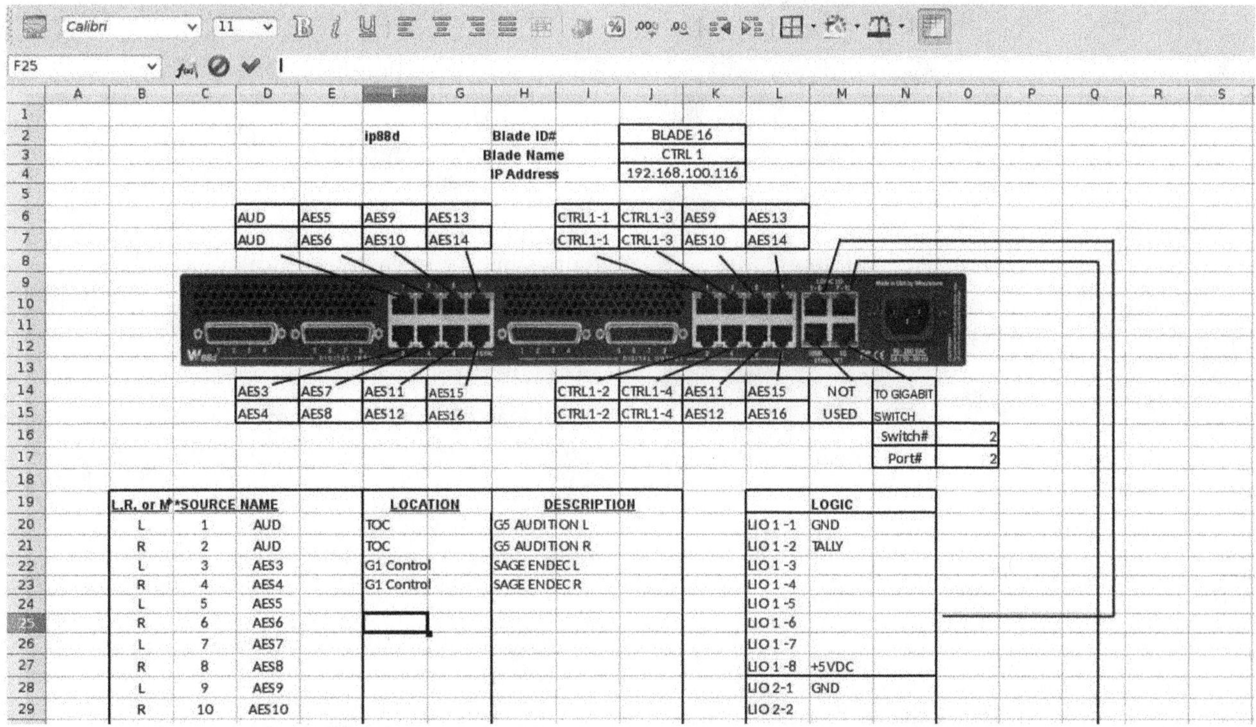

FIGURE 4.2-5 A sample worksheet for part of an AoIP system

is essential to have a detailed list of each unit, the IP address(es) to be assigned and the plain-English identifying names that will be used for quick recognition of sources, destinations and nodes (refer again to the worksheet of Figure 4.2-5).

Because the vendor is likely to specify a "smart" network switch with individual port configuration(s), it's not a bad idea to even plan which switch ports will be used. Checklists should be made and then used throughout the process.

Choose a Standard Sample Rate

Professional digital equipment will usually permit you to select a sample rate, typically 44.1 KHz or 48 KHz. Choose a standard for the facility and add this to the checklists; the configuration for each source will need to be confirmed. Mismatched sample rates can cause an extremely odd problem: the result may be audible, but it will sound as though it is at the wrong "speed." This can be quite baffling to someone who has never experienced it.

PLANNING STEP SIX: INSTALLATION

While we cannot detail every possibility in this chapter, we can provide some guidelines. If the studio will have analog or discrete AES wiring, more thought should be given to cable routing. If AoIP will be used, inter-studio wiring is vastly simplified. In each studio, analog lines may be needed for microphones, CD players, monitor speakers and other discrete non-digital items. All other interconnections will be done via the Ethernet underlying the AoIP system.

If a new facility is being built, the new facility should have all new wiring. It is never good engineering practice to reuse old cabling unless absolutely necessary, and not without first testing the wiring to be reused. If possible, the old, abandoned wiring should be removed after the new facility is put on the air.

Labeling and Color Codes

Following the checklists and formal wiring plan, it is good practice to use different colors for each type of cabling and apply cable identification labels at each termination of each cable, to allow quick visual identification (see Figure 4.2-6). Choose a cable color and labelling scheme in advance and ensure that every member of the engineering team follows it faithfully. It is also good practice, if the facility uses wire numbers, to make a list of the wire numbers for cross-reference, and keep it with the rest of the cabling documentation.

Ethernet (for AoIP)

Generally, with an AoIP system, the Ethernet should be put in place first. Use a cable network analyzer to confirm link quality. If such test equipment isn't available, at least use a laptop on each connection to check connectivity. *Remember to label all network lines as well.* It's good practice to label the network switch(es) with the IP address of the switch itself, its function, and so on.

FIGURE 4.2-6 Selected wire colors, neat routing and labeling are essential

Courtesy Amanda Hopp, KLZ, Denver, CO

The idea of using different colors for the cable(s) especially applies to the several networks that will typically be used within a radio station facility. For example, the engineering team might plan on orange CAT6 for AoIP, blue CAT6 for the station automation network and white CAT6 for the office network.

Additional Wiring

For other connections, using the installation manuals as a guideline (another reason why it's a good idea to have these in advance), make up the needed cables *and label them*. If this is a new facility, these new lines may be prerouted through furniture and fixtures. If these will be put in place during the upgrade of an existing facility, the entire bundle of new cables for each room should be loosely wrapped or tied together, then clearly marked for the intended room.

To provide an idea, in the production room example shown in Figure 4.2-3, the most time-consuming part of the upgrade was cutting a hole in the furniture for the new control surface. This particular studio had two microphones and an Internet computer for the retrieval of audio. The cables for these analog sources had been prepared in advance, so the total time for cable installation was less than an hour.

The creation of wiring and equipment documentation is critical on any project, however small. It is essential to keep detailed documentation and records, and to label everything. A good label maker and several boxes of labeling tape should be among the very first purchases, and then used continuously, without exception throughout the project. Electrical outlets should be labeled with the circuit number. Phone and network jacks should be labeled clearly to prevent confusion. *Label everything*. Some label makers are equipped with serial or USB ports that can attach to a laptop with software to turn even the longest wire list into a set of neat labels. If cable numbers are used, use number series that make sense: for example, 1000 series numbers for AoIP, 2000 series numbers for analog audio, 3000 series number for digital audio, and so on.

As each room's wiring is completed and checked off, double-check the names that will appear for sources and destinations on the AoIP network. Carefully check for duplicate IDs and/or IP addresses. Check the sample rates and other settings. Give each source and destination a meaningful name that will identify it to the staff. Often, asking end users what names they want to use for various sources and destinations will help them remember them. Keep a list of source and destination names handy in each technical area and control room as an aid for staff that is learning the new equipment.

Final Testing

Finally, as each room is completed, final commissioning and a full test of all functions must be made. Levels should be calibrated and a default set of sources and destinations should be programmed into the control surfaces. This test should include sending audio to each expected destination and checking the result. Obviously, the goal is to uncover problems *before* facing the stress of making the new facilities operational. Especially critical are legal practices such as EAS testing.

PLANNING STEP SEVEN: TRAINING

At this point, the new facilities would be nearing completion. If these are in a new location, staff should be brought in and trained on the equipment. If this is an upgrade from analog to AoIP, emphasize the similarities (sources are still sources, faders will still "fade," etc.). Then emphasize the new features, such as "connect once, use anywhere." Whether the facility is moving or simply expanding in place, plan to temporarily set up a small AoIP system in a convenient space for staff training and experimentation. Connect a control surface and a couple of nodes together with a small network switch, with sources, monitor speakers and headphone amplifiers to allow them to hear the results.

A typical control surface can be daunting at first glance, so the creative staff will need sufficient hands-on time to be become familiar with it. They should soon feel comfortable, once they realize that its operation is much the same as that of a classic analog mixer.

It is probably not practical to drag out the drawings and explain all of the system intricacies to the staff, but they should be aware of available sources and how to select them. Alternatively, they should be cautioned as to what *not* to do; many AoIP systems will allow engineering to "lock" certain selections; this should be considered for critical functions. Notebooks of specific functions, and "quick start" guides should be available in all work areas so that first-time users can have aids to refresh their knowledge if needed.

PLANNING STEP EIGHT: MOVING DAY

The remodel is essentially complete; all studios have now been wired. The equipment has been installed and thoroughly tested. Levels have been set, key source and destination assignments have been made.

The staff has been trained. In sum, it is time to start bringing rooms online.

Clearly, a radio station will need to remain on air during this process. This will require careful planning and everything should be mapped out in advance. Tools, materials and personnel should be in place as well. Detailed checklists should be distributed to key personnel to ensure that nothing is overlooked.

Ideally, if the move is to be made with minimal disruption, the new equipment will be installed and configured prior to relocation. Operations may continue on the old equipment, at the old location, until just prior to the move. In that case, it will be a (relatively) simple matter of bringing each new studio "live," one at a time.

If this isn't possible, alternatives may need to be considered. For a radio station, it may be possible to directly connect a PC workstation to the STL, or directly to the transmitter, while the move is being made. In any event, careful planning will be needed to keep each station on air.

SUMMARY

This chapter has been based on the author's personal experience. An engineer facing his or her first big studio rebuild project might view a large upgrade as just "one studio at a time"—the same as usual, only with more "quantity." That is not the case. The larger the facility for audio production and distribution, the more advanced planning will be required. A failure to do this will inevitably result in a badly delayed project that runs over budget and may be compromised upon launch.

Each proposed facility will be different; no two are alike. It should also be pointed out that there is actually a great deal of overlap between the discrete steps outlined above; they are guidelines only. Other times, several of these steps may be iterative; the planners of an audio facility may make several back-and-forth trips between the architect, contractor, various trades and station personnel to work out details, even after the equipment is ordered or construction has begun.

Plans will need to be modified once they are exposed to the real world. The unforeseen will occur. But by planning as much as possible in advance, the team will be able to smoothly pivot onto these unexpected problems and deal with them as a coordinated unit.

ACKNOWLEDGEMENTS

Special thanks go to the author's supervisor, Cris Alexander, Director of Engineering for Crawford Broadcasting Company, who has overseen most of the projects upon which this material is based. Amanda Hopp, Chief Engineer for Crawford Broadcasting in Denver, also provided photos and advice. The author's assistant at Crawford Broadcasting in Birmingham, Todd Dixon, helped in too many ways to name in this limited space. Finally, thanks go to Jay Tyler with Wheatstone Corporation, who has made "teaching" AoIP a passion and whose student the author has been.

CHAPTER

4.3

Microphones for Broadcast Applications

TY FORD

GREG SILSBY

INTRODUCTION

Improving a broadcast facility's sound by choosing the right microphone is the direct result of knowing enough about microphones to make the right decisions. The combination of studio acoustics, type of microphone, directional characteristics, and preamplifier plays a role in the selection of the microphone that sounds best for any given situation. A microphone is an electro-acoustic transducer which, when activated by acoustic energy from a sound source, converts or transduces that energy to another form—an electric current. In the microphone, acoustical energy (sound waves impinging on the diaphragm) is converted to a varying voltage that is the electrical analog of the sound. This chapter describes the basic types of microphones, their construction, and how their different characteristics can be used in various applications for best results.

ACOUSTIC ENVIRONMENT

The successful use of any microphone depends on the acoustic environment in which it is used and the particular voice or instrument on which the microphone is used. The problems of poor acoustic environments cannot always be overcome by using more expensive microphones. The end result might be technically excellent audio and a great performance in what can be heard as an obviously bad acoustical environment. Too many large glass windows and other hard reflective surfaces create a difficult acoustical environment in which no mics perform well. In addition, glass windows are often mounted with the glass presenting a downward angle to sound. While this works in large music studios, it doesn't work well in small rooms because the sound usually bounces back down into the rear or side of a microphone. Minimizing the amount of glass, using window treatments such as curtains, and angling the glass to reflect the sound upward and away from the microphone are much better strategies.

In addition to acoustic issues, individual voices vary greatly. Historically, part of the reason some men and women were chosen for on-air work was for the quality of their voices. This is less frequently the case now. There is only so much compensation a mic and preamp can do to improve speech quality. Great sounding audio, then, is the result of a great source, a great environment, and the right microphone and microphone preamplifier.

MICROPHONE CHOICES

Of the many microphones available, broadcasters usually choose handheld, boom-mounted, and lavaliere styles. The handheld microphone is often used for on-camera and electronic newsgathering (ENG). Boom-mounted microphones are usually found in the broadcast studio, and some handheld microphones also are boom mounted. The lavaliere microphone is usually used in TV news, where a handheld, desktop, or boom-mounted microphone would be obtrusive. A miniature microphone can also be mounted on a headset device, with or without headphones. Lavaliere microphones may be useful in some radio talk show programming where the guest may be unfamiliar with good microphone technique or intimidated by a large boom-mounted microphone.

MICROPHONE PREAMPLIFIER

Because of the low signal level from a microphone, preamplifiers are often used to boost the level to that needed for the mixing bus in an audio console. The preamplifier may be located in the console or externally in an equipment rack. From a technical perspective, the input impedance of the preamplifier will affect how well a microphone will match with a preamplifier. However, there is no hard data as to the best impedance, or combination of resistance, capacitive reactance, and inductive reactance. Some microphones are relatively immune to these variables, whereas others are not. Quality and price are not necessarily factors. High-quality microphones and preamps can substantially improve the sound of a station. The same high-quality microphone can be compromised by connecting it to a low-quality microphone preamplifier.

One solution, albeit a costly one, has been to design preamplifiers with tunable capacitive and inductive front ends, as in some phonograph preamps. These preamps allow the user to vary the input impedance of the preamplifier and, thus, the sound of the microphone/amplifier combination. On a more practical level, when one is considering both budget and application, there are usually a number of suitable choices.

MICROPHONE ATTRIBUTES

The four attributes of a microphone most broadcasters need to consider are *self-noise, directionality, sensitivity*, and *ruggedness*.

Self-Noise

Self-noise is normally perceived as a hiss. It is the electronic noise generated by the passive or active circuitry of the microphone itself. Condenser microphones use a vacuum tube or field effect transistor (FET) to convert the high impedance of the condenser capsule to a lower impedance, more suited for the input of a microphone preamp. Vacuum tube condenser microphones are generally noisier than condenser microphones that use FETs. However, careful selection of the vacuum tube can result in a tube microphone being quieter than its FET counterpart is.

Self-noise is an important factor in studio recordings, where every effort is made to eliminate extraneous noise. Self-noise is less important in applications in which the ambient noise level is considerably higher. The ambient noise level in a broadcast air studio may be quite low, equaling that of a well-designed music studio. On the other hand, there may be noise from HVAC systems, tape machines, computer hard drives and fans, and sound transferred through the structure from the surrounding environment. If the combined noise from any or all of these sources cannot be reduced, spending extra money for a microphone with exceptionally low self-noise is not the best use of that money. Finding the best-sounding directional microphone and positioning it to reject as many of the ambient noises as possible may be a better approach.

The self-noise of dynamic or ribbon microphones is usually much less because these microphones do not use active electronics. However, the output of a dynamic or ribbon microphone is usually lower than that of a condenser. Because of this, more gain is required of the microphone preamplifier to bring the dynamic and ribbon microphones up to the appropriate operating level. Less expensive or poorly designed microphone preamps also generate noise, also usually perceived as hiss.

Directionality

Omnidirectional and *directional* microphones each have their place in broadcasting. Omnidirectional microphones pick up most sounds from all directions equally. However, they are usually directional at high frequencies. Omnidirectional microphones are often used in newsgathering interviews in which one or more voices from different directions are to be picked up. Their wide pattern is useful because the position of the microphone does not have to be moved as accurately from person to person. Most omnidirectional microphones are also less susceptible to popping and wind noise than are more directional microphones. However, some small-diaphragm omnidirectional condenser microphones are quite sensitive to popping. If there is high ambient noise, or if only one voice is needed, a directional microphone may be more effective.

Directional microphones with *cardioid* (heart-shaped), *hypercardioid*, and *supercardioid* patterns are used in broadcasting to reduce ambient noises. *Shotgun* microphones, sometimes referred to as supercardioids, offer high directionality at upper frequencies, but are much more omnidirectional at middle and lower frequencies. While their "reach," or ability to hear sounds at a distance, is greater than hypercardioid patterns, they are best used on a large well-damped soundstage or in quiet exterior spaces.

Hypercardioid patterns perform much better than do shotgun mics in tight studio environments. Because they are more directional than a shotgun, they don't pick up as much sound reflected from the ceiling, walls, floor, and other hard, flat surfaces.

Sensitivity

Output level, or the sensitivity of a microphone, is important because all microphone preamps add a certain amount of noise, especially at the upper end of their operating range. The higher the output of the microphone, the less amplification will be required from the microphone preamp. Condenser microphones with their internal amplifier are more sensitive than dynamic and ribbon microphones.

Ruggedness

Ruggedness is more important in ENG (electronic newsgathering) and EFP (electronic field production) recording than in controlled studio situations. Microphones for these applications must withstand rough handling and harsh elements (cold and hot weather and rain and very dry conditions). Ribbon microphones should not be used in the field because of the fragile ribbon element. Condenser microphones designed for recording studio use are better suited for more controlled environments because they are sensitive to humidity, temperature changes, and wind. However, some new condenser microphones are almost as rugged as dynamic microphones.

High-Frequency Response

Condenser microphones are typically more sensitive to high frequencies than are dynamic microphones. Special care must be taken in choosing condenser microphones if the intended environment is small, is noisy, and has many reflective surfaces. Their sensitivity to high frequencies in that environment will result in an increase in unwanted room sound.

Hot Spot

The *hot spot* (sometimes called a *sweet spot*) of a microphone is that area within its pickup pattern in which a particular sound source sounds best. In a quiet and well-damped environment, the hot spot may be larger. However, if the sound source is thin and would benefit from more bass response, the hot spot for a cardioid microphone might be within the range of its "proximity effect," usually two to four inches from the microphone.

If the sound source, an announcer for example, moves around too much, a hypercardioid pattern may provide too narrow a hot spot, resulting in noticeably uneven levels and sense of presence. If the announcer cannot be trained to stay within the hot spot, a broader cardioid pattern may be a better choice.

MICROPHONE TYPES

Microphones can be typed according to their electric-generating element (transducer):

- Ribbon
- Dynamic
- Condenser

Ribbon Microphone

The *ribbon*, or *velocity*, microphone utilizes a very thin, corrugated metallic foil ribbon suspended within the flux field of a strong permanent magnet. While the ends of the ribbon are held in place, the rest of it is allowed to move freely back and forth in a sympathetically induced mechanical recreation of the amplitude and frequency of the sound presented to it.

As the metallic ribbon is moved across the magnetic flux lines, it induces a small AC current through the ribbon. Wires from the ends of the ribbon connect it to a step-up transformer, which converts the low impedance of the ribbon (approximately 1 ohm) to a value between 50 and 500 ohms more suitable for matching the input to the preamplifier. Ribbon microphones are available in a variety of fixed or variable patterns and are known for delivering a very warm sound, due to the ribbon's sensitivity to low frequencies. Blowing into a ribbon microphone or using it in a windy environment can destroy it. Even rapid panning on a studio boom has caused ribbon failure. Newer designs, however, have provided considerable improvements in durability and a lower failure rate.

Some ribbon microphones should not be connected to a microphone preamp with active phantom or A/B powering, because the voltage from these supplies can damage the ribbon.

Dynamic Moving Coil Microphones

Although the ribbon microphone is a type of dynamic microphone, in common usage, the term *dynamic* microphone usually refers to a microphone with a moving coil. The dynamic microphone has a diaphragm attached to a voice coil, as illustrated in Figure 4.3-1. This lightweight coil of wire is suspended in a magnetic field supplied by a permanent magnet structure. The ends of the voice coil are brought out to stronger leads, which connect to either a transformer or the microphone's output connector.

Sound waves reaching the diaphragm cause it to move back and forth. The attached voice coil cuts the lines of flux in the magnetic field, causing a small AC current. This signal closely emulates the sound waves in frequency and amplitude. The diaphragm must be highly compliant to allow effortless excursion at all frequencies of interest. In addition, this movement must be accomplished with maximum linearity and a minimum of break-up modes. Break-up modes occur when a portion of the diaphragm resonates independently of the rest of the surface. Phase cancellation results and, with it, response anomalies occur.

The design and construction of a high-quality dynamic microphone suitable for broadcasting use blend science and art. As is true in other areas of engineering, design trade-offs are numerous and the laws of physics tend to win in the end.

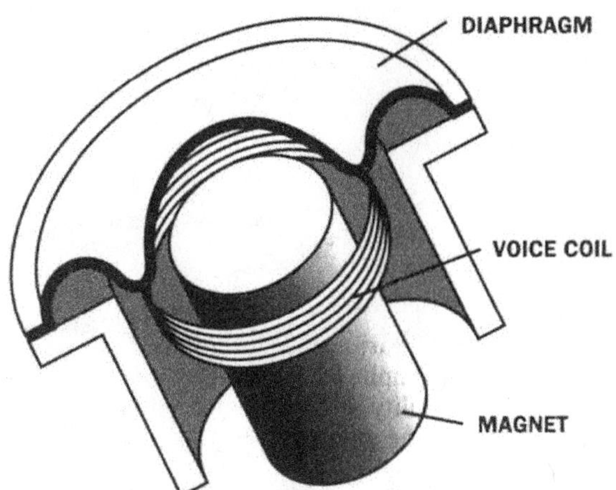

FIGURE 4.3-1 Dynamic moving coil element

Courtesy of Shure, Inc.

Size plays an important role in the performance of the dynamic microphone. Small dynamic mechanisms tend to have low acoustic sensitivity and high mechanical sensitivity. The result may be a poor system signal-to-noise ratio (SNR) and excessive handling noise or noise transmitted through the microphone stand. Internal shock-mount systems may be used to reduce the mechanical excitation, but the design goal of small size may then be defeated. Small size usually means sacrificed low-frequency response in dynamic microphones. This is not to say that large dynamic microphones will always have an extended low-frequency response.

Another physical characteristic of the dynamic microphone that affects its performance is the mass of the diaphragm/voice coil assembly. The greater the mass, the more limited will be the high-frequency response. Common design practice includes the use of Helmholtz resonators[1] immediately in front of the diaphragm, as illustrated in Figure 4.3-2, that creates peaks and effectively extends high-frequency response beyond the normal limits of the system. Advances in metallurgy, specifically the use of stronger neodymium magnets, have resulted in dynamic microphones with higher outputs.

Most broadcast-quality dynamic low impedance (Z) microphones exhibit an impedance that is a function of the number of turns and gauge of the voice coil wire. Some older, more public-address-oriented microphones employ a transformer within the housing to correct for design trade-offs in the voice coil. The transformer adds to the microphone's cost and may also restrict performance, if it is not a high-quality unit, by limiting the frequency response and possibly increasing distortion. Properly designed dynamic microphones can be the most rugged of the high-quality transducer types. Some have truly become legendary for their capability to provide high-quality broadcast audio with virtual bullet-proof construction.

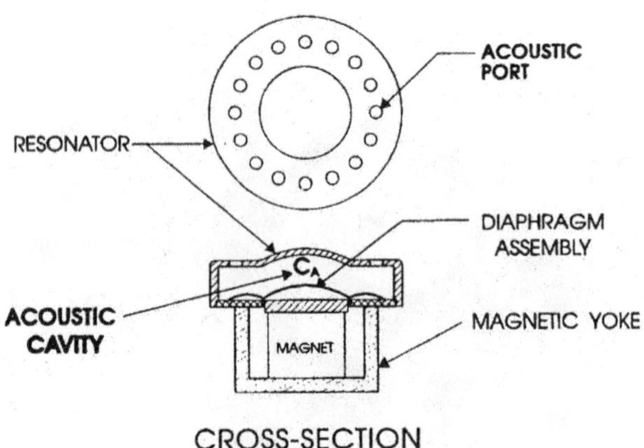

FIGURE 4.3-2 Cutaway drawing of a dynamic microphone element showing Helmholtz resonators

Courtesy of Electro-Voice, Inc.

Condenser Microphone

In the *condenser* microphone, a capacitor forms the generating element, as illustrated in Figure 4.3-3. One side of the capacitor is the diaphragm; the other is the fixed backplate. Air between these two plates acts as a dielectric. The capacitor, of course, must possess a positive electrical charge on one plate and a negative charge on the other. The conventional or discrete condenser receives this polarizing or bias voltage from an external DC power supply. *Phantom* power is used in most conventional condenser systems to deliver the required DC voltage to the microphone over the same cable used to carry the audio signal.

Upon activation of the power supply, a voltage is quickly built up on the surface of the diaphragm or backplate of the capsule. This causes an electrical current to flow through the resistor until the surface of the backplate finally receives an opposite charge of equal value. As sound waves (air pressure changes) strike the diaphragm, causing it to move back and forth, the distance between the two plates rapidly increases and decreases. This causes proportional changes in the capacity of the condenser. The result is an AC current flow in the resistor and a voltage across the resistor that corresponds to the excursion of the diaphragm.

While this voltage effectively represents the output voltage of the microphone, the source impedance is too high to be carried for any distance over microphone cable. This output signal, then, is presented to an impedance converter circuit, usually a vacuum tube or FET inside the microphone. Power for the impedance converter is derived from the same source that provides the polarizing voltage for the element. The impedance converter delivers a low impedance output that can be fed through long microphone cables with minimal loss.

Electret Condenser

The *electret* condenser microphone utilizes a material that has the capability to hold a charge applied during the manufacturing process. Most high-quality electrets apply this material to the fixed backplate of the capacitor, as shown in Figure 4.3-4. Some designs employ a charged diaphragm instead. Lowering the

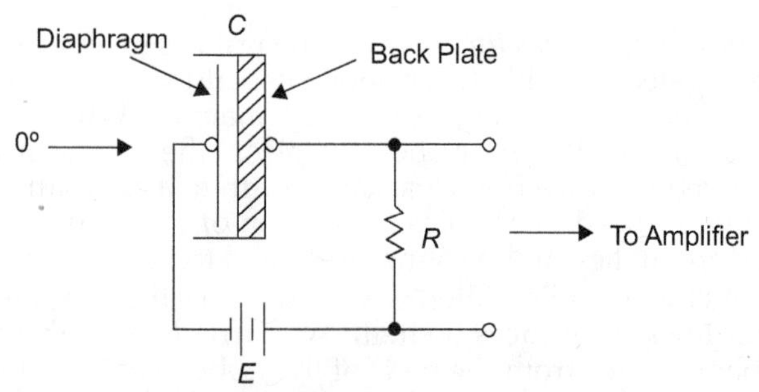

FIGURE 4.3-3 Conventional capacitor microphone system

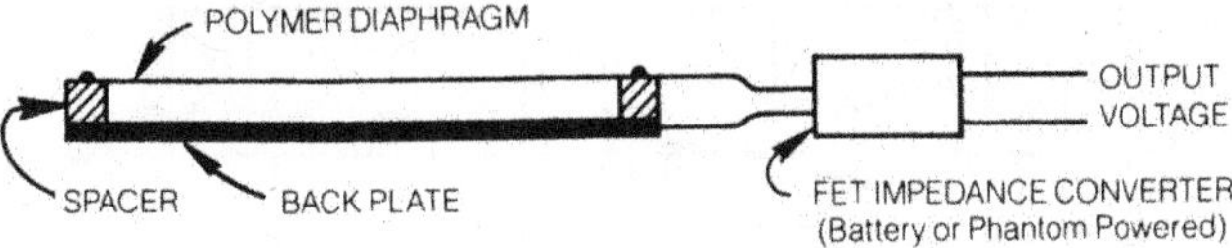

FIGURE 4.3-4 Electret condenser element

Courtesy of Audio-Technica U.S., Inc.

weight of the diaphragm, by moving the electret material to the backplate, results in lower handling noise, extended frequency response, and improved transient response.

Although the electret functions much like the discrete condenser, but produces its output voltage without the need for an external high-voltage DC supply, an impedance converter is still required. The low voltage needed to power it may be derived from internal or external batteries or an external AC-powered supply.

PHANTOM POWER

Phantom power, or *simplex power*, provides one means for remotely powering condenser microphones and may range from 9 to 52 VDC from an external supply, usually the audio console or mixer. While many electrets will operate over a wide range of voltage, most modern discrete condenser designs require 48 volts. The phantom supply voltage for nonelectret condenser microphones is often stepped up by an internal circuit to provide a sufficient capacitor-polarizing charge for good signal-to-noise figures.

The amount of current delivered by the power supply is also a factor. While some condenser microphones can operate with less than 1 mA, others require up to 2.5 mA or more. If insufficient current is provided, the level of the microphone is reduced and the signal is distorted.

In a phantom power circuit, the positive side of the DC supply is applied equally to both of the signal-conducting leads of a balanced microphone line and the negative to the shield. This is accomplished by means of either a build-out of matched precision resistors or a center-tapped transformer. In each case, the return path is the shield. In the microphone, the DC may be similarly tapped via the resistor or center-tapped transformer method to provide the power it needs. The DC is prevented from appearing at the impedance converter output by DC-blocking capacitors or the internal center-tapped transformer.

If a balanced-output dynamic microphone is connected to a line with phantom power present, performance should not be altered, nor should damage occur to the dynamic element. The voice coil or output transformer winding connects across the two signal leads and should see no potential difference between them. Because there is no connection between either lead and the shield (the DC return path), there is no current flow. If an unbalanced dynamic microphone is connected to a phantom supply, the DC current will pass through the voice coil and may destroy it. A less common powering system called *A/B* or *T* powering is not compatible with dynamic or phantom-powered microphones. A/B power puts the positive side of the DC on one signal lead and the negative on the other. This will damage even a balanced output dynamic microphone.

MICROPHONE CABLES

Regardless of the style or pattern of microphones chosen, the best results require the use of a high-quality, well-shielded microphone cable. Quality cable is more important here than anywhere in the audio chain because of the high amount of gain applied by the microphone preamplifier. Poorly shielded or low-quality cable may not reject electromagnetic (from nearby AC power lines) or radio frequency (from nearby transmitters) induced noise. As a result, that low-level noise is amplified by the microphone preamplifier and can become audible.

MICROPHONE SPECIFICATIONS

Some microphones come with individual graphs for frequency response and polar pattern. While specifications tell part of the story, there is nothing better than trying a microphone in a specific application. Note that the same microphone may sound different when amplified through different preamps.

Frequency Response

One of the first specifications considered on a microphone data sheet is the frequency response range. Often, more attention is given to the response limits than to how the microphone actually sounds in its intended application. Nonlinearities of the response often contribute more to the listener's subjective impression of sound transmitted by the microphone than the response range. A specification that reads "frequency response 40 to 18,000 Hz," by itself, says little about a microphone's actual sound in use. Adding to that some amplitude limits, such as ± 3.0 dB, improves the specification but, depending on other characteristics of the microphone, the result may still be unsatisfactory.

The shape of the frequency response and polar pattern characteristics contribute to the character or personality of the microphone's sound. Response nonlinearities can create acoustic feedback in sound reinforcement, a nasality sound, poor intelligibility, excessive sibilance, muffled sound, or any of a variety of other acoustic problems. On the other hand, a microphone's response may be deliberately tailored by the design engineer to solve problems rather than create them. A rolled-off low-frequency response and a rising high-frequency response may be employed in a microphone that is intended to reduce the effects of unwanted low-frequency information, such as traffic noise or the rumble of air-handling systems.

Some microphones that are intended to be worn on the body exhibit a response that compensates for the chest cavity resonance, which they tend to pick up in a lapel or clip-on mounting position. A rolled-off low-end response may help considerably in attenuating handling or stand-borne noise, as well as wind noise or the breath blasts of plosives in speech.

Published specifications can serve as only one guide in understanding a microphone. Even curves from well-respected manufacturers may be difficult to compare due to the variety of test procedures and standards. For example, the frequency response chart is an X-Y graph that compares output to frequency. If that chart is compressed vertically or stretched horizontally, the response curve may appear to be more linear. Such variables as the recording speed, damping, or even the direction of the tone sweep (low to high or high to low) may result in substantially different curves.

Directional microphones present a phenomenon known as the *proximity effect*, which results in a bass-boosted output when used close to a small sound source. Proximity effect is neither good nor bad; its value depends on the intended application. Various designs will differ in the amount of proximity effect that is possible to attain. The response curve of a particular directional microphone, tested at a specific (or perhaps even unknown) distance, may be of little value to someone who wishes to use the microphone at another distance. Ideally, directional microphone data should include close and distant curves, as shown in Figure 4.3-5.

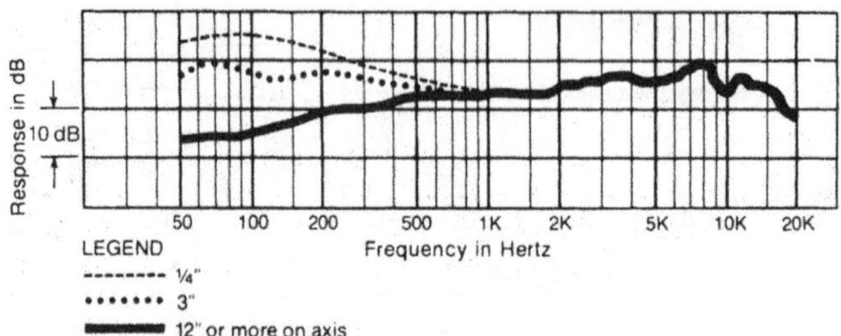

FIGURE 4.3-5 Influence of proximity effect on a directional microphone response

Courtesy of Audio-Technica U.S., Inc.

Polar Pattern

As difficult as it might seem at times to pick up a desired audio signal, the real problem often lies in eliminating unwanted sounds. Microphones with various directional patterns may be often used to improve the ratio of desired signal-to-ambient noise or other unwanted sounds. Ambient noise, leakage from other instruments in a band or orchestra, room reverberation, and feedback potential from public address systems or monitor speakers are some of the reasons why it is important to know the off-axis response of a microphone. The best view of the microphone's off-axis response is obtained by examining several different polar plots. These should be drawn at low, mid-band, and high frequencies. Overlaid, these plots should reveal how well the microphone maintains its directionality at each frequency, as illustrated in Figure 4.3-6.

There are several broad categories of polar patterns to which most microphones' directional characteristics conform to some extent or another:

- Omnidirectional
- Bidirectional
- Cardioid
- Supercardioid

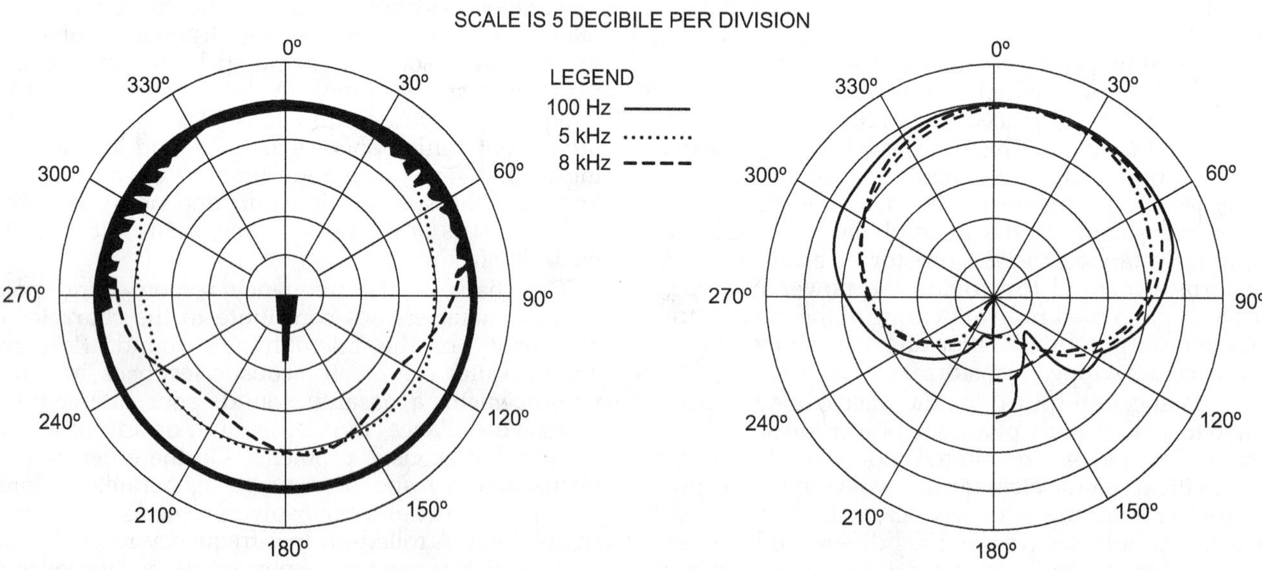

FIGURE 4.3-6 Polar patterns drawn at several frequencies

Courtesy of Audio-Technica U.S., Inc.

Hypercardioid

Figure 4.3-7 shows the directional patterns obtained from mathematical models representing the perfect polar characteristic for each example. Actual microphone patterns may vary from near perfection to close resemblance. In practice, the microphone design engineer must go beyond math equations to accomplish a desired axial response and sensitivity while maintaining polar uniformity. The chart in Figure 4.3-7 shows how the various patterns should relate to the reference omni in their ability to reject unwanted energy arriving from various points off-axis.

Polar Scaling and Range

It is important when reading polar patterns to observe the several variables in the way that they are represented, as illustrated in Figure 4.3-8.

First, determine whether the scale is logarithmic or linear. A log scale (the most commonly used) will show a fairly modest inward curve of the cardioid pattern at 90°, indicating a 6-dB drop in level. The linear scale for the same microphone will show a polar pattern that appears much more directional. The outside circle of the linear polar represents 100% output, whereas the center of the circle equals zero output. Because a 6-dB loss is equal to a 50% drop in voltage, on a linear scale the polar curve at 90° sweeps to half the distance between the outside of the circle and its center.

Second, determine if the graduations between concentric circles are 5 dB or 10 dB apart. Finally, take note of the range of the polar pattern. This may be determined by counting graduation lines inward from the point where the polar crosses 0°, in 5- or 10-dB steps (as marked) to the smallest inner circle. Patterns may be found in most any range, with 25, 30, and 40 dB all being common. These differences will also alter the shape of a polar pattern.

Note that the polar pattern represents a cross-sectional, two-dimensional diagram of a three-dimensional function. The 131° pickup arc, for example, in the chart in Figure 4.3-7, that is described by the 3-dB down points on either side of the axis of the cardioid microphone, can really best be thought of as a conical area within which the microphone is virtually uniformly sensitive. This area is often referred to as the microphone's *angle of acceptance* or *included angle*.

Omnidirectional

The omnidirectional microphone consists of a diaphragm and generating element backed by a totally sealed case. When placed in a sound pressure field,

CHARACTERISTIC	OMNI-DIRECTIONAL	CARDIOID	SUPER-CARDIOID	HYPER-CARDIOID	BIDIRECTIONAL
Polar response pattern	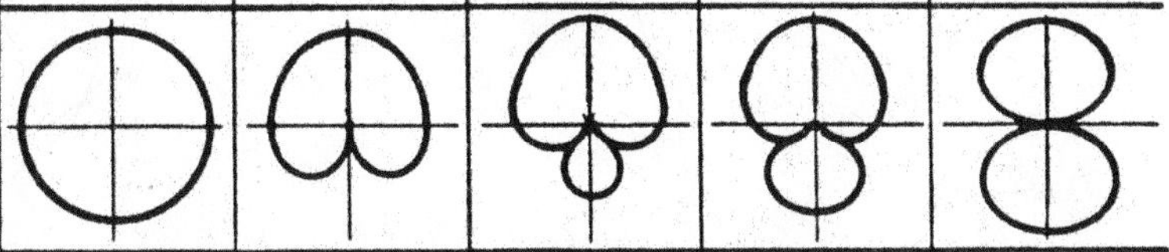				
Polar equation	1	.5 + .5 cos θ	.375 + .625 cos θ	.25 + .75 cos θ	cos θ
Pickup ARC 3 dB down (1)	—	131°	115°	105°	90°
Pickup ARC 6dB down	—	180°	156°	141°	120°
Relative output at 90° dB	0	−6	−8.6	−12	−∞
Relative output at 180° dB	0	−∞	−11.7	−6	0
Angle at which output = 0	—	180°	126°	110°	90°
Random energy efficiency (REE)	1 0dB	.333 −4.8dB	.268 −5.7dB (2)	.250 −6.0dB (3)	.333 −4.8dB
Distance factor (DF)	1	1.7	1.9	2	1.7

NOTE:
1 = Drawn shaded on polar pattern
2 = Maximum front-to-total random energy efficiency for a first order cardioid
3 = Minimum random energy efficiency for a first order cardioid

FIGURE 4.3-7 Microphone polar patterns and characteristics

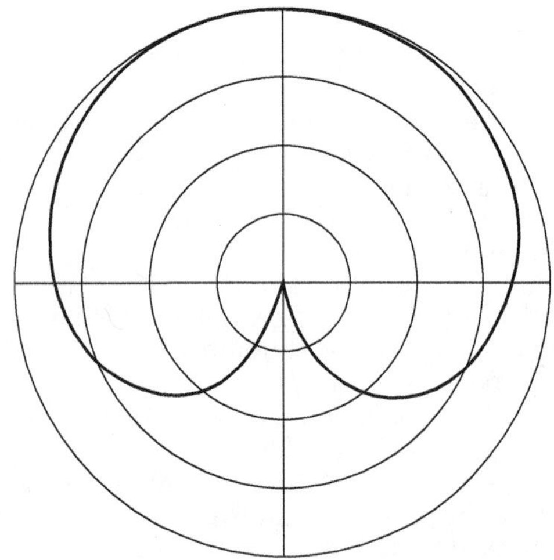

FIGURE 4.3-8A Cardioid log scale polar. Scale is 10 dB per division.

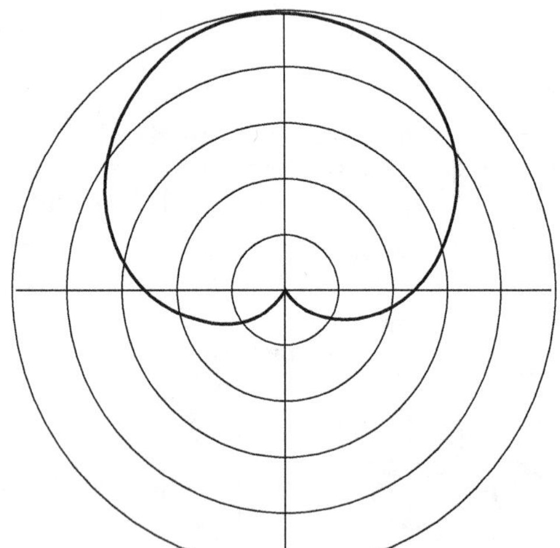

FIGURE 4.3-8B Cardioid linear scale polar

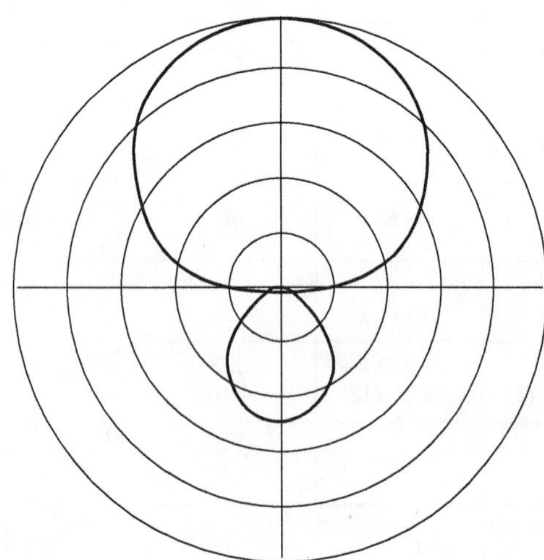

FIGURE 4.3-8C Hypercardioid linear scale polar. Dynamic range is 50 dB.

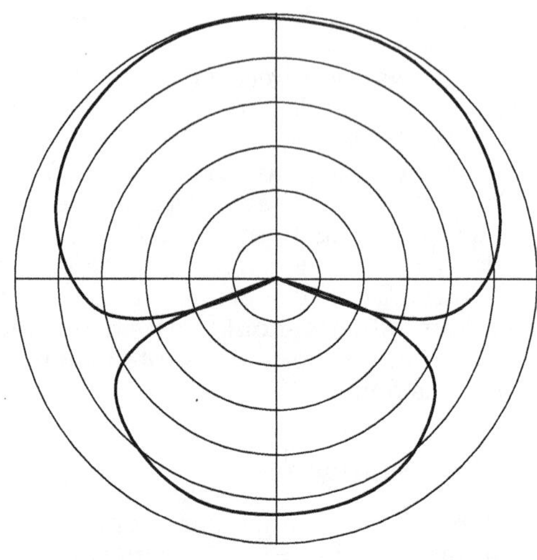

FIGURE 4.3-8D Hypercardioid log scale polar. Scale is 10 dB per division. Dynamic range is 40 dB.

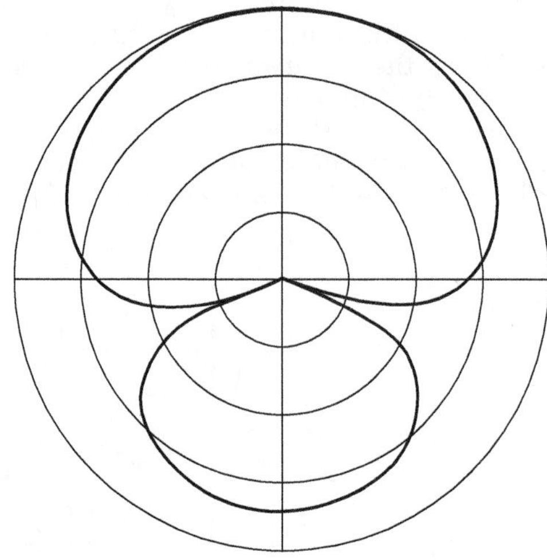

FIGURE 4.3-8E Hypercardioid log scale polar. Scale is 10 dB per division. Dynamic range is 60 dB.

the perfect omni disregards the direction of the sound's origin. A positive pressure (air expanding) at the diaphragm, for example, causes the diaphragm to move inward regardless of the sound's point of origin, as illustrated in Figure 4.3-9. Such a microphone may also be referred to as a *pressure microphone*.

Most omnidirectional microphones, however, are not truly omnidirectional. The case of the microphone represents a barrier to higher frequencies arriving from off-axis. Because of this case effect, most omni mics are increasingly directional at higher frequencies. The smaller the omni mic, the more truly omnidirectional it may be. In addition to the case effect, energy arriving at the diaphragm from on-axis is reinforced at those frequencies to which the size of this *baffle area* is significant. The baffle effect causes a rise in the microphone's high-frequency output, but only with respect to energy arriving on-axis.

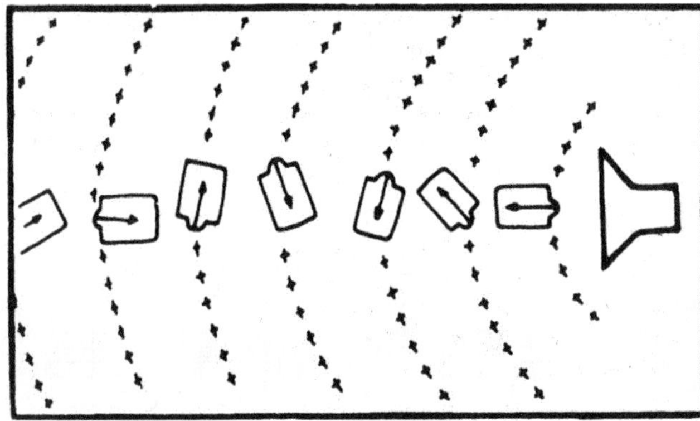

FIGURE 4.3-9 Omnidirectional microphone principle

Cardioid

Directional microphones employ a damped porting system in their element design that allows sound waves to act upon the rear of the diaphragm as well as the front. The design introduces varying amounts of phase shift for sound arriving from off-axis, resulting in cancellation. The rear entry ports of most directional microphones are spaced at a single distance or "D" from the diaphragm, illustrated in Figure 4.3-10. Multiple port systems are also available and are designed to reduce the proximity effect.

When using any directional microphone, avoid obstructing the ports with the microphone clip, hand, clothing, gaffer's tape, or logo flags. Covering any of the ports results in serious degradation of the microphone's directional characteristics and overall sound quality.

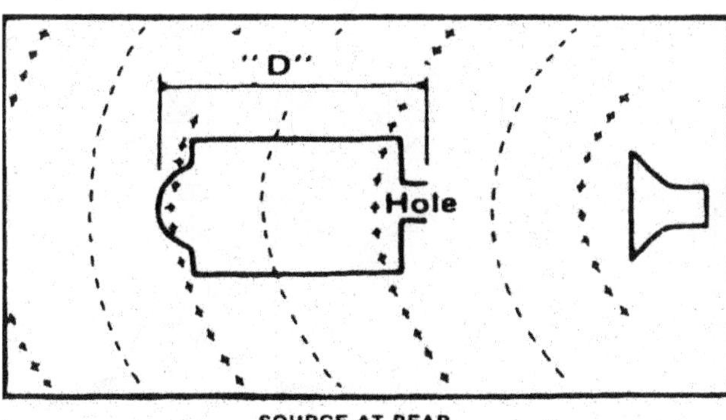

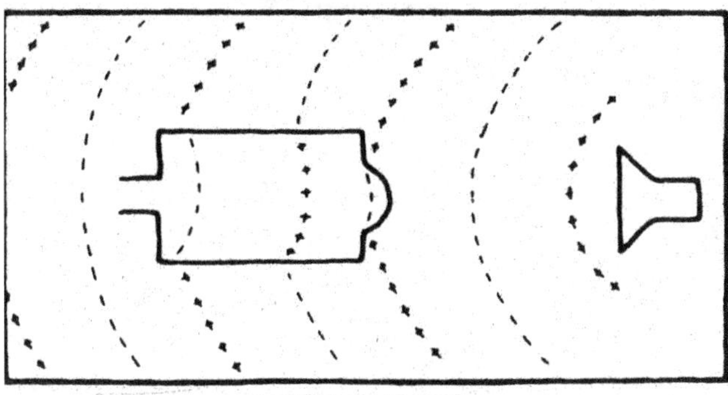

FIGURE 4.3-10 Single-D cardioid microphone operating principle

A sound source that delivers, for example, 60-dB sound pressure level (SPL) to a cardioid microphone on-axis from one foot away will drop 6 dB when the microphone is rotated to position the sound source at 90° off-axis. Here, a properly designed cardioid emulates well its mathematical model. At 180° off-axis, however, the cardioid cannot live up to the equation. The chart indicates zero output. In reality, well-designed cardioids are capable of something only on the order of a 20-dB differential. That is equivalent to moving the sound source to ten times its actual distance from the microphone.

The 180° response curves (back curve) of many cardioid microphones show their tendency to more closely resemble an omni mic at both the low and high frequencies. The much more impressive cancellation in the midrange sometimes causes a manufacturer to release a data sheet that shows only one polar pattern and for an unknown frequency. One of the most beneficial performance advantages to look for in a well-designed microphone is off-axis linearity, as illustrated in Figure 4.3-11. Note the comparatively smooth back response of the higher quality condenser microphone in (b), compared to that of the dynamic microphone in (a).

Omni Versus Cardioid

While the omni picks up sound from all angles, the cardioid reduces the pickup of ambient noise and reverberant energy from behind the microphone. The comparison chart (Figure 4.3-7) shows that the *random energy efficiency* (REE) of the cardioid is 0.333 compared to an REE of 1 for the omni.

The random energy efficiency is a measurement that compares a microphone's sensitivity to random (or reverberant) energy to its on-axis sensitivity. While this shows the cardioid to be one-third as sensitive to random ambient noise as the omni, note that discrete sound sources positioned at the null of the cardioid will be attenuated to a much greater extent. This will prove true in outdoor performances where sounds arrive at the microphone directly with minimal reflections. Indoors, the advantage of the cardioid's deep rear null is appreciated only when the microphone is situated within *critical distance* of the offending sound source. Within the critical distance, the direct sound is greater in intensity than the reflected energy. After that point, the two remain approximately equal. Other characteristics of the cardioid mic include

- It is more susceptible to the problems of "pop" (the blast of plosives from words that begin with "B," "P," and "T" in speech), wind noise, and handling or mechanical noise.
- It is more complex to design and construct and generally costs more than an omni of apparent equal audio quality.
- More complex construction makes it generally less rugged than an omni mic.
- It has greater resistance to feedback in most sound reinforcement applications due to its lower REE and is further aided by the proximity effect.

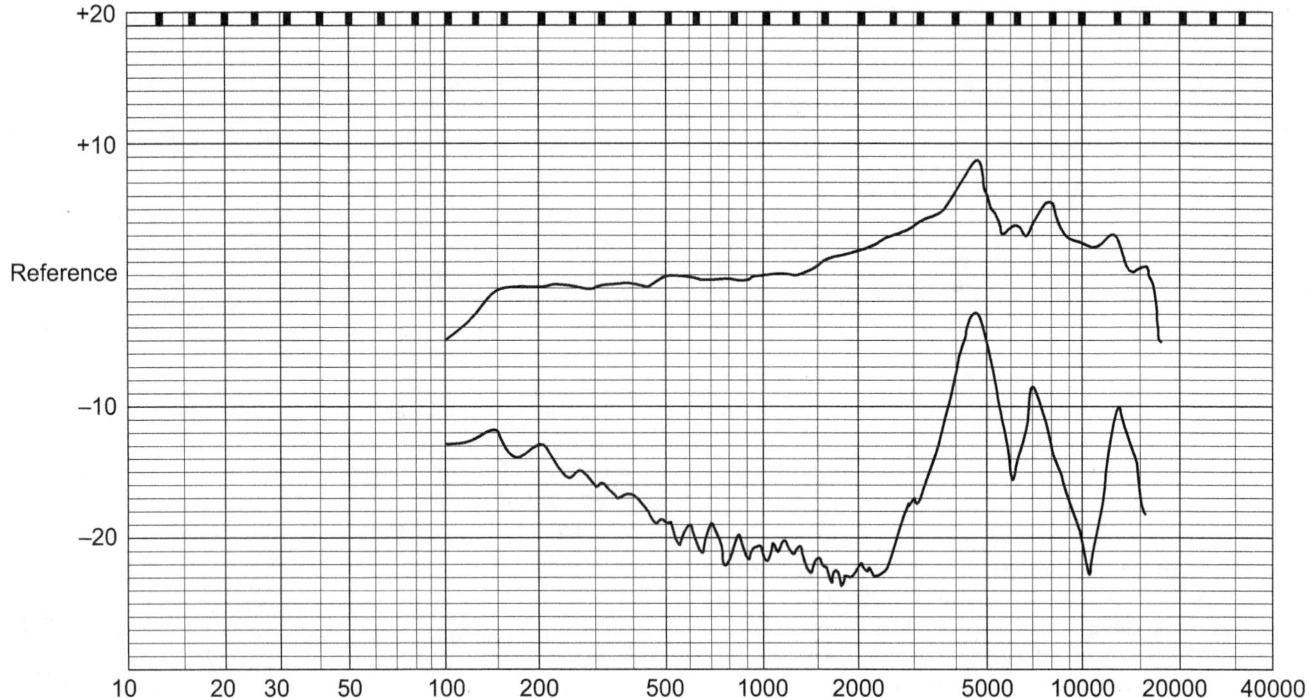

FIGURE 4.3-11A Front and back curves of a typical cardioid dynamic handheld vocal microphone. Vertical scale is in dB; horizontal scale is in Hz.

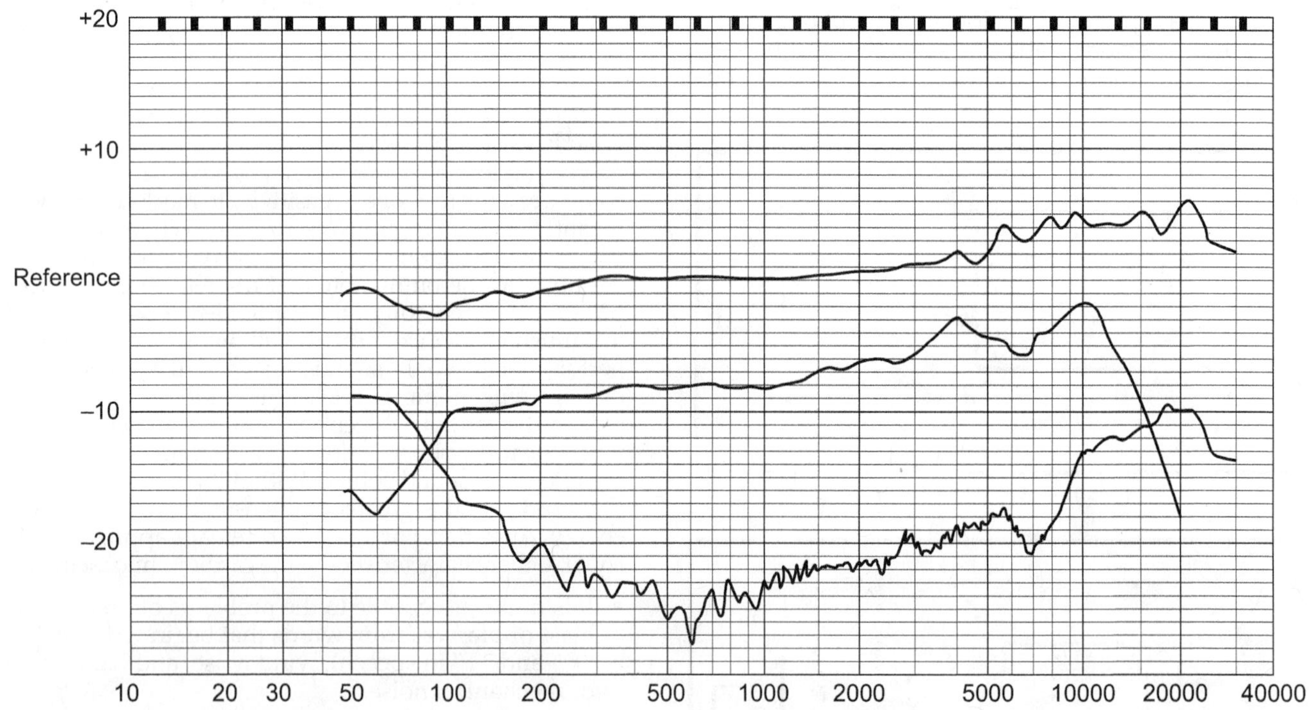

FIGURE 4.3-11B Front, side, and back curves of a high-quality cardioid condenser microphone. Vertical scale is in dB; horizontal scale is in Hz.

- It increases the effective working distance. From Figure 4.3-7, note the distance factor (DF) of 1.7 for the cardioid mic, meaning that it has a working distance advantage over the omni of 1.7:1. This factor is calculated on the assumption of a perfect cardioid, in a totally diffuse noise field. An ideal cardioid, then, could be used at a distance of 1.7 times that of the omni for a given ratio of desired, on-axis signal to ambient noise.
- The nonlinear polar response and the inability of cardioid microphones to achieve total cancellation

at their null would seem to reduce the cardioid's working distance advantage. However, increasing the working distance often has more to do with attenuating a single, offending noise source than with overcoming a diffuse noise field. Directing the deep null of a good cardioid microphone at an offending noise source in the field or studio may offer more than a 1.7:1 working distance advantage over an omni.

- The cardioid exhibits proximity effect. While some designs are quite low in proximity effect, all exhibit some bass-boost effect when used close to the source. Although this may be considered an enhancement in many close-mic applications, it is important to avoid preamp input overload or loss of intelligibility that may result from excessive proximity effect.

Other Patterns

Again referring to Figure 4.3-7, note how the three other polar patterns compare to the omni and the cardioid. The hypercardioid, for example, combines a tight acceptance angle with superior side rejection and offers the lowest REE. The bidirectional pattern offers the best side rejection, but with no advantage over the cardioid in REE. Bidirectional (also called *figure of eight*) microphones are typically ribbons or dual-diaphragm condensers. They are useful in eliminating unwanted sounds from the side and for picking up two sound sources (such as two people opposite each other) while incurring no phase problems.

Output Impedance

The impedance (Z) of a microphone is a measurement of its AC resistance looking back into the transducer. Broadcast microphones should be low impedance, ranging typically from 50 to 600 ohms. Dynamic moving-coil microphones achieve their low impedance by either a low-Z voice coil winding or a transformer. Condenser microphones use an impedance converter circuit to step down the capacitor's high-Z output.

Low impedance offers the advantages of low susceptibility to hum and electrical noise pickup and the capability to use relatively long cables with a minimal loss of level or high frequencies. Unlike matching power amplifier impedances to speaker systems, which may be desired for best power transfer, microphones need load impedances on the order of ten times their internal impedance. This assures maximum voltage transfer. A microphone that looks like a resistive source of 150 ohms, looking into a load resistor of 150 ohms, for example, will suffer a 6-dB voltage drop compared to an open-circuit connection.

Dynamic Range

The difference between a microphone's own self-noise and the maximum sound pressure level it can handle before distortion is the *dynamic range*. In many field applications, ambient noise provides sufficient masking to make the self-noise specification of minor interest. The importance of this specification increases as greater working distances are demanded or ambient noise levels are lowered.

The impedance converter of condenser microphones, like any active circuit, will create some noise, which will vary from one design to another. The impedance converter design also determines the headroom or maximum SPL that the microphone can handle. A maximum SPL of as high as 141 dB is achieved in several high-quality condensers.

Dynamic microphones contribute virtually no self-noise. When they are greatly amplified, only the noise of the thermal agitation of air molecules is detected. While this is very low in level, the dynamic microphone does not automatically rank as the first choice in a low-noise system. Because the output level of the dynamic is often lower than that of a condenser system, the user may end up working into the noise floor at the upper extremes of the preamplifier in order to provide sufficient system gain.

Some new dynamic microphones employ powerful rare-earth magnets to increase the efficiency of their motor mechanisms. Their higher output, while still not as high as many condenser microphones, can provide a considerable S/N advantage over earlier designs.

WORKING DISTANCE

Sometimes it is not sufficient to reduce ambient noise merely by using a polar pattern that offers the lowest REE. In very noisy environments (for example, in an aircraft, in a factory, or at a sporting event), it may be desirable to differentiate between close sound (an announcer, for example) and distant sound. Microphones that offer considerable proximity effect may be used to advantage in these situations by having the announcer work the microphone very close and roll off the low end as needed to flatten the response.

Noise-Canceling Microphone

In extreme situations, a noise-canceling (differential) microphone may be required. Because of the special design (rear ports and back damping systems) of the differential microphone, sound arriving from a distance strikes both sides of the diaphragm with equal intensity and in phase. A positive pressure on the front, for example, would encounter a positive pressure on the rear of the diaphragm, causing the signal to be canceled. A combination of inverse square law and port damping causes sound that originates very close to the front of the noise-canceling microphone to be lower in intensity and to exhibit some phase error by the time it arrives at the rear of the diaphragm. The noise-canceling microphone is able, therefore, to differentiate between close and distant sound sources. The audio quality of such systems normally limits them to voice communication applications.

Inverse Square Law

The easiest, and certainly the least expensive, way to limit the apparent working distance of a microphone is by positioning the microphone very close to the sound source. Inverse square law shows that decreasing the distance between the microphone and the sound source by one-half (for example, from 8 feet to 4 feet) increases sound intensity at the microphone by a factor of four, or 6 dB, as shown in Figure 4.3-12. As the input sensitivity control of the mixer or recorder is lowered to compensate for the additional 6 dB available from the now-closer sound source, the microphone, in effect, becomes less sensitive to distant sounds.

Working at a Distance

In applications in which the sound source is at a significant distance, the effective maximum working distance may be determined by the electronic signal-to-noise ratio of both the microphone and subsequent amplifiers. For example, the selection of an ideal boom microphone for picking up dialogue in a quiet environment, with no reverberation problems, may have little to do with polar patterns. Instead, the desirable microphone would have a high output and low self-noise. Most often, electronic signal-to-noise and signal-to-ambient-noise ratios are the main concern.

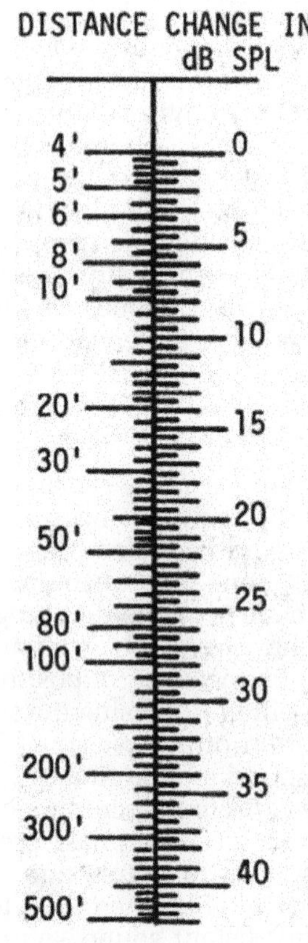

FIGURE 4.3-12 Inverse square law

HEADSET MICROPHONES

Headset microphones provide benefits gained from always being a fixed distance from the speaker's mouth. Background noise is reduced (because of the inverse square law) and levels remain consistent. Omni, cardioid, and differential elements are available in headset systems. Cardioids offer the best combination of ambient noise suppression and acceptable broadcast quality.

SHOTGUN MICROPHONES

Effective working distances beyond those afforded by cardioid, supercardioid, or hypercardioid systems may be realized through the use of a shotgun microphone, which uses a long, slotted interference tube ahead of the element to provide a high degree of cancellation at the sides. Sound waves arriving on-axis are essentially unaffected by the tube. Sound arriving from slightly off-axis, however, is forced to turn and travel down the tube to the element. This results in numerous out-of-phase conditions being set up in the tube, with cancellation increasing as the microphone is rotated to 90°.

Newer shotgun microphone designs from several manufacturers provide superior pattern control using shorter interference tubes than those required by older standards. The new generation of shorter, lighter products are much easier to handle in boom applications.

Some shotgun microphones are much less uniform in off-axis response than are conventional hypercardioids, as shown in Figure 4.3-13. Even with their multilobed polar patterns, however, their increased reach over hypercardioids' reach may make them a logical choice in the right acoustical environment.

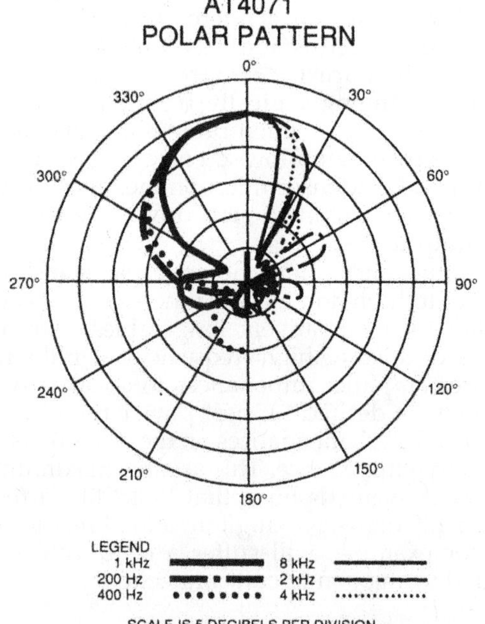

FIGURE 4.3-13 Shotgun polar patterns

Courtesy of Audio-Technica U.S., Inc.

Shotguns work best outdoors and in controlled acoustic environments, such as well-designed studios. Distant miking down a hallway will not be assisted greatly by the use of a shotgun microphone. A shotgun microphone pointed upward toward an actor wearing a wide-brimmed hat may work better than one boomed from above, unless there is also an HVAC duct in the path above the actor.

ACOUSTIC GAIN DEVICES

While shotgun microphones increase working distance by rejecting off-axis sound energy, thereby narrowing the acceptance angle, some devices increase working distance by providing *acoustical gain.* The most commonly used acoustic gain device is the *parabola* or *parabolic reflector,* as illustrated in Figure 4.3-14. The parabolic reflector is shaped so that sound is reflected onto a focal point a short distance in front of the center point of the dish. An omnidirectional microphone placed at this point receives multiple reflected sound waves in phase, which add to produce significant gain. The response of such systems is ragged and limited. Low-frequency response is extended as the dish diameter is increased. While unacceptable for most broadcast applications, the audio quality achieved with the parabolic microphone is often deemed adequate for sound effects pickup such as at sporting events.

A second type of acoustic gain device is the *horn.* Low cost re-entrant horns are often used for talk-back in paging systems and are quite directional and sensitive. Installed on the side of a building, the small horn is virtually as inconspicuous as a light fixture and is seldom thought of as a microphone. Some horns are built with 45-ohm voice coils, providing higher output signals to microphone inputs.

Acoustic gain is also realized by using a microphone in the very close vicinity of a large, hard, reflective surface. Omnidirectional microphones may be flush mounted into the barrier, facing out. In this position, the microphone is in a half-space environment, or looking into only half the world. The output for sound arriving on-axis is increased by 6 dB. As the sound source is rotated off-axis, however, the microphone output drops. At 90° off-axis, the output is down 6 dB, or equal to the omni in free space. The resulting polar pattern resembles a cross-section of a cardioid cut through the microphone at 90°.

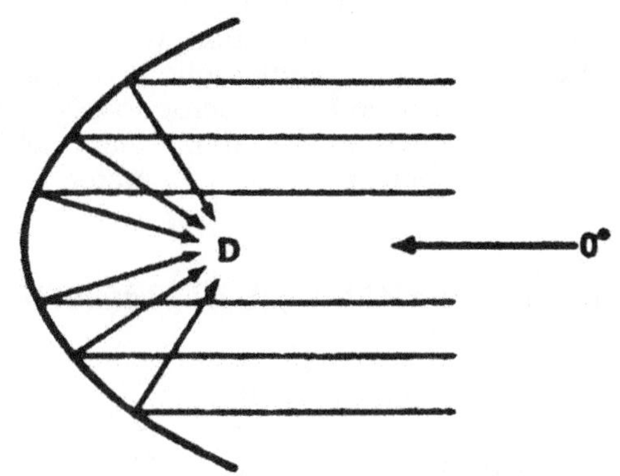

FIGURE 4.3-14 Parabolic microphone principle

Frequency Response and Distant Miking

Distant miking may result in noticeable or even dramatic changes in the spectrum of the sound being recorded. High frequencies, attenuated by the air, may require boosting to restore a normal sound. Similarly, high-pass (low-cut) filters may prove helpful in reducing low-frequency room reverberation or background noise, thereby extending the useful working distance.

SENSITIVITY RATINGS

Microphone sensitivity is rated in several ways, including open-circuit output voltage or power level into a given impedance with a specific sound pressure level. Sensitivity is specified according to a test procedure in which a 1-kHz tone at 94-dB SPL applied at the input of the microphone element produces an open-circuit voltage that is measured or the power into a specified impedance is measured. A 94-dB SPL is also described as the pressure of 1.0 Pa (pascal) or 10 dynes/cm^2, and the output voltage is measured in terms of mV/Pa (millivolts per pascal) according to standards IEC 60268-4 and EIA SE-105.

Open-Circuit Output Voltage

Microphones are often specified as having a particular output voltage when looking into an open circuit. In most modern equipment, microphone preamplifier inputs are at least ten times the measured impedance of the microphone and may be regarded as an open circuit. Specifications may be given as an actual output voltage or as decibels below one volt at a sound pressure level of 74 dB (1 dyne/cm^2 or 0.1 Pa) or 20 dB below the standard procedure. These ratings are referred to as *the open-circuit output voltage* rating or *open-circuit sensitivity.*

The open-circuit sensitivity may be expressed in dB by means of the following formula:

$$V_{oc} = 20\log E_0 - SPL + 74$$

where

V_{OC} = Open-circuit voltage in dB (referenced to 1 V/0.1 Pa)

E_0 = Microphone output in volts

SPL = Actual SPL at the microphone

The nomograph shown in Figure 4.3-15 provides an easy method of calculating the open-circuit voltage.

Power Level

Microphone sensitivity may also be specified in terms of its output power level. This equivalent power-level rating takes into consideration the open-circuit rating and

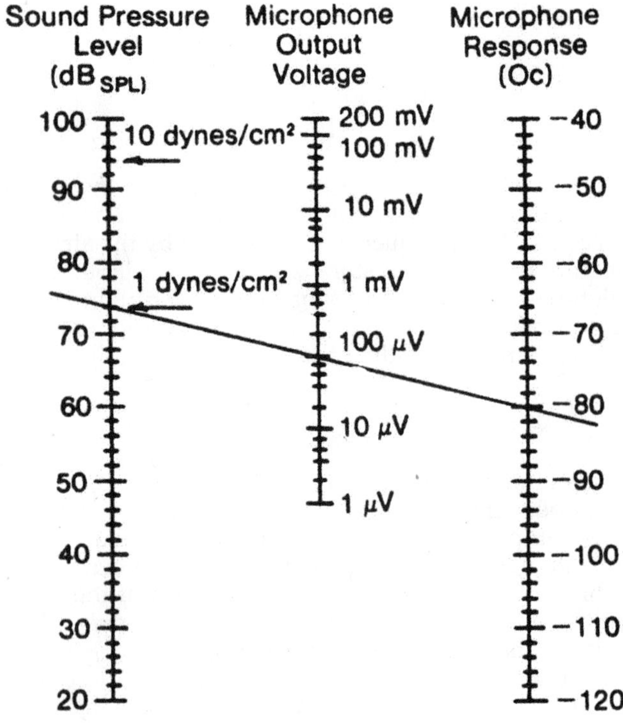

FIGURE 4.3-15 Nomograph: open-circuit voltage rating

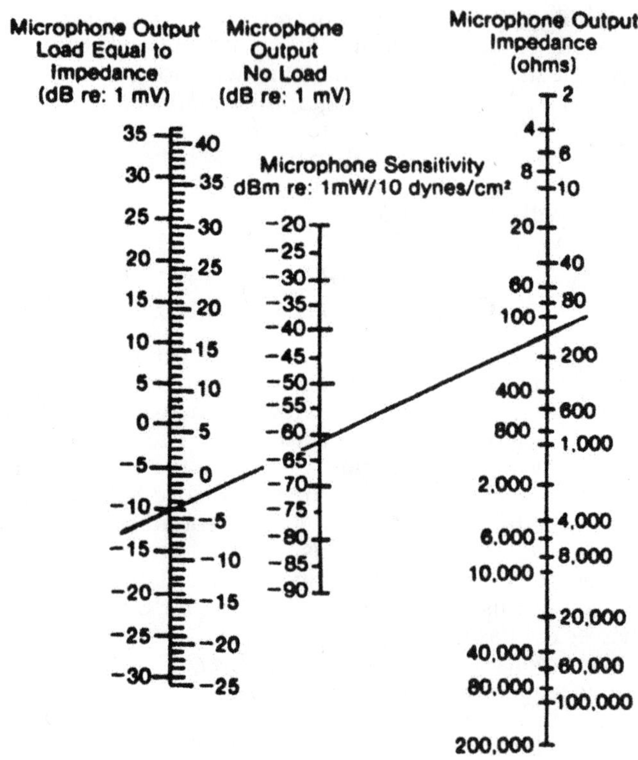

FIGURE 4.3-16 Nomograph: equivalent power rating

either the actual impedance of the microphone or the rated impedance. Specifications given would be in dBm (or just dB) referenced to 0 dB = 1 mW / 10 dynes / cm² or 0 dB = 1 mW / Pa.

The following formula is used for calculating the power-level rating:

$$P_E = V_{\propto} - 10\log_{10} Z + 44 \text{ (dB)}$$

where

P_E = Equivalent power level

Z = Impedance of the microphone

A nomograph for determining the equivalent power rating is shown in Figure 4.3-16.

EIA Sensitivity Rating

The EIA sensitivity rating is sometimes specified but rarely used. The formula for determining EIA sensitivity is

$$ESR = V_{\propto} - 10\log_{10} R_{MR} - 50 \text{ (dB)}$$

where

ESR = EIA sensitivity rating

V_{OC} = Open-circuit voltage in dB

R_{MR} = Center value of the nominal impedance range

Table 4.3-1 may be used for determining R_{MR}. An EIA sensitivity rating nomograph is given in Figure 4.3-17.

TABLE 4.3-1
Center Value for Nominal Impedance Ranges

Range (ohms)	Center Value (ohms)
20–80	38
80–300	150
300–1250	600
1250–4500	2400
4500–20,000	9600
20,000–70,000	40,000

LINE-LEVEL MICROPHONES

For situations involving long mic cables, using a portable preamp/limiter combination will reduce loss and hum pickup. Line-level microphones designed for remote use may incorporate a microphone, preamplifier, limiter, and power supply in one handheld, or a separate unit with belt clip, battery-operated package.

HIGH ACOUSTIC LEVEL APPLICATIONS

Properly designed dynamic microphones are difficult to drive to audible distortion. The distortion heard when a dynamic microphone is subjected to the lips-touching proximity of a very loud rock and roll vocalist is usually caused by the peak clipping in the

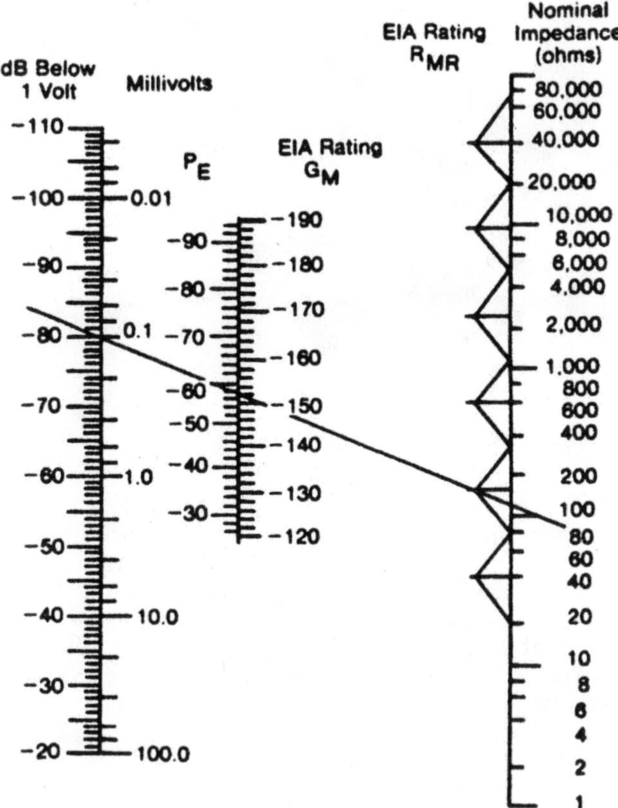

FIGURE 4.3-17 Nomograph: EIA sensitivity rating

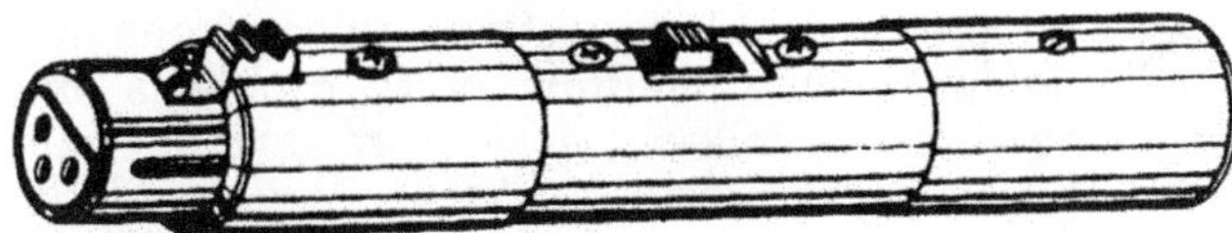

FIGURE 4.3-18 Microphone in-line attenuator

Courtesy of Audio-Technica U.S., Inc.

electronics following the microphone. Outputs of 1 volt or more may actually be delivered in such applications as rock and roll music. Preamp or amplifier clipping may be avoided by attenuating the microphone output or adjusting the trim (gain adjustment) of the specific input on the mixer. Note that many mixers offer adjustment only after a gain stage or transformer, either of which may distort before any control is possible. In-line attenuators, or pads, are commercially available that allow the selection of 10, 20, or 30 dB of attenuation and plug directly into the microphone line at the input of the mixer. Before using any in-line device, verify that it is compatible with the powering system being used for condenser microphones. A typical in-line attenuator is shown in Figure 4.3-18.

AVOIDING NOISE PROBLEMS

Unwanted signals include wind noise, "P-pop," mechanical or handling noise, AC hum, and radio frequency interference (RFI). The reduction or elimination of each of these can be handled both through microphone design and user technique.

RFI problems can usually be traced to a point in the low-level circuitry at which the signal leads are unbalanced, high-Z, or both. Condenser microphones, for example, may sometimes be sensitive to RFI at or around their impedance converter. If RFI is a problem, the manufacturer should be consulted for low-pass modifications or information.

P-pops due to the plosives from words beginning with "B," "T," and "P" in speech may be reduced in several ways:

- Use an omni microphone. Directional microphones are much more prone to P-pop problems than are omnidirectional ones.
- Position the microphone out of the area of the breath blast. In an announce application, speak across the microphone with the microphone 45° left or right of the person speaking. Stand-ups and handheld interview miking should be done with the microphone capsule below the axis of the mouth.
- Use a pop filter. This is often the same as the manufacturer's windscreen. Test the combination carefully for frequency response and directional characteristics before putting it into service. Windscreen/pop-filter foam is specially designed reticulated foam that comes in a variety of densities. Even very acceptable open-cell foam may be too thick for use on some microphones. Nonreticulated foam (such as "Nerf balls") rolls off high-frequency response and alters the polar patterns of directional microphones.
- Fashion a pop filter. For radio and other off-camera miking, a piece of silk may be suspended a short distance in front of the microphone diaphragm. A number of systems are commercially available that use a frame made of wood or plastic, and some allow the insertion of multiple screens. Other systems use perforated metal discs. Some of these pop filters install directly on the microphone suspension mount. Others clamp onto the mic boom.
- Use a high-pass filter. Most of the disturbing plosive energy of a P-pop is very low in frequency. Try using a very abrupt high-pass (low-cut) filter in the microphone line. Rolling this energy off before it gets to the mixing board or recorder input will further reduce distortion in the audible range.

Wind Noise

While wind noise may be dealt with similarly to P-pop, superior results with shotgun microphones can be attained by using a well-engineered fabric/mesh cylindrical screen that provides an air space between the material and the microphone, as shown in Figure 4.3-19. To handle severe cases (gale-force winds), special

FIGURE 4.3-19 Zeppelin-type windscreens for shotgun microphones

Courtesy of Audio-Technica U.S., Inc.

fur-like socks are available to wrap around the tubular windscreen. While using these materials will result in some performance trade-offs, recordings made under such conditions are typically not intended for critical listening. Windscreens must cover all openings to the element—front and rear.

Use a High-Pass Filter

A microphone with a limited low-frequency response will help minimize wind noise. Extended-response condenser microphones can produce very high outputs of infrasonic energy when moved or when air around them is moved by wind or air-handling systems. The result may be preamp overload or undesirable compressor or limiter action. Again, windscreens and/or high-pass filters may solve these problems.

Handling or Mechanical Noise

The problem of mechanical, nonacoustic noise is one that is encountered by the user whether the microphone is handheld, body worn, or hardware mounted. The reduction of a transducer's sensitivity to such noise, or the improvement of its acoustic-to-mechanical noise sensitivity ratio, starts with the basic element design.

Microphone elements are often internally shock mounted by the manufacturer to avoid the transfer of noise from the case to the element. Lowest noise is achieved through the combining of omni or omni condenser systems with internal shock mounts, as illustrated in Figure 4.3-20.

External shock mounts are often employed in stand or boom-mounted microphone applications. Properly designed shock mounts allow excursion on-axis, or perpendicular to the diaphragm plane. The combination of mechanical isolation and an internally shock-mounted microphone provides the best results.

Another method of reducing mechanical noise is to raise the resonant frequency of the mechanical drive system. An example would be that of bracing wooden platforms, tables, or lecterns to eliminate the very audible, drum-like sound produced when they are struck. The use of high-density materials for microphone support systems will result in a higher resonant frequency. A microphone stand set onto concrete or into sand also takes advantage of this density.

Mechanical Cable Noise

Mechanical noise transfer to the diaphragm may also be reduced through decoupling the diaphragm from tensile forces, converting them to lateral forces. This may be demonstrated by selecting a microphone that has some noticeable handling noise problems and plugging it into a talk-out system, raising the gain until the handling noise is evident. Now hold the microphone face up (diaphragm horizontal), with the cable hanging straight down, and tap on the cable. This should produce a low thump. Next, rotate the microphone 90° so that the cable is hanging at a right angle to the microphone axis. Tap the cable again, and the thump should be all but gone. This effect may be applied to custom hardware designs and the dressing of cables on body mics or as they enter stand or boom-mounted microphones. A loop of cable or a small coiled cord lowers mechanical noise transfer by this method.

AC Hum Rejection

Microphones may also be sensitive to noise induced in the element and cables by electromagnetic or electrostatic radiation, which may be the result of proximity to power transformers, fluorescent ballasts, high-voltage AC lines, and SCR dimmers, as examples.

Ensure that lines are balanced and low impedance. The higher the impedance of the microphone, the greater the voltage of electrostatically induced hum. Balanced lines ensure that nearly equal hum will be induced on each conductor. Little differential is seen at the amplifier input, resulting in common mode rejection of the hum.

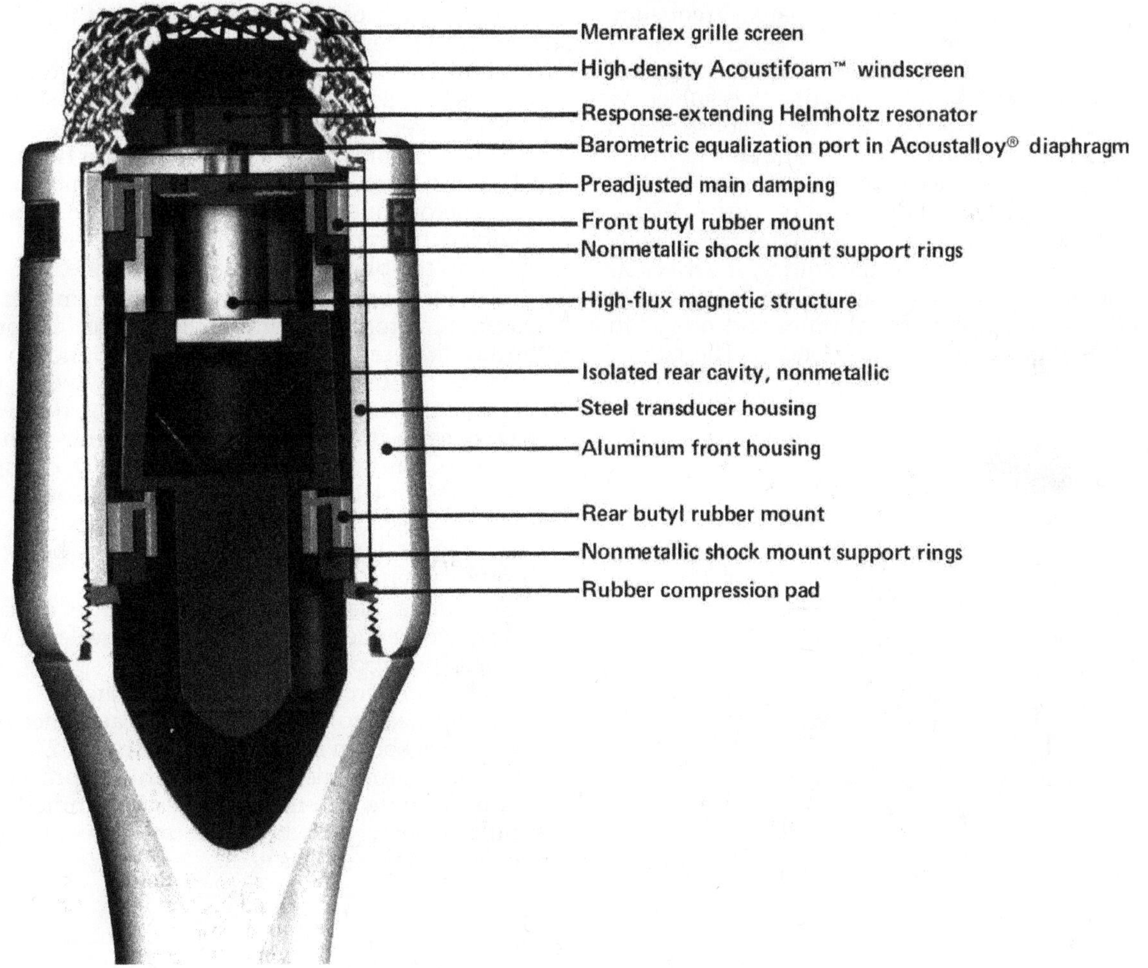

FIGURE 4.3-20 Cutaway view of shock-mounted omni

Courtesy of Electro-Voice, Inc.

Route cables with caution. Avoid running low-level signal cables long distances near AC power cables. When such cables must cross, they should do so at right angles. If more than one AC line must cross microphone lines, separate the AC lines so that they cross at different points.

Use twisted-pair cable. Leads should be twisted inside the microphone and out. The virtually identical positioning this provides for the two conductors within the hum field and the fact that they are out of phase with each other will further reduce induced hum.

Use well-shielded cable. Installed cable may use a foil shield, as flexibility and low-flex memory are not factors. Good stage and field cable normally has a braided shield. Cables that offer the best combinations of flexibility and good shielding use conductive cloth or conductive vinyl under the braided shield.

Follow good grounding practices to avoid ground loops.

In general, dynamic microphones are much more sensitive to induced hum than are condenser microphones. The voice coil is a very effective inductor. Hum-buck coils are employed in some designs that lower electromagnetic hum sensitivity by about 20 dB.

The hum-buck coil is wrapped around the outside of the motor mechanism and wired in series with the voice coil but out of phase. When both coils are placed into an electromagnetic field, equal energy is induced onto each. Because they are out of phase with each other, the offending signal is canceled.

Transformers located within microphones should be avoided if electromagnetic hum is a possible problem. Some transformers are constructed with humbucking characteristics that reduce hum induction.

In severe problem situations, operation at line level rather than microphone level may be required.

ACOUSTIC PHASE INTERFERENCE

Another miking problem relates not to sounds that are added to the output, such as popping or hum, but to portions of the spectrum that are greatly attenuated.

This change in the microphone's apparent frequency response is the result of acoustic phase cancellation that may occur when two or more microphones are mixed, or even when a single microphone is subjected to an overdose of reflected sound. The response charts in Figure 4.3-21(b) through (e) show the severe phase cancellation problems that can result from incorrect use of multiple microphones laid out as in Figure 4.3-21a.

Although sound arriving at each microphone is identical, originating at the same source, it arrives at the two microphones by paths of varying lengths. This causes a difference in the arrival times and results in phase cancellation at certain frequencies. The curves given for each of the examples are fast Fourier transform (FFT)–derived displays of the actual frequency response of the two microphones combined, with respect to a sound source positioned as shown. The FFT analyzer and its companion microprocessor were used to compare the combined output of two matched, calibrated microphones to the output of one of the two microphones by itself. If no phase cancellation occurred, no trace variations would appear on the X/Y plot, and the plot would be a straight line.

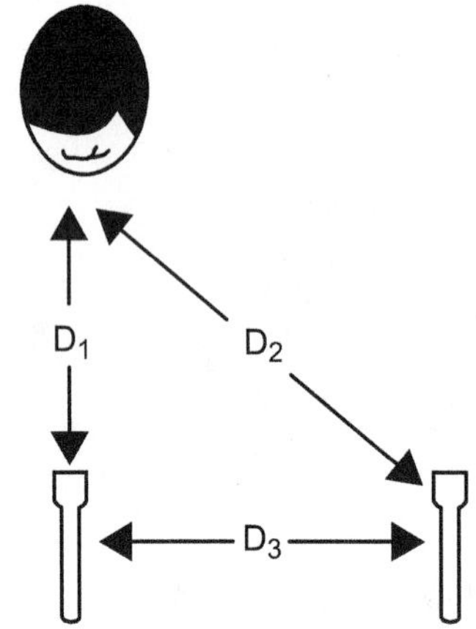

FIGURE 4.3-21A Phase cancellation with multiple microphones

The response charts show that no matter which way the microphones and sound source are oriented, the summed response of the two microphones is poor. These experiments reveal graphically what the ear often perceives as a comb-filter or notch-filter effect that sweeps up and down in frequency (and even changes Q) as the variables D_1, D_2, and D_3 change with the movement of the microphones or sound source. In more subjective terms, the resultant sound may be described as hollow, as if the sound is being forced through an empty cardboard tube.

Situations that cause acoustic phase cancellation arise quite frequently. One classic example occurs with a pair of microphones on a podium, spaced apart to provide on-microphone coverage as the speaker turns his or her head to address all of the audience in front. The curves shown in Figures 4.3-21b and 21d are typical of the problems caused by this approach. If the output of these two microphones is summed and fed simultaneously to a house sound system, feedback problems may occur as well.

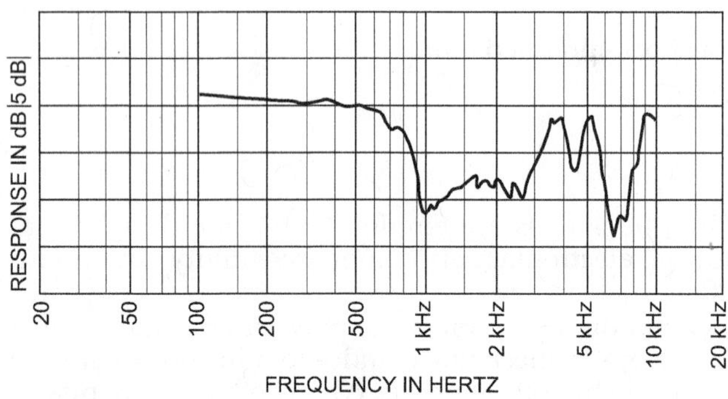

FIGURE 4.3-21B D1 = 12″, D2 = 21.6″, D3 = 18″

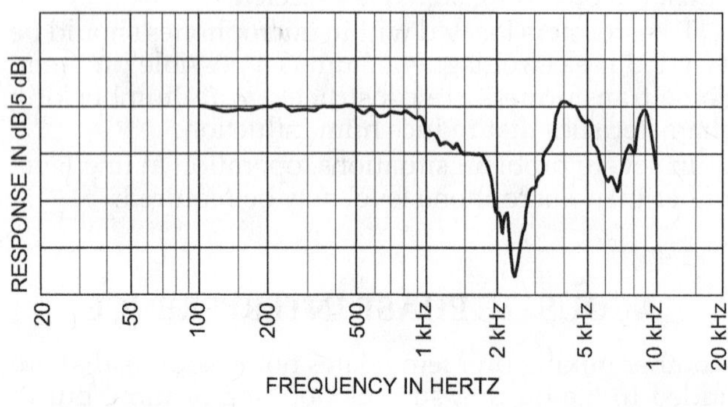

FIGURE 4.3-21C D1 = 18″, D2 = 21.6″, D3 = 12″

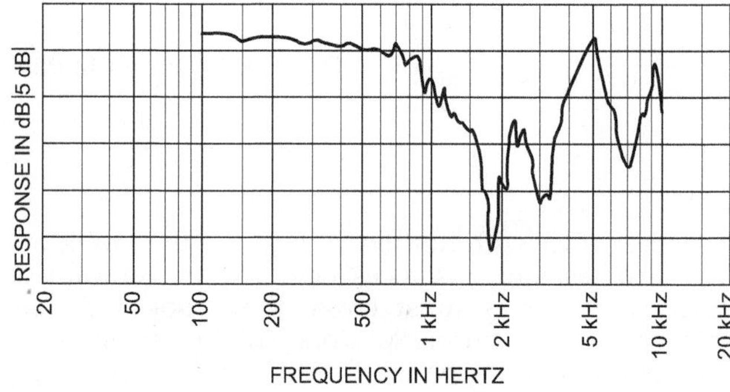

FIGURE 4.3-21D D1 = 24″, D2 = 30″, D3 = 18″

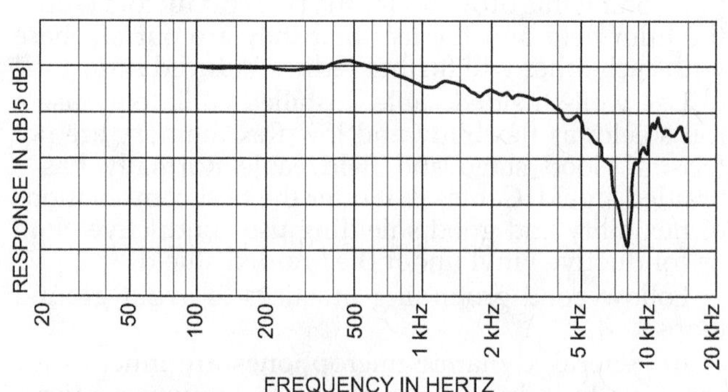

FIGURE 4.3-21E D1 = 5.6″, D2 = 6″, D3 = 2″

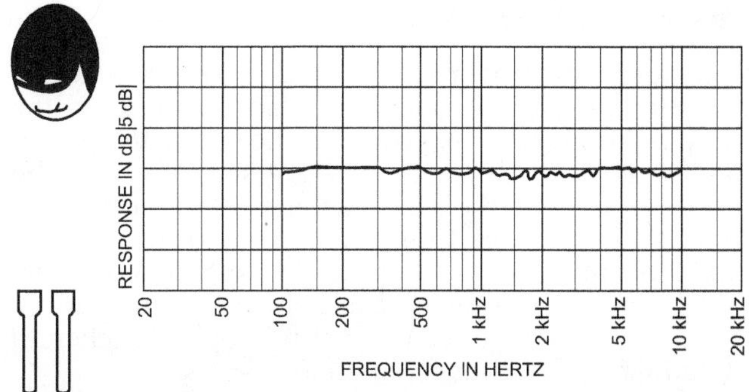

FIGURE 4.3-22 Response of two microphones when co-located for redundancy

Courtesy of Shure Inc.

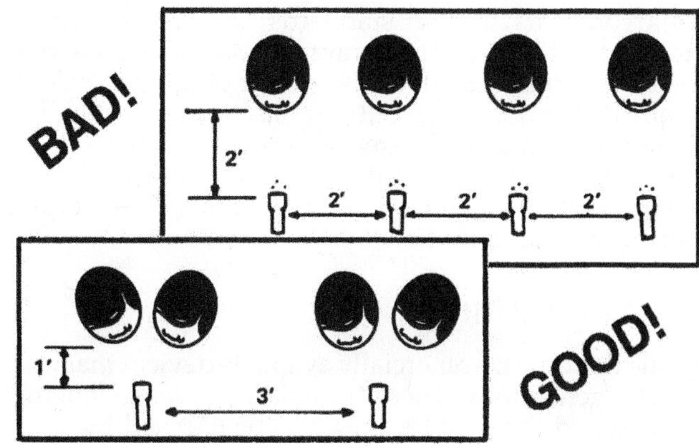

FIGURE 4.3-23 Obeying and violating the 3:1 ratio rule

The simplest solution to the problems caused by this spaced-pair podium miking technique is to use one microphone only, placing it in front of the person speaking and toward the center of the podium. When redundancy miking is desired for critical applications, the two microphones should be placed immediately adjacent to each other, preferably one above the other. Figure 4.3-22 shows the response of two adjacent microphones. Normally, only one of these microphones would be open at a time; the second is strictly a backup. Sometimes multiple microphones are used to feed separate systems, such as for house PA, government agencies, and broadcast. Each may still be used as a backup.

Adjacent pairs of cardioid microphones may at times be angled inwardly with their axes crossed and their diaphragms closely spaced. This arrangement will broaden the acceptance angle of the two microphones while still maintaining some cancellation at the rear. The close proximity allows their diaphragms to occupy nearly the same point in space, thus reducing sonic time path differences. This ensures that negligible phase cancellation will occur should their output be summed. The same formation is often used as a two-microphone stereo pickup technique and has the added benefit of good mono compatibility.

There are occasions when the outputs of two or more open microphones must be mixed and, in this case, phase cancellation must be avoided. The problems shown earlier in Figure 4.3-21 can be substantially reduced by employing the 3:1 ratio rule, where D_3 must always be at least three times D_1. Figure 4.3-23 shows examples of compliance with and violation of the 3:1 ratio rule. Subjective tests have shown that an amplitude difference of at least 9 dB between the two signals will reduce phase cancellation to an inaudible level. The 3:1 ratio rule is a means by which this 9-dB minimum difference may be quickly approximated in most multiple-microphone setups.

The amplitude variance desired may also be obtained through the mixer's gain or fader controls. In general, only microphones in actual use should be opened to their normal operating levels; others should be lowered in level or preferably off.

Acoustic Phase Cancellation With a Single Microphone

Acoustic phase cancellation can also occur in a single microphone system when reflected energy from a nearby barrier such as a music stand, podium, table, or floor is introduced at the microphone's diaphragm at a sound pressure level within 9 dB of the direct sound. Such problems may be avoided in several ways:

- Increase the reflected path length
- Shorten the direct sound path length
- Reduce the reflectivity of the barrier—it may be possible to cover the barrier with an acoustically absorptive material or construct it out of an acoustically transparent material. For example, use an acoustically transparent, visually opaque screen in chroma keying to eliminate reflections into a weather person's lavaliere mic.

MICROPHONE POLARITY REVERSAL

Phase cancellation will also occur if the outputs of two microphones, positioned in the same sound field, are combined with their polarities reversed (which could occur if pins 2 and 3 are reversed at one end of one cable). The sound energy from the two microphones will cancel, and the degree to which cancellation will occur depends on how far apart the microphones are spaced, how closely matched their frequency responses are, and the relative levels of the two mixed signals. Note that the terms *phase* and *polarity* do not mean the same thing. Phase refers to a difference in the relative timing of two signals. Polarity refers to the wiring of a microphone or connectors in its circuit and, when reversed, results in a shift of 180° in the phase of the signal. Having noted this distinction, in common usage, the terms *in-phase* and *out-of-phase* are often used to refer to matters of polarity. Most microphones will be wired to what is sometimes called the *XLR pin count,* which is:

Pin #1 shield

Pin #2 high (+)

Pin #3 low (−)

This conforms to IEC standards 268-12 and 268-4. Refer also to EIA-221 (paragraph 3.3) in which the in-phase terminal shall be the red (or other than black) conductor and that the out-of-phase terminal shall be the black conductor. The terms *in-phase terminal* or *high* (pin #2) indicate the terminal that has a positive voltage present when a positive pressure is applied to the microphone diaphragm.

Checking Polarity

While there are commercially available devices that use a pulse generator to check for polarity reversal, microphones and their cables may also be checked by simply bringing them together and summing their output while speaking into them from a foot or so away. Two microphones that are "in-phase" will deliver a higher output under such a test; if one is reversed in polarity, the output should drop noticeably.

Polarity Reversal as a Tool

While inadvertent polarity reversal in a microphone line can result in some very bad audio, deliberate polarity reversal is sometimes employed as a problem solver.

Reducing Background Noise

A pair of microphones may be reversed in polarity to reduce the pickup of ambient noise. This technique is sometimes employed with two microphones in fixed locations, such as in a press box at a sporting event. If these microphones are brought together, a noise-canceling or differential microphone is created. The speaker must now talk into one of the microphones only, virtually in a lip-touching position. Because of the inverse square law, the amplitude of the voice at that microphone will be much greater than at the other, resulting in a reasonable output level. Distant sound will be picked up equally well, however, by both elements and canceled.

STEREO MIC TECHNIQUES AND MONO COMPATIBILITY

A goal of providing high-quality stereo audio is to do so without compromising monophonic audio quality. Broadcasters must maintain compatibility with monaural receivers (or stereo receivers operating in the monaural mode and FM receivers that blend to mono as the received signal drops off, etc.).

Spaced-Pair Microphones

The need for mono compatibility normally excludes the use of spaced-pair microphone techniques involving omnidirectional or cardioid microphones. Spaced microphones depend on a combination of amplitude and timing (phase) differences to provide stereo separation. They do not sum well for mono, as the very phase differences that aid in separation result in multiple comb filter effects in the mono mix.

Coincident Microphone Techniques

Coincident microphone techniques utilize two microphones whose diaphragms are placed as near to the same point in space as possible. They offer the potential for good stereo without adversely affecting the mono signal with the phase anomalies introduced by spaced microphones. Coincident microphones depend only on amplitude differences for stereo separation and imaging and provide excellent mono compatibility. There are several coincident microphone schemes including X-Y, M-S, and Blumlein.

X-Y Microphone Technique

The simplest of the coincident techniques is called *XY*, which crosses two directional microphones so that their patterns meet at their 3-dB down points, as shown in Figure 4.3-24a. The two microphones should be positioned so that one capsule is directly above the other and on the same vertical axis as shown in Figure 4.3-24b. This minimizes any reflection or shadowing of high frequencies that each might contribute to sound arriving on the horizontal plane.

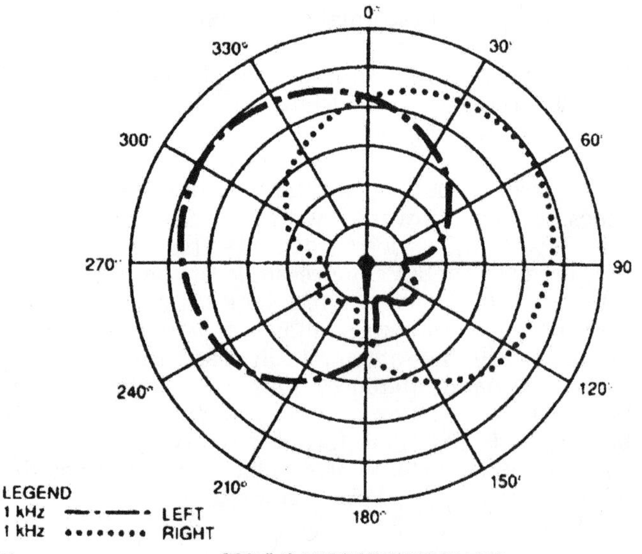

FIGURE 4.3-24A X-Y pattern orientation

Courtesy of Audio-Technica U.S., Inc.

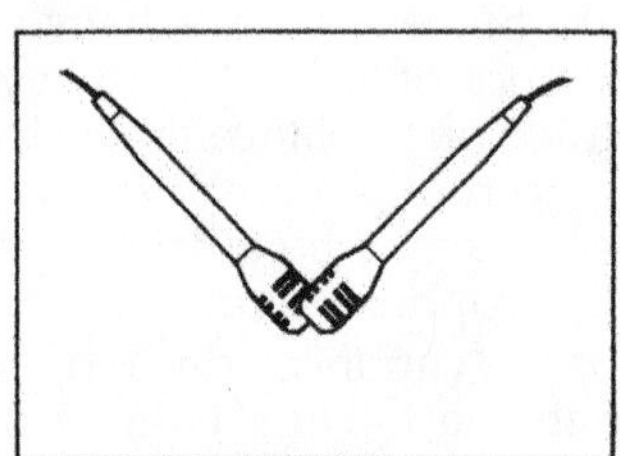

FIGURE 4.3-24B X-Y capsule orientation

Courtesy of Audio-Technica U.S., Inc.

An ideal cardioid microphone would have an acceptance angle of 131° and so would be 3 dB down at 65.5° off-axis. If the angle is too great, sound sources at the center of the stereo image are placed farther off-axis of each microphone and are thereby attenuated, making them sound as if they are farther away. Similarly, too narrow an angle results in near-center sources sounding louder, or appearing to be closer.

Crossing the patterns to overlap at their 3-dB down points ensures that sound arriving from the center of the stereo stage will be summed such as to provide uniform sensitivity from left through center to right.

While a 131° angle would be correct for ideal cardioid capsules, the optimum angle for real microphones will likely be somewhat less. It is important for the microphones used in the X-Y technique to have uniform polar patterns because the patterns of most cardioid microphones tend to collapse at higher frequencies, suggesting that a better X-Y positioning would be 90°. This narrow spacing, however, often results in too much overlap of the patterns. Stereo separation suffers and center-channel information tends to be brought forward of where it should lie in the stereo image.

The optimum angle for many cardioid elements will be approximately 120°. Experimentation and a thorough knowledge of the patterns of the microphones chosen to use will obtain the best X-Y results. Even highly directional shotgun microphones may be used successfully in X-Y, particularly if some of the newer models are selected that have greatly improved polar pattern uniformity. Cross the microphones at the elements, not the ends of the microphones.

Several X-Y stereo microphones are available that integrate two directional elements into one housing that greatly simplifies microphone placement. It may seem obvious that in most cases the axes of the left and right microphones should be near horizontal. Some X-Y microphones, however, hide their capsules in round housings or windscreens that do not permit a quick visual indication of just what is horizontal. It is not good practice to use an X-Y microphone for a close-up announcer or reporter application. Even a slight side-to-side head movement can cause the voice to shift dramatically from one channel to the other.

Testing Mics Outdoors

A quick check outdoors can be made of the angle adjustment of an X-Y pair using the following procedure. Sum the outputs of the two microphones into a mono audio monitor and provide equal gain for each. Feed pink noise into a small powered speaker about 5 to 6 ft. in front of the microphones.

Rotate the X-Y pair horizontally at their capsules and monitor the pink noise from far left channel to far right. Pay particular attention to the amplitude at center. There should be a smooth transition from left to right channel. If there appears to be a hole in the middle, the angle is too great. If the noise seems suddenly closer at the center, try increasing the angle.

Mid-Side Coincident Microphones

The most versatile of the coincident microphone types for stereo broadcasting is the M-S, or "mid-side," microphone, whose polar pattern is shown in Figure 4.3-25. The M-S microphone is a combination of a *middle microphone*, typically a cardioid or hypercardioid, and a bidirectional *side microphone*. The capsules of the two are placed as close together as possible. An M-S pair may be constructed using mid and side microphones plus a matrixing system. A matrixing network combines their outputs and decodes them as left and right channel information, as illustrated in Figure 4.3-26. The information derived from the matrix is nearly identical to that delivered by an X-Y pair, but with some important control advantages.

Sound originating from directly on-axis of the M-S microphone will be picked up by the mid element and delivered equally to left and right channels through the matrix. The side microphone, with one lobe facing

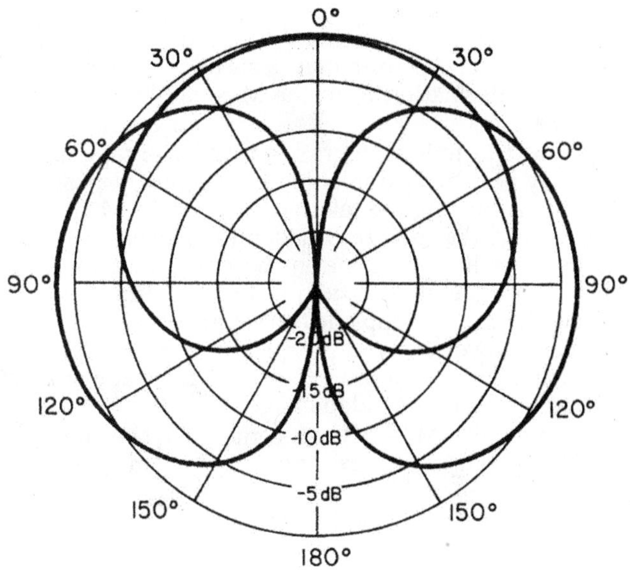

FIGURE 4.3-25 M-S microphone pattern orientation

Courtesy of Shure Inc.

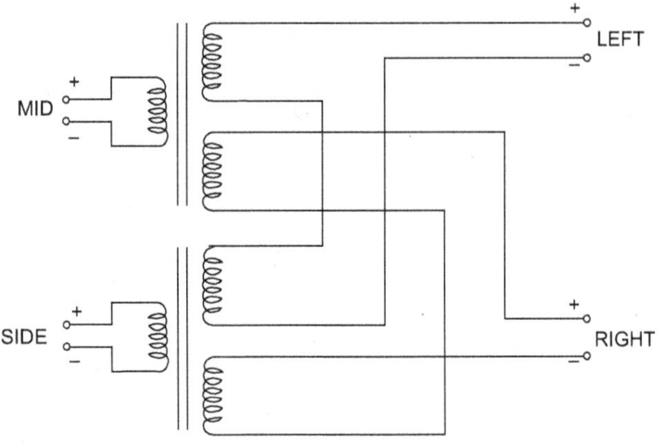

FIGURE 4.3-26 A passive M-S matrix

left, and the other lobe facing right, is insensitive to sound arriving from the center of the stereo stage, as the sound is arriving at 90° off-axis, where the null is deepest. It is, of course, sensitive to sound arriving from each side—a figure eight pattern.

This is part of the process by which the M-S microphone derives directional cues. Sounds arriving from the left are picked up by the mid and side elements and, because they are in-phase, are summed and sent to the left channel. Because the rear of the side element is out of phase with the mid microphone, their sum cannot be used to produce right channel information. Instead, an inverted-polarity version of the side microphone output is mixed with the mid microphone and delivered to the right channel. This processing happens in the sum-and-difference matrix according to the following equations:

Left = mid + side

Right = mid − side

Commercially available M-S systems offer well-matched capsules, easy operation, and considerable control flexibility such as a choice of outputs: mid and side or stereo. A mixer may also be used for deriving L/R information from the M-S pair, as shown in Figure 4.3-27.

The M-S technique offers several control capabilities that may be exercised in either production or postproduction. Adjusting the relative levels of the M and the S signals will narrow or broaden the perceived stereo image. This may be done in the field using the M-S microphone's matrix system or by recording the outputs of the M and S capsules on separate tracks, saving the matrixing of them for postproduction. Matrixing in post will allow the audio perspective to be adjusted to make sense with the video image.

Panning the M signal off-center may be done to deliberately shift the stereo image. For example, crowd noise at a sporting event may be shifted to appear more closely balanced left and right of a microphone position without moving either the microphone or several thousand fans.

Substituting various patterns, from omni to hypercardioid, for the mid microphone will affect the apparent microphone-to-sound source distance as well as the signal-to-ambient or reverberant noise ratio.

Blumlein Technique

The Blumlein miking technique[2] employs coincident crossed bidirectional elements and, like the M-S, responds to amplitude differences to achieve stereo separation. The sound source is placed between the 90° arc of the front capsules of the two bidirectional elements. The stereo sound achieved by this approach can be quite natural, and mono integrity is well maintained. The Blumlein technique is more sensitive to ambient noise and reverberation than the M-S, and placement is critical.

MAINTENANCE

Microphones require a certain amount of care in their handling and storage. Misuse, or even some attempts to service or clean the microphone, could affect some manufacturers' warranties. When in doubt, ask or return microphones to the manufacturer's recommended service organization for maintenance.

Use a windscreen or pop filter to protect the microphone if it is to be subjected to airborne contaminants, such as dust or smoke.

A foam windscreen will also protect a microphone from exposure to rain or snow. Over time, the cells will fill with water, resulting in high-frequency loss and level drop. The foam may be squeezed to reduce

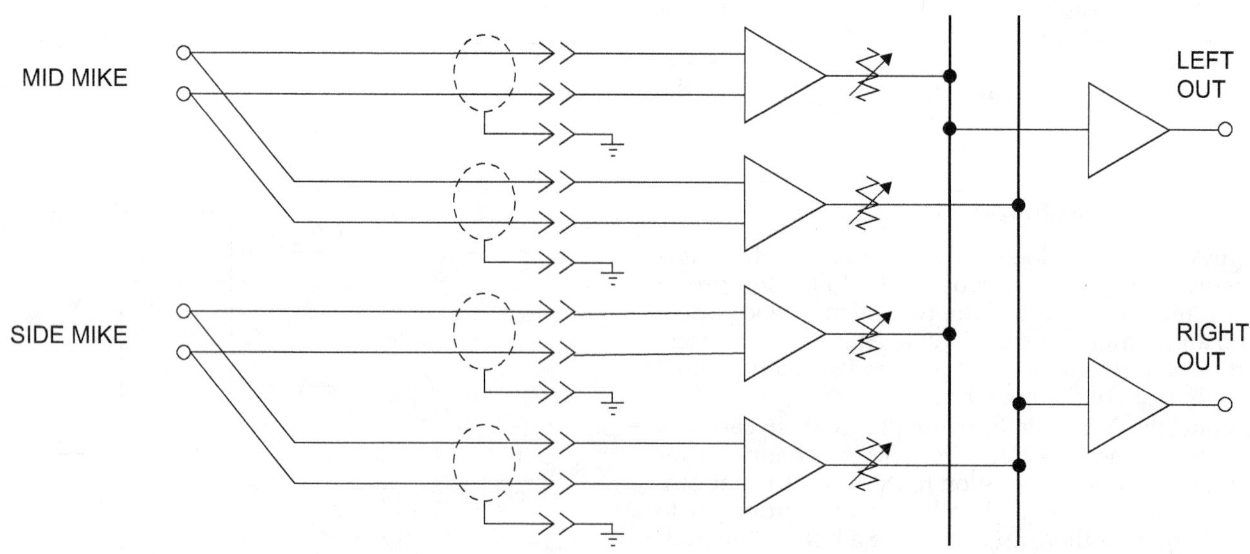

FIGURE 4.3-27 Mixer used as an M-S matrix

the moisture content or a dry screen substituted as required.

Foam windscreens will accumulate deposits of dust and other contaminants. The result will be a deterioration of frequency response and, perhaps, even altered polar response. Foam may be cleaned with soap and water. Rinse well to remove all residue. Nondetergent soaps work well.

Many microphones may be carefully opened to remove a foam pop filter and sometimes a cloth insert. Do so only in a very clean environment. These filters should be cleaned as detailed previously.

Avoid placing microphones on workbenches or other areas where metal particles or metallic dust may be attracted to their internal magnet structures. Very small metal particles can work their way onto the diaphragm and alter the response greatly. In some cases, the dynamic microphone can be opened to reveal the diaphragm for examination. Metallic particles may be very carefully removed onto the magnetized tip of a screwdriver. The screwdriver shaft should be steadied on the edge of the microphone case and the tip very carefully lowered to attract particles, which would likely be held immediately above the voice coil gap.

Avoid subjecting electret condenser microphones to high temperatures, such as in the trunk or glove compartment of a car left in the sun on a hot day or on a boom very close to hot lights. The result may be a loss of charge on the capacitor element and a drop in level.

Avoid moisture with all microphones but especially with condenser microphones.

If given a choice between using mercury or alkaline batteries to power a microphone, note that mercury cells die much more suddenly than do alkalines. The gradual drop in level with an alkaline battery may save a production. Mercury batteries also drop in output level in cold weather and may give off a gas that can corrode the contacts.

Avoid unnecessary mechanical shocks. Store in clean, padded enclosures.

Moving a condenser microphone from a cold environment to a warm one may cause noise problems from condensation.

Avoid moisture in cables and connectors, particularly where phantom power is being used.

Notes

[1] Named for Hermann von Helmholtz, German physicist who explained certain acoustic and psychoacoustic principles. For more information, see http://newt.phys.unsw.edu.au/jw/Helmholtz.html.

[2] The Blumlein difference technique employs two microphones aimed forward and separated by the width of the human head with an acoustic absorption material in between. It was invented by British sound engineer Alan Blumlein in 1931. Blumlein is also credited with the invention of stereo sound and many other audio and recording techniques.

Bibliography

Abbagnaro, Louis. "Microphones: An Anthology of Articles on Microphones." *Journal of the Audio Engineering Society*, AES, New York, 1953–1979.

Alexander, Robert Charles. *The Inventor of Stereo, the Life and Works, of Alan Dower Blumlein*, Focal Press, Waltham, MA, 1999.

Burroughs, Lou. *Microphones: Design and Application*, Sagamore Publishing Company, Inc., Urbana, IL, 1974.

Davis, Don, and Carolyn Davis. *Sound System Engineering*, Second edition, Focal Press, Waltham, MA, 1997.

Eargle, John. *Microphone Book*, Focal Press, Waltham, MA, 2004.

Eargle, John. *Sound Recording*, Van Nostrand Reinhold Co., New York, 1976.

Ford, Ty. *Advanced Audio Production Techniques*, Focal Press, Waltham, MA, 1993.

Ford, Ty. *Audio Bootcamp Field Guide*, Second edition, www.tyford.com.

Long, James. "Layman's Guide to Microphone Specifications," *Audio Magazine*, August, 1969.

Long, James. *The Microphone Handbook*, John Eargle, ed., Elar Publishing Company, Inc., 1982.

Olson, Harry. *Modern Sound Reproduction*, Van Nostrand Reinhold Co., New York, 1982.

Sank, Jon R. "Microphones," *Journal of the Audio Engineering Society*, AES, New York, July/August 1985, vol. 33, no. 7/8, 514–547.

Tremaine, Howard M. *Audio Cyclopedia*, Howard W. Sams & Co., Indianapolis, IN, 1974.

Woram, John M. *The Recording Studio Handbook*, Elar Publishing Company, Inc., 1982.

CHAPTER

4.4

In-Studio Audio Recording for Radio and TV

RICH RAREY

Rareworks LLC
Washington, DC

JEFF BRUGGER

Turner Studios
Atlanta, GA

This chapter covers techniques for in-studio recording for radio and television broadcasting. Some techniques are similar or identical between the radio and television mediums, but in other cases, specific methods vary. Because of this differentiation, studio recording techniques for the two mediums are covered separately and sequentially here, with radio presented first.

IN-STUDIO AUDIO RECORDING FOR RADIO

Since the beginning of broadcasting, there has been a need for in-studio production. Whether it was a daytime AM station using its main control room for commercial production after signoff, or a full-time AM/FM facility that needed multiple control rooms for commercial, program, and promotion production, these stations all share the need to create and prepare audio elements for presentation to the audience at a later time. Because the human voice is by far the most common production element, this chapter will begin by discussing the selection and use of single and multi-microphone arrangements in the production room. Since technology advances rapidly, we will review the general signal flow and processes within the production room rather than dwell on specific branded pieces of equipment. (For more on studio microphones, see Chapter 4.3.)

In addition, we will examine how a modern station's production room accomplishes its task in common monaural and stereo formats; explore some time-proven production tips and fundamentals that are as old as broadcasting itself; and offer the reader a sense of how aesthetic *taste* and *approach* are as important to a successful production as the equipment used to produce it.

The Human Voice: The Most Heard Sound in Broadcasting

We begin with the notion that station management would like the audience to hear a consistent voice quality. While every individual's voice on the air is distinct and unique, the audience shouldn't perceive any particular voice having unusual or unpleasant audio characteristics like boominess, muddiness, or sibilance. Two fundamental factors will help achieve a consistent sound: Carefully selecting the microphone model and type, and the use of a high-quality preamplifier connected to the microphone.

Essential Microphone Technology

The first step is for station engineering and management to select a single microphone model and type for use throughout the station: The on-air studio's mic model and type should match those used in the production facility. Choosing a station microphone must take into account reliability versus cost, keeping in mind that a higher-cost professional microphone will likely save the station money in the long run, due to lower maintenance costs and less frequent need for replacement. It is also recommended that the production control room be of a similar size and shape as the on-air control room; if an unused closet is converted to a production room, there will be acoustic differences between the rooms that a "station-standard microphone" selection cannot solve.

Exceptions to the station-standard microphone model are those used on remotes, especially outdoor and sports remotes, as these microphones will be used in noisy environments, may have humidity or temperature variations, and must withstand being transported to and from the remote location. (Again, see Chapter 4.3 for more on remote microphones.)

Because a microphone's natural electrical output is so low (a few millivolts), the signal must be *preamplified* to bring its level up to match other station audio sources, such as workstations, media playback devices, and the like. A preamplifier is designed to amplify the small microphone signals without adding significant noise, while also providing high headroom before clipping. The effect is that there won't be electrical hiss and noise when listening to a quiet voice, nor will an extremely loud voice sound distorted. Fortunately, modern professional audio consoles generally have good mic preamplifiers built in, making this less of a concern than in the past. Newer console systems, such as those based on audio over IP (AoIP), greatly simplify microphone connections further by locating a mic module near the microphone, which immediately preamplifies and converts the microphone signals to a networked data stream for transport throughout the facility without risk of additional noise or distortion being added.

In other cases, microphones may be connected directly to computer workstations, sometimes using the computer's (or laptop's) built-in microphone input. Doing this limits the station's microphone options; because these inputs are intended for lower quality mic sources, they do not offer phantom power to run the electronics within higher quality microphones, and they are not designed with a high dynamic range. An easy solution is to use a high-quality external USB mic preamplifier with these workstations. These devices have one or more quality mic preamplifiers with phantom power and a physical level control. Installation is straightforward: Plug the device into an unused USB port and install any manufacturer software onto the computer (this is becoming less necessary as more computer operating systems recognize the device and automatically begin using it).

Prior to the introduction of external USB preamplifiers, computer workstation users typically bought professional sound cards, which were installed into a desktop or "tower" computer. The sound card was connected to the manufacturer's multi-pair umbilical cable that exposed the card's microphone inputs on standard 3-pin microphone connectors. The umbilical cable also presented line-level analog and digital inputs and outputs for incorporation into the station's physical plant. While this approach may still be found in some facilities, the external USB approach is now dominant and preferred.

The Single-Microphone Technique

Using proper studio microphone technique during recording will save time later in "fixing" a poorly miked production. For voice recording in the typical studio, place the microphone three to five inches from the person speaking and at an angle, so the mic is pointing to the edge of the person's mouth from slightly above. This position has the following benefits:

1. Microphone overload from spoken plosives are minimized. In the English language, a person speaking a "p" or "t" sound (saying "Peter" and "tough," for instance) emits a little puff of air—called a plosive, or "P-Pop." This puff of air has the power to rattle a microphone element, creating an unpleasant sound distortion. Spoken plosives are naturally directed as a narrow column of air, either straight outward from the center of the mouth (for "p"), or slightly downward (for "t," "k," and "f"). You can test this yourself by holding the palm of your hand an inch in front of your mouth and reciting, "Peter Piper picked. . . ." You'll feel the plosive on the palm of your hand with each "p" utterance. Move your palm slightly downward, and try the t, k, and f sounds. Then notice when your hand is slightly above your lips and a bit to either side, you no longer feel the plosives when pronouncing any of these sounds. Although avoiding plosives may seem ridiculously obvious, it's not uncommon to hear plosives on the air, indicating that improper mic placement was used.
2. Assuming a directional microphone is being used (as recommended), the microphone output should include a pleasing *proximity effect*. All directional microphones exhibit some degree of low-frequency boost when positioned close to a sound source (in this case, a person's mouth). The proximity effect is caused because directional microphones act less directionally (i.e., cancel less sound arriving from off axis) at lower frequencies, and therefore seem to add bass to any closely miked sources. The closer the positioning, the more bass boosting is evident due to proximity effect on most directional microphones. Also, in general, the smaller the directional microphone, the more proximity effect it may exhibit. Proximity effect has been used creatively to impart "gravitas" to the spoken word. Excessive use simply makes the speaking person sound unpleasantly "muddy" or "boomy," however. Many modern microphones have a "roll-off" switch that reduces the low frequencies, and thereby can reduce the impact of proximity effect. Depending on the microphone, the usage techniques, and the station's preference, the low-frequency roll-off switch may be used, but for a consistent station sound, this switch setting should be set the same way on all station microphones.
3. The voice talent can easily see scripts and ad copy when the microphone is slightly above the mouth and slightly off to either side. Microphone positioning should be simple, but this topic has troubled the talent who are trying to read, see the console meters, operate equipment while speaking, and still sound good on the air.

Directional Microphones and Headset Microphones

Figure 4.4-1 shows three common microphone pickup patterns: omnidirectional, bidirectional ("Figure 8") and cardioid. Cardioid- or hypercardioid-patterned

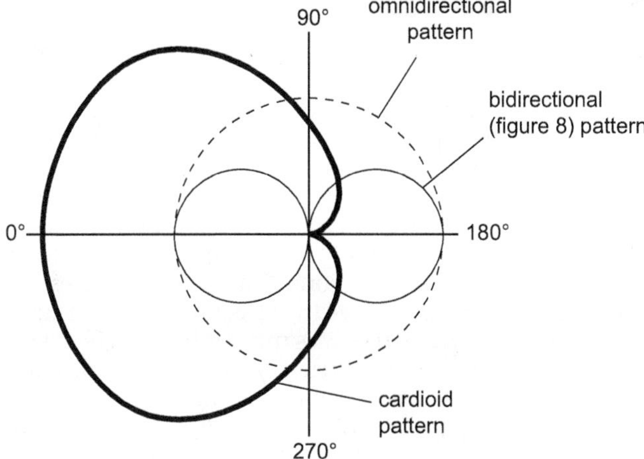

FIGURE 4.4-1 Three common microphone pickup patterns: omnidirectional, bidirectional, and cardioid. The zero degree axis represents the front of the microphone.

microphones are good choices for the production room. These microphones employ a series of small baffles to acoustically reduce sound coming from behind the microphone, and as their name suggests, the pickup pattern on paper looks like an upside-down valentine heart. In real life, the pickup pattern approximates a three-dimensional heart-shaped *sphere* that extends from the front of the microphone. These patterns are not noise cancelling, but rather noise *reducing*. Do not confuse these so-called "unidirectional" cardioid and hypercardioid microphone patterns with that of a microphone in a parabolic dish found on the sidelines at sporting events. Those dish mics physically block noise from the rear and focus the incoming sound to a tight beam, while in contrast, using a cardioid or hypercardioid pattern microphone in a noisy area will reduce sound from the back and side (mostly at middle and high frequencies only), but not eliminate it.

We briefly note the headset microphone—an assembly that mates single- or double-muff headphones with a microphone mounted on a small boom with a flexible extension. These are not commonly used in the radio production studio, but are generally relegated to remote work such as sporting events, where a headset mic takes the place of headphones, a mic, a mic stand, and connecting cables, allowing the talent to move to follow the action without going "off-mike." The headset microphone element is typically smaller than most studio microphones, and it is prone to P-pops, therefore requiring proper placement. (More on headset microphones in the TV section below).

Listening to a Room

If you have wondered what a room would sound like on the air before a microphone has been plugged in, you can use the following simple technique: Assuming you have normal hearing in both ears, carefully place an earplug in *one* ear. What you will experience is the "single point pickup" sound, in which your unplugged ear acts like a single microphone. You may become aware of background noises that you hadn't noticed before—the whine of a computer fan, the buzz of florescent lights, whistling and chuffing from the air handler. This exercise is helpful to identify and hopefully remove and reduce the noise-producing devices in the room.

Closing your ear with a finger is less effective than using an earplug, because you will hear the low frequency sounds of your muscles through your fingertip.

Microphone Techniques for Two or More Voices

Over many decades of radio broadcasting, some useful recommendations have evolved for multi-microphone recording. The general recommendations can be distilled down to the following:

- Place a microphone on each separate source of sound. This is the "one-person, one-microphone" guideline (see Figure 4.4-2).
- Avoid using omnidirectional microphones in the studio (at least for voices).
- Avoid using figure-8-pattern microphones with a person on each active side.
- Mute all speaker monitors in the studio when the mics are live; if participants need to hear other sources during the production besides others studio participants, headphones should be provided.
- If there are more than four or five voices, the audience will have trouble discerning who is speaking. A subtle reintroduction of the person speaking will help the audience recognize the voice ("We're talking with Joe Flabeets of Universal Rundle today. . . .").
- Use the "3-to-1 Rule," in which adjacent microphones are three times farther away from each other than they are from their respective sources. For example, microphone 1 is three inches from person 1. Any adjacent microphones must be at least nine inches away from microphone 1. While this is easily achieved in most studio productions, it becomes more important when more participants are involved, as in dramatic performances (see Figure 4.4-3).
- For productions of dramatic or staged performances, the performers should stand rather than sit, and if possible, a separate mic should be provided for each performer, using floor or boom stands. Standing will improve the performer's projection and allow them to easily move closer and farther away as the performance requires. Other tips for such productions: If possible, arrange the mic stands in a circle, so the performers can see each other without moving away from the microphone's pickup pattern. Note also that if an audience is present, performers may instinctively turn toward and play to the audience, sometimes forgetting the microphone

entirely. Providing headphones can help in this regard, but performers may prefer not to use them.

- In seated, multi-mic productions (such as group interviews), place people across the table from one another. People talking crave eye contact, and their body language will naturally cause them to shift to face the speaking person—and sometimes away from the microphone's pickup pattern.
- Riding the gain on multiple microphones helps reduce the room reflection on an individual's voice, especially if the room has little or no acoustic treatment (e.g., bare drywall and no carpeting). This is performed by reducing (but not turning off) the level of all microphones except for the person speaking. This requires the concentration of the person operating the mixing console to keep up with the conversation and quickly turn up the correct microphone at the correct time.
- Avoid panning microphones to the extreme left or right in a stereo production. It is a matter of taste during a stereo production to use the console's pan control to "position" a microphone to the left or right in the stereo soundscape. If it is desired to use panning as an effect, keep the sound panorama fairly conservative; the pan control should be around the 10 o'clock position of the control (left pan) or around the 2 o'clock position of the control (right pan). Panning a source to hard-left channel or hard-right channel can result in unmatched volumes among voices for listeners hearing the program in mono.

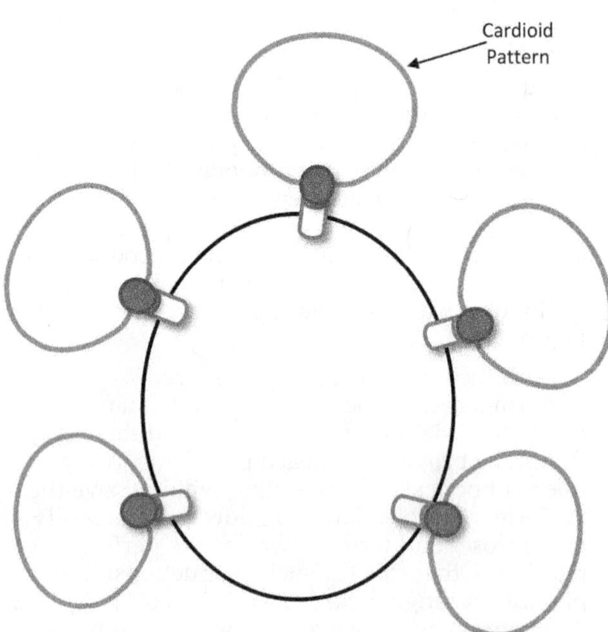

FIGURE 4.4-2 A five-microphone roundtable conversation, showing the cardioid mic pattern around each mic

Non-Voice Recordings

While numerous sound effects libraries are available commercially, there may be a time when you desire to record your own sound effect, such as recording the sound of a station's door opening and closing. For recording sharp, brief sounds, like a handclap or a dog bark, the mic position and room acoustic reflection become part of the "art." Since we don't usually put our ears directly next to clapping hands or a dog's snout, placing the microphone at a short distance from the source will help achieve the effect we expect to hear. An example of where this was misapplied can be heard in reruns of the old TV series *Lassie*; every dog bark, whine, and vocalization has the background acoustics of a sound stage, with the effect that the viewer sees Lassie on screen barking in a living room or outside, but your ears tell you the bark is taking place somewhere else entirely.

Some sound designers who record loud sound effects for films[1] prefer to use condenser microphones over the more rugged dynamic mics and prefer to record, on separate tracks, several microphones each placed farther away from the loud source but still pointed *at* the sound source. Turning the microphone's "dead" (non-pickup pattern) area *toward* the sound source can sometimes be as effective as moving the microphone away from the sound source.

For quiet sound effects, the electronic noise floor seems to be the limiting factor; with the console gain at maximum, you may hear "electrical hiss" from the preamplifier or the electronics within the microphone itself. This means filtering out the noise later or using an audio workstation "restoration" filter that dynamically removes the noise.

In all cases, it is critical that the session has approximately 30 seconds of room ambience recorded in the

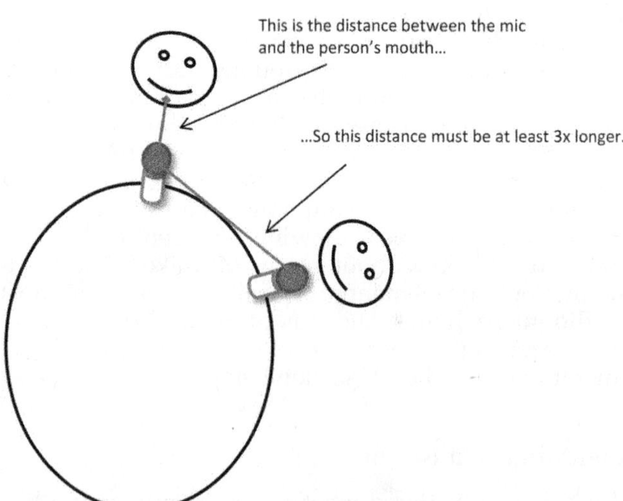

FIGURE 4.4-3 The 3-to-1 Rule. The nearest adjacent microphone must be three times farther away than the microphone the person is speaking into.

clear without anything else happening in the studio, and recorded with the same settings as used in recording the rest of the session. (This is often called "room tone.") If there is noise in the recording session, this ambience in the clear can be used as a reference for the workstation processing, and can also be used as a base audio track to smooth edits in the final production.

Recording Musical Performances for Radio

Studio music performances can take either of two miking approaches: multi-microphone or overall/ambient microphone pickup.

The multi-microphone approach is best when you wish to have complete control over the level and processing on every instrument. Each instrument and voice is provided a microphone, and drum kits have a group of microphones pointing at individual elements within the kit (cymbals, toms, snares, kick drum), and possibly a mic for picking up percussion groups (cowbell, wood block, etc.). This setup requires a well-stocked mic closet with plenty of stands and cabling. The mixing console must have an input for each microphone and preferably the ability to electronically subgroup microphones together. The subgrouping feature allows the engineer to set a tasteful balance of a *group* of microphones—the drum kit microphones, for instance—and then assign the output of those faders to a subgroup master fader. The overall drum kit level can then be easily adjusted without changing the balance among the individual drum microphones. This type of production session works well with multitrack recording equipment, where each microphone (or a group) is routed to a track (or channel) on the multitrack recorder. The recording can be mixed down later. For immediate broadcast use, a two-channel (stereo) mixdown of all the microphone channels can be made on the console and routed to a conventional stereo recorder and/or directly to broadcast.

The other technique for capturing musical performance is the "overall/ambient" setup. This technique is frequently used in concert and ensemble performances where it is impractical to place a mic for each instrument because there are too many performers, or the ensemble sound is intended to be heard "as a group" rather than "a collection of individual performers." The overall/ambient technique uses two or more microphones and is straightforward in setup, but it offers less control over the sound than does the multi-mic technique. Generally, one or more pairs of microphones are positioned at some distance from the ensemble so they gather the overall sound, including the room's acoustic reflections. This approach is generally only successful in a large studio or proper performance space; the typical radio production studio is too acoustically "dead" (i.e., absorbent and nonreflective) for overall miking to produce a pleasant result.

If a soloist is performing, an additional microphone can be provided and panned appropriately in the stereo image (i.e., if the soloist mic is center stage, then the microphone would be panned to the center.) Some engineers will hang additional microphones over an ensemble's groups, such as strings or brass, to give the recording more "clarity" and "presence," or provide a measure of control over the overall balance, without resorting to placing a mic on every instrument as in the multi-mic technique. (Note that there are numerous other books covering the topic of studio music recording in much greater detail.)

Live Broadcast or Recording Sessions With an Audience

One of the most complex studio production scenarios involves using additional microphones to capture audience reaction (applause, laughter, etc.) for the broadcast. In addition, the room may require sound reinforcement ("P.A.") for the audience to hear all parts of the ensemble performance adequately. This sound reinforcement mix is called the "house sound" or "house mix."

An additional complication occurs when the band or ensemble requires a separate sound reinforcement "stage" or "monitor" mix. This mix feeds on-stage monitor speakers facing the performers or earpieces they wear, so they can hear themselves adequately during their performance.

For one person to mix the performance for broadcast *and* simultaneously try to mix two good sound reinforcement feeds is laborious and fraught with disappointment. It is far preferable to split those tasks and have one engineer manage the broadcast mix and another engineer manage the "house," and perhaps yet another handling "stage" sound reinforcement mixes.

Splitting Microphone Feeds

When providing separate mixes from the same set of microphones, a mic splitter box is used. The splitter box provides two or more isolated microphone level outputs from each microphone—for example, one output for the broadcast mix, and another for the house mix. The microphone outputs are isolated from each other so that hum, noise, and accidental cable short circuits on one system cannot affect the other system. The microphone's output "sees" the correct impedance and the two outputs present the correct impedance to the respective consoles so the microphone's level and frequency response is not affected.

There are several types of microphone splitters available:

Transformer Balanced

This splitter has its first output directly connected to the microphone and to the input ("primary") of an internal high-quality audio transformer. The output ("secondary") of the transformer is connected to the splitter's second, isolated output for that microphone. Some split boxes provide a ground-lift switch that disconnects the second output's ground from the microphone's ground. This is useful for eliminating noise-causing "ground loops." These mic splitters can be bought from professional broadcast equipment vendors, or even homemade using high-quality audio transformers, XLR microphone connectors, and some passive components. Figure 4.4-4 shows a one-channel microphone splitter schematic based on a Jensen model JT-MB-D transformer.

Audio Over IP (AoIP) Node

As AoIP becomes more prevalent in the broadcast facility (and performance venues), traditional mic splitter boxes may be replaced by AoIP mic boxes that convert each microphone's signals to an AoIP data stream. Instead of bulky multi-pair microphone cables running from stage to mix position, a single Ethernet cable (a conventional CAT 5 or CAT 6 type) returns all the microphone signals to the AoIP data network. This means every microphone connected into the AoIP data network is available to every AoIP console connected to that data network.

Mult Boxes: Feeding One Output to Many

At events that cause journalists to gather—a press conference or speech, for example—a *mult box* is usually provided. This device splits the podium microphone signal into many isolated, microphone-level outputs so that each reporter can plug into a mult box output to record the event. At large events where there may be musical performances as well as speeches, a separate mult box for each may be provided.

The Audio Workstation

While most stations still have a separate production room with an audio mixing console, multiple microphones, and outboard processing equipment (which can serve as a backup to the main on-air studio), the rapid rise of computer-based audio workstations in broadcasting has changed the notion of a production "room." For simple voice and news production, the production room often devolves into a small booth (hopefully not a repurposed broom closet) equipped with a computer running audio workstation software, an external USB microphone preamplifier, a microphone on a boom stand, a pair of headphones, and perhaps monitor speakers and an external telephone interface or ISDN/IP/phone codec. These small facilities are part of a station's newsgathering operation and have found favor in stations, because several of these booths can be installed in the same space as a traditional control room. Meanwhile, traditional production rooms that remain are also outfitted with an audio workstation because of its convenience and its ability to replace a rack full of processing equipment.

But what is an audio "workstation"? At its simplest form, the digital audio workstation (DAW) is an off-the-shelf computer with an audio editing software application installed. The computer provides the hardware interface, video interface, data network connectivity, local storage, and the underlying operating system, and the audio production software becomes just another application running on the computer. The audio software has the means to ingest audio into the computer from analog and digital sources and can record audio from these sources. It provides a method of editing and processing the audio and saving the finished audio product (an audio file) to the station's media playback system. The audio editing software can be open source and freely downloaded, like the popular Audacity® for Windows, Linux, and OSX platforms, as seen in Figure 4.4-5.

Some audio editing software is provided in commercially sold and supported packages, such as Adobe Audition®, Avid Technology's Pro Tools®, Hindenburg Systems' *Journalist* and *Broadcaster*, to name but a few.

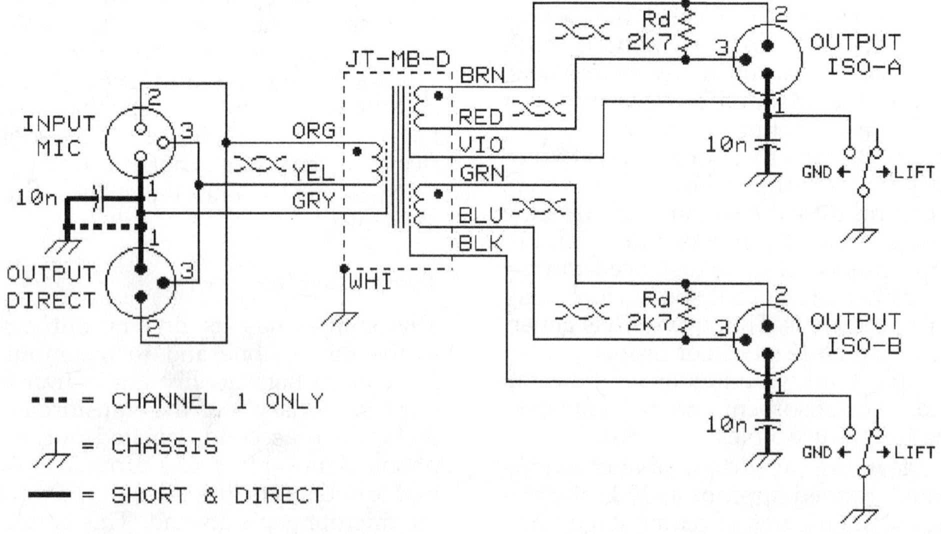

FIGURE 4.4-4 A one-channel microphone splitter using the Jensen model JT-MB-D transformer. If building multiple channels within the same enclosure, only one channel has its input grounded to the chassis.

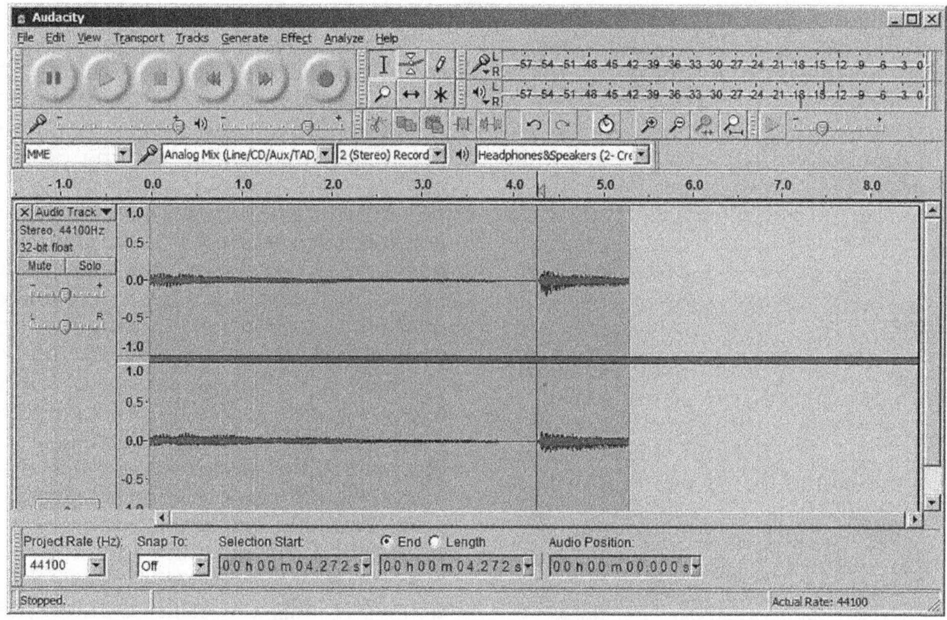

FIGURE 4.4-5 Audacity is a free, open-source audio editing package. Its software license allows a user significant flexibility, and it is popular among audio enthusiasts and professionals.

All audio software products list the minimum computing hardware processing power and hard-drive specifications required for proper operation. To entice users to buy their product, some manufacturers allow a user to download and install a free trial version of the fully featured commercial software, which will only operate for a limited time after installation. The caveat here is that commercial software can be so powerful and complicated that a trial period may not be enough time to fully learn and use its features. Indeed, a robust third-party publishing business has emerged, offering books on "how to" use a particular audio editing software tool.

Physical and Virtual Workstations

Traditionally, the audio workstation uses a physical computer to perform all of its tasks. More recently, however, the physical computer has become "virtualized," meaning that a powerful, centralized computing system runs multiple instances of the operating system and workstation software and presents one instance for each user. This bigger system manages the hardware, monitors the "health" of each instance, and shares memory, physical audio In/Out, and storage among all instances. To each user, it appears that they are interacting with a physical computer using a traditional keyboard, mouse, video monitor, and speakers. The benefits of virtualizing the computers include lower maintenance, higher reliability, and sharing of hard drive, memory, and audio resources. In a virtualized computing environment, there are typically two identical systems that provide a primary and a backup system. When the primary fails, the backup system seamlessly takes over and alerts the system administrator.

Although an audio workstation could operate "standalone" and remain unconnected to the station's data network, the workstation is most useful when the user can easily copy its audio files to the station's media playback system, such as an automation or a "jock live-assist" system. In larger stations and in station groups, a media asset management (MAM) system may supplement the digital audio editing software. In a MAM system, not only is the audio file cataloged, but also additional information about the audio (metadata) can be virtually bundled with the audio file. Metadata can include scripts, pictures, artist/title/album fields, and other relevant information. This bundled collection can be curated by the system for automatically publishing information to the station's website and for transmitting real-time RBDS artist/title, HD Radio program service data, and HD Radio artist experience (i.e., album art) to the listener, as well as be incorporated into the station's Internet audio stream. (For more on MAM systems, see Chapter 5.9.)

Workflow

For our purposes here, we will discuss the workflow elements common to all brands of audio workstation software, which include loading or recording audio into the computer ("ingest"), editing the audio file as a single-track or multitrack project, processing the audio or portions of the audio, mixing the audio elements, and finally exporting the finished audio production.

When choosing the material to load, obtain the highest quality source material as possible. You should assume the finished production will be *transcoded* (that is, converted from one audio format to another) as it is transmitted over a studio-to-transmitter link, fed into an HD Radio™ encoder and Internet streaming encoder, or fed by satellite uplink to remote stations. Every time audio is transcoded, especially using "lossy" audio compression algorithms, the audio quality is degraded. While the degradation may not be easy to detect at the first transcoding, additional transcodings may begin to produce audible artifacts. For this reason, the best production source formats are lossless: WAV files, AIFF files, ripped audio from compact disc, FLAC[2] files, and other formats.

Loading the Elements

After the material has been chosen, load the audio elements into the workstation. Some audio editing applications call the area where these elements exist a "media pool" or "library," but it's only a fancy name for a directory on the computer where the application expects to find audio files. The sources of these elements are many and varied: files downloaded from the Internet or music service; lossless files made by ripping a CD track; analog tape or disc audio fed at real-time into the workstation; perhaps one or more microphones in the production room for capturing live audio; and so forth. Figure 4.4-6 shows a simple overview of sources. For your sanity, allow the workstation software to place these elements in the same hard drive location, either in a local or shared directory. There are few things more annoying than time wasted searching for "lost" or misplaced audio files.

Single-Track and Multitrack Productions

After the audio materials have been chosen and loaded into the workstation, one may choose between two work methods: a single-track production or multitrack production. Single-track production involves placing every audio element in series, one after another, and creating a rapid crossfade between each element. This method is useful when you are working with one element, say, an interview, and you only need to cut out the unwanted sections. Seasoned broadcasters will liken single-track editing to the old razor blade and splicing tape editing performed on reel-to-reel tape; the edits are made *destructively*. That is, the original audio file is changed by the edits made in the single-track session. Once the file has been saved, the parts of the audio that have been cut out are gone and cannot be recovered. Some users make a backup copy of the audio file before starting this kind of session, in case important audio is inadvertently deleted.

Modern digital workstations offer features that allow the user to easily undo an edit, even in a single-track editing session. The workstation may keep temporary files that contain the edited changes, even after the audio file has been saved, which gives the user more confidence to freely edit the audio as needed, because the changes can be undone.

In multitrack sessions, a workstation typically doesn't change the original audio file; this is known as *nonlinear* or *nondestructive* editing. When the user edits the audio, the workstation builds up a collection of internal markers, called an edit decision list (EDL), that "point" to the location of each edit within the audio file. There may be other information in the EDL, such as the length of the crossfade between the audio sections, level changes, and the like. During playback, the workstation simply plays the original audio files, but only the sections pointed to by the EDL. This type of editing was first applied toward (non-real-time) video editing and filmmaking, and became practicable for real-time audio uses when desktop computer processing power increased to such a degree that the rapid reading and reliable playback of an EDL was seamless. Figure 4.4-7 shows a multitrack session with two monaural tracks and one stereo track.

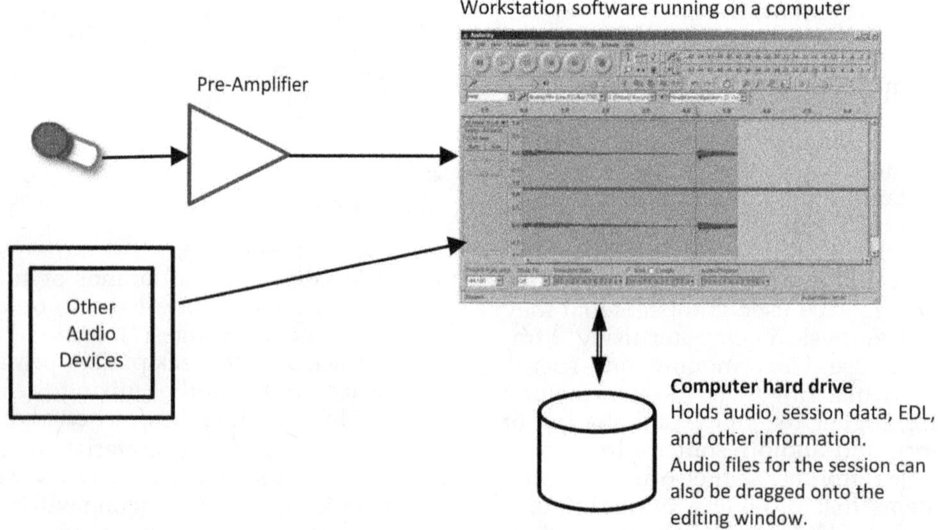

FIGURE 4.4-6 Audio elements can be recorded into the workstation in real time, or audio files can be dragged and dropped onto the editing screen

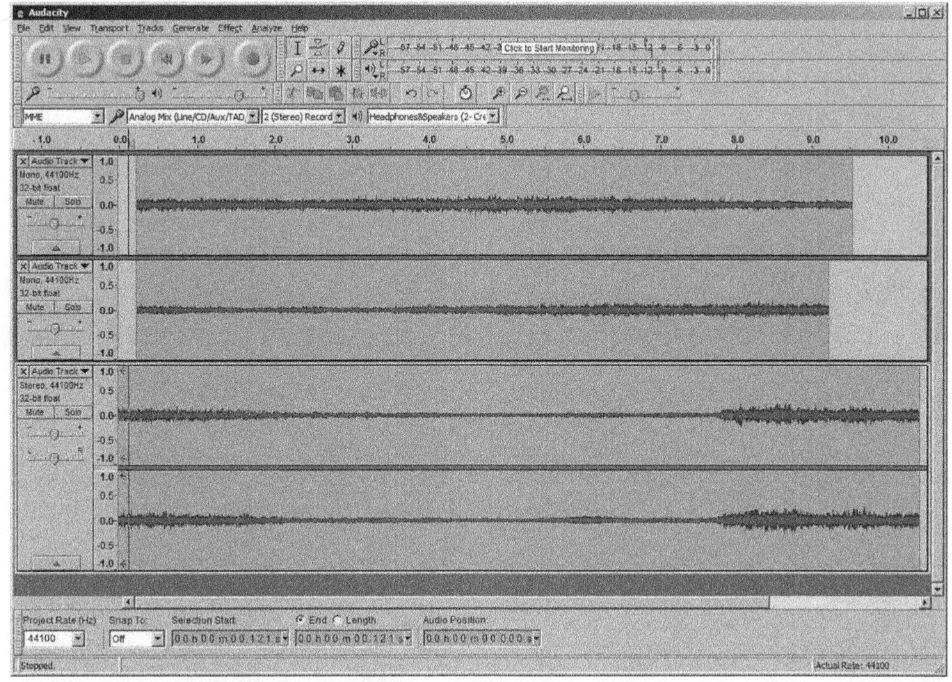

FIGURE 4.4-7 A multitrack session with one stereo track and two monaural tracks

The EDL is saved as a hard disk file, typically in a manufacturer-proprietary format. Some workstations call the file a "session file," which, like any computer file, can be archived (saved together in a .zip, .7-Zip, or other compression utility and stored for later use). When the session file and all the audio elements are archived together, the user has the ability to easily recreate the session at a much later date by "unzipping" the files to the local hard drive. Note that the audio files must be saved in the archive as well as the session files, or else the edit points in the EDL are meaningless and the user will get "file not found" errors when attempting to recreate the session.

Nonlinear editing gives tremendous creative flexibility and confidence after the session that every audio element can be moved, altered, or recut if a "tweak" is needed to satisfy the client, the manager, or the sponsor.

Multitrack production on a digital audio workstation is fairly straightforward. The user either drags and drops prepared audio elements or records them through the workstation's audio inputs, each element in its own *track*, onto the workstation multitrack panel. Each track, unlike analog multitrack production, may contain stereo or monaural audio. This process is called *layup*. A layup is made by, for example, placing a voice track on the first track (recorded onto that track directly or dropped in from an audio file), music on a second track, sound effects on a third track, another music element on the fourth track, and so on. The object of the layup is to choose and load the audio onto their respective tracks to get a sense of the timing and position of the elements; none of the elements will be synchronized to other elements yet.

Working With the Audio Elements

Once the layup is completed, the user can begin working through the production elements in sequence from first to last. It is here the power of a digital audio workstation is realized; the ease of moving, editing, and changing elements as needed makes audio production more efficient and simple. For this discussion, we will illustrate the steps to build a simple multitrack production, such as a commercial or promo. The script calls for music, voice, different music, voice, sound effect, voice, and music up and out at the end.

First, drag the music element to the beginning of the multitrack timeline. Next, position the voice track audio at the approximate location within the track where it is to be heard. Since the workstation is using an EDL, you have the flexibility to move the audio within its timeline—sooner or later in the production—and literally bracket the desired section of that audio. Move the end bracket of the voice track to the end of the paragraph. Drag the second music element to coincide with the end of the first voice track. Now drag and drop another instance of the voice audio file onto the voice track and position it to start at the desired location in the second music element. Choose the needed section of this voice track by bracketing the audio. Drag the sound effect to coincide with the end of this new voice track. As before, drag and drop another instance of the voice audio file onto the voice track and position it to start at the end of the sound effect, and as before, bracket the desired section of this new voice track. On the music track, drag and drop a new instance of the music selection onto this track and adjust the brackets so the music ends at the desired time, say, 30 seconds from the beginning of audio.

Mixing the Elements

If you play your production at this point, you will hear all elements in proper sequence, but as a cacophony of sound—the audio level for each element hasn't been set. Depending on the workstation configuration, you may have two methods to adjust the levels: through the console in real-time—each track appearing on a console fader—or by changing the levels of each track in the workstation. The former method gives an experienced producer an immediate, tactile mix, while the latter method becomes part of the session and is saved with the EDL—and thus is fully adjustable later. Some workstation manufacturers have realized that using physical faders for a mix can be more efficient for humans than using a mouse to laboriously tweak each level, and therefore offer a mixing "surface" that has physical faders, the movement of which directly integrates into the workstation software. As the user changes the fader level, the track level changes and becomes a recorded part of the saved session—the best of both worlds.

Processing Audio

What if there are more spoken words than can comfortably fit into the production? Modern workstations offer a time shift or *time compression* feature, where the audio is speeded up but its pitch remains the same, thus more words can fit into a smaller amount of time. Technically speaking, the digital signal processing feature uses pitch-synchronous overlap and add (PSOLA), where the audio is divided up into small overlapping segments. To speed up the audio without changing the pitch, some segments are discarded and the remaining segments are recombined. This technique is often used where legal requirements dictate a commercial or announcement has a "disclaimer" at the end (e.g., "The side effects of this product include . . ." or "Your mileage is based on . . ."). The text of the disclaimer may, at normal speed, consume precious seconds of the production, so the disclaimer is processed to be rapid but intelligible.

One audio processor that is universally liked and *mis*used is the compressor. This processor reduces an audio track's dynamic range—the difference from the softest to loudest sound—the effect of which is to make the track sound "louder" without increasing its maximum electrical level. While this tool can help manage extreme variations in a track's level (e.g., a talent shouting in excitement), overuse makes a production sound "squashed," and in a longer length production, the audio can become fatiguing to the listener. Excessive compression is not usually desired in a production, because most radio stations have a compressor/limiter at the transmitter input that creates a station's signature "sound," so highly compressing a production's audio dynamic range in the workstation will make an undesirable listener experience when the production is heard on the air, after passing through further compression in the station's air chain.

Monitoring the Production

How does the audience consume your audio? Listening on ear buds, on headphones, in the car, through one or more speakers, all of the above? Trying to tailor a production sound and mix for every possible listening environment can be a frustrating exercise, but it doesn't have to be; in absence of hard demographic listener data on your station, you can make some educated guesses about where your audience hears your work and plan accordingly. For instance, podcasts are typically consumed on headphones, on earbuds, and in the car. Productions aired in drive time are most likely to be heard in the car over speakers, or for some listeners, with earbuds while riding mass transit.

Before beginning the production process, take a moment to let your hearing adjust to the quieter studio-like ambience, especially if you are coming inside from a noisy city street, driving a convertible with the top down and the wind blowing into your ears, or have just been listening to music in headphones at a high volume. It won't take long for your hearing to adjust to the quieter environment. Set the monitor volume at the beginning of the session and don't change it during the session.

Hopefully, your production space has a pair of well-matched speakers powered by a quality amplifier, and these components have no internal audio processing and are acoustically neutral. Although a misplaced notion circulates around the industry that "speakers are no longer necessary because everyone listens on headphones," speakers remain a necessary component in the production room.

The point is, speakers give the producer the freedom to listen to the audio in free space—the area between the speakers and the ears—and allow the producer to more easily detect distortion, phase anomalies, mismatched levels, and excessive processing and to determine if the audio production has achieved the producer's goal. So, a good rule is to mix while listening through speakers first, and then listen to the audio again with earbuds/headphone/car speakers as needed.

The Finished Radio Product

The audio during your session has been recorded, processed, laid up on multiple tracks, level adjusted, and then mixed down to stereo, and everything seems in place. At this point, you know that the *content* is correct—no words in the script have been left out, the music swells and dips where it should, and the sound effects enhance the message. What is left to check?

Quality: Does the audio sound complete and correct? If at any point during the playback, your ears prick up at a sound, then there's something amiss.

Phase: Does the production sound as expected in monaural, that is, when the left and right channels are summed together? There are still many monaural radio receivers in the world, and an out-of-phase production will sound strangely attenuated on these receivers. Remember also that even on an FM stereo radio, blending to mono may occur when the received signal strength is low.

Timing: Does the production fit within the allotted time?

Metadata: If the production is for an automation system, does metadata, such as artist/title/album, or a trigger, which starts another event at the conclusion of the playback, need to be added?

IN-STUDIO AUDIO RECORDING FOR TELEVISION

The need to accommodate studio talent on camera for television necessitates some significantly different audio recording techniques from those of radio production. The remainder of this chapter will address audio techniques in the television studio.

General Considerations for Recording

As with radio production, the first step in the process is to learn the purpose of the end product. The session might be live to air, or it could be recorded and assembled in post. Not every project will start with this information, but it will be a guide in deciding how to make the best recording. A good rule of thumb is to choose the best microphones and equipment possible. Because your recording will be in front of cameras, the microphone selections will be limited to ones that are unobtrusive or have a particular look. Otherwise, the remainder of the signal chain should be of the highest quality possible, as noted in the radio production section above.

Miking the Talent

When working on a television set, the acoustic environment is often a very live, reverberant space, with a relatively high ambient noise floor. Microphone choices and locations should be chosen to isolate the talent from any ambient noise and be visually acceptable to the producer. Headsets are often used in sports settings or in other noisy environments, because they offer excellent isolation and intrinsically include audio monitoring for the talent. They are not suitable for situations requiring a more subtle visual approach, however, such as on a studio news set. In such cases, a well-placed lavalier (a clip-on "lapel" microphone) will likely be the first choice. Lavalier mics should be positioned at the center of the talent's chest when possible; avoid placing the mic under the neck or off to the side. This will minimize changes in sound quality when the talent's head moves.

As described in the radio section above, proximity effect is the low-frequency build-up found in directional microphones when the talent works them closely. This is rarely a problem on the television set for several reasons. The first is that most lavaliers are omnidirectional and therefore without proximity effect. Headset microphones are meant to be worked closely and if they have any proximity effect, it is well controlled in their design.

The use of headsets eliminates many miking problems in difficult acoustical conditions. Their use is no longer a sonic compromise, because current designs have overcome past shortcomings. Headsets fix the mic in a position that does not change, making for consistent sound quality and repeatability.

Multi-Voice Techniques

For nonmusic recordings, the best results are achieved when each person has their own mic. When possible, all the talent should have the same type of microphone, so there is a consistent sound quality and a consistent look. Omnidirectional microphones are the norm in television and will work well, except when the talent

FIGURE 4.4-8 A typical television sound stage presents challenging acoustical conditions

work very close together. Since omni mics are sensitive to sound arriving from all directions, close voices can leak onto a nearby microphone easily. In these situations, you may need to consider other options, like directional microphones, typically cardioid-patterned microphones. Cardioid-patterned mics reject sound behind them, making situations where talent are close together more workable, since there is less bleed due to the directional pattern of the mic.

Sometimes on-camera talent will use a handheld microphone. The main difficulty in using a handheld mic is that the position is not consistent. If the talent does not have the awareness of where they are positioning the microphone, the quality of the sound will change. The most obvious example of this is when the on-camera talent is interviewing someone with only a single handheld mic and must try to keep up with rapid-fire questions and answers. This is rarely completely successful. A better solution is the "one voice, one mic" approach, which in this case can be accomplished by putting a lavalier or a headset on the talent, allowing talent to use the handheld mic exclusively on the interviewee, which should result in a more balanced presentation.

Multiple talent will also benefit from the use of headset microphones because they can turn their head, looking at different people in the group, without affecting the quality of the pickup of their voice. If multiple on-air talent are present and headset mics are not available or desired, then individual microphones should be utilized. These can be lavaliers or handheld. If multiple people are miked in a noisy environment, a directional microphone will be helpful with noise rejection and isolation.

For dramatic or staged readings, headset microphones without large, over-the-ear headphones can be used, because they are small and less visible to the audience. Alternatively, wireless lavaliers can be hidden in costumes or headpieces. Some lavaliers are very small or can be tinted with makeup in order to better hide them. These choices can make the microphones invisible to the audience as well as the camera. You may also want to consider the use of overhead boom microphones. Booms will need to be kept out of the shots, which requires planning or knowledge of the location and movement of the talent on stage. If the boom fits into the production constraints, it can be utilized to cover a larger number of performers with a single microphone output.

As with radio production, when using multiple microphones the quality of the mix is improved by riding the gain of each mic. This not only balances the mix, but reducing the level of mics not in immediate use improves the overall audio quality—but care must be taken to carefully follow the discussion in order to have each mic up full when in use, and microphones not in current use should never be turned completely down or off.

Locating Voices in the Sound Field

Microphones used for on-camera talent or off-camera voiceovers (VO) should always be panned to the center of a stereo recording, or front-center in a 5.1-channel surround recording. If this is a stereo recording or mix, panning voices to the center will assure that if upmixed to surround, the voices will also end up in the front-center channel of the surround mix. Conversely, if this is a surround production, all voices should appear only in the center channel. This will ensure a usable downmix from 5.1 to stereo. Additionally, if all dialogue is contained within the center channel, it will be easier to generate mix-minuses that can be used in repurposing a production for another language.

Ambience and Audience Mics

Ambience or crowd microphones should always be placed in pairs. This allows for hard panning (full left or full right) in a stereo image. These stereo pairs are then more easily placed into a surround mix in the front or rear channels. When a PA system is used in the production, the recorded sound quality can be heavily compromised, because the PA is often loud enough to interfere with capturing isolated crowd noise through the production microphones. To minimize crosstalk, you must take special care in selecting the microphones and their pickup patterns and placement in order to maximize quality. In addition, discussion of house and stage monitor speaker locations with the PA operator prior to the recording is vital in order to maximize the recording quality.

Recording Music Performances for TV

Most of the concepts presented in the radio section above for music recording applies similarly to TV, with the possible exception of a heightened concern by producers for minimizing the visibility of microphones on the set or soundstage. Otherwise, the use of mic splitters and mult boxes is identical to that described in the radio section above.

Recording Systems and Audio Workstations

Although recording on magnetic tape is now rare, the paradigm of feeding signals to a multitrack tape deck tape machine remains the same regardless of what the actual recording device is. Typically, you will be routing your mixing console outputs to multiple locations or devices for recording. There is typically a device recording the "line cut" or video output from the video switcher. This primary record device will also record the program mix and possibly any "mix-minus" channels required. The recording device might also be a digital recording device capable of recording many video channels and audio channels simultaneously, requiring many individual audio channels to be planned, processed and routed to the appropriate record channels. This allows for "iso" (isolated) records of individual cameras, or sometimes switched programming. Each of these unique recording paths may require a unique audio mix. Careful planning will be needed to make sure the appropriate audio is sent to the correct devices.

FIGURE 4.4-9 Screen shot of a DAW post-production session showing tracks synchronized to video, with equalization processor and loudness monitoring windows at lower right

Like radio, TV audio systems also use digital audio workstations. This is more likely to happen in a production recording for a program going to subsequent post-production than in live scenarios. For TV, however, the DAW may possibly include a video interface for recording reference video. The DAW also will require audio and sync sources from the facility in order to function correctly and in proper synchronization with the video. Many separate audio channels can be recorded at the same time. When the recording is completed, there are multiple options for editing and processing of the audio. This is generally done in parallel with the video editing during the post-production process.

Good Practice

Always start with the best-quality choices possible. Select the best microphones available and maintain a clean, distortion-free path throughout. Set up all processing in the mixing console or in the DAW prior to the session to avoid introducing noise and muting in the audio path during the recording session.

Carefully set up the output of the audio console so that all of the mixes, mix-minuses, and isolated components have their own outputs and are clearly labeled.

Prior to the actual session, make test recordings to ensure that microphones and mixes are reaching their intended targets. Confirm that your outputs match the labels they carry and fulfill the requirements of the producer or deliverable specifications.

After Recording

In the case of a live shoot, there is little opportunity to make changes after the recording has completed. If the recording is intended to go to post-production, however, there is much more that can be done to the audio downstream. Prior to delivery to post, the DAW session file should be properly labelled and backed up. Any notes made during the recording should be passed along to the post personnel along with the session audio.

During the post-production process, the sound designer will coordinate closely with the post editor to match edits, time-shift B-roll, etc. The sound designer might also need to record additional VO or narrator tracks. There is often significant cleanup of the dialog tracks, as well as sound effects (SFX) added to enhance or replace sounds recorded during the event. Sometimes, the dialog or VO recorded during the shoot needs to be replaced. When this is required of an on-camera voice, it is referred to as dialogue replacement, usually abbreviated ADR.[3] This is a separate skill that is not covered in depth in this chapter, but is useful to keep in mind if corrections are required to the script.

During the mixing process, there are significant amounts of audio processing and effects that can be applied to the recorded audio. Some processes involve controlling the dynamic range with limiting or compression. Tonality or timbre can also be changed by applying equalization. Both of these types of processing can be applied to correct sounds or make particular tracks (or microphones) fit better into the final mix. There is also time-domain processing that can stretch or shorten time to make elements fit the program's timeframe. Reverb or echo can be added—and sometimes subtracted—in order to make elements fit together in an acoustically pleasing way. All of these processes are combined to deliver the best representation of the audio within its given delivery method.

Loudness

Since the CALM Act was enacted in 2012, along with similar protocols in other countries, television broadcast has a standard loudness in most parts of the world. The particular specifications may be left to the network or particular country, but they add to the workload of live events. Not only is the audio operator trying to make a clean, distortion-free recording, they are also trying to hit the required loudness measurements so that further processing is not needed downstream. (For more on this important aspect of television audio, see Chapter 5.14.)

Recording Live Events

The live audio mixing environment is a chaotic one. Not only is the A1 (i.e., principal audio mixer) trying to record all of the microphones that have been set up, there is the need to also control audio sent to many different recording devices, often with unique audio on each path. This takes significant amounts of planning prior to the shoot. During live events, A1s not only need to output pleasing audio mixes, they must also listen to the director's calls, the camera operator's chatter, and almost every other communication going on among the crew. The mixes they are producing are often the least important audio they monitor during the recording, because the timing and cues from the director cannot be missed during a live event.

During sessions where the audio is recorded for post-production, the session is no less chaotic, but it usually has the opportunity to record more than one take if the performance or technical aspects are not exactly right.

Control-Room Recommendations

The audio control room should be isolated acoustically from the other production areas so that the audio mixer can make an accurate judgment of the mix and the quality of the audio coming into, and out of, the mixing console.

The speakers should be full range on every channel. This will allow the audio mixer to discern quality issues quickly. Mixes should be checked in all formats in which a program is likely to be heard by viewers. Many programs are only produced in stereo, so these should be checked in mono as well. Other programs are produced in 5.1 surround, so the control room should be properly set up for surround monitoring, and program mixes should also be checked in stereo (played over left and right speakers only) and mono (played through center speaker only). Increasingly, television programs are viewed on mobile devices, so checking mixes in stereo on headphones is also a good idea.

In the future, immersive (or "3D") sound mixes may require control rooms with up to 24 channels of audio monitoring, along with rendering engines for audio objects and scene-based formats like higher-order ambisonics (HOA). Like today's surround formats, these mixes will be backward compatible to legacy formats via downmixing, so mixes produced in these advanced formats will need to be checked with proper monitoring in 5.1 surround, stereo, and mono. (For more on TV audio delivery formats, including immersive audio technology, see Chapter 5.18.)

The monitoring volume in the control room should remain consistent. If sound mixers leave their monitor volume controls alone, they will learn how loud things are by using their ears and will have less dependence on the meters. Constantly changing the monitoring

FIGURE 4.4-10 A live audio control room capable of monitoring and mixing up to 5.1 surround audio

volume can result in an inconsistent output level and should be avoided. (Specific recommended practices on monitoring for consistent loudness are presented in Chapter 5.14.)

The monitoring environment should be of the highest quality possible, including the room acoustics and any ancillary monitoring equipment. Audio mixers should be familiar with the sound of their control rooms so they have confidence in what they are hearing. An accurate monitoring environment gives the audio mixer the basis for making accurate judgment calls throughout the recording.

In closing, remember to listen and trust your ears more than anything else. With practice, your listening skills will be trained to deliver high-quality mixes, with consistent output levels and as little distortion as possible.

Notes

[1] "How to Record Extremely Loud Things—Randy Coppinger Interviews Chuck Russom and Michael Raphael," Posted February 10, 2015 by Asbjoern Andersen in Recording, SFX interview, Sound Design Guides, www.asoundeffect.com/how-to-record-extremely-loud-things/.

[2] Free lossless audio coding, an open, royalty-free audio data compression format.

[3] Automatic dialog replacement, also called "looping," in which the actor can repeatedly rehearse one line or portion of a line at a time—watching the screen and hearing the field-recorded dialog audio—until good lip-sync is achieved.

CHAPTER

4.5

Telephone Network Interfacing

KIRK HARNACK AND JOE TALBOT

Telos Systems
Cleveland, OH

INTRODUCTION

Compelling, thought-provoking programming brings talent, listeners, experts, and events together. The result is content—the "Content Is King" kind of content. Creating thoughtful content often implies using voices from outside the studio. Expert guests, listener-callers, interviewees, reporters, and witnesses in situ—all must be quickly mixed or routed on-air in context with the instant programming.

Phone hybrids, talk show systems, and remote (outside) broadcast codecs are a few of the tools for creating compelling content.

More than ever before, the equipment used to connect people outside broadcast studios to those inside is the telephone. As of early 2016, there were over seven times more mobile phone subscribers than fixed, landline subscribers around the world.[1] Indeed, there are about as many mobile phone subscriptions as there are people on Earth.[2]

Whether mobile or landline, a telephone enables anyone anywhere to go immediately live on-air. Broadband-connected phones allow something else as well—a high-quality voice audio connection over IP.

Despite being commonplace in broadcast facilities, interfacing telephone and IP networks to broadcast studio equipment can still be challenging. It's not so difficult to attach a commercial phone hybrid between a POTS (plain old telephone service) line and the studio mixing console. But what if the line is noisy or the level is too low or varies too much from caller to caller? Then, knowing how the phone network works will be an advantage. For advanced operations such as required for call-in shows or for sharing lines among studios, broadcast vendors offer sophisticated and specialized telephone-to-broadcast interface equipment. Knowing how this gear works is essential to getting the most from it. Mobile phones and Internet connections present opportunities, but working knowledge is the key to applying these technologies.

Today, voice calls may be delivered by analog lines POTS, ISDN (integrated services digital network), or voice over IP (VoIP). Many engineers have worked with two or even all three of these transport technologies. Most recently, though, VoIP connections are edging the others out.

This chapter is designed to not only provide a reference for legacy technologies—POTS and ISDN—but to provide a working knowledge and best practice reference for IP-based remote connections.

THE PUBLIC SWITCHED TELEPHONE NETWORK (PSTN)

Most broadcast facilities have made the transition to digital systems—and many to audio over IP (AoIP) networked routing and mixing. Note that the public switched telephone network (PSTN) has been almost entirely digitized for decades, beginning in the early 1960s with the first T1 circuits.[3] Telephone engineers appreciate digital technology for the same reason broadcasters do: reduced susceptibility to noise and other disturbances and improved ability to switch, monitor, and maintain the circuits.[4]

Although the worldwide PSTN has been a significant achievement, for decades it was mostly made from a simple ubiquitous element: digital circuit-switched channels of 64 kbps each. Circuit-switched means that the channel is connected end to end with the entire capacity available for the duration of the call. This is in contrast to packet-switched systems, such as the public Internet, where capacity is shared among users and there is often no guaranteed bandwidth or time of arrival. Internally, telephone companies (telcos) have been moving to packet-switched networking between cities and continents. These networks are usually private or, at least, offer guaranteed throughput performance. At the PSTN's edges, subscribers will get the performance implied by 64-kbps circuit-switched connections.

Speech Coding/Standard Codecs Used in the Telephone Network

Full-spectrum, linear audio, represented digitally, requires about 1.5 to 2.4 megabits per second (Mbps) of data. These are the data rates typically used in linear CD audio, AoIP systems, and other audio transport and storage media where quality is imperative and bandwidth is cheap. Such high data rates are still not practical for worldwide use in telephony at this time; as such, telecom technologies include data bitrate reduction techniques. These techniques are interchangeably referred to as "data reduction," "data compression," "audio encoding," or "compression."

Older audio encoding techniques employed mathematical formulae and transforms to effect significant reductions in the amount of data required to transport voice audio. The G.711 and G.722 codecs are examples of simple but effective voice codecs. These make use of adaptive differential pulse-code modulation (ADPCM) to reduce the full-bandwidth, linear sampled audio. The key trade-offs are audio bandwidth and signal-to-noise ratio.[5]

More recent audio encoding uses psychoacoustic modeling to determine the precision to which some given audio must be encoded in order to suit the communications purpose.[6] The adaptive multi-rate wideband (AMR-WB), and MPEG family (MPEG-1 Layer 2, MPEG-1 Layer 3, and AAC family) of audio codecs are examples of commonly used psychoacoustic encoders. These can code medium to wideband audio at a wide variety of bit rates.

G.711 (A-law/u-law)

While the telephony world makes forward strides in technology, it must remain backward compatible with the billions of wired and wireless phones extant across the globe. As such, audio bitrate reduction is and will remain an integral part of any telephone conversation.

G.711, introduced in 1972, is a narrowband audio codec that provides toll-quality audio at 56 or 64 kbps.[7] G.711 first filters incoming audio to the range of 300–3400 Hz, then samples the audio at of 8,000 samples per second (8-kHz sampling). It then applies logarithmic quantization with 8 bits used to represent each sample, resulting in a 64-kbps bit rate.[8] There are two variants of G.711—A-law and μ-law. μ-law is used with the T1 standard in North America and Japan. A-law coding is used with the E1 standard in the rest of the world. The difference is in the analog sampling method. In both schemes, the signal is not sampled linearly, but in a logarithmic fashion. A-law provides a bit more dynamic range than does μ-law.

As G.711 is a simple companding application and not a psychoacoustic codec; its inherent coding delay is quite small—only 0.125 milliseconds.[9]

What's important to note about G.711 is its familiarity and comfort to our hearing. It's by no means high or even medium fidelity, but it is devoid of audible annoyances. Speech coded with G.711 is generally clear and understandable.

G.722

The G.722 audio codec offers 7-kHz audio (considered "wideband" in the telephony world), at a data rate of 56 or 64 kbps.[10] This is useful for voice over IP applications, and offers a significant improvement in speech quality over narrowband codecs such as G.711, with little increase in implementation complexity. Although G.722 is not implemented across the PSTN, it's often used within campus or corporate telephone systems, especially those using VoIP. G.722 has also been widely used by radio broadcasters for sending commentary-grade audio over a single 56- or 64-kbps ISDN B-channel, and more recently over IP connections, including over the public Internet.

With G.722, the input audio is passed through a digital filter that divides the audio into 0-Hz-to-4-kHz and 4-kHz-to-8-kHz audio bands.[11] These sub-bands are then each encoded using sub-band ADPCM encoding. Most voice energy is concentrated in the lower audio band (0–4 kHz), so 48 kbps of the bandwidth is dedicated there. The other 16 kbps is allocated to the higher sub-band.[12]

G.729

G.729 is another speech algorithm that compresses digital voice audio. It is officially described as "Coding of speech at 8 kbit/s using code-excited linear predictive speech coding" (CS-ACELP).[13]

Because of its low bandwidth requirements, G.729 is mostly used in VoIP applications where bandwidth must be conserved,[14] such as conference calls. Standard G.729 operates at a bit rate of 8 kbps, but there are extensions, which provide rates of 6.4 kbps and 11.8 kbps, offering worse or better speech quality, respectively.[15]

G.729 is sometimes offered by VoIP/SIP providers as the default end-user delivery codec. It allows a fivefold increase in the number of calls over a given IP bandwidth. However, this is done with a substantial cost to audio quality, particularly if the call includes other perceptual algorithms.[16] A transmission path that includes two or more audio encode/decode cycles implies "cascading" of coding algorithms. Such cascading can result in garbled, unintelligible speech reproduction, especially when one or more of the codecs is of a very low bit rate.

Complex Perceptual Codecs

Environments where bandwidth is more constrained may prefer one of the more bitrate-efficient codecs, such as G.722.1 (Siren7) or G.722.2 (AMR-WB).[17]

G.722.1 is a completely different coding algorithm, typically used within a campus or corporate phone system. It's not widely supported outside such systems.

G.722.2 is the ITU designation for adaptive multi-rate wideband. This codec passes audio up to 7 kHz, but can do so at well under half the bit rate of G.722. AMR-WB also works, at a slightly reduced high-end cap, down to 6.60 kbps. Ranging automatically from 6.60 to 23.85 kbps, AMR-WB is the most advanced low-bitrate telephony codec that provides a reasonable listening experience. Mobile network operators, including T-Mobile (USA), WIND (Canada), Orange (UK), France Orange, plus carriers in Taiwan and Ukraine, are supporting AMR-WB within their own networks.

Other mobile network operators have agreed to implement AMR-WB, both in the handset hardware they sell, and compatibility within their networks. Over the long term, these mobile network operators will be able—and should eventually be willing—to interoperate with AMR-WB and other, better audio codecs. They should also be able to provide SIP connections to broadcasters, allowing AMR-WB–encoded calls directly into broadcast facilities.

The POTS Analog Line

Overview/Availability Outlook

The plain old telephone service lines provided by telcos are known officially as subscriber loops, trunks, or simply CO (central office) lines. Trunks are usually lines destined for private branch exchange (PBX) systems and sometimes include special signaling as well. In current generation central office switches, two-wire circuits use a two-to-four-wire converter (also called a hybrid) to interface the analog lines to its internal four-wire system. This process takes place on a line card, which is also responsible for digitization, talk battery insertion, off-hook detection, and ring generation.

Basic Analog Line Audio Characteristics

In the days of an all-analog PSTN, frequency response, noise, and distortion were quite variable due to connection distance and customer equipment at each end. Since digitization of the PSTN's infrastructure, audio characteristics have been consistent per specifications of the digital-to-analog and analog-to-digital converters.

Audio bandwidth for POTS calls for decades has been strictly limited to a 3.4-kHz bandwidth by the sharp low-pass filters required for proper digitization.[18] The phone network's 8-kHz sampling rate permits a theoretical Nyquist upper frequency limit of 4 kHz, but a 600-Hz transition band is necessary for anti-aliasing and simple reconstruction filtering.

The public switched telephone network, nationwide in the United States, has an average measured conversation level of −16 dBm. Because of variations in line length and the arrangement of systems at the CO, the actual level arriving at a subscriber's home or at a broadcaster's phone interface may have a range of−40 to −4 dBm, as illustrated here.

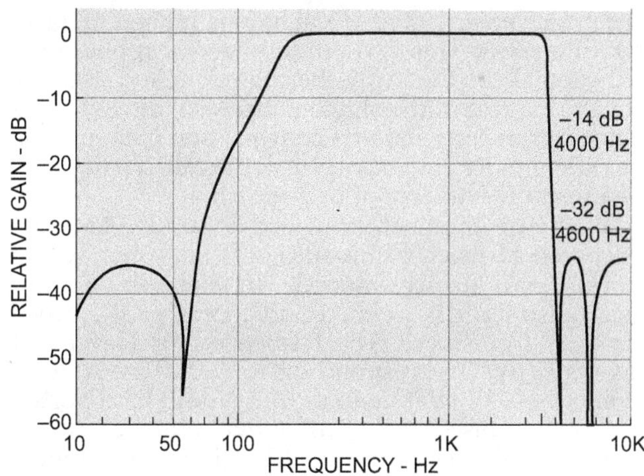

FIGURE 4.5-1 Frequency response of the POTS system to permit 8-kbps sampling and remain within the 4-kHz Nyquist frequency limit

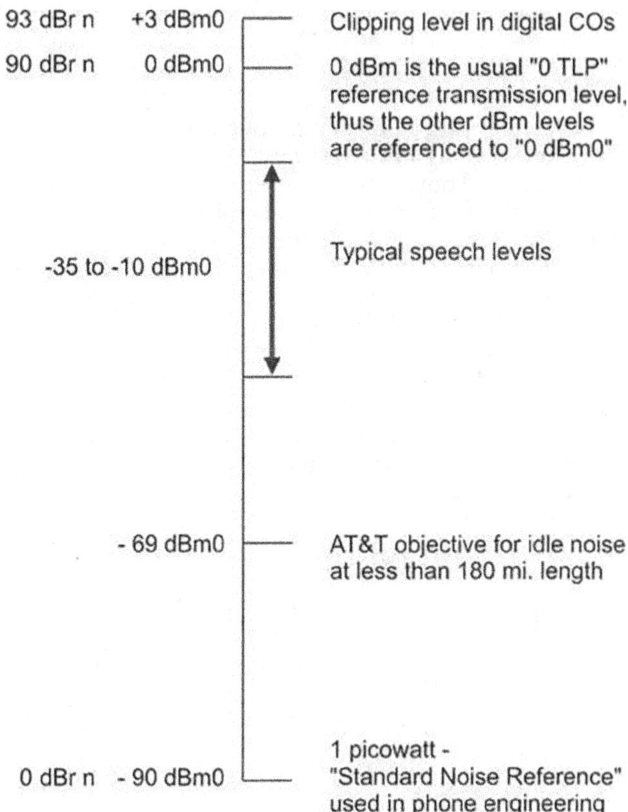

FIGURE 4.5-2 Signal and noise level references used in telephone engineering

Audio sent into the telephone line must be limited to the average −9 dBm as specified in Part 68.308 of the Federal Communications Commission (FCC) Rules.[19]

Audio loss on any given local loop is limited by tariff to 8 dB or less. This loss limit, however, applies only to the loop from the CO to the subscriber and does not include the rest of the signal path. Also, the 8-dB loss may occur at each end of a conversation path, once at the calling party end and again at the called party end, for a possible total loss of up to 16 dB.

Telephone engineers measure noise upside-down compared to the way broadcasters and other audio professionals are accustomed.[20] In telephony, a reference noise floor is defined and then signal is measured *up* from there. The reference noise level is 1 picowatt (pW), corresponding to −90 dBm. A noise level of −60 dB relative to 0 dBm would be reported as 30 dBrn noise (dBrn = dB above reference noise). Again, upside down from conventional audio measurements, the higher this number, the worse the noise.

Note that idle channel noise used in telephone engineering is not the same as the familiar signal-to-noise ratio used in professional audio, because the presence of a signal causes noise in a digital audio system to increase. This effect is called modulation or quantization noise and is primarily dependent upon the number of bits per sample.

A C-weighting filter is employed when measuring the phone line signal-to-noise ratio (S/N). The C-message curve as shown in Figure 4.5-3 was developed to simulate the frequency response of an old-style telephone earpiece and, accordingly, it has considerable low-frequency roll-off. This means that a telephone line can have significant hum and other low-frequency noise and still meet the noise specifications. Since broadcast telephone interfaces often have good low-frequency response, such noise can be heard on the air. The noise meters used by telephone company technicians do have a "flat" setting, which can be used to detect problems of this kind.

Basic Signaling

Dual-tone multiple-frequency (DTMF) dialing uses four high-group frequencies, one for each keypad column, and four low-group frequencies, one for each row, as illustrated in Figure 4.5-4; thus, each button press generates two tones simultaneously. They are transmitted at a level between −10 and −6 dBm. The typical combined level is about −2 dBm. Tones in the high group are transmitted with a 2-dB greater level in order to compensate for high-frequency roll-off in the phone line. These DTMF frequencies were carefully chosen to avoid problems with harmonic distortion causing incorrect detection.[21] Frequency tolerance is ±1.5% at the transmit side and ±2% for the digit receiver. The digit receiver is designed to avoid "talk-off," which is the tone detector accidentally misinterpreting speech or noise as a key-press tone. The time required to recognize any digit tone is 50 ms, with a minimum interdigit interval of 50 ms. If a valid tone pair is accompanied by significant additional audio energy, a well-engineered DTMF receiver assumes noise is present and suppresses detection.

Foreign Exchange (FX) Loops/Designed Circuits/ Pair Gain

Foreign exchange (FX) provides local telephone service from a central office that is foreign to the subscriber's exchange area. If a station is located in the suburbs and the choke network CO is downtown, FX loops will be necessary to connect the lines to the station. FX circuits have also been popular to provide a local phone number presence for stations with coverage areas that spanned across a telephone provider's local access transport area (LATA). These scenarios often include stations with coverage across state lines, and even national borders.

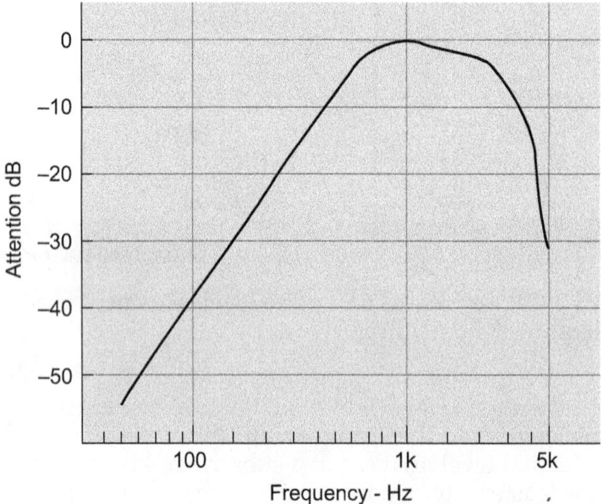

FIGURE 4.5-3 The C-message noise-weighting curve used when measuring noise on telephone company channels

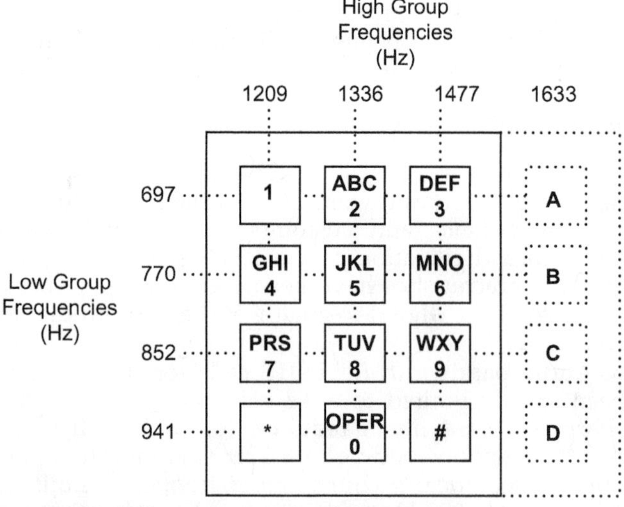

FIGURE 4.5-4 DTMF tone keypad frequency assignments. The four tone pairs in the last column (ABCD) are for special applications.

In recent years the virtualization of certain aspects of the PSTN, and how competing local exchange carriers (CLECs) work with that, the need for FX service is greatly diminished. Local, regional, and national CLECs are viable options for broadcasters seeking lower telephony costs and/or more flexible and transparent service. Radio and TV stations seeking high-profile phone numbers are choosing a single toll-free number, or a single, easily memorized 10-digit number for on-air promotion. As many, or even most, callers are using mobile phones, the concept of a toll call is rapidly vanishing; all calls—at least within a country's borders—are free or nearly so. The changing landscape of toll calls implies that broadcasters likely no longer need FX or other special telephone lines in order to broadly serve their listeners.

Choke Networks

In past years, many broadcast stations were required by their incumbent local exchange carrier (ILEC) to use special, high-volume exchanges for their contest and request lines. These lines are referred to as public response calling service (PRCS) or high-volume call-in (HVCI) network lines. Broadcasters simply call these "choke lines."

The ILECs' requirement to use choke lines may originate from the days when aggressive program directors desired the publicity that overloading a central office would generate. The choke network works by diverting calls beginning with the unique choke prefix around the local serving CO's usual interoffice trunking and instead sending them directly to the choke switching exchange, usually located downtown, via dedicated trunks as illustrated. The usual switching and routing path is bypassed.

Today's mass-volume call scenario is lessened due to several factors.

1. More radio stations with generally less market share each
2. More telephone companies through which listeners are calling
3. More capacity and dynamic routing provided by these diverse telcos

Considering the foregoing factors, the need for choke lines is largely gone. Public service and other priority PSTN users are not at particular risk of blockage by any given broadcaster's call-in promotion.

One unfortunate outcome of choke networks' design is an inability or unwillingness of the ILEC to "port" a popular choke network phone number to a CLEC. This results in forcing a broadcaster to choose between keeping a popular and expensive request line phone number with the ILEC, or choosing a new public phone number from a CLEC that can provide an ISDN PRI or VoIP/SIP connection for it. Recent experience has demonstrated that the perceived value in a broadcaster's legacy phone-in numbers may be overrated; the listening public quickly adapts to using a new phone number, should the station be forced to change to a new one.

Caller ID

Caller ID (CID) allows the called party to know the phone number of the calling party. This capability is useful for call-in shows, where it might be desirable to deny access to problem callers.

There are several incompatible standards around the world for caller ID. These include Bellcore FSK, ETSI FSK, DTMF, and V23 FSK.

Bellcore FSK is widely implemented. Between the first and second ring of an incoming call, the CID information is sent in a packet using a 1200-baud modem. This is the same modulation scheme used in computer modems operating at this rate. Customer equipment normally suppresses the first ring, so the answering user does not take the call before the CID information is fully transmitted.

ISDN and VoIP/SIP telephone connections have more standardized CID protocols, as this feature was built in from the beginning and widely recognized.

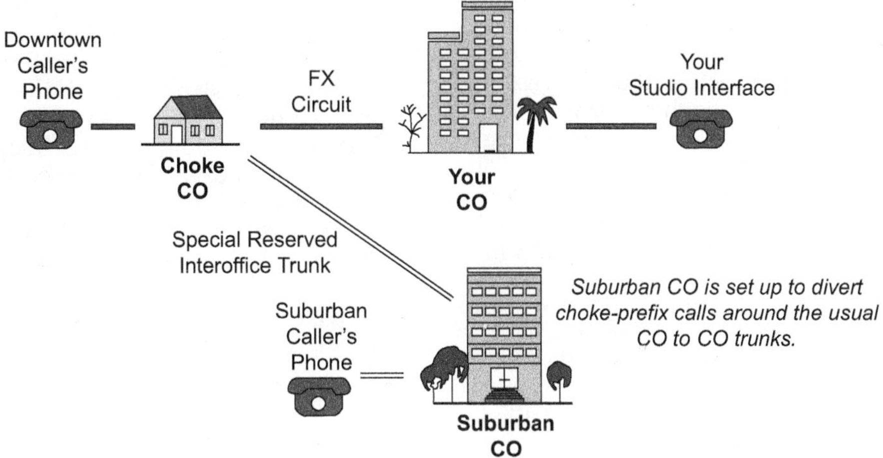

FIGURE 4.5-5 Choke exchange setup used for radio station high-volume applications

PBX Line Equipment

A private branch exchange—or PBX—is becoming more commonplace in broadcast operations. Once only found in large corporations with hundreds or thousands of telephone extensions, the PBX is useful, especially to broadcasters who need to connect several studios to a variety of incoming phone line technologies.

Many broadcasters wish to continue using some legacy phone lines, such as high-volume "choke" lines, as well as newer incoming services like ISDN PRI and VoIP. A local or remotely hosted PBX, such as one using the Asterisk phone software system, is finding its way into broadcast operations. Asterisk is a software implementation of a telephone PBX.[22] It allows attached telephones—including on-air telephone hybrids and talk show systems—to make calls to one another and to connect to other telephone services, such as the PSTN, via POTS or ISDN, and/or via VoIP/SIP services.

One of the most useful aspects of implementing a VoIP/SIP PBX in-house is the ability to equip station personnel with smartphone apps, allowing them to remotely call in to newsrooms and studios using a higher-quality audio codec from the field. A 7-kHz audio quality broadcast can originate from any location with 3G or 4G mobile data coverage or over WiFi. In practice, each smartphone app is assigned an extension number off the SIP PBX. This avoids using the PSTN and allows a direct connection between the smartphone and the station's SIP phone system, via the SIP PBX.

ISDN Basic Rate Interface (BRI)

History, Overview, and Availability Outlook

By the mid-1980s, much of the telephone network was already using digital voice channels. In addition, the amount of data traffic being sent over the switched voice network was growing; data exceeded voice traffic in the late 1990s, and by 2003 data traffic exceeded voice by ten times. Translating data into modem tones and using 100-year-old transmission and signaling techniques were necessary only because customer access was still via analog connections. This approach meant that 64-kbps network channels were being used to convey at most about 33 kbps of user data, and reliability was poor. Providing direct customer access to the digital network would be a much better way. This is what ISDN was designed to do. ISDN uses the digital switched telephone backbone for either voice or data service. Because ISDN provides customer connections directly in digital form to the telephone network, it often improves voice quality on normal calls. A characteristic of ISDN important to broadcasters is that the B channels are true four-wire full duplex, with no crosstalk between the send and receive signal paths. More importantly, users may bypass the normal POTS speech-coding methods and supply their own algorithms, such as those standardized by Moving Pictures Expert Group (MPEG). There are two ISDN services: the *Basic Rate Interface* (BRI), which consists of two-bearer channels of 64 kbps each, and the *Primary Rate Interface* (PRI), which has 23 bearer channels.

Physical Layer—the "S/T" Interface

The "S/T interface" is the common 4-wire user connection for ISDN BRI devices. It's used primarily in Europe and in some PBXs. This can be a multi-device interface point. It supports two 64-kbps bearer channels and a single 16-kbps data channel that is used for call setup.

Physical Layer—the "U" Interface

The "U" interface is the 2-wire connection used by ISDN BRI devices primarily in North America. It provides the same services as the S/T interface point but cannot support multiple devices.

ISDN Primary Rate Interface

Overview/Availability Outlook

While ISDN BRI service is being deprecated by telcos in much of the world, the ISDN PRI interface will be with us for a while; though it's likely that you'll see more "emulated" PRIs in the field. While the interface point and protocol will remain the same, the PRI will be locally generated by a router or integrated access device (IAD) and the actual traffic delivered from the provider via IP.

The important take away here is that circuit-switched data is generally not possible with an emulated or encapsulated PRI. Circuit-switched data is used by all popular ISDN codecs in the broadcast industry to carry the compressed high-quality audio between studio units and field units. This implies that an emulated PRI service cannot be demultiplexed into BRI circuits that support circuit-switched data, as is required by ISDN audio codec connections.

At the time of this writing, it appears that SIP lines and trunks delivered via IP will replace ISDN completely in the near future.

"T1" Interface

The common "T1 line" is actually the physical interface for a digital circuit.[23] It may carry data, channelized telephone traffic, or an ISDN PRI. It is commonly used in North America and Japan (called a "J1" there). Most T1 circuits today are HDSL over one or two pairs and are terminated with a network interface unit (NIU) that is used to regenerate the line signal and provide a remote "loopback" functionality to the provider.[24] The loopback may be activated and deactivated remotely from a test center, and the line may easily be checked for digital errors, in both directions. A channelized T1 may carry twenty-four 56–64-kbps channels. PRIs use channel 24 as a data channel for call setup for the remaining 23 bearer channels.[25]

"E1" Interface

The E1 interface is common outside the United States and carries thirty-two 64-kbps channels. A PRI on an E1 is typically 30b + 2d; that is, 30 bearer channels and two data channels that are used for call setup.[26]

VoIP Options and SIP

For the past few years, we've seen newer standards emerging for IP telephony or voice over IP. While many concepts remain the same, new issues have emerged, mainly from the world of IT—information technology. Previously, we never had to worry about authentication, encryption, lost packets, or jitter; those issues either don't or barely exist with time division multiplex (TDM) technology that has been the backbone of the telecom world.

VoIP and SIP

"VoIP" stands for Voice Over Internet Protocol. This broad term covers any phone calls originated or terminated over a packet-switched network, as opposed to traditional telephone lines, otherwise known as the PSTN. Other terms are used interchangeably with VoIP, such as IP telephony, Internet telephony, voice over broadband, broadband telephony, IP communications, and broadband phone service. They all describe the fact that a packet-switched network—private or public, such as the public Internet—is used to digitally transmit voice-signaling data to another telephone or endpoint.[27]

VoIP is being adopted at a swift pace around the globe.[28] The term encompasses a group of protocol technologies, including proprietary ones such as Skype, and open standards, such as Session Initiation Protocol, or "SIP." SIP is the most popular connection protocol in the VoIP realm. The terms VoIP and SIP are used almost interchangeably when discussing telephony services over IP. Neither term necessarily implies whether the services are delivered over the public Internet, private WAN, or dedicated facilities such as DSL, point-to-point IP radio, or fiber connections.

SIP or VoIP Lines

In the SIP realm, "lines" or "extensions" are connected to telephones, often called "endpoints" or "user agents" in the IT world. They have a feature set that is useful for end users and are provisioned differently than so-called "trunks" because of this. The SIP VoIP phone on your desk is served as an extension by a SIP PBX, either local or remote. That PBX then connects your phone extension to the outside world via a trunk after you dial the number to which you want to connect.[29]

SIP Trunks

Trunks connect switching equipment, such as PBX, to a larger network or to a PSTN gateway via a service provider. Broadcasters in particular should make sure to insist that trunk providers use only the G.711 (A-law or μ-law) codec for their connection to the PSTN. Some Internet telephone service providers (ITSPs) default to the much lower quality, highly compressed G.729 codec.[30] Use of this codec in trunks, in endpoints, or anywhere in a call's path will dramatically reduce the high-frequency response, clarity, and overall intelligibility of that call. When service is ordered and provisioned, a broadcast customer should verify that the provider agrees that any use of the G.729 codec is not acceptable.

Common Provider Practices

Many larger businesses have direct and dedicated connectivity to their traditional telco or service provider, frequently delivered over dedicated T1 circuits or increasingly via IP over fiber. As opposed to using the public Internet, this direct and dedicated connection is the preferred method of delivery, providing better reliability and security.[31] While VoIP service may be provided over the public Internet, the audio quality and consistency will vary if any part of the path is underprovisioned for the peak traffic. The public Internet's architecture isn't optimized for real-time voice traffic and is delivered on a "best effort" basis.[32] While some prioritization can be implemented at or near the endpoints, the effectiveness of this can vary. This is not to say that the public Internet cannot or should not be used for VoIP delivery. Rather, it's important that broadcast engineers are aware of VoIP delivery options and tradeoffs.

With dedicated VoIP service, a provider typically will install their own managed router or integrated access device at your facility. Data transport will be via the most desirable delivery media such as fiber, cable TV (DOCSIS modem), microwave radio, or high-capacity (HICAP) telephone lines. HICAP lines are most often high-bitrate digital subscriber lines (HDSL) provisioned as T1s.

Using the Public Internet for Voice Service Delivery

While the public Internet has been steadily improving, VoIP customers should consider whether their organization could tolerate occasional dropouts or jitter-induced impairments, such as out-of-order packets that produce a stuttering effect on the recovered voice channel. Such impairments would likely occur using today's public Internet. The Internet is considered vital for a business today. Still, Internet providers and most IT professionals still don't treat the voice network as telephone providers historically have in the past in terms of down time, repair responsiveness, and concerns about maintenance work affecting customers.

The public Internet can be great for telephony, but every location and Internet service provider is different. Often there are busy periods when many are using it to deliver television using "over the top" (OTT) TV services such as Netflix, YouTube, or Hulu. Using the public Internet for telephony can be very attractive due to low costs and widespread availability. At the time of this writing, it's recommended that you do testing of your own before committing to VoIP service delivered via the Internet. Hints on testing for VoIP viability appear near the end of this chapter in the Troubleshooting section.

Services available from ITSPs range from basic endpoint station service and SIP trunks to hosted PBX solutions. Costs of these Internet-delivered services vary widely, so careful shopping is encouraged.

These offerings have developed quickly, and may change just as quickly. In 2016, there are many smaller providers, often merging with larger ones. This has created a very dynamic and often turbulent business climate. Mergers can lead to service, billing, or support problems, where there were none before. "Old line" telephone providers with a utility history have also been slow to adopt and understand VoIP technology. They are now fully engaged in migrating to it themselves. As VoIP has developed, the newer technologies and their advantages were quickly understood and adopted by the smaller, hungrier providers. We are now seeing the leading smaller providers being acquired by the larger ones, with varying degrees of success (from a customer standpoint).

Broadcast users should be concerned with the level of technical savvy and the stability of a potential VoIP service provider. At this time, delivery of VoIP service over the public Internet should be very carefully scrutinized from a technical and support standpoint. Many of the organizations promoting hosted PBX service are marketing-centric organizations or simply resellers of a larger provider's services. Generally, at this time, hosted PBX services tend to be "cookie cutter" implementations that may have a difficult time serving broadcasters well. Broadcasters must be concerned about things like call audio quality, high-volume calling (resulting from contesting), and large hunt groups that are not typically encountered by other businesses. Many broadcasters are using such services successfully and lowering their monthly expenses significantly. However, comparing, testing, and planning are critical to implementing reliable VoIP telephony in a broadcast facility.

Provider Options

LECs

LEC stands for local exchange carrier; often the term "ILEC" is used, which means almost the same thing, but stands for "incumbent local exchange carrier." In plain English, this means the original old phone company—most often a former Bell or independent company (AT&T, Verizon, Centurylink, Frontier, and others).

LECs, at the time of this writing, don't generally offer SIP endpoint services, but do offer trunk services. This is still new to them, and they have a limited number of staff who are trained for provisioning and troubleshooting SIP circuits. This will certainly change, but this is the case in 2016.

CLECs

Competitive local exchange carriers, or "the feisty new guys," were first to adopt and offer VoIP services. Some CLECs started out life as long distance companies. CLECs tend to be leaner and more technically driven operations that embrace the newer technologies earlier. They tend to be more flexible technically and lower in cost. At this time, CLECs seem to understand VoIP better than their former utility counterparts. CLECs also tend to operate regionally rather than nationally, and therefore must be compared and considered on a location-by-location basis.[33] Countries outside of North America each have their own situations, though many seem to follow the North American model described here.

Long Distance Providers

The larger long distance companies often act as CLECs or may provide local services. They can be good choices for broadcasters and can offer many advantages, including a near-national footprint. This includes companies like AT&T, Verizon (MCI), Sprint (though they may be dropping some of their long distance service offerings), XO, CenturyLink, Windstream, Level 3, and others. When shopping for a long distance provider, beware of Internet search results. Many of these are only "lead generation services," lacking in first-hand technical knowledge. Wherever possible, speak to colleagues or friends about their experiences with telco and VoIP providers. Avoid term contracts that extend beyond two or three years, and make sure that you are able to maintain some kind of leverage in the event of carrier nonperformance.

Toll-Free Numbers

The use of toll-free or "800" numbers may have peaked in popularity due to flat-rate long distance and cell phones. If your broadcast coverage area spans multiple area codes and you wish to only advertise a single (possibly vanity) telephone number, then a toll-free number may be a good choice for you. Toll-free numbers in the North American Numbering Plan (NANPA) include the area codes 800, 888, 877, 866, 855, and 844.

One of the reasons that we now have so many toll-free area codes is that numbers are generally not "recycled." There are several companies acting as long distance providers, but they actually exist to control as many numbers as possible, either to sell at a premium or to promote "services" of dubious value and high expense. If you have a toll-free number that is important to you, be careful not to allow that service to lapse, as these unscrupulous providers may take your lapsed number as their own, and you'll never get it back. Callers to your old number may reach a recording, often promoting such things as "adult services."

Broadcasters should consider that they will pay by the minute for each answered call-in to their toll-free number. A talk radio station, where many callers spend a lot of time on hold, will want to seriously consider those costs. As the popularity of toll-free numbers appears to have leveled or is declining, some providers are more competitive with pricing; you might be offered a deal that makes the change to a toll-free number worthwhile or even attractive.

One key advantage to having toll-free numbers is automatic number identification (ANI). ANI is like caller ID, but it cannot be blocked by the caller. As such, the ANI number delivered to you is very likely to be accurate.

Also available from many providers are ANI "II digits." These are extra digits not dialed by the customer that indicate the type of service the caller is using to reach you. This includes typical POTS phones, cell phones, pay or charge-a-call phones, prison phones, and various other special arrangements. A current list of II digits is available online from NANPA.[34]

It should be noted that your PBX equipment must be capable of utilizing ANI II digits if they are made available to you and you want to use the information.

Additionally, inbound toll-free calls can be limited to certain geographic areas, implying that calls could be limited to the broadcaster's service area. Many providers offer schedule-controlled or web-accessed call routing changes. This feature is perfect for broadcasters that change origination studios or cities.

Internet Telephone Service Providers

ITSPs offer SIP VoIP services delivered over the public Internet. They sell everything from single lines, with a vendor-provided and -managed analog terminal adapter (ATA) such as Vonage, to trunks and direct inward dial (DID) telephone numbers. With most providers, telephone numbers are available from most exchange locations throughout North America and from many countries throughout the world.

Many ITSPs offer a wide variety of often creative services, even extending your PBX extension onto mobile devices. Some current examples of popular ITSPs include Broadvoice, Vonage, IP-Com, Vitelity, CallCentric, Voicepulse, Flowroute, RingCentral, 8x8, and even the old PBX maker Mitel.

When shopping, broadcasters should consider and verify that the ITSP offers services that use the G.711 codec. Ensure that the use of the G.729 codec is not required or used in that provider's trunking to the PSTN. Other things to watch for are requirements that exclusive hardware be used. In the telecom industry, user "lock in" has been used for decades to exclude the use of competitor's hardware under the guise of stability. Now that telecom users finally have open standards, it's best to choose companies and products that support those standards and allow the user to choose the best hybrid solution for their needs. This is particularly important in an evolving market, where one company might go out of business or be merged with another, leaving users stranded or with limited or expensive options.

Private Networks

At one time, only the largest companies had private voice networks, but now many companies have branch offices around the country or around the world. As they need data connections for administrative purposes, many organizations have chosen to use some of that bandwidth to carry their own voice traffic. This has allowed them to share operators and network resources (voicemail and other systems) across multiple locations. For broadcasters, an advantage of private networks is the ability to provision SIP phones and audio devices to use the G.722 codec. G.722 is sometimes called "HD voice" because of its higher-quality, 7-kHz audio bandwidth. As long as calls remain on the private network, the G.722 codec may be used, and it delivers much-better-sounding audio on internal calls.

Broadcasters can take advantage of VoIP connections across these same private networks. News, weather, and traffic reporters, as well as other on-air talent, can quickly and easily call to the on-air VoIP phone systems of other connected facilities, allowing good-quality HD voice for such reports and updates.

Mobile Networks

Mobile networks are very much in transition right now; perhaps it's more accurate to say that they have always been in transition since their creation. The technologies have been evolving at a rapid rate since the introduction of wireless telephony. Call audio quality to or from mobile devices can vary widely. Low-end phones typically offer no ambient noise-cancelling capability. Calls traversing the PSTN from a given carrier, or handed off from one carrier to another, often suffer from cascading low-bitrate codecs. Experience has shown that some of these calls are virtually unusable on-air due to overwhelming audio coding artifacts.

However, some improvements are showing up in the wireless business at the time of this writing. There is more use by carriers of the AMR-WB codec within their own networks. Data transmission speeds are growing dramatically in many locales. Mobile reliability is generally better overall, with fewer dropped calls.[35] And higher-end phones often have 2- or 3-mic systems, allowing for intelligent ambient noise reduction, which is shown to be very effective. Such noise reduction from the mic system improves mobile calls in two ways. It dramatically reduces nonvocalized audio, and it presents the audio codec with a far simpler coding task, improving audio quality by allowing most of the bits to be dedicated to speech coding, rather than ambient noise coding.

AMR-WB delivers 7-kHz audio voice transmission using what the carriers consider an acceptably low bit rate. T-Mobile was first to implement it in the United States and has been followed by some others. Calls between smartphones running the AMR-WB codec can sound exceptionally good. Unfortunately, when they connect to the PSTN, there is no way to bridge level of quality to other devices. This is because the PSTN doesn't support codecs other than G.711, and most other phone devices don't or can't support others either.

The mobile data networks have improved speed by popular demand. Often this data connection can be used by broadcasters to bypass the voice network and PSTN altogether, using third-party apps or devices to move higher-quality audio or even video using whatever codec the user desires.

Selecting Carriers

Choosing a carrier is limited by what's available at your site—this means the utility phone company (LEC or ILEC) and hopefully some other facilities-based provider using cable or a metropolitan area network (MAN). When choosing a new studio location, part of the selection process should include noting the available providers in the building or area.

Many office parks and high-rise buildings have service from multiple fiber-based, cable, or fixed microwave providers. Broadcasters may also consider leasing "dark fiber" that terminates in a co-location facility (CoLo) where many carriers are present. Since data and voice network services are constantly changing and evolving, it's wise to have as many options as possible.

Choose a carrier that is comfortable providing the array of services that you require. If you want your voice traffic delivered TDM (that is, time division multiplex, channelized, ISDN PRI) most carriers can provide that without difficulty. If you need circuit-switched data to run ISDN codecs or ISDN-based network encryption devices, you'll need a carrier that understands those services and plans to carry that type of (synchronous) traffic for some time. At the time of this writing, these services are being "sunsetted" by many providers. Others that still offer these services have few trained technicians, making repair and new installations clumsy and difficult.

Brokers exist that can ease the shopping process. If you choose to work with one, have a candid conversation about their preferences of one company over another. Most brokers have your long-term satisfaction (and business) in mind. Note that brokers are sometimes commissioned by carriers.

Multiple physical routes and separate facilities should be considered if very high reliability is required. Often a fixed microwave or fiber provider used in conjunction with an LEC will be sufficient to prevent complete communications failure.

Choke Networks/High Volume Calling/Contesting

TV and radio influence people to do things on a daily basis. Broadcast advertising and programming cause people to buy products, vote for candidates, go to events, etc., en mass. The same thing happens when a TV or radio station runs a contest or talk program with a provocative topic. In the past, the high calling volume has created problems for telephone companies and even public safety to some extent. Improvements in the capacity and robustness of the transmission and switching fabric have largely made higher calling volumes less of a concern. Though in some telephone companies, fear of "the smoking a central office" remains.

Legacy Choke Networks

Legacy choke networks and "high-volume calling prefixes" are typically only encountered in metropolitan areas that are/were served by a "Bell System" ILEC, although there are very few exceptions to this rule.

Choke networks were a solution to a shortage of analog trunks, typically from nearby residential areas to a city's center, where most of the broadcast facilities were located in the 50s, 60s, and 70s. Mechanical switchgear and analog trunks were the rule back then, and the telco's approach was to establish a "special prefix" with limited and separate trunking for the radio and TV stations. When a viewer or listener called the station, the call would take special routing to the station because of the special prefix. This prevented contests from blocking regular traffic—especially high-priority calls to police, fire, and medical services.

If your radio or TV station was not located in the city center, you generally had to pay extra to have the choke lines extended to your studio location. This usually degraded the audio, caused signaling problems, and was expensive.

Unless you're in New York City, give away cars, houses, or huge sums of money to the 901st caller, you probably won't be creating traffic that reaches potentially dangerous levels on today's modern telecom networks.

Provider Choking

With POTS service, choking may only be done by the provider. Their central office switch will bear the brunt of 100% of the call attempts.

With PRI or SIP service, choking can still be done by the provider, and that is the ideal case. It's generally best to choke traffic as far upstream as possible to avoid unnecessarily sending high volumes of traffic to the destination. This protects your own on-site or remote PBX switch.

If you have toll-free numbers, the long distance carrier would be the best place to choke. There is no reason to allow more calls to a telephone number than it can accept. If you have a six-line hunt group, ask the carrier to allow no more than six calls through at a time. Some won't do this or don't understand the requirement; it may be a matter of carrier policy. Generally, long distance carriers can do this.

PBX Choking

With SIP or PRIs, choking can be done at the customer's site, using the customer's PBX, but you will be protecting only the endpoint equipment (the on-air phone system itself). This is less desirable, since all traffic will "hit" the PBX. Ideally, in this case, the PBX should be a hardened Asterisk PBX or something capable of withstanding many call setup messages in large and potentially frequent traffic "spikes."

TELEPHONE-TO-BROADCAST INTERFACING

One-Way Interfacing Using Couplers

Despite being decades-old technology, a "QKT" or "telephone coupler" is still a valid method for extracting (listening to) audio from a POTS phone line or impressing audio (sending) to such a phone line.

The "telco" side of the QKT is attached across a POTS line or to terminals inside a special telephone. Such a special telephone will usually have an "exclusion" switch to inhibit audio to and from the phone's handset, preventing room audio from appearing in the audio provided by the QKT. Note that the QKT cannot itself terminate a POTS phone line; it is bridged across a phone line that is otherwise terminated by a telephone.

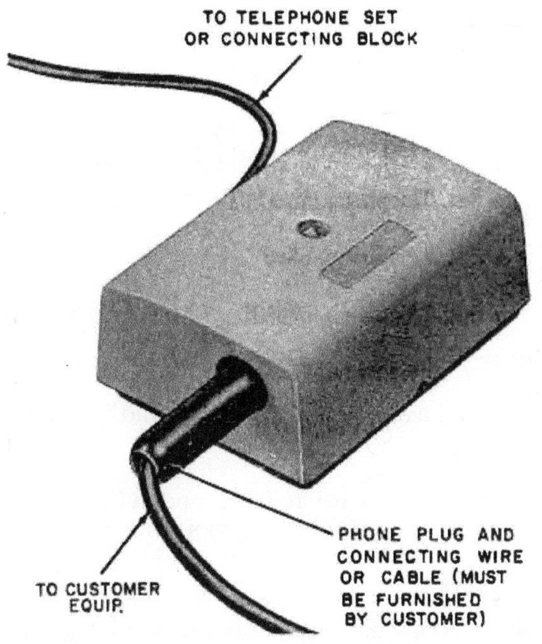

FIGURE 4.5-6 QKT or "30-Type" voice coupler

"Bell System Practices," AT&T Co. Standard, Issue 4, September 1972

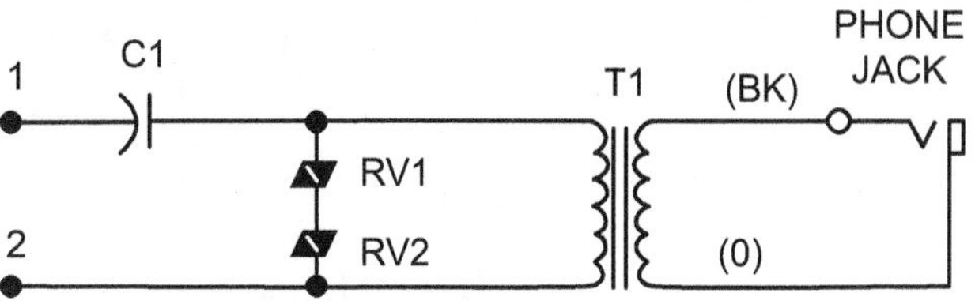

FIGURE 4.5-7 Schematic diagram of QKT or "30-Type" schematic diagram

The QKT's 1/4-inch phone jack provides the audio output, allowing connection to an audio mixing console or other audio destination. The audio level is typically rather low—about −30 dBm—so some gain is typically needed before or in the audio equipment.

The QKT may alternatively be used to send audio to the phone line. An analog audio output from a mixer or other audio device may be wired to the QKT's phone jack, sending that audio to the other end of the POTS connection.

Two-Way Interfacing Using Hybrids

A telephone hybrid converts between the two-wire "last mile" of POTS phone lines and four-wire audio connections suitable for use in professional audio applications. Used in broadcast facilities to enable the airing of telephone callers, a broadcast-quality telephone hybrid provides not only this basic interfacing, but also dynamically "nulls" the telco-side impedance connection to provide the clearest audio in both directions (send and receive) with no echo.

The need for hybrids comes from the nature of analog POTS home or small business telephone lines, where the two audio directions are combined on a single two-wire pair.[36] Within the telephone network, switching and transmission are done using four-wire paths, with the two signals being separated. In older analog networks, conversion to four-wire was required so that repeater amplifiers could be inserted in long distance links. In today's digital systems, each speech direction must be processed and transported independently.[37]

The hybrids in place within the telco infrastructure are of limited performance, allowing a certain level of "leakage" between send audio and receive audio at each hybrid along the way. For normal conversation with handsets, this is not problematic. However, for

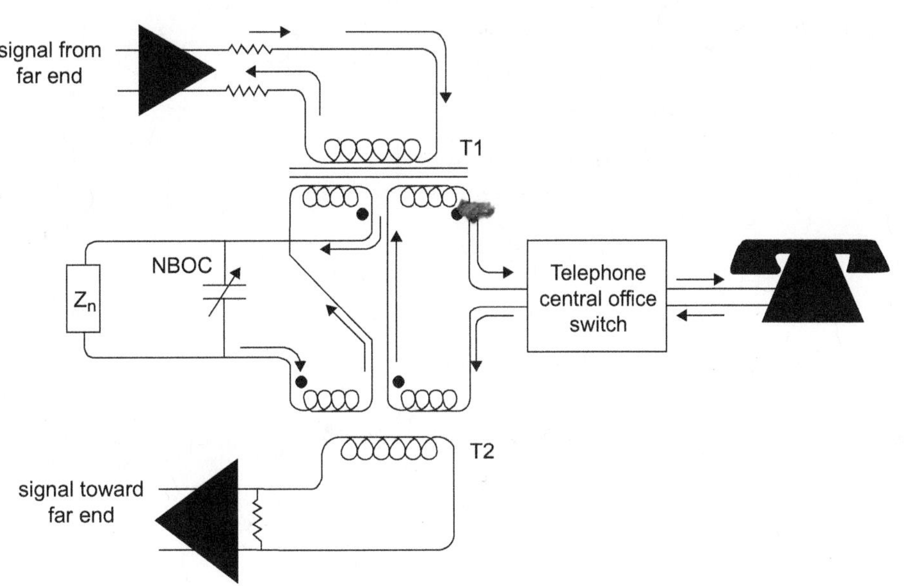

FIGURE 4.5-8 A telephone hybrid—a four-wire to two-wire converter with nulling circuit

Wikipedia User Constant314 (Creative Commons 0 license—Public Domain)

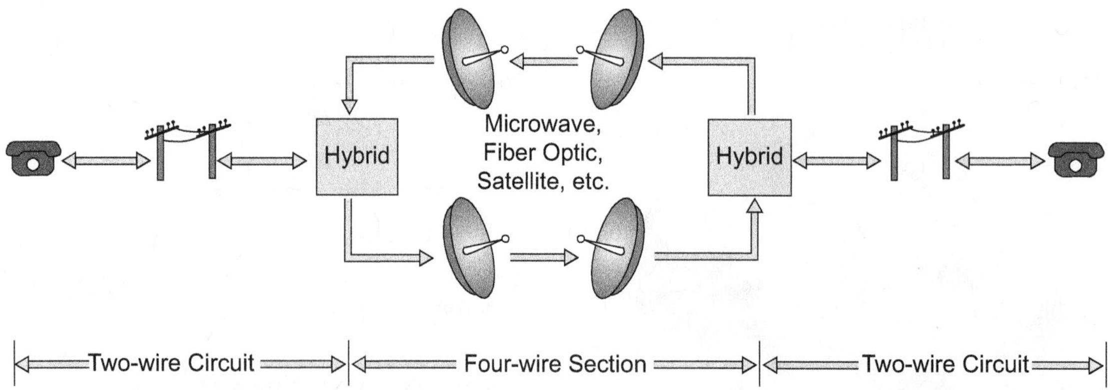

FIGURE 4.5-9 Including the users' telephones, up to four hybrids may be present on a landline-to-landline phone call

a clean, two-way broadcast conversation, it's important to eliminate and send audio from appearing in the hybrid's receive output. Older, analog broadcast hybrids were limited in their performance. However, designs from the past three decades have improved this performance greatly.

Digital Signal Processing Hybrids

Digital signal processing technology became available to audio equipment manufacturers in the early 1980s. DSP—using appropriate dynamic algorithms—found its way into the broadcast telephone hybrid, significantly improving the audio performance. Adding some audio processing, equalization, caller ducking, and pitch shifting on the send audio also improved hybrid performance.

These improvements led to caller audio being used more frequently on both radio and TV programs, as callers could be heard reliably and without echo or phasing effects allowed by previous hybrid designs.

ISDN Hybrids

Telephone companies began offering the integrated services digital network, or ISDN, in the late 1980s. ISDN allowed the digital audio signals, normally confined to the telco infrastructure, to be extended to the customer's premises. ISDN service can bridge between an analog (POTS lines) caller and an ISDN subscriber. This allows a broadcaster to bring ISDN lines into the facility and still take phone calls from those with POTS or mobile phones.

The key benefit to broadcasters of ISDN has been the elimination of two hybrid instances in a typical call to or from a remote POTS line. Also eliminated is the two-wire connection between the nearest telco central office and the broadcaster. Using ISDN at the broadcast station implies that once a caller's voice is digitized at

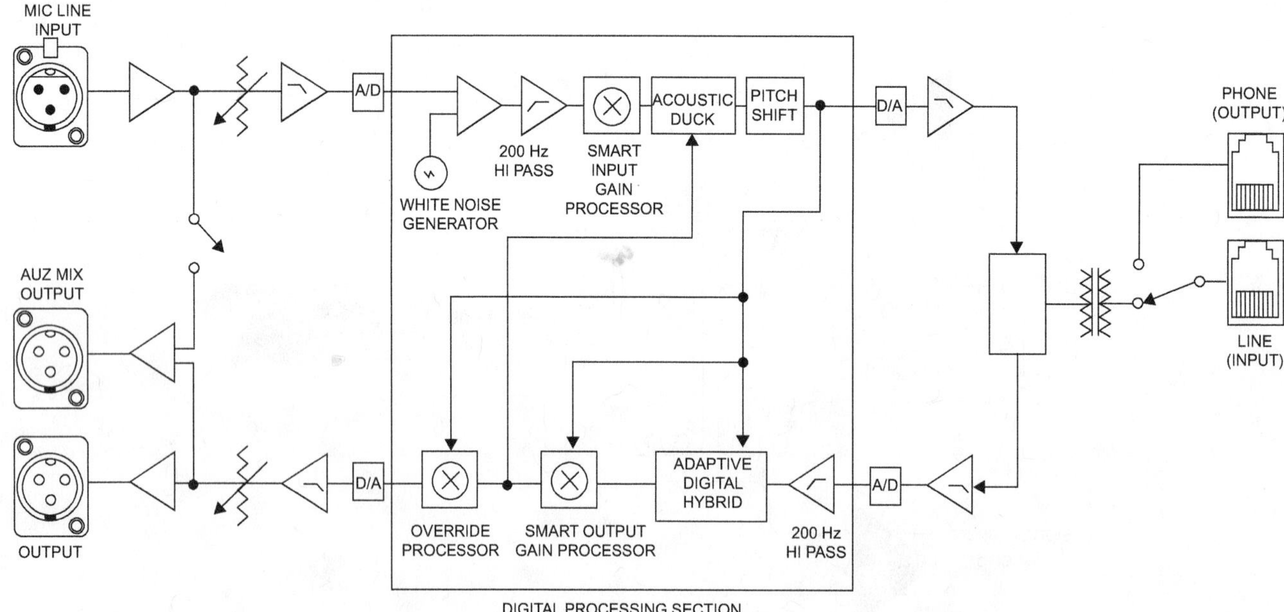

FIGURE 4.5-10 DSP-based adaptive digital hybrid added to a transformer-based phone hybrid

Telos ONE manual, ©TLS Corp., used with permission.

the telco equipment near their end of the connection, audio remains digitally transported all the way to the broadcaster's phone interface.

A hybrid functionality is still desired at the broadcaster's end. This is to reduce or eliminate any echo from the hybrid in the caller's phone or the hybrid at the caller's nearest central office.

In the United States and many other countries, ISDN service is being curtailed in preparation to eliminate it altogether eventually. IP-based infrastructure is the replacement technology. Indeed, some ISDN, as well as POTS, circuits are being emulated over IP with the termination equipment providing an ISDN or POTS interface for the end user.

Note that it is ISDN BRI (basic rate interface) that is being phased out in many areas. The multi-line ISDN PRI (primary rate interface) is thought to have some years left before end-of-life is announced. Managed IP services can properly replace either ISDN BRI or PRI service with voice over IP, either now or in the future.

VoIP Hybrids

A VoIP hybrid may be designed and sized to handle one call at a time or manage multiple calls across several studios.

VoIP lines typically originate in a VoIP PBX, either on-site or off-site. All "lines"—no matter their source—are delivered to the VoIP hybrid over a single Ethernet cable. Within the VoIP hybrid, each line is terminated by a separate "user agent"—VoIP terminology for "soft phone."

A single- or dual-line VoIP hybrid will have a SIP configuration similar to those found in VoIP desk phones. This configuration, shown in Figure 4.5-11, will specify the IP address of the desired SIP server, as well as login credentials for each SIP extension or phone number.

A multi-line VoIP phone system will have one or more configuration pages for SIP lines, as shown in Figure 4.5-12. A multi-line VoIP system may be able to register multiple lines with several PBXs—local or remotely hosted—to make available all the desired connections to each studio.

Whether a single line or multiple lines, a VoIP hybrid will bridge IP telephone connections outside the broadcast plant with networks and equipment inside the plant. Indeed, a multi-line VoIP talk show system will likely have two network ports—one for the VoIP/SIP network and one for the local LAN providing control of lines and shows as depicted in Figure 4.5-13.

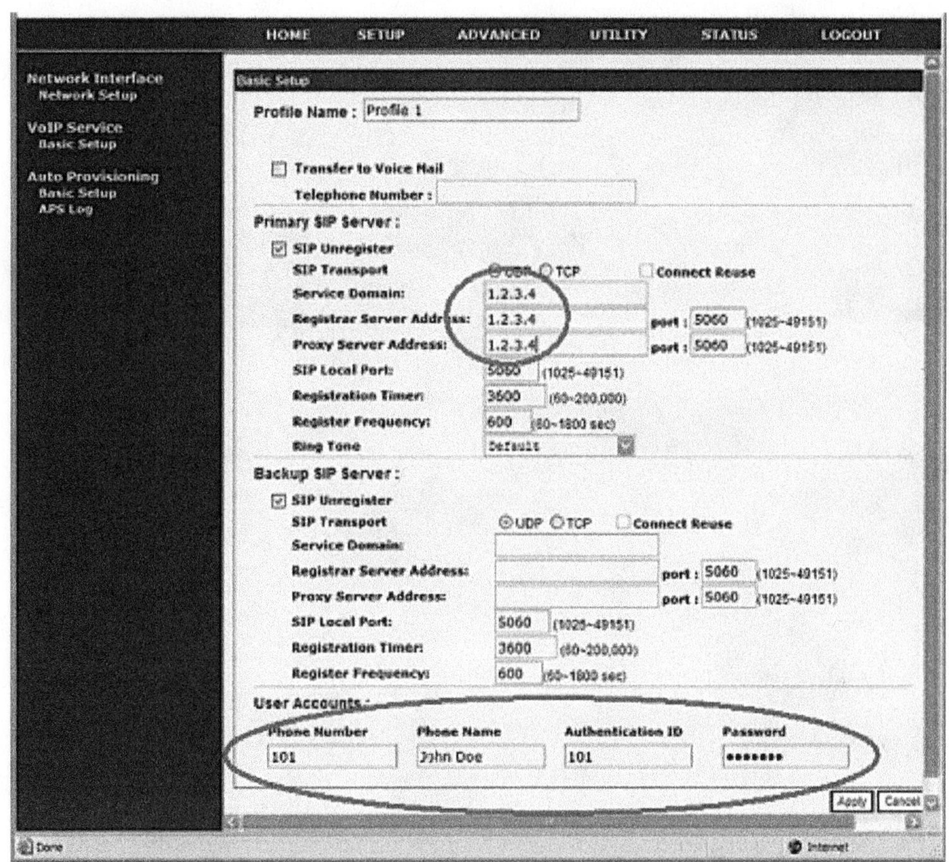

FIGURE 4.5-11 Typical VoIP phone configuration page

Kirk Harnack

IncrediblePi

General Settings:

SIP Server:	192.168.0.16
Name:	IncrediblePi
External IP:	
VX Domain:	

Server is in use by studio(s) "Lab Desk" (show "KIRK Radio"), "Sports Studio" (show "Morning Mayhem")

Lines:

Extension	Register	Expires	Auth User	Auth Password	Shows (Studios)
102	✓		102	••••••••	**KIRK Radio** (Lab Desk), **Morning Mayhem** (Sports
201	✓		201	••••••••	**Morning Mayhem** (Sports Studio)
202	✓		202	••••••••	**Morning Mayhem** (Sports Studio)
203	✓		203	••••••••	**Morning Mayhem** (Sports Studio)
204	✓		204	••••••••	**Morning Mayhem** (Sports Studio)

FIGURE 4.5-12 Multi-line SIP registration configuration on a broadcast talk show system
Kirk Harnack

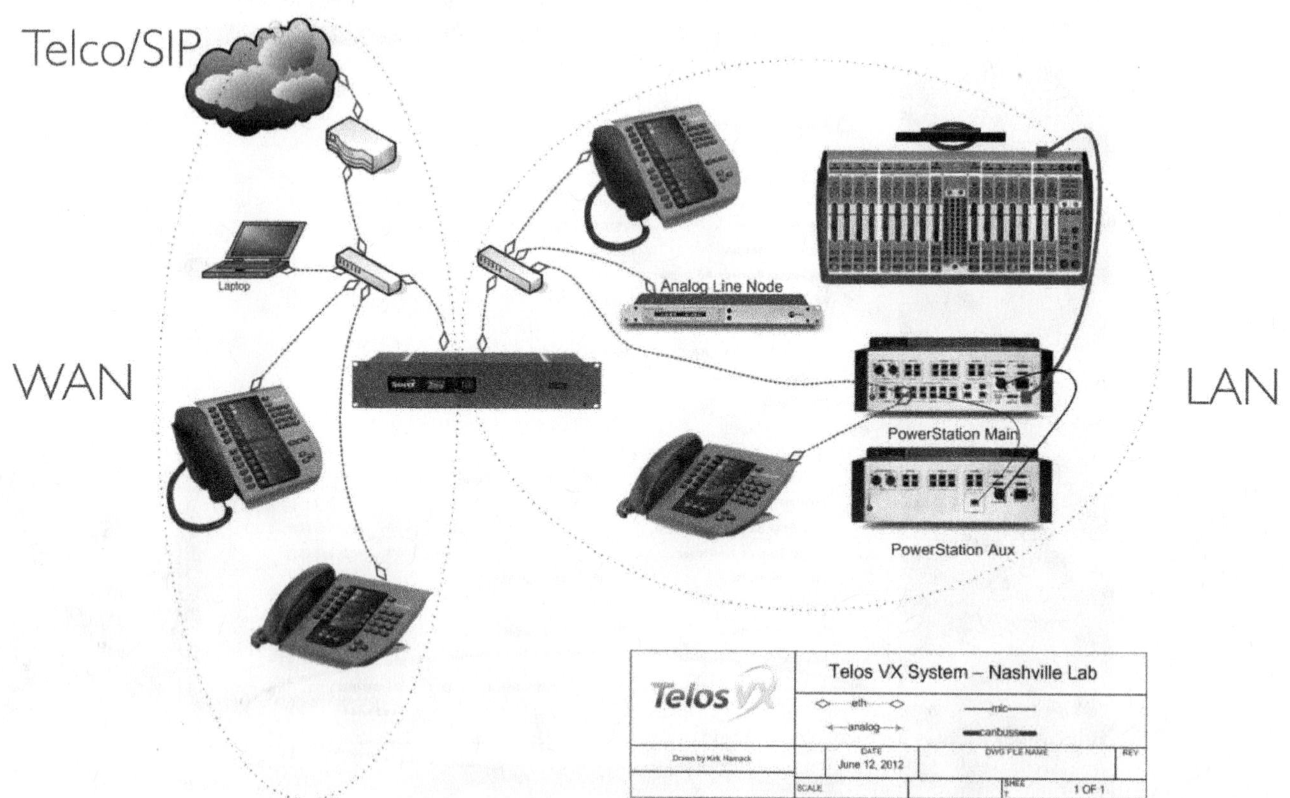

FIGURE 4.5-13 Network and equipment layout for a VoIP + AoIP on-air phone system
Kirk Harnack

Unlike a POTS connection, a VoIP connection is being updated every few seconds. The user agent (software phone) in the VoIP hybrid initiates a handshake to the PBX with which it's registered.[38] And unlike both POTS and ISDN talk show systems, where lines are switched as desired to one or two hybrids, a multi-line VoIP talk show system terminates each line or extension to a full-time user agent. For example, a VoIP talk show system with 50 incoming VoIP lines will have 50 full-time IP connections to 50 user agents. The data rate for each termination is very low until a call is placed or received. Typical data usage for calls is about 90 kbps. For example, 11 simultaneous phone calls will consume about 1 Mbps of data in each direction.

Interfacing to PBX Systems

In a radio or TV studio, there is business beyond talking with callers or interviewees on the air. Coordinating station business is important, and "private" lines or PBX extensions are helpful to this end. Rather than have a separate air and business phone in each studio, it's often a good choice to put all of those lines, public and private, on your on-air phone system. Also, using direct inward dial (DID) numbers for hotlines, warmlines, and special lines is far cheaper than POTS service and usually sounds better.

With older analog POTS-based phone systems, any PBX lines would need to be provisioned as POTS stations and would simply be assigned as a line on the on-air phone system.

If a PBX uses proprietary phones only, there is no real way to integrate them with on-air phone systems.

With a SIP based on-air phone system, it's easier to fully integrate with a PBX than ever before. Most business PBXs support the SIP protocol these days, and it's generally just a matter of configuration. Many of the commercial PBXs available require licensing of stations, or trunks. You pay for what you use. The open source Asterisk PBX does not have this limitation, though some commercial releases of it do use a licensing model.

When selecting a PBX, make sure that the system supports "third-party SIP endpoints" and that any limitations of the PBX's licensing model or hardware compatibility do not preclude you from accomplishing your goals. Some products have limitations on the number of line appearances or endpoints from a particular IP address. It's often better to simply install an Asterisk system and connect it to the business PBX via a trunk between the systems.

Multi-Line Studio Telephone Interface Systems

Multi-line broadcast telephone systems are a great way to handle all telephone needs within one system. The workflow of most of these systems emulates the key telephone, with multiple lines and a hold button. Most systems are POTS, but systems are available that directly support ISDN BRI, SIP, and even mobile networks, directly or through gateway devices.

There is usually a control surface that selects the line to be connected to a 2 wire to 4 wire (separate transmit and receive) hybrid. Mix-minus audio is sent to the caller, and the receive audio goes to the console or router. The feed to the caller is "minus" the phone output itself, so that you aren't feeding the phone to itself, which is confusing to the caller.

Many of these systems have some audio processing, including leveling and even EQ.

The systems can interface to computers for call screening and database functionality.

Using VoIP in a Broadcast Facility

It's likely that any PBX that is bought at this time will be a VoIP-based PBX. Using VoIP within an office or studio facility is simple and has little risk. Best practices include keeping the voice and data networks separate or using deterministic, managed switches to reduce jitter or packet loss caused by other users' network traffic.

Voice over IP is a "real-time" process. It's important that VoIP data packets be transferred from end-to-end consistently and reliably, so that the resulting decoded voice audio is smooth and continuous. Packet buffering is used in endpoint devices (IP telephones) as a basic technique to smooth out small variations in packet delivery, also known as "jitter." Jitter on a network causes audio dropouts, ticks and pops, extra delay, and even misaligned syllables when packets come in out of order. Buffer sizes, measured in terms of time, are typically 100 to 200 milliseconds. Buffer times longer than 200 ms result in an unacceptable delay in communications between those talking.

As VoIP engineers and users, we have less control over packet transfer characteristics outside our premises. However, inside our on-premise network it's important to handle these real-time VoIP packets expeditiously.

Considerations for External VoIP

We'll define "external" as any use of VoIP outside of the broadcasting facility, and on the "exiting side" of the firewall. Most often, this means SIP providers (ITSPs) delivering service over the public Internet. "External VoIP" also includes PBX extensions located "off premises." This could include at news bureaus, sales offices, the owner's home, or the transmitter sites. Where possible, do not expose your VoIP system or extensions to the outside world. If you must, expect to be reviewing logs and applying security patches continuously. The best approach is to block everything and allow only known traffic into your system. Never expose ports 5060–5082 (SIP) or Ports 443 and 80 (HTTPS and HTTP) directly to the outside world. Your PBX—or even endpoints—will be hacked by "bots" to the point of uselessness. Hackers and bots may never "get in," but the endless dictionary attacks will become, effectively, a denial of service attack.

VPN tunneling is one good solution to prevent hacking and even hacking attempts.[39] Individual remote IP phones may be configured to first establish a VPN connection to the VoIP PBX's network before sending login credentials. A VPN connection common to a remote office or facility provides a more convenient approach when both ends are under common IT governance. In this case, all remote IP phones, as well as other hosts, share a common, secure tunnel to the main location's network.

Another approach to remote VoIP connection security includes access control lists (ACLs) combined with DNS and fully qualified domain name (FQDN) specificity from the login client. This technique offers security at the IP firewall, preventing "script hackers" from using simple IP address and port scanning to identify their marks.

The rule set for your firewall is of critical importance. Never put a device "naked" on the Internet, or use a router's "DMZ" feature to avoid a proper remote access setup.

Transmitter sites are frequently fed by digital IP radios that can effectively extend your network out to the site. These work well to carry VoIP telephones, video cameras, program audio, and remote control to your site.

Connecting fixed, remote facilities together with reliable IP voice and other audio is becoming more commonplace within broadcast organizations. The public Internet may not provide the packet-by-packet desired reliability. Many ISPs offer non-public Internet connectivity where MPLS is an option. Multi-protocol label switching (MPLS) is a technique for prioritizing packets over ISP's networks. MPLS operates at a layer that is generally considered to lie between traditional definitions of OSI Layer 2 (data link layer) and Layer 3 (network layer), and thus is often referred to as a layer 2.5 protocol. It was designed to provide a unified data-carrying service for both circuit-based clients and packet-switching clients that provide a datagram service model. It can be used to carry many different kinds of traffic, including IP packets, as well as native ATM, SONET, and Ethernet frames.[40] MPLS networks are engineered to support high-quality voice and video service with some quality of service (QoS) guarantee. As such, broadcasters have the option to specify MPLS to ensure high-priority packet handling over an ISP's network, even when handed off to intermediate carriers. MPLS service is useful to broadcasters needing packet reliability for remote audio codecs.

Considerations for Internal VoIP

Excellent performance from internal data networks is relatively easy to attain and maintain by adhering to good IT engineering practices where they're concerned.

Keep voice traffic prioritized over other data traffic. This is ideally done using a physically separate Ethernet network for IP telephony. Such a separate network is often desirable anyway, as power over Ethernet (PoE) switches may be employed for powering the IP telephones, thereby reducing office wiring and under-desk clutter.

Power Over Ethernet

PoE switches provide DC power to VoIP telephones over the same Ethernet cable carrying network data. However, PoE switches cost more than non-PoE units. As such, for best economy they are most economically used only where their power-providing capabilities are needed on a majority of their ports. This bolsters the case for employing physically separate IP networks for VoIP and general office data. Using PoE switches also simplifies AC power wiring and outlet use at each telephone's location, as a "wall wart" to power each phone locally is not needed.

PoE Ethernet switches, as their non-PoE counterparts, may be of the "managed" or "unmanaged" variety. Managed PoE switches will likely offer the engineer extra monitoring tools and alerting mechanisms. For example, current (or power) draw from each Ethernet port may be monitored with alarms available for changes or disruptions. This allows easy checking on any IP phone's presence and power drawing status.

A good network will provide audio quality and signaling reliability that is indistinguishable from a TDM network.

Interconnecting External and Internal VoIP Systems With Broadcast Gear

Whether all telephone service in a broadcast facility is via VoIP or not, the technology provides a valuable way to interconnect equipment and route calls within studio facilities. An important benefit is the possibility to integrate all of a facility's networks into one infrastructure. Most stations now have an Ethernet data network, a phone PBX system, and AES3 and analog audio connections. A professional audio-over-IP (AoIP) system, such as Axia's Livewire+ or AES67, allows audio and data to share one network. VoIP is the final piece that consolidates all the different pieces into one unified system. Another benefit is the possibility to have tighter integration of the office PBX system and the on-air phone system, so lines can be shared among studios, calls can be easily transferred between the studios and offices, etc. This has always been a problem with traditional PBXs due to their closed design and the limitations of analog connection signaling. VoIP solves this with its open standards and rich signaling protocol.

A variety of audio codecs can be used with VoIP. Some have better quality than traditional voice codecs, and these have the potential to enhance on-air quality for phone segments.

An important distinction between VoIP on local networks and over WANs that carry calls offsite is that LANs have plenty of bandwidth and are completely under local control, so high quality is achieved with a bit of care. Transmitting a call over the public Internet is quite another matter. There is no guarantee of any kind with the Internet, so good quality may occur on one day and annoying dropouts the next.

Broadcast on-air systems may use VoIP in a number of ways. Simple hybrids could interface to VoIP PBXs.

Elaborate multi-studio setups could use VoIP as their internal interconnect, and IP could be used as a transport for offsite connections to remote announcers and intercoms.

On-Site VoIP PBX

The elements needed to assemble a VoIP system are summarized below:

- Gateway—The element that interfaces the PSTN to the IP network. It converts the audio from either analog POTS or ISDN to the IP packet format. It also translates the signaling from one to the other—typically SIP. The gateway is only needed during the transition period when phone calls are coming to the studio via the traditional circuit-switched network. When a facility transitions fully to SIP trunking or a hosted SIP PBX, gateway functions are no longer needed. It's also possible to use gateway functionality contained in a business-use VoIP PBX to handle studio lines. Popular recently has been to replace an older, traditional PBX with an Asterisk-based PBX. The Asterisk PBX can contain line cards to accommodate POTS, ISDN PRI, or similar, traditional telco services, while integrating Ethernet-delivered SIP services as well. The Asterisk PBX can unify these sources and provide PBX services to both business and studio lines internally.
- Softswitch—The controller for the call-processing functions; for example, it listens to the keystrokes from IP telephone sets and initiates calls by sending commands to the gateway. Even though it is called a "switch," no audio passes through it. The actual interconnection of audio is done in the Ethernet switch. This offers an important benefit—the "soft" in the name emphasizes that its characteristics can be much more easily changed than in the traditional system with the control and switching tightly bound in a single device. Some vendors refer to their softswitch products as call servers or communications servers.

PBX Functionality

PBX application server—An appliance or PC-based device that provides functions beyond basic switching. Voicemail is the most common, but other sophisticated tasks such as might be required by a high-volume call center would be handled here. Any number of these may be attached to the network to perform a range of tasks. The gateway, softswitch, and application servers could be independent physical boxes, each individually connected to the Ethernet, but most products for small businesses have these functions integrated into a single package. Products in this category are provided by many of the well-known PBX manufacturers such as Nortel, Siemens, Avaya, and Mitel. Vendors from the computer-networking world such as Cisco and 3Com also have popular products. More recently, there are many more offerings from firms that started up after or just as VoIP telephony has become popular. Grandstream, Audiocodes, Digium, Elastix, Rhino, and Sangoma are a few common and respected VoIP PBX manufacturers. Some engineers—interested in the technology—have built their own VoIP PBX systems using desktop or server-quality PC hardware and one of several implementations of Asterisk PBX.

IP router/firewall—The device typically used for connecting Ethernet to the Internet or private company WANs. This device examines the packets on the network to determine which should be sent outside or kept in the local network. It performs the same function in the other direction, usually with a firewall service that prevents unwanted traffic from entering the broadcaster's network. Phone calls that arrive at the studio via IP will pass into the network via this box rather than the gateway. Sometimes the router and firewall are included as part of the gateway and softswitch.

Broadcast interface—A specialized device made by a broadcast equipment vendor to support studio telephone operations. It converts the audio from the phone format to the professional audio format necessary to connect to studio mixers, including sample rate conversion. It provides the hybrid function to cancel echo as well as the broadcast-specific switching and conferencing.

Audio nodes—Required to interface to legacy analog and AES-digital studio equipment. When studio equipment based on AoIP is being used, there is no need for this interface because the packets can flow directly to the audio routing and mixing elements.

IP phones—Connect to the Ethernet, but otherwise work like usual business telephone sets. They range from simple single-line sets to ones that have multiple features and line buttons, as well as an LCD screen that can even display rich graphics.

Studio console—May be equipped with a call controller module that talks to the broadcast interface to select lines for on-air transmission. This generally will be a control-only device, with no audio passing through it.

IP softphones—Run on PCs, providing similar functions as the dedicated hardware phone sets. A headset/microphone plugged into the soundcard or USB jack is used in place of the usual handset.

Producer/screener applications—These are similar to the call screener packages described earlier, but because VoIP systems put the phone audio on the network, the PC can serve as a softphone as well as a controller, or it can work in conjunction with an associated VoIP telephone set.

One of the advantages of VoIP is that it can readily scale from serving a single studio to accommodating the demands of sophisticated multiple studio facilities. Ethernet switch ports are numerous and low cost. Plugging in more phones, controllers, and PCs for expansion is straightforward.

Interfacing With a Commercial VoIP PBX

The Session Initiation Protocol—SIP—defines a standard framework for different VoIP systems to communicate with each other. SIP allows VoIP telephones from one manufacturer to work seamlessly with a VoIP PBX from another manufacturer. As long as each manufacturer supports SIP, then basic functionality should be assured.

Some makers of VoIP PBXs would prefer their customers to buy into their entire ecosystem. In such cases, generic SIP devices may not work properly with their VoIP PBX products.

Broadcast-oriented SIP VoIP equipment is typically nonproprietary in its SIP implementation. Indeed, current broadcast industry VoIP offerings are designed to work with Asterisk PBX, which is an open-source software PBX system.[41]

More than a few broadcasters have been surprised to learn their business "VoIP PBX" is actually a proprietary system, unable to integrate with third-party equipment, or to discover that "SIP extensions" on their PBX require annual licensing fees to enable standard SIP functionality with third-party devices.

Fortunately, most VoIP PBXs will interconnect just fine with third-party VoIP phones. This interoperability is the touchstone that implies compatibility with broadcast-oriented VoIP equipment, such as intercoms, phone hybrids, and talk show systems. Indeed, a good question to ask your PBX vendor is, "Will this VoIP PBX work with 'generic' VoIP phones?" A positive answer implies that it should work fine with broadcast-oriented SIP VoIP equipment.

The physical interconnection between a VoIP PBX and, for example, a VoIP broadcast talk show system is an IP network. The VoIP talk show system will appear as some number of SIP extensions off the VoIP PBX. Each extension represents a "line" available for incoming or outgoing calls. A modern, multi-line VoIP talk show system is typically capable of accepting extensions (or "lines") from several VoIP PBXs. This implies that a broadcast VoIP system can have some lines coming from the broadcaster's business phone system, while other lines come from a VoIP PBX that's dedicated to high-volume call-in numbers.

Considerations for a Dedicated Broadcast VoIP PBX

A VoIP PBX for broadcast use should be extraordinarily robust, especially from a hardware standpoint. It will be ignored and abused but must continue to provide service without interruption or intervention. When selecting or specifying VoIP PBX hardware for on-air use, consider the following attributes:

Redundant power and a RAID 1 (mirrored) hard disk set are a good idea. Adequate cooling is vital for long hard drive and component life overall. Smaller rack mount machines tend to have tiny screaming fans just to move enough air through them to keep the machine running. They also usually have special smaller form factor motherboards, which can be difficult to replace in a hurry. A larger rack mount case will accommodate just about any hardware and permit the use of larger, lower RPM (quieter) fans.

An adequate number of Ethernet interfaces is necessary for a VoIP PBX. Most server motherboards have two onboard gigabit ports at the time of this writing.

The system should have enough CPU horsepower and RAM to be able to process large numbers of call setup requests during periods of heavy calling for contests and the like. A server that would run a typical office, such as an Atom-based system, should be considered inadequate for broadcast use.

Also important for a VoIP server is the number and type of free interface "slots." You will need a machine capable of accommodating "long expansion cards" if analog POTS interfaces and expansion cards for things like T1s and PRIs are to be used. In the near future, it is expected that the need for these types of interfaces will fade and the mere mention of this requirement will amuse the reader.

Improving Phone Audio Quality

POTS, ISDN, and T1/E1 Audio Optimization

Whether extracted from analog or digital lines, audio that originates from a traditional telephone set is hardly studio quality. Until everyone has a high fidelity VoIP set, techniques to improve inherent phone audio need to be considered.

One important quality limitation results from the anti-aliasing and reconstruction filters in PSTN G.711 codecs. These filters usually have an ultimate roll-off of around 35 dB. Audio above the 4 kHz Nyquist frequency will alias and appear in the 300-Hz to 3.4-kHz band as distortion; thus, typical codecs have distortion of 2% to 3% caused by aliasing. The raspy noise that seems correlated with the speech sometimes heard on a telephone circuit is a result of the effects of this kind of distortion combined with audible quantization errors. Older codec filters used switched-capacitor technology, which can be noisy.

An ISDN or VoIP connection at the studio solves half of the problem, because at least one of the telco codecs is bypassed. Still, the caller's phone set and connection are unknown and out of the station's control. Fortunately, there are some remediation possibilities. Filtering, equalization, gating, and dynamic compression are the primary tools. Most commercial interfaces have some or all of these processes built in:

- Filtering—On a dial-up phone line, there is very little audio above 3.4 kHz, but there are noise and digital reconstruction distortions. Thus, a filter with a steep roll-off above the telephone passband will reduce phone line noise significantly without affecting conversation audio. Low-frequency hum is often a problem, as well. This is usually 60 Hz mixed with its second harmonic, 120 Hz. It is often a good idea, then, to have a sharp roll-off starting at about 200 Hz.
- Equalization—An equalizer (EQ) used to shape the frequency response of the phone line within its audio bandwidth can result in marked improvements in perceived quality. A typical phone line has an excess of energy at around 400 Hz and considerable roll-off at both the top and bottom ends of its passband, so the idea is to compensate by adding gain at both. Boosts at

2.5 kHz and 250 Hz and attenuation at 400 to 500 Hz with a parametric equalizer will help achieve better sound.

When it is not possible or practical to make custom adjustments, an adaptive multiband EQ can be an effective tool. Some advanced hybrids include this function. The principle is much the same as implemented in broadcast transmission processors. Audio is filtered into multiple bands, and an automatic gain adjustment is performed on each spectral segment. Given the limited frequency range of telephone calls, three bands are sufficient.

- Noise gating—Another effective processing device is the expander or noise gate. These devices may be used to reduce gain between the words of a conversation, thus making phone line noise less objectionable. On extremely noisy lines, however, the gating action can make noise more distracting by causing it to come and go with the words. In such cases, it might sound better to leave the gate off and let the noise remain present at a constant level.
- Dynamic compression—Levels on phone calls vary widely. A compressor helps to smooth this out. An AGC that maintains a constant compression ratio regardless of average gain reduction produces more consistency. Freeze gating is also important so gain does not increase during caller speech pauses.

VoIP Audio Optimization

Audio quality on VoIP talk show systems is typically an improvement over the equipment and connections it replaces. However, mobile calls still represent a broad range of equipment, codecs, environment, and signal conditions. Some calls from mobile users may be nearly unusable, especially if the caller is using a Bluetooth-connected mic and earpiece, or an inline earbud microphone. Such callers also sound bad on traditional POTS hybrids, so this is no fault of VoIP technology.

One of the first steps to better caller audio quality is to disallow callers from getting on-air when they're using a mic that's not near the mouth. This includes wireless mic and earpiece sets, inline earbud mics, or the speakerphone function of their mobile device. Callers should be in a reasonably quiet place with the phone or a wired headset close to the mouth. In most cases, simply asking the caller to handhold the phone to their ear and cheek can make a significant improvement.

The balance of improving VoIP audio quality involves optimizing the VoIP packet delivery into the hybrid or talk show system. If the SIP call connection is good, with packets timely and reliably delivered, then the same audio optimization techniques described above apply to VoIP connections. Perhaps the only difference would be in the case of a SIP connection using G.722 for 7-kHz audio. In this case, the narrowband filtering used to reduce aliasing distortion should be automatically switched to accommodate the wider HD voice codec.

Because IP remote broadcast equipment uses the same underlying technology as VoIP, the same network performance issues apply as for VoIP. Notably, VoIP telephony and IP remotes may well eventually converge to use the same equipment.

Broadcast codecs are usually single-box solutions that include an ISDN TA and/or IP interface and a number of selectable coding algorithms. Most are full duplex, with provision for transmitting and receiving simultaneously. Some offer a feature to allow connecting to POTS phones for voice communications or to POTS codecs for higher-than-phone fidelity over analog lines. End-to-end, parallel contact closures offered by many codecs may be used to control playout automation systems and other devices. Some portable units include a mixer for multiple audio inputs and outputs. Many include a receive-side mixer to combine the mix-minus signal from the studio with the local audio.

HD Voice Connections

HD voice, using a codec such as G.722 or AMR-WB, is possible between a data-connected smartphone (or tablet) and VoIP broadcast phone system. This connection, however, cannot traverse the PSTN. In practice, the remote smartphone will run a "softphone" app, such as Linphone, Acrobits Softphone, Grandstream Wave, or CSipSimple. This app is configured to register with a VoIP PBX, typically at the broadcast facility. Now the G.722 smartphone app is registered on the same PBX as is the VoIP talk show system at the studio. A simple extension-to-extension call is placed, and an HD voice SIP connection is negotiated.

From anywhere in the world, then, a good-quality remote audio connection is established and may be placed on-air. The mic and audio subsystem in a smartphone is typically as good as or better than the G.722 connection. An external mic or even a mixer may be connected to the smartphone or tablet for more sophisticated audio production.

BROADCAST CODECS: HIGH-FIDELITY REMOTES OVER POTS, ISDN, AND IP

Audio Coding

Audio coding in the context of VoIP has been discussed. A new class of codecs has been designed for better fidelity than those made for VoIP telephony application. High-fidelity audio coders are able to reduce the bit rate of a digital audio signal to as little as 5% of the original size, while preserving fidelity very close to the original. There are a number of these audio coding methods, each with advantages and drawbacks.

ADPCM Coding

Adaptive differential pulse-code modulation (ADPCM) predates MPEG perceptual coding and is much simpler than the perceptual methods but has much poorer performance.[42] It achieves data reduction by transmitting only the difference between successive samples. G.722 dates from the late 1970s.

It is a sub-band ADPCM codec that has a frequency response extending to 7 kHz at 56- or 64-kbps bit rate.[43] Unless there is no alternative, it should be used only for voice feeds, as music transmitted via G.722 has a distinct fuzzy quality. It is good also for cueing and intercom channels. Only 2 bits are allocated per sample for audio frequencies above 4 kHz—sufficient for conveying the sibilance in voice signals but not adequate for intricate musical sounds. Also, the predictor model used to determine the step size in the adaptive function is designed only for speech. G.722 has the lowest delay of all popular coding methods, about 20 ms.[44] For this reason, it is often used as a return channel so round trip delay is reduced, even when a higher fidelity method is used for the on-air feed. G.722 uses a procedure called statistical recovery timing (SRT) or statistical framing to lock the decoder to the data stream. The process usually happens instantaneously but can take up to 30 seconds. The locking can be sensitive to audio present on the G.722 path, as it relies upon the properties of the audio bit stream itself. Some audio material and tones can prevent lock altogether. Low-level noise is the most reliable signal for locking, and undistorted voice is usually acceptable. There can be problems with sine-wave tones and distorted voice or music signals, in which case removing or lowering the audio by about 12 dB or so for a few seconds will generally cause lock to occur. In rare cases, it may be necessary to disconnect and redial. Tones and noises may be present before locking occurs, and some continuous audio tones may cause momentary unlocking.

Perceptual Coding: MPEG

By far, the most popular perceptual coders rely upon techniques developed under the MPEG umbrella. The MPEG process is open and competitive. A committee of industry representatives and researchers meets to determine goals for target bit rate, quality levels, application areas, and other parameters. Interested organizations that have something to contribute are invited to submit their best work. A careful, double-blind listening test series is then conducted to determine which technology delivers the highest performance. The subjective listening evaluations are done at various volunteer organizations around the world that have access to both experienced and inexperienced test subjects. Finally, results are tabulated, a report is drafted, and a standard is issued. In 1992, this process resulted in the selection of three popular related audio coding methods, each targeted to different bit rates and applications. In 1997, another algorithm, advanced audio coding (AAC), was added to the MPEG suite.[45]

All of the MPEG codecs rely upon the acoustic masking principle—a unique property of the human aural perception system.[46] When a tone is presented at a particular frequency, we are unable to perceive audio at nearby frequencies that are sufficiently low in volume. As a result, it is not necessary to use precious bits to encode these inaudible, masked frequencies. In perceptual coders, a filter bank divides the audio into multiple bands. When audio in a particular band falls below the masking threshold, few or no bits are devoted to encoding that signal, resulting in a conservation of digital bandwidth that can then be used for the bands where it is needed. Masking also occurs in the time domain, with low-level signals occurring shortly before or after a masking event being inaudible.[47]

Predominant MPEG coding technologies include the following:

- MPEG Layer 2, which is widely used on satellite links and high-capacity terrestrial paths such as primary ISDN or T1 channels. Layer 2 is the standard for European Eureka 147 terrestrial digital broadcasting.
- MPEG Layer 3 (MP3), which is perfectly matched to the bit rates available on ISDN BRI lines, permitting full FM broadcast quality. Full-fidelity 15-kHz mono is possible on a single ISDN B channel, and near-CD-quality 20-kHz stereo is achievable using both ISDN B channels. It is widely supported in broadcast codec equipment from a number of manufacturers.
- MPEG AAC is a very powerful audio coding method. According to careful tests, it achieves quality indistinguishable from the original at 64 kbps per mono channel and has approximately 100% more coding power than Layer 2 and 30% more power than the former MPEG performance leader, Layer 3.[48]
- MPEG High-Efficiency AAC, previously called aacPlus, which uses spectral band replication (SBR) to add yet more coding power to AAC.[49] It has reasonably good fidelity down to 24-kbps mono and 48-kbps stereo. The codec used for the U.S. HD Radio™ system also uses SBR and is similar to this one.
- MPEG AAC-LD, which is a low-delay variant of AAC. It is not quite as powerful as AAC but has about a third of the delay.
- MPEG AAC-ELD is the high-efficiency version of AAC-LD. This codec offers both low-bitrate efficiency and low-delay encoding and decoding.
- MPEG xHE-AAC saw its first commercial implementations in 2015. This codec addresses the differences in coding both speech and music at very low bit rates. It quickly adapts to the dominant audio type—speech or music—to provide very low coding artifacts even at low bit rates.

Cascading Codecs

When multiple codecs are used in sequence, there is the chance that audio could be impaired by the additive effects of the combined coding processes. Discovering which combinations are satisfactory and which create audible problems is often a matter of trial and error. As a general rule, the goal is to get as much coding headroom as possible at each stage. This is achieved by using the highest possible bit rate at each stage of the signal chain or using the most powerful coding method of those available at each stage.

ISDN Codecs

ISDN has been the premium high-fidelity transport service for over 20 years. It offers fully guaranteed bandwidth, very low delay, and sufficient bit rate for studio-grade audio when used with an audio codec such as MPEG Layer 3 or AAC. These codecs are a near perfect fit to the ISDN capacity: One B channel is enough for mono, while both are combined for stereo operation. ISDN is well established, and many if not most radio stations have at least one ISDN codec on hand.

The chief downside in the United States and many other countries is that new ISDN BRI service is no longer being offered in most locations.[50] Moreover, existing ISDN BRI service has seen triple-digit rate increases since major telcos began announcing an end to ISDN expansion. Assuming ISDN BRI is available, the time-honored downsides have included long distance that are per minute, and the time-based costs can add up. ISDN is not widely used for either general telephony or Internet access, so it is not nearly as ubiquitous as POTS or IP connections. It has to be specially ordered well in advance of need.

Broadcast codec equipment is used much like a telephone. In the United States, the ISDN basic rate interface telco line is connected via the U interface. In Europe, the telco provides the NT1 device and connection is via the S interface. Most codecs accommodate both. Calls are dialed using familiar telephone numbers on a standard keypad. Users select from among the various provided codecs, with care that both ends are operating with the same algorithm. Often the receive-side codec can automatically detect and adapt to the incoming signal.

IP High-Fidelity Codecs

IP connectivity can be provided by wired Internet connections, available WiFi, or cellular wireless data such as 4G-LTE. As noted earlier, the public Internet is attractive due to its ubiquity and low cost, but it offers no performance guarantees; however, it may be useful for noncritical applications, especially if adaptive technology is used to adjust the codec optimally to network conditions. IP service-level technologies are available, such as MPLS, which provide guaranteed—or prioritized—service levels. These are intended for VoIP application, but can certainly be used for broadcast remotes, point-to-point backhaul, and wider-area IP-audio distribution.

The core technology for high-quality IP audio is the same as used for VoIP telephone systems, but better codecs enable higher fidelity. Not all high-quality codecs are useful for IP applications, though. A suitable codec must have effective mechanisms for concealing the audible effects of dropped packets that occasionally occur on almost all IP links. Some otherwise useful coding algorithms have no such concealment, and depend on perfect data transfer. Delay is another concern. Because IP networks will have more delay than ISDN or POTS, codecs with low delay are preferred for two-way interactive applications to keep the total delay within acceptable range.

When IP codecs are to be used on nonguaranteed networks such as the public Internet, adaptive buffers are an important part of the receiver system. These detect the jitter in the network and can dynamically adjust for optimum performance. When jitter is low, the buffers can be small, so delay can be minimized. When jitter is high, the buffer is extended so dropouts are reduced. Of course, this comes at the expense of longer delay. A time-stretching/contracting algorithm allows the buffer to adjust inaudibly while the program is ongoing. A further refinement makes the codec bit rate variable to adapt to available network bandwidth.

IP codecs may sometimes have to be used behind firewalls or network address translation (NAT) devices—IP routers. A "rendezvous server" or "transversal server" placed outside the firewall or NAT allows remote codecs to call and connect to ones inside the firewalled network. It works on the principle that firewalls and NATS will usually open an incoming path in response to an outgoing one being initiated. When it is powered up, the inside codec makes an outgoing connection to the transversal server, which then opens a return path and keeps it open with occasional pings. The remote codec can then ask the transversal server to mediate a connection request to the inside codec. Name servers are often integrated with the transversal server to allow text names, rather than IP numbers, to be used for dialing.

MIX-MINUS

Both phone hybrids and broadcast codecs require mix-minus in order to provide a meaningful backfeed to talent in typical studio/remote setups. An example of a mix-minus setup is shown in Figure 4.5-14. All perceptual audio codecs have too much delay for on-air talent on remote to hear themselves via a round-trip loop; therefore, a mix-minus arrangement is required so the remote on-air talent's voice is not heard via the studio return. Instead, the remote talent's microphone is mixed locally with a studio feed that has a mix of everything minus the remote audio—thus the "mix-minus" designation. European broadcasters call this "M-1."

A second problem is caused by hybrid leakage. The mix-minus sent to the remote site will contain residual leakage of the announcer audio being sent to the caller. Because the codec causes this to be delayed, the announcer experiences this as a low-level echo in the headphone feed. Because the longer the delay, the more noticeable this echo, choosing coding algorithms with reduced delay is part of the solution. A digital hybrid with maximum trans-hybrid loss is required. If it has variable override (caller ducking), the amount can be increased when remotes are in progress to enhance the effective trans-hybrid loss. To achieve the best performance, use an ISDN or a VoIP hybrid.

With regard to delay, it's the round-trip that counts—the sum of both send and return path delays. This can be reduced in a typical remote broadcast by

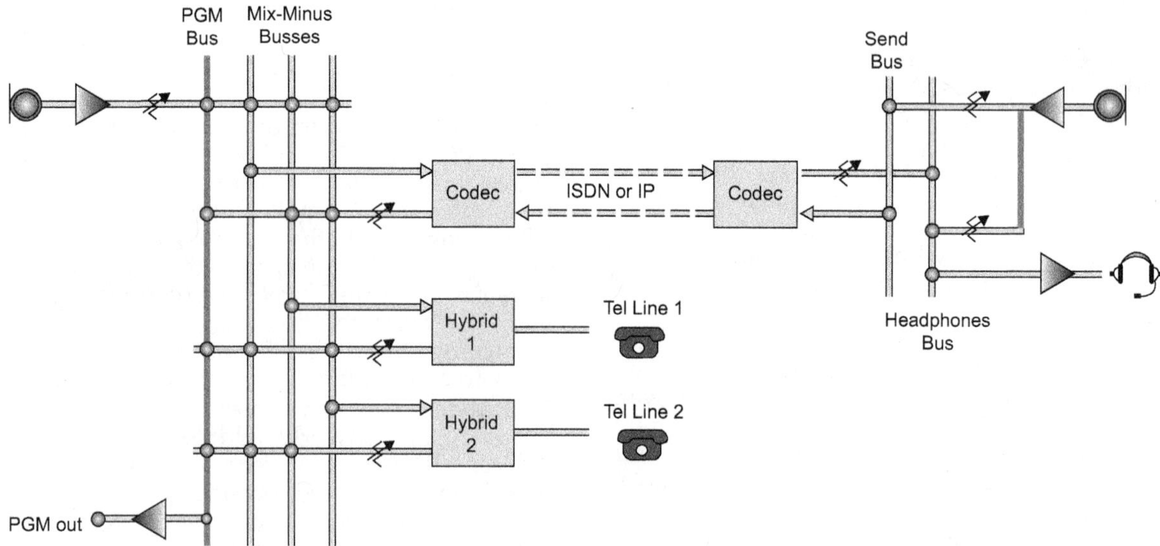

FIGURE 4.5-14 Simplified block diagram shows the mix-minus required for hybrids and codecs

using a low-delay algorithm such as G.722 for the return cueing path.

Even without codecs in the picture, hybrids require mix-minus feeds. Without them, hybrids will generate feedback when forced to "chase their tails." A multiple mix-minus setup is also essential for conferencing multiple lines.

Most modern consoles have built-in mix-minus, often with provision for selective switching of sources into the feed. With codecs and hybrids playing such an important role in modern broadcasting, the trend is for consoles to have a mix-minus return capability associated with each source and fader.

Mix-minus busses may fed be pre- or post-fader. Post-fader is preferred, so callers and remote sites hear the on-air mix; however, phone hybrids are often used as fancy speakerphones for off-air conversations or recording. In this case, microphones need to feed the phone even when their mix channels are off and faders down. Some consoles are able to automatically switch between the two modes depending on whether the phone channel is active to air. If the console does not have a dedicated mix-minus, aux sends can be configured to create mix-minus busses. This is common in TV audio, where hybrids, codecs, and satellite feeds are routine.

INTERFACING PRODUCTION INTERCOM SYSTEMS

To aid communication with the field crew during remote broadcast projects, connecting the production intercom system to dial-up telephone lines is often desired. Smooth integration of live news remote feeds, for example, requires that production personnel at all locations be able to communicate with each other in a simple, trouble-free fashion. This is especially true when multiple remote sites are involved, such as for election coverage, major sporting events, and telethons. Ideally, crews at each location would use the intercom system without regard for the distances involved. Most often, access to the dial-up phone network is available by wire or cellular, so an interconnection of the intercom system to the telephone network is a common solution. Intercom system manufacturers typically offer hardware modules that interface to a POTS or ISDN line. More recently, IP network interfaces are becoming available for VoIP system connection to intercom systems.

Four-Wire Intercom Systems

Four-wire systems are those in which the two speech directions are kept separated in the switching and distribution process. Although it would be possible to use special four-wire telco circuits (or two standard loops) to maintain independent signal paths to remote sites, it is more economical and convenient to be able to use a single phone line. To accomplish this, an effective conversion between the two-wire phone line and a port on the four-wire intercom matrix is needed. It will be necessary to separate the send and receive speech signals on the phone line with a hybrid.

Trans-hybrid loss performance will be important when intercom stations with open loudspeakers and mics are to be used and when conferencing of multiple telephone lines is desired. In the first case, the acoustic coupling between the speaker and mic completes a feedback path that includes the hybrid. Clearly, the better the hybrid's isolation, the higher the feedback margin will be. In the second case, a feedback path exists from each active hybrid through all of the others that are conferenced to it. When the total gain exceeds unity, feedback results. The goal is to have the best possible trans-hybrid loss so the maximum line-to-line gain may be achieved.

An auto-answer and disconnect function may be required for unattended operation. This circuit responds to a phone line ringing signal by activating the hybrid and deactivates the hybrid when the calling party hangs up. A dial-tone detector may be necessary to ensure reliable operation. The tone detectors are connected to respond to signals on the hybrid's separated telco receive audio signal. Were this not the case, and the detector was merely connected across the phone line, there would be a major problem when multiple lines are used together in a conference. This is because the tones would be conveyed to each line in use (through the intercom-switching matrix) from every other line, causing all of the detectors to respond to the tones from all of the other lines as well as their own. When one line's interface receives a "disconnect," all of the others would turn off as well. Therefore, there is a requirement in this setup that trans-hybrid loss must be sufficient to be certain that any cross coupling is below the threshold of the tone detectors. The same situation applies with any DTMF detection that is used on a per-line basis.

Two-Wire Systems

Two-wire systems are the popular party-line systems. Here, the interface requires two hybrids connected back-to-back so the intercom hybrid receive output is fed to the phone line hybrid send input and vice versa. Appropriate gain and processing stages are inserted in the four-wire path. This system is what telephone engineers refer to as a two-wire-to-two-wire repeater. High-quality hybrids are required to prevent feedback. The signals can feed around the loop and feedback could build up. This happens when the combined trans-hybrid loss of the hybrids is not at least as great as the gain in the two amplifiers. As telephone circuits have widely varying and unpredictable end-to-end transmission characteristics, interfacing intercom systems to phone lines without gain and AGC is not likely to work very well.

VoIP for Intercoms

Because digital ISDN and VoIP connections are inherently four-wire, they are perfect for the intercom application. Used with a four-wire intercom system, speech paths may be kept separated end to end. Applied to a two-wire intercom system, the problem of maintaining sufficient hybrid balance is eased. ISDN and VoIP lines are less expensive and easier to get than the special analog four-wire lines sometimes used in the past for intercom interconnection. With ISDN, two channels are available so that production and talent busses may be kept separate. Codecs that combine good fidelity and low delay such as G.722, AAC-LD, or AMR-WB are good choices for the intercom application.

TROUBLESHOOTING

When it involves the phone network, troubleshooting can be challenging. Engineers who are comfortable fixing 50-kW transmitters may be lost when dealing with telcos, especially new phone service providers existing only "in the cloud." Perhaps the main source of frustration is that, unlike the rest of the facility, telephone services are not under the engineer's full control. The authors are in a unique position to understand, having been involved for many years with assisting station engineers and phone companies to effectively connect. The first round of troubleshooting should involve simple substitution, such as trying a different line or trying different equipment on the line in question. For IP-based systems, initial troubleshooting involves checking for ping responses from the relevant equipment and service providers. If these do not quickly point out the problem, a more sophisticated approach may be necessary.

About Problem Solving

The principles here apply whether solving problems at the system level or at the component level. At the different levels, the actions required can be very different, but the process is the same. The basic philosophy of troubleshooting any system is to follow these steps:

- Observe the behavior to find the apparent problem.
- Observe collateral behavior to gain as much information as possible about the problem.
- Look at previously troublesome equipment.
- Generate a hypothesis about the problem.
- Generate an experiment to test the hypothesis.
- Fix the problem.

Repeat, if necessary, to attack additional problems. The step-by-step troubleshooting sequence in detail is:

- Step 1—Observe the behavior to find the apparent bug. In other words, determine the bug's symptoms. Remember that many problems are subtle and reveal themselves via a confusing set of symptoms.
- Step 2—Observe collateral behavior to gain as much information as possible about the problem. Does the problem only occur with a specific phone line or piece of gear? Does the audio problem correlate to an alarm LED or odd behavior in some other portion of the system? Try to avoid studying a problem in isolation, but also be wary of trying to fix too many symptoms at the same time. Assume that there is only one problem and work to solve it. Once it is solved, tackle the remaining symptoms, if any.
- Step 3—Look at previously troublesome equipment. At the system level, always suspect the configuration settings, the cables, the telco line setup, the punch-blocks, etc.
- Step 4—Generate a hypothesis. Before changing things, formulate a hypothesis about the cause of the problem. To generate the hypothesis, more data on the problem may be needed. Sometimes you will have no clue as to what the problem might be. Start with the basics of looking for a bad connection or cable, loose connector, or other seemingly unrelated cause.

- Step 5—Generate an experiment to test the hypothesis. Change the ISDN connection to a known good line; call a known good phone or hybrid at the other end. If long distance does not work, try a local call. Plan the tests to eliminate 50% of the possible problems in one test, if possible. Keep track of what is being done to determine what has been eliminated. Keeping notes will also make it easier to run the scenario by a fellow engineer or to discuss the problem with the equipment manufacturer's tech support staff to get a second opinion and reality check.
- Step 6—Fix the problem. Constantly apply sanity checks; for example, just because the telco line was checked last night and it was fine does not mean that it is okay now. Remember that the system worked well at one time and likely will work well again.

Tools and Test Equipment

Given the importance of various telephone and networking to our industry, it is important to have the appropriate tools and test equipment on hand to quickly solve problems. The importance of the various types of lines to the broadcast facility will determine how extensive the collection of telephone-related test equipment should be. In some cases, test equipment must be on hand for immediate use; in other cases, a particular test item may only be needed on rare occasions and could be shared among a number of stations. Online auction sites have made it much easier to purchase test equipment at reasonable prices, in anticipation of future needs. Typical useful test equipment for telephone technology troubleshooting may include:

- For POTS—A *subscriber loop analyzer* is a simple line tester that allows measurement of loop current, ring voltage, and on- and off-hook loop voltage, as well as resistance and capacitance. A *transmission test set* is basically a variable tone generator and a dBm meter in one device. Some units only cover the voice band, while the better units are intended for testing program channels and therefore can operate at a wider bandwidth. Typically, one is placed on each end of a circuit to be tested, with one sending tones and the other receiving. When in the receiving mode, the meter can be used to measure noise and includes the appropriate bandpass filters to give the same measurements used by the telcos. When connected to a milliwatt-tone number at the local CO, it can determine circuit loss. Most transmission test sets do not include the functions of a loop analyzer, although some loop analyzers do include selected functions of a transmission test set. It is important that the correct termination method be used to get accurate level and noise measurements. If the user equipment is left connected to the line (and off-hook), then the transmission test set should be set to *bridge* the line, whereas if the test set replaces the user equipment, it must be set to *terminate* the line with the appropriate termination selected (usually 600 ohms).
- For digital circuits—Different digital technologies offer differing low-level signals, so different test equipment will be required for different types of lines (some offer interchangeable modules to cover more than one type of interface). For ISDN BRI, a tester that can handle both the four-wire (AMI) S interface as well as the two-wire (2B1Q) U interface is desirable. It is quite useful if the unit can emulate an NT1. The ability to act as a telephone is typical and can give a quick nonquantitative test of basic function. An important function that most ISDN BRI testers offer is the ability to do a bit error rate test (BERT). Most can do so by dialing from B1 to B2 (thereby testing both B channels simultaneously); however, some require dialing into a piece of equipment that is able to loop back the bit stream. The BERT function sends a test pattern and then examines the bits coming back to determine if errors have been introduced.
- For T1—In T1 (and low-level ISDN PRI) testing, a similar test set is used. Typically, these can be placed across the incoming or outgoing half of the T1 and can be used to passively monitor the framing bits for improper patterns. They can also determine the level of the incoming signal from the telco, or channel service unit (CSU) if placed after it. If framing errors or bipolar violations are detected, this is indicative of a problem. Most also permit monitoring audio and the robbed-bit signaling bits. A more complete test is to put the T1 tester on the line in place of the user's equipment. In this case, end-to-end bit error rate testing can be done by placing another tester at the far end or by placing the far end into loopback mode. Most T1 test sets include digital transmission test set functions that work just like an analog transmission test set but over a specified channel of the T1. Some T1 testers can emulate a CSU for substitution testing.
- For ISDN PRI—ISDN PRI test sets include some, or all, of the functionality of a T1 tester, but they also have the ability to emulate a piece of PRI user equipment. In this mode, it can make or place calls just as the PBX or telephone system would. Some units can emulate the network, allowing connection of user equipment in the absence of a working PRI. Basic Layer 3 protocol monitoring is often included.

Protocol analyzers are available for both ISDN BRI and PRI (some units handle both). These typically do not include other test functions but do allow viewing the D channel protocol transaction in detail. These are only rarely needed on telco-provided ISDN but are quite useful for troubleshooting compatibility problems with PBX-generated ISDN circuits. User equipment such as broadcast codecs may have built-in protocol analyzers, and some basic ISDN test sets offer limited protocol monitoring.

It's important to note that many of today's digital delivery circuits are terminated in customer-provided equipment (CPE). Whether a stand-alone appliance or a PC card, this equipment typically offers a user interface via a web browser. Most, if not all, troubleshooting information needed can be gleaned from web pages offering status and troubleshooting routines. Familiarity with proper status indications is helpful here. The authors suggest making screenshots of GUI-accessible interfaces when they're working properly, and save for comparison when there's trouble.

Some equipment is controlled and diagnosed using a command-line interface, and plenty of documentation is available in online manuals. These status and diagnostic resources can help an engineer quickly determine if the trouble is with the CPE or with the service provider's line or central office equipment.

Troubleshooting POTS

The most common POTS line issues tend to be with noise—crackle, "static," or hum—or with an unintended disconnect. Another may be when the far end hangs up, but the line remains unavailable to make or receive another call.

For problems with calls hanging up while in use, check the loop current by opening up one side of the line and inserting an ammeter in series with it, or by using a loop analyzer. The loop current should be at least 23 mA but is typically 30 mA or greater. Problems where calls fail to clear when the far end disconnects may be due to lack of calling party control (CPC) or may possibly be related to the user equipment's CPC detection threshold. A differential oscilloscope (or battery-operated unit without a ground reference) will allow for easy measurement of the duration of interruption in loop current present when the far end hangs up, if this occurs.

Noise problems can be difficult to isolate. Because a typical telephone instrument has considerable loss at the high and low ends of the spectrum, often the line in question will sound okay on a phone. However, when placing that same line on-air via a good-quality hybrid, equalizer, and added audio processing, noise from the line may become quite apparent. Testing with a transmission test set (see above) will yield useful information. If the difference in noise (measured across the line) is more than a few decibels different between the 3 kHz and the C-message filters, then the problem may be circuit balance. This is particularly likely if the complaint is a 60-Hz hum. Even if the C-message noise is barely within the telco tariffs (usually 20 dBrnC), the telco may assist if it can be proved that the "longitudinal balance" is not up to specs. To do so, measure the noise to ground (sometimes called "power influence"), which generally should be 80 dBrnC or less. Subtracting the tip-to-ring number (called "metallic noise") from the noise-to-ground number will give the figure for balance, which should be 60 dB or less.

Troubleshooting ISDN BRI

ISDN BRI circuits involve two basic classes of troubleshooting. The first is the physical layer—the S or U interface, as appropriate. Equipment designed for use on a U interface has a built-in NT1, which normally has an LED that shows NT1 status. If the U interface of the NT1 remains in the initializing state (or will not stay initialized for very long), then this indicates the line itself is bad or marginal. However, the NT1 could be malfunctioning. A quick check is to connect an analog telephone to the ISDN jack; if white noise or clicking is heard, then it is probably ISDN, whereas if a dial tone is heard it is not. The trouble may be a bad cord, jack, or inside wire. Try connecting the ISDN device directly to the network interface or demarc point.

If the S interface is used, the NT1 will have an LED indicating the interface's status. The S interface requires two pairs and is usually in the form of an eight-pin/eight-position miniature modular jack (RJ-45 style). There are several different allowable wiring configurations, and these have different termination resistor requirements. A direct connection of a single device to an NT1 using a short 1-meter cord should always work, no matter the wiring configuration.

If the U and S (if present) interfaces are both synchronized and remain so for 60 seconds or more, then the equipment should be able to establish Layers 2 and 3 and perform basic dialing functions. If the ISDN phone or codec in question supports multiple calls—one on each B channel—then calling from one B channel to the other is a good basic test. If the equipment supports both data and voice calls, try both types of calls, as appropriate.

If the physical layer comes up and seems stable but will not dial or receive incoming calls, then the problem may be a Layer 2/3 problem. In this case, first check the equipment configuration to determine that the ISDN protocol setting is matched to the line and, if using one of the U.S. protocols that require SPIDs, make sure they have been entered correctly. Check that the directory numbers (DNs) are correct and entered in the same order as the SPIDs. It may be necessary to reboot the equipment after changing any of these parameters. For non-U.S. ISDN, multiple subscriber numbers (MSNs) are optional. If these are incorrect, they will prevent receiving incoming calls, so remove them while troubleshooting.

If you can only place voice or data calls, but not both, then the line may have been provisioned for just one type of service. The correct provisioning that includes both is called "alternate voice data service." In many cases, problems are due to advanced features being enabled when they are not needed. Most broadcast equipment does not need anything but the most basic ISDN features.

If the problem is only with long distance calls, there may not be a long distance carrier (IXC) or the chosen IXC may not handle calls. This can be determined by attempting both types of calls. Also, check that each type can be received, as an IXC is not required for incoming calls.

Troubleshooting T1

The first place to check when troubleshooting a T1 circuit is the user device and the CSU. These will typically offer a *loss-of-signal* alarm that will illuminate if the incoming signal can no longer be detected. The CSU will generally have such an LED for both the user side as well as the network side. CSUs have the ability to maintain an error log that shows the number of various errors for each 15-minute period for at least 24 hours. If an ESF T1 is being used, the telco can poll the CSU for this data. If the CSU has a serial port or LCD, it may be possible to access this important information on the instrument. The smart jack or HDSL/HDSL2 transceiver may have error LEDs or a serial port that, if the cabinet is unlocked, can yield useful information (try 9600 bps). Remember to configure the CSU and user equipment to accommodate the amount of cable between them.

Once all of the above sources of data have been checked, the next step would be to bridge a T1 tester into the incoming T1 signal before the CSU. Check for level and for a lack of CRC errors (ESF only) and bipolar violations (primarily for SF). When this group of tests has been completed, move the tester to other points on both sides of the CSU (four points total). If all else fails, for a dedicated T1, place one end into loopback (a simple adapter can be made, or the equipment may have this function built in), and do a bit error rate test. Be sure to test for at least 15 minutes with each of the test patterns supported by the tester.

Troubleshooting IP Networks

IP and VoIP Debugging

Troubleshooting VoIP telephony issues is similar to doing so with other technologies. As engineers, we realize that it's important to become familiar with systems when they're operating correctly. We must also understand both the system as a whole and how to get the parts to interact. A good way to begin an understanding of VoIP telephony is to personally experiment with the technology. Sign up for a free or low-cost VoIP phone number. Then purchase and install a low-cost IP telephone to configure with the service. This exercise will not only yield a new working phone service, it will force the engineer to become familiar with the concepts and protocols germane to VoIP.

The IP phone acquired for experimentation and learning will also come in handy for troubleshooting in the broadcast plant. Such an IP phone may be substituted for portions of a larger VoIP talk show system, in case there are troublesome extensions.

As VoIP telephony depends on proper IP network performance, it's essential that engineers and/or IT technicians apply their skills to VoIP troubleshooting. Good IP network performance and configuration will lead to good VoIP performance.

Once in a while, though, some serious skills and tools are needed. Debugging a SIP trunk or station may ultimately require that you be able to see all of the packets used for the call, including call setup and the RTP (media) stream. A free, useful tool to do this is Wireshark. Wireshark is a packet capture and logging program. It will clearly show and record details of the call set-up (or attempted call set-up) and even the audio of the call itself.

Another useful tool is called MTR or "My Trace Route." It'll show you the route that a call took and any impairments (like packet loss) that it can see. You can leave it running to catch intermittent problems as well.

For VoIP calls that traverse the public Internet, it's desirable to characterize each end—or at least the studio end—of the IP connection prior to use, to see if there are any configuration issues or impairments to smooth packet transmission and reception. Experience is showing that many—if not most—"Internet connection" troubles occur in the "last mile" and the local router. This implies that characterizing each end of a proposed IP connection is a valid way to prequalify that connection. It's less important to check end to end and more important to characterize each end's connectivity with the public Internet. The public Internet itself sometimes, but not often, turns out to be a source of connection trouble.

While a speed test is a good place to start, there's more to characterizing an IP connection than simply measuring upload and download speed. Packet jitter, especially at various times of the day, is important to know. A measure of dropped packets is important, too. Additionally, an understanding of the local router's behavior can point out issues before they affect an on-air broadcast. Many of the available online IP path tests require Java to be running as a plugin with your web browser. As Java is falling into some disfavor due to persistent known security issues, there are a few tests that do not require Java.

Several free, online tests are available to pre-qualify or perform very basic bandwidth and jitter tests on Internet connections used to carry VoIP traffic. As of this writing, some of these online tests require Java to run in the browser.

Interpreting Test Results

What should we look for when analyzing these results? Certainly sheer speed or bandwidth is critical. If we have, say, 1 Mbps of upload bandwidth, then there's room for about 10 simultaneous VoIP calls, but with little or no margin for any other data.[51] There's no magic ratio of anticipated bandwidth usage to available bandwidth, but a reasonable figure is about a 100% margin, not including other users' data. Indeed, the most reasonable Internet connection scenario is to have a dedicated connection for VoIP and other real-time data. Internet data cannot be prioritized—at least not in the incoming direction—though some quality of service (QoS) effort may be configured in many Ethernet routers. Such QoS configuration can have a limited benefit for ingress and egress of data between the router and the public Internet.[52] Nevertheless, establishing a dedicated Internet path for real-time data of known maximum bandwidth is the best practice. Other non-real-time data, such as web browsing, file transfers, video and audio downloads, and e-mail, should use a separate Internet feed if at all possible.

Two network tests that bear some commonality are for round-trip time and jitter. Round-trip time (RTT) measures the time required for a packet to travel to a remote server and back again. If that remote server happens to be your hosted VoIP PBX, then this figure becomes important. We want to see it as low as reasonably possible, and we also need to be fairly consistent. Large variations in RTT imply a bottleneck of Internet traffic somewhere between your location and the remote server. If more consistent RTTs are found from, perhaps, closer servers, then the difficulty may well be with your ISP and/or its gateways with other ISPs or Internet backbone providers.

Jitter is a measure of inconsistency in the periodicity of packet arrivals, presuming they were sent on a perfectly periodic basis from the sending server. What's desired is jitter of less than about 20 ms, which represents about 1 IP audio packet in time. Higher jitter usually means there are other packets contending for attention in the upload or download path. The receive buffer on any VoIP device will need to be set to cover the worst-case jitter figure. In other words, we cannot allow the jitter buffer on a VoIP endpoint to become empty, waiting for the next packet to arrive, or else an audio silence will occur.

In VoIP telephony, it's common for the jitter buffer on any "user agent" (VoIP telephone) to be set at about 100 milliseconds. A 100-ms delay, added to the delays already inherent through typical network connections and the PSTN, generally remains within an acceptable range for easy conversation. Higher buffer times may be configured in VoIP devices to force a longer wait for any given packet to arrive, but this is at the expense of natural conversational flow. If audio dropouts are heard, and increasing the buffer time isn't helping, then it's likely that there is actual packet loss somewhere in the IP path.

The Netalyzr test from berkeley.edu shows a worst-case network buffer measurement for both downlink and uplink buffers. This is the worst case that the test was able to cause, but could be higher under real-world conditions. Again, higher buffer numbers are usually caused by too much other Internet traffic in and out of the router's WAN port. Lower numbers here are better. Various routers will behave differently on this test and may indicate that a different router should be tried if the buffer times are uncomfortably high.

Remember, when using any of these tests, they reflect the IP path between your connection and the (assumed) highbandwidth server connection at the other end. They may serve as a proxy for expected performance with other remote devices, such as a hosted VoIP PBX.

SUMMARY

Broadcast and communication technology—as most other tech—is ever changing. It seems no sooner are we fully versed and comfortable with one technology that a new approach is upon us. These changes typically hold the promise of better performance and lower cost, but only if engineers—those of us implementing new tech in our workplaces—understand the key concepts and are wary of potential pitfalls.

It's commonplace these days to find three or more different telecom technologies in broadcast plants. From POTS and T1, to ISDN, to VoIP delivery and interconnections, to wireless IP systems both private and carrier-based, engineers are challenged with interfacing these various technologies with broadcast gear. We're also committed to doing so with economy in mind, as well as reliability, all while getting the highest possible audio quality from each.

Engineers can fully exploit the benefits of each technology by having the foregoing technical information and experienced advice at their fingertips.

Notes

[1] "The world in 2015 — ITU." 2015. 3 Aug. 2016 <www.itu.int/en/ITU-D/Statistics/Documents/facts/ICTFactsFigures2015.pdf>

[2] "There are now more gadgets on Earth than people—CNET." 2014. 3 Aug. 2016 <www.cnet.com/news/there-are-now-more-gadgets-on-earth-than-people/>

[3] "Public switched telephone network." 2014. 29 Jul. 2016 <ftp://ftp.ulim.md/Ingenerie/2007-2008/Eremeev%20V/Telecommunications%20systems/Lectures/PSTN.doc>

[4] "Telephone network interfacing—The Telos Alliance." 2015. 29 Jul. 2016 <www.telosalliance.com/support/Telephone-Network-Interfacing>

[5] Karapantazis, S. "VoIP: A comprehensive survey on a promising technology." 2009. <www.utdallas.edu/~kxs028100/Papers/VoIP.Survey.pdf>

[6] Brandenburg, Karlheinz, and Marina Bosi. "Overview of MPEG audio: Current and future standards for low bit-rate audio coding." *Journal of the Audio Engineering Society* 45.1/2 (1997): 4–21.

[7] "G.711—Expand your mind—Revolvy." 2015. 29 Jul. 2016 <www.revolvy.com/main/index.php?s=G.711>

[8] "G.711—Wikipedia, the free encyclopedia." 2011. 29 Jul. 2016 <https://en.wikipedia.org/wiki/G.711>

[9] Apostolopoulos, John, and Nikil Jayant. "Broadband in the last mile: Current and future applications." *Broadband Last Mile: Access Technologies for Multimedia Communications* (2005): 1.

[10] Hanzo, Lajos L., Clare Somerville, and Jason Woodard. *Voice and Audio Compression for Wireless Communications*. New York: John Wiley & Sons, 2008.

[11] Ritz, Christian. "Decomposition and interpolation techniques for very low bit rate wideband speech coding." (2003).

[12] "G.722—Wikipedia, the free encyclopedia." 2011. 29 Jul. 2016 <https://en.wikipedia.org/wiki/G.722>

[13] "G.729—Wikipedia, the free encyclopedia." 2011. 29 Jul. 2016 <https://en.wikipedia.org/wiki/G.729>

[14] "How to decode G.729—CiscoZine." 2014. 29 Jul. 2016 <www.ciscozine.com/how-to-decode-g-729/>

[15] Shafi, N.A., Farrok, O., and Ali, M. "An analysis of VoIP communication and overview of Bangladesh Practical Field, *International Journal of Computer Theory and Engineering* (2013), Vol. 5, No. 3, 574–577. <www.ijcte.org/papers/752-T183.pdf>

[16] Gardner, William G. "Reverberation algorithms." *Applications of Digital Signal Processing to Audio and Acoustics* (2002): Vol. 437, 85–131.

[17] "G.722[1] is an ITU-T standard 7 kHz Wideband audio codec operating . . ." 2016. 29 Jul. 2016 <www.theinfolist.com/php/SummaryGet.php?FindGo=G.722>

[18] "3-10—Scribd." 2016. 29 Jul. 2016 <www.scribd.com/document/46128144/3-10>

[19] "Telephone network interfacing—The Telos Alliance." 2015. 29 Jul. 2016 <www.telosalliance.com/support/Telephone-Network-Interfacing>

[20] Church, Steve, and Rolf Taylor. "Telephone network interfacing." In Edmund A. Williams (Ed.), *NAB Engineering Handbook* (2007): 609–644, Milton Park, Abingdon, UK: Taylor & Francis.

[21] Messerschmitt, David G. "Digital terminations and digital signal processing." In John C. McDonald (Ed.), *Fundamentals of Digital Switching* (1990): 285–322. New York: Springer Science+Business Media.

[22] "View sample PDF—IRMA-International.org." 2011. 13 Aug. 2016 <www.irma-international.org/viewtitle/62756/>

[23] "All you wanted to know about T1 but were afraid to ask." 29 Jul. 2016 <www.dcbnet.com/notes/9611t1.html>

[24] "HDSL basics—OoCities." 2011. 13 Aug. 2016 <www.oocities.org/~katmci/pdfs/hdsl.pdf>

[25] "Configuring ISDN PRI—Cisco." 2013. 13 Aug. 2016 <www.cisco.com/c/en/us/td/docs/ios/12_2/dial/configuration/guide/fdial_c/dafchant.pdf>

[26] "Primary rate interface—Wikipedia, the free encyclopedia." 2011. 13 Aug. 2016 <https://en.wikipedia.org/wiki/Primary_Rate_Interface>

[27] "SIP vs. VoIP, What' the Difference—IP telephony explained—SIP Blog." 2013. 29 Jul. 2016 <www.sip.us/sip-vs-voip-whats-difference/>

[28] "Federal communications commission DA 14–95 before the federal . . ." 2015. 13 Aug. 2016 <https://apps.fcc.gov/edocs_public/attachmatch/DA-14-95A1_Rcd.pdf>

[29] "SIP trunking vs. VoIP: What's the difference?" 2012. 29 Jul. 2016 <http://searchunifiedcommunications.techtarget.com/answer/SIP-trunking-vs-VoIP-Whats-the-difference>

[30] Kulathumani, Vinodkrishnan. "Voice over IP: Products, services and issues." 23 Nov. 1999.

[31] "What is ethernet-dedicated Internet?—SearchNetworking—TechTarget." 2013. 2 Aug. 2016 <http://searchnetworking.techtarget.com/feature/What-is-Ethernet-dedicated-Internet>

[32] Mohsin, Mansoor, Winnie Wong, and Yogesh Bhatt. "Support for real-time traffic in the Internet, and QoS issues. Paper published by the Department of Computer Science, The University of Texas at Dallas, Richardson, TX (2001). <http://citeseerx.ist.psu.edu/viewdoc/download?doi=10.1.1.10.8075&rep=rep1&type=pdf>

[33] Grubesic, Tony H., and Alan T. Murray. "Constructing the divide: Spatial disparities in broadband access." *Papers in Regional Science* 81.2 (2002): 197–221.

[34] "NANPA: ANI II digits—View Assignments." 2013. 29 Jul. 2016 <www.nationalnanpa.com/number_resource_info/ani_ii_assignments.html>

[35] "Mobile network performance in the US: A special RootMetrics first . . ." 2015. 29 Jul. 2016 <www.rootmetrics.com/us/blog/special-reports/2015-1h-national-us>

[36] "Telephone hybrid—Expand your mind—Revolvy." 2015. 29 Jul. 2016 <www.revolvy.com/main/index.php?s=Telephone%20hybrid>

[37] "Telephone network interfacing—The Telos Alliance." 2015. 29 Jul. 2016 <www.telosalliance.com/support/Telephone-Network-Interfacing>

[38] Jaber, Aws N., Chen-Wei Tan, Selvakumar Manickam, and Ali A Khudher. "Session initiation protocol security: A brief review." *Journal of Computer Science* 8.3 (2012): 348.

[39] "Firewalls—TechNet—Microsoft." 2012. 13 Aug. 2016 <https://technet.microsoft.com/en-us/library/cc700820.aspx>

[40] "Multiprotocol label switching—Wikipedia, the free encyclopedia." 2011. 13 Aug. 2016 <https://en.wikipedia.org/wiki/Multiprotocol_Label_Switching>

[41] "What is asterisk?—Getting started—Open Source Asterisk." 2012. 3 Aug. 2016 <www.asterisk.org/get-started>

[42] "DPCM." 2003. 14 Aug. 2016 <http://einstein.informatik.uni-oldenburg.de/rechnernetze/dpcm.htm>

[43] "ITU-T Rec. G.722 (11/88) 7 kHz audio-coding within 64 kbit/s." 2013. 14 Aug. 2016 <www.itu.int/rec/dologin_pub.asp?lang=e&id=T-REC-G.722-198811-S!!PDF-E&type=items>

[44] Gibson, Jerry D. "Speech coding methods, standards, and applications—ViVoNets Lab." 2010. <http://vivonets.ece.ucsb.edu/casmagarticlefinal.pdf>

[45] Brandenburg, Karlheinz. "Low bitrate audio coding state of the art, challenges and future directions." 2005. Ilmenau Technical University & Fraunhofer IIS Arbeitsgruppe Elektronische Medientechnologie Ilmenau, Germany. <www.mp3-tech.org/programmer/docs/icct482.pdf>

[46] "On beer and audio coding—The Telos Alliance." 2015. 14 Aug. 2016 <www.telosalliance.com/support/On-Beer-and-Audio-Coding>

[47] Zwicker, Eberhard, and Hugo Fastl. *Psychoacoustics: Facts and Models*. Berlin, Germany: Springer Science & Business Media, 2013.

[48] "On beer and audio coding:—The Telos Alliance." 2015. 14 Aug. 2016 <www.telosalliance.com/support/On-Beer-and-Audio-Coding>

[49] "MPEG-4 HE-AAc v2—audio coding for today's media world—Ebu.ch." 2011. 14 Aug. 2016 <https://tech.ebu.ch/docs/techreview/trev_305-moser.pdf>

[50] "Radio world: The grand transformations of IP audio." 2013. 14 Aug. 2016 <www.radioworld.com/article/the-grand-transformations-of-ip-audio/221929>

[51] "Determine how much Internet bandwidth you need for SIP trunking . . ." 2014. 2 Aug. 2016 <www.sip.us/determine-much-internet-bandwidth-need-sip-trunking-3-steps/>

[52] "How to use Quality of Service (QoS) to get faster Internet when you . . ." 2011. 2 Aug. 2016 <www.howtogeek.com/75660/the-beginners-guide-to-qos-on-your-router/>

CHAPTER

4.6

Radio Remote Broadcasts

LORNA WHITE

NPR

Chicago, IL

Radio remote broadcasts, both live and recorded, are effective ways to connect your station with your listening audience and sponsors. Station events provide a sense of immediacy and inclusiveness, which can be very popular in this era of instant, on-demand content. Broadcast remotes are also a great opportunity for the public to see and interact with the staff. While the electronics have become more compact and user friendly, remote planning remains complex, particularly for the larger ones. The size and scope of remotes varies from covering a small news conference with a single microphone to a large public event with multiple microphones and a remote truck. The basic audio principles apply to every remote, and this chapter will cover some of the practical aspects.

THE REPORTER KIT

Digital audio recorders are great improvements over previous analog recorders. Gone are the days of 30-pound boxes with a shoulder strap and a delightful bruise to show for your work. Solid-state flash recorders are small and versatile, and most feature balanced XLR inputs for external microphones. These handheld devices are robust and are excellent choices for field reporting. Paired with a good dynamic microphone and headphones, these units are excellent tools for any reporter. Some of these devices' microphone inputs provide phantom power, which allows a broader choice of microphones. A spare XLR cable, extra CF or SD cards, and spare batteries are fundamental items to include in the reporter bag as well.

There are options for recording directly to a laptop. Computer audio interfaces will connect your microphones and provide level control while feeding audio to your hard drive. Software programs provide the tools for editing and manipulating the tracks into a final mix on the laptop.

One indispensable tool that most of us carry every day is the smartphone. It can act as a backup computer, audio recorder, note taker, and video camera, and it fits in the palm of your hand. Dictation apps allow greater efficiency for taking notes while on the move. There are recording apps that allow a reporter to record audio and then upload to a station server. Those files can then be retrieved by a producer, edited, and packaged for air.

Take the opportunity to gather the remote equipment in one place. Consider making a master checklist that can be referred to in the future. It's common for shared equipment to become misplaced or lost. The kit is only as good as the last person who packed it. The checklist could look similar to this basic reporter kit.

NEWS REMOTES

Familiarize yourself with the equipment prior to the remote so you are at ease with its operation. Hook it up, turn it on, and make a test recording. Confirm the settings, record a short audio clip, and listen back to make sure the components are in good working order. Double-check the amount of recording time left on your device. There's nothing quite like running out of audio storage a few minutes into an interview. Delete the old sound files and reformat the media if possible.

Basic Reporter Kit

- Audio recorder
- Power supply and/or batteries
- Extension cord
- CF or SD cards
- Microphone(s)
- Windscreen
- XLR Cables
- Headphones
- Rowi-style C-clamp
- Gooseneck
- Mic clip
- Desk stand
- Monopod (also called a fish pole)
- Line-to-mic level barrel adapter
- Other in-line adapters, as needed
- Durable bag

FIGURE 4.6-1 Suggested equipment list for a radio remote reporter's kit

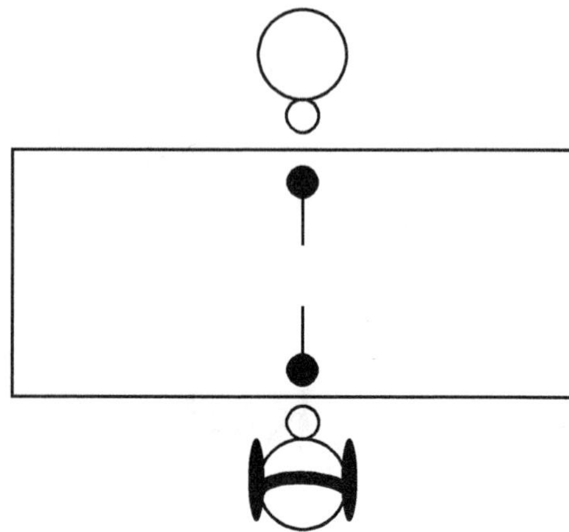

FIGURE 4.6-2 Recording guest and reporter with two cardioid microphones

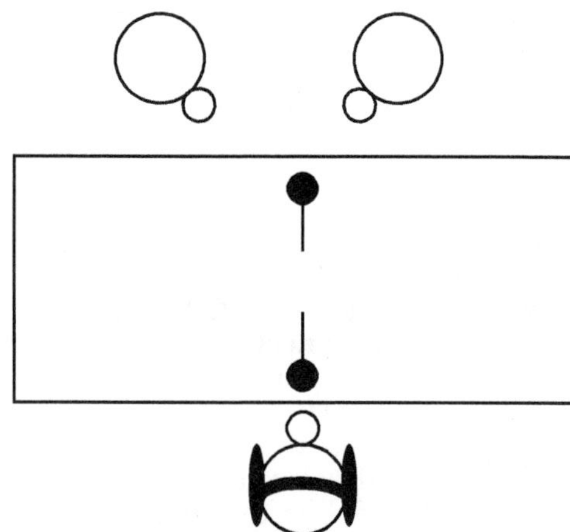

FIGURE 4.6-3 Recording two guests and reporter with two cardioid microphones

Determine how you can best provide power to your equipment. If it's a small remote and you need to be mobile, batteries are the best choice, so have plenty on hand. If you will be in one place feeding live for long periods, you should opt for a more reliable source of power. Consider bringing your own extension cord and power strips so you do not have to depend on someone else to provide them for you.

A common remote arrangement is a face-to-face interview with one person. An office or conference room provides an adequate recording environment. Air handlers can be temporarily shut off to eliminate fan noise. Overhead fluorescent lights can be turned off if they make noise. Draperies can be drawn to reduce outdoor noise and reduce hard reflective surfaces.

If two interview subjects will be recorded at the same time, it can be done with the same microphone attached to a desk stand and placed between the two subjects. A cardioid microphone is a good choice.

The microphone should be no farther than 8 inches from the mouths of the interview subjects. This distance can be estimated by stretching out the thumb and pinkie fingers on your hand. Place the microphones so that the interviewees' heads are oriented directly toward the microphone rather than having them talk across the table and away from the microphone. Attention should be paid to any noise from tapping or bumping the table. (This may only be noticeable if you are listening on headphones, which is essential.) If it is not possible to sit at a table, a microphone on an extendable monopod, also known as a "fish pole," held in front of the interview subjects will work as well.

If the interviews take place outdoors, an omnidirectional dynamic microphone will minimize some effects of wind. A good option for a more directional microphone is a pistol-grip zeppelin and windsock, which offers some protection from windier conditions. Even with these precautions, it may be necessary to shield the microphone with your body. You can also move to the sheltered side of a building to reduce the intensity of the wind. Again, determining whether the amount of wind noise is tolerable can only be determined by monitoring on headphones.

Whether you are recording outdoors or inside an office building, it is useful to capture a minute or so of location ambience. Even though the interior of a room seems quiet, there will be some type of undesirable sound, for instance, HVAC noise, nearby automotive traffic, or workplace chatter. Recording an additional minute or so of the location will aid in post-production by providing a bed that can be later added to the mix to smooth over edits. A good practice is to record the ambience with the microphone in the same place as your interview so there will be little difference in the sound. Take note if you hear trains, airplanes, lawn mowers, and any other distinct sounds that will need to be captured for possible later use. This also serves as a good test to make sure your recording setup is working as expected.

In some cases, radio frequency (RF) interference may be introduced into your recording equipment. This may sound like a buzz, hum, or actual audio leakage. If this occurs, move to a different location while monitoring on headphones to see if the interference changes. In many cases, moving just a few feet will greatly reduce or eliminate the problem. If you are recording very close to a high-powered transmitter, however, you may need to take more remedial action, such as changing microphones or recorders, using a shorter cable, switching from AC to battery power, or moving to a below-ground location.

News stories covered in the field often present other unique challenges. There may not be a suitable space to conduct the interview. A parked car with the engine turned off can provide a controlled space to interview someone or record reporter voice tracks. Be sure to park away from other people and machines. In a hotel, avoid large conference rooms with only hard reflective surfaces, which can affect the sound quality. Moving to a corner of a large room with ceiling-to-floor drapes (fully closed) can also help. Many journalists have reported stories from under a blanket or an open umbrella while sitting on a bed in their hotel. Two sofa cushions propped up at a 90-degree angle on a table can provide some sound isolation for a mic on a desk stand. A hotel room closet with hanging clothes or hanging towels can also work. Creative use of sound dampening materials will benefit your audio tracks recorded on the road.

In case of natural disasters where buildings are destroyed and cellphone towers have been damaged, it could be necessary to travel to a location with available WiFi to upload stories—perhaps a fast-food restaurant, coffee shop, or hotel lobby. Other journalists will inevitably have the same idea in such situations, and a number of reporters filing online in the same location at the same time can have a negative effect on available bandwidth.

COVERING EVENTS

If a scheduled news event you are covering is an announcement by public officials, a press conference, or the like, the event organizers may supply a "mult box" for members of the press. This is a metal box that has one input and multiple outputs. (In some cases, one broadcaster or news organization may provide this for the benefit of other members of the press; this is called a "pool feed.") A single microphone input from a lectern is split into multiple outputs, which avoids the need for a forest of microphones to be clustered on the lectern. The mult-box feed to press users normally appears as a male XLR jack. It is important to ask if the mult box feed is a line-level or a mic-level signal. If it is line level, confirm the input setting on your recorder is also line level. If your recorder input is set to mic level, the audio will likely be distorted and not usable. In such cases, either use the line-level input on your recorder, or a line-to-mic-level in-line barrel pad that attenuates the signal by approximately 50 dB. Of course, none of this guarantees a clean feed. It is good practice to listen to your input and change to another output jack on the mult box if you find one with hum or noise.

If a mult box is not made available, the opportunity may exist to clamp a microphone onto the lectern. A C-clamp, gooseneck, and microphone clip can be assembled and attached to the edge of the lectern. A longer microphone cable will be required to run from the lectern location to your recording device. Sometimes a fellow member of the press may offer a feed off their equipment. It is also possible to record audio from a public address mixer or by miking a PA speaker, provided the PA audio is clean without additional buzz or hum from the system itself. Hold or place the microphone close to the PA speaker to minimize ambient noise. Make sure to watch your levels and use a mic pad if necessary. This isn't the best way to gather audio, but it can work in a pinch. There is always the risk of picking up chatter from nearby audience members, of course, so keep this in mind when choosing a microphone location.

COMMERCIAL REMOTES

A station might choose to broadcast from a commercial business in the local community—perhaps a bookstore or a car dealership—in order to promote something like a charity event or festival. During load-in, scout out a power source to run to the equipment. A table and a few chairs may be needed to set up a small broadcast position. When selecting a spot, think about where the power and audio cables need to be run so you can avoid heavy foot traffic. A few strips of carpet and gaffers tape can secure the area from any trip hazards.

At a bookstore, a position near bookcases will not only provide a visually appealing backdrop, but the books will help diffuse reflected sound. Avoid setting up in a corner with hard reflective surfaces and near large flat windows. If there is little choice as to where you can set up, pipe and drape can be rented and deployed to delineate the broadcast position. Should the decision be made to set up outside, a portable canopy is useful to shield the staff and the equipment from the hot sun. A few tarps in the remote kit are essential in case of rain.

A small mixer is sufficient for this type of event. Two microphones will handle audio for the talent and a guest with inputs left over for a playback device and backhaul audio from the studio. The talent can wear headphones. A cellphone can be used for real-time talking or texting to the station. A laptop is practical for instant messaging and script changes, as long as the Internet connection is reliable. If there is sufficient time, order a dedicated drop from the local Internet service provider to make sure there is dedicated high-speed bandwidth for the broadcast.

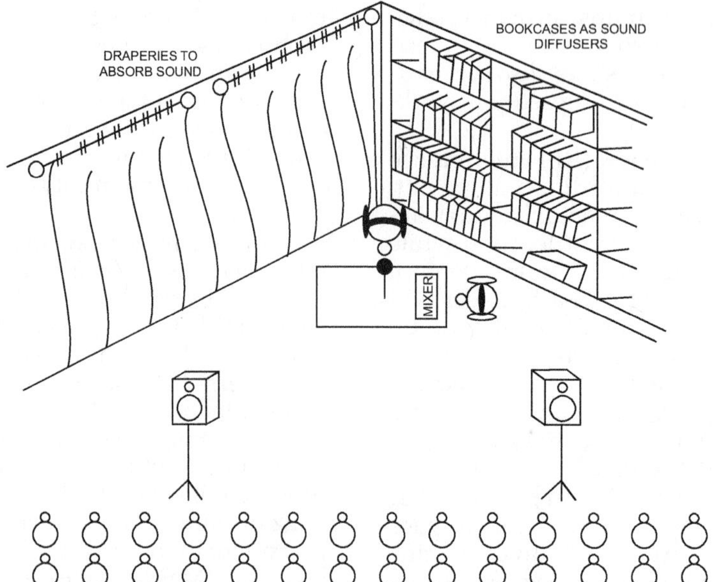

FIGURE 4.6-4 Bookcases make a good acoustical location

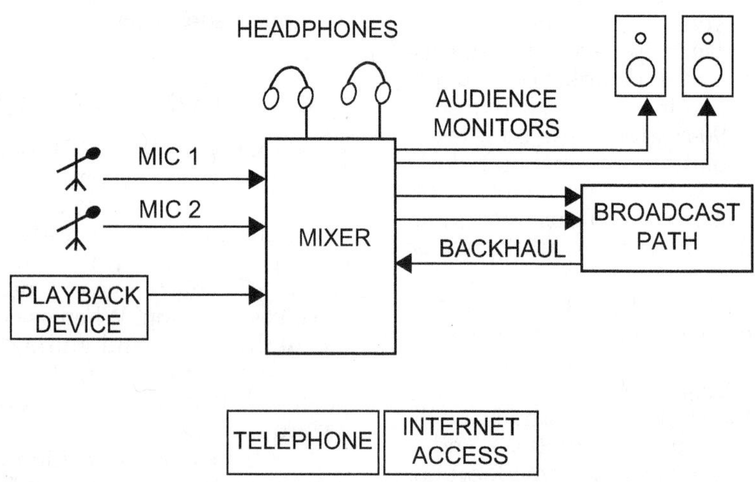

FIGURE 4.6-5 Small remote setup

If an audience will be gathering to watch the remote, they will need to hear as well. A pair of self-powered audio speakers on stands will provide coverage to the immediate area. Place them pointed away from the broadcast position and keep the levels at a comfortable level to avoid PA sound being picked up by the microphones.

COMPLEX REMOTES

Larger remotes require greater lead-time to plan. Determine what the requirements will be by asking the producers what they have in mind. No question is too extreme when it comes to this type of production.

An advance site visit is recommended to become familiar with the space, as well as to meet the venue staff. If temporary phone lines and Internet are needed, begin making arrangements six weeks ahead of the remote date. If you are sending audio via IP codec (see Chapter 6.1), it is worth the investment to order a dedicated line from a service provider so you are not sharing your bandwidth with everyone else in the facility. Always use hard-wired network connections and avoid use of WiFi or other wireless IP service whenever possible, even for supportive services such as broadcast staff computer connections. (WiFi may work well during testing, but can fail during the actual event due to much heavier usage when the venue is fully functional and crowded with other users.) Sometimes the venue can provide you these services in-house. Otherwise, you will need to contact the local phone and Internet service providers and set up a small business account. Give them the address of the location for

Producer Questions

1. When is the event? (This is always the first question.)
2. Where will it be?
3. Is it a live broadcast or a recording session?
4. How big is the venue?
5. Will there be an audience?
6. If there is an audience, when will they be admitted to the venue? (This will determine the end of set-up/rehearsal time.)
7. Does the production require lighting and sound reinforcement?
8. Who are the people involved in coordinating the event?
9. Is it a union house? This is important when planning the crew call.
10. Will Internet access be required and for what purpose? How much bandwidth?
11. Are phone lines necessary?
12. Will a workspace or other preparation area for talent or crew be required?
13. Should a video feed be made available?
14. Will members of the press request access to cover this event?
15. What are the load-in arrangements (parking, loading dock, any stairs to climb, etc.)?

FIGURE 4.6-6 Sample questions to ask when planning a complex remote broadcast

installation, as well as a contact name and phone number of someone who will be present at the appointed time. It never hurts to mention it is for a live broadcast, even if your remote is a recording session. A sense of urgency can be beneficial. Pick an installation date a week out from the remote and test the service immediately. If there are any problems with the service, this will allow time to correct the issue. Remember to disconnect the lines the week after the broadcast if you no longer require them. Keep all correspondence in case questions about billing arise.

The venue will ask for a work schedule so enough stagehands are present to help with the production. Break down your setup into tasks and estimate how much time is needed to complete each one. Ask about the work rules concerning coffee breaks and meal times in order to avoid scheduling penalties. Some houses allow work to continue on stage during breaks, while others insist that the stage "go dark" during breaks. Respect the rules and the stagehands, and coordinate the schedule accordingly.

If your remote requires professional lighting and additional sound reinforcement, and the infrastructure is not in place, the venue should be able to recommend companies familiar with the space. These companies will need to load in and set up before you bring your broadcast gear in, and load-in paths or docks can become crowded. The performance area must be clear to freely move light trusses, road cases, sound rigging, and the like. It is common for venues to use boom lifts to adjust lights, so ask first before running your cables so they won't be run over and damaged. Stage crews aren't always careful with where they roll things, particularly if they feel you are impeding their work.

Sound reinforcement for speech is different than sound reinforcement for music. Conversational spoken word can become muddy and lost in a large room without well-placed speakers. Large theatre venues will probably have a center cluster or arrays hung stage left and right to cover the floor and balcony sections. Subwoofers in the system can be muted or turned off, as they are not necessary for spoken word. A moderate sound level with uniform coverage of the hall is ideal to meet the needs of the recording/broadcast and the audience. Small fill speakers can be added along the lip of the stage to reinforce the audio for the first few rows. If the venue is a music hall with highly reflective surfaces, distributed speakers with digital delay can be spaced along the side walls to maintain coverage and intelligibility, while keeping PA sound levels relatively low and manageable for the broadcast. During the sound check, walk around the hall and listen to each PA speaker to make sure all the speakers are working properly. Never turn the PA volume up any louder than necessary, but bear in mind that an empty hall will require less volume than when it is filled with an audience.

Whether the remote is a music performance or a panel discussion, you will be planning a multiple microphone setup. Rugs on the deck of the stage will eliminate the sound of people walking and reduce unwanted acoustical reflections. Microphones on stands are easy to position. They can be placed to pick up various audio sources—for example, soloists, orchestra sections, or a narrator. If the talent sits in fixed positions on stage throughout the program, the microphones can be hard-wired with long XLR cables run to the mixer. Another option is a head-worn microphone system. These small, unobtrusive microphones provide comfort for the talent and offer great mic placement options for the audio engineer. They are available for rental, and suppliers offer them in different colors to match skin tones and make them less visible. If talent is required to move around the stage, wireless transmitters and receivers can be used, along with the proper microphone adapters, to eliminate cables on the stage. The talent wears a small transmitter ("body pack") clipped to the waist or tucked in a pocket. Be sure to coordinate the wireless frequencies you are using with the house sound tech to avoid interference problems. Microphones that can be used with a wireless transmitter include handheld, lavalier, or the head-worn type mentioned above. (See Chapter 4.3 for more on wireless microphones.)

Once the number of microphones has been determined, add up the other audio sources that will be required and choose a mixing console that will support that number of inputs, plus a few more. Alternatively, a second console can be operated as a stage "sub-mixer," controlling the individual microphone inputs and focusing on the balance of the performance mix. The output of that console is then sent to the broadcast production mixer, where additional, broadcast-specific inputs—such as on-air talent microphone(s) and broadcast playback devices—can be added. This allows the stage mix engineer to concentrate on the performance feed alone and not worry about broadcast timing, cues, off-air monitoring, etc.

You may need to provide a program feed with IFB (interruptible fold-back) to the talent on stage. This allows a broadcast program's director to talk to the talent during the program, providing warnings, cues, and other direction on top of the program audio fed to the talent headphones. If the console does not have a built-in feature, one can be made using one of the console's auxiliary buses in post-fader mode.

At the mix position, monitoring the program can be done with either loudspeakers or headphones. Both have their pros and cons. Self-powered speakers are a good choice, since they avoid the need to bring a power amplifier. If using headphones, a headphone amp with multiple outputs will be helpful. If the broadcast mix is taking place in a room away from the action, take time to listen to the speakers' sound in that space. Play audio material that you like and are familiar with, and listen closely to how it sounds in the room. Allow yourself time to get used to it so you can later mix in the room with some confidence. If permitted by the venue, you may be allowed to add temporary sonic treatment, like foam tiles or moving blankets, to the walls and even the ceiling. A closed-circuit video feed from the stage can also be helpful in such an arrangement, so the mixer can see what's happening on stage.

For musical performances with live audiences, a separate sound-reinforcement mix should be anticipated. The need for a house mix position in the performance space is critical. Ideally, a single set of stage microphones will be run through a mic splitter to provide independent control of the same mics by the house sound and the broadcast mixers. Adjustments in microphone levels and other processing for the house sound can be made independent of the broadcast mix. The location of the house mix position is often flexible. Larger venues have dedicated audio booths but should be agreeable to moving a console into the house if that is your preference. Being in the house gives the engineer a better feel for what the audience is or is not hearing.

Remote trucks are a welcome convenience and take the worry away from setting up a broadcast mix position in the venue. They provide a controlled environment for monitoring and mixing, and allow the producer and engineer to communicate freely during the

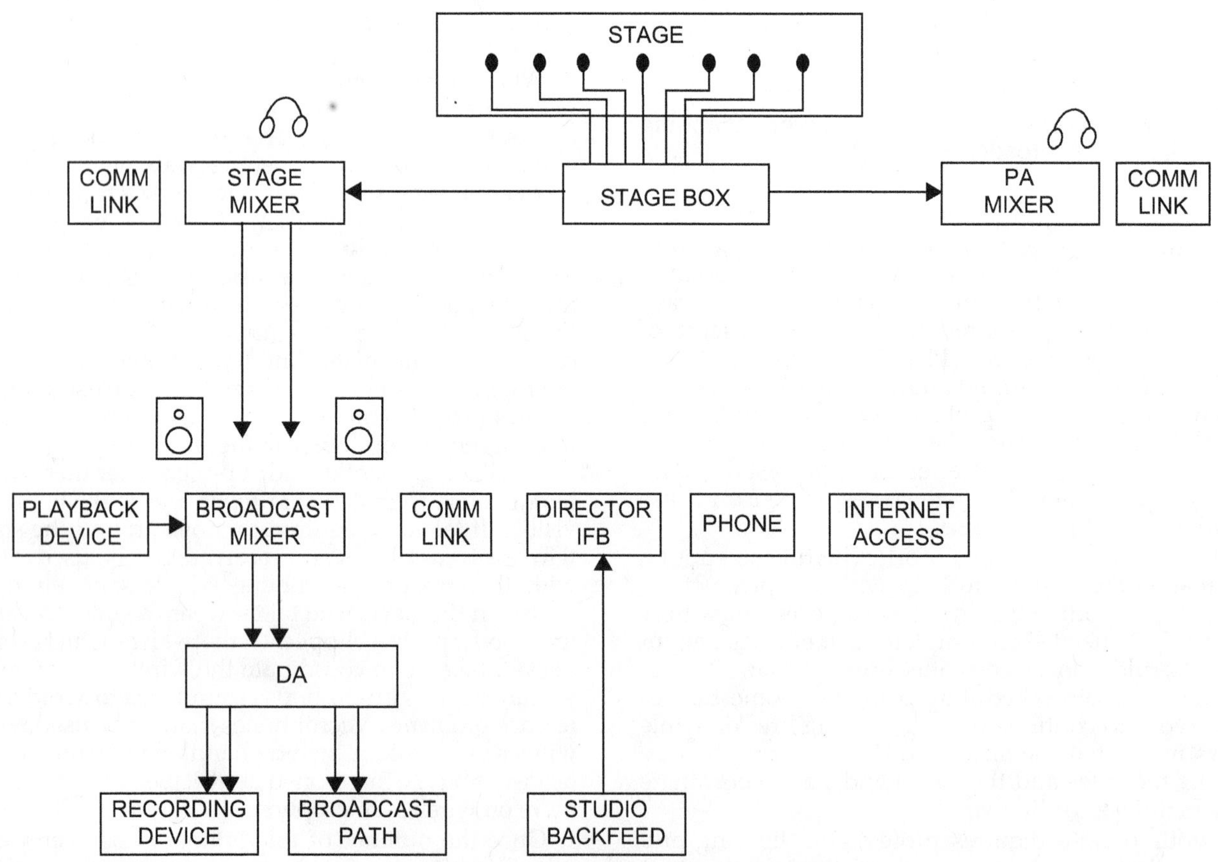

FIGURE 4.6-7 Large remote mixer

event. They are also useful when the venue is too small for an inside mix position. Arrange for parking of the vehicle with the venue ahead of time. Venue management may be able to provide access to reserved space. In some cases, the remote truck operator may need to apply for a permit from local/municipal authorities to park on a public street. You will have to run one or more multi-pair cables ("snakes") to the truck, so plan the cable run as safely as possible.

Remote gear needs to be organized and contained. This is helpful both before and after the event. The stage clearing and loading out of equipment is a concerted effort by all crew to tear it down and pack it up efficiently. Overtime and penalties from a slow strike can eat a large part of your budget. For broadcast remote gear, clear plastic containers with lids are convenient and inexpensive, and allow you to see what is inside. The larger polyethylene containers are moisture resistant and stackable. They can be labeled and color-coded according to use.

If your station is investing in a large amount of remote gear, factor road cases into the budget. They are an additional expense but will protect the gear for years to come. Consider wheeled, plywood cases with foam-padded interiors. Smaller cases can be stacked on top, making it easy to move equipment around. Instead of wrapping electronics in bubble wrap after each remote, opt for custom foam cutouts that allow you to drop the electronic units into dedicated slots. (This also makes it easy to see if any equipment items are missing when packing up.) Cases with rack mounts are available and are quick to set up. Judicious planning is necessary, however, as they can be heavy when the dimensions are too large, and the racks mounts can be damaged if the case is dropped. There are waterproof hard cases that are quite durable. Options include inserts with compartments similar to a camera case. These are useful for smaller pieces of gear, office supplies, batteries, tools, and other miscellaneous items. Each case can be numbered for organizational purposes.

As a broadcaster obtains remote gear, it is helpful to compile a master list or spreadsheet with the model, serial number, and inventory number of all devices. You could add in the cost as well. It helps to know what each case contains for accurate inventory and insurance purposes.

Radio remotes provide an excellent way for broadcasters to connect with their communities and provide listeners with uniquely engaging content. They do require special effort, however. The key to successful remotes includes properly selected equipment, well-trained personnel, and above all, adequate planning and preparation.

CHAPTER

4.7

Radio Station Automation, Networks, and Audio Storage

EUGENE NOVACEK, P.E.
ENCO Systems
Southfield, MI

INTRODUCTION

The term "radio station automation" is a very broad concept. It generally refers to the use of devices, processes and system interconnections designed to make a broadcast station (radio or TV) run a series of scheduled events automatically—without operator intervention. This includes every scheme ever devised, from massive reel-to-reel tapes and interconnected cart machines to all-encompassing control systems that steer the directional array, turn on the transmitter, play the audio, control the console cross-fades and generate transmitter logs and billing information.

While the individual pieces of hardware involved in broadcasting have evolved to include powerful automatic functions and interconnect capabilities, nothing has advanced the state-of-the-art of station automation more than the personal computer. The purpose of this chapter is to educate the reader on the current best practices of station automation. To do this, we will explore some background on how personal computers and their powerful data manipulation and communications capabilities have provided the necessary "glue" to bond all of a station's subsystems together. Additionally, how these strengths coupled with powerful and intuitive user-interface features have created a complete automation solution. In this process, we will examine the evolution of practical network architectures and cabling standards, as well as the current "favorites." We will discuss network bandwidth requirements, storage capacity and speed requirements, multi-user performance issues and data redundancy options. We wrap up with a look at some current storage architectures that fulfill these necessities, and point out a few popular methods of protecting these systems from catastrophic failure.

The majority of the details discussed in this chapter come from practical radio station automation systems; however, the concepts also apply to television station automation.

THE PC HAS REVOLUTIONIZED STATION AUTOMATION

It wasn't too long ago that station automation referred to a 15-inch reel of tape playing at 1-7/8 inches per second on an auto-reversing tape deck. This automation arrangement could provide most of a day's programming with minimal operator intervention, but the audio quality was mediocre due to the slow tape speed, and it required production personnel to invest a great deal of time assembling the contents of the tapes. Once the tapes were assembled, it was very difficult to make any changes, especially if the tape was playing on-air.

A more flexible arrangement used a stack of cart machines with the secondary cue (SEC) output of one wired to the start input of the next, and so on, with the last machine connected back to the first. This allowed the operator to load several songs and commercials to play automatically in sequence, freeing up short amounts of time for the operator to handle other tasks. This form of automation was followed by numerous variations of mechanical beasts (I affectionately referred to them as "Cartasauruses" or "carousels") designed to hold and cycle dozens and sometimes hundreds of carts through multiple playback decks. These electromechanical marvels were capable of automatically sequencing several days' worth of programming.

These machines were incredible achievements in station automation and have faithfully served the broadcast industry for many years. However, their dependency on extensive mechanical transports made them high-maintenance devices and limited their flexibility. Even the audio quality was hard to maintain due to the number of capstans and tape heads involved. Broadcasters needed newer, higher quality, easier to maintain, more versatile automation systems.

As CD players and DAT tape machines became commonplace in the home and in the studio, audience demand for *CD-quality* broadcast audio also grew. Along with the improved audio, this new generation of equipment brought improved control features that allowed broadcasters to elevate automation to a new level. In addition to the standard START and STOP functions, a much more complete set of instructions, including shuttling and indexing, were now available, usually through a serial data protocol. This enabled a single electronic controller to "talk" to multiple devices, directing them to cue up then play individual tracks from CDs and tapes containing multiple tracks. With *CD jukebox* devices available that could hold and play tracks from over 300 CDs and DAT tape machines, cuing and playing hours' worth of programming, a station could now automate for days at a time, if the controller was programmed properly.

While several manufacturers developed automation systems for their own equipment, these were mostly proprietary hardware and not able to communicate with devices from other manufacturers. These units often employed special keypads for data entry and ran ROM-based programs that were difficult to upgrade or modify. This is where the personal computer made its biggest mark on station automation. PCs were already being used in business and at home to run database applications, perfect for storing a log of scheduled events. PCs had excellent serial (and parallel) communications capability and could be configured/programmed to "speak" any protocol required. PCs used a standard user interface (keyboard and monitor). PC-based programs were stored on disk and run from RAM and were therefore relatively simple to modify and reload. The personal computer was an excellent platform for station automation controllers.

A number of manufacturers created custom software packages using standard PC hardware to automate CD jukeboxes, DAT players and fire relays to play standard cart machines. These were quite effective and could provide a good level of *walk away* automation, but some of the stations' most important material, e.g. commercials, IDs and promos, were left to the weakest link—the *cart machine*. It became increasingly obvious that a more sophisticated CD-quality version of the cart machine was needed to play spots, IDs, jingles, etc.

As digital cart machines began appearing as direct hardware replacements for the existing analog units, some clever computer people were realizing that everything needed for a digital cart machine was already in a PC, except the analog-to-digital and digital-to-analog conversion electronics. This could easily be added by building a custom signal-processing circuit board designed to plug into one of the existing PC expansion slots. Since PCs were already driving the development of the hard disk storage technology, they had access to the capacity required to store many hours of digital audio. This could be divided into any number of any size pieces, allowing an extensive inventory; and unlike tape or floppy disk based systems, PC-based systems could provide random access to any and all of the material they contained.

Adding these powerful audio capabilities to the automation strengths of the PC has created the perfect platform for the continued evolution of station automation.

KEY FEATURES OF AN AUTOMATION SYSTEM

The personal computer is capable of providing all of the key functions required for a powerful station automation system in a single box: random access high-quality audio, database operations, communications capabilities, programmability, user-friendly operator interface and file redundancy. But to be an effective automation system, the hardware and software must be designed to provide certain key features.

Professional (CD) Quality Audio and Sound Cards

When the compact disk (CD) came out, it set a new standard of audio quality that became a widely used benchmark when discussing audio performance. The term "CD quality" implies the following basic specifications:

Frequency response	+/− 0.5 dB (20 Hz–20 kHz)
Dynamic Range	>90 dB
Signal to Noise	>90 dB
THD + N	<0.05%
Phase error between L-R channels	<1 degree @ 15 kHz

These specifications are very close to what signal processing theory would predict for a digital system using 16-bit *quantization*. Current multi-gigahertz PC platforms are quite capable of manipulating this (and considerably higher resolution) data while maintaining its purity. Therefore, a determining factor in the overall audio quality of a PC-based digital audio system is the quality of the electronics used to convert between the analog and digital worlds. In a PC-based system, this process occurs in what is generally referred to as the *sound card* or *audio card*.

These devices include the analog input stage, the input *anti-aliasing filter*, the analog-to-digital (*A/D*) and digital-to-analog (*D/A*) converter devices, the output *reconstruction filter* and the analog output stage for each audio channel. It is very important that the analog stages have very low noise, a wide dynamic range and be very linear. The input low-pass filter must have a steep enough response to preserve the maximum high-frequency information for a given sample rate, minimal pass-band ripple, minimal ringing and a linear phase response. The analog-to-digital converter must have very accurate quantization levels (high linearity) and must complete each conversion very quickly (short conversion time). The digital-to-analog converter and reconstruction filter should use higher resolution (more bits) than the A/D to keep from raising the noise floor and limiting the headroom of the output signal.

Another option is to directly interface to AES digital audio, which combines two channels of audio into a single digital data stream. This is then routed into or out of the sound card on a single shielded twisted-pair cable and XLR connector, or in some cases on a fiber-optic link. Most professional sound cards are available with either analog, digital, or both analog and digital inputs and outputs. Utilizing AES digital audio reduces the number of A/D and D/A conversions in the audio chain, which can improve the sonic quality of audio that has passed through many "nodes" from source to destination.

It is often challenging to obtain these specific details without reviewing extensive technical documentation. To complicate things further, many manufacturers will measure these parameters and report the resulting specifications using differing terminology and references. But fortunately for the user, these components have advanced to the point that almost every PC-based sound system available today will provide CD-quality audio when configured and connected correctly. The type of connections required will generally determine the type of sound card required.

Sound cards exist in several forms. Typically, they are plug-in cards designed to fit into a free expansion slot inside the computer. Many contain the A/D and D/A electronics on the card itself, and this has proven to be a successful format for many years. Some manufacturers have questioned the engineering soundness (no pun intended) of putting highly sensitive analog electronics inside a computer's "noisy" digital environment. They have chosen to design their computer cards to contain only the digital electronics while moving all analog functions to an external chassis. This concept not only benefits audio purity, but the flexibility in this design allows the use of different audio interfaces designed for different standards (e.g. analog/digital, balanced/unbalanced). Another variation of external audio interface connects to the computer via a USB port. These are available in both "consumer" and "professional" grade as well.

Similar to this is the relatively new concept of the remote audio card. Since audio signals inside a PC are really just streams of data, this design uses a software driver to convert these audio streams to packetized data and transmit them to the audio hardware via an *Ethernet*-based network. Some designs use a dedicated audio network architecture (CobraNet, WheatNet, Virtual Audio Cable) while others use standard Internet Protocol (IP) over a conventional switched Ethernet network (Axia LiveWire, Ravenna, AES67, etc.). These "audio over IP" networks not only allow the analog or AES digital audio electronics to "live" in a properly conditioned environment, but the inputs and outputs can exist in different physical locations throughout a facility. This concept provides traditional sound card functions and can also serve as the primary audio routing system and provide audio console functions as well.

In addition to these "add-on" solutions, many current PCs include audio hardware as a part of the main board (motherboard). This hardware usually falls into the "consumer grade" or "multimedia" category. While these can provide good audio performance, consumer units typically have more high frequency roll-off for a given sample rate, more pass-band ripple, more inter-channel phase error and higher distortion due to less expensive A/D and D/A converters. Generally speaking, products intended for the professional market (radio, TV, recording studio) use very high-quality components and provide measurably better audio quality than consumer-grade multimedia units. Pro units also usually provide balanced analog and digital audio connections, whereas consumer units do not. And most pro units provide onboard DSP "horsepower" to perform real-time data compression and audio effects, whereas consumer units do not. But with PC processing power growing by leaps and bounds, many of these functions are being provided in software (plug-ins) and off-loaded to the main *CPU*. This architecture can provide very cost-effective audio functionality, so these consumer devices have their place.

Thus, there are a number of ways to provide high-quality audio from a PC-based system, and a key feature of a good automation system is to be flexible enough to make the best use of the different audio resources available to it.

Databases Keep It Organized

With the ability to record and play high-quality audio established, the next most important feature an automation system must possess is a way to organize its inventory. This is where the PC's database capabilities are utilized. There are many different types of database formats. The system should be able to hold numerous fields of information for each piece of audio recorded into the system. Things like the title, album, artist, length, genre, record date, start date and kill date (to name just a few) should be stored along with a unique identifier (numeric or alphanumeric) for every recording. The system should be able to sort the database by any field and offer search functions able to find data from any of the fields. It should be easy to change, update or remove data as well as to generate electronic and/or hard copy reports of the contents. More advanced systems might also include features like storing segue points, auxiliary cue points, back timer points, number of plays counters and the date and time of each cut's last use. Most importantly, the information and tools to manipulate it must be presented clearly and logically to the user through the software without the need to manipulate data in the database directly. This is a function of the graphic user interface, which will be discussed a bit later.

From this overall database of the complete system inventory, an automation system must be able to create (or import from an external source) sub-databases that include the elements that will be used (played) in succession. These *playlists* might contain elements for a single show or might contain an entire day's programming and thus may represent the station's daily program schedule or log. In addition to the actual audio elements, these playlists need to include automation information that instructs the system how to sequence from one event to the next automatically. The ability to program lists to cover multiple days or the ability for one list to sequence to another list is also very important.

Since these playlists might typically hold hundreds or possibly thousands of elements, scheduling each and every event and programming the automation manually would be very tedious. Again, the power of the PC comes to the rescue. There are many PC-based programs in existence that use database functions combined with user-defined sets of rules to manage and schedule commercials/promos (traffic) and music. Each category is handled somewhat differently. Programs that schedule commercials generally take sales department information (orders) and apply rules based on airtime rates, time of day, time of year and adjacent programming to generate a daily schedule or log. These programs are usually also tied to the billing department and are often able to accept *as-played logs* from the automation system and reconcile them with the original schedule to generate the appropriate billing invoices. Music scheduling programs, on the other hand, must choose cuts based on the station's format, time of day, music category and artist, and may apply rules such as prohibiting two artists from the same music category from playing back to back or limiting specific cuts from playing more than once per hour or once per day.

While some automation systems provide these scheduling features as part of an overall comprehensive automation package, it is not necessarily a great advantage to obtain everything from a single vendor. With all the excellent scheduling software that is already available from a multitude of vendors, often the most flexible approach is to choose each package separately based on their individual merits (possibly already familiar or in use), and then select an automation system that can interface with these systems to import and merge the traffic and music schedules and provide seamless integration of scheduling and playback, as well as maintain library data. The automation system then exports an as-played log representing what actually aired to the traffic system, which uses the data to expedite the billing process as described above. Immediate reconciliation is also possible through direct communication between applications. In short, to be a good automation system, it is not necessary to possess scheduling features, but to be able to use the schedules generated by other applications and then report back to them the on-air results.

Sequencing, Cross-Fades, and "Play While Record"

A professional automation system will include several options for controlling when and how the system automates playout from event to event. Features to preset and/or execute transitions at a specific time of day are important for cueing *stopsets* and starting programs. Transitions that tie the end of one event to the beginning of the next are useful for connecting multiple spots into a single stopset. Advanced systems often provide the ability to overlap events to provide segues between audio elements and to add voiceover announcements over the transitions. The ability to perform these overlaps with dynamic level control (e.g. cross-fades, ducking) all within a single stereo playback channel is highly desirable. Larger facilities that take in many networked programs, news feeds, actualities, etc., will dedicate a "net catch" workstation. This workstation lives on the automation network and is configured to simultaneously record multiple inputs from various sources, then automatically trim and ingest the content into the automation system for playback. Advanced operations will allow for greater recording flexibility by controlling audio switchers to bring in various audio inputs from external sources (i.e. satellite, webstream). Various styles of broadcast quality audio cards will have models geared towards higher numbers of recording inputs and outputs.

Another desirable automation system feature is the ability to record into the device while simultaneously playing audio. This allows production personnel to create new inventory without having to wait for specific breaks in the normal on-air programming. Most systems provide this function by offering multiple record and/or play audio channels. Some systems are scalable, allowing the channels to be added as needed by inserting additional audio devices.

An important extension of this function is the ability to play a file while it is still being recorded. One excellent use of this feature is to perform short interval time shifts. Once a program is being recorded, you may start playback of that recording at any time, even before the program (and recording) is completed. This allows long-form programs to be delayed by as little as a few minutes or even seconds. Another powerful use is the ability to edit and play excerpts of a recording while the recording is still in progress. News feeds provide an excellent example. Sound bites can be trimmed and used on air before the originating news feed recording is complete. While most automation systems require a recording be in progress for a specified amount of time before the cut may be accessed for edit or playback, this interval is typically only a few seconds making this feature very handy.

Peripheral Interfaces

An automation system rarely contains everything necessary for complete walk-away operation. For it to be truly effective, a system needs to be able to interface with various other broadcast devices, like audio switchers, audio consoles, satellite receivers, CD players, DAT players, station clocks, and more. These interfaces are created in two parts: the hardware interface, which provides proper electronic interconnection and the software interface that communicates specific commands to and from the peripheral.

Systems based on personal computers can take advantage of a multitude of *off-the-shelf* hardware (plug-in I/O boards) that can provide any mix of RS-232, RS-422, RS-485 and large numbers of opto-isolated contact closure inputs and dry contact relay outputs. Systems that are able to address these generic I/O products are more flexible than those that offer their own custom I/O hardware that may offer a mix of these features.

In addition to these "traditional" communication interfaces, many new consoles and routing switchers feature network interface ports and are able to communicate via IP over standard Ethernet networks. This is a perfect match for the network hardware of a PC platform and can greatly simplify the physical interconnections for systems that are capable of communicating through this channel. This can allow PC-based automation systems to control input routing, fader labeling and bus assignment and even automate mixes and move motorized faders through a simple network connection.

With the hardware connections becoming more standardized and commonplace, we see that the real strength of an automation system is in the software portion of this interface—how it uses the hardware to communicate with the various devices external to the PC. A wide variation of command syntax and order exists among the various peripheral devices an automation system might connect to. Command strings, string lengths and even the use of carriage return and/or line feed vary from device to device. Even simple contact closure-controlled devices can often require specific closure sequences to perform a given function. It is therefore very important that the automation system software support the creation and transmission of custom protocols, be capable of storing and running *control macros* that execute specific command sequences and provide these features through a user programmable mechanism so that each station's unique compliment of equipment can be administered and maintained.

Many system manufacturers provide software modules (drivers) that perform the data translations necessary to allow the automation system to communicate with other popular devices. In many cases, that is a simple and adequate way to provide interoperability, as long as the configuration never changes. But things will change. An automation system that is able to utilize industry standard methods of communication and provides a mechanism to "tweak" communication details is more flexible and able to adapt to future developments.

Data communication abilities of automation systems are now extending beyond peripheral control and being used to create a better user experience through ancillary data services. An automation system can send database information about its audio (often referred to as metadata) to external data channels that can provide relevant "Now Playing" information to the end user. Examples of this include RDS encoders feeding conventional FM transmitters to display title, artist and station ID text strings on users' receivers and roadside billboards. IBOC HD Radio includes an enhanced version of this feature through its program-associated data (PAD) channel. Metadata over IP is now allowing broadcasters to tightly integrate their on-air product with their station websites and streaming services. By and large, these services are employing standardized information exchange formats like eXtensible Mark-up Language (*XML*) and Javascript Object Notation (JSON), which can provide a very powerful and flexible framework for future development.

Automation-Production System Integration

This communication flexibility and standardization is also providing a path toward tighter integration of automation and production systems. Many facilities utilize multitrack production systems to create content because of advanced production tools available in those environments. But completed projects then need to be moved to the automation system to benefit from their more advanced play-to-air sequencing and control features. Often times these systems run on different hardware platforms and employ different operating systems to run their application software. In the past, this created a significant roadblock to data sharing and required the transfer to be accomplished in real time by dubbing from one system to another. This also required additional time and effort to manually re-enter the metadata into the destination system's database.

Today, networking technology allows us to connect even cross-platform devices, allowing them to share data at the file system level. Most systems have adopted standardized audio file formats, such as WAV and AIFF, allowing direct file transfers from one system to another. Even if the standards are different, many production and automation applications have file conversion applets that allow non-native files to be reformatted during transfer. And if not, there are a number of generic format converter applications available that can make each side of a connection happy.

But in addition to the audio, we need to be able to share the metadata. This has been somewhat problematic in the past due to the many types of databases in use and the specific fields of information stored and manipulated by differing systems. But more and more system vendors are offering application programming interfaces (APIs) to their systems that allow an outside application to extract and inject relevant information (metadata) to and from their native database structure. This allows the vendor of one system to create an interface module to another vendor's application that allows it to access, manipulate and return updated information to the host application. When properly implemented, this can create integration so tight that the user does not realize they are changing environments. An example of this is clicking an edit button inside an automation system and having the audio file with its associated metadata open inside a different vendor's multi-track editor. Upon completing the edit session, the user clicks Save in the multi-track editor and returns to the automation system's screen with the edited audio and metadata instantly available.

Even if specific integration modules are not available, more and more systems are starting to support generic ways of moving metadata. A simple approach to this is to place key information into the file name of the exported audio file using a special character as the delimiter. At the destination, an application like ENCO Systems' NewDB can be configured to parse this information from the file name and use it to populate the associated metadata fields in its system's database.

Some systems go a step further and import and export metadata in XML files that are linked to the audio file. XML is well suited to this purpose since data tags are defined within the data itself, making the structure flexible enough to handle any type and size of metadata.

Metadata can be contained within the audio file itself. Many system vendors are supporting extensions to standard file formats such as the Broadcast Wave (*BWF*), *Cart Chunk* and Extensible Metadata Platform (XMP) formats. These are nonproprietary standards that allow metadata to be contained within a "chunk" that is embedded into a standard WAV file, such that "aware" applications see it and nonaware applications ignore it. Systems supporting MP3 files are making use of their *ID3* tags, which serve a similar metadata function but are only applicable to MP3 audio files. Other file exchange formats that combine media and metadata, such as Open Media Framework Interchange (*OMFI*), Media eXchange Format (*MXF*) and Advanced Authoring Format (*AAF*), are becoming more widespread as well.

Graphic User Interface (*GUI*)

Even if an automation system possesses all of the key features described thus far, it *must* have a clear, well-organized, intuitive operator interface or much of its power will go unused. Many of the early automation systems were quite powerful but displayed busy text-based screens and required multiple-key keyboard commands to operate. The "computer-phobic" shied away from these units, making them useful to only the more computer-literate station personnel. Graphic user interfaces (GUI) like Mac O/S, IBM O/S2, Unix X-Windows, and various versions of Microsoft's Windows have made it possible to display screens that provide more visual clues as to function and use. The necessary text-based data remains, but it is surrounded with buttons and scroll boxes that allow the operator to navigate the screen and operate the system with a simple pointing device like a touchscreen, mouse or trackball (see Figure 4.7-1).

Touch-screen technology coordinates a touch sensitive x-y matrix placed over the display screen with the graphics/text displayed through it by the program. The operator can "press" a button on the screen and get the illusion that they are operating a physical button due to the program's resultant visual clues, like color change and shadow reversal. This type of interface has engendered PC-based automation systems into common use by even the most computer-phobic staff members and has enhanced the role of the automation system as an on-air live-assist device. Screens that present lists of cuts to be played sequentially along with their *hit times*, transition information, a countdown clock, a time-of-day clock and audition functions work well for tightly formatted live-assist periods. Screens that present panels of programmable playback buttons provide the instantaneous playback capabilities needed for spontaneous live shows. Script-based

FIGURE 4.7-1 PC user interface combining audio and database functions

programs like newscasts can benefit from systems that allow the integration of text and audio to provide sound bite playback from buttons that are embedded in and scroll with the script. Examples such as these illustrate how a well-designed human-machine interface can broaden an automation system's usability and enhance its effectiveness (see Figure 4.7-2).

With operators handling multiple tasks at once, it is important that the automation system have a user-friendly interface, providing as little work for the operator as possible. Many automation systems provide usable touch-screen interfaces that take the place of the keyboard and mouse. There are two types of commonly used touchscreens—capacitive and resistive. A capacitive touchscreen consists primarily of a piece of glass coated with a conductive material. When a person touches the screen's surface, they act as an electrical conductor, creating a change in the electrostatic field. This charge is measured internally to determine the location of the touch, resulting in a process of information on the screen. The downfall to this type of screen is that it relies on the human body to create its process. When a person wears a glove or uses an object to touch the screen, in most cases that information will not be processed. Resistive forms of touchscreen rely on several layers of thin conductive flat material. When the screen is pressed, these materials are pressed together resulting in a conductive change, measuring the location based on the point of connection. This form of screen allows the user to be able to use different objects to press on the screen to create the same process (stylus, credit card, gloves, etc.). Costs for this type of screen are relatively lower than other types, and it is less likely to create accidental "presses" from static charge. However, this type of screen is more susceptible to screen damage if it is pressed upon too hard or dented with a sharp object. It is also not as bright as a capacitive touchscreen due to the many layers of material in front of the display.

NETWORKED AUTOMATION SYSTEMS

An automation environment is far more efficient when each workstation is referring to the same databases and sharing a common set of resources. Networked systems keep files organized and in a central location for all users to access in accordance with their permission level. A single, stand-alone workstation with the features described thus far can do an excellent job providing the basic production, live-assist and automation features required for a single station. If you need to automate multiple stations, you have two choices: try to squeeze the additional resources from a single workstation or add additional workstations to handle the additional stations. Using a single workstation simplifies inventory management and allows the production and scheduling to be shared; however, as you add additional processes to a single workstation, the unit's capabilities will be spread thinner and thinner until the overall performance is no longer satisfactory. If you add more independent units, you maintain each system's performance level but add the burden of managing multiple inventories, multiple production facilities and multiple scheduling systems. To achieve the better of these two scenarios, we would need to interconnect the workstations so they could share audio inventory, production elements and scheduling resources while they each provided high-performance live-assist and/or automation features to their individual stations. The sharing of these resources reduces the fragmentation of data among the workstations and the overall need for storage space. This kind of interconnection is exactly what PC networks are designed to provide.

FIGURE 4.7-2 Sample of touch-screen-controlled live-assist graphic user interface

Networking the individual workstations together enables them to move audio, automation schedules, as-play logs, scripts and wire copy text, configuration and control data and software updates from unit to unit without having to physically transport the data via removable media. Eliminating the time and effort associated with such a *"sneakernet"* is usually by itself enough to justify the expense of the network hardware and software. But in addition to providing a conduit to expedite the exchange of information between one unit and another, networking can actually increase the overall efficiency of a group of workstations by eliminating unnecessary replication of resources and by allowing individual workstations to specialize. For instance, it is usually more efficient in both equipment and manpower to equip a single workstation with the production facilities to create spots for *all* of the stations than it is to perform production at *each* of the on-air workstations. This also makes the on-air workstations more efficient by allowing them to devote their resources to their primary on-air functions. The same is true of other group-wide functions, such as wire service integration, news feed collection, commercial/music scheduling and overall inventory management. There are further benefits to be realized from specific network interconnections, and therefore some discussion of network architecture is in order.

NETWORK ARCHITECTURES

A number of computer network architectures are presently being utilized for broadcast applications. Each has certain advantages and disadvantages. The three most common configurations are the *peer-to-peer network*, the *audio (video) server*-based network, and dedicated *file server*-based network.

Peer-to-Peer

A peer-to-peer network is a simple and inexpensive way to link two or more workstations at the file system level, permitting file transfer between them. This arrangement has achieved reasonable popularity since it only requires the addition of network interface cards (*NIC*) to each workstation and a basic network switch. If only two workstations are networked, then a direct Ethernet cable connection can be used, which eliminates the need for a network switch. In the past, this required a special "cross-over" Ethernet cable, but most current NIC technology will auto-sense how it is connected and can use a standard "straight-through" cable. The primary drawback of not using a network switch is that there is no easy way to add additional workstations or computers.

Peer-to-peer networks allow each workstation to access the contents of another workstation's drive. In early generations of PC hardware, a workstation's capabilities could easily be taxed by the overhead required to service a network client trying to access its hard drive. This would often affect the host workstation's on-air performance. For this reason, audio was usually not shared directly, but was copied at a throttled rate across the network to the client's local hard drive before it could be used. This added a considerable delay between production and playback, especially if the cut to be transferred was quite large. This method of file sharing also usually dictated that each peer run from its own separate database, which required considerable file maintenance to keep the multiple separate inventories organized.

Current PC workstations with multi-core 3+ GHz processors, SATA or solid-state (SSD) disks and 100 Mbps or 1000 Mbps (gigabit) LAN hardware are much more capable of "serving" while maintaining high-performance local operations. This has made effective peer-to-peer configurations very practical and common. However, a workstation is not a server, and these configurations should be limited to just a few workstations. This limitation is also reinforced by most operating systems, which usually limit client (peer) connections to ten or less. Typically, a limit of five is suggested, as performance may suffer in a larger peer-to-peer environment. The disk configuration can also affect performance, so it is generally recommended that the database and audio files be stored on a drive that is separate from the operating system and applications. This drive should have a 7200 RPM speed rating or higher (or SSD), and if possible be a RAID 1 configuration (two drives mirrored) to provide a degree of fault tolerance and faster read performance.

Audio Server

An audio server-based network usually consists of one or more workstations communicating with a dedicated and often proprietary computer, which houses both the audio-processing hardware (DSP boards) and audio file hard-disk storage. The workstations act as control interfaces that command the central audio server to perform all record and play functions. A nice feature of the arrangement is that all the audio data is stored centrally, making it instantly accessible by any workstation. Audio recorded by a production workstation can be immediately played by an on-air workstation without requiring any file copying to the local hard drive. A serious disadvantage to this network configuration is that all audio processing is performed by the audio server, which has a physical limit to the number of audio channels it can contain. This limits the number of workstations that can be supported by the system and makes it difficult and expensive to expand. Also, since all audio inputs and outputs occur at the audio server, audio interconnect wiring is required to and from each studio, adding to the installation complexity and potential points of failure. And since the workstations contain no audio hardware themselves, they have no functionality in the event of an audio server or network failure.

File Server

A dedicated file server-based network combines the best features of the first two architectures presented. The network generally consists of one or more workstations linked to a central high performance, nonproprietary computer, which runs a standard high-performance operating system like Windows Server or a Linux/Unix-based server capable of SMB protocol (to support Windows workstations). The server presents itself to the rest of the system as a large shared disk drive (or drives) and manages the storage of all audio and system data. As with the audio server concept, the audio is stored centrally and is therefore instantly accessible by any of the workstations. The major advantage this configuration has over the audio server configuration is that the audio-processing hardware (sound device or driver) is located in each of the individual workstations. This places the audio inputs and outputs at the workstation's location with the network used to transfer packets of digitized audio data between the workstation and server. Not only does this allow all workstations to share the centrally stored audio, but it also provides *audio* distribution via the *data* network. An audio source local to a workstation at one end of a facility can be recorded and then immediately played by a workstation at the other end of the facility. A practical example of this is a workstation installed in an equipment room with the station's satellite receivers and programmed to automatically record multiple news feeds throughout the day. Workstations anywhere within the facility can access, edit and play portions of these feeds without having any direct audio feed from the receivers themselves. A significant benefit to this design is that the complete interconnect for distributing the data and audio to and from a workstation is via inexpensive CAT5e, CAT6 or CAT7 cables.

The fact that the file server contains no audio hardware is also very significant to server functionality, because this means it is not bound by any physical number of channels. The server is designed to service large numbers of data streams bound only by performance and capacity of its disk I/O subsystem and the speed of the network. With proper server design, any number of workstations/audio channels can be supported, making this type of network very scalable.

Another excellent benefit of the dedicated file server-based network is the various *data redundancy* (backup) options available. With a number of workstations relying on the file server to store all of their work, it is very important that this storage remain online at all times and be protected in the event of catastrophic hardware failure, such as a hard-disk failure. Features like disk mirroring, server mirroring (clustering) and RAID (described later in this section) are available, along with support for a considerable range of industry-standard hardware components designed to provide performance and redundancy. Also, the dedicated file-server architecture is able to provide a level of redundancy via the workstation. The primary mode of operation is for each workstation to record and play directly to and from the server; however, since each workstation contains its own audio-processing hardware, local hard-disk storage can be can be included for emergency use in the event of file server or network maintenance or failure.

Recent developments in audio over IP networks (AoIP) and virtualization have made a file-server environment even more advantageous. A file server (or server cluster) with enough "horsepower" and capacity could contain not only the database and audio files, but would also contain any number of "virtual" workstations. These workstations would function as normal audio and control devices and would be accessed via "thin clients" located in the individual studios. Audio would be routed via AoIP protocol data from the server to the individual network "nodes" located as needed to interface with external audio sources and destinations. This environment provides an added degree of redundancy and fault tolerance, as well as reducing the hardware and rack space requirements for a larger facility. An entire facility with dual-redundant fileservers and 10 or 15 virtual workstations could be contained in a single 4RU chassis.

Each of these network architectures provides useful connectivity between workstations and has its place in station automation. But, all things considered, the dedicated server-based network offers the best collection of features from which to build a multi-station automation system.

Now let us examine how these network architectures are physically accomplished.

NETWORK TOPOLOGY

The way cables are arranged to provide interconnection between devices in a network is known as its *physical topology* or shape. The way the signals are actually routed through the network is known as its *logical topology*. The physical topology and logical topology may not necessarily be the same, as we will see shortly.

Early network design utilized bus-link physical topology. The bus topology uses a single cable that each workstation and server attaches to and shares. When any node (connected devices are often referred to nodes) transmits, it broadcasts over this transmission line with every node hearing every transmission. Communications would become hopelessly confused if multiple workstations tried to transmit on this shared line simultaneously; therefore, bus networks must employ a control mechanism to make sure that only one transmitter is active at a time. This media access control mechanism will generally lower the network's efficiency well below its theoretical value. The bus topology can be easily expanded by making the bus cable longer and attaching more workstations up to the electrical limitations of the cable. It is also possible to expand beyond this limit by using repeater electronics and multiple cable segments. Figure 4.7-3a represents the basic concept of the bus topology.

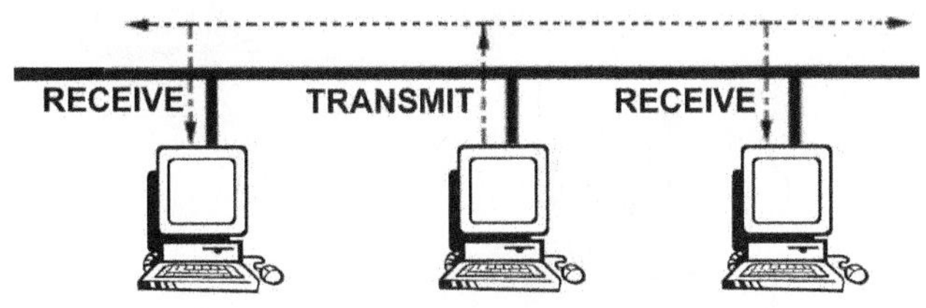

FIGURE 4.7-3A Bus topology network

A more modern method is point-to-point topology that does not broadcast on a shared cable, but as the name implies, transmits to exactly one receiver and receives from exactly one transmitter. Before the development of managed network switches, as Figure 4.7-3b illustrates, the transmitter of one workstation was connected directly to the receiver of another with the transmitter of that workstation connected back to the receiver of the first. Since transmission and reception occur over separate wires, this topology had the advantage of supporting duplex operation (simultaneous transmit and receive), which doubled communication speed. However, point-to point networks are somewhat more complicated to expand. One way to add workstations is to provide a point-to-point connection between every pair of workstations in the network, as shown in Figure 4.7-3c. Known as a *mesh*, this can provide very efficient point-to-point communications, but it is obvious that these connections would become overly complex for even a small group of computers.

Most modern networks are an implementation of the point-to-point link network called *star topology*. In this arrangement, all workstations and servers have a point-to-point connection with a centrally located device that provides the signal routing between the nodes. Early configurations of this topology utilized a network hub, but almost all current networks use a managed or unmanaged network switch (see the Switched Ethernet section below). A star network offers some distinct advantages over other topologies. Point-to-point connections can provide more efficient *full duplex (FDX)* communications. Centralized connections provide for easy expansion and reorganization of the network. Connection ports in the switch typically include signal conditioning electronics to clean and reshape waveforms, which improve the performance, reliability and range of the network. In addition, the conditioning electronics allow the switch to ignore a disconnected or defective port or cable, making the network less sensitive to a single-point failure. The network switch also contains the intelligence needed to route the data coming in only where it is intended to go, thus reducing unnecessary network traffic. And most switches also include traffic monitoring and management functions, which allow system administrators to analyze and fine tune their networks.

How the central device (internally) connects the nodes determines how the network communications take place; a.k.a. the logical topology of the network.

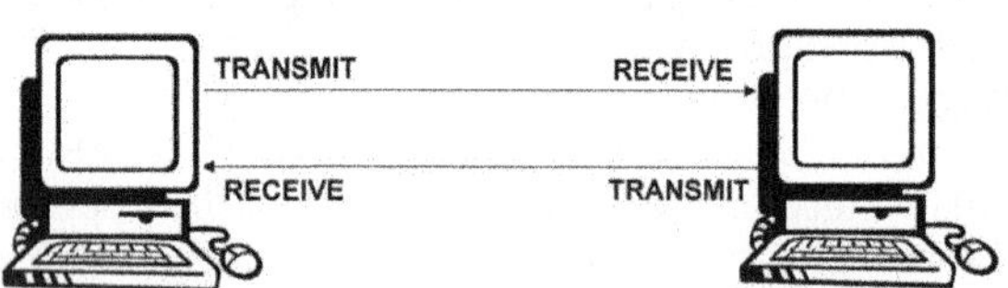

FIGURE 4.7-3B Simple peer-to-peer network

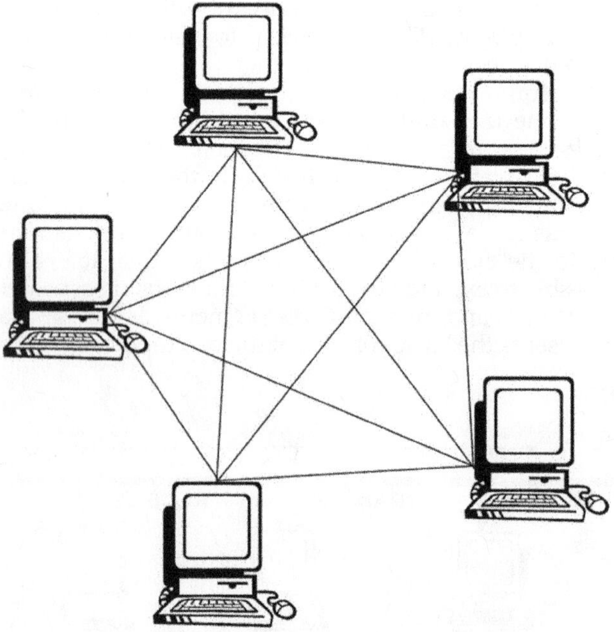

FIGURE 4.7-3C Mesh peer-to-peer network

Switched Ethernet

The current state of the art in local area networking (LAN) is *switched Ethernet*. In this type of star-connected network, the central connection device, called a *switch*, provides high-speed port-to-port packet-switching functions that create a virtual mesh logical topology. As described earlier, a mesh network provides very efficient point-to-point communications, exactly what we need for multi-workstation-to-server and workstation-to-workstation transfers.

This complicated work requires a sophisticated device. A switch is "smart" in that it decodes the addressing of each packet it receives and routes it to the specific destination without "bothering" parts of the network that are not involved in the exchange. This effectively makes each port look like a separate segment, which isolates the traffic (*collisions*) and provides full Ethernet bandwidth on each port.

To accomplish this, the switch must have electronics to do the decoding (processor), packet buffers (RAM) to store packets until forwarded, memory (more RAM) to store the address table of destinations and a fast matrix (backplane) to move the packet to the desired port. Several different algorithms can be used to decode, check and process the packets before they are sent out.

Each feature described above is important to the overall performance of a switch. The faster the processor and the higher the backplane speed, the higher the throughput. The greater the onboard memory, the larger the packet buffer size and the larger the address table storage.

These individual specifications are not always provided by the manufacturer. Fortunately, all of these combine to determine the real benchmark of interest, *latency*. This is a measure of how long it takes a packet to get through the switch and on to its destination. An automation system requires packets to arrive in a very timely manner to prevent its audio and control functions from being interrupted. High latency can reduce a workstation's ability to keep its audio buffers full and can make the user interface too sluggish to use effectively.

Therefore, a very important part of creating and maintaining a successful station automation system is to build it upon a backbone with very low latency. Fortunately, today even "economy" switches will generally have latency under 100 microseconds and will work quite well in small to moderate size systems. High-quality switches will have latency figures in the low microsecond range and should be used in large systems and/or if high data throughput (such as multiple PCM audio streams) are required.

Another important note is that even high-quality switches with great throughput and low latency specs can still degrade system performance if not properly configured. A *managed switch* allows a user to control and program its features through a serial or IP connection, using a telnet session or web-style interface. These switches often contain various additional software algorithms that can be valuable in general but should be disabled in some specific situations. Features to limit broadcast traffic, Spanning Tree Protocol (*STP*) to find and block redundant links, *port trunking/teaming/link aggregation* to provide higher bandwidth to critical devices and the ability to manage *multicast* traffic are all powerful and useful features but can sometimes cause severe interactions with features of an automation system. Typically, the factory default settings for a switch will provide proper performance, but be aware of the potential complications and contact your system vendor if in doubt.

NETWORK CABLING AND HARDWARE

So what does all this mean to the station engineer who needs to install and connect the station automation system? To get to the real-world nuts and bolts of networking, we must discuss cabling standards.

Ethernet is by far the most favored network standard in use. Ethernet uses *baseband* signaling, which is indicated by the inclusion of "base" in each of the specific designations. This means that only a single Ethernet signal occupies a transmit or receive path at a time.

The most common implementations of Ethernet today are 100BaseTX (fast Ethernet) and 1000BaseT (Gigabit Ethernet). There have been other variations in the past but in most cases they are no longer used (*10Base2*, *10Base5* and *10BaseT*, which are 10-Mbps technologies, and *100BaseT4*, *100BaseTX* and *100BaseFX*, which are 100-Mbps technologies). There are also Gigabit Ethernet over fiber optic standards (*1000BaseLX* and *1000BaseSX*). The ultimate (so far) Ethernet standard that has recently become commonly used is 10-GigabitEthernet over fiber (10Gbe) and CAT6 or CAT7 twisted-pair copper (10GBase-T). These newer and faster copper technologies have limitations as to length of the cabling and may require shielded cabling.

ETHERNET HISTORY

10Base5 and 10Base2 were coaxial cable-based networks that have largely disappeared from the networking landscape. 10Base5 used 0.5-inch diameter RG-8 style backbone and was dubbed *thicknet* due to the wire size. The trunk had a maximum length of 500 meters, hence "5" in the IEEE designation. 10Base2 used a 0.2-inch RG-58 style trunk and was called *thinnet* due to the thinner cable. The trunk could extend approximately 200 meters, thus the "2." These coaxial bus networks were popular in their day, but the fact that a single-point failure could kill the entire network caused most installations to move to twisted-pair infrastructures.

10BaseT stands for 10 Mbps over twisted-pair wiring. This was the popular standard until 100BaseT (Fast Ethernet) emerged with two different cabling standards: 100BaseTX and 100BaseT4. More recently, 1000BaseT, known as Gigabit, has become widely available. All of these employ a star physical topology and utilize unshielded twisted-pair (*UTP*) cabling to make point-to-point connections between each workstation or server and a central switch. Twisted-pair networks use affordable, easily attached *RJ-45* connectors. Individual cable runs can be up to 100 meters (328 feet). Each connection port is isolated, which reduces the sensitivity to single-point failure and provides convenient patch points that can ease expansion and load balancing. The low cost of cabling and the ease of installing, connecting and maintaining these networks has made the twisted-pair configuration the most popular cabling arrangement for almost all modern LANs.

The quality of UTP cable in use is vital to system performance and must meet certain specifications based on the Ethernet standard in use. UTP cable includes four separate twisted pairs in a single sheath, and although seemingly simple in structure, many factors influence its electrical quality. Things like the insulating material, the spacing of the wires in a pair and whether the wires in a pair are bonded together all have an effect. Each pair is twisted at a different rate per length to minimize crosstalk between the pairs, but this can also lead to a skew in timing between the pairs. These and many other parameters determine the frequencies and data rates a cable can reliably support.

To help standardize cabling practices, the Telecommunications Industry Association (TIA), which represents the communications sector of the Electronic Industries Alliance (EIA), has developed a series of three standards for designing and implementing cabling systems. These are formally known as ANSI/TIA/EIA-568-B.1-2001, -B.2-2001, and -B.3-2001, but are often collectively referred to as just TIA/EIA-568-B. Within this standard, several categories (abbreviated CAT) of performance are defined to qualify twisted-pair cabling. Ratings commonly found in the field include *CAT3*, *CAT5*, *CAT5e* and *CAT6*. *CAT6a* and *CAT7* are also starting to be utilized in high-end installations.

10BaseT connections require only two of the pairs (one transmit, one receive) and require UTP cable rated Category 3 or higher. 100BaseTX also only uses two of the pairs but requires Category 5 or higher. 100BaseTX links are limited to 100 meters (328 feet) with an overall segment length limit of 250 meters (820 feet). 100BaseT4 uses all four twisted pairs and can use Category 3, 4 or 5. 100BaseT4 was originally developed to allow a simple upgrade path to 100 megabit for existing 10BaseT installations that utilized Category 3 wiring. But this 100-Mbps derivative never really caught on and this hardware is not widely available. So, for CAT3-based infrastructures, rewiring is the only realistic way to upgrade.

Currently, 100BaseTX using Category 5 or 5e wiring is still the most common network infrastructure, but 1000BaseT has become the new standard. Gigabit Ethernet uses echo cancellation and a 5-level pulse amplitude modulation (PAM-5) scheme to put 250 MHz of data bandwidth on each of the four pairs, and thus requires a higher grade of cable. CAT5e is the "enhanced" specification capable of handling 1000BaseT, but any new facility would be wise to use CAT6 or the highest standard available when all new wiring is to be installed.

Ethernet can also utilize optical fiber. 100BaseFX was an early 100-Mbps technology that has given way to two gigabit specifications: 1000BaseSX and 1000BaseLX. All use a star physical topology with each connection using two strands of fiber-optic cable, one for transmit and one for receive. 1000BaseSX uses short-wave (0.85 micrometer) LASER light over multi-mode fiber to provide link distances up to 550 meters. 1000BaseLX uses long-wave LASER light, which can provide the same 550-meter link when transmitted over 50-micron multi-mode fiber, but can also provide long distance links of up to 5000 meters when sent over 9-micron single-mode optical fiber.

Figure 4.7-4 presents a quick comparison of these common Ethernet configurations.

STORAGE REQUIREMENTS FOR A DIGITAL AUDIO SYSTEM

Now that we have a better idea of what a station automation system is and how individual units can be networked to enhance efficiency, we need to decide how to store all this data. The first thing we need to know is how much data there is.

Data Size

Audio is encoded for digital storage through a modulation technique known as *sampling*. This is sort of like taking a snapshot of the audio waveform at regular intervals. Sampling a continuous waveform produces a stream of pulses, and is thus a form of pulse modulation. The amplitude of each pulse is then converted into a digital number that can be stored as data, which is known as *coding*. Therefore, the overall process of sampling an analog signal into a stream of digital data is known as *pulse-coded modulation (PCM)* and is often referred to as *linear PCM* to emphasize that no data compression or reduction schemes have been applied. The amount of data generated by this process depends on how fast we sample, how many bits are devoted to each sample and how many signals we sample simultaneously (mono, stereo or multi-channel).

In order to completely describe the incoming audio waveform, the *Nyquist Sampling Theorem* states that the sampling must occur at a rate at least twice as fast as the highest frequency in the incoming audio. Due to the limitations of real-world filtering required to keep samples from be contaminated (known as anti-aliasing filters), the sample rate must actually be slightly higher than twice the highest input frequency expected. 32-kHz sampling has a reasonable frequency response up to 15 kHz, 44.1-kHz sampling is good to about 20 kHz and 48-kHz sampling can produce frequencies up to 22 kHz or so. In recent years, much higher sample rates like 96 kHz and even 192 kHz have been added by the audio hardware manufacturers. These high sample rates provide more accurate high-frequency reproduction with less coloration by allowing the anti-aliasing filters to roll off smoothly, farther up the spectrum from the audio frequencies of concern.

The number of bits used to describe each sample is known as the level of quantization and determines the resolution or accuracy of the sample. The number of bits also determines the quantization (digital) noise floor, which relates directly to the overall dynamic range and signal-to-noise ratio of the digitized signal. 16-bit quantization provides a theoretical dynamic

Network Media	Ethernet Specification	Speed (Mbps)	Cable	Description
Twisted Pair	100BaseTX	100	CAT 5	"Fast Ethernet", uses 2 pair, max length 328'
	1000BaseT	1000	CAT 5E CAT 6	"Gigabit" or "GigE", uses 4 pair, max length 328'
	10GBaseT	10000	CAT 6A CAT 7	"XGbE" or "10GbE", max length 328' on CAT 7.
Fiber Optics	100BaseFX	100	Multi-mode, Single-mode	Fiber-based Fast Ethernet, 1.2 mile (2 KM) on multimode, 6 mile (10 KM) on singlemode fiber
	1000BaseLX	1000	Multi-mode, Single-mode	long wave laser, 1800' (550 M) on multimode, 3 mile (5 KM) on single mode
	1000BaseSX	1000	Multi-mode	short wave laser, 1800' (550 M) on multimode fiber
	10GBase SR/LR/ER	10000	Multi-mode, Single-mode	Fiber-based 10Gb, 1300' (400 M) to 24 miles (40 KM)

FIGURE 4.7-4 A summary of common Ethernet networks

range of 98 dB, which is very near the limit of human perception. While this is adequate for most broadcast purposes, 24-bit quantization with its 146 dB of dynamic range is becoming more commonplace, with some applications even using 32-bit quantization.

As for channels, this number is pretty simple—stereo produces twice as much data as mono. And there are now multi-channel formats to be considered. Six- and eight-channel files are being used to store multilingual programs and store surround sound material. These files contain six and eight times as much data as mono or three and four times as much as stereo audio files.

Putting these numbers together, we arrive at the first equation in Table 4.7-1. This shows us the rate at which data is created as a function of the number of channels, the sample rate and the quantization level. Applying some basic conversions and rearranging this equation, we obtain two more useful relationships for analyzing storage requirements. The first gives the number of megabytes required to store a minute of audio, while the second yields the number of hours of audio that can be stored per gigabyte of storage capacity. Evaluating these equations for common 44.1-kHz-sampled 16-bit audio provides us with 10.1 MB/Min and 1.69 GB/hour respectively.

Let's do a sample calculation. Station X has a basic inventory of 1000 songs with an average length of 3 minutes. It also maintains 100 commercials averaging 60 seconds and 100 station promotions averaging 30 seconds. It also carries five separate 1-hour-long syndicated shows that are recorded and played back weekly. All audio is stored as 16-bit linear PCM sampled at 44.1 kHz. This adds up to:

$$(1000 \times 3) + (100 \times 1) + (100 \times 0.5) + (5 \times 60) =$$
$$3450 \text{ minutes of stereo audio}$$

At a sample rate of 44.1 kHz, Table 4.7-2 shows we require 10.1 MB/Min for a total of 34823 MB, which is about 35 GB of total storage. By itself, this is not very large compared to current storage standards. But combine several stations together, as is most common today, and provide production and archive space and it quickly starts to add up.

Data Rate

Knowledge of the *data rate*, as defined in the first equation of Table 4.7-1, is important to determine: (1) how many simultaneous streams of audio our network can handle and (2) how fast must our storage system be to keep up with the reads and writes associated with this amount of system-wide I/O. Again, using common numbers: A stereo signal sampled 44,100 times per second and quantized to 16 bits generates a data stream of 1,411,200 bits per second (bps) or about 1.35 Mbps. Dividing by 8, this represents 176,400 Bytes per second (Bps) or about 172 KB/s.

In the past, when network speeds topped out at 10Mbps and SCSI disk buses could only handle 5 MBps, these data rates posed a significant challenge. A single network segment would only support about seven streams with no other traffic, and a single disk drive channel would not even support 30 streams. To build a useful, networked digital audio system, these data rates and storage needs needed to be reduced.

Compression

This brings us to the issue of *data compression*. This term generally refers to a mechanism, usually a digital algorithm, that is used to reduce the amount of data needed to accurately represent an object like a data file or a digitized signal. The amount of reduction is usually expressed as a ratio (known as the compression ratio) and is a factor that can be applied to the formulas of Table 4.7-2 to obtain the effective data rate or required storage capacity. Table 4.7-2 compares linear PCM, *Dolby AC2*, *Dolby AC3* and several *MPEG*-compressed

TABLE 4.7-1
Digital Audio Data Rate and Storage Requirement Calculations

$$\text{Data Rate (Bytes/Sec)} = \text{Channels} \times \frac{\text{Samples}}{\text{Second}} \times \frac{\text{bits}}{\text{Sample}} \times \frac{\text{1 Byte}}{\text{8 bits}}$$

For a stereo, 44.1 kHz sampled, 16-bit audio stream: **Data Rate = 176,400 Bytes/Second**

Adding some conversions we get a more useful number for determining capacity requirements:

$$\text{Data Rate (MB/Min)} = \text{Channels} \times \frac{\text{Samples}}{\text{Second}} \times \frac{\text{bits}}{\text{Sample}} \times \frac{\text{1 Byte}}{\text{8 bits}} \times \frac{\text{1 MegaByte}}{\text{10242 Bytes}} \times \frac{\text{60 Sec}}{\text{1 Min}}$$

For a stereo, 44.1 kHz sampled, 16-bit audio stream this becomes: **Data Rate = 10.1 MB/Min**

Another useful way arrange this equation yields the amount of audio that can be stored per unit of storage:

$$\text{Audio Storage (Hours/GB)} = \frac{1}{\text{Channels}} \times \frac{\text{Seconds}}{\text{Sample}} \times \frac{\text{Samples}}{\text{bits}} \times \frac{\text{bits}}{\text{Bytes}} \times \frac{1024^3 \text{ Bytes}}{\text{1 GigaByte}} \times \frac{\text{1 Hour}}{\text{3600 Sec}}$$

For a stereo, 44.1 kHz sampled, 16-bit audio stream this becomes: **Audio Storage = 1.69 Hours/GB**

TABLE 4.7-2
Disk Storage Capacity Requirements

16-bit Quantization	Sampling Rate kHz	Data Rate Kbps/ch	Compression Ratio	MB per stereo minute Divide by 2 for Mono	Stereo hours per GB Multiply by 2 for Mono
Linear PCM	32.0	512	N/A	7.32	2.33
	44.1	705.6	N/A	10.09	1.69
	48.0	768	N/A	10.99	1.55
Dolby AC2	32.0	N/A	6.0	1.22	13.98
	44.1	N/A	6.0	1.68	10.15
	48.0	N/A	6.0	1.83	9.32
Dolby AC3	32.0	64	8.0	0.92	18.64
	32.0	80	6.4	1.14	14.91
	32.0	96	5.3	1.38	12.35
	44.1	64	11.0	0.92	18.60
	44.1	80	8.8	1.15	14.88
	44.1	96	7.4	1.36	12.51
	44.1	112	6.3	1.60	10.65
	44.1	128	5.5	1.84	9.30
	44.1	160	4.4	2.29	7.44
	44.1	192	3.7	2.73	6.25
	48.0	64	12.0	0.92	18.64
	48.0	80	9.6	1.14	14.91
	48.0	96	8.0	1.37	12.43
	48.0	112	6.9	1.59	10.72
	48.0	128	6.0	1.83	9.32
	48.0	160	4.8	2.29	7.46
	48.0	192	4.0	2.75	6.21
ISO/MPEG Layer 1	32.0	32	16.0	0.46	37.28
	32.0	64	8.0	0.92	18.64
	32.0	96	5.3	1.38	12.35
	44.1	32	22.1	0.46	37.37
	44.1	64	11.0	0.92	18.60
	44.1	96	7.4	1.36	12.51
	44.1	128	5.5	1.84	9.30
	48.0	32	24.0	0.46	37.28
	48.0	64	12.0	0.92	18.64
	48.0	96	8.0	1.37	12.43
	48.0	128	6.0	1.83	9.32
	48.0	160	4.8	2.29	7.46
ISO/MPEG Layer 2	32.0	32	16.0	0.46	37.28
	32.0	48	10.7	0.68	24.93
	32.0	56	9.1	0.80	21.20
	32.0	64	8.0	0.92	18.64
	32.0	80	6.4	1.14	14.91
	32.0	96	5.3	1.38	12.35

TABLE 4.7-2 *(continued)*
Disk Storage Capacity Requirements

16-bit Quantization	Sampling Rate kHz	Data Rate Kbps/ch	Compression Ratio	MB per stereo minute Divide by 2 for Mono	Stereo hours per GB Multiply by 2 for Mono
	44.1	32	22.1	0.46	37.37
	44.1	48	14.7	0.69	24.86
	44.1	56	12.6	0.80	21.30
	44.1	64	11.0	0.92	18.60
	44.1	80	8.8	1.15	14.88
	44.1	96	7.4	1.36	12.51
	44.1	112	6.3	1.60	10.65
	44.1	128	5.5	1.84	9.30
	48.0	32	24.0	0.46	37.28
	48.0	48	16	0.69	24.86
	48.0	56	13.7	0.80	21.28
	48.0	64	12.0	0.92	18.64
	48.0	80	9.6	1.14	14.91
	48.0	96	8.0	1.37	12.43
	48.0	112	6.9	1.59	10.72
	48.0	128	6.0	1.83	9.32
	48.0	160	4.8	2.29	7.46
ISO/MPEG-1/2 Layer 3	32.0	32	32.0	0.23	74.56
	32.0	48	21.3	0.34	49.63
and	32.0	56	18.3	0.40	42.64
	32.0	64	16.0	0.46	37.28
ISO/MPEG-2/4 AAC	32.0	80	12.8	0.57	29.82
	32.0	96	10.7	0.68	24.93
	44.1	32	44.1	0.23	74.57
	44.1	48	29.4	0.34	49.71
	44.1	56	25.2	0.40	42.61
	44.1	64	22.1	0.46	37.37
	44.1	80	17.6	0.57	29.76
	44.1	96	14.7	0.69	24.86
	44.1	112	12.6	0.80	21.30
	44.1	128	11.0	0.92	18.60
	44.1	160	8.8	1.15	14.88
	44.1	192	7.4	1.36	12.51
	44.1	224	6.3	1.60	10.65
	44.1	256	5.5	1.83	9.30
	44.1	320	4.4	2.29	7.44
	48.0	32	48.0	0.23	74.57
	48.0	48	32.0	0.34	49.71
	48.0	56	27.4	0.40	42.56
	48.0	64	24.0	0.46	37.28
	48.0	80	19.2	0.57	29.83

TABLE 4.7-2 *(continued)*
Disk Storage Capacity Requirements

16-bit Quantization	Sampling Rate kHz	Data Rate Kbps/ch	Compression Ratio	MB per stereo minute Divide by 2 for Mono	Stereo hours per GB Multiply by 2 for Mono
	48.0	96	16.0	0.69	24.86
	48.0	112	13.7	0.80	21.28
	48.0	128	12.0	0.92	18.64
	48.0	160	9.6	1.14	14.91
	48.0	192	8.0	1.37	12.43
	48.0	224	6.9	1.59	10.72
	48.0	256	6.0	1.83	9.32
	48.0	320	4.8	2.29	7.46

formats in terms of effective compression ratios and required storage capacity for the sample rates and data rates commonly used by automation systems.

The details of specific data compression algorithms are quite complex and beyond the scope of this discussion; however, a few observations are in order. In the data world, compression algorithms (e.g. WinZip) work by eliminating *redundant* information. This information is "extra" and can be removed without affecting the inverse algorithm's ability to completely restore each and every bit of the original source. This is called *lossless compression*. Unfortunately, sampled audio has very little redundant information and does not compress with these utilities. For audio, a different class of algorithms has been developed that remove the *irrelevant* information—meaning information not important to how a human perceives sound. Since these algorithms throw away data that cannot be brought back by an inverse operation, they are referred to as *lossy compression*.

This does not mean these processes are flawed. Through a single encode/decode cycle, lossy algorithms, which include Dolby AC2, AC3, and the whole MPEG family, are quite transparent and imperceptible to all but the highly trained ear. At a time when network bandwidth and hard-disk capacity was very costly, these compression schemes made great economic sense and most systems employed them.

But some information *is* sacrificed in the process, and unfortunately the automation system is not the only place compression exists. Internet streaming (SHOUTcast and Icecast for example), satellite downlinks, studio transmitter links (STL) and even satellite (SiriusXM) and terrestrial (Ibiquity HD Radio) delivery mechanisms all employ some form of data compression. There has always been concern that audio quality might not hold up through multiple encode/decode cycles or through a cascade of different lossy compression schemes. As we move forward in the digital age, this is becoming the primary concern, and many automation system users are migrating back to full bandwidth PCM to keep the source material as pure as possible.

The good news is that in the last decade, network bandwidth and hard-drive capacity have both increased by two orders of magnitude (100X) while the basic unit cost of each has dropped roughly in half. In the same period, disk drive transfer rates have doubled six times! Compression is no longer a necessary evil—at least not for storing the source audio material.

Bandwidth Considerations

Let us look at the some numbers: a dedicated 1000TX gigabit network link in a full-duplex switched environment can carry 1000 Mbps in both directions simultaneously. This represents 1000 Mbps / 1.35 Mbps = 740 simultaneous stereo streams both in and out. A 100TX link will only handle a tenth of this, or about 74 streams, in both directions. Current day workstations often include four or more physical outputs and each may layer up to three streams at a time during overlap sequences. Only ten channels, all playing three streams simultaneously, represents half of the available bandwidth of a 100TX link, leaving the rest for database transactions and other miscellaneous I/O.

As automation software continues to advance and become more interconnected, I/O traffic on the network also increases. A single 100TX connection to a central storage device can become a potential bottleneck, particularly if all workstations are reading and writing to and from this same device. Just three full-featured workstations as described above could overload a single 100TX connection to a server. Most modern managed network switches and server NICs support link aggregation, which allows multiple physical links to be combined to effectively multiply the bandwidth (and additionally provide fault tolerance) between the two devices, but this can only help a 100TX network so much. Obviously, gigabit connections are very beneficial for server connections, and when combined with link aggregation, dozens, even hundreds, of workstations can be supported.

Destination: Disk Drives

So it appears we have no problem moving full bandwidth PCM audio to and from our network. Now, how and where do we store this sizeable volume of data? As described earlier under Network Architectures, we can choose to provide storage in each workstation, in a centralized server, or in both. But no matter where we choose to store the data, the current state of the art dictates that it will end up on hard-disk drives. This technology has made huge advances in capacity, speed and reliability, which has helped propel digital computing systems into the mainstream of nearly all broadcast operations. But due to their core mechanical properties, hard-disk drives remain the Achilles heel of all automation systems and therefore deserve some special understanding.

I have found by experience that the greatest point of failure in any computer-based system is the hard-disk drive. The issue is not *if* a disk drive will fail, but *when* it will fail. Current premium-quality, high-performance drives are quoting figures of 1 million hours mean time between failure (*MTBF*), which is over 100 years! However, a large percentage of the high-capacity drives being used for multimedia systems are failing far short of this mark—many failing within the first six months of operation. There appear to be several reasons for this. A few stem from manufacturing difficulties, whereas most are related to environmental conditions such as heat, shock and vibration.

Although the drive manufacturers are continuously improving the reliability of disk drives by reducing the internal parts count and using higher levels of integration in the on-board electronics, the ultimate storage mechanism is mechanical, and therefore subject to mechanical failure.

In the very early releases of what were considered "high-capacity" drives, primarily the 4- and 9-GB sizes, several manufacturers had problems with the bearings used to support the spindle and its associated media platters. The bearings would wear prematurely, causing the spindle and platters to shift and wobble, thereby degrading track alignment and causing loss of data. It was not uncommon to see drives wear so badly that they would rumble loudly then come to a horrible grinding halt.

While this sort of problem has largely disappeared through improved design and manufacturing processes, mechanical problems associated with handling remain.

Disk drives are extremely sensitive to shock and vibration. A short fall to a workbench or floor is often enough to destroy a drive. The manufacturers have addressed these kinds of problems through better head parking designs and improved packaging and handling procedures, but what shipping companies and stock room personnel do with the drive before it gets to you is another story. The best advice is to only deal with reputable system vendors and distributors. There is probably a good reason for those "super deals" you can find on the Internet.

One popular alternative to conventional hard-disk drives that has been gaining traction over the years are solid-state drives (*SSD*). Instead of featuring mechanical parts that wear out over time, these use solid-state flash memory with a microcontroller to manage incoming data and distribute load. At one point, these drives had far lower capacities and much higher costs associated with them. Now they are nearly matching conventional hard-disk drives in terms of capacity and have come down to more reasonable costs and offer a viable alternative. They are much faster for reading from and writing to, do not suffer mechanical wear, are more resistant to drops and falls, operate using less power and generally improve performance over their conventional counterparts.

However, SSDs do have a few negatives that are worth mentioning. First is the higher cost. While prices have come down dramatically, they still cannot match traditional HDDs in terms of cost. Also, the flash memory used in them has a limited number of times that it can be programmed. This is mitigated by the memory controller chip on the SSD to balance the load. This puts the MTBF of a modern SSD effectively around that of a regular HDD. So they can still potentially fail, even if not mechanically. Also, SSDs tend to suffer rapid write performance drop as they approach maximum capacity due to the nature of how they must write and erase pages of memory. Modern operating systems such as Windows 7 and higher help to mitigate this by use of a feature called TRIM that marks pages for deletion by the drive's internal garbage collection algorithm. Also, many modern drives leave a fraction of the drive unformatted so that they do not encounter this situation. This is called "over provisioning" and the amount can range from 10% to 25% of the maximum capacity. This can be identified by how much space the drive advertises as available. All SSDs are in capacities that are powers of two (64 GB, 128 GB, 256 GB, 512 GB, 1024 GB, and so on). If an SSD lists less than that maximum, then it normally has already been over provisioned from the factory. If an SSD lists the maximum possible capacity, then it may be wise to manually leave some unformatted space on the drive when partitioning and formatting it. It also means that just looking for the drive with the highest capacity may not be the best decision, as they may be the same internally.

One more interesting technology that has come from the development of HDDs and SSDs are solid-state hybrid drives (*SSHD*). These are mostly conventional HDDs with some flash memory added as a large buffer. Files that are regularly used are written to the flash memory to speed up performance. This flash memory is handled completely on the drive and is not managed by any sort of optimizations in the OS. The drive behaves as a conventional HDD otherwise. Positives to this technology is that it has potential to speed up access to commonly used files and is only marginally more expensive that standard HDDs of similar capacity. Negatives are the same as a traditional HDD, as it requires accessing the data a few times before it gets added to the flash memory, the rotation speed of the drives can be lower than HDDs of the same capacity, and the capacities of flash memory can vary and are usually low. Suffice it to say, SSHDs are more of a stopgap measure until SSDs reach complete parity with

HDDs in terms of capacity and cost. However, they do offer another possible solution in the meantime for those that need high capacity and performance, but at a lower cost of a full SSD.

Once the drive has made it to the final destination and is up and running, the biggest environmental factor is *heat*. Disk-drive cabinets must have adequate cooling fans, and the fan filters must be kept clean. Airflow through the enclosure should be free flowing and unimpeded by doors or covers. Some cabinets offer temperature-sensitive fans that will increase flow as the temperature rises. Others provide temperature-monitoring devices and alarms to alert maintenance personnel of a problem. Even though most drives are rated to operate in temperatures up to 50 degrees C (122 degrees F), the cooler they are, the longer they will run. This also applies to SSDs, as they are just as susceptible to heat as the chips on the motherboard.

Unfortunately, as CPU clock speeds go up and as drives spin faster, they consume more power and generate more heat. And as more and more equipment is packed into "rack rooms," it usually becomes necessary to employ additional air conditioning equipment to remove the heat from the room and keep the computers and their disk drives cool. So the pertinent question at his point is, "how much cooling do I need?"

A good estimate of the heat load produced by a system can be determined by estimating the total amount of AC power it will consume and then assume that all of this power is turned into heat. Obviously the units will draw less than the maximum (probably closer to half the max), but designing for the max leaves headroom to cover all situations.

To estimate power consumption, count the total number of PC power supplies that will be operating within the space being considered. Multiply the number of supplies by the max power rating of the supply (e.g. 600 watts). Then add the total number of monitors involved multiplied by an average of 150 watts per monitor. Add the power of any other major power-consuming devices to get the total power in watts. Then use the conversion:

1.0 KWH = 3.4 KBTU, which yields the following relationship:

Heat Load (BTU/hour) = Total Power (Watts) × 3.4 (BTU/WH)

An *HVAC* professional can take this heat load and determine how many tons of air conditioning are required and whether the current HVAC system has the needed capacity or if additional AC equipment will need be necessary.

Disks Fail—Plan for It

In an expected lifetime of 100 years, failures within six months to a year could be considered infant mortality—certainly within the hardware warranty. But a radio or TV station can become very dependent on such a system in that time frame and will trust it with a considerable amount of irreplaceable data. It is therefore imperative that the data storage system be able to handle disk failures while continuing to function. This is achieved through disk redundancy.

The basic concept of disk redundancy is to store extra (redundant) data along with the target data and arrange it among multiple hard drives, such that this extra data can be used to replace the original data should one of the disk drives fail. This can be achieved several ways. The most basic method is called *disk mirroring*. In this configuration, each disk drive is paired with an identical disk drive. The data on the first drive is automatically replicated onto the second drive, either by the operating system or via a special disk controller (as in *RAID* technology, to be explained later). If either disk should fail, the operation will be unaffected because the system will just use the other disk for all reads and writes. It is also possible to mirror more than two disks for additional redundancy; however, this is usually not very cost effective.

Disk mirroring can also enhance system performance. Since two (or more) copies of the data are available on independent drives, the controller can pull from both, using whichever drive is better positioned to service a particular request. This effectively increases the number of heads available to gather data.

To increase disk performance even further, a disk configuration called *spanning* or *striping* is often employed. The idea behind this scheme is to subdivide the data between two or more drives. The operating system or a specialized disk controller takes the data stream and divides into smaller units then writes one unit to each drive. Since each drive is able to write (or read) simultaneously with the other drives, the disk I/O throughput is increased by a factor equal to the number of drives. This also means the total capacity is equal to the sum of all the drives.

However, this configuration offers *no* redundancy whatsoever and actually puts a facility at greater risk of overall failure. If any of the drives in the span fail, *all* of the data on *all* of the drives is lost. Since more drives increases the total parts count, the probability of failure due to one of the components increases. For this reason, I would never recommend using spanning or striping by itself.

A great solution is to combine mirroring with spanning/striping. In this configuration, you span a set of drives, each of which has its own associated mirror drive. This gives you the increased performance and the additive capacity of spanning/striping, as well as the redundancy and performance advantages of mirroring. In fact, this arrangement performs so well that it is possible for it to experience multiple drive failures without significantly affecting system performance. This makes it very important to diligently monitor the status of your drive subsystem and correct any failures as soon as possible. This configuration offers great scalability with both performance and redundancy.

The obvious downside of disk mirroring is that for a given capacity, the disk drive requirements are doubled. For systems requiring large capacities, this can be very costly, not to mention the added complexity, space, power and heat concerns. These issues, along with advances in disk controller technology, have caused the simple but bulky mirrored disk arrangements to be replaced by a more complex, but more efficient, redundant disk configuration know as RAID.

RAID

Although RAID is often perceived to be a relatively new innovation, this technology was first described in 1987 by Berkeley authors David Paterson, Garth Gibson and Randy Katz. RAID is an acronym that stands for *redundant array of inexpensive disks*, but when you price a unit worthy of multimedia storage requirements, you might wonder where they got the "inexpensive" part. Because of this misnomer, manufacturers have adopted a slightly modified acronym: redundant array of *independent* disks. Either way, the important point is that the technology has matured, has become very dependable and offers great value.

These disk subsystems are usually comprised of a specialized multi-channel disk controller combined with three or more disk drives. The reason for their popularity is that they are capable of providing high capacity, excellent performance and data redundancy with fewer disks than any other configurations.

RAID units can function in several modes, referred to as RAID levels. The various levels offer different combinations of capacity, performance and redundancy. *RAID Level 0* is also known as disk striping and provides scalable capacity and high performance but offers *no* redundancy, as described earlier. *RAID Level 1* is also known as disk mirroring and provides redundancy but doubles the number of disks required for a given capacity. RAID controllers can also provide a combination of RAID 0 + RAID 1 (often referred to as RAID 10), which provides scalable capacity, excellent performance and redundancy, but still gives away one-half of the capacity to accomplish the redundancy. RAID Level 2 uses multiple check disks to allow error correction without having to completely duplicate all data. This can provide fully redundant storage with fewer than double the number of disks, but since higher RAID levels are even more efficient, Level 2 is rarely used. RAID Level 3 provides redundancy by using nine synchronized drives with a bit of each byte written to the first eight, along with a parity bit that gets written to a ninth dedicated parity drive. Level 3 can provide excellent performance, but storing parity at the byte level requires a large number of synchronized drives, which add considerable expense of the system.

The real efficiency comes with RAID Levels 4 and 5. In both of these modes, the data stream is subdivided into smaller blocks (which can be many bytes), with each block being written to a separate disk—just as in disk striping (Level 0). In addition, the blocks written to each drive are compared by an algorithm that generates parity data that is written to another drive. If any drive in the stripe fails, this parity information is combined with the other good drives to recreate the data from the missing drive. When the failed drive is replaced, a similar process is used to rebuild the new disk to the original data structure.

The main difference between the two levels is that Level 4 always stores the parity information for every stripe on a single drive, while Level 5 stores parity information for each stripe on a different drive. RAID Level 4 can only access one stripe at a time, limiting its effective I/O for random transfers; however, its parity calculations are somewhat simpler, making it effective for long sustained sequential operations, as in single-user multimedia applications. *RAID Level 5* is able to process multiple stripes simultaneously, making it better suited for random scattered disk requests, as in multi-user centralized storage environments. The parity structure of RAID Level 5 is a bit more complicated than other levels, but given the proper horsepower and caching to offset this overhead, it is the best choice for a wide variety of file I/O operations.

The complexity of RAID 5 operations can often impose a limitation on the maximum size of a single RAID 5 array. To overcome this limitation, most controllers will allow multiple RAID 5 arrays to be combined under a RAID 0 stripe to create an array with the aggregate capacity. This is commonly referred to as RAID 50.

With *RAID Level 6*, like RAID 5, the read data transactions are very fast. With RAID 6, if two drives fail you still have access to all data, even while the drives are being replaced and rebuilt, providing for greater data security compared to RAID 5. Read performance is similar to RAID 5, while write performance decreases due to increased parity. RAID 6 is a good all-around system that provides efficient storage with excellent security and good performance. It is preferable to RAID 5 in file and application servers that use many large drives for data storage.

Figure 4.7-5 summarizes the features of the various RAID levels.

STORAGE ARCHITECTURES

Disk arrays can be built with the capacity and performance necessary to handle just about any size automation system, but the array cannot be accessed directly from the network. Important functions like rights management and file locking need to be provided by a layer in between the actual storage and the network users. These functions are traditionally performed by a server.

The File Server

Currently, the most common storage architecture for automation systems is to use a centralized file server. This is basically just another networked PC with a lot of storage capacity. However, server hardware is usually more powerful, with multiple high-speed CPUs, multiple gigabytes of RAM, multiple gigabit network-interface connections and a large RAID disk array. The operating system is also usually a specialized "server" version that optimizes and prioritizes file I/O functions over user-interface features and is designed to handle many simultaneous client connections to its resources. Network operating systems (NOS) like Windows Server 2008 and 2012 are the most common, but various forms of UNIX and Linux have their share of application as well.

RAID Level	Name	Description	Min. Disks	Performance	Fault Tolerance
0	Striping	Data stream is divided into blocks that are distributed equally between all disks increasing the overall I/O transfer speed above that of the individual disks. No data redundancy: loss of a single component can lead to unrecoverable date loss.	2	Very high read and write I/O with virtually no overhead	None
1	Disk Mirroring	Each disk is paired with a duplicate disk resulting in twice the cost per Byte. All data is 100 percent duplicated on the mirrored disk.	2	Writes equal to single disk, reads 2x better	Excellent Highest of all levels
0+1	Striping and Mirroring (Mirrored Stripes)	Combination of RAID levels 0 and 1. Data is striped across several physical disks. Each stripe has a mirror stripe generated and written across the same set of disks. This level provides performance through striping and redundancy through mirroring.	3	High read and write data rates	Good Same as RAID 5
2	Disk Array	Hamming code error checks across the disks. Data can be recovered without complete duplication of all data, although several check disks are needed. All disks in a group must be accessed, even for transfers, and the slowest must finish before the transfer is complete.	3	High data rates possible with more check disks Transaction rate equal to single disk	Very Good Between RAID 5 and RAID 1
3	Disk Array	Data block is divided and striped across data disks. Redundancy is accomplished through stripe parity data that is generated and written to a single parity disk. Disk spindles are typically synchronized. All disks must be accessed and the slowest must complete the process before a transfer is complete.	3	Very high read and write data rates but transaction rate equal to single disk	Good Handles single drive failure
4	Disk Array (Striping with Fixed Parity)	Full data block written to each data disk with block level parity stored on a single parity disk. This results in faster individual disk reads for small transfer but the parity check disk becomes a throughput bottleneck on writes.	3	Very high read transaction and transfer rates. Very low write transaction and transfer rates	Good Handles single drive failure
5	Disk Array (Striping with Floating Parity)	Data blocks and parity data generated as in RAID 4, but parity information is spiraled across all data disks. This eliminates the parity disk throughput bottleneck problem of Level 4. Distributed parity increases write performance but introduces high overhead to track the location of parity addresses.	3	High read transaction rate Medium write transaction rate Good I/O data transfer rates	Good Handles single drive failure
6	Disk Array (Block Level Striping)	RAID 6 extends RAID 5 by adding another parity block; thus, it uses block level striping with two parity blocks distributed across all member disks.	4	Decrease in write performance vs. Raid 5 due to increased parity	Excellent Handles two drive failures
10	Striping Mirrored Disks	Combination of RAID levels 1 and 0. Each physical disk is paired with a mirror disk. Data is striped across the mirrored pairs. This level provides redundancy through mirroring and increased performance through striping.	4	Very high read and write data rates with highest redundancy	Excellent Same as RAID 1
50	Striping multiple Disk Arrays	Combination of RAID levels 5 and 0. Multiple RAID 5 arrays are striped together to combine their capacities. Striping can also improve overall performance above that of the individual RAID 5 arrays.	6	High read/write transfer rates and high I/O data rates for small requests	Very Good > RAID 5 unless 2 failures in one array
JBOD	Just a Bunch Of Disks	Each disk is operated independently like a normal disk controller, or multiple disks can be spanned and seen as a single large disk. No data redundancy.	1	Just equal to individual disk performance	None

FIGURE 4.7-5 RAID (redundant array of independent disks) features

The most essential component of any NOS is to provide access control and rights management. The important data you will be entrusting to your storage system needs to be protected from nonauthorized users. Clients must login and authenticate to the server NOS to gain access to the files they need. Even authorized clients must be controlled as to which files they can read, which they can execute, which they can modify and which they can write or delete. And it even gets trickier when one client needs to read from one part of a file while another client needs to write to the same file. Sophisticated file locking at the byte level coupled with caching and transaction-tracking mechanisms are required for this level of interaction to be successful. These are the functions accomplished by a good NOS. Without it, there would be data anarchy and chaos.

Servers can also provide more than just file services. Servers can also serve applications to the clients. This means the application software resides on the server and a client pulls the components necessary to run the application across the network and loads them into memory for each session. This reduces the local storage requirements and makes network-wide software updates much easier to manage, since the applications reside in a single centralized place.

But all of these features tend to make servers somewhat complex to set up and maintain. Highly trained administrators are usually required to make changes and keep system configurations up to date. This may not be an issue if your facility has its own information technology (*IT*) department to provide this needed expertise. For the moderate- to large-sized facilities where computing systems are involved in almost every general business application, in-house IT makes good business sense. But for small installations, this is usually not the case, and server administration can be intimidating and costly. Unfortunately, this often leads to complacency and to the server being ignored, which can be very dangerous. Many automation systems don't need anything more than basic file services. If the I/O needs are moderate and the automation system runs in an isolated environment where complicated access control is not necessary, there is a somewhat simpler solution.

NAS

This acronym stands for network-attached storage. This basically means a "black box" that you connect (attach) to your network that acts like a big disk drive that anyone can use for storage. The structure of a *NAS* is similar to that of a server, but on a simpler, more appliance-like level. Hardware-wise, it is a specialized PC with good network connectivity and a large disk array. But the NAS concept is all about simplicity: connect it to the network, plug it in and turn on. It just works! NAS is designed to work with very little interaction and generally doesn't even have a keyboard or monitor attached. Configuration is typically done through a simple web page accessed via a standard web browser on any PC on the network. Since little user interface is desired, NAS devices generally utilize compact efficient operating systems like Linux that can maximize file system performance by running through a simple command line interface that lends itself to telnet or web server control.

A NAS device can generally be used in conjunction with an existing server. Set up is a breeze with a configuration option that allows it to follow the access control list and rights set up on the existing server or domain. This is the beauty of NAS: a simple way to add more drive space and locate it where it the network needs it.

But a professional-grade NAS can also be used as the primary (and only) storage device. In this role, the main difference between a file server and a NAS is in the granularity of control. NAS devices are simple to configure because they have fewer options, but this also means there are fewer aspects you can control. Access control is simpler, but that can make the file system less secure. Rights management is less tiered, but fewer choices makes it more of an "all or nothing" sort of assignment.

Simpler configuration can impact performance as well. Many NAS units only allow RAID 1 (mirrored) configurations. Even those that provide RAID 5 do not provide much flexibility in block sizes, partitioning or hot spare configurations. Just pick the RAID level, the file system and say go. This certainly makes things simple and can be completely acceptable with the high-performance drives available today. But this is another area in which NAS generally provides a compromise.

While most servers use *SAS* (serial attached *SCSI)* disk drives, most NAS units employ arrays of less expensive serial ATA (*SATA*) disk drives. This makes the units more affordable but can affect the level of performance. Many systems now use SATA II and SATA III drives that provide data transfer rates of 3 Gbps and 6 Gbps respectively, and these have provided respectable results. However, even SATA drives currently have lower rotational velocities and lower bus transfer rates than the SAS drives typically used in high-performance servers. This means you need more drives in the array to achieve the same level of performance. But, more drives also means more potential for drive failure. Also, SATA drives offer lower MTBF figures and are generally regarded as being less robust than SAS. One potential alternative would be to use SSDs instead of mechanical HDDs to avoid mechanical failure, but there is still an issue with capacity and cost with those. Furthermore, the currently available multi-channel SATA RAID controllers are not as mature or as full-featured as the SAS offerings. Fortunately, many NAS systems retain the drive failure alerting mechanisms that we have come to expect with servers.

Overall, the future of NAS looks bright. There are currently high-end SATA-based NAS devices that easily handle 50 linear PCM streams in *and* out. This indicates that NAS and SATA are up to the challenge. How these units will hold up to the rigors of 24 × 7 × 365 digital audio abuse remains to be seen, but we all know that SAS drives fail too. The bottom line is the performance/price value that SATA-based NAS is able to deliver is impressive and hard to beat. For many applications, it can adequately handle the job right now, and it will only get better. For these reasons, I not only expect to see the application of SATA-based NAS grow in the future, but also expect to see greater application of SATA-based storage with file servers as well.

SAN

Another acronym that should be explained at this point is *SAN*. It often gets confused with NAS since it is the same three letters, but SAN stands for storage area network and is something quite different.

A storage area network is a specialized high-speed network that lives between a server and its storage disks. Clients cannot access this network directly; they access the server as usual via a standard Ethernet network then the server uses the SAN to communicate with its disk drives. High-speed, high-bandwidth connections like *Fibre Channel*, *SSA* and *SAS* are utilized to provide similar data rates as if the disks were directly attached, but being a network, SAN allows multiple servers to connect to and share a pool of storage disks. This improves system administration by allowing a facility to manage all of its disk drives as a single resource instead of having to manage multiple isolated arrays distributed amongst multiple independent servers.

SANs are very scalable and provide efficient, redundant storage. SAN management software allows disk drives to be added or removed at any time and merged in and out of redundant RAID configurations. The aggregate storage of the combined disks can be partitioned and assigned to individual servers as needed. With all disk drives managed as a single entity, disk maintenance and backup operations are much easier to schedule and control. And routine backups take far less time to complete, because backup devices can connect directly to the SAN and move data at SAN speed without interfering with client/server operations.

SAN does not necessarily offer any performance advantage to a server over directly attached disk drives (often referred to by yet another acronym, *DAS* for direct attached storage). And the hardware and management software needed to build and maintain a SAN can be very expensive. The main advantage of a SAN is the ability to consolidate all of the storage needs for every in-house data system (accounting, automation, email, graphics, news, sales, traffic, etc.) and to supply the overall required storage from a single array of standardized drives managed by easy-to-reconfigure software. SAN does not make much sense for small facilities due to its cost, but in large facilities with many systems to share the price tag, SAN can be an excellent and economic solution.

BACKUP OPTIONS FOR A DIGITAL AUDIO SYSTEM

Although all of these redundant systems provide excellent fault tolerance, they usually require user intervention after a failure to return the system to full capability. It is important that the station personnel responsible for these systems be aware of potential failures and be prepared to take corrective action as soon as possible to guarantee data safety.

But what if the unthinkable happens and all your data is destroyed? This is a remote but real possibility. For instance, in a RAID 5–based storage system, all it takes is for two disk drives to fail at the same time to lose 100% of the data. These catastrophic events can happen and must be planned for.

Tape

There are several ways to protect data from catastrophic failure. The most basic approach is to use a tape backup device and to keep complete data backups in a secure location. Tapes are available in a number of different formats: QIC, Travan, DAT, 8mm, Mammoth, AIT, DLT, Super DLT, ADR, LTO-1 through LTO-6 currently with LTO-7 in development, and VXA. The "right" format for a particular installation depends on many factors, including capacity per tape, transfer rate, and whether a single tape drive or an auto-loader is required to do a complete backup. All of these factors affect the overall price and must be considered carefully. It is beyond the scope of this section to lay out a complete and comprehensive backup plan, but a brief discussion of some key components and backup architectures is in order.

Tape systems use backup software to stream the hard-disk data onto the tape, which can then be removed and stored safely in another location until needed. The data stored on tape is a snapshot of the files at the time of the backup, so it is only as current as the last backup. While many backup software packages allow you to schedule periodic backup sessions, these will always lag behind the rapid changes made daily in a broadcast automation system.

Once the software has been configured, this process can usually be unattended unless it takes multiple tapes to contain the full backup. In this case, an operator must switch tapes when instructed by the backup application, or an expensive tape changer or autoloader must be employed to perform these changes automatically. Currently, LTO-6 cartridges have the greatest capacity at 2.5 TB native and 6.25 TB compressed, with systems scalable using multiple tape cartridges to greater than petabyte capacity.

But the biggest problem with tape devices is the data transfer rate. This has improved over time, with LTO-6 now having a data transfer rate of 160 MBps native and 400 MBps compressed. Transfer rate must be considered in any workflow for both backup and restoration. If the primary data is destroyed, the system is generally useless until all the data can be restored from tape. This can lead to considerable down time. Even if only a partial restoration is needed, the data on the tape must be accessed linearly and sequentially, so restoring even a single file can be time consuming.

With these facts in mind, it should be obvious that tape-based backups cannot be considered an "online" resource. It takes good planning, diligent execution, and lots of time to collect and safely store tape backups. However, they are worth their weight in gold if they are ever needed.

Disk-to-Disk Backups

A strategy that has recently become popular is to use disk drives as the backup media. With large 2-, 3- and 4- up to 8-TB SATA drives now available, it is becoming very affordable to keep data backups on disk drives. Portable disk enclosures that connect via "plug and play" USB 3.0 or 1394 FireWire and eSATA interfaces makes it a simple task to plug in and back up entire workstations. Server storage can be divided and written to several disks, with the backup drives easily disconnected and stored in a secure location. And transfer rates are at very high disk-to-disk speeds, which greatly reduces backup and restore times.

Another useful form of this is to include large disk drives in each of the workstations and use them to back up the relevant portions of the server content. This could be done either manually or by using an application designed to identify and copy the correct elements by following a user defined set of rules. An example of such an application is shown in Figure 4.7-6.

In a system with a server holding data for several separate stations, each on-air workstation could back up the content necessary for its own operation. This way, each station would remain operational should the server or network go down. And by combining the data from all the workstations, there would be a complete copy of all the data should the server ever need to be restored.

A secondary or backup server is another popular disk-to-disk architecture. In this arrangement, another server with capacity greater or equal to the primary server holds the backup copy of the data. Some systems utilize an intelligent application (like the one in

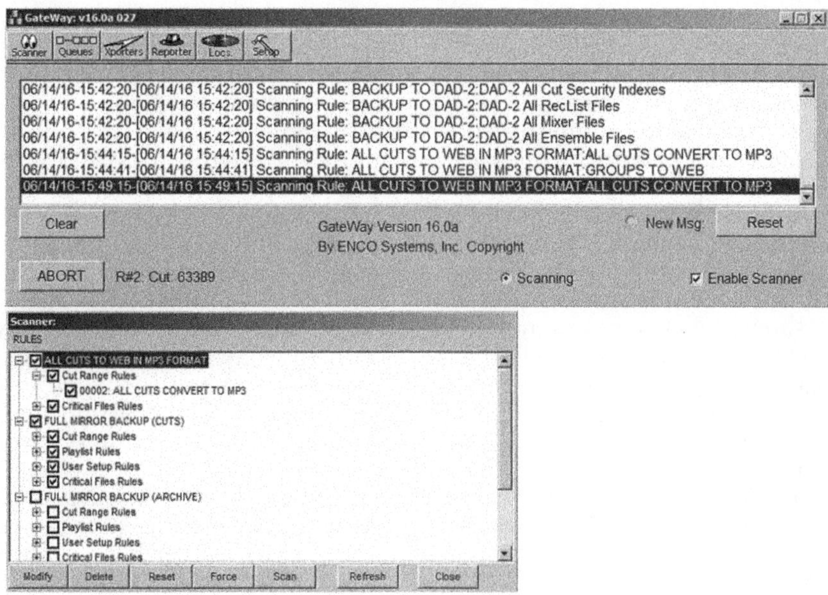

FIGURE 4.7-6 Example of a rule-based file synchronizer application used for disk-to-disk backups via LAN or WAN

Figure 4.7-6) running on the backup server that is constantly scanning for changes and actively synchronizes the file systems of the two servers. Other systems make use of server clustering algorithms that are implemented at the operating system level. Generally, the client workstations can access data from either server, making this a truly "online" solution.

There is also an extremely interesting version of this mechanism available in some SAN-based systems. Using disk-imaging algorithms, a "snapshot" of the contents of a disk array can be written to another array with the same SAN. This can be completed in a fraction of the time it takes to actually copy the data. Hundreds of gigabytes can be captured in minutes! This is the epitome of disk-to-disk backup schemes, but it is very expensive.

Disk-to-Disk-to-Tape Backups

One more extension of this architecture that deserves mention is disk-to-disk-to-tape backup. One of the most time-consuming attributes of tape is that it can only read or write a single data stream at a time. If several sources need to be copied, they must be done sequentially, adding considerable time to the procedure. By adding a disk-to-disk layer ahead of the tape process, the best attributes of disks, speed and random access can be combined with the best attributes of tape, portability, long shelf life and zero-power off-site storage. With the cost of large disk drives dropping (see NAS above), this arrangement is often more cost effective than upgrading to newer, faster tape drives, and certainly adds to the flexibility. Software packages are available that can manage the entire process and utilize almost any arrangement of existing hardware.

As a basic example of this process, let's examine a dual server system with a tape drive installed in the backup server. The disk-to-disk copy occurs between the servers as described above and proceeds quickly due to the speed of the disk arrays. Then backup software running on the backup server streams the server's data from its disks to its locally connected tape drive. Since this part of the process if fully contained within the backup server, it can occur at native tape drive speeds without placing any burden on the primary server or its disks. The overall process results in both an online backup (backup server) and a tape backup that can be moved off site and stored in a secure place in case of catastrophe.

This principle can also be applied within a SAN to provide extremely efficient backups. The snapshot mechanism explained above can perform the disk-to-disk image in minutes, and with a tape system connected directly to the SAN, the disk-to-tape process can proceed at full tape speed without burdening any servers or clients.

CLOUD-BASED CONTENT DISTRIBUTION

Advances in technology have allowed station operators to be able to control automation systems without ever having to be in front of the actual system. In this section, we will cover many of the new conveniences of smartphone and cloud-based solutions to create seamless operations of the automation system from wherever you are in the world. Due to many of the unique workflows and fast-paced operations, there are various cloud-based ways to accomplish the same end result.

Web-based library management/voice tracking allows users to be able to access their automation database from anywhere in the world via the web. A web server is created at that station that allows users to be able log in via an http site. From here, users can access their database libraries, preview and download cuts, add/remove cuts, create playlists and voice track. Voice tracking is done using a visual interface allowing the operator to utilize their on-board soundcard and microphone. Users are sent parts of the audio cuts and are able to track and create segues over them. The convenience of this form of remote management allows any user to be given login credentials and access to the system. It eliminates the need to actually remote into the computer, potentially creating unnecessary latency or accidental mistakes with the automation system controls. Figure 4.7-7 shows what a common web-based library management screen looks like.

Remote voice-tracking server/client is a remote voice-tracking solution that allows the user to utilize a local version of their station's automation system at their home or remote studio. This style of tracking gives the user the true look and feel, as well as editing functions, of their main automation system. This method allows voice tracks to be transferred from a remote facility to the main audio server by use of an FTP site. A utility is housed on each end checking for changes to playlists. When playlists are added and updated from the station side, they automatically are uploaded to an ftp server along with partial audio from the cuts that precede and follow the voice track. This gives the operator a sense of live announcing by being able to hear the intro and outro of the cut on each side of the voice track. When the remote operator is ready to voice track their show, they load their remote system, and the program will connect to the ftp site and download any updates. Once the operator completes his or her session, the process is automatically reversed and the updates to the playlists are sent to the ftp site and then automatically uploaded and synced with the station playlist. Figure 4.7-8 outlines the process in which the remote playlists communicate between the automation server and remote client.

In today's society, the immediate need for communication with the world at our fingertips is critical. The "smartphone" and "tablet" have given humans instant access to email, social media, texting, banking and even controlling broadcast transmitter levels. The need for the ability to interact with and control an automation system has now become a norm. Reporters in the field have come a long way from the days of transmitting their broadcast via microwave or the costly ISDN lines. With a simple app on an IOS or Android device, users can prerecord audio clips and interviews, edit them and send them to the automation system from anywhere in the world. This has given the user complete control over their remote broadcast or reports. Special on-scene coverage has never been easier. As long as the user has a cellular or WiFi connection, transmissions of audio clips are picked up and ingested into the automation system, ready for playback at any time. Recent developments have now allowed users to be able stream their audio reports live from their smartphones directly to the station, where the automation control can output their stream live on the air. With an operator in the studio and reporter into the field, the studio operator can allow access for the reporter to "go live" and speak directly to the listeners with minimal delay and with clear audio quality. These advancements have increased workflow efficiencies for reporters and announcers by cutting equipment costs and setup time, as well as providing them with

FIGURE 4.7-7 A common web-based library management screen

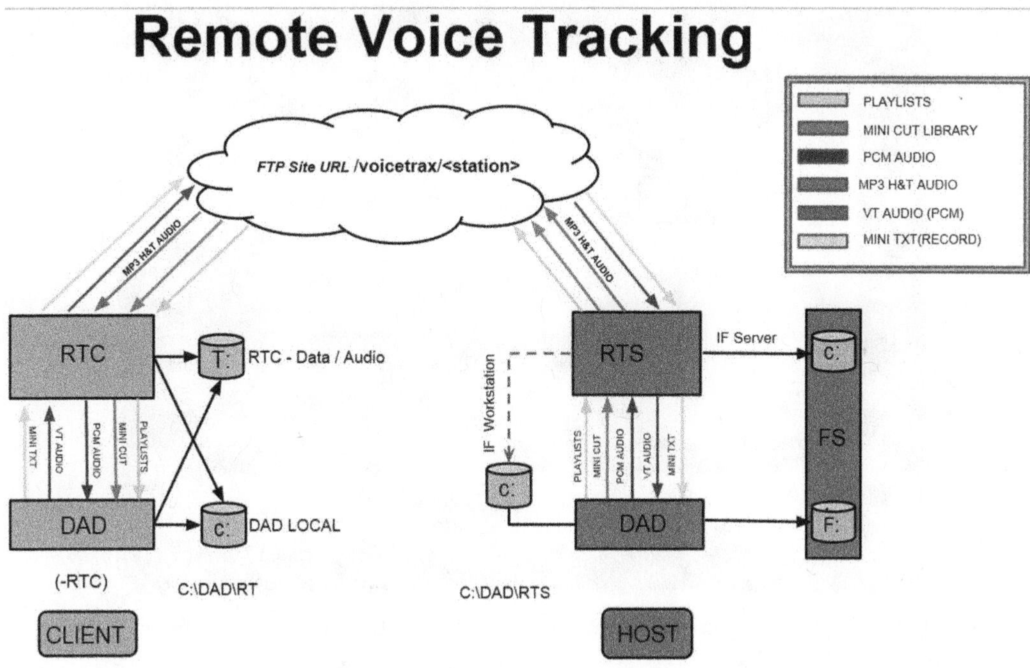

FIGURE 4.7-8 Remote playlist communication process

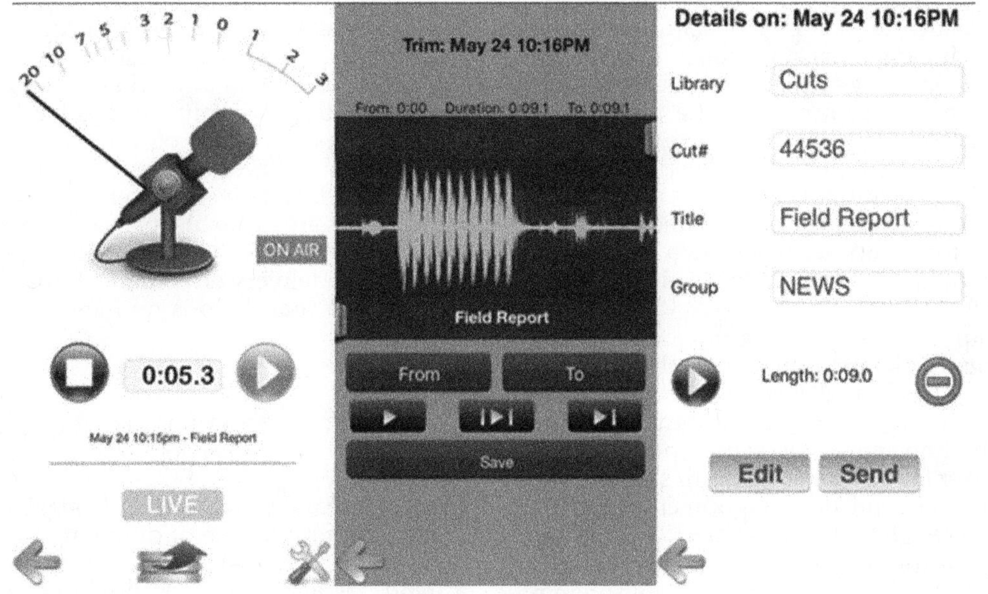

FIGURE 4.7-9 Simple user interface for remote audio recording and editing on an IOS or Android device

useful and easy-to-understand application controls on a device that they are familiar with and rely on every day. Figure 4.7-9 outlines the simple user interface for an iOS or Android device.

Modern station automation systems have taken the ability to transmit audio from a mobile device a step further. They are now creating macros that allow full control for the automation system from an iOS or Android tablet, as shown in Figure 4.7-10. This type of control allows the operator to be able run the station from any remote location, make changes to playlists, start and stop the automation system as well as voice track. The days of having to drive into the station in the middle of the night to make an important change are over. With two-way communication to and from that automation system, the operator can simply open the control app, sync the automation screen to real time, make playlist/content changes, add voice tracks, etc.

FIGURE 4.7-10 User interface for remote access to automation system from iOS or Android tablet

DISASTER RECOVERY

Natural disasters do happen and, unlike most other industries, a broadcaster's ability to continue operations during such a disaster is of utmost importance to public safety. One way to plan for a disaster is by having an automation computer at the transmitter. While many stations may already have something like this in place, it is important to keep the libraries and audio files synchronized with the automation system at the studio. By synchronizing the necessary files, you allow the computer to run as a standalone workstation and play out content to a switcher at the transmitter site.

It is also important to interface your automation system with your hardware equipment at the transmitter site. This would include silence detectors, switchers and backup RDS encoders. Setting up the automation computer at the transmitter site can help in cases of STL failure, power failure of the studio (if studio is not on generator power) and other equipment malfunctions that would result in loss of signal.

For simple control over an automation system, it may be possible to send commands to an automation computer using a web browser, smartphone or tablet. This will enable the engineer to load playlists, start and stop playback and even communicate with hardware equipment, such as running an RWT, controlling a switcher or even setting levels on an AoIP console.

The most critical component to disaster recovery is having and adhering to a plan. It should be written in a station's manual how offsite backups are to be performed, how to control transmitter-side automation computers and how to transition to alternative hardware when a disaster threatens the continuation of operations.

FAULT-TOLERANT SERVERS

One method of reducing downtime and providing a seamless transition to a backup server is to use fault-tolerant servers. A fault-tolerant server is usually a pair of servers or a "cluster" of servers that are interconnected with software and hardware, so that if the "main" server goes off-line, the "backup" server takes over the server operations with no interruption in audio or data delivery. Several types of systems and applications are available that can accomplish this, but it is important to know that any system or application in an audio-delivery environment must be specifically designed so that there is no interruption in the data stream.

VIRTUALIZATION IN BROADCAST AUTOMATION

In recent years, virtualization has gained in popularity among a multitude of different industries because of its versatility and lower cost of ownership. Such a trend has also caught on in the broadcast industry, where broadcasters are constantly looking for ways to do more with less.

For those unfamiliar with the concept of virtualization, imagine an extensive network of KVM extenders to connect workstations to the monitor, mouse and keyboard in the studio, except imagine replacing the individual workstations with one single hardware unit that hosts all of the individual operating systems. The result is a virtual environment.

The current model of using standalone workstations requires each computer to have its own drive for storage, processor and RAM. With the virtual computing model, however, hardware resources are shared

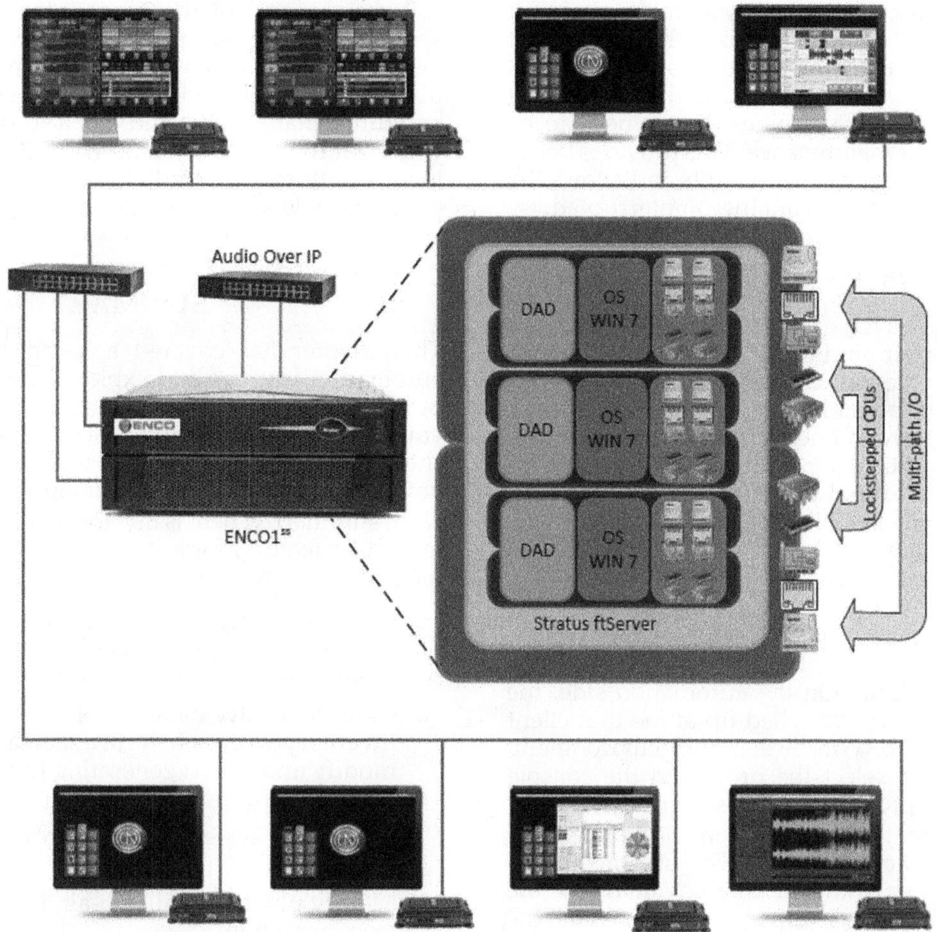

FIGURE 4.7-11 Virtualized automation network, in which a single server connects via KVM extenders to a number of user terminals, and hosts each terminal with its own independent OS

among the different virtual workstations. This is made possible because a number of virtual workstations can run on a single hardware unit. The many advantages provided by a virtual environment have led to this technology catching on in the broadcast industry.

One of the principle advantages of virtualization is the reduction in the total cost of ownership. While the initial capital outlay may be higher than for traditional automation workstations, the cost savings are realized in lower maintenance costs and hardware upgrades. This is due, in large part, to the fact that the centralized hardware is a shared resource utilized by every virtual workstation and file server on the network. Additionally, the cost of powering and cooling the system is drastically lower than powering and cooling the multiple computers of a traditional system, especially considering the power drawn when systems are in standby.

The initial deployment of a virtual system is easier than deployment of traditional workstations, because everything is in one box and all control points are connected by simple network cabling. An existing AoIP network provides further ease when deploying a virtual automation system. These attributes also make relocating a virtualized system very simple. Furthermore, when station personnel decide it is time to upgrade the hardware, the upgrades can be done quickly with less downtime.

With a virtual system, any studio can access any workstation session using the thin clients. This enables operators to use studios as a multipurpose space. Operators can go live to air from any studio or can use any of the available studios for production work.

Finally, a virtual system protects your facility against data loss because of the ability to keep redundant copies of audio assets and databases. This redundant environment allows for easy virus removal and cloning of virtual workstations to scale upward when your facility's needs change.

Reduced Cost of Maintenance

When all virtual workstations share a common set of hardware components, maintenance of the system is far less expensive than replacing or upgrading the same components on each hardware workstation. Along with being cost-effective to replace or upgrade hardware components, a virtual system also makes it easy, as there is only one centralized hardware unit on which to perform maintenance. Many of today's hardware units allow components and drives to be hot swapped, so the virtual machines do not need to be shut down to render maintenance. This provides better uptime over traditional hardware units and allows the on-air program to continue running uninterrupted.

Studio Space Is Not Fixed in Its Functionality

Traditionally, studios have been functionally fixed. An air studio was an air studio and a production studio was used for production. Creating multipurpose studios was certainly possible but not always practical. Within a virtual environment, workstations are not assigned to any particular control point, thus any client can control any virtual workstation. With this level of versatility and the flexibility of an AoIP routing system, any studio can be used for any purpose regardless of its original intended use.

Being able to create these multipurpose studios is tremendously efficient, especially during busy times, such as pledge drives, when talent might be waiting on a production studio. On the automation side, the appropriate session can be called up at the thin client located in the studio. Within the AoIP environment, the operator would select the profile on the console that matches their production needs.

A virtual automation environment has some clear advantages over the traditional workstation model, and stations are beginning to take advantage of the ease and versatility. It is important to evaluate the needs of a station and determine how a virtual environment can improve the workflow, data security and maintenance requirements.

Visual Radio

In the past, delivery of "audio programming" through radio services was dedicated to "over the air" on a radio frequency broadcast channel. The growing availability of Internet services has allowed for audio delivery through the Internet. While "OTA" transmission still continues today, increased Internet availability has led to streaming audio from many radio stations. This allows access to everyone with data services to receive their favorite radio programming on a variety of mobile handheld devices anywhere, thus extending the reach of any radio station. Through Internet enhancements, software, hardware development and cost reductions, it is now within the reach of every audio provider to very easily include any video content to accompany any audio service.

Visual radio can be integrated with the radio station's own legacy playout system, so the interactive visual channel created by the radio station's content producers is synchronized with the audio programming. The "radio automation" system can be a fully automated software solution that integrates and automatically switches playout servers to video, graphics or cameras by analyzing audio signals and XML data from the radio station's automation system. Camera switching can be automated with the use of VOX software as part of the voice-automated production system. During "studio live" interviews, the video may be camera video of the host and guest. During playback of radio audio assets, the video assets may be a direct match to the audio or a substitute video that you wish to run at any time with the audio. Integrating and automating "radio" with cameras, videos and graphics means that "radio" can now also be "visual."

SUMMARY

This chapter has covered a broad range of station automation features and explored the wide variety of options available to network, store, share and protect your digital audio assets. These general discussions should provide a framework with which to evaluate your own specific station automation needs.

To summarize, here is my Top Ten (actually 17) list of important features for a state-of-the-art networked digital audio automation system (in no particular order):

1. CD-quality audio with balanced analog and AES/EBU digital I/O. High sample rates, 24-bit quantization and multi-channel audio are a plus
2. Comprehensive database of entire network inventory with easy-to-use search, sort, edit, modify and report generating functions
3. Automated synchronization of logs from scheduling system of choice to automation playlists
4. Modify playlists while on-air, right up to airtime from anywhere
5. Automated reporting of as-played data back to scheduling and billing systems
6. Voice tracking and highly adjustable segues for live sound during automation
7. Automatically record news feeds independent of on-air playback
8. Accept closures (real or virtual) from satellite receiver to play stopsets, IDs and jingle/liner rotations
9. Control audio routing, console ON/OFF keys, telephone systems
10. Send "Now Playing" metadata to HD transmitter, RDS encoder and/or web site via IP
11. Easy to learn and use operator interface with touchscreen operation for live assist
12. Integration of third-party multi-track production equipment
13. Integration of wire services, text and audio
14. Dedicated file server-based network architecture
15. Switched Ethernet gigabit network using UTP
16. Multi-processor, multi-NIC file server or NAS with integrated disk array or SAN
17. High-speed (15K RPM), high-capacity (300 GB), high-transfer-rate (>300 MBps) disk drives configured as a RAID Level 5 array.

Station automation is a continuously evolving entity. With the personal computer at the heart of it, this evolution will only accelerate due to the rapid advances being made in computing platforms, networks and the software applications that run on them. One thing the first decade of these systems has shown us is that advancements occur so quickly that what is rocket science today will surely be obsolete long before it wears out. It is therefore wise, when building or upgrading a system, to use the most advanced, stable and reliable technology available at that time.

I hope this chapter has given the reader some insight as to the state of this technology in 2016 and provided enough history to help the user prepare for what emerging technology may bring in the future.

COMMONLY USED TERMS

1000BaseLX—1000-Mbps fiber-optic Ethernet network using long-wave laser

1000BaseSX—1000-Mbps fiber-optic Ethernet network using short-wave laser

1000BaseT—1000-Mbps twisted-pair Ethernet network (four pairs of CAT5E or CAT6)

100BaseFX—100-Mbps fiber-optic Ethernet network

100BaseT4—100-Mbps twisted-pair Ethernet network (four pairs of CAT3, CAT4 or CAT5)

100BaseTX—100-Mbps twisted-pair Ethernet network (two pairs of CAT5)

100BaseVGAnyLAN—100-Mbps star-wired network using *DPP media access control* mechanism over four twisted pairs of CAT3 or higher

10Base2—10-Mbps RG-58-style coax bus-cabled Ethernet network

10Base5—10-Mbps RG-8-style coax bus-cabled Ethernet network

10BaseT—10-Mbps twisted-pair Ethernet network (two pairs of CAT3)

10GBaseT—10,000-Mbps twisted-pair Ethernet network (CAT6a or CAT7)

10GbE—10-Gigabit Ethernet

AAF—Advanced Authoring Format; an interchange file format developed by the AAFAssociation, Pro MPEG Forum and SMPTE, primarily for post-production interchange of media and metadata

A/D—analog-to-digital conversion

AES—Audio Engineering Society

AES/EBU—Audio Engineering Society/European Broadcast Union; used to refer to the accepted professional format for serial digital audio

anti-aliasing filter—a low-pass filter used to limit input frequencies to an A/D converter to one-half the sampling rate to satisfy *Nyquist* criteria and prevent sample contamination

as-played log—a log file containing date, audio file ID, start and stop time information for every element aired

ATA—AT attachment specification that defines the IDE interface; AT refers to the IBM PC/AT, which was the first computer to use the IDE interface

audio server—a file server that contains both the audio file hard-disk storage and the audio-processing hardware (DSP boards) to enable it to record and play audio directly

baseband—communication in which only a single signal occupies a transmit or receive path at a time

bus link—a communications link where every node shares a single communication medium (e.g. cable)

BWF—Broadcast Wave Format; an extension of the WAV file format developed by the *EBU* that embeds metadata into a broadcast extension (BEXT) "*chunk*" contained within the audio file; described in EBU Document Tech. 3285 and its supplements

Cart Chunk—Cart Chunk format; an extension of the WAV file format developed by D. Pierce and G. Steadman with input from the radio broadcast community that embeds metadata into a cart "*chunk*" contained within the audio file; a public domain, license-free interchange standard ratified by the AES as AES46-2002

cart machine—traditional analog tape machine that used an endless loop "cart" ridge

CAT3—Category 3 voice-grade unshielded twisted-pair cable (*UTP*); frequencies up to 16 MHz; used for telephone and for data up to 10 Mbps

CAT5—Category 5 data-grade *UTP* cable; frequencies up to 100 MHz; used for up to 100 Mbps

CAT5e—Category 5 enhanced data-grade *UTP* cable; frequencies up to 125 MHz; used for up to 1000 Mbps

CAT6—Category 6 data-grade *UTP* cable; frequencies up to 250 MHz; used for up to 1000 Mbps

CAT6a—Category 6a data-grade *UTP* cable; frequencies up to 500 MHz; used for up to 10 Gbps

CAT7—more formally known as ISO/IEC 11801 Class F cable; has four individually shielded twisted pairs (ScTP) inside an outer shield; frequencies up to 600 MHz; used for 10 Gbps

CD jukebox—a CD player that housed hundreds of compact disks and used robotics to move the CDs from storage to an active CD transport for playback

CD quality—frequency response flat from 20 Hz to 20 kHz, dynamic range > 90 dB, signal-to-noise ratio > 90 dB

cheapernet—slang name for 10Base2; comes from low cable cost

chunk—the basic data building block of a file based on the Microsoft Resource Interchange File Format (RIFF), such as a WAV file

coding—the process of converting signal amplitude to a digital word *collision*—when two or more transmitters access a shared bus simultaneously and their data packets interfere with each other

control macros—sets of control functions executed in sequence as a single command

CPU—central processing unit

CSMA/CD—carrier sense multiple access with carrier detection; contention-based network control mechanism used by Ethernet networks

D/A—digital-to-analog conversion

data compression—the process of reducing the amount of data needed to accurately represent an object

data rate—the speed at which data is produced or consumed

data redundancy—storing extra data such that original data can be recovered in the event of a storage component failure

DAS—direct attached storage; refers to disk drives connected directly to a server or workstation as opposed to network-based architectures like *SAN* and *NAS*

disk mirroring—data written to two (or more) identical disk drives for redundancy

Dolby AC2/AC3—digital audio data reduction algorithm developed by Dolby Labs

DPP—Demand Priority Protocol; network access mechanism used by 100BaseVG

DSP—digital signal processing; refers to digital processing hardware used to manipulate audio data and execute data reduction algorithms

duplex communications—the ability to transmit and receive simultaneously

EBU—European Broadcast Union

Ethernet—the first and still most popular LAN cabling standard that provides shared access through the *CSMA/CD media access control* mechanism; includes 10Base2, 10Base5, 10BaseT, 100BaseFX, 100BaseT4, 100BaseTX, 1000BaseLX, 1000Base SX and 1000BaseT specifications

Fibre Channel—a high-speed transport mechanism used primarily to connect devices in a *SAN*; most often implemented with fiber-optic connections, although three types of electrical (copper) connections exist

file server—A computer with an integrated network operating system that presents itself to the rest of the system as a large shared disk drive and manages the storage of all audio and system data

FDX—*full duplex*; communication that provides simultaneous transmit and receive

GUI—graphic user interface; computer screens designed for point-and-click operation as opposed to keyboard-only operation

hit times—time of day that a program element is scheduled to air

hub—a central wiring cabinet or device used in star-wired networks

HVAC—heating, ventilation and air conditioning

IEEE—International Electrical and Electronic Engineers standards body

ID3—IDdentify an mp3; a metadata standard for MP3 files developed in 1996 by Eric Kemp; current version 2 is known as ID3v2

IP—Internet Protocol

IT—information technology; generally refers to all computer and network equipment and the departments and personnel that manage and maintain it

LAN—local area network; computer network existing in one geographical location

latency—time delay between the request for data and the receipt of data

linear PCM—uncompressed *PCM*; sampled, quantized full-bandwidth digital audio

link aggregation—the process of combining individual network connections (links) into a single logical link that offers higher bandwidth and/or link redundancy

logical topology—how a network functions

lossless compression—algorithms that eliminate redundant information and can completely restore each and every bit of the original data through an inverse algorithm, because no unique data is discarded

lossy compression—algorithms that remove irrelevant information but cannot completely restore each and every bit of the original data through an inverse algorithm, because some minimally useful data has been discarded

macros—a set of functions programmed to execute in sequence for singular effect

managed switch—an Ethernet switch that provides a user interface (usually serial, telnet or web based) to a menu of internal configuration parameters

Mbps—megabits per second; millions of data bits per second

mesh topology—a network where each node has a point-to-point connection with every other node

media access control—how a network limits transmission to a single node at a time

MPEG—Motion Picture Experts Group; used to refer to the digital audio data reduction algorithms developed by the group

MTBF—mean time between failures; statistic relating to longevity of a device

multicast—a part of the IP protocol suite that allows a single-source packet to be delivered to many destination addresses

MXF—Media eXchange Format; an interchange file format developed by the AAF Association,

Pro MPEG Forum and SMPTE, primarily for the interchange of media and metadata between storage, broadcast and play-out systems

NAS—network attached storage

NIC—network interface card; a plug-in card that attaches a computer to the network wiring; also used to refer to integrated main board electronics that serves the same purpose

node—any device that communicates on a network

NOS—network operating system; software that manages a shared file system

Nyquist—mathematician who developed sampling theory that states a signal must be sampled at more than twice its highest frequency

off the shelf—nonproprietary computer hardware that is available from common computer suppliers

OMFI—Open Media Framework Interchange; an open interchange standard pioneered in 1992 by Avid Technology, Inc., along with industry partners to provide a platform independent method of exchanging media and metadata

packet—the basic unit of information exchanged between nodes on a network; consists of data packaged with address information

PATA—Parallel *ATA*; the original IDE interface introduced in 1986

PCM—*pulse-coded modulation*; sampled, quantized full-bandwidth digital audio

peer-to-peer network—a basic network link between two or more workstations at the file system level that permit file sharing and transfer between them

physical topology—how a network is connected; also known as its shape

playlist—a database representing elements to be played in sequence

point-to-point link—a communications link where exactly one transmitter communicates with exactly one receiver

port trunking—another name for *link aggregation* often used in Ethernet switches

quantization—the process of breaking continuous analog levels into discrete digital levels or steps

RAID—redundant array of inexpensive (independent) disks—a disk technology that provides data redundancy through additional error correction disks

RAID Level 0—also known as *disk striping*

RAID Level 1—also known as *disk mirroring*

RAID Level 5—a disk array where redundancy is provided by a single additional disk

RAID Level 6—a disk array that allows for the loss of up to two drives simultaneously

RAM—random access memory; fast rewritable memory

RJ-45—the eight-position modular connector used in twisted-pair Ethernet wiring

reconstruction filter—a filter used to smooth the pulsed output of a D/A converter back to a continuous analog waveform

ring topology—an arrangement of point-to-point links that connect in a circle or ring

ROM—read-only memory; slow, nonvolatile memory

sampling—the process of periodically measuring a signal's value

sample rate—the frequency at which a signal is sampled

SAN—storage area network

SAS—serial attached *SCSI*; designed as a serial communication-based replacement for parallel (traditional) SCSI; it offers simpler interconnects and higher speed data transfers; uses standard SCSI command set easing integration and is also compatible with *SATA* technology

SATA—serial version of the ATA (originally defined as parallel) IDE interface

SCSI—small computer systems interface; refers to a standard interconnection and command set used to transfer data between computers and peripheral devices; most commonly used for connecting workstations and servers to disk and tape drives, but is designed for device independence and can be used with other devices such as printers, scanners, CD and DVD burners; SCSI has evolved through several generations using 8- and 16-bit-wide parallel data bus to provide data transfer rates from 5 MBps to 320 MBps

sneakernet—slang term for transferring data from one computer to another by running (in sneakers) a floppy disk between the two

sound card—a plug-in card that contains the electronics necessary to allow a PC to receive as input and send as output audio signals; also used to refer to integrated main board electronics that serves the same purpose

spanning—combining multiple disk drives to act as a single drive with the sum capacity of all the individual disks

SSA—serial storage architecture; a serial transport protocol developed by IBM to provide redundant data paths to up to 192 disk drives in direct attached storage configurations; utilized *SCSI* command set and was adopted as an ANSI X3T10.1 standard; promoted by the SSA Industry Association as an open standard, but lost favor to the more widely adopted *Fibre Channel* protocol

SSD—solid-state disk; an alternative to a traditional HDD, using flash memory instead of moving mechanical parts

SSHD—solid-state hybrid drive; combination of a traditional HDD with a small amount of flash memory added to improve performance for frequently accessed files *standalone*—refers to a single computer that functions all by itself

star topology—a network where each node has a point-to-point connection with a central hub

station automation —the use of devices, processes and interconnections to make a broadcast facility function automatically

stopset—a set of audio elements that play in succession after a program segment has completed (stopped); usually refers to a group of commercials or other announcements placed between program segments

STP—shielded twisted-pair cable

striping—dividing and writing a data stream to multiple drives to improve performance

switched Ethernet—an Ethernet network in which the central connection device, called a *switch*, provides high-speed port-to-port packet switching functions that create a logical *mesh topology*

teaming—another name for *link aggregation* often used by *NIC* manufacturers

thicknet—another name for 10Base5; comes from thick RG-8-style cable

thinnet—another name for *10Base2*; comes from thin RG-58-style cable

UTP—unshielded twisted-pair cable

walk away—station automation so complete, the operators can leave the equipment unattended

visual radio—the integration of video assets with audio assets of a radio station

XML—eXtensible Mark-up Language; a standardized, self-describing information exchange format that allows very flexible organization and description of the data payload; this simplifies communicating complex data between applications

XGbE—10-Gigabit Ethernet

References

AAF and MXF—A Complementary Pair. (n.d.). Retrieved September 15, 2006, from the AAF Association Web site: www.aafassociation.org/

Bird, Robert. (1991, August). Hard Disk Recording for Broadcast Use. *Broadcast Engineering*, p. 60.

Cabot, Richard. (1990, July). Performance Aspects of Digital Oversampling. *Broadcast Engineering*, pp. 26–93.

Carlson, Bruce. (1975). *Communication Systems* (2nd edition, pp. 294–326). New York, NY: McGraw-Hill, Inc.

Carter, R. Scott and Stephanae Ann Stevenson. (1994, July/August). Strike Up the Bandwidth. *NetWare Connection*, pp. 17–27.

Category 6 Cabling Overview, FAQs and Whitepapers. (n.d.). Retrieved September 14, 2006, from the Telecommunications Industry Association Web site: www.tiaonline.org/standards/technology/cat6/faq.cfm

Category 7 Cable. (n.d.). Retrieved September 13, 2006, from Wikipedia, The Free Encyclopedia Web site: http://en.wikipedia.org/wiki/Category_7_cable

Chalmers, R. (1997, Winter). The Broadcast Wave Format—an introduction. *EBU Technical Review No. 274*, pp. 16.

Definition of: Gigabit Ethernet. (n.d.). Retrieved December 1, 2005, from the PC Magazine.com Encyclopedia Web site: www.pcmag.com/encyclopedia_term/0,2542,t=Gigabit+Ethernet&i=43779,00.asp

Gigabit Ethernet. (n.d.). Retrieved December 1, 2005, from the TechWeb TechEncyclopedia Web site: www.techweb.com/encyclopedia/defineterm.jhtml?term=Gigabit+Ethernet

Conover, Joel. (1995, November). ATM Fast, Fast Ethernet, Fibre Channel Fast. *Network Computing*, pp. 4660.

Direct Attached Storage. (n.d.). Retrieved September 13, 2006, from Wikipedia, The Free Encyclopedia Web site: http://en.wikipedia.org/wiki/Direct_Attached_Storage

EBU Document Tech. 3285 (July, 2001). *Specification of the Broadcast Wave Format—Version 1.* Retrieved September 14, 2006, from the EBU Web site: www.ebu.ch/CMSimages/en/tec_doc_t3285_tcm6-10544.pdf

ENCO Systems, Inc. (2001). *DAD$_{PRO32}$ Digital Audio Delivery System Reference Manual* (Spring 2001 Revision, pp. 12-5 to 12-17). Southfield, MI: ENCO Systems, Inc.

Heywood, Drew, CNE, Dulaney, E., Homer, B., Niedermiller-Chaffins, D., Orr, S., Stevens, S., & Stone, H. (1995). *Inside NetWare 3.12* (Fifth edition, pp. 61797). Indianapolis, IN: New Riders Publishing,

Introduction. (n.d.). Retrieved September 14, 2006, from the Cart Chunk Organization Web site: http://cartchunk.org:8080/introduc.htm

Katron Technologies, Inc. (1996). *Installation Guide: Fast Ethernet PCI Adapter* (pp. 1–19) [Manual]. Miami, FL: Katron Technologies, Inc.

Marks, Howard. (2005, September 14). Review: Disk-to-Disk-to-Tape Software, Escape the Tape. *Network Computing*, pp. 67–72.

Meggyesi, Zoltan. (1994, August 15). *Fibre Channel Overview.* Retrieved January 11, 2006, from the CERN High Speed Interconnect Pages Web site: http://hsi.web.cern.ch/HSI/fcs/spec/overview.htm

PC Technology Guide. (2003, November 15). *Tape Storage.* Retrieved December 21, 2005, from the PCTechGuide Web site: www.pctechguide.com/15tape.htm

Pizzi, Skip. (1991, August). Digital Audio Workstations Diversify. *Broadcast Engineering*, pp. 56–66.

Pohlmann, Ken. (1989). *Principles of Digital Audio* (2nd edition, pp. 41–98). Carmel, IN: SAMS.

RAID Tutorial. (2005). Retrieved September 13, 2006, from the Advanced Computer and Network Corporation Web site: www.acnc.com/04_01_00.html

Rodgers, Adam. (1992, September). Understanding RAID Technology. *Inside NetWare*, Vol. 1, No. 9.

Serial Attached SCSI. (n.d.). Retrieved September 13, 2006, from Wikipedia, The Free Encyclopedia Web site: http://en.wikipedia.org/wiki/Serial_Attached_SCSI

Smyth, Stephen. (1992, February). Digital Audio Data Compression. *Broadcast Engineering*, pp. 52–60.

Storage Computer Corporation. (1991). *RAID Aid: A Taxonomic Extension of the Berkeley Disk Array Schema* (p. 14). Nashua, NH: Storage Computer Corporation.

Tagging Introduction. (n.d.). Retrieved September 14, 2006, from the ID3 Organization Web site: www.id3.org/intro.html

Whitmann, Art. (1996, May). Hey, Hewlett-Packard! Let's Stop The Insanity! *Network Computing*, pp. 103–105.

CHAPTER

4.8

Digital Audio Compression Technologies

SCHUYLER QUACKENBUSH

Audio Research Labs
Scotch Plains, NJ

FRED WYLIE

Audio Processing Technology, Ltd.
Belfast, Northern Ireland

INTRODUCTION

Virtually all applications of digital audio deal with enormous amounts of data. Even though modern computer networks have a capacity that may eliminate the need for audio signal compression, there remain many channels, including nearly all wireless communications channels, for which signal compression is essential for commercial viability of applications. Because of this, compression is an integral part of all digital radio and digital television systems, such as

- the Advanced Television Systems Committee (ATSC) digital television broadcast system, including the ATSC 3.0 system,
- the DTS (formerly iBiquity Digital Corporation) HD Radio System (standardized by the National Radio Systems Committee, NRSC),
- commercial satellite radio broadcast systems (such as SiriusXM Satellite Radio),
- other terrestrial DAB systems (such as Digital Radio Mondiale, Eureka-147, and ISDB-T) and cellular streaming audio systems.

Audio compression involves taking the "full" digital representation of an audio signal (usually obtained by digitally sampling an analog audio signal from a microphone or other audio source) and sending it through an *encoder*, which removes information from this full representation in a prescribed manner. This encoded signal is sent through some medium (for example, a broadcast channel), then on the receive end is sent through a *decoder*, which either restores the audio signal to its original quality or in some cases, by design, to a signal of lesser quality as a result of some limitation of the system (for example, severely restricted transmission bandwidth).

Many existing and proposed audio compression systems employ a variety of processing techniques. Any scheme that becomes widely adopted can enjoy economies of scale and virtually guaranteed interoperability. Timing, however, is critical to market acceptance of any scheme. If a scheme is selected well ahead of market demand, more cost-effective or higher-performance approaches may become available before the market takes off. On the other hand, any particular scheme may be merely academic if it is established after alternative schemes already have become well entrenched in a specific marketplace.

These forces are shaping the audio technology of the future. Numerous scenarios have been postulated for the hardware and software systems that will drive the digital audio production and transmission facility in the 21st century. However, one thing is certain: it will revolve around audio compression.

The professional audio industry continues to demand ever more complex equipment for the capture, storage, post-production, exchange, distribution and transmission of high quality audio, whether it is mono, stereo, multichannel or immersive sound formats. This demand is being driven by end users, broadcasters, filmmakers and the recording industry, all of which are converging to an all-digital environment. Continuing advances in digital signal processing (DSP) technology make it possible to create low-cost, real-time hardware for digital audio signal compression that implement complex and powerful signal processing algorithms. Such systems can significantly lower the bandwidth and storage requirements for the transmission, distribution and exchange of high-quality audio.

The introduction in 1982 of the compact disc (CD) digital audio format set a quality benchmark that the manufacturers of professional audio equipment had to meet or exceed. The introduction in 1998 of the DVD and later DVD-Audio and Super Audio CD (SACD) raised this quality benchmark by providing 5.1 channels of high-resolution audio (higher sampling frequency and greater sample word length than those used in the CD). Some consumers expect the same quality from radio and television receivers, which represents a potential challenge for the broadcaster.

The infrastructure required to support the shift from analog to digital audio could be expensive. It is a rather complex technical exercise to fully implement a digital pulse-code modulation (PCM) infrastructure. To demonstrate the advantages of distributing compressed digital audio over wired or wireless systems and networks, consider again the CD format as a reference. The CD uses a 16-bit linear PCM representation, which has excellent sonic qualities, but with the handicap of the amount of bandwidth the digital signal occupies in a transmission system. A stereo CD transfers data at 1.411 Mbps, which would require a bandwidth of approximately 700 kHz to avoid distortion of the digital signal. In practice, additional bits are added to the signal for channel coding, synchronization and error correction, which increases the bandwidth further. The commonly quoted bandwidth requirement is 1.5 MHz, such that a circuit is capable of carrying a CD or similarly coded linear PCM digital stereo signal. This can be compared with the 20 kHz needed for each of two circuits to distribute the same stereo audio in the analog format, a 75-fold increase in bandwidth requirements.

This chapter may use two-channel stereo signals as a baseline, but 5.1 channels are now ubiquitous and immersive formats are envisioned in next-generation audio systems. This is evident in audio for today's high definition television, and there is interest on how broadcast might be extended to deliver more immersive sound using next-generation audio codecs. In the case of radio, there is the issue of making a transition to stereo digital radio, and then a further transition to multichannel digital radio. Obviously, the amount of audio data in a multichannel broadcast is significantly larger than in a stereo broadcast, so it seems certain that new audio compression technology will play an important role in making multichannel digital radio a practical reality.

AUDIO BIT RATE REDUCTION

Lossless and Lossy Systems

Audio compression systems can be either lossless or lossy. A lossless system is able to reconstruct an output signal that is bit-identical to its input signal. Therefore, a lossless system that compresses a signal in CD format can truly be said to have CD quality. By comparison, a lossy system reconstructs an output signal that is not identical to its input, but if the system does not use overly aggressive compression, it is still *perceived* by listeners to be identical to its input. Lossless compression systems typically remove 50% of the input data, while lossy compression systems typically remove 90% (or more) of the input data.

Redundancy and Irrelevancy

An audio signal is very complex and contains a great deal of information. Some aspects of the information are highly predictable and are therefore redundant. Such signal components can be removed in the encoder and replaced in the decoder, a process often referred to as *statistical signal compression*. Other aspects of the information are not perceived by the listener, and are therefore *irrelevant*. Such signal components, once removed in the encoder, do not need to be replaced in the decoder (since they are irrelevant). They are irretrievably lost in the compression process, but are not *perceived* to be missing, a process referred to as *perceptual signal compression*. Coders that use this process are perceptual coders, and if their compression is not too aggressive, can result in a perceptually lossless compression system. Hence, lossless compression systems remove only the redundancy in a signal, while lossy compression systems remove both redundancy and irrelevancy.

Some signals, such as pure tones, are high in redundancy and low in irrelevancy. They can be compressed quite effectively, almost totally as a statistical compression process. Conversely, other signals, such as complex audio or noisy signals, are low in redundancy and high in irrelevancy. They also can be compressed quite effectively, but almost entirely as a perceptual compression process.

The Human Auditory System

A variety of phenomena exhibited by the human auditory system has been studied and forms the basis of what is called the *psychoacoustic model*. The sensitivity of the human ear is greatest at the lower end of the audible frequency spectrum, around 3 kHz. At 20 Hz, the bottom end of the spectrum, and 17 kHz, the top end, the sensitivity of the ear is reduced by approximately 50 dB relative to its sensitivity at 3 kHz. The curve shown in Figure 4.8-1 represents the typical limit of human hearing, such that sounds having a loudness below the threshold are not audible.

Another aspect of the hearing process is that a loud sound (or tone) at a given frequency will mask a quieter sound at a nearby frequency. This effect is called *simultaneous masking*. For a given separation in frequency between loud and quieter tones, it is more pronounced when the loud tone has a frequency that is lower than the quieter tone. Furthermore, it is more pronounced as the loud tone increases in frequency. For example, as illustrated in Figure 4.8-2, with a 2-kHz tone at a level of 70-dB SPL, tones at 1.5 kHz or 3.4 kHz would require levels greater than 40-dB SPL to be heard.

Finally, the ear is also subject to *temporal masking*, where a strong sound will mask a weaker sound that precedes or follows the stronger sound. This effect is strongest when the weaker sound follows the stronger one.

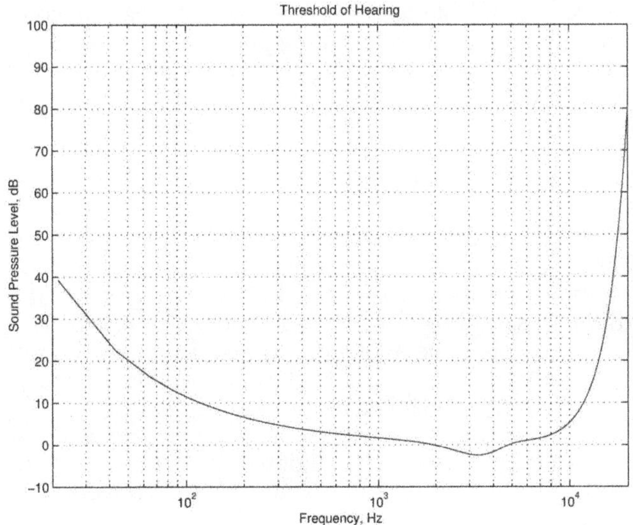

FIGURE 4.8-1 Typical threshold of hearing curve

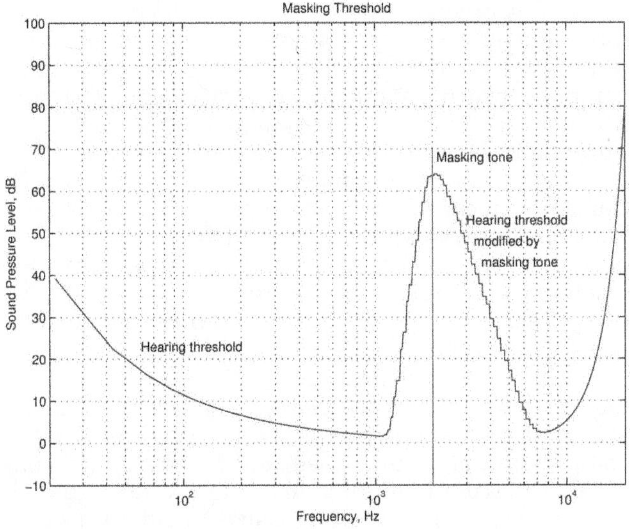

FIGURE 4.8-2 Masking effect of a loud tone at 2 kHz (shown as vertical line in figure)

By exploiting these three aspects of the human auditory system—threshold of audibility, simultaneous masking and temporal masking—an audio compression system can identify the irrelevant components of the input signal that the ear is unable to hear, not code or transmit those components, and thereby achieve data compression. The maximum tolerable level of these irrelevant components that can be removed, across the range of frequencies and for a given segment of the signal, is called the *noise masking threshold*.

Sampling and Quantization

The analog signals that comprise the audio and visual stimuli that we perceive in the world around us are continuous-time and continuous-valued functions, which is to say that they take on a continuously varying range of values over a continuously varying range of time. Sampling is the process of converting a continuous-time signal into a discrete-time signal, which is done by sampling the continuous function at a sequence of discrete, typically uniform, instants in time. Quantization is the process of converting a continuous-valued function into a discrete-valued function, which is done by mapping a given interval of values to a single value (typically the interval's midpoint). Pulse-code modulation is one of the most common forms of quantization, in which a continuous value is represented by an N-bit digital word.

An analog-to-digital (A/D) converter performs simultaneous sampling and quantization, so that it can convert an analog signal to its representative digital format. For example, the CD represents an analog signal that has been converted into a sequence of 16-bit PCM values uniformly sampled at 44.1 kHz. The level of each audio sample, therefore, is one of 65,536 discrete levels or steps. A complementary process, that of digital-to-analog (D/A) conversion, converts a digital audio signal back into analog form suitable for human listening.

The process of sampling and quantization typically entail some loss of information. While a continuous-time variable can have a theoretically infinite range of frequency content, an important rule of information theory known as Nyquist's theorem states that the highest frequency component in a signal that is to be sampled must be no greater than one-half of the sampling frequency. The reason is that the process of sampling creates copies, or *aliases*, of the original signal's spectrum at integer multiples of the sampling frequency. If the lower sideband of the first alias (which has its DC or zero frequency component located at 1 times the sampling frequency) overlaps with the upper region of the baseband audio, audible and objectionable aliasing effects occur. Hence, signals that are to be sampled must first be filtered to ensure that they do not contain frequencies greater than half the sampling rate. In practice, the sampling rate is set to slightly above the highest desired frequency, making such "anti-aliasing" filters much less expensive to produce. This is the reason that the CD sampling rate is 44.1 kHz, so that all audible frequencies (20 Hz to 20 kHz) are preserved while permitting some margin for a practical anti-aliasing filter to transition from the passband at 20 kHz to the stopband at 22.05 kHz.

Quantization is a nonlinear process, in which a continuous-valued input is mapped to a discrete-valued output, as is shown in Figure 4.8-3, where vertical lines indicate the quantizer decision boundaries and the "X"

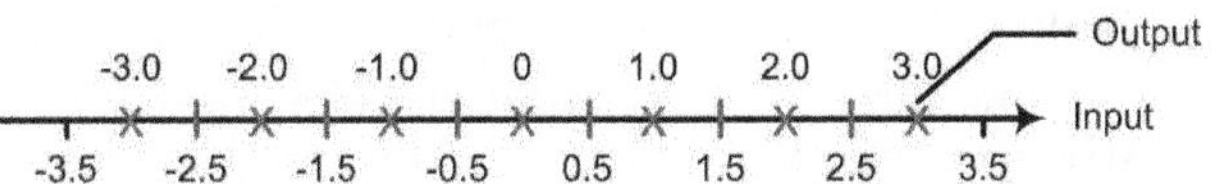

FIGURE 4.8-3 Seven-level uniform quantizer function for converting a continuous input value to a discrete output value. Horizontal axis is input range, vertical strokes show input decision boundaries and "X" indicates quantized values.

indicates the set of discrete output values. For example, an input value anywhere in the range of 1.5 to 2.5 is mapped to the output value 2.0. As in the case of the seven-level quantizer illustrated in the figure, the quantization process results in an error in the quantized output value relative to the unquantized input value. The error is referred to as quantization noise. Quantization can be modeled as a process that adds quantization noise to the signal to be quantized.

It is good operating practice to limit the number of A/D and D/A conversions in the audio processing chain, and similarly it is good practice to limit the number of compression stages in the audio chain. Typical lossless or lossy compression systems entail more than one quantizer, each of which injects quantization noise. Such noise can be minimized by operating the compression systems at as high a bit rate (i.e., as low a compression ratio) as is practical. Otherwise, after a number of A/D and D/A conversions and passes of compression coding, the accumulation of quantization noise and other unpredictable signal degradations inevitably will exceed the noise-masking threshold and be perceived by the listener.

Bit Rate and Compression Ratio

The bit rate, B, of a digital signal is defined as

$$B = \text{sampling frequency} \times \text{word length} \times \text{number of audio channels.}$$

Hence, the bit rate of a CD is

$$B = 44{,}100 \text{ samples/sec} \times 16 \text{ bits/sample} \times 2 \text{ audio channels} = 1.4112 \text{ Mbps}$$

for a two-channel stereo signal. The compression ratio, R, is defined as the ratio of the bit rate of the input signal to the bit rate of the output signal of the compression system

$$R = B_{in} / B_{out}$$

For example, if a CD signal is compressed to 128 kbps, the resulting compression ratio is

$$R = 1.4112 \text{ Mbps} / 128 \text{ kbps} = 11.025$$

Current industry practice suggests the following example bit rates and codecs for the various links in the audio delivery chain:

- For production infrastructure where further compression is anticipated: systems that use MPEG-1 Layer 2 should use at least 128 kbps per mono channel (256 kbps for stereo) as a minimum bit rate; systems that use Dolby E should use at least 150 kbps per mono channel (768 kbps for 5.1 channel signal) as a minimum bit rate.
- For emission coding, systems that use MPEG-4 Advanced Audio Coding should use 64 kbps per mono channel (320 kbps for 5.1 channel signal); systems that use Dolby AC-3 should use 76 kbps per mono channel (384 for 5.1 channel signal).

These industry practices speak to the fact that high levels of compression early in the processing chain can lead to problems if any subsequent stages of compression are required or anticipated. With successive stages of compression, the noise floor and the audio bandwidth will be set by the stage operating at the lowest bit rate. Therefore, it is worth noting that when processing chains include such a low bit-rate stage, the result cannot be subsequently improved by a stage operating at a higher bit rate.

A stage of compression may well be followed in the audio chain by another digital stage, of either compression or other processing, but which operates at a different sampling frequency. If tandem D/A and A/D conversion is to be avoided, a digital sample-rate converter must be used. This can be a stand-alone unit or it may already be installed as a module in existing equipment. If the stages have the same sampling frequencies, a direct PCM or AES/EBU digital link can be made, thus avoiding the need for sample rate or D/A conversion.[1]

Processing Delay and Computational Complexity

A signal is processed by a system in which there is an input and an output. The processing delay, or throughput *latency*, is the time between the arrival of a signal component at the input of the system and its reproduction at the output of the system. For the purpose of throughput latency, an audio compression system can be viewed as a single module consisting of encoder, transmission channel and decoder, since the signal, as a waveform, only exists at the input of the encoder and the output of the decoder. Although one can assign a time to a frame of compressed data produced by the encoder, it is primarily the processing delay of the encoder-channel-decoder system that is of practical interest.

The processing delay of an audio compression system is composed of two elements: algorithmic delay and implementation delay. The former depends on the compression algorithm used and is greater for coders that process long intervals of the input signal all at once (i.e., using long block sizes), and such delay is unavoidable given the compression algorithm. The latter depends on how the compression algorithm is implemented, such as the size of input or output signal buffers, or the extent of pipelining in the calculations of the algorithm. A more powerful DSP chip may reduce the implementation delay, hence the computational complexity of a compression algorithm may be a consideration, not only due to the cost of such chips, but also due to its potential effect on processing delay. The processing delay of an audio compression system can range from a few milliseconds (msec) to tens and even hundreds of milliseconds, and must take into account the additional delay imposed by the communications channel that connects encoder to decoder.

Most audio compression systems operate in real time, which is to say that signals can be processed "on the fly." However, due to algorithmic and implementation delay, all systems have some processing delay that introduces a measurable delay into the audio chain. The amount of delay will be important if the equipment is to be used in an interactive or two-way application. As a rule of thumb, more than 40 msec of round-trip delay in a two-way audio exchange is problematic, and in some cases, the delay due to transmission of the compressed data over the channel may be non-negligible. While a 1000-km, full duplex, optical telecommunications link has a propagation delay of approximately 5 msec in each direction, the propagation delay in satellite and long terrestrial circuits is considerably longer. Hence, the processing delay of a codec and transmission delay of the associated channel must be taken into consideration when designing or implementing a two-way interactive audio system.

Editing Compressed Data

The minimum temporal resolution of a compressed data format may or may not be adequate to allow direct editing in the compressed domain (i.e., direct editing of the compressed audio signal waveform). The minimum set of audio samples that can be manipulated via editing of the coded signal is determined by the size of the time block associated with a frame in the compressed domain and directly corresponds to the number of samples in a block in the case of block-processing compression algorithms (see the discussion of transform coders in the next section). The larger the time block, the more temporal granularity occurs in the editing process.

FILTER BANK TIME-FREQUENCY ANALYSIS

The Time-Frequency Domain

Most audio coders that have a compression ratio of 4:1 or more are perceptual coders, which achieve compression by the removal of both redundancy and irrelevancy. Since irrelevancy is primarily due to masking, and masking principles are best applied in the frequency domain, most perceptual coders involve a transformation of the signal from the time domain to the frequency domain. This typically involves using a filter bank that divides the audio signals' spectrum into from 4 to 1024 (or more) subbands.

Historically, audio coders employing such filter banks were divided into two classes: subband coders and transform coders. Both employ filter banks, and from a mathematical point of view they are not different. *Subband coders* typically have fewer subbands (e.g., 32) with sharper rejection of frequency components in adjacent subbands due to their relatively long filter lengths. *Transform coders* typically have more subbands (e.g., 1024) with less rejection of frequency components in adjacent subbands due to their relatively short filter lengths. Most subband coders have filter lengths that are several times longer than the number of subbands (e.g., four times), while transform coders have filter lengths that are at most a few times longer than the number of subbands (e.g., two times).

Examples of subband coders are G.722, an ITU-T standard introduced in the mid-1970s, and MPEG-1 Layer 1 and 2, standardized in 1993. Examples of transform coders are Dolby AC-3, a proprietary algorithm introduced in the early 1990s and MPEG-2 advanced audio coding, standardized in 1997. In the remainder of this chapter, the term subband coder will refer to both subband and transform coders.

Subband Filtering—Redundancy Reduction

The algorithms mentioned above process the PCM signal by splitting it into a number of frequency subbands, in one case as few as two (G.722) or as many as 1024 (AAC). MPEG-1 Layer 1, with 4:1 compression, has 32 frequency subbands.

Subband filtering enables the frequency domain redundancies within the audio signals to be exploited, such that the coded bit rate can be reduced relative to PCM while maintaining the same signal fidelity. Spectral redundancies are present whenever the signal energies in the various frequency bands are unequal at any instant in time. By altering the bit allocation for each subband, typically by dynamically adapting it according to the energy of the signal in each subband, the quantization noise can be reduced across all bands.

On its own, subband coding, incorporating PCM in each band, is capable of providing compression, or *subband gain,* compared to that of full-band PCM coding. This is due to the reduction in redundancy provided by the subband coding gain. Subband coding gain is defined as the improvement in the signal-to-noise ratio (SNR) when using subband coding as compared to full-band PCM coding, given that the bit rate in each domain is constant. This can be quantitatively expressed as the ratio of the variance of the quantization errors when coding in the full-band PCM domain and the variance of the quantization errors when coding in the subband domain, again with the constraint that the bit rate in each domain is constant. Subband coding gain increases as the number of subbands increase, although the complexity of the compression algorithm also increases.

Figure 4.8-4 charts subband gain as a function of the number of subbands for four essentially stationary,[2]

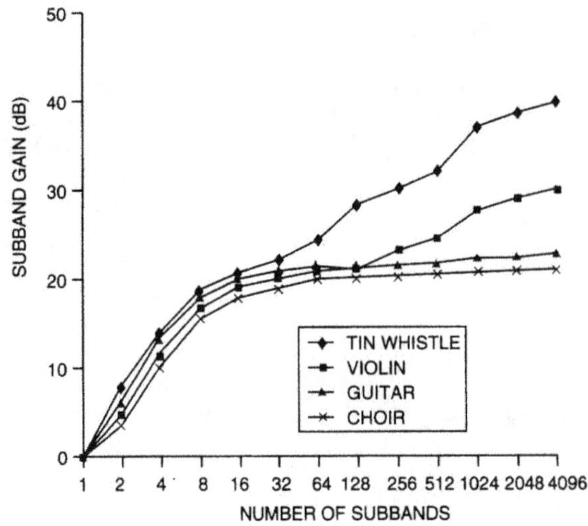

FIGURE 4.8-4 Subband gain as a function of number of subbands

but differing, complex audio signals. This figure suggests that subband coders will deliver greater compression as the number of bands is increased, even to 4096 bands or more. However, the fallacy in this is that the subband gain shown is possible only for stationary signals. A large number of subbands typically require long filter lengths, and if, during a time interval comparable to the filter length, the character of a signal varies in a way that is not predictable, the redundancy in the signal is sharply decreased. Hence, the subband coding gain is sharply reduced.

Subband Filtering—Irrelevancy Reduction

The most important contribution to compression in perceptual coders comes from irrelevancy reduction. As stated previously, this comes from exploiting the three types of auditory masking—threshold of hearing, simultaneous masking and temporal masking. The human auditory system itself can be modeled as a subband filter bank, but one with nonuniform bandwidths that are narrow at low frequency and wide at high frequency. This is equivalent to having low temporal resolution at low frequency and high temporal resolution at high frequency. Simultaneous masking is best exploited with a high-frequency resolution filter bank, while temporal masking is best exploited with a high-time resolution filter bank. Although a nonuniform bandwidth subband filter bank would seem to be the obvious solution, they are typically not used, as they can result in reduced signal compression and increased computational complexity.

In the end, the number of subbands in a coding system reflects the trade-off between higher frequency resolution that permits greater redundancy reduction and higher temporal resolution that prevents errors in irrelevancy reduction. Many audio codecs, such as AC-3, MPEG-1 Layer 3 and MPEG-4 AAC address this trade-off by using filter banks that can seamlessly switch between a high-frequency resolution mode and a high-time resolution mode.

Subband Quantization and Coding

Exploiting the irrelevancy properties of the auditory system requires computing an estimate of the short-time spectrum of the signal, since masking models are best manipulated in the frequency domain. A common method of doing this is to use a fast Fourier transform (FFT) that is aligned in time and frequency with the subband filter bank. This means that the FFT frequency bins are aligned with the band edges of the subband filter bank, so that the power in the FFT is an estimate of the short-time power spectrum of the subband signals flowing out of the encoder filter bank.

The short-time power spectrum is converted from Hz to *Bark* (at any given frequency, unit intervals on the *Bark scale* correspond to a width of one auditory *critical band*). The advantage of the Bark scale is that simultaneous masking curves are, to first order, identical at every frequency. Based on the power spectrum, along with notions as to whether a frequency region is tone-like or noise-like, a masking threshold is computed. This threshold, which specifies the "just noticeable" quantization noise at each frequency, can be used to directly set the quantizer step sizes within each subband. These step sizes are typically transmitted as side information. Note, however, that maintaining a level of "coding margin," which keeps the actual quantization noise some amount below the estimated masking threshold, is important if further compression is contemplated as part of any additional post-production or transmission steps.

Different coding algorithms may use different strategies to code the quantized subband values. For example, subband values can be coded in vectors, and entropy (i.e., lossless or noiseless) coding can be used to further compress the codes assigned to these vectors.

Finally, some systems exploit the significant redundancy between the two channels (left and right) in a two-channel stereo (or more for multi-channel) signal by using some form of joint channel coding. For stereo signals, this might result in coding regions of the stereo spectrum as a single amplitude-panned spectrum, or computing left, right, sum and difference spectra and then, for each region of the spectrum, to select the domain (left/right or sum/difference) that provides the greatest coding gain.

DIGITAL AUDIO COMPRESSION ALGORITHMS

This section presents the technical details of several audio compression systems that are widely adopted in the marketplace, or are anticipated to have a significant impact on the marketplace.

Algorithms and Compression

- *Dolby AC-3 (also known as Dolby Digital).* Standardized in 1996, this audio compression scheme is used in the ATSC DTV system and is included in the DVB digital TV standard. It can encode five surround audio channels plus a low-frequency effects channel at a (typical) total bit rate of 384 kbps. This configuration is referred to as 5.1 channel surround sound.
- *ISO/MPEG-1 Layer 2.* Standardized in 1993, this audio compression scheme is widely used as a storage format in audio production, as it permits signals to undergo many stages of compression without suffering perceivable degradation. It is also the audio compression format used in the Eureka 147 DAB system.
- *ISO/MPEG-1 Layer 3 (MP3).* Standardized in 1993, this audio compression format is used in the majority of portable music players worldwide, and has significant presence in Internet download and streaming music applications and services.
- *ISO/MPEG-4 Advanced Audio Coding (AAC).* Standardized in 1997, this audio compression format is used in portable music players and in music download and streaming services. It is also the audio

compression format used by the Integrated Services Digital Broadcasting (ISDB) service in Japan.

- *ISO/MPEG-4 High Efficiency Advanced Audio Coding (HE-AAC).* Standardized in 2003, this audio compression scheme is used in satellite-delivered DAB in the United States, and in Digital Radio Mondiale. The core technology in HE-AAC (spectral band replication) is used in HD Radio, the audio compression scheme used by the in-band/on-channel digital audio broadcast system for AM and FM in the United States. It is also used in many music-streaming services.
- *HD Radio Audio Compression.* The "HD codec" (HDC) was developed by iBiquity Digital Corporation and is used in their HD Radio in-band/on-channel (IBOC) AM and FM digital radio systems.[3]
- *Dolby E-AC-3 (Enhanced AC-3, also known as Dolby Digital Plus).* This audio compression technology was incorporated into ATSC Standard A/52B in June 2005. It provides additional compression relative to AC-3 by incorporating a high-frequency-resolution transform, spectral extension, enhanced coupling channel and "pre-echo" noise suppression tools, while retaining compatibility with AC-3 at the level of data framing and meta-data.
- *Next Generation Audio (NGA).* This label covers an emerging set of audio compression algorithms (including AC-4 and MPEG-H) that support immersive audio presentations with rich metadata that provides new functionality at the audio decoder. Immersive is characterized by having loudspeakers at the listener's ear level (as with 5.1 channel layouts) and also above and below ear level.

Table 4.8.1 shows the typical operating parameters used by each of these audio coding algorithms when coding a stereo signal.

The following sections will discuss each of these audio compression algorithms.[4]

AC-3

The AC-3 audio compression algorithm specified in the ATSC DTV system can encode from 1 to 5.1 channels of PCM audio into a serial bit stream at data rates ranging from 32 kbps to 640 kbps.[5]

A typical application of this compression algorithm entails the following steps. A 5.1-channel audio program is converted from a PCM representation requiring more than 4 Mbps (6 channels × 48 kHz × 168 bits = 4.608 Mbps) into a 384-kbps serial bit stream by the AC-3 encoder. This may be multiplexed with other data (e.g., compressed video) and have timing, synchronization and transport information added to the stream. Radio frequency (RF) transmission equipment converts this stream into a modulated waveform that is applied to a terrestrial broadcast signal. The amount of bandwidth required for the transmission of the audio information has been reduced by a factor of 12 by the AC-3 digital compression system. The received signal is demodulated and demultiplexed to recover the 384-kbps serial bit stream, which is decoded by the AC-3 decoder. The result is the reconstructed 5.1-channel audio program.

AC-3 Encoding

The AC-3 encoder accepts PCM audio and produces the encoded audio bit stream. The AC-3 algorithm achieves high coding gain by coarsely quantizing a frequency domain representation of the audio signal. A block diagram of this process is shown in Figure 4.8-5. The first step in the encoding chain is to transform the representation of audio from a sequence of PCM time samples

TABLE 4.8-1
Typical Audio Compression Algorithm Operational Parameters From Compressing a 2-Channel Signal

Coding System	Typical Compression Ratio	Subbands	Bit Rate, kbps (stereo)[1]	Typical Implementation Delay, ms[2]
MPEG-1 Layer 1	4:1	32	384	19
MPEG-1 Layer 2	6:1	32	256	40
AC-3	7:1	256	192	50
MPEG-1 Layer 3	9:1	576	160	120
MPEG-4 AAC	11:1	1024	128	130
E-AC-3	15:1	256/1536	96	50–150
HDC	15:1	(not disclosed)	96	340
MPEG-4 HE-AAC	29:1	64/1024	48	340
NGA (incl. AC-4 and MPEG-H)	15:1	1024	96[3]	*TBD*

[1] Typical values shown; each system is capable of operating at various bit rates.
[2] The total one-way delay (encoder input to decoder output), neglecting any additional transmission channel delay.
[3] However, NGA is primarily aimed as immersive programs with e.g., 11 audio channels.

into a sequence of blocks of frequency coefficients, which is done by the analysis filter bank. Overlapping blocks of 512 time samples are multiplied by a time window and transformed into the frequency domain. These PCM blocks have 50% overlap, so that each new block contains only 256 new samples. However, because the filter bank is *critically sampled*, the 512 PCM samples processed by the transform result in only 256 frequency coefficients, so that there is no growth in data rate (i.e., 256 new PCM samples per block result in 256 frequency coefficients). Because of the overlapping blocks, each PCM input sample is represented in two adjacent transformed blocks.

The individual frequency coefficients are represented in block floating point, with a common exponent for a set of coefficients and a mantissa for each coefficient in the set. The exponents are encoded into a coarse representation of the signal spectrum, referred to as the spectral envelope. This spectral envelope is used by the core bit allocation routine, which determines how many bits should be used to encode each individual mantissa. The spectral envelope and the coarsely quantized mantissas for six audio blocks (1536 audio samples) are formatted into an AC-3 frame. The AC-3 bit stream is a sequence of AC-3 frames.

The actual AC-3 encoder is more complex than shown in the simplified system of Figure 4.8-5. Most notably, the following functions also are supported in the encoder:

- An AC-3 frame has a header that contains a synchronization word and other information (bit rate, sample rate, number of encoded channels and additional data) such that the decoder can synchronize to and decode the bit stream.
- Error-detection codes are inserted to allow the decoder to verify that a received frame of data is error-free.
- The analysis filter bank spectral resolution may be dynamically altered to better match the time/frequency characteristic of each audio block.
- The spectral envelope may be encoded with variable time/frequency resolution.
- A more complex bit allocation may be performed, with parameters of the core bit allocation routine modified to produce an optimum bit allocation.
- The channels may be coupled at high frequencies to achieve higher coding gain for operation at lower bit rates.

In the two-channel mode, a rematrixing process may be selectively performed to provide additional coding gain and to allow improved results to be obtained in the event that the two-channel signal is decoded with a matrix surround decoder.

AC-3 Decoding

The decoding process is essentially the inverse of the encoding process. The basic decoder, shown in Figure 4.8-6, must synchronize to the encoded bit stream, check for errors and deformat the various types of data (the encoded spectral envelope and the quantized mantissas). The bit allocation routine is run, and the results are used to unpack and dequantize the mantissas. The spectral envelope is decoded to produce the exponents. The exponents and mantissas are transformed back into the time domain to produce the decoded PCM time samples. Additional steps in the audio decoding process include:

- Error concealment or muting may be applied in the event a data error is detected.
- Channels with high-frequency content coupled must be decoupled.
- Dematrixing must be applied (in the two-channel mode) whenever the channels have been rematrixed.
- The synthesis filter bank resolution must be dynamically altered in the same manner as the encoder analysis filter bank was altered during the encoding process.

AC-3 in the ATSC Audio System

A simplified block diagram of the audio subsystem of the ATSC DTV standard is shown in Figure 4.8-7. It comprises the audio encoding and decoding functions that connect the audio inputs and outputs to the transport subsystem. The audio encoder is responsible for generating the audio elementary stream, which is an encoded representation of the baseband audio input

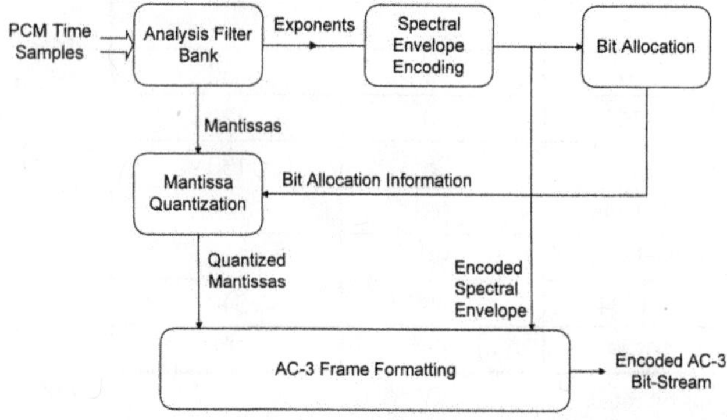

FIGURE 4.8-5 Block diagram of AC-3 encoder

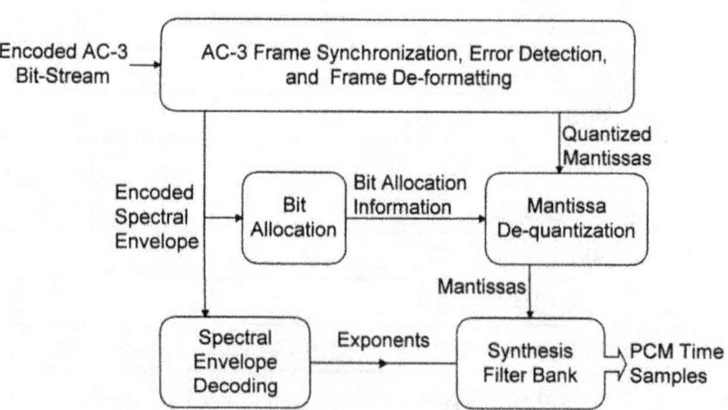

FIGURE 4.8-6 Block diagram of AC-3 decoder

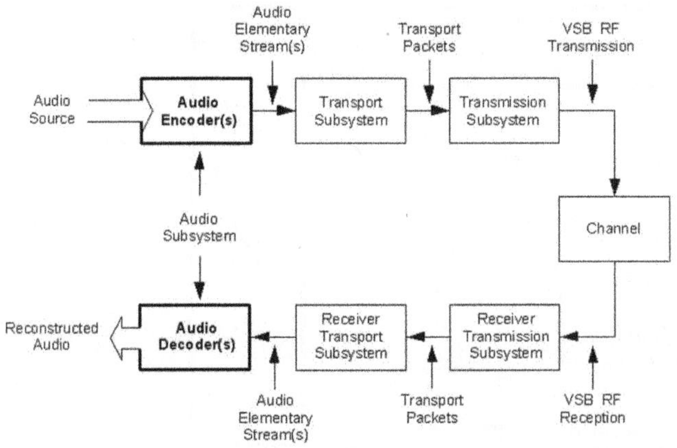

FIGURE 4.8-7 Block diagram of ATSC audio subsystem

signals. (Note that more than one audio encoder may be used in a system.) The transport subsystem formats the audio elementary stream into packetized elementary stream (PES) packets, which are then further formatted into transport packets. The transmission subsystem converts the transport packets into a modulated RF signal for transmission to the receiver. At the receiver, the signal is demodulated by the receiver transmission subsystem, and the receiver transport subsystem converts the received audio packets back into an audio elementary stream. The flexibility of the transport system allows multiple audio elementary streams to be delivered to the receiver. At the receiver, the transport subsystem is responsible for selecting which audio streams to deliver to the audio decoder subsystem. The appropriate audio decoder is then responsible for decoding the audio elementary stream back into baseband audio.

The partitioning shown in Figure 4.8-7 is conceptual, and practical implementations may differ. For example, the transport processing may be broken into two blocks; the first would perform PES packetization and the second would perform transport packetization. Or, some of the transport functionality may be included in either the audio coder or the transmission subsystem.

AC-3 System Timing Issues

The AC-3 system conveys digital audio sampled at a frequency of 48 kHz, and this sampling clock must be phase locked to the 27-MHz system clock referenced in the ATSC DTV standard. If analog signal inputs are employed, the A/D converters should sample at 48 kHz. If digital inputs are employed, the input sampling rate should be 48 kHz, or the audio encoder should contain sample rate converters that translate the sampling rate to 48 kHz. Obviously, the sampling rate at both the input to the audio encoder and the output of the audio decoder is locked to the 27-MHz system clock, but they also must be locked to the video clock to ensure proper synchronization of the audio and video subsystems.

In general, input signals should be quantized to at least 16-bit resolution, although the AC-3 audio compression system can convey audio signals having up to 24-bit resolution.

MPEG-1 Layer 2

A block diagram of the MPEG-1 Layer 2 encoding and decoding systems is shown in Figure 4.8-8.

MPEG-1 Layer 2 Encoder

In the encoder, the incoming linear PCM signal is formed into blocks of 1152 samples and filtered by a polyphase analysis filter bank to divide the input signal into 32 equal-bandwidth subband signals of 36 samples per subband. *Polyphase* refers to a specific implementation of a filter bank that permits significant computational efficiency. At a 48-kHz sampling rate, the duration of each input block is 24 msec and the bandwidth of each subband is 750 Hz. The filter bank, which displays moderate delay and minimal complexity, has sufficiently high frequency resolution to effectively exploit simultaneous masking and sufficiently high time resolution to assure that temporal masking is modeled correctly. Multiple quantizers can be used for the subband samples, each of which have an adaptive step size. Furthermore, the encoder has an adaptive bit allocation among the set of subbands. In the decoder, after demultiplexing and decoding of the audio data and side information in a frame, the synthesis filter bank reconstructs the output block of 1152 PCM samples.

The PCM signal is fed both to the filter bank and to an FFT, which computes the short-time power spectrum of the input signal. This information is used by the psychoacoustic model to determine the masking threshold for the set of subband data. This, in turn, determines the quantizer step sizes such that the quantization noise is below the masking threshold in each of the subbands. More precisely, it determines signal-to-mask ratio (SMR) threshold, which is the difference between the signal power (in dB) and masking threshold (in dB), computed in each subband. The actual number of levels for each quantizer is determined by the bit allocation.

MPEG standards only specify the format of a bit stream of compressed audio data and the process for decoding the data. An MPEG encoder is not specified other than to be a process that produces a standard-conforming bit stream. Since the psychoacoustic model is found only in the coder, it can evolve over time to permit better estimates of the masking threshold and hence better signal compression. For example, in a broadcasting system that utilizes this method of audio coding, as advances in coder technology are made, only the encoder portion of the system need be upgraded. The decoder portion (in receivers) does not need to be changed to experience this improvement. This is an important feature of the technology, since typically there are few encoders but many receivers in use.

MPEG-1 Layer 2 Scale Factors and Side Information

A scale factor is determined for the set of 12 samples in each subband. The maximum of the absolute values of these 12 samples determines a scale factor word

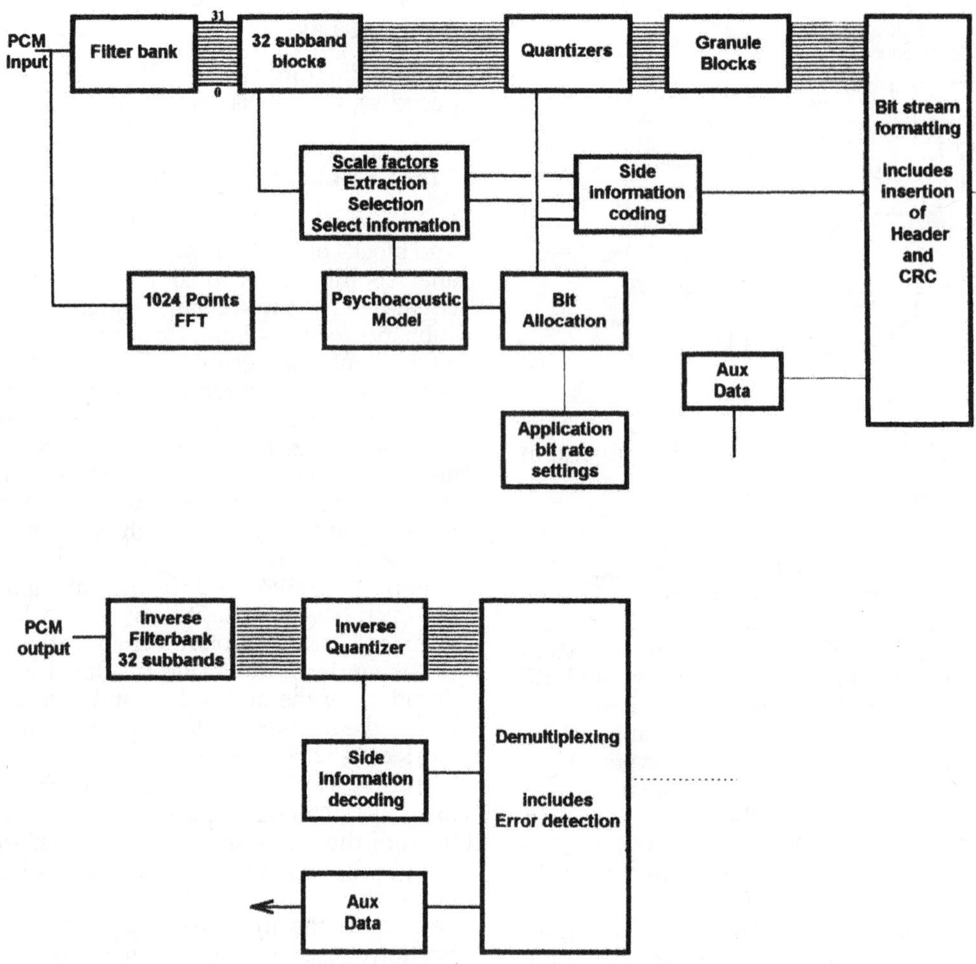

FIGURE 4.8-8 Block diagram of MPEG-1 Layer 2 encoder (top) and decoder (bottom)

consisting of 6 bits. With a quantization step size of 2 dB, this results in a dynamic range of more than 120 dB. Because each frame of audio data in Layer 2 corresponds to 36 samples in each of the 32 subbands, this process will generate three scale factors per subband per frame. However, the transmitted data rate for these scale factors can be reduced by exploiting redundancy in the data. Three successive subband scale factors are analyzed and a pattern is determined. This pattern, which is obviously related to the nature of the audio signal, will decide whether one, two or all three scale factors are required. The four possibilities of the two additional scale factors being present or absent is signaled by the scale factor select information (SCFSI) data word of 2 bits.

In the case of a fairly stationary tonal-type sound, there will be very little change in the scale factors and only the largest one of the three is transmitted, so that the corresponding data rate will be (2 + 6) or 8 scale factor bits per subband. However, in a complex sound with rapid changes in content, the transmission of two or even three scale factors may be required, producing a maximum scale factor bit rate demand of (2 + 6 + 6 + 6) or 20 bits per subband.

The number of data bits allocated to a frame's bit pool is determined by key operating parameters, those being sampling frequency, compression ratio, and where applicable, limitations imposed by the transmission medium. In the case of 20-kHz stereo being transmitted at a data rate of 384 kbps, if the signal is sampled at 48 kHz, then this necessitates a compression ratio of 4:1.

After the number of side information bits required for scale factors, bit allocation codes, cyclic redundancy check (CRC) and other functions has been determined, the remaining bits left in the pool are used for coding of the audio subband samples. The allocation of bits for the audio is determined by calculating the SMR, via the FFT, for each of the 12 subband sample blocks. The bit allocation algorithm then selects from 15 available quantizers, such that the overall bit rate limitations are met and the quantization noise is masked, or at least masked to the greatest extent possible. Note that if the signal is identically zero, no bits are allocated so that no quantizer needs to be signaled. If the composition of the audio signal is such that there are not enough bits in the pool to adequately code the subband samples, then the quantizer step sizes are increased to obtain a best-fit solution that should minimize the impact of the quantization noise on the subjective quality of the decoded audio signal.

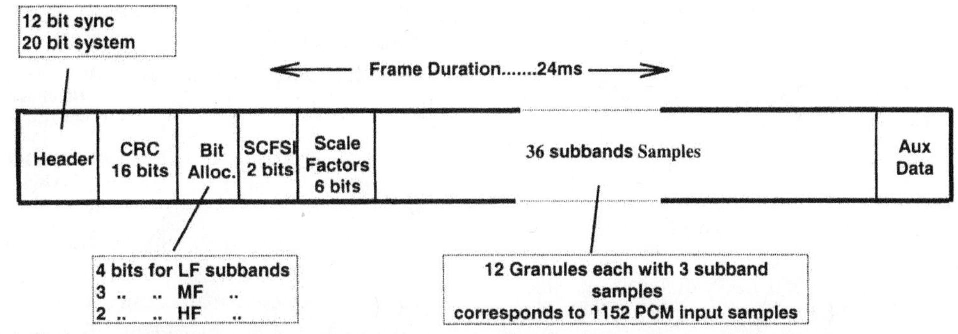

FIGURE 4.8-9 MPEG-1 Layer 2 bit stream frame structure

If the signal block being processed lies in the lower one-third of the 32 subbands, a 4-bit code word is generated to signal the selected low-subband quantizers, with this word carried as side information in the main data frame. Similarly, a 3-bit word signals the selected mid-subband quantizer and a 2-bit word signals the high-subband quantizer. This allows for at least 15, 7 and 3 different sets of quantizer ranges, respectively, in each of the three subband groupings.

As with the scale factor data, some further redundancy can be exploited in the coding of quantization levels. For the lowest quantizer ranges (3, 5 and 9 levels), three successive subband sample blocks (three groups of 12 subband samples) are grouped into a *granule* and a single quantizer range can be specified for this entire set of samples using a single code word. This is particularly effective in the higher frequency subbands where the quantizer ranges are invariably set to the lower end of the scale.

MPEG-1 Layer 2 Framing

Error detection information can be relayed to the decoder by inserting a 16-bit cyclic redundancy check word in each data frame. This parity check word allows for the detection of up to three single bit errors or a burst of errors of up to 16 bits in length. A codec detecting an uncorrectable set of errors can mute the signal, or if it incorporates an error concealment strategy, it might replace the impaired data with a previous, error-free data frame or some other suitable estimate of the damaged signal block. The typical data frame structure for MPEG-1 Layer 2 audio is given in Figure 4.8-9.

MPEG-1 Layer 3 (MP3)

MPEG-1 Layer 3, referred to more commonly as MP3, offers significantly greater compression than that provided by Layer 2. It is typically used to compress a stereo signal sampled at 44.1 kHz to a data rate of 128 kbps to 160 kbps. It can compress to a fixed bit rate, so that it can operate over fixed-rate channels or in variable rate mode, appropriate for operating over a packet network channel. In the latter case, the encoder attempts to maintain a constant signal-to-mask ratio for each signal block and permits the length of the encoded frame of data to vary frame by frame.

A block diagram of the Layer 3 encoder algorithm is shown in Figure 4.8-10.

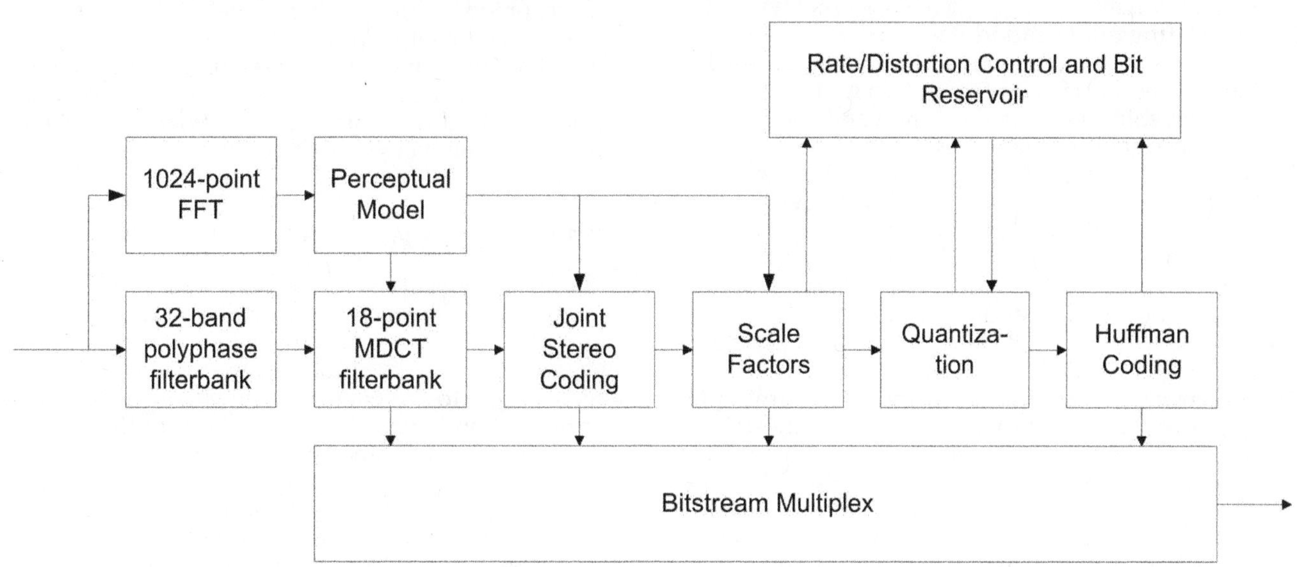

FIGURE 4.8-10 Block diagram of the MPEG-1 Layer 3 (MP3) encoder

MP3 Encoder

The filter bank used in MPEG Layer 3 is a hybrid filter bank that consists of a polyphase filter bank and a modified discrete cosine transform (MDCT). This hybrid form provides a common framework with Layer 1 and Layer 2. The further subdivision of the 32-band polyphase filter bank by the 18-band MDCT filter bank produces a division into 576 subbands, providing a greater opportunity to exploit signal redundancy as a means to gain compression. It also permits a finer tracking of the masking threshold at lower frequencies, thus permitting the coder to better exploit signal irrelevancy. In the case that the normal "high-frequency-resolution" mode of the MDCT filter bank would have insufficient time resolution to avoid poor coding performance due to temporal masking issues (the so-called "pre-echo" problem), the MDC filter bank can switch to a "high-time-resolution" mode, with a time resolution of 4 msec by limiting the division of the signal into only 192 subbands.

The perceptual model is the principal component that determines the quality of a given Layer 3 encoder implementation. It typically uses a FFT to compute the short-time energy of the signal, which is then used to calculate the masking threshold. It is important that this energy estimate be aligned in time and frequency with the subband signal that is being coded. The output of the perceptual model consists of a signal-to-mask ratio, or the allowed quantization noise in each subband. If the bit rate is sufficient to keep the quantization noise below the masking threshold, then the reconstructed signal should be perceptually indistinguishable from the original.

Joint stereo coding takes advantage of the fact that left and right channels of a stereo channel pair often contain nearly the same information. These stereophonic irrelevancies and redundancies are exploited to reduce the total bit rate. Joint stereo coding is used in cases where both high compression and stereo output is desired.

Quantization and coding to a fixed bit rate typically involves setting quantization step sizes and determining the size of the coded frame, including bits allocated from the bit reservoir. A system of two nested loops is the common solution for quantization and coding in a Layer 3 encoder. The quantization process effectively allocates noise to each subband based on the signal-to-mask ratio. Since some signal blocks are harder to code than are others (based on the instantaneous statistics of the signal), they will inherently require more bits to meet the requirements posed by the ratio. Rather than suffer inconsistent audio quality under the constraint of constant bit rate, Layer 3 solves this problem with the concept of a bit reservoir. With this structure, the number of bits comprising a bit stream frame is allowed to vary based on the requirements imposed by the masking threshold. When a frame requires greater than the nominal number of bits, they are supplied by the bit reservoir, when it requires fewer, the surplus is returned to the reservoir. While this has the advantage of raising the audio quality of the most difficult to code signal blocks, it has the tradeoff of imposing an additional throughput delay proportional to the size of the bit reservoir.

Quantization is done via a nonuniform power-law quantizer. In this way, larger values are automatically coded with less accuracy so that a mild degree of noise shaping is built into the quantization process. The quantized values are entropy coded using Huffman codes, which is a lossless process. Since this does not inject any noise into the signal, it is also referred to as noiseless coding.

To find the optimum gain and scale factors for a given signal block, the bit rate and signal-to-mask ratio threshold is usually calculated using two nested iteration loops in an analysis-by-synthesis manner:

- *Inner iteration loop (rate loop):* the Huffman code tables assign shorter code words to more frequently occurring (which are typically smaller) quantized values. If the number of bits resulting from the coding operation exceeds the number of bits available to code a given block of data, this can be corrected by adjusting the global gain, resulting in a larger quantization step sizes, leading to smaller quantized values. This operation is repeated until the resulting bit demand from Huffman coding fits within the current frame's bit budget. This loop is called the rate loop because it modifies the overall coder bit rate until it is within the desired range.
- *Outer iteration loop (noise control/distortion loop):* to shape the quantization noise according to the masking threshold, scale factors are used to normalize the subband signal within each scale factor band. Initially the scale factors have a default factor of 1.0 for each band. If the quantization noise in a given band is found to exceed the masking threshold in a given band, the scale factor for that band is adjusted to reduce the quantization noise. Since achieving a smaller quantization noise requires a larger number of quantization steps and thus a higher bit rate, the rate adjustment loop has to be repeated every time there are new scale factors. In other words, the rate loop is nested within the noise control loop. The outer (noise control) loop is executed until the actual noise (computed from the difference between the original spectral values and the quantized spectral values) is below the masking threshold for every scale factor band.

MP3 Bit Stream Framing

In the context of transmission channels, Layer 3 can operate over a constant-rate isochronous link, and has constant-rate headers (as do Layers 1 and 2). However, Layer 3 is an instantaneously variable-rate coder, which adapts to the constant-rate channel by using a "bit buffer" and "back pointers." Each of the headers signal the start of another block of audio signal. However, due to the Layer 3 syntax, the frame of data associated with that next block of audio signal may be in a prior segment of the bit stream, pointed to by the back pointer. See Figure 4.8-11 and specifically the curved arrows pointing to main_data_begin.

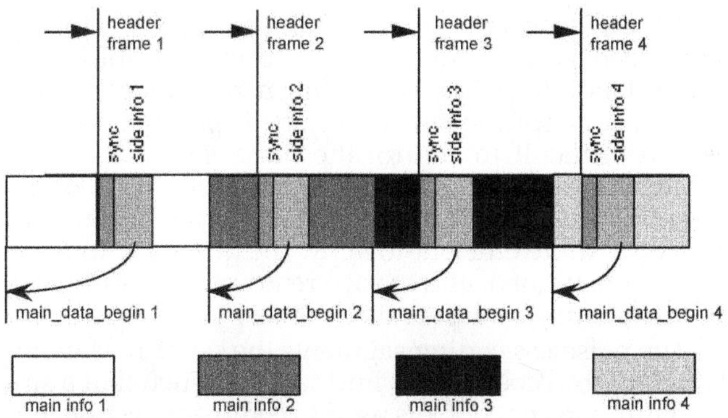

FIGURE 4.8-11 MPEG-1 Layer 3 frame structure during active use of bit buffer

MPEG-4 Advanced Audio Coding

MPEG-4 advanced audio coding is a sophisticated perceptual coder that builds upon the principles of MPEG-1 Layer 3 in a number of ways to increase coding efficiency. First, it has a filter bank that provides higher frequency resolution (1024 subbands compared to 576 in Layer 3) to better exploit signal redundancy and signal irrelevancy at low frequencies. Second, it has improved joint stereo coding, in that coding decisions can be made on a group-of-subbands basis rather than for the entire block of subbands (as in Layer 3). Third, it has improved Huffman coding, in that it uses more codebooks and better represents a range of signal statistics. Finally, it incorporates new coding tools: temporal noise shaping (TNS) and perceptual noise substitution (PNS). With its coding efficiency, an AAC system operating at 96 kbps produces the same sound quality as ISO/MPEG-1 Layer 2 operating at 192 kbps—a 2:1 reduction in bit rate. A block diagram of the AAC system is shown in Figure 4.8-12.

AAC was first standardized in 1997 as part of MPEG-2. MPEG-4 AAC is equivalent to MPEG-2 Low Complexity AAC (the most widely used version of MPEG-2 AAC), with the addition of the PNS tool and the adoption of the MPEG-4 system layer and file format.

MPEG-4 AAC provides the capability of up to 48 main audio channels, 16 low frequency effects channels, 16 overdub/multilingual channels, and 10 data streams. By comparison, ISO/MPEG-1 Layer 1 provides two channels and MPEG-2 Layer 2 provides 5.1 channels (maximum). AAC is not backward compatible with the Layer 1 and Layer 2 codecs.

MPEG-4 AAC Encoder

Each of the blocks in the encoder is described.

- *Filter bank:* AAC uses a resolution-switching filter bank that can switch between a high-frequency resolution mode of 1024 bands (for maximum statistical gain during intervals of signal stationarity), and a high-time resolution mode of 128 bands (for maximum control of time-domain coding error during intervals of signal non-stationarity). The modified discrete cosine transform is used to implement the filter bank, which has the properties of perfect reconstruction of the output (when no quantization is present), critical sampling (so that the number of new PCM samples processed by the transform exactly equals the number of time/frequency coefficients produced), and 50% overlapping transform windows (which provide signal smoothing on reconstruction).
- *Temporal noise shaping (TNS):* The TNS tool modifies the filter bank characteristics so that the combination of the two tools is better able to adapt to the time/frequency characteristics of the input signal. It shapes the quantization noise in the time

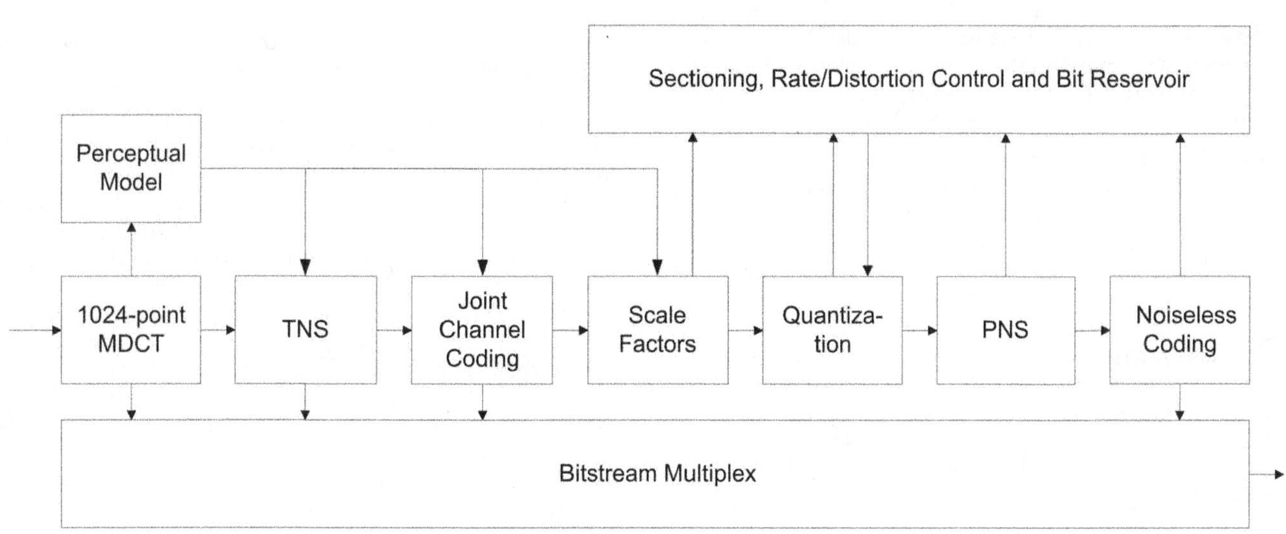

FIGURE 4.8-12 Block diagram of the MPEG-4 AAC encoder

domain by doing an open loop linear prediction in the frequency domain, so that it permits the coder to exercise control over the temporal structure of the quantization noise within a filter bank window. TNS is a technique which has proved very effective for improving the quality of signals, such as speech, when coded at low bit rates.

- *Perceptual model:* The perceptual model estimates the masking threshold, which is the level of noise that is subjectively just noticeable given the current input signal. Because models of auditory masking are primarily based on frequency domain measurements, these calculations typically are based on the short-term power spectrum of the input signal, as calculated by an FFT whose bins align with the MDCT bins. Threshold values are adapted to the time/frequency resolution of the filter bank outputs. The threshold of masking is calculated relative to each frequency coefficient for each audio channel for each frame of input signal, so that it is signal-dependent in both time and frequency.
- *Joint channel coding:* This block actually comprises three tools, intensity coding, mid/side (M/S) stereo coding (also known as "sum/difference coding") and coupling channel coding, all of which seek to protect the stereo or multichannel signal from noise imaging, while achieving coding gain based on exploiting the correlation between two or more channels of the input signal. M/S stereo coding, intensity stereo coding, and L/R (independent) coding can be combined by selectively applying them to different frequency regions, and by using these tools, it is possible to avoid expensive overcoding when using binaural masking level depression to correctly account for noise imaging, and very frequently to achieve a significant saving in bit rate.
- *Perceptual noise substitution (PNS):* This tool identifies segments of spectral coefficients that appear to be noise-like and codes them as random noise. It is extremely efficient in that, for the segment, all that needs be transmitted is a flag indicating that PNS is used and a value indicating the average power of the noise. The decoder reconstructs an estimate of the coefficients using a pseudo-random noise generator weighted by the signaled scale factor power value.
- *Scale factors, quantization, coding and rate/distortion control:* The spectral coefficients are coded using one quantizer per scale factor band, which is a division of the spectrum roughly equal to one-third Bark. The psychoacoustic model specifies the quantizer step size (inverse of the scale factor) per scale factor band. As with MPEG-1 Layer 3, AAC is an instantaneously variable rate coder that similarly uses a bit reservoir. If the coded audio is to be transmitted over a constant rate channel, then the rate/distortion module adjusts the step sizes and number of quantization levels so that a constant rate is achieved.

The quantization and coding processes work together. The first quantizes the spectral components and the second applies Huffman coding to vectors of quantized coefficients in order to extract additional redundancy from the nonuniform probability of the quantizer output levels. In any perceptual encoder, it is very difficult to control the noise level accurately, while at the same time achieving an "optimum quantizer" (in the minimum mean square error sense). It is, however, quite efficient to allow the quantizer to operate unconstrained, and to then remove the redundancy in the quantizer outputs using entropy coding.

The noiseless coding segments the set of 1024 quantized spectral coefficients into sections, such that a single Huffman codebook is used to code each section. For reasons of coding efficiency, section boundaries can only be at scale factor band boundaries so that for each section of the spectrum, one must transmit the length of the section, in scale factor bands, and the Huffman codebook number used for the section. Sectioning is dynamic and typically varies from block to block, such that the number of bits needed to represent the full set of quantized spectral coefficients is minimized.

The rate/distortion tool adjusts the scale factors such that more (or less) noise is permitted in the quantized representation of the signal, which, in turn, requires fewer (or more) bits. Using this mechanism, the rate/distortion control tool can adjust the number of bits used to code each audio frame and hence adjust the overall bit rate of the coder.

- *Bit stream multiplexer:* The multiplexer assembles the various tokens to form a coded frame, or access unit. An access unit contains all data necessary to reconstruct the corresponding time-domain signal block. The MPEG-4 system layer specifies how to carry the sequence of access units over a channel or store them in a file (using the MPEG-4 file format).

AAC has a flexible bit stream syntax for a coded frame that permits up to 48 main channels and up to 16 LFE channels to be carried in an access unit, but in a manner that does not incur any overhead for the additional channels. In this respect, it is as efficient for mono, stereo and 5.1 channel representations.

MPEG-4 High Efficiency AAC

The MPEG-4 High Efficiency AAC (HE-AAC) audio compression algorithm consists of an MPEG-4 AAC *core coder* augmented with the MPEG-4 spectral band replication (SBR) tool. The encoder SBR tool is a pre-processor for the core encoder, and the decoder SBR tool is a post-processor for the core decoder, as shown in Figure 4.8-13. The SBR tool essentially converts a signal at a given sampling rate and bandwidth into a signal at half the sampling rate and bandwidth, passes the low-bandwidth signal to the core codec, and codes the high-bandwidth signal using a compact parametric representation. The lowband signal is coded by the core coder, and the lowband compressed data and the highband parametric data are transmitted over the channel. The core decoder reconstructs the lowband signal and the SBR decoder uses the parametric data to reconstruct the highband data, thus recovering the full-bandwidth signal. This combination provides a significant improvement in performance relative to that of the core coder by itself, which can be used to either lower the bit rate or improve the audio quality.

MPEG-4 HE-AAC SBR Principle

A perceptual audio coder, such as MPEG-4 AAC, provides coding gain by shaping the quantization noise such that it is always below the masking threshold. However, if the bit rate is not sufficiently high, the masking threshold will be violated, permitting coding artifacts to become audible. The usual solution adopted by perceptual coders in this case is to reduce the bandwidth of the coded signal, thus effectively increasing the available bits per sample to be coded. The result will be a cleaner sound, but duller due to the absence of high-frequency components.

The SBR tool gives perceptual coders an additional coding strategy (other than bandwidth reduction) when faced with severe bit rate restrictions. It exploits the human auditory system's reduced acuity to high-frequency spectral detail to permit it to parametrically code the high-frequency region of the signal. When using the SBR tool, the lower frequency components of the signal (typically from 0 to between 5 kHz and 13 kHz) are coded using the core codec. Since the signal bandwidth is reduced, the core coder will be able to code this signal without violating the masking threshold. The high-frequency components of the signal are reconstructed as a transposition of the low-frequency components followed by an adjustment of the spectral envelope. In this way, a significant bit rate reduction is achieved while maintaining the same audio quality, or alternatively an improved audio quality is achieved while maintaining the same bit rate.

MPEG-4 HE-AAC SBR Technology

Using SBR, the missing high-frequency region of a low-pass filtered signal can be recovered based on the existing low-pass signal and a small amount of side information, or control data. The required control data is estimated in the encoder based on the original wide-band signal. The combination of SBR with a core coder (in this case MPEG-4 AAC) is a dual rate system, where the underlying AAC encoder/decoder is operated at half the sampling rate of the SBR encoder/decoder. A block diagram of the HE-AAC compression system, consisting of SBR encoder and its submodules, AAC core encoder/decoder and SBR decoder and its submodules is shown in Figure 4.8-13.

A major module in the SBR tool is a 64-band pseudo-quadrature mirror analysis/synthesis filter bank (QMF). Each block of 2048 PCM input samples processed by the analysis filter bank results in 32 subband samples in each of 64 equal-width subbands. The SBR encoder contains a 32-band synthesis filter bank whose inputs are the lower 32 bands of the 64 subbands and whose output is simply a band-limited (to one-half the input bandwidth) and half-sampling rate version of the input signal. Actual implementations may use more efficient means to accomplish this, but the illustrated means provides a clearer framework for understanding how SBR works.

The key aspect of the SBR technology is that the SBR encoder searches for the best match between the signal in the lower subbands and those in the higher subbands (indicated by the curved arrow in the Time/Freq. Coef. box in the figure), such that the high subbands can be reconstructed by transposing the low subband signals up to the high subbands. This transposition mapping is coded as SBR control data and sent over the channel. Additional control parameters are estimated in order to ensure that the high-frequency reconstruction results in a highband that is as perceptually similar as possible to the original highband. The majority of the control data is used for a spectral envelope representation. The spectral envelope information has varying time and frequency resolution such that it can control the SBR process in a perceptually relevant manner while using as small a side information rate as possible. Additionally, information on whether additional

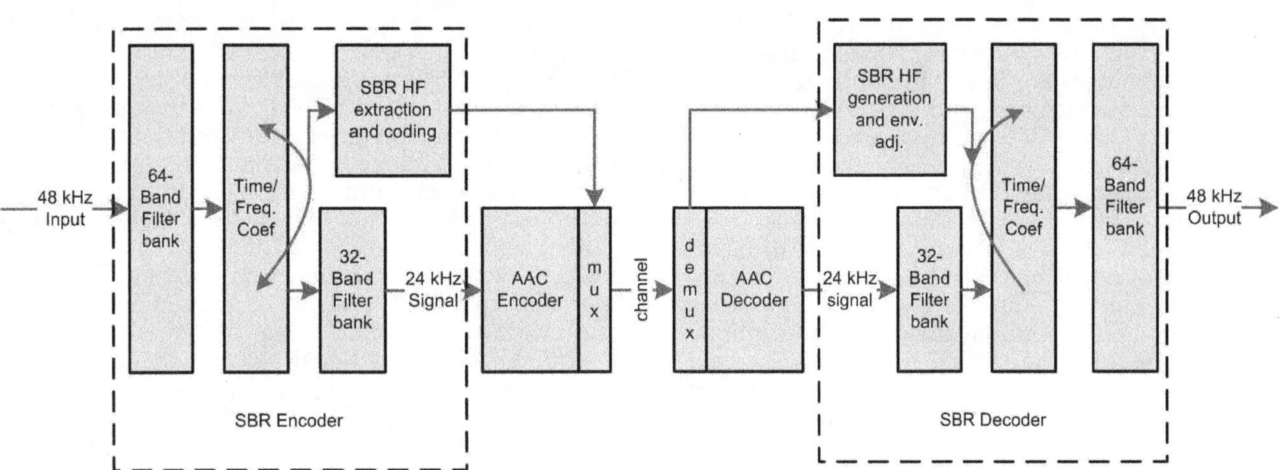

FIGURE 4.8-13 Block diagram of HE-AAC encoder and decoder

components such as noise and sinusoids are needed as part of the highband reconstruction is coded as side information. This side information is multiplexed into the AAC bit stream (in a backward-compatible way).

In the HE-AAC decoder, the bit stream is demultiplexed, the SBR side information is routed to the SBR decoder and the AAC information is decoded by the AAC decoder to obtain the half-sampling rate signal. This signal is filtered by a 32-band QMF analysis filter bank to obtain the low 32-subbands of the desired time/frequency coefficients. The SBR decoder then decodes the SBR side information and maps the lowband signal up to the high subbands, adjusts its envelope, and adds additional noise and sinusoids if needed. The final step of the decoder is to reconstruct the output block of 2048 time-domain samples using a 64-band QMF synthesis filter bank, whose inputs are the 32 low subbands resulting from processing the AAC decoder output and 32 high subbands resulting from the SBR reconstruction. This results in an up-sampling by a factor of two.

MPEG-4 HE-AAC Target Applications

The technology is suited for any application where the full audio bandwidth cannot be sufficiently well coded by a waveform coder. This makes it an excellent tool for applications such as

- Digital radio, such as Digital Radio Mondiale,
- Digital TV transmission, such as DVB, and
- Mobile music services such as streaming and music download services.

HDC

The HD codec is a subsystem of the DTS (formerly iBiquity Digital Corporation) HD Radio in-band/on-channel digital radio system. As such, all the bit rates described here refer to the bit rates made available to the codec and its associated transport and does not include other overheads associated with other parts of the HD Radio system.

HDC incorporates many techniques to maximize audio quality at lower bit rates. Three of the more important techniques are filter bank optimization, bit allocation and quantization, and spectral band replication. Filter bank optimization is used to improve encoding of transients in the audio signal (e.g., sharp attacks in castanets, triangles, drums, etc.). Distortion of attacks (abrupt transitions in level and response) create noticeable artifacts at lower bit rates. In HDC, a signal-adaptive switched filter bank, which switches between two modified discrete cosine transforms, is employed for analysis and synthesis. This technique substantially reduces the artifacts normally associated with audio attacks and allows for higher frequency resolution and transient control. Sophisticated bit allocation and quantization strategies, which rely upon nonuniform quantization and analysis-by-synthesis and entropy coding, are introduced to allow reduced bit rates and improved quality. Perhaps the single most important technique is the use of SBR, a method for highly efficient coding of high frequencies that can be used with nearly all audio codecs. By efficient encoding of the high-frequency information, the underlying perceptual codec is only responsible for accurately reproducing the lower frequencies, which results in the significant reduction of noticeable artifacts, especially at the lower bit rates.

An important feature of HDC is that it has been optimized for use in the HD Radio digital radio system. It incorporates powerful error concealment techniques for mitigating the effects of channel errors. It also offers the ability to split the bit stream into a core bit stream, which is independently decodable, and an enhanced bit stream, providing the optimal audio quality for the prevailing channel conditions and/or interference scenarios. Moreover, HDC supports all bit rates (in 8-bps increments) from 96 kbps down to 11.5 kbps.

Perceptual Coding and HDC

In the bit reduction of audio material, human perception plays a key role. This is related to the situation where source coding, a method used very successfully in speech signal compression, does not work nearly as well for music. Recent U.S. and international audio standards' work (HDTV, DAB, MPEG-1, MPEG-2, ITU-R), therefore, has centered on a class of audio compression algorithms known as perceptual coders. Rather than minimizing analytic measures of distortion, such as signal-to-noise ratio, perceptual coders attempt to minimize perceived distortion. Implicit in this approach is the idea that the signal fidelity perceived by humans is a better quality measure than the signal fidelity computed by traditional distortion measures. Perceptual coders define compact disc quality to mean perceptually indistinguishable from compact disc audio.

HDC employs source coding techniques to remove signal redundancy and perceptual coding techniques to eliminate signal irrelevancy. Furthermore, HDC employs SBR. The basic principle of SBR relies on the fact that the higher one or two octaves are psychoacoustically less relevant but still require a significant part of the overall bit consumption when coded with a waveform perceptual codec. SBR is an efficient method to code the high frequency part of the spectrum by extracting guidance information of the original high-frequency characteristic, whereas only the low-frequency information will be coded by the perceptual coder. In the decoder, the high-frequency content will be reconstructed through guided transposition of the low-frequency portion to match the original frequency response. The amount of SBR guidance information is small when compared with what is required for conventional perceptual coding. With SBR, an increase in efficiency can be achieved while maintaining the original sonic quality.

Combined, these methods yield a high compression ratio while ensuring maximum quality in the decoded signal. The result is a high-quality, high-compression-rate coding algorithm for audio signals.

HDC Structure

Figure 4.8-14 shows a more detailed block diagram of the HDC encoding algorithm, and illustrates the flow of data between the algorithmic blocks.

Enhanced AC-3

Enhanced AC-3 (E-AC-3) extends and enhances the ATSC AC-3 standard. E-AC-3 provides superior compression as compared to AC-3, is backwards compatible such that all E-AC-3 decoders can also decode AC-3 bit streams, and was designed so that E-AC-3 bit streams can be easily converted into AC-3 bit streams. This last feature provides a low-complexity means for enhanced services that use the E-AC-3 compression format to support the installed base of nearly 20 million A/V receivers containing 5.1-channel Dolby Digital (AC-3) decoders.

Conversion of bit streams from E-AC-3 format to AC-3 format utilizes a special form of transcoder that minimizes quality degradations resulting from tandem coding. This is possible due to E-AC-3 and AC-3 sharing the same input signal framing, modified discrete cosine transform and bit allocation strategies. This permits the use of a frame-synchronous partial E-AC-3 decoder and partial AC-3 encoder that are designed to minimize changes in the original quantization of mantissas, and thus provides significantly less distortion as compared to an asynchronous tandem coding, in which the signal is decoded to baseband PCM as an interim step.

An example application of this special form of tandem coding, or *transcoding*, is "fallback" audio for terrestrial broadcasting. In this application, it is desired to have a more robust simulcast of the audio program than provided for in the main channel, for example, by including a lower bit rate signal in conjunction with a more robust, but lower throughput, modem signal constellation, along with the main channel signal constellation. Whenever transmission channel conditions deteriorate such that the main signal is received with an unacceptable error rate, the receiver "falls back" to the robust, low-rate simulcast signal. If AC-3 is used for the regular audio program, and E-AC-3 is used for the fallback program, then the receiver can synchronously switch from AC-3 to E-AC-3 (since their signal

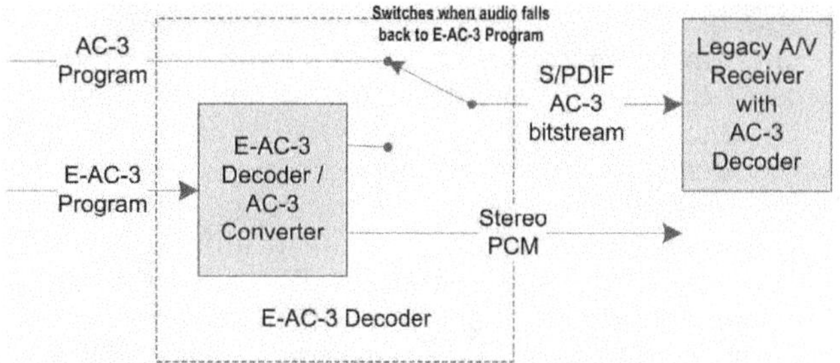

FIGURE 4.8-15 Broadcast receiver that can "fall back" from AC-3 "normal" transmission to E-AC-3 "robust" transmission

framing is identical). Also, if the receiver audio is nominally being routed (as an uncoded AC-3 bit stream) to an A/V receiver, then during the fallback mode the E-AC-3 decoder can transcode the robust E-AC-3 signal to AC-3, and hence the A/V receiver need not have an E-AC-3 decoder to accommodate the fallback mode of transmission. This is illustrated in Figure 4.8-15.

New Technology in E-AC-3

E-AC-3 incorporates a number of new coding tools that give greater coding efficiency and increased flexibility. It preserves the framing, metadata and filter bank structure found in AC-3, and adds the following:

- *Adaptive hybrid transform (AHT):* AC-3 uses a modified discrete cosine transform that produces 256 time/frequency samples. When standardized in 1996, this provided a reasonable tradeoff between audio coding gain and decoder implementation cost, in terms of computation and memory requirements. However, it is well known that a higher spectral resolution can provide increased compression for stationary signals. Therefore, E-AC-3 incorporates a second transform, a Type II discrete cosine transform (DCT). The two transforms in cascade form the AHT, which takes six sets of 256 MDCT coefficients and transforms them into a single set of 1536 time/frequency coefficients. The bit stream provides signaling such that the AHT can be adaptively enabled or disabled in response to signal statistics.

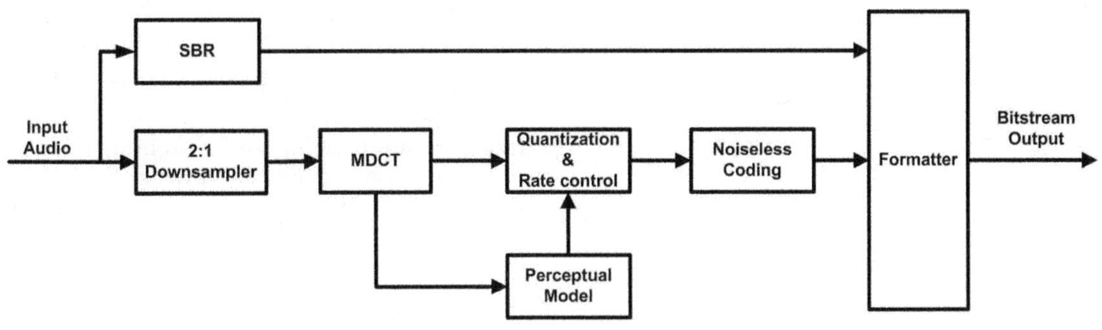

FIGURE 4.8-14 Block diagram of the basic SBR-enhanced HDC audio encoding structure

- *Spectral extension:* This coding tool synthesizes high-frequency transform coefficients without the need to actually code and transmit them, which is done by copying regions, or bands, of low-frequency coefficients to high-frequency regions. The energy envelopes of the transposed coefficients are adjusted to match that of the original spectrum. A shaped noise spectrum can optionally be added, since in audio signals high frequencies tend to be more "noise-like" than low frequencies. The spectral extension technique is schematically illustrated in Figure 4.8-16. The frequency region labeled "A" is the copy region, which is repeatedly transposed, or copied, to the synthesized high-frequency region (indicated by the arrows). Next, the high-frequency envelope is adjusted (indicated by the decreasing envelope with increasing frequency for the copied regions). Finally, a noise spectrum is added (indicated by the hatched regions).
- *Enhanced coupling:* This tool improves the imaging properties of coupled channels by adding phase compensation and decorrelation to the amplitude-envelope processing of conventional coupling. Prior to downmixing the coupled channels to a single composite signal, the encoder derives both amplitude and phase information for each coupled channel. Phase information is obtained by using a modified discrete sine transform in conjunction with the existing MDCT. The decoder uses the information to recover the coupled signals from the composite signal using a combination of amplitude scaling, phase rotation and decorrelation. The result is an improvement in soundstage imaging as compared to conventional coupling, which allows the technique to be used at lower frequencies than conventional coupling, thus improving coding efficiency

  Because it may be desirable in some circumstances to use channel coupling for a mid-range portion of the frequency spectrum and spectral extension for the higher-range portion of the frequency spectrum, spectral extension is fully compatible with channel coupling, with both tools being enabled at the same time, but for different portions of the frequency spectrum.
- *Transient pre-noise processing* : Perceptual coders must balance the requirements of high-frequency resolution for maximum redundancy removal during stationary signal intervals against high-time resolution for minimum errors in irrelevancy reduction during nonstationary signal intervals. E-AC-3 can disable the AHT and use the 256-length window "short-block" mode of the AC-3 transform during signal nonstationarity. However, at low bit rates even this level of time resolution may not be sufficient to guard against transient "pre-echo" coding artifacts.

  The transient pre-noise tool detects this condition and, for each audio channel, replaces the noise with a portion of the signal from the prior transform block. This replacement is done in the time domain by extending the signal from the prior transform block into the block with the transient up to the point of the onset of the transient. The extension is done using aspects of time-scale modification, and the exact channels and signal intervals that receive the pre-noise processing is determined using aspects of auditory scene analysis.
- *Increased number of audio channels and increased bit rate range:* Encoding of up to 13.1-channel audio programs is supported, which easily encompasses the 6.1- or 7.1-channel audio programs that are envisioned today. The bit stream can range in rate from 32 kbps to 6.144 Mbps. The higher rate supports not only the larger number of audio channels, but also a higher rate per channel, as is beneficial in applications that serve legacy equipment by transcoding from E-AC-3 to AC-3.

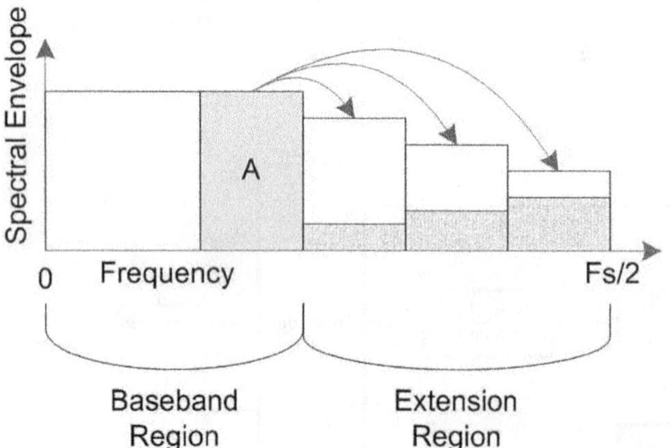

FIGURE 4.8-16 Illustration of spectral extension in E-AC-3

Next Generation Audio (NGA)

Next generation video codecs will support video resolutions beyond high definition (HD, typically 1920 × 1080 pixels), and such high-resolution displays typically will be larger in size and so give viewers a more immersive viewing experience. Next generation audio codecs are being designed to give the listener a similarly more immersive listening experience. This is done by moving beyond 5.1 loudspeaker layouts to add loudspeakers above and below the listener's head to create a more enveloping listening experience. Immersive layouts can range from a relatively straightforward extension of 5.1 to add two front high speakers to the 22.2-channel loudspeaker layout pioneered by NHK in Japan. However, an immersive format that promises to be widely adopted is 7.1 speakers at the listener's ear level plus four elevated speakers, two front and two back. Immersive loudspeaker layouts are often designated as x.y.z, where x is number of loudspeakers in the mid (ear level) plane, y is number of low frequency effects speakers and z is number of speakers above the mid plane.

While immersive audio presentations are typically related to loudspeaker layouts in a home TV setting, NGA systems also support mobile and handheld platforms. These may be personal tablet TVs with frame-mounted loudspeakers or smartphones with an NGA binauralization engine that permits spatially rich audio programs to be presented via headphones.

NGA Content

In addition to more immersive audio presentation, NGA is characterized by support for more flexible audio content formats. Current generation audio codecs almost exclusively process channel-based audio content. NGA codecs will support audio content formats that are a combination of channels, objects and scenes:

- *Channels*: The conventional method of presenting an audio program, in which the production engineers have (more or less) exact knowledge of the locations of the reproduction loudspeakers. NGA sometimes refers to this as a channel bed, as it can be the complete audio program or a "foundation" to which audio objects are added.
- *Objects*: The notion of *sound objects* with associated (possibly dynamic) positional metadata is a very important part of NGA. Objects can give compression efficiency (e.g., dialog in several different languages can be transmitted as sound objects, with one selected object mixed onto a single transmitted channel bed), or support user interactivity (e.g., adjust dialog level versus background level).
- *Scenes*: Some NGA codecs are able to compress and reproduce *higher order ambisonic* (HOA) signals. These represent a 3D sound pressure field as a set of "coefficient signals" that describe the sound field at all angles and distances. Such a representation can be recorded using a single multi-capsule microphone or a set of spot microphones. The format supports simple rendering to any loudspeaker layout.

NGA Metadata

Whereas current codecs that compress stereo or 5.1-channel signals may require minimal metadata other than a channel mapping, NGA supports rich metadata that supports very flexible presentation of the audio program. As already mentioned, metadata may relate to the dynamic position of sound objects. However, metadata can also be associated with

- *Program loudness control and compliance:* Loudness processing and control is mandated in many regions of the world and metadata is the key to compliant program loudness control. Metadata can carry program loudness for the production playout format and also for downmix formats, e.g., the pervasive stereo presentation.
- *Dynamic range control (DRC):* Users can consume NGA programs in various scenarios, as different as a home theatre setting and a portable device. This may require different audio dynamic range to enjoy the program given constraints of, for example, background noise. Furthermore, even a home setting may require constrained dynamic range (e.g., "night mode") for programs produced with dramatic sonic effects.
- *Language selection:* Languages can be represented as multiple dialog objects. Metadata identifies the objects and languages such that the user can easily select the desired language. Alternate languages can be broadcast in a program multiplex or delivered via alternate means, e.g., an Internet connection. Such metadata and user interaction is not limited to language selection—it can encompass accessibility functions such as video description, or personalization functions such as selecting the home team or away team announcer in a sports program.
- *Dialog level enhancement:* Dialog captured as audio objects can be easily manipulated by the user, for example, to boost the dialog level relative to the program background level. Metadata in the bit stream identifies the dialog objects and metadata at a user-to-decoder interface specifies an amount of boost.

The ATSC has specified two NGA systems in its ATSC 3.0 family of specifications (each has *Candidate Standard* status at the time of this writing). As important examples of NGA technology, they are briefly described in the following sections. Both standards provide advanced metadata functionality as outlined above. It is expected that each of these Candidate Standard NGA systems will achieve "Excellent" subjective audio performance for immersive audio content, e.g., 7.1.4 at 384 kbps, as measured using the BS.1534-3 MUSHRA test methodology.

MPEG-H 3D Audio

MPEG-H 3D Audio can support any combination of the three content formats mentioned above: channels, objects and HOA. This permits great flexibility, in terms of near-term market adoption (with channel-based content), the user personalization that channels and objects provide, and the compelling scene-based capture and rendering that HOA provides. In particular, HOA holds great promise for capture of immersive sound in live recordings such as sports venues and for playback of virtual reality experiences.

MPEG-H 3D Audio specifies both a compression engine and a rendering engine, and so provides a fully specified path from bit stream input to rendered audio output. The MPEG-H 3D Audio codec is standardized as ISO/IEC 23008-3. The standard and its associated amendments specify the 3D Audio technology and the 3D Audio Low Complexity Profile of the technology, which is particularly suited to broadcast applications. A block diagram of a MPEG-H 3D Audio Low Complexity Profile decoder is shown below in Figure 4.8-17.

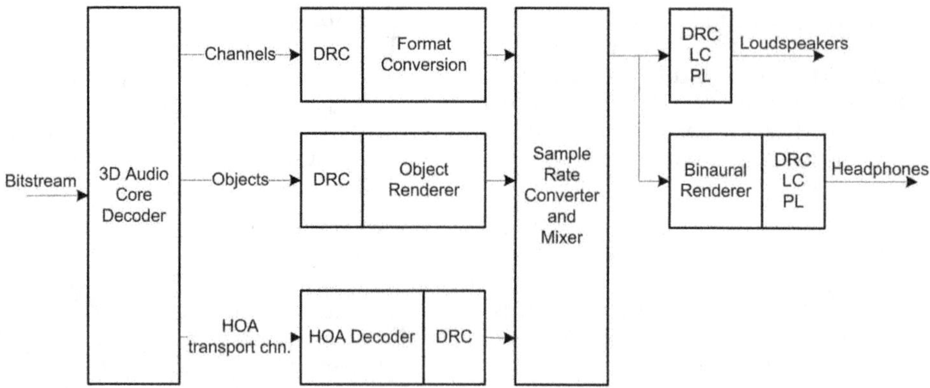

FIGURE 4.8-17 Block diagram of MPEG-H 3D Audio Low Complexity Profile decoder, showing main processing blocks and dynamic range control (DRC), loudness control (LC) and peak limiter (PL) blocks

The 3D Audio system has a core audio coder that extends and enhances the MPEG Unified Speech and Audio Coding (USAC) technology. This coder was specifically designed to provide equally good performance for all types of input content, especially the important content types of speech and music. 3D Audio core audio coder has a dynamically selectable set of coding tools, the most important of which are described here:

- *Modified discrete cosine transform*: As used in MPEG-AAC. In fact, USAC incorporates virtually all of the AAC coding tools, but replaces Huffman entropy coding with the more efficient arithmetic coding.
- *Linear prediction tools*: For somewhat speech-like signals, linear prediction is used to model the spectral envelope, and the MDCT is used to code the residual signal. The AAC toolbox is also used in this mode. For very speech-like signals, a long-term linear prediction filter can be used to model the human speech production mechanism, which is coupled with adaptive and fixed excitation codebooks such that the 3D core coder achieves the performance of state-of-the-art wideband speech codecs.
- *Multichannel-coding tool (MCT):* This extends the MDCT domain USAC stereo coding tools to apply to multichannel signals. It exploits inter-channel correlations in a flexible, signal adaptive way.
- *Intelligent gap filling (IGF):* Applies parametric coding techniques to fill spectral holes introduced by the quantizer in the encoder due to bit rate constraints. IGF is utilized as both a high bit rate noise-filling technique and a low bit rate, parametric high-frequency signal reproduction tool.

MPEG-H 3D Audio provides for fully specified, normative and very flexible rendering of the decoded channel/object/HOA signals from the core audio decoder to the output loudspeakers. As playback systems move from stereo to immersive, it becomes more likely that the consumer's loudspeaker layout is not an exact match to the layout that was assumed during production. It could be that the user has a layout that is different from the production layout (e.g., 5.1.2 as opposed to 7.1.4), or (perhaps more commonly) the user may have the right number of loudspeakers, but have them in the wrong locations. Flexible rendering is needed to accommodate both of these issues. MPEG-H 3D Audio incorporates flexible rendering engines for each content type:

- *Channels*: If the transmitted number of channels is greater than that in the user's actual loudspeaker layout, a *format converter* selects downmix matrices that best match production speaker layout to rendering speaker layout. Furthermore, the matrices are not static, but rather the downmix is actively monitored and adaptively modified to avoid artifacts, such as signal cancellation or comb filtering, that can occur when mixing partially or largely correlated signals.
- *Objects*: Objects have associated metadata that specifies the object's possibly dynamic location in the sound stage. The object render uses vector base amplitude panning to render objects to a given location, which automatically identifies the appropriate triangle of output loudspeakers within which the object should be located.
- *HOA*: The 3D Audio core coder transports a coded form of HOA coefficients. The coded coefficients represents HOA signals that are decomposed into *ambient sounds*, without a strongly directional component, and *predominant sounds* that have a strongly directional component. This coding permits a significant reduction in the number of transport channel as compared to the original HOA signal. These coded transport channel signals must be decoded to recover the HOA coefficient signals. At this point, the HOA signal can be rendered to the target loudspeaker layout via a simple matrix multiply.

In addition, MPEG-H 3D Audio specifies an engine to binauralize the audio program for presentation on headphones, which is particularly important for mobile use cases. Such rendering maintains the spatial detail and envelopment of the audio program. In addition, a decoder-side interface is defined such that the user can load a custom, user-specific head-related transfer function (HRTF) to obtain a more personalized and realistic audio presentation.

AC-4

Like all NGA systems, AC-4 incorporates features that go beyond traditional audio coding algorithms. As mentioned above in the overview of NGA, AC-4 supports immersive audio programs, personalized presentations, advanced loudness management and interactive dialog enhancement. In addition, the structure of AC-4 provides for video-frame-synchronous coding.

AC-4 is standardized as ETSI TS 103 190. It is capable of processing any combination of channel-based and object-based content. The decoder architecture has the following elements:

- *MDCT and audio spectral front-end*: The input audio signal is processed by an MDCT using one of five different window lengths. This permits an optimal tradeoff between coding gain and control of temporal artifacts. A bit reservoir permits a variable allocation of bits per coding frame.
- *MDCT and speech spectral front-end*: Similar to the audio front-end, the input audio signal is processed by an MDCT. However, the input block size is reduced to approximately 5 ms, permitting the transform coefficients to capture the transient details typical in speech signals. Aspects of the spectral envelope, which is associated with speech formants, are coded and a predictor is used in the MDCT coefficient domain to provide coding gain associated with speech pitch production.
- *Pseudo-QMF domain processing*: This tool provides an oversampled, complex-valued time/frequency representation of the input signal as processed by either the audio or speech front-ends. It supports advanced spectral extension, which is similar to MPEG-4 SBR, but provides for a very flexible combination of waveform-encoded spectral regions and parametrically encoded SBR regions. In addition, the QMF domain processing permits use of the companding tool. In the encoder, QMF-domain signals are compressed to reduce their dynamic range. After quantization, coding, transmission and decoding, the signal is expanded in the decoder to restore it to its original dynamic range. At the same time, quantization noise is also expanded such that higher noise levels are associated with higher signal levels, thus providing a noise shaping that follows the signal's temporal envelope.
- *Advanced joint channel coding*: This tool represents an immersive audio signal as a five-channel downmix plus very low bit rate side information to guide an upmix. In the encoder, the downmix is dynamic and signal-adaptive, and in the decoder, the upmix can use both "dry" and "wet" parameters to control the level of decorrelation in the upmix.
- *Advanced joint object coding*: Objects play an important role in NGA, and this tool permits objects to be conveyed in a downmix with associated side information to guide in reconstructing the original audio objects. Side information primarily indicates the time- and frequency-varying elements of the downmix to be used in creating an upmix to reconstruct the component object signals.

In AC-4, rendering of channel-based output signals is required when the production loudspeaker configuration does not match the user's presentation loudspeaker configuration. Rendering could be done via either downmixing or upmixing to achieve a match of production and consumption loudspeaker configurations. The general form of such mixing matrices is specified, and can be modified via parameters transmitted in the bit stream. Alternatively, an external rendering engine could be used. For object-based signals, rendering is always required, and it is envisioned that this will be done by an external rendering engine. If it is desired to render an immersive program for presentation via headphones (e.g., using HRTF), it is also expected that an external rendering engine will be used.

QUALITY MEASUREMENTS

Perceptual audio coding has revolutionized the processing and distribution of digital audio signals. However, one aspect of this technology that is rarely discussed is how difficult it is to determine the quality of perceptually coded signals. Audio professionals could benefit from a simple yet accurate method for signal characterization, in that it would provide a reliable means to check the audio quality within a given facility.

Most often quality assessment involves subjective evaluations of audio quality, in which groups of listeners compare reference audio material to coded audio material and then judge the level of impairment caused by the coding process. It is widely acknowledged that whenever an audio processing system involves perceptual audio coding, subjective evaluation is the most reliable means of assessing audio quality. However, such a procedure is both time-consuming and costly. An alternative is to use fully objective assessment methods. Although these are not considered as accurate as subjective methods (when perceptual coding is involved), they are fast and inexpensive, and can be performed on a personal computer in real time. Both subjective and objective assessment methods will be reviewed in the following sections.

Subjective Quality Measurements

In subjective quality measurement, human judgment determines the audio quality rating. Typically, a set of relevant audio test items (informally referred to as *audio clips* or *audio cuts*) is identified, the systems to be evaluated process the test items, and a set of subjects listen to each processed item and judge its audio quality. For a given audio processing system, the average score of all listeners is the score given to that system. This is intuitively satisfying, in that humans, after all, are the ultimate consumers of audio programs. However, humans demonstrate both "noise" in their judgments and also personal differences in taste, which lead to random and systematic differences in listener judgments, respectively. Accurate assessment is obtained only by using a relatively large number of subjects, such that these differences are reduced by averaging over the set of subject responses.

A procedure for the subjective assessment of very-high-quality systems has been standardized in ITU-R Recommendation BS.1116-1, "Methods for the Subjective Assessment of Small Impairments in Audio Systems Including Multichannel Sound." This standard specifies the test environment, the test procedure and the analysis process. Concerning environment, it is paramount that the tests be conducted in an acoustically conditioned listening room or sound booth. The listening space must have little to no background noise or reverberation, so that potential audio processing distortions are not masked by the listening environment itself. It also specifies the configuration of loudspeakers and the position of the listener.

The procedure requires that the listener hear the unprocessed audio ("reference") followed by two additional audio samples ("A" and "B"), one of which is the reference and one of which is processed. For this reason, it is also called the "triple-stimulus hidden reference" method. Using this method of presentation, both the processed and reference items are in the listener's short-term aural memory, permitting the assessment of very small differences in the processed item relative to the reference item. The listener grades both A and B, and one must receive a score of 5.0 indicating that it is judged to be the hidden reference. In order to eliminate any systematic errors in assessment, the location of the hidden reference (A or B) and the order of presentation of test items is randomized, preferably with different randomizations for each listener. Any difference between the reference and the processed audio is to be regarded as a distortion or impairment, which listeners rate using the following ITU-R five-grade impairment scale shown in Figure 4.8-18.

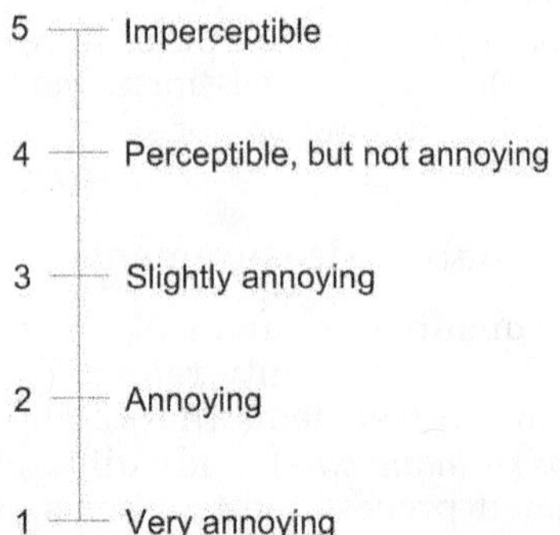

FIGURE 4.8-18 ITU-R five-grade impairment scale

The results are analyzed by computing the sample mean and 95% confidence interval of the mean score associated with a given processing system as pooled, for example, over all listeners or all listeners and all test items. When presenting the results of the assessment *diffgrades* are used, which are equal to 5.0 minus the average subjective score for a coding system, so that the diffgrade is always zero or negative. Systems with scores that approach zero are of very high quality. Figure 4.8-19 shows an example of a subjective test result using the BS.1116 methodology. The 95% confidence interval is indicated by the vertical stroke with end caps, and the sample mean is indicated by the marker graphic.

In many broadcast satellite or terrestrial digital radio applications, the target bit rate for delivery of audio programs is such that near-transparent audio quality is not possible. A procedure for the subjective assessment of intermediate quality systems has been standardized in ITU-R Recommendation BS.1534-1 (3/2001), "Method for the Subjective Assessment of Intermediate Audio Quality." It is also referred to as Multi-Stimulus with Hidden Reference and Anchor (MUSHRA). This relies on BS-1116 for the specification of acoustic environment and other aspects of the testing procedure. As with BS-1116, any difference between the reference and the processed audio is regarded as a distortion or degradation, which listeners rate using the 100-point scale shown in Figure 4.8-20.

The procedure requires that, for each test item, the subject be able to listen to the unprocessed test

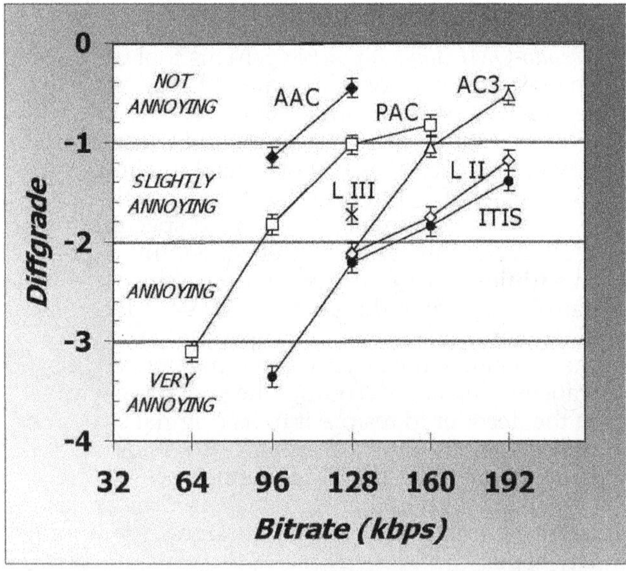

FIGURE 4.8-19 Example of BS.1116 Subjective Assessment (data taken from March 1998 *Journal of the Audio Engineering Society*). The processed audio material is stereo. In the graph, AAC is MPEG-2 Advanced Audio Coding; PAC is Lucent Perceptual Audio Coding; AC3 is Dolby AC-3; L II and L III are software implementations of MPEG-1 Layer 2 and 3, respectively; and ITIS is a hardware implementation of MPEG-1 Layer II.

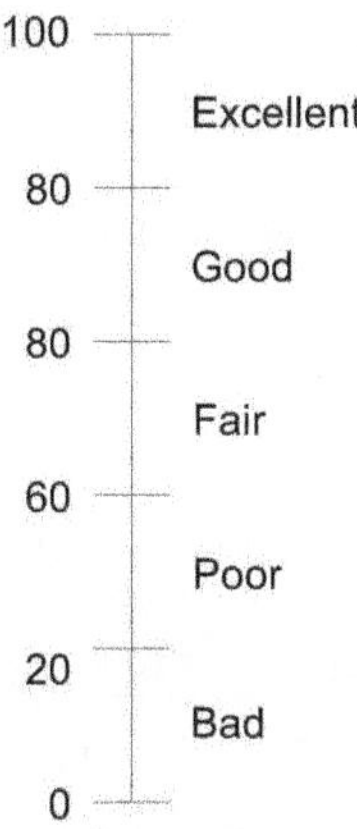

FIGURE 4.8-20 100-point scale used in MUSRA testing

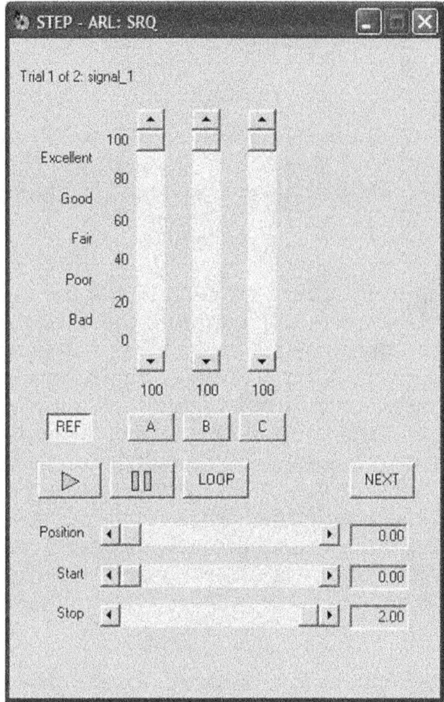

FIGURE 4.8-21 Example of BS-1534 MUSHRA graphical user interface (from Audio Research Labs' software subjective evaluation tool STEP)

item (the "open reference") or any of the processed test items and to grade all processed items before proceeding to the next item. The test subject can listen to the reference and processed items in any order and as often as desired in the course of making the evaluations. This is only possible because of an innovation in the MUSHRA methodology, which recommends that computer-controlled presentation and scoring be used. An example of a MUSHRA graphical user interface is shown in Figure 4.8-21 (from the Audio Research Labs' subjective quality assessment software tool STEP).

The systems to be evaluated using the MUSHRA procedure are often a considerable distance from the open reference, in terms of points on the grading scale. While the open reference represents the high end of the grading scale (i.e., transparency), a second, lower-quality reference called the *low anchor* is mandated in the procedure to give a point of reference at the low end of the scale. Specifically, a 3.5-kHz and a 7.0-kHz low-pass filtered version of the reference are used as low anchors. Additionally, a hidden reference is included, which is often used as a means to test listener reliability.

For each test item, the listener must rate all of the processed items, which are denoted by letters (A, B, C, etc.) underneath the vertical scroll bars, while the reference is indicated by the button REF to the left of the letter buttons. The scroll bars are used to designate the score for each item. For the example test interface shown in Figure 4.8-21, clicking on any of the buttons instantly switches the audio to that processed item. In this way, the listener can repeatedly compare and contrast the processed items as a means to converge on a final set of scores. When the listener is satisfied with the scoring, the NEXT button is pressed to proceed to the next trial.

Figure 4.8-22 shows an example of a subjective test result using the MUSHRA methodology. The graph in this figure shows results aggregated over all listeners and all audio items in the test. The 95% confidence interval (a statistical measure of reliability) is indicated by the vertical stroke, and the sample mean (i.e., the average value of scores for a system) is indicated by the horizontal tick mark.

Quality measurements made with properly executed subjective evaluations are widely accepted and have been used for a variety of purposes, from determining which of a group of perceptual coders performs best to assessing the overall performance of an audio broadcasting system.

Objective Quality Measurements

Traditional objective measures of audio performance, such as signal-to-noise ratio or total harmonic distortion, are not reliable measures of the audio quality delivered by a perceptual audio coder. To remedy this situation, the ITU-R developed Recommendation BS.1387-1, "Method for Objective Measurement of Perceived Audio Quality" (PEAQ). The method requires feeding the original audio signal and the processed audio signal into the PEAQ measurement device. For each block of the signals, PEAQ calculates the degree to which noise in the processed signal is above a masking threshold derived from the original signal and accumulates parameters based on that impairment over the duration of the signal. From the impairment parameters, an estimated total signal quality is computed.

PEAQ can be realized in software or hardware. It offers a basic version, which is designed to support real-time implementations, and an advanced version optimized for the highest accuracy but not necessarily in real time. The primary applications for PEAQ are summarized in Table 4.8-2.

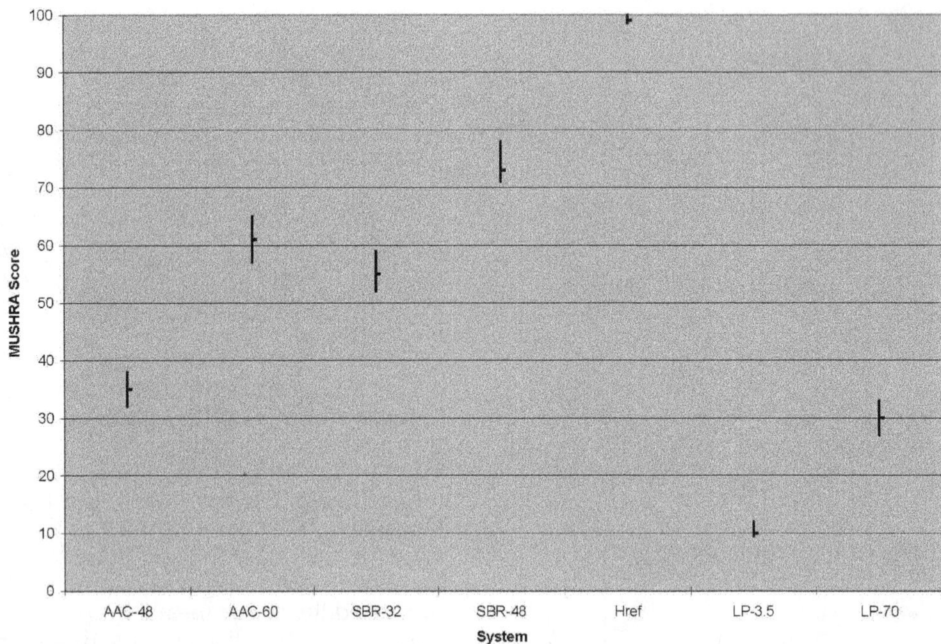

FIGURE 4.8-22 Example MUSHRA subjective assessment. The processed audio material is stereo, and in the graph Sys1_32s is at 32 kbps, Sys1_48 is at 48 kbps, Sys1_64 is at 64 kbps and Sys1_96 is at 96 kbps; Href is the hidden reference; LP-35 and LP-70 are 3.5 kHz and 7.0 kHz low-pass filtered versions of the reference, respectively.

TABLE 4.8-2
Target Applications for ITU-R Rec. BS.1116 PEAQ

Category	Application	Version
Development	Network planning	Both
	Codec development	Advanced
Diagnostic	Assessment of implementations	Both
	Equipment or connection status	Advanced
	Codec identification	Both
Operational	Codec operating mode	Basic
	Online monitoring	Basic

SUMMARY

In designing an audio processing chain, a balance must be struck between the degree of compression selected and the level of distortion that can be tolerated. This is true whether the processing chain is the result of a single coding pass or the result of a number of passes, such as in a complex audio production and delivery system. There have been many outstanding successes for digital audio data compression in communications and storage, and as long as the limitations of the various compression systems are fully understood, the number of successful implementations will continue to grow in number.

Looking to the future, it is clear that the processing power of DSP chips will continue to increase, even as their price continues to drop. This will permit ever more complex audio compression algorithms to be considered for practical implementations. However, it is expected that the value that might be provided to the user by "more channels" (as a result of greater compression) will be complemented by the value provided by new delivery paradigms (such as channels and objects) with rich metadata that supports a new and compelling level of user customization and user interactivity.

Notes

[1] AES/EBU link refers to a standard digital audio interface commonly used in broadcast audio equipment. For additional information see "AES-3id-2001: AES information document for digital audio engineering—Transmission of AES3 formatted data by unbalanced coaxial cable," Audio Engineering Society, www.aes.org.

[2] In this context, "stationary" refers to a random process (in this case, an audio signal) where its statistical properties do not vary with time.

[3] The HD Radio system was standardized in 2005 by the National Radio Systems Committee as NRSC-5-A, "In-band/on-channel Digital Radio Broadcasting Standard," however the audio codec is not specified in this standard.

[4] MPEG-1 Layer 1 is not discussed separately as it is largely obsolete and superseded in use by MPEG-1 Layer 2 or Layer 3.

[5] See ATSC Standard A/52B.

Bibliography

[1] ATSC Standard A/52:2015: *Digital Audio Compression (AC-3, E-AC-3)*. Advanced Television Systems Committee, Washington, DC, Doc. November 24, 2015.

[2] ATSC Candidate Standard A/342: *Part 1: Audio Common Elements; Part 2: AC-4 System; Part 3: MPEG-H System*. Advanced Television Systems Committee, Washington, DC, Doc. May 3, 2016.

[3] Bosi, M., Brandenberg, K., Quackenbush, S., Fielder, L., Akagiri, H., Fuchs, H., Dietz, M., Herre, J., Davidson, G., and Oikawa, Y., "ISO/IEC MPEG-2 Advanced Audio Coding," *J. Audio Eng. Soc.*, Vol. 45, No. 10, October 1997, pp. 789–814.

[4] Brandenburg, K., and Gerhard Stoll, "ISO-MPEG-1 Audio: A Generic Standard for Coding of High Quality Digital Audio," 92nd AES Convention Proceedings, Audio Engineering Society, New York, NY, 1992, revised 1994.

[5] Brandenburg, K., "Introduction to perceptual coding," in *Collected Papers on Digital Audio Bit-Rate Reduction*, N. Gilchrist and C. Grewin, Eds., New York: Audio Engineering Society, 1996, pp. 23–30.

[6] Dietz, M., Liljeryd, L., Kjörling, K., and Kunz, O., "Spectral Band Replication, a Novel Approach in Audio Coding," 112th AES Convention Proceedings, Audio Engineering Society, Munich, Germany, May 10–13, 2002. Preprint 5553.

[7] ETSI TS 103 190 (2014-04), "Digital Audio Compression (AC-4) Standard."

[8] Fielder, Louis D., Andersen, Robert L., Crockett, Brett G., Davidson, Grant A., Davis, Mark F., Turner, Stephen C., Vinton, Mark S., and Williams, Phillip A., "Introduction to Dolby Digital Plus, an Enhancement to the Dolby Digital Coding System," 117th AES Convention Proceedings, Audio Engineering Society, San Francisco, CA, October 28–31, 2004. Preprint 6196.

[9] Herre, Jürgen, Johannes Hilpert, Achim Kuntz, and Jan Plogsties, "MPEG-H Audio—The New Standard for Universal Spatial/3D Audio Coding," *J. Audio Eng. Soc.*, Vol. 62, No. 12, December 2014, pp. 821–830.

[10] ISO/IEC JTC1/SC29/WG11 (MPEG), "Information technology—Coding of moving pictures and associated audio for digital storage media at up to about 1.5 Mbit/s—Part 3: Audio," 11172-3:1993 (MPEG-1 Audio).

[11] ISO/IEC JTC1/SC29/WG11 (MPEG), "Information technology—Generic coding of moving pictures and associated audio information—Part 7: Advanced Audio Coding (AAC)," 13818-7:2006 (MPEG-2 AAC).

[12] ISO/IEC JTC1/SC29/WG11 (MPEG), "Information technology—Coding of audio-visual objects—Part 3: Audio," 14496-3:2006 (MPEG-4 Audio).

[13] ISO/IEC JTC1/SC29/WG11 (MPEG), "Information technology—MPEG audio technologies—Part 3: Unified speech and audio coding," 23003-3:2012.

[14] ISO/IEC JTC1/SC29/WG11 (MPEG), "Information technology—High efficiency coding and media delivery—Part 3: 3D audio," 23008-3:2015.

[15] ITU-R Recommendation BS.1116-3 (02/2015), Methods for the subjective assessment of small impairments in audio systems.

[16] ITU-R Recommendation BS.1387-1 (11/2001), "Method for objective measurements of perceived audio quality."

[17] ITU-R Recommendation BS.1534-3 (10/2015), "Method for the subjective assessment of intermediate quality levels of coding systems."

[18] Jayant, N. S., and Noll, P., *Digital Coding of Waveforms*. Englewood Cliffs, NJ: Prentice Hall, 1984.

[19] Kjörling, K., et al., "AC-4—The Next Generation Audio Codec," 140th AES Convention, Paris, France, June 2016. Preprint 9491.

[20] Moore, B. C. J., "Masking in the human auditory system," in *Collected Papers on Digital Audio Bit-Rate Reduction*, N. Gilchrist and C. Grewin, Eds., New York: Audio Engineering Society, 1996, pp. 9–19.

[21] National Radio Systems Committee, *NRSC-5-A In-Band/On-Channel Digital Radio Broadcasting Standard*, Arlington, VA: Consumer Electronics Association and Washington, DC: National Association of Broadcasters, September 2005.

[22] Neuendorf, M., et al., "The ISO/MPEG Unified Speech and Audio Coding Standard—Consistent High Quality for All Content Types and at All Bit Rates," *J. Audio Eng. Soc.*, Vol. 61, No. 12, December 2013, pp. 956–977.

[23] Painter, T., and Spanias, A., "Perceptual Coding of Digital Audio," *Proc. of the IEEE*, Vol. 88, No. 4, April 2000, pp. 451–513.

[23] Todd, C., Davidson, G. A., Davis, M. F., Fielder, L. D., Link, B. D., and Vernon, S., "AC-3: Flexible Perceptual Coding for Audio Transmission and Storage," 96th AES Convention Proceedings, Audio Engineering Society, NY, February 1994. Preprint 3796.

[24] Zwicker, E., and Fastl, H., *Psychoacoustics Facts and Models*. Berlin, Germany: Springer-Ferlag, 1990.

SECTION

5

VIDEO PRODUCTION AND TV STUDIO SYSTEMS

CHAPTER

5.1

Principles of Light, Vision, and Photometry

JERRY WHITAKER

Advanced Television Systems Committee
Washington, D.C.

Based on text by W. Lyle Brewer, Robert A. Morris, Donald G. Fink[1] and Alan R. Robertson, Joseph F. Fisher, and Jerry C. Whitaker[2]

INTRODUCTION

Television images originate as light coming from a scene being viewed by a camera. The television system ends with the presentation of those images on a display, which generates light that is picked up by the eye and perceived by the human brain as a representation of the original scene. The key to the whole process is the science of light and how the human visual system detects light and processes the signals that it receives. This chapter introduces some of the scientific principles of light, vision, and photometry that make television possible.

PRINCIPLES OF LIGHT

Sources of Illumination

Light reaches an observer directly from a light source or after being reflected from some object. The original source of such energy typically is radiation from molecules or atoms resulting from internal (atomic) changes. The exact type of emission is determined by:

- The manner in which the atoms or molecules are supplied with energy to replace that which they radiate
- The physical state of the substance, whether solid, liquid, or gaseous

The most common source of radiant energy is the thermal excitation of atoms in the solid or gaseous state, although light is also generated by other mechanisms.

The Spectrum

The electromagnetic spectrum comprises radiation of a wide range of wavelengths. As shown in Figure 5.1-1, light constitutes a small section in the range of electromagnetic radiation, extending in wavelength from about 400 to 700 nanometers (nm) or billionths (10^{-9}) of a meter. The wavelength of the light determines its color. When a beam of light traveling in air falls upon a glass surface at an angle, it is *refracted*, or bent. The amount of refraction depends on the wavelength, its variation with wavelength being known as *dispersion*. Similarly, when the beam, traveling in glass, emerges into air, it is refracted again (with dispersion). A glass prism provides a refracting system of this type. Because different wavelengths are refracted by different amounts, an incident white beam is split up into a number of beams corresponding to the many wavelengths contained in the composite white beam—thus is obtained the spectrum of many colors. If a spectrum is allowed to fall upon a narrow slit arranged parallel to the edge of the prism, a narrow band of wavelengths passes through the slit. Obviously, the narrower the slit, the narrower the band of wavelengths, or the sharper the spectral line. Also, more dispersion in the prism will cause a wider spectrum to be produced, and a narrower spectral line will be obtained for a given slit width. Note that purples are not included in the *spectral colors*. The purples belong to a special class of colors; they can be produced by mixing the light from two spectral lines, one in the red end of the spectrum, the other in the blue end. Purple (magenta is a more scientific name) is therefore referred to as a *nonspectral color*.

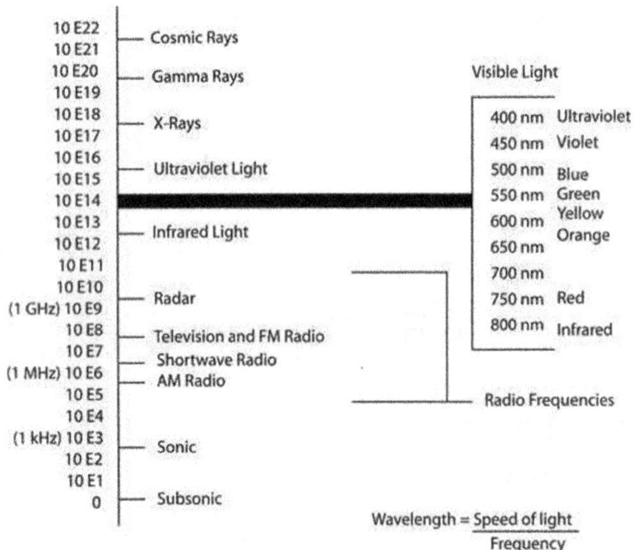

FIGURE 5.1-1 The electromagnetic spectrum.

Spectral Distribution

A plot of the power distribution of a source of light is indicative of the watts radiated at each wavelength per nanometer of wavelength. It is common to refer to such a graph as an *energy distribution curve*. Individual narrow bands of wavelengths of light are seen as strongly colored elements. Increasingly broader bandwidths retain the appearance of color but with decreasing purity, as if white light had been added to them.

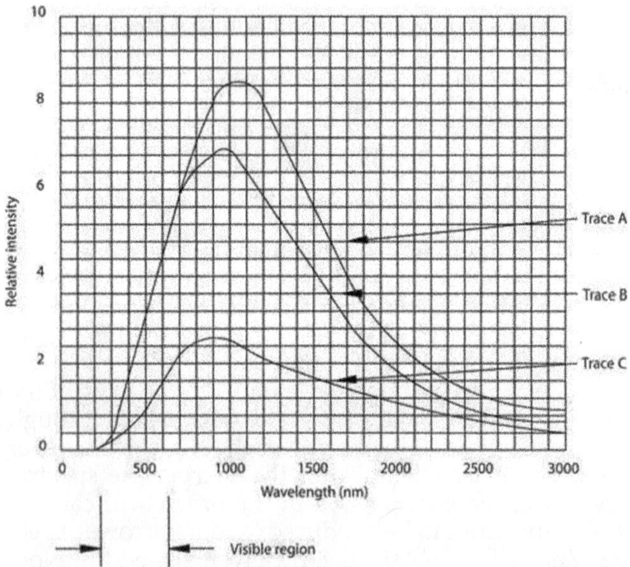

FIGURE 5.1-2 The radiating characteristics of tungsten: (Trace A) Radiant flux from 1 cm^2 of a blackbody at 3000°K. (Trace B) Radiant flux from 1 cm^2 of tungsten at 3000°K. (Trace C) Radiant flux from 2.27 cm^2 of tungsten at 3000°K (equal to curve A in the visible region).

Adapted from *IES Lighting Handbook*, Illuminating Engineering Society of North America, New York, 1981

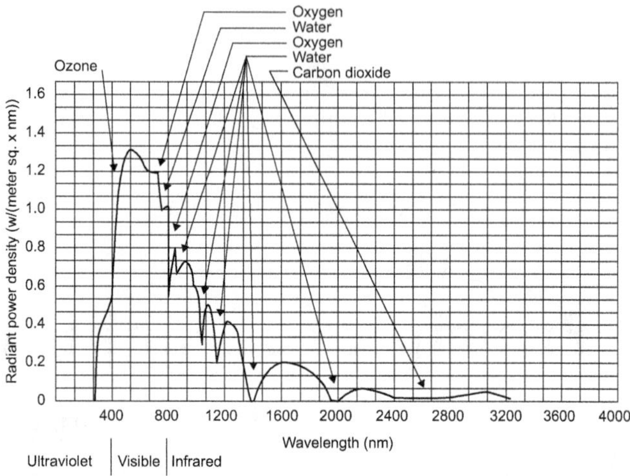

FIGURE 5.1-3 Spectral distribution of solar radiant power density at sea level, showing the ozone, oxygen, and carbon dioxide absorption bands.

Adapted from *IES Lighting Handbook*, Illuminating Engineering Society of North America, New York, 1981

A very broad band extending generally throughout the visible spectrum is perceived as white light. Many white light sources are of this type, such as the familiar tungsten-filament electric light bulb for which the radiating characteristics are shown in Figure 5.1-2. Daylight also has a broad band of radiation, as illustrated in Figure 5.1-3. The energy distributions shown in Figures 5.1-2 and 5.1-3 are quite different and, if the corresponding sets of radiation were seen side by side, would be different in appearance. The light bulb would appear to have a yellow-orange tint, while the daylight would have a bluish tint. Either one, particularly if seen alone, however, would represent a very acceptable white. A sensation of white light can also be induced by light sources that do not have a uniform energy distribution. Among these is fluorescent lighting, which exhibits sharp peaks of energy through the visible spectrum. Similarly, the light from a monochrome (black-and-white) video cathode ray tube is not uniform within the visible spectrum, generally exhibiting peaks in the yellow and blue regions of the spectrum, as shown in Figure 5.1-4, yet it appears as an acceptable white.

Additive Color Mixing

It can easily be demonstrated using light sources of different colors that two or more colors can be added together to produce a mixture of light, which is perceived as a new color. Using the three colors of red, green, and blue, nearly all other colors can be reproduced. As shown in Figure 5.1-5 (when reproduced in color), mixing red and blue together produces magenta; red and green produce yellow; and blue and green produce cyan. Furthermore, when the correct proportions of red, green, and blue are mixed together, a white light is produced. The three colors red, green, and blue (RGB) are known as *primary* colors. The colors magenta,

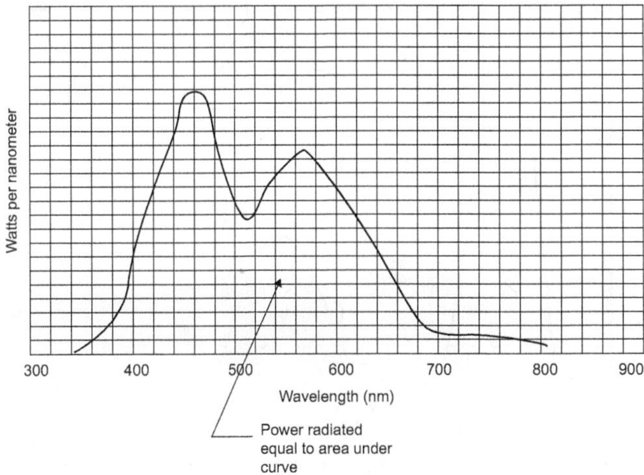

FIGURE 5.1-4 Power distribution of a monochrome video picture tube light source.

Adapted from Fink, D.G., *Television Engineering*, 2nd ed., McGraw-Hill, New York, 1952

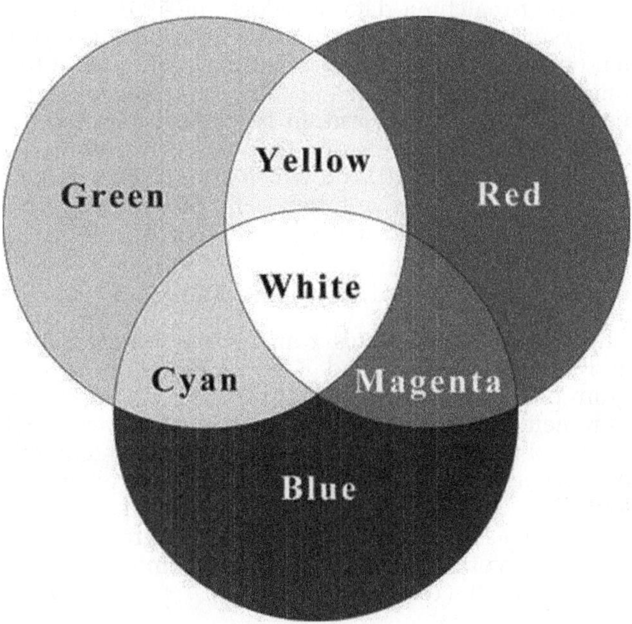

FIGURE 5.1-5 Additive color mixing.

yellow, and cyan are known as *secondary colors*. One of the characteristics of primary colors is that two of them cannot be mixed to produce the third. Color television systems can only work because of this fundamental property of light and vision. It allows almost any color from a scene to be analyzed into values of red, green, and blue light by a camera pickup device and also enables those values to be added together on a display device to reproduce the original color.

CIE COLOR SYSTEM

The International Commission on Illumination (CIE, the initials of its French name) is the international organization dealing with light, illumination, and color. In 1931, the CIE defined a set of color-matching functions and a coordinate system that have remained the predominate international standard method of specifying color since (see www.cie.co.at/).

RGB Color Matching

The color-matching functions of the initial CIE effort were based on experimental data from many observers measured by Wright [1] and Guild [2]. Observers had to produce a match for a series of test colors by mixing the appropriate amounts of red, green, and blue primary color light. Wright and Guild used different sets of primaries, but the results were transformed to a single set—namely, monochromatic stimuli of wavelengths 700.0, 546.1, and 435.8 nm [3]. The units of the stimuli were chosen so equal amounts were needed to match an *equienergy stimulus* (constant radiant power per unit wavelength throughout the visible spectrum). These tests resulted in a set of RGB color-matching functions, defining what is known as the "1931 CIE Standard Observer." One characteristic of the RGB functions was that they required some negative values in order to reproduce the full gamut of colors (that is, a primary color had to be added to change the test color before a match could be produced).

Figure 5.1-6 shows a chromaticity diagram based on these color-matching functions plotted on r and g axes with parameters that are calculated from R, G, and B. When these spectral chromaticity coordinates are

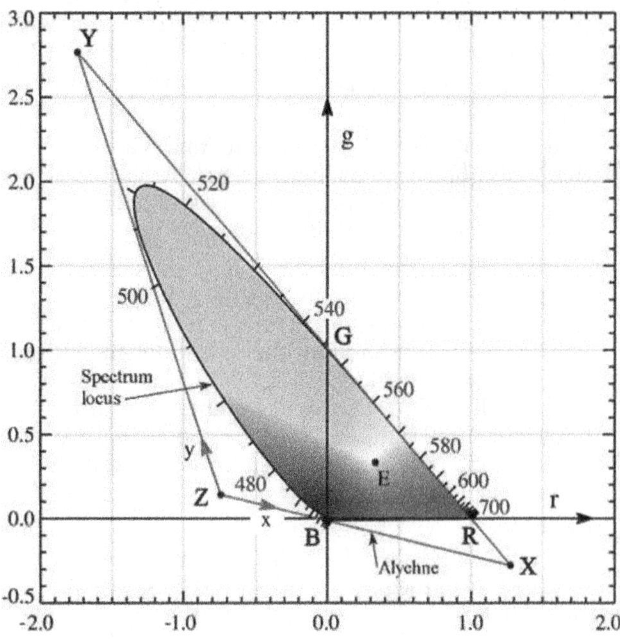

FIGURE 5.1-6 Spectrum locus and alychne of the CIE 1931 Standard Observer plotted in a chromaticity diagram based on matching stimuli of wavelengths 700.0, 546.1, and 435.8 nm. The locations of the CIE R, G, and B primaries and the imaginary primaries **X**, **Y**, and **Z** are shown.

plotted, as shown, they lie along a horseshoe-shaped curve called the spectrum locus. The extremities of the curve correspond to the extremities of the visible spectrum—approximately 380 nm for the blue end and 780 nm for the red end. The straight line joining the extremities is called the purple boundary and is the locus of the most saturated purples obtainable.

XYZ Color Space

At the same time it adopted these color-matching functions as a standard, the CIE also introduced and standardized another color space with a set of primaries that would encompass all the visible colors and allow the color-matching functions to be everywhere greater than or equal to zero. This involved some ingenious concepts. The set of real physical primaries was replaced by a new set of imaginary nonphysical primaries with special characteristics. These new primaries are referred to as **X**, **Y**, and **Z**, and the corresponding tristimulus values are X, Y, and Z. The chromaticity of the new **X**, **Y**, and **Z** primaries in the RGB system are shown in Figure 5.1-6.

Primaries **X** and **Z** have zero luminance, and all the luminance in a mixture of these three primaries is contributed by **Y**. This convenient property depends only on the decision to locate **X** and **Z** on the alychne, which is defined as the locus of colors with zero luminosity. It still leaves a wide choice of locations for all three primaries. The actual locations chosen by the CIE were based on the following additional considerations:

- The spectrum locus lies entirely within the triangle **XYZ**. This means that negative amounts of the primaries are never needed to match real colors. The color-matching functions $\bar{X}(\lambda)$, $\bar{Y}(\lambda)$, and $\bar{Z}(\lambda)$, are as shown in Figure 5.1-7. Therefore all positive at all wavelengths.
- The line $Z = 0$ (the line from **X** to **Y**) lies along the straight portion of the spectrum locus. Z is effectively zero for spectral colors with wavelengths greater than about 560 nm.
- The line $X = 0$ (the line from **Y** to **Z**) was chosen to minimize (approximately) the area of the **XYZ** triangle outside the spectrum locus. This choice led to a bimodal shape for the $\bar{X}(\lambda)$ color-matching function because the spectrum locus curves away from the line $X = 0$ at low wavelengths. A different choice of $X = 0$ (tangential to the spectrum locus at about 450 nm) would have eliminated the secondary lobe of $\bar{X}(\lambda)$ but would have pushed **Y** much further from the spectrum locus.
- The units of X, Y, and Z were chosen so that the tristimulus values X, Y, and Z would be equal to each other for an equienergy stimulus.

This coordinate system and the set of color-matching functions that go with it are known as the *CIE 1931 Standard Observer Colorimetric System*. The color-matching functions are shown in Figure 5.1-7, where relative tristimulus values are plotted against wavelength (λ).

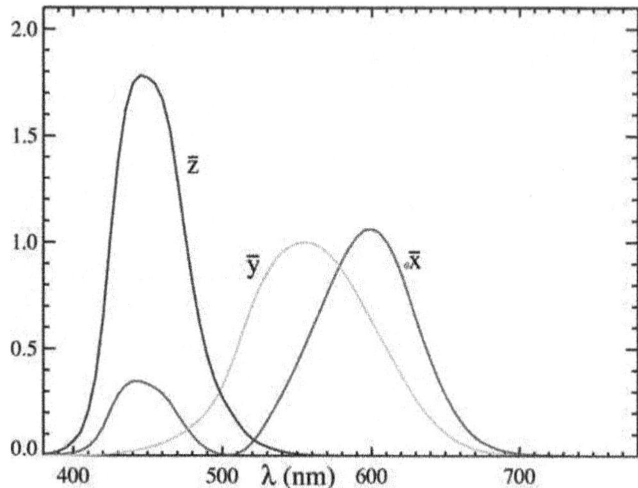

FIGURE 5.1-7 CIE 1931 XYZ color-matching functions.

CIE 1964 Supplementary Standard Observer

The color-matching data on which the 1931 Standard Observer is based were obtained with a visual field subtending 2° at the eye. Because of the slight nonuniformities of the retina, color-matching functions for larger fields are slightly different. In 1964, this prompted the CIE to recommend a second Standard Observer based on a visual field of 10°, known as the *CIE 1964 Supplementary Standard Observer*, for use in colorimetric calculations when the field size is greater than 4°.

Chromaticity Coordinates and Diagram

Chromaticity coordinates x and y may be calculated as functions of X, Y, and Z and a chromaticity diagram plotted as shown in Figure 5.1-8. It is important to remember that the CIE chromaticity diagram is not intended to illustrate appearance. The CIE system tells only whether two stimuli match in color, not what they look like. Appearance depends on many factors not taken into account in the chromaticity diagram. Nevertheless, it is often useful to know approximately where colors lie on the diagram.

Standard Illuminants

The CIE has recommended a number of standard illuminants, $E(\lambda)$, for use in evaluating the tristimulus values of reflecting and transmitting objects. Originally, in 1931, it recommended three—known as A, B, and C. These illuminants are specified by tables of relative spectral distribution and were chosen so they could be reproduced by real physical sources. (CIE terminology distinguishes between *illuminants*, which are tables of numbers, and *sources*, which are physical emitters of light.) The sources are defined as follows:

Source A—A tungsten filament lamp operating at a color temperature of about 2856°K. Its chromaticity coordinates are $x = 0.4476$ and $y = 0.4074$. Source A represents incandescent light.

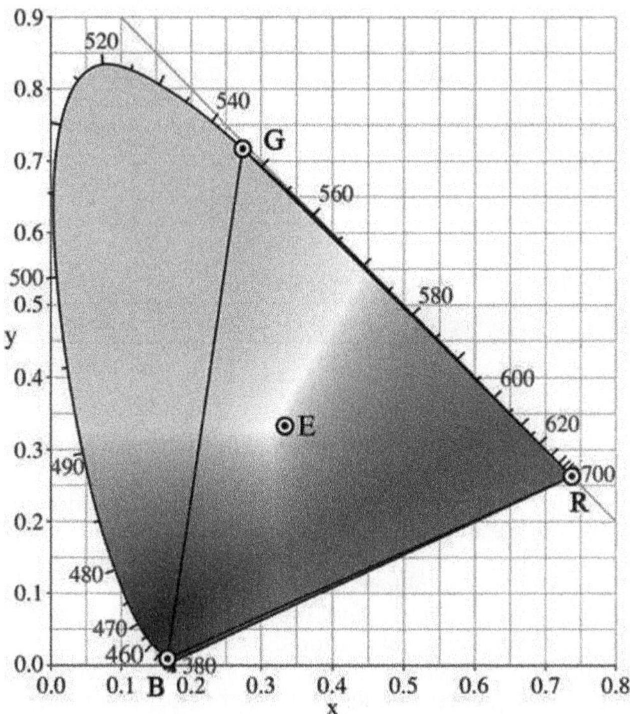

FIGURE 5.1-8 CIE 1931 Chromaticity diagram, showing RGB primaries.

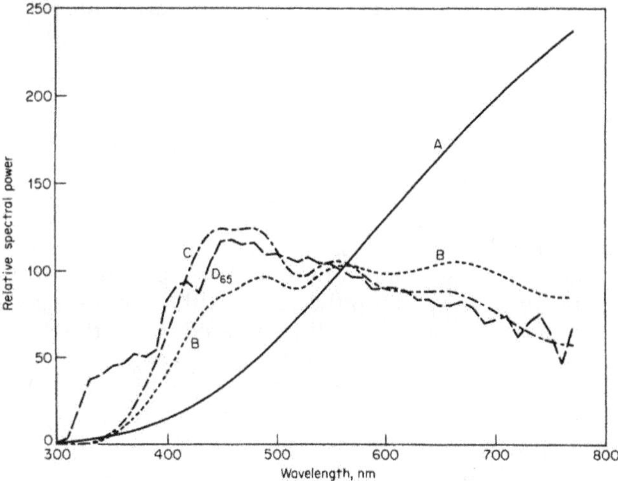

FIGURE 5.1-9 The relative spectral power distributions of CIE standard illuminants A, B, C, and D_{65}.

From Robertson, A. R., et al., in *Standard Handbook of Video and Television Engineering*, 4th ed., Whitaker, J. C., Ed., McGraw-Hill, New York, 2003

Source B—A source with a composite filter made of two liquid filters of specified chemical composition [4]. The chromaticity coordinates of source B are $x = 0.3484$ and $y = 0.3516$. Source B represents noon sunlight.

Source C—This source is also produced by source A with two liquid filters [4]. Its chromaticity coordinates are $x = 0.3101$ and $y = 0.3162$. Source C represents average daylight according to information available in 1931.

In 1971, the CIE introduced a new series of standard illuminants that represented daylight more accurately than illuminants B and C [5]. The improvement is particularly marked in the ultraviolet part of the spectrum, which is important for fluorescent samples. The most important of the D illuminants is D_{65} (sometimes written D6500), which has chromaticity coordinates of $x = 0.3127$ and $y = 0.3290$. The relative spectral power distributions of illuminants A, B, C, and D_{65} are given in Figure 5.1-9.

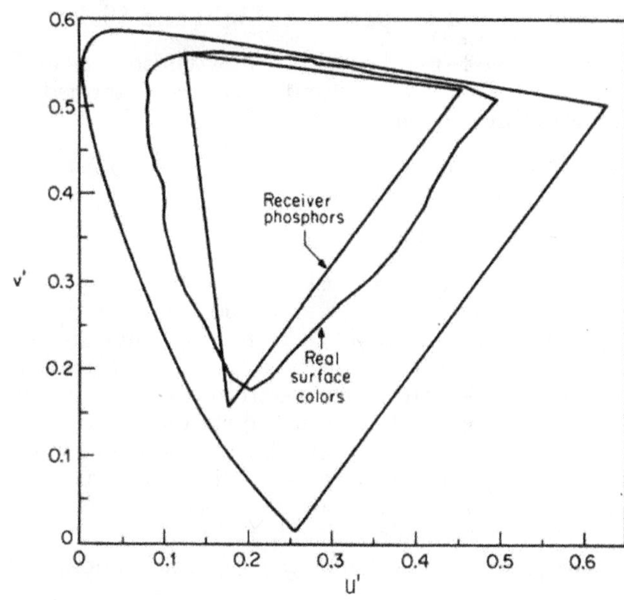

FIGURE 5.1-10 The color triangle defined by a set of color television receiver phosphors compared with the maximum real color gamut on a u',v' chromaticity diagram.

Adapted from Pointer, M. R., *Color Res. Appl.*, 5, 145–155, 1980

Gamut of Reproducible Colors

In a system that seeks to match or reproduce colors with a set of three primaries, only those colors can be reproduced that lie inside the triangle of primaries. Colors outside the triangle cannot be reproduced because they would require negative amounts of one or two of the primaries. In a color-reproducing system, it is important to have a triangle of primaries that is sufficiently large to permit a satisfactory gamut of colors to be reproduced. To illustrate the kinds of requirements that must be met, Figure 5.1-10 shows the maximum color gamut for real surface colors and the triangle of typical color television receiver phosphors used in CRT-based consumer sets. These are shown in the CIE 1976 u',v' chromaticity diagram in which the perceptual spacing of colors is more uniform than in the x,y diagram. High-purity blue-green and purple colors cannot be reproduced by these phosphors, whereas the blue phosphor is actually of slightly higher purity than any real surface colors.

THE HUMAN VISUAL SYSTEM

The human visual system is powerful and exceeds the performance of artificial visual systems in almost all areas of comparison. Vision results from stimulation of the eye by light and consequent interaction through connecting nerves with the brain. Although the range of human vision is small compared with the total energy spectrum, human discrimination—the ability to detect differences in intensity or quality—is excellent. Under ideal conditions, the human visual system can detect:

- Wavelength differences of 1 millimicron (10 Å; 1 Angstrom unit = 10^{-10} cm)
- Intensity differences of as little as 1%
- Forms subtending an angle at the eye of 1 arc minute, and often smaller objects

Sensation and Properties of Light and Vision

The perceived (psychological) color vision sensation associated with a light stimulus can be described in terms of three characteristics:

- Hue
- Saturation
- Brightness

The spectrum contains most of the principal hues: red, orange, yellow, green, blue, and violet. Additional hues are obtained from mixtures of red and blue light—these constitute the purple colors. *Saturation* pertains to the strength of the hue. Spectrum colors are highly saturated. White and grays have no hue and, therefore, have zero saturation. Pastel colors are of low or intermediate saturation. *Brightness* pertains to the intensity of the stimulation. If a stimulus has high intensity, regardless of its hue, it is said to be bright.

The psychophysical analogs of hue, saturation, and brightness are:

- Dominant wavelength
- Excitation purity
- Luminance

These equivalent properties are listed in Table 5.1-1.

By means of definitions and standard response functions, which have received international acceptance through the CIE, the dominant wavelength, purity, and luminance of any stimulus of known spectral energy distribution may be determined by simple computations. Although roughly analogous to their psychophysical counterparts, the psychological attributes of hue, saturation, and brightness pertain to observer responses to light stimuli and are not subject to calculation. These sensation characteristics as applied to any given stimulus depend in part on other visual stimuli in the field of view and upon the immediately preceding stimulations.

TABLE 5.1-1
Psychophysical and Psychological Characteristics of Color

Psychophysical Properties	Psychological Properties
Dominant wavelength	Hue
Excitation purity	Saturation
Luminance	Brightness
Luminous transmittance	Lightness
Luminous reflectance	Lightness

Color sensations arise directly from the action of light on the eye. They are normally associated, however, with objects in the field of view from which the light comes. The objects themselves are therefore said to have color. *Object colors* may be described in terms of their hues and saturations, as is the case for light stimuli. The intensity aspect is usually referred to in terms of lightness, rather than brightness. The psychophysical analogs of lightness are *luminous reflectance* for reflecting objects and *luminous transmittance* for transmitting objects.

Scotopic and Photopic Vision

Light entering the eye passes through a lens that focuses the image onto the *retina*, which has two types of light-sensitive elements known as *rods* and *cones*. Cones are able to respond to different colors of light, whereas rods respond only to luminance levels. At low levels of illumination, objects may differ from one another in their lightness appearances but give rise to no sensation of hue or saturation. All objects then appear to be of different shades of gray. Vision at low levels of illumination is called *scotopic vision*, as distinct from *photopic vision*, which takes place at higher levels of illumination. Table 5.1-2 compares the luminosity values for photopic and scotopic vision. Only the rods of the retina are involved in scotopic vision; the cones play no part. As the fovea centralis (at the center of the retina) is free of rods, scotopic vision takes place outside the fovea. Visual acuity of scotopic vision is low compared with photopic vision. At high levels of illumination, where cone vision predominates, all vision is color vision and visual acuity is high.

Trichromatic Color Vision

Color vision processing in the human visual system starts with the absorption of light by the light-sensitive cones. Based on research carried out during the 1970s and 1980s, three different classes of cones were identified, each containing a different type of photosensitive pigment. These are short-wavelength sensitive (S-cones), middle-wavelength sensitive (M-cones), and long-wavelength sensitive (L-cones), all having different but overlapping spectral sensitivities. The spectral sensitivity of S-cones peaks at approximately 440 nm, M-cones peak at 545 nm, and L-cones peak at 565 nm, although different researchers have found slightly different results. Interactions between at least two types of cone are necessary to produce the ability to perceive color. In this way, the brain can compare the signals from each type and determine both the intensity and color of the light. Based on this *trichromacy* of color vision, it is apparent that many different physical stimuli, or mixtures of light, can evoke the same sensation of color. All that is required for two stimuli to be equivalent is that they should each cause the same number of light photons to be absorbed by any given class of cone. In that case, the message to the brain, and thus the color sensations generated, will be the same even though the stimuli are physically different. This characteristic of the human visual system is fundamental to the science of colorimetry, and without it color television as we know it could not exist.

MEASUREMENTS AND PERFORMANCE

This section discusses some further aspects of vision and the visual system that can be quantified and measured.

Photometric Measurements

Photometry is the measurement of the properties of light. Evaluation of a radiant energy stimulus in terms of its brightness-producing capacity is a photometric measurement. An instrument for making such measurements is called a *photometer*. In visual photometers, used in obtaining basic photometric measurements, the two stimuli to be compared are normally directed into small adjacent parts of a viewing field. The stimulus to be evaluated is presented in the *test field*; the stimulus against which it is compared is presented in the *comparison field*. For most high-precision measurements, the total size of the combined test and comparison fields is kept small, subtending about 2° at the eye. The area outside these fields is called the *surround*. Although the surround does not enter directly into the measurements, it has adaptation effects on the retina and thus affects the appearances of the test and comparison fields. It also influences the precision of measurement.

Luminosity Curves

A *luminosity curve* is a plot indicative of the relative brightness of spectrum colors of different wavelength or frequency. To a normal observer, the brightest part of a spectrum consisting of equal amounts of radiant flux per unit wavelength interval is at about 555 nm. Luminosity curves are therefore commonly normalized to have a value of *unity* at 555 nm. If, at some other wavelength, twice as much radiant flux as at 555 nm is required to obtain brightness equality with radiant flux at 555 nm, the luminosity at this wavelength is 0.5. The luminosity at any wavelength λ is, therefore, defined as the ratio $P_{555}/P\lambda$, where $P\lambda$ denotes the amount of radiant flux at the wavelength λ, which is equal in brightness to a radiant flux of P_{555}. The

TABLE 5.1-2
Relative Luminosity Values for Photopic and Scotopic Vision

Wavelength (nm)	Photopic Vision	Scotopic Vision
390	0.00012	0.0022
400	0.0004	0.0093
410	0.0012	0.0348
420	0.0040	0.0966
430	0.0116	0.1998
440	0.023	0.3281
450	0.038	0.4550
460	0.060	0.5670
470	0.091	0.6760
480	0.139	0.7930
490	0.208	0.9040
500	0.323	0.9820
510	0.503	0.9970
520	0.710	0.9350
530	0.862	0.8110
540	0.954	0.6500
550	0.995	0.4810
560	0.995	0.3288
570	0.952	0.2076
580	0.870	0.1212
590	0.757	0.0655
600	0.631	0.0332
610	0.503	0.0159
620	0.381	0.0074
630	0.265	0.0033
640	0.175	0.0015
650	0.107	0.0007
660	0.061	0.0003
670	0.032	0.0001
680	0.017	0.0001
690	0.0082	—
700	0.0041	—
710	0.0021	—
720	0.00105	—
730	0.00052	—
740	0.00025	—
750	0.00012	—
760	0.00006	—

luminosity function that has been accepted as standard for photopic vision is given in Figure 5.1-11. Tabulated values at 10 nm intervals are given in Table 5.1-2. This function was agreed upon by the CIE in 1924. It is based upon considerable experimental work that was conducted over a number of years. Chief reliance in arriving at this function was based on the step-by-step equality-of-brightness method. Flicker photometry provided additional data.

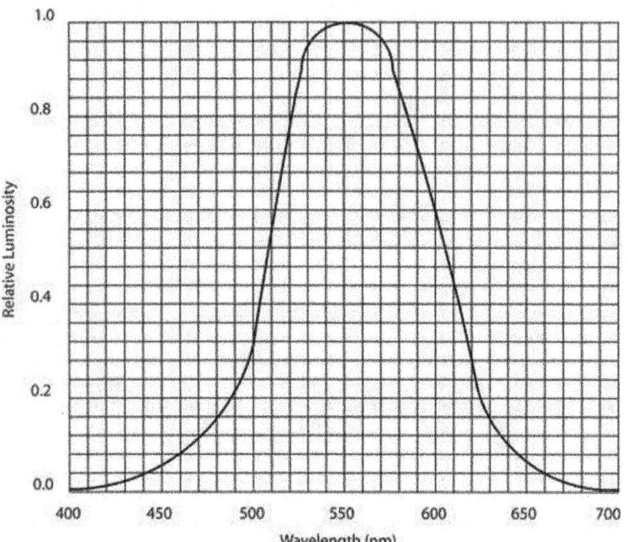

FIGURE 5.1-11 The photopic luminosity function.

Adapted from Fink, D.G., *Television Engineering*, 2nd ed., McGraw-Hill, New York, 1952

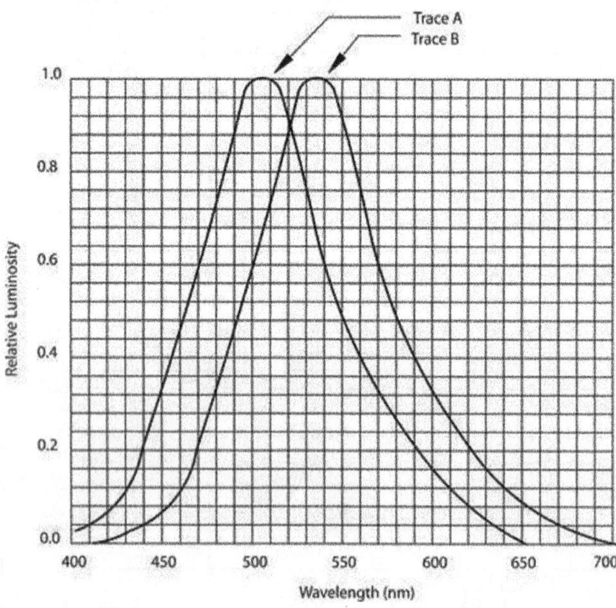

FIGURE 5.1-12 Scotopic luminosity function (Trace A) as compared with photopic luminosity function (Trace B).

Adapted from Fink, D.G., *Television Engineering*, 2nd ed., McGraw-Hill, New York, 1952

In the scotopic range of intensities, the luminosity function is somewhat different from that of the photopic range. The two curves are compared in Figure 5.1-12 and the relative values are listed in Table 5.1-2. The two curves are similar in shape, but there is a shift for the scotopic curve of about 40 nm to the shorter wavelengths.

Measurement of Luminosity

Measurements of luminosity in the scotopic range are usually made by the *threshold-of-vision* method. A single stimulus in a dark surround is used. The stimulus is presented to the observer at each of a number of different intensities, ranging from well below the threshold to intensities sufficiently high to be definitely visible. Determinations are made of the amount of energy, at each chosen wavelength, that is reported visible by the observer a certain percentage of the time, such as 50%. The reciprocal of this amount of energy determines the relative luminosity at the given wavelength. The wavelength plot is normalized to have a maximum value of 1.00 to give the scotopic luminosity function. In the intensity region between scotopic and photopic vision, called the *Purkinje* or *mesopic region*, the measured luminosity function takes on sets of values intermediate between those obtained for scotopic and photopic vision. Relative luminosities of colors within the mesopic region will therefore vary, depending on the particular intensity level at which the viewing takes place. Reds tend to become darker in approaching scotopic levels; greens and blues tend to become relatively lighter.

Luminance

Brightness is a term used to describe one of the characteristics of appearance of a source of radiant flux or of an object from which radiant flux is being reflected or transmitted. Brightness specifications of two or more sources of radiant flux should be indicative of their actual relative appearances. These appearances will depend in large part upon the viewing conditions, including the state of adaptation of the observer's eye. Luminance, as indicated previously, is a psychophysical analog of brightness. It is subject to physical determination, independent of particular viewing and adaptation conditions. Because it is an analog of brightness, however, it is defined in such a way as to relate as closely as possible to brightness.

The luminosity function is the best-established measure of the relative brightness of different spectral stimuli. In evaluating the luminance of a source of radiant flux consisting of many wavelengths of light, the amounts of radiant flux at the different wavelengths are weighted by the luminosity function, which converts radiant flux to luminous flux. As used in photometry, the term *luminance* is applied only to extended sources of light, not to point sources. For a given amount (and quality) of radiant flux reaching the eye, brightness will vary inversely with the effective area of the source.

Luminance is described in terms of luminous flux per unit projected area of the source. The greater the concentration of flux in the angle of view of a source, the brighter it appears. Luminance is therefore expressed in terms of amounts of flux per unit solid angle or *steradian*.

In considering the relative luminance of various objects of a scene to be captured and reproduced by a video system, it is convenient to normalize the luminance values so the *white* in the region of principal illumination has a relative luminance value of 1.00. The relative luminance of any other object then becomes the ratio of its luminance to that of the white. This white is an object of highly diffusing surface with high and uniform reflectance throughout the visible spectrum. For purposes of computation, it may be idealized to have 100% reflectance and perfect diffusion.

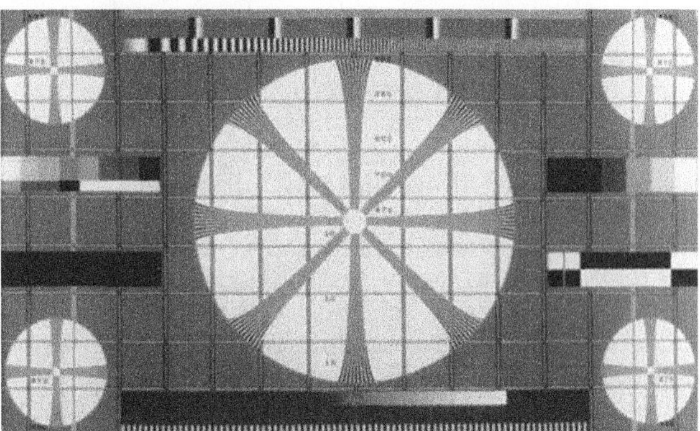

FIGURE 5.1-13 Test chart for checking resolution, geometry, bandwidth, and color reproduction.

Courtesy of Tektronix

Perception of Fine Detail

Detail is seen in an image because of brightness differences between small adjacent areas in a monochrome display or because of brightness, hue, or saturation differences in a color display. Visibility of detail in a picture is important because it determines the extent to which small or distant objects of a scene are visible and because of its relationship to the *sharpness* appearance of the edges of objects. *Picture definition* is probably the most acceptable term for describing the general characteristic of crispness, sharpness, or image-detail visibility in a picture. Picture definition depends on characteristics of the eye, such as visual acuity, and upon a variety of characteristics of the picture-image medium, including its resolving power, luminance range, contrast, and image edge gradients.

Visual Acuity Measurements

Visual acuity may be measured in terms of the visual angle subtended by the smallest detail in an object that is visible. The *Landolt ring* is one type of test object. The ring, which has a segment cut from it, is shown in any one of four orientations, with the opening at the top or bottom or on the right or left side. The observer identifies the location of this opening. The visual angle subtended by the opening that can be properly located 50% of the time is a measure of visual acuity. Test object illuminance, contrast between the test object and its background, time of viewing, and other factors greatly affect visual acuity measurements. Up to a visual distance of about 20 ft, acuity is partially a function of distance, because of changes in shape of the eye lens in focusing. Beyond 20 ft, it remains relatively constant. Visual acuity is highest for foveal vision, on the center of the retina, dropping off rapidly for retinal areas outside the fovea. Normal vision, corresponding to a Snellen 20/20 rating, represents an angular discrimination of about 1 minute. Separations between adjacent cones in the fovea and resolving power limitations of the eye lens give theoretical visual acuity values of about this same magnitude.

Resolution

The extent to which a picture medium, such as a photographic or a video system, can reproduce fine detail is expressed in terms of *resolving power* or *resolution*. Resolution is a measure of the distance between two fine lines in the reproduced image that are visually distinct. The image is examined under the best possible conditions of viewing, including magnification. Two types of test charts are commonly employed in determining resolving power: either a wedge of radial lines or groups of parallel lines at different pitches for each group. For either type of chart, the spaces between pairs of lines usually are made equal to the line widths. Figure 5.1-13 shows a test signal generated by a video measuring test set.

Resolution in conventional photography is usually expressed as the maximum number of lines (counting only the black ones or only the white ones) per millimeter that can be distinguished from one another. Measured values of resolving power depend upon a number of factors in addition to the photographic material itself.

Sharpness of focus depends upon the general quality of the focusing lens, image, and object distances from the lens, as well as the part of the projected field in which the image lies. In determining the resolving power of a photographic negative or positive material, a test chart is generally employed that has a high density difference, such as 3.0, between the black-and-white lines. A high-quality lens is used, the projected field is limited, and focusing is critically adjusted. Under these conditions, ordinary black-and-white photographic materials generally have resolving powers in the range of 30 to 200 line pairs/mm. Special photographic materials are available with resolving powers greater than 1000 line pairs/mm.

Resolution in a video system is expressed in terms of the maximum number of lines, counting both black and white, that are discernible in viewing a test chart. The value of horizontal (vertical lines) or vertical (horizontal lines) resolution is the number of lines equal to the vertical dimension of the raster.

Sharpness

The appearance evaluation of a picture image in terms of the edge characteristics of objects is called *sharpness*. The more clearly defined the line that separates dark areas from lighter ones, the greater the sharpness of the picture. Sharpness is, therefore, related to the transient curve in the image across an edge. The average gradient and the total density difference appear to be the most important characteristics. No physical measure has been devised, however, that in all cases will predict the sharpness (appearance) of an image. Picture resolution and sharpness are to some extent interrelated, but they are by no means perfectly correlated.

Pictures ranked according to resolution measures may be rated somewhat differently on the basis of sharpness. Both resolution and sharpness are related to the more general characteristic of picture definition. For pictures in which, under particular viewing conditions, effective resolution is limited by the visual acuity of the eye rather than by picture resolution, sharpness is probably a good indication of picture definition. If visual acuity is not the limiting factor, however, picture definition depends to an appreciable extent on both resolution and sharpness.

A Model for Image Quality

Researchers have studied the human visual system extensively to ascertain the most efficient and effective methods of communicating information to the eye. An important component of this work has been the development of models of how humans see, in an effort to improve image processing systems. The classic approach to image quality assessment involves the presentation to a group of test subjects visual test material for evaluation and rating. The test material may include side-by-side display comparisons or a variety of perception threshold presentations. One common visual comparison technique is the *pair-comparison* method. A number of observers are asked to view a specified number of images at two or more distances. At each distance, the subjects are asked to rank the order of the images in terms of overall quality, clearness, and personal preference. An image acquisition, storage, transmission, and display system need not present more visual information to the viewer than the viewer can process. For this reason, image quality assessment is an important element in the development of any new video system. Evaluation by human subjects, while an important part of this process, is also expensive and time consuming. Numerous efforts have been made to reduce the human visual system and its interaction with a display device to one or more mathematical models. (Some early research is cited in References 6 through 10.) After the system or algorithm has successfully passed the minimum criteria established by the model, it can be subjected to human evaluation. The model simulation requires the selection of many interrelated parameters.

VIDEO SYSTEM REQUIREMENTS

The objective in any type of visual reproduction system is to present to the viewer a combination of visual stimuli that can be readily interpreted as representing, or having close association with, a real viewing situation. In order to achieve that, it is by no means necessary that the light stimuli from the original scene be duplicated precisely. There are certain characteristics in the reproduced image, however, that are necessary and others that are highly desirable. Only a general qualitative discussion of such characteristics as they relate to video systems will be given here.

Resolution

In monochrome video, images of objects are distinguished from one another and from their backgrounds as a result of luminance differences. In order that detail in the picture is visible and that objects have clear, sharp edges, it is necessary that the video system be capable of rapid transitions from areas of one luminance level to another. This degree of resolution need not necessarily match that possible in the eye itself, but too low an effective resolution results in pictures with a fuzzy appearance and lacking fineness of detail. Matching the resolving capabilities of the human eye is a commendable engineering goal, but a difficult proposition in any practical imaging system. The move from standard-definition imaging (in particular, the constraints imposed by encoding systems such as NTSC, PAL, and SECAM) to high-definition imaging led to enormous improvement in overall picture quality. This improvement has stretched from the camera through the entire transmission system to the display device. This trend will no doubt continue as imaging technologies continue to improve, particularly cameras and displays. Historically, improvements in one area tend to push improvements in another.

Dynamic Range

Luminance range and the transfer characteristic associated with luminance reproduction are likewise of importance in a video system. Objects seen as white usually have minimum reflectances of approximately 80%. Black objects have reflectances of approximately 4%. This gives a luminance ratio of 20:1 in the range from white to black. To obtain the total luminance range in a scene, the reflectance range must be multiplied by the illumination range. In outdoor scenes, the illumination ratio between full sunlight and shadow may be as high as 100:1 [11]. The full luminance ranges involved with objects in such scenes could not be reproduced in conventional video reproduction equipment. Modern imaging systems, with advanced cameras and displays, have made great strides in approximating this luminance range.

Monochrome video transmits only luminance information, and the relative luminance of the images should correspond at least roughly to the relative luminance of the original objects. Red objects, for example, should not be reproduced markedly darker than objects of other hues but of the same luminance. Exact luminance reproduction, however, is by no means a necessity. Considerable distortion as a function of hue is acceptable in many applications. Luminance reproduction is probably of primary consequence only if detail in some hues becomes lost.

Depending on the camera pickup element, the dominant wavelength and purity of the light may be of consequence. Most video pickup elements exhibit sensitivity throughout the visible spectrum and, consequently, marked distortions in luminance as a function of dominant wavelength and purity are not encountered. Their spectral sensitivities seldom conform exactly to that of the human observer, however, so some brightness distortions do exist.

Scanning Frequency

Images in a video system are transmitted one point, or small area, at a time. The complete picture image is repeatedly scanned at frequent intervals. If the frequency of scan is not sufficiently high, the picture appears to flicker. At frequencies above a *critical frequency*, no flicker is apparent. The critical frequency changes as a function of luminance, being higher for higher luminance. The basic requirement for video is that the *field frequency* (the rate at which images are presented) be above the critical frequency for the highest image luminance.

Color Reproduction

Images of objects in color video are distinguished from one another by luminance differences or by differences in hue or saturation. A number of factors may contribute to color rendition in a practical system, categorized as follows:

- *Exact color reproduction*, where the reproduction is a metameric match to the original. Exact color reproduction will result in equality of appearance only if the viewing conditions for the picture and the original scene are identical. These conditions include the angular subtense of the picture, the luminance and chromaticity of the surround, and glare. In practice, exact color reproduction often cannot be achieved because of limitations on the maximum luminance that can be produced on a color monitor or display.
- *Colorimetric color reproduction*, a variant of exact color reproduction in which the tristimulus values are proportional to those in the original scene. In other words, the chromaticity coordinates are reproduced exactly, but the luminances all are reduced by a constant factor. Traditionally, color video systems have been designed and evaluated for colorimetric color reproduction. If the original and the reproduced reference whites have the same chromaticity, if the viewing conditions are the same, and if the system has an overall gamma of unity, then colorimetric color reproduction is indeed a useful criterion.
- *Corresponding color reproduction*, a compromise by which colors in the reproduction have the same appearance that colors in the original would have had if they had been illuminated to produce the same average luminance level and the same reference white chromaticity as that of the reproduction. For most purposes, corresponding color reproduction is a suitable objective of a color video system.
- *Preferred color reproduction*, a departure from the preceding categories that recognizes the preferences of the viewer. It is sometimes argued that corresponding color reproduction is not the ultimate aim for some display systems, such as video games or even television programs, and that it should be taken into account that people prefer some colors to be different from their actual appearance. For example, suntanned skin color is preferred to average real skin color, and sky is preferred bluer and foliage greener than they really are.

Even if corresponding color reproduction is accepted as the target, some colors are more important than others. For example, flesh tones must be acceptable—not obviously reddish, greenish, purplish, or otherwise incorrectly rendered. Likewise, the sky must be blue and the clouds white, within the viewer's range of acceptance. Similar conditions apply to other well-known colors of common experience. The imaging system improvements brought into practical usage by high-definition video have given program producers considerable flexibility in using the video system for highly accurate picture capture and display, or for using the system creatively to convey a mood or effect.

SUMMARY

This chapter has introduced many of the basic principles of light and vision upon which color television systems rely. Other chapters in this book build on this foundation. For more in-depth coverage of the subject, other sources of information are listed in the further information, bibliography, and references sections that follow.

ACKNOWLEDGMENTS

Portions of this chapter were adapted from Whitaker, J. C., and Benson, K. B., Eds., *Standard Handbook of Video and Television Engineering*, 4th ed., McGraw-Hill, New York, 2003. Used with permission.

DEFINING TERMS

- *Brightness:* A term used to describe one of the characteristics of appearance of a source of radiant flux or of an object from which radiant flux is being reflected or transmitted.
- *Critical frequency:* The rate of picture presentation, as in a video system or motion picture display, above which the presented image ceases to give the appearance of flickering. The critical frequency changes as a function of luminance, being higher for higher luminance.
- *Dispersion:* The variation of refraction as a function of wavelength.

- *Energy distribution curve:* A plot of the power distribution of a source of light giving the energy radiated at each wavelength per nanometer of wavelength.
- *Field frequency:* The rate at which images in an electronic imaging system are presented. For an interlace system, two fields are presented each second to make up one *frame* (a complete picture).
- *Refraction:* The bending of light as it passes from one medium to another, such as from air to glass, at an angle.
- *Resolution:* The extent to which an imaging system, such as a photographic or a video system, can reproduce fine detail.
- *Sharpness:* The appearance evaluation of a picture image in terms of the edge characteristics of objects contained therein. The more clearly defined the lines separating dark areas from light ones, the greater the sharpness of the picture.

Further Information

The International Society for Optical Engineering (SPIE; Bellingham, WA, see www.spie.org/) offers a number of publications examining the characteristics of the human visual system. The organization also conducts technical seminars on various topics relating to optics and the application of optical technologies.

Notes

[1] Adapted from: Brewer, W. L., R. A. Morris, and D. G. Fink: "Light and the Visual Mechanism," and "Photometric Quantities," in *Standard Handbook of Video and Television Engineering*, 4th edition, Jerry C. Whitaker and K. Blair Benson (eds.), McGraw-Hill, New York, 2003, pp. 1-1 to 1-33.

[2] Adapted from: Robertson, Alan R., Joseph F. Fisher, and Jerry C. Whitaker: "The CIE Color System," in *Standard Handbook of Video and Television Engineering*, 4th edition, Jerry C. Whitaker and K. Blair Benson (eds.), McGraw-Hill, New York, 2003, pp. 1-1 to 2-50.

Bibliography

Boynton, R. M., *Human Color Vision*, Holt, New York, 1979.

Committee on Colorimetry, *The Science of Color*, Optical Society of America, New York, 1953.

Davson, H., *Physiology of the Eye*, 4th ed., Academic Press, New York, 1980.

Evans, R. M., Hanson, Jr., W. T., and Brewer, W. L., *Principles of Color Photography*, Wiley, New York, 1953.

Fink, D. G., *Television Engineering*, 2nd ed., McGraw-Hill, New York, 1952.

IES, *IES Lighting Handbook*, Illuminating Engineering Society of North America, New York, 1981.

Kingslake, R., Ed., *Applied Optics and Optical Engineering*, Vol. 1, Academic Press, New York, 1965.

Pointer, M. R., The gamut of real surface colours, *Color Res. Appl.*, 5, 145–155, 1980.

Polysak, S. L., *The Retina*, University of Chicago Press, Chicago, IL, 1941.

Richards, C. J., *Electronic Displays and Data Systems: Constructional Practice*, McGraw-Hill, New York, 1973.

Schade, O. H., Electro-optical characteristics of television systems, *RCA Rev.*, 9, 5–37, 245–286, 490–530, 653–686, 1948.

Whitaker, J. C., *Video Display Engineering*, McGraw-Hill, New York, 2001.

Whitaker, J. C., and Benson, K. B., Eds., *Standard Handbook of Video and Television Engineering*, 4th ed., McGraw-Hill, New York, 2003.

Wright, W. D., *The Measurement of Colour*, 4th ed., Adam Hilger, London, 1969.

Wright, W. D., *Researches on Normal and Defective Colour Vision*, Mosby, St. Louis, 1947.

References

[1] Wright, W. D., A redetermination of the trichromatic coefficients of the spectral colours, *Trans. Opt. Soc.*, 30, 141–164, 1928/1929.

[2] Guild, J., The colorimetric properties of the spectrum, *Phil. Trans. Roy. Soc. A*, 230, 149–187, 1931.

[3] Robertson, A. R., Fisher, J. F., and Whitaker, J. C., The CIE color system, in *Standard Handbook of Video and Television Engineering*, 4th ed., Whitaker, J. C., Ed., McGraw-Hill, New York, 2003 pp. 2-27 to 2-50.

[4] Judd, D. B., and Wyszencki, G., *Color in Business, Science, and Industry*, 3rd ed., Wiley, New York, 1975, pp. 44–45.

[5] CIE, *Colorimetry*, Publ. No. 15, CIE, Paris, 1971.

[6] Grogan, T. A., Image evaluation with a contour-based perceptual model, in *Human Vision, Visual Processing, and Digital Display III*, Rogowitz, B. E., Ed., SPIE, Bellingham, WA, 1992, pp. 188–197.

[7] Barten, P.G.J., Physical model for the contrast sensitivity of the human eye, in *Human Vision, Visual Processing, and Digital Display III*, Rogowitz, B. E., Ed., SPIE, Bellingham, WA, 1992, pp. 57–72.

[8] Daly, S., The visible differences predictor: an algorithm for the assessment of image fidelity, in *Human Vision, Visual Processing, and Digital Display III*, Rogowitz, B. E., Ed., SPIE, Bellingham, WA, 1992, pp. 2–15.

[9] Reese, G., Enhancing images with intensity-dependent spread functions, in *Human Vision, Visual Processing, and Digital Display III*, Rogowitz, B. E., Ed., SPIE, Bellingham, WA, 1992, pp. 253–261.

[10] Martin, R. A., Ahumanda, Jr., A. J., and Larimer, J. O., Color matrix display simulation based upon luminance and chromatic contrast sensitivity of early vision, in *Human Vision, Visual Processing, and Digital Display III*, Rogowitz, B. E., Ed., SPIE, Bellingham, WA, 1992, pp. 336–342.

[11] Brewer, X., Lyle, W., Morris, R. A., and Fink, D. G., Light and the visual mechanism, in *Standard Handbook of Video and Television Engineering*, 4th ed., Whitaker, J. C., Ed., McGraw-Hill, New York, 2003, pp. 1–13.

CHAPTER

5.2

Planning a Video Production and TV Studio Facility

Systems, Principles, Design, Architecture

RALPH S. BLACKMAN

Rees Associates Inc.
Dallas, Texas

Updated for the 11th Edition by

DAVE GUERRERO

Chief Operating Officer
WLVT-TV, Bethlehem, PA

INTRODUCTION

This chapter offers an overview of the considerations needed when planning a broadcast media production facility. The information presented is solely intended to provide guidance to the broadcast engineer in understanding requirements of building the "average, non-news" broadcast facility.

For any substantial project, it is highly recommended that specialists in system integration, acoustics, and building construction be engaged to ensure the success of the new facility. As with most engineering projects, a project plan, project manager, and dedicated team are imperative in overall project success. From the initial concept through the finished product, attention to detail is the key attribute guaranteeing success.

Broadcast media production facilities have particular requirements that are similar albeit different from other modern technical facilities. Issues of sound control, HVAC, and lighting require significant consideration. Media production requires space definitions not found in any other facility: television studio, voice-over booths, edit rooms, client screening rooms, and so forth.

Yes, the media production facility is quite unique, and no two are exactly the same. As most businesses have unique clients and employees, so do production houses. The client base and employees' requirements drive the demand for physical space and therefore the overall design.

GENERAL PLANNING CONSIDERATIONS

Where to Build

Gone are the days of colocating the media production facility at or near the transmission facility, far removed from "downtown." Traditionally the media production facility was built adjacent to the transmission facility due to the economics of leasing/using common real estate. This has proven to be a poor investment; although the land and taxes are cheaper on top of a mountain, the practical side is that no one wants to spend a few days producing popular media in the woods . . . unless they are making a reality show (which, by the way, does not require a studio).

Today the production facility needs to be near the clients. The media production facility should be located in some suitable proximity to the region's popular shopping/entertainment district. Locating the media production facility in the "heart of the town" enables a very positive result in attracting and maintaining an active clientele. (See Figure 5.2-1.)

Let's Get Started

Plan, plan, plan, and then plan some more.

One can never review the details of any significant project enough. It is amazing how many times I've heard a technician comment on the poor planning that a seasoned systems integrator has made.

The hired experts are only as good as the quality of the information that they are given . . . period!

When in planning meetings or evaluating project details, please review this line daily, and have everyone on your team review it daily!

Proper facility planning starts with listening!

The project manager is your best friend. Every project detail needs to be collected, evaluated by the *entire* project team, and prioritized. Good project managers are indispensable; they will have their hands on every level of the project and every detail required.

The Team

The project team needs to be cross-functional, with representation from every department within the station. An example of key personnel comprising this team and meeting no less than weekly: CEO, COO (or GM), CE, CFO (or Controller), Dir. Sales, Production Manager, Exec. Producer, Dir. Marketing, Chief Operator (if so), IT Lead, and Facility Manager. All staff should be available for monthly planning/progress meetings; it is advisable that everyone currently touched by this transition be included in discussions. For early planning, it is imperative that individual work groups are created to discuss even the smallest of details. (Have you ever seen a master control room with no or very little access to simple 120 VAC power receptacles for connecting a vacuum cleaner? . . . I have.)

Initial Planning: You Need to Ask Questions to Acquire Details

1. Why are we building a new facility?
   a. More space
   b. Closer to the community
   c. Old facility outdated and renovation extremely expensive
   d. Lease of old facility runs out
2. What (if any) are issues with the old facility?
3. Is the location of the new building critical?

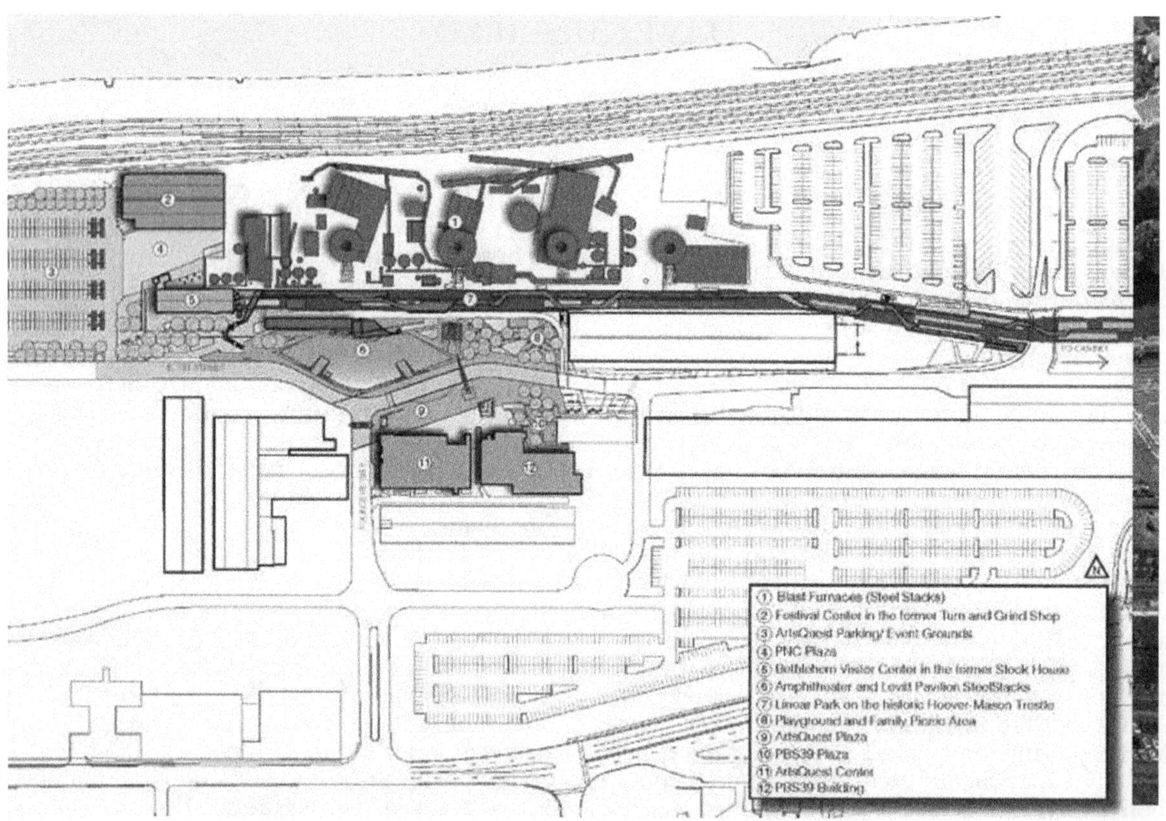

FIGURE 5.2-1 This is a rendering of the location plan of the TV station currently located in the redevelopment district of Bethlehem, PA. The new SteelStacks campus currently includes WLVT-TV, a new arts center, Chamber-funded visitor center, and the ruins of the Bethlehem Steel Corp. The campus is used nearly daily for arts and entertainment and includes the Levitt Pavilion, which offers live music free to the community Thursday through Sunday every week in the summer as well as numerous other events. Locating the station on this campus has afforded a huge amount of community support of the station's programming and events.

Courtesy of WRT

4. Will all services transfer directly from the old facility?
   a. Are we creating a new technical infrastructure?
   b. Has our business model changed?
   c. Are there old or new partnerships to consider?
5. Will all staff move into the new facility?
   a. Plans to relocate sales staff to a better location
   b. Plans to move a segment of technical staff to a different location
   c. Plans to outsource services, changing the composition of the team
6. How will connectivity to various distribution networks be affected, and can these be updated?
   a. Broadcast affiliate
   b. DirecTV
   c. DISH Network
   d. Cable TV headends
   e. Transmitter
   f. Internet
   g. E-commerce
7. To what level will the new facility be designed for enhanced building security and monitoring?
   a. Staff access
   b. General public access
   c. Special event considerations
8. Parking access on-site?
   a. Staff
   b. Visitors
   c. Events
   d. Availability of nearby public parking

Answering these questions will give the team a good start in planning. Every facility has different requirements; however, starting fresh with an open mind will create a new environment that is future proofed and will last indefinitely. (See Figure 5.2-2.)

> *A facility built to current technical requirements, current staffing, and current business model will usually be outdated, maybe the day it is commissioned. . . .*

With that in mind, all spaces should be designed using open concepts, technical closets/cores/IT booths should be designed for no less than 100% expansion, flexible office space should be provided, and finally any space dedicated to a specific task needs to be reviewed by the entire team to the smallest detail.

FIGURE 5.2-2 This is a bird's-eye view of the campus as it is today. Note the 5.6M uplink on the roof of the WLVT building as a reference to scale of the image.

Christenson Photography for WRT

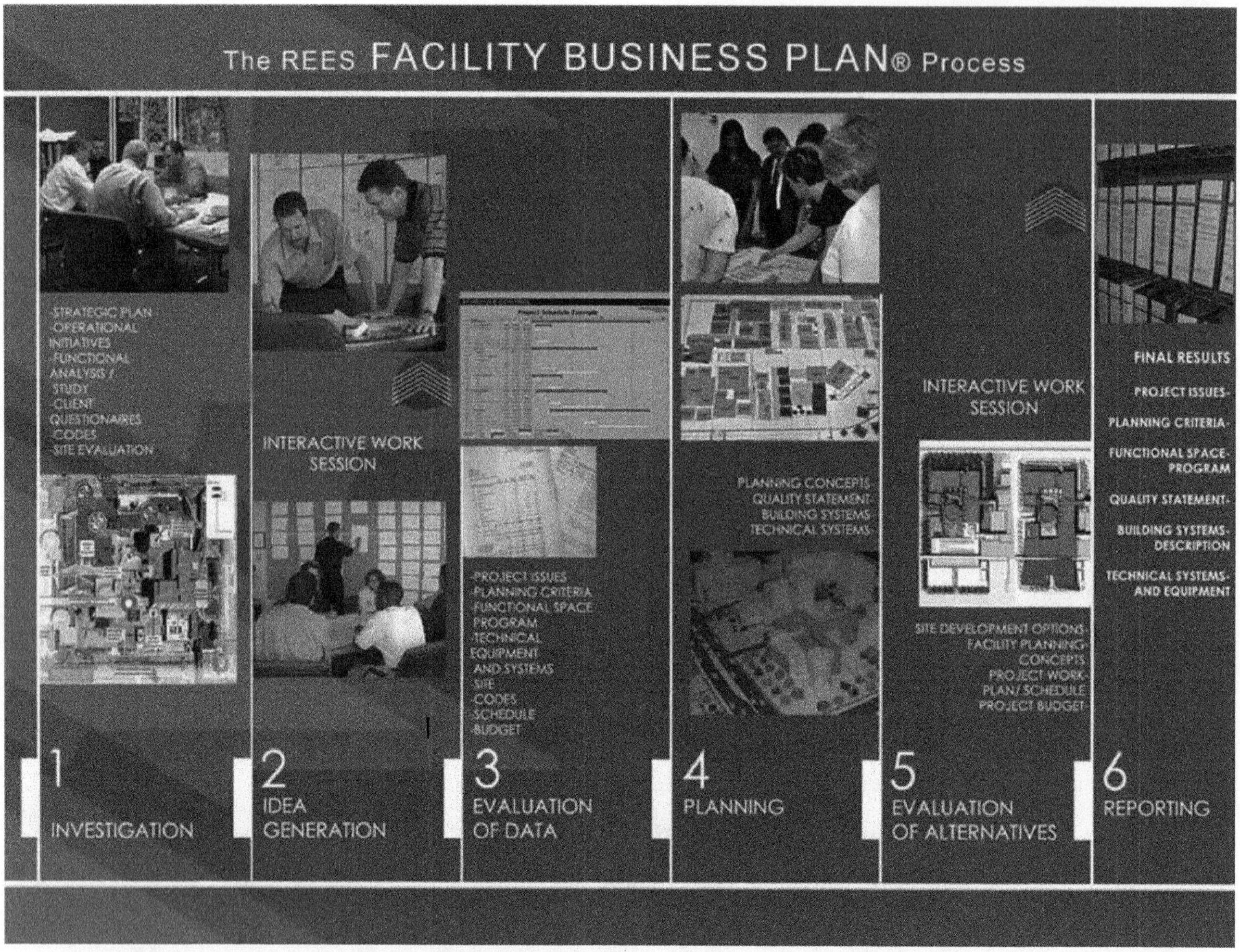

FIGURE 5.2-3 Facility Business Plan process.

Courtesy of Rees Associates

Space Planning

Some of the issues that should be discussed and considered when planning the building layout include:

- Relationship of sound-sensitive areas to other areas of the facility;
- Relationship of studios to storage (primarily for sets and properties);
- Relationship of control and edit facilities to equipment and other technical areas;
- Relationship of technical areas to any and all areas that require technical support or cabling;
- Relationship of visitor and employee areas;
- Departmental relationships (Sales to Traffic to Business, Master Control to Engineering, etc.);
- Security (now a very important consideration);
- Duopolies (multiple stations operating out of one facility).

(See Figure 5.2-3.)

DESIGN CONSIDERATIONS

Studios

Size and Height

Modern studios should be able to handle television production with wide-screen pictures. Cameras that shoot 16:9 have a wider aspect ratio than the traditional 4:3, resulting in a need for wider sets, which ultimately can require wider studios. It might appear possible to reduce the height of the studio for a given horizontal angle of view when shooting 16:9. However, other considerations, such as the height of suspended lights and air conditioning ducts at high level, also come into play, as well as the need for shooting low-angle shots. These factors indicate that the height of studios should not be reduced just because wide-screen television is being introduced.

It should be noted that for high-definition (HD) television, sets must be well defined and more detailed than in the past because HD will show up any and every imperfection of the set.

Floors

All efforts should be made to ensure that a super-flat concrete floor slab is placed. ASTM "F" numbers of F_L 55 (levelness) and F_F 35 (flatness) should be target levels.[1] When met, these targeted levels are more than capable of allowing the operation of pedestal-mounted cameras without visible camera-shake.

Some studio floors are finished with a self-leveling epoxy-type surface treatment but, with a well-finished slab, all that may be required is a good concrete seal. This allows the production unit to paint the floor or add whatever surface is required for particular production needs.

Air Conditioning

Air should be delivered to the space with high volume and low velocity to provide the cooling required for the thermal load of lights and equipment with noise levels to meet the acoustic criteria for the studio.

Acoustics

Background noise levels and reverberation should be appropriate for the studio application. While all studios need to be quiet, exactly how quiet depends on the application, as does the amount of reverberation. Not all studios need to be totally dead.

Target acoustical noise criteria for different types of studios and methods of controlling reverberation with sound absorbers and diffusers are mentioned later in this chapter and discussed in detail in Chapter 4.1, "Principles of Acoustics and Noise Control for Broadcast Applications."

Doors

Doors should be sized and of suitable construction (based on acoustic criteria) according to the ultimate usage of the studio. Typically, wider and higher than normal doors are provided for the movement of large scenery and set pieces. The opening itself is usually a minimum of 10 ft wide and high. Doors into studios, edit facilities, and other sound-sensitive areas should be coordinated with the sound transmission class (STC) rating of the surrounding construction (see Chapter 4.1).

Sound Vestibules

To reduce noise ingress into a video production studio, *sound vestibules* or *sound locks* should be included in the facility design, as shown in Figure 5.2-4. These vestibules should be the main entrance into the studio for all audience members and, in some cases, crew, talent, and cast. In most cases, the doors will need to open out, as part of the egress requirements from the space.

Studio Control Rooms

Direct line of sight from the control room into the production studio is not as necessary as it may have once been. In fact, when control rooms are constructed with windows into the studio, they are often eventually obstructed with other materials or props. It is not usually necessary to provide windows for direct line of sight unless there is an overriding functional reason for the productions being planned for the studio.

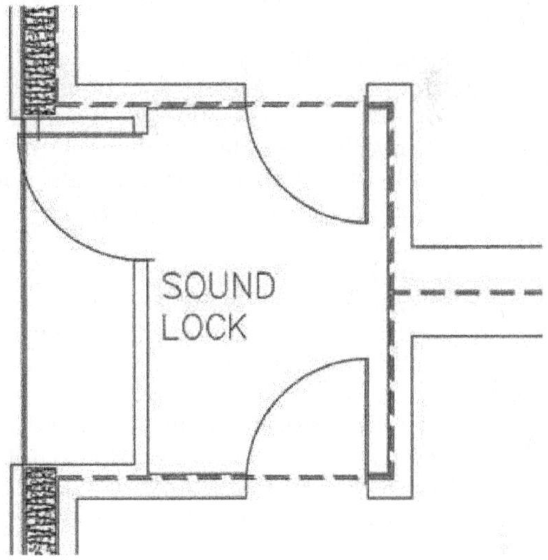

FIGURE 5.2-4 Studio vestibule/sound lock.

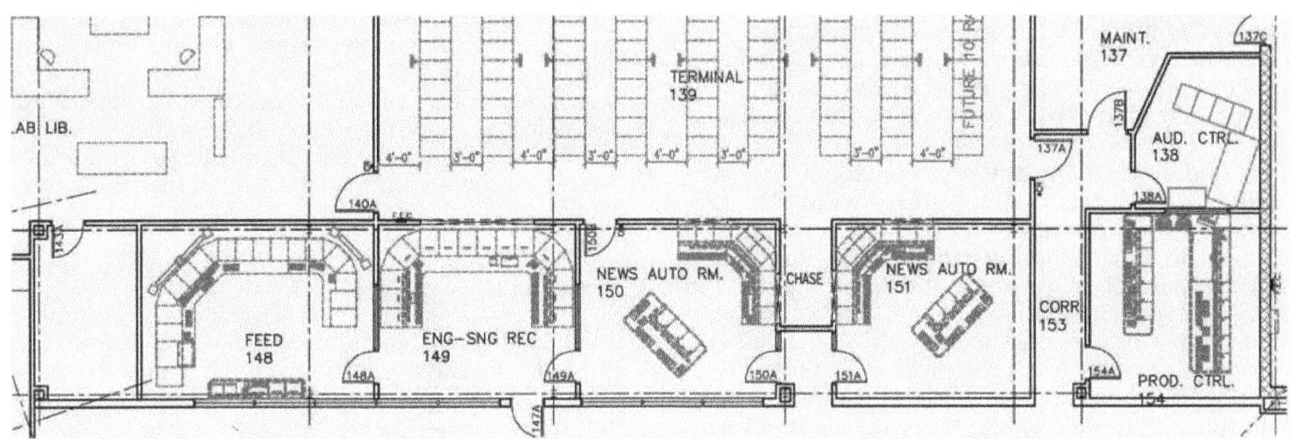

FIGURE 5.2-5 Plan of a typical television control room suite.

One valid reason for providing windows into control rooms, however, is to allow visibility for visitors touring the facility. In this case, observation windows can be strategically placed to allow viewing from an adjacent area in a manner that will not disturb the control room operations.

Inside the control room suite, it is still desirable to provide a visual link from the audio engineer to the remainder of the control. Often audio control will be in a separate room to provide acoustic isolation, and a window between the two areas will be needed.

Figure 5.2-5 shows the layout of a typical suite of control rooms for a studio, with other rooms for news operations and feed control.

EFFECTS OF DIGITAL TECHNOLOGY

Changes to the "Tech Core"

"Tech core" for the purpose of this discussion includes all areas such as ingest control rooms, edit rooms, master control, production control, news control, audio control, electronic rack rooms, server rooms, IT, and engineering areas.

A good design usually will address these areas by grouping rooms with related functions as close together as possible, which is generally good for station operations. In addition, placing as many technical rooms as possible on raised modular access flooring in turn addresses the primary cabling issues of most facilities.

The switch to digital technology has changed the size and quantity of racks in the tech core. There tends to be more racks that are more densely populated. While it is true that most pieces of digital equipment are getting smaller for similar or increased functionality, more equipment is often being populated within the racks. This has resulted in a deeper rack (36-inch depth is commonly used), with more power dissipation and a need for more air conditioning to keep the equipment cool.

Air Conditioning

The tech core requires air conditioning systems that maintain humidity and temperature within close tolerances. These conditioning systems require redundancy in order to prevent equipment from shutting down or being damaged due to adverse temperature conditions if an air conditioning unit fails.

Some digital equipment is more temperature sensitive than analog, and temperatures within the tech core should not rise above 85° F. Systems should be designed to provide uninterrupted temperature and humidity control. One approach that is employed to respond to this 365/24/7 criteria is the use of computer-room-type air-conditioning (CRAC) units with built-in redundancy (dual compressors or with chilled water with direct expansion backup). Effective temperature and humidity control is easily achieved with these units when used in conjunction with a properly placed vapor barrier.

Technical Power

Just as digital equipment is sensitive to temperature, it is also power sensitive and tends to be unforgiving when it comes to power disturbances. Clean and reliable power is critical to a digital facility.

A site with two different sources of power is definitely a plus, and this is a consideration when selecting a site for a new facility. Two power supply lines from the power company (if truly independent) would allow for a redundant system design with either an automatic transfer switch or a conveniently placed manual transfer switch. However, when this is not available, some type of generator backup is usually essential. The generator should be sized to back up all critical loads of the facility, including those systems that serve the tech core.

An uninterruptible power supply (UPS) or electromechanical flywheel may be required to bridge the gap between power failure and the time when the generator comes online. In any case, given the sensitivity of modern equipment to power line disturbances, a large UPS supplying all tech core electronics is highly desirable. The alternative arrangement, of many smaller units distributed with individual items of equipment, is much less satisfactory and difficult to monitor and maintain.

Changes to Control Rooms, Edit Suites, and Post Suites

The advent of smaller equipment for digital production switchers and editing systems has reduced the spatial requirements for equipment within the design of these particular rooms, but the number of people wanting to occupy these rooms has not changed radically. So these areas should be designed around the number of people involved in the action, with less consideration needed for the amount of equipment than in the past.

The bulky consoles and huge walls of monitors are now largely obsolete. Today, many consoles are replaced with simple, functionally tiered desks (see Figure 5.2-6). Monitor walls are usually composed of a series of flat panel displays or a projection system that allows multiple-source viewing.

Totally networked facilities can reduce the number of traditional viewing and editing bays. Initial video browsing and editing can now be performed at the desktop. This capability is becoming increasingly cost-effective and therefore more staff may require quicker and easier access to these features. This implies a need for more working spaces with these facilities in offices.

Cable Management

Most facilities still require some traditional coaxial and twisted pair cable interconnections. However, nearly all modern equipment requires networking, and copper- or fiber-based networks are now part of the landscape in digital facilities. Wireless network systems may be used in some limited applications.

FIGURE 5.2-6 Edit room.

Courtesy of WJLA, Allbritton Communications, Arlington, VA

A properly designed facility will allow for multiple changes in technology, which inevitably will require change in equipment locations and replacement of wiring. In order to respond to these changes, a complete and cohesive cable management system is required, including all or some of:

- Modular access floor systems (where appropriate for the room);
- Overhead cable trays;
- In-floor duct banks and conduits.

Where multi-floor technical facilities cannot be avoided, fully accessible cable risers with cable support systems must be provided. How and where each of these methods for handling cables is incorporated into the design depends on the type and location of each room, the constraints of the overall building design, and, to some extent, how the end user sees the facility ultimately functioning.

Whichever methods are selected for running cables, part of the management system design is to separate different categories of cables in order to minimize interference between cables carrying different types of signals.

Audio

Quality acoustically treated spaces must be provided for audio production and monitoring. In the past, quality sound was not always a priority for television broadcasters. Today's viewer demands a higher quality of sound. Along with the great improvements in video, broadcasters and production facilities are expected to provide similar results in audio. Provision will need to be made for high-quality monitoring of stereo and, increasingly, 5.1-channel surround sound.

Technical Furniture

In today's environment, the more flexibility that can be designed into a facility the better. The use of manufactured movable furniture instead of typically built-in millwork is one approach to ensure flexibility of space. Areas such as edit bays, engineering technical benches, and audio consoles, to name a few, are areas to consider for movable modular furniture such as that shown in Figure 5.2-7. This will allow areas to be easily repurposed when the need arises. The use of premanufactured instead of built-in furniture will, however, have an impact on the budget of the facility.

(a)

(b)

FIGURE 5.2-7 Modern modular premanufactured furniture.

Courtesy of Winsted Corporation

Facility Monitoring Systems

Many facility support systems now have computer-based control systems, providing the ability for centralized monitoring. The most commonly monitored systems are those for energy management, security, phones, networks, and fire and safety. Discussions with the many different vendors regarding plans for using such monitoring will assist in selection of the appropriate manufacturer and system for each application. The level of monitoring of facility systems will also have a direct impact on the project budget.

Tape Storage

Servers are being used increasingly throughout the broadcast and production industries. Many in the industry believe that ultimately such disk-based systems will replace tape-based technology completely. This may eventually happen, but until that time there is still a need for tape archives and storage space.

Most facilities will need some tape storage for the foreseeable future, both for legacy analog and digital tape material. Other digital media such as CDs and DVDs are also still going to need a home. The use of high-density storage systems, as shown in Figure 5.2-8, should be considered. Such systems may appear at first to be expensive but, when compared to construction cost, they may be cost-effective, especially for a large library. High-density storage will also have an impact on the budget.

BUDGET DEVELOPMENT

When project budgets are developed, they should so far as is possible include all "probable" or "potential" project costs. These costs may include, but are not limited to, the following:

- Land cost;
- Land surveys;
- General construction of the facility based on a gross square foot cost. General construction costs cover all exterior and interior building costs for the facility. Renovation will be different from new construction. Gross square footage should include all the net square footage of assigned spaces in the facility, with a gross-up factor, plus percentages for circulation, mechanical, electrical, plumbing, and building envelope;
- Moving or adding additional satellite dishes;
- Construction of a new tower, if needed;
- Any special construction required (such as dry pipe fire suppression, hurricane protection, etc.);
- Furniture, furnishings, and accessories;
- Fees for architecture, engineering, technical reports (such as geo-technical and environmental reports, etc.);
- Any radio frequency (RF) studies, if required;
- Any terrestrial studies, if required;
- Communication systems;
- Computer systems;
- Security system;
- Broadcast equipment systems;
- Bid escalation for time that the project maybe on hold;
- Design and construction contingencies.

Timing of construction is increasingly a major cost concern in the construction industry. Costs for new or renovated construction have escalated in recent years and, at the same time, the skilled labor force has dwindled. So the earlier the process can be locked down, the better chance there is of controlling the overall cost of the project.

The U.S. Green Building Council has introduced a Leadership in Energy and Environmental Design (LEED) "green" building rating system.[2] LEED-friendly design and/or certification should be discussed in the initial stages of the project, primarily because of the cost implications required to accomplish true LEED certification. If LEED certification is desired, it must start at the beginning of the project because of the documentation required from initial design through the construction process. Buildings can incorporate green design techniques if desired and not have to be certified. A few of the green techniques that can be employed include:

- Gray water irrigation;
- Lights-on motion sensors;
- Recycled materials and programs;

(a)

(b)

FIGURE 5.2-8 High-density storage systems.

Courtesy of Russ Bessett Corporation

- Energy management systems that monitor carbon dioxide levels and that allow for economizer cycles. These allow fresh air to condition the space;
- Waterless waste technologies.

DESIGN AND CONSTRUCTION PROCESS

There are several different scenarios that may be used in delivering a project, including:

- Design, bid, build;
- Design/build;
- Fast track, or negotiated.

The following sections describe the process for each scenario.

Design, Bid, Build

The phases of a typical *Design, Bid, Build* process are as follows.

Schematic Design

During the schematic design (SD) phase, the architect will interpret the functional space program that is usually provided by the owner and reflect this interpretation in a floor plan as scaled representation of the new or renovated facility. During this time, aesthetics, materials, finishes, schedule, and budget will begin to be identified and discussed.

Design Development

After approval of the schematics, the design team will move into design development (DD). During this phase the project is further defined in terms of actual wall thickness, interior and exterior window systems, and selection of all interior and exterior finish systems and materials; exterior elevations and building sections will be generated to assist in defining the building vertically as well as horizontally. All design decisions must be made prior to finishing this phase of the project. Also, at this point 3D sketches will begin to help clarify the overall aesthetic of the facility. Budget and schedule will again be discussed and updated.

Construction Documents

After approval of the design drawings, the design team will move into preparing the construction documents (CD). With all decisions made in DD, the project team now can build the project on paper. This is the technical exercise that allows the team to do all the final designing and documentation before putting the project out for bid.

Quality Assurance

Before any project goes out for bids, a third-party review or quality assurance (QA) review should be performed. This effort is the final step of coordination of all the disciplines—civil, structural, architectural, mechanical, electrical, and plumbing—that needs to take place before the CDs are given to contractors for bid.

Bid/Negotiation

With a bid/negotiation (B/N) type of project, this is the stage in which several preselected and prequalified contractors, usually no more than three, will be provided with the CDs and asked to formulate their bids (price) for the project that is defined in the documentation. Typically, a three- to four-week period is allocated for the bid process, and, at the end of this time, bids are submitted. From the bid process, a contractor is selected and a contract for construction is negotiated. Once the contract for construction is executed, construction will usually start.

Construction Phase

Two points regarding the management of the construction process should be noted:

- *Point of Contact:* When construction starts, the owner should assign one individual as the main project contact. This individual will be the contact point between the design team and the contractor from day one through the end of construction. In fact, it is best if this individual is assigned during the initial planning or SD phase of the project. Direct contact by others with the contractor during this time should be discouraged and not allowed.
- *Change Orders:* No design team is perfect, and there will inevitably be change orders in most construction projects. However, change orders add to the cost of the project, and the key is to minimize them, if at all possible. Allowing a project to proceed in a methodical and logical progression is one of the steps that will help in the minimization of change orders. Well-coordinated and documented CDs will also minimize change orders.

The design process outlined in the preceding sections represents a conventional design approach for a "hard" or "competitive" bid project. While it is methodical and logical, the downside is that the contractor is not a design team member. The contractor comes in only at the B/N phase and then proceeds with construction. Waiting this late to bring a contractor into a project may create a more adversarial project environment.

Design/Build and Negotiated

An alternative approach for project implementation is known as *Design/Build*. In this arrangement, the contractor and architect, as a team, design and construct the project. A single contract is executed between the owner and the design/build team with an agreed price for the project. The team then proceeds to deliver a complete project within the budget and on schedule.

Another delivery method is that of the *Negotiated* approach for the construction part of the project. In this arrangement, the owner has already selected the architect, and the owner and architect together preselect a group of general contractors for interviews. These firms may be selected by a list of predetermined qualifications such as previous experience, personnel, and/or cost. The successful contractor is then included in the design team from the beginning of the project and through construction. The contractor firm is then available to monitor cost estimates and consult on constructability. Many customers have found this approach fits a broadcast facility project well and has the best track record of keeping a project within the initial budget and on schedule.

Selecting a General Contractor

The general contractor's expertise is a necessary part of the project's success, with skills that should include project scheduling, keeping within the budgeted dollars, material selections, and constructability for the particular region of the country. This key player must also have a good knowledge of technical facilities and understand the importance of flexibility for the future.

The selection of the appropriate general contractor is one of the most crucial decisions a broadcast facility owner has to make when planning and constructing a broadcast facility. There can be serious challenges from the first day if the selected contractor is not suitably qualified.

When selecting a general contractor, there are four main areas to evaluate:

- *History:* A consistent management philosophy should have been established if the firm has been in business for at least ten years. This should be prevalent in the interviews of personnel during the selection process.
- *Reputation:* This consideration includes not only the firm, but its leaders and its relationships with other subcontractors. Does the reputation coincide with your company standards? Does the reputation reinforce the management philosophy? Does the reputation hold with subcontractors?
- *Financial stability:* The selected firm's net worth should be worth at least half of the projected project cost of the planned project.
- *Record:* What is the record of the firm concerning technical facilities? As part of the qualifications, request a list of projects the firm has completed, with total construction costs, sorted by year. Determine the bonding limit of the contracting firm and make sure this dollar amount is at least twice the construction budget. This guarantees the performance of the contracting firm, and the insurance company can provide valuable information on the company. Ask for detailed information of the last three technical facilities projects the firm has completed. The information should include each project's location, construction budget, completed cost, the year built, and a person to contact regarding the project at the facility. If there are no technical facilities, then the contractor is not qualified.

CONSTRUCTION

Exterior Construction

The building exterior should be designed to adequately isolate extraneous noise at the site. In some cases, interior sound isolation construction should be resiliently decoupled from the exterior construction. Roofs over sensitive spaces should be designed to adequately attenuate rainfall impact noise and other environmental noise.

Selection of exterior cladding materials should take account of the area. For instance, the use of brick in northern areas is often undesirable because materials may be hard to come by and installation may be difficult due to the climatic conditions. In that case, a premanufactured panel of some type may be better value for money.

Selecting the proper building structure is critical to several areas of the project. This selection will have an impact on cost, schedule, and expandability of the facility. Whatever form of construction is chosen should allow for ease of reconfiguration or future modifications. Steel or concrete framing will usually allow this. A load-bearing wall structure is much less flexible and does not easily reconfigure. Obviously, when remodeling an existing building there is not as much flexibility with the structure, so the interior planning must allow for ease of reconfiguration.

Interior Construction

Walls

Interior construction should usually be primarily based on partition walls built with steel studs. This type of construction allows for easy reconfiguration as the partitions are easily moved or removed and can be insulated for acoustic isolation when required. In areas such as studios, with critical acoustic requirements, grout-filled concrete block walls are commonly used if dead load is not an issue. Otherwise, various designs of stud walls with multiple layers of cladding and insulation may be used as discussed in a later section.

Ceilings

Areas without special acoustic requirements may use conventional suspended ceilings with lay-in ceiling tiles. However, utilities such as water, electrical, fire suppression, wiring, communications, and data are often run in the ceiling voids of corridors and other areas. In some areas, technical cabling is also run in overhead cable trays. To allow access to these overhead utilities and cabling, it is recommended that the use of suspended ceilings be eliminated wherever possible. Clearly aesthetic considerations may require the use of ceiling tiles in some areas such as offices, but their use should be kept to a minimum.

Areas that require acoustic isolation may require special sound-isolating ceilings as discussed in a later section.

MECHANICAL, ELECTRICAL, PLUMBING, AND FIRE PROTECTION ENGINEERING

Main Service

A new main electrical service, typically fed from a utility company pad-mounted transformer, will be required. Transformer secondary distribution voltage should be 480Y/277 volts, three-phase, four-wire with a solid ground. The main switchboard and each level of distribution for the technical, mechanical, and studio lighting power systems should be equipped with separate transient voltage surge suppression devices in accordance with IEEE standards.

Secondary Services

Reliability of the secondary electrical distribution system should be a major design consideration to eliminate single points of failure in the critical power path. Separate panel boards should be provided for emergency, technical, mechanical, studio lighting, and normal power.

Emergency Backup Generator

A standby diesel-driven generator set should be provided, either located in a heated acoustical weatherproof housing or in an acoustically treated room within the building, with either a skid-mounted or external fuel tank sized for a minimum 24-hour run period.

UPS and Flywheel

All technical power systems that serve broadcast processing equipment, control equipment, technical equipment, and non-broadcast-related LAN equipment should be served by a central three-phase static or rotary UPS system. UPS systems should be backed up either by lead acid batteries (sealed or flooded cell) with a minimum of 5 minutes supply duration or a rotary flywheel for 20 seconds duration, with a backup 24-volt supply from the UPS to start the standby generator.

Air Conditioning

A distribution of conditioned air should be provided to all occupied spaces in the building, with the exception of mechanical rooms, using either packaged rooftop units or air handling units located in mechanical rooms. Cooling may accomplished by central chiller units with a chilled water distribution around the building or by unitized direct expansion equipment located where required. Associated ductwork, terminal units, diffusers, registers, return fans, and associated controls are normally required for a complete system. Each rooftop unit should include air filters, hot water preheat coil, direct-expansion or chilled water cooling system, centrifugal supply and return fans, double-wall casing, stainless steel double-sloped drain pan, and vibration isolation roof curbs. Where required for noise considerations, duct silencers should be provided.

Each studio, control room, or other acoustically sensitive space should be provided with a dedicated constant-volume terminal unit served from a variable volume air handling unit. Care must be taken in the location of the rooftop units, terminal units, selection of fans, and sizing of ductwork, to achieve the required acoustic and thermal goals for each space.

Technical equipment areas on raised access flooring should be served by cooling-only units. Air may be ducted from above ceiling to precision registers or through the raised floor air grilles. Redundant air conditioning units for these areas should be taken into account for both maintenance and failures. The technical area spaces should also be furnished with a humidification system to allow constant humidity to be maintained.

Technical Grounding

The facility should be provided with a low-resistance grounding system consisting of ground rods around the building perimeter. The grounding system should be connected to structural steel and to the lightning protection system. An isolated signal reference ground system should be provided for all broadcast and technical equipment, with a single point ground, as discussed in Chapter 11.2, "Facility Grounding Practice and Lightning Protection," and in Chapter 4.2, "Planning a Radio Studio or Audio Production Facility."

Energy Management

A central energy management control system should be used to monitor and control the air conditioning and heating systems. The system should utilize direct digital controls (DDCs), with variable volume terminals equipped with DDC controllers. This central system should monitor and alarm all critical temperature and humidity levels in technical broadcast and information technology equipment spaces. Such systems utilize remote paging options to alert personnel of approaching potentially serious conditions.

Fire Suppression

The facility should be furnished with a complete fire protection system that meets building code requirements. Sprinkler heads should be furnished throughout the facility to accomplish total coverage. The technical areas should be provided with a pre-action double interlock sprinkler system actuated by any two detectors (connected to the fire alarm system).

An addressable fire alarm system should be provided for a new facility. Smoke detectors should be provided throughout the building and also manual pull stations at egress points as required.

In the technical equipment spaces, an FM 200 or FE-25 gaseous extinguishing system should be considered for the areas above or below the raised floor.

ACOUSTICAL DESIGN CRITERIA

The following criteria can be used as a guide for the required maximum background noise levels for mechanical and electrical systems, in terms of Noise Criterion (NC) ratings. These criteria can also be used to characterize the sensitivity of each space to outside noise intrusion.

TV/Production Facilities	
Large production studio	NC 20–25
Small production studio	NC 20–25
Production control rooms	NC 25–30
Audio control rooms	NC 20–25
Announce booths	NC 20–25
Master control rooms	NC 30–35
Instructional studios	NC 25–30
Editing suites	NC 25–30
Client editing suites	NC 20–25
Audio sweetening rooms	NC 20–25
Radio	
Performance studio	NC 20–25
Announce booths	NC 20–25
Control room	NC 20–25
Editing rooms	NC 25–30
Miscellaneous	
Offices	NC 35–40
Conference rooms	NC 33–35
Technical workrooms	NC 35–40
Data center/core equipment rooms	NC 40–45

ACOUSTIC SEPARATION OF SPACES

Requirements for acoustic separation of the various program spaces should be established based on the sensitivity of the spaces, as defined in preceding sections, the sound-generating characteristics of each space, and the specific space adjacencies as developed in the building layout.

To the maximum extent possible, the building layout should take advantage of opportunities to minimize or eliminate expensive sound isolation construction—for example, decoupling certain elements of the building structure, providing buffer space such as corridors and storage rooms between sensitive spaces, locating noisy shop spaces remote from studios, and avoiding vertical adjacencies between the most sensitive spaces.

Most facilities have two general levels of physical acoustic separation required in critical spaces, with requirements as follows:

Level I

Floor: Typically, there are no special isolation requirements for on-grade floors, unless there are special ambient circumstances such as an adjacent railroad or holding area for trucks. Above-grade floors may possibly require heavier than normal construction for insulation and isolation of airborne noise from vertically adjacent spaces.

Walls: Typical sound walls are of multiple layers of gypsum board on both sides of heavy-gauge steel framing, extending full height, and sealed to the structure above, with full-depth batt insulation in the stud cavities. Construction details are provided in Chapter 4.1.

Ceiling: Typical suspended acoustic tile ceiling, or no special requirement.

Level II

Floor: An isolated floor may be required within each room, consisting of a 4- to 6-inch concrete slab supported by a mat of fiberglass insulation with resilient isolators spaced on 12- to 16-inch centers atop the structural slab, as illustrated in Figure 5.2-9. The isolated floor should be decoupled from all other elements of the building construction.

The structural slab may require depression to resolve floor elevations between isolated and non-isolated spaces, and cabling pathways will need to be incorporated into the isolated floor design as required. If the actual structural floor slabs between adjacent sensitive spaces can themselves be resiliently decoupled, for example, with expansion joints, then the isolated concrete floor may not be required.

Walls: Walls should consist of separate multiple layer gypsum board assemblies for each space being isolated, supported on the respective isolated floors and decoupled from the structure above with resilient angle brackets. Construction details are provided in Chapter 4.1.

Ceilings: Ceilings should consist of multiple layer gypsum board assemblies supported with combination spring and neoprene isolation hangers, as shown in Figure 5.2-10, with all services surface mounted or suspended below the sound barrier. The ceilings should be decoupled from all wall and other surrounding constructions. When double ceilings are required for additional isolation, one ceiling will be of gypsum board construction as shown in the figure and one will

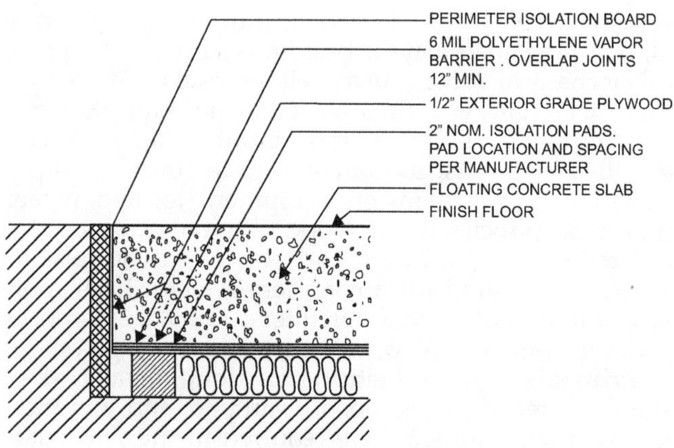

FIGURE 5.2-9 Sound isolation floor.

be lay-in. The lay-in ceiling is 1- to 1.5-inch thick glass fiber with a canvas face or other finish—not the typical lay-in ceiling found in offices.

Other acoustical treatment items for Level II areas include resilient details and seals at all duct, pipe, electrical/other wiring infrastructure, door, and window penetrations.

Doors

Three types of sound-reducing doors are typically used in sound-critical areas, depending on the use and acoustic isolation required.

- *Pre-engineered sound-rated door and frame assemblies.* These have integral doors, frames, and acoustical seals. They comprise heavy gauge, stiffened and insulated door panels, heavy gauge frames, magnetic head and jamb seals, cam-lift hinges, and fixed bottom seals. Glazed panels incorporate insulated laminated glass.
- *Standard doors with field-installed acoustical seals.* These assemblies typically have insulated door panels, standard frames, and field-installed adjustable perimeter gaskets, including an automatic door bottom seal.
- *Sliding glass doors.* These assemblies will typically be exterior weatherproof type with insulated glass assemblies incorporating laminated glass. In some cases, double sliding doors may be required.

Isolation for the most acoustically sensitive spaces will require sound locks—that is, two sets of doors separated by a vestibule. For sound locks, the inner door should be the sound-rated type and the outer door may be either a sound-rated door or a standard door with field-installed acoustical seals.

Acoustical Room Finishes

Substantial portions of the wall and ceiling surfaces in most of technical/production spaces will require sound absorption treatment of some type. Sound absorptive treatment may consist of rigid acoustic insulation between 1 and 6 inches thick, covered with sound transparent facing such as fabric, either in premanufactured panels or a field-fabricated system. Some spaces may be provided with special suspended ceilings as discussed previously, depending on the sound isolation requirements of the area.

Broadcast production studios should have sound absorptive acoustical treatment on the full extent of the wall, to grid height, and also on ceiling surfaces. The treatment can consist of 2- to 4-inch rigid acoustic insulation with protective sound transparent facing such as metal mesh or fabric on nailing strips, as shown in Figure 5.2-11. Studios will typically also require special units for absorbing low frequency sound and for sound diffusion. Further information on sound absorption and diffusion is provided in Chapter 4.1. Television studio floors are almost always left with a hard level surface finish to allow for smooth camera movements.

Critical audio listening and mixing rooms will require acoustic treatment with sound absorption materials and usually also require prefabricated sound-diffusing elements on the walls and/or ceiling surfaces.

Floors in control room, master control, and technical operation areas are typically high-pressure laminate on computer access flooring systems. Although sometimes used in control rooms to assist with acoustic absorption, carpets should generally be avoided in these areas. Even with "static-resistant" carpet, there may be some static discharge potential over the life of the material that can give rise to problems with equipment. When planned correctly, these rooms should be located in close proximity so as to limit the amount of access flooring required and to minimize cable runs.

Office areas are treated very much in the manner of typical office space, with lay-in ceilings, paint wall finishes, and carpet or vinyl tile flooring. However, in sound-critical environments the use of hard reflective surfaces should be minimized. For example, there should not be a hard floor surface next to an edit bay or control room without allowing for proper acoustic isolation. The treatment would need to isolate the potential noise-generating surface from the interior of the room, to minimize any sound transference between spaces.

Mechanical Systems Noise Control

The noise and vibration control design for the building mechanical systems should anticipate and plan for the following features:

- Physical location of major equipment remote from sound-sensitive spaces, on grade where possible;
- Selection of quiet type (low pressure) air distribution systems to serve the most sensitive spaces;
- Selection of quiet type equipment;

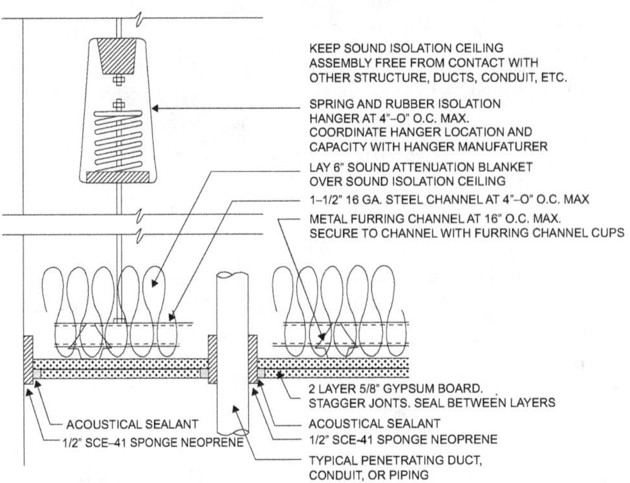

FIGURE 5.2-10 Sound isolation ceiling.

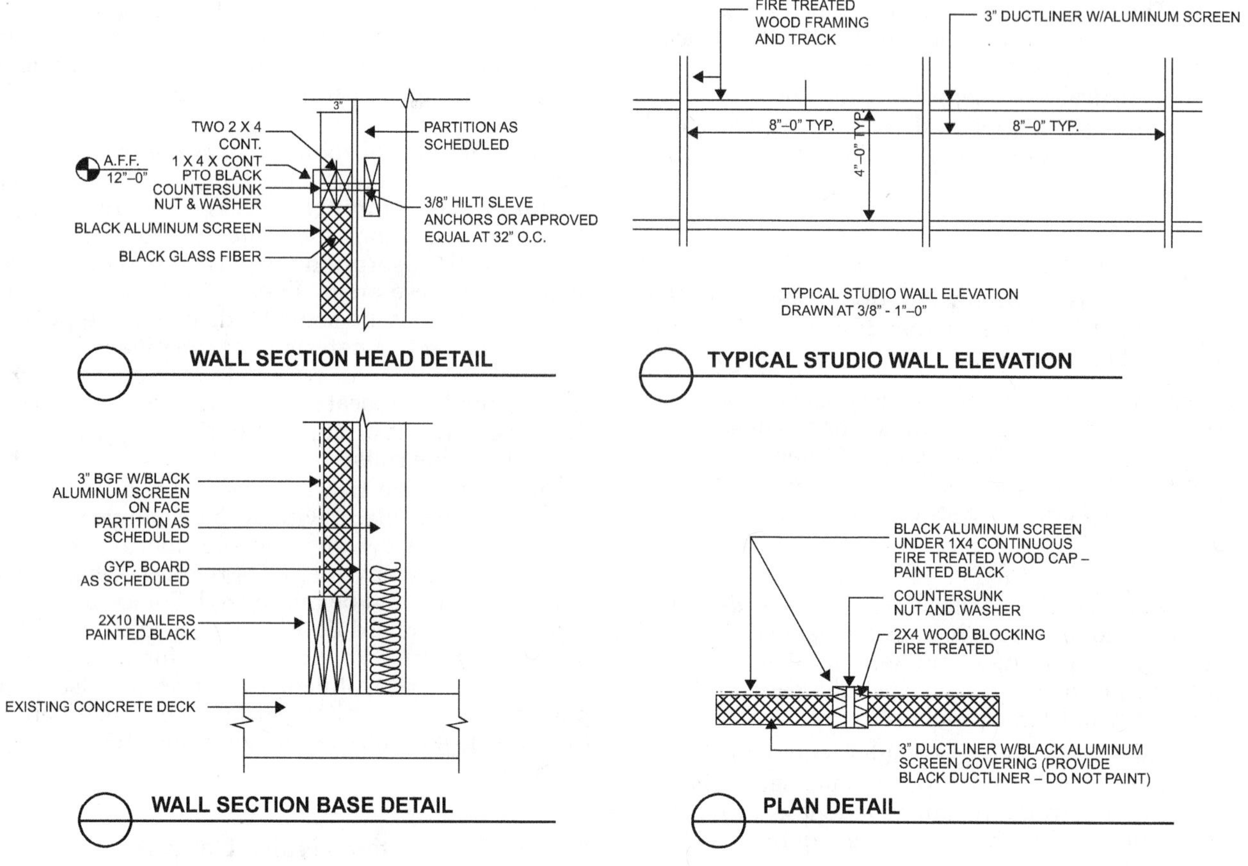

FIGURE 5.2-11 Wall acoustic treatment detail.

- Long lengths (30 to 60 ft) of duct run between air handling equipment and sensitive spaces, to allow dissipation of low-frequency noise;
- All ductwork serving critical spaces to have 1-inch or 2-inch internal acoustical lining;
- Ductwork to be sized and routed to minimize regenerated noise due to air turbulence;
- Ductwork to be designed according to guidelines for maximum air velocities, which in the most sensitive spaces will be as low as 300 or 600 feet per minute;
- All above-grade equipment to be vibration isolated from the building structure to minimize structure-borne noise and vibration. To the extent necessary, the pump and piping systems to be vibration isolated, and air handling units to incorporate flexible duct connections;
- Selection of quiet studio exhaust system equipment, with remote location to allow natural dissipation of noise in the ductwork.

Electrical Systems Noise Control

Power

Electrical power systems, including major substations and transformers, should be reviewed with respect to potential structure-borne noise excitation. Power transformers may require vibration isolation from the building structure.

Studio Lighting

Lighting systems should be reviewed with respect to the applicable background noise criteria for the studio and the noise characteristics of the fixture and control systems. Lamps and dimmer systems should be selected accordingly.

Unusual Ambient Noise Conditions

Sometimes a site is selected for reasons outside the control of the design team and they are left with site conditions that are less than desirable. An example would be a site that is directly in the flight path of an airport, or with an active railroad less than 30 feet away, or perhaps an active rock quarry not far away. Site conditions such as these clearly have major challenges for acoustic isolation, and the project will require the expertise of an experienced acoustician. An acoustic consultant should be selected who is familiar with the requirements of a broadcast or production facility. The acoustician should be brought into the design process early and made familiar with the application of the facility and how it will be operated. Sometimes the smallest detail can lead to substantial cost savings.

SUMMARY

The intent of this chapter has been to give an overview of the design and planning process that should be undertaken once the decision is made to build a new video facility or renovate an existing one.

To ensure success, it is imperative that a team of experienced individuals be assembled in the early stages of the project to work through the details of programming, planning, designing, engineering, scheduling, and estimating for the new facility and its technical systems. The team should be composed of an architect, engineers, contractor(s), and system integrators who have all worked together on a broadcast or production facility or, at a minimum, have several members with experience in this type of facility.

ACKNOWLEDGMENT

Thanks are extended to Julian Rachman of DFW Consulting Engineers, Irving, TX, for his help and advice in preparing this chapter.

Notes

[1] ASTM E 1155–96, "Standard Test Method for Determining FF Floor Flatness and FL Floor Levelness Numbers," see http://astm.org/

[2] U.S. Green Building Council, "LEED," see www.usgbc.org/

Bibliography

Gregg, Walter L. Preparing Your Building for Digital Television, *Broadcast Engineering*, September 1998, pp. 66–72.

CHAPTER

5.3

Lighting for Television

BILL MARSHALL

Updated for the 11th Edition by

CINDY HUTTER CAVELL

Cavell, Mertz & Associates, Inc.
Manassas, VA

INTRODUCTION

Lighting is a fundamental, critical component of the television process, requiring a unique blend of optics, mechanical, electrical, and electronic engineering, which are then coupled with the necessary artistic elements. Lighting systems, both for television studios and theatrical stage, have increased greatly in complexity and sophistication over the years as new technologies continue to evolve in light sources, robotics, and lighting control systems. This chapter discusses the basic requirements for lighting systems, including lighting fixtures, suspension systems, dimmers and control systems, as well as new developments in fixtures and controls.

Television lighting techniques are as equally complex and carefully crafted as those used in theater and film productions. The lower light level requirements of modern studio cameras, coupled with new developments in fixtures, lamps, and sophisticated control systems, allows complex multi-camera shoots to be produced in real time—something that would have been more difficult to undertake and endure in the early days of television.

Television is still a two-dimensional medium; therefore lighting techniques are critical in order to suggest three dimensions. Even the impressive resolution of UHD camera systems can only replicate an image as it is revealed by light and shadow. Good lighting will model the form of the subject and its surroundings while maintaining its relative balance of intensity to the rest of the image. As such, lighting design can be most accurately defined as *constructive* use of controlled light for a predetermined objective. With the advent of virtual sets, subject modeling and good use of light across a seamless background are even more critical.

BASIC PRINCIPLES OF LIGHTING

Many of the images created in television "shots" are close-ups; therefore, well-composed lighting of faces is the single most important task to be accomplished. The techniques for modeling faces with light, as used in still photography and later adopted for motion pictures and television, are simply a modern adaptation of the same approach used by portrait painters for centuries.

Three-Point Lighting System

The basic arrangement of lights for television lighting is commonly referred to as the *three-point system*. This arrangement endures, no matter what type of lamp is in the fixture.

Key Light

The first point is the *key light*, which is the principal illumination of the subject's face. It is placed on an axis or slightly to one side of the camera and generally at an elevation of about 30° above the horizontal line between the subject's face and the camera lens. The elevation should be adjusted relative to the particular structural elements of each subject's face such as length of nose or depth of eyes relative to the brow. The location in practical terms will obviously be determined by where or how the fixture can be supported. In studio work, this means the key light is very often suspended much lower than the general grid height of the rest of the studio fixtures to gain precise positioning. Typically, a "spot" light (light types are discussed below) is used as a key light.

Backlight

The second point of this lighting approach is *backlight*. Located above and to the rear of the subject, the backlight creates a glow on the hair and a highlight on shoulders, separating the subject from the background. The "spot" type of fixture is also used as a backlight, often with a set of barn doors to keep the light off of objects other than the subject's hair and shoulders. Light control devices such as barn doors are also discussed further on in the chapter.

The combination of a camera-mounted key light and stand-mounted backlight is the most prevalent arrangement used in electronic news gathering (ENG) and electronic field production (EFP) situations. The use of the camera-mounted key light often creates a somewhat less than ideal picture, but it has come to have direct association with realism.

Fill Light

The third element of the classic three-point lighting system is the *fill light*. A single key light can often provide modeling that is too severe in the absence of other ambient light sources, so fill lights are used to soften and blend the back and key light accents while maintaining their purpose of highlighting and separating the form.

Production Complexities

While a setup such as described here is relatively easy with one camera and one subject, the reality of television production is that there are generally multiple camera angles, more than one subject, and a complex background. The lighting director will need to expand the basics of the three-point system to cover all the subjects and all of the camera angles. This often necessitates clever fixture arrangements that require that a single fixture has more than one use throughout the course of the production. For example, in a typical two-person interview, the lighting design might be arranged such that one subject's key is simultaneously serving as the other subject's backlight.

The lighting director must also carefully consider the control of the shadow the light sources will create. For each camera angle, the key light and backlight relationship must continue as well as control of the shadow they create. Other common shadow problems, which are generally within the lighting director's control, are elongated nose and chin shadows that occur from improperly positioned key or backlight fixtures.

The use of shadows, shapes, or light patterns can be some of the most powerful techniques available to a lighting director in suggesting time, place, mood, or just pure background decoration. The character and quality of light are often more apparent from the shadows it casts. Shadows can be hard, soft, transparent, or even a different color than the apparent source. Because shadows are so important, major lighting arrangements should be carefully planned and drawn to scale. Section views in scale will help to predetermine where the shadows will fall, if there will be any scenic conflicts, and most of all, this preplanning will save time in the studio or on location.

Once the lighting of the faces is established, the lighting director plans the background lighting. While a TV program can occur in almost any conceivable background, the lighting of faces continues to determine the relative brightness of the backgrounds. The reflective value of a face ranges between 28% and 41%, while a white wall can be 96% reflective. A lighting director must therefore be careful to use lighting to maintain a proper balance since the viewer's attention will be naturally drawn to the brightest spot of a television picture. Lighting intensity, as well as scenic element selection and decoration, is critical to controlling the viewer's focus. A background can easily be over-lit, and then in contrast, it can overpower and appear to darken the skin tones. Elaborate scenic treatment may require all types of built-in specialty lighting and a whole assortment of fixtures for each unique requirement.

Lighting Levels

There is no one proper level of light, except as required to allow a particular camera to make a good picture under specific conditions. The level required will vary by camera type, lens, type of action, quality of teleprompter glass (mounted in front of the camera lens), existing ambient lighting conditions, and other purely aesthetic considerations. The lighting director must control the balance of the lighting for every given situation whether it is in brilliant sunlight or in a dark dramatic stage setting.

Cameras that have been developed in recent years for HD and UHD utilize improved pickup devices. Even for cameras with lenses that are not top of the line, light requirements for a modern camera are significantly less than they have historically been.

Color Temperature

For accurate color reproduction, television pictures need to be maintained within a consistent range of shades of color, called *color temperature*. Color temperature is measured in degrees Kelvin (°K) and varies dramatically between types of artificial light, time of day, and atmospheric conditions. A television camera must be "white-balanced" under the same lighting conditions as the scene that will be shot. Fortunately, television systems are actually quite flexible, permitting color temperature swings of ± 300°K without major effect to visible color rendition.

Outside light, such as incidental light from exterior windows into a set with artificial lighting, can present the greatest color temperature problem, since exterior daylight is at a substantially different color temperature and is extremely bright relative to the artificial interior lighting. Large sheet filters that correct both color temperature and intensity are available for this purpose; however, if the window area is extremely large and daylight dominates the scene, it is generally easier to light the set with daylight color temperature light sources rather than incandescent. The color temperatures that most studio or remote productions will deal with are incandescent (3200°K), daylight (5600°K), and architectural fluorescent (4700°K), which is not easily matched.

LIGHTING FIXTURES

The type of lighting equipment that is used in the studio differs considerably from that used "on location" outside of the controlled studio environment. On location, equipment is chosen for its light weight, portability, and efficiency, yet the on-location lighting designer must cope with varying natural light conditions from full sunlight to little or no light. Conversely, studios are designed and equipped to give the lighting designer complete control over all aspects of lighting: intensity, color, and placement.

There are two basic categories of lighting fixtures: *spotlights*, with directional, hard-edged light beams that produce distinct shadows; and a range of "other" fixtures that produce a softer light with diffuse shadows. There are numerous varieties of each type of fixture, and the focus of the light and area of illumination can usually be adjusted, sometimes by moving the internal position of the lamp or lens and always by using external accessories for particular applications.

Spotlights

In a studio, the most common and useful fixture is the *Fresnel lens spotlight*, as shown in Figures 5.3-1 and 5.3-2. The light from a Fresnel lens[1] is a diffuse but directional beam of light and has a smooth, even field. When the position of the lamp is adjusted behind the lens, the fixture will produce either a narrow spot or wide flood beam of light. Equipped with *barn doors*, this beam can be further shaped to virtually any pattern. Slots in front of the lens allow *color filter*, *diffusion media*, or *screens* to be inserted for intensity control. A complement of single and double thickness and half- and full-frame screens should be provided for all Fresnel lights even when dimmers are available. Fresnel lights are commonly used for key and backlights or anywhere direct lighting is required.

For most 14 ft high grid studios, Fresnel spots in 4", 5", 6", and 8" sizes, with wattages from 300 to 1000 watts, are the most common units. In smaller studios or tight applications, smaller fixtures are often employed. These units enclose the same wattage lamps in small housings. Many lighting directors prefer these "baby-size" units for their easier handling and somewhat different optical characteristics. In larger studios with grids higher than 14 ft, fixtures with higher wattages may be required. In those cases, the standard complement of fixtures might be 2- and 5-kilowatt units or even up to 10 kilowatts. Conversely, Fresnel lens units are also available in sizes down to a 2-inch lens with a 100-watt lamp.

FIGURE 5.3-1 Studio-quality Fresnel spotlights are more durable and have superior optical systems compared to some inexpensive theatrical units.

Courtesy of De Sisti Lighting

FIGURE 5.3-2 Studio-quality Fresnel spotlights must have long-leaf barn doors as an essential accessory.

Courtesy of Strand Lighting

In selecting a line of Fresnels, the studio lighting designers should evaluate all of the optical and mechanical features including:

- *Stability*. Poorly made fixtures will not focus properly once they are hot.
- *Balance*. Fixtures should be well balanced, even with barn doors attached, to remain in focus and not change position.
- *Available accessories*. These features include barn doors, screen sets, diffusion frames, and stand and grid mounting hardware.

Soft Lights

Soft lights are the second most common studio lighting fixture, as shown in Figures 5.3-3 and 5.3-4. They provide a controllable diffuse base or fill light. In this fixture the light sources are totally concealed and the light is reflected in an indirect manner. These units are not particularly efficient but are unequaled in providing a shadow-free light, which can be controlled with *egg-crate*-like attachments, and they are available in a variety of sizes. Typical units in the 1000 to 2000 watt range are popular as fill lights because they are easy on the talent's eyes. With lower light level requirements, soft lights have become the preferred broad base light source. Color-corrected fluorescent fixtures are often used in similar applications and offer large power and heat savings.

Area Lighting

Incandescent scoop or "broad" floodlights were once very commonly used in studios to build up high levels of

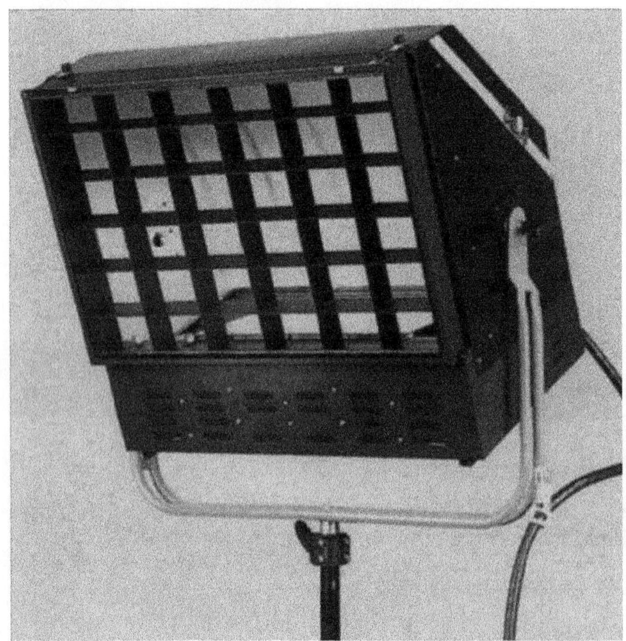

FIGURE 5.3-3 Spotlights fitted with egg crate louvers offer controlled, smooth, and even lighting.

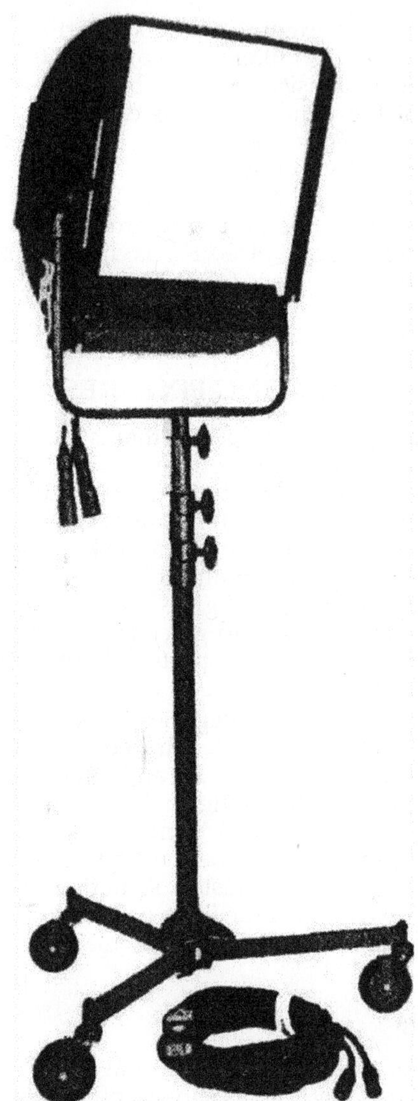

FIGURE 5.3-4 Soft lights produce a diffuse source and have switches to select as many lamps as necessary.

Courtesy of Mole-Richardson

base light that is no longer required for current-technology cameras. While the majority of lighting can be done with Fresnels and some type of diffuse source, various fixtures produce a different quality of light for special applications. For a harder, rectangular field of fill light, a "broad" may be used; this is similar to a scoop in concept, except its housing/reflector creates a somewhat less diffuse light. It too is always used with diffusion media. These fixtures are now most commonly limited to lighting backgrounds.

Other Fixtures

A popular and efficient incandescent fixture used in television studio lighting is the *PAR light* (parabolic aluminized reflector), as shown in Figure 5.3-5. The lamp itself contains both the reflector and the lens,

FIGURE 5.3-5 The Source 4 PAR provides similar output as a standard PAR 64 but at a lower wattage. It has interchangeable lenses for a variety of beam spreads.

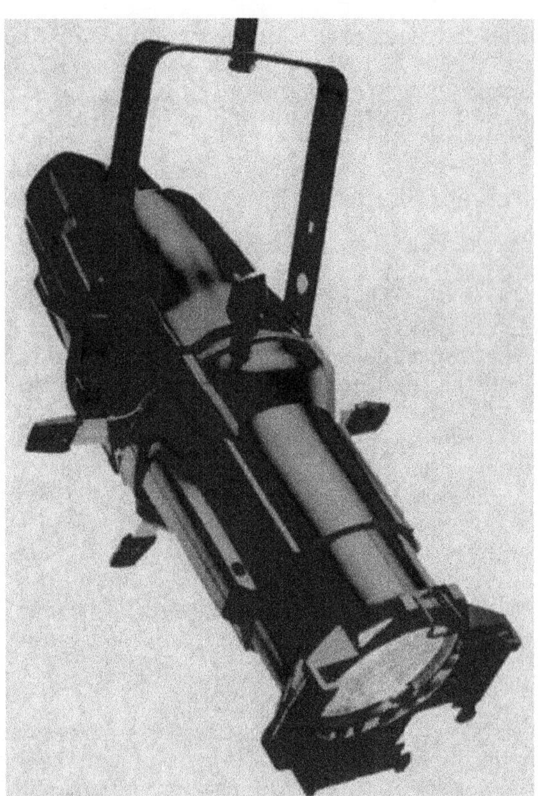

FIGURE 5.3-6 The Source 4 ellipsoidal lighting instrument can produce sharp-edged patterns for a variety of special effects.

which is available in five different beam spreads. The PAR is an inexpensive and powerful tool for the lighting director, and while it has little in the way of adjustments, the application of barn doors and holders for color filters makes this fixture popular for key and backlighting. A new variation of this project, called the ParNel, allows adjustment of the beam spread without changing lenses, acting much like a Fresnel.

Ellipsoidal (or "Leko") fixtures are a purpose-built spotlight used for providing a round source of light. They are often used with a colored gel or metal pattern "cookie" which can be placed in front of the lamp in a metal holder, as seen in Figure 5.3-6. The Leko projects the light through the pattern or gel to great textural or colored backgrounds onto a solid or drapery background, called a *cyclorama*, discussed below.

Fixture Lamps

While all of these types of fixtures were originally designed to accommodate tungsten-halogen lamps (commonly called "quartz" lamps), there are many manufacturers that make fluorescent soft lights. Recently, LED lighting is becoming popular and there is a wide array of both LED spotlights and soft lights and some of the tungsten fixture manufacturers are offering LED lamp replacement kits for the tungsten fixtures. Tungsten lamps operate at 3200°K, making them ideal for studio lighting where there is no incident daylight or need to color match architectural fluorescents. Studio fluorescent lamps were developed in the 1990s, primarily for use as a soft or fill light. These fixtures were more energy efficient than the quartz lamps, and were good for lighting large areas evenly. Early broad light spectrum fluorescent fixtures were a bit odd-colored, but the technology has developed well, and 3200°K fluorescent fixtures are now commonplace in most TV studios. HMI lighting fixtures are large, high-wattage daylight (5600°K) lights that are good at filling in shadows outside. LED lamps can be made to work with a broad range of color temperatures, from 3200°K to 5600°K. All of these lamps will be discussed and compared in more detail throughout this chapter.

Robotic Fixtures

Robotic fixtures have become commonplace in theatrical and television lighting. These lights fall into two categories. The first group can position a light fixture by means of a motorized yoke. The second group of robotic fixtures utilizes a motorized mirror that reflects the output beam of light into the correct position. Within these two mechanical solutions, there are *wash* lights with a soft-edge beam of light and *hard* lights, which project a defined beam of light and can project many types of patterns. These lights are primarily used for concert lighting and theatrical productions where the movement of the lights and changing colors provide dramatic effects under computer or manual control. These lights are rarely used for face lighting unless they present the only practical alternative because of the useful remote focus capabilities.

Fluorescent Fixtures

Fluorescent fixtures are popular in many studio installations. Utilizing the improved color temperature of compact fluorescent lamps, several manufacturers have developed a variety of useful fixtures. These systems offer the compelling features of low energy consumption and minimal heat production. They can provide soft or fill light but are somewhat limited in throw distances. Fluorescents are most commonly employed in news and other continuously on-air forms of production to take advantage of their energy efficiency. Fluorescents are another tool for the lighting director and can be effectively used in conjunction with standard incandescent fixtures. Figure 5.3-7 shows fluorescent soft lights being used to light a wide area of a news set.

LED Fixtures

LED light fixtures have opened up an entire universe of possibilities for lighting. These fixtures make it possible to shoot in any color temperature light with the same fixtures by equipping the fixture with a combination of daylight and tungsten LEDs. Traditional studio tungsten fixture manufacturers are now providing LED conversion kits for legacy tungsten Fresnel fixtures.

LED fixtures emulate the various "traditional" tungsten device types—they are available in a Fresnel-type fixture, as well as soft light or "broad" and PAR-style fixtures.

There are numerous benefits to LED lighting:

- Tremendous efficiency per watt and electrical cost savings. A 5000-watt LED "Fresnel" uses 900 watts of power.

FIGURE 5.3-7 Fluorescent soft lights being used on a news set.

Courtesy KMIR-TV

- Simplified studio electrical wiring. The LED lamps can be plugged into standard 15-amp wall outlets.
- The lights last much longer than tungsten or fluorescent lamps. Some manufacturers indicate that LEDs will last as much as 50,000 hours, compared to an average of 1000 for tungsten or 5000 for fluorescent.[2]
- HVAC savings can be substantial, as LED lighting fixtures generate very little heat.
- LED fixtures can be dimmed with no change in color temperature, and many LED fixtures have built-in dimmers, eliminating the need for large dimmer banks or individual dimmer packs. This ability allows light level to be raised or lowered without the need for scrims.
- A single fixture, when outfitted with a combination of LEDs, can serve to match a wide range of color temperatures, which are handy for lighting in a non-studio environment.
- Entire TV news sets can be lit using 3200°K daylight lighting, negating the need to color-balance all of the monitors and large expanses of exterior windows that are increasingly popular on news sets.
- Many LED light fixtures can be fitted with traditional barn doors for additional light control.
- LED arrays can be added together for extremely large areas to be lit.
- LED fixtures are very suitable for coloring white cycloramas ("cycs"), as noted in the Cyclorama portion of this chapter.
- LED lights are also excellent for use as set decoration lights due to their longevity, color stability, and minimal power consumption.
- LEDs are also becoming popular for battery-operated field lighting, as shown in Figure 5.3-8 due to their light weight, relative durability, low energy consumption, and color temperature versatility.

While it might seem that LED lights are becoming a "Holy Grail" of fixtures, there are still a few drawbacks:

- The initial cost of ownership is still substantially higher than for tungsten or fluorescent fixtures.
- Even though LED fixtures are much improved over the originals for light output, the tungsten fixtures provide a higher light output ratio than LEDs. In situations where there are light stands on the floor that would get in the camera shot, it is more practical to light using a smaller number of tungsten fixtures, as the light is "thrown" a greater distance than with LEDs. LED lights have to be placed closer to the subject in order to produce enough light to properly light it.
- Not all LED fixtures are the same. Different manufacturers use different colors and types of LEDs to produce a particular color temperature. Some LED luminaires are single color-temperature

FIGURE 5.3-8 ENG Camera with a Battery-Operated LED Camera Light.

Courtesy John Bullard/Lip Flap Productions

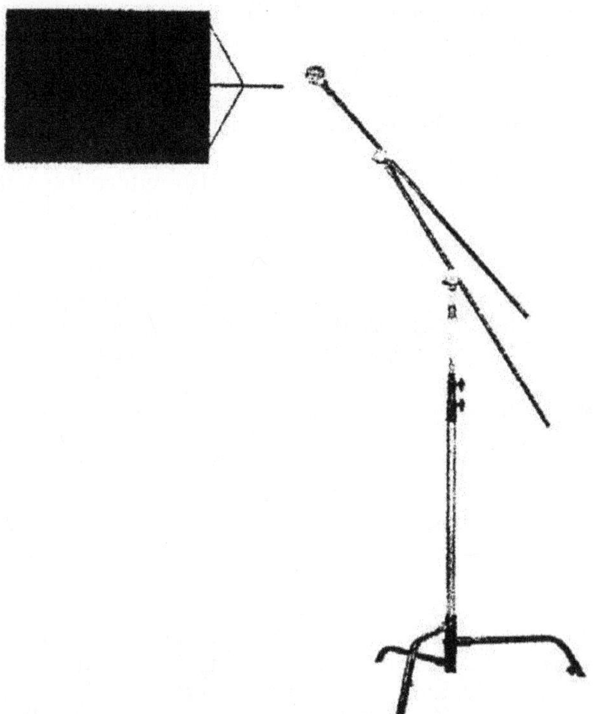

FIGURE 5.3-9 The three level legs allow the "C-stand" to fold flat for storage. The head can grasp flags, cutters, cookies, and other accessories.

only; others can change color temperature with the flip of a switch. Not all LEDs used in fixtures are of the same quality.

The studio lighting plan will be fundamentally the same whether the lighting director uses incandescent, fluorescent, or LED fixtures. The same principles of lighting described in the early portion of this chapter still apply no matter which type of "lamp" the lighting fixture uses.

Lighting Fixture Accessories

There are a number of fixtures to assist a lighting director in positioning a light fixture and controlling the light output of each fixture to create the necessary lighting effects.

Lighting Stands

Rolling floor stands can position lighting fixtures in locations that can't be reached from the lights on the grid.

Extension rods, either *straight* or *telescopic*, allow fixtures to be hung at a lower level than on the fixed grid. For example, a key light at a 30° angle (elevation) above the talent could be moved much closer to the talent and provide more light if it were hung lower. Counterbalanced devices such as pantographs and spring-load telescoping hangers are not recommended because they often cannot be locked in place.

The *Century stand*, shown in Figure 5.3-9, has multiple applications around the studio. It can hold and position materials such as *flags*, *cutters*, and *cookies*, and can also support reflectors. Century stand use is limited only by the lighting designer's imagination. No matter how many are available, they will all be put to good use.

In addition, many types of *clamps* and general grip equipment are available for special mounting situations.

Barn doors are used to shield subjects and backgrounds from unwanted light spill from a spotlight fixture. Barn door attachments usually have four "leaves" and attach onto the front of a Fresnel fixture. Typically, barn door attachments also have slots that will accommodate frames that can hold scrims or colored gels.

Egg crates are typically used with soft lights. They also control the unwanted spill of light, and have a smaller footprint than a barn door. Egg crates come in various beam angles to narrow the light beam from nearly 180 degrees for a soft light down to as little as 40 degrees. Because the egg crate is mounted in front of the fixture's lamp it acts like a *scrim* to reduce the light output, so an egg crate can be used instead of a scrim if they are in short supply, and works better than a dimmer, which will alter the color temperature of a tungsten lamp.

Flags can also be used to block unwanted light.

A *cookaloris* (or "cookie") is a pattern cut into a metal disc, which is installed into a holder on the front of an ellipsoidal light. Cookies are used for projecting a light pattern onto a background. The holes in the disc allow the light through.

Many devices are available to the lighting director to control the levels of illumination, including a wide variety of television studio and location lights. The distance between the lighting fixture and the subject will directly affect intensity. Intensity can also be controlled by *scrim*, which is actually a wire screen inserted in front of the lamp. The greater the density of the screen, the more the light level is reduced. A wide variety of available sheet *diffusion* materials will also reduce intensity and reshape the pattern and quality of the light. Diffusion materials soften the light and shadows produced.

Color filters or "gels" not only produce colored (or color-correcting) light, but also reduce light output relative to the coefficient of transmission of the particular color. For example, a blue filter passes very little light, whereas a yellow-green filter will reduce the output light level very little. Blue gels can be used with an incandescent fixture for lighting outside, as the color temperature of daylight is very blue as compared to incandescent light.

The most precise and convenient level control of a lighting fixture is *dimming*. Compact portable dimmers are also used outside a studio or theatrical environment, but do not necessarily eliminate mechanical means of level control. While dimming does change the color temperature of an incandescent lamp, it can still be used effectively. Dimming can be used effectively with a fluorescent fixture if the fixture has an incorporated ballast. Dimming is most effective with LED fixtures, and many LED fixtures have built-in dimming capability.

Efficient Use of Lighting Fixtures

Ultimately, the proper quantity of fixtures, accessories, and other components of a lighting system for a studio of a given size can vary widely according to the requirements of a specific situation. For example, a studio with a fixed grid and a tight production schedule will function more efficiently with a wide range of fixtures available in the grid. With a large quantity of fixtures, constant relocation of the units is minimized. When sophisticated motorized grid systems are available, the ease of relocating fixtures will reduce the total quantity of fixtures for the same size studio. The choice of either of these two approaches is also dramatically affected by labor costs within a particular facility. The expense of a few extra fixtures is not significant compared to the additional time and labor required to fully utilize a minimal complement of equipment or the cost of elaborate rigging. Final discussions on equipment purchase require a careful analysis of a station's specific production requirements in coordination with all the financial ramifications of various approaches.

For any production, close cooperation between the lighting designer for the show and the "video shader" in charge of the cameras is important. Great television pictures are the result of careful teamwork by the lighting designer and the video shader (see Chapter 5.5, "Television Camera Systems").

HMI Lighting—On Location

In some instances, good lighting design may not even require the use of lighting fixtures per se. In many daylight on-location (remote) situations, a lighting director may choose to simply use reflectors and scrims to achieve the desired effect. Movement of the sun is predictable, so a survey of the site must be made at the same time of day of the event shoot in order to select workable camera positions and to determine the proper equipment.

The sun as a key source is harsh and requires intense fill light for the talent. Reflectors are very effective for this purpose. They require no electrical power but must be attended by a person at all times to keep them oriented correctly to the moving sun. Lightweight folding reflectors are convenient to pack and set up but do not present the necessary hard, stable surface needed to provide a smooth field of light. The heavy solid panels are less convenient but are stable.

Another way the harsh light of the sun can be controlled in intensity is by using large silks or nets to soften or to shadow the talent. Nets and silks under 6 ft × 6 ft attached to frames are called *butterflies*, and those up to 20 ft × 20 ft are called *overheads*. Wind is an obvious factor in the decision to use these tools.

Small-Scale Setup

For smaller-scale remote setups, the lighting director's selection of fixtures is often governed by weight and portability relative to lumen output. A wide variety of ingenuous and lightweight fixtures are available in kits that have an assortment of stands and grip hardware, as shown in Figure 5.3-10, to accommodate any situation which may occur while on location. Many of these kits are designed to incorporate filters to match the 3200°K incandescent lighting to the 5600°K daylight. This filter reduces the light output quite noticeably.

Large-Scale Setup

For larger location work involving daylight conditions, *HMI lighting* has become the standard of both the film and television industries. HMI is a registered trademark of OSRAM, although it is often used (or misused) in referring to lamps of other manufacturers. The name refers to the basic elements that are combined in the lamp's quartz envelope to create the unique color temperature: "H" for mercury (Hg), "M" for the various metal halide rare earths, and "I" for the halogen iodine and bromine.

These high-efficiency ballasted arc source lamps come in a wide variety of sizes from 200 watts to 20,000 watts. They provide about 80+ lumens-per-watt at 5500°K (which is four times the output of an incandescent light). HMI lighting is often the best solution for interior situations with substantial window exposure. As in any exterior situation, the lighting director must carefully monitor the changing color temperatures of daylight and remember that the light of a clear sky late afternoon is quite a different color and character than mid-morning when the sky is overcast. Figure 5.3-11 shows a typical HMI device.

Dealing with Architectural Fluorescent Lights

Often a lighting director has to shoot in places where the existing lighting is a low ceiling covered with

FIGURE 5.3-10 For remote lighting, compact kits containing a large assortment of fixtures, stands, reflectors, and grip hardware are available.

Courtesy of Lowel-Light Manufacturing, Inc.

FIGURE 5.3-11 HMI lamps are most often used in Fresnel lens housings and are available in a wide variety of sizes. The associated lamp ballast is often a separate unit from the lamp housing for high wattage fixtures.

Courtesy of Arriflex Corporation

fluorescent fixtures, and they must be used because they characterize the space. Common architectural fluorescent lights do not provide a full spectrum of light, and the most common type, cool white, is very short of red energy. While it is possible to color correct each fluorescent lamp using a minus green color correction *gel*, it is often impractical. By color balancing the camera under a representative mix of the fluorescent and the 3200°K talent's lights, the lighting director can produce acceptable skin tone while the background may still appear greenish. Today, a wide variety of fluorescent lamps are available, and some offer quite a high color rendering index. Fluorescent studio fixtures have become very popular but are somewhat limited in control capability due to the relatively large size of the light source and the diffuse nature of the output.

Nighttime Location Lighting

"Exteriors" at night for TV require great simplicity to look real. Large HMIs are very useful because one large source can supply the basic illumination for a wide area. Within an exterior scene there are always elements supposedly lit by artificial sources, either seen or imagined. With the camera balanced for 3200°K, quartz lights can be used for the "people areas" while the overall HMIs at 5500°K will seem very blue. If the HMI light is too blue, it can be corrected slightly until it seems to be the proper gray-blue of moonlight. Using smaller HMIs as back and rim light will further enhance the moonlight effect.

In attempting to provide realism, lighting must suggest the correct mood. Lighting is essential in creating the feeling or mood the production team envisions.

STUDIO DESIGN

In the design of a television studio, the lighting system is a major consideration because it is so closely interrelated to the physical size and shape of the room. While the television industry has seen major technological improvements throughout its evolution, the basic physics of light and its ability to describe three-dimensional forms cannot change. Consequently, while many older studios have upgraded virtually every piece of electronic equipment, the basic components of an older, well-designed lighting system may still function extremely well.

Studio Size

Ideally, a studio provides the optimum environment for any type of production. In practice, however, studios of various sizes tend to function best for particular types of production. The typical broadcast plant requires several sizes of studios to most effectively service its programming schedule. Typically, a small studio (1200 sq ft) is dedicated to news, interviews, and public affairs, whereas a larger (2400 sq ft) studio is the most common size of small program production facility. For general production, a studio of 5000 sq ft or more will offer fewer limitations. The size of the studio must be carefully determined by existing and future programming requirements, keeping in mind that the larger the studio, the fewer the limitations (see Figure 5.3-12). These are important decisions, since the lighting system must be planned relative to the size and specific requirements of the studio.

Studio Height

When determining the height of a new studio, the lighting suspension system must be considered. The grid height for small and medium-size studios is a function of the TV aspect ratio and the normal wide-angle zoom lens. Most zoom lenses can cover approximately a 45° field. When the actual width of this maximum horizontal dimension is calculated, the height of the suspended lights can be determined by applying the 16:9 aspect ratio.

However, television in the studio is primarily a close-up medium, and limitations in grid height can often be overcome by various camera angles and special fixture-mounting systems.

The suspension system for the studio lighting fixtures is an important factor in determining studio height. The system must allow lighting fixtures to be hung anywhere over the entire studio, and several arrangements are possible, ranging from simple fixed pipe grids to systems with adjustable hangers and manual and motorized hoist systems, sometimes with several different types or combinations.

Fixed Grid

A *fixed grid* as shown in Figures 5.3-13 and 5.3-14(a) is the most common and least expensive system for mounting lighting fixtures. The pipes are generally laid out with a 4 ft × 4 ft grid spacing that provides adequate flexibility in hanging positions. The most common fixed grid height is 14 ft because it offers a good compromise between easy ladder reach and adequate clearance for wide shots. A fixed grid that is much higher than 14 ft increases the amount of labor needed to install and adjust fixtures. Bi-level fixed grids have been utilized to create a higher apparent background without additional cost and to keep the major portion of the grid at a reasonable height.

Fixtures may be attached directly to the pipes of a fixed grid, but when lights are needed at a lower

FIGURE 5.3-12 This large multipurpose studio has a flexible lighting system equipped with motorized battens and hoists that allow fixtures to be easily adjusted and located for different productions.

Courtesy of Bavaria Studios/De Sisti Lighting

FIGURE 5.3-13 This fixed grid studio employs two levels to create a higher apparent background. Multiple pigtail outlets are positioned for cyc lighting units.

Courtesy of WFAA-TV

level for particular effects, and then various fixed- or adjustable-length suspension rods or poles and sometimes spring-loaded adjustable *pantographs* are used. Self-powered man lifts can be used to provide reasonable access to higher fixed grids, but in a large multipurpose studio, a fixed grid can be a limitation. Several other systems have been developed that are more flexible and efficient to use, albeit at a higher initial cost.

Catwalks

Catwalks, as shown in Figure 5.3-14(b), are generally also built in large studios at a fixed level above the studio floor and need sufficient height above them to allow personnel access. Catwalks provide a measure of increased efficiency over a fixed grid, but they are considerably more expensive. They allow studio electricians to work on lighting at the same time carpenters are handling the scenery. This can be an important time savings on a tight production schedule. Catwalks are usually arranged to create a fixed grid utilizing handrails in conjunction with an extension rod for suspending each lighting fixture. When hung too low, catwalks create the same height limitation as a fixed grid, but generally they are employed at heights greater than would be reasonable for a fixed grid. This arrangement primarily saves time in setup and strike and offers the lighting crew a safe work platform above the studio. In recent practice their high cost has prevented many installations. Most shows set up and stay in a studio for their entire run. With only one major setup per production, the cost of catwalks can be hard to justify.

Adjustable Hoist Systems

The larger and higher the studios become, the more complex the lighting and its suspension system become, and various alternatives to fixed lighting grids may be used. Figure 5.3-14(c) illustrates a plan for suspended battens.

In the development of a fully integrated lighting system, it is essential to coordinate all the building's structural, electrical, and mechanical systems in relationship to each other and the grid. Whereas in normal construction many of the mechanical elements in a ceiling are placed where convenient, it cannot be overemphasized that, in a TV studio, improperly planned or installed air conditioning ductwork and electrical conduit runs can be a hindrance to production.

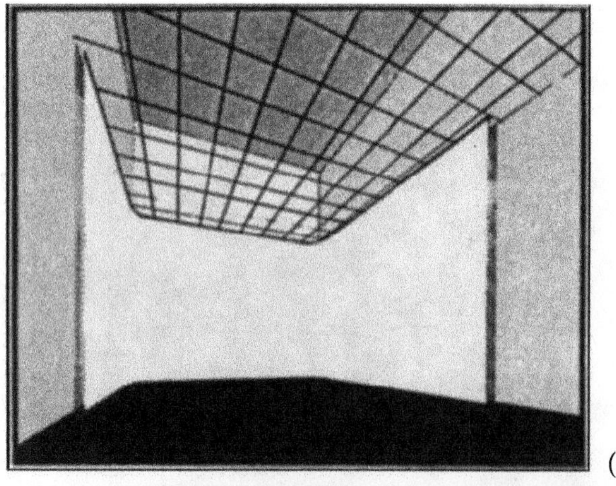
(a)

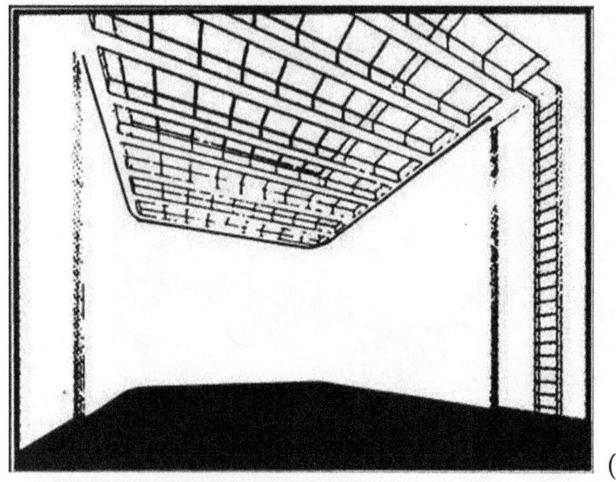
(b)

(c)

FIGURE 5.3-14 (a) Fixed grid plan. (b) Catwalk grid plan. (c) Batten plan.

Pole-Operated Fixtures

Whichever system of suspension is used, light fixtures have to be adjusted for direction and focus. In studios with fixed sets such as news, adjustments may be rare, but for some general-purpose production studios, frequent adjustments may be needed. To avoid having to

FIGURE 5.3-15 When a long pole-mounted crank is inserted into the adjustment cups, this fixture can be panned, tilted, or spot/flood focused from the floor.

Courtesy of Mole-Richardson

use a ladder or man lift to reach a fixture to adjust it, most studio fixture manufacturers offer pole-operated yokes for their instruments. Such fixtures have sockets to which a pole crank can be attached, with adjustments for each of the basic functions of pan, tilt, and spot-flood (for adjustable spotlights), as shown in Figure 5.3-15. There is a reasonable limit to how long a pole can be easily manipulated, but this feature considerably enhances most types of suspended light fixtures, whether attached direct to a fixed grid or on a hanger. This feature is especially useful for adjusting otherwise unreachable fixtures blocked by scenery or other obstructions and, although more expensive, is seriously worth considering when purchasing light fixtures.

CYCLORAMA ("CYC")

Designing the cyclorama used for backgrounds is integral in planning the studio lighting and grid systems. The grid or suspension system must provide a mounting position for the cyc lights located at the proper relationship to the cyc. Also, the total area of cyclorama will affect the calculation of the studio power service because of the number of lights needed to illuminate it evenly.

Hard Cyc

Small studios often incorporate *hard* cycloramas that are smooth, plastered surfaces that actually blend flush into the floor. They provide the ideal infinity effect that draperies cannot equally simulate. Hard cycloramas can be easily painted any color as needed and are especially effective for certain Chroma key techniques that can be spoiled by even the most invisible cloth seams. Hard cycloramas are generally limited to small studios since their hard surface area invariably creates acoustical problems. Several manufacturers now make precast hard cyc pieces. Put together, these preformed pieces can create a nearly infinite number of shapes for a hard cyc, most often for "virtual set" applications.

Drapery Cyc

For most studios, a *drapery cyclorama* is the most convenient solution. This seamless drape is hung on carriers that roll along a track and allow the cyc to be positioned anywhere around the perimeter of the studio. Two parallel tracks permit another type of background to be pulled in front of the stretched cyc, as shown in Figure 5.3-16. Switches on the track system allow the draperies to be easily transferred to the front or rear tracks.

In large studios where experienced lighting directors are available, true white cycloramas are used to achieve greater color intensity on the cyc. However, in smaller studios, where the talent occasionally must work very close to the background or where limited control equipment is available, a 60% TV white cyc should be used for better control of contrast.

Drapery cycloramas are generally furnished with jack chain weights. Removable pipe weights bent to match the shape of the track should also be provided to create a wrinkle-free background. One of the most common errors in cyclorama design is an insufficiently large radius at the corners of the cyc. No matter what material the cyc is made of, the larger the radius, the easier it is to light evenly and accomplish the desired effect.

FIGURE 5.3-16 Cycloramas and draperies can be shifted to various track configurations by utilizing transfer switches (CBS-NY).

Courtesy of Peter Albrecht Corporation

Chroma Key or "Green Screen" Wall

A news studio weather set or talent "standup" set is usually the application in which a Chroma key set is used the most often. Typically the green screen is a rigid wall, but it can also be a curtain or piece of material stretched on an aluminum frame for smaller setups. The background is most often a particular color of green or blue that will electronically "key out" when processed by a production switcher's Chroma key keying software, and can be replaced with a graphic, live video, or other picture.

Cyclorama Lighting

Generally, the arrangement of doors into the studio will define the most functional area for the cyc to be positioned. In the basic design of the grid or suspension system, the type of cyc lighting system should be predetermined and the proper hanging system provided. There are several types of cyclorama lighting fixtures, including strip lights and a fixture commonly known as a *Far Cyc*, shown in Figure 5.3-17. Strip lights are continuous rows of quartz halogen, MR-16, or PAR lamps, which, for an 18 ft high cyclorama, should be mounted 5 to 6 ft from the cyc, as illustrated in Figure 5.3-18, whereas the Far Cyc units should be mounted 7 to 8 ft from the cyc. Another type of cyc fixture is a version of a strip light made up of individual cells that can be placed to match any curve or straight surface. The entire suspension system should be designed around these dimensions. Far Cycs generally will light the cyc

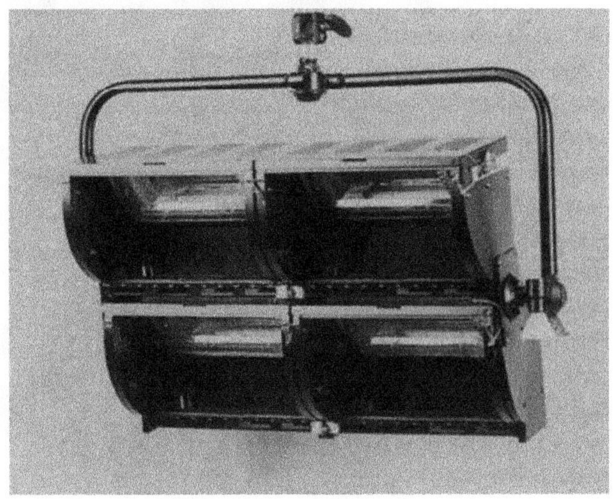

FIGURE 5.3-17 Four-unit "Far Cyc" fixtures are very effective in lighting tall cycloramas.

Courtesy of Colortran

as evenly as strip lights but with less wattage. Because Far Cycs are mounted a greater distance from the cyc, they force the talent farther from the background. Generally, strip lights should be limited to three colors; otherwise, the separation between alternately colored lamps is too great to provide even coverage.

Chroma key lighting is typically done with "broad" lights, whether fluorescent strip lighting, incandescent fill lighting or LED panels set up to replicate a broad light pattern. Care must be taken to make sure that the Chroma key background is evenly lit. Proper training of the studio talent is also critical, as the closer the talent stands to the wall, the more difficult it will be for the production switcher operator (usually the director or the technical director) to adjust the keyer software in the switcher to eliminate keying anomalies caused by the talent's shadow on the wall. Additionally, the talent needs to be lit in such a way that he or she is visually separated from the green screen; the use of proper key, fill, and especially backlights is necessary. Figure 5.3-19 features a weathercaster standing in front of a lit green screen, and Figure 5.3-20 demonstrates how the electronic keyer "replaces" the green color on the wall with weather graphics.

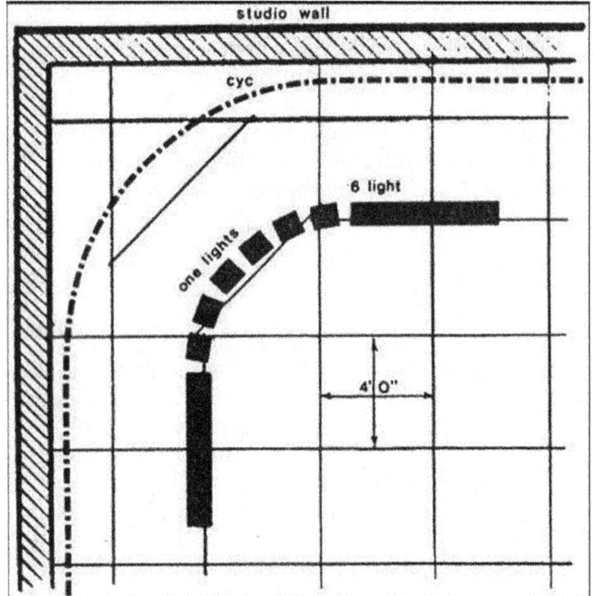

FIGURE 5.3-18 Cyc light layout: typical arrangement of strip cyc lights relative to the cyc curve.

If the green hard cyc or material extends down onto the floor for a more "virtual set" application, care has to be taken to light the floor evenly as well as the wall. If the floor has less light than the wall, the talent standing in front of the camera may have green shadows under their chin from green light reflecting upward from the floor. The point at which the wall meets the floor must be lit carefully in order to keep a dark horizontal line from forming along the join.

ARCHITECTURAL CRITERIA

Electrical Power Requirements

With the basic studio size determined and the net production area (NPA) defined by the cyclorama, it is possible to determine the power requirements for any given studio. The power requirements for studio

FIGURE 5.3-19 Weathercaster in studio in front of a green Chroma key wall.

Courtesy of KMIR-TV

FIGURE 5.3-20 The finished Chroma key in the production control room.

Courtesy of KMIR-TV

lighting are a direct function of area and the required level of illumination. This power requirement remains consistent regardless of the grid height. For a lower grid, a greater quantity of smaller wattage fixtures is used, whereas a higher grid will require fewer fixtures of increased wattage. In either case, the total watts per square foot will remain roughly the same.

For tungsten fixtures, an average of 55 watts per square foot of NPA has been proven in production to provide sufficient power for any normal television lighting requirement. This method of calculating the studio load will provide sufficient power for virtually any situation. This is generally more power than is actually required for the average studio with typical television camera equipment, but it allows for over-lighting by novices and higher levels necessary for special situations. It also provides sufficient power for an average cyclorama as well as lighting the entire studio as required for audience participation.

This calculated power service describes a real maximum probable electrical load, and the feeders must be able to supply this full amount of power. Only certain limited productions will ever require this full amount. Also, note that for dimmer-per-circuit systems, the dimming capacity will be far greater than this calculated power service. The full dimming capacity need not be fully serviced, since the larger number of dimmers is a matter of convenience and will never be fully loaded beyond the maximum probable load. Large production studios will require a variety of additional power services in various voltages for special effects, rigging, and motorized lights.

When adding in fluorescent and LED fixtures into the lighting system, electrical needs can take into account that fluorescent lights draw roughly 25% of the power that incandescent fixtures do for the same wattage, and LEDs roughly one-tenth of the power. Keep this in mind when applying the steps below for HVAC capacity; LED light fixtures give off almost no heat, even though they produce several kilowatts of light.

Calculating Air Conditioning and Electrical Power

Because the maximum lighting load seldom occurs, it is unnecessary to use that maximum load as the basis for the air conditioning capacity. Production practice has shown that a diversity of 60% can be applied to the maximum load and still provide sufficient capacity for full period shooting. Of course, any other heat-generating devices and the population of the studio should be included in the air conditioning calculations. A properly designed air conditioning system will require very large ducts to meet stringent acoustical requirements. These large ducts can often interfere with a grid system and should be closely coordinated. Determining studio power and air conditioning requirements according to the factors discussed earlier:

- Start with the net production area (NPA), which is the usable studio area in square feet minus the area behind the cyc and other areas that are unusable for production.
- To determine the number of outlets required at the patch panel, divide NPA by 18 (one outlet for each 18 sq ft).
- Then divide the outlets into 20 amp and 50 amp circuits to determine how many dimmers will be needed of each variety.
- To determine the Studio Lighting Load, multiply the NPA by 55 watts to find the maximum possible load in watts on the air conditioning system.
- To determine the Studio Lighting Power Service required, divide the maximum load power in watts by 120 volts to obtain the maximum current in amps.
- To determine the Studio Lighting Power Service, divide the total amps by 3 (for three-phase, four-wire, 120/208 volts) to find the maximum load in amps per phase leg and then round off to the next larger standard panel size.

Electrical Distribution

For a studio to be flexible, lighting equipment power must be distributed uniformly throughout the studio. At the grid level, power is commonly distributed through prewired plugging strips. These strips are mounted directly to the grid, catwalks, or on fly-in-and-out battens or movable grid sections.

Each circuit terminates in the studio in a pigtail or outlet. For large incandescent lighting fixtures, there are two types of connectors in common use: *stage pin connectors* and *twist lock*. Stage pin connectors are less expensive and more common in rental equipment. If additional fixtures are rented on occasion, this may be an important consideration. In addition, the cost savings of stage pin connectors recur with each fixture and cable purchase. Twist lock connectors have a positive locking feature. The final choice should be based on the studio's specific requirements. For LED and fluorescent fixtures, standard 15A plugs are more typical.

Wall-mounted outlet boxes should be provided at 30 inches above floor level, around the perimeter of the studio, as shown in Figure 5.3-21. Generally, the governing factor for their placement is relative to a layout of floor-mounted cyc strips. Otherwise, these outlets are used for miscellaneous lights on floor stands and practical lighting fixtures on the set.

The number of circuits and their capacities are also related to studio size. The actual number of circuits is based initially on the square footage (approximately one outlet every 18 sq ft of the NPA) and then altered as necessary to conform to the particular grid system and the cyclorama layout, as illustrated in Figure 5.3-22. Dedicated circuits for the cyclorama are often overlooked. The cyc lights require a large number of circuits, and once they are hung in place, they will seldom, if ever, be moved. Because of the rather wide spacing of the cyc units, it is sometimes more efficient to feed these lights from individual grid-mounted junction boxes rather than a plugging strip.

Additional circuits should be located around the outer perimeter of the studio at grid level. This is a natural backlight position for a set facing away from the wall. Properly located circuits save considerable time in running jumper cables.

For most studios the majority of circuits will be rated 20 amps. However, the larger the studio, the greater the quantity of 50 amp circuits needed. In a medium-size studio (3500–5000 sq ft), 50 amp circuits are generally located in a regular pattern throughout the center area of the grid and slightly more frequently around the perimeter backlight position. In larger and higher studios the density of 50 amp circuits must be increased, although the overall outlet density should not be significantly decreased. On adjustable-height grids, a full complement of both 20 amp and 50 amp outlets should be provided for use of appropriate fixtures at varying heights. Large studios (over 10,000 sq ft) with high grids must be furnished with 100 amp circuits.

FIGURE 5.3-21 Wall-mounted outlet boxes are generally located around the perimeter of the studio.

Courtesy of Electronic Theater Controls

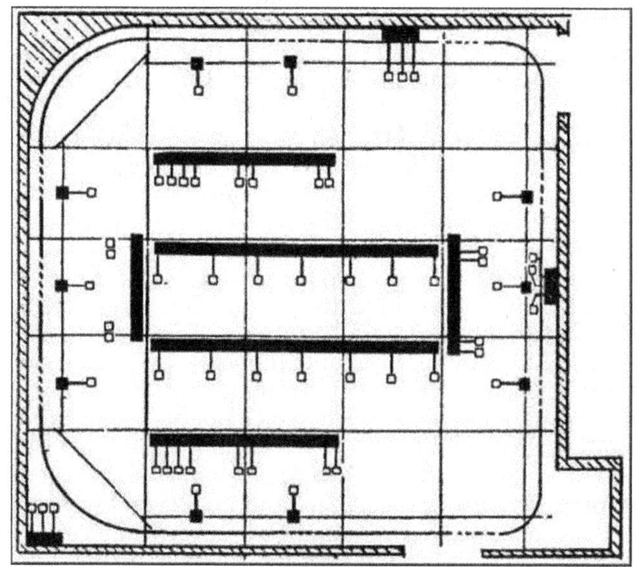

FIGURE 5.3-22 Typical studio lighting outlet distribution plan for a small industrial studio.

Lighting Control

The dimming and control system is an important part of the lighting system. Dimmers allow easy control of numerous fixtures, balancing and recording of levels, and the blending of colors. A dimmer system frees the lighting director of the unnecessary burden of calculating and controlling the loads through more labor-intensive mechanical methods. Electronic dimming and control allow the execution of complex lighting cues, which are a very effective production element. Modern LED and some fluorescent fixtures may have built-in dimming as an option.

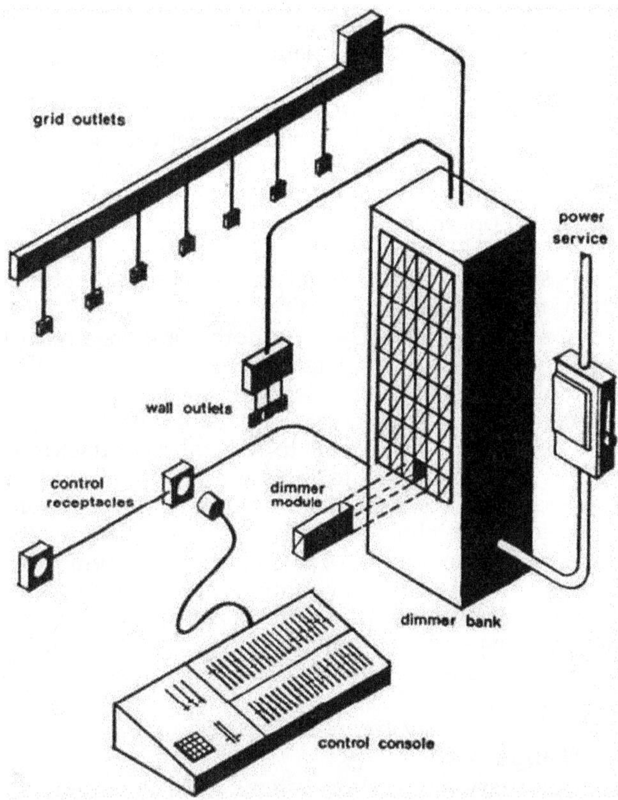

FIGURE 5.3-23 Diagram showing the basic components of a studio lighting control and dimmer system.

Most studios are outfitted on a basis of one dimmer per circuit in which every circuit terminates in its own dimmer, with the integral circuit breaker protecting both the dimmer and the circuit. The dimmer-per-circuit system gives the lighting director individual control of each individual fixture, or group of fixtures, plugged into that circuit. Normally, a studio utilizing mostly quartz fixtures is outfitted with 20 amp (2.4 kW) and 50 amp (6 kW) dimmers. Figure 5.3-23 illustrates a typical studio lighting control dimmer system arrangement. Many fluorescent and most LED fixtures come with dimmers already built into the fixtures, so large external dimming systems are not necessarily needed.

Dimmer Bank

For older incandescent systems, individual dimmer units are used. They generally plug into electronic equipment racks to form the dimmer bank. This modular system also allows quick plug-in substitution of faulty dimmer modules. Depending on the manufacturer, up to 96 individual 2.4 kW dimmer modules can fit into a single rack. From the 1970s until recently, most studio-quality electronic dimmers have utilized silicon controlled rectifiers (SCRs), although a few have used MOSFET. These units are reliable and are universally available in 2.4 kW, 6.0 kW, and 12.0 kW (20 amp, 50 amp, and 100 amp, respectively, at 120 V) ratings. Only dimmers that have sufficient filtering to prevent unwanted RF interference and excessive filament vibration should be considered for use in television studios. Toroidal chokes are utilized to control the rise time of the dimmer and the associated electrical noise. More recently, sine wave dimming has become commercially available. These dimmers, described in the next section, are more expensive but have several advantages compared to normal SCR dimmers.

The location of the dimmer bank is an essential part of the initial studio space planning. The dimmer room should be centrally located to minimize the length of all the wiring to avoid voltage drop and excessive installation cost. This room should be sized to allow sufficient space for required conduit radii, access to the feeder lugs, and adequate front clearance as specified by the local code.

Most SCR dimmers are 95% to 98% efficient. Therefore, they could create heat perhaps up to 5% of the energized lighting load. Since the maximum lighting load is an infrequent occurrence, the dimmer room cooling should more reasonably be based on 5% of the diversified load on which the studio air conditioning system is based.

Centralized and Distributed Dimming

SCR dimmers tend to be both acoustically and electrically noisy, which is one of the reasons why in the past they have always been located in a separate dimmer room—an arrangement known as *centralized dimming*. Dimmers such as the *sine wave, or pulse width modulation (PWM) dimmers* described below, are acoustically silent and do not create electrical interference, enabling an alternative *distributed dimming* arrangement to be used for all or some of the studio lighting channels. Because these dimmers can be located in sensitive areas, they can be located closer to the fixtures and loads, on the lighting grid, or elsewhere in the studio at a convenient location. This capability provides increased flexibility in design and can simplify installation, which may be useful, particularly when upgrading or expanding an existing installation.

Sine Wave Dimmers[3]

Although SCR dimmer designs have improved over the years, they have several flaws due to the basic characteristics of their design; they tend to generate acoustical and electrical noise, and may also induce noise in the lights that they control.

SCR dimmers work by varying the switch-on point of the lamp current each half cycle. These dimmers slice the waveform in half and, using chokes, increase the rise time and use the inertia of the filament to smooth out the switching change. This causes the filaments in the lamp to buzz (creating noise and shortening lamp life). In addition, the electronic spikes that occur in these dimmers create radio frequency interference. The high frequencies created by the SCR switching waveforms may also be radiated or carried through electrical power circuits. The unfortunate results of this cross-interference may be picked up by other electronics (such as amplifiers and wireless devices). Although lighting equipment manufacturers have improved these negative artifacts with advanced technology and high-quality chokes, it is impossible to completely silence these dimmers.

In noise-sensitive environments such as television studios, modern SCR dimmers may just be too noisy. If complete silence from the lighting system is desired, sine wave dimming may be the solution. Unlike SCRs, sine wave dimmers produce a pure sine wave output with variable amplitude to control lighting levels. Through the use of transistors to slice the mains into pulses, these dimmers vary the current using pulse-width modulation, average the result, and produce a continuous, variable-amplitude smooth sine wave. This, in turn, eliminates the noise to the tungsten filament. In addition, this technology lowers the operating cost as it uses less reactive power (produced by harmonics in SCR dimmers), lowers the maintenance cost as lamps last longer, and eliminates radio cross-interference.

Sine wave dimming also has the advantage of allowing control of almost any kind of load. This includes neon, HMI, and LED lights, and even some motors, which cannot be controlled with conventional SCR dimmers.

Control Consoles

In electronic dimming systems, each dimmer circuit represents a channel that can be programmed to retain a particular dimmer level setting. A group of preset levels is called a *scene*. While manual control panels with two scenes are sufficient for some productions, even a relatively simple production could require many more. While the physical size of manual multi-scene control panels can be cumbersome, numerous methods of storing and recalling the level settings for each preset have been developed.

Even the most modest control consoles offer enough control channels for individual settings of every dimmer, although it is possible to electronically patch dimmer circuits to a smaller number of control channels. Although the simplest multi-scene systems physically resemble a standard manual two-scene preset system, they have virtually all the capabilities of the largest systems. Each slider can represent any channel or group of channels, or act as a sub-master rather than a hardwired individual dimmer. The specific number of control channels is a matter of the physical electronic design of the console. Obviously, more channels afford greater control within a single scene.

In a dimmer-per-circuit system, it is not unusual for a medium-size studio to have in excess of 200 dimmers, and large studios may have a thousand or more. To control this large quantity of channels, it is possible to patch the dimmers to control channels. The patching function allows any dimmer to be controlled by any channel. For example, the blue cyc lights load may require 12 separate dimmers. If they are patched into the same channel, they will operate together in perfect unison. All this patching occurs within the console and does not require any cords, plugs, or diode pins. For small to medium-size studios, consoles with over 100 miniature potentiometers in two or three presets are available to provide the operational convenience and simplicity of manual systems.

After the dimmers are patched to channels and the desired levels are set, the entire preset can be stored in a memory by assigning a memory number. This preset can then be recalled by keying the appropriate number or operating the slider to which the preset has been assigned. This type of system is very economical and is suitable for most small and medium-size studio situations.

For studios that encounter more complex production requirements, there are a number of systems that resemble a personal computer with specialized keypad and controls with much greater capabilities. One type of system that is common to several manufacturers utilizes a video display to show the various functions. In operation, the screen displays the channel numbers, and below each channel is a two-digit number for the intensity. At the bottom of this field of numbers is the *cue sheet*, which displays various operational functions such as cue numbers and fade times. The most complex systems employ dual displays to actively display more information. These systems offer many features and provide an indication if the lamp is burned out. Examples of control consoles are shown in Figures 5.3-24 and 5.3-25.

Computerized lighting systems store their memory on one of several types of removable storage media for

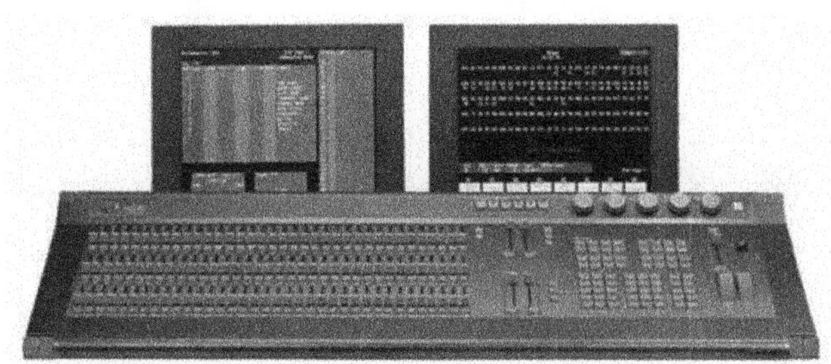

FIGURE 5.3-24 For medium-size studios, lighting consoles offer both manual and memory control.

Courtesy of Electronic Theater Controls

FIGURE 5.3-25 The most sophisticated consoles can control up to thousands of channels, with advanced sequencing and memory capabilities for all types of robotic fixtures, using both tactile and touch screen control user interfaces.

Courtesy of Electronic Theater Controls

reuse. This permits a complete copy of all the settings in memory. Larger systems are available with varying degrees of internal backup memory systems and can be networked to interconnect various types of remote controls and peripherals. The operational software is permanently stored in ROM within the machine.

While many computer lighting control systems were designed for theatrical shows, where the easy daily repetition of very complex multipart cues is the primary goal, some have modified their programming to be more sympathetic to television's less predictable demands. In selecting a system with this caliber of sophistication, the lighting director must carefully determine which features are really necessary.

Computerized lighting systems have greatly simplified the installation of the control wiring. While analog systems required at least one wire for every dimmer, these computerized systems use standard computer network wiring. An advantage to this simplified cabling is that the console can be easily relocated to any of the plug-in stations. In addition, when the main computer is located outside the studio, a small remote control the size of a handheld calculator can be used in the studio to activate fixtures as necessary for focusing or other simple operations. The technology of computer networking is now used on lighting control systems to add greater flexibility and multiple control locations. This allows the console to be located in the most convenient position for a particular phase of the production. It also allows several studios to share a more sophisticated system as required, since the largest systems can plug into the same control wiring as the smallest.

Computerized control has made sophisticated remotely controlled (robotic) fixtures possible. Most consoles now have the capability to control moving lights as well as the dimmer system. A single console can be used to program and record the intensity, position, speed of movement, iris, internal pattern selection, color mixing, pattern rotation, as well as soft or hard beam patterns. The console records and replays all these functions as desired by each cue programmed.

Control Protocols and Standards

Early lighting control systems used analog control signals, with multicore cables and one wire per channel to connect control consoles with lighting dimmers. As systems became more sophisticated, most manufacturers developed their own dimmer control protocols with coded signals to eliminate the bulky multicore cables, but protocols were proprietary and incompatible with other manufacturers' equipment. As the industry grew and systems became more complex, the need for cross-manufacturer compatibility became critical.

In response to the need for standardization, in 1986 the United States Institute for Theater Technology (USITT) published the DMX512 Digital Data Transmission Standard for Dimmers and Controllers (usually known simply as DMX), later revised in 1990. This standard describes a method of digital data transmission between controllers and lighting equipment and accessories. DMX (Digital MultipleX) covers electrical characteristics, data format, data protocol, and connector type. It is intended to provide for interoperability at both communication and mechanical levels with controllers made by different manufacturers. In 1998, maintenance of the DMX512 standard was transferred to the Entertainment Services and Technology Association (ESTA). Subsequently, a further revision of DMX was adopted as ANSI standard E.1.11-2004, Asynchronous Serial Digital Data Transmission Standard for Controlling Lighting Equipment and Accessories. DMX has been widely adopted throughout the entertainment industry, and the majority of current control systems and dimmers use this system.

ESTA coordinated development of further standards for control of integrated systems for the entertainment industry, and a new standard, ANSI E1.17-2006, Entertainment Technology—Architecture for Control Networks, better known as ACN, has recently been published. ACN will not replace DMX but is a suite of documents that specifies an architecture, including protocols and language, which may be configured and combined with other standard protocols to form flexible, networked audio, lighting, or other control systems. Wireless implementation of DMX (W-DMX) is possible and has been implemented by various companies; a standard has been proposed but is not finalized.

Sources of information on DMX, W-DMX, ACN, and other standards for studio and staging systems are listed in the Resources section.

SUMMARY

This chapter has introduced the subject of television lighting, describing the various parts of the system with some guidelines on their use, including basic requirements, lighting fixtures, suspension systems, dimmers, and control systems. New technologies and techniques will no doubt continue to be introduced, and the reader is encouraged to refer to the resources provided here for more information.

Most television production professionals have the opportunity to be involved in planning a new studio only once or twice in their entire careers. Even years of experience in studio production are not necessarily the best preparation for coordinating studio requirements into a construction process. Often, a new studio provides an opportunity to acquire a complement of new equipment that is more sophisticated than the existing staff's level of experience. In this situation, an experienced lighting designer should be consulted to assist while evaluating the requirements and in the preparation of the equipment purchase orders. While manufacturers are sometimes helpful in this area, they are still primarily interested in selling their product, and no single manufacturer offers a full line of suitable equipment in every area. Working with a consultant with experience in the planning and design of the lighting system for a television studio will provide a better perspective and a more reasoned approach to designing the overall lighting system.

RESOURCES

Contact information is given below for a selection of resources related to studio lighting:

- United States Institute for Theater Technology: www.usitt.org
- Information on DMX512 and other standards: http://old.usitt.org/DMX512.aspx
- Summary of DMX512: www.ubasics.com/DMX-512
- Entertainment Services and Technology Association (ESTA): www.esta.org
- The ACN standard is available from the ESTA Foundation website: http://tsp.esta.org/tsp/about/index.html
- Information about wireless DMX solutions, with numerous downloadable fact sheets, is available at: www.wirelessDMX.com
- Information on lighting fixtures, grid and suspension systems, control systems, and dimmers is available from various companies, including:

  Electronic Theater Controls (ETC): www.etcconnect.com

  Entertainment Technology: www.usa.lighting.philips.com/products/lighting-controls

  Leviton Lighting Management Systems: http://lms.leviton.com

  Strand Lighting: www.strandlighting.com

  Lowel Light Manufacturing: www.lowel.com

  Arriflex Corporation: www.arri.com

  De Sisti Lighting: www.desisti.it

  Mole-Richardson: www.mole.com

Notes

[1] French physicist Augustine Fresnel in 1822 invented a lens consisting of concentric rings of glass prisms to make a more efficient lighthouse beacon.

[2] The BBC's Low Energy Lighting Guide for TV Productions (2011)

[3] This section is adapted from "Modern Dimming: Choosing the Right Technology" by David Martin Jacques, published in *Church Production Magazine*, November/December 2006. Used with permission.

C H A P T E R

5.4

Television Playout and Content Delivery

PETER WHARTON

Vice President, Technology and Business Development, BroadStream
Atlanta, Georgia

INTRODUCTION

The entire function of a master control system is to assemble content for broadcast—program segments and live feeds, commercials, promotional clips and station IDs. Today, master control must do much more than simply stitch together content. It needs to also insert informational and promotional graphics and branding, including emergency alert messaging, sports news and weather data and credit roll squeeze-backs. It needs to handle multiple file formats and video resolutions, multichannel audio and multiple languages, subtitles and captions. It needs to insert watermarked ratings metadata, normalize audio levels and manage video aspect ratios. Frequently, master control is expected to provide near real-time display of social media content, which is something that wasn't even contemplated when the last issue of this handbook was written. All of these challenges compound the complexity of master control.

In the United States and other regions, ATSC 3.0 is expected to become the new terrestrial broadcast standard, moving broadcast from linear video transport streams to IP-based delivery and opening up a host of new possibilities. Internet-only linear television channels are being launched. Availability of UHD-TV content and delivery is increasing, and by the time you read this is likely to be ubiquitous. Immersive audio is also gaining popularity. Seamlessly integrating Web streaming, Video on Demand (VOD) and over-the-top (OTT) delivery into master control operations and workflows is becoming essential, especially for multichannel operations. In the last edition, this chapter was titled "Television Master Control Systems and Network Distribution." But with all of the additional burdens noted above, the term master control has become an anachronism, and "playout" is now used as a more encompassing term. Playout operations have become content preparation and delivery factories, where agility and adaptability are the prime concerns as the future becomes increasingly harder to predict and change occurs at a faster pace all while competitive pressures constantly increase.

In this section we'll use the term "linear playout" to define the creation of a real-time output stream, a signal that is typically broadcast over the air or fed as a cable or satellite channel, while "nonlinear playout" is used to describe the workflow that creates similar programming for delivery as files for Video on Demand and Internet delivery.

PLAYOUT FUNDAMENTALS

This section looks at playout from two perspectives: technology and operational models. Playout technologies have evolved from traditional, discrete master control systems with multiple individual hardware components performing specific functions to integrated, software-based solutions that can even be implemented in virtualized and cloud environments. Operational models include everything from single-channel facilities to large, multichannel operations and centralized playout facilities.

Master Control and Playout Technologies

Traditional Discrete Master Control

Initially, master control was a combination of a multi-source audio and video routing switcher and an audio and video switcher (or "mixer") that provided the ability to seamlessly switch between sources. (See Figure 5.4-1.) Unlike a production switcher, the master control switcher switches audio along with the video (typically called "audio follow video"). In some countries, the master control switcher is called a vision or presentation mixer.

By the mid-1990s, automation and video servers were becoming prevalent in master control systems as evolutionary replacements for videotape machines and cassette robots. These automation systems were a technological breakthrough at the time, but were complex systems to set up and operate. Numerous external devices including video servers, logo inserters, audio cart machines, character generators (CGs) and other graphics-generating devices were needed to create a complete program output. These were typically purpose-built hardware-based components. When a number of these devices are used to provide master control functionality, these systems are often referred to as "discrete master controls," as each function is provided by a discrete device. Each device had its own control protocol, typically serial, which varied from manufacturer to manufacturer. Likewise, each device had its own, often proprietary, user interface, making operations more difficult. At some point, machine controls became more standardized, which allowed facility managers to choose "best-of-breed" devices for each discrete function in the master control facility. It was especially important to have this flexibility when, for example, a particular channel required a complex graphic design that one specific product could generate, or there was a large library of content that was proprietary to a specific device.

Streamlined Discrete Master Control

A decade later, master control became more streamlined. Most playout facilities had then become multichannel facilities and managed the playout of a number of channels or program streams simultaneously. In this streamlined model, automation was (and still is) used to manage the master control systems, allowing one operator to easily run many channels or to run a single channel with complex requirements such as live sports integration. The continued growth of multichannel playout systems requires master control systems that are more scalable. To provide scalability, a modern discrete master control switcher typically integrates a number of devices including an audio clip player, logo inserters and graphics capabilities into the switcher itself. The source routing is typically handled by the main facility routing switcher with destinations to the master control switcher for digital video effects (DVE) sources and external graphics devices. Additional purpose-built hardware devices are being added into the chain to handle newer requirements—such as audio watermarking and SCTE trigger insertion. The automation controls these external devices as well, each device still sometimes with its own proprietary control protocols. With technology improvements and fewer connections, the master control switcher signal

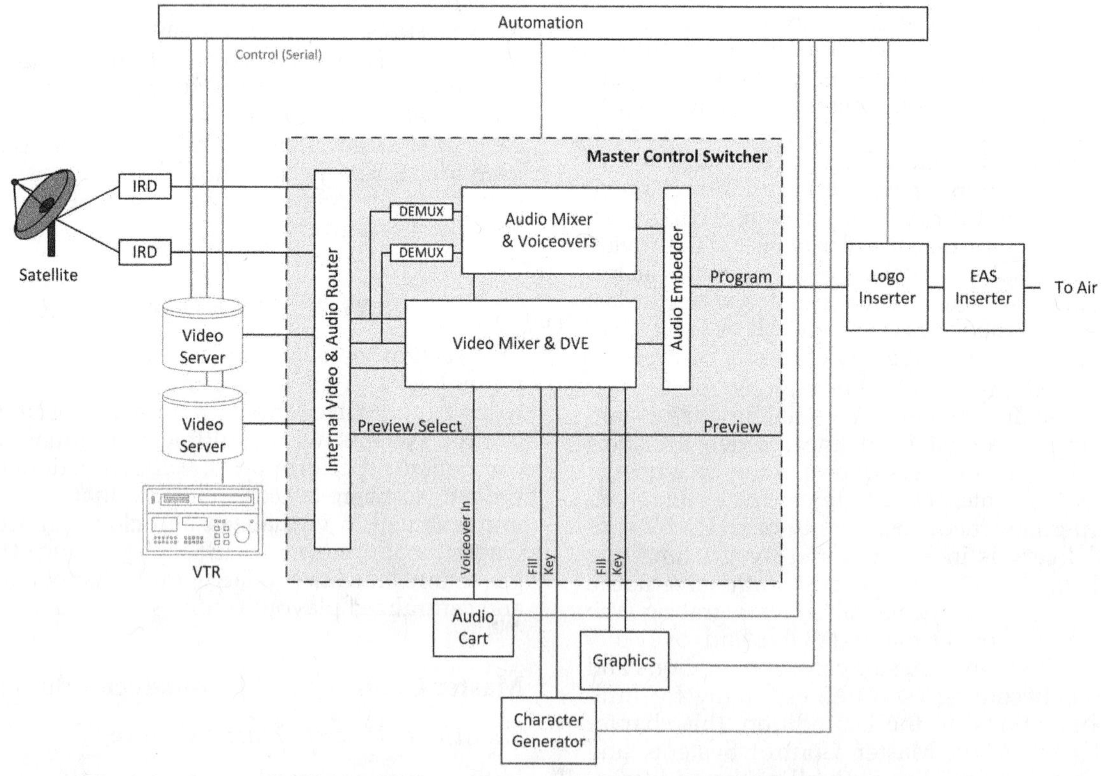

FIGURE 5.4-1 Traditional Discrete Master Control.

processing chassis shrank from many rack spaces to as little as one rack unit (or less with some card-based modular versions), making it possible to automate and play out a large number of channels in very little rack space. Full-featured modern discrete master controls are complex systems to implement and maintain, but for master control facilities that demand best-of-breed capabilities such as sports networks with specific graphics requirements, they often remain the preferred solution. At the same time, additional demands were being placed on master control—support for hybrid SD and HD simulcast channels, multichannel audio and even more advanced graphics. As a result, even though many of the previously required external components were now integrated into the master control processor, the additional requirements meant more external hardware was still needed to create a complete playback channel. (See Figure 5.4-2.)

A principal characteristic of the discrete master control approach is that the functions of the master control system are mostly performed by dedicated, purpose-built hardware. The hardware of a discrete master control provides a high level of performance and reliability, but at a steep price. As the needs change in a broadcast facility and new demands are placed on master control, these systems are not able to easily adapt.

Software-Based Integrated Playout

An alternative to hardware-based master control systems is software-based "integrated playout systems." These systems run specialized (often proprietary) software and perform most or all of the required functions of master control on commercial off-the-shelf ("COTS") IT server hardware.

Running a channel in a software environment provides many advantages. (See Figure 5.4-3.) The systems are inherently more compact, more power efficient and more scalable than hardware-based master control solutions, and are usually less expensive. Since they are software based, they can readily adapt to changing requirements without requiring additional hardware. Discrete master control systems were implemented in an era where equipment had a service life of 10 to 20 years. With the rapid rate of change in the television industry brought on by technology advancements via the IT industry, television facilities must be able to easily adapt to new requirements; moving from hardware-based to software-based solutions enables this adaptation. With multichannel systems, each channel can have its own software configuration to meet its specific requirements while still having the same physical platform as all of the other channels. Finally, redundancy and disaster control are simplified.

Not all functions of master control are integrated into a single server with integrated playout solutions. Many systems require a shared set of core services to provide channel and media management, often running from a cluster of redundant database servers. Some systems integrate the video server, master control switcher and graphics functions into a single server, but run the automation engine in a separate server. In the instance where all master control and playout functions can be run in a single standalone server, the functionality is often called "Channel-In-a-Box" (CIaB).

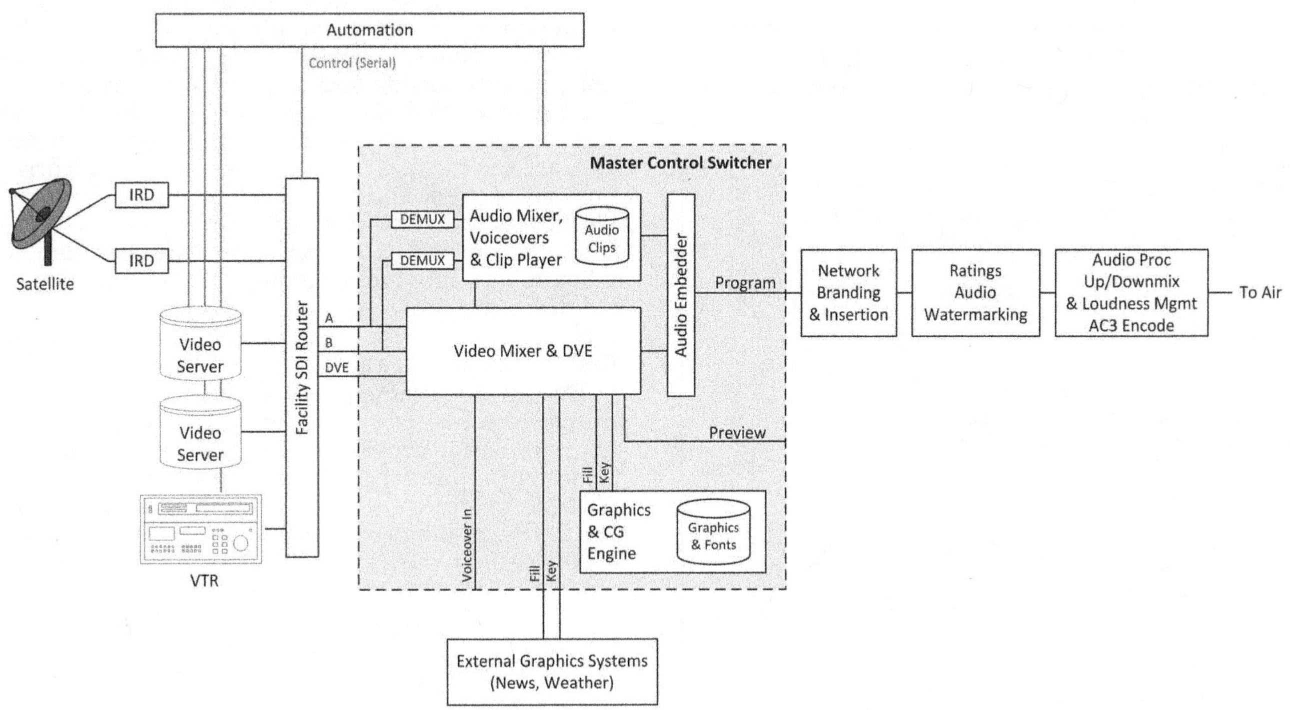

FIGURE 5.4-2 Streamlined Discrete Master Control.

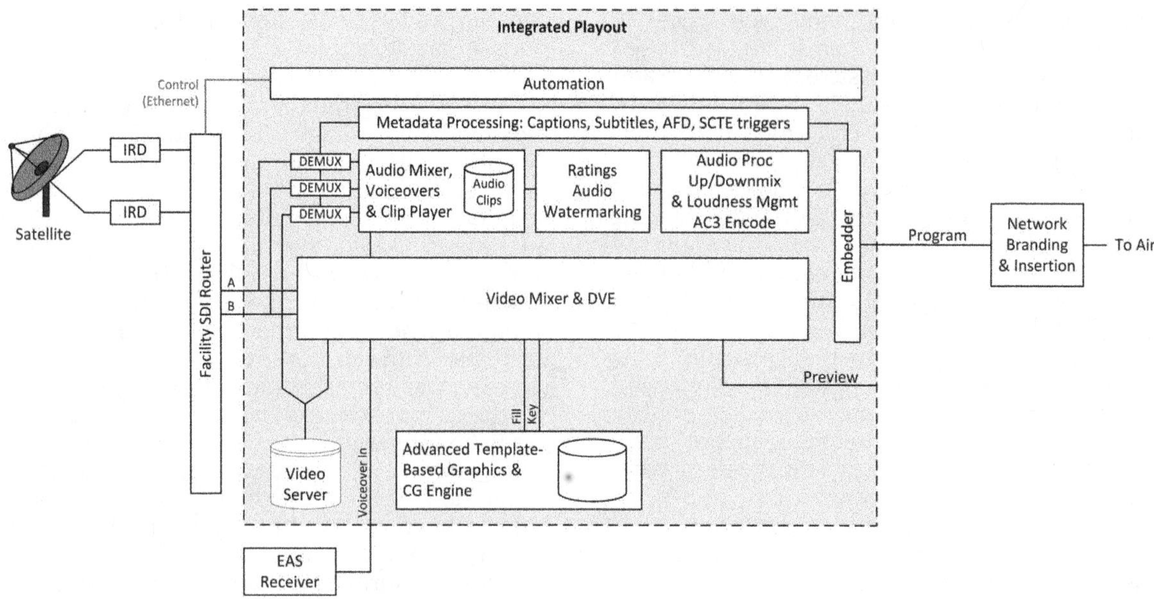

FIGURE 5.4-3 Software-Based Integrated Playout.

Integrated playout systems can often support not only SDI I/O, but also streaming IP inputs and outputs, with the streaming decoder and encoders integrated into the software. For SDI I/O, a video card is installed in the server. With some systems, the video card is simply a frame buffer and SDI I/O; on other systems the video card provides additional processing such as video format conversion, DVE and de-embedding and embedding of audio and VANC (vertical ancillary data space) metadata.

A software-based playout system incorporates a wide range of capabilities as shown below.

Typical features of an integrated playout solution:

• Video server with wide range of file formats supported • Media management, including missing media alerts • Master control switcher • Router control • Active format description (AFD) management • Dolby decoding and encoding • Audio level normalization—both file-based and live inputs	• Multichannel automation with control of all capabilities • Live input switching • Logo insertion • DVE • Animated graphics and branding • Character generator • Closed captions • V-chip insertion • SCTE-35/SCTE-104 receive and inject • IP inputs and outputs	• EAS alerts • Audio mixing and voice-overs • Audio clip playback • Video up, down and cross-conversion • Audio up/down mixing • Nielsen ratings watermarking • Integrated redundancy management

A software-based integrated playout system has a number of advantages over discrete master control solutions. A software-based solution allows the system to adapt to changing demands and add new features and incremental upgrades without hardware "forklift" upgrades or external devices. Use of COTS hardware provides simpler support and upgrades with greater protection from obsolescence. Tightly coupled automation and playout engine assures that all playout capabilities are controllable from automation. With discrete master control systems, the playout system often has inherent capabilities that the selected automation does not support, making such capabilities unavailable to the use. Integrated playout systems typically offer end-to-end metadata integration from files and incoming streams, and traffic and playlist metadata, which drives graphics, control and output parameters, simplifying workflows that would be difficult if not impossible to implement with a discrete master control. Simpler integration, with a limited requirement to control external devices typically controlled via Ethernet instead of serial data, allows for easier operations as all capabilities are provided in a single, unified interface.

Despite all of these advantages, integrated playout can have some shortcomings when compared with discrete master control systems: As all functions are typically performed in the server and software, it is possible to create events so complex—such as multiple DVEs, graphics, file playbacks and IP decodes/encodes—that the system becomes resource-starved and fails. Hardware-based discrete master control systems have defined capabilities that avoid these resource concerns, but are more finite. Discrete master control systems typically run on embedded real-time operating systems similar to those used to run automobile computers and other mission-critical systems. Software-based playout typically runs in Windows environments or occasionally Linux, making them less stable and requiring greater IT security and redundancy schemes. Discrete master control and automation systems typically support a wider range of external devices such as advanced graphics systems and hard-panel master control surfaces. This often makes them the preferred choice for complex news and sports channels with extensive live switching and specific graphics requirements. Software-based integrated playout is more dependent on a well-engineered and maintained IT network than a discrete master control solution, and requires an engineering staff fluent in modern networking technologies and IT systems.

The Next Generation of Playout

It is possible to build playout facilities today using software-based integrated playout that are completely IP based and have no SDI video. An all-IP facility provides additional benefits:

- Unified wiring and routing infrastructure—the same wiring and routing technologies used across the enterprise for both business and engineering applications.
- Greater flexibility, scalability and protection from hardware obsolescence.
- An infrastructure that is agnostic with regard to frame rates, video formats and resolutions.

While integrated playout delivers numerous benefits, it still requires dedicated hardware for each channel, suitable for 24/7 linear broadcast, cable and satellite television channels. But newer delivery systems such as content channels streamed over the Internet or to mobile devices may make it desirable to operate some channels for a short, defined period such as during an election or sports season. For these "pop-up" channels, dedicated hardware that might sit idle for long periods could make them uneconomical or impossible to launch in a short period of time. They need a different approach.

With proper software architecture, an all-IP software-based solution can run on virtual machine platforms. A virtualized playout system, by definition, is an all-IP solution; any conversion from IP to SDI is done by edge devices (devices that sit on the receiving end of the channel) instead of by video cards in the playout server. This methodology provides a highly scalable solution where channels can easily be added or dropped as needed. Virtualization further abstracts playout from the hardware, providing economies and greater flexibility to adapt to changing demands, and each channel consumes just the resources needed by its configuration. Redundancy is more economical in a virtualized environment as spare resources can be allocated to other tasks until needed for redundancy. The virtual machine playout concept can be extended to "private cloud playout," a type of master control centralization in which an organization runs their software-based playout on dedicated virtual machines in a private data center that they own or lease that is shared by all of the operations centers in the enterprise. This provides even greater economies and scalability.

With the appropriate management, security and connectivity, playout can also be run on virtual machines in the public cloud, providing additional benefits. Public cloud playout eliminates most capital expenses (CAPEX), making it more economical for occasional playout applications. It allows organizations to focus on content creation and delivery instead of technical expertise and infrastructure. A public cloud environment can provide greater scalability and redundancy than the private cloud model while offloading more responsibilities onto the cloud service provider. Playout in a public cloud can also reduce costs as both content providers and distributors may have direct connections into the same cloud environment. An ideal public cloud implementation would allow for content to be acquired, produced, managed and delivered all without ever leaving the cloud environment, creating substantial efficiencies and economies. At the time of publication of this *NAB Handbook*, several major cable network channels are expected to be originating from public cloud playout solutions.

In addition to linear playout, master control facilities are tasked with nonlinear content delivery for Internet "over-the-top" and Video on Demand (VOD) services. A unified approach to providing linear and nonlinear content delivery offers compelling economies.

When implementing an all-IP solution, whether on-site in dedicated hardware or as a virtualized system, it's essential to make sure that all of the services and components that are needed for the particular master control are actually available in the software; individual requirements may still drive the need for a small amount of external hardware. For example, in the software-based integrated playout solution drawing above (see Figure 5.4-3), the network branding and insertion is still provided by a downstream device provided by the affiliate network with SDI I/O.

Other Master Control Systems

Alternative playout models to the above technology also exist. These typically serve specific applications such as local station or commercial insertion into network feeds and are covered in greater detail below under "Regionalization and Localization."

Playout Operational Models

Playout systems can be implemented in several ways. A playout system can be provided "on premise," where the solution is implemented in a television station or facility. Organizations will often elect to centralize the playout for multiple stations, facilities or regions into a hub for efficiencies and operational benefits. Playout can also be outsourced to third-party hosting facilities that run centralized playout operations for multiple customers. Any of the previous master control and playout technologies can be used to support any of these operational models.

Playout can also be implemented in a private or public cloud using appropriately designed software-based playout.

On-Premise Playout and Control

The simplest form of playout is one where the playout hardware/software and operators are colocated in the same facility, and the channels that originate from this facility are distributed in the same region or fed to a satellite uplink. (See Figure 5.4-4.) Many such facilities operate multiple channels; larger commercial cable origination facilities may operate more than 100 channels in such facilities. Typically each operator manages anywhere from one to as many as twenty channels simultaneously, as determined by the channel complexity and operational requirements. Live events requiring manual operations typically constrain an operator to managing a single channel.

The operator typically runs the channel from a computer screen that displays the automation system interface. A larger, wall-mounted display is fed from a multiviewer that shows the various source signals, channel output signals and off-air signals for confidence monitoring. Additional screens and interfaces may be provided for facility monitoring and control, transmitter monitoring, traffic and media management. If a physical master control panel is used, it might also have a control screen.

Centralized Playout

Centralized playout operations have been used since the late 1990s as a way to reduce operational costs while improving efficiency and performance. A centralized playout facility is one in which the operator manages multiple channels from a common location, but the broadcast or distribution of those channels occurs in one or more separate regions. (See Figure 5.4-5.) This means that the operator cannot directly monitor the off-air or off-satellite feed, and a provision must be made to backhaul the off-air monitoring signal from the region in which it is broadcast back to the operator's location.

Benefits of Centralization

Centralization can bring numerous benefits to an organization or group of organizations. The most obvious are cost savings—typically a centralization project allows an organization to reduce significantly the number of content prep and master control operators and redundant equipment in the enterprise, which may offset the initial capitalization costs and ongoing connectivity costs. In a centralized playout facility, a single operator can run a large number of channels simultaneously, aided by automation and advanced monitoring by exception.

Less obvious, but just as beneficial, are the operational improvements. Centralization generally leads to fewer on-air errors and shorter outages resulting from the capabilities of the monitoring and control system. Operator errors are also reduced as they now work in a more collaborative environment with greater oversight and centralized training. Elusive faults are easier to diagnose when there are a number of identical systems running concurrently, and centralization also benefits from "fix it once, fix it everywhere" versus each station troubleshooting the same issue independently. Centralization can make new initiatives affordable, such as improved branding, promotions and an overall improvement in the on-air product, leading to greater audience retention.

In addition to centralizing playout, other operations across the enterprise can also benefit from centralization, including:

- Graphics and promotions
- News operations
- Incoming content record and prep
- Traffic

Centralization Models

There are two main models for centralized playout. In the first model, the entire playout system is located at the operator's control hub location—the operator workstation, the playout channels and the monitoring and control system. High-bandwidth circuits feed the channel output to the in-region transmission systems. In the second model, the actual playout system along

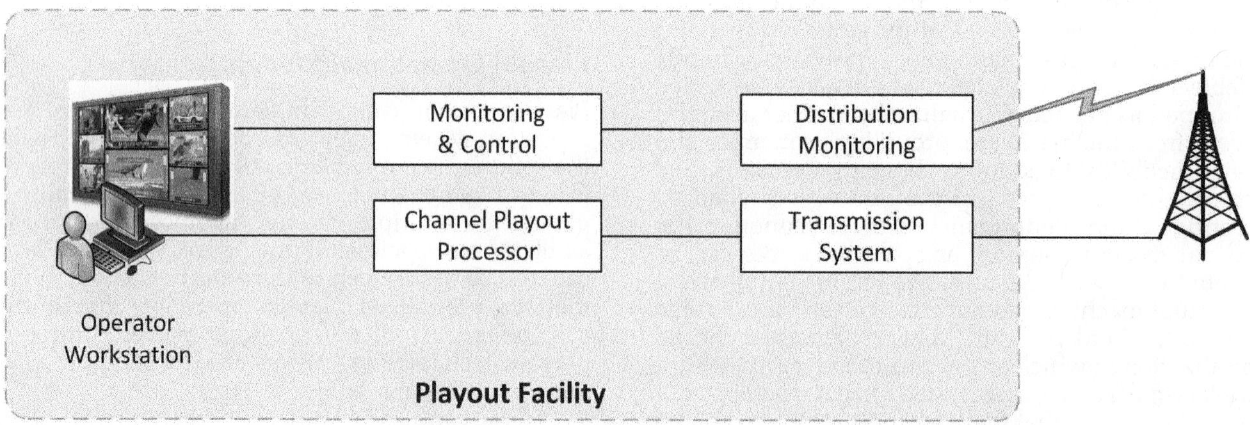

FIGURE 5.4-4 On-Premise Playout Facility.

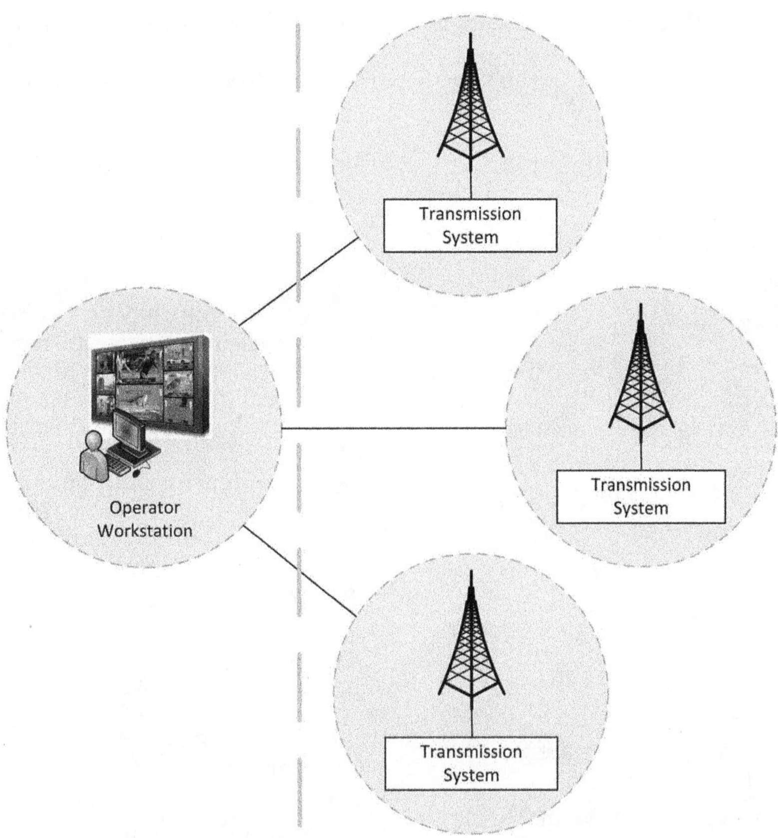

FIGURE 5.4-5 Centralization Hub and Spoke.

with the monitoring and control system are colocated in region with the transmission system, and only the operator workstation, monitor wall and other operator interfaces are extended to the control hub. Each model has benefits, and some larger deployments use a mix of both models.

Centralized Control and Playout

In centralized control and playout, the channel playout and processing are located at the control hub with the operators. (See Figure 5.4-6.) The channel outputs are fed over high-bandwidth circuits such as DS-3 circuits (45 Mbps), MPLS Ethernet circuits (typically 100 Mbps), SONET ring networks or leased video circuits to the remote transmission locations. An advantage of Ethernet circuits is that the return bandwidth is the same as the outbound bandwidth; this allows for very high-quality return monitoring—sometimes the full transmitted transport stream.

The content stream that is fed from the playout facility to the transmission facility can be at a mezzanine level of compression; this is typically done when it is recompressed and multiplexed with other signals at

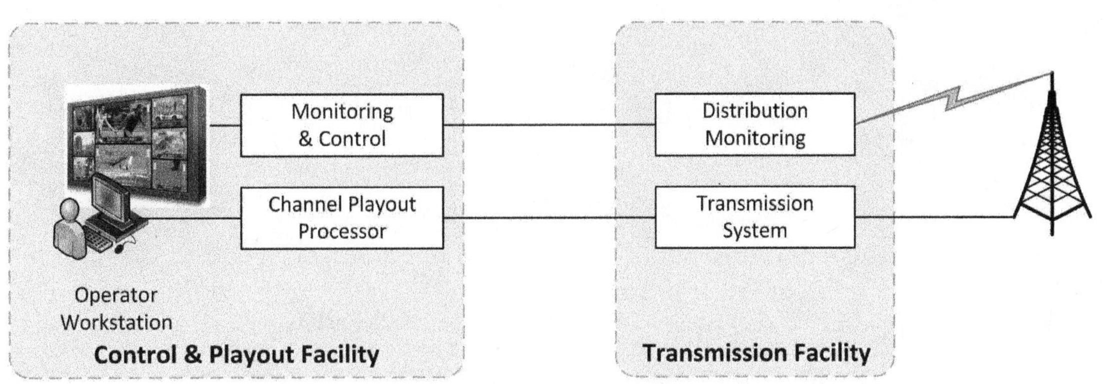

FIGURE 5.4-6 Centralized Control and Playout.

the transmission facility. Another configuration is to multiplex and package the entire signal at the playout facility, and the feed from the hub arrives at the transmitter directly.

Advantages of Centralized Control and Playout Model

All playout equipment is at the hub, allowing for:

- Simpler maintenance and upgrades
- The availability of on-site spares
- Colocated operations and engineering, common core including IT network, HVAC and UPS
- Shared media for all channels, reducing media duplication
- Efficiencies of common ingest and media prep
- Affordability of highly trained operations and engineering staff

Disadvantages of Centralized Control and Playout Model

Disadvantages include:

- Recurring interconnection circuit costs are expensive.
- High-bandwidth circuits may not be available in all locations, and last-mile construction can be very expensive.
- Playout or connection redundancy can be expensive: duplication of circuits, SONET ring network, satellite backup or duplicate redundant playout equipment and operations staff at the transmission facility.
- Stations can change ownership; moving playout operations out of the prior owner's hub can be complicated.
- Local content—commercials, programs, graphics—all needs to be sent to the hub for integration into the channel stream(s).
- Live local content from the distant stations, such as news, must be delivered back to the hub in real time to be integrated into the final program stream.

Centralized Control with Local Playout

In the centralized control with local playout, the actual playout systems are located at each local transmission facility and are remote from the operator location and hub. (See Figure 5.4-7.) The hub houses only the monitoring and control surfaces, and all of the processing is done in-region. The control side is very low bandwidth, typically under 1.0 Mbps. The return bandwidth can also be very low; these systems have been operated over a single T-1 circuit (1.5 Mbps). At this low bit rate, the return monitoring is only confidence quality in order to have a low enough latency to support live switching. From an operator's perspective, both systems appear practically identical; the only difference is in the quality of the monitor wall images.

Advantages of the Centralized Control with Local Playout Model

All playout equipment is at the transmission facility, allowing:

- Simpler playout system redundancy (N+1 at transmission facility), mitigating the risk of catastrophic hub failure.
- Simple, inexpensive hub redundancy and transfer of control between hubs.
- Local channel control at transmission facility for redundancy, local insertion and local live event switching. In an emergency, the station can be controlled remotely from any PC with network connectivity.
- Simpler insertion of local content and multiplexing.
- Fewer complications in the case of acquisition or divestiture of stations, since each station can be operated locally.

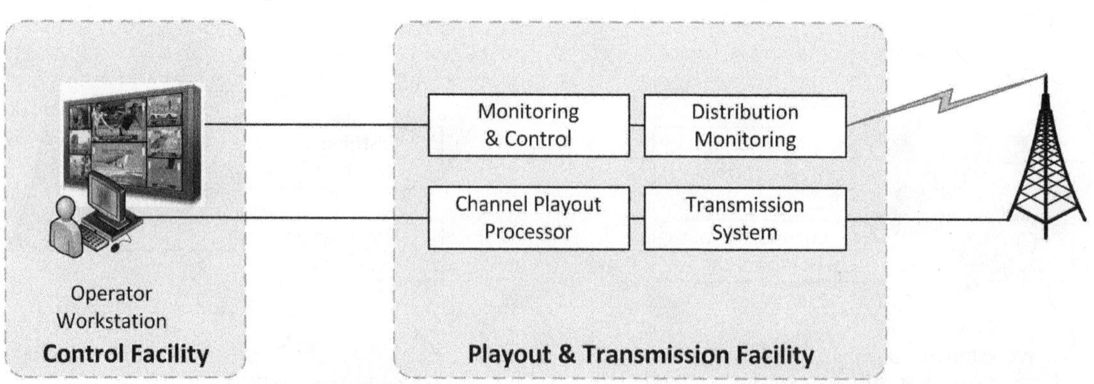

FIGURE 5.4-7 Centralized Control With Local Playout.

- Usage of low-cost, low-bandwidth and readily available interconnections from the station to the hub.

Disadvantages of Centralized Control with Local Playout Model

Disadvantages include:

- Low-quality monitoring prevents true image and audio quality analysis at hub.
- Occasionally requires on-site engineering support for some diagnostics, repairs and upgrades, although most work can be done remotely.
- Duplication of media at all distant stations, although storage costs continue to drop, making this less of a factor.

Network Considerations for Centralized Operations

When planning a centralized operation, careful attention must be paid to the design of the network that interconnects the central hub facility to the in-market stations, including the following parameters.

Bandwidth

The network must have sufficient bandwidth to adequately support the remotely controlled operations, monitoring and distribution. The on-air video and playout control signals must be prioritized over less important network traffic such as file transfers and enterprise business activities. This can be done by providing dedicated circuits with fixed bandwidths. More often, shared Ethernet networks are used with carved-out prioritized and guaranteed bandwidth for these functions using MPLS (multi-protocol level switching) classes of service.

Latency

When used for centralized control with local playout models, network latency is critical for real-time switching of live events. Latency both affects the control of the local playout system from the central operations hub and introduces additional delays in the monitoring. The encoding used for monitoring the local playout system also needs to introduce minimal latency; typically all I-frame encoding is used.

Redundancy

For fully centralized playout operations, redundant networks are often deployed; SONET networks provide redundancy coupled with self-healing. For local playout models, the network redundancy is less critical, especially where real-time live switching is not typical. The local playout system should continue to play while the network is down as long as there are playlists loaded and content cached, and local staff can step in to handle manual switching if needed.

Outsourced Versus Internal

In addition to centralizing to a hub or hubs owned by the broadcaster, there are numerous companies that offer hosted channel playout from commercially operated hubs. Some hubs specialize in cable networks, others only host PBS stations, while some handle a wide mix of clientele and channel types. With hosted playout, the level of service can range from simply hosting the equipment (platform as a service [PaaS]), providing and maintaining the equipment (infrastructure as a service [IaaS]), providing the software and channel management (software as a service [SaaS]) to providing the operators and completely running the channel (fully outsourced). Some third-party providers are even willing (for a price) to perform all of the traffic functions as well. Another form of outsourcing is cloud-based playout where the service is used as a subscription model.

Should you consider outsourcing? For some enterprises, planning and executing a centralization project or running a hub may require more effort than they're prepared to commit to. For smaller enterprises, the efficiencies that make hub operations attractive to larger centralization projects may be unobtainable. Whatever the reason, for enterprises looking to centralize but unwilling to embark on this course internally, outsourcing is an option.

Figure 5.4-8 illustrates the various service models for playout.

Third-party hosting facilities may offer a range of advantages. They have broad experience in hosting centralized playout. Because of their larger size, they can operate with greater efficiency. They have economies of scale as well as greater leverage in purchasing solutions. They may be better located on a fiber backbone or in a more suitable location. Their facility may include features such as generators, UPS and satellite backup for the fiber, providing greater resilience than a dedicated hub might be able to provide. They may be able to offer additional services that can be test driven, such as enhanced graphics, secondary and mobile DTV channels, VOD and Web services—all without the expense and commitment that an internal launch would require. By outsourcing playout to a hosting facility, enterprises are able to focus instead on their real mission—providing compelling program content.

Conversely, third-party facilities run at a profit, potentially diminishing the intended ROI. They usually operate as centralized control and playout models and own the equipment, making it harder to remove the stations from the hub if they are sold. The service provider may not fully understand a company's business or the needs of a particular operation. Typically, hosted playout includes a service-level agreement (SLA) that guarantees a level of performance.

Successful Centralization

Some centralization projects are highly successful, yielding excellent return on investment and operational improvements. Other similar projects have failed and

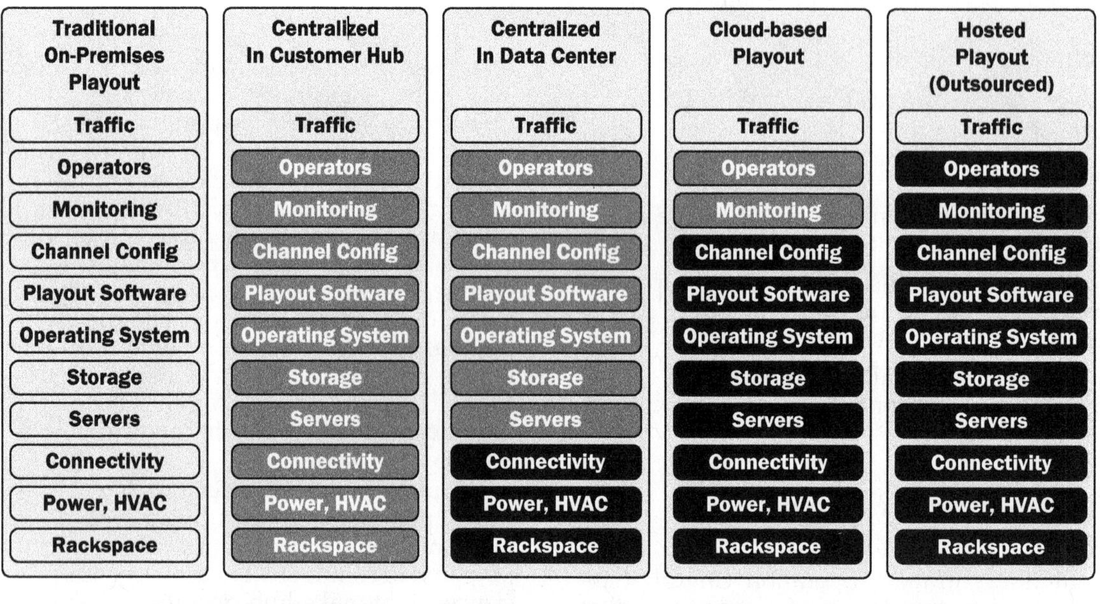

FIGURE 5.4-8 Playout Service Models.

the stations have returned to on-premise playout. In most cases the difference was not the technology; it was the discipline, determination and consistency in execution that made the difference. The most challenging part of a centralization project is rarely the technology; it's getting an existing operation to give up control of their playout to a hub and become cooperative partners in supporting a unified approach to playout operations.

AUTOMATION

Automation is essential in modern facilities to efficiently control playout channels. It manages a schedule while controlling the devices and processes required for creation of a finished air signal. Even in playout operations that handle substantial live events and their associated manual operation, automation is essential in executing the complex synchronized actions required during manually triggered events. Automation provides the efficiencies that allow for a single operator to manage a large number of channels simultaneously or for a station to run a program stream "lights out" without any operators.

A playout system consists of two main components—the control system and the signal processing. The control system typically includes an automation schedule to run a list of program events according to the time of day, with manual control capability for intervention or live events, and to log what was output for reconciliation. The signal processing system includes the audio, video, graphics and metadata processing that creates the final output for distribution.

The automation system uses a playlist to provide a schedule-based control of all of the devices and functions in a playout system. As shown in Figure 5.4-9, the operator sees the playlist on a graphical user interface (GUI), which allows the operator to modify the playlist and manually override it. Without automation, we could have neither the complexity of playout that exists today nor the efficiency of multiple channels managed by a single operator.

Automation Event Types

Automation provides scheduled playout of a playlist of events. A playlist typically consists of two main event types: primary and secondary events.

Primary events are the main scheduled program events. They include program segments, commercials and other full-screen events. Primary events can have *hard start* times, starting precisely at the time of day that they are scheduled, *floating times* where they begin when the preceding event finishes or *manual start times* where the playback is initiated manually. Hard start times are common for the first event in a program, as this time needs to match the published schedule and electronic program guide. Subsequent program segments that are separated by commercials use floating times when the program and commercials are all played out locally. When a recorded program comes from a network feed, the start times for commercial breaks are usually published in advance and entered into the playlist as hard start times. For live events, manual start times are used and the operator triggers the manual event using the *"take next"* function in the automation software GUI. Other commands an automation system might utilize in manual operation include *cue, skip next* and *hold*.

If an event with a hard start time begins before the prior event is complete, it will truncate the prior event, typically with a warning in the playlist GUI. If an event ends before the next event with a hard start begins, a

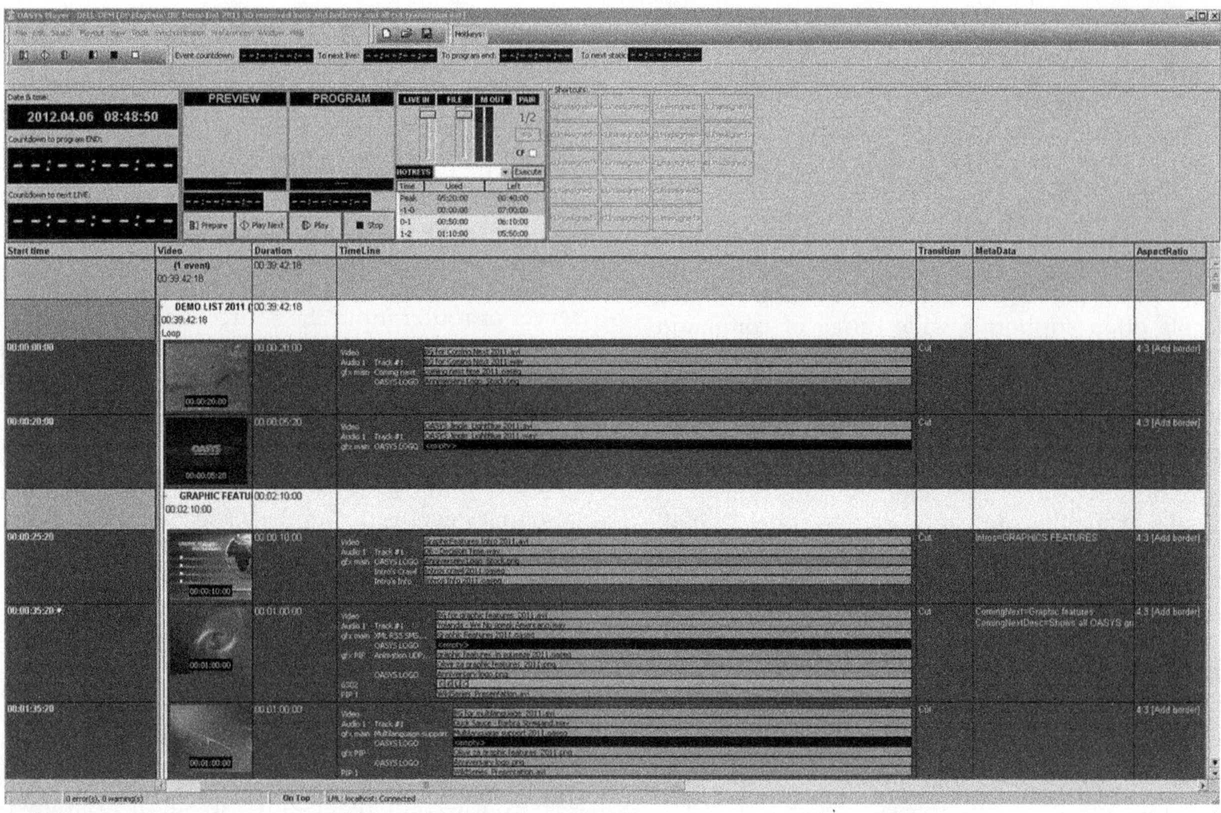

FIGURE 5.4-9 Typical Automation Operator Interface (GUI).

Courtesy of BroadStream Solutions

gap in the playlist will occur with a warning notification in the playlist GUI. Some automation systems are able to automatically fill these gaps with additional material of the correct duration.

Some automation systems support the concept of *grouped events*, where an entire set of events can be treated as a single group. A group could be a local commercial block, an entire program including program segments and commercials, an alternative playlist in case a live event runs long or for breaking news. Actions and properties can be assigned to an entire group, such as automatically make each breakaway from network for local commercials a manually triggered event.

Secondary events are usually some type of overlay over a primary event and are timed relative to the associated primary event. Examples of secondary events include graphics such as logos, on-screen clocks, news crawls, audio voice-overs and DVE functions. Secondary events such as DTMF and SCTE triggers are not apparent on-screen because they are utilized to embed commands in the output signal to control downstream devices. Most of the complexity in automation comes from the handling of these secondary events.

Automation Control

Automation controls all of the processes required for master control and playout, including any necessary external devices. Control of external devices via automation was originally done via EIA RS-422 serial connections, often with nonstandard pinouts. Ethernet is rapidly replacing serial control as the preferred method.

In a discrete master control, automation must support the desired features of each controlled device utilizing the supported control protocol of that device. For example, when controlling a graphics device, the control protocol of the graphics device may be extensive, and the automation system driver may only support a small subset of those commands. With integrated playout this is seldom an issue, as the integrated playout automation is often the only means for controlling the playout functions, and the functionality originally dedicated to discrete graphics engines is built into the software of the integrated device.

Some common protocols used in discrete master control environments include the following:

VDCP Server Control

VDCP is a common serial control protocol for video servers that supports ingest, playout, and some media management. VDCP is often augmented with Ethernet-based API control for non-real-time activities including more extensive media management.

Master Control Switcher Protocols

Whereas VDCP is widely supported and nonproprietary, master control switcher protocols are typically proprietary and are based on legacy protocols with additional, unique extensions and customizations that have been added over the years to support advanced features. In North America, most manufacturers have used the Grass Valley M-21 protocol as a reference, while in Europe the ProBel TX-220 and Oxtel Imagestore protocols form the underlying command set. The Oxtel Imagestore protocol also supports Ethernet-based control. Some contemporary products support multiple selectable protocols based on these original protocols. The extended versions of these protocols often support a wide range of commands including channel management, graphics and CG and advanced audio management; the protocol documents are often several hundred pages long. It's not unusual for a vendor to offer several products, each utilizing a different subset of a protocol's commands. Automation vendors typically only support a fraction of the protocol's available commands. It's essential to make sure the functions needed in a master control deployment are supported by the protocol commands implemented by the automation vendor.

Graphics and CG Protocols

Like master control switchers, graphics and character generator ("CG") control protocols are typically proprietary and require specific automation support for full utility. The Chyron Intelligent Interface (CII) protocol is a common protocol often supported to provide basic functionality in graphics devices when a proprietary protocol is not supported.

GPI/O

Not really a "protocol," general purpose interfaces (GPIs) are used to interface two disparate systems when they lack a serial or Ethernet protocol for communication. A GPI is simply a single-wire input trigger that is held "high" internally on the receiving device with a pull-up resistor to a voltage bus (typically 5V or 12V), and pulled to ground by the sending device (most often with a toggle switch or push button) to trigger an action. The sending device output is usually called a GPO (general purpose output). Some GPOs use relays and GPIs use opto-isolators or relays to provide greater isolation between devices. A GPI can be momentary, where it triggers on the change of state (from high to low) with sequential triggers to toggle the action or uses separate GPIs for on and off actions. Alternatively, a GPI can be latching, triggering the on action when low and off when high. A GPI is useful for simple actions such as running an emergency alert message on a channel, but becomes unwieldy when used to implement a large number of actions. And a GPI can only pass an on-off state, whereas a control protocol can include file names, on-screen data and similar complex information.

A playout system has a time reference input to allow it to run the schedule with frame-accurate precision. Typically this is timecode using LTC or VITC inputs; network time protocol (NTP) is also supported. Newer systems support Precision Time Protocol (PTP) per SMPTE 2059/IEEE-1588.

Traffic and Schedule Creation

Traffic systems are used to create schedules for playout, providing a list of programs, commercials, interstitials and other content used to create the air schedule. The traffic system output is processed by the automation system to create a playlist that drives the channel on air. This schedule from traffic typically contains all of the main scheduled program events and commercials, event types, start times and durations. When the content comes from an external source such as a video server or router input the source is also defined.

A traffic system creates a schedule based on a set of business rules. These rules might include how many commercials to run in a break, or how to prevent conflicts between content elements such as two competitors' commercials back-to-back.

Many traffic systems also support the scheduling of secondary events. Examples of secondary events automatically scheduled by traffic might include:

- Closed-caption and V-chip ratings graphics for the first 5 seconds of each program segment. The appropriate V-chip logo is defined by the program file metadata.
- The specific TV station logo beginning 5 seconds after the beginning of each program segment and ending 2 seconds before end of each segment.
- Next program lower-third "tease" at 1 minute prior to end of last program segment.

Traffic Systems

The output schedule log from the traffic system is imported into the playout system to create the on-air playlist. The output is typically a file structure and protocol that is proprietary to the traffic vendor, using text, CSV or XML formatting. An SMPTE-standardized file format and protocol, BXF (SMPTE 2021), is increasingly used instead of the proprietary formats.

Codes or tokens are typically used to represent various functions that must be translated to automation playlist functions by the traffic schedule import process. The schedules provided by traffic are often incomplete and lack secondary events; these can be automatically added in by the schedule importer using rules or manually added by operators as part of the playlist approval process. A typical rule might be to add a secondary event to turn on the station logo at the top of each program segment at 5 seconds from the start of the segment with a duration of 10 seconds. The insertion of graphics for caption and V-chip IDs is often automated by the schedule importer.

In some smaller playout systems, no traffic system is used and schedules are created manually via a GUI using text or spreadsheet formats and imported into the automation system.

Schedules are typically imported into automation daily, either as a manual process or using a watch folder to automate the process. Schedule updates can be implemented the same way.

As-Run Logs

The playout system creates an "as-run" log file as it manages the playout channel. This is a list of all of the events that were performed by the playout system. It can be created in a custom format or a BXF-compliant format. The as-run log is used to reconcile what actually aired with what traffic originally scheduled; combined with compliance recordings, they provide the evidence needed to bill advertisers for commercials that were run and also prove compliance with regulatory requirements.

Traffic systems typically create a "reconciliation key" for each event scheduled, which is also passed to automation and inserted into the as-run log with the corresponding event. This allows for each instance of the same content, such as a commercial that repeats in the same program, to be properly reconciled.

BXF Protocol

Given the wide variation in traffic system schedule formats and capabilities, the process of interfacing between traffic and automation systems can cause problems when implementing a new playout system or adding capabilities. To simplify the interface and improve interoperability, the SMPTE BXF protocol (SMPTE ST 2021) was developed to standardize the interface between traffic and automation. BXF has the potential to allow faster implementation of playout systems while expanding the capabilities that the traffic system can support. BXF is a broad standard and encompasses many capabilities; unfortunately, seldom have two different traffic vendors implemented the standard identically, so additional work may still be required to provide interoperability between traffic and automation.

BXF defines an XML schema (XSD) for the exchange of schedule information and metadata. BXF also supports near real-time updates, allowing incremental updates to the schedule and the passing of as-run information back to the traffic system immediately after airing. This allows a traffic system to take corrective action such as scheduling additional commercials for a live event that runs longer than scheduled.

BXF supports:

- Traffic schedule and as-run logs
- Primary events
- Secondary events including "hot starts"
- Content metadata
- Media movement instructions
- Near real-time two-way communications between traffic and automation
- Information exchange with advertising agencies for media buys

Electronic Program Guides

Electronic program guides (EPG) are the channel and program schedules presented to users on their viewing device—typically a set-top box or mobile device—that allows them to see a menu of what's airing on a given day and select a show to watch or record. While the device in master control that embeds the EPG into the program stream only processes the EPG on a 24-hour basis, the traffic system often can publish the EPG far in advance of the air date. Ten days to two weeks in advance are common.

Automation systems can also support real-time updates of some EPG processing systems. This allows late changes made in the automation playlist to be reflected in the user's EPG.

Live Event Handling

Playout systems that require handling and switching of live events need specialized capabilities; this is especially true for channels with live news and sports where the operator must manually trigger events including graphics and live cut-ins. Live event capabilities include the following:

Hard Panel

The operator needs to have manual control of a channel to trigger the next event. A touch screen or mouse is usually insufficient, as it requires the operator to look away from the monitor wall to confirm where the mouse pointer is on the automation screen. A hardware control panel allows the operator to maintain eye contact with the live event program on the monitor wall while performing accurate and timely actions on the channel. A control panel could be as simple as one with a few programmable buttons for router sources, graphics on/off and playlist "take next," or a more complex panel that mimics the functions of a traditional master control panel including multiple playlist control functions.

Ganged Operation

An operator might need to perform live event handling on multiple channels simultaneously. An example would be a network morning news program fed through a centralization hub that inserts local programming and commercials for several stations simultaneously. The operator would be able to select which channels are ganged, and a "take next" would then trigger the commercial breakaway across all of the ganged channels. Some systems allow multiple operators to control the same channels, providing additional flexibility with combined station and hub or operator and supervisor control.

Live Event Scheduling

Live events might not fit exactly into an allocated time slot in the schedule. If they run short, a "filler" set of events can be inserted into the playlist to fill the time needed until the next scheduled event starts with a hard start time. Often, a live sports or entertainment event will run long and overrun the next scheduled event, or breaking news will interrupt scheduled playout. In these cases, there are two ways the original schedule can be managed:

Join in Progress (JIP)

When the live event runs over into the next scheduled event, the scheduled event does not air completely. When the live event ends, the original schedule resumes exactly where it would have been had the live event not preempted it. This is also called "roll-under," as the original schedule continues to roll even though the live event is still on the air—even if it is not visible to the operator. Some systems allow the commercials that would have been overridden to be automatically appended to the end of an interrupting live event, maintaining revenue while sacrificing more of the preempted programming.

Schedule Delay

In this mode, the scheduled event start times are delayed instead of being preempted. When the live event ends, the schedule picks up exactly where it left off when it was overridden. At some point the schedule needs to be brought back onto the original schedule, especially if there are hard start times in the upcoming playlist. Often this is done by cutting short other live events such as live news programs. Nonessential schedule items such as promos and PSAs can also be dropped to make up time; some systems can prioritize events and automate this function. Finally, some systems will allow the automatic recording and delaying of incoming live events such as network programming, delaying up to an entire evening of scheduled programming until the late evening news, which then may be cut short to make up the lost time.

Clean Feed and Secondary Recording

It is often necessary to record a "clean" feed of a live program for later replay or other distribution such as VOD. A clean feed is one where some or all of the graphics inserted in playout are omitted—for example, a clean feed of a sports event might include the graphics related to the event itself such as the score and game clock, but omit other graphics that would be stale during a replay such as scores from other games that were occurring at the same time or sponsor graphics where VOD has a different sponsor. This clean feed could be recorded in the playout system itself, a feature called "secondary recording," or externally. The recording can be automatically segmented in the system for secondary recording when the commercial breaks are manually inserted in the original live program playout, allowing it to be easily reused for later replay or creation of VOD assets without additional editing.

Profanity Delay

Live events may require profanity delay and blanking in playout operations. This delay may be simply a fixed delay that allows the operator enough time to insert alternate audio and video to override the profanity.

Live Caption Insertion

Closed captions must be inserted into live programming. This is covered below in the Captions and Subtitles section under Metadata.

Latencies and Monitoring

All master control systems have latencies, from system throughput, source synchronization and graphics/DVE processing to encoding and distribution. Virtualized, centralized and cloud-based playout solutions with their compressed IP video connections introduce additional delays as described above in the section entitled Network Considerations for Centralized Operations. When live switching is a requirement for playout, attention must be paid to the accumulated latencies in the architecture and their impact on the operator's ability to perform timely switches in the programming.

VIDEO PROCESSING

The master control or playout solution includes video processing. Typical processing includes:

Transitions

A transition is the ability to mix between two sources using a variety of fade and cut effects. Simple cuts or fade-through-black "V-fades" are most common; occasionally a "U-fade" is utilized when content does not have black frames at the beginning and end and the server leaves the last or first frame in cue to prevent seeing this still frame on air. More advanced transitions include wipes, digital video effects, fade-through-color and the use of animation file overlays to provide graphical transitions.

Video and Audio Routing

Video and audio routing involves internal routing or control of an external upstream router to select the sources for the video mixer.

Internal Graphics Overlays

Internal graphics overlays provide still and animated graphics for logos, ratings information and other on-screen graphics coupled with internal storage.

Character Generation

Character generation ("CG") capability is used to create on-screen crawls and text including clocks and data-driven graphics. The CG also provides the capability for EAS alert text crawls and the system interfaces to external EAS receivers.

Digital Video Effects (DVE)

This involves processing to provide picture-in-picture effects. DVEs can be 2D, providing simple rectangular "squeeze-back" effects; 2.5D, where trapezoidal squeezes are supported; and 3D, which provides effects such as page curls. Multiple DVE windows can be on the screen simultaneously to squeeze back the credit roll of a terminating program to accommodate "coming up next" video clips. Some broadcasters use multiple DVEs for "hot starts," which involves playing the beginning of the next show alongside the credit roll of the previous show, allowing additional time for commercials in the next show. Sources for the DVE are most often routed video inputs and video server ports in a discrete master control or from additional video playback engines in an integrated playout system.

Preset Output Monitoring

A preset output allows the operator to preview the next primary event source, often with a separate graphics overlay so that graphics can also be previewed.

Graphics Keying

Graphics keying involves the process of providing external fill and key video inputs with internal keyers to support the keying of external graphics systems over video sources. More advanced broadcasters such as news and sports channels often require more additional advanced graphic layers than are possible with internal graphics capabilities, and such systems may utilize multiple external CG and graphics systems.

Frame Synchronization

Master control and playout systems typically require synchronized video feeds and rarely have built-in frame synchronizers. Where unsynchronized sources are present, frame synchronizers should be installed in front of the playout system. On most systems, the output timing can be adjusted to align it with other router inputs.

Up/Down/Cross Conversion

Video format converters may also be needed in mixed-format facilities if the playout engine cannot handle mixed formats internally for live inputs. Many integrated playout and software-based systems include the ability to handle mixed formats and unsynchronized sources. Outputs may also require up or down converters to allow both SD and HD outputs from a channel.

Aspect Ratio Conversion

When up- or down-converting an input or output, an aspect ratio converter manages the transition between 4:3 and 16:9 aspect ratios. This is covered in detail in the Aspect Ratio Management and AFD section under Metadata.

File Formats

Television facilities use a wide range of file formats for video, audio and metadata. A playout system needs to be able to support ingest and playback of the supplied files, whether through native capability or transcoding. Older video servers are typically limited to one or a few internally supported file formats, and outboard transcoding of incoming material is required for any unsupported formats. Newer video servers and integrated playout solutions support a wide range of formats natively and require less external transcoding. Even with a wide range of supported file types, in cases where a greater variety of files are ingested, the likelihood increases that a file will fail to play back as intended. Most facilities will therefore define one or a few "house" formats that will be used in playback and transcode nonconforming files to one of these preferred formats. Often the house formats will include the format used by the prior playout system to ensure backward compatibility and simplify the transition from one system to another.

Smaller organizations and playout facilities that accept content from a number of sources typically have to accept any file formats provided, whether by having a playout system that can natively support the formats or via external transcoding. Larger organizations, most notably content aggregators and content delivery services, often provide file delivery specifications that content creators and providers who send content to the delivery services must adhere to. These specifications typically include:

- Video resolutions, formats, frame rates and bit depths
- Aspect ratio and active format description (AFD)
- Color space and gamut
- Video encoding such as XDCAM or AVC-I
- Audio channels and coding, metadata for track tagging, language mapping, Dolby metadata and loudness
- File wrappers such as MXF OP1a AS-11
- Segmentation
- Captions and subtitles
- Timecode
- Watermarking
- File-naming convention

An example of this is the Digital Production Partnership (DPP) in the UK, which publishes specifications for program delivery. To make it easier for their clients, content delivery services will often ask their clients what their "house" format is, and deliver a personalized set of content already conformed to that house format.

Video and audio file formats are detailed in other chapters of this handbook. For playout, file formats should provide the desired balance between content quality and file size. More efficient codecs require more computational resources in the playout engine, which may limit other functionalities or channel density per integrated automation server.

The playout format needs to support the necessary number of audio tracks. Metadata is a critical component of playout; for this reason, MXF is usually the preferred file wrapper.

AUDIO PROCESSING

The playout system processes audio from files and live sources, de-embedding audio from video sources, decoding compressed audio, shuffling audio tracks, mixing audio and voice-overs, managing audio levels and loudness, re-embedding the audio on the video outputs and compressing the audio outputs when necessary. Audio may also need to be up-mixed from stereo to multichannel surround or down-mixed from surround sound to stereo.

Most playout systems use embedded audio for SDI sources and distribution; this simplifies routing and avoids the inadvertent introduction of audio-to-video timing errors that result in many lip sync problems. For IP inputs and outputs, audio can be embedded with video, as with SMPTE 2022-6, or carried as separate essences, as in MPEG transport streams (with the same concerns for timing errors).

The audio processor de-embeds the audio from the video input and processes it separately internally in the system. Some file and live sources have compressed audio, such as MPEG, Dolby-E and Dolby Digital (Dolby-D or AC3). As these compression systems introduce delays, the audio and video processing need to account for the separate stream compression and processing delays to maintain proper synchronization. Files and compressed audio sources often also have associated metadata that needs to be extracted to manage audio tracks and levels.

The audio mixer typically operates in an "audio-follow-video" mode, where the audio source and the video source are the same, and the video transition (cut, cross-fade, fade through black or U-fade) determines the audio transition as well. Audio transitions can also lead or lag the video transition if the video and audio transitions occur at separate times.

A program may have multiple audio streams:

- Stereo two-channel (2.0) audio
- A separate surround mix (typically 5.1 channels, although 7.1 audio is becoming more common on cable movie channels, and immersive, object-based audio is forthcoming with ATSC 3.0)
- Secondary Audio Program (SAP) for alternate language
- Descriptive Video Service (DVS) for the visually impaired

In some regions, several iterations of each stream may be necessary in order to support several languages. To support this myriad of audio streams and channel lineups, metadata describing the incoming signals is required to understand which streams and languages are on each channel so that they can be shuffled and mixed accordingly. Playout systems also typically have audio up and down mix capabilities so that stereo audio signals can be utilized in a 5.1 mix and vice versa. In some cases it may be preferable to simply insert the 2.0 stereo audio into a 5.1 program stream with appropriate metadata identifying it as 2.0 in a 5.1 stream, especially if the audio was not produced for surround-sound or is expected to be down-mixed to 2.0 in downstream equipment.

Voice-overs allow another audio source to be superimposed over the program audio mix. Typical playout voice-over applications include sound effects, promotional announcements and station IDs and EAS alert audio messages. The voice-over source can be from an internal audio file, external discrete audio input, or de-embedded from another video source. The program audio is reduced in volume or "ducked" when the voice-over audio is played to maintain a consistent overall volume; in some cases, such as EAS or other full-screen graphics, the underlying program mix is completely muted.

Audio events can also be part of complex graphic event sequences such as credit roll squeeze-backs with coming next show promotions. More complex audio mixing is required when these DVE-type effects are used and there are multiple video sources on-screen simultaneously; in this case the audio mix instructions are normally part of the graphics event sequence. These sequences may include timelines that control the audio mix and playback of voice-over files with separate audio level control tracks for each audio source.

Audio signals include metadata that may need to be used, passed and regenerated. Typical audio metadata may include information about the audio tracks (channels, position and language) as well as loudness (dialnorm).

On playout, audio loudness processing is nearly always required. Audio levels can be managed using simple automated gain and limiter functions, but these do not correlate to the loudness a user perceives; more advanced loudness management is required that controls the perceived loudness of a program based on content and dialog levels integrated over a period of time. In the United States, broadcasters are required to manage loudness according to the ATSC A/85 standard referenced by the Commercial Advertisement Loudness Mitigation Act (CALM) of Congress H.R. 1048, S. 2847. A/85 requires loudness management that integrates both short-term and long-term loudness levels using an algorithm based on the ITU-R BS.1770 specification. The output signal contains dialnorm metadata, a value between 0 and –32, which indicates the overall average of a program. This value determines an offsetting audio level adjustment made in the consumer's playback system and is intended to maximize dynamic range. While dialnorm values can change throughout a program and between programs, most organizations today prefer to normalize all content and playout to a fixed value of –24 LKFS as recommended by the ATSC and the NAB.

With proper loudness management and use of dialnorm, perceived loudness can be consistent between programs and commercials and even between channels. But loudness processing on the outbound signal also reduces the overall dynamic range and interferes with the creative intent of the program creator. Solutions also exist to perform loudness processing on an entire file; it allows the audio level of the entire file to be adjusted as a single value to provide loudness normalization while having no impact on the dynamic range. Network feeds and other sources can likewise be normalized at the source instead of the output; when fed through the playout system a mechanism is employed to reduce the loudness processing applied to simply be protective, maintaining loudness compliance while maximizing dynamic range.

MEDIA MANAGEMENT

Media management is an essential component of a playout automation solution. The playout system automation utilizes media management to verify the availability of the required media when a new playlist is loaded, to move required media to the playout locations and to indicate on the playlist when media is unavailable or cannot be located.

Playout systems may also interface to or include the capabilities of a media asset management (MAM) system, which provides a more comprehensive set of media management functions for near-line and online content management as well as an archive management system to manage long-term off-line content. Functions of a MAM and archive management system include:

- Interfacing to a number of storage locations to locate and retrieve media
- Retrieving media and caching on playout cache location
- Transcoding of incoming media to preferred playout formats
- Automating quality control (QC) of content
- Managing off-line archives including tape drives, robotic drives and cloud-based storage
- Duplicating media for redundancy and mirrored storage systems
- Managing available space on playout caches and online storage locations
- Deleting and archiving content based on a set of rules and conditions
- Maintaining missing media information and alerts to master control operators and traffic personnel

Workflows That Integrate OTT/VOD

In addition to linear playout, facilities are also tasked with nonlinear content delivery for Internet "over-the-top" and Video on Demand (VOD) services. A unified approach to providing linear and nonlinear content delivery offers improved workflows with compelling economies. A unified playout and OTT workflow would allow the same playout engine workflows to also create files for VOD, inserting graphics and triggers for downstream dynamic ad insertion (DAI). The output files would be rendered in a variety of formats and bit rates to comply with the requirements of each content distribution network (CDN) and user device along with the additional metadata and packaging that is required. For live events, the secondary recording function would automate fast turnaround of live events as VOD assets. Tight integration between traffic, MAM and playout can enable these efficient OTT workflows.

GRAPHICS AND BRANDING

In addition to the scheduled playout of programming and commercials, a playout system is expected to insert a wide range of graphics. Graphics can be informational, such as news and weather information, or to meet regulatory requirements, such as ratings and captioning information. Graphics can also provide branding and promotional information, defining a channel's identity in a crowded market and improving audience retention.

Examples of graphic types include:

- *Still images:* Full-screen slides for station ID or semitransparent network and station logos (bugs) in the corner of the screen to identify the channel. U.S. broadcasters are also required to display the program rating (V-chip) graphic and captioning logo at the beginning of each program segment.
- *Animations:* Animated versions of the still images and promotional graphics displayed on the bottom of the screen ("lower thirds") that promote upcoming shows, often called "now-next" graphics or in-program messages (IPMs).
- *Character generator (CG) text:* Informational news, sports and weather crawls or stock tickers that display information from external sources in a programmatic way, either as a continuous crawl of data or as fields that constantly change with updated information. Required CG crawls in the United States are Emergency Alert System (EAS) messages that warn viewers of possible or pending emergencies.

Often it is desired to integrate multiple graphics along with video clips and DVE moves as well as audio voice-overs to create a complex branding event. This can be done by scheduling a number of secondary events, one for each element, action and duration. This is difficult to implement and support as the playlist becomes cluttered with a large number of secondary events. Some systems use macros to simplify the scheduling of repeated events with multiple actions.

More advanced systems support template-based graphics, where a graphical template can reference a large number of individual elements and actions with a timeline-based sequence—a mini-playlist within the playlist. The timings can be relative to the primary event they are attached to, such as, "begin 5 seconds after the start of the program segment and end 5 seconds before the end," or actions that are relative to the internal content, such as a video clip used in the template. The template can include graphical elements, external data sources, DVE image positioning and audio mix information. As a result, such templates are easily tested, highly repeatable and reusable across channels.

In some cases an external graphics system is used, either because of its advanced capabilities that exceed the integrated graphics capabilities of the playout system or because the user has invested in an existing workflow and the graphics template library that would be difficult to transition to the new system. The selected playout system will likely have fill and key signal input capability in order to support this external graphics system as well as an automation interface to control it.

Pre-Rendering Just-in-Time Graphics

An integrated playout system often has limited ability to render the complex graphics that some channels require. Multiple DVEs and advanced CG rendering effects can exceed what a real-time system can generate (or at minimum, significantly raise the cost of the playout platform). An alternative is to pre-render the graphics in the graphics department, typically slower than real-time, and air them as "flattened" single-element animation files. This simplifies playout and the requirements of the playout engine, and allows very complex graphics to be produced, all while using the graphics authoring tools the artists are most familiar with. This technique also creates other workflow issues:

- The programming that may be needed for DVE moves might not be available off-line in production.
- The graphics are pre-rendered, making last-minute changes difficult.
- The graphics cannot be easily previewed in the actual air schedule with the scheduled program.

One solution is to have the playout system actually control the graphics authoring software rendering. The graphics software project template is scheduled in the playlist, and the playlist also provides the metadata needed by the project to create each desired instance. When the playlist is loaded, the rendering is scheduled and the resulting flattened output files are queued for air. Last-minute changes to the metadata in the playlist cause the change to be re-rendered with the updated data. This provides just-in-time rendering of advanced graphics without requiring a powerful graphics engine in the playout system. Since this solution still does not function in real time, it's not suitable for quickly needed news, sports and weather information, but it does provide for sophisticated "now-next" style graphics driven directly from the playlist schedule.

Workflows That Integrate Social Media

It is common for playout graphical requirements to integrate with social media, providing real-time user feedback on air from social media forums such as Twitter and Facebook. The integration for this information requires connectivity to the social media platforms as well as tools to collect the user feedback and moderate the feeds to prevent inappropriate content from airing.

METADATA

Playout facilities need to be efficient, scalable and adaptable. To achieve these objectives, signal management, file workflows and operational practices must be carefully planned. Underlying all of these, metadata is the enabler, allowing disparate systems from different vendors to work harmoniously together. Metadata is the "data that describes the data," providing additional information about the content or "essences."

An example of metadata in everyday life is the packaging that surrounds a food product such as a carton of orange juice. The picture of an orange on the carton and the orange color allows the consumer to readily find the juice on a crowded grocery store shelf. In media workflows, a thumbnail image provides the same capability, whether in operations to quickly find the correct file or on the consumer's media device. The carton also contains additional information—ingredient listings, nutritional statistics, descriptive and marketing information. The orange juice carton also has the ubiquitous bar code that allows for fast and accurate checkout. Without this metadata, grocery shopping is much more cumbersome. Consumers need to bag their own oranges when selected in the grocery department. Checkout is much slower as the clerk asks what kind of oranges they are and then tries to look up the appropriate code; somehow those little produce code labels never survive the trip from the produce bin to the cash register. Finally, the consumers must create the juice themselves when they get home, with uncertain results, as oranges can spoil without any indication on the outside. An orange juice carton provides packaging and associated metadata that transforms the grocery store workflow, provides consistent quality and an improved customer experience all while reducing cost and effort for the consumer.

Content needs similar information to facilitate efficient workflows. In the past, media metadata was attached to the physical media. Metadata was written on the label on a videotape cassette and on the tape's library sleeve, and sometimes on a separate sheet inside the cassette case. More extensive metadata was kept in a separate database where it could be searched and facilitate rundowns and playlists. Many systems are built around such centralized databases, typically relational SQL databases that contain and manage the metadata. For some metadata this centralized workflow is essential, much as a grocery store needs a centralized database to manage inventory and perform price lookups and changes. A media asset management system (MAM) lives at the heart of a media facility and manages this database. A MAM enables fast searches and locating of media. But other metadata provides greater efficiency and accuracy when it is attached directly to the media itself, much like the videotape label, and travels with the media throughout its journey. It avoids constant lookups from a MAM to enable the metadata-derived workflows, lookups that can slow down operations and impair system resilience. It also avoids metadata becoming disassociated or orphaned, especially when media is transferred outside the facility network to systems that cannot access the MAM.

Metadata can be attached to media directly through several methods. For SDI video signals, metadata is embedded, and stored in the HANC and VANC ancillary data spaces. Many file formats support metadata, whether embedded with the essences themselves or contained in the file header. For file types that don't support sufficient metadata, options include transcoding or rewrapping the file to a more metadata-rich format, adding a metadata sidecar file to the original file, or storing the additional metadata in the centralized MAM database. Traffic systems that provide schedules for automation and playout can also provide metadata that is used in the playout process or inserted into the output.

Below are some examples of metadata instrumental in the playout workflow. Their applications are described in greater detail in this chapter.

- *AFD* (active format descriptor) is used to define the aspect ratio of the media format and the image contained within, and is essential for driving up- and down-conversion processes that alter the aspect ratio.
- *Audio tags* define languages and audio channels to allow correct languages to be mapped to the output channels and up- and down-mixed. Audio dialnorm defines the target loudness for a program stream.
- *Captions and subtitles* are both metadata and essences as they describe the audio content.
- *SCTE-35 and SCTE-104* triggers control downstream program and ad insertion as well as content availability and blackouts.
- *Program metadata* is used to provide real-time updates of electronic program guides (EPG) and also to provide the data for on-screen "now-next-later" graphics and schedule boards.

When designing playout systems, it is advisable to plan for the capabilities and preservation of metadata throughout the workflow. For example, an incoming SD feed could have specific AFD metadata embedded in the VANC that must be captured during the recording process and then used during playout to control SD to HD up-conversion, DVE and graphics formatting. In playout the AFD could be regenerated to reflect the changes made to the content and then re-embedded at the output to manage downstream down-converters, whether done at a cable headend or in the consumer set-top box or display.

In analog video, metadata exists in the vertical blanking interval (VBI). Metadata that is frequently carried in the vertical blanking interval includes (NTSC line number shown):

- Timecode (typically repeated on two nonadjacent lines; 12 and 14, 14 and 16 or 16 and 18 in NTSC, or lines 19 and 21 in PAL) per SMPTE 12M
- CGMS-A copy protection flag (lines 20, 283, 284) per CEA-608 and IEC-61880
- Closed captions (line 21) per EIA-608
- XDS data (line 21) per CEA-608:
  - V-chip program ratings
  - Time of day
  - Program information
  - Station ID
- Teletext (line 22)

Line 19 in NSTC is reserved for the ghost-canceling signal per ATSC A/49.

In digital video (SDI) there is room for much more metadata outside of the picture area than there is in analog video. The ancillary (ANC) space in an SDI signal includes both the horizontal ancillary data space (HANC) and vertical ancillary data space (VANC) per SMPTE 291 and SMPTE 334. Audio, up to 16 channels in SD and HD, or 32 channels in 3G-HD, is stored in the HANC. Common VANC data includes:

- Audio metadata per ATSC A/85 and SMPTE 2020
- Active format description per SMPTE 2016
- Closed captions per CEA-608-B, CEA-708-B and ATSC A/53
- Program description
- Data broadcast
- MXF file VBI metadata per SMPTE 436
- MPEG2 user data
- Digital program insertion (DPI) triggers per SCTE-104 and SMPTE-2010
- Text tags to indicate the source of the video
- V-ISAN content identifier
- Generic and user data

It should be noted that some types of metadata can exist in multiple locations simultaneously. An AFD value can be in the wrapper or header of a file while also embedded in the VANC in each frame of video. Likewise, captions can exist in multiple places not only in a file but also in an SDI stream. This becomes especially problematic when a file or signal is processed in such a way that the metadata is changed in only one location; the overarching system needs to know which metadata source to use for subsequent processing. Examples would include a file being edited where new content is spliced in that lacks the embedded captions or AFD, or the captions are retimed for accuracy in a sidecar caption file but not in the original embedded captions.

Captions and Subtitles

Closed captions are used to provide on-screen text to assist the hearing impaired when watching television. Subtitles similarly provide text to the viewer and are more often used to translate the program audio into alternative languages.

For both captions and subtitles, the transmission can be "closed" as in closed captions; the text data is carried as metadata and is rendered and displayed on-screen only if selected by the viewer. Most televisions and set-top boxes have a caption or subtitle button on the remote to turn this feature on and off. Open captions are "burned-in"; the text is pre-rendered on-screen during playout as part of the video image and is seen by all viewers. In addition to transcribing or translating the spoken dialog in the program, captions and subtitles also describe other audio including sound effects, ambient sounds and music including lyrics.

In NTSC analog television systems, closed captions were carried on line 21 of the video signal per the EIA-608 standard. This was the first active line of the signal, which was meant to insure that captioning survived the recording and transmission process. Because the CRTs originally used to view television over-scanned the image, line 21 was not visible on most sets. The data rate of the line 21 service is only 120 cps (960 baud), and the same service is also used to transmit XDS data.

With the advent of digital television, captions were also carried as digital metadata streams in the SDI signal per CEA-708-B, SMPTE ST 334 and EG 43, and broadcast per ATSC A/53 and A/54 standards. The digital captions support up to 9600 baud data rate and allow for multiple caption streams as well as enhanced font, positioning and character sets in the caption distribution packets (CDP). The simpler EIA-608 captions are also included in the SDI signal's CEA-708-B caption distribution packets, primarily to allow subsequent digital-to-analog conversion of the video signal. When the digital 608 captions are authored with the 708 captions they are referred to as "compatibility bytes." When they are converted from legacy analog 608 line 21 captions they are referred to as "translated" or "up-converted" captions. A second 608 caption stream can also be included in the SDI signal as 608 data, although this is not recommended per SMPTE EG 43; the preferred location for 608 caption data is inside the 708 captions. In an SD-SDI signal, the line 21 analog captions can also remain.

Captions can be carried in several places depending on the file format, including as embedded VANC data, in the file as a separate essence or in a sidecar caption file.

With multiple locations for captions, problems can arise when captions are changed in one location and not another. This is especially true when captions are retimed as required when re-airing a program that originally aired live and has since had the captions corrected and retimed—which set of captions was corrected? These problems are resolved in the file domain when using MXF-wrapped files with SMPTE ST 346 ANC carriage, which provides for metadata essences. For SDI signals, using only 708 captions with 608 compatibility bytes also resolves this issue.

Live Captions

When airing live programs, live captioning must be provided. This is typically provided by an outside service that monitors the feed and sends caption data via a phone line or Internet connection to the broadcaster for insertion in the master control signal. The playout system must have the capability to insert these live captions during live programming, whether as an integral function or as a downstream inserter. The insertion of these captions must also not overwrite captions already present in pre-produced content such as commercials.

Live captions inherently lag the video, and this delay is exacerbated by the delay observed by the caption author caused by the compression and delivery system of the programming. As a result, live captions are often several seconds behind the actual programming. When recording live programming for later re-airing, the captions must be retimed to remove this delay.

Caption Identification

At the beginning of each program, a graphic such as the one in Figure 5.4-10 must be superimposed on the program video to indicate that the program contains closed captions.

Watermarking and Ratings

Broadcasters include a signal in their broadcast to allow the reception by viewers to be tracked and the audience measured by a ratings company. This signal, Audience Measurements of Lineups (AMOL), is done by inserting a watermark in the audio using a proprietary algorithm. This allows a "people meter," a device that listens to the broadcast audio, knows and tracks which program was being watched and what broadcaster aired the program. The same mechanism is also used to track the viewing of recorded programming and VOD/OTT programming. The playout system needs to include the ability to insert this watermarking in the broadcast signal as either a native capability or using an external encoder.

Aspect Ratio Management and AFD

It is not unusual to handle a combination of SD and HD content on a channel as well as content that has 4:3 and 16:9 aspect ratios. The up and down conversion of video that occurs in master control and subsequently in the distribution chain, set-top box and display can cause undesirable image aspect ratio manipulation.

FIGURE 5.4-10 Closed Caption Symbol.

Courtesy of WGBH

Examples of improper aspect ratio management include tall images when a 16:9 image is squeezed into a 4:3 raster, wide images when a 4:3 image is stretched to fill a 16:9 raster and, perhaps worst of all, postage stamp images where the video image only fills a portion of the raster, with black bands on all four sides, as the result of sequential up and down conversion.

To remedy this, metadata that defines the picture inside the raster is used to control downstream scaling and prevent inappropriate aspect ratio conversion. Active format description (SMPTE 2016 and ATSC A/53 part 4) defines both the raster (coded frame) aspect ratio and the image aspect ratio. A signal can be defined as having a raster of 4:3 or 16:9 (14:9 is also supported in the standard but rarely implemented.) Within a 4:3 raster, the image can be full-frame or letterboxed. Within a 16:9 raster, the image can be 16:9 full-frame protected, 16:9 with a 4:3 center-protected area, or 4:3 pillarboxed. The most commonly used AFD codes are highlighted in bold in Table 5.4-1.

An SD image with a 4:3 raster (coded frame) and image would have an AFD of 1000. When converted to HD with a 16:9 raster, the image will retain its 4:3 aspect ratio and be bounded by pillarboxes on either side, which could be black or a graphical image, and the AFD would become 1001. If the same signal was later down-converted back to SD with a 4:3 raster, the AFD 1001 value would tell the down-converter to crop out the pillarboxes in the 16:9 image and display the resulting 4:3 image full-frame.

With the increasing predominance of 16:9 displays, many broadcasters have started to also transmit their SD signals in 16:9 full-raster, sometimes referred to as anamorphic 16:9. Broadcasters are also using SD 16:9 full-raster for news acquisition. In this case, the coded frame indicates that the raster is 16:9 (bit 2 = 1) and the AFD code is 1010 (16:9 full frame protected) or 1111 (16:9 image with 4:3 center protected).

AFD is also used to manage graphics inserted in playout. A 16:9 image with a 4:3 center protect (AFD

TABLE 5.4-1
AFD Codes (from SMPTE 2016-1-2009)

Active Format Description		**AFD Code**
In a 4:3 coded frame (bit 2 = 0)	**In a 16:9 coded frame (bit 2 = 1)**	**a3, a2, a1, a0 (bits 6–3)**
Undefined	Undefined	0000
Reserved	Reserved	0001
Letterbox 16:9 image, at top of the coded frame (not common)	Full-frame 16:9 image, the same as the coded frame (not common)	0010
Letterbox 14:9 image, at top of the coded frame (not common)	Pillarbox 14:9 image, horizontally centered in the coded frame (not common)	0011
Letterbox image with an aspect ratio greater than 16:9, vertically centered in the coded frame (not common)	Letterbox image with an aspect ratio greater than 16:9, vertically centered in the coded frame (not common)	0100
Reserved	Reserved	0101
Reserved	Reserved	0110
Reserved	Reserved	0111
Full-frame 4:3 image, the same as the coded frame	**Full-frame 16:9 image, the same as the coded frame**	**1000**
Full-frame 4:3 image, the same as the coded frame (same as 1000, which is preferred)	**Pillarbox 4:3 image, horizontally centered in the coded frame**	**1001**
Letterbox 16:9 image, vertically centered in the coded frame with all image areas protected	**Full-frame 16:9 image, with all image areas protected**	**1010**
Letterbox 14:9 image, vertically centered in the coded frame (not common)	Pillarbox 14:9 image, horizontally centered in the coded frame (not common)	1011
Reserved	Reserved	1100
Full-frame 4:3 image, with alternative 14:9 center (not common)	Pillarbox 4:3 image, with alternative 14:9 center (not common)	1101
Letterbox 16:9 image, with alternative 14:9 center (not common)	Full-frame 16:9 image, with alternative 14:9 center (not common)	1110
Letterbox 16:9 image, with alternative 4:3 center (center-only protect, not common)	**Full-frame 16:9 image, with alternative 4:3 center (center-only protect)**	**1111**

Courtesy of SMPTE

1111) would require that the inserted graphics support being cropped to 4:3. If the playout system were to change the image by performing a DVE squeeze with an L-bar graphic during a morning news show, it would also need to change the AFD to 1010 to make sure the L-bar was not cropped in a downstream down conversion.

The AFD metadata is inserted on a frame-by-frame basis in SDI and compressed video signals. Some file formats also support AFD in the file header or metadata, which can cause erroneous results when the embedded AFD conflicts or the AFD value changes during the file.

XDS and V-Chip

Extended Data Services originated with line 21 closed captions and are carried in digital video as 608 captions in 708 caption distribution packets. XDS includes program information including V-chip content ratings. A playout system may need to insert V-chip data on playout when the original file or live content lacks this. Broadcasters are also required to insert on-screen a V-chip logo with the ratings information for a program at the beginning of each program segment.

Cueing Signals

Playout systems may be required to insert triggers into their program output and OTT/VOD files to control the insertion of commercials in downstream devices at cable headends, local TV stations and VOD/OTT delivery systems. These triggers can be as simple as a white line on the first line of active video to DTMF tones in an unused audio channel or SCTE-35/SCTE-104 cueing signals. These are covered below under "Regionalization and Localization."

Audio Metadata

SDI signals and media files can carry a wide range of audio essences for which metadata is essential in order to properly identify the audio and arrange it during playout. In an SDI signal, audio metadata includes:

- Channel mode (acmod)—whether the audio stream is stereo (2.0) or surround (5.1)
- Dialog level (dialnorm)—defines the dialog level and loudness of the program
- Surround parameters
- Down-mix parameters
- Dynamic range control (DRC)
- Program description including language

Likewise, similar information is included in media files to identify the audio tracks and mixes. Different file formats use different methodologies, and the MAM or playout system needs to be able to properly read the audio track tags to play out the audio correctly.

REGIONALIZATION AND LOCALIZATION

Sometimes a television channel is tailored to individual markets or regions by using the main feed to create sub-feeds that have different interstitials and/or graphics. Some applications for sub-feeds include:

- Insertion of local EAS alerts at cable headends and remote translators
- Insertion of network branding and graphics at each affiliate station separately
- Different branding for each regional transmitter in a statewide network
- Insertion of different commercials in each cable and direct broadcast satellite (DBS) feed from a station; for example, a station that covers a three-state area could run different political campaign ads for local elections in each cable system within the region
- Networks that provide local availability for content insertion and commercials
- Alternate programming to cover sports events blacked out in a particular region
- Insertion of commercials at the set-top box or player device targeted to individual users or households

Inserters and Bit Splicers

To provide this capability, a local inserter switches between the main programming feed and local content. Triggers are inserted in the main program feed to indicate to the inserter when to break away for local insertion and when to return to the main feed. The management of the local content to be inserted is usually done by the inserter at the distant end, which runs a mini-playlist during the available window; however, it is also possible to have the main feed include the information about what content is to be inserted.

Typically the main program feed in which the local insertion is being performed is a compressed signal. Many inserters uncompress this signal to baseband, perform the needed insertion the same way a master control system would, and then recompress the signal. To preserve picture quality and reduce impairments caused by compression concatenation, some local inserters act as bit splicers. A bit splicer switches between the main program and inserted content in the compressed domain as a transport stream switch. Bit splicers also can insert graphics overlays by decompressing just that area of the picture where the overlay occurs, compositing the new image, recompressing it and integrating it back into the original signal.

Bit splicers can preserve picture quality, especially when the main program signal is already highly compressed for final transmission. For networks that distribute their content to stations as 19.34 Mbps ATSC-ready bitstreams, a bit splicer can preserve image quality. For other networks that distribute at a higher contribution or mezzanine bit rate, the impairments caused by concatenated compression may be less noticeable. Utilizing bit splicers will limit the capabilities of the local inserter. Graphics are confined to a small region of the screen. The switching between compressed feeds also needs to be done on the MPEG I-frame, and feeds using long-GOP 15-frame compression need a half second of black at both in and out points to allow for clean transitions.

Cueing Systems

In order to control downstream insertion devices, the main program feed contains cue signals to trigger the insertion and then the return back to the main feed. The period when local insertion is allowed is typically referred to as a local avail (short for availability). The common methods of cueing signals include:

Out of Band

The signal can be sent as an "out of band" trigger—one not carried by the SDI signal. This could be as simple as a GPI that could be carried over an MPEG transport stream or an Ethernet-based signaling and control system.

White Line Trigger

The first line of active video is replaced by a white line while in a local avail.

DTMF Tones

A sequence of DTMF tones (telephone touch-tones) are sent over an audio channel. Typically these are sent over an unused audio channel and not one with programming. The tones are typically four digits plus a # or * to indicate on and off, providing a level of addressability.

Unused VANC Data

Signaling can be sent using unused or user-defined VANC data such as the user data in teletext messaging or captions/XDS. Some U.S. networks use this method to trigger their affiliate insertion devices.

SCTE-35 and SCTE-104 Messages

These SCTE protocols define a messaging system for controlling dynamic ad insertion (DAI) or digital program insertion (DPI) systems. SCTE-35 defines the messaging in an MPEG transport stream, while SCTE-104 defines the messaging in an SDI signal. SCTE-104 messages are carried in the SDI VANC per SMPTE 2010. These SCTE messaging protocols provide a rich array of commands for managing regionalization and localization of feeds, including:

- Immediate triggers for live events
- Pre-roll triggers for scheduled insertion
- Defined durations for avails
- Multiple methods for addressing receive devices individually or as groups
- Carriage of DTMF tone data to support legacy systems
- Schedule download
- Regional blackout flag
- Web delivery allowed flag
- User-defined messages

The usage of SCTE-35 and SCTE-104 commands is increasing, especially as they are used extensively in OTT and VOD distribution to manage individual commercial insertion for each viewer.

QUALITY CONTROL FOR PLAYOUT

"Five nines" reliability ("5–9s") is a term one often hears in broadcasting—meaning 99.999% uptime. It's a term that first originated in the telephone companies as a measure of the reliability of their systems. You can imagine the disruption that occurs in a telephone company—and the economy—if the telephone network fails. In order to achieve 5–9s, the system must not incur problems for more than 5 minutes 15 seconds per year in total. That's a difficult requirement to meet, especially since a single on-air failure could easily cause more than 6 minutes of disruption. For many stations or hosting facilities with SLAs (service-level agreements) with their customers, a missed or wrong commercial or missing graphics count against the uptime measurement.

The first challenge in achieving 5–9s is to quickly identify a problem when it occurs, or even better, *before* it occurs! Traditional master control rooms already have some of this capability. Even though these systems are run under automation, they still utilize source and preview monitors, which allow operators to see these signals before they go to air and potentially correct problems in advance. This monitoring has its limitations, because some signals, such as server video clips, are not visible until they play to air, and few station operators are able to listen carefully to 5.1 audio. Compound this with the need to monitor several off-air signals (transmitter, cable and satellite at a minimum), and the odds of recognizing a problem in 6 minutes are substantially reduced. What if that same operator also has to oversee ingest and maybe answer the front door during off-hours? In such a scenario, how many *days* might it take before someone notices that on the down-converted and down-mixed SD cable feed the center channel wasn't being mixed in properly? Without a system that monitors these signals and logs any errors, how would a station even have the metrics that would allow the personnel to determine how close to 5–9s they were performing?

Playout systems designed to deliver high quality and reliability include monitoring and QC components that assist in:

- Input and source material quality analysis
- Providing extensive monitoring capabilities coupled with monitoring by exception

- Redundancy and self-healing
- Detailed logging and analysis tools that provide accurate metrics and enable continuous improvement

A monitoring and control system oversees the health of the devices in the broadcast chain as well as that of the signals passing through them. An effective monitoring and control system can improve quality considerably by reducing the time it takes to discover and remedy faults. The monitoring system saves operators time trying to discover faults and instead allows them to focus on fixing them. Some monitoring and control systems can take protective action automatically when a fault is detected, from simply switching to a backup source, signal path or device to automatically provisioning and switching over to another system entirely.

Quality Control and Monitoring Components

File Analysis and Media Alerts

Incoming files need to be checked before playout to validate their integrity. File checking systems can validate files using two methods: *Detailed File Inspection* and *Validation*. These file QC tools inspect a file with great detail, analyzing the header, metadata and essences for any errors, omissions or anomalies. The video and audio essences can be inspected to verify their integrity and levels; compression artifacts can be measured; and systems can also check the timing of captions against the video program and against the audio for accuracy. These systems can provide absolute assurance that a file meets the desired specification. Conversely, they require careful configuration to minimize false alarms for errors that, while indeed present, would be masked or corrected during playout.

In some playout systems, files can be checked using the same engine as is used for playout. Files can be run through this file check system, which provides the assurance that the file is actually playable by the system that is preparing to play it out. A perfect file could pass the previous file QC tools yet not play on the playout system because it is in an incompatible format; the playout system file check tools keep such situations from occurring. In addition to warning about errors, the file management system also provides "missing media" alerts to the operators in the case that some of the media that is needed in a playlist is not ready and prepared for playback.

Multiviewers

A monitoring and control system is essential for playout operations. A typical monitoring system includes two components—a multiviewer that allows the operator to view a number of video signals simultaneously and a control system to assess and repair the health of the facility and components.

A multiviewer is typically a card-based or integrated I/O matrix, rack-mounted chassis system that is used to display a number of video and audio signals onto one or more screens in front of an operator position as a mosaic of video images. Multiviewer cards commonly support a mix of SDI and IP inputs. These signals include live sources and the in-house inputs to the playout system, the program output, a preview or preset output and off-air return signals for confidence monitoring. The preset output typically shows the next source cued for air, while a preview output also shows upcoming graphics actions.

In addition to displaying the video signals, the multiviewer also can include audio meters and alarms for signal quality and external conditions. These alarms help draw the operator's focus to the signals requiring attention.

Device Monitoring and SNMP

Device monitoring can use proprietary or native protocols to communicate with components that report signal and health status to the monitoring system. SNMP is an Internet-standard Ethernet-based protocol that is commonly used to provide health and status monitoring. A management information base (MIB) defines the capabilities of a monitored device so that the monitoring system knows what parameters are available. The monitored device sends unsolicited requests called "traps" to update the management system of any changes. Management systems often have "MIB browsers" that allow the user to select which traps to listen for and report to the user. The downside of utilizing SNMP traps is that there is no verification of the unsolicited requests, so some SNMP messages "fall through the cracks" and aren't picked up until the message is re-sent. Sometimes, the problem may be quite apparent by the time another message is sent.

Content Verification and Lip Sync

With both signal and device monitoring, a monitoring system is able to make sure that signals are present and that the infrastructure is healthy. However, neither type of monitoring will recognize whether the signal is the correct signal. It is possible for a signal to be misrouted, especially when this routing occurs outside of the managed facility. It is possible for a cable headend to switch away to an alternate feed for blackout coverage and fail to return, or insert a commercial at the wrong time. To validate that the correct content is being received at all locations, fingerprinting technology is in development and being standardized (SMPTE ST 2064). This technology takes a small sample of the video and audio from a signal—a fingerprint—and compares the fingerprints from multiple signal locations. From this comparison the monitoring system is able to determine if the content is identical and whether proper lip sync is maintained throughout the transmission and distribution chain.

Monitoring and Control System

In addition to the multiviewer, most playout systems employ a monitoring and control (M&C) system. The M&C system receives all of the status and alarm signals from the various devices and components and displays them for the operator. An M&C system can collate multiple alarms into a single status, making it simpler for the operator to assess system health and locate faults. A monitoring and control system can also interface to a multiviewer, using the multiviewer to probe signals and sending status information to the multiviewer display. Other devices can report their health to the system using standard SNMP interfaces.

By virtue of the design of an M&C system, an operator should be able to determine the cause of a problem by quickly seeing a correlation across stations or signals within a station. If a server shows an alarm at a station's server, the other downstream devices will also be in alarm, and it will be readily apparent to the operator that the server is the highest upstream device in error and the likely cause of the system failure. Likewise, a failure in the network satellite signal will show up as alarms on the network integrated receivers/decoders (IRDs) for all of the stations. When more detailed analysis is required, an operator can use the GUI to click on the alarm and bring up an active signal flow diagram showing the alarms in a schematic diagram and bring up a configuration panel for a device. This correlated-alarm and drill-down approach provides rapid analysis and resolution of problems. It is extremely important when evaluating M&C systems that the system is designed to not remove visibility of the overall status alarms for all channels, paths and devices being monitored while drilling down to analyze a particular problem. The M&C system should also provide complete end-to-end monitoring of the playout and transmission signal path.

Monitoring by Exception

Larger systems with sometimes hundreds or even thousands of signals to monitor can become unwieldy; with so many signals, an operator needs assistance in focusing on the ones with faults to be able to quickly evaluate the root cause. A concept known as "monitoring by exception" uses the monitoring and control system to monitor the health of all of the system components and signals, and then only display the ones in error on the display monitors. With monitoring by exception, operators no longer focus on trying to discover *if* there are problems; instead the alarms immediately bring problems to their attention and they focus on taking corrective action. Now operators no longer need to constantly look for signal errors; instead the system reports such exceptions to them. The system will even monitor all of the distribution paths—cable, satellite, repeaters—validating the signals to all of these customers, even when such signals cannot be received directly at the station.

Redundancy and Resilience

Systems designed for high quality use multiple levels of redundancy to achieve the resilience required by the owners. At the most elemental level, redundant power supplies are common in broadcast equipment. Redundant components, signal paths and changeover switches provide protection from component failure. Many systems employ completely redundant playout systems to achieve even greater resilience, while maintaining an off-site backup or disaster recovery facility assures maximum resilience and business continuity. This section looks at some of the strategies being used to provide highly resilient playout.

Basic Redundancy

For playout systems, redundant power supplies are typically standard in all components. Devices that operate in the signal chain typically have internal bypass relays that maintain the signal through the device in case of failure.

Playout Channel Redundancy

In traditional discrete master control systems, small routers (10 × 1) are used to bypass portions of the signal path during failure and take signals directly to air. Integrated playout systems typically employ spare channels for redundancy, using 1 + 1 mirrored and synchronized playout servers and automatic 2 × 1 changeover switches to provide nonstop operation. A more economical approach is to use fewer backup playout servers than main channel servers in either an N + 1 or N + M configuration and a facility router to select which server feeds the path to air. A downside to the N + 1 approach versus the 1 + 1 configuration is the time it takes for the standby server to be reconfigured and synchronized with the server it is protecting. The automatic changeover switch in the 1 + 1 configuration can also provide additional probing and monitoring of the server outputs and can potentially detect faults not recognized by the playout server itself.

Disaster Recovery

For high-value channels, geographic diversity and redundancy are essential; for this reason, off-site disaster recovery (DR) facilities are employed. Additional instances of the playout system are operated here to provide backup for the main facility in case of catastrophic failure. The DR facility may also be used when the main facility system is being upgraded to provide protection.

Even with these additional benefits, a DR facility remains a significant financial burden to both launch and maintain, especially when a well-engineered and managed main facility should have a downtime of less than 6 minutes per year. Alternatively, public cloud-based playout solutions offer a strong economic option for use as DR applications. A cloud DR solution would only be used during the time when backup is needed. If media storage and connectivity costs can be managed and a cloud DR system rapidly provisioned when there is a main system failure, cloud DR has the potential to make DR affordable for more than premier high-value channels.

Resilience for Channels without Live Content

A number of television channels rarely air live content, mostly playing prerecorded shows back-to-back. For these channels, the majority of issues that arise might be playlist and media issues. One way to reduce these is to play out the channel several hours in advance of the actual airtime, with the output then feeding a corresponding delay server before airtime. The backup or redundant channel would play out at the normal airtime, and the redundancy changeover switch would select between the advanced channel's delay server output and the real-time backup channel.

When a problem is noticed on the advanced channel output, the operator will have several hours to correct the issue. The fix can be tested on the real-time channel, and once approved, the operator would switch to the real-time channel before the fault appears at the output of the advanced channel's delayed output.

Channels without live content can also use on-site and off-site delay servers to provide redundancy and disaster recovery. The delay server typically runs a 3- or 7-day delay and records the output of the main program. If the main program feed should fail, the delayed output is put on air. While this would not likely be the same exact content as what would have aired from the main server, it's likely to be the correct program, just the wrong episode, which may be sufficient to maintain viewership while the technical difficulties are being resolved.

Self-Healing

Merely reporting errors is not enough to achieve 5–9s; one must also be able to quickly restore the impaired signal. The issue isn't just getting operators to discover problems and their cause; it is also the difficulty in getting them to take corrective action out of fear of making the problem worse or because their attention is diverted elsewhere. Any delay in taking needed corrective action reduces the likelihood of achieving the desired level of reliability. To reduce this delay, many systems are programmed to automatically take corrective action. This could be as simple as an automatic changeover card that switches between redundant signal paths and as complex as provisioning a complete backup path and then switching over to it. By implementing such self-healing systems, downtime is greatly reduced.

Logging and Reporting

Monitoring systems maintain detailed logs of all conditions monitored; these are often essential in troubleshooting failures and determining the root cause. In many instances, an upstream failure will cause a cascade of failures to be reported from downstream systems. Other failures may affect a component common to a large number of signals. More advanced systems can narrow down an underlying cause from a number of failure conditions and reduce an avalanche of error warnings to the primary root cause. Operator error is often the cause of on-air faults; detailed and accurate logs help to quickly pinpoint whether the cause was technical or operational and move towards resolution.

CHAPTER

5.5

Television Camera Systems

MICHAEL BERGERON

STEVEN MAHRER
Panasonic
Newark, NJ

Updated for the 11th Edition by

STEVEN MAHRER
Panasonic
Newark, NJ

IDIN ROSHAN
Idin Roshan Productions

MARTIN MARIETTA
Producer, Camera Operator, and Robotic Camera Movement Specialist
New York, NY

INTRODUCTION

Since the *NAB Engineering Handbook*'s 10th edition, the television landscape has continued to evolve and change. NTSC analog camera systems have all but vanished. Typically the only place NTSC analog technology exists is in the systems that are used to play and digitize legacy content that is stored in that format. Most broadcasters are working diligently to convert and transfer any content of long-term value to digital files. Even digital standard definition (SD) is no longer seriously considered when broadcasters are looking at camera systems. High definition is now the minimum requirement, and many broadcasters are looking at 4K and UHD camera systems for capture as a means of creating high-quality content that is future resistant and can be downgraded easily to lower resolutions such as HD or SD.

The topics that follow in this chapter explore and explain some of the capabilities, challenges, and components that are part of the rapidly changing camera landscape. Some of these features appear to be paradoxical in nature, as higher resolution pushes for larger imagers, while creative content producers simultaneously look for smaller-sized cameras that can be placed in locations that cannot accommodate a standard-sized camera, and higher-resolution images with greater depth of field and dynamic range push television camera operators toward being more cinematographer and less videographer.

DIGITAL VIDEO STANDARDS

Before continuing a discussion on SD, HD, and UHD digital video, we should first review what the standards are.

The chart in Figure 5.5-1, based on Table A3 of Advanced Television Systems Committee (ATSC) A/53[1],[2] outlines the currently defined ATSC digital television formats for emission. The bottom two rows represent SD formats, with 480 × 640 being closest in resolution to the quality of analog NTSC video, although this resolution actually corresponds to the computer VGA standard and not a video production format. The 480 × 704-line digital video formats are defined for production by the ITU 601 and Society of Motion Picture and Television Engineers (SMPTE) 125M[3] and SMPTE 293M[4] standards (where they actually have 720 pixels per line). ITU 601 defines spatial and temporal characteristics, colorimetry and transfer characteristics, and the sampling parameters for the images, while the SMPTE standards further define other aspects of particular formats. Transfer characteristics define the *gamma correction* applied to compensate for the nonlinear characteristics of traditional cathode ray tube display tubes. The serial digital interface (SDI) for these formats is defined in SMPTE 259M.[5]

The top two rows of the chart represent high-definition formats. The 1080-line digital video formats are defined for production by the SMPTE 274M standard,[6] and the 720-line formats are defined by the SMPTE 296M standard.[7] These standards define the spatial and temporal characteristics and the sampling parameters for the images. The colorimetry and transfer characteristics for the HD formats are defined by the ITU-R BT 709 specification[8] (referred to as "ITU 709" in this chapter), which also defines the spatial characteristics for international program exchange. The high-definition serial digital interface (HD-SDI) for these formats is defined in SMPTE 292M.[9],[10]

Scanning Lines	Horizontal Pixels	Aspect Ratio Pixel Shape	Picture Rate	Horizontal Frequency
1080 Active (1125 Total)	1920 Active	16:9 Square	60i, 30p, 24p	33.74 KHz (60i)
720 Active (750 Total)	1280 Active	16:9 Square	60p, 30p, 24p	45 kHz (60p)
480 Active (525 Total)	704 Active	16:9, 4:3	60i, 60p, 30p, 24p	15.734 kHz (30p/60i) 31.5 kHz (60p)
480 Active (525 Total)	640 Active	4:3 Square	60i, 60p, 30p, 24p	15.75 kHz (60i)

FIGURE 5.5-1 ATSC-defined video emission formats.

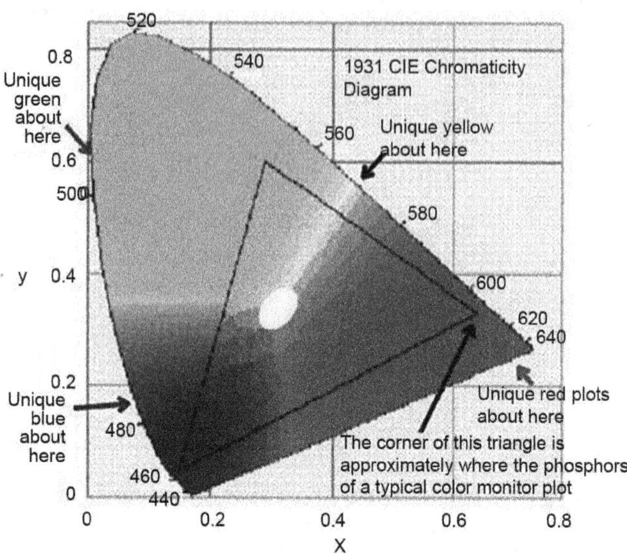

FIGURE 5.5-2 CIE chromaticity diagram.

To be clear, the above formats and standards describe digital video signals and are not specifications for imagers or signal processors. The purpose of the standards is to ensure that camera systems and other video equipment can work together; many cameras, however, tend to follow the standards internally from the beginning of the chain to simplify signal processing.

Regarding the frame rates of the above formats, although the chart shows 24, 30, and 60 frames per second (fps) and 60 fields per second, all the formats have equivalent versions with these frame and field rates divided by 1.001. These alternative rates are needed for compatibility with NTSC (analog) standard video, for which the frame rate had to be adjusted by that factor when color was introduced many years ago. This use of non-integer frame rates will be necessary as long as NTSC and DTV material coexists in the same plant and conversions between the NTSC and DTV video standards are required. For the alternative rates, 30 fps becomes 30/1.001 fps, which is usually referred to as 29.97 fps, and 60 fields per second becomes 59.94 fields per second. Similarly, 24 fps becomes 23.98 fps.

COLOR VIDEO

Color Perception

The human visual system is capable of not only interpreting varying intensities of light as an image but also analyzing the frequency characteristics of the light within the range of what is known as the *visible spectrum*. This spectral analysis is perceived as color. Laser or other coherent light sources produce a single wavelength; however, most naturally occurring radiated or reflected light sources emit a spectral response that includes components of many wavelengths (colors). The human vision system interprets the spectral makeup by analyzing it into three specific color components. The retina contains light receptors known as cones that have sensitivities centered on red–orange (580 nm), yellow–green (545 nm), and blue (440 nm).

A study conducted in 1931 by the International Commission on Illumination (CIE) analyzed and plotted the entire range of human color perception. One result of this work is the CIE chromaticity diagram shown in Figure 5.5-2. The range of all possible colors is referred to as the *gamut*. For a *tri-stimulus* (three-color) system, the gamut of colors that can be produced by any particular system is defined by the triangle whose vertices are the location of the three *primaries* on the CIE chart. Color video systems define color using red, green, and blue primaries of slightly different wavelengths depending on the particular standard being used. Like the human eye, color video cameras are tristimulus systems and they must separate the three primaries in order to capture a color image.

Color Space

For any video standard, the precise color (or placement on the CIE chart) of the primaries must be part of the specification. This is part of what we refer to as *color space*. The RGB primaries defined by the CIE set red at 700.00 nm, green at 546.1 nm, and blue at 435.8 nm. Because the human eye adds and subtracts colors in its system, pure red, green, and blue primaries are not suitable for a video system, and primaries must be chosen that can be mathematically combined more simply. These primaries are what define the gamut of each video standard, and the primaries defined for NTSC, ITU 601, and ITU 709 are all slightly different.

Red, green, and blue primaries are not the only three components that can be used to define a color image. Like any three-dimensional coordinate system, it is possible to transform RGB into an infinite number of three-variable bases, and these can be derived mathematically by using a 3 × 3 matrix to transform the components as follows.

$$[Y, Pb, Pr] = \begin{vmatrix} a & b & c \\ d & e & f \\ h & i & j \end{vmatrix} \times [G, B, R]$$

This matrix equation is a representation of the three equations that determine the values of Y, Pb, and Pr:

$$Y = aG + dB + gR$$

$$Pb = bG + eB + hR$$

$$Pr = cG + fB + iR$$

where Y is the *luminance* and Pb and Pr are the *color difference* signals. These transforms form the basis of component video as it is usually managed in a broadcast environment. The actual values of the matrix terms/equation coefficients (*a*, *b*, *c*, *d*, *e*, *f*, *g*, *h*, and *i*) are specific to each video standard, and these parameters (along with the values of the RGB primaries) define the color space. It should be noted that the color space is likely to change when video is changed from one format to another, such as when it is down-converted from HD to SD.

Component Video

Cameras deal with color image acquisition in RGB; most monitors display images in RGB, and color imaging on most information technology (IT) platforms, including film postproduction, works in RGB space, so why do component digital video systems transform RGB to YPbPr? The answer is part legacy "baggage" and part smart engineering.

As background, the NTSC color video standard was originally created with the mandate that earlier black-and-white equipment had to be compatible, so a transform was created such that one of the three primaries would be the black-and-white video signal. This component was, and still is, often (imprecisely) referred to as *luminance* after the CIE-defined term denoted as "Y." Two other *color difference* parameter signals were defined to carry the color information. These produced the "U" and "V" signals that were processed further, bandwidth reduced, and encoded on the color subcarrier, where the color information was referred to as *chroma* or *chrominance*. Those terms are not used in component video systems. The NTSC luminance signal by itself will provide a good monochrome image to a black-and-white monitor. Color difference signals are needed with the luminance in order to produce a color picture.

Because the luminance parameter used in video signals is *gamma corrected*, its value is not identical to the well-defined CIE term luminance, and it is correctly referred to as *luma*, denoted with the term "Y′." In this chapter we will use the precise term *luma*. As shown in the matrix equations above, luminance can be derived from R, G, and B. These equations relate to luma when the signals are gamma corrected.

It is the luma and color difference signals that are used in analog component video. As mentioned, the definition of luma is specific to the video standard. Looking at the ITU 601 specification, it can be seen that luma is defined as:

$$Y' = 0.299R' + 0.587G' + 0.114B'$$

and the color difference signals are defined as:

$$R' - Y' = 0.701R' - 0.587G' - 0.114B'$$

$$B' - Y' = -0.299R' - 0.587G' + 0.886B'$$

The terms used for analog color difference channels are Pb and Pr, which are directly derived from the R − Y and B − Y signals (indicating that they are essentially the remaining red and blue channels after the luma has been removed). Strictly speaking, digital color difference signals should be referred to as Cb and Cr. These conventions are not always followed, so if a schematic indicates Pb and Pr this does not necessarily mean that the signal is not digital; it may just be careless labeling on the part of the designer or writer.

Color Subsampling

The human visual system perceives significantly more detail in luminance (luminance is correct, because we are talking about human vision) than it does in color information. In fact, half of the color resolution in a video picture can be dropped without a perceptible difference in sharpness, as long as luma resolution is maintained. Based on this information, the horizontal sampling rates for digital video were chosen for digital video standards with a two-to-one ratio of luma to color information. This is commonly referred to as color subsampling and is denoted by 4:2:2. The arrangement used in most component digital interface systems is 4:2:2. Other varieties of color subsampling, such as 4:2:0 and 4:1:1, are possible.

Y′ Cb Cr signals are subsampled as part of the digitization process, including those carried on SDI interfaces. While the human eye may not notice the change brought about by color subsampling, it affects the robustness of the signal if it is processed downstream. To reduce the effects of color subsampling, image processing should wherever possible be carried out before the signal is transformed to Y′ Cb Cr component digital, which is one reason why the 4:4:4 format is almost exclusively used with RGB components.

In a video camera system design, it is up to the engineers to decide where in the signal processing chain to transform the RGB video signal into component Y′ Cb Cr, a process known as *matrixing*. Some signal processing takes advantage of the nature of luma and therefore can only be done after matrixing. Contouring and detail enhancement, for example, are applied only to the luma signal, as they seek to enhance the perceived sharpness, which is almost entirely a function of the luma signal.

VIDEO CAMERA SYSTEMS OVERVIEW

Turning Photons into Electrons

The human visual system is capable of capturing and interpreting an image focused on the image plane of the retina. Video cameras capture images similarly. Both must use some kind of lens to focus light bouncing from an object on to a light-sensitive surface. The light pattern, or image, must then be captured and interpreted. To electronically capture an image, it is necessary to represent varying light levels with analogous voltage levels. The voltage levels can then create a representation of the image focused on the image plane. Image capture is, in essence, an analog process, regardless of the technology involved. Early video cameras used an electron beam to scan lines on the imaging surface of a vacuum tube, which allowed detection of varying light levels on that surface. The resulting voltage variations were amplified and processed directly into an analog video signal.

Imagers

Modern cameras use solid-state *imagers*, which are manufactured similarly to other semiconductor devices and as such are referred to casually as *chips*. Imaging chips comprise an array of *photoreceptors* or *photosites*, each converting photons into an electrical charge. When the charges from the individual receptors or *charge wells* are extracted sequentially from the imager, this creates a variable voltage, much like the variable voltage generated by the scanning in a tube camera. Although the solid-state imager architecture is based on individual pixel sensors, they are still analog devices. In most cases, the imager produces an analog signal, which must then be sampled. When the voltage values of the samples are converted to binary code words, the signal can be considered digital video.

Traditionally, in most solid-state imagers, the number of photoreceptors corresponded directly to the number of pixels required to produce a given video image format, so the photoreceptors on a solid-state imager are also referred to as pixels. When an imager's pixels match the final video format pixels one to one and are scanned similarly, this is referred to as *native capture*. Because of the need for cameras with flexible output formats, as well as the necessity to optimize the relationship between pixel size and pixel density, many current cameras do not employ native capture. Having a greater density of photoreceptors than output pixels allows *oversampling* of the image with potential quality improvements. Having fewer pixels than is required for native resolution is often a necessary compromise for smaller imagers that would otherwise lack sufficient sensitivity due to smaller pixel sizes.

System Architecture

A video camera system needs to accomplish a specific task, which is to capture an image and deliver the image data in a form compatible with downstream systems. Regardless of the technologies involved or the specific applications for the camera, this task is accomplished by the following components:

- *Optics*—Including lenses, filters, and prism block;
- *Imager*—Charge-coupled device (CCD) or complementary metal-oxide semiconductor (CMOS); three-chip or single-chip; converts light images to video signals;
- *Processing*—Analog processing and digital processing;
- *Control circuitry*—Access to processing to adjust camera parameters and images;
- *Interfaces*—Necessary for reference, input, output, and control signals and essential for multiple cameras to work together;
- *Output or recording*—Camera must output a standard format.

Camera Components

The interrelation between these components is determined by the camera application. Components are combined in some systems and split in others to allow some components to be moved to a dedicated operation area. A camera in a studio application usually has the camera optics and lens with its *zoom* and *focus* controls and the imagers and electronics in a *camera head* run by the camera operator, while control electronics, most interfaces, and often a good deal of processing are in a *camera control unit* (CCU), which is installed in racks in an equipment room so it can be easily interconnected with other systems. *Lens iris* and camera *black level* and color adjustment (*paint*) controls are sent to a remote panel so an engineer or *shader* in a control room can watch exposure and other settings and match cameras. Fine image adjustments might be done in yet another area with another remote control unit.

Some point-of-view (POV) cameras have even more circuitry away from the imager to reduce the size of the camera head; while an electronic news gathering (ENG) camcorder will pack everything but the lens into a single housing. Cameras without interchangeable lenses are also available at the professional level, and these offer a true one-piece system.

Figure 5.5-3 shows a diagram of a typical studio camera system configuration and Figure 5.5-4 shows a typical ENG camera with a built-in recorder and also a radio link for a wireless microphone.

Camera and Imager Sizes

Historically, imager size has also been determined by the application. This is still a factor, but the lines have been blurred thanks to the ability of new solid-state devices to deliver good quality with smaller imagers. Full-size studio cameras are often referred to as *hard cameras*.

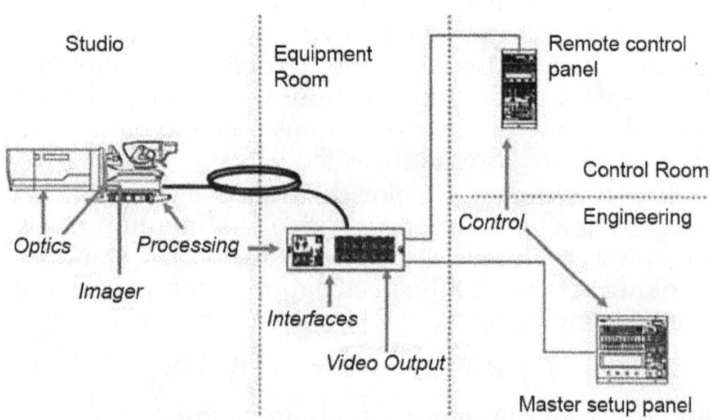

FIGURE 5.5-3 Studio camera system.

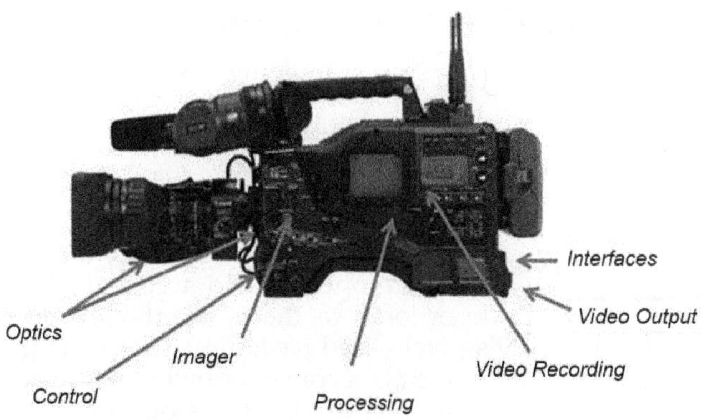

FIGURE 5.5-4 ENG camera system.

They have fewer size and weight constraints than portable cameras, so camera designers would in the past often include larger imagers and more advanced processing in the camera. Currently available studio hard cameras, however, all use the same 2/3-inch three-CCD image blocks found in ENG or electronic field production (EFP) cameras. The size of a hard camera is now driven by the need to include studio features and the ability to mount large box-type lenses, rather than the size of the imager, as the lens mount on the smaller cameras cannot support these heavy lenses.

Other than the potential for more sophisticated lenses, image quality available from even quite modest camcorders has largely caught up to that produced by the largest and most expensive studio cameras. Although the camera sizes still vary, nearly every broadcast application is currently employing 2/3-inch imagers. Twenty years ago, imagers smaller than the 2/3-inch format were relegated to industrial applications; however, three CCD imagers of 1/2 inch and 1/3 inch have begun to find a place in some broadcast applications. Most of these cameras are chosen based on the significant cost savings and an expectation that the image quality will be good enough for the particular application.

Achieving high performance with small-format optics can be difficult, and many small-format cameras mitigate this by integrating fixed optics into the design. This dictates a fixed lens rather than the traditional professional design that allows for alternative lenses. That is not to say that the lens itself is fixed; zoom lenses are virtually universal for modern television cameras (although not for electronic cinema, where cameras that mount 35 mm film camera prime lenses are also available). Most producers have approached smaller imager cameras with the attitude that this allows the production to use more cameras for the same price and provide more content, albeit of slightly less quality.

The requirements of UHD production have led to larger than 2/3-inch imager cameras making a comeback. These have primarily been introduced into production where the potential for better performance is combined with, and enhanced by, the desire to shoot with cameras more similar in operation to 35 mm film cameras. However, as price points decline and consumer expectations rise, it is quite possible that these large imager cameras may become the mainstay just as HD cameras did in the past.

Background: Pixel-Based Focal Plane Imaging Arrays

In 1985, the world's first solid-state imager-based camera made its debut in the world of television broadcasting. That early camera with its prototypical comparatively low-resolution CCD imagers, although primitive by today's standards, paved the way for a sea change in the methodology of television image capture. The benefits of solid-state imagers compared to tube-based cameras quickly became apparent to all. They were simpler and more robust, and they had no tube (target) burn, no highlight comet tail problems, and no requirements for tube registration and other complex alignment before use. Today, tube-based cameras are archaic and quaint, a distant relic of a bygone age. Indeed, many of today's broadcast engineers may have never seen one, let alone had to use or maintain one.

The advent of tubeless cameras was not, however, without a struggle. Device physics, analog signal processing, and technology limitations all conspired against the CCD pioneers; however, the relentless march of technology and the advent of digital computing power made it possible to overcome the problems. Today, CCD and CMOS video cameras are ubiquitous. They are used in broadcast and production cameras, consumer camcorders, smartphones, tablets, point-and-shoot digital cameras, single lens reflex cameras (SLRs), spy satellites, robotic vision, webcams, laptop computers, security cameras . . . the list is almost endless. They range from marginal to ultra-exotic in performance and from almost free to hugely expensive in cost. State-of-the-art imagers used in specialized applications such as satellite and medical imagery contain arrays of 100 megapixels or more. Obviously, these systems used for these types of specialized applications are hugely expensive. But consider that an average smartphone will have a 12-megapixel imager capable of capturing 4K motion pictures. How those imagers are designed, manufactured, and integrated into a device sets the performance of the entire system.

CCDs were invented in the late 1960s; their initial application was to provide an analog delay function for computer applications. Their ability to move an electric charge in a controlled fashion soon led to other applications, such as audio "bucket-brigade" delay lines and as video imaging/storage devices. Bell Labs produced a prototypical CCD-based television camera in 1970. Device technology rapidly improved to the point that by 1975 a more advanced CCD camera was considered almost broadcast quality.

CCD Image Development

Figure 5.5-5 shows some examples of CCDs from the past 30 years of camera design. They range in size from an early 2/3-inch RCA Frame Transfer CCD to a 1/4-inch Interline Transfer device suitable for ultra-compact point-of-view and closed-circuit television cameras.

It should be noted that the nominal size of a solid-state imaging device is based on the diameter of an equivalent legacy camera pickup tube. The actual diagonal size of the active imaging area is approximately 16 mm for a 1-inch device, 11 mm for a 2/3-inch device, 8 mm for a 1/2-inch device, 6 mm for a 1/3-inch device, and 4.5 mm for a 1/4-inch device. Early CCDs suffered from extremely low manufacturing yields, often less than 0.01% for broadcast-worthy devices, and were thus very expensive. As semiconductor production technology improved, it became possible to build CCDs with improved yields, or smaller CCDs for less demanding applications (e.g., the consumer and business/industrial markets). Those markets are considerably more tolerant of defects and appreciate the user benefits of less expensive, more reliable, and easier to operate cameras that the new technologies bring. The enormous consumer camcorder market sector would not have been possible without solid-state imagers.

Current Imagers

Today's solid-state imagers are based on two technologies: CCDs and CMOS. Both are intended to fulfill the same basic function, namely to capture light on a pixel-based focal plane array and then to convert that light into a video signal that can be processed for later image display. Both device families have their own unique strengths and weaknesses, and the user should choose the best device for the intended application. As semiconductor physics and manufacturing processes for both device families improve, newer devices with improved resolutions or other specific parameters may become viable. Imagers are like most things in life in that there is no "one size fits all" solution.

FIGURE 5.5-5 A collection of CCDs from 1983 to 2006. The top left device is a 1/4-inch CCD from an inexpensive "lipstick" camera; the lower right is a 2/3-inch high-definition 16:9 progressive HD imager.

CHARGE-COUPLED DEVICES

CCDs are a family of similar imaging devices:

- Frame transfer (FT) CCDs;
- Frame interline transfer (FIT) CCDs;
- Interline transfer (IT) CCDs.

The FT, FIT, and IT imagers are basically analog devices. They all capture photons and accumulate an *electric charge* proportional to the incident light falling on the CCD's pixel-based *photosites*. Those charge packets are then moved in a conveyer-belt type of process, one packet at a time, to the CCD's output stage. Here the minuscule electric charge from each pixel is converted to a voltage proportional to the charge; this voltage is the output signal of the device. The CCD structure is fabricated by means of diffusion and doping of the silicon and by the addition of metal and polysilicon conductive bus wiring to provide the defined pixels areas with their interconnect and drive requirements. Although the basic imaging array works in a similar general fashion, the three device types transfer and store the imaged signal charge in different ways, which are now explained.

Frame Transfer CCD

The first CCDs were frame transfer (FT) devices. FT CCDs for broadcast applications comprise two basic sections. In an imaging array, light is converted by means of the *photoelectric effect* to an electrical charge. Light hits the silicon of the imager, and, in efficient devices, nearly every photon generates an *electron–hole* pair. The electron is captured under each pixel in a tiny electrically generated temporary storage area called a *charge well*. The CCD's photosensitive pixel array is normally exposed to light for the majority of the television field or frame period (1/60 sec), usually ~16 msec. The electrons present in each well represent the cumulative charge generated by the light falling on that pixel during that 16 msec period. CCDs are extremely linear devices; thus, more light equals more charge and thus more signal output. In FT CCDs, this signal charge is rapidly moved from the imaging area to a similarly structured storage area during the TV signal's vertical blanking time. The only real difference between the imaging and storage areas is that the storage area is covered by an opaque mask of chrome or other such metal. This is necessary to avoid unwanted light from generating extraneous charge and corrupting the signal collected from the imaging array. Figure 5.5-6 details the basic structure of a frame transfer CCD. It is not necessary for the video engineer to be an expert on such devices; however, an appreciation of the differences may prove helpful in determining their best use in differing applications.

The white shaded imaging array (IA) consists of contiguous pixels; each pixel accumulates charge during the active integration period. During the vertical blanking interval, that charge is clocked at high speed to the chrome-masked storage array (SA). It is interesting to note that the FT imager uses the actual

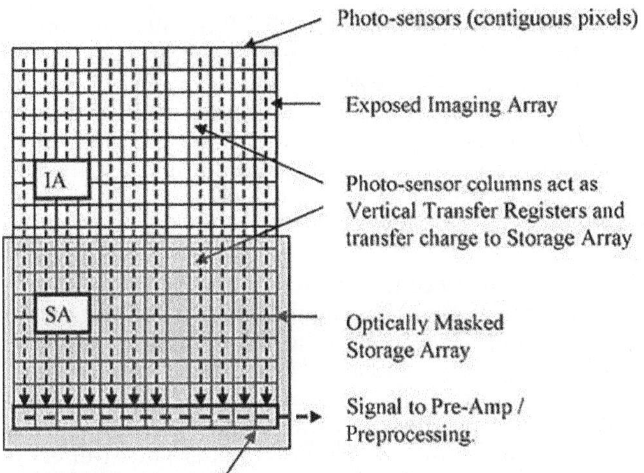

FIGURE 5.5-6 The typical frame transfer CCD is composed of a light-sensitive imaging array and an optically masked storage area. Charge is accumulated in the imaging array and then quickly transferred to the storage array during the vertical blanking interval. The stored charge is then clocked out of the storage array in real time by means of the vertical and horizontal transfer registers for subsequent video processing.

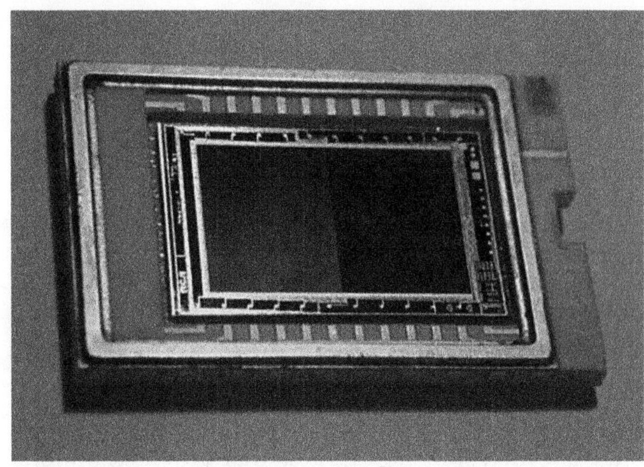

FIGURE 5.5-7 An early RCA 535 × 480 pixel frame transfer CCD (circa 1984). Note the clearly demarked imaging and storage areas.

photosensitive pixels as vertical transfer columns. This is efficient and simplifies manufacturing but requires that the camera use a physical shutter to block light from impacting the CCD during this transfer period. If an optical shutter is not used, or if the shutter is inoperative or mistimed, transfer streaking will occur. This is a visible effect caused by extraneous light polluting the stored charge during its transfer to the masked storage area. The effect resembles a bright vertical highlight smearing artifact in the image.

The use of a mechanical shutter is fairly simple but may have undesired and objectionable visual artifacts when viewing alternating current discharge lighting in a television scene, on electronic scoreboards, or any devices that use scanned display arrays. Such devices may exhibit a strobing effect, even on still shots.

In the FT CCD, the stored charge is clocked out of the storage area by the horizontal shift register in a *real-time* line-by-line fashion; the charge for each pixel is then fed to the output stage of the CCD. This important part of the CCD is a tiny sample-and-hold circuit that converts each pixel's tiny electrical charge to a voltage proportional to that charge. It should be noted that the electric charge per pixel is extremely small, usually ranging from about 5 to 10 electrons for black (no light) to possibly 20,000 electrons for peak white. As can be imagined, such tiny charges and such sensitive circuitry require very specialized circuitry, which is critical to the performance of the CCD. Techniques such as *correlated double sampling* are used to minimize various noise sources; for example, the N-MOS source followers in the CCD output stages have a significant flicker noise component. The processing at this stage is analog and determines the noise floor of the device; extreme care should be taken to provide the cleanest signal for subsequent video processing.

An example of a frame transfer CCD can be seen in Figure 5.5-7. Although an early device, it clearly shows the different imaging and storage areas of the imager. It is interesting to note that this apparently 16:9 widescreen imager is actually comprised of the two 4:3 imaging and storage sections sitting side by side. The lighter gray area is the chrome-covered storage area and the darker area is the imaging array.

Frame Interline Transfer CCD

The next type of CCD to consider is the frame interline transfer (FIT) device. This CCD was devised as a means of simplifying the structure and complexity of the FT imager and to obviate the need for a mechanical light-obscuring shutter. The FIT has the usual imaging array of pixels to capture the light and form the signal charge; however, in the FIT, optically obscured vertical transfer registers are added between the columns of pixels in the imaging array.

After the imaging area's light integration period, the accumulated charge is quickly moved from the imaging section to these adjacent vertical transfer registers, then to the storage area. The use of the metal-masked vertical transfer registers means that the CCD does not require a mechanical shutter during the transfer period, as the chromed vertical transfer registers are masked from seeing any extraneous light.

The lack of a physical shutter means that the imager may be shuttered electronically and the shutter duration adjusted to eliminate any visual beating effects from strobing light sources. This feature is a significant benefit for both production and ENG cameras and is often referred to as "Synchro-Scan" or "Clear-Scan." It is used to eliminate strobing when shooting computer screens, televisions, and other non-constant light sources.

Figure 5.5-8 shows the basic structure of the FIT CCD. The imaging array is similar to the FT, but the readout/transfer is accomplished by the vertical transfer registers (shaded gray). The transfer from the imaging area to the storage area is very fast, usually less than 1% of the frame period. The fast transfer and chrome-masked transfer registers/storage array give the FIT CCD a transfer smear specification of about −130 dB, meaning that the classic transfer smear, visible as a vertical white line on highlights, is virtually eliminated.

There are some custom variations of the FIT CCD; one example is the M-FIT, a Panasonic CCD device in the mid-1990s. This particular device was a 480-line 60 frames per second progressive 16:9 aspect ratio imager with a split storage area that provided options for clocking the image out as either an interlace or progressive signal. It was used in the AJ-PDW900P, a DVCPRO-50P 480p/480i camcorder.

Interline Transfer CCD

The third type of CCD imager is the interline transfer (IT) device. This CCD is of much simpler construction and therefore both easier and less expensive to manufacture than the FT and FIT devices. The IT device imaging array is very similar to that of the FIT CCD and, as in the FIT, vertical transfer registers are located between each column of the array. The IT CCD, however, does not use a separate storage area, as the vertical transfer registers perform both the charge transfer and storage functions. Figure 5.5-9 details the simple structure of the IT device. The imaging array accumulates charge during the active field or frame period; the charge is then moved laterally to the vertical transfer registers. These, as in the FIT device, are masked to eliminate extraneous light from contaminating the signal. During readout the charge is clocked in real time, both down and sideways, to the output section of the device.

The simplicity of the IT imager and low cost of production make this device ideal for consumer video cameras. Early IT devices, however, had an undesirable highlight overload effect that caused an ugly vertical streaking artifact. This effect is widely seen on early ENG news cameras when shooting bright lights, most especially at night. The classic transfer smear effect is seen as a bright vertical line on picture highlights; in extreme cases it can be full picture height. This phenomenon gave IT imagers a bad name and relegated them to ENG and other low-end applications where such effects could be tolerated. The FIT imager with its much-improved highlight handling performance was usually specified for more serious EFP and production work. As with most technological problems, time and advances in process technology came to rescue the lowly IT devices from the stigma of being suitable only for news applications.

Current IT devices feature much more efficient blooming drains and very much improved highlight handling, so much so that current devices are equal to the more complicated and expensive FIT CCDs. An example can be seen in the Panasonic 1280 × 720 progressive HD imager, which has a transfer smear specification of −135 dB, basically on a par with current FIT devices. A basic rule of thumb is that an HD FIT imager usually costs at least twice as much as a similar

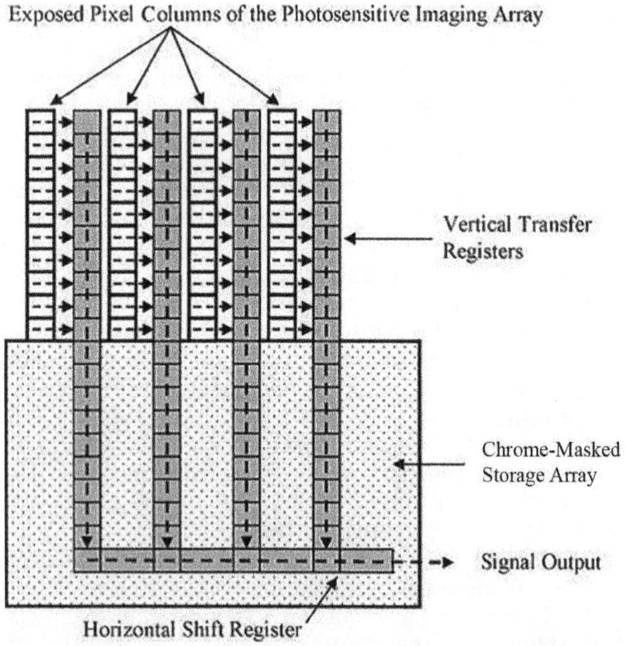

Exposed Pixel Columns of the Photosensitive Imaging Array

Vertical Transfer Registers

Chrome-Masked Storage Array

Signal Output

Horizontal Shift Register

FIGURE 5.5-8 Frame interline transfer CCD basic structure shows an imaging array with a vertical shift register feeding the optically masked storage area. The signal charge is then clocked out of the storage area in real time.

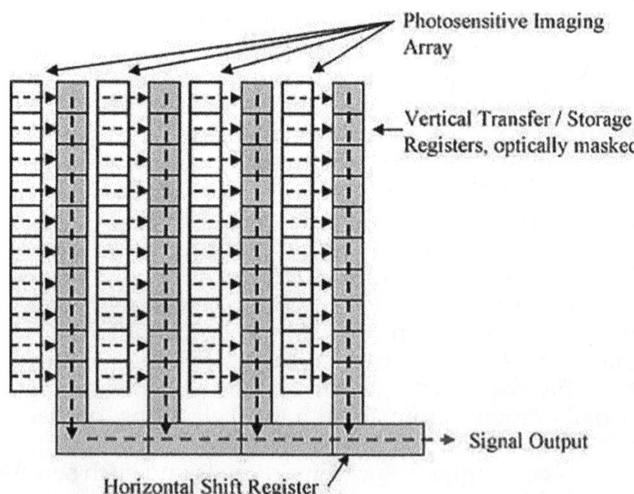

FIGURE 5.5-9 Interline transfer CCD basic structure shows an imaging array with vertical shift registers acting as both transfer registers and local storage for the charge from the imaging area. These dual-purpose registers then feed the horizontal shift register/output stage.

HD IT imager to produce, so if an IT device performs equivalently then the savings realized from using an IT imager mean more affordable products and a wider product choice for the end user.

The era of IT cameras being subperformers is over; this can be seen in the marketplace today where most of the high-end production, studio, and EFP cameras feature IT imagers. The FIT imager is no longer worth the cost and most development work on CCD imagers is very much centered on IT devices.

COMPLEMENTARY METAL-OXIDE SEMICONDUCTOR IMAGERS

Complementary metal-oxide semiconductor (CMOS) imagers, like CCDs, have variants and range in complexity from quite simple to amazingly complex. CMOS technology has the innate ability to mix both analog and digital processing functions on the same device. Thus, unlike CCDs, CMOS devices can place their digital clock generators, row and column drivers, and even complex analog-to-digital converters on the same chip as the analog photosensitive imaging array.

Passive Pixel CMOS

An example of a simple CMOS imager can be seen in Figure 5.5-10. This device, known as a passive pixel CMOS imager, is the least complex of any of the CMOS family. It is basically an array of photodiodes connected to MOS switches, which in turn connect to a series of horizontal and vertical bus lines that transport the pixel charge to the sample-and-hold circuit output stage. In passive pixel devices and other similar basic CMOS devices, the lack of any on-pixel buffering causes the tiny electric charge to encounter problems of the significant stray capacitance of the switched bus matrix.

This results in noise problems and poor performance. It is, however, straightforward in construction and is often used for simple cameras (e.g., cell phones), where performance is less of a factor than cost and complexity. Figure 5.5-11 details enhancements to the pulsed power supply (PPS) CMOS imager with the addition of buffering on the column busses; this helps reduce noise and improves performance. The next logical step in this process would be the addition of *on-pixel* sample-and-hold circuitry along with individual buffering. This raises the complexity of the device considerably but results in marked picture quality improvement.

Active Pixel CMOS

Figure 5.5-12 shows details of a considerably more complex, but much improved, active pixel sensor CMOS imager. Although similar to the simpler passive pixel device, each pixel has its own sample-and-hold circuit and buffer amplifier to reduce signal deterioration when switched by the column/row switch matrix to the analog-to-digital converter. The introduction of a per-pixel sample-and-hold circuit and buffer can, however, lead to fixed pattern noise problems due to the individual pixel sample and hold circuits not being 100% identical.

The fixed pattern noise comes from differences in both the electrical gain as well as DC offsets in the sample-and-hold circuits for each pixel. This gain/DC error signal causes a fixed pattern static modulation of the imager's signal and is often quite visible. Fixed pattern noise is not so much of a problem for digital still cameras, as it resembles random film grain in appearance; however, in television applications the human eye is very good at correlating this fixed pattern noise across time, and even small brightness differences (often less than ~1%) tend to become very visible.

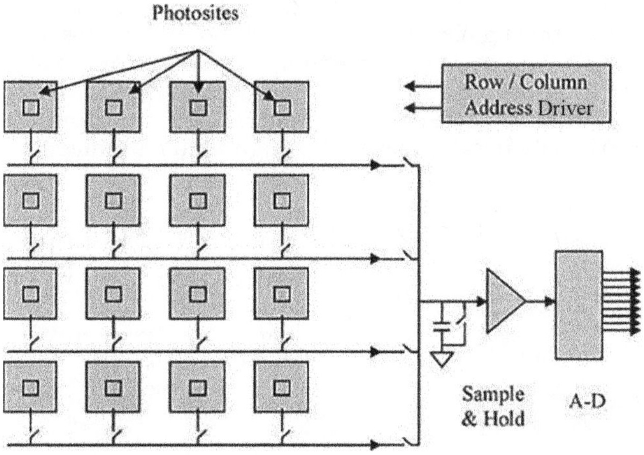

FIGURE 5.5-10 Block diagram of simple passive pixel CMOS imager. Row and column switches route charge to a sample-and-hold circuit followed by an analog-to digital converter. This simple imager structure is used in applications such as cell phone cameras.

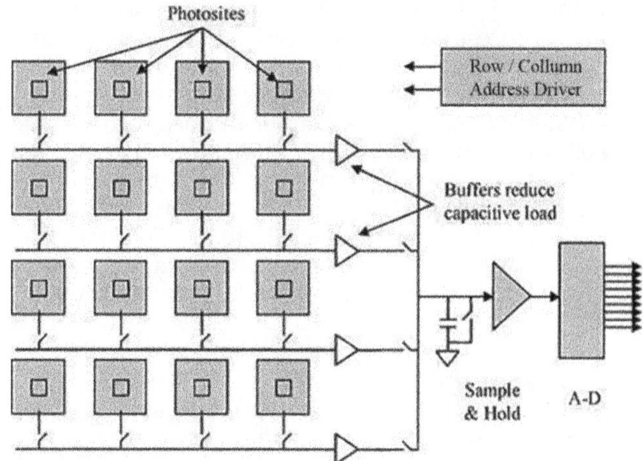

FIGURE 5.5-11 Block diagram of slightly more complex passive pixel CMOS imager with buffering on the column busses to reduce the stray capacitance. It increases the complexity but is worth the improvement in performance.

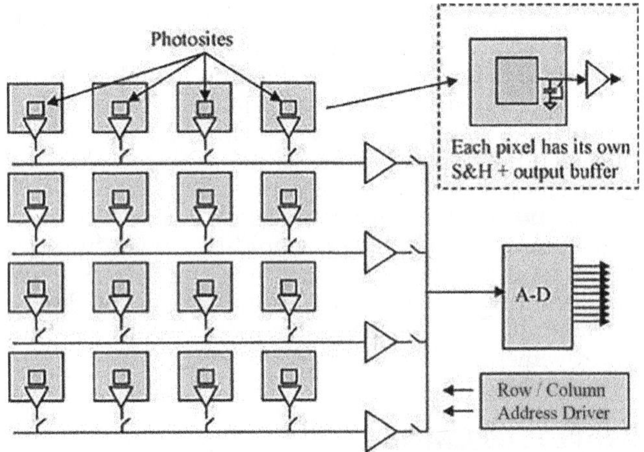

FIGURE 5.5-12 Block diagram of more complex active pixel sensor CMOS imager. In this device, each pixel has its own sample-and-hold and buffer amplifier; however, as each pixel's sample-and-hold circuit and buffer amplifier are unique and not 100% matched to adjacent pixels, fixed pattern noise can become an issue.

CMOS Developments

CMOS imagers can be as simple as the passive pixel device or as complex as each pixel having its own sample-and-hold amplifier, buffer, *and* analog-to-digital converter. Such complexity, however, causes problems of cost, heat (from the many active devices on the wafer), and reduced fill factor of the pixel, due to the addition of more metalizing and polysilicon to provide power and signal distribution busses. More metal means less silicon is available to catch the all-important photons, and devices become less and less sensitive.

The pixels of CMOS imagers are addressed on an individual basis. Although this may initially be considered an inconvenience, it does permit the user to map the imager and download only the area of interest. This process has uses for applications where it may be desired to read only a certain area or zoom into an area of interest. CMOS imagers are often used for high-frame-rate cameras; for example, an imager may have a 1000 × 1000 pixel matrix and a maximum frame rate of 100 fps. It is possible, however, to address only the center 100 × 100 pixels of that imager and read them out at a much higher frame rate (e.g., 1000 fps). The same 10,000,000 pixels per second are being processed, so there is no issue of speed of the A/D or sample-and-hold circuitry. The user is simply choosing temporal over spatial resolution for the viewing application.

In broadcast applications, some 1920 × 1080 pixel progressive CMOS imagers have the ability to be addressed as native 1280 × 720 imagers. In this mode, the user is choosing to address only the center portion of the imager. This is often done to permit operation at a higher frame rate or to switch the imager between the two main HD imaging formats (1920 × 1080 and 1280 × 720). This sub-mapping process imposes a very high requirement for lens performance. It also has the additional unwanted side effect of the native 2/3-inch imager becoming windowed to the equivalent size of a 1/2-inch imager. This has an unfortunate twofold effect of increasing the effective focal length of the lens by about 22% and losing about 1 f-stop of sensitivity. In news applications, this can reduce a $20,000 8 mm wide-angle HD lens to a not-so-wide 11 mm lens. The modulation transfer function (MTF) requirements are also vastly increased as the effective 1/2-inch center portion of the lens is expected to provide the full resolution of the 2/3-inch format.

FILL FACTOR AND CCD MICRO-LENSES

Fill factor is the term used to specify how much of the surface area of a given pixel or imager is actually photosensitive and exposed to the incident light. By way of example, if an imager has a contiguous array of 10 nm square pixels, yet only the center 5 nm square of each is light sensitive, the fill factor for the pixel is 25 nm^2/100 nm^2, or 25%. This simple example shows that only 25% of the light actually hits the photosensitive portion of the pixels. This means that the imager will be some 2 f-stops less sensitive than an equivalent imager with a 100% fill factor. To mitigate that loss of sensitivity requires either a faster lens (usually very expensive) or having to add 12 dB of electronic gain to make up for the light loss; 12 dB of electronic gain is a significant amount of gain and will make any camera noisy.

Low pixel fill factors are not easy to correct. The surface of the CCD or CMOS imager is usually covered with numerous metalized busses, with additional metal plugs and via holes that move power or signals around the inside of the imager. The more complex the imager is, the more circuitry it is likely to have, and consequently the more likely it is to have opaque metalizing that contributes to the lower fill factor. One means of correcting a poor fill factor to use a *micro-lens* to focus the light onto the photosensitive portion of the pixel. Figure 5.5-13 shows the concept of the micro-lens approach. Pixel-sized lenses above each active pixel focus the light onto the active photosite to greatly improve the sensitivity. Typically, micro-lens systems provide an increase in the light sensitivity of 200%–300% (some 1 to 1.5 f-stops) and can bring the sensitivity back on par with a high fill factor imager.

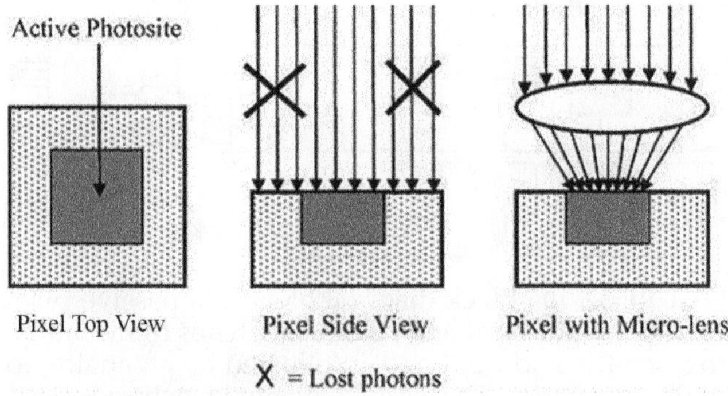

FIGURE 5.5-13 CCD imager with micro-lenses.

SENSITIVITY AND DYNAMIC RANGE

Both CMOS and CCD imagers share the same basic imaging process and are covered by similar device physics. Figure 5.5-14 details large and small pixel devices. Usually, the larger the pixel is, the more photons it will catch, thus the more sensitive the imager is. These large pixels also have large charge wells and therefore are capable of holding a larger charge. An imager with large pixels at peak exposure may hold ~50,000 electrons in its charge well.

That same imager with no incident light still may hold 25 electrons; the majority of these are caused by thermal activity and by chemical impurities causing leakage in the bulk silicon. Interestingly, one or two of those electrons may be caused by high-energy photons from cosmic rays and even radioactive isotope emissions from the CCD's ceramic packaging. Examining the minimum to maximum charge levels (50,000/25 or 2000:1), we can say that the imager has a useful dynamic range of about 2000:1, which represents some 11 f-stops.

Now consider a 2/3-inch high-definition imager with 2 million pixels. In this case, the active pixel size will be about 5 nm square. Unfortunately, electrons do not change size just because we use smaller imagers; hence, the much smaller charge well of the HD imager may only hold 10,000 electrons at peak well level, so if the noise floor of the pixel is 10 electrons then we now have only a 1000:1 dynamic range (10 f-stops). This smaller HD pixel is also less sensitive compared to the larger pixel, as the photosensitive portion of the pixel may only be 1/4 the size of the larger pixel and thus able to intercept only 1/4 of the light.

In summary, the HD imager is 2 f-stops less sensitive and has half the dynamic range of the device with larger pixels. This limitation can be seen with the current crop of digital point-and-shoot consumer camcorders; these usually utilize 1/4-inch or smaller imagers with around 5 million pixels. They have severe limitations with both sensitivity and dynamic range, and testing of these cameras has indicated only about 7 or 8 f-stops of latitude and poor low-light performance—a testament to the fact that more is not always better!

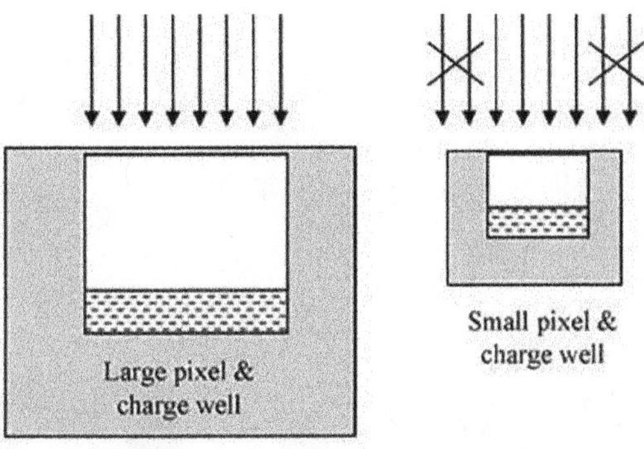

Large pixels are more sensitive due to their larger active surface area, they simply catch more photons. The charge well size is also large and can hold more electrons before the onset of blooming.

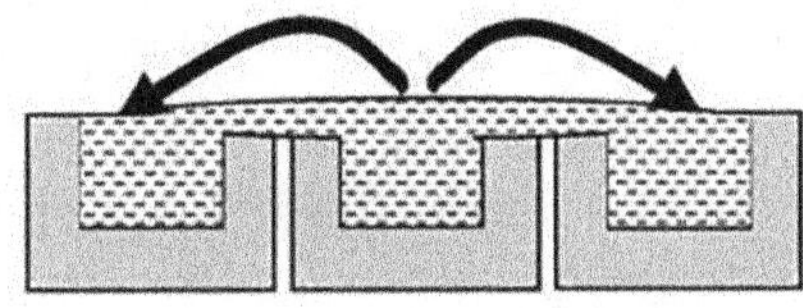

Small pixels are less sensitive and have smaller charge wells. When a pixel is overexposed, its charge well overflows and spreads the charge to the adjacent pixels in a process known as highlight blooming.

FIGURE 5.5-14 CCD imager with micro-lenses.

INTERLACE AND PROGRESSIVE SIGNAL GENERATION

Most CCD and CMOS imagers have an inherently progressive imaging array; that is, the light hits the uniform array of pixels all at the same time. If the imager is required to output signals suitable for interlaced television applications, the stored image data will require processing to derive the required interlaced signal structure. This process is usually accomplished by a technique called *row–pair summation*.

Figure 5.5-15 shows how the odd and even field signals are derived from the inherently progressive imaged data. The charge data from row 1 and 2 of the CCD is summed in the CCD sample-and-hold output circuit to provide a new pseudo-pixel of the required TV field. This row–pair summation process has two interesting benefits. First, as the charge from the two pixels is summed, the signal value is doubled. This doubling increases the effective device sensitivity by 6 dB, or 1 stop. The second benefit of the row–pair summation is an increase in the device's signal-to-noise

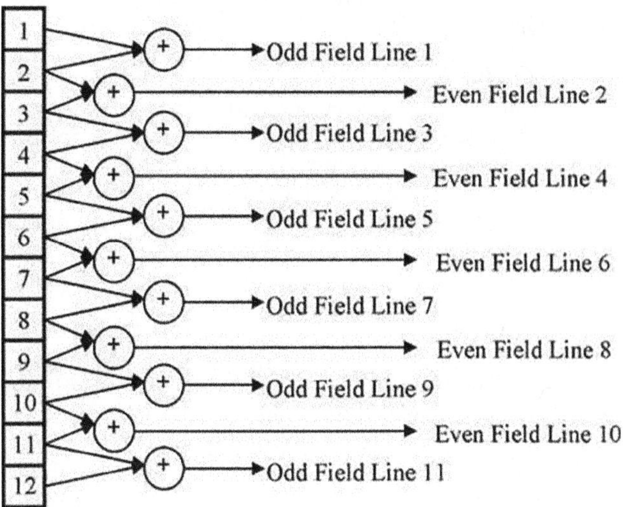

FIGURE 5.5-15 CCD row–pair summation generates interlace signals from progressive device.

ratio by 3 dB. This occurs in the summing of the noncoherent random noise of each pixel; the differing noise of each pixel tends to slightly cancel the other, resulting in a useful 3 dB improvement in the signal-to-noise ratio.

The row–pair summation process of mixing charge from adjacent lines (i.e., 1 + 2, 3 + 4, and so on) generates a new single pseudo-sample (pixel) sited between the two original pixel rows, thus generating the half line vertical offset required to simulate the odd TV fields composed of lines 1, 3, 5, . . . In the next field, pixels are summed from rows 2 + 3, 4 + 5, . . . to create the even TV field composed of lines 2, 4, 6, . . .

Any imager used to create an interlaced signal must use a carefully optimized optical low-pass filter to limit the vertical resolution of the device/optics. If not, the row–pair summation process will result in unpleasant and very visible vertical alias artifacts in the video. This filtering may impose limitations of low vertical resolution upon the system if the camera is to be switched between interlace and progressive imaging modes. If optimally optically filtered for interlace, any progressive signals may appear as *soft*, or lacking in vertical resolution. Conversely, if optimized for progressive mode, the interlace signal may exhibit severe interlace *twitter*.

With the increases in large-scale integration (LSI) technology, a more recent approach to the problem of switchable interlace/progressive cameras is to generate a full-resolution progressive signal from the imager, then by means of filtering and spatial image interpolation derive an optimally filtered interlaced signal electronically. This approach, however, does lose an f-stop of sensitivity and the 3 dB reduction in the signal-to-noise ratio. As is usual in most things in life, there is no free lunch, and compromises abound.

DICHROIC PRISM BEAM-SPLITTING ASSEMBLIES

Almost all ENG and production television cameras utilize a prism to split the light from the lens into the three constituent primary colors of red, green, and blue. This system is efficient in that none of the light is wasted by color absorption filters; instead, a three-channel R, G, and B prism beam splitter is used to channel the spectrally selected light to three dedicated imagers. Figure 5.5-16 details the structure and ray paths of the typical three-port beam-splitting prism. The prism is assembled of three glass blocks glued together. *Dichroic coatings* on the prism faces are used to generate two *optical filters*, one low-pass (LPF) and one high-pass (HPF). Light enters the front prism port and proceeds to the dichroic coating on the back face of the first prism, which forms the high-pass filter; this filter permits green and blue light to pass through. The low-frequency red light is reflected and bounces to exit the lower prism exit port to the red imager. The back of the second glass prism block is coated with another dichroic layer to form a low-pass filter; this reflects blue light up and out of the top prism port to the blue imager. Green light is not affected by either the low- or high-pass filters and passes straight through the prism assembly to exit the rear port to the green imager. Figure 5.5-17 shows an optical low-pass filter and infrared cut filter on the front of the prism front port.

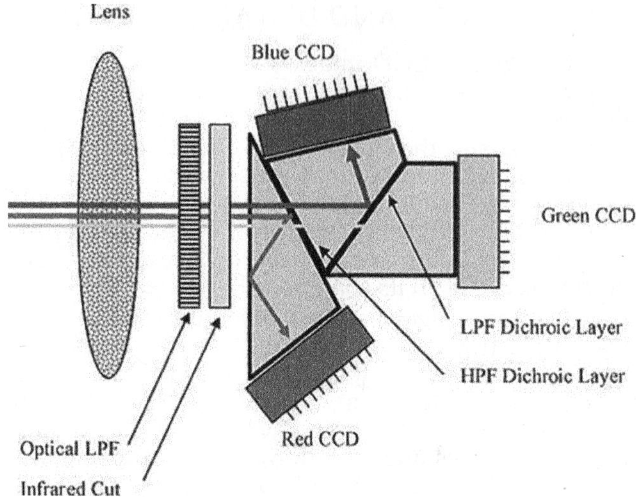

FIGURE 5.5-16 Basic three-port RGB prism and CCD optical assembly.

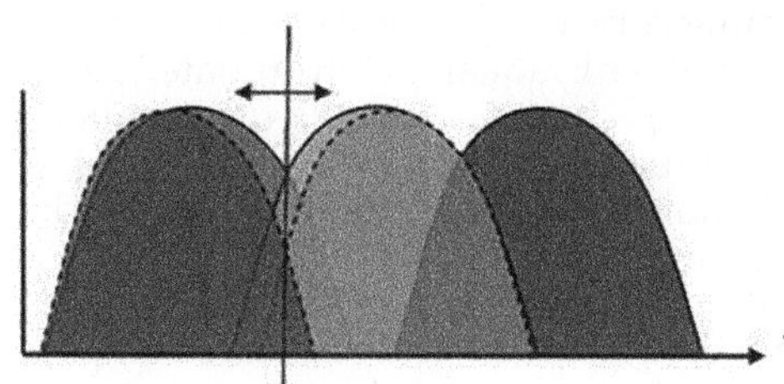

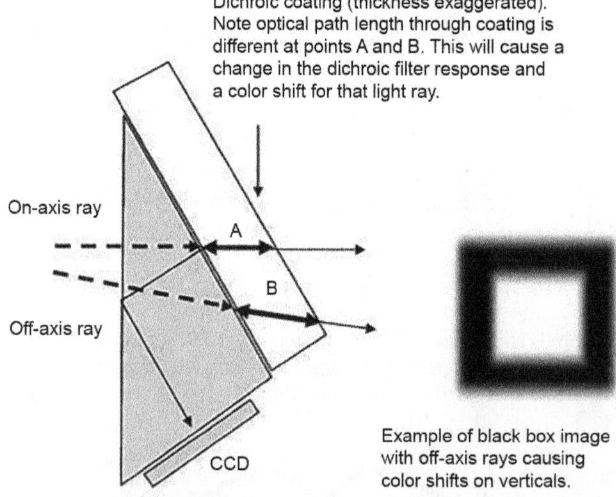

FIGURE 5.5-17 Off-axis light hitting the dichroic coating causes a shift in the low-pass filter and high-pass filter, thus causing a color shift.

The infrared (IR) cut filter is required to severely curtail the IR response of the optical path to avoid the red CCD imager from seeing this IR portion of the light. Infrared light will show up as a strange color shift in some colors due to some IR component of the subject's spectral emission or due to the different focus point of the lens to IR, as with an out-of-focus halo on streetlights. CCDs are still relatively sensitive to IR light even at wavelengths approaching 1.0 nm. The IR cut usually starts at about 625 to 650 nm and will have response of less than 1% at 750 to 800 nm.

It is of particular importance that the ray path of the lens's exit pupil be parallel. If an optical mismatch occurs (e.g., due to the wrong lens on the camera, a misaligned rear extender, or the iris diaphragm opened too wide), there is a chance that the exit pupil light cone may diverge. This will result in an unsightly magenta/green shading effect on out-of-focus objects in the image.

The prism's dichroic filter is made by depositing finely controlled layers of metal oxide coatings that are adjusted in thickness to match the wavelength of the light being filtered. Figure 5.5-17 shows how an off-axis ray can cause a filter shift as it enters the dichroic filter, thus incurring a transit path length change. Off-axis light entering the dichroic coating at an angle will shift the filter response and thus cause a color shift for that ray. This problem is very obvious when using a long lens with a wide-open iris diaphragm. An external lens extender will exacerbate the effect.

Port Prism/CCD Assembly

An interesting variant of the common three-port prism is the four-port version shown in Figure 5.5-18. In this device, there are two green ports on the prism. This permits the attachment of two green imagers, G1 and G2, as discussed in the next section.

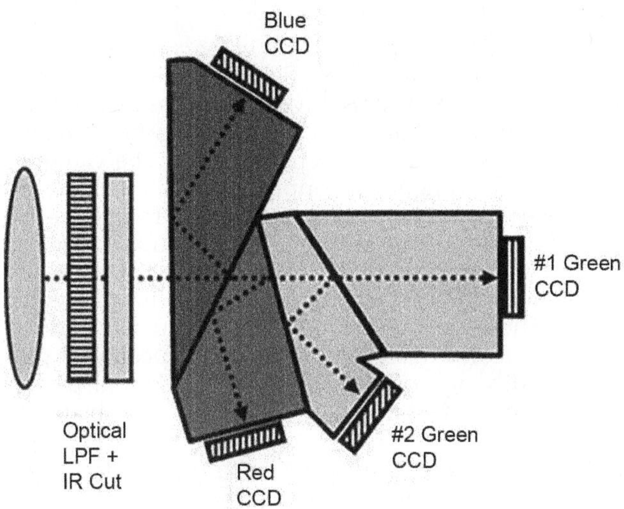

FIGURE 5.5-18 A four-port prism assembly (G1 + G2 + R + B).

THE USE OF PRECISION SPATIAL OFFSET OF IMAGERS

Two Offset Green Imagers

If the alignment of the two green images is very carefully offset so there is a half pixel horizontal and vertical offset between the imagers, it is possible to generate a double-resolution image by merging the two green signals. CCDs and CMOS imagers have a defined pixel matrix for their imaging array. That X–Y matrix of pixels will determine the Nyquist limit for the imager in both the horizontal and vertical directions; for example, if an imager has a 1000 × 1000-pixel count, the Nyquist limit for both directions will be 1000/2, or 500 cycles. Any image resolution above that of the 500-cycle limited imager will result in an alias effect. It is possible to build imagers with higher pixel counts to improve resolution, but this often results in a loss of sensitivity and also lowers manufacturing yields. The spatial offsetting technique with two imagers is often used as a practical means of increasing the resolution of an imaging system; for example, using a four-port prism assembly like that detailed in Figure 5.5-18 with two green channels. If these two imagers are aligned with a half pixel offset, it is possible to combine the signals from two green channels and double their combined resolution.

Figure 5.5-19 shows two CCD imagers with a half-pixel offset sampling a high-frequency sine wave chart. The imaged test chart is exactly at the Nyquist limit of each CCD imager, so if the imagers have a fairly narrow sampling aperture they will each integrate the image scene to a DC value of white or black depending on the phase of the pixel to that of the image. In a correlated double sampling system, the pixel samples from the two imagers are interleaved as CCD1, then CCD2, then CCD1, etc.

Steady-state video values for CCD1 + CCD2 =
"white" DC levels + "black" DC levels

Correlated sampling of CCD1 + CCD2 provides an accurate reproduction of high-frequency image detail and doubles the pixel resolution of a single imager. Because green has by far the greatest contribution to luma, using two, offset, green imagers effectively doubles the system resolution.

Spatial offsetting can work in horizontal, vertical, and combined horizontal and vertical directions. This technique is becoming more common on higher resolution 4K and 8K HD cameras. An individual 4K or

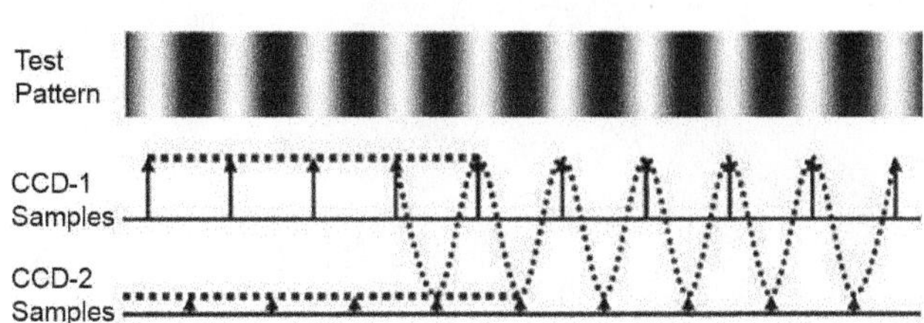

FIGURE 5.5-19 Spatial offsetting of imagers to increase resolution.

8K imager would be prohibitively expensive and quite insensitive. The use of two 2K or 4K imagers, quincunx (biaxially) spatially offset, will produce the resolution of a 4K or 8K imager at much lower cost and with more practicality.

Offset Green with Red and Blue Imagers

A more common approach that uses spatial offset is a conventional three-port prism camera with individual red, green, and blue imagers. Here, however, a half pixel horizontal offset is added to the green imager relative to the red and blue imagers. This spatial offsetting generates a "super-sampling" effect that occurs in the luma matrix when the red, green, and blue channels are combined. The ITU 709 luma matrix is described as Y′ = 0.701G′ + 0.212R′ + 0.089, and it is interesting to note that the 30% increase in effective luma response from this approach agrees closely with the approximate 30% contribution of the combined red and blue channel signals in the Y′ matrix. With the offset pixel imagers, the increase in resolution will only occur when there is coherent signal content in the green and one or more of the color channels. Should the system be imaging a primarily green image, there will be little or no red or blue signal content in the image and thus little or no additional sampling content to be added to the green, hence no increase in effective luma resolution. An example of the benefits of the use of spatial offset in three-CCD imaging systems can be seen in Figure 5.5-20.

This image was taken from the Panasonic AG-HVX200, a small P2 memory-based SD/HD camcorder. This camera uses three 1/3-inch CCD imagers in the usual R, G, and B prism configuration; however, the use of 1/3-inch HD CCDs has negative consequences for sensitivity and dynamic range if the pixel density is too high. It was therefore decided that this camera would use a novel approach of biaxial spatial offsetting in combination with a new 1/3-inch 16:9 960 × 540 pixel progressive CCD. The choice of the imager resolution was a careful compromise between resolution and sensitivity/dynamic range. As the AG-HVX200 was intended for HD production applications, dynamic range and sensitivity were critical constraints. The use of a quincunx (biaxial horizontal and vertical) spatial offset greatly helps in raising the system resolution to about that of a native 1280 × 720 camera, while also maintaining the sensitivity and dynamic range benefits of the large pixel imagers. In the figure, note the marked increase in resolution, comparing the upper left corner (effectively that of a single 960 × 540 imager) with the rest of the image, which is effectively 1280 × 720. The significant increase in resolution results in a clean 600 TV lines per picture height resolution being quite visible in the combined luminance signal.

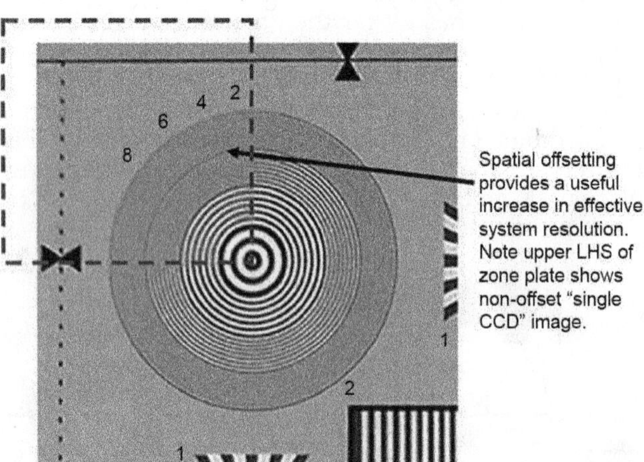

FIGURE 5.5-20 Results of biaxial spatial offsetting of imagers.

Single Imager Color Systems Versus Three-Imager Systems

If a 35 mm film camera is considered as a model for a still or movie camera, it at first appears quite simple in concept. Light enters a lens system, passes through a shutter mechanism, and then falls onto the film plane. In this model, both electronic and film cameras can be considered pretty much identical. Both capture light and create an image at the focal plane imager, be it a piece of film or an electronic CCD or CMOS imager. There is no easy electronic analog to a piece of film; that simple square of film with three ultra-thin light-sensitive emulsions is still a wondrous achievement.

It is possible to make cameras using a single combined color imager that are smaller, simpler, and lighter than three-CCD systems but suffer from being less sensitive and, depending on the type of imager/color separation filter system used, may have issues of resolution. Cameras with beam-splitting prisms and three or more CCDs or CMOS sensors do not absorb light by filtration as do single imagers; thus, they are more sensitive but also more costly, more complex, and physically larger. The relative merits of both approaches should be considered when choosing a camera system for a particular application.

Efforts to make single imager color cameras go back many years. Ignoring CBS's field sequential system of the early 1950s, which involved a large spinning color wheel, as being impractical, a more realistic approach was to use striped optical filters on a tube camera's pickup tube. Diagonally opposed, thinly striped gel filters were carefully placed on the tube faceplate. The striped filters, when scanned by the tube's electron beam, would provide color information automatically modulated on subcarriers embedded in the video signal. These subcarriers were demodulated and the respective color difference signals produced and processed. Although of low color resolution and performance, they served well in millions of early tube-based VHS camcorders.

A simple yet elegant solution to the single imager requirement was developed for the digital still camera of today, of which hundreds of millions are sold annually. The most common approach to a single imager color sensor is the Bayer color mask, which is composed of individual red, green, and blue color filters overlaid onto the CCD or CMOS imager's pixels, as shown in Figure 5.5-21. The interleaved R, G, B, G pattern has twice as many green pixels, because green is the color most sensitive to the human visual system. It is possible by means of a "de-Bayering" process to interpolate the Bayer pattern to individual red, green, and blue signal components. The Bayer mask approach is useful but results in a loss of light due to the absorptive optical filtering. It is obviously important that the filters are of the correct density and color response and also accurately matched to adjacent pixels, to avoid blotchy colored areas in the picture. It is a tribute to modern manufacturing technology that a tiny 1/4-inch CCD 5 million pixel imager can be covered with a Bayer mask and also with the tiny micro-lenses that are used to regain some of the lost sensitivity.

Large-Format Imagers

CCD and CMOS imagers range in size from tiny 1/10-inch or smaller imagers for cell phone cameras to the larger than 4-inch wafers for specialist large-field cameras, such as the Hubble Space Telescope cameras. Several companies are now producing 35 mm sized imagers for use in film-style production cameras. Most of these cameras use a single imager, with some form of Bayer or color mask. Very large imagers like these are difficult to make and are expensive, but such cameras are made in modest quantities, usually in the tens to one hundreds, so limited imager yields are not an insurmountable problem. The large-format imager means that the camera can make use of the large range of very high-quality 35 mm lenses used in motion picture film production. One interesting benefit is that large-format cameras are more tolerant of less-than-perfect lens specifications. A 35 mm sized imager may require a lens with only about 70 lines/mm response to provide a very high quality image. A much smaller 2/3-inch CCD requires a lens with almost three times that performance to provide equal quality, a very much more difficult task for the lens.

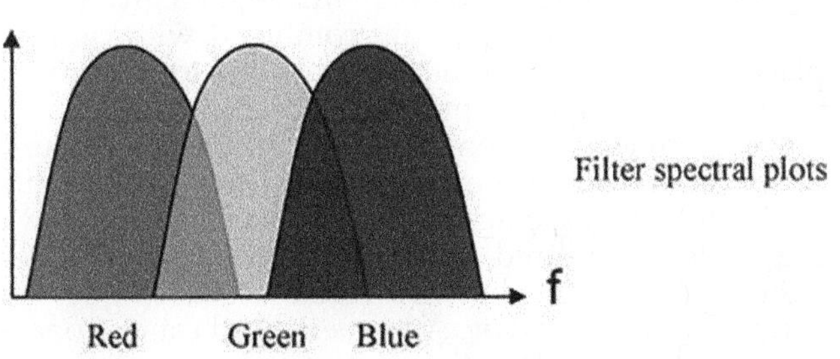

G	R	G	R	G	R	G	R	G	R
B	G	B	G	B	G	B	G	B	G
G	R	G	R	G	R	G	R	G	R
B	G	B	G	B	G	B	G	B	G
G	R	G	R	G	R	G	R	G	R
B	G	B	G	B	G	B	G	B	G
G	R	G	R	G	R	G	R	G	R

FIGURE 5.5-21 Bayer mask with R, G, B, G pattern. The mask absorbs about 67% of the light, and only 33% of incident light is captured.

CCD Versus CMOS Quality

As of the time of this writing, CCD imagers have the advantage of superior image quality, especially in the broadcast size of 2/3 inch. CMOS, although an interesting technology, is still probably a few years away from viability in mainstream broadcast cameras. While offering some advantages, particularly lower power consumption, CMOS technology suffers from lower sensitivity, fixed pattern noise, and production manufacturing issues; little can be done to easily improve these parameters, although perhaps in due course they will be overcome. Figure 5.5-22 is an interesting testament to the relentless advances in technology, as it is an image taken by a current 1280 × 1024 pixel cell phone camera. Although optically challenged by cost and size constraints and lens physics, it has about twice the resolution of the 1984 RCA CCD1 professional ENG news camera with its 1/2-inch frame transfer 403 × 480 pixel imagers. This cell phone camera has a lens that is about the size of the letter "O" printed here.

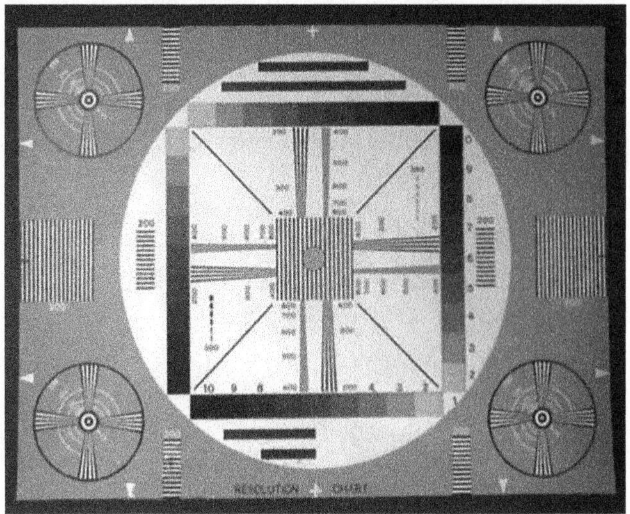

FIGURE 5.5-22 Test chart capture from a 1.3 megapixel cell phone camera with a 1280 × 1024 pixel native imager.

OPTICS

The optical system of every camera can be thought of as everything affecting the image before the light waves (photons) change to voltages (electrons). The optical systems of cameras possessing radically different image capture technologies often have a great deal in common. The core of the optical system is the lens; however, before the light reaches the lens, there are accessories such as lens hoods and optical filters that can improve the image by taking account of conditions in the scene, such as point light sources and other stray light, which may degrade the performance of the lens itself. As shown in Figure 5.5-23, behind the lens the optical system primarily involves built-in optical filters and the color separation system. Dichroic prism beam splitting with color separation was discussed earlier.

Optical Filters

Filters can be placed either in front of or behind the lens. Behind-the-lens filtering is common practice in video cameras because it keeps systems streamlined and there is less danger of a misaligned filter interfering with the imager. Filters behind the lens are usually of two kinds: permanent filters, which are part of the *optical block*, or those placed on a rotating *filter wheel*.

Common fixed filters include:

- Infrared cut (IR) filters, which protect the imagers from infrared waves, to which solid-state imagers are very sensitive;
- Anti-aliasing filters, which prevent high-frequency details that might cause aliasing in the video signal from being picked up.

Other behind-the-lens filters are normally placed in a filter wheel. The filter wheel makes changing of filters easier. It is important to remember that placing any plane of glass behind the lens will change the back focus, so most filter wheels include one clear filter to maintain consistent back focus when filters are removed. Filters on filter wheels include:

- *Neutral density* (ND) filters for adjusting the exposure without setting iris or gain;
- *Color correction* filters for shooting outdoors with tungsten balanced imagers;
- Effects filters including *cross* or *star* filters as well as *diffusion*.

Filtering is also used for color separation in the imager architecture, but this should not be confused with general optical filtering, which is no different than it has always been in motion picture or still-frame film photography.

Lenses

In principle, a lens can be as simple as a pinhole but it can also have almost unlimited complexity. Television lenses are always comprised of multiple *elements*, and the best-quality lenses can easily outstrip the price of even an expensive HDTV or digital cinema camera.

The two key parameters for a lens are its *focal length* (fixed or variable) and its *aperture*.

Focal Length

Lenses with a fixed focal length are referred to by this parameter (see Figure 5.5-24). Variable focal length *zoom* lenses are referred to by their maximum and minimum focal lengths or by the minimum value and the *zoom ratio*. The actual measurement of the focal length (typically in millimeters) is the distance from the principle plane (optical center) of the lens system to the focal point (the point at which light rays converge when the lens is focused on a subject at infinity).

The focal length determines what viewing angle of the scene will be focused on an imager of a given size, so the choice of focal length, combined with camera placement, determines the point of view for the scene. Shorter focal lengths capture a *wider angle* of view, and longer focal lengths (tending toward *telephoto*) capture a narrower field of view. Increasing the focal length seems to bring the scene closer, as a smaller viewing angle is captured by the same size image plane. Adjusting focal length with a variable focal length lens is referred to as *zooming*. An experienced camera operator will have an intuitive feel for what portion of a scene a given focal length captures; however, when the imager size is changed the relationship between focal length and field of view is changed. A smaller imager is looking at a fraction of the image a larger imager would capture. When these images are shown on the

FIGURE 5.5-23 Camera optical system.

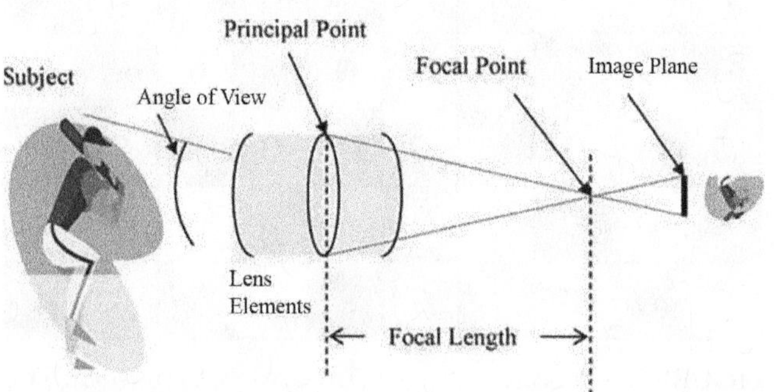

FIGURE 5.5-24 Focal length.

same size monitor, the smaller imager picture will capture a viewing angle similar to what the larger format system might have produced with a longer focal length lens.

Aperture

The lens aperture determines the amount of light that passes through the lens and is a fundamental characteristic of the lens design based on the size and arrangement of the lens elements. In most lenses, the amount of light passed can be controlled with an adjustable *iris diaphragm* and is referenced by an *f-number* or *f-stop*. The term *t-number*, with slightly different meaning, is used on many lenses intended for electronic cinematography, as t-numbers are usually used in the movie industry. This adjustment is referred to as *aperture, iris,* or *exposure*. The f-numbers indicated on the lens are normally specified in one stop increments, and each change from one f-number to the next represents halving or doubling the amount of light passing through the lens. The smaller the f-number, the more light reaches the imager. When the lens is set to its lowest f-number, this is referred to as *full aperture* or *wide open*. Increasing the f-number by one increment (one stop) cuts the light in half. This is equivalent to reducing the camera gain by 6 dB or applying 0.6 of neutral density (ND 0.6) filtering. Camera sensitivity is often indicated by the f-number; in this case, the specification will normally specify the lux rating (light intensity), such as 2000 lux, that is needed for the camera to produce good-quality pictures at that f-number.

Front and Back Focus

Front focus is the operational focus adjustment for the lens and is used to bring the subject matter in the scene into sharp focus. This means the proper image plane in the scene is being focused on the imager plane in the camera. For this to occur the image plane has to be the correct distance from the lens and proper focus on the imager itself is set with the back focus, which is a setup adjustment (see below). When the back focus is set properly the front focus should track throughout the zoom range and the focus marks (the distance indicators on the lens) should represent real distances from the camera to the subject.

Depth of Field

Whereas front focus will set the focal plane, the *depth of field* determines if a given object will be in sharp focus. Depth of field is the range of distances from the camera that are simultaneously in focus. A shallow depth of field means that small ranges of distance, before and behind the focal plane (perhaps even a fraction of an inch), are in sharp focus. The depth of field is proportional to the f-number (F_{NO}) as well as the object distance (l), and is inversely proportional to the focal length (f). The formula for depth of field is a bit complicated for practical field use:

$$[\delta * F_{NO} * l^2] / [f^2 - \delta * F_{NO} * l] + [\delta * F_{NO} * l^2] / [f^2 + \delta * F_{NO} * l]$$

where δ refers to the "circle of confusion," which is determined by the smallest circle that can resolve as a point, which is important to know in terms of the near limit and far limit in relation to the object distance. The formulae for the far and near limits of depth of field are as follows:

$$D1 = [\delta * F_{NO} * l^2] / [f^2 - \delta * F_{NO} * l]$$

$$D2 = [\delta * F_{NO} * l^2] / [f^2 + \delta * F_{NO} * l]$$

Depth of field calculations in the field, if needed, are typically made using tables. If more depth of field is required, then it is necessary to shoot with a higher f-number or a wider focal length; if a shallower depth of field is required, then the iris must be opened if possible or a longer focal length selected. Iris settings can often be compensated for by adjusting gain, light levels, or neutral density filters; different focal lengths can achieve the desired framing by changing the position of the camera. It is worth remembering that, because different size imagers capture different angles of view with like focal lengths, a similarly framed shot will use a different focal length and therefore exhibit a different depth of field for a given f-number. For this reason, it may be difficult to achieve a shallow depth of field with a small imager.

Back Focus Adjustment

Back focus for film camera lenses has traditionally been set in the shop using a collimator, and these lenses are also designed to maintain consistent back focus, as it cannot be checked in the field. Video lenses, on the other hand, have a user-adjustable back focus so they can be checked and adjusted in the field. The back focus adjustment technique normally used takes advantage of the difference in depth of field and depth of focus as focal lengths are changed. At longer focal lengths, front focus is critical due to the shallow depth of field; at wider focal lengths, back focus is more critical.

A long focal length with a shallow depth of field affords a more precise front focus setting and does not require precise back focus. A wide-angle focal length has a deeper depth of field but it requires a precise back focus setting. To set back focus in the field, the iris should be wide open to minimize the depth of field. After setting front focus zoomed in, the lens should then be zoomed out to a widest position. If the back focus is off, the image will go soft and it should be adjusted on the lens. At a wide angle, the back focus becomes critical, so a fine adjustment can be made. It is often necessary to repeat the process of focusing the front of the lens zoomed in and the back focus with the lens wide to achieve a good back focus. ENG cameras require frequent back focus checks because they are often handled roughly, but studio cameras also require a resetting of back focus from time to time, especially if they experience changes in temperature. A *Siemens star* (see Figure 5.5-25) is a useful tool for setting back focus because sharp focus can be easily judged at many focal lengths. Any chart or object that has a variety of object sizes can be used for back focus as long as it provides multiple-sized high-contrast figures on the same focal plane.

FIGURE 5.5-25 Siemens star test pattern used for adjusting back focus.

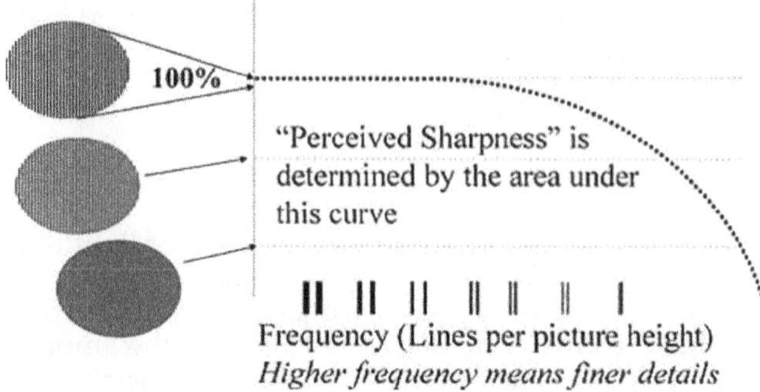

FIGURE 5.5-26 Modulation transfer function.

Comparing Lens Quality

Camera lenses price ranges are a function of lens speed (or the wide-open f-number) wide-angle capability, zoom ratio, MTF, and overall image and construction quality. A lens capable of transmitting more light is clearly of more value, as is a lens capable of delivering a greater range of focal lengths. Achieving high specifications for all of these parameters simultaneously becomes increasingly difficult for lens designers, and small improvements can lead to large price changes. Comparing available lenses it is easy to see how the key lens specifications impact the cost. With the introduction of HD cameras, some confusion arises because lenses with similar focal length and speed specifications may vary greatly in price. This is a function of lens image quality, which is somewhat more difficult to measure.

Modulation Transfer Function

The *modulation transfer function* (MTF) (Figure 5.5-26) is a measure of an optical system's perceived sharpness. It is a more comprehensive parameter than resolving power, which is measured optically as line pairs per millimeter or as lines per picture height on a video signal, both of which deal with the maximum frequency signal that can be resolved. MTF takes into account the fact that, even if an imager is capturing a high-frequency image, it may not be capturing it at its true contrast level. In other words, a series of black-and-white stripes might be resolved in a given system but it may be resolving as dark gray and light gray rather than black and white. The percentage of full contrast level for a given frequency is known as the modulation depth.

Modulation transfer function plots the modulation depth as a function of the frequency (size of the objects to be resolved). This parameter is called *frequency* because it can describe the frequency of a periodic signal on a horizontal scanning line imaging black-and-white line pairs. As is shown in the figure, the MTF exhibits a roll-off as the Nyquist frequency is approached. Sharpness is the measure of what a *human observer* will describe as *sharp* when viewing an image, and this is actually a function of the square of the area under the MTF curve. That means that the perceived sharpness of an image is determined by both the resolution and the contrast as combined in the MTF. The overall MTF of a displayed image is determined as an aggregate of the entire system, including the lens, camera, video format, recording algorithm, and display technology; therefore a high-quality lens can have a positive effect on any video system, but it will have diminishing returns on a system where the imager recording format or display has a limited MTF.

Lens Performance

The stated MTF for a lens may not be useful for comparison unless something is known about the consistency of the lens's performance. Lens designers will always have more difficulty achieving good MTF at the extremes of the lens parameters, such as iris and focal length. At the extreme wide, long, or wide-open settings, lenses may experience a reduced MTF, as well as a lesser degree of consistency across the image. Diminished MTF at the corners is a typical problem with zoom lenses, particularly at the widest focal lengths. The number and severity of lens artifacts are perhaps more important than MTF when determining lens performance in most applications. As with MTF, the artifacts will become better or worse depending on the focal length and iris setting, so time and care should be taken when looking for them.

There are five basic lens flaws known as aberrations (*Seidel's five aberrations*). *Spherical*, *chromatic*, and *astigmatism* aberrations appear as zones of the image that may exhibit color fringes or are not completely focused at some lens settings. *Curvature of field* means that the focal plane is not flat, and *distortion* is exhibited when an image that ought to be rectangular exhibits curved edges so as to appear like a *pincushion* (concave lines) or *barrel* (convex lines). All of these have to be corrected or reduced through complex lens design, and the level of success helps determine the quality of the lens. Failures contribute to lens inconsistencies and chromatic aberrations, which are observed as color fringing on contrasty images. Chromatic aberrations tend to occur near the edges of the image and at extreme settings. Another optical problem is known as *flare*, which is a catchall term for unwanted light reflected off surfaces inside the optical system. Flare can prevent an imaging system from producing a true black, and it adds a light bias because the reflected light typically has some tint.

In evaluating a particular lens, care should be taken to test it at focal lengths and exposures where it is most likely to be used. Some aberrations at very wide angles might be tolerable if the lens performs well in other circumstances. Lenses for ENG use can tolerate flaws at extreme settings (such as an inconsistent f-stop at long focal lengths) in order to have access to more focal lengths and exposure settings. Digital cinematography lenses may limit the zoom range to maintain consistent exposure throughout the range.

It may seem counterintuitive at first, but the larger the image format, the easier it is to manufacture a good-quality lens. This is because the large imager can function within wider optical tolerances. Line pairs per millimeter can only be compared for two lenses used on the same size imager, as a smaller imager will necessarily be enlarged more to be shown on the same display, so any flaw will also be blown up. Small image format cameras typically use lower-cost lenses, so it should come as no surprise that lens aberrations will be far more common in these cameras.

TELEVISION PRODUCTION CAMERAS: THE MOVE TO S35MM LARGE-FORMAT SINGLE IMAGER CAMERAS

In the late 1990s, studio cameras evolved from using 2/3″ CCD imagers with standard definition pixel arrays numbering in the 800–900 (h) × 480 (v) to the new high-definition 2/3″ 1280 × 720 or 1920 × 1080 format native imagers. The new 2/3″ HD imagers doubled the spatial resolution of the older SD imagers, resulting in smaller, more densely packed photosites with the attendant issues of reduced sensitivity and dynamic range, both of which can, to some extent, be mitigated. The quartering of a given pixel's photosensitive area results in the loss of two stops of sensitivity; likewise, the photosite's storage well size/depth is also diminished, reducing the dynamic range by some two stops. Two stops of sensitivity loss is a huge burden to overcome, requiring four times the light for the same resultant signal levels. Four times the light in a studio environment means massively increased lighting loads and similar HVAC increases. Given time, however, technology improves; pixel-based micro-lenses and higher fill factors help with sensitivity, while silicon doping profile changes to deepen the pixel storage "well" help offset dynamic range deficits. Today's 2/3″ HD cameras offer quite acceptable sensitivity and sufficient dynamic range for most applications.

The shift from the 2000s-era 2/3″ HD studio and EFP production cameras, with their 2.2M pixel 1920 × 1080 HD native imagers, to today's larger s35mm single imager sensor-based cameras has brought many changes, challenges, and benefits for production. One immediate result came from the use of a native s35mm imager, much larger than that of a 2/3″ studio camera; this provided the much admired shallow depth of field, and used the same PL 35 mm lenses from earlier film productions.

Currently, the majority of the cameras used for "prime-time television production" employ large-format imagers, much of this driven by the aesthetic of the s35mm format's shallow depth of field/subject isolation, as used for cinematic production and narrative storytelling. In reality this move is a simple reinforcement of the subjective appeal of the 35 mm film acquired programming of previous years. Many aspects of the "film look" are attributable to the large-format image capture and 24 fps acquisition frame rate. It's a learned, if somewhat nostalgic, attribute, but nevertheless one that's respected and deemed crucial for all serious productions.

The technology for manufacturing large-format camera imagers involved the maturation of many aspects of imager design and large-scale wafer/device fabrication, much of this coming from the LSI evolution and manufacturing technology trends and the experience gleaned from making large silicon wafers. At the same time, the digital single lens reflex (DSLR) industry was evolving and rapidly replacing the traditional 35 mm film-based SLRs. Those early DSLRs often came with crude "video" acquisition modes or options. Some ten years later those cameras and their associated video capture and recording modes have greatly improved, and today DSLRs play a significant role in 4K television episodic and cinematic production.

Figure 5.5-27 shows the imager size comparisons for an Arriflex 65 mm imager (OG), a Canon 5D Mk3 full frame 35 mm 1.85:1, the VariCam s35mm 4K imager, and the "standard" 2/3″ HD 16 × 9 sized imager. As the imager format size changes, so do the optics. S35mm PL lenses are significantly larger and heavier, and bring limitations to everyday expectations like focus ease or

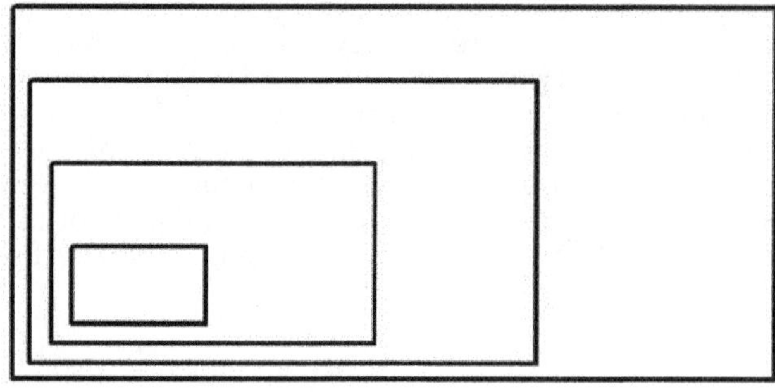

FIGURE 5.5-27 Comparative sizes of 65 mm, 35 mm full frame 1.85:1, s35mm (3perf) and 2/3″ HD imagers.

zooming; 65 mm lenses are few in number and fill a very specialized niche market. (See Figure 5.5-28.) 2/3″ HD studio cameras often use 20–40x box zoom lenses; sports productions like the huge 101x zooms because of their "reach." Meanwhile, 35 mm PL zoom production lenses are often limited by weight, cost, and physics to less than a 4 or 5 times zoom ratio. Examine the examples of the s35mm PL 24–290 mm lens, and its much smaller 22x 2/3″ ENG/EFP cousin. PL lenses are large and costly, and few have any integral servo drives for zoom or focus. There are no 50x1 PL zoom lenses; cost and physics will help that remain the case.

Today 4K and even higher resolution s35mm imagers are now common; the photosites of these imagers, even at 4K resolutions, are larger than the photosites of the 2/3″ imagers used in the 2/3″ HD cameras. The 5 µm square pixels of a 2/3″ imager from a 2000-era EFP/studio camera could be expected to provide around 10 stops of dynamic range and effective ISOs (International Standards Organization) in the 500–800 range. Today's s35mm MOS imagers designed for 4K production can achieve an impressive 14 stops of dynamic range or more, and native ISOs range from 500 up to an amazing 5000 (as seen in the VariCam 35's 800 and 5000 dual native ISO imager). Please note the emphasis on *native* ISO; electronic gain is not being used as an ISO boost.

Modern s35mm large-format MOS imagers are designed for high spatial resolution and wide dynamic range image capture. The performance of the current crop of 4K imagers now rivals what was possible only a few years ago with the best 35 mm film negative stocks. Despite some dubious marketing claims, large-format imagers are still subject to the laws of physics, as various performance factors often conflict. Some hard choices have to be made as to which particular parameter is the most important for a specific application or camera product.

Large imagers are generally more sensitive. They also have a higher dynamic range and suffer less from diffraction. However, their lens and optics choices are more limiting, complex, and expensive. Smaller imagers by comparison are usually less sensitive, have a lesser dynamic range, and suffer more from diffraction, yet their lens choices are much wider, more affordable, and practical. Imager format size also determines aesthetics like the depth of field of an image, the bokeh of an iris pattern on the out-of-focus highlights, or the look of an older "portrait" lens. S35mm is wonderful for selective focus and subject to background isolation, but it is not so good for sports where exact focus is required at all times.

Examine Figure 5.5-29 for some examples of the pixel size and the light-sensitive surface area of pixels from various format modern camera imagers. Given the same light levels, and assuming the same quantum and pixel readout efficiencies, the surface area of the pixel basically determines the sensitivity. Double the pixel size and you quadruple the sensitivity, simply because the pixel catches four times the light. The small 1.5 µm square pixels in the iPhone 6's 4K camera work fairly well in bright light, but in challenging lighting situations, we all know the limitations of such devices. The much larger 36 sq. micron area of the 4K s35mm VariCam imager means that it is 16 times more sensitive than the iPhone 6; 16 times is exactly 4 stops.

To overcome the low sensitivity of some imagers, electronic gain (amplification) is used; this artificially boosts the sensitivity, but also adds noise. Dynamic range can be improved by HDR techniques such as those used in DSLRs, namely the blending of several bracketed exposures. Such techniques work fairly well for still images; however, two or more different exposures taken at disparate time intervals when combined can generate motion trail issues; imagine nighttime fireworks shot this way! The useful dynamic range performance of a cell phone camera may only be 5–6 stops; add the small photosite surface area/very low relative sensitivity, and all sorts of processing is required to provide a usable image. Few of those processing tricks are applicable or even acceptable in today's production cameras.

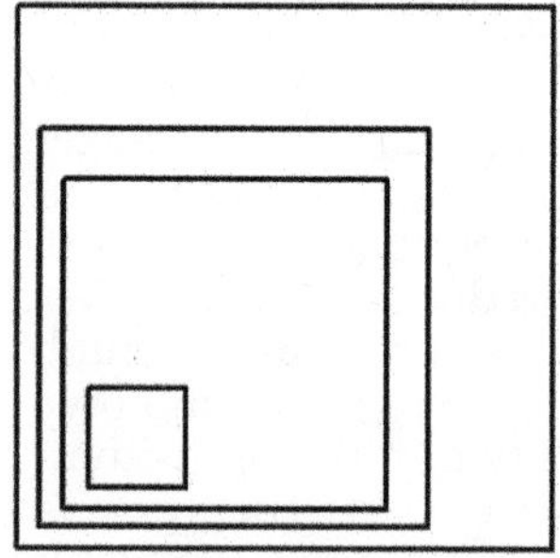

FIGURE 5.5-29 Comparative pixel sizes for various common imager formats.

FIGURE 5.5-28 Angenieux Optimo 24–290 mm 35 mm PL zoom (left). Fujinon 22 x ZA22x7.6 2/3″ EFP/ENG lens (right). Interestingly, the size and cost are similarly scaled!

Large-format production cameras are used to create imagery for high-end television and cinematic productions. As such, their performance needs to be exemplary in most of the following image attributes: high sensitivity, wide dynamic range, low noise, high spatial resolution, high frame rates, accurate rendition, and low imager smear/rolling shutter artifacts. Additionally, careful attention is needed for the optical low pass and infrared filtration. The imager is part of a carefully designed and developed optical system; any deficiencies in the imager design or fabrication will manifest as unwanted artifacts later. Technical requirements like high dynamic range raise these specifications to new levels. A 14 stop capable imager has the technical capability of discriminating 16,384 levels of grayscale tonality.

New Imagers, New Workflows

Large-format 4K s35mm cameras/imagers bring changes in signal formats and workflows. An imager capable of 14 stops of dynamic range does not fit well into live television, as TV's Rec. 709 EOTF provides only 7–8 stops of display system dynamic range. So workflows much like shooting 35 mm film are now employed. The two basic variations for these new workflows are a logarithmic (log) based wide dynamic range/wide gamut approach followed by mild mezzanine-level compression. The alternative is a RAW data workflow where the imager's wide dynamic range signal values are collected and processed individually, effectively reading and storing the individual RGGB pixel values from the imager's Bayer CFA quads.

Often, to reduce the signal processing throughput and data storage sizes, some form of perceptual encoding is used to reduce the 14- or 16-bit linear A/D conversion data from each pixel to a more pragmatic 12- or 10-bit form. That perceptual encoding usually takes the form of a "log" style OETF (Opto-Electric Transfer functions). Twelve-bit "log" encoding processing provides the equivalent of at least 14-bit linear, so the process is quite transparent and greatly improves the subsequent recording efficiency. Imagers with greater than 14-bit A/Ds offer very few performance benefits over 14-bit devices due to the influences of various natural noise sources outside of the cameras' design influence. Shot noise is an interesting such example.

Imager Noise Sources

Shot noise, flicker noise, and other random noise sources are often natural in origin and not easily correctable by technology advances. These noise sources set the low-light/dynamic range performance of any given camera. One interesting example is shot noise (a.k.a. photon noise); this is a natural phenomenon, a statistical fluctuation in the number of photons arriving on the imager in any one-time interval. This noise is statistical in nature and follows an approximate square root function. For example, a million photons per frame would have a square root of 10^6, or 10^3, photon randomicity; 1000/1,000,000 is 0.1% "noise," very low and not visible. At much lower light levels, for example, 1000 photons/frame, the noise will be the square root of 1000, or ±32 photons; that's a significant noise component. Near the bottom of the imager's sensing capability, but at a still detectable and discernible 10 photons/frame, the shot noise component will be ±3 photons, or 66% noise. That's very visible. Shot noise is a significant issue where wide dynamic range cameras are used at low light levels. Choices need to be made as to how and when to filter, remove, or reduce that noise. In-camera filtering/noise reduction may have negative impacts on the image quality; 3D recursive filtering will add motion blur and smear to adjacent frames, while 2D filtering will usually soften imagery if used in excess. In many such cases the choice is often to not process the noise in-camera, but to use noise reduction later in postproduction where it is more controlled and can be best optimized for the scene content.

Figure 5.5-30 shows the very wide dynamic range capabilities of the VariCam 35; note the 16-stop tone mapping capability of the V-Log OETF. The camera's 4K s35mm imager is easily capable of in excess of 14+ stops of capture range, which is more than 16,384 steps of discernible grayscale tonality.

RAW workflows used in today's production cameras vary due to the philosophy and design characteristics of the imager and workflow. Some would argue that the RAW data from an imager is best kept untouched and not modified in any respect—the suggestion being that as de-Bayering techniques and noise reduction capabilities improve with time and research, the camera's original RAW content could once again be processed, and this time additional image information can be recovered. That is laudable, but in many cases not the optimal approach. Imagine a camera exposed to dimmed tungsten lighting, for example, 2800°K. The camera's imager's color filter array (CFA)/native color temperature may be 5600°K (often a Bayer CFA design choice), so exposed to 2800°K light, the red pixels of the Bayer CFA may well be overexposed (too much red in dim lighting) and the blue Bayer pixels would be very much underexposed . . . so a significant color channel imbalance has occurred and will be passed through the color pipeline in post. Hopefully the director of photography or digital imaging technician will correct the exposure to avoid any channel clipping—often a problem with the newer LED-based lighting with its narrow and spike-like spectral emissions. That exposure correction lowers the underexposed blue channel even more.

Some manufacturers process the RAW signal to correct for the white point of the image (best RGB channel match) and also for ISO/gain processing, the premise being to modify the imager data in an uncompressed linear light/full signal level environment before any compression or truncation to lesser word lengths is performed.

Should that 12-bit log-encoded Bayer RAW A/D data be clipped in the red channel and only fill half the data values in the blue, it then poses a problem for later color grading and processing. RAW workflows use metadata extensively to control the signal processing in post/grading. However, if the original RAW data is compromised, as in the case of a severe lighting/

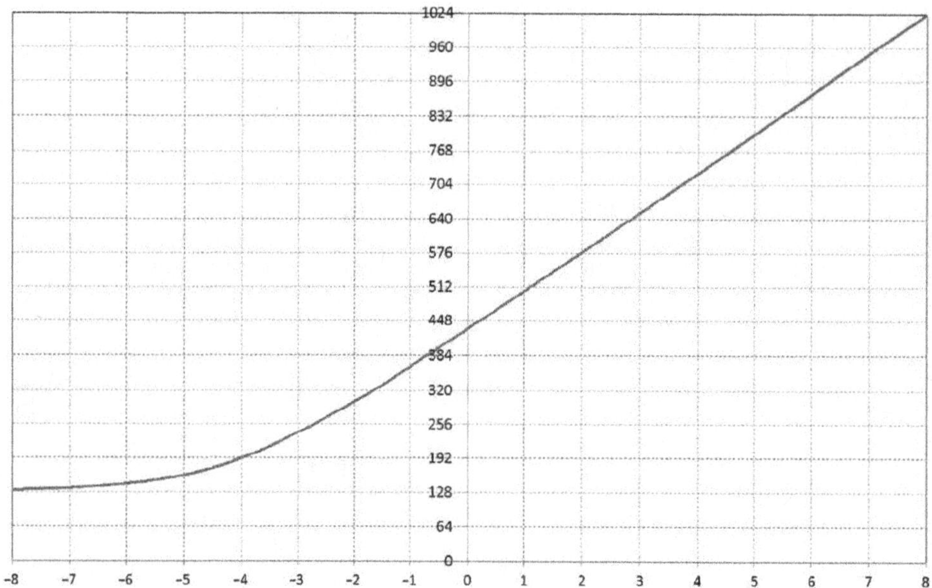

FIGURE 5.5-30 Panasonic V-Log OETF as used in the VariCam 35 camera series.

imager color mismatch, some channels may clip whilst others are underutilized. That limits grading and color correction capabilities in later workflows.

RAW workflows are metadata rich, and vary in complexity. In addition to perceptual encoding, RAW signals are in some cases compressed to make RAW workflows and large data volumes/processing more practical. Log-encoded 12-bit uncompressed RAW is very transparent, but comes at a price: 4096 × 2160 × 24 × 12 = 2.548 Gb. Some choose to apply compression technology to reduce the RAW data bit rates to more manageable levels; schemes vary as to the transforms used for this compression, often JPEG 2000 or MP4, or more recently H.264. Regardless, the compression used in such examples is lossy, and at the 14 stop levels of signals involved, much care and attention is needed to avoid unwanted artifacts appearing. JPEG 2000 will exhibit areas where detail/intra-frame noise is locally softened/blurred. MP4 or H.264 discrete cosine transform (DCT) processing may add blocky artifacts if stressed by severe image content. (See Figure 5.5-31.)

Testing of any camera, its signals, and post/grading workflow is advised in *all* production situations and should use representative content.

Allied with the huge increase in dynamic range, modern imagers, by means of a carefully optimized matrix, can have a very wide taking gamut. Figure 5.5-32 shows the classic CIE 1931 chromaticity space. Superimposed are color gamut maps for ACES, ITU-R BT.709, P3DCI, ITU-R BT.2020, and VariCam's V-Gamut. Note that V-Gamut (the native taking gamut of the VariCam 4K cameras) is larger than Rec. 2020. This wide taking gamut covers a very large portion of the human visual color sensitivity, and almost all of the Pointer Surface Colors; indeed, V-Gamut is about 200% larger than the well-known ITU-R BT.709 color space.

Modern postproduction takes advantage of the large gamut and dynamic range of the captured image, and, by the careful use of 3D LUTs and human-optimized grading controls, can adjust the color space, scene contrast, and colorimetry to produce the imagery desired in a wide variety of formats. This process is very efficient, as many differing deliverable masters can be produced by a simple render process to suit various applications. Examples are UHD HDR (ST 2084), "good old-fashioned" HD in Rec. 709, or a cinema DCP in P3. The flexibility of a mastering nondestructive render process is a welcomed improvement over the film transfer and telecine color correction process of yesteryear.

With the ability to make whatever deliverable is required from the log, wide gamut master, and wide dynamic range camera master, it was not long before on-set workflows started to evolve to include on-set preview capabilities by means of adding a LUT to a camera's HD monitor output. Adding a 3D look-up table (LUT) and any chosen color decision list (CDL) trims would provide an accurate means of previsualizing the equivalent of a fully graded deliverable, all the while on set and without modifying the all-important camera's original content. This very efficient and logical process can save considerable time and money in the grading room. Metadata derived from the in-camera preview process is embedded in the camera's recordings and later used in the grading process to add the required "look" to the content. (See Figure 5.5-33.)

This preview grading process has now evolved to the extent that, in the VariCam 35 family of cameras, full wireless control of the in-camera previsualization process of the finished grade is possible on set. The associated metadata—file names, time code (TC), and so forth—is always embedded in the camera's recordings. To even further enhance the workflow options,

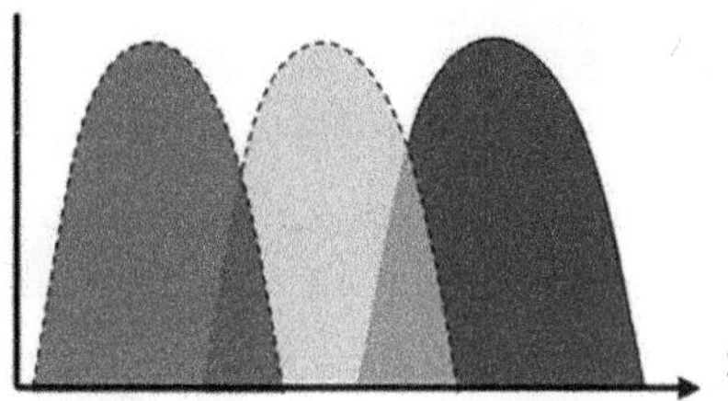

f

Optimal imager / light
color temp match
Balanced RGB levels

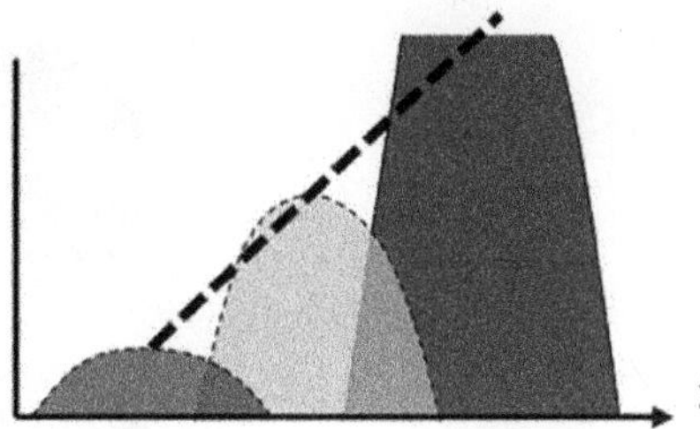

f

Channel imbalance due
to high light color temp
Clipped blue, low red

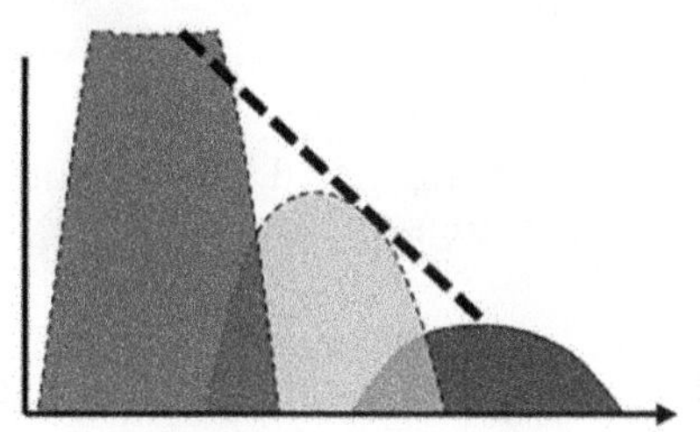

f

Channel imbalance due
to low light color temp
Clipped red, low blue

FIGURE 5.5-31 Examples of mismatch of RGB data values in extreme lighting.

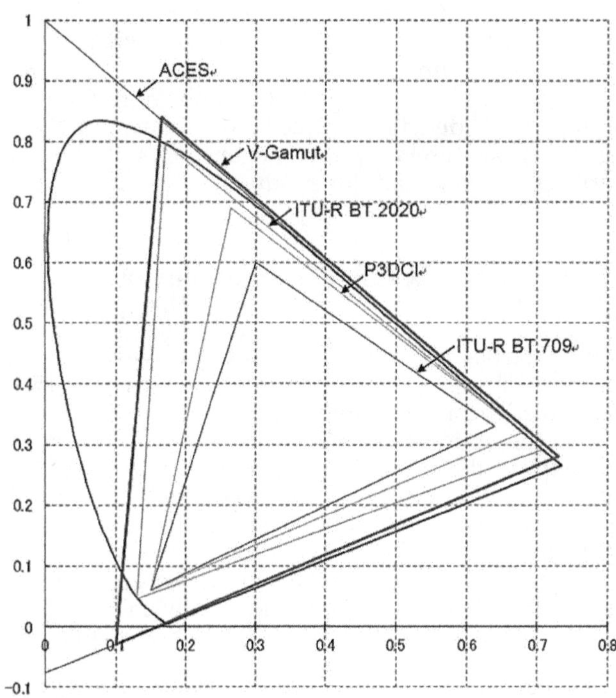

FIGURE 5.5-32 CIE 1931 with overlays for ACES, ITU-R BT.709, P3DCI, ITU-R BT.2020, and Panasonic V-Gamut.

a sub-recording, usually 2K or HD, is generated from the 4K/UHD camera content. That 2K sub-master can have the desired 3D LUT and CDLs baked in if required. Editorial now has instant access to a 2K sub-master with baked-in grading for final color and edit. An additional low-resolution but very efficient MOV proxy is generated for remote viewing, even via wireless FTP.

Lastly, to add the full benefit of a hierarchical workflow, it is possible to also simultaneously record uncompressed V-RAW in addition to the 4K, 2K, and associated proxy. Workflow choices are just that, choices, but more options add flexibility to any production system.

Imager design cannot hope to accommodate let alone excel in all aspects of performance; imagers are improving as to sensitivity, dynamic range, and even higher frame rate operation. In some cases, a newer technology provides improvement for specific artifacts, one example being the thorny issue of "rolling shutter" versus "global shutter." The majority of imagers in use today use a rolling shutter readout process; this simplifies the pixel structure electronically, and increases the fill factor and thus sensitivity. The downside, however, is that the readout of the image data is real time, and any camera motion during the readout process will result in a spatial shift in image content of the frame. (See Figure 5.5-34.) Examples are

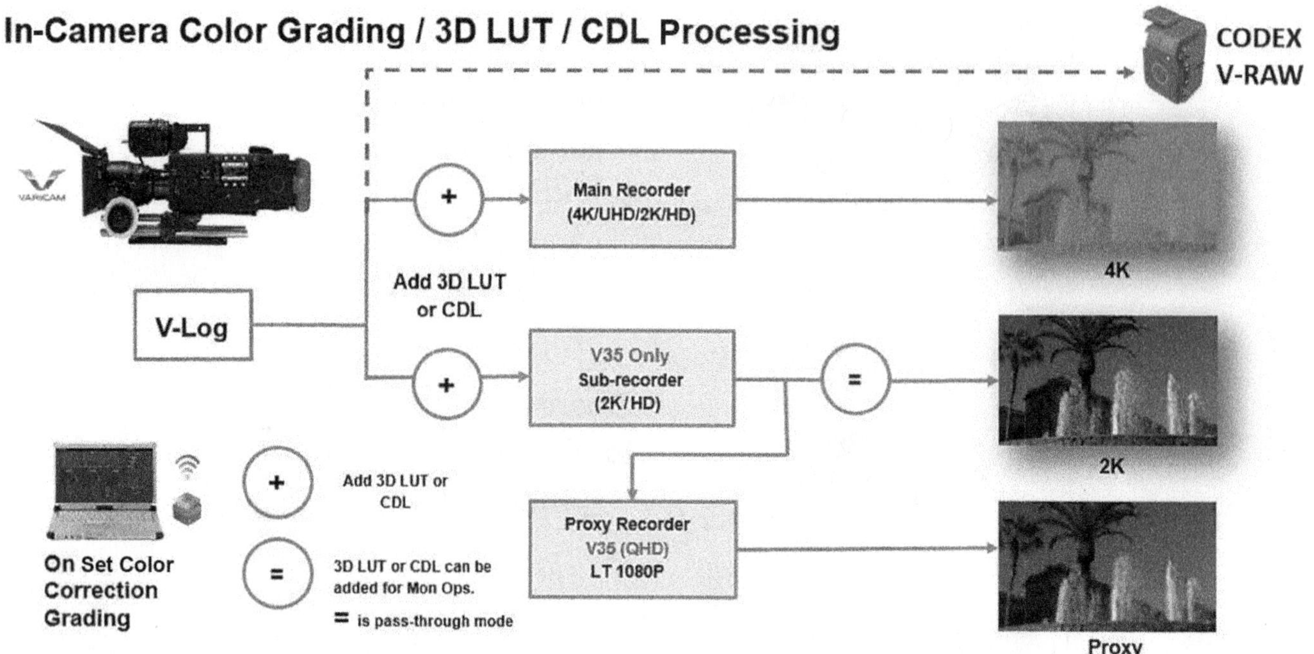

FIGURE 5.5-33 VariCam 35's parallel 4K RAW, 4K mezzanine, 2K mezzanine, and low-resolution proxy recordings. (Note the option to bake in 3D LUT/CDLs to the 2K and proxy recordings, or not.)

FIGURE 5.5-34 Example of partial frame flash illumination during readout phase of rolling shutter imager.

well known as tilting verticals, with the most apparent being that partial frame illumination caused by use with strobe lights/flash guns active during a portion of the imager's readout time.

Global shutter based imagers are available, and improving in performance. To date, however, their performance, while exemplary in terms of full frame buffered readout and motion-related skew elimination, is not quite matched by certain parameters deemed more critical for production. The additional electronic components required for the image storage and readout on each pixel can result in 1 or even 2 stops loss of dynamic range; that extra storage can also lead to fixed pattern noise (FPN) or highlight smear issues. The FPN is very much a concern with the huge dynamic range of current production cameras. A small FPN of 0.01% may be quite invisible on an ENG camera working in HD and Rec. 709's limited contrast and display capabilities. However, for a 15 stop capable production camera, that 0.01% FPN is a serious detriment and may not be acceptable, even after in-camera black level shading correction, as grading of the linear or log-based imagery will greatly stretch the image contrast and make the FPN more visible.

Time and technology will come to the rescue; in ten years, global shutter imagers will be the norm and the performance deficiencies rectified. Today, however, most manufacturers use the imager most suited to the particular application; for production, absolute image quality is the predominant quest.

Small-Format 4K Cameras, the Irony of Diffraction

Diffraction is a phenomenon whereby a wave (in the case of optics, light) is "bent" around a structure or sharply defined edge, or some similar mechanical abrupt transition in the optical path. Diffraction causes interference wave fronts that result in the cancellation and blurring of the image detail. The interference effects cause small details in the image to soften and reduce in amplitude; this is the result of the Airy disk impulse spreading and crippling of the system MTF. Diffraction occurs in all optics; the effect is, however, markedly increased on smaller formats such as 1/3″ or 2/3″ imagers and with any increase in native system resolution.

The recent move to large-format 4K production cameras has set the expectations for 4K production and eventually 4K television. Today, 4K cameras are shooting exquisite imagery with enormous dynamic range and a huge color palette, which is then graded to become the programming and movies you see and appreciate daily. An s35mm 4K camera actually places relatively low demands on the optical path, and combined with the accepted "gentle look" accomplished with the use of contrast-reducing diffusion filtration, the hero or star of the movie looks great, and the filmic experience is maintained.

By contrast, 4K ultra-sharp sports at 60 frames per second is almost the opposite of the prime-time experience. It is sharp and detailed and very unforgiving; it's hyper-real! The problem for 4K production comes with the need to generate 4K 60p live for end-user viewing. Sports needs large zoom lenses capable of fast shock zooms and snap focus to deliver the "game" experience. S35mm lenses do not and cannot provide that performance, so what to do? One solution is to use B4 mount 2/3″ HD lenses with some sort of PL to B4 mount and optical format conversion to match the s35mm camera with the lens required.

As can be seen from Figure 5.5-36, the optical footprint of 2/3″ 16 × 9 optics is far from that required to fill the 35 mm imager. Any adaptor will be required to expand the 2/3″ B4 lens exit pupil from a mere 0.378″ × 0.212″ to the native 0.907″ × 0.49″ s35mm imager of the 4K camera. That optical projection system is at best a 2.2:1 expansion, and with it comes 2.6 T-stops of light loss. This is a loss of 80% of light, and the optical performance of the format adaptor and 2/3″ HD lens is often sadly lacking for 4K camera systems. The results are a poor compromise at best. It's not 4K, and the optical light losses are a big problem in low light levels, such as in nighttime sports.

The sports experience of fast action, fast zoom and focus, and the close following of the action across the field is tough to change. It's a 30-year learned expectation/experience. Is our solution 4K cameras using a native 2/3″ imager format? This sounds like a very logical solution—same zoom and fast focus, same depth of field, huge zoom ratios, everything the same as HD 2/3″. Alas, here diffraction enters and spoils things. A 4K native 2/3″ camera system places huge demands on the optics and physics involved, and although possible to achieve, certain basic issues arise.

A 4K native 2/3″ single imager system comprised of a single-chip 4K imager (Bayer CFA) and a more advanced, yet costly, three-chip 4K 2/3″ prism beam splitter system are both possible. The single imager 2/3″ 4K imager would be heavily compromised as to sensitivity and dynamic range. The 3 times 2/3″ 4K imager beam splitter array is interesting; it promises 3 × 4096 × 2160 RGB (4:4:4) processing coming with co-sited pixels. However, the small optical format suffers from a serious 2-stop sensitivity loss, as well as limited dynamic range. The 2/3″ 4K optical system also runs into serious diffraction loss problems.

Diffraction, as mentioned, is optical format and lens aperture linked. Comparisons can be made to recent HD ENG formats using 1/3″ HD optics and 1920 × 1080 imager systems. The cameras would perform superbly at lens apertures of f1.7 to f5.6, delivering textbook HD resolutions and MTF. Above f5.6, however, things change. At f8 the MTF (modulation transfer function) would drop appreciably, by about 50%; by f11 the HD camera would resemble that of a standard-definition camera. Using the same cameras at f16, focusing is almost impossible to achieve, as the MTF is so low as to negate camera operator/lens/viewfinder (VF) focus control feedback. The images following are from a 1/3″ HD three-chip camera; these image/resolution examples can be linearly extrapolated to performance of a 4K 2/3″ camera of similar design—1920 HD to 3840 UHD is a simple 2 times interpolation exercise.

The WFM screen captures shown in Figure 5.5-35 show the huge MTF hit when the same camera/lens

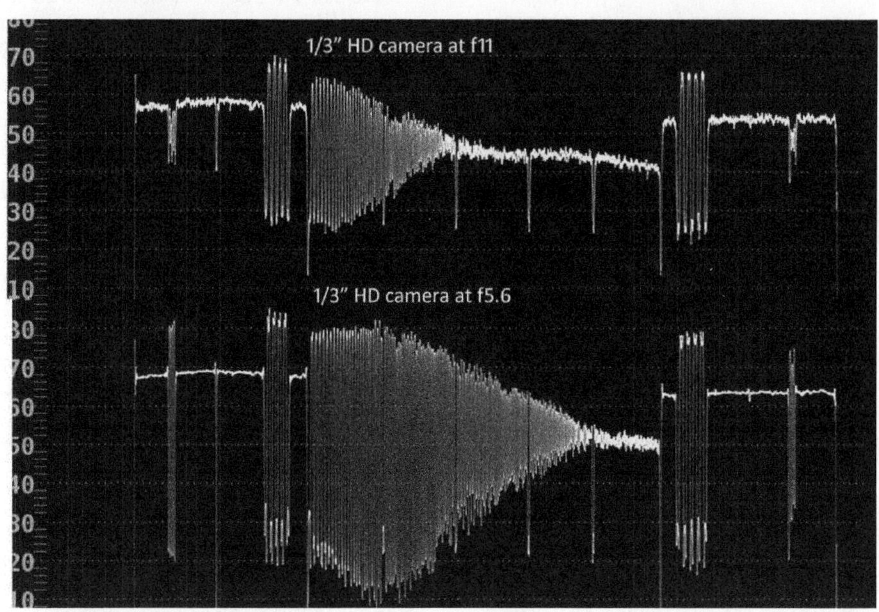

FIGURE 5.5-35 MTF versus lens aperture = diffraction losses.

is operated at f11 versus f5.6. The shift from f5.6 to f11 has reduced the 1920 × 1080p imagery to less than SD resolution. This is diffraction at work, and these results easily translate to the use of UHD (3840 × 2160) 2/3″ imagers on 2/3″ 4K cameras. Interestingly, the 2/3″ lens performance may even make the example of Figure 5.5-35 worse, as 2/3″ HD lenses are rarely 4K performance capable, so the 4K 2/3″ MTF results may be worse than the above.

Any 4K small-format (2/3″) camera will need to be operated at a very wide iris setting to avoid severe loss of detail and associated focusing issues. This is manageable for smaller lenses, such as those used for news and EFP. However, the large sports-style box lenses may ramp severely with zoom, often dropping the lens f-stop appreciably and thus suffering diffractive losses. New operational disciplines will need to be developed to ensure that any 4K small-format production does not run into diffraction-based limitations. Maybe some form of lens to camera feedback/automatic gain control (AGC) to maintain wider iris?

Large-format production cameras have revolutionized television and cinema production. The inherent image quality combined with advanced RAW or log-based mezzanine compressed workflows enable the creative choices to be made without any field acquisition compromise. Like film capture over the past 50 years, the camera originals have additional value for later retransferring or regrading, especially with the multitude of new display formats being added. High dynamic range remastering from the camera-original wide dynamic range content is a simplified process; so will be advances in wide color gamut. Users should, however, be aware of the camera performance basics, and understand their inherent limitations. Today's production community is awash with camera and workflow options. RAW workflows, once the dream of a few moviemakers, are now everyday occurrences. Choose carefully and knowingly.

4K AND ULTRA HIGH DEFINITION (UHD)

This section will discuss the 4K resolution standard from operational, evolutionary, and human visual system vantages. It is not intended as a reference for signal standards specifications; these can be accessed through available industry white papers by the appropriate standard-setting organization. It is also not intended as a source for 4K workflow or technical infrastructure design recommendations, which are beyond the scope of its current purpose.

The next widely adopted stage in a resolution standard will be 4K, generally denoting 4000 horizontal lines of effective sensor resolution. The transition is analogous to the one from standard definition to high definition in the early 2000s.

The benefits of 4K are fundamental. It is a major step forward in bringing the reproduction of motion imaging closer to emulating the biological sensory experience of the world around us.

The primary factors serving as impetus for 4K market saturation are consumer TV set prices and availability of content. Industry challenges for satisfying an emerging need for content are the need for efficient data infrastructures to process 4K footage, implementation of a broadcast transmission standard, and possible retooling of transmission infrastructure.

As is common to major innovations in history, the consumer delivery implementation of 4K is a slightly downgraded format from true 4K or DCI 4K at 4096 horizontal lines of resolution. This has generally become referred to as ultra high definition (UHD) at 3840 horizontal lines of resolution.

The 4K Standard

Although there are slight variances dependent on production, the two most prevalent 4K formats can be defined primarily by these resolutions:

DCI 4K 4096 × 2160

UHD 4K 3840 × 2160

Digital Cinema Initiatives (DCI) is an industry working group composed of six major film studios in the United States—Disney, Fox, Paramount, Sony, Universal, and Warner Bros. DCI's mission is to self-regulate various industry standards for digital cinema production. The inventor of 4K capture resolution may be regarded as Teledyne Dalsa. Produced in 2003 by a group of imaging veterans with significant film backgrounds, in the form of the Dalsa Origin camera, the sensor resolution was 4096 × 2048. UHD display technology surfaced prior to 2003 beginning with IBM and later NHK.

The term "UHD" is broad and includes 4K and 8K resolutions as well as high dynamic range (HDR), wide color gamut (WCG), and high frame rate (HFR) capabilities. Video cameras have been manufactured for professional use with UHD sensor resolution. Acquiring footage in UHD may provide increased efficiency in postproduction by skipping the DCI conversion to UHD process.

As 4K TV set sales continue to increase, the broader capabilities of UHD may become more familiar as a consumer specification. Going forward, the areas discussed as 4K will also generally apply to UHD from an imaging standpoint.

Benefits and Challenges for Production

Depth of Field

Operational focus will be a common and expected difficulty for 4K production during acquisition. This is not only due to larger sensors with less depth of field but also to increased resolution. With a 4K camera using a 2/3″ chip, focus sensitivity can still be estimated to increase by a factor of four versus an HD camera using the same size chip. With HD the occurrence of softness in shots increased dramatically from SD. An additional reason is the lack of focus calibration being performed by the operator in the camera viewfinder, where often peaking, contrast, and brightness are erroneously set, unknowingly producing an artificial expansion to the depth of field. This compounds the problem of detecting focus in higher resolutions.

Dynamic Range

An increase in pixel density and photosites can play a part in increasing dynamic range given the capability of the camera's electronics to process the information.

Resolution

An increase in planar resolution results in an increase in the depth of the working visual palette. Two effects result:

1. Higher planar visibility: This includes artifacts, digital processing anomalies in the imaging chain, optical conversion properties, and colorimetric construction.
2. Higher resolution adjustment: The transition from SD to HD brought with it more precision in the optical and colorimetric sculpting of the image. If we can compare the SD canvas to painting with broad brushes, with the HD canvas the brushes became much finer, that is, there is an increase in the resolution of image adjustments. The importance of highly subtle adjustments thus becomes pronounced. The same occurs with 4K colorimetry.

As when Technicolor introduced a new world of chroma resolution, a resurgence will occur in chroma detail as understanding and crafting of the BT.2020 space progresses. The audience was unaware of the possibilities HD could reveal over SD, such as seeing individual strands of hair on an actor's head or pores on an actor's face. BT.2020 will achieve similar results for color.

Another way to think about this increase is in the form of an operational Z axis added to the X and Y axes of a two-dimensional screen. One of the descriptions by early adopters of HD's exceedingly drastic sharpness over SD was to liken it to a window producing the illusion of three-dimensionality. From the standpoint of an image analyst and image craftsperson this can be treated as a Z axis, comparably nonexistent in SD. Learning to visually process a picture using this dimension can allow the brain to layer high-resolution adjustments with better balance. Adjustments such as shadow saturation, matrix circuit correction, gamma curve customization, highlight detail, spectral spike detection, lighting balance, contrast ratios, detail creation, and skin correction can all benefit quite well from this perceptual technique.

Modulation Transfer Function (MTF)

The transition from light to dark in a high-contrast portion of a 4K image or in a subject's contour against background can be much more fine and subtle compared to HD. The importance of MTF and how it is treated through the imaging pipeline should be stressed. The physics of MTF management for capture and post treatment, about which there is some documentation, remain the same as lower resolutions except with a deepening of the previously mentioned Z axis as resolution increases. Without sufficient attention paid to crafting MTF, the signature visual impact of spatial detail in 4K can noticeably decrease.

Oversampling

Production operations can reap the resolution benefit of 4K in some form today without a requirement to have in place 4K delivery. When footage acquired using 4K sensors is oversampled or downscaled to HD, the results are several perceptual effects:

- Improved MTF—also referred to as detail creation or contrast, the perceived sharpness of the image as originally converted through the optical path of lens to sensor and through the camera's digital signal processing (DSP) is substantially improved compared to HD acquisition.
- Chroma Depth and Precision—perceptually, hues will be represented with increased accuracy. There can be improved separation and gradation, dependent on encoding bit depth maintained between acquisition and delivery.
- Refined Gamma and Highlight Rendition—increase in resolution requires escalation of pixel counts, which increases photon capture and subsequently image data. With adequate DSP implementation, density of data can be proportional to gamma and highlight transitions.

Perceptual benefits refer to the human visual system. The above are inherent to well-captured 4K footage. For HD oversampling, a working understanding of contrast ratios, lighting informed by sensor dynamic range and color gamut limitations of not only the acquisition device but the post and delivery process is assumed. Detailed familiarity with and quality of the down-scaler technology employed is essential to superior result.

It may seem implausible to scientifically measure such benefits when downscaling from a high-resolution format to a low-resolution one. The history of workflow methodology in various forms of electronic media reproduction shows otherwise. A common practice for full-page photographic advertisements has traditionally been capture using large-format 120 mm still cameras rather than the smaller 35 mm format. Although the size of the final delivery medium remains the same at 8 × 10, the larger format camera delivers better contrast and color sharpness. Frame rate considerations aside, the same knowledge was the basis for programming and TV advertisements shot on 35 mm film and distributed on SD television or VHS. In motion pictures, the occasional studio film acquired in 70 mm, despite primary distribution in 35 mm, showed improved color reproduction sharpmess and contrast rendition when projected in 35 mm. In music production, also a medium converting the analog world to electricity, the highest-resolution microphones and magnetic tape machines produced a better listening experience on low-fidelity end-user audio cassette than using acquisition equipment of lower sonic resolution.

A scientific analysis of this effect would track the quantifiable integrity of the image's building blocks through every conversion stage beginning in the analog world—the acquisition machinery, format, storage, and codec; the post pipeline, delivery, and transcoding—while correlating this data at every stage with the effect on the human vision system's processing of it. The simplified and meaningful explanation of the basis for this effect has been ingrained in the toolset of seasoned imaging practitioners for decades—the more one begins with, the more one ends with. Technically, the subtractive process produces better results when originating real-world analog data than an electronic, additive process attempting to emulate analog data. In the end, perceptual benefits to the human vision system outweigh numerical analyses.

Cropping

A popular benefit of 4K acquisition for HD delivery has been the ability to recompose the image in post due to four times the availability of image data. For specific applications, where one may want to overcome shooting constraints for example, this is a useful tool. Protecting or specifically composing for HD while shooting 4K, the same as we would protect for SD 4:3 when shooting HD, is also a manner in which 4K acquisition can be quickly prepared for HD delivery by cropping to the predetermined HD composition in post. The most significant benefit of these methods is future repurposing of the program for full 4K broadcast and distribution when transmission and delivery ecosystems have matured.

In overview of the direction towards the next prevalent standard being 4K, these methods do tend to slow the adoption of 4K. Although the sensor and DSP capture attributes of a 4K camera will be partially on display with a one-quarter crop of the full resolution, they do not display the entire visual story of 4K acquisition as an oversampled HD program would. Additionally, as shooting technique these methods have a short life, becoming less viable as 4K use and demand for full resolution increases. In a sense, they may also decrease some measure of discipline in craft by relegating the art of photographic composition to post.

Infrastructure, Delivery, and Transmission

Secondary operating challenges of working with 4K material are data throughput, compression, and post infrastructure.

The first challenge of 4K has been data management. The amount of storage required to transfer, back up, and store camera acquired footage was prohibitive at the introduction stage of 4K cameras. Being the oldest hurdle, there has been much time to improve on this, with storage costs perpetually dropping and transfer speeds of drives continually increasing. The popularity of solid-state disk (SSD) technology has aided this. For facilities with infrastructures spanning multiple locations, full 4K implementation is in many cases still on the drawing board.

Using the limited-scale system as the basic model, once storage started becoming manageable, CPU and graphics processing speeds in nonlinear editing (NLE) workstations and the software design to handle multiple real-time 4K streams on a timeline showed need for improvement. This stage is still settling into a stabilization phase in comparison with the same level of efficiency and speed that existed with HD systems across different platforms.

One might conclude that realizing UHD transmission in the majority of households may depend on the same path of government legislation that enacted the conversion from SD to HD. The difference this time is the prevalence of Internet connectivity. Web services such as YouTube, Netflix, and Hulu are today streaming movies, on-demand TV programs, and mixed genre short-form videos in full resolution UHD, which can be superior in the previously mentioned image quality parameters compared to any HD content available, including Blu-ray DVD sourced HD. As the popularity of these services increase, it's possible that the viability of traditional delivery platforms may come into question for the consumer.

Adoption

The year 2015 marked a significant rise in the number of 4K TV sets. This was directly driven by an intentional reduction in the retail price point. If the adoption history of the HDTV set was an indication, sales will continue to increase until market saturation is achieved, at which point a 4K TV set will be considered as commonplace as an HD set is today. HDTV in turn will be considered equivalent in obsolescence to SDTV. This pattern began in part prior to television, as we observed 16 mm film, at a resolution that is roughly half of 35 mm, on a progressive decline in usage against various formats driven fundamentally by resolution and resolving factor.

In the United States, 4K programming is being distributed via domestic pay satellite services and Internet streaming services. At the acquisition stage, 4K production activity is considerable. The delivery, however, is often downscaled to HD to accommodate distribution.

Determinants

It is theorized here that technologies such as 3D TV technology did not achieve lasting market saturation because, as technological innovation, they did not adequately complement the human drive for perceptual realism.

In this regard, 4K is a repeat of the SD to HD transition. As a new generation of people is raised watching 4K TVs, HDTV will become an unfamiliar technology. Just as film was unknown to the video generation, SD is unknown to the current generation. The process will continue and repeat with 8K.

BEYOND 4K: RESOLUTION AND HUMAN PERCEPTION

4K contains 8 megapixels of image data; 8K capture and transmission is currently in development. Going forward, DSLR megapixel resolutions of today will be surpassed with the arrival of a potential 32K television standard.

Surveying the historical landscape of image reproduction technology, we see that resolution prevails as the technology of consistent progression. Plotting the trajectory from cave paintings to the planned Large Synoptic Survey Telescope camera at 3200 megapixels to be used for photographing anomalies in outer space, resolution is the primary driver of reproducing what we perceive to be objective reality. The reason—the natural human drive to faithfully relive an organic experience. Currently, we:

- restrict the visual span of a story within a two-dimensional rectangular aspect frame;
- maintain a physical space between the storytelling medium and the spectator.

This serves a purpose for information processing by the human brain. Imagine a machine that allows the user to experience a story by interfacing directly with the optic nerve. It could immerse spectators into the story as if they were physically present in the scene. The illusion of limitless engagement is produced. In contrast, the two-dimensional screen reduces information to a readily digestible box, physically separating the audience from the world of the story. Although interactivity may still occur at the cognitive level, a physical detachment creates the sense of safety, increased information processing efficiency, and, by virtue of these limitations, quicker assessments of the story.

As long as the planar format exists, resolution will increase up to the threshold of the human eye, currently estimated at approximately 500 megapixels. HD is approximately twice the resolution of SD, 2K is twice that of HD, 4K is double the amount of 2K, and so on.

At the current rate of doubling resolution, the human vision system will be exceeded at 256K or an increase of 64 times the resolution of 4K.

HIGH DYNAMIC RANGE CAPTURE AND HDR DISPLAY

The recent interest in high dynamic range (HDR) signaling and display systems is unprecedented in expectation. The primary advantage that drives interest in HDR imaging is that HDR is very easily perceived by everyone who sees it. The result of a side-by-side comparison of HDR versus conventional standard dynamic range (SDR) systems is very often described as stunning, making the viewer acutely aware of the limitations of legacy SDR TV systems. HDR provides a huge leap forward in perceived image quality as it removes the traditional barriers of black level detail, 8-bit image quantizing, reduction and clipping of volume in highlights, and significantly higher image dynamic range. HDR is often judged more beneficial to resultant image quality than the HD to UHD doubling of the horizontal and vertical spatial resolution. Most agree that HDR provides the biggest improvement in perceived image quality in recent times, being compared to the revolution that color TV transmission brought over black and white.

A new generation of television displays makes HDR viewing possible. These displays take advantage of the increased brightness capabilities of recent consumer (and professional) displays, include 10-bit accuracy, and have a wider color gamut that includes near P3 cinema performance and in some cases exceeds P3.

Postproduction signal processing for HDR and delivery systems is evolving to process the camera's scene capture capabilities to provide the display with signals capable of displaying a much wider dynamic range than current TV systems allow.

A simple analogy to high dynamic range video may be found in audio systems. Audio is adjusted to a specific working level for normal tones, but audio systems have an inbuilt "headroom" margin (often +12 or 20 db) to handle the peaks and transients that preserve subtleties of audible signals. Until the advent of HDR, video signals have not enjoyed any equivalent overload margin or overload capability. Scene-referred white levels represented the maximum signal levels; any content above was harshly truncated by hard clipping. With the advent of HDR video systems, there is provision for the inclusion and processing of camera-captured "video peaks" (e.g., small area highlights and speculars) and scenic volume (the ability for fully saturated colors to be displayed in bright images). These image dynamics are often many times brighter than the previously permitted "peak white" of traditional TV transmission systems.

Until now, TV displays have been limited by the current performance standard known as ITU-BT Rec. 709. In this system, the characteristics of the video system's gamma functions, and the signal fed them, determine the placement of the scene reference white level near peak exposure at approximately 95%–100% overall capability. These standards were adequate for the 100-nit brightness (a performance level referenced on the CRT-based displays of yesteryear); today's LCD and OLED-based displays can offer SDR performance brightness levels of 200–300+ nits (a gain of 2 or 3 times). HDR capture mandates a performance level that could be as much as 10 times more light output, which conventional gamma-based EOTFs do not serve in performance.

HDR signaling presupposes that the HDR display will be capable of much higher brightness and therefore higher contrast and the ability to render higher volume. Current SDR high-definition "Rec. 709" displays provide a useful dynamic range of little more than 6–7 stops, or about 200 or so levels of gray scale. This is not so much based upon display limitations but upon the gamma-based limited dynamic range signal processing and EOTF based on 8-bit digital signals. Scene content above the reference white level of 95%–100% is severely clipped, and any brightly colored content is desaturated and blends into white as exposure approaches 100%.

HDR promises a minimum of 10 stops of useful dynamic display range in a 10-bit system, thus offering a minimum of 1024 levels of gray scale, which enables a significant visual payoff. However, this is cleverly configured to provide significant highlight detail and color above the normal scene-referred white level, which is set at 50% of the operating level. Importantly, as the white point is moved down to 50%, HDR permits much expanded color volume; saturated colors do not hit the clipper and desaturate to white. HDR scenic content places the scene reference white levels at similar levels to older CRT/Rec. 709 displays, around 100 nits; however, their extra brightness permits modulation and display of highlights exceeding those levels by 8 times, around 800 nits, some 3 stops higher. With HDR it is possible to see small area highlights, specular metal reflection, stars at night, car headlights, and so forth in a much more natural and lifelike way than possible with Rec. 709 video signals.

High Dynamic Range Signaling Systems

There are two outstanding proposals for HDR signaling systems that are based on open industry standards. The offerings of HDR systems are *Hybrid Log-Gamma (HLG)* and *SMPTE 2084* (also known as perceptual quantizing or simply PQ). Both of these systems offer a new EOTF (electrical optical transfer function), which determines how an electrical video signal is converted to a light output value on a display screen. In July of 2016, the ITU published BT.2100: Image Parameter Values for High Dynamic Range Television for Use in Production and International Program Exchange. This document is suggested reading for specifics.

Hybrid Log-Gamma

Hybrid Log-Gamma, as its name suggests, is a system designed to marry the lower half of a conventional gamma-based "709-like" EOTF curve with a new "log" encoded highlight curve.

Figure 5.5-37 shows the EOTF for the proposed HLG system; for comparison purposes, this figure also shows a conventional "709" gamma curve, and the same "709" gamma curve modified with an added linear knee for highlight control, often found in many of today's HD cameras. Working in HLG, and by placing the HLG scene reference white level at 50% (X), this approach will provide sufficient contrast and dynamic range for it to be viewed on a conventional 709 SDR display. The gamma-based EOTF of the SDR display will show somewhat compressed highlights, rather similar to a display showing camera content with highlight knee processing. The results are normal scene contrast with useful, if limited, highlight viewing capabilities.

When displayed on an HDR display with the matching HLG EOTF, the upper log encoded portion of the HLG EOTF will be expanded to produce more

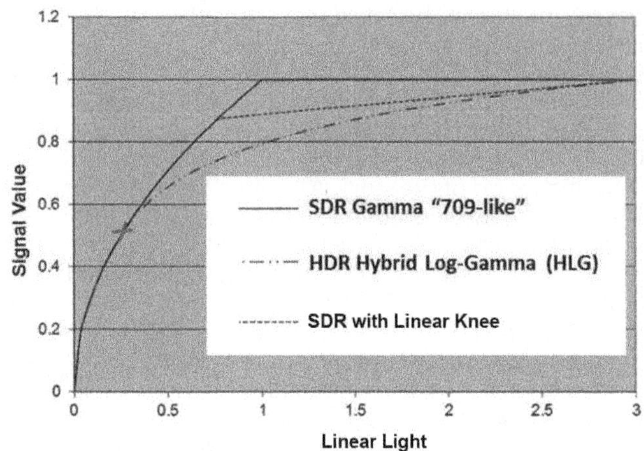

FIGURE 5.5-37 EOTF for the proposed HLG system.

FIGURE 5.5-36 SDR and HDR displays. Note the brightness levels in nits between the two displays.

Note: With traditional displays, such SDR/HDR comparative examples often do not show the significant image differences.

highlight contrast, and thus more dynamic range. Those extra 3 stops of highlight range result in a dramatically more realistic image; peak levels of the highlights are now 600–800 nits.

HLG is a relative newcomer to the HDR arena. Although more recent, its biggest advantage is its inherent compatibility with existing display technologies. Current TV display characteristics can elegantly handle the high brightness without harsh distortions. Current broadcast, cable and satellite, and over-the-top technologies need little or no upgrading to handle the signal for this proposed system. Testing and standardization is ongoing and making quick progress as the BBC, NHK, and some U.S. broadcasters are firm advocates of HLG. Work is close to the final industry standardization of HLG.

HLG, while elegant and fairly compatible, has a complication in that the Hybrid Log nature of its EOTF makes transcoding to other HDR systems difficult, so HLG is currently seen as a useful but closed production system, good for broadcast or live TV, but not so useful for later processing or remastering to other HDR systems, most of which use SMPTE ST 2084 EOTFs. Time may provide solutions. The world of color science transformation is moving ahead rapidly. Currently, HLG is seemingly preferred by broadcasters seeking a simple and "compatible" system for HDR purposes. HLG is likely to be an "output" menu option on some products, for example, studio cameras. Interestingly, the studios and content providers, however, favor ST 2084 (PQ). Time will tell which will prevail; more likely, both will find uses in applications they suit best.

ST 2084

Another HDR approach uses a completely new EOTF, standardized by SMPTE as ST 2084. Sometimes referred to as PQ, or perceptual quantization, ST 2084 is not based upon conventional gamma-based processing for the display EOTF; ST 2084 provides a much better "fit" to the human visual system, with better gray scale tracking and lessened luminance error visibility. ST 2084 provides the ability for much higher brightness and wider dynamic range displays than conventional gamma-driven displays used in today's SD and HD systems. The downside is that the ST 2084 EOTF is not compatible with current "gamma" based video processing and display technologies. ST 2084 is display "output" referred, thus absolute light levels are defined by code word values on the EOTF. The ST 2084 OETF handles light levels of up to 10,000 nits; however, recent testing has shown a more real-world approach is to limit this to about 1000 nits, representing the upper end of available HDR displays.

Figure 5.5-38 shows the ST 2084 EOTF and an overlay of the current ITU-R BT.1886 (which includes Rec. 709) gamma profile. Note the limited dynamic range/output levels of the legacy HD/BT.1886 curve. The ST 2084 EOTF was carefully constructed to provide maximum dynamic range for various applications. The lower half of the 10-bit HDR10 EOTF produces the current SDR (100 nits) capable brightness points. It

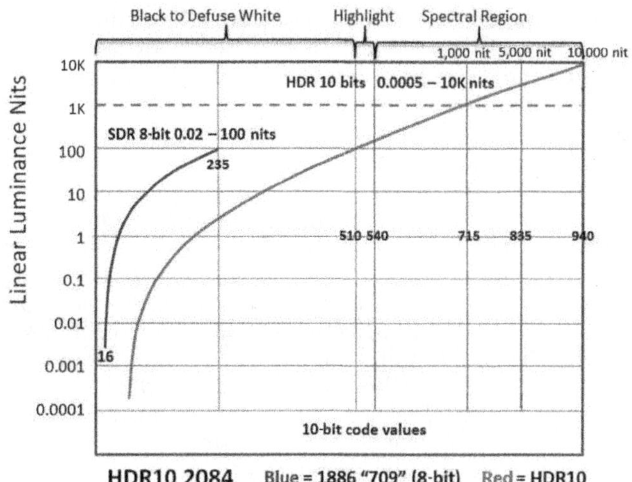

FIGURE 5.5-38 ST 2084 EOTF—HDR10.

does this using 500+ code values; this is twice the number of code values as 8-bit 709. The current generation of 1000-nit TV display levels are indicated by a dotted line. These are currently available and have found acceptance as being not too bright or tiring to watch. Note that the ST 2084 EOTF only uses approximately 200 code values for display brightness of 1000 nits to 10,000 nits; therefore, it's highly efficient and clearly human vision correlated.

HDR Camera Capture and Workflow Considerations

HDR will require careful attention to exposure, and new HDR-capable monitors to replace the existing "709" displays for review and grading purposes. HDR requires that the capture camera have a dynamic range well in excess of the 10+ stops of display signal range. The good news is that many of today's production cameras, mostly large-format s35mm (super 35 mm) cameras, have dynamic capture ranges of some 12 to 14 stops. These cameras use a lower contrast log-based OETF to permit recording of the 12–14+ stops of dynamic range without signal clipping. In most cases this is accomplished by means of a "log" based recording scheme. Log-C is fairly common, and most of the production cameras today utilize some form of log-based recording scheme. Panasonic's VariCam 35 uses V-Log (V for VariCam), an internal OETF chosen to optimize dynamic range yet not compromise resolution or add noise. V-Log is capable of accurately encoding 16 stops of scenic contrast.

Production and Image Capture for High Dynamic Range: "Start with More"

True 4K/UHD capture resolution combined with HDR production requires cameras with dynamic range capture of greater than 10 stops (hopefully more), and in this area many of today's s35mm cameras qualify, with 14 or more stops. Log-based encoding allows easy recording of that entire dynamic range for later postproduction, be it SDR or HDR. The important aspect here is to faithfully capture more resolution and scenic dynamic range than is required for the postproduction process of choice.

The concept of wide dynamic range capture for display at much lesser range is not new; 35 mm film shot in 1980 for prime-time TV shows captured a wide dynamic range as well as greater spatial resolution than required for the deliverable. Postproduction involved color correction of the scanned film and careful grading to optimize the captured image to that best for display on the limited CRT-based TVs of the day. Typically, television in the 1980s through today was set for an approximate +2/−4 stops of deliverable contrast. This allowed for enough image contrast to present a "good and representative" viewed image. With HDR, we have added at least 3 stops of display dynamic range, so our captured image must be greater than that (after grading/color correction) to saturate the display.

Today's production technique of shooting wide latitude imagery, often with the attendant high spatial resolution, enables more choice in post. Much like film shot for TV, the ability to modify the scene contrast at will is helpful, and the oversampling by the original capture format enables useful techniques like cropping or reframing of the image with little or no visible degradation. Much of today's production is geared toward these types of post processes. So if the required delivery format is HD or UHD, SDR or HDR, having a very high dynamic range, spatially oversampled camera master is helpful. Color correction of HD and conventional SDR delivery is easy: use the EOTF to match the Rec. 709 display, and grade/control the image parameters as required.

For HDR mastering, the process is very similar; the wide dynamic range "log" camera master is viewed through a LUT (look-up table, a nondestructive way to transform one visual signal into another) to replicate the inverse of the EOTF of the HDR display device, in this case HLG or ST 2084. Now that the display matches the content, normal grading takes place, and contrast is added where needed to create interest and the desired look. The highlights in HDR are, however, no longer crushed or clipped. Those extra 3 or more stops of highlight dynamic range detail have the ability to add "volume" to the image, to create contrast depth, to enable saturated colors even in high lighting conditions, and to allow specular highlights to really shine (pun intended).

HDR and Live Production

The majority of television in the United States is postproduced and mastered for prime-time viewing; little live TV is normal, with the exception of news or sports. Live TV, however, has a large appeal, and combined with certain types of content, it can be very compelling. Imagine the Olympics "live" in HDR. . . . Live HDR is possible, and that indeed was the intent of the BBC and NHK with their development and introduction of the HLG, Hybrid Log-Gamma, HDR system.

Designed to enable easy mixing and production of live content, HLG promises to bring HDR to live TV. Sports programs require very different production techniques to the more traditional prime-time TV scripted content. Sports camera coverage implies live action, fast-paced motion, the need to zoom into the action from across a large playing field or stadium. As such, the more traditional s35mm production cameras with their prime lenses, or at best small large format zooms, cannot hope to compete with smaller (2/3″) HD/UHD cameras and large servo-driven 50:1 or even 100:1 zoom lenses.

Several companies have introduced 2/3″ production and studio cameras in both HD and UHD. These cameras have advanced image sensors that can provide the 12+ stops of useable dynamic range required for HDR processing. Work is underway to determine what types of HDR production tools and workflows will be needed, HLG, or possibly other EOTFs. These cameras will require some sort of 3D LUT to provide camera operators with sufficient image contrast so that they can track fast-moving sports plays and enable critical focus and exposure control. Live matching of multiple HDR cameras is not that much different to normal 709 HD.

BOX-STYLE CAMERAS

Traditional broadcast cameras have a familiar form factor. They can be operated handheld or mounted on a tripod, and all of their controls are easily accessible. Quick-release plates allow for simple mounting onto many sorts of camera supports. There is, however, another, more simplified form of camera that is referred to as "box-style." Many of the major manufacturers, such as Panasonic, Ikegami, and Hitachi, are producing this style of camera now. Local camera controls have been eliminated.

These cameras are much smaller and lighter, but they have the same imagers and camera circuitry as their comparable full-size versions. For example, the well-known Sony HDC-1000 and HDC-1500 broadcast cameras have a box-style version, which is called the HDC-P1. By having matching internal components, these two cameras can be easily used on the same multi-camera production, while capturing comparable images and maintaining the same workflows. The same lenses can also be used.

Box-style cameras are sometimes referred to as POV cameras because their small size allows them to be utilized where a full-size camera would be too large or too heavy. These lightweight camera bodies may only weigh 3–4 pounds, making them perfectly suitable for robotic camera movement. When motors are used to control camera placement, less mass can help to provide for smoother and more precise outcomes. Additionally, the small size of a box-style camera means that smaller devices can be employed in their support and movement. Reducing the size of camera rigs can have many benefits. They can be easier to transport and assemble. Their limited size makes them more discreet, and at live events the audience could be less distracted, and their views may be less obscured. These attributes can save a production both time and expense.

Box-style broadcast cameras are approximately 4.5″ wide by 5.5″ high and 8″ long. They are often the best choice for remotely operated applications both in the studio and on location. They have been deployed on remote heads on cranes, jibs, remote dollies, robotic pedestals, telescoping columns, and other vertical motion rigs, as well as cable and pulley systems, such as Skycam. They are also quite useful as fixed cameras capturing unique shots in out-of-the-way locations. Mounted on a remote head, these small cameras can produce a view that would otherwise be impossible. Some common examples are goalpost and net shots in professional sports and close-up keyboard and drummer views at live music events. Entertainment venues can also benefit from their application on a small remote dolly mounted just off the lip of the stage, or on a tower that can provide dynamic aerial views. They are also increasingly applied to innovative use in reality and episodic television shows. Maintaining world-class production values while solving complex physical challenges is a prime benefit of these cameras.

One of the newest of iterations is the Sony HDC-P43. It is suitable for a 4K resolution workflow. Additionally, it allows for both high dynamic range supporting ITU-R BT.2020 space and high frame rate content at 2x speed. At regular HD output it can capture slow-motion footage at 8x speed. This camera is the smaller form factor version of the full-size HDC-4300. (See Figures 5.5-39 and 5.5-40.) It is equipped with a three-chip, 2/3″ 4K image sensor, which helps to provide traditional television depth of field characteristics. It interfaces with its camera control unit using a single-mode optical fiber connecting directly to the camera head. This fiber cable can be extended up to 1.5 miles for control of the camera's necessary functions.

Small form factor box-style broadcast cameras are being utilized in many of the newer up-and-coming production technologies. Stadium sports can now utilize 360-degree systems featuring dozens of fixed-position box-style cameras capturing the entire playing field. By digitally stitching multiple views, virtual perspectives can be created, allowing for points of view that are similar to a player's, and rendering slow-motion replays with the ability to rotate around the action. There is a trend for these fully functional box-style cameras to become even lighter and more compact, and we

FIGURE 5.5-39 HDC-4300 full-size camera.

Courtesy of Sony

FIGURE 5.5-40 HDC-P43 box-style camera.

Courtesy of Sony

will see them used on more and more projects, where traditional full-size units are no longer necessary.

CONCLUSION

The challenge in creating a meaningful chapter on television camera systems is recognizing the fundamental change that has taken place. In the early days of television broadcasting, the technology that made up a "broadcast quality" camera was very good, and that was enough. The quality of the mass-produced consumer display and the analog distribution technologies that were available to get the "broadcast quality" content to the viewer guaranteed that virtually any imperfection at the creation end would be masked by the lower-quality consumer display and torturous path that the content had to traverse. The advent of digital technologies and displays has completely negated this concept, and it is quite possible that the consumers' systems may have capabilities that exceed the content creators'. In this environment, broadcasters must ensure that they create content at the highest quality possible and do no harm to the content that passes through their systems.

Notes

[1] ITU-R Recommendation BT 601-5, Studio Encoding Parameters of Digital Television for Standard 4:3 and Wide-Screen 16:9 Aspect Ratios, International Telecommunications Union.

[2] ATSC A/53E, ATSC Digital Television Standard, Revision E, with Amendments No. 1 and 2, Advanced Television Systems Committee, 2006.

[3] SMPTE 125M-1995, Television—Component Video Signal 4:2:2—Bit-Parallel Digital Interface, Society of Motion Picture and Television Engineers.

[4] SMPTE 293M-2003, Television—720 × 483 Active Line at 59.94-Hz Progressive Scan Production—Digital Representation, Society of Motion Picture and Television Engineers.

[5] SMPTE 259M-1997, Television—10-Bit 4:2:2 Component and 4fsc Composite Digital Signals—Serial Digital Interface, Society of Motion Picture and Television Engineers.

[6] SMPTE 274M-2003, Television—1920 × 1080 Image Sample Structure, Digital Representation and Digital Timing Reference Sequences for Multiple Picture Rates, Society of Motion Picture and Television Engineers.

[7] SMPTE 296M-2001, Television—1280 × 720 Progressive Image Sample Structure—Analog and Digital Representation and Analog Interface, Society of Motion Picture and Television Engineers.

[8] ITU-R BT 709-5 Recommendation, Parameter Values for the HDTV Standards for Production and International Programme Exchange, International Telecommunications Union.

[9] SMPTE 292M-1998, Television—Bit-Serial Digital Interface for High-Definition Television Systems, Society of Motion Picture and Television Engineers.

[10] Poynton, C. *A Technical Introduction to Digital Video*. New York: Wiley 1996. Chapter 6, "Gamma," is available on line at: www.poynton.com/PDFs/TIDV/Gamma.pdf.

CHAPTER

5.6

Camera Supports

MARTIN MARIETTA

Producer, Camera Operator, and Robotic Camera Movement Specialist
New York, NY

Motion imaging is an art form dependent on technology. The success of this aesthetic relies largely on science and engineering. Mechanical and electronic devices, in the form of camera supports, permit much of the visual magic to take place. Modern advances in video camera technology are allowing higher-quality images to be captured with smaller and smaller equipment. This is enabling artists to utilize a new range of tools, which will continue to grow in their ability to create unique and beautiful moving pictures. In the same vein, both traditional and robotic camera supports are becoming more useful and easier to implement for the average video production. This path towards more powerful and often simpler use of both manual camera supports and robotic movement machines is certainly progressive. This chapter will explore some of the more obscure current methods in this discussion of video camera usage, and will also touch on a broad scope of production tools, including those used in television studio productions, such as news, talk, and game/talent shows, and also certain supports relevant to episodics, award shows, commercials, and sports. Understanding the full scope of what is possible in modern production is the professional standard of this industry.

Camera support has everything to do with how human beings work with a camera: holding a camera in the hand, attaching it to the body, supporting it on engineered devices, mounting it on controlled motion machines, achieving free-flowing, three-dimensional movement, and even allowing it to be operated by a robot itself.

Cameras can now be mounted anywhere. Various rigging systems help us. Irregular and softer surfaces can be made firm and squared-off in a variety of ways. Structures, bracketry, and many types of hardware all contribute to holding a camera steady. Capturing a static image without any movement is often critical. It is also important to be able to move a camera any number of ways without introducing any distracting disturbances to the image. When a viewer sees motion in an image, the element of time becomes more entwined into the visual space. A moving camera provides a continuous flow of composition, adding dynamic visual complexity.

The craft of camera movement has become quite technical in recent years, as robotics have become commonplace. Interestingly, however, creating unique and sometimes mission-critical components using handcrafting techniques is common and necessary. In this era, individual technicians can design and then make or commission custom solutions quickly. The opportunity to devise new gear, and its connected methodology, is very much alive in this industry. This is substantially because of the evolving technology of robots that are revolutionizing camera supports. There are now automated ways to place a camera at one location in three-dimensional space, and then very quickly if necessary place it in some other location. A number of these robocams, with their remote operator interfaces, provide the possibility of repetitively and accurately accomplishing very complicated on-air moving shots. Today often a single camera operator can manipulate many different cameras simultaneously. It is a time of innovation, striving to find the most effective and creative ways to use these amazing new tools in production locations worldwide.

What are the most elegant and expedient mounts and methods? This paradigm changes constantly, but the classics remain, as the future unfolds.

MOVEMENT DETERMINES SUPPORT

Static Shots

Static shots are often used in video production. Sometimes camera support is only used to keep the framed image from shifting and to make the camera safe. If a fixed camera is delegated to one specific shot then it might be unmanned. If a specialty shot requires the camera to be mounted in some out-of-the-way location, then it will be set up in advance and may only require simple positioning bracketry and a safety cable.

Some productions might only require an operator to reposition a camera off air, reframing the image for another lock-off shot. In these instances, camera operating is less intensive, and it may be possible to use the most basic simplified supports. Most commonly, however, productions want to perform on-air camera moves, making it necessary to utilize more complicated supports, which must be finely tuned and balanced, so that the motion is smooth and easy to control.

Camera Moves

Moving a camera, the image created often mimics human experience. Panning a camera simulates turning your eyes from side to side. When a person surveys his or her view of a scene, it is natural to scan across the visible panorama. To pivot the lens of the camera to the right or the left, while keeping the horizon line level, is to *pan*. You cannot "pan up" or "pan down," even though the phrase may be spoken accidentally, or naively. When an operator pivots the camera's lens directionally upward or downward, similar to nodding your head, it is called a *tilt*. A tilt is also performed while keeping the horizon line level. When a camera makes a compound move, panning and tilting simultaneously, the point of view moves diagonally. While this may be accomplished fairly easily by an operator with a well-balanced manual camera rig, high-end motorized remote heads should also be able to move in any direction by compounding a pan and a tilt without pausing or stuttering. Stepping can also be a problem, when the camera movement does not create a smooth diagonal, but instead incorrectly alternates between panning and tilting.

The more rarely used *roll* axis allows the horizon line to be canted diagonally on dry land, or corrected and leveled when a camera is mounted on a sailboat, which heels or rolls on the water. The technique of intentionally setting the shot where the horizon is not level is known as a *Dutch* angle. It is generally for some special dramatic effect, or to create a sense of imbalance or uneasiness. This type of shot has been held effectively at every angle, from a few degrees cant, all the way up to an oblique angle. If the roll is dynamic, it can take an extreme in one direction, then roll in the other direction, for amplified effect. Though in fact, however, most camera supports quite usefully help the operator by keeping this axis from changing. Everything is level, and it stays level.

To *pedestal* a camera is to simulate raising or lowering the height of a person's head. When an operator hears the command "ped up," it is an instruction to raise the camera straight up, usually while keeping the subject positioned in the field of view, often necessitating a small amount of downward tilt. Raising and lowering the camera can be quite useful in establishing the viewer's perspective. For example, when capturing the image of a person, is the eye line level, high, or low? As with any camera placement or motion, what is the artistic perspective?

When a camera is moved towards or away from the subject, it is called a *dolly*. "Dolly in" or "dolly out" might refer to walking the camera forwards or backwards handheld or with a Steadicam rig. It also might refer to the use of an actual dolly on tracks, a remote rail system, any wheeled rig, or even the expansion/contraction of a telescoping crane. A similar move across the ground that is perpendicular to the orientation of a dolly move is called a *truck*. This takes place when the lens is pointed forward and the camera support travels to the left ("truck left") or to the right ("truck right"). The difference between a dolly move and a truck move can be especially useful when operating pedestals, because they move in any direction.

A tracking shot moves side to side and parallel to the subject, but not necessarily straight. One could track around something in a circle, or follow a curving trajectory. To *track* is to maintain generally the same distance from a subject, often matching its movement lest it leave your frame. When operating a remote dolly that is rigged at the front of a stage, creating a smooth reveal of the action and following at will, you would say, "Track left, track right." Although technically tracking and trucking have specific meanings, when a dolly is used to make these moves, it is common to hear the terms "dolly left" and "dolly right." These terms may be used interchangeably, and often are.

David Smith, Emmy Award-winning Camera Operator, recommends,

> When performing a tracking shot, it is important to consider the spatial proximities of the lens, the background, the subject, and any other objects. As the camera moves, relative angles visibly shift. The viewer will see a continually changing relationship between the subject and the background. If the camera is too far away, this dynamic will be less effective. Perspective changes are different for every element in the frame, based on its distance from the lens. Foreground elements can encourage the perception of movement, as the camera appears to travel past them at a faster rate. This is compounded by the speed and distance of the camera's actual move. These visual cues of motion can create a more dramatic and dimensional feel to the shot.

A *zoom*, interestingly, is not natural to the human eye, nor is it actually a camera move. It is a shortcut, produced by a trick of optical physics inside of the lens. It is accomplished by changing the focal length, which has to do with the converging and diverging of light, by changing the distances between the glass optics of the lens and the focal plane. The focal plane is at the surface of the sensor in the camera, where the light is focused to create a sharp two-dimensional image, using the illumination that has reflected off of the three-dimensional scene. A longer focal length creates magnification of the subject, which proportionally narrows the field of view. The opposite is also true, in that a shorter focal length allows for a wider field of view with a subject that appears smaller. In the lingo of camera operating, to "tighten" or to "push" means to zoom in, and to "widen" or to "pull" means to zoom out. When a lens zooms in, everything that remains within the frame is magnified equally. This is different than a dolly move, where foreground elements become magnified as the dolly approaches, and elements in the distant background appear generally the same. It is possible for the viewer to notice a potentially distracting zoom, so they are often used sparingly. A *dolly zoom* is when a zoom is performed in juxtaposition to a dolly move, and a very unsettling effect is created. When you dolly forward while zooming out, or vice versa, the unusual nature of a zoom becomes obvious. The background shifts in an unnatural way in relation to foreground elements, causing an effect that is considered disturbing. This is often referred to as the Hitchcock, because of his use of the technique in the 1958 film *Vertigo*. Certain *floating camera* moves, like booming a long arm or flying, are another example of visual movement that humans do not normally experience. These sorts of shots are often best used subtly, artistically, or when motivated by the subject, so as not to interrupt viewers by making them consider your choice of camera movement.

There are many advanced camera moves that can often trick the audience into thinking something is happening that is not. Ted Erne, Director of Photography, describes how to create the sensation of movement by using compound camera moves:

> A formulation of panning, tilting, and zooming can create the illusion of movement with a stationary camera. Studied composition of start and end framing makes this possible. For example, a camera positioned above the head height of the talent, initially framed with copious headroom, performing a simultaneous execution of proportionate tilt and zoom, ending with normal headroom, while maintaining the bottom of the frame as it relates to the subject, will give the impression of a descending camera. Adding a sweeping pan can promote this illusion.

OPERATOR-SUPPORTED CAMERAS

Soft Supports

Video cameras can be damaged fairly easily. They have irregularly shaped bodies with electronic controls. Attached to a camera is an expensive glass lens that has moving parts. All of this adds up to a piece of equipment that must be handled carefully. It is important to protect a camera from the natural world. *Padded covers* are conventionally thought of as *camera protection*, similar to a road case, but they can actually provide additional options for camera placement. They are made to fit perfectly over many professional cameras. This is especially useful for "run and gun," ENG (electronic news gathering) type work, which may necessitate that the camera rig will be set upon, leaned against, or in contact with damaging surfaces. These covers are meant to remain on the camera while it is being used in the field, and have access panels that allow the user to control the camera's functions. These covers can extend the life of a camera by protecting it from physical contact, and they are also good at limiting exposure to heat and cold. Padded covers offer some protection from moisture, but when necessary, it is important to use a waterproof *rain cover*. Some companies sell specific rain covers for their gear, and there are specialists who design and craft custom rain covers by commission. Often, operators, technicians, or camera assistants will create water protection for their various rigs on the spot, using plastic sheeting and trick line, clamps, or tape.

One of the simplest forms of camera support is the *beanbag*. A hard or irregular surface can be transformed into a soft cradle to hold the camera. These can be purchased commercially and come in a variety of sizes to fit any camera. Some beanbags are constructed with a mounting plate for a camera head, but most are simple bags, sometimes with a special shape, possibly with a few rings or straps. Many times these bags can be made by hand, and they are filled with Styrofoam or plastic beads, rice, buckwheat seeds, or actual beans. In a pinch, a sandbag can be used, but it will not conform as well, and will clearly weigh a lot more.

Handheld

Fortunately, one of the best camera supports is the one that is always accessible, the human body. Not only is it very adjustable on the fly, but it can also react to outside forces quickly and often gracefully. In fact, humans are so good at stabilizing their own vision system, that excessive camera shake can become quite disturbing to watch. Of course, one of the compromises of operator-supported cameras is that they must be continuously attended by the operator during use, and they cannot create a perfectly stable image. Creating great images with a handheld or a shoulder-mounted camera is a difficult skill, which takes a lot of practice to perfect.

Mark Whitman, renowned Handheld Camera Operator, reveals:

> You have to control your breathing, and sometimes not breathing at all helps. There are times when I will add a slow zoom to hide a bit of the shake. Maintaining a high level of physical fitness and staying away from caffeine is important. Athletics can help with keeping your balance. After practicing martial arts for ten years, I find that it very much supports this camera specialty.

Shoulder mounts are meant to conform to the shape of an operator's shoulder. They allow for a stable base to balance a handheld camera. Full-size video cameras usually have built-in versions of shoulder mounts. As cameras are made smaller, the shoulder mount becomes an accessory, and it can be part of a more elaborate system that is customized for a particular kit. Many built-up systems have a shoulder pad that attaches to the camera rods. Ergonomic versions may take into account the natural slope of an operator's shoulder, utilizing asymmetrical shapes to help in keeping a good horizon line. Often there is a tendency for a camera operator to tense the shoulder muscles in an attempt to steady the camera and maintain a level image. The best practice is to allow the weight of the rig to settle into the necessary position, as the operator maintains a fully relaxed stance that can be held for longer periods of time. In situations where there is flexibility to choose shots, continuously alternating between various positions will allow for prolonged shooting that is athletic and sustainable. Having a zoom rocker on the top handle helps with shooting from the hip or near the ground on one knee. Balancing the camera on convenient surfaces for a shot can provide a momentary respite as an operator moves through the potential stances.

There are configurations available to build up cameras, which allow for the addition of many different *handles*, *grips*, and mounts, depending on how the assembly will be used. An ergonomically shaped solid wood handle for an operator's dominant hand can make long hours of shooting more effective. Camera controls, such as start/stop or zoom, can be added to these grips. Smaller cameras, when held on the shoulder, can be improved by adding a pair of handles extended in front of the rig, and perhaps even a counterweight to the rear to allow for better stability and balance. Sometimes, dual shoulder supports are used. Certain rigs also add a front bar support, which attaches to a chest harness or belt, in order to assist in maintaining a level and shake-free camera.

A popular form factor for a handheld camera buildout is sometimes called a *Fig Rig*, named after the original designer, film director Mike Figgis. This steering wheel type cage allows the operator's arms to be spread apart and extended in a very supporting way. The posture automatically helps with stability by giving the operator easy and relaxed control, allowing the natural shock-absorbing ability of the body to work efficiently. It is commonly necessary to extend the camera's focus and zoom controls to the handles. There are many contemporary iterations of this form that utilize a similar cage style.

A *butt dolly* is an adjustable-height rolling stool, with several high-end casters, meant for a handheld camera operator. On a reasonably smooth surface, it can glide quietly in any direction. This creeper dolly allows for quick and flexible camera movements from a lower angle. When needed, the camera may be placed directly on the seat as a soft platform. Also, the seat and post can be removed allowing for a direct mount to the base, creating a simple multidirectional caster dolly with a very low center of gravity.

POV Mounts

From the beginning of motion imaging, there have been subjective camera views. These shots represent the point of view (POV) of a character or subject. Objective shots, on the other hand, are looking in on the action like an unseen observer. While both types of camera angles are clearly points of view, POV has come to generally mean small cameras or cabled camera heads, often *action cameras* mounted to a person, as shown in Figure 5.6-1. These worn mounts can provide

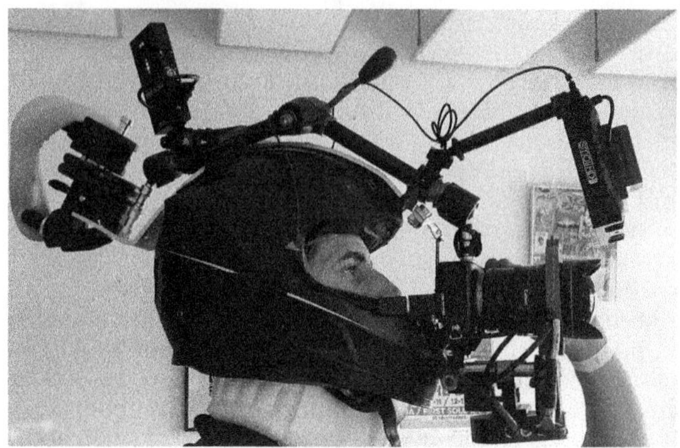

FIGURE 5.6-1 POV Headgear

Courtesy of Arnaud Ly Van Manh, director, and Jonathan Grimault, director of photography, in collaboration with HK Corp

a camera angle that simulates the wearer's vision. Miniaturization technology has created a prototype, which could be available in the future, of a digital contact lens, which can capture video, allowing for the ultimate POV camera position. Until then, however, we attach small cameras to our bodies and our gear any way that we can. Helmet mounts can include pan/tilt capability, the hardware to support high-end optics with lens control, viewfinder monitors, counterbalance weights, and even electromechanical camera stabilization.

Cameras may also be mounted to parts of the body other than the head, or rigged onto extensions, allowing the image to capture the wearer. One interesting shot is accomplished by mounting the camera in front of the subject and looking back. When it stays in perfect synchronization with the talent as they move, it has an unusual effect on the background. It creates the counterintuitive effect of the subject remaining still, while the background moves. This form of mount is often referred to as a Snorricam or simply a body mount. Professional versions of this rig allow for the framing to have high, low, or level eye lines, and can even mount the camera behind the subject for a unique over-the-shoulder perspective.

Only the imagination limits the ways to build both simple and complex rigs mounted to body points and other unusual locations. POV mounts and action cameras might also be used anywhere it would be impossible to mount a regular video camera, or when there is the possibility of damage to a high-end camera. Possibly the cheapest way to get a quick and irregular aerial shot is to use an action cam parachute. A waterproof POV camera attached to a small semi-rigid parachute is thrown high in the air, and on the way down it can capture a wide-angle view of any scene.

Action cameras get mounted to surfboards, skis, and other sports equipment, or hidden various places, peeking out to capture the action. A POV camera can provide a special viewpoint, which helps to show a subject in a unique way. Often, these rigs must be weatherproof and durable enough to sustain likely impacts. Headgear, eyewear, chest harnesses, belts, suction-cups, and sewn-in, strapped-on hardware are all part of the potential POV trick bag. The action sports market has enabled the introduction of a huge variety of these mounts. The possibilities are endless.

Selfie Sticks

These are short poles with a handle on one end and a camera mount on the other. They are generally used with smartphones or other small POV/action cameras, and enable users to point the lens back at themselves, in order to capture video framed as a self-portrait. Because this short pole camera is essentially handheld, it takes some practice to keep the image framed correctly. In the field, some organizations, especially when in dangerous areas or when a low budget is a priority, have begun using this tool to support the capture of their regular news reporting. With the onset of simplified video streaming from mobile devices, a single journalist can report from the field, using a microphone, a smartphone, and a selfie stick. These simple supports are used for ENG-style reporting, walk-and-talk type capture, and even fast-moving action sports.

Monopods

With only one leg, they are not self-supporting. Additional stability is usually provided by the camera operator, but occasionally clamping a monopod in a piece of grip hardware may be useful. Some monopods have short spreader legs and a balance point that can be used for momentary static shots. They are generally compact, versatile, and easy to reposition quickly, being rather unobtrusive in crowded places and tight spaces. Monopods only provide support vertically, therefore the horizon line can easily tilt, making them difficult to stabilize and sometimes limiting their usefulness for serious video capture. Of course, taking advantage of the support's instability could be a good thing. For example, an operator could lean into a close-up with a preset point of focus, bringing the subject from blur into clarity with a smooth and precise movement. Certain methods can allow a camera operator to be more effective with a monopod and to be possibly quick and nimble, with a bit of practice. Keeping the rig close to the body helps. Bracing the base with your instep and the shaft with your thigh can be a good method. A belt or chest harness can also provide an additional point of support for a monopod. Creative uses in the field allow you to mimic the effects achieved by a jib or a slider.

String Monopod

Much of camera support has to do with ways to stabilize a camera. Certain lenses and cameras have internal image stabilization capabilities, which can help, but the majority of stabilization techniques rely on external camera support. An old trick that is rarely utilized today can teach us something about forces that stabilize movement. A string monopod is made with a piece of cord. A simple weight that you can stand on, like a fender washer, is tied to one end. A metal eyebolt is threaded into the camera bottom, and the other end of the cord is tied to it. After correcting the length of the cord for lens height, stepping on the weighted end and lifting the camera will tension the cord. This force, opposing the downward pull of gravity and limited by the length of cord, creates a semi-rigid structure, which forms one axis of stability. Conveniently, the entire rig fits easily into a pocket.

Aerostat, Bungee, and Easyrig

A balloon filled with a gas that is lighter than air has a tendency to float upwards. It has buoyancy. If a camera rig is lifted by a force in proportion to its mass as affected by gravity, a beautiful floating balance can be achieved. The main defining axis of force and movement on this planet points directly at its core. When lift is equal to weight, there is an equilibrium. In this state, it takes very little force to control the object's movements. This is what a camera operator wants. Of course, if the upward force were actually a balloon, the wind might just as easily take control of the feather-light camera rig. Although a few interesting videos have been made with a handheld camera rigged under a large envelope of helium, much of the current work in this area has the balloon and the camera way up in the sky capturing aerial shots.

A bungee cam is a flexible rigging method that builds a highly customizable piece of gear. There are many possible ways to rig bungee tethers, which help to create the sought-after equilibrium of weightlessness. The most common design has four vertical bungees supporting the corners of a camera-sized cheese plate platform with two pan bars for control. Many custom bungee rigs have been made over the years, especially for isolating the camera from a vehicle's or a dolly's movement.

Perhaps the ultimate portable bungee cam is what is known as the Easyrig. This device hangs the camera from a cable suspended from an overhead rod with a shock absorber. The rod is supported by a body harness, which transfers the weight of the rig to the operator's strong core and leg muscles. By matching the proper rig, each with a different spring force, to the weight of the camera package, it becomes easier to operate handheld within a limited range of motion. The most advanced version of this product allows for adjustment to the cable tension, so that the optimum lift can be found for each configuration.

Gyroscopes

The classic stabilizing gyroscope is a spinning metal wheel mounted in a restrained gimbal, which can change its orientation only in a limited way. The force created by its fast spin resists motion by trying to keep its axle, or spin axis, pointing in the same direction. These electronic gyros can be used as effective image stabilizers when there is random motion that needs to be damped. Perhaps the most well-known provider of these gyros is Kenyon Laboratories. They make a two-axis gyro that resists movement of both panning and tilting (pitch and yaw) when it is mounted in line with the lens, utilizing two counter-rotating metal wheels spinning at 20,000 rpm. Another version has four flywheels and stabilizes three axes, including roll. Each version comes in several different sizes, which can be matched to the weight of a camera package. Generally, a three-axis stabilizer is best suited for video; however, when size and weight are important considerations, like with a handheld or shoulder-mounted rig, a two-axis gyro may be the best choice. These systems can be used to steady a camera in many creative ways, from a mount high up on a pole, to a rig flying through the air on a cable.

Tyler Camera Systems' Minigyro and Gyrobot are complete systems for stabilizing a camera package up to 32 pounds. These systems contain two battery-operated KS-8 Kenyon Gyros in a lightweight body with two reconfigurable handles. It can be mounted under a camera and used handheld, or in combination with its shock-absorber column base. This base can be rested on an unstable surface if necessary, although gyros perform best without contact. On a moving platform, it must be loosely tethered to a hard mount for safety, where it will reduce the effect of unwanted motion from a vehicle, boat, or helicopter. It can be used in conjunction with an Easyrig or a bungee cam. The spinning gyros need 20 minutes to reach operating speed, and unfortunately, they create a fair amount of noise. This sort of gyro cannot be swung too quickly for a move, or it will create an undesirable percussion, due to an action called gimbal lock. The larger the gyro, the faster it can be panned.

When working with gyros, it is important to consider the forces that need to be stabilized. When there is a fair amount of movement or heavy winds, a larger system will provide better stability. The operation of these systems takes a bit of practice. It is best to use a light touch when operating, allowing the device to do its work.

A different kind of gyroscope is used in the brushless gimbal systems, which will be covered in the next section. They are MEMS gyroscopes (microelectromechanical systems), specifically, vibrating structure gyroscopes, which are small electronic components used to detect movement.

Handheld and Body-Worn Stabilizers

The simplest form of this device is the pivot-style, lowered center of gravity, mechanical stabilizer. On top of an ergonomic handle, there is a pivot, which is actually a gimbal with a limited range of motion. This type of handheld rig features a handgrip on a swiveling ball-joint, which attaches to the camera's baseplate. An offset arm curves from the base, around the handle, hanging an adjustable weight at a point below the grip. This, in effect, lowers the center of gravity, and allows the rig to maintain balance dynamically. The pendulum effect causes the camera to automatically return to a central and plumb position. This increases the camera's inertial stability. By carefully balancing and using the correct amount of counterweight, unwanted movements can be regulated fairly well. It can be used with good effect, and it is relatively inexpensive.

Another type of mechanical stabilizer has a three-axis gimbal and a straight column between the camera and the lower counterweight. The handle is offset and connects to the gimbal on the column. This connection point can be adjusted along the length of the column, in order to gain the best-balanced placement. Additionally, it is possible to extend or retract the length of the weighted column, in order to establish the most controllable physics. This style of mechanical stabilizer can be easily mounted by its handle to a stand, for hands-free balancing and weighting. The handle also allows for mounting to a rigid arm on a body harness, which transfers the weight of the system from the operator's arms to the larger muscles of the body during prolonged use. Of course, when the skilled damping movements of the operator's arms are removed from the equation, the stabilizing effect is diminished.

One of the most well-known and extensively used camera stabilizing rigs of this sort is called the Steadicam. It is used to support a camera by an operator for extended periods of time, without the shake that comes from fatigued muscles. Even though the weighted column is attached to a custom-fitted exoskeleton vest, it isolates any movement from the operator's body by using a spring-loaded, shock-absorbing, articulated arm. This mechanical arm, when tuned and balanced completely, does a much better job of correcting for the body's motion than a camera operator can accomplish with only his own arms and a handheld rig. Since its introduction to the industry, the Steadicam has revolutionized what is possible, creating smooth, continuously traveling, handheld shots, even while running at a fast pace.

> Images from a Steadicam cut well with other shots, whether they are handheld and shifting, or precisely framed and firmly supported. The subtle, almost organic, stabilization of the Steadicam can create a close approximation of our vision system's natural view. This aesthetic has become an indelible part of the language of film and video production.
>
> (Mark Sparrough, SOC, DP, Steadicam owner/operator)

A Steadicam can be used in the traditional high mode, where the camera sits on top of the weighted column. It can also be configured in low mode, in order to capture low-angle shots. There is also a horizontal mode, where the column is balanced to be stabilized sideways, so that the camera can be outrigged over a lateral surface, such as a table or a fence. Finally, there are ways to rig a Steadicam arm onto a rigid support, using a hard mount, or a *vehicle mount* adapter, as shown in Figure 5.6-2. This is commonly utilized for moving vehicle shots for safety, rigging directly onto speed rail or onto a Mitchell mount with the necessary adapter.

> A Steadicam allows a great amount of flexibility. Moving the camera in certain ways can elicit feelings of anxiety, fear, excitement, or a sense of calm. In this photograph, I am hard-mounted on a *Rickshaw Rig*, matching the speed of joggers, and creating a smooth gliding motion for a fitness commercial.
>
> (Matthew Fleischmann, Steadicam owner/operator)

Self-balancing personal movement devices like the Segway have been modified and used in conjunction with a Steadicam rig, usually with a custom

FIGURE 5.6-2 Steadicam Vehicle Mount

Courtesy of Matthew Fleischmann

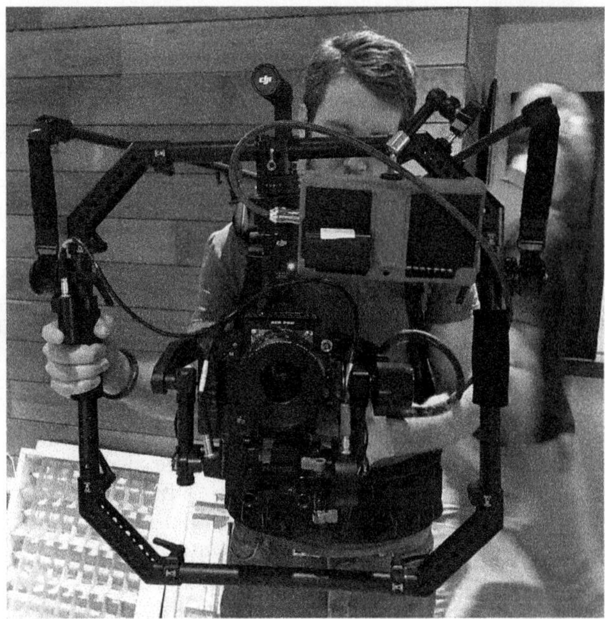

FIGURE 5.6-3 Fully Rigged DJI Gimbal System with Harness Support and Lens Control

Courtesy of Redrock Micro and Hamad Almuftah

hard-mount assembly, to create versatile dynamic stabilized shots. This combination is commonly called a *Steadiseg*, and it enables an operator to perform complex free-roaming movements at higher speeds and without the fatigue of supporting the rig.

Another newer type of stabilizer, called a *gimbal system*, shown in Figure 5.6-3, can also enhance the ability to shoot smoothly while using a handheld camera. There are varieties suitable for the smallest POV cameras, and versions capable of holding fully built-up rigs. These electronic, gyroscopic, three-axis, motorized gimbal stabilizers are found on set regularly. They utilize brushless DC motors and sophisticated software algorithms, in order to counter any negative effects of movement. These algorithms help the device to permit smooth and even camera moves, working to eliminate unwanted shake. Firmware updates are continually improving the effectiveness of these systems and adding new control features. This small lightweight gear is changing the way many productions work. It is increasing the speed of capture for many shots that would have required additional time to set up other more painstaking equipment. Almost completely silent, any motor noise should be below normally recordable levels. A gimbal system may be operated simply handheld out in front of the body for short periods of time, or it is commonly suspended from two support rods and a chest harness. Another form of support, specifically designed to assist in holding the weight of a gimbal system, uses two adjustable-tension spring arms and an exoskeleton vest. Gimbal systems are a necessary part of image stabilization for drones, and they are mounted on many other compatible camera motion projects that need additional reduction of random movements and vibrations. It is important to note that these systems automatically adjust for a level horizon, which is a skill that Steadicam operators must develop through practice. Brushless gimbal systems can be operated solo, or in tandem with another operator, who watches a handheld wireless video monitor and remotely controls the pan, tilt, and roll axes. With the addition of wireless focus, iris, and zoom motors, full lens control can be accomplished. Additionally, a high-precision *pan bar controller* has been developed, which remotely operates the pan and tilt axes, using a hard camera form factor with which most operators are familiar.

RIGID SUPPORTS

Connections

Most cameras connect with their camera supports using one, or more, threaded holes found on their bottom. These holes are most commonly threaded SAE 3/8–16 (UNC) for full-size cameras and SAE 1/4–20 (UNC) for smaller units. Some form of *camera plate* is usually mounted to the camera using machine screws. This could be a simple dovetail-style quick-release plate for easy mounting and dismounting from a pan/tilt head of some sort. There are heavier plates that slide fully onto the head before locking, and there are V-shaped plates that hold some of the heaviest cameras. It could also be part of a camera platform, which might include bases, grips, shoe mounts, or other mounting points for various accessories, such as lights, microphones, monitors/viewfinders, matte boxes, or a rod system for lens support and control. As cameras accessorize, there is a need to interconnect a system of supports and handles in ways that protect, enhance, and simplify the use of a complex group of delicate equipment. Rod systems connect with various simple bridges or full bridge plates. Many types of *cages* can be built around cameras. These may be full or partial cages, and they are often made up of custom metal bracketry with many threaded holes for mounting and customization. *Cheese plates* are often useful both for building up camera systems and also for rigging cameras. They are multipurpose mounting plates that have a pattern of either threaded holes or thru-holes for 3/8″ or 1/4″ connections. A common standard matrix pattern spaces the holes 1″ apart, center to center. Some holes may be strategically countersunk for mounting hardware flush to the surface of the plate.

On full-size broadcast cameras, there is a heavy-duty lever-released baseplate, which has a series of threaded holes on the bottom that allows for the best placement of the quick-release plate for general weight balancing on a pan/tilt head. This baseplate essentially covers the built-in shoulder mount, and allows for a quick conversion from hard-mounted to handheld modes. Sometimes known as *convertible cameras*, they can be used portably in the field using the contoured shoulder pad, or in a traditional studio configuration, or attached to a moving camera support and used as a remote camera.

Fixed Cameras

When a camera has a special location, it might be rigged in unusual ways. Once mounted, the camera may be left safely without continuous operator support. Many different types of hardware can be used, hopefully creating a completely stable and shake-free image. Solid metal parts have structural rigidity, and they can be paired with hardened steel rated bolts and attached to other load-rated metal pieces, such as lighting grids, truss structures, or speed rail pipes. When in doubt, always use the highest-grade hardware available. A grade 8 bolt has a tensile strength of 150,000 lbs per square inch. This is relevant to its preload strength as it is tightened, because it is a connector's threads that ultimately support the union. Of course, components that thread together should always have matched hardness levels, so that a softer set of threads cannot be damaged easily by over-torquing the connection.

Truss, as it applies to video production, is generally aluminum 12″ × 12″ plated box truss. It comes in standard lengths of 5′, 8′, and 10′. Plated truss refers to metal plates on the ends, with mounting holes for bolted unions. When attaching to a truss structure, the main supports are pipes with a 2″ outside diameter. There are other sizes and shapes available, as well as light, medium, heavy, and extra heavy-duty ratings. Available in 20′ lengths, *speed rail*, otherwise known as 1–1/4″ schedule 40 aluminum pipe, can be cut to any size. The actual dimensions of speed rail pipe are 1.38″ inside diameter and 1.66″ outside diameter. Speed rail stock can be found in a heavier 1–1/2″ size, which can also be used for *lighting grid* pipes. In conjunction with the many matching connectors, corners, flanges, fittings, and pipe clamps, both fixed and swivel, almost any supporting structure might be built. A *cheeseboro clamp* is commonly used to make connections to a pipe. Fixed cheeseboros are set at a right angle, and swivel cheeseboros adjust to any angle, connecting two pipes together. Other cheeseboros come in narrow profile, or have just one clamp and specialty ends, like posts, eyebolts, and hooks. Additionally, various grip mounts, some normally used for lighting, can help to mount cameras and remote heads in unusual locations. Whether building the structural foundation for a camera on a simple leveling head or an advanced robotic movement device, working in conjunction with an experienced *rigger* is always a good idea.

There is one thing to always be sure of. When objects are rigged overhead, safeties must be used. There cannot be any metal parts or heavy objects overhead that could loosen or fail to hold. Use redundant clamps when possible. Load-rated carabiners and climber's webbing tethers (Dyneema slings) are very flexible, and can be especially useful for devices that move, like remote heads. They should be reserved for outdoor use, not in studios, because they are not fireproof. The most common safeties are made from steel aircraft cables. Every item added to a lighting grid, for example, will have a metal safety cable.

Hi-Hats

These are simple, yet invaluable, camera supports. The simplest hi-hats are short, rigid, metal risers with mounting holes on the bottom that allow for a simple wooden base or for bolting onto cheese plate or other rigging. The top may be designed to fit one of the standard head interfaces: Mitchell, 75 mm ball, 100 mm ball, 150 mm ball, or flat mount. More advanced hi-hats are made in the form of very short tripods, and allow for some adjustability in height and level. An ultra-short version of a hi-hat might be referred to as a *low-hat*. Hi-hats can be useful as a desktop mount, perhaps for building and testing a camera before mounting it on a location rig or onto a robot. They allow for the use of a traditional pan/tilt head where you have very little room to work. They can facilitate low-angle shots or custom mounts in tight or unusual spaces.

Camera Risers

These metal risers come in various shapes and sizes. They can be configured to extend a hi-hat or to change the camera height on any number of hard-mounted rigs and dollies. The traditional metal pipe style comes in several stackable lengths: 4″, 6″, 12″, 24″, 36″, and customizable versions that utilize lengths of speed rail, which may be cut to any specification.

An *offset* is a heavy metal bracket, which shifts the camera mount position laterally. They come in many sizes, and can allow for space restrictions, or when the base of a rig cannot be directly under the camera, often helping with the interaction of the camera operator. Sometimes referred to as a *side arm*, they are often used for placing the camera off the side of a dolly. There are rotating versions, with firm locking mechanisms, that provide fast adjustability. Camera risers and offsets generally connect via Mitchell mount or Euro mount.

A *bazooka* is another metal column form of riser for elevating your camera package. Most connect using the international *Elemack* standard, also known as the *Euro mount*. This 80 mm system consists of a male Euro-adapter that fits into a female Euro-clamp collar, tightened with a convenient handle. It is a dual-safety coupling, with a redundant bolt that locks the mount connection through the center. Various adapters are used to conform to other head mount types. Similarly, there are adapters available to convert bowl mounts and Mitchell mounts to Elemack. Bazookas are quite stable on a moving support. An *Elemack mounting plate* can be placed strategically onto a dolly platform to attach a bazooka or to start directly with an offset for "low mode" configurations.

These support pillars may also be used underslung, perhaps hanging from a grid, thus eliminating the need for any ground footprint altogether. Bazookas come in three main types, two of which are height-adjustable on the fly. This ability to adjust elevation quickly off-shot can be another prime benefit of using a bazooka. These different sizes of adjustable risers can provide a range of height changes, from a few inches to a couple of feet. There are pneumatic gas-column risers and mechanical spindle hand-cranked risers. Fixed-length bazookas come in 3.5″, 6″, 8″, 12″, 16″, 18″, 20″, and so forth.

Bazookas can also be used freestanding in conjunction with a *bazooka base*, which is low to the ground and adjustable for level. This configuration is sometimes referred to as a *camera stand*. Convenient multi-configuration *Elemack center plates* can be built out with three to six legs. A base with three equal legs and adjustable feet is balanced and sturdy. With the addition of studio wheels and an offset seat, a simple caster dolly is formed. Four equal legs, skateboard wheel brackets, metal track, and an operator's seat create a simple and effective track dolly. Various combinations of different length legs can suit any space requirement. Customized camera mounts can be constructed using legs that attach to 1–1/2″ speed rail. Cut-to-length vertical pipe-ready leg mounts can conform to multiple height levels or uneven surfaces. Horizontal pipe attachments allow for safe and sturdy rigging. Hinging variable tilt elbow risers can fix any angle, and rotating risers quickly provide for any pivot orientation.

Leveling Mounts

These short camera mounts have the ability to adjust their top plate for perfect level. There are several different versions of common leveling heads. They use threaded bolt supports with turn knobs, offering multiple points of adjustment. There are *four-way* and *three-way* types, with three and four points of adjustment respectively. Some have top and bottom Mitchell mounts or Elemack bases; others have cheese plate bases and slotted top plates. A *ball-leveling mount* adjusts the level of the top plate with a ball in socket form of swivel, which locks firmly in place at the turn of a knob. Leveling mounts are necessary for leveling on camera cars, dollies, bazookas, risers, offsets, and various fixed mounts, both for manual and robotic configurations. A *wedge plate* is a pair of hinged camera plates, providing 0 to 90 degrees of variable, locked-off tilt for changing the necessary orientation of a fixed camera or a camera head. This is not to be confused with the quick-release camera plate that is sometimes also referred to as a wedge plate. Also in use, there are versatile mounting plates with adjustable threaded rod leveling feet, meant for a variety of quick placements directly on surfaces and for unusual rigging setups. They can mount a camera directly, or utilize hi-hats and risers for height.

Hard Cameras

The term "hard camera" can denote different configurations. The general description is a full-size broadcast camera on a stationary tripod with a studio viewfinder. A hard camera should be placed on a stable surface at a location to cover the action. The operator will control the rig with extended lens controls mounted onto the pan bars. Variations include different pan/tilt heads and lens sizes. Another term to describe this type of setup could be studio configuration.

When the position of a hard camera is far away from the subject, or when the highest-quality image is needed, then the largest of broadcast lenses, which are telephoto box lenses, must be used. These studio/field box lenses generally provide the best high-definition performance for contrast and sharpness, both at the lenses' center and at their edges. Their multiple large glass elements cause these units to weigh a lot, often over 50 pounds. In order to safely merge a box lens and a full-size broadcast camera, they are built onto a metal platform called a *sled*. The purpose of the sled is to provide the stiffest possible connection. To this end, there is a thick metal bracket for the lens to hang on, keeping all of the weight off of the camera connection point. There are several different versions, some having cage-like grab bars for lifting the finished rig, and some containing additional built-in support electronics. The sled buildup configuration is used widely for live events, stage shows, and studio productions, when a hard camera position necessitates the best or the largest and longest of telephoto lenses. These heavy camera packages rely on heavy-duty pan/tilt heads and tripods. The connection uses a V-shaped plate that is mounted with machine screws to the bottom of the sled, allowing it to be locked firmly onto the head.

Camera Platforms

It is often necessary to elevate a hard camera buildup and its operator higher than a tripod will allow, in order to create optimal sightlines. This problem might be solved by utilizing two height-adjustable 4′ × 4′

staging risers. One is for the tripod and camera, and the other is for the operator. A small gap between the two platforms keeps any movement from the operator's platform from reaching the camera. Platform legs might be adjustable for height, or they could be pipe cut to custom length. When building this type of setup, there is a compromise between more height and more instability created by the riser. When necessary, use a heavy-duty ratchet strap around all four legs of each platform, like a belt, so that there is a moderate inward force helping to connect the legs and their bracing, uniting the entire platform structurally. Also for consideration, there are a variety of lightweight portable kits from different manufacturers in many shapes, sizes, and levels of stability. Probably the best known of these kits raises the camera an additional 23" and is called a Spider Pod.

Tripods

Commonly called *sticks*, these three-legged collapsible devices are probably the first piece of camera support that most people learn about. They provide a portable and easy to manipulate structure that can be leveled and adjusted for various heights. The most stable setup for a tripod, if the ground is level, has each of the three legs extended equally, and each foot equidistant from an imaginary point on the ground, centered below the head. When necessary, level the tripod itself, by adjusting each leg. Once the tripod is perfectly level, the pan/tilt head can then be leveled to match.

There are both heavy-duty and lightweight tripods. Heavier camera packages, with longer lenses and more accessories, need a bigger tripod. When in the field, it is convenient to have lighter tools, but not at the expense of the ability to operate your camera well. Stronger tripods are more rigid, so they will deform less when moving force is applied to them. Even the most minor twisting or bending is bad, because it will affect the stability of the shot. Even worse, the stored energy from the structural distortion will rebound, causing a backlash effect that can be visible in the image. When in doubt, use a heavier tripod.

Tripod legs are generally manufactured from either aluminum or a plastic composite reinforced with carbon fiber. Both composites and aluminum have a high strength-to-weight ratio. Aluminum is potentially less likely to sustain damage from an impact, but also provides very little in the way of vibration damping. Carbon fiber is the most expensive, but it is a bit lighter, damps vibration better, and does not feel as cold in low temperatures.

Spreaders connect the three legs and add form to the tripod, while keeping each leg from shifting position. A *ground spreader* adds the most stability and performs well on reasonably flat hard surfaces. A *midlevel spreader* is slightly less stable, but allows for adjustment on uneven surfaces. There are different types of feet. There are foot ends that fit well into the ends of the ground spreader, rubber feet that resist slipping on smooth surfaces, and spiked feet that can be pushed into the ground. Some tripods have all three types of feet on a swivel array at the end of the legs for easy swapping.

There are many functions and features to consider. What is the effective range of lens heights that can be produced with the setup that you are building? Are the extension locks on the legs adjustable for tension, so that they can be maintained effectively? What tools are needed for repair and adjustment? What kind of spreader will be needed? What kind of feet will be used? Are wheels needed for easy repositioning? What pan/tilt head is best optimized for the weight of the camera package, and which type of mounting connection does it have?

There are a few specialty tripods on the market that can be quite useful in certain circumstances. Although rarely seen, there are professional tripods, suitable for heavy camera packages, featuring an adjustable-height center column that provides fast elevation changes using a simple hand crank. Another heavy-duty tripod on the market features a built-in three-way leveling Mitchell mount, which allows leg heights to be optimized without regard for the level of the Mitchell casting, making it an excellent choice for heavy remote heads, jib arms, or just a full-size camera and a box lens.

A basic tripod dolly with caster wheels, called a *skid*, can be useful for quick repositioning or even careful on-air moves on a smooth floor. These wheels attach to the tripod feet in the form of pairs that pivot at the end of three arms, jointed at a central point, like a spreader. When removed, they fold and lock together, for easy handling. In a studio, the best wheels tend to be solid rubber and firmer, for control and stability on hard floors. In the field, OB (outside broadcast) skids have heavy-duty arms, with taller wheels and treaded pneumatic tires made of softer rubber, in order to perform over irregular ground surfaces. Features like cable guards and positive wheel locks with easy engagement can be important, because fast changes in camera position can be necessary. Locking a caster that might swivel unexpectedly is a good idea, so that the force of a panning head will not inadvertently move the tripod. If live event action dictates constant repositioning, a well-placed operator's foot can help to steady a caster. The wheels of some skids can be fixed directionally in 60-degree increments, locking out the swivel action, so that various smooth and defined camera dolly trajectories can be made without track.

For even more precise moves, a tripod and skid can be mounted on a pair of traditional metal dolly tracks. There are many systems that are capable of building these simple tripod dollies. There are skids that function as floor/track hybrids, allowing for use of flexible dolly track made of smooth rubber rope, which can be quickly installed and easily transported.

MANUAL CAMERA MOVEMENT

Pan/Tilt Heads

Most manually operated camera heads control two axes: pan and tilt. The most common type of professional-grade manual pan/tilt head is called a *fluid head*. Mechanical components inside of the head are forced to move in a controlled way, because they are surrounded by a viscous fluid. Continuously variable or stepped settings change the way that the fluid resists, allowing for adjustable motion damping, providing the preferred amount of drag to the head's rotational movements. The camera moves are performed using *pan bars*. Two-handed control is the easiest way to get exact moves, one pan bar on each side of the head. Often, the zoom and focus controls are extended to the handles, and a viewfinder monitor is rigged onto the camera to allow the operator a large, easy-to-see video image. If the camera is mounted onto a pan/tilt head ENG-style, only one pan bar is used, and the operator looks through the viewfinder eyepiece, placing his or her free hand directly on the lens for control.

Once the camera is fully built, it must be balanced front to back, so that the tilt can be worked effectively. Different heads have various mechanisms for this. Basically, the camera plate will slide by either manually pushing or pulling, or by turning a dial. There is a friction lock that must be opened before adjustment and closed afterwards. Before balancing the camera package, if there is a counterbalance setting, start at the lowest setting. Also, make sure that there is no drag on the tilt axis. This will allow the camera to tilt freely, in order to observe its balance. If the likely range of tilt for a camera on a production is known, then it can be a simple matter of adjusting the camera plate so that the camera will remain stationary in the middle of the tilt range, which is usually level and horizontal.

The next consideration is counterbalance. Most manual pan/tilt heads do not allow for the vertical adjustment of the camera base, and the camera sits on top above the pivot of the tilt. If the axis of rotation for tilt could be matched to the camera's center of gravity, then the camera could hold its position freely anywhere in the complete tilt range. Instead, the camera tends to fall off of the horizontal plane once any tilt movement is initiated. To solve this, we use *internal counterbalance* to keep the tilt from becoming increasingly weighted the farther it gets from the horizontal, as it tilts backward and up, or forward and down. Counterbalance may be engineered into a head in several different ways. Seldom does it work perfectly, but a good compromise can usually be found. If too much counterbalance is added, the camera will fight to stay level, making it harder to tilt. Often, the best solution may be to set the counterbalance to the lowest level that helps to keep the tilt from moving due to gravity alone. If the counterbalance adjustment is continuously variable, it may be dialed in carefully. Many systems use stepped settings, and the amount closest to your need must be chosen.

Both the tilt and pan axes should have the necessary range of adjustable *drag* that is appropriate for the weight of the camera package. This resistive force is crucial. It creates a level of damping, which smoothes out every deliberate motion. If the head's movements were frictionless, it would be impossible to start and stop a move gracefully without wavering. Once the proper amount of continuous drag is set according to the operator's preference, a move can start gently and stop confidently, staying at rest until the next move begins. A common way to set drag is to find a comfortable level for the pan, and then use a slightly firmer setting for the tilt. This allows for easy and level panning without accidentally engaging in tilt. Both the pan and tilt axes have a friction lock in order to avoid accidental movement. Heavier heads will usually also have a mechanical hard lock as an additional safety for the tilt axis, which holds any unbalanced weight against the force of gravity.

A *geared head*, or "gear head," is considered to be the optimum head for holding a camera rock steady, and then creating even, repeatable, and precise pans and tilts. It utilizes a mechanical advantage of metal gearing, separating the action of the pan from the tilt, and is controlled with two hand wheels. The design of these heads enables the performance of very controlled, cinematic, sweeping movements without backlash or exaggerated motions. The tilt gearing is in the shape of an arc, which is a portion of the circumference of an imaginary circle. The camera package's center of gravity could be placed at the center point of that imaginary circle, and also aligned with the center of the pan axis, in order to create a perfect balance of stationary weight. However, the alignment of the lens' *optical center* at this point, for reasons of perfecting the visual geometries during a move, might take precedence over perfect weight balance. The addition of a counterweight along the tilt axis can help correct for a misaligned center of gravity, if necessary.

A clear advantage of a geared head is that the camera will not move unless the operator moves a hand wheel. Also, when coming to a sudden stop, there is no visible wobble, even when the image is magnified on a big screen. Another feature is the ability to repeat moves for exact placement by keeping track of the number of revolutions that each wheel will make during a shot. Some of the more well-known classic gear heads in the industry are manufactured by Panavision and Arri. They generally have three selectable gears, allowing for slow, medium, and fast speed ratios. Many feature an adjustable wedge plate for facilitating more angled tilts past the range limit of the tilt gear's normal arc. Traditionally these heads have been used to move large and heavy camera packages, but recently there are new smaller versions on the market, designed for the weight and size of modern cameras.

Fluid heads, for many operators, are more instinctual to use and can usually allow for quick starts and instantaneous reactions. Long lens work is especially suited to a fluid head. The pan and tilt motions can be used in combination, for a subtle float-like movement. One thing to keep in mind is that fluid heads have a reputation for drifting, sometimes very slowly and continuously, when you least expect it. Correct use of the settings, and the right kind of practice, can allow a great operator to perform world-class camera moves.

Once a video image has been captured, the playback and viewing of that work of art represents a manufactured point of view. It is the created point of view of an observer, formed by the camera. Where a lens points, forms a directional point of view, but additionally the choice of location in space in relation to other objects, is the basis of emotional and practical thoughts about what is seen. Inside the lens, the optical center dictates the location that the "viewer" is placed, and from what point in space the viewer's eyes will see. In other words, the placement of the lens' optical center is fully relevant to the perspective from which the audience interprets the video and personally identifies with what they see. Furthermore, in order for video to appear completely natural, this "nodal" point should be aligned to match the rotational axes of the head upon which it is affected. When this alignment is perfected, rotation of the camera on an axis does not cause near and far objects to shift position relevant to each other, as they appear in the image. When objects in the lens' field of view appear to remain at the same locations in relation to each other as the camera moves, the axis is aligned with the optical center and is considered nodal.

Geared heads, certain fluid heads, and even specially designed modern friction-based pan/tilt heads will allow the nodal placement of a lens' optical center. These are called *nodal heads* (see Figure 5.6-4). These heads allow for the camera plate to slide backwards farther than normal, and have the ability to raise and lower the baseplate. There is some commonly misunderstood theory surrounding their use. There is one optical center. There are two actual nodal points in a lens, which help us to understand and locate the optical center. These nodal points, or principle points, are axial points, but are technically different than the optical center. Light moves straight through these points, following the optical axis, or principal axis, of the lens. The optical center is a point on this axis, where all of the light from the lens intersects, on its way to the focal plane, to form the upside-down image of what the lens sees. It is the vertex of the lens' angle of view.

When the pan and tilt axes rotate around the optical center of the lens, it most closely replicates the action of an eyeball rotating in its socket. Consider the human visual system. As the human eye pans, the optical center does not move. It is centered inside the eyeball. Now consider the arc that a non-nodal camera and lens creates. Imitate this arc by moving your head as if it is a satellite orbiting around an imaginary point. Do you notice the unnatural difference in what you are seeing? When objects in view shift in relation to each other, it is known as *parallax error*. Some studies show that human perception does not always notice these details, but it is quite likely that what is not consciously recognized is still subconsciously registered to some effect.

Even the roll axis can benefit from "nodal" lens alignment. This is how a perfect 360-degree spinning image is created. Of course, in practice there are often more important considerations for camera work, such as nodally balancing a camera's center of gravity. Also, when using the zoom feature on a multi-element lens system, the optical center will shift, changing any precise alignment. So, while a nodal head can be a wonderful tool for image making, it is certainly a specialized one. Close-up tabletop work and special effects traditionally benefit greatly from aligning a lens' optical center, but there is a growing consideration for its use in more standard areas of image capture. The Cartoni Lambda head has special sliding counterweight rods that can be perfectly adjusted, so that both center of gravity and the lens' optical center can be placed in line with the rotational axes.

FIGURE 5.6-4 Lambda 50 Three-Axis Nodal Head on a Heavy-Duty Three-Way Leveling Tripod

Courtesy of Cartoni

Pan/tilt heads and tripods have a few different types of connection methods. There are a few proprietary designs, which are not likely to be encountered. The most common mounting systems are discussed below. The simplest base on a head is a *flat base*. It attaches with either one or four 3/8–16 mounting bolts, and it can be the lowest profile for specialty rigging. A *ball base* utilizes a socket-type mount, and is made in a range of sizes: 60 mm, 75 mm, 100 mm, and 150 mm. Larger diameter balls provide more stability for heavier camera packages. Ball mounts allow for the easiest head leveling, because they can adjust for 360 degrees of horizon simultaneously and lock simply by turning the bottom mounting knob. The most robust type of mount is the *Mitchell base*. This mount has a large, approximately 2–7/8" diameter center hole that fits a 2–3/4" threaded column insert. One or more 3/8" wide keyways accept a registration node in the base to avoid unwanted rotation. The mount is firmly locked in place with a special *castle nut*. It is important for the castle nut to use all of the available threading so that there is no movement in the mount. If it will not thread completely on, a custom 1/2" thick spacer washer can be used to take up the slack. The castle nut should always be stored attached to the Mitchell base when not in use, in order to protect the threads. Damaged threads can make assembly difficult, or even impossible. If necessary, they might be carefully repaired with a triangle file. Having a specialty wrench, matched to the particular shape of the castle nut that is being used, can be quite helpful. There are ring-shaped adapter plates available to convert a Mitchell mount to a ball mount, and 150 mm ball mount to 100 mm, or 100 mm to 75 mm, as needed. Other adapters are available to convert a flat base to a Mitchell base, or a ball base, and vice versa. Adapters have their own ratings for load and may lower the overall weight capacity of a rig.

Three-Axis Camera Heads

Although less frequently utilized, there are manual camera heads that control the roll axis as well as pan and tilt. The Cartoni Lambda head, as discussed above, has the ability to provide a Dutch angle, using its third axis roll feature. The older and possibly less convenient Weaver-Steadman three-axis fluid head is still very much in use in the industry. It is less expensive to rent, quite rigid, and has the ability to expand to a very large configuration. Another industry-favored nodal head is the Ronford-Baker Atlas 7 fluid head, which is highly adjustable using various sizes of platform blocks and fine-tunable adjustments. Like the Lambda head, it utilizes a steering wheel in order to create roll, pan, and tilt simultaneously. Similarly, adjustable counterweights allow for perfect weight balancing. This head provides for a 360-degree tilt by reconfiguring its components into a side mount for unrestricted tilt action, with an image field of view that is unobstructed by the head or the camera riser. There is also a "fourth axis" configuration that allows the camera to point straight down, and then pan right or left with 360 degrees of movement.

An interesting alternative to a three-axis head is a swiveling wedge plate, which can be used under a traditional fluid head to create manual roll effects or to extend a tilt for straight-down shots. The Matthews CAM Tank, shown in Figure 5.6-5, allows for these effects and can also create a very low-angled shooting platform with the ability to pan and tilt, or pan and roll.

FIGURE 5.6-5 CAM Tank

Courtesy of Matthews Studio Equipment

Perhaps the most recognized third axis add-on device is the Tango Swing Head. It is a low-profile (2.3") manual roll mechanism that can be used with any traditional pan/tilt head. It is a proven tool, sturdy and silent, with 90 degrees of available motion. Because the camera resides inside of the mechanical roll apparatus, the lens' optical center may be aligned to the axis of its movement, allowing for a nodal configuration. For various reasons, it is not always possible to position the camera to roll around the lens' optical axis, thus creating a less natural motion. With an uneasy, awkward movement of the image often being the desired outcome for a Dutch angle, this traditional deficiency might be used to good effect.

Dollies

Dollies have been used in motion imaging since the early days of the craft. Placing the camera on a rolling platform has great advantages. The camera can now be smoothly moved laterally, side to side, or forward and back. Many dollies are convertible between free-roaming steerable devices, which can roll on any smooth surface, and *track dollies*, which glide on different styles of tracks. Track profiles are made in many custom shapes but are sometimes simple metal pipe, such as speed rail. Tracks must be carefully leveled in order to promote the best vibration-free effect from the dolly itself. Bubble levels and laser levels used with wedges and cribbing (flat dimensional lumber pieces) are common solutions.

Often, fully integrated smaller track dollies are called *sliders*. Some slide on a frictionless surface instead of using wheels, hence the name. Most often operated manually, they are generally lightweight and portable. They can be placed directly on a surface, or elevated with one or more tripods or heavy-duty stands. Some sliders can mount vertically, so that the camera can glide up and down, or even diagonally on an incline/decline, depending on the unit's necessary orientation. In order for a slider to maintain balance as it moves off-axis to horizontal, it must be counterbalanced by a sliding weight that is relative to the mass and position of the camera package. Some of the most functional slider designs maintain their balance on top of a pivoting head on a single stand or a tripod, allowing for the addition of a jib-like swing as well.

On location, a slider can accomplish moves similar to a traditional dolly, or a jib, when there isn't time or space for a full-size rig. Smaller cameras can use smaller sliders, some of which can be carried all as one unit, or even slung over the shoulder by a strap. There are motorized sliders, which can be used like a primitive remote dolly system. Also, there are versions that are programmable and can repeat a move exactly for multiple takes, special effects, or for incremental moves when creating a time-lapse.

A simple and effective slider dolly base of 12″ × 12″ × 3/8″ aluminum cheese plate with Mitchell mount allows for full-size pan/tilt heads and cameras. Mounted onto the base are four sets of wheels that roll on a sturdy track built from custom lengths of speed rail and adjustable speed rail brackets. Different sized hi-hats can be used on this slider to raise camera height or to allow for camera heads with different mounts. Traditionally, everything is rigged on top of the slider's base, which is on top of the track. This is called *overslung*. When necessary, it can be configured in *underslung* mode, where the head and camera are mounted hanging under the base, sliding through the rails of the track itself.

A very simple dolly, created by *dolly grip* James Saldutti, is called the Dutti Dolly. (See Figure 5.6-6). It is a low and stable yet very compact and portable 20″ wide metal plate, with a Mitchell mount and four sets of skate wheels. It can be used on track, or directly on the floor, to create super low-angle shots, or it can be built up like any dolly, with bazookas, risers, and so forth. At less than 2″ off the ground, it can be used for gliding step-off shots by a stabilized handheld operator.

A slightly more complex style, shown in Figure 5.6-7, is called a *doorway dolly*. Less than 29″ wide, these narrow platform dollies have pneumatic tires, a T-bar for steering, and the option of an *operator seat*. A *turret* can also be utilized. This allows both the seat and the camera mount to swing 360 degrees together on a strong and balanced rotating support beam, which can then be locked into position. When necessary, a second seat can be added for a camera assistant/focus puller. Traditionally rolled on the ground, a doorway dolly has the ability to adapt to rail use with the addition of track-wheel brackets. When these dollies need to be wider, sideboards can be added and the axles can

FIGURE 5.6-6 Dutti Dolly with a Cartoni Fluid Head

Courtesy of Matthews Studio Equipment

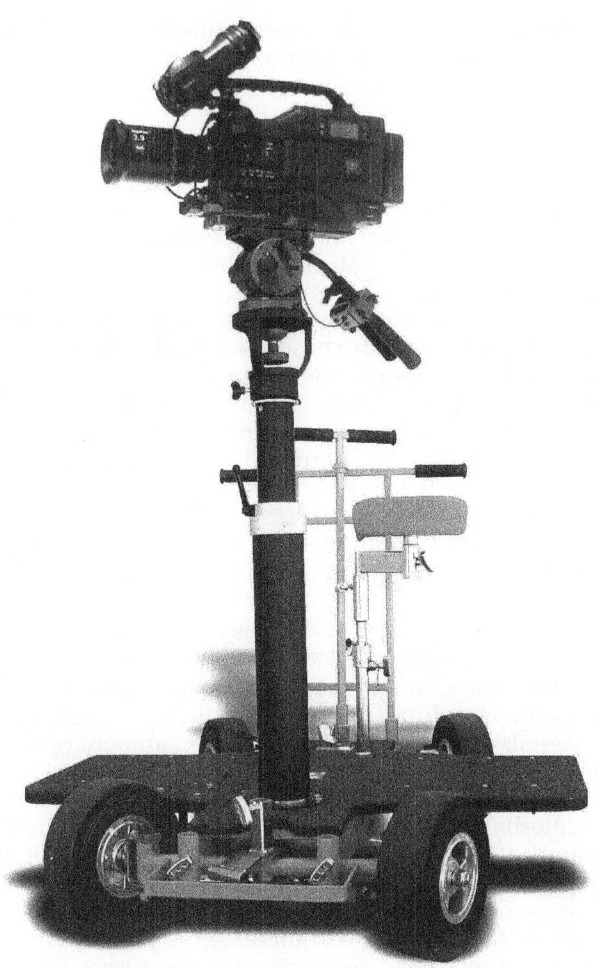

FIGURE 5.6-7 Doorway Dolly with Adjustable Bazooka and Operator's Seat

Courtesy of Matthews Studio Equipment

be extended for dual ground wheels, which greatly increase stability. A *western dolly* is a larger and heavier version of the doorway dolly. With wider tires and a longer wheelbase, it is smoother for travel across the ground, often useful for hauling equipment.

Renowned Italian engineer Sante Zelli began producing his Elemack Spyder dolly in the 1960s. It gained popularity due to its versatile features. The Spyder dolly has a single hydraulic column, which is operated with a foot pedal. The wheels' spread can be changed from narrow to wide, with one set for smooth surfaces, and another set of wheels for tubular tracks. This dolly can change between two-wheel steering and four-wheel crab. Advanced versions added boom arms for additional height and reach. The newer Cricket dolly has made even more improvements to the system, such as an electric-powered hydraulic column with deceleration at the stops. Traditional cine-style dollies have a built-in boom arm, or a telescoping column (See Figures 5.6-8 & 5.6-9), which provides fast and easy adjustment for lens height.

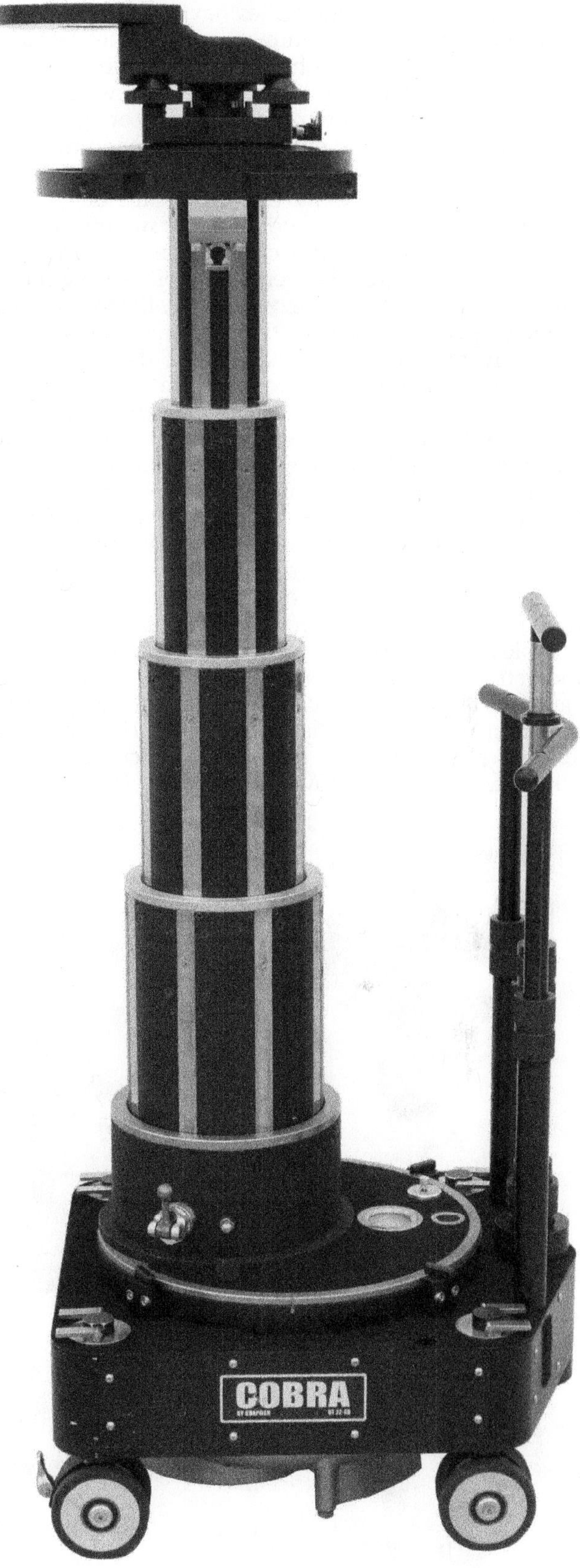

FIGURE 5.6-8 Cobra Dolly

Courtesy of Chapman/Leonard Studio Equipment

Traditional cine-style dollies have a built-in boom arm, or a telescoping column, which provides fast and easy adjustment for lens height. Often, a primary feature of the lift action is a straight up and down motion, as opposed to the arcing trajectory of a jib or crane. The movement is quiet and either hydraulic, pneumatic, or electromechanical in design. (See Figure 5.6-9.) Compound operating of dolly travel with lift or descent can accomplish complex and perfectly smooth camera movements. These dollies can change their configuration between three different types of steering. First, there is rear-wheel steering, where only the rear wheels move closest to the dolly grip, who pushes and pulls the dolly forward and back. Second, there is round steering, where both the front and rear wheels turn in opposition to each other, causing a curved trajectory or to match the arc of a curved track. Third, there is crab steering, where all four wheels turn in tandem, allowing the dolly to move in any direction. When a smooth floor is needed, often a layer of 3/4″ plywood and then a layer of 1/4″ PVC or other smooth sheeting is laid. *Dance floor moves* are when a dolly is moved freestyle in more than one direction. Once mastered, experienced dolly grips can perform multi-point moves with complex booms that are perfectly synchronized with the action. When *dolly track* is used, the entire dolly can be lifted onto *skateboard dolly channels*, one set for the front wheels and one for the back. This creates a smooth repeatable ride on straight or curved track, without having to remove the ground wheels. Often, dollies have swappable wheels: different types of soft or hard, pneumatic or studio, and combinations with integrated *track wheels*.

FIGURE 5.6-9 Hybrid IV Dolly

Courtesy of Chapman/Leonard Studio Equipment

Pedestals

Pedestals can easily and quietly move a camera in any direction. They usually have a base with three steerable double-wheel sets that are guided with a horizontal steering ring on the top of their column. The simplest pedestals have pivoting wheels with foot-activated locks on each caster, but more advanced pedestals have two steering methods. There is single-wheel steering, where one wheel is controlled by the steering ring and the other two wheels follow, and crab steering, where all three wheels are controlled in tandem. There are different wheels for different surfaces. In a studio or on any smooth floor, a pedestal will have solid smaller (~4″–6″) hard rubber wheels, but on location for outside broadcast, the wheels will likely be air-filled soft rubber and taller (~5″–8″). Cable guards help to push cables out of the way, so that they do not get run over, disturbing the smoothness of a moving shot.

The most distinguishing feature of a pedestal is its telescoping column with pneumatically assisted lift. Pedestals are charged with the correct amount of air or pressurized nitrogen and trimmed with small weights to balance the mass of the camera package. When counterbalanced properly, the camera will remain stationary at any point along the operational stroke of the column. It is important to note, however, that temperature changes will affect this equilibrium. A rise in the ambient temperature will increase the pressure inside of the column, and a decreasing temperature will lower the pressure. Small adjustments to the gas charge or to the trim weights will solve this issue. Depending on the range of movement needed, between two and four stages of column slide smoothly within each other to allow easy on-shot elevation or descent for the camera. High-end units are produced with exact tolerances between the stages and have almost no twist or sway. There are a range of sizes, column strokes, portability, and other features available for different uses and budgets.

Local Versus Remote

When the camera is out of the reach of the operator, it must be controlled remotely. Even when it is mounted on a basic studio head on a tripod, and it is only a few feet away, the zoom and focus controls are extended electronically or mechanically to the handles of the pan bars. Many times the distances are farther away, perhaps at the end of a jib, or even farther away and out of sight of the operator altogether. In these instances, a *remote head* is used to create camera moves while the operator watches the results on a video monitor. Remote and robotic movement devices that enable cameras to move through space are also controlled from a distance whenever necessary. Remote operation is enabled by the use of an electronic network, which sends and receives control data.

When using a remote head with electronic controls and viewing a monitor, the operator has a more detached observational view of the scene. This creates its own type of interaction, possibly allowing for a more concentrated focus on specific elements of the creative process. Only viewing a monitor limits how much of the set is available to see, so all efforts are contained to interpreting and manipulating the image at hand. Framing and borders of the image become a point of hyperfocus. Conversely, when the operator moves with the camera on a dolly, a platform crane, or in certain handheld and body-worn configurations, there is a more instinctual, all-encompassing, hands-on, whole-body way of feeling the dynamic as it takes place. A broader, more general, yet emotionally driven response might be the result. Also, the entire scope of the set is visible, and exact cues are available as to when and how elements and action will enter the frame. These are two clearly different approaches to the immersive control of motion for the video art form.

If the action of a movement device or settings on a lens need to be observed back at the remote operator's control position, small locked-focus *witness cameras* can be mounted for observation. Custom sensor-driven devices can be very effective in maintaining awareness of the exact location of a moving camera. This author is known as the creator of the LEDA System: an operator's real-time see-in-the-dark visual location prompter for remotely operated camera movement. Other alert systems exist to effectively warn an operator of a pending automated deceleration or end-stop limit.

Sometimes the camera operator can only reasonably handle the camera's movement, and a camera assistant/focus puller will remotely adjust the lens controls. One of the most common appliances for this type of work is the "fiz," or FI+Z, which is short for "focus, iris, and zoom," made classically by Preston Cinema Systems. Lens rod mounted, external, digital motors drive geared lens rings, by receiving wireless commands at the motor driving transceiver unit. Advanced hand unit controllers display a specific lens' calibrated distance marks for exact focus. Other types of remote lens control are widely available on the market now. For many remote and robotic installations, broadcast-quality tri-servo zoom lenses get their commands directly from a remote control desk, greatly simplifying the task of lens control.

Platform Cranes

One of the iconic scenarios, often talked about in motion imaging, is now less frequently deployed than ever: the camera operator riding on a platform crane. A ride-on platform crane can be operated completely locally. No remote head is necessary. This ride-on crane is swung by a dolly grip, and the platform at the end of the crane seats a camera operator, and possibly the director, the director of photography, or a camera assistant. Two people can sit at the end of a platform crane, but clearly serious precautions must be taken to ensure the safety of the crew. There are cranes that can be configured as either a rideable platform crane, or a remote crane, as needed.

An operator moving on the end of a crane can feel the physical movement of the shot itself, and has perfect immediacy of control. Being fully immersed in the environment of the set at the same point in space where the viewpoint of the audience ultimately will be placed has its benefits. These cranes can also ride on dolly track, using accessory track wheels. Platform cranes are often highly reconfigurable. Certain models can even mount the *crane platform* directly onto the *crane base*, eliminating the *crane arm* itself, to form a traditional ride-on dolly.

One interesting use of the platform crane is to place an operator with a Steadicam high up on the platform and perform a graceful descending boom of the arm. Just as the platform reaches the ground, the operator can step off and continue an uninterrupted take onto the set. This move is known as the famous introduction of the Steadicam to the industry in the 1976 film *Bound for Glory*.

Remote Cranes and Jibs

Kelly Richardson of Atomic Dolly and Crane Rentals concisely explains the popularity of jibs:

> Productions like to work with a jib, because it is a very cost-effective tool for creating many unique camera angles. Much of a jib's work can be done without repositioning the base, creating a large amount of production value in a short amount of time.

A camera jib is a boom arm mounted asymmetrically in its middle onto a two-axis joint. It has controls at the operating end, and a camera at the other. In order to control the camera, at the end of the arm, a remote head is used to pan, tilt, and sometimes roll. This makes most jibs and cranes hybrids of both local and remotely operated camera movement.

John Kosmaczewski, Jib Operator extraordinaire, shares his experience:

> It is important to create a mental map of a set's usable air space, as it relates to physical objects and sources of light. By working with the lighting director, the clean space can be determined, where your jib will not interfere with necessary light, or cause shadows. Where and how the jib can be swung will dictate the length of arm to be built. From an operational standpoint, executing moves with an exact start frame and end frame come down to the confidence and strength of your basic technique. Minimizing your body movement behind the controls is a main ingredient for a beautiful camera move. If given a chance to rehearse, I try to do it without changing the position of my upper torso. Any act of stepping, or chasing of the controls, can change the pressure that your hands exert on the joystick and the zoom rocker. The result isn't optimal. Where possible on a longer move, I take only one step across a pivot foot, extending my arms at the start and end of the swing. With faith in the way that I am physically doing the move, I practice forward and reverse several times, complete with arm, pan/tilt, zoom, and focus. Amazingly, muscle memory takes over quite quickly.

The tail of a jib is short and holds added weights, which counter the mass of the head and the camera. The head end of a jib, being longer and holding a heavy camera package, has the physics of leverage to deal with. In fact, the longer the jib is, the more counterweights that must be added to create a perfect equilibrium of balance. This state of balance is critical for easy operation, but also to avoid potential hazards. These booms usually have either a simple tripod base, a wheeled dolly base, or a track wheel configuration. Different sizes of jibs are measured by the length of their reach, from the arm's pivot point to the camera head. They can be quite short or as long as 40′. The longer the arm, the more dramatic the swing is, as it reaches out into three-dimensional space. Of course, as a jib's configuration is extended, it needs additional structural support in order to stay rigid.

Cranes tend to be sturdier in every way than a jib, and their wheeled bases are proportionately more heavy-duty. Camera crane dolly bases often have hand-cranked leg supports, which lift the wheels off of the ground for quick leveling on uneven surfaces. These support legs are often outrigged, forming a wider and safer base than the wheels alone. Some systems have a center column, allowing adjustment for height. They may utilize a self-propelled mobile platform with a post mount for moving shots or location changes.

Remote Handheld and Harness-Mounted Booms

An ultralight camera pole can be mounted to a chest harness, allowing an operator to control a small jib while maintaining the flexibility and mobility of a body-worn rig. The company Polecam has largely spearheaded this form factor. These super maneuverable rigs can place a controllable camera in unusual and difficult places simply and easily. Similarly, a very long monopod with a lightweight camera can be crafted from a heavy-duty expandable pole, allowing for quick handheld low-altitude aerials. RodCam has created a professional version that utilizes a telescopic fishing pole form and a wireless miniature stabilizing head with an action camera, to reach into areas as far as 7 meters away. Many innovative shots can be made on the fly, without disturbing the shooting environment or taking the time to build more traditional equipment.

REMOTELY OPERATED AND ROBOTIC CAMERA MOVEMENT

Manual Versus Robotically Assisted

> Throughout the history of imaging, we have communicated in two visual techniques—real vision and surreal vision. A medium shot on a tripod is real vision. A telescopic jib, a drone, or a cable cam can give us surreal vision. A greater potential of robotic camera movement is the careful melding of live-action realism with imaginative surreal technique in a way that is effectively processed by the human visual system.
>
> (Idin Roshan, Image Scientist)

The term "remotely operated" denotes remote control by a human operator, who has the ability to visually observe the actions of the device, either by direct line of sight, by the use of video feeds, or via other types of sensors. "Robotic" in general means to perform automatically, either autonomously or by way of advance programming. The term "robotic" is sometimes used to describe a device that is operated by remote control, but in order to be specific, it can be useful to understand the differences.

Drones, rolling robots, perfectly geared remote heads, and zero-latency remote lens control can all be operated from a desk while watching monitors. It is the real-time image that is being controlled.

In order to operate a rig from a distance, we need the assistance of user interfaces and robotic automation. Many different groups of engineers and technicians have tried their hand at developing one-off solutions to the problem of manipulating a camera from a distance. Some of these handcrafted robots have been quite successful at creating very watchable entertainment. We are now entering a new phase of this very specialized technology, however. Well-known and popular corporations in the industry, as well as niche up-and-comers, are focusing their considerable R&D budgets on the design and production of world-class robotic camera movement devices. The results are installed in television studios across the world. It is likely that every major market, especially in the TV news category, has studios that have begun the process of automating their daily camera operating duties.

In the field of live events, moderate to high-end video productions are utilizing specialists to install and operate a variety of remote motion devices, which not only provide dynamically interesting images, but often help the show attendees by obstructing fewer views. Ron Washburn, Technical Supervisor for Television at the Metropolitan Opera, tells us:

> It is a challenge trying to capture a live event, such as a classical music performance, creatively but with a minimum of distraction to the paying audience. It is often impossible to put a full-size camera and operator in the position that would give the best angle for a shot. With new equipment now available, we can place small remotely operated cameras in locations that have never been possible. New camera angles, and cameras tracking through the middle of an orchestra or along the lip of the stage, allow a director to work much more creatively than in the past.

Remote and robotic camera movement is all about the details. When installing and tuning a robocam, it is critical to conscientiously try for the highest standards and tolerances, sometimes to pixel-point precision. This ensures that the machine will perform at its absolute best, move after move. Imperfect motion in a video image can be both consciously and even subconsciously recognized by the viewer. Therefore, it becomes necessary to exceed any expectation of quality, whenever possible. By giving up shortcuts and maintaining strict installation protocols, the equipment will provide optimum performance. Similarly, it is important to not be tricked into thinking that there is only one adjustment that is a likely solution to motion deficiencies. It is by correcting each and every parameter, by tuning and balancing at every level, that beautiful motions are created. When all forces combine in their best form, the very best movements of the camera can happen.

Robotic camera movement devices should have the ability to set acceleration/deceleration ramps, using kinetic damping algorithms for each individual axis of motion. Small details in tuning will affect the entire system's performance. For example, if a system's mass is imperfectly balanced, there will be a tendency to add more damping to make up for the lower-quality movement. Whether the damping is digital or mechanical, too much will make it harder for the operator to perform precise camera movements, causing mistakes, such as not landing a move directly on target. This causes the operator to use less ramping, which maintains some wobble. Also, it is important for the director, camera operator, and robo tech to know and agree on *practical zoom limits*. When an image is magnified, any small vibration or motion error is multiplied proportionately.

It is necessary to understand that the contributing architecture of any mounting system involved in a camera build must provide for ultimate stiffness. For example, if there is any amount of space between neighboring components, such as talent monitors, show clocks, or prompters, it must be eliminated, along with any ability of the components to move independently. It is critical to interrupt the potential for any minor wobble or oscillation, which would cause visible movement in the video image. A firm, yet pliable, damping material of rubber or dense foam should be sandwiched in between such parts, and then tightly held together by a completely rigid bracketry. Reengineer any seesaw physics that stem from a large camera buildup, which might be troubled with outrigged pendulum-like components and other attachments. This will stop off-axis vibrations during camera moves from amplifying into harmonic sway due to the inertial energy from a heavy payload. Heavier camera packages are inherently more difficult for machines to control, because they have more mass and more inertia, which encourages motion to continue, necessitating more counterforce, leading to backlash at the end of a move.

Range of motion limits must be set in order to confidently operate a remote movement device. Setting limits on each axis will allow for uninterrupted camera operating, safety for the crew and spectators, and a reduced risk of damaging the equipment. It is also important that motion systems have a well-placed emergency shutdown button.

Remote Heads

Whenever a camera operator cannot be in direct contact with the pan/tilt/roll mechanisms, a remote head can be utilized. They come in many shapes and sizes, depending on the camera to be used and the specific needs of the production. Marks of a high-quality remote head that is suitable for on-air moves include: clean diagonal movement without any stepping, smooth starts and stops, feathered ramping of acceleration and deceleration, no inherent backlash at the end of moves, no jitter or shake during moves, and the ability to mount cables properly without risk of interference to the necessary movements.

Whenever a motor-driven mechanical system needs to pivot smoothly and evenly, the first step is to be extremely critical of the balance. A majority of production technicians are familiar with the concept of sliding the camera package forwards and backwards on its mount for weight balance, so that it is neither front heavy nor back heavy. It is also necessary, however, to adjust vertically, so that the camera package's center of gravity is aligned with the tilt axis. If this is not corrected and the camera is too low, gravity will affect the camera by consistently trying to bring it back towards level while the operator is tilting up or down. If the camera is too high, gravity will encourage a tilt to continue forwards or backwards, as the center of gravity rides over the top of the tilt axis. When all drag has been removed from the system, the camera must be able to stop and remain still at any point in the tilt's range of motion. Because a camera package's center of gravity is not visible, matching it to the tilt's pivot axis must be done systematically. The first step is to correct the front-to-back adjustment by moving the camera plate, so that when the camera is level, it does not move. Next, check for vertical adjustment by manually tilting the camera to see how it reacts. Adjust the height of the *camera cradle* vertically, and note any improvement. Now recheck the front-to-back adjustment, as it can now be better refined. Continue to alternate adjusting between the two axes until the system is absolutely perfect. When the camera can stop anywhere in the tilt range without any rebound movement, then the center of gravity is matched to the tilt axis, and the remote head's motors will perform at their very best.

Another consideration for camera placement on a remote head is placement on the camera cradle. Many remote heads utilize an L-shaped camera cradle, which is only supported on one side. While these cradles are generally stiff and strong, the heavier the camera package is, and the farther the camera is from the support side, the more there is the possibility of a slight cradle deflection. If the cradle is deflected, the camera will lean very slightly to the side, but even more importantly, cradle deflection acts as a spring, storing a small amount of energy, which can result in motion error. Whenever possible, keep the camera close to the supported side of the camera cradle.

An advanced feature on some heads is integrated 360-degree *slip rings*. This allows the video signal, control data, and power to pass through the head, even as it moves continuously. Cable ports underneath the pan section transport signals past the rotating pan axis to matched cable ports at the camera on the tilt section. This provides a neater and safer setup, while allowing the head's movement to continue indefinitely without having to return back to a central position to unwind cables after an extended move. Without slip rings, cable management on remote heads can be a bit tricky, because a snagged cable can be the cause of bad movement or equipment damage. Even cable tension from stiffness can have a small influence on the quality of a move. Some heads without slip rings have pass-through holes in both the pan and tilt axes, so that cables are less likely to become over-twisted.

Full-size remote heads are the most common and are designed to carry a full-size camera. *Micro remote heads* have the smallest profile. They are great at disguising their position, but generally can only carry a smaller payload, such as a box-style camera. *Stabilized remote heads* use gyroscopically referenced adjustments to maintain smooth shake-free images, even as their supporting structure moves or vibrates.

Controllers and Control Desks

Remote operator interfaces come in a variety of shapes and sizes, generally with some consideration for ergonomics. Wireless controllers are designed to be handheld and portable, but many can also be used functionally while sitting at a control station. When the operator sits at a table, every opportunity should be taken to create a perfected and easy to control workspace. *Foot pedals* can be an excellent feature for controlling motions other than those of the remote head. Side-to-side dolly moves and up and down elevations are prime candidates for foot pedal controls.

The ergonomics of different *control desks* vary quite a bit. Camera operators become familiar with a certain style of operating, and often create strong preferences. A common comparison in shooting styles is the pilot versus pointer designation. One group of operators likes to move a joystick back or down, like an airplane pilot, to make a remote head tilt up. Another dedicated group, pointers, pushes the stick forward or up, to tilt up. For this sort of reason, it is always a good idea for remote controllers to have the ability to change the directionality of any control mechanism. It is also quite necessary to be able to fine-tune the sensitivity of each control, in order to suit an operator's style and habits. Additionally, any moving axis should have precise velocity limits, which can be set at a fixed level, and also a variable speed setting that can be adjusted on the fly.

Pan bar controllers can provide an operator with the fully functional control of a remote head, while maintaining the same form factor of a traditional pan/tilt head buildup. A special digital encoder head, with traditional pan bar lens controls, sends commands to the remote head, which emulates the operator's motions perfectly, allowing for intuitive tracking shots and precise detailed movements.

On a jib, cables are run from the remote head to the interfaces for pan, tilt, zoom, and focus, which are mounted on the weight bar, so that the arm can be swung while simultaneously operating the camera. On telescoping cranes, the control for telescopic extension is often a cable-tethered handheld controller, commonly referred to as the "pickle." This allows an assistant to operate the arm length, or it can be mounted at the back of the crane for the primary operator to control. Often, the remote head on a crane will be controlled by another operator from a seated position using a control desk.

When there is some system failure or malfunction, consider a few common possibilities. An "air-gap" is when some power cable or control cable is either fully or partially unplugged. Another likely problem might be a bad cable, or a cable that is too long for the perfect transmission of control data. Too many junctions or barrel connectors may diminish a cable run's effectiveness. Whenever possible, make a "home run," using a single cable. Consider running backup cables, in case of a problem. Another potential cause of control errors is EMI, or electromagnetic interference, which could come from any electronic device, even a cell phone or a walkie-talkie. This radio frequency disturbance can negatively affect an electrical circuit, causing unexpected results. If there is incompatibility between devices, operation could be disrupted or become erratic. If the EMI source can be identified, simply separating the devices physically may stop the problem. Additionally, there are shielding solutions using conductive mesh to stop incoming interference and outgoing electromagnetic emissions.

Remote and Robotic Dolly Systems

These systems are commonly known as railcams, and they can be one of the most practical additions to a technically advanced video production when the goal is to create subtle and interesting camera movement. (See Figure 5.6-10.) They often use straight dolly tracks, but the many systems can utilize various radii of curved tracks, or even a combination of straight and curved sections, if necessary to conform to staging requirements. While dolly systems are traditionally configured overslung, built onto a flat stable surface, certain systems may be underslung and suspended.

FIGURE 5.6-10 TrackRunner Remote Dolly with LiftArm and Stab-C Compact Stabilized Remote Head

Courtesy of Atlantic Cine Equipment

Building a system onto a lighting grid or a truss structure can provide overhead shots that fly smoothly and silently throughout a set without interfering in the action. A self-driven remote dolly with an onboard motor is sometimes simpler and easier to install, and it can drive very long distances on both straight and curved track, but a rope-driven railcam on a straight track does not have to consider traction issues, so it can pull a heavier umbilical *drag cable* with ease, making it potentially more reliable in certain situations. The addition of a *telescoping lift column* gives a production the ability to make remote lens height adjustments. It can provide smooth elevation to capture perfect eye lines whenever needed. An interesting alternative to a lift column is the *lift arm*. Keeping a fairly low profile, these devices can be used unobtrusively to lift a camera to the proper perspective for a special shot, and then lower it back out of sight.

Remote dollies are meant to be operated live, providing the ability to carefully remotely control every movement of the device, including the settings of the lens. At the foot of a stage, a rail system can follow and complement the action in an intimate way, while dramatically revealing details to the viewer. A buildout that has a narrow track and a low-profile remote dolly with a short unobtrusive remote head holding a box-style camera can be the least distracting form factor during a live show. One of the best combinations to date is the A.C.E. RailRunner with the Nettmann MicroPlus remote head, as shown in Figure 5.6-11.

On a sports field, these rail systems can match fast athletes' performances, keeping up with their every step. Advanced features for a remote dolly may include programmable acceleration and deceleration (also known as ramping), foot pedal drive control, initiation of automatic deceleration as the dolly approaches the mechanical shock-absorbing end stops, and emergency stop button.

Robotic dollies, such as that shown in Figure 5.6-12, have the ability to record and program the robotic placement of the camera and lens positions. In a studio setting, where talent knows to stay on their marks, and production

FIGURE 5.6-11 RailRunner Remote Dolly with MicroPlus Remote Head

Courtesy of Atlantic Cine Equipment

FIGURE 5.6-12 Furio Robotic Dolly

Courtesy of Ross Video

standards give all participants advanced knowledge of the action, entire shows can be structured so that robotic cameras make time-saving moves with perfect efficiency. Absolute positioning systems can enable one talented operator to control many cameras simultaneously, once all of the detailed programming has been accomplished. Advanced user interfaces, such as that shown in Figure 5.6-13, are now available to organize countless variations of camera positions in a quickly accessible format.

A fairly new iteration of robotic dolly is the automated parabolic slider, as shown in Figure 5.6-14. This piece of gear can create dolly moves at any focal length to capture additional dynamic footage in a multi-camera shoot without adding another operator. The curved rail is designed to keep a stationary subject framed and in focus. Speed control, acceleration and deceleration ramping for direction changes, and time-lapse modes make for a versatile piece of gear.

Robotic Pedestals

Much like a robotic dolly, a robotic pedestal, such as that shown in Figure 5.6-15, can accomplish the most complex placements of camera and lens at the push of a button. These amazing free-roaming devices are installed in smooth-floored television studios around the world. Optical targets and complex encoder

FIGURE 5.6-14 One Man Crew Director Automated Parabolic Slider

Courtesy of Redrock Micro

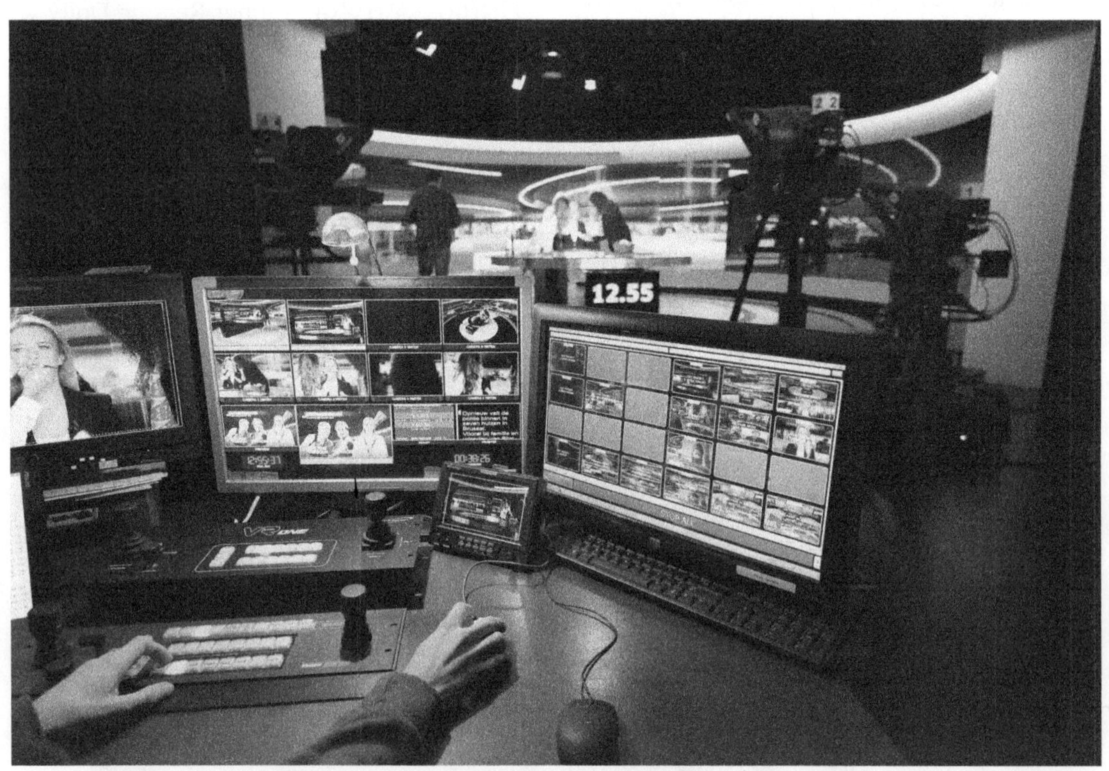

FIGURE 5.6-13 Touch Screen Shot Matrix Interface

Courtesy of Ross Video

FIGURE 5.6-15 CamBot 700XY Robotic Pedestal

Courtesy of Ross Video

systems allow these pedestals to track their exact position as they recreate rehearsed camera moves. In order to get top-level results, it is important to follow a prescription of standard operating procedures and strict studio protocols. However, an excellent outcome for repeatable multi-camera formulas creates a favorable economic situation for daily productions.

Arguably, the task of accomplishing the cleanest result for most robotic movement devices can be aided by limiting the weight of the camera package and its accessories, but the best robotic pedestals can carry full-size cameras with heavy box lenses and viewfinder monitors, prompters, talent monitors, clocks, and other necessary items. An advanced feature of these devices is the ability to switch between remote and local controls. When necessary, an operator can manipulate the device using traditional pan bars attached to the head, and can drive the robotic base and elevation column using a locally mounted joystick.

Telescoping Jibs and Cranes

Telescoping jibs and cranes add an additional degree of freedom to camera movement with their ability to quietly and evenly extend and retract their boom arms. This is accomplished by proportionately shifting stacked counterweights toward the crane's pivot point as the arm retracts, so that balance is maintained. When the arm extends, the weights move back towards the tail of the rig. Possibly the best-known telescoping crane is the Technocrane. When introduced, these cranes became the catalyst for a vast array of new and innovative camera movement possibilities. They are now used throughout the world for every category of world-class production. Like many cranes, they can ride on track for even more complicated moves. The SuperTechno crane comes in several sizes: 15′, 22′, 30′, 50′, 75′, and 100′. The 100′ version is massive with a reach of over 97′ and a maximum lens height of almost 105′. It creates grand possibilities for incredible lateral swings and booming that rivals many aerial shots, yet it can execute perfectly controlled floating camera moves at ground level with an arm that can smoothly extend almost 82′. The minimum lens height, off the side of a building or dropping down below the ground level of the crane base, is almost −63′.

The lightweight Techno-Jib is becoming very popular for modern television production studios. Like a traditional jib, every aspect of a camera move can be performed by a single operator. There are two versions of these telescoping boom arms, 15′ and 24′.

Scott Buckler, creator of the 45′ PowerScopic, shown in Figure 5.6-16, lends us his insight:

> A subtle approach is important when operating camera movement devices. Using the limits of your gear arbitrarily and without motivation is not good storytelling. When I work with telescoping cranes, I focus on the ability to travel laterally, rather than on gaining altitude. The telescoping feature is perceived as natural, in a way that a zooming lens is not. The crane is usually positioned at a central point, with the huge potential of automation-assisted moves in any direction, from low to the ground to up in the air. A rule that I like to keep in mind is to limit camera height to 18′, unless directed otherwise. This is three times the normal height of an actor, and it makes for dramatic, yet realistic, shooting.

A very long and very unique amphibious telescoping camera crane is made by Chapman/Leonard. The 73′ Hydrascope, shown in Figure 5.6-17, can extend its arm for a travel distance of over 60′. When its custom 5′ extension is utilized, it can provide a reach of over 75′ and an overslung lens height of over 80′. This crane is completely waterproof and can perform underwater. It has an automated telescoping adjustable-height center post and a four-point leveling system for operation on uneven ground.

FIGURE 5.6-16 PowerScopic 45′ Telescoping Crane

Courtesy of Riggin Design, Inc.

FIGURE 5.6-17 73′ Hydrascope Telescoping Crane on Ultra Maverick Base

Courtesy of Chapman/Leonard Studio Equipment

Not all telescoping cranes are large. The Miniscope 5, shown in Figure 5.6-18, has a compact design for use in small spaces with the ability to travel through narrow doorways. When necessary, it can be operated by one person, like a jib. It has a maximum reach of 8.5′, and mounted on the Hybrid IV dolly, it can provide a lens height of over 15′. This crane has a standard Mitchell mounting base, making it a highly versatile tool for the industry.

Remotely Operated Cranes and Jibs

Even the boom arms of these devices are swung remotely. There are a few different incarnations of remotely operated camera cranes that are designed for mounting onto camera cars and other moving transports, as we will discuss in the next section. There are also a number of *motion control rigs* in the industry that have robotic arms and can be used for the exact duplication of programmed shots. This type of apparatus is commonly used in the creation of advanced special effects.

The *JitaCam*, shown in Figure 5.6-19, is a 360-degree remotely operated overhead jib system that is suspended from firm rigging points, such as trusses. It can be used as a stand-alone device with a reach of up to 30′, or it can perform in conjunction with a underslung remote dolly and up to 80′ of track, covering an area of practically 60′ × 140′. When space is at a premium, the jib can be configured in very short lengths as well. It is built up in 2.5′ increments, and uses no floor space at all. This versatility and the unique perspectives provided by this device have made it popular for live events, ring sports and indoor action sports, concerts and award shows, as well as scripted productions. JitaCam also produces a robotically automated version of their jib system, which can recall preset moves at the touch of a button. These systems are permanently installed in several advanced technology television production studios.

FIGURE 5.6-18 Miniscope 5 on a Hybrid IV Dolly Base

Courtesy of Chapman/Leonard Studio Equipment

Camera Cars

There are many types of camera cars that are used to support cameras for moving shots. On-camera prop vehicles, commonly called *picture cars*, or sometimes *hero cars*, are rigged with cameras in many configurations using specialty hardware such as: *limpet mounts* (vacuum cups), grip heads and arms, speed rail and ratchet straps, cheese plates, and ball leveling mounts. Innovative car-mounted rigs might include a miniature curved rail system configured as a *hood mount*, capturing the classic reverse-angle through-the-windshield shot with the ability to dolly between views of the driver and passenger. Similar rails have been mounted to travel from the outside of the vehicle through a window for interior close-up shots, taking the place of the standard *hostess tray* door-rigged side mount.

Professional camera cars are used to provide movement for the cameras with their supports and the crew. These vehicles are rented in conjunction with an experienced driver/rigger. Commonly referred to as an *insert car*, features may include: quiet onboard AC power generators, upgraded suspensions, specialized seating with three-point safety harnesses, and custom build-out configurations to suit any need. An operator with a tripod can be safely rigged onto a deck. Remote heads can be mounted anywhere, and controlled from within the cabin.

A full-size pickup truck is a common choice. They can be built out with diamond plate, 3/8″ matrix cheese plate, speed rail receivers, and scaffolds to suit many different uses. Platforms are mounted in front of the engine compartment (front lower deck),

FIGURE 5.6-19 Remotely Operated Jib

Courtesy of JitaCam

above the cab (upper deck), in place of the rear bed (main deck), behind the rear bed (rear lower deck), or even on the sides. These trucks can accommodate many different types of camera cranes, and allow for endless possibilities of camera mounting options. They can be used in conjunction with a *tow dolly* or a *process trailer* to tow on-camera vehicles for all sorts of moving shots. Process trailers are designed to be very low to the ground, in order to provide realistic camera angles for shooting the vehicle that is being towed. They may have additional side platforms, which provide enough room for dolly track or other camera rigging on every side. These trailers are traditionally towed behind the insert car, but special rigs allow for side towing, which provides many creative camera angles.

All-terrain utility "gators" can be a good choice for a smaller cargo box or flatbed form of vehicle. Speed rail scaffolds allow for high or low mounts from all four sides. Pan/tilt heads, turrets, and crane arms can be attached to built-in *Mitchell mount base* configurations, along with rigid arm mounts and post supports, allowing for versatile camera placement, rigged either overslung or underslung. Local camera heads can be operated by hand, or stabilized remote heads can be controlled from a seated control desk position. Steadicam mounts might be supported on the rear deck, allowing for direct operator-controlled stabilized positioning of the camera. A spinning gyroscope can help to keep the framed image from shifting as the vehicle moves at speed.

A *chase car* is meant for tracking fast action, usually shots of a moving on-camera vehicle. These are often high-end SUVs from Porsche or Mercedes, or even a sports coupe for the fastest work. Modified off-road vehicles are used for the most serious terrains. These chase cars are often outfitted with *remotely operated camera cranes*, which are an extremely versatile feature, allowing for a range of motion that is hard to match. These fully remote cranes are known as *remote arms*, and even the crane's arm can swing and boom remotely. One of the most popular versions is called the Ultimate Arm. It can be built between 9′ and 17′ long, with a lens height of over 20′, and can be operated at speeds in excess of 100 mph. These rigs are used in conjunction with any of several compatible stabilized remote heads to capture smooth dynamic action shots.

A *spring mount* is a mechanical stabilizer arm, which utilizes spring tension and pneumatic struts in order to reduce any major jolts that a vehicle's suspension doesn't smooth out, so that the stabilized remote head that it holds can eliminate the remaining smaller vibrations. This form of *vertical axis stabilizer* can be quite effective on any form of transport on land, sea, or air. While it doesn't move like a remote arm, it is much less expensive to rent.

Another useful device for reducing impacts prior to reaching the stabilized head is the *vibration isolator*. These passive isolation systems can come in many forms, but the most common type is the wire rope isolator. Compact and durable, stranded steel cable is looped to create a spring stiffness, which causes unwanted energies to dissipate within the designed damping frequency of the system.

There are electric versions of camera vehicles on the market that are quiet and simple to operate, accelerating quickly and evenly. Electric utility vehicles, and even electric motorcycles, can be very effective camera cars. *Electric tracking vehicles*, such as that shown in Figure 5.6-20, are a common sight on professional sports fields and are designed to be safe and low impact on ground surfaces.

A highly innovative new option for creating automotive content has been created by The Mill. The BLACKBIRD is an electric vehicle platform used in conjunction with CG renderings. Its chassis is reconfigurable to create the dimensions of virtually any car, and it is programmable, emulating different drive performance characteristics. Using camera arrays, advanced sensors, and 3D laser scanning, the images and data that are captured allow for virtual renderings that are photo-realistic, without ever having a picture car itself.

FIGURE 5.6-20 Olympian III Electric Tracking Vehicle

Courtesy of Chapman/Leonard Studio Equipment

REMOTELY OPERATED GROUND AND UNDERWATER VEHICLES

Ground Vehicles

Unmanned ground vehicles (UGV) are now being used in conjunction with various stabilization technologies to move a camera remotely, when it is less desirable or impossible to have a human operator at the camera's position. The Panaconda 4 × 4 all-terrain mobile crane base from Panavision can move a 4400 lb. payload, while under full wireless control. This machine can mount a SuperTechno 50′ Telescoping Crane on its adjustable center column for a 48′ lens height.

The Mantis by Motion Impossible (see Figure 5.6-21) is a new type of camera dolly in the form of a stabilized wirelessly operated mini camera car. It comes in two versions. There is a slow smooth studio version and a high-speed model. Before adding a modern brushless gimbal system remote head, there are two options for pre-damping with a vertical axis stabilizer. One is low to the ground, and the other has an adjustable post for a camera height up to 63″. Both stabilizers create a shock-absorbing suspension, similar to a spring mount on a full-size camera car.

Freefly Systems has also created a small-scale remotely operated vehicle for camera movement, as shown in Figure 5.6-22. They mount their popular three-axis stabilizer, the MōVI gimbal system, onto the TERO, a 31″ long, 18″ wide, and 10″ tall high-performance mobile platform. They use tensioned wire rope vibration isolators, which are specifically matched to this application, to smooth out some of the larger hits that the vehicle might encounter on uneven terrain. This allows for a sliding path of low-angle camera motion with a steerable trajectory, even at high speeds, up to 30 mph or so. It can be used as a chase car or to place a camera remotely, where it would be unsafe for an operator. This new type of remote camera motion allows for many creative possibilities.

Submersibles

A fairly new frontier in video drones, the submersible, is also known as an underwater remotely operated vehicle, or an unmanned underwater vehicle. With the ability to go deeper, for longer periods of time, they are capturing footage never before possible. There are autonomous robots and remotely operated versions.

FIGURE 5.6-21 Mantis Remotely Operated Ground Vehicle with Ethos Head

Courtesy of Motion Impossible and DEFY

FIGURE 5.6-22 TERO Remotely Operated Ground Vehicle

Courtesy of Freefly Systems

Some of these underwater robots can follow a diver, swimming in formation around a wearable beacon. Some have sonar for obstacle avoidance, capabilities of object recognition, and the ability to automatically frame a subject.

AERIAL PLATFORMS

Classic

Aerial shots have been created since the start of motion imaging, by finding a high camera position with the necessary perspective. Sometimes, in order to get more of a bird's-eye view, a camera might be placed up on scaffolding, or even onto a construction lift. Some amount of sway will likely be present. The longer the focal length, or tighter the zoom, the more important a stable image becomes. Wide *cover shots* are the most forgiving, and can hide a small amount of random motion. Jibs and cranes have made their contribution to high-angle shots with their portable and versatile qualities. Now, there are other innovative devices, which are giving productions the opportunity to capture dramatic aerial views in new and wonderful ways.

Remote Telescoping Columns, Vertical Rail Systems, Descender Rigs

In the camera department, one issue is always in full consideration, the very important question of lens height. It is an indefatigable opportunity. So many forms of camera support allow for vertical adjustment. When there is the possibility of a remotely automated vertical move, so much the better.

Vertical motions can be very energetic. A smooth camera rise can be a great tool for showing a reveal or for defining the intensity of a shot. Gracefully descending upon a subject can be quite dramatic, as well.

Remote telescoping columns can do a lot in a very small footprint. A Towercam can elevate a lens height to over 33′ above the ground at a maximum speed of around 2.5′ per second. SuperTower produces a version that travels at around 5′ per second. Some telescoping columns can be inverted, allowing cameras to descend from a high mount for a shot and then retreat out of the way. There are several different lengths. Shorter columns can be mounted onto dollies, or peek out of small openings. Taller towers can be effective jib replacements, when space limitations mandate. A single operator can run this sort of rig, using foot pedals to control the column, and a control desk for the remote head.

A vertical rail system is a remotely operated camera mover that can perform on-shot elevations and descents. They are often monorails, rather than two-rail tracks, and are generally between 1 and 6 meters long. Certain versions are meant to be rigged onto a camera car in conjunction with adjustable shock absorbers and a stabilized remote head. Others can be installed into studio sets, taking advantage of their rather slim profile.

Descender rigs are vertical cable systems, much like the lateral line cams that we will discuss next. They are winch-operated cable and pulley systems, which can quickly lower a stabilized remote head down from above to capture a shot, and then return to an out-of-the-way position until needed again. They can be used to great effect for many live events or strategically for scripted productions of all types.

Remote Cable and Pulley Systems

There are many incarnations of cable-suspended rigs, called line cams, or cable systems. These systems are installed to custom dimensions on location. Longer cable lengths require more tension and safer rigging points. It is important to be able to qualify a structure for rigging lateral load. Riggers are specially trained for this. Cable thickness is proportionate to its weight, and also to its ability to carry a tensioned load. Considerations for the angle of the line's descent can be made as well. The key is to find the best formula for the weight of the camera package, the rigging details, and the necessary speed and trajectory of the shots that are desired. Several possible systems could be used.

Single-axis, point-to-point, zip-line type systems have advanced substantially. They are now often single line cables with motorized trollies, wireless control, and stabilization provided by control arms, gyroscopes, and gimbal systems. (See Figure 5.6-23.) A cable is mounted between two firm load-bearing anchor points. These devices are commonly used as tracking systems, keeping up with a moving subject for long distances. They can be especially convenient for events that are going to be repeated continuously in the same location. Often manipulated by a single operator, they can maintain a level horizon while flying with smooth accelerations and gentle decelerations, as the operator pans and tilts to frame the perfect shot. Modern technical features can include: RF video, wireless operation, chassis self-drive via onboard motor, battery-powered simplicity, regenerative braking, adjustable vibration isolation, travel speeds from very slow to over 35 mph, start and stop ramping, and automatic end-stop safeties.

Two-axis systems are also horizontal, A to B, two-point systems. However, they can additionally control the vertical height of the camera, by manipulating the amount of sag that the camera produces on the suspension cable. Three-axis systems allow for full coverage of a large area, such as a stadium, arena, or outside space. These systems use four existing anchor points, or when necessary compatible aerial lifts or construction cranes. They allow the suspended camera platform to move laterally in any direction and vertically as well. Powerful three-phase AC-powered winches safely and precisely place the camera anywhere in the fly zone. Commonly used for sporting events and concerts, they are also used in many dramatic and creative ways for special shots throughout the world. Within the defined three-dimensional space, the camera can move freely up to 25 mph. Three-axis stabilized remote heads are used to maintain a level and smooth image. Operation can be manually remote, or it can utilize motion control software for programmed and repeated shots. During control, the computer software automates the action of the winches, translating operator commands into positional coordinates tracked by the system.

Helicopters

Helicopters have been used for decades to capture moving images from up in the air. A camera operator

FIGURE 5.6-23 Dactylcam Pro Remote Cable System with Ethos 3-Axis Gimbal Remote Head

Courtesy of DEFY

can shoot from an open door, while protected from a fall by a safety harness, or a stabilized remote head can be controlled from within the craft, while it is mounted on the outside. There are specialized rigging platforms that allow three-axis heads to be mounted to the front, or the side, of the helicopter. Fully enclosed housings protect the head, camera, and lens from damage. The pilot has a major influence on the success of this sort of shoot, and strategies for getting the necessary shots are worked out in advance. When working with a helicopter, time is always an important factor, so efficiency is critical. The environment is quite noisy, and all parties communicate via headset.

Remotely Piloted Aircraft Systems

There are many different types of RPASs, also known as UAVs (unmanned aerial vehicles), UASs (unmanned aircraft systems), or civil drones. The most common form factor is known as a multicopter, which generally has between four and sixteen propellers to provide lift. Figure 5.6-24 provides an example. The size of these devices, compared with a traditional helicopter, is quite small, so issues such as air disturbance, ground shadows, and cost of use are minimized. RPASs can also be used indoors and in smaller spaces, where traditional aircraft cannot. These aerial robots may be fully or partially autonomous in their operation. Generally, in video production use, they are controlled by a *remote pilot*, accompanied by a remote camera operator and a spotter/technician. The most modern versions of these devices will have safety features such as flight stabilization, sense and avoid, GPS position and altitude hold, operational redundancy, auto-launching parachutes, and automatic return to base in case of low power level or loss of wireless control. It is also good to remember that, unlike most camera supports used in video production, propellers are quite noisy, and will affect live sound capture.

Cameras can now be flown from ground level to high in the air. They can travel through openings, taking their camera support with them. The ability to create uninterrupted shots that change from a dolly's to a crane's to a helicopter's point of view is revolutionizing what is possible. We are now seeing dramatic aerial images in new and safer ways, at a lower cost, and it is only the beginning. The ultimate bird's-eye view can now give watchers access to many sights before unseen. Even during a hover, small floating movements are automatically part of the video capture, giving a dreamy sensation to the footage. These are new and incredible sights, opening a form of creative possibilities like never before. Sweeping vistas and dynamic dramatic camera moves, when motivated by the story, provide excitement, while allowing the viewer to literally see the bigger picture.

Cameras moving almost limitlessly in a three-dimensional space is the newest possibility, and all of this can be utilized as a very affordable system; however, the use of such systems often requires approval by a country's aviation authority. In the United States, the Federal Aviation Administration (FAA) is the final authority regarding when, where, and how these systems may be used. For most television and film purposes, drone use is considered to be a commercial operation by the FAA, and therefore will be subject to certain rules and regulations. For instance, commercial drones must be registered, and used only for allowable types of missions. An FAA-licensed UAS pilot must be involved and that person will have an FAA certificate, obtained after taking training, passing an exam, and receiving a UAS pilot certification. Operationally, there are many considerations, such as: safety conditions, wind and weather, maximum allowable altitude, and where you are allowed to fly. There are no-fly zones that exist, such as near airports, and even some entire cities are off limits. Temporary flight restrictions (TFRs) are put in place by the FAA, sometimes without notice, for areas where VIPs may be traveling, or during special events. In some instances, special waivers may be issued by the FAA by following a process, although the lead time can be as long as 90 days. Due to the growing recognition of the utility of drone aircraft in many industries, the FAA's rules and policies have evolved and will likely continue to change in the coming years. As such, it is important to become conversant with the latest FAA guidance and rules before using a UAS. Advance planning is paramount. Local governments may also have restrictions in place limiting the use of RPASs. Insurance, in case of damage or liability, is an issue unto itself. Additionally, it may be relevant to consider whether or not public opinion needs be taken into account. The current use of these devices is commonly misunderstood, causing excitement or possibly leading to some negative reactions from casual observers in the field. Local police as well are likely to be suspicious of use in their jurisdiction, so it is a given that a production must be able to provide evidence of their compliance with all existing regulations at any time. Be aware that this is new technology, and that the governmental agencies are still deciding what the rules are.

There are large multirotor RPASs capable of handling full-size production cameras, and there are micro drones, which often have built-in cameras. Some of the more advanced aerial drones can fly themselves, based on preprogrammed parameters. There is an enormous range of products currently on the market, each with their own capabilities and restrictions. One style of aerial robot is labeled VTOL for vertical take-off and landing. These drones look more like an airplane than a helicopter, but they still launch vertically, repositioning their rotors for lateral movement once they are in the air. Another type of remotely piloted aircraft system is the *tethered drone*. There are several benefits to having an umbilical cable that connects your aircraft to the control station. The very nature of a tether limits a drone's movement, and may allow for safer operation. Eliminating wireless communication can be more reliable. Additionally, in circumstances where battery power does not provide a long enough flight time, cabled power can be continuous. Video over cable is also more reliable and does not have the latency issues found with some lower-end wireless transmitters. When a wireless video signal is used by operators in order to gauge how they control a remote

FIGURE 5.6-24 ALTA 8 Remotely Piloted Aircraft System with MōVI M15 Gimbal System

Courtesy of Freefly Systems

device, the smallest fraction of a second of delay will cause diminished reaction times. Another iteration is the *LTA platform*. These lighter-than-air platforms are often shaped like a blimp, and they may be free-flying or tethered. Japan's NHK has developed a gyro-stabilized, remotely controlled powered aerostat, which can operate in winds up to 15 mph. Panasonic is currently in development of a very promising entry into this category, called ballooncam.

In summation, motion imaging is an art, a science, and also a craft. The craft of utilizing supports to place a camera anywhere in a space is a detailed discipline, which carries with it many important traditions. The science of camera movement is continuing to evolve in many exciting new ways. Never before have there been so many choices for making beautiful images in video. The ability to enhance a viewer's experience with moving camera perspectives is making great contributions to the art. As this technology becomes more and more advanced, it will continue to be a great adventure for us all.

CHAPTER

5.7

Closed Caption Services

CHRIS HOMER
Public Broadcasting Service
Arlington, VA

INTRODUCTION

Captions provide a written representation of spoken audio along with pertinent information on other audible actions (e.g., gunshot, or creaky door). Subtitles are simply representation of a spoken word or as a translation for alternate language (e.g., Spanish, French). Outside the United States and Canada the term subtitles is used to describe both use cases.

Closed and open captions for television are considered access services. Open captions are burned into the video and cannot be disabled. Closed captioning is the ability to present caption information that is hidden until requested by the viewer. This method is preferred over open captions, which can be distracting to persons who do not require captions.

The audience for television open and closed captions is as varied as non-access service. It may be important to understand the definition of hearing impaired, which begins to identify this diversity.

HEARING IMPAIRED

"Hearing impaired" encompasses a range of impairments that can be characterized as anything from a person's inability to hear clearly to complete deafness. The latter is called *anacusis*. The first language of a person not born deaf (or not deafened very early) is a spoken language such as English. For most deaf people who are born deaf, their first language is sign language. For both groups the ability to read what is being spoken can allow the enjoyment of televised program content to a much greater extent than the picture only.

Even those who are hard of hearing also benefit from caption information to help fill in what has been missed by the viewer. Others who may find captions useful are those in a noisy environment or watching a program in another language.

CAPTION HISTORY

Open Captions

Open captioning first appeared in 1972 on PBS's *The French Chef* with Julia Child. WGBH's Caption Center was the first captioning agency and began open captioning of programs such as *Zoom*, *ABC Evening News*, *Masterpiece Theatre*, and *Once Upon a Classic* shortly thereafter.

Closed Captions

Closed captioning was demonstrated for the first time at the First National Conference on Television for the Hearing Impaired in Nashville, Tennessee, in 1971. A second demonstration of closed captioning was held at Gallaudet College (now Gallaudet University) on February 15, 1972, where ABC and the National Bureau of Standards demonstrated closed captions embedded within a normal broadcast of *The Mod Squad*.

The deaf community was excited about these demonstrations, and with assistance from the National Association of Broadcasters deliberated on how to move forward with a true captioning service. The federal government funded development and testing of this evolving technology and in 1973, the engineering department of the Public Broadcasting Service (PBS) started the project under contract to the Bureau of Education for the Handicapped of the Department of Health, Education and Welfare (HEW).

PBS engineers then developed the caption-editing system that would be used to provide closed captions for prerecorded programs. Later on, participants including PBS realized that securing the cooperation of commercial television networks required a single-purpose organization with a specific mission to perform this captioning, so in 1979, HEW initiated the National Captioning Institute (NCI). A device called the TeleCaption from NCI was used to decode the first closed captions for the television audience. There were two units originally available: the TeleCaption "adapter," which could be hooked up to any TV, and the TeleCaption "integrated receiver," a TV set with the adapter built in. The units were sold at Sears, Roebuck & Co.

Real-time captioning, used for captioning live broadcasts, was later developed by NCI in 1982. In the initial real-time captioning system, trained personnel used a "stenotype" device to write at speeds of over 225 words per minute, giving television viewers immediate access to live programs including news, sports and live entertainment. As a result, the viewer sees the captions within seconds of the words being spoken.

CAPTION TECHNOLOGY

CTA 608

The standard for creating, transmitting and decoding analog captions is Consumer Technology Association CTA Standard CTA-608. In NTSC analog television for the United States, text from captions is translated to data and sent over the airwaves embedded in the analog video signal within the vertical blanking interval (line 21, field 1 and field 2). The caption data channels are assigned as shown in Table 5.7-1. The waveform for the signal is shown in Figure 5.7-1.

TABLE 5.7-1
CTA-608 Caption Data Channels

Name	Field	Channel	Description
CC1	Field 1	Channel 1	Primary closed captioning—synchronized
CC2	Field 1	Channel 2	Special use captioning—non-synchronized
CC3	Field 2	Channel 1	Secondary closed captioning—synchronized
CC4	Field 2	Channel 2	Another special use captioning—non-synchronized
Text 1	Field 1	Channel 1	Primary text service
Text 2	Field 1	Channel 2	Secondary text service
Text 3	Field 2	Channel 1	Alternate text channel
Text 4	Field 2	Channel 2	Alternate text channel
XDS	Field 2	Channel 3	Extended data service

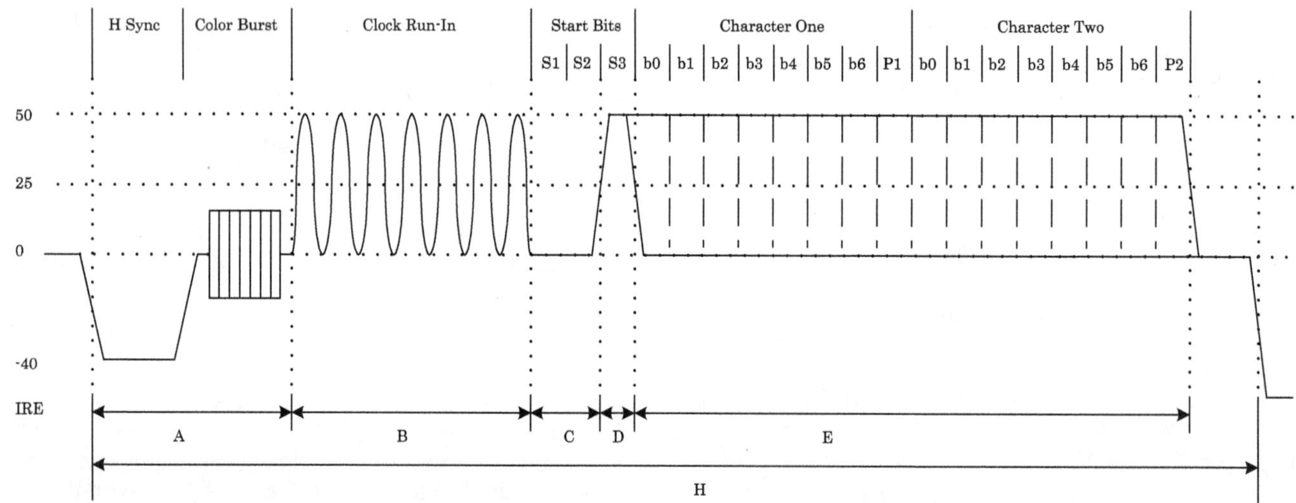

FIGURE 5.7-1 Line 21 Waveform Diagram

Courtesy of CTA

NOTES: 1) All interval measurements are made from the midpoints (half amplitude) on all edges.

2) If only Field 1 is being encoded, Field 2 may be 0 IRE, 7.5 IRE (setup) or clock run in with two null characters. If only Field 2 is being encoded, Field 1 shall be clock run in with two null characters.

3) If both fields are to be encoded, each field shall be encoded within the tolerances specified in Table 1. The transmission channel between the encoder output and the decoder input shall not change the difference in the amplitudes of corresponding signal parameters between the two fields by more than 5 IRE.

4) Deviations introduced anywhere in the signal due to source generated switching transients shall be no greater than ± 2 IRE. However, decoders should accept larger deviations, which may be introduced by passing the signal through a transmission channel and/or signal security system.

5) No setup level allowed.

6) Line 21, Field 2 is also known as line 284.

Primary and secondary closed captions are typically used for main and secondary languages respectively and are synchronized to the audio channel. Text 1 and Text 2 are for special use only and need not be synchronized. Text 3 and Text 4 are discouraged for use, and finally XDS provides for program data such as program name, V-chip data, channel and time of day.

A decoder within a stand-alone device or the television receiver decodes the data from line 21 and displays the words and sentences in a prescribed location determined by the caption provider.

CTA 708

When broadcasting moved to digital with ATSC transmission, an improved design was created to provide closed caption services for digital transmission called Digital Television Closed Captioning (DTVCC). This new delivery standard for creating, transmitting and decoding captions is Consumer Technology Association CTA Standard CTA-708.

This new signal was formatted and provided to the vertical ancillary data (VANC) as a digital bitstream according to SMPTE ST-334 within a logical data channel that would not be compressed. The DTVCC format is agnostic of the video format.

The caption data channel consists of five layers following the Open Systems Interconnect (OSI) model for layered protocol, as shown in Table 5.7-2.

Carriage of CTA-608 data stream in DTVCC is provided to support TVs and converter boxes with NTSC outputs. The entire CTA-608 data stream is preserved during the delivery process.

The components for the DTVCC service are carried in various locations within the DTV bitstream. Caption service data is carried in the Program Map Table (PMT) and Event Information Table (EIT). The video user data bits carry the DTVCC Transport Channel that includes caption service data such as caption text, window commands and so forth as well as NTSC 608 captions if present. (See Figure 5.7-2.)

TABLE 5.7-2 CTA-708 OSI LAYER DEFINITION

Layer	Definition
Transport Layer	Extraction of DTV closed caption data or CTA-608 data stream
Packet Layer	Detect and resync of 708 data stream
Service Layer	De-multiplexing of 708 caption service
Coding Layer	Parsing of 708 syntax
Interpretation Layer	708 window appearance and placement and interpretation of display video

Caption Attributes

Roll-up or scroll-up captions appear from left to right, up to one line at a time. When a line is filled in roll-up mode, the whole line scrolls up to make way for a new line, and the line on top is erased. Typically two or three lines are displayed at a time, so when a new bottom line causes the replacement of the top line, the whole display doesn't disappear. This method is used when captioning video in real time such as for live events, where a sequential word-by-word captioning is needed or a pre-made file isn't available.

The lines typically appear at the bottom of the screen but can actually be placed on any of the 14 predetermined screen rows to avoid covering faces, graphics or action. This method is signaled on CTA-608 by a two-byte caption command replacing rows for a roll-up effect and duplicating rows for a paint-on effect. This allows for real-time editing of words from right to left in the event a word needs to be erased and replaced.

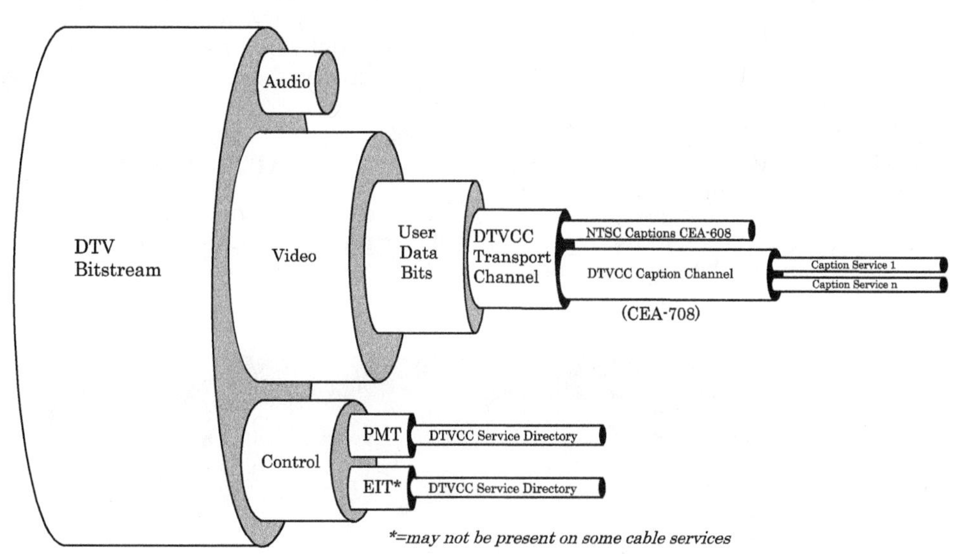

FIGURE 5.7-2 DTVCC Data in DTV Bitstream

Courtesy of CTA

Pop-up captioning is provided by loading the captions into a non-visible memory and swapping the memory so they appear to pop onto the screen image. Pop-up captions can work well for prerecorded programs. In pop-up captions, effort is made to keep sentences together unless two sentences are very short or a single word, in which case they can be combined. Pop-up captions can be positioned horizontally on the screen left to right to indicate who is speaking.

Paint-on captions are a combination of both roll-up and pop-up and are seldom used.

CAPTION WORKFLOW

Real-Time Captioning

Real-time captioning requires captioners to keep up with the spoken word, which can reach up to 250 words per minute (wpm). The best typist on a keyboard might be able to reach 100 wpm, which is why stenography, with capability to do up to 300 wpm with great accuracy, was chosen as the technology for live captioning of television programs.

Stenographers use what are called phonemes, which are the smallest unit of sound that make up a word. So stenographers create words by how they sound rather than how they are spelled. In addition they can hit several keys at the same time, like chords on a piano, to produce a whole syllable or even a whole word. Stenographers can also produce what are called "briefs," which are abbreviations for words they expect to come up during the course of the program. This provides the speed required to keep up with the spoken word in a courtroom as well as a live television broadcast. Research shows that comprehension can be achieved at 220 wpm or higher regardless of age, sex, education or severity of hearing disability.

A relatively new caption service called "re-speak" or "voice writing" has emerged in which highly trained captioners listen to program audio and repeat each word into a proprietary system that converts the speech to the proper caption text and caption attributes. This process allows for either live or short turnaround for faster delivery of programs.

Semi-live captioning follows the same workflow as live captioning but has the added advantage of loading a verbatim script into the caption system ahead of the live broadcast, whereas only the missing dialogue needs to be added. This is possible in a news or magazine program where many of the stories and scripts are preproduced before air. This might be less commonly used, as there may be more cost associated with this process.

The interface to the caption encoder is typically a serial connection either directly into the unit or sent via a modem so the caption author need not be located at the broadcast studio. Therefore, caption authors can be located either at the caption facility or even at their home while the caption encoder can be placed at the studio or master control facility closer to the air chain. As technology has evolved, caption encoder interfaces have moved to Internet protocol (IP) since IP offers a more reliable interface, and the use of "plain old telephone service" (POTS) lines has become rare. If a POTS line becomes unstable, it will result in disconnects and garbled captions. As more telco carriers eliminate POTS services, IP caption encoders will become the norm. An example is shown in Figure 5.7-3.

Offline Captioning

Offline captioning has the advantage of improved quality as the author has time to consider position, timing and formatting. Offline captions are typically pop-up captions, and relative position typically indicates who is speaking. Offline captions can be created as linear or nonlinear.

Linear captioning requires the use of a timecoded videotape dub of the content so the author can see the video in order to place captions in the most appropriate part of the screen and sync the caption appear and disappear times with the video to match the spoken word. The timecoded caption files can then be sent to a facility to be encoded into the videotape.

Nonlinear captions skip this last step, as the timecoded video is sent as a file, usually as an MPEG file, rather than a videotape. The caption system can then read the timecode directly off the file and produce a caption file that can be imported back at the editing facility directly into a nonlinear editing system. As long as no changes have been made to the program, the caption file will match frame accurately to the master program timecode. The caption file can then be embedded into the video or delivered as a separate file to the broadcaster depending on the broadcaster's delivery requirements.

CAPTION REGULATION

The Americans with Disabilities Act (ADA) brought about many changes for the disabled, including the hearing impaired. Though the act does not address

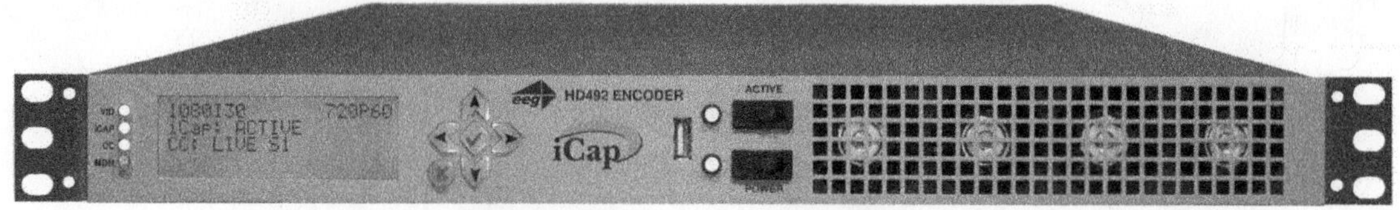

FIGURE 5.7-3 IP Caption Encoder

Courtesy of EEG Corp.

closed captioning wholly, it does reference public service announcements (PSAs) that are federally funded, as stated below.

Sec. 611. Closed-captioning of public service announcements

Any television public service announcement that is produced or funded in whole or in part by any agency or instrumentality of Federal Government shall include closed captioning of the verbal content of such announcement. A television broadcast station licensee

(1) shall not be required to supply closed captioning for any such announcement that fails to include it; and
(2) shall not be liable for broadcasting any such announcement without transmitting a closed caption unless the licensee intentionally fails to transmit the closed caption that was included with the announcement.

The TeleCaption decoder provided the means for the hearing impaired to decode programs with closed captions but could still cost a couple hundred dollars to purchase. The Television Decoder Circuitry Act of 1990 required any TV set with a screen size 13 inches or greater to have a built-in closed caption decoder chip. This reduced cost for the consumer and stirred an increase in closed caption content to be produced.

The Telecommunications Act of 1996 brought about a phased-in approach to provide closed captioning for both English and Spanish language programming. The Federal Communications Commission (FCC), in order to reduce the burden on broadcasters, only required programs airing after January 1, 1998, and up to 450 hours per quarter starting on January 1, 2000, for English and January 1, 2001, for Spanish. The act calls for 100% compliance by January 1, 2006, for English and January 1, 2010, for Spanish. There are exceptions, however, including program providers with less than three million dollars a year in revenue, commercials, programs produced in languages other than English or Spanish, and a few others.

In 2010 the Congress enacted the Twenty-First Century Communications and Video Accessibility Act (commonly known as the CVAA), roughly 20 years after the ADA, which included wide-ranging provisions to improve overall accessibility of audio and visual materials, impacting broadcasters, device manufacturers (including TVs, smartphones, computers, video game consoles), Web site operators and others. This launched the rulemaking for closed captioning of video delivered via Internet protocol, which went into effect in 2012, for video description in 2011, and for emergency crawl information in 2013, among many others.

Finally, in February 2014 the FCC adopted new rules for closed caption monitoring, record keeping and quality. The quality rules deal with caption accuracy, synchronicity, completeness and placement. The ruling also amends the exemptions from previous rulings. It is always good to consult with a station's FCC attorney for any questions regarding closed caption compliance.

RECOMMENDATIONS

The delivery of closed captions requires the entire station ecosystem to be considered. This includes program acquisition, program ingest, quality control as well as production, postproduction, broadcast playout and monitoring. Paying close attention to caption quality along with audio and video is in everyone's best interest. Caption complaints can be time-consuming and costly. Another recommendation includes maintaining close relationships with local cable companies and national satellite service providers that distribute station content to subscribers. Finally, consider providing closed captions as an opportunity to reach a wider audience to engage and entertain rather than simply as a chore.

LOOKING FORWARD

With the upcoming changes that can be afforded by ATSC 3.0 there will be many new opportunities for accessibility and closed captions. Caption technology for ATSC 3.0 leverages a profile of the W3C Timed Text Markup Language (TTML) called Internet Media Subtitles and Captions 1.0 or IMSC1. IMSC1 is substantively equivalent to SMPTE Timed Text. TTML in general has been used widely for the Internet and over-the-top ("OTT") media delivery for high-value content. There are two profiles defined in IMSC1—text and image. These profiles allow IMSC1 to truly be a worldwide format supporting nearly unlimited languages.

Another feature in ATSC 3.0 is to decouple the caption file as a separate component (i.e., not "tunneled" in the video component). This will allow the broadcaster to send one caption file over broadcast and another in an alternate language over broadcast or broadband that can be received by the main receiver or a companion device. This also has advantages in the production process similar to maintaining separate audio and video tracks that do not get combined until delivered to the receiver. With alignment with Internet technology, the opportunity allows for new possibilities and streamlines workflow when servicing viewers on both broadcast and broadband.

Bibliography

Americans with Disabilities Act of 1990, ada.gov, 1990
Consumer Technology Association ANSI/CTA CEA-708-E Digital Television DTV Closed Captioning, 2013
EBU Report on Access Services, EBU-UBR, 2004
FCC Closed Captioning Report and Order, Declaratory Rulemaking, and Further Notice of Proposed Rulemaking, February 20, 2014
Jensema, Dr. Carl J., and Burch, Dr. Robb, *Caption Speed and Viewer Comprehension of Television Programs Final Report*, Institute for Disabilities Research and Training, Inc., Silver Spring, MD, 1999
Robson, Gary D., *The Closed Caption Handbook*, Focal Press, Boston, MA, 2004

CHAPTER

5.8

Workflow and Professional Media Networks

BRAD GILMER

Gilmer & Associates
Atlanta, GA

At a high level, workflow-based solutions focus on what the business wants to achieve rather than focusing on a hardwired technical facility. This chapter presents some of the benefits, challenges and potential new directions for broadcasters considering the deployment of workflow-based solutions.

To aid in the understanding of these systems, let's start by describing a small portion of a traditional, hardwired master control facility. (See Figure 5.8-1.) In this facility, we receive commercials via a "catch server," move selected commercials to an on-air server and then play them out according to a predetermined schedule.

This system is efficient, stable and well understood by operators, and it delivers good performance day-in and day-out. It is built upon well-proven and time-tested technical designs, and it fulfills the master control function quite adequately.

In the past, changes to this facility's processes were relatively rare, so its hardwired nature was not seen as a constraint. These days, however, changes to the master control scenario seem to arise frequently, resulting in costly reconfigurations. To illustrate this point, imagine that in the scenario above, a new commercial "catch server" is installed. This server provides files in a format that is incompatible with the current master control playback equipment. A transcoder device is purchased and installed in the master control area to handle the format mismatch. Shortly thereafter, another system is installed that provides recording, transcoding and storage of news stories for later viewing by the general public via the Web. See Figure 5.8-2.

We will use this example to illustrate the difference between hardwired technical facilities and facilities designed using a workflow-driven approach (as follows).

FIGURE 5.8-1 Simplified traditional master control facility.

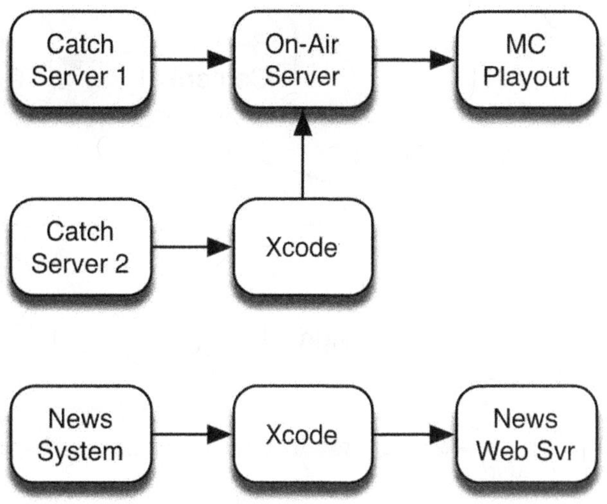

FIGURE 5.8-2 Master control with added equipment.

DESIGNING WORKFLOW-BASED SYSTEMS: BASIC CONCEPTS

Compare the traditional design process described above to a workflow-driven approach. Using this approach, when we add on to the original facility, we start by asking, "What is the business trying to achieve with this new workflow?" In the case above, we have two new business objectives. First, we are playing back commercials received from a new advertising distribution company so we can reduce the cost of ingesting videotapes manually. Second, we are making news content available on the Web after a live broadcast has concluded so we can create new Web-based advertising revenue. It is important to note that in both cases, we state the objective to be achieved, but we also clearly acknowledge the value to the business if we achieve this objective. This is important because, as engineers, this keeps us focused on what the business is actually trying to accomplish, rather than immediately dropping down into a lot of technical detail. Once we understand the two business objectives, we begin to model the workflow.

A critical question is how much detail is required in the model. Large professional media companies have invested millions of dollars, collectively, in modeling their workflows. In a typical operation it is not necessary to go to that extent. It is also not necessary to invest in one of the many expensive modeling software programs available in order to complete this task. Frankly, a sharp pencil and a piece of paper is all you need for the initial modeling pass.

The first task in the modeling process is to sketch out the steps in any given workflow. For example, Figure 5.8-3 shows a high-level model of the workflow involved in the new commercial delivery process.

You can see that this first pass at the model is quite coarse, and contains what are called *epic functions*. These are "rolled up" processes that consist of a number of smaller *atomic functions*. Modeling workflows with epic functions is problematic because it may oversimplify a workflow that is actually quite complex, and also because it may hide opportunities to reuse common functions in other workflows.

FIGURE 5.8-3 High-level model of commercial delivery workflow.

For example, the function *Move File* is an atomic function. By contrast, the function *Ingest Content* is an epic function, because that function consists of a number of atomic functions, such as *AssignID*, *Transcode*, *QC* and so on. A better workflow model (although reduced in complexity for space considerations here) is shown in Figure 5.8-4.

The exercise of developing these models is sometimes called *business process modeling* or BPM, although in practice, BPM involves a set of specific tools and uses a specific vocabulary to describe the different elements used in the models. (You should be aware that people who do process modeling for a living use a very specific set of symbols, usually on specialized, graphical drag-and-drop software systems, with the ultimate goal of "compiling" BPM into code that can be executed in a computing environment.) While it is quite possible that the professional media industry may evolve to this level of computer automated integration, at this point we are simply borrowing some techniques and vocabulary from the world of BPM. A simple model drawn with a pencil and consisting of boxes without any fancy symbols will get you a very long way.

Going back to our two examples, Figure 5.8-5 shows a workflow model for the news capture and Web replay system (again, at a very high level, for the purposes of illustration).

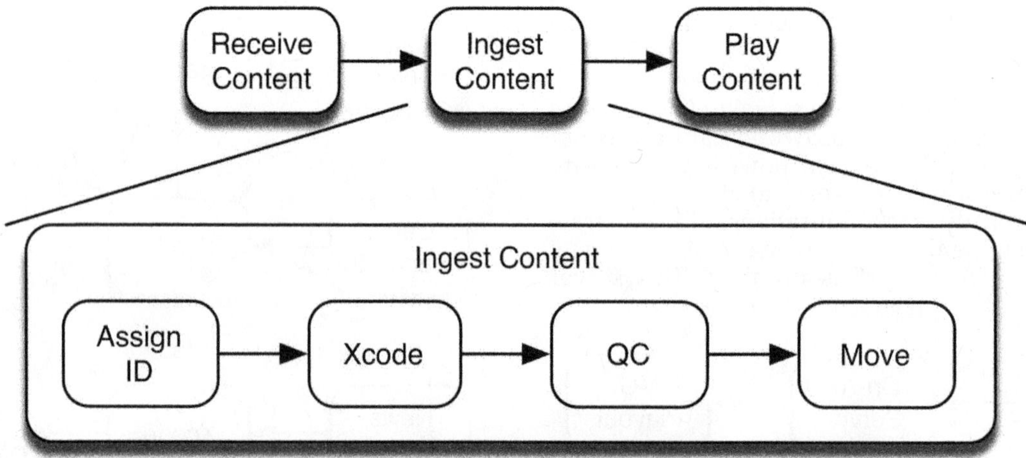

FIGURE 5.8-4 Expanded commercial delivery workflow model.

FIGURE 5.8-5 News story capture and Web replay workflow model.

REFINING THE DESIGN

Once the initial modeling work is done, there is an opportunity to look for places where workflows may be improved, and where processes may be duplicated. In our examples, note that there are places where the atomic functions are identical—*Transcode* and *Move*, for example. Whether these represent actual opportunities for improvement depends on a number of factors, but this example proves the point that without a workflow-based approach these duplications might not be revealed and analyzed.

The promise of a workflow-based approach goes beyond simply identifying potential duplication, however. Let's examine further how increased agility is enabled.

In the hardwired master control example above, content moves from one place to another along a well-established path. Errors are handled as they come up (e.g., poor-quality audio is resolved by contacting the commercial delivery company and requesting delivery of another version of the commercial). Changes in workflow are possible, but they require sending memos to people in different departments, staff training, and the rewiring of equipment. Steps in this workflow are executed in a serial manner. (See Figure 5.8-6.)

In the workflow-driven approach, there is typically a system (usually called an *orchestrator*) that executes the steps in a given workflow using a set of rules that have been established in advance based upon the workflow modeling process described above. The workflow moves through a predetermined set of states, which may be described in a state diagram or state view. A particular atomic function, for example, *Move*, may be employed to move all sorts of files between many different locations simply by passing the *Move* function the information it needs in order to accomplish its task. Each atomic function stands alone, and it may be executed before, or after, any other function. The practical effect of this architecture is that workflows may be modified by altering the rules in the orchestrator. This is also how a single atomic function can be reused in more than one workflow. As Figure 5.8-7 shows, complex workflows can be executed in any order using a variety of atomic functions available to the orchestrator.

Note in Figure 5.8-7 that the workflow can end in an error state. Recall that in the traditional, manual workflow, errors are handled by operators as they come up. One advantage of well-designed workflow systems is that they anticipate, from the beginning, that a workflow may enter an error state at particular points. Sometimes, as humans, we prefer to avoid thinking about what to do if something goes wrong and only think about the "happy path." A well-defined workflow-based approach not only lays out the path of expected behavior if all of the steps succeed; it also defines error conditions, and the operations that should be taken in order to initiate recovery from these errors.

As Figure 5.8-7 shows, the orchestration client drives the workflow by sending a request to a specific service and taking action based upon the response from that service. In our initial example we added a new commercial delivery service that required a transcoder in order to be used in our facility. If at a later time the delivery service adds a new video format compatible with our playback equipment, we can eliminate the transcode step by simply changing the rules in the workflow client.

CHALLENGES TO IMPLEMENTING WORKFLOW-BASED SYSTEMS

If this all sounds too good to be true, let's also address some of the challenges related to a workflow-driven approach. For now, in the real world of the professional broadcast engineer, the reality is that there are a number of vendors offering different, incompatible workflow-based solutions.

For example, you may have a newsroom system, a transcoding system, and an on-air automation system, all designed around workflow, all promising to allow you to be more agile, but in reality you may be unable to achieve the promised benefits because the systems do not work well together. A more holistic approach to workflow-based technologies is required, and may develop. Some success in this area has been achieved by the FIMS[1] group of the Advanced Media Workflow Association[2] (AMWA); however, at this writing, adoption of FIMS technology including interoperable interfaces has been spotty.

Another issue that broadcast engineers need to be aware of in workflow systems is the serial reliability problem. This problem states that the reliability of any given workflow is the product of the reliability of each

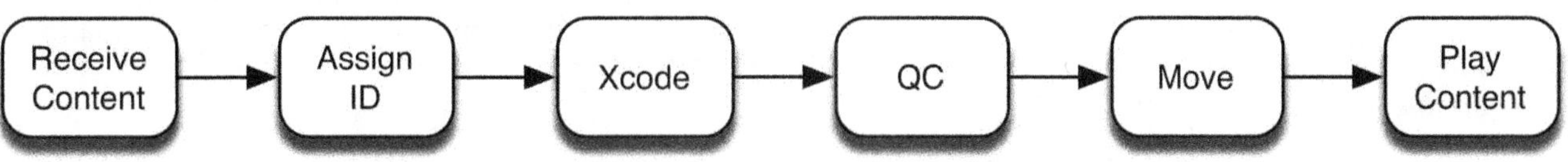

FIGURE 5.8-6 Simple serial workflow.

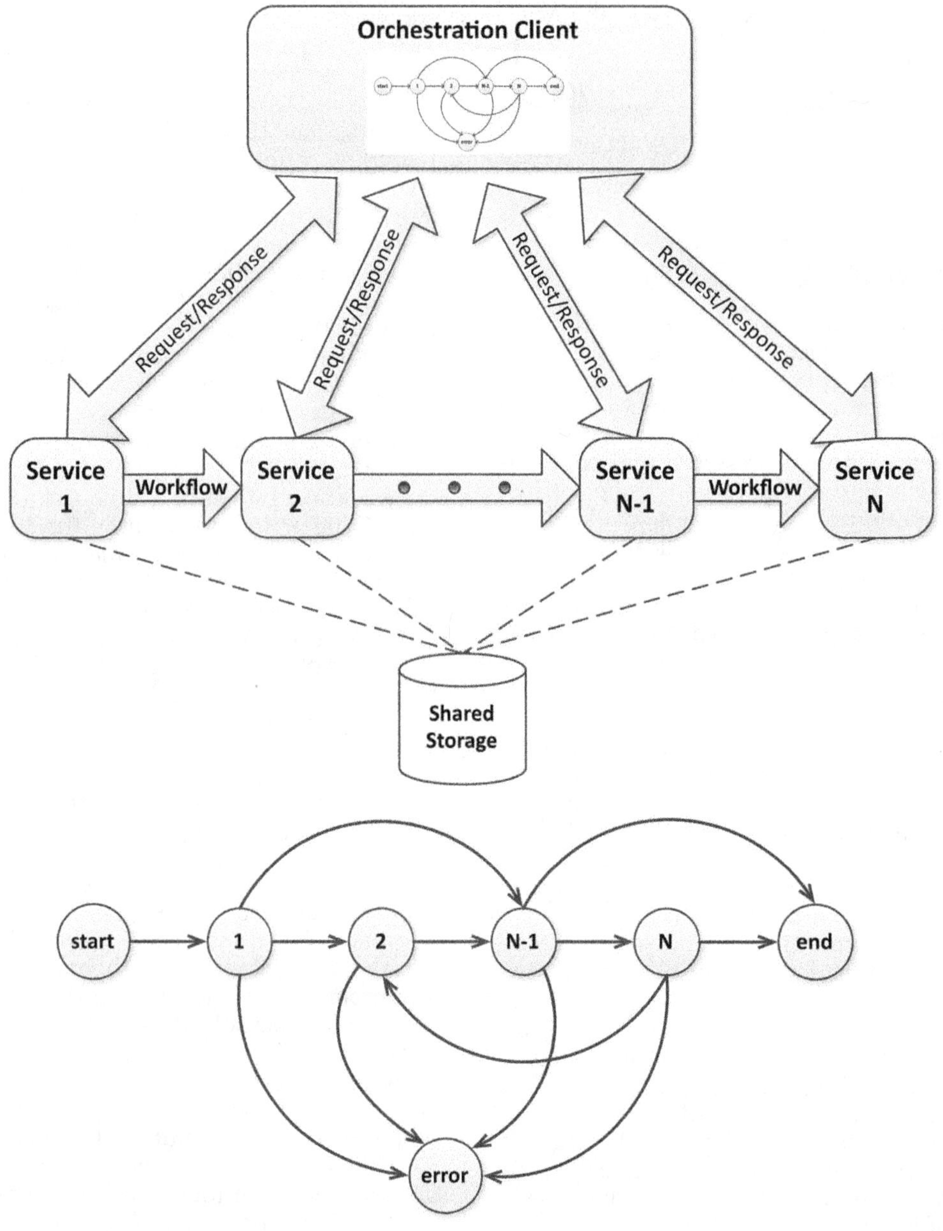

FIGURE 5.8-7 Orchestrator driving dynamic workflows.

Courtesy of Jim Trainor, Gilmer & Associates, Inc.

individual atomic service composing the workflow. Mathematically, the serial reliability problem is stated as:

$$R(N) = \prod_{i=1}^{N} R(i)$$

R(i) = individual reliability of process step i
R(N) = overall reliability of the workflow

For example, a manufacturer may say that their workflow is 96% reliable, meaning that it fails four times in one hundred operations. If a series of five steps with 96% reliability are chained together, then the overall reliability of the workflow is 0.96 * 0.96 * 0.96 * 0.96 * 0.96 = 81.5%, meaning that the overall workflow fails a little less than twenty times for every one hundred times it is initiated. This may or may not be what you were expecting, but in any case, it is a good formula to tuck into your back pocket when thinking about workflows.

NEW DIRECTIONS

It is important to acknowledge that alongside this movement toward workflow-based systems is the growing trend toward Internet-driven technologies affecting the entertainment industry as a whole. This phenomenon will continue to have an impact on workflow-based system design in broadcast production facilities, as described below.

In 2013, the Joint Task Force on Networked Media[3] (JT-NM) was formed by the European Broadcasting Union[4] (EBU), the Society of Motion Picture and Television Engineers[5] (SMPTE) and the Video Services Forum[6] (VSF) to assess the impacts of the adoption of Internet technology on professional media networks. The goal of the group was, and remains, to aid the industry by increasing interoperability in the transition from traditional SDI[7]-based technology to systems based upon computer science and the IT industry. While this effort addresses the question of an "SDI replacement," it also extends to facilitating the establishment of critical fundamental frameworks, standards and best practices for the broadcast industry going forward. The reason to include this information in a chapter on workflow is that the work of the JT-NM will have a profound effect on the workflows, facilities and technologies used by broadcasters in the future.

Work of the JT-NM to Date

The JT-NM began its work by collecting over 150 business-driven user stories regarding the functions and capabilities required in professional media networks. These user stories were condensed down to sixteen use cases. The JT-NM next issued a Request For Technology (RFT), looking for technologies that could meet the requirements enumerated in the use cases. The good news is that technology existed to address all of the requirements. The bad news is that there was no way to make a coherent system from all of the different submitted technologies. In short, there was no interoperability among the submissions, and no clear path on how to achieve the group's overall goals. The challenge appeared to be especially daunting in the area of *live* television production.

Because live television in a professional networked media environment presents a number of unique challenges to conventional IT technology, the JT-NM decided to develop what was considered to be a Minimum Viable System to support a typical, live half-time sports studio production scenario. The requirements of such a scenario were captured in the JT-NM Minimum Viable System (MVS) report.[8]

Significantly, this report, along with the activities of the Studio Video over IP (SVIP) group within the VSF, ultimately led to the creation of the first specification for elementary video, audio and data streams over IP networks, VSF Technical Recommendation 03 (TR-03).[9] TR-03 captures many of the essential thoughts contained in the JT-NM recommendations, and specifically within the JT-NM Reference Architecture. Importantly, VSF TR-03 forms the majority of the basis for standards[10] for future professional media networks.

The JT-NM decided to create a Reference Architecture (RA)[11]—a collection of best practices, successful patterns, fundamental frameworks and standards that, taken together, could potentially lead to broad interoperability in the industry. The JT-NM Reference Architecture covers a lot of ground, but there are three key concepts that every broadcaster should understand. These are capabilities, fundamental frameworks, and the JT-NM conceptual model.

Capabilities

Figure 5.8-8 presents a high-level system view diagram of the JT-NM Reference Architecture. At the top of the figure is the operation layer. The operation layer describes the high-level media and business operations that a media company performs. Examples of these operations include live studio production and corporate accounting; these are operations performed by people in order to accomplish the overall goals of the media company. People performing media and business operations use media and business applications. These applications, such as video editing or corporate accounting, require certain capabilities in order to perform their higher-level functions. These applications are shown in Figure 5.8-8 in the application layer.

At the bottom of the figure is the infrastructure layer, which consists of the networks used to move professional media, and above it, the data center and specialty hardware required by the broadcast industry.

The platform layer rests on top of the infrastructure layer. Capabilities provide the connection between the platform layer and the application layer.

Several key points result from this architecture:

1. It is the job of the platform layer to deliver up latent capabilities available in the infrastructure layer to meet the requirements of the application layer. (By latent we mean capabilities that exist but that are not yet being consumed.)
2. Once engaged, a latent capability becomes a part of the workflow chain for the consuming application.
3. Once an application has finished using a capability, that capability is released and again becomes a latent capability and is available to be consumed by the same or another application in the facility.

In addition to the concept of capabilities, there are four fundamental frameworks introduced in the Reference Architecture that are important for broadcast engineers to be aware of. These frameworks are *Identity, Timing, Discovery and Registration*, and *Connection Management*. It is the belief of the JT-NM RA authors that the appropriate standards and best practices in these key areas result in a stable, interoperable base layer for future professional media systems. This base layer establishes interoperability in IT-based systems, while leaving room for vendors to innovate and differentiate themselves from others. These fundamental frameworks and why they are so critical are described as follows.

Identity

In part, the JT-NM says, "Internet-style Unique Resource Identifiers (Internet Standard STD 66 and

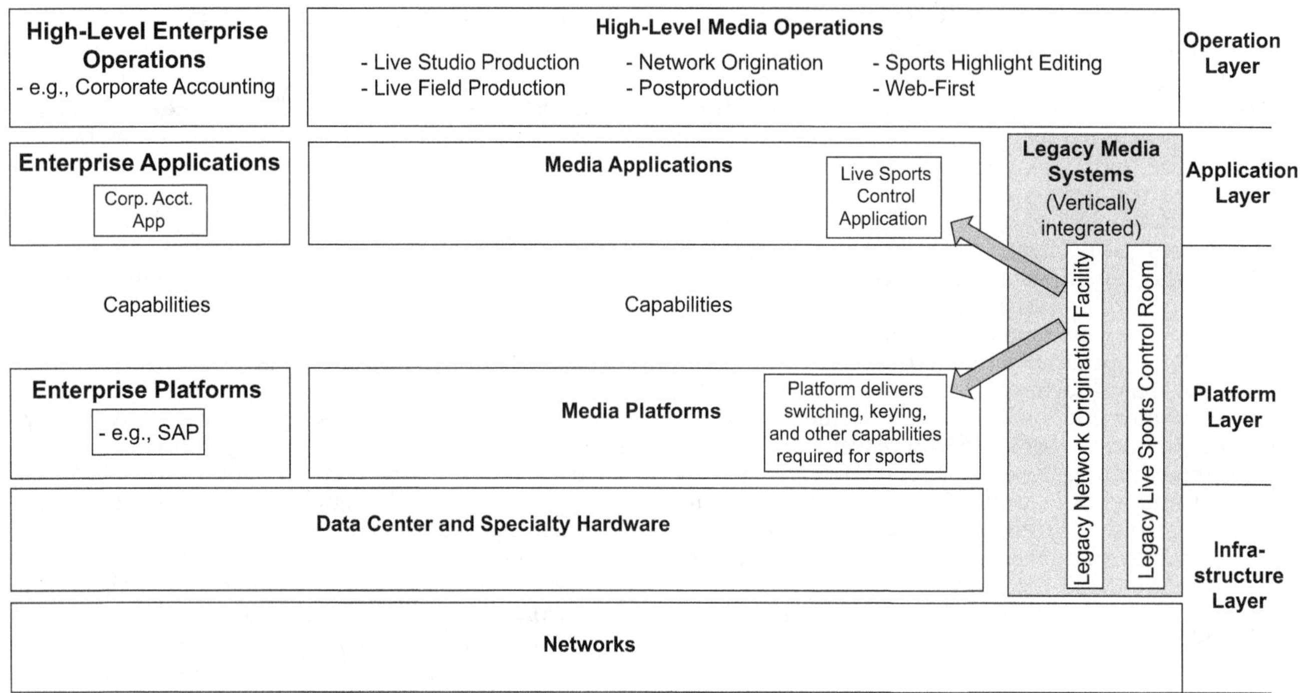

FIGURE 5.8-8 High-level system view diagram of the JT-NM Reference Architecture.

RFC 3986) shall be used to identify the resources of networked media." This text, along with the other parts of this section in the RA, basically say, "Identify professional media resources the same way we identify other things that use Internet technology." Unique resource identifiers (URIs) can be generated locally, the domains of the identifiers can be scoped, and URIs are used natively in many applications. As an additional benefit, URIs are extremely well understood in the software and computer science domains. Using URIs as described in the RA brings the broadcast industry in line with the rest of the computing world.

Timing

It is well known that since the beginning of television, timing has been critical to its operations. This does not change just because we move to an IT-based infrastructure. That said, it would be rather narrow-sighted to assume that the professional media industry is the only industry with a need for accurate timing in a networked environment. Surely financial markets, aviation, medical, military and other fields have as stringent, if not more stringent, requirements. In fact, this was the assumption of the JT-NM when it looked into this area, and the assumption turned out to be correct.

The IEEE has developed a standard called *Precision Time Protocol*, or PTP, in which a precision timing signal is distributed over a network (IEEE 1588).[12] Because of its wide use in other fields, both the Audio Engineering Society (AES) and SMPTE use IEEE 1588 as the basis for the development of profiles in the professional video and audio domains. These two profiles have been harmonized, resulting in a technology for our industry that is actually more accurate than the traditional "Black Burst," has new features beyond time code, and can be distributed over a network with sufficient accuracy to synchronize video, audio and other components.

Discovery and Registration

In an SDI world, knowing what is available on the router is not difficult. Engineers connect a device to a specific port, and then they either make appropriate label entries like "CAM1" in the router software, or they reference the physical input and output in order to make a connection (e.g., "Connect Input 10 to Output 2"). In an IP environment, this is no longer the case.

Generally speaking, you can plug a device into any port on an IP router and it will be connected to the network. But who keeps track of the network details of all these devices? In a small IT network this is generally not a problem. Addresses of devices are manually configured, and operators know about every single device connected to the network.

However, writing down the specifics of every device, its network configuration, functionality, et cetera, is simply not feasible, even in a relatively small television production facility. Discovery and registration provides a standardized way for a user to plug in a camera, have it discovered on the network, and have its details recorded automatically in a registry for later use. This technology addresses a fundamental problem in networked broadcast facilities that is otherwise unmanageable.

Connection Management

The easiest way to explain connection management is to consider a situation where a technician who maintains remote trucks has a small SDI router on his workbench. The technician retrieves a camera from a remote truck, brings it to his workbench and plugs it in to his router. He selects the camera on the router, and its output is displayed on a monitor above his bench—simple.

Now consider this scenario in a professional networked media world. The task to be performed is the same, but this time, the technician brings in the camera and plugs it into an IP router. A monitor is also plugged into the same router. Where is the "control panel"? How does the technician view the output of the camera on the monitor? This issue is addressed by a combination of discovery and registration (which has noted the existence of the camera and monitor in the registry described above) and connection management (which instructs devices to join the appropriate multicast stream). In effect, the camera is like a little TV station, transmitting on a specific frequency, or in this case, sending out a stream with a specific multicast address. The monitor, acting a little like a television set, is instructed, through connection management, to tune to the appropriate channel (i.e., join the appropriate multicast) in order to view the output of the camera.

Potential Impact of JT-NM on Future Facilities

Returning to the design of broadcast facilities, the vision of the JT-NM is that broadcasters in the not-too-distant future will be able to build scalable facilities, which can ramp up or scale back based upon demand. New workflows can be quickly accommodated by using the appropriate orchestration tools, and by adding capabilities if the new workflow requires a new feature that is not currently present in the facility. If the facility requires expansion, the broadcaster can purchase more of the needed latent capability from the appropriate vendors. Facilities become much more elastic, able to respond to changes in demand, or even changes in the underlying business.

If all of this sounds farfetched, you should know that this is the way Internet technology works. Today. The future described here is absolutely possible if video, audio and metadata become first-class citizens in an Internet technology environment.

The Cloud and Virtualization

If you are in any technically related field, it is almost impossible to avoid the topic of the cloud or cloud computing. One reason for this is because a lot of businesses are "moving to the cloud." Another reason is that there are some very significant marketing efforts being put into the topic of the cloud.

The cloud is nothing more than servers and networking equipment with some "dashboard" user interfaces, and a business model that lets users lease, rather than buy, the hardware involved. The one thing that is impactful about the cloud is its scale. Computing, storage and networking capacities are tremendous, and growing at an amazing rate.

One way to think of how the cloud could benefit broadcasters is to think about buying JT-NM capabilities "by the pound." Storage, transcoding and video processing are just a few examples of things that broadcasters can already purchase on a monthly basis.

Virtualization is a concept that could revolutionize the broadcast industry, especially when coupled with JT-NM capabilities and the fundamental frameworks. Virtualization allows an engineer to "spin up" servers that have been pre-configured with software, which are capable of performing media operations such as transcoding or content playback. This relates back to the term "latent capability" used earlier in this chapter. Assuming that computer resources are available, and assuming that the appropriate software licensing terms are in place, a broadcast facility has at its fingertips the ability to launch instances of software applications that can perform important steps in a media workflow. While the resources are available and the licensing is in place, the capabilities provided by the server are "latent," meaning that they are available to be used. Once they are instantiated and are being used, they are actual capabilities that are part of a media workflow.

Note that the actual hardware creating virtual servers can be anywhere—inside a broadcaster's four walls, across town at a postproduction facility, or in the cloud. Over time, however, it may be that the majority of the media services consumed by broadcasters are in the cloud. Fortunately, the architectures and best practices described here allow broadcasters to make the appropriate choice about where services are hosted based upon issues such as security, service agreements (SAs) with cloud vendors and applicable tax laws.

Virtualization has come a long way, to the point that—depending upon the function to be performed—a single piece of hardware may be able to act like five, ten or more individual servers. In the past, one would have needed one piece of hardware per each copy of server code, but this is no longer the case. We are now at a point where, with the proper advances in vendor products, and with well-engineered and architecturally designed plants, broadcasters can realize truly scalable, flexible and robust media facilities.

SUMMARY

This chapter started by exploring some topics around workflow-based systems and the difference between them and traditional hardwired facilities. It then introduced the concepts in the JT-NM Reference Architecture around capabilities, the fundamental frameworks of identity, timing, discovery and registration, and connection management. It then explained that one way to think about the cloud is that it is simply a collection of servers that provide capabilities you can buy "by the pound."

All of these elements will increasingly define broadcast facilities and fundamentally change the nature of broadcast operations in ways that help it to flourish in the networked digital economy. Broadcast engineers should therefore strive to understand these basic principles, and work toward implementing them in future facility designs.

Notes

[1] More information at www.fims.tv
[2] More information at www.amwa.tv
[3] www.jt-nm.org
[4] www.ebu.ch
[5] www.smpte.org
[6] www.videoservicesforum.org
[7] SDI (serial digital interface): A common, SMPTE-standardized interface and interconnection method for professional video equipment.
[8] The MVS report is available at www.videoservicesforum.org/jt-nm/documents/JT-NM MVS Report.pdf
[9] www.videoservicesforum.org/download/technical_recommendations/VSF_TR-03_DRAFT_2015-10-19.pdf
[10] The SMPTE ST2110 Standards are under development at this writing.
[11] The JT-NM RA v1.0 is available at www.jt-nm.org/RA-1.0/index.shtml
[12] https://standards.ieee.org/findstds/standard/1588-2008.html

CHAPTER

5.9

Digital Asset Management

WENDY ALLEN
21st Century Fox
Los Angeles, CA

INTRODUCTION

In the late 1980s, less than 1% of the world's stored information was in digital form. By 2007, it had increased to 94%, and was over 99% by 2014.[1] It is estimated that the world's capacity to store information has increased from 2.6 exabytes (optimally compressed) in 1986, to some 5,000 exabytes in 2014 (5 zettabytes).[2] Cloud computing entered the mainstream by 2015, and now with multi-device consumer behavior at the forefront, more and more data is being tracked, managed and stored on-premise and in the cloud. With the amount of data stored in the cloud expected to double by 2020, it is imperative that broadcasters have digital asset management (DAM) systems in place to search using metadata, to improve decision making and to promote their businesses' viability.

Broadcasters have moved steadily toward digital assets and DAM systems since the early 1990s, as it became faster and more cost-effective to work within the digital domain for managing video and audio media. Meanwhile, the traditional broadcasting model has continued to evolve. Expensive, traditional and dedicated hardware-based systems like editing suites moved to computer-based nonlinear editing, not just because of the benefits of working in a nonlinear mode, but due to reduced infrastructure costs, which became increasingly compelling as the technology improved. Similarly, traditional broadcast supply-chain infrastructures for distribution have moved components to the cloud, and eventually entire supply chains will be supported. This technology movement has not just produced a new generation of hardware, but has resulted in development of agile software with new feature sets, resiliency and faster processing.

In conjunction with these technical advances, broadcasters' businesses have engaged in more media deals, with the media deliverables increasing at a staggering rate to remain viable in a rapidly changing market. These new broadcast distribution models include multichannel video programming distributor (MVPD), over-the-top (OTT), live streaming, over-the-air (OTA) and 4G or Long Term Evolution (LTE) wireless ventures, to name a few. With the move toward more distribution paths at lower costs, the sheer number of channels has increased as well. Broadcasters need to identify specific markets, deliver content with pinpoint accuracy to niche sectors, and prove to advertisers that the targeted audience was reached. This implies a need for greater knowledge about the audience, their behavior patterns and metadata associated with accessible assets.

As a result of increasing distribution paths and models, there is potential for an increase in organizational pain points and even in the loss of revenue. Some organizations attribute the loss of revenue to the lack of metadata management. An example would be the inability to propagate a unique identifier throughout the entire supply chain, as the adaptation of ID standardization remains challenging. DAM systems are capable of solving these types of problems through better metadata management processes.

A DAM system drives behaviors and manages accessibility, visibility and workflow decisions using metadata. Metadata management is fundamental when searching and retrieving assets within a DAM deployment. It is estimated that creative staff search for media files 80 to 100 times a week, and that in as many as one-third of those attempts, they fail to find what they are looking for due to insufficient metadata.

TABLE 5.9-1
Changing market conditions and opportunities in broadcasting.

1970 1980	1990 2000	2010 2020
Free TV dominates	Subscription/pay-per-view significantly increases	Targeted and valued-based revenues
Single-channel broadcasting	Multichannel broadcasting	Multi-platform broadcasting, narrowcasting
Ratings-based advertising revenues	Pay-per-view, channel bundling, viewer "packages"	Pay per pull, multi-service plans, micropayment, push service payments
Broadcast and cable TV distribution	Satellite, cable, DTT	Cable download, trickle feeds, live event real-time feeds
Cinema, TV, VHS compete for viewers' attention	Digital cinema, DVD, Internet, video games, mobile phones	HDTV, mobile devices, Internet/set-top box/online gaming
		OTT, live streaming, and programmatic AD delivery

ANALOGUE → DIGITAL

DAM systems address the fundamental search, locate and restore functionalities required, and can improve and automate the increased workflow demands. A digital asset consists of the core content (*essence*) plus additional metadata. Program video and audio files, low-resolution copies, scripts, subtitle files, audio files, press photos, logos and lower-third *supers* are examples of broadcast digital assets.

DAM solutions come in various sizes and capabilities. They provide asset management functionality with decision-driven capabilities, based on ingest, cataloguing, storage and distribution metadata. As the media distribution model continues to evolve, the need for digital management, data ontology and data accessibility becomes more and more relevant for the broadcast industry. This chapter explores options for consideration when deciding to implement a DAM system.

DAM AWARENESS AND PLANNING

DAM discovery and planning is critical in the process when considering what system is best for your organization. DAM systems can be built for specific business goals and objectives. They can be sized for different levels of engagement from small or medium to an enterprise solution model. There are a number of factors to consider when assessing the system of choice, including the following: organizational change and impact, digital asset management types, digital asset management characteristics, digital asset management project management and return on investment (ROI) justification.

Organizational Change and Impact

Organizational change is inevitable when implementing a DAM system. The ability to list and access assets from anywhere within the organization and from any geographical location is empowering to an organization's staff, but it can also invoke risk and a sense of loss of control by those who once manually managed (or micromanaged) these assets within their own disparate systems or siloed environments.

When implementing a DAM, it is critical to develop a group of stakeholders and dedicated resources across the organization that can commit to the design process and messaging. These selected staff are typically the same resource pool performing the day-to-day work of the organization because they know the workflows inside and out and are eager to eliminate their pain points. They are also the people with the least spare time, however. Not having adequate project resource allocation is one area where DAM deployments can fail both on delivery of expectations and timeline. Therefore it is important to have a small group of operationally knowledgeable stakeholders who can provide continuity throughout the project. In addition, providing periodic updates for the managers and the key resources that will eventually take ownership of the system will lead to a smoother transition.

The saying "measure twice and cut once" applies well to the DAM design phase. At times this phase can seem overwhelming, but while upfront preparation is time-consuming, the more upfront work that is done the more beneficial it will prove in the long term. While going through this process, job positions and resources will need to be realigned. Changing workflows and/or automating them can lead to new organizational layers for exception handling and monitoring. Too often organizational changes occur as an afterthought or as a result of a painful rollout and risk diversion.

No matter how strong a communication plan is, it is never enough. Creating learning materials, service-level agreements (SLA), process and workflow diagrams, and user adaptation are recommended. DAM solutions inherently provide media visibility and accessibility, which for a media organization is their most valuable asset. Unfortunately this brings inherent risk. Well-constructed policies, rules, SLAs and permission oversight and management need to be clearly documented, communicated and developed as the project evolves. See Figure 5.9-1 for a conceptual overview of DAM processes.

FIGURE 5.9-1 DAM systems integrate services, apps and storage systems with expanded capability and functionality.

Digital Asset Management Types

There are several ways to define the types of DAM systems. They can be categorized on deployment size (small workgroup or enterprise), location (Software-as-a-Service [SaaS], on-premise, off-premise or hybrid) or by business need or niche (print, branding, etc.). All DAM systems regardless of purpose can range in scale, configuration complexities and security levels. It also is important to consider all factors that weigh into a DAM selection, including the organization's ability to sustain and support the selected system.

Off-premise (SaaS model): A cloud-based digital asset management system, that is, SaaS, is a software package solution accessed and delivered over the Internet. The system is scalable and can be accessed anytime by anyone in the organization (subject to permissions), using an Internet connection. SaaS DAM systems provide elasticity, and typically the product and platform upgrades are the provider's responsibility. SaaS pricing models can vary based on number of transfers (upload and download), amount of storage, data transfer rate and number of transactions. Benefits can include faster provisioning and setup, scaling up/down with business demands, eliminating the hardware maintenance and sustaining pain points. Over time, pricing models can become more predictable based on analytics and pay-per-use models, especially when periodic events require peak-usage increases.

SaaS deployments require organizations to move from a capital expense model to an operational expense model. Costs can be incurred every time an asset is touched, but also in other activities, such as when upgrades are performed (additional support and SLA terms are required), when additional business criticality and continuity are required and when real-time metrics and monitoring are essential. Also there can be limited ability to address customer-specific enhancements in this model.

On-premise: Broadcast DAM systems have traditionally been purchased and installed on infrastructure hosted within an organization's existing facilities. The expense of in-house hardware, storage, power and equipment can be costly and time-consuming. Additional, dedicated staffing is usually required for on-premise solutions as well. Benefits include in-house control and management. Organizations that produce and distribute live content, operating 24/7 with fast turnaround to market and five nines or better reliability may require an on-premise solution. This solution may also serve special security needs for proprietary information. Full system control is usually shared by internal engineering and IT teams. Sustaining and managing updates, new releases, version control, storage watermarks, security and so forth can become resource intensive for the organization.

Hybrid: As a result of the above analysis, organizations often explore options for a hybrid DAM model. This model could be implemented based on contractual or regulatory restrictions for high-valued content, specific speed and storage connectivity requirements, legacy workflows and/or time to market. An example could be financial documents or sports events in which production and postproduction such as editing require quick turnaround time to air. Also, there are DAM deployments that are open text hosted in the cloud—but the storage is located on premise. Regardless of where the software and hardware is hosted, users on the organization's staff shouldn't know the difference; from the user interface (UI) perspective, they are working with a single, unified DAM system.

Open source: Many DAM systems use open source software, with source code available for modification and enhancement by anyone. Not all commercial DAM vendors make their source code available. Typically, if the software is open source, your options for manipulating the software are either to build the software development experience in-house, via the open-source community, or via third-party "partners" with expertise on the technology platform. Vendor support for open source may vary. For example, an Enterprise Service Buss (ESB) vendor that is based on open-source technology can also provide additional features not part of the core open-source product offering. Advantages of open source include increased flexibility and creative solutions when developing new feature sets, but this can become very expensive and require expanding an organization's software development and networking teams, who will manage and sustain the solution.

Security is another area where DAM systems can range from small and easier deployments to more agile systems with greater security oversight. Table 5.9-2 is a basic chart showing the progression of IT management levels.

Digital asset management systems can be defined as limited scope with a specialty focus for deployment, or as systems that span an entire enterprise. Smaller or limited-scope DAM systems tend to concentrate on a reduced range of benefits. Typically, they remain within a single working group that shares or accesses common material and works within a particular operational context. An example of this type of deployment would be a promotional department DAM, in which there are a limited number of users, a defined range of assets to be managed, and usually a small set of operational groups. Limited-scope DAMs tend to be straightforward to implement, since they have a smaller range of requirements. In the promotional department example, video, audio, graphics, documents, scripts, schedules and workflows are usually well defined, and the required DAM capabilities might be reduced to Ingest, Describe, Search and Edit. Table 5.9-3 presents more detail.

Limited-scope projects have advantages, particularly in the case where niche suppliers have a specialty focus. These applications usually require fewer professional customization services for successful implementation, primarily because there are limited workflow changes necessary. Commonly, it becomes possible to share material with faster access to newly received content, which is particularly useful in a news environment, for example, where several editing suites may be working on the same story or different packages

TABLE 5.9-3
Examples of desired business benefits to be achieved in limited-scope DAM systems.

Limited-Scope DAM Benefits	
Project	• Easy-to-define scope based on dealing with a single business objective • Focused set of user stories to address and personas of typical users • Limited interaction with adjoining departments • User roles clearly understood
Price	• Predictable • Usually very little custom development (which is expensive) • Cost of scaling well known • Training costs low
Implementation	• Whole project duration predictable • Limited impact on physical installation • Training is usually short due to limited functional scope • Benefits easy to identify and quantify

TABLE 5.9-2
Progression of IT management levels for DAM systems.

Basic	Standard	Advanced	Federated
• Access control at resource level • Impersonation • Authentication as authorization	• *n*-tiered application architecture • Impersonation • Some role-based access control authorization • Access control at application level • Roles and permissions abstracted from resources	• Some service-oriented architecture • Trusted subsystem model • Role-based • Authorization gateway • Primary authorization interface on ESB • Cross-platform authorization available • Global roles available	• Wide adoption of service-oriented architecture (SOA) • Authorization gateway based on federated identities • Supports claim-based tokens • Authorization token-based • Support for Web Services (WS) authorization tokens • Authorization based on federated identities

for major breaking news. Workgroup solutions rarely address many of the changing business requirements that broadcasters face, however. To tackle these issues and to make significant economy-of-scale changes, an enterprise-solution approach to digital asset management is recommended.

Enterprise DAM systems can be complex and expensive—a disincentive for introduction at a time when many broadcasters are already making significant investments for the digital transition across much of their material-gathering and infrastructure capabilities. With the increase of channels and platforms and the fragmentation of the market (often resulting in declining viewership per channel), broadcasters typically seek out methods of achieving "more with less."

An enterprise DAM solution is able to use metadata to guide and initiate processes, such as records, access control rules for input, playout decoders, transcoders and channels. Enterprise solutions especially become important when integrating with rights management, trafficking, third-party applications and hierarchical storage management systems, where function and performance have to be aligned. In addition, defining migration, distribution and purge rules, as well as establishing retention policies, all lead to complexities but can also generate long-term cost savings.

DAM scalability is critical when considering enterprise DAM solutions. Importing and exporting, storage, proxies and derivatives are just a few of the areas where the lack of proper architecture planning (scaling, I/O, etc.) can result in untenable issues and costs. This is especially evident with sports. When an event is recording and simultaneous users are accessing, previewing, editing and manipulating a high-valued asset simultaneously, performance and functionality can suffer if not properly scaled. Continuing with the sports example, there may be multiple logs and data sources coming in and updating or affecting metadata. These transactions can cause performance issues and need to be appropriately scoped.

Enterprise solutions are highly configurable with feature richness, global accessibility, infinite scalability, built-in redundancy, geo-replication, and enhanced integration capabilities that span a wider range of applications and enterprise security oversight. These types of DAM systems can be deployed on-premise, hosted in the cloud, or implemented as a hybrid solution.

DAM Characteristics

The following list of digital asset management system characteristics have been defined by the DAM Foundation Board. Systems that wish to gain a certification from the DAM Foundation as possessing all ten characteristics must apply to have their systems evaluated.

1. DAM systems *ingest* assets individually or in mass sets, and allow for the manipulation of those assets and their metadata individually or with mass actions. This is accomplished in part by assigning a unique identifier to each asset on ingest.
2. DAM systems *secure* the assets they contain. Security in a DAM extends to defining access control lists (ACLs) for assets and defining roles for users accessing the system.
3. DAM systems *store* assets as both binaries and metadata. A DAM system can store multiple file types, and allows for the customization of metadata fields and the metadata in those fields attached to the stored files.
4. DAM systems *render/transform* assets on ingest into new forms, such as thumbnails or proxy files. The new forms generated on asset ingest via transformation should all be stored as asset parts of the original file uploaded.
5. DAM systems *enrich* assets through the extension of metadata and metrics regarding the use and reuse of the asset throughout its lifecycle.
6. DAM systems *relate* assets by tracking the relationships between and among an original asset and versions/variants of the original. Versioning and version-control tools are central to an asset's life in a DAM system.
7. DAM systems regulate a structured *process* in the management, creation and review of assets with workflow tools. Via programmed workflows, DAMs allow for a decentralized workforce to collaborate together in a centralized system.
8. DAM systems allow for users to *find* assets and to retrieve those assets by facilitating search through metadata, collections, workflows and access control tools. By increasing the discovery of assets that may not have been easily accessible before ingest, a DAM assists workers in leveraging existing content for maximum work potential.
9. DAM systems have a *preview* function that allows users to view assets before downloading or opening a file on their own device. By allowing users to take a look at assets in search quickly, without download, DAM systems reduce the amount of time users must spend in search.
10. DAM systems *produce/publish* content by providing methods whereby assets may be shared, linked to or otherwise be distributed outside the system. This DAM function may be as simple as generating a URL on ingest or as complex as allowing users to build collections of items for sharing with a workgroup.

DAM Project Management

A common error when exploring an enterprise solution is to start looking in the wrong place. Suppliers often hear that a broadcaster client seeks a technology solution that will "support and distribute all its digital assets." While not a surprising viewpoint—because this may be the primary problem that triggers the desire for a DAM solution—the result may be less than perfect.

Broadcast Projects International[3] has developed a four-layer model that has been helpful for a number

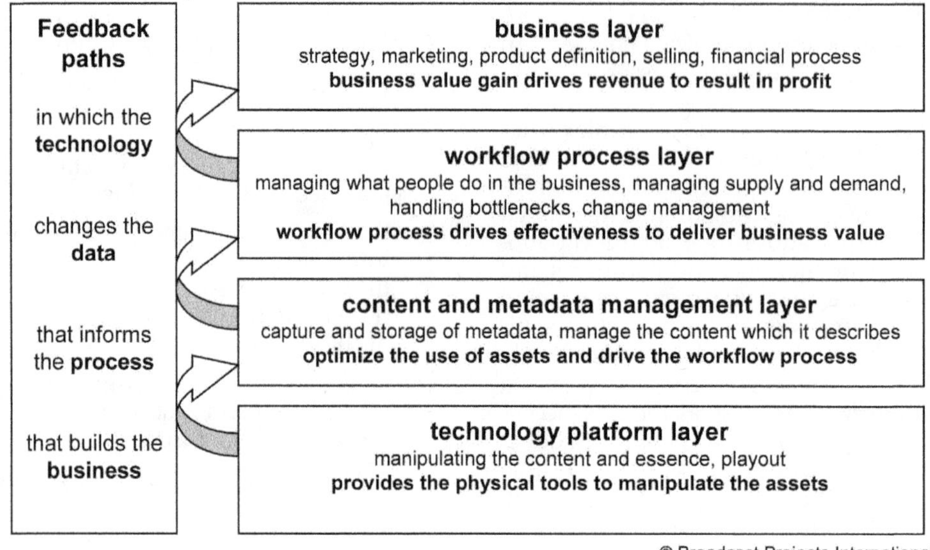

FIGURE 5.9-2 Four-layer DAM model.

Courtesy of Broadcast Projects International

of clients looking to achieve a step change in efficiency and capability. This model, shown in Figure 5.9-2, helps clients think first about who they are and where they really want to be. What are the services they intend to supply? How will they make money? Who are their markets? What is the compelling content that will support their long-term business proposition? Understanding at an early stage how revenues will be realized and processed is vital before designing enterprise systems, and will help identify applications and skill sets necessary to achieve the outcome.

When examining the business layer in more detail, a number of benefits can be achieved using a DAM system. They should go well beyond achieving workflow efficiencies of current production processes. Table 5.9-4 outlines some examples of the benefits that broadcasters may expect to realize.

The advantage of thinking top-down is that it helps define a system that delivers the overall business objectives, many of which reach beyond single department views or technology benefits.

Content that is already owned by a client can be reused either to create additional content (new programs) or repurposed for different platforms. For this to work, content must be thoroughly described and managed within the DAM system. One of the fundamental concepts behind DAM systems is the mantra, "If it can't be found, the content has no value."

Content is also of value to other content providers. If another program maker can reduce its production costs by using high-quality material that already exists, its cost and time to delivery are reduced. Therefore, visibility to the external world is obviously important, as is an efficient and secure process to deliver the material with usage rights. Reduced program acquisition costs benefit all broadcasters, who are constantly looking to drive down their programming production costs.

TABLE 5.9-4
Examples of desired business benefits to be achieved with a DAM system.

Business Benefits to Be Achieved	
Revenues	Increased by reselling of content to other content suppliers
	Increased by providing new platform services that provide compelling value to consumer
	Reduced by improved workflows on production processes
Costs	Reduced by effective repurposing and reuse of content on in-house production
	Ability to respond to new distribution opportunities
Flexibility	Ability to reach targeted market
	Ability to integrate additional revenue applications (e-commerce)
Competitive Advantage	Integrated systems working across multiple revenue streams enable more agile response to revenue opportunities

Newer delivery platforms are a potential revenue source for personalized services. Perhaps the most obvious example is sports highlights, where consumers can request rapid delivery of sports action from a broad range of sports, or perhaps highlights of their own team. With the increase of the range of mobile devices and the quality of the consumable product, this has been identified as a major opportunity for rights holders. News and reality shows are also useful revenue streams when personalized services can be implemented in a cost-effective fashion.

Although recent years have seen rapid technology changes, with technologies and applications continuing to grow and systems coming in and out of the market, it is certain that there is more change on the way. This drives the requirements of the DAM system toward open architecture using industry standards since additional applications will almost certainly need to be integrated over time.

Digital Asset Management ROI Justification

When preparing for a DAM implementation, creating the ROI can be daunting. Often people try to assess the average time saved on everyday tasks. Unfortunately, this usually is a subjective measurement and often can't be measured because of multitasking, interruptions throughout the day and overall biased opinions of how long it should take to perform a task. Also, the rewards of expediting or automating routine tasks through a DAM system won't necessarily be realized until enough time on the system has been established. Developing realistic and methodical productivity measurements to assess DAM benefits and success is important, along with consideration for risk analysis. Addressing these concerns in the DAM design stage not only will prevent potential issues but also will lead to maximizing ROI. An organization's assets are its livelihood. Being able to ingest, locate, access, restore, distribute and back them up quickly and efficiently is crucial to being profitable. Time to market is one of the most important benefits of a DAM system.

The following are some benefits of a well-designed DAM:

- Saving time when performing a comprehensive search and accessing assets
- Faster retrieval of assets, "time to market"
- Identification markers
- Version tracking
- Metadata tags that can promote automated processes
- Easier and more complete backups
- Faster and easier distribution of assets to stakeholders
- Saving time in retrieval and executing for distributions
- Eliminating duplication-based data limitations or lost work
- Eliminating historical manual, one-off knowledge sources
- Improving organizational knowledge through a central, searchable repository for assets that drive engagement and power marketing efforts
- Improving employee efficiencies and increasing ROI through the reuse and repurposing of assets
- Enabling better cross-departmental practices
- Enabling consistent brand culture growth

DAM CONFIGURATION

Metadata

DAMs provide organizations the ability to manage files throughout every step of the digital asset lifecycle. The stages in a typical digital asset lifecycle are Creation, Ingest, Manage, Distribute, Archive and Restore, as shown in Figure 5.9-3.

Creation stems from the asset's origination—for example, a camera shoot in the field or studio plus graphics. Ingest encompasses how the digital asset is transferred into a system. This includes legacy migration and new assets. Management of an asset spans from version control, permission, policies, review and approvals. Distribution can involve sharing assets internally between departments and locations, or sending assets to external destinations via established distribution channels and paths, as well as through external distributors, multiple platforms or via social media. Archive in the traditional sense is the preservation of the asset, but many broadcasters' archives are used more like a near-line storage system with Restores performed for reuse purposes. Every stage of the lifecycle process carries valuable metadata, which is often referred to as "data about data."

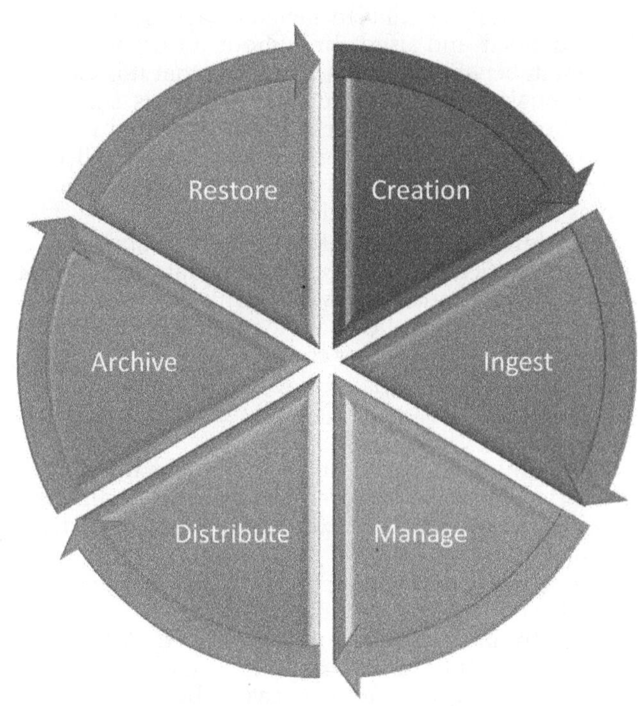

FIGURE 5.9-3 Asset lifecycle.

There are two main types of metadata: *structural metadata* and *descriptive metadata*. Structural metadata is data about the containers of data. Descriptive metadata describes individual instances of application data or the data content.[4]

Metadata can help organize, identify, associate and locate assets. Metadata searches will aid in the discovery of relevant content or system information describing the asset.

Metadata can be stored in either human-readable or binary form. Broadcasters typically use XML or sidecars that accompany the media to describe their technical information or the intellectual properties of the object such as segment description and timecode. Descriptive metadata describes a resource for discovery and identification. It can include elements such as title, abstract, author and keywords for search and locate. Often metadata tagging and keywords facilitate a wider range of search returns that is applied to the asset during cataloguing and ingest.

People also refer to "adaptive metadata," which is a model for deciding metadata attributes, such as time-based information like duration. Adaptive metadata also can be referred to as "class-oriented metadata" to highlight the relationship with object-oriented software engineering principles. Classes can have hierarchical relationships that include associations, such as permissions or taxonomy information.

Administrative metadata is information about the content produced, such as the type of equipment, software, date and location and when an asset has been modified—for example, by whom and when an asset was ingested, or device activity such as an LTO[5] tape ID.

Metadata has two main properties, *fixed* and *flexible*, which are important to understand, along with the benefits of each. Fixed metadata is a schema that is predefined and used throughout an organization. The main benefit of a fixed schema is that it is easy to apply business rules, since specific data is stored in predefined places. In other words, applications know exactly where the data is stored and can be collected via fixed metadata. Fixed metadata schemas define tables built into relational databases.

The downside with the fixed metadata approach is that it is difficult to change the metadata schema for additions over time. This is especially true when business models change and adapt to new platforms, transactional methods, or other changes that come along.

Many developers have moved away from a relational database to a NoSQL[6] solution. This model is referred to as "flexible" or "extensible" metadata. As business models change and evolve over time, flexible metadata allows different assets to have different metadata. A user can add extra fields when needed, for example.

There are many different structures that can be used for organizing metadata, and many standards have been and continue to be published. Adherence to industry-standard schemas makes the exchange of assets from external suppliers easier, but there are no standards yet that comprehensively cover all the needs

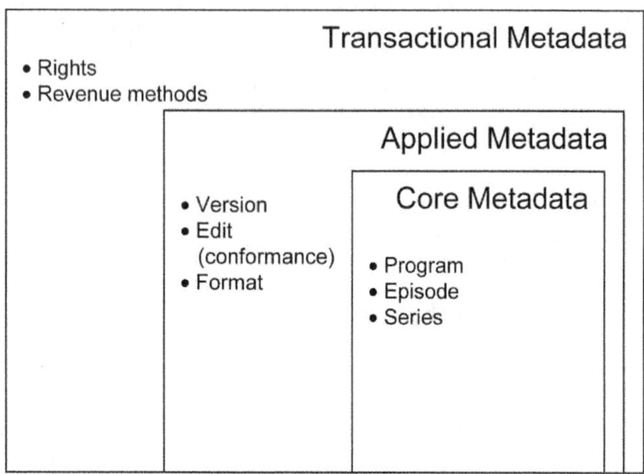

FIGURE 5.9-4 Simplified metadata model.

of broadcasters. A simplified way of viewing the organization of metadata is shown in Figure 5.9-4.

Organizing metadata in different business areas helps in understanding the purpose of the data. For example, a broadcaster may buy a program's rights but the program's core metadata won't change since it describes the essence itself. Additional metadata may be applied for more specificity. For example, there may be an edit list to create a version of the program that is suitable for family viewing, or programs may be broken out into segments to allow for commercial playout. All of these are applied to the core essence to make the revised or applied version.

At a business level, all transactional data must be tracked and used to trigger various processes. Broadcasters may have the Video on Demand (VOD) rights for a particular *release window*, which is an available date range for consumer viewing. The transactional rights, which may be VOD with window conditions, are carried with the content and provide a download capability window for the consumer. When the window ends and rights expire, the material becomes unavailable for end users.

Metadata is not inherently complicated, but a structure is required for entering the information. From a user point of view, this can either be a list of fields that can be filled in or, if the DAM system offers a hierarchical structure, a selection of major headings with subheadings and details. This simplifies the organizing of descriptive data about different types of content.

Opening a heading in a hierarchical system shows the subheadings appropriate for that branch, and so on. An example of a hierarchical and "flexible" metadata structure is as follows: Suppose there is a library that catalogues sports feeds. If it receives a baseball game, then the metadata might have a structure that looks like Figure 5.9-5. This example is obviously a very basic version of the type of data that may need to be logged. Typically, this would be set up to match the types of searches that a sports department might run. For example, "Show me all the games where winning home runs occurred in the ninth inning." This structure

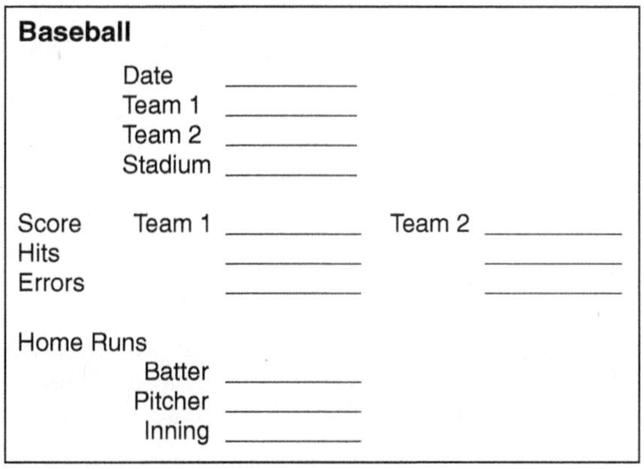

Baseball

Date ________
Team 1 ________
Team 2 ________
Stadium ________

Score	Team 1	________	Team 2	________
Hits		________		________
Errors		________		________

Home Runs
Batter ________
Pitcher ________
Inning ________

FIGURE 5.9-5 Simplified example of a subset of metadata structure.

is fine, until the library starts receiving material from a completely different sport. In the case of NASCAR the returned value would require a completely different data structure, allowing the entry of data about the drivers, the final results, or major crashes, for example. Being able to create new branches at any point in the descriptive metadata structure without having to start over with additional development is essential in this situation, and needs to be supported from a business-rules point of view.

It is also easy to generate XML files that show all the data that has been stored, even if it is not yet understood by any other application on day one.

Numerous different types of data can be entered into metadata fields, and the type is usually part of the definition of the field. Using a controlled vocabulary eliminates many of the problems that arise from the most common "freeform" vocabulary, in which any letter, number, or character can be used. Most users are familiar with the standard range, and limiting the data to specific types can have great benefits to an organization over the long term. Freeform is used to index large descriptions, but this format should be used sparingly. Wherever possible, fixed vocabulary lists should be used. An example would be when entering names of team members in data describing a baseball game. If all the team names have been entered into a list (a purely administrative task), then users can select the one they want (from a drop-down list) and prevent spelling errors, abbreviations, and other unidentified search words in the team name field, which could hinder rapid searches.

Boolean values are usually available, which make it easy for business rules to be applied from other applications, or for workflow control. Fixed-date data generated from calendars should be used to ensure adherence to proper date formats. Broadcasting uses standardized dates and timecodes in particular formats. Dates must work across the world for multi-geographical organizations, and different timecode types (e.g., dropframe vs. non-dropframe) in particular must be handled carefully and validated upon entry. Another obvious complication is the wide range of recording or file formats used by different organizations or in different parts of the world. Sensible labeling disciplines also need to be applied. If a new field description of digital content is "New Baseball Graphics," this will inevitably cause confusion in years to come. Descriptions must be specific and, if at all possible, unique.

Unique identifiers (ID) are a critical piece of program metadata. As systems scale, assets grow, integration increases and delivery models expand, unique identifiers become increasingly important because they link to other descriptive data that provides context about an asset. Identifiers provide the ability to link key elements in a system with other key application elements.

Being able to track and count assets immediately upon delivery is a key to success and optimized DAM system ROI. Unfortunately, this is an area that broadcasters are struggling with because IDs are not being propagated through the supply chain and to all delivery end points. As the number of applications, cloud services, distribution chains and end points increase, a DAM can better manage them, but standardized IDs are required.

Descriptive and enriching metadata entries serve many purposes for broadcasters. In the traditional library function, metadata must be added to content to locate it efficiently. With hundreds of thousands of digital assets, cataloguing of material must be specific and done in a method that will allow those searching the library to find what they want without having to wade through hundreds of unusable returns on a query.

Major cost efficiencies can be achieved using automatic data enrichment. The most obvious example is storyboarding. When material is ingested, a low-resolution *proxy* copy becomes available for use in a Web-browsing environment. This proxy copy may use a variety of formats, including MP4, H.264 and Windows Media. However, when users are searching or browsing, content storyboards (i.e., selected key frames) can be developed to show the sequence of shots that forms the material rather than searching complete proxies. The number of storyboard frames can be chosen either by fixed times (for example, one every five seconds) or automatically by detecting shot changes. After a storyboard is created, each frame can have additional data entered describing major items of interest like a close-up of a celebrity in the audience, or an aesthetic shot, such as the sunset, to facilitate very rapid searching. DAM systems themselves cannot necessarily do this sort of automatic data enrichment, but they can be used to send material to applications that have these capabilities, and then store the resultant data along with descriptive metadata, enabling reuse. When multiple technologies are integrated, such as voice and image recognition, there is potential for significant time and cost savings. Another example is automating the creation of secondary language subtitles and searching caption files. When using a workflow engine in conjunction with a DAM system, these tasks as well as others would traditionally have taken several staff many days to achieve, but these tasks now can be completed in a matter of minutes. Major savings can be achieved when the majority of the work is performed through automation, allowing staff to manage exceptions or areas in which technology sophistication isn't yet adequate to perform tasks automatically. These types of automatic data enrichment are particularly useful when large quantities of poorly catalogued material exist. Also, metadata enhancement applications like face and voice recognition processors can run across large quantities of data to enable expensively acquired assets to be optimally reused.

Program automation ("playout") systems have been using ingest processes since the early 1990s that have always used some method of attaching metadata to the video files. While some suppliers chose to wrap this metadata into the file itself (typically in the file header), others preferred to use separate databases to hold the additional data. There are different benefits to each method. Simple databases are easy to change, whereas data wrapped in file headers is more complex to open and edit. Having metadata wrapped into the essence means the metadata can never be lost, however, avoiding unidentified video pieces in the library.

Workflow

Workflow is the interaction of processes and users with assets. The goal of any workflow is to improve efficiencies, mitigate the individual "hand-holding" management of a process, prevent bottlenecks, track the asset, and ensure that a task is performed within the time required. If the workflow runs smoothly then everyone is happy, but if problems occur, difficulties can arise with staff, clients or management. The more efficient the workflow is, the higher the probability of ensuring long-term success and better return on investment. Workflows can be short, point-to-point interactions, or longer-term, end-to-end processes, with various tasks that make up the entire chain. DAM workflows typically consist of tasks such as Retrieve file, Catalog file, Define metadata, Share file, View and approve action, Route, Push, and Publish/Distribute a file. When defining an organization's workflow, consider the following:

- Identify the use cases to address
- Identify the current pain points within those use cases
- Define the goals and the objectives of the new workflows
- Prioritize tasks that need to be performed
- Identify roles and groups of ownership for workflow
- Identify how to measure the workflow success both at the component (task) level and across the full end-to-end chain
- Ensure any new workflow can scale and adapt to change

DAM workflows, when integrated with a Business Process Management (BPM), can provide additional ROI. Through BPM's process orchestration, workflows can be automated to increase efficiencies for the business units, operations and engineering/IT. Supply-chain orchestration includes automating the review and approvals process; messaging handoffs, escalations and alerts; and progressing a job from one department to another through the system with ease. Business units can design workflows more easily by using existing behaviors, profiles and extended libraries as new processes are added.

When designing workflows and scoping a DAM system, it is best to determine what flexibility is required for the business unit to manage its own workflow, rather than relying on IT or professional services.

Also be aware that larger media files can lead to trouble. Consider the following: How do they get to the DAM? Where do they get stored, retrieved from, and so forth? How do they get to the end user, along with the networking concerns that go along with it? Is public Internet upload/download viable? What is the network capacity at source/destination? Will browser-based viewing be used, or will access be "assisted" with a managed file transfer?

When answering these types of questions and deploying automation in a new environment, it is important to have good reporting and monitoring along with alerts and exception-handling options. Adequate upfront time and deep stakeholder involvement when designing such workflows can lead to a successful DAM implementation—including significant cost savings.

Search

A core component of a DAM system is the search engine. Searching vast quantities of data has driven the need for a range of search tools. Complex searches can be built up, using ranges across a series of metadata fields such as catalogue roots, subheadings, dates, descriptors and so on, to help users come up with a sensible range of results. Some search systems are proprietary by the DAM vendor, whereas other suppliers prefer to use powerful "off-the-shelf" engines that are integrated into the DAM system. The capabilities of these search engines vary greatly, and the suitability of different systems depends on the quantity of data, the number of users, and the range and formats of the data. Search capabilities vary greatly across different DAM systems, with some providing more advanced searching techniques. Search parameters include searches on keywords, criteria, controlled metadata values, numeric ranges, file type, asset status, collections categories, text, file formats and folders, to name a few. When searching on keywords, many users can craft advanced search syntax to return precise date ranges, file formats or metadata values.

An asset's file name is just as valuable as the keywords attached to it. Consistency with the naming convention can prove to be beneficial. Best practices for digital asset naming is to not load them with keywords and keep the length under 50 characters.

Metadata is critical to search in any form, and the more complete or rich a content library's metadata is, the more it will benefit everyone in the organization. The speed and performance of DAM searches can vary, nevertheless. DAM systems that search databases directly can be slow as the database size increases and the data set expands. Most DAM systems do index searches, however, and in larger DAM deployments it is common to have search engines that can be decoupled from the main application server. In these deployments, dedicated resource allocations for indexing and searching tasks provide greater performance and can effectively broaden their search capabilities. Search engine optimization uses keywords or keyword phrases that are used in HTML or structural code such as tags, headlines, body text, descriptions, and other characteristics to broaden their search capabilities. Search engines can provide a highly optimized index for all asset metadata. Semantic search[7] is becoming increasingly mainstream.

Analytics

Business models are evolving and changing daily as more media platforms and technology drivers for video consumption emerge. Demands for solution integrations, business intelligence and analytics within a DAM are becoming a greater focus. By leveraging the vertical and horizontal layers of the organization, DAM systems become the central component to any media company. Connectivity and interoperability through integrated solutions, such as application programming interfaces (APIs) and connectors, must be advanced. If not, the company could suffer loss of opportunity.

Because a DAM system is the core component, incorporating business intelligence (BI) and analytics tools will help drive decisions across the organization—whether from a management, creative, operational and administrative level, or simply based on how media is accessed and used. Advanced visualization tools for dashboards and reporting provide easier ways to assess information, and they can lead to faster response time and better decision making when business or technology changes occur. Combining supply-chain metrics along with customer behavior information—such as the number of downloads, posts to social networks, views of share links and video assets' performance—all roll up and can provide a more complete picture of DAM system ROI.

Some DAMs offer a range of visualization and analytic toolsets, which may be helpful to broadcasters as they attempt the vertical and horizontal integration of disparate systems. Although DAM systems can bring efficiencies and improved value to some of these matters, DAMs still need to support cloud services and improve on integration and interoperability in order to serve the entire enterprise's digital ecosystem.

Figure 5.9-6 provides a high-level view of an enterprise-wide DAM architecture.

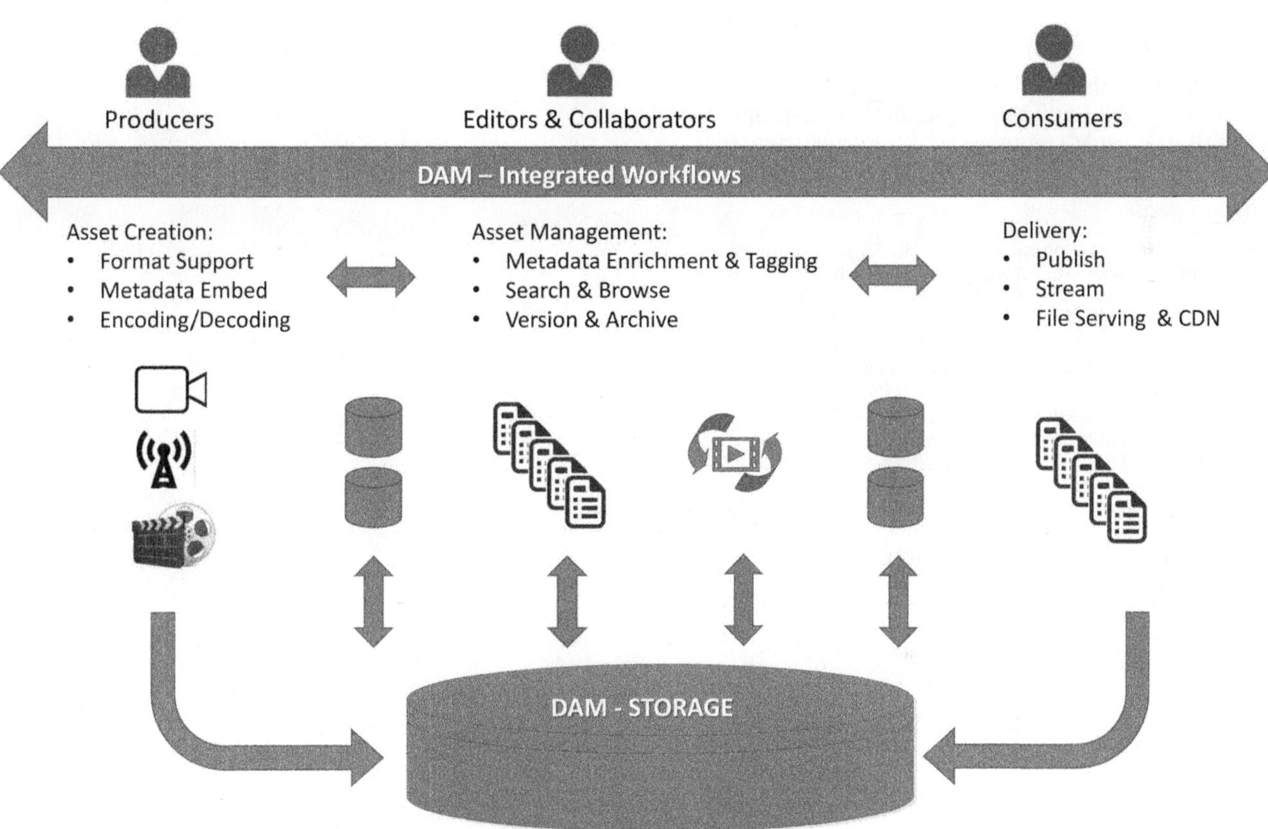

FIGURE 5.9-6 Enterprise digital asset management.

SUMMARY

Digital assets are a company's most valuable commodity. As new technologies and distribution models emerge, broadcasters must harness, access and manage their digital assets and enrich their metadata during all stages of the supply chain to the consumer.

DAM systems manage and enable process efficiencies and improvements, increase productivity, enable reuse of content and repurposing of assets, and make assets accessible throughout the digital ecosystem, which lead to significant cost savings and faster time to market. Digital asset management implementation requires a considerable amount of planning and commitment to implement successfully.

Most successful projects are carried out in a multiphase approach, limiting disruption and building confidence and support as the project progresses. Organizational changes will inevitably need to be addressed; and because training is central to any technology change, developing learning materials and building in-house DAM expertise—along with incorporating the analytics required to measure success into your deployment—will lead to clear expectations and deliverables.

A successful enterprise DAM implementation not only will create significant cost savings but can also position broadcasters to benefit from new revenue streams and provide new capabilities to readily respond to new opportunities.

ACKNOWLEDGEMENT

Some content in this chapter was adapted from the 10th edition of the *NAB Engineering Handbook*'s chapter entitled "Digital Asset Management," by Rod Fairweather.

Notes

[1] "The World's Technological Capacity to Store, Communicate, and Compute Information," especially supporting online material, Hilbert, Martin, and López, Priscila (2011), *Science*, 332(6025), 60–65, www.martinhilbert.net/worldinfocapacity-html/.

[2] "Information in the Biosphere: Biological and Digital Worlds," doi:10.1016/j.tree.2015.12.013, Gillings, M.R., Hilbert, M., and Kemp, D.J. (2016), *Trends in Ecology & Evolution*, 31(3), 180–189.

[3] www.bpi.uk.com/home.asp

[4] National Information Standards Organization (NISO) (2001), *Understanding Metadata* (PDF), NISO Press, Baltimore, MD, ISBN 1-880124-62-9.

[5] Linear Tape-Open, an open-standard magnetic tape storage format.

[6] *NoSQL* is defined as referring to "non-SQL," "non-relational" or "not only SQL" databases.

[7] Semantic search is a technique for improving search accuracy by understanding the searcher's intent and determining the contextual meaning of the words and terms the searcher is using.

CHAPTER

5.10

ATSC 1.0 Encoding, Transport, and PSIP Systems

RICHARD CHERNOCK, ScD
Triveni Digital
Princeton, NJ

MATTHEW S. GOLDMAN
Ericsson
Bedford, NH

CHRIS LENNON
MediAnswers
Monument, CO

INTRODUCTION

This chapter covers the implementation of equipment and systems for producing compressed digital television (DTV) bit streams in accordance with the original ATSC Digital Television standard [1]. In light of subsequently developed ATSC standards suites—called ATSC 2.0 and 3.0—the subjects covered in this chapter are now retrospectively referred to under the general title of "ATSC 1.0." (Hereafter in this chapter, however, they are simply referred to as "ATSC.")

The first part of the chapter covers the video encoder, the transport multiplexer, system solutions, and video coding technology of ATSC. Next, there is discussion of issues related to the implementation of Program and System Information Protocol (PSIP), followed by the use of Programming Metadata Communication Protocol (PMCP) and the role of the various sources of PSIP metadata. The chapter ends with a summary of the PSIP requirements stated in the FCC rules.

Other issues related to DTV encoding and transport systems are covered elsewhere in this handbook.

OVERVIEW

Encoding and multiplexing are central parts of the ATSC digital television (DTV) system, which is described in the ATSC Digital Television Standard, A/53 [1]. A generic block diagram of the encoding and multiplexing subsystem in a typical broadcast station is shown in Figure 5.10-1, with basic components comprising the video and audio encoders, a transport stream multiplexer, a PSIP generator, and (optionally) a data server. The output of the multiplexer is fed to the 8-VSB modulator that is associated with the over-the-air transmitter.

The main interface between encoders and the transport multiplexer conforms to the ATSC transport layer as defined in A/53. The ATSC *transport layer* is based on the MPEG-2 *Transport Stream* (TS) format, as defined by the MPEG-2 Systems standard [2]. The link and physical layers are defined by the *DVB Asynchronous Serial Interface* (ASI) [3]. In typical implementations,

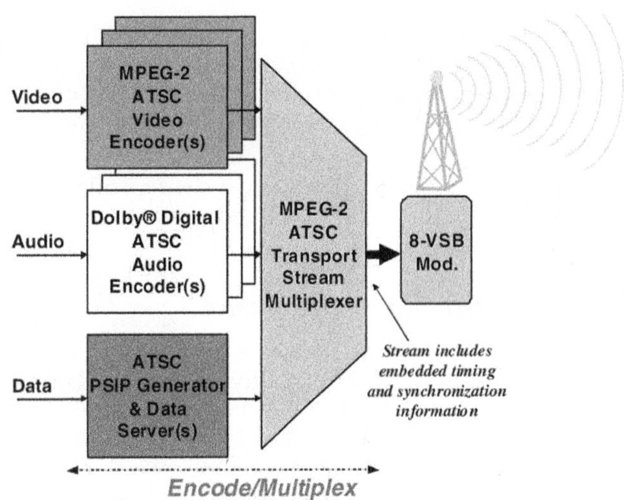

FIGURE 5.10-1 Encoding and multiplexing subsystem in a typical broadcast station.

the video and audio encoders and the ATSC transport multiplexer each create output bit streams in the TS format. The TS format provides a mechanism to encapsulate and multiplex coded video, coded audio, and generic data into a unified bit stream. It includes timing information in the form of *time stamps* in order to enable the real-time reproduction and precise synchronization of video, audio, and data (as necessary).

In order to facilitate parsing of the information contained within the bit stream, in-band control information, known as *program specific information* (PSI), is also defined. PSIP expands upon the PSI to provide comprehensive channel "tuning" and program guide information as discussed later in this chapter. The TS format also was designed to facilitate real-time transmission and reception of DTV over error-prone physical transmission paths, in particular, over-the-air broadcasting.

For a detailed explanation of the theory behind the ATSC transport and compression layers, see *Proceedings of the IEEE, Special Issue on Global Digital Television*, January 2006 [5].

VIDEO ENCODER

When specifying or reviewing an MPEG-2 video encoder, there are many factors to consider:

- What is the image quality, particularly at the lower bounds of practical bit-rate usage?
- Are the necessary input and output interfaces supported?
- Is multichannel audio encoding supported internally or is an external audio encoder required?
- How many audio services are supported (to be encoded or passed-through from an external audio encoder)?
- Will the encoder be used in a single broadcast service solution or will multiple services be broadcast in the same 6 MHz transmission channel?
- If multiple services will be broadcast, does the encoder support efficient bandwidth utilization? In particular, will the encoder operate as part of a statistical multiplexing system (see section later in the chapter)?
- Does the encoder support the required ancillary data carriage (for instance, closed captions)?

Video Compression Performance

Arguably the most important factor when choosing a video encoder is the picture quality it produces. Unlike reviewing a checklist of, for example, input/output interfaces, the factors that impact video compression performance are often the least understood and, to many people, the most mysterious. Particular areas to consider are preprocessing (including noise reduction), motion estimation, and rate control [4].

Since bandwidth is a limited commodity, the performance of an MPEG-2 encoder is usually tested at the lowest possible bit rate. While it is true that some encoders fail in more noticeable ways than others at the low end of practical data rates, differences between encoders may show up under a number of different test conditions. Some encoders might have difficulty with particular types of motion, others with noise, others with scene cuts, and so on.

The picture quality produced by an MPEG-2 encoder depends on many factors that include:

- The quality of the original video source material;
- Preprocessing operations;
- Encoding architectures and algorithms;
- System parameters (pixel rate, bit rate, and number of services in the transmission channel, in the case of multiservice systems);
- Size and quality of the display device.

The encoder, however, has to produce the highest picture quality of the compressed video bit stream for any set of operating parameters and for any given source material.

Preprocessing

The preprocessing functions of an encoder typically include:

- Picture resizing;
- Noise reduction;
- Noise level detection;
- Film mode (or 2:3 pull-down) detection;
- Forward analysis.

Picture Resizing

At low bit rates, reducing the detail (pixel rate) of the source is a commonly used mechanism for trading between image sharpness that contains visible MPEG *coding artifacts* and a softer picture that reduces or eliminates visible MPEG coding artifacts. The most common MPEG-2 coding artifact that occurs when there are insufficient bits to encode the detail is known as *blockiness* or *macroblocking* (see Figure 5.10-2).

Most encoders provide a large number of resolution subsampling options. Common picture resizing is to reduce the horizontal resolution by one-quarter or one-third, so:

- 1920 × 1080 (1080i) → 1440 × 1080 or 1280 × 1080
- 1280 × 720 (720p) → 960 × 720
- 720 × 480 or 704 × 480 (480i/p) → 544 × 480 or 528 × 480

If too much detail is removed, the picture will lack contouring and edge detail, commonly known as being "too soft."

FIGURE 5.10-2 An example of the macroblocking artifact.

As such, the *coding efficiency*—picture quality versus bit rate used—is one of the most important parameters that needs to be tested and established for an encoder.

Noise Reduction

Most video sources contain a small amount of noise. While it may not be visually very annoying, even a low level of source noise may add a significant encoding overhead that increases the required bit rates. This is because noise, by its nature, is random and usually contains substantial high-frequency components that do not compress well. When an encoder does not recognize noise as such, it will treat it as a high-detail component of the image and will waste valuable bandwidth by attempting to encode it properly. This behavior becomes even more significant in cases where the compressed bit stream is statistically multiplexed with other compressed video sources, as discussed later in the chapter. In that case, instead of reducing its bit rate for input sequences that should be simple to code (say, with little motion and low detail), an encoder with a noisy source will continue to require relatively high bit rates even if the picture material itself (without the noise) could have been compressed further.

Measurements have shown that even low levels of source noise can increase the bit rate demand of encoders almost by a factor of two [6]. Figure 5.10-3 shows the bit rate demand of a noisy source (30 dB SNR) with and without noise reduction as a probability density function. In this example, noise reduction reduces the most likely bit rate from 5 to 3 Mbps.

Properly designed noise reduction algorithms reduce the visibility and the effects of noise without introducing picture artifacts themselves. As a

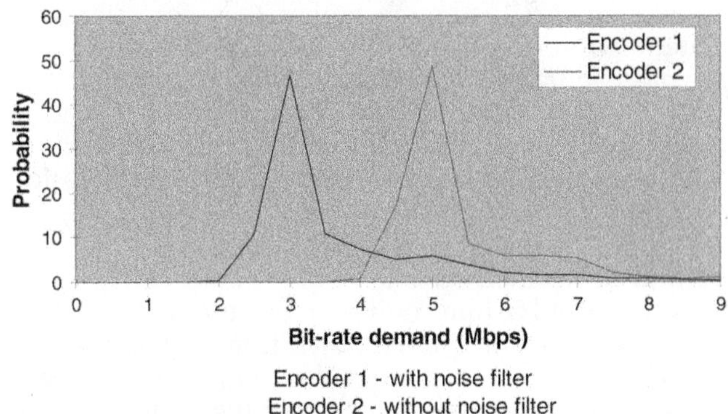

FIGURE 5.10-3 Bit rate required with and without noise reduction for a noisy source.

preprocessing function to compression, it is particularly important that the processes of noise reduction, followed by compression, do not in themselves lead to undesirable artifacts. Examples of such artifacts are *motion blur* or *double imaging* (both commonly caused by excessive temporal noise reduction). In the case where the source material contains high levels of noise, it is generally better to add an element of spatial noise reduction while keeping the amount of temporal noise reduction at a moderate level. If such a filter is overapplied, the image will become noticeably soft and lacking in detail.

Noise Level Detection

As noted, an encoder must be adaptive in how it applies the noise filters or poor image quality will result. Therefore, even more important than the basic capability to reduce the effects of source noise is the requirement to measure the amount of noise present on the source material so the proper type and amount of filtering may be applied. Noise levels may vary over a wide dynamic range, from extremely low levels of noise (typically expected at the output of a digital studio) to relatively high levels (such as from older, archived material or material that has already undergone a series of recordings or compression/decompression cycles).

A studio or broadcast center typically contains or passes content material from a myriad of different sources. Most noise-reduction algorithms, when applied to clean sources in full strength, will reduce picture quality. Since, in real-time broadcast operation, it is not practical for an operator to manually change the configuration of an encoder for each input source, the encoder itself must be adaptive and vary the level of noise reduction in accordance with the level of source noise present at the input. The flexibility and effectiveness of noise processing are therefore very important functions to test and verify.

Film Mode Detection

A considerable proportion of broadcast material originates from film. Since the compression efficiency for film material is much higher than for interlaced material (due to the lower frame rate, the fact that progressive frames are easier to code than interlaced, and for 60 Hz systems, the elimination of duplicate video fields), it is important to detect the presence of film-originated input and change the compression algorithm accordingly.

For film transfer to 60 Hz video, the well-known 2:3 pull-down method is used to generate 60 Hz video fields from 24 Hz film frames. In this method, two and then three fields are pulled down from successive film frames, thus generating a video sequence that contains two repeated fields for every five video frames. By detecting and eliminating the repeated fields, the encoder is able to reconstruct and compress the original film sequence, resulting in a significant improvement in coding efficiency. Measurements have shown that coding efficiency may be improved by more than the simple estimate of 20% due to the higher coding efficiency on progressive frames.

In reality, however, recognizing the 2:3 pull-down sequence is not always straightforward. Once the film source has been transferred to video, the material is usually subjected to further editing in the video domain. As a result, the 2:3 pull-down sequence, or *cadence*, which would otherwise be readily detectable, is disrupted and encoders must implement a more involved algorithm to reliably detect the sequence. An encoder's ability to detect and eliminate repeated fields in heavily edited film-mode material is therefore another aspect to test and verify.

Forward Analysis

In addition to the noise level and film mode, there are a number of other parameters that may be automatically measured at the input of the encoder to help with the compression process. Statistical analysis of the input signal assists the compression engine in selecting the most appropriate coding modes for a particular piece of content. A superior encoder will measure field dominance as well as field and frame picture activity to select between field and frame picture coding. It will also detect scene cuts and fades to select coding modes and to give advance warning to the rate control algorithm.

An extreme case of forward analysis involves doing a full encoding pass, measuring the efficiency of the encoding, and passing parameters such as image complexity and picture type to a subsequent compression engine. Typical industry terminology for this is "look-ahead processing" or "multi-pass encoding." This is not only helpful for single service encoding, but also greatly improves the efficiency of statistical multiplexing for multiservice encoding.

Motion Estimation

One of the most crucial components that determine the overall quality of a real-time MPEG-2 encoder is its motion estimation engine. Performance criteria of motion estimation algorithms encompass more than just the total search range explored for the motion displacement vectors, and the ability to track true motion is less significant than one might expect. What is important is the reliability of the motion vector and prediction mode to produce an overall (global) minimum prediction error.

Motion estimation is computationally intensive. In order to achieve a large enough search range, therefore, motion estimation is often carried out in a hierarchical manner using a down-sampled, lower-resolution picture. Down-sampling reduces computational complexity because pixel displacements in the down-sampled image correspond to larger displacements (and therefore a larger search range) in the original image. Hierarchical motion estimation is not without its disadvantages, but the details are beyond the scope of this book.

Exhaustive motion estimation is an alternative to the hierarchical approach, while still keeping computational complexity in check. A smaller search range is used but assuming the search range is still adequate for all prediction modes, exhaustive motion estimation results in the best possible prediction as it always finds the global minimum of prediction errors. The search range required depends on the specific content, with high-motion sports programming having the largest potential range. However, the vast majority of motion vectors has been shown to be in a relatively small search area [7]. Therefore, the reliability of the motion estimation is more important than the size of the search area. Note also that fast motion is typically less difficult to compress because of camera blur and the human visual system's lack of fine detail response during fast motion.

An encoder that does a good job at motion estimation will maximize picture quality for a given bit rate; hence, this is another important factor that impacts the overall performance of an MPEG-2 encoder.

Rate Control

Without rate control, the bit rate generated by an MPEG-2 encoder may vary over more than three orders of magnitude. Black frames require less than 70 kbps to code whereas white noise may generate instantaneous bit rates exceeding 100 Mbps. Rate control is required to convert this very large variation into a compliant constant bit rate (CBR) or variable bit rate (VBR) bit stream, with a known (and often restrictive) bit rate cap. To achieve this, the rate control algorithm must react quickly on both ends of the scale. At low bit rates, stuffing bits must be put into the bit stream to prevent buffer underflow. At the maximum allotted bandwidth, the rate control algorithm must be able to reduce the bit rate such that buffer overflow is prevented under all conditions.

In practice, the large variation in bit rate is due to a number of factors, including:

1. The type of picture material (film, sports, etc.)
2. Temporal correlation and predictability (fades, scene cuts, etc.)
3. Global scene criticality
4. Type of motion (transversal, zoom, rotation, etc.)
5. Picture type (I, P, or B frame)
6. Spatial complexity variation within pictures
7. Macroblock coding type (intra, forward predicted, etc.)

To achieve a fast response to changes in the video signal, the encoder needs to analyze the input signal at the earliest possible stage and forewarn the rate control about significant statistical changes in picture criticality, noise level, and other parameters. This is often done in high-end encoders by using dedicated forward analysis processing. The most accurate results are achieved by using a full-function "look-ahead" or multi-pass encoder before the actual encoding of the video signal. Information obtained from both the initial forward analysis of the video signal and the "look-ahead" encoder is passed on to the final encoder in order to optimize compression parameters and to obtain an accurate estimate of the required bit rate.

Audio Encoding

Two-channel (stereo) Dolby AC-3 audio compression is typically included within a video encoder unit because it greatly simplifies system configuration and audio/video sync. In most cases, multiple stereo services (for example, multiple languages associated with the same video service) may be supported through the addition of optional additional audio cards. Some manufacturers are also beginning to offer 5.1-channel AC-3 audio compression as an option in an integrated video/audio encoder unit.

In most cases, however, where local 5.1-channel audio encoding is needed, an external multichannel audio encoder is used. In such situations, the audio encoder must be locked to the video encoder in order for A/V sync to be maintained. This is typically done by sending an AES/EBU clock reference from the video encoder or the station clock source to the external audio encoder (note that this is separate from locking the video encoder to the station reference clock, which is often done as well). In addition, the delay through the external audio encoder must be compensated for in the video encoder unit so that A/V sync will be aligned. Typically, both the external audio encoder and the video encoder have configuration adjustments so that proper lip sync may be obtained.

Interfaces

The following is a summary of typical interfaces on a video encoder:

- *Video Input*
  - Serial digital interface (SDI) (SMPTE 259M)
  - High-definition serial digital interface (HD-SDI) (SMPTE 292M)
  - Analog component/composite
- *Audio Input*
  - Analog balanced
  - Analog unbalanced
  - AES-3 digital
  - Note: Audio may be embedded in the video SDI/HD-SDI (as per SMPTE 291M)
- *Caption Input* (CEA 608 and CEA 708)
  - Serial SMPTE 333M
    - Note: Captions may be embedded in the video SDI/HD-SDI (SMPTE 334M)
    - Note: Analog SD captions (CEA 608) may also be carried in video Line 21
- *Other Inputs*
  - H-Sync (for synchronization with a station master clock)
  - Asynchronous Serial Interface (ASI) (ETSI EN 50083-9 Annex B), if an internal remultiplexer is included
  - Serial interface (RS-232 or similar) for control system
  - 10/100Base-T Ethernet or similar for control system
  - Voltage-level sensitive general purpose interface (GPI) for control system
  - Feedback control channel for statistical multiplexing (typically Ethernet)
- *Outputs*
  - Asynchronous Serial Interface (ASI) (ETSI EN 50083-9 Annex B) (typically more than one for redundancy support and confidence monitoring)
  - AES/EBU clock reference (for connection to external multichannel audio encoder)
  - Synchronous Serial Interface (SSI) (SMPTE 310M)

- IP/Ethernet (for direct connection to an IP network, for scenarios that do not use ASI)
- Control channel for statistical multiplexing (typically Ethernet)

TRANSPORT MULTIPLEXER

When the ATSC system was first envisioned, it was assumed that the DTV channel would comprise a single high-definition (HD) service. However, standard-definition formats were included in the standard, and their reduced bit rate requirements allow multiple SD services to be carried simultaneously. Today, various combinations of HD and SD (and also data) services are being delivered by broadcasters in a single DTV channel. This is made possible with the transport multiplexer.

The *transport multiplexer* (also known as a *mux*) is responsible for receiving the compressed video bit stream, compressed audio bit streams, PSIP bit stream, associated data (synchronous, asynchronous, and synchronized) bit streams, and independent data bit streams, packaging them up into a single multi-stream transport, and ensuring that streams that require synchronization among them are correctly aligned and do not exceed established bit rate or buffer size parameters, and then deliver (in a real-time stream of bits) the multi-stream transport to a channel modulator or other network interface.

The transport multiplexer's role can be best illustrated by understanding the ATSC transport subsystem. Figure 5.10-4, from ATSC standard A/53, illustrates that the ATSC transport subsystem resides between the video and audio encoders (application encoders in the figure) and the video and audio decoders (application decoders in the figure), at both the transmission site and the consumer receiver location. The transmit site's transport subsystem is responsible for formatting the coded bit streams and multiplexing the different components of the program for transmission. The receiver's transport subsystem does the inverse function, recovering the coded bit streams to pass them to the appropriate decoder and for the corresponding error signaling.

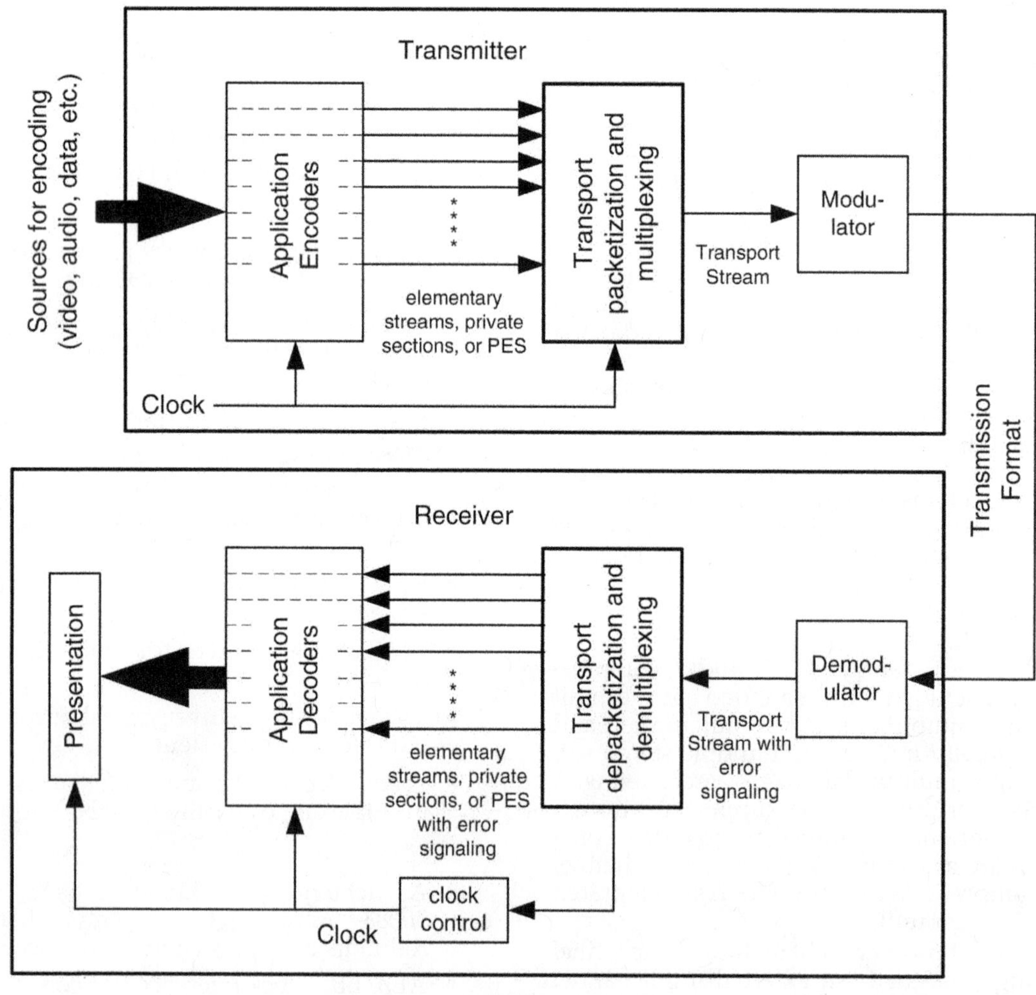

FIGURE 5.10-4 A functional overview of the ATSC transport subsystem.

Courtesy of ATSC

The transport multiplexer is a physical device that incorporates the functions depicted as transport and packetization multiplexing in the figure. Note, however, that transport packetization and multiplexing are also implemented in a typical video encoder because the TS format is the de facto standard for interconnections between encoders and transport multiplexers. The difference is that the transport multiplexer receives TS-formatted bit streams from many sources. In essence, a transport multiplexer is both a multiplexer of data to be put into the TS format and a remultiplexer/resynchronizer of TS-formatted data from encoders and other devices that choose to use the TS format as an interconnect.

In a simple system, a remultiplexer function may be implemented on a single card and that card may be placed inside a video encoder. In such cases, the video encoder may perform almost all the functions of a "station in a box": video encoding, multiple audio encoding, and remultiplexer for data and PSIP.

Two-Layer Multiplexing

One way to visualize the TS multiplexing function is to consider it as a combination of multiplexing at two different layers. In the first layer, a *single service Transport Stream* is formed by multiplexing packets from audio, video, and in-band control information, known as *program specific information* (PSI). This typically occurs within a video encoder. In the second layer, one or more single service Transport Streams are combined to form a *service multiplex* of programs (also known as a Multi-Program Transport Stream [MPTS] in the MPEG-2 Systems standard, and a Digital Television Standard *multiplexed bit stream* in the ATSC standard). This occurs in the transport multiplexer (re-multiplexing).

Data streams—such as a data application in its own right or PSIP data—may enter the transport mux in one of two ways: either as real-time streaming bit streams (already packetized into the TS format by the data server or PSIP generator) or as a typical file transfer. In the latter case, the transport multiplexer is responsible for packetizing, multiplexing, and synchronizing with other services within the multiservice TS, if applicable. In the former case, the data stream is re-multiplexed into the multiservice TS in a similar fashion as done for video/audio services entering the transport multiplexer. The transport multiplexer is responsible for handling timing/synchronization of packets, the correctness of the stream syntax (including in many cases the content of the PSI, although in some cases this is done by the PSIP generator), and the overall output rate control.

Conceptually, the TS may be represented as a large communications pipe containing one or more smaller pipes, as discussed in [5]. Each smaller pipe represents a single service (such as a DTV minor channel—see PSIP description later in the chapter). Figure 5.10-5 shows an example where the TS carries a single service,

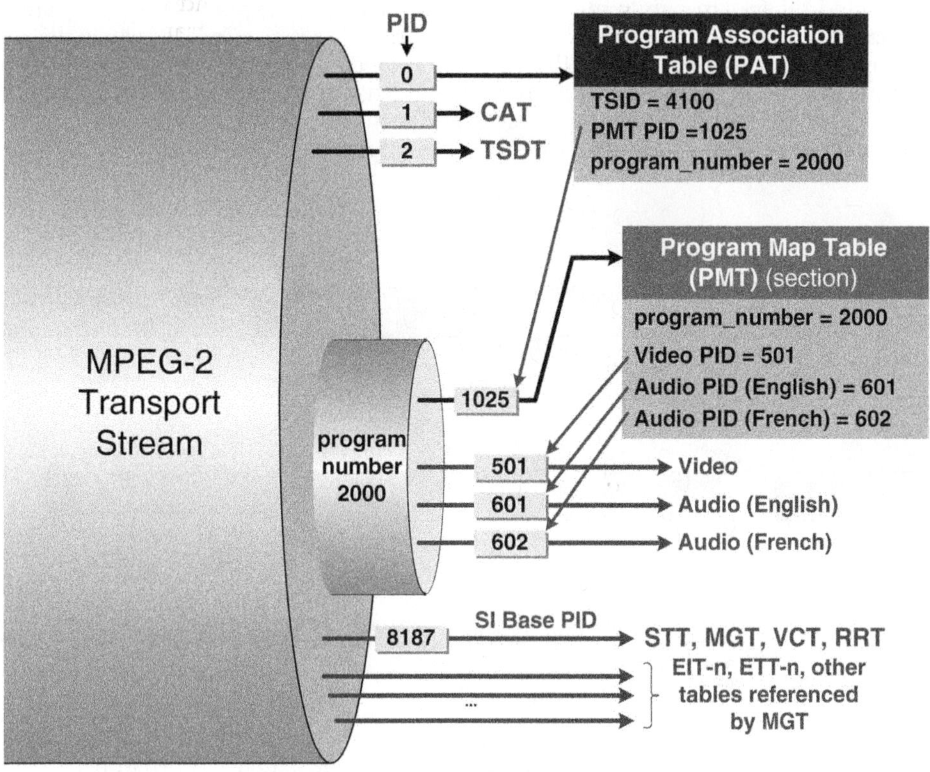

FIGURE 5.10-5 An illustration of a Transport Stream showing two-layer multiplexing.

Courtesy of IEEE

but this concept can easily be extended to multiple services within the TS by adding additional smaller pipes (each one with a different program number from program number 2000 shown in the figure). Each service (MPEG-2 program) comprises one or more elements, which may include video, one or more audio (such as multiple languages), and data streams.

The program specific information (PSI) provides information about the contents of the Transport Stream. For ATSC streams, the PSI comprises the Program Association Table (PAT), which lists all the programs (services) in the multiplex; the Program Map Table (PMT), which identifies the elements that make up each program; and the Conditional Access Table (CAT), which provides information relating to scrambled programs.

Basic Transport Stream Structure

Figure 5.10-6 illustrates the packet structure hierarchy of an MPEG-2 Transport Stream. A video encoder that encodes both video and several audio services produces a set of elementary streams, one for each video and audio. Each elementary stream is segmented into a series of packetized elementary stream (PES) packets as shown in the figure. The PES packets, in turn, are further segmented into fixed-length TS packets to facilitate multiplexing and transmission in real time. Multiplexing of an MPEG-2 Program is the process of interleaving the TS packets of all the elementary streams (and other section data produced by the encoder, such as PSI) that make up the MPEG-2 Program into a single unified bit stream, while maintaining timing synchronization of each elementary stream. At the next layer, the multiplexing of multiple MPEG-2 Programs into a single multiservice multiplex is the process of interleaving the TS packets of more than one MPEG-2 Program into a unified bit stream, called a multi-program TS (MPTS). The MPTS construct enables the deployment of practical, bandwidth-efficient digital broadcasting systems, with each service capable of being delivered at an independent, variable bit rate from other services within the overall fixed bit rate of the MPTS.

The packet identifier (PID), contained in the header of each TS packet, is the key to sorting out the components or elements in the TS. The PID is used to locate the TS packets of a particular component stream within the service multiplex in order to facilitate the reassembly of the payload of each TS packet back into its higher-level constructs; that is, TS packets into PES packets and PES packets into an elementary stream. A series of TS packets containing the same PID include either a single program element (for instance, a video elementary stream), or descriptive information about one or more program elements (for instance, a PSI table).

An MPEG-2 Transport Stream is a continuous series of TS packets as shown in Figure 5.10-6. A TS packet is 188 bytes in length and always begins with a 4-byte (including the synchronization byte) TS packet header. The remaining 184 bytes are available to carry up to 184 bytes of TS packet payload.

A transport multiplexer may be configured for a fixed set of services (MPEG-2 Programs), with a fixed set of streams for each service. However, in practice, encoders and multiplexers are controlled by a sophisticated control and management system, so that high-availability solutions (automatic switching to backup or redundant components in case of a failure) and service flexibility (such as the changing of one HD service to multiple SD services) are possible. In addition, in the case of multiservice systems, there could be hundreds of parameters that need to be provisioned, and using a control and management system to handle these operations reduces human error and eases changes in configurations.

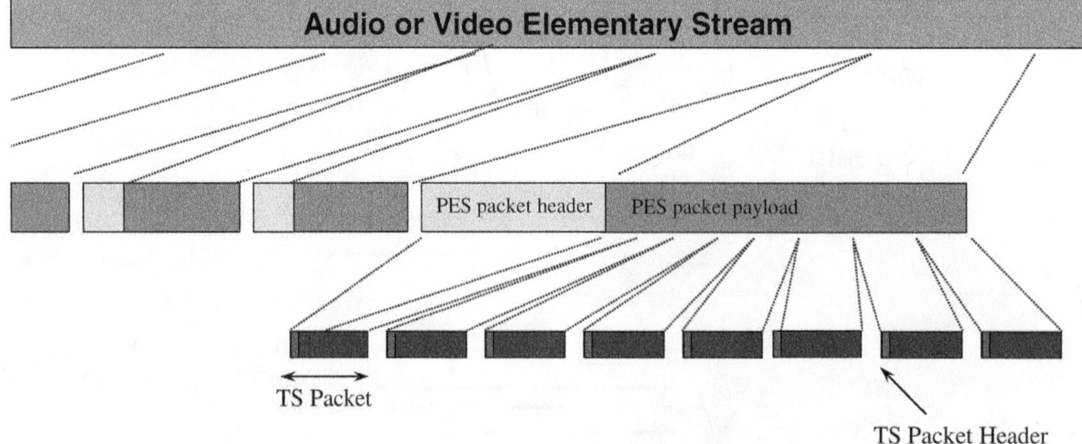

FIGURE 5.10-6 Packet structure hierarchy of a Transport Stream.

Courtesy of IEEE

Statistical Multiplexing

The ATSC transport is limited to 19.39 Mbps in a 6 MHz transmission channel, which was originally intended to provide sufficient bandwidth for an HDTV service. However, today many stations are either broadcasting or planning to broadcast multiple services in a single transmission channel—either using multiple SD services or a combination of HD and SD. As such, a mechanism is needed that will use the available bandwidth in the most efficient manner. Closed-loop *statistical multiplexing* is the most efficient mechanism for combining multiple services into a single transport (refer to Figure 5.10-7).

The quality measure is a very important part of a statistical multiplexing system. It should reflect the picture quality that will be perceived by the viewer, and this can be difficult to achieve. The quality measurement in the encoder and the bit rate allocation algorithm in the multiplexer form a closed-loop system. This must be capable of responding quickly to changes in the criticality of the video input, for example, at a scene change. Using a variety of parameters, each encoder computes the video quality and complexity and forwards this information to the transport multiplexer. The multiplexer, in turn, informs each encoder how much bit rate is available to it over a specific time period (typically adjusted on a video frame-by-frame basis) based on an evaluation of each encoder's parameters and service priorities defined. Each encoder then sets its output bit rate to the exact amount allocated for exactly the period allocated. The multiplexer then combines all the streams together to create an MPTS that, in the case of an ATSC 6 MHz transmission channel, is 19.39 Mbps.

Data services may be combined with the TV services in the same manner. Both "best effort" (*opportunistic*) and fixed data rates may be assigned. See Figure 5.10-8 for an example of a multiservice statistical multiplex that includes four video services, and opportunistic and fixed data services.

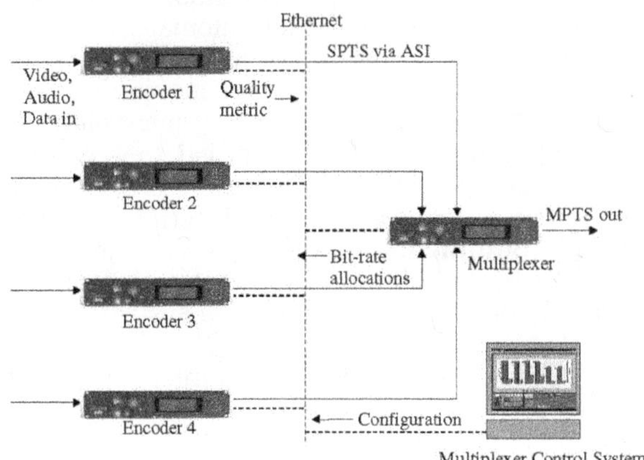

FIGURE 5.10-7 A four-channel statistical multiplexing system. SPTS = Single Program (service) Transport Stream. MPTS = Multi-Program (service) Transport Stream.

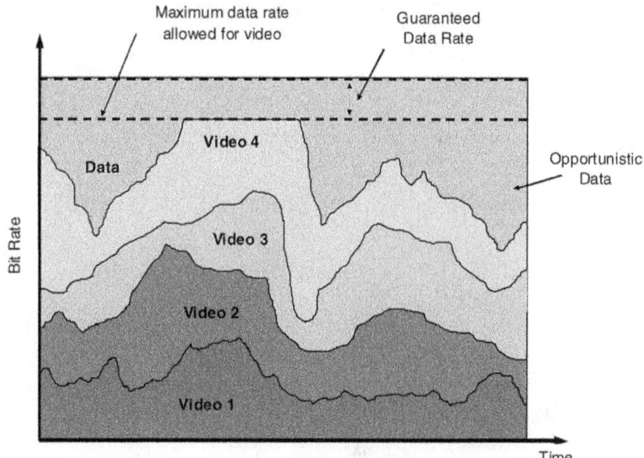

FIGURE 5.10-8 An example of statistical multiplexing within an MPEG-2/ATSC transport multiplex.

Courtesy of IEEE

Statistical multiplexing works because MPEG-compressed video, by its nature, is variable bit rate for each frame. In addition, different bit rates are required to encode different types of content. Mostly still pictures that consist of large areas with relatively little motion and/or little detail (such as scenes of landscapes with blue sky or news commentators) are easiest to code, whereas fast sports material is hardest. There are also short-term variations due to scene changes. By statistically taking advantage of the mixture of low complexity and high complexity content, instantaneous bandwidth may be reallocated to the service that needs it. Statistical multiplexing systems typically also allow priority to be placed on certain content to prevent high complexity in a secondary service from "stealing" bandwidth from a primary service when both call for additional bit rate.

Statistical multiplexing is difficult to do well. If multiservice systems are highly desired or a requirement, the resultant picture quality of all the encoders in a statistical multiplex should be evaluated closely.

Interfaces

The following is a summary of typical interfaces on a transport multiplexer:

- *Inputs*
  - Asynchronous Serial Interface (ASI) (ETSI EN50083-9 Annex B) (multiple)

    From one or more video encoders

    From one or more data servers

    From an external PSIP generator

- 10/100Base-T Ethernet or similar
  From one or more data servers
  From an external PSIP generator
- Serial interface (RS-232 or similar) for control system
- 10/100Base-T Ethernet or similar for control system
- Control channel for statistical multiplexing (typically Ethernet)

- *Outputs*
  - Asynchronous Serial Interface (ASI) (ETSI EN 50083-9 Annex B) (typically more than one for redundancy support and confidence monitoring)
  - Synchronous Serial Interface (SSI) (SMPTE 310M)
  - IP/Ethernet (for direct connection to an IP network, for scenarios that do not use ASI or SSI)
  - Feedback control channel for statistical multiplexing (typically Ethernet)

ENCODING AND TRANSPORT SYSTEM SOLUTIONS

Broadcast requirements differ from station to station, and there are many possible encoding and transport solutions or system architectures to address those needs. In general, however, these may be grouped together into a few common system solutions that will address most of the station's needs. Variations on these common scenarios are possible. The three main types of system architectures are:

- *Basic:* Single HD service, no redundancy, simple control and management
- *Mid-level:* Multiservice, automated redundancy, dedicated control and management system
- *Advanced:* Multiservice with statistical multiplexing, automated redundancy, and flexible/high-function control and management system

The basic architecture covers the simplest of broadcast scenarios, where there is no requirement for system redundancy (controlled fail-over to backup hot standby components) and there is a strong desire to keep the system simple to operate and maintain.

The mid-level architecture is typically used by small or mid-sized stations that want to broadcast a combination of HD and SD services, and have a requirement for system redundancy (automated fail-over to a hot standby component).

The advanced architecture is typically used by large stations or those that want the highest level of flexibility and performance. It supports a combination of HD and SD services, automated redundancy for high availability, and closed-loop statistical multiplexing for the most efficient use of service bandwidth possible.

BASIC SYSTEM ARCHITECTURE

The basic system architecture, as shown in Figure 5.10-9, is for the simplest scenario where there is no requirement for system redundancy or statistical multiplexing. Configuration is handled using front-panel controls or a simple Web browser loaded on a colocated personal computer (PC); a dedicated control and management platform is not needed or desired in this scenario.

The system solution supports one encoder (HD or SD) with a built-in multiplexer card that supports an external PSIP generator. Optionally, two additional SD encoders can be supported as well since the multiplexer card has multiple inputs. A SMPTE 310M (SSI) output is also available in addition to ASI.

An Ethernet switch is used to interconnect all the devices to a local PC so that users can launch the graphical user interface (GUI) for any Web browser available on each device.

A separate AC-3 encoder has been included in the diagram to represent the case where the video encoder does not include an integral multichannel (5.1 surround) audio encoder and local multichannel encoding is required. The audio encoder is fed into the audio input interface of the HD encoder.

For closed caption handling, a translator unit is shown. This is to illustrate the required configurations at the encoder level for closed captions handling in its possible standards and formats, when such a device is required; that is, for CEA-708 DTV captions via a separate SMPTE 333M or "Grand Alliance" interface, or in any case where captions are not embedded in the HD-SDI or SDI using SMPTE 334M.

Mid-Level System Architecture

The mid-level system architecture, as shown in Figure 5.10-10, is designed for a small or mid-sized station that wants to broadcast a combination of HD and SD channels, has a requirement for automated redundancy, but does not need statistical multiplexing. It provides easy-to-use configuration and alarm monitoring via GUIs using a dedicated control and management platform.

The system solution supports 1+1 (one online and one hot standby) HD encoder, 2+1 SD encoders, 1+1 transport multiplexer with PSIP insertion and SMPTE 310 (SSI) output, and associated redundancy switching. A dedicated control and management system is used to configure and monitor the system, and automatically controls redundancy.

An Ethernet switch is used to interconnect all the devices to a local PC. The dedicated control system GUIs are launched from the PC.

A separate AC-3 encoder has been included in the diagram to represent the case where the video encoder does not include an integral multichannel (5.1 surround) audio encoder and local multichannel encoding is required. The audio encoder is fed into the audio input interface of the HD encoder.

For closed caption handling, a translator unit is shown. This is to illustrate the required configurations

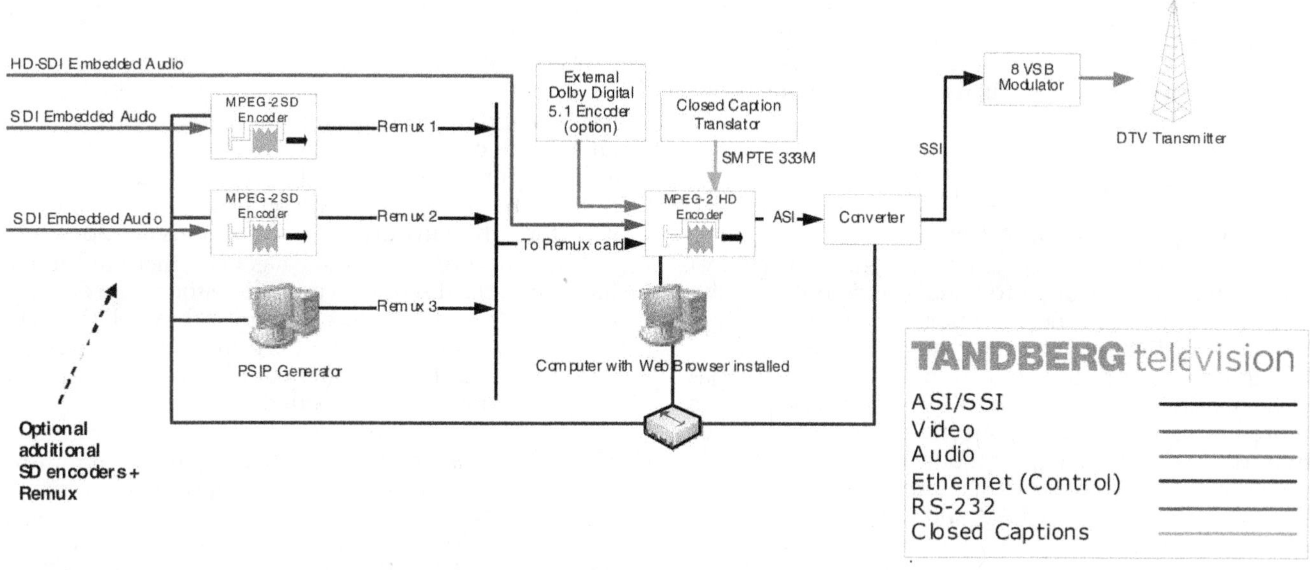

FIGURE 5.10-9 Basic encoding/transport system architecture. For color figure detail, please refer to the companion website provided by the publisher.

Courtesy of Ericsson

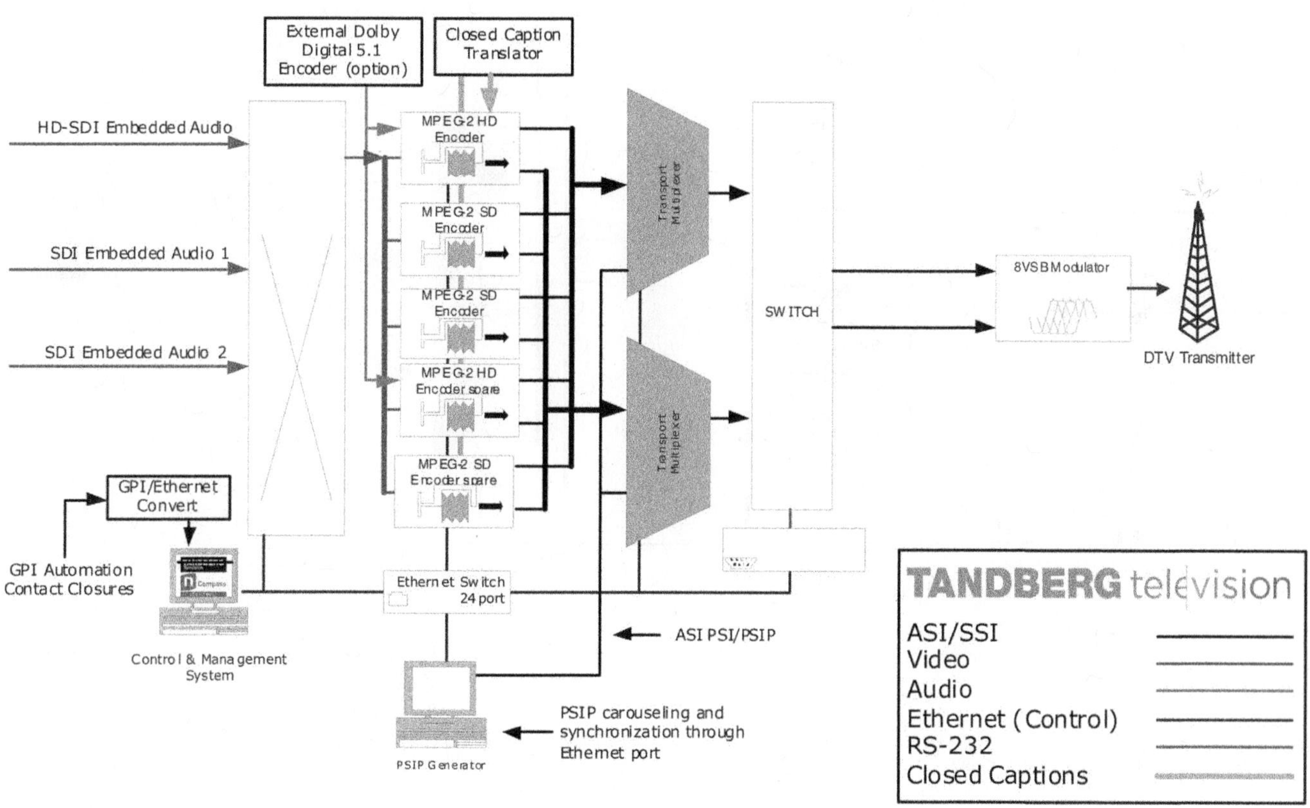

FIGURE 5.10-10 Mid-level encoding/transport system architecture. For color figure detail, please refer to the companion website provided by the publisher.

Courtesy of Ericsson

at the encoder level for closed captions handling in its possible standards and formats, when such a device is required; that is, for CEA-708 DTV captions via a separate SMPTE 333M interface, or in any case where captions are not embedded in the HD-SDI or SDI using SMPTE 334M.

Advanced System Architecture

The advanced system architecture, as shown in Figure 5.10-11, uses a dedicated comprehensive control and management system to handle and configure the devices involved in the Transport Stream processing. An advanced control system typically provides intuitive interfaces, sophisticated "n+m" redundancy capabilities, easy-to-download and backup physical maps representing equipment interconnections, profile files that represent TS configurations (for example, PID assignments, stream types contained within the TS, and other configuration data), and the ability to run the configurations live (online) or in a test/demo mode.

The advanced system architecture includes the highest-end solution for video compression, which may include features such as multi-pass or "look-ahead" encoding and statistical multiplexing of the services/channels in order to optimize picture quality per bit rate across all services.

The inputs to the encoders are handled by a router that is both HD-SDI and SDI compatible with embedded SDI audio. However, in situations in which external digital or analog audios are a requirement, an audio router could be included to handle their input to the system. The encoders are set up as 1+1 HD and 2+1 SD, although many other configurations are possible.

A separate AC-3 encoder has been included in the diagram to represent the case where the video encoder does not include an integral multichannel (5.1 surround) audio encoder and local multichannel encoding is required. This encoder could also be the input to a digital audio matrix if needed.

For closed caption handling, a translator unit is shown. This is to illustrate the required configurations at encoder level for closed captions handling in its possible standards and formats, when this device is required; that is, for CEA-708 DTV captions via a separate SMPTE 333M interface, or in any case where captions are not embedded in the HD-SDI or SDI using SMPTE 334M.

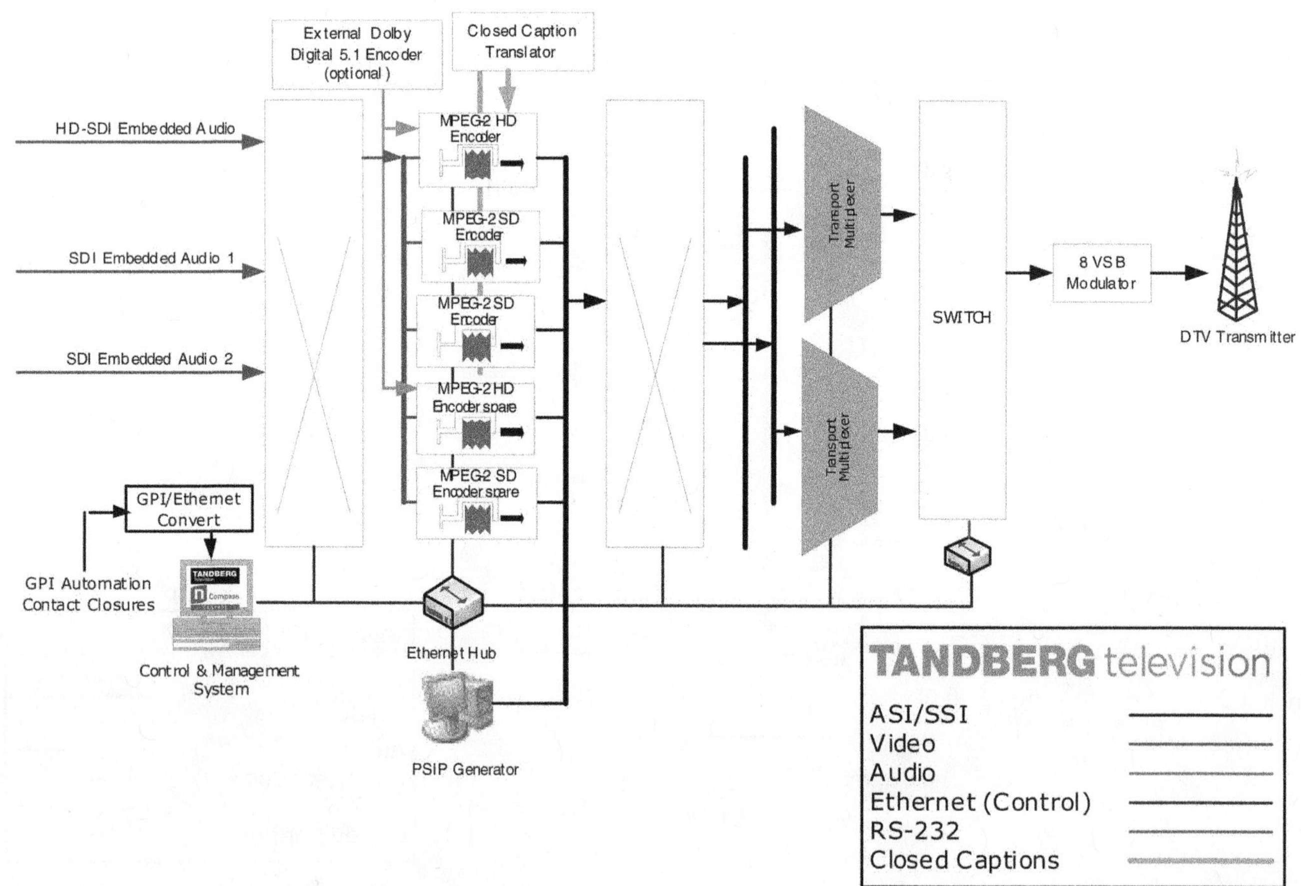

FIGURE 5.10-11 Advanced encoding/transport system architecture. For color figure detail, please refer to the companion website provided by the publisher.

Courtesy of Ericsson

For multiplexing purposes, two transport multiplexers are shown in a 1+1 redundant configuration. Each transport multiplexer contains two multiport ASI input cards and one SMPTE 310M SSI output card.

The PSIP generator could either feed the multiplexers via Ethernet or via one of the ASI ports of each multiplexer.

At the output stage, a switch is used to control the SMPTE 310M Transport Stream outputs to the RF system section.

The Ethernet switch keeps all the devices connected in the same network, and the translator server converts the RS-232 control protocols into TCP/IP for those devices that require it.

NEXT-GENERATION VIDEO COMPRESSION

The MPEG-2 Video standard [8] is the ubiquitous video compression technology used for the first generation of digital television systems, including digital terrestrial broadcasting, digital cable, direct-to-home satellite, and digital versatile disc for video (DVD video). The standard was finalized in October 1994, and was the state of the art at that time when memory was approximately $50 per megabyte, silicon wafer fabrication technology was greater than 1 micron, and typical microprocessor clock rates were well below 500 MHz. While many of the mathematical algorithms used were known years earlier, the technology for practical encoder signal processing and compact, cost-effective receiver implementation limited what could be achieved.

More than two decades have passed and the state of technology has leapt ahead. For the first time since MPEG-2 Video, three systems are now available that significantly decrease the amount of bandwidth needed for broadcast-quality digital video: MPEG-4 AVC [9], SMPTE VC-1 [10], and HEVC.

Application and Usage

At the time of writing, the new codec technologies are not approved by the FCC for use in the main, free, over-the-air television broadcast service in the United States. MPEG-2 Video is still required because existing digital TV receivers do not all support the next-generation compression technologies. However, there is interest in using next-generation coding for backhaul, contribution, alternative pay TV services, and mobile applications. As such, future digital TV receivers may include these new compression technologies.

Because of the different organizations involved, the industry is using various terms to describe the same technology standards in some cases. Figure 5.10-12 explains many of the names in use. For further details on video coding standards, see Chapter 5.11.

Improvements over MPEG-2 Video

In a similar way to MPEG-2 Video, successor codecs also are organized into Profiles and Levels to define specific

• **MPEG-2** • **H.262**	The ubiquitous video codec standard used digital television today – Terrestrial Broadcasting, Cable, DBS, DVD-V
• **MPEG-4 Part 2** • **MPEG-4 SP/ASP**	A follow-on video codec standard – not widely used for DTV (*therefore not discussed further here!*)
• **MPEG-4 AVC** • **H.264** • **MPEG-4 Part 10** • **AVC** • **"JVT"**	Advanced Video Coding (AVC) – A next generation video codec standard jointly developed by ISO/IEC MPEG and ITU-T VCEG
• **SMPTE 421M** • **SMPTE VC-1** • **Windows Media™ Video 9 (WMV9)**	A next generation video codec initially developed by Microsoft and now a SMPTE standard
• **HEVC ISO/MPEG** • **H.265**	High Efficiency Video Coding, the most advanced system developed to date, likely to be widely used in next-generation digital television systems; also known as ISO/MPEG-H Part 2 The ITU nomenclature for HEVC

FIGURE 5.10-12 Terminology primer for various video compression technologies.

Application	*MPEG-2 Video (H.262)*	*MPEG-4 AVC (H.264)*	*SMPTE VC-1 (Windows Media™ Video 9)*
Mobile Devices **Video Conferencing** **Internet Streaming**	-	**Baseline**	**Simple**
Broadcast Quality (TV, Cinema, IPTV)	Main Profile SD: Main Level HD: High Level	Main Profile SD: Level 3 HD: Level 4	**Main (progressive displays only)**
		High Profile SD: Level 3 HD: Level 4	Advanced Profile SD: Level 1 HD: Levels 2-3

FIGURE 5.10-13 Profiles and Levels versus application for video codecs.

Courtesy of IEEE

interoperability points. A Profile is used to specify the exact set (or subset) of algorithmic tools used and a Level defines constraints on those tools. As Figure 5.10-13 shows, only certain Profiles are applicable for broadcast-quality video such as that required for terrestrial and satellite broadcasting and cable television. Both the terminology itself and the Profile usage have caused some industry confusion as potential users attempt to compare video quality of what they believe are encodings made by the same technology but in fact are not. Examples include digital cameras and World Wide Web video streaming applications to PCs.

First-generation implementations of both MPEG-4 AVC and SMPTE VC-1 have realized 30% to 50% compression efficiency gains over MPEG-2 Video. As with MPEG-2 Video, continual refinements of real-time implementations will almost certainly occur over the next few years. In 1994, the state of the art for real-time full ITU-R SD resolutions was 8 to 8.5 Mbps (see Figure 5.10-14). With refinements in algorithmic

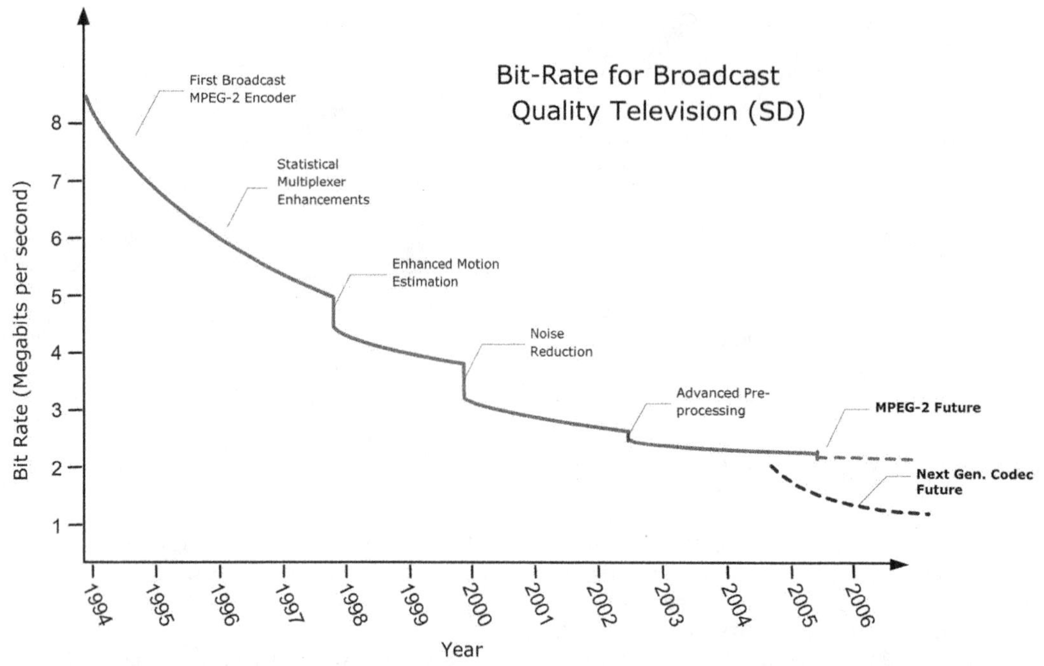

FIGURE 5.10-14 Comparison of SDTV picture quality for MPEG-2 and some successor codecs.

Courtesy of IEEE

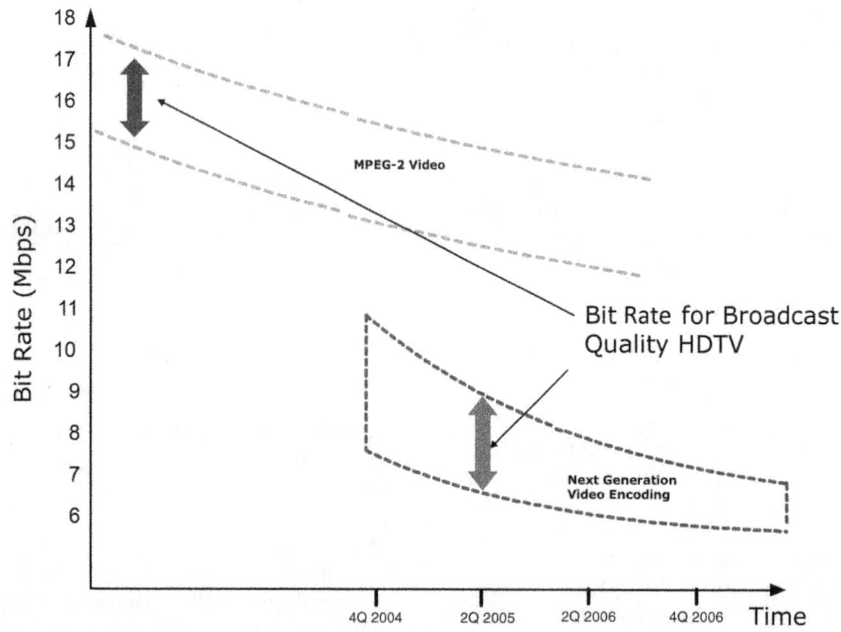

FIGURE 5.10-15 Comparison of HDTV picture quality for MPEG-2 and some successor codecs.

Courtesy of IEEE

implementations, advanced preprocessing, technology advances, and statistical multiplexing, this has been reduced to under 3 Mbps for the same picture quality. Most experts believe that the MPEG-2 Video improvement curve is near its asymptotic theoretical minimum. For next-generation compression technologies, SD rates for similar picture quality are near 1 Mbps and will drop further within the next few years.

With HD content, bit rate reduction is even more dramatic as an amount of consumed bandwidth per service. Only a few years ago, HD content required over 19 Mbps. While today's MPEG-2 HD content is being compressed at rates between 12 to 18 Mbps, next-generation compression rates, at similar picture quality, are approximately 6 to 10 Mbps today, and will likely be reduced further in the next few years (see Figure 5.10-15).

Tool	*MPEG-2 Video* (H.262)	*MPEG-4 AVC* (H.264)	*SMPTE VC-1* (Windows Media™ Video 9)
Intra Prediction	-None: MB encoded -DC predictors	- 4x4 Spatial - 16x16 Spatial - I_PCM	- Frequency domain Coefficient
Picture Coding Type	- Frame - Field - Picture AFF	- Frame - Field - Picture AFF - MB AFF	- Frame - Field - Picture AFF - MB AFF
Motion Compensation Block Size	- 16x16 - 16x8, 8x16	- 16x16 - 16x8, 8x16 - 8x8 - 8x4, 4x8 - 4x4	- 16x16 - 8x8
Motion Vector Precision	- Full Pel - Half Pel	- Full Pel - Half Pel - Quarter Pel	- Full Pel - Half Pel - Quarter Pel
P Frame Feature	- Single Reference	- Single Reference - Multiple Reference	- Single Reference - Intensity Compensation
B Frame Feature	- 1 Reference Each Way	- 1 Reference Each Way - Multiple Reference - Direct & Spatial Direct Modes - Weighted Prediction	- 1 Reference Each Way
In-Loop Filters	- None	- De-Blocking	- De-Blocking - Overlap Transform
Entropy Coding	- VLC	- CAVLC - CABAC	- Adaptive VLC
Transform	- 8x8 DCT	- 4x4 Integer "DCT" - 8x8 Integer "DCT"	- 4x4 Integer "DCT" - 8x4,4x8 Int "DCT" - 8x8 Integer "DCT"
Other	- Quantization Scaling Matrices	- Quantization Scaling Matrices	- Range Reduction - In-Stream Post Processing Control

FIGURE 5.10-16 Comparison of algorithmic tools used in MPEG-2 Video, MPEG-4 AVC, and VC-1.

As with MPEG-2 Video, obtainable bit rates for a particular overall picture quality vary greatly with content, with real-time encoded high-motion sports being one of the most difficult classes to code.

The coding gains come from the ability to perform more parallel processing and select better matches (that is, better results on the rate-distortion curve) in real time in the signal processing stages, a more efficient bit stream syntax, and more computationally intensive entropy coding resulting in fewer bits used in the stream processing stage.

Figure 5.10-16 contains a summary of the algorithmic tool differences among MPEG-2 Video, MPEG-4 AVC, and SMPTE VC-1.

For more information on video compression systems, see Chapter 5.11.

WHAT IS PSIP?

Program and System Information Protocol (PSIP) is an essential set of broadcast metadata that is required under the FCC rules. It has three functions. First, it allows end users to access the DTV broadcast in a manner similar to the way they are used to with analog broadcasts. Second, it provides critical information for the DTV receiver so it can correctly decode the signal. Finally, it provides promotional information, which allows the station to brand itself and advertise present and future services.

PSIP helps to make DTV as simple to access as analog TV, hiding the complex interrelationships between digital components of the system from the end user. PSIP provides capabilities familiar to analog TV users such as channel numbering, up and down tuning, and on-screen program guide.

In addition, PSIP provides the information necessary to tune and decode DTV signals. Decoders depend on the PSIP to help them differentiate different services in the digital stream. While MPEG defines digital services (with the PSI tables mentioned earlier in the chapter), it is the PSIP that defines *virtual channel numbers*, and then associates that information with a station name.

The PSIP in a DTV stream also serves an important marketing function. PSIP allows the DTV broadcaster to brand its signal with promotional information such as the *electronic program guide* (EPG). At its simplest, PSIP provides signaling information about what is playing now, in the future, and where the user has to go in order to find the virtual channel he or she is looking for.

PSIP Generator

The PSIP information is collected and processed in a device known as a *PSIP generator*, which typically is a dedicated personal computer running a specialized software package that provides the required functionality. The PSIP generator has network and other interfaces to allow it to communicate with other systems for receiving source information and for feeding the PSIP output to the multiplexer.

PSIP IN THE DTV STREAM

PSIP consists of a number of MPEG-2 tables, whose structure and usage are defined in ATSC standard A/65C [11]. The tables are carried in MPEG-2 transport packets in the DTV Transport Stream. To provide for random access tuning, the PSIP tables are repeatedly cycled and updated according to a specific schedule, also defined in A/65C. The following is a list and brief description of the most important tables.

- *Master Guide Table* (MGT): The MGT serves as a directory of all PSIP tables. It lists the locations, sizes, and versions of each table so the receiver can extract them from the stream.
- *Virtual Channel Table* (VCT): The VCT defines each of the virtual channels within the DTV stream, defining major and minor channel numbers as well as which program elements to assemble to reconstruct the television program.
- *System Time Table* (STT): The STT carries current time information from the broadcaster to the receiver.
- *Event Information Table* (EIT): The EITs carry the main program guide information (event titles, times, and information about captioning, ratings, audio, and broadcast flag).
- *Extended Text Table* (ETT): ETTs carry extended text descriptions of individual events within the EITs.
- *Rating Region Table* (RRT): The RRT defines the rating parameters used for the region.

For a more detailed explanation of the PSIP tables, see *Proceedings of the IEEE, Special Issue on Global Digital Television*, January 2006 [5].

PLANNING FOR PSIP

Some of the information that the PSIP generator needs is relatively static and should be configured at installation. Static information includes items such as the station's Transport Stream Identifier (TSID, identified by the FCC), virtual channel number, and packet IDs (PIDs) associated with audio and video streams.

To facilitate a smooth and quick transition when implementing a PSIP system at the station, it is important to generate a list of this information before beginning installation. The information should be collected in a *Digital Site Survey* document (DSS).

Preparing a Digital Site Survey

The DSS is a crucial step that should be pursued prior to implementing PSIP in the TV station. The information to be collected is listed in Table 5.10-1.

The finished site survey may look something like Table 5.10-2.

Many stations undergo changes in the virtual channel configuration during the broadcast day, known as *day part changes*. Day part changes are transitions from multiple standard-definition virtual channels to a single high-definition channel and back. Because the virtual channel configuration is signaled by PSIP, the PSIP data must be modified to match the actual configuration dynamically, as the day part change takes place. If the station has implemented or plans to implement day part changes, the site survey should contain information for both time periods (see Table 5.10-3).

Sources of Schedule Data

In addition to the information required at startup, the PSIP generator requires a great deal of information or *metadata* to create the EPG. This information includes:

- Branding;
- Characteristics: ratings, captioning, broadcast flag (at the service level);
- Announcement;
- What programs are coming up and when;
- Descriptive information;
- Characteristics: ratings, captioning, broadcast flag (at the program level).

All of this information is dynamic; it requires updating on a regular basis to remain current. Dynamic information is available within systems already in use in broadcast operations:

- *Listing services (traditionally provide guide information for print and other media):* Channel name/number, program name, time, descriptions, CC, ratings, audio types.
- *Traffic/program management (traditionally used for business operations—ensuring that content is available when needed and advertising revenue):* Program name,

TABLE 5.10-1
Digital Site Survey for PSIP Implementation

Type of Information	Where to Get This Information
Transport Stream ID	The unique identifier for a specific station's stream, provided by the FCC
Packet IDs (PIDs) for all audio and video streams	Mux user interface, or from the encoder
Program number for all digital services	Comes from the source for the station MPEG tables. See the next section for more details
Major channel number	The channel that users will use to enter into their remote control in order to tune to the station's DTV signal. These numbers are typically between 2 and 69 and are usually (but not always) based on the station's historical NTSC channel number
Minor channel numbers	For stations that broadcast multiple services, the minor number differentiates the different virtual channels. For example, if a station is broadcasting one high-definition service and one standard-definition service: 39-1 High-definition service 39-2 Standard-definition service The minor channel number is the -1 and -2

TABLE 5.10-2
Digital Site Survey Results Example

Station TSID:	1432					
Number of Services	3					
	Service Name	**Virtual Channel**	**Program ID**	**Video PID**	**Audio PID**	**Audio PID 2**
Service 1	WEXA-HD	31-1	3	49	52	53
Service 2	WEXA-SD1	31-2	4	65	68	69
Service 3	WEATHER	31-3	5	81	84	85

TABLE 5.10-3
Digital Site Survey Results with Day Part Changes

Station TSID:	1432					
Number of Services	Daytime: 3	Evening: 1				
	Service Name	**Virtual Channel**	**Program ID**	**Video PID**	**Audio PID**	**Audio PID 2**
4:00 AM—11:00 PM						
Service 1	WEXA-HD	31-1	3	49	52	53
Service 2	WEXA-SD1	31-2	4	65	68	69
Service 3	WEATHER	31-3	5	81	84	85
11:01 PM—3:59 AM						
Service 1	WEXA-NTL	31-1	3	49	52	53

time (optionally: channel name/number, program descriptions, CC, ratings, audio types).

- *Automation (controls actual playout of content and switching between feeds):* Program ID, accurate time (optionally: channel name/number, program name, descriptions, CC, ratings, audio types).
- *Human operator (ultimate control of operations):* Subset or all of the above.

As one moves down the list of sources above, the information becomes more chronologically accurate but less informative. At one end of the spectrum, the listing services have a considerable amount of detail for each event (such as the title, full description, actors, and other details that are useful). However, due to the long lead times characteristic of these services, the schedule information might be inaccurate. At the other end, with the automation system the actual event timing is exact (because the automation system is responsible for starting and stopping the playout of content). However, the automation system typically does not have much detail about the events themselves and in many cases does not differentiate between shows and interstitials. In order to formulate the most accurate and informative guide, it is necessary to draw information from multiple sources over time and intelligently merge the information.

When selecting an input source for the scheduling information, cost and time both play a role. A listing service may provide accurate and detailed information, but it also typically requires a yearly subscription fee. Furthermore, listing services typically require an Internet connection to the PSIP generator. If the PSIP generator is located at a transmitter site, it may not have an Internet connection. When evaluating vendors for PSIP generators, it is important to check that the device supports the schedule import method that is being planned to use. More information on the ATSC Programming Metadata Communication Protocol (PMCP) standardized method of communicating this information is provided later in the chapter.

PSIP INSERTION

Once the programming information is encoded into the binary PSIP tables, the PSIP data needs to be injected into the broadcast transport through the station multiplexer. Either the PSIP generator will have an ASI output, which will feed directly to the mux, or it will communicate with the mux via IP over a network connection. The actual configuration depends on the interoperability between the mux and the particular PSIP generator.

ASI PSIP Insertion into the Transport Stream

The simplest method for doing the PSIP data injection is to utilize an ASI connection from the PSIP generator into one of the input ports of the multiplexer, in a similar manner to the connection from an encoder, as shown in Figure 5.10-9. In this arrangement, the PSIP generator creates an MPEG-2 packetized stream with all of the PSIP tables encoded and scheduled to meet the required (or desired) cycle times. The multiplexer is provisioned to allow sufficient bandwidth to accommodate the PSIP stream (typically, less than 250 kbps is necessary).

PSIP Insertion Using Mux Carouseling

Some muxes have the capability of storing encapsulated PSIP tables in internal memory and playing them out according to a predetermined schedule; this is known as carouseling. For carouseling, an Ethernet connection is used between the PSIP generator and the mux. The PSIP generator uploads the encapsulated PSIP tables into the mux using User Datagram Protocol (UDP) over an IP connection, along with instructions on how to schedule the playout. The tables only need to be refreshed when there is an update, typically at the three-hour EIT boundaries or when an operator manually changes schedule information. Carouseling offers a form of error resiliency in case of communication problems between the PSIP generator and mux. If the PSIP generator fails to communicate, the mux will continue to carousel the information it has indefinitely.

Multiplexer Configuration Synchronization

Some PSIP generators can synchronize directly with the encoder/multiplexer systems to automatically download static data and deal with day part changes. If day part changes are being planned (see above), then a mux/PSIP generator pair should be used, which automatically synchronizes this data. Otherwise, manual intervention will be required every day to switch between the two different program lineups.

PSIP AND PSI CONSIDERATIONS

One point that is often overlooked is the coordination of MPEG PSI (program specific information) and ATSC PSIP. Some of the information carried in the PSIP is also signaled in the PSI. In order to guarantee that services will be decoded by receivers, the MPEG-2 program information (from the PSI) and the PSIP virtual channel information (from the PSIP) must be consistent. Besides the necessary matching between the MPEG-2 programs and the PSIP virtual channels, some program-specific signaling information is carried in both PSIP and PSI structures (examples include the *caption service descriptor* and *redistribution control descriptor*). Only the PSIP generator has access to the schedule data required for proper insertion of these descriptors.

The links between PSI and PSIP present a potential problem for broadcasters. If the PSI information does not match the information generated by PSIP, there is no guarantee that decoders will function. Furthermore, if more than one source of PSIP and/or PSI resides on the system, they may collide, resulting in intermittent or total decode failure.

PSIP and PSI conflicts remain among the most common problems encountered when implementing PSIP in a digital television system. To help avoid these problems, two important rules should be followed:

1. Use only *one* source for PSI information into the multiplexer. All other PSI sources should be blocked or disabled. The same is true of PSIP information—only *one* device in the system should supply encoded PSIP data to the multiplexer.
2. Regardless of the device that generates the PSI tables, the configuration of the PSIP generator must *match* the information in the PSI tables.

Rule 2 is automatically satisfied if PSI information is generated in the PSIP generator, and blocked or disabled elsewhere on the system. For this reason, it is recommended that PSI information come directly from the PSIP generator.

Time Information

Accurate time information is critical for PSIP. The system time table (STT) carries the broadcaster's notion of current time to the receiver, where it is used for the EPG and increasingly for the start and stop of recording through digital video recorders (DVRs). ATSC standard A/65 requires that the time carried in the STT be within 1 second of GPS time to avoid confusion during channel tuning.

It is relatively straightforward to broadcast an accurate STT. Many stations utilize SMPTE time code, carrying time of day, in their operation. It is a simple matter to equip a PSIP generator with a SMPTE time code reader, which will automatically synchronize to the station time. The station clock should itself be synchronized to a standard time reference—typically GPS. Obtaining time reference direct from a GPS receiver is another relatively low-cost option and easy to implement with a connection into the PSIP generator. If neither of these references is available, accurate time information can be obtained from the Internet using the Network Time Protocol (NTP). NTP servers are available on many office networks or on the Internet.

If the station in question utilizes a UDP connection from the PSIP generator to the mux rather than an ASI input, the STT timing information is usually the responsibility of the mux. In this case, the mux generates the STT based on an internal clock, which should be synchronized to an external source. Specific arrangements about SMPTE, GPS, or NTP time sync are usually determined in consultation with the PSIP generator and mux vendors.

PSIP ARCHITECTURES

There is a wide range of PSIP architectures in use for broadcast television stations, groups, and networks. The simplest (and most common) configuration is for each station to be considered a separate entity that gathers all of the information required for PSIP and then injects the PSIP data tables into the broadcast Transport Stream, within the station. At the other end of the spectrum are "central casting" operations—large station groups that may be geographically distributed but with a centralized location for managing and creating PSIP.

Individual Stations

For an individual station, the architecture will resemble Figure 5.10-17; however, typically only some of the schedule sources will be involved. Most often, the PSIP generator and the encoding/multiplexing equipment will be collocated, allowing direct connections from the PSIP generator.

Central Casting

For large station groups, economies of scale can be achieved via central casting—especially in terms of operations. Commonly, schedule management and PSIP-related operations are performed at a central location, with the resultant information carried to the edges for broadcast. In some cases (as discussed in Chapter 5.4 on master control systems), master control has also been centralized. Central casting PSIP arrangements involve a single PSIP generator at the central location (network operations center or NOC), which feeds the individual remote stations/transmitters. Schedule aggregation and PSIP table generation takes place at the NOC. The method for carrying the PSIP tables to the network edges for broadcast varies depending upon the central casting design, as discussed below.

The architecture used for central casting depends upon a number of factors:

- Common programming at all stations versus varied programming: Can the same PSIP information be used at different stations with only changes to certain fields (TSID, major/minor channel numbers, and others), or does the PSIP information need to be generated for each station?
- Content distributed in compressed (MPEG) form to remote stations (remote encoding) versus analog form: Can encoded PSIP be carried in-band (part of the transmission) versus out-of-band, typically over a wide area network (WAN)?
- Network connectivity to remote stations: How is the PSIP data carried to the remote station—over a WAN? Over POTS (plain old telephone service)? As part of a digital (MPEG) transmission?

Figure 5.10-18 shows a generalized architecture for central casting, with three possible connections illustrated. Path A illustrates a connection for remote emission sites with broadband WAN connectivity. While it is possible to stream the MPEG-2 packetized PSIP tables across the WAN connection, uncertainty in packet loss and latency suggests that a carousel connection is best for this situation. Path B illustrates a similar situation (remote emission sites), but with no WAN connectivity available. For this situation, carouseling is the only

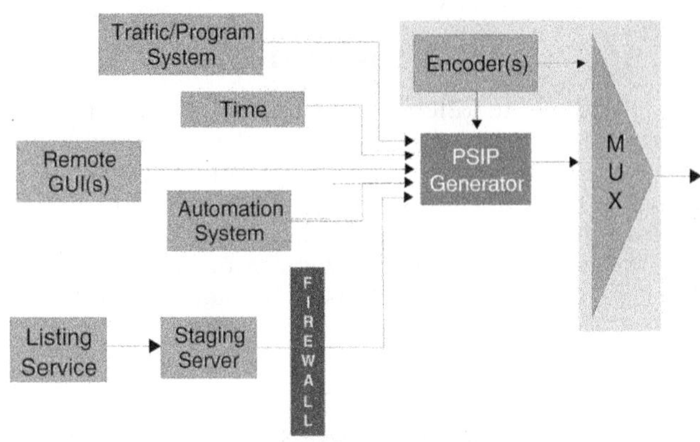

FIGURE 5.10-17 PSIP data flows.

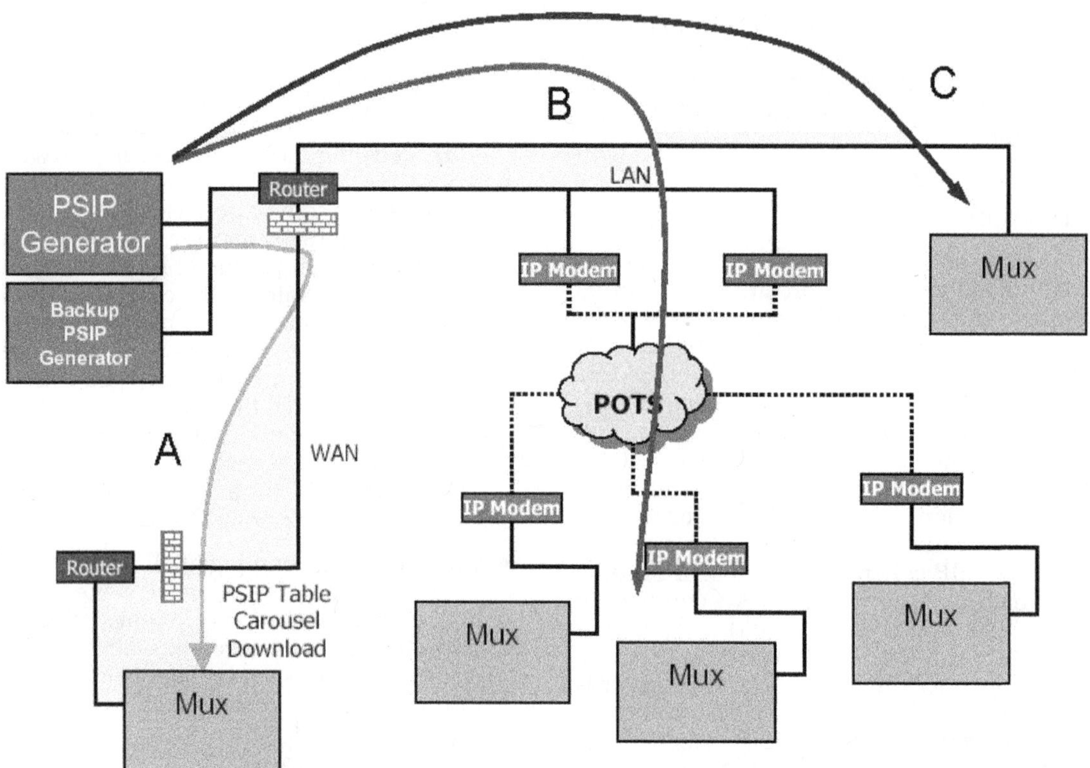

FIGURE 5.10-18 Example PSIP central casting architecture.

reasonable approach, using IP modems across POTS connections. Path C illustrates connections to local multiplexers, where LAN connectivity is available. In this situation, it is reasonable to either stream MPEG-2 packetized PSIP tables or to utilize carouseling. In many instances, the actual solution will involve a mixture of these types of connections.

Central casting typically requires close coordination between the station information technology (IT) and engineering departments, as well as consideration of a number of IT policies: security (firewall, antivirus, VPNs, etc.), services that are allowed/disallowed (for example, SNMP auto-discovery), and naming and addressing schemes such as static IP addresses and dynamic domain name servers (DDNS).

PSIP AND PMCP

As previously mentioned, it is necessary to draw programming information, or metadata, from multiple sources and intelligently merge the information to generate the PSIP tables. To try to avoid a multitude of proprietary interfaces to achieve this, the ATSC developed the Programming Metadata Communication Protocol (PMCP), published in 2004 as ATSC standard A/76.

What Is PMCP?

PMCP is a common language that systems can use when exchanging PSIP-related metadata. It is based on XML (eXtensible Markup Language), and supports exchanges of data via files or socket-based connections.

PMCP is used for communication between the PSIP generator and other sources of PSIP-related information, which may also communicate among themselves using PMCP. Data transfer will typically take place over a network, although any means of transport may be used for file transfer.

What Are the Benefits?

PMCP is the main means of connection between manufacturers of systems that are involved in the PSIP production chain. A manufacturer can build a single PMCP interface and use it for communications with multiple partners' systems. Reduced development time, reduced costs, improved reliability, and simpler implementation are just a few of the benefits.

Broadcasters also benefit from the adoption of PMCP, with an assurance of compatibility between systems that need to exchange PSIP data. If any of these systems in a facility is replaced at a later date, providing the new system also includes PMCP compliance, there should be few problems when connecting with other systems already in place at the station.

There are other benefits. PMCP is based on XML, and is designed to be quite human-readable, even by those without an extensive technical background. XML support is easily incorporated into products using most modern development environments, and is already well known to most of the manufacturers of broadcast systems. It is by its very nature extensible, which means that it can be extended and updated as needs evolve. In addition, provision has been made for carriage of private information throughout the schema so that systems can enhance their PMCP interfaces with heretofore unthought-of data elements.

Finally, much of the PMCP structure has been adopted by the highly successful and widely deployed Broadcast eXchange Format (BXF). This means that many broadcasters and broadcast system vendors have a comfort level with the structure used by PMCP.

Static Versus Dynamic PSIP

In the first years of digital television in the United States, many over-the-air broadcasters sent out *static PSIP* (or, in some cases, no PSIP data at all). Static PSIP provided enough information for receivers to tune to the broadcast signal, but not much else. FCC rules have required transmission of *dynamic PSIP* for some time now, which is of far more use to viewers. Dynamic PSIP provides viewers with an accurate listing of what is scheduled to air on each channel, permitting them to tune to what they want to watch at the time the program will actually air, and to know how long it will be on air.

Dynamic PSIP is important for other reasons. The most obvious are recording devices that use PSIP, such as DVRs. In order for such devices to record what is intended, PSIP must accurately reflect what is going to air, and at exactly what time. Less obvious, but still important is rating (V-chip) information. If PSIP for the next program comes on before the program begins, or if incorrect PSIP is sent, and the rating differs from what it should be, viewers who have set their receiver to prevent viewing of programs with certain ratings may be unable to view an otherwise suitable program.

Typical Systems Involved

One of the things that make PMCP so helpful is that it can be used to connect a wide variety of systems.

Because PSIP data typically resides on several different systems, in order to have accurate, dynamic PSIP, this data must often be obtained from multiple sources.

A typical broadcast station may obtain PSIP data from some or all of the following systems:

- *Listing service:* A few companies provide services that aggregate program listing information for stations across the country. Similar services are also available outside of the United States. Unlike the other systems involved in PMCP, these systems are typically not hosted in the broadcast facility.
- *Program management system:* Broadcasters who themselves originate a good percentage of their programming lineup may have a formal program management system. These typically manage program contracts, rights, and program schedules, as well as financial aspects of programming.
- *Traffic system:* Virtually every broadcaster has a traffic system. These systems manage spot sales contracts, commercial material management, assembly of the complete broadcast schedule, reconciliation of what ran, and invoicing of spots, among other things.
- *Automation system:* Since the early 1990s, broadcasters have increasingly moved to systems that manage their on-air operations. These systems control the ingest of broadcast material, management of material and devices, on-air playback of that material, and a variety of other tasks.

Challenges: Data Ownership and Granularity

When PSIP data resides in multiple systems, it may be asked, why not just choose the one with the highest quality and quantity of data, and use that as the single data source for the PSIP generator? That would be wonderful if such a system existed, but it does not.

The PSIP paradox is that as accuracy of PSIP data increases, the richness of that data decreases, as illustrated in Figure 5.10-19.

A system-by-system examination of the problem follows.

Listing Service

Because the job of these services is to maintain rich viewer-friendly program listings, the depth of the data found here is excellent. Many details for the event information table (EIT), as well as much information for the extended text table (ETT), can be found here. These systems typically have databases with full and

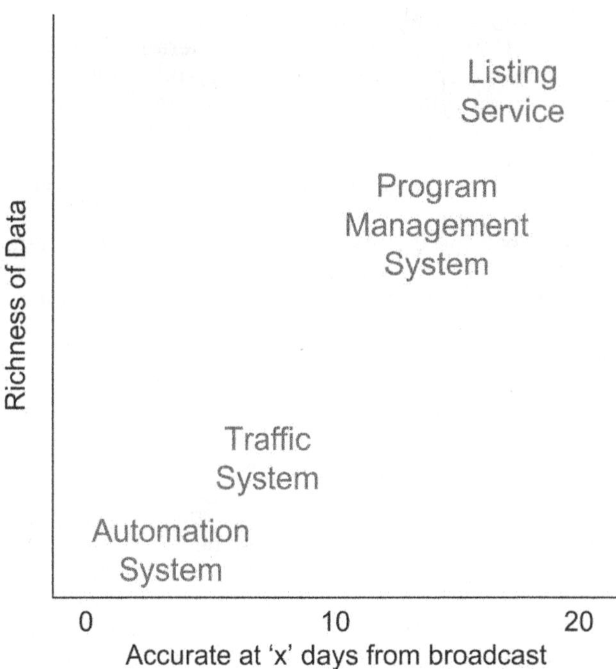

FIGURE 5.10-19 Richness of data versus accuracy at 'x' days from airtime.

consistent information for every program that could be scheduled, removing any worry about typos or incomplete information. The problem lies in the fact that listing services' view of the program schedule is typically frozen in time about 2 weeks prior to air. This means that any program schedule changes occurring within this window are not reflected in these systems.

Program Management System

In terms of depth of information, in many cases these systems can challenge listing services. Program management systems have the capability to track virtually every detail concerning programs. However, there are some issues. The first point is that data must be entered into the system. Unlike listing services, which have centralized databases containing details relating to virtually every program ever produced, program management systems typically rely on personnel at the station or station group to either hand-enter this information, or capture it from external source(s). As a result, information held here can be incomplete and is more likely to contain typographical errors than that which is found in listing services. However, program management systems have one significant advantage. Being in-house systems, they are typically updated as the schedule gets closer to air, and often contain accurate representations of the on-air schedule up until a day or two before broadcast.

Traffic System

At this point the depth of program-related metadata often degrades significantly. Because traffic systems are primarily concerned with the management of commercials, they do not require the depth of programming metadata that the other systems mentioned thus far do. The typical depth of information that can be expected from a traffic system is generally limited to date, time, channel, show name, episode/title name, and show duration. In some cases, it may be possible to get more, but those will be the exception. Although the data is not rich, it is normally far more accurate than that found in either the program management system or listing service. Traffic systems have an accurate and complete view of the entire broadcast schedule up until a few days, or in some cases, a few hours prior to air. Changes to the program schedule made in this time window will normally be reflected in the traffic system.

Automation System

In terms of richness of metadata, automation is not typically any better than traffic. In fact, in many cases, automation may know even less than traffic when it comes to program material. It normally does not have visibility of the program schedule more than 1 to 3 days in advance. So, what makes it an essential PMCP source? In a word, *accuracy*. Automation's primary responsibility is to control what airs, and at precisely what time it is to air. So, if accurate PSIP is the goal—to reflect exactly what is airing and at what time it is airing—the information must come from the system responsible for the on-air product, which is automation. In addition, automation is probably the only source of PSIP data that is truly mission-critical. As such, it is typically built with a great deal of redundancy and reliability, in most cases far more so than any of these other potential PMCP sources.

Clearly, ownership of PSIP data changes as a function of time. Although there can be multiple sources, the authoritative source of this data will vary as the time gets closer to broadcast.

Which system should be relied upon for PSIP data? Ultimately, the decision is up to the broadcaster, and different broadcasters will make different choices. However, two factors are key in determining from which systems to get PMCP data:

1. How close to air does your program schedule generally get "locked in?"
2. How accurate do you want or need your PSIP to be?

Another Challenge: Accurate Names

As discussed, it may be desirable to have the ability to update PSIP from the traffic system, and possibly from the automation system. As outlined earlier, although the information available in these systems is very basic (typically, start time, duration, and show name), the accuracy is very good. An important thing to consider before putting in place any PMCP interface from either of these systems is that the show names in use in these systems may not be in a form that is acceptable for display to viewers. Some broadcasters have program names such as "Friends (Barter Version)" or "Late Night Movie (Weekend)" in their traffic and automation systems. These names are not appropriate for the EPG. So, before putting these interfaces in place, it is good practice to spend some time considering how to clean up these names and make them presentable for the public.

PMCP: A Use Case

To better understand the system interactions involved, a use case is presented.

1. Months in advance, a blockbuster movie is scheduled. This is done in the program management system.
2. Two weeks prior to broadcast, the program management system sends this information to the listing service.
3. Shortly thereafter, the program management system sends this information to the traffic system.
4. The day prior to air, the traffic system sends details of this movie to the automation system.
5. On the day of air, a breaking news event preempts the movie at the last minute. Automation must communicate this to the PSIP generator.
6. Eventually, the live event ends, and the movie is started later than expected. Again, the automation

system must communicate this to the PSIP generator, as well as any shifts of programming that follow the movie and whose start times are affected.

This is a fairly simple, common, and straightforward illustration, and it alone has six exchanges of PSIP data taking place. It can be far more involved than this. Because of this, it becomes clear that standardized communications are of great benefit in helping to solve this problem.

Recommended Implementations

How PMCP is implemented in a particular facility is up to the broadcaster. Hopefully, the information provided here is helpful in understanding the topic well enough to make the right decision for each particular situation. As a guide, Table 5.10-4 offers some possible implementations based on different types of programming.

SUMMARY OF THE FCC RULES ON PSIP

The entire Report and Order issued by the FCC in September 2004 concerning PSIP [12] is quite lengthy, but perhaps the most important section relating to PSIP and PMCP is the following.

> We conclude that adoption of ATSC A/65B (PSIP) into our broadcast transmission standards will serve the public interest. As pointed out by commenters, during the development of PSIP, the ATSC carefully considered which elements of PSIP should be mandatory and which should be optional. . . . We therefore require that broadcasters fully implement PSIP to the extent that ATSC A/65B requires. In order to give broadcasters adequate time to come into compliance, this requirement shall take effect 120 days after publication in the Federal Register. We expect broadcasters to populate the required tables and descriptors with the proper information to help receivers assemble functioning guides. *All tables and descriptors that require one time setup should be set correctly, including TSID, Short Channel Name, Service Type, Modulation Mode, Source ID, and Service Location Descriptor. ATSC A/65B also requires that broadcasters send populated EITs covering at least a 12 hour period. These EITs should be populated with the correct information, so that the user knows what programs are on for this 12 hour period.* Also, we expect that manufacturers will have every incentive to build equipment that looks to PSIP for its basic functionality, but we will revisit the issue if necessary. Standardized use of the data transmitted through PSIP will ensure that the full benefits and innovations of the new digital system will be available to the public.

A few portions of this are particularly important, and those have been highlighted. An examination of these passages is helpful in understanding the core of the rule.

> *All tables and descriptors that require one time setup should be set correctly, including TSID, Short Channel Name, Service Type, Modulation Mode, Source ID, and Service Location Descriptor.*

The implication here is that all of these attributes must be set up correctly in the PSIP generator. Since these are "static" elements of PSIP, they should only have to be set up once, saved, and rarely revisited.

TABLE 5.10-4
PSIP Metadata Sources

Program Schedule Is Locked Down	Is Programming Mainly Local Origination?	Suggested PMCP Sources
2+ weeks out	Yes	Listing service/ program management
2+ weeks out	No	Listing service
1 week out	Yes	Listing service, then program management
1 week out	No	Listing service, then traffic
A few days out	Yes	Listing service, then program management, then traffic
A few days out	No	Listing service, then traffic
Never—lots of live programming	Yes	Listing service, program management, traffic, then automation
Never—lots of live programming	No	Listing service, traffic, then automation

This is the easy part. The rest of what the FCC requires implies "dynamic PSIP," which is not quite so straightforward.

> *ATSC A/65B also requires that broadcasters send populated EITs covering at least a 12 hour period.*

EITs contain the basic program guide information that allows the consumer to see what programming currently is being shown and what is coming up. These tables include information such as show name, start time, and duration of the entire program. ATSC A/65B actually requires that the first four EITs (EIT-0 through EIT-3) are sent, each containing 3 hours of events. However, as programs are completed in the current EIT (EIT-0), it will reduce from 3 hours, eventually down to zero, meaning that right at the end of EIT-0, only 9 hours of EITs (basically EIT-1 through EIT-3) will be sent out. Although the FCC Report and Order specifies a minimum of 12 hours, the actual A/65B standard that they cite really only requires this 9- to 12-hour range covered by EIT-0 through EIT-3.

> *These EITs should be populated with the correct information, so that the user knows what programs are on for this 12 hour period.*

The key word in this sentence is "should," which may be considered as meaning, "we highly recommend you do this, and if you don't, we'll make you." So, it would be prudent to make sure that the information sent out in the EITs is accurate.

Required Accuracy

If EITs are based on information from a listing service that might be 2 weeks old, this may not be accurate enough. This leads to the question of where to get more accurate information regarding program names, start times, and durations. Traffic and automation systems certainly have this information.

If the goal is to have all of this information as accurate as possible, an interface from the automation system to the PSIP generator is required. However, some believe that getting this information from traffic may be good enough. To resolve this, the following question must be answered.

How Many Hours of EIT Information Should Be Sent?

Although the minimum requirement in A/65B is the next four EITs, covering 9 to 12 hours of programming, will that be sufficient? To give viewers a good experience, probably not. The ideal amount of EIT information to send out may be more along the lines of a week's worth. As mentioned in the Static versus Dynamic PSIP section, there are several reasons for this, relating to DVR use, rating information, and the viewer experience.

However, there is another compelling reason to send out this much EIT data. PSIP is sent out by each broadcaster individually, and it will typically only provide information about their own services. This means that a receiver, when tuned to a particular channel, must acquire the PSIP data relating to that channel, and make it available to the viewer. This may take several seconds. To permit quicker tuning from channel to channel, some receivers perform a scan of PSIP available from all available services periodically when not in use (perhaps in the middle of the night). This information is stored and displayed to viewers very quickly when they tune from channel to channel, giving a channel surfing experience similar to that available in the analog world, without a potential delay of a second or two while changing channels. This helps provide an improved viewing experience.

Rating (V-Chip) Information

The FCC order also requires that rating information be included in PSIP, and be accurate. The pertinent passages from the FCC Report and Order regarding V-chip are as follows:

> . . . the Event Information Tables ("EITs") defined within PSIP will contain any available Content Advisory Descriptors ("CADs") for broadcast programming.

> . . . we believe it is reasonable to provide an 18 month transition period. After the transition period, all digital television receivers will be required to provide v-chip functions following the regulations that we adopt in this proceeding.

This means V-chip data must be provided to the PSIP generator. Traffic and automation systems typically either do not track this information, or if they do, they often do so in nonstandard ways. One broadcaster may include this information on the schedule as secondary events, another may do so in text-based comments, while yet another may find other ways in which to include this information. As a result, there is no single straightforward way in which to extract this information from traffic and automation systems for the purposes of providing it to the PSIP generator.

For this reason, it may make more sense for PSIP generators to get this information from systems such as program management and listing services, which are more geared toward dealing with this parameter.

Looking Ahead

Many have asked, "How does PMCP apply to ATSC 3.0?" This is an excellent question. PMCP was authored over a decade ago, specifically targeted at the terrestrial broadcast environment we now refer to as "ATSC 1.0." ATSC 3.0 introduces many changes—some quite revolutionary—into the mix, and is specifically not backward compatible with ATSC 1.0 (see Chapter 8.4). For this reason, it is likely that a new method for exchanging "guide information" like PSIP will be developed. The form of this PMCP replacement is yet to be determined at the time of writing of this book. However, it is fully expected that meeting this requirement is on the radar of those involved in ATSC 3.0's development, and it certainly will be addressed.

SUMMARY

Encoding, multiplexing, and the resultant Transport Stream are fundamental parts of any DTV system. This chapter has discussed the principles of operation of the equipment that broadcasters need to use for this vital part of their operations. The discussion about PSIP should help ensure that broadcast stations comply with the FCC rules in what is often a confusing area. PSIP compliance is made easier via the PMCP standard for metadata communications and its close cousin, BXF. While MPEG-2 encoding will be with us for many years to come, it is expected that the new advanced codecs discussed earlier in this chapter and in Chapter 5.11 will inevitably become a part of the toolkit that broadcasters have to draw upon to provide high-quality yet highly efficient coding for broadcast programming.

ACKNOWLEDGMENTS

Figure 5.10-4 is taken from the ATSC Digital Television Standard A/53 published by the ATSC and used with permission. Figures 5.10-5, 5.10-6, 5.10-8, 5.10-14, and 5.10-15 are based on figures in "ATSC Video and Audio Coding" and "The ATSC Transport Layer, Including Program and System Information Protocol (PSIP)," first published in *Proceedings of the IEEE, Special Issue on Global Digital Television*, January 2006, ©2006 IEEE, used with permission of the publisher.

References

[1] ATSC Digital Television Standard, A/53 Revision E with Amendments No. 1 and No. 2, September 2006. Advanced Television Systems Committee. See www.atsc.org/.

[2] ITU-T Rec. H.222.0 | ISO/IEC 13818-1:2000 Coding, Information Technology—Generic Coding of Moving Pictures and Associated Audio. Part 1: Systems. Information: Systems, December 2000.

[3] ETSI EN 50083-9: Cabled Distribution Systems for Television, Sound and Interactive Multimedia Signals. Part 9: Interfaces for CATV/SMATV Headends and Similar Professional Equipment for DVB/MPEG-2 Transport Streams, (DVB Blue Book A010), Annex B, Asynchronous Serial Interface.

[4] Bock, A., "What Factors Affect the Coding Performance of MPEG-2 Video Encoders?," DVB '99 Conference Proceedings, March 1999.

[5] *Proceedings of the IEEE, Special Issue on Global Digital Television,* January 2006.

[6] Jordan, J., and Bock, A., "Analysis, Modelling and Performance Prediction of Digital Videostatistical Multiplexing," *Proceedings of 1997 International Broadcasting Convention*, September 1997, Amsterdam, The Netherlands, pp. 553–559.

[7] Bolender, S., Hackett, A., Heimburger, C., and Knee, M., "Motion Content of Transmitted TV Images: A Statistical Survey," *Proceedings of 1994 International Broadcasting Convention*, September 1994, pp. 405–410.

[8] ITU-T Rec. H.262 | ISO/IEC 13818-1:2000 Coding, Information Technology—Generic Coding of Moving Pictures and Associated Audio. Part 2: Video, February 2000.

[9] ITU-T Rec. H.264 | ISO/IEC 14496-10 |:2005, Information Technology—Coding of Moving Pictures and Associated Audio-Visual Objects. Part 10: Advanced Video Coding, December 2005.

[10] SMPTE 421M-2006, VC-1 Compressed Video Coding 1. Bitstream Format and Decoding Process, February 2006.

[11] ATSC Standard A/65:2013, Program and System Information Protocol for Terrestrial Broadcast and Cable, http://atsc.org/standard/a652013-program-and-system-information-protocol-for-terrestrial-broadcast-and-cable/

[12] FCC 04–192 Second Periodic Review of the Commission's Rules and Policies Affecting the Conversion to Digital Television. Federal Communications Commission, September 7, 2004.

CHAPTER

5.11

Video Compression

MATTHEW S. GOLDMAN

Ericsson
Bedford, NH

PETER SYMES

Consultant
Grass Valley, CA

INTRODUCTION

The electronic transmission or recording of moving images has always presented challenges because of the sheer volume of information that must be handled. From the early days it has been necessary to find compromises between bandwidth and quality. The challenge has always been to find the techniques that provide the greatest savings in bandwidth for the least loss in delivered quality.

In the analog world, the first major development in this direction was interlaced scanning. For a given picture size and static resolution, and a chosen repetition rate sufficient to prevent large-area flicker, interlace permits halving the bandwidth that would otherwise be required. But, there is a cost; segments of vertical and temporal spectra are overlapped. This results in a loss of vertical resolution on moving objects and interlace artifacts caused by vertical energy appearing as small-area flicker.

The other technology that provides dramatic saving of bandwidth is the National Television System Committee (NTSC) color system. This brilliant development interleaved color information into the luminance spectrum so as to permit color transmissions in the same bandwidth as monochrome NTSC (and maintained monochrome backward compatibility). Again, there were consequences: some degradation of the monochrome signal, cross-channel artifacts, and a barely adequate chroma bandwidth. However, few would argue today that the compromise was a bad one or that a better solution was available at the time.

Digital video brings many benefits, but, in the formats used for production, it requires vast quantities of data. Standard-definition 525 line component 4:2:2 video at 29.94 frames/sec, as used for NTSC coding, is carried in the studio at a rate of 270 megabits per second (Mbps). This is more than needed for transmission—about 23% is dedicated to blanking and carries no picture information. Also, for delivery to the home, 10-bit precision is unnecessary, and 4:1:1 or 4:2:0 coding would be adequate. Taking all of these factors into account, it would be possible to save about half of the original 270 Mbps; however, even the remainder would require ten or more 6 MHz broadcast channels using conventional transmission systems.

High definition (HD) is, of course, much more demanding. In the studio, HD signals are transmitted at almost 1.5 gigabits per second (Gbps), five and a half times the standard-definition rate.

Today, digital transmission of high-definition signals is accomplished in many countries within a single 6 MHz channel.[1] Advanced modulation techniques provide one part of the solution, but the most important enabling technology is video compression. MPEG-2, the near ubiquitous compression standard used for almost all digital video broadcast systems (from the mid-1990s to the mid/late 2000s), can provide acceptable picture quality while reducing the required bit rate by a factor of between 50:1 and 100:1. The latest compression systems can provide even higher ratios.

Delivery to the home is the obvious imperative for video compression—without this technology, digital television transmission would be totally impractical—but compression is also a vital technology for storage of digital video. A two-hour movie, even in standard definition, represents well over 100 GB but can be compressed to less than 9 GB for a DVD. Personal video recorders (PVRs) use compression to store many hours of video on standard PC hard drives at very low cost. In the professional areas of production and program storage, lower compression ratios are used, but video compression is still the enabling technology for all current digital videotape formats, nonlinear editors, playout servers and digital archives.

Compression systems really are *systems*: complex combinations of a variety of tools that operate in quite different ways. This chapter will discuss the various approaches to compression, the tools that implement these, and the ways that tools are combined into complete systems. In most applications, video compression is used to provide an appropriate balance between compression ratio and picture quality. The term "appropriate" is very important, as diverse applications and the various steps in the production/delivery chain call for very different approaches to compression. In all cases, the objective is the same as discussed at the beginning of this introduction, that is, maximum savings in bandwidth/bit rate for a loss of quality that is acceptable in the specific application.

OVERVIEW

Terminology

In a practical application, there are three major elements to a system that uses compression. The *encoder* takes a standard video signal as its input and provides a compressed *bit stream* that is sent to a transmission or storage system. The bit stream is then sent to a *decoder* that reconstructs the video. The video from the decoder will likely be of lower quality than the input to the encoder. Terminology varies, but most people differentiate between picture *losses*, such as a loss of sharpness, and *artifacts* that are visible effects in the output picture that were contributed by the compression system. The most obvious example of an artifact is the "blockiness" often seen in stressed MPEG systems.

Lossy and Lossless Compression

There are two fundamentally different parts of practical compression systems. The first is to ensure that only necessary information is transmitted (or stored); the second is to code that information in the most efficient way possible. In the video world, compression is usually thought of as a lossy process; it is expected that the output of an encoder will not be an exact match to the input. Lossy compression works by approximating the image in a way that can be transmitted with fewer bits while minimizing the loss *as perceived by the human visual system* (HVS). This is a critical factor. An image compression system must be designed to produce the desired quality as perceived by a viewer under a particular set of assumed viewing conditions. This is the first step described above—ensuring that only necessary information is transmitted. This part of the process may also be described as *eliminating visual redundancy*.

Lossless compression refers to techniques that reduce the number of bits required by coding the information in more efficient ways but always guaranteeing that the output of the lossless decoder will be absolutely identical to the input to the lossless encoder. Lossless compressors can themselves be systems employing a number of tools. Well-known examples are the *PK Zip* and *StuffIt* programs used on personal computers. This process of lossless compression, also known as *entropy coding*, can also be viewed as eliminating redundancy. It substitutes nonredundant coding (or less redundant coding) for inefficient (redundant) coding of information; however, within this article the term "redundancy" will be used to refer to visual elements; improving coding efficiency will be referred to as just *lossless compression*.

Lossless compression is rarely used alone for video because it can provide only a relatively small degree of compression (2:1 is regarded as typical); however, after lossy techniques have been used, lossless compression is generally applied to the resulting bit stream to improve the efficiency of the overall system. The distinction is very important. As described above, the first step is to remove, as far as possible, all redundancy in the representation of the image. The necessary principles and techniques will be discussed later in this chapter. However, if this step is performed efficiently, any subsequent error will remove nonredundant information with unpredictable results, so it is essential that the following steps be, so far as is practical, lossless.

Spatial and Temporal Compression

Video is a sequence of fields or frames, each of which is a still image. Spatial compression treats each image separately and compresses it without reference to any other image. This is straightforward, relatively simple to implement, and the preferred solution for some applications. However, video is not a sequence of unrelated images—in any real video sequence there is obviously a great deal of similarity between adjacent or nearby frames. This means that a spatial-only system will necessarily send similar or identical information many times and cannot be the most efficient approach. There is still redundancy in the signal. Temporal compression takes advantage of the real-world similarity of frame sequences to improve the compression ratio.

Image Types

Compression systems used for video are based on the assumption that the information to be coded represents real-world "photographic" images. This is important in a number of ways. As explained below, the initial conversion from analog to digital must be performed correctly to avoid aliasing. In fact, all processing steps must be designed carefully to prevent artifacts such as excessive overshoots and clipping. Noise is inherently unpredictable and, therefore, generally uncompressible. High-quality encoders will often include sophisticated processing to remove as much noise as possible prior to compression. Graphics systems must be selected carefully. Simple systems can produce very fast edges that can exceed the bandwidth of the television system. Filtering these edges can result in clipped overshoots that can substantially impair the performance of a compression system.

The Human Visual System

The human visual system, or HVS, is the term used to describe the combination of eye, nerves, and brain that allows us to see. It is a remarkable and complex system (still not fully understood), but it does have quantifiable limitations, and these are exploited by image compression systems. The most obvious parameter is limited acuity. There is a limit to the smallest details we can see—generally about 20 cycles/degree; however, the design of television systems already takes this into account. The number of lines in a television system is chosen so the line structure is not visible at the designed viewing distance; for example, NTSC is designed for viewing at a distance of six to seven times picture height, and the 480 active lines correspond to about 25 cycles/degree at this distance.

Compression systems use another, related, limitation of the HVS. The figure quoted above, 20 cycles/degree, represents (approximately) the absolute limit of vision. These spatial frequencies can be resolved only under optimum conditions and at high contrast. The HVS has a characteristic known as the *contrast sensitivity function*, shown in Figure 5.11-1. The peak of HVS sensitivity is around 4 cycles/degree. Below this value, sensitivity decreases somewhat, but this effect is not generally useful for compression systems. Sensitivity falls off rapidly at higher spatial frequencies, and this effect is the most important element of lossy image compression systems. This characteristic of the HVS means that image elements that have high spatial frequency and low contrast do not need to be transmitted, because they will not be perceived by the viewer. More importantly, the decreasing sensitivity with increasing frequency means that the higher the spatial frequency the lower the precision that needs to be transmitted. Less precision means fewer bits and more efficient transmission.

TOOLS FOR SPATIAL COMPRESSION

Spatial compression is the compression of the data representing a single static image. Originally developed for still images (see JPEG section below), the techniques may be used to compress each single field or frame of a video sequence. Simple video compression systems, such as Motion JPEG, use only spatial compression.

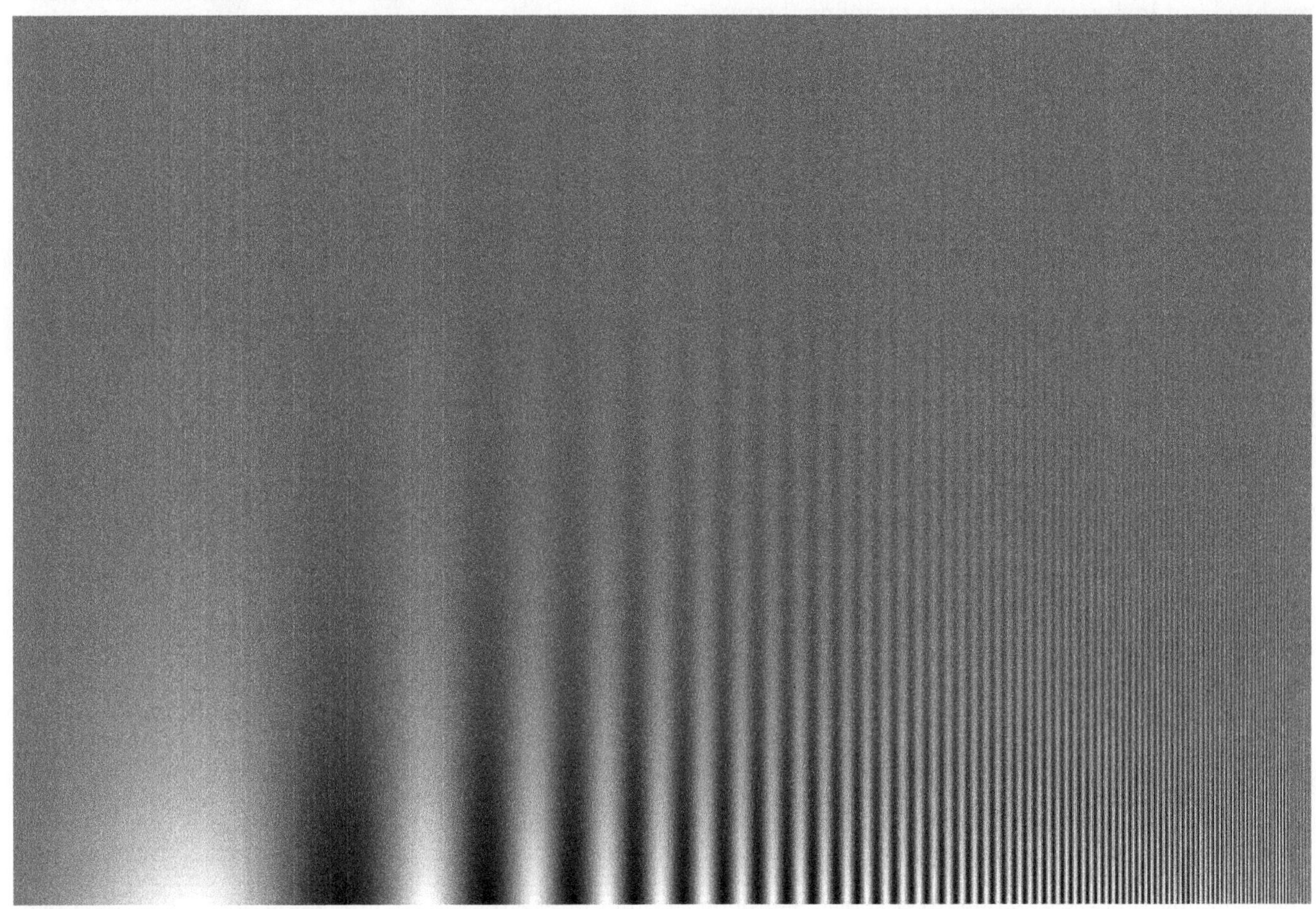

FIGURE 5.11-1 This illustration shows the increasing spatial frequency (left to right) and decreasing contrast (bottom to top) to demonstrate the contrast sensitivity function of the human visual system.

Sampling

Sampling is not really part of the compression process but is closely linked to it. Sampling is the conversion of a continuous analog signal into a sequence of discrete values, each representing an instantaneous measure of the signal. Sampling can be performed spatially and/or temporally; for example, a film movie camera samples a complete scene temporally by imaging it onto a photosensitive emulsion 24 times per second.

The sampling mechanisms are not always simple or immediately obvious. A non-shuttered camera tube produces an almost continuous analog signal that approximately repeats every frame. If this signal is sampled for digital conversion, each sample represents a unique point in space–time. In a shuttered charge-coupled device (CCD) camera, the shutter samples the scene temporally into a sequence of frames. The physical cell structure of the CCD samples each frame horizontally and vertically, but all these samples represent the same instant of time. The cell charges are then read out sequentially from the sensor in the form of an analog signal that is then sampled in an analog-to-digital converter!

Sampling is a critically important process because, if it is performed improperly, errors and artifacts will be created in the sampled signal that can never be removed. An important concept in sampling theory is the Nyquist theorem, which can be stated: "To ensure that a signal can be recovered from a series of samples, the sampling frequency must be more than double the highest frequency in the signal." This statement suggests that frequencies higher than half the sampling frequency (known as the Nyquist limit) cannot be recovered. This is true, but the problem is far worse than that. If a signal is sampled that contains frequencies above the Nyquist limit, this energy will reappear as different, lower frequencies when the signal is recovered from the samples. This effect is known as *aliasing*. A classic example of temporal aliasing is filmed wagon wheels—rotation that is too fast (frequency too high) for the sampling is reproduced as slow rotation in the opposite direction. Spatial aliasing can often be observed as moiré-like patterns in the image.

The relevance of this discussion is that the compression process will start with a sampled signal. The design of the compression system will assume that the signal has been correctly sampled; any aliasing present will, like any other "improper" signal content, impact performance. Particular care must be taken to ensure that signals to be compressed have been properly filtered prior to sampling.

Quantization

Quantization is the process of setting a value to one of a predefined set of possible values and is important in a number of ways. First, quantization is the tool used with sampling to perform analog-to-digital conversion. The sampling process was described above. Sampling alone produces a series of analog values—the values are constrained only by the precision of the sampling process; however, to generate a digital signal it is necessary to limit the possible values to a set that can be represented by the chosen number of bits. For example, with an 8-bit analog-to-digital converter, the output value of each sample must be represented by a single 8-bit word that can represent one of 256 possible values. Quantization in this example is the process of choosing one of these 256 possible values for each sample. Usually the quantizer will choose the value closest to the actual analog value measured by the sampler. So, quantization is fundamental to creating any digital signal. The number of bits used determines the precision with which the original signal can be represented. It also determines the best possible signal-to-noise ratio of the digital signal, because the quantization process allocates values that are not identical to the original signal values. This deviation is a form of noise known as *quantization noise*.

In the discussion of the human visual system, it was stated that high spatial frequencies could be represented at lower precision than the frequencies to which the HVS is most sensitive, without significant loss to perceived picture quality. Quantization is the tool used for this purpose. The section on transforms will describe how images may be represented in terms of their spatial frequency content. Quantization is the tool that permits transmission of the various frequency components at differing precisions; for example, it may be determined that it is adequate to transmit only eight possible values of a particular frequency component, so this component can be transmitted using only three bits.

Prediction

Prediction is one of the most powerful tools available to the compression system designer. It is used in many different ways but always plays a substantial role. The technique of prediction is to use some information that is already known and coded to estimate a value for the next item to be coded. As a very simple example, a prediction algorithm might estimate the value of a pixel to be the same as the value of the pixel immediately to its left, or immediately above, or some combination of these values. Sometimes the algorithm will predict correctly; more typically, the estimate will be wrong to some degree. The estimate must be compared with the actual value to determine the prediction error; however, if the prediction algorithm is a good one, it will be more efficient to code the prediction errors than to code the actual values.

Some spatial predictors are very simple, as described above. Others are more complex, using adaptive algorithms that choose a predictor based on the context created by the surrounding image. Modern codecs such as Advanced Video Coding (AVC) provide multiple predictors, and it is possible for the encoder to test all of the possibilities and choose the one that represents the lowest number of bits for the image segment in question. Most spatial predictors operate on actual pixel brightness values, but some operate on coefficient values after a transform. Prediction is also the key tool in providing temporal compression, but this application will be discussed separately.

Transforms

As discussed above, a key element of compression is the exploitation of the limitations of the human visual system, but this requires knowledge of the spatial frequency content of the image. This may be derived by examining the brightness values of multiple pixels, but individually the pixel values provide no spatial frequency information.

A *transform* is a mechanism for changing the representation of information. The most familiar transform to television engineers is the conversion between RGB and YUV color spaces. Each representation consists of three values that together specify a single color. RGB and YUV are two different representations of the same thing, and with sufficient precision the transform is reversible to any chosen degree of accuracy; however, the two representations have substantially different properties. The UV signals may be frequency limited without substantial damage to the image, and this is a fundamental requirement for color systems such as NTSC and Phase Alternating Line (PAL).

For video compression, there is a need to represent image information in a form permitting exploitation of the contrast sensitivity function of the HVS. The transform must take pixel brightness values for all or part of an image and transform the representation of that information into a set of coefficients that represent the horizontal and vertical spatial frequencies in the image. In this case, a block of brightness values is transformed into a block of frequency coefficients.

Most compression systems use some version of the *discrete cosine transform* (DCT). The principal exception is the JPEG 2000 Standard, which uses the *discrete wavelet transform* (DWT), and this transform will be discussed in the JPEG 2000 section. DCT is a derivation of the Fourier transform, long used for frequency analysis. The DCT used by compression systems is a two-dimensional transform, usually operating on an 8 × 8 block of pixels. The transform takes 64 pixel brightness values and returns 64 spatial-frequency coefficients. Each coefficient represents the value of one combination of horizontal and vertical frequencies. The possible combinations are shown in Figure 5.11-2. The top-left component in the illustration is used to represent the DC (average) value of the pixel block. To the right of this are seven components representing various horizontal frequencies ranging from 1/4 cycle to about 3–1/2 cycles across the block of pixels. Down the left-hand side of the illustration the same vertical frequency components appear. All of the rest of the frequency components represent some combination of horizontal and vertical frequency.

The transform is reversible. Each of the spatial frequency components, known as *basis functions*, can be represented by a block of 64 pixel values forming a pattern; for example, all 64 values for the top-left DC component are the same; the 64 values of the bottom-right form a checkerboard pattern. The coefficients generated by performing the DCT transform represent the quantity (positive or negative) of each of these patterns that must be added to get the original 8 × 8 pattern of brightness values.

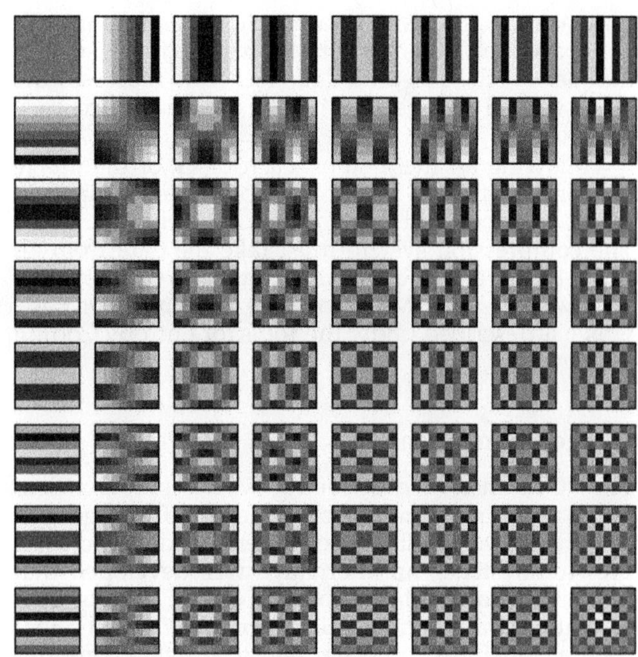

FIGURE 5.11-2 A representation of the 64 basis functions of the DCT transform.

As stated above, a transform creates a new representation of the same data. The DCT transform takes 64 brightness values and returns 64 coefficients. With adequate precision, these 64 coefficient values may be reverse-transformed to yield the original set of brightness values. "Adequate precision" means generally that the coefficients must be represented with about three more bits than the brightness values; transforming 8-bit brightness values requires 11-bit coefficients for reversibility. So, the transform has actually created more data! However, when the coefficients resulting from the transform of a real image are examined, it is found that many of the coefficients are zero, or very close to zero, and can be discarded without significant impact. Because of the HVS contrast sensitivity function (CSF), low values of high-frequency coefficients can also be discarded because these represent low-contrast, high-frequency information that is below the CSF threshold. The higher the spatial frequency represented by the coefficient, the higher the threshold below which that coefficient may be discarded. Even for coefficient values that cannot be discarded, many may be quantized more coarsely, thus using fewer bits.

In practice, these steps are usually combined. In one implementation, a weighting factor is assigned to each basis function; small values are used for very significant frequencies, large values for frequencies where the HVS is less sensitive. Each coefficient is then divided by its weighting factor and the result rounded down. This results in small coefficients being zeroed and others reduced to a value within a small range that is represented by a low number of bits.

Irrespective of any weighting, an overall quantization factor (an additional divider, uniform across all the coefficients in the block) may be applied. If the quantization level is increased, more coefficients will be zeroed, and others will be represented by a smaller number of bits. The total data needed to represent the block will be decreased but at the expense of accuracy and quality.

Following all these steps, each block of the image is represented by a block of 64 coefficients, many of which are zero and some others coarsely quantized. The compression system must take the data from all of the blocks and code the information in the most efficient way possible. Many tools are used in this process, which is described under the Tools for Entropy Coding section.

Integer Transforms

The DCT transform described above is a complex calculation requiring many cosine values and floating-point arithmetic. During the development of advanced compression systems, an alternative was developed that implements a rough approximation to DCT but uses just integer values in the calculations. It was found that such a transform could be used to provide compression efficiency very similar to that of true DCT. Integer transforms are incorporated into the AVC and VC-1 systems (see below).

TOOLS FOR TEMPORAL COMPRESSION

Prediction

As discussed above, video is not a sequence of unrelated images. The content of a frame is usually very similar to the preceding frames. If one frame of a sequence has been compressed and transmitted, this frame is already present at the compression decoder when we send subsequent frames. It should be possible to predict the content of some areas of a subsequent frame and to use the image content already transmitted. For static scenes, this is simple, but it is more complex if objects in the scene move or if the camera moves, thus changing the position in the image of background content. Even when the camera pans and objects or actors move, however, there is typically considerable content in common, albeit in a different location in the image. It should still be possible to benefit from prediction if we can gather information about where content has moved. This is the role of *motion estimation.*

Frame Types

Most compression systems that perform temporal prediction use *reference frames*, or *intra-frames*. An intra-frame is one that is coded without reference to any other image, using just the spatial compression tools described above. Reference frames may be intra-frames, but predicted frames may also be used for reference (see below). A *predicted frame* or *inter-frame* is coded, so far as possible, by prediction from one or more reference frames. The reference frame is transmitted, and at the receiving end it is decoded and stored. The encoder also incorporates a decoder so it can store a reference identical to that stored in the decoder (not the same as the original image because the compression is lossy). The new frame to be coded is compared to the stored reference frame. When a valid prediction is found, the decoder is instructed to use the appropriate data from its copy of the reference frame.

Motion Estimation

The most obvious case where prediction is possible is in a sequence where the camera has not moved so the background is static from frame to frame. In this case, the encoder can instruct the decoder to extract unchanged blocks from the stored reference frame and place them in the same position in the new frame. More generally, of course, there may be content common to the reference frame and the frame to be predicted, but likely not in the same position in the image. Objects in the scene may have moved, or the camera may have moved. Prediction using these elements is much more difficult but is essential for efficient temporal compression.

Motion estimation is the process of searching a reference frame for image content that matches the content in the predicted frame that needs to be encoded. This is the most complex part of an encoder, and the range and accuracy of motion estimation are major factors determining encoder performance.

Motion estimation is generally based on the *macroblock*—usually 16 × 16 pixels. A motion estimator takes a macroblock and attempts to find an area in the reference frame that is a close match in content. When a match is found, a *motion vector* is generated. The motion vector is a pair of values representing horizontal and vertical offsets; it is used to instruct the decoder where the matching content is to be found in the stored reference frame. Although the content that is being searched *for* is determined by the macroblock boundaries on (usually) a 16 × 16 pixel grid, the matching content generally is not found on this grid. For efficient motion estimation, the system must search to very high precision, varying from 1/2 pixel resolution in most systems to 1/4 pixel in more advanced systems. An exhaustive search is, therefore, an enormous computational task.

The performance of an encoder is closely linked to the efficiency of its motion estimation. It is obvious that the more content that can be matched to image information already at the decoder, and the more accurate the match, the better the system will perform. Less obvious is the fact that it is important to find the right match. There may be many places in the stored reference frame that are a good match for the macroblock being coded, particularly for image content such as blue sky; however, once the match is found, the motion vector has to be transmitted to the decoder. Each motion vector is a pair of numbers, each perhaps 10 or 12 bits long, so these must be coded in an efficient manner or much of the benefit of temporal compression will be lost.

In image areas where an object is moving (or the background is apparently moving because of a camera pan or tilt), there should be a high correlation between motion vectors of adjacent macroblocks; in fact, for linear motion, the motion vectors will be identical. This is an ideal situation for predictive coding (again). After initialization, instead of sending the motion vector values, the coder will send just the differential from the previous macroblock, quite likely zero. However, if the motion estimator were to select suitable matches (for blue sky, for example) from anywhere in the reference frame, all correlation would be lost, and the predictive coding would not be effective.

Residual Coding

In practice, the motion estimator will usually find excellent matches for some macroblocks, marginal matches for others, and no useful match for the remainder. Where the match is sufficiently good, only the motion vector needs to be transmitted; where there is no useful match, that macroblock has to be coded using just the spatial tools. When the match is marginal, some benefit can still be obtained by subtracting, pixel by pixel, the reference macroblock from the macroblock being coded. This array of error values is known as the *residuals* and can be encoded using the same spatial tools. If the match is reasonably good, it is likely that the total bits required for the motion vector and residuals will be less than required to code the entire macroblock spatially.

I-, B- and P-Frames and the GOP

Temporal compression requires image content to be stored at the decoder to be used as a reference. Usually complete frames are used for reference. A system generally will use three types of frame. The simplest type of frame is the I-frame, or *intra-frame*. Intra-frames are coded using only spatial tools and the information within that frame; no use is made of any information outside the frame being coded. An I-frame will generally be used as the first frame of a sequence, providing a reference frame. I-frames are also inserted, usually at regular intervals, to prevent the propagation of coding errors.

A P-frame, or *predicted frame*, is coded using predictions from a previous I-frame or P-frame. Macroblocks that cannot be predicted are coded using the spatial tools. B-frames are coded using two reference frames, one from earlier in the sequence and one from later in the sequence. This can only be implemented if both reference frames have been transmitted to the decoder, so the use of B-frames requires reordering of the frames for transmission and introduces significant delay. The reference frames may be I-frames or P-frames; in most systems, B-frames are not used as reference frames.

The *group of pictures* (GOP) represents the structure or cadence of the coded sequence. In typical applications, a GOP may have a length from a few frames to a few tens of frames. For transmission channels such as terrestrial broadcast with a relatively high error rate, a GOP length of about 1/2 second is common. A typical GOP structure for such an application might be as follows:

I-B-B-B-P-B-B-B-P-B-B-B-P-B-B-B-[I-B-B-B-P-. . .]

Such a sequence is described as *long GOP*. Systems may use a *closed GOP* structure, where all references (for temporal compression) are within the same GOP, or *open GOP*, where references are permitted across GOP boundaries. GOP structures are usually flexible; the chosen structure is used when there is no reason to change it, but an event such as a scene change may cause the encoder to insert an extra I-frame or restart the GOP cadence (see also the Rate Control section below).

TOOLS FOR ENTROPY CODING

The final step in a compression encoder is to take the result of all the processes described above and to reduce this information to the minimal number of bits for transmission. This process has to be lossless; otherwise, the data fed to the decoder would be meaningless. The entropy of a set of data is a measure of its lack of predictability. If the coding is to some degree inefficient, this means that there is some predictability (or redundancy). The coding will use more bits (per pixel, per frame, for example) than the calculated entropy of the data. In general, entropy is expressed in bits per symbol, where a *symbol* is any one of the possible values that might be transmitted (for example, for an alphanumeric data stream, a letter or number would be a symbol). Shannon, the mathematician responsible for much of modern communication theory, stated that data cannot be represented in any coding scheme by fewer bits than its entropy. The job of the entropy encoder is to approach this limit as closely as possible, within appropriate constraints of cost, delay and computational resources.

Run-Length Coding

This is the simplest tool. If a value is repeated many times, it is obviously more efficient to code this information by a scheme that says, in effect, "Take value *m* and repeat *n* times." Sometimes the data must be manipulated to maximize the gain available from run-length coding. During the development of the original JPEG system, it was found that, after quantization, the transform coefficients of a typical block included a large number of zeroes. It was discovered that reading out these coefficients in a diagonal pattern (rather than horizontally or vertically) resulted in a stream that usually contained long runs of adjacent zeroes—suitable for run-length coding.

Variable-Length Coding

Many coding schemes such as ASCII use a fixed number of bits for each item to be coded. Variable-length coding is used when the statistics of a data set are known; it uses small codes for common values and longer codes for infrequent values. A simple example of a variable-length coding scheme is Morse code, which is (crudely) optimized for English language text. Common letters such as *e* and *t* are represented by short codes; less frequently used letters such as *q* and *w* are allocated much longer codes. Variable-length coding is particularly useful following prediction. In a predictive coding system, the prediction error (rather than the actual value) is transmitted. If the prediction is good, there will be many more small errors than large ones, and the most common errors can be represented by very short codes. A system used in several standards that allocates variable-length codes in an efficient manner is referred to as *Huffman coding*.

Arithmetic Coding

Techniques such as Huffman coding can usually compress in lossless fashion to a number of bits close to the calculated entropy of the data, but are limited by the fact that each coded symbol has to be transmitted using a whole number of bits. Some systems mitigate this effect by coding combinations of symbols, but at the expense of much higher complexity.

Arithmetic coding is a family of technologies originally developed by IBM for facsimile transmission. Like Huffman coding, the techniques are based on an analysis of the statistics of the data and the probability of occurrence of each of the possible different symbols. Arithmetic encoding transmits a code stream that represents a binary fraction of the interval between zero and one. The operation is recursive: When each symbol is received, the precision of the fraction is increased by adding more bits to the code stream, effectively coding the cumulative probability of the input sequence. Decoding is also recursive and is performed by analyzing what the encoder must have received to generate the successively finer granulation of the interval.

The principal advantage of arithmetic coding is that the techniques effectively permit fractional bits to encode symbols. Also, arithmetic encoders can adapt on the fly to the statistics of the data, rather than relying on a prebuilt table of code words. The principal disadvantage is complexity; the techniques are relatively demanding in hardware implementations and are difficult to implement efficiently with general-purpose processors.

RATE CONTROL

Rate control is the science and art of using all of the tools discussed above to achieve the best compromise of bit rate *versus* quality. In most systems, the actual control mechanism is very simple—it is just the quantization level applied to each block of data being encoded. The algorithms used to decide the appropriate quantization at any point, however, may be extremely complex. The objective is generally to compress the input video to a given number of bits per second while maintaining the highest possible subjective quality. The term "subjective" is very important; heavily compressed images will contain artifacts, but a good compression system will ensure that the artifacts are as invisible as possible.

Many factors affect visibility of artifacts. Many artifacts are less visible in busy parts of the picture—areas with considerable detail or of rapid motion. The human visual system is very good at integrating over time, and it is generally possible to quantize B-frames more coarsely than reference frames without significant perceived impairment. Sophisticated encoders will use "look ahead" techniques to determine the complexity of upcoming frames in advance of starting the actual compression process. This allows the encoder to use fewer bits on simple frames, reserving more for the complex frames.

Statistical Multiplexing

Conventional rate control mechanisms aim to achieve the best possible quality within a given permitted bit rate. In television distribution and delivery, the bit streams of several programs are commonly combined into a *multiplex*. Early systems used a fixed allocation of bits from the total stream for each program, but efficiency can be improved to a remarkable degree by statistical multiplexing. Video programs vary enormously in their bit-rate requirements. "Talking head" static shots contain little information and may be heavily compressed, whereas a sports event such as basketball contains a great deal of complex motion and is a severe challenge to compression systems. Bit-rate requirements vary not only by program genre but also with time in a single program. Statistical multiplexing uses these variances to fit more programs into a multiplex while maintaining adequate quality. It relies on the fact that not all programs need their maximum number of bits at the same time. In a statistical multiplexing (*statmux*) system, one encoder is provided for each program; each encoder measures the complexity of its video and requests an appropriate bit rate from the statmux controller. The controller aggregates the requests, applies any priority rules that may be in force, and tells each encoder the actual number of bits it will be allowed.

PUTTING IT ALL TOGETHER

Figure 5.11-3 shows the elements of a spatial encoder and decoder. Only the luminance signal path is shown. Each frame is split into blocks, and the blocks are transformed, producing transform coefficients. These are weighted and quantized, and the resulting sparse set of coefficients is scanned, run-length encoded, then passed to the (lossless) entropy coder. The stream is monitored by the rate controller, which determines the appropriate level of quantization.

In the decoder, the entropy encoding and run-length coding are reversed, generating a set of values identical to that output by the quantizer. The coefficients are then de-quantized. (Note that this creates an approximation to the original magnitudes, but the precision discarded by the quantizer cannot be recovered.) The recovered coefficients are then subject to the reverse transform, resulting in a close approximation to the original image.

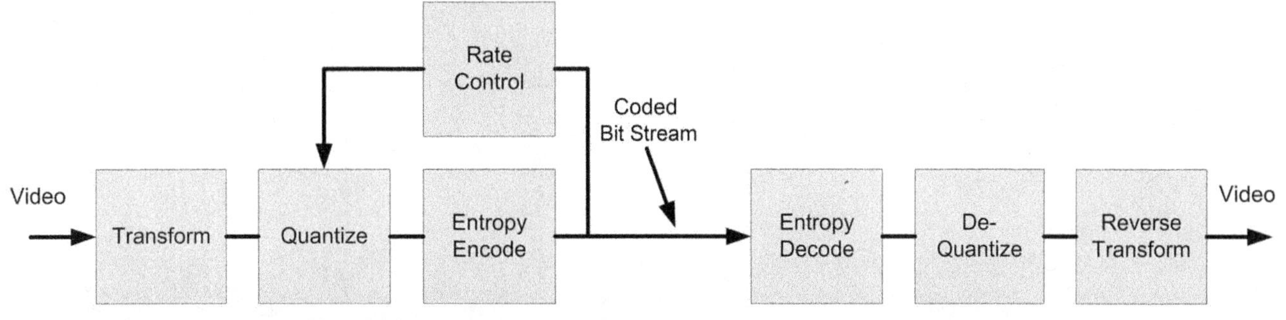

FIGURE 5.11-3 A simple spatial encoder and decoder.

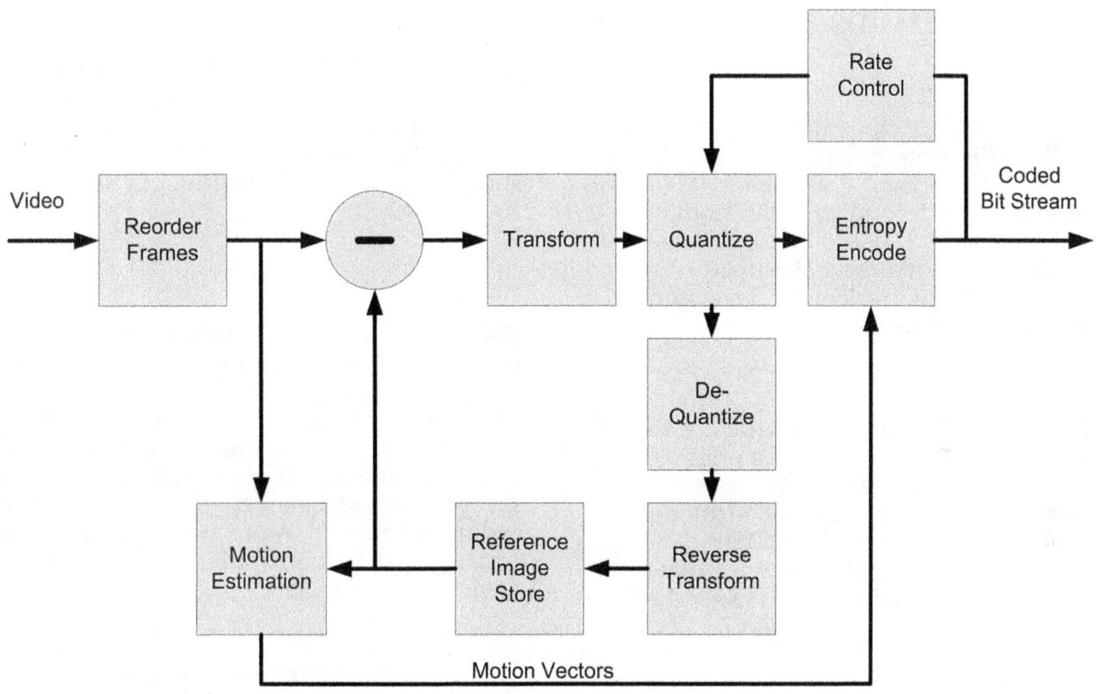

FIGURE 5.11-4 A simplified encoder with motion estimation and prediction.

Figure 5.11-4 shows a simplified diagram showing the basic elements of an encoder with motion estimation and temporal compression, such as an MPEG-2 encoder.

Many of the elements are the same as for the spatial encoder, but a prediction loop is added. Quantized coefficients are de-quantized and reverse transformed. This generates the same image (an approximation to the original) as will be generated within the decoder. This reference image is stored, and is available for comparison when subsequent images are coded. When a match is found, the reference block is subtracted from the block being coded, and any residuals are transformed and quantized. A motion vector is generated to point to the block used for prediction, and the motion vectors are multiplexed with the rest of the encoded bit stream. Where no useful prediction is found, blocks are encoded spatially, just as in the previous description. In the decoder (Figure 5.11-5), reference frames are stored, and motion vectors are used to fetch predicted areas from the store. Where there is no motion vector, the transmitted values are treated as spatially encoded, just as in the spatial decoder.

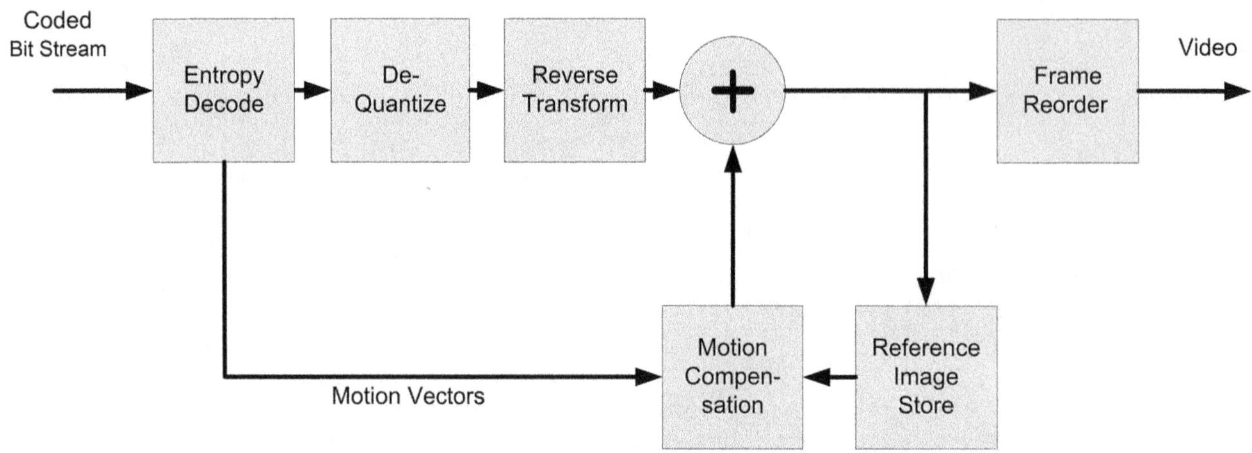

FIGURE 5.11-5 The simplified decoder corresponding to the encoder of Figure 5.11-4.

THE STANDARDS BODIES

ISO/IEC

The International Standards Organization (ISO) and the International Electrotechnical Commission (IEC) formed a Joint Technology Committee (JTC-1) to develop standards for information technology. One of the subcommittees of JTC-1 is SC29, charged with image technology. Two Working Groups (WG1 and WG11) are responsible for much of the development of modern compression systems.

JPEG

JTC-1/SC29/WG1 is the Joint Photographic Experts Group (JPEG). The development of the original JPEG standard for still pictures created the basis for most modern image compression schemes. Although the ISO version of JPEG was not published until 1994 (one year after MPEG-1), the early work begun around 1983 created the spatial tools used in both JPEG and MPEG standards.

MPEG

JTC-1/SC29/WG11 is the Moving Picture Experts Group (MPEG). This group was established in 1988, and the MPEG-1 standard was published in 1993.

ITU

The International Telecommunication Union (ITU) is a treaty organization covering all aspects of telecommunications; it is divided into ITU-R for radio (wireless) communications, and ITU-T covering (generally) wired communications. Part of ITU-T is the Video Coding Experts Group (VCEG), which has developed video compression systems for video conferencing and so forth. ITU-T has also published standards such as JPEG and MPEG-2 as ITU-T Recommendations.

JVT

In 2002, ITU-T and ISO/IEC agreed to formalize the existing cooperation and formed the Joint Video Team (JVT) to develop compression standards for publication by both organizations. JVT developed the AVC standard known as both Rec. ITU-T H.264 and ISO/IEC 14496-2 MPEG-4 Part 10.

SMPTE

The Society of Motion Picture and Television Engineers (SMPTE) is a body that focuses on standards for moving images. SMPTE does not generally engage in research to optimize compression systems but ensures that coding specifications are accurate and complete to permit interoperability. SMPTE has standardized several mezzanine-level (lightly compressed) video coding schemes that are useful for contribution and primary distribution applications, under the common names VC-1 to VC-5.

COMPRESSION SYSTEMS

JPEG

JPEG (now often referred to as JPEG-1 to distinguish it from JPEG 2000) was intended for static images. It is a block-based system using DCT and Huffman variable-length coding.

Motion JPEG

The compression provided by JPEG was more efficient than anything else available at the time, and a number of manufacturers recognized the benefit of coding video as a sequence of static images, each JPEG coded. This technique became known as *Motion JPEG*. However, the JPEG committee did not provide any mechanism for rate control in this application or any other tools for coding a sequence. Unfortunately, this resulted in each manufacturer developing a proprietary scheme, and Motion JPEG streams are not generally interchangeable.

DV

This compression system was developed by a consortium of manufacturers for use in consumer camcorders. It uses the DCT transform followed by quantization and a complex system of priorities and ordering to ensure that standard-definition video can be encoded at 25 Mbps, with a constant number of bits per frame, suitable for recording on tape. Encoding and decoding are performed within the same single chip. The performance of DV exceeded expectations, and it has been widely applied in professional environments, particularly in camcorders for news broadcasts. A professional variant was created, using two chips and 50 Mbps, for applications requiring higher quality.

MPEG-1

The first compression standard specifically intended for video was developed between 1988 and 1992 and was aimed specifically at the Video CD application. MPEG-1 encodes CIF Video (352 × 240) at 30 Hz and two-channel audio all within a bit rate of 1.5 Mbps. MPEG-1 uses motion prediction and then DCT for transforming blocks that could not be predicted, or residuals of predicted blocks. After quantization, Huffman coding is used for entropy coding. MPEG-1 also established a new standardization philosophy. The standard defines the *syntax* used in the bit stream, and exactly what the decoder must do in response to the bit stream. The standard does *not* specify the encoder—anything that produces a legal bit stream that can be decoded according to the standard is a legal encoder. This approach permits a great deal of evolution in encoder design and substantial improvements in performance, without changing the standard or the decoders in the field.

The importance of MPEG-1 should not be underestimated. Although designed for an application that had only a short life, it established the foundation for the video compression systems in use today. It also included the MP3 audio compression scheme that is still dominant in many fields. The syntax created for MPEG-1 was so robust that, prior to the availability of MPEG-2, it was extrapolated for use in satellite television and in an early high-definition system.

MPEG-2

Work started on MPEG-2 even before completion of MPEG-1. MPEG-2 video was a joint development between MPEG and VCEG. It was realized that there were important applications for compression in broadcast television, but that these applications required coding of larger pictures and provision for coding interlace with reasonable efficiency. The standard is based heavily on the concepts of MPEG-1, but with many extensions and improvements. MPEG-2 introduced *profiles* and *levels*. Profiles define a set of tools that may be used—more precisely, syntax elements that a compliant decoder must be able to process. A level defines a set of parameters, such as maximum bit rate, decoder memory requirements, and so forth, that size a decoder for a specific application. A *conformance point* is a combination of a profile and a level. In MPEG-2, a decoder for a particular conformance point must decode bit streams from all lower profiles and levels. For example, a common conformance point is the *Main Profile at Main Level* (expressed as MP@ML), suitable for standard-definition television and used in DVD players. The *Main Profile at High Level* (expressed as MP@HL) may be used for high-definition television, as in the ATSC television system. An MP@HL decoder must be able to decode MP@ML bit streams, but not vice versa.

MPEG-2 was the dominant video compression scheme from the mid-1990s to nearly the end of the 2000s, and it is still widely used at the time of the writing of this publication. It is the basis of the current digital television transmission systems (ATSC, DVB, and ISDB) and is used for satellite and digital cable television distribution. It is also used for video distribution on DVDs, and a decoder is present in almost every personal computer. There are well over 1 billion MPEG-2 decoders, MP@ML or higher, in the field.

The MPEG-2 standards also included very robust systems-level tools that have been used by later standards and diverse applications. The *MPEG-2 Transport Stream* is the most widely used mechanism for transmitting synchronized audio, video, and ancillary data in a single bit stream.

MPEG-4

Many people consider that MPEG-4 is where the committee lost its way. The set of standards included an enormous array of techniques such as facial and body animation schemes, but none of these has achieved widespread employment. MPEG-4 audio has been adopted in some areas, most notably in the Apple iPod. MPEG-4 did include a range of new video compression schemes, collectively known as *MPEG-4 Part 2*, and offered some performance improvement over MPEG-2 (perhaps 15%), but in most applications this improvement was not sufficient to overcome the benefit of the large number of MPEG-2 decoders already deployed.

AVC (MPEG-4 Part 10, H.264)

Both MPEG and VCEG recognized that there was a need for a video compression system substantially more efficient than MPEG-2 and that they should focus their efforts in this direction. The two organizations formed the Joint Video Team (JVT) to produce such a standard, to be published by both organizations. This accounts for the many names of the resulting standard. Officially, it is both Recommendation ITU-T H.264 and ISO/IEC 14496-2 MPEG-4 Part 10.

AVC is very efficient. It is able to achieve up to approximately twice the coding efficiency of a sophisticated MPEG-2 encoder. This is a sufficient advantage to make it worthwhile commercially to upgrade, when the possibility arises to change decoders, or to use as the initial codec in "greenfield" deployments, where no legacy exists. One additional factor must be considered for its use, however, for terrestrial broadcasting: such usage is regulated by governments. For example, in the United States, MPEG-2 MP@HL decoders are mandated for digital TV transmission (so called ATSC 1.0) and it would take a change by the Federal Communications Commission or Congress to migrate to a new format. In countries such as Australia, where the Digital Video Broadcasting (DVB) transmission system has been adopted for high definition, using MPEG-2, the same applies. In Europe, however, existing DVB decoders are standard-definition only. As such, AVC has been adopted for high-definition services, requiring a new generation of receivers and decoders. Brazil has adopted the Integrated Services Digital Broadcasting (ISDB) transmission system (originally in Japan only) but will use it in a unique mode, employing AVC for compression.

The initial version of AVC published in 2003 (Main Profile) initially faced a serious challenge from VC-1 (see below) because the latter appeared to be more efficient with encoding large images such as HDTV. However, the JVT produced an amendment in 2004 that added *Fidelity Range Extensions* (FRExt), which provided better tools for large images and again is becoming the preferred choice for most applications, including a new High Profile for direct-to-consumer applications that improved coding efficiency on par with or exceeding VC-1. AVC has since been adopted in many applications, including Blu-ray Discs and "over-the-top" Internet-based streaming services.

AVC uses tools similar to MPEG-2, but with many refinements and improvements. The transform is an integer approximation to DCT, and spatial compression is aided by a sophisticated spatial predictor. Motion estimation is much more sophisticated. One major improvement is the ability to define a global motion vector, improving efficiency considerably for camera pans and tilts. The basic unit for prediction is still the macroblock, but to track accurately the edges of complex moving objects a motion vector may be applied to blocks as small as 4 × 4. AVC improves on the prediction of B-frames and can use multiple reference pictures for even more capable prediction. The system can use an entropy encoder called *Context-Adaptive Variable-Length Coding* (CAVLC), an enhancement of Huffman coding, but, for greatest efficiency, an alternative entropy encoder, *Context-Adaptive Binary Arithmetic Coding* (CABAC), is employed. CAVLC is a significant improvement on Huffman coding, but CABAC improves by a further 10% to 15% in many applications. Finally, AVC includes a deblocking filter that masks the very obvious blocking artifacts seen when MPEG-2 systems begin to fail.

The greatest challenge of AVC is its computational complexity. Decoding is about three times more complex than for MPEG-2. Encoding, as with any MPEG standard, is greatly variable, but a high-efficiency AVC encoder may be perhaps ten times as complex as a good MPEG-2 encoder.

VC-1

In 2003, Microsoft Corporation decided to freeze the bit stream definition for the video coding in Windows Media 9 and publish this as a standard to permit implementation by multiple vendors. It was standardized in SMPTE as VC-1. VC-1 is similar in many ways to AVC, except that the tool set is less comprehensive and was chosen to reduce computational complexity. Like AVC, VC-1 uses an integer transform. It has a deblocking filter, but it is of a simpler design than in AVC, and it uses variable-length coding rather than an arithmetic coder. The result is a compression system that is not quite as efficient as AVC but with a decoder that is much easier to implement in software on a general-purpose processor (which was a serious concern at that time based on the capabilities of general-purpose processor technology).

VC-2

SMPTE VC-2 was originally a proposal from the British Broadcasting Corporation for a simple compression system (known as Dirac, in honor of theoretical physicist Paul Dirac) specifically designed to compress 1920 × 1080 50 or 60 Hz progressive (requiring 3 Gbps uncompressed) by a factor of 2:1, permitting transmission over a standard 1.5 Gbps high-definition serial digital interface. The original integer DCT-based design has been abandoned, and the latest proposed variant uses a wavelet transform. The full proposal now includes a wide range of capabilities, including the combination of block-based motion prediction with a wavelet transform. The intent is to have an "open source" (and believed royalty-free) compression system suitable for use throughout the production and delivery chain.

VC-3

SMPTE VC-3 is the compression system originally proposed by Avid Technology, Inc., based on its DNxHD (Digital Nonlinear Extensible High Definition) compression schemes for use in editing environments. It uses a lot of techniques similar to JPEG, including DC prediction and variable-length codes with run-length encoding and independently encoded frames only; however, it is based on the DCT transform as opposed to JPEG's wavelets.

VC-5

SMPTE VC-5 is the standardization of the Cineform codec, which was designed for intermediate workflows for film and TV applications. It is based on an integer reversible wavelet compression, with a nonlinear quantizer to achieve higher compression. Compression data rates typically range from 10:1 to 3.5:1.

JPEG 2000

JPEG 2000 is the successor to the original JPEG standard and is also intended for still images; however, this time provision has been made for a standardized, interoperable mechanism for rate-controlled compression of a video sequence.

This system used a discrete wavelet transform (DWT) rather than a transform from the DCT family. The wavelet transform is essentially a set of complementary high- and low-pass filters that operate on the image, both horizontally and vertically. This produces high (H) and low (L) horizontal frequencies, and high and low vertical frequencies. The filtered image is separated into four parts or quadrants, one with just high vertical and horizontal frequencies (HH) and the other combinations (HL, LH, and LL). Because of the filtering, alternate samples from each of these parts may be discarded, so each of the four quadrants has 1/4 of the number of pixels of the original image.

Three of the image parts (HH, HL, and LH) are wavelet transform coefficients and are quantized in a manner similar to that used in DCT systems. The LL image is treated differently. It is a quarter-sized (1/2 × 1/2) version of the original and can be filtered again by the same filters, resulting in four new quadrants. Again, the resulting LL is a smaller version of the original (now 1/4 × 1/4 size). This process, called *decomposition*, may be repeated—a typical JPEG 2000 compression may use five iterations. The JPEG 2000 coder uses sophisticated techniques to predict values in one iteration from those in others and a complex arithmetic encoder as a final step.

JPEG 2000 is remarkable in that it is scalable in multiple dimensions. The bit stream that represents the compressed version of the full-size, full-quality image may be ordered in many ways. Using these techniques, it is possible to decode many potential subsets of the complete image. Using data from less than the full number of decompositions will yield a smaller image. Other techniques may give just chosen areas of the picture, or a monochrome image, or lower signal to noise. Often these techniques are used in a progressive manner to provide a small, low-quality image immediately, building to a large, high-quality image as more data is decoded. JPEG 2000 has been adopted by the Digital Cinema Initiatives (DCI) and SMPTE for digital cinema and has been used in a number of television products.

HEVC (MPEG-H Part 2, or ITU-T H.265)

The most recent video compression system standardized at this writing is High Efficiency Video Coding (HEVC). HEVC was also a product of a joint development between MPEG and VCEG, in this case called the Joint Collaborative Team for Video Coding (JCT-VC). In ITU-T it is known as Recommendation ITU-T H.265. In ISO/IEC it is known as ISO/IEC 23008–2 (MPEG-H Part 2).

After AVC was first standardized, some previous trends sharply accelerated and totally new trends appeared:

- HD content is increasing. That content needs storing, moving and delivering in a multiscreen environment. Multiscreen households now have multiple HD-capable TVs, computers, tablets and mobile phones. Consumers also expect the best picture quality possible on all their media devices.
- Telcos are now deeply involved in video content delivery and want to increase service reach in IPTV over DSL.
- Cable operators in turn are increasingly offering telephony and want to increase the number of video services in their available bandwidth.
- Terrestrial broadcasters, who have long been dealing with severe bandwidth constraints, want to offer more services, including mobile TV.
- Interest in 1080p60/50 workflow is growing, and a cost-effective business model is needed.
- The 3DTV business model continues to be debated, and there are calls for full HD 3DTV delivery economics to be improved.
- In 2003 mobile phones were still primarily traditional phones, for voice conversations. Now they are enormously more powerful and evolving fast.
- Possibly most significant is the emerging need to deliver Ultra HDTV to broadcast audiences.

The HEVC target goal was to improve coding efficiency compared to AVC by up to 50%—effectively halving the required bit rate—and this has wide ramifications for many different industries that need to get high-quality images across constrained links. See Figure 5.11-6.

HEVC became an international standard in 2013, and it has been amended with additional features in the years that followed. The HEVC codec contains a set of advanced coding tools that together make HEVC significantly more compression efficient compared to AVC. Figure 5.11-7 depicts a high-level tool comparison between AVC and HEVC.

Some key differences between HEVC and AVC include the following:

- A coding unit with block sizes ranging from 8 × 8 to 64 × 64
- Hierarchical block coding, in a tree structure, with quad-tree partitioning
- Independent transform unit and prediction unit quad-trees, with minimum block size of 4 × 4
- Non-square transforms (NSQT)

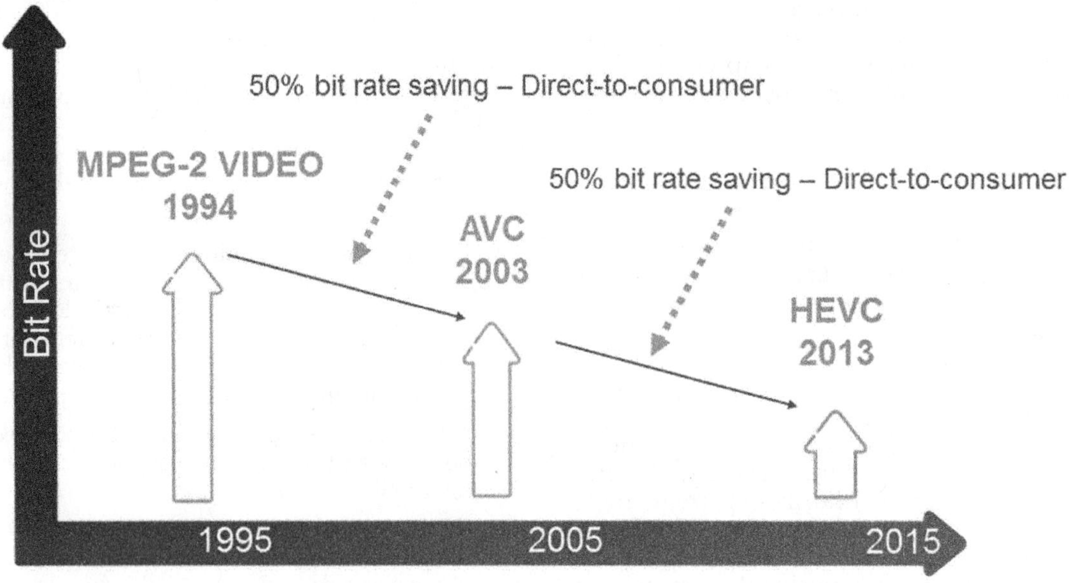

FIGURE 5.11-6 Video compression bit rate trends.

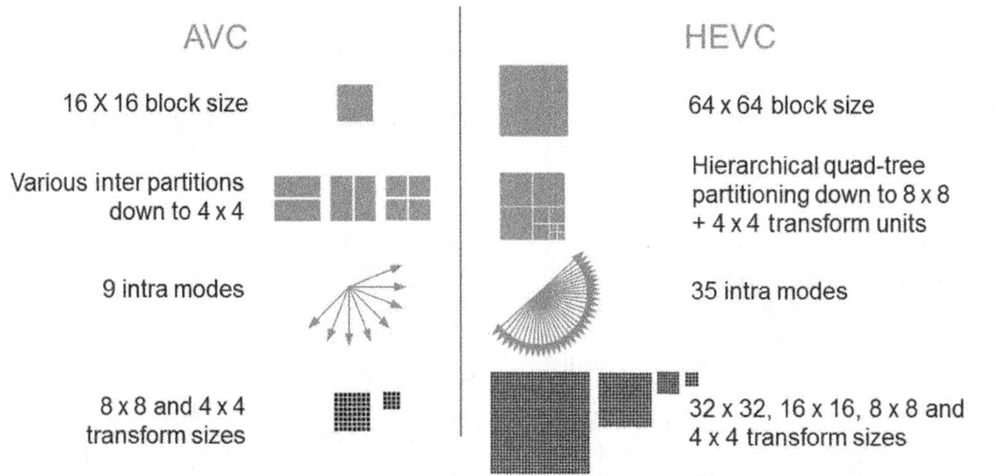

FIGURE 5.11-7 A high-level tool comparison between AVC and HEVC.

- 35 intra prediction modes
- Asymmetric motion prediction (AMP) (non-square partitions)
- Intra chroma prediction from luma
- 8-tap luma/4-tap chroma interpolation filters
- Advanced motion vector prediction
- 10-bit precision storage and output
- Sample adaptive offset (SAO) filter and adaptive loop filter (ALF)
- Parallel processing added to Context-Adaptive Binary Arithmetic Coding (CABAC)
- Additional parallel processing tools: tiles (picture partitioning into rectangular non-overlapping regions), wavefront coding and entropy slices

Table 5.11-1 lists some of the major differences between AVC and HEVC. Note that the two common profiles of HEVC for direct-to-consumer video are the Main Profile and the Main 10 Profile. The latter is a significant development in video compression because it is the first profile designed for direct-to-consumer usage that uses 10-bit sample bit depth (i.e., using 10-bit coefficients); the two HEVC profiles are otherwise identical.

One of the marvels of an HEVC coded video sequence is how little data from the original content is present. What the viewer sees is almost entirely a construct. Up to an estimated ten times more encode processing power and up to estimated three times more decode processing power than AVC is needed to perform the calculations required by the new tools, as well as to execute many of the calculations in parallel.

TABLE 5.11-1
Tool comparison of AVC versus HEVC, by coding-tool configuration.

AVC High Profile (8-bit)	HEVC Main Profile (8-bit)/ Main 10 Profile (10-bit)
16 × 16 macroblock	Coding unit quad-tree structure, 64 × 64 down to 8 × 8
Partitions 16 × 16 to 4 × 4	Prediction units, 64 × 64 to 8 × 8
Square + non-square (inter)	Square + non-square (inter) + asymmetric (inter)
8 × 8 and 4 × 4 transforms	Transform units, 32 × 32 to 4 × 4, 4 × 4 skip
Intra prediction (8 directions + DC)	Intra prediction (33 directions + DC + planar)
Inter prediction luma 6-tap + 2-tap, to % pel	Inter prediction luma 8-tap, to % pel
Inter prediction chroma bilinear interpolation	Inter prediction chroma 4-tap, to 1/8 pel
Motion vector prediction	Advanced motion vector prediction (spatial + temporal)
In-loop deblocking filter	In-loop deblocking filter and sample adaptive offset (SAO) filter
CABAC or CAVLC	CABAC using parallel operations

CABAC—Context-Adaptive Binary Arithmetic Coding
CAVLC—Context-Adaptive Variable-Length Coding

While the Main Profile and Main 10 Profile were defined to address the initial set of applications, additional profiles were developed as well for different deployment environments, ranging from newsgathering and point-to-point studio links to delivery to consumer handheld devices.

The following subsections explore some of major areas of coding improvements.

HEVC Techniques: Block Structure

As video resolution increases, larger blocks can provide increased compression efficiency. A traditional MPEG macroblock (the basic image processing unit; an N × N array of samples or transform coefficients) consists of 16 × 16 pixels, but up to a 64 × 64 pixel block size is defined for HEVC; early testing has shown significant efficiency improvements versus 16 × 16 for resolutions from 720p videoconferencing to 4K TV resolutions, and beyond.

To enable the use of large blocks while keeping the coding performance of small detailed areas in the same image, HEVC uses a hierarchical block coding partition method, with the basic image processing unit referred to as the largest coding unit (LCU) or treeblock

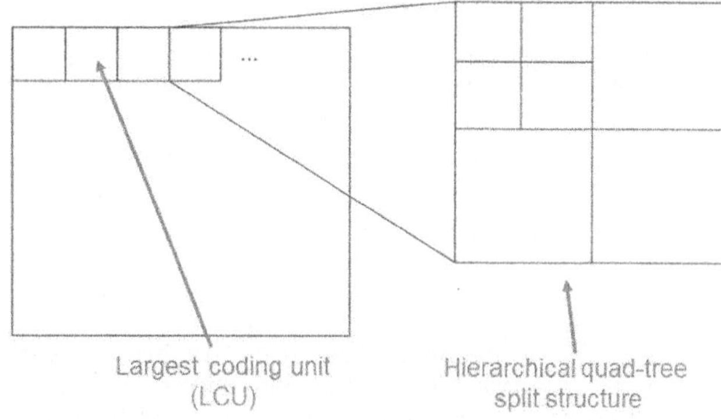

FIGURE 5.11-8 Example of HEVC picture structure.

(analogous to the macroblock). A block is referred to as a coding unit (CU).

LCUs are scanned left to right in the same way as macroblocks in AVC. Each LCU may be split, and the resulting blocks (CUs) may be split again hierarchically in a quad-tree fashion. There is also a smallest size defined, for which no further split is allowed; this block is called the smallest coding unit (SCU).

Figure 5.11-8 shows how a picture is divided into a number of LCU blocks, and one example of how an LCU can be split in a quad-tree fashion into smaller blocks.

Each CU may be further quad-tree split into two additional independent quad-tree structures, the prediction unit (PU) quad-tree (containing motion vectors) and the transform unit (TU) quad-tree. In the HE10 configuration, non-square transforms and predictions (asymmetric motion prediction, AMP) are also supported.

HEVC Techniques: Spatial Intra Prediction

For increased Intra compression efficiency, up to 35 prediction modes may be used (Intra planar, Intra DC, 33 directions, and a chroma mode using luma samples). In AVC there are up to 9 Intra prediction modes. This means that spatial prediction can be done with much finer granularity in HEVC compared with AVC.

HEVC Techniques: Motion Interpolation Filter

Both HEVC and AVC use motion vectors with 1/4 pel resolution. For HEVC luma space, an 8-tap separable filter is defined while the corresponding AVC filter has 6 taps. HEVC chroma space uses a 4-tap filter while the AVC chroma uses bilinear interpolation. These changes relative to AVC provide increased compression efficiency for HEVC.

HEVC Techniques: In-Loop Filters

HEVC uses a deblocking filter similar to that of AVC, but after the deblocking filter has been applied, two additional in-loop filters are used in HEVC, the sample adaptive offset (SAO) filter and the adaptive loop filter (ALF). Both these filters differ from a deblocking filter by not being constrained only to filter block boundaries; they are able to filter any pixel within a picture. Both in-loop filters are controlled by the encoder, and their syntax elements are sent in the bit stream. The ALF is not used in the Main Profile.

Ultra HDTV—the "Killer App" for HEVC?

With the seemingly inexhaustible consumer desire to have artificial reality match reality and the entertainment industries' desire to deliver compelling content to audiences, Ultra HD (UHD) televisions, projectors and camcorders were first shown to the public at the 2012 Consumer Electronics Show. Subsequently, the format continued to improve through the incorporation of other features, starting with its initial capability of 4K spatial resolution (3840 × 2160 pixels), then adding high dynamic range (HDR) video, wider color gamut (WCG, generally referenced to Rec. ITU-R BT.2020), and high frame rates (HFR, up to 120 Hz).

UHD delivery to the home is one of the prime drivers behind HEVC. 4K resolution has been in use for many years in digital cinema. There is a large existing infrastructure of 4K digital cameras, 4K film scanners and telecines, 4K postproduction systems, 4K playback servers and 4K display devices. Much of today's HD content is actually shot at 4K and then subsampled ("down-rezzed") for delivery. This means that there is a growing library of 4K content that could be the basis of future film, TV drama and TV documentary channels to the home.

4K digital cinema content can require 48 MB (or more) to store a single non-compressed frame at 4096 × 3112 pixels,[2] using 4:4:4 10-bit sampling (or higher). These daunting figures might seem to rule out practical 4K TV for the foreseeable future, but, in fact, the reality is much more plausible.

4K TV only involves handling *four* times more data than HDTV, 3840 × 2160 pixels at 4:2:2 10-bit—a slightly more modest 20 MB per frame, non-compressed. See Figure 5.11-9.

Early HEVC tests have shown that original non-compressed 4K TV can be shown at 18 Mbps compressed and still show stunning results. This figure effectively means 4K TV can be shown at bit rates previously used for MPEG-2 video-based HDTV services. See Figure 5.11-10.

Beyond 4K resolution, using HEVC to deliver other UHD attributes (HDR, WCG and HFR) requires little additional data. (HDR utilizes HEVC's ability to efficiently deliver 10-bit video, thereby reducing the visible "banding" effects often seen with 8-bit video in earlier systems.) Table 5.11-2 shows the relative amount of additional bit rate required by all the UHD video attributes after HEVC compression (relative to traditional digital HDTV delivery).

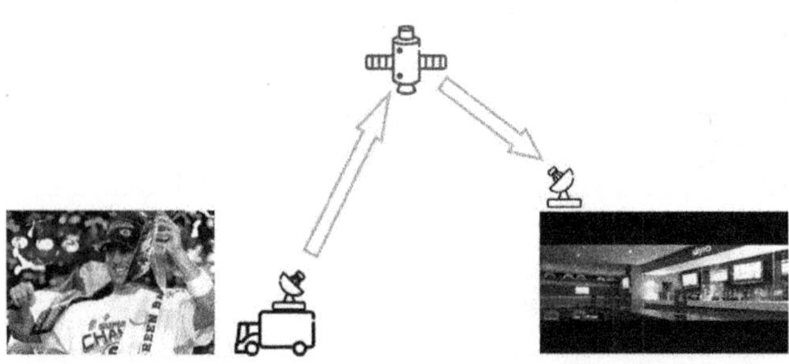

FIGURE 5.11-10 HEVC enables live UHDTV to be delivered to sports bars and the home.

TABLE 5.11-2
Relative impact to transmitted bit rate of various ultra HD features when delivered via HEVC.

UHD Feature	Uncompressed	Compressed (consumer-grade)
4K (2160p) vs. HD (1080p)	400%	about 250%
High dynamic range + wide color gamut + 10-bit sample depth vs. conventional 8-bit video	25–30%	about 0–20%
High frame rate (100–120 fps) vs. conventional frame rate (50–60 fps)	200%	about 30%

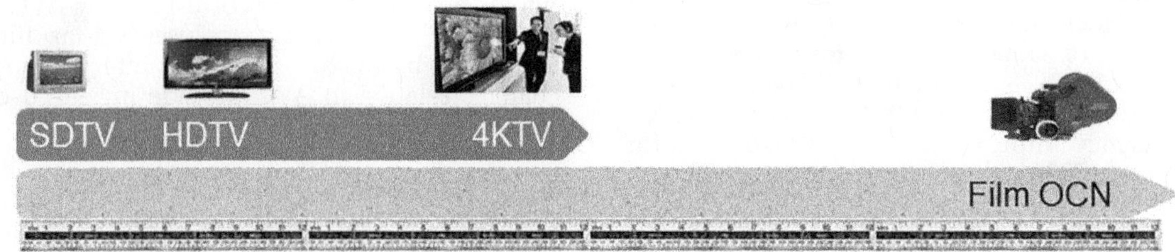

FIGURE 5.11-9 4K TV is four times more data than HD.

INTELLECTUAL PROPERTY AND COMPRESSION

Compression systems depend on very sophisticated technology, resulting from very expensive research in many companies around the world. This expertise is brought together in committees such as MPEG, but the contributing companies need to see some return on their research investment. This results in the requirement that most compression systems may be legally implemented only with the appropriate patent licenses.

MPEG-2 resulted in a new concept—the patent pool. Each company that believes it has patents essential to implementing the standard may submit the patents for examination by independent experts. If a patent is found to be essential, it may be included in the pool. Companies wishing to build systems may take one license for all the patents in the pool, and the royalties are shared among the patent owners.

This system has been controversial, particularly with MPEG-4 and later systems. The original MPEG-4 license required payment of royalties, not only on equipment but also on each actual use of the compression (for example, a television transmission). When AVC was first completed, the proposed license terms were similar to MPEG-4, and there was a similar reluctance to adopt the standard. Since then, the license terms have been modified substantially and are now widely accepted. However, similar uncertainty is occurring today, at the time of this writing, for the licensing of HEVC.

JPEG 2000 is unusual in that the committee adopted a policy that the only technologies adopted within Part 1 of the standard would be those that were contributed on a royalty-free basis. It is believed, therefore, that no royalties are payable for basic JPEG 2000 implementations—one of the reasons that it was attractive to the Hollywood movie studios and a factor in its adoption for digital cinema.

SUMMARY

Compression is a factor in almost every aspect of the television system. The technology has evolved rapidly over the last 20 years—today the most efficient systems can transmit good-quality color images at an average of about 1/10 of a bit per pixel. This is made possible partly by the development of flexible and extensible standards, with contributions from hundreds of experts from all parts of the world. The other major factor is the continuing effort in many companies to exploit these standards and the availability of more and more computational power, to provide ever-improving performance, often beyond the expectations of those who created the standards. Like interlace and NTSC in earlier times, compression always represents a compromise between bandwidth and performance. Like many compromises, it can produce disastrous results if applied incorrectly or without understanding. Compression is an art as well as a science.

Notes

[1] In other countries, 7 MHz or 8 MHz channel bandwidth is used.

[2] A convention based on full aperture scanning of 35 mm film OCN (original camera negative).

Bibliography

Books on Video Compression

Hubbard, B. B., *The World According to Wavelets*, 2nd ed., A. K. Peters, Ltd., Wellesley, MA, 1998.

Mitchell, J. L., Pennebaker, W. B., Fogg, C. E., and LeGall, D. J., *MPEG Video Compression Standard*, Chapman & Hall, New York, 1996.

Pereira, F., and Ebrahimi, T., *The MPEG-4 Book*, Prentice Hall, Upper Saddle River, NJ, 2002.

Rabbani, M., and Jones, P. W., *Digital Image Compression Techniques*, SPIE Optical Engineering Press, Bellingham, WA, 1991.

Richardson, I.E.G., *H.264 and MPEG-4 Video Compression*, John Wiley & Sons, New York, 2003.

Symes, P. D., *Digital Video Compression*, McGraw-Hill, New York, 2004.

Useful Websites

http://mpeg.chiariglione.org/
http://www.jpeg.org/
http://www.symes.tv/

CHAPTER

5.12

Video Recording, Servers, and Storage

KARL PAULSEN
Diversified
Pittsburgh, PA

INTRODUCTION

Video servers are just the tip of what has become a huge shift in infrastructure. Video server technologies, which have essentially replaced traditional magnetic recording on videotape, now encompass not just recording and playout, but also the processes associated with file-based and streaming workflows, which include image capture and field recording, image processing, live video encoding, storage of the assets, networking for signal transport and storage, distribution to myriad end points and lastly the archiving of the content for preservation and protection purposes.

Major shifts in technology since the previous edition of this handbook have taken video servers, storage solutions, and asset management activities outside the facility and up into the domain of the cloud. Many, if not all, of the technologies discussed in this 2017 update are now ported or could be moved to a cloud-based environment—whether that cloud be public, private or hybrid. This fundamental shift in video-applications may once again change the way we look at video servers, content storage, file-based workflows and content management.

From Film to Tape to Disk

Video recording's evolution is marked by significant changes in the recording media itself; highlighted in particular by the migration from film to videotape to disk. The most notable medium captures sequences of individual images onto a stock made of transparent celluloid, acetate or polyester base, which is coated with emulsion containing light-sensitive chemicals. The coated base is exposed in a camera and later developed to produce what is otherwise known as *film*.

Film has a long, rich history that continues to be revived and rejuvenated as motion picture artists bring new life to legacy image capturing and recording. Video recording has further changed the complexion of image capture and retention, having seen dozens of technology transitions since its commercial introduction in the early 1950s.

Two-Inch Quad to Half-Inch Cassette

The first video recording demonstrations, presented by Bing Crosby Enterprises' electronics division, were shown in Los Angeles on November 11, 1951.[1] Commercial electronic recording for broadcast purposes began at the end of 1953.[2] First relegated to analog video recordings on videotape in monochrome (black and white), color recordings commenced only a short time later in 1954. Finally, the quadruplex videotape machine (using 2-inch videotape) was introduced by Ampex on April 14, 1956, at the Chicago NAB convention; this was followed by a color machine by Ampex produced in 1958. See Figure 5.12-1.

Cassette formats were introduced in 1969 when Sony produced the prototype that would become the U-matic format, an analog composite product that was standardized as 3/4-inch (1.905 cm) and commercially produced in 1971. A successor format using 1/2-inch (1.27 cm) tape, also from Sony, emerged as Betacam in 1982; with Panasonic producing a competing format M-I ("m-one"), which was replaced by an updated version called M-II. Panasonic's M-format never reached the commercial success of Betacam and dwindled from

FIGURE 5.12-1 First VTR from AMPEX Corporation.

Courtesy of AMPEX

FIGURE 5.12-2 Sony Component Digital D-1 format VTR from the early 1990s primarily in postproduction.

Courtesy of Sony

production before the time that digital recording technologies emerged.

Replacing Analog with Digital

The path to digital was inevitable, given the physics and limitations of analog recording technologies. Video recording was transformed to digital by taking the three component analog "color difference" components (Y, B-Y, R-Y), sampling them at 8 bits and conforming those bit sets into a component digital signal, per the specifications of what was then the Recommendations of the CCIR (Consultative Committee on International Radio) and which would later become ITU-R Recommendation CCIR-601. As a result of the standard originally issued in 1982, analog to digital and the means of sampling the color difference signals interleaved with the luminance (Y) signals paved the way to modern digital coding technology and is representative of methods still in use some 35 years later.

The "Consultative Committee on International Radio" (CCIR) would be renamed the "International Telecommunication Union—Radiocommunication" (ITU-R) sector. The ITU-R (previously CCIR) standards would guide and define technical methods for encoding analog video signals into a digital video format. This would be referred to as "component digital" and for videotape best known as the "D-1" tape format. The digital video samples would be quantized as 8-bit or 10-bit pulse-code modulated (PCM) bits derived from YCbCr analog component signals.

From the specifications of the D-1 format came a composite version otherwise called "D-2" or "composite digital," represented as a four-times subcarrier ($4f_{sc}$) sampling of the single channel analog video signal. D-1 and D-2 were essentially the last of the uncompressed, "high-quality" video recording forms before compressed video formats (e.g., a compressed composite D-3) were introduced. High-definition video recording, which came initially on open-reel videotape transport devices, and then later on cassette and optical mediums, was recorded in either an uncompressed (open reel) or variations of compressed formats to conserve data space on the medium itself. See Figures 5.12-2 and 5.12-3.

FIGURE 5.12-3 Sony Composite Digital D-2 format VTR used throughout the 1990s.

Courtesy of Sony

Digital Video Protocol

Standard-definition (SD) protocols for broadcast digital video were fairly straightforward in the beginning. For standard-definition video, ITU-R Recommendation BT.656 (originally called CCIR-656) described a digital video protocol for *streaming* uncompressed NTSC (525 SD) and PAL (625 lines) signals. The interchange recommendation was built upon the 4:2:2 digital video encoding parameters defined in ITU-R Rec. BT.601—the process for interlaced (field-based) video data streaming that uses the YCbCr color space with a 13.5 MHz sampling frequency.

These two protocols (BT.601 and BT.656) with their associated SMPTE standards set the tone for the modern era of digital video even before compressed video became achievable and long before digital high definition was developed. These fundamentals for the encoding of video have endured up through present-day high-definition (HD) and ultra-high-definition (UHD) television systems. However, the recording of these digital video protocols and how video was adapted for compression is a much deeper subject that is covered in greater detail throughout this handbook.

Bits on Disks

In close time coincidence with the launch of compression-based nonlinear editing (NLE) systems, postproduction equipment manufacturers adapted those high-quality component digital formats used for standard definition (SD) to spinning magnetic media (hard disk drives, or HDDs). However, there was no truly *standardized* method for saving the bits to actual physical disk media. Real-time interoperability between systems or devices was extremely limited except when converted back to the serial digital interface (SDI) transport streams necessary for playout/recording interconnections. For other interchange, such as in non-real-time (NRT) file-based workflows, video could be exchanged either on physical media or across a network connection. In its infancy, consistency across the formats was limited and differences in color video acuity or even pixel dimensions were widespread.

Since the evolution of digital recording technologies, nearly all forms of information have been retained on either magnetic or optical media. Real-time record/playout in various compressed video formats were created and were supported by the manufacturers of disk-based digital video recorders and play-to-air video servers. As solid-state media technologies evolved, like other disk-based systems, so did the ability to freely interchange digital video recordings without the necessity to depend upon linear videotape.

To support the growth in nonlinear editing as file-based production, and for the recording and playback of that content, new and expanded capabilities for storage would be needed.

What's in Storage?

Rapid development of disk- and file-based technologies, in conjunction with audio/video compression, drove industry away from linear videotape (and film) toward what we know today as the routine way of capturing moving images and sound to a storage medium. Nonetheless, with all this new media-centric information, methods for long-term preservation were needed. Preserving the content, through archival means, of the ever-expanding dimensions in digital video media would soon require alternative technologies.

Preserving Back to Film

Throughout the course of finding a means of preserving content and prior to the widespread acceptance of using digitization for that preservation, some organizations sought an archive methodology that was less risky and could provide for a higher-quality recovery much later in time. Since deterioration of original film content was inevitable, many studios looked to take a more *permanent preservation* approach—which included rescanning tri-group (color separation images as RGB) back to film-material with each color held as luminance-only values. By removing the color properties found in a composite image, high-quality, higher-resolution scanning could then capture and later recover the individual channels. The three images would be recombined with each color channel properly set during the reprinting process, thus making an accurate—as close to the original as possible—full-color reproduction. Before digitization, the methods offered a better-controlled system without necessarily depending upon what could otherwise become an obsolete procedure.

The idea of preserving three discrete color image channels is not unlike approaches taken for preserving digitally scanned images, where filters would remove the color information and retain the primary individual (red-blue-green) color channels as light level "negative-like" information. The "physical" approach to archive for film content is similar to today; we use color difference encoding with a weighted scale for the R-Y (Cr) and B-Y (Cb) channels.

Others have looked beyond physical media archive technologies, taking other approaches including techniques as described next.

DOTS

In more recent times, as an alternative for long-term permanent archive, a technology using data deposited to a metal-alloy tape that then uses a human-readable format/coding methodology was developed. Pioneered by Group 47, Inc., who acquired the technology from Kodak, DOTS (which stands for "digital optical technology system") creates data that is written at a density in excess of 8000 dots across the width of 1/2-inch media. The data dots can be "eye-readable" using polarized light. An archival longevity of 100 years,[3] according to Group 47, is possible "without loss due to magnetic failure." The process "is chemically inert, is immune to electromagnetic pulse (EMP) dangers and can be stored in normal offices from 15 to 150 degrees Fahrenheit."

DNA

Other still developing experimental storage technologies used to retain imaging as digital data include both holographic forms and the encoding of content into artificial, "non-biological" DNA.[4] See Figure 5.12-4.

DNA storage was publicly demonstrated during the 2016 Las Vegas Consumer Electronics Show (CES),[5] produced by the former Consumer Electronics Association

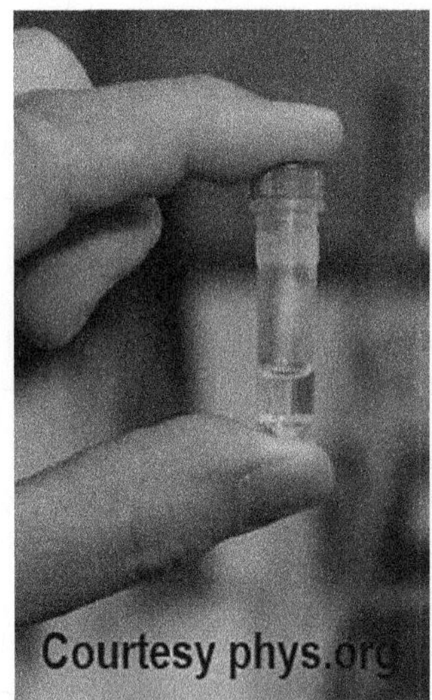

FIGURE 5.12-4 Vial of DNA "data." One droplet of the solution could contain one million copies of an old movie encoded onto DNA.

Courtesy of phys.org, from "Technicolor Stores Hollywood History in a Bottle," by Frankie Taggart, April 5, 2016

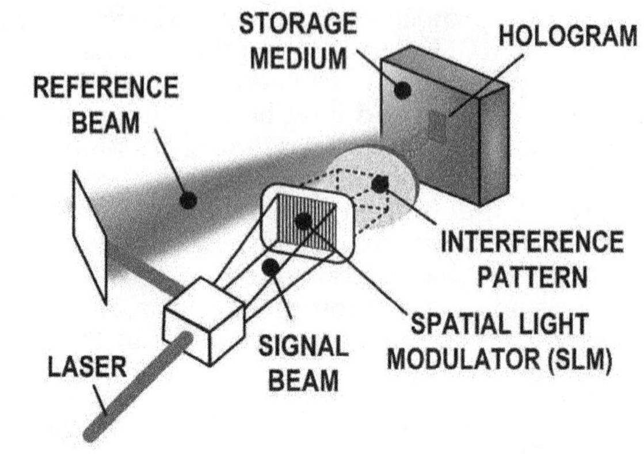

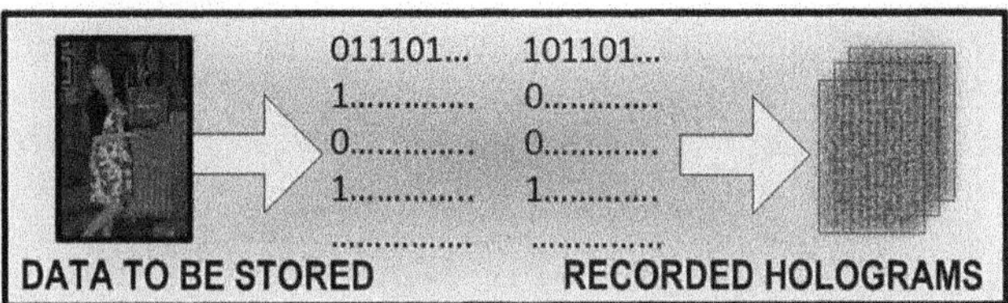

FIGURE 5.12-5 Holographic image capture—hologram write function with reference beam and signal beam combining to form the hologram.

(CEA), which is now known as the Consumer Technology Association (CTA). Technicolor—in collaboration with Harvard Medical School—displayed actual vials of liquid containing a portion of a motion picture that was converted into non-biologic DNA strands and emulsified in liquid. Observers were allowed to "taste" the motion picture by sampling a few drops from the vial that contained the DNA archive material.

Holography

One other method for long-term preservation used a technique where the data set was embedded into a holographic "solid" medium that could be scanned by a precision laser and decoded. The holographic storage prototype was shown at the 2005 NAB convention at Maxell Corporation of America. In April 2009, GE Global Research demonstrated its own 500 GB holographic storage material employing discs that would utilize similar read mechanisms as those found on Blu-ray Disc players.

Holographic storage technologies prompted another prospective manufacturer to prototype a video recorder/player that used the same concept, but the product was never really brought to market. See Figures 5.12-5, 5.12-6 and 5.12-7.

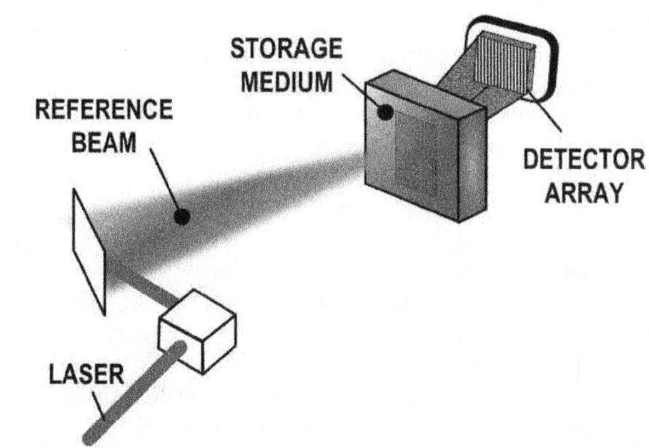

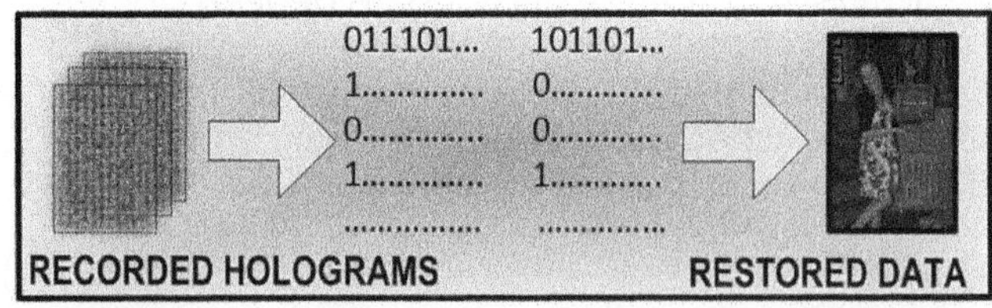

FIGURE 5.12-6 Holographic data recovery uses only the single reference beam and the detector array.

DISK RECORDING AND STORAGE

Topics that have been updated or expanded from the previous handbook include how video servers and storage technologies have matured over the past decade. This section outlines how the transition from linear recording to nonlinear recording and playback have become well-accepted technologies, and in essence the new foundation for today's video recording, storage and server systems.

The foremost changes in recording and playback of video media over the past decade have come principally from:

- the *flexibility* and *modularity* that is available in software-based encoding/decoding

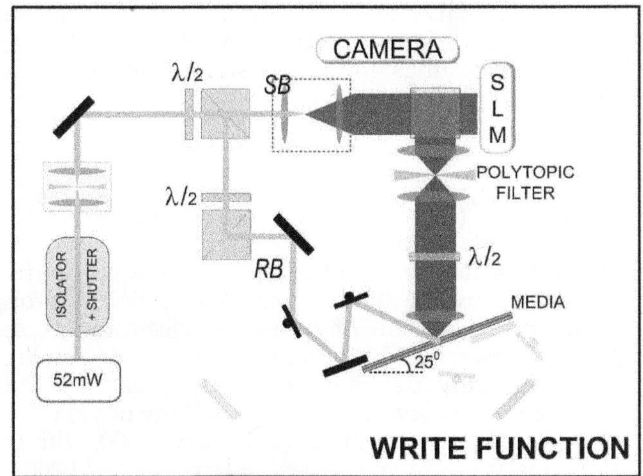

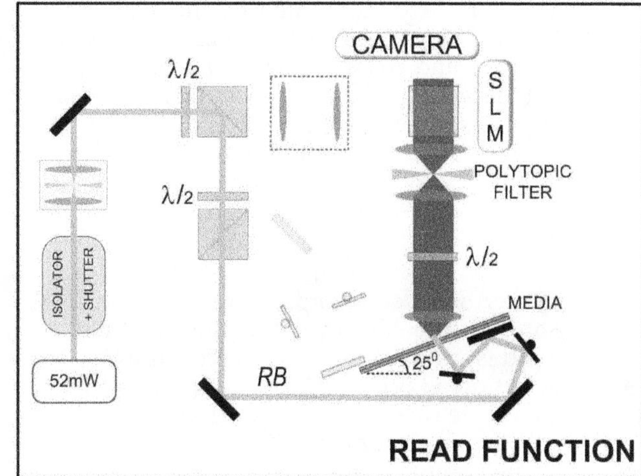

SLM = SPATIAL LIGHT MODULATOR
READING → REFERENCE BEAM (RB) ONLY IS REQUIRED FOR READING
WRITING → REFERENCE BEAM (RB) + SIGNAL BEAM (SB) BOTH USED FOR WRITING

FIGURE 5.12-7 Hologram write and read schematic function to (write) and from (read) the physical media.

- the enormous impact that *high-capacity, high-performance storage* for media-centric data has had on the broadcast media and entertainment industries
- how *compression* and *networking technologies* have changed the overall nature of media management across the entire ecosystem and most likely impact every user of technology and media today

Analog Optical (Video) Disc

Developing features sets for modern video technologies can be reflected upon by examining older, legacy technologies. One of the now long-forgotten consumer/industrial formats used for home video, theme park and museum presentations, corporate training and education, plus many other applications was in the analog optical videodisc format—a technology offshoot that began with an optical video recording patent in 1961. The first format developments, known as "transparent recording discs," were introduced by Philips in 1969 and became available commercially in December 1978.

In opposition to the VHS and Betamax videocassette systems made available two years prior, the optical disc format never really took off. Pioneer Electronics purchased a majority interest in the optical disc format and then, for marketing reasons, renamed the format to LaserVision and the brand name became LaserDisc. A significant library of play-only content (motion pictures) was generated and was said to be of superior quality to VHS. Audio quality added a much-needed bonus unavailable in many VHS players. Educational institutions and corporations invested in the technologies, finding it quite useful for interactive training and presentations. See Figure 5.12-8.

The ability to stop the playback, freeze the image, step-rewind or skip-forward (a.k.a. "stunt play") was a driving factor for many owners and literally helped pioneer the many concepts we have grown to expect in all video recorder/playout feature sets.

FIGURE 5.12-8 The first commercial optical disc, the Pioneer LaserDisc was an analog recording technology for both home entertainment and business and industry.

This, and other legacy technologies, has certainly helped to shape the modern era of video recording using disk- (and disc-) based mediums. Yet while products like the LaserDisc had their day, other recording technologies have also endured and evolved.

End of an Era

The once predicted and now certain demise of linear cassette-based videotape as a *primary* recording medium was seriously accelerated by the availability of solid-state media for field recording, mobile devices, studio recording and mass storage systems. Helping support this acceleration were higher (areal) density rotating magnetic media for storage and workflows coupled with faster network interfaces.

The combination of these forces has allowed massive amounts of data (sometimes called "Big Data") to be recorded, processed, managed and distributed on a global basis.

The majority of video recording since around 2010 is now captured directly to solid-state (flash) media; and that trend is still increasing. In the consumer space it is harder to find a DVD or a VHS-deck. In mid-2016, Japan's Funai Electric company announced that "July [2016] would be the last month that a [VHS] cassette deck would be manufactured."[6]

Physical media such as CDs and DVDs continue to show declining sales as online media, in the cloud storage, and streaming takes over.

Mobile devices manufactured in the last five years all utilize flash memory for media and image storage. Many laptops and full desktop computers have switched to or are integrating solid-state drives and hybrid SSD/HDDs for some or all of their storage. Flash memory is being embedded on processors and hard drives to aid in buffering and handling continual calls or routines essential to the basic functions of the computer system.

For the professional broadcast production industry, there still remain exceptions to the "demise of videotape" prognostication; mainly when needing to record high bit depth, wider bandwidth high-definition content or when a particular workflow is intended for non-file-based applications. Ironically, SD and HD professional HDV and DVCAM videotape decks and camcorders are still available, as well—but with most of those recordings going to flash memory instead of linear videotape.

Blu-ray

Blu-ray Disc (BD) was aimed at superseding the DVD format; with a target to significantly increase the amount of data that could be stored on a single disc. Developed by the Blu-ray Disc Association, a group representing makers of consumer electronics, computer hardware and motion pictures, it is technology used frequently to store motion pictures, data and field capture images from video cameras. Sony unveiled the first Blu-ray Disc prototypes in October 2000, with the first prototype player released in Japan in April 2003.

The "Blu" in Blu-ray refers to its blue laser (actually a violet laser), which is used to read the disc. The longer wavelength (650 nm) red laser used for DVDs could not produce the needed data density; thus the blue laser (wavelength between 360 and 488 nm) was developed, which produces a much smaller spot beam yielding a much higher data density. See Figure 5.12-9.

A BD plastic disc is 120 mm in diameter and 1.2 mm thick. Conventional BD can contain 25 GB per layer, and dual-layer discs will store 50 GB. Triple-layer discs will store 100 GB and quadruple-layers store 128 GB, which is available for BD-XL re-writer drives and is typically used for data (and also by Sony in the XDCAM products) but seldom found in the consumer or movie presentation market.

BD is capable of storing full-length high-definition motion pictures, making it ideal for consumer systems that want a higher presentation quality than what

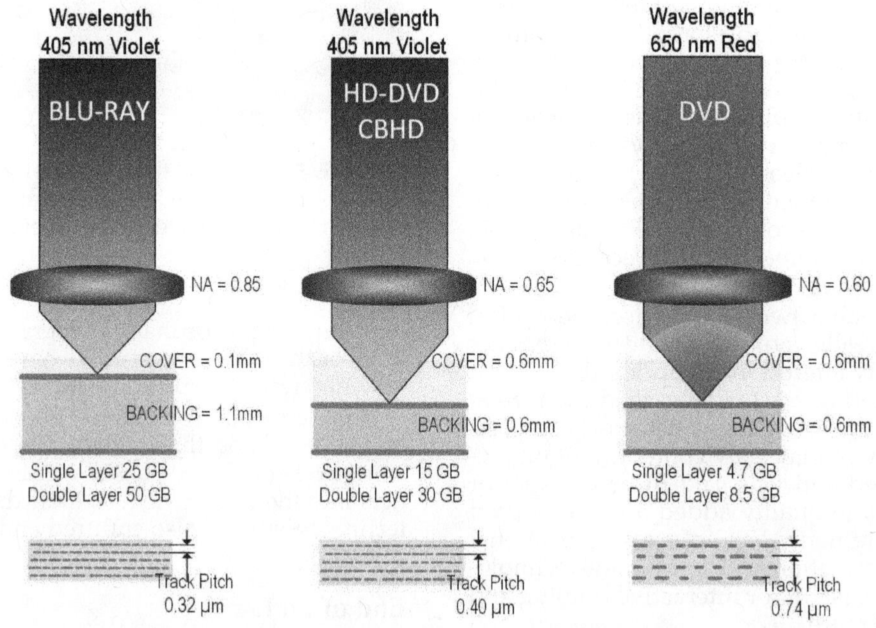

FIGURE 5.12-9 Optical disc format technical specifications showing track pitch, numerical aperture, wavelength and disc capacity per layer.

DVD, or cable television, can provide. Blu-ray Disc titles were first released on June 20, 2006.

Sony developed its preformatted XDCAM 50 GB dual-layer professional disc (PFD50DLA) and the 128 GB quad-layer (PFD128QLW) from the same blue-laser technology.

Last Expected VTR Format

Introduced as the next-generation, high-end progression beyond HDCAM (which is also known as D-11 and standardized as SMPTE ST 367M), Sony's HDCAM SR format may indeed be the last linear videotape format developed for broadcast content capture, production and postproduction activities.

HDCAM SR is offered exclusively by Sony. The format, introduced circa 2003, was standardized as SMPTE ST 409M-2005. HDCAM SR provides high-quality, high data rate recording capabilities, which are available in a linear videotape transport with cassette-based media and in a similar device which uses solid-state media for both acquisition and editing. See Figures 5.12-10, 5.12-11 and 5.12-12.

Throughout its nearly 15-year life cycle, many productions have started in the field on this format and

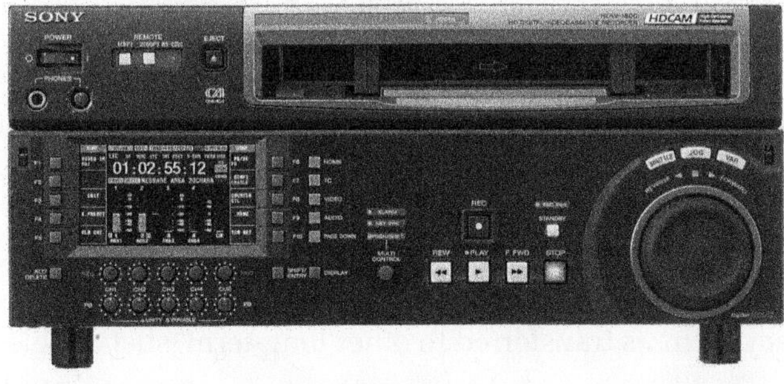

FIGURE 5.12-10 Sony HDCAM (Model HDW-1800), one of the last videotape transports to be produced before the transition to solid-state memory devices, e.g., the Sony SRR1000.

FIGURE 5.12-11 At the time of this writing, this is the latest generation Sony, Model SRR1000 (SRMASTER Storage Unit). It uses solid-state memory cards, e.g., SR1TS25. This flash memory–based unit replaced the HDCAM SR units, which relied on magnetic videotape.

FIGURE 5.12-12 Sony 1 TB (SR-1TS25) ultra-high-speed flash memory card for use in the HDCAM SRMASTER series. It has transfer speeds of up to 2.5 Gbps and uses advanced error correction and concealment technologies.

followed through editing and to production master, and then ultimately released for distribution as DVD, Blu-ray or through a content distributor.

Its 440–600 Mbps payload offers a 10-bit video depth in 4:2:2 or 4:4:4 RGB. Employing MPEG-4 Part 2 Simple Studio Profile for compression allowed for some features such as dual-stream recording for stereoscopic/3D. Sony increased the available audio channels with up to twelve at 48 kHz/24-bit. HDCAM tape transport models can offer a 2x-mode, increasing the video data recording rate to 880 Mbps.

In 2011, the industry experienced what some felt might become the "presumptive demise" of the SR tape format. Influenced, in part, when the SR tape manufacturing plant was rendered unusable following the March 11, 2011, magnitude 9 earthquake that shook northeastern Japan and unleashed a devastating tsunami. At that time, many companies that depended upon the HDCAM SR videotape format for their production mastering became concerned and started to look into alternatives for this very successful eight-year-old tape-based medium.

When the ability to acquire raw tape stock was thought to be challenged, many organizations rethought their videotape strategy and began the move toward alternative means of recording, editing, postproducing, mastering and preserving their media assets. Fortunately, the SMPTE 409M standard endured and the shift toward solid-state media was accepted, and today, HDCAM SR capabilities are still a valued form of image capture and mastering.

For electronic digital cinema camera capture, the storage of higher bit rate, high-volume files directly to solid-state media packages is still one of the predominant means of image capture from high-definition video cameras. Other production work, such as that with existing SR tape inventories, still use this videotape recording format.

With digital capture one of the more cost-effective means of production and changes to higher-resolution images including UHD, 4K and beyond, it is not uncommon to find the master images captured in the native camera manufacturer format. Many times a secondary backup may be created on linear data tape (e.g., LTO) so the files can be protected and/or migrated onto another platform for other uses, including archiving. Depending upon the workflow and production setups, HDCAM SR may be used to capture a parallel higher-resolution version as a down-converted production daily in 1920 × 1080p60.

Migration to Non-Tape Alternatives

Sony HDCAM SR products are still carried online, in catalogs and by distributors. However, for both technology's sake and due to a limited (tape) media availability becoming an inevitable possibility, a migration to SRMASTER A/V recording and storage using solid-state SRMemory Cards drove videotape still closer to obsolesce.

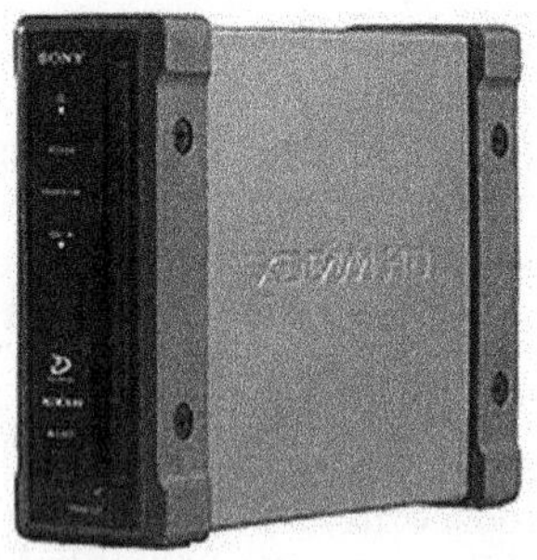

FIGURE 5.12-13 Sony XDCAM optical recording media based on Blu-ray Disc (BD) technologies and player unit model PDW-U1.

Sony and Panasonic were instrumental in developing broadcast media specific, high-capacity solid-state storage in card-based form factors. The technology allowed camcorders to flourish and professional production to continue without the need for bulky, mechanically burdened videotape transports. Sony had also created a non-flash-based storage medium using optical media, known as the "professional disc" and found in the XDCAM format. See Figure 5.12-13.

Others offered ruggedized portable magnetic spinning disk products, but found the reliability and durability of lugging hard drives attached to camera bodies was beyond sustainability.

Sony's XDCAM product line was also introduced in 2003, becoming the foundation for what would be called the "XDCAM Professional Disc" system. The format and product lines offered digital recording capabilities in four different lines: XDCAM SD, XDCAM HD, XDCAM EX and XDCAM HD422. Other companies up through around 2012 were developing products based on alliances with Sony to produce XDCAM or included the XDCAM codec in original equipment manufactured directly.

A Sony software utility, known as XDCAM drive software, allowed a computer to connect to a PDW-U1/U2 Professional Disc Drive Unit and convert from XDCAM disc to files for use in editing on Sony (and other) production editorial products. See Figure 5.12-14.

Nonetheless, these new recording paradigms are continually shifting away from tape due in part to the relatively declining availability of tape-based media and because other alternatives provide for a better total cost of ownership (TCO). Still, some workflows remain dependent upon on videotape recorders because of legacy tape-based content libraries, for other incentives or for financial reasons. Getting legacy formats transferred to other long-term, sustainable

FORMAT NAME	VIDEO CODING	CONTAINER	COLOR SAMPLING	FRAME SIZE	FRAME RATE SCANNING	BIT RATE (bit depth = 8-bits)	AUDIO CODING
DVCAM	DV	MXF DV-AVI	4:2:0	720 x 576	25i, 25p	25 Mbps - CBR	PCM 4-channel 16-bit 48 kHz
			4:1:1	720 x 480	29.97i, 29.97p		
MPEG IMX	MPEG2 422P@ML	MXF	4:2:2	720 x 576	25i, 25p	30 Mbps - CBR 40 Mbps - CBR 50 Mbps - CBR	PCM 8-channel 16-bit 48 kHz PCM 4-channel 24-bit 48 kHz
				720 x 480	29.97i, 29.97p, 23.98p		
MPEG HD	MPEG2 MP@H14/HL	MXF MP4	4:2:0	1920 x 1080	25i, 25p, 29.97i, 23.98p	35 Mbps - VBR	PCM 4-channel 16-bit 48 kHz
				1440 x 1080	29.97i, 29.97p, 23.98p, 25p	18 Mbps - VBR 25 Mbps - CBR 35 Mbps - VBR	
				1280 x 720	59.94p, 29.97p, 23.98p, 50p, 25p	25 Mbps - CBR 35 Mbps - VBR 19 Mbps - CBR	
MPEG HD422	MPEG2 422P@HL	MXF	4:2:2	1920 x 1080	29.97i, 25i, 29.97p, 25p, 23.98p	50 Mbps - CBR	PCM 8-channel 24-bit 48 kHz PCM 4-channel 24-bit 48 kHz
				1280 x 720	59.94p, 23.98p, 50p		

FIGURE 5.12-14 Various formats found that are deployed on the XDCAM platform.

media storage can be a costly process; as such, most users are still looking for the most readily available and best cost-effective options that will decrease their dependency on videotape.

Overtaken by Flash Memory

The new storage front-runner for the origination of video—across consumer and professional applications—is flash-based memory. Historically, the more recent attempts to bring ruggedized hard disks to market never effectively caught on, and the success of optical media for field capture has now evolved to a more reliable and much higher density medium. Solid-state or flash-memory data density has increased to proportions that optical media cannot achieve. The flexible capacities and continued cost decreases in flash have steadily outweighed the previous value propositions for nearly all of the non-solid-state storage devices—at least for acquisition. See Figure 5.12-15.

The long running stretch of XDCAM technologies, built upon blue-laser technologies for optical disc media, was quite successful. However, the optical media form factor for field recording has since faded in favor of multi-terabyte solid-state devices, commonly known as SSDs (solid-state drives). Optical media is making a renewed comeback for archival purposes, especially in the write-once-read-many (WORM) environments and when the environment for linear digital tape poses both a short-term and a long-term risk.

ExpressCard for Video Media

Sony also developed a form factor flash memory media card with high-speed data writing called "SxS"

FIGURE 5.12-15 Flash memory variations utilized in consumer and professional applications.

(pronounced ES-by-ES). The card is compliant to the SanDisk/Sony created ExpressCard standard.

ExpressCard is a technical standard that specifies the design of slots (usually built into the computer) and of the expansion cards that insert in the slots. ExpressCards include electronic circuits and, in some implementations, connectors for external devices. ExpressCard replaces the PC Card (also known as PCMCIA) standard and is maintained by the USB Implementers Forum (USB-IF).

Files and Streams

In what is now a recurring post-1990s theme, recording moving images to videotape has effectively now moved on. Originating on linear videotape has eroded to the point that file-based capture and production for much of the broadcast and production industry is the norm. Before that transition could be successful, a long steady set of proprietary and standardized adoptions of file-based media would be experienced.

For many years, content that began on videotape would be digitized into a purpose-built production/editorial device, edited and then printed (playout) to a videotape recorder (VTR). The quality degradation that occurred was termed a "generation loss"; and sometimes that was concatenated multiple times, to the point the images were not satisfactory for what was at that time "broadcast quality." Improvements continued that would involve several technological advancements, particularly focused in higher-quality video compression that bridged the Motion JPEG formats to MPEG formats suitable for retention on disk-based storage systems.

Sequential JPEG Files

The migration from videotape to disk storage technologies began with the storing of discrete sequential images, as "files," to rotating magnetic disk drives through a process referred to as "digitizing." The digital image files derived from the videotape were usually JPEG image formats, which could be successively played back at 30 (29.97) frames per second for NTSC (25 fps for PAL). Since JPEG was a "tool kit" of various formats, interchange of JPEG files was limited to like/proprietary systems.

Some of the high-quality videodisk production devices would use other methodologies, sometimes via proprietary digital video interfaces; yet each device would for the most part store each individual frame discretely.

Workflow

There were no true "file-based workflows" or "streaming media" in these early days. Workflows involved playing a VTR in real time to a receiving system consisting of a converter and a storage device that "encoded" the linear images into the discrete files that were then organized on the storage device so they could be replayed sequentially at 30 (or 25) fps, on command, to a "decoder" that would reverse the process and produce a real-time lineal video that would be recorded by an external VTR. Since these workflows were prior to MPEG encoding technologies, many manufacturers essentially created a closed system whose only interchange was videotape or through a coax containing the video stream itself.

Real Time Only

Even though video would be stored in a sometimes proprietary digital imaging format, most of the workflow for ingestion and processing was at best a real-time function. Other than the ability to seek from one group of images (frames) to another almost instantaneously, all ingestion and playout functions were not via files, but were actually real-time playbacks to and from the encoding/decoding and storage system.

Although the video content may have originally been captured (recorded) to digital videotape, ingesting that content to the videodisc recorder platform would require a real-time transfer—usually as analog video—from tape to encoder to storage. Until digital video interfaces (i.e., SDI) were standardized, all broadcast-quality video storage was essentially derived from an analog interface.

Linear Video to Disk Recorders

Around the end of the 1980s, recording video to hard disk drives had stepped into reality, but it was not commonplace. Although video could now be stored to a disk drive, it needed something more. Systems were created that could transfer from one drive platform (videodisk recorder) to another, and could segment the longer clips into short clips that would be organized on the drive to linear playout. Using similar techniques to those on VTRs, assemble edits could be strung together to create a sequence of frames that were now continuous. But that was just the beginning. A system would need to be developed that displayed similar functionality to that of the CMS-editing systems used for program editing in a production control room environment. See Figure 5.12-16.

Off-line and Online Editing

From the late 1980s through the mid-1990s, *nonlinear "off-line" edit systems* (the predecessor to desktop editors) were developed and used to produce an edit decision list (EDL), which would later be used to *conform* the original material (residing on videotape) into finished products in what was termed the *"online" editing room.*

Online editors (e.g., CMX, Sony and Grass Valley) were essentially electronic control units operated by human editors that would manage linear videotape transports (VTRs) and other production tools—video switchers/vision mixers, graphics devices, effects units (DVE), digital disk recorders and audio consoles—to assemble selected video and audio segments into a program—complete with effects, transitions, graphics and sound.

Nonlinear Editing

With the support of video compression technologies and hard disk drive storage, the off-line editing platform would evolve from rough cutting poor-quality

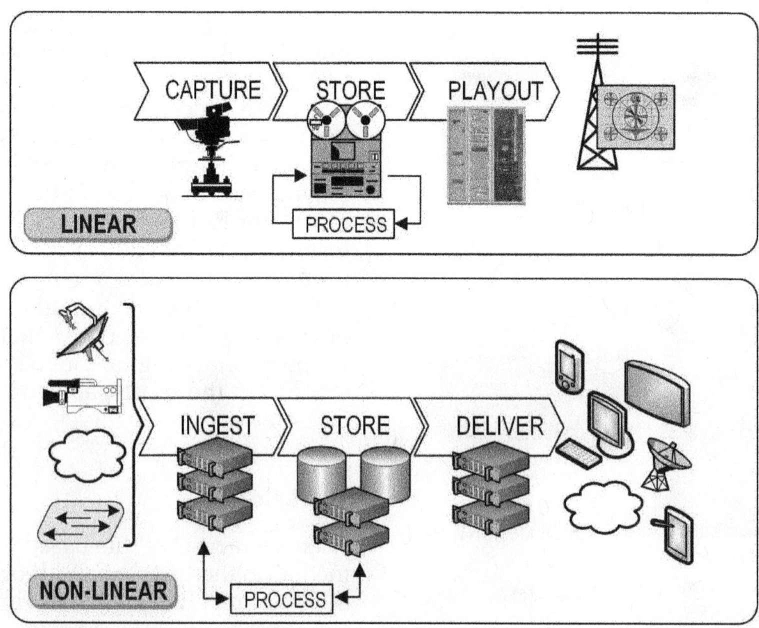

FIGURE 5.12-16 Linear workflows (top) were single threaded and contained. File-based nonlinear workflows (bottom) are multi-threaded and unconstrained.

video into a full resolution, full quality system that could produce completed programs without the need for exporting an EDL to a conforming session in the online editing room. Eventually these rudimentary nonlinear editing (NLE) systems would gain sufficient sophistication that they would replace the online rooms entirely—but that would take almost another decade, up into the early 2000s, before the online room would become a dinosaur.

Videodisk Recorders Develop

File-based workflows did not truly begin until video server technologies and nonlinear editing systems developed to the point they produced content that could be exchanged as digital streams (not digital files) and a form of file-structure could be agreed upon that would not require decoding to baseband video for interchange.

Standard definition—serial digital interface (SD-SDI) technologies (e.g., SMPTE 259), which were progressing rapidly in the early 1990s, added the ability to reduce the general degradation of the video image as it moved from device to device in real time. Yet prior to SDI, if one wanted the editorial work to be delivered to another system, it would be played out from the NLE in either single-wire analog composite (NTSC or PAL) format or three-wire component analog (Y, R-Y, B-Y or similar format) and captured by VTRs.

Before there were any actual "video server" systems as stand-alone, purpose-built encoding/storage/decoding platforms, a lengthy development period would ensue. Beginning with the storing of multiple individual images as files and then replaying those images in sequence to create a contiguous video "stream," video servers combined multiple record and playout channels into a single unit that could derive files from a common integral storage system.

First Uses for Videodisk Storage

The first applicable uses in a professional environment likely began in the postproduction industry. Adding digital disk recording (DDR) capabilities for production purposes spawned the era of non-optical video effects (VFX) for commercial and motion picture content. The new capabilities to store an endless frame of video as a static image; to instantly search, play or record to either a single frame or a series of frames; or to play video slower, faster, backwards and forwards (as in the instant replay) were features not previously available to the creative community.

Clip playback devices, graphics still stores, animated character generation and short-term video caching all became viable and usable features for broadcast, news, computer animation and such. As computer server technologies developed, their features (including file sharing, networking and file management) were all adapted to the videodisk system, and almost overnight the dawn of a new era in video recording was launched.

From the DDR to a Video Server

The evolution of migrating video from its origins (sources) on any form of media and then bringing it to a mobile device or into the cloud is now almost second nature; that is, today, video is just another *commodity*-based technology that is *expected* to work regardless of how the content was created or where it is to be delivered and consumed.

As expressed earlier, the video server wasn't developed overnight. Instead, video servers, for professional uses, were an evolutionary development with a steady progression that combined multiple technologies into feature sets that are now deployed almost everywhere and on almost any platform.

Initially the architecture of a video server consisted of multiple "DDR-like" components connected (see Figure 5.12-17) by varying interfaces and attached to a well-controlled, dedicated storage configuration—often with each of the components built into a single chassis configuration. The concept worked reasonably well until technology changed faster than the base product could be modified. As video compression

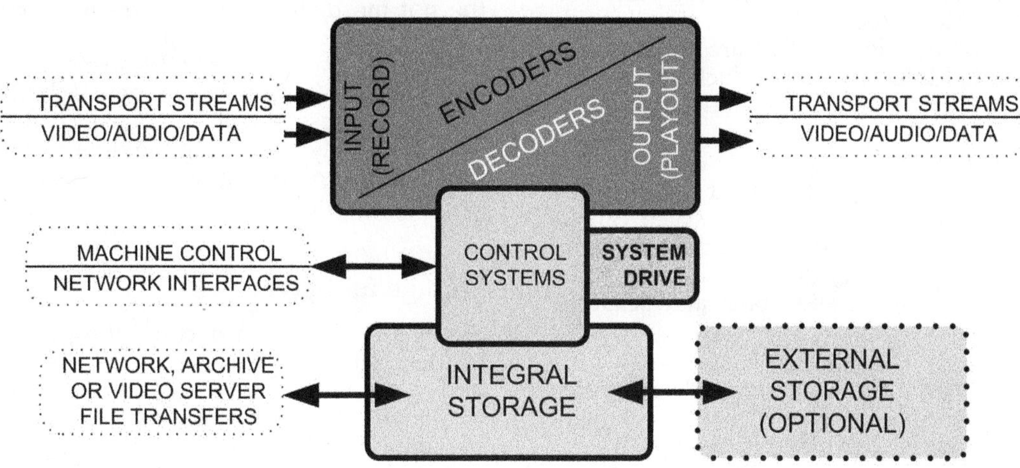

FIGURE 5.12-17 Fundamental basic blocks of a video server system.

evolved it improved exponentially, and the need for more codecs came became a natural progression. And that led to storage as one of the first sets of components that needed change. There simply wasn't enough space on a video server to hold the growing content requirements. The IT and computing sectors drove the developments in hard drives. The professional video server manufacturers leveraged that development and sought a means to add external storage solutions—which rapidly moved the entire industry to the shared storage model we're using today.

Next, additional and varying video compression formats became more popular. This drove the fundamental way each video server operated, forcing the inner workings to either be modified or replaced. Early codecs were set into hard-coded silicon. Changing those features meant replacing circuit boards—something not easily accomplished in the field. Eventually the video server moved to modular codecs on interchangeable sleds. When a new codec came out, it could be added to a new plug-in module and the features were now brought up to date. Servers could now be configured and updated in much the same way one might have configured a component home theatre or stereo system.

The development of software codecs and software-based RAID controllers changed everything. As processor speeds improved, so did the ability to do more in the core. Hard-coded silicon could migrate to flexible field-programmable gate arrays (FPGAs) or similar form factors. Disk drives could be upgraded without having to necessarily replace the controller or the chassis. The video server had migrated to the point where you could update certain features and capabilities without a wholesale forklift upgrade.

Commodity-Level Solutions in Two Decades or Less

Professional broadcast video servers now come in a multitude of form factors and configurations. Video playout and recording devices range from card-based sleds with SSD devices to common/commercial-off-the-shelf (COTS) servers with external third-party I/O cards (e.g., from Matrox and AJA), and they include purpose-built modular devices that are either standalone or embedded into various products from video production switchers to on-air "channel in a box" broadcast air chains.

Within just two decades, the concept of a video (or media) server has matured to the point that they are nearly a "commodity level" solution, which can be used for production, playout, time zone delays, graphics, security and surveillance, and at all levels of users from consumers through professional and scientific applications.

Applications for video server technologies now extend to full-service graphics systems. They stretch functionality and terminology to encompass conventional data servers with video playout; to "clip-servers" for all types of short-form playout; and to ingest platforms that take real-time video into a storage platform. For news and postproduction systems, multiple video servers running on COTS hardware provide a very high degree of sophistication including real-time layering or compositing effects, full-function media asset management and archiving capabilities.

Video servers now simultaneously generate multiformat files that include one to many high bit rate (HBR) masters and can add low bit rate (LBR) versions for browse or proxy or other streaming media purposes. Video servers can be configured to automatically generate or add technical or user metadata. They allow user tagging for file indexing and content identification. Video servers can store content to integral drives, to networked drives, to the cloud or to huge arrays of shared content stores that can be replicated across continents or campuses.

Video servers are capable of being controlled by any number of peripheral control interfaces, natively over IP, by apps, remotely over the Internet, directly via established control protocols (e.g., VDCP, ProBel, etc.) or by media asset management (MAM) and nonlinear editing for news or production devices. Yet that's not the end of the string of emerging uses for video recording devices.

Building in Software

Employing the concept of modular building blocks for broadcast servers is a model that continues. However, like all technologies, advancements in core services and approaches to architectures often redirect their uses and applications, or drive entirely new models for implementations. Video servers are no different.

SDI to IP

Within the last decade, products that were once purpose-built into "black boxes" and sold as a dedicated product are shifting to products that are more software-centric. Depending upon the services needed, purpose-built black boxes are gradually shifting to COTS (common/commercial-off-the-shelf) servers with SDI input-output cards and BNC connectors. In the not too distant future, these may be augmented with or replaced entirely by IP-based network technologies that are taking SDI-transport technologies to the domain of packetized video technologies. These various processes tend to drive costs down while simultaneously improving the time to market for new products or for adding new processes and features to existing products.

Storage for Servers

Storage for media and computing activities can be resident either on the actual (COTS) server or connected via a network to a storage system that consists of a larger "pool" of storage devices. In many broadcast equipment rooms you now find multiple sets of one- and two-rack unit "pizza box" servers configured as entire systems, servicing SDI ingest, SDI playout and media management. The servers are connected to external storage arrays of stunning proportions, reaching multiples of petabytes and interconnected via either Fibre Channel or 10 GbE (Gigabit Ethernet) network topologies.

Virtualization

For most technologies, change is inevitable. Next-generation video server architectures appear headed down yet another path. Much like "cloud" configurations, video server technologies are expected to move to a far more virtualized environment as well. Virtualization isn't actually new, but it is certainly making headways into the professional and broadcast services marketplace in noticeable way.

By virtualizing processes, entire sets of servers can be administered from a central control point, modified as workflows change, spun up to provide additional services and then spun down or reconfigured to perform other tasks once the previous services are no longer needed. Blade servers, a cost-effective "plug-in" like card-based server implementation, are an example of how virtualization may be used to support multiple services on an as-needed basis. See Figure 5.12-18.

Virtualization for broadcast (e.g., local commercial television station) servers may not necessarily make sense for mainstream broadcast station program playout, commercials or interstitial material simply because most station activities are for over-the-air (OTA) purposes and remain fairly well fixed. However, as broadcasters continue to offer more ancillary services, including secondary OTT or video on demand, virtualization of streaming services for content distribution makes perfect sense.

Instead of using a group of specifically configured servers for each of the streaming media preparation tasks, a pool of servers would be created that could easily (in software) be configured and then reconfigured for each of the workflow applications. The workflow would recall files from the news system and send them to this set of virtualized servers. The servers would load up the appropriate workflow task and perform the function as directed. Since this pool of servers was provisioned with varying software services, the same servers would perform a set of pre-stage functions (e.g., transcoding); and once those functions were complete, they would reconfigure themselves in software for a different task. Ultimately these same servers would provide delivery to the content distribution network or act themselves as an origin server and deliver directly to the hosting service or as necessary.

During the streaming processes, or as demands change, these servers would be throttled up to do just streaming or just transcoding or some other tasks (closed captioning, graphics overlay, etc.), all based upon demand by the devices requesting that content. The amount of physical resources becomes less, the utilization of the hardware increases, the number of errors is reduced, and the ability to maintain a single type of server instead of multiple types is increased.

How systems utilize the same physical servers for multiple purposes will ultimately change how systems are designed and will function, which is precisely the model used when those services are ported to the cloud.

OTT and Streaming Media Servers

The evolution of video playout has led the industry to a relatively new, and exceedingly popular, means of supplying content to a wider mixture of devices and consumer audiences. Yet, this requires that content must be prepared in a more diverse set of formats to meet the needs of the receiving devices (tablets, phones, etc.). Industry has named this service "over-the-top" (OTT) delivery. OTT is fundamentally based on technologies that utilize differing techniques depending upon how the content (files) are sourced and to which device(s) the content will be delivered.

Video servers once purpose-built to deliver only SDI video would not be the recommended means for preparing and delivering OTT video. Streaming media servers with specific software apps have become the mainstream component that is utilized for these relatively new kinds of services.

FIGURE 5.12-18 Blade servers are a cost-effective alternative when virtualizing operations where servers can provide individual or pooled services depending upon the operational needs at the moment.

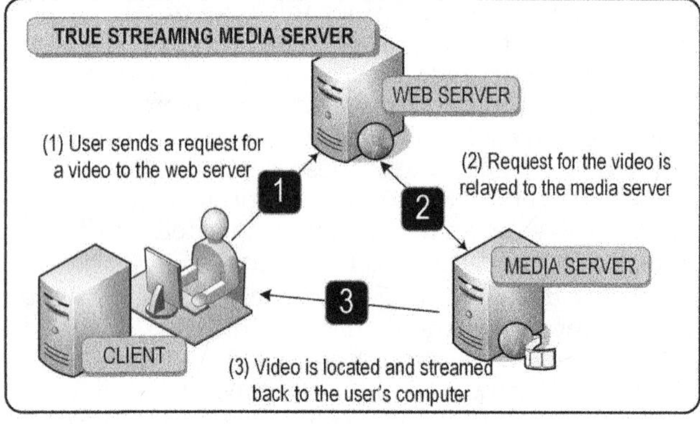

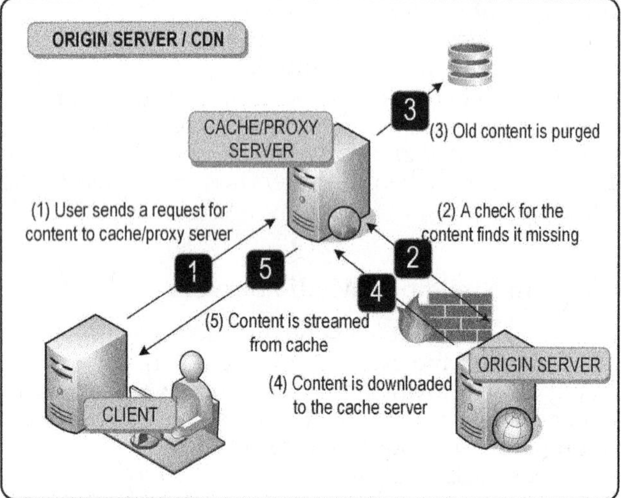

FIGURE 5.12-19 A "true streaming" media server concept (top) and the "origin server" concept with cache/proxy server (bottom) for media delivery.

In the United States, a distributor of video that provides video content by means of the Internet (or via other Internet protocol [IP] based transmission) is referred to as an "online video distributor" (OVD)—according to and as defined by FCC 13–99. The content distributed is considered "over-the-top" content and is delivered via streaming media servers and other conforming devices that shape the streaming files into IP streams according to the specifications of the receiving (often mobile) device. See Figure 5.12-19.

OTT video content is different from over-the-top messaging (e.g., "instant messages") in that the former is intended to deliver video content while the latter is intended for delivering short-form messages sometimes including video, audio or just text. OTT video content delivery will typically employ content servers (which store the original content) to feed programming files to the conforming devices and encoders; for example, as in HTTP Live Streaming (HLS) or adaptive bit rate (ABR) encoding. Streams are "packaged" into properly conformed groups, which are then sent over the Internet to those devices requesting the files—in that specific format.

The growth in OTT and streaming media has seriously altered the once pristine single methodology for getting content to viewers over broadcast systems. The technology components—that is, the servers, schedulers, transcoders and so forth—needed to properly conform the content so that mobile and fixed players can receive and decode the content is now part of the everyday system makeup for many broadcasters or their associated content delivery providers.

MAGNETIC RECORDING

Magnetic recording was first proposed in 1888, some 43 years after the discovery of rotational magnetic polarization. Yet it was not until the early 1950s before any true commercial development for storing computer-generated data in a semipermanent format began in earnest.

The first evidence of storing data magnetically began with Danish telephone engineer Valdemar Poulsen, who worked at the Copenhagen Telephone Company in 1893. Poulsen began experimentation with magnetism to record telephone messages. He built and patented the first working magnetic recorder, called the Telegraphone. See Figure 5.12-20.

Commercial data storage development was first demonstrated by depositing magnetic materials on cylindrical drums in patterns that emulated a data code—similar to binary codes we still use today. Recovery of that information was made possible by a device that would become what we now know today as the "magnetic head."

First-generation disk drives had a severely limited life span because the heads physically needed to contact the surface of the disk itself. As development progressed, IBM engineers found a means to float the head on a cushion of air—a fundamental principle that would become mainstay magnetic disk recording technology.

RAMAC Debut

Manufactured hard disk drives for commercial uses (usually internal to IBM products) were introduced on September 13, 1956. IBM's 305 RAMAC (which stood for Random Access Method of Accounting and Control) then had a storage capacity of five million characters. The RAMAC used fifty 24-inch diameter disks with an areal density of 2 kilobits per square inch. At a transfer rate of only 8.8 kilobits per second, the IBM model 355-2 used only a single head and cost $74,800 in 1956 dollars—an equivalent of about $6,233 per megabyte.

IBM's direct access storage product manager, Alan Shugart, assigned David I. Noble the task of designing a cheap and simple device that could load operating code into computers. To be labeled the *Initial Control Program Load* (ICPL), it was to cost only $5 and have a capacity of 256 KB. Noble experimented with tape cartridges, RCA 45 rpm records, dictating belts, and a magnetic disk with grooves developed by Telefunken before, in

HALF CENTURY OF SPINNING DISK DRIVES

1956	FIRST DISK DRIVE	IBM model 350 "RAMAC"	5 MB capacity, Consisting of fifty 24-inch disks
1961	FIRST USE OF ZONED RECORDING	BRYANT model 4240	90 MB capacity, Consisting of twenty-four 39-inch disks
1962	FIRST DRIVE TO USE AIR BEARING HEADS	IBM model 1301 "Advanced Disk File"	28 MB capacity, Consisting of twenty-five 24-inch disks
1963	FIRST REMOVABLE DISK PACK	IBM model 1311 "Low Cost File"	2.69 MB capacity, Consisting of six 14-inch disks
1965	FIRST DRIVE TO USE VOICE COIL ACTUATOR	IBM model 2310 "RAMKIT" - Single Disk Cartridge	1.04 MB capacity, Single 14-inch disk
1966	FIRST DRIVE WITH FERRITE CORE HEADS	IBM model 2314	29.17 MB capacity, Consisting of eleven 14-inch disks
1971	FIRST TRACK FOLLOWING SERVO SYSTEM	IBM model 3330-1 "MERLIN"	100 MB capacity, Consisting of eleven 14-inch disks
1971	FIRST FLEXIBLE DISK MEDIA (READ ONLY)	IBM model 23FD "MINNOW"	0.0816 MB capacity, Single 8-inch disk
1973	FIRST FLEXIBLE DISK MEDIA (READ/WRITE)	IBM model 33FD "IGAR" [sets industry standard for 8"]	0.156 MB capacity, Single 8-inch disk
1973	FIRST LOW MASS HEADS, LUBRICATED DISK	IBM model 3340 "WINCHESTER" [first sealed assembly]	35 or 70 MB capacity, two or four 14-inch disks
1976	FIRST 5.25-INCH FLEXIBLE DISK DRIVE	SHUGART ASSOCIATIONS model SA400	0.2188 MB capacity, Single 5.25-inch disk
1979	FIRST 8-INCH RIGID DISK DRIVE	IBM model 62PC "PICCOLO"	64.5 MB capacity, Consists of six 8-inch disks
1980	FIRST 5.25-INCH RIGID DISK DRIVE	SEAGATE model ST506	5 MB capacity, Consisting of four 5.25-inch disks
1981	FIRST PERSONAL COMPUTER	IBM model 5150 "IBM PC" with Floppy or Cassette only	[first IBM desktop model 5100, introduced in 1975]
1981	FIRST 10.5-INCH RIGID DISK DRIVE	FUJITSU model F6421 "EAGLE"	446 MB capacity, Consisting of six 10.5-inch disks
1981	FIRST 3.5-INCH FLEXIBLE DISK	SONY model OA-D30V	0.4375 MB capacity, Single 3.5-inch disk
1982	FIRST 9-INCH RIGID DISK DRIVE	CONTROL DATA model 9715-160 "FSD"	150 MB capacity, Consisting of six 9-inch disks
1983	FIRST 3.5-INCH RIGID DISK DRIVE	RODIME model RO 352	10 MB capacity, Consists of two 3.5-inch disks
1985	FIRST CARD-MOUNTED DISK DRIVE	QUANTUM model "PLUS HARDCARD"	10 MB capacity, Single 3.5-inch disk [disk plus controller]
1988	FIRST ONE-INCH FORM FACTOR 3.5-INCH DISK	CONNOR PERIPHERALS model CP3022	21 MB capacity, Single 3.5-inch disk
1990	FIRST DISK DRIVE WITH PRML ENCODING	IBM model 0681 "REDWING"	857 MB capacity, Consists of twelve 5.25-inch disks
1991	FIRST USE OF MAGNETORESISTIVE HEADS	IBM model 0663 "CORSAIR"	2.139 GB capacity, Consists of ten 3.5-inch disks
1993	FIRST 7200 RPM DISK DRIVE	SEAGATE TECHNOLOGY model ST 12550 "BARRACUDA"	2.139 GB capacity, Consists of ten 3.5-inch disks
1997	FIRST 10,000 RPM DISK DRIVE	SEAGATE TECHNOLOGY model ST19101 "CHEETAH 9"	9.1 GB capacity, Consists of eight 3.25-inch disks
1999	FIRST ONE-INCH DISK DRIVE	IBM "MICRODRIVE"	340 MB capacity, Single 1-inch disk
2000	FIRST 15,000 RPM DISK DRIVE	SEAGATE TECHNOLOGY model ST318451 "CHEETAH X15"	18.350 GB capacity, Consists of three 2.5-inch disks

FIGURE 5.12-20 Historical benchmarks for spinning (magnetic) disk drives.

1968, creating his own solution—called the "memory disk"[7]—which eventually became the *floppy disk*.

Noble's memory disk creation was nicknamed the "Minnow." It was an 8-inch diameter plastic disk, 1.5 mm thick, coated on one side with iron oxide and attached to a foam pad. The read-only disk rotated on a turntable driven by an idler wheel. The Minnow's capacity was only 81.6 KB. A read-only magnetic head was moved over the disk by solenoids that read prerecorded data tracks on a disk with a density of 1100 bits per inch. The disk was *hard-sectored*, meaning the disk was punched with eight holes around the center that marked the start of each data sector.

Single Sided Goes Double

By February 1969, the floppy was coated on both sides and had doubled in thickness to a plastic base of 3 mm. In June of 1969, the Minnow was added to the IBM System 370. Its name was changed to *Igar* (c. 1970). Two short years later, in 1971, the Igar disk became the 360 rpm model 33FD—marking it as the "first commercial 8-inch floppy disk."

Touting an access time of 50 milliseconds, the 33FD was dubbed the "Type 1 diskette." Its eight hard sector marking holes were later replaced by a single index hole, creating the first "soft sector diskette." The floppy had 77 tracks, which became known as "IBM sectoring." In 1976, the first double-density, frequency-modulated 1200-kilobit model 53FD floppy was introduced. It was followed by the 43FD dual-head disk drive, which permitted both read and write capability on both sides of the two-sided diskette.

The IBM floppy disk ironically emerged almost simultaneously with the Intel's first microprocessor, the 4004, released in 1971. It would take another ten years, August of 1981, before the first personal computer, IBM's "PC," would incorporate removable floppy disk drive technology for data storage and the infant microprocessor. These sets of components and their respective products would change the way the world worked forever.

Disk Drives in the 1960s, '70s and '80s

By 1962 IBM had introduced its first commercially available 28 MB disk drive with air-bearing flying heads. The IBM 1301's heads rode above the surface at 250 microinches, a decrease from the previous spacing of 800 microinches. A removable disk pack came into production in 1965 and remained popular through the mid-1970s. A year later, ferrite core heads became available in IBM's model 2314, which would later accompany the first modern PCs.

IBM's Winchester drive (introduced in 1973) bore the internal project name of the 30-30 Winchester rifle and employed the first sealed internal mechanics. The 3340 Winchester drive had both a removable and permanent spindle version, each with a capacity of 30 MB. Flying head heights had now been reduced from the previous 250 microinches to 17 microinches.

A 5–1/4-inch form factor drive, Seagate's ST-506, was introduced in 1980. The drive featured four heads with 5 MB capacity. IBM later introduced the PC/XT with an available 10 MB model ST-412 drive. The 5–1/4-inch hard drive and 3–1/2-inch floppy configuration set the standard for many of the PC-compatible products for the next few years.

In 1983, Rodime introduced its 3–1/2-inch form factor model RO352. The 3–1/2-inch drive and mechanical interfaces remained the "universal size" for most of the HDDs until the 2–1/2-inch form factor was introduced for portable applications in 1988. See Figure 5.12-21.

Early Video Recording—Disk and Tape

Linear (tape) and rotating magnetic storage surface video recordings had a similar and parallel development. In step with the development of magnetic recording tape, random-access video on demand (first predicted in 1921) was demonstrated in principle as early as 1950.

Not long after, John Mullin in 1951 at Bing Crosby Enterprises demonstrated an experimental 12-head VTR that ran at 100 ips (inches per second). The concept of recording video onto a spinning platter was demonstrated shortly before in the late 1950s, at about the same time the first disk drive and NTSC television were introduced. Black-and-white NTSC was first developed in 1941, but without a provision for color. A second NTSC version, with backward compatibility, was adopted in December 1953.

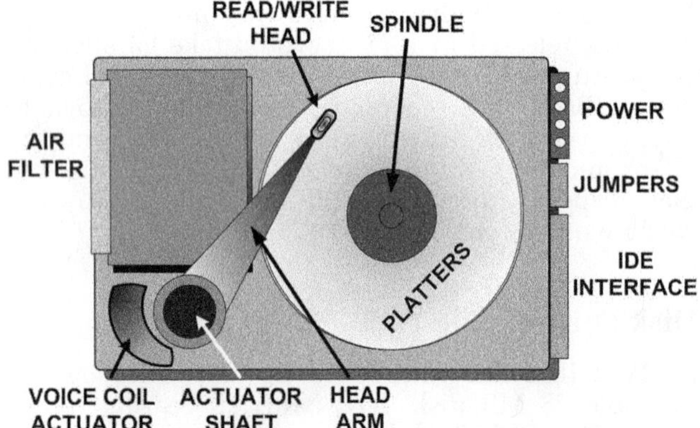

FIGURE 5.12-21 Components of a legacy IDE magnetic hard disk.

In 1957, Italian Antonio Rubbiani demonstrated a rudimentary plastic videodisk at the Salone Internazionale della Tecnica. Inspired in part by Rubbiani's demonstration, a few years later, technologists at CBS developed a technology that would become a videodisk recorder.

Videotape Recording

Developed by a team of engineers from Ampex Corporation (founded in 1944) and led by Charles Ginsburg, work began on the videotape recorder (VTR) in 1951. Ampex demonstrated the first three-head system in November 1952. In March 1953, a second system, using four heads, was shown. However, problems known as the "Venetian blinds" effect persisted due to discontinuous recording from one head to the next.

By 1954, Charles Anderson and the Ampex team, including Shelby Henderson, Fred Pfost, and Alex Maxey, were working on an FM circuit that debuted in February 1955. Ray Dolby later designed a multivibrator modulator, Maxey discovered how to vary tape tension, and Pfost developed a new sandwich-type magnetic head.

This team would go down in history as the fathers of videotape recording. The National Academy of Television Arts and Sciences (NATAS) would award Ampex its first Emmy in 1957 for its development. Ampex received a total of 12 Emmys for its technical video achievements. And a half century after the debut of the "quad," the original team was awarded, some posthumously, a Lifetime Achievement in Technology Emmy for their contributions.

First Public Demonstrations and Videotape Is Trademarked

In anticipation of the first public demonstration of video recording for the April 16, 1956, Chicago convention of the National Association of Radio and Television Broadcasters (NARTB), an improved VTR, which would later become the Ampex Mark IV, was shown to Bill Lodge of CBS and others. The Mark IV, later renamed the VRX-1000, used 2-inch-wide videotape running at 15 inches per second past a transverse track, rotating head assembly. Circuitry included frequency modulation (FM) for the video and amplitude modulation (AM) for the sound recording. Ampex trademarked the name *videotape* for its recorder.

In 1959 the first color videotape recording debuted during the Nixon-Khrushchev Kitchen Debate in Moscow. See Figure 5.12-22.

From Tape to Disk

Renowned Minnesota Mining and Manufacturing (3M) Company, which produced the first 2-inch videotape for the Ampex VRX-1000, showed a noise-plagued videodisk, publicly demonstrating in 1964 that "this new *disk* format had a future." Although less than a year before the first showing of a random access,

HISTORY OF RECORDING AND MEDIA

1877 – Edison makes the first recording of a human voice on the first tinfoil cylinder phonograph.

1898 – Valdemar Poulsen (Denmark) patents the "Telegraphone," the first magnetic recorder that uses steel wire.

1925 – The first electrically recorded discs and Orthophonic phonographs go on sale.

1928 – Dr. Fritz Pfleumer patent in Germany for application of magnetic powders to strip of paper or film.

1944 – The Minnesota Mining and Manufacturing (3M Company) began coating experiments for tape in under Ralph J. Oace.

1948 – The first Ampex Model 200 tape recorders employed for *The Bing Crosby Show #27*, recorded on 3M Scotch 111 gamma ferric-oxide coated acetate tape.

1951 – The Ampex team, led by Charles Ginsburg, begins work on a videotape recorder in October. Bing Crosby Enterprises demonstrates its experimental 100 inches per second, 12-head VTR.

1953 – Vladimir K. Zworykin and RCA Labs demonstrate a longitudinal three head VTR running at 360 inches per second that uses amplitude modulated (AM) sound

1956 – Ampex demonstrates the first practical quadruplex VTR at the Chicago National Association of Broadcasters (NAB) exhibition on April 14. The "quad" uses 2-inch-wide 3M tape running at 15 ips over a rotating head assembly recording at a slant on tape surface with AM sound. Over the next four years, 600 units at $75,000 each were sold, mostly to the television network stations.

1959 – Toshiba demonstrates a prototype single head helical scan VTR using 2-inch tape running at 15 ips. Following the demonstration, Sony begins to develop its helical scan VTR.

1963 – Sony markets the first open reel 1/2-inch helical scan VTR for home use at a retail price of $995.

1967 – The first instant replay is shown using the Ampex HS-100 color video disk recorder for ABC's "World Series of Skiing" in Vail, Colorado.

1969 – Sony introduces the 3/4-inch U-Matic one-hour the first videocassette; which is available in the U.S. by 1971.

1975 – Sony introduces the Betamax consumer VCR. The console sold for $2295, with a one-hour 1/2-inch tape cassette selling for $15.95.

1976 – JVC introduces to Japan the 'Video Home System' (VHS) format videocassette recorder (VCR) for US$885.

1978 – Pioneer develops the LaserDisc, first used by General Motors to train its Cadillac sales staff. Pioneer began selling the home version LaserDisc players in 1980.

1984 – Digital Equipment Corporation (DEC) introduces Digital Linear Tape (DLT), which was formerly called CompacTape; and the TK50 drive for MicroVAX II and PDP-11.

1985 – Magneto-optical (MO) disc drive is introduced.

1985 – The standard for the Compact Disc Read Only Memory (CD-ROM) computer discs (produced by Sony and Philips) that use the same laser technology as the audio CD is introduced.

1987 – Digital Audio Tape (DAT) players are introduced.

1991 – First ADAT (Alesis Digital Audio Tape) recorders are shipped in March.

1992 – Digital Compact Cassette (DCC) by Philips and Matsushita (Panasonic) created. As a rival, Sony develops the MiniDisc.

1996 – DVD players are first introduced, and only available in Japan.

1997 – The first DVD players available in the United States.

1998 – The first DVD players are available in Europe.

2000 – Sony introduces the PlayStation 3 console in Japan, with capabilities of playing DVD movies.

2000 – September, first commercial LTO-1 tape format with 100 GB storage and 20 MB/second data transfer speeds introduced.

2002 – The Blu-ray Disc Association is founded. Sony, on February 19, 2002, in part with nine companies, unveil the plans.

2002 – On August 29, Toshiba and NEC propose to the DVD Forum, a next generation optical disc format which will become the HD DVD.

2004 – January, Sony introduces the Hi-MD, a further development of the MiniDisc format, which would later support MP3 as well as other computer data formats.

2004 – Toshiba unveils first prototype HD DVD player at the Consumer Electronics Show, to be backward compatible with the DVD.

2005 – Sony and Toshiba begin discussions of a possible single format HD version of the DVD, which end up nowhere.

2006 – The first Blu-ray players are sold, with the first title released on June 20, 2006. Microsoft announces it will support the HD DVD for the Xbox 360 gaming platform, including movie playback.

2007 – Combination HD DVD and Blu-ray players as dual format are introduced. It is almost too late, the studios are starting to announce it is stopping support for HD DVD in favor of Blu-ray Disc.

2008 – Through out the year, HD DVD production and retailers stop carrying or producing content – Blu-ray Disc wins the format battle.

2010 – January, first LTO-5 specification is announced, products become available in Q2-2010.

FIGURE 5.12-22 Historical records of recording and media through 2010. By this point, most of the recording processes were file-based and utilized a media that might not tangibly be touchable or physically transportable (with the exception of digital data tape, such as LTO). The digital age had, by now, transformed an industry forever.

still-image generating, recording, and playback device, the demonstration portrayed disk-based recording technologies that would change the future of recording in a profound way.

Instant Replay History

The infancy of videodisk recording is probably best associated with live sports in the form of the "instant replay." According to historical recollection, the first ever use of an instant replay came during a Canadian hockey game in 1950 that utilized film as the recording medium. Toronto director George Retzlaff used a "hot processor" to develop kinescope footage of an ice hockey goal for replay within 30 seconds. Without an approval for Retzlaff's experiment, and no other technology to replicate it, the Canadian Broadcasting Corporation (CBC) that produced the game prevented Retzlaff from reusing it.

Fifteen years later at the July 1965 SMPTE conference in San Francisco, MVR Corporation showed a 600-frame (20 second), black-and-white video recorder, the model VDR-210CF, that recorded individual frames. Around August 1965, CBS used the technology as a freeze-action videodisk. The MVR, with its shiny aluminum, nickel-cobalt–coated magnetic disk, was used in football telecasts to instantly play back short action sequences in normal motion with the capability to freeze the motion on a single frame.

Taking a different approach, Ampex created an 1800 RPM spinning metal disk with a series of stepper motor-driven recording heads that moved radially across the platters, creating 30 seconds of normal video using analog recording technologies.[8] The device recorded 30 video tracks per second, with each track holding one NTSC frame, giving a total of 1800 NTSC fields. The heads could be rapidly moved to any location on the disk for replay at normal speed or, when the head stepping was slowed and the same frame repeated in multiple sets, slow motion playback was possible. When the playback stepper heads were stopped (with the platter continuing to spin), the same frames would be repeated to produce a freeze-frame image display. Broadcast-quality video from a disk recorder was tested during a live July 8, 1965, CBS Television football broadcast.[9]

FIGURE 5.12-23 Videodisk recorder's platter from the Ampex HS-100.

AMPEX Corporation Press Release

A commercially available videodisk recorder with true slow and stop motion, the Ampex HS-100, was officially placed into broadcast service in March 1967. The "video effects" device enabled "quick turnaround playback in normal, slow, or stop action." With the World Series of Skiing program (ABC's *Wide World of Sports*, March 18, 1967[10]) from the U.S. Ski Championships in Vail, Colorado, the disk recorder marked the dawn of instant replay for television broadcast. See Figure 5.12-23.

Disk storage and disk recording technologies would change dramatically between 1965 and 1984, when the concept of a "video serving device" would emerge, a migration that would take its roots from the early instant replay of television sports.

Disk-Based Video Production

Early applications for (digital) disk-based video storage and image reproduction came in the form of graphic arts applications. Development for recording multiple static images to spinning disks continued as technologies for disk drives advanced. From these foundations, the emergence of *digital* storage was born.

In 1981, Quantel introduced its *Paintbox*, the premiere video graphical composition tool that became the benchmark for television art and graphics. *Paintbox* laid the groundwork for what would ultimately change television production, video graphics and composition forever. Like with other devices that needed a means to store work, Quantel needed to figure out how to store digitally generated graphic compositions. Without a digital video (tape) recorder available for data storage, digital representations of images from the *Paintbox* were off-loaded onto either 8-inch floppies or their Freestyle Dash (FSD) removable hard drive, a transportable hard disk drive from Control Data/Hitachi/NEC, permitting exchange between proprietary Quantel systems only.

In 1986, Quantel's *Harry* became the first high-quality, integrated, and true nonlinear editing (NLE) system to use disk-based digital technologies to create multilayered moving video and effects. *Harry* remained the standard for television special effects and graphics (which included *Henry* and *Hal*, circa 1992). With Quantel's 1993 introduction of their Dylan "fault-tolerant drives," the company set the stage for RAID-like centralized and protected video storage going forward, albeit entirely for their proprietary-dedicated systems.

Following these developments by Quantel, the need for a simpler, stand-alone disk recording device with component digital quality surfaced. NTSC (composite) recording, onto less than full-resolution rigid and semi-rigid media, was already available. However, these recordings were constrained to short segment recording and playback with simple stunt features such as reverse or stop motion, and faster than real-time playback. For general-purpose storage of graphic images, manufacturers began to produce electronic still stores (ESS)—devices that could store hundreds to thousands of full-frame (or less) video images or graphics, and then recall them live/on demand using an integrated database or library-like "asset management" system. The storage systems utilized the same types of data storage drives and removable disk-packs that the mainframe computer industry used.

Abekas Video Systems, co-founded by Junaid Sheikh and Phil Bennett in 1982, created the A62/A60/A64 series of disk-based production systems. Their innovations pioneered a new trend for production-quality digital video recording. Available in a single disk chassis, or as pairs of disk sets with compositing or layering engines, Abekas digital disk recorders worked with both PAL and NTSC standards in composite analog, composite digital or component digital video. Animation, graphics special effects and early computer graphics (for animation and 3D rendering) drove single-frame recording and short segment clips. However, another important and significant technology also emerged from the concepts of the Abekas designs—the foundation of *file-based interchange* of digital images over a simple network interface.

File-Based Interchange and Control

Abekas disk recorders accepted and produced files that were data representations that could be transported (via TCP/IP) to other devices, in particular 3D and 2D computer-generated imaging (CGI) platforms. Abekas files were produced in two formats: 24-bit RGB and 16-bit YCbCr. Both image formats were 720 pixels wide by 486 pixels high—almost precisely what the first component digital video format would become for television moving images. Pixels from the products were not square, so an aspect ratio conversion factor of either 0.9 or 1.111 needed to be applied depending on whether squaring or unsquaring the pixel was needed for the proper resolution and scanning rates of finished video.

Abekas products featured control and transfer capabilities, such as Telnet, allowing files to be exchanged between other computer-centric devices. Simple single-line ASCII codes told the devices to select or play a single frame or a series of frames identified by numerical indexing specific to the drive. The Abekas A60 stored 750 PAL frames (for 25 fps) or 750 NTSC frames (30 fps) of broadcast standard video in the digital storage format specified by (at that time) the CCIR-601 specification. The disk format's native YCbCr format also had encode or decode firmware that processed 3-byte red-green-blue (RGB) images on the fly, permitting TCP/IP file transfers to and from external devices.

These concepts became the de facto process for 3D computer graphics and animation, given there were, at the time, no standards for file formats or interoperability between platforms. Graphics and post facilities would develop their own processes for integrating the individual frames as discrete files. A precise video storage format would create files that had no options, no header and were always 720 pixels per line, 576 lines per image.

VIDEO AND MEDIA SERVERS—COMPONENTS

When deployed for professional broadcast uses, the video server is usually construed as "a nonlinear recording and playout device dedicated to the tasks of delivering video in real time." In broader terms, this *computer-based storage and encoding device* is user configurable and can perform a multitude of functions that can be deployed in a number of differing platforms and form factors.

Sometimes the term "media server" is used in a similar context; however, the media server is, by definition, a computer-based device that simply stores and shares digital media as (compressed) video files. Media servers may, depending upon what software (or firmware) options are added, do more than just serve media-centric files; yet they are not usually associated with real-time, broadcast-specific SDI images. Instead, the media server, per se, is better associated with streaming media, such as the distribution of compressed video streams over IP-based networks.

Wider Uses

Servers built for video or media services now encompass a much wider selection of applications and uses, particularly with the rapid growth of Internet services including OTT, multiscreens, video on demand (VOD) and not to forget the widespread inclusion of user-generated video on social media. For the purpose of this section, the descriptions and technologies are intentionally focused on the resources typically available for professional broadcast "video server" applications, those deployed for long-form (program length) content and commercial (short-form) or interstitial (promotional) content.

With that statement, we now see the tides of server development and applications are changing rapidly. Within just this past decade the traditional (broadcast) video server has evolved from a number of low-input ports/low-output ports to much larger scales, which now include dozens or more I/O ports and interfaces to large-scale storage systems for content management. Video servers are deployed in almost every content delivery model used by broadcasters today. Services utilized to deliver video and the devices that now view those videos have driven the traditional, original concepts for the video server to new dimensions.

While videotape recorders (VTRs) essentially had a limited set of functions (record, edit, playback), video servers have opened the door to many more video-centric tasks. This next section looks briefly at the variances available in devices that "serve video" for broadcast, non-broadcast and emerging applications.

Types and Tasks

In computing, a server is a computer program or a device that provides functionality for other programs or devices (usually called *clients*). There may be client–server models, single computational models or IT models where a computer program provides services to other computer programs or devices. Computational services may be distributed across multiple processes or devices, and, as in a Web server, the computer program serves requested HTML pages (or files) to clients or users.

A video server is essentially a computer-based device that is tailored to deliver video and audio content.

Functionally, the tasks that the video server can be used for include:

- ingesting of video content from one to many sources
- the retention of video files either locally or to a shared network storage system
- the management of those files and their associated metadata
- the consolidation, editing and assembly of the files into video clips of varying lengths
- the transfer and delivery of the files (as clips or long-form programs) to additional resources including other video servers, storage systems (including the cloud and archives) or for playout of the files as real-time video streams for broadcast purposes

Media Server Differences

The term "media server" is somewhat vague, as these devices are more closely associated with non-real-time media, video on demand, networked delivery platforms and similar workflows.

Typically, when thinking about a media server (versus a video server), it is sometimes in the context of devices such as personal (digital) video recorders (PVRs and DVRs), home theatre or home media content servers, streaming media for personal and professional uses, and even origin, upstream and edge servers that specialize in video on demand (VOD) and over-the-top (OTT) delivery and potentially are integrated into upcoming ATSC 3.0 devices.

Media servers are sometimes viewed as "file servers" that are tailored to moving image and sound content. They often exclude the real-time ingest from baseband codecs and will generally only send those files to software systems that have players that do the "heavy lifting" of changing media files into real-time viewable/hearable content. Whereas the traditional broadcast video server generally incorporates an end-to-end systemization—which can take baseband (digital) video in and store it for playout in the same fixture—media servers manage the content at the file level and distribute it to external devices, typically over a network, through a content delivery network (or "CDN"), or to another device such as an origin server (a device that "originates" the file delivery), as prescribed by the platform(s) it supports.

Origin[11] and Upstream Servers

In computer networking, an "upstream server" implies a server that provides service to another server. The upstream server is located higher up in the "hierarchy of servers"—that is, the varying levels of services that are provided by two or more servers in a network.

The highest server in the hierarchy is sometimes referred to as an "origin server." The inverse term, downstream server, is rarely used, but it might imply a device which is lower or lowest on the server hierarchy. The terms upstream and downstream are exclusively used in contexts where requests and responses move in opposite ways, somewhat like "northbound and southbound" is referenced in a network topology.

An origin server can be configured to support a wide range of multiscreen video services delivered across public and private content delivery networks or CDNs. Intelligent caching and dynamic content adaptation features enable support for streaming media demands—which differ from what is typically done for broadcast video servers. Origin servers provide support and delivery of stream-based (and sometimes file-based) media to an ever-increasing array of consumer viewing devices over connected and wireless topologies.

Origin servers can support live, on-demand and network DVR services on a single platform. Through variable processes, sometimes referred to as "dynamic content adaptation" (DCA), content is customized (formatted) at the point in time when it is streamed. This difference ensures compatibility with the consumer device on which the content is intended to be viewed. Features such as just-in-time repackaging, digital rights management and session-based policy controls enable services to reach more devices, thus reducing the time to market and managing storage costs more efficiently.

Origin servers may be configured to support integrated caching and storage with internal, external or third-party video storage solutions. With the dynamics of broadcast changing as rapidly as technology (and ideas) grow, origin server software features are capable of supporting live broadcast, VOD, catch-up TV and similar network single-copy and shared-copy digital video recorder (DVR) applications or workflows. The platform it serves must seamlessly integrate with public and private CDNs in order to deliver reliable video streams across multiple devices and at multiple bit rates and formats.

Edge Devices and Edge Servers

The term "edge" is used to describe what resides at the far end point of a network or a connection. Thus sometimes there is confusion or overlap in defining the functionality and the location of edge-based services, devices or servers. See Figure 5.12-24.

An edge *device* generally provides the entry point into an enterprise or service provider's core network. When thinking in networking terms, these points might include routers, routing switches, integrated access devices (IADs), multiplexers and a variety of metropolitan area network (MAN) or wide area network (WAN) access devices or points.

Edge devices can further provide connections into carrier and service provider networks, implying that the end point of one's internal network—the point in the central equipment room where the signals are about to leave or enter the premises—reaches out to (or in from) the edge (i.e., the origin point) of a carrier's network.

Two examples might lead to the terminology confusion. An incoming feed from a carrier may land in an edge device (or edge server) that resides "on-premise" where it waits to be brought into the local area network of the facility. Conversely, an outbound feed may be cached to an edge device, where it waits to enter the

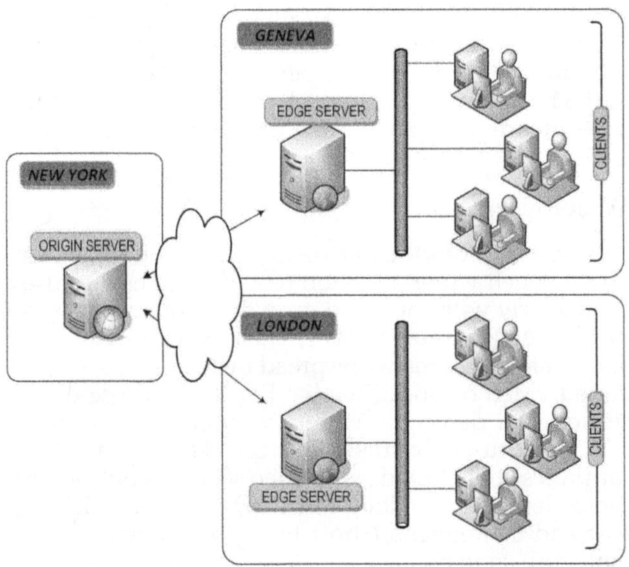

FIGURE 5.12-24 Edge servers deployed at local sites and served by an origin (primary) server.

carrier network before it actually "enters" that outbound network.

An edge *server*, in a network system administration context, is any server that resides on the "edge" between two networks—typically a private network and the Internet. Edge servers can serve different purposes depending on the context of the functionality in question.

Thus the edge device will have a minimum of two interfaces, one facing an internal or known network and the other facing the Internet or a foreign network. Usually in network implementations, "edge" implies the device separates a private network from the Internet—that is, unless the network has its own autonomous system (AS) number and the network employs an exterior gateway protocol designed to exchange routing and reachability information, as in Border Gateway Protocol (BGP).

Video and Media Server Capabilities

Video servers and media servers each play important parts in today's digital world. Whether as part of a news editorial and playout system, a broadcast program ingest and on-air playout system, a component in a live video production control room, or as a streaming media engine distributing content to a content delivery method, the concepts and technologies employed in video/media server technologies have shaped how our content is collected, stored and distributed in profound ways.

The following topics cover the foundational principles, components and basic operational architectures utilized primarily in video servers designed fundamentally for broadcast applications.

Scalability

When resources are added, performance should not decrease. Broadcast video server products are generally designed with a modular approach that allows for expansion of the structural hardware that supports its operation.

Scalability is the measure for a system to increase performance when an increase in loading occurs, as when resources (services and hardware) are added. A system whose performance improves when adding hardware proportionally to the capacity added is said to be a *scalable system.*

Expansion capabilities may include (but are not limited to):

- adding more input or output ports
- increasing the number of streams being ingested or played back
- increasing the ability to simultaneously increase services—such as adding FTP bandwidth or doing file-based ingest while performing all other required operational tasks
- incrementally increasing storage capacity without excessive reconfiguration

Extensibility

A system design principle where the implementation takes into consideration future growth, *extensibility* is a systemic measure of the ability to (a) extend a system and (b) manage the level of effort required to implement that extension. Obsolescence, especially in technology-driven systems, is a fact. For a product to be extensible infers that the product can continue to be of value and will not be rendered useless, so that feature sets can be added without a complete replacement of the system, all while minimizing the impact to existing system functions and operability.

In general, the concept means that the devices used in the current server implementation (which were built before later definitions were established) will accept and process variations in the materials that they have been designed to handle. A measure of extensibility includes the ease with which a system or component can be modified to fit the area it addresses now and in the future.

Efficient Interoperability

Video server designs, even in their infancy, had the ability to provide services to, exchange services between, and accept services from other systems; that is, they were intended to be interoperable. Those services include, at a minimum:

- the ability to exchange standard digital (or analog, if so equipped) video signals between the video server and other video systems (video routing, video monitors, etc.)
- the ability to exchange file-based data between like systems (e.g., to FTP files between video servers from the same manufacturer)
- the ability to be controlled through established mechanisms with a minimal set of functionality (i.e., to start, stop, cue a clip for playout, remove that clip for a queue, etc.)

For a video server to be most efficient, services and data need to be exchanged in a fashion that will enable them to operate effectively together.

Ideally, video servers need to exchange information transparently between systems, regardless of the native format of the video server itself. For a server to be extensible, or "future-proofed"—from an interoperability perspective—it should be capable of conveying its information over interfaces currently used and be adaptable (within limits) to those that have not yet been defined.

At the boundaries of these capabilities—excluding file-based interchange, which came later in server development—is the concept that a video server should have no less functionality than its predecessor, the videotape transport. Indeed, the video server should also offer new features that were previously unobtainable in a videotape-based environment.

File Interchange

The movement from a common baseband video interchange to one that was file-based was no easy task.

As was evidenced early on, the lack of interchange on most of the origination file formats available from any given manufacturer (an example being Motion JPEG) was crippling to interoperability. The industry, and the marketplace, recognized that video servers needed to strive for a common set of standardized interchange formats and protocols with which to carry and exchange media and control data from manufacturer X to manufacturer Y. Even after over two decades of video server product implementation, this process is still not without certain issues, and in many cases it requires external services (e.g., transcoding) to be included in the overall systemization mix.

Developments in the Material eXchange Format (MXF) opened the door to a solution, but this has still not proven to be an overarching, universal and without modification interchange, due to many factors. As new implementations unfold, existing standards evolve and new protocols for transport emerge. The ability to interchange content at a file or stream level amongst dissimilar video server products may still require "gateway" type devices that can modify and/or adapt a file from one server to another server.

As the broadcast industry moves into the domain of Internet Protocol Video (IP Video), the adaptations necessary for the real-time transport and interchange of packetized digital video—uncompressed, as in the developing SMPTE ST 2110 suite of standards, or as encapsulated SDI, presently available in the SMPTE ST 2022 suite—must become less complex than those that we have in place today.

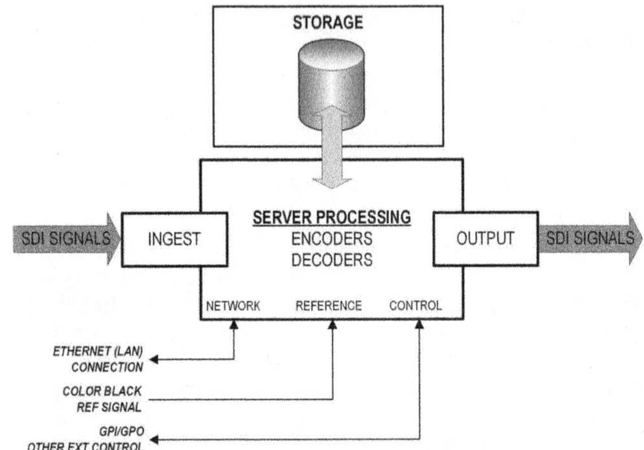

FIGURE 5.12-25 Fundamental video server block diagram (native SDI I/O, external storage, network interface).

Basic Video Server Components

Broadcast video servers are fundamentally composed of two system components: (a) the record/play and/or network systems that draw from and/or deliver to a media storage component and (b) the media storage system that warehouses the content that moves to/from the record/play or network component.

Sometimes, for smaller configurations, both of these components will be integrated together into a single system in a single platform form factor. For larger broadcast- or network-level systems, station playout or enterprise systems, the components are physically separate devices connected on a network and may be managed by discrete sets of programming and/or application interfaces controlled via a dedicated system manager.

As illustrated in Figure 5.12-25, server inputs and outputs (I/O) link to encoders that accept real-time video and compress the signal to a user-selected format (e.g., MPEG, DV or AVC-Intra) and store it on the HDD system (or in some devices, SSDs). On playout, decoders convert files and produce real-time video streams, usually in a digital (SDI) format.

The server's machine control interface may be via videodisk *control* protocol—sometimes shown as videodisk *communications* protocol (VDCP)—that is asynchronously communicated at 38.4 Kbps transfer rate to controller systems conforming to the traditional EIA RS-422A or another "native" network-based protocol. Based originally upon Louth automation systems' control instructions, VDCP allows the outside world to issue commands such as start, stop, cue, send to archive, and several other functions.

Content may be stored integrally with the video server (as part of the same chassis supporting the server) or in a separate data storage system external to the video server itself. External storage may be classified as a dedicated storage product designed to match the server components (e.g., a SAN that is part of a production editing system); as a storage system that is directly connected as a network-attached storage (NAS) package; or as a shared storage system provided by a third-party vendor (i.e., a SAN or a NAS tailored with the appropriate drives and interfaces) that may be part of a much larger central storage supporting much or all of the media in a facility.

One other important component in the server ecosystem is the *system manager*. This, like the inner workings of a personal computer, will generally house one or more operating system drives, file system management, control interfaces, diagnostic systems, a database for external integration, and those applications necessary to make the servers and storage systems work together. The system manager may also be the interface to third-party devices or to an external media asset management system either via application programming interface (API) or via direct communication services.

System managers play important roles in the overall architecture of a system. They may perform functions like load balancing and resource allocation. Each server manufacturer will configure their version of the system manager and tailor it to the codecs, servers and storage systems that they provide. In some cases, as when deploying servers coupled to external or third-party storage platforms, the system manager will aid in regulating how data is transferred to/from the various components in the system.

Channels and Engines

A "video engine," or "channel" as they are sometimes referred to, consists of the record/play unit, the codecs for converting video to files and vice versa, the physical layer I/O (connectors such as BNCs for SDI video) and other elements described in the following sections (see the illustrations in Figures 5.12-26 and 5.12-27). Depending upon the manufacturer's specific product, I/O and control components are sometimes combined to a single codec board that permits port reconfiguration when the system changes the channel's function from an ingest (encoding) device to a playout (decoding) device. This is sometimes referred to as a "bidirectional" channel and is often used in the quality assurance process for reviewing content that was just recorded.

Video Signal Input-Output

Video inputs and outputs are usually physical 75-ohm BNC connectors that accept digital video as SDI, per these and other SMPTE standards:

- SMPTE ST 259M-2008 for standard definition (SDTV 10-bit at 143/270/360 Mbps)
- SMPTE ST 292–1:2012 for high definition (HDTV nominally at 1.5 Gbps)
- SMPTE ST 424:2012 for payloads of 2.97 Gbps or 2.97/1.001 Gbps (formerly SMPTE 424M-2006)

Over the course of technology's evolution, legacy signals including analog composite (NTSC/PAL) and three-wire component analog (Y, Pb, Pr) video have almost disappeared from professional broadcast video server products. Users are now expected to provide external format conversions for these legacy inputs or outputs since those analog signals for most facilities have moved on.

Most servers produced since the early 2000s include embedded audio using the AES standard (for digital audio) over a serial transmission format as two-channel linearly represented digital audio data. One or more AES serial streams may be carried in a single SDI signal transport. Some servers may still offer discrete (non-embedded) AES capabilities, but with lessening availability as technology moves further from discrete transports of video.

Signal I/O interfaces often accompany the actual codec card, which may be mounted on a "sled," allowing for changes in the server configuration to be made

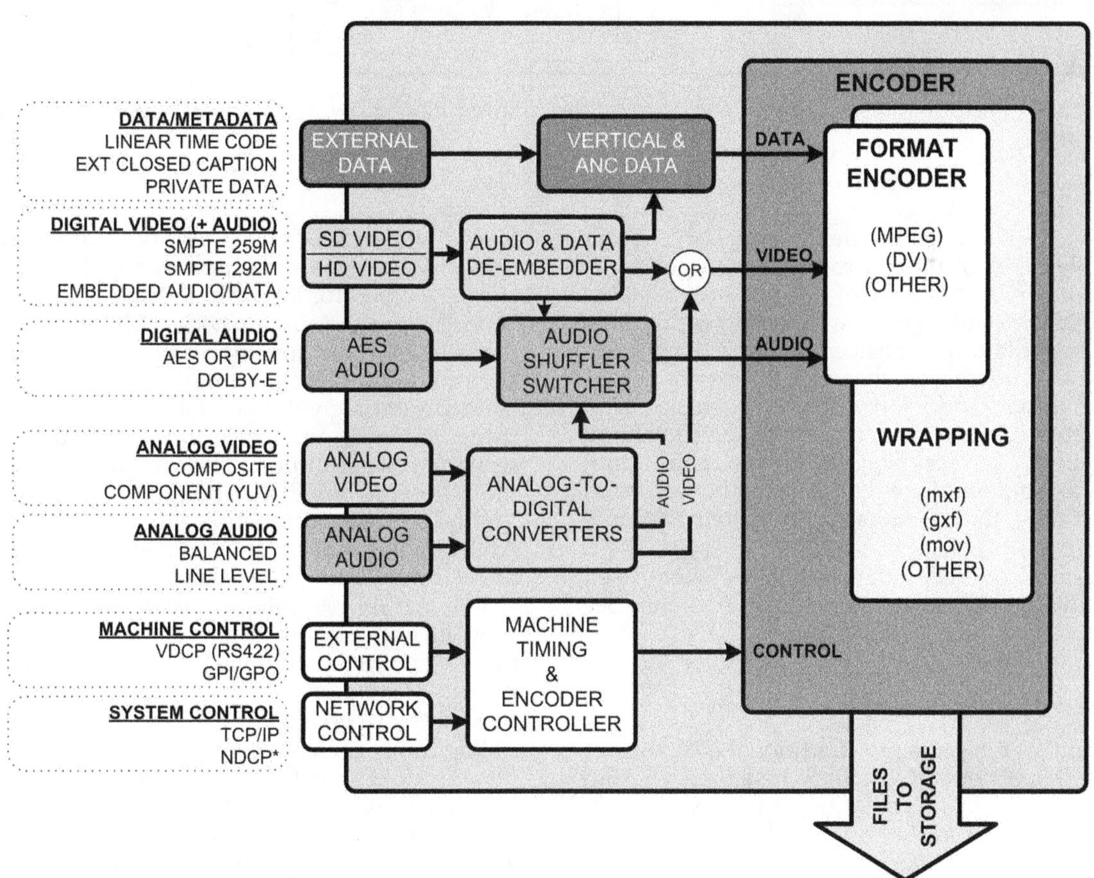

FIGURE 5.12-26 Input functions of a server channel showing analog, digital, control and other ancillary data (time code, metadata, closed captioning). Many of these signals in today's technologies are no longer "discrete" or individual connections and instead are carried on an IP network over Ethernet or are embedded in the SDI transport.

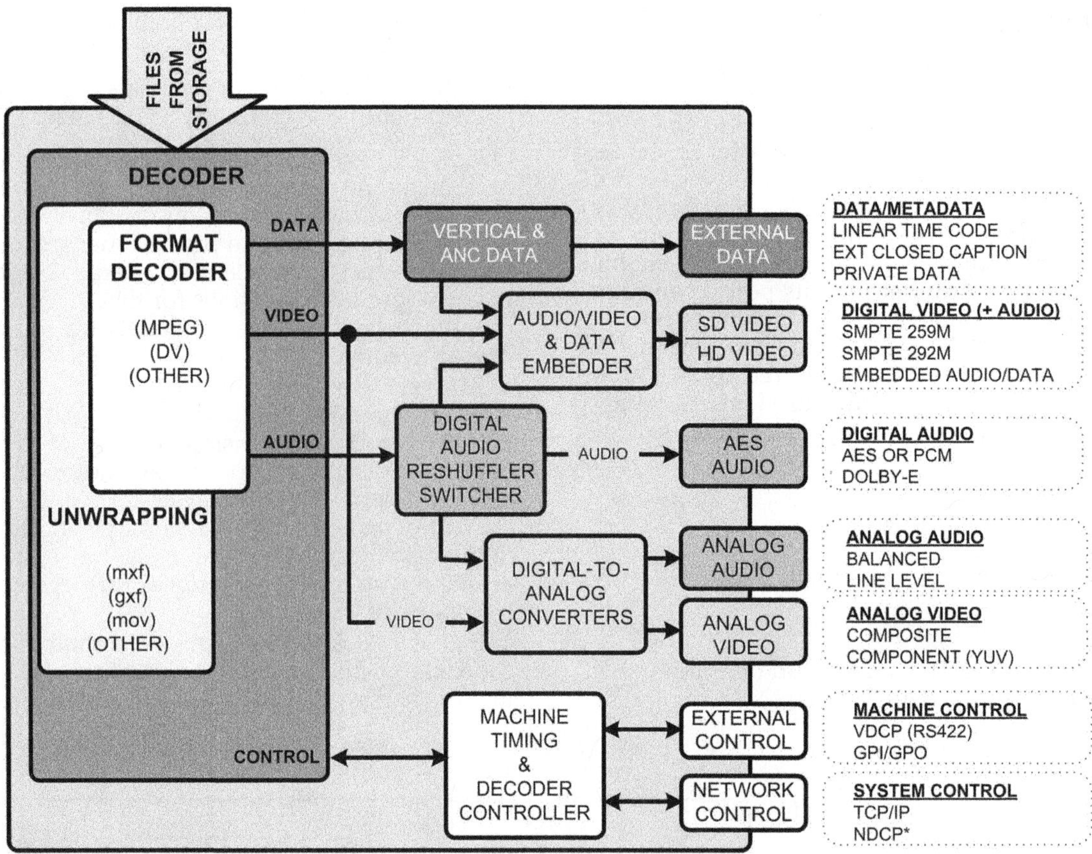

FIGURE 5.12-27 Output (playout) functions of a server channel also showing analog, digital, control and other ancillary data.

by the user. Modular components for video servers use multiple types of video processor or codec cards and may come with a variety of options and available video formats. Modularity allows users to configure a server chassis to the specific needs of their operations. For example, one server chassis may be dedicated strictly to ingest; therefore, it would be packaged with video input BNCs and encoders, with the likelihood that at least one decoder (for confirmation monitoring) be included. Another server chassis may be dedicated for playout only; that is, no record functions and only with decoders configured.

Most servers offered today utilize software-based codecs, allowing for even greater system flexibility. See Figure 5.12-28.

Transport Streams Input–Output

Occasionally users need a non-traditional I/O, such as MPEG-2 transport streams (TS), to be stored without needing to be decoded. TS recording and playout capability was necessary in the infancy of high-definition terrestrial broadcast (circa 1998–2000) because some of the network signals were sent as compressed video wrapped in an ASI (asynchronous serial interface) transport stream. When an MPEG-2 encoded HD signal needed to be temporarily stored, it could be cached to this type of specialized I/O port on the server. Signals did not need to be decoded if they were simply being "stored" for a later broadcast or time zone shift.

Video servers may provide ASI inputs and outputs that allow for the capture and storage of compressed bit streams. In this configuration, the high-definition baseband digital signal (1.485 Gbps) was precompressed using external encoding and ingested through the server's ASI input port. The idea was to directly store the compressed MPEG-2 HD signal as either single or multiple program transport streams (SPTS or MPTS). An equivalent decoder would convert the MPTS or SPTS signals back to baseband video if necessary.

PBS would use this process to send HD programs from its network operations center to its member stations as "ATSC-ready" HD signals that could then be directly turned around and sent to the local station's transmitter as an ATSC-compliant bit stream.

Beginning on November 1, 1998, the ABC Television Network began its HD broadcasts sending their 720p HD signal as a 45-megabit SPTS stream. The origination signal was stored as uncompressed HD video in the New York HD Release Center. For transmission, ABC encoded the HD video to a 45 Mbps MPEG-2 TS and sent it over satellite to the affiliates. Affiliates in the Mountain and Central time zones would then record (usually to D5 videotape) and then time shift

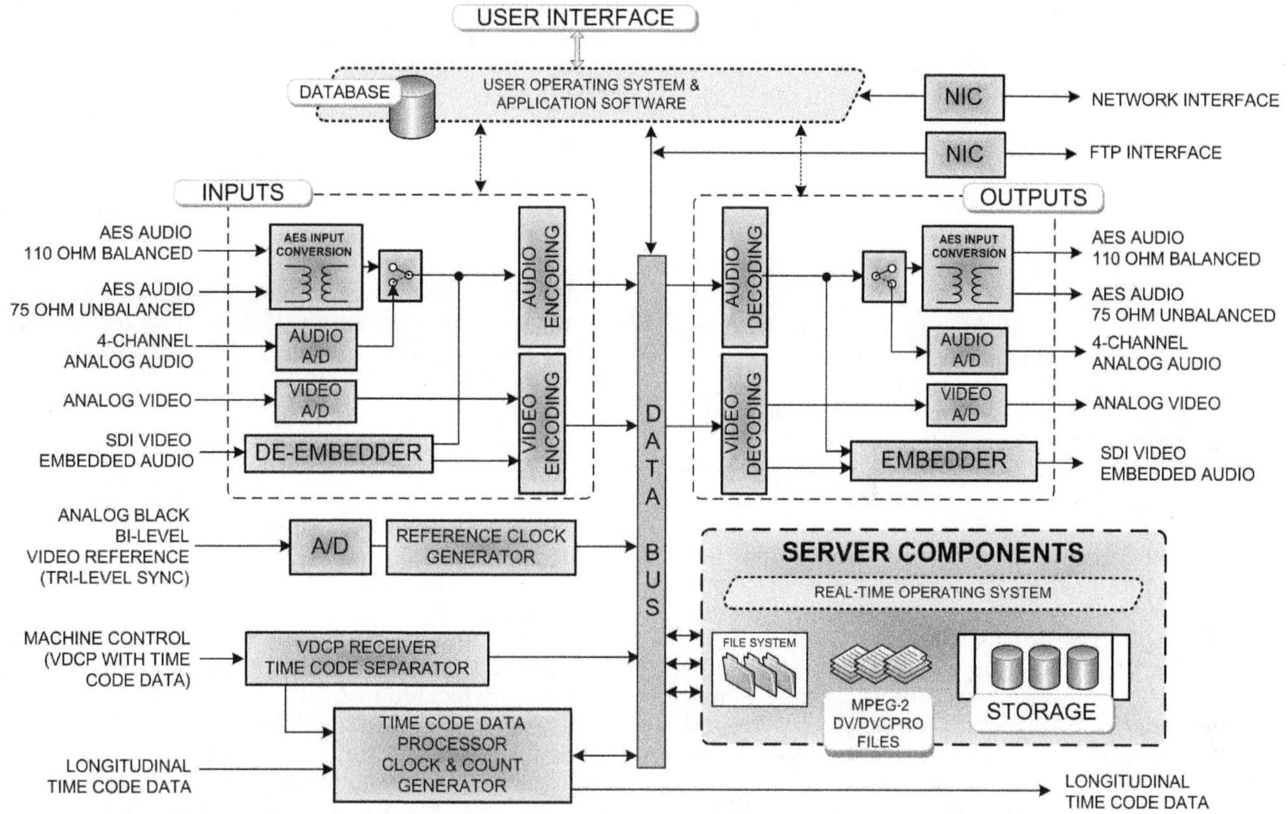

FIGURE 5.12-28 Detailed fundamental operational systems found in a "traditional" video server system. Manufacturers often combine several signal conversion functions (which might be shown individually in this diagram) and integrate them either internally or externally. Storage components, media management or analog-to-digital/digital-to-analog components may be peripheral components supplied by third parties.

"Moving Media Storage Technologies," K. Paulsen

the HD signal to compensate for the local time zone. Some of the ABC affiliates used a purpose-built ASI/TS server for the turnaround.

Fox Television had a similar plan utilizing a live splicing system to allow program switching or regional commercials to be inserted directly at the station. Fox did not provide time zone shifting via a localized server.

Multichannel Audio Inputs

Matched to each video input would be a minimum of two pairs of PCM digital audio carried as per AES standards and specifications (i.e., as an AES3 transport). Individual physical audio interfaces included either 75-ohm BNC for unbalanced (coaxial) signals that will to carry a single stereo pair per BNC, or a pair of three-pin XLR connectors when a 110-ohm balanced signal was preferred.

Professional broadcast servers nominally accept 48 kHz PCM sampling at 16-, 20-, or 24-bit resolution. Multichannel/multilingual and surround sound drove the AES audio I/O upwards to account for at least six and often eight discrete audio tracks.

Most servers can handle four AES inputs as either discrete physical inputs or embedded in an SDI transport. Some server units may only have two external physical BNCs and expect any additional channels to be carried as an SDI-embedded format.

Broadcast servers—especially those carrying multilingual and surround sound content—need to preserve certain compressed (non-PCM) audio sets, especially Dolby E and occasionally Dolby Digital. This configuration is universal on most professional broadcast video servers and allows for the ability to capture, retain and play back many more audio tracks in the same data set than a single AES stream would consume.

Few servers provide analog audio inputs any longer, and, as in those legacy analog video sources, these audio I/Os would expect the ADC and DAC processes to occur externally.

Timing or System Reference

Video servers, like most broadcast equipment, expect to see an external video reference signal. Typically, synchronized analog video black burst is fed from the facility's reference generator to the server's gen-lock or reference input. Users usually select between external reference or internal reference (free run), or they may lock the server to one of the video inputs (either digital or analog, if available). The latter method is rarely used unless the server stands by itself and does not necessarily require synchronization (for transition purposes) with other video signals.

Control

Compared to ten years ago, there are numerous options for controlling the video server. The most typical, and universal, is via the EIA RS-422 protocol—the signal protocol originally used by broadcast for videotape machine control. RS-422 emulation of common VTR modes, such as the BVW-75 protocol, is used to provide similar functionality for most video servers.

Most servers will provide at least one RS-422 machine control port (using the 9-pin DSUB connector) for each channel of the server. Here the term "channel" refers to either an input (encode) record function or an output (decode) playout function. Occasionally one might find that the 9-pin interface would support RS-232 serial communications, but those implementations are now rare and would be more likely found on industrial servers that control video messaging or other types of digital signaling applications.

The RS-422 interface usually employed the VDCP protocol, which is commonly known as the Louth protocol. Many adaptations of VDCP evolved as server manufacturers added their own feature sets to the servers.

As networked systems continue to dominate current installations, and as media asset management systems and multichannel automation systems become more prevalent, the trend is for Ethernet-based "native" interfaces to be the control and management methodology for server port configurations and transport feature functionality. There are a variety of network-based control protocols in use, many of them developed by specific server or automation companies.

Native, Ethernet-based interfaces also provide several additional services that VDCP over RS-422 could not provide. Metadata, clip numbering, sequencing and other operational features are available, as well as the routine functions necessary for record and playback.

Another protocol, the Network Device Communications Protocol (NDCP), developed by Harris Broadcast prior to its acquisition of Louth (in 2000), was effectively a proprietary protocol that was used in only limited implementations. NDCP was never fully adopted by the industry and was replaced as network sophistication continued to grow.

Services and System Managers

Most video server products utilize some form of service device often known as a *system manager* to oversee the video server system at many levels. Nearly all server products have 1 gig or 10 gig Ethernet ports for communications and network services, which are apart from those services that might be provided for file transfer and data movement between storage and server I/O components.

Older units had RS-232 terminal ports and/or USB ports—sometimes connected as part of a KVM (keyboard/video/mouse) solution—that would allow for connection to a dedicated PC terminal used for initial setup, interrogation and diagnostic monitoring of the servers.

In a networked environment, this information may now be real-time communications or it might be transmitted via TCP/IP or similar protocols direct to the facility's broadcast LAN or other network.

System Drives

System drives store lower-level starting sequences that contain instructions and applications necessary to initialize the various server components. If there is a system manager (i.e., a dedicated system server) for the product, the operating system drive may also reside there. If the video server system is composed of COTS server-based hardware, system startup drives will generally be located in each of those server chassis.

System drives, for reliability and root-level diagnostic purposes, should run autonomously from the applications software that controls the media-centric functions. System drives may be dedicated to a different operating system (e.g., Linux) and are likely to use an SSD for better reliability, faster performance and isolation. System drives may be mirrored (i.e., duplicated in RAID 1) for the protection of the critical system functions and consequently never hold actual program content, essence or metadata, as content is not contained in a discrete or complete form on any individual disk, but rather the data is distributed across multiple drives.

Alarms and Monitoring

Connections for alarms may be added as GPO contact closures or other independent ports. Users can configure alarm signaling measures as they desire and may trigger external alarms, which could include primary power supply failure, fan or airflow failure, over temperature, or a fault in the computer or compression engines themselves.

More often, in today's "lights out" operating centers, more sophisticated monitoring is expected—generally over a network connection. For servers, video servers, storage systems and networking components, it's not unusual to find a centralized monitoring system (e.g., DataMiner from Skyline) that continually reviews multiple parameters across the entire hardware complement.

GPI/GPO

General-purpose input (GPI) or general-purpose output (GPO) interfaces usually have at least the same number of inputs and outputs as there are channels.

GPI/GPO functions may be dedicated or "hard-wire configured," or shifted into software where the functions are assigned by the user via API or server-provided apps.

GPIs allow server operations without a facility automation system to control start, stop, recue, or advance clips during playout, provided the server has the capability to generate its own playlist internally or it could be directed by another means, such as a MAM.

Time and Control Code

SMPTE ST 12M-1 (as revised in 2014 and previously in 2008) and SMPTE ST 12M-2 (formerly SMPTE RP 188) are considered "standard" for use in systems at 29.97, 30, 25, and 24 frames per second. Video servers will typically take in SMPTE time code information upon video ingest, store it in the server as pulse codes or metadata, and then add it back to the signal during playout.

SMPTE time and control code can be presented, in synchronization, with the video content, and it is delivered in a variety of ways and formats—depending upon your system design.

Time code, generally referenced to the facility house clock (master sync system), or when specific to the actual video clip, is often carried in harmony with the source video (e.g., a VTR) via the RS-422 serial control interface corresponding with the VDCP control signaling. When time code is unavailable on the RS-422 interface from the VTR or other source, it may be ingested through a physical time code connector (XLR or BNC) using data that originates from the master house time code generator.

In some systems, vertical interval time code (VITC) or D-VITC may be employed instead of a linear analog time code stream. Details related to the specific numbering of video frames, in reference to the original time code captured, varies between manufacturers. Specific frame numbers of the actual video frames are (typically) not retained in the SMPTE 12M formats and instead are generated from counters and metadata integral to the video server. Part of this rationale has to do with video compression and segmenting of clips, which for the most part might destroy the original individuality of the time code numbering.

Files delivered to the video server may indeed carry time code, labeling or other control code specific information depending upon if it were added during the file preparation process or if a transcoder or the MAM made those data sets available to the server itself. Work continues on how to standardize specific time code and labeling of numeric data to MXF files, but that work is incomplete as of this writing.

Thus, time code numbering, when referenced, is generally non-absolute frame numbering that is then conformed to the video frame structure upon clip playout. The video server's internal database manages that metadata, aligns it with the locations of the actual file data on the drives, is carried through to the codecs and then is applied appropriately depending upon the specific purpose that data is used for.

Ancillary Data and Blanking Information in Digital Servers

Broadcasters have requirements for conveying certain metadata that is typically placed in the VBI during transmission/emission, such as closed caption data, teletext, ratings and content validation or audience measurement information. When converting (or ingesting) legacy analog content to digital, not all the data originally held in the VBI (from some source material) may be preserved.

A digitally encoded signal no longer needs to consume the nearly 8% of a video frame that is used for synchronization (i.e., "blanking") in the analog domain. Since digital video does not need the traditional *analog* blanking information necessary in analog television, some digitization processes may obliterate portions of those VBI data sets. That leaves the digitization process and the video server with the dilemma of how to retain the information originally contained in that portion of the signal.

Video servers use various product-specific methods for capturing and retaining the data originally carried in the VBI. For example, when ingesting analog video, the CEA 608 closed captioning originally set on analog video line 21 must be preserved as the content converts from analog to a compressed digital format for storage in the server. Typically, this type of information is stored as separate files (ancillary metadata) that are linked through the server database so that information may be recovered and reproduced upon delivery.

Digital video equivalents of ancillary data (e.g., CEA 708 captioning, Dolby E metadata or even rights management) may be carried in the ancillary data portions of the digital transport signal, per standards. Each manufacturer's approach to capturing and storing this information *internally* (within the server) is different; however, the output results should comply with standards-based delivery requirements, which returns that data to its proper location and functionality during playout or file delivery.

To reconstruct the vertical blanking interval (VBI) portion of the picture, signaling bits (i.e., data in the digital bit stream) trigger the decoder to place the equivalent blanking data back into the serial digital signal—at the time of decode and presentation to the SDI transport (i.e., the physical BNC server output connector). It is at this point where the VBI information is returned to its digital equivalency.

Early server implementations were faced with several data-reconstruction and insertion issues—most of which are now resolved. Nonetheless, if the legacy content were exported to, for example, an archive system, the return of that data to a more current video server may find this data is no longer appropriate for their server implementation, or may not be present at all.

Video Server Storage and Network Connections

As alluded to earlier in this chapter, storage options now vary significantly for video servers.

Essentially, server architectures now accommodate almost any storage structure, from integral storage onboard, to closed group/shared storage via a storage area network (SAN) or network-attached storage (NAS), or via another connectivity platform, for example, InfiniBand or Fibre Channel, which integrate with third-party storage from a variety of providers (e.g., Quantum, NetApp, Dot Hill/Seagate, and many others).

Increases in network transfer speeds, compute power and peripheral LAN-based storage systems now help to enable many heretofore difficult storage solution capabilities. Most servers can attach to most any tier or structure of storage. When system parameters—such as latency and transfer bandwidths—are appropriately managed, video server products and applications can then make use of suitable external storage.

Caution is urged, however, as some features—such as "editing in place" on the storage or support for "growing files"—may not be achievable using conventional storage management techniques and applications. Furthermore, as collaborative workloads and various streaming needs expand, users must be observant that the draw on any network storage system will eventually reach the bottleneck point where systems slow down or may possibly collapse.

When the organization expects to regularly and routinely use workflows that involve multiple functions, collectively and simultaneously—for example ingest, preview/review, proxy generation, editing and packaging in a collaborative environment—the system design should strive to employ specialized media-purpose-built storage systems designed to address those specific demands. This is the point where attempts to "build it yourself" from COTS storage or server components may be unwise, as the needs of these high-performance environments can become quite costly, unwieldy and even disastrous if not built properly.

Gateways

Server manufacturers often provide various means to augment or supplement the transfer of files between servers, storage and other sites. When high-speed file transfers require additional bandwidth to deliver files using standard Ethernet connectivity, a gateway may be necessary to manage the protocol differences and throttle the transfers based upon needs or priorities.

Gateway technologies help support the overall systemization, especially when large amounts of data must be transferred, as when migrating media from server to server or between storage tiers—such as near-line storage or archive systems. High-bandwidth and high-speed transfers now expect to see at least 10 GbE (Gigabit Ethernet) connectivity, which can support transfer performance of 1500 to 4000 Mbps (i.e., 200 to 599 megabytes per second). Gateways will also be utilized when support for multiple (e.g., greater than 10 to 20) users is necessary.

Gateways are also used to attach one form of shared storage (e.g., Fibre Channel) to another (e.g., Ethernet) and will generally provide redundant paths and links for file transfer protocol (FTP) delivery to or from external resources, including the cloud. Some gateways may include protected storage (RAID 6 or similar) so that buffering of data at these high speeds will not impact the shared storage being utilized for other purposes.

Gateways are also available with SSDs (flash memory, in total or as a supplement to the HDD storage) to support higher input-output operations (IOPS) and faster caching. SSD gateways can be significantly faster than HDD storage and may provide for many simultaneous requests (threads) for transfers or other operations. Broadcast-purposed gateways will usually be tailored for media awareness; that is, they will accept and properly manage not just FTP transfers, but also file types that may be native to the particular servers as well as industry-recognized media wrappers and file types such as MXF, DV, AVI, MOV and so forth.

Server Types and Applications

With the generalization of the term "server" come many new sets of applications residing on all forms of technical hardware. Many more applications are being ported directly into software, yielding capabilities and adaptabilities that continue to rise. This is certainly a far-reaching change from when the video server was first introduced in the mid-1990s.

COTS Servers and Interfaces

Using conventional, IT-based servers from HP, IBM or Dell as foundational elements for media-centric video solutions is now without question the wave of the future. Of course, for the applications manufacturer, the utilization of COTS hardware changes the model for development and product release (time to market) tremendously. This further opens the door even wider for deployment of products that reside "in the cloud"—a trend that is growing in popularity and acceptance.

Thin, one rack unit (1RU) "pizza box'" servers are nothing new to the IT industry. For the broadcaster, they are becoming commonplace for many installations. COTS devices can be configured into many different purpose-built and virtualized solutions, and they offer a large variety of "standard" components, including:

- single to multiple processors
- memory from a few gigabytes to hundreds of gigabytes
- internal storage in most any RAID form factor
- internal storage with HDD or SSD high-capacity memory
- removable media (DVD or Blu-Ray)
- slots for additional third-party cards (I/O cards, FC interface cards, etc.)
- network connections (1 Gb, 10 Gb, and beyond)
- additional card slots for peripheral devices (RS-422 cards, SDI cards, etc.)

Use of these servers is continuing to grow. To attempt to define where and how these COTS servers are employed would be impractical. However, some recent examples might suggest that the terms "video server," "video file server" or "media server" should be rethought. The crossover point between a dedicated "video server" product and a "full function product" (such as the Channel in a Box) is narrowing with every year.

Some of the more noted and selected applications for COTS-like video/media/file servers include:

- FIFO (first-in, first-out) loop recorders that continuously record hours to days of video "just in case" it might be useful (satellite feeds, surveillance, network feeds monitoring, program verification, competitor analysis)
- Replay systems for sports or other reviewable activities
- Clip servers for stadiums, arenas and similar entertainment venues
- Clip servers for high-resolution, high frame rate (e.g., UHD) video used in theme parks and museums or parks and recreation sites
- News system and production servers supporting media asset management (MAM) systems
- Virtualization as in 2D and 3D virtual sets and augmented reality
- Play-to-air services as main and/or redundant services, including remote "edge servers" for disaster recovery (DR) or business continuity protection (BCP)

Servers built on COTS hardware are changing the way on-premise broadcast equipment centers are being designed, built and utilized. It is difficult to predict what the next stage of development will be for servers, but if the recent (2016) NAB Exhibition was any indication, it appears that virtualization (the sharing of resources dependent upon load and need) may shift these "pizza box" systems in one or more directions. Expect to see the flat-profile 1RU and 2RU server systems shift to "blade servers," that is, multiple servers in vertical card slot configurations in a single chassis. The concept of blades reduces the number of dedicated power supplies, operating or management systems and such by considerable proportions. Services deployed on blades can be spun up and spun down as needed; even more likely, many of the services typically heretofore placed on-premise will move to the cloud, where they are virtualized on a global basis.

Storage Interfaces

Depending on their drive structure, physical storage drives may be mounted internally with the server chassis or in secondary enclosures. Most drive systems use some level of RAID protection; note that this changed historically when early video servers and clip servers were deployed with very simple direct-attached storage (DAS). Some clip servers made today still use this approach, and some have moved entirely to solid-state implementations.

Storage interfaces have moved away from the days when the server's drives were connected via a multi-pin SCSI connector in a chain (or loop) configuration. It is doubtful those external small computer systems interface (SCSI) connections are even available any longer; however, some transport interfaces still utilize SCSI principles but in an IP technology methodology (e.g., iSCSI). Other systems utilize USB (e.g., USB 3C) to connect from host (server) to other devices, including fixed or transportable storage.

Now we find nearly all external storage arrays are connected via some form of high-speed interface, such as Fibre Channel or Gigabit Ethernet or a backbone-like interface such as InfiniBand. As Gigabit Ethernet interfaces and network systems reach 40- to 100-gigabit wire speeds, storage systems are expected to follow—which may change the complexion of the storage and server industry once again.

VIDEO SERVER STORAGE

Recent and ongoing developments in servers, storage and networking are forever changing the nature of how media and associated image-based data is captured, stored, protected, distributed, replicated and utilized.

The growth in unstructured data has accelerated the need for storage capacity increases and is driving new storage management strategies that include how, and on what kinds of storage devices, that data is to be kept. When the prospects of ultra high definition as UHDTV-1 (or 4K) and UHDTV-2 (or 8K) are added, the industry storage requirements are expected to become larger still.

New Dimensions in Storage Systems

Unlike legacy linear videotape, which has a finite recording period based upon the physical tape length and whose primary metric is measured in "hours and minutes," storage systems are essentially huge bit buckets. Storage systems come in many form factors and can contain myriad forms of information (a.k.a. "data"). They can perform a nearly infinite number of functions, and they, like most other compute-sensitive products, can benefit from protection, interface through some form of networking, are best architected for resiliency and redundancy, and in most instances can (and will) fail or will be expected to need upgrading over the useful life of their investment.

Some of the items are seldom thought about by the users; others are certainly just taken for granted. Nonetheless, storage systems are the future of this industry, and all the users must depend upon them, whether they are in their mobile devices, their computers, their entertainment systems or their workplaces.

Storage Ecosystem

The following sections bring forward key topics related to the storage ecosystem. The information is not necessarily provided for the purpose of developing any particular design model; yet it is intended to give the reader insight into the technologies that are presently available, or are on the cusp of changing how media, data and information is stored, protected and distributed in today's broadcast and media entertainment industries.

Networked, Intelligent Storage

Usually most storage discussions begin by describing the various forms of storage, such as JBOD (just a bunch of disks), DAS (direct-attached storage) and such. In the youthful days of computer storage systems, each vendor usually packaged their own flavor of media storage integrally with the compute-product. Certainly with video and broadcast manufacturers who offered editing systems (e.g., CMX), character generators (e.g., Telemation, Chyron), or graphics composition and still stores (e.g., Harris, Abekas), specific storage technologies were needed to keep the data stored under strict requirements and specifications that could be controlled by the devices themselves. For most first-generation systems, this meant borrowing mainframe computer storage technologies and adapting them for the applications with which they were associated.

It was generally impossible in the 1970s to utilize the same storage platform for anything more than one focused application. Character generators used one form of disk drive, still stores or still-frame libraries needed another type, and editing systems only stored database-type information to a floppy drive or even paper punch tape. Hard disk drives, in either 5–1/4-inch or later 3–1/2-inch form factors, were not commercially viable, and were unlikely to be found in any of these devices until after the birth of the IBM PC in 1981.

Storage systems quickly evolved in many dimensions. Numerous technological advancements drove storage ecosystems at a time when Moore's law was allowing computer technologies to explode exponentially.

This section begins by reviewing the historical perspectives of the storage evolution; details the three fundamental types of drive structure configurations (DAS, NAS and SAN); and then describes how various subsystems that manage the physical storage medium are applied to the drive configurations.

Early Uses for Disk Storage

Early uses of disk drive storage in professional broadcast applications were centered on storing still graphic images and generating character overlays for use in television news. The drives utilized were adaptations from larger mainframe computer storage systems, like those in an International Computers Limited ICL 2966 disk drive with EDS 200 disk packs.

Early digital disk recorders and some basic video server deployments were configured with simple dedicated direct-attached storage (DAS). When compressed motion imaging was limited to Motion JPEG file formats, the limitations on run times for these devices became directly proportionate to the physical storage capacity of the drives.

As multichannel video server products were introduced, their storage systems' capacities grew up to the limit of the RAID controller or the physical size of the chassis that contained them. Physical and electronic technologies or protocols then governed how many and which type of hard drives could be added to the storage subsystem. Video server performance was limited by the level of continuous I/O and data transfer activities and restricted by the bandwidth available for transfers between the encoders and storage (as in ingest) and the decoders and storage (as in playout). Any overhead activity required for operational purposes, including external file transfers to archive devices and to/from other servers' storage subsystems, was secondary to getting media into or out of the server.

At these early stages of deployment (late 1980s through mid-1990s), video servers generally had just one or two inputs with one or two output complements available. As uses continued, more I/O was required, and more storage to support the additional I/O was required. To facilitate the sharing of the data across many I/O ports, the concept of a *centralized storage system* was recognized. Only a select few video server manufacturers fully embraced the concept of true centralized or *shared* storage. However, the growth of Fibre Channel technologies supported high-bandwidth data transfers between multiple server chassis, storage arrays and even other devices. This permitted the desired shared storage and in turn allowed Fibre Channel to grow into the predominant storage technology for video servers, where it still remains de facto in several storage solutions available in the second decade of the 2000s.

Sizes, bit depths and the types of media files continue to grow—despite serious improvements in compression technologies that include H.264 AVC and more recently H.265 High Efficiency Video Coding (HEVC). High-definition and ultra-high-definition video and compressed multi-program transport bit streams, alongside increased ancillary requirements for proxy generation, near-line archive management, and external transfers of files via FTP or other file acceleration technologies from companies like Signiant and Aspera—an IBM company—all necessitate more advanced means of addressing storage and a continual movement toward higher bandwidth systems and network topologies.

Video media server systems can now process massive amounts of contiguous data. To satisfy the management of that data, a higher degree of intelligence is required from the storage architecture. Sophisticated storage systems must appear to function like any of the other elements in a common network environment.

Smart Storage

Storage intelligence, as it applies to networked storage, grew out of the increasing requirements for high availability, security and a measurable quality of service (QoS). Network storage architectures, like those shown in Figures 5.12-29 through 5.12-31 (or their variations), provide a path for disaster recovery and aid in the prevention of data loss. Network storage systems enable consolidated backups, the archiving of data assets and the ability to provide for business continuity protection (BCP) by geographically locating redundant and connected groups of storage around the globe. Employing networked storage to a media server system also offers the consolidation of storage with improved capacity, utilization and a unification of storage management.

Early video server systems either handled all the media storage in the same chassis where the codecs and I/Os were, or they were directly attached to storage consisting of a single or small group of disk arrays connected via SCSI or Fibre Channel (FC) interface. As server systems grew to address the tapeless working environments of modern production, editorial systems and play-to-air, a network-based solution for storage also began to evolve.

Overall, the influence of network-centric systems has brought three terminologies to defining storage architecture: *direct-attached storage* (DAS), *network-attached storage* (NAS) and *storage area networks* (SAN).

Direct-Attached Storage

Direct-attached storage (DAS) was the first, and remains the most common, approach for storing data on magnetic spinning media. Evolving from the 5 megabyte to present-day multi-terabyte hard disk drives (HDDs), DAS is the easiest and simplest of storage architectures still employed. Direct-attached storage describes a storage device that is directly connected to a host system. DAS devices housed in external chassis, such as the early SCSI-transportable storage devices, have subsided, only to be rapidly replaced by USB (universal serial bus) solutions offering many hundred times faster data transfer speeds than SCSI—and without any of the complications of cumbersome connectors, bulky cables and limited strings of drives in a link. See Figure 5.12-29.

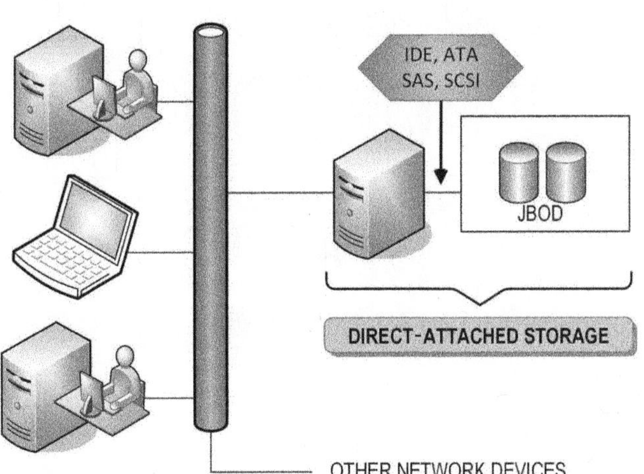

FIGURE 5.12-29 Direct-attached storage (DAS) is the simplest form of storage.

When DAS utilized IDE, ATA and other forms of SCSI drives and interfaces, the storage system experienced many limited capabilities and performance governed by controllers that had to interface directly to the bus and the drive. There were a finite number of devices that could be attached to the server/computer bus, and expansion was seriously limited by the type of connector or cabling, the number of command channels available to the operating system, and the actual operating system itself.

DAS drive systems were, and still are, generally found on smaller, dedicated, stand-alone Windows or Mac computers that don't necessarily require external high performance, high bandwidth, or high throughput performance. In the past, DAS drives might have been used only as external storage or for transferring larger files between different workstations. Today, the portable USB drives (which reach capacities in excess of a few terabytes) are routinely used not just for external storage but for working projects including editing, graphics and file transfers.

Network-Attached Storage

Network-attached storage (NAS) detaches storage access and management from the primary server, as illustrated in Figure 5.12-30. By definition, NAS is a server device dedicated to file sharing. The NAS head consists of both the file system and the drives themselves. NAS is typically simple to implement and is used for editing systems and video server systems depending on the applications of the server.

The choice to use a NAS versus another storage form is that of either the server manufacturer or the user. Factors including cost, throughput, performance and bandwidth requirements are part of decision to use a NAS.

A NAS solution will consist of hard disk or solid-state storage, including multi-disk RAID systems; software for configuring and mapping file locations

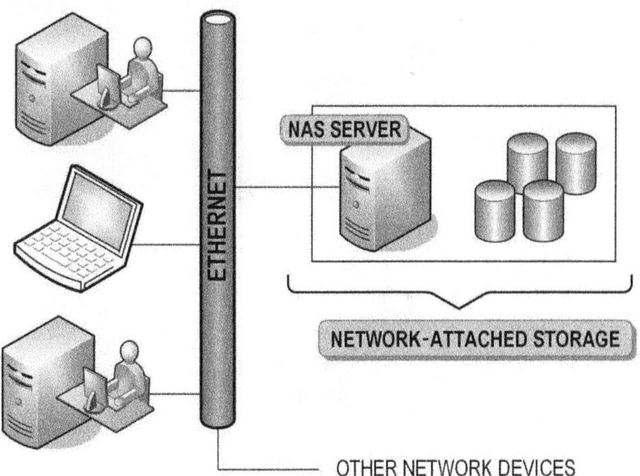

FIGURE 5.12-30 Simplified network-attached storage (NAS) with clients attached via Ethernet to a NAS server that administers the storage and file management (usually supports both CIFS and NFS protocols).

to network-attached devices; and its own network address. System-wise, NAS will follow a traditional client/server design. In a NAS environment, storage and computer processing no longer compete for the same processor resources, so by removing storage management from the main processor's activities, applications run independent of the main server, resulting in both file management and run time applications that can be served faster.

Network-attached storage devices are typically attached to the local area network (LAN) via Ethernet. Requests for files are mapped by the main server to and from the NAS file server.

A NAS does not physically need to reside with the server. NAS allows for the incremental addition of storage without the requirement for shutdown or rebooting. Network-attached storage can be included or may be seen as part of a SAN.

The NAS device (or "NAS head") will generally contain its own operating and file system. Most NAS systems communicate over TCP/IP, with I/O requests supported by common file sharing protocols, such as the Unix-based Sun Network File System (NFS), Samba, and the original IBM/Microsoft-developed file sharing support for DOS, formerly known as Server Message Block (SMB), which has since become Common Internet File System (CIFS) for Windows machines. By design, a NAS system will allow many computers to access the same set of files over the network.

The NAS system identifies data by file name and byte offsets, transfers file data or file metadata (i.e., the file's owner, permissions, creation dates, etc.) and handles security, user authentication and provisions for file locking.

Storage Area Networks

The counterpart to network-attached storage is called a storage area network (SAN) or sometimes simply a *storage network*. Storage networks are distinguished from other forms of network storage by the low-level access methods that they use. Most storage networks will communicate between servers and devices using SCSI protocol, although the low-level physical interfaces (e.g., the legacy parallel cabling used in early implementation between drive devices) will not be used.

SANs employ block storage, where it is the server that initiates a transfer request for specific blocks (referred to as data segments) from specific drives, addressing that data by disk block number, and then transfers raw disk blocks of data. See Figure 5.12-31.

In a SAN, storage administration is simplified, adding flexibility because physical cables and storage devices are not moved from one server to another. However, with the exception of SAN file systems and clustered computing, each logical unit number (LUN) on the SAN is owned by a single computer host, called the initiator, and thus storage in a SAN is still on a one-to-one relationship. By contrast, NAS allows many computers to access the same set of files over a network.

Types of SAN

SANs tend to provide faster and more reliable access than the high-level protocols employed in a NAS, because the SAN will generally be built on a specifically designed infrastructure that handles storage communications only. Historically, the SAN has undergone

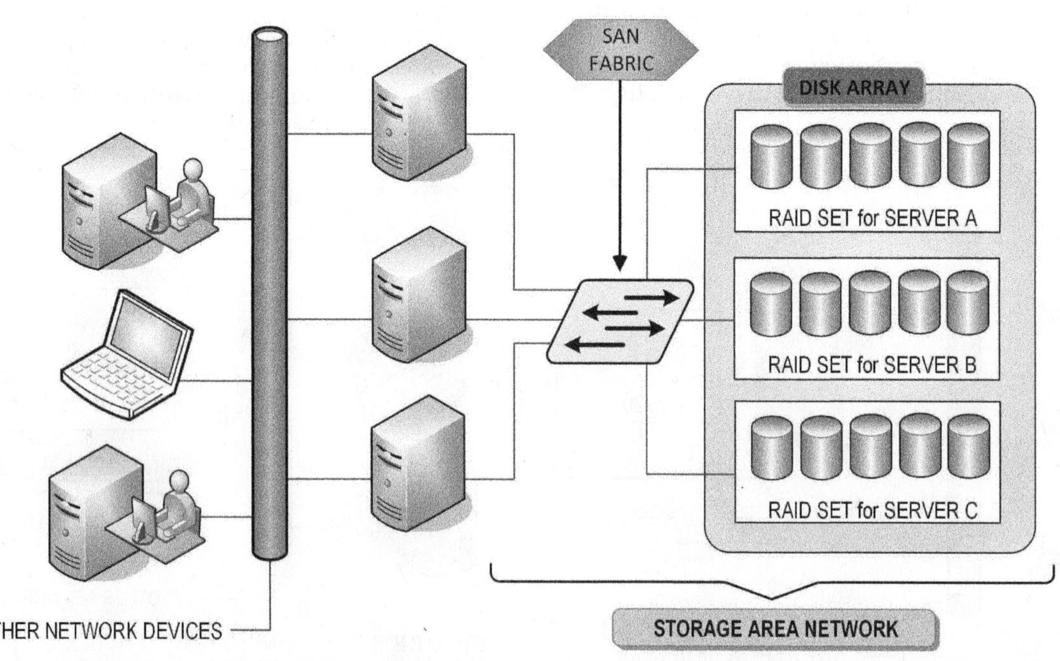

FIGURE 5.12-31 Storage area network shown with the SAN fabric as either Fibre Channel or Gigabit Ethernet. Other storage structures, explained in the following Types of SAN section, are possible.

the highest degree of architectural change, from a mostly exclusive to a Fibre Channel system to a hybrid of iSCSI, and now in Gigabit Ethernet topologies.

Fibre Channel (FC) networking still remains the most common SAN deployment technology, whereby the SAN is comprised of a number of Fibre Channel switches connected in a fabric or network structure that attaches to a large number of spindles (HDDs) that fuel capacity and bandwidth.

A SAN may have one or more server hosts and one or more storage structures. Storage may be comprised of disk arrays, tape libraries or optical-based storage devices. When large contiguous blocks of data must be transferred at high throughput data rates, such as in a video server or a large-scale video production editing system, the SAN can provide greater bandwidth and thus exchange data between the store and the server with much-improved performance.

Early Issues in SANs

During the development of SAN technologies, and partly because Fibre Channel was rushed to market without thorough compatibility testing, SANs began to acquire a poor reputation due to a lack of compatibility testing between network components, servers, operating systems and the storage arrays themselves. Another significant drawback was that FC fabrics and associated drives were more expensive than Ethernet switches and the conventional SCSI-only drive components of a NAS.

Interoperability problems still surfaced even though the basic FC storage protocols were formally standardized. Some higher-level functions would not work between differing devices (e.g., drives, switches and adaptors) despite meeting then current standards compliance.

The early incompatibility issues have since been resolved, but for these types of reasons and during initial introductions to SANs and Fibre Channel topologies, most mission-critical broadcast video server manufacturers would seldom provide or warrant storage systems that have not been thoroughly qualified in their labs and by the manufacturer. Essentially, most major video server manufacturers would only provide SAN (or NAS) storage directly.

Control, Management and SAN File Systems

A SAN includes a set of control and management systems that collectively provide for the connections, data transfers and other block-based services. In this context, services are defined as the input and output operations for data movement between servers (hosts) and storage.

A SAN may provide file-based services; however, for media-centric applications, SANs are generally configured to meet the demands of large continuous blocks of data that move predictably and efficiently between I/O and storage. There are cases when a "SAN file system" may be configured and will enable a different set of interchange amongst multiple servers that may have different operating systems.

A storage area network *file system* is programming that enables the exchange and sharing of the same *copies* of files stored on common storage media. Although different servers may share common storage media (i.e., through methods of virtualization), without a SAN file system, the servers are unable to share the *same* files. Thus the SAN file system streamlines SAN management and simplifies operations. The SAN file system optimizes the use of storage resources, allows network components to be scaled individually, minimizes storage and retrieval time and eliminates the necessity for storage redundancy when in a mixed file sharing environment.

iSCSI Protocol

Another variation of SAN technology, not based upon a fabric network SAN protocol, is also available. Referred to as iSCSI (for Internet Small Computer System Interface), which is a transport layer protocol that describes how (SCSI) packets should be transported over a TCP/IP network, it took the SCSI command sets and implemented them (typically over Ethernet) using Internet technology protocols, thus creating a SAN using Ethernet switches.

SAN Elements

There are specific functional elements required for the storage area network.

The first element is the *device*, otherwise considered a collection of storage *elements* or storage *systems*. Storage devices may be a single JBOD (just a bunch of disks) array, a series of RAID chassis, or a massive array of FC drives arranged, for example, in a split-bus configuration providing high system bandwidth and high availability storage capacity. The device might be an archive system driven by a gateway server that buffers data flow or other inhibitors, such as interruptions from the tape drive mechanics, while continuing to maintain a high throughput, constant performance (transfers to the tape system) or restorations (the transfer back from data tape to the storage or server system).

The second element is *connectivity*, consisting of components such as routing, switching, the physical media (copper or optical cabling) and the appropriate protocols for the exchange and transport of data between those components. Connectivity interfaces are at the physical (PHY) layer and are administered through specific and usually standardized protocols that provide for compatibility between media and the physical elements on the network.

The third element is *control*, the management of the data paths, transfers, resources associated with those devices (e.g., storage arrays) and the regulation of actual data within the SAN. For the control element, network management is the process by which a stable transport of data across a network's infrastructure is maintained. In the case of the SAN, control must be maintained such that peak limits are obfuscated, that server requests are handled according to their preset hierarchy for delivery, and that backup or protection paths are enabled and ready to take over when or if needed. Additional elements of control include volume management, data resource and data backup management, file access, and the reliable transport of data between storage elements and servers when called upon.

Metadata Controllers

A *metadata controller* (MDC), sometimes called a *metadata appliance*, is SAN technology employed in managing file locking, space allocation and access authorization. Most high-performance SAN solutions include the MDC in their complement of equipment.

Physically, the MDC appears like a server with associated (integral) disk drives. Sometimes the device is a separate chassis, and other times it is integral to the drive tray system. The MDC device is what manages the activities associated with accessibility, not just from users, but also to/from the file transfer/file system components. One functional component is *file locking*—an important element that manages and controls file accessibility. File locking is what prevents unauthorized access to the drive array data.

Each operating system and/or manufacturer will have its own file-locking scheme, just like most editing systems, which have their own bin-locking or access control, as well. File locks (in Unix-like systems and Mac OS X, for example) apply to the actual file itself and not to the name assigned to that file.

It is important to recognize that the metadata contained in the MDC for a SAN is different from what is generically termed "metadata" in a file-based workflow or *media asset management* (MAM) product. In the former, the MDC is an integral part of the SAN and the file system itself; in the latter, the metadata describes "the bits about the bits"—that is, the file components' elements, origination, date and size of the file, and myriad "user-defined" information important to categorizing and cataloging the assets (files) in the storage system.

MDCs vary by the manufacturer and can be quite complex due to what they do, how they manage the files, how rapidly they access the files and much more. MDCs can provide services such as the linking between high-resolution files and their proxy-level equivalents. MDCs can manage file hierarchy and track additions or deletions to the SAN, as well as control "snapshots" of the system in the event that certain levels of corruption are detected.

MDCs are considered key components—the "secret sauce" of the storage management system—because of how they distinguish functionality and feature sets between storage in the SAN and the systemization of the entire SAN solution.

Comparing NAS and SAN

Traditional SANs differ from NAS in several ways, even though on the surface they might appear quite similar. Both SANs and NAS generally employ RAID-protection schemes for storage, are connected in a network-like environment and, at a high level, can serve the same purposes, yet there are important differences that can affect the way data is utilized, as listed in Figure 5.12-32.

NAS	SAN
Almost any machine that can connect to the LAN (or is interconnected to the LAN through a WAN) can use NFS, CIFS or HTTP protocol to connect to a NAS and share files	Only server class devices with SCSI Fibre Channel can connect to the SAN. The Fibre Channel of the SAN has a limit of around 10 km at best
A NAS identifies data by file name and byte offsets, transfers data or file metadata (e.g., owner, permissions, creation, description), and handles security, authentication, file locking	A SAN addresses data by disk block number and transfers raw disk blocks
A NAS allows greater sharing of information, especially between disparate operating systems such as Unix and NT	File sharing is operating system dependent and does not exist in many operating systems
File system managed by NAS head	File system is managed by servers
Backups and mirrors are done on files, not blocks, for a savings in time and bandwidth. A snapshot can be very small compared to its source volume	Backups and mirrors require a block by block copy, even if the blocks are empty. An entire mirror machine must be equal to or greater in capacity compared to the source volume

FIGURE 5.12-32 Principal differences between NAS and SAN implementations.

From a connection perspective, a NAS employs TCP/IP networking, including Ethernet, FDDI, and even now legacy ATM. SANs can use Fibre Channel (FC) high-speed networking, which commonly runs at 1, 2, 4, 8, 16, 32, and 128 gigabit per second rates. FC is primarily used to connect computer data storage to servers. SANs can also be built on conventional Gigabit Ethernet topologies.

NAS protocols use TCP/IP and NFS/CIFS/HTTP, with a SAN using encapsulated SCSI (information is available under the "SCSI Encapsulation Protocol" [SEP] drafted through the IETF in May 2000) or Fibre Channel over IP (FCIP).

Fibre Channel over IP is technology for linking Fibre Channel storage area networks. FCIP and iSCSI are complementary solutions that enable a wider access to storage, such as in a campus environment. FCIP is used to interconnect Fibre Channel (FC) SAN islands over IP networks, while iSCSI allows IP-connected hosts to access iSCSI or FC-connected storage.

iSCSI and FCIP are typically used for different purposes. In iSCSI implementations, SCSI commands and data frames are encapsulated in IP to support I/O disk access over an IP network. With FCIP, Fibre Channel frames are encapsulated in IP so that both SCSI and non-SCSI frames can be transported over an IP network. The technologies are useful applications for remote backup and replication, in addition to enabling Fibre Channel I/O communication.

The lines between NAS and SAN are blurring, as evidenced by the crossover in transport, topologies and even networks. Some enterprise systems may actually share both SAN and NAS technologies, depending upon the degree of existing hardware, as evidenced in Figure 5.12-33.

Scaling Storage

Frameworks for standard storage arrays have, for many years, employed two controller heads for redundancy and multiple sets of disk trays, jammed with hundreds or more spindles for bandwidth and capacity. Controllers are connected to the SAN to provide storage to the servers. Each of spindles in each of the trays is connected to storage controllers so that the servers can access the disks through the two controllers.

Scaling systems can become a triangle-like balancing act. Users attempt to maximize each of the elements (capacity, bandwidth, and resiliency) while managing capital or operational costs, physical size, and maintenance requirements. Figure 5.12-34 depicts these elements in terms of the overall impact to storage management, and Figure 5.12-35 in terms of resiliency and performance.

Increasing performance (i.e., bandwidth) while increasing capacity often involves adding more trays of disks, which are then added to the same two controllers. This methodology is what is referred to as

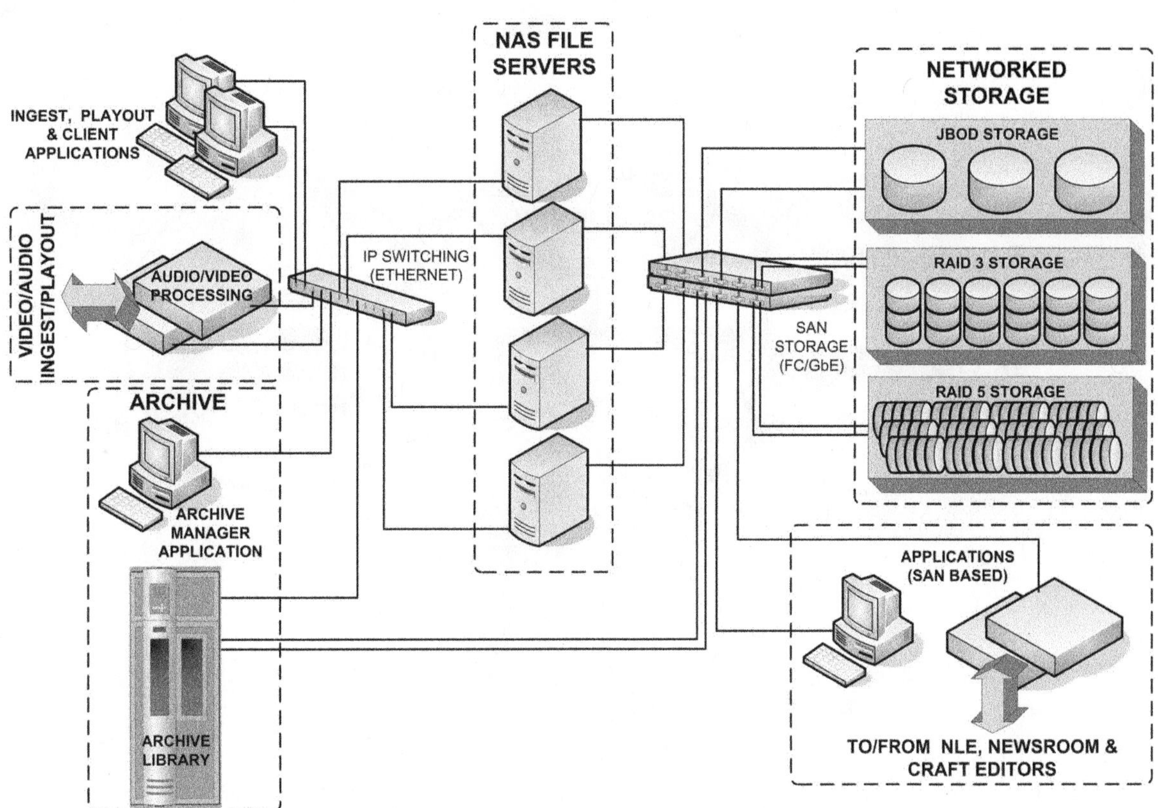

FIGURE 5.12-33 A complete server, client/application, storage and archive system that might be deployed in a broadcast facility or an enterprise, campus or production company.

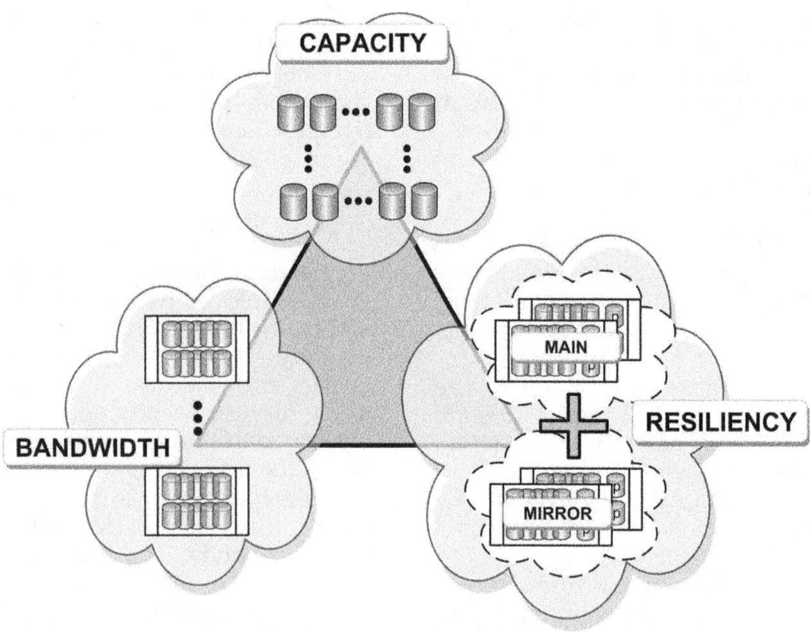

FIGURE 5.12-34 Representative "triangle of balance" in storage management. The three elements shown are methodologies for improving performance—and in turn impact each other in varying ways.

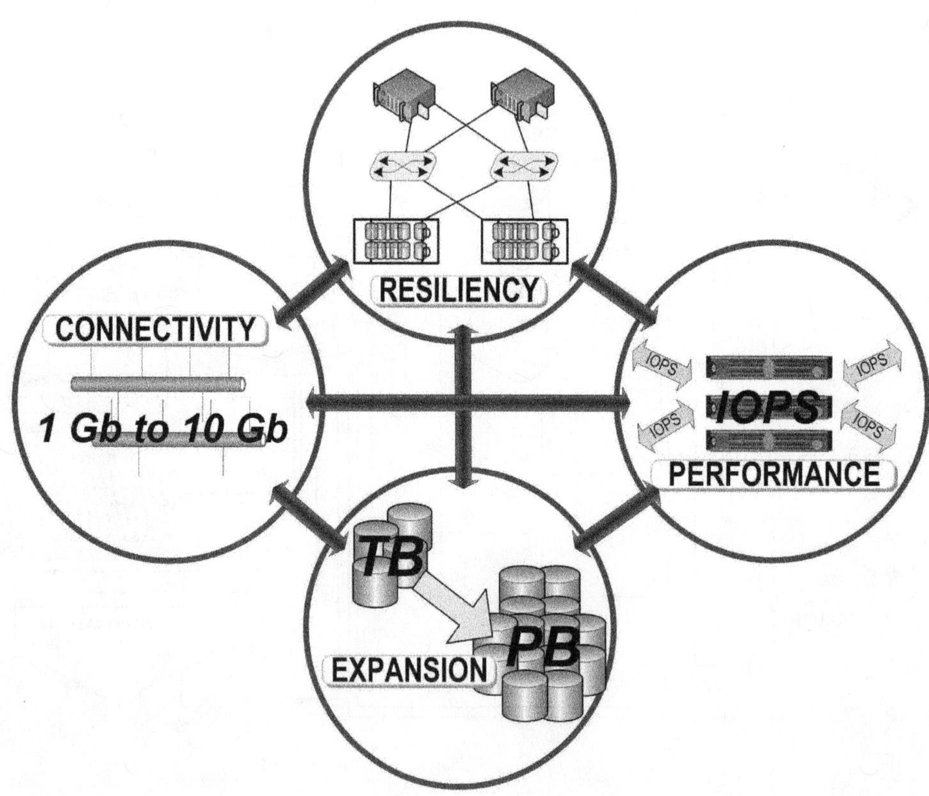

FIGURE 5.12-35 Systems growth can be accomplished by several means, as shown by the upscaling of systems for resiliency and performance—including connectivity (networking), which may be necessary to accomplish certain improvements, such as IOPS increases. The architecture (and in turn performance) of any storage system depends upon properly balancing these—and the elements shown in the previous figure—into a harmonious system that meets the needs of the users.

"scale-up" (sometimes called "vertical") expansion. Some storage systems grow using this methodology, while others employ what is called "scale-out" (sometimes called "horizontal") expansion.

Scaling Up

Scaling up is addressed by increasing the quantity of drives in the system. Many SANs have provisions to add drive "trays," which can contain tens to hundreds of physical, usually 3–1/2-inch, form factor drives, ranging from sub-terabytes to multiple terabytes per drive. The drives are arranged in LUNs that communicate, in these groups, to a controller system (usually two for protection and monitoring purposes).

It was first thought that by adding more disks the system actually improves overall performance. This was believed to be true because disk throughput was probably the limiting factor from a performance standpoint. This also increased capacity, also seen as a good thing. However, as the load on the array is driven upward and disks are added to accommodate the concern, a bottleneck begins to brew because of the limitations now placed on the two controllers.

Adding more drives requires more and more processing to handle the additional RAID calculations of the SAN or the NAS. As this situation continues to increase, the controllers become saturated to the point they can do no more.

Adding faster disks to an already overloaded controller pair only places more stress on the controllers themselves. The resolution to this dilemma forces the user to buy another complete new array with another paired set of controllers and then move a portion of the workload onto this new array. Adding in the new controller/array components is often disruptive to the workflows and can result in risks that are unavoidable.

Scaling Out or Storage Clustering

An alternative solution, one that alleviates the problems described in "scale-up" storage, is to leverage the addition of servers rather than finding a new set of controllers or customizing the controller platform to match the additional drive trays and spindle counts. The new concept is called *storage clustering*.

A group of servers working together forms a clustered storage array and provides LUNs (or file shares) connected over a network in the same operational structure as a traditional array. The core architecture behind this network-centric storage/server concept, which can use conventional Gigabit Ethernet topology, is that the servers in the storage array now all communicate efficiently to balance the loading amongst the ever-growing number of spindles and trays.

The addition of nodes of disks to a scale-out array adds drives plus additional servers to the cluster. Doing so adds more network ports, more CPU and more memory (RAM). When the capacity of a scale-out array increases, its performance increases as well. Adding nodes—in most cases—results in linear performance improvements. When extremely large volumes of data must be managed, spreading the management out over servers and storage nodes increases efficiency without reaching controller limitations that would otherwise become unwieldy to address.

Adding nodes is largely nondisruptive. A normal maintenance window can therefore be used to add capacity and performance to the array, rather than a full system outage. Some storage systems are engineered to add nodes or even upgrade drives without any detrimental impacts to the system or workflows.

Scaling Performance

Adding nodes adds usable capacity, but it in turn also increases IOPS. Upping the IOPS increases the performance, so this becomes a "win-win" objective. But of course there are risks in this scale-out architecture.

Upgrades with servers may show software incompatibility issues with existing servers in the system—particularly when server technology changes or obsolescence gets in the way. Cluster administration and management can be complex, as code upgrades may necessitate down time or server rebooting. In systems with a large number of nodes, making the upgrades across tens to dozens of nodes (servers) can be time-consuming—so plan your upgrades accordingly.

The goal is to eliminate scalability and performance bottlenecks while allowing the disk arrays to grow as the load is being applied. As with any technology change, being aware of the limitations helps you avoid trouble. Be certain you are conscious of the way the scale-out storage expands and where its scale limits are.

PERFORMANCE, RESILIENCY AND PROTECTION

Information, as data, as assets or however the information may be referenced—that is, the bits and bytes that make up digital assets—would be valueless if it were not properly stored, capable of being accessed and, in many cases, provided with protective measures to guarantee the integrity and sustainability of that data. Storage technologies have developed several methods to provide data protection and system resiliency. Most of those methods, by their very nature, provide many additional benefits and duties, which in turn increase overall performance based upon how the protection scheme is engineered and implemented.

Since the mid-1980s, when spinning magnetic disks were elevated to a level that they could provide reliable storage for information, computer users and data-related industries that use these various storage platforms have looked for safety measures that both retain the data captured during failures of devices and continually increase the performance of their storage ecosystem investment.

This sections looks into the technologies and prominent methodologies for providing high performance, maintaining maximum resiliency and for protecting those assets in the event of component failures at the storage level of the ecosystem. Most of these means are used in nearly every storage architecture in use today, whether in your mobile device, your laptops, the enterprise or in a cloud environment—whether that cloud is on premises, off premises or in a hybrid combination of both.

Defining RAID

The term "Redundant Array of Independent Disks" (RAID) was originally called a redundant array of *inexpensive* disks and was intended to fundamentally provide a means to store the same data in *diverse* places. Developed in 1988 by David Patterson, Randy Katz, and Garth Gibson of the University of California at Berkeley, RAID technologies were described in terms of five basic data protection methods applicable to arrays of disk drives, that is, groups of hard disk drives assembled to function as a coherent unit.

The methods were aimed to guard against data destruction due to drive failure in the grouping. RAID described techniques for mapping (or striping) data across arrays and designated their performance benefits accordingly. Commercial RAID development began in 1989 when the first non-mirrored version was announced by Compaq. By the mid-1990s the concept had become a virtual household word.

The word "inexpensive" was later modified to become "independent," presumably because the word *inexpensive* is relative, especially as storage costs have continually become less expensive in similar context to Moore's law analogies that began two decades prior in 1965. The numeric annotation (also called a *level*) was an outgrowth of the first RAID publication. RAID levels were adapted over time to indicate certain types of protection and data mapping schemes. Since the researchers' first publication, various protection equations were added, some of them by specific storage manufacturers to indicate their formula for their particular implementation.

Principle advantages to RAID come from the concept of data distribution. Data is spread across a number of individual disks in order to provide additional security for that data versus storage of the same data on a single disk. Data protection is provided by a methodology whereby redundancy data is stored separately from the primary data. The redundant data scheme may be a complete copy of the data (i.e., a *mirror*) onto another drive, or through the addition of information that can be used to reconstruct the primary data should some element of the hardware fail. In this latter case, the information used to reconstruct a bad drive is called *parity data*.

RAID technologies may be applied to stand-alone independent storage systems used in computers, imaging devices or on-air playout servers as well as other systems designed for all forms of content used in streaming media, video on demand or cloud systems. Since the dawn of RAID technologies, many other means and forms for data storage have surfaced, including flash and object-based storage. Each has helped support the evolution of data storage, protection, resiliency and high-performance data processing—irrespective of whether that information is for short-term data analysis or as an alternative to other storage mediums including optical (archive) and digital linear tape.

Cost-Effectiveness

With the cost-effective availability of small form factor disk drives, promoters of RAID claimed significant improvements in reliability and maintainability. RAID also provides a conduit for higher-data bandwidths and improved performance capabilities necessary for environments that include complex data analysis (e.g., medical imaging or oil and gas exploration), graphics processing and rendering, video production editorial systems or large groups of collaborative users who must access large files such as those used in the news, media and entertainment production areas.

Early opposition to the concept stated that RAID "can be expensive and unnecessary," and claimed there are "alternative methods to similar performance at a better cost point." Such arguments subdued significantly as the storage (areal) densities for 3.5-inch form factor drives continually increased and other factors, such as flash media, were added to compensate for physical issues that constrained or limited data storage.

RAID by the Numbers

RAID *levels* eventually transcended to become simply RAID *numbers*, signifying an *architecture* for the storage of data on magnetic spinning disks. The well-recognized yet often misunderstood term was deeply entrenched in the computer data industry long before disk arrays were used for video recording and storage for playout servers or nonlinear editing systems. RAID terminology, when used in the broadcast video server context, may still remain a source of confusion brought on by both marketing hype or the perceptual fear of failure resulting in the loss of any data.

During the early years of video servers, the term RAID was both widely used and sometimes abused. RAID numbering was quite possibly used as much as a marketing tool as it was a technology explanation. This further resulted in a misconception that some users perceived as "the higher the RAID number, the better the performance." That myth is easily disposed of when one understands the functionality of the RAID system, and where it is best placed in service.

Unfortunately, this RAID numbering misconception continued for several years irrespective of any explanation and implementation. Today, RAID levels are used mainly in fundamental disk array arrangement and may still be found as part of the surrounding subsystems integrated into a total storage array or solution (e.g., in DAS, NAS, SAN or object storage).

RAID System Architecture

Architecture, in the technology sense, describes a "physical structure of components or elements in a system." Such components become a subsystem within an overall storage system, since storage arrays are generally composed of many sets or logical units of hard disk drive (HDD) groups. In computer storage, a *logical unit number* (LUN) is a unique identifier that identifies either an individual or collection of physical or virtual (storage) devices. The devices execute input/output (I/O) commands with a host computer per protocols established by the industry. The *logical unit* is that device that is addressed by the small computer systems interface (SCSI) or storage area network (SAN) protocols, which encapsulate the data when employed in Fibre Channel or iSCSI services.

The term *subsystem* is used because, for the most part, RAID is intended to be a configuration element of the system, which is comprised of:

- the host (typically a computer and system drive)
- an input/output engine (a video encoder or decoder)
- the storage media

To distinguish between differing RAID architectures, a numbering scheme was developed that was intended to identify the various combinations of drives, how the data was striped across those drives and how the parity information was developed and carried within the array. As reflected previously, in early RAID development some expressed concern that the use of the term RAID *x*, sometimes referred to as Level *x* (where *x* is generally a number from 0 to 5) was an attempt to capitalize on the incorrect notion that the higher the RAID level, the higher the reliability or performance.

RAID levels recognized by the original Berkeley papers and the *RAID Advisory Board* (RAB) have been categorized from RAID Level 1 to Level 6. Other levels recognized by the Berkeley papers, but not by the RAB, include Level 0 (disk striping), where data is mapped in stripes across the entire array.

A storage array cannot simply be created by connecting a series of drives to a controller. Arrays generally consist of drive electronics that format, code and distribute data in some specifically structured format across all of the drives. Specialized instructions operate in conjunction with the various drives to perform protective and fault-tolerance functions necessary to meet the level of RAID designated. Some video server manufacturers developed their RAID architecture in software, thus eliminating the dependency on a physical set of outboard circuitry to perform the specific RAID operations. Today, nearly all RAID control architectures are carried in software, which is made possible by chipsets that contain the controller firmware for the HDDs. These chipsets allow groups of HDDs to communicate amongst each other at a metadata/control level, thereby allowing groups of drives to be configured to certain RAID levels in firmware.

Software-Based RAID

RAID in software is best used for large block applications such as data warehousing or video streaming. When the servers have the available CPU cycles to manage the I/O-intensive operations that certain RAID levels require, a software implementation can be quite cost-effective.

A drawback is that all the RAID functions are handled by the host CPU, which can severely tax its ability to perform other computations. The hard disk drive management component only needs a standard controller. Workstations only needing a RAID 1 mirroring support with only a couple of HDDs will generally be implemented in software.

Hardware-Based RAID

Hardware-based controllers are best used for small block applications such as transaction-oriented databases and Web servers. In concept, the processor-intensive RAID operations are off-loaded from the host CPU to the hardware controller in order to enhance performance, primarily because the controller is specifically focused on disk management, leaving the CPU to handle the database-related activities.

Including a battery backed-up write-back cache (as a solid-state flash memory device) can dramatically increase performance without adding to the risk of data loss.

RAID Differentiation

The techniques used to provide redundancy in a RAID array are a primary differentiator between RAID levels. Redundancy is provided in most RAID levels through the use of mirroring, parity and striping. Combining RAID levels, sometimes referred to as *hybrid RAIDing* or *nesting*, is exemplified in RAID 53—which combines disk striping with the parity features found in RAID 3, which in turn yields a much higher transaction rate. Another example, RAID 10, mixes RAID 0 disk striping and RAID 1 (mirroring), resulting in excellent I/O performance (through striping) combined with RAID 1's data resiliency in mirroring.

The original *RAIDbook* and the RAB suggested that "RAID levels be chosen with numbers that impart meaning, not confusion." Nonetheless, the confusion surrounding RAID nomenclature continued, and by 1997, the RAB had recognized nine RAID implementation levels. Of those nine levels, five conformed to the original 1988 Berkeley RAID researcher's efforts, with four other RAID terms later used and acknowledged by the RAB.

RAID 0, RAID 6, RAID 10 and RAID 53 were developed through committee work anchored by manufacturers, suppliers and consumers. As with any technology, there are advantages to some forms of RAID that can yield certain disadvantages as well. A careful understanding of both applications and performance are necessary when selecting one storage configuration versus another, whether it be in RAID level choices, DAS or SAN/NAS applications.

RAID Level 0

Also referred to as *striping*, this process distributes data blocks randomly across an array of disks without providing redundant information or parity. RAID 0 requires at least two drives.

The striping of an array, although enhancing performance especially in high transfer rate environments, has one critical downside. Since no redundant (or parity) information is created, and there are no redundant disks, the failure of any one drive results in the complete loss of all data on the drive array.

The term RAID 0, by itself, is seldom used despite the principle being commonly practiced. Initially RAID 0 referred to "the absence of any array technology." The RAB states that the term implies *data (or disk) striping*—"a means to evenly distribute data so that during the read request, the blocks can be rapidly recovered at random with a minimum of latency."

Disk striping without parity is not *fault tolerant.* Disk striping without RAID may be used for temporary data, scratch space, or in situations where a master copy of the data is easily recoverable from another storage device.

To address the potential for data loss with RAID 0, a RAID set typically uses at least one stripe for parity. The parity information is commonly calculated by using the binary exclusive (XOR) function and is stored on a physical drive in the RAID set. If a storage drive in the striped RAID set fails, the data is recoverable from the remaining drives and the parity stripe.

For a data set with n drives, the data might be striped on drive n through *n-minus-1* ($n - 1$), and the nth drive would be reserved for parity. For example, in a RAID set with ten drives, data could be striped to nine drives, leaving the tenth drive reserved for parity.

Disk striping with RAID provides redundancy and reliability. RAID 4 and RAID 5 protect against a single drive failure. RAID 6 can use two drives for parity and protects against two drive failures; or it may duplex (copy the parity information) across all the drives. Data protection can be extended beyond two storage device failures using a technology called erasure coding (explained later).

A disadvantage of disk striping with parity is there exists a performance penalty for small random writes, as the system must access all the stripe units in the striped RAID set.

RAID Level 1

Level 1 is the simplest, most reliable and easiest of the RAID architectures to implement and understand. In this RAID architecture, data is continually duplicated and managed on a secondary drive, as a redundant copy, as shown in Figure 5.12-36.

RAID 1 is referred to as disk mirroring, shadowing, or duplexing. When data is stored on two or more separate drives, total redundancy is achieved. The approach provides for a continual backup of all the data, thus improving reliability while still providing a high level of availability, especially when reading data. If one disk component fails or is replaced, all of the information can be restored without any down time.

RAID 1 writes as fast as a single disk. Systems are typically single I/O and non-scalable, which means that performance is not increased by adding disk elements to the array. RAID 1 is not prominent in servers designed for video purposes.

System Note

Mirroring, for video servers and recording systems, is typically accomplished by providing two complete and independent video server systems consisting of I/O, processing engines, controllers and storage. Mirrored video server systems nearly always run in parallel, with a failure of any component in the primary system resulting in a switch over to the mirrored system until the primary system can be repaired. Mirrored server systems often operate under control of an automation system that controls the recording or playback functionality and can also monitor the health of the overall system.

For highly reliable systems, the RAID 1 concept is sometimes extended to optical archives, data tape systems, and other large-scale spinning disk arrays that keep content online, near-line, or off-line. A recent and upcoming alternative to on-premise mirroring is the

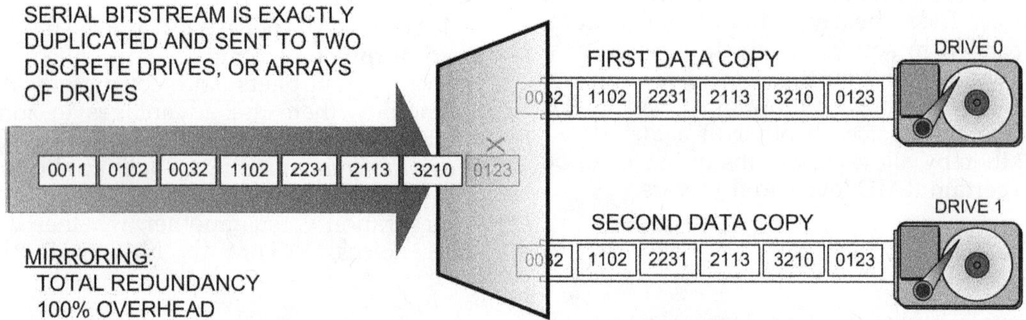

FIGURE 5.12-36 Bit mapping of data string to drive 0 and its "mirror" drive 1 is referred to as RAID 1 or mirroring. All data is duplicated for 100% redundancy. RAID 1 provides good reliability for data, does not require a complicated RAID controller and is relatively inexpensive to implement.

cloud, whereby the data in the *on-premise* system is replicated via a network connection to a cloud service provider, be it public or private.

Disk striping (RAID 0) can be combined with disk mirroring (RAID 1) to speed performance and expand capacity by striping data across multiple sets of mirrored drives. The disadvantage of disk striping with mirroring is the 50% overhead inherent in using half the capacity to make an exact copy of the data for protection.

RAID 1E

Some RAID 1 implementations will treat arrays with more than two disks differently, creating a *non-standard* RAID level, in this case known as RAID 1E. In this arrangement, data striping is combined with mirroring, accomplished by mirroring each written stripe to one of the remaining disks in the array. See Figure 5.12-37.

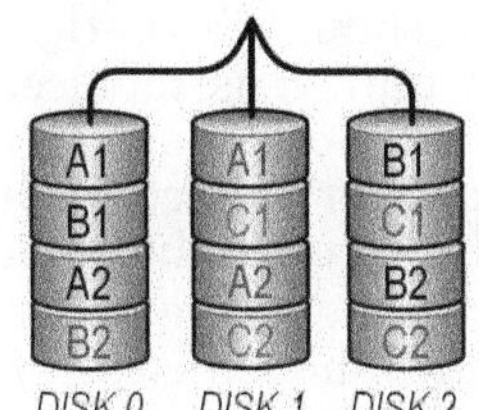

FIGURE 5.12-37 RAID 1E, a non-standardized (non-RAB recognized) configuration.

RAID Level 2

As the only RAID level that won't use one or more of the techniques of mirroring, block striping and/or parity, RAID 2 was developed for the early, large diameter disk sets that operated in 100% synchronization. In those early implementations, the huge storage devices were prone to constant disk errors, so RAID 2 employed error correction coding (ECC), known later as Hamming codes, whereby ECC data would be stored on an independent set of drives, adjacent to and mathematically connected to the primary data drive sets. In the 1950s, such data errors might have been corrected by encoding four bits of data into seven, using the additional three parity bits stored on the other drive set; thus, it was called a [7,4] Hamming code.

RAID 2 interleaves bits or blocks of data. The redundant (parity) information helps to detect read cycle errors, to correct single bit errors and detect double bit errors. ECC algorithms determined the number of disks required in the array. Later technologies replaced the extra disk drives and used RAM to detect and correct errors.

RAID 2 was quite expensive, with virtually no commercially available implementations, and has all but evaporated from all storage environments.

RAID Level 3

Level 3 is an array of *n* drives with *byte* striping (or sometimes *bit* striping) across all but one of the drives, as shown in Figure 5.12-38. A single dedicated parity drive contains redundancy data that is used to mathematically reconstruct the total data set should another member drive fail.

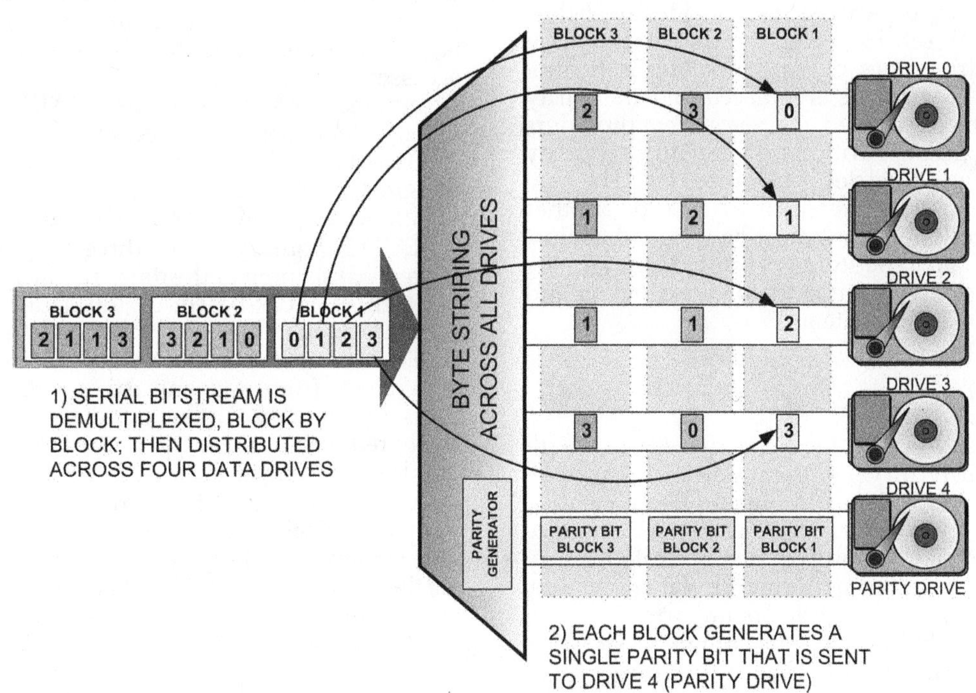

FIGURE 5.12-38 In RAID 3, data is distributed across all four primary drives. A parity bit is built from that block data and stored on the fifth drive.

RAID 3 allows for parallel data access, giving it the advantage of high data rates and increasing performance over single-disk structures. Each disk in the entire array is synchronized and used for each portion of the read/write operation. Both read and write performance cannot be hampered. Each logical record from the data stream is broken up and interleaved amongst all the data drives. Each time that new data is written to the data drives, a new parity is calculated and then rewritten to the parity drive.

RAID 3, when standing by itself, is one of the preferred choices when large blocks of sequential transfers are required. Applications in imaging, CAD/CAM and digital video or media servers may select RAID 3 because of its ability to handle large data block transfers. For streaming video applications, contiguous blocks of data are written to the array so that during playback only a minimal amount of searching is necessary, thus increasing efficiency and throughput. Sequential reads and writes will have good performance; however, for random read/write operations, RAID 3 yields the worst performance.

Another disadvantage—which is common to all striped arrays—is its poorer level of write performance when compared with single or duplex/mirrored drives. This drawback can be controlled by proper buffering and sectoring during the write process.

By definition, the entire RAID 3 array can execute only one I/O request at a time, referred to as *single-threaded I/O*, which may or may not be important, depending on the application. Some controllers and smart arrays have minimized this impact by providing intelligent algorithms and larger disk caches (including solid-state media) to buffer data temporarily while being written to the drive.

RAID 3's discrete parity drive provides resiliency to the entire disk set. However, the drawback is that when a data drive goes down, protection is temporarily lost until that drive is replaced and the parity information is reconstructed. Storage system providers may configure mission-critical arrays with two parity drives, giving a second level of protection. The cost of this implementation (versus a RAID 6 or another hybrid-RAID configuration) must be weighed against the volume of data to be stored per unit dollar and the amount of physical space it takes to house, power and cool this kind of RAID system.

RAID Level 4

Level 4 is characterized by *block-level striping* with a dedicated parity drive. A minimum of three disks are required, two for data striping and one for parity. Block striping improves random access performance compared to RAID 3; however, its dedicated parity disk remains a negative performance (latency) issue, especially during random write operations.

A *block* is a contiguous set of bits or bytes that forms an identifiable unit of data. Disk drives must establish parameters for block sizes and use metadata (the "physical record") to locate where those blocks can be found on the physical disk media (i.e., the *tracks* and *sectors* of the individual HDD *platters*). Block structuring is discussed in greater detail later in this section.

RAID 4 lets the individual member disks work independently of one another. Benefits include good input and output performance for large data transfers and good read performance, fault tolerance, format efficiency and other attributes similar to those found in RAID 3 and RAID 5. A significant difference between RAID 4 and RAID 3 is that RAID 4 uses blocks instead of bytes for striping. RAID 4 is like RAID 5, except that RAID 4 uses dedicated parity instead of the distributed parity in RAID 5.

As with most striping, write times are extended because the data is dispersed across several drives in segments. The method provides good random reads due to drive independence, but suffers from random writes resulting from having to write the parity data to a separate drive each time. With RAID 4, if a block on a disk goes bad, the parity disk need only rebuild the data on that specific drive.

A drawback to RAID 4 is in those extra steps that become necessary to update check data and user data. More significantly, should the parity disk fail, all data protection is lost until it is replaced and the parity drive data is rebuilt. Sometimes an additional parity disk might be added for protection, but this raises similar concerns as discussed in the section on RAID 3.

RAID 4 storage systems are not commonly used in storage architectures; nonetheless, at least one video media storage company utilizes RAID 4 in its storage product.

RAID Level 5

Level 5 is one of the more widely utilized RAID formats, especially for media-centric storage activities and when high transaction throughput and multiple concurrent accesses to data is necessary (such as video on demand or nonlinear editing systems). RAID 5 is employed when independent data access with high read-to-write ratios is required. Typically, the more disks in the array, the greater the availability of independent access, thus the higher bandwidth (or throughput) levels.

RAID 5 requires at least three disks and uses block (or record) striping of the data stream, as shown in Figure 5.12-39. Block-based, digital data is spread across all of the disks in the array. Independent access is available because it is possible to extract the entire specific data block from any one drive without necessarily accessing another. Latency and seek times are effectively reduced, resulting in performance increases and good redundancy.

As each block of data is written to the array, a rotating parity block is calculated and inserted in the serial data stream. The parity block is interleaved throughout all the disks, allowing that data to be recovered from any of the drives at any time. When parity information is rotated across all of the drives, the I/O bottleneck of accessing the single parity disk (as in RAID 3) when concurrent accesses are requested is significantly reduced.

When a single drive in a RAID 5 array fails, read and write operations will continue because data will

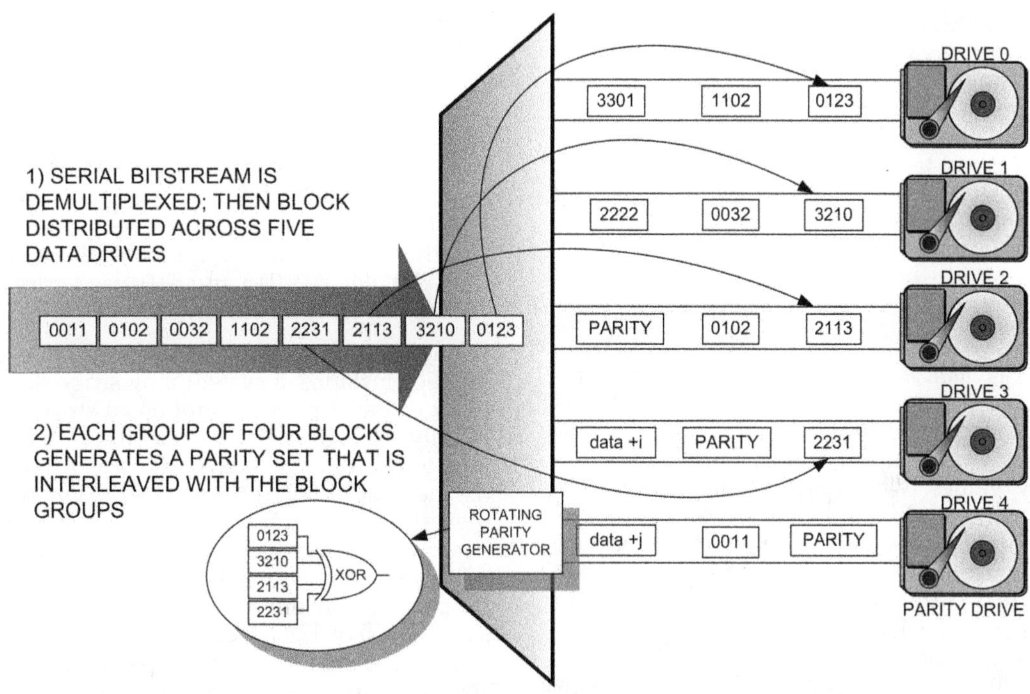

FIGURE 5.12-39 In a RAID Level 5 configuration, should any single drive fail, the read and write operations may continue. The parity bits generated are interleaved throughout the entire set of disks, allowing for continued operations and rapid recovery. RAID 5 (like RAID 3) is used in high-performance applications.

or has been block striped across all the drives. During a read cycle, as long as the data to be read resides on an operational drive, there are no problems. During a write cycle, data that is to be written to the array is protected because the controller simply inhibits writing to the failed drive. Parity data is still constructed and then distributed across the remaining drives until the failed drive is replaced and the RAID rebuild process is completed.

Data that resides on a failed drive is still available because the read process will use the remaining parity information, which is interleaved throughout the remaining drives, to reconstruct the missing data. Algorithms determine where the data resides and are statistically tailored to allow nearly transparent operations during failure modes. The host never sees the failed drive, and the entire array remains transparent.

RAID 5 "Enhanced"

Some manufacturers have developed enhanced "E" versions of RAID arrays (see RAID 1E above) in order to provide additional protection through the inclusion of hot spare drives. While the terminology isn't necessarily standard-issue by any means, the marketing concepts and added layer of resiliency have improved sales and reduced risks in at least a few general instances.

RAID 5E

RAID 5E is a RAID 5 array with an active hot spare drive that utilized in the actual array operations. In the conventional RAID 5 configuration with an added "hot spare," the drive sits inactively next to the array in standby and waiting for a drive to fail. When a failure occurs, the hot spare becomes available and the controller then rebuilds the data set with the new HDD.

Advantages to this operational method include that the user/administrator knows in fact that the drive intended to be used as a hot spare is in working order. Also, the additional drive is indeed included in the array, thus further distributing the array's I/O load. When more spindles are active, it equals better performance in most cases. Arguably, RAID 5E *could* perform better than typical RAID 5.

There are a few disadvantages accompanying RAID 5E as well. First, controller support for RAID 5E is limited. Second, a hot spare drive cannot be shared between arrays. Third, rebuild time can be slow.

RAID 5E storage capacity is effectively the same as the raw storage capacity of a RAID 5 array that contains a hot spare.

RAID 5EE

RAID 5EE is quite similar to RAID 5E, but with one significant difference: the hot spare's capacity is integrated into the stripe set. In contrast, under RAID 5E, all of the "empty space" is kept at the end of the array. As a result of interleaving empty space throughout the array, RAID 5EE provides for a faster rebuild time than in a RAID 5E configuration.

RAID 5EE has all of the same advantages as RAID 5E, with the enhancement of a faster rebuild time than any of the other RAID 5 configurations. On the downside, RAID 5EE has the same cons. It is like RAID 5E, with the main negative point being that controller support is minimal.

Note that the descriptions for enhanced "E" drives were originally developed in the 2007–2012 time frames. Implementation and terminologies will change, sometimes dramatically, as technologies improve, and the rationale for this type of enhancement may be superseded by other forms of configurations.

RAID Level 6

Known also as *double-parity* RAID, this particular implementation over the course of the past decade has gained significant popularity, especially for media-centric and mission-critical applications. RAID 6 uses two sets of parity values, which are calculated for each parcel of information. Like RAID 5, the active data is block-level striped across all the drives. The difference is that "dual-parity" blocks are created for each block of data, and are distributed across the entire drive set.

The real value in RAID 6 over others (such as RAID 5) is that up to two disks may fail and the storage system will still function. Even with two drives out of service, data may still be reconstructed without having to go to an outside data backup. This integral "second tier of protection" makes RAID 6 ideal for mission-critical playback and record operations and is popular when deployed for archive purposes.

Drawbacks to RAID 6 include the addition of extra drives necessary to account for the dual-parity configuration. Another drawback comes with every write operation requiring two parity blocks be written; thus write performance is lower, yet the read performance of RAID 6 still keeps on a par with RAID 5.

Mirroring and Striping

RAID is essentially based upon two fundamental principles whereby data is either completely *duplicated* or it is *distributed* across sets of drives to increase resiliency and increase performance. In many storage systems, combinations of mirroring and striping add further value to the resiliency/performance equations.

Mirroring

The process where multiple copies of a given RAID level are employed is referred to as *mirroring*, and sometimes may include or add the term *duplexing*. Examples include RAID Levels 0+1 and 1+0, which represent a layering of the two elementary levels of RAID. In the example, RAID 10 combines data striping (as RAID 0) along with equal data splitting (RAID 1) across multiple drive sets. Duplexing is a variant of RAID 1 that includes mirroring of the disk controller as well as the disk.

Disk Striping

Disk striping is the process of dividing a set of data into blocks and distributing those data blocks across multiple storage devices—in this example hard disks, yet it can apply to solid-state drives (SSDs) as well.

- A *stripe* consists of the data divided across the storage devices.
- A *striped unit* (or strip) references the data slice on an individual drive.
- *Striping with parity* is where single RAID Levels 2 through 7, and multiple RAID Levels 0+3 (e.g., RAID 53), 3+0, 0+5 and 5+0, use parity with striping for data redundancy.

Storage systems vary in the way they perform data striping. For instance, a system may stripe data at the byte, block or partition level, and it can stripe data across all or only some of the disks in a cluster. A storage system with ten hard disks might stripe a 64 KB (kilobyte) block on the first, second, third, fourth and fifth disks and then start over again at the sixth through tenth disks. Another system might stripe 1 MB (megabyte) on each of its ten disks before returning to the first disk to repeat the process. Still another system might change the block size to 4 KB or 128 KB or anywhere in between.

How or in what way a particular drive array is striped is up to the supplier or user's configuration. The reasoning depends entirely upon the application. In applications that manage proxy images of a size 256 Kbits to 1 Mbit, the block stripe size may be midway between the low- and high-end dimension—since much of the data is relatively small. For metadata chunks, usually 2 to 10 Kbits, a smaller 64 KB block size should suffice. However, when handling much larger image files (e.g., 10 MB) or entire streams of contiguous images (hundreds of megabytes), then a 4 KB block size may be more advantageous in terms of access and seek latencies.

Block Sizing

A *block* is an addressable unit of storage with a specified size. Hard drive efficiency depends on a relative way of addressing trillions of bits, and the block (or block numbers) refers to an address that contains *x* number of bits. This is why, for example, if you save a file that is 1 KB (kilobyte) on a hard drive with a 16 KB block size, the file will appear to be 16 KB. If the file were 16.1 KB, it would appear to be 32 KB (which wastes data space), and so forth. If a computer had to address each bit individually, significant bottlenecks would be introduced; thus the rationale behind slicing the drive data space into blocks.

Block sizes can be elusive, as there is sometimes both a RAID block size (more commonly known as *chunk size*) plus a separate *file system block size*, with various file system products using differing techniques. In this scenario, the file system may write a 4 KB file system block to hard drive, which uses a 256 KB chunk size. This means that additional metadata is kept that maps the 4 KB block of data into multiple 256 KB chunks for distribution on the HDD or SSD.

Managing the block size and chunk size is typically the job of the file system manager, which means that the drive array itself must be compliant with the file system, file system block size and other services that can impact the performance and value proposition of all the components.

Dynamic Allocations for Varying Block Sizes

Other types of systems may use different striping techniques. One technique, called *dynamic striping*, allows various block sizes to be used across various stripe layouts. In *static striping* (the typical means of striping a disk array), data is striped across disks in chunks of a certain size. Stripe layouts are statically determined at stripe group creation time; however, storage allocation workloads can change dynamically. Flexibility can be gained by striping the same data more than once with different widths, but even with partial coalescence of the duplicated data, the storage overhead can be more than 50%, which is quite prohibitive.

Static striping also causes the performance of the entire stripe group to be limited by the performance of the slowest disk in the group. This can inhibit, for example, streaming media full quality playout, since the I/O cannot complete faster than the slowest disk can read or write the blocks stored on it. Here the overall performance gets degraded to the lowest common denominator in the disk pool.

Storage Pool Allocation

Dynamic striping leverages that a file system, designed to incorporate dynamic striping, can write any block to any device in a storage pool that is managed by the storage pool allocator (SPA) program. The system would need to incorporate "copy-on-write, write-anywhere" file system capabilities. That is, each write to a block command then drives that block to be written in a new location.

When the SPA is presented with a sequence of blocks to write out, it automatically fans out the writes across all the available devices in the storage pool, allowing each block in the file to be written to any location on any disk. Dynamic striping requires no administrator configuration beyond the creating of the storage pool allocation and a few short code words.

Dynamic striping further allows a general-purpose file system to handle heterogeneous (differing) devices, due to its ability to allocate blocks from disks in proportion to their performance. Adding one slower device to the storage pool does not degrade performance because it basically then writes fewer blocks to the slower device versus to those of faster devices.[12]

Block Type Contrasts

In these previous sections, while related to data storage at the hard drive level, the information should be placed in contrast to other instances where the term "block" is used. While there are some corollaries between the terms, their uses must be understood in the context in which they are presented. For example, an *operating system block* is the minimum unit of data that the operating system can read or write. By contrast, an *Oracle block* is a logical storage structure whose size and structure are not known to the operating system.[13]

Pros and Cons of Disk Striping

The other principal advantage of disk striping is that it yields higher performance across the entire disk storage system. For example, striping data across three hard disks would provide three times the bandwidth of a single drive, less some additional overhead. For example, when a pool of three disk drives each runs at 200 input/output operations per second (IOPS), by employing disk striping up to 600 IOPS is made available for data reads and writes.

A disadvantage to disk striping is lower resiliency. The failure of any physical drive in the striped disk set results in the loss of the data on the striped unit, and consequently, the loss of the entire data set stored across the set of striped hard disks. This is why parity information, which is created from each data block and per each location of the block, is absolutely necessary to prevent total loss of data from any single given disk drive.

Hybrid Multilevel RAID

When RAID configurations are *layered* or *nested*, they become *hybrid storage systems* that can then improve protection, performance, access time and overall system *bandwidth* and *I/O throughput*. Nested arrays combine single-digit RAID array components, also referred to as *segments*, with the features of other types of segments to achieve these goals.

Some of the "common nomenclatures" for multiple hybrid RAID levels are identified in Figure 5.12-40, although there is no specific industry recognition for some of the terms.

Most of the nested RAID configurations are sanctioned by industry groups, yet some are made by storage system vendors who are looking for that next best device to be provided and marketed and in turn are not necessarily standardized or thought to be consistent across other vendor's offerings.

The follow sections describe in text and diagram how the commonly accepted "single-digit" RAID configurations are shaped into hybrid or nested form factors.

RAID 01

Known also as RAID 0+1 or RAID 0/1, this level uses a mirror of stripes, achieving both replication (duplicative copying) and sharing of data between disks. A minimum of four disks are required to create a RAID 01 configuration. The usable capacity of a RAID 01 array is the same as in a RAID 1 array made of the same drives, in which one half of the drives are used to mirror the other half. See Figure 5.12-41.

HYBRID / NESTED RAID LEVEL	ALTERNATIVE / EXAMPLE NOMENCLATURES	DESCRIPTIONS & NOTES
RAID x + y	example: RAID 01 alternatively: RAID 0+1 common name: "mirrored stripes"	*Each set is a RAID 0 array,* *which is then mirrored across both RAID 0 sets* *(ex: eight disks divided as two mirrored sets of four drives each)*
RAID y + x	example: RAID 10 alternatively: RAID 1+0 common name: "striped mirrors"	*Each set is a RAID 1, mirrored array,* *which is then striped across all five RAID 1 sets* *(ex: eight disks divided as four sets of two drives each)*
RAID 0+1	RAID 01 or RAID 0/1	*RAID 01 is a mirrored configuration of two striped drive sets*
RAID 1+0	RAID 10 or RAID 1/0	*RAID 10 is a strip across a number of mirrored drive sets*
	Note: Some designers use RAID 01 and RAID 10 interchangeably	
RAID 0+3	RAID 03	*Formed by putting into a RAID 3 array a number of striped RAID 0 arrays*
RAID 3+0	RAID 30	*Formed by striping across a set of RAID 3 sub-arrays*
	Note: Combines byte striping, parity, and block striping; and sometimes seen as "RAID 53"	
RAID 0+5	RAID 05	*Formed from a RAID 5 array comprised of a number of striped RAID 0 arrays*
RAID 5+0	RAID 50	*Formed from a RAID 0 array striped across RAID 5 elements* *(this is an uncommon configuration)*

FIGURE 5.12-40 Common nomenclatures for multiple (hybrid or nested) RAID levels, generally provided for data protection and/or resiliency, as well as bandwidth management improvements.

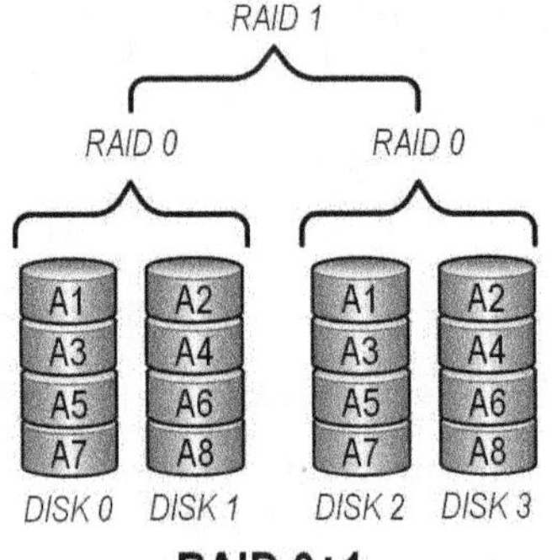

FIGURE 5.12-41 RAID 0+1 achieves both replication and sharing of data between disks.

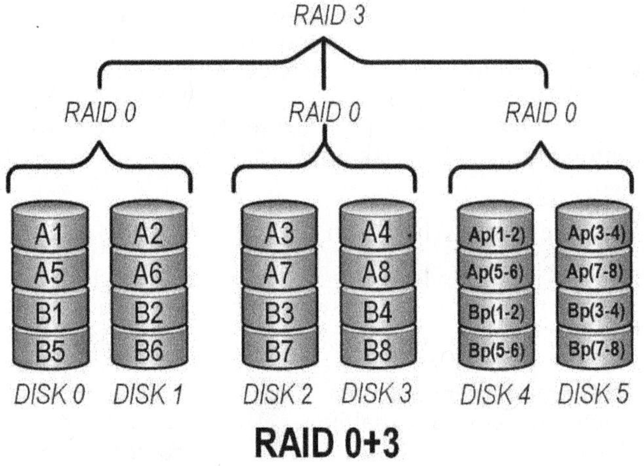

FIGURE 5.12-42 RAID 0+3 is similar to RAID 0+5 and 5+0.

RAID 03

Known also as RAID 0+3, RAID 03 is like RAID 01 except that the data (like in RAID 3) uses byte striping with dedicated parity combined with block striping.

This set is probably the most confusing of the various multilevel RAID configurations. These levels can be found in combinations of RAID 03 (= RAID 0+3) or RAID 30 (= RAID 3+0), and as RAID 53 (often implemented as RAID 30—but not as RAID 03).

Characteristics in RAID 0+3 and 3+0 are found to be similar to those in RAID 0+5 and 5+0. RAID 30 and 03 tend to be better for large files than RAID 50 and 05. See Figure 5.12-42.

RAID 10

Also known as RAID 1+0 and sometimes seen as RAID 1/0, implementation takes a minimum of four disks and combines disk mirroring and disk striping across mirrored drive sets. As long as one disk in each mirrored pair is functional, data can be retrieved.

RAID 10 leverages redundancy and performance and is one of the better choices for I/O-intensive applications. A disadvantage is that only 50% of the total raw capacity of the drives is usable due to the mirroring.

According to manufacturer specifications and independent benchmarks, RAID 10 provides better throughput and latency than all other RAID levels except RAID 0, which excels in the throughput category.

RAID 10 is ideal for applications such as database, email, Web servers and other uses requiring high disk performance. See Figure 5.12-43.

RAID 50

RAID 50, also seen as RAID 5+0, might better be called RAID 03 because it is implemented as a striped (RAID 0) array whose segments are essentially RAID 3 arrays. Layout-wise, the array is a mirrored pair of RAID 5 striped sets, which includes distributed parity across each RAID 5 segment.

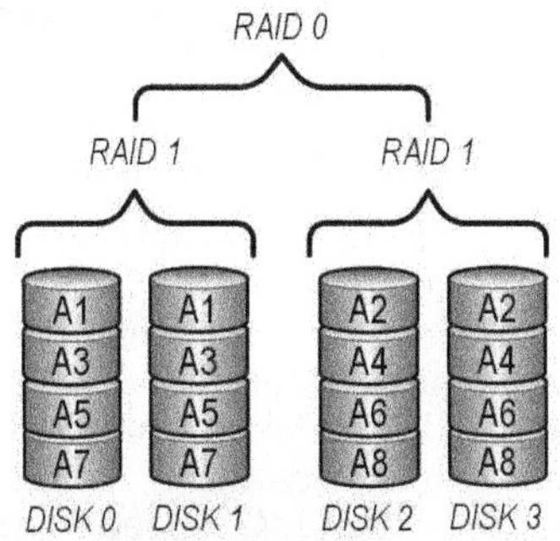

FIGURE 5.12-43 Sometimes known as RAID 1/0, this requires a minimum of four disks in a striping and mirroring configuration.

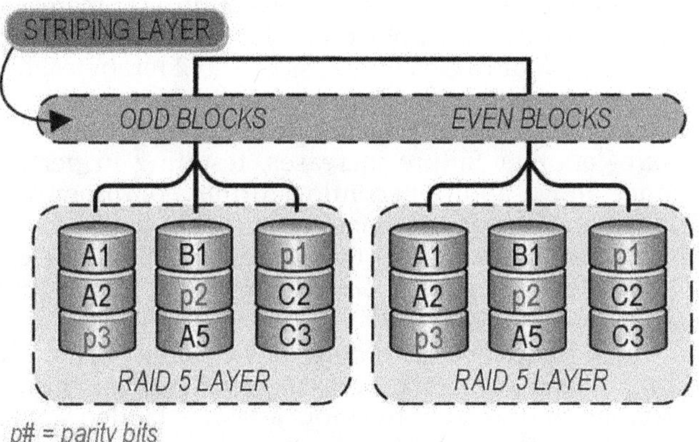

FIGURE 5.12-44 Sometimes known as RAID 5+0, the array is a mirrored pair of RAID 5 striped sets and includes distributed parity across each RAID 5 segment. This is an uncommon implementation.

Its introduction gained favor in the mid-1990s, but it was found to be very expensive to implement. Like RAID 3, the array requires that the disks be synchronized to gain access to the data, which limits drive selection choices.

RAID 50 has better fault tolerance than RAID 5, but it also requires twice the parity overhead. The RAID 5 segments provide for high data transfer rates. For small requests, the I/O rates are high due to its RAID 0 striping.

A failure of two drives in any one of the RAID 5 segments renders the entire array useless. See Figure 5.12-44.

RAID 53

This RAID level is very similar to RAID 50 and should probably be labeled RAID 03. RAID 53 is implemented as a striped (RAID 0) array whose segments are RAID 3 arrays. RAID 53 has the same fault tolerance as RAID 3 and the same fault tolerance overhead.

High data transfer rates are achieved due to its RAID 3 array segments, and high I/O rates (for small requests) because of its RAID 0 striping.

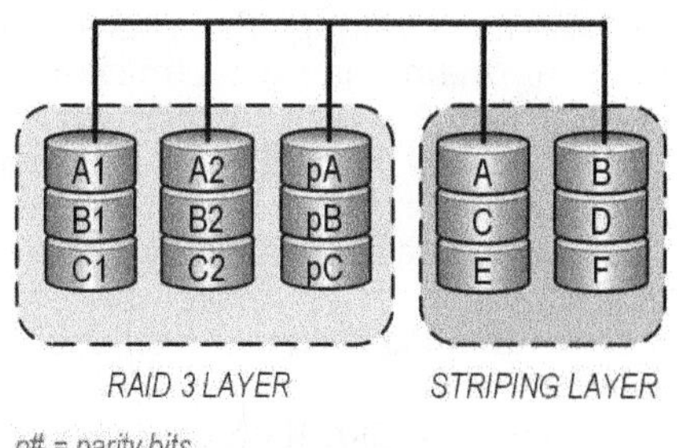

FIGURE 5.12-45 Similar to RAID 50, and better called RAID 03, the array achieves high transfer rates but requires spindle synchronization—a costly approach due to the uniqueness of the specialty drives necessary.

Disadvantages include a high cost to implement, spindle synchronization (which in turn limits the choices of drives) and byte striping (which results in poor format capacity utilization). See Figure 5.12-45.

RAID 7

RAID 7 is not an open industry standard but a trademarked marketing term from a computer storage vendor, used to describe their proprietary RAID design. The vendor describes RAID 7 (RAID 3 + RAID 4) as a "hybrid of RAID 3 with a write cache and RAID 4."

RAID Rebuilding

Anyone with any type of high-performance storage system for a video playout server, a play-to-air system or a nonlinear editing solution of any scale has probably experienced this. One of their disk drives fails completely or they get that error warning of "imminent drive failure—change drive ooooX3Hd IMMEDIATELY!"

Some may procrastinate and risk the certain impact. Others take heed and elect to change the failing drive. Still another group sits back, knowing they'd planned ahead and bought an extra hot spare and that the drive controller system will take over without human intervention.

Resiliency to Data Loss

When a hard drive in RAID configuration fails, there is a period whereby the system fault tolerance or resiliency to data loss decreases. Depending upon the RAID level employed or the protection scheme in place, the risks may range from moderate to serious. The most significant and primary concern to the system is if another drive in the same array fails. Such a loss compromises the entire storage system and renders all the data in that LUN, array or possibly the system useless. When the protective element in either the dedicated parity drive (as in RAID 3 or RAID 4) or the secondary protective parity set (as in dual-parity RAID 6) is no longer available, the period between then and when the array is rebuilt and back online, at 100% service level, can be dicey at best.

When a failed, or failing, drive is detected, the administrator or maintenance technician must first replace the bad drive, which in turn triggers a process called the "RAID rebuild." RAID rebuilding is the data reconstruction process that mathematically reconstructs all the data and its parity complement, so that full protection (with fault tolerance) is restored, in essence returning the system's resiliency back to a "normal" state.

Sometimes when certain data checks or errors are detected by the controller, the drive array may go into a reconfirmation period whereby checksums and/or parity algorithms perform a track-by-track, sector-by-sector, block-by-block analysis on each drive. Ultimately, parity or checksums are all compared and/or rewritten, and the array is then requalified to a stable, active state.

Steps to Recovery

RAID fault tolerance involves a number of steps. In one example, should a disk fail, the RAID controller attempts to copy the resilient data to a spare drive while the failed one is replaced. Using parity data and RAID algorithms, which vary depending upon the RAID level, parity data is then reassembled back to either the dedicated parity drive (as in RAID 3 or RAID 4) or is distributed across all the drives, as in RAID 5 or RAID 6. See Figure 5.12-46.

For other RAID levels, should one of the main data drives fail and no active hot spare is available, a new HDD must then be installed, at which time data from the other remaining drives is reconstructed back onto the new drive using data extracted from a dedicated parity drive or from the parity blocks distributed across the array. Either way, the risks during the rebuild time are elevated until the new drive is brought up online with all the reconstructed data and parity elements having been restored.

Large-scale drive arrays, those with hundreds of spindles (HDDs), usually have sufficient overhead, intelligence and processing bandwidth to compensate for certain fault issues. Those that employ intelligent RAID controllers can also be proactive. Should the RAID controller suspect or detect that a hard drive is about to fail, the proactive controller may either begin the process of RAID rebuild to a (hot) standby drive or signal the user to replace the failing drive or add in another drive in an available slot so that the RAID rebuild process can be kick-started before an actual failure occurs.

Failures during the Rebuild Time

One of the drawbacks to RAID—when in a failure mode or during a rebuild process—is that the performance of certain applications or processes may be impacted due to system latency. System throughput—otherwise known as bandwidth—may be reduced because: (a) not all the drives are functioning; and (b) the rebuild process takes away the I/O speed while it rapidly moves blocks of data sets from the remaining active drives onto the new/replacement drive. The reduced performance can be especially noticeable when the array is relatively small, that is, when the number of spindles (drives) is low or when the individual HDDs are very large.

As hard disk storage capacities continue to increase, rebuild process times will take longer and longer. In some cases, for drives in excess of 1 to 2 terabytes, the rebuild process can last from several hours to several days. Of course, during this period, latency and the risk of another failure increases, resulting in performance and usability becoming further compromised. This is one reason, amongst others, that high-performance storage solutions may tend to use smaller-capacity HDDs (e.g., 300 GB to 750 GB) and may put many more drives into a single chassis or array.

These issues and considerations become part of the selection and decision process regarding how to choose a storage solution. For the more advanced video server or media-centric products, those built for mission-critical operations, many manufacturers have already taken these conditions into account and provide sufficient fault tolerance or resiliency to "ride through" most of the more commonplace maintenance, upkeep and failure situations.

Flash for Resiliency

Recent technology solutions used to supplement the resiliency process involve flash memory or, more appropriately, solid-state drives (SSDs) to either supplement the array or replace the hard disk drive completely. Another of the possibilities, becoming more prevalent in solutions for archive or long-term storage, is to employ another form of storage architecture that doesn't utilize RAID principles but instead employs a

FEATURES	RAID 0	RAID 1	RAID 1E	RAID 5	RAID 5EE	RAID 6	RAID 10
MINIMUM DRIVE COUNT	2	2	3	3	4	4	4
READ PERFORMANCE	HIGH	HIGH	HIGH	HIGH	HIGH	HIGH	HIGH
WRITE PERFORMANCE	HIGH	MODERATE	MODERATE	LOW	LOW	LOW	MODERATE
DEGRADED READ PERFORMANCE	N/A	MODERATE	HIGH	LOW	LOW	LOW	HIGH
DEGRADED WRITE PERFORMANCE	N/A	HIGH	HIGH	LOW	LOW	LOW	HIGH
UTILIZATION (% CAPACITY)	100%	50%	50%	67–94%	50–88%	50–88%	50%
DATA PROTECTION	NONE	1 DRIVE FAILURE	1 DRIVE FAILURE	1 DRIVE FAILURE	1 DRIVE FAILURE	2 DRIVE FAILURE	UP TO 1 DRIVE PER SUB-ARRAY

FIGURE 5.12-46 Comparison of selected RAID configurations.

technology called "erasure coding" to distribute data across significantly more drive spindles with algorithms that allow the data to be reconstructed should one to many (several) drives cease to function.

RAID Assisted by Solid-State Devices

Solid-state storage devices (SSDs) provide users with additional levels of performance, such as much higher *input/output operations per second* (IOPS) and less latency versus HDDs. Some of the RAID performance enhancements typically found for spinning magnetic disks become somewhat moot, which, in turn, puts a different focus on data availability and protection advantages for SSD.

Many modern flash memory chips have some basic RAID functions already built into their structure. The features are designed to increase redundancy and extend longevity. Other functions inherent to flash and SSDs are not necessarily RAID related, yet overall will add performance improvements to system functionality. The value proposition in utilizing SSDs is in how the devices are used in the overall system.

Tier or Cache

How solid-state storage is used, as a tier or as a cache, becomes a key consideration, depending upon what performance goals are expected to be achieved.

Understanding how, what type, or when to employ SSDs in a tier or a cache can be a confusing and even misleading proposition. For example, when used in a cache configuration, some SSD/HDD implementations may require a lower-tier spinning disk confirmation prior to confirming a write operation. Furthermore, many cache operations, beyond read-only activities, are unlikely to offer the immediate data protection expected.

Utilizing an SSD cache (or tier) to support a storage system with these impacts can indeed affect performance and do not much more than add unnecessary cost to the system. Validate the functionality of the entire solution and understand how a prospective vendor might address this type of implementation.

SSD to Replace the HDD

Many believe that using an SSD device as a direct replacement for the HDD is a practical and viable method to increase performance overall. That analogy may be accurate, depending upon what the current storage composition is made of.

When the SSDs are used simply to replace some HDDs in a system, then the same RAID configuration may be applied. However, when adding in the components that form a RAID group, invariably you will need the same RAID to be applied within and across all the RAID groups to allow for the tiering and movement flexibility expected from the same HDD configuration.

For high-end systems, RAID 5 or RAID 10 are likely sufficient, but for added security the double-parity provisioning from RAID 6 is likely a superior choice.

The raw reliability of solid-state storage (flash memory) is entirely relevant, as SSD manufacturers now claim "x years" for a given number of *full daily writes*, which is considerably better than the equivalent specifications provided by HDDs.

RAID with SSD—Why?

Mitigating single points of failure, especially in a storage system, is one of the principal reasons that RAID is employed for mechanically based spinning magnetic storage (HDD). Looking at storage from a solid-state perspective can add new dimensions to a long-standing approach for performance, protection and resiliency.

As an alternative in server storage implementation, placing a solid-state storage PCIe card in a server is a popular approach aimed at boosting application and storage performance. The implementation is effectively still a single unit, direct-attached storage (DAS), which then translates into similar single point of failure (SPOF) risk factors. A straightforward RAID 1 (mirrored) pair of flash cards could be an applicable data loss protective alternative, despite the fact it is an expensive solution. Frequently these cards are implemented as *read-only cache*, allowing the protection to be more economically achieved at the HDD level.

Flash Relevancy to RAID

Flash drives (a.k.a. SSDs) have no moving parts and consequently are not subject to mechanical failures like HDDs. Solid-state devices implement wear leveling, a process that extends longevity of the drive material, since there is a limited life span for the cycling of the SSD's memory cells between erase-write and read. Wear leveling processes monitor where data is placed in the memory cell structure and how often those data cells are modified. These algorithms distribute, over time, the write I/O functions, which if left uncontrolled would cause these devices to fail prematurely. See Figure 5.12-47.

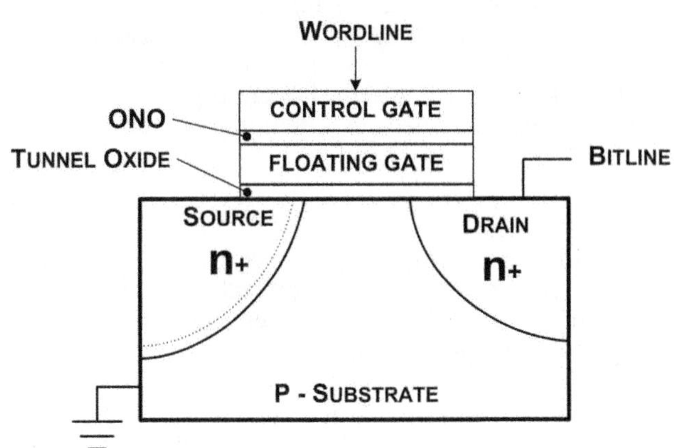

FIGURE 5.12-47 Flash memory cell construction.

Like any other component, electronic or mechanical, there is always a risk of failure—flash memory/SSDs are no different. Component failures in an SSD could, for example, include issues with the device controller or with a weak memory cell that might be unable to hold a steady state. Eventually an SSD will fail due to its reduced (weakened) write I/O capacity; that is, too many cells are weakened, rendering data mapping unobtainable to those cells.

All-flash storage systems can typically implement system-wide redundancy. This happens by evenly distributed write I/O functions across the entire array, which in turn gains the maximum life expectancy from the solid-state components. Some manufacturers have products that are tagged as "virtual RAID" whereby the I/O load is distributed across all components, ensuring the normal erase cycle encountered when writing to SSD does not affect the performance of other I/O host traffic.

Some manufacturers have addressed the erase cycle impact for read I/O by creating proprietary RAID in software solutions that treat read I/O delays, on a single flash unit, as a device failure. The troubled device reads the data by rebuilding the read request from other devices within the same parity group. It is the high performance and consistent response times inherent in SSD devices that allow this process to be successful.

Options in Solid-State Drives

In the early years, before modern RAID and SSDs, most storage components were principally ranked by the individual raw capacity of each storage device or, in the case of "just a bunch of disks" (JBOD), an array or group of disks. As technologies evolved, the very earliest of broadcast video servers used hard drives that barely approached 4 GB (e.g., the Tektronix PDR-100 of the early 1990s).

Professional video servers needed protected storage, and a Fibre Channel RAID configuration offered that solution. Once HDDs became more common and more affordable, storage descriptions shifted to include a RAID level numeric specification.

Initially the RAID level number didn't mean too much to the user. The mere fact that the storage on a (PDR-100) video server was limited to about 12 hours meant that only a limited number of commercials or promos could be stored on the drives, so besides the fact that there was some fundamental protection, a lot of the backup remained on videotape.

High-Speed Card-Based Memory

As magnetic spinning hard disks reached or exceeded the 10 TB range, in a 3.5-inch form factor, and field camera capture and mobile devices began to proliferate, a greater need for a smaller form factor and removable memory evolved. Most of the solid-state memory in mobile devices and tablets was not removable. Mobile device manufacturers didn't want to add the additional ports and power necessary to support USB, so it has remained "rare" except for a few mobile platforms.

For desktops or laptops, consumers have looked toward USB external connections, but also at USB "jump drives" (flash drive sticks with USB connectors) or other flash memory devices (e.g., compact flash) to support consumer and commercial storage needs. Collectively, these USB, SD (secure digital) and similar extensions use generic *flash memory*, which is generally comprised of NAND flash memory components.

In support of HD and 4K resolution video, these SD-card memory devices come with published storage at upwards of 128 GB. The companion device, which is usually integral to the capture device or the computer, is the solid-state drive, mostly in 2.5-inch form factors, which have reached half a terabyte or above.

The memory management on a conventional SSD is built on the assumption it will be a companion or replacement to a spinning magnetic disk. This makes the products suitable for computer work, versus a similar size data disk that lacks the operational features of an HDD in a solid-state form factor.

Space Availability

Understanding how much actual storage is available on a system, whether for RAID or SSDs, is not as easy as it might seem. Regardless of the media type, the available space *reported* to a user by an operating system with an SSD or other flash memory device almost never matches the physical (raw) storage device capacity listed on the nameplate. The reasons can be confusing and misunderstood, sometimes causing users to roll the dice on what they're really getting for their precious "pennies on the gigabyte."

In all cases, some of the storage memory cells for any flash memory product will be utilized for device formatting during manufacturing. Subsequently, additional cells will be consumed during operating system (OS) formatting, and more are used for administrative computing operations either for computing inside the card or computing within a host device, for example, the camera or Windows or Mac OS. The so-called hidden space is unavailable to the user for data storage. However, it does serve to improve the performance and/or the reliability of the storage medium when used with host devices.

Controller firmware, defined as "the code that runs inside the storage device," is an allocation set aside for updates and other internal administrative functions. In some applications, memory cells are reserved to administer copy protection or other security feature requirements found in Secure Digital (SD) cards.

Function-wise, the hidden space may also be allocated for such things as testing the memory cells to identify usable versus unusable cells. In this case, the system reserves (or deactivates) the unusable cells to avoid reliability issues or performance degradation in the user experience. These cells cannot be written to and remain inactive for any user data.

File Tracking

During the OS configuration for a device, which varies for Mac versus Windows operating systems, the device-formatting operation will allocate space to track such things as user file locations, labels for dates and times, and other metadata such as filenames, types or sizes. The device also stores administration records describing used versus free file system space, which is then reported to users through the operating system. For example, on a Windows OS (like an HDD), a memory card or SSD will (usually) create a file allocation table (FAT) with root directory and subdirectory elements carving out some portion of the storage as unavailable for actual user data.

Flash memory devices are known to "age" (or "wear"); that is, they deteriorate based upon the number of read-erase/write-erase cycles they are subjected to. Spare storage cells are allocated in the NAND for administrative purposes, which will extend over the life of the card or drive. The "spare cells" are available for use as substitute cells when an in-use memory cell should fail to store (or successfully read) a user data item. If a nonfunctioning cell is detected, the cell is effectively "retired." Then, a spare cell from the reserved pool is activated and the user is allowed to place data there—without ever knowing the previous cell was deactivated. Overall, the available user space remains unchanged; as the spare cell is activated it equalizes storage capacity as the nonfunctioning cell goes dormant (deactivated).

Card-Based COTS Memory with Ultra-High-Speed Performance

The advent of streaming media, including higher resolution/higher bit rate imaging (both for moving and still images), is bringing with it an increase in performance requirements for flash and other high-speed memory components. More recent nomenclature for flash Secure Device memory now includes designations such as Ultra High Speed (UHS) II and UHS Speed Class 3 (U3) (see Figure 5.12-48). Read speeds, at the time of this writing, are now reaching upwards of 300 MB per second, with write speeds up to 260 MB per second.

UHS-II memory cards (left) have additional contacts for data transmission and width improvements.

Ultra High Speed II   Ultra Plus SD

FIGURE 5.12-48 High-performance solid-state (nonvolatile) Secure Digital (SD) memory cards have additional contacts to support the much higher transfer speeds.

With transfer and read/write speeds in these dimensions, memory device manufacturers may recommend specifically developed high-speed readers that enable the high-volume transfer of data (i.e., RAW photo images and 4K video) between the card/device and a Mac or PC. Adding this component can dramatically accelerate workflow and offer users performance improvements straight out of the package.

Not unexpectedly, as soon as a target is set for one device, others emerge that drive the performance capabilities even higher. As evidence, formats such as SD, SDHC, or SDXC are now superseded by a new XQD card format. Sony's G series XQD Format Version 2 technology states transfer speeds of 400 MB per second (read) and 350 MB per second (write). Transfer speeds will also vary depending upon the interface and other influences revolving around the host devices and/or the receivers.

When employing other interface technologies (e.g., PCI Express Gen. 2 and USB 3.0 interfaces) coupled with these UHS (ultra-high-speed) cards, support is now available for stable recording at up to 600 Mbits (= 75 MB) per second. Depending on the device (i.e., the camera/camcorder being used) and the card model being utilized, maximum read speed rates of 1.44 Gbits (180 MB) per second are achievable, enabling faster ingest performance as recorders move away from proprietary card solutions to commodity-off-the-shelf (COTS) memory cards.

Where once the industry specified storage only by capacity, users now must factor in speed—with both read and write consideration being given to workflow performance. UHDTV/4K production continues to drive the thirst for both capacity and performance, with accessibility to the professional as well as the casual shooter.

Beyond RAID Protection

Since 1988, when RAID was first developed, six basic RAID levels have endured, with many more hybrid versions now available from most storage array providers. Deciding the appropriate RAID technology for a user's application situation requires a sometimes complicated blend of equipment and technology costs, performance requirements and bandwidth (or throughput) data availability needs, coupled with an accurate assessment of capacity needs balanced with organizational goals including utilization and growth.

RAID protection for storage is solid, enduring technology—whether the data is contained on spinning magnetic disks, in optical archive drives, on flash/solid-state memory or even digital linear tape. As alternatives continue to be developed or utilization and storage tier levels (including the cloud) evolve, the presence of RAID in servers, storage vaults or home computers is unlikely to diminish for quite some time.

Hybrid Drives

Hard disks have utilized spinning magnetic media accompanied with integrated solid-state (flash) memory for many years. The name often applied to this technology is *hybrid drive*. Hybrid drives have been gaining in popularity as the industry contemplates a migration from spinning media to fixed, solid-state media.

Hybrid Versus Solid-State Storage

When a hard disk drive contains some amount of NAND flash memory, it is likely considered a "hybrid HDD." The addition of the NAND flash memory serves as a nonvolatile cache, allowing the spinning magnetic media component to possess some of the properties of solid-state storage (SSS). The boundaries of SSS may be found in solid-state storage modules or as components embedded into the control electronics of an individual HDD. Hybrid drives should not currently be considered *solid-state drives* (SSD), as the latter will essentially have no moving parts.

Interfaces for SSS devices include SAS, SATA, PCIe and Fibre Channel. The devices can be used in numerous applications and at various performance levels. The nonvolatile cache in an SSS device allows data to be retained through a power cycle, and allows for improved performance by reducing read or write time periods generally set aside as routine operations.

A main value point in adding SSS devices to a spinning disk (or "rotating magnetic media") is to improve access time and reduce latency.

Adding Intelligence to the Drive

Conventional HDDs employ software algorithms that learn the patterns of where certain data is located on the drive tracks and how to predict when that data needs to be accessed. When a flash memory (SSS) component is added to the HDD, onboard firmware now will use such learned access patterns to optimize access performance. This in turn reduces access time or latency. Firmware can also store frequently accessed data (e.g., system files and some routine applications) into the NAND flash cache that is now resident on the hybrid HDD. Now, instead of the system waiting for the drive to spin around to the track and sector containing system files and for the actuator arm to pick up that data, the controller recognizes the file purpose and memory location and moves it from the SSS memory, instead of from the HDD itself.

Pairing and Trimming

Alternatively, rather than employing a hybrid HDD, some systems will use "storage pairing"—a combination of a lower-cost SSS and a high-capacity HDD. The SSS provides for fast access to commonly needed files and allows for a much faster boot time. An SSS device also performs *garbage collection*, a background process where blocks of data marked for deletion are accumulated and then a "whole block erasure" occurs on the garbage data, reclaiming that space for upcoming write operations. Garbage collection, when employed on SSS, improves write performance by eliminating individual erasures of no longer needed block data prior to each write—instead of making the individual cell erasures happen at the time of each write function.

Another feature of the hybrid-HDD integration or stand-alone SSS devices is called the *trim command;* a predictive erasure command and an operating system function that informs the NAND flash which data blocks are no longer needed and pre-issues an erase command just for those blocks. Normally, at the time of a typical write process, firmware must first issue the erase command for a data block area and then perform the actual write-to command. By predictively erasing the unused (or no longer needed) blocks ahead of a write command, the write new data process is accelerated.

Wear Leveling

Endurance of the SSS device is of paramount importance to users, especially as costs and performance for the SSS devices approach the price of HDD or physical (DRAM and similar) memory. Of concern is the life expectancy of a NAND flash memory, which is determined by the number of erasures across the flash cells. This aging is equated to *wear leveling*, the term applied to a set of firmware algorithms that the flash controller uses to distribute erases and write commands across the entire array of flash cells in the SSS. The wear leveling algorithm's goal is to prolong the useful life of the SSS (i.e., in flash-based storage) by delaying individual cell wear-out.

Integrity and Security

Data integrity and data security are becoming accelerated issues, especially for mobile devices and for systems that depend upon the Internet for various identification or location-based tasks. Since more users depend upon mobile devices (phones, tablets, SSD-only laptops, etc.), the need and ability to protect data should the device be compromised is a growing concern.

Data integrity is an assurance that the data is valid, is not destroyed and remains *unaltered* through the course of any unauthorized process, activity or manner of use. From a security perspective, *data encryption* is the process used to protect user data from unauthorized access. The integrity and security methodologies utilized are subjects of both concern and of contention in this rapidly developing SSS domain—and have become the subject of controversy in certain terrorist cases involving cell phone data, which is essentially "self-encrypted" and accessible by no one without the proper key and access codes, plus a password.

Drive Encryption

Another innovation that continues to expand in availability involves a means to encrypt drive information that can only be unlocked by a key, which is known only to the individual who sets that key and enables the encryption functionality for that specific drive. The drives are better known as self-encrypting drives (SEDs), despite the fact users have no control or say in how the encryption is created.

SEDs are relatively recent additions to the rotating magnetic memory marketplace, which are now joining the ranks of the SSS space en masse. SEDs embed encryption logic directly into the drive's silicon, yielding a coding script that is permanent and unalterable. An SED will automatically encrypt data as it is written to the device or drive and then decrypt all the data from the media automatically.

Extremely strong passwords, up to 32 bytes, are permitted as per recommendations of the Trusted Computing Group (www.trustedcomputinggroup.org/). TCG is the international vendor-neutral standards group that publicizes specifications and uses membership implementation as examples to protect business-critical data and systems.

Password Losses Become Drawbacks

Obviously a serious drawback to the SED is that users must rely on their password for access. The loss of a password essentially becomes the demise and failure point of the SSS or HDD. Data recovery companies are addressing this issue more frequently as users discover their own lack of human memory in password-remembrance processes. Further encumbrances occur, for example, when an employee/user is dismissed from a company whose computer data was stored on an SED. Should that user's password be changed, or never registered, the data stored may not be recoverable without great cost and risk, if at all.

Yet there still remains only marginal interest in self-encrypting drives (SSS or HDD), due in part to the poor-performance perceptions related to early SED-based storage. For newer SSDs, SEDs do not measurably impact performance; however, managing the keys (otherwise called *key management*) can be complex and burdensome for large multiple-drive systems. In larger systems the loss of one password could mean the loss of entire data sets; thus, new and broader schemes for data storage are evolving.

Users of computers (laptops and desktops) and many server systems have likely been using self-encrypting and/or hybrid hard disk drives for years, yet they may not have been aware of it. Both types of storage devices are relatively new to the general-purpose computing industry; however, they've been around for some time, although found in different form factors. According to the Storage Networking Industry Association (SNIA) and the Drive Trust Alliance (DTA), nearly all HDDs manufactured in recent years (circa 2010 to present) are SEDs; yet the lion's share of those drives do not enable the SED functionality.

The technologies and these issues related to security, encryption and key management are becoming everyday issues as fraud and thefts increase. As demands for increased performance grow, knowing the options and architectures of storage systems just becomes a little more complex and a lot more variable, demanding that new perspectives be understood and enacted to ensure integrity and improve workflows.

File Systems

The data structures and methods that an operating system uses to keep track of files on a storage device (e.g., an HDD or a partition of that device) are referred to as the *filesystem* (also written as two words, as in *file system*). Filesystems manage the methodology in which data files are organized on storage devices. The term can be used to refer to a partition or disk drive that is used to store the files, or it may be used to describe the type of the filesystem itself.

Filesystems can be optimized for the type of storage or its functionality. The primary purpose is to control how data is stored and how it is retrieved. Filesystems might include built-in fault tolerance (e.g., ZFS originally from Sun Microsystems or ReFS from Microsoft). Others are organized for flash memory or solid-state media, or as record-oriented file systems for storing collections of records or shared-disk file systems (also known as shared-storage file systems or SAN or clustered filesystems).

Distributed filesystems, also known as "network filesystems," are located independent of the primary systems and will often be managed by access control lists (ACLs). These may be found in cloud services (e.g., Amazon S3) or in secure files systems (SFS) used in highly controlled or sensitive environments. Distributed filesystems may incorporate fault tolerance and be parallel (i.e., data is striped over multiple servers) for high-performance applications—or a combination of any of the above, such as in a "distributed, parallel fault-tolerant filesystem," including Google File System (GFS), EMC/Isilon's OneFS or Panasas ActiveScale File System (PanFS), amongst others.

Examples of disk file systems, which are usually block-oriented and fully random access based, include:

- CP/M—Control Program for Microcomputers (released in 1974)
- FAT—File Allocation Table, used in the Microsoft Windows operating system; for example, FAT12 (for floppy disks), FAT16, FAT32
- NTFS—New Technology File Systems (used on Microsoft Windows NTFS-based computers)
- HPFS—High Performance File System (used on OS/2)
- ISO 9660—Used on CD-ROM and DVD-ROM discs
- JFS—Journaling File System (from IBM)
- LTFS—Linear Tape File System (for LTO tape)

Differentiation between a disk or partition and the filesystem that it contains is critical. Some programs (e.g., programs that create the filesystems) operate directly on the raw sectors of a disk, solid-state media or a partition. If there exists another (foreign) filesystem in that space, it could be seriously corrupted or even destroyed. Programs must generally operate on a filesystem, and are unlikely to work on a partition that doesn't contain an active and correct (appropriate) filesystem.

The process of "creating the filesystem" must occur before any partition or disk can be used. This process requires that the drive be initialized and the data management structures be written to the disk. Once the filesystem is created, software then will communicate between storage and the operating system to make the exchange of data possible.

Many other different forms and form factors exist for filesystems that are dependent upon the physical medium they are applied to and the applications used by that medium for storing files. Some simple file system examples include:

- Disk file systems, including optical discs and shared disk file systems
- Tape file systems, such as LTFS for LTO tape greater than Generation 5
- Database file systems, those files identified by characteristics
- Transactional file systems, designed to update multiple files all at once
- Network file systems, including those with remote access protocols
- Device file systems, including I/O devices and pseudo devices

Coding for Archive and Objects

Founded even before RAID, a different flavor of data coding technology—initially used in 1950s satellite data communications—was found to be potentially relevant for use as a storage platform. In more recent times, the coding technology has come back to life for storage applications. Some of the features of the coding principles, called *erasure coding* or *fountain encoding*, are moving into the media and entertainment industry, implemented in part for archive and long-term preservation on spinning disks.

The epic proliferation of the cloud combined with the inherent risks of lengthy rebuild times associated with large-capacity disk drive components are again changing long-term storage strategies. Thus, the elements of this seemingly forgotten technology are now being applied as a component of *object-based storage* (OBS), taking disk-based media archiving to a renewed perspective.

Driven by the continued motivations for archiving material content, users are finding that depositing all their content onto digital linear tape adds certain operational complexities and expenditures to the total cost of ownership (TCO). One alternative to tape-based archiving places long-term storage on spinning nonlinear media (magnetic or optical) or sends the data out to a commercial cloud service provider.

For some, storing valuable assets in the cloud may not yet be a desired solution given some of the issues cited by owners and users, concerns that include security, control, time of accessibility and long-term unknown retention or recovery costs. With tape-based archiving, the cost of technology migration as well as potential unforeseen risks (damage, media failure, etc.), continue to challenge long-term storage practices.

Paramount to selecting a storage solution is making sure users determine the application that needs to be fulfilled before choosing a particular form of storage technology. What works well for online high-bandwidth editorial functions is probably not as applicable for near-term, long-term or archive purposes, and vice versa. One would not put a huge investment into a Fibre Channel SAN simply to store data that is not routinely accessed by more than a few attached clients. With that perspective, the methodology for data protection (RAID versus non-RAID alternatives) provides justification for just how a specific storage system is modeled.

Crossing the Terabyte Storage Boundary

When disk drives crossed the 2 TB capacity threshold, users as well as administrators of digital media storage systems started to realize there might be potential operating inefficiencies when employing traditional RAID types of storage protection. Of particular concern was during drive failures, that time between when a drive failed and when it was replaced, data was restored and full operations were returned. Risks during this period include the possibility that a second drive in a RAID 5 array (or third in a RAID 6 configuration) might fail—forcing a catastrophic loss of data—or that when write errors occur during the rebuild/restoration process, those errors wouldn't be found until much later, it ever.

Object Awareness

When long-term storage and archive or where high bandwidth requirements for multi-threaded operations (editing, ingest and playout) are not factors, the use of disk-based storage continues to gain popularity as drive capacities increase and file sets get larger. As such, the storage format for archive applications appears to be drifting away from RAID and toward object-based storage architectures.

Object-based storage shows great potential with renewed promise, especially as more data storage systems approach petabyte scales and beyond. OBS is still relatively new for media and entertainment applications, but that landscape is continuing to change.

Fountain Erasure Coding

Object-based storage systems may employ a coding theory called *fountain erasure code,* which is designed for use with large objects, as compared to the familiar Reed-Solomon erasure code, designed for much smaller objects. Fountain codes, also known as *rateless erasure codes,* were first used for recovering data signals from deep space satellites and are now found to be a reliable means of multimedia multicasting.

The coding property allows a potentially limitless sequence of encoding symbols, which can be generated from a given set of source symbols such that the original source symbols can ideally be recovered from any subset of the encoding symbols of size equal to or only slightly larger than the number of source symbols. Information theory defines *erasure code* as a forward error correction (FEC) code for a binary erasure channel. The coding will transform a message of k symbols into a longer message (i.e., the "code word") with n symbols such that the original message can be fully recovered from a subset of the n symbols. The *fountain* term comes from the metaphorical fountain that produces an endless supply of water drip (i.e., "encoded packets"). Fountain encoding refers to the fact that the codes do not exhibit a fixed code rate and can be recovered (extracted) if enough packets (drops) are collected and analyzed.

OBS with fountain codes works by dividing a data set into many smaller sets of data, erasure coding each block and then interleaving those encoded blocks with the data found in the transmission.

When the data is transformed for distribution across multiple drives and nodes, the data can still be recovered if one or more of the nodes (up to a single remaining node) is lost or disabled. Generally, a system requires a minimum of at least three nodes before the system becomes both cost-effective and fully reliable. The nodes are generally distributed around a facility's data center, across portions of a campus or globally around the world for high reliability and risk mitigation.

Figure 5.12-49 shows an oversimplification of how a set of algorithm-based equations is assigned to each chunk. Fountain coding equations, which are in reality much larger polynomials necessary to represent thousands of bytes of data in chunks, can resolve missing data when distributed across multiple drives in multiple nodes and in multiple geographic locations. Should a drive (or node) fail, through the mathematical solving of many simultaneous equations, the algorithm will then recover the missing data and rebuild that drive or node. In typical systems with 10 to 20 drives per node, as many as four to eight drives could be lost and the entire data set would still be recoverable.

Distributed data chunks minimize the risk of multiple drive losses causing a total collapse of the storage platform, with resilience far above that of conventional storage protection methods, such as RAID. By not having to immediately replace a single drive in a RAID set, the reliability, integrity and maintainability of the storage system becomes much greater while the risk of data losses is reduced considerably.

In sharp contrast to RAID, OBS systems may use erasure coding to spread smaller data chunks across many sets of drives or nodes. Each data object knows enough about where the other data objects are to allow

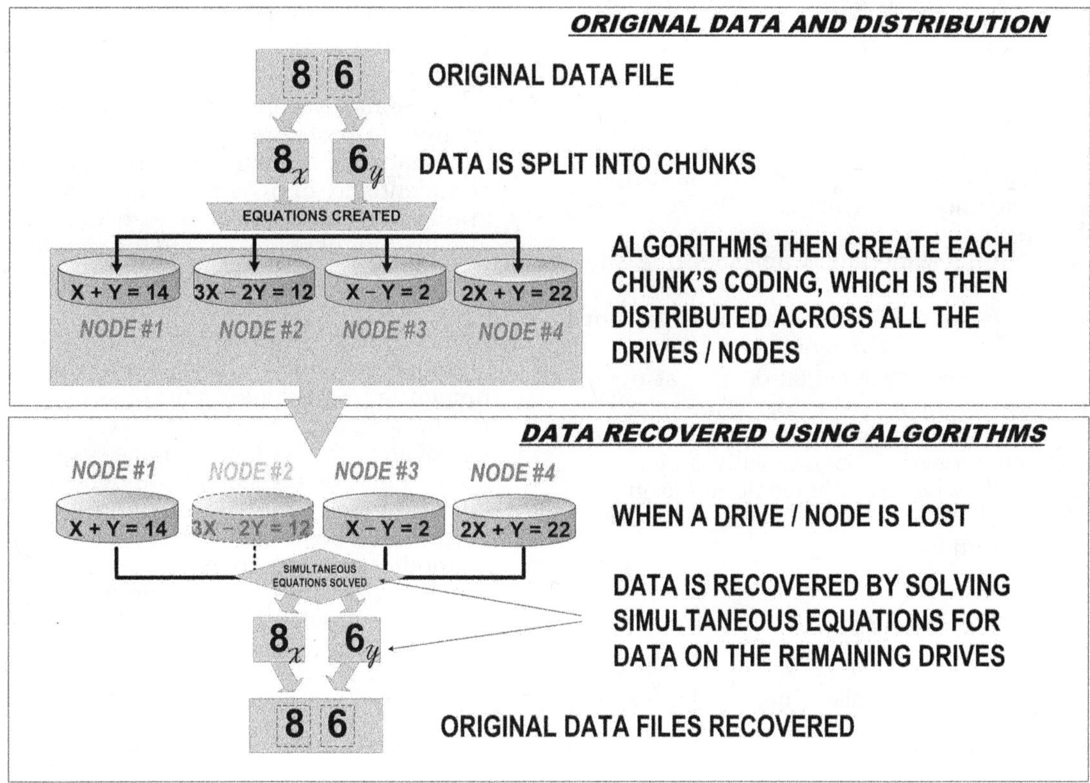

FIGURE 5.12-49 Basic fountain coding data recovery principles—noting that the actual algorithms are much larger and provide for much better "resolution" when applied to large data sets.

TV Technology, August 2013, by Karl Paulsen

a complete set of data to be recovered should as much as 40% of the drives (nodes) become unavailable. This is accomplished without striping, parity bits or dedicated parity drives.

Systems employing erasure coding and/or OBS have the capability for individual drives, groups of drives or entire nodes to employ a power-down mode, further decreasing the power load, reducing cooling requirements and extending the life of the drives. When systems utilize these technologies they can provide for the highest level of data durability, exceeding "10 to 15 nines" or beyond.

Note that "object-based storage (OBS)" is different from "objects" (as containers) themselves, which are addressed shortly.

Active Archive

A recent archive model (possibly employing OBS) allows organizations to build "active archives" on a single tier of reliable, highly efficient and easily accessible disks. If necessary, they can still keep a tertiary backup set on physically removable media (tape or optical disc) if desired. However, users must recognize that this is an archive model and not a high-bandwidth multi-dozen workstation "edit in place" model that one might find in a file-based workflow environment with many users accessing the storage system simultaneously. For the latter form of storage, production will need high spindle counts working in parallel, which still remain the norm and should not be confused with the need for an archive.

Data-Aware Fetching

Systems employing OBS may include benefits such as data-aware prefetching and caching, and shared access by multiple clients. OBS allows for mixing varying disk drive sizes in the same unit. Capacity can easily be scaled upward as needs increase.

At the storage layer, intelligent space management and reliable security are added benefits.

A drawback to OBS implementation is that overall bandwidth is less than when dozens of spindles are churning out data from a large striped RAID set. As previously mentioned, OBS is ideally suited for archiving, especially when the storage nodes (as in the cloud) are geographically dispersed across campuses, regions or throughout the globe.

Objects as Data Containers

As discussed earlier, one of the hotter topics in recent times is that of *object storage*, also simply "object storage" or "objects."

Despite some abstract perceptions, object storage is *not* necessarily an emerging technology—it is just getting renewed attention. Data stored as *trillions of objects* has likely already exceeded the exabyte scale (i.e., 1000 petabytes). Companies like Amazon (with S3) provide object storage via a public cloud platform at mammoth scale. Data storage as objects can also be implemented in the local media data center using open technology or vendor-provided solution sets. To understand the fundamental concepts of object storage, one needs to comprehend the definition and functionality of an object in a more general sense.

Containers and Wrappers

An *object* may be thought of as a *container* or a *wrapper*. Not exactly like the structure of the Material eXchange Format (MXF), but similar in that each is a device that encapsulates a set of information and produces a compact method for keeping relevant properties such as data, payload, metadata, ancillary data and so forth together as those components are stored, transferred or migrated throughout the workflow.

Object storage (not 'object-*based* storage') is therefore a storage *architecture* that manages data as objects, versus other storage architectures (i.e., file systems) that manage data as a *file hierarchy* and *block storage*, which manages data as blocks within sectors and tracks (in HDDs) or within cells of solid-state flash memory (SSDs).

Object Composition

Objects can be composed of the following:

- The *data* (sometimes called the payload) unto itself. Data can be anything from a single photo to a movie to the complete script for a motion picture production.
- The *metadata* (sometimes referred to ancillary or user data). Metadata may be defined by the creator of the object. It often contains contextual information about what the data is, its rights, confidentiality or ownership data, uses or applications for the data, or essentially anything else that might be relevant to the way in which the data is used or other mechanisms which help to support the use of that data.
- A *globally unique identifier (GUID)*, sometimes called a unique material ID (UMID) or any other such nomenclature that allows the object to be addressed in a system. In some instances—especially for global and/or very large storage systems—the ID applied may be registered (through a registration authority such as SMPTE or ISO) so that the object could be found over a distributed system, in another geolocation, or in the cloud. With a properly conformed or registered *unique* identifier, it becomes possible to find the data without having to know the physical location of the data.

Blocks, Objects and Metadata

Differentiating between the widespread uses and definitions of block versus object storage is not terribly complicated, especially if you understand the basics of, for example, file-based workflows or media asset management systems.

Block Storage

In block storage, files are split into evenly sized blocks of data, each with its own address but with no additional user information (i.e., metadata) that would otherwise provide more context for what that block of data is. It is expected that users will encounter block storage in the majority of enterprise workloads. Block storage has a wide variety of uses, as evidenced by the popularity of storage area network (SAN) arrays.

Object Storage

Object storage will not split files up into raw blocks of data. Instead, collected sets of data are stored in a unique "container" that houses the data and metadata and is labeled by a unique identifier. Object storage is quite powerful and can be customizable to the needs of the user, the project, the storage medium or the purpose.

The principle of objects is a founding component behind the Archive eXchange Format (AXF) standardized by the SMPTE as ST 2034–1:2014. AXF is a means to archive data or content of any dimensions regardless of the media type (linear or nonlinear format), the medium itself (tape, optical, magnetic or other form factor) and if there is a file system or not.

Object Metadata

Object storage, in principle, places no limit on the amount or types of metadata that can be contained in any one or all of the objects. Metadata can include anything from the security classification of the file within the object to the importance of the application associated with the information.

Given the ever-increasing use of social media, those who have stored a picture on Facebook or listened to a song on Spotify have used object storage without necessarily having any notion of it. These social media sites utilize enormous amounts of metadata that describe, link or convey information between the payload data contained in the object and the "tag" information contained in other databases that link friends, "likes" or number of plays.

In an enterprise data center, object storage is used for similar types of storage needs. The concepts behind object storage aid in making the information (payloads, metadata and IDs) highly available and highly durable. Thus, most cloud-based storage solutions utilize object-based storage, and more storage solution providers are making object storage a part of their offerings.

Manipulating Objects

Object storage may not necessarily provide the ability to incrementally manipulate (edit) one part of a file, which remains a rationale for block-based storage and why block storage continues to prevail for high-performance collaborative workflows such as nonlinear editing.

Objects need to be manipulated as a unit, requiring that the object be accessed, updated and then rewritten in its entirety. Implementations such as AXF provide *standardized* methodologies for making changes to objects (or components within the object), which in turn yields interoperability across any file system or any storage medium.

The ability to maintain an object allows objects that are created later in time to reference objects already stored, irrespective of where those objects reside.

Often secondary applications, such as an object storage manager, will utilize vendor-provided interfaces or APIs to compose, read, modify and write the payload or metadata contained in a specific object. If the objects are AXF compliant, the interface would be universal to any vendor who provides read/write capabilities complying with SMPTE ST 2034–1:2014.

Block storage can be directly accessed by the operating system as a mounted drive volume. Object storage may not do this without potential degradation to performance. Corruption of an object should not occur because the operating system directed a block storage movement that failed or improperly associated the file allocation table to a corrupted block structure. Object storage is well suited to archive or preservation platforms, where accessibility and modifications are not necessarily routine operational functions. The tradeoff here is that with object storage, the storage management overhead of block storage (such as remapping volumes) is relatively nonexistent.

Resolving Data Growth Issues

Data growth, especially in unstructured data—which makes up the largest share of data for video, audio and images—is an increasing problem, which can no longer be ignored. Of paramount importance is the *management* of that data. Simply expanding physical storage capacities is no longer the answer. Data management must be addressed collectively with the volume of storage alongside the issues of accessibility, containment, security and searchability.

Object storage is an ideal component for addressing data growth issues. As more data is generated, storage systems are required to grow. Sometimes the cloud can be a viable solution set. However, for many enterprises, especially those in broadcast or motion picture production where their assets require specific control and access, on-premise storage with total control remains a requirement—at least for the primary set of data. This infers that segregating the storage into active and inactive storage may become more necessary, shifting the trend from totally on-premise to hybrid on-premise/in-cloud solutions.

Treating all storage equally is no longer possible, due to media costs (drives), physical space (where can one expand to) and ongoing operational expenses (power, cooling, network access and security).

Expanding a block-based storage system to multiple petabytes or beyond requires that durability issues be addressed. Storage medium migration (updating and replacing aging or worn-out HDDs) is only one part of the problem. Hard limitations to the storage infrastructure currently in place may force a complete changeover, as in a "forklift upgrade," to the storage platform. Budgets for management turn into more than just overhead, with the next storage solution forcing the capital budget process to another dimension.

Managing to Scale

Provisioning management becomes evident whenever storage expansion, at any scale, is necessary. This is one area where object storage excels.

Object storage can support many differing asset forms, such as static Web content, data backup or archives. Object-based storage architectures (for tier-two and tier-three storage) are easily scaled out, that is, capacity expanded and managed, by merely adding additional nodes.

The fundamentals in object storage, including flat name space organization of data in combination with its expandable metadata functionality, facilitate and ease the continual growth support requirements that affect the enterprise. In essence, object storage is not constrained by the physical issues of block-based storage.

Responding to Resiliency

Another advantage of object-based storage is in its responsiveness to resiliency while still controlling the costs of growth. Objects can remain protected by storing multiple copies (i.e., replication) of data over a distributed, node-based system. If one or more nodes fail, data can still be made available, in most cases, without the application or the end user being impacted.

The architecture of an object-based storage system suggests that at least three copies of every file be stored. These *replicated* files may be in geographically distributed locations and connected by a network that continually monitors any changes, additions or deletions to the storage system. By employing these concepts, common issues such as drive failures, bit rot, server failures and power outages are managed without disruption to the accessibility or reliability of the data.

Distributed storage, as described above, is designed for high availability and allows for less-expensive commodity (COTS) hardware to be used, primarily because the data protection scheme is automatically built into the object architecture.

Advantages and Tradeoffs to Object Storage and OBS

Overall, there are advantages and compromises in using *object storage* deploying *object-based storage* solutions. Advantages include:

- reducing infrastructure costs by allowing organizations to employ less-expensive COTS hardware
- reducing management time through the ease of scalability
- reducing complexity through flexibility in addressing all types of storage needs in an easy to deploy, geographically dispersed environment

Object storage solutions become one of the available "storage tiers," which is better suited as an alternative to those presented by conventional linear tape archives.

On the other hand, object storage is not the answer for every storage need.

- Object storage should not be used for transactional data such as financial databases, ATM machines or other structured data.
- Object-based storage was not created as a general replacement for NAS file access and sharing.
- Object storage is not intended to support the locking and sharing mechanisms needed to maintain a single accurately updated version of a file.

Block storage is a better fit when many simultaneous accesses to noncontiguous unstructured data is necessary, as in broadcast news, video production or nonlinear editing. For those applications, having hundreds of spindles accessing data that is striped across many storage drives will secure the needs of streaming-media files better than having those files distributed in varying locations for protection purposes.

Consistency

A primary decision point in determining whether to employ object or block storage is consistency. Object storage systems are *eventually consistent*; while block storage systems are *strongly consistent*. Users need to decide which type of architectural storage is appropriate for their needs, balancing the requirements for a scalable storage solution while providing resilience and performance.

Eventual Consistency

Object storage has the capability to provide virtually unlimited scalability, ensuring high availability for data that needs to be durably stored while remaining relatively static. Data should not be expected to change much, if at all. Storing photos, video and other unstructured data is an ideal use case for object storage systems. This kind of essence does not need to be constantly altered. The downside to eventual consistency is that there is no guarantee that a read request returns the most recent version of the data.

Strong Consistency

Block storage is needed for real-time systems, such as transactional databases, which are constantly being written to. Block storage should provide limited scalability and reduced availability as a result of hardware failures. Such scalability becomes more difficult within a geographically distributed system. Strong consistency is a requirement, however, whenever a read request must return the most updated version of the data.

Workloads for Object Versus Block Storage

Object storage works very well for unstructured data sets where data is generally read but not written to.

Data that is best stored as objects include static Web content, data backups, archival images, and multimedia (videos, pictures or music) files. Databases in an object storage environment ideally have data sets that are unstructured, where use cases suggest data that will not require a large number of writes or incremental updates.

RAID on Tape as RAIT

Introduced in the late 1990s and early part of 2000, Redundant Array of Inexpensive Tapes (RAIT) is a means of storing data as a collection of N + 1 tapes that are aggregated to act as a single virtual tape. Typically, data files are simultaneously written in blocks to N tapes in stripes, with parity information written in bitwise exclusive-OR fashion stored on from one to many additional tapes. RAIT will have a higher performance compared to systems that only store duplicate copies of information.

One drawback to RAIT is, because data must be striped across multiple tapes, to read a particular set of data requires that all the tapes storing that data must be mounted and then simultaneously and synchronously read in order to reconstruct any of the files stored on the tape. Even though RAIT utilizes parity information for the benefit of reconstructing data lost when a tape has gone missing or has unrecoverable data, should one or more tapes be found unreadable, the system must wait until all of the remaining tapes can be read and parity information can be used to reconstruct the data set.

Since the introduction of digital tape for archiving, manufacturers of tape-based storage systems looked at means to protect tape-based information in archives and backup systems in a similar fashion to RAID protection for disks. The concepts included extending high data rate throughput, robustness and virtualization. Through the use of striping, the quality of service (QoS) would be improved by combined the number of data stripes (N) with the number of parity stripes (P). Defined as "scratch mount time," N + P sets would define the QoS for a particular configuration.

By implementing parity, much like that in a disk RAID set (e.g., RAID 5), the number of virtual-volume reads necessary before data loss is managed. For example, by prescribing an array of striped tapes in a RAIT configuration—composed of a set of eight striped tapes plus from one to four parity tapes—the tolerance profile for the number of virtual-volume reads before data loss (as when one or two tapes go missing out of the set) is elevated, and is further described in Figure 5.12-50.

RAIT essentially works the same as RAID, but with linear data tape instead of spinning media. RAIT is ranked by six increasing levels of protection (or security) ranging from RAIT Level 0 to RAIT Level 5.

Mirroring—RAIT Level 1

Known as disk mirroring or disk duplexing, RAIT 1 is where all data is written to two separate drives simultaneously. Should a drive fail for any reason, the other drive continues operation unaffected.

REDUNDANT ARRAY OF INEXPENSIVE TAPES (RAIT)

VIRTUAL-VOLUME READS
NUMBER OF 160 GB VIRTUAL VOLUME READS BEFORE INCURRING A LOSS OF DATA

PROTECTION SCHEME	TOTAL TAPES (*'x' + parity*)	QUANTITY OF MISSING TAPES		
		0 MISSING	1 MISSING	2 MISSING
NONE	8	10^2	1	1
1 PARITY	8 + 1p	10^9	10^2	1
2 PARITY	8 + 2p	10^{16}	10^9	10^2
3 PARITY	8 + 3p	10^{23}	10^{16}	10^9
4 PARITY	8 + 4p	10^{30}	10^{23}	10^{16}
MIRRORING	8 + 8	10^{10}	10^3	15

RAIT is sometimes called "Redundant Array of Independent Tapes"

FIGURE 5.12-50 Redundant Array of Inexpensive Tapes (RAIT) protection schemes are based upon the volume of tapes and how parity is dispersed. Note that "Inexpensive" may be found written as "Independent."

Data striping—RAIT Level 3

This employs data striping, in which the data stream is divided into equal parts, depending on how many drives there are in the array. In RAIT 3, there is a drive dedicated to maintaining parity. A parity drive can potentially cause a bottleneck because all the read/write requests must first go through the parity drive. In the event a drive should fail, the parity information is then used to reconstruct the lost data.

Data striping—RAIT Level 5

The parity information is divided equally among all drives. Level 5 RAIT provides maximum data integrity; however, it is often at the expense of data throughput.

RAIT—Not for Random Access

RAIT technologies are not a substitute for random access to data. The concepts were more theoretical than practical. In the data archiving and backup world, RAIT has not been widely accepted. One manufacturer (at the time StorageTek in 2001) once stated a commitment to COTS RAIT; however, the development of virtual tape library technologies never seemed to take off.

Limited information can be found past around mid-2006, and there remains less than a marginal commitment to RAIT at either the network or the architectural level.

DEVELOPING A STORAGE STRATEGY

To put the storage equation into a global perspective—continual estimates and studies routinely increase their own prognostications as to the amount of data generated and stored on a worldwide basis. We entered the "zettabyte era" somewhere between 2007 and 2009. A zettabyte (ZB) is a trillion gigabytes (1,000,000,000,000 gigabytes) or 10^{21} bytes. See Figure 5.12-51.

Some examples:

- Research by scientists at the University of Southern California reported in 2011 that "human kind has stored more than 295 billion gigabytes [equivalent to 295 exabytes] of data since 1986."[14]
- Scientists also concluded that "2002 should be considered the beginning of the digital age," pointing out that it was the first year digital storage capacity overtook total analog capacity worldwide.
- In 2007, humankind successfully sent 1.9 zettabytes of information through broadcast technology such as televisions and GPS.[15]
- By 2020, 35 zettabytes (ZB) are expected to have been generated, with one-third of that data "living in or passing through the cloud."[16] Data production at that time will be 44 times greater than it was in 2009 (estimated then at 0.79 ZB).

NAME	NUMBER OF BYTES	EXPONENTIAL NOTATION (BYTES)	RANGE OF COMPARATIVE ANALOGIES
BYTE	8 BITS	1.00E+00	One single decision to a telegram
KILOBYTE	1,000 BYTES	1.00E+03	Very short novel to a large box of punch cards
MEGABYTE	1,000,000 BYTES	1.00E+06	3.5-inch floppy disk to a CD-ROM of data
GIGABYTE	1,000,000,000 BYTES	1.00E+09	Pickup bed of paper to a floor of academic journals
TERABYTE	1,000,000,000,000 BYTES	1.00E+12	50,000 trees made into paper to the print in the Library of Congress (2012)
PETABYTE	1,000,000,000,000,000 BYTES	1.00E+15	Most of the U.S. acedemic research libraries (2012) to amount of digital magnetic tape in 1995
EXABYTE	1,000,000,000,000,000,000 BYTES	1.00E+18	All the words spoken by human beings to the amount of global IP traffic predicted by Cisco (=20 exabytes/month by 2020)
ZETTABYTE	1,000,000,000,000,000,000,000 BYTES	1.00E+21	Per IDC: the total amount of global data expected in 2012 was 2.7 zettabytes
YOTTABYTE	1,000,000,000,000,000,000,000,000 BYTES	1.00E+24	Likely what is stored by the NSA (unsubstantiated) in 2014
XENOTTABYTE	1,000,000,000,000,000,000,000,000,000 BYTES	1.00E+27	A very big number
SHILENTNOBYTE	1,000,000,000,000,000,000,000,000,000,000 BYTES	1.00E+30	An even bigger number
DOMEGEMEGROTTEBYTE	1,000,000,000,000,000,000,000,000,000,000,000 BYTES	1.00E+33	A really, really big number

FIGURE 5.12-51 Comparative data sizes for today and into our future (with selected analogies of representative unsubstantiated figures).

Content Clutter

Data, as a set of qualitative or quantitative values, is generally classified (or modeled) as either *structured* or *unstructured*.

Structured data is information that can be easily categorized or cataloged, as in the information in a relational database; for example, ATM transactions, online order entries, banking or database operations or household accounting activities.

Unstructured data has no specific or rational categorization or cataloging. It has no predefined data model and is not organized in a predefined manner. Video and audio media are considered unstructured, giving way to inconsistency in content, and effectively are non-numerical by nature. See Figure 5.12-52.

To put facts to perspectives, in 2014 (according to the Data Never Sleeps 2.0 Infographic),[17] every minute the following sets of information were generated:

- Facebook—users share nearly 2.5 million pieces of content
- Twitter—users tweet nearly 300,000 times
- Instagram—users post nearly 220,000 new photos
- YouTube—users upload 72 hours of new video content
- Apple—users download nearly 50,000 apps
- Email—users send over 200 million messages

Each of these entities utilize tens of thousands to hundreds of thousands of servers to collect, place, store and distribute their content—many (e.g., Facebook, YouTube or Instagram) as *unstructured* video content that incorporates some degree of "video server" in the process. The video server may be a core component in their host server architecture or a playout application on a local PC or mobile client.

Regardless of what the storage management solution makeup is, as "hardware + software," much of the technology that allows the ingest and playout of the media content for such services came from the core video server technologies discussed in this chapter. The balance of the system, the physical storage components, consists of myriad variations of mechanical, electrical and solid-state devices that collect and store "data" in both structured and unstructured forms.

Storage Management Processes

If one accepts that traditional videotape recording technologies are essentially gone from most modern operations, and that media content that is collected in the field, as files, migrates from one physical medium to another depending upon the operational functions (i.e., the "workflows") that utilize that data, then one can quickly understand why developing a "storage strategy" is important to fulfilling the requirement and processes that are supported by that media content.

Process (or workflow) wise, most raw content moves from capture to production editing and is ultimately passed to storage or distribution (playout). These processes now virtually eliminate all *human handling* of the physical media, with the possible exception of moving from the camcorder SSD to the ingest platform.

When users first tiptoed into file-based workflows, each, unknowingly, created their own independent storage management practices. Most users, at that time, were far from developing a "strategy-like" approach to storage, simply because there was so little continuity between platforms, vendors or technologies. Many users relied upon videotape as their long-term storage medium for finished projects; some used optical media (CD-R, DVD-R and later Blu-ray), and for some those archiving models may still remain.

oox2315	oox4175	ooxa5403
oox2316	oox4176	ooxa5404
oox2317	oox4177	ooxa5405
oox2318	oox4178	ooxa5406
oox2319	oox4179	ooxa5407
oox2320	oox4180	ooxa5408
oox2321	oox4181	ooxa5409
oox2322	oox4182	ooxa5410
oox2323	oox4183	ooxa5411
oox2324	oox4184	ooxa5412
oox2325	oox4185	ooxa5413
oox2326	oox4186	ooxa5414
oox2327	oox4187	ooxa5415
oox2328	oox4188	ooxa5416

STRUCTURED DATA SET

oox2315	oox4175	ooxa5403
oox4179	oox2319	oox2328
oox2317	oox4177	ooxa5405
ooxa5404	ooxa5408	oox4176
ooxa5412	ooxa9813	ooxa5407
oox2320	oox4180	oox4186
ooxa5415	oox4183	ooxa5409
oox2322	ooxa5410	oox2321
oox4188	ooxa9831	ooxa5411
oox2324	oox2326	oox2318
oox4182	oox4185	ooxa5413
oox2323	ooxa5406	oox4181
oox2327	oox4187	oox4184
oox4178	ooxa5414	ooxa5416

UNSTRUCTURED DATA SET

FIGURE 5.12-52 Structured (ordered, sequential numbering, consistent organization) versus unstructured (no order, random numbers, random colors) data. Video and most media are considered "unstructured data," whereas accounting, ATM transactions or databases are "structured data."

In today's models, many are retaining content that never existed on physical or transportable media. So, for long-term preservation they are moving their assets to other forms of physical media, including tangible media, for example, digital lineal tape platforms (such as LTO), or intangible forms, such as the cloud.

Users recognize there is an increased value to retaining more of their content, whether for archival purposes, content repurposing or legal matters. As the unit cost of storage continues to decrease and compression technologies afford more images being stored in fewer bits, users are now able to retain more of their content in varying formats including full resolution, proxy resolution and mezzanine resolution without necessarily burdening their overall storage infrastructure.

Today, high-definition imaging and production editing systems are everywhere. HD is sending standard-definition imaging to almost a legacy format domain. Capturing compressed HD video at 35 to 50 Mbps means that more data can be originated, stored, copied for protection and transcoded to proxies for preview and other editorial purposes. Once that story is completed, a master format is then transcoded into multiple release formats. Full-bandwidth, uncompressed HD video (1.485 Gbps) is seldom utilized except in live production at the studio or in the commercial production and motion picture industries.

Individualized Workflows Stress Storage

Master- and mezzanine-level storage are putting enormously different demands on storage, production workflows and archive systems. Until only recently, each release version had been stored individually, in some location on some form of media and for some undefined period of time. Dozens to tens of thousands of versions of the same basic content (story line) must now be produced for every major piece of content. These unwieldy requirements drove the development of both a new workflow strategy and a supportive storage strategy at each and every organization and for each version of the file-based workflow.

Having a different strategy for every version of the product becomes impractical, so a means that would allow a master set of content to be modified by deploying standards-based versioning was created to address not just the varying workflow practices, but to mitigate the unnecessary burden of managing the storage required to hold each individual version.

The Interoperable Master Format (IMF), otherwise known as the SMPTE ST 2067 group of standards, enables this new workflow and indirectly this new storage strategy. In IMF, only the discrete elements that compose a version need to be stored. When it is time for a release to distribution, the IMF procedures are used to assemble each version without having to make and retain an inordinate number of individual composite files. No longer do separated discrete versions of every release need to be postproduced and stored individually.

Storage Strategy Variances

Storage strategies will vary between organizations irrespective of the entity's size or location. Several factors can affect how the organization uses a storage platform. Departments, workgroups or entire enterprises may deploy differing storage strategies based upon the workgroup's size, the length of the project and the workflows. The age, type and size of the storage systems employed in the respective groups can impact the storage strategy for that group and for the overall organization.

Some organizations already divide storage components and storage strategies into segments so as to improve network management, isolate burdensome functions, address security, or in some of the larger organizations address geographic locational differences from campus-wide to international global dimensions. Other groups, especially those with numerous legacy and current storage systems, use storage management tools that consolidate or virtualize storage to extend its useful life or to increase shared storage capacity.

Defining a storage strategy may not necessarily be vital to small-scale systems until such time as the operations are expanded or merged with other workflows involving digital media. Island workgroups or systems that consist of a modest set of PC or Mac workstations (maybe three to five in a group) may choose integral direct-attached storage (DAS) on a local level. Workgroups may also attach to a NAS-head that drives a common pool of RAID-protected spinning magnetic media. In the latter case, the NAS solution creates a central storage system that can be routinely backed up or centrally managed by a media asset management (MAM) solution.

At mid-scale, such as in a studio facility that produces a continuing set of programs, small-scale storage management concepts may be expanded to include peripheral components, such as a common ingest platform that uses both a video server ingest module to capture content at the baseband level and file-based ingest through transfers from an outside network. Assets might be shared on a storage area network (SAN) that is configured with capability and bandwidth to perform editorial processes directly on the storage system.

In a broadcast facility, such production/editorial models may incorporate video server playout features with mirrored, protected storage for finished air-ready content. Organizations that prepare content for multiple network origination often may have several "production islands" or units with varying forms of localized storage systems coupled through a server-based editorial system. The production units may be connected to a centralized SAN that allows for the sharing of content and assets across the entire operation.

Strategy Justification

Storage strategies and their associated components come into play when new subsystems must be added to increase productivity or expand storage capacity.

Without a storage strategy plan, adding another program to the production schedule, replacing obsolete equipment that must be retired (e.g., legacy components that can no longer be supported or are unavailable) or updating the overall system because the bandwidth is insufficient to meets the needs of the organization becomes complicated and often quite expensive.

With a storage strategy plan in place, the organization is better prepared to adjust for change. A strategy lets the user look far beyond just addressing the immediate need of insufficient capacity or a performance degradation because too many users are accessing the storage system. To develop a storage strategy, users should analyze each of the current systems and workflows independently, and then look at them in an aggregate perspective to make decisions that support short-term and long-term needs or issues.

For instance, suppose the read/write access loading from some production islands is deemed insufficient to support the next required production workflow. An option might be to expand the bandwidth of the local SAN by adding more drives or restriping the system to increase the availability of those drives. Another option would be to look at migrating away from the independent "production island" concept to a centralized storage platform (a NAS or a SAN) that not only allows common access to files for all the connected clients but also provides greater accessibility for each workstation connected on the storage system itself.

The previous example is conceptually simple. However, the actual implementation could be far more demanding—physically and economically. Bridging production islands into a common shared storage configuration is useful when the geographic locations of the islands are in relative close proximity and can share a common local area network (LAN). When this model crosses the boundaries to a campus or an entire enterprise, a larger-scale storage strategy will not only make better sense, it may be required.

Resolving Short-Term Issues

Hasty decisions aimed at resolving short-term isolated problems are unwise and sometimes end up in disaster—functionally and economically. Users and administrators need to understand the big picture, making sure there aren't other influences (e.g., network connectivity) that would still affect productivity despite fixing other issues such as storage capacity or workflow throughput.

A *workflow management tool set* can help to maintain control by monitoring all the activities that impact operations. The resolution to low-bandwidth productivity might be as simple as throttling the "back office" like activities, or changing processes to time slots that aren't during peak production editorial periods. A workflow management *dashboard* lets administrators see what's really going on across the entire system, not just at a local workstation or storage island. By shuffling tasks amongst workstations, storage-related issues might be mitigated and may allow the organization to extend services and uses of the equipment even further.

Options Available for Change

The digital media production industry has never before been afforded so many choices and options in managing their enterprise no matter how small or how large. When the enterprise adds in concerns such as security, accessibility, archive and redundancy, the storage strategy bar is raised even higher.

Employ experts in the fields you are addressing. The "IT department" that doesn't routinely work with unstructured media workflows or operations is unlikely to possess the skills needed to address the paramount issues surrounding a true media-centric storage strategy.

Realize there will be many options from every potential corner of the storage and media asset vendor community. When looking to implement change, ask the hard questions of the vendors, for example: "How do you manage growing files?" or "At what point does the read or write process begin to deteriorate?" or "What alternatives to the SAN configuration do you offer?" If you're unsure what to ask, invest in a consultant (and a firm) who has a reputable track record in the industry you work within.

Map out and document your plans. Know all the numbers (reads, writes, threads, churn rate and many more). Network with your peers. Check with users who've employed the suggested tools, storage systems, media asset management services or other strategies.

Understand the big picture before you simply add another NAS or put more disks in the chassis in order to resolve yet another hidden set of items.

CONVERGENCE: A WAVE OF THE FUTURE

The past several years, starting around 2010, have seen media management technologies being refreshed to support content creation and delivery. These new directions took off with enormous popularity and interest from users, by owners and managers, and in various media market vertical segments—such as sports, digital signage, fan experience venues, and for a plethora of personal and professional entertainment possibilities.

The most obvious change is the migration from dedicated *purpose-built hardware* to flexible software-based systems running on COTS components (servers, storage and networking). These IT systems and Internet protocol (IP) technologies now play key roles in facilitating this change, a fact we expect to persistently increase as the needs for new services grow and the time to deploy those services shrinks.

Harmonization

Hardware-based systems, once primary components in many media systems, are fading—giving way to the convergence of IT-based components on network-centric architectures. The harmonization of software-defined storage and networking is aiding the virtualization of these components and subsystems, allowing the facility to expand with limitless boundaries. This so called "convergence" is leading to faster times to market and less costly efforts to adjust when changes in or new services are needed.

One reason for the move away from hardware-based systems is development time. In the past, developing a new product could take years from concept to first release. If the products were built on new technologies, it would take considerable effort to develop, test, debug and release those systems to the industry. When those products required integration with third-party components, it could easily take another year or more for entire systems to reach maturity and be accepted by mainstream users. By that time, the industry's need for the product could have changed substantially or even been replaced by a competing technology offering a better solution for less cost.

The Paradigm Shift

As solutions shift from purpose-built dedicated hardware to hybridized server and software-oriented systems, new and more programming services begin to emerge. This is a topic the general broadcaster had only marginal experience with a decade ago. The ability to rapidly add services or create new revenue streams was boosted by changing the way manufacturers approached the technologies necessary to assemble, test or reconfigure file-based workflows for program generation and distribution.

When videotape recorders were the primary means to reproduce content, it was often costly to add another transport and its supporting frame sync, audio and video distribution, monitors and so on—just to add another "stream." So the concept of producing another revenue stream for the broadcaster was a quagmire lost in capital budgeting, engineering design, physical installation, space and so forth.

Servers changed this model, since the ability to add another channel or to store more content was relatively straightforward and not nearly as costly as adding a VTR and associated support gear.

Resource Independence

In today's rapidly changing "Internet" marketplace, businesses can no longer afford to wait six to nine months to spin up a new service. Stakeholders often address new business plans that necessitate bringing on a new service in a matter of weeks, not months. The effort to bring up new competitive services can no longer be driven by hardware development cycles or burdened by unpredictable capital cost or budget restrictions. Systems have and are becoming more flexible and less costly to deploy, having less impact on operations if the service needs to be modified or even shut off after an unsuccessful run.

Furthermore, these systems now offer resource independence. Servers no longer need to be dedicated as a single resource to a single solution—they are now capable of serving multiple functions that can adjust based upon demand or need.

Software Defined

Media management, content creation and program distribution now depend upon a new dimension called "software defined" networking, storage or services. The emergence of this approach reduces the "time to market" and in turn changes the landscape of technological development forever. And these systems can be deployed in the cloud, in a private colocated data center or on-premise within the footprint of the enterprise.

To support this paradigm shift, product requirements and system architectures now leverage IT functionality more than ever before. This is why IP, Ethernet and networking in general are found in nearly every aspect of the video media industry today.

Systems that once consisted of dedicated purpose-built boxes are now enabled on commodity (or common) off-the-shelf (COTS) components. Systems that once utilized a few PCs to interface hardware amongst systems or were employed strictly as administrative tools now utilize multiple servers as core components to orchestrate and manage multiple tasks simultaneously.

Servers have essentially taken the place of the hardware-based devices, allowing for increased flexibility and service modifications without a forklift change or a major new capital procurement process.

Software Defining Flexibility

The practice of deploying multiple servers for a set of specific subsystems, such as a transcode farm, encoding farm or archive management system, is only about a decade old. The transition began about the time networks moved from 10/100 Mbps to gigabit transport speeds, and when compression technologies moved from strictly MPEG-2 to Advanced Video Coding (AVC) such as H.264 and beyond. These two enablers changed the dynamics of what could be built and what could be transported within the plant and outside it.

While multiple dedicated server concepts continue in many of today's system architectures, that perspective is likely to change as the next generation of IT-based technology for media evolves. The professional broadcast industry is already seeing this as Studio Video over IP (SVIP) takes shape amongst the manufacturers and as the users of SDI-only equipment and systems take the next steps.

Integrating Playout

To illustrate the foundation of this transition, look back at how the early video server (circa the mid-1990s) has transformed the way play-to-air and master control operations have changed. Early video servers were assembled in similar fashion to how component stereos were outfitted in the '70s (see Figure 5.12-53 "A"). The next generation of the video server's functionality is now being integrated directly into the same physical chassis as the broadcast playout channel chain (see Figure 5.12-53 "B"). Between compression advances,

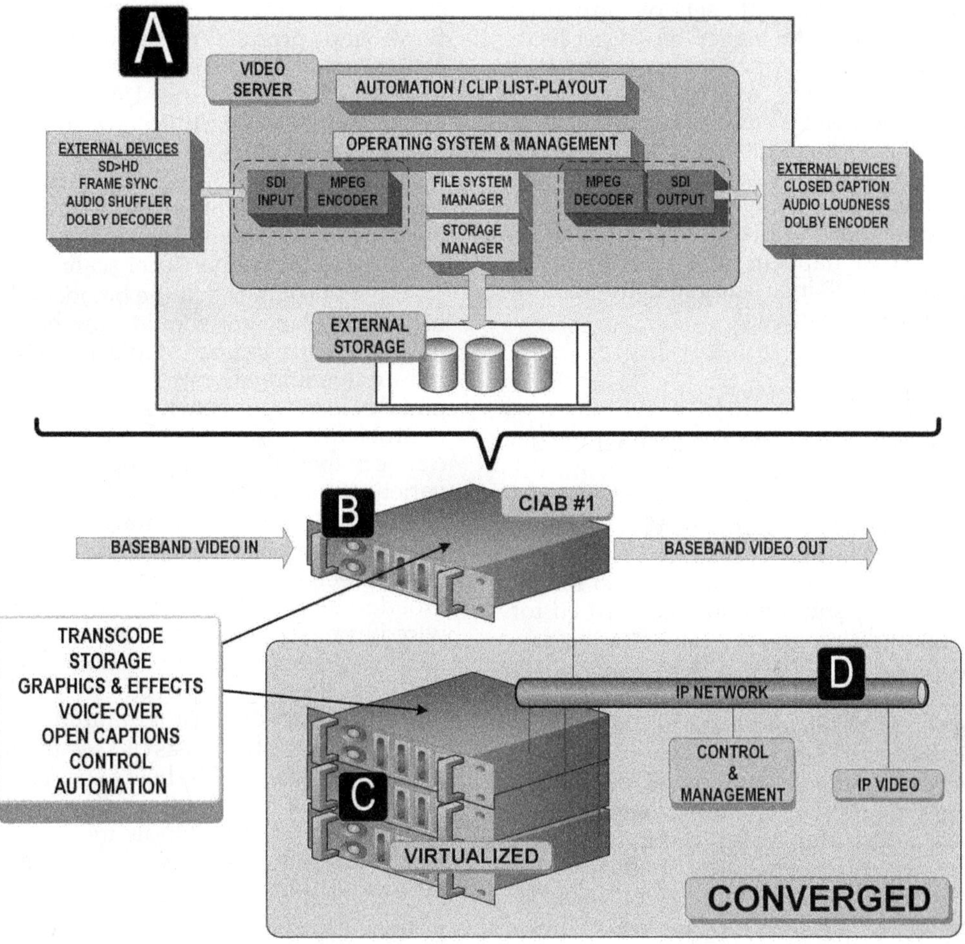

FIGURE 5.12-53 The discrete components found in legacy video server playout systems (A) are now integrated into the Channel in a Box system shown in (B). Future IP-only systems are virtualized (C) with various services distributed "as needed" and managed over the network (D).

storage capacity increases and network control and content transportability, the need for multiple chassis devices with independent functionality managed by a central automation system has decreased significantly.

The functionality and physicality of the equipment is now integrated into a single system, sometimes in a single-unit chassis. These "integrated playout systems" now include file-based transcode, software decode and IP playout featuring various codecs, graphics and audio overlay, near-term storage, automation and internal SDI conversion. Some or all of these services are wrapped into a single channel and deployed on a dedicated server. The system is now entirely software enabled—meaning users don't need to buy another box or wire another component to gain current or future features.

Channel in a Box

A term coined in the late 1990s, "Channel in a Box" (CIAB), now describes the next generation of playout systems. CIABs operate fundamentally like their discrete legacy systems, but with more functionality, less complexity and certainly less cost. Manufacturers can now assemble discrete sets of software routines—once allocated to dedicated boxes—into a single powerful computer-based server running as IT.

The CIAB model is likely to remain the traditional broadcaster's operational model for the relative near future. However, for those service providers that address dozens of channels, provide for thousands of streams or routinely need to change program delivery methods, this CIAB model is unlikely to be their ideal solution.

Even with the CIAB model fairly well established, the handling of many hundreds of streams requires a different set of parameters. Traditionally—like in the data center domain—when a new application was to be presented, the IT department would deploy a new server, expand or add new storage and add another pair of network switches. This is a costly, time-consuming task that often prohibits experimentation by the programmer.

Even today, conventional broadcast plants are built out using a hundred or more COTS servers, dozens of network switches and multiple sets of individual storage systems. They, like the legacy broadcast facility, require a main (and backup) central core network switch, redundant firewalls, aggregation switches and other network components. Cabling is hardly ever reduced, because the systems now need dual home runs to multiple sets of 24- to 48-port switches. The dangers come when equipment, power and cooling overrun the physical space available to contain those components. Once that happens, the scaling model becomes constrained and increasing the number of expandable services is jeopardized.

Convergence—Change on the Horizon

Another significant change is emerging either on the horizon or, in some cases, already above it. As the media industry moves forward, systems will be configured using virtualized environments. Virtualization enables new services to be spun up, placed online and manipulated. If (or when) the services are no longer needed, they are simply shut off and reconfigured for another service or application.

Conceptually, Figure 5.12-53 "C" depicts an all-IP-based virtualized environment providing various services on an as-needed basis, operating entirely over a network (Figure 5.12-53 "D").

The management framework allowing these software-based infrastructures to operate is orchestrated as though it were addressing an entire platform on a global level. The new framework, like its data center counterpart, is referred to as a "converged infrastructure"—also known as a *converged system* or, in some circles, a *unified computing system* (UCS).

Automation, Orchestration and Optimization

Relatively new technologies, changing business trends and continually adjusting service demands are shaping media services to look more like the information technology world than the traditional broadcast world.

Virtualization is simplifying networks and components, thereby allowing broadcasters and multichannel video programming distributors (MVPD) to take advantage of the expanding 10 gig and 40 gig pipes, software-defined systems and IT-based technologies for many of their service infrastructures.

Automation is reducing the manual intervention needed to facilitate new services. What were once fixed, finite methods of generating a program schedule have now become core feed points into orchestration layers that facilitate the management of all the delivery chain components.

Bottlenecks Becoming Smaller

Among the many storage and networking components involved in the change is solid-state storage, which is relieving storage performance bottlenecks by improving speed and accessibility. Distributed and shared storage alleviates discrete islands of incompatible storage, opening the doors to more beneficial workflows and production opportunities—that is, more content with fewer steps to release.

Optimization of the LAN/WAN coupled with enhancements in caching and improvements in data transmission all contribute to improvements in moving content between global locations. And cloud services are allowing the enterprise to store data or add or burst up services with improved accessibility, offering greater flexibility and easier scaling with a capital cost.

The end result is that the broadcast and program distribution businesses should now be able to streamline management functions, deploy new services faster and reduce operational costs. With integrated systems composed of fewer components, troubleshooting connections or problem areas becomes less complicated. Components can be managed by a single operator console instead of a variety of incompatible management systems.

Convergence is becoming the answer to implementing the future of multiple services such as OTT, multiscreen and directed program content delivery for broadcasters (i.e., those intending to use ATSC 3.0) and wired or mobile consumers on multiple platforms and display systems. These concepts can, and most likely will over time, be deployed universally—even for the traditional broadcast programming service provider.

It is now relatively easy to see the broadcast central equipment room (CER) being shaped into a virtualized central data center that can live in almost any space and be controlled to handle most any set of programming requirements in a rapid, cost-effective and expeditious fashion.

PRESERVATION AND BACKUP

Throughout these storage topics and discussions, the subject of protection and backup have alluded to the discussion of archive and content management.

Archiving is a process that can begin at the capture point of an asset or happen at the end of the workflow. When archiving begins from the capture point, it becomes what is sometimes referred to as a "digital birth certificate" for the asset and can follow the production or replication of the asset for protection and preservation purposes through eternity. When at the end of the workflow, it is usually for library or simply a means to preservation for legal purposes or historical purposes.

Archiving can serve several concepts, such as preservation, backup or as a searchable library solution controlled by a media asset management system.

The archiving of the valuable assets as media content becomes a complex and site-specific and/or enterprise-specific and policy-based activity. Sometimes the purpose of the archive is strictly for preservation of a historical nature. Other times the archive is an extension of the storage (another "tier" level), which augments and extends the facility's storage capacities.

Archives per se may not be for every operation. Determining what policies are suitable for the enterprise (or the individual user) is a decision that is sometimes difficult for the user to make. Deciding on an archive involves assessment of the operation, workflow needs, market size, value of the assets and the labor necessary to initiate and manage that archive for the long term.

Many organizations have historians or archivists whose purpose is to catalog and manage the media within their department or across the entire enterprise. This was a relatively simple and easy to administer process 20 or 30 years ago; that has all changed since the world has become "born-digital" in nature. With the millions of files already being created in the larger organizations, a human being can no longer physically (or emotionally) manage all the digital assets in an unassisted manner.

Archives sometimes consist only of backup videotapes, or they may involve linear data tapes, or optical storage on DVD-R. Some organizations have outsourced the digitization of their assets and then placed them in the cloud (e.g., Amazon's Glacier), where they will presumably stay until eternity. Some organizations may only use a solution like Carbonite or even Dropbox. There are no "one-size-fits-all" models for archives or their workflows.

Some early video server platforms provided the flexibility to transfer native files from the storage platform to other media. The methodologies surrounding those techniques and technologies varied immensely. Organizations sometimes wouldn't even attempt archiving unless they owned a sophisticated hierarchical media asset management (MAM) or digital asset management (DAM) solution. Others simply duplicated the files by playing them out to videotape and storing them in an office closet or someone's basement.

Video server solutions and third-party products now have interfacing capabilities between server/automation solutions and physical robotic disk or digital tape libraries. These solutions are generically referred to as *archive managers* and have varying capabilities, management processes and physical or network interfaces associated with them. Archive managers are often tightly integrated into each of the asset management components in the overall system. They can operate by themselves, or they can operate under the umbrella of a MAM or DAM.

Besides the controls and interfaces, the archive manager becomes the handler of the assets, making certain that the assets are cataloged, protected, secured, backed up and ready for use with sufficient notice such that the content appears "always available." These products usually monitor the condition of the media (tape) and the robots and drives within their library. They routinely will prepare secondary copies (duplicates) of the tape assets or create protection copies when the number of passes of the tape across the heads reaches levels that become risky to the sustainability of the physical tape asset itself.

CONCLUSION

The evolution of recording has seen many changes and forged many paths. We have crossed the boundary of videotape as a principal recording medium and essentially left it behind. We have nearly outgrown the storage of all digital assets on premises. We are currently witnessing a paradigm shift in the way images are prepared, transported, preserved and manipulated.

We can say with some sensible rationality that video server and storage technologies have changed the way we create, hold or view audio/video and static images. Many take "video" and "video recording" (in any form) as second nature (still calling it "filming" in some circles), much like many of today's youth have never held a telephone handset in their palms or taken a photograph with a film-medium camera.

The volume of data being collected will at some point reach a limit, which none of us can predict. At what point do these assets (the video, audio or pictures we generate) become of no additional value, and are in turn left as forgotten? There will, at some time, be a point where storing all this data becomes unmanageable at any level.

No one knows for sure what the next quantum leap in imaging, video recording or server technologies will bring—whether at the professional, broadcast or consumer levels. One thing for certain, however, is we're beginning to see that the importance level for *quality* and *value* seems to be changing, driven in part by technologies including mobile, with its "any time/any place" user attitude and an insatiable desire for accessibility and instantaneousness.

Notes

[1] "Tape Recording Used by Filmless 'Camera'," *The New York Times*, Nov. 12, 1951, p. 21.

[2] "Magnetic Tape Used By RCA to Photograph Television Program," *The Wall Street Journal*, Dec. 2, 1953, p. 1.

[3] Adrian Pennington, "DOTS Lands Hollywood Studio for Long Term Archive," www.thebroadcastbridge.com/content/entry/978/dots-lands-hollywood-studio-for-long-term-archive and www.group47.com (headquartered at 5344 Don Pio Drive, Woodland Hills, CA).

[4] Frankie Taggart, "Technicolor Stores Hollywood History in a Bottle," http://phys.org/news/2016-04-technicolor-hollywood-history-bottle.html

[5] "CES Day Two: Technicolor's VR Strides & Digital Life Take the Show Floor," www.technicolor.com/en/who-we-are/press-news-center/news/ces-day-two-technicolor-s-vr-strides-digital-life-take-show-floor

[6] Forbes online, www.forbes.com/sites/brittanyhodak/2016/07/23/rip-vhs-worlds-last-vcr-to-be-made-this-month/#2c3636fe70a2, July 23, 2016.

[7] Georgi Dalakov, "Floppy Disk," http://history-computer.com/ModernComputer/Basis/floppy_disk.html

[8] Steven E. Schoenherr, "Television Instant Replay," www.aes.org/aeshc/docs/recording.technology.history/television8.html

[9] Tom Howe, "Ampex Instant Replay Disk Recorder," www.cedmagic.com/history/instant-replay-hs-100-deck.html

[10] AMPEX Corporation, Press Release, March 22, 1967, www.digitrakcom.com/literature/HS-100%20Press.pdf

[11] "Client to Edge to Servers to Origin," https://developer.akamai.com/stuff/Overview/Client_Edge_Servers_Origin.html

[12] Val Henon, Matt Ahrens and Jeff Bonwick, "Automatic Performance Tuning in the Zettabye File System," Sun Microsystems, Inc., Menlo Park, CA, 2003.

[13] For more information, see Lance Ashdown and Tom Kyte (primary authors), "Oracle Database Concepts: Logical Storage Structures," https://docs.oracle.com/cd/E11882_01/server.112/e40540/logical.htm#CNCPT89139

[14] Lucas Mearian, "Scientists Calculate Total Data Stored to Date: 295+ Exabytes," www.computerworld.com/article/2513110/data-center/scientists-calculate-total-data-stored-to-date--295--exabytes.html
[15] Suzanne Wu, "How Much Information Is There in the World?" http://news.usc.edu/29360/how-much-information-is-there-in-the-world/
[16] "The Rapid Growth of Global Data" (Infographic), www.csc.com/insights/flxwd/78931-big_data_universe_beginning_to_explode
[17] Susan Gunelius, "The Data Explosion in 2014 Minute by Minute – Infographic," https://aci.info/2014/07/12/the-data-explosion-in-2014-minute-by-minute-infographic/

CHAPTER

5.13

Format and Standards Conversion

PAOLA HOBSON, PhD
InSync Technology Ltd
Petersfield, UK

INTRODUCTION

Since the earliest days of TV broadcasting, global viewers have wanted to watch programs from other countries—live breaking news, international sporting events, entertainment and cultural programming. However, simple international program exchange has never been possible due to the huge differences in TV standards around the world. Similarly, transfer of movies to formats suitable for home TV viewing inherently requires both format and frame rate conversion, for audiences in all regions.

With the transition of services from analog to digital and from SD to HD, and now to UHD, as well as proliferation of mobile and Internet content services, there have never been so many different formats and standards for video content available to global viewers.

Format and standards converters are needed to ensure every audience gets the most faithful reproduction of the originally produced content, at least to the best approximation available on their display device.

BACKGROUND

Video pictures are described in shorthand by their number of lines and their frame rate in pictures per second (usually abbreviated to Hz—which is used to refer to either frames or fields per second). The spatial representation is illustrated in Figure 5.13-1.

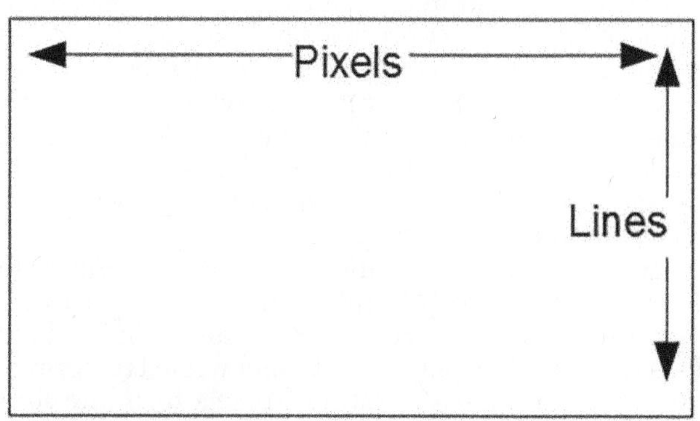

FIGURE 5.13-1 Image size (resolution)

A TV frame may be composed of two "interlaced" fields or may be "progressive" frames. "Interlaced" (i) means that there are two fields that comprise a frame:

- odd lines of the frame are scanned in field 1 at a time T
- even lines of the frame are scanned in field 2 at a time T + field interval

The scanning pattern may also be even lines of a frame scanned at time T to create a field 1 and odd lines scanned at time T + field interval to create a field 2.

An example of an interlaced standard is standard-definition (SD) PAL signals, which are produced at 50 fields per second—that is, 25 frames per second (field interval of 20 ms). It is the refresh frequency of the

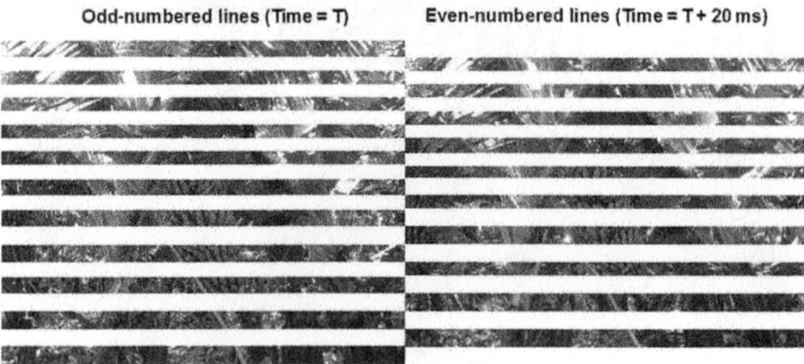

FIGURE 5.13-2 Interlacing example

pictures that gives the viewer the impression of continuous motion, even though each field is just a "snapshot" of a scene at a given time. A conceptual view of interlacing is illustrated in Figure 5.13-2. Note that real interlaced pictures comprise single interleaved lines, not blocks of image as shown in Figure 5.13-2.

A progressive frame (p) is one where all lines are present in all frames—that is, one complete picture is available to the display at every frame interval. The special case of progressive segmented frames (psf) is discussed later in this chapter.

The "display aspect ratio" of a picture defines the correct relative width-to-height ratio when presented on a display device. Typically standard-definition formats have a 4:3 display aspect ratio, while HD formats are 16:9. Note that the ratio of pixels to frame lines need not match the display aspect ratio for formats with a non-square pixel aspect ratio.

The number of pixels per line is derived from the standard and the expected aspect ratio. Some commonly used standards and their abbreviated descriptions are shown in Table 5.13-1. Note that Table 5.13-1 refers to "active" pixels and lines. The complete image for transmission historically included areas known as horizontal blanking interval (HBI) and vertical blanking interval (VBI), which applied per line and per frame (field) respectively. These blanking regions were needed to allow for analog TV displays to correctly position the picture, but in digital TV systems addressing modern displays, such intervals are not needed. Instead, provision is made for ancillary data—namely, horizontal ancillary data (HANC), located in non-picture portions of horizontal scan lines, and vertical ancillary data (VANC), located in non-picture regions of the frame or field. Standards with 1080 active interlaced or progressive lines per frame, or 720 progressive lines per frame, are considered high definition (HD). Ultra-high-definition pictures can be UHDTV1 (3840 pixels × 2160 lines) or UHDTV2 (7680 × 4320), progressive only [1].

Note that Table 5.13-1 is a very short summary of the range of production, transmission and display standards in use internationally. The use of the abbreviations PAL and NTSC for SD pictures is common, but not technically correct. Strictly speaking, PAL and NTSC refer to composite analog transmission formats, whereas most content is now sourced and transmitted digitally. Throughout this chapter, shorthand notation has been used for fractional frame rates. Where 59 and 59.94 Hz are used, these actually refer to 60/1.001. Similarly, 29 and 29.97 Hz mean 30/1.001, and 23.976 Hz is used to refer to 24/1.001.

Introduction to Format Conversion

Format conversion refers to any operation applied to a video sequence, without a change of frame rate. Format conversion includes aspect ratio conversion (ARC), where the aspect ratio is altered to suit the user's display; deinterlacing, for example, where 1080i material is converted to 720p or 1080p at the same frame rate; and up- or downconversion, where the content size is altered to suit a specific transmission mechanism, for example, 625 50i to 1080 50i conversion.

One example of format conversion could be where content was originally sourced at an aspect ratio of 4:3 (common in SD services) but the broadcaster's service is targeting viewers with 16:9 displays. Presentation of a 4:3 picture without any ARC would lead to distorted objects, as shown in Figure 5.13-3(a). With aspect ratio

TABLE 5.13-1
Some typical broadcast standards

Active pixels	Active lines	Interlaced or progressive	Frame or field rate	Display aspect ratio	Abbreviation
720	576	i	50 Hz	4:3	PAL (625 50i)
720	484	i	59.94 Hz	4:3	NTSC (525 59i)
1920	1080	i	50 Hz	16:9	1080 50i
1920	1080	p	25 Hz	16:9	1080 25p
1280	720	p	59.94 Hz	16:9	720 59p
3840	2160	p	23.976 Hz	16:9	2160 23p

conversion, the broadcaster can choose to take a 16:9 segment and zoom it to fill the 16:9 screen, as in Figure 5.13-3(b), or can add black bands to the sides of the picture, as in 5.13-Figure 3(c) ("pillarbox").

Another example of format conversion is where a broadcaster wishes to integrate a segment of standard-definition (SD) material into a high-definition (HD) program for simultaneous SD and HD services within one country. This requires an "upconversion"—that is, the number of pixels and lines per frame need to be increased, but there is no change in frame rate.

Introduction to Frame Rate Conversion (Standards Conversion)

Standards conversion or frame rate conversion refers to any operation applied to a video sequence where the frame rate is changed. Standards conversion may also include a format conversion at the same time.

An example of frame rate (standards) conversion could be a TV program sourced in the UK at 1080 50i (i.e., 1080 lines per frame, 50 Hz interlaced) that is to be transmitted to U.S. viewers at 720 59p (i.e., 720 lines per frame, 59.94 Hz progressive).

To understand standards conversion, we need to visualize the video content in both the spatial and temporal domains, as illustrated in Figure 5.13-4.

In order to format and frame rate convert the content, we need to create new video pixels and lines, at new spatial positions and new temporal intervals. For example, a 50 Hz to 59.94 Hz standards conversion requires creating new output fields every 16.67 ms, from input fields spaced 20 ms apart. Simply retiming the fields would not work, as we would be creating a video that runs faster than it was filmed, making the motion look completely unrealistic.

APPLICATIONS

Standards Conversion Applications

One of the most commonly applied standards conversions is for international program exchange, where

Original 4:3 image

(a) Presentation of 4:3 source on 16:9 display without ARC

(b) 4:3 source on 16:9 display as center cut with zoom

(c) 4:3 source on 16:9 display with black bands

FIGURE 5.13-3 Aspect ratio example

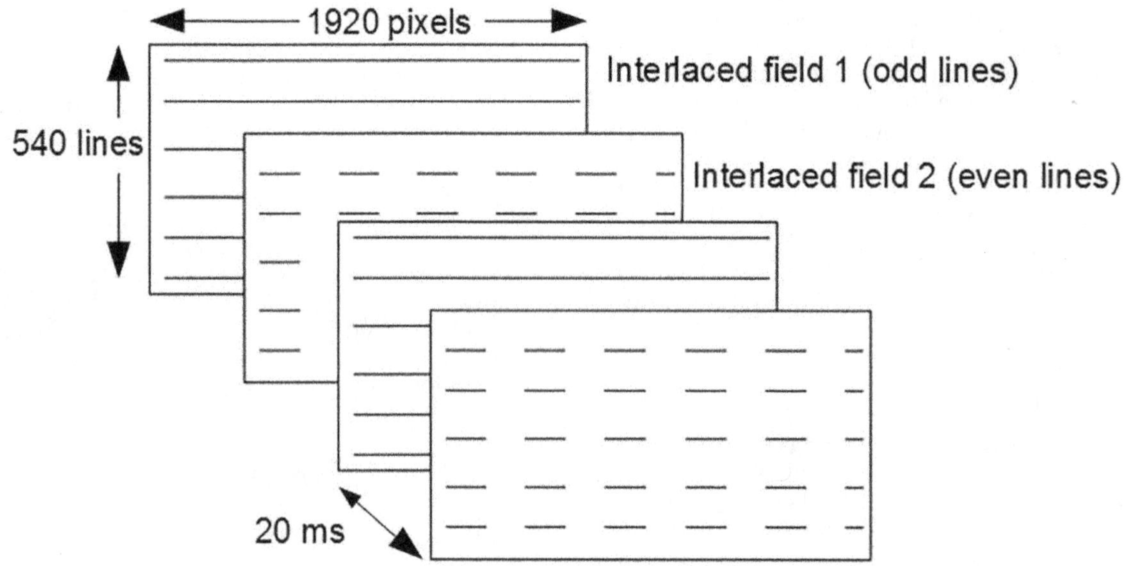

FIGURE 5.13-4 Illustration of an HD video program shown field by field

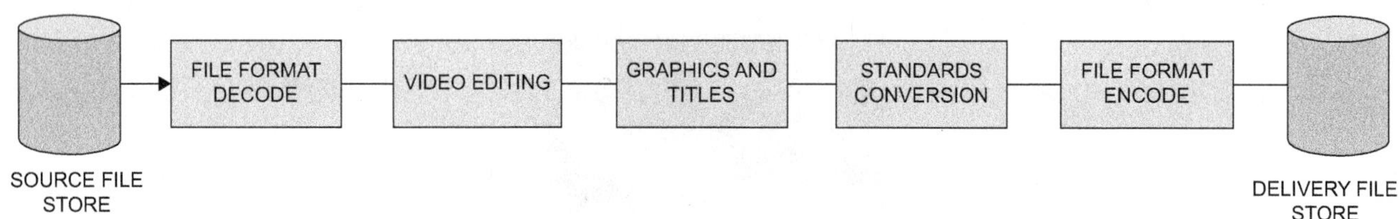

FIGURE 5.13-5 Standards conversion in a simplified edit workflow

material acquired in one country is broadcast in another. A typical example is for international sports distribution, where a major event, such as the FIFA World Cup, is simultaneously broadcast around the world. Cameras at the event will usually be operating at the standard of the home country (e.g., 1080 59.94i), so any services at 50 Hz (e.g., for European or Australian viewers) will need to apply frame rate conversion. Such conversions will typically be required live—that is, directly in the broadcast chain—as viewers will want to watch the action as it takes place. Historically, such live conversions have been provided by standards conversion hardware, due to the complexity of the processing and the high data throughputs required.

Conversion of other TV material is also required on a daily basis, where popular TV shows produced in one region are broadcast in another. Commonly, after a successful run in the home country, a TV series is repurposed for global audiences. Standards conversion will usually be performed on the entire series as a single project, and the conversion step can take place before or after editing for localization. Using conversion hardware, standards conversion would be carried out as a tape-to-tape process. The introduction of software standards converters has enabled the conversion process to be more tightly integrated into the edit workflow, as illustrated in the simplified workflow of Figure 5.13-5.

Standards conversion, including format conversion, is required when transferring movies to formats suitable for TV transmission. Movie material originally produced on film will be transferred to a digital file format using a high-resolution scanner. Typically, films may be scanned to UHDTV1 resolutions or above, but scanning directly to HD resolutions is also common. Since film is captured at 24 frames per second, frame rate conversion to 50 Hz or 59.94 Hz is always required, and the additional step of format conversion is also required if UHDTV scanning was performed. Digital film production often uses UHDTV and higher resolutions (e.g., "5K" 5120 × 3840, "6K" 6144 × 3160 and above), so format and frame rate conversion are always required for transmission on TV services.

Film conversions generally require aspect ratio conversion in addition to scaling and frame rate conversion. Common movie aspect ratios include 1.85:1 and 2.35:1. Conversion to 16:9 (1.78:1) for TV transmission may require addition of black bands ("letterboxing") to preserve the entire original content (Figure 5.13-6a), or a suitable portion of the frame may be extracted via a manual edit process known as pan-scan, where an operator chooses the 16:9 portion to be transmitted (Figure 5.13-6b), which is subsequently zoomed to fill the screen. Whereas letterbox formats may be desirable since the director's original artistic intent is preserved, some viewers find the black (or colored) bands

FIGURE 5.13-6A Framing a 2.35:1 movie in a 1.78:1 transmission

FIGURE 5.13-6B Extraction and zoom of a 1.78:1 portion of a 2.35:1 movie

distracting and prefer to compromise on content in order to watch a full-screen presentation. Note that with the advent of aspect ratio signaling (see later in this chapter) and suitably enabled displays, presentation of letterboxed or zoomed content may be chosen by the viewer.

Format Conversion Applications

Format conversions—where the size, interlacing or aspect ratio of a video frame is altered but there is no requirement for frame rate conversion—are needed where content produced in one format is converted for services using a different format. As the broadcast industry evolves, and higher-resolution programming becomes available, there is always a need for format conversions due to coexistence of different production sources and transmission requirements. For example, during format transitions, such as from SD to HD and from HD to UHD, it is not commercially viable for broadcasters to completely reequip their production facilities in a short space of time, so multiple production formats may be in use. Similarly, with home viewers possessing a range of TV sets (SD, HD and UHD), simultaneous services at different formats are required. A typical production environment might therefore require both spatial upconversion (e.g., HD to UHD) and downconversion (e.g., UHD to HD) to meet all viewer requirements, as illustrated in the simplified diagram of Figure 5.13-7.

Format conversion is a highly important process in monetization of content assets. Reuse of archived material in new services may require the content to be upconverted to a higher resolution. For example, integration of SD archive material into an HD service needs extremely high-quality upconversion processing so that viewers remain satisfied with their HD viewing experience.

Format conversion is also sometimes needed for some Internet and mobile content services, where limitations in transmission bandwidths or display size may restrict the viewer's ability to receive the highest-resolution version of the program. Similarly, where user-generated content acquired from consumer devices is integrated into news or other professionally produced programs, format conversion may be required.

FUNDAMENTAL PROCESSES IN FORMAT AND STANDARDS CONVERSION

Deinterlacing

Typical standards conversions require steps of deinterlacing, temporal rate conversion and rescaling. As deinterlacing is the first step in the conversion, it is essential that this be of the highest possible quality. The deinterlacer must preserve the maximum possible picture resolution and introduce zero (or, in practice, a minimal level of) artifacts. Since any later rescaling stage to a higher resolution will effectively magnify any tiny picture defects, introduction of undesirable effects at the deinterlacing stage could have a huge impact on visual quality of the output.

In recent years, deinterlacing technology has been improved through application of a number of techniques, mainly focusing on nonlinear, adaptive methods. More computationally intensive methods such as full motion compensation can be applied to the problem, but it is not always possible to use such methods affordably and robustly. In practical TV systems, source content may be noisy, grainy or unsteady, especially if it came from legacy sources. The material may also contain artifacts due to previous stages of compression encoding and decoding, so adaptive deinterlacing methods relying on identification of moving versus static areas need to be robust and repeatable.

A good deinterlacer will apply different processing in moving and stationary areas in order to obtain the maximum possible output resolution. Problems can arise if the deinterlacer fails to identify a picture area correctly—for example, a very slowly moving area is identified as static, or a stationary area is identified as moving because it is close to an area in motion. Weak analysis can also generate errors in periodic structures (such as posts in a fence or stripes in wallpaper) where a moving periodic area is classified as static.

Typical problems with deinterlacers include over-filtering, leading to a very soft output, ringing, and poor static area/moving area adaption. Other problems include jagged diagonals, loss of picture detail and edge artifacts (ringing). Any downstream processing of picture areas containing such errors can generate extremely visible picture defects. In particular, if the deinterlacing stage is before a motion analysis process,

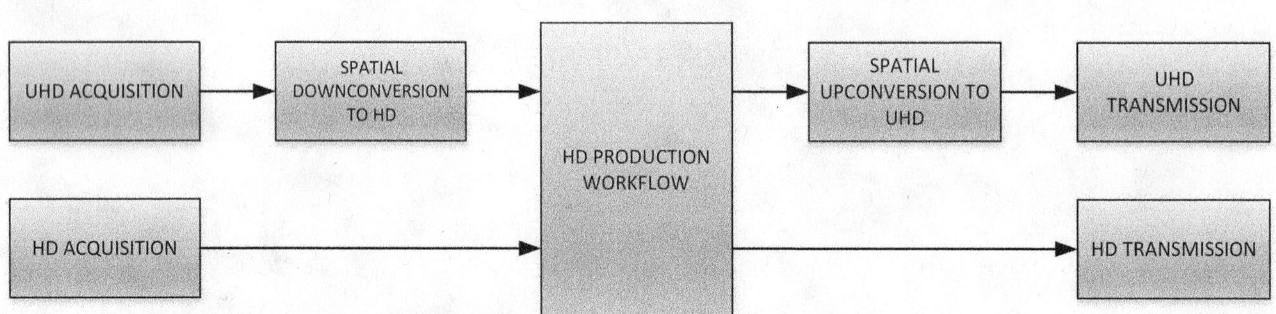

FIGURE 5.13-7 Hybrid HD and UHD production environment

artifacts such as line twitter can affect the reliability of motion vector generation.

Historically, deinterlacers also had problems when processing film-originated content originally sourced at 23.976 Hz contained in a 59.94 Hz 2:3 cadence sequence. In such content, incorrect cadence detection can actually lead to generation of more picture defects than if the processing ignored the source cadence. Modern deinterlacers are generally cadence-aware, so such problems are rare, but they are occasionally encountered in less sophisticated converters that do not have modes to manage mixed cadence material, for example, where 2:3 material is inserted into part of a video sequence, or where video captions and graphics are overlaid on 2:3 content.

Sometimes, deinterlacing errors can be masked by other processing in the transmission chain, such as compression, so they are less likely to be visible in HD material when presented on modern domestic, flat-screen monitors. However, when rescaled to UHDTV resolutions, any small defect becomes immediately more visible. An example is shown in Figure 5.13-8a, where poor adaptation has led to distortion of the edges of the captions in an HD sequence. This error has arisen due to moving text being interpreted as part of the stationary background. When upconverted to UHDTV, the distortions are very visible, as shown in Figure 5.13-8b; for example, see the horizontal edges of the letter E.

Typical methods for detecting stationary areas or image similarity use field-to-field (or frame-to-frame) absolute differences, with simple noise coring. Such

FIGURE 5.13-8A Deinterlacing artifact: incorrect moving/static adaptation

FIGURE 5.13-8B Magnification of Figure 5.13-8a when upconverting

methods have a number of weaknesses, which include errors in areas of high vertical frequency due to aliasing, over enhanced sources causing resolution variation that is falsely interpreted as motion or where there are illumination changes between subsequent images. Improved solutions are robust to noise and aliasing, and independent of lighting and other typical scene effects. A high-quality solution will contain specific algorithm refinements to directly mitigate away the substantial false negatives that would otherwise be obtained when attempting to identify still areas in the presence of interlace alias. A robust solution should be equally capable of rejecting the frequent false positives obtained when attempting to identify still areas in the presence of spatially repetitive moving image elements. However, such processing must be highly efficient, as very fast processing times are essential in live video applications.

Other filtering and interpolation artifacts such as ringing may be avoided, at no loss of resolution, through the use of nonlinear analysis and filtering techniques. Such methods can offer significant advances over linear filtering methods, and often lead to a higher *perceived* resolution—that is, a heightened impression of picture sharpness. Suitable nonlinear methods include frequency-selective processing, where different processing may be applied in different picture areas depending on the picture content—for example, presence and orientation of edges, or related to the image texture.

Temporal Rate Conversion

Frame rate conversion requires creation of new picture content in temporal locations that do not necessarily exist in the source. Conversion of material from a lower frame rate such as 23.976 Hz or 50 Hz to a higher frame rate such as 100 Hz or 120 Hz requires interpolation of new frames based on the available source frames. Clearly, the lower the source frame rate, the larger the nominal amount of motion between frames, and the harder it becomes to accurately predict the content in intermediate frames.

The simplest methods for frame rate conversion are generally unacceptable due to their poor motion performance. Retiming a sequence of images to play back at a different frame rate will affect the speed of motion of the objects in the image, and will distort the audio associated with the video. In practice, such "off-speeding" is usually only applied to very small frame rate changes such as 23.976 Hz to 25 Hz, in situations where audio pitch can be more easily corrected with simple processing and where the broadcaster is prepared to accept a 4% change in program length.

Conversion of 59.94 Hz to 50 Hz represents a 16% change, so off-speeding cannot be used as it would create grossly unnatural motion, highly significant audio pitch alterations and a dramatic change in program length. For such conversions, very simple frame rate converters simply copy frames from input to output where the input and output presentation positions would be closely related in time, and skip (or duplicate) frames to maintain the required overall frame rate. This is illustrated in Figure 5.13-9, where simple copying of a 50 Hz sequence to 59.94 Hz moves content to the wrong temporal position and requires periodic duplication of frames (e.g., frames F3 and F8 are duplicated).

Frame copying with drop/repeat has many obvious drawbacks, including unnatural and discontinuous motion, audio artifacts (where audio packets are lost or repeated at the frame skip/repeat) and corruption of metadata (e.g., closed caption packets would be lost or repeated). Therefore it is not recommended in professional video applications.

Linear Interpolation

One method to create new output pixels and lines from input sources is to linearly interpolate from the existing picture content, as illustrated in Figure 5.13-10, which is the basis of low-complexity, linear

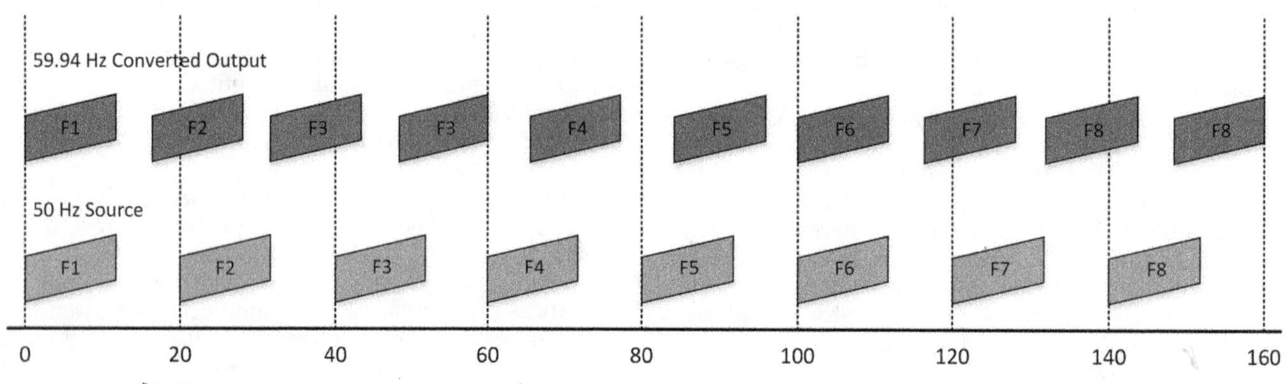

FIGURE 5.13-9 Frame copying with duplication

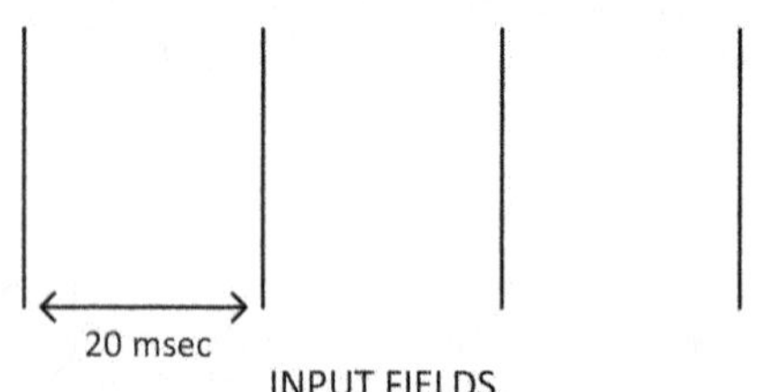

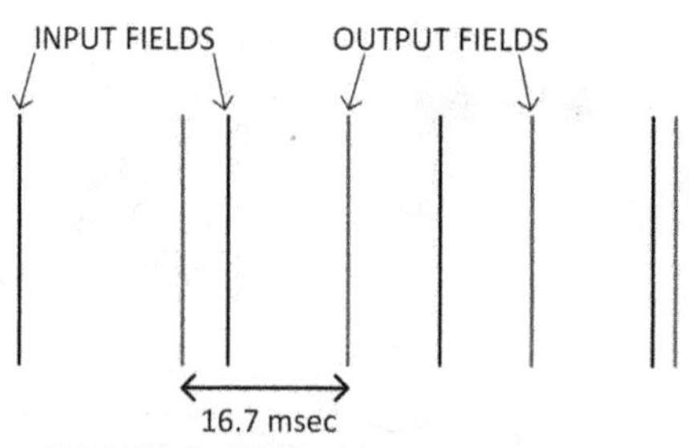

FIGURE 5.13-10 Temporal linear interpolation between fields

frame rate converters that are used in many low-cost applications. Figure 5.13-10 shows a simple case where pixels from two input fields are used to create the pixels in a new output field. Simple linear interpolation using a weighted sum of existing pixels to generate new output pixels gets round some of the problems associated with frame copying but can cause picture quality problems such as blur, loss of resolution and unnatural movement of objects, for example, judder.

An illustration is shown in Figure 5.13-11, where a still frame from a sequence converted using a linear converter is presented. As can be seen in Figure 5.13-11, the image suffers from blurring in moving areas and an overall lack of perceived sharpness. This is because content from different points in time are being combined, without taking into account the movement of objects in those temporal intervals.

Advances in linear frame rate conversion generally allow for motion adaptation. Motion adaptive methods apply different processing in stationary and moving areas of the image. In theory, where an area is stationary, input frame pixels can be copied into output frames at new time intervals, without loss of picture quality. In format conversion, moving areas are interpolated using intra-field processing, which is a spatial-only process, but in frame rate conversion, there will always be a temporal interpolation as well, with inevitable double imaging effects and blurring. In practice, real-life images are rarely entirely stationary, and simple effects such as lighting changes, image noise and aliasing can all contribute to false detection of motion, which inevitably reduces quality in the output image. Thresholding to account for small picture perturbations can lead to disturbing motion artifacts where moving areas are treated as static.

For applications with slow-moving content, or in situations where the source content suffers very obvious picture defects (e.g., low-resolution footage originating from electronic newsgathering equipment), linear standards converters may be acceptable, despite the motion artifacts. Linear converters are also widely used in disaster recovery or backup applications, where the converter is idle for most of the time, so a low-cost solution is required.

Motion-Compensated Frame Rate Conversion

Motion-compensated processing overcomes many of the drawbacks associated with linear standards converters. A motion-compensated frame rate converter calculates the motion between frames in the content, and works out where to move objects to when creating new frames in between, as illustrated in Figure 5.13-12. In Figure 5.13-12, if we can calculate the change in the object's position between frames 1 and 2, and we know the time interval between those frames, and if we assume that the object moves at a constant speed, we can work out where the object should be at any other

FIGURE 5.13-11 Illustration of linear frame rate conversion

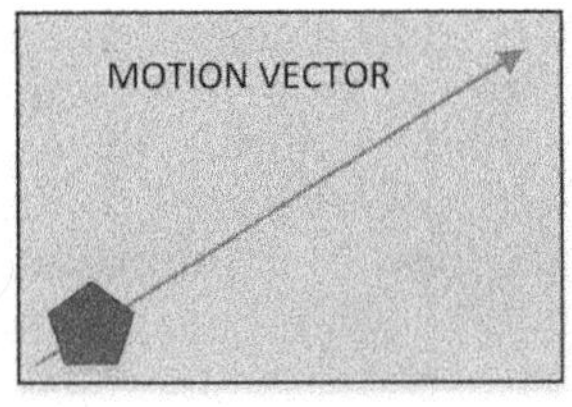

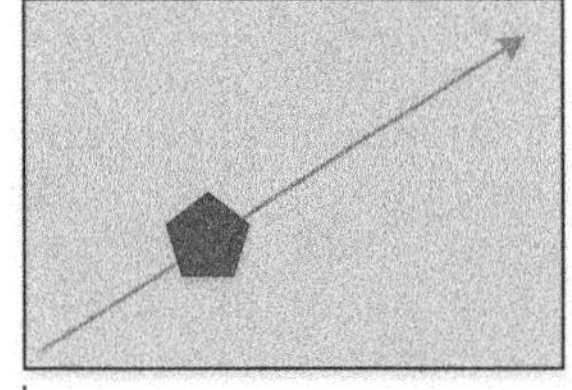
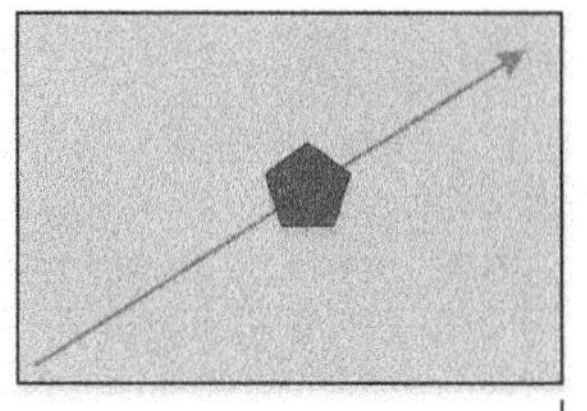

FIGURE 5.13-12 Illustration of motion-compensated frame rate conversion

FIGURE 5.13-13 Illustration of different types of motion in a typical scene

time interval. A motion-compensated converter can thus reproduce the object within any interpolated or retimed frames. In this way, all picture objects remain sharp and in focus, and their motion is smoothly portrayed without any judder or irregular movement.

Although motion-compensated conversion has the capability of delivering outstanding quality conversion, it relies on highly sophisticated processing to ensure consistent results. Specifically, creating new pictures at different temporal intervals and different resolutions relies on extremely accurate calculation of the motion vectors: In the simple case illustrated in Figure 5.13-12, a single object is moving at a constant speed without any change of scale. In practice, real TV programs contain multiple objects, moving at various speeds, occluding each other, moving into and out of the scene, and moving into and out of the camera field of view. Even the assumption of constant speed is a huge simplification.

An example illustrating some typical motion is shown in Figure 5.13-13, where object motions are shown as arrows of various lengths (illustrating different speeds and directions of motion). In Figure 5.13-13, each person is walking in a different direction at different speeds. People cross each other, they come into and out of the shot, some walk towards the camera so appear to get larger and some walk away from the camera so appear to get smaller. Furthermore, part of the flexible sign in the top right of the picture is stationary, but the lower part will be moving with irregular motion (which is wind dependent) and will appear to be warping, and although we cannot see this in a still image, the camera could be panning across the shot, meaning that there is global as well as local motion. Typical scenes also include rotating objects, camera zoom, and special effects. Further complications arise from moving graphics superimposed over the picture (e.g., titles and credits).

In addition to the need to deal with multiple, complex motions in a scene, the motion estimator has to handle cases where, although a human can easily work out that something is moving, a non-human processing device cannot. For example, if a camera is panning across the fence shown in Figure 5.13-13 at a speed where the camera horizontal movement equates to exactly one fence post per frame, it could appear to a motion estimator that there is no motion.

Failure to accurately measure motion in a scene leads to objectionable defects in the converted program such as irregular movement of objects, blurring, and insertion of erroneous picture information projected from elsewhere in the sequence. Therefore, motion-compensated frame rate converters generally use complex, multi-stage processing to get the best results.

As illustrated in Figure 5.13-14, motion-compensated standards conversion preserves the picture detail and ensures that motion portrayal is realistic and accurate. A comparison of Figure 5.13-14 with the linear conversion of Figure 5.13-11 very clearly shows the benefits of motion compensation in standards conversion.

Motion Estimation

Motion estimation is used in many image processing applications including video compression codecs. A method in use for many years is termed "block matching." In block matching methods, areas of the image in one frame (usually rectangular image blocks)

FIGURE 5.13-14 Illustration of motion-compensated frame rate conversion

are projected backwards in time into previous frames or forwards in time into subsequent frames, according to a set of candidate motion vectors. If the position of the projected block matches the image content in the appropriate frame, the motion vector is considered valid. Much has been written on the subject of matching criteria, since it is unlikely that every pixel in the projected block will identically match that in the previous or next image, due to noise, illumination changes and picture distortions.

Block matching methods have been refined over a number of years to improve efficiency and accuracy. Using small block sizes and overlapping blocks can improve motion estimation accuracy, but at the expense of higher computational requirements, especially in the search stages. There is also a higher chance of a false match if the block size is too small (in the limit a single pixel), but this can be improved by spatial consistency checking, that is, whether all the motion vectors in a given image spatial area are closely related. Simple block matching methods do not work well when motion is not translational (e.g., content rotation), and they do not cope well with zoomed, uncovered or occluded objects. Unless totally exhaustive searches are used (which is impractical in real applications), block matching methods may fail for objects in very fast motion, if the motion falls outside the search area.

Although block matching is useful for compression codecs that use decomposition into blocks as the basis of their compression algorithms, a block-based model does not suit standards conversion applications where picture areas do not generally move in neatly ordered block-sized segments. Gradient methods for motion estimation, which include optical flow analysis, get round the block paradigm by applying computations to large image areas to estimate global motion, with successive refinement stages to estimate local motion. Optical flow methods are based on the assumption that image intensity is conserved along a spatiotemporal path along the motion. Such methods work well for small amounts of motion, but are less reliable for flat textures and where objects have fast, complex motion.

Many motion-compensated standards converters use phase correlation [2] as the basis of the motion estimation. In the phase correlation method, image data are transformed into the frequency domain using the Fourier transform. Once transformed, the position of picture detail is encoded by the frequencies' phase information. Frequency magnitudes are adjusted so that they contribute equally. The size and direction of motions present may then be derived by subtracting the phases obtained from two sequential frames and transforming the result back to the spatial domain.

Phase correlation is insufficient to perform motion compensation by itself. Significantly, the method identifies motions present within the image but does not define which areas in the image have that motion. Therefore, the measured motions must somehow be associated with specific regions within the image, which typically represent separate real-world objects. Content properties such as shot changes, concealed and revealed picture elements, brightness changes and the presence of abrupt picture boundaries must also be addressed. These are reasons why different converters using this method of motion estimation exhibit differing results.

All motion estimation techniques may be impaired by the presence of noise, low resolution or blurred content. Extreme motion speed or motion complexity can be causes of failure. Similarly, the presence of repetitive structures such as fences or grilles are a well-known source of error. Hence, converters also differ in performance by how they preprocess content prior to applying the phase correlation method, how the method is refined to make it accurate and resilient in the presence of real-world content challenges, and how the results are post-processed to avoid incorrect motion compensation.

Transparency can be particularly challenging, such as where there are objects with one direction of motion overlaid with semitransparent foreground or background objects that have a different motion (e.g., a person running behind a fountain). Figure 5.13-15 illustrates this type of error, encountered with a typical motion-compensated converter that uses phase correlation. In Figure 5.13-15, it can be seen that the motion analysis has been unable to differentiate areas that appear very similar—for example, the highlights on the child's head are difficult to distinguish from the flowing water. This has led to incorrect reconstruction of the head, as well as defects in the flowing water, in the output frame.

Standards conversion of film-rate material (e.g., 23.976 Hz and 24 Hz sources) can be particularly problematic, especially when frame rate upconverting to a higher frame rate where the source contains fast movement. There are two important considerations when converting film-rate content with fast movement. First, longer shutter duration results in a high degree of "film blur" along the path of motion, which increases the difficulty of deriving accurate motion vectors. Second, objects will have relatively more spatial displacement from frame to frame, requiring much larger search areas for valid vectors. When frame rate upconverting to high frame rates such as 100 Hz or 120 Hz, many more frames need to be interpolated, so there are more chances of possible visible artifacts and interpolation errors. Therefore, most standards converters will include a fallback "safe" processing method such that a non-damaging interpolation will be applied if there is insufficient confidence in the reliability of a motion vector selected for a particular image region.

FIGURE 5.13-15 Detail from conversion with error due to incorrect motion analysis

Scaling and Aspect Ratio Conversion

Where the standards or format conversion requires a change in picture size (number of lines and/or number of pixels per line), scaling or resizing is needed. In the simplest concept of rescaling, one could create new output pixels and lines from the input sources via linear interpolation, possibly within a field, as illustrated by the downconversion of Figure 5.13-16, or inter-field where the content is static. In Figure 5.13-16 we show how an SD output could be created by interpolation from an HD source. The solid black points represent input pixels, and the pale points represent output pixels.

Where an output pixel position is co-sited with, or close to, an input pixel position, the pixel may be copied from input to output. Where the output pixel position is between two input pixel locations, an interpolation is made. A linear interpolation applies a spatial filter within a field (or frame) to create the output pixels. For example, on the top line, we illustrate a 2-tap filter where a weighted average of two input pixels is used to create an output pixel. At the bottom of the frame, we show an example of where four input pixels are used to create an output pixel. The benefit of this method is that it is low complexity, requiring low data storage and few calculations to create the output picture. The disadvantage is that such simple spatial filtering decreases the picture resolution, so areas of interpolated pixels become blurred. Where some pixels are interpolated and some are copied, depending on the relationship between input and output pixel positions, there will be significant differences in subjective appearance, which may be quite objectionable.

In practical conversion applications, scaling ratios may not always be simple 2:1 upconversions or downconversions. For example, when converting content between SD and HD, the acquisition and display aspect ratios need to be taken into account. Overall display aspect ratio (DAR) is the product of pixel aspect ratio (PAR) and storage aspect ratio (SAR). Although the ratio of number of pixels to number of lines is informally referred to as aspect ratio, this is generally incorrect, as both 720 × 576 and 720 × 486 format pictures are presented to the viewer as SD 4:3. For more on aspect ratio conversions, see later in this chapter.

Cadence Management

Cadence in video processing is generally understood as the pattern of a frame or field structure in a video sequence. For example, interlaced video content is referred to as 1:1 cadence, as there is a regularly repeating pattern of one field comprising the odd-numbered

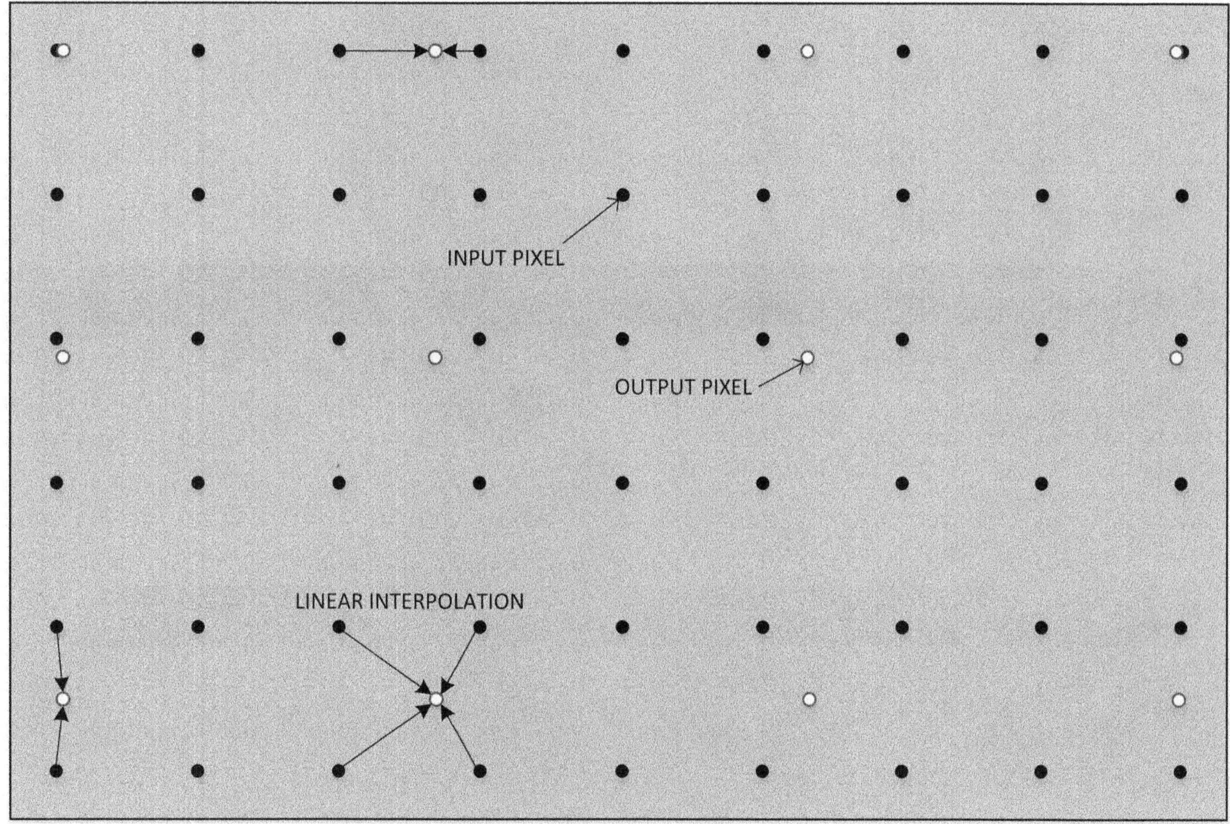

FIGURE 5.13-16 Spatial linear interpolation within a field

video lines followed by one field comprising the even-numbered video lines making up one video frame.

To avoid an interpolation step in conversion of film-rate material to 59.94 Hz video for TV transmission, a process known as 2:3 pull-down was adopted, which is illustrated in Figure 5.13-17. For each film frame (either scanned or from telecine) S_i, a series of intermediate interlaced fields are created, f_{ij}, where two or three interlaced fields are generated from each source frame S_i in a consistent, alternating 2:3 pattern. For example, source frame S_0 is used to generate two intermediate interlaced fields (f_{01} and f_{02}), whereby source frame S_0 is optically scanned or digitally sampled with appropriate filtering using a field 1 sample grid to generate the intermediate field f_{01}, and the same source frame S_0 is then sampled again using a field 2 sample grid to generate the intermediate field f_{02}.

The second source frame, S_1, is used to generate three intermediate interlaced fields (f_{11}, f_{12} and f_{11}), whereby image content S_1 is sampled using the field 1 sample grid, generating intermediate field f_{11}, then S_1 is sampled again using the field 2 sample grid, to create f_{12}, and S_1 is again sampled using the field 1 sample grid, generating intermediate field f_{11}.

The output frame sequence is then created using the intermediate fields. Irrespective of how many fields are generated from a given source frame S_i, the sampled intermediate fields f_i must always alternate between a field 1 and a field 2 as per normal interlaced video. Interlaced video is recorded and broadcast as a whole number of interleaved frames comprising a field 1 and a subsequent field 2. For example, as illustrated in Figure 5.13-17, output fields R_{11} and R_{12} comprise one 29.97 Hz output frame.

When 59.94 Hz material with 2:3 cadence is subsequently frame rate converted to 50 Hz, care is needed to ensure that fields are paired correctly according to their temporal sampling position. Problems can occur if the 2:3 sequence has been incorrectly edited on a field boundary, leaving an "orphan" field. For example, in Figure 5.13-17, if the editor removes a small segment {R_{21}, R_{22}}, the resulting sequence now has a discontinuous cadence 2:2:1:3:2:3. The standards converter needs to have sophisticated cadence detection tools to ensure that such anomalies are identified and the correct processing applied.

Good cadence detection requires careful analysis since there are many cadence patterns in use, including 2:3:3:2 and 2:2:2:4, which are common cadences derived from camcorders, and 5:5, 6:4 and 8:7, which are common animation cadences.

A cadence pattern commonly used in TV productions is the progressive segmented frame (psf) format, which is often referred to as 2:2. Unlike true progressive format video where each video frame represents the entire image at that time interval, a psf frame comprises two fields ("segments"), each of which contains half the lines in the frame, scanned alternately odd and even lines. The two fields are transmitted separated in

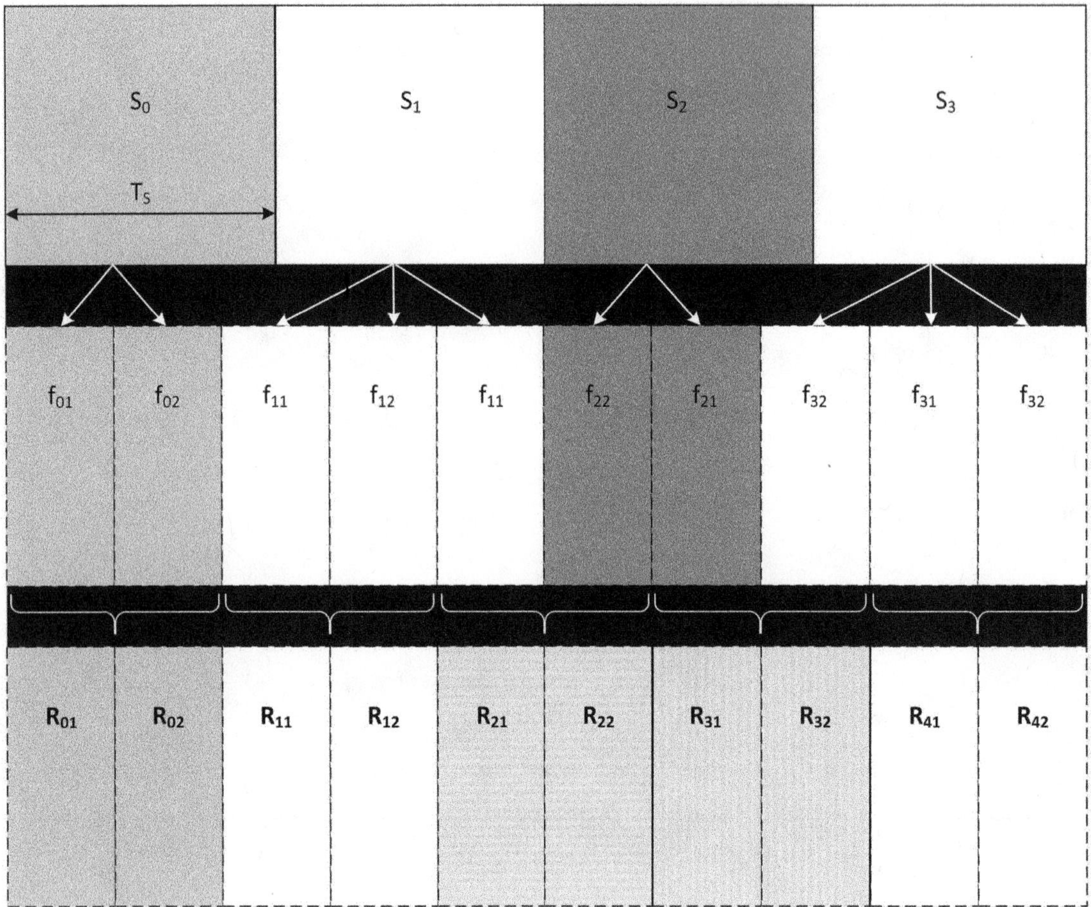

FIGURE 5.13-17 Illustration of 2:3 pull-down for conversion of 24 Hz progressive content to 59.94 Hz interlaced output

time by half a frame interval, so that they can be carried within a transport designed for interlaced video. The receiving device must be told (e.g., via metadata) that the content is psf so that the fields can be stored and reassembled correctly. Since there is no motion between the fields in a psf frame, interpretation of a psf source as interlaced fields by a standards converter would result in severe motion artifacts. Where a live source contains SMPTE ST 352 payload identification, or a video file contains i/p/psf metadata, the standards converter will recognize and correctly convert psf material.

One difficulty presented to standards converters is where a source contains mixed cadence. This can occur where a program has been produced using a film-rate camera (or has been scanned from movie film) in a psf format, and then video graphics are edited over the program. An example might be a 25psf drama, over which 50i closing titles are edited. In these situations, the user has to make a subjective decision whether to process the content as entirely interlaced, thus preserving the caption movement, or entirely psf, thus favoring the background. More complex systems where different processing may be applied to different regions of the image can be constructed, but such systems require expert user intervention, which is generally not practical in cost-conscious, time-sensitive conversion applications.

FORMAT AND STANDARDS CONVERSION PRODUCT REQUIREMENTS

Live Conversion Products

Format and standards conversions are used in live broadcast applications where action taking place in one region (e.g., breaking news or major sporting or musical events) is to be viewed in another, where the remote and local production standards are different. Of critical importance to the broadcaster are real-time processing, excellent picture quality, high equipment reliability and low delay in the chain from remote camera to local production studio. Further considerations include equipment cost, power consumption, form factor and ease of integration into the overall system.

For example, consider a simplified outside broadcast application at a major sporting event, Figure 5.13-18. The production is made at 1080 59.94 Hz for delivery to clients (usually fixed studios) in both the home country and internationally. The outside broadcast production company will supply either a clean feed or a feed with their logo plus simple titles and graphics (e.g., team names, latest score, etc.) at both 1080 59.94 Hz and 1080 50 Hz. The clients add commentary, audio description, closed captions (subtitles) and local titles in their own language to create a final program for their audience. In practice, there could be a requirement for more client feeds, for example, direct from some of the cameras, and at more than just two output standards.

In Figure 5.13-18, the cameras and processing equipment may all be close to the venue where the action is taking place, possibly all colocated in an outside broadcast (OB) truck. Such OB trucks will have limited space available for equipment, as they need to be small enough to be driven on a public road and also must accommodate the production crew. Equipment will be mounted in 19" racks and is expected to have SDI connectivity. The standards converter must be controllable by simple front panel operation or via a remote control system over Ethernet. Since not all OB operations require standards or format conversion, the converter may not be part of the truck's fixed equipment. It is possible that the converter may be rented for a specific event and will be operated by a non-expert; therefore, simple, unambiguous controls are essential. In particular, it must be possible for the standards converter to auto-detect the input standard, so that the user can set the converter to generate the required output standard, with no intervention needed in the case that the input standard changes.

Advances in processing technology, combined with very efficient motion estimation methods, have enabled motion-compensated standards converters to be available in single height form factors, suitable for compact 1U rack mounting. Further refinements have reduced the form factor down to a single circuit board, suitable for use in modular installations, which are popular in OB trucks due to their compact footprint.

In the live workflow of Figure 5.13-18, the standards converter is located after the final program production. This poses additional requirements on the converter, as it will need to deal with content that may contain extremes of high-frequency components, due to upstream graphics insertion. Titles and effects with high image frequency edges, overlaid on pictures that may be noisy or suffer compression codec artifacts, cause particular problems for motion adaptive format converters and motion-compensated standards converters, since noise and high-frequency edges can be misinterpreted as motion. The solution is to manage the process at the earliest deinterlacing stage of the converter, as described in the Deinterlacing section presented earlier in this chapter.

An alternative live workflow is shown in Figure 5.13-19, where the standards or format converter is

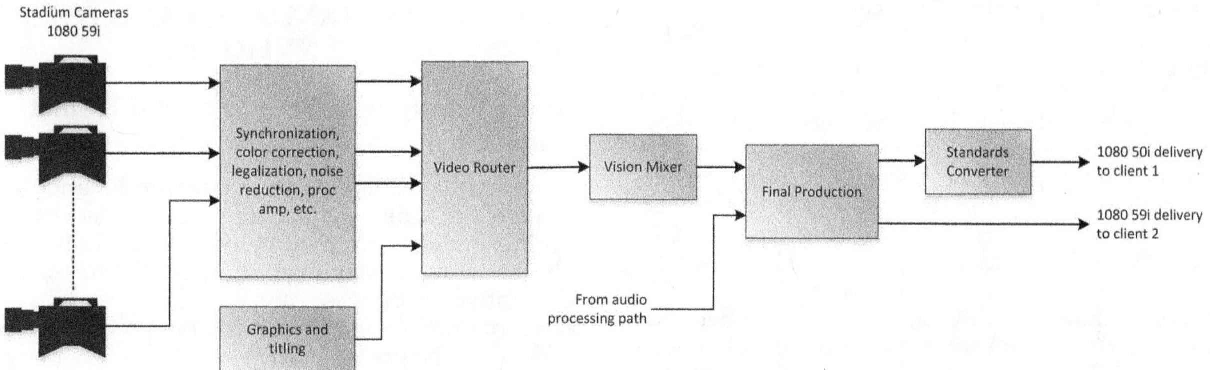

FIGURE 5.13-18 Example outside broadcasting of a live event

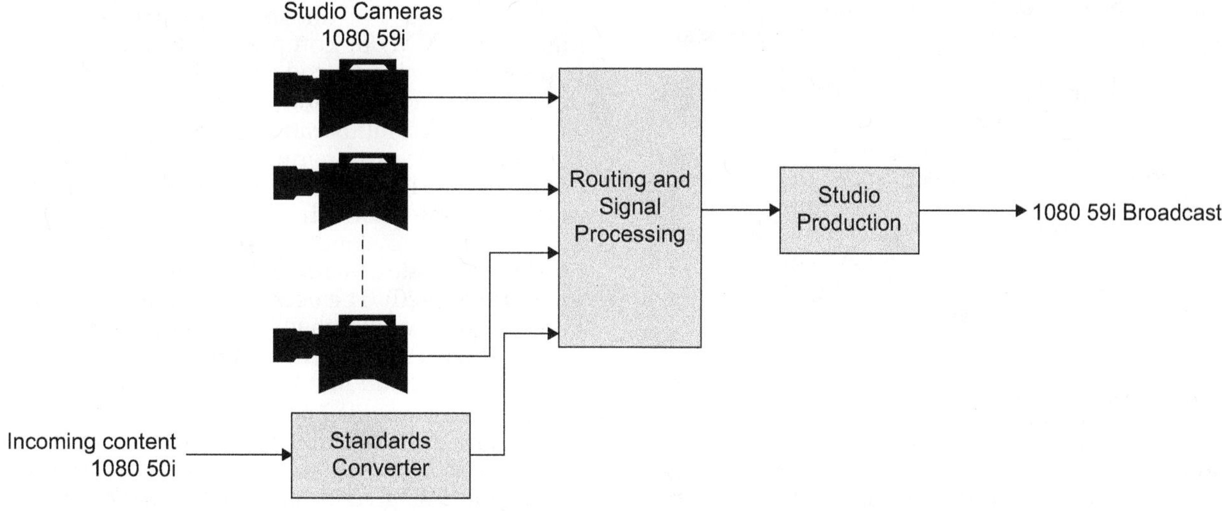

FIGURE 5.13-19 Example studio production

located at the ingest point of the workflow. This is typical in applications where content from field acquisition is used within a studio production, for example, news feeds or live sports coverage.

In the live application of Figure 5.13-19, the standards converter may be expected to perform a wide range of signal processing operations, in addition to frame rate and format conversion. For example, the converter may incorporate frame synchronization, audio de-embedding, legalization, noise reduction, picture enhancement, metadata management (closed caption and timecode), audio and video compensating delay, audio gain adjustment, audio track shuffling, logo insertion, sidebar keying and other video processing operations.

In the live workflow, standards converters typically support SDI input and output. With the advent of UHDTV production, requirements for 6G and 12G SDI have arisen. Because of limitations on cable lengths at the higher frequencies, alternative interfaces such as fiber and Ethernet have been introduced. SMPTE ST 2036–3 [3] provides for mapping into single-link or multi-link 10 Gb/s optical SDI, but this requires dual-link for UHDTV1 content at 50 and 60 Hz, and octa-link for UHDTV2. Higher frame rates and word lengths of 12 bits and above would call for higher-order multi-link connectivity, which is generally impractical. Proposed SMPTE standard ST 2062 [4] supports up to UHDTV2 at 60 Hz on single-link, but it is a costly solution.

An alternative is to use real-time streaming protocols to convey compressed UHD content to and from standards converters. 10 GbE IP network hardware is now readily available (40 GbE and higher are also no longer uncommon), which enables multiple 3G SDI signals to be routed around a studio and therefore to be available to the standards converter. The simplest mapping of an SDI payload into an IP transport is provided by SMPTE standards 2022–6 [5]. A solution that can be more bandwidth efficient is the Technical Recommendation TR-03 from the Video Services Forum [6], which provides a standard for Video over IP without SDI encapsulation. A further refinement, VSF TR-04 [7], defines SMPTE ST 2022–6 as a payload type in the context of a VSF TR-03 system. While SDI using conventional BNCs remains the most popular interface to the standards converter for live HD applications, IP interfaces may prove to be more practical for UHD services.

One particular advantage of using IP connectivity is the opportunity to implement the standards converter as software on generic computer hardware such as a high-performance server. Such software implementations can be feasible if the processing power of the server is sufficient to enable real-time conversion. More efficient conversion architectures and application of graphics processing units (GPUs) also help reach real-time performance in software.

File-to-File Conversion Products

Non-live workflows that typically need standards conversion include postproduction applications, archive services, content distribution and content repurposing. Common examples are where a popular TV show is converted for international distribution, or transfer of a movie from its original cinematic format to a format suitable for small-screen TV broadcast. The simplified edit workflow of Figure 5.13-5 would be applicable for such applications, where content originates and terminates as media files.

File-based conversion workflows are conveniently supported by software standards converters, where key functionality and processing, originally designed into proprietary hardware, is ported to a software implementation. Software standards converters offer more flexibility for users, as they may be deployed in Windows or Linux platforms, to match the user's overall operational requirements. Of particular advantage is that the software may be deployed in virtual machine (VM), server farm and cloud implementations. These allow the user to trade processing resource usage for speed of conversion, where high-priority jobs can be serviced using multiple cores of the server, whereas low-priority conversions can run (e.g., overnight) on any remaining free cores. This can be particularly important when a content distributor has the rights to rebroadcast certain material for a short period of time, and any time lost in conversion means lost revenue.

PRACTICAL STANDARDS AND FORMAT CONVERTERS

Audio Handling in Standards and Format Conversion

In practical deployment of standards and format conversion, video signals are accompanied by audio. Multichannel audio may be present, including possibly stereo main language, multiple mono (or stereo) alternative languages, music and effects, and audio description. Sixteen or more channels of uncompressed audio may be present, which in live productions may be embedded with the video in an SDI transport. Dolby-encoded audio may also be present, Dolby 5.1 being commonly used within HD productions. In file-based applications, audio may be multiplexed with the video information as uncompressed (pulse-code modulation, or PCM) or compressed using one of many popular compression codecs such as AAC. For more discussion of audio formats, see Section 4.

Figure 5.13-20 illustrates a live conversion application, with audio processing. Although embedded audio is common in TV production environments, audio will have a separate production chain to the video. In some applications, the standards or format converter may need to integrate other sources of audio (as illustrated in Figure 5.13-20). The user may also wish to mute certain incoming audio channels if they are not needed, and may wish to insert new audio, for example, when dubbing a program into the local language.

In frame rate conversion, output video pictures will be presented at different time intervals to the originating source presentation times. Furthermore, any video processing associated with format and frame

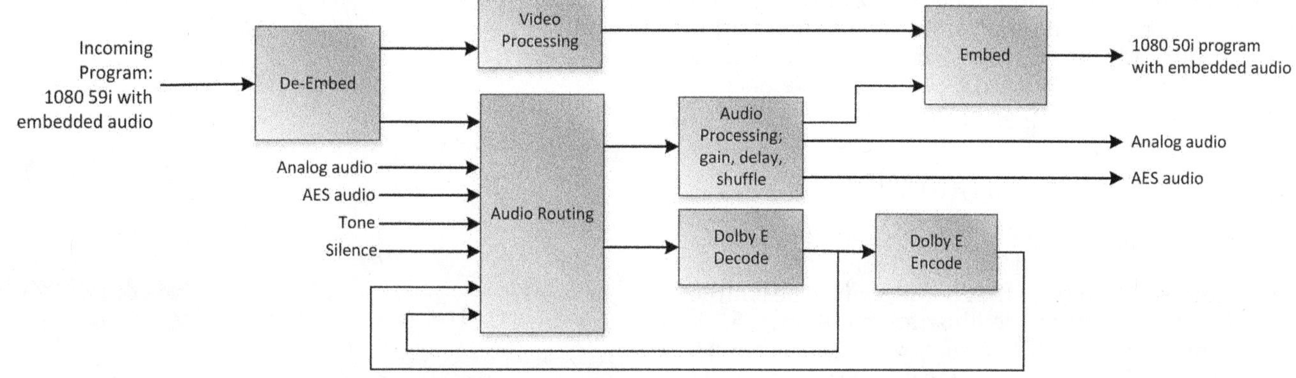

FIGURE 5.13-20 Conversion application with audio processing

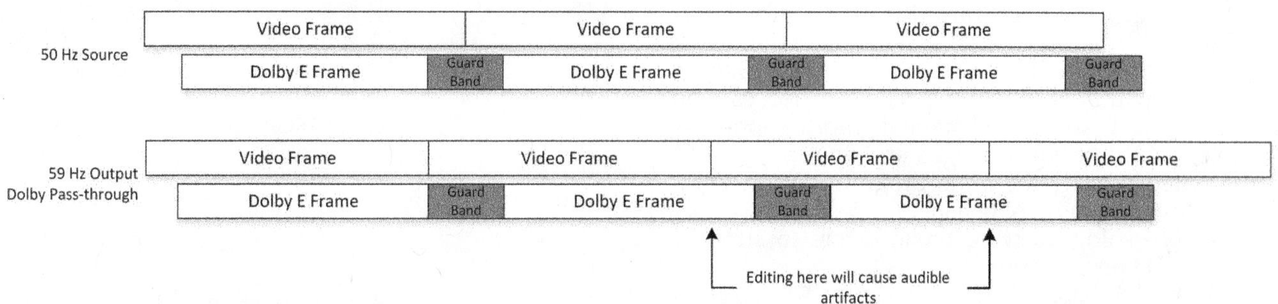

FIGURE 5.13-21 Dolby E misalignment when using a pass-through mode when standards converting

rate conversion will inevitably take a certain amount of time, and it is important for any associated audio to remain co-timed with the video. Given that the audio and video processing stages will take differing amounts of time depending on the conversion required, compensating audio delay will be needed in order to keep the audio and video in step. Timing discrepancies are especially noticeable when people are speaking ("lip-sync"), and it is widely recognized that images presented early with respect to audio are significantly more preferable to images presented late with respect to audio for the simple reason that the former never occurs in normal real-world experience, whereas, due to the finite and relative speeds of light and sound, the latter occurs frequently in real-world experience.

The standards converter is an ideal place to carry out small audio adjustments that would otherwise require separate processing hardware. For example, if English is on audio channel 1 on the input, but Spanish is required to be audio channel 1 on the output, the standards converter can include audio mapping functionality to shuffle the audio channels. Similarly, if an upstream error has advanced or attenuated one of the incoming audio channels, additional delay or gain can be provided by the converter.

More complex audio handling is needed if the incoming audio is compressed and encoded using Dolby E. Dolby E divides audio into frames at a rate aligned with the associated video. A special feature of Dolby E audio is that complete Dolby E packets can be lost or repeated with minimal audio distortion. As illustrated in Figure 5.13-21, Dolby E frames are slightly shorter than video frames to allow for switching time in routers and mixers. If a Dolby E frame is truncated, the audio will not be correctly decoded, and audible artifacts will be heard.

In the simplest format and frame rate conversion applications, Dolby E data can be passed through the converter from input to output without any intervening processing. This enables Dolby E data to be correctly decoded if there is no downstream processing. However, as illustrated in Figure 5.13-21, the Dolby E frames will no longer be aligned with the video frames, so if the program output is passed to a synchronizer or other processing where Dolby E packets may be partially wrapped, dropped or repeated, or undergoes editing where edits are made on the video frame boundaries, corruption of the Dolby E data will occur.

In format conversion where input and output video are the same frame rate, for many applications, it is adequate for the format converter to realign the Dolby E guard band on the output side, which will be set automatically, with respect to the output standard. The user may also add delay or advance to match downstream processing requirements.

To guarantee correct guard band alignment when input and output frame rates are different, it is necessary to decode the incoming Dolby E audio, and re-encode on the output side at the new frame rate.

Some standards converters include a Dolby E option that enables this processing (as shown in Figure 5.13-20), and any additional delay or advance can thus be incorporated into the Dolby E decode/re-encode process.

Aspect Ratio Conversion (ARC)

As already mentioned in this chapter, aspect ratio conversion (ARC) is needed in cases where the output aspect ratio is different to that of the input. The simple cases shown in the example illustration of Figure 5.13-3 were common in SD/HD conversion applications, where SD content sourced at 4:3 was to be upconverted for presentation as 16:9 HD, or content produced at 16:9 HD was to be downconverted for broadcast to 4:3 SD monitors.

During the SD to HD transition, some broadcasters adopted 14:9 aspect ratios to avoid excessive black banding in up- and downconversion. Positioning of the picture within different aspect ratio frames was also introduced, so that additional information—for example, subtitles, tickers, logos and so forth—could be presented outside the main program picture area. SD monitors with a display aspect ratio (DAR) of 16:9 were widely sold, enabling broadcasters to support full-screen SD presentations.

While broadcasters continue to offer SD services and while SD material continues to be incorporated into HD programming, aspect ratio conversions will be needed. Situations may occur where a broadcaster receives an incoming feed where the actual picture presentation is not known, and may change during the daily schedule, depending on the origin of the source. This is illustrated in the downconversion application of Figure 5.13-22, where the incoming source may be (1) HD full frame 16:9, or (2) HD 4:3 in a 16:9 frame (pillarbox). The SD viewer wants to see the whole picture on their SD TV set, so in Case 1, the presentation on the 4:3 SD display is letterboxed 16:9, but in Case 2, the presentation takes the 4:3 center from the 16:9 pillarboxed frame and displays it on the SD monitor (center cut). Thus the converter will have to apply different processing for each of Case 1 and 2. If a center cut is taken in Case 1, the viewer misses the sides of the picture. If letterboxing is applied in Case 2, the viewer sees a very small picture with excessive black bands. This is illustrated in Figure 5.13-23.

Applying the correct ARC processing without manual intervention requires a method of signaling to the converter the format of the source. The user can then set the converter to obey a set of rules in order to achieve the desired result automatically. Signaling standards include Wide Screen Signaling (WSS) [8], Video Index (VI) [9] and Active Format Description (AFD) [10].

To use AFD as an example, the 16:9 full frame picture of Figure 5.13-22 Case 1 would be 16:9 AFD8, and the 4:3 center picture of Case 2 would be 16:9 AFD9. The user can apply rules such as "fit to width" or "fit to height" such that the standards converter applies the appropriate aspect ratio conversion based on the

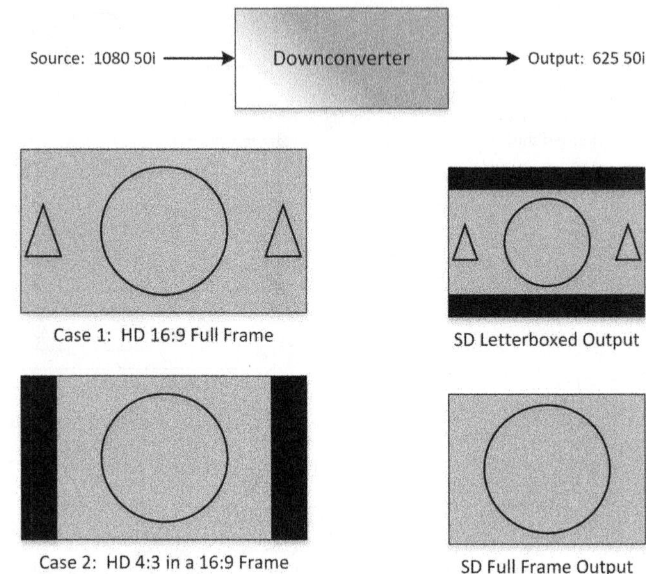

FIGURE 5.13-22 Downconversion ARC illustration—desired operation

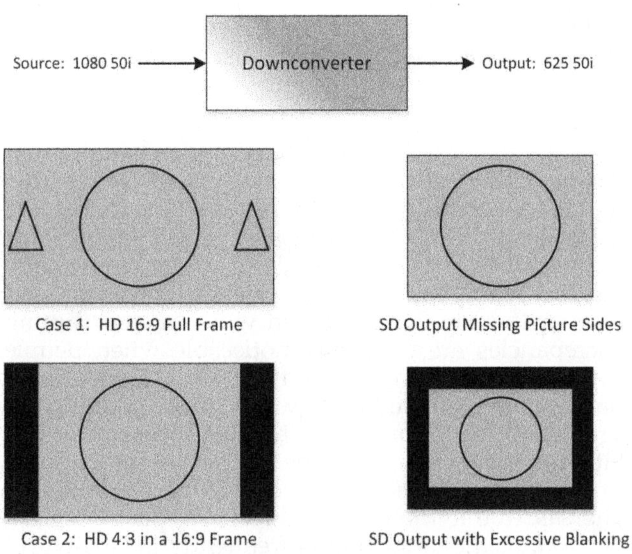

FIGURE 5.13-23 Downconversion—undesirable aspect ratio conversions

AFD. The converter also adds the correct AFD to the outgoing picture, so that downstream equipment and consumer displays know how to present the image. For example, in Case 1 of Figure 5.13-22, the image is 4:3 AFD10 (16:9 vertically centered letterbox in a 4:3 frame), and for Case 2 it is 4:3 AFD8 (full frame). In all, 16 AFD codes are described for each of 16:9 and 4:3 framing in SMPTE ST 2016–1:2009 [10].

Metadata Management—Timecode

Timecode is defined by SMPTE [11] as a mechanism for indicating the location in time that media items such as video and audio can be found. SMPTE timecodes contain binary-coded decimal hour:minute:second:frame identification, for example, 01:00:00:00 might indicate a program starting at exactly one hour. In some cases, the actual studio "wall clock" time may be used for timecode, but equally timecode may start from an arbitrary time.

Where integer frame rates are in use, such as 50 Hz, timecode time will align with actual time; for example, timecode 01:10:03:00 means that 10 minutes and 3 seconds have elapsed since the start of program (assuming 01:00:00:00 start of program). When non-integer frame rates are used, such as 59.94 Hz (actually 60/1.001), there will not be a whole number of frames in a second, so timecode counts 60 frames per second. This means that an "hour of timecode" at a nominal frame rate of 59.94 Hz is actually longer than an hour of wall clock time by 3.59 seconds, leading to an error of almost a minute and a half over a day. To deal with this, drop-frame timecode is defined, in which timecode counting skips 2 frame counts every minute, except every tenth minute.

When frame rate converting, it is often required that the output signal has a correct timecode, which should be generated with respect to the output frame rate, and taking into account whether the user wants drop-frame or non-drop-frame timecode in the case of 59.94 Hz output. Generally, users will want to specify the output timecode start time, as conventions for start of program vary; for example, 10:00:00:00 is commonly used as start of program in the UK. Another useful feature of timecode in format and frame rate converters is its use as a trigger to start conversion. A complete program may include run-up material before the start of program, for example, black bars, clock and so forth. Often such material is referenced by timecode before the start of program—for example, starting at 00:58:00:00. The standards converter can be set to start conversion triggered by receiving timecode 01:00:00:00, thus converting exactly from the start of the actual show, that is, the first frame of the output video sequence is exactly start of program.

Metadata Management—Closed Captions/Subtitles

Many countries mandate the inclusion of closed captioning or subtitles for hearing impaired people. Different standards exist for the presentation of such captions, ranging from simple presentation of text, positioned bottom left of the screen, which replicates dialog, to more complex representations including different colors for different speakers, positioning to indicate which person on screen is talking, and special characters for music and effects. Ideally, closed captions (called subtitles in Europe) should be co-timed with the program audio, and some countries mandate the maximum delay between audio and caption presentation.

In format conversion, where there is no change of frame rate, subtitles and other metadata can be passed from the input to the output without losing any frame association. There may be a need to re-encode the closed caption or subtitle metadata when up- or down-converting in situations where the caption encapsulation standard differs between SD and HD services. For example, if the SD service uses CTA-608 captions, the caption information will need to be encoded into the compatibility space of CTA-708 captions when upconverting to HD.

Closed captions and subtitles for hearing impaired people can be passed from input to output when frame rate converting, but the resulting data will have errors when decoded. This is because the data capacity of the vertical ancillary data space, in which such caption or subtitle data are carried, is dependent on the frame rate. A higher capacity measured in bits/second is available in 59.94 Hz services compared to 50 Hz services. So, for example, when frame rate converting from 1080 59i to 1080 50i, caption packets would have to be dropped, and when converting from 50i to 59i, packets would have to be duplicated. This would inevitably lead to incorrect decoding of the caption packets, leading to erroneous text presented on-screen. The only solution in the live chain is to de-embed the caption data before frame rate conversion, decode, buffer and re-encode the data to match the new required frame rate, then reinsert the data after conversion. In file-based conversions, this problem can be circumvented if caption data are conveyed as separate files that are reintegrated with the main audio and video program after standards conversion.

Color Space Conversion

When converting between SD and HD standards, it is important to note that different color spaces are used, and a format or standards converter must take this into account when creating the output picture. SD production typically used BT.601 color space [12], whereas HD production uses BT.709 [13] and UHD production uses BT.2020 [14]. The full explanation of color space appears in Chapter 5.1. For standards and format conversion, we note that the appropriate mappings will need to be made between the different color standards if there is a format conversion between SD, HD and UHD.

OTHER APPLICATIONS OF FRAME RATE CONVERSION

Program Length Changing

When needing to adjust the duration of a program or clip to fit a transmission slot, or to enable insertion of additional commercials, two methods were historically used. The first was simply to play back a videotape at a higher or lower frame rate than it had been recorded. This might be possible for very small program length changes, but is inappropriate for large changes, as viewers would notice the unrealistic motion and audio pitch changes.

The second method, which is also generally unacceptable, is to remove (or repeat) program content. Such editing runs counter to the original production intent and is very time-consuming. In an extension to the edit method, more recent manual or semi-automated methods look for suitable frames to drop (in the case of program length reduction) or repeat (in the case of program length extension), in the expectation that there would be enough blank or stationary frames to make adjustments without their effect being visible. This method fails for all but the smallest amount of program length change, and when used for large program length adjustments (e.g., 2 minutes reduction in a 28-minute program), it represents a destructive video process, whilst also requiring complex audio editing.

To enable nondestructive program length adjustment, it is possible to use frame interpolation methods to recreate a sequence at effectively a new frame rate, which, when played back at the target broadcast frame rate, would occupy a longer or shorter period of time.

Extending the playout duration of a clip effectively means stretching the original "story" to occupy a longer period of time. For example, if we have a 30-second clip, produced at 29.97 frames per second (fps), we have 899 source frames. To stretch the clip to 33 seconds means we need to generate 908 output frames (occupying 33 seconds), which will be played back at the original frame rate of 29.97 fps. Conceptually, the start and end frames are the same in the source and destination—it is just that more frames need to be generated to occupy the desired playout length.

As an example, consider 10 input frames. To make the 10% program length extension, we need to generate 11 output frames, which will still be played back at 29.97 fps, so will therefore extend the playback by 10%. This is illustrated in Figure 5.13-24.

Similarly, to shrink a clip duration, we need to tell the original "story" in a shorter period of time. For example, the complete contents of our 30-second clip, consisting of 899 frames produced at 29.97 fps, needs to be repackaged into 809 frames. When we play those 809 frames back at 29.97 fps, they will occupy 27 seconds, that is, a 10% reduction in clip length. Note that we have not deleted any of the content in the clip—we have just adjusted the number of frames so that the whole story can be played out in the desired duration.

As an example, consider 10 input frames. To make the 10% program length reduction, we need to generate 9 output frames, which will still be played back at 29.97 fps, so will therefore reduce the playback by 10%. This is illustrated in Figure 5.13-25.

As discussed above, motion-compensated processing can be used very effectively to carry out very high-quality frame interpolation, thereby enabling program length increase or decrease. It should be noted that audio processing as a separate process may be needed to match the video processing—for example, pitch adjustment.

Synthetic Slow Motion

Synthetic slow motion systems are an attractive proposition for broadcasters wishing to offer slow motion effects in a variety of programs without the expense of dedicated high-speed cameras and specialized slow motion equipment. Synthetic slow motion effects can be applied to content captured from ordinary production cameras, which reduces the need for special on-site equipment, or from archived content that was produced at standard frame rates, allowing slow motion to be applied to any type of content. An application at a sports venue where synthetic slow motion is incorporated into a normal motion production is shown in Figure 5.13-26.

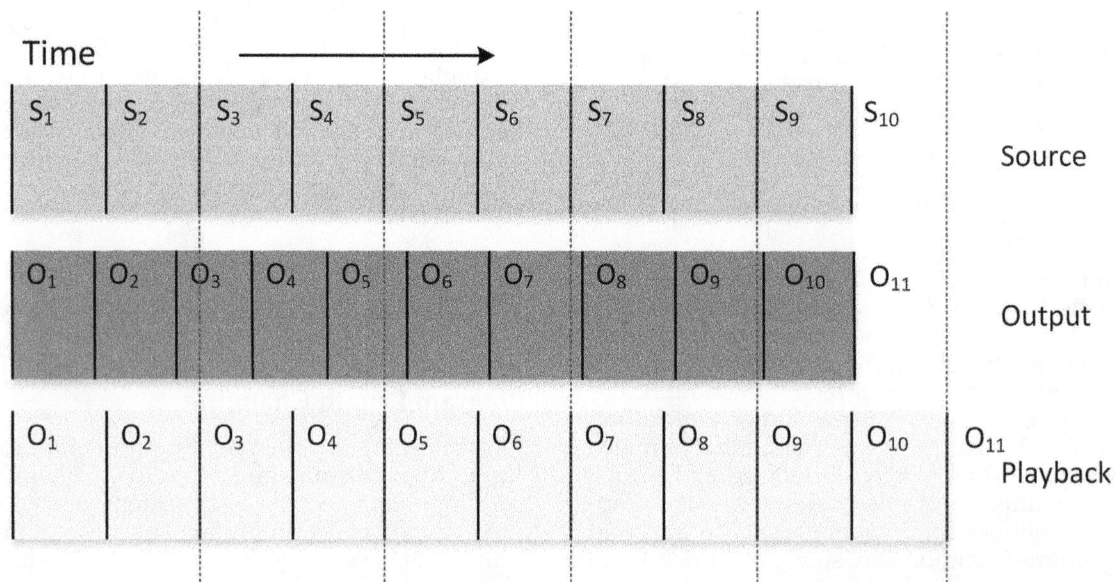

FIGURE 5.13-24 10% program length extension example

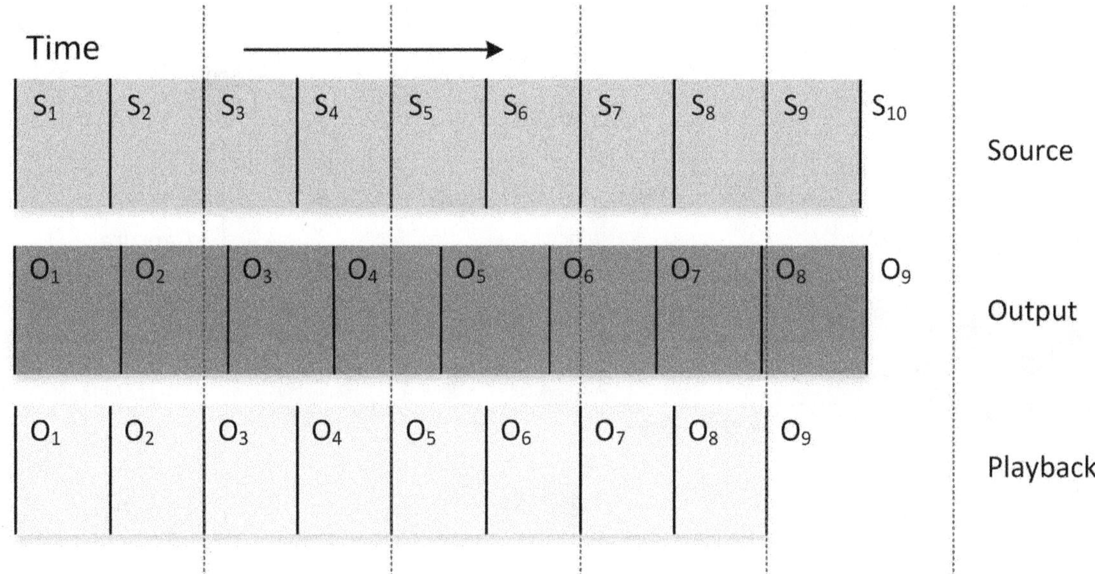

FIGURE 5.13-25 10% program length reduction example

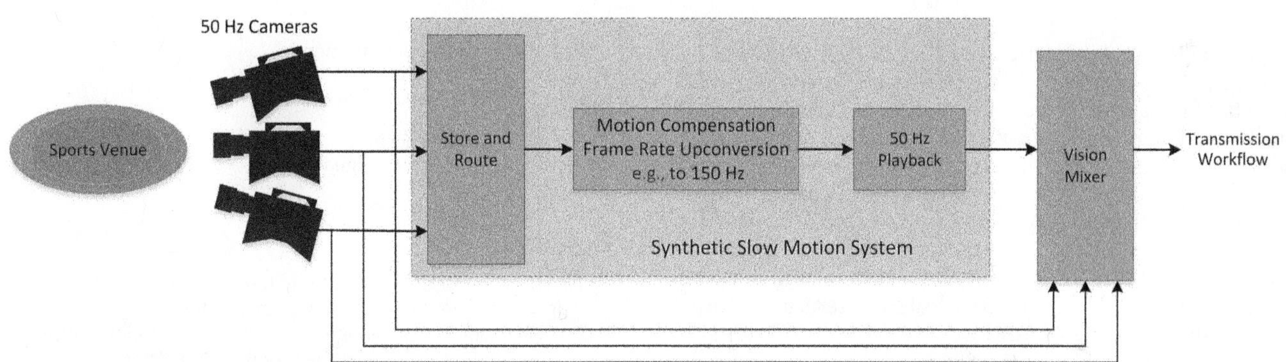

FIGURE 5.13-26 Use of synthetic slow motion in a sports application

A synthetic slow motion system is one in which material shot at standard broadcast frame rates—for example, 50 Hz or 59.94 Hz—is used to create slow motion effects by interpolation of frames representing the slower (playback) rate. For example, if the director wishes to play back at one-third speed some action that has been captured at 50 Hz, the synthetic slow motion system will frame rate upconvert to a 150 Hz frame rate, and then the file system will play back the sequence at 50 Hz. This is illustrated in Figure 5.13-27, where source frames are captured at 50 Hz (S_i) then are frame rate upconverted to three times the rate (U_i), that is, 150 Hz. If the frame rate upconverted sequence is now played out at 50 Hz, the effect will be motion at one-third speed.

Synthetic slow motion is an extremely demanding application of standards conversion, as any artifacts introduced by the processing will be very visible to the viewer, since the eye can track objects more easily when they are moving slowly. In particular, slow motion effects are often applied to sports content, to enable viewers to see more clearly action that was in real time too fast for the human brain to interpret, for example, a ski jump or a tennis shot. Content with fast movement is already challenging for motion analysis, and to increase the accuracy of synthetic slow motion results, it is recommended to carry out analysis in multiple orientations to generate a set of results that would then be post-processed to obtain a consistent decision.

FUTURE STANDARDS CONVERSION

Over the 90 years since the BBC made their first TV broadcasts, the range of standards has continued to increase. The number of possible conversions needed in a hardware standards converter already exceeds 1,000. If the range of formats and frame rates now needed for mobile and Internet distribution are included, the requirement for conversion escalates. New production formats at higher frame rates, high dynamic range (HDR) and wide color gamut (WCG) also add to the number of different conversions needed.

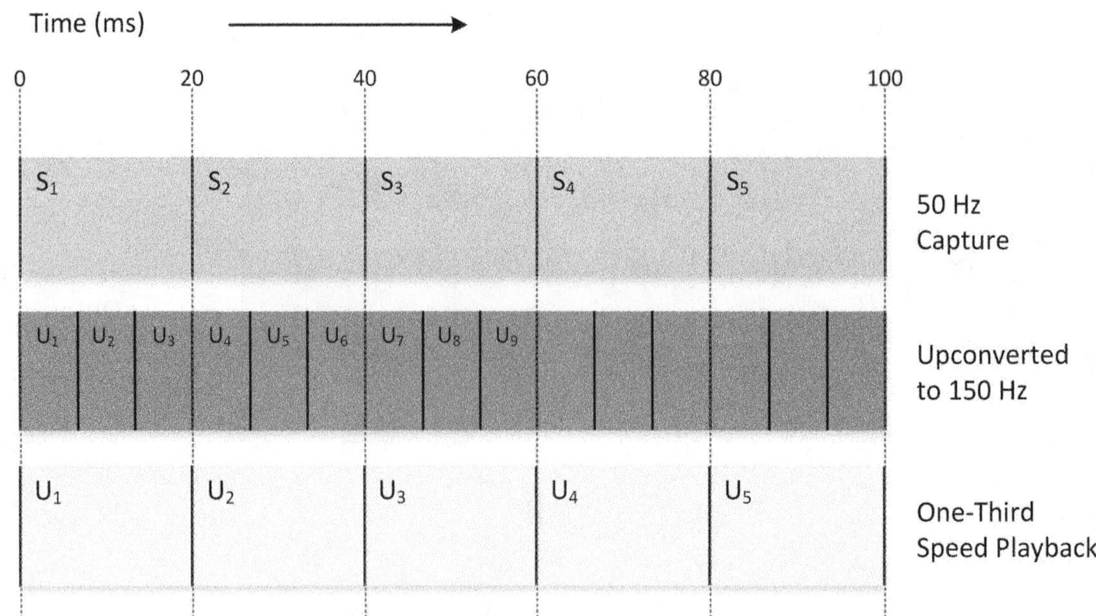

FIGURE 5.13-27 Illustration of synthetic slow motion using 50 Hz source

However, an opposing trend of IP-distributed content targeting software players decreases the demand for standards conversion. Software players can generally accommodate a range of different frame rates, when combined with multi-rate displays, and it would be expected that such players would, in future, incorporate additional processing to overcome motion artifacts that are present where there is a mismatch between playback rate and display refresh rate. Thus, content owners would be able to distribute content at a single frame rate for domestic and international audiences.

It seems likely that standards and format conversion will remain important for some years to come, especially as existing SD and HD content now needs to be reformatted for UHD services. However, future converters are likely to take advantage of new processing technology and advanced architectures, such that more flexible software-based converters are likely to slowly replace fixed hardware solutions.

Glossary

AAC	Advanced Audio Coding
AFD	Active Format Description
ARC	Aspect ratio conversion
DAR	Display aspect ratio
fps	Frames per second
HANC	Horizontal ancillary data
HBI	Horizontal blanking interval
HD	High definition
HDR	High dynamic range
i	Interlaced
OB	Outside broadcast
p	Progressive
PAR	Pixel aspect ratio
PCM	Pulse-code modulation
psf	Progressive segmented frame
SAR	Storage aspect ratio
SD	Standard definition
SDI	Serial digital interface
UHD	Ultra-high definition
VANC	Vertical ancillary data
VBI	Vertical blanking interval
VI	Video Index
VM	Virtual machine
WSS	Wide Screen Signaling

References

[1] SMPTE ST 2036–1:2009 "Ultra High Definition Television—Image Parameter Values for Program Production."

[2] Watkinson, J., "The Engineer's Guide to Motion Compensation," Snell and Wilcox Handbook Series, 1994, Published by Snell & Wilcox Ltd.

[3] SMPTE ST 2036–3:2010 "Ultra High Definition Television—Mapping Into Single-Link or Multi-Link 10 Gb/s Serial Signal/Data Interface."

[4] SMPTE Draft ST 2062:2012 "25 Gb/s Serial Signal/Data Interface."

[5] SMPTE ST 2022–6:2012 "Transport of High Bit Rate Media Signals Over IP Networks."

[6] Video Services Forum (VSF) Technical Recommendation TR-03 "Transport of Uncompressed Elementary Stream Media Over IP," November 2015 (being standardized as SMPTE ST 2110).

[7] Video Services Forum (VSF) Technical Recommendation TR-04 "Utilization of ST-2022-6 Media Flows Within a VSF TR-03 Environment," November 2015.

[8] ETSI ETS 300 294 "Television Systems; 625-Line Television Wide Screen Signalling (WSS)," 2003.

[9] SMPTE RP 186–2008 "Video Index Information Coding for 525- and 625-Line Television Systems," 2008.

[10] SMPTE ST 2016–1:2009 "Format for Active Format Description and Bar Data."

[11] SMPTE ST 12–1:2008 "Television—Time and Control Code."

[12] ITU Recommendation BT.601 "Studio Encoding Parameters of Digital Television for Standard 4:3 and Wide Screen 16:9 Aspect Ratios," 2008.

[13] ITU Recommendation BT.709 "Parameter Values for the HDTV Standards for Production and International Programme Exchange," 2015.

[14] ITU Recommendation BT.2020 "Parameter Values for Ultra-High Definition Television Systems for Production and International Programme Exchange," 2015.

CHAPTER

5.14

Digital Television Audio Loudness Management

JIM STARZYNSKI
NBCUniversal
New York, NY

BACKGROUND AND HISTORY

Along with high-definition (HD) pictures, digital television (DTV) introduced cinema-quality surround sound that had a tremendous increase in audio dynamic range compared to its analog predecessor. With this increase in dynamic range came the unintentional consequence of severe loudness variation when loudness was not managed correctly. Initially, the home listening DTV audience often felt the negative impact of the new audio system in the form of extreme loudness variation between commercials and programs on a given channel and loudness variation when changing channels across the dial. Viewers were not accustomed to the inconvenience of constantly adjusting the volume with their remote controls, but were now burdened with having to do so to enjoy digital HDTV.

At the onset of the analog-to-digital transition and at the crux of the problem, TV station engineers, whose priority and typical focus is on keeping signals on the air, were now, among many other pressing demands, presented with DTV's entirely new Dolby AC-3 audio system. The new Advanced Television System Committee (ATSC) DTV standard used Dolby AC-3 digital audio encoding to reduce the bit rate needed to deliver up to 5.1 channels of high-bandwidth, high dynamic range surround sound. This new AC-3 digital encoding system was a completely new and relatively complicated technology, and it shared little similarity with loudness-control techniques previously used with analog NTSC transmission. Station engineers still needed to become familiar with requirements for correct operation, including spending the time to master the system's enhanced capabilities and establishing matching program and advertising content delivery specifications.

In the meantime, suppliers were continuing to mix their digitally recorded soundtracks to specifications that were intended for use with analog transmission systems that would constrain range by design, with no regard for DTV's extreme range capabilities. No new rules were in place for correctly delivering DTV mixes, and no guidelines existed to help understand how to effectively use DTV's expanded range for creative advantage. In this environment, uncontrolled soundtracks compounded by incorrectly set station encoding often resulted in loudness swings of 10 dB or more.

Meanwhile, in the mid-2000s, affordable rear-projection and flat-panel televisions had entered the market. As a result the DTV audience was growing significantly, and so was viewer annoyance with poor DTV loudness control. As the situation worsened and viewer complaints grew, television engineers became as frustrated as the audience.

By 2006, the ATSC had recognized the problem, and in 2007 it took action by creating an audio group composed of industry experts to identify technical issues and to write guidelines for solving the problem. Shortly thereafter (and unaware of the ATSC's work in progress), the U.S. Congress responded to the American public's dissatisfaction and initiated HR 1084, *The Commercial Advertisement Loudness Mitigation Act* ("CALM Act"). The original bill required the audio of TV advertisements that accompanied programming on U.S. television stations to not be "noisy or strident" and "not be presented at modulation levels higher than the program material they accompany." In 2010, a revised version of the CALM Act became law, mandating the use of newly released ATSC's loudness practices by the U.S. television broadcasting industry.

This chapter on digital television loudness management provides background and explains proven, effective and *required* practices for mastering loudness across the television audio ecosystem, with particular focus on television engineering and operations.

The Analog-to-Digital Transition: A New DTV Audio System and the Makings of a Growing Problem

At the commencement of the analog-to-digital television transition, an additional 6 MHz television channel was assigned to each broadcaster to facilitate the transition from the legacy NTSC analog service to a new, forthcoming and mandated digital television service using the ATSC standard. The latter would eventually remain as the broadcaster's only transmitted channel once the date for the analog shutoff was reached. This situation posed the interim operation of an additional digital broadcast plant, with content that, for the most part, mirrored the output of the analog plant (i.e., it was a simulcast transmission). To accomplish this, standard-definition (SD) digital video signals were upconverted in many cases to high definition, and encoded for emission (using MPEG-2 video compression) using the ATSC standard on the station's DTV channel.

Analog audio signals were digitized and encoded to the ATSC's standard using the Dolby AC-3 system (branded as *Dolby Digital* in the consumer electronics world), requiring a unique set of essential parameters in the encoder to describe the audio in a completely new DTV audio metadata authoring process. One such parameter was *channel mode*. At the commencement of DTV, most if not all audio soundtracks were encoded in 2/0 (stereo), given that the source of the audio was content from the analog, dual-channel mono or stereo-only NTSC channel used by North American (and other) broadcasters. This is where DTV audio's similarity with analog ends.

The Introductiovn of Digital Audio Encoding

At the time, DTV audio encoding was performed by an onboard audio encoder that resided alongside the video section of the ATSC broadcast encoder, or by a stand-alone, stereo Dolby DP567 digital audio encoder. The output stream of the device was connected to the ATSC encoder in audio pass-through mode, enabling the 567's AC-3 audio stream to remain intact during emission encoding. The connection was straightforward, but the setup for both types of encoders could be puzzling due to new metadata authoring requirements, which bore little or no resemblance to the analog TV audio emission setup. The DP567 stereo encoder was quickly supplanted by the DP569, a 5.1 audio channel capable unit, which was for many years thereafter the only device available to encode digital, discrete surround sound for DTV.

As seen in Figure 5.14-1, the DP569 bears little physical resemblance to earlier analog audio equipment, and its setup was equally new to the user. As part of the metadata authoring process, numerous parameters required examination of their default settings by the broadcaster. Some settings were for informational purposes, but many directly controlled the audio being encoded.

As listed in Table 5.14-1, "dialog level" (which would come to be known as "dialnorm" or "dialog normalization") tops the list of metadata control parameters. This item would become the focal point of loudness control for digital television for content creation, broadcast emission and, most importantly, the listener experience, and it remains so today. That being just one of twenty-seven new parameters, it's easy to understand how the broadcast operator could become consumed with many new settings yet still need to investigate the overall workings of DTV loudness control using this very important dialnorm parameter. This all took place in the context of those engineers having to focus on the revenue-producing analog NTSC channel while establishing a new ATSC DTV channel, as well.

In NTSC analog television, audio loudness was controlled to ensure that 100% modulation of the frequency-modulated FM audio carrier was not exceeded. Analog TV broadcasting is capable of ~50 dB of range, with dialog typically 17 dB below 100% modulation.[1] ATSC AC-3 audio is capable of >100 dB of dynamic range; perceptually, this equates to 32 times the dynamic range of analog broadcast audio. See Figure 5.14-2.

NTSC audio was frequently processed with dynamic range control ahead of the studio-transmitter-link or at the transmitter. This was often performed by a stand-alone device, or with an optional card in or alongside the stereo generator at the transmitter. This device was relied upon for protection against overmodulation and, as a by-product, it was effective against loudness shifts while evolving to create pleasant although *altered* sound for the listener. As a result, this process changes the sound from what the content creator intended, but it was necessary to fit the audio within the modulation constraints of the analog transmission. See Figure 5.14-3.

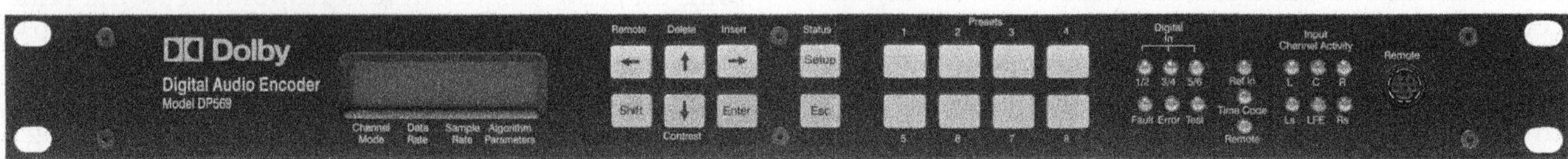

FIGURE 5.14-1 The Dolby DP 569 Digital Audio Encoder, the first device capable of encoding 5.1 surround sound for ATSC Digital Television.

TABLE 5.14-1
The Metadata Parameter table from the Dolby DP569 Digital Audio Encoder Operator's Manual. (Extended Bitstream Information parameters are in italics.)

Metadata Parameter	Informational	Control
Dialogue Level		×
Channel Mode		×
LFE Channel		×
Bitstream Mode	×	
Line Mode Compression		×
RF Mode Compression		×
RF Overmodulation Protection		×
Center Downmix Level		×
Surround Downmix Level		×
Dolby Surround Mode		×
Audio Production Information	×	
Mix Level	×	
Room Type	×	
Copyright Bit	×	
Original Bitstream	×	
Preferred Stereo Downmix		×
Lt/Rt Center Downmix Level		×
Lt/Rt Surround Downmix Level		×
Lo/Ro Center Downmix Level		×
Lo/Ro Surround Downmix Level		×
Dolby Surround EX Mode		×
A/D Converter Type	×	
DC Filter		×
Lowpass Filter		×
LFE Lowpass Filter		×
Surround 3 dB Attenuation		×
Surround Phase Shift		×

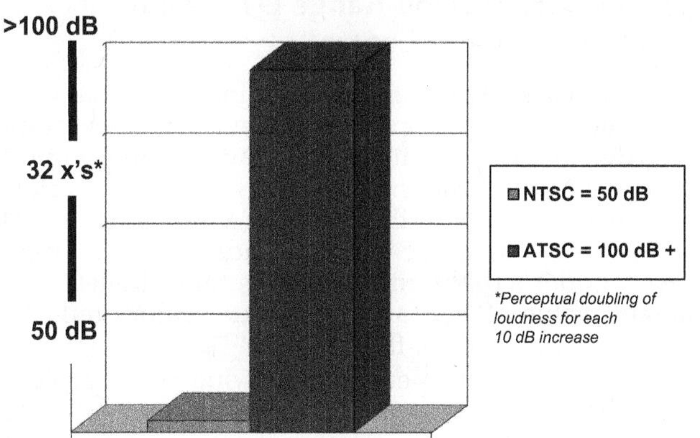

FIGURE 5.14-2 Analog versus digital TV audio dynamic range.[1,2]

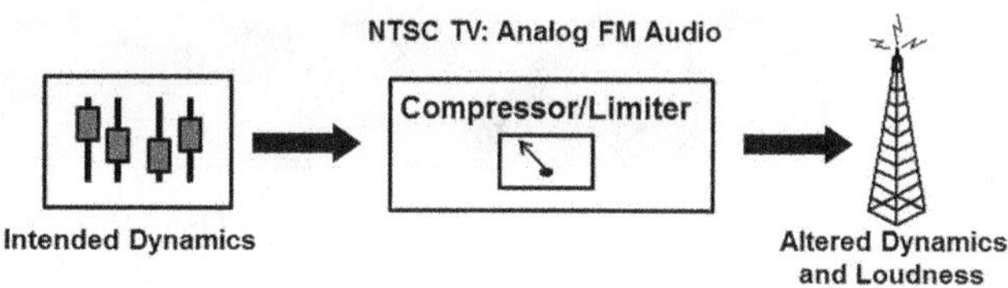

FIGURE 5.14-3 Typical NTSC audio loudness control.

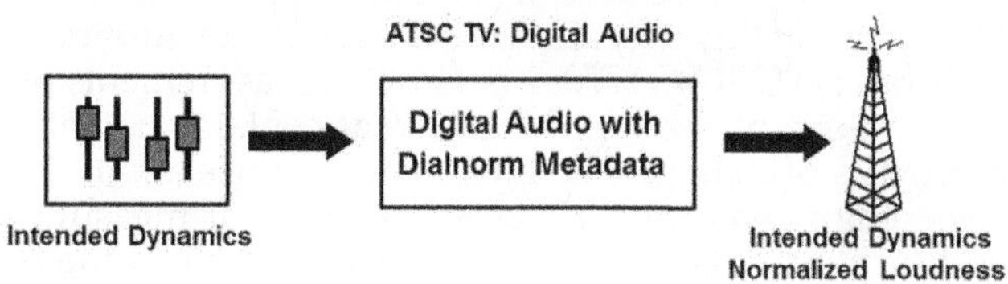

FIGURE 5.14-4 ATSC audio loudness control, in which intended audio dynamics are preserved in the broadcast via dialog normalization.

As mentioned earlier, the AC-3 audio system defined in the ATSC digital television standard uses metadata to control loudness and other audio parameters more effectively without permanently altering the dynamic range of the content.[3] The *dialnorm* parameter is used to normalize the loudness of the content. As required in the standard, for correct operation of the loudness normalization system, the value of the dialnorm parameter must match the measured loudness of the content. This has come to be called "the Golden Rule."[4] See Figure 5.14-4.

Analog TV audio measurement techniques did not focus on the actual *loudness* of the content. Familiar volume unit (VU) and peak program meters (PPM) measured electrical (not perceived loudness) levels; they were used to protect equipment against overload and to establish a nominal operating point above the noise floor while maintaining adequate headroom below clipping. In contrast, two key elements for audio measurement in the digital TV system are *true peak* and *loudness* levels. True peak measure helps the operator protect against clipping, and a useable loudness measurement is needed to avoid unwanted wide variations in what the listener hears.[5] The difference between the familiar, available measurement techniques used for analog TV and the new ones required for DTV added yet another challenge for the operator.

Introduction of the Broadcast Loudness Meter

This DTV loudness paradigm included the development of a new digital television broadcast loudness meter, whose first of many incarnations would be the Dolby LM100 (see Figure 5.14-5). At its introduction, the device read in units of *Leq(A)*, an available legacy technique effective for measurement of motion picture soundtracks. The meter lacked true peak measurement capability at the time. Leq(A) would soon be supplanted by ITU-R BS.1770: "Algorithms to Measure Audio Program Loudness and True Peak Audio Level."

FIGURE 5.14-5 The Dolby LM100 Broadcast Loudness Meter.

This seminal document is described in Chapter 5.18 and will be further discussed later in this chapter.

The meter introduced a selectable algorithm called Dialogue Intelligence, allowing the LM100 to automatically base ITU-R BS.1770–1 or Leq(A) measurements on the portions of the input signal that contain the characteristics of dialog.[6] The LM100 did not use a familiar sweeping needle or LED cluster display. It measured and displayed its readout as a single numeric value of the audio loudness, measured over a period of time. The operator could choose either "short-term," the report of an updated average of the content loudness for the previous 10-second period, effective for measurement for live mixing in real time, or "infinite," an ongoing average that could be controlled by pause/reset selections on the front panel, and effective for measuring recorded content. In addition, the LM100 was capable of reading the dialnorm parameter from an AC-3 bitstream, and it displayed that value next to the measured audio loudness reading for easy comparison. The LM100 was limited to reporting loudness of 2 channels of PCM sources and up to 5.1 channels of Dolby AC-3. It could also measure up to 8 channels of Dolby E, which was at the time a new, mezzanine-level[7] audio codec[8] intended for professional postproduction and distribution use.

Equipped with a broadcast loudness meter, the operator could now measure the standardized loudness of content at production, postproduction and ingest, as well as the loudness of the digital television station output and its dialnorm value at transmission.

Content Loudness and Dialnorm

Equipped with loudness metering but with no new content delivery recommendations for audio loudness in place, operators quickly realized that programs and commercials were being recorded and delivered at widely variable loudness levels. To fulfill DTV's promise of theater-like sound, no compressor-limiter-type control devices were in the path to manage useable range and normalize the loudness at the station output. Instead, it was intended that the operator rely on dialnorm to normalize the loudness. To do so as designed, dialnorm metadata, reflecting the measured loudness of the audio, needed to accompany each piece of content, travelling from content creation, through distribution and to emission encoding. If present, a single dialnorm value could dynamically change within the AC-3 stream to match the measured average loudness of complete shows or complete commercials it accompanied. Decoders would apply the specific dialnorm value at content boundaries, attenuating the audio at the output of the receiver by an offset value needed to achieve the system normalization reference value of –31.

For example, content encoded with a –20 dialnorm value would be attenuated 11 dB, while content encoded with a –31 value would not be attenuated. If dialnorm always matched actual content loudness (the Golden Rule), the audience would be presented with smooth loudness transitions at content boundaries and channel changes, all normalized to –31 in a process that was seamless, while maintaining the content creator's artistic intent.

Encoding Dialnorm Metadata with Content

The means to include dialnorm metadata with the audio essence at this time was limited to encoding the audio at postproduction with Dolby E—or in a subsequently developed process, as vertical ancillary (VANC) data—inside the serial digital interface (SDI) video and audio signal.[9]

The Dolby E codec was developed to expand the number of audio channels available at production from the limits of four-channel VTRs and servers to eight channels, therefore facilitating DTV's new 5.1 surround format and more.[10] The codec included metadata authoring compatible with AC-3. This data and audio was efficiently designed to be distributed over a single AES3 pair,[11] either alone or embedded in SDI. Unfortunately, this would add another layer to DTV audio challenges by adding two frames of delay and requiring special monitoring gear. Though very effective when operated in a managed system, and with some success in postproduction situations, terrestrial broadcasters shied away from Dolby E's universal usage. VANC metadata insertion practices came later on, but existed under two different standards. They were rejected by the Hollywood postproduction audio community due to the risk posed to a show master when assembling the data into the video portion of a tape, an unfamiliar technique for audio operators.

The Results of Wide-Range DTV Audio without Loudness Control

In many cases, with no valid loudness metadata and no analog loudness control type device in place, audio was delivered to the audience unaltered and without loudness normalization, exactly as supplied by various content creators, with the ability to access the full >100 dB of DTV's dynamic range. In addition, the broadcast encoder's dialnorm value was typically left at its default value of –27, and this value seldom reflected the actual loudness of the content.

As evidence of operators' growing dissatisfaction with the performance of the system, a number of stations changed their static dialnorm value to –31, essentially "turning off" the DTV loudness normalization system. Exacerbating the problem, this now created excessive loudness variations on channel changes by users, in addition to shifts at content boundaries within a channel (e.g., programs vs. commercials). In addition, stations attempting to apply the Golden Rule were now penalized in some markets, because a station adhering to the ATSC standard and the Golden Rule had its audio attenuated to AC-3's −31 loudness reference, while stations turning off the system and broadcasting audio at a typical −20 dBFS were transmitting audio 11 dB louder than compliant stations.

As affordable flat-panel displays hit the market in the mid-2000s and the DTV audience grew, so did annoyance with television audio. The ineffectiveness of the system irked the listener with its loudness variations often exceeding 10 dB, and it plagued the broadcaster in the form of growing user complaints and impending government regulation.

The need for a reevaluation of the system or a mutually agreed-upon industry recommended practice was never more apparent.

ATSC Takes Action

With growing dissatisfaction over the use and performance of the DTV audio loudness system, television station engineers referred to the ATSC standard for guidance on exactly what was required to manage audio loudness.

What they found in the original ATSC A/53 Audio Standard was the following:

> *Dialogue Level*
>
> The value of the dialnorm parameter in the AC-3 elementary bit stream shall indicate the level of average spoken dialogue within the encoded audio program. Dialogue level may be measured by means of an "A" weighted integrated measurement (LAeq). (Receivers use the value of dialnorm to adjust the reproduced audio level so as to normalize the dialogue level.)[12]

TV engineers questioned the wording of the standard. Why was average spoken dialog used as the reference? What if content had no dialog? Was Leq(A) (corrected abbreviation for "LAeq" that appeared in the standard) adequate for the measurement of TV sound?

Fortunately, work was underway at this time in the ITU-R (International Telecommunications Union—Radiocommunication Sector) on development of a contemporary loudness measurement recommendation. Aside from that, however, there was no other industry-wide recognized and suitable information, and no other standards or official recommended or best practices that could be used at the time. The section on dialog level of the ATSC standard seemed ambiguous and unfamiliar. In the meantime, consumers did not expect and were annoyed by large changes in audio loudness from program to interstitials, and from channel to channel. Clearly, a proper solution was required.[13]

Formation of ATSC S6–3 Audio Subgroup

In December 2006, the ATSC Board of Directors recognized the loudness problem and assigned a work item to the responsible Technology Group. In April of 2007, ATSC S6–3 formed an ad hoc audio subgroup to develop a scope of work to "investigate issues relating to variations in reproduced audio levels within the ATSC Digital Television System," with an objective to have their work "ultimately result in the drafting of a Recommended Practice on Audio Loudness."[14]

The proper expertise and industry representation was needed in order for the group's work to be effective and for its efforts to achieve consensus and approval by the ATSC membership and by the industry at large. With this in mind, a group of engaged and assertive audio experts from commercial TV and cable networks, public television, professional audio equipment manufacturers, academia and other standards organizations was assembled. The group's expertise included broadcast system and laboratory engineering, audio production and postproduction mixing, system and circuit design, motion picture sound, and cable, satellite and telco television distribution, along with input from other standard development and industry trade organizations. With these key stakeholders in place, the group took on the task of solving the DTV audio loudness problem.

The first face-to-face meeting of the group was a "Digital Television Audio Loudness Summit" held in July 2007 at the University of Southern California (USC), attended by over 60 participants. Its intent was to create a heightened awareness of the issues involved. The outcome of the meeting concluded that DTV loudness was indeed a serious problem shared by many parts of the industry along with the TV audience.

ATSC S6–3 then determined that a thorough and lengthy industry-wide effort would be required to create effective guidelines that would be both sanctioned by the ATSC and effectively practiced by the entire audio ecosystem, from content creation to emission. They immediately began their work.

An Initial Step: Using ITU-R BS.1770

In order to create effective loudness guidelines, a uniform measurement method was deemed essential, and considered to be the cornerstone of any loudness normalization solution. Given that ITU-R BS.1770 had been approved and published in July 2006 (following years of development), the group soon agreed to base its solution on such a worldwide, common reference for loudness measurement. This recommendation introduced a value identified in units of *LKFS* (**L**evel integrated over the time of the segment, using **K** frequency weighting—a new equalization curve developed for loudness measurement, reflecting human perception—and relative to **F**ull **S**cale digital level). A detailed description of ITU-R BS.1770 is presented in Chapter 5.18.

An experiment at USC was conducted in the fall of 2007 by using typical broadcast content and having a group of six award-winning mix engineers "normalize" the clips by ear in a calibrated listening environment. Fader information was collected from these adjustments and compared to the BS.1770 measured values for the same content. The maximum deviation between objective measurement data (via a BS.1770

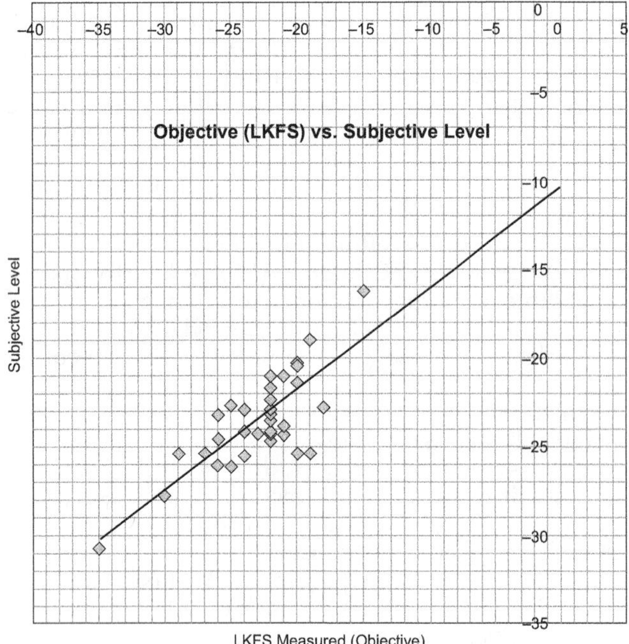

FIGURE 5.14-6 ATSC S6–3 objective versus subjective loudness measurement experiment results.

loudness meter) and subjective data (expert listener reaction) was less than +2 dB.[15] (See Figure 5.14-6.) With these firsthand results, the group was satisfied with the performance of the measurement recommendation and could now move on to consider additional essential elements of study for their main objective—a recommended practice on loudness.

ATSC A/85: Techniques for Establishing and Maintaining Audio Loudness for Digital Television

With group acceptance of the new ITU loudness measurement recommendation secured, S6–3 focused its work on the practices engineers would need to master and implement in order to solve the loudness problem. The initial draft for the group's recommended practice, to be called "ATSC A/85: Techniques for Establishing and Maintaining Audio Loudness for Digital Television," contained sections on loudness measurement, audio monitoring environment, metadata management considerations and dynamic range management. As their work progressed over the period from April 2007 to the initial release of A/85 in November 2009, followed by updates in 2011 and 2013, this list would expand as the group continuously tracked the loudness problem by reacting to ongoing technical activity in the industry and impending government regulation.

A/85 and the CALM Act

In 2008, the Commercial Advertisement Loudness Mitigation (CALM) Act was introduced in Congress as a reaction to consumer annoyance with the loudness of commercial advertising on television. In June of 2009, Congressional hearings were held and Congressional staff attended a technical demonstration on the progress of the ATSC's work, where it was shown that an industry-wide answer was near completion and its release was imminent. In October, a markup of the bill was created that would later become law in December 2010. The signed bill stated:

> Rulemaking Required. Within 1 year after the date of enactment of this Act, the Federal Communications Commission shall prescribe pursuant to the Communications Act of 1934 (47 U.S.C. 151 et seq.) a regulation that is limited to incorporating by reference and making mandatory (subject to any waivers the Commission may grant) the 'Recommended Practice: Techniques for Establishing and Maintaining Audio Loudness for Digital Television' (A/85), and any successor thereto, approved by the Advanced Television Systems Committee, only insofar as such recommended practice concerns the transmission of commercial advertisements by a television broadcast station, cable operator, or other multichannel video programming distributor.[16]

The revised bill reflected a key shift away from subjective terms such as "strident" and "modulation" in favor of contemporary solutions, and this language ultimately became law. ATSC A/85's modern recommendations had now been validated as the solution to the irking problem of loud commercials on TV. These ATSC industry guidelines became government-mandated practices that TV broadcast stations and cable, satellite and telco TV operators must follow.

Six years later, after its effectiveness had been proven, ATSC A/85 was honored with a 2015 Academy of Television Arts and Sciences Engineering Emmy Award for dramatically improving consumer satisfaction.[17] See Figure 5.14-7.

FIGURE 5.14-7 Members of the ATSC Leadership and S6–3, receiving A/85's 2015 Academy of Television Arts and Sciences Engineering Emmy Award for dramatically improving consumer satisfaction, at the Bellagio Hotel, Las Vegas, NV, January 8, 2015. Left to right: Jerry Whitaker, VP, ATSC; Glenn Reitmeier, ATSC Board Chair; J. Patrick Waddell, ATSC S6 Chair; Jim Starzynski, ATSC S6–3 Chair; Mark Richer, ATSC President; Rich Chernock, ATSC TG-3 Chair; Greg Coppa, ATSC S6–3; Craig Todd, ATSC S6–3.

The next sections of this chapter will explain the mastering of A/85's vital technical practices for digital television audio loudness management when using the ATSC AC-3 audio system, including references to current government DTV loudness regulations and enforcement at this writing.

EFFECTIVE LOUDNESS NORMALIZATION USING ATSC A/85

Just like other important technical facets of television, the mastering of DTV loudness requires the harmonization of operational and engineering practices by the operator. This applies to the entire audio ecosystem from content creation, distribution and emission. There are a number of essential elements that must be considered that, when effectively coordinated and practiced, will yield an easily manageable loudness workflow for the professional, resulting in an enjoyable listening experience for the consumer.

These practices are documented and explained in ATSC A/85: Techniques for Establishing and Maintaining Audio Loudness for Digital Television. They are the basis of effective and compliant loudness control, mandated by the federal government and reflected in the FCC rules.

Applying the "Golden Rule"

The Golden Rule of the ATSC recommended practice is that the transmitted dialnorm value must correctly identify the loudness of the content it accompanies in order to prevent excessive loudness variation during content transitions on a channel (e.g., TV program to commercial) or when changing channels.[18] Implementation and consistency with this fundamental is the starting point and key to ensuring success and a good DTV loudness consumer experience.

To effectively apply the "Golden Rule," an operator must consider the following:

1. The importance of dialnorm.
2. The use of agile or fixed metadata during emission.
3. If the operator chooses a fixed metadata system that matches the chosen dialnorm value selected in the emission encoder, a content target loudness value must be universally implemented across the entire operation.
4. A content delivery specification should be written and distributed to all internal and external content suppliers explaining technical audio and loudness requirements and the expected characteristics of delivered content based on these needs. This specification should also describe any audio loudness practices used downstream by the recipient operator during content ingest, distribution and/or emission for reference by the supplier.
5. Good loudness metering practices ensuring the creation of compliant content must be employed at ingest, during production and postproduction, through distribution and emission, and for monitoring of short-form and long-form content. ATSC A/85 defines specific practices for these different operations.
6. Content creation, editing and monitoring facilities should be designed and equipped to adequately support operator loudness judgments in all environments.
7. The operator should analyze all content delivery paths and ensure a suitable loudness control method is employed for each, preventing the emission of non-normalized, noncompliant content to the audience, with a goal of maintaining audio quality.

The next part of this section will explain how these important points contribute to effective loudness management. ATSC A/85 documents the necessary processes. An explanation of each element and how they work together follows.

Dialnorm

During the initial design of the AC-3 system in the late 1980s and early 1990s, those responsible for its specification and features were very familiar with both current analog practices for loudness control and television audio soundtrack creation. As explained earlier, an analog broadcast processor protected against distortion and overmodulation, smoothed overall loudness for the listener and was effective at doing so over a broad range of content loudness at its input. As a consequence, however, the dynamic range of the content as delivered by the supplier was altered and no longer represented the full intent of the content creator. At the time, content came from many suppliers working in different genres at numerous facilities. Although reference setup tones on TV content were delivered at −20 dBFS for the most part, mixes were not delivered under specific dialog loudness standards. Those developing AC-3 recognized this and did not believe that this practice could change.

With this in mind, the AC-3 system was designed to manage loudness and dynamic range in a completely different manner than its analog predecessor. With a goal of carrying the content creator's intent all the way through to the listening audience, a means would need to be developed that could make use of digital's wide-range capabilities but still yield a pleasant listening experience for the audience. Dialnorm was invented to accomplish these goals.

The dialnorm concept:

- A mix engineer should have the latitude to choose how loud the dialog ("anchor") element is in a mix, placing it appropriately in the overall available dynamic range of the soundtrack, with genre-based considerations for soft passages, and allowance of headroom for loud effects and music.
- The *anchor element* is the perceptual loudness reference point or element around which other elements are balanced in producing the final mix

of the content, or that a reasonable viewer would focus on when setting the volume control.[19]

- Dialog forms the anchor element in the majority of television programs.[20] This is exemplified by a listener's choice to set his or her listening level to the storyline of the content. (A/85 describes proxies to be used as anchor elements if no dialog is present in the content.)
- With one common point identified as a "reference" or "anchor" section in any soundtrack, this point is loudness-measured and labeled with dialnorm metadata.
- Then, at content boundaries, *all elements* of every soundtrack are raised or lowered in level *by the system* once, so the anchors match each other in loudness during decoding at the listener's receiver.

Using this audio loudness management process makes the loudness of each soundtrack anchor perceptually the same, thereby normalizing loudness *and* meeting the goal of maintaining full creative dynamic range. This is essentially identical to a listener physically moving a volume control up or down at program and commercial boundaries to maintain a comfortable loudness, except this action is now happening automatically by using a continuous flow of changing dialnorm metadata in the bitstream to control the loudness of the receiver. Because *all elements* of the mix get louder and softer by the same increment as the anchor is adjusted up or down, dynamic range is unaltered.

Metadata System Options: Agile, Preset or Fixed

As an essential first step in an audio loudness management plan, the digital television operator must determine a metadata operating mode for the audio encoder. A/85 lists three possibilities: *agile metadata, preset metadata* and *fixed metadata*. The dialnorm parameter is the basis of the loudness system and is impacted by the choice of metadata mode selected. Many factors contribute to an operator's decision to choose one over the other. An explanation of how the dialnorm parameter is impacted by this choice follows.

Agile Metadata System

An agile metadata system (see Figure 5.14-8) is required to achieve the full capabilities of dialnorm. Because metadata must be present with the content and maintained at all points in the digital TV ecosystem, the following practices and conditions apply when a broadcaster chooses to pursue end-to-end dialnorm loudness implementation by means of an agile metadata system:

- Content is specified to contain a dialnorm metadata value encoded in the deliverable. The loudness value is left to the discretion of the content creator.
- The content recipient (operator) specifies how the dialnorm metadata will be encoded in the deliverable based on the systems the recipient has in place for ingesting and distribution of the content. A/85 specifies these techniques in its section 7.5.1. Two such methods are standardized by SMPTE.[21] The Dolby E codec is another method of delivery that also uses metadata encoding.
- The entire broadcast plant infrastructure must be compatible with synchronous and uninterrupted passage of the metadata stream.
- Monitoring equipment in the path must react to the dialnorm parameter in order to present an accurate representation of the content loudness that will be presented to the listener.
- The broadcast audio encoder must be capable of receiving the active dialnorm metadata and changing the dialnorm parameter in real time, under metadata stream direction.

Meeting the conditions needed for an agile system presents a difficult challenge to the broadcaster. This is based on the many different parts of the ecosystem that must work in perfect concert to rule out potential risk to content. If the flow of metadata becomes corrupted or interrupted for any reason, the operator should be aware that a protection or reversion mode engages in the broadcast encoder that must be provisioned to use a set of parameters suited to guarantee an acceptable (though not ideal) presentation of any supplied content.

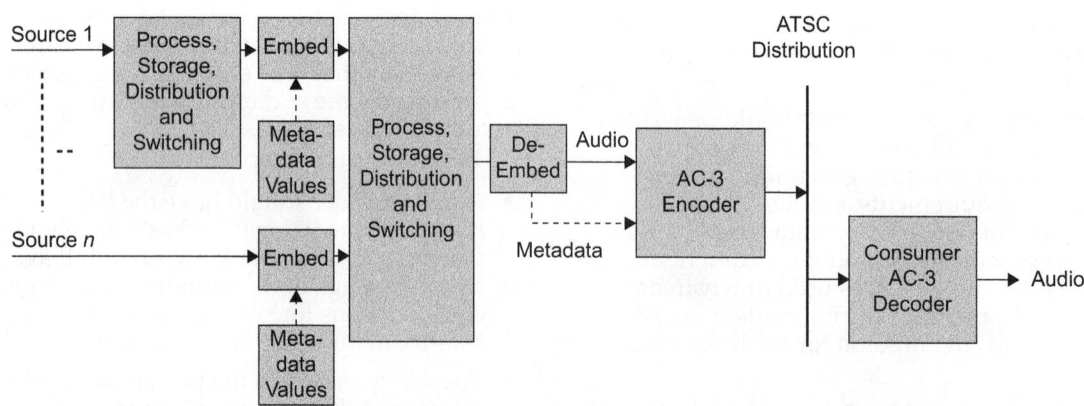

FIGURE 5.14-8 Simple diagram of an agile metadata system from ATSC A/85:2013.

As explained earlier, dialnorm is one of many metadata "control" parameters in the AC-3 system. Under fault conditions, an inaccurate dialnorm value of content loudness will make content soft or loud, but for the most part still understandable. However, if 5.1-channel surround sound content is mislabeled with incorrect channel metadata (also encoded as "control metadata" and carried in the same metadata stream as dialnorm), for example, 5.1 (3/2L mode) content is mistakenly labeled with stereo (2/0 mode) metadata, only the left and right channels of the 5.1 content will be heard. These channels usually contain only music and effects and in most cases do not contain dialog. In this situation without dialog, the listener is presented with an unacceptable condition, under any circumstance.

The conditions above, particularly along with the lack of interest by Hollywood in encoding metadata into a show master and with a resistance to the use of Dolby E by terrestrial broadcasters, make it difficult for a broadcaster to accept the risk associated with the complexity of an agile metadata system regardless of the benefit. Therefore, A/85 offers its other two alternatives, preset and fixed metadata.

Preset Metadata

The preset metadata system (see Figure 5.14-9) is based on engaging encoder presets loaded with different dialnorm values. This method is useable only when unique triggers are available from the plant automation system or by general-purpose interface (GPI) from master control.

If a category of content uses the same dialnorm value and plays out with a unique trigger, a preset can be engaged in the audio encoder with dialnorm metadata matching the loudness of this content. For example, a disc playback device stores long-form program inventory, all recorded at the same −27 LKFS dialog loudness, and when this device is selected by automation to go to air, the audio encoder receives a trigger from the automation system to engage a preset with a −27 dialnorm setting. The same station uses a different device to store short-form commercial inventory all recorded at the same −22 LKFS loudness. When this device is selected by automation to go to air, the audio encoder receives a trigger from the automation system to engage a different preset with a −22 dialnorm setting.

This technique allows for a change in dialnorm value based on content loudness but does not require the complexity of a fully agile system. It affords the operator some latitude in content loudness while reducing major risk. This method can also be used to switch other metadata parameters, for example, channel mode, potentially switching between 5.1 and two-channel stereo content. However, caution is required, since preset mode presents the possibility for an audible glitch in the receiver during switching, and therefore an AC-3 frame synchronizer should be considered to mitigate this problem.

Preset metadata mode was suitable early in the DTV transition, when content loudness and channel mode could be easily categorized. Programs were frequently delivered at the same dialog loudness and the same value worked well in many cases for the majority of two-channel commercials and promos. Considering the possibility for varied loudness across all content and its susceptibility to audible glitches, preset mode was limited in its usefulness, however.

Fixed Metadata

If one is able to imagine the notion of taking the agile metadata system and working in reverse, this would create the basis of the fixed metadata system (see Figure 5.14-10).

As described in the agile system, to fulfill the Golden Rule, dialnorm is set to match content loudness. In the fixed metadata system, to fulfill the Golden Rule, the opposite process is applied, by which *content loudness* is adjusted to match the dialnorm value. A single dialnorm value is set or "fixed" at the audio emission encoder, and all content is adjusted in level to meet the loudness indicated by this dialnorm setting.

To make this effective, content creators and broadcasters both must agree on a deliverable loudness recommendation that suits their needs. Broadcasters' main interest is minimizing risk to the content. A fixed system poses the least amount of such risk, having the audio encoder simply set to a value that is never changed. Content creators did not embrace the encoding of dialnorm metadata in their soundtracks, so they

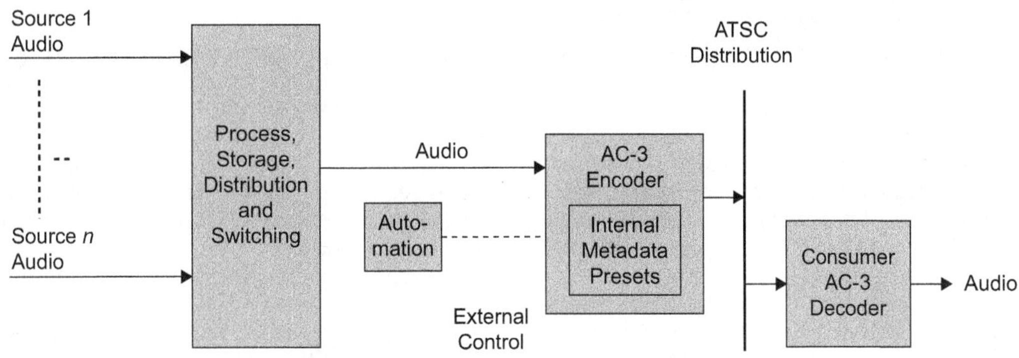

FIGURE 5.14-9 Simple diagram of a preset metadata system from ATSC A/85:2013.

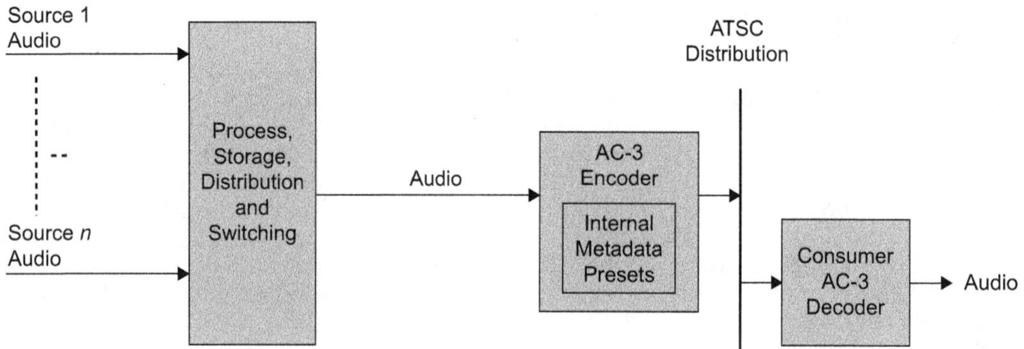

FIGURE 5.14-10 Simple diagram of a fixed metadata system from ATSC A/85:2013.

came to understand where the dialog loudness "sweet spot" for television could reside.

The Fixed Metadata System and ATSC Target Loudness Recommendation

The experts in ATSC S6–3, in conjunction with their content suppliers, agreed on the following industry recommendation for content dialog loudness, which was intended to work under a fixed metadata system:

> For delivery or exchange of content without metadata (and where there is no prior arrangement by the parties regarding loudness), the Target Loudness value should be −24 LKFS.[22]

As noted earlier, the creators of the AC-3 system never foresaw the possibility of this type of agreement between content creators and operators and therefore created the dialnorm system and the agile metadata concept in pursuit of a wide-range listener experience and creative flexibility during content creation. Nevertheless, a universal target loudness recommendation and the fixed metadata system became the preferred answer when operators looked for an alternative to agile dialnorm metadata.

The selected value of −24 LKFS was close to the figure for similar work going on in Europe by the European Broadcast Union (EBU), which had chosen −23 LUFS (Loudness Units Full Scale, an identical measure to LKFS). ATSC compromised at −24 LKFS, a figure that was also within reasonable reach of −27 LKFS, the average dialog loudness of typical episodic content at the time.

Given this agreed-upon recommendation, operators set their dialnorm values uniformly to −24 and specified content to match the dialnorm value, therefore fulfilling the Golden Rule and minimizing risk to content by reducing the complexity of the content delivery and broadcast distribution system. This undertaking resulted in a gradual improvement in program-to-interstitial transitions and established a uniform loudness across the dial. Broadcasters adopted the recommendation, and a significant improvement to TV loudness was presented to the listener.

Metadata System Choice

For all the reasons stated above, broadcasters should give serious consideration to developing individual workflows that support target loudness normalization of content upstream of the broadcast encoder, and the choice of a fixed metadata and dialnorm value. Use of a hierarchy of these workflows preserves quality and will be described later in this chapter. However, A/85 suggests the option of a loudness-control processor at the end of the signal chain to force target loudness at the output of the channel if upstream normalization of content is not practical. This choice should only be considered if no other means are available, as it limits dynamic range and alters the content, similar to applying range-altering NTSC techniques to DTV.

In most cases, the ATSC recommendation of −24 LKFS will be suitable to content creators and operators for exchange and emission of the long-form programs and events, short-form commercials, promotional material and public service announcements that make up most broadcast content.

Note that a fixed metadata system can work effectively using any value appropriate to the content as long as the Golden Rule of content loudness and dialnorm matching is fulfilled. Creators and distributors of "premium" content may choose a different average dialog loudness and dialnorm value, for example, −27 LKFS, to permit additional headroom for wider dynamic-range soundtracks.

CONTENT DELIVERY SPECIFICATIONS

A *content delivery specification* is required to identify the elements of good loudness practices.

Documenting and distributing the necessary technical requirements and recommendations for content delivery are important but often overlooked steps to managing audio at any facility. The purpose of the specification is to inform both *internal* and *external* content suppliers of the necessary steps to be taken when mixing audio, so the delivered soundtrack is compatible with systems used by the operator, and the soundtrack will present well to the audience. The specification strives to ensure that the content will meet the expectations of the creator, the operator and the audience.

ATSC A/85 is clear on a number of important guidelines that should be conveyed to those creating content and specified by the operator receiving the content. The operator should list these guidelines for the supplier, making certain to include any specifics that pertain to how the operator's handling of the content will impact the audio delivered to the listener.

Though other technical parameters for delivery should be included in the specification, this section describes the content delivery specification only as it applies to DTV loudness. Note that this information assumes the operator is employing a fixed metadata/dialnorm system. The specification should include the following:

- A reference to *ATSC A/85, Techniques for Establishing and Maintaining Audio Loudness for Digital Television*, for information on creating, measuring and distributing audio for DTV. It is a good practice to include a link to the ATSC website for access to the document, but not necessarily a direct link to the document, as the version may change and a direct link may become outdated.
- The use of ITU-R BS.1770–3 as the measurement method for determination of content loudness. (This Recommendation's *-3 revision* is required per ATSC A/85:2013 and FCC 14–71.)
- A Long-Form Content Loudness Specification: Specify the required *average* dialog loudness for soundtracks of long-form content. This value must match the dialnorm value that is carried in the operator's bitstream transmitted to the audience. It should be referred to in units of LKFS and be clear about any tolerance around the expected value, for example, "Program suppliers must provide a measured average of –24 LKFS (±2 dB) dialog loudness, not whispered or shouted." Note the extremes of the range are not to be targeted. Also include any special provisions that may be used by the broadcaster to correctly normalize any noncompliant content after it has been received. This may involve subjecting the content to file-based loudness scaling or real-time loudness processing, described later in this section, with its use at the discretion of the operator.
- A Short-Form Content Loudness Specification: Specify the required *average* loudness of the full program mix (not just the dialog element) for soundtracks of short-form content. This value must match the dialnorm value that is carried in the operator's bitstream transmitted to the audience. It should be referred to in units of LKFS, and be clear about any tolerance around the expected value, for example, "Program suppliers must provide a measured average of –24 LKFS (±2 dB) of the full program mix." Note the extremes of the range are not to be targeted. Include any special provisions that may be used by the broadcaster to correctly normalize any noncompliant content after it has been received. This may involve subjecting the content to file-based loudness scaling or real-time loudness processing, described later in this section, with its use at the discretion of the operator.
- Guidance on target value: Explain the correct interpretation of the numeric loudness value, and that the displayed value *will deviate* from the target average value due to the nature of audio dynamics.[23]
- An explanation on the use of wide dynamic-range, short-form content: State that the delivery of wide dynamic-range, short-form content may pose unintentional loudness attenuation of the anchor element within this type of content.[24]
- A True-Peak Specification: Specify a maximum true peak level for soundtracks.[25]
- A note on Metadata Authoring: State that the authoring of all DTV metadata will be by the operator/recipient. The operator should include a table specifying the in-use metadata parameters that will be transmitted with the content to audience receivers.
- An example of the loudness comfort zone: Consider including information pertaining to the loudness comfort zone (see Figure 5.14-11), and how use of the comfort zone impacts acceptable perceived loudness when transitioning between content items.[26]
- An explanation of downmixing: For operators accepting 5.1-channel soundtracks, note to the supplier that any 5.1-delivered soundtracks will be downmixed at time of broadcast for the stereo-listening audience in the viewer's receiver. The operator should note how the mixed elements of a 5.1 soundtrack will impact the stereo downmixed version with special consideration to loudness of the downmixed version.[27] (See Figure 5.14-12.)

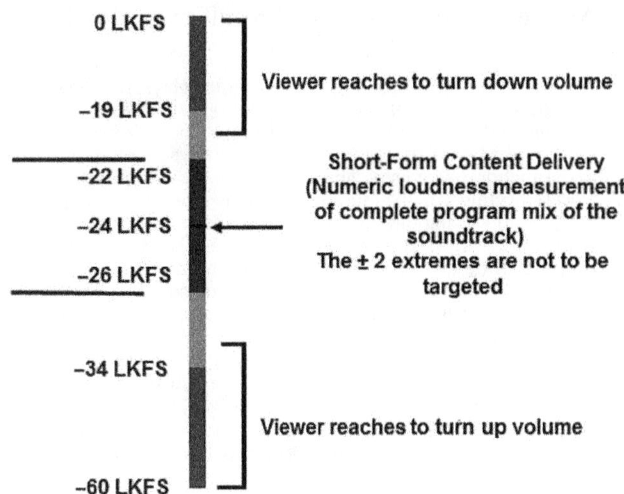

FIGURE 5.14-11 An interpretation of the loudness comfort zone using a –24 LKFS target for short-form content.[28]

Understanding Elements in the Content Delivery Specification

Several key elements within the content delivery specification warrant fuller description for a thorough understanding by broadcasters:

1. *Target loudness and additional information on long- and short-form deliverables, their measurement, and how loudness is perceived:*

As described previously, when a fixed metadata system is chosen, a target loudness value must be established and enforced. This target must match the fixed dialnorm value in the metadata stream, thereby fulfilling the Golden Rule. There is a significant difference in measurement of long-form and short-form content described in A/85. Nevertheless, under a fixed system, a single loudness target is used effectively for both types of content for the following reasons:

- For long-form programming, a listener will set the volume control to the storyline of the content, which is usually normally spoken dialog—that is, not whispered or shouted—and this becomes the common "anchor element." Though the soundtrack may contain soft passages and loud effects, a listener will be tolerant of reasonable changes in range as long as they are of acceptable duration, and therefore listeners will not move the volume control under these conditions.
- Short-form content—for example, commercial advertising, promos, PSAs and other interstitial announcements—often focuses on overall perceptual impact to the listener, creating an overall anchor element (i.e., the entire mix serves as the anchor element), not a dialog/storyline-based anchor element. To accomplish this, these short-form elements are often produced with narrow dynamic range. Therefore, a full program mix measurement was chosen as the loudness determinant for these categories of content in A/85.

An example of the intended result presented to the listener:

> When a correctly mixed long-form program plays, the dialog level (not whispered or shouted) will be consistent from scene to scene and from act to act.[29] Loud effects and soft passages will be mixed with acceptable range and duration, referenced to the anchor dialog level. The listener sets the volume control to the storyline's dialog loudness, with the intention of setting it once and leaving it there. Next, a short-form advertisement with the identical average target value as the surrounding program (but measured using its full-program mix) plays during a commercial break in the program. With the volume control previously set for the *dialog* loudness of the program's storyline, the commercial's soundtrack transitions smoothly, based on the typical narrow dynamic-range characteristics of most ads and promos, and the use of the full program mix measurement.

2. *Wide-range short-form content poses a problem under the scenario described above.*

With a typical movie trailer as an example, if the trailer content is recorded with dialog (as many are), the wide-range music and effects will influence a significant portion of the average reported by the full program mix measurement. This will therefore reduce the perceptual loudness of any dialog in the trailer (due to loud music and effects), resulting in a poor transition from program to trailer. Because A/85 specifies a full program mix measurement for all short-form content, A/85 cautions the content producer on delivering wide dynamic-range content, and details the loudness results when doing so.[30]

An example of an unintentional result presented to the listener in such cases follows:

> When a correctly mixed long-form program plays, the dialog level will be consistent from scene to scene and act to act. Loud effects and soft passages will be mixed with acceptable range and duration, referenced to the anchor *dialog* level. The listener sets the volume control to the storyline's dialog loudness, with the intention of setting it once and leaving it there. Next, a short-form advertisement with a dialog element plays during the commercial break of the program. The short-form content has a wide overall dynamic-range containing effects and music that are loud relative to the dialog for a significant portion of the content. Although the short-form content has been properly delivered at the specified LKFS target *full program mix* value, the perceived loudness of the ad's *dialog* has been reduced well below the LKFS value of the adjacent program. The listener perceives the advertisement as softer than the program, due to the loudness mismatch of the dialog anchor loudness of program compared to the dialog loudness of the advertisement and its full program mix anchor.

Therefore, short-form content providers should be cautious when delivering DTV content with wide or extreme dynamic range. Mixes should always be influenced by knowledge of the features and the characteristics of the DTV system, as documented in A/85.

3. *The downmixing process and its impact on content loudness are sometimes misunderstood by content suppliers.*

When 5.1 content is specified and transmitted, the soundtrack presented to the portion of the audience that listens to the main soundtrack *in stereo* is created at the time of broadcast by the receiver, from the elements of the 5.1 soundtrack, under direction of operator-transmitted downmix control metadata accompanying the audio. (This excludes any separate alternate language and video description presentations.) See Figure 5.14-12.

If 5.1 content is mixed with identical phase and amplitudes of the same elements on multiple channels of the soundtrack, the resultant stereo downmix of the 5.1 content can be louder than the original 5.1 soundtrack.[31]

Therefore, content suppliers providing 5.1 channel soundtracks should measure both the loudness of the 5.1 version and the downmixed stereo version as created and monitored using the identical 5.1-to-stereo downmix metadata parameters used by the operator. The 5.1 mix should be adjusted to meet the required loudness for both versions.

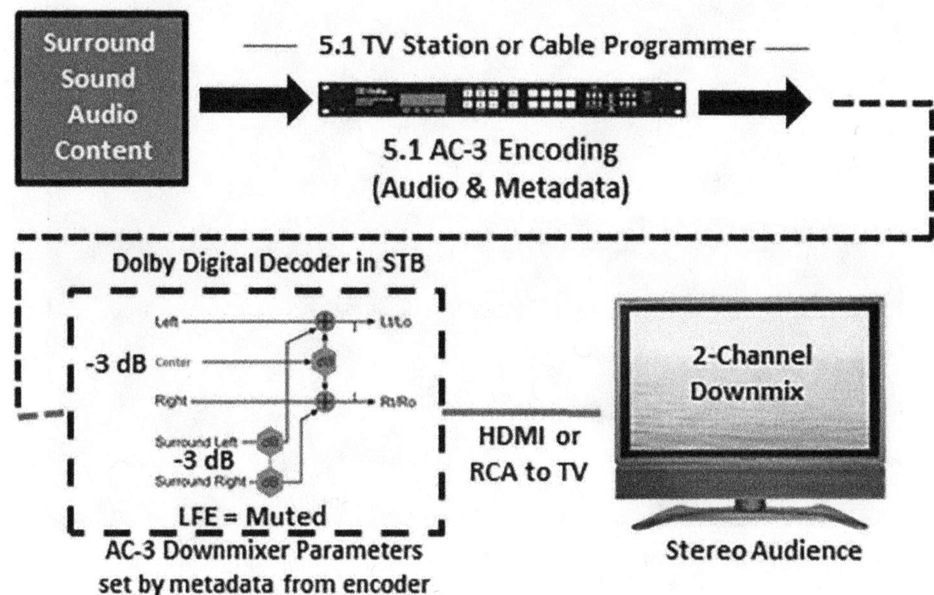

FIGURE 5.14-12 Typical ATSC 5.1-to-stereo downmix process.

A/85'S METHODS OF CONTROLLING PROGRAM-TO-INTERSTITIAL LOUDNESS

ATSC A/85's methods to control program-to-interstitial loudness are the key to all loudness normalization techniques. This section lists the choices for loudness normalization that apply to all types of content. Though this section may seem limited to loudness practices for content transitioning, the recommendations can apply to every situation for normalizing loudness at any step throughout the entire audio ecosystem. Therefore, these methods apply to workflows for live local and remote content creation, long- and short-form content ingest, postproduction and transmission operations.

With a fixed metadata system, accepted measurement practices and a target loudness recommendation in place that fulfills the Golden Rule, the operator can choose the management of loudness for all types and paths of content, one by one. These recommendations will be referred to in all parts of this section as they apply to various content creation and monitoring situations.

When audio quality preservation is a goal, these three methods of loudness normalization apply in this order of preference:

1. *Ensure that all content meets the target loudness, and that long-term loudness matches the dialnorm value.*
2. *Employ a file-based scaling device to match long-term loudness of non-conformant file-based content to the target value.*
3. *Employ a real-time loudness processing device to match the loudness of non-conformant real-time content to the target value.*[32]

The following two subsections detail each of the above methods.

Specifying and Verifying Loudness of Content as Supplied

When matching loudness to a target value is performed by the content creator, the highest-quality audio can be delivered to the audience by leaving loudness decisions in the hands of the supplier, thereby moving control of content loudness as far upstream as possible.

When working to a content delivery specification, normalizing audio as far up in the signal chain as possible, during the initial mix, permits content creators to establish the relationship of elements in the soundtrack based on their creative decisions, eliminating the need for further downstream normalization by the operator. Once the content supplier understands the necessary target required, content can be mixed and auditioned to meet expectations of the supplier and the recipient. (See Figure 5.14-13.) This practice will yield the highest-quality soundtrack to the audience if the distribution and transmission signal path is unaltered from creator to listener.

For an unaltered distribution path to be practical and for an operator to meet compliance, every item of content (e.g., locally and remotely originated live and recorded programs, ads, promos and postproduced material) must all be mixed and delivered meeting the loudness target.

If all content is mixed and delivered to a content loudness target value specification that is enforced

FIGURE 5.14-13 A contemporary audio console displaying a loudness histogram and an LKFS target value during content creation.

to match the dialnorm value, no other practices are required downstream to meet loudness compliance. Soundtracks presenting the supplier's unaltered creative intent can be delivered directly to the audience via a clean, process-free path. From a quality perspective, this is a best-case scenario.

The details of the necessary practices for this method of normalization will be described later in this chapter.

Taking Control of Content Loudness after Delivery

In many cases, especially for short-form content, some type of normalization may be required after delivery to the operator to meet compliance. A/85 specifies two means to do so, in this order of preference:

1. *Employ a file-based scaling device to match long-term loudness of non-conformant file-based content to the target value.*[33]

This method uses a transcoder type device that incorporates LKFS loudness normalization consisting of a "measure" and "scale" process. (See Figure 5.14-14.) The operator enters the target loudness value in the transcoder. On an automated first pass, applied file-based content is analyzed and a measured loudness value is determined. The device compares the measured loudness to the target loudness. Only if a difference is detected, during a second pass the scaler adjusts the overall loudness of all elements of the content to create a match.

Because the transcoder applies the loudness "scale" to all elements in a linear fashion, the dynamic range of the content is not altered, as long as there is enough headroom to scale louder (i.e., increasing gain) to the target without going into peak clipping. If there isn't, a limiter is required to prevent clipping. Loudness scaling that normalizes content softer (i.e., reducing gain)—which is typical for most short-form content—is simplistic, as it is reducing loudness with no need for limiting.

Because this operation is file based, it can be performed faster than real time (e.g., faster than playing linear tape), but the length of the process will vary based on content and manufacturer implementations.

Scaling is an excellent choice for the normalization of file-based content, as it is automated, precise and in most use cases will not alter the dynamic range or quality of the content. Most scaling devices work in the PCM domain and scale audio loudness of content to match the target. Other implementations may be available that can change the dialnorm value to match content loudness within AC-3 bitstreams, for use when agile and preset metadata systems might be chosen.

The focus of the loudness measurement during the "measure" and "scale" process must follow the rules of ATSC A/85. Remember that long-form content measurement is dialog based, while short-form content measurement is of the full program mix. The normalization section of the transcoding device must be specified and set up to perform the correct action on the content type. Some devices may include dialog recognition for long-form measure and scale, but some may not. In a rare instance, A/85:2013 footnotes the acceptable action of using ITU-R BS.1770–1 (formula 2) for automated devices.[34]

It should be noted that a manual scaling process is also possible. An operator can measure and scale content using an LKFS meter and adjust overall gain of the measured content to match a target. Because a unit of LKFS is the same size as a decibel, a loudness offset

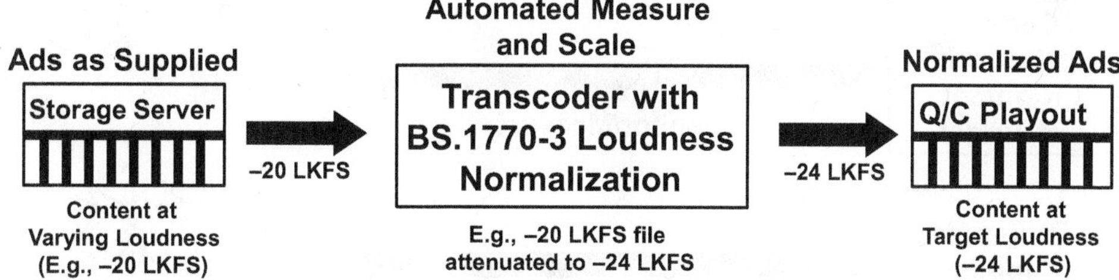

FIGURE 5.14-14 Simple diagram of automated measure and scale workflow.

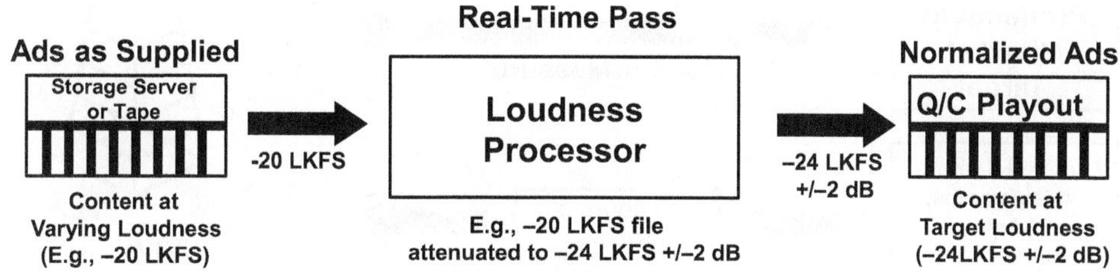

FIGURE 5.14-15 Simple diagram of loudness processing.

adjustment on the player device can be done in units of dB. Once the offset is determined and a unity gain copy of the content is made using the loudness offset applied at the player, the resulting tape or file will be correctly normalized.

2. *Employ a real-time loudness processing device to match the loudness of non-conformant real-time content to the target value.*[35]

This method uses a loudness-control processor to control range and fit content inside of a loudness target window. It is very effective for achieving content compliance by constraining the range of the applied audio in a similar fashion to the way an NTSC processor reduced range (and therefore altering the creative intent of the content) to prevent overmodulation, and as a by-product, making loudness transitions smoother by doing so.

If content either was not mixed to a target, or cannot be scaled (either with a file-based device or manually), the operator can employ a loudness processor to meet compliance. (See Figure 5.14-15.)

Note that the operator is well served by having a manufacturer's expert consult on the setup of the processor for every unique implementation. Though it may seem that a device can be set up quickly by choosing a preset named to fulfill a need, like "General Broadcast," a good match of processor to operation requires the setup be tailored to the type of content being normalized by the processor and the desired target. A processor with settings that are not well matched to content and loudness target can severely degrade the quality of the audio applied to it. This setup is best left to an expert on the specific device.

The simplest approach to overall audio loudness normalization is to apply a "one size fits all" processor downstream in transmission, at the output of the channel. However, if no normalization is enforced upstream, the unpredictability of the loudness levels of all types of content will require the processor to work very hard to place the varying loudness within a target window. Undesirable audible artifacts will most likely be created when doing so, which can be noticed by the audience and especially by content mixers and producers.

Hybrid Loudness Normalization Approach

With audio quality preservation as a goal, consideration should be given to establishing specific normalization workflows starting as far up in the signal path as possible. As it's unlikely that all content will be received or locally mixed to the loudness target, in many cases a plant-wide operation can benefit from use of a quality-driven, "hybrid" loudness normalization approach, using a combination of the methods described above. (See Figure 5.14-16.)

The benefit of using the hybrid approach is that the operator can make certain that quality is maintained whenever possible by considering the most preferred method of normalization, first based on content origin and available workflows, and moving down in preference to quality-altering methods only if necessary. (See Figure 5.14-17.)

For long-form content of various genres, enforcement of the content delivery specification should allow the broadcaster to use these programs with no need for further normalization, leaving loudness and creative decisions in the hands of the supplier. This applies to both externally supplied entertainment-type content as well as internally supplied news-type content, as examples. Working down the normalization

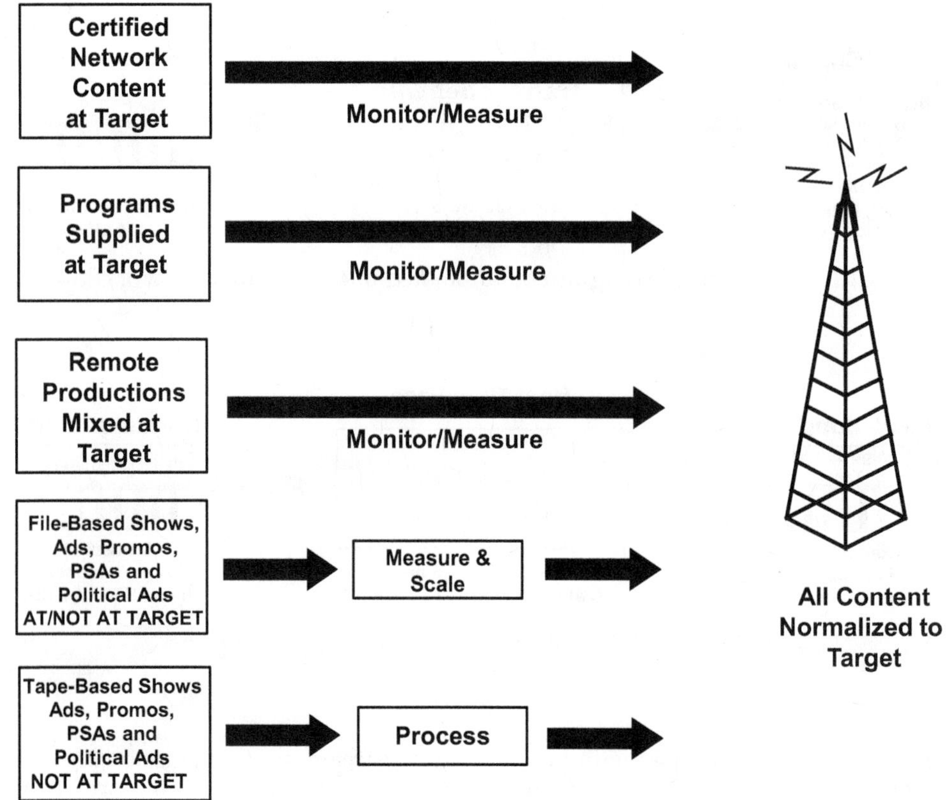

FIGURE 5.14-16 Simple diagram of the elements of a hybrid loudness normalization approach.

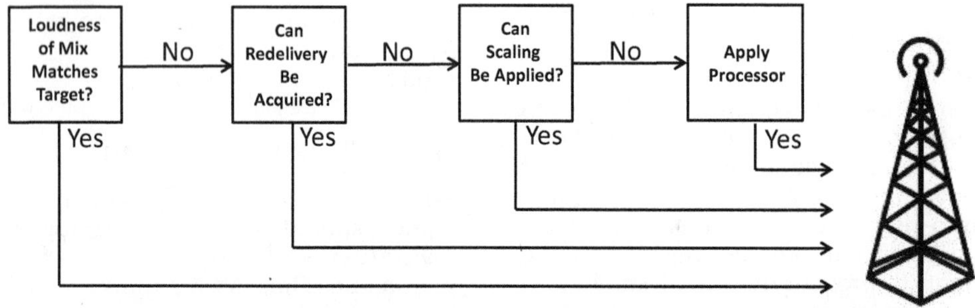

FIGURE 5.14-17 Order of preference for an audio-quality-driven loudness compliance process.

preference list, if this content cannot be relied on for loudness matching the target value and a replacement copy is not an option, the next best normalization choice is a file-based scaler with dialog detection for correcting a content loudness and target mismatch. If this is unavailable, next in preference, manual scaling can be performed. Last in preference, if manual scaling is not workable, the content can be normalized by a real-time processor.

For short-form content, enforcement of the content delivery specification *should* allow the broadcaster to air ads, promos, PSAs and other interstitials with no need for further normalization; however, it is unlikely that this process can always be relied upon. Advertisers and marketing departments strive to stand out among their competitors, frequently attempting to use a loudness advantage to do so. An operator can try to enforce the content delivery specification and ask for replacements when needed. However, the broadcast schedule, the sheer number of ads received and a delicate customer relationship may make this impractical. Based on this, a mandatory workflow that applies loudness scaling to all short-form, file-based content is a prudent step that should be considered by the operator.

If the operator sources file-based commercial advertising via a third-party aggregator, it is possible the aggregator may apply loudness normalization to content as a service to the recipient operator. The operator needs to determine if this is sufficient. If short-form content is acquired from other sources as well, it isn't practical to solely rely on the aggregator, and therefore it makes sense to normalize all short-form content through a scaling workflow managed by the operator.

If file-based scaling is not possible, or short-form content is received on tape, a manual scaling process can be used. If this is impractical, the ad or promo can be normalized by a processor.

With use of an effective hierarchical hybrid approach, the operator strives for a clean, unaltered path for compliant content to be presented to the audience without alteration. This is possible because the quality-altering processor is left only in content workflows that could not be normalized by more preferred methods upstream.

LOUDNESS MANAGEMENT DURING CONTENT CREATION

The first critical steps to identify when starting a project requiring the operator to meet a loudness target value are:

- Identify the content being mixed as long- or short-form type.
- Mix content focusing on average *dialog* loudness for long-form and on average *full program mix* loudness for short-form content.
- Read the content delivery specification and know the target loudness that's required. Content must be delivered to this loudness. The operator is obliged to have this specified loudness match the dialnorm value encoded in the AC-3 bitstream, therefore fulfilling the Golden Rule.

The following steps have been found to be effective for creating compliant mixes that should not require any further normalization by the recipient operator:

- Verify that the ITU-R BS.1770–3 loudness metering is set up correctly and in plain sight:
    - If available, set the meter up to "ATSC" if offered. The meter will now label loudness values in units of LKFS (ATSC) and not LUFS (EBU), though derived values for each are identical.
    - Set up the term. Many engineers use a very effective 10-second rolling average.
    - Next, make sure all channels of the soundtrack are being measured. If a 5.1 soundtrack is being mixed, some meters, for example, the Dolby LM100 mentioned earlier, will not be able to read all channels of audio without the content being encoded to 5.1 Dolby Digital or Dolby E, and the encoder connected to the input of the meter.
    - Do not engage dialog detection.
- Verify the monitoring system has been calibrated as described in a later section of this chapter. If a monitoring-level sound pressure level (SPL) that corresponds to the room size cannot be verified, at least make certain the main loudspeakers are each calibrated for equal amplitude at the mix engineer's listening position, and the low-frequency effects (LFE) channel has been set 4 dB louder than the other speakers.[36]

Live and Postproduced Long-Form Content

- Bring up the talent vocal on the console fader while the talent is speaking normally, not whispering or shouting. Raise or lower the fader, observing the metered loudness value. Park the fader where the loudness target is achieved.
- If mixing in an SPL-calibrated room, the monitor pot will be fixed at a predetermined volume and should be left there.
- If mixing in a non-SPL-calibrated room, the monitor pot should be adjusted while observing the target value on the loudness meter while talent is speaking normally and set to a comfortable volume for the mix engineer and not moved once set.
- With talent vocal at the target loudness and monitors set and fixed, the mix engineer blends the remaining elements of the soundtrack by ear, verifying the isolated dialog's loudness periodically and confirming the LKFS reading matches the target value and the mixer's perceived loudness. ATSC A/85 notes the use of an acceptable ±2 dB tolerance around this value, but it is clearly stated that *the extremes of this range are not to be targeted*.[37]
- The dialog loudness of correctly mixed content will be consistent within and across the acts of a program and from actor to actor.[38]
- Mostly by ear, the mix engineer determines the loudness relationship of the elements of the soundtrack, making certain they blend well against the loudness of the dialog anchor, noting that dialog intelligibility should never be inadvertently compromised by other elements of the mix.
- If creating 5.1 content, the two-channel downmix derived using the same audio encoder metadata parameters as the operator (typically –3 center, –3 surround and mute LFE) should be checked frequently during the mix to confirm mix and loudness compatibility of both versions.
- The mix engineer can refer to the loudness comfort zone for guidance in understanding when a listener will reach for the volume control to adjust for varying loudness.[39]
- If the program does not contain a dialog anchor, for example, in the case of a musical concert, or some types of documentary programming, a surrounding announcer vocal or a portion representing what a typical listener would set his or her volume control to, along with the loudness comfort zone, can be used as a reference to establish the loudness anchor of the non-dialog-based program.

These practices will yield good results that can be confirmed (if needed) if the live program is recorded and subsequently measured with a meter or software as follows:

- Using a manual approach, an operator can isolate the portions of the content where dialog (not whispered or shouted) is the only element present and take measurements in a spot-checking process by reviewing a few different sections of the content. The results should reveal that a correctly mixed program will have consistent dialog loudness that meets the target.
- In an automated approach, if the content is recorded as a file, use of a software program with dialog detection can scan the content faster than real time and report the average integrated dialog loudness value of the entire program.

Postproduced Short-Form Content

- Decide on the anchor element for the soundtrack, keeping in mind the measurement of the full program mix of short-form content will be used to determine the loudness value of all elements of the soundtrack, not just dialog.
- Bring up the anchor element (or multiple elements making up the anchor) on the console fader(s). Raise or lower the fader(s), observing the metered loudness value. Park the fader(s) where the loudness target is achieved.
- If mixing in an SPL-calibrated room, the monitor pot will be fixed at a predetermined volume and should be left there.
- If mixing in a non-SPL-calibrated room, the monitor pot should be adjusted while observing the target value on the loudness meter while the anchor is present and set to a comfortable volume for the mix engineer, and not moved once set.
- The mix engineer determines the dynamic range of the overall soundtrack and the loudness relationship of the elements within the soundtrack by ear. If dialog is included in the soundtrack, the listener will expect to be able to hear dialog normally. Like long-form content mixing, dialog intelligibility should never be inadvertently compromised by other elements of the mix.
- With the anchor element at the target loudness and monitors set and fixed, the mix engineer blends any remaining elements of the soundtrack by ear, verifying loudness periodically and confirming the LKFS reading matches the average target value and the mixer's perceived loudness.
- ATSC A/85 notes the use of an acceptable ±2 dB tolerance around the target.[40] However, it also notes that *the high and low side of this range should not be targeted*. If short-form content is delivered with an average value at the upper or lower side of this range, and the content recipient employs a scaling device to ensure loudness compliance, average loudness will be precisely readjusted to the recipient's target value specified in the scaling device. In these cases, delivering a value other than that specified by the content recipient may yield an unexpected, undesirable change in the overall perceptual loudness and composition of the soundtrack during air.
- To achieve desired results downstream, short-form content should be auditioned at the target loudness prior to delivery to the recipient. Mixing should be fine-tuned to make certain the full program mix meeting the target loudness value also meets the creative expectations of those producing the content.
- It is advisable that short-form content intended to play out directly after or before a program break and in a group with other short-form content (e.g., a commercial break composed of multiple commercials and promos) be auditioned playing side by side with dialog at the target loudness and with other typical short-form content at the target loudness, to hear the spot to be delivered in context.
- Noted above in this chapter, A/85 cautions about the use of wide dynamic range short-form content. Given its importance, that guidance is repeated here:

> Those choosing to create and deliver wide dynamic range short-form content should note that the louder elements of this type of material will increase the loudness measured with a long-term integrated method, and consequently reduce the perceived Anchor Element loudness after normalization. This can cause an unacceptable match to long-form material measured with an anchor-based method.[41]

- If creating 5.1 content, the two-channel downmix derived using the same audio encoder metadata parameters as the operator (typically −3 dB center, −3 dB surround and mute LFE) should be checked frequently during the mix to confirm mix and loudness compatibility of both versions.

VERIFYING LOUDNESS

These practices will yield good results, but it is highly encouraged that the supplier verifies the loudness target of short-form content by subsequent measurement with a meter or software as follows:

- Using a manual approach, choose "infinite" or "integrated" measurement (continuous measurement, not short-term). Ensure the meter can measure all audio channels of the content. Disable any dialog detection. Clear the meter value, and pause the measurement. Play the content at the very beginning and precisely start the meter. Stop the meter at the exact end of the content. The resulting value is the average, integrated loudness value of the content used to report loudness per A/85.
- In an automated approach, if the content is recorded as a file, use of a software program measuring the full program mix can scan the content in faster than real time and report the average loudness value.

Ingest

For ingest practices, the operator can use similar methods to those described above to confirm that the loudness of content matches the target loudness, repeated here:

For Long-Form Content:

- Using a manual approach, an operator can isolate the portions of the content where dialog (not whispered or shouted) is the only element present and take measurements in a spot-checking process by reviewing a few sections of the content. The results should reveal that a correctly mixed program will have consistent dialog loudness and, if mixed as described above, that it meets the target.
- In an automated approach, if the content is recorded as a file, use of a software program with dialog detection can scan the content faster than real time and report the average integrated dialog loudness value.
- If a correctly set up automated scaler is employed, the scaler ensures that all average integrated content loudness meets the target value.

For Short-Form Content (these practices can be used to confirm commercial advertising loudness):

- Using a manual approach, choose "infinite" or "integrated" measurement (continuous, not short-term). Ensure the meter is measuring all audio channels of the content. Disable any dialog detection. Clear the meter value, and pause the measurement. Play the content at the very beginning and precisely start the meter. Stop the meter at the exact end of the content. The resulting value is the average, integrated loudness value of the content used to report loudness per A/85.
- In an automated approach, if the content is recorded as a file, use of a software program able to measure the full program mix can scan the content faster than real time and report the average loudness value.
- If a correctly set up automated scaler is employed, the scaler ensures that all content loudness meets the target value.

Transmission or Master Control

The transmission engineer can verify the loudness of incoming programs from the studio or remote locations similar to the way other engineers verify the loudness of long-form content:

- Verify that the ITU-R BS.1770–3 loudness metering is set up correctly and in plain sight:
  - If available, set the meter up to "ATSC" if offered. The meter will now label loudness values in units of LKFS (ATSC) and not LUFS (EBU), though derived values for each are identical.
  - Set up the term. Many engineers use a very effective 10-second rolling average.
  - Next, make sure all channels of the soundtrack are being measured. If a 5.1 soundtrack is being mixed, some meters, for example, the Dolby LM100 mentioned earlier, will not be able to read all channels of audio without the content being encoded to 5.1 Dolby Digital or Dolby E, and the encoder connected to the input of the meter.
  - Do not engage dialog detection.
- Using conventional VU and not loudness metering, in a pre-program tech-fax check, have the studio or remote engineer send a 1 kHz tone to validate unity gain between locations and adjust as necessary. Acquisition of a steady –20 dBFS is typically used for this process.
- Next, have the studio or remote engineer send dialog (not whispered or shouted; spoken live by talent, a production assistant or from a recording) through the mix and distribution path that will be used for the program or event.
- Ensure the engineer confirms that the target loudness set on the mix console has been verified by loudness meter at the mix location that is set up in a similar fashion to the transmission engineer's loudness meter.
- The transmission engineer measures the content being sent by the studio or remote mix engineer, comparing values. If the distribution path is at unity gain and metering is set up correctly on both sides, the readings will be similar and the means to establish good loudness has been confirmed.
- Situations can vary and common sense should be used, but in cases where downstream monitoring has revealed a problem or discrepancy upstream, an accepted practice is to have the responsibility for correcting the problem as far upstream as possible. Once fixed upstream at the source, this reduces the potential for inadvertent re-correction of the problem downstream, only making matters worse.

The Monitoring Environment

An essential practice for making good loudness judgments and overall good audio mixing is working in a room with quality monitoring that has been calibrated for correct monitor gain, established in reference to room size. ATSC A/85 does a thorough job of explaining the requirements and the practices used to establish reliable environments capable of content creation meeting expectations and yielding interoperability across facilities.

Three approaches for room calibration are documented in A/85. The operator should choose the most comprehensive method practical for the operation in this order of preference:

1. The first approach is intended for planners, design engineers, installation engineers, maintenance engineers and users.[42] This is a comprehensive guide to the characteristics of rooms and spaces,

installation of the system and setup method to achieve calibration of monitoring. The key elements contained are two tables categorizing audio room types typically found in broadcast environments (see Table 5.14-2) and appropriate sound pressure levels for rooms based on room volume in cubic feet (see Table 5.14-3). Also included is a step-by-step method to calibrate a room in the correct fashion based on conditions from the two aforementioned tables.

2. The second approach is a quick reference on monitoring setup for television.[43] Its purpose is to describe just the necessary steps to ensure unity gain through the mix console along with appropriate test noise that can be played and monitors set for SPL, per the included room volume table.
3. The third approach is a really quick reference guide and is clever, simply instructing the user to set unity gain in the system and compare and adjust mixed content to a known item of content at −24 LKFS, the example target loudness, by ear.[44]

It should be noted that the embedded links in A/85 may not connect to available audio test content as documented in this section of the Recommended Practice. A/85 specifies the type of test content needed, and the operator should acquire suitable test material elsewhere if not available from ATSC.

In addition to room calibration, A/85 also gives guidance on room acoustics and loudspeaker placement, describing room layout and construction along with suitable acoustical treatments.[45] A section on room correction is included as well, explaining to the reader

TABLE 5.14-2
Categories of audio control rooms and spaces used in TV production (from A/85:2013).

Category	Characteristics
I	Principal audio monitoring control rooms with specialized acoustics and sound systems. Channel range up to 5.1 (3 front/2 surround/0.1 low-frequency enhancement). Well isolated from other operations. Widest frequency and dynamic ranges equal to best home cinemas properly aligned. This type of room may be used for quality control at the network level, for example, checking program material for conformance to delivery requirements when a question arises at ingest stations. Sound monitor quality dominates over production requirements in this category of room. Broadcast organizations might be expected to have only a small number of such rooms.
II	Audio-mostly production spaces with equipment needs and placement supplanting absolute audio monitoring conditions, although audio monitoring is still expected to be good. Channel number equal to highest number used for material originating in the room. Good isolation from other operations. This type of room may be used for program origination, with its output occasionally subject to check in a Category I room. Low-frequency range and headroom may be somewhat restricted compared to a Category I room.
III	Audio editing spaces, premix and prelay rooms, and other spaces the output of which is typically expected to be integrated into programs in a Category II room or better. If used for final mixing, apply the level and equalization recommended practice herein.
IV	Trucks and booths for program mixing. These spaces have special considerations due to their small room volume, high background noise level, high level of early reflections, and communication needs in a production environment.
V	Headphone monitoring systems recommendations. Used for ingest stations in crowded environments, quality control in machine rooms, and the like.

TABLE 5.14-3
Reference sound pressure level (SPL) recommendations for mixing rooms of varying size (from A/85:2013).

Categories	Room Volume in Cubic Feet	SPL in dB re 20 μN/m^2
I, II	> 20,000	85*
	10,000 < 19,999	82
	5,000 < 9,999	80
	1,500 < 4,999	78
	< 1,499	76
III	Depends on room usage. For editing purposes, may be controlled by the editor for use with the material at hand. For final program mixing, follow the recommendations for categories I, II above.	
IV	< 1,500	76
V		Use 2 cc coupler and set 440 Hz level to 74 dB.

* Per SMPTE RP 200 [6]

both effective and ineffective methods to achieve good results.[46]

Practical room considerations:

- If possible, the audio mix room should be designed as an isolated acoustical space with high-quality loudspeakers that are suitable for the content being mixed and are comfortable and non-fatiguing for the mix engineer to work with.
- Most importantly, however, the monitor system, especially one for 5.1 surround sound, must be calibrated in some fashion, making it a reliable reference and not a handicap to accurate content creation. At minimum, using pink noise, the main amps/speakers should be set to equal amplitude at the listening position with the LFE set 4 dB louder, using a simple SPL meter (slow response, C-weighting) or at least by ear.
- A correctly set up broadcast loudness meter should be connected to the audio system, able to measure all channels of the content being mixed or monitored. The meter must be installed in clear sight of the mix engineer or operator.
- Large rooms with high reference SPL yield wide dynamic range mixes. Smaller rooms using a lower SPL reference yield more range-constrained mixes. Though the AC-3 system is technically capable of cinema-like sound with >100 dB of range, realistic, good-sounding DTV mixes have been done in smaller rooms with ~78 SPL. Typical episodic content for digital television is frequently mixed with these considerations in mind.
- If applicable, the content lay-back session (the process of rejoining audio to video after audio postproduction mixing) is an excellent opportunity to double-check loudness for matching the target value and the mix for compatibility and quality. A small room with lower SPL, equipped with a small surround system and/or stereo speakers, the speakers in a flat-panel TV, and a loudness meter or loudness software are perfect tools for examination of the 5.1 and derived two-channel mix before dub and delivery.
- In a transmission room situation where audio monitoring is limited to a two-channel, rack-mounted speaker-panel device, an operator can get a good sense of continuity and presence of elements in a mix. In these cases, however, surround sound content will be downmixed to two-channel for listening. Presence of signal can usually be observed on LED bar graphs to determine level, and some units may include LKFS loudness monitoring. Though not able to monitor full surround sound, the panel is useful for spot-checking and quality control (QC), and it is very effective for isolating and critiquing channels of audio on speakers and especially with headphones.

THE CALM ACT AND FCC RULES

This section identifies important considerations for establishing a loudness normalization plan as specified in the 2011 FCC Report and Order based on the 2009 and 2011 releases of ATSC A/85. The full FCC Report and Order should be referred to, as this section does not include all the information and footnotes from that document. A later part of this section will list minor changes to the rules from the FCC's Second Report and Order issued in response the 2013 release of ATSC A/85. Again, the full FCC Report and Order should be referred to, as this section also does not include all the information and footnotes from the Second R&O.

When researching the necessary steps to complete loudness normalization that complies with FCC rules, the broadcaster or operator is strongly encouraged to consult legal counsel for an interpretation of the law and for guidance with the execution of a suitable operating plan that fulfills all mandated requirements.

For an interpretation of how these rules apply to a TV station, an NAB report from December 19, 2011, is included for reference.

This section lists:

- What the CALM Act requires of the FCC
- FCC documents and their current location
- The starting date of CALM Act enforcement
- The FCC's response to the CALM Act
- FCC loudness management rules
- NAB TV TechCheck, December 19, 2011
- FCC response to the 2013 release of ATSC A/85
- FCC loudness management rules (2014)

CALM Act Requirements

Reacting to the DTV loudness problem and identifying ATSC A/85 as a solution as described here and in an earlier section of this chapter, the Commercial Advertisement Loudness Mitigation Act that became U.S. law in December 2010 and enforceable in 2012 required the following:

> Within 1 year after the date of enactment of this Act, the Federal Communications Commission shall prescribe pursuant to the Communications Act of 1934 (47 U.S.C. 151 et seq.) a regulation that is limited to incorporating by reference and making mandatory (subject to any waivers the Commission may grant) the 'Recommended Practice: Techniques for Establishing and Maintaining Audio Loudness for Digital Television' (A/85), and any successor thereto, approved by the Advanced Television Systems Committee, only insofar as such recommended practice concerns the transmission of commercial advertisements by a television broadcast station, cable operator, or other multichannel video programming distributor.[47]

The FCC responded to the signing of the CALM Act into law by adopting a Notice of Proposed Rulemaking (www.gpo.gov/fdsys/pkg/FR-2011-06-03/html/2011-13822.htm). Comments on the proposal were addressed, and ultimately a first Report and Order (https://apps.fcc.gov/edocs_public/attachmatch/FCC-11-182A1.pdf) was issued on December 13, 2011.

A second Report and Order (https://apps.fcc.gov/edocs_public/attachmatch/FCC-14-71A1.pdf) based on the 2013 release of ATSC A/85 was released on June 4, 2014.

The First Report and Order became enforceable on December 13, 2012.[48]

The Second Report and Order became enforceable on June 4, 2015.[49]

In the First Report and Order, the FCC summarizes the CALM Act's requirements of the Commission as follows:

> The CALM Act directs the Commission to incorporate into its rules by reference and make mandatory a technical standard, developed by an industry standards development body, that is designed to prevent digital television commercial advertisements from being transmitted at louder volumes than the program material they accompany. As mandated by the statute, the rules apply to digital TV broadcasters, digital cable operators, and other digital multichannel video programming distributors ("MVPDs").
>
> The rules [we adopt today] are designed to protect viewers from excessively loud commercials and, at the same time, permit broadcasters and MVPDs to implement their obligations in a minimally burdensome manner. As described below, we will require broadcast stations and MVPDs to ensure that all commercials are transmitted to consumers at the appropriate loudness level in accordance with the industry standard.
>
> In the event of a pattern or trend of complaints, stations and MVPDs will be deemed in compliance with regard to their locally inserted commercials if they demonstrate that they use certain equipment in the ordinary course of business. For the embedded commercials that stations and MVPDs pass through from programmers, we also establish a "safe harbor" to demonstrate compliance through certifications and periodic testing [See Figure 5.14-18]. This regime will make compliance less burdensome for the industry while ensuring appropriate loudness for all commercials.[50]

The Report and Order also notes the following:

> The "Golden Rule" of the RP is that the dialnorm value must correctly identify the loudness of the content it accompanies in order to prevent excessive loudness variation during content transitions on a channel (e.g., TV program to commercial) or when changing channels. If the dialnorm value is correctly encoded—if it matches the loudness of the content, which depends in turn on accurate loudness measurements—the consumer's receiver will adjust the volume automatically to avoid spikes in loudness.
>
> In addition to requiring the Commission to incorporate the RP by reference, the CALM Act requires the Commission to incorporate by reference "any successor thereto."
>
> In addition, we adopt our tentative conclusion that "all stations/MVPDs and not only those using AC-3 audio systems" are subject to our rules.
>
> We conclude that the statute makes each station/MVPD responsible for compliance with the RP as incorporated by reference in our rules with regard to all commercials it transmits to consumers, including both those it inserts and those that are "embedded" in programming it receives from program suppliers.[51]

The NAB followed with this report on the CALM Act and the FCC Report and Order:

> On December 13, 2011, the Federal Communications Commission adopted an Order required by the Commercial Advertisement Loudness Mitigation (CALM) Act designed to prevent commercials from being louder than programs surrounding them.
>
> (see July 11, 2011, issue of *TV TechCheck*)
>
> The Order incorporates the entire ATSC Recommended Practice on *Techniques for Establishing and Maintaining Audio Loudness for Digital Television* (ATSC A/85) into the regulations and makes A/85 mandatory for managing relative loudness of commercials and programs. The FCC plans enforcement to be based upon patterns of public complaints, instead of an audit program.
>
> First and foremost, all television broadcast stations (and MVPDs) are deemed ultimately responsible for controlling the loudness of all commercials they transmit (with respect to the program segments before and after them). Stations are expected to directly (or indirectly) follow the recommendations in ATSC A/85 to measure and thereby control the loudness of the audio segments fed to the service's AC-3 encoder (regardless of its physical location). There are multiple approaches available to broadcasters that can mitigate their exposure to a Notice of Apparent Liability for loudness violations.
>
> Most stations will find that operating at a fixed loudness (as measured per ATSC A/85) and leaving the "dialnorm" setting constant in the AC-3 encoder will be the most practical implementation approach.
>
> Stations technically can choose a different fixed operating loudness value for each program source (virtual channel); however, the loudness of the network feed will establish that virtual channel's operating point. Further, since program providers have many distribution outlets, and typically do not want to deliver different levels to different outlets, expectations are that most sources will be delivered to stations at the

CALM ACT CERTIFICATION

This is to certify that:

1. As required by Section 76.607 of Title 47 of the Code of Federal Regulations, all commercial advertisements embedded in programs carried on **This Television Network** are in compliance with the loudness control practices contained in Advanced Television Systems Committee (ATSC) A/85: Recommended Practice: Techniques for Establishing and Maintaining Audio Loudness for Digital Television ("ATSC A/85 Recommended Practice") at the point of distribution by **This Television Network** to authorized reception equipment of downstream Television Stations.
2. Compliance with the ATSC A/85 Recommended Practice is determined by **This Television Network** through the use of equipment and associated software that is installed, utilized and maintained in a commercially reasonable manner.

Executed this X day of X
By: Senior Vice President, Broadcast Operations

FIGURE 5.14-18 Example of a CALM Act Certificate.

recommended operating loudness of –24 LKFS (as measured with the ITU BS.1770–1 method). This value is then entered into the AC-3 encoder's "dialnorm" setting.

The rules for broadcasters will be in Part 73.682(e), and include six major sections. The first section covers compliance with ATSC A/85, as briefly described above. The remaining sections cover the following:

1. Commercials inserted by stations
2. Embedded commercials—safe harbor
3. Use of a real-time processor
4. Commercials locally inserted by a station's agent—safe harbor
5. Demonstrating actual compliance

For commercials inserted by stations (1, above), the rule reads:

A television broadcast station that installs, utilizes and maintains in a commercially reasonable manner the equipment and associated software to comply with ATSC A/85 shall be deemed in compliance with respect to locally inserted commercials, which for the purposes of this provision are commercial advertisements added to a programming stream by a station prior to or at the time of transmission to viewers. In order to be considered to have installed, utilized and maintained the equipment and associated software in a commercially reasonable manner, a television broadcast station must:

(i) install, maintain and utilize equipment to properly measure the loudness of the content and to ensure that the dialnorm metadata value correctly matches the loudness of the content when encoding the audio into AC-3 for transmitting the content to the consumer;
(ii) provide records showing the consistent and ongoing use of this equipment in the regular course of business and demonstrating that the equipment has undergone commercially reasonable periodic maintenance and testing to ensure its continued proper operation;
(iii) certify that it either has no actual knowledge of a violation of the ATSC A/85 RP, or that any violation of which it has become aware has been corrected promptly upon becoming aware of such a violation; and
(iv) certify that its own transmission equipment is not at fault for any pattern or trend of complaints.

For the "embedded commercials—safe harbor" situation (2, above), the rule is much more complex. The Order presents the FCC's determination that the CALM Act establishes stations as ultimately responsible for the loudness of all commercials broadcast—even those inserted upstream. In general, it calls for stations with annual receipts of more than $14 million to either obtain a certification from the upstream provider that it is A/85-compliant, or perform at least two annual tests on that provider's content to verify its loudness compliance. Either of these alternatives establishes a degree of protection ("safe harbor") against fines to the station in the event that excessively loud commercials are inserted upstream and cause complaints.

If the spot-check approach is taken, the first check must be completed by December 13, 2012. (Spot-checking is defined to include 24 hours of measurement and analysis of the audio loudness transmitted by the broadcast station. The Order includes various suggestions and requirements on how to implement the spot-checking process.) This section also includes a requirement to perform a 24-hour spot-check after a "pattern of complaints" result in an FCC inquiry, with a progressive escalation process that can lead to fines in the event of continued non-compliance.

The Order also provides the option (3, above) for a station's use of a properly maintained real-time audio processor, with record-keeping requirements for demonstrating its consistent and ongoing use.

Also contained in the Order (4, above) is the option to establish a safe harbor for the special case where commercials are locally inserted by a station's agent. Stations may demonstrate compliance by relying on the third-party local inserter's certification of compliance with ATSC A/85, conditional upon meeting the terms detailed in the Order's relevant subsection.

Finally, a station also may document actual compliance with ATSC A/85 with regard to any commercial advertisements that may become the subject of an inquiry, and certify that its own transmission equipment is not at fault for any such pattern or trend of complaints (5, above).

The Order further explains that if after a broadcaster informs an upstream source of a loudness issue, and it is not fixed in a timely fashion, that each station carrying that source's content will be subject to liability if the problem persists. The progressive test and report process outlined in the Order should provide incentive to the source to fix the problem, given that some stations might stop carrying the source's content to avoid financial exposure.

A streamlined financial hardship waiver for some of the above processes is available to small broadcast stations. A "small broadcast station" is defined for purposes of the streamlined waiver as either a station with no more than $14 million in annual receipts, or one that is located in television markets 150 to 210. Small broadcast stations must file for such waivers by 60 days prior to the effective date of the rules.

The following are excerpts from the FCC Second Report and Order on CALM, issued June 4, 2014:

2. The Commission's rules implementing the CALM Act, adopted December 13, 2011, require digital TV broadcasters, digital cable operators, satellite TV providers, and other digital MVPDs to ensure that the commercials they transmit to viewers comply with the television industry's 2011 ATSC A/85 Recommended Practice (RP), which describes how the industry can monitor and control the loudness level of digital TV programming. As mandated by the statute, the Commission incorporated into its rules by reference and made mandatory the 2011 ATSC A/85 RP. The rules took effect on December 13, 2012.

3. Section 2(a) of the CALM Act mandates that the Commission's rules incorporate by reference and make mandatory "any successor" to the RP. On March 12, 2013, the ATSC published a successor document to its 2011 A/85 RP. As described by the ATSC, the Successor RP applies an improved loudness measurement algorithm to conform to the International Telecommunication Union's (ITU) updated BS.1770 measurement algorithm, "BS.1770–3." BS.1770–3 employs "gating" that will exclude very quiet or silent passages of a commercial when calculating the average loudness of that commercial. Use of the new algorithm may reduce the volume of some commercials in certain circumstances. The Successor RP also contains other minor changes that do not affect our rules.

Additionally:

5. As required by the statute, we adopt the Successor RP and will incorporate it by reference into our rules. We also find, as we tentatively concluded in the *FNPRM*, that the only substantive change created by the Successor RP as it relates to our rules is the change to the measurement algorithm to conform to BS.1770–3. This finding is consistent with the ATSC's description of the Successor RP and is not disputed in the record. As a practical matter, this change seems to be designed to prevent advertisers from using silent passages to offset excessively loud passages when calculating the average loudness of program material. Thus, once this Successor RP is implemented, consumers may notice a modest decrease in the perceived loudness of certain commercials. This change is consistent with the type of updates that we believe Congress intended the Commission to incorporate in its rules by specifying in the CALM Act that the Commission shall make mandatory successor versions of the RP.

6. We adopt the proposal in the *FNPRM* to make the Successor RP mandatory as of June 4, 2015, one year from the release date of this Second Report and Order. NAB, the only commenter on this issue, supports this approach.

CONCLUSION

Confronted with unintentional consequences of a new high dynamic range audio technology, industry and government came together, recognizing DTV's early loudness problem and establishing government regulation that codified an industry-developed solution. With the timely development of the effective techniques explained in this chapter, a group of audio experts became uniquely empowered to influence quality-driven rules, ultimately solving the problem to the satisfaction of government, industry and, most importantly, the television audience.

This chapter describes the story, solutions and successes of the first generation of digital television loudness control. Next-generation audio systems are already beginning to benefit from the pioneering efforts described here, leveraging and perfecting these practices that laid the groundwork.

ACKNOWLEDGEMENTS

The author wishes to thank his colleagues Margaret Tobey, Tomlinson Holman and Glenn Reitmeier for their assistance in compiling this chapter.

Notes

[1] Tim Carroll and Jeffrey Riedmiller, *NAB Engineering Handbook*, 10th edition, "Audio for Digital Television" (p. 1329), 2007.
[2] Bertram Scharf and Rhona Hellman, "Using Magnitude Estimation to Teach About Loudness," *Journal of the Acoustic Society of America*, Volume 58, S1, November 1975.
[3] ATSC A/85:2013, Section 1.
[4] FCC 11-182A1, Report and Order, December 13, 2011.
[5] ATSC A/85:2013, Section 4.
[6] Dolby LM100 Broadcast Loudness Meter User's Manual, Issue 5.
[7] Mezzanine-level compression is used for efficient distribution in applications prior to final emission coding.
[8] Codec = Coder/Decoder.
[9] SMPTE ST 2020 Parts 1, 2, 3.
[10] Dolby E's eight channels could be assigned with six channels for a 5.1-channel surround mix, and the remaining two channels used for a synchronized stereo mix.
[11] AES3–2009, "AES standard for digital audio: Digital input-output interfacing—Serial transmission format for two-channel linearly-represented digital audio data," Audio Engineering Society, New York, reaffirmed 2014. www.aes.org/publications/standards/search.cfm?docID=13
[12] ATSC A/53 Part D 2005.
[13] ATSC A/85:2013, Section 10.1.
[14] ATSC S6–3–001r0.
[15] ATSC S6–3 Loudness Experiment Report, September 2009.
[16] S. 2847 (111th), CALM Act.
[17] ATAS Press Release, October 14, 2014.
[18] Federal Register, "Rules and Regulations," Vol. No. 131 (p. 40279), July 9, 2012.
[19] ATSC A/85:2013, Section 3.4.
[20] ATSC A/85:2013, Section 5.2.
[21] SMPTE ST 2020, Parts 1, 2 and 3.
[22] ATSC A/85:2013, Section 6.
[23] ATSC A/85:2013, Section 5.2.3.
[24] ATSC A/85:2013, Section 5.2.4.
[25] ATSC A/85:2013, Section 6.
[26] ATSC A/85:2013, Annex E.
[27] ATSC A/85:2013, Section 5.2.5.
[28] Riedmiller, J., Lyman, S., and Robinson, C, "Intelligent Program Loudness Measurement and Control: What Satisfies Listeners?," AES Pre-print, 115th Convention, October 2003, Audio Engineering Society, New York, NY.
[29] DTV Audio Group Meeting, AES Convention, Los Angeles, October 1, 2016.
[30] ATSC A/85:2013, Section 5.2.4.
[31] ATSC A/85:2013, Section 5.2.4.
[32] ATSC A/85:2013, Section 8.1.1.
[33] ATSC A/85:2013, Section 8.1.1.
[34] ATSC A/85:2013, Page 16, Footnote 3.
[35] ATSC A/85:2013, Section 8.1.1.
[36] ATSC A/85:2013, Section 10.3.
[37] ATSC A/85:2013, Section 6.
[38] DTV Audio Group Meeting, AES Convention, Los Angeles, October 1, 2016.
[39] ATSC A/85:2013, Annex E.
[40] ATSC A/85:2013, Section 6.
[41] ATSC A/85:2013, Section 5.2.4.
[42] ATSC A/85:2013, Section 10.
[43] ATSC A/85:2013, Annex D Section 1.
[44] ATSC A/85:2013, Annex D Section 2.
[45] ATSC A/85:2013, Annex B.
[46] ATSC A/85:2013, Annex C.
[47] S. 2847 (111th), CALM Act.
[48] FCC 11–182 December 13, 2011.
[49] FCC 14–71 June 4, 2014.
[50] FCC 11–182 December 13, 2011.
[51] FCC 11–182 December 13, 2011.

CHAPTER

5.15

Weather Radar Systems

BARON SERVICES, INC.

Huntsville, AL

INTRODUCTION

As early as 1915, a British scientist named Robert Watson-Watt used radio technology to detect thunderstorms in England. Radar, which is derived from the words radio detection and ranging, was developed by British (and other) militaries to detect enemy aircraft during World War II. However, it was quickly realized that radar also detected precipitation. It was initially seen as a nuisance. However, after the war was over, the development of weather radar began almost immediately (e.g., Sim 2014; Duda 2015).

During the 1940s and 1950s, the U.S. Weather Bureau, in cooperation with universities and research laboratories, produced the first true weather radar systems. The Weather Surveillance Radar (WSR-57) was developed in 1957, and a network of these radars was deployed across the United States. These radars were capable of showing where precipitation was occurring, and how intense it was. Also, during the 1950s and 1960s, early Doppler radar was developed (Sim 2014). Doppler radar can detect the motion of precipitation toward or away from the radar site, allowing for detection of winds and severe storms, including tornadoes. Doppler radar technology advanced greatly through the 1970s. The WSR-88D, also known as NEXRAD (NEXt generation weather RADar), a Doppler weather radar system, was deployed nationwide during the 1990s by the National Weather Service (NWS). Recently, the WSR-88D network was modified so that all of the radars have dual-polarization capabilities. This has improved many aspects of the radar system, including tornado and hail detection, and delineation of rain versus snow in cold air.

The overall mechanics of a weather radar are simple. A beam of radio frequency (RF) energy is transmitted from the radar site. As it travels outward from the radar, some of the energy is reflected back to the radar as the beam encounters objects such as precipitation. The radar can then determine, knowing the speed of light and the time of return of the energy, how far away the precipitation is. Repeating this process in every direction from the radar provides an aerial map of where precipitation is occurring. Additional techniques are applied to determine the intensity of the reflecting precipitation, the movement of it toward or away from the radar, and the type of precipitation.

In this chapter, we start out in the Weather Radar Theory and Operation section discussing the basic theory and operation of a weather radar, including the transmission of the outgoing energy and the reception of reflected energy. This section includes explanations on how the intensity of precipitation is determined, and the Doppler effect. Other issues, including attenuation of the radar beam by heavy precipitation, and how the radar beam propagates in the atmosphere, are also covered. The Weather Radar Hardware and Design section addresses the hardware and design of the radar. This includes the types of transmitters that generate the radar beam and their different components, the receiver that detects reflected energy from precipitation, the antenna that focuses the outgoing radar beam and receives the incoming radar energy reflected from precipitation. The Radar Meteorology section presents the basics of radar meteorology such as radar reflectivity and certain patterns therein that indicate certain weather phenomena; Doppler velocity and its many uses, including wind patterns and tornado detection; dual-polarization radar variables and how they differentiate precipitation types and detect tornadoes; and derived radar products. The Radar Procurement section contains some important information television stations must consider when deciding to buy a weather radar.

WEATHER RADAR THEORY AND OPERATION

Transmission of Outgoing Energy

A weather radar detects the intensity and location of precipitation by transmitting an antenna-focused beam of radio frequency (RF) energy, detecting what portion of energy is returned to the radar, and how long it takes for the energy to return (see Figure 5.15-1). The RF energy is in the form of sinusoidal waves. However, there are multiple sources of RF waves in the atmosphere, including radio and television transmissions, satellite communications, as well as the sun. Due to these other sources of RF energy, most radars transmit RF energy at a single frequency and only process incoming energy at that frequency. Additionally, the Federal Communications Commission (FCC) regulates the frequencies of RF radiation that may be used for different purposes; weather radars are allotted a very specific band of frequencies. Most weather radars operate in the microwave band of frequencies between 3 and 10 GHz, or at wavelengths between 10 and 3 cm, respectively. The S-band frequency is allocated to the NWS radar network. Broadcast television station radars have historically been C-band, with a recent move to high S-band. Table 5.15-1 shows the four most common weather radar frequencies.

A weather radar emits pulses of RF energy in a focused, conical beam for the detection of precipitation, and at least two pulses in a given direction are required to determine the Doppler phase shift. The number of such pulses emitted per second is known as the pulse repetition frequency (PRF). Outgoing pulses of energy are emitted by a radar over a very short time period known as the *pulse width*. Given that the speed of light is 3×10^8 meters per second, a pulse width of 1.57 microseconds (typical of a NEXRAD radar) will produce a pulse of RF energy with a physical *pulse length* of 471 meters. The pulse length determines the radial resolution (usually 1/2 the pulse length, or in this case, 235 m).

TABLE 5.15-1
Common Radar Frequencies and Wavelengths*

Radar band	Frequency	Wavelength
X	9.5 GHz	3 cm
C	5.5 GHz	5.5 cm
High S	3.5 GHZ	8.5 cm
S	3 GHz	10 cm

* values represent median bandwidth

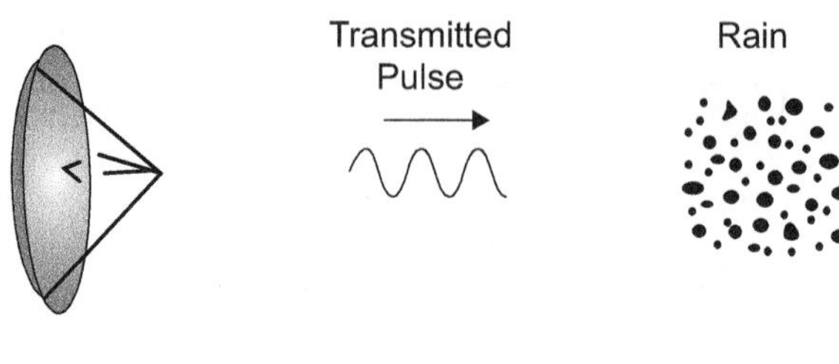

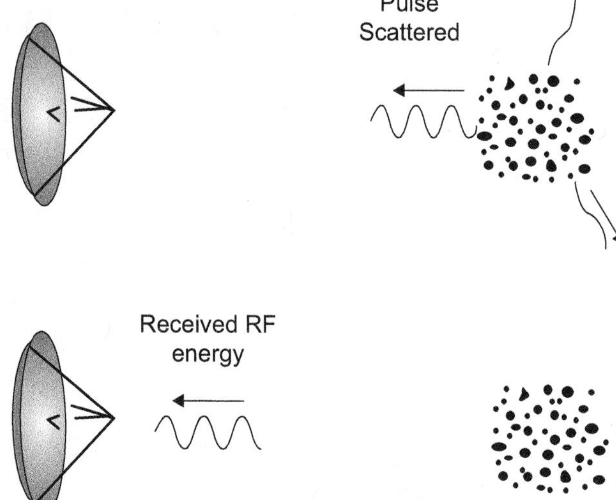

FIGURE 5.15-1 Transmission and Scattering of RF Energy

In modern, dual-polarization radars, two types of RF pulses are emitted. Some are horizontally polarized (meaning the electric field in the RF wave oscillates horizontally), while others are vertically polarized.

The conical beam carrying the transmitted energy has a certain *beam width,* usually given in degrees. The weather radar beam width is typically on the order of 1°. The transmitted radar beam is typically Gaussian, meaning its energy is strongest in the center and gets weaker as it moves outward away from the center of the beam. The beam width is normally the width across the radar beam between the two points where the beam's power drops off to half the maximum power. The radar examines *pulse volumes*, also known as *gates*, which are essentially a volume equal to half the pulse length multiplied by the cross-sectional area of the radar beam at a given range. Since the radar beam is conical, it widens as it travels further from the radar. Therefore, resolution decreases with increasing distance from the radar.

Received Energy Reflected from Targets

Azimuth and Range

Most of the RF energy transmitted in a beam by a radar never returns to the radar. However, certain targets, including precipitation (and aircraft, trees, birds, and insects), scatter the RF energy that hits them. Some of the energy is scattered sideways or forward, but some part of the energy is scattered, or reflected, back toward the radar (Figure 5.15-1). The focusing radar antenna detects this returned energy. Such returned energy is called a *radar echo.*

When the radar is in scanning mode, known as planned position indicator (PPI), it transmits a beam of RF energy at a given *elevation angle* (the angle above horizontal), then receives and processes radar echoes. The radar repeats this process in many different compass directions, known as *azimuth angles*, where 0° is due north, 90° is east, 180° is south, and 270° is west. Typically, a radar in PPI mode will transmit and receive in 360 directions, or every 1° of azimuthal direction. Radars sometimes go through PPI scans at multiple elevation angles, forming what is known as a *volume scan*, or a 3D examination of the precipitation around the radar.

For each azimuthal direction, the distance between the radar and any reflecting targets causing radar echoes must be calculated. This distance of the echoes from the radar is known as the *range*. The distance of a target from the radar may be determined using the known speed of light, and the time between the transmitted pulse and the received pulse. One may calculate the distance an RF pulse travels using the simple equation:

$$d = ct$$

where d is the distance traveled by the EM pulse, c is the speed of light (3×10^8 meters per second) and t is the time in seconds. However, *the radar energy must travel out to the target and back*, so the range (r) of a target from the radar is actually

$$r = \frac{ct}{2}$$

where t is the time between the transmitted pulse and the received pulse.

The radar may receive echoes from multiple targets at multiple ranges in any azimuthal direction. The energy from each echo will return to the radar at a given time, allowing for the determination of its range. An example of such a PPI scan is shown in Figure 5.15-2.

The radar is in the center, the rings are at certain range intervals (in this case 50 km), and the spokes are at certain azimuth intervals (in this case 45°). The area of precipitation, some of it heavy, is indicated primarily to the west and northwest of the radar.

Range Folding

As previously discussed, a radar transmits multiple pulses of energy in each azimuthal direction at a given frequency, or PRF. Once a second pulse has been transmitted, the radar cannot differentiate between energy returning from the first transmitted pulse and the

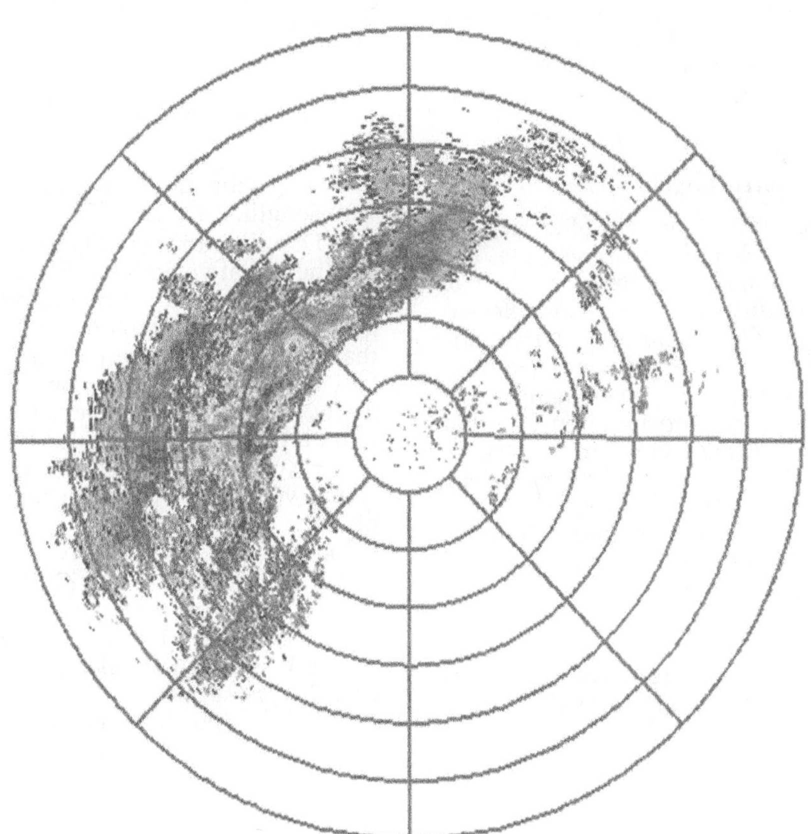

FIGURE 5.15-2 PPI Reflectivity Scan

second transmitted pulse. Due to this factor, if pulse 1 is transmitted at time t_1, and pulse 2 is transmitted at time t_2, the time between the transmission of those two pulses is $t_2 - t_1$ and is inversely proportional to the PRF. This may be written as

$$t_2 - t_1 = \frac{1}{PRF}$$

If a target is far enough from the radar that its returned energy from pulse 1 does not make it back to the radar before the next pulse is transmitted, the radar will assume that the returned energy was actually from pulse 2, not pulse 1. Since range is determined using the time between the transmitted pulse and the received pulse, the range of the distant target will be severely underestimated (because the radar assumes it is reflected from the later pulse 2). This situation is known as *range folding*. The radar echo from the distant target is displayed by the radar at a much closer range than it really is, and is known as a *second trip echo*. Range folding will only occur with targets beyond the range of the radar where the time it takes for the energy to return to the radar is greater than $t_2 - t_1$. This makes a radar's maximum "unambiguous" range, based on the previous two equations:

$$r_{max} = \frac{c}{2(PRF)}$$

Any targets beyond r_{max} will be incorrectly displayed as second trip echoes. One may increase a radar's maximum unambiguous range by decreasing the PRF. Decreasing the PRF allows for more time ($t_2 - t_1$) between pulses and, therefore, more time for a pulse to travel out to a greater distance and back to the radar before the next pulse is generated.

Reflectivity from Distributed Targets

According to Rinehart (1997), there are often at least 1 billion raindrops in a single radar pulse volume. The energy received by the radar from a large number of small targets, known as *distributed targets*, is affected by the number of targets present per unit volume, the average diameter of the targets, and their range from the radar (but the radar adjusts for power changes due to range). The intensity of the precipitation (number and diameter of raindrops) affects how much energy is reflected back to the radar, and is called the *reflectivity* of the distributed targets. Radar reflectivity (z) is normally reported on a logarithmic scale, since it varies greatly based on the diameter of the raindrops. This radar reflectivity factor, Z, is given by

$$Z = 10log_{10}z$$

Keeping this equation in mind, a reflectivity factor of 30 dBZ is actually ten times as much reflectivity as 20 dBZ, 40 dBZ is actually ten times as much reflectivity as 30 dBZ, and so forth.

Doppler Radial Velocity

In 1842, Christian Doppler provided an explanation for the perceived change in color of stars due to their motion. Essentially, as waves of any frequency (sound waves, electromagnetic waves) are being emitted by an object approaching an observer, their frequency of arrival at the observer is increased above their actual frequency. If the object is moving away from the observer, the frequency of arrival will be decreased below the actual frequency. This is known as the "Doppler effect," and is easily observable as a car blowing its horn passes by a stationary person. Radars can use a similar principle. A Doppler radar detects not only the reflectivity of distributed precipitation targets, but also the *radial velocity*, or the component of the velocity of the precipitation targets toward or away from the radar. Positive radial velocities are away from the radar, and negative radial velocities are toward the radar. A Doppler radar sends out multiple pulses of energy at the same distributed target. If the target is moving toward or away from the radar, there will be an *extremely small* difference in the time between transmission and reception from one pulse to the next, because the target will have moved slightly between the two pulses. This time difference causes a shift in the phase of the two returned EM waves (Figure 5.15-3), which results in a frequency shift.

In Figure 5.15-3, Pulse 2 is 45 degrees "behind," or later than, Pulse 1. This is a phase shift of +45°, indicating that the target is moving away from the radar. Analyzing this data, it becomes clear that the second pulse took longer to return to the radar than the first one.

The phase and frequency shift are related to the radial velocity, and the time between pulses is known (through the PRF). The change in frequency (Δf) is given by

$$\Delta f = \frac{2V_r}{\lambda}$$

where V_r is the radial velocity of the target and λ is the wavelength.

A Doppler radar in PPI mode can therefore also map out the radial velocity of the targets around the radar. This radial velocity, when applied to distributed targets like precipitation, implies a wind speed, assuming that the raindrops are moving with the wind. Recall that the radial velocity is only the *component of the velocity toward or away from the radar*. Therefore, unless the wind is blowing directly along the radar beam (toward or away), the radial velocity will be smaller than the actual velocity of the raindrops. And, if the wind is blowing perpendicular to the radar beam, the radial velocity will be zero (Figure 5.15-4).

Figure 5.15-4 shows an illustration of radial velocity (V_r) measured by radar given three actual wind velocities (V). In cases 1 and 2, V is the same speed, but in case 1 V_r is much smaller and inbound (negative). In case 2, V_r is almost as large in magnitude as V, and it is outbound (positive). In case 3, V is perpendicular to the radar beam, so $V_r = 0$.

Doppler Spectrum

The millions of raindrops in a radar pulse volume generally have a variety of radial velocities. Turbulence,

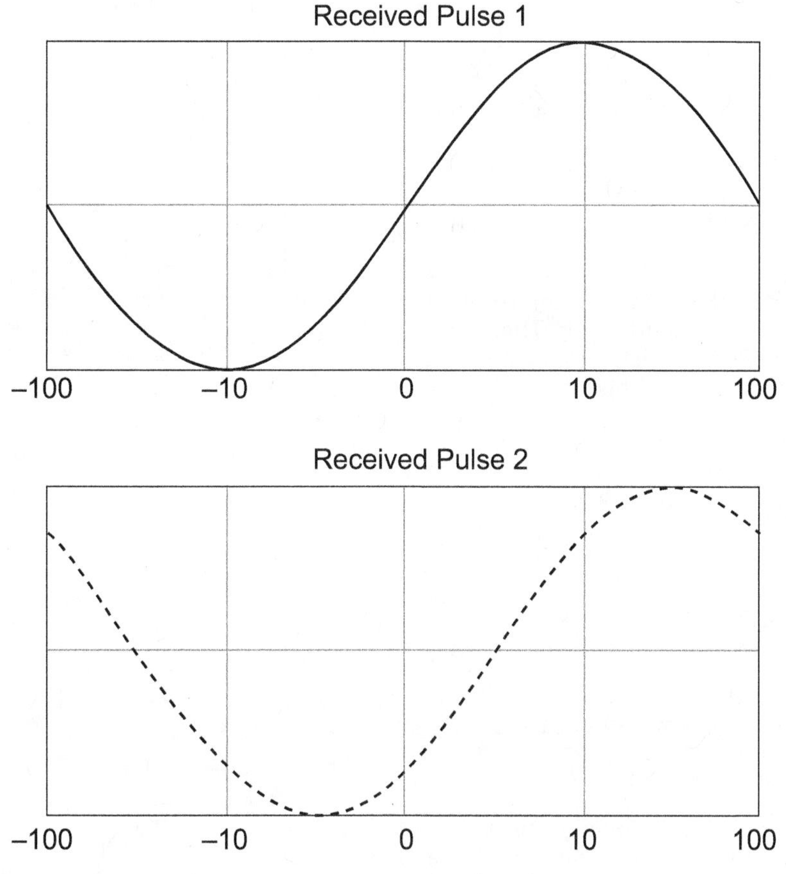

FIGURE 5.15-3 Received Pulse Phase Shift

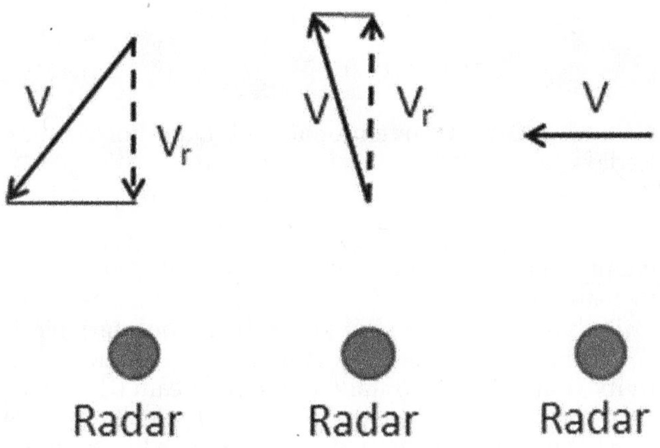

FIGURE 5.15-4 Illustration of Three Radial Velocities

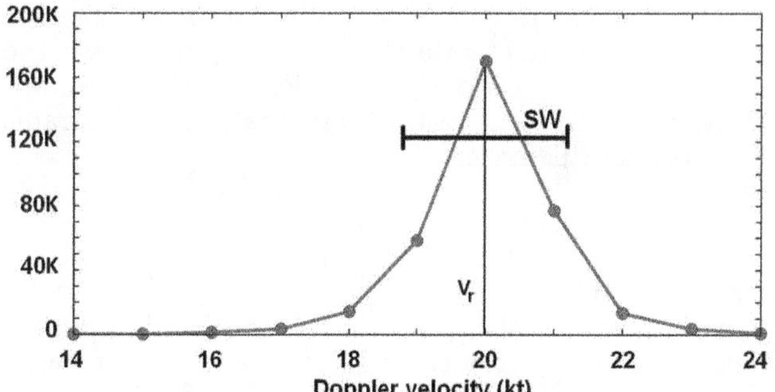

FIGURE 5.15-5 Single Pulse Volume Doppler Velocity Spectrum

raindrop size, and other factors affect the velocity of individual raindrops, and the radar will detect millions of different radial velocities in a pulse volume. This variety of radial velocities is known as the *Doppler spectrum*. The spectrum of velocities is usually a normal distribution forming a bell curve. The median velocity of all the targets is reported as the actual radial velocity in that volume, and the standard deviation of all the radial velocities in the volume is known as the *spectrum width* (Figure 5.15-5). Large values of spectrum width may indicate turbulence, or it may indicate that two different types of small-scale wind flow are overlapping within the pulse volume.

In Figure 5.15-5, the *x*-axis shows *V*r, and the *y*-axis shows the returned power (dB) of reflectors in that volume at that radial velocity V_r. There are reflectors with V_r ranging from +14 knots to + 24 knots. The median is 20 knots, so that would be reported as V_r. The standard deviation of the velocity is 1.2 knots (either side of 20 knots), so that is the spectrum width (SW).

Velocity Folding

The phase shift between two received pulses provides the information necessary to determine the frequency shift and, therefore, the radial velocity. A sinusoidal wave of energy goes through 360 degrees of phase in one wavelength. The phase shift is usually indicated between −180° and +180°. If the radial velocity of a target is large enough, the phase shift may be larger than 180°. On a sinusoidal wave, +190° is at exactly the same phase as −170°, +200° is at the same phase as −160°, and so on. However, the radar will interpret all phase shifts to be between −180° and +180°. Therefore, an actual phase shift of +200°, which should imply a large outbound radial velocity, will be interpreted by the radar as a phase shift of −160°, which would imply a large inbound radial velocity. This is known as *velocity folding*. Therefore, a Doppler radar has a maximum unambiguous velocity that it can determine, both inbound and outbound, before velocity folding occurs.

This is known as the *Nyquist velocity*, and is given by

$$V_{max} = \frac{\lambda(PRF)}{4}$$

This indicates that a radar's Nyquist velocity is increased by longer wavelength or higher PRF.

The Doppler Dilemma

A large PRF provides a larger unambiguous velocity, or Nyquist velocity. However, as shown earlier, a small PRF produces a larger unambiguous range for the radar. Depending upon the data required, a PRF frequency must be chosen to either unambiguously detect large radial velocities or unambiguously detect targets at long range. This is known as the *Doppler dilemma*. As shown by Rinehart (1997), if one solves the equations for r_{max} and V_{max} for PRF and then equates them, the following result mathematically illustrates the Doppler dilemma:

$$V_{max} r_{max} = \frac{c\lambda}{8}$$

For a given wavelength λ, increasing V_{max} implies a decrease in r_{max}, and vice versa. It should also be noted that, regardless of PRF, a radar with a longer wavelength λ will allow for larger unambiguous ranges and/or larger Nyquist velocities than a radar with a shorter wavelength. This is one of the reasons why radars with longer wavelengths, such as S-band radars, are a better choice overall than shorter wavelength radars, such as C-band (Figure 5.15-6).

Dual-Polarization Variables

As discussed in the Transmission of Outgoing Energy section, many modern radars are dual-polarization radars. These radars emit RF pulses where the electric field is horizontally polarized, and other RF pulses where the electric field is vertically polarized (see Figure 5.15-7). Differences and correlations between the received horizontally and vertically polarized RF

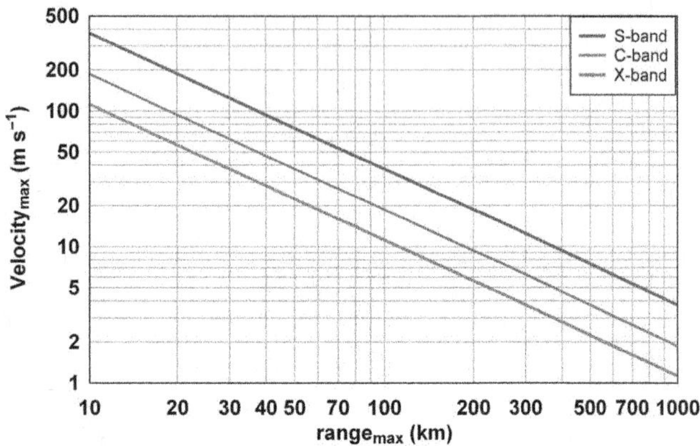

FIGURE 5.15-6 V_{max} versus r_{max} for S-, C-, and X-Band Radar

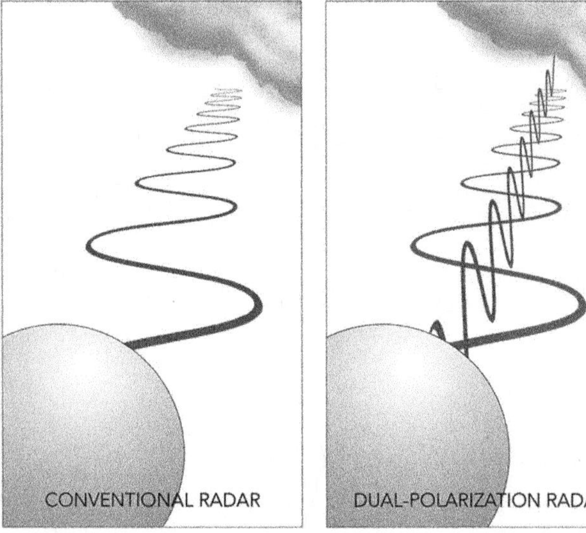

FIGURE 5.15-7 Conventional and Dual-Polarization Radars

energy may be used in several ways to improve radar measurements.

Perhaps the most well-known dual-pol variable is *differential reflectivity* (Z_{DR}). It is the ratio of the reflectivity from the horizontally polarized beam (Z_H) to the vertically polarized beam (Z_V). It is typically written logarithmically and expressed in dB, as in the equation below:

$$Z_{DR} = 10\log\left(\frac{Z_H}{Z_V}\right)$$

One example of the use of Z_{DR} is in identifying precipitation type. Since raindrops are generally oblate and fall with the long axis to the ground, Z_H is often larger than Z_V, so $Z_{DR} > 0$ dB. Hailstones, on the other hand, are more spherical, and may even fall with a slightly vertical orientation, so Z_{DR} in hail is either closer to 0 dB, or in some cases may be negative (meaning $Z_V > Z_H$).

Another common dual-polarization variable is *differential phase*, or Φ_{dp}. As both horizontally and vertically polarized RF energy passes through precipitation, it is slowed down slightly, due to the increased refractive index of water (or ice). This change in the speed of the RF waves will cause a slight change in the phase at which they return to the radar. Differential phase is the difference between the phase of the received horizontally polarized RF energy Φ_H and the phase of the received vertically polarized RF energy Φ_V from each given range, or approximately $\Phi_H - \Phi_V$. Since most weather targets have a $Z_{DR} > 0$ ($Z_H > Z_V$), the horizontally polarized energy will be slowed down more than the vertically polarized energy as it passes through weather targets, and therefore will have a larger change in phase. So, Φ_{dp} almost always increases with range from the radar. The rate of increase of Φ_{dp} with increasing range from the radar, $\Delta\Phi_{dp}/\Delta r$, is known as the *specific differential phase* K_{dp}. This quantity provides another estimate of rainfall rate, and it is not affected by attenuation.

The third main dual-pol variable is called the correlation coefficient (ρ_{HV}). It indicates the correlation between the horizontally polarized and the vertically polarized received energy from a given range gate. If all of the targets in a range gate are exactly spherical (all have Z_{DR} = 0 dB), or even exactly the same shape of raindrop (all have Z_{DR} = 3 dB), then the correlation is perfect, and ρ_{HV} = 1. It is often between 0.95 and 1 in rain. However, if there is a mixture of targets within a range gate with different shapes, in other words if Z_{DR} varies within the range gate, then the correlation will not be as high. Low values of ρ_{HV} may indicate a mixture of rain and hail, large snow aggregates, ground clutter, bioflyers, or tornado debris. The latter has become one of the most important uses of ρ_{HV} over the past few years. Correlation coefficient, in addition to the other dual-polarization variables, will be discussed in more detail in the Radar Meteorology section.

Other Considerations with Weather Radars

Weather radar is extremely complex, and a thorough discussion of every detail or potential problem that may occur with Doppler radar is beyond the scope of this section. However, several issues are particularly important and will be addressed here.

Attenuation

It must be considered that as a radar beam goes through one area of precipitation, some of the RF energy is scattered away and some is absorbed by the raindrops and so forth. Most of the radar beam continues onward beyond the initial precipitation it encounters and can detect additional precipitation at further ranges from the radar. However, some of the beam's energy is lost through scattering and absorption in each area of precipitation it encounters; this is *attenuation*. Since the radar beam is less energetic once it is attenuated, there will be less energy to illuminate, and therefore to reflect back from, targets beyond where the attenuation occurs. This causes radar echoes in an area where the beam is attenuated to appear weaker than they really are.

Attenuation is more pronounced at higher frequencies that have shorter wavelengths. This is why the NWS NEXRAD system uses the longest wavelength, S-band (10 cm). Using equations from Burrows and Attwood (1949) and Doviak and Zrnic (1993), and interpolation, the two-way attenuation (the signal gets attenuated going away from the radar and coming back, going through rain both times) may be calculated in dB of attenuation for a given distance traveled in a certain intensity of rain. Using a rain rate of 1 inch per hour (about 45 dBZ under typical conditions) at a temperature of 64°F, the following two-way attenuation occurs at each radar wavelength over a distance of 5 miles of radar beam propagation: X-band 7.8 dB, C-band 0.8 dB, high S-band 0.4 dB, S-band 0.1 dB. So, a C-band radar signal is attenuated quite a bit more in significant rain than S-band. At high S-band, newly available to television stations, the attenuation is only half of that at C-band, the traditional television radar.

Beam Propagation

Another major issue with radar is the propagation pattern of the beam away from the radar. In a nutshell, the radar beam generally bends toward the earth somewhat due to refraction, but the earth curves out from under the beam even more quickly. So in most cases, the net effect is a radar beam that gains altitude above the ground with increasing range. As illustrated in Figure 5.15-8, both horizontal and refracted beams propagate relative to the earth's curved surface. Both gain altitude relative to the surface as the range increases.

If only the earth's curvature is considered, ignoring refraction, a perfectly horizontal radar beam gets to a fairly high altitude above the earth's surface in a fairly short range (see Figure 5.15-8(a)). However, the speed of light is only the standard 3×10^8 meters per second in a pure vacuum. In real media such as the atmosphere, RF energy is slowed down just slightly below that. A radar beam travels more slowly in a denser atmosphere (mainly related to temperature and pressure). When RF energy, such as a radar beam, encounters the real atmosphere, where the air density varies with height, it is usually refracted. *Standard refraction*, or a given profile of the change of the speed of light with height due to changes

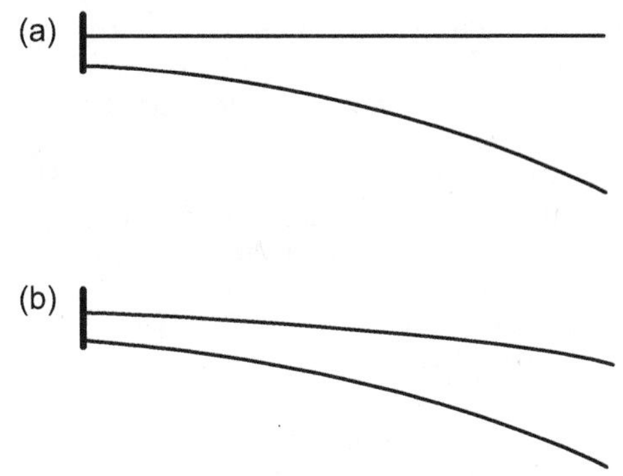

FIGURE 5.15-8 Horizontal and Refracted Beams

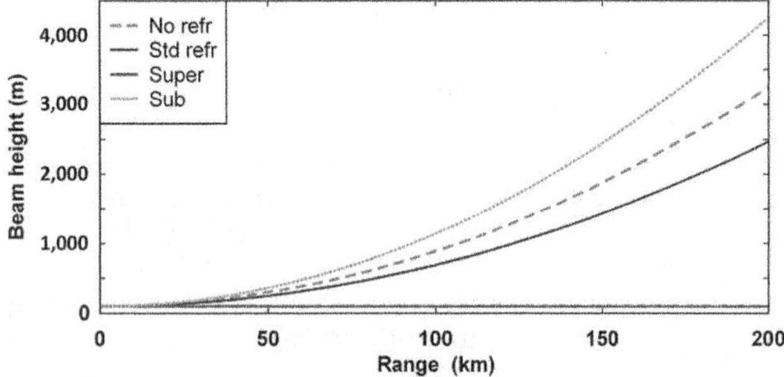

FIGURE 5.15-9 Radar Refractions

in the atmosphere, is typically assumed. In the case of standard refraction, the radar beam is bent downward toward the earth's surface somewhat, but not enough to offset the earth's curvature out from under it. So, the radar beam slowly gains altitude with increasing range (see Figures 5.15-8(b) and 5.15-9).

Sometimes, in the case of an extremely rapid decrease in density with height, the radar beam bends downward toward the ground more quickly than is normal, a situation called super-refraction (Figure 5.15-9).

If super-refraction becomes intense enough to bend the radar beam to the point that it remains essentially parallel to the ground, it is called anomalous propagation (AP). Super-refraction and AP occur most often at night, when a temperature inversion is in place. Especially in cases of super-refraction, some of the radar beam may reflect back to the radar from non-meteorological targets such as trees, buildings, insects, and aerosols including pollen and pollution. These are known as *ground clutter*.

In some rare cases, atmospheric density may increase with height, causing sub-refraction. However, this is rare enough to require no further study herein.

Beam Blockage

Even under conditions of standard propagation, solid non-meteorological objects such as trees, buildings, and mountains may block part of the radar beam, preventing it from propagating beyond the location of these objects. Such radar *beam blockage* may be partial, in that only part of the radar beam is blocked from continuing to propagate to ranges beyond the object(s) causing the blocking. In some cases, especially in the mountainous areas in the western United States, the beam may be fully blocked in some directions.

When the radar beam is fully blocked at any azimuth and range, it cannot detect any precipitation beyond the range of the blocking at that azimuth. When the beam is only partially blocked, then some percentage of the radar beam is blocked, but the remainder of it continues on beyond the range of the blocking. This reduces the energy in the beam that has been partially blocked, so if that beam encounters precipitation, the amount of energy hitting the precipitation is reduced, and therefore the energy returned to the radar is reduced, and that precipitation appears to the radar as being lighter than it really is.

A simplified example of partial beam blocking is shown in Figure 5.15-10*. One half of the radar beam is blocked by mountains at a range of 50 km, between azimuths 315° and 360°.

This results in the loss of half of the radar beam beyond the blocking terrain, and reduces the energy returned to the radar by any rain beyond 50 km (in those directions) by half, reducing the reflectivity by 3 dBZ (recall that reflectivity is logarithmic).

Volume Control Patterns (VCP)

As previously discussed, weather radars can go through multiple PPI scans at various elevation angles, producing a three-dimensional examination of the precipitation around the radar. The various types of 2D or 3D *volume scans* produced by a radar are controlled by the user-selected VCP (Figure 5.15-11).

The National Weather Service, during widespread severe weather, often uses a VCP for its NEXRAD radars that combines 14 PPI scans, each at different elevation angles (from 0.5° above the horizon to 19.5°). This VCP produces a 3D view of storms, and takes about 4.5 minutes to complete (each PPI scan lasts about 20 seconds). During good weather, the NWS often uses a "clear-air" VCP that performs only five low-elevation PPI scans, and the radar moves much more slowly to increase its sensitivity. During severe weather, television stations that own their own radars often use a VCP with only one to three low-elevation PPI scans that can be completed in less than 30 seconds. This provides an update on the intensity and location of storms much more frequently than any VCP used by NWS for their radars, and is often used as a competitive advantage and marketing tool for the TV station radar.

One problem encountered with the use of any feasible VCP is that the highest elevation angle is usually

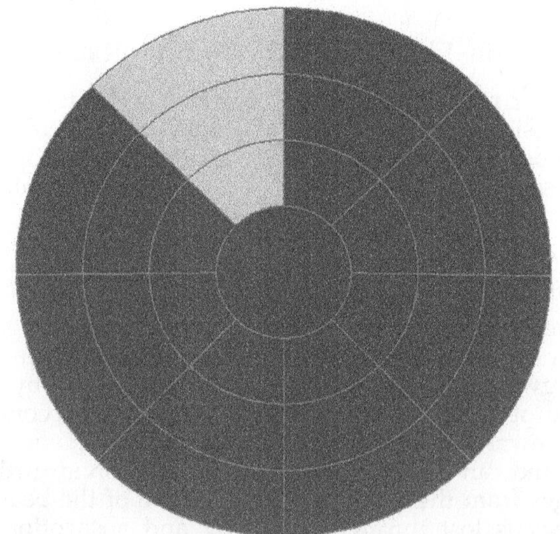

FIGURE 5.15-10 Partial Beam Blocking

***Dark gray indicates reflectivities of 30 dBZ or more, and light gray indicates reflectivities of 29 dBZ or less in this image. Range rings are every 50 km, and azimuth spokes are every 45 degrees.**

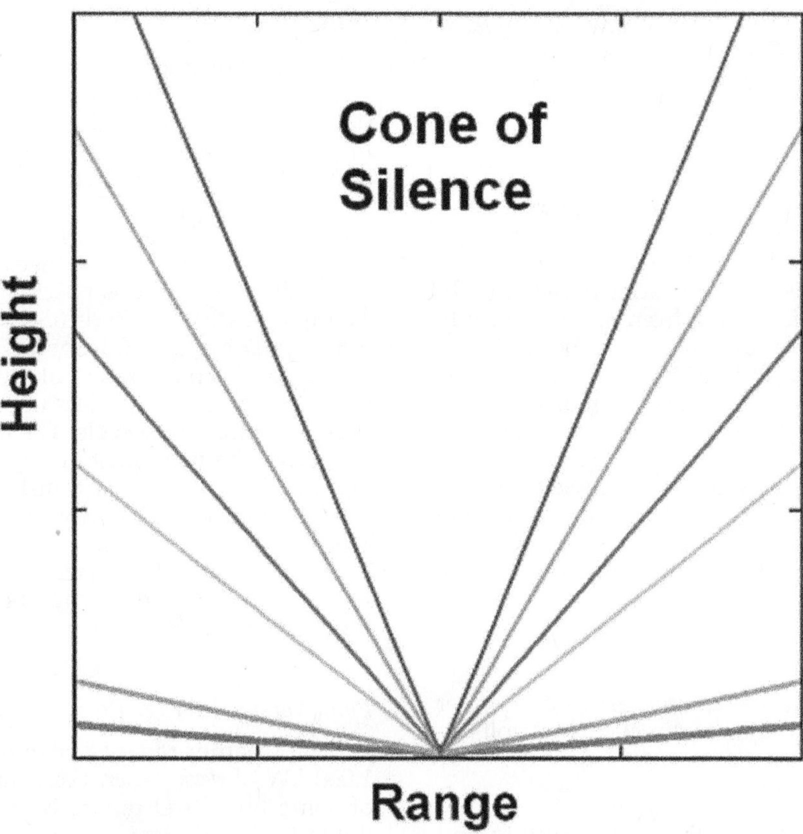

FIGURE 5.15-11 Typical VCP

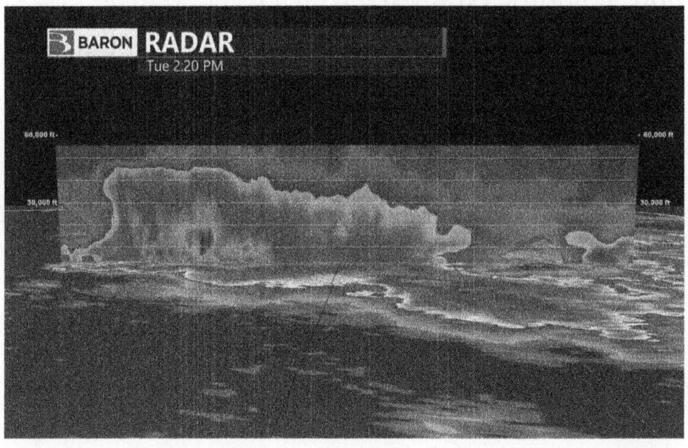

FIGURE 5.15-12 RHI Scan of Thunderstorms at Range

below 20°. For storms very close to the radar, even at the highest elevation angles, the radar may not be able to scan the top parts of the storms. This volume around the radar where no PPI scan has a high enough elevation to examine the environment is known as the *cone of silence*. Within the cone of silence, there are no radar PPI scans and, therefore, no reflectivity or velocity detection.

Occasionally, meteorologists wish to examine a storm's vertical structure in great detail. So, instead of using a PPI scan, a *range height indicator* (RHI) scan examines elevation angles up to 60° or more in a fixed azimuthal direction. An example of an RHI scan is shown in Figure 5.15-12.

WEATHER RADAR HARDWARE AND DESIGN

A weather radar is composed of a complex set of components that are finely tuned. This is due to the fact that extremely small changes in returned power and frequency from targets have a large effect on data being gathered. The main radar components, some of which have multiple subcomponents, are listed below:

1. The *transmitter* that generates the RF pulse.
2. The *duplexer*, a switch that changes the radar from transmission to reception in millisecond-range time intervals. It also protects sensitive components.
3. The *antenna*, which focuses the outgoing RF pulses into a conical beam. It also "amplifies" the incoming reflected energy.
4. The *pedestal*, which supports and rotates the antenna.
5. The *receiver*, which processes returned energy from targets.

In this section, each of these components and their subcomponents are covered. Additionally, the importance of radar calibration, the computer that generates the radar products meteorologists use (known as the *radar product generator* [RPG]), and various output sources for the radar products are discussed.

Transmitter

There are multiple types of weather radars in use today that transmit RF energy at frequency f_T. The type of radar primarily depends on the type of transmitter. A *magnetron* is the oldest type of radar transmitter. In a magnetron, an electric current from a power supply is sent through the center of a vacuum tube. The tube contains fixed magnets cut into shapes that, when the electric current passes through the center, produce resonance and EM waves. However, the waves produced by a magnetron are not at a stable phase or frequency, so the changing frequency must be processed through additional components. Since it has a more complicated structure than a *klystron* radar or a *solid-state* radar, it will be used to demonstrate the functions of the various radar components.

Power Supply

Every radar must have a stable, uninterruptible power supply, producing thousands of direct current volts.

Modulator

A modulator is connected to the power supply and essentially turns the radar transmitter on and off to provide the correct pulse width (the time period at which that energy is being transmitted). The modulator also serves a similar role as a capacitor, in that it stores up electric charge from the continuous power supply between transmitter pulses.

Duty Cycle

When talking about transmitter pulses, one usually hears the terms pulse width, pulse repetition frequency (PRF), and duty cycle. The pulse width is the period of time the transmitter pulse is present (typically 400 ns to 2 μs for weather radar). The PRF is the frequency of the transmitter pulses, typically 250 to 3000 Hz, and the pulse repetition time (PRT) is the reciprocal of the PRF. Transmitters are specified with a maximum *duty cycle*. The duty cycle is specified as a percentage. For example, a duty cycle of 0.15% states that the transmitter can only transmit for 0.15% of the time without damage. The combination of the pulse width and PRF must yield a duty cycle less than the transmitter maximum specified duty cycle. The longer the pulse width, the lower the PRF must be. In the example shown in Figure 5.15-13, the radar transmits its peak power (750 kW) for a very short period of time (the pulse width). This is done once during each pulse repetition time. The duty cycle is the pulse width divided by the pulse repetition time. The average power is the duty cycle times the peak power.

Peak Versus Average Power

Typical weather radars transmit between 250 kW and 1,000 kW of *peak power*. For example, the peak power of some NEXRAD radars is 750 kW. However, since the radar is only transmitting a very small percentage of the time (the duty cycle), the *average power* is much smaller than the peak power. The average power may be computed by multiplying the duty cycle by peak power. For the NEXRAD, with a duty cycle of 0.157% and a peak power of 750 kW, the average power is only 1178 W. The radar produces very short bursts of

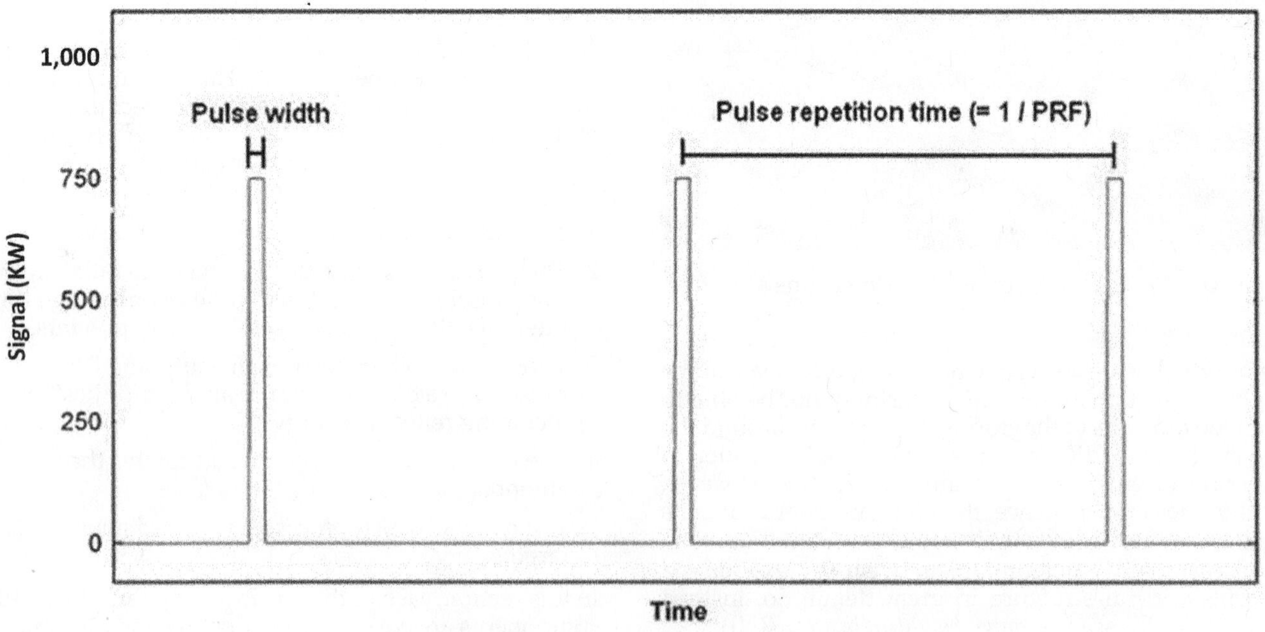

FIGURE 5.15-13 Duty Cycle Graph

enormous power by storing up power in the modulator during the time between pulses.

Waveguide

Given the extremely high frequencies and power associated with the RF pulses, a coaxial cable or other simple conductor is not sufficient to transmit the signal produced by the transmitter to the radar antenna. Instead, hollow, usually rectangular, metal tubes are used. These metallic tubes not only act as a Faraday cage, which keeps external RF radiation out, but they also have significantly lower power loss than a cable. The size of the waveguide depends on the frequency (and wavelength) of the radar.

The waveguide for transmission of the radar beam generally exits the transmitter and goes through the duplexer and then to the radar antenna through the antenna feed, as shown in Figure 5.15-14.

Antenna Feed

An *antenna feed* (also known as a feedhorn), or horn-shaped end to the waveguide (Figure 5.15-14), greatly broadens the physical size of the stream of emission of RF radiation as compared to the waveguide alone. The antenna feed allows the transmitted EM radiation to spread across the radar's parabolic antenna. The antenna then focuses the radar beam into the desired beam width.

Duplexer

The *duplexer*, also known as the *transmit/receive switch*, is located between the radar transmitter and the antenna (Figure 5.15-15).

Since the same antenna and section of waveguide are used to transmit the outgoing radar beam and to receive the reflected energy from targets, the duplexer must be in place to prevent the high-powered transmitted energy pulse from traveling down the receiving waveguide. The receiver components are not designed to be exposed to such large amounts of energy. When the radar is transmitting energy, the duplexer closes the portion of the waveguide that feeds into the receiver. While the radar is not transmitting, it switches, closing the waveguide that leads from the antenna to the transmitter and opening the waveguide from the antenna to the receiver.

Antenna

Parabolic Reflector

The radar antenna is usually a parabolic reflector. It focuses the energy transmitted to it by the antenna feed into a conical beam of RF energy of desired beam width. It is then transmitted away from the radar. The reflector also receives and focuses the RF energy reflected from targets, such as precipitation. The ratio of focused received power from the antenna to that of an isotropic receiver is known as antenna gain.

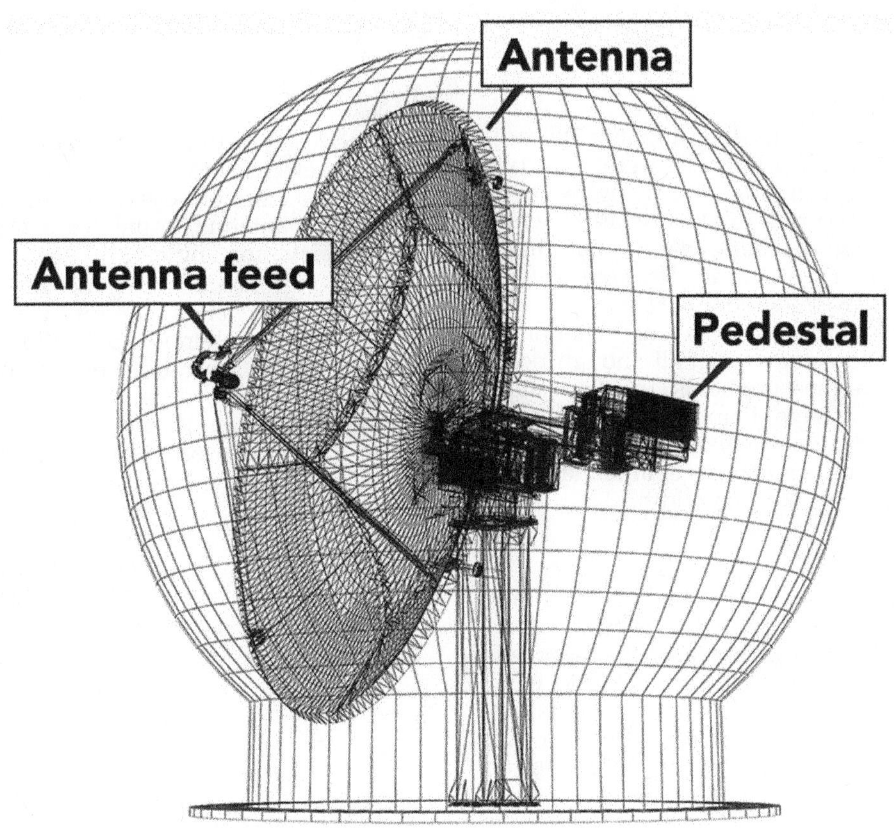

FIGURE 5.15-14 Pedestal-Mounted Radar Antenna and Feed

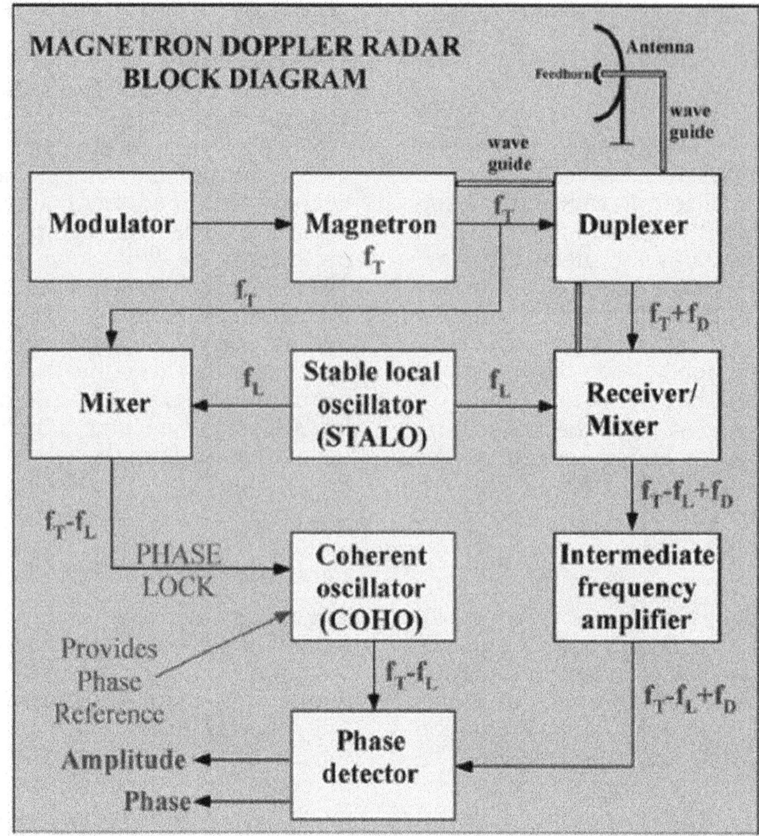

FIGURE 5.15-15 Magnetron Doppler Radar Block Diagram

Size Versus Wavelength

The parabolic antenna focuses the outgoing RF radiation into a conical beam. A larger reflector (i.e., one with a larger diameter) can more effectively focus the energy into a smaller beam. A smaller beam is usually more desirable for weather radars, as it provides higher-resolution measurements. The angular *beam width* is proportional to the wavelength of the radar's RF radiation and inversely proportional to the diameter of the antenna. Therefore, to achieve a small enough beam width for precise radar measurements (1° is fairly standard), one must use a large diameter antenna, especially for radars with longer wavelengths. For example, an S-band radar requires a much larger antenna than a C-band radar to achieve a 1° beam width.

Radome

The antenna of most weather radars is contained within a radome, a spherical dome that is built around the antenna's entire sweeping domain. This radome is necessary to prevent errors in accuracy due to wind loading, and to protect the radar from precipitation. The radome is typically coated with a hydrophobic coating that repels water, so that it will not retain too much moisture that would cause excessive attenuation of the radar beam, along with the returned radar energy from precipitation targets.

Pedestal

The pedestal supports the radar antenna and contains the electric motors that rotate it. The rotation can pivot on a horizontal axis, changing the azimuth of the radar's antenna, or on a vertical axis, changing the radar's elevation angle.

Key components of the pedestal are the azimuth and elevation *rotary joints*. The rotary joints typically have one waveguide channel and optionally coax channels depending on the radar configuration. These rotary joints are capable of transferring the high-powered RF pulses through the moving pedestal to the antenna feed. They can also be used to transfer the low-level signal returns from the antenna feed back to the signal processor.

For electrical and communications, a *slip ring* package is used. These rotate in azimuth and contain multiple signal paths to pass electrical power and communications signals to the elevation section of the pedestal.

Receiver

The receiver contains several components involved in converting and analyzing the energy returned to the radar.

Low-Noise Amplifier (LNA)

The signal returned to the radar from precipitation targets is very weak, even after the gain from the antenna. Therefore, a low-noise amplifier (LNA) is used to boost the power of the signal. The LNA has a low *noise figure*, in that it introduces very little additional noise to the signal it is amplifying beyond what noise was already in the signal before it came through the amplifier. This helps maintain a high *signal-to-noise ratio* that is important for radar system accuracy.

Downconverter and Coherence of Signal

The RF energy received is at a frequency $f_T + f_D$, where f_T is the transmitted frequency and f_D is the Doppler frequency. The received RF energy is then mixed with some much lower, but very stable frequency f_L, produced by a STAble Local Oscillator (STALO). This lowers the overall frequency of the received radar energy (see Figure 5.15-13). The radar energy, at microwave frequencies above 3 GHz, is converted to intermediate frequencies (IF, typically 50–100 MHz). This new, mixed signal at a lower frequency ($f_T - f_L + f_D$) is transmitted to an analog *intermediate frequency amplifier*. The combination of a *mixer*, stable local oscillator, and IF amplifier are, together, known as a *downconverter*. A radar signal with a lower frequency is more easily processed (e.g., Skolnik 1980; Rinehart 1997).

As a Doppler radar uses the change in phase to determine Doppler radial velocity, the radar must be coherent; the Doppler frequency f_D from targets must be determined even though a magnetron may not produce a stable frequency from one pulse to the next. To accomplish this, the energy from the STALO is not only mixed with the received energy from the antenna, but also with the transmitted RF energy in a separate mixer. This mix produces energy at a frequency $f_T - f_L$. It then goes through a *coherent oscillator* (COHO) that maintains the frequency $f_T - f_L$, and also maintains the phase relationship between the mixed signal and the transmitted signal. The mixed, phase-locked frequency $f_T - f_L$ is then sent to the phase detector (see Figure 5.15-15).

The frequency coming from the COHO into the phase detector is $f_T - f_L$, and the frequency coming from the downconverter into the phase detector is $f_T + f_D - f_L$. This being the case, the frequency $f_T - f_L$ may be cancelled out. f_D, which is the accurate Doppler frequency of targets, is determined even though the magnetron may not be emitting the same f_T in each pulse. The actual Doppler frequency f_D of two successive pulses may then be determined, and the difference in frequencies between two pulses provides the Doppler velocity (see the Received Energy Reflected from Targets section).

IF Digitizer

The signal coming from the phase detector contains amplitude information (proportional to the amount of returned energy and, therefore, the radar reflectivity) and phase/frequency information, allowing for determination of Doppler velocity. This analog information is then passed through a digitizer, so that computers may analyze the information and determine the base moments. Modern IF digitizers (IFD) typically have 16 bits of resolution.

Signal Processor

Information on amplitude and frequency shift is analyzed by the signal processor. The signal processor is sometimes considered part of the receiver. It produces the *base products*, which include reflectivity, radial velocity, and spectrum width. For dual-polarization radars, differential reflectivity, differential phase, and correlation coefficient are also computed. Ground clutter suppression is often performed by the signal processor.

In state-of-the-art radars, the receiver and most of the signal processor are mounted to the antenna inside the radome. This is done to minimize signal loss, whether to a waveguide, cable, or the transmitter at the base of the building.

Calibration

Value of Calibration

A weather radar must be calibrated properly in order to provide accurate measurements of both the location and intensity of precipitation. A radar may become uncalibrated slowly through a variety of processes; these include, but are not limited to, a slight change in output power, a change in the hydrophobic coating on the radome, a change in antenna gain, and changes in temperature causing electronic misfires. Due to these factors, the radar must be periodically calibrated so that true reflectivity values are received. In addition, the accuracy of the radar's azimuth and elevation must be maintained. Otherwise, there will be geometric errors in the location of radar echoes.

Impact of Errors in Calibration

Small errors can affect numerous radar-derived parameters. For example, if the radar is underestimating reflectivity by 3 dBZ, the following errors may occur: radar-derived rainfall rates may be underestimated by 30% or more at high rainfall rates, vertically integrated liquid may be severely underestimated, and estimated hail size may be underestimated. Errors become much larger at higher reflectivities, since reflectivity is logarithmic. Such errors may lead to the failure to detect flash flooding, failure to detect severe hail, and other issues.

Once dual-polarization radar is considered, the Z_{DR} bias must be determined. This is the error or bias induced by imbalances between the horizontal and vertical transmit and receive paths. Z_{DR} accuracies of better than ±0.2 dB are required to make accurate Z_{DR}-based measurements.

Finally, if the indicated azimuth angle of the radar has a 1° error, this translates into an error of 3.5 km (about 2 miles) at a range of 200 km (125 miles). The elevation angle is even more critical, as radar propagation patterns (see the Other Considerations with Weather Radars section) are assumed given atmospheric conditions, and an error of only 0.1° in elevation may result in an error of around 400 meters in elevation above ground level (AGL) at a range of 200 km.

Calibration Techniques

There are several techniques for the calibration of weather radars. Some of these work in real time, while others must be done periodically.

For periodic calibrations, several methods can be used. One common technique is called the *bird bath* calibration. This calibration ensures that the horizontally and vertically polarized beams are correlated with each other. Here, a radar is pointed vertically, at 90 degrees elevation. As will be discussed in the Radar Meteorology section, raindrops fall as oblate ellipsoids; their horizontally polarized reflectivity is larger than their vertically polarized reflectivity. However, if the radar is pointed vertically in rain (so that the radar antenna looks like a bird bath), both the "horizontal" and "vertical" radar beams will be along the same axis of the raindrops, so there should be little or no difference between horizontal and vertical reflectivity. If there is a large difference, then there is a calibration problem with one or both of the perpendicular scans. The problem with this calibration method is that it can only be performed in certain weather conditions. After the radar has changed temperature, enough Z_{DR} bias is typically created to cause significant Z_{DR} errors.

Solar scans are used to calibrate the accuracy of the radar's elevation and angle measurements and its measurements of reflectivity. Given easily accessible azimuth and elevation locations for the sun at any given time and location, the radar may be pointed in the general direction of where the sun is supposed to be, and scan around that region in a small window of azimuth and elevation. If the maximum power from the sun is indicated at the correct azimuth and elevation angle, and an exponential decrease occurs as the radar is pointed toward azimuths and elevations slightly away from the direct sunlight, then the radar is calibrated correctly for its azimuth and elevation (see Figure 5.15-16). In addition, the radar may be aimed directly at the sun, and moved slightly back and forth across the sun's center to calculate the antenna gain, and therefore the received power, for the radar. It is compared to known values of incoming energy from the sun, available on many Internet sites (see Figure 5.15-16).

Real-Time Calibration

Numerous methods of real-time calibration of radars may be employed. Test signals are generated after each volume scan, sent through the LNA, STALO, and IF digitizer, then examined by the receiver. If the receiver determines that both the horizontally and vertically polarized returned power is within tolerance, the radar is correctly calibrated. If not, adjustments must be made.

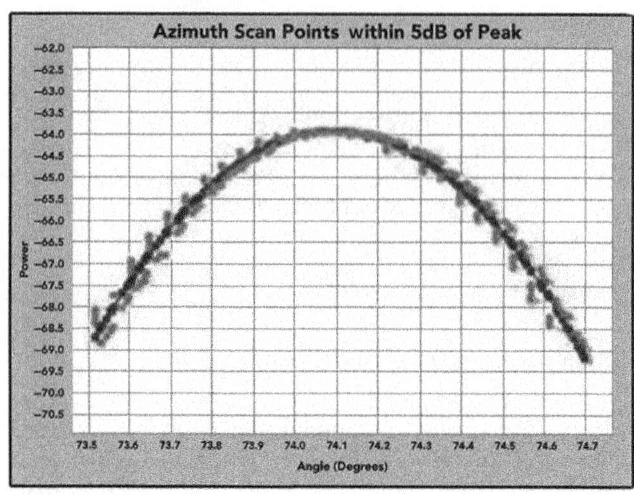

FIGURE 5.15-16 Solar Scan Calibration

Radar Product Generator (RPG)

The RPG is a computer system that receives base data (reflectivity, radial velocity, spectrum width, and dual-pol base variables) and generates radar products. These products include simple PPI maps of base data at different elevation scans. However, they also include *derived products* calculated using base data. These include products such as vertically integrated liquid (VIL), storm-relative velocity (SRM), composite reflectivity, rainfall/snowfall rate and accumulation, hail probability and size, and mesocyclone and tornado vortex signature products. The most sophisticated RPGs will create value-added products, such as shear rate, shear markers, tornado debris signatures, and storm tracks, to help on-air meteorologists convey weather threats to their viewers easily.

Output Data Sources

Basic Displays

Data from modern radars is displayed in a variety of ways, depending on the requirements of the user. For example, if the radar data is being used by National Weather Service meteorologists, basic maps of radar reflectivity, Doppler velocity, spectrum width, and derived products may be superimposed over a simple geographical map, allowing forecasters to determine the location of precipitation, storms, or wind data. Research meteorologists may use more complex displays of the same variables, with less geographical information.

Television Station Graphics Systems

In television stations, some computer systems are used exclusively for the display of radar data, either from the TV station radar or an NWS radar. These systems must show the reflectivity and velocity data on precise and aesthetically pleasing maps. The maps must be detailed enough to allow for large zoom factors; identification of small surface roads; small towns; the location of venues such as schools, churches, and sports stadiums; as well as TV station storm spotter locations. In addition, derived parameters such as mesocyclone locations, tornado potential, and future storm track (including arrival times for downstream towns) are typically options in the TV station radar display.

Also, most TV graphics systems have the ability to create *composite images*, or the reflectivity from multiple radars across a region or the entire United States. This radar data is sometimes overlaid with satellite data.

Raw Data

TV station radar data may be disseminated in a raw format so that it can be displayed on the TV station's websites or its own weather apps in real time.

Notes on Klystron and Solid-State Radars

It should be pointed out here that many radars use klystron transmitters, which emit RF energy at a very stable frequency from pulse to pulse. These radars still require a STALO to lower the frequency of received energy to values that are better managed by computers and other components. However, they do not require the STALO to transmit to a separate mixer with the transmitted frequency and then to a phase-locked COHO (see Figure 5.15-15), since their frequency is stable.

Some radars, generally radars that are either examining close-range targets or large targets, use solid-state transmitters. Current state-of-the-art solid-state transmitters are typically limited to a few hundred or a few thousand watts as compared to the hundreds of kilowatts of the klystron radars. While some techniques may be used to improve range and sensitivity for these lower-powered radars, there are always trade-offs. For a typical broadcast radar customer, the cost of a high-power solid-state system is prohibitive.

RADAR METEOROLOGY

All of the technology and equipment that make up weather radar, discussed in Sections 1 and 2, are for the purpose of improving the detection and forecasting of weather events. These two main tasks encompass multiple weather events, problems, advantages, and equipment. Those subjects are covered in this section, with specific examples listed below:

- Hazardous and nonhazardous weather phenomena including rainfall, flooding, snowfall, ice, wind profiles, severe windstorms, large hail, and tornadoes.
- Common problems with weather radars and how to recognize them, including bright bands, ground clutter, and attenuation.
- The advantages of dual-polarization radars, such as precipitation type determination (rain vs. hail, rain vs. snow, etc.); improvement in radar-estimated rainfall; and tornado debris detection.
- Radar products, including vertically integrated liquid, composite reflectivity, and many others.

Radar Reflectivity

As discussed in the Received Energy Reflected from Targets section, the reflectivity from a pulse volume, or range gate, is proportional to the number of reflectors in that volume and the diameter of those reflectors. Usually, those reflectors are raindrops, but they may also be cloud droplets, snowflakes, ice pellets, or non-meteorological targets such as insects, trees, and so forth. Higher reflectivities generally indicate heavier precipitation.

Precipitation Location and Intensity

Weather radar is generally run in PPI mode, meaning the radar scans a cylindrical volume of the atmosphere (or a circular area, as shown in plan view), at a constant elevation angle. This allows for a snapshot of the location and intensity of precipitation for each PPI scan. This information is often placed on top of a map of the radar domain that may contain counties, cities, roads, and so forth, to make placement of the precipitation easier (Figure 5.15-17). If volume scans are being used, the vertical extent of precipitation echoes may be determined, providing a 3D picture. Using volume scan information, the radar operator may either examine high elevation PPI scans to determine the vertical extent of precipitation features (Figure 5.15-17—right side), or cut user-defined cross-sections through the volumetric data (Figure 5.15-18).

Figures 5.15-17 and 5.15-18 show a line of heavy rain (presumably thunderstorms) to the northwest of the Atlanta, Georgia, radar. The left panel in Figure 5.15-17 shows a PPI reflectivity scan at 0.5°, indicating a long northwest-southeast oriented line of very heavy rain, about 75 miles northwest of the radar. Additional areas of lighter rain are indicated to the southwest of the radar. If meteorologists wanted to determine the height of the 50 dBZ echo in the heavy rain, representative of the vertical extent of the heavy rain and possibly hail, they could continue to look at PPI scans at higher elevations until there was no 50 dBZ detected. However, many software packages allow for vertical cross-sections to be taken through volume scan data, and this was done here along the Alabama/Georgia border in Figure 5.15-17. This vertical cross-section is shown in Figure 5.15-18. Note that echoes above 50 dBZ are mainly confined below 3,000 m AGL.

Movement of Rain and Short-Term Forecasting

Although the synoptic-scale or mesoscale systems that produce rain and other types of precipitation evolve over time, successive reflectivity images can often be used to forecast the future time and location of precipitation.

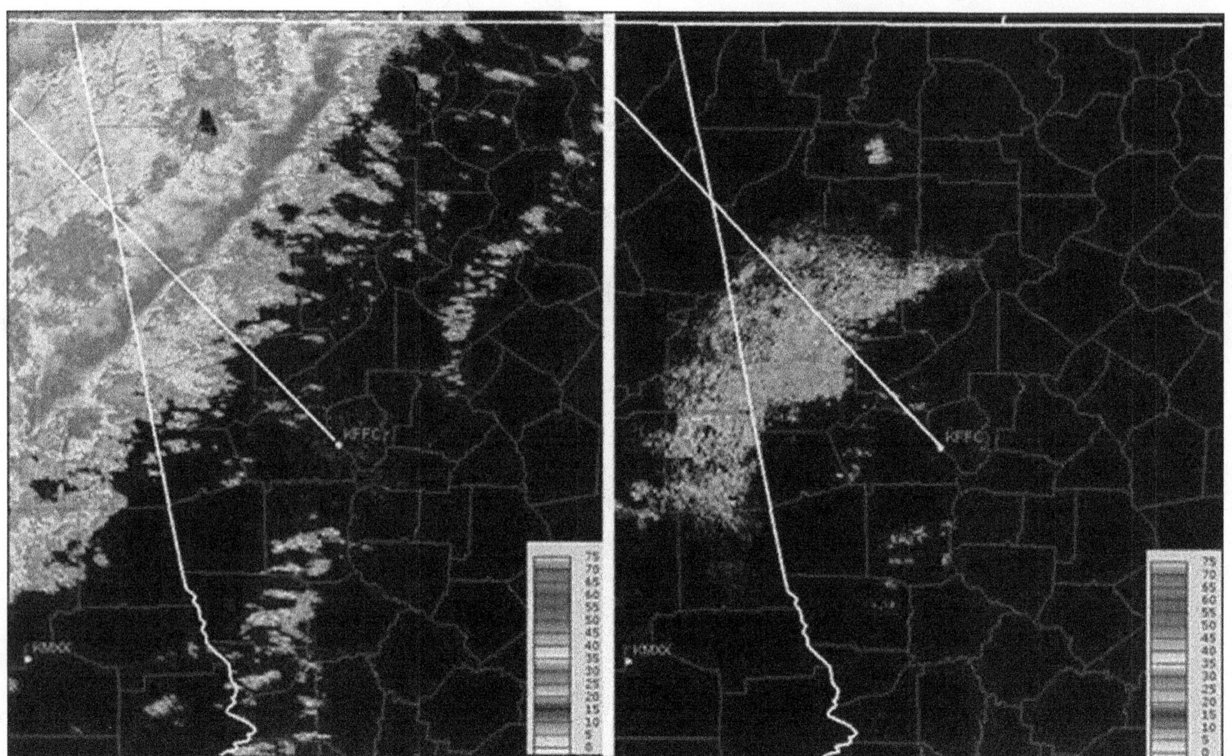

FIGURE 5.15-17 PPI Reflectivity Scans (dBZ) 0.5° (left) and 4.0° (right) Elevation

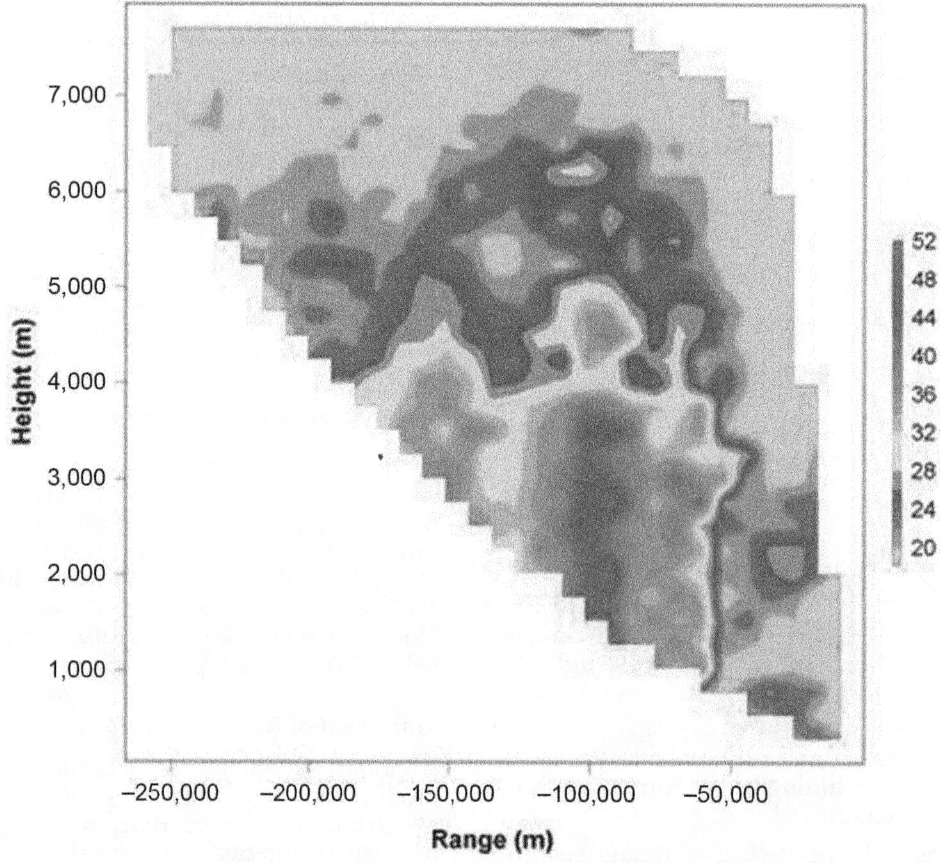

FIGURE 5.15-18 Cross-Section of Reflectivity along the Vertical Beam in Figure 5.15-17

For example, a wide band of rain may be moving across a large area, such as a state. In the case shown in Figure 5.15-19, an area of mainly light to moderate rain is shown at 1205 GMT (abbreviated as "Z"), and at 1303 GMT.

The area of rain is clearly expanding and/or moving north-northeastward, as evidenced by its change in position over the 58-minute interval between radar scans (the range rings in Figure 5.15-19 are at 10-mile intervals). The front edge of the precipitation is approximated at 1205 GMT using a straight line. At its closest point to the radar, the range of the leading edge of the area of rain was 62 miles. Fifty-eight minutes later, at 1303 GMT, the leading edge of the rain was at a range of 33 miles from the radar. So, a rough estimate is that the rain area is moving NNE at a rate of 29 miles every 58 minutes (or 30 miles per hour). Assuming the rain area continues to move in that same direction at that same speed (usually a good assumption on small time scales, v with large rain areas), one may forecast that it will move another 29 miles NNE every 58 minutes. At that rate, it will have reached just north of Tuscaloosa (KTCL), and be just south of the radar site at KEET, by 1401 GMT (shown by the line segments in Figure 5.15-19). Going forward another time step, it may be assumed that the rain area will be just south of Birmingham (KBHM) by 1459 GMT. It should be emphasized that such linear extrapolation is subject to multiple potential errors due to a change in synoptic-scale dynamics, mesoscale changes, drying or moistening aloft, and so forth.

This type of forecasting, using direction and speed derived from past radar scans to predict the future location of precipitation, can be more difficult with a single, 3D entity like a severe storm. In Figure 5.15-20, a tornadic supercell storm is shown northwest of Indianapolis, Indiana, during the November 17, 2013, tornado outbreak.

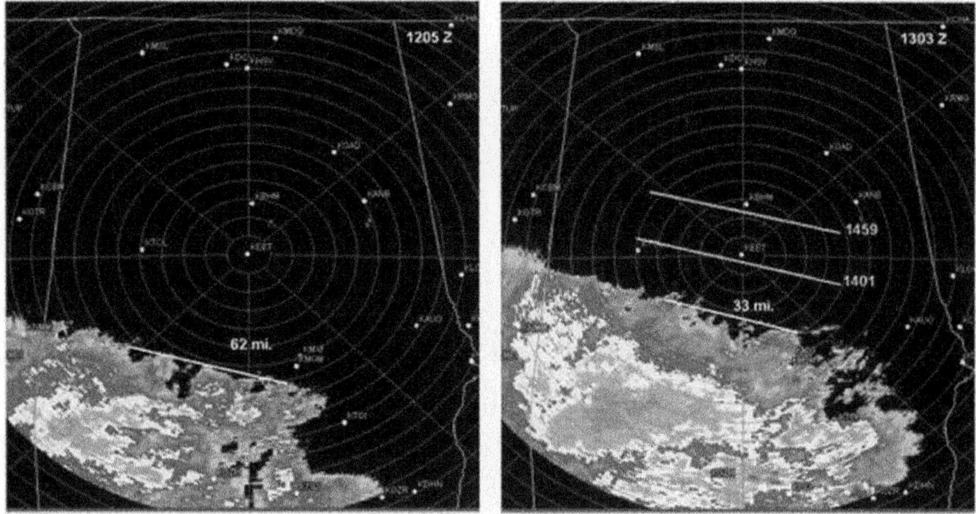

FIGURE 5.15-19 PPI Reflectivity Scans from NEXRAD Radar, at 1205 Z and 1303 Z

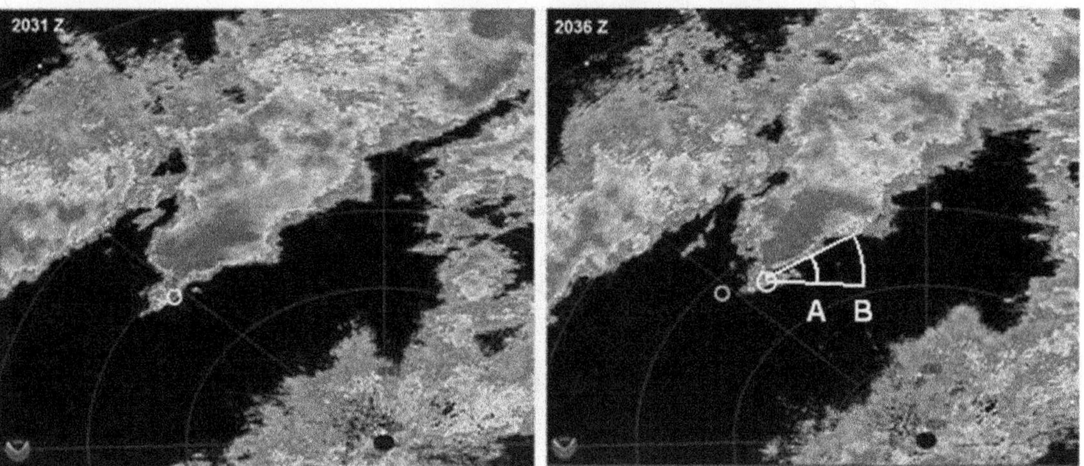

FIGURE 5.15-20 PPI Reflectivity Scans at 2031 Z and 2036 Z

Note: Circles in Figure 5.15-20 indicate locations of radar-indicated tornado at 2031 and 2036 Z. Cone and curves inside indicate likely locations of radar-indicated tornado in the future.

Given the motion of the storm over the five-minute period between 2031 and 2036 Z, the likely location of the radar-indicated tornado may be forecast at 2041 and 2046 Z (somewhere along the curves A and B in Figure 5.15-20). However, since supercell storms and tornadoes often take right turns when intensifying, or left turns when dying, the direction could also change. That is why the area of possible location is a cone, with arrival times in five minutes along curve A, and in ten minutes along curve B.

Severe Storm Identification

One of the most critical uses of weather radar is identifying severe storms that may cause bodily injury or property damage. Advancements in radar have driven most of the improvements in U.S. tornado warnings over the past 60 years, resulting in a greatly decreased fatality rate due to storms like tornadoes (e.g., Coleman et al. 2011).

Many reflectivity signatures are excellent indicators by themselves, without any additional data (radial velocity, etc.). *The importance of these reflectivity signatures in severe storm identification is often not realized*, given the wide availability of Doppler radial velocity data.

One easily recognizable reflectivity pattern considered a strong indicator of damaging straight-line winds is a *bow echo*. Strong surface winds often occur along the leading edge of the cold pool in a quasi-linear convective system (QLCS), also known as a squall line. This cold pool is helping to generate new updrafts and maintaining the intensity of the QLCS. When surface winds become very strong in a certain part of the QLCS, the cold pool is pushed forward faster in that part of the QLCS relative to other sections, causing the updrafts and associated heavy rain to move faster also. When one part of the QLCS begins to move faster than the remaining part, the high radar reflectivity associated with the updrafts in the QLCS begins to form an arc shape, or a bow. When this radar reflectivity pattern is noted, enhanced straight-line winds and wind damage are more likely in the area of the bow echo. Bow echoes may occur in only a small part of a QLCS, in a part of the squall line as small as 25 miles or less (Figure 5.15-21(a)), or may be as large as 100 miles long and produce widespread wind damage (Figure 5.15-21(b)).

Two more examples of bow echoes are shown in Figure 5.15-22. One is a radar reflectivity image from a large bow echo that caused widespread wind damage

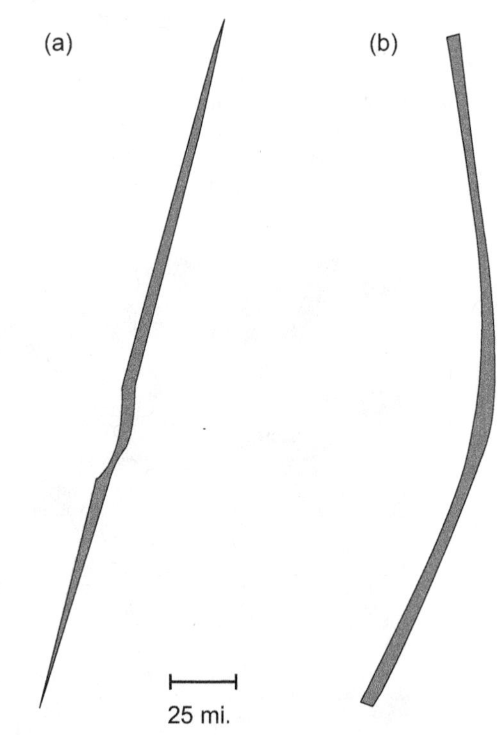

FIGURE 5.15-21 Schematic of Bow Echo Radar Reflectivity Area Where Z > 50 dBZ

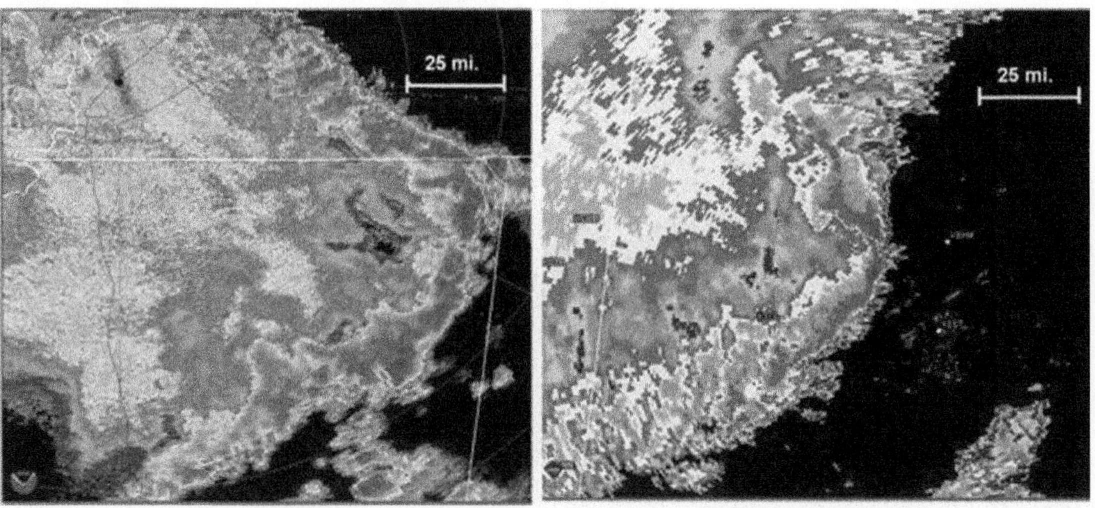

FIGURE 5.15-22 PPI Reflectivity Scans with Bow Echoes

NOAA; Corfidi, S. F., J. S. Evans, and R. H. Johns, cited 2015: About Derechos: Serial Derechos.

(and wind gusts over 75 mph) over central Alabama on February 16, 2001 (right). The other is from a *derecho*, or long-lived, steady bow echo that caused widespread wind damage from eastern Oklahoma across Arkansas, southwestern Tennessee, northern Mississippi, and northern Alabama on June 12, 2009 (left).

Another radar reflectivity pattern that is sometimes seen in QLCSs is the *line echo wave pattern* (LEWP). These typically occur in environments with high wind shear. Often, as shown in Figure 5.15-23, a bow echo will develop small bow echoes/bulges, resembling waves.

At the center of these bulges, the risk for straight-line winds is increased. However, given the natural area of horizontal vorticity (vertical shear) along the leading edge of a QLCS, the "kinks" in the QLCS where two bulges meet are sometimes associated with enhanced convergence and associated large updrafts.

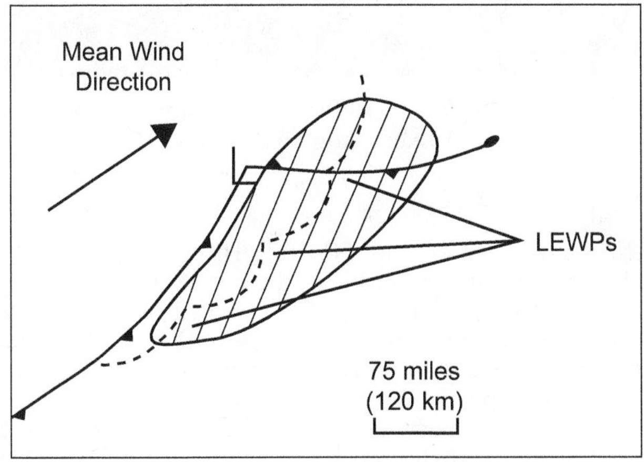

FIGURE 5.15-23 Core of Highest Radar Reflectivity QLCS Area Containing LEWPs

Corfidi et al., 2015.

These regions may encourage the tilting and stretching of vorticity, resulting in storm rotation and an increased risk of tornado formation near the kinks in the QLCS (Figure 5.15-24*).

Perhaps one of the best-known radar reflectivity signatures is the *hook echo*, an indicator that a storm is rotating and may produce a tornado. Three hook echoes were detected on radar and associated with tornadoes as early as 1953: one in Texas, one in Illinois, and one in Massachusetts (e.g., Coleman et al. 2011). Many tornadoes, and almost all violent tornadoes, are associated with a long-lived, rotating thunderstorm known as a supercell. In almost all supercells, the updraft is located on the southwest side of the storm, and the precipitation is dropped northeast of the updraft due to wind shear. Often in a supercell, the updraft is so strong that precipitation cannot form quickly enough to fill the lower portions of the updraft, leading to a *weak echo region*, traditionally known as overhang. However, if the updraft is rotating significantly, some of the precipitation falling out of the northeastern part of the storm may wrap around the western and even southern side of the precipitation-free updraft at low levels, producing the classic hook echo.

Examples of hook echoes can be seen in Figure 5.15-25. These images were associated with violent tornadoes in El Reno, Oklahoma, on May 31, 2013 (top), and in Birmingham, Alabama, on April 8, 1998 (bottom). Occasionally, instead of a full appendage in reflectivity, only "part" of the hook echo can be seen, where rotation has only slightly distorted the rainfall pattern. This may appear as an *inflow notch*.

Not all supercell storms with hook echoes produce tornadoes. However, one excellent indicator that a rotating supercell storm with a hook echo is actually producing a tornado is the presence of a *debris ball* in radar reflectivity. The raw reflectivity received by the radar from a range gate is proportional to the number

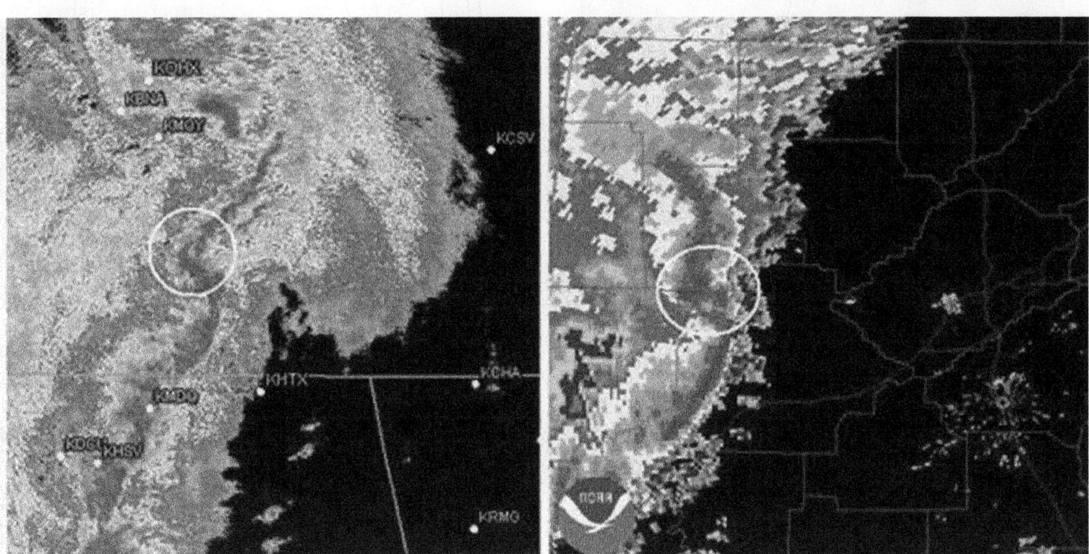

FIGURE 5.15-24 PPI Reflectivity Scans

***White circles indicate the area of LEWP.**

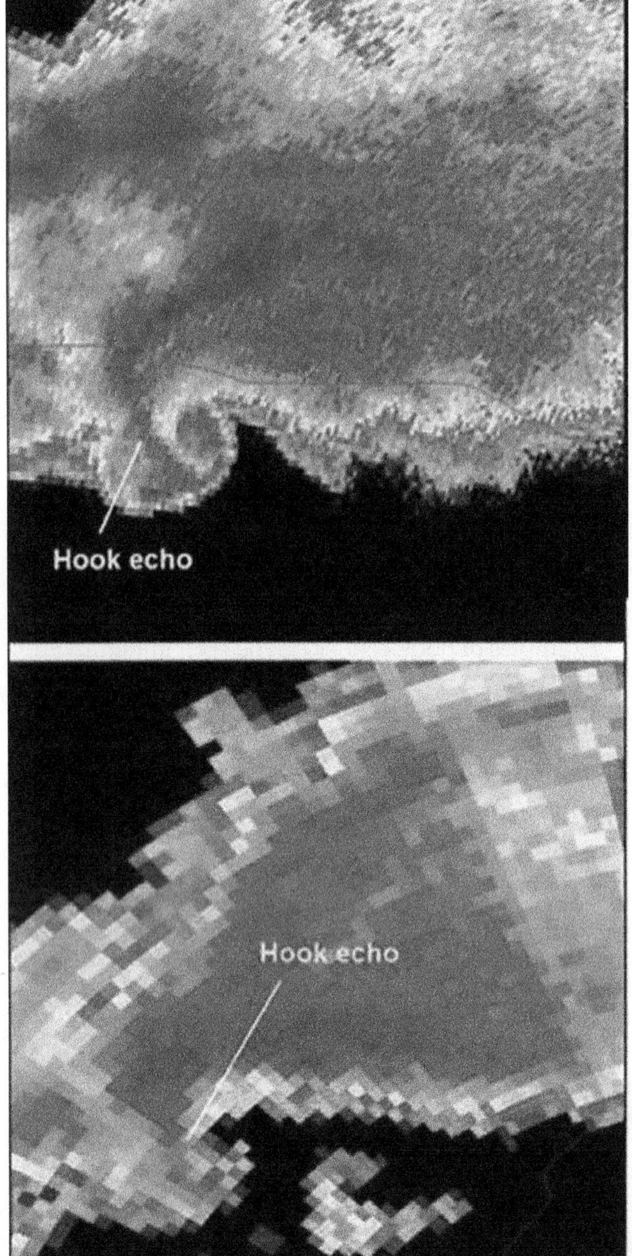

FIGURE 5.15-25 Two Hook Echoes Associated with EF-5 Tornadoes

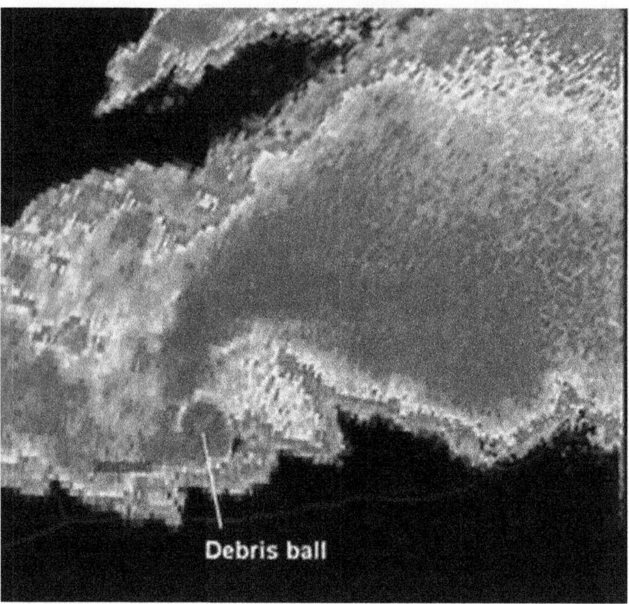

FIGURE 5.15-26 Radar Reflectivity Image of Supercell Storm Debris Ball Exceeding 65 dBZ

of targets present at each diameter times the diameter of the targets to the sixth power. Most raindrops have diameters of 1–5 mm, and even most hailstones have diameters smaller than 50 mm; reflectivities above 50 dBZ are usually only seen in heavy rain or hail. However, even though it may take several hundred raindrops with diameters around 5 mm within a given volume to produce a reflectivity of 50 dBZ, it may only take one or two large objects, such as pieces of metal roofing, vinyl siding, or lumber, with equivalent diameters up to 1,000 mm, to produce a reflectivity greater than 50 dBZ. Tornado debris that is lifted into the air (and into the radar beam) is often composed of such objects. One would not normally expect to see especially heavy rainfall or hail near a hook echo, so when extremely high reflectivities are found there, particularly when the area of high radar reflectivities is very circular, it must be suspected that tornado debris may be causing the high radar return. In long-track tornadoes, debris balls often grow in size and intensity during and immediately after a tornado passes over a populated area, as was the case with the Tuscaloosa tornado on April 27, 2011 (Figure 5.15-26).

Hail detection may also be accomplished using reflectivity, although examination of dual-pol parameters makes confidence much higher. Very high reflectivities, greater than 60 dBZ, are almost always associated with hail, because raindrops simply do not often achieve the required diameter or number to produce such values. However, one must also consider the temperature profile in the atmosphere when attempting to detect hail using reflectivity. For example, if one applies the National Severe Storms Laboratory (NSSL) algorithm for the probability of hail at the ground, the difference in height between the highest altitude where 45 dBZ reflectivity occurs and the height of the freezing level is calculated and applied to a probability curve (Figure 5.15-27).

Another indicator of hail in reflectivity is a *hail spike*. When a storm contains large hailstones (usually 1 inch in diameter or larger), some radar energy may initially scatter off the hailstone toward the ground. Some of this energy may then reflect back up into the hailstones in the storm, and some of this energy is then returned to the radar. This process is shown in Figure 5.15-29, and is also known as *three-body scatter*.

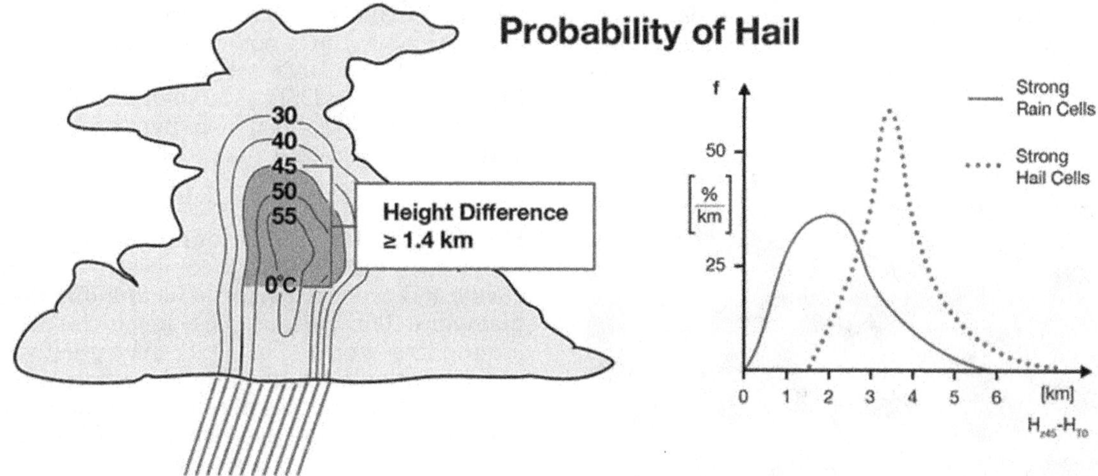

FIGURE 5.15-27 Thunderstorm Reflectivity Cross-Section and Probability Curve for Rain Versus Hail

Cross-sections of storms are usually best, in order to locate the highest altitude of large reflectivities (Figure 5.15-28).

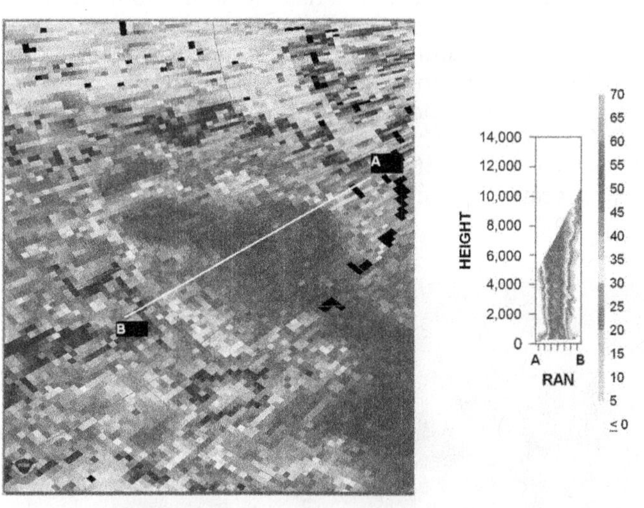

FIGURE 5.15-28 PPI Scan (left) and Cross-Section along Line AB (right) of Reflectivity (dBZ)

Note: In the cross-section, 50 dBZ echoes extend all the way up to the highest elevation (and likely into the cone of silence), near 8,000 m (24,000 feet MSL). Soundings (not shown) indicate that the freezing level was near 13,000 feet MSL.

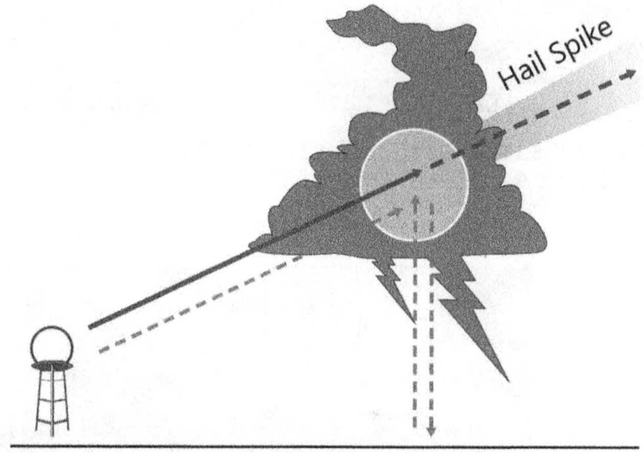

FIGURE 5.15-29 Three-Body Scatter Leading to Hail Spike

The light dashed lines represent the actual propagation path of the radar beam (from the radar to the hail, to the ground, back to the hail, then back to the radar). However, the radar determines the range of an echo based purely on its time of arrival, and since the process of the RF energy going to the ground and back takes extra time, the radar will simply assume that this energy came from some farther distance from the radar than it actually did (darker dashed line). This assumed farther distance causes the radar to show a thin line of light reflectivity directly behind the storm (relative to the radar). This process of a significant amount of RF energy striking three targets (hail, ground, and hail again), instead of just one, and still returning to the radar, requires fairly large hail, and is an excellent indicator of it. An actual hail spike is shown in Figure 5.15-30.

A *wake low* is another type of weather with the potential for damaging winds that can be difficult to detect; a wake low is an area of low pressure at the rear of a mesoscale convective system (MCS) caused by descent of air from aloft and rapid warming. This descent of air and warming often dries up precipitation very abruptly along the back side of the MCS, producing a distinctive large gradient in reflectivity there (Figure 5.15-31*).

Rainfall Accumulation Rate

Radar reflectivity has been used for decades to determine rainfall rates (typically in inches per hour) and total rainfall accumulation over time periods from one hour up to one week. If the distribution of raindrop sizes is known (drop size distribution, or DSD), the basic definition of raw reflectivity can be used, related to the number of drops and the diameter of those drops, to develop a relationship between the raw reflectivity (Z) and the rainfall rate (R). These Z-R relationships are typically of the form

$$Z = aR^b$$

FIGURE 5.15-30 Hail Spike Northwest of the KBMX Radar on April 1, 2012, in PPI View

where Z is the raw reflectivity, R is the rain rate (in mm per hour), a is some coefficient and b is some exponent. If raindrops are distributed according to the Marshall-Palmer DSD (Marshall and Palmer 1948), a coefficient $a = 200$ and an exponent $b = 1.6$ are obtained, and the following Z-R relationship:

$$Z = 200R^{1.6}$$

However, Battan (1973) found that 69 Z-R relationships, each with different combinations of the coefficient a and exponent b, could be found in the literature that worked for various DSDs in various climatic areas around the world. The DSD may greatly vary at a given location with time, due to different types of air masses, different types of rainstorms, and so forth. The one most commonly used by the NWS is given by

$$Z = 300R^{1.4}$$

This relationship changes frequently. Despite calibration of the radar, no single Z-R relationship may be trusted in all situations even at a single location. Advanced radar software has the ability to make real-time corrections to the Z-R relationship to produce more reliable rainfall estimates.

To determine the rainfall over an entire storm event, a radar uses some given Z-R relationship to convert reflectivity at each volume scan time to a rainfall rate, assumes that rainfall rate continues until the next volume scan, then multiplies the rainfall rate for each

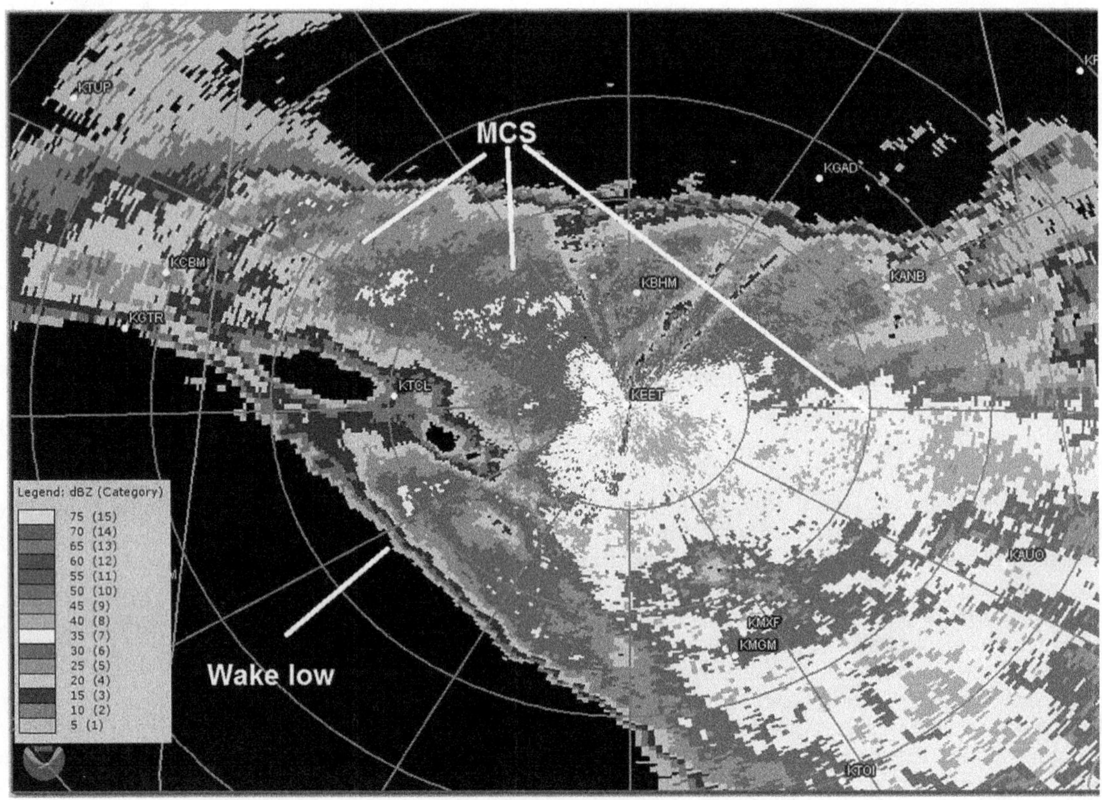

FIGURE 5.15-31 Classic Wake Low Signature in Reflectivity

***Range rings are at 25-mile intervals.**

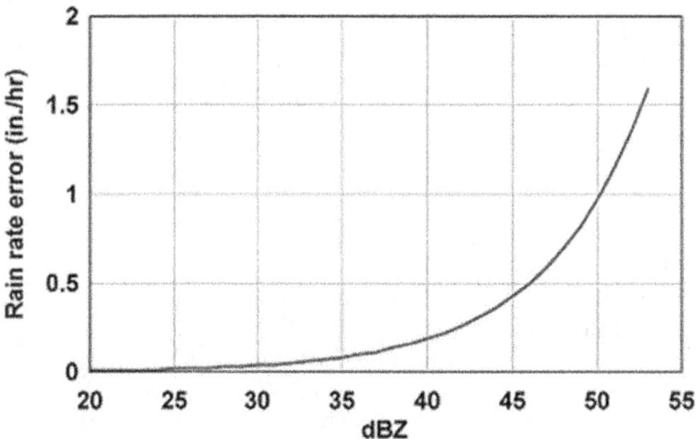

FIGURE 5.15-32 Underestimation of Rainfall Rate

volume scan by the time between volume scans. These individual rainfall totals for each volume scan are then summed at each location to determine a storm total precipitation. One-hour and three-hour precipitation amounts are often calculated in the same way, and are used in predicting flash flooding.

In addition to changing *Z-R* relationships, another problem in rainfall estimation from radar is associated with partial beam blockage (see the Other Considerations with Weather Radars section). When the radar beam is partially blocked, the reflectivity will be underestimated. Suppose that 50% of the radar beam is blocked (Figure 5.15-32). This results in a loss of 3 dBZ in radar reflectivity. This may not cause a significant problem at very light rainfall rates, but at high rainfall rates it may cause large errors.

For example, as shown in Figure 5.15-32, at 25 dBZ the rainfall rate error is less than 0.02 inches per hour even with 50% beam blockage. However, at 50 dBZ, the same beam blockage causes an error of 0.97 inches per hour. In a long-term heavy rain event, rainfall in areas with a partially blocked beam may be significantly underestimated. Advanced radar software may correct for these errors due to partial radar beam blockage.

Hail may also cause severe overestimation of rainfall rates, given its large diameter and subsequently large effect on radar reflectivity. Standard practice has historically been to truncate all reflectivities at 53 dBZ, assuming that any values above that are affected by hail. However, dual-pol radar can identify hail, and this is a classic situation where dual-pol radar can improve precipitation estimates.

Problems to Recognize

There are a number of potential problems to be aware of when using radar reflectivity data. For example, primarily in large areas of cool stratiform rain, the rain starts as snow aloft and then melts on the way down. When the fairly large snowflakes start to melt, they develop a coating of liquid water, which is much more reflective than the ice crystals themselves. These melting snowflakes aloft produce a layer of artificially high reflectivity called a *bright band*. The bright band typically begins to form around the freezing level and may extend as far 1,500 feet below it.

Therefore, on a PPI radar scan, the bright band will generally appear as a doughnut-shaped area of enhanced reflectivity, with the range from the radar depending on the height of the bright band. These bright bands may be misleading as to the rain intensity in the bright band area. More importantly, they may produce a significant false maximum, also doughnut-shaped, in storm total rainfall calculated by the radar, as shown in Figure 5.15-33, where there appears to be anomalously high rainfall totals at ranges from 50 to 75 miles (top). This is associated with a bright band.

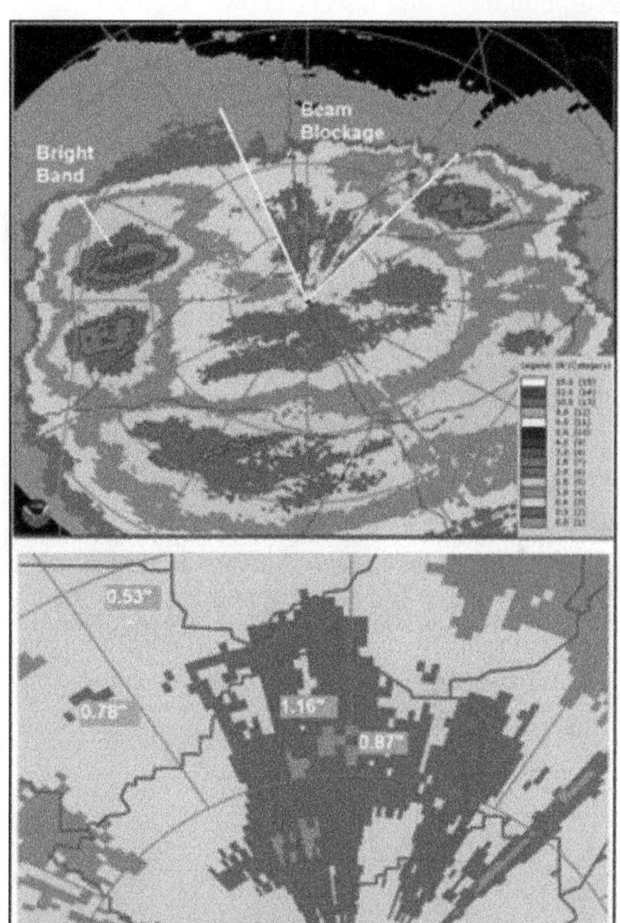

FIGURE 5.15-33 Bright Band Effect

Coleman, T. A., and J. A. Westland, 2010: Underestimation of QPE in a Flash Flood Situation Due to Partial Beam Blocking and Attenuation: Correction Using the BREAM model. 24th Conf. on Hydrology. Note: The distinct minimum to the NNW, N, and NNE of the radar site is caused by partial beam blocking just north of the radar. In the zoomed-in area north of the radar (bottom), measured rainfall totals are included. Notice that for areas north of the radar with partial beam blockage, where rainfall amounts were near 1", radar indicated less than 0.6". Yet, in areas without beam blockage northwest of the radar, rainfall was estimated correctly, between 0.6" and 1.0".

As discussed in the Received Energy Reflected from Targets section, sometimes rain occurs beyond the maximum unambiguous range of the radar, and the radar interprets the rain to have come from the second radar pulse instead of the first radar pulse, greatly underestimating its range. This is known as range folding, and produces faulty *second-trip echoes*. Precipitation that is far away from the radar shows up as much closer to the radar than it actually is.

The azimuth and intensity are shown correctly, but the range is not. As shown in Figure 5.15-34, the second trip echoes are elongated along a constant azimuth and do not look natural.

The radar may also detect ground targets like trees, buildings, and insects, known as *ground clutter*. Ground clutter is made worse during situations where super-refraction occurs, because more ground targets are detected (the radar beam is lower to the ground). The worst ground clutter occurs in situations of *anomalous propagation* (see the Other Considerations with Weather Radars section).

Some ground clutter, if it is large enough, tall enough, and close enough to the radar, is detectable during almost any type of propagation conditions. Recall that the change in temperature with height plays a huge role in the refraction of the radar beam. During the daytime, standard propagation or sometimes sub-refraction occurs, and ground clutter is minimized. However, on clear nights when a temperature inversion forms, the radar beam becomes super-refracted, and may even have AP. Figure 5.15-35* shows a series of radar images, in late afternoon, an hour after sunset, and three hours after sunset, from the Evansville, Indiana, radar. At 2207 GMT (top, afternoon) some ground clutter is detected within 30 miles of the radar. At 0055, ground clutter is detected as far as the 90-mile range. At 0304, there is more ground clutter, and some extends beyond 90 miles.

After sunset, the ground clutter around the radar is greatly increased. Some of this may be due to the emergence of more insects after sunset, but on the date of the radar image (March 18, 2015), there should not be a *large* insect population around Evansville yet because it is still in the cold season. Much of the increase is due to the temperature inversion and subsequent super-refraction. Advanced weather radars have various clutter suppression techniques, and these are important to determine which echoes are real and which ones are due to ground clutter.

A radar beam also suffers *attenuation* as it goes through rain (see the Other Considerations with Weather Radars section). This may lead to the false indication of rain being lighter than it is, or suddenly weakening when rain passes over the radar. Attenuation is worse when rain is occurring at the radar site and more pronounced at high frequencies (such as C-band), as opposed to S-band and high S-band.

One type of attenuation that can be troublesome for meteorologists occurs in QLCSs or other linear storms, because when the line is over the radar, the radar beam is moving through a long stretch of high reflectivity and being attenuated significantly (see Figures 5.15-36 and 5.15-37). Due to this phenomenon, it appears that the line has weakened slightly (lower overall reflectivities due to attenuation), when in fact its intensity has changed little or none.

Doppler Radial Velocity

Wind Speed and Kinematics

A Doppler radar is able to detect the component of wind speed along the radar beam. Therefore, wind speed (and direction) may be inferred from Doppler velocity data (see Figure 5.15-4). In addition, certain

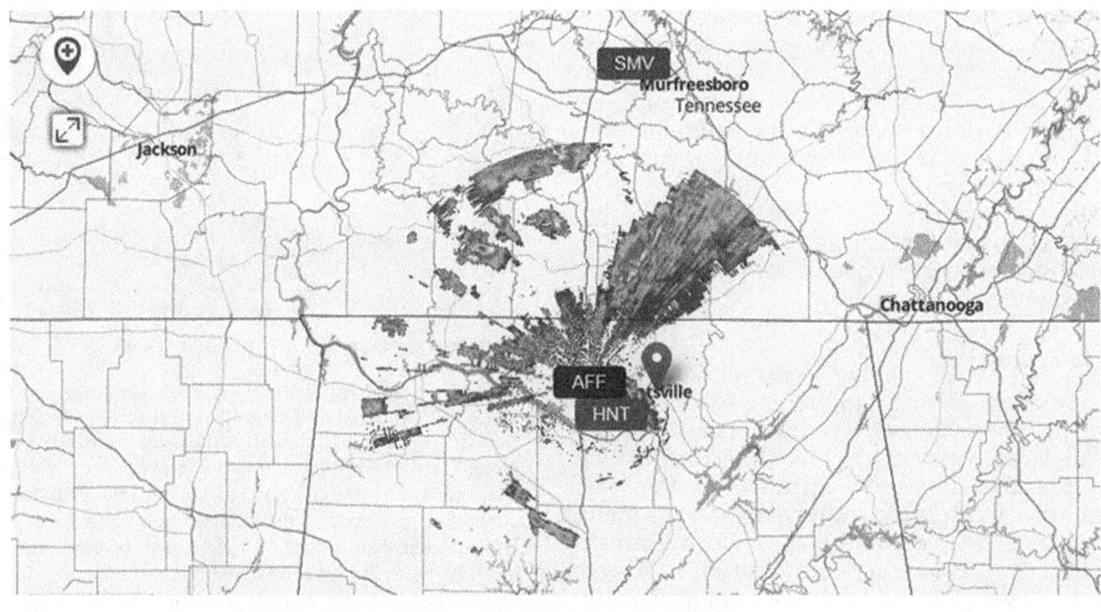

FIGURE 5.15-34 Second-Trip Echoes

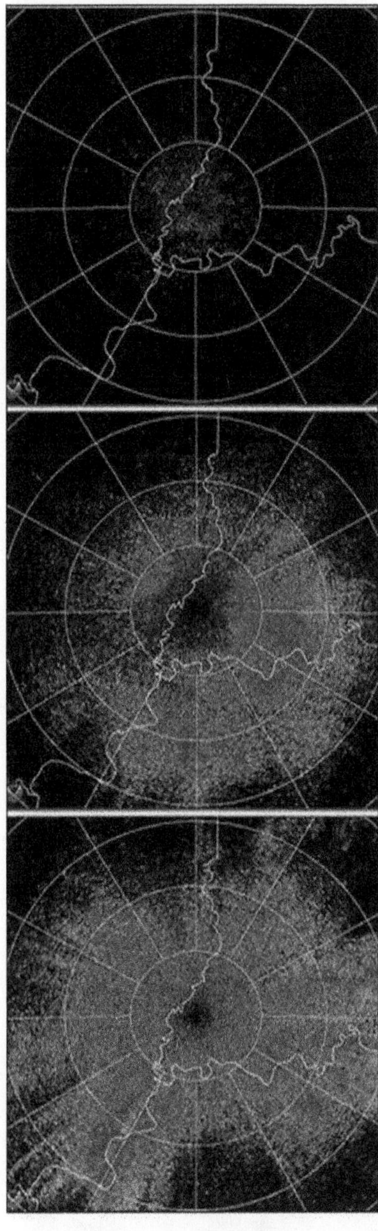

FIGURE 5.15-35 Reflectivity Images with Ground Clutter

*Range rings are at 30-mile intervals.

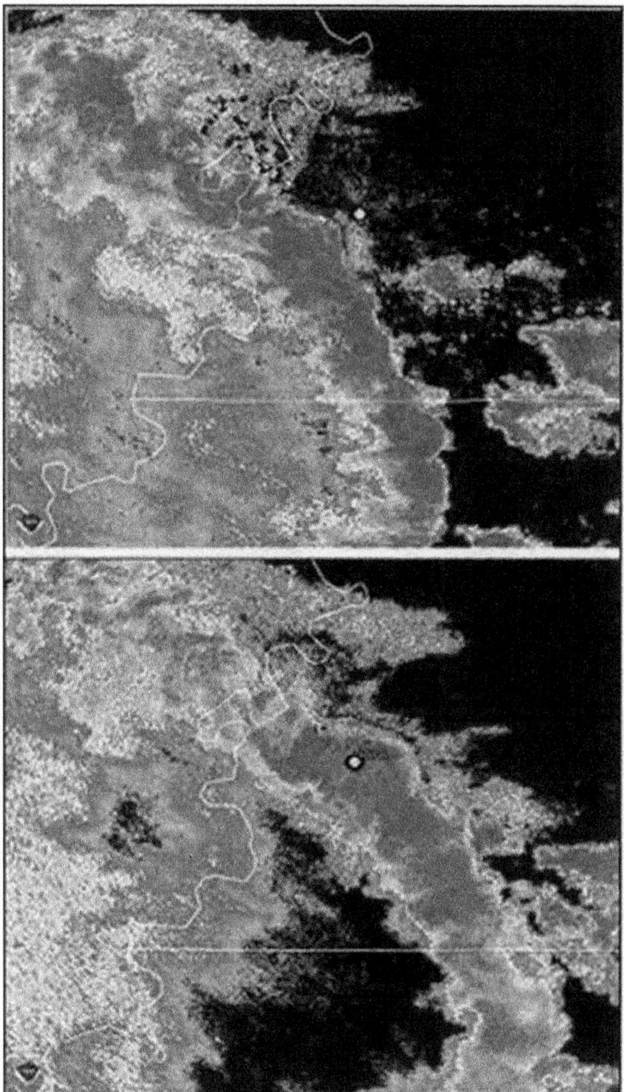

FIGURE 5.15-36 Reflectivity Images before and during Attenuation

patterns of Doppler velocities indicate storm-scale motion of different types.

It should be pointed out that, on radar displays, inbound velocities are negative by convention and are generally displayed using cool colors, such as blues and greens. Outbound velocities are positive, and are indicated by warm colors such as red and pink.

If outbound and inbound velocities are adjacent to one another along a constant azimuth (along the radar beam), this indicates convergence or divergence. If outbound and inbound velocities are adjacent to one another at a constant radius (perpendicular to the radar beam), this indicates rotation. This is shown in Figure 5.15-38, as it would be detected by a Doppler radar positioned to the south of the divergence or rotation (see top panel). In both cases, actual flow is shown on the left, and Doppler radial velocities are shown on the right. Note that the radar only detects the component of the wind toward or away from the radar. So, just north and south of the center of rotation in the bottom panel, the radial velocity is zero, even though the actual velocity is just as high there as it is on the east and west sides of the rotation.

When outbound radial velocities and inbound radial velocities are found adjacent to one another on a PPI scan, convergence/divergence and/or rotation are indicated.

There are many uses for the detection of divergence on radar. For example, if one observes strong divergence *at low levels* in a thunderstorm, it indicates that a microburst may be occurring, with winds spreading

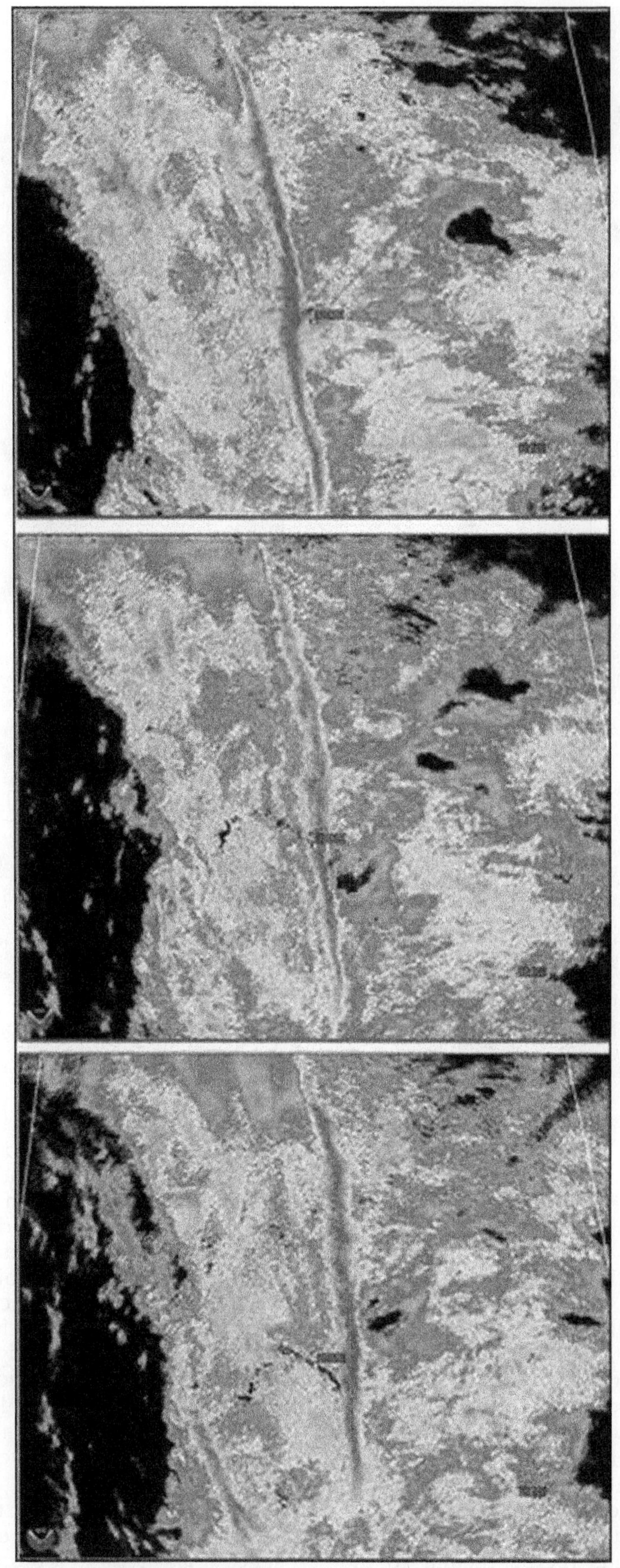

FIGURE 5.15-37 Reflectivity before, during, and after Passage of an Atmospheric Wave over a Radar

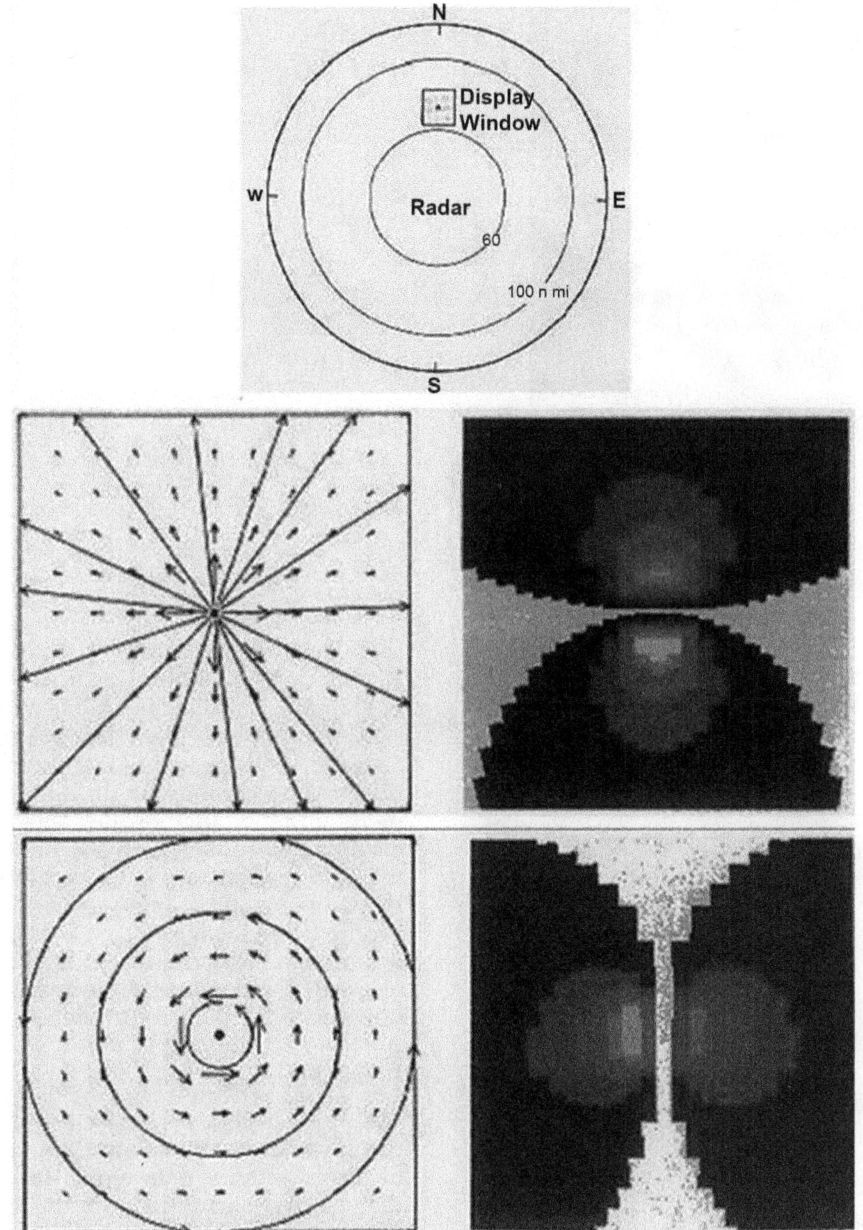

FIGURE 5.15-38 Schematic Showing Pure Divergence (middle) and Pure Rotation (bottom)
NOAA; Brown, R. A., and V. T. Wood, 2007: *A Guide for Interpreting Doppler Velocity Patterns: Northern Hemisphere Edition.* NOAA/NSSL, Norman, OK, 2nd ed.

Note: The direction of the radar must be known from the pattern in radial velocity to understand what type of motion (convergence or rotation) is being observed. See the top of Figure 5.15-38.

out in all directions (see Figure 5.15-39). This is known as a *microburst signature*. Also, when a storm exhibits extreme divergence near its top, it indicates a very strong updraft and the possibility that a storm is particularly intense. Storms with *storm-top divergence* and strong updrafts are more likely to produce large hail.

VAD Wind Profiles

Assuming that the wind flow is fairly uniform near the radar site, as is usual, Doppler radial velocities around the radar site itself can be used to determine the wind direction and speed at different altitudes. This is known as a *velocity azimuth display* (VAD). For each given altitude, some elevation angle and range from the radar are selected, providing a certain altitude. In Figure 5.15-40, the radial velocities at a range of 5 miles and an elevation of 2.4° in the top panel (PPI display of radial velocity) are plotted in the bottom panel, producing a VAD. Using a best-fit sine wave, the wind speed and direction may be approximated at that height. In this case, it is from 170° at 38 knots. Note

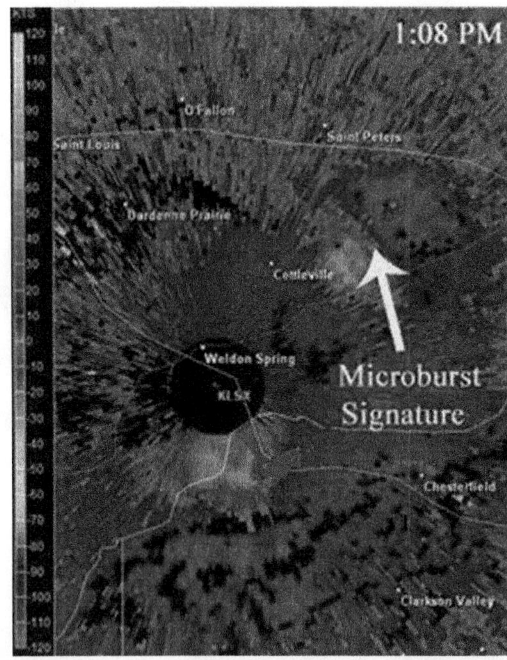

FIGURE 5.15-39 Microburst Signature Detected by the St. Louis, Missouri, Radar

Courtesy of NOAA

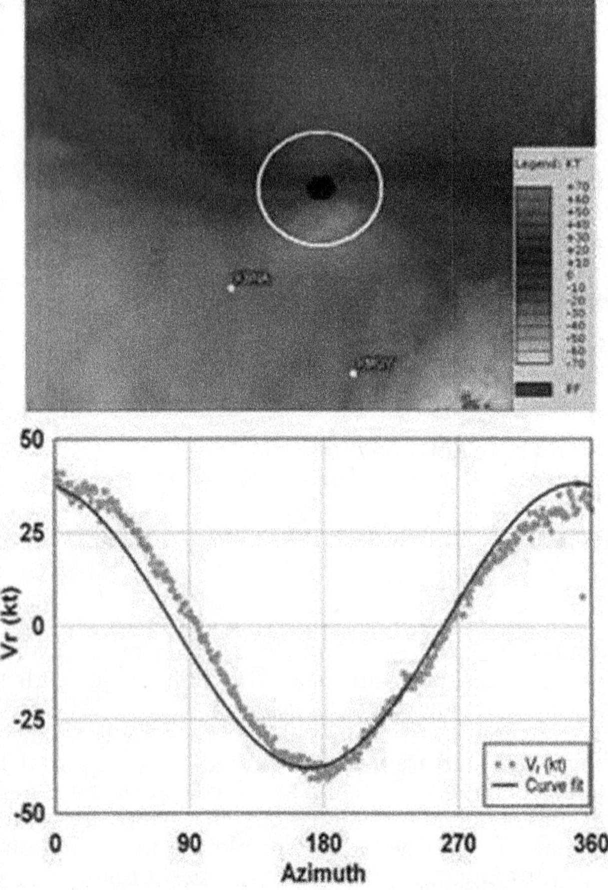

FIGURE 5.15-40 Velocity Azimuth Display

ALT KFT	
	ND
10	ND
9	ND
8	
7	
6	
5	
4	
3	
2	
1	

FIGURE 5.15-41 VAD Wind Profile

Note: "ND" indicates that there was not enough data at those altitudes to determine wind speed and direction accurately.

that the strongest inbound velocities are near 170°, and the strongest outbound velocities are near 350°. Also note that radial velocities are near zero where the wind is perpendicular to the radar beam, at 80° and 260°.

If the wind speed and direction are calculated using this VAD technique at multiple altitudes, by either increasing the range or the elevation, a *VAD wind profile* may be produced. This wind profile provides similar information to what a weather balloon does, but in almost real time. Figure 5.15-41 shows a VAD wind profile from Raleigh, North Carolina, during the large tornado outbreak of April 15, 2011. The different shades of gray (or colors in the companion website) indicate how much the data is scattered from the sine wave.

Severe Storm Identification

Multiple types of severe local storms may be detected using radial velocity data that may not be detectable using reflectivity data alone. In other cases, velocity data increases confidence in the implications of a particular reflectivity signature.

Detection of storm rotation is probably the most common and well-known use of Doppler radar, because it has greatly improved the detection of tornadoes. The rotation may be broad, indicating rotation of the entire updraft. This type of rotation is known as a *mesocyclone*. Two examples of mesocyclones are found, one in a supercell, shown in Figure 5.15-42, and one in a QLCS. Typically, mesocyclones are fairly deep (detectable at multiple altitudes). The key to differentiating a mesocyclone from convergence or divergence is that the inbound radial velocities are on the left side of the inbound/outbound velocity signature, as viewed by the radar, and the outbound radial velocities are on the right side. In other words, *the velocity difference is at a fairly constant range from the radar.* The intensity of a mesocyclone is often measured by the difference between the maximum inbound and maximum outbound velocities. For example, if the

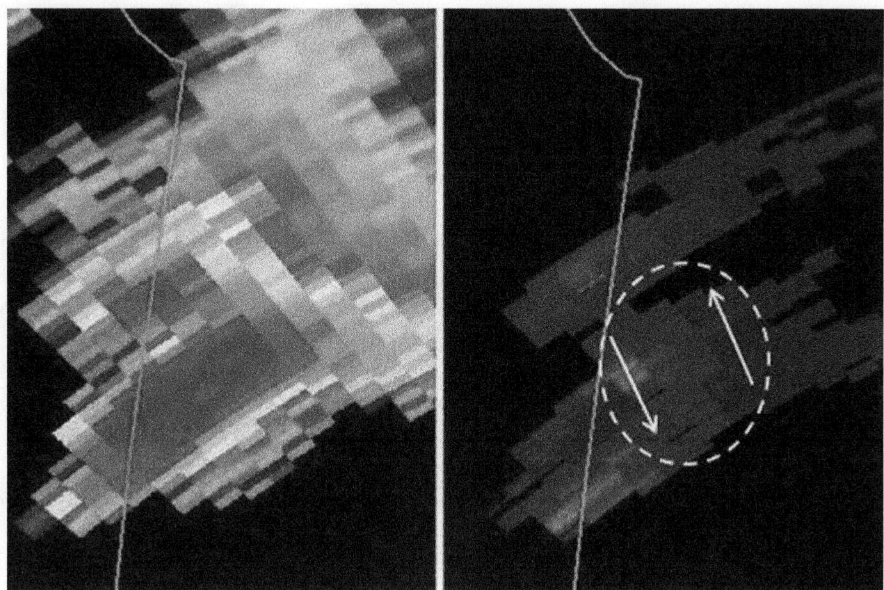

FIGURE 5.15-42 A Broad Mesocyclone within a Supercell—Reflectivity (left) and Radial Velocity (right)

maximum inbound velocity is −40 knots, and the maximum outbound velocity is +30 knots, then the intensity or shear of the mesocyclone is 70 knots. A stronger mesocyclone is more likely associated with a tornado.

Given that the translational motion of the parent storm is sometimes as large as the rotational velocities in a mesocyclone, a mesocyclone moving toward the radar may show very large inbound velocities and near zero outbound velocities, while a mesocyclone moving away from the radar may show large outbound velocities and near zero inbound. For cases like this, *storm-relative velocity* is often utilized. The radar calculates the average storm motion and vectorially subtracts it out to show the radial velocities within a storm *relative to storm motion*. It allows for easier detection of mesocyclones in storms that are moving with a large component toward or away from the radar.

Indication of very intense rotation, with large inbound and outbound velocities, a large difference between the two, and these vastly different velocities occurring in adjacent (or nearly adjacent) range gates, is known as a *tornado vortex signature (TVS)*. Normally, a TVS is only detectable within about 50 miles of the radar, and indicates that a tornado is likely on the ground or forming. However, a TVS is still not a guarantee that a tornado is on the ground. An example of the TVS associated with the Tuscaloosa to Birmingham tornado on April 27, 2011, is shown in Figure 5.15-43 where the strong inbound and strong outbound velocities are gate to gate, with no separation between them. In this case, the maximum inbound velocity was −139 knots, and the maximum outbound velocity was +110 knots, for a shear of 249 knots.

In addition to tornadoes, straight-line winds, associated with bow echoes, derechos, and microbursts, are potentially damaging wind events that can be detected using Doppler radial velocity data. Not all straight-line wind events in QLCSs are associated with bow echoes

FIGURE 5.15-43 Radial Velocity Image of the Tuscaloosa-Birmingham EF-4 Tornado on April 27, 2011

or derechos, and it is important to be able to examine the radial velocity associated with the winds to determine the approximate wind speed and the potential for wind damage.

Figure 5.15-44 shows the radial velocities associated with a large QLCS approaching Milwaukee, Wisconsin, on June 30, 2014. There is a fairly large area of radial velocities between 40 and 50 knots, and given their range from the radar, they are only 500–1,000 feet above the ground. One must consider that the wind speeds will generally be lower at the ground than at the level of the radar beam, especially with increasing range from the radar. However, it is also important to point out that the radar can only detect the component

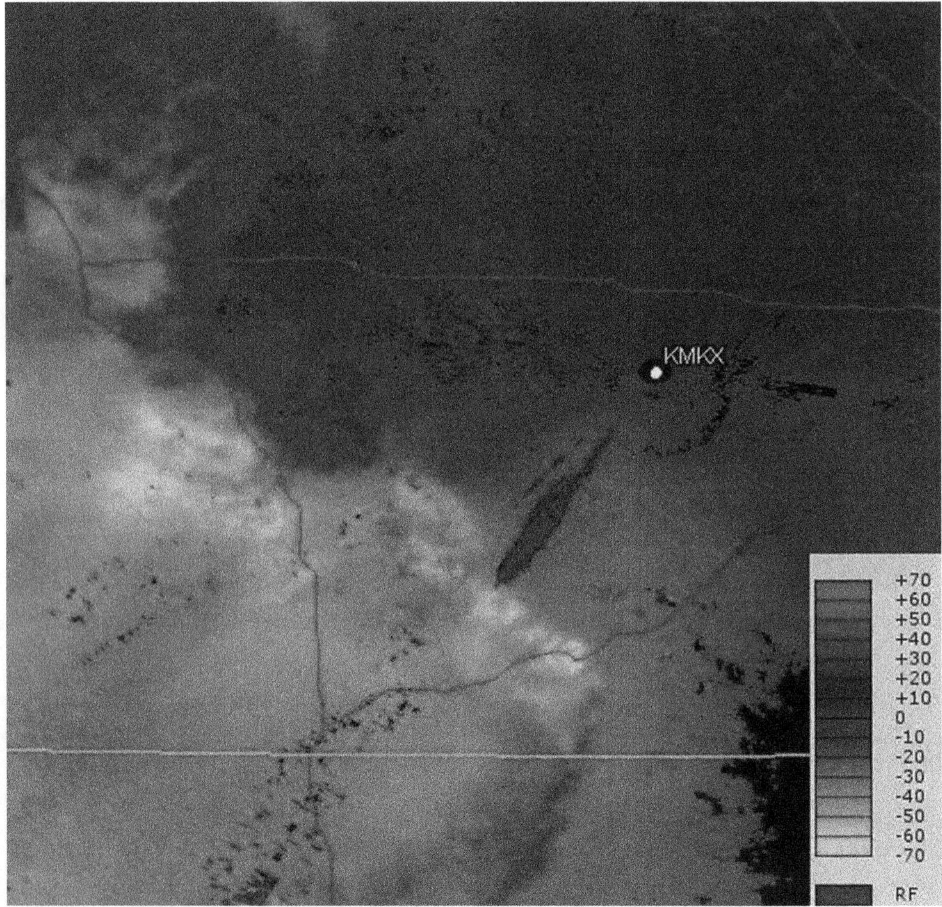

FIGURE 5.15-44 Radial Velocity Image Associated with a QLCS

of the wind toward or away from the radar, so the radial velocities of 40–50 knots may actually represent higher wind speeds, depending on the angle between the wind direction and the radar beam.

Advanced weather radar solutions will provide value-added products, such as shear markers and shear rate, that process the radial velocity to produce easily identifiable indicators of rotation and straight-line wind events to help with situational awareness.

Velocity Folding

A radar has a certain Nyquist velocity, or maximum unambiguous velocity that can be detected based on the wavelength and the PRF (see the Received Energy Reflected from Targets section). If a radial velocity is beyond the Nyquist velocity, then the phase shift is more than ±180 degrees, and the radar folds the velocity around to the other end of the Nyquist interval. This happens frequently in high-wind events. A classic indication of velocity folding is when an area of wind, with its maximum at the center, shows a radial velocity pattern where the sign of the velocities (inbound vs. outbound) suddenly changes and starts getting weaker. Modern radar systems have multiple algorithms for the detection and removal of velocity folding.

Dual-Polarization Radar Variables

Dual-pol radar is on the cutting edge of weather radar technology, and many of these products are still being researched to determine how they can best benefit the meteorological community. This subsection covers the meteorological uses of the dual-pol radar variables introduced in the Received Energy Reflected from Targets section, including differential reflectivity Z_{DR}, specific differential phase K_{dp}, and correlation coefficient ρ_{HV}.

Differential Reflectivity

Differential reflectivity (Z_{DR}) is simply the ratio of the returned horizontally polarized RF energy (Z_H) to the returned vertically polarized RF energy (Z_V), and is usually written logarithmically as follows:

$$Z_{DR} = 10\log\left(\frac{Z_H}{Z_V}\right)$$

One useful advantage of dual-polarization radar, and specifically Z_{DR}, is in *identifying the type of precipitation that is falling*. Raindrops are generally oblate, especially at larger diameters, and fall with their largest axis being horizontal, as shown in Figure 5.15-45.

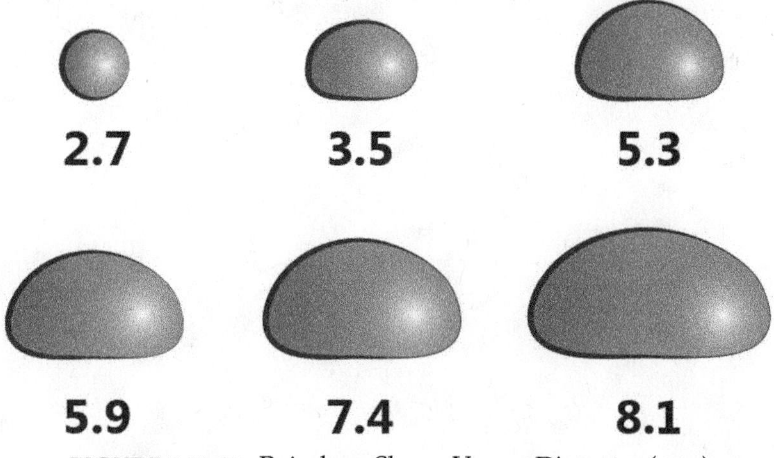

FIGURE 5.15-45 Raindrop Shape Versus Diameter (mm)

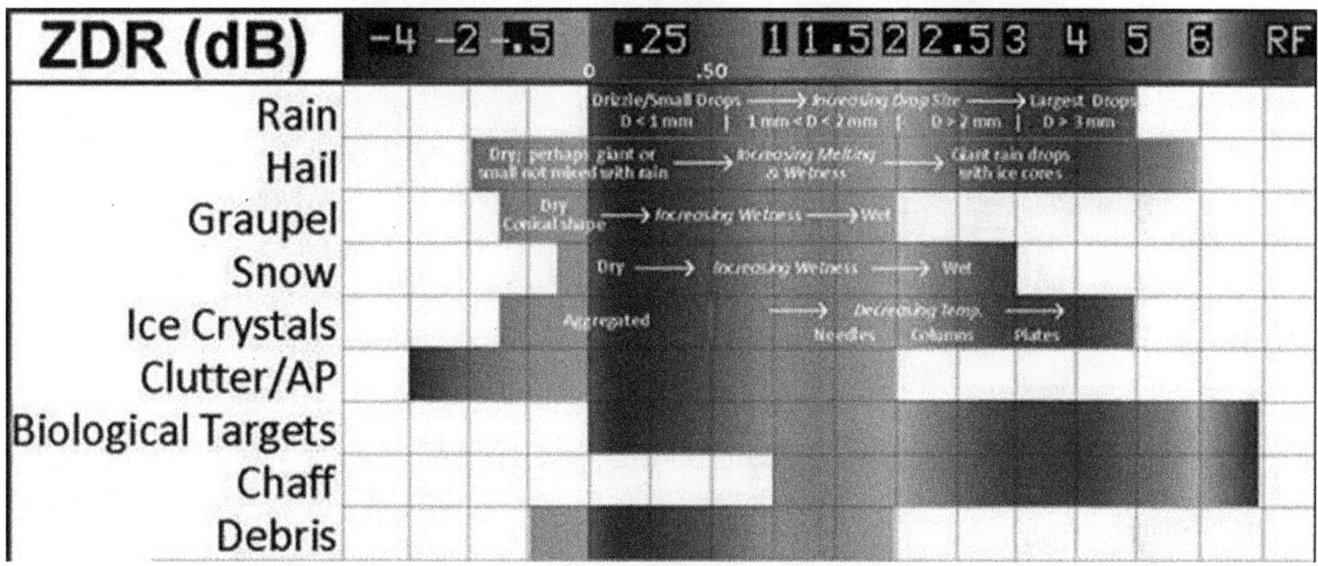

FIGURE 5.15-46 Typical Values of Differential Reflectivity Using an S-Band Dual-Polarization Radar (WDTB 2015)

Courtesy of NOAA

Therefore, for most raindrops, $Z_H > Z_V$, so $Z_{DR} > 0$ dB, and gets larger as the raindrops get larger. In other words, since most raindrops fall as shown in Figure 5.15-45, the horizontally polarized beam will be reflected more by the horizontally oriented raindrops than the vertically polarized beam will. Hailstones, on the other hand, are more spherical, and extremely large hailstones may even fall with a vertical orientation. For hail, Z_{DR} is generally smaller than rain, and may be less than 0 dB for large hail. Snow also usually has a lower Z_{DR} than rain, but as snow becomes wet and multiple flakes aggregate together, it also falls with a more horizontal orientation, so Z_{DR} may be large in wet snow, often near the rain/snow line. A table showing the general Z_{DR} characteristics of most hydrometeors and some other objects using an S-band dual-polarization radar is shown in Figure 5.15-46.

The use of Z_{DR} in identifying the rain/snow line in a heavy winter precipitation is shown in Figure 5.15-47, over north Alabama on February 25, 2015. Z_{DR} is generally low, between 0 and 0.7 dB, over northern Alabama, where heavy snow was falling. Along an east-west line through Birmingham (KBHM), a band of high Z_{DR} shows the rain/snow changeover line. Large, wet snowflakes fell in this area, mixed with rain, and produced the very high Z_{DR} values of 1 to 3 dB. Z_{DR}, just like reflectivity, is shown at the altitude of the radar beam. In this case, at a range of 35 miles from the radar, the radar beam is about 2,000 feet AGL. This means that the rain/snow line at the surface is likely a little bit farther north.

Differential reflectivity may also be used to identify hail, especially large hail. In the example shown in Figure 5.15-48, large hail, some greater than 2 inches in diameter, fell in northwest Alabama on March 31, 2015. Recall that hail has a lower Z_{DR} than rain because it falls more spherically, so Z_H is similar to Z_V. In very large hail, Z_{DR} is often near or below 0 dB, as large hailstones

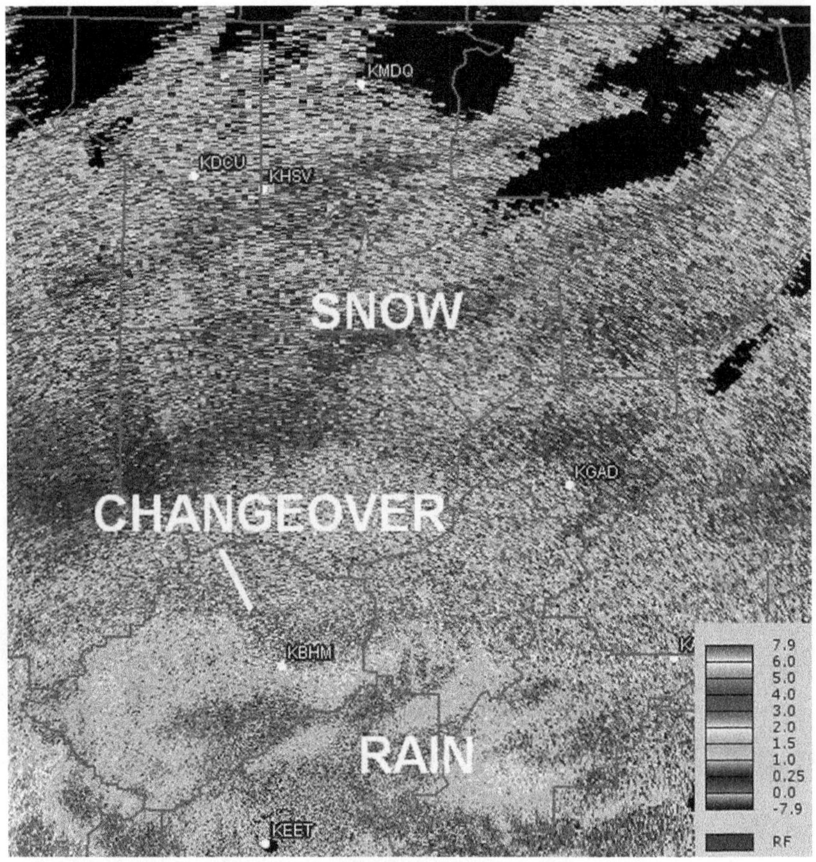

FIGURE 5.15-47 Z_{DR} in a Winter Weather Event with a Sharp Rain/Snow Line and Heavy Precipitation

tumble as they fall and may become slightly vertically oriented due to their weight. However, in identifying hail, one must find areas where high reflectivity values are colocated with low Z_{DR} values; two of these locations are shown in Figure 5.15-48. In the southeastern storm, reflectivity values were above 65 dBZ, and differential reflectivity Z_{DR} was actually below 0 dB over several areas, indicating the very large hail. Note that there is an area of low Z_{DR} to the northwest of the southeastern hailstorm. However, in that area, the reflectivity is only 15 to 25 dBZ, so hail is unlikely in that area.

Similarly to the rain/snow line discussed above, at the *melting level*, or the altitude where falling snow is melting and becoming rain, the melting snow becomes more horizontally oriented, producing large values of Z_{DR}. If this melting level is fairly uniform around the radar, it will produce a ring of high Z_{DR} at some range from the radar, depending on the height of the melting level. This can be particularly important in events where the melting level is slowly descending due to latent cooling and the meteorologist is concerned about the rain changing over to snow at the ground. As the ring of high Z_{DR} gets closer to the radar, this indicates a lower melting level (see Figure 5.15-49).

For many years, rainfall accumulation rates have been estimated using radar reflectivity Z through Z-R relationships. Most Z-R relationships have been of the form $Z = aR^b$, where R is the rainfall rate (typically in mm/hr), and a and b are constants for a particular type of rainfall event, depending on the drop size distribution (DSD).

For example, a widely used Z-R relationship in the NEXRAD network is given by

$$Z = 300R^{1.4}$$

As the DSD is different in almost every event, the standard Z-R relationship rarely provides extremely accurate rainfall estimates by itself. However, since larger raindrops are more horizontally oriented than smaller raindrops (see Figure 5.15-45), larger raindrops also have a higher differential reflectivity Z_{DR}. Z_{DR} may improve the radar-estimated rainfall rate in many situations because it provides some information on the drop size distribution.

Many relationships between reflectivity, differential reflectivity, and combinations of the two with rainfall rate have been developed. One reported by Fabry (2015) used both reflectivity and differential reflectivity, and is written as

$$R(Z, Z_{DR}) = 0.0067Z^{0.927}Z_{DR}^{-3.43}$$

Research is ongoing in the area of radar estimates of rainfall amounts.

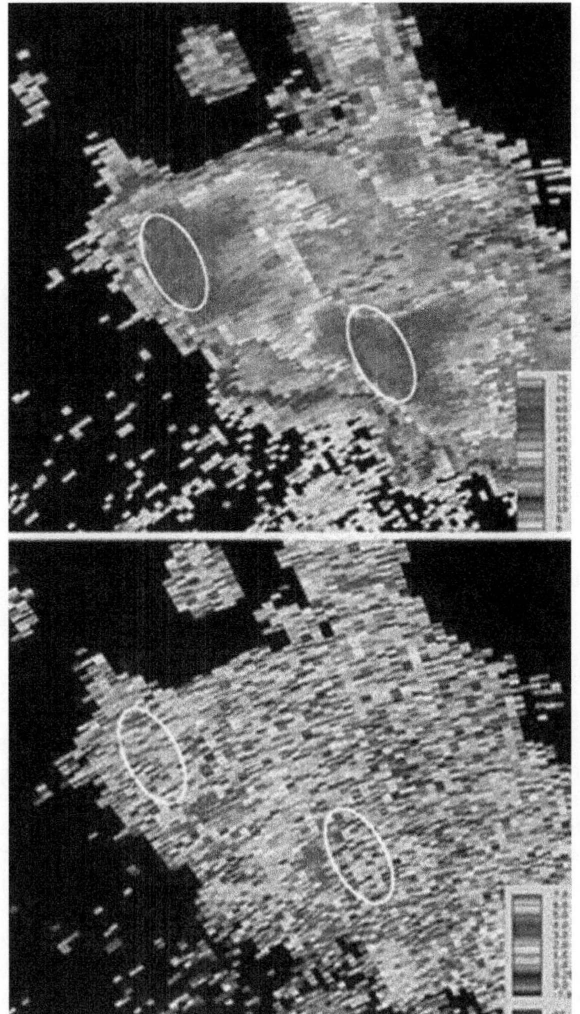

FIGURE 5.15-48 Reflectivity (top) and Z_{DR} (bottom) for Two Hailstorms over Northwest Alabama on March 31, 2015

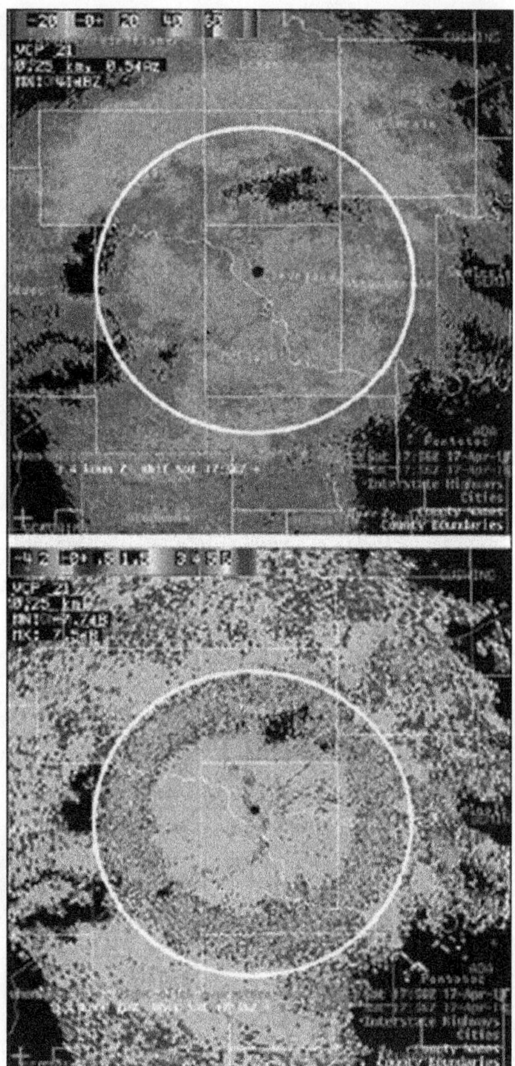

FIGURE 5.15-49 Reflectivity (top) and Z_{DR} (bottom) for Showing the Melting Layer in Stratiform Precipitation

Courtesy of NOAA

Correlation Coefficient

The correlation coefficient ρ_{HV} indicates the correlation between the horizontally polarized and the vertically polarized received energy from a given range gate (see the Received Energy Reflected from Targets section). If all of the targets in a range gate have the same shape (and therefore the same Z_{DR}), then the correlation is perfect, and $\rho_{HV} = 1$. If various types of objects are in a range gate, there will be less correlation between the horizontally polarized and the vertically polarized energy; the more that Z_{DR} varies within a range gate, the lower ρ_{HV} will be. A chart showing typical values for ρ_{HV} for various targets (again, for S-band radars) is shown in Figure 5.15-50.

Rain generally has a high value of ρ_{HV}, generally between 0.95 and 1 (see Figure 5.15-50). As hail gets larger, it becomes more variable in its size and orientation within a range gate. Also, if hail gets very large, depending on the wavelength of the radar, one may transition from the more typical Rayleigh scattering to Mie scattering. The important point is that as hail gets larger, its ρ_{HV} decreases; however, ρ_{HV} even in large hail is generally above 0.8. Correlation coefficient ρ_{HV} also tends to be smaller in wet snow, because there will typically be a range of types of hydrometeors, including some snowflakes, some rain, and some water-coated ice. This also occurs near a rain/snow line or at the melting layer.

However, one particular phenomenon stands out in Figure 5.15-50. *Non-meteorological echoes* such as ground clutter, insects, birds, chaff, and tornado debris tend to have low correlation coefficients. Insects and birds have various Z_{DR} because of their different shapes and various flying angles relative to the radar beam while chaff is quite random in how it falls. Given the varying Z_{DR} within these non-meteorological echoes, they tend to have low values of correlation coefficient ρ_{HV}. This can be helpful when determining whether or not an echo on radar is actually precipitation or simply a non-meteorological echo. In Figure 5.15-51, an evening with a

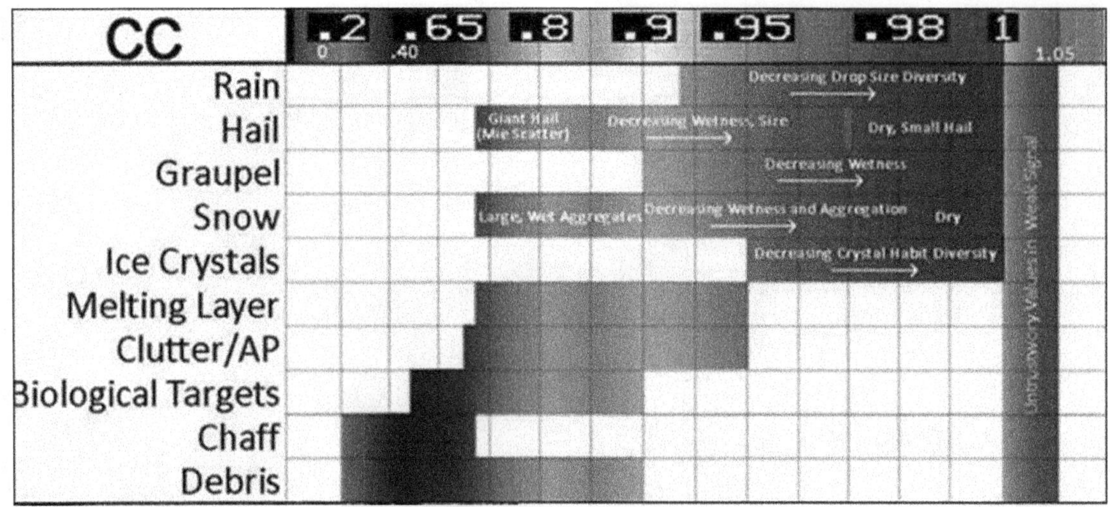

FIGURE 5.15-50 Typical Values of Correlation Coefficient for Various Types of Objects

Courtesy of NOAA

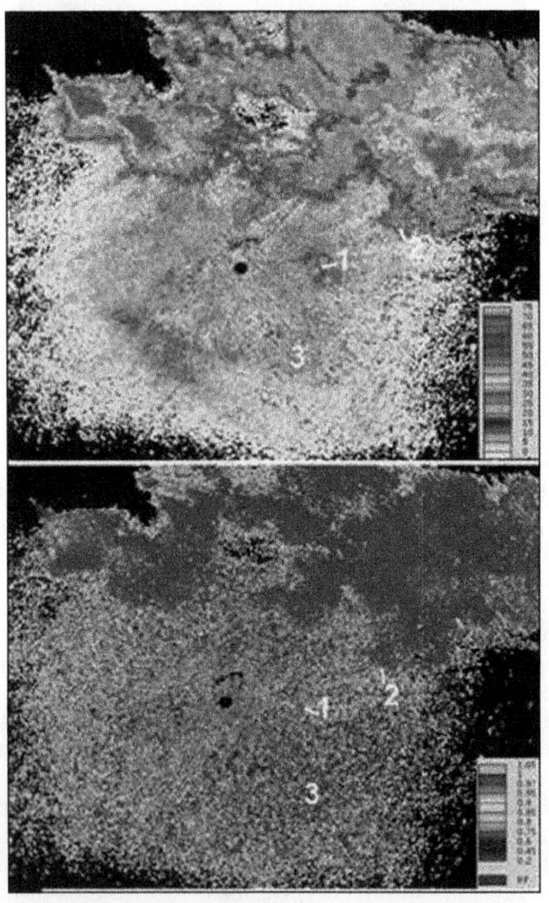

FIGURE 5.15-51 Reflectivity (top) and ρ_{HV} (bottom) with Ground Clutter, Insect Bloom, and Precipitation

temperature inversion near the radar site is causing a large amount of ground clutter and insect return around the radar site. However, there are thunderstorms moving in from the north. Some of the areas of precipitation are obvious (areas with reflectivities greater than 30 dBZ, for example). However, a couple of areas within the ground clutter/insect area may be questionable as to whether or not they are actually precipitation echoes. Since ground clutter and insects have low values of ρ_{HV}, but rain typically has ρ_{HV} near 1, it is clear that areas 1 and 2 are indeed precipitation, as their ρ_{HV} values are between 0.94 and 1. However, area 3 is entirely ground clutter and insects, with ρ_{HV} varying widely, and many areas having ρ_{HV} less than 0.5.

Once the NEXRAD system went to dual-polarization and low ρ_{HV} values associated with tornado debris were examined in more detail, the *tornado debris signature (TDS)* became well known. Detecting tornadoes with radar has always been nearly impossible because they occur at and near the ground. In the past, meteorologists had to rely on reflectivity patterns like hook echoes and LEWPs, and velocity patterns like mesocyclones and TVSs, to predict the possibility that a tornado would touch down. Even though it cannot predict a tornado before it forms like some of these signatures can, a TDS can verify to a high degree of certainty that a tornado is indeed on the ground doing damage. This is because the TDS shows an area of low ρ_{HV}, or lofted debris, in an area where some other indication of storm rotation is detected in reflectivity or velocity products. The TDS is especially useful in situations where marginal reflectivity or velocity signatures do not indicate the presence of tornadoes. According to the NWS and Schultz et al. (2012), a TDS contains reflectivity larger than 30 dBZ, a strong indication of rotation in Doppler velocity, ρ_{HV} less than 0.8, and a Z_{DR} near 0.

Additional research is now focusing on the correlation between tornado intensity and 1) the maximum height and width of the TDS, and 2) how low ρ_{HV} is within the TDS (e.g., Bodine et al. 2013; Entremont and Lamb 2014). These studies may allow forecasters to use TDS parameters to predict tornado intensity with some degree of objectivity for the first time. Examples of TDSs are shown in Figures 5.15-52 and 5.15-53.

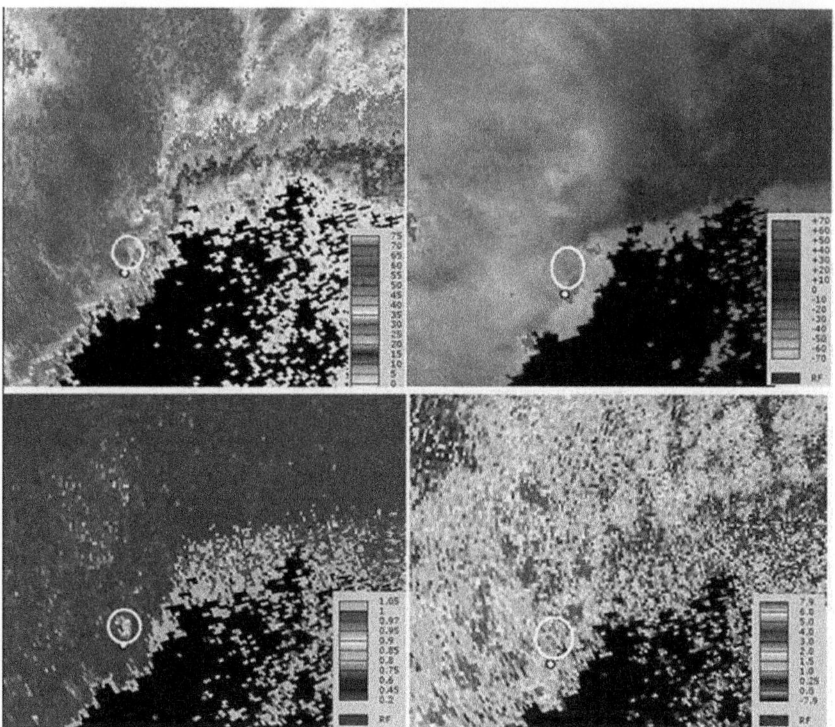

FIGURE 5.15-52 Clockwise from top left: Reflectivity, Velocity, Differential Reflectivity, and Correlation Coefficient Showing a Tornado Debris Signature (TDS) West of Green Bay, Wisconsin, on August 7, 2013

The white circles indicate the area of tornado debris. Note that reflectivities are fairly high, there is a strong mesocyclone signature, Z_{DR} values drop to near zero, and ρ_{HV} is as low as 0.61.

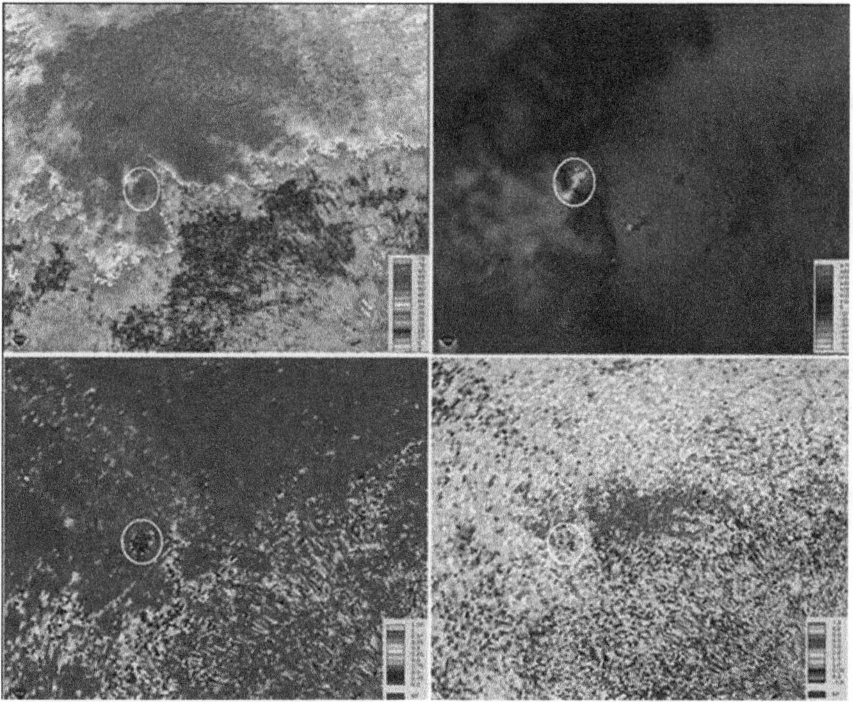

FIGURE 5.15-53 Clockwise from top left: Reflectivity, Velocity, Differential Reflectivity, and Correlation Coefficient Showing a Tornado Debris Signature (TDS) near Bessemer, Alabama, on April 28, 2014

The white circles indicate the area of tornado debris. Note that reflectivities are very high (debris ball), there is a strong mesocyclone signature or possibly a TVS, Z_{DR} values drop to near zero, and ρ_{HV} is as low as 0.21.

Advanced weather radar solutions provide processing that automatically identifies tornado debris signatures using the various radar variables.

As mentioned earlier, a ring of minimum values of ρ_{HV} around the radar often indicate the *melting layer*, because Z_{DR} will vary greatly in that layer as snowflakes melt and aggregate (see Figure 5.15-54).

Most radar return is caused by Rayleigh scattering. Mie scattering occurs when the radius of the reflector is near to or larger than the wavelength of the radar. In cases of *extremely large hail* (generally 2 inches in diameter or larger), due to the transition to Mie scattering, Z_{DR} may vary greatly and cause a low ρ_{HV} (see Figure 5.15-55).

Specific Differential Phase

RF energy slows down very slightly as it goes through rain, snow, hail, and so forth, and this causes a very small change in phase of the returned RF pulse to the radar. The differential phase Φ_{dp} is the difference in phase of the received horizontally and vertically polarized RF energy. Since most weather targets have a $Z_{DR} > 0$ ($Z_H > Z_V$), the differential phase is almost always positive and gets larger as the radar beam goes through more precipitation (at greater ranges from the radar).

However, since Φ_{dp} almost always increases with range from the radar as the beam goes through precipitation, and then stays basically the same once the beam is no longer going through precipitation, it is difficult to interpret physically. The rate of increase of Φ_{dp} with increasing range from the radar, $\Delta\Phi_{dp} / \Delta r$, known as the *specific differential phase* K_{dp}, is easier to interpret, because it is only significant in areas where precipitation is slowing down the radar beam and Φ_{dp} is increasing. Figure 5.15-56 from WDTB shows a schematic of how Φ_{dp} and K_{dp} produce different results; after passing through one area of rain 5 km wide, Φ_{dp} increases to 20° and remains 20° until it encounters another area of rain, then it increases to 35° and remains at 35° from that range forward. K_{dp} in the first rain area encountered is 2 degrees per km (it may seem like it should be 4 degrees/km, but recall that Φ_{dp} is affected by the rain twice, once on the way out to the rain and again on the way back to the radar). K_{dp} is 1.5 degrees/km in the second rain area. So, the values of 20° to 35° for Φ_{dp} in the second rain area mean little in terms of

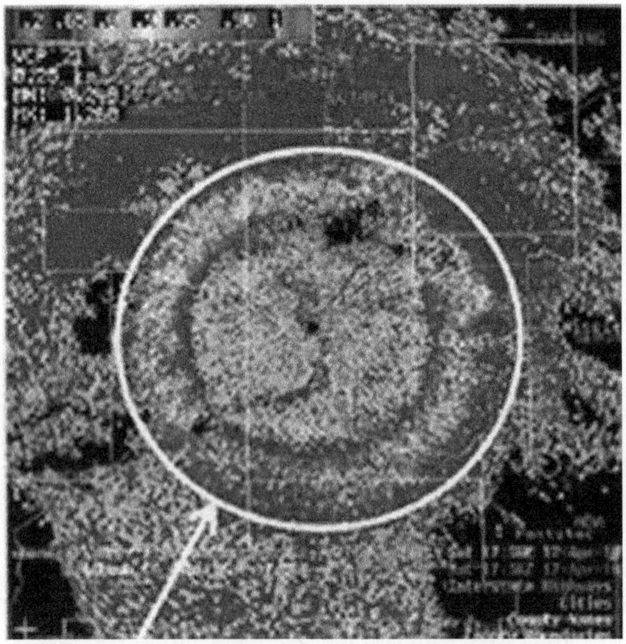

FIGURE 5.15-54 Band of Low ρ_{HV} Radar Showing Melting Layer (WDTB 2015)

Courtesy of NOAA

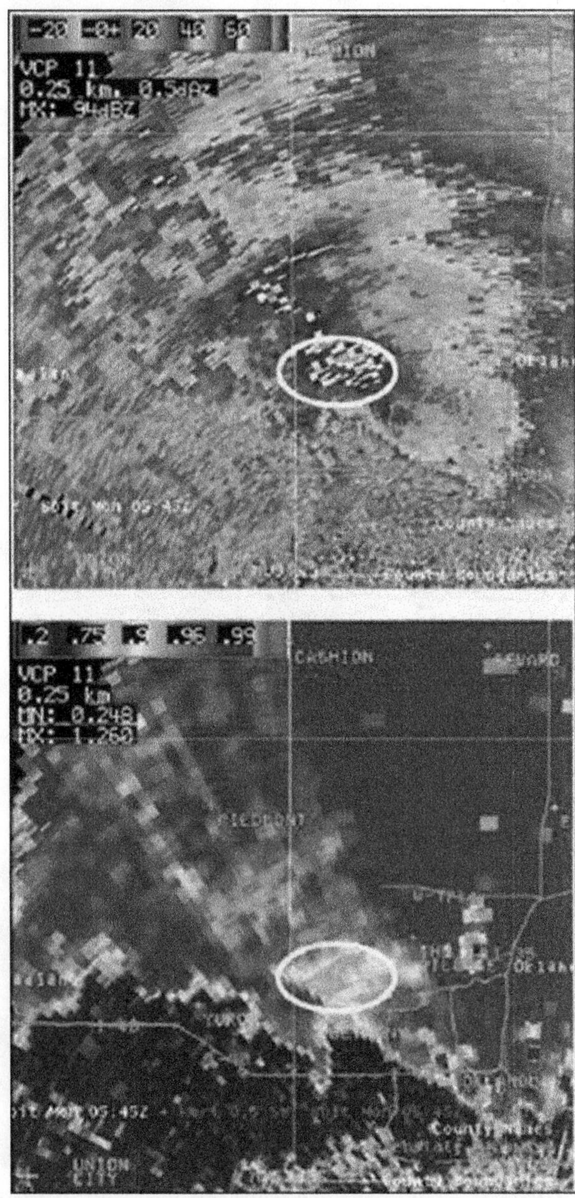

FIGURE 5.15-55 Reflectivity (top) and ρ_{HV} (bottom) Showing Area of Very Large Hail inside White Circle

Courtesy of NOAA

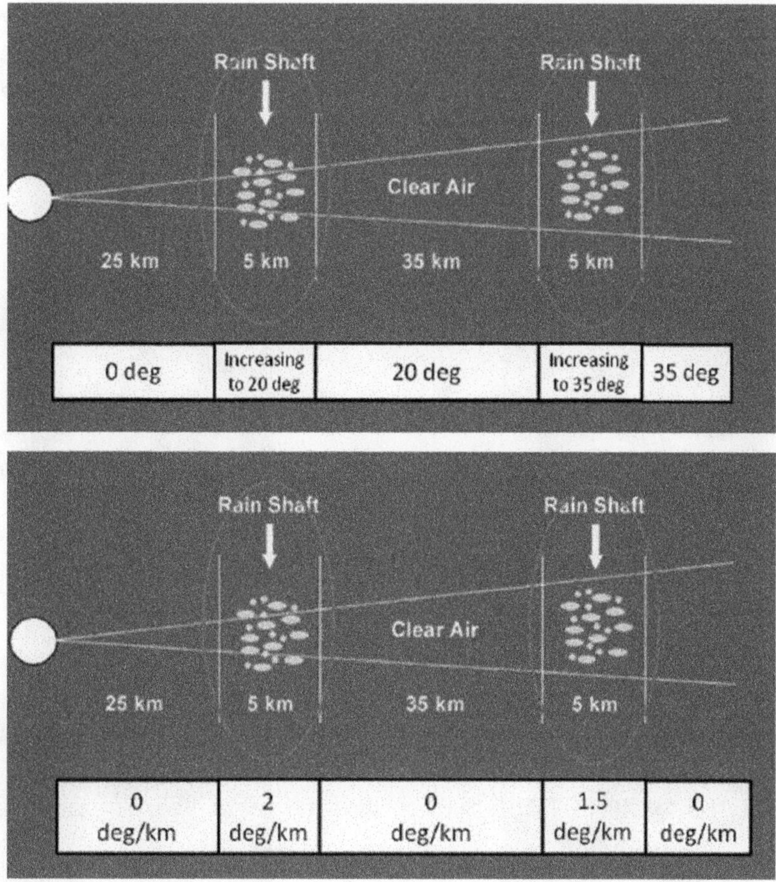

FIGURE 5.15-56 Differential Phase (top) and Specific Differential Phase (bottom)

Courtesy of NOAA

rain intensity there, only about the rain intensity at all points between there and the radar combined. In this case, the K_{dp} value of 1.5 degrees/km does provide information on how much the radar beam is being slowed down in that particular rain area.

Figure 5.15-57* shows a comparison of Φ_{dp} versus K_{dp} at a specific time during the precipitation event that caused the historic floods in South Carolina in October 2015. There is a large area of rain over South Carolina, but the heaviest band of rain at the time of this radar scan is to the northeast and east of the radar site, as indicated by reflectivity. Differential phase Φ_{dp} increases with increasing range from the radar, and along 60° azimuth, it increases from 80° at a range of about 15 miles to 110° at 50 miles. In areas east of the radar where the beam goes through more rain, Φ_{dp} is as large as 200°. However, the Φ_{dp} data is streaky and difficult to interpret. Specific differential phase K_{dp}, on the other hand, is primarily maximized where the highest reflectivities are located. Along the same 60° azimuth, K_{dp} is 2.5°/km at 15 miles and 0.1°/km at 50 miles (the same two points referenced in Φ_{dp}).

As K_{dp} is roughly proportionate to reflectivity, it is also related to rainfall rate. However, since K_{dp} is based on a phase difference as opposed to returned power (like reflectivity), it is not affected by attenuation. This makes K_{dp} another valuable dual-polarization variable in determining rainfall rate and then storm total rainfall. Research is ongoing to find ways to use K_{dp} either alone, or in conjunction with reflectivity and Z_{DR}, to improve radar estimates of rainfall.

Derived Products

There are dozens of derived products which the radar product generator (RPG), or even value-added software from radar data vendors, can compute. These products may be used for a variety of meteorological analyses using the raw radar data. A few common derived products are discussed below.

Storm Cell Identification and Tracking (SCIT)

SCITs are assigned to specific storms by radars so that the storm's motion may be tracked and other attributes of the storm may be evaluated over time. The radar uses scans of the storm at various elevation angles to determine the 3D mass centroid of the storm, then gives it an ID (usually a letter and a number, like T4). The radar will then track the motion of the storm and note trends in its attributes (mesocyclone intensity and depth, maximum reflectivity, VIL, etc.).

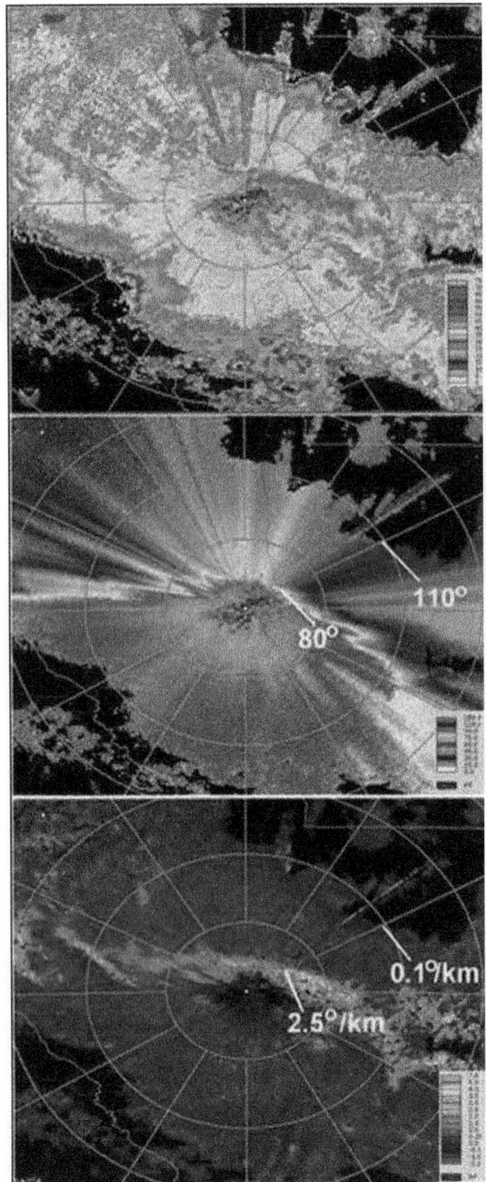

FIGURE 5.15-57 Reflectivity (top), Differential Phase (middle), and Specific Differential Phase (bottom)

***Range rings are at 25-mile intervals.**

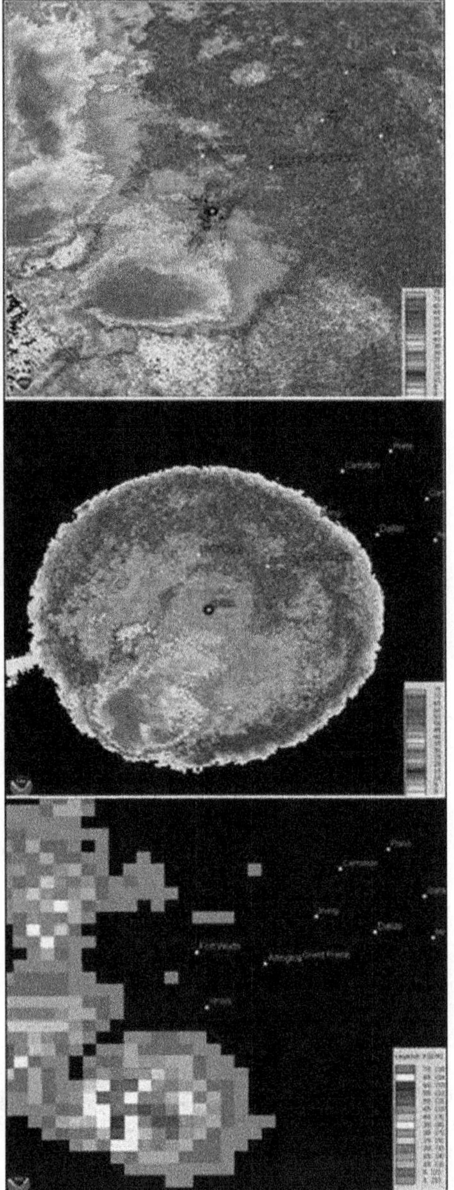

FIGURE 5.15-58 Reflectivity at 0.5° (top), 15.6° Elevation (middle), and VIL (bottom) for a Storm That Produced Large Hail near Dallas, Texas, on April 26, 2015

Vertically Integrated Liquid (VIL)

Given a value of reflectivity and assuming some drop size distribution, the radar can estimate the amount of liquid water in kilograms per cubic meter of air. Within a storm, the reflectivity is determined at numerous altitudes using various elevation angles. By determining the amount of liquid water in kg/m^3 at each elevation angle, the radar can interpolate between elevation angles. It then integrates the amount of water in each m^3 from the ground all the way to the highest elevation scan, determining the total amount of liquid water (in kg/m^2) above any square meter of ground surface. This is called the vertically integrated liquid, VIL.

VIL provides a good approximation for the depth of a storm, and specifically the depth of high reflectivity values in a storm. Deep, intense storms are more likely to produce large hail and damaging winds (see Figure 5.15-58). VIL may be biased upward due to hail contamination (a raindrop size distribution is assumed, so hailstones inflate VIL), but forecasters often use this hail contamination in VIL to diagnose which storms, especially in a low-shear pulse storm environment, will produce large hail. Once a hail report has been received and correlated with a VIL value, forecasters may determine the "VIL of the day," and all storms with VIL above that value are closely monitored for the possible issuance of a Severe Thunderstorm Warning.

Composite Reflectivity

The composite reflectivity at a given location is the maximum reflectivity observed at any altitude at that location (by any elevation angle of the radar). It is particularly useful when evaluating storms that are high-based, or storms that have high reflectivity cores aloft, as they often do during the developmental stage. In storms that are tilted due to wind shear, the composite reflectivity provides an analysis of the trend in overall storm intensity, as shown in Figure 5.15-59. In this storm, there is obviously a large overhang, as the radar elevation scan at 5.1° (about 25,000 feet) shows high reflectivities above areas with little or no significant reflectivity at 0.5° elevation (about 4,000 feet AGL). The storm is tilted downshear, so the area of reflectivities greater than 50 and greater than 60 dBZ shown in the composite reflectivity image are much larger than those areas in either the 0.5° or the 5.1° elevation scans. Composite reflectivity may be used to identify storms with very high reflectivity somewhere aloft, scanning at various elevation angles to locate its altitude.

FIGURE 5.15-59 Reflectivity at 0.5° (top), 5.1° Elevation (middle), and Composite Reflectivity (bottom) for a Sheared Storm in Western Alabama on April 11, 2015

Hail Probability and Size

A radar system uses the height of various reflectivities, along with the temperature profile, specifically the heights of the freezing level and −20°C, to determine not only the probability of hail (or severe hail), but also the maximum hail size. The NSSL algorithm for the probability of hail at the ground uses the difference in height between the highest altitude where 45 dBZ reflectivity occurs, and the height of the freezing level (see Figure 5.15-27). The NSSL algorithm for determining the probability of severe hail uses the reflectivity between the heights of the 0°C and −20°C isotherm (Holleman 2001). The temperature and reflectivity data are also used to determine the maximum hail size. Modern radar systems incorporate dual-polarization variables in hail algorithms, including Z_{DR}, ρ_{HV}, and even K_{dp}.

RADAR PROCUREMENT

There are many competitive advantages for a television station (and its news and weather operations) that has its own weather radar. First of all, it is very promotable, and can set one TV station apart from the others in the market when it comes to weather; weather has consistently been shown to be the primary reason people watch local news in most markets in the United States. Also, the meteorologists at the TV station have full control of the radar, in terms of the PRF (for range vs. Nyquist velocity), the volume control pattern, and RHI scans. For example, PPI scans could be restricted to only the area where storms are occurring, giving more frequent updates than NEXRAD, which updates only every 4.5 minutes. The TV meteorologist may also show RHI scans on the air, showing a unique perspective of storms to viewers. TV station radars also enhance public safety, with more data for TV meteorologists to look at, more value-added products from the RPG, and more data to share with the public.

Perhaps the largest advantage from a meteorological point of view, and one that a TV station can definitely promote, is that a TV station radar *does not have to complete volume scans* at all. Instead, it can make PPI scans at a constant, low-elevation scan and get updated reflectivity, velocity, and dual-polarization information every 10 seconds to 1 minute, as opposed to every 4.5 minutes for NEXRAD data. NWS has implemented SAILS (Supplemental Adaptive Intra-Volume Low-Level Scan) using NEXRAD radar, but that only provides one to three extra low-level PPI scans during the 4.5 minute VCP (ROC 2015). A TV radar can be set to get ten or more low-level updated PPI scans in 4.5 minutes, providing important updates during severe weather. The extra minutes of lead time on rapidly changing storms can save lives, and the greater number of scans over any period of time produces much more fluid, beautiful radar data animations.

Trade-Offs

There are many types of commercially available radars for television stations to choose from. The starting point is usually the budget. This typically determines the magnetron/klystron and the output power decisions.

Klystron Versus Magnetron Transmitter

As pointed out in the Transmitter section, a magnetron transmitter does not transmit a stable frequency, requiring the use of a STALO and COHO to calibrate the frequency and Doppler shift in real time. A klystron transmitter transmits a very stable frequency, and a COHO and frequency-stabilizing equipment are not necessary. A klystron transmitter is usually more expensive up front than a magnetron transmitter at the same peak power with all other factors being equal; however, a klystron transmitter typically has a longer life span than a magnetron and offers savings over the lifetime of the radar system. A klystron radar is generally considered superior to a magnetron radar, and its more stable frequency provides better clutter rejection and velocity data.

Sensitivity

A radar's sensitivity is basically the minimum signal the radar can detect. It depends on factors such as beam width, antenna size, wavelength, and maximum transmitted power. A shorter wavelength, higher-frequency radar (such as C-band) has a higher sensitivity to meteorological targets than high S-band or S-band radars, and dual-polarization radars are slightly less sensitive than single-polarization radars. However, experiencing the lower attenuation at high S-band and having dual-polarization capability are both usually worth the loss in sensitivity unless a TV station is attempting to detect very light radar echoes (drizzle, light snow, etc.) at long ranges from the radar.

Also, if one places the radar antenna on a high tower, ground clutter may actually increase due to radar sidelobes (transmitted energy outside the main beam), and sensitivity may decrease. An ideal location for a radar is in a small bowl-shaped valley, where the sidelobes and associated ground clutter will be eliminated and sensitivity will be enhanced.

Frequency

As discussed in the Other Considerations with Weather Radars section, low-frequency, long-wavelength radars are less attenuated by rainfall than high-frequency, short-wavelength radars. Historically, NEXRAD radars have been S-band with a frequency of 3 GHz, and TV station radars in the United States have been limited to C-band (5.5 GHz) by NOAA and the FCC. However, very recently, the high S-band frequency (about 3.5 GHz) has been made available to commercial vendors by the U.S. government.

A high S-band radar is superior to a C-band radar in most weather situations due to its much smaller attenuation. However, the beam width is proportional to the wavelength and inversely proportional to the diameter of the antenna (see the Antenna section). In order to produce a reasonably small beam width, the antenna dish must be larger for a high S-band radar than for a C-band radar. For example, the NEXRAD system has an approximately 1-degree beam width at S-band (10 cm wavelength), and the antenna must have a diameter of 8.5 m (about 28 feet). For a high S-band radar (8.5 cm wavelength), the antenna must be 7.2 m in diameter (about 24 feet) to maintain the same beam width as NEXRAD. A C-band radar only requires a 4.5 m (15 feet) diameter antenna. So, although a high S-band radar has advantages over most TV station radars currently in place, a high S-band radar requires a larger antenna, and therefore a larger radome, and is more expensive than a C-band radar.

Siting of Radar

A television station must consider several advantages and disadvantages when determining where to place their new radar. The radar should be placed away from the center of the largest population area, to keep that area out of the cone of silence and to limit ground clutter issues. Ideally, the radar would be placed on the opposite side of the populated area away from the direction the most common severe weather originates. This keeps the radar tower out of the heavy precipitation at critical times. There are a number of other issues that must be considered.

Avoidance of Beam Blockage

As discussed in the Other Considerations with Weather Radars section,, any objects (trees, buildings, mountains) that block the radar beam will cause errors in radar reflectivity at all points beyond the range of the blocking (see Figure 5.15-60). The beam rises fairly quickly with height, but one must place the radar in an area where it is not within 1 or 2 miles of any obstructions (trees taller than the radar tower, or buildings) and where it is not too close to high terrain. Many radar vendors can help calculate the beam propagation pattern and potential beam blockage at any location.

Avoidance of Interference

The radar must be placed in an area where it will not receive interference, or false signals, from outside sources of microwave or RF energy in the vicinity of the radar. The sources could include high-voltage power transmission lines (that emit an electromagnetic field including RF energy), television towers, or communications equipment.

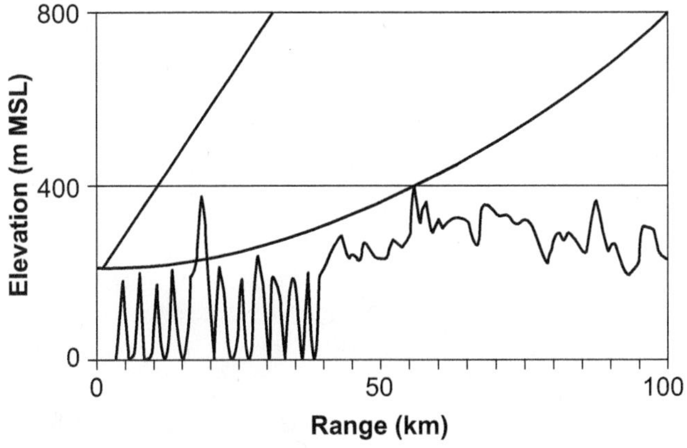

FIGURE 5.15-60 50% Beam Blockage (3 dBZ) Due to High Terrain

Coleman, T.A., and J.A. Westland, 2010: Underestimation of QPE in a Flash Flood Situation Due to Partial Beam Blocking and Attenuation: Correction Using the BREAM model. 24th Conf. on Hydrology.

Power and Data Access

When selecting a location for a radar, access to commercial power lines and high-speed, reliable data lines must be considered. Some remote locations would be excellent for a radar, but without access to the power grid or the communications grid, these sites could be unusable, or at the very least cost much more money, because the TV station may have to pay utility companies to take power and data lines to the site.

Radiation Hazards

There should be verification that the RF energy being emitted by the weather radar will not produce a health hazard to the surrounding community (e.g., Leone et al. 1989).

Physical Accessibility (Road)

It is important that the radar site is accessible by a road, or close enough to an existing road that a small service road, even gravel, can be built. Trucks must have access to the radar site to deliver fuel for the backup power supply, and be able to drive other vehicles there to service the radar. The road should not be dirt, as these roads sometimes become impassible after heavy rain.

Permitting

Frequency (FCC)

In addition to the need to avoid interference with the radar from outside RF sources, the radar must also be placed in an area where *it will not interfere with other RF signals* such as television, communication, or other radars. The FCC must approve the installation of all high-energy transmitters, especially weather radars.

FAA

The radar tower and radome must be colored (for daytime) and lit (for nighttime) appropriately according to FAA standards to avoid any potential hazards to aviation.

Construction

Construction permits will be required to build the radar tower, the building containing the transmitter, the antenna and radome, and any service roads. TV stations must make sure that the land they want to place their radar on is zoned properly also.

References

Battan, L.J., 1973: *Radar Observation of the Atmosphere*. Chicago: The University of Chicago Press.

Bellon, A., and F. Fabry, 2014: Real-Time Radar Reflectivity Calibration From Differential Phase Measurements. *J. Atmos. Oceanic Technol.*, **31**, 1089–1097.

Bodine, D.J., M.R. Kumjian, R.D. Palmer, P.L. Heinselman, and A.V. Ryzhkov, 2013: Tornado Damage Estimation Using Polarimetric Radar. *Wea. Forecasting*, **28**, 139–158.

Burrows, D.R., and Attwood, S.S., 1949: *Radio Wave Propagation*. New York: Academic Press.

Cohen, L., 2015: RADAR Transmitter Overview Tube and Solid State. Naval Research Laboratory. URL: www.its.bldrdoc.gov/media/31069/CohenRadarTxOverviewISART2011.pdf.

Coleman, T.A., K.R. Knupp, J. Spann, J.B. Elliott, and B.E. Peters, 2011: The History (and Future) of Tornado Warning Dissemination in the United States. *Bulletin of the American Meteorological Society*, May 2011.

Colorado State University (CSU), 2015: X-Band Attenuation Example: 30 July 2012. URL: www.chill.colostate.edu/w/X-Band_attenuation_example:_30_July_2012.

Corfidi, S.F., J.S. Evans, and R.H. Johns, 2015: About Derechos. URL: www.spc.noaa.gov/misc/AbtDerechos/derechofacts.htm#types.

Doviak, R.J., and D.S. Zrnic, 1993: *Doppler Radar and Weather Observations*. 2nd ed. San Diego, CA: Academic Press, 562 pp.

Duda, J., 2015: A History of Radar Meteorology: People, Technology, and Theory. Iowa State University. URL: www.meteor.iastate.edu/~jdduda/portfolio/HistoryPPT.pdf

Entremont, C., and D. Lamb, 2014: Relationship Between Tornado Debris Signature (TDS) Height and Tornado Intensity. National Weather Service.

Eumetcal, 2015: URL: www.eumetcal.org/intralibrary/open_virtual_file_path/i408n169839t/co/.

Fabry, F., 2015: *Radar Meteorology Principles and Practice*. Montréal and New York: McGill University and Cambridge University Press.

Holleman, I., 2001: Hail detection using single-polarization radar. Scientific report 2001/01, Royal Netherlands Meteorological Institute (KNMI).

Ice, R.L., J.G. Cunningham, A.K. Heck, W.D. Zittel, and R.R. Lee, 2015: Polarimetric Weather Radar Antenna Calibration Using Solar Scans. URL: www.roc.noaa.gov/wsr88d/PublicDocs/Publications/PolarmetcirWXRadar_Cal_Using_SolarScans_AMTA2014_final.pdf

Kessinger, C.J., E.A. Brandes, and J.W. Smith, 1995: A Comparison of the NEXRAD and NSSL Hail Detection Algorithms. Preprints, *27th Conference on Radar Meteorology*, American Meteorological Society. 603–605.

Lee, L., 2008: The Kings Mountain Tornado of 28 April 2008. National Weather Service.

Leone, D.A., R.M. Endlich, J. Petriceks, R.T. Collis, and J.R. Porter, 1989: Meteorological Considerations Used in Planning the NEXRAD Network. *Bull. Amer. Meteor. Soc.*, **70**, 4–13.

Marshall, J.S., and W.M. Palmer, 1948: The Distribution of Raindrops with Size. *J. Meteor.* **5**, 165–166.

National Weather Service (NWS), 2015: Microbursts. URL: www.srh.noaa.gov/bmx/?n=outreach_microbursts

Prupachher, H.R., and K.V. Beard, 1970: A Wind Tunnel Investigation of the Internal Circulation and Shape of Water Drops Falling at Terminal Velocity in Air. *Q. J. R. Meteorol. Soc.*, **96**, 247–256.

Radar Operations Center (ROC), 2015: Multiple Elevation Scan Option for SAILS. URL: www.roc.noaa.gov/wsr88d/NewRadarTechnology/NewTechDefault.aspx.

Rinehart, R.E., 1997: *Radar for Meteorologists*. 3rd ed. Grand Forks, ND: Rinehart Publishing, 428 pp.

Schultz, C.J., S.E. Nelson, L.D. Carey, L. Belanger, B.C. Carcione, C.B. Darden, T. Johnstone, A.L. Molthan, G.J. Jedlovec, E.V. Schultz, C.C. Crowe, and K.R. Knupp, 2012: Dual-Polarization Tornadic Debris Signatures Part II: Comparisons and Caveats. *Electronic J. Operational Meteor.*, **13**, 138–150.

Sim, P., 2014: Making Waves: Robert Watson-Watt, the Pioneer of Radar. British Broadcasting Company. URL: www.bbc.com/news/uk-scotland-tayside-central-27393558

Skolnik, M.L., 1980: *Introduction to Radar Systems*. New York: McGraw Hill, 581 pp.

Waldvogel, A., B. Federer, and P. Grimm, 1979: Criteria for the Detection of Hail Cells. *J. Appl. Meteor.*, **18**, 1521–1525.

Warning Decision Training Branch (WDTB), 2015: Dual-Polarization Radar Training. National Weather Service. URL: www.wdtb.noaa.gov/courses/dualpol/

CHAPTER

5.16

Television Graphics

KEVIN ENTREKIN, CHRISTOPHER KELLY, AND PAUL LACOMBE

Ross Video Limited
Iroquois, Ontario, Canada

INTRODUCTION

Television graphics are how we generate synthetic pictures and video for use in broadcast. At a high level we will discuss online graphics, which include character generators (CG), clip players, branding devices, and augmented reality and virtual studio generators (AR/VS); and off-line graphics, which are graphics rendered in software to a file and used through an existing online device or in a postproduction environment.

As Moore's law describes in short: processing power doubles every two years. And there are few technologies that have realized the benefits of this like television graphic systems. The gap in abilities between off-line rendered graphics and online rendered graphics is closing. What took proprietary and expensive rendering devices can now be performed on more easily sourced and mass-produced professional workstations. In some instances consumer-grade computers have been used to create graphics that air worldwide.

For most broadcasters, the hardware is only as important as the software's ability to leverage the hardware's power. That is common between off-line and online graphic generation. So if given the choice of hardware platforms, it is important to consult the software's recommended hardware complement.

In this chapter, we will describe at a macro level the differences between online and off-line graphic generation and at a micro level the differentiation between workflows in online graphic generation.

TYPES OF GRAPHIC SYSTEMS

As described above, there are two primary types of graphic systems: online and off-line. Inside each type, there are various use cases and workflows that differentiate them further.

Off-line Devices

These devices do not directly output to a baseband hardware output, and are generally a software-only solution. The software takes the images created and writes them to a file or bitstream to be used in a nonlinear editing solution, or eventually in another online system. Where online devices need to provide a new frame of video in real time, 16 ms or less at 60 frames per second (fps), or 20 ms or less at 50 fps, off-line creation gives the creator more flexibility to render more complex graphics to be used in a postproduction environment.

The output of the graphics comes in many flavors and with many parameters. Resolution, pixel aspect ratio, color depth, alpha channels: all must be taken into consideration, as the file will eventually be displayed alongside other video content and should match. As many postproduction workflows use compression to economize on storage, compression of graphics should also be taken into consideration. Compression adds many factors, including artifacting, file size, and processing. One compression algorithm may be lossless visually with little to no artifacting, and small file sizes, but requires much more processing power to unpack the images to baseband video. Another algorithm may require less processing power, but comes at the cost of either more artifacts or bigger file sizes.

These files are then transported as various file types, also with their own unique properties. Image sequences, like Targa or PNG, are where the individual frames of video are their own file. See Table 5.16-1 for a listing of common still image file types. One second of 60 fps graphics would be 60 files, each with its index number appended to its filename. The resolution, compression, and so forth would be dictated by the capability of the image format. One weakness of this file format is audio is separate from images. A strength is most if not all graphic systems accept image sequences as there are rarely any proprietary instructions for the editing platforms to interpret.

Audio Video Interleave (AVI) and QuickTime (MOV) files are other commonly accepted formats. These file types are explicitly video formats where all of the frames are "wrapped" inside one file, again with the resolution, compression, and so forth dependent on the capabilities of the format. In addition, many times the audio is embedded in the file. While these particular file types are widely accepted, some systems are incompatible with one or the other due to proprietary instruction sets. Table 5.16-2 lists the types of common video files.

Be it an image sequence or a video file, these files are flattened and not editable once they are in the editing solution. If a change needs to be made, one must go back to the creation software to adjust and export or save a new file.

Most editing solutions have their own simple graphic tool to create a simple CG, or to add text to an image or video created outside of the editing tool. This can save time and space, but it is more limiting in how the graphic will look. In addition, some editing solutions have a plug-in architecture that allows for third-party middleware to create specialized graphics, saving time and space, without compromising much of the rich graphic capabilities of the more advanced graphics systems. In both situations, rarely do they offer rich graphic creation from scratch in the editing solution; rather, they use templated graphics from dedicated graphics software and allow for editing of text and simple images, all from the same editing user interface.

TABLE 5.16-1
Common Still-Image File Types

Extension	File Type
.jpg	JPEG Image
.png	Portable Network Graphics File
.psd	Photoshop Image File
.tga	Targa Image File
.tif	Tagged Image Format File

Online Devices

Online graphic systems take pre-rendered and real-time rendered graphics from the central processing unit (CPU) and in most cases a graphics processing unit (GPU) and output them from a hardware baseband video output such as analog composite, analog component, serial digital interface (SDI), high-definition serial digital interface (HD-SDI), and high-definition multimedia interface (HDMI). A discussion at greater depth on hardware outputs can be found below in the Output Hardware section.

The online device can link into the broadcast chain in many places, simply grouped into two categories: upstream and downstream. When a graphic system is upstream, it is connected to a switcher or keying device upstream from the broadcast signal path. When a graphic system is downstream, it is in line with the broadcast signal path, and the layering is done by the same hardware that produces the rendered graphics.

Upstream Devices

There are two kinds of upstream devices we will discuss. The most common is insert graphics in a production control room. More recently, graphic devices are used in conjunction with camera tracking-capable hardware and software to provide graphics that adjust viewpoints based on the camera's viewpoint.

Most people associate television graphics with the first type: upstream insert graphics in a production control room. In this instance, two signals are produced: the fill channel, which is the full-color representation of the rendered graphic, and the key channel, which is the alpha channel of the rendered graphic. The alpha channel is a luminance-only signal that is an instruction set for how the graphic is to be layered, or keyed, over another signal. This channel must be matched frame accurately with its paired fill channel in order for it to be displayed correctly. With few exceptions, the graphic device must have two hardware outputs for every keyed channel of graphics.

When connected to a switcher or other hardware keying device, the timing of the signals must match, in

TABLE 5.16-2
Common Video File Types

Extension	File Type
.avi	Audio Video Interleave—Wrapper developed by Microsoft for video applications. Can contain many different types of video content, like Radius Cinepack, MPEG1, MPEG2, MPEG4, Uncompressed, and many more.
.mov	QuickTime—Wrapper developed by Apple for video applications. Can contain many different types of video content, like Radius Cinepack, MPEG1, MPEG2, MPEG4, Uncompressed, and many more.

order to line up the top line of the graphics with the top line of the other sources in the keying device.

Camera-tracked graphics are another type of upstream device. The fundamentals are very much the same as with traditional insert graphics. However, because the graphics are paired with a very specific camera and its movement, the keying is usually done upstream of the broadcast production switcher. If the graphics are providing a full virtual studio, they will be composited in an external chroma keyer, and the finished product will be sent to the switcher flattened. If the graphics are augmented reality, graphics are layered over the top of the camera, but not behind, and only need a linear keyer.

Downstream Devices

Downstream graphics are where the keying and layering of the graphics are done internally by the graphic device. There are two methods in which this keying and layering can be done. In both instances, there is an input to the framebuffer device. The broadcast signal is routed into the input, where the keying and layering of graphics is performed. The difference between the two methods happens here.

On most framebuffer devices, the keying can be performed in place on the card. The graphics are rendered, layered, keyed, and passed immediately to the output of the card. The broadcast signal never passes through the local bus of the computer. For many applications this is the preferred method as there is little to no signal delay through the card. More importantly, the broadcast signal, and its associated ancillary data, is often untouched with the exception of the graphics being introduced.

The input source, however, can be made part of the graphic composition, in which case the input is captured, processed, rendered, and output in near real time. This allows for the broadcast signal to be processed as a part of the overall graphic rendering, giving digital video effect (DVE) like capabilities. This method does come at a cost, however. Because the image is being captured and processed, it is susceptible to image quality loss and stripping of ancillary data. In addition, the signal can be delayed a number of frames.

Downstream graphics are typically used for channel branding of the broadcast, and/or constant information like a weather or news tickers. If either branding or tickers were done upstream, that switcher or keying device would need to be left untouched until the graphics were taken off-line.

ARCHITECTURE AND FUNDAMENTALS

Hardware

Most modern online graphics systems are comprised of the same basic components:

1. CPU/Motherboard
2. Graphics Processing Unit (GPU)
3. Framebuffer Input/Output Device
4. Other Workstation/Server Accessories (Storage, Network Interface, etc.)

The components listed above, when tied together with CG software, are the foundation for the modern CG, and often referred to as render engines. In this section, we're going to discuss in greater detail the hardware components of the render engine, as well as the software requirements to produce broadcast graphics.

Rendering Hardware

The two most critical components of the CG are the central processing unit (CPU), and the graphics processing unit (GPU). Think of the CPU as the operations manager for the system. The CG software is reliant on a nonproprietary operating system (like Windows or Linux), so much of the basic UI functionality is performed within the bounds and limitations of the OS, as well as some ancillary tasks as needed. Think of the GPU as the artist and creative director. The GPU, with the aid of the OS's application programming interface (API), performs most if not all of the actual image rendering.

After the operating system is started, and the CG application launched, the software initializes resources in the GPU to start writing frames. Where those frames are rendered to is dependent on the type of system. For a traditional online system, the frames will be created in the GPU and sent to one of a few possible locations in the workstation:

1. Framebuffer Device
2. IP Device
3. File

With a framebuffer device, output of the images comes from a dedicated internal or external device, connected to the local computer bus. The graphics software generates uncompressed frames of video as well as packets of audio, and writes that material to dedicated memory locations on the framebuffer device. The software then tells the framebuffer device in what order the frames are to be displayed, and when a buffer location is no longer in need of a particular frame the software writes over that frame with a new frame.

As described in the discussion on upstream/downstream devices, most often there are two compositions sent to the framebuffer for every actual frame. One is the fill channel, which is the full-color 4:2:2 image. The other is the key channel, also referenced as the alpha channel. The key channel is a luminance-only picture designed to be the instruction set for the keyer downstream of the graphics device *or* for the internal keyer in the framebuffer device. The keyer takes that luminance-only image and, based on the luminance value of a particular pixel, determines the opacity of that pixel as it keys over the top of another video source. Framebuffer boards are not always simply input/output devices. Many offer more complex operations to be processed on their own hardware layer. It is up to the CG software to implement and initialize those operations, like up/down/cross conversion and edge softening. Many devices also offer simple black output when frames are not written to the device, or test patterns for setup and diagnostic purposes.

While the GPU is drawing image frames, the CPU is keeping the operation in order. Still and video images are decompressed and stored on the overall system RAM as well as the GPU RAM cache for faster recall. It is also the liaison between the operator and the renderer. When an operator requests for a graphic to be output, the CPU instructs the GPU what to draw and how it should be displayed.

Output Hardware

There are three types of output we will discuss in this section: baseband video, IP, and as a file. For most users, CG will eventually come out of a baseband video device. Even rendered files will likely playout via baseband video.

As described above, when rendering to baseband video, there is a hardware component with enough frame buffering capability to store a finite number of video and audio frames for output. Once those frames are rendered and stored, it is up to the hardware device to output in the proper format. The software CG engine uses the framebuffer's SDK (software development kit) and drivers to set the appropriate output modes, with frame rate, resolution, and whether it's interlaced or progressive. Then the board will take the frames and output them to the device's interface or interfaces. Some framebuffer boards offer multiple interfaces out, for both audio and video. Most broadcasters will use serial digital, regardless of whether it's standard definition or high definition. Some framebuffer boards can simultaneously output to serial digital as well as analog composite or high-definition multimedia interface. For every channel of graphics, there will need to be two physical outputs, one each for the fill channel and the key channel. So what is commonly referred to as a two-channel device is really four individual outputs.

When discussing IP, it is important to note that IP delivery of graphics is much like the rest of the IP discussion: unresolved. But there are a few basic principles with IP that make it different than a standard baseband video delivery. First and foremost, at this moment there is no framebuffer device for IP. We are reliant on the internal PC architecture to deliver the frames to the network adapter. So the frames, when buffered, need to be buffered back to the system's RAM. The frames can then be sent as packets to the appropriate IP router and output device.

Lastly, many CG systems can render the frames to a file. Much like the off-line creation software systems available, the online systems can render what was supposed to be streamed in real time, and instead write the frames to a file. As discussed in the Off-line Devices section, there are many formats and many codecs. CGs vary widely on what formats they support, so it will be important to check with your CG of choice to determine what formats it can output to.

Keying Types

There are two types of ways a graphic can be layered on top of video: shaped and unshaped. When a graphic is processed by a keyer, the keyer needs instructions on how the graphic was intended to be keyed.

A shaped key has the alpha channel pre-applied to the video, or fill channel. When viewing the fill channel by itself, the image will look like it was keyed appropriately over black.

With an unshaped key, the alpha channel is *not* pre-applied to the video or fill channel. When viewing the fill channel by itself, many times the image has no gradients to the black background. In addition, there will be bleed to the edges, leaving the key channel to perform the full cutout over the top of video.

In either condition, most keying devices can take a shaped or unshaped key and display it where once it's keyed it would display properly. When previewing a graphic on a monitor or multi-display device, an unshaped key will not as closely match the output because the multiplication of the fill and key will not yet have been performed.

Software

CG systems require a system-level rendering application interface to the hardware and GPU. Specifically, there are two APIs that most CGs use: OpenGL and DirectX. In both instances, the API gives the CG direct access to the 3D rendering hardware. By offloading this processing to the GPU, it greatly reduces the rendering time of the graphics and reduces the overhead on the more multipurpose CPU. There are substantial differences between the two, however, and we will discuss those below.

OpenGL

OpenGL is an open standard API, developed by Khronos Group. OpenGL API calls are *only* for graphic rendering. There are no functions that handle audio, networking, or other multimedia applications. For many CG systems, they need to come up with another API to handle those parameters in conjunction with the graphics layer. Like its competitor, DirectX, OpenGL is used primarily in the 3D rendering world for not only broadcast graphics, but also for things like video games and CAD software. The API is implemented in the GPU drivers on a vendor-by-vendor basis, so it is possible for different hardware to exhibit different behaviors for similar API calls. OpenGL is a cross-platform API, which means that an application can be ported between many operating systems, including Windows, Mac OS X, and the many variations of Linux. When new versions of OpenGL are introduced, new functions and libraries are added, where existing libraries and functions usually remain the same. This means upgrading the version of OpenGL should not affect how the developed CG operates.

DirectX

DirectX is a platform developed by Microsoft and is only available in the broadcast environment through a Windows OS. DirectX has libraries for not only graphic rendering, but also audio, networking, and other multimedia applications. This means from a development standpoint the development team needs only be familiar with one family of API libraries. With each new version of DirectX, the functions and libraries change. So developers of a CG system using DirectX will typically standardize on a version and will thoroughly test existing known features in the API when making the upgrade to a later version. Hardware manufacturers implement the DirectX API in their drivers through a common runtime. Therefore, there should be little variation between GPU performance and behavior.

Design and Templating

When a design is implemented into a CG system, there are many factors on how a CG graphic can be produced using different systems. In this section, we will discuss a few of the parameters that affect how a graphic looks and is rendered.

2D/3D Space

Most CGs built today rely on 3D space to render their graphics. This means that creation of 3D graphics are based on 3D objects placed in XYZ space. Three-dimensional graphics require a camera viewpoint in order to render. So the XYZ space of an object is relative to the viewpoint of the camera. And an object closer to the viewpoint of the camera will render over the top of another object that is further away from the viewpoint of the camera.

Objects are invisible without two parameters: materials and lighting. The object itself provides the structure of the visuals as a flat plane, cube, sphere, cylinder, and more. The material represents the wrapping of the object. At a simple level, a red material combined with a cylinder would result in a red cylinder. Objects can have multiple faces to which different materials can be applied. So one end of the cylinder could have a green face, and the other end could have a blue face. Materials can also have shaders, which are visuals more complex than simple RGB colors. Shaders, usually a function of an implementation of the GPU's API, allow the CG to display pre-rendered still and video graphics, as well as internally generated patterns and visuals like gradients.

In addition, lighting affects how a material is viewed, based on the position of the lighting relative to the object and the camera viewpoint. Material parameters like diffuse color and emissive color dictate how light reacts with the material as a whole—diffuse influencing how light reflects from the material, and emissive how light emits from the material.

Some CGs still rely on 2D space for their graphic representation. In this instance, graphic element positions are represented in XY space, and what objects render over others is based on a simple layer order represented in a list format. Three-dimensional looks can be represented in a 2D design environment. However, usually the objects are in a single layer, and do not react with one another in 3D space.

Pre-Rendered Movie Files

While real-time 3D rendering has improved over many years, there are still some GPU and CPU intense effects that are difficult to render in enough time to sustain the frames per second we as broadcasters work in. To that end, many designers will pre-render effects to short video clips, and use the video clips in conjunction with the real-time rendering from the GPU. Because a video clip is really no more than a sequence of images, the video files can be displayed one of two ways. Some CG systems will decode the video with the CPU, and then pass the raw video frames to the GPU for processing as a material. Another method commonly used is a hardware-based clip player included in the frame of the CG alongside the GPU and CPU. Both video signals are passed to the framebuffer board and mixed/keyed on the output.

Because raw HD frames of video require a great deal of sustained bandwidth to transfer across the main bus, there is a finite amount of video clip rendering that can be a part of a real-time composition, and it varies from system to system and generation to generation of hardware. In some applications, it is possible to render video clips to nonstandard broadcast resolutions. For example, if one is designing something with a logo animation that is intended to reside in a small corner of the screen, it may not be necessary to render a full 1920 × 1080 or 1280 × 720 clip for imagery that will only occupy 200 × 200 of space. Techniques like this can conserve rendering capabilities when creating the most visually complicated compositions.

Single-Frame Images

In addition to video clips, it is possible to create visually appealing graphics using still-frame images and the power of the CG to animate those items in the composition. Many designers still mock up the original design in products like Adobe Photoshop. And the elements that are a result of that mock-up are used in the assembly in the CG system. For some of the elements, a designer can use 3D primitives native to the CG system, or model new 3D geometry to match the flattened mock-up of the composition. But under the right conditions, it can be simpler and more efficient to take the still mock-up and assemble in the CG using flat objects in either a 2D or 3D composition.

In a simple graphic where there will be little to no animation, many will simply export the flattened image with any text removed that is intended to be replaceable in the composition. Then they can place the image in the CG composition, with a replaceable text field where the text is supposed to appear. Most CGs support TrueType fonts and OpenType fonts, so the graphic can match the look of the mock-up.

In a more complicated graphic, it is advantageous to export a still composition from a product like Photoshop into individual layers. This allows the person assembling the graphics to have full control over how the individual parts of the graphic layer, and then allows them to independently animate using keyframing and interpolation to create a motion effect out of a collection of still parts.

3D Geometry

Most CG systems are now capable of taking 3D geometry and rendering it in real time. It's the cornerstone of their rendering platform. What differentiates standard 2D images and 3D geometry is the precision of the pixel. In an image, the each pixel is accounted for and lives in only 2D space, X and Y. If that image is scaled beyond 100% of its actual resolution, you will see more imperfections in the edges and definition. Three-dimensional objects can have structure in 3D space, and not every pixel is accounted for. Three-dimensional graphics rely on points in XYZ space called vertices. A series of vertices are connected to create polygons. And the polygons connect as tiles to create structure and shapes in 3D. So what looks to be a sphere, in reality is a series of very small tiles. What looks to be a curve is a series of very short lines that bend slightly at each vertex. In many applications the number or "resolution" of the 3D object is referred to as tessellation. The greater the tessellation, the more precise or smooth the curves become. However, the greater the tessellation, the greater the number of polygons are included in the object and more processing the GPU will need to do to render the object visually.

There are two ways in which 3D geometry can be introduced into a CG system, as a primitive or as a file modeled in an external 3D modeling application. Some CG systems even have simple modeling capabilities natively built in.

When using the primitives you are at the mercy of what your CG system provides. Most CG systems provide a simple 2D plane to apply imagery to. Also often included are shapes like spheres, cubes, and cylinders. Combined, these objects can create more complex shapes. Primitives, however, don't often give you complete control over an individual vertex or polygon.

To have that level of control, you must work in a 3D modeling program. Most CGs have the ability to render 3D models from 3D modeling applications. Those models are created and saved out to one of the many model formats. That model then becomes an object itself in the composition, with many of the properties of a primitive.

Keyframes and Interpolation

In order to create motion in a CG system, most use keyframing to interpolate a positional parameter over time. This is specifically important when trying to create animations in and out of a particular graphic. The process is fairly simple, in that a minimum of two keyframes are needed to create a motion. The first keyframe sets the starting position of a particular object. As an example, at frame 0 in time, a keyframe is placed with a value of 0. At frame 4 in time, another keyframe is placed with a value of 10. The CG system will mathematically determine that at frame 1 the object would need to be at 2.5, and at 2 it would be at 5, and at 3 it would be at 7.5, and at 4 it would be at the keyframed value of 10. There would be no need to set more keyframes for frames 1, 2, or 3. This is based on a linear graph. Many CG systems give you the ability to adjust tension (the amount of curve), continuity (the direction of the vector at the point of the keyframe), and bias (a relative time differential of the curve at the moment of the keyframe). This can prevent motions from looking mechanical as they "ease in" and "ease out" of a movement.

Character Encoding

As data is entered into graphics, the characters ultimately are stored as bits in RAM and on disk. The original and most recognizable character set is the American Standard Code for Information Interchange or ASCII for short. The original ASCII was designed to be seven bit, with the eighth bit of the overall byte reserved for parity. With seven bits, this allowed for 128 codes, or slots for a specific character. This was enough for the 26 letters of the alphabet, multiplied by 2 for lower and upper case, numerals, punctuation, and special characters suitable for the English language. Extended ASCII character encoding enabled the eighth bit, which doubled the number of slots to 256, used to support a broader character set with dots, accents, and other special characters found in many European and Latin American countries.

However, even 256 positions are not enough for many languages, especially languages that originate in Asia. Many two-byte character sets were developed to support more than 64,000 characters in a set. To aid in unifying these different sets, a standard named Unicode was developed to make all fonts compatible with the same byte structure. The three Unicode types are UTF-8, UTF-16, and UTF-32, where the number represents the minimum number of bits. UTF-8 is the most common and compact, as a minimum single-byte character set that can expand to two, three, or four bytes. This is the most backwards-compatible version of UTF for ASCII. For most applications in the CG environment, UTF-8 is sufficient.

Graphic Playout and Storage

Most CGs work on the basic principle of template and page (or take item). Generally speaking, a template is a page, but it's not intended to be the actual playout page. It's what you recall to store new data into. CGs vary on how they accomplish this task, however.

The most known method is to store a page or a file at a location on disk by number. This workflow goes back to the early years of CG, where graphics were saved in a proprietary storage location. Number recall is more preferable because it makes graphic recall more simple using a standard ten-numeral keypad, though in most modern systems the file is stored into a standard

Windows file system and can be named anything. In the numbered environment the file name is the number, and there is a common extension for that particular CG system. In most applications, the page itself is just an instruction set for the CG rather than the whole graphic and associated media. The file points to the media and fonts on disk by file path and name. In the page-based workflow, there is generally a page dedicated to be the template, and it is stored at another number close to other template pages. That page is recalled, data is entered, and then it is usually saved to a new numbered location on the same render engine. Rarely, though, do these pages link, where if the overall look of a template is made . . . the changes do not ripple to the pages that were made based on the original template.

A project-based workflow is a newer workflow where one project file can encompass all of the templates and the derivative pages for a particular broadcast or look. Much the way the page-based system works, rarely is the associated media stored within the project file. Rather, it's stored as it would be on disk, and the project points towards the media. A project-based workflow relies on template graphics being created, and those templates are put into a sequence for playout. Multiple versions of the same template can be in the sequence, with certain fields of text or images on objects replaced to make derivative graphics. Many times the sequence items are given ID numbers for quick recall, similar to the workflow of the page-based system.

Other systems use a more complex but powerful workflow, where everything connects back to a central database via a network. Templates, take items or pages, and even the assets used to create the graphics are seamlessly delivered to the individual rendering machines.

External Control

Many graphics systems have an external control interface. With production automation, master control automation, and other third-party playout controllers, every CG has some sort of interface to the rest of the broadcast world.

The most widely known and adopted interface is what's called Intelligent Interface, or II for short. This interface is a standard serial text interface that can also be delivered via IP, delimited by forward slashes (\) and terminated by double forward slashes (\\). A common II string might look like V\5\3\1\1\2000\1\\. For short, this command for many CGs would recall page 2000 to output 1, based on the fourth and fifth element of the string after the V. Different CG systems prefer different commands to suit how their CG works best. But the same basic principles apply in that there is usually a letter command to start the string and a double forward slash to end it, and the parameters between are separated by a single forward slash. This protocol is still one of the widest adopted for CGs as it's the least proprietary and most understood.

Intelligent Interface is also limited because it is not proprietary. The protocol has to be something to every CG. For more integrated capabilities, some CG systems will have their own proprietary protocol and/or a complete application programming interface, or API. By using a library of programming objects and functions, a developer has access to many more parameters to evaluate which gives them tighter control over the CG and its operation. It does, however, require some programming experience and a foundation in the language that the CG system relies on. For some it's a variant of C, Visual Basic, or Java, to name a few.

Graphics Workflow

While CG render engines may be the same, how graphics output is dependent on the application. News graphics, sports graphics, special event graphics, even branding graphics . . . how they're recalled to playout is unique to the application. Graphics are reliant on data populated into their fields. What that data is and where it is retrieved from helps to determine the workflows. In addition, in what order, or lack thereof, will determine how and where graphics are accessible.

There are two workflows that best illustrate the wide range of applications: scripted and unscripted. News graphics are a good example of a scripted workflow, as they are generally very ordered, and the data comes from the editorial team. To that end there, the toolsets are designed to drive the CG render engines in a way that's most efficient for playback. The sports environment is an excellent example of unscripted workflow, as they are generally disordered. Graphics playout is based on the tone and trends of the game. There is little order or sequence. In addition, the data can be manually entered or linked to many common data sources. Because of this, there are toolsets and workflows designed to retrieve the most data as accurately and as quickly as possible.

Scripted Graphics Workflow

The link between the news department and graphics has always been critical in order to get relevant graphics on air. In the past, the workflow included printing raw text to a printer for CG operators to manually type in their graphics. In later years, light automation was introduced using Intelligent Interface so the editorial team could type the graphic into their editorial system and it would download to the CG. The limitation was that it was only really useful for text replacement.

The most commonly used interface now for scripted shows and news is through the MOS protocol. MOS, short for Media Object Server communications protocol, is a method in which CG instructions are embedded in a script and rundown in a newsroom computer system. Those messages, stored and processed as XML, are currently embedded using an ActiveX plug-in that lives on each newsroom client, but which communicates back to a central CG MOS server. When a newsroom rundown is made active, the newsroom system sends the messages back to the CG MOS server so it knows what graphics need to be prepared, but also knows what order they are to appear in. When a story is removed or moved inside of a rundown that is active, the newsroom system sends a command back to the CG MOS system to reorder the sequence of graphics. Because the protocol allows flexibility from vendor to vendor, the CG system can send more parameters than just text replacement, like paths to imagery, for example. Using the plug-in, the editorial staff can get more immediate feedback on what the graphic will look like. This allows for a richer user experience and saves a lot of redundant work.

Unscripted Graphics Workflow

In an unscripted environment, often the operator doesn't know what graphic will be needed until it is requested to take on air. In order to get graphics on the air quicker, the graphics need to be built in a way that limits the focus of the graphic to only what needs to be changed. With some CGs, they use what are called tab fields. Tab fields are locations on the graphic where a text value is editable. The operator can fill in the data, press the tab button on the keyboard, and move to the next value. In a similar operation, some text objects are their own objects as opposed to one big text object. In that case, the objects are "published" or made editable and ordered in such a fashion that the workflow is similar to how tab fields work.

In addition, external files and data sources can be used to retrieve data from known and existing places. In many stadiums and arenas, the scoreboard vendor offers a data interface that can be used with all CG systems. That feed will deliver the clock time, score, and other parameters that are displayed during an event. This data then can be pushed into a graphic more easily and is refreshed often enough to keep up with a running clock. This frees the operator to focus on other more descriptive data and graphics. But even that can be a challenge to retrieve and keep track of. Many stadiums and arenas keep stats on a computer system in real time with the game. Those applications often provide a raw data file in a format like XML, which would allow the CG to parse the data from the elements and attributes of the XML file. That way operators only have to know what the player uniform number is, and what statistics they want to show.

As long as the data is in a predictable and repeatable location, a CG system can often retrieve that same data repeatedly. To that end, customers have used files as simple as a text file (comma separated value or CSV), Microsoft Excel, XML, and server-based data like SQL queries and Rich Site Summary data (RSS feeds). With these methods, the viewer at home enjoys more statistical data being shown during the broadcast, and usually more reliable data.

Augmented Reality and Virtual Set

This section covers the technical aspects of designing and building augmented reality (AR) graphics and virtual sets (VS), leveraging new technology while avoiding the pitfalls. The importance of thinking through the lens during the design process, and working with designers, directors and producers, will produce the best results.

The same process used for traditional computer graphics production (scan-line rendering) holds true for real-time graphics, for the most part. A scene is modeled, textured, and lit; however, for real-time graphics it is more important to understand the on-air computing hardware in order to maintain real-time performance. The director typically determines the camera positions, angles, and field of view. Therefore the VS/AR needs to hold up from all anticipated angles and levels of detail, especially for the close-ups, which account for the majority of shots in typical studio productions. We refer to this as "thinking through the lens," "previsualization," or "shot blocking," an integral part of the design process, done prior to being on set.

How many polygons per second and how much texture memory is available for the scene? These are the two reoccurring questions that need to be asked every few months due to the rapid advancement in graphics hardware, thanks to the booming gaming industry. Frames or fields rendered at 60 Hz equate to 1/60 sec = 16.67 ms, and 50 Hz equates to 1/50 sec = 20.0 ms, (16.67 ms NTSC or 20.0 ms PAL). If render time exceeds 16.67 ms (NTSC), the graphics will not be synchronized with the composited video and production will be halted. Tools to measure and monitor resources are mandatory, and trade-offs in design and production need to be understood and managed carefully.

Maya, 3DStudio Max and Cinema 4D account for the majority of 3D modeling tools in current broadcast use; in addition, files from AutoCAD and SketchUp, used by traditional set designers, may be imported to save time and money. Once textures are applied and mapped to the model, the lighting process begins. Techniques such as ambient occlusion (AO) and global illumination (GI) are important to achieve photorealistic results. Details on these techniques are beyond the scope of this chapter; interested readers are encouraged to research these topics on the Web, where the information and video examples are readily available. Once the scene is lit, a process of "baking" is begun; this technique bakes the lighting with the texture to create a new texture that is remapped onto the geometry as a fixed lighting setup.

The 3D scenes are optimized to maintain real-time requirements of the hardware and then imported into the real-time 3D engine where further optimization and visual enhancements are done.

Photos are often used to create set extensions from real to virtual, and for out the window cityscape backgrounds. Looping video clips are even more effective as backgrounds, adding life to the set, provided they are high resolution supporting narrow field of view close-ups. Care must be taken to monitor the use of videos; efficient decoding varies amongst codecs and 3D engines. Monitor walls typically viewed in news sets often use quad splits and map the various quarter-resolution videos to different monitors. This works remarkably well provided you do not zoom in full frame on a virtual monitor.

Video I/O plays a big role in VS and AR. Virtual monitors use live video input from the router or switcher AUX bus for on-air transitions. Video input may also be used to take the camera feed into the system in order to key internally. This approach eliminates the need for an external keyer at the expense of using inputs. Chroma key shaders also use resources and should be evaluated carefully. Video output requires both fill and key signals that are sent to an external chroma (VS) or linear (AR) keyer. Both approaches have their pros and cons with quality and costs. Trackless virtual sets use video input from a locked-off camera feed of the talent standing still in a green- or bluescreen. This video input is applied to a rectangular surface in the virtual set. The surface and set are then manipulated to simulate limited camera motion. Billboarding orients the surface towards the camera as it moves, maintaining the illusion of 3D for small camera movements.

One of the challenges with augmented reality is to blend computer graphics into a real-world environment. Applying the lighting and reflections from the real world onto the materials may be accomplished using panoramic photos. Shadows cast from the AR onto the real-world surfaces are semitransparent layers rendered from the key lights perspective.

Virtual set design and preproduction should begin early in the project to best leverage the technology and achieve the desired results. For augmented reality, previsualization should be done using panoramic photos or photogrammetry to create composites for approval prior to being on set or on location.

Computer graphics continue to progress rapidly, largely due to the multibillion-dollar gaming industry pushing the quality higher, lowering costs, and improving the ease of use, making AR and VS commonplace.

CONCLUSION

Graphic production and CG are growing more and more important in broadcast. To that end, the workflow from creation through fulfillment and playout is critical to the success of the broadcast. And with most if not all CG systems, the parameters and handles are there to customize the product to suit your individual needs.

CHAPTER

5.17

ENG, SNG, and Remote Video Production

TOM JENNINGS
Accelerated Media Technologies, Inc.
Auburn, MA

STEPHEN ALHART
NEP
Pittsburgh, PA

MOBILE VIDEO PRODUCTION

Mobile broadcast productions can be produced using several different platforms. Three in particular are:

- Traditional 53′ self-contained semi-trailer systems
- Hybrid production/uplink trailers
- Outside broadcast controlled from client home base

Traditional 53′ Trailers

Many traditional 53′ trailers for outside broadcasts remain as a single expanding trailer facility with an adequate amount of equipment to meet client production requirements. See Figure 5.17-1. A single-unit remote broadcast trailer could be equipped with twelve camera control units (CCUs), six multichannel record/playback servers, graphics machines, a suitably powerful audio console, a matrix-based communications system, audio and video processing, a routing switcher, multiviewer, terminal gear and operating positions for the production personnel and technical crew. Several of the larger outside broadcast trailers are built utilizing two 53′ trailers, and one existing mobile broadcast facility is made up of five 53′ trailers.

These multiple trailer facilities may contain upwards of 36 CCUs, 12 or more multichannel record/playback servers, several audio consoles that are networked together, a matrix communications system capable of hundreds of ports (stations), nonlinear edit facilities, multiple graphics machines, extensive fiber-optic patching, fiber-optic routing, and a huge routing switcher and multiviewer. Multiviewer technology

FIGURE 5.17-1 TV Compound at the PGA Championships

has replaced individual small monitors throughout the industry, as the flexibility afforded by multiviewer technology is almost without bounds. Over the last few years, multiviewer technology has improved greatly. One of the most useful traits of multiviewer technology in mobile units is the ability to use a few, lighter weight flat panel screens for a large number of inputs (many of which can be routed via the routing switcher). The weight savings in a mobile unit can be substantial.

A side note about the useful life of a production trailer: A production trailer will most likely have several iterations while the chassis is on the road. As technology has advanced from analog through digital and now HD, 4K and beyond, trailers have been repurposed for other functions. One example is the former ABC-1, which was the last mobile platform built and commissioned by ABC-TV for ABC Sports. When ABC finally sold their outside broadcast fleet, the ABC-1 production trailer was purchased by NEP and was rebuilt into NEP ESU, a full-service HD video routing distribution system with extensive analog, AES and MADI audio facilities, routing and distribution and other space for add-on services as requested by a client. As the chassis and enclosures for large tractor-trailer mobile units (not to mention the customization for power, air-conditioning and electronics) can have a long life if well maintained, and are extremely expensive, it is almost always cost-effective to repurpose them instead of building new trailers.

Fiber-Optic Infrastructure

Many of these multiple trailer facilities interconnect with a 12-strand single-mode fiber-optic cable, known in the industry as "TAC-12" (12-fiber tactical) cable. See Figure 5.17-2. Interconnections between trailers utilize a 12-position hermaphroditic connector for speedy, reliable connections. Fiber-optic technology has greatly simplified many aspects of a typical mobile broadcast setup.

FIGURE 5.17-2 DT-12 Interface Panel

A typical golf event, covering all 18 holes, may have a single TAC-12 fiber-optic cable deployed to each of the 18 greens, and an additional TAC-12 run to both the media center and the "rain room." The principal announce booth location will usually have two TAC-12 cables to handle all the cameras, the ancillary video feeds (for the announcers to see video of the various greens and other locations) and audio.

Cameras

Most modern cameras connect in an inside studio configuration to a camera control unit (CCU) by means of a SMPTE optical cable system. Current-generation HD and 4K cameras, depending on optional configurations, can connect with triax or SMPTE optical cable. Triax has distance limitations, primarily from the attenuation of the RF carriers and secondarily from the voltage drop through the length of the cable. Most sports venues have existing triax infrastructure still in place and functional from the analog days of Ikegami 355 and BVP-900 cameras. Triax systems can utilize repeaters to extend the cable length, though the use of repeaters is obsolescing rapidly.

There are two different optical connection formats available for mobile broadcast needs. SMPTE cable allows connection, including CCU power, between the outside broadcast unit and the camera. The construction of a SMPTE camera cable is two single-mode fiber strands, two control (sense) conductors, two pairs of power conductors and an overall copper shield for grounding and cable strength. The power conductors are doubled up, lowering the end-to-end resistance while maintaining the cable in a round format. Many sports venues have some installed SMPTE cable infrastructure, with more venues upgrading as time and budgets allow. An interesting note to SMPTE cable and termination: the cable jacket has markings every few feet to show which end attaches to the camera. Not only will this help the client's utility crew, but also the cable construction crew, as the fiber strands in the cable are pre-positioned for termination without bending or twisting the strands, preventing fractures and failure during the connector installation process.

Still, in some instances (golf in particular), the cameras may be thousands of yards from the production trailers, which is practically impossible for a standard SMPTE optical system connection, and definitely impossible for a triax-connected system. The other optical connection option is to utilize two strands of single-mode fiber. With properly engineered tactical fiber-optic link budgets, the distance between the truck and camera can be miles. To interface the standard type of SMPTE optical camera system to a TAC-12 fiber strand requires interface equipment at both the camera end and the CCU end of the fiber to convert the standard camera control system for use over long-distance tactical fiber. This interface system, in particular the one sold by Belden, uses a "SHED" (SMPTE Hybrid Elimination Device), which is connected to the specific CCU that will control the camera. The purpose of the SHED is to "trick" the CCU into believing a camera is connected at the distant end of the fiber strands. Similarly, an "HDX" is the distant-end interface for the camera. The function of the HDX is to interface the two long runs of single-mode fiber strands to the SMPTE cable and apply AC power to the camera via the SMPTE cable. Two hundred meters of SMPTE cable can be connected between the HDX and camera. The HDX requires nearby AC power at the HDX to operate the camera head and peripherals.

The latest generation of Sony HD camera also has a provision for a data trunk between the camera head and CCU. The data trunk can be used for digital data streams for production enhancement tools such as Protracer technology, which graphically highlights a golf ball's trajectory for the viewer at home.

The form factor of the current generation of HD camera is designed as a handheld portable camera. For an outside production facility to best utilize its camera assets, camera manufacturers offer what is known in the industry as a sled or large lens adapter. This structure allows a handheld camera to be securely mounted to it, and a large lens to hang off the structure at the front. This sled configuration mounts on an industry standard pan head and field tripod or pedestal. All connections between the sled and camera are completed automatically when the camera is mounted. Connection to the lens is unchanged from past generations of lenses. The 36-contact micro-ribbon connector remains the standard.

Camera viewfinder technology has moved well beyond old CRT-based monochrome units. Color viewfinders are a great assist to the camera operator, especially in sports, with all of the colorful team uniforms. HD-resolution OLED color viewfinders are now available as an option on Sony cameras.

Record/Replay

The videotape recorder and disc- or tape-based "slo-mo" recorder formerly were a vital aspect of any outside broadcast. This author remembers (through the years) remote broadcast trailers equipped with 2" quad, 1" Type C, Betacam, HDCAM and Ampex HS-100 slo-mo videotape recorders. Now, video servers handle all the record, replay and slo-mo functions. Current designs of these servers in a mobile unit are usually deployed so that up to some combination of 12 record and replay channels are available, all in HD, with multiple channels of audio embedded with each video input or output.

Depending on how the server is configured for remote control operation, it is possible for several operators to control separate channels of the multichannel server. These multichannel servers, while not being lightweight, do make the best use of valuable rack space in the technical spaces of the production trailer, as they are more compact than their tape- and disc-based predecessors. One of the drawbacks of these servers is the heat they generate, combined with the fact that they are typically installed tightly together in the "tape bay" racks. Proper attention to air-conditioning within the trailer and management of waste heat is critical during the design phase and physical layout of the trailer and the placement of equipment racks. As with any piece of electronic equipment, a plan for periodic cleaning and replacement of the filters in all mobile unit equipment is required. Excessive heat will severely impact equipment performance.

ENG Field File Acquisition for Remote Production

Field ENG (electronic news gathering) material storage has also moved away from videotape. Optical discs, flash cards and solid-state drives are now integrated into single-piece camcorders. These camcorders utilize various compression schemes to achieve good results with very portable systems. Recording bitrate and standard will vary with the remote production or client.

Audio

Gone are the analog audio consoles made by SSL, Neve, Mackie and Yamaha. Enter the digital world of Calrec, DiGiCo, Lawo and Studer. Digital audio consoles and high-density audio connectivity have revolutionized audio production and distribution. Physically, the control surface of a digital audio console is simply linear and rotary encoders, OLED displays and pushbuttons, giving it the "feel" of a legacy audio console. See Figure 5.17-3.

All the I/O, processing and routing within the desk is accomplished in the digital sound processing (DSP) chassis. The operational power of the DSP chassis is significant, providing extensive EQ, multitrack routing, dynamics and, important in today's HD environment, individual assignable and variable delay of any signal routing in the chassis. This is important to compensate for delays in HD video processing originating from RF camera systems; a video delay of several frames to as much as one second is typical with the current digital video compression formats in the

FIGURE 5.17-3 Audio Sub-Mix Control Surface

RF camera transmitters. If the talent being shot by the RF camera is speaking into a hardwired microphone connected to the outside broadcast trailer, the audio necessarily must be delayed to match the video from the wireless camera.

Fiber-optic cable plays a significant new role in remote production audio. Fading into history is the well-known DT12 cable and its boxes and fan-outs. DT12 cable runs have been replaced with fiber strands that connect to either stand-alone analog interfaces or analog interfaces that now directly network to the audio consoles as direct fiber-optic connections. Calrec's HYDRA technology is a prime example of networked audio connectivity and interface. Other bulk audio connection technology includes Multichannel Audio Digital Interface (MADI) and Digital Audio Network Through Ethernet (Dante). MADI technology allows the transmission of 56 or 64 channels of digital audio data at up to 48 kHz, or 28 to 32 channels of digital audio data at up to 96 kHz in one direction. Two MADI streams in a bidirectional connection are equivalent to a 128-channel audio snake. The connection medium can either be over coax for reasonable distances or via single-mode fiber for many miles, which is often need in stadiums or golf courses.

The ability to network several audio console "surfaces" is very beneficial in large outside broadcast events. In a networked IP audio environment, the power of having hundreds, if not thousands of facility audio sources available to each work surface at the click of a mouse or the drag of a finger on a touch screen, is a bonus for the sound engineer, the maintenance engineer and the design of the facility. See Figure 5.17-4. There is currently a MADI router available with 1024 × 1024 routing capabilities with the signal interface as 32 MADI streams, either on coax or fiber.

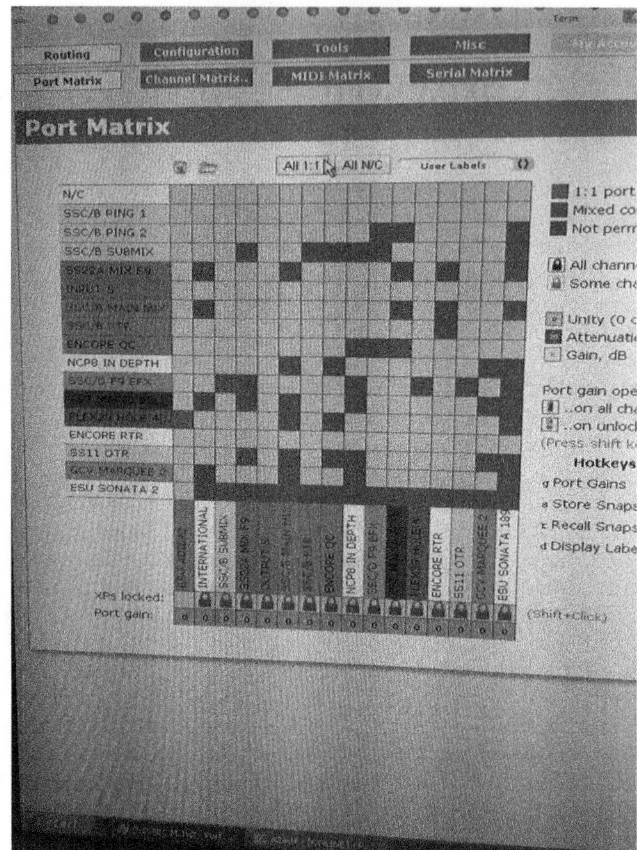

FIGURE 5.17-4 MADI Router Bulk Control GUI

A major advancement in audio transmission has been the growth and expansion of embedding and de-embedding technology. Embedding is the marriage of up to 16 audio channels into the HD video signal stream (or even more into a UHD stream). The current technology is several generations advanced from the early days of embedding audio onto a video signal for transmission. "Sound-in-Sync" or FM subcarrier channels were the earliest form of embedding. This progressed to four and then eight channels embedded in a standard-definition digital video stream, and now sixteen channels in the HD video stream.

Communications

Large communications (or intercom) matrices have become the standard in outside broadcast facilities. The flexibility of the communications matrix allows all aspects of comms to be integrated into one system. Options for current communication matrices include MADI and Dante interfaces, IP-connected stations and distant matrices. The stage manager in the announce booth and the producer in the mobile unit are connected by powerful technology.

In the past, communications and audio were two separate ecosystems. With the current and emerging technology, communications and audio are blended into one system, linked through MADI, Dante or other TDM (time division multiplexing) technology.

Video/Audio Routing Matrices

Signal distribution in a mobile production unit environment has evolved from signal distribution utilizing distribution amplifiers to a router-centric design. Smaller mobile facilities might have 128 × 128 video matrices, while larger facilities have video matrices could utilize 256 × 256. Some of the largest facilities have video matrices on the order of 1024 × 1024. At the time of this writing, SDI routing matrices have started to include internal capabilities to embed and de-embed 16 audio channels to or from each HD video source or destination. This technology advancement further diminishes the need for external embedders and de-embedders, and eliminates the rack space they currently occupy. Many video routing matrices have moved away from the industry standard BNC connector in favor of the 1.0/2.3 DIN connector for physical space considerations on the rear panel. These matrices embed and de-embed, but also have the capability to ingest, construct, deconstruct and distribute MADI streams. Utilizing this technology, a switching matrix can have an audio switching capacity of several thousand sources and destinations. In the matrix-centric facility, the matrix performs all the audio routes, from transmission to monitor speakers in the control room.

Currently most of the routing matrices in remote broadcast facilities are crosspoint-based designs. Emerging matrix technology is based on IP routing, which offers scalable matrix configurations. There are pluses and minuses with this new IP-based routing technology. There is a four-line penalty associated with one of the system designs. This penalty is related to the IP "encapsulation" and "de-encapsulation" process within the matrix. Sometimes the newest technology is not always the most useful.

Hybrid Production/Uplink Mobile Units

To fill a market for cost-conscious clients, the hybrid production/uplink mobile facility offers a viable alternative to the traditional fleet of 53′ outside broadcast trailers. See Figure 5.17-5. Facilities integrated in the hybrid trailer are downsized, but offer full HD production capabilities with the added ability to perform Ku uplink and downlink connectivity.

Remote Broadcasts Controlled from a Fixed Control Room

Through the rapid growth of point-to-point fiber-optic connectivity and fiber availability into many venues, regional switching centers in or near metropolitan areas and broadcast facilities, the concept of "home-base production" is becoming more attractive. At the origination venue, the broadcaster has a small technical facility that may fly to the production site in "road cases" with cameras, CCUs, audio gear and preamps, embedders for transmission of isolated audio channels and interfaces for tally lights. See Figure 5.17-6.

These sources are encoded and multiplexed for transmission to the home base production center. High-bandwidth bidirectional fiber-optic connections via any number of national fiber-optic "telco" carriers make this production technique possible. One significant positive of the at-home production plans is cost savings in crew travel, lodgings and shipping of graphics and playback/record equipment.

At the home base production center, these feeds are ingested, de-multiplexed, decoded and treated as traditional sources feeding video production switchers, audio consoles and communications systems. Viewers at home are probably not aware of the separation between venue and production control room.

The concept of transmitting isolated camera and audio feeds from a remote location to a distant control room is not new. It has been in practice since the 1980s, at a time when connectivity was based on AT&T discrete video and audio (VANDA) circuits. One VANDA circuit was used for each analog video/camera feed, with two analog audio signals each. Often, the audio portion of the VANDA had a different signal path than the video signal. A return VANDA from the home control room carried return video (program video) for the on-air talent along with talent cueing (IFB) so that producers in the home control room could talk to the on-air talent at the remote location. Early IFB and intercom communications were accomplished through POTS (Plain Old Telephone Service) lines equipped with phones and headsets. Additional information about video distribution can be found in Chapter 6.2.

FIGURE 5.17-5 Uplink/Production Vehicles

FIGURE 5.17-6 Remote Equipment in Road Cases

IP Networking and Routing

Every piece of equipment in a mobile production unit today has (for better or worse) a network connection. Gone are the days of a broadcast engineer being a master of component- and system-level electronics, of being able to design a circuit and knowing how to work around equipment failures. Now, all broadcast facilities need a networking specialist (and great patience) to manage all the networks within a broadcast facility.

ENG AND SNG

Television stations started migrating to digital microwave transmission in 2004 and began to utilize the increased data capacity that QPSK (quadrature phase shift keying) modulation schemes had to offer. Not long after, smartphones, LTE and Wi-Fi transmission revolutionized mobile newsgathering. All of a sudden, not every live shot required a mobile unit. Soon after pioneers began using these technologies for live shots, however, they realized the downside of using public utilities for live newsgathering. Some intriguing (and often competing) technologies filled the void.

The Technological Revolutions Begin

Beginning in 2008, manufacturers began working on satellite systems that could be used while in motion. Dubbed "Comms on the Move" or COTM by military organizations, these satellite systems utilized creative ways to track satellites in real time while in motion. It was technology already in use for data acquisition on shipboard applications, but it was being adapted for use on land by several of the key antenna providers for the purpose of supporting two-way VoIP, data and streaming video, as well as tactical military and commercial radio extension, over the toughest terrain. Typical applications for the technology included Command and Control (C2) emergency and first responders, border patrols and civil patrols, and it was the intent to add news media and broadcasting to the list of users. The terminals were (and still are) available in multiple aperture sizes in Ku-, Ka- and X-bands. The technology tested exceptionally well throughout the world and had a proven track record in the Middle East and in countless tactical engagements where it was invaluable in providing critical communications where no terrestrial infrastructure existed, and it remains a key tool in military operations. The adaptation to broadcast applications was bit more challenging. First, there was the cost of the hardware. An inexpensive system started in the mid $60,000 range. More rugged units, that could take the constant abuse of city streets and near constant use a news crew would subject it to, could cost nearly $150,000. The Achilles heel to the technology for commercial use, however, was the cost of bandwidth and lack of spacecraft. Since the technology was designed to be used in motion, systems generally used a phased-array satellite transmission mode, which allows for instantaneous electronic scanning and multiple simultaneous beam capabilities. Unfortunately, that translated into large amounts of bandwidth being consumed for relatively low throughput, which again translated into very expensive spacecraft time. As an example, it was reported that a leading news station wanting 24/7 access of approximately 2 Mb of data exceeded $35,000 per month. Additionally, the relatively small antenna aperture size, when compared to conventional satellite antennas, made the gain characteristics of these antennas too low for reliable links to be established in urban city environments. The added complication to these systems was the tracking requirements. The United States stacks satellites very close together with minimal 3-degree spacing. The complexity of the tracking systems required to follow a satellite while driving 60 MPH, hitting potholes and turning corners proved to be more of a challenge than these systems could handle reliably. The technology worked best on linearly moving vessels such as railcars, aircraft and in some cases marine vessels and in wide-open spaces. This technology is still being developed in the hopes it will reach a price/reliability point that would make it practical for broadcasters to implement for truly mobile newsgathering.

Cellular Bonding Enters the Scene

Covering live news events via traditional ENG/SNG technologies has, to this point, been an expensive and logistically difficult business. It relied on specialized and expensive hardware and trained/licensed professionals to operate. However, with the advent of cellular bonding technology, broadcasters and online video professionals, with minimal exposure to specialized training, could suddenly, easily and cost-effectively report live from any location using a field-proven method—cell phones.

Bonding combines multiple cellular and wireless networks such as 2.5G, 3G, 4G LTE, Wi-Fi and WiMAX to obtain a reliable, robust video uplink capable of transmitting HD video.

Cellular bonding takes compressed video (H.264) and transmits some of the packets over each of the multiple cellular modems. (Typical systems use three to ten individual cell cards). Instead of relying on a single, potentially unreliable cellular link with a single point of failure, the systems bond several cellular links together, which minimizes the inherent risks while achieving the desired or greater performance. These systems continuously, and in real time, monitor all of the links in use and dynamically adapt to use each of them efficiently. These systems also dynamically adapt the video encoder according to the momentary total available bandwidth in all the links. The bonding systems also compensate for, and recover from, any losses, and are able to interact with the operator to best manage individual needs. On the receive side, which can be anywhere in the world, software and servers are installed to receive those multiple packet streams, aggregate them and reconstruct the video.

For roughly $25,000 per transmit package, broadcasters can potentially replace all ENG and digital satellite news gathering (DSNG) assets and equip every photographer and newscaster with a small, low-complexity cellular bonding device, which ranges in form factor from a single rack-mounted device to small cases/backpacks to transmit live news from anywhere in the world.

Without question, cellular bonding allowed broadcasters greater flexibility than ever before. It took them to regions where conventional technology would never have been able to access. The inconspicuous form factor and high reliability were adopted by many newsrooms in the United States and around the globe. In the early days of cellular plans, it was common for carriers to offer unlimited data plans for a low set price. Broadcasters took advantage of these plans—wireless bandwidth was cheap, reasonably reliable and readily available. It required no infrastructure to maintain or install and it was virtually everywhere, vastly expanding the typical 40-mile coverage of electronic news gathering (ENG) microwave and a fraction of the cost of DSNG transmission.

Unfortunately, frequencies and bandwidth are not unlimited, and the capacity of cellular infrastructure is somewhat limited (albeit impressive). As popularity of these systems grew, some actual physical limitations became apparent. Crews began to run into cellular bandwidth problems when multiple competing broadcast entities would arrive on scene, especially at a breaking news story where the existing cellular systems may be already at full capacity with high cell phone use. The cellular systems, designed to handle hundreds, if not thousands, of calls simultaneously, can suddenly be choked by the emergence of bonding technologies attempting to push large packets of streaming data on a priority basis. Additionally, newsrooms began expanding the use of these technologies to cover sporting events and other remote locations that suffer from issues ranging from oversaturated cellular towers or no cellular towers. As a result, the technology began to develop a reputation as being unreliable or at least unpredictable in certain situations.

IP Satellite to the Rescue

Launched October 19, 2011, aboard a Proton rocket, ViaSat-1 holds the Guinness record for the world's highest-capacity communications satellite, with a total capacity in excess of 140 Gbps, *more than all the satellites covering North America combined*, at the time of its launch.

ViaSat-1 is capable of two-way communications with small dish antennas (0.75 M to 1.2 M) at higher speeds and a lower cost-per-bit than any satellite before. It was part of a new satellite system architecture to create a better satellite broadband user experience, making satellite competitive with DSL and wireless broadband alternatives for the first time. It operates on a new Ka-band frequency previously not utilized by broadcasters for mainstream transmission.

While the initial intent of ViaSat-1 was to provide broadband Internet access to every household in the United States, regardless of location, ViaSat-1 quickly became a commercial powerhouse providing Internet to rail systems and airline clients alike.

In order to understand how cellular bonding might benefit from this technology, it is important to understand how and why this Ka-band satellite was a good fit. During a typical broadband Internet query, a small 256–512 Kb inquiry is transmitted over the satellite while a 5–10 Mbps *download* is received in response. Broadcasters sought to use the system in the opposite fashion from the way it was essentially conceived; *uploading* 5–10 Mbps and receiving smaller 1 Mb return data. The capacity existed on the satellite, and it was a partnership between Accelerated Media Technologies Inc. in Auburn, Massachusetts, and Dejero Systems in Waterloo, Ontario, that first presented and tested a *hybrid* cell/Ka bonded system using ViaSat's "Exede" service.

The system works by overlaying bonded-cellular with Ka-band satellite, thereby providing "on-demand, dynamic" bandwidth. To the bonded IP encoder, the Ka-band satellite signal looked like any other wireless bandwidth. The cellular systems could aggregate between cell and satellite bandwidth dynamically to minimize satellite bandwidth costs, and also take advantage of Ka-band satellite when cell signals become suppressed. Likewise, the aggregation works in reverse when using cellular to back up a satellite link in rain-fade situations. They are very complementary technologies and create a very robust and economical pipeline.

Generally it requires less than 1 Mbps of cellular bandwidth to correct for any data packet loss that can occur on the Ka-band link, caused by weather, low cloud cover and rain. Rain fade continues to be a major concern with Ka-band transmission, but the marriage to cellular bonding equipment has made this a non-issue for broadcasters.

When coupled with bonded-cellular, Ka-band satellite gives broadcasters a multipronged cost saving approach that allows smaller, more mobile and fuel-efficient trucks, lower hardware costs and significantly lower costs of operation. Ka-band terminals cost about one-third the price of comparable IP-over-Ku-band equipment, and the cost of Ka-band bandwidth is easily 30%–50% less, which makes it very attractive to fleets nationwide. As an added bonus, Ka-band does not require scheduling or coordination; acquisition is an automated one-button deploy. These systems were designed to be operated by reporters or photographers with less technical background than a conventional ENG or SNG truck operator.

The use of Ka-band technology drastically reduced the unpredictability of cellular-based solutions and moved the technology into a new category by moving it from an "in addition to" technology to an "instead of" technology replacing legacy Ku satellite trucks in news fleets with smaller, more diverse fleets of SUV-based hybrid Ka-band/cellular technologies.

Ka-band differs from Ku-band. Remotes are downlinked at ViaSat Gateways and not at the station. Distributed gateways are tied together by a fiber ring to a ViaSat Core Node. See Figure 5.17-7. Stations link to the Core Node via the Internet (all open standards IP).

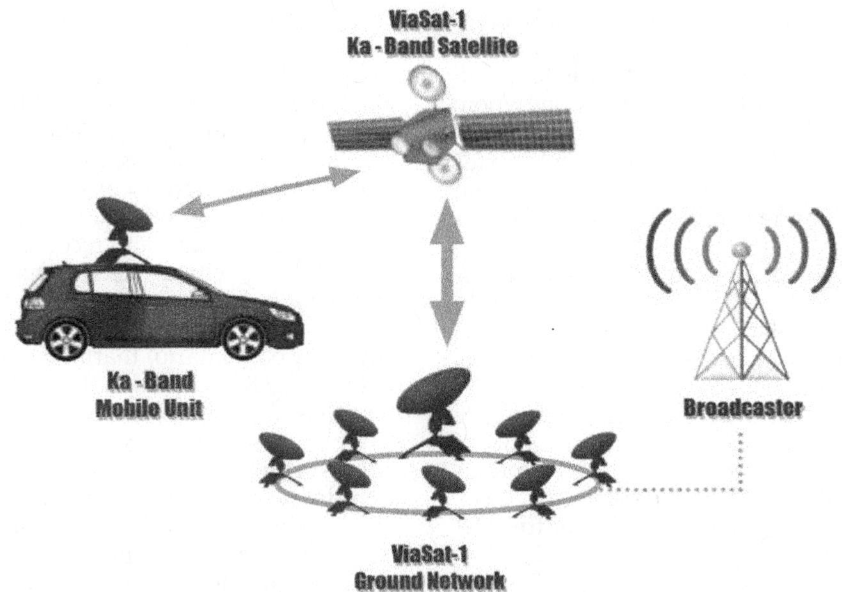

FIGURE 5.17-7 ViaSat Ka-Band System

Courtesy of ViaSat

FIGURE 5.17-8 Next Generation Ka/Cell Bonded SUV Vehicle

Courtesy of Accelerated Media Technologies/Telemundo 47 New York

The ViaSat Ka-band system:

- Two-way IP connectivity provides editorial efficiency
- Dedicated and prioritized upload speeds for SNG
- Utilizes lightweight, small antennas that fit on smaller, less expensive vehicles like what photographers drive every day
- Solves safety issues with masts and weight restrictions
- Can drastically simplify electrical/generator requirements
- Expands coverage area beyond microwave limits and wireless over capacity
- Has lower bandwidth costs than Ku-band

Next Generation Ka-Band/Bonded Cellular Vehicles

An example of a "next generation" Ka-band/bonded cellular vehicle is shown in Figure 5.17-8. Figure 5.17-9 shows the interior of a typical general use vehicle, while Figures 5.17-10 and 5.17-11 show a specialized "weather chaser" unit.

FIGURE 5.17-9 Ka/Cellular Bonded SUV Interior Typical

Courtesy of Accelerated Media Technologies/Meredith Broadcasting WFSB Hartford

FIGURE 5.17-10 Ka/Cellular Bonded Weather Chaser

Courtesy of Accelerated Media Technologies/NBC Hartford WVIT

Evolutions in Microwave Technology

Cellular bonded/Ka solutions continue to dominate modern newsrooms and the combined cell/Ka-band technologies have shifted where TV stations spend their money. Instead of capital equipment and personnel costs, stations have increased operating costs for subscriptions and satellite airtime. While significantly lower than legacy solutions, these operating costs can be difficult to budget and even more difficult for station management to entertain as cellular data rates increase without warning and news is unpredictable. Additionally, a portion of the technology relies on public bandwidth, which can be variable depending on public use and cellular infrastructure in a given area. The legacy microwave technology is still a good asset in the newsgathering garage, and it is on the verge of its own revolution that will give news stations an even larger and more dedicated pipeline that is licensed for exclusive use by broadcasters in the 2 GHz spectrum.

The move from a COFDM, single-direction microwave technology to a more useful LTE bidirectional infrastructure is under way, and it offers the spectral

FIGURE 5.17-11 Weather Chaser LTE Microwave, Ka-Band with Cellular Bonding

Courtesy of Accelerated Media Technologies

efficiency and flexibility of a private cellular IP network over a licensed link.

The original bidirectional microwave technology was pioneered by General Dynamics (GD) for the purpose of providing long range bidirectional communications links in remote, war-torn regions where communications systems were nonexistent. Using 2, 7 and 13 GHz as well as military bands, the company invested significant time and resources to making the system able to handle bandwidths in excess of 75 Mbps using easy to deploy omnidirectional antennas in the worst possible conditions. Not to be confused with short-range mesh networks that require hundreds of nodes to cover a large geographic area, this technology is a high-power bidirectional system designed to transmit and receive video, audio and communications data up to 80 km away using simple omnidirectional antennas on the mobile units and panel antennas on the tower sites, drastically simplifying installations and limiting ongoing costs to maintain and operate.

As is often the case with wartime technologies, once the major conflicts ended in the Middle East, these previously classified technologies began to emerge into the commercial segment. Once the technology was declassified, a collaborative effort to adapt the technology to the broadcast segment began between GD, Accelerated Media Technologies and several leading broadcast entities, and ENGenesis was developed.

Given the performance of a legacy COFDM single-input receiver, which typically reaches receive threshold on or about −88 dBm, as shown in Table 5.17-1, or a COFDM diversity receiver, which typically reaches threshold on or about −93 dBm, the ENGenesis system surpasses COFDM's threshold down to −126 dBm.

TABLE 5.17-1
Receiver Threshold of COFDM Versus LTE

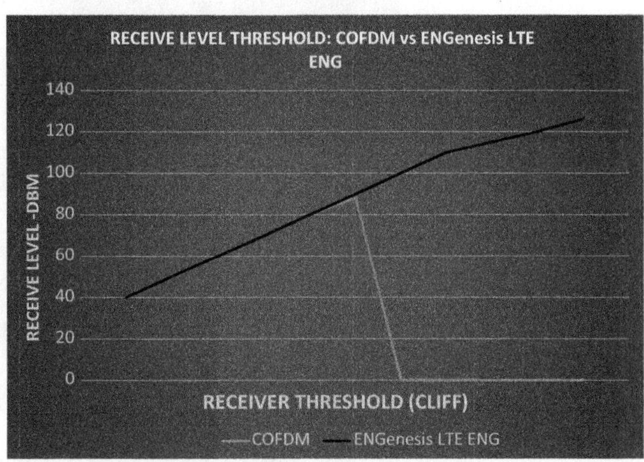

Courtesy of Accelerated Media Technologies

The superior front-end system gain combined with a worldwide industry standard LTE modulation scheme allows users to plan and deploy a private network to cover their market without the need of having to "tune in" a shot. Wide area coverage is achieved through proper placement of receive sites, each consisting of three sectors, combined for 75 Mbps of total throughput per site. Three vehicles using MPEG4 video codecs can operate in each single sector simultaneously using

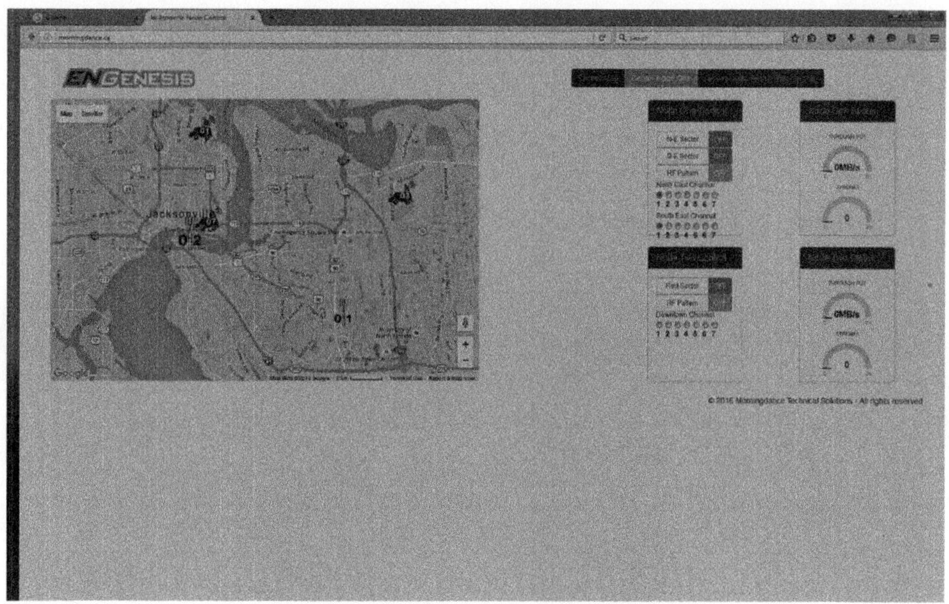

FIGURE 5.17-12 ENGenesis GUI Interface

Courtesy of Accelerated Media Technologies

the same frequency. With the addition of the two other sectors, it is theoretically possible to have nine mobile assets operating into a single receive site.

Additionally, the ENGenesis technology allows full IP communication to and from the studio and the dynamic user interface allows the station to track all field unit locations, individual data usage and a number of remote control capabilities (see Figure 5.17-12). Any IP-capable gear in the field can be monitored, configured and controlled from the studio or anywhere a station's IT department has connections into the ENGenesis private network.

CHAPTER

5.18

Television Audio
Analog and Digital Systems

TIM CARROLL AND JEFFREY RIEDMILLER
Dolby Laboratories
San Francisco, CA

INTRODUCTION

The field of audio for digital television (DTV) has expanded greatly over the last decade. Audio quality and features have advanced significantly. The transition to over-the-air (OTA) digital television (DTV) is an accomplished fact, and at this writing the industry is deploying a significant next-generation version of the system. Simultaneously, content is increasingly consumed from the Internet. One thing that has not changed is that audio remains an area causing apprehension with many TV engineers and managers, especially as the industry has already begun distribution of personalized and immersive audio.

Next-generation audio (NGA) systems promise to deliver significantly more sophistication, artistic creativity, greater processing power, and higher quality than their predecessor formats. This chapter provides a fast review of analog television audio, and a review of first-generation DTV audio practice (including how one of its largest problems—loudness control—has largely been solved), with the remainder devoted to an introduction to next-generation audio systems. By no means an exhaustive discussion of what is involved, the material presented here is intended to document the current state of the industry and focus on what an NGA system delivers.

TELEVISION AUDIO SYSTEM OVERVIEW

Analog Audio

Television audio was analog for many more years than it has yet been digital. It began as a single monaural frequency-modulated carrier at 4.5 MHz in the 6 MHz TV channel, as defined by the NTSC specifications. Stereo was superimposed on this carrier by the BTSC standard in much the same way as stereo was introduced to FM radio.

Compared to methods of delivering audio to consumers by nearly any digital means, even the most carefully aligned BTSC system is not comparable in quality. Some of the issues with analog TV audio included the side effects of high-frequency pre-emphasis, limited frequency response and dynamic range, low channel separation, and elevated levels of noise and distortion. For subcarrier channels that carry auxiliary programs, such as second audio program (SAP) or descriptive video service (DVS), the problems are worse since these are monaural channels, limited to 7.5 kHz frequency response, and noisy due to limits on modulation.

Digital Audio

When the process for standardizing a digital television system began, one of the requirements was that the audio system should be enhanced well beyond the capabilities of the BTSC system. It was decided that stereo, then later multichannel, audio would be delivered digitally. To conserve data bandwidth, audio data rate reduction, or data compression, would be employed, with the objective of maintaining performance that was as measurably and audibly as close to the original as practical.

Multichannel Sound

The first actions toward achieving this goal focused on MPEG-compressed delivery of two-channel audio.

NAB ENGINEERING HANDBOOK

The need for supporting multichannel surround sound was introduced because a substantial quantity of theatrical motion pictures had been mixed with four or more channels of audio: left front, right front, center, low-frequency effects (for driving the subwoofer), left surround, and right surround, commonly abbreviated Lf, Rf, C, LFE, Ls, and Rs.

In the early 1990s, Dolby Laboratories introduced a new system called AC-3, also known as Dolby Digital, which delivered discrete (i.e., non-matrixed) 5.1 channel audio via a composite bitstream that required precisely the same bandwidth as stereo MPEG-compressed audio of the day. Today, Dolby Digital (AC-3) and its subsequent extension called E-AC-3 (or Dolby Digital Plus) are integral parts of broadcast standards around the world.

AC-3 Audio

ATSC standard A/52:2015[1] describes in detail the AC-3 audio coding system, and readers are encouraged to download a copy for reference purposes. This document also describes E-AC-3, an enhanced version of AC-3 with more features and better efficiency. Broadcasters should be aware that OTA transmission of a standard AC-3 stream is still required to maintain compatibility with the large installed base of ATSC receivers. E-AC-3 is widely used for delivery of over-the-top (OTT) and Video on Demand (VOD) content.

AC-3 is an efficient audio coding system capable of carrying from 1 to 5.1 channels of audio, with high quality for 5.1 channels at data rates as low as 384 kbps. It also conforms with the broadcast television production environment, which uses an audio sample rate of 48 kHz and up to 24-bit audio resolution. Importantly, the AC-3 stream carries a parallel audio control-data path called *metadata* (detailed below), which provides additional key features used by decoders to optimize reproduction of the audio program by the receiver in a given listening environment. The system is designed such that a single encoded bitstream can be reproduced by any consumer decoder, regardless of whether the source content is encoded as 5.1 channels, stereo, or mono, and regardless of whether the decoder is capable of reproducing 5.1 channels or only a single channel. Programs of any number of channels can be reproduced by *any* AC-3 decoder. Figure 5.18-1 shows an example of how a single bitstream can serve all types of decoders.

Note that multichannel decoders also contain a matrix decoder of some type (usually Pro Logic or Pro Logic II) to handle content that is delivered in two channels. This audio will be matrix-decoded using level and phase information and will then have the capability of providing audio to all speakers when appropriate.

Audio that is transmitted using the ATSC or OpenCable standards must at least be AC-3 encoded and may also be E-AC-3 encoded. Pulse code modulation (PCM) audio is fed to the AC-3 encoder and a packetized elementary stream (PES) results. It is this PES data that will be multiplexed with video and other data into an MPEG-2 transport stream for transmission to consumers.

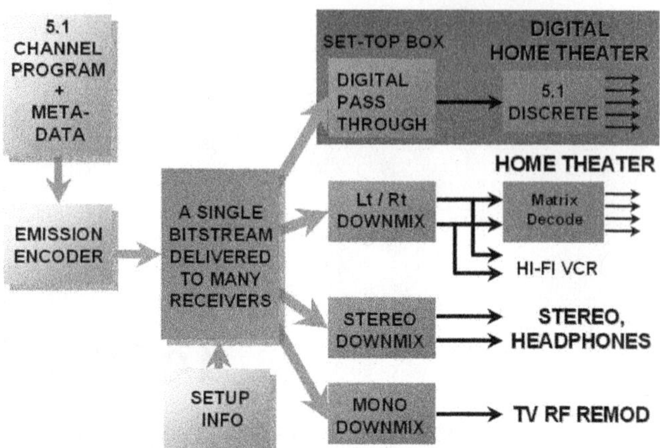

FIGURE 5.18-1 Single bitstream delivered to many different types of decoders; any bitstream can be decoded by any decoder under the guidance of metadata.

LOUDNESS MEASUREMENT

This section briefly discusses and provides a high-level overview of audio loudness measurement devices and practices. It by no means attempts to give the reader a full treatise on this subject. There has been and continues to be research in the area of loudness estimation and devices to measure this subjective quantity, and the reader is encouraged to follow the publications of professional societies and standards bodies that work in this area.

Historical Milestones of Loudness Meters

The former CBS Laboratories performed extensive research into and developed a Loudness Level Monitor (meter) for use in broadcasting in the 1960s. In the early 1980s the original CBS meter was improved (based on further psychoacoustic studies). The CBS loudness meter described functions by dividing the signal into eight bands, which are each weighted according to a 70-phon equal loudness contour [2]. This equal loudness contour is similar to the Fletcher-Munson contour below 1 kHz and the Robinson and Dadson contour above 1 kHz. According to research, the output of each of the filters should respond with a 10 ms attack and 200 ms decay time constants. The filter outputs are then summed and applied to a 200 ms time constant before being displayed. During listener (validation) experiments, CBS found that the meter performed very well and the indicator was never more than 2.5 dB from the median listener judgment. In summary, the work of CBS Laboratories continues to be applicable today, as well as various derivatives of it being utilized in multiband processors to control program dynamics and overall level.

Another method is ISO 532, which specifies two methods for computing the loudness of complex groupings of sounds. Method A (Stevens) utilizes an analysis of the spectrum in one octave bands, where method B (Zwicker) uses 1/3 octave bands [3]. The 1/3 octave bands are utilized to approximate to the critical bands. Zwicker's method also takes into account level dependence, nonlinear frequency response, and the masking effects of the human auditory system. The procedure for calculating loudness found in method B involves three steps based on a set of graphs included in ISO 532. The graphs themselves provide the user a manual means of plotting and combining the physical spectrum (in 1/3 octave bands) of the sound being measured and then converting it to yield the total loudness. Different graphs are provided for frontal sounds and diffuse fields. While ISO 532 showed promise, it was considerably more complex and difficult to implement onto real-time hardware platforms that were available during the mid- to late 1970s.

VU and PPM

VU (volume unit) meters and PPMs (peak program meters) were originally developed for measuring signal level in order to best match the signal to audio production, reproduction, or transmission equipment. This is a different problem than matching audio levels to a desired perceived loudness. Nonetheless, these and other "signal level" meters are often used to estimate subjective loudness or to level programs to a reference level.

The VU meter has considerably slower ballistics than the PPM and will indicate somewhere between the average and peak values of a complex waveform. Moreover, the VU meter only approximates momentary loudness changes in program material and can indicate moment-to-moment level differences that are greater than what our ears perceive. The VU meter also incorporates a relatively flat frequency response over the entire audio spectrum and, therefore, does not address the nonlinear nature of the human auditory system. This can result in large meter deflections that do not correlate with a change in perceived loudness. Perhaps most important, these types of devices can lead to subjective interpretation errors among operators.

Even to the experienced operator, the VU meter is often very difficult to interpret due to its dynamic characteristics and small useable dynamic range. The useable dynamic range is approximately 13 dB, whereby the top 6 dB of this range is dedicated to 50% of the meter's overall scale. Thus, with uncompressed material the indicator tends to fluctuate more than the perceived loudness change, making it difficult for this type of device to assess the subjective loudness of broadcast programming among multiple operators. Experienced operators generally use their ears to balance the different elements of a program and to set the overall loudness, while watching the PPM or VU meter to make sure that signal peaks are not overloading the equipment.

As implied by the name, a peak program meter (PPM) responds very quickly to changes in the signal level. It was designed to help identify peaks or transients that may exceed distortion limits in a device being used. However, the human ear is not particularly sensitive to instantaneous peaks in signal level. While peaks of short duration may be present in a signal, the perceived loudness of the overall signal is typically not significantly affected. This is why a PPM is less effective in indicating loudness. Psychoacoustic experiments have shown that for short intervals of time, perceived loudness is less for shorter sounds, but that at some time interval, somewhere around 100–200 ms, increasing the duration of a sound doesn't make it any louder to the listener.

Equivalent Loudness

The textbook definition of equivalent loudness (Leq) can be defined as the level of a constant sound, which in a given time period has the same energy as does a time-varying sound. The equivalent loudness measure itself can be and is often coupled with a frequency-weighting network (a filter) to better approximate the frequency sensitivity of human hearing at different loudness levels. Common (and standardized) filters include A, B, and C weighting, which are meant to approximate the equal loudness contours at increasing loudness (playback) levels, where A weighting approximates the 30 phon equal loudness contour, B weighting approximates the 70 phon contour, and the C weighting approximates the 100 phon contour. Equivalent loudness measurements utilizing one of these weighting networks is often referred to as Leq(A), Leq(B), and Leq(C), respectively.

In mathematical terms, Leq is computed as the mean-square energy within the measurement interval where x_w is the frequency-weighted signal and x_{ref} is a reference level, as follows:

$$\mathrm{Leq}(w) = 10\log_{10}\left[\frac{1}{T}\int_0^T \frac{x_w^2}{x_{\mathrm{ref}}^2}\,dt\right], \mathrm{dB}$$

The label (w), if present, represents the type of frequency-weighting network utilized for the measurement. For example, a measurement expressed in Leq(A) indicates that the signal being measured was passed through an A-weighting filter, as shown in Figure 5.18-2, before the energy was summed and the average over time was computed. Therefore, any equivalent loudness method (e.g., Leq, Leq(A), Leq(B), Leq(RLB), etc.) provides the operator with a single loudness value that represents the entire measurement period (program) on a long-term average basis.

Loudness Meters

Leq(A) has been a standardized measure since the mid-1980s and is defined in IEC 60804—Integrating-Averaging Sound Level Meters. The use of Leq(A) is also referenced in several other standards, including ATSC A/53E and CEA-CEB11—NTSC/ATSC Loudness Matching. In ATSC A/53E, Leq(A) is recommended for use in measuring the average level of spoken dialog within broadcast programming to determine the proper dialog normalization (dialnorm) value within the AC-3 bitstream for a given program. This is an important and often overlooked step in the encoding

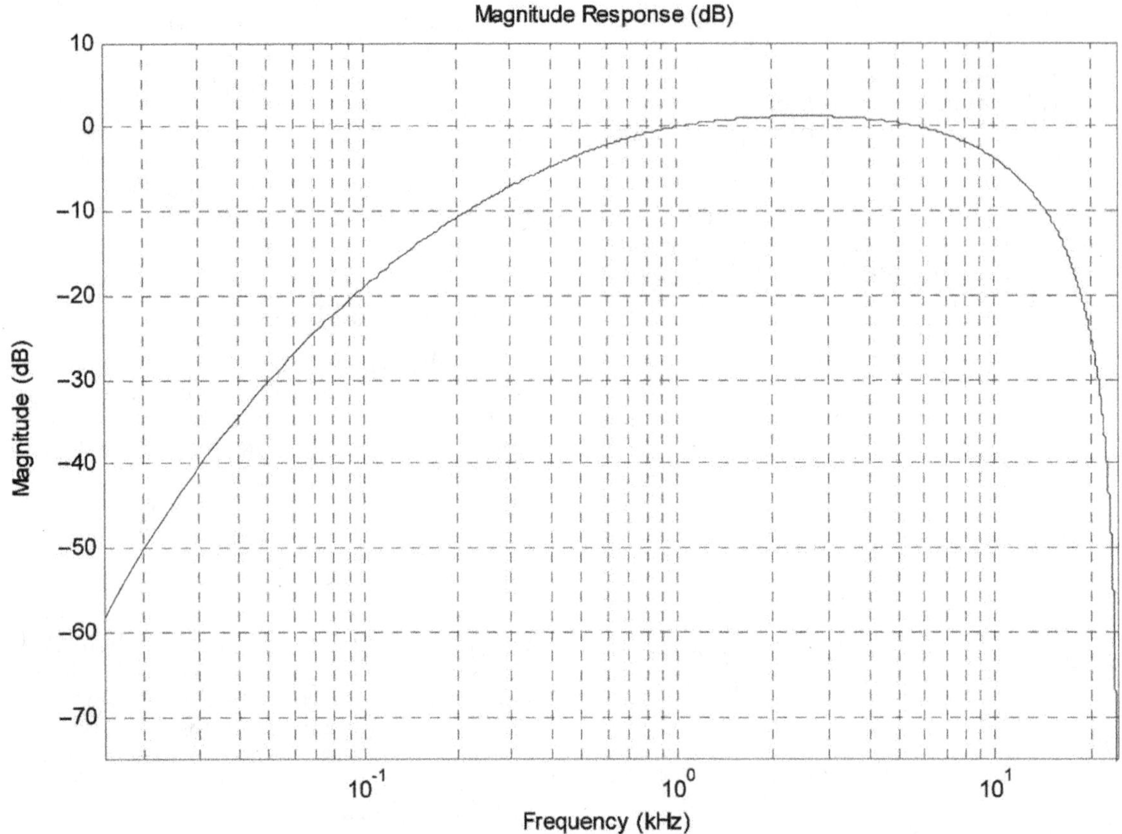

FIGURE 5.18-2 A-weighting filter contour utilized in Leq(A).

process that contributes directly to the reproduced level of the decoded stream. The CEA-CEB11 document provides guidance to set-top box manufacturers on maintaining uniform audio loudness (via the internal gain structure relationships) between NTSC audio services and ATSC audio services while preserving the dynamic range of the ATSC audio service. Importantly, Leq(A) is the metric utilized to quantify the dialog levels throughout the CEB11 document as well.

In July 2006, the ITU-R approved and published a new recommendation for estimating the loudness of broadcast programs, ITU-R Rec. BS.1770, "Algorithms to Measure Audio Programme Loudness and True-Peak Audio Level." It is the intent of this algorithm to estimate the overall loudness of an audio program by computing the frequency-weighted energy average over time. Hence, it also provides (as does Leq(A)) a single measurement value that represents the overall loudness of an entire program.

The ITU-R Rec. BS.1770 algorithm was derived from, and based on, the Leq(RLB) algorithm described by Soulodre [4] to support mono, stereo, multichannel, and advanced immersive audio signals while retaining a low computational complexity. This in turn allows it to be easily implemented and/or adopted by many equipment manufacturers at a low cost. During the ITU-R study, subjective testing showed that the BS.1770 algorithm yielded the best performance among several other methods including algorithms based on psychoacoustic models. Figure 5.18-3 shows a high-level functional block diagram of the ITU-R meter taken from ITU-R Rec. BS.1770.

The BS.1770 algorithm can accommodate any number of channels for multichannel audio programs. In its revision BS.1770–2, the gating block (shown at right in Figure 5.18-3) was added to provide absolute gating of silence, and a relative gate to improve accuracy. Later, BS.1770–3 removed unused filtering, thus simplifying and improving accuracy further. At this writing, the latest revision, BS.1770–4, includes additional input channels to support advanced immersive audio systems. [5] Note that the LFE channel is not included in any of the measurements.

Within the BS.1770 algorithm, each of the individual audio channels being measured is first passed through two filters in cascade. The prefilter, which has a shelving characteristic (shown in Figure 5.18-4), is to account for the acoustical effects of the human head; it is followed by the RLB (Revised Low-frequency B) filter (shown in Figure 5.18-5), which is a modified version of the standard B-weighting curve, and has a low-frequency response that falls between the C- and B-weighting curves. Together these combine to provide the so-called "K-weighting" filter used by the BS.1770 algorithm (as shown at left in Figure 5.18-3 above).

Once the input signal for each channel is filtered, the mean-square energy for the channel is computed for the measurement interval (time). The individual

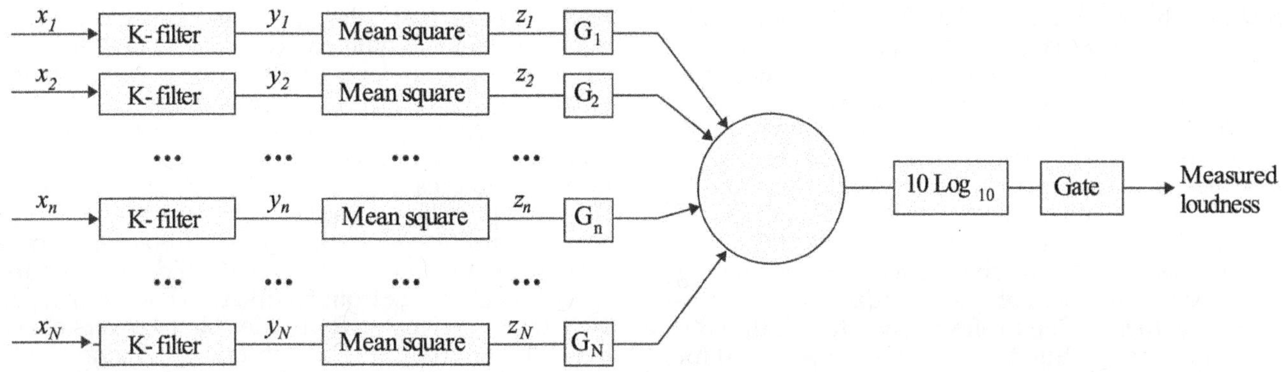

FIGURE 5.18-3 ITU-R loudness meter/algorithm functional block diagram, as specified in ITU-R Rec. BS.1770.

channel loudness results are then weighted in accordance to the angle of arrival (refer to Table 3 in ITU-R Rec. BS.1770-4 [5]) and then linearly summed to provide the overall (final) loudness value as follows:

$$\text{ITU Loudness} = -0.691 + 10\log_{10} \sum_{i}^{N} G_i \bullet z_i, \text{ in dB}$$

The weighting blocks (G_i in Figure 5.18-3 above) are always 1.0 (0 dB) for left, center, and right channels, while the weightings for the left surround and right surround channels are always 1.41 (+1.5 dB) each. The emphasis on the surround channels acknowledges the fact (based on the research of the ITU-R group) that sounds arriving from behind the listener could conceivably be perceived as being louder relative to those arriving from the frontal direction.

Note that the constant value (–0.691) in the equation above is a calibration constant that addresses the combined effects of the prefilter and RLB filter at 1 kHz (note the filter *gain* at 1 kHz in Figure 5.18-4). Further, in support of advanced immersive audio systems, the weighting blocks (*Gi*) now include azimuth (θ) and elevation (φ) angles. Unchanged with respect to legacy formats, the newest recommendation should be consulted for specific values in immersive applications.

A final important note: The definition for ITU-R Rec. BS.1770 utilizes a full-scale *square wave* as the 0 dBFS reference point, but it is common broadcast practice that meters utilize a full-scale *sine wave* as the 0 dBFS reference point. Hence, a 1 kHz sine wave in either left, right, or in the center channel of a program at 0 dBFS (i.e., the peaks of which reach full scale) will measure and display –3 dBFS with ITU-R Rec. BS.1770 meters, whereas they may measure and display 0 dBFS on legacy Leq(A) meter types. Modern meters, which may support both Leq(A) and BS.1770 measurement modes, will exhibit this 3 dB difference when switching modes.[2]

It is important to note, however, that this difference only affects how the measured value is displayed. Consider an operator checking a 1 kHz level of a single channel per SMPTE RP-155, which specifies that the reference signal shall be a 1 kHz *sine* wave at 20

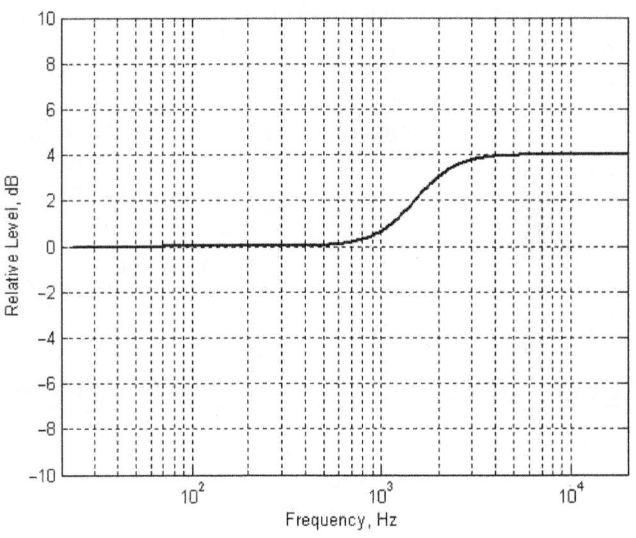

FIGURE 5.18-4 ITU-R Rec. BS.1770 prefilter characteristics.

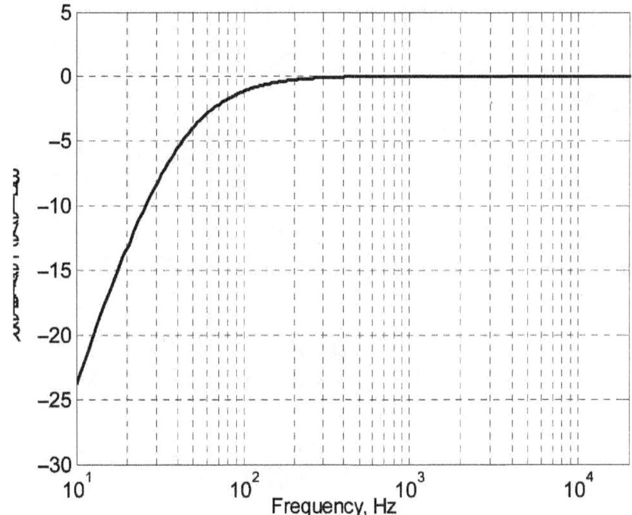

FIGURE 5.18-5 ITU-R Rec. BS.1770 RLB filter characteristics.

dB below the system limit of 0 dBFS. A signal of this type and a display that follows this convention would tell the operator that 20 dB of headroom exists. If the measurement and display device follows ITU-R Rec. BS.1770 (which is calibrated to a *square* wave), however, the display will correctly read –23 dB LKFS with the same (*sine* wave) signal source. Critically, only 20 dB—not 23 dB—of headroom truly exists.

Note that while both ITU-R Rec. BS.1770 and speech-based Leq(A) have been effective in estimating the subjective loudness of typical broadcast programming, they do *not* claim to effectively estimate the subjective loudness of pure tones. Such tones are used for meter-alignment purposes only, not for their loudness measurement. It is important to remain aware of this distinction.

Leq(A) and ITU-R Rec. BS.1770—Dialnorm Measurement

This section briefly discusses the importance of speech-based measurement and its application within the North American DTV system. Also provided is a high-level comparison between Leq(A) and the BS.1770 algorithms and their application/use for provisioning the dialog normalization value within the North American DTV system. However, this section is also relevant to DTV systems outside of North America as a guideline for normalizing the loudness of broadcast programming based on speech (speech-based normalization).

It is useful to measure speech loudness levels. Studies have found that listeners are generally more satisfied when program leveling is based on the dialog segments within programs. For many television programs, speech, as the information-bearing portion of the signal, is most important to the listener. Evidence also suggests that television listeners make adjustments to their volume controls in an effort to create consistent speech levels within their own listening environments. One study conducted by Benjamin [6] concluded that television viewers in a typical living room environment preferred the dialog level to be at a mean sound pressure level of 57.7 dB(A), where listeners in a typical home theater setting preferred a dialog level (mean) of approximately 64.8 dB(A). In another independent study, Pearsons et al. [7] determined that speech levels range from 55 dB(A) to 66 dB(A) at conversational distances. Thus, television viewers choose to set the listening levels such that the dialog level is consistent and that the dialog level chosen is closely related with ordinary conversation levels.

A second benefit of loudness normalization based on dialog is that measuring within a specific content type yields a higher correlation among listeners. Listeners more closely agree on relative loudness when comparing two dialog segments than when comparing arbitrary audio signals. This is not necessarily the case for general audio content.

Figure 5.18-6 is a histogram from a study [8] that shows the frequency of leveling errors between speech and nonspeech broadcast programming. The gray bars indicate the frequency of errors made when the listeners were comparing speech to speech, and the white bars indicate error frequency when they were comparing nonspeech to speech. As the figure shows, errors when comparing speech to speech are more closely clustered around 0 dB, with approximately one-third of the listeners agreeing to within 0.5 dB.

The correlation histogram in Figure 5.18-7, taken from "Intelligent Program Loudness Measurement and Control: What Satisfies Listeners?" [9], compares the results of only two specific samples evaluated by the listener panel. One of the samples primarily contained dialog, and the other contained a portion of a program where only footsteps were heard (i.e., a scene from a movie). According to the correlation histogram, it is apparent that there was general agreement among the listeners when they leveled the dialog item, where 19 out of the 21 listeners agree with each other to within

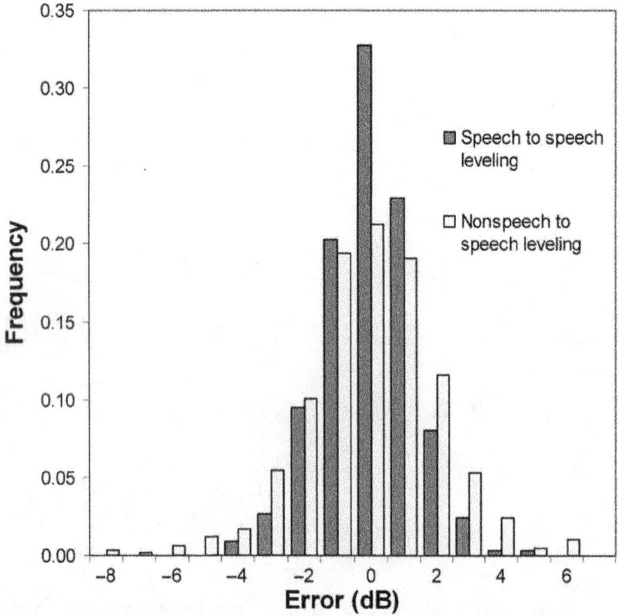

FIGURE 5.18-6 Frequency of listener leveling error.

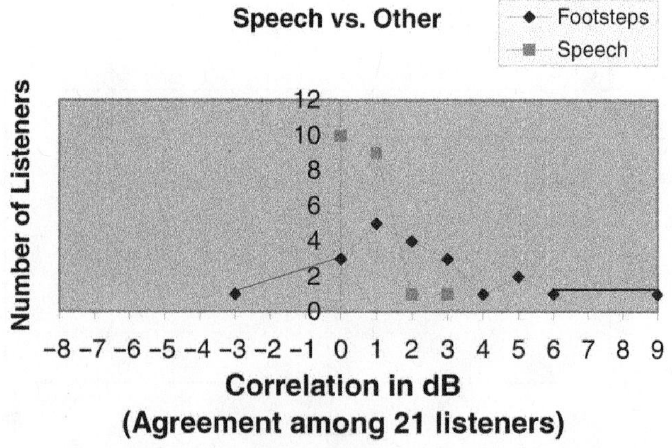

FIGURE 5.18-7 Agreement among listeners when evaluating speech items versus other signal types.

1 dB. However, there was a large disagreement within the group when they attempted to level the "footsteps" piece to the reference. Indeed, one person indicated a need to adjust the footsteps up by 3 dB, while another indicated a decrease of 9 dB to make it agree to the reference.

Given the results of the studies outlined above, focusing loudness measurements on the speech portions of the content has been successfully utilized to improve the performance of objective loudness measurements—even using simple methods such as weighted and non-weighted Leq types—for use in broadcast applications, especially with long-form programming.

AUDIO METADATA

An essential part of digital television audio is its metadata, or data about the audio data. Metadata conveys information such as the loudness of a program, how many channels have been encoded, how to downmix those channels if the bitstream is decoded by a two-channel decoder, and dynamic range control values to help match the audio to the listening environment.

Dialog Loudness (dialnorm)

Based on research performed by Bell Laboratories, the BBC, Dolby Laboratories, and others, it has been determined that dialog provides the most common perceived-loudness "anchor" of a program. It is what most listeners will use to judge the relative loudness of one program versus another. The *dialog loudness* (also known as *dialnorm*) metadata parameter is used to indicate the long-term, A-weighted loudness of a given program with respect to 0 dBFS (full scale). Each program requires its own measurement and the assignment of a unique dialog loudness value. This value directly controls a 1 dB per-step attenuator present in all AC-3 decoders, allowing programs to be scaled to the internal target for AC-3 of –31 dBFS.

Figure 5.18-8 shows how a typical program would be analyzed to determine a dialog loudness parameter, and the results of applying this value after decoding.

The usefulness of this approach is illustrated in Figure 5.18-9(a), where multiple programs having different dialog-loudness values are applied to the encoder, and upon decoding, they are then scaled by the proper amount as shown in Figure 5.18-9(b).

It should be immediately apparent that while the average dialog loudness of each program in Figure 5.18-9 is matched, the signal peaks have not been affected. Audio metadata delivers the unique ability to separate loudness matching from dynamic range control. This means that it is possible to more closely preserve content as originally produced, and leads to a discussion of the *requirement* for some sort of dynamic range control that supports this concept.

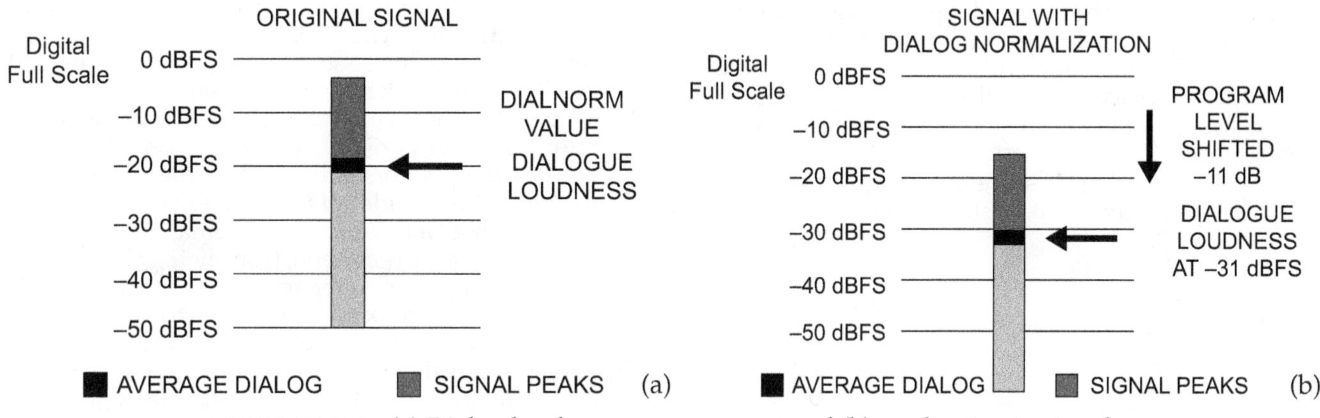

FIGURE 5.18-8 (a) Dialog loudness measurement, and (b) application to signal.

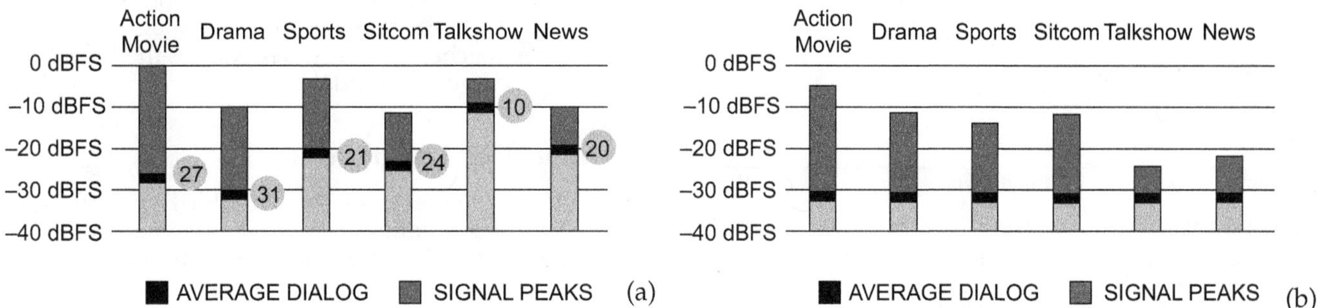

FIGURE 5.18-9 (a) Dialog loudness assignments for each program prior to encoding and transmission, and (b) results after reception and decoding.

Dynamic Range Control

Television audio has traditionally kept dynamic range tightly controlled to ensure that loudness is consistent and that programs are intelligible in as many listening environments as possible. Unfortunately, this has meant that processing has been adjusted for the lowest common denominator, the 3-inch speaker on the side or front of a TV set, to the detriment of larger, higher-quality reproduction systems. With digital audio, loudness can now be matched without severely impacting a program's dynamic range. It also should be apparent that far less dynamic range control is necessary by the broadcaster, and it can be applied using a different approach.

Audio metadata contains dynamic range control (DRC) gain words that can be generated by the AC-3 encoder and applied to the audio at the time of *decoding*—the original, transmitted audio is not affected. Many Dolby Digital decoders offer the consumer the option of defeating the DRC metadata, but some do not. Decoders with six discrete channel outputs (full 5.1-channel capability) typically offer this option. Decoders with stereo, mono, or RF-remodulated outputs, such as those found on DVD players and set-top boxes, often do not. In these cases, the decoder automatically applies the default DRC metadata associated with the decoder's selected operating mode.

The system relies completely on the dialog loudness parameter being measured and set correctly by the broadcaster, or else inappropriate DRC values will be generated (which will result in level control problems). Also, while the control loop used for generating these values is frequency weighted, the gain control is wideband and is applied to all channels simultaneously. This places practical limits on the degree to which dynamic range can be controlled without causing objectionable audible artifacts.

Figure 5.18-10 illustrates the transfer function for the DRC subsystem of AC-3. It shows that although it is

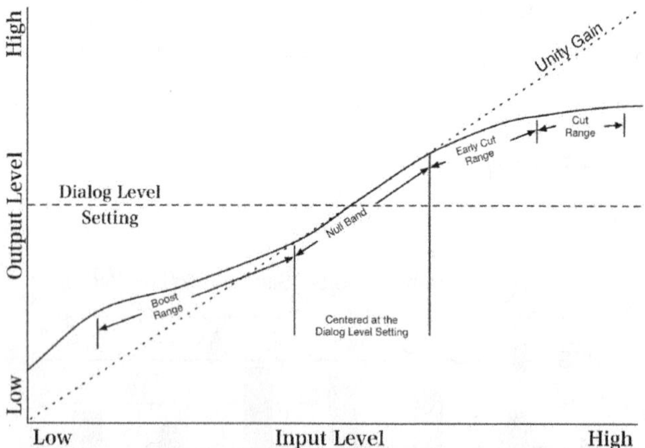

FIGURE 5.18-10 Transfer function of the DRC system in AC-3. Note that the minimal action null zone is roughly centered around the dialog level. No knowledge of transfer functions is required for broadcasters to generate dynamic range control parameters. Presets available in all AC-3 encoders make the task straightforward.

constrained by the fact that is it wideband, the sophisticated structure of the compressor can yield reasonable results, again assuming that the dialog level parameter has been correctly set. Due to inherent natural delays in the audio encoding process, look-ahead processing is possible, which further improves the audible performance of the DRC system.

Additional information on audio processing and loudness can be found in Chapters 5.14, "Digital Television Audio Loudness Management," and 7.2, "Transmission Audio Processing."

Monitoring

Although its full examination is beyond the scope of this chapter, acoustical room alignment is critical for creating a reference monitoring environment, and therefore a brief overview will be presented here.[3] It is impossible to know if a problem truly exists if the monitoring system is uncalibrated. This is best done with a real-time analyzer (RTA) and calibrated microphone, but can also be accomplished with an SPL (sound pressure level) meter.[4] All channels have an individual speaker, including a subwoofer for the LFE channel.

Once the speakers have been physically aligned, electrical alignment is next. Set the real-time analyzer or SPL meter to apply a C-weighting curve and a slow response. A reference listening level of 79 dB/C/Slow is recommended as it most accurately matches the average listening level for most viewers. Generate pink noise at the reference level as shown on the meters of a console or other metering device used during mixing (i.e., set the level of the pink noise so that it averages around –20 dBFS). Route the pink noise to the center channel and adjust the monitor volume control on the console or the monitor controller until the SPL meter reads 79 dB/C/Slow. Mark the position of this volume control and keep it there for the remainder of the calibration. Apply pink noise individually to each of the remaining speakers except for the subwoofer and trim their gain (not the master gain that is set at reference) until each reads 79 dB/C/Slow. When pink noise is panned to each speaker, they should all reproduce at the same level of 79 dB/C/Slow with no need to adjust any levels.

The subwoofer requires a slightly different alignment, as it needs an additional 10 dB of gain compared to the other main channels in order to match consumer reproduction standards. The net result of this extra subwoofer gain is that it will cause operators to mix the sounds to the LFE channel 10 dB quieter. When this audio then reaches the consumer system, the LFE channel is boosted by 10 dB and the net of the process is unity.

Channel Configuration

SMPTE ST 2035 replaces the older SMPTE 320M and specifies track layouts for multichannel audio media. Case 11d is one of several commonly observed layouts and matches what was previously described in 320M:

- Track 1: Left Front
- Track 2: Right Front

- Track 3: Center
- Track 4: LFE (Low-Frequency Effects)
- Track 5: Left Surround
- Track 6: Right Surround
- Track 7: Lt or Lo (Left total or Left only)
- Track 8: Rt or Ro (Right total or Right only)

It is helpful if the channels are specified in this order or one covered by ST 2035. As the layout could be one of many, a facility and its operators must be prepared to shuffle individual channels.

DISTRIBUTION

While ATSC specifications fully define how to send audio from the broadcaster to the consumer, an area subject to wide variability is network distribution, or routing the signal from the program distributor to the broadcaster. There are several approaches that can be used, and a logical combination of different techniques may yield the best results.

Baseband Methodology

Keeping audio in the uncompressed PCM domain allows straightforward access to the audio and perhaps easier control of audio/video synchronization. Audio can either be separate AES3 pairs or can be embedded along with video in an SDI stream.[5] Note that a separate time-aligned path for audio metadata storage and routing is required for baseband systems.

One distribution approach inserts, among other things, the audio metadata into the vertical ancillary data (VANC) space of an SDI (HD or SD) signal.[6] Done properly, this method allows for video, audio, and all associated metadata to be stored, routed, and switched within one signal path, and synchronization between the audio and its metadata can be accurately maintained—at least in theory.

Practice has proved this method to be unreliable due to the lack of consistent preservation of the vertical ancillary space by equipment like contribution and distribution encoders and IRDs (integrated receiver/decoders).

Compressed Methodology

Using audio mezzanine data compression systems such as Dolby E[7] allows multiple channels of audio to be carried along with metadata over a single, bit-accurate path. Noting that the status bits of the AES3 frame are often not preserved in the recording process, the data rate is confined to the audio and aux portions of the frame, equaling approximately 1.92 Mbps. Although compressed systems impart some amount of latency during decoding, they also guarantee that audio and metadata remain tightly synchronized.

When dealing with compressed audio, it is important to choose equipment that has known timing and performance characteristics. While Dolby E is video frame based, different equipment can cause timing shifts. Just like video, this timing must be known when designing a system. Modern audio test equipment from vendors such as Linear Acoustic, Phabrix, Tektronix, and others enables verification of timing and bitstream compatibility.

Transport Stream

At least one U.S. terrestrial broadcast network has chosen to send a pre-compressed, ready-to-air ATSC transport stream to its affiliate or member stations. This was done initially as a cost-saving method in the early days of DTV, which allowed local stations to simply feed the transport stream direct to the transmitter, getting them on the air with no need to purchase an expensive local ATSC encoder. Of course, this meant a lack of local programming or branding, but it did help stations obey FCC rules while simultaneously getting SD and HD content on the air.

There are two choices that a local station user has with transport stream distribution: decode to baseband audio and video, or splice in the compressed domain. Decoding to baseband audio and video produces signals that can be routed and switched with locally generated audio and video signals, then re-encoded for transmission. The usually preferred method, however, is splicing the transport stream, in which case original quality can be maintained all the way to the consumer.

Technologies available for processing and splicing transport streams have overcome most of the early drawbacks of transport stream distribution and allow many operations, such as local logo insertion, to be performed in the compressed domain. They also allow the output of local ATSC encoders to be seamlessly spliced in place of the network transport stream.

It is also possible to apply splicing techniques directly to audio that has been AC-3 encoded. The benefit of allowing pre-encoded content to pass through a facility and straight to transmission to the consumer is that it eliminates all other coding steps, and the ultimate quality is maintained with no need for local personnel to worry about metadata or other settings. This is useful also in situations that require the video to be decoded to baseband for local processing and logo insertion through traditional means, while still allowing the encoded audio to pass through to consumers except when local content is switched in. The AC-3 bitstream is particularly well suited to this type of operation and responds well to splicing so long as timing constraints are obeyed. Techniques and products to support all types of local station needs exist, such as voice-over and program insertion, and are successfully employed by many stations.

LOUDNESS

Soon after the transition to digital television broadcasting, issues arose regarding maintenance of consistent audio loudness. Subsequently, the industry began down the path of solving these issues. After several years of work, the result was ATSC Recommended Practice: Techniques for Establishing and Maintaining Audio Loudness for Digital Television (A/85:2013), which provides guidance to broadcasters and creators of audio content. It recommends production, distribution, and transmission practices to provide the highest-quality audio soundtracks to the digital television audience. It focuses on audio measurement, production, and postproduction monitoring techniques, and methods to effectively control loudness for content delivery and exchange. It further recommends methods to effectively control program-to-interstitial loudness, discusses metadata systems and use, and describes modern dynamic range control. A/85 provided the basis of Report and Order FCC 11–182A1, a regulation that implemented the U.S. Congress's Commercial Advertisement Loudness Mitigation (CALM) Act, which became law in 2012. Over this period of time, there were also revisions to the original ITU-R BS.1770 loudness metering recommendation, which is a fundamental part of A/85.

For more on this topic, see Chapter 5.14, "Digital Television Audio Loudness Management."

AUDIO PROCESSING FOR DTV STATIONS

Audio processing for NTSC television evolved from a necessity to guard against overmodulation, to guarding against annoying loudness shifts, to a way to create a so-called "signature sound" for a television station, to combinations of all of these things. The typical net result has been a permanent reduction of the peak-to-average ratio of audio signals, and has led directly to higher and denser modulation levels.

With the advent of digital television and the uncoupling of loudness and dynamic range controls, the elusive goal of transmitting audio as it was intended by content creators while simultaneously protecting viewers from overly dynamic or objectionably loud programs seemed to be within reach.

The promise was that if loudness metadata was set correctly and program producers mixed their programs with appropriate dynamic range, then the system would work as designed and viewers would have a pleasant experience. If metadata was incorrect or simply run in default on any channel or on any program, however, all of this falls apart. It must be correct everywhere, or else unacceptable loudness shifts are guaranteed to creep in, and that is exactly what happened in the early DTV days. Old techniques have gradually been replaced by modern measurement and control tools, however, and as of this writing, digital TV audio quality and consistency has been remarkably transformed. Proper technique has found its way into every stage of the content ecosystem, starting with production. Most critically, heavy audio processing at the very end of the broadcast chain is no longer seen as an acceptable solution.

NEXT-GENERATION AUDIO (NGA) SYSTEMS

Next-generation audio systems give the industry tools and technology to enable the creation and delivery of content with increased accessibility, personalization, and immersiveness, resulting in dramatically improved consumer experiences. New systems and codecs with increased efficiency have been created to deliver these new experiences and adapt them to different playback environments. Creating next-generation audio content can be accomplished alongside existing stereo and multichannel workflows, expanding incrementally and only as needed.

Object-Based Audio and Rendering

In traditional channel-based audio mixing, sound elements are mixed together and mapped to individual, fixed, speaker channels, for example, left, right, center, left surround, and right surround. In object-based audio mixing, individual sound elements and their position in three-dimensional space are delivered to the playback device, where they are rendered based on the speaker layout in use, including the number and position of the speakers in either two or three dimensions. The benefit of this approach is that by decoupling audio mixing from a fixed channel-based layout, the mix can be reproduced optimally for any speaker layout in use today or in the future. In other words, the creative process mixes for the "space" rather than to a fixed speaker location that is assumed to be identical in every playback environment.

One commercially deployed system called Dolby Atmos provides a mix between traditional channel-based and object-based elements [34]. Diffuse, textural sounds such as ambience, crowd noise, and music are delivered using a traditional channel-based mix referred to as an "audio bed." Audio bed channels are mapped one-to-one to speaker channels.

In practice, sound mixers and directors use a combination of channel-based audio beds and audio objects to create their sound mix. The spatial accuracy, homogeneity, and resolution possible with object-based audio enables new possibilities for the sound designer, mixer, and director, and when coupled with flexible rendering, it can satisfy listeners on any device in any environment.

Accessibility

Next-generation audio systems can provide higher-quality experiences that cater to non-native speaking, visually and hearing impaired audiences by leveraging a fundamentally different approach with how programming is distributed to consumers. Instead of distributing *complete* mixes of each type (e.g., hearing impaired, visually impaired, etc.), next-generation systems can independently distribute program *elements* as discrete objects containing music and effects, main dialog, alternate dialog, and dialog descriptions to the receiver for combining during playback and under the consumer's control (a.k.a. personalization). This approach improves efficiency and the quality of experience for non-native speakers and the hearing/visually impaired by providing each of them with the ability to enjoy a complete mix.

Immersive Audio

Immersive audio is an extension of traditional surround sound reproduction that includes higher spatial resolution and full 3D audio rendering techniques including ear-level and overhead speakers to produce more realistic and natural auditory cues for the listener. Immersive audio also includes advanced headphone playback or reproduction techniques commonly referred to as "binaural" that can achieve a more natural and pleasing experience for a headphone listener.

Personalized Audio

Audio personalization enables multiple producer- and broadcaster-controlled presentations tailored to viewers' needs and preferences to be delivered. It can provide a deeper level of control over the viewers' audio experience and provides a more engaging connection to the program.

In traditional live audio mixing for television, multiple audio elements from an event (crowd mics, announcer mics, music, sound effects) arrive at an audio mixing console where an audio engineer mixes these sources into a stereo or 5.1-channel mix for delivery. This single mix is delivered to consumers, where it is reproduced on television speakers, a home theater system, or a mobile device. Basic metadata provides the capability to downmix or upmix the audio program optimized to the user's playback configuration.

A limitation is that this singular mix isn't tailored to listeners with different preferences or needs. Sports fans may want to listen to a commentator that favors a particular team, or they may want to switch off the commentator altogether. Someone with impaired hearing may want to increase the volume of the commentator and turn down the stadium crowd or ambience, and someone with visual impairment may want to enable an audio description track while preserving the rest of the surround sound mix. Object-based audio satisfies these needs by providing the ability to send individual audio program elements from the mixing console all the way to consumers, thus offering individual consumers the ability to personalize what they experience. Further, audio objects enable content creators to deliver an audio experience that can be optimized for all types of playback environments and applications.

Sending individual sound elements to a playback device allows the audio program to be personalized based on individual consumer preferences and/or needs. Personalized sound elements can be included, excluded, volume-adjusted, or repositioned in the sound field. This provides the viewer with an unprecedented number of audio experience choices. An example of a set of audio elements that may accompany a live broadcast of a sporting event, along with how they may be combined (under user and broadcaster control) to form a number of different presentations, is shown in Figure 5.18-11. These choices could be presented to the program's viewers via a set of on-screen icons, allowing users to select the audio presentation of their choice.

NEXT-GENERATION AUDIO PRODUCTION

If next-generation audio (NGA) production is to be commercially viable, particularly for live events, it is important that any additional sound capture and mixing work be minimized. Existing microphone plans and mixing desk configurations that have developed over many years of operational practice can form the basis of these new productions.

In the initial phases of deployment, object-based audio production is simply an add-on to the current channel-based production operation. The add-on model means that both existing channel-based mixes and object audio mixes will be generated simultaneously.

Immersive Audio Production

Film and television mixing has become increasingly complex, time-consuming, and expensive as content creators strive to get more from the sound. Any new mixing technology should enable new creative options, but it must also integrate into existing post-production workflows without adding excessive time

	Ambience	Main commentary	Home team biased commentary	Away team biased commentary
Default mix	•	•		
Just the action	•			
Clear commentary		•		
Home team	•		•	
Away team	•			•

FIGURE 5.18-11 Audio elements and presentations.

or cost to the process. A hybrid model of channel- and object-based audio therefore allows most sound design, editing, premixing, and final mixing practices to be performed in the same manner as they are today. Object-based audio and metadata are recorded in the session in preparation for the pre- and final-mix stages.

Metadata generation is integrated into the console surface, allowing the individual fader channels, panning, and audio processing to work with channels and channel groups ("stems") as well as audio objects. The metadata can be edited using either the console surface or via a workstation user interface, and the program is monitored using an external renderer, such as the Dolby Rendering and Mastering Unit (RMU). This external hardware device typically has a large number of inputs and outputs (>128) to support core rendering and monitoring on large speaker configurations. Another key function of this renderer is to provide simultaneous re-rendering capabilities to streamline (reducing time/cost) the creation of legacy (channel-based) deliverables. For example, a native object-based (immersive) mix can be monitored on 32 loudspeakers while rerecording simultaneous 7.1/5.1/2.0 channel re-renders of the full object-based mix, or as separate stems using the remaining output channels.

Plug-in applications for digital audio workstations supporting immersive audio creation allow existing panning techniques to remain unchanged. In this way, it is possible to record both channels and objects within the workstation using traditional 5.1- and 7.1-equipped editing rooms then transfer the automation to the full object renderer in the dubbing stage. Alternatively, an integrated "software-renderer" plug-in can be used to monitor object-based audio automation in smaller listening environments.

Figure 5.18-12 illustrates a typical mixing and authoring workflow using a hybrid channel and object format, such as Dolby Atmos. The source audio objects and channels are rerecorded in a traditional manner, but existing audio paths are complemented by metadata exchange over IP for object rerecording. During authoring and mastering, a rendering unit is responsible for monitoring via the room's physical loudspeaker layout, as well as controlling the re-rendering output options (e.g., 5.1/7.1) and generating the deliverables. See Figure 5.18-13.

Specific tools implemented as plug-ins or natively supported in mixing consoles and workstations provide the mixers with the ability to author object-specific metadata alongside and simultaneously with the normal creation of 5.1/7.1 content. See Figure 5.18-14.

Personalized Audio Production

One goal for next-generation audio is to limit or even reduce the operational and financial impact to existing workflows while providing more flexible and engaging content. New tools have been developed to augment existing production practices and facilitate creation of personalized, immersive object-based audio using existing facilities. Object-based production will initially be an add-on to the current channel-based production. The add-on model means that both existing channel-based mixes and object audio mixes will be contributed simultaneously.

Live-broadcast mixers take all of the microphone feeds from a live event and mix them down to a high-quality 5.1 or stereo program, while production mixers create different sub-mixes (e.g., ambience/effects, music, dialog) before mixing down to a final program. With object-based audio, these sub-mixes comprise the audio beds and objects, which when sent discretely to the playback device can be personalized and rendered as desired. For example, a music and

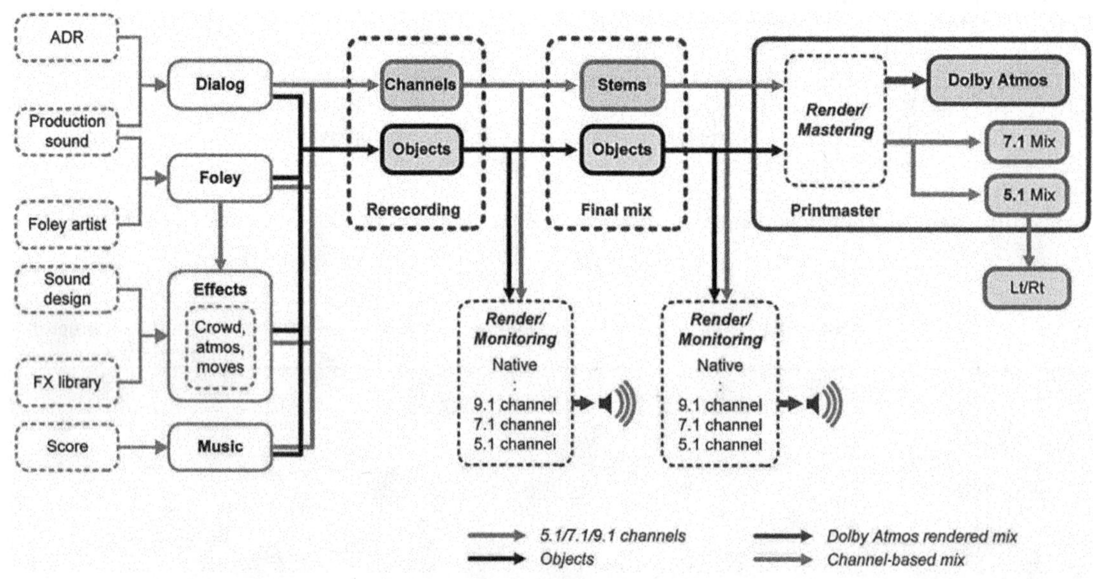

FIGURE 5.18-12 Authoring workflow, showing combination of channels and objects for theatrical or episodic production.

FIGURE 5.18-13 Immersive speaker layouts in production. Dolby Atmos dub stage with 42 channels (left), mobile unit with 5.1.2 monitoring (center), and control room with 5.1.4 monitoring (right). ITU-R BS.2051 [31] defines suitable loudspeaker configurations for mixing in broadcast and smaller control room environments.

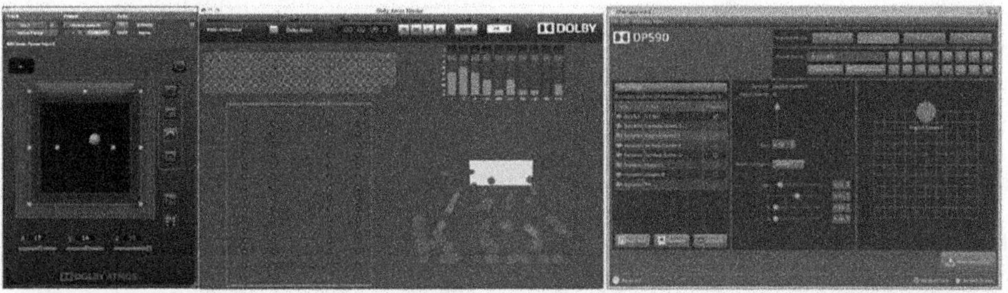

FIGURE 5.18-14 Screenshots of Dolby immersive audio production tools. At left is the Dolby Atmos Panner plug-in UI for Avid Pro Tools. At center is the Dolby Atmos rendering and mastering unit monitoring application, showing speaker feeds and object motion. At right is the Dolby DP590 Personalized Audio authoring tool for defining personalizable objects, audio beds, and presentations.

effects sub-mix could be represented by a 5.1 channel audio bed, and each of the multiple dialog sub-mixes (e.g., for multiple commentators or languages) could be represented by an audio object.

Creative reuse of existing microphones can easily form the basis of next-generation audio mixes for live broadcast events. Additional microphones can be added to provide a more immersive audio experience.

A channel-based live audio production today is comprised of a number of individual elements, which may include the following:

- Crowd sound—constructed from a number of sources
- Spot sounds, for example, ball kick, basketball bounce
- Offscreen dialog—commentary or announcer
- On-screen dialog—Studio links and RF camera interviews to camera
- Audio effects associated with on-screen graphic transitions
- Prerecorded material (e.g., a highlights package or replay elements)
- Synthesized "fill sounds" such as a helicopter or garage sounds in a pit lane

An object audio production is made up of exactly the same elements, but instead of combining them into a finished mix, they are sent to the receiver as a number of different audio *presentations*, and then the selected *presentation* is rendered in the receiver to the final speaker configuration.

In a hybrid "bed plus object" production, each audio element can be mixed into the bed in the conventional manner or handled as a separate object. There is a trade-off in terms of the number of additional audio tracks that need to be delivered versus the final production flexibility.

NEXT-GENERATION AUDIO METADATA AND RENDERING

There are several metadata categories necessary to describe different aspects of next-generation audio:

- Immersive program metadata—metadata utilized for object-based audio rendering. Includes parameters such as position, and speaker-dependent metadata.
- Personalization metadata—metadata to specify audio presentations and define the relationships between objects for consumer user interfaces. Includes parameters such as object enable/disable, alternate object position, 5.1/stereo-aware render metadata, and inter-object dependencies (e.g., only allowing one dialog object to be active at a given time).

- Intelligent loudness—metadata defined per ETSI TS 102 366 Annex H payload_id 0x1.
- Program synchronization—metadata to allow IP and other sources/streams to be synchronized with the primary (emitted) presentation with frame-based accuracy.
- Legacy metadata—traditional metadata including dialnorm, DRC, downmixing for channel-based audio, and so forth.

Immersive Program Metadata

Immersive programming requires generating and delivering dynamic metadata[8] to playback devices. For immersive programming, object position and rendering control metadata is essential for enabling the optimum set of experiences regardless of playback device or application.

Personalized Program Metadata

Object-based audio metadata defines how audio objects are reproduced in a sound field, while an additional layer of metadata defines any personalization aspects that might be enabled. This personalization metadata serves two purposes: to define a set of unique audio "presentations" that a consumer could select, and to define constraints to ensure that a personalized presentation always sounds optimal.

Presentation Metadata

An audio *presentation* specifies which object elements/groups should be active along with their position and their volume levels. Defining a default audio presentation ensures that audio is always output for a given program. Producers and sound mixers can define multiple audio presentations for a program to allow users to switch easily between several optimally predefined audio configurations. For example, in a sports event, a sound mixer could define a default sound mix for general audiences, sound mixes for supporters of each team that emphasize their crowd and favorite commentators, and a commentator-free mix. Presentation metadata defines the details that create these different sound experiences.

Presentation metadata can also provide conditional rendering instructions that specify different audio object placement/volume for different speaker configurations. For example, a dialog object's playback gain may be specified at a higher level when reproduced on a mobile device as opposed to an AVR (A/V receiver).

Presentation metadata typically will not vary on a frame-by-frame basis. However, it may change throughout the course of a program. For example, the number of presentations available may be different during live game play but may change during a half-time presentation.

Interactive Metadata

Users may want complete control over personalizing an audio program. To ensure that every customization results in an optimal sounding mix, interactive metadata provides a set of rendering rules (governed by the content creator or broadcaster) to use for personalization only. Interactive metadata can specify object parameter min/max values, inter-object mutual-exclusion, inter-object position/volume/ducking rules, and overall mix rules.

The interactive metadata is typically leveraged by a consumer user interface to prevent the creation of a non-ideal rendering (mix). For example, if both English and Spanish dialog objects are present in the audio stream, the interactive metadata would prevent a user from enabling both objects simultaneously.

METADATA CARRIAGE

Different methods enable the carriage of metadata described above within file-based and real-time (HD-SDI) contribution/distribution workflows to address a wide range of industry needs related to interoperability and reliability necessary for day-to-day operations.

File-Based Carriage of Object/Channel-Based Audio with Metadata

With the growing interest across the worldwide broadcast industry in enabling delivery of both immersive and personalized experiences, metadata must coexist with audio essence data to fully describe these new experiences and paradigms. The EBU Audio Definition Model [35] has provided the foundation for the ongoing development of an international recommendation within ITU-R WP6B.

The ITU-R Audio Definition Model (ADM) specifies how XML data can be generated to provide definitions of tracks and associated metadata within Broadcast Wave Format (BWF), RF64 files, or as a separate file that references associated essence files. In general, ADM describes the associated audio program as two parts via XML. The *content* part describes what is contained in the audio (e.g., language, loudness, etc.), while the *format* part describes the technical detail of the underlying audio to drive either decoding and/or rendering properly—including the rendering of object-based audio as well as signaling of compressed audio formats in addition to linear PCM.

The ADM is envisioned to be incorporated into the Broadcast Wave and RF64 file formats, as described in Recommendation ITU-R BR.1352, "File Format for the Exchange of Audio Program Materials with Metadata on Information Technology Media," allowing this ubiquitously supported audio file format to carry numerous audio program representations including object-based immersive, along with audio programming containing elements that are intended to be used for personalization.

Real-Time Carriage of Object/Channel-Based Audio with Metadata

The reliable carriage of audio metadata across real-time interfaces and workflows within HD-SDI has been a long-standing challenge for the industry. Moreover, the existing method(s) could only describe a limited number of "channel-based" audio programs along with limitations in terms of extensibility to support future needs. For next-generation audio systems, a new framework and accompanying bitstream format has been proposed, which would be carried across any AES3 channel pair within HD-SDI and audio over IP signals. This framework/bitstream is a SMPTE 337 formatted derivative of the Extensible Format for the Delivery of Metadata (EMDF) defined in ETSI TS 102 366 Annex H,[9] shown in Figure 5.18-15. The EMDF specifies the carriage of metadata in a serialized (and efficient) form made up of "payloads," each with a unique ID. Payload IDs can signal the carriage of several types of metadata (and associated protection information) necessary for next-generation audio, including immersive (object), personalized, intelligent loudness (i.e., per ETSI TS 102 366 Annex H payload_id 0x1), second-stream synchronization, and so on, on a frame-by-frame basis, matching the intended (downstream) codec frame rate, if applicable. It is also understood that tools to translate to/from the metadata formatting defined in the ITU-R BWF/ADM referenced earlier are necessary—including any frame rate conversion(s) required by emission systems. Open standardization of the EMDF bitstream and associated payloads has been proposed, to enable efficient interchange of immersive (object), personalized, intelligent loudness, second-stream sync metadata within the SMPTE ST 337 family of standards, for use in HD-SDI and future environments including IP interconnects. Note that the EMDF bitstream, framework, and associated payloads can easily be migrated for use across any of the standardized IP-based interconnects/workflows such as AES67, and such standardization work is in progress at this writing.

ESSENTIAL METADATA REQUIRED FOR NEXT-GENERATION BROADCAST

This section provides a high-level overview of the most essential metadata parameters required for enabling next-generation audio experiences. It should be noted that this is not an exhaustive listing of parameters required, and only intends to provide the reader with a general understanding of the basics.

Intelligent Loudness Metadata

Intelligent loudness metadata provides the foundation for enabling automatic dynamic bypass of cascaded (real-time or file-based) loudness and dynamic range processing commonly found throughout distribution and delivery today. It can also provide instant and frame-accurate indications of program loudness, which aids compliance monitoring and verification. Intelligent loudness metadata is supported for both channel- and object-based audio.

Dialog Normalization Level—This parameter indicates how far the average dialog level is below 0 LKFS, identical to legacy dialnorm.

Loudness Practice Type— This parameter indicates which recommended practice was followed when the content was authored or corrected. For example, a value of "0x1" indicates ATSC A/85. A value of "0x2" indicates EBU R 128. A value of "0x0" signifies "not indicated."

Loudness Correction Dialog Gating Flag— This parameter indicates whether or not dialog gating was used as part of the loudness measurement when the content was authored or corrected.

Dialog Gating Practice Type— This parameter indicates what dialog gating practice was followed when the content was authored or corrected. Typically 0x02—"Automated Left, Center and/or Right Channel(s)"; other values indicate manual selection of dialog or other channel combinations.

Loudness Correction Type— This parameter indicates whether a program was corrected using a file-based correction process (e.g., measure and scale) or a real-time loudness processor.

Program Loudness, Relative Gated— This parameter indicates the overall program loudness per ITU-R BS.1770. In ATSC countries, this parameter would typically be −24.0 LKFS for short-form content, per ATSC A/85.

Program Loudness, Speech Gated—This parameter indicates the speech-gated program loudness. In ATSC geographies, this parameter would typically be −24.0 LKFS for long-form content per ATSC A/85.

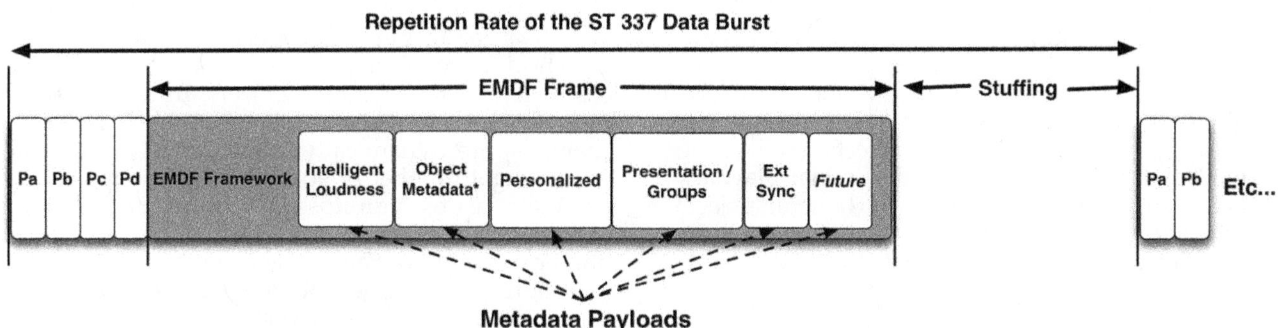

FIGURE 5.18-15 Serialized EMDF frame formatted per SMPTE ST 337.

max_loudstrm3s—This parameter indicates the maximum short-term loudness of the audio program measured according to ITU-R BS.1771.

max_truepk—This parameter indicates the maximum true peak value for the audio program measured according to ITU-R BS.1770.

loro_dmx_loud_corr—This parameter is used to calibrate the downmix loudness (if applicable), per the Lo/Ro coefficients specified in the associated metadata and/or emission bitstream, to match the original (source) program loudness.

ltrt_dmx_loud_corr—This parameter is used to calibrate the downmix loudness (if applicable), per the Lt/Rt coefficients specified in the associated metadata and/or emission bitstream, to match the original (source) program loudness.

Personalized Metadata

The essential metadata parameters necessary for personalized audio include:

- Number of presentations
- Default presentation value
- Per presentation information:
  - Number of substreams per presentation
  - Substream indices
- Per substream information:
  - Content type (for example, dialog, music & effects)
  - Language
  - Gain

Nonessential metadata parameters for personalized audio are:

- Unique program identifier
- Per substream information:
  - Is mutable
  - Minimum gain value
  - Maximum gain value
  - Positional constraints

Object Audio Metadata

Object-based audio consists of one or more audio signals individually described with metadata. Object-based audio can contain bed objects (similar to channel-based audio) with a fixed nominal playback position in three-dimensional space, dynamic objects with explicit positional metadata that can change with time and/or in an intermediate spatial format. Object-based audio is closely linked to auditory image position rather than presumed loudspeaker positions. The object audio metadata contains information used for rendering an audio object. The primary purpose of the object audio metadata is to:

- Describe the composition of the object-based audio program
- Describe how objects should be rendered
- Describe the properties of each object (e.g., position, size, etc.)

Object audio metadata is essential to provide the best audio experience and to ensure that the original artistic intent is preserved. Some essential parameters must always be carried, else there will be no way to render the objects. The remaining nonessential metadata fields are used for enhanced playback applications or aiding in the transmission and playback of the program content.

ENABLING INTERCHANGE OF OBJECT-BASED IMMERSIVE PROGRAMMING

Delivering object-based immersive audio programming made up of a large number of independent audio tracks creates challenging interchange problems. A novel immersive interchange translation process has been developed to overcome this challenge and minimize the bandwidth required—in the form of required audio tracks/channels—so that it can easily flow through broadcast, over-the-top (OTT), and consumer delivery pipelines and consumer device interconnections respectively.

Intelligently grouping objects into what are called spatial object groups, a smaller number of PCM channels plus metadata can be interchanged and, when rendered to any speaker layout, sounds identical to the original, higher channel-count object-based mix. The grouping process is primarily based on perceptual loudness and spatial distortion metrics that work to minimize spatial error when compared to the original object-based mix. Spatial object groups are effectively an aggregate set of the original audio objects. Generating this immersive object interchange format is motivated by the reduced spatial resolution requirements for typical home theater loudspeaker layouts. Keep in mind that there are approximately 7–20 speakers in a typical "immersive" home theater layout, as opposed to approximately 30–64 speakers in a typical immersive cinema layout.

As a simplified example, the left side of Figure 5.18-16 shows the original object-based presentation (i.e., a hybrid channel/object mix), including nine bed channels and ten dynamic objects. In the translation process, individual source objects (including the bed channels) are dynamically and optimally aggregated into a target number of spatial object groups [24, 25]. Some spatial object groups can be comprised of several of the original source objects (downmixed to a monophonic element), or can be the combination of an original source bed channel and one or more objects, while some objects may also be redistributed amongst multiple groups. In this example, 11 spatial object groups +

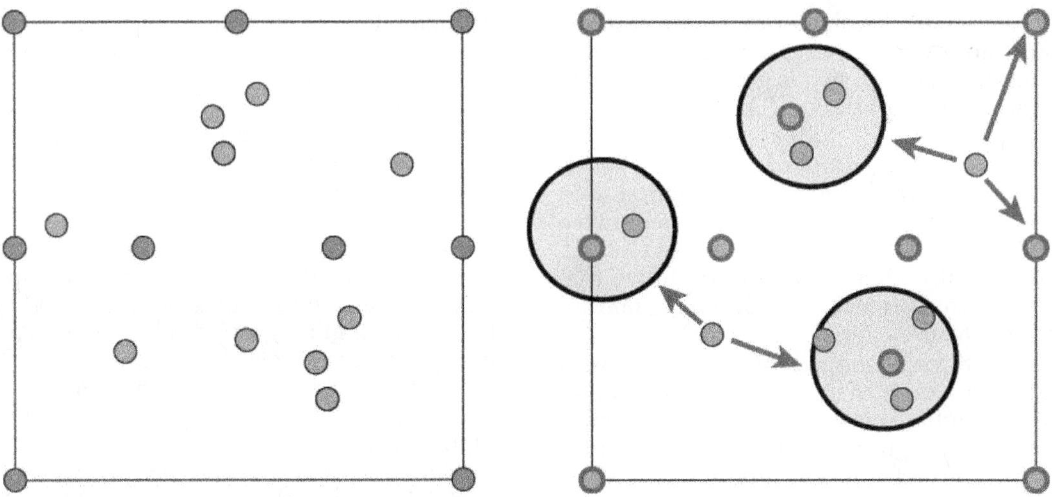

FIGURE 5.18-16 Object audio "pre" (at left) and "post" (at right) interchange translation.

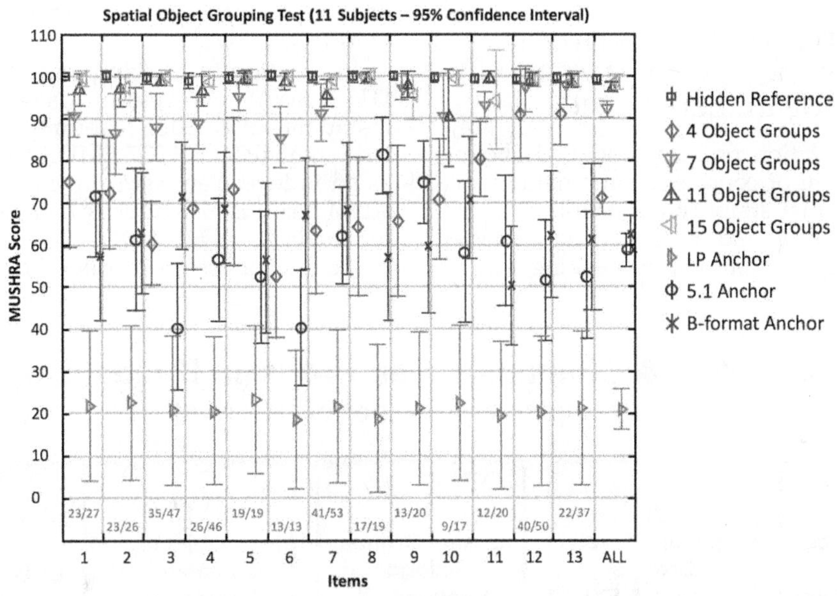

FIGURE 5.18-17 Perceptual testing results for the performance of the translation process for different output spatial object groups rendered on a 7.1.4 loudspeaker system. The numbers at the bottom of the graph indicate the average and peak number of active objects in the original content. It can be seen that 11 to 15 clusters provide perceptually lossless results for these items.

metadata are generated from the original source containing 19 objects + metadata using this technique. Each of the spatial object groups can be carried over a single AES3 channel/track in linear PCM format. Thus, the 11 spatial object groups generated by the translation process would require 11 AES3 channels/tracks (i.e., 6 AES3 pairs/tracks) while the metadata can be formatted and carried per BWAV (ADM/ITU) and/or SMPTE 337 EMDF as described above.

It is important to note that the representative translation process never discards any of the original source objects. Instead, it analyzes the spatial information contained in the original mix to create a version of the program that can be more easily delivered across existing broadcast and OTT workflows and ultimately encoded for consumer delivery. Moreover, the translated output still contains all of the original audio object information and associated metadata, and it can be rendered to arbitrary speaker layouts, including those with many more speakers than the number of spatial object groups carried within the interchange format (e.g., a file/stream with 15 channels/tracks in the representative immersive object interchange format can also play back on 22.2-channel systems).

Although not reversible, the translation process is a pure, linear, wideband combination, without any waveform or perceptual audio compression applied. The end result can therefore be delivered in a lossless manner, similar to how a 5.1 mix is generated from an original digital audio workstation (DAW) based mixing session.

Figure 5.18-17 shows the result of a perceptual evaluation test comparing the source object material with a number of translation scenarios using different numbers of spatial object groups. Note that both 11 and 15 groups are statistically identical to the full home theater mix (the hidden reference), which can contain up to 128 objects/channels.

The speaker configuration used for the test followed ITU-R BS.2051–0 Sound System G with the exception that the left and right "screen" channels were omitted, resulting in a configuration commonly known as "7.1.4"[10] and includes left, center, right, left surround, right surround, left rear surround, right rear surround, left top front, right top front, left top rear, right top rear, and low-frequency effects channels.

SPATIAL AND CHANNEL GROUPS: THE BUILDING BLOCKS FOR IMMERSIVE AND PERSONALIZED AUDIO

The interchange translation process described above enables the interchange of countless combinations of spatial object groups that simplify the enablement of immersive higher-quality and personalized experiences for native-speaking, non-native-speaking, and/or visually impaired viewers. The audio essence and associated metadata for each of the program building blocks is envisioned to be carried in the ITU-R BWAV/ADM file format either as PCM or in a mezzanine compressed format. Within each of the blocks (A–C) in Figure 5.18-18, "N" indicates that the number of tracks/channels (each carrying a spatial object group and metadata) is flexible to address a wide range of workflow capabilities and compatibility needs. This approach also allows users to trade off spatial resolution to meet their business and/or operational needs. Common spatial groups along with channel-based group program elements envisioned for immersive and/or personalized program interchange and consumer delivery are represented in blocks A–I in Figure 5.18-18.

Note that program building blocks G and I represent 5.1 channel and 11.1[11] channel-based immersive music and FX elements (i.e., no audio objects used). These types of building blocks may be found in day-to-day operations that require little or no dependency on supporting dynamic metadata throughout their workflow.

In addition to carrying metadata formatted per the ITU-R BWAV/ADM format, a serialized version of the metadata suitable for real-time applications is formatted per SMPTE ST 337 EMDF and can also be carried as an independent set of tracks in the ITU-R BWAV/ADM file. It is also directly compatible with and ready for carriage via HD-SDI interfaces, including delivery to the final emission encoder if necessary.

Spatial Object & Channel-Based Program Building Blocks
(examples only — not exhaustive)

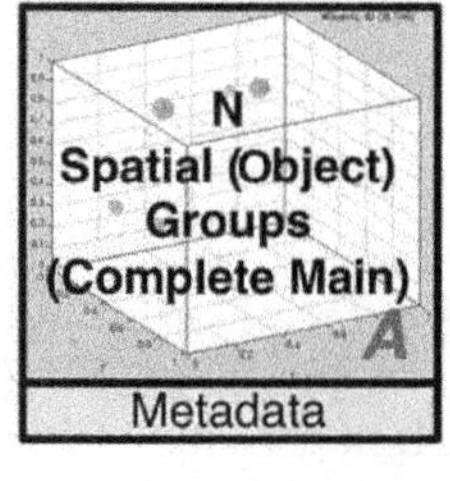

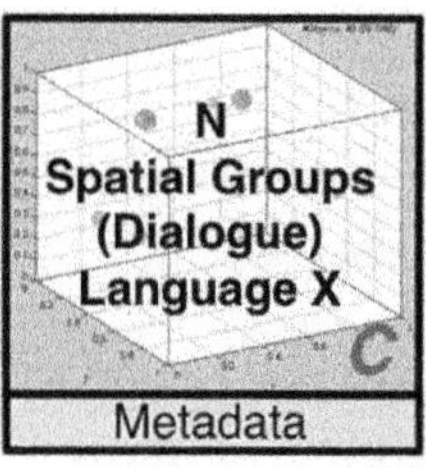

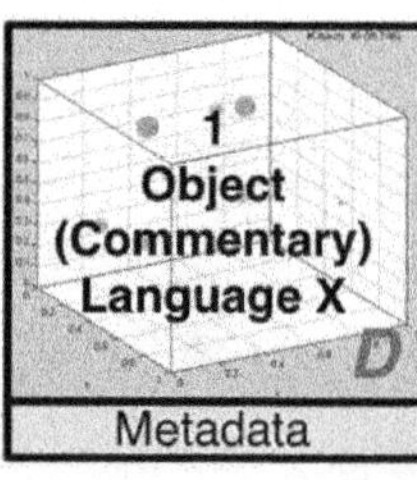

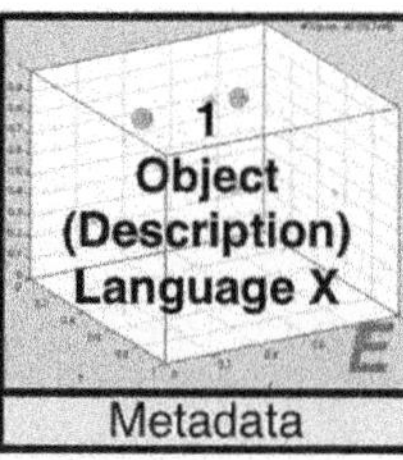

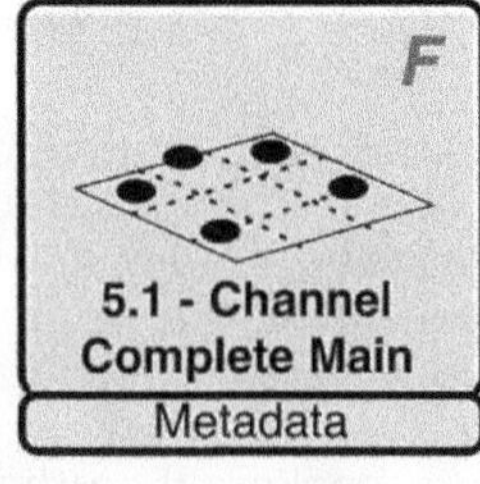

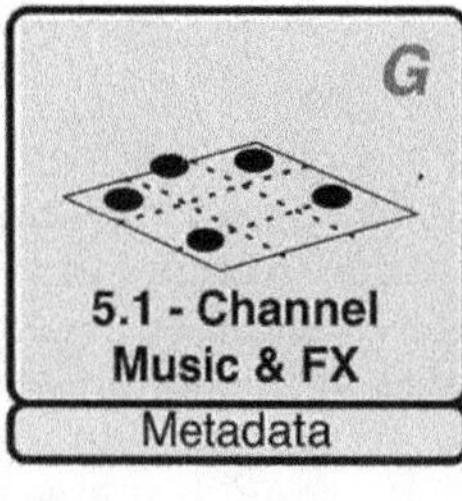

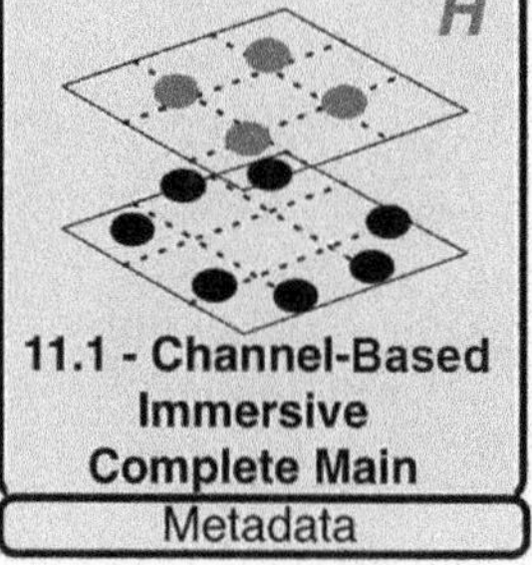

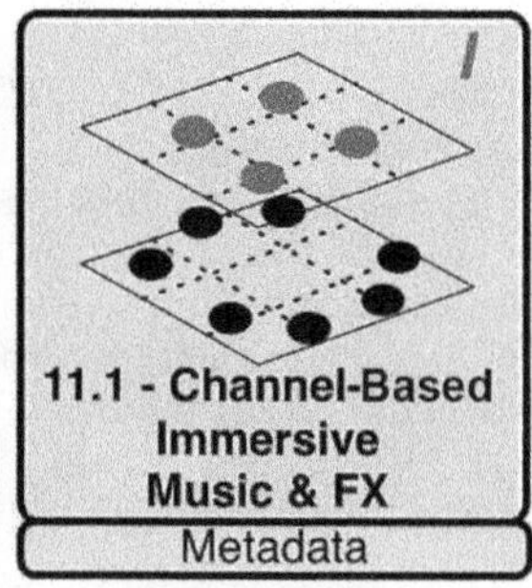

Metadata -> File carriage via BWF(ADM/ITU)
-> Serial (real-time) carriage via AES3/SMPTE 337 or embedded in ED2

FIGURE 5.18-18 Example surround and immersive programs with personalized program interchange elements.

FILE-BASED IMMERSIVE/INTERACTIVE PROGRAM INTERCHANGE, QC, AND DELIVERY PREP

This section describes a representative set of ingest, QC, and content preparation processes in an immersive/interactive audio system. Figure 5.18-19 shows how these processes offer a high degree of flexibility to accommodate a wide variety of day-to-day broadcast, programmer, and OTT operational practices, workflows, and infrastructure constraints. Most importantly, this set of tools can enable a smooth transition from the traditional channel-based paradigm to object and/or hybrid channel/object paradigms. First the blocks are described, followed by a few example use cases for the processes.

Spatial Object Group(s) Combiner [block #3]

This functional block combines or recombines multiple spatially grouped immersive sources like those shown in Figure 5.18-18 with additional channels to generate new combinations of spatial object groups. This spatial combining step can be automated through the use of embedded *presentation* metadata carried within the source files or via presets. For example, spatial object groups 1a (containing N^{12} spatial object groups plus metadata representing the program's Music & FX elements) and 1c (containing N^{13} spatial object groups + metadata containing the spatialized dialog elements) can be automatically combined in a dynamic manner into a single complete main presentation with appropriate dynamic metadata as shown in 1d in Figure 5.18-19.

Adaptive Channel/Object Loudness Estimation and Correction [block #4]

This functional block provides automated loudness estimation and correction from both channel and/or object-based audio presentations contained in the source files. For example, immersive program files at location 1d in Figure 5.18-19 will have their individual loudness estimated and dynamically corrected (if necessary) per regional regulatory requirements. This block also generates or updates the intelligent loudness descriptors defined in Annex H of ETSI TS 102 366 (payload_id 0x1) carried in the metadata per the ITU-R BWAV/ADM file specification discussed above. Importantly, the intelligent loudness descriptors defined in Annex H of ETSI TS 102 366 [18] are utilized by downstream emission, redistribution, and playback systems to perform a number of functions, including those targeted to eliminate quality degradations from cascaded loudness and dynamic range processes by automating dynamic bypass of processing.

Metadata/Preset-Driven Dynamic Rendering [block #5]

This functional block provides automated rendering and/or pass-through of immersive presentations (via user-controlled templates/presets) presented to the

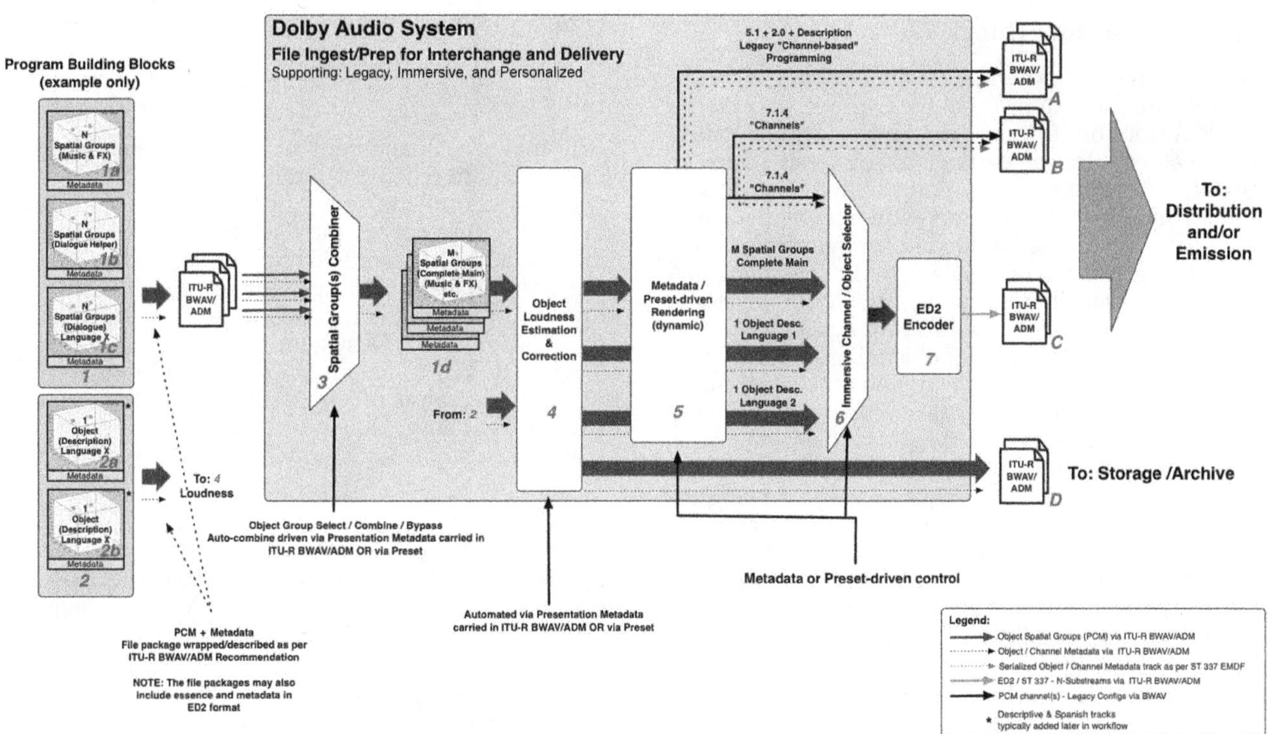

FIGURE 5.18-19 Workflow for immersive and personalized file ingest and preparation for distribution and delivery.

input stage. Rendering at this point in the workflow provides the broadcaster/programmer/OTT-provider with the tools (and flexibility) to automatically generate channel-based file formats (from the immersive sources) to seamlessly feed the existing legacy or new (channel-based) distribution and/or delivery paths. Additionally, the broadcaster/programmer may also choose to offer immersive programming in a pre-rendered channel-based format (e.g., 7.1.4). This approach is likely most suitable for workflows that have additional capacity via HD-SDI and a desire to operate their emission audio codec in a static configuration (i.e., eliminating the need for dynamic metadata to be carried through the entire workflow/plant). Moreover, the channel-based outputs from the renderer can also include an additional audio track containing serialized metadata per SMPTE 337 EMDF discussed earlier. Metadata in this form is ready for linear playout via HD-SDI. It is also important to note here that while the output files A and B in Figure 5.18-19 contain metadata in various forms, some statically configured/provisioned workflows (including emission audio encoders) may choose to ignore the metadata, similar to many traditional broadcast audio operations.

Immersive Channel/Object Selector [block #6]

This functional block simply allows the user to select which type of immersive source would be presented to the ED2 encoder if applicable (e.g., either the spatially grouped objects, or in a pre-rendered immersive channel configuration for distribution and/or delivery to consumers).

Dolby ED2 Encoder [block #7]

This functional block employs the use of a backwards-compatible extension to the Dolby E mezzanine audio distribution format. This new format, called Dolby ED2, enables the carriage of immersive and/or personalized audio essence, along with synchronized metadata in a bit-efficient and video-aligned format. It is applicable to file-based and/or real-time interchange. Extending the original Dolby E format ensures that existing contribution and distribution systems (e.g., encoders, decoders, monitoring equipment, etc.) are able to support the flow of object-based audio plus metadata throughout a wide range of existing infrastructures. The core ED2 format is supported by legacy Dolby E decoder products, thereby addressing confidence monitoring needs. Currently, the ED2 system can support up to four ED2 substreams—each carrying up to eight audio channels plus metadata. Each channel can be an object, a spatial group, or any combination. Each ED2 substream generates a SMPTE 337 formatted bitstream for carriage over an AES3 channel. ED2 substreams can also be linked and synchronized (across multiple AES3 pairs) to enable robust distribution/interchange of channels, objects, or spatial object groups in multiples of eight. The ED2 format has one other critical benefit: it eliminates the challenges of keeping dynamic object metadata sample-aligned through a PCM-based HD-SDI workflow. Future deployments of IP-based interconnects replacing AES and HD-SDI infrastructures are also being designed to handle many audio channels along with tightly coupled metadata, and these will reduce the dependency on ED2 or other mezzanine formats in use.

CONSUMER DELIVERY OF NEXT-GENERATION AUDIO

While all of the features described above might seem attractive and desirable for over-the-air (OTA) or over-the-top (OTT) delivery, immersive audio and personalization requires delivery of all audio components (audio and metadata) necessary to support all of the desired presentations in the audio bitstream. To enable this, extensions to the E-AC-3 format [7] have been developed to provide efficient delivery of immersive and personalized audio to the consumers. These experiences can be delivered in a backwards-compatible manner to allow legacy E-AC-3 decoders to reproduce a default audio presentation. Audio objects can then be discretely encoded to allow for a full range of personalization.

The Dolby AC-4 system [19] is one of the audio delivery methods specified in next-generation DTV systems such as ATSC 3.0. It provides efficient delivery of immersive and personalized audio experiences. The AC-4 system also incorporates a number of system-related features that address many of the day-to-day challenges broadcasters and multichannel video programming distributors (MVPDs) face with maintaining quality and consistency across a wide range of consumer devices and applications. See Figure 5.18-20.

Loudness

Automated management of loudness and advanced dynamic range control must be aligned with worldwide standards and regulations. This is accomplished via native support for the EMDF framework and payload metadata described earlier. If the metadata is present and valid, the content can be directly reproduced. If the metadata is missing or the source cannot be authenticated, the system runs a real-time loudness leveler (RTLL) to generate an ITU-R (loudness)-compliant gain value for transmission in the AC-4 bitstream. That gain value is automatically applied in the playback system. See Figure 5.18-21.

Frame Alignment

Bitstreams are frame aligned with the associated video frame rate. This simplifies common downstream processes including frame synchronization, splicing for ad or program insertion, or when performing bitstream transitions for adaptive streaming applications. By avoiding unnecessary decode/re-encode, frame alignment also helps avoid audible artifacts and reduces complexity. See Figure 5.18-22.

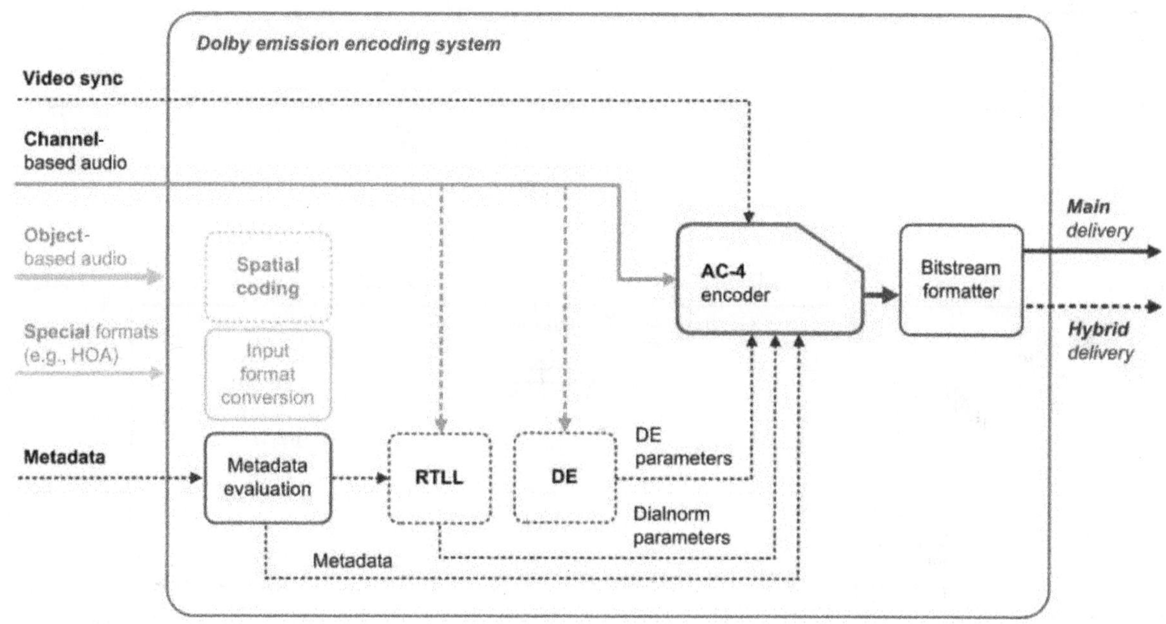

FIGURE 5.18-20 Channel-based AC-4 emission encoding system processing channel-based audio.

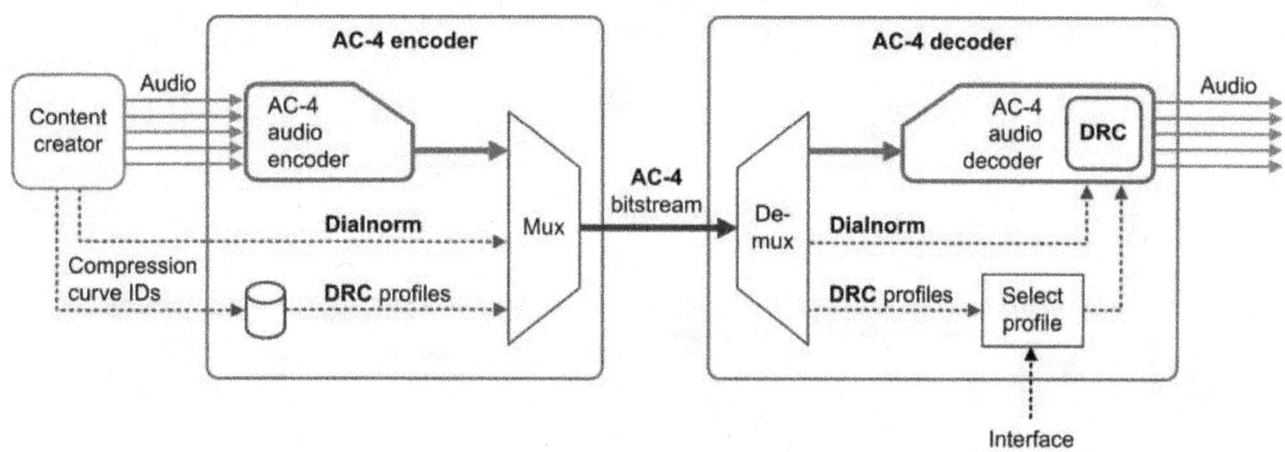

FIGURE 5.18-21 DRC is generated at the decode side, enabling greater customization and loudness control.

Dynamic Range Control

The AC-4 codec supports a range of playback device types that are extensible to address new device types in the future. No longer "one size fits all," the system uses profiles with optimized loudness and dynamic range settings—called target reference loudness (TRL)—for home theater AVRs, flat panel TVs, portable/mobile speakers, and headphones. See Figure 5.18-23.

For dynamic range control, the AC-4 codec and bitstream syntax generate and carry one or more compression profiles to the decoder. In addition to the standard profiles, content creators, broadcasters, or operators can define their own compression profiles for any type of playback device and/or content. The approach minimizes bitstream overhead while still supporting a more typical and desirable multiband DRC system that can be applied as needed to the final rendered result. This again eliminates the need for "one size fits all" upstream processing that would likely be unable to effectively process audio that may be as simple as stereo or 5.1 channels or as complex as immersive and personalized.

The AC-4 bitstream is organized into substreams as described in ETSI TS 103 190 [19]. A single substream can carry a single audio channel/object-element or a group of channels/object-elements. Substreams can further be organized into presentations where a single presentation can contain one or multiple substreams. Additionally, the same substream may also be shared across multiple presentations.

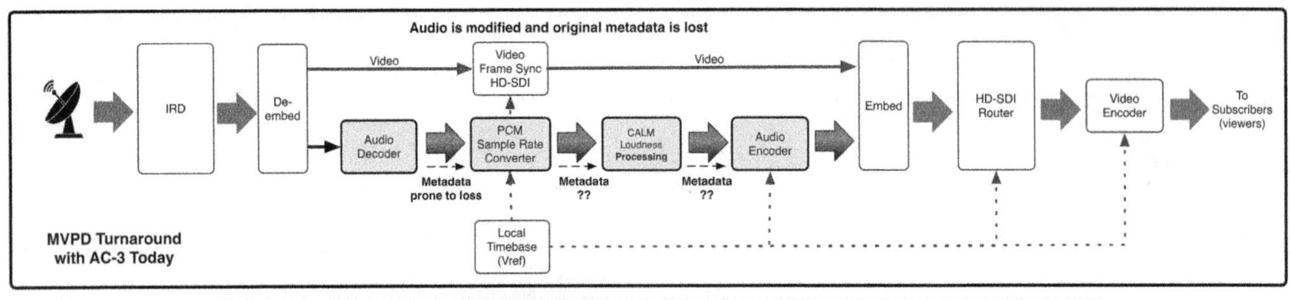

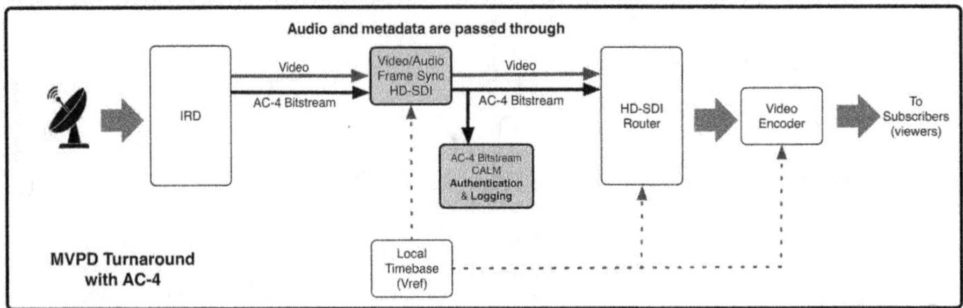

FIGURE 5.18-22 Comparison of AC-3 (at top) and AC-4 (at bottom) systems, showing reduced integration complexity and increased quality with protected metadata in the AC-4 system.

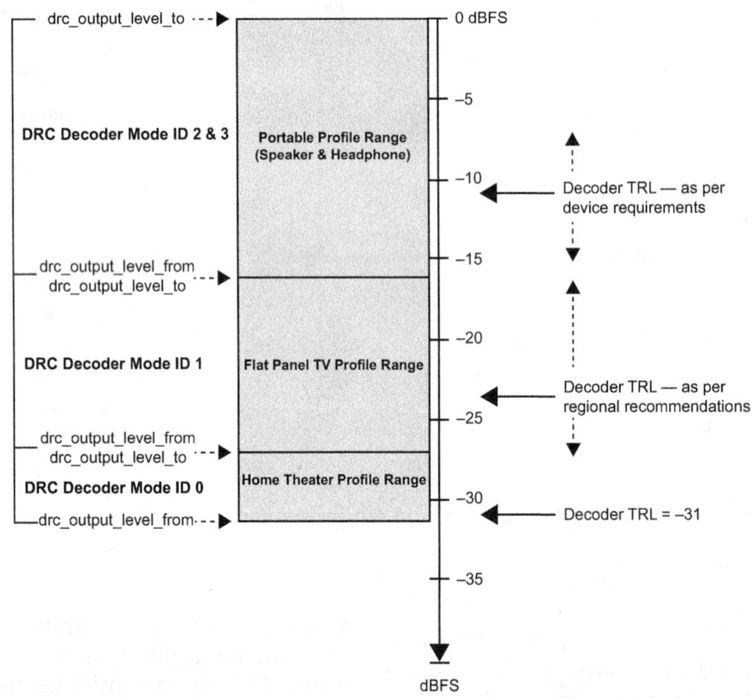

FIGURE 5.18-23 Multiple target reference loudness (TRL) supports different devices.

IMMERSIVE PLAYBACK AND RENDERING

Speaker Layouts for Immersive Audio Playback

Commercial cinema formats [14] as well as ITU recommendation BS.2051 [33] define suitable loudspeaker layouts for immersive audio playback. For backwards compatibility with current channel-based cinema sound systems, as well as providing a predictable output when up-/downscaling loudspeaker configurations, the speaker layouts can be organized around a concept of zones. Each zone corresponds to a group of loudspeakers. This organization in zones is also fundamental to the concept of artistic rendering control, as previously discussed in the Object Audio Metadata section. See Figure 5.18-24.

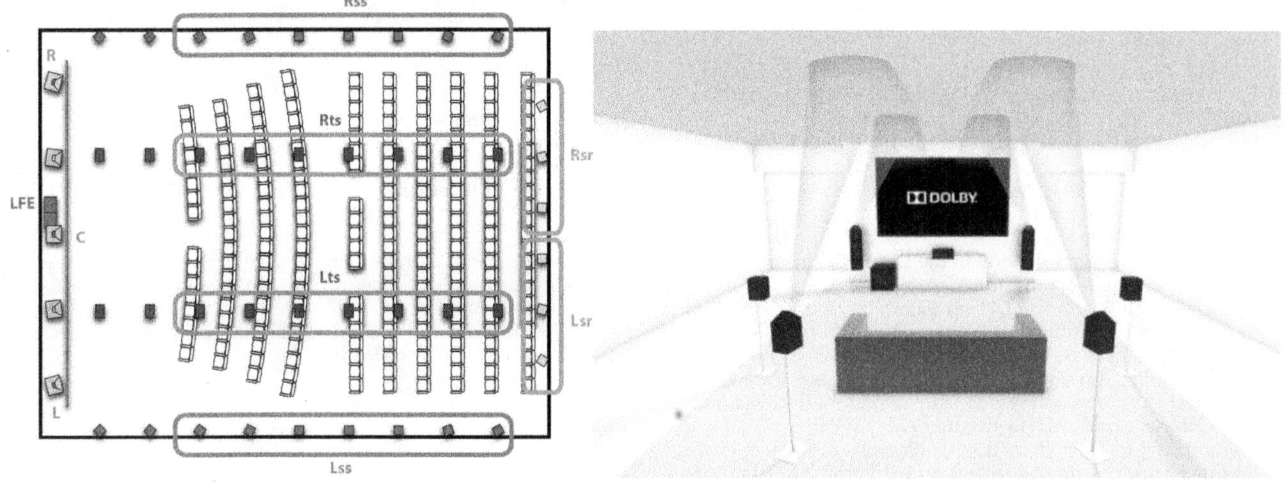

FIGURE 5.18-24 Illustration of different immersive speaker layouts in the cinema (left) and in the home (right). In the cinema, each zone includes a large number of speakers. In the home, most implementations will generally use one speaker per zone. As the diagram depicts, in some cases overhead sounds are reflected off the ceiling and do not require ceiling-mounted loudspeakers.

Three main zones are defined—screen, sides, and back—following the traditional layout of loudspeakers used for surround sound production, with the addition of a new overhead zone. These four zones are addressable through corresponding speaker feeds/channels in the hybrid spatial audio description model. Depending on the room size and geometry, each zone may contain one or more speakers. Channels may address multiple speakers at a time in each zone. In addition, each individual speaker is also addressable through the object-based model.

SUMMARY

Audio for digital television has advanced significantly since first appearing in the 10th edition of this handbook. At that time, the industry was amidst a confusing and stressful transition away from an analog system that was over 60 years old. The move was as much analog to digital as it was standard-definition to high-definition pictures and sound, and from stereo or mono to 5.1 channels of sound at that. The industry has not only met the challenge, it has exceeded expectations with a demand for audio quality and features so great, it has resulted in another transition to the second generation of digital television broadcasting. There is a continuously growing desire for personalized and immersive experiences among consumers, and delivering these experiences requires new thinking, technology, and careful consideration at every point in the broadcast workflow. Audio systems are no exception, as described in this chapter.

Finally, the authors would like to thank those whose work contributed significantly to this chapter, including Sripal Mehta, Scott Norcross, Nicolas Tsingos, Prinyar Boon, Tim Onders, and the greater sound team at Dolby Laboratories.

Notes

[1] ATSC A/52:2015: Digital Audio Compression (AC-3) (E-AC-3) Standard, available at http://atsc.org/standards/atsc-standards/.

[2] An example is the Dolby Laboratories LM100 Broadcast Loudness Meter with Dialog Intelligence.

[3] ATSC A/85:2013, available at http://atsc.org/standards/recommended-practices/, has an excellent section describing proper audio monitor setup and calibration.

[4] There are several applications for iOS and Android devices that when coupled with an inexpensive calibrated microphone result in a useful, highly portable RTA and/or SPL meter. In a pinch, the microphone built into the device could be even used for SPL approximation.

[5] Described in SMPTE 259M/272M (SD) and 292M/299M (HD).

[6] Described in SMPTE 334M.

[7] Dolby E carries up to eight channels of PCM audio and metadata over a single 48 kHz/20-bit AES3 pair.

[8] Dynamic metadata is mandatory for delivering object-based audio experiences.

[9] Note: The EMDF framework described in Annex H is also known as the Evolution Framework (EVO). Dynamic object metadata carried via this method embedded in HD-SDI is required to maintain sync to within ~ ±40 AES samples (@ 48kHz) with the associated audio object channel(s)/track(s).

[10] In this nomenclature, the first numeral indicates the number of speaker channels on the listening plane, with the second number indicating low-frequency effects (LFE, i.e., subwoofer) channels, and the third numeral indicates the number of height channels (above the listening position). In some documentation, an alternate nomenclature such as "7.1+4" or "7.1+4h" is used.

[11] In 7.1.4 format.

[12] A typical number of spatial object groups representing M&E is 15. However, there are no restrictions on group count.

[13] A typical number of spatial object groups representing Dialog is 6. However, there are no restrictions on group count.

References and Bibliography

[1] Moore, B. *An Introduction to the Psychology of Hearing*. San Diego: Academic Press, 1997.

[2] Jones, B.L., and Torick, E.L. "A New Loudness Indicator for Use in Broadcasting," *SMPTE Journal*, pp. 772–777, Sept. 1981.
[3] ISO 532, *Acoustics: Method for Calculating Loudness Level*. Geneva, Switzerland: International Organization for Standardization, 1975.
[4] Soulodre, G.A., and Norcross, S.G. "Objective Measures of Loudness," 115th AES Convention, Oct. 2003.
[5] ITU-R Rec. BS.1770–4, "Algorithms to Measure Audio Programme Loudness and True-Peak Audio Level."
[6] Benjamin, E. "Preferred Listening Levels and Acceptance Windows for Dialog Reproduction in the Domestic Environment," 117th AES Convention, Oct. 2004.
[7] Pearsons, K.S., Bennert, R.L., and Fidell, S. "Speech Levels in Various Noise Environments," Report No. EPA-600/1–77–025, Environmental Protection Agency, Washington, DC, 1977.
[8] Riedmiller, J.C., Robinson, C.Q., Seefeldt, A., and Vinton, M., "Practical Program Loudness Measurement for Effective Loudness Control," 118th AES Convention, May 2005.
[9] Riedmiller, J.C., Lyman, S.B., and Robinson, C. "Intelligent Program Loudness Measurement and Control: What Satisfies Listeners?" 115th AES Convention, Oct. 2003.
[10] Vinton, M., and Robinson, C.Q. "Automated Speech/Other Discrimination for Loudness Monitoring," 118th AES Convention, May 2005.
[11] CEA CEB11, NTSC/ATSC Loudness Matching.
[12] ETSI TS 101 154 V1.11.1, "Digital Video Broadcasting (DVB); Specification for the Use of Video and Audio Coding in Broadcasting Applications based on the MPEG-2 Transport Stream."
[13] "Object-Based Audio Applied to Football Broadcasts—The 5 Live Football Experiment," BBC R&D White Paper WHP272, 2013.
[14] Robinson, C. "Scalable Format and Tools to Extend the Possibilities of Cinema Audio", *SMPTE Motion Imaging Journal*, vol. 121, pp. 63–69, 2012.
[15] Dolby Laboratories, "Dolby E and PCM Solutions for Broadcast," www.dolby.com/us/en/professional/technology/broadcast/dolby-e.html
[16] SMPTE ST 337:2008, "Format for Non-PCM Audio and Data in an AES3 Serial Digital Audio Interface."
[17] SMPTE ST 2020–1:2008, "Format for Non-PCM Audio and Data in an AES3 Serial Digital Audio Interface Transport."
[18] ETSI TS 102 366 V1.3.1, "Digital Audio Compression (AC-3, Enhanced AC-3) Standard."
[19] ETSI TS 103 190 V1.1.1, "Digital Audio Compression (AC-4) Standard."
[20] Robinson, C., N. Tsingos, and S. Mehta, "Scalable Format and Tools to Extend the Possibilities of Cinema Audio," *SMPTE Motion Imaging Journal*, pp. 63–69, Nov. 2012.
[21] Rumsey, F., *Spatial Audio*. Oxfordshire, UK: Taylor & Francis, 2001.
[22] Begault, D., and F. Rumsey, *An Anthology of Articles On Spatial Sound Techniques: Part 2—Multichannel Audio Technologies*. New York: Audio Engineering Society, 2004.
[23] de Vries, D., *Wave Field Synthesis*. AES Monograph, 2009.
[24] Furness, R.K., "Ambisonics—An Overview," AES 8th International Conference, Washington, DC, 1990.
[25] Lossius, T., P. Baltazar, and T. de la Hogue, "DBAP—Distance-Based Amplitude Panning," International Conference on Computer Music (ICMC), Montreal, 2009.
[26] Dickins, G., M. Flax, A. McKeag, and D. McGrath, "Optimal 3D-Speaker Panning," Proceedings of the AES 16th International Conference, Spatial Sound Reproduction, Rovaniemi, Finland, pp. 421–426, Apr. 1999.
[27] Pulkki, V., "Virtual Sound Source Positioning Using Vector Base Amplitude Panning," *Journal of the Audio Engineering Society*, vol. 45, no. 6, pp. 456–466, June 1997.
[28] Kostadinov, D., J.D. Reiss, and V. Mladenov, "Evaluation of Distance Based Amplitude Panning for Spatial Audio," Proceedings of ICASSP 2010, pp. 285–288, 2010.
[29] Tsingos, N., C.Q. Robinson, D.P. Darcy, and P.A. Crum, "Evaluation of Panning Algorithms for Theatrical Applications," Proceedings of the 2nd International Conference on Spatial Audio (ICSA), Erlangen, Germany, Feb. 2014.
[30] Robinson, C., and N. Tsingos, "Cinematic Sound Scene Description and Rendering Control," SMPTE 2014 Annual Technical Conference.
[31] ITU-R BS.2051, Advanced Sound System for Programme Production, www.itu.int/rec/R-REC-BS.2051–0–201402-I.
[32] "Dolby Atmos Specifications," www.dolby.com/us/en/technologies/dolby-atmos/dolby-atmos-specifications.pdf.
[33] Tsingos, N., E. Gallo, and G. Drettakis, "Perceptual Audio Rendering of Complex Virtual Environments," SIGGRAPH 2004, Aug. 2004.
[34] Moeck, T., N. Bonneel, N. Tsingos, G. Drettakis, I. Viaud-Delmond, and D. Alloza, "Progressive Perceptual Audio Rendering of Complex Scenes," I3D, Symposium on Interactive 3D Graphics and Games, 2007.
[35] EBU Tech 3364, "Audio Definition Model," https://tech.ebu.ch/docs/tech/tech3364.pdf.

CHAPTER

5.19

Intercom and IFB Systems

VINNIE MACRI AND KARI EYTHORSSON
Clear-Com, LLC
Alameda, CA

INTRODUCTION

"Intercom" is a general term to describe a stand-alone voice communication system or devices that allow people in different locations to speak to each other operating independently of the public telephone network. There are several types of production intercom subsystems that will be examined in this chapter. They are: wired party-line systems, wireless party-line systems, matrix systems, system interfaces, virtual communications and communication subsystems such as IFB and commentary.

Within the production environment, some of the more common highlights of intercom include the following:

- Can be expanded to hundreds if not thousands of intercom stations
- Can carry out multiple conversations between stations at the same time
- Can route calls to specific stations
- Can forward calls if station is busy or if call goes unanswered
- Can operate auxiliary relays to unlock doors, turn on lights, key radios and so forth
- Can use plain text to identify stations within operating menus
- Can interface with analog or digital or IP telephone systems
- Can interface with analog or digital two-way radio systems
- Can interface with other analog or digital two-wire or four-wire or IP communication systems
- Can interface with security management systems and automation systems
- Have ability to leverage IT infrastructure and work within IP networks
- Have ability to be configured and monitored using a computer

NOMENCLATURE

Understanding intercom functionality and operational capability may sometimes be confusing because of the use of identical terms that have different meaning in different contexts. Here are some of the more popular and generic intercom terms:

Point-to-Point (PtP): A point-to-point connection allows a one-to-one communication path. Within intercoms, a point-to-point is a direct, one-way, simplex communication between two intercom stations or between stations and interfaces whereby what is said by one caller can only be heard by the other. It is the basis of communication in matrix intercom systems. Conversation is limited to those selected by the originator of the call. This connection normally includes a "tally" subsystem; see Tally description below. Most connections are full duplex.

Point-to-Multipoint: A point-to-point connection allows a one-to-many communication path.

Typically, conversation is limited to members of an assigned group, but can be restricted in software. An example would be a call from the stage manager to two-way radios and carpenter PL arranged as a group. This system normally includes a "tally" subsystem (call signal). Most connections are full duplex.

Fixed Group/Group Call: A fixed group is a set of participants or devices such as controls, interfaces, key panels and party lines that can be treated as a single item. A user who has a talk key programmed for a fixed group is able to simultaneously talk to everyone who is part of that fixed group. A fixed group differs from a party line in that the group membership is set by the configuration of the intercom system (not changed dynamically). A fixed group is intended to be used for one-to-many or point-to-multipoint type communication. The group call function allows for multiple participants in a special group communication circuit. For example, a group call can be an "all" call for a director to get the attention of everyone involved in the production.

Channels: A "channel" is one individual circuit of communication. An example would be a party-line channel for camera operators.

Party Line (PL): The name came from the original telephone systems where everyone shared the same line and could hear and join all conversations at once. A conference or PL is classically used when several users, such as belt packs, are active in a common task and they must communicate with each other all the time in a full-duplex mode. A party line is intended for use as a conference with a significant amount of back and forth communication between users. Each station must activate the listen key to the desired party line to listen and the talk key to talk. Stations are dynamically added and subtracted from a party line as users activate talk and listen keys.

IFB: IFB stands for interruptible fold-back and, in the simplest application, allows a user to hear one audio source, which may then be interrupted by another audio source. These systems are generally used to cue talent. IFB systems are also called interrupt feedback, program interrupt, director interrupt, director cue, director interrupt, or interrupted return feed. They can be viewed as one-way intercom systems. With the rise of matrix systems as the dominant type used in broadcast, IFB systems are no longer considered as separate systems but are an integral part of the main intercom. A typical broadcasting IFB application is when talent listens to program audio in an earpiece and a producer can interrupt that audio to offer the talent information and instructions. The program can be muted or dimmed (attenuated) to a predetermined level. An IFB can be provided to people other than the designated talent, and it can also feed a loudspeaker as well as an earpiece.

In an IFB system, program audio is fed to the talent (usually through an in-the-ear earphone). The program feed is generally a feed of the program mix without the talent's voice and is therefore referred to as a mix-minus feed. These mix-minus feeds are sometimes called "clean feeds" or "confidence feeds."

Mix-Minus: Mix-minus is an audio feed where all the audio elements are present except the talker's own voice. This is necessary when the latency from a remote venue to where the audio is mixed, and then back again to the venue, is such that talkers will hear themselves delayed enough to cause distraction. Mix-minus, when used, is fed to the talent's IFB feed. Often when you see reporters pull their IFB earpiece out during a stand-up it means that they are not receiving mix-minus and are also hearing themselves. Most modern broadcast consoles make some provision for at least one mix-minus.

Creating mix-minus circuits is important during the installation of a broadcast-to-telephone interface. You want to feed the caller an audio signal so that he or she can hear what is going on in the studio. Most mix-minus circuits feed the entire on-air mix except, of course, the caller. This allows the caller to hear everything over-the-air listeners hear. When installing more than one hybrid, one needs to create a separate mix-minus feed for each hybrid. Keep in mind that you want each caller to hear all of the other callers. The program mix should treat the hybrid outputs as simply additional audio sources. See Figure 5.19-1.

Some modern digital audio consoles have an N – 1 (mix-minus) function that allows for setting up complex conference calls. All subscribers who are off-air (fader down) can talk to each other. If one of them is put on air (fader up), the corresponding path will be removed from the conference. When the fader is closed that person is once again added to the conference.

The program audio can be interrupted or isolated by an operator (usually the director or producer) to give verbal cues to the talent. Typical systems have control panels (or keys on the intercom panel appropriately programmed) for the producer, director, assistant director, and the audio operator. IFB systems can be quite extensive, involving multiple program inputs and links to off-premises locations. Their heaviest use is in sports and news operations.

ISO: ISO stands for "isolated" and is an intercom function that permits a user on a conference or party line to be isolated from that conference for a private conversation. A good example for the ISO function is used by video engineers to isolate a camera from the camera PL. The video engineer can then work directly and privately with the camera operator on camera setup without disturbing the other camera operators on the camera PL.

There are specialized ISOs, such as an IFB ISO, which is a function of some matrix intercoms and can be a very useful feature in busy environ-

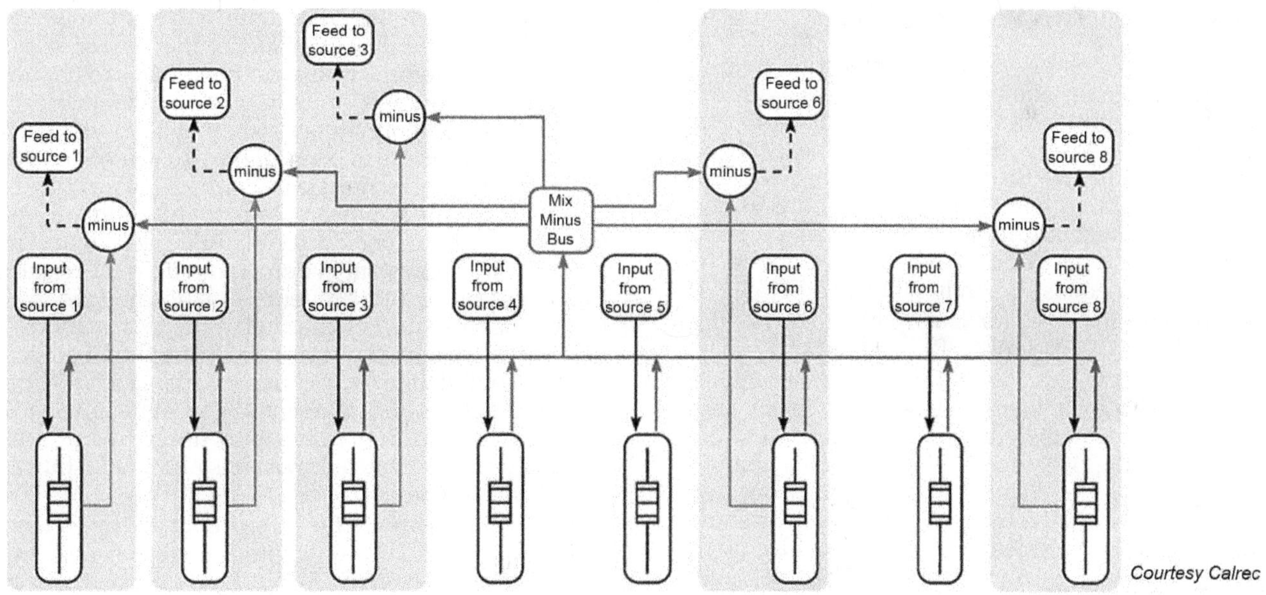

FIGURE 5.19-1 Mix-Minus Audio Console Flow

Courtesy of Calrec Audio

ments. This basically allows a user, normally the producer, who may have numerous talk latched assignments, to ISO an IFB, thus removing their talk from any latched talk other than the ISO IFB. This basically eliminates others from hearing conversation meant for the IFB.

Tally: Tally is a signal sent to indicate the status of a particular function. A telephone ringer can be described as a tally. In an intercom with multiple channels, it can be a visual signal to indicate to which station a calling voice belongs. It can be used to indicate that a particular function is not available. Tally can also indicate a call waiting. Call lights used in party-line intercoms are a form of tally (blinking call light) to get the attention of a user. The call light signal is superimposed over the signals representing "intercom channel" audio. This signal is not heard because it is a 20 kHz tone in RTS TW systems and a power interrupt in Clear-Com PL systems. In these cases, the voice audio and the call light signal are multiplexed. Within some matrix intercoms the call signal (tally) is embedded in the serial data.

ANALOG PARTY-LINE SYSTEMS

The name PL (party line) came from the original telephone systems where everyone shared the same line and could hear and join all conversations. It is often called a conference system or TW, which stands for two-wire (see the Two-Wire User Stations section). A party-line system allows a group of people to intercommunicate. A party line is classically used when several users, such as belt packs, are active in a common task and they must communicate with each other all the time.

Characteristics of party lines:

- When the party line is already in use, if any of the other users to that line pick up a headset, turn it on and listen, they can hear and participate in the conversation
- Completely non-private lines
- Systems are created from building block components to correspond to the demand of the event

Party-line systems can be distributed or centralized. Most of the systems are distributed conference systems. Distributed means that a station can be plugged in at any arbitrary point along the bus or channel.

A party-line intercom is known as a two-wire system. From a technical point of view, an analog party line or two-wire system is a communications system where the path is the same for both talk and listen. These systems are also referred to as "talkback" systems.

A two-wire system refers to an intercom system that carries both intercom audio and DC voltage for powering belt packs and speaker stations over a single twisted audio pair. The term "two-wire" refers to the two wires that make up the audio conductors of a twisted pair. Bidirectional audio and DC power are carried over the two wires plus shield in a two-wire system.

Clear-Com developed the Partyline and RTS developed the TW (Two-Wire) systems that are the most common analog two-wire intercom systems. Differences between them are explained later in this chapter. PL is an abbreviation for "party line," but in the vernacular can refer to an intercom system, a user station on an intercom system, as well as any kind of communication conference (including a teleconference).

Party Line (PL)/Two-Wire Intercom

Two-wire intercoms provide communication by offering a conference on one or more channels, with each channel functioning over a single pair of wires. This connection is typically used to carry the party-line voice in one wire and the station operating power on another. There is no point-to-point communication in a two-wire system. Multiple channels can be implemented so that a number of different party-line conferences may be used, as shown in Figure 5.19-2.

Thus, there may be, for example, a camera PL, a director PL, and a production PL.

Both the Clear-Com and RTS systems offer bidirectional audio over a single shielded twisted audio pair terminated with standard three-pin XLR connector. The RTS TW system allows for two channels of audio, with audio channel 1 on the pin 2 conductor and audio channel 2 on the pin 3 conductor, as shown in Figure 5.19-3. The shielded twisted pair also carries 30 volts DC that powers end points such as belt packs and remote speaker stations. Clear-Com's Partyline system and the RTS TW system puts this DC voltage between pin 2 and pin 1 (shield).

Both Clear-Com's Partyline and RTS's TW operate on an unbalanced line where the audio signal is carried between a signal conductor and ground. Party-line and TW intercom systems are typically designed around a power supply or (as is often the case with Clear-Com) a main station that provides the DC voltage applied to the pin 2 audio conductor.

The most important standard developed by these manufacturers was the use of a fixed line impedance, usually around 200 Ω, with the individual intercom stations bridging the line with a high impedance, generally higher than 10,000 Ω. The power supply includes a system termination for the audio channel. In some RTS models an impedance generator exists with DC at the source, and in other RTS and Clear-Com systems, there is a simple resistor and capacitor in series as the termination circuit. This audio termination circuit is what allows many intercom user stations to be connected to a single party line. Each channel of the system would have its own terminator. The termination prevents drastic changes in the impedance of the PL channel if remote stations or belt packs are added or removed from the PL line. An un-terminated line will cause excessive levels, possible oscillation of line drivers, and squealing in the headsets. An intercom line with double or multiple terminations will cause low levels and the inability to null the headsets.

User stations bridge the intercom line at impedances of greater than 10,000 ohms and, therefore, only minimally load the system. Audio levels remain constant as stations (within the specification of the system) are added to or subtracted from the intercom line. Two-wire systems are referred to as distributed amplifier systems since they do not have central electronics as does a matrix intercom system. Advantages of these systems include the ability to power belt packs and speaker stations without external AC power supplies or batteries and the ability to use microphone cable for interconnect between power supplies, user stations, belt packs, and speaker stations.

Party-line intercom systems are well suited for remote television production since they are relatively easy to set up and tear down. Providing power to belt

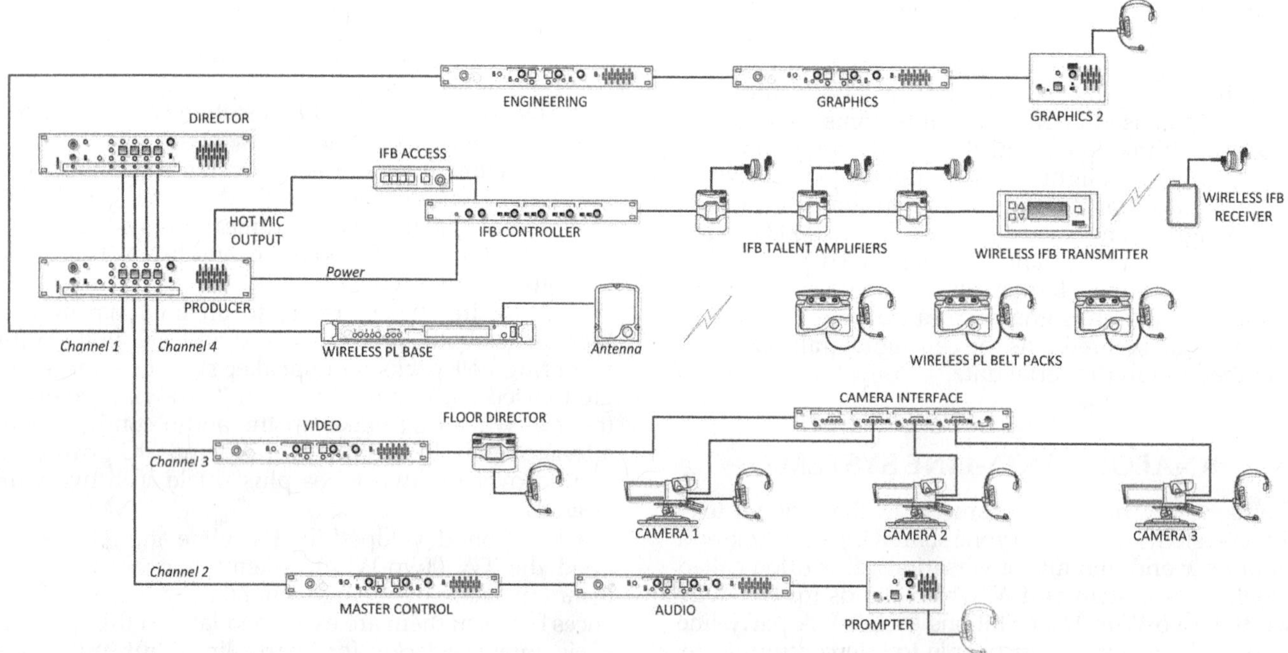

FIGURE 5.19-2 Four-Channel Party-Line TV Station Intercom

Courtesy of Clear-Com

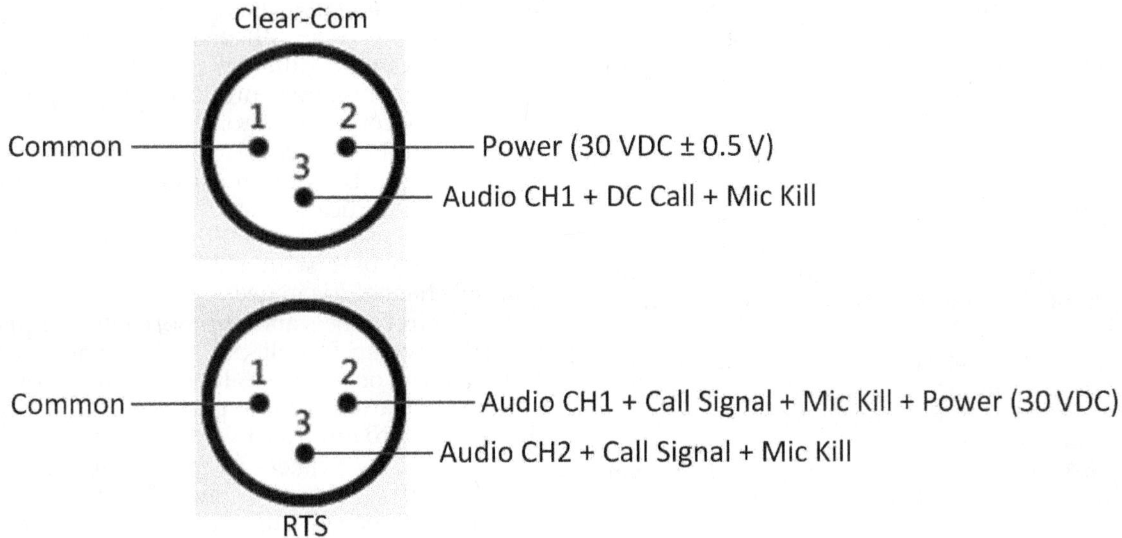

FIGURE 5.19-3 PL/TW XLR Pinouts

packs over the intercom line is an important advantage in this type of situation.

Each user station in a party-line system has its own electronics and a power supply for powering microphone preamps, headset and speaker amplifiers, as well as for providing power to signaling electronics such as call lights.

Two-Wire (PL) User Stations

Two-wire user stations use headsets and gooseneck microphones. They can include internal loudspeakers, hot mic outputs, program inputs, relay outputs and logic inputs. Wired party-line intercoms can be described as a distributed amplifier. The distributed amplifier is built into the various user stations and offers each user listening level adjustment. User stations also include a microphone preamplifier, volume control(s), talk switch(es) and call switch(es). Some user stations may have visual signal circuitry, such as call light indicators (e.g., LEDs to signify power on the line plus power fault), and channel selectors.

Since a two-wire system is a distributed system, it does not have a central intercom unit. Some two-wire intercom panels can display mnemonics, but these are created in the panel. Tally information is transmitted as a call light from a calling panel to a called panel.

Signaling

Some signaling can be provided but usually not to the extent that every operator can selectively signal every other operator. "Call" lights allow user stations to generate visual indications for cueing purposes or to show that someone is trying to talk to a particular station or belt pack. Clear-Com uses a DC voltage to activate call lights; RTS uses a 20 kHz tone. Call signals can also be used to trigger external closures such as relays to two-way radios and other devices. Devices are available that will detect both types of call light signals and activate a relay closure.

Remote Mic Kill

Remote mic kill is a useful feature of many two-wire intercom systems. An open, unattended microphone is an annoying and disruptive aspect to an intercom system. Extraneous noise picked up by open, unattended microphones can be heard by everybody on the intercom channel and can make communication difficult.

Clear-Com systems accomplish remote mic kill by momentarily interrupting the DC on an intercom line. RTS TW systems use a 24 kHz tone to signal user stations to shut off their microphones.

Source Assignment

Source assignment panels function as a means of routing two-wire intercom channels to intercom buses. Source assignment panels are, by and large, manually operated matrix switches. A source assign panel is usually a separate piece of hardware. These are used in more complex systems that have many party-line intercom channels. For example, a two-wire system with 12 intercom channels and 20 intercom buses would use a source assignment panel to route any of the intercom channels to any of the intercom buses. Inputs to the source assign panel are sensibly called "Sources" and are typically the party-line channels from a main or master station or party-line interfaces from a matrix or similar system. The "Destinations" are the party-line circuits or buses themselves, that is, cabled outputs from the source assign panel that are "wet" (have supply voltage 24–30 VDC). Remote intercom stations, belt packs, or other interfaces that can also be groups of such devices connected together are patched to various circuits/buses. The source assignment panel offers flexibility in assigning intercom channels to intercom buses and provides the ability to easily reassign channels to buses on the fly.

Two-Wire IFB Systems

Two-wire IFB systems generally require a central electronics control unit to allow multiple users to select one of many IFB channels to talk to different talent. An example of a two-wire IFB system is shown in Figure 5.19-4.

The circuitry in the central electronics unit allows for the selection of program sources and the switching of those sources to an IFB channel. When a user talks on an IFB channel, that interrupt audio is mixed with the program source audio, which is dipped in level according to a user-defined setting.

The IFB control unit also superimposes a DC voltage on the IFB line to drive the talent receivers. These IFB receivers can, according to model, be either a two-channel model that can provide non-interrupt audio in one ear and program with interrupt (IFB) audio in the other ear, or a single-channel model that provides only interrupt audio.

The interrupt audio is generated at IFB control panels that are interfaced to the IFB controller or from a user station. User stations can also generate a local IFB where the program audio that is fed into the user station is interrupted by the talk at the user station and is sent down an intercom channel. A locally generated IFB does not have the flexibility of a centrally based IFB system, since only that user station can talk on a given IFB channel. With a central system any number of users can potentially cue talent on the same IFB channel. In a two-wire system, IFB priorities can be set so that one station has priority over another station for talking on an IFB channel. User panels can be programmed with different priority levels to ensure that the most important producer in a production will always have communications access to talent.

The central electronics unit provides tally information so that an IFB button assigned to a channel will illuminate to indicate when that channel is in use. For example, if a user talks on IFB channel 1, the IFB 1 button on every user station will illuminate to indicate that IFB channel 1 is in use.

IFB control panels and appropriately equipped user stations have an IFB all call button that will allow a user to speak on all IFB channels simultaneously. This is a handy feature when a producer needs to provide the same cue to multiple talent on a set.

The monitor speakers in control areas can be connected to the IFB system such that they are muted or dimmed when the IFB is active. This will prevent acoustic leakage of control room monitor audio into the IFB during an interrupt.

Commentators/Announcers

A commentator is a radio or television individual who commentates on the events and actions of sport or news. Other common terms are announcer, newscaster, presenter or reporter. Within remote sports productions they are referred to as sportscasters, analysts and play-by-play and color commentators. Some sporting events require numerous inputs from specialists, statisticians and other external cueing sources.

Commentators may work from the event itself, or in some cases "off tube" via the host signal while based in their home nation. "Off tube" refers to when commentary on an event is produced in a studio whilst

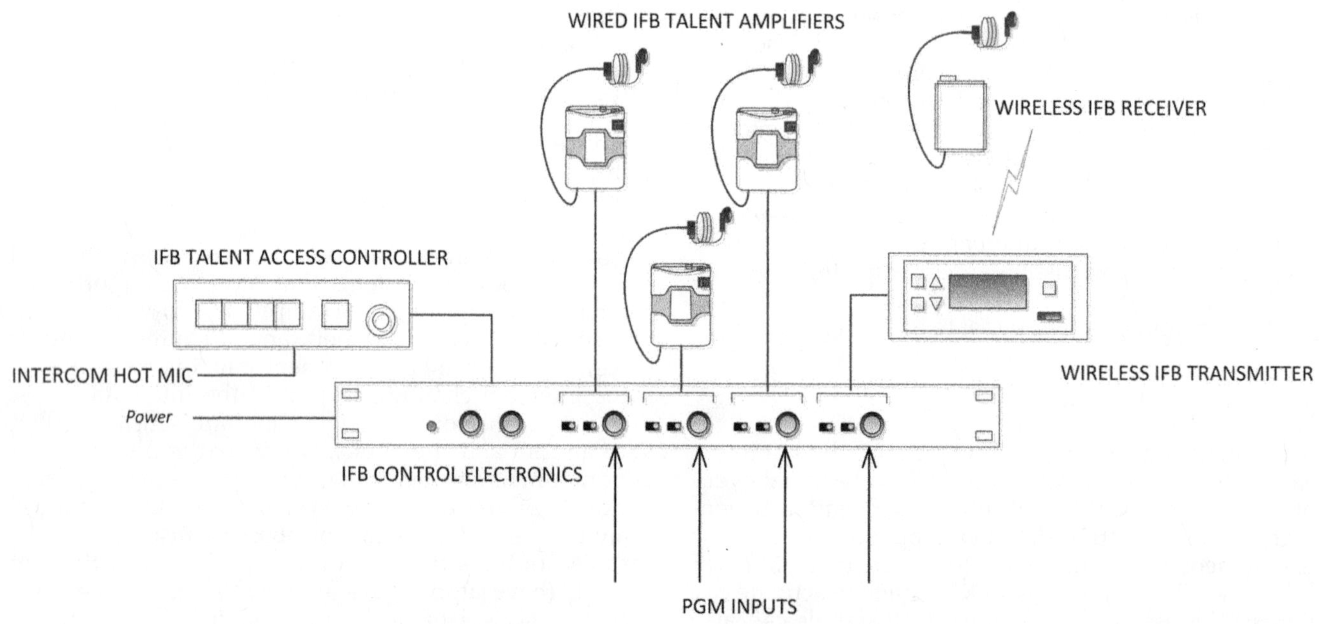

FIGURE 5.19-4 Example of Party-Line IFB

watching the event on a monitor. In the simplest operation a commentator may create the program mix on the commentary unit itself, and a program output is ready for broadcast.

Commentary/Announce Box

Commentators connect their headphones and microphone to the commentary/announce box. In-the-ear earpieces are the most commonly used listening device, but communication headsets with high audio quality boom microphones are often used for sportscaster applications.

These devices allow announcers to manage their own microphone's on/off button and monitor a variety of sources. This on/off button is typically called a "cough button" and is settable as a momentary action button that turns off the microphone input only when it is held down. In addition, a commentary/announce box will feature a coordination or TB (talkback) circuit. This is the main off-air communication channel between the commentators, the engineer, sometimes the audio A1, and the coordinator. The coordinator is an editorial position responsible for the commentators, usually a producer or director. Line-level outputs are provided, often with a mixed output of all mics.

For applications that require complex reporting, commentator user stations with multiple-input headphone mixers are used. Some units provide up to two or three commentary positions, allowing them to set up individual headset mixes. In these applications a commentator may receive the program audio feed as broadcast, the international sound of the stadium crowd, advice on mic levels or popping from the engineer, a local input from a sports statistician and of course off-air instruction from the coordinator/producer. Most units are capable of split-ear operation, individual level and pan controls for each source and four-wire connections to intercom systems and external line devices.

DIGITAL PARTY-LINE SYSTEMS

Various intercom manufacturers provide a digital party-line system. They offer hardware with similar attributes to an analog system, that is, belt packs, speaker stations (biscuits) and remote stations. Of these, the human interfaces are also familiar—headsets and gooseneck microphones with loudspeakers.

Similar in concept to the Partyline and TW systems, one system offers interconnections via standard shielded twisted pair mic cable, maintaining the familiar deployment and workflows of Partyline and TW systems. Instead of a single DC voltage applied to pin 2, one implementation applies a 30- to 48-volt DC phantom power to both pins 2 and 3 (with the shield as return) to power belt packs and speaker stations. The DC voltage is generated from a power supply or a master station. Audio is transported as a modified AES3 format that is capable of bidirectional transmission on a single cable, and all signaling, such as remote mic kill and call signaling, is transmitted as user bits in the modified AES3 data stream. The digital audio is two channels each at 16-bit resolution sampled at 48 kHz. An AES3-based party-line system requires specific cabling considerations and active infrastructure devices.

Another intercom manufacturer transports the intercom audio and signaling as packetized Ethernet using CSMA/CA (carrier sense multiple access with collision avoidance) power line technology, where the Ethernet packages are modulated onto the power line. This allows Ethernet-based devices to be connected using standard twisted pair shielded three-pin XLR cables using the same topologies utilized in analog party-line systems, including passive splits, daisy chains and extended cable distances. Ethernet-based systems allow for interconnection of devices and system control with standard Ethernet links as well as the capability of being powered from standard Power over Ethernet (PoE) switches.

Digital party-line systems offer improved audio quality and resiliency to external interference that the unbalanced party-line systems are susceptible to, as well as minimal cross-talk between adjacent channels.

WIRELESS INTERCOM

In most applications, the wireless intercom is an extension of the wired system, used by those staff members who require mobility for safety or convenience.

Wireless intercoms are typically used in a group/party-line communication workflow as detailed in the Analog Party-Line Systems section above. They are seldom used as point-to-point communications; however, there are wireless systems available that have these capabilities.

Systems are generally comprised of a "base station" and a limited number of user wireless belt pack/headset stations that can work with and are "paired" to the base station.

Wireless intercom communication is full duplex; however, some systems offer different operational modes whereby the system can be used in a half-duplex manner, allowing more wireless users on the system.

Wireless Intercom Architecture

Wireless intercom systems are commonly constructed around a base station that includes internal or external radio modules to communicate to users wearing wireless intercom belt packs as well as to provide interfaces to other intercom systems and peripherals.

Wireless intercom systems are commonly equipped on a per-channel basis with a two-wire loop-thru (three-pin XLRM&F) interface to analog and digital party-line systems and four-wire I/O (typically an RJ-45) for connection to matrix intercoms, additional wireless intercom base stations and other third-party audio devices. The two-wire interfaces generally are compatible to both RTS TW and Clear-Com analog party-line systems and have manual or automatic nulling control.

A dedicated program audio input is generally available at the wireless base station to feed analog audio to the wireless belt packs.

Stage announce (SA) is a system feature similar to the wired party-line systems presenting an audio output. These outputs are always analog line level and often do not have an output-level trim control. In operation, the audio from a belt pack, which is the headset microphone signal, or audio from the base station microphone appears on an output XLR3 male connector on the base station. Audio from this connector is typically wired to a paging system or series of paging systems for the benefit of anyone on a studio set, scene shop or other work areas. A relay is often associated with the SA output so that the activation of the SA can key a two-way radio or turn on an on-air light.

Analog Wireless Party Line

Analog wireless party-line intercom systems have utilized VHF and UHF frequency bands with FM or AM transmission modulation to construct wireless party-line systems. In these systems each individual user's wireless belt pack transmits on a discrete frequency its talk to a central base station. The base station mixes all of the belt pack talk paths with its external two-wire and four-wire inputs and transmits the mixed audio to all belt packs on a separate, shared transmission frequency, as shown in Figure 5.19-5, as well as to the external two-wire and four-wire outputs. The belt packs' sidetone is generated in the base station and allows the user to verify that the transmission is being made.

Dual-listen wireless intercom systems are common in television production, as they allow the user to monitor the production's main channel as well as having a channel dedicated to the user's department, each with individual listen-level controls. This requires a base station with two radios for transmission of the discrete channels to all belt packs.

ISO (isolation) or WTA (wireless talk around) talk options are provided to the belt pack operators in order to be able to talk discreetly with other wireless users on their base station without it being heard on any of the base station's external two-wire and four-wire interfaces.

Analog wireless intercom systems can be expanded in size by adding base stations to allow for more duplex wireless belt packs. Ordinarily, different departments within the production may be grouped together within a single or multiple base stations, depending on the number of users required. The department's intercom channel(s) is then shared between the base stations, and one base station is configured to transmit the combined channel(s) to all belt packs in the department.

Up to six RF carriers can be used by a four-user, full-duplex, dual-listen analog wireless intercom system. In some systems, belt pack talk paths can be shared with a mechanism that momentarily locks another belt pack from transmitting to cut down on use of RF carriers and base stations.

Digital Wireless Party Line

Digital wireless party-line intercom systems follow most of the same principles of the analog wireless party-line systems, yet make for efficient use of the available RF spectrum by using unlicensed spectrum with common technologies and components that allow the systems to coexist with other users.

A common transmission scheme used for digital wireless intercom is spread spectrum frequency hopping (SSFH), a technology originally developed for military guidance systems to make them insensitive

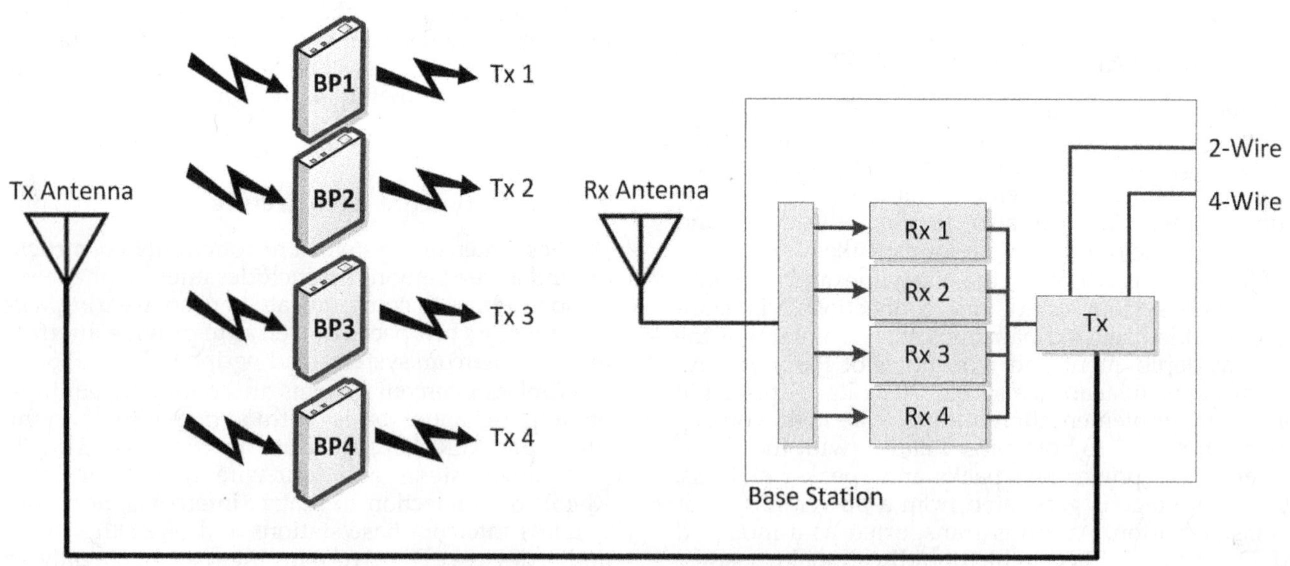

FIGURE 5.19-5 Analog Wireless Concept

to frequency jamming. SSFH-based wireless intercom systems transmit digitized audio samples for all intercom channels and belt pack return talk paths in a single transmission burst serially on a narrow-band RF carrier. This is then repeated on various frequencies in a defined pattern known by the base station and the belt pack. Depending on the implementation, SSFH may be used in conjunction with redundant transmission where a second transmission pattern is utilized in parallel, or redundant transmission of the digitized audio on the same carrier after the primary transmission. The most recent versions additionally include a mechanism that validates the next frequency in the SSFH pattern before using it for transmission or moving onwards to the next frequency. Frequency bands that are commonly used for SSFH include 908–928 MHz (U.S. only), 2400–2480 MHz and the 5 GHz range.

Another common digital transmission scheme is the DECT (digital-enhanced cordless telecommunications) standard. Originally developed for cordless phones, the DECT multicarrier TDMA (time division multiple access) technology allows for very efficient spectrum utilization over two to ten common carriers in the 1880–1930 MHz range depending on regional allotments. The DECT protocol automatically assigns timeslots and carriers to individual devices.

Digital wireless systems require a mechanism to pair the wireless belt packs to the base station of the system; this is normally done by assigning credentials to either device by the manufacturer or the user to ensure that the devices only operate with the system they belong to. The credentials can be used as well to establish keys for encrypted communication.

Digital wireless party-line systems often allow for the belt pack talk paths to be shared between multiple belt packs. This is where more belt packs can be used with a base station than there are belt pack talk paths. The talk paths are then taken up by belt packs on a first come, first served basis.

Digital wireless party-line systems exhibit audible latency due to the conversion, processing, buffering and transmission of the digital audio. For this reason, the belt pack user's sidetone is generated locally in the belt pack.

Digital Wireless Matrix

Digital wireless matrix intercom systems utilize the RF transmission technologies for digital party-line systems with a discrete audio send and receive for every single belt pack. This allows the base station to become an intercom matrix, or a card in a larger hybrid wired and wireless intercom matrix. In these systems, the belt pack controls cross points, or groups of cross points on the base station's router according to the configuration of the system. As every belt pack receives a discrete audio transmission from the base station (as opposed to a "shared" transmission, as in party-line systems), point-to-point communications can be established between different wireless devices, or wired I/O. Various configurations of party lines, fixed groups, IFBs and point-to-point communications can be realized with this system architecture using software that allows for real-time changes to the system over the air. See Figure 5.19-6.

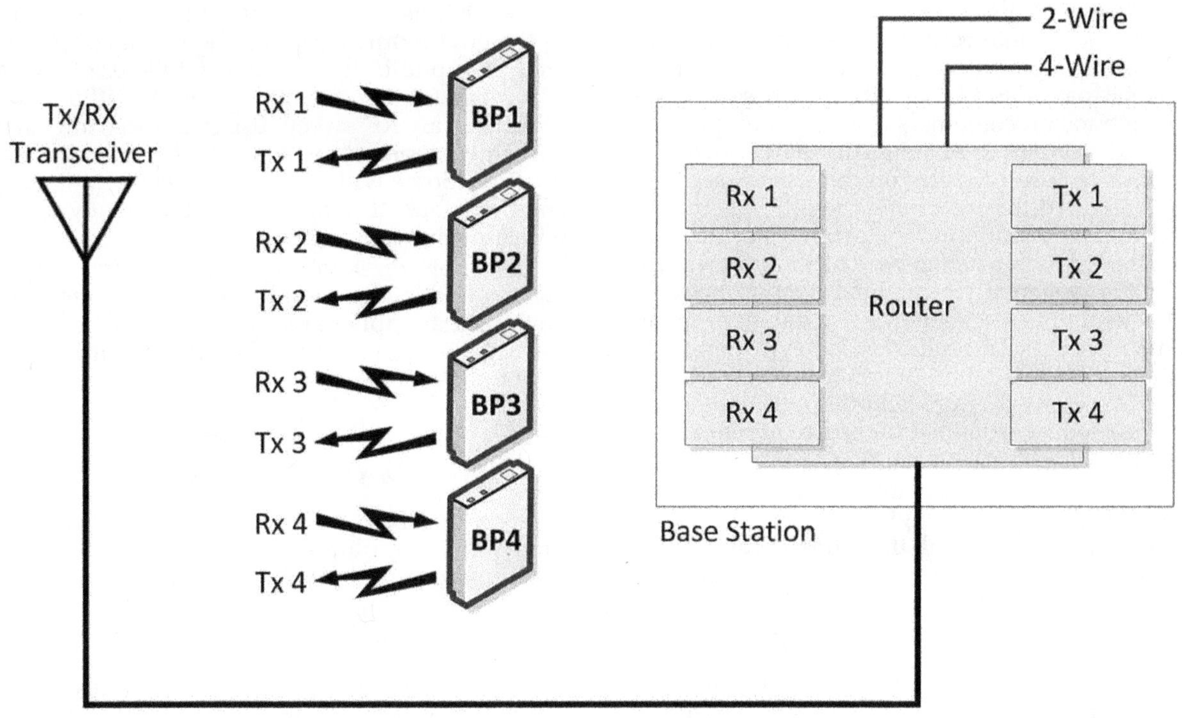

FIGURE 5.19-6 Digital Wireless Matrix Concept

Wireless IFB

Occasionally within a fixed facility and within some remote field applications, wireless talent cueing methods are chosen over wired systems. Traditionally, a wireless IFB system utilizes a VHF or UHF transmitter whose source audio is the IFB output of a matrix or two-wire intercom system where the IFB is created and actioned. The talent wears a battery-powered wireless receiver belt pack with a volume control to hear the IFB, via a replicable acoustical tube that connects to a cable-mounted transducer.

Alternative wireless IFB receivers include miniature earpieces with VHF receivers or induction loop receivers. These devices are popular with many producers and talent due to their low profile, yet they require care with battery management, sealing to the ear canal and transmitter location.

Duplex communication with talent is often realized by using a pre-fader send of the talent's microphone to the matrix key panel of anyone who has to communicate with the talent.

Currently many remote field applications rely on IP-capable communication devices like smartphones and tablets for talent connections to the producer and director. See the IP User Stations section below.

Extended Wireless Coverage

Wireless systems, be they analog or digital, have a finite range that will be heavily dependent on the environment the system is being deployed in. At times, it can be impossible to make a single system cover all of the areas where coverage is needed for a production, therefore requiring the deployment of an auxiliary distribution system for RF.

Analog wireless intercom systems may need distribution of the discrete wireless send and receive signals to/from belt packs and base stations. These systems will utilize transmit combiners from the multiple base stations and splitters to antennas in the various coverage zones required for the production. A separate system is required for combining the multiple receive antennas and distribution to the receivers of the base station. These kinds of antenna distribution systems are normally designed for minimal overlap between zones to ensure phase coherence of combined signals whilst maintaining unity gain at each device and keeping system noise to a minimum. Fiber converters that modulate analog RF to/from light for optical transmission can be used for extended cabling requirements.

Digital wireless intercom systems can use similar distribution for multiple antennas to extend coverage. In most digital systems, a single antenna is used both for send and receive, making the infrastructure different. Minimal overlap between antennas and unity gain is very critical in antenna distribution systems for digital wireless intercom systems to ensure high-coherence reception.

Various digital wireless intercom systems allow for an alternative distribution of expansion of wireless system coverage; this can include allowing the radio transceiver of the system to be remoted from the base station to allow it to be optimally located for coverage. The interconnectivity between base stations can be digital on TDM transmission protocols such as E1/T1 or via packetized Ethernet.

Some digital wireless intercom systems allow the belt packs to automatically switch between different transceivers that the belt pack considers to be of the same system. The belt pack will be able to monitor the signal strength and quality of the transmission from any of the transceivers that it is registered to and able to lock onto. Once the belt pack makes the determination to switch between transceivers, the switch will break the first connection and establish the second one to the second transceiver.

Another manufacturer of digital wireless intercom systems allows for automatic make-before-break switching between transceivers, where the belt pack requests its audio data stream from the transceiver it is about to switch to, while still being connected to the first transceiver. During the switch, the transmission to and from the belt pack is duplicated by both transceivers and the received audio from both is mixed together by the belt pack. Once the belt pack has successfully completed the switch, the first transceiver will tear down its connection and its slot will be available for another belt pack.

Wireless Coordination and Considerations

Wireless transmission of different systems should be coordinated to avoid potential conflicts and to work with environmental and regulatory considerations.

Intermodulation calculations should be carried out for VHF and UHF analog transmissions for the frequency bands of the equipment being used, as well as the environment. Environmental data can be gathered from publically available databases of allocations of various frequencies by regulatory authorities or by capturing an RF sweep using a spectrum analyzer. Software programs are available for this kind of coordination. Some will import RF sweeps from analyzers and the public RF databases and potentially control the wireless devices being used.

Users of wireless equipment should always be aware of national, federal and local regulatory considerations that apply to use of wireless devices and only use legally approved equipment within its specified limits.

MATRIX INTERCOMS

Today's matrix production intercom systems utilize time division multiplexing (TDM) technology to provide interconnection of high-performance intercom stations, including the ability to provide custom-mixed intercom signals to specific intercom stations and other connected devices within the system. TDM is a form of multiplexing many signals, contributed by multiple TDM-connected components combined into a single data frame that is shared once per audio sample amongst all devices sharing the TDM bus. The utilization of TDM technology permits relatively large, high-performance intercom systems to be constructed. Today, nearly all matrix intercoms are based on TDM or similar technology.

Without question, the best method to understanding matrix production intercom is from an audio perspective. These platforms are referred to as full summing, non-blocking matrices when stripped to the core. This refers to the capabilities that any input or number of inputs could be routed to any output or any number of outputs with full cross point level control/adjustment, as shown far right in Figure 5.19-7.

The term "cross point" refers to a one-way audio path from one port's input to another port's output. Cross points exist between every pair of ports, virtual or otherwise, in the system and are connected and disconnected as needed to provide communication paths between system ports on the matrix backplane. The outstanding feature here is the cross point level control/adjustment, which is not typical of traditional audio routers. When you have many people talking, the ability to adjust listening levels (cross points) is critical to efficient production cueing and directions. This becomes essential with multi-talent IFB work and group/conference communication.

The basic hardware elements of matrix intercoms are:

- Interface frames, which are used to host interface modules.
  - Because the intercom is considered "critical communications," each matrix frame typically operates with two independent power supplies. These must be connected to a main and backup power source for redundancy. If one of these power supplies fails, the second supply automatically takes over.
  - Before one can interconnect any user stations or external communications, one must know that the heart of any matrix intercom must have a central processing unit (CPU). The central processing unit provides the serial data and an Ethernet connection for PC control. The CPU stores complete system configurations in its memory. Each matrix system contains at least one central processing unit, located either on its own removable CPU card or in internal circuitry. Most if not all matrices are supplied with two CPU cards (in a master and slave relationship), ensuring fail-safe operation. In the event of a complete power failure, a matrix may automatically restore itself to a complete functioning system by retrieving configuration information from its nonvolatile memory.
- Various circuit cards, sometimes called client cards, used for I/O.
  - There are a wide variety of different client cards available for matrix systems. These client cards are the I/O and interfacing connections to various user intercom stations and third-party sources and destinations. They can be used for "trunking," that is, matrix-to-matrix connections, described in the Trunk Systems section below. These client cards offer various audio format conversions as well. Client cards are described in terms of "ports." Ports refer to the number of connections available to external devices to and from the matrix. Typically, a port is assigned a function within the configuration

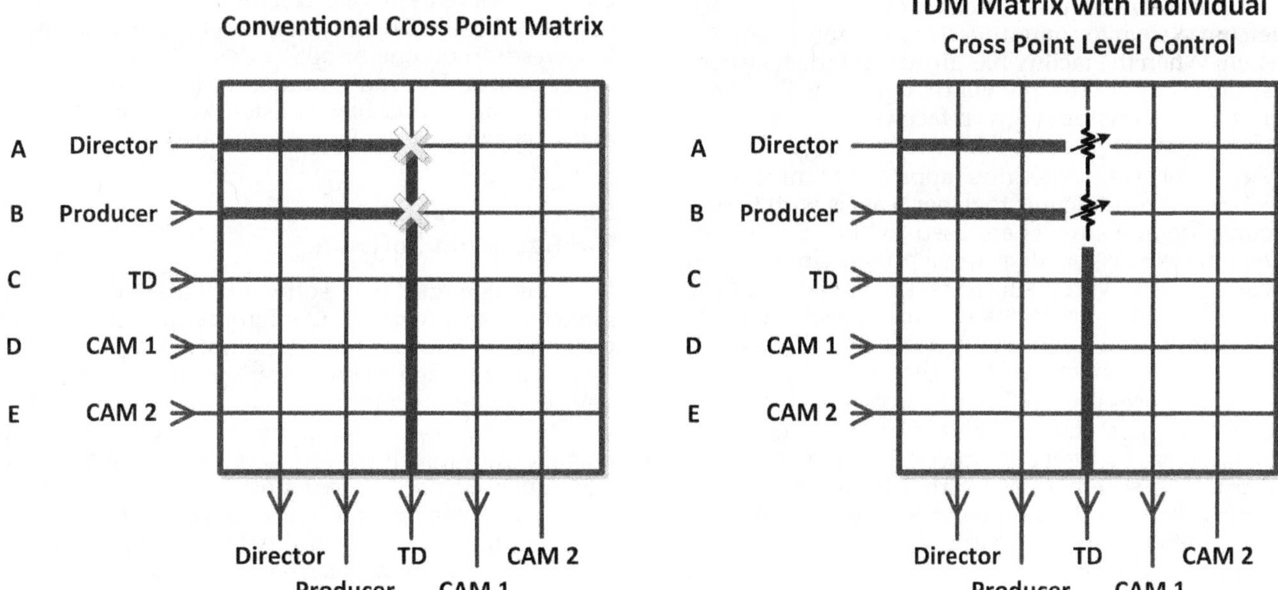

FIGURE 5.19-7 Conventional Matrix and TDM Matrix
Taken from the Telex *Handbook of Intercom Systems Engineering*

software. This function specifies what type of device is connected to the port, such as, but not limited to, a user key panel, an interface, a four-wire or a trunk. Software is used to set the parameters of the function for that port. These include analog client cards (typically referred to as four-wire ports), AES client cards' E1/T1 connections and MADI (Multichannel Audio Digital Interface) client cards and audio over IP cards.

  - Rear-panel connectors. A matrix connects to remote devices such as user key panels, interfaces, general-purpose inputs and outputs, local area networks (LANs) and other matrices through its rear-panel connectors. RJ-45 connectors are found on analog four-wire, IP circuit cards, and Ethernet for PC control, while digital connections may include BNC or fiber or both for MADI I/O and BNC for AES I/O. Typically D-sub connectors are used for externals functions like GPIOs.
- User control panels.
- An external computer (PC), which hosts the configuration software.

A feature of audio matrices, production intercom matrices included, is the function of audio conversion. Because the I/O to modern intercom matrices includes various analog and digital audio formats, the circuit I/O cards convert whatever audio format is being generated into one of the many digital formats used by other circuit cards. You may have analog program audio input put on a MADI stream to feed the audio console, as an example.

Trunk Systems

There inevitably comes the need for users in one matrix intercom system to communicate with users in another system. When the facility has grown and thus the need arises to connect multiple matrix intercoms that serve different areas within a given facility or separated by countries, we employ trunking methodology.

An older yet sometimes appropriate means for linking matrix intercoms together was/is with tie-line circuits. Tie-line circuits are used to "tie" two matrix intercom systems together over a private circuit. These circuits come in various formats, such as copper four-wire or digital connections or, nowadays, IP circuits. These simple one-to-one connections are surpassed by what is better known as "intelligent trunking."

Trunking uses proprietary protocols and algorithms, and the support for trunking is tightly integrated into all manufacturers' matrices. Most manufacturers' trunking schemes are referred to as "intelligent," meaning they function automatically, establishing the trunk, continuously monitoring and reporting on status of trunk utilization, and finally releasing the trunk when the conversation is over. Multiple calls to one destination use only one trunk line. Also, calls from one source to multiple destinations will use only one trunk line. This enables an efficient use of available audio bandwidth. Trunk lines may not only support point-to-point calls but also conference calls and group calls.

Most matrices allow trunks to be indirectly routed to the final destination when there is no direct link. In some systems, if a direct trunk line connection exists, then this is used. If there is no direct connection, the software knows that there is a possible path using another matrix as a hub connection.

Several intercom manufacturers have implemented a number of methods for interconnecting multiple intercoms using AES and MADI digital circuit cards, E1/T1 with E and M signaling, IP circuit cards, and optical fiber circuit cards in order to provide varying degrees of fault tolerance and redundancy. Some of the high-density circuit cards like MADI and IP may be partitioned to allow multiuse, such as X number of ports for user stations and X number of ports for trunk circuits, allowing flexibility in design and affordability.

In general, as one plans for a system that requires a multiple trunked system, a maximum of as many trunk lines are needed as there will be different conversations.

Matrix Intercoms Operations

One of the most basic and powerful operations provided by a matrix intercom is point-to-point communication. Keys on intercom panels can be programmed either from the intercom configuration computer or from the panel itself to allow direct communication with any other user on the system. In the case of trunked intercoms, users from one intercom can be programmed to appear on keys on intercom panels attached to a different system. This simply and easily enables communication between users who might be as much as half a world apart.

It is worth noting that point-to-point communication may not be as private as it might seem. Other users of intercom panels on a system can turn on a listen key to eavesdrop on one or both sides of a point-to-point conversation. The configuration software may offer a privacy feature that blocks listen key access to ports that have this privacy feature activated.

Matrix Control Systems—Configuration Software

It is the configuration software that makes matrix intercoms so powerful. Configuration software provides online and off-line configuration capabilities to create and manage communication pathways between devices including matrices, interface cards, interface modules and user panels. It gives you the ability to manage multiple intercom systems and assign and reassign users to different ports, as well as dynamically add intercom hardware to your system. Basic functions include the capability to load preset configuration files, which means saved configurations can be uploaded to the live application at any time. The control software also offers extensive diagnostic and maintenance capabilities, including the ability to monitor real-time performance of intercom hardware including

I/O cards, controller cards and intercom panels. It also gives a user the ability to monitor cross point status and to enable and disable cross points from the control system itself. The intercom control software can also set audio levels for specific cross points and intercom ports as well as for party-line conferences.

Add-on software packages are offered by many manufactures to control subsystem functions like trunk management, program logic statements or to monitor and control the distribution of conferences and IFBs and audio from multiple platforms, such as tablets, PCs and intercom panels. Nowadays users can drag and drop graphical elements to add ports to conferences and add sources to IFB destinations as well as monitor return feeds, move ports from one conference to another, route four-wires and change labels. By using the scroll wheel or on-screen fader, users can easily change IFB dim and source input and output levels to IFBs.

Matrix Intercom User/Subscriber Control Key Panels (Traditional)

User control panels are varied and so complex in feature sets that it is beyond the scope of this chapter to explain every programmable feature from every manufacturer. We shall only describe most of the common attributes of user control key panels.

The most visible part of the intercom, and the one piece of equipment that most users associate with an intercom, is the intercom user control panel, examples of which are shown in Figure 5.19-8.

User control panels are the human interface hardware in intercom systems used by production staff to communicate while performing their function (creating a show). In most applications it is the only piece of equipment that ties together production staff. Control key panels are populated with communication paths to the various individual personnel and/or groups of people with similar tasks (like camera operators). These communication paths are accessed on the user panels by the use of "keys."

Keys and Buttons

User control panels are fully compatible with digital matrix systems but not interoperable between manufacturers. They are available in 8-key, 12-key, 16-key, 24-key, and 32-key pushbutton, rotary, and lever key formats. Keys refer to the number of control positions on the face of the unit. A typical "key" would include the following:

- A display (sometimes LCD, sometimes OLED, sometimes TFT, depending on the manufacturer) and panel series from a manufacturer showing a destination or function.
- A talk button or lever function (usually down for talk); a listen button, encoder or lever function (usually up for listen). Some user control panels have separate listen and talk keys, and some user control panels use keys that share talk and listen functions. Some user control panels use the display itself as the talk/listen button. These buttons/levers also activate other functions such as GPIO assignments/logic assignments/audio routing, which may be independent or in combination with voice communication. Keys can be configured for momentary or latching action.
- A listen-level control (sometimes a rotary encoder or up/down button). Sometimes a lever key is also the listen volume control, and/or a rotary listen-level control may be the listen on/off button.

Most of these panels can be programmed either from the intercom control system or at the panel itself. Panels typically use LED, LCD, electroluminescent, or fluorescent displays to show intercom mnemonics and tallies.

Status LEDs (Tallies)

Status LEDs (tallies), which may be the display itself, indicate the status of a key, audio route, or menu

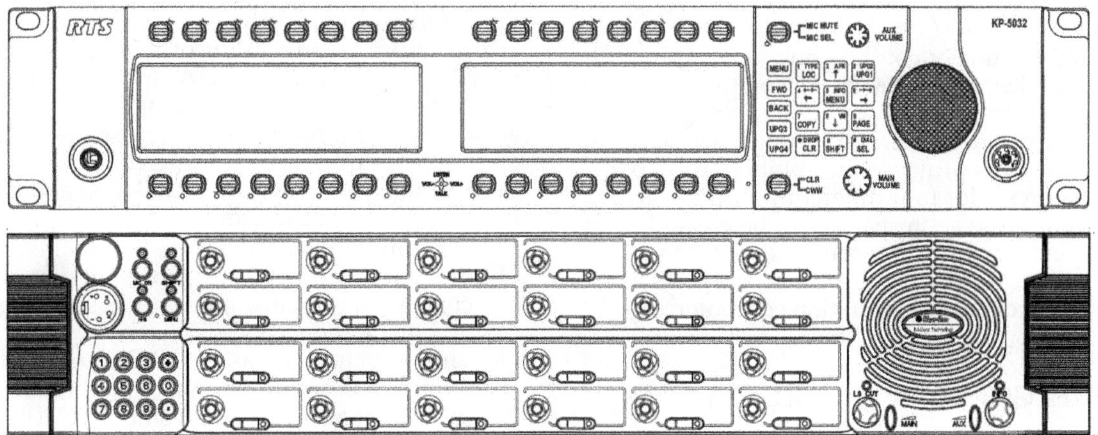

FIGURE 5.19-8 Examples of Matrix User Control Key Panels

Courtesy of Clear-Com and RTS

option. Depending on the manufacturer, these indicators can be multicolored.

Indications include but are not limited to the following:

- A listen path (audio route) is active.
- A talk and listen path (audio route) is active.
- Key is either not configured or the menu option is not selectable.
- Either user action is required, or there is an incoming call or call signal.
- Key is configured as a talk key, or a menu option is selectable.
- Key is configured as a listen key, or menu option is selectable.
- Key is configured as a talk and listen key, or menu option is selectable.
- Menu mode is active.

Control Key Panel Audio Inputs and Outputs

Intercom user control panels typically have connections on the rear panel that provide access to a variety of features. These include auxiliary (*aux*) audio inputs and outputs, an external speaker output, an external headset connector, a footswitch input, GPI inputs, relay outputs and a microphone preamp output (referred to as *hot mic*). Typically, the hot mic audio from intercom panels feeds a facility/studio IFB system for talent cueing.

Intercom panels offer the ability to switch between the panel microphone and panel speaker and a headset, with the ability to mute the panel microphone and mute/dim the loudspeaker.

Setup

Matrix intercom user control panels have many properties that are programmable, which include basic panel options, global options, menu options, audible alert options, tally options, fast key assign options, DTMF options, gain options, logic input options, voice operated switch (VOX) options, auxiliary relay options, and so forth.

Matrix intercom user control panels can store multiple pages of key assignments. A 16-button control panel may, for example, have four setup pages. Each button on the panel on each setup page can be given an assignment. This provides the ability to have panel assignments preset for four different productions. In a matrix system these setup pages can be created from either the panel or from the configuration computer. Similarly, a given setup page can be activated from either the panel or the configuration computer.

INTERFACES TO TWO-WIRE SYSTEMS

Interfacing generally involves interconnection of separate production communications systems. The most common interface method to wired analog party-line systems is to convert the two-wire signals to four wires. The challenges with interfacing intercoms are conversion, level matching and signaling.

The best means for conversion is the two-wire-to-four-wire converter, also known as a "hybrid."

Hybrids/2W-to-4W Interface

The term "hybrid" refers to a device that converts two-wire to four-wire audio and vice versa. Analog hybrids initially used transformers, as shown in Figure 5.19-9, and later used op-amps to convert between two-wire and four-wire audio. Digital hybrids with digital signal processing (DSP) chips are most commonly used to perform two-wire/four-wire conversion.

One problem with hybrids is poor trans-hybrid loss. This is a measure of the loss or isolation between the transmit and the receive ports on the four-wire side of the circuit. Trans-hybrid loss depends on signal cancellation accomplished by defining the line impedance and mirroring it in a balance network. The variations in impedance presented on the two-wire side of the hybrid make balancing difficult and often result in poor trans-hybrid loss. See Figure 5.19-10.

Hybrids are also used in intercom systems as interfaces to telephone networks (see the Telephone Hybrid section) and as interfaces between two-wire and four-wire intercom systems. The inherent problems associated with two-wire to four-wire conversion justify why it is better to use the four-wire interface that is available on most camera control units (CCUs) instead of a two-wire interface when connecting CCUs to a matrix intercom system (more on this under Cameras below).

Null

Nulling refers to adjustments made in a balancing network to achieve greater trans-hybrid loss by making resistive, capacitive and inductive adjustments to match the impedance of the two-wire side of the circuit. Inductive, resistive and capacitive nulling affect low-, mid- and high-frequency bands. Modern hybrids are digital and are capable of auto-nulling.

A hybrid's ability to isolate the transmit signal from the receive signal in the two-wire-to-four-wire interface is critical. The quality of this isolation is technically known as return-loss. A sidetone nulling control fine-tunes the circuitry to best match the devices to the acoustic conditions near the intercom, as well as to the electronic conditions on the intercom line. They should be set at the time of system installation and adjusted as is comfortable for the user.

Four-wire audio is more or less defined as a pair of conductors carrying an input/receive signal and a second pair carrying the output/send signal. This hybrid circuit connects the four-wire audio to the single wire in such a way as to variably restrict the user's reception of his or her own voice on the intercom line, which is often referred to as "sidetone." There should be no component of the talk signal in the listen signal. The basis to remove the talk signal is to add an inverse polarity copy of the talk to the listen. The level of the

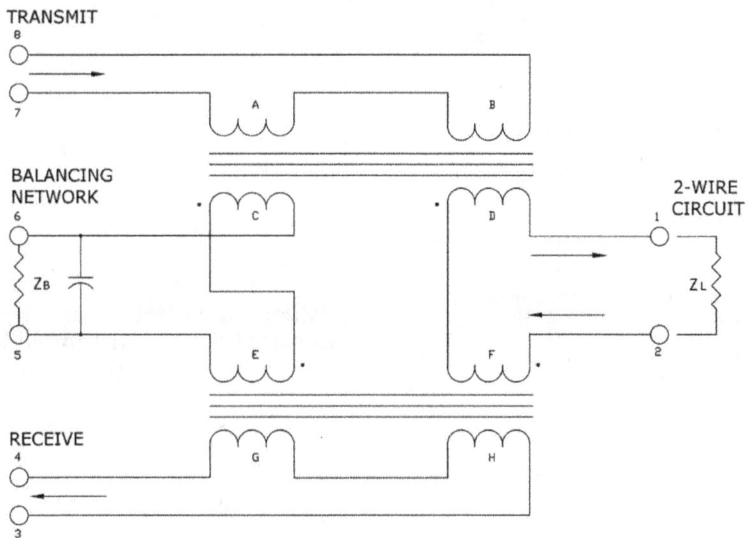

FIGURE 5.19-9 Transformer Analog Hybrid

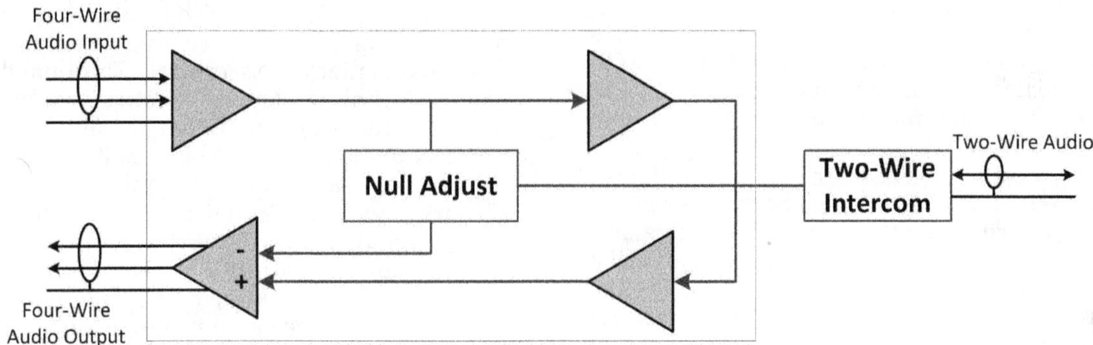

FIGURE 5.19-10 Two-Wire-to-Four-Wire Interface with L/C/R Null Circuit

inverted talk signal must be exactly the same amplitude as the two-wire circuit. High gain between the send and receive poses a risk of oscillation or "howling" resulting from acoustic and/or electronic coupling within a headset or between a speaker and a microphone.

With manual nulling devices, there are the following accessible controls: Separate "R" (resistance), "L" (inductance) and "C" (capacitance) controls compensate for each component of the line's impedance, providing the best null possible. Ideally, there should be no portion of the talk signal in the listen signal. The variation of the two-wire line phase coherency is affected by the cable capacitance (length of wire and gauge) and inductive elements of the line. The "C" control compensates for cable capacitance; the setting depends on the length of the line. The "L" control compensates for the low-frequency inductive and capacitive elements the wiring of the external party line presents to the line.

The importance of two-wire termination (as explained above), lack of, or double termination also influences the null result.

Every time communication goes through a two-to-four-wire conversion (hybrid), the signal is degraded somewhat.

Cameras

One of the reasons that interfacing with camera intercoms is such a problem is the lack of standardization. There is no standard design for camera intercoms among different camera manufacturers. In addition, some manufacturers don't even use the same intercom design within their own line of cameras.

The majority of current broadcast-grade cameras have built-in intercom capability that is acceptable for most applications. If we are designing with a matrix, a direct connection is usually possible with high-end cameras, as they generally have a four-wire connection available. If we are designing in the two-wire world, it may be more difficult, as interfaces may be needed. Several camera manufacturers are designing their circuitry so that it will connect directly to the currently popular two-wire systems.

If the cameras are not compatible with the chosen intercom system or if the built-in facilities are inadequate, we must use interface devices or ignore the camera's built-in facilities completely and use external belt packs at the camera head. This can get a bit messy, and a full discussion is beyond the scope of this chapter.

The majority of large installations and virtually all mobile units use camera isolate systems. A camera isolate system allows an operator (usually the video operator or technical director) to have private communications with any of the camera operators for setup or maintenance purposes without interfering with ongoing communication between the other cameras and the director. In matrix systems the required functionality can be easily programmed.

Cameras and Analog PL systems

The best way to interface a camera intercom to analog party-line systems is, without exception, to access the CCU intercom circuit as a four-wire circuit and then use a two-to-four-wire interface to convert to party line. A four-wire connection is frequently not available on lower-cost cameras.

If a four-wire connection to the camera intercom is not available, almost always the next best choice is to interface to the headset connection on the CCU/DCU. (The headset connection must be a three- or four-wire circuit.)

It is always best to interface each camera individually. Trying to combine the two-wire party-line circuits the CCU/DCU provides into a single party line and then use an interface to convert the camera party line to Clear-Com or RTS party line is the least effective way to interface multiple cameras. There are several reasons.

- First, the more cameras you connect in parallel, the worse the party-line impedance characteristics become, therefore a good null is difficult, if not impossible, to obtain.
- Second, if the impedance characteristics of a single camera are worse than the other cameras (which is not uncommon), the overall party-line impedance characteristics are degraded to the level of the worst camera.

NOTE: Many camera CCUs/DCUs have the intercom connection on a three-pin connector, or on three pins of a multi-pin connector (frequently along with the "tally" controls). The pin-out is frequently identified as +/hot, −/cold and common. Many people assume that this indicates a three-wire intercom circuit, with one pin "send," one pin "receive" and one pin "common" to both. This is not true. The +/hot and −/cold pins are a two-wire party line, with the "common" simply being a shield/ground. There are no known exceptions to this. These problems can be eliminated completely through bypassing the camera altogether. Run a microphone cable with the camera cable, and plug a belt pack in at the end. This is commonly done at remote productions such as a news shoot or sporting event when there is a set on the field.

Analog Telephone Interface

Telephone interfaces allow dial-up telephone system access to an intercom and typically are designed to work on standard (wet) telephone lines. Wet lines are typical tip/ring telephone lines that carry ring voltage, talk battery, and full duplex, balanced audio over two wires to an analog telephone. In the real world, when you send your voice down the telephone line it has a tendency to bleed over into the caller's audio. The hybrid must adapt to the phone line in order to properly separate transmit and receive audio. These interfaces are typically integrated into the intercom control system, allowing a phone call to be initiated from and received by an intercom panel. Telephone interfaces typically offer auto-answer capability, password protection and the ability to remotely signal specific ports via DTMF (dual-tone multi-frequency) tones. Telephone interfaces are often used to communicate with a talent at a remote location. An example circuit is shown in Figure 5.19-11.

The Telephone Hybrid

As mentioned before, a hybrid's ability to isolate the transmit signal from the receive signal in the two-wire-to-four-wire interface is critical. The quality of this isolation, technically known as return-loss or trans-hybrid loss, is measured in dB. A high value is important, especially in applications where multiple two-wire-to-four-wire interfaces are used together.

Hybrids were invented to separate the send and receive signals from the common two-way phone pair. Nowadays, most hybrids are made with active components and are known as active hybrids. Both circuit types use the same principle and achieve the same effect. The balancing network is a circuit consisting of capacitance, resistance and sometimes inductance, forming an impedance network. Depending on the hybrid's application, this circuit can be very simple or it can be comprised of a large number of components and have a very complex impedance characteristic.

If the phone line and balancing network are pure resistances, then, clearly, the phone line and the balancing network must have the same value in order for the signals at the differential amplifier to have the same amplitude and for complete cancellation to occur. The phone line, however, is not purely resistive, but rather is complex impedance, causing both the amplitude and phase to vary as the send signal frequency varies. Two-to-four-wire converters, transformers, repeaters, T-carrier systems and other telco systems are responsible for significant impedance bumps. Loading coils also usually have a deleterious effect on the performance of hybrid interfaces since the coils can create resonant peaks and phase anomalies in the phone line's impedance curve that are difficult to null out. Only when the impedance of the balancing network is the same as the phone line, and the signals at the differential amplifier are matched in both amplitude and phase, will full cancellation of the send signal be achieved. Otherwise, leakage results—the scourge of hybrids.

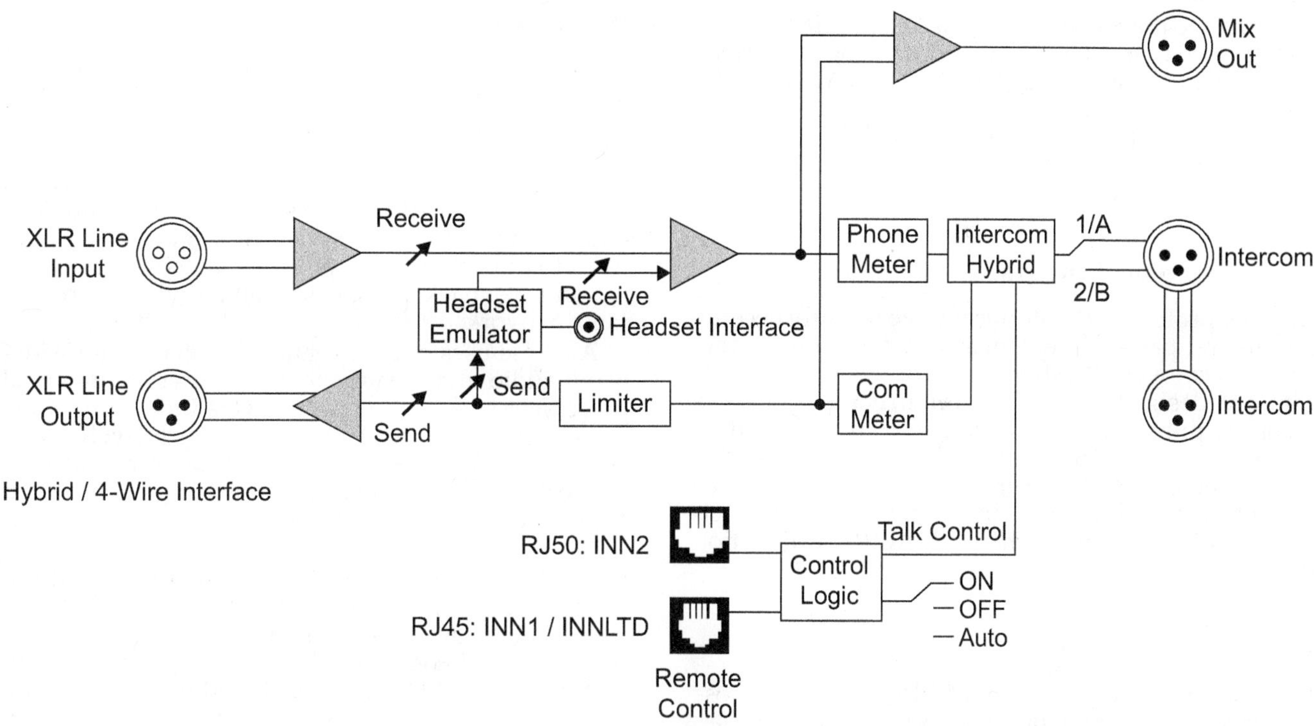

FIGURE 5.19-11 Telephone Hybrid Circuit

Courtesy of JK Audio, Inc.

Analog phone systems can connect to VoIP trunks via multi-port ATA adapters. An analog telephone adapter (ATA) is a device for connecting traditional analog telephones, fax machines and similar customer-premises devices to a digital telephone system or a voice over IP telephony network. In an enterprise setting, an ATA usually has multiple telephone jacks and an RJ-45 connection to a 10/100BaseT Ethernet hub or switch and is used to connect to a local area network (LAN). Such an ATA digitizes voice data and uses protocols such as H.323 or SIP (Session Initiation Protocol) to communicate directly with a VoIP server so that a softphone is not required. An ATA that connects telephones to a LAN is sometimes called a VoIP gateway.

Interfacing to PBX Phones

It is usually possible to interface to PBX phones for on-air use. However, this is best reserved for casual phone use such as for the occasional request or contest winner call. For applications where phone calls are a significant programming element, it is usually better to consider the specialized on-air systems from broadcast-oriented manufacturers.

One reason is that the hybrid interface cannot determine when a new call is selected, so it cannot adjust its null to the new line before the conversation starts. (However, since the hybrid can null on voice during conversation, null will be achieved in perhaps four seconds. This is acceptable if only a portion of the call is to be aired, as is common with on-air requests, contest winner calls and the like.) Another shortcoming of the direct-to-electronic phone approach is that the line switching clunk is not muted, although this is not a problem when calls are not aired directly and sequentially.

Another potential problem is audio quality. The primary weakness is usually noise, most often the result of the data signals cross-talking into the audio. Often, frequency response is limited by too small line coupling transformers or from other causes. Poorly designed digital systems may suffer from quantization and aliasing noise and distortion. Few PBX manufacturers publish specs on audio performance. Since, clearly, this is of importance to those of us who need to get decent quality from phones for on-air use, we'll want to make sure that the audio is at least reasonable. When choosing a new PBX, ask the phone system dealer for audio performance data or arrange to conduct at least a few simple tests yourself.

AES Interfaces

AES3 and AES10 digital audio interfaces have become important as digital audio systems have entered broadcast facilities. Matrix intercom systems often have a substantial number of ports directly connected to digital production audio consoles for program audio and mix-minus feeds. In addition, many operations use the intercom as a monitoring device with remote feeds, talent microphones and program feeds serving as sources to the intercom. Digital AES interfaces eliminate the need for external digital-to-analog converters. It is worth noting that digital AES3 and AES10 audio distribution is currently migrating to IP audio network connectivity. IP audio networks will be discussed in detail in Chapter 4.2, "Planning a Radio Studio or Audio Production Facility."

Internet Protocol Interfaces

Internet protocol (IP) interfaces have recently become an integral part of intercom systems. Manufacturers offer voice over IP (VoIP) and audio over IP cards that are designed to work over private, and more recently, public IP networks. Intercom control data and voice traffic can be sent over these IP networks, which is the basis for user control key panel device connectivity and the trunking of matrices across networks. (More on this in the IoIP/AoIP/VoIP section below.)

GPI Control Interface

Intercom systems typically have GPI (general-purpose interface) inputs and outputs for control purposes. A common use of an output is the triggering of a two-way radio base station. GPI inputs are used to trigger an intercom event from an external contact closure. An example of the use of a GPI input would be an external button push that triggers the intercom to change the program source of an IFB.

INTERCOM OVER IP/AUDIO OVER IP/VOICE OVER IP

This section is not meant to be a study and/or tutorial for the Internet protocol suite (often referred to as TCP/UDP IP). We are aware that the Internet protocol (IP) is "the" method by which data is sent from one network device to another on the Internet. We also understand that audio networking is based on standard Layer 2 Ethernet and Layer 3 IP making it routable across networks. Because of this, it will work with commonly available and existing network infrastructure and devices rather than requiring specific format-compliant switches. IP capabilities can allow for a very effective deployment that can, at times, reduce infrastructure requirements cabling as well as adding abilities that would otherwise be impossible to realize from the standpoint of practical engineering and budgetary constraints.

Intercom over IP can be considered to be an additional method to interconnect a traditional intercom system with a system-specific control mechanism, as well as being a gateway for third-party connectivity requirements that includes audio and more basic logic control. These two main interface categories include the transmission of the fundamental building block of an intercom system, four-wire connectivity, over IP.

IP Gateways

IP gateways are commonly used to incorporate audio in and out of intercom systems. These include wideband PCM audio from routers and mixing consoles for mix-minus feeds, program audio, pre-fade microphone feeds, sources that need to be quality controlled and such. Using IP gateways for this application can additionally allow the system's administrators and users to utilize them for recording, monitoring, metering and playback using compatible IP devices on the same network.

As these uses often demand low latency and wideband audio that require high bandwidth, minimal processing and low IP frame payload to maintain the signal's synchronicity, they are often exclusively maintained on LAN-based IP networks often confined to networks, or VLANs, with a minimal number of switch hops for the best performance.

Commonly used interconnectivity IP standards for wideband, low latency audio gateway in and out of intercom systems currently include AES67, Dante and AVB. In systems without an IP gateway, these sources would commonly be brought into the system with analog or AES standards-based digital four-wire connections.

An intercom system may require IP gateways that are able to connect to lower bandwidth devices on a WAN or over the public Internet. These applications may include four-wire connectivity to/from commentators, connectivity to remotely located two-way radios and to telephone systems, including SIP-based telephony systems.

For these applications, a higher priority may be placed on the resiliency of the connection and the bandwidth used by the circuit than latency and wideband audio. This is accommodated by data compression and buffering mechanisms that commonly utilize industry standard codecs such as G.711, G.722, G.723, G.729 and Opus, which can have user adjustments that allow them to be optimized for the connectivity route of the signal.

In addition to being able to compress the audio to a lower bandwidth signal, add error correction information, adjust buffer sizes to improve utilization of the payload and silence suppression of dataflow, the receive side can also vary buffer sizes to preserve the audio as IP packages are received out of order as well as provide an error correction mechanism to deal with dropped or corrupted packages.

Sometimes limited control and monitoring capabilities are incorporated in these systems as well as stream reservation and control mechanisms, either standard as a part of the codec used or using the manufacturer's proprietary protocol. These methods are essential to allow use of the lower-cost remote connectivity that can be delivered with the public Internet.

SIP (Session Initiation Protocol)-based connectivity adds a standards-based, IP-based control mechanism for telephony on top of the compressed and buffered four-wire audio stream. The standard includes defined command language to dial, establish connections of IP audio streams and to tear them down, allowing the POTS—plain old telephone system—to migrate to IP-based networks.

IP System Interconnect

IP connectivity can be utilized to interconnect the various devices that construct an intercom system. These might include user stations, be they physical or soft user interfaces, antennas for wireless intercom systems, system-specific components and modules as well as intelligent trunking between systems. This normally includes the methods and protocols used for IP gateways, depending on the application, and adds a dedicated control layer for the system and is normally proprietary to the system.

IP User Stations

In addition to the use of IP-enabled conventional user stations, such as key panel control stations, the use of IP opens up a new world of possible user station types. Most IP-capable communication devices could theoretically be deployed to be user stations for an IP-capable intercom system. A non-fictional example would be to convert devices that end users of intercom systems are intimately familiar with and use in their daily life into intercom user stations. Various manufacturers make "soft" user stations in the form of mobile devices and computers that can connect to the intercom system via IP. A very popular application in use today is feeding IFB program and cue to the remote talent's cell phone.

IP Alternative System Architecture

This chapter describes how IP is utilized in a traditional intercom system topology that is built around a central router or mixing engine, which is also the controller of the system. Other IP intercom topologies can also be realized, and those may include the use of IP stream reservation based on the user's channel selection, wherein the incoming audio streams that the user station is subscribed to are mixed in the user station, allowing for a decentralized system architecture to be formed.

IP System Engineering Considerations

IP was developed for packetized delivery of data, where packages can be missed and delivered upon a second request or arrive at the receiver out of order, even via separate routes. This makes it inherently very robust and effective for transmission of data that does not have to arrive on time with other feeds. Synchronous media such as audio and intercom are very time critical, as the human ear is sensitive to gaps and misalignments in audio due to latency and lack of network quality.

As has been mentioned in this chapter, many methods and technologies have been devised to work around this limitation. These include methods to deal with non-optimal network environments as well as the ability to provide system synchronicity and quality of service on networks that need to be understood, specified and maintained for the application at hand as a part of a broadcast infrastructure.

It should be taken into account when specifying an intercom system that the IP infrastructure does add an additional variable, and a potential point of failure, to the system. The risk-benefit ratio of IP interconnectivity should always be considered when specifying a system. At times, direct connections in parts of the system can be worth considering.

IP Future

Adding IP capability is obviously a tremendously powerful addition to any intercom system. IP has been developed to route data and, due to it being agnostic to the data it routes, an immense amount of development is ongoing to make it more capable and easier to use. Intercom concepts and workflows having been inherently devised around routing audio leads to immense synergies for using IP for intercom.

Remote connectivity that was previously impossible due to economics can today be established with the assistance of tethering data from a standard cell phone network with full control and manageability. Inside a broadcast plant, inexpensive IP-based cabling, switching and routing infrastructure that is also used for other applications can be utilized for interconnect between user stations, third-party gateways and the system's components instead of a mission-dedicated infrastructure. Some of this is depicted in Figure 5.19-12.

The use of IP will lead to flexibility that administrators and operators of intercom systems couldn't have imagined years ago, and the broadcast industry can certainly expect the intercom system of the future will further embrace IP for system backbone and interconnectivity.

SUMMARY

Digital intercom systems have become increasingly powerful, flexible and complex. This is particularly true for matrix and wireless intercoms. The size and features of modern intercom systems allow more users on a communications system, and the high-quality communication between those users can take place from disparate and faraway locations. The end result is worth the added effort, as these communication systems have helped make increasingly complex broadcast productions possible. The melding of IP technologies with production intercom has yielded a variety of solutions. They include server-based systems that can set up a local or global intercom network on either private networks or via cloud-based servers. Such systems connect digital matrix frames to intercom key panels

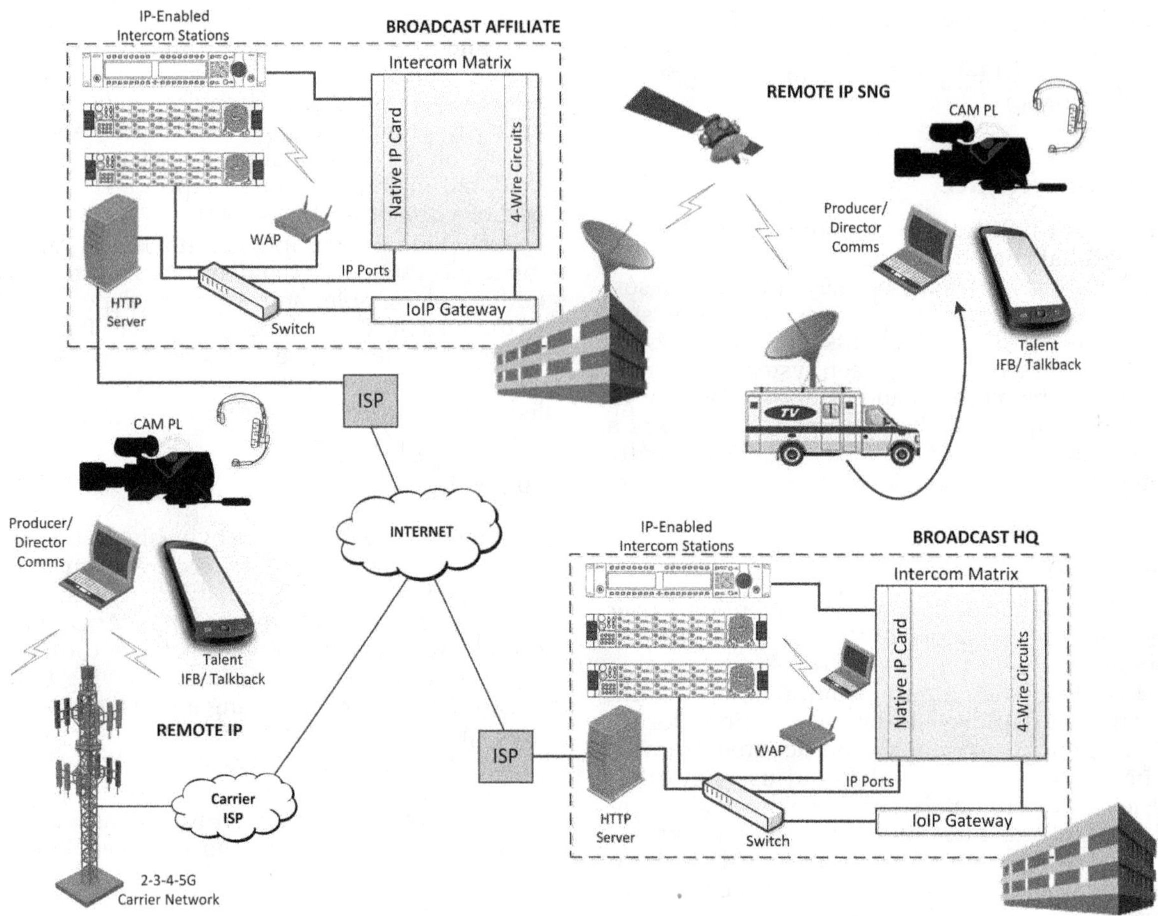

FIGURE 5.19-12 Intercom over IP Circuits

via standard Ethernet networks and to laptops, tablet computers and smartphones with a cellular or Wi-Fi/ IP connection, allowing them to become remote virtual intercom stations and join any desired point-to-point or party-line communications.

Bibliography

The following publications were referred to by the authors in preparing this chapter and may provide the reader with additional useful information:

Clear-Com Partyline and Digital Matrix System Installation Manuals at www.clearcom.com/support/user-manuals.

Telephone Network Interfacing by JK Audio, Inc. at www.jkaudio.com/downloads/datasheets/JK-Audio-Interchange-Datasheet.pdf.

Telex Handbook of Intercom Systems Engineering and RTS Basic Intercom Application Guideline at www.rtsintercoms.com/us/rts/file?i=97125&lg=eng.

CHAPTER

5.20

Ultra-High-Definition Television

MATTHEW S. GOLDMAN

Ericsson
Bedford, NH

At the 2012 International Consumer Electronics Show, consumer television took a leap forward when 4K Ultra-High-Definition TV (UHDTV) displays were introduced. The spatial resolution of 4K UHDTV is 3840 × 2160 pixels, four times the spatial resolution of HDTV (1920 × 1080). While the emphasis at that time was on the greater spatial resolution of the display, the great advantage of UHD has to do with its immersive TV viewing experience or realism. Spatial resolution is just one aspect of several; other enhancements to the content—such as higher frame rates, increased dynamic range, increased color representations, and deeper sample bit depth—greatly increase the realistic viewing experience, and are all components of Ultra HD television.

PROPER VIEWING DISTANCE

An important aspect to achieve the more realistic viewing experience of UHD is to understand the proper viewing distance. The proper TV viewing distance decreases as the display resolution increases. Since the mathematics of human visual acuity are complex, with angular resolution (both vertical and horizontal fields of view) included in the computation of optimal viewing distance, this is outside the scope of this book. A common, more basic rule of thumb (and therefore a bit more inaccurate) is to estimate the approximate proper viewing distance in terms of the number of picture heights. Using this form of measure, a distance of approximately six picture heights from the front of the screen is recommended for standard definition TV (SDTV) and three picture heights for HDTV. This number halves again for 4K UHDTV to one and one-half picture heights. Regardless of how the proper viewing distance is derived, if one sits closer than the recommended minimum viewing distance, then the pixels themselves will begin to be recognized individually as opposed to the human visual system processing the entire picture as a whole, and most likely the viewer will feel uncomfortably close. In addition, if one sits further away than the maximum recommended viewing distance, then the increased resolution of the screen is not discernible and the greater image detail provided by the higher screen resolution is not observed. The proper viewing distance is not a singular number, but rather is in a range of distances between the minimum and maximum stated above.

The decreased viewing distance coupled with the increased screen sizes offered with UHDTV results in the TV screen occupying more of the human central field of vision, which is approximately 90–100 degrees. Figure 5.20-1 demonstrates this, with HDTV

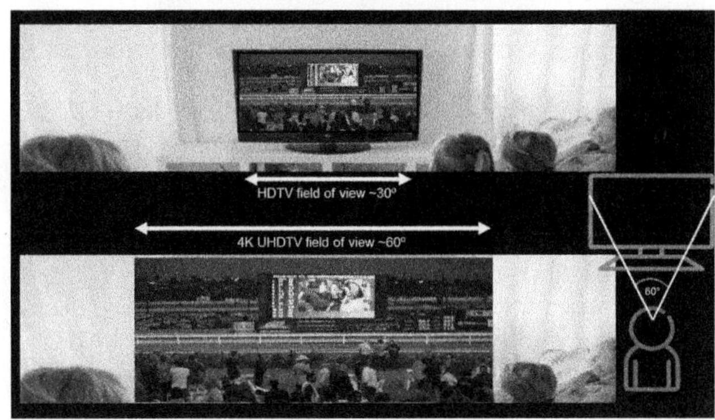

FIGURE 5.20-1 Comparison of 4K UHDTV and HDTV field of view within human central field of vision.

occupying approximately a 30-degree field of view and 4K UHDTV occupying approximately 60 degrees [1] of the human central field of vision. This wider view angle, in combination with the higher spatial resolution, provides a greater sense of presence [2] or sensation of reality and hence a more immersive viewing experience [3].

THE MORE IMMERSIVE TV VIEWING EXPERIENCE

Since perception of spatial resolution is strongly linked to screen size and viewing distance, research suggests that increasing resolution alone will have limited consumer impact on typical TV sizes (40″–65″ in 2016) viewed at typical viewing distances. While early emphasis of Ultra HD was on the greater spatial resolution of the display, other enhancements to improve the viewing experience have since been increasing in importance. These include standardizing on progressively scanned 50 and 60 frames per second (fps) and possibly adding higher frame rates of 100 and 120 fps, to improve motion representation, and a wider color gamut, to more accurately reflect the capabilities of the human visual system (HVS). Over the past two years, however, one aspect arguably has stood out above all others as having the largest impact on advancing the viewing experience or TV realism, and that is high dynamic range (HDR).

The following subsections describe new technologies to television that enable a more immersive viewing experience.

Higher Frame Rates

Because with 4K UHDTV the TV screen size now occupies more of the human central field of vision when viewed at the proper distance, images that move across the display cover a wider arc (greater angular change) in the field of view and therefore the human eye will be more sensitive to motion behavior. As such, higher frame rates than those used in SDTV or 1080i HDTV may be needed to represent fast motion across the greater angular change, or visual artifacts will be observed. The impact of higher frame rates on the overall viewing experience is highly content dependent; however, higher frame rates can reduce temporal problems. For very high motion content on larger screens, such as some sports or nature documentaries, the human visual system can perceive artifacts. Using higher frame rates than those used in 1080i HDTV more accurately represents fast motion across the greater angular change, reducing visual artifacts.

When the camera capture frame rate is too low, motion blur may be introduced (assuming a correspondingly appropriate open shutter time). Motion blur becomes more noticeable with larger screen sizes. Figure 5.20-2 illustrates an example of motion blur.

FIGURE 5.20-2 Example of motion blur due to low frame rate.

Motion blur may also occur due to motion tracking by the HVS. The HVS tracks a moving object continuously, but a TV screen displays the moving object in discrete frames. As such, the projection of the object on the retina will move during the display of each frame, causing a motion blur [4].

FIGURE 5.20-3 Example of motion judder due to low frame rate.

Strobing or motion judder artifacts are introduced when the position of objects in sequential frames are not where they should be based on the natural motion assumed by the human visual system. In these cases, the human visual system perceives the motion of an object to stutter as it traverses as opposed to being smooth and continuous. Motion judder artifacts are observed commonly today when motion picture content (most of which is produced at 24 frames per second) is converted to video through a process called 2:3 pull-down. For example, 24 progressive frames per second of a motion picture in the cadence AA BB CC DD converted to 60/1.001 interlaced fields per second of video results in the cadence AA BB BC CD DD.

Figure 5.20-3 illustrates an example of strobing or motion judder. Motion judder is difficult to represent using a single still image; the judder in Figure 5.20-3 is simulated for illustration purposes using a strobing effect.

Motion judder will be recognized more readily with the larger screen sizes of UHDTV due to the greater angular change of cross-screen motion, although the effect with motion pictures is somewhat masked due to wider angle filming, which means that the angular velocity does not increase in the same ratio as the resolution. Notwithstanding, live sports—which is often considered premium content—is very susceptible to motion judder at lower frame rates, and as such, today's 25 and 30 frames per second (fps) (50 and 60/1.001 fps if interlaced) frame rates are far too low. This issue has been under debate in several standards organizations and industry forums, with frame rates at least as high as 120 fps being vetted, with the weighing of the increase in picture quality to its complexity (cost) to implement. Some TV-related standards already have been updated to include 100 and 120 fps (for 50 Hz and 60 Hz regions respectively).

What is clear from observations today is that a minimum of 50–60 fps is required for high motion content. For other types of content, there may be little, if any, noticeable difference.

Sample Bit Depth

All direct-to-consumer TV production and transmission/delivery today represent digital data values with 8-bit precision (sample bit depth). This results in a maximum of 2^8 or 256 levels for both luminance (brightness) and chrominance (color). With the relatively closer viewing distances of larger displays, another artifact that will be noticed more readily is image banding, also known as *posterization*. Image banding is commonly seen in TV today, particularly in large areas of backgrounds where there is a shallow ramp in chrominance/luminance accompanied by low or still motion, as well as during fades. In Figure 5.20-4, notice the circular banding in the rays of the sunset on the left side. This is caused by not having a sufficient number of different levels for chrominance/luminance to represent smooth gradients; clearly, quantization steps are noticeable. On the right side, however, 10-bit sample depth data precision is used—with 2^{10} or up to 1,024 levels of gradient—resulting in much smoother contouring of the sun's rays.

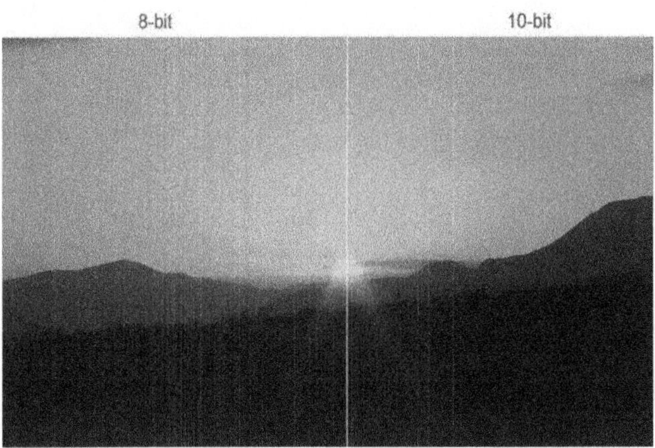

FIGURE 5.20-4 A simulated example of banding artifacts due to insufficient data precision. On the left, 8-bit data values cause banding to appear. On the right, smoother contouring with 10-bit data values eliminates the banding effect.

UHDTV specifications, such as SMPTE ST 2036-1 [5], define 10-bit sample bit depths as the minimum data precision values. Existing production workflows use 10-bit sample bit depths throughout, and it is only when the direct-to-consumer format is created that the sample bit depth is truncated to 8 bits during final stage compression. The Joint Collaborative Team for Video Coding—consisting of the ISO/IEC Moving Picture Experts Group (MPEG) and the ITU-T Video Coding Experts Group (VCEG)—created a Main 10 Profile in the first version of the High Efficiency Video Coding (HEVC) standard [6] for this reason. While previous compressed digital video standards defined profiles that included 10-bit sample bit depths, these were for professional applications only. HEVC is the first to include a profile that supports 10-bit precision for consumer direct-to-home applications.

Wider Color Gamut

UHDTV also has an expanded color gamut, or space, compared to that defined for HDTV [7]. In Figure 5.20-5, note the greatly expanded region of greens and the inclusion of more reds. This enables UHDTV to have a more realistic color palette than HDTV, which in turn would take advantage of the 10-bit precision described previously. Current UHDTVs do not support the full BT.2020 color gamut, but improvements in color space have occurred with each new generation of UHDTVs.

High Dynamic Range: The Ultimate Experience?

High dynamic range (HDR) video could offer consumers a much-improved viewing experience compared to current broadcast video. The HVS is much more sensitive to luminance (brightness) than it is to chrominance (color), with an extremely wide dynamic

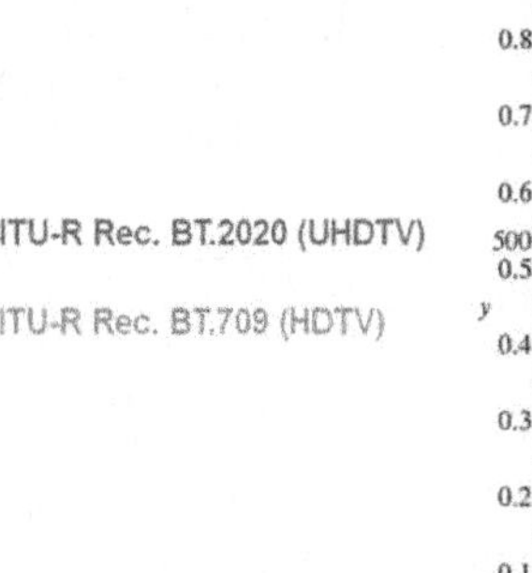

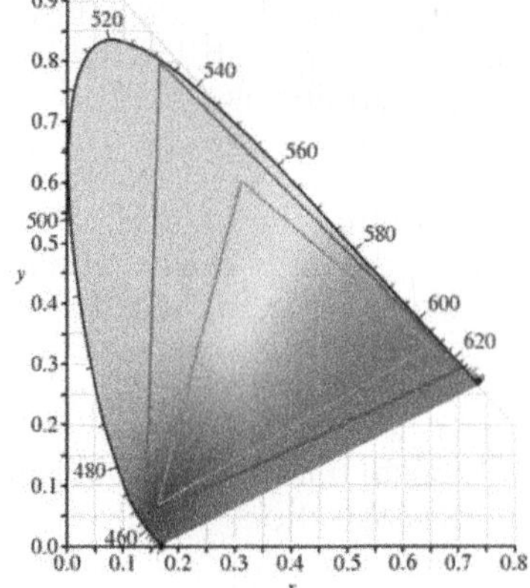

FIGURE 5.20-5 The expanded color gamut of UHDTV.

Adapted from Commission internationale de l'Eclairage proceedings, 1931. Cambridge University Press, Cambridge, UK.

range of approximately one million to one (a contrast ratio of the luminance of the brightest object to that of the darkest) [8]. The HVS dynamic range can discern from 10^5 cd/m^2 (candelas per square meter, or "nits" of luminance) bright sunlight to 10^{-4} cd/m^2 dim starlight. It is highly complex, adaptive and not fully understood in terms of television viewing. Unlike increased resolution, which works best on a very large screen, increasing the dynamic range that a viewer can see is equally applicable to any screen size, as well as with resolutions less than UHD.

The production standard for consumer video, however, has not been changed since the physics of cathode ray tubes (CRTs) were first documented in the 1930s, including setting the peak white level to 100 nits. Although modern video cameras have been able to capture a very wide dynamic range for several years now and the latest HDTVs claim maximum peak output in the range of 400–1,200 nits, TV production standards are just in the process of being updated to support HDR at this writing.

Reduced dynamic range translates to the inability to see both lowlights (e.g., details in deep shadows) and highlights (e.g., clouds in a bright sunny day) simultaneously; one or the other will be "lost." For example, think of watching a football game on a conventional, standard dynamic range (SDR) TV, with the stadium partially in shadow and partially in sunlight. The impact of reduced dynamic range particularly is noticeable for specular reflections, such as sunlight reflecting off of the surface of water or metal. In real life, such light usually causes a physiological response in the viewer ("feeling" the brightness of the highlights, including squinting of the eyes, for example). For an HDR TV, your eyes may not squint, but you will still sense the isolated brightness of the highlighted area, giving the overall image a much more realistic viewing experience.

The impact of HDR is difficult to represent without a high-output active light source such as a studio reference monitor designed for this purpose. However, the concept can be expressed by simulating HDR through the combination of multiple exposures of the same image (albeit without the dramatic impact of specular reflections). Figure 5.20-6a shows a scene of the St. Louis Gateway Arch with lowlight exposure. Note that the central building is "crushed" (i.e., overexposed, such that all detail in that part of the image is lost). In Figure 5.20-6b, the exposure is adjusted for proper highlights, so that the building is reproduced properly, but now the detail on the side buildings and in the sky are no longer visible. In each of these cases, SDR is not capable of reproducing both dark and light areas simultaneously. Figure 5.20-6c simulates how an HDR image would be represented, which more closely resembles the realism that the human visual system provides.

Demonstrations of HDR TV over the past several years have convinced standards development organizations to specify this new dimension of immersion into the TV viewing experience. Significant benefits have been shown for not only UHD but also HD resolutions, resulting in many believing that HDR is arguably *the* most important new development for TV.

FIGURE 5.20-6A Example of lowlight standard dynamic range.

FIGURE 5.20-6B Example of highlight standard dynamic range.

FIGURE 5.20-6C Example of high dynamic range, simulated by tone mapping.

High Dynamic Range and the Display Technology

The existing broadcast chain nominally assumes a peak brightness of 100 nits [10]. Typical consumer televisions can achieve a greater brightness than this, and continuing developments in display technologies seek to increase both the maximum brightness and the possible contrast ratio (dynamic range).

The most commonly used technology today is an LCD panel illuminated by an LED backlight. Typical panels of the last few years achieve a light output varying from about 0.1 nits to about 300 nits [11]. Typical LED backlights use blue LEDs combined with a yellow phosphor, which emits a relatively wide spectrum, to produce "white" light which is then filtered to give red, green and blue in the LCD panel [12]. A trade-off exists between the brightness and the width of color gamut achievable. A color filter that more accurately filters the incoming light to give a single saturated color will filter out more light and give a lower brightness. Recent, more sophisticated LCD displays can achieve high brightness and high contrast ratio by using "local dimming," where the brightness of the backlight can be locally varied across the display, to a black level less than that possible by the LCD shutter itself. As backlight technology improves, the ability to localize dimming over a smaller area of the display will increase, resulting in better-localized high contrast ratios than possible in a conventional LCD display.

Recently, various manufacturers have announced LCD displays that include the use of backlights utilizing nanocrystals (sometimes referred to as "quantum dots") where a blue LED is used to excite red and green quantum dot phosphors [13]. These backlights can emit narrow bands of red, green and blue light, enabling a wider color gamut and either a higher brightness or better efficiency, as less light is absorbed in the LCD color filters. Displays using this technology announced at the 2015 International Consumer Electronics Show claimed a maximum brightness of up to 1,200 nits, with black levels significantly lower than possible with LCD panels.

While LCDs work by filtering light from a backlight, in directly emissive display technologies, none of the light produced is absorbed within the display and so they offer the possibility of obtaining even higher brightness with much lower black levels. Individual pixels can also be turned fully off, and so these displays could give greater contrast ratio. OLED (organic light emitting diode) based consumer televisions are available where each pixel is formed by red, green and blue LEDs (with sometimes an additional white LED or a white LED with color filters). These offer a very dark black level, but some do not offer a significantly greater peak brightness [14]; however, levels of peak brightness of up to about 800 nits are now being claimed [15]. This gives a higher contrast ratio than most LCD screens, though the extent to which the darker black level can be perceived is dependent on the reflectivity of the screen and the ambient light level in the room where it is viewed. Display technology continues to develop; as previously mentioned, a technology challenging OLED panels is direct emissive quantum dot LED displays [16]. These would offer increased brightness and improved color gamut compared to OLED.

High Dynamic Range in Standardization

At the time of this writing, the industry is still in the midst of defining HDR standards. The Society of Motion Picture and Television Engineers (SMPTE) completed the Study Group Report on HDR Imaging Ecosystem in October 2015 [17], which articulates the issues that need to be addressed and standardized to complete the end-to-end production and delivery workflows. SMPTE also has recently created new standards in this area, including an HDR electro-optical transfer function (see the next section below for more details), known as perceptual quantization, for mastering reference displays [18] and mastering display color volume metadata supporting high luminance and wide color gamut images [19]. Dynamic metadata-based HDR systems are still a work in progress at the time of this writing (draft SMPTE ST 2094-x series).

In July 2016, the ITU-R published a recommendation for HDR-TV image parameters for use in production and international program exchange, ITU-R BT.2100 [20], and an accompanying report, ITU-R BT.2390 [21]. This includes two transfer functions, Perceptual Quantization (PQ, SMPTE ST 2084) and Hybrid Log-Gamma (HLG) [20]. The Digital Video Broadcast project (DVB) updated its standards to include UHD-1 Phase 1 (4K resolution at up to 60 fps) in 2014. UHD-1 Phase 2 adds HDR and was published on November 17, 2016, for commercial deployments in 2017–2018. The HDMI Forum and the Consumer Technology Association defined an extension to HDMI to support a static HDR metadata system known as the HDR10 Media Profile (CTA-861.3)—which includes signaling support for SMPTE ST 2084 PQ and ST 2086 metadata—that will update HDMI to support the signaling of HLG and upcoming dynamic metadata systems (draft SMPTE ST 2094-x series) (CTA-861-G). Meanwhile, two industry groups are addressing Ultra HD interoperability: the Ultra HD Forum and the UHD Alliance. Standards work on Ultra HD in progress at this writing includes the Advanced Television Systems Committee (ATSC) and the Society of Cable Telecommunications Engineers (SCTE).

The impact of HDR on the TV viewing experience, for HD and UHD, is well appreciated by consumers, scientists and engineers alike [22].

ELECTRO-OPTICAL TRANSFER FUNCTION AND OPTO-ELECTRICAL TRANSFER FUNCTION

The electro-optical transfer function (EOTF) describes how to convert different coded levels of information (either analog voltage levels or digital numbers) into visible light, such as how a TV renders and displays its luminance output levels.

Production artists make creative decisions on the proper "look" of content using a reference monitor. This typically is done at the camera control station in the case of a live production (such as live sports) or at a postproduction grading desk in the case of off-line productions (such as motion pictures). This process maps a wider dynamic range capture environment to a narrower dynamic range consumer display environment in whatever way achieves the desired artistic intent. A defined EOTF is required to obtain consistent image reproduction. Note that for consumer televisions this defined EOTF gives a baseline; actual consumer televisions can, and do, adjust the EOTF in order to try to give the most pleasing result to the consumer.

While used in practice for decades due to the physical properties of a CRT display, the EOTF was not standardized in full until 2011, when Rec. ITU-R BT.1886 [10] was created for flat panel displays (FPD). Since it is based on the CRT, it ensured consistency of rendering content for SDR displays. Not being standardized until 2011 is somewhat ironic, since this codified the response curve of a CRT (which is virtually extinct), but this was needed in order to standardize the baseline so that derivations of existing practice could be made to address HDR. A conventional CRT has a power-law response to voltage; that is, luminance produced at the face of the display is approximately proportional to the applied voltage raised to the power of 2.5. The numerical value of the exponent of this power function is colloquially known as *gamma* [23]. This nonlinearity must be compensated for in order to achieve correct reproduction of luminance.

The inverse function, the opto-electrical transfer function (OETF), is used for image capture. It was derived from the EOTF and includes a rendering "intent" in its definition. This process has been colloquially known as *gamma correction* where linear *R,G,B* tristimulus values are transformed into nonlinear *R'G'B'* quantities. This process allowed reduction in the bit depths required. While the correction was aimed to compensate for the characteristics of CRT displays, it also follows the HVS's lightness response, therefore ensuring perceptual uniformity of encoded video signal [24]. The OETF for HD is included in Recommendation ITU-R BT.709 [25] and the same transfer function was included for UHD in Recommendation ITU-R BT.2020 [7].

The combination of OETF and EOTF results in an overall transfer characteristic often called *system gamma*. For SDR, the system gamma is not equal to 1 but rather 1.2 and can be adjusted through display brightness. For digital cinema systems, the gamma used is 1.5, which is due to dark surroundings. Conversely, for bright surroundings, the system gamma is closer to 1.

Transfer Functions for High Dynamic Range Systems

As noted in the previous section, two new transfer functions have been defined to support HDR, namely PQ and HLG. Ideally, the EOTF should be defined by the human visual response (the current *gamma*, which is the inverse of a CRT's nonlinearity, coincidentally works well perceptually to the HVS, but only at relatively low luminance levels and small dynamic ranges). Increasing display brightness relates to the increase in distance between rendered luminance values for adjacent code words, leading to visible artifacts known as banding or contouring. According to the Weber-Fechner model [26], two luminance levels cannot be distinguished if the difference is smaller than 1%. The contrast sensitivity function of the HVS varies depending on factors such as display brightness or field of view [27]. These factors were included in Barten's model [28], which has been used for predictions of the minimum detectable contrast [29].

The predictions from the model show that if the industry chose to standardize the peak brightness of the reference HDR display as 10,000 nits (as defined by SMPTE ST 2084 PQ), then the current ITU-R BT.709/BT.1886 transfer functions would be found insufficient: There are too few levels available to represent low luminance levels and, consequently, bits would be *wasted* for high luminance values where they are not required. This could lead to visible banding artifacts. One way to reduce the artifacts would be to increase the sample bit depth of the video signal (see the preceding section on Sample Bit Depth). However, for synthetic signals such as a dark shallow ramp, the banding would be visible at bit depths beyond 10 bits per sample. This could be deemed impractical for broadcast systems. If, on the other hand, the peak brightness of the reference HDR display were set to be in the range of 2,000–3,000 nits, the current ITU-R BT.709/BT.1886 transfer functions for a 10-bit signal would provide equivalent detectable contrast as the current 8-bit SDR systems (i.e., the 10-bit variant of ITU-R BT.1886 EOTF stretched to peak brightness of about 2,800 nits gives equivalent detectable contrast as the 8-bit variant of ITU-R BT.1886 at peak brightness of 100 nits).

It is due to the aforementioned limitations of the current ITU-R BT.709/BT.1886 transfer functions that alternative transfer functions have been defined for encoding and rendering HDR signals. The EOTF (and its inverse OETF) standardized in SMPTE ST 2084 PQ is defined for the absolute brightness of 10,000 nits and aims to follow predictions of the Barten's model. HLG was designed for peak brightness <10,000 nits.

In summary, the focus for the HDR transfer functions has been to avoid banding artifacts due to increased peak brightness of HDR displays. In consequence, these new transfer functions exhibit much higher nonlinearity compared to the transfer functions used in today's SDR systems (see Figure 5.20-7a and b). This has several implications on the existing broadcast chain, such as a possible increase in bitrates required for emission of HDR services, modifications to graphics and video processing equipment, and the incompatibility of HDR services with existing SDR displays.

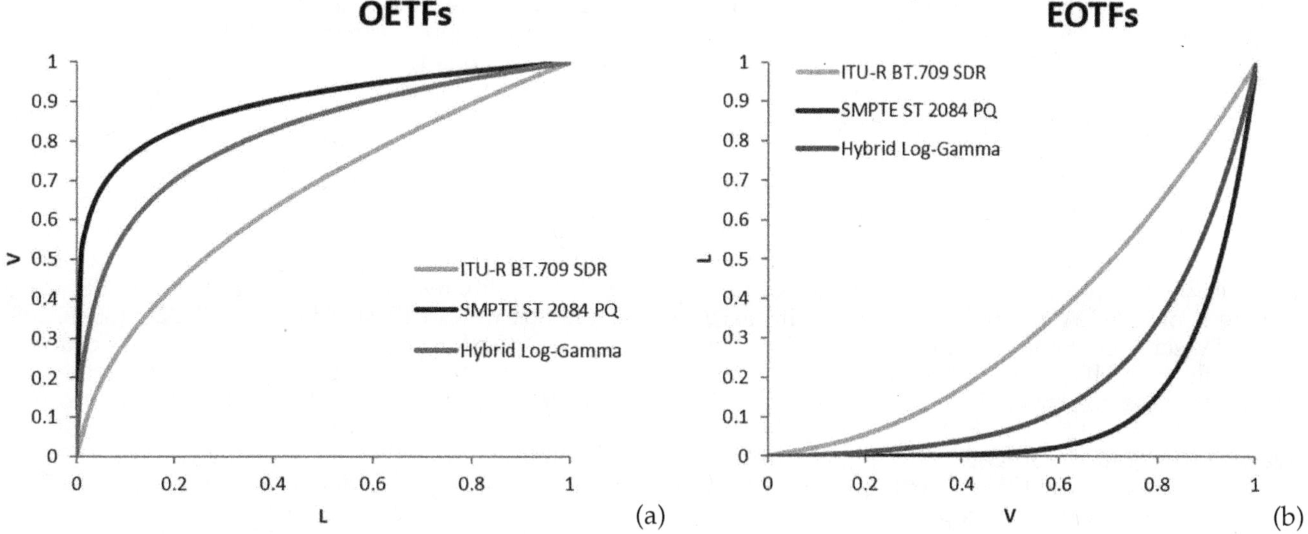

FIGURE 5.20-7 The higher nonlinearity of Perceptual Quantization and Hybrid Log-Gamma HDR transfer functions versus the BT.709 standard dynamic range transfer function.

THE ENABLER: HIGH EFFICIENCY VIDEO CODING

Due to both economic and technical bandwidth constraints of direct-to-consumer networks, a more highly efficient video coding technology was needed to enable UHD content delivery to the home and for more widely used video delivery to mobile handheld devices. In January 2013, the first release of HEVC [6] was completed.

HEVC, officially known as ITU-T Rec. H.265 and ISO/IEC 23008–2 MPEG-H Part 2, includes the Main Profile and Main 10 Profile to address direct-to-consumer applications, such as direct broadcast satellite, cable television, telecom television (IPTV), terrestrial broadcasting, Internet-delivered video (TV over IP, or "over-the-top" TV) and consumer-packaged media (e.g., UHD Blu-ray Disc). The HEVC codec contains a set of advanced coding tools that together make HEVC significantly more compression-efficient than prior generations of video codecs [30]. The coding efficiency improvement of HEVC compared to Advanced Video Coding (AVC, also known as Rec. ITU-T H.264 and ISO/IEC 14496–10 MPEG-4 Part 10)—the most bitrate efficient standardized codec prior to HEVC—is up to 50%, effectively halving the required bitrate, or up to four times as bitrate efficient (one-fourth the bandwidth) as MPEG-2 Video (also known as Rec. ITU-T H.262 and ISO/IEC 13818–2). See Figure 5.20-8. For 4K UHDTV, HEVC's bitrate reduction over AVC may exceed 50% [31]. This has major ramifications for many different industries that need to deliver high-quality

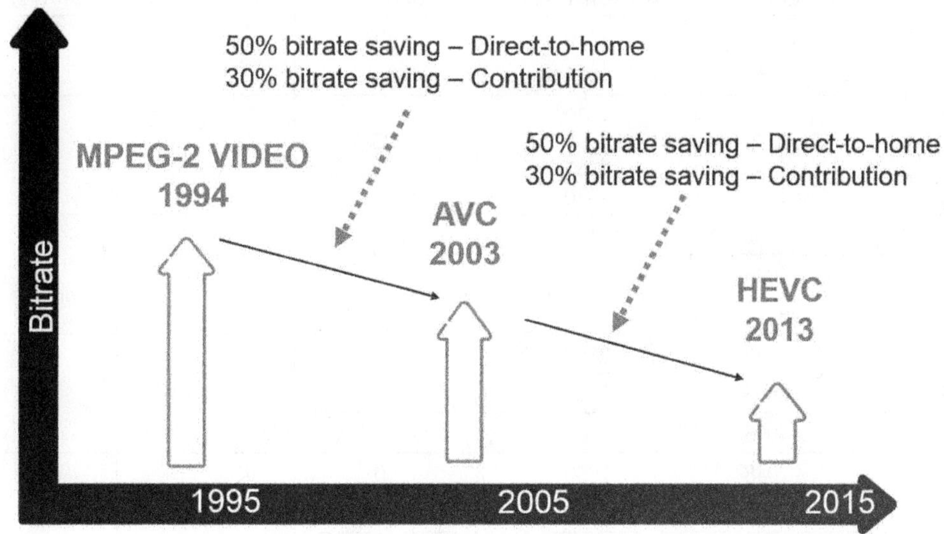

FIGURE 5.20-8 Compression bandwidth efficiency trends.

images across constrained links, or for more efficient storage applications.

ULTRA HD INTERFACES

The uncompressed video bitrate for 4K UHDTV equates to approximately 12 gigabits per second (Gbps). In comparison, uncompressed 720p and 1080i HDTV requires 1.5 Gbps, while "full" HDTV 1080p60 (or p50 in 25 Hz parts of the world) requires 3 Gbps.

There is already a "bandwidth crunch" with existing HDTV services. Requiring four to eight times the bandwidth over HDTV to deliver uncompressed 4K UHDTV services presents a challenge. Many broadcast facilities today use 1.5 Gbps HD-SDI [33], while some have been upgraded to 3G-SDI to support 1080p60 [34]. Although 12G-SDI [35] is starting to appear, the most common way to interconnect 4K UHD today is to use four 3G-SDI linked together to carry a single live uncompressed 4K UHDTV signal. As shown in Table 5.20-1, the amount of data for transporting UHDTV can be significant.

Many in the broadcast industry are questioning how far traditional SDI architectures will be practical above 12 Gbps. Alternatives such as lightly compressed "mezzanine-level" encoding are one option. For many usage scenarios, mezzanine formats cause little picture quality degradation while conserving significant bandwidth. There are many mezzanine compression formats available, so defining interoperability points within a subset of those standards likely will garner the most industry support.

Another alternative is to map a "UHDTV-SDI" frame into an Internet protocol (IP) stack and use the more ubiquitous Ethernet data link and physical layers. In this fashion, as higher bitrate Ethernet-based transports become viable, uncompressed video can ride the Ethernet economy of scale.

There is a major shift occurring in the professional video infrastructure ecosystem, where more facilities are looking to leverage their IT infrastructure to move video to an IP-based workflow. To aid in the delivery of high-quality real-time video over IP networks, there is a family of SMPTE standards, ST 2022-x [36, 37, 38, 39]. Defining forward error correction (FEC) codes to improve the underlying quality of service of the network, these standards also specify the transport format for various video compression algorithms encapsulated in the near ubiquitous MPEG-2 Transport Stream format [40]. They also define the carriage of uncompressed video streams (referred to as high bitrate media transport), from SD (270 Mbps) to 1080p HD (3 Gbps). In all cases, RTP/UDP/IP protocol layers are required, as they provide the standard header for the data essence and FEC streams.

SMPTE extended the ST 2022 series of standards with Part 7, a standard for "hitless switching" [41]. ST 2022–7 defines the seamless reconstruction of a stream of SMPTE ST 2022 RTP datagrams, based on transmission of two streams of identical content over potentially diverse paths.

For uncompressed video, SMPTE ST 2020–6 maps the video and embedded audio into an SDI frame first, which is in turn mapped in IP. The latest scheme in the transformation of broadcast facilities to "all IP"

TABLE 5.20-1
Formats and interfaces for real-time uncompressed digital video.

System Nomenclature	Horizontal Pixels	Vertical Pixels	Frames per Second	Total Payload	
				10-bit 4:2:0 10-bit 4:2:2	12-bit 4:2:0 12-bit 4:2:2 12-bit 4:4:4 10-bit 4:4:4
4320p60 / 59.94	7680	4320	60	48 Gbps	96 Gbps
4320p50	7680	4320	50		
4320p30 / 29.97	7680	4320	30	24 Gbps	48 Gbps
4320p25	7680	4320	25		
4320p24 / 23.98	7680	4320	24		
2160p60 / 59.94	**3840 / 4096**	**2160**	**60**	**12 Gbps**	24 Gbps
2160p50	3840 / 4096	2160	50		
2160p48	4096	2160	48		
2160p30 / 29.97	3840 / 4096	2160	30	6 Gbps	12 Gbps
2160p25	3840 / 4096	2160	25		
2160p24 / 23.98	3840 / 4096	2160	24		
1080p60 / 59.94	1920 / 2048	1080	60	3 Gbps	6 Gbps
1080p50	1920 / 2048	1080	50		
1080p48	2048	1080	48		

involves the separate mapping of video, audio and data essence directly into IP protocol layers. As of this writing, this work effort is still in process, and is known as draft SMPTE ST 2110. The bulk of the preliminary work was defined in the Video Services Forum's Technical Recommendation TR-03 [42].

SUMMARY

Ultra HD video typically consists of at least four times the spatial resolution of HDTV, at least 10-bit sample depth precision for data values (vs. the 8-bit values used today for consumer direct-to-home HDTV applications), an expanded color space, high dynamic range, and can include at least two times the temporal resolution of 1920 × 1080 HDTV. All these factors create a more immersive viewer experience, providing a greater sense of presence and sensation of reality than possible with conventional HDTV. High dynamic range video is arguably the most important new development for television. It enhances greatly the realism of the TV viewing experience, both for HD and UHD.

References and Bibliography

[1] Sugawara, M., Mitani, K., Kanazawa, M., Okano, F., and Nishida, Y., *Future Prospects of HDTV—Samsung Trends Toward 1080p*, 2005.

[2] Bracken, C.C., and Botta, R.A., "Presence and Television: Form versus Content", *The Fifth International Workshop on Presence*, 2002.

[3] Sugawara, M., Masaoka, K., Emoto, M., Matsuo, Y., and Nojiri, Y., "Research on Human Factors in Ultrahigh-Definition Television (UHDTV) to Determine Its Specifications", *SMPTE Motion Imaging Journal*, pages 23–29, April 2008.

[4] *Temporal Rate Conversion*, accessed January 2015, https://msdn.microsoft.com/en-us/library/windows/hardware/dn642112(v=vs.85).aspx

[5] SMPTE ST 2036–1–2014, *Ultra High Definition Television—Image Parameter Values for Program Production*, revised October 2014.

[6] Rec. ITU-T H.265, *High Efficiency Video Coding*, revised October 2014.

[7] Rec. ITU-R BT.2020–1, *Parameter Values for Ultra-High Definition Television Systems for Production and International Programme Exchange*, revised June 2014.

[8] Barton, H., and Byrne, K., "Introduction to Human Vision, Visual Defects & Eye Tests," in *Proceeding of the Joint Technical Meeting ILE/BOHS*, Salford, Inglaterra, page 22, March 2007.

[9] McCoy, K., *StLouisArchMultExpEV+4.09*. Licensed under CC BY-SA 3.0 via Wikimedia Commons, accessed January 2015, http://commons.wikimedia.org/wiki/File:StLouisArchMultExpEV%2B4.09.JPG#mediaviewer/File:StLouisArchMultExpEV%2B4.09.JPG

[10] Rec. ITU-R BT.1886, *Reference Electro-Optical Transfer Function for Flat Panel Displays Used in HDTV Studio Production*, March 2011.

[11] *Plasma vs LED vs LCD TV*, accessed January 2015, www.rtings.com/info/lcd-vs-led-vs-plasma

[12] *Quantum Dots Unleash High Color Gamut Performance in LED-Backlit Displays*, accessed January 2015, http://dot-color.com/2011/10/18/quantum-dots-unleash-high-color-gamut-performance-in-led-backlit-displays/

[13] *CES 2015: Placing Bets on the New TV Technologies*, accessed January 2015, http://spectrum.ieee.org/tech-talk/consumer-electronics/audiovideo/ces-2015-placing-bets-on-the-new-tv-technologies

[14] *LG OLED TV Lab Measurements and Technical Analysis*, accessed January 2015, www.displaymate.com/LG_OLED_TV_ShootOut_1.htm#Lab_Tests

[15] *LG's OLED HDR TV Shines Brighter (Hands-on)*, accessed January 2015, www.cnet.com/news/lgs-oled-hdr-tv-shines-brighter/

[16] *Nanocrystal Display*, accessed January 2015, www.technologyreview.com/demo/405755/nanocrystal-displays/

[17] Society of Motion Picture & Television Engineers (SMPTE), www.smpte.org/standards/reports

[18] SMPTE ST 2084:2014, *High Dynamic Range Electro-Optical Transfer Function of Mastering Reference Displays*, August 2014.

[19] SMPTE ST 2086:2014, *Mastering Display Color Volume Metadata Supporting High Luminance and Wide Color Gamut Images.*

[20] Rec. ITU-R BT.2100–0, *Image Parameter Values for High Dynamic Range Television for Use in Production and International Programme Exchange*, July 2016.

[21] Report ITU-R BT.2390–0, *High Dynamic Range Television for Production and International Programme Exchange*, February 2016.

[22] Hannart, P., Korshunov, P., Ebrahimi, T., Thomas, Y., Hoffmann, H., "Subjective Quality Evaluation of High Dynamic Range Video and Display for Future TV", *IBC 2014 Conference Proceedings*, September 2014.

[23] Poynton, C., *Frequently Asked Questions About Gamma*, January 1998, www.poynton.com/notes/colour_and_gamma/GammaFAQ.html

[24] Poynton, C., and Funt, B., "Perceptual Uniformity in Digital Image Representation and Display", *Color Research & Application*, Volume 39, Issue 1, pages 6–15, February 2014.

[25] Rec. ITU-R BT.709–5, *Parameter Values for the HDTV Standards for Production and International Programme Exchange*, revised April 2002.

[26] Hecht, S., "The Visual Discrimination of Intensity and the Weber-Fechner Law", *Journal of General Physiology*, Volume 7, Issue 2, pages 235–267, 1924.

[27] Report ITU-R BT.2246–3, The Present State of Ultra-High Definition Television, March 2014.

[28] Barten, P.G.J., *Contrast Sensitivity of the Human Eye and Its Effects on Image Quality*, SPIE Optical Engineering Press, Bellingham, WA, 1999.

[29] Cowan, M., Kennel, G., Maier, T., and Walker, B., "Contrast Sensitivity Experiment to Determine the Bit Depth for Digital Cinema", *SMPTE Motion Imaging Journal*, pages 281–292, September 2004.

[30] Goldman, M., "High Efficiency Video Coding (HEVC)—The Next Generation Compression Technology", *Society of Motion Picture and Television Engineers Technical Conference Proceedings*, October 2011.

[31] Goldman, M., and Horton, M., "High Efficiency Video Coding: Next Generation Compression Technology Driving New Business Models for Television", *NAB Broadcast Engineering Conference Proceedings*, April 2012.

[32] Goldman, M., Baumann, O., and Litwic, L., "The Future of 4K Ultra High Definition TV—Examining Methods to Acquire, Exchange and Distribute Content", *NAB Broadcast Engineering Conference Proceedings*, April 2014.

[33] SMPTE ST 292–1:2012, *1.5 Gb/s Signal/Data Serial Interface.*

[34] SMPTE ST 424:2012, *3Gb/s Signal/Data Serial Interface*, October 2012.

[35] SMPTE ST 2082–1:2015, *12 Gb/s Signal/Data Serial Interface—Electrical.*

[36] SMPTE ST 2022–1:2007, *Forward Error Correction for Real-Time Video/Audio Transport Over IP Networks*, May 2007.

[37] SMPTE ST 2022–2:2007, *Unidirectional Transport of Constant Bit Rate MPEG-2 Transport Streams on IP Networks*, May 2007.

[38] SMPTE ST 2022–5:2012, *Forward Error Correction for High Bit Rate Media Transport Over IP Networks*, October 2012.

[39] SMPTE ST 2022–6:2012, *Transport of High Bit Rate Media Signals Over IP Networks (HBRMT)*, October 2012.

[40] ITU-T Rec. H.222.0 | ISO/IEC 13818–1, *Information Technology—Generic Coding of Moving Pictures and Associated Audio Information: Systems*, revised June 2012.

[41] *SMPTE ST 2022–7–2013, Seamless Protection Switching of SMPTE ST 2022 IP Datagrams*, October 2013.

[42] *Video Services Forum TR-03, Transport of Uncompressed Elementary Stream Media Over IP*, November 2015.

SECTION

6

FACILITY INTERCONNECTION

CHAPTER

6.1

Audio Contribution and Distribution Methods

SKIP PIZZI
NAB
Washington, DC

JOHN KEAN
Cavell Mertz & Associates, Inc.
Manassas, VA

INTRODUCTION

This chapter deals with the transmission of high-quality audio by broadcasters prior to, or "upstream" of, actual broadcast—in other words, any point-to-point transmission of audio signals involved in the production of broadcast content. This chapter does not cover specialized, licensed radio frequency (RF) transmission systems owned and operated by broadcasters for local remote backhaul (Remote Pickups [RPUs]), or studio to transmission-site links (STLs), that are covered in Chapter 6.4. (Some of the same audio and data transmission technologies covered in this chapter may be applied on those links, however.) Interconnection by voice-grade telephone is covered in Chapter 4.5.

Note also that although most current connectivity in this chapter uses some form of digital encoding and transport, there still may be some analog audio circuits available to broadcasters in any given locale. Because much of these analog legacy services are becoming increasingly unavailable for a growing number of locations across the United States, discussion of these older systems was reduced for this edition.

Terminology

In recent years, providers and users of audio transmission circuits have developed a nomenclature that establishes a hierarchy of signal pathway types, namely contribution, distribution, and emission channels:

- Contribution channels are those used typically for transmission from program origination sites to broadcast assembly centers. They may be called "backhaul" circuits, and common examples are the signal paths used from a remote site to a broadcast station or from a sports venue to a broadcast network operations center.
- Distribution channels are typically those used by broadcast networks to send programming content from operation centers to broadcast stations or transmission centers. Satellite distribution is a common means for television and radio networks to send content to affiliate stations.
- Emission channels are those used in the "last mile" path from stations' transmitters to consumer receivers.

The quality metrics of contribution channels are higher than those of distribution channels, that are in turn higher than those of emission channels. Such a hierarchy allows each channel to be optimized to the audio content it is likely to encounter (e.g., a contribution channel may precede remixing, dynamic processing and encoding cycles, that requires higher quality than the following distribution or emission channels).

Contribution and distribution channels are generally point-to-point in nature, whereas emission channels, as the name suggests, are point-to-multipoint broadcast transmissions. Contribution and distribution channels used for remote backhaul and network interconnections of audio content require special attention when they are not owned, controlled, or operated by broadcasters who use them, but by third parties—typically telephone companies or other telecommunications carriers ("telcos").

For clarity, the term "circuit" will designate a telco audio path, replacing old (and potentially confusing) synonyms of "loop," "private line," "leased line," "program circuit," and the like. As well, "telco" will refer generically to telecommunications companies that provide local audio and data carrier service, both wired and wireless. "LDS" (long-distance service) will refer to public and private telecommunications companies that provide interconnection service between local telcos.

Dealing With Telco

Broadcasters and telcos have been associated since the earliest days of both industries and will likely remain so. This chapter considers the telco audio interconnections that may be available to broadcasters and examines how they can implement these hookups.

To minimize problems with a telco, the station engineer should establish a good working relationship with the appropriate personnel and understand as much as possible about the company's services and operation. If the station's staff comes across as friendly and knowledgeable, but also professionally firm and businesslike, things should go well. If possible, keep the station's liaison to the telco limited to one staff member, and try to always deal with the same person at telco as well. The latter may be challenging, as local and regional telcos merge and change their business structures.

For ordering digital circuits, another option may be a data-line brokerage service. Here, the station gives its time, place, and quality service requirements to a third party, who books the line for the station at no charge and receives a commission from the telcos whose circuits they book. This service can be especially helpful in long-distance applications, where two local telcos (one foreign to the station) and an LDS are involved.

More recently, telcos have offered Internet service provider (ISP) service, which can provide data-interconnection consultancy for stations. In these cases, digital interconnection services for both digital audio interconnection and computer data-networking needs may be bundled into a single agreement with a local service provider, for increased cost-effectiveness and stable operations.

This transition has moved much of what broadcasters need from telco for audio transmission into the consumer domain. While this may considerably decrease costs for broadcasters, some service packages may also reduce the level of telco customer service from what broadcasters formerly enjoyed as special clients to that of ordinary retail customers.

An Abundance of Options

There have never been more choices of technologies and pathways available for broadcasters' transport of audio programming. Many years ago, the only choice for broadcasters were standard bandwidth analog circuits from a single service provider. Today the options are myriad, including the last vestiges of such dedicated analog paths, to dedicated digital paths (e.g., T-1), to switched digital circuits (e.g., ISDN), to packet-based networks (telco IP and "bandwidth on demand" services), to broadband access to the packet-based "cloud" of the Internet (e.g., DSL, DOCSIS), to other wireless broadband access (e.g., Wi-Fi, 3G/4G/LTE cellular, WiMAX, satellite broadband global area networks, etc.). In some cases, first-generation digital services such as ISDN—widely used by radio broadcasters for remote backhaul—have been themselves replaced by IP-based services.

This chapter will address all of the above, in the order of their emergence. Naturally, because most of these technologies are implemented as local terrestrial services, new systems are introduced—and older systems phased out—gradually, region by region. Not all options will be available in all areas at any given time, although most portions of the United States already have at least some choice among multiple service providers today. Over time, emerging technologies will become broadly available in all but the most remote areas. Consult your local telecommunications services providers for currently available options. (Readers who have no access to or interest in legacy analog audio interconnection are advised to skip ahead to the "Digital Audio Circuits" section.)

ANALOG CIRCUITS

While most analog audio service has been phased out, in some locations, analog circuits may still be offered. The standard classes of service offer maximum bandwidths of 3.5 kHz, 5 kHz, 8 kHz, and 15 kHz. Most local telcos have by now replaced analog circuits with digital substitutes, while LDSs have long since phased out analog services for their long-haul paths.

Where they are still offered, costs for both service and installation of analog program circuits continue to increase. Installation of an analog program circuit generally requires several hours of an experienced technician's work to equalize the line, as opposed to digital services, which take much less time set up and pass spec at installation.

Analog circuits might provide wider bandwidth than what was ordered, whereas digital services typically cut off exactly as specified. But analog services that pass between telco central offices or switching centers will be converted to a digital stream and bundled on an interoffice digital carrier.

Obtaining and Testing Analog Telco Circuits

Analog circuits may be available in more than one bandwidth and under temporary or permanent status. Check with the local telco to determine which, if any, are available and at what cost. Installation charges for permanent lines may be much higher than those for temporary service, because the telco may want to actually install new wiring rather than permanently occupy any pairs on its existing network cables. In an increasing number of areas, analog audio circuits will not be offered at all, in which case the reader may skip the remainder of this section. For those legacy users of analog audio program circuits, however, the following is provided.

The 15 kHz circuits may be ordered as a stereo pair, incurring a one-time installation surcharge for stereo conditioning. This ensures that both lines are routed together throughout their runs, so that inter-channel phase differences will be minimized. A third line can be ordered for backup, and should be included in the stereo conditioning. Actual routing of the circuits may be much longer than billed routing, increasing the opportunity for phase differences. (An approximate 5 μsec time difference occurs for every mile of path length difference.)

Lines should be ordered well in advance of the telco's lead time. The start date should be at least one business day earlier than the actual requirement to allow time for station tests on the lines to be performed. Check frequency response, signal-to-noise ratio (S/N), distortion, and headroom. For stereo pairs, check relative phase response and relative polarity.

Frequency response should be at least within ±3 dB of what was ordered. Be sure to check outside the passband, because response may not roll off but instead rise beyond the cutoff frequency. Use caution with any complementary analog noise reduction (NR) system used on a telco circuit. As a general rule, be sure that the audio fed to the NR encoder is pre-filtered to match the line's response so that the audio passband seen by the encoder is relatively similar to that seen be the decoder after the audio has passed through the circuit. If the decoder sees a substantially narrower bandwidth than the encoder, it will not decode the noise reduction process in a complementary fashion.

The proper procedure for calibrating levels on analog program circuits employs sinewave test tones of 400 Hz or lower, fed at the telco program operating level (POL) of +8 dBm. Frequency response tests and tones above 400 Hz must be sent at the telco test level of 0 dBm to minimize cross talk into other circuits via capacitive coupling of higher frequencies.

For measuring noise, telcos use a unique approach. They consider a noise level of −90 dBm to be absolute quiet or noise free, and measure noise from that reference point. The unit used is 0 dBRN (RN for reference noise). Therefore, a −50 dBm noise level would be called 40 dBRN by telco. If the telco's specified audio reference level of +8 dBm is used by the customer on this circuit, this results in a 58 dB SNR (signal-to-noise ratio).

The greater the distance an analog circuit travels, the noisier it becomes. The wider a circuit's bandwidth, the quieter it needs to be. Although specs vary between telcos, noise specifications generally reflect those observations. A typical noise level for local 15 kHz circuits is 33 dBRN or lower, providing 65 dB or better SNR. Again, check with the local telco for circuit specifications, and when connected always verify that those specifications are met.

Continuous random noise, such as hiss, is far less objectionable than coherent noises sometimes found on telco circuits. Capacitive coupling between adjacent pairs in multipair cable can produce clicks and other switching noises. Circuits should be carefully auditioned at the receive-end after installation, without any audio on the line, and checked again prior to each on-air use. Report any cross talk or impulse noise problems to telco at the first sign of trouble.

Although there are usually no published specs for distortion or phase response on telco circuits, total harmonic distortion (THD) should be < 0.25% on 15 kHz lines. For stereo pairs, relative phase response should be within 30° across the passband. Widely divergent frequency response between the two circuits in a stereo pair is a tip-off to check phase response carefully.

The maximum level guaranteed on telco circuits is +18 dBm, which provides only 10 dB of headroom above the +8 dBm reference level. A sensible alternative is to use +4 dBm as a reference level (most professional audio hardware uses this level anyway), thus allowing a more realistic 14 dB of headroom, at the expense of 4 dB less SNR—generally a worthwhile trade-off.

Interfacing Procedures

Figures 6.1-1 through 6.1-5 illustrate some do's and don'ts of audio interfacing to analog telco circuits. This is where impedance matching is still important, because telcos equalize their circuits for flat response and measure signal levels based on a 600-ohm source and 600-ohm termination. Due to the reactive components in long wired paths, varying these impedances will affect the frequency response and levels of audio

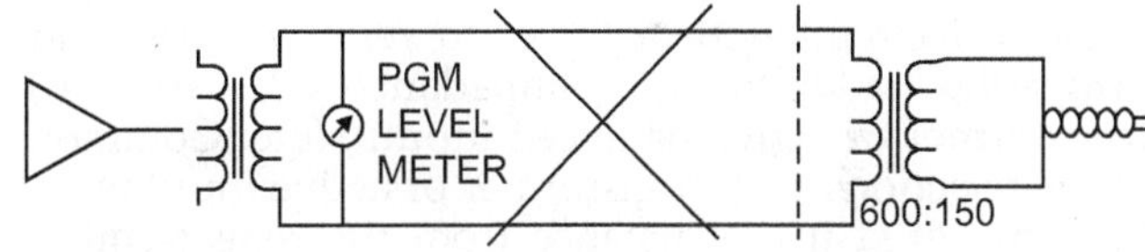

FIGURE 6.1-1 Don't meter the signal across the input to the repeat coil (transformer at right), and don't feed the repeat coil directly from another transformer, if possible.

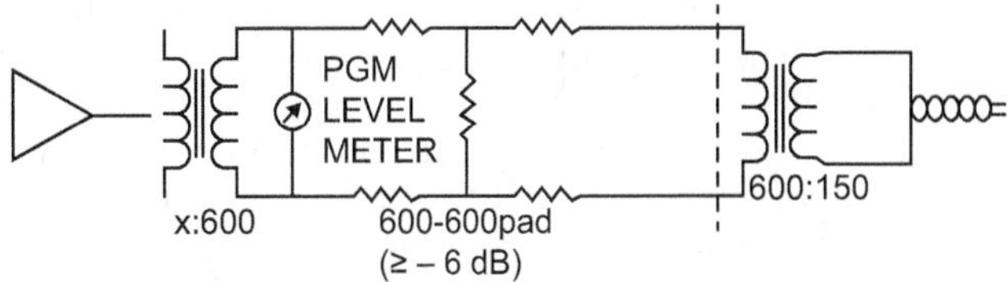

FIGURE 6.1-2 Do place a resistive pad between the output device's transformer and the repeat coil, and do place a level meter before the pad, calibrated for the voltage across the pad's output when terminated with a 600-ohm resistor in place of the repeat coil. The resistive elements of the "H" must be sized to provide a 600-ohm source to the repeat coil, taking into account the impedance of the output transformer.

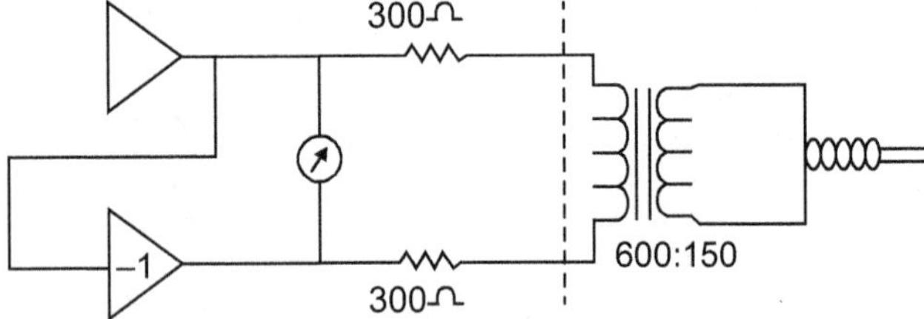

FIGURE 6.1-3 Do feed low-impedance sources (e.g., op-amps) through a 600-ohm differential balanced pad, and do place the level meter before the pad, calibrated as in Figure 6.1-2.

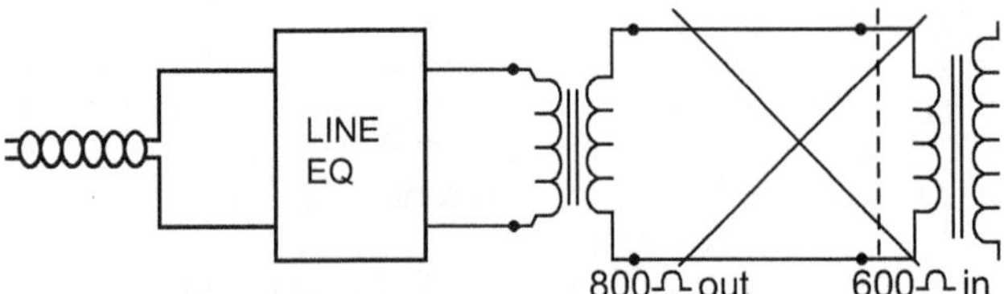

FIGURE 6.1-4 Don't terminate the receive-end repeat coil (transformer at right) with another transformer, if possible. Its loading may vary with frequency from the true 600-ohm resistive termination used in line-up, causing level and frequency response variations.

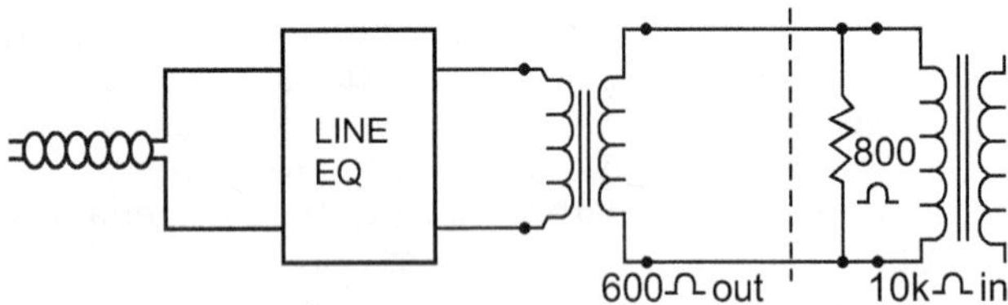

FIGURE 6.1-5 Do terminate the receive-end of an equalized circuit with a 600-ohm resistive load and bridge the load with a high-impedance (e.g., 10k ohm) transformer or active balanced input. Note: Common mode rejection of the transformer or active input must be considered if the distance from the repeat coil is great or is near other lines that induce cross talk.

carried on the circuit. Contemporary audio equipment have solid state amplifiers that are transformerless, resulting in output impedances much lower than 600 ohms, and input impedances much higher than 600 ohms. These amplifiers require interface circuits to work with equalized lines, as described later.

The secondary of the repeat coil should be resistively terminated. Instead of a bridging transformer, an active, balanced input circuit works nicely, provided any RF is bypassed before the first stage of amplification. An alternative, often seen on the input of broadcast line amps and modulation limiters, is a 600-ohm H-pad, effectively the reverse of Figure 6.1-2. It should be noted that many audio mixers are designed with input and output circuits that are not suitable for equalized lines. Following the aforementioned interface steps, however, will ensure that transmission loss and frequency response closely match those the telco set up.

DIGITAL AUDIO CIRCUITS

As with much of the technological progress in our industry, the digital audio revolution has dramatically affected the audio contribution and distribution world that began with analog telco. The transmission of data is nothing new to telco, but as discussed in this section, advancements in technology have offered ways to reduce the costs and offer new options for digital transport.

Audio program circuits are often ordered from telcos not as audio services per se, but simply as digital circuits with appropriate bandwidth (i.e., data rate) for the transmission task at hand. Unlike dedicated audio circuits, the actual signal(s) passing through the line are of little consequence; a data channel on a digital circuit cares not whether the content it carries is a symphony or a spreadsheet.

Data Rates

Like their analog counterparts, digital circuits come in various bandwidths. But rather than specifying cutoff frequencies of the audio passband, telcos specify digital circuits in terms of their data rates. Broadcasters now need to consider bandwidth requirements in two dimensions: that of the audio signal itself and the data-transmission path. The former is specified directly (in Hz), whereas the latter is specified as a data rate (in bps). The resultant audio quality through a digital transmission path is a function of the bandwidth of the digital circuit and the performance of the coding algorithm at that data rate. For example, linear PCM (pulse code modulation) will provide only telephone-like results at 64 kbps, but ISO/MPEG-1 Layer III Audio Coding can provide near CD-quality mono audio on the same 64 kbps circuit.

Therefore, broadcasters must develop acuity for the appraisal of various data rates' capabilities, both with and without data compression. In the linear PCM (uncompressed) mode, the data rate requirement of a given audio signal is determined by multiplying its sampling frequency (in Hz) by its resolution (in bits). In Table 6.1-1, three examples are given. This is only a partial list, as sampling rates, bit resolution, and numbers of channels may vary from these examples.

The resulting bit rates shown could be costly for leased telco circuits. Fortunately, data compression (or bit-rate reduction) systems have made possible transmission paths that previously were considered useful only for computer interconnection. Reductions from linear digital audio transmission of 4:1 are commonplace, and compression ratios of 12:1 or more without significant audible penalty are now frequently used. (For more detail on using data compression see the section "Considerations for Planning and Operating Digital Circuits.")

Telephone company installations and tariffs provide a variety of services, with more new services continually being deployed in many cities. Table 6.1-2 shows some common wired, point-to-point data services and their capabilities.

TABLE 6.1-1
Some commonly used digital audio data rates

Application Examples	Sampling Rate		Resolution (Bits/Sample)		Primary Bit Rate		Audio Channels		Resulting Bit Rate
CD Audio	44.1 kHz	×	16	=	705.6 Kbit/s	×	2 (stereo)	=	1.4112 Mbit/s
Professional Recording, Studio Mixing	48.0 kHz	×	24	=	1.152 Mbit/s	×	2 (stereo)	=	2.304 Mbit/s
Recording, Mixing, Distribution	48.0 kHz	×	16	=	768.0 Kbit/s	×	2 (stereo)	=	1.536 Mbit/s

TABLE 6.1-2
Some U.S. digital telco data services and their characteristics*

Designation	Line Rate	Mode/Carrier
Switched 56	56 kbit/s	Switched/copper
ISDN-BRI	128 kbit/s	
ISDN-PRI	1.472 Mbit/s	
DS 0	64 kbit/s	Dedicated/copper
DS1 (T-1)	1.544 Mbit/s	
DS2	6.312 Mbit/s	
DS3	44.736 Mbit/s	
OC1	51.84 Mbit/s	Dedicated/optical
OC3	155 Mbit/s	
OC12	622 Mbit/s	
OC48	2500 Mbit/s	
OC192	9600 Mbit/s	

*The ISDN data rates shown are for B-channels only.

One of the most important differences to the broadcaster between analog and digital circuits is that like dial-up, but unlike analog program circuits, digital services are almost always provided bidirectionally. This fact should not go unnoticed when making cost comparisons to analog circuits. Although interfacing hardware for return path channels must still be provided, their circuits require no separate costs or orders, as they likely would have with analog paths.

Telcos have always had their own insider vocabulary and set of acronyms, but since the introduction of digital services, this has expanded dramatically. Many of these are explained in the "Glossary of Telco Terms" at the end of this chapter.

DS1 and T-1 Service

Digital audio transmission on Digital Signal 1 (DS1) or T-1 lines has become widely available and is often cheaper than standard analog circuits in both service and installation charges. (See the "Glossary of Telco Terms" for the distinction between DS1 and T-1 nomenclature.) DS1 is a bidirectional 1.544 Mbps serial data link. The previous data rate calculations show how DS1 can carry a single, linear PCM stereo audio signal or several such compressed channels.

DS1 service is extremely reliable. Its bit error rate (BER) of 10^{-9} (the specification of error that no more than one erroneous bit in a billion is carried by the circuit) is among the lowest available. By comparison, IEEE and ITU-T standards establish 10^{-6} as the BER required for data customers.

A DS1 circuit is actually a multiplex of 24 data channels, or slots, of 64 kbps each, called DS0 channels. For standard telco T carrier use, each DS0 carries a digital voice grade circuit, using an encoding algorithm called µ-Law, which is a nonlinear companding (compression and expanding) scheme to get more dynamic range than is possible from linear coding with 8-bit samples at an 8 kHz sample rate. When a customer leases a DS1 circuit, it can be configured to carry any bandwidth channel for which DS1 hardware is available for (3.5 kHz, 5 kHz, 7.5 kHz, 15 kHz) in any combination, up to the customer's payload data limit. When a customer leases a full DS1, the telco may use one DS0 slot for framing and other overhead, in addition to the 8 kbps synchronization slot, leaving around 1.4 Mbps for customer data. Check with the telco if it is important to know the exact rate.

A rack of coding and multiplexing hardware appears on each end of the DS1 line, usually as customer-provided equipment (CPE), and the circuit can be reconfigured simply by changing the appropriate cards in the proper slots in the racks at both ends. The customer can perform these reconfigurations at any time, without telco involvement or notification.

Unlike the labor-intensive installation and equalization of an analog circuit, putting in a T-1 circuit has become nearly as routine as a standard dial-up telephone service installation. This and the excess capacity in some areas continue to lower costs for DS1 service, although other newer services—also cheaper than DS1 in most cases—are also being aggressively marketed in some regions. Customers' use of digital compression systems on these channels can also increase this economy. Whereas a 15 kHz (mono) audio channel had originally required six DS0 slots, today the use of perceptual coding can reduce this to two (or even a single) DS0 slots.

For overseas links, rough equivalents to each of the domestic services shown in Table 6.1-2 do exist outside the United States, but their actual data rates differ. Format conversions are therefore required for international transmissions, but most LDSs can handle this for the broadcaster. For example, European telcos offer an E-1 rate of 2.0 Mbps, in lieu of the North American T-1 rate of 1.5 Mbps.

Fractional T-1

In some areas, Fractional T-1 service is available, generally only for local intra-LATA applications (LATA refers to telephone communications within a "local access and transport area"). This service allows a customer to lease only the number of DS0 slots on a DS1 circuit that are needed for a particular application. Although installation charges will be about the same as a full DS1, service charges may be substantially reduced for many remote audio applications.

Switched 56

Another telco digital offering that has been used by broadcasters is the Switched 56. This digital service became available in the 1980s in many metropolitan areas in the United States, and from some long-distance carriers, but it is now largely obsolete. It provided a full-duplex (bidirectional) 56 kbps data path between dial-up terminals up to 5500 meters from the CO over standard 26-gauge copper twisted pair wiring. Like POTS, a monthly service/network access fee was charged (which often included at least some free local calling), with long-distance calls billed by the minute, at rates similar to regular dial-up long-distance service. In some cases, local calls were also billed for connect time. An installation fee was also typically charged for initiation of service.

Some telcos may still offer a similar unswitched service, in which a single DS0 channel can be leased on a monthly basis. For heavy point-to-point users, this may be more economical than a switched approach. As single DS0s, these circuits operate at 64 kbps (but one bit per byte is used for band signaling, leaving 56 Kbps for data), and their terminal hardware is less expensive because it need not accommodate switch signaling.

ISDN Service

During the 1990s, integrated services digital network (ISDN) became widely available in most U.S. urban centers and eventually became almost universally deployed. The most common variety of ISDN service is the basic rate interface (BRI), which provides two 64 kbps paths (bearer or B-channels) and one 16 kbps D-channel. BRI-ISDN is often referred to as 2B+D. Like Switched 56, ISDN is carried over lines between the customer and the telco's CO switch.

Bearer channels carry customer data, whereas data channels carry signaling and call routing data. This feature is a significant departure from POTS and Switched 56 service, in which the call signaling data is routed via the same data path as the program audio or data. (This is why DTMF (dual tone multiple frequency) tones and call-waiting beeps or pulses are audible on a POTS call, for example.) Such interruptions can cause problems to data communications, which benefit from continuous, bidirectional connections. ISDN's use of such out-of-band signaling provides significant improvement in

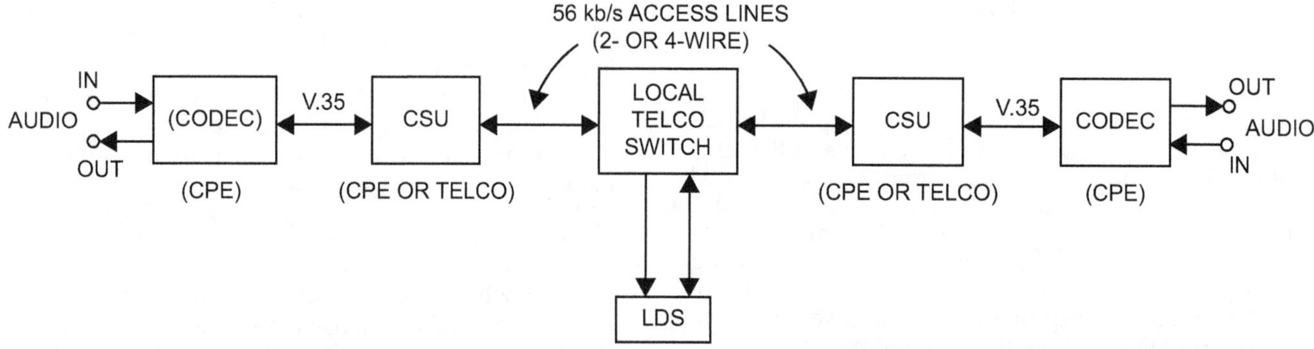

FIGURE 6.1-6 Block diagram of Switched 56 circuit path. (See "Glossary of Telco Terms" at end of chapter for acronym definitions.)

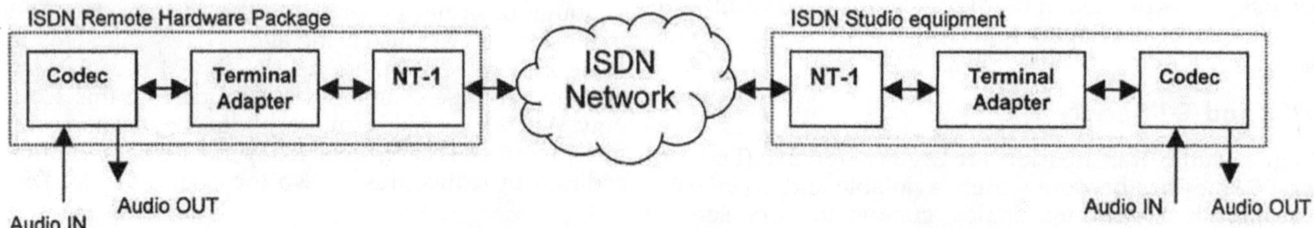

FIGURE 6.1-7 Block diagram of typical BRI-ISDN service used for a radio remote

connection speed, throughput, and robustness for switched data communications.

BRI-ISDN is intended as a standard residential service, while business service is provided by ISDN's primary rate interface (PRI). This service provides 23 64-kbps B-channels and 1 64-kbps D-channel (23 B+D). (The D-channel in PRI-ISDN operates at a higher rate than the BRI-ISDN service's D-channel because it has many more B-channels to manage.) By using a bonding protocol, the twenty-three B-channels can be combined to form a single 1.472 Mbps data channel, which is nearly equivalent to the capacity of a T-1 circuit (1.536 Mbps). (Some PRI-ISDN systems in the United States offer users only 56 kbps of payload data on their B-channels, for a net capacity of 1.288 Mbps.)

ISDN is a bidirectional, customer-switched service, operating as a dial-up, billed minute data network, allowing both circuit-switched and packet-switched operations. Its multichannel nature allows simultaneous voice and data to separate destinations (like multiline POTS service) or other applications in which multiple B-channels are combined to provide a single, higher-bandwidth call. This is done via a process called inverse multiplexing (IMUX). A typical IMUX might be capable of inverse multiplexing together four BRI-ISDN lines to provide a throughput of 512 Kbps. This saves the expense of having to purchase or lease equipment to individually terminate each ISDN line, and it provides an efficient way to increase network bandwidth.

The ISDN service is provided by a terminal adapter (TA). A terminal adapter connects using an RJ-45 connector to the ISDN line at the customer premises and provides electrical termination for this line. The other interface on the terminal adapter is typically an RS-232, RS-366, RS-530, or V.35 serial interface such as DB25.

By the late 1990s, BRI-ISDN had become the delivery method of choice for most radio remotes, offering mono or stereo 15 kHz broadcast audio plus communications to/from a remote site. Often a single B-channel is used with one of several audio codecs capable of providing high-quality mono audio transport at a 56 or 64 kbps rate. In other cases, two B-channels are IMUXed for high-quality stereo at 112 or 128 kbps. Numerous devices are available for such applications, and most include an integrated ISDN terminal adapter and codec, plus machine-control closures (see Figure 6.1-7). These devices may offer multiple codecs, including ISO/MPEG-1 Audio Layers II (AAC) and III (MP3), APT x-100, Microsoft Windows Media Audio Professional (WMA Pro), Apple QuickTime, MPEG-4 AAC-LD (optimized for low delay), or another with extended frequency response, called High-Efficiency AAC (HE-AAC, also known as AAC+).

At the time of this writing, many telcos are no longer installing new ISDN service, and some are phasing out or have eliminated the service, to the dismay of many radio broadcasters who had happily used the service for many years.

DSL Service

One of two popular methods being deployed for broadband Internet connectivity (although it can be used for other purposes, such as IPTV), and which played a role in broadcast audio transport, is the digital subscriber line (DSL) service. It provides multimegabit connectivity over the telcos' existing twisted-pair infrastructures, which is accomplished by implementing adaptive digital signal processing (DSP) at both ends of the signal path, thereby compensating for the distortion inherent in long copper lines. Path lengths for DSL are limited, but the most widely deployed systems are designed to accommodate the typical nonrural telco customer-to-CO runs of 12,000 to 18,000 feet. DSL is provided by telcos as a data plus voice service on the same pair of wires, with a standard switched POTS service combined with dedicated broadband data service. The two services are split at either end of the line. In many areas, stand-alone or dry DSL service is also offered (often called "Naked DSL"), which provides the broadband data only without POTS.

DSL paths are dedicated, point-to-point services, and intermediate repeater amplification is not used. Two different transmission technologies were originally employed, Discrete Multitone (DMT, developed by Amati Communications), and Carrierless Amplitude / Phase modulation (CAP, from AT&T), but most deployments today use DMT, which is essentially similar to orthogonal frequency division multiplexing (OFDM). Importantly, DSL's data service is not switched (so it is not a direct substitute for ISDN), and it generally is installed between a customer's service location and a telco central office, not from service location to service location (so it is not a direct substitute for T-1). Nevertheless, it may provide an economical alternative to either service for transmission of audio files via the Internet, or in some cases, for real-time audio backhaul (see the later section "Audio Over IP").

DSL service comes in a range of data rates (typically from 128 kbps to 24 Mbps), and is offered in both symmetrical (SDSL) and asymmetrical (ADSL) forms. The latter offers a higher data rate from the telco CO to the customer's service location ("downstream") than in the opposite direction (from customer to CO, or "upstream"). Actual speeds of any service are affected by physical wiring path lengths between the service location and termination at the telco CO, but it is not uncommon for ADSL service to be limited to an upstream data rate of 64 kbps. Because the upstream path is what a broadcaster would use when feeding audio from a DSL-connected remote site, it is the more important data rate for such applications, although telco typically identifies and markets the service by its higher downstream rate. So take care when ordering the service to specify the minimum upstream data rate from the service location. If this service is shared with other networked users at the originating site, audio quality will be further constrained, so it is recommended that any such circuit be exclusively dedicated to audio backhaul when used for broadcast applications.

More appropriate to broadcasters may be the high bit rate (or very high bit rate) digital subscriber line (HDSL or VDSL), which can be configured to provide symmetrical high-speed capacity, although over shorter distances than noted previously. For example, a typical VDSL can provide around 10 Mbps in each direction across a 1-mile path. Lower speeds may be possible over longer path lengths. In some cases, multiple physical DSL lines can be bonded to achieve higher aggregate bandwidths.

If voice telephony service is otherwise provided at the site, Naked DSL may be an appropriate choice for broadcasters who wish to use it simply for broadband Internet access at a given location.

DOCSIS Cable Service

The other primary wired method of broadband Internet access to consumers is over cable television systems' coaxial cable, typically via the Data Over Cable Service Interface Specification (DOCSIS). At this writing, the DOCSIS 2.0 standard has been widely deployed, which is better suited for broadcast applications (again, given the previous caveats regarding use of the public Internet), due to the 2.0 spec's higher upstream speeds and improved QoS capabilities over the predecessor DOCSIS 1.0 and 1.1 standards. In some areas, a successor format, DOCSIS 3.0, is emerging at the time of this writing.

DOCSIS uses quadrature amplitude modulation (QAM) (from 16-QAM to 256-QAM are specified), with a maximum upstream throughput of 30 Mbps per channel and a maximum downstream capacity of 42 Mbps. A "channel" is carried on a 6 MHz-wide TV channel of the cable system infrastructure, and each of these is typically shared by multiple cable-modem customers. DOCSIS 2.0 supports one channel in each direction, while DOCSIS 3.0 can bond any number of upstream or downstream channels to multiply throughput.

To reduce congestion during heavier usage, cable operators set limits on each customer's upload and download rates, typically with lower limits for uploads, which is the direction broadcast audio transport needs most. Because the DOCSIS capacity is shared among multiple users, the throughput may be lower than the advertised rate in some dayparts and can drop steeply during moments of high demand. To avoid an unexpected shortage in throughput, customers should order a substantially higher rate than the audio transport requires, and test the audio thoroughly for dropouts before the live event. Despite the apparent excess capacity, to avoid momentary shortages it is advisable to order a separate service for other data, such as file downloads, Internet browsing, and so forth.

WIRELESS BROADBAND LINKS

Wireless broadband for digital audio transport has grown to deserve its own section of this chapter. It has two distinct categories:

- The IEEE 802.11x ("Wi-Fi") family of standards for fixed local area networks, which are unlicensed, and its descendants, IEEE 802.16x and 802.20x, which may operate in frequency bands requiring an FCC license by the operator.
- The use of wireless links over mobile (cellular) data networks that are operated by telecommunications carriers.

With few exceptions, all these services are simply first/last mile access routes to/from the Internet, so they are subject to the same caveats noted previously on use of the public Internet for broadcast audio transport applications.

The 802.11x family has adequate bandwidth for even uncompressed broadcast audio application, and is generally provided as a private, short-range service, in the unlicensed spread spectrum bands of 2.4 or 5 GHz. (See Table 6.1-3 for details.)

The maximum data rates specified are achieved only under optimal signal conditions; when signal strength or quality fades, data rates are reduced in steps until the link fails. For example, the 802.11b standard provides 11 Mbps bidirectional connectivity under optimal conditions, shifting down to 5.5 Mbps, then to 2.0 Mbps, and then to 1.0 Mbps, as signal strength or quality decreases. All signal ranges quoted assume indoor transmission and reception, using omnidirectional

TABLE 6.1-3
Selections from the IEEE 802.11x family of wireless LAN standards and their basic characteristics

Standard	Release Date	Operating Freq. Band	Data Rate (Typical)	Data Rate (Max)	Range	Comments
802.11	1997	2.4 GHz	1.0 Mbps	2.0 Mbps	?	Legacy
802.11a	1999	5 GHz	25 Mbps	54 Mbps	30 m	Less interference than 2.4 GHz but more propagation concerns
802.11b	1999	2.4 GHz	6.5 Mbps	11 Mbps	30 m	Original "Wi-Fi"
802.11g	2003	2.4 GHz	25 Mbps	54 Mbps	30 m	Backward compatible to 802.11b
802.11n	2009	2.4 GHz & 5GHz	200 Mbps	540 Mbps	50 m	Adds MIMO antennas
802.11ac	2013	5GHz	500 Mbps	7 Gbps	50 m	Wider channel bandwidths and higher-order modulation

antennas. Somewhat longer ranges can be expected outdoors or with directional antennas. (MIMO = multiple input/multiple output antennas; see "Glossary of Telco Terms.")

Note also that the data overhead of 802.11x limits the available throughput for streaming content to about 60% of available bandwidth for TCP and about 70% for UDP. Thus, on a fully available, best-case 802.11b link, for example, maximum throughput for audio streaming is never more than 7 Mbps or less, and often lower. Nevertheless, this is still adequate for a single, uncompressed stereo audio feed.

More importantly, however, the availability and reliability of 802.11x service are variable and, on the whole, less than wireless services provided by actual wireless data service providers (telcos), due to their generally private and highly localized ("hotspot") deployments. Note also that even if an 802.11 service is available, it may be blocked for usage by the access point owner. So "Wi-Fi" is useful for broadcast audio transmission only in certain very specific, predetermined, short-haul cases.

At this writing, it is expected that 802.16x (WiMAX) systems may be of somewhat greater utility to broadcasters, given their anticipated greater coverage. While 802.11x is considered a wireless local area network (WLAN) standard, 802.16x is a wireless metropolitan area network (WMAN) standard, which is anticipated to operate in a "mesh" fashion, covering a large area from multiple antenna sites, with customer devices that have an operating range up to several miles. It will operate at various spectral locations between 2 GHz and 66 GHz (typically 2–11 GHz), offering 40 Mbps or greater bidirectional customer connectivity. The 802.16e variant will offer mobile reception capability (dynamic handoffs between nodes, like cellular telephony), but initial deployments are expected to be for fixed usage only. If licensed services are deployed by communications services operators, these may provide adequate reliability for broadcasters to consider using for Internet access, particularly for quick availability from random remote sites around their metro areas.

Meanwhile, most mobile telephone companies are developing so-called 3G (third-generation) wireless data services, as extensions to their existing voice and low-speed data networks. These are almost all packet-based, asymmetrical formats designed for Internet page browsing on handheld terminals with small, low-resolution screens, so they are not particularly applicable to real-time broadcast backhaul use—particularly given their low upstream capacity. One proposed service might prove to be an exception, however. It is called EV-DO Rev A, for Revision A to the Evolution-Data Optimized standard (also called 1xEV-DO Rev A), which will offer upstream capacity of up to 1.8 Mbps, along with a wideband audio element optimized for real-time, low-latency streaming.

Broadband Global Area Network (BGAN)

While the wireless broadband systems discussed in the preceding sections are all terrestrial services, the international satellite telecom service provider Inmarsat has developed its own form of broadband connectivity called BGAN. This service provides voice plus data, with adequate upstream bandwidth (about 256 kbps) for real-time audio up-linking via the Internet from nearly anywhere on earth. BGAN may be useful to broadcasters for "extreme" remotes where no other form of connectivity is available.

DIGITAL NETWORK TECHNOLOGY FOR AUDIO CONTRIBUTION

A new generation of digital audio solutions has begun. This new phase moves away from systems that packetize and directly modulate carriers using traditional telco data transmission formats to a method that encapsulates digital audio packets into the Internet protocol (IP) for networked transmission.

Audio Over IP

Audio over IP (AoIP) is the distribution of digital audio across an Internet protocol (IP) network. It is being used increasingly to provide high-quality audio feeds over any distance. A contribution over IP offers additional transmission efficiencies, and leverages the cost-effectiveness of the IP networking architecture, which can be carried over a variety of transmission paths, including Ethernet, ISDN, ATM, FireWire and DSL. As ISDN line availability declines, AoIP has increasingly become the solution for audio transport of all types, from field reporters to major sporting events to music concerts with multichannel sound.

The fundamentals of AoIP networking are beyond the scope of this chapter, but are explained in many other widely available general texts on the subject (see "References" section). More details on audio over IP networks are also available in Chapter 4.2 of this handbook. Some specifics of AoIP interfacing are worth noting here, however.

Whether working with a small internal network or one that expands to the planet, digital networking architecture should be kept in mind, which was originally defined by a seven-level stack formulated by the ISO, called the open systems interconnect (OSI) model. In today's digital networking (including the Internet), the OSI model has been condensed into five layers, as shown in Figure 6.1-8.

While the network layer (IP) is ultimately responsible for delivering data packets from device to device via packet-address management, the transport layer (TCP or UDP) is charged with protecting the integrity of the data thus delivered. This protection is done on a "best-efforts" basis, meaning that accurate delivery is not guaranteed in all cases.[1]

IP networking allows both point-to-point ("unicast") and point-to-multipoint ("multicast") transmission styles. When the latter approach is used, the UDP (rather than TCP) transport is often engaged. UDP/IP offers less robust error recovery, but it provides a simpler and more direct way to send and receive data

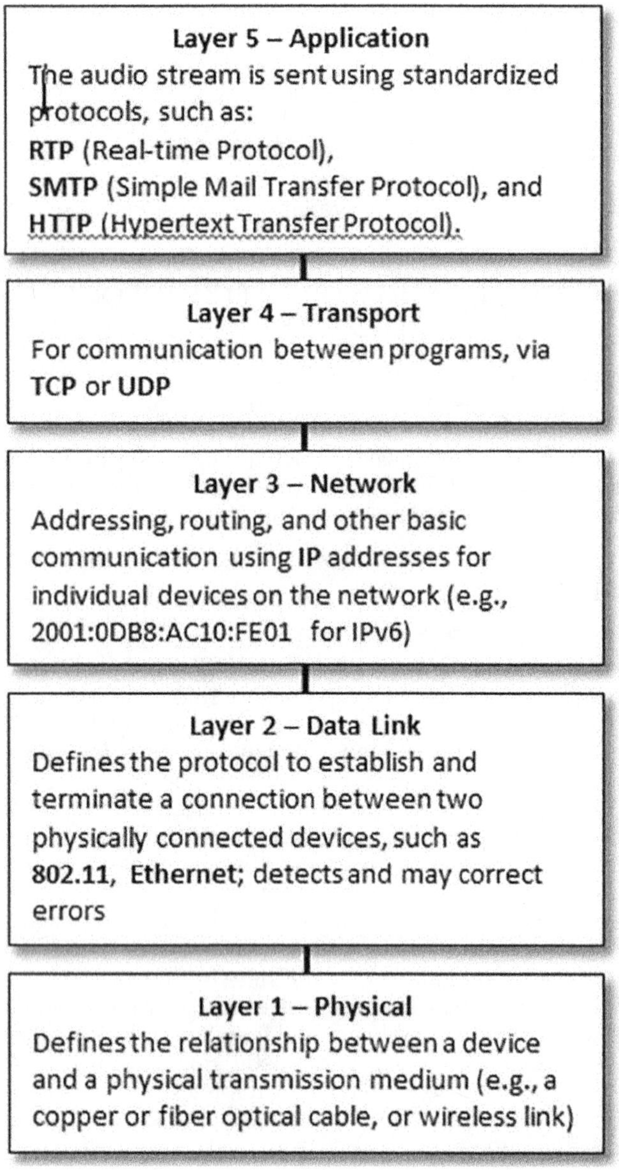

Layer 5 – Application
The audio stream is sent using standardized protocols, such as:
RTP (Real-time Protocol),
SMTP (Simple Mail Transfer Protocol), and
HTTP (Hypertext Transfer Protocol).

Layer 4 – Transport
For communication between programs, via **TCP** or **UDP**

Layer 3 – Network
Addressing, routing, and other basic communication using **IP** addresses for individual devices on the network (e.g., 2001:0DB8:AC10:FE01 for IPv6)

Layer 2 – Data Link
Defines the protocol to establish and terminate a connection between two physically connected devices, such as **802.11, Ethernet**; detects and may correct errors

Layer 1 – Physical
Defines the relationship between a device and a physical transmission medium (e.g., a copper or fiber optical cable, or wireless link)

FIGURE 6.1-8 Typical digital network layers

packets. Figure 6.1-9 illustrates the transmission of audio packets generated by application "A" to application "B." A combination of networking elements is used to prepare and send the packets over the IP network. (Note that the data packets appear to flow in one direction, but the network is bidirectional and some handshake data flows from application B to A. Some networks are fully bidirectional.)

AoIP equipment working within the hardware of the link layer will suffice for small internal networks, but AoIP networks of greater size must rely on the network layer. To facilitate interoperability between previously competing AoIP systems, in 2013 the Audio Engineering Society developed the AES67 Standard.[2] Since its publication this Layer-3 protocol suite has been implemented by several IP-based audio networking manufacturers including RAVENNA, Livewire,[3] Q-LAN, WheatNet-IP, and Dante.

Audio-Specific Techniques

Digital networking is primarily designed for the transfer of data files, which is inherently not a real-time process. Part of basic network functionality is the mitigation of data collisions for optimum use of the network by many simultaneous users. In the interest of minimizing file-transfer time and the possibility of collisions, increased networking speeds are constantly sought after. Today's state-of-the-art for LAN applications of IP is gigabit Ethernet architecture, providing approximately 1 Gbps of network bandwidth on CAT-5 wiring.

These networking speeds can manage faster-than-real-time transfer of very large audio files, but with proper care, they can also provide reliable real-time, multichannel audio streaming on LANs. Various network applications have been developed to provide this kind of network management, such as CobraNet. Originally developed by Peak Audio, CobraNet is now implemented by several manufacturers for Ethernet-based digital audio routing.

Protocols such as CobraNet and others address the specific needs of real-time audio transmission over a digital network, in particular the reduction of TCP-and UDP-related latencies associated with high-bandwidth (uncompressed audio) streams and improved quality of service (QoS) management. (See the "Latency" section in this chapter for more information on this topic.)

Thus, the IP Audio approach formats digital audio signals into a structure that can easily pass over any IP-enabled network, whether it is an intra-facility LAN, a dedicated inter-facility network, or the public Internet.

Of course, the use of the Internet for a critical broadcast signal path presents potential problems, due to the uncertainties of network conditions over time. The Internet's "best efforts" service implies that no particular quality of service (QoS) is specified or guaranteed. This stands in stark contrast to more traditional digital telco transmission services, where robustness is high and a minimum BER is typically specified.

This does not imply that IP Audio is inherently unreliable, however. It is important to maintain a clear distinction between IP and the Internet. The latter is simply one of the physical networks (link layer) that utilize IP as a networking protocol. Many applications of IP Audio do not use the Internet, but travel on more robust broadcaster-operated networks, which may even have dedicated bandwidth assigned for a critical audio signal path.

For the applications considered in this chapter, however, it is implicit that transmission across non-broadcaster-operated networks (i.e., telco networks) will often be required. IP Audio is the format of choice for an increasing number of these data transmission offerings.

IP Audio Applications

There are three basic methods available to broadcasters today for inter-facility transmission of their audio

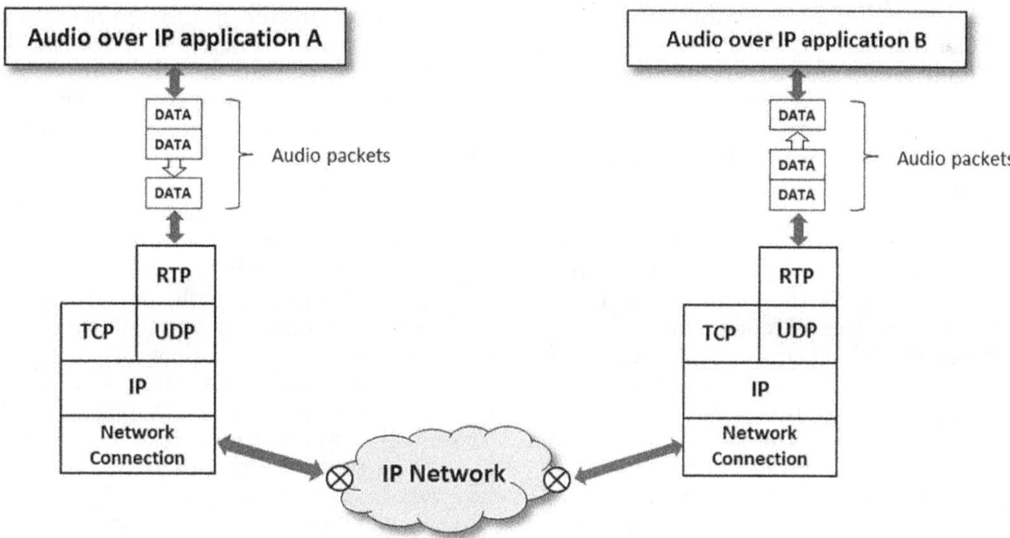

FIGURE 6.1-9 Sending AoIP from one application to another

content via IP. The first is the use of traditional dedicated digital transmission services such as T-1, which can be repurposed for IP transmission with the appropriate terminal hardware at each end.

The second approach is to simply purchase IP bandwidth from telco. The actual signal-transport architecture used will be determined by telco. This may be a more cost-effective approach if the requirements on a given path are only occasional, or particularly if bandwidth needs vary widely over time, in which case a "bandwidth on demand" option may be provided by telco.

The third method is the use of broadband Internet service, with each involved facility connected via broadband Internet access of some form, such as a telco digital subscriber line (DSL) or a cable modem (DOCSIS), both described later.

Of particular interest to broadcasters for remotes is an emerging variant on this option, which offers *wireless* broadband Internet access, via a wide range of technologies, each offered by different wireless service providers (see "Wireless Interconnect" later in this chapter).

Using the Public Internet for Broadcast Audio

The first two methods described in the preceding section require lead time for ordering and dedicated installation, plus relatively high cost for CPE hardware and/or service, but they provide high reliability and guaranteed QoS.

The third method—in all its variants—offers the option of easy access with no dedicated-line installation, and if Internet access is already available on site, no lead time or service orders at all. Connection hardware is also generally inexpensive and off the shelf. This flexibility is highly attractive to broadcasters, but connections via the Internet do not offer guaranteed availability or quality of service. As a result, broadcasters typically have avoided this option for live-to-air or other critical audio transport applications. Yet some other emerging technologies have reduced the risks involved with use of the public Internet, as have overall—and continuing—improvements in robustness, availability, and bandwidth of the Internet and its access paths.

These devices include generic data as well as audio-specific devices. Among the former are LAN-contention and prioritization devices (sometimes called broadband boosters or accelerators) that can provide improved robustness for a particular connection requiring continuous, real-time connection within the upstream data on a local area network. These devices are targeted at the growing voice-over-IP (VoIP) telephony and online gaming markets, but can be applied for any real-time signal usage over a shared Internet connection. These devices work by monitoring all the traffic on a LAN and determining which applications within it require real-time priority. Of course, if an Internet connection path is used wholly for a broadcast audio application, without traveling through a LAN sharing the Internet access, these devices are of no value.

At this writing, an example of an audio-specific device for this application is a system developed by Comrex Corporation called Broadcast Reliable Internet Codec (BRIC), which allows adjustable and adaptive redundant packetization of audio, such that a stable audio connection can be maintained across an Internet connection with minimum latency (typically 100 to 250 ms). Such processing is likely to substantially improve results on any type of broadband Internet connection, although guaranteed service over the Internet can still never be given.

Other basic networking techniques can also be used to improve robustness over the Internet, such as the use of fixed IP addresses and, in some cases, connection via virtual private networking (VPN).

Another issue to consider is the asymmetry of most broadband Internet connections. Upstream data rates are generally far lower than downstream rates. This implies that the upstream speed of the broadband service at the originating source location of an audio signal path will be the limiting factor in any Internet-based contribution channel.

Regardless of the technology or access method employed, using the Internet for real-time audio transport should always be considered as a second choice to dedicated point-to-point service due to the inherently lower reliability of Internet connections. But the differential between dedicated service and broadband Internet service is continually shrinking, and when considered with cost/benefit ratios and availability issues, the use of a broadband Internet connection for broadcast audio contribution or distribution may be an appropriate choice in an increasing number of cases.

Good recommended practice for using the public Internet therefore includes the following:

- Choose the most appropriate, efficient, and fault-resistant codec.
- Set the codec's output to the lowest tolerable bit rate for the application.
- Fortify the stream and set up the network appropriately at the sending end (as described earlier).
- Perform adequate testing.
- For live broadcasts, have a hot-standby (typically POTS or POTS-codec) backup.

Finally, many radio remotes may require Internet connectivity for other reasons, such as reporters' need for access to news sites and so on. For this reason, the use of a broadband Internet connection can serve double duty for both audio backhaul and a staff research resource. On the other hand, remember that the connection's bandwidth is shared by all users, so any critical audio signal paths traveling over the broadband Internet link should take priority, with adequate bandwidth reserved for their use whenever required.

CONSIDERATIONS FOR PLANNING AND OPERATING DIGITAL CIRCUITS

Wireless Interconnect

Wireless packet-based networks are now a common transport for digital audio, ranging from the local elimination of interconnection cables, as in WiFi and Bluetooth, to long-distance microwave links spanning tens of miles, using IEEE 802.16 (WiMAX) or dedicated radio hardware. For more detail on the types of wireless transport, protocols, typical bit rate values, and so forth, the reader may refer to the sections earlier in this chapter.

Latency

All digital transmission links suffer from some degree of throughput delay, or "latency." These delays—if longer than a few hundred milliseconds—can cause complications when the circuit is used for real-time broadcast applications, because it will require mix-minus to be used for monitor return feeds from the studio to the remote site. It may also result in pauses or interruptions when studio talent engages in conversation with remote talent.

The digital audio codecs used for digital audio transport can add significant delay—on the order of 100 ms in each direction in some cases—in the contribution (backhaul) path, as can studio-to-transmitter links and other components of the air chain. The cumulative delay from these devices can cause monitoring problems for talent listening to their own voices or in conversation with others. For this reason, dedicated analog communications lines may be useful even when air monitoring or digital return is possible.

When the Internet is involved, latency can become extreme—extending to several seconds—as well as quickly variable and unpredictable. Wireless broadband links add yet another layer of high and variable latency. Thus, broadcasters should be prepared to face the reality that even if reliable, high-fidelity connectivity can be achieved over a digital circuit that involves wired or wireless Internet connections, real-time two-way communication suitable for live broadcast may not be possible.

Data Compression

While the straightforward nature of linear PCM may have been helpful in making the transition from analog systems, especially where bandwidth was cheap and available, it is an inefficient method for encoding digital audio. The performance of today's linear PCM systems often exceeds the actual needs of most listeners, and significant reductions in transmitted data can be achieved by applying data compression algorithms to the data streams that linear PCM conversion produces. That being said, linear PCM with as high a resolution (bit depth) as economically feasible is still a good idea for the original conversion of analog signals to the digital domain, as well as for any digital production and signal processing. But for signal delivery systems (and in some cases, for long-term storage), data compression is an appropriate tool.

Earlier compression systems (again pioneered by telcos) used statistical methods to reduce a data stream's coding redundancies. These so-called lossless systems could be used to reduce the data rate of any kind of data transmission and by definition provide a data stream after decoding that is an exact, bit-for-bit representation of the original bit stream. Audio waveforms have limited redundancy, therefore this technique only provides a reduction of about one-third in bit rate. More recent "lossy" data compression systems designed for audio signals exploit the limitations of human aural perception to achieve much higher data compression ratios than lossless systems. Thus, they are generally referred to as "perceptual coders."

Unlike the lossless coder's statistical analysis of the bit stream itself, perceptual coders analyze the audio waveform and allow a recoding of the linear digital audio signal in such a way that far fewer bits are required to represent it, without substantially noticeable aural impairment in the process. Also, unlike lossless systems, this reduction is permanent, in that the original linear signal is never regenerated exactly as it was. But a good perceptual coder will arrange the signal on this lesser number of bits in such a way that the listener is not aware of the difference. Figure 6.1-10 is a chart showing subjective listening scores for different codecs having efficiencies (i.e., bit reduction ratios, relative to linear PCM) from approximately 1:3 to 1:23. These MUSHRA[4] tests give a score of 50 a rating of "fair," a score of 70 is "good," and 90 is "excellent." (For more on perceptual audio coding, see Chapter 4.8.)

Degradation From Cascading and Transcoding

As discussed earlier, digital audio compression introduces new and more cost-efficient ways to transport audio. With modern audio coding, the audio quality of each stage may be excellent. Unfortunately, broadcasters sometimes experience significant problems when using lossy audio compression in audio contribution circuits. This is often because lossy audio compression may be used not once, but two or more successive times, in contribution circuits, network distribution, and as the "last mile" in digital audio broadcasting and Internet audio streaming. The result may be audible artifacts in the audio, described as "grittiness," "watery" sound, and "swishy" or distorted sibilants.

This process is referred to as cascading, when the same audio coding is used, or transcoding when different audio coding is in the chain. Despite the likelihood of this occurring, there has been little research to date into quantifying the effect or determining what audio coding combines better or worse with the same or other audio coding.

One of the few studies using controlled listener testing was performed in 2005.[5] That study was performed by the BBC, which evaluated combinations of five codecs as they would be used in stages for acquisition, contribution, studio mixing, distribution, and broadcast. The choices of coders and bit rates was oriented to the television networks of the BBC, IRT (Institut für Rundfunktechnik in Germany), Radio France, and NRK (Norsk Rikskringkasting in Norway), which is less applicable to North American broadcasting, especially radio. Nevertheless, the testing generally indicated that combinations of MPEG-1 Layer II (MP2) resulted in higher scores, in isolation or in combination with other codecs. (MP2 is a perceptual audio coder for 1- or 2-channel content at typical bit rates of 128–384 kbit/s. The MP2's relatively low complexity coding may make it more resistant to cascading degradation, although it is less efficient than recently developed codecs.)

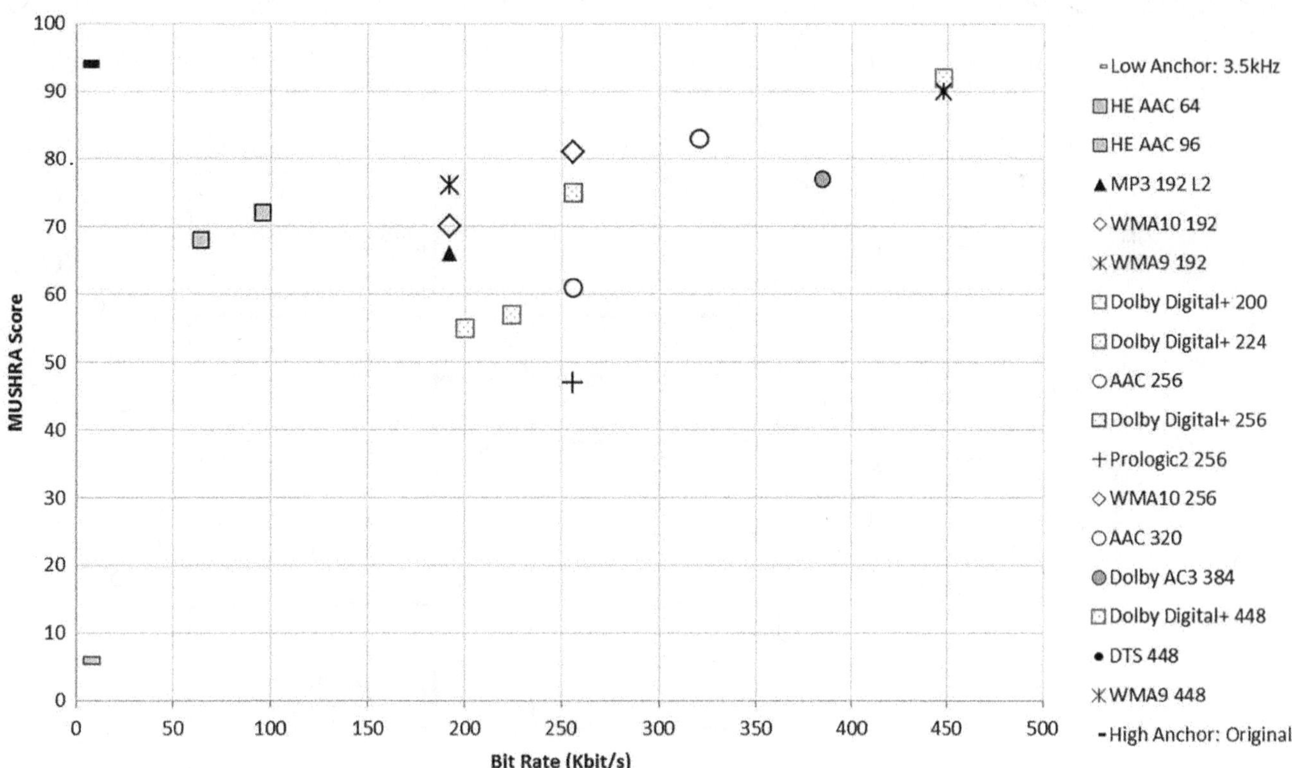

FIGURE 6.1-10 Subjective listening scores of various audio codecs and bit rates, as reported by the European Broadcasting Union (EBU)

Data courtesy EBU

Today's audio content production and broadcast are technically complex operations. The audio signal may pass through multiple stages of encoding before it is broadcast or streamed. It may be difficult to determine where lossy compression is introduced and specifically what codecs are in use. There are some guidelines that will help choose audio codecs and reduce the artifacts of cascading and transcoding, as shown in Figure 6.1-11.

On the right side, audio contribution occurs from the remote site. As the first stage of digital compression, recommendations for the choice of codec are shown below: one with low coding efficiency (bit rate reduction), which results in a relatively high bit rate. In the middle, audio may go through mixing, recording and distribution. This stage in the chain should also use high bit rates with low coding efficiency for minimal cascading degradation. On the left side, it is likely that the "last mile" codecs must have high efficiency to support low bit rates for streaming or broadcast. Their design rules out cascading, and while there is little flexibility in these codecs, the overall system will perform best with low-efficiency, high bit rate codecs on all earlier generations.

Cost of Services

For local service, there is normally only one provider of telco circuits, analog or digital, which is, of course, the local telephone operating company. So there is little competitive choice in the matter for intra-LATA service pricing. Check with the local telco frequently to determine which services are more economical, for either permanent or temporary applications. Analog services continue to increase in price in most areas (if they are even available), whereas the cost of digital services declines at varying rates around the country—although some telcos are also moving away from point-to-point services entirely in favor of their broadband Internet businesses. Therefore, it is worthwhile to keep a close eye on these changing rates and service offerings.

For long-distance paths, there is significant competition for inter-LATA digital service, so prices are kept low and continue to drop (inter-LATA refers to telephone communications between "local access and transport areas," or LATA; there is little or no inter-LATA analog service available anymore). Shop around for the best deal on inter-LATA digital service, or use a data line broker. Secondary services may also be offered, such as switching and monitoring.

Backup is also an important issue, particularly for live broadcasts. Many broadcasters establish primary and secondary backhaul and communications paths, using different services for each (for example, ISDN primary and POTS secondary services). This adds a level of redundancy because the services are typically routed differently. Costs of each service are an important factor, particularly when broadcasters are paying for something (the secondary service) that they hope they never have to use. The use of a lower-cost service for backup therefore makes both technical and economic sense, and it is a highly recommended practice.

For those broadcasters who still book a significant number of traditional point-to-point telco program circuits (or for historical reference), it also may be helpful to understand how telco billing works for these services. Figure 6.1-12 shows the traditional route-billing concepts followed by local and long-distance common

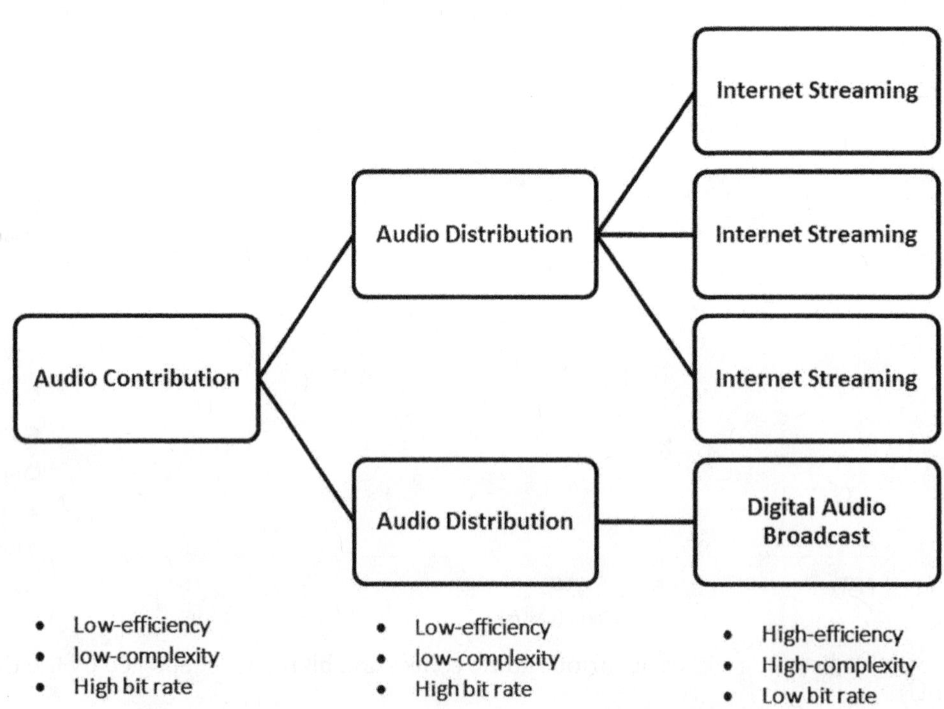

FIGURE 6.1-11 Considerations for minimizing cascading and transcoding effects

LEGEND

CM — Channel Mileage (per air-mile billing basis)
CO — Central Office (LEC switching center; also called SWC—Serving Wire Center)
CT — Channel Termination (flat rate billing)
IOC — Interoffice connection on LDS network (distance-sensitive)
LATA — Local Access and Transport Area (telco service zone)
LDS — Long-Distance Service (common carrier)
LEC — Local Exchange Company (local telco)
POP — Point of Presence [LDS's office in each LEC; also called SO (Serving Office)]

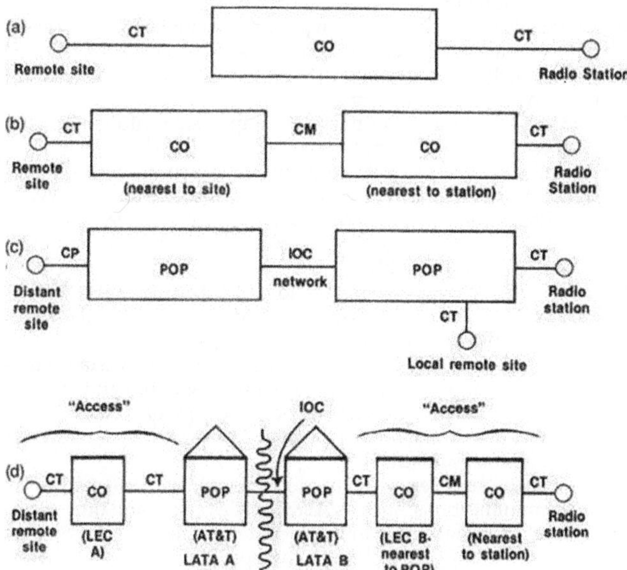

FIGURE 6.1-12 Traditional billing methodology for dedicated telco circuits

carriers for dedicated (nonswitched, point-to-point) lines. In (a), the two ends of a local circuit are served by the same central office, or "rate center." In (b), two different central offices are involved. In (c) and (d), long-distance service is depicted. LEC connections to LDS are referred to as "access." Access shown in LATA B will be more expensive because the radio station is in a different rate center of LEC B than the long-distance carrier's point of presence (POP).

On the other hand, costs for today's broadband Internet connectivity from telcos are generally quite straightforward, with flat monthly rates for a given bandwidth service being the norm. Remember that many services are asymmetrical, so *two* rates—upstream and downstream—are being quoted. Telcos are used to stressing the downstream rate to their general customers, but as noted previously, broadcast customers using Internet connectivity for remote backhaul are more concerned with the upstream rate. It is also important to understand that the actual access bandwidth may vary over time. While this is particularly true for DOCSIS and wireless services, even a DSL connection's bandwidth can vary with overall network load conditions. Here again, *upstream* rate consistency is the more critical consideration. Note that the quoted bandwidths refer only to first/last mile access to/from the ISP; throughput bandwidth and latency over the Internet backbone are "best effort" only and never guaranteed.

Numerous Internet sites and applications are available to test or verify the bandwidth of a given connection. They should be used by broadcasters whenever new service is established or when any change in service or recovery from an outage takes place. Note that many of these services test only the downstream rate, so be sure to use a test that considers upstream performance as well. Bandwidth test results should reach at least 90% of the specified rate (in both directions). Contact the service provider if results are significantly and consistently below this value.

The importance to broadcast radio of interconnection, immediacy, and fidelity underscores the importance of a long-term relationship between broadcasters and telcos. Good awareness of and rapport with communications service providers are essential to the daily work of the broadcaster. In today's context, that also means keeping abreast of the changes in telecommunications that affect broadcasting as the digital revolution continues to roll along.

GLOSSARY OF TELCO TERMS

1xEV-DO: See *EV-DO*.

3G: General name applied to third-generation wireless telephony, which includes significant broadband data capability.

802.11x: IEEE standards for wireless LAN (local area network), often called "Wi-Fi." Uses unlicensed spread spectrum in 2.4 and 5 GHz bands, offering nominal data rates of 11 Mbps and above (minimum 1 Mbps under impaired signal conditions), with nominal range of 30 meters (100 feet) or more. Includes numerous variants (802.11a, 802.11b, etc.); entire family referred to as 802.11x, not to be confused with 802.1X, which is a remote authentication protocol for wired or wireless Ethernet networking.

802.16x: Developing IEEE standards for wireless MAN (metropolitan area network), often called "WiMAX," is expected to offer 40 Mbps or higher data rates over cellular-type network of metro-area connectivity, from fixed or mobile terminals, in either unlicensed or licensed forms, in various bands of spectrum located between 2 GHz and 66 GHz.

Access point: The name used for an IEEE 802.11x ("Wi-Fi") transceiver, which connects the wireless service area to a wired network.

ADPCM: Adaptive differential pulse code modulation. A form of digital coding more efficient than linear PCM because it codes only the difference between one sample and the next, instead of assigning a fully discrete value to each sample. It also adapts its coding to the signal values currently under process. Considered a form of statistical data compression.

ADSL: Asymmetric digital subscriber line. A high-speed digital service running on telcos' existing copper (twisted-pair) infra-structure, which provides data at a higher rate from the telco CO to the customer's premises ("downstream") than in the opposite direction ("upstream"). For example, service may run at 512 kbps from the telco CO to the customer, but only at 64 kbps from customer to telco. Typically used for broadband Internet access or IPTV service by telcos.

AMI: Alternate mark inversion. The binary modulation code used by the telephone company for data and digital voice transmission. It uses RZ coding in an alternate bipolar scheme, with logical 0's corresponding to zero volts, and logical 1's alternating between +3V and −3V. (The first logical 1 produces a +3V output, the next 1 produces −3V, the next +3V, and so on.) Self-synchronization is possible with this approach, but the number of continuous zeros must be limited.

Baud: Symbols per second.

B-channel: In ISDN service, a channel designated for customer data transmission, uninterrupted by any signaling data.

Bellcore: Bell Communications Research. The R&D firm that feeds technology and standards to the RBOCs, and is funded by them. Formerly Bell Labs.

Bonding: The generic term for any of several protocols (such as Multilink-PPP) that allows a data signal to be inverse multiplexed (IMUXed) over several physical circuits, to increase the effective bandwidth of a signal path. (Originally, BONDING was an acronym for a specific IMUX protocol, but this protocol is no longer widely used, and the term has become generically applied to all IMUX applications.)

BRIC: Broadcast-Reliable Internet Codec. A proprietary system from Comrex that fortifies an audio streaming signal for more reliable transport over the public Internet.

Carrier: In telco parlance, a multiplexed digital interoffice signal, containing many individual calls or signals in a single cable or fiber.

CCITT: Consultative Committee of International Telephone and Telegraphy. The international standards setting organization for telephone systems established by the United Nations.

CO: Central office. The generic name given to a telco's switching and service center, where all the telco circuits in a given physical area are terminated.

Codec: Coder/decoder. Any device that includes digital transmission/encoding and reception/decoding circuitry in the same chassis.

Contribution channel: A digital audio channel used between an outside origination site and the first permanent broadcaster facility the signal reaches; in other words, the circuit used from a remote site to a broadcast operations center or radio station's studio.

CPE: Customer-provided equipment. Any network interface hardware not provided by telco.

CSU: Channel service unit. Terminal hardware for a telco data line, either CPE or telco provided. Also referred to as CSU/DSU (DSU = data service unit) in T-1 applications. Interfaces unipolar NRZ computer style data streams to the RZ bipolar (AMI) telco data format. A switched CSU includes a keypad for call direction and other switch control.

D-channel: In ISDN service, a channel designated for signaling data only.

DDS: Dataphone Digital Service. The first telco data service in the United States, originated in the mid-'70s by AT&T.

Distribution channel: A digital audio channel used to deliver finished programs from a broadcast operations facility to transmission facilities that will deliver it to listeners, such as a satellite feed from a broadcast network to its affiliates, or an STL.

DOCSIS: Data Over Cable Service Interface Specification (v1.0, 1.1, 2.0, 3.0), an asymmetrical, high-speed data service deployed by cable TV operators to offer broadband Internet access over their coaxial cable networks.

DS0: Digital Service 0. A 64 kbps data channel.

DS1: Digital Service 1. A 1.544 Mbps data service usually configured as 24 DS0 channels plus an 8 kbps sync channel.

DS2: Digital Service 2. Four DS1 channels multiplexed together for transmission. Generally reserved for telco interoffice transmission, and not offered to customers directly.

DS3: Digital Service 3. Twenty-eight DS1 channels multiplexed together with additional control data, providing a data rate of 44.736 Mbps (generally quoted as 45 Mbps). Used for compressed NTSC and high-definition video distribution.

DSL: Digital subscriber line. See also *xDSL*.

Emission channel: A digital audio channel used to deliver a broadcast signal to end users such as IBOC, DAB, or SDARS services.

EV-DO: Evolution-Data Optimized (or Evolution-Data-Only). A format for broadband wireless data service via PCS. (Also known as 1xEV-DO.) The Revision A (EV-DO Rev A) of this format includes broadband upstream capacity and added robustness for high-quality streaming media delivery.

First-mile: The signal path between a program's origination site and its entry point to a common carrier's network or a private satellite uplink. Usually a terrestrial RF link or a local telco circuit.

G.722: A CCITT standard for audio data compression. It uses two sub-band ADPCM coding to put 7.5 kHz audio into 64 kbps.

HDSL: High bit rate digital subscriber line. A digital service that uses telcos' existing copper (twisted-pair) infrastructure to provide high-speed bidirectional connectivity over limited distance. Adaptive terminal hardware allows higher speed than previous equipment was capable of on the same physical paths. Bidirectional rates of 6 Mbps or higher are possible. The smart hardware allows quick installation without extensive testing and tweaking of lines. HDSL replaced T-1 service in some areas.

IMUX: Inverse multiplexing. See also *Bonding*.

InterLATA: Telco service or rates between LATAs, or long-distance service.

IntraLATA: Telco service or rates within a LATA, or local service.

Inverse multiplexing: See *Bonding*.

IP: Internet protocol. The networking layer used in the Internet; serves as an intermediate layer between the data transport layer (such as TCP or UDP) and the network link (physical) layer (such as Ethernet or Wi-Fi).

IP Audio: Streaming audio packets encapsulated into an Internet protocol (IP) data stream.

ISDN: Integrated services digital network. A new telco service designed to eventually replace POTS with flexible digital service. It will be offered in basic rate (2 B+D) service, intended for home use, and primary rate (23 B+D) service for business customers.

ISP: Internet service provider.

J.41: A CCITT standard for digital audio encoding. Using 1411 PCM encoding (14 bits for lower-level signals, 11 bits for higher-level signals), it places 15 kHz audio on 384 kbps.

Last-mile: The short haul signal path between a long-distance network terminal point (or private satellite downlink) and the customer's receive point. Usually a local telco circuit.

LATA: Local access and transport area. The service area of a local exchange company (LEC).

LDS: Long-distance service. A carrier of long-distance (InterLATA) telecommunications, such as AT&T, MCI, Sprint, and others.

LEC: Local exchange company. A local telco. Each RBOC contains one or more LECs. Also refers to independent, non-Bell local telcos.

Mark: The telco term for high-level data pulse, usually corresponding to logical 1. (See also *Space*).

MIMO: Multiple input/multiple output. Antenna technology allowing increased data throughput and extended range on wireless links via use of multiple antennas for spatial diversity and spatial multiplexing.

Naked DSL: A DSL line provided with broadband data only, and without the POTS service included in telco's standard DSL offerings. Also called dry or standalone DSL.

NRZ: Non-return zero. The most basic form of binary modulation coding, in which logical 1's and 0's are directly represented by high and low levels, respectively. Because no level transition occurs between continuous strings of like logical values, this form of modulation is not self-synchronizing and requires an external bit clock output for synchronous operation.

OC: Optical carrier. Specifies the speed of fiber optic networks in the SONET standard. Base rate (OC-1) is 51.84 Mbps. Common multiples used by telcos internationally include OC3, -12, -48, and -192.

Packet Switching: A sort of data party line, in which data is transmitted in addressed bursts or packets, occupying the transmission channel only for the duration of the packet, after which the channel is free for other packets to or from the same or other users. Many users can be interconnected to the same line, but data can be sent discretely to each destination.

PDN: Public data network. Telco data services, including both switched and leased lines.

POTS: Plain old telephone service. Refers to the public switched telephone network (PSTN).

PSTN: Public switched telephone network. The standard dial-up phone system.

QoS: Quality of service. The ability of a network to guarantee a specified level of accuracy and/or timeliness in its signal delivery.

RBOC (or BOC): (Regional) Bell Operating Company. The seven *Baby Bells* created when AT&T divested itself of its local telephone operations.

RZ: Return zero. A form of digital modulation coding in which logical 1's and 0's are directly represented by high and low levels, respectively, but where coding output returns to low level following each high pulse.

SDSL: Symmetrical digital subscriber line. A form of DSL in which upstream and downstream paths operate at the same data rate (see also *ADSL*).

Slot (or time-slot): Generally refers to a DS0 channel within a DS1 signal.

SONET: Synchronous optical network. An international standard for connecting fiber optic transmission systems. SONET establishes optical carrier (OC) levels from 51.8 Mbps (OC-1) to 9.95 Gbps (OC-192).

Space: The telco term for low-level data pulse, usually corresponding to logical 0. (See also *Mark*.)

Switch: Generic name for any telco call routing and connection hardware.

Switched 56: A switched digital service offering 56 kbps data service on a dial-up network, generally no longer available in most areas, having been largely replaced by ISDN.

T-1: The copper network and hardware used to carry DS1 service.

TA: terminal adapter. The terminal equipment ("telephone") used at the customer's premises in ISDN service.

T-carrier: See *Carrier*.

Tariff: A schedule of services and their prices that a telco will provide to a given service area, subject to approval by the appropriate regulatory agency.

TCP: Transmission control protocol. One of the two most commonly used transport layer protocols in the Internet. Serves as an Intermediate layer between the IP networking layer and the application layer, such as HTTP. (See also *UDP*.)

UDP: Universal datagram protocol. One of the two most commonly used transport layer protocols on the Internet. Serves as an intermediate layer between the IP networking layer and the application layer, such as RTP. UDP uses less overhead than the other common IP transport layer, TCP. Therefore, UDP provides less reliability than TCP, but faster operation for time-critical applications, so UDP is preferred for streaming media transport.

V.35: An older CCITT telco standard for low-speed data I/O to a CSU, with a unique multipin connector.

VDSL: Very high bit rate digital subscriber line. A digital service that can be configured either symmetrically or asymmetrically; in the latter form it can provide a maximum of 52 Mbps down-stream and 12 Mbps upstream over a single twisted pair of wires. (See also *HDSL*.)

VSAT: Very small aperture terminal. Refers to Ku-Band satellite earth stations for fixed or portable use with dish diameters on the order of 1.5 m or less.

Wi-Fi: See *802.11x*. WiMAX: See *802.16x*.

xDSL: Generic terminology for any of the variety of digital subscriber line services offered by telcos, which provide high-speed connectivity across limited distances on the existing copper (twisted-pair) infrastructure. (See also *ADSL, HDSL, SDSL, VDSL*.)

Notes

[1] L. Johnson and M. Coinchon, "Streaming audio contributions over IP—a new EBU standard," EBU Technical Review, Q1 2008, https://tech.ebu.ch/docs/techreview/trev_2008-Q1_coinchon.pdf

[2] "AES67–2013: AES standard for audio applications of networks—high-performance streaming audio-over-IP interoperability," Audio Engineering Society. 2013-09-11, www.aes.org/publications/standards/search.cfm?docID=96.

[3] Steve Church and Skip Pizzi, *Audio over IP—Building pro AoIP systems with Livewire*, Focal Press, 2010.

[4] Multi-Stimulus High Resolution Audio test, a subjective listening test paradigm specified in ITU-R BS.1534.

[5] A.J. Mason and D. Marston, "Cascaded audio coding," BBC R&D White Paper WHP 118, September 2005.

CHAPTER

6.2

Video Contribution and Distribution Methods

CINDY HUTTER CAVELL

Cavell, Mertz & Associates, Inc.
Manassas, VA

Video contribution and distribution started at the simplest level with the film industry. Filmmakers made films; from the early 1900s, the distribution of these films proved to be quite lucrative and competitive. One of the distribution methodologies saw producers selling copies of their films to a local salesperson in a number of territories. Several salespeople across several territories would schedule the films into as many theaters and nickelodeons as they could until the copies physically fell apart.[1]

Prior to the advent of digital media, video contribution and distribution can be characterized in one of two ways: recorded video contribution and distribution, and live (or scheduled) video contribution and distribution. For the purposes of this chapter, videotape and disc distribution will be discussed, as film distribution to the broadcast industry and to the public has long since faded into history.

RECORDED VIDEO CONTRIBUTION AND DISTRIBUTION

In the 1950s, West Coast U.S. television network affiliates received programming live from the networks' New York studios earlier in the afternoon in the Pacific Time Zone than the programs were broadcast on the East Coast. Station managers on the West Coast wanted to air the programs in the same day-parts as their East Coast counterparts, and network time-shifting was born. This early recording was done via kinescope film, which required a film camera specially adapted to shoot a video monitor. The film was then sent off to processing. Often, it took all of the three hours of delay to get the film shot, out of the processor and wound on a reel to be ready for air.

Early videotape was confined to broadcasting. The 2-inch quadruplex tape machine was developed in 1956 by the Ampex Corporation.

"Quad tape," as it was called, was the first practical analog videotape recording format. The tape machines were roughly the same size as the medium-powered FM transmitters of the same era. The equipment and parts and the tape medium were originally fabulously expensive (a new quad machine approached $400,000) and the setup of the machine was a long, painstaking process that had to be repeated several times during the day to create a stable recording. Often, early videotapes recorded on one machine would not play back on another tape machine of the same model. Videotapes that were created on an early version Ampex quadruplex machine would almost never play on an early RCA quadruplex machine. Nonetheless, through the 1960s, the technology and ease of recording and playback on these machines improved greatly. They and their successors in professional broadcast format—1-inch helical scan, 3/4-inch helical scan (commonly known as U-Matic, as branded by the Sony Corporation, which developed the tape machines), and 1/2-inch helical scan—were the workhorses in broadcast facilities across the world for decades. Recorded videotape affected all aspects of a television plant:

Commercial tapes: Commercial advertising agencies began transferring commercial film to videotape for distribution to television stations. The ad agencies continued to shoot their commercials in film for a long time because the finished film

FIGURE 6.2-1 Ampex VR-1000 2-inch videotape machine

Courtesy of www.Big13.com

FIGURE 6.2-2 ENG soundperson with portable 1-inch VTR

Photo Courtesy of Stephen Alhart

product had higher resolution than footage shot directly onto videotape with a vintage video camera—not until digital video cameras matured did the ad agencies start creating directly onto videotape or disc. The quad tape was cut to fit the length of the commercial, usually 30 seconds to 2 minutes; wound onto a small, inexpensive plastic tape reel; and put in a box and mailed to the networks or local TV stations. Sometimes each television market received just one tape per commercial, because the quad tape was still a rather precious commodity—the first station to receive it would record it onto an in-house videotape and send the "original" to the other stations in town, often by bicycle courier—hence, the concept of "bicycling tape" from one station to another.

Program tapes: Videotape took over as the program tape delay medium for the U.S. West Coast in one of two ways: either the network themselves would record a time-shifted version of the East Coast content, and play it out three hours later via telephone company or satellite (see next section, "Real-time Contribution and Distribution"), or the stations themselves would record the original content and then play it out at the appropriate time. Soon, time-shifting was occurring for stations in the Central Time Zone; stations in the Rocky Mountain Time Zone, Alaska and Hawaii still record and time shift network content today.

News footage: The early 1970s saw the "miniaturization" of videotape machines and cameras for news use. The portable versions of the cumbersome studio equipment made "portable" would change the face of news reporting by being able to cover more news "up close and personal," and bring news stories to air faster, as videotape didn't need developing. Videotape could be edited electronically, maintaining the originals as they were shot, and creating an edited version that could be changed to air on subsequent newscasts or updated as more current news came in. Chapter 5.12 (Video Recording, Storage, and Servers) and Chapter 5.17 (ENG, SNG, and Remote Video Production) cover these topics in more detail.

Videotape served well as a contribution tool as well as a distribution tool. Television stations were able to create long-form programming that could be shot in the studio—previously, only live shows could be done this way—recorded onto videotape and be edited for rebroadcast, either by the individual television station or by a network. PBS was a pioneer in this sort of contribution. Major market television stations in New York, Pittsburgh, Washington, and San Francisco could produce programs for distribution by the network to the member stations.

REAL-TIME CONTRIBUTION AND DISTRIBUTION—MICROWAVE AND VIDEO CIRCUITS

As the TV networks broadened their reach by adding affiliates across the United States, they needed a way to distribute live programming to them. AT&T had developed a nationwide coaxial system to carry telephone "long line" traffic. Additional coaxial cables along the same routes increased the capacity of the system. Early "radio relay" systems designed by the engineers at Philco operated in VHF at 200 MHz and carried TV programming from New York to Boston and Philadelphia. After World War II, the system was migrated to 1200 MHz. In 1949, the FCC set aside three "common carrier" microwave bands: 4000 MHz, 6000 MHz, and 11,000 MHz. Initial AT&T "radio relays" were built in the 4 GHz band (known then as the TD-2 system) with towers roughly 30 miles apart. Narrow beamwidth horn/reflector antennas allowed more efficient use of the radio spectrum to yield more channels. Heterodyne repeaters were developed so that microwave signals could be carried from tower to tower without demodulation and remodulation at each tower. To further increase the number of available audio, video and voice channels, AT&T soon began overbuilding the system at 6 GHz (the "TH system"), which could utilize the same tower routes and add channels without interference, as the two sets of frequencies were far apart from one another. Some of the iconic AT&T towers with their crown of horn antennas still survive today.

Coaxial cables were apportioned into channels of 2.7 MHz width for video transmission—at the time, wider-band channels required repeaters or amplifier nodes that were too close together for a practical system. Coaxial systems were initially more expensive than microwave, and considering purchasing the right-of-way, cable burial and the cost of the cable itself (coaxial cables for video transmission were bundled in groups of eight) could cost as much as $50,000 per mile in 1950. Microwave was wider bandwidth for the channels apportioned in 1949—the microwave systems of the day could provide channel widths of 4.5 to 6 MHz.

AT&T completed the transcontinental system in 1951. The coaxial cable plus microwave interconnection and switching system covered most regions of the country, as shown in Figure 6.2-4.

FIGURE 6.2-3 AT&T Long Lines Microwave Facility in Kingston, New York

Photo Courtesy of Paul Thurst

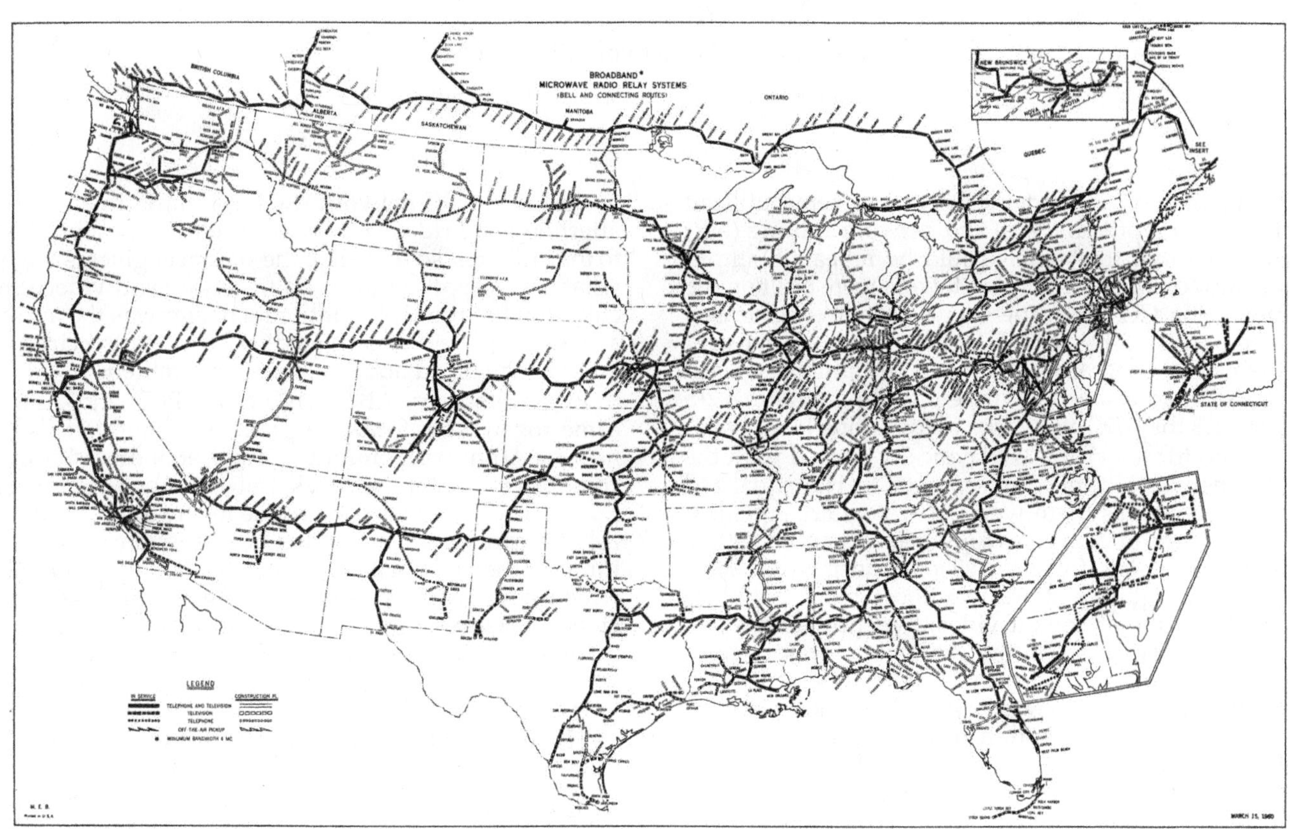

FIGURE 6.2-4 AT&T National Long Lines System, c.1960

Courtesy of AT&T Archives and History Center

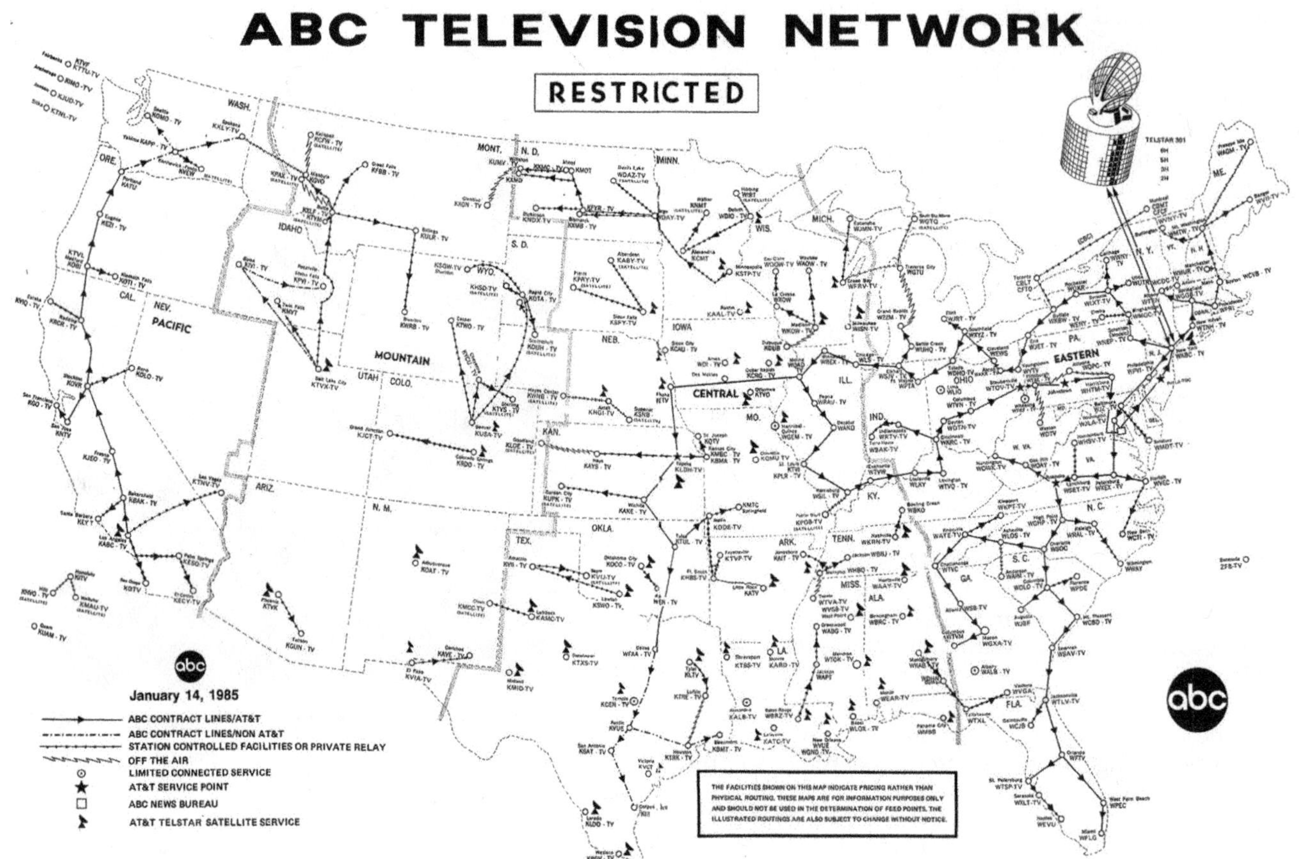

FIGURE 6.2-5 ABC network diagram (1985)

Courtesy of American Broadcasting Company

Both the coaxial and microwave systems utilized AT&T central offices (COs) and network routing (NR) centers so that the signals could be repeated, amplified, and/or equalized. Many towers were built on the pre-existing CO sites. The television networks leased a good deal of the AT&T system for their network systems for program delivery to the affiliates. Each company had a slightly different routing. Figure 6.2-5 illustrates the ABC Network video routing as of 1985.

Audio for television was typically carried with the radio network traffic on separate 5KHz equalized audio carriers, as there were separate gain, equalization, and termination rules for video and audio. Often, the audio signals for a television transmission were handled by equipment that was floors, blocks, or miles away from the attendant video. Synchronizing them and switching them at the same time was complicated; sometimes when circuits were switched either the video or the audio was switched, but not the other. Many times, network transmission engineers were confronted with a program in which the audio didn't match the video. Weekend afternoons when different sporting events were being played in different regions of the country could be fraught with frustration for television network and AT&T engineers making sure that the routing for all audio and video switches were correct. The ABC network used the "dark time" in the afternoon on Fridays to feed a closed-circuit broadcast to the affiliates, so that all of the station engineers knew what was supposed to be on the audio and video circuits at what time. Not until much later was the audio actually diplexed (modulated) onto the video circuits and the equalized audio circuits were retired.

Engineers from ABC, CBS, NBC, PBS, and AT&T came together to form the Video Transmission Engineering Advisory Committee; a subset of this advisory committee was the Network Transmission Committee (NTC). The NTC was responsible for creating and publishing various analog video transmission standards through the years, culminating in NTC Report No. 7,[2] which addressed recommended procedures; parameters, measuring them and performance objectives; interconnection requirements and processing errors.

THE NEXT ITERATION—THE SPACE RACE AND COMMUNICATIONS SATELLITES

As commercial network affiliates and PBS member stations grew in numbers, distributing several hours per day of programing to a large number of affiliates was becoming expensive over "land lines." Networks needed a better one-to-many method of distribution. Further, local affiliates were creating their own programming, and desired to show network programs at a different time or date. TV stations in the Rocky Mountains and the West Coast that were now able to record and time-shift network distributed programs and air them "at the correct time," instead of whenever the network sent them (8 p.m., which was the start of "prime time" in the Eastern Time Zone, was 6 p.m. in Denver, right when those stations wanted to air their evening news). Commercial communications satellite technology had developed, just as the coaxial infrastructure did, to carry telephone signals and audio signals. Additional satellite "payload" offered enough bandwidth so that the cost of delivering video programming to affiliates via satellite approached the cost of delivering over Telephone Company ("telco") video circuits, with the added advantage of being nearly infinitely scalable on the receive side—all that was needed to add another member or affiliate to the feed was to install a satellite downlink antenna and attendant receiver at the station. Typically, networks bore the cost of the satellite downlink installations, with the added benefit that what the stations received and recorded could be controlled by the network. Cable-only networks added encryption codes to their programming for delivery to cable head-ends in the early 1980s so that viewership could be controlled.

The first domestic commercial geosynchronous satellite, Westar I, was launched in 1974. C-Band telecommunications satellites were the first used for analog video distribution. This spectrum was the first to be assigned to commercial telecommunications satellites; it was a known quantity, as it employed 5.850 to 6.425 GHz for the uplink frequencies and 3.625 to 4.2 GHz for the downlink frequencies similar to the frequencies used for the terrestrial telephone company microwave systems (Russian C-Band frequencies are slightly different, and there are frequencies in the "extended C-Band" as well). Specifics of communications satellite technology are discussed in Chapter 6.3 in depth. C-Band telecommunications satellites have 24 transponders on board; each is 36 MHz wide and are spaced 40 MHz apart. To minimize interference between spacecraft, the C-Band satellites are spaced 2 degrees apart, and the Ku-Band satellites are spaced 1 degree apart in orbital position. Every other transponder is the opposite polarity, to keep interference between transponders on each spacecraft to a minimum. Like their terrestrial counterparts, satellites are designed as relay stations; they receive an incoming signal, amplify it, and translate the incoming frequency to the downlink frequency, and send it earthward.

Original communications satellites were only able to provide a small amount of signal gain, so satellite uplink transmitters were necessarily powerful, and both transmit and receive antennas were relatively large to keep signal viable over the 25,000-mile round-trip to the geostationary satellite and back. Contribution from the major PBS member stations and network-owned stations was provided by fixed satellite uplink installations, which were usually also implemented at the networks' expense.

As newer satellites with improved on-board signal amplification technology were launched, smaller antennas and/or less powerful transmitters could be employed, making way for transportable live C-Band satellite program contribution. While the VYVX fiber optic system (covered later in this chapter) terminated at a large number of professional sports stadiums, there were a number that were still not connected to telco routes. With the entry of sports-centric cable networks in the 1980s, mobile satellite trucks were a way to broadcast from the "unwired" sports venues, and bring a larger array of live sporting events to the viewing public.

News, as well, benefitted from C-Band satellites. Live news from overseas or across the country was now possible, both from a distribution and contribution standpoint. Live or videotaped material could be fed by either a stationary or mobile C-Band uplink from anywhere in the world to the network news centers in the United States.

Again utilizing familiar telco frequency bands, Ku-Band satellites (11 GHz uplink and 13 GHz downlink) with improved gain capability and 32 transponders started being launched in 1990. The higher frequency signals are more susceptible to rain fade, and receive amplifiers and antenna sizes necessary for reliable (99.999%) reception are larger in areas with higher yearly rainfall. Ku-Band is still extremely popular for "occasional" news feeds from transportable and mobile units. Transponders, or portions of transponders, are made available for contribution (individual news trucks) or distribution entities such as ABC's NewsOne news feed to affiliates in increments of as short as 5 minutes, making the cost reasonable for even the smallest of stations to cover news far from home.

For analog video, satellite transmission was typically single program per transponder. With the advent of digital video and video compression algorithms, the DVB-S and DVB-S2 standards were developed. DVB-S uses the MPEG-2 digital video standard, and was capable of single-program-per-carrier (SCPC) as well as multiple-program-per-carrier (MCPC). MCPC signals are several transport streams multiplexed together. It is primarily used for the transport of standard definition (SD) signals. The DVB-S2 standard modulations schemes encourage roughly 30% more performance gain; with the addition of the more advanced MPEG-4 (AVC) compression standard, an uplinker can effectively provide twice the content in the same transponder bandwidth as they were previously capable of.

The television networks also made the switch to Ku-Band satellites in the early 1990s, as the cost of C-Band transponder leasing for ever-increasing program content was also increasing. By taking advantage of the increased payload and efficiency, networks and program providers were able to provide either the same number of services in half the bandwidth, or an increased number of programs and services in the same bandwidth. PBS still carries one C-Band "in the clear" feed by Congressional mandate for rural viewers with legacy backyard mesh-type C-Band antennas left over from the late 1980s.

Most PBS and commercial network downlink systems at the affiliate have become very complex. Affiliates/member stations have two or three receive antennas of varying sizes, often with four antenna feeds in each. Networks ship several pre-configured racks to each station, complete with IRDs (integrated receiver-decoders), and complete automated remote systems, which enable the network control center to switch between satellites for special sectional broadcasts, sun outage periods or maintenance windows. The Fox network distribution system includes a bit splicer, as the switching is done on the ASI, rather than baseband, level. Some of the networks provide simple graphics equipment either for network logo information insertion, or for affiliates to insert the local time and temperature over the morning news and entertainment shows. Additional feeds to affiliates may include a 24/7 news clip feed (CNN also provides this to local affiliates), or a "rain-fade" or disaster recovery (DR).

Many television network and CNN affiliates use the network system in reverse to contribute news and entertainment content back to the network for distribution. These stations are outfitted with stationary satellite transmit facilities (uplinks) for the contribution. NBC outfits a number of affiliates with a "PuP" (portable uplink), controllable either locally or by the network for stations to feed special events or newsworthy clips to the network system. Many of the larger PBS member stations have large production facilities that produce weekly or long-form programs that are either fed live (PBS NewsHour) or carefully produced (NOVA) and fed to the network for later distribution.

With the DVB-S2 technology, commercial third-party uplink services were able to offer their "occasional" or small bandwidth clients smaller slivers of transponders, often inserting a single transport stream into a larger MCPC "bundle" and lowering the cost for the client. As costs for satellite services went down, more clients were able to use the services, providing a good business model for the third-party uplinkers.

As for additional satellite-driven businesses, the nature of the terrestrial ATSC television transport stream has allowed television stations to broadcast multiple digital transport streams in one 6 MHz television channel, and stations are interested in additional content to monetize using this technology. Into the breach came a number of "diginets"—companies that purchase the right to use classic television content and distribute it to television stations to carry on sub-channels. Ku-Band satellite distribution of an MCPC made up of several SD program transport streams for stations to downlink and air is reasonably priced and simple to manage. The stations are given (or purchase at a discount) the required satellite receiver and commercial insertion equipment, and add a feed and LNB (low-noise block downconverter) to an existing antenna pointed to the same satellite.

Program syndicators and commercial agencies were also seeking a distribution method other than the slow and cumbersome "bicycle" method or expensive telco or fiber landline distribution method; the one-to-many delivery of telecommunications satellites was tailor-made for companies wishing to deliver the same program or set of commercials to 500 TV stations in various cities across the country. This business model has given birth to a number of companies that serve as a third-party distribution mechanism for program syndication companies. Full-length programs are "prepped" (See "Television Playout and Content Delivery—Master Control Systems and Automation," Chapter 5.4), each segment is timed, black spaces inserted for local commercial content cover, and "barter" spots inserted. Program formats are provided to stations in advance.

PitchBlue handles only long-form programming, pushed to member stations, either for live linear playback or to be recorded for future airing. The company is a consortium between Deluxe/GDMX (which serves Warner Brothers), 21st Century Fox, and CBS, with the operating managed by CBS. The system utilizes C-Band satellite only, and transponder space is provided by the clients. Distribution methodology over satellite is multiple program transport stream (MPTS). Each program provider is assigned several program streams, each carrying different programming. End-user television stations are provided with an IRD/Server, with a graphics card providing baseband video outputs and the ability to "lock" the server to the station facility timing system. Each server will hold close to a week's worth of programming with an Internet connection from the server to the PitchBlue operations center that reports when the server is getting full enough to not have sufficient space to store additional programming. The network operations center (NOC) can address the individual IRD/servers at all times. The Internet connection also serves to confirm whether or not the content has been recorded properly in the individual servers, requesting additional "pushes" of content to the server until confirmation of good content is confirmed back to the NOC. The system utilizes DVB-S2 with MPEG-4 encoding with an 8PSK constellation (again, see Chapter 6.3 for explanation and additional content).

Content is delivered to the television station in the "house" format the station desires, which requires a good deal of transcoding to be done from the original formats internally before the individual copies are pushed to client stations. There may be hundreds of different formats that are transcoded from a single program for distribution. Delivery can be adapted to either single-station use or for hub-and-spoke master control models as described in Chapter 5.4. Live content is available through the system, and stations are able to record content while simultaneously playing back content live. They system works on standard master control system protocols, so that a station doesn't need to have a master control standing by to take the recorded content out of the PitchBlue server and transfer the file over to the master control playout server system. If enabled, the content will automatically be transferred.

Extreme Reach offers another delivery service of syndicated material via a combination of satellite and

Internet delivery. Predominantly short-form commercials, promos, news packages are sent as files to stations with XML metadata "side-cars"; these are also custom-transcoded by Extreme Reach to match a station's "house" format. CNN and ABC's NewsOne service use Extreme Reach to deliver news clips and packages to member stations. While having a fairly wide Internet pipe is preferable, the company indicates that it pushes non-real-time (NRT) content to stations with Internet connections of as little as 3 Mbps. Arguably, it takes more time for the content to transfer with a narrower-bandwidth connection, but most content transfers occur in the overnight period. Longer-format infomercials and syndicated content are also sent as files via satellite, so that they can take advantage of the wider bandwidth to send large files more quickly.

PBS, in addition to uplinking twelve channels of SCPC satellite content from the main uplink facility, pushes some 50% of its NRT content files to approximately 120 member stations via the Internet. A typical Internet connection used at the station level is 100 Mbps to download PBS-generated material, which is archived at the standard rate of 50 Mbps. Member stations in Washington, New York, and Boston are connected to the PBS hub for both contribution and distribution via OC-3 fiber. Others are connected to the hub for contribution purposes via satellite uplink, either their own or that of a leased third-party facility.

Most recently, with the advent of bonded-cellular and WiFi ENG video transport, many ENG vehicle manufacturers have begun using Ka-Band (18 GHz) satellite spectrum, which is normally used to carry "bursty" data, to fill gaps when the local cell service isn't able to handle the amount of traffic caused by increased cell phone use during a local event, or in the case of cell system collapse during a natural disaster. More on this use of the technology is available in Chapter 5.17—ENG, SNG, and Remote Video Production.

THE RETURN TO TERRESTRIAL DISTRIBUTION—FIBER OPTIC CIRCUITS AND THE INTERNET

The Williams Communications Group, Inc. (WilTel) began installing fiber optic cable on the pipeline rights-of-way owned by its parent company, Williams Energy. The intent was to pioneer a multiservice broadband network that would accommodate voice, audio/video, and eventually Internet services. Williams Network Services was set up in 1990 to provide the switching of private high-bandwidth fiber optic circuits. They packaged the fiber service with on-premise transmitting and receiving equipment for customers. Interconnections were run to major-market sports facilities—football stadiums, basketball arenas, and baseball stadiums. Fueled by the rise of television and cable network sports content, Williams Network Services (the video transport system now called Williams VYVX) supplied private, switched circuits across the country that, like occasional satellite services, were ordered up on a scheduled basis, utilized for the duration of the program, and then "goodnighted." This enabled content providers to leave the expensive and maintenance-prone satellite uplink vehicles at home and use them for shorter-duration events at locations where there was still no VYVX network. At the beginning, connection between the stadiums and the VYVX local technical operations center (TOC) was usually made by the local phone company on existing balanced video copper circuits with two channels of audio or some local common carrier using early fiber optic technology.

Eventually VYVX installed or leased dark fiber to connect stadiums, production houses, TV stations and corporate offices to their TOCs. This all-fiber network enabled the advancement to allow uncompressed HD connectivity between content producer and broadcaster.

The 1990s and early 2000s saw all of the other large carriers—AT&T, MCI, and Sprint—overbuilding their telco networks with fiber optic systems for broadband delivery. In recent years, the fiber optic technology stretches to other continents via large, high-capacity submarine cable.

Today, private, IP packet-switched networks share a fiber backbone with Metro-E circuits and VPN circuits. For video contribution and distribution, there are service level tradeoffs. Security and reliability can be traded for cost reduction if real time delivery is not a factor.

IP was originally developed to send short, "bursty" data. FTP is an excellent and inexpensive method of sending a video file on a non-real-time basis from one place to another for replay at a later time. Most non-real-time video transfer is done this way. Log onto an FTP server, give it the proper credentials, and the process begins. The time to download the file is related to the size of the file, the bandwidth of the transmitting facility's connection, the bandwidth of the network system, and the bandwidth of the receiving facility's connection. Security can be an issue when downloading files via the Internet. Companies that deliver content via satellite maintain that a satellite link, especially an encrypted one, is still a secure private "network," no matter how many entities the uplinker is delivering to, the delivered content will not suffer packet loss, hacking, or "backhoe fade."

Real-time video delivery of MPEG content over IP (MPEG/IP) networks is only successful if the decoder is able to lock to the program stream via a program clock reference (PCR). Ideally, a private circuit (VPN) with a minimum amount of switching is the best method. It is, however, usually the most expensive. Packets sent over a network are collected in the integrated receiver/decoder (IRD) in a buffer. "Network jitter" is the measure of the deviation between the time a packet is expected at the distant end and the time it actually arrives. Too much deviation and the decoder may not be able to recover the PCR. Because the stability of the picture is based on the receiver being able to lock to the clock reference and enough packets in the IRD buffer in the right sequence to spit out a correct stream, too much network jitter could cause the IRD buffer to not be able to piece enough packets together in order to send out as a baseband stream. "PCR jitter" is a combination of overall network jitter and any inherent accuracy errors in the encoder. IP networks that distribute real-time video can be monitored at intervals to evaluate network quality. Often, it is possible to establish a service level agreement (SLA) with a carrier on a less private service—a metro-ethernet service, for example, which switches private circuits and Internet data all together—which establishes a maximum PCR jitter. If the IRD can't recover the PCR, the resulting video is not viewable due to the amount of resulting macro-blocking.

While more expensive, private circuits nearly all come with a very restrictive (on the provider) SLA, they can be very pricey. A few companies offer video delivery over what is usually called "managed Internet service," which usually includes encoders and decoders as well as an "appliance" that offers some additional forward error correction (FEC) as well as additional buffering on the receive end to overcome as much PCR jitter as possible so that packets have every advantage to get to the distant end. This can be an improvement over plain Internet service, and less costly than a private circuit.

Nearly all Internet-delivered video is MPEG-4 encoded (AVC or H.264) or HEVC (H.265) object-coded, as more aggressive encoding is needed to push live, HD (or more) video over an Internet connection.

"Cloud" Internet-based delivery is a platform as a service (PaaS), which may include off-site content storage for simple file retrieval via Internet or full play-to-air capability of recorded media content. Cloud providers offer other services as well, such as media asset management (MAM) services, media "prep" for air and master control automation. It is yet another outsource mechanism for video delivery that takes the burden off of the end user.

Direct-to-Consumer Contribution and Distribution—Cable, DBS, Video on Demand, and Streaming

Strictly speaking, cable television services and direct-to-home satellite services are content providers even though the content is made up of "channels" of pre-packaged content and commercial interstitials. Cable entered the content provider field in 1980 to provide 24/7 "niche" viewing such as sports, weather, news, and movies. Cable networks such as CNN, ESPN, the Weather Channel and Discovery were developed to provide alternate programming to that which was delivered over the air by the traditional television networks. New content was created and packaged into additional networks to answer a growing demand for different types of content. It was the beginning of the concept of content consumers having to pay to receive content.

Cable companies wired facilities direct to the home; the more populated the area was, the easier it was to reach all or most of the homes that they cable system passed. The direct-to-home small aperture satellite services such as DirecTV and Dish Network took advantage of the fact that cable providers didn't bother to serve the more far-flung neighborhoods and rural areas, as it was too expensive to invest in infrastructure that served so few. Again, the perfect application of the one-to-many nature of satellite delivery began in the mid-1990s. Special Ku-Band satellites were launched to serve the portions of the country unserved by cable systems. In-home IRDs were small and easy to set up, and typically included a built-in VCR, which have been replaced by video servers in recent years, which hold a good deal of content before they have to be purged for additional recording.

The development of the home-use VCR and cable and satellite delivery of content allowed consumers to do two things that have completely changed the landscape of the heretofore time-based appointment viewing of content: record it for viewing at another time, and skip over the commercial breaks.

YouTube is by far the most widespread video contribution/distribution site in the world today. The platform enables individual users to create and upload video content and enables other users to download recorded videos they want to watch. Content providers worldwide are combining streaming video with their more traditional content delivery methods. Television stations will stream their news while broadcasting it over the air at the same time. Traditional television networks and cable nets include the ability to download and/or stream content, usually after it has aired over the network. How-to videos and video user guides are available for download from all manner of companies from how to install hardwood flooring to how to use label-maker software. Companies like Hulu and Netflix have become content aggregators of a different sort, not only gathering the rights to stream movies and other network content that has already been "aired," and now having original content created just for their use.

There are three types of video streaming in use at time of this writing:

- *Progressive download*: This type is mostly considered "fake streaming." Previously, the entire file would have to download (typically from an HTTP server) in its entirety before it could be played back. Currently, files that are progressively downloaded only have to fill a playback buffer before playout can begin, but the user doesn't have full control over the video until it has completely downloaded. Until the download is complete, fast-forward or location-seek controls can't be used. For example, the user can't click on the end of the video to see a clip of it before that portion has downloaded completely. The best use is for "linear" viewers that want to start a video and view it linearly without jumping around while it downloads. Progressive downloads are delivered to all users at a single, fixed bit rate (the bit rate at which the content was encoded), no matter the end user's connection speed. This type of video file-sharing is the least secure, the least efficient and the least adaptable to the broad universe of different end user devices.

- *Real time streaming protocol*: Instead of being fed from an HTTP site, real time streaming video is sent from a streaming server using real time streaming protocol (RTSP), delivered via UDP, or real time messaging protocol (RTMP), delivered via TCP. The video is not downloaded onto the user's computer; rather the streaming server sends a portion of the video (shorter and gets to the consumer more quickly) to the user's PC or other device, depending on which portion of the video the user requests with their individual controls. Once the consumer has viewed the video, the receiving device discards it. It is more efficient for the provider as well. In a progressive download, the streaming site is occupied until the content has completed streaming—an unnecessary waste if a user abandons the video partway through watching it. This type of streaming requires not only a streaming server, but a player or application on the user end.
- *Adaptive bit-rate streaming*: Current streaming servers allow for *multi-bit-rate* encoding, which enables them to encode at a number of different bit rates to reach different devices at different resolutions, depending on the device being used and Internet connection restrictions. Adaptive bit-rate streaming is progressive download streaming via HTTP packets. This model will act like streaming by delivering small chunks—typically 4 seconds or less—of content via an app or player, but it is also utilizing progressive download so that the end user retains a copy of the downloaded material *if desired*. The idea of adaptive bit rate streaming, given the number of connection speeds and devices that want to download and/or play video, is to keep the incoming bit rate of the stream close to the bit rate that the connection or the resolution that the device can handle so there is no video stuttering or "buffering" notices while the video is frozen on the screen. Types of adaptive streaming are Apple's HLS Live Streaming, Adobe's HTTP Dynamic Streaming, and the MPEG group's MPEG-DASH.

Even though the inherent hardware and software have evolved over time, and the way that people consume content worldwide is also changing, all three of the original methods of video contribution and distribution—terrestrial (Internet), microwave (wireless), and satellite still remain as effective platforms.

ACKNOWLEDGMENTS

Many thanks to Marc Drazin at PitchBlue, Fred Cunha at Extreme Reach, Stephen Alhart at NEP, Frank Graybill at WNET, and Tom Crowe at PBS. Special thanks to all of the retired transmission engineers at ABC for their deep knowledge of all things AT&T and VYVX.

Notes

[1] The Society of Independent Motion Picture Producers. "The Early Film Business."

[2] *"NTC Report No. 7—Video Facility Testing—Technical Performance Objectives"* (revised 1976), a joint effort of the television networks and the common carriers.

CHAPTER

6.3

Satellite Uplinks and Downlinks for Contribution and Distribution

SIDNEY M. SKJEI

Skjei Telecom, Inc.
Falls Church, VA

INTRODUCTION

The use of satellite technology by radio and television broadcasters is highly developed and in widespread use throughout the world. Satellites are routinely used for broadcast contribution, backhaul and distribution services, and they provide low cost, reliable operation. These and many other services are made possible due to the unique characteristics of satellites placed in geosynchronous satellite orbit. The capability of a single quasi-stationary repeater in the sky, visible to large, contiguous regions, offers a wide variety of distribution capabilities for terrestrial broadcast service providers as well as the public in general. Satellite communications are also particularly useful for mobile services, long-distance communication services, and for services across difficult terrain. This chapter addresses system and equipment aspects of uplinks and downlinks that are used for television contribution and distribution.

Satellites in the 22,300 mile high geosynchronous orbit rotate from west to east. They appear fixed in space to earth stations on the ground because they orbit in synchronism with the earth's rotation. A satellite that is closer to the earth orbits faster and one that is beyond synchronous orbit rotates slower than the earth. Satellites located in the geosynchronous orbit have direct lines of sight to almost half of the surface of the earth, as shown in Figure 6.3-1.

Communication by satellite was made possible by parallel advances in space technology and electronics. In 1945, Sir Arthur C. Clarke, the noted British scientist and science fiction writer, proposed relay stations in geostationary orbit for satellite communications. By 1963, due to advances in technology, solid-state electronics, and the thrust capability of rockets, the ability to place a satellite into a stationary orbit was achieved.

Frequency Bands

Communication satellites today operate at many different microwave frequencies. Excluding satellites intended primarily for mobile communications or direct to home satellites, the frequencies primarily used by U.S. broadcasters are shown in Figure 6.3-2. In the United States, the domestic commercial communications satellite networks operate in the fixed satellite services (FSS) frequency bands as defined by the Federal Communications Commission (FCC). Most of the domestic systems operate in either the C-band (6 and 4 GHz), Ku-band (14 and 12 GHz), and Ka-band (30 and 20 GHz) frequency ranges. For completeness, it should also be noted that L-band (approximately 1.6 GHz) satellite frequencies for the Mobile Satellite Service and S-band (approximately 2.5 GHz) satellite frequencies for the Digital Audio Radio Service are also used by broadcasters. This chapter deals primarily with the C-band, Ku-band, and Ka-band frequency FSS satellites, as those are most commonly used by broadcasters today.

International satellite systems provide services on a global basis and also on a regional basis to countries visible from a single orbit location. Intelsat, SES, and Eutelsat are examples of this type of system, which typically use both the C-band and Ku-band and less frequently, Ka-band. The international satellite communication frequency bands are similar to the U.S. frequencies at C-band (except that some locations have

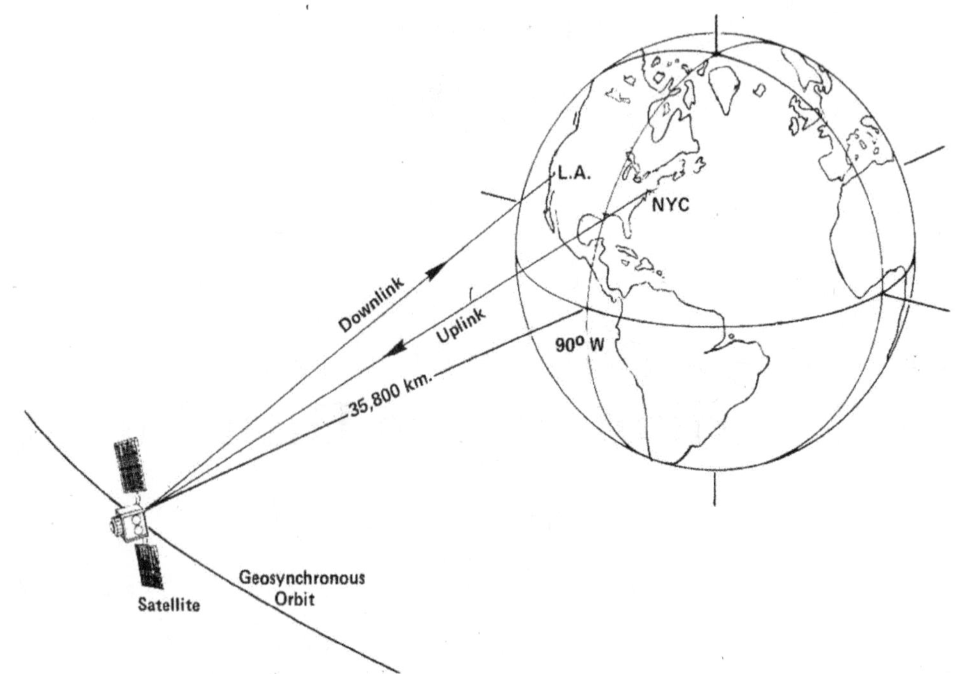

FIGURE 6.3-1 Satellite in geosynchronous orbit

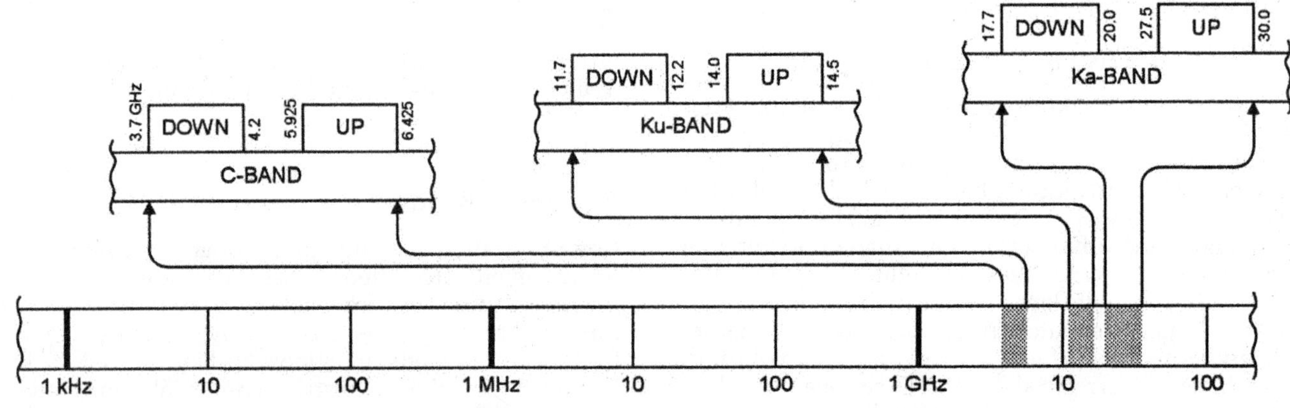

FIGURE 6.3-2 Frequencies in the microwave range of communications satellites

extended this band) but can be somewhat different at Ku-band and Ka-band, and can vary on a geographical region basis. The frequency assignments are determined in joint negotiations by the countries of the world through the auspices of the Radio Communication Sector of the International Telecommunication Union (ITU-R).

Satellite Stationkeeping

It was stated earlier that the synchronous satellite appears stationary in space. Actually, a synchronous satellite is never perfectly stationary, because a number of forces, including the pull of the sun and the moon, perturb its orbit. If left alone, the satellite would eventually drift out of orbit. To overcome this, the position of the satellite is continuously monitored by an earth station, called a telemetry, tracking, and control (TT&C) station, and small jets of propellant such as hydrazine or xenon are used to keep it in position within a prescribed station-keeping box. The station-keeping box is typically a square ±0.05° or less on each side and oriented with the sides parallel and perpendicular to the orbital plane. Sufficient rocket propellant must be carried on board to last for the satellite's predicted life, usually from 15 to 20 years. This life can and sometimes is extended by cessation of "north-south" stationkeeping and allowing the satellite's orbital inclination to increasingly exceed the station-keeping box, normally at a rate of 0.9° per year (the "COMSAT" maneuver). This type of operation requires FCC approval, is termed "inclined orbit operation," and can extend the remaining life of a satellite by a factor of 12. Of course, to use this type of operation for any significant period of time, tracking earth stations are required on the earth.

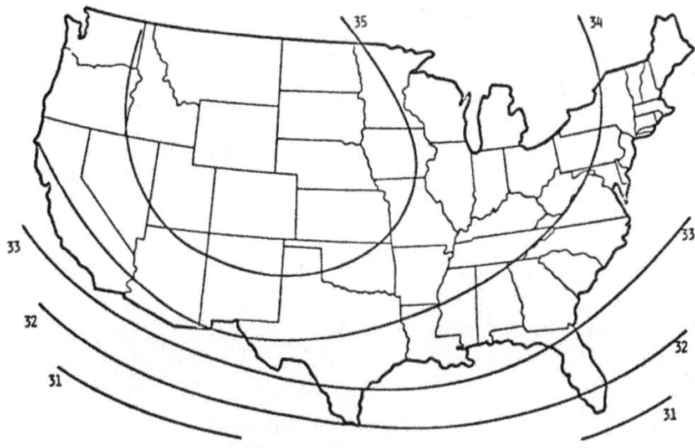

FIGURE 6.3-3 Satellite footprint. Lines with numbers indicated varying antenna gain and consequently varying levels of downlink power (EIRP) or uplink receive system performance (G/T or SFD).

Satellite Footprint

The transmitting and receiving antennas on a satellite are designed to cover only specific desired regions of the earth's surface. This has several purposes: it concentrates the power radiated from the satellite into the desired direction, increases the sensitivity of its receiving antennas, and helps prevent interference with signals from and to other satellites. The part of the earth's surface covered by a satellite is called the satellite's footprint. The footprint may cover one or more relatively localized regions of the earth or almost a complete hemisphere. A typical footprint is shown in Figure 6.3-3.

Polarization

The term "polarization" refers to the spatial orientation of either the transmitted signal or the receive (or transmit) antenna. Electromagnetic waves (signals) and antennas are always polarized in some manner. The polarization may be linear or (approximately) circular. Linear polarizations and circular polarizations are aligned in space as shown in Figure 6.3-4. A linearly polarized antenna receives maximum power from an incident linearly polarized wave if the tilt angles (also called the polarization angles) of the wave and the antenna polarizations are aligned in space as shown in Figure 6.3-4(a). The wave is then said to be co-polarized. As the tilt angle of the wave or antenna rotates from co-polarization, the received power decreases. When the tilt angles are 90° apart, as shown in Figure 6.3-4(b), the antenna is said to be cross-polarized to the wave and receives (ideally) no power from it. The antenna and the wave then have orthogonal polarizations. A given satellite can employ two orthogonal polarizations that exist simultaneously and carry different information without interference. This principle, frequency reuse, is used to increase the information capacity of satellites and the geosynchronous orbit.

Circular polarizations have either right hand (RHC) or left hand (LHC) senses. RHC and LHC polarizations are orthogonal. A circularly polarized satellite and a circularly polarized earth station are co-polarized if they have the same senses and are cross-polarized if they have the opposite senses. The relative tilt angles of circular polarized antennas and waves are of no consequence and are not defined. This represents an advantage of circular polarization over linear polarization, because the tilt angle of the earth station does have to be adjusted for a particular satellite.

SATELLITE SYSTEM CHARACTERISTICS

The design of a satellite communication system is an intricate process involving trade-offs among many variables to obtain maximum performance at a reasonable cost. The major cost and complexity trade-off occur between satellite and earth stations or, more generically, between the space segment and ground segment. The dominating design factors in both segments for systems using geostationary satellites include the following.

Space Segment:

- Weight and size of satellite
- DC power generated on board
- Propulsion system technology
- Dimensions and complexity of satellite antennas
- Requirements of the communications payload.

Ground Segment:

- Allocated frequency bands
- Earth stations' antenna size and radiofrequency (RF) requirements and capabilities
- Earth stations' multiple access and signal processing techniques.

The weight of the satellite is limited by the high cost of launching a spacecraft into geostationary orbit, a cost that generally increases with the weight of the satellite. Satellites are transitioning to electric propulsion and away from chemical reactive propulsion. This has a twofold effect: it saves mass on the satellite and allows for additional communications payload or solar power; it also results in better stationkeeping due to the continuous operation of the electric thrusters versus regular stationkeeping burns with the chemical system.

For a satellite of limited weight and size, a limited number of solar cells can be deployed, which defines an upper limit on the DC power available for the communication transponders. The size and power limitations translate into the fact that the spacecraft has limited RF output power, which then must be transmitted onto particular areas of the earth (e.g., the continental

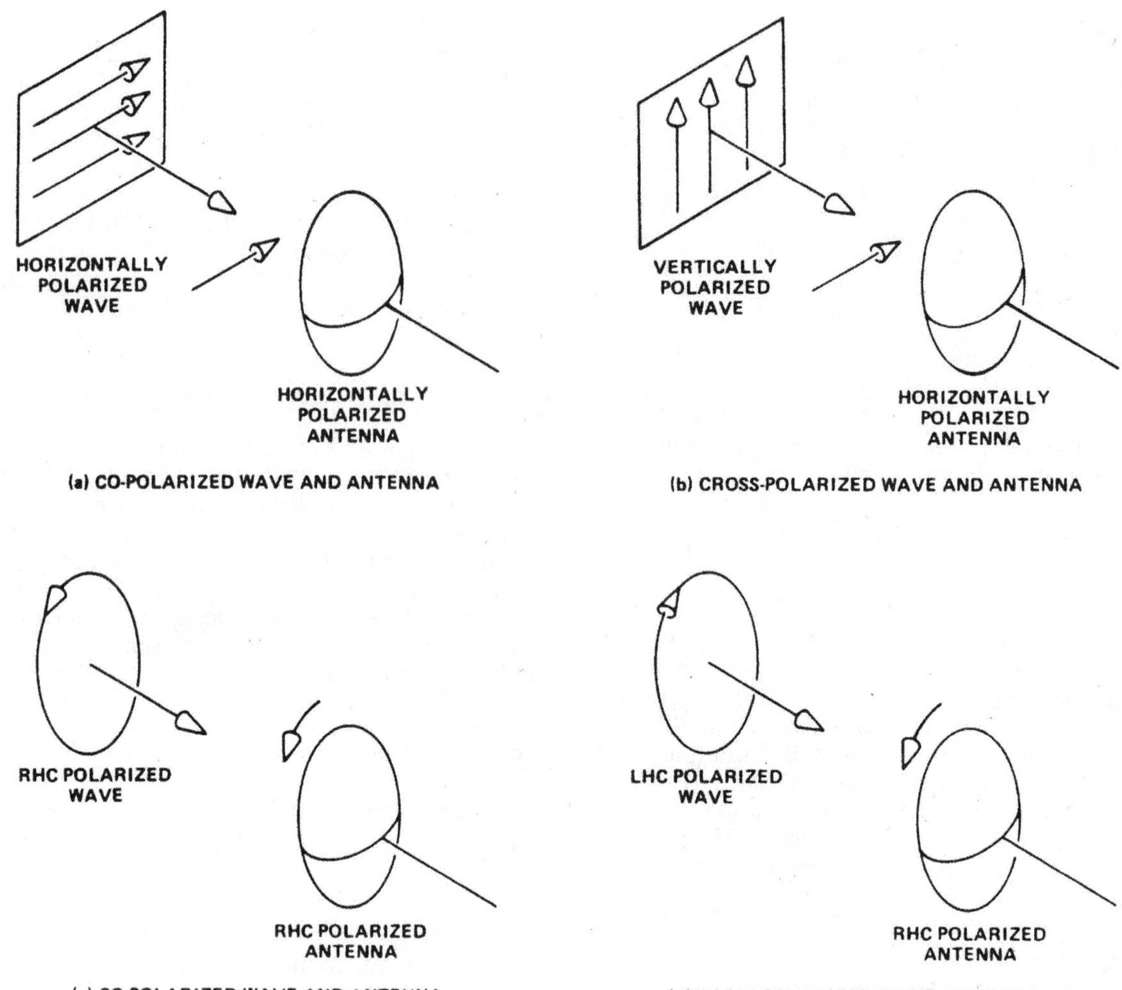

FIGURE 6.3-4 Linear and circular polarizations; (a) linear co-polarized wave and antenna; (b) linear cross-polarized wave and antenna; (c) circular co-polarized wave and antenna; (d) circular cross-polarized wave and antenna

United States). Furthermore, power densities over the earth's surface are limited, depending on operating frequency bands, to allow interference free coexistence with other communications systems operating at the same frequencies.

Multiple access and multiple destinations are distinctive virtues of satellite communications. The methods by which a large number of earth stations share one satellite or one transponder providing the required connectivity (multiple access techniques) also have a significant impact on system design. The multiple access can be achieved by sharing the transponder bandwidth in separate frequency slots—frequency-division multiple access (FDMA)—or the transponder availability in discrete time slots—time-division multiple access (TDMA). A third but less commonly used technique, code-division multiple access (CDMA) or spread spectrum, shares the transponders by allowing coded signals to overlap in time and frequency.

A satellite communication system must be designed to meet certain minimum performance standards, within limitations of the transmitted power, RF bandwidth, and antenna sizes. The most important performance criterion is the bit error rate (BER) of the demodulated signal. This BER depends on a number of factors, such as the pre--detection carrier-to-noise density ratio (C/No), the type of modulation, and the information data rate.

Satellite Transmission Modes for Television and Radio

Analog and digital modulation methods have both been used for the transmission of television and radio signals via satellite. Historically, analog techniques were predominant until technological developments in digital compression and modulation systems resulted in their greater efficiency and cost-effectiveness. Consequently, analog techniques are rarely used today and will not be discussed in this chapter except as regards earth station architecture. Information on analog techniques is available in previous editions of this handbook.

Television and Radio Transmission

A typical digital transmission uplink consists of a digital encoder, a modulator, upconversion equipment, and a power amplifier. The digital encoder converts the analog source content to a digital signal, and this invariably involves use of digital compression techniques to reduce the signal bandwidth (e.g., MPEG encoding). The modulator then uses the digital signal to modulate a continuous wave (CW) carrier, producing a digitally modulated carrier information to and from a modulated carrier at an intermediate frequency (IF), typically 70 MHz or 140 MHz. The upconversion equipment converts the modulated carrier at IF to a satellite frequency corresponding to the transponder of the satellite, and provides this signal to the power amplifier that boosts the signal power and sends it to the antenna for RF transmission. More commonly today, the upconversion is distributed between the modem and a block up-converter (BUC) that combines the functions of upconversion and HPA.

Taking a closer look at the process, analog content is digitally encoded and applied to the FEC section of the modulator for forward error correction (FEC) encoding. This process adds redundancy to the bit stream by appending additional bits to the original information to provide error detection and correction capability. Increasingly, two separate FEC systems (e.g., a block FEC and a convolutional FEC) are utilized, with the applications sometimes separated by an interleaver that acts to break up error bursts to permit the FEC to be more effective. The signal is then scrambled using a standard algorithm to produce random-like data for a signal spectrum that will be noise-like in nature.

The aggregate data (original data plus error correction bits) is applied to the modulator for modulation onto an IF carrier. The IF carrier is selectable, typically in the range of 50 to 90 MHz (for 70 MHz operation), 100 to 180 MHz (for 140 MHz operation) or 950–2000 MHz. The center frequency of the modem modulator is tuned to position the signal within the satellite transponder.

A satellite digital transmission system is characterized by:

- Data rate
- Data interface
- Code rate
- Modulation scheme.

Data rate refers either to the number of output bits per second transmitted by the modulator (transmitted data rate) or to the data rate of the digital input stream provided to the modulator for transmission (information data rate). The information data rate is typically front-panel selectable. Modulators typically support a number of data interfaces. The data interface refers to the physical connector and signal levels and protocols.

Code rate refers to the FEC encoding scheme. In most modulators, the code rate is selectable. The code rate configuration is referred to as m/n, where m is the number of original bits per block of transmitted bits and n is the number of original bits plus error correction bits per block of transmitted bits. Thus, a code rate of 3/4 means that for every three data bits input four data bits are transmitted; thus, a 1024 kbps modem operating with a code rate of 3/4 would transmit 1365 kbps over the satellite channel.

The most widespread FEC coding technique currently in use is concatenated coding, in which two different coding methods are successively applied to the information to be transmitted, sometime separated by an interleaver that acts to distribute burst errors (on the receive side) so the outer coding system (the one applied first) can correct errors not corrected by the inner system (the one applied second). Frequently, convolutional codes are used for the outer code and block codes are used for the inner code, but block codes may be used for both.

The modulation scheme refers to the method of carrying the data bits on the RF carrier. Two common digital modulation schemes employed in satellite transmission systems are quadrature phase-shift keying (QPSK) and eight-phase-shift keying (8PSK). These modulation schemes generate periodic phase shifts in the RF carrier referred to as symbols. The desired symbol rate (the number of symbols per second) and data transmission rate (which depends upon the information source itself as well as the amount of FEC coding employed) determine the amount of bandwidth required in the channel. At Ka-band, higher order modulations like 16APSK or similar are now also being used.

In QPSK, four phase shifts are used to represent two unique states, so each transmitted symbol represents two transmission bits. For this case, the symbol rate is equal to half the data transmission rate. The 8PSK scheme uses eight phase shifts, thus transmitting three bits per symbol. For 8PSK, the symbol rate is equal to one-third the transmission rate. For a given data transmission rate, 8PSK requires less bandwidth than QPSK, but it requires increased performance (carrier-to-noise ratio, C/N) from the channel to achieve an equivalent BER. Neither QPSK nor 8PSK employs an amplitude modulation component (all modulation is phase modulation). This permits them to be employed in a saturated (maximum power) satellite transponder when utilized in single carrier per transponder mode.

C-Band Satellites

C-band was initially favored for communications satellites because of the favorable propagation characteristics at these frequencies. The specific bands in most common use are the 5925 to 6425 MHz (uplink) and the 3700 to 4200 MHz (downlink) band pair. U.S. domestic FSS requires the use of 36 MHz bandwidth channels placed on 40 MHz centers. A satellite using a single polarization can provide 12 such transponders, although all satellites are mandated to employ frequency reuse and provide 24 such transponders. Frequency reuse is implemented by the use of orthogonal polarizations and by staggering the center frequency of opposing transponders.

As an example of a typical satellite, the transmit and receive frequency plans of a C-band satellite are shown in Figure 6.3-5. The numbered brackets represent each channel. The bandwidth of the channel is represented by the width of the bracket. The carrier frequency, shown above the channel number, is centered on each channel. The signals of alternate transponders in the frequency plan of Figure 6.3-5 are nominally orthogonal. If they were exactly orthogonal and the associated earth stations were ideal (with respect to polarization), there would be no interference caused by the overlapping sideband energy of adjacent transponders. In practice, the polarizations of the antennas of the satellite and earth stations are not ideal. Some small amount of interference occurs, but the combination of nearly orthogonal polarizations and the use of the staggered frequency plan provides for high-quality transmission under almost all weather conditions.

C-band FSS share frequencies with terrestrial microwave systems. This places constraints on the location of C-band earth stations and limits the permissible power density. Prior to licensing a C-band transmit antenna, a frequency coordination process must be performed (such frequency coordination is not required at the Ku-band).

Ku-Band Satellites

The first systems using the 14.0 to 14.5 GHz (uplink) band and the 11.7 to 12.2 GHz (downlink) band were launched in 1976. The higher propagation loss characteristics at these frequencies require higher spacecraft equivalent isotropic radiated power (EIRP) to achieve the same transmission performance as C-band frequencies, and this is obtained by a variety of methods, including the use of greater spacecraft antenna gains, readily achievable at the higher frequencies. Because in the United States the Ku-band frequencies are not shared with terrestrial systems, the power flux density (PFD) limitation is much less stringent, and there is no requirement for coordination with terrestrial microwave systems; consequently, Ku-band satellites employ higher power satellite amplifiers than do C-band satellites—as much as an order of magnitude higher. The high powers permit the use of very small earth station antennas at or near the user's premises. This provides an important economic advantage for many services and makes the use of this frequency band very attractive; however, a good part of the higher satellite power achievable is necessary to offset the additional attenuation that is experienced at these frequencies during heavy rain conditions.

There is no mandated frequency plan for transponders in this frequency band, although typical transponder bandwidths today are 36 MHz. Because the bandwidth is the same as C-band, it is possible to have a similar 24 transponder, 36 MHz frequency plan with 40 MHz channel spacing when frequency reuse is utilized.

Ka-Band Satellites

The commercial application of Ka-band satellites was first introduced by NASA's experimental Advanced Communications Technology Satellite (ACTS), which demonstrated successful use of the 17.5 to 22.5 GHz downlink and 29.5 to 34.5 GHz bands for various services. Use of this band for direct broadcast satellite (DBS) video and Internet service is a direct result of those experiments. It is likely that these bands will be increasingly used for video applications, with considerable interest in narrow casting and point-to-point applications.

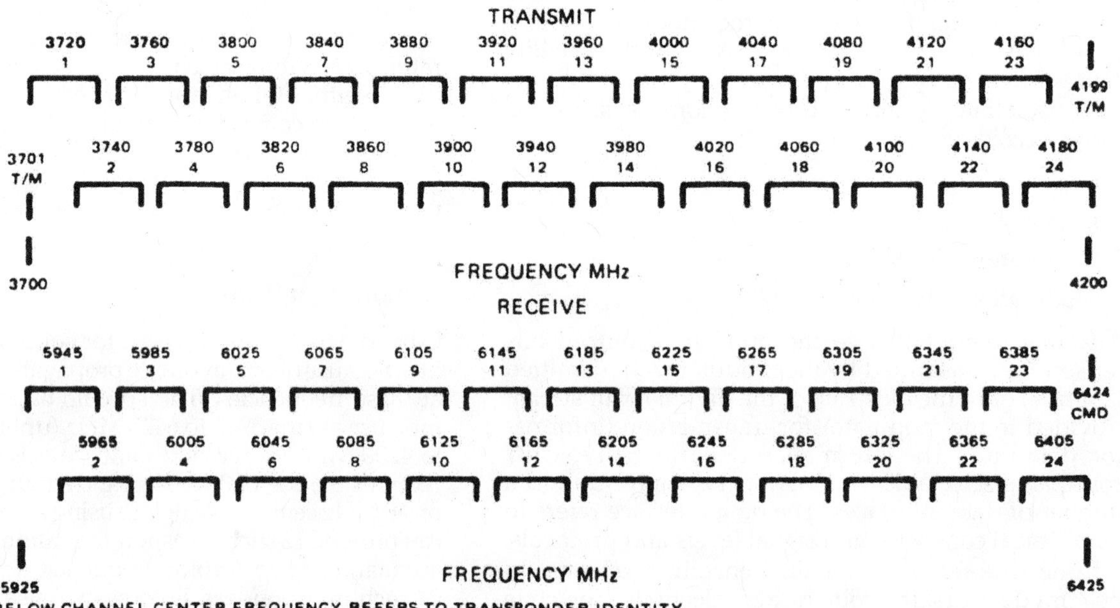

FIGURE 6.3-5 Typical C-band satellite frequency plan

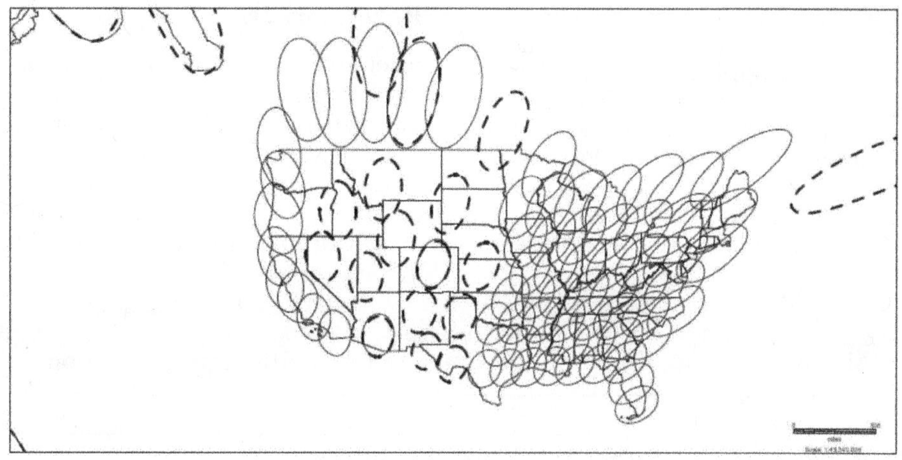

FIGURE 6.3-6 Beam pattern for representative Ka-band Satellite (Viasat-1). Gateway beams are shown in dashed lines.

Drawing courtesy of Viasat

Due to the higher frequency range, Ka-band satellites generally include many small spot beams in lieu of one large contiguous U.S. (CONUS) or geographically large beam, as is shown in Figure 6.3-6. This significantly improves frequency reuse and overall satellite capacity but at the expense of connectivity, such as the ability of an uplink to simultaneously broadcast to many geographically dispersed downlinks. Ka-band signals suffer from greater attenuation due to the presence of rain and atmospheric oxygen than do C-band and Ku-band services. This is illustrated in Figure 6.3-7, where the approximate center of each band is marked by a dashed line. This attenuation problem has historically made Ka-band rather unattractive for satellite communications, but the scarcity of spectrum in other bands and desirability of small spot beams for some applications, as well as advances in satellite communications technology that greatly increase capacity available at a given orbital location, appear to override this disadvantage.

In addition to core satellite voice, data, and video offerings, Ka-band licensees will provide many low-cost, broadband interactive services such as:

- Direct-to-home video
- Internet access.

At the time of this writing, DirecTV is using Ka-band satellites to provide DBS video service. Although some of these satellites were originally designed for on-board switching, they are being used in a "bent pipe" configuration for local-into-local video, including HDTV. Another current use of Ka-band is for home and small office Internet service, which was initially implemented by the ViaSat Wild Blue satellite system (for Internet applications). In 2011, ViaSat launched the ViaSat-1 satellite with a claimed capacity of more than 130 Gbit/s, more capacity than all C-band, Ku-band, and Ka-band satellites serving the United States at that time. ViaSat-1 satellite operates in the 28.1–29.1 GHz and 29.5–30 GHz uplink bands and 18.3–19.3 GHz and 19.7–20.2 GHz downlink bands; employs 72 user spot beams (63 United States and 9 Canada) and 20 gateway spot beams (17 United States and 3 Canada); and provides 12/3 Mbit/s Internet service to more than 600,000 users. ViaSat-2 spacecraft scheduled to be in service in 2017 with more than 300 Gbit/s of throughput and service to over 40 countries in the Americas. At the time of

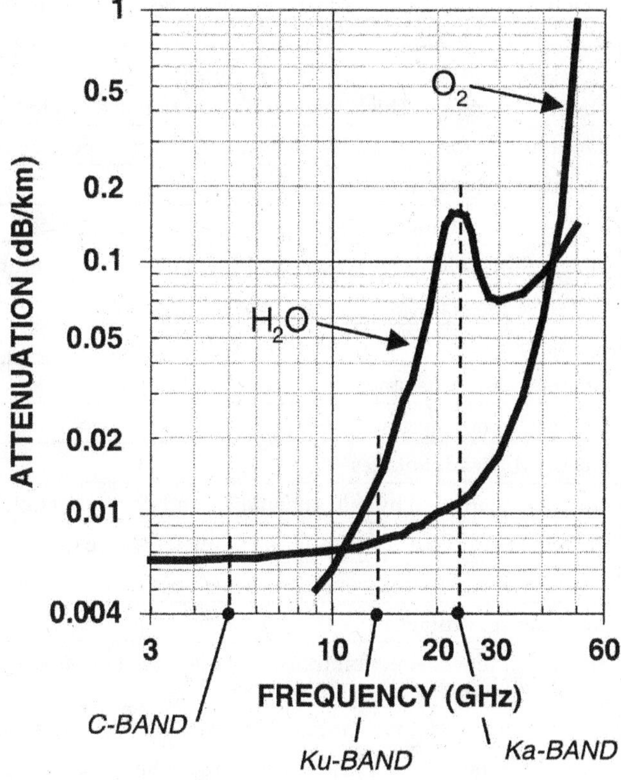

FIGURE 6.3-7 Signal attenuation versus frequency due to atmospheric oxygen and water vapor

writing, little or no Ka-band trunking or backhaul is taking place, but this type of point-to-point application would seem a natural application of Ka-band systems, particularly in drier regions of the country. Echostar, formerly Hughes Network Systems, is also using Ka-band satellites for home and business small dish Internet service.

The advantages and disadvantages of C-band, Ku-band and Ka-band are summarized in Table 6.3-1.

Regulatory Issues

Satellite communication systems are governed by the FCC in the United States and by the ITU-R on the international level. The governing agencies assign frequency bands of operation, satellite performance characteristics, and orbit location and provide technical specifications of radiated power density and radiation gain patterns for the earth stations. The FCC is

TABLE 6.3-1
Merits of C-band, Ku-band, and Ka-band for satellite communications

C-Band Advantages
C-band is usually the most reliable due to less susceptibility to rain outages.
The C-band space segment is normally less expensive than other FSS space segments.
C-Band Disadvantages
The frequency band is congested because it is shared with terrestrial microwave, making frequency coordination a requirement.
C-band requires relatively large antennas because of low satellite EIRP levels and the necessity of narrow half-power beamwidth to allow 2° spaced satellites.
Avoiding terrestrial interference can make site selection a difficult process.
The use of artificial shielding to block interference can increase total system cost.
Faraday rotation of polarization can affect system performance.
Ku-Band Advantages
The frequency band is used only for satellite communication.
Smaller antennas may be used because of higher gain and higher satellite EIRP.
Site selection is easier because of the smaller size of antenna and reduced terrestrial interference.
Ku-band is suitable for direct-to-home application.
Channel plan is flexible.
Ku-band is not affected by Faraday rotation.
Ku-Band Disadvantages
Ku-band is affected by rain attenuation and depolarization.
Waveguide and coaxial transmission line losses are quite high.
Interference can occur from radar detectors located in passing automobiles; site surveys should assess.
Ka-Band Advantages
Space segment cost per bit transmitted is lowest because satellites have much greater capacity due to increased frequency reuse via spot beams and gateways.
Smaller antennas (than Ku-band) can be used in many areas where atmospheric attenuation is low.
Higher frequency is more amenable to spot beam use and facilitates small dish uplinking and point-to point applications.
Smaller downlink beams can be tailored to the coverage area and exclude reception elsewhere.
Ka-Band Disadvantages
Increased atmospheric attenuation requires a greater link margin and can reduce reliability when fixed link margins are used.*
All-spot beam operation results in significant disadvantage for broadcast (point-to-multipoint) service where broad geographical distribution of the same signal is required.

*It should be noted, however, that many two-way Ka-band networks such as contribution links use adaptive coding and modulation (ACM) and operate with little or no margin. This allows for very high data throughput rates during clear sky conditions; as conditions degrade, the mod-code point changes within a range to use higher gain codes and less complex modulation to maintain the link at the expense of reduced throughput rates. Thus overall reliability within an operating range is as good as or better than fixed margin implementations, and because higher throughput is achieved during (95%) clear sky conditions, the overall capacity is significantly higher with ACM.

the licensing body for all transmit earth stations in the United States and licenses receive-only earth stations at the owner's request. Such receive antenna licensing ensures that the antenna technical characteristics exist in the master FCC database and must therefore be considered and analyzed from a potential interference standpoint by anyone considering licensing a transmit antenna in its vicinity. The FCC Rules and Regulations, CFR 47 Part 25, form the basis of the applicable documents that must be followed for the planning and implementation of any FSS band satellite communication system.

The FCC amends and interprets the rules as the technology and the requirements of satellites change through amendments, decisions, and declaratory orders; therefore, it is necessary to review the current FCC Rules prior to planning a new satellite project.

The FCC previously had separate 25.209 antenna gain and 25.134 input power density rules, both of which needed to be met independently. The FCC now has implemented EIRP density masks for the various services that essentially combine the 25.209 and 25.134 rule parts, allowing the applicant to comply only with resulting EIRP density as a function of off-beam angle or "theta." This allows the input EIRP density to be traded for shortfalls in antenna gain performance and with low power density input waveforms allow the use of much smaller antennas than would normally be allowed with the 25.209 rule alone.

The following sections present, in tabular form, the transmitted EIRP density allowable for earth station antenna routine licensing. Other densities may be permitted but require special showings. One such special showing would exist if the satellite carrier coordinated density levels exceeded those assumed by the FCC in its rules, which is frequently the case.

DIGITAL EARTH STATION OPERATION IN THE CONVENTIONAL OR EXTENDED C-BANDS

1. For co-polarized transmissions in the plane tangent to the GSO arc:

$26.3-25\log_{10}\theta$	dBW/4 kHz	for	$1.5° \leq \theta \leq 7°$
5.3	dBW/4 kHz	for	$7° < \theta \leq 9.2°$
$29.3-25\log_{10}\theta$	dBW/4 kHz	for	$9.2° < \theta \leq 48°$
−12.7	dBW/4 kHz	for	$48° < \theta \leq 180°$

The EIRP density levels specified for $\theta > 7°$ may be exceeded by up to 3 dB in up to 10% of the range of theta (θ) angles from ±7 to 180°, and by up to 6 dB in the region of main reflector spillover energy.

2. For co-polarized transmissions in the plane perpendicular to the GSO arc:

$29.3-25\log_{10}\theta$	dBW/4 kHz	for	$3° \leq \theta \leq 48°$
−12.7	dBW/4 kHz	for	$48° < \theta \leq 180°$

These EIRP density levels may be exceeded by up to 6 dB in the region of main reflector spillover energy and in up to 10% of the range of θ angles not included in that region, on each side of the line from the earth station to the target satellite.

3. For cross-polarized transmissions in the plane tangent to the GSO arc and in the plane perpendicular to the GSO arc:

$16.3-25\log_{10}\theta$	dBW/4 kHz	for	$1.5° \leq \theta \leq 7°$

4. A license application for earth station operation in a network using variable power density control of earth stations transmitting simultaneously in shared frequencies to the same target satellite receiving beam may be routinely processed if the applicant certifies that the aggregate off-axis EIRP density from all co-frequency earth stations transmitting simultaneously to the same target satellite receiving beam, not resulting from colliding data bursts transmitted pursuant to a contention protocol, will not exceed the off-axis EIRP density limits permissible for a single earth station.

DIGITAL EARTH STATION OPERATION IN THE CONVENTIONAL KU-BAND

1. For co-polarized transmissions in the plane tangent to the GSO arc:

$15-25\log_{10}\theta$	dBW/4 kHz	for	$1.5° \leq \theta \leq 7°$
−6	dBW/4 kHz	for	$7° < \theta \leq 9.2°$
$18-25\log_{10}\theta$	dBW/4 kHz	for	$9.2° < \theta \leq 19.1°$
−14	dBW/4 kHz	for	$19.1° < \theta \leq 180°$

The EIRP density levels specified for $\theta > 7°$ may be exceeded by up to 3 dB in up to 10% of the range of theta (θ) angles from ±7 to 180°, and by up to 6 dB in the region of main reflector spillover energy.

2. For co-polarized transmissions in the plane perpendicular to the GSO arc:

$18-25\log_{10}\theta$	dBW/4 kHz	for	$3° \leq \theta \leq 19.1°$
−14	dBW/4 kHz	for	$19.1° < \theta \leq 180°$

These EIRP density levels may be exceeded by up to 6 dB in the region of main reflector spillover energy and in up to 10% of the range of θ angles not included in that region, on each side of the line from the earth station to the target satellite.

3. For cross-polarized transmissions in the plane tangent to the GSO arc and in the plane perpendicular to the GSO arc:

$5-25\log_{10}\theta$	dBW/4 kHz	for	$1.5° \leq \theta \leq 7°$

4. A license application for earth station operation in a network using variable power density control of

earth stations transmitting simultaneously in shared frequencies to the same target satellite receiving beam may be routinely processed if the applicant certifies that the aggregate off-axis EIRP density from all co-frequency earth stations transmitting simultaneously to the same target satellite receiving beam, not resulting from colliding data bursts transmitted pursuant to a contention protocol, will not exceed the off-axis EIRP density limits permissible for a single earth station.

25.138 LICENSING REQUIREMENTS FOR GSO FSS EARTH STATIONS IN THE CONVENTIONAL KA-BAND

Applications for earth station licenses in the GSO FSS in the conventional Ka-band that indicate that the following requirements will be met and include the information required by relevant provisions in 25.115 and 25.130 may be routinely processed:

1. The EIRP density of co-polarized signals in the plane tangent to the GSO arc will not exceed the following values under clear sky conditions:

32.5–25log(θ)	dBW/MHz	for	$2.0° \le \theta \le 7°$
11.5	dBW/MHz	for	$7° \le \theta \le 9.2°$
35.5–25log(θ)	dBW/MHz	for	$9.2° \le \theta \le 19.1°$
3.5	dBW/MHz	for	$19.1° < \theta \le 180°$

where θ is the angle in degrees from a line from the earth station antenna to the assigned orbital location of the target satellite.

2. In the plane perpendicular to the GSO arc, as defined in §25.103, the EIRP density of co-polarized signals will not exceed the following values under clear sky conditions:

35.5–25log(θ)	dBW/MHz	for	$3.5° \le \theta \le 7°$
14.4	dBW/MHz	for	$7° < \theta \le 9.2°$
38.5–25log(θ)	dBW/MHz	for	$9.2° < \theta \le 19.1°$
6.5	dBW/MHz	for	$19.1° < \theta \le 180°$

3. The EIRP density levels specified in paragraphs (a)(1) and (2) of this section may be exceeded by up to 3 dB, for values of $\theta > 7°$, over 10% of the range of theta (θ) angles from 7° to 180° on each side of the line from the earth station to the target satellite.
4. The EIRP density of cross-polarized signals will not exceed the following values in the plane tangent to the GSO arc or in the plane perpendicular to the GSO arc under clear sky conditions:

22.5–25log(θ)	dBW/MHz	for	$2.0° < \theta \le 7.0°$

A license application for earth station operation in a network using variable power density control of earth stations transmitting simultaneously in shared frequencies to the same target satellite receiving beam may be routinely processed if the applicant certifies that the aggregate off-axis EIRP density from all co-frequency earth stations transmitting simultaneously to the same target satellite receiving beam, not resulting from colliding data bursts transmitted pursuant to a contention protocol, will not exceed the off-axis EIRP density limits permissible for a single earth station, as specified in paragraphs (a)(1) of §25.138.

Any antenna licensed for reception of radio transmission from a space station in the FSS shall be protected from radio interference caused by other space stations only to the degree to which harmful interference would not be expected to be caused to an earth station employing an antenna conforming to the standards defined in paragraphs (a) and (b) of §25.138.

FCC License

The FCC requires licensing of transmitting earth stations and permits licensing of receive-only (RO) earth stations. It is desirable for a broadcaster to license a C-band RO earth station, as licensing protects the station from future interference from domestic microwave systems. This is because a licensed antenna is included in the FCC earth station database and anyone wanting to license a new transmitter needs to frequency coordinate with all licensed and identified antennas, including receive antennas. Because the FCC Rules and Regulations are ever evolving, it is necessary to review the latest rules before filing satellite applications. Licensing information is available on the Internet at www.fcc.gov and the FCC Rules are available on the Internet at www.ecfr.gov.

SYSTEM PERFORMANCE ANALYSIS

Considering an RF link as illustrated in Figure 6.3-8 with transmit power P_t and transmit gain G_t, the effective isotropic radiated power, EIRP, for the station along the main beam of the antenna is the product

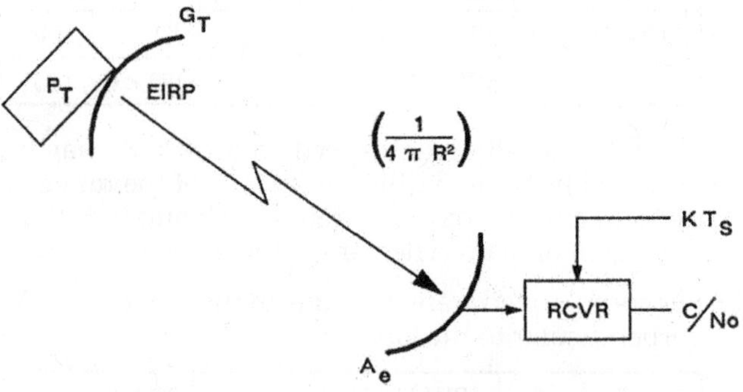

FIGURE 6.3-8 RF link diagram

$G_t \times P_t$. At a distance R meters from the transmitter, the radiated flux density, S, becomes:

$$S = (G_t P_t)\frac{1}{4\pi R^2} k_a \text{ Watts} / m^2 \quad (1)$$

where:

k_a = atmospheric attenuation factor < 1.

If an antenna with an effective area in square meters, A_e, is receiving this flux density, the received carrier level, C, at the antenna output is:

$$C = S A_e = (P_t G_t A_e)\frac{1}{4\pi R_2} k_a \text{ Watts} \quad (2)$$

At the same antenna output point, the effective noise power density, N_o is given by:

$$N_o = k \times T_s \text{ watts/hertz} \quad (3)$$

where:

k = Boltzmann's constant = 1.38×10^{-23} Joules/K or −228.6 dB

T_s = System noise temperature.

Consequently, received carrier-to-noise density becomes

$$\frac{C}{N_o} = (P_t G_t A_e)\frac{1}{4\pi R^2}\frac{1}{k T_s} k_a \quad (4)$$

A fundamental relationship in antenna theory is that the gain, G_r and the effective area of an antenna, A_e, are related by:

$$A_e = G_r \frac{\lambda^2}{4\pi} \text{ meter}^2 \quad (5)$$

Substituting this relation into the expression for C/N_o,

$$\frac{C}{N_o} = (P_t G_t G_r)(\lambda/4\pi R)^2 \frac{1}{kT_s} k_a \quad (6)$$

or

$$\frac{C}{N_o} = \text{EIRP}\frac{G_r}{T_s}(\lambda/4\pi R)^2 \frac{1}{k} k_a \quad (7)$$

The factor $(\lambda / 4\pi R)^2$ is often inverted and defined as the spreading loss or space loss factor. This spreading loss can also be expressed as:

$$L_s = (4\pi Rf/c)^2 \quad (8)$$

where:

c = Speed of light = 3×10^8 meter/second

f = Frequency in hertz.

Link calculations are usually carried out in dB rather than directly from the relations mentioned earlier because of ease of working in common logarithms. For clear sky conditions, C/No in dB can be calculated by:

$$(C/N_o)\text{ dB} = 10 \times \log (C/N_o)$$

$$(C/N_o)\text{ dB} = G_t \times P_t - L_s + (G/T) + 228.6 - k_a \quad (9)$$

where:

EIRP = $10 \times \log (G_t \times P_t)$ dBW.

$$L_s = 20 \times \log (4\pi Rf/c)\text{ dB} \quad (10)$$

$$= 92.45 + 20 \times \log R \text{ (R in km)} + 20 \times \log f \text{ (f in GHz)}$$

$$(G/T) = 10 \times \log (G_r/T_s)\text{ dB/K} \quad (11)$$

Alternately, C/N_o can be expressed in terms of flux density, S, as:

$$(C/N_o) = S + (G/T) - A_i + 228.6 - k_a \text{ dBHz} \quad (12)$$

$$S = \text{EIRP} - L_s + A_i \text{ dBW/m}^2 \quad (13)$$

where A_i is the effective aperture of an isotropic radiator in dB:

$$A_i = 10\log(4\pi/\lambda^2) \quad (14)$$

Equation 9 is a fundamental tool for characterizing space link performance. It will be utilized later when calculating overall satellite link performance.

Earth Station Receive Figure of Merit: G/T

G/T is the figure of merit of a receive system. It is primarily a function of the gain of the antenna along with the antenna noise temperature, first amplifier noise temperature and losses located between the antenna and the first amplifier. The importance of the term G/T in Equation 9 and Equation 12 cannot be overstated. Examination of the C/N_o expression shows that for a given available transmitting power and information format (and thus bandwidth), the only available method of controlling the received signal quality that can be used by the downlink operator is through the system G/T. Note that the G/T provides a direct dB relationship with C/N_o.

Figure 6.3-9 shows a block diagram of a typical receive system. Each device in the RF path has an associated gain or loss and a noise temperature. These contributions are combined to reflect the noise power weighted by the gain distribution through the chain. The earth station G/T is given by:

$$(G/T) = G_a - 10 \times \log(T_s)\text{ dB/K} \quad (15)$$

where:

G_a = Antenna gain referenced to LNA or LNB input (dBi)

T_s = System noise temperature referenced to LNA or LNB input (K).

The system noise temperature, T_s, referenced to the LNA or LNB input can be calculated by adding as noise powers the equivalent noise temperatures of all

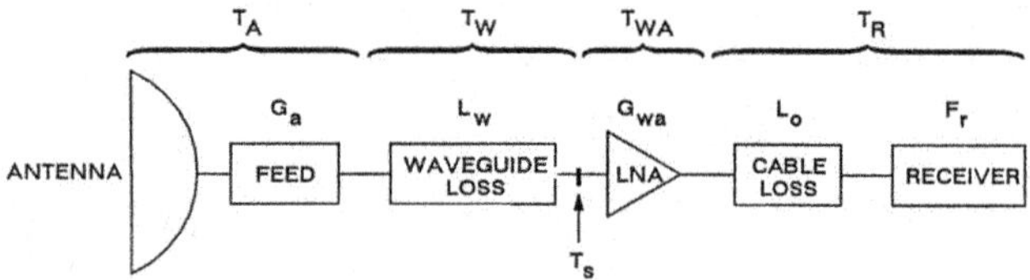

FIGURE 6.3-9 G/T system diagram: receive-only earth station

noise contributors, weighted by the net gain between the point in which that noise is been added and the LNA or LNB input, that is:

$$T_s = (T_a/L_w) + T_o \times (L_w - 1) / L_w + T_{vswr} + T_{lna} + [(L_t - 1) + L_t \times (F_r - 1)] \times T_1/G \quad (16)$$

where:

- T_a = Antenna noise temperature (K)
- L_w = Waveguide loss between antenna and LNA or LNB (linear power ratio)
- L_t = Transmission loss between LNA or LNB and receiver (line power ratio)
- T_o = Ambient temperature (K)
- T_1 = 290 K
- T_{lna} = LNA or LNB noise temperature (K)
- T_{vswr} = LNA or LNB-Antenna impedance mismatch noise temp. (K)
- F_r = Receiver noise figure (linear power ratio)
- G = Net gain between LNA or LNB input and receiver input (linear, includes interconnect cable loss).

The antenna temperature is usually minimum at zenith, typically 15° to 25° for a low loss, C-band antenna with low wide-angle sidelobes. As the elevation angle decreases, the antenna temperature increases because more of the higher level sidelobes look at the earth that has a temperature of about 290 K. A typical curve of the variation of noise temperature with elevation angle is illustrated in Figure 6.3-10. Similarly, Figures 6.3-11 shows typical G/T system performance for different C-band and Ku-band antenna diameters as a function of elevation angle.

Satellite Transponder

The orbiting spacecraft provides a one-hop carrier relay over a wide geographic area. In C-band systems, the uplink signal is transmitted near 6 GHz, received by the satellite, amplified, translated in frequency, filtered, and retransmitted near 4 GHz. Likewise, in Ku-band systems, the uplink occurs in the 14 GHz range and the downlink in the 12 GHz range; in Ka-band systems, the uplink occurs in the 30 GHz range, the downlink is offset by 9.8 GHz and is in the 20 GHz range (see Figure 6.3-2).

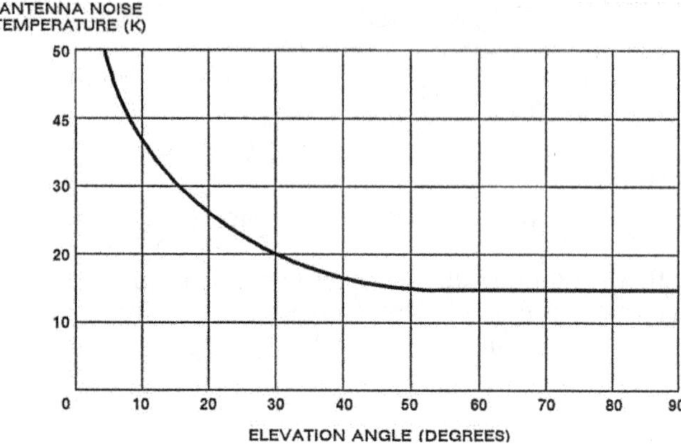

FIGURE 6.3-10 Typical antenna noise temperature variations with elevation angle

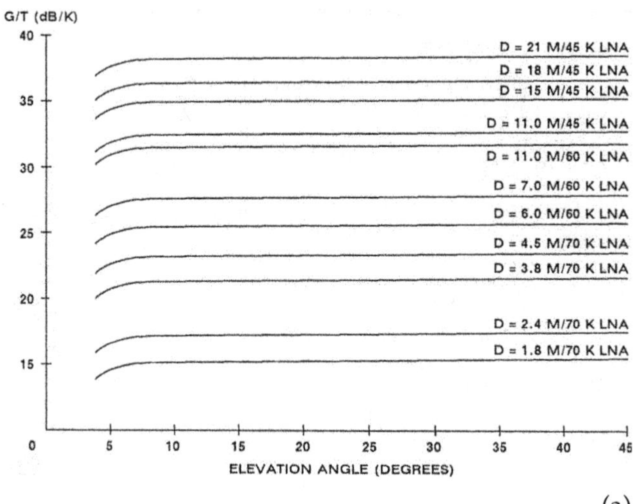

(a)

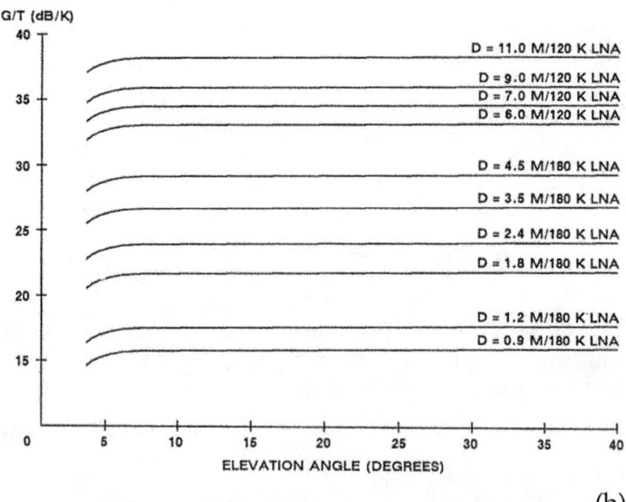

(b)

FIGURE 6.3-11 (a) Typical C-band G/T system performance versus elevation angle for different commonly used antenna diameters. (b) Typical Ku-band G/T system performance versus elevation angle for different commonly used antenna diameters.

Because the satellite serves as a transmit/receive station, it must be characterized by a G/T for the uplink side and by saturated EIRP for the downlink side. To couple the uplink and downlink signal strengths and as a definition of the transponder sensitivity, the uplink RF spectral flux density required at the satellite to saturate the transponder is also specified (saturation flux density, SFD). These three satellite parameters vary with geographic location. Contour maps, called footprints, are usually available for assessing these variations. Typical footprints for C-band and Ku-band satellites are shown in Figures 6.3-12 and 6.3-13, respectively. The footprint for a Ka-band satellite was previously shown as Figure 6.3-6.

Another important parameter that characterizes the transponder performance is the input/output power transfer and the intermodulation response. Both performance parameters are normally specified in terms of input back-off (BO_i) and output back-off (BO_o), as a function of the power reduction (expressed in dB) with respect to saturation. Figures 6.3-14 and 6.3-15 show typical transponder responses for a satellite equipped

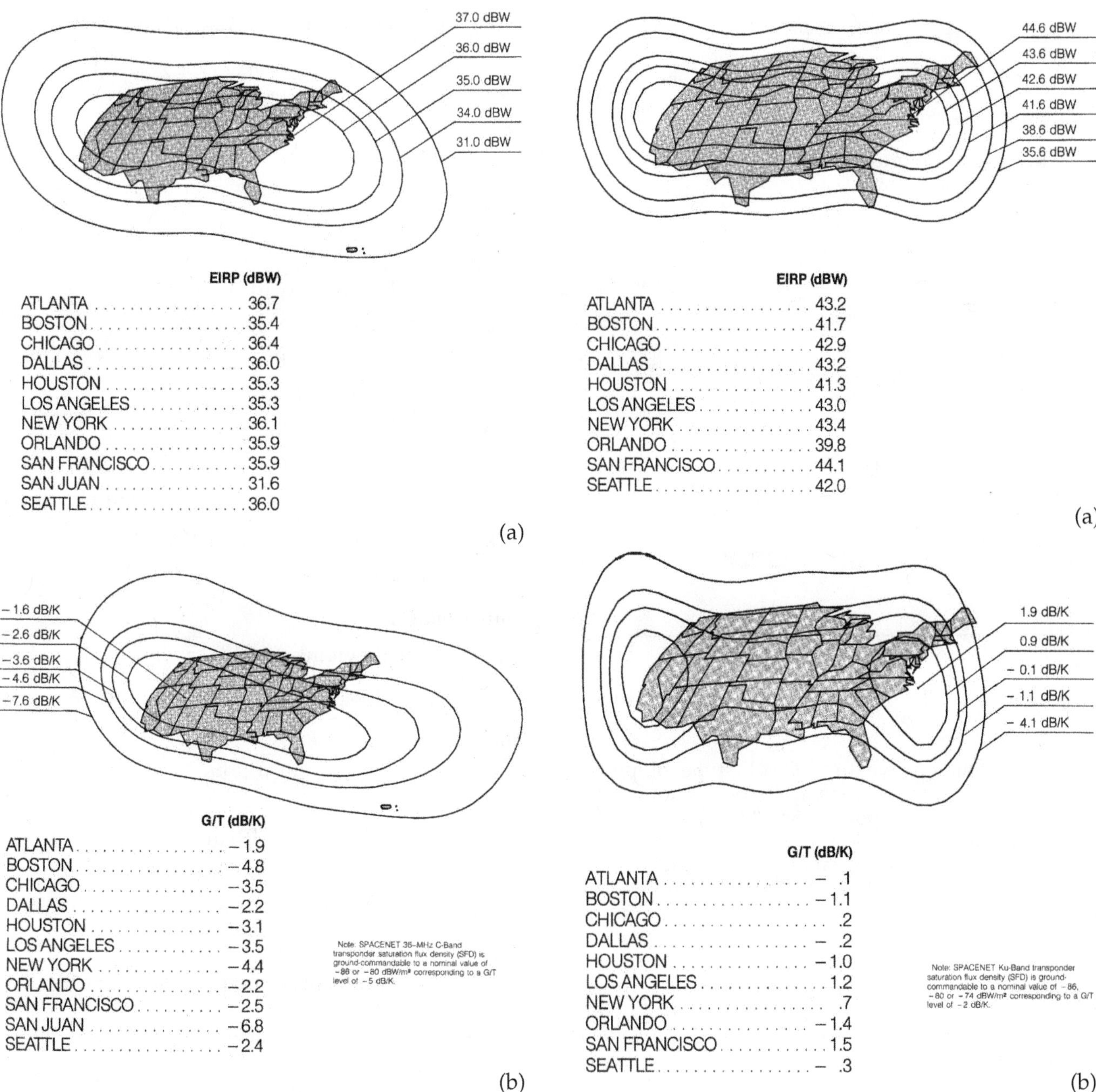

FIGURE 6.3-12 (a) C-band satellite EIRP footprint; (b) C-band satellite G/T footprint

FIGURE 6.3-13 (a) Ku-band satellite EIRP footprint; (b) Ku-band satellite G/T footprint

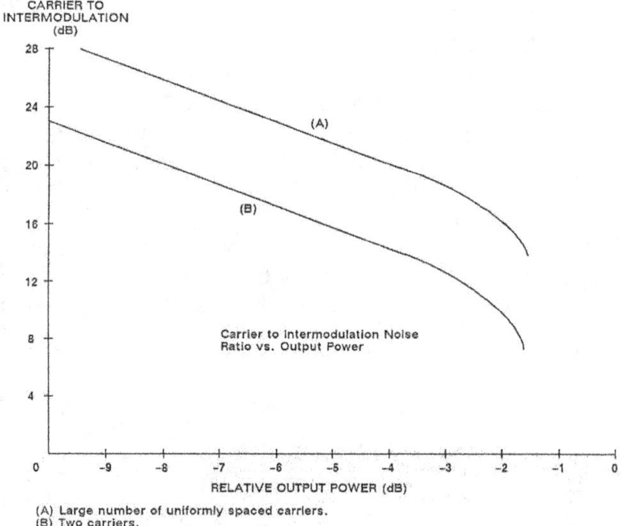

FIGURE 6.3-14 Output power normalized to single carrier saturation point

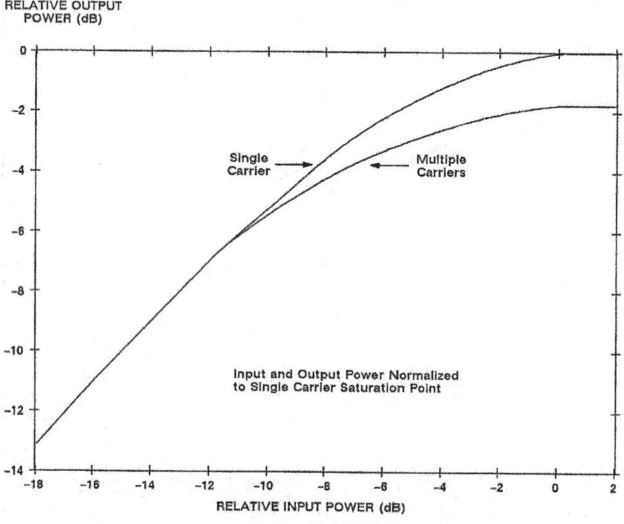

FIGURE 6.3-15 TWT power transfer characteristics

with a traveling wave tube (TWT) type of power amplifier. Solid-state power amplifiers (SSPAs) are also used as satellite transponder amplifiers, particularly as driver amplifiers and at C-band. Their improved linearity and increased reliability (longer device life) recommend them highly, but they cannot be used in all applications, particularly where higher powers are required.

Satellite Link Analysis

With the preliminary procedures and formulations described previously, link calculations can be conducted. First, the distance or slant range from the satellite to the earth station needs to be determined so that the space loss may be calculated. From orbit geometry and Equation 10, the space loss expressed in dB is found to be:

$$L_s = 185.05 + 10 \times \log [1 - 0.295 \cos(H) \cos(\Delta L)] + 20 \times \log f \quad (17)$$

where:
- H = Latitude of earth station
- ΔL = Difference in longitude for earth station and satellite
- f = Frequency in GHz.

The overall satellite link can now be calculated.

Uplink C/N

From Equation 9, the uplink $(C/N_o)_u$ becomes:

$$(C/N_o)_u = EIRP_u - L_u + (G/T)_s + 228.6 - k_a \quad (18)$$

or

$$(C/N_o)_u = S - A_i + (G/T)_s + 228.6 \quad (19)$$

and

$$S = SFD - Bo_i \quad (20)$$

where:
- S = Flux density (dBW/m²)
- L_u = Uplink space loss (dB)
- $EIRP_u$ = Uplink EIRP (dBW)
- $(G/T)_s$ = Satellite G/T (dB/K)
- A_i = $21.5 + 20 \times \log f$ (GHz) (dB/m²)
- SFD = Saturation flux density (dBW/m²)
- BO_i = Transponder input back-off (dB)
- k_a = atmospheric attenuation.

Downlink C/N

Likewise, the downlink $(C/N_o)_d$ can be calculated by:

$$(C/N_o)_d = EIRP_d - L_d + (G/T)_{e.s} + 228.6 - k_a \quad (21)$$

and

$$EIRP_d = EIRP_s - Bo_o \quad (22)$$

where:
- $EIRP_d$ = Downlink EIRP (dBW)
- $EIRP_s$ = Saturated EIRP (dBW)
- L_d = Downlink space loss (dB)
- $(G/T)_{e.s}$ = Earth station G/T (dB/K)
- BO_o = Transponder output back-off (dB).

It is important to note that Equations 20 and 22 are related by the nonlinear power transfer function of the transponder, therefore, for transponder operation below saturation the input and output relationship needs to be resolved graphically with the aid of Figure 6.3-15 or its equivalent.

Once uplink and downlink noise contributions are determined the composite link performance in terms of total carrier to noise density ratio $(C/N_o)_t$ can be readily obtained by simple noise power addition because the uplink and downlink contributions are incoherent. This yields:

$$(C/N_o)_t = \{(C/N_o)_u^{-1} + (C/N_o)_d^{-1}\}^{-1} \quad (23)$$

This equation represents a simplified situation in that only thermal noise is added to the carriers. In actuality there are other sources of perturbations and interference, transponder nonlinearity is the cause of some of the more important ones. As shown in Figure 6.3-15, operating the transponder near maximum power for better efficiency implies that compression, due to the instantaneous nonlinear transfer characteristic of the amplifier relative to the signal amplitude, becomes more significant. Under this condition, when more than one frequency is amplified, interaction between the signals occurs and consequently a spectrum of spurious frequencies or intermodulation products is generated.

Particularly, the so called third order intermodulation product of the form $(2f_1 - f_2)$, a consequence of the third order nonlinearity of the transponder, constitutes a significant interfering signal because it is the largest product and it falls in the same operating bandwidth of the information signal. Figure 6.3-14 shows how carrier-to-intermodulation ratio varies as a very sensitive function of the transponder operating output back-off.

System C/N

Figure 6.3-16 depicts a complete satellite link. Other sources of interference have been added, such as uplink interference due to off beam radiation from other earth stations and uplink cross-polarization isolation, combined, represented by the quantity $(C/I_o)_u$. Similarly, in the downlink the quantity $(C/I_o)_d$ represents the combined effects of downlink cross-polarization isolation and adjacent satellite interference. When all these terms are considered the total link $(C/N_o)_t$ can be calculated by:

$$(C/N_o)_t = \{(C/N_o)_u^{-1} + (C/I_o)_u^{-1} + (C/N_o)_d^{-1} + (C/I_o)_d^{-1}\}^{-1} \quad (24)$$

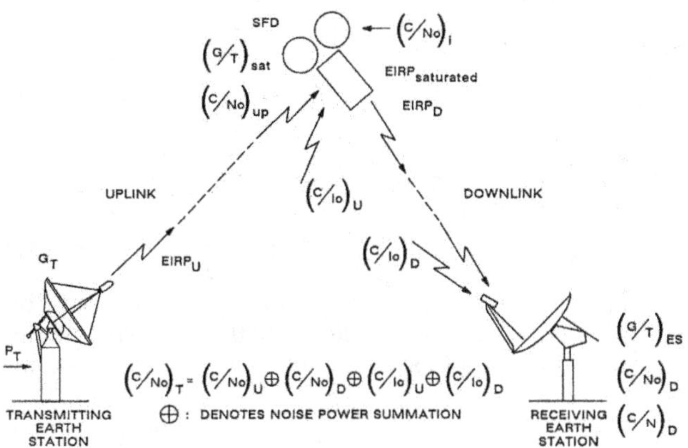

FIGURE 6.3-16 Satellite link model

Figure 6.3-17 shows the typical interaction of the different terms in Equation 24 as a function of transponder input back-off and in the presence of thermal and intermodulation noise. The total $(C/N_o)_t$ can be maximized by reducing the transponder input drive and adjusting the transponder gain. Backing off the TWT reduces $(C/N_o)_\mu$ and $(C/N_o)_d$ (through the input/output relationship of the transponder), but as $(C/N_o)_I$ increases rapidly when the input drive is reduced, an optimum value of $(C/N_o)_t$ is obtained at a specific back-off level. Interference noise can be kept down by proper antenna design, transponder sensitivity and frequency coordination of satellite services.

Rain Effects on System Performance

Rain is the dominant factor in satellite propagation for frequencies above 10 GHz. Rain propagation has been studied intensively since the late 1960s and only a brief discussion will be presented here. Due to the basic interaction of electromagnetic waves with water in liquid form, raindrops cause absorption, scattering and depolarization phenomena. Absorption and scattering result in signal attenuation and an increase in sky noise temperature, with the consequent degradation of the received C/N_o. Depolarization has an effect on dual polarization systems and creates interference between cross-polarized signals.

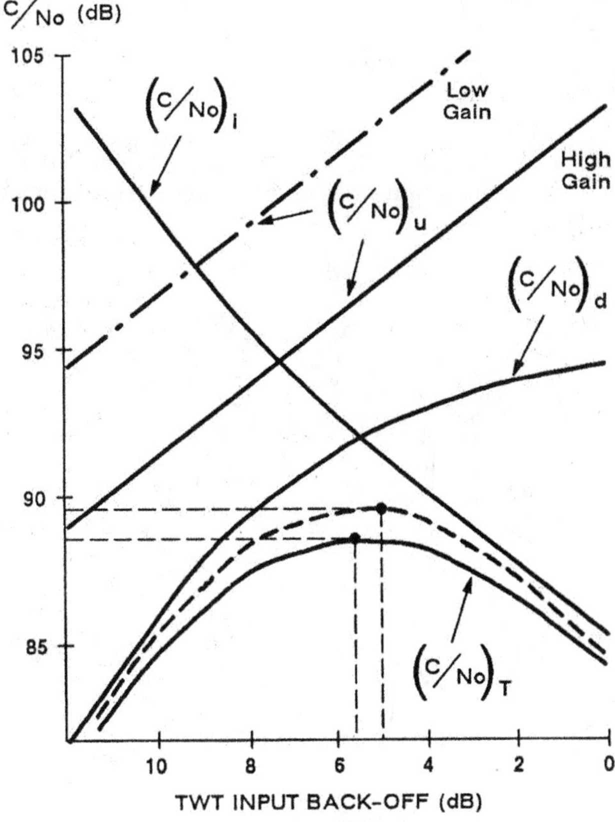

FIGURE 6.3-17 Optimum TWT operation

Signal Attenuation

The amount of attenuation depends fundamentally on the rain intensity or rain rate and the signal path length in rain. Rainfall data are available for most parts of the world; different types of climates have been defined and boundaries of their regions identified. Figures 6.3-18 through 6.3-20 show the NASA rain rate climate regions. The long-term behavior of rain is described by the cumulative probability distribution or exceedance curve. This gives the percentage of time that the rain rate exceeds a given value. Table 6.3-2 gives the rain-rate distribution values versus per cent of year for the various rain climate regions of Figures 6.3-18 through 6.3-21. Figure 6.3-21(a) and (b) plot the rain rate cumulative probability distributions for the regions presented on the previous maps.

The calculation of the rain attenuation involves two basic steps. The first step is to determine the rain rate in mm/hr as a function of the cumulative probability of occurrence. This probability will be defined by the grade of service or availability of the link to be provided. The second step consists of the calculation of the actual rain attenuation associated with the rain rate that was exceeded with such probability.

The attenuation per unit of length (specific attenuation), λ_r(dB/km), is tied to the rain rate R (mm/hr), by the empirically derived relationship:

$$\lambda_r = a(f) \times R^{b(f)} \text{ dB/km} \quad (25)$$

where a(f) and b(f) are frequency dependent coefficients. For the frequency range between 8.5 and 25 GHz, Equation 25 becomes:

$$\lambda_r = 4.21 \times 10^{-5} f^{2.42} \times 1.41 \times f^{-0.0779} R \text{ dB/km} \quad (26)$$

The attenuation per unit length is heavily frequency dependent, Figure 6.3-22 shows frequency dependence of λ_r for various rain rates. Introducing the concept of equivalent path length, $L_e(R)$, the total rain attenuation in decibels is simply:

$$Ar = \lambda_r \times L_e(R) \text{ dB} \quad (27)$$

Equivalent path length is primarily determined by the height of the freezing level or 0° isotherm, which depends on latitude, season and rain rate, the cosecant of the elevation angle, and site altitude. For latitudes within +30°, the freezing level is at 4.8 km. Curves of equivalent path lengths versus elevation angle and for different rain rates are shown in Figure 6.3-23.

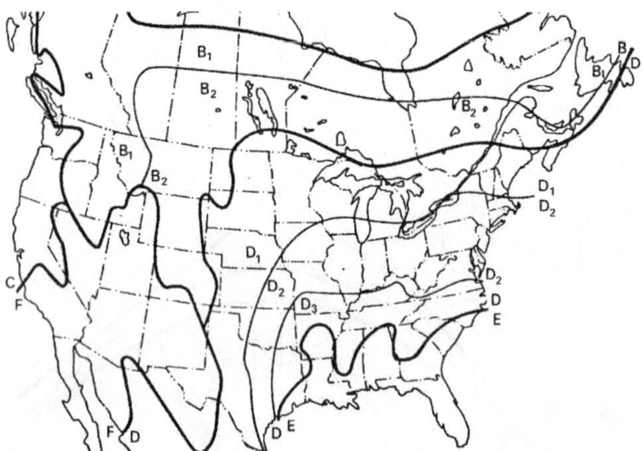

FIGURE 6.3-18 Rain rate climate regions for the continental United States showing the subdivision of Region D

NASA Propagation Effects Handbook for Satellite System Design, ORI TR 1679

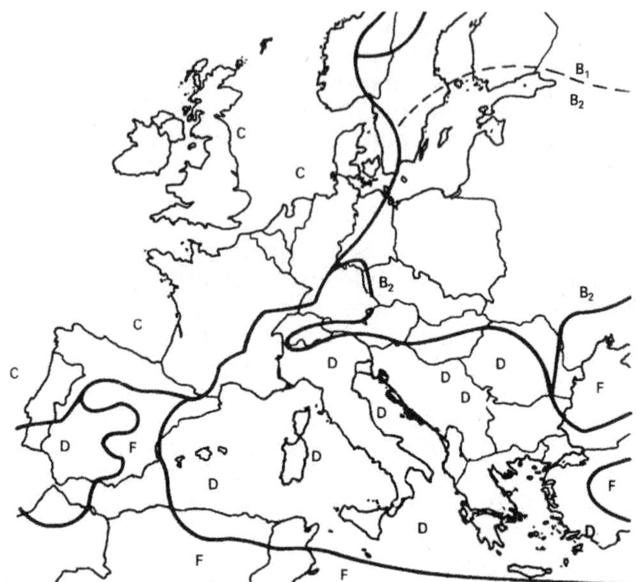

FIGURE 6.3-19 Rain rate climate regions for Europe

NASA Propagation Effects Handbook for Satellite System Design, ORI TR 1679

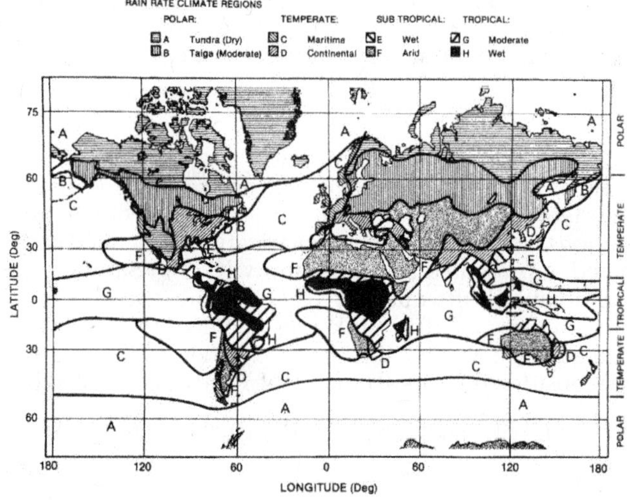

FIGURE 6.3-20 Global rain rate climate regions, including the ocean areas

NASA Propagation Effects Handbook for Satellite System Design, ORI TR 1679

TABLE 6.3-2
Point rain rate distribution values (millimeters per hour) vs. percent of the year rain rate is exceeded

Percent of Year	Rain Climate Region										Minutes Per Year	Hours Per Year
	A	B	C	D1	D2	D3	E	F	G	H		
0.001	28.0	54.0	80.0	90.01	02.0	127.0	164.0	66.01	29.0	2512.0	5.3	0.09
0.002	24.0	40.0	62.0	72.0	86.0	107.0	144.0	51.0	109.0	220.0	10.5	0.18
0.005	19.0	26.0	41.0	50.0	64.0	81.0	117.0	34.0	85.0	178.0	26.0	0.44
0.01	15.0	19.0	28.0	37.0	49.0	63.0	98.0	23.0	67.0	147.0	53.0	0.88
0.02	12.0	14.0	18.0	27.0	35.0	48.0	77.0	14.0	51.0	115.0	105.0	1.75
0.05	8.0	9.5	11.0	16.0	22.0	31.0	52.0	8.0	33.0	77.0	263.0	4.38
0.1	6.5	6.8	7.2	11.0	15.0	22.0	35.0	5.5	22.0	51.0	526.0	8.77
0.2	4.0	4.8	4.8	7.5	9.5	14.0	21.0	3.8	14.0	31.0	1052.0	17.50
0.5	2.5	2.7	2.8	4.0	5.2	7.0	8.5	2.4	7.0	13.0	2630.0	43.80
1.0	1.7	1.8	1.9	2.2	3.0	4.0	4.0	1.7	3.7	6.3	5260.0	87.66
2.0	1.1	.2	1.2	1.3	1.8	2.5	2.0	1.1	.6	2.8	10,520.0	175.30

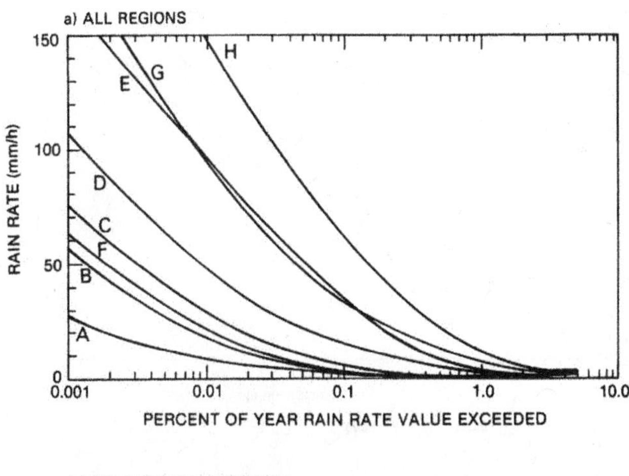

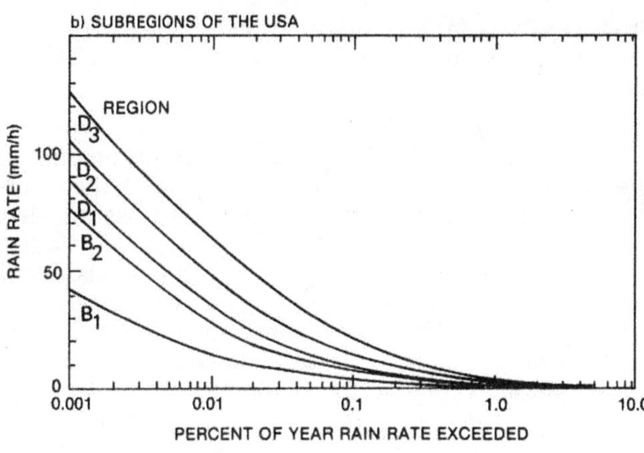

FIGURE 6.3-21 Rain rate cumulative probability distributions for the regions presented on the previous maps

Reprinted from Louis J. Ippolito, R. D. Kaul, and R. G. Wallace, *Propagation Effects Handbook for Satellite Systems Design* [NASA Reference Publication 1082(03)], National Aeronautics and Space Administration, Washington, DC, June 1983. Courtesy of NASA.

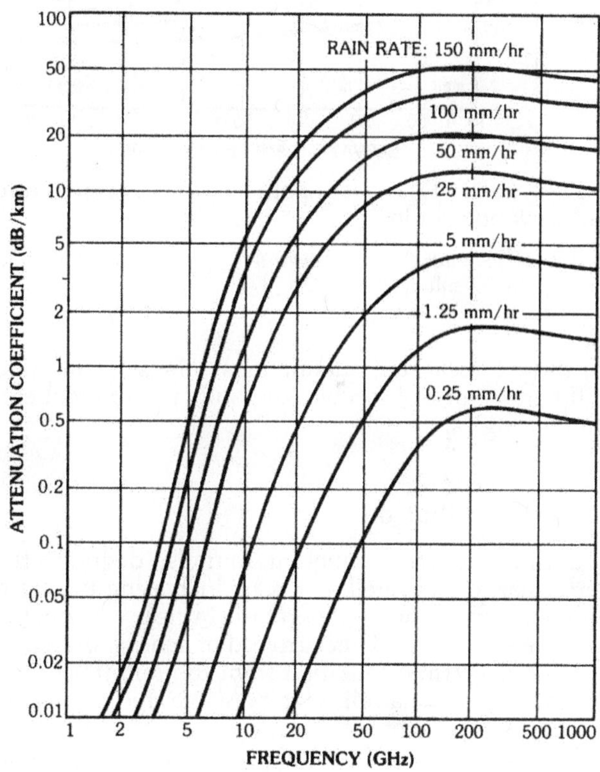

FIGURE 6.3-22 Attenuation per unit length versus frequency and rain rate

K. Miya, ed., *Satellite Communications Technology*. Tokyo: KDD Engineering and Consulting, Inc., 1982.

The rain attenuation is required to be added to the satellite link as a margin to allow the specified availability under fading conditions. Figures 6.11-24(a) and

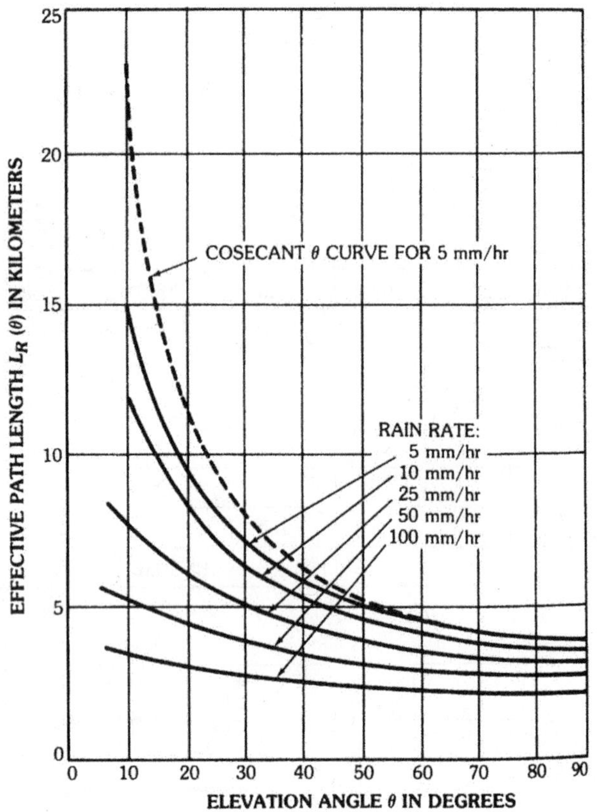

FIGURE 6.3-23 Equivalent path length versus rain rate and elevation angle

K. Miya, ed., *Satellite Communications Technology*. Tokyo: KDD Engineering and Consulting, Inc., 1982.

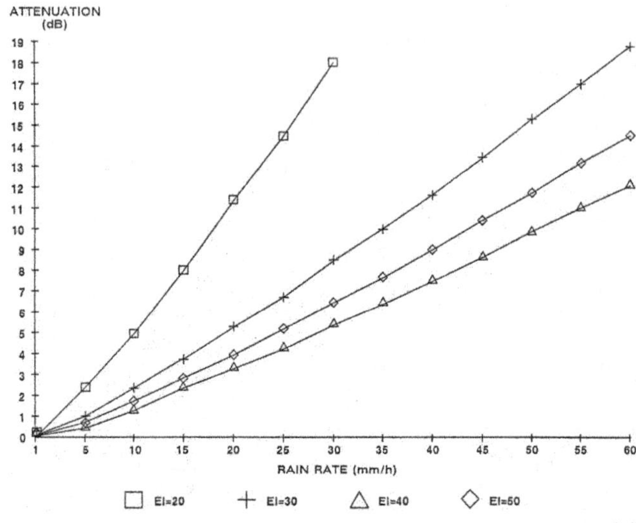

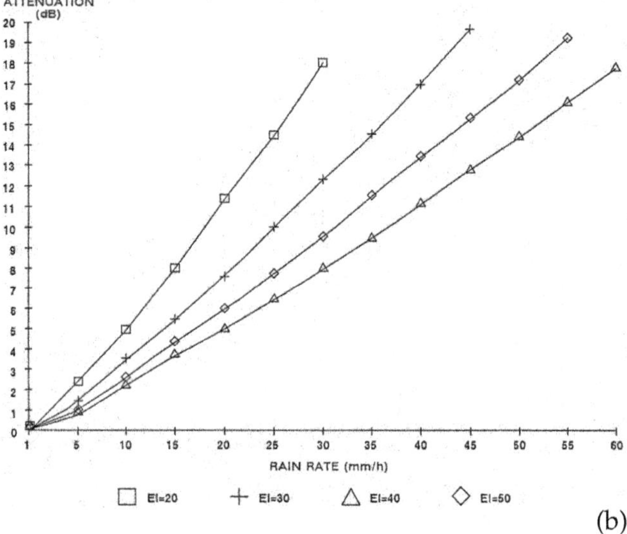

FIGURE 6.3-24 (a) Rain attenuation (11.95 GHz) 4.8 km zero-degree isotherm; (b) Rain attenuation (14.25 GHz) 4.8 km zero-degree isotherm

(b) show typical rain attenuations versus rainfall rate in the transmit and receive Ku-bands for different elevations angles.

Noise Contribution

In addition to the attenuation, rain also degrades the performance of a satellite link by increasing the earth station antenna noise temperature. In clear weather the antenna sees the cold background of space, but in rain it receives thermal radiation from the raindrops. The increase in antenna noise temperature due to rain, T_r, may be estimated by:

$$T_r = 280 \times (1 - 10^{-A/10})\ \text{K} \qquad (28)$$

where A is the rain attenuation in decibels. Figure 6.3-25 shows the impact of the rain contribution of noise temperature on the normal clear sky G/T for different clear sky system temperatures. The G/T degradation corresponding to the rain attenuation for the stipulated link availability also must be added to the satellite downlink. This is to provide sufficient margin to compensate for the combined rain effect of signal attenuation and noise increase.

The allocations of rain fade margins in the uplink and downlink can be done independently, corresponding to specific availability requirements of the uplink or downlink and consistently with the availability requirement of the total link. The assumption is that, due to the localized nature of the rain fades, the uplink fade and downlink fade can be considered as two statistically independent processes. Therefore, total link availability can be obtained as the reciprocal of the summation of the uplink and downlink outages calculated as if they occurred independently and one at a time.

Example of a System Link Calculation

In digital satellite systems, the most common modulation technique is phase-shift keying (PSK). Variations of this technique are known as binary PSK (BPSK) when logic symbol zeros and ones are mapped into RF signals 180° apart in phase and as quadrature PSK

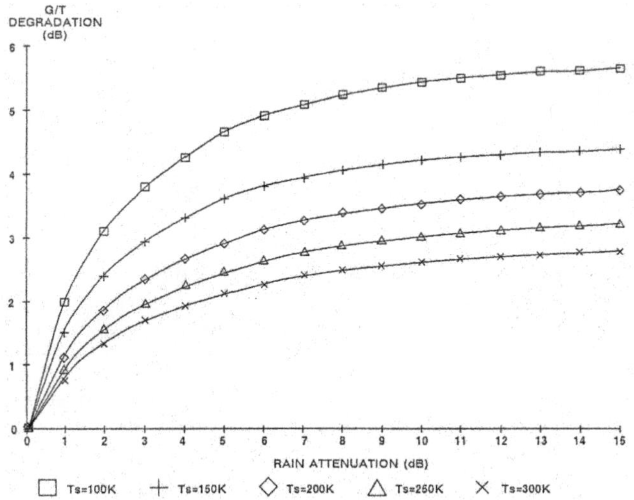

FIGURE 6.3-25 G/T rain degradation

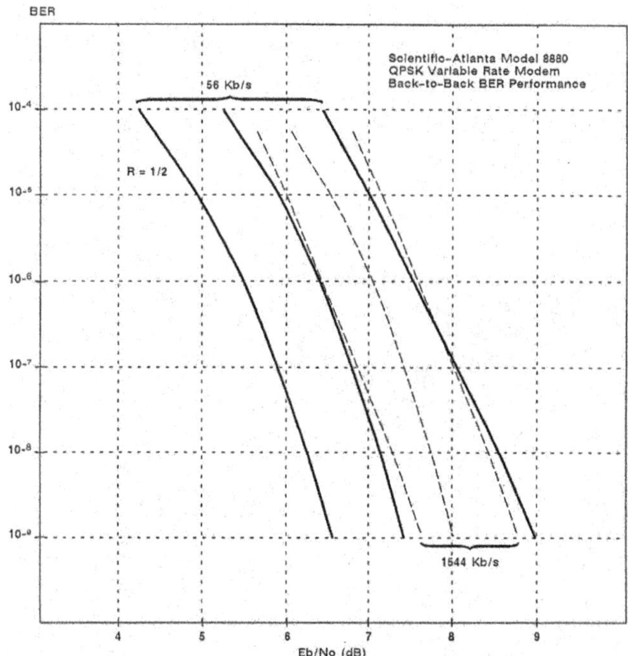

FIGURE 6.3-26 QPSK modem BER performance versus bit energy over noise density ratio

(QPSK) when the phases are 90° apart. In recent years, 8PSK, using 45° phases, has also come into widespread use. The BER performance of these systems is evaluated as a function of the energy per bit of information transmitted versus noise density (E_b/N_o). Sophisticated digital coding and decoding techniques exist that, by adding error control bits to the information data stream, allow substantial improvements in BER that can translate into transmit power reductions of up to 5 dB. The ratio between the uncoded data rate and the coded one is the coding rate (R). Figure 6.3-26 shows the performance of a typical QPSK modulator and demodulator for different data and coding rates.

Table 6.3-3 shows a typical link budget for a digital video signal without uplink power control. Rate 3/4 FEC and 188/204 Reed-Solomon block coding are applied. Performance is in accordance with the DVB-S specification.

Earth Station Block Diagram

The block diagram of Figure 6.3-27 depicts an earth station capable of providing uplink services for analog and digital video and digital data in the vertical and horizontal polarizations. All subsystems are redundant for maximum reliability. A computer-based monitor and control system, by means of a serial control bus, offers centralized operation of the complete earth station with the ability of monitoring all status and controlling all variable parameters of every subsystem from the local or remote terminals.

Figure 6.3-28 shows a simplified block diagram of a digital radio uplink. C-band or Ku-band could be used. Reduced cost of C-band transponders and rain fade margin is causing many radio broadcasters who converted from C-band to Ku-band several years ago to take advantage of smaller uplinks and simplified licensing, to convert back to C-band.

Figure 6.3-29 shows the block diagram of the corresponding dual polarization receive-only terminal. This low-cost earth station, with an L-band (950 to 1450 MHz) interfacility link (IFL), can provide simultaneous reception of analog and digital video and audio. Data capability (not shown) can be readily added. For simplicity, analog reception is shown on one polarity and digital on another, but in an actual terminal either polarity can receive and process either or both types of video, as well as multiple signals.

Interference Considerations

The consideration of interference in a satellite communication system is important, not only from the standpoint of interference to the desired satellite signal but also as it applies to the satellite's signals generating interference into other systems. The FCC requires a proposal for a satellite transmit system in the United States to include a coordination filing with an interference analysis. This analysis must show the impact of the proposed system on existing operational systems and must satisfy the allowable interference requirements of the FCC.

Antenna Characteristics

The primary characteristic of the antenna that affects the interference analysis is the angular discrimination, defined as the gain differential between the on-axis gain and the gain of an off-axis angle for the interfering source. Both receive and transmit antenna characteristics should be considered. Off-beam cross-polarization isolation of the antenna should also be considered, and in this regard it should be noted that during periods of heavy rain the polarization of the incoming and

TABLE 6.3-3
Link budget for digital Ku-band video satellite link

Satellite	SES Americom AMC-3			
Beam type	Conus (Continental United States)			
Type of service	Digital SCPC			
Transmit/receive connectivity	5.0 m New York to 3.0 m Dallas			
Information rate	3.0 Mbps			
Combined forward error correction (FEC)	0.69, using rate 3/4 convolutional coding and 204/188 Reed Solomon block coding			
Transponder bandwidth	36.0 MHz			
Parameter	**Values**			
I. UPLINK NOISE	Clear Sky	Uplink Fade	Downlink Fade	Units
Earth station EIRP per carrier	55.3	55.3	55.3	dBW
Pointing losses	0.7	0.7	0.7	dB
Path loss	207.2	207.2	207.2	dB
Isotropic antenna area	44.5	44.5	44.5	dBW/m²
Saturation flux density	–80.8	–80.8	–80.8	dBW/m²
Rain attenuation	0.0	5.3	0.0	dB
G/T including footprint advantage	4.3	4.3	4.3	dB/K
Input back-off per carrier	18.7	24.0	18.7	dB
Uplink thermal C/N	15.5	10.2	15.5	dB
Co-channel interference	26.1	20.8	26.1	dB
Off-beam emissions interference	18.6	13.4	18.6	dB
Total uplink C/(N + I)	13.6	8.2	13.6	dB
UPLINK AVAILABILITY 99.93%				
II. INTERMODULATION NOISE	21.7	16.3	21.7	dB
III. DOWNLINK NOISE	46.0	46.0	46.0	dBW
Satellite saturation EIRP	4.0	4.3	4.0	dB
Transponder output back-off	31.6	26.3	31.6	dBW
EIRP per carrier	205.5	205.5	205.5	dB
Path loss	0.0	0.0	6.0	dB
Rain attenuation	0.6	0.6	0.6	dB
Pointing losses	26.7	26.7	26.7	dB/K
Earth station G/T	0.0	0.0	3.5	dB
G/T degradation	16.1	10.8	6.6	dB
Downlink thermal C/N	21.1	15.8	18.1	dB
Co-channel interference	14.4	9.6	6.3	dB
Total downlink C/(N + I)				
DOWNLINK AVAILABILITY 99.93%				
IV. TOTAL C/(N + I) NOISE	10.8	5.5	5.5	dB
Total C/(N + I)	64.8	64.8	64.8	dB – Hz
Occupied bandwidth per carrier	75.6	70.3	70.3	dB – Hz
$C/(N_o + I_o)$ total	5.5	5.5	5.5	dB
Required E_b/N_o	5.3	00.0	0.0	dB
Margin				
TOTAL LINK AVAILABILITY 99.86%				

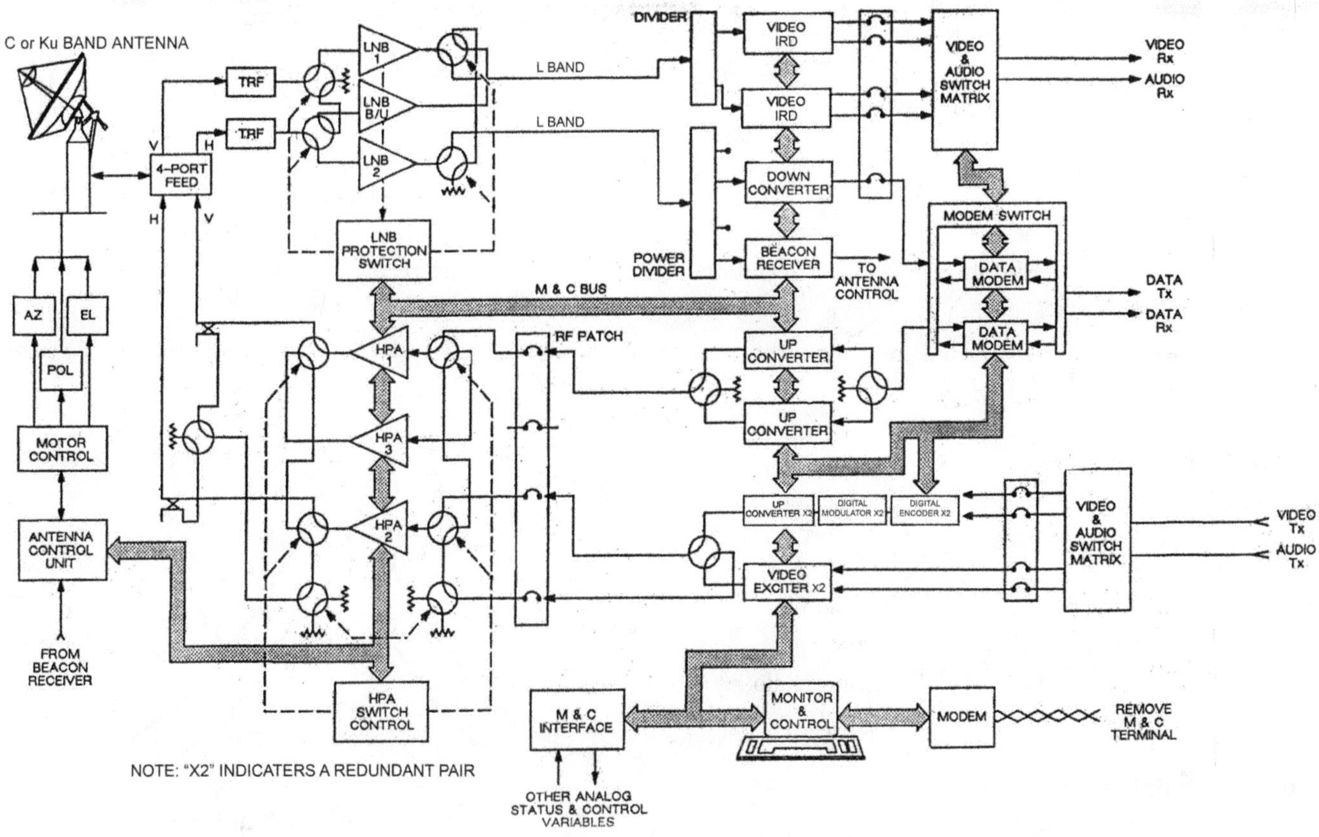

FIGURE 6.3-27 Digital video and data broadcast uplink earth station block diagram

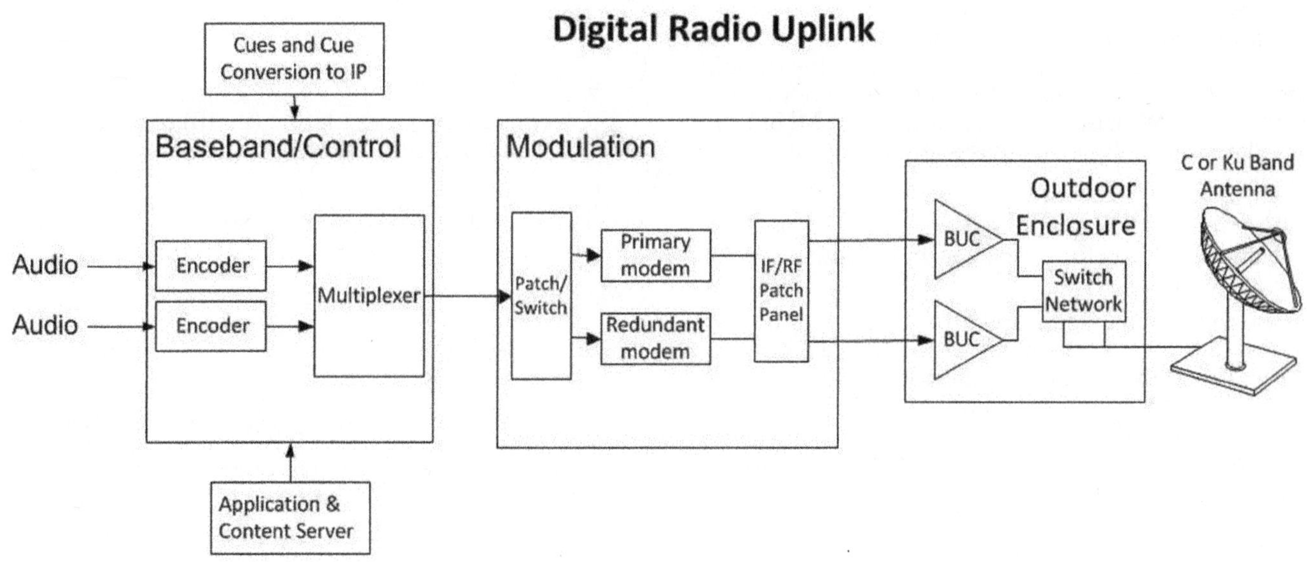

FIGURE 6.3-28 Typical digital radio uplink earth station

Courtesy Kayou Communications

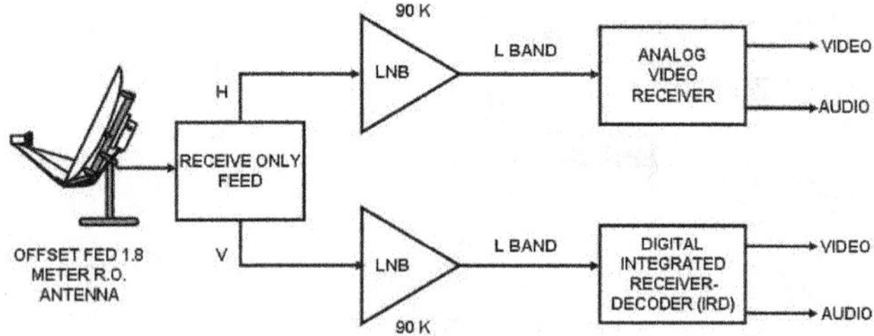

FIGURE 6.3-29 Video/audio receive-only earth station

outgoing signals may be affected such that the full cross-polarization isolation is not realized.

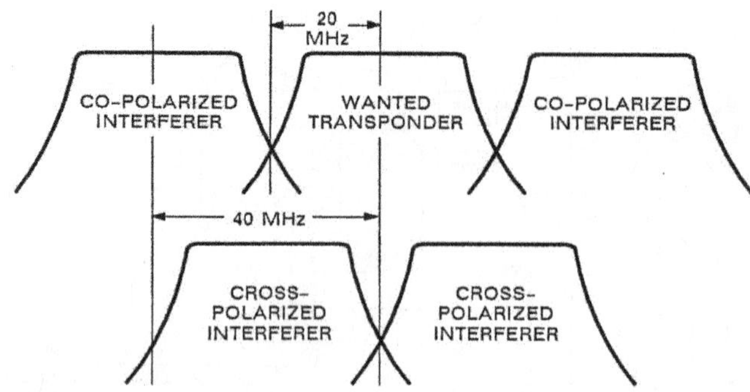

FIGURE 6.3-30 40 MHz transponder frequency reuse transponder plan

Sources of Interference

Interference into a geostationary satellite communication system can originate from several sources, including:

- Adjacent satellite uplink or downlink signals
- Cross-polarization signals
- Terrestrial microwave or radar detector signals.

Adjacent Satellite Interference

Interference from adjacent satellites occurs in two ways: (1) uplink interference from earth stations transmitting to adjacent satellites and (2) downlink interference from adjacent satellite transmission into the desired earth station. The interference in both the uplink and downlink consists of many possible sources, but it is primarily caused by the co-frequency channels/transponders and the two half transponder bandwidth offset-channels/transponders in a frequency reuse system. The particular interferers for a 40 MHz transponder bandwidth system as shown in Figure 6.3-30 are the following:

- The two 20 MHz offset-frequency, co-polarized transponders on the first adjacent satellite on either side.
- The co-frequency, cross-polarized transponders on the first adjacent satellite on each side (note: as shown in Figure 6.3-31, even though the cross-polarized antenna characteristic is nominally 10 dB less than the co-polarized antenna characteristic, this specification is not always met in practice, particularly on receive).
- The co-frequency, co-polarized transponders on the second adjacent satellite on either side.
- The two 20 MHz offset-frequency, cross-polarized transponders on the second adjacent satellite on either side.

The contribution to interference from satellites at orbital positions greater than 4° from the desired satellite tends to be noise like in that it is the result of a number of small, relatively noncoherent signals.

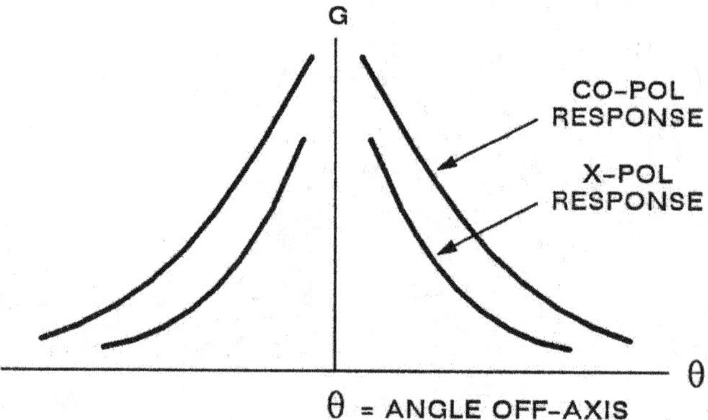

FIGURE 6.3-31 Earth station antenna radiation characteristics

The equations for calculation (in dB) of the adjacent satellite interference are given below:

$$(C/I)_u = \sum_{i=1}^{N} \oplus \{(EIRP)_I - (G_i - G(\theta_i)) + F_i + P_i\} \quad (28)$$

where:

$\sum_{i=1}^{N} \oplus$ = Series power summation

$(EIRP)_{es}$ = Earth station radiated power in dBW

$(EIRP)_I$ = Effective radiated power of interfering earth station in dBW

G_i = Peak gain of the interfering earth station in dBi

$G(\theta)_I$ = Gain of the interfering earth station in direction (θ) in dBi

F_i = Frequency discrimination factor for ith earth station

P_i = Polarization discrimination factor for ith earth station

$$(C/I)_d = (EIRP)_{sat} + G_{es} - \sum_{i=1}^{N} \oplus \{(EIRP)_I - (G_{es}(\theta_i) + F_i + P_i\} \quad (29)$$

where:

$(EIRP)_{sat}$ = Effective radiated power of satellite in the direction of the receive earth station in dBW

G_{es} = Gain of the receive earth station in dBi

$G_{es}(\theta_i)$ = Gain of the receive earth station in the direction θ_i in dBi

F_i = requency discrimination factor for the ith earth station

P_i = Polarization discrimination factor for the ith earth station

N = Number of transponders considered ($N \geq 3$).

The total adjacent satellite interference is then calculated by combining the uplink and downlink contributions in a power summation manner:

$$(C/I)_{adj.sat} = (C/I)_u \oplus (C/I)_d \quad (30)$$

The polarization discrimination factor in the previous equations is the system discrimination rather than that of the receive or transmit antenna alone. A well-designed dual linearly polarized antenna can achieve excellent cross-polarization discrimination on or near the main beam axis (greater than 35 dB relative to the co-polarized energy) and reasonable rejection of the cross-polarized signals in the close-in sidelobe regions. The adjacent satellite signals are received through the sidelobes of the earth station, and the [19–25 × log(θ)] envelope is normally assumed. This assumption, rather than being conservative, may be optimistic when one considers various factors such as the interactions of the ionosphere and atmosphere on the transmitted signals (from the earth station and/or the satellites) and actual off-axis cross polarization antenna performance. A more conservative analysis may assume a slightly reduced discrimination of perhaps [21–25 × log(θ)] for the cross-polarized sidelobe energy in the off-axis regions.

The frequency discrimination factor is related to the spectra of the desired and undesired signals. This factor can range from 0 to 30 dB depending upon the interfering power from the different services.

Crosspol Interference

The internal interference in a satellite system utilizing 40 MHz transponders is primarily due to the crosspolarized 20 MHz offset transponders. The interfering power from different services is calculated by convolving the power spectra of the individual services or by addition on a power basis. This data, taken together with the appropriate polarization discrimination term, determines the amount of interference. Crosspol interference is also dependent upon transponder backoff and transponder attenuator (gain step) or "pad" setting.

Terrestrial Interference

Terrestrial microwave carriers are centered on frequencies offset by 10 MHz from the satellite carriers. To analyze the effects of terrestrial carriers, it is necessary to determine the power level of the interfering signal and the spillover of the terrestrial carrier spectra into the passband of the receiver. The first factor involves site details, such as angular discrimination and distance to the interfering transmitter. The second factor can be computed from the spectral distribution projected for the terrestrial carrier and the filter characteristic of the receiver. Other sources of terrestrial interference to satellite circuits can include:

- Radio altimeters (C-band)
- Military radars (C-band)
- Vehicle-borne radar detectors (Ku-band)
- Non-geostationary satellite uplinks (Ku-band and Ka-band)
- Ground to air Internet (Ku-band).

Sun Transits and Eclipses

Communications satellite systems experience predictable service interruptions involving the sun. A sun transit outage occurs when the pointing angles from a receiving earth station to a satellite and to the sun so nearly coincide that the additional noise power presented by the sun renders transmission unusable. A solar eclipse occurs when the earth shadows the sun from the satellite. The eclipse event is not as serious as the sun transit event because the satellite has battery backup systems to augment the solar primary power.

Daily sun transits of all geostationary satellites serving an earth station occur during one week in the spring and again in the fall. The exact dates depend primarily on the latitude of the receiving earth station. The geometry and duration associated with a sun transit are controlled by the off-axis gain of the earth station antenna, the receiving system noise temperature, the solar noise power profile, and the minimum acceptable S/N ratio. In late February or early March, short daily outages affect earth station systems situated near the U.S.-Canadian border. Two or three days later, these systems experience maximum outages lasting 5 minutes or more, depending on transmission parameters and permissible S/N. Outages at these earth station locations end after an additional 2 to 3 days, and the sun transit outage paths progress southward at a rate of about 3° latitude per day. All outages affecting U.S. earth station antenna systems above north latitude 26° cease prior to mid-March. Conversely, in the fall, the daily outages progress from south to north, affecting southern U.S. earth stations beginning October 1 and ending in the north about mid-October.

Eclipses of geostationary satellites can be expected for a total of about 90 evenings per year in the spring and fall. Eclipses occur near apparent midnight of the time zone at each satellite's longitude, beginning in late February or early March and ending mid-April. Fall events begin about September 1 and end mid-October. Eclipses of about 70 minutes duration occur on the dates of the spring and fall equinoxes. Communication satellites are provided with batteries to prevent circuit outages and to maintain pointing, attitude control, stationkeeping, telemetry, and command capabilities during eclipses.

EQUIPMENT CHARACTERISTICS

An earth station system is made up of four major subsystems:

- Antenna subsystem
- Transmitting subsystem
- Receiving subsystem
- Monitor and control subsystem.

Antenna

The antenna provides the means of transmitting signals to the satellite and collecting the signal transmitted by the satellite. The antenna must not only provide the gain necessary to allow proper transmission and reception but must also have radiation characteristics that discriminate against unwanted signals and minimize interference into other satellite or terrestrial systems. A further function of the antenna is to provide the means of polarization discrimination of unwanted signals. The individual communication system operational parameters dictate to the antenna designer the necessary electromagnetic, structural, and environmental specifications necessary for the antenna. Antenna requirements can be grouped into several major categories: electrical or RF, control systems, pointing and tracking accuracies, and environmental and miscellaneous requirements such as radiation hazard and primary power distribution. Table 6.3-4 summarizes many of the more important parameters of an earth station antenna.

Electrical Performance

The primary electrical specifications of an earth station antenna are gain, noise temperature, voltage standing wave ratio (VSWR), power rating, receive/transmit group delay, radiation pattern, polarization, axial ratio, isolation, and G/T. All of the parameters except the radiation pattern are determined by the system requirements. The radiation pattern should meet the minimum requirements set by the FCC and/or the ITU-R. Earth stations that operate in a regulated environment in the U.S. domestic system must meet the requirements set forth in the FCC regulations for earth station antennas pertaining to antenna aperture diameter, sidelobes, and/or radiated power density. The desired radiation properties to satisfy the communication system design dictate the choice of the type of antenna to be employed as an earth station. The three most important radiation properties are gain, sidelobe performance, and noise temperature. Most earth station antennas are designed to maximize gain and minimize noise, thereby maximizing G/T. These two criteria have led to the predominance of reflector-type antennas for earth station applications, although other types of antennas such as arrays and horns have been used.

Types of Earth Station Antennas

Several types of earth station antennas are in use in the United States and abroad. These antennas can be grouped into two broad categories: single beam and multiple beam. A single beam earth station antenna is defined as an antenna that generates a single beam pointed toward a satellite by means of a positioning system. A multiple beam earth station antenna is defined as an antenna that generates multiple beams by employing a common reflector aperture with multiple feeds illuminating that aperture. The axes of the beams are determined by the location of the feeds. The individual beam identified with a feed is pointed toward a satellite by positioning the feed without

TABLE 6.3-4
General considerations for earth station antenna design

Electrical Performance	Mechanical Performance	System Considerations
Frequency (Bandwidth)	Angular Travel	Operational Function
Gain	Drive Speed and Acceleration	Local and/or Remote Operation
Noise Temperature	Pointing and Tracking Accuracies	Availability and Maintainability
Radiation Pattern	Compatibility and Environmental Conditions	Design Lifetime
Polarization	Reflector Surface Accuracy	Interface Conditions with Other Subsystems
Axial Ratio	Physical Dimensions	Space Needed in Antenna Hub for Electronics
VSWR	Weight	
Power Handling Capability	Ability to Survive in Strong Winds	
Port-to-Port Isolation		
Out-of-Band Emissions		

moving the reflector. The dual-frequency antennas may be considered another class of antennas as they produce two coincident simultaneous beams and as such are categorized as single beam antennas.

Single Beam Antennas

The majority of the earth station antennas in use are single beam antennas. Single beam antenna types used as earth stations are paraboloidal reflectors with focal point feeds (prime focus antenna), dual reflector antennas such as the Cassegrain and Gregorian configurations, horn reflector antennas, offset-fed paraboloidal antennas, and offset-fed multiple-reflector antennas. Each of these antenna types has its own unique characteristics, advantages, and disadvantages to be considered when choosing them for a particular application.

Axisymmetric Dual-Reflector Antennas

The predominant choice of many system operators has been the dual-reflector Cassegrain antenna. Cassegrain antennas can be divided into three primary types. The classical Cassegrain geometry employing a paraboloidal contour for the main reflector and a hyperboloidal contour for the subreflector (Figure 6.3-32). The paraboloidal reflector is a point focus device with diameter D_p and focal length f_p. The hyperboloidal subreflector has two foci. For proper operation, one of the two foci is the real focal point of the system and is located coincident with the phase center of the feed; the other focus, the virtual focal point, is located coincident with the focal point of the main reflector.

A geometry consisting of a paraboloidal main reflector and special-shaped, quasi-hyperboloidal subreflector shown in Figure 6.3-33, is appropriate for describing this antenna. The main difference between the classical Cassegrain and this antenna is that the subreflector has been designed such that the overall efficiency of the antenna has been enhanced, thereby yielding improved gain performance. This technique is especially useful with antenna diameters of approximately 30 to 100 wavelengths—for example, a 5 m antenna in the 6/4 GHz frequency band.

A generalization of the Cassegrain geometry consists of a specially shaped, quasi-paraboloidal main reflector and a specially shaped, quasi-hyperboloidal subreflector. The subreflector is shaped to redistribute its incident energy such that the illumination of the main reflector is optimized for high gain and desired radiation pattern. The main reflector is then shaped to correct the phase of the aperture field such that it is in phase. The feed must have a high beam efficiency, and its radiation pattern should be circular symmetric. This technique allows the antenna designer to synthesize the surfaces to achieve an arbitrary aperture distribution. The dual reflector antenna offers excellent gain performance, and for aperture sizes larger than approximately 75 wavelengths the sidelobe performance can meet the FCC pattern requirements. Dual

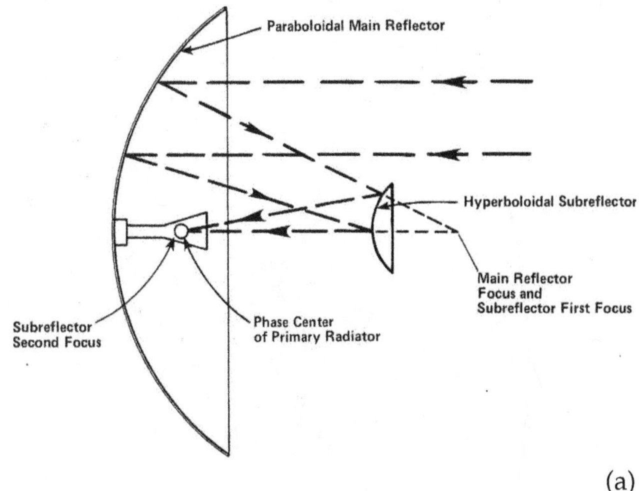

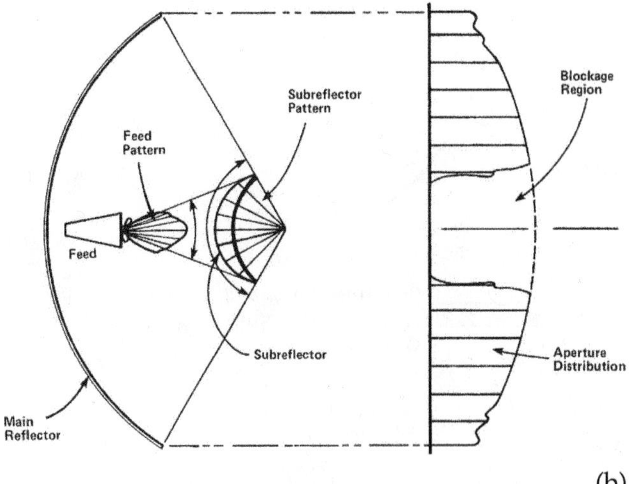

FIGURE 6.3-32 (a) Cassegrain antenna geometry; (b) aperture distribution of a Cassegrain antenna

reflector designs are employed for earth station antennas for apertures as small as 50 wavelengths to as large as 500 wavelengths.

Prime Focus-Fed Paraboloidal Antennas

The prime focus-fed paraboloidal (PFFP) antenna is another of the most often employed antennas for earth stations. This type of antenna can have excellent sidelobe performance in all angular regions except the spillover region around the edge of the reflector, but even in this region the pattern requirements of the FCC can be met. This antenna configuration has a lower cost than dual reflector antennas and offers a good compromise choice between gain and sidelobes. Its basic limitations are its location of the feed for transmit applications and, for aperture sizes less than approximately 30 wavelengths, the blockage of the feed and the feed support structure raises the sidelobes with respect to the main beam such that it becomes exceedingly difficult to meet the FCC sidelobe requirements. The PFFP

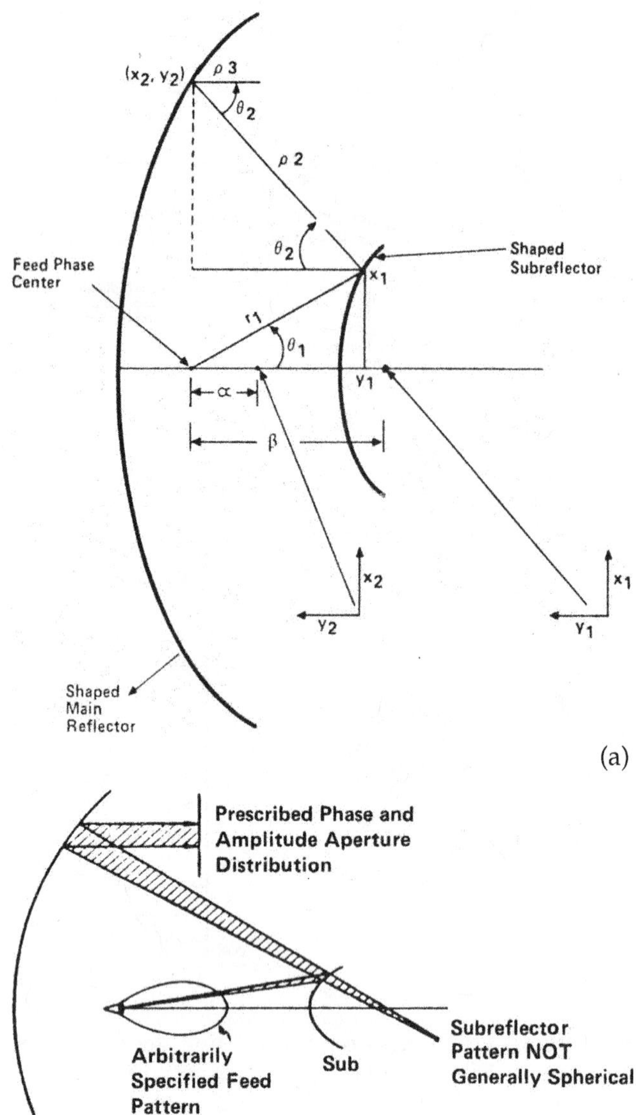

FIGURE 6.3-33 (a) Dual-shaped reflector geometry; (b) circularly symmetric dual-shaped reflectors

antenna is used for many receive-only earth station antennas as well as for transmit/receive applications when only one transmit polarization is required.

Offset-Fed Reflector Antennas

The offset-fed reflector antenna (Figure 6.3-34) was originally used primarily in small-aperture antennas for VSAT applications but is now used for larger antennas as well. The offset-fed reflector antenna can employ a single main reflector or multiple reflectors, with two reflectors being the more prevalent of the multiple reflector designs. The offset, front-fed

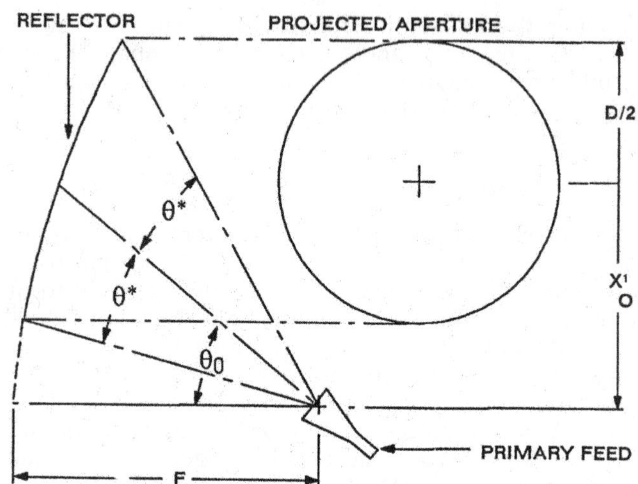

FIGURE 6.3-34 Offset single-reflector antenna

reflector, consisting of a section of a paraboloidal surface, eliminates the direct aperture blockage from the feed and feed supports and minimizes diffraction scattering by removing the feed and feed support structure from direct illumination of the aperture current distribution. Limitations of the offset-fed single-reflector antenna can include its polarization performance, reduced cross-polarization performance off-axis for linear polarizations, and beam squints in opposite directions for two orthogonal circular polarizations. The actual offset feed can be located above or below the antenna.

These antennas are sometimes selected for use to avoid the need for deicing systems in some locations where the need for deicing is not great. This is because an offset-fed reflector is oriented more toward the vertical than a prime focus-fed antenna. Use of an offset-fed antenna and either a cover or a hydrophobic coating can obviate the initial and recurring cost of antenna deicing systems.

The offset dual-reflector antenna (Figure 6.3-35) can be designed to have all the desirable characteristics of an axisymmetric antenna with increased gain and lower sidelobes. The polarization problems sometimes associated with the offset single-reflector design can also be compensated for with a two-reflector antenna design. The only disadvantages of the offset dual-reflector antenna are its cost of manufacturing for large apertures consisting of multiple sections and the complexity of its mount geometry and associated cost.

Multiple Beam Antennas

Several multiple-beam antenna (MBA) configurations are used for earth station applications (see Figure 6.3-36). These include the spherical reflector, the torus antenna, and a class of offset-fed Cassegrain antennas. All of these configurations employ multiple feeds to generate the multiple beams. The multiple feeds must be physically small such that the individual beams may be pointed at desired satellites. When the desired

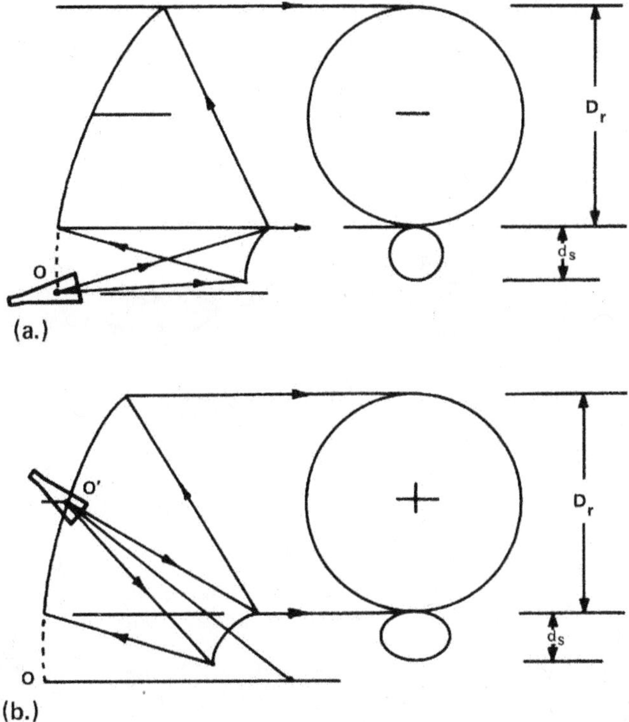

FIGURE 6.3-35 Offset dual-reflector geometries: (a) double-offset geometry (feed phase center and paraboloidal vertex at 0°); (b) open Cassegrainian geometry (feed phase center located at 0°; paraboloidal vertex at 0°).

satellites are spaced as close as 2° apart, the MBA may not be practical. The obvious advantage of the MBA is that a single antenna installation can transmit or receive signals to several satellites simultaneously, such as the case in DBS receive applications where three LNBs are used to receive three satellites with one antenna. Multiple-beam antennas can be particularly attractive where space is limited and many simultaneous feeds are required or when high reliability is required and antennas cannot easily be repointed in the event of a satellite failure. The disadvantages are the complexity of the feed arrangements for maintaining pointing at several satellites at the same time when the primary antenna aperture, the main reflector, remains fixed with respect to the earth's coordinates and the stringent requirements for the initial installation of the antenna system. Another disadvantage is that the MBA is not easily steerable in the dimension perpendicular to the orbital arc and can therefore be difficult to use with inclined orbit satellites.

Mechanical Performance

The mechanical design of an earth station antenna must provide the structural integrity to accurately point the antenna beam toward the desired satellite and to maintain the pointing accuracy within the environmental conditions for the locale. Further, the mechanical design of the antenna must ensure the required tolerance of the radiating surface such that the radiation performance of the antenna is not compromised. The antenna pedestal must also provide the means to steer the antenna beam to the satellites of interest.

The location and size of an earth station antenna system (antenna, pedestal or mount, electronics, and control housing) usually make it subject to local building codes. The code that is almost universally accepted is Minimum Design Loads in Buildings and Other Structures (ASCE 7, formerly ANSI A58.1), which requires that buildings or other structures and all parts thereof be designed and constructed to support safely all loads, including dead loads, without exceeding the allowable stresses (or ultimate strengths when appropriate load factors are applied) for the materials of construction in the structural members and connections. When both wind and earthquake loads are present, only that one that produces the greater stresses needs to be considered, and both need not be assumed to act simultaneously. The loads that must be safely supported by an earth station antenna system are the weight of the antenna and the attached equipment, the expected ice and snow load, earthquake load, and the wind load. Of these, the wind load is usually the largest single contributor to the stress and deflection of the structure.

Earth station antennas have a specification that is variously called maximum wind, survival wind, or withstand wind. These terms should be considered synonymous. At the manufacturer's specified survival wind, the system must be safely supported without exceeding the allowable stresses for the materials. Survival wind, as defined herein, when combined with ice and dead weight results in the design load as defined in Standard EIA-222C. In addition to survival wind, two other sets of wind conditions are usually specified: the operational wind velocity and the drive-to-stow wind velocity. The operational wind velocity is the maximum value at which the antenna system fully meets the performance specifications. The drive-to-stow wind velocity is the maximum value the antenna may be driven through the azimuth and elevation actuators to the prescribed stow position (usually zenith).

Pointing and Tracking

The pointing and tracking accuracy are two very important considerations for an earth station antenna system. Pointing accuracy is defined as the precision with which an antenna can be held (for a fixed-position antenna) or steered under the specified operating conditions. The pointing error is a measure of pointing accuracy and is defined as the space angle difference between the command vector and the actual position of the antenna communication RF axis. Pointing error is usually specified to less than 0.2 of the half-power beamwidth (HPBW) of the antenna in the transmit frequency band. Tracking accuracy is the precision with which an antenna can track a source under specified operating conditions. The tracking error is a measure of tracking accuracy and is defined as the space angle difference between the communication RF axis of the

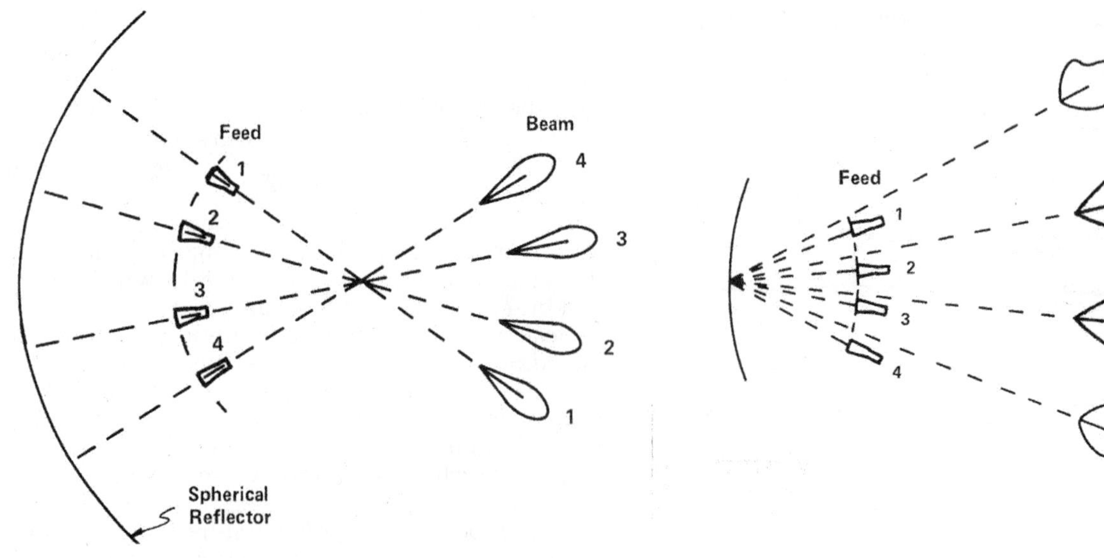

Conventional spherical multibeam antenna using extended reflector and multiple feeds

Alternative spherical multibeam antenna using minimum reflector aperture with scanned beam feeds

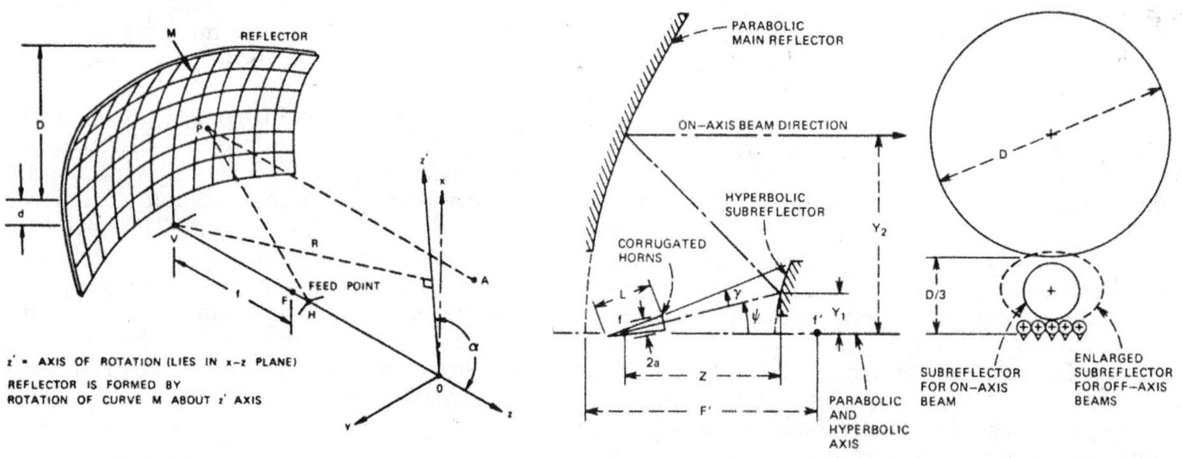

Torus–antenna geometry
(Copyright 1974, *COMSAT Technical Review*.
Reprinted by permission.)

Geometry of the offset–fed multibeam Cassegrain antenna.
(Copyright, 1974, *American Telephone and Telegraph Company*.
Reprinted by permission.)

FIGURE 6.3-36 Multiple-beam configurations

antenna and the vector to the RF source. Tracking error is usually specified to be less than 0.1 of the HPBW. Many earth station systems operate in the point mode; that is, there is no requirement for automatic tracking of the satellite. This condition exists when the satellite orbital location is maintained within a small fraction of a degree (< 0.1°) and when the earth station antenna HPBW is sufficiently broad (> 0.5°). Automatic tracking may become necessary as the antenna becomes large in terms of wavelengths (very narrow RF beam) or if the satellite is allowed to transverse an inclined orbit. The complexity of the tracking system is determined by the overall system accuracy requirements and the allowance in EIRP and G/T that is budgeted for impaired operation.

A hierarchy of pointing and tracking systems is as follows:

1. Initial fixed pointing is satisfactory (receive-only).
2. Repointing of the antenna is required to switch between various satellites or to correct for satellite motion.

3, Tracking is required to correct for satellite drift. Satellite position versus time is known and program track is satisfactory.
4. Automatic tracking is necessary but can be satisfied by a simple step-track system.
5. Full automatic tracking is necessary (extended inclined orbits).

The simple step-track system is satisfactory for most satellite communication applications when automatic tracking is required. The step-track systems generate tracking information by moving the RF beam in several steps, comparing the signal level, deciding the proper direction to move for the next step, and then continuing this process until the RF signal is maximized. The steptrack system uses a very low frequency servo loop and therefore will not track out such disturbances as wind. Step-track can be susceptible to fade conditions unless the sampling circuitry is pre-set to cut off when a large signal loss is evident. Step-track may also be augmented with a program track mode whereby the satellite movement is memorized and then followed by a memory command circuit.

Fully automatic tracking systems are typically used for TT&C earth stations or for those earth stations operating under extreme conditions with very narrow RF beamwidths. The automatic tracking configurations include conical scan, electronic beam scanning, single-channel monopulse, and three-channel monopulse. The electronic scanning and three channel monopulse techniques offer the advantage of providing a data channel and a transmit channel without tracking modulation superimposed on the signals. This is not possible with the conical scanning technique.

Transmit Electronics

The transmit subsystem for digital transmission consists of equipment from baseband to the high power RF amplifier. Depending on the application digital encoders, video and audio signals, modulators, upconverters, and high-power RF amplifiers may be employed.

Digital Video Encoders

A system used for the transmission of digital video uses a perceptual video codec (coder/decoder) such as MPEG to sample the applied video and audio waveforms and generate a digital representation of the input signals. The encoder performs compression to reduce the bit rate of the signal. Digital video systems are described elsewhere in this handbook, but in general the encoder will include an integrated (for Single Channel Per Carrier service) or external (for Multiple Channel Per Carrier service) MPEG transport stream multiplexer. This combination of encoder/ multiplexer can accommodate a wide variety of input signals and quality levels, including not only video and audio but also encapsulated data signals. As video varies widely in its encoding requirements, the systems should be capable of different data rates. MPEG-2 4:2:0 or 4:2:2 video encoding may be used or newer encoding systems such as MPEG-4 part 10 AVC, SMPTE-VC-1 (Windows Media 9), H-265, JPEG-2000, or newer encoding systems being developed all the time, depending on the application.

Radio Service

Audio (e.g., radio) signals are transmitted by satellite in both analog and digital form. Most of the domestic U.S. nationally distributed audio material is delivered in digital format. The satellite distribution service encodes program material in digital form at the source and distributes the information in that form.

At the time of writing, the most popular digital audio encoding rates are 64 kbps, 128 kbps, 192 kbps, and 256 kbps (MPEG or AAC). Typically the 64 kbps audio channel is closer to a 10 kHz audio response. The stereo 192/256 kbps audio response is CD quality at 20 kHz.

Digital Modem

The compressed digital audio or video signal is modulated onto the carrier using a digital modem or modulator/demodulator. The output frequency is typically 70 MHz. The modem will normally accommodate a broad selection of input digital data rates and will offer a wide variety of FEC techniques such as block codes, convolutional codes, turbo codes, and low-density parity check (LDPC) codes. Often a combination of two of these FEC techniques, sometime separated by an interleaver, is used to permit a greater range of error performance. Although QPSK predominates at the time of this writing, use of 8PSK is increasing due to its increased bandwidth efficiency. Other modulation systems such as 16 QAM (or APSK) or 32 QAM/APSK or even forms of minimum-shift keying (MSK) will likely see increased use in the future because of their bandwidth efficiency.

High-Power Amplifier

The high-power amplifier (HPA) amplifies the RF output signal from the upconverter to the required power level for transmission to the satellite. Amplifiers for satellite video applications are typically sized in the range from 1 W to 3 kW. Amplifiers in the 1–200 W range are available in solid-state power amplifier (SSPA) configurations. Traveling wave tube amplifiers (TWTAs) are available in configurations up to approximately 3000 W, although lower power TWTAs (approximately 200 to 400 watts) are more common.

For power levels above 750 W, klystron tube amplifiers are frequently used, but these are typically limited to an output frequency range of less than 140 MHz without retuning, whereas SSPAs and TWTAs have a 500 MHz output range. The lower power SSPAs restrict their application, particularly for multicarrier operation where the amplifier must be operated in the linear region to prevent the generation of intermodulation product interference; however, for applications where SSPAs can be used, they are generally preferred because of greater reliability and linearity.

The HPA usually contains bandpass filters to reject harmonics and power sampling circuits for monitoring the output transmit power and the reflected power from the antenna. Often, protection circuitry is added to turn off the HPA when the reflected power exceeds some predetermined level.

Receive Electronics

The receive electronics are similar in scope to the transmit subsystem but operate in the reverse order. The incoming RF signal is filtered, amplified, downconverted (optional, depending on the frequency band of operation), and passed to the receiver, where the signal is further downconverted, amplified, and demodulated and decoded to baseband video and audio.

Low-noise amplifiers/low-noise block converters The first active signal processing of a downlinked satellite signal occurs at the low-noise amplifier (LNA) or low-noise block converter (LNB). Traditional C-band broadcast applications use an LNA or LNB mounted at the antenna and connected to the indoor electronics through a length of coaxial cable. Typical Ku-band systems use an LNB at the antenna that amplifies and downconverts the signal to L-band (950 MHz to 1450 MHz).

The LNA or LNB provides:

- High gain and low noise to establish a high system G/T.
- Transition from antenna waveguide to coaxial cable to eliminate long expensive waveguide runs.
- Adequate mechanical strength to be mounted directly to the antenna waveguide and to connect a coaxial cable to go to the receiver.
- Radiofrequency interference (RFI)/ electro-magnetic interference (EMI) tight weatherproof housing for the sensitive amplifier circuitry.

While some LNAs are still in use in an ever-decreasing number of locations, LNBs have mostly replaced them due to their lower cost and higher reliability. LNBs convert the RF signal block of frequencies to a standard L-band intermediate frequency of 950 to 1450 MHz (Figure 6.3-37). This conversion may be accomplished with or without the use of an external reference signal and with or without a phase-locked loop stage in the LNB. Due to the normal selection of the oscillator frequencies (5150 MHz at C-band and 10,750 MHz at the U.S. FSS Ku-band), spectral inversion occurs for the C-band signal but not for the Ku-band signal.

Integrated Receiver-Decoder (IRD)

For digital video signals, an IRD is typically used. The IRD takes an L-band input from the LNB, demodulates, de-interleaves, FEC decodes, and decrypts (as

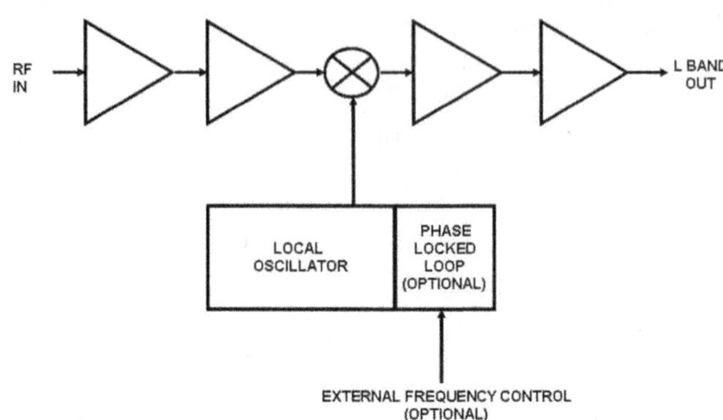

FIGURE 6.3-37 LNB block diagram

needed) the compressed video and audio signals and provides them to the MPEG decoder, which digitally decodes the signals and provides video and audio outputs, as well as associated data and closed-captioning signals if included. Figure 6.3-38 shows a basic block diagram for an IRD that includes a control processor, as a wide variety of data rates, FEC, and digital video compression options must be accommodated.

Protection Switching

With the exception of receive-only systems, most satellite transmission systems contain redundant subsystems to meet high availability specifications. Satellite teleports advertise availability specifications as high as 99.995%. Protection switching is used to implement automatic subsystem redundancy. The protection switch monitors one or more online subsystems for failure. Upon detecting a failure in the subsystem, the protection switch switches to the backup unit and configures the backup to the configuration of the failed online system.

The configuration of the subsystems protected by a protection switch is referred to as m:n, where m is the number of backups available and n is the number of online units monitored and protected by the protection switch. The simplest configuration is 1:1—a single backup is available to replace a single online unit.

Operation of the 1:1 configuration is simplified in that the backup may be tuned to the same operating parameters as the online unit; switching to the backup merely requires switching the source of the input and output signals from the online unit to the backup. An example of a larger configuration is 2:6, where two backup units protect six online units; this configuration is often used with large multiple-channel uplinks. The larger configurations become more cost effective as the number of online channels increases. Earth station monitoring and control systems, described below, are increasingly being used to implement protection switching, often eliminating the need for dedicated (separate) redundancy switchover logic devices.

Monitor and Control Systems

Monitor and control refers to systems used to monitor earth station components for failures and provide

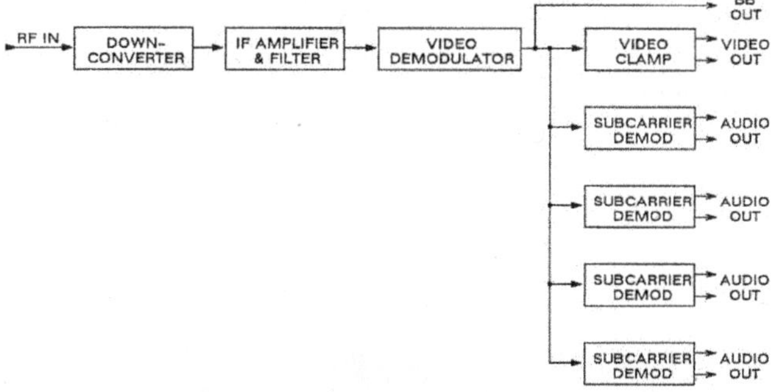

FIGURE 6.3-38 Integrated receiver-decoder (IRD)

manual and automatic control of the components. These systems are widely used for a variety of reasons. Although most earth station components provide front panel monitor and control functions, there are generally too many earth station components to monitor from the front panels of the respective components. A monitor and control system provides a single point of monitoring and control for the operator, thus easing the operator workload and allowing the operator to handle more transmissions. Because many earth stations are located away from the studio or master control room, a monitor and control system permits remote operation of the earth station. Remote operation can employ the Internet or a low-speed data circuit between the earth station equipment and the monitor and control computer in the studio or master control. This circuit can be a fiber optic or wireless channel, a satellite data circuit, a data subcarrier on a microwave radio channel or another data carrier, a dial-up phone circuit with computer modems, or a dedicated EIA-422 hardwire connection (for distances less than 1000 m).

The state of the art of M&C systems has advanced to the point that they are replacing dedicated protection switching in many applications.

Earth Station Interface

The M&C system computer interfaces to the earth station components through a number of interfaces, including serial, contact closures, and other customer supplied interfaces:

- Serial interfaces. The most common interface is a serial ASCII-protocol-based interface. The interface is usually an asynchronous-character-oriented scheme utilizing EIA-232C or EIA-422 signal levels. EIA-232C interfaces are used for short cable distances (computer to device) less than 5 m or for connection to a modem. EIA-422 interfaces may be used for cable lengths of up to 1000 m and are used in multidrop mode, thus allowing many devices to share a single interface port. The protocols are usually ASCII based because there is no satellite communication standard and most vendors use their own version of an interface protocol. Thus, earth station computers must offer a number of serial equipment interface ports and also support a variety of protocols on those ports.
- Contact closure interfaces. Older components typically provided contact closure interfaces and offered no serial interface functions. Many of these components are still in use today. In addition, components such as waveguide switches and shelter alarms (intrusion, air conditioner, emergency generator, etc.) offer only a contact closure interface; for example, a waveguide switch may provide two status points and two control points. Thus, earth station computers must provide some method of accommodating contact closure controls. Some vendors offer systems that connect the contact closure directly to the computer. Others offer general-purpose interfaces (GPIs) that reside in the earth station and interface to the earth station computer through a serial interface. Status inputs to the earth station computer are usually optically isolated. The earth station computer supplies the optical isolator. The monitored device sinks current through the isolator to indicate one of two states. The alternate state is no current flow. Control outputs are of two types. The most flexible interface is the Form C output. The Form C output provides a common connection and a normally opened (NO) and normally closed (NC) connection. The second type of control output is the open collector output. This interface provides a connection to the collector of a transistor to sink current. One control state is with the transistor on, thus sinking current from the controlled device. The alternate state is with the transistor off.
- Vendor-supplied computer interfaces. A vendor-supplied, device-specific contact closure adaptor. Most TWT amplifiers provide contact-closure-based remote monitor and control interfaces. Additionally, forward and reflected power indications are provided by a signal with voltage level proportional to power. Some TWT amplifiers require an analog current signal to control the power level attenuator; however, most TWT amplifier vendors offer an interface adaptor that converts the contact closures to a serial-protocol-based EIA-232C or EIA-485 interface. These adaptors reside in the rack with the TWT amplifiers and connect to the earth station computer through a serial connection.

SUMMARY

There are many aspects of the design, installation, operation, and maintenance of an earth station antenna system that have not been discussed in this chapter. Site selection and preparation are, in particular, critical to the successful operation of the system as well as the foundation design. Details of this aspect of the earth station design should be accomplished with the assistance of experienced engineers and frequency coordination experts. Although the operations building and equipment houses should be in proximity to the earth station, remote operation is both possible and practical. The power requirements for the earth station should also be carefully planned to provide adequate, reliable, and conditioned power for the electronics, including the transmitter equipment and any power required for antenna deicing where applicable.

Satellite communications used by broadcasters has over recent years been diminished by alternative use of terrestrial interconnections, especially fiber optic. The advent of high throughput satellites, however, has the potential to reverse this trend for point-to-point links and point-to-multipoint links where the destinations are not numerous. Increasing coding gains due to new encoding systems such as H.265 also reduce satellite bandwidth and consequently cost. Some broadcasters have converted to terrestrial distribution only to find decreased overall reliability in these links due to outages and high mean time to repair (MTTR) associated with some terrestrial transmission. Consequently, use of satellite for broadcasting purposes is alive and well and still a fundamental transmission component for television and radio broadcasting.

ACKNOWLEDGMENT

Daryl Hunter of Viasat and Mark Johnson of KAYOU Communications were very helpful in the updating of this document and in providing new material. Much of the material in this chapter is from the version in the 10th Edition of the *NAB Engineering Handbook*, co-authored by James H. Cook Jr., Scientific-Atlanta, Inc., Atlanta, Georgia.

Bibliography

Cook, J.H., Jr. and Hollis, S., Eds., *Communications Symposium '83 Notebook*, Scientific-Atlanta, Atlanta, GA, 1983.

DVB-EN 300 421, Framing Structure, Channel Coding and Modulation for 11/12 GHz.

DVB-EN 302 307, Second-Generation Framing Structure, Channel Coding and Modulation Systems for Broadcasting, Interactive Services, News Gathering and Other Broadband Satellite Applications (DVBS2).

DVB-TR 102 154, Implementation Guidelines for the Use of MPEG-2 Systems, Video and Audio in Contribution Applications.

DVB-TSI-TR 101 154 V1.4.1, Digital Video Broadcasting (DVB): Implementation Guidelines for the Use of MPEG-2 Systems, Video and Audio in Satellite, Cable and Terrestrial Broadcasting Applications, Annex B, July, 2000.

Inglis, A.F., Ed., *Electronic Communications Handbook*, McGraw-Hill, New York, chapters 2, 5, 6, 17, 1988.

Ippolito, L.J., Kaul, R.D., and Wallace, R.G., *Propagations Effects Handbook for Satellite Systems Design*, NASA Ref. Publ. 1082, National Aeronautics and Space Administration, Washington, DC, 1981.

ISO/IEC 13818–1, International Standard, Information Technology: Generic Coding of Moving Pictures and Associated Audio Information: Systems.

ISO/IEC IS 13818–2, International Standard, MPEG-2 Video, 1996.

Johnson, R.C. and Jasik, H., Earth station antennas, in *Antenna Engineering Handbook*, Cook, J.H., Jr., Ed., McGraw-Hill, New York, 1993, chapter 36.

Jordan, E.C., *Reference Data for Engineers: Radio Electronics, Computers and Communications*, Howard W. Sams, Indianapolis, IN, 1985.

Martin, J., *Communications Satellite Systems*, Prentice-Hall, Englewood Cliffs, NJ, 1978.

Morgan, W.L. and Gordon, G.D., *Communications Satellite Handbook*, Wiley-Interscience, New York, 1989.

Pratt, T. and Bostian, C.W., *Satellite Communications*, John Wiley & Sons, New York, 2003.

Pritchard, W.L. and Sciulli, J.A., *Satellite Communication Systems Engineering*, Prentice-Hall, Englewood Cliffs, NJ, 1986.

CHAPTER

6.4

STL Systems for Radio and TV

ERNEST M. HICKIN JAMES H. ROONEY III

Updated for the 11th Edition by

DOUG IRWIN

Technical Editor, Radio Magazine
Los Angeles, CA

INTRODUCTION

For many years, and for both radio and television broadcasters, the typical studio-to-transmitter link (STL) has been a privately owned microwave system, with leased telephone lines being a popular option for radio stations. The first decade of the 21st century saw a marked increase in alternative choices, with fiber optic and digital carrier circuits gaining steadily in popularity. The dawn of digital broadcasting has spawned the need for even greater diversity in connecting the studio to the transmitter. In some cases, radio and TV broadcasters have turned to unlicensed, spread spectrum microwave for STLs or data links.

The conversion by TV broadcasters in the United States to the Advanced Television Systems Committee (ATSC) digital television (DTV) system created a need to support multiple video streams and increased ancillary data. (The coming ATSC 3.0 standard presents different requirements that are under development as of this writing and not covered in this chapter.)

In AM and FM radio broadcasting, the development of the DTS (formerly iBiquity Digital) Corporation's HD Radio system (standardized by the National Radio Systems Committee, NRSC, as the NRSC-5 Standard) for in-band/on-channel (IBOC) digital radio, has created a similar situation in that additional STL capacity is needed to support both multicasting (in digital) as well as simulcast digital and analog programming.

Frequency modulation (FM) techniques have been the mainstay in microwave STL communications for decades, but are clearly on the decline as the digital transition increases the complexity of STLs and transmitter-to-studio links (TSLs). While microwave radio still accounts for the largest proportion of STL systems in use, this chapter will also examine the alternatives so as to provide information that will help an engineer decide which is best for a particular situation.

MICROWAVE PROPAGATION

With certain exceptions, microwave radio communication is a line-of-site method, and is well suited to broadcast STLs and intercity relay applications. However, microwave paths are subject to numerous atmospheric anomalies, most of which can adversely affect reliability. These effects may be random in nature, or they may be seasonal. They may be very short in duration, or they may last for hours, even days.

Free Space Loss

Free space path loss for any electromagnetic wave arises from the spreading of the wave front as it propagates from its source. A simple way to visualize this effect is to observe a flashlight beam, which closely resembles a microwave signal in most respects. At its origin, the light beam is strong and narrowly concentrated. As the distance from the light source increases, the beam spreads out and grows weaker.

After passing beyond the near-field radius of its transmitting antenna, a radio signal will lose 6 dB (a 4:1 power ratio) every time the distance from the transmitter doubles (at microwave frequencies where the wavelengths are short, the near field is usually a fraction of a mile). Because a 2:1 increase in range leads to a 2^2:1 reduction in power, this relationship is referred to as the inverse square law.

The formula for free space path attenuation (A) at microwave frequencies (F) over a distance (D), which is independent of ground or atmospheric effects, is given by:

$$A\,(dB) = 96.6 + 20\log F(GHz) + 20\log D\,(miles),\ or$$

$$A\,(dB) = 92.45 + 20\log F(GHz) + 20\log D\,(km)$$

This is the loss between isotropic antennas (theoretical antennas that radiate or receive equally in all directions).

The gain (G) of a microwave antenna is then expressed in dBi (gain relative to an isotropic antenna). For a parabolic reflector antenna, diameter d (in feet), with an efficiency of 55% (which is typical of all but the smallest antennas), gain is given by:

$$G(dBi) = 20\log d(ft.) + 20\log F(MHz) - 52.6$$

Strictly speaking, an antenna being a passive device cannot possess power gain. The value G is the amount by which the radiation in a desired direction has been increased by redirecting energy, which would have been radiated in unwanted directions by an isotropic antenna. It should be noted that below 1 GHz it is common to express gain relative to a dipole where 0 dBd = 2.2 dBi; thus, the gain of a VHF or UHF TV transmitting antenna is most likely to be quoted in dBd.

The ratio of received power to transmitted power between two correctly aligned antennas of gains G1 and G2, when D miles apart, will be given by:

$$\frac{Pr}{Pt} = G1 - 36.6 - 20\log F - 20\log D + G2 \text{ (units of dB).}$$

This number will be negative; the numerical value (i.e., absolute value) is referred to as path loss.

As examples, if the antennas are 6 feet in diameter and the path is 30 miles long, then at 2 GHz:

$$\frac{Pr}{Pt} = 28.5 - 36.6 - 66.0 - 29.5 + 28.5$$
$$= -75.1 \text{ or path loss} = 75.1 \text{ dB.}$$

and at 7 GHz:

$$\frac{Pr}{Pt} = 39.8 - 36.6 - 76.9 - 29.5 + 39.8$$
$$= -63.4 \text{ or path loss} = 63.4 \text{ dB.}$$

Note that in these equations doubling the frequency increases the total antenna gain by 12 dB (2 × 6) while the path loss increases by only 6 dB. This means that lower-frequency systems require larger antennas or more power, or both, for a given received carrier level.

Atmospheric Bending and k Factor

The pressure and hence the density of the atmosphere surrounding the earth vary with height, lessening as the height increases and the weight of the air above decreases. As a result the dielectric constant also decreases with height and this has a prismatic effect causing microwaves (and light waves) to bend toward the earth. Under normal conditions, the bending is less than the curvature of the earth but nonetheless microwaves will go farther than simple geometry would suggest. A convenient way to allow for this when drawing profiles is to increase the radius of the earth until the microwaves appear to be traveling in straight lines.

The ratio of this effective earth's radius to the true earth's radius is called k and its value is approximately 4/3 or 1.33 for over 90% of the time in most parts of the world. However, there are times when k can be anything from infinity to as low as 0.45. When k is equal to infinity, the earth appears to be flat; it is a condition where mirages are seen and radar echoes are received from hundreds of miles away. Values of k between 1 and 0.45 can occur for a few percent of the hours in a year and it is necessary to allow for this if a reliable link is to be established. Figure 6.4-1 is a map of the continental United States showing contours of equal minimum k factor; this is based on refractive index measurements made by the Central Radio Propagation Laboratory.

Path Clearance Requirements

One of the major tasks required to engineer an STL system is the path analysis between the STL transmitter at the studio and the STL receiver location. To determine what constitutes a clear path, the concept of Fresnel zones for optical theory is applied to radio waves. Most of the electromagnetic energy at a receiving point is concentrated in an elliptical volume that is a function of the distance between the transmit and receive points and the wavelength. The energy outside this volume either cancels or reinforces the energy within the volume, depending on whether the distance that the energy travels to the receive point is longer by

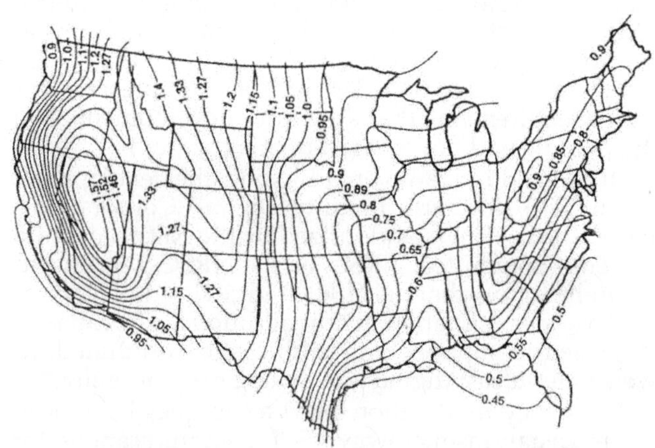

FIGURE 6.4-1 Map of the continental United States showing contours of equal estimated minimum k factor [2]

an even or odd number of one-quarter wavelengths. Even distances result in radio wave cancellations; odd distances result in radio wave reinforcement (see Figure 6.4-2).

The radius of the first Fresnel zone, which defines the boundary of the elliptical volume, is given by the following:

$$F_1 = 72.1\sqrt{\frac{d1 \times d2}{f \times D}},$$

where:

F_1 = the first Fresnel zone radius in feet

d1 = the distance from the transmitting antenna to the obstruction in miles

D = the total path length in miles

d2 = (D – d1) in miles

F = the frequency in GHz

H = the distance from the top of the obstruction to the radio path (as shown in Figure 6.4-2).

The significance of the first Fresnel zone is that 96% of the transmitted power is contained within it, thus requiring adequate first-zone clearance to be maintained over the entire path. For reliable operation, obstructions should not project into the area thus defined. Empirical studies, however, have shown that performance is substantially the same for H greater than 0.6 F_1.

Figure 6.4-3 shows the extent to which a microwave signal is attenuated when it passes close to an obstruction. Clearance is stated as a fraction or multiple of the first Fresnel zone, F1. This figure illustrates that the attenuation with a Fresnel zone clearance (FZC) of 0.6 is equal to the free space attenuation. However, as noted earlier, k-factor variations will mean that more than this clearance will have to be built in to allow for values less than 4/3, and a typical design parameter is to plan for 0.3 FZC for the lowest value of k expected on the path, as derived from Figure 6.4-1. While such a clearance will introduce 2–8 dB of loss, this is well within the fade margin of a well-designed link.

Fade Margin

The excess of signal over the minimum required for satisfactory service is called the fade margin. Systems are typically designed to have fade margins in the range 26–46 dB and (ideally) larger for higher frequencies, longer paths, and over water or similar difficult situations.

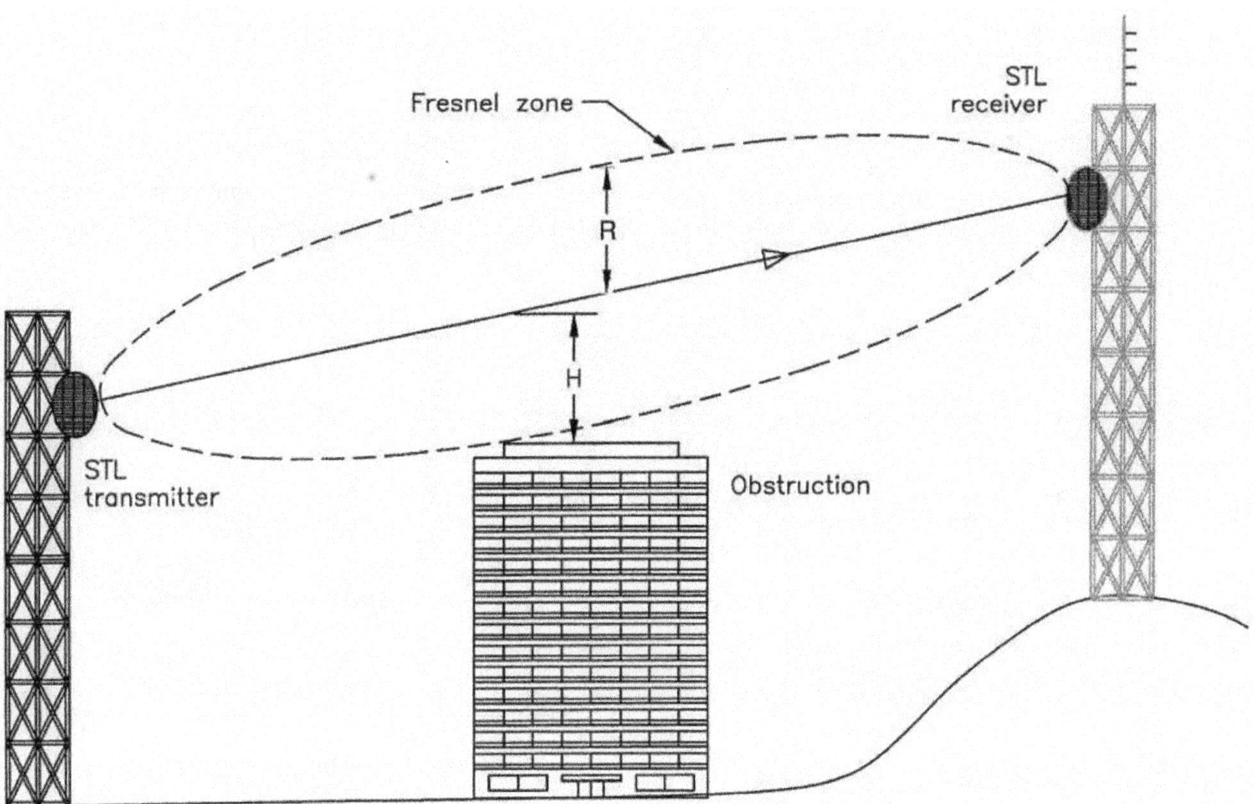

FIGURE 6.4-2 Fresnel zone clearance for an STL path [2]

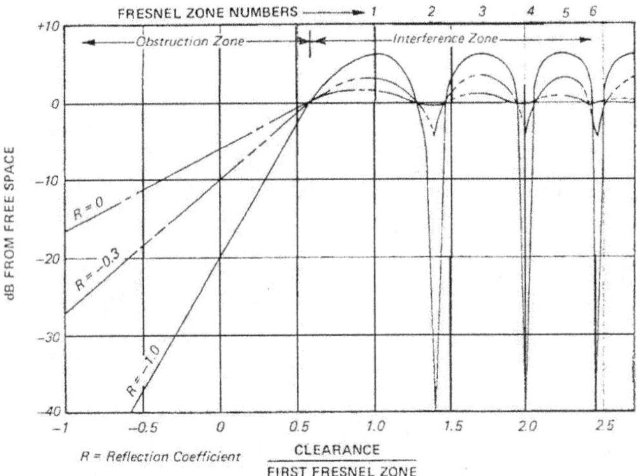

FIGURE 6.4-3 Attenuation versus path clearance for various types of obstruction [2]

The choice of fade margin may involve compromises, for example, when the need to use existing towers or masts may limit the size of antenna. It will also be influenced by the environment, requiring a higher than normal margin in humid, flat country, whereas in dry, mountainous regions, a lower than normal margin may be used.

Often times, a microwave signal will reach its destination not just by the direct path, but by one or more indirect paths as well. Depending on its phase relationship to the direct signal, a signal reflection can add to or cancel the direct signal, causing increases of up to 6 dB or reductions of more than 50 dB. Multipath fading is defined as the fading caused by the desired signal arriving by multiple paths, caused by reflections from water, hills, buildings, or atmospheric discontinuities. Although multipath reflections are typically atmospheric, or ground based (including water), reflections may also be caused by internal system discontinuities. Radio frequency (RF) channel filters, band filters, circulators, power dividers, transmission line, connectors, and the antenna itself are all potential sources of reflections, and should be carefully evaluated when problems arise.

There are multiple factors that contribute to a system's fade margin, including:

- Thermal fade margin (TFM): A reduction in the signal level reaching the receiver; not frequency selective. Virtually all path performance and fade margin calculations are based on thermal margin, unless otherwise specified.
- Dispersive fade margin (DFM): A frequency-selective fade, usually caused by a multipath reflection that manifests itself as a notch moving through the receiver passband. Multipath can alter the ratio of the carrier to the sidebands in an FM system, which modifies the effective deviation and therefore the level of the demodulated signal. In a digital system, multipath can introduce inter-symbol interference to a level that the demodulator cannot tolerate, causing temporary path failure.
- External interference fade margin (EIFM): The receiver threshold degradation due to external system interference (independent of thermal noise).
- Adjacent channel interference fade margin (AIFM): The receiver threshold degradation due to interference from adjacent channel transmitters in one's own system, or a neighboring system. This is caused by closely spaced systems at the same site, on an opposite polarization, or a different antenna.

The ability of a receiver to perform in a fading environment is determined by the type of modulation and the effectiveness of the adaptive equalizer (if present). When digital systems are involved, manufacturers typically characterize receiver fading performance by moving a phase delayed 6.3 ns notch across the intermediate frequency (IF) passband. As the notch frequency is changed, the depth of the notch is adjusted to produce a 10E-6 bit error rate (BER, a measure of digital system performance). The result of a typical radio response is shown in Figure 6.4-4. The response shown is referred to as the equipment signature, M or W curves.

These four fade margins are power added to derive a composite fade margin (CFM) as follows:

$$CFM = 10\log(10^{-TFM/10}\ 10^{-EIFM/10}\ 10^{-AIFM/10}).$$

The outage time due to multipath fading is calculated by:

$$T = (rT_o \times 10^{-(CFM/10)}) / I_o$$

where:

- T = outage time in seconds
- r = fade occurrence
- T_o = (t / 50)(8 × 106) = length of fade season in seconds

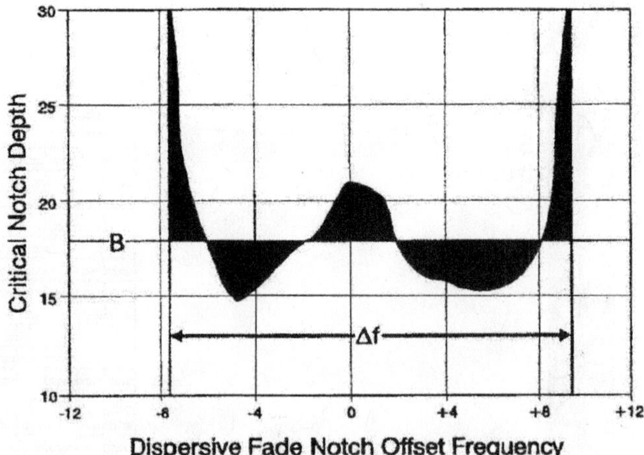

FIGURE 6.4-4 Dispersive fade margin measurement—W curve [2]

t = the average annual temperature in degrees Fahrenheit

CFM = composite fade margin

I_o = the space diversity improvement factor; factor = 1 for no diversity; ≥ 1 for space diversity.

The fade occurrence factor, *r*, is calculated from the basic outage equation for atmospheric multipath fading:

$$r = c(f/4)(D/1.6)^3 \times 10^{-5} \text{ (English)}$$

$$r = c(f/4)D^3 \times 10^{-5} \text{ (metric)}$$

where:

c = climate or C factor (see Figures 6.4-5 and 6.4-6)

f = frequency (GHz)

D = path length (miles or km).

A close look at the formula reveals that path length has an exponential effect on outage, which better explains the need for higher power or space diversity antennas on longer paths.

System Availability

The outage time due to propagation effects is usually given in seconds or minutes per month or per year. System availability, which is the reciprocal of unavailability, is typically expressed as a percentage of a year or a month; based on the fact that there are 525,600 minutes in a 365-day year. For example, if the annual outage time has been calculated as 9.5 minutes, express this as a percentage:

Unavailability is 9.5 / 525,600 × 100 = 0.00181% of a year

If there were no outages ever, the availability would be 100%, so availability may be calculated subtracting unavailability from the ideal, or 100% minus 0.00181% = 99.99819%. There is a strong tendency among engineers to strive for availability numbers that are close to 99.99999%, however, this may not be an economically achievable goal.

Digital STL systems require a significantly higher degree of RF linearity, which results in lower system gain figures, and a lower fade margin than comparative analog systems. The inherent lower system gain of a digital system may require larger antennas or space diversity to maintain the same reliability. The spacing for diversity antennas is based largely on empirical data, but there is broad agreement among microwave engineers that 30–40 feet of vertical separation is a good rule of thumb, while some point to 40 wavelengths as being the proper spacing. Field experience has shown that spacing as low as 15–20 feet can provide a valuable improvement factor.

The improvement factor of space-diversity receiving antennas may be more accurately predicted by using a formula developed by Arvid Vigants of Bell Laboratories in the 1970s:

$$Isd = (7.0 \times 10^{-5} \times f \times s^2 \times 10^{F2/10}) \;/\; D,$$

where:

Isd = improvement factor

F = frequency in GHz

s = the vertical antenna spacing in feet, center to center

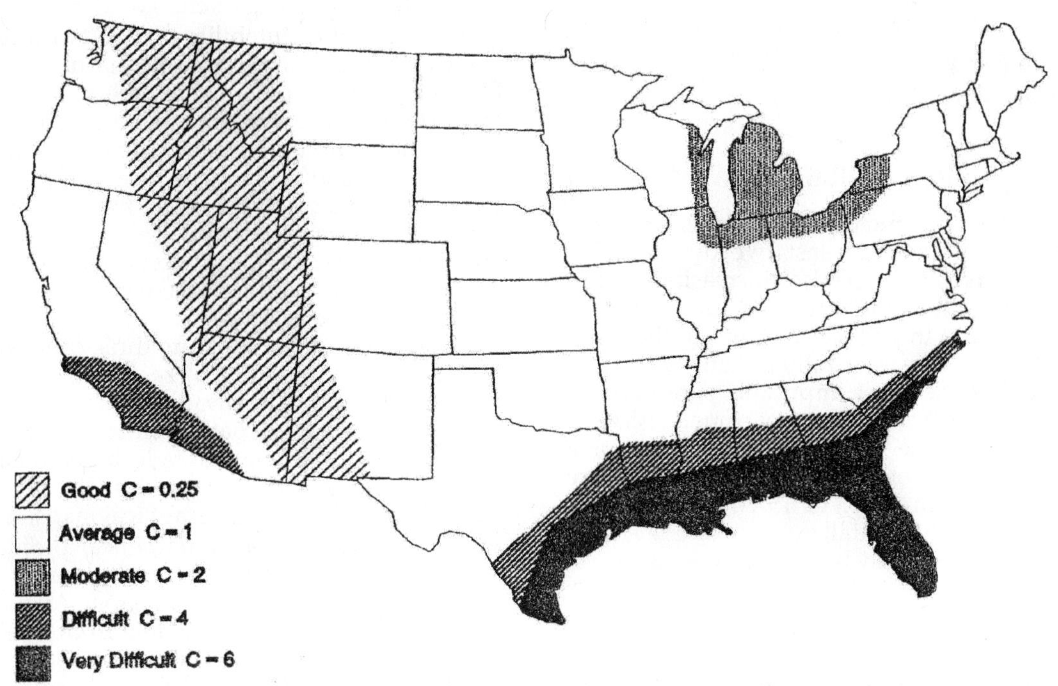

FIGURE 6.4-5 Values of climate factor for the continental United States [2]

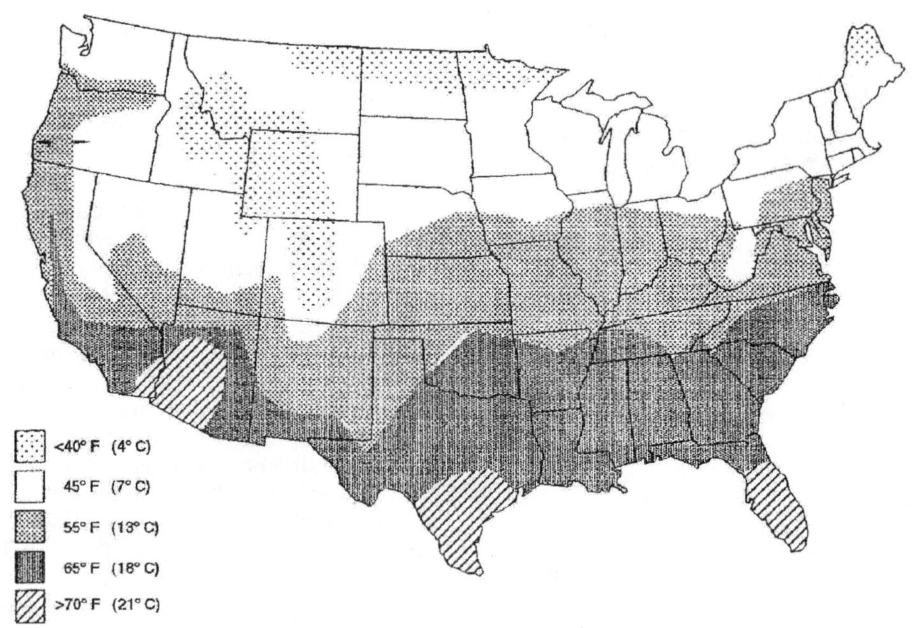

FIGURE 6.4-6 Average annual temperature for the United States [2]

D = path length in miles

F2 = the fade margin in dB of the antenna with the lower fade margin (the antenna with the higher fade margin should be used to calculate U).

Isd as calculated here represents a ratio, for example, if Isd is calculated to be equal to 43, the improvement factor is 43 to 1. To calculate the total unavailability for a space diversity system:

$$Usd = \mu / Isd,$$

Usd is unavailability with space diversity

U is unavailability, nondiversity [1]

Isd is improvement factor

Moisture Absorption and Rainfall

The impact of rain on microwave systems has been studied increasingly over the last two decades. The principal finding is that rain causes increasing attenuation at frequencies above 10 GHz, and should always be taken into account in path calculations at these frequencies. The peak rain rate rather than the annual average rainfall is the most important parameter to analyze. A day-long drizzle and a short thundershower may precipitate the same volume of water, however, a dense cell in a thunderstorm will cause the most severe outage. On longer paths at 13 or 15 GHz, and for all paths above 18 GHz, the rainfall outage dominates and multipath effects can usually be ignored.

Path Profiles

Microwave terminals are likely to be located at studios, earth stations, or TV transmitters, all of which are known, fixed sites. Should a new electronic news gathering (ENG) site or a microwave repeater site be required, the choice may be constrained by availability and access. A new access road and utility service connections are often the most costly part of building a new site. As such, it may prove more economical to avoid the highest point of ground and use a taller tower or other support structure located closer to an existing road and utility connections. In some cases, an existing tower may be close enough to the desired path or coverage area to warrant approaching the owner about a leasing arrangement.

Assessing the suitability of two or more sites, and determining the required antenna heights for establishing point-to-point microwave paths, requires an accurate knowledge of the intervening terrain and any obstructions along the path. The standard method of gathering this information is to record the vertical elevation data between the sites, from which a vertical earth profile may be drawn. This can be done quickly and easily with the aid of economical computer programs that draw profiles based on U.S. Geological Survey (USGS) topographical survey data. Some programs now include aerial photos and three-dimensional mapping that allow the user to "see" in all directions from ground level at any spot on the map. Still, others provide a complete set of path performance and reliability calculations, including rain fade, based on inputs from the user. Manually drawn profiles may be done by taking data from USGS topographical maps, if available.

Most programs draw a flat, linear profile on a rectangular format, much like the manual method of drawing on rectangular graph paper. At one time, special k-factor paper was available from a few vendors, but was limited to one k factor only. Flat profiles have the distinct advantage of allowing the engineer to analyze multiple k factors, and plot reflection points, all on the same page with allowances included for average tree heights and known manmade obstructions at obvious critical points along the way. A critical point is defined as one that has the potential for blocking the microwave beam, if sufficient antenna height cannot be achieved at one or both ends of the path.

Once a decision has been made to pursue a particular path, an on-site examination of key path elements is recommended to ensure that tree heights are correct and no new buildings have been added in the path since the data was compiled, or since the maps were printed. The typical criteria used in programs and in manual calculations are as follows:

$1.0F_1 + k = 4/3$ for "normal" clearance, and

$0.3F_1 + k = 2/3$ for "worst case" clearance

where F1 is the Fresnel zone radius in feet. Although a clearance of 0.6F1 is adequate, it is viewed as the absolute minimum, and leaves no room for future tree growth or topographical survey errors.

ANTENNA SYSTEMS

For point-to-point microwave systems, the usual antenna choice is a parabolic reflector type. The parabolic antenna consists of a frequency-sensitive feed assembly at the focus of a reflector; which may be placed in the aperture for minimum side-lobes, or in front of the aperture for maximum gain.

In the 950 MHz aural broadcast auxiliary service (BAS) band, the sectional parabolic is favored in a majority of STL systems, followed by the grid parabolic, and finally the solid parabolic. Yagi arrays are still being used in some AM and FM STLs but are not recommended for new installations, and should be replaced in existing installations due to their poor side-lobe patterns and low front-to-back ratios.

At frequencies up to 3.7 GHz, the feed is typically a dipole with a subreflector and a solid main reflector. At frequencies above 3.7 GHz, waveguide feeds and solid reflectors are required to maintain the same level of efficiency. Grid-style main reflectors may be used up to 2.7 GHz to reduce wind load, but are limited to a single polarization.

The gain ratings of microwave antennas are referenced to an isotropic radiator, as noted by the reference to gain in dBi. An isotropic radiator is a hypothetical antenna with a radiation pattern that is perfectly spherical. Directional antennas are designed to enhance the radiation in a given direction at the expense of other directions, which has the effect of concentrating the signal in the desired direction and reducing the signal in other directions.

Antenna Types and Applications

The range of typical antenna types in use today for aural and television service include the following:

- Standard Parabolic. For many years, the workhorse antenna of microwave communications has been the standard parabolic, which offers a good compromise between gain, side-lobe performance, and front-to-back ratio (Figure 6.4-7). Standard antennas may be single or dual polarized, and include both solid and grid reflectors, although grid antennas cannot be dual polarized. Low VSWR versions are available in most models, and in some cases may be the only type available.
- Sectional Parabolic. Although based on a parabolic design, the truncated sectional parabolic does not employ the typical round reflector, but only a portion of the reflector. In practice, these antennas employ grid reflectors (Figure 6.4-8), and have elliptical radiation patterns as shown in Figure 6.4-9. A sectional antenna may be successfully employed in uncontested areas, where potential interference is minimal. The only drawback to a sectional antenna stems from its elliptical radiation pattern and the fact that the optimum side-lobe suppression is realized only when the antenna is vertically polarized.

  While the Federal Communications Commission (FCC) rules do not currently specify a minimum beam width for the 950 MHz aural band, they do enforce a de facto standard that may preclude use of a paraflector.
- Grid Antennas. Grid-style reflectors were developed to reduce wind load, and are built with grid elements spaced close enough to act like a solid reflector in the specified band of interest. The grids must be aligned in parallel with the polarization of the feed, or the gain will be reduced at least 20 dB. Over the years, they have been very popular for the 950 MHz aural band, and the 2

FIGURE 6.4-7 Standard performance parabolic antenna

Photo courtesy Andrew Corp.

GHz TV STL band. Grids are effective to around 2.7 GHz, and are restricted to one polarization only. Above 2.7 GHz, the grid spacing is reduced to the point that it no longer provides a wind load advantage, and the feed also becomes less efficient.

- Enhanced Performance Parabolic. There are enhanced designs have come on the market to fill the gap between standard performance and high-performance antennas. The enhanced-performance parabolic usually includes a deeper than normal reflector, and is sometimes called a "deep-dish" antenna (Figure 6.4-10). The deep-dish designs may also include special feed assemblies, and selective shield attachments at the edge of the reflector. The main advantage of enhanced designs is lower side-lobe radiation that may meet FCC category A compliance where a standard antenna of the same size might only meet category B compliance. Some enhanced designs may sacrifice a small amount of gain as part of the tradeoff for cleaner side-lobe performance. The significance of categories A and B is based on side-lobe suppression, where category A is more stringent than category B. Part 101.115 of the FCC's rules discusses exact details of category A and category B.

FIGURE 6.4-8 PR-950 "Paraflector" sectional parabolic antenna

Photo courtesy Kathrein-Scala

- High-Performance Antennas. The term high performance with regard to a parabolic antenna refers to an antenna that has been purpose-built to suppress the side-lobes to a much higher degree than a standard antenna of equal size (Figure 6.4-11). It is not unusual to find, for example, that a 6 foot high-performance antenna will provide better adjacent channel protection than does an 8 foot standard-performance antenna, and with less wind load. Vendors typically offer several models of high-performance antennas, with increasingly tighter patterns or higher polarization discrimination in order to support difficult frequency coordination requirements in a crowded environment.
- Angle Diversity Antennas. The angle diversity antenna is a highly specialized design, intended to help reduce multipath fading on paths that require space diversity, but are limited to one antenna at each site. The antenna is built with two separate

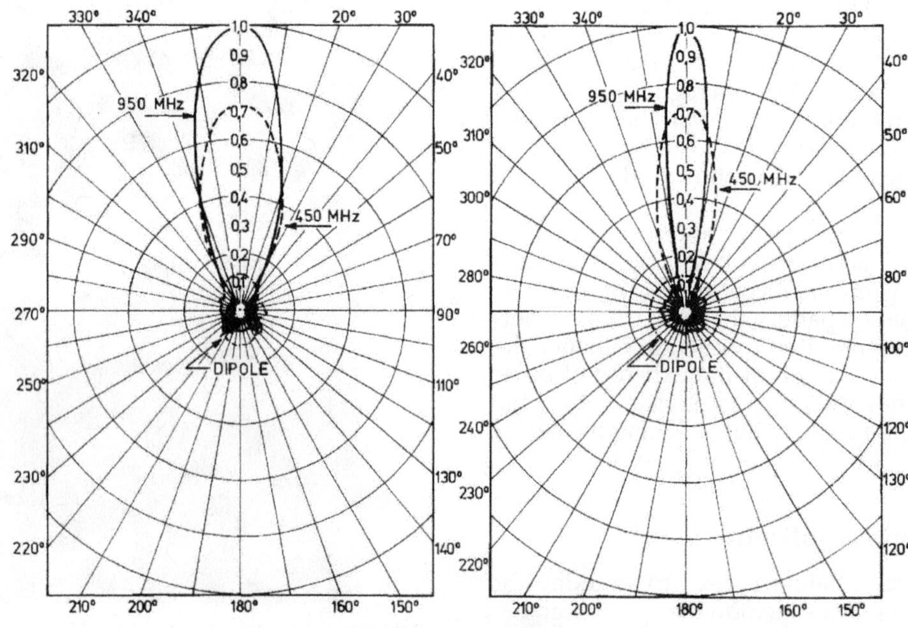

FIGURE 6.4-9 Radiation plots sectional antennas at 450 and 950 MHz; (a) is the pattern for a horizontally polarized antenna, and (b) is the pattern for a vertically polarized pattern [2]

FIGURE 6.4-10 Enhanced performance "deep dish" antenna

Courtesy Andrew Corp.

FIGURE 6.4-11 Typical high-performance, shrouded antenna

Courtesy Andrew Corp.

feed assemblies that propagate identical beams with a small angular displacement. Theoretically, the multipath fade mechanism will affect each beam differently, and effectively reduce the fading.

Radiation Patterns

In order to comply with FCC and European Telecommunications Standards Institute (ETSI) standards, the radiation pattern for every variation of microwave antenna that is offered for sale in fixed link applications must be measured, recorded, and made available to customers and frequency coordinators.

A typical radiation pattern is shown in Figure 6.4-12. The pattern shown provides information about the typical gain and side-lobe performance of an 8 foot dual-polarized antenna for the 7 GHz broadcast auxiliary service (BAS) STL band.

Radomes and Loss

Depending on the manufacturer and antenna type, a radome may or may not be included with a specific antenna. Whether included or optional, the addition of a radome to any microwave antenna provides several benefits. Protection from weather extremes and wind-loading reduction are at the top of the list. A radome-equipped antenna is also better protected against falling ice and the potential for feed horn damage as a result.

For standard antennas, the radome is usually convex in shape and made of molded fiberglass or ABS plastic. Radome options include embedded heating elements, extra high-strength versions, and custom colors to blend better with the environment. High-performance shrouded antennas always include flexible unheated radomes made with rubber-coated nylon or polymer-coated fiberglass sheeting. The flexibility is an advantage in windy conditions, as the movement tends to shed any ice buildup.

While the loss of a radome is included in all high performance antenna gain specifications, it is generally not included in standard antenna gain figures, and must be accounted for when performing path calculations. Table 6.4-1 provides average radome loss for typical standard antennas, interpolated from catalog values for the broadcast auxiliary service bands. Exact values can be obtained from the specific manufacturer.

It can be seen that the loss is almost negligible in the 2 GHz band, but rather substantial in the 13 GHz band. When making path calculations, the manufacturer's stated loss figures should always be used. Losses have been omitted at 950 MHz, as they are negligible, and most aural STL installations do not include radomes. In the 18 GHz and 23 GHz bands, most antennas are of the high-performance type, and the losses are included in the gain figures.

Transmission Lines, Tuning, and Pressurization

Below 3.7 GHz, antennas are normally fed with coaxial cable that is rated for microwave use. The cable may use a foam dielectric between the inner and outer conductors, or an air dielectric construction that includes a spiral wound center-to-outer spacer. Air dielectric cables exhibit the lowest transmission loss, however, foam dielectric cables do not require dehydration equipment and are often completely adequate in spite of higher losses.

Above 3.7 GHz (and also below if the cable loss is excessive), elliptical waveguide is the typical choice for transmission line. In cases where the loss of elliptical

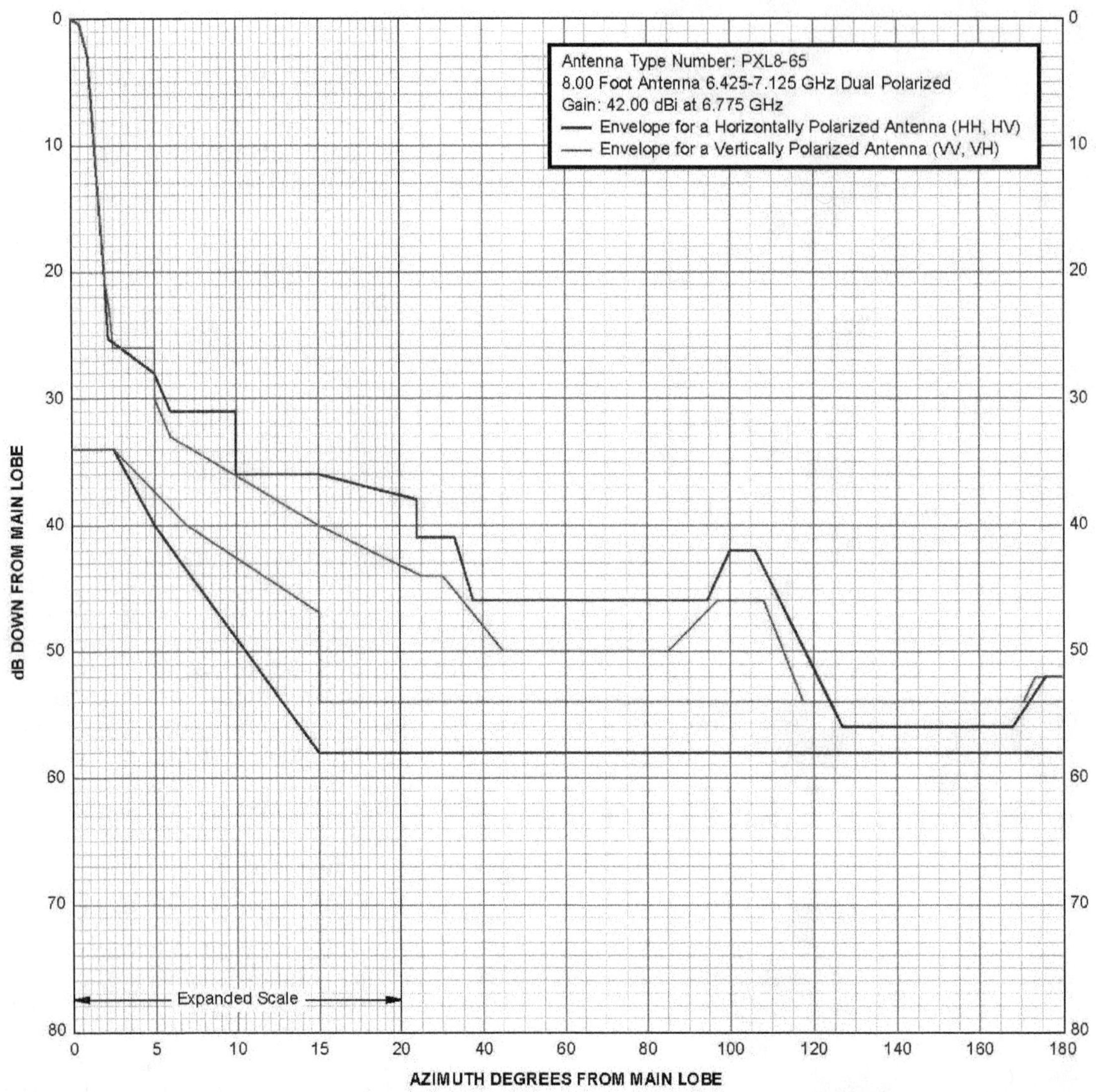

FIGURE 6.4-12 Radiation pattern envelope for an 8 foot standard parabolic antenna at 7 GHz, with a dual-polarized feed

Courtesy of Andrew Corp.

TABLE 6.4-1
Typical radome loss for standard parabolic antennas [2]

Frequency Band	Antenna Size		
	6ft	8ft	10 ft
2 GHz	0.1 dB	0.1 dB	0.2 dB
7 GHz	0.7 dB	0.8 dB	1.1 dB
13 GHz	1.7 dB	1.8 dB	2.1 dB

waveguide is a problem, circular waveguide may be deployed. In all cases, microwave transmission line is delicate in nature, and must be installed to manufacturers' specifications for bending radius, support points, and pressurization.

Connectors for both cable and waveguide need careful attention during installation to avoid mismatch, air leaks, and the ensuing possibility of moisture ingress over time. Special flanging and attachment tools can be purchased or rented from manufacturers to aid in the correct and consistent assembly of connectors.

As the broadcast world converts to digital, and as more radio channels are combined on a single path, tuning the transmission line becomes much more important. RF reflections can reduce the effective range of adaptive equalizers in digital radios by stressing them needlessly, or they can cause excessive distortion at spot frequencies in analog radios. Sweep tuning the waveguide during installation, with occasional sweep tests at periodic intervals thereafter, will assure that system performance is at its best when fades do occur, or when new RF channels are added.

Installations that employ waveguide or air dielectric cables must be maintained at a positive pressure (typically 0.5–5.0 lb/in^2) above atmospheric with dry air or nitrogen. A wide range of dehydrators and accessories are available to facilitate proper pressurization and control.

Dehydrators can be fully automatic (where the desiccant is automatically dried out by the unit) or semi-automatic (where the desiccant must be periodically dried out or replaced). For remote sites, automatic dehydrators are strongly recommended to avoid the possibility of moisture ingress and damage associated with desiccant exhaustion. Proper sizing of the dehydrator is also important for reliability. An undersized dehydrator may operate at a high duty cycle and shorten the compressor life. To avoid rapid cycling of the compressor, the volume of air in the system should not be less than 1.5 cubic feet; this can be achieved in small systems by adding a regulating tank.

In systems with multiple transmission lines, it is good practice to install a pressurization manifold with shutoff valves and metering for each transmission line. Should one of the lines develop a leak, it can be isolated for repairs, while pressure is maintained on the others.

Commscope, ERI, Radio Frequency Systems (RFS), and Kathrein-Scala each provide a wealth of information on their websites and in their catalogs relative to antennas, transmission lines, connectors, and all of the accessories needed for installation.

Antenna Alignment Techniques

The highly directive nature of a parabolic antenna can present quite a challenge when attempting to align a microwave path. At lower frequencies, such as the 950 MHz band, the job is made easier by a comparatively broad main lobe; however, at 7 GHz and above, the narrower beam widths can make it difficult to find the signal when performing an initial alignment. Knowing a few tricks of the trade can be very helpful when the tower crew arrives and needs direction in the alignment process. More often than not, the crew has an idea of the basics, but lacks technical savvy. Although this section focuses on the alignment of a new installation, a realignment is essentially the same after the point that a signal has been acquired.

Pre-align the Antenna

First and foremost, it is vitally important to know the exact pointing azimuth for each antenna, and to pre-align the antenna as closely as possible to its final position. If antennas or their mounting structures can be seen with the naked eye or with a pair of binoculars from one end of the path to the other, the job should be straightforward. If the other end of the path is not visible, another way must be found to pre-align the antennas. Google Earth (www.google.com/earth/) is an excellent resource for determining the azimuth from one site to the other.

The first step in pre-aligning the elevation is usually a matter of setting the peak of the main lobe on the horizon (zero degrees elevation referenced to the horizon), unless there is a great difference in antenna height at opposite ends of the path. A short path will exaggerate the effect, while a long path reduces it. For extremely narrow beam widths, in most cases an "eyeball" elevation alignment is not sufficient, and a leveling tool must be used. The simplest tool is an inexpensive angle finder held against the structural mounting ring on the back of the antenna. An angle finder is a handheld device that uses a weighted indicator to measure the angle of any surface with respect to gravity, and can be found in most hardware stores. The scale is usually calibrated in 0.5 degree increments, making it an easy matter to adjust the antenna bore sight elevation to the specific path requirements.

Pre-aligning the azimuth is a bit more complicated. USGS topographical maps and virtually all path programs (including Google Earth) use true north as a reference, however, the most common direction-finding tool is a magnetic compass. Let's say, for example, that a crew in downtown San Francisco must align an antenna to an azimuth of 115 degrees. In San Francisco, the magnetic declination, or the difference between true north and magnetic north is +14.7 degrees. In order to find the correct direction referenced to true north, the declination must be subtracted from the compass reading, which we now know is 14.7 degrees beyond where the antenna should point. So, 115 degrees minus 14.7 degrees yields a magnetic azimuth of 100.3 degrees. On the East Coast, the declinations are negative, and must be added to the compass reading.

Path performance calculations should always be done in advance, to provide an accurate estimate of the expected receive carrier level.

It is extremely important to verify that the antennas were assembled according to the manufacturers' instructions, especially the positioning of the feed assembly. Failure to do so can degrade the performance and distort the pattern, causing lower than expected signal levels and increased susceptibility to interference. For example, if the antennas are cross-polarized at the start, the signal will be at least 20 dB down from where it should be.

Also, the transmission lines should be checked carefully for any kinks, dents, or extreme bends, and the connectors should be checked to verify that they have been correctly installed. It is a good idea to tune adjustable connectors and sweep the transmission lines connected to the antenna using a microwave network analyzer and a return loss bridge, to benchmark the performance immediately after installation.

Alignment Hints

First and foremost, it is very important that the polarization is identical at both ends of the path, and that it matches the polarization that the path is licensed for. Coaxial and waveguide feeds usually have arrows near the input connector showing the polarization of the feed that can be checked for verification purposes.

When the tower crew is in place, and the antennas are pre-aligned, check to see if there is any signal at all in the receiver. If not, have the crew on one end of the path start to pan the azimuth slowly about 10 degrees to each side of the pre-set center. If the first attempt at receiving a signal is not successful, return the antenna to its pre-set azimuth, and try the same procedure at the other end of the path.

Once a signal is acquired, there are numerous ways to coach the tower crew into a perfect alignment. To begin with, everyone involved in the antenna alignment should be in voice communications. VHF and UHF handy talkies are useful for such communications.

One successful method of directing an alignment is to call out automatic gain control (AGC) readings to the person that is panning the antenna, after explaining the significance of the changes. The tower crew should be instructed in advance on the plan, in particular what to expect with regard to the feedback that they will be receiving while listening to the AGC readouts. The antennas should be moved very slowly during this process or the signal can be missed completely. Many crews have done this type of work before, but it helps to set ground rules that everyone can understand.

Many microwave radio systems have complete network access and a user-interface that can be reached via Internet protocol (IP) communications. It is especially helpful during any alignment process to be able to "see" the far end (specifically, the received signal strength indication) so that directions can be given to the tower crew as to what adjustment is most effective in finding the peak level. If the far-end site has IP communications capability, even at a minimal level, it can be very helpful to configure it ahead of time, so that a remote terminal connection to the far end receiver can be made, even before the alignment process starts.

AURAL STL RADIO SYSTEMS

While a substantial number of analog STL systems are still feeding analog AM and FM exciters, the landscape is changing, and the rate of change is accelerating. The majority of new and upgraded AM and FM radio STL systems employ digital modulation formats that support the AES/EBU audio interface in addition to legacy analog inputs. Access to IP communications has become commonplace at transmitter sites and its bidirectional nature has led to a substantial increase in the number of devices located there.

Digital AM and FM exciters and newer AM and FM transmitters all support IP communications. The most current exciters are capable of supporting standard analog and the NRSC-5 digital format, also known as IBOC or by the trade name of HD Radio. All-digital STLs and exciters still continue to generate analog transmissions; however, many are capable of adding IBOC/HD Radio support when the station wishes to do so.

The principal frequency range used by aural STL systems in the United States is between 945.0 MHz and 952.0 MHz (inclusive). Use of private operational fixed point-to-point microwave services—otherwise known as OFS—is specified in Part 101, Subparts C and H, of the FCC's Rules. Using OFS it is possible for radio stations to license spectrum in the 952–960 MHz band, as well as parts of the 6 and 11 GHz bands. Stations with very short paths have sometimes chosen 18 or 23 GHz, while some have ventured into the unlicensed ISM bands at 900 MHz, 2.4 GHz, and 5.8 GHz. Radio systems operating in these bands typically use digital modulation techniques (described later in this chapter) and carry payloads based on time division multiplexing (TDM) or IP, in a bidirectional fashion.

CONVENTIONAL MULTIPLEX SIGNAL GENERATION

In many FM and AM broadcast operations, the various components of the final signal are still sent from studio to transmitter site on separate paths, or generated on-site. FM-modulated monaural STL systems still have their place for both AM and FM broadcasting, though their use has diminished. Prior to the development of multiplex systems, stations often used dual-parallel systems for stereo or dual-mono applications with a common antenna at the transmitter and receiver.

For many years, analog devices have been utilized by broadcasters to generate the FM multiplex (MPX) signal. By nature of the FM technology, the final stereo multiplex is an analog signal feeding the analog FM transmitter. This FM multiplex signal is synthesized by the stereo generator and, as a minimum will consist of the mono baseband audio (L+R, 30Hz to 15 kHz), a 19 kHz pilot tone, and the difference signal (L-R). The difference signal (L-R) signal modulates a 38 kHz carrier, producing a double-sideband, suppressed carrier signal occupying the baseband range of 23 to 53 kHz.

Furthermore, various radio data systems can occupy higher ranges of the MPX spectrum as shown in Figure 6.4-13. Systems differ by country but the Radio Data System (RDS) digital data subcarrier is the most commonly used.

The MPX signal (see Figure 6.4-13) is an analog signal carrying a frequency spectrum of at least 60 kHz or even 80 or up to 100 kHz (57 kHz carrier for RDS). The technical recommendation is to provide a data path of 80–100 kHz including the direct band spectrum and more. This is the main reason why the MPX signal generator (the stereo coder) has typically been placed as close as possible to the transmitter equipment; keeping the cable lengths as short as possible helps maintain the quality of the signal. Poor frequency response or poor phase response in the MPX signal interface between the generator and the exciter will compromise the ultimate stereo performance.

The 19 kHz pilot tone, which is phase-coherent with the 38 kHz L-R modulator, is transmitted as part of the overall multiplex signal and allows the FM receiver to accurately demodulate the L-R signals, and thus reconstruct both left and right channels. It is therefore

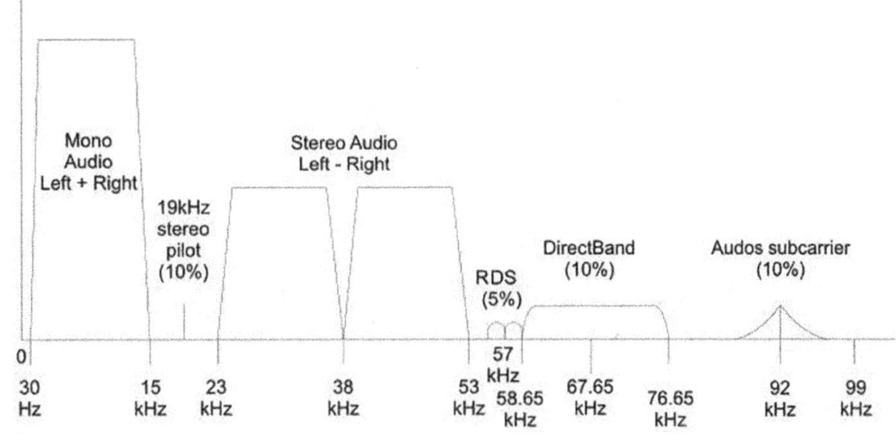

FIGURE 6.4-13 The complete multiplex frequency spectrum for analog FM transmission

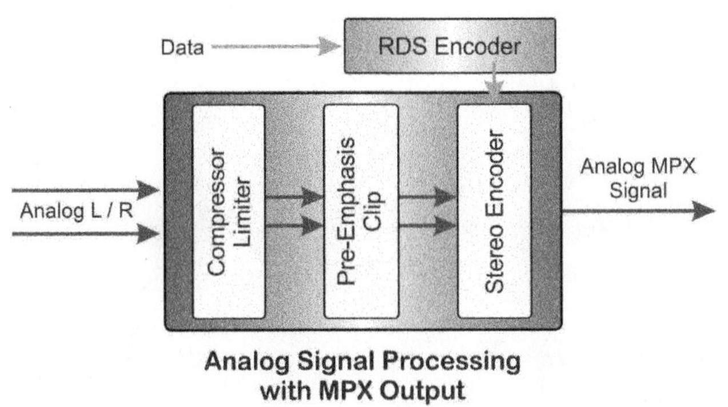

FIGURE 6.4-14 The principle of analog multiplex generation

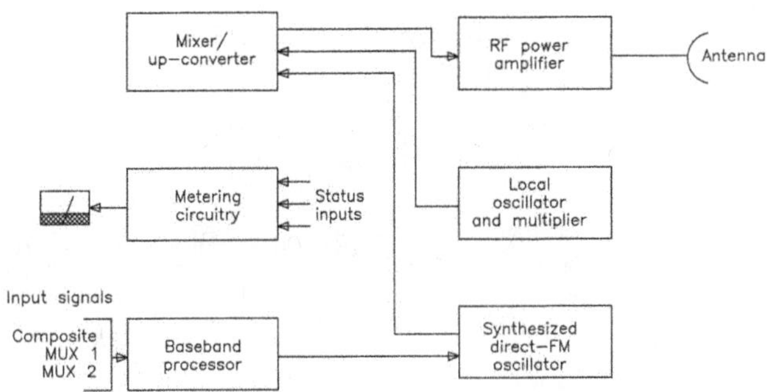

FIGURE 6.4-15 Block diagram of a typical composite STL transmitter [2]

extremely important for proper performance of the stereo broadcast that the phase transition of the 19 kHz pilot is not manipulated, affected, or corrupted by the transmission chain. As a result, the MPX signal is very sensitive, and its quality is of the utmost importance to the quality of the FM modulation.

The diagram of Figure 6.4-14 represents a typical signal processing chain for a single program on a single transmitter site. As occurs in many sites, it shows a baseband audio signal that is carried by an STL to the transmitter site. The location of the various components within the signal chain can have significant effect on the performance of the system overall. As was previously mentioned, the frequency and phase response needed for accurate representation of a stereo is such that system performance overall is often optimized by having the stereo generator co-located with the exciter.

Even so, many stations opt to use composite aural STLs. They may be found in a number of hardware variations, all designed to accomplish the same goal: a low-distortion relay of a multiplex signal from the studio stereo generator to the remotely located exciter and transmitter. The multiplex signal from the STL receiver is fed directly into the FM exciter. A block diagram of a representative composite STL transmitter is shown in Figure 6.4-15.

A block diagram of the companion composite STL receiver is shown in Figure 6.4-16.

A later variation in STL systems came in to common use after the industry acceptance of the AES/EBU standard for digital audio transmission (see Figure 6.4-17). Left and right audio were processed at the studio end of the STL, and were further processed, and converted to the stereo multiplex signal at the far end.

DIGITAL STLS FOR RADIO BROADCAST

As the conversion to digital radio broadcasting moves forward, so does the sophistication of aural digital STLs. By its nature, digital transmission offers more robust path performance during fades, up to the point of reaching a threshold at which point the digital system performance drops significantly. As long as the received signal strength and carrier-to-noise exceed a defined minimum value, the bit error rate remains relatively constant, and the decoded baseband signal is an exact duplicate of the input signal at the studio (Figure 6.4-18). With adequate error correction, signals can be received error free. Another advantage can be found where multiple microwave hops are required; digital STLs regenerate the digital stream at each repeater, eliminating additive signal degradations found in analog systems.

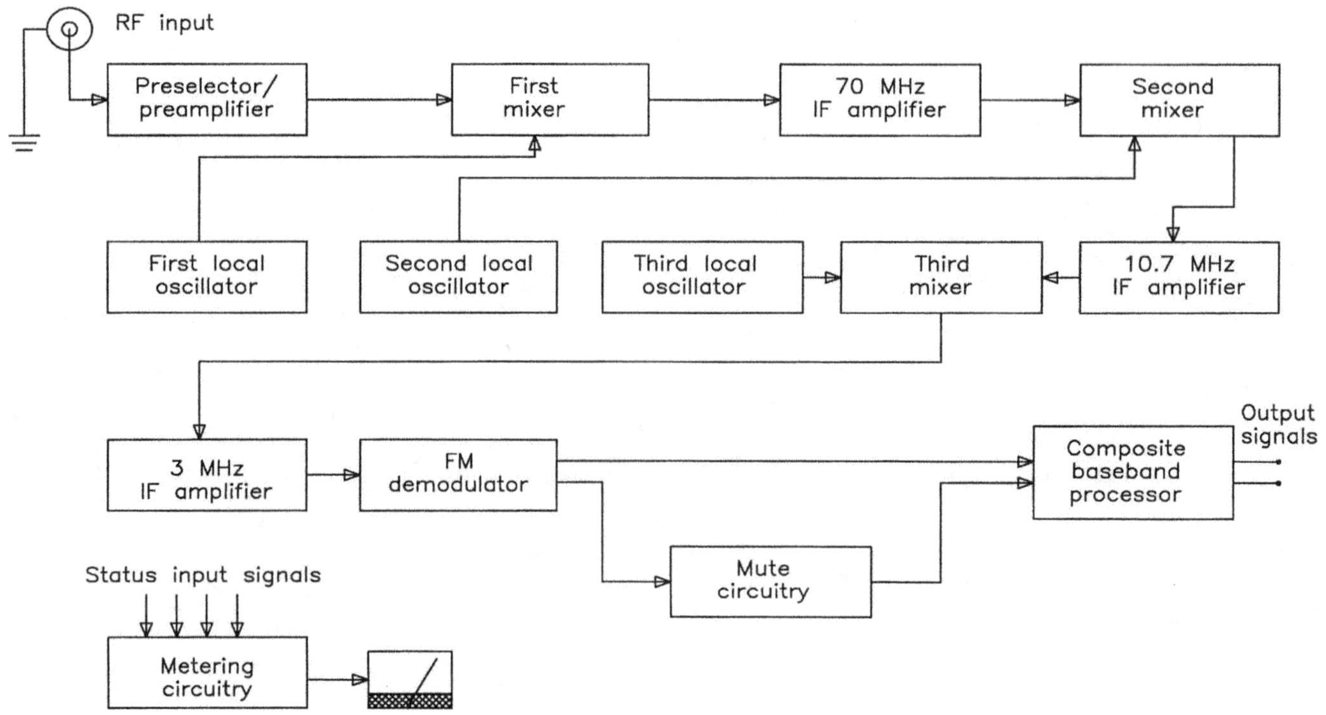

FIGURE 6.4-16 Block diagram of a typical composite STL receiver

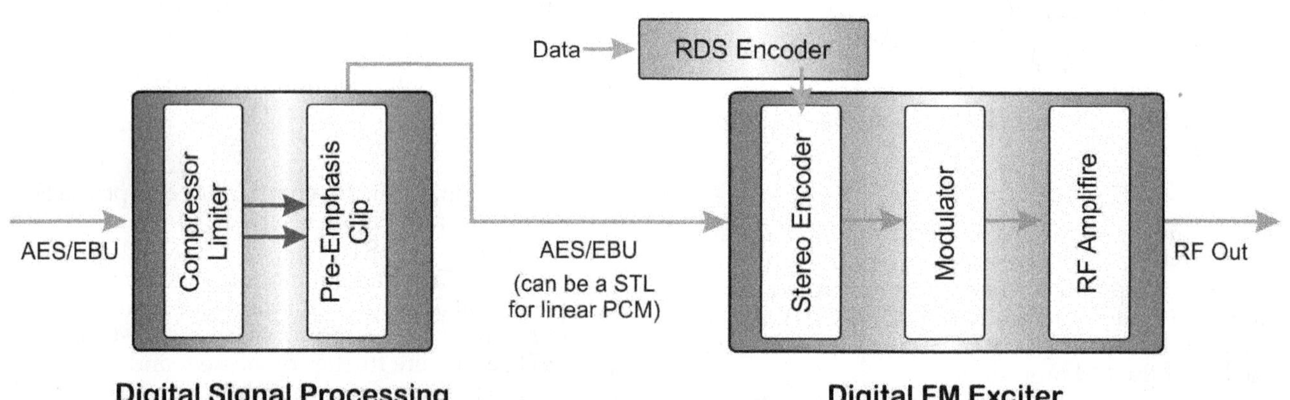

FIGURE 6.4-17 STL system with intermediate AES/EBU connection

It is fair to point out that digitization of the audio signal always brings with it degradation in the form of quantization errors. However, the high sampling rates typically used for professional audio applications reduce such degradation to extremely low levels that are virtually inaudible.

The process of analog-to-digital conversion has an output that is defined in terms of its sample rate and word length. The sample rate determines the upper frequency limit (also known as the Nyquist limit) of the signal being sampled. The bit rate defines the overall dynamic range that can be represented in the subsequent conversion from digital back to analog:

Dynamic range = $20 \log 2^n$ where
n = word length in bits

In practice the dynamic range will be somewhat less because dither noise is sometimes added to the analog output to mask the quantization noise that remains from the digital to analog conversion process.

Multiplication of the bit rate times the word length gives the basic data rate, before consideration of any overhead. A properly operating transmission channel can be assumed to provide error-free throughput; if that is indeed the case, the digital signal can be regenerated at the receiving point as an exact duplicate of the input waveform, quantization errors excepted.

As previously mentioned, the benefits of a digital STL for radio broadcasters can best be appreciated by comparing the performance of a digital system and an analog system (see Figure 6.4-18). A digital

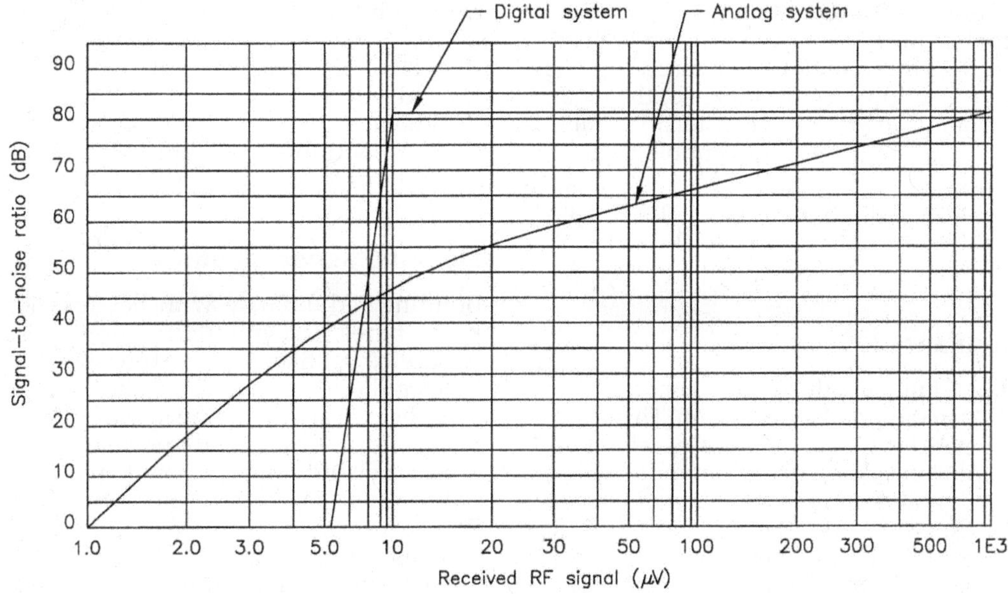

FIGURE 6.4-18 The benefits of digital versus analog STL systems in terms of S/N and received RF level [2]

STL typically permits broadcasters to extend the fade margin of an existing analog link by 20 dB or more. Furthermore, audio signal-to-noise (S/N) improvements of at least 10 dB can be expected for a given RF signal strength. Alternatively, for the same S/N, the maximum possible path distance of a given composite STL transmitter and receiver can be extended. These features could, in some cases, mean the difference between a one-hop system and a two-hop system.

The spectrum-efficiency of a digital STL is of great importance today in highly congested markets. The system may, for example, be capable of relaying four program channels and two voice-grade channels. The use of digital coding also makes the signals more tolerant of co-channel interference than a comparable analog STL.

Combining 950 MHz Systems With IP for System Hardening

The 950 MHz RF STL, whether analog or digital, is inherently subject to degradation due to environmental interference. Legacy STLs are one-way radio links operating in the studio-to-transmitter direction and historically it has not been possible for the system to adapt dynamically to conditions that can reduce system throughput. The advent of ubiquitous IP networks presents the opportunity to add two-way communications to these STLs, enabling not only real-time adaptation of the RF modem parameters to changing environmental conditions, but also potentially, the ability to completely back up programming from RF to IP.

An STL system encodes program content at a studio site and transports this content wirelessly over a licensed frequency to a far-end transmitter site. A modern digital RF STL system can encode and multiplex multiple services onto the same radio link, including high quality audio, lower bandwidth voice quality audio, and data services such as IP and RS-232. As with many communication systems, link degradation or failure can occur due to equipment failure or environmental conditions.

STL Data Bandwidth

For the main analog FM broadcast-type program audio, an uncompressed linear stereo with 16-bit samples and 32 kHz sample rate (typical for a 15 kHz FM signal) requires just over 1 Mbps of data bandwidth. Increasing the sample rate to 48 kHz for 20 kHz audio and the requirement jumps to 1.5 Mbps. Audio compression algorithms such as AAC and apt-X can reduce these data requirements considerably but also compromise the audio quality. Depending on the HD Radio signal requirements, traffic such as Importer-to-Exporter (I2E) or Exporter-to-Exgine (E2X) data, the STL will need to transport up to 330 kbps of IP data traffic. Add to this the transport of auxiliary audio channels such as subsidiary communications authority (SCA) and RDS, and the total data bandwidth requirements can exceed 2 Mbps.

Three basic factors affect the amount of data that can be sent across the RF STL link: the amount of RF bandwidth available, the modulation scheme used, and the RF gain of the link. The first of these is simply the amount of RF bandwidth allowed by the Part 74 license. For FM stations this is usually 300 kHz but can be as little as 200 kHz or as much as 500 kHz. The amount of data that can be carried in this RF bandwidth depends on the type of modulation chosen. Quadrature amplitude modulation (QAM) is a modulation scheme widely used in STLs and many other communication applications. Common QAM forms are 16-QAM, 64-QAM, 128-QAM and 256-QAM. By moving to a higher order symbols or constellations, it is possible to transmit more bits per second in the same RF bandwidth. However, in doing so, the mean energy between points in the QAM constellation must be closer together and thus more susceptible to noise and other corruption. A higher order QAM requires a higher signal-to-noise Ratio (SNR) than a lower order QAM to achieve the same bit error rate (BER) performance [1].

RF Link Performance

Prior to building a link, a path study is conducted to determine the total required RF system gain. This study factors in parameters such as antenna gain, cable loss, transmit power, distance, fade margin, required BER, and so forth. Design and optimization of a digital RF STL is a tradeoff between spectral efficiency and BER. The difference in required SNR when moving between different QAM orders is around 3–4 dB per step. So, for example, 128-QAM will require 3– 4 dB more signal to achieve the same BER performance as 64 QAM. Figure 6.4-19 shows the SNR required to achieve 10^{-6} BER performance versus spectral efficiency. Note that some forward error correction (FEC) coding information is carried in the RF data that greatly improves performance and as a practical matter is always implemented.

Redundancy Using Multiple Data Paths

The RF STL is a unidirectional data transmission path; however with the advent of inexpensive and ubiquitous IP links it is possible to introduce an IP-based data return path from the transmitter site back to the studio site. An IP return path enables the transmitter site unit (the receiver) to send information regarding the performance of the forward path back to the studio site unit (the transmitter). Because IP links are for the most part bidirectional, the IP link can also serve as a forward data path for program audio material from the studio to the transmitter site. Such an IP STL data path can serve as a backup path to the primary RF STL link, or as the primary path itself (with the RF STL as the backup path), or the two data paths, RF and IP, can augment each other to provide greater data bandwidth than one alone can provide [2]. The GatesAir Intraplex HD Link was a system designed around this idea.

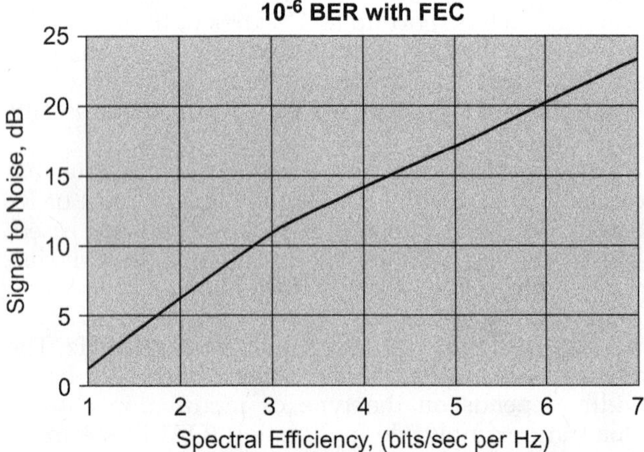

FIGURE 6.4-19 Signal-to-noise versus spectral efficiency

Defining Service Profiles

Building redundancy with heterogeneous STL data paths types, RF and IP, requires the configuration of service profiles for each one. Each service profile defines the attributes of all the audio and data services being carried on the system. These attributes in turn factor into the total information rate, that is, the total data bandwidth required on the STL path. A service profile can be associated with a particular path. For purposes of describing redundancy, the service profiles can be called main and backup. The main and backup service profiles are independent. How to configure each profile depends on how much bandwidth (data carrying capacity) is available on the RF and IP paths. However, the backup profile would normally be set up to use less bandwidth than the main, as its purpose is to enable some or all of the services to remain on-the-air in some format when throughput on the primary path is impaired, or when switching to a backup path with lower capacity than the primary.

On the RF path, the data bandwidth is determined by a combination of the RF bandwidth that is available (200 kHz to 500 kHz), the QAM rate, and the amount of FEC carried as overhead data. For example a 500 kHz channel operating at 256 QAM has about 3.1 Mbps of user data throughput. On the IP path, the available bandwidth is determined by the nature and technology of the path being implemented. This may range from relatively low-speed paths such as DSLs, up to gigabit-per-second type networks. To properly configure an STL system, it is necessary to know what the minimum guaranteed bandwidth will be on your IP path.

RF-Only, IP Return Path Is Available

If an IP path is available but cannot be used for service transport due to bandwidth limitations or unreliability, it may be of use still as a means of performance feedback. The RF STL transmitter can adjust its RF parameters based upon performance; for example, if the RF SNR becomes degraded below a user defined threshold, the transmitter can automatically throttle down to a lower QAM order resulting in a lower data bandwidth thru the STL. Doing so results in improved RF performance because a lower QAM order requires less RF SNR to achieve the same BER performance as a higher QAM order. The RF STL receiver can report degradation of the RF performance to the RF STL transmitter, enabling the RF STL units at both ends to switch to a lower bandwidth backup profile when the RF path degrades.

As an example, the bandwidth for the backup profile can be reduced compared to the main profile by switching to a lower audio sampling rate (32 ksps vs. 48 ksps), or by reducing services (removing voice grade channels used for SCA). In another example, assume that the Ethernet data stream is normally running at 512 kbps, carrying the HD Radio E2X signal at 320 kbps bandwidth, with the remainder used for control and LAN data. The backup profile can be configured to carry only the critical 320 kbps of HD Radio data, and drop the rest. Note this requires the STL system to prioritize and shape the Ethernet traffic so that low-priority traffic does not interfere with critical HD Radio traffic.

Figure 6.4-20 illustrates how this type of fallback operation using main and backup profiles could work. The switch from main to backup service profile can be made to revert. If, for example, the RF performance, as measured by the receive SNR improves above a user defined threshold, the system can automatically switch from the backup service profile back to the main profile. Note that in this scenario, the services being fed into the RF STL link system at the studio does not change during activation of the backup profile; all the changes take place inside the two RF STL link units, nothing external needs to be configured.

In the next scenario, the backup path has at least as much bandwidth available as the primary path so that all services can fail over to the backup path without the need to use different service profiles or individual service path mapping for reconfiguration. The most common situation here is when you have a licensed RF STL primary path along with a wideband IP path for backup. In this case, all services can failover from the primary path to the backup path. The primary path can be either RF or IP.

Figure 6.4-21 shows a scenario where RF is the primary path but a wideband IP path is available. All services are backed up to IP in case of an RF path failure. This could also be the reverse, where all services are normally on IP and switch to RF upon IP failure. The criteria for determination of when to switch services away from the RF STL path can be the same as the previous scenario of RF only with IP return path—receive SNR. The criteria for switching services off the IP path can be multidimensional, but as a practical matter, loss of service packets at the IP receiver is usually sufficient.

In another scenario, the primary path (RF or IP) has enough bandwidth to carry all the desired services during normal operation, but the backup path has more limited bandwidth and as such is able to carry only the most critical services when the primary path is down. In this case, if the primary path is RF, then the critical services are set to use RF, with failover to IP. If the IP path has sufficient bandwidth to support the services on at all times, then critical services can be set to use RF and IP simultaneously. The noncritical services are set to use RF only. Alternatively if the primary path is IP, then the critical services are set to use IP, with failover to RF. The noncritical services are set to use IP only. Figure 6.4-22 shows a scenario where the primary path is IP.

The RF link has less bandwidth than the IP, so only the most critical services are carried during backup. The two main audio programs are carried in their normal configuration but the two auxiliary audio programs and the Ethernet data stream are dropped. This could also happen in reverse, where RF is the primary path and only the most critical services are backed up to IP during RF path failure.

The transmission robustness of the audio and data services transported over an RF STL system can be greatly enhanced whenever an IP path is available between the studio and the transmitter site. Using both the RF and IP paths allows the STL system to adapt to

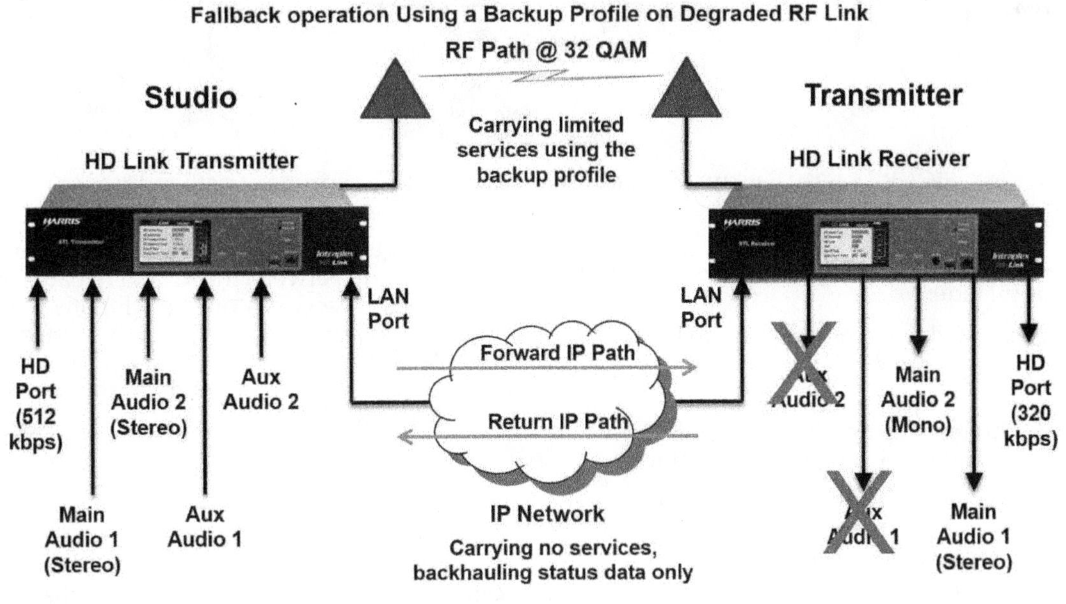

FIGURE 6.4-20 Fallback operation using a backup profile on degraded RF link

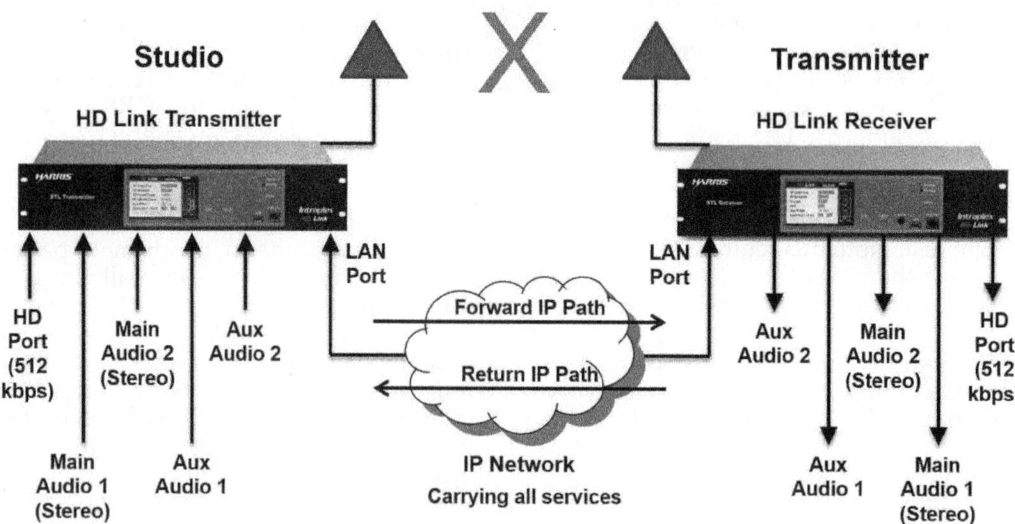

FIGURE 6.4-21 Backup of all services during primary path failure

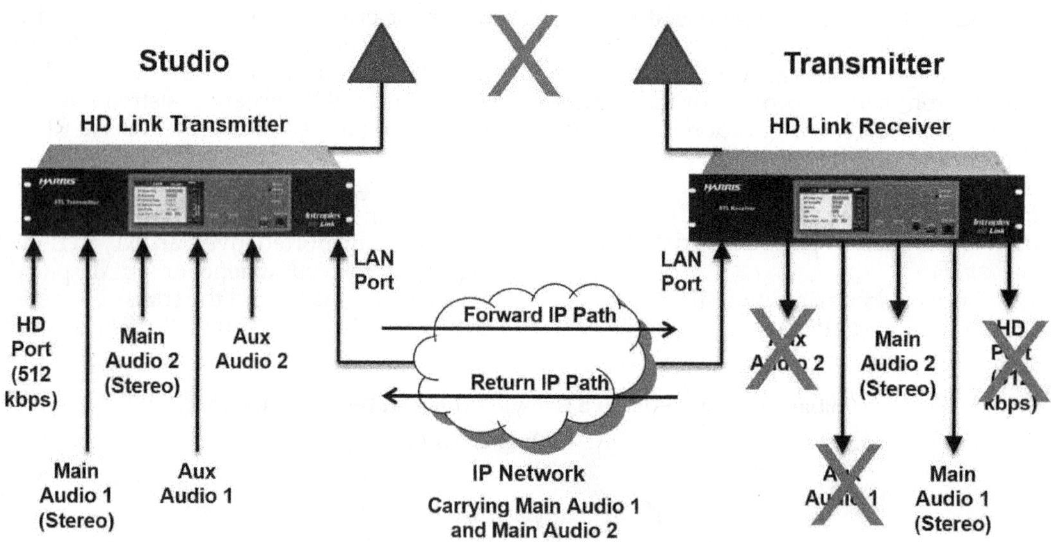

FIGURE 6.4-22 Backup of critical services only during primary path failure

changing network conditions on either path. Adaptation scenarios include:

- Automatic path failover from primary to backup, where the primary can be either the RF or the IP path.
- Support of different capabilities on primary and backup paths using main and backup profiles.
- Adaptive RF modem attributes to compensate for a degraded RF path.
- Independent path mapping for each of the supported services.

Taken together, these adaptations can provide automatic backup between the RF and IP paths or automatic scale back of services on the RF path when environmental or other factors degrade the performance of the RF STL signal at the transmitter. They also enable the STL system to provide load sharing between the RF and IP paths.

The IP-Based STL

For many years synchronous technologies have been considered by broadcasters as the backbone technologies for broadcast audio delivery. Balanced analog copper circuits, microwave, and synchronous-based systems have been the traditional choice for audio contribution and distribution. These synchronous links have been readily and affordably available from all major service providers. The suitability of such links to the application of delivering broadcast grade audio is due to the fact that they offer dedicated, reliable point-to-point and bidirectional communication at guaranteed data and error rates.

Although not without a distinct subset of issues and problems, IP technology is now challenging these legacy synchronous links as the benefits over existing synchronous networks are increasingly proving too persuasive for broadcasters to ignore. It is clear to see that the use of IP networks for audio delivery enables broadcasters both to leverage their existing infrastructure while achieving greater flexibility in terms of content sharing and network configuration. IP is fast replacing traditional RF and synchronous type links (E1, T1, ISDN) for both STLs and studio-to-studio communications.

Today the use of IP and various Ethernet standards to transmit audio from point to point is an important capability in broadcast and the use of IP networks for audio transmission is increasing. If the IP network being used is some kind of private network between facilities (virtual private networks [VPNs] or multi-protocol label switching [MPLS], etc.), the network can be managed—established routes can be defined between the various nodes on the network, and quality of service (QoS) controls can be enforced that will regulate the traffic on any network segment, with (presumably) the highest priority given to the audio packets, so that they may arrive at the destination on time and in sequence. However, these types of "managed" networks can be expensive to deploy and maintain, particularly when sending audio over some distance, such as between cities or to different parts of a region or country.

While a managed IP network presents few challenges to a broadcast engineer, there are some broadcasters that cannot afford this level of dedicated service and, indeed for applications such as ad hoc remote broadcasts (a typical domain for ISDN), it is simply not practicable. For these types of broadcast links, the public Internet offers an inexpensive means to deliver audio. The benefits of the Internet as an audio transport backbone are many and some are obvious:

- It is available locally nearly everywhere worldwide.
- The Internet payload capacity is increasing yearly.
- The Internet allows communication across virtually any distance.
- It can be leased for short periods of time.
- It is also self-healing to a degree, which offers some protection against the failure of a single segment in a communications link.

However, the public Internet is not without its challenges, primarily the fact that it is "unmanaged." The lack of IP network management such as QoS mechanisms, traffic balancing, or any intercommunication between packet source and packet destination often results in unreliable and interrupted data streams with large variations of data transmission quality. The types of issues encountered on unmanaged IP networks include:

- Jitter: A high and variable rate of packet delay caused by bottlenecks on backbone routers, particularly if they operate very close to their overload limit.
- Packet Loss: Loss of single audio packets due to network inconsistencies.
- Loss of Packet Cluster: Loss of a cluster of audio packets, sometimes called a "burst" loss.
- LOC: Loss of connection, temporary network failure.
- Sequence Error: Packets arrive out of sequence due to variations in network latency.

The definition of an unmanaged network is that it operates without management. On unmanaged networks no corrective action will take place should the traffic on any given segment exceed the nominal load capacity. The unmanaged network will react with an overload condition and start to drop packets due to buffer overloads on router queues, and the packet latency will increase significantly. Conditions like this can be seen as the worst case scenario for real-time media content streamed over the network.

The audio over IP (AoIP) industry has proposed several mechanisms to cope with these conditions when a network fails after reaching a certain level of error rate in the transmission chain. The intelligence and control provided by protocols such as QoS can help to assure a secure transmission if the network supports these mechanisms. However, QoS is not available on the public Internet.

Recently, a number of technologies have emerged that aim to enable broadcast audio via the open Internet, utilizing dynamic parameter manipulations:

- Dynamic jitter buffer: This technology is the most obvious approach for coping with increasing delay jitter. This technology dynamically expands or reduces the size of the delay jitter buffer. As a result, the overall transmission delay will vary significantly throughout the broadcast.
- Bit rate adaption: This technology dynamically changes the data rate of the signal, reducing the packet rate. Should the bandwidth available be restricted, the audio quality will be scaled back in order to maintain continuity of service. The implications of this approach are that the audio quality may vary considerably during the transmission and a duplex connection is required.
- Packet resend: There are different approaches to resending packets available on the market. The most obvious transmission protocol for this is

TCP/IP. TCP/IP always asks for packets to be resent and therefore will always require a duplex link. As a result, the shortest possible audio delay with a TCP/IP connection will always be significantly higher than that of a UDP connection. There are also some proprietary protocols that enable packets to be resent using the UDP protocol.

In order to effectively counteract the drawbacks of what is otherwise a very attractive and cost-effective network for professional audio transmission, the sending of broadcast quality audio over public Internet connections will require a different approach.

Redundant IP Packet Streaming

Due to the nature of an IP network, a sender is not physically connected to a data receiver as would typically be the case in synchronized networks. The lack of a direct connection results in situations where two program streams could be directed on different routes from the same source location to the same destination. One stream could travel via route A, and the other through a very different route B. Even individual packets from the same program stream could be sent via different routes.

Recognizing the fact that two data streams from the same source to the same destination can show significantly different performance characteristics, it follows that by using both routes A and B for each of the program streams could increase the reliability of both program feeds. Figure 6.4-23 shows the principle of a real-world configuration.

The inherent nature of the redundant streaming approach means that twice the content is sent, and therefore twice the bandwidth will be required. However, given that bandwidth is becoming an increasingly affordable commodity this is unlikely to represent a true disadvantage for the majority of broadcasters.

Redundant packet streaming technology is an inherently scalable technology, meaning that it has applications applicable to major market broadcasters and small market broadcasters alike. It is also suitably generic, so it can be applied to any audio coding algorithm or none (linear PCM) or to any physical transport topology that is presented as IP. This makes it suitable for use on satellite links, and licensed or unlicensed microwave links. It can essentially involve anything from an expensive managed IP network through to an inexpensive, consumer-grade xDSL link or wireless 3G/4G connection. As such, the redundant streaming approach by itself plays a significant role in increasing the reliability, quality, and consistency of an audio link over the public Internet.

Additional measures can be taken to safeguard the delivery of the audio content. For example, statistical diversities can be attributed to each stream to ensure greater variation in the routes taken. More advanced resequencing techniques can be put in place on the decoder side to enable seamless playout with an unwavering level of delay. These measures elevate the increased reliability of redundant streaming to a level that is comparable (and often superior) to the Five Nines service offered by traditional synchronous links.

The following are various sample configurations of redundant streaming, specific to the types of networks used:

- Redundant managed networks. Managed networks are usually far less susceptible to packet losses, sequence errors or traffic overload conditions, but they are not immune to a complete failure on an individual link. Utilizing redundant streaming ensures permanent redundancy without the need to switch between two sources at the receiver decoder. Figure 6.4-24 shows the principle on a redundant IP network with two access/ingress points and two exit/egress points.

  The specific encoder generates the redundant streams, which are routed through the network and each reach the exit/egress points having traversed different paths. Given their divergent routes, typically, the arrival times of stream A and stream B will differ. In order to maintain playout, the receive buffer size must be larger than the time difference of stream A and B.
- Redundant network with managed network main feed and XDSL redundancy. This example describes a network topology that transmits the audio program chiefly on a managed network with an inexpensive DSL link deployed as a redundant connection. As in the last example, two identical

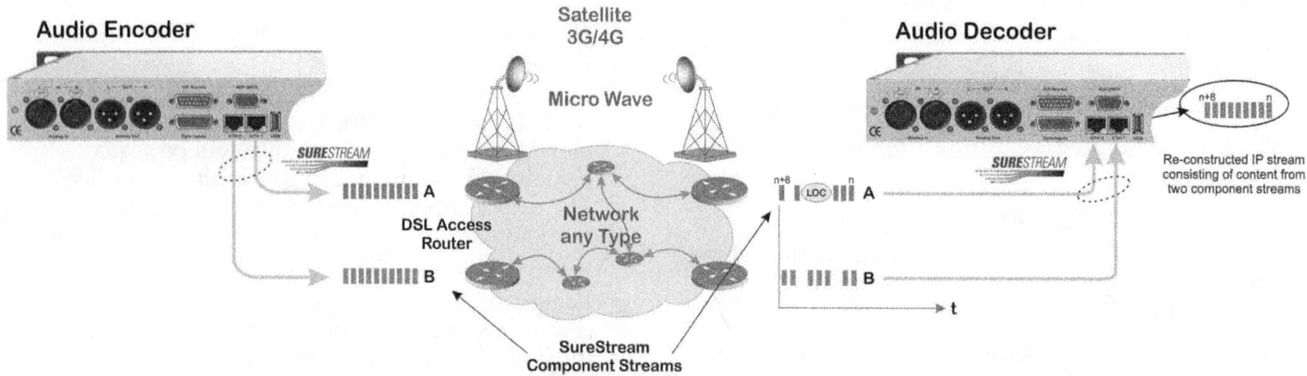

FIGURE 6.4-23 Redundant streams could take very different routes through the public Internet

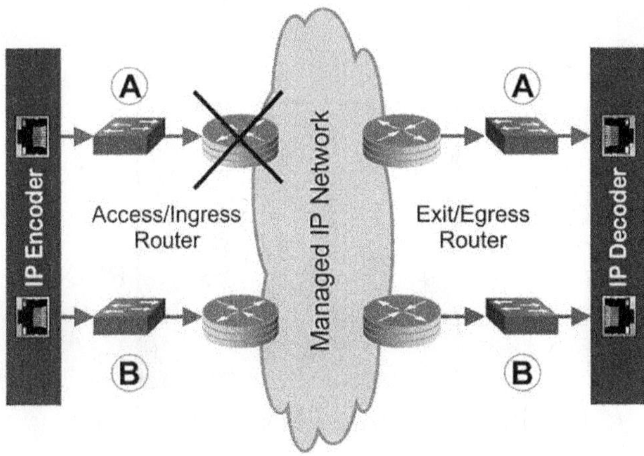

FIGURE 6.4-24 Typical redundant managed network

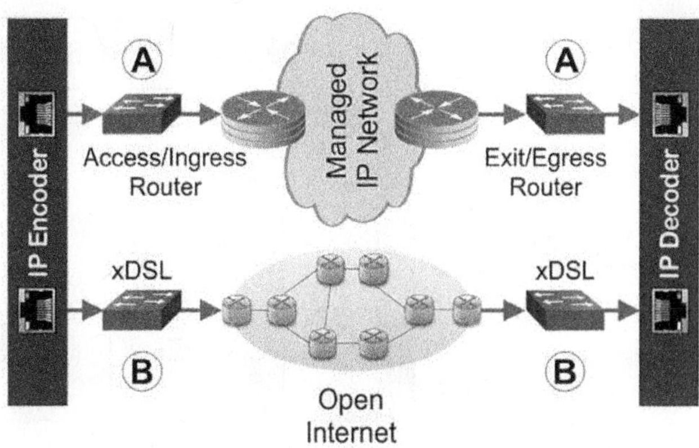

FIGURE 6.4-25 Managed network as main, XDSL as alternate

component streams are generated. However, in this instance the characteristics of these links are significantly different. The managed network has a much lower delay jitter and statistically the higher availability while the DSL link may suffer from congestion conditions and higher delay jitter figures. Utilizing both links in a redundant streaming application will create a slightly compromised performance for the overall link in terms of average latency when compared to the managed link used in isolation. Nevertheless, the redundant stream protects the broadcast link from a complete failure and any burst errors on the link, and therefore increases the reliability of the link as a whole. Figure 6.4-25 below shows the principle of this type of network design. As described earlier, the loss of one complete component stream (LOC) or the loss of packets of either of the streams does not affect the seamless audio play out. The receive buffer size must be larger than the time difference of stream A and B.

- Redundant packet streaming via dual ADSL. Using the open Internet for broadcast-grade audio transmissions presents the most challenges. The Internet was not designed for real-time audio distribution using RTP over UDP and, as detailed previously, its network behavior is therefore not wholly appropriate for it. Although all protocols required for audio transmission are supported, the Internet does not provide any protection mechanism for UDP streams. Redundant packet streaming, as seen below in Figure 6.4.26, has proven itself a reliable solution in this environment.

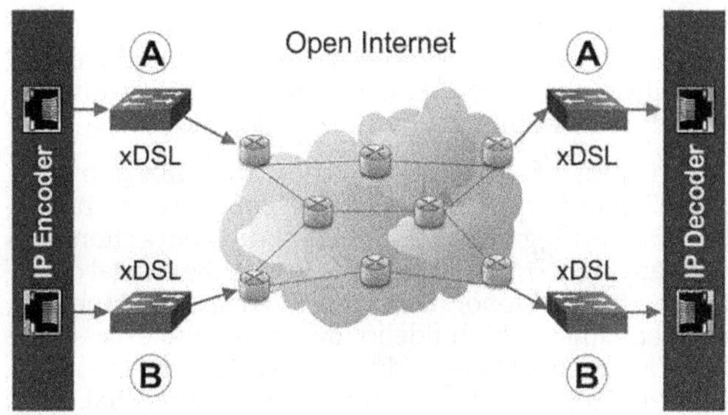

FIGURE 6.4-26 Redundant streaming over the open Internet

Transmission of FM Multiplex via IP

The total digital transmission path is capable of providing outstanding performance results and there are no reasons why an all-digital broadcast facility that provides superior quality broadcasting via either an MPX over AES, or MPX over IP connection, cannot exist today.

Why couldn't the digital audio processor, which already has the MPX encoder inside, be able to connect its digitally generated baseband signal directly to the digital modulator of the exciter? This would be analogous to the analog composite input on any exciter. Today, firmware is capable of operating at a very high sampling rate, as would be the case with 192 kHz. The economy of scale allows manufacturing of high speed sampling at a fraction of the cost.

Audio connectivity standards as defined by the Audio Engineering Society (AES) now offer interoperability up to 192 kHz sampling, and this is more than sufficient for transporting the FM-Stereo MPX signal. Connectivity is made available between the stereo generator's output of the audio processor to the input of the exciter by an AES connection and utilizing 192 kHz sampling. An example of this is shown in Figure 6.4-27.

This method provides the best possible interface to the exciter, and the performance benefits are significant. Aside from the obvious benefit of a complete digital transmission system, there are a couple of advantages this technique offers:

- Zero overshoots due to added emphasis, coding, or sample rate converters.
- Point-to-multipoint MPX connectivity. This is especially beneficial when the desire or need is to

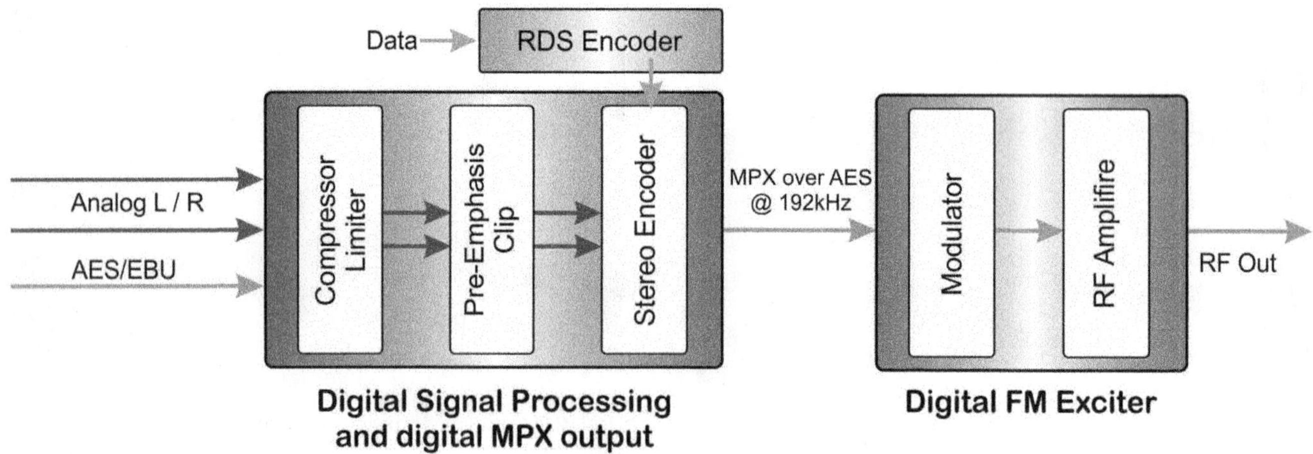

FIGURE 6.4-27 Full digital link to a transmitter site

utilize one master processing system, and have the ability to transport a mirror image of the composite output to numerous transmitter locations.

The advent of AoIP technologies, formalized by the AES67 standard, enables extreme flexibility for broadcasters with regards to common audio connections. As with any networked system, payload is the critical determining factor. Mono, stereo, and even multichannel signals (streams) of high fidelity do not require excessively large amounts of bandwidth, but even so, the challenge is not so much the technical aspects, but financial. Data links of this size are not inexpensive, and are not everywhere commonly available to be used for an STL link. Still, many stations will find the means by which this technique can be used within their budgetary parameters.

Two Approaches to Digital MPX

There are two approaches to the transmission of the MPX via digital means. One is by providing a backwards-compatible interface for an analog MPX signal; the analog signal is digitized in the MPX codec. The analog-to-digital converter (ADC) in the encoder is responsible for the digital signal quality, and the user trusts the encoder device with total control of the signal in the FM system. The converted MPX signal is then sent over an IP network to the far end.

In the second approach, the audio processor/stereo generator device is provisioned with an AES output with a 192 KHz sample rate. Once the digital MPX signal is generated the encoder will transmit it over an IP network. This signal chain provides a simplified interconnection to the exciter with the promise of excellent performance. Lab tests have shown that a digital MPX signal with 16-bit resolution is ample to deliver acceptable quality, but almost all manufacturers offer a 24-bit resolution.

Network Considerations for an IP-Based Multiplex STL

The data stream to be sent over an IP network is a digital representation of the MPX signal, sampled at 192 kHz and likely using 24-bit word length. It will be transmitted as RTP over UDP packets or encapsulated in an MPEG transport stream. Here are the major points to consider when planning to the deployment of digital MPX links over an IP network:

- Network capacity. An AES signal with 192 kHz Fs and a 24-bit sample size has a mean bit rate of 4608 kbps. For signals originating from 16-bit samples the bit rate is 3072 kbps. With a packet size of 2ms, the overall bit rate required from the network is ~4750 kbps for 24-bit and ~3260 kbps for a 16-bit MPX signal.
- Network type. The digital MPX signal is a robust signal by nature. Nevertheless it is the final product of the entire production chain and represents the station's sound, reliability, image, and technical quality. It is highly recommended to use managed networks only. The network should provide support Quality of Service (QoS) flags in order to allow prioritization of the audio stream.

As discussed earlier in this chapter, transmission of audio over an IP network is subject to errors associated with jitter, packet loss, complete loss of connection, and sequence errors. There are various protection mechanisms available to cope with these potential network errors: Delay jitter and sequence errors can be compensated on the decoder site by the de-jitter buffer. Packet loss and loss of packet clusters can be avoided by a utilizing an FEC scheme (with certain limitations). Adding an error concealment mechanism can also be of great benefit. If the MPX signal is affected by the transmission chain, the risk of overshoots increases. FEC and error concealment are recommended on links with limited bandwidths. A complete loss of connection can be avoided only by establishing at least one other fully complete and independent stream in order to build redundancy into the entire STL system.

A managed private IP network usually does not suffer from a large number of packet losses, sequence errors, or traffic overload conditions. But it is not immune to those problems, and managed networks can still experience a complete failure on an individual link. Utilizing redundant streaming ensures permanent redundancy without switching between two sources at the receiver decoder.

Due to the nature of the packetized network, a sender is not physically connected to a data receiver as would typically be the case in synchronized networks. The lack of a direct connection results in the capability to direct two or more program streams on different routes from the same source location to the same destination. One stream could travel via route A, and the other through a very different route B. Even individual packets from the same program stream could be sent via different routes.

Use of more than one private IP network makes it easy to route two streams via different routes to the same destination. Figure 6.4-28 illustrates the principle of a redundant IP network with two access/ingress points and two exit/egress points. Redundant IP streams are produced for the two network access/ingress points. The streams are routed through the network and each reaches the exit/egress points having traversed different paths. Given their divergent routes, typically, the arrival times of stream A and stream B will differ slightly. An algorithm recombines the playout stream from packets of both component streams. The redundant packets will be discarded.

Multipoint Distribution of Digital MPX

The nature of IP connectivity means that digital MPX signal from the stereo encoder can be sent to one transmitter or many transmitter sites. The central signal processing of the digital MPX signal allows for great consistency across all sites of a broadcast network. While local programs or those with many regional insertions may be less suitable for central processing than programs with the same content in a large region, even individual stations can benefit from the improved quality and signal control that digital MPX transport can offer. Advantages of centralized MPX generation and distribution include:

- Centralized audio processing allows the station sound is exactly the same from each transmitter site. Digital MPX distribution allows the FM system to benefit from a completely digital processing system; less sample rate conversion, and no overshoots.
- Only one stereo encoder (or a redundant pair) is necessary for each program, regardless of the number of transmitter sites.

Single-frequency networks could also benefit from this type of MPX distribution, because the modulation encoder would be identical on each outgoing feed, and thus on each transmitting node.

Multicast Distribution

Being able to deliver a single centrally processed signal to multiple locations over IP has other advantages as well. Multicast distribution can be used. A point-to-multipoint distribution can be done with a wired network or via a satellite. The distribution principle is the same: the centrally processed FM signal is fed to a central point and distributed unchanged to many locations. In IP networks, multicast is a bandwidth-saving method at the entry point, because the network multiplies the streams. Before considering such as system, make sure your network provider can support IP-multicast. Figure 6.4-29 shows the basic configuration.

System Redundancy and Self-Healing Networks

Whether for the transmission of television or radio, high-speed IP connectivity between a studio location and a transmitter sites provides many benefits. Ethernet and IP communications links are easily made to be self-healing. This not only provides an obvious advantage in keeping a broadcast station on the air, but it also frees those in charge from having to do the "healing" on an emergency basis. Systems can be designed and then configured to do automated switching of data pathways, letting those in charge know after the fact by way of e-mail (or other remote control functionality).

While what follows may seem like more of an IT topic, it's important to discuss it (at least conceptually) in order to form the basis of the idea of self-healing. At least two links will need to be established, and, after establishment of the links, there are multiple choices in how to configure the self-healing aspect:

- Load balancing at layer-2. Both ends of both links will terminate in a layer-2 switch. The

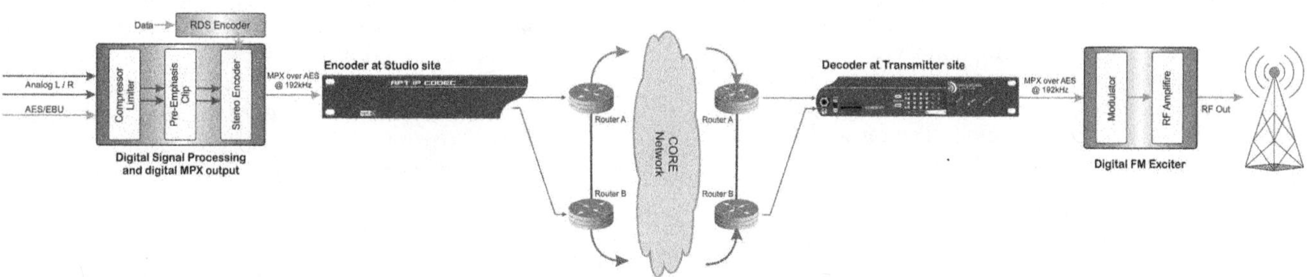

FIGURE 6.4-28 The principle of redundant MPX stream

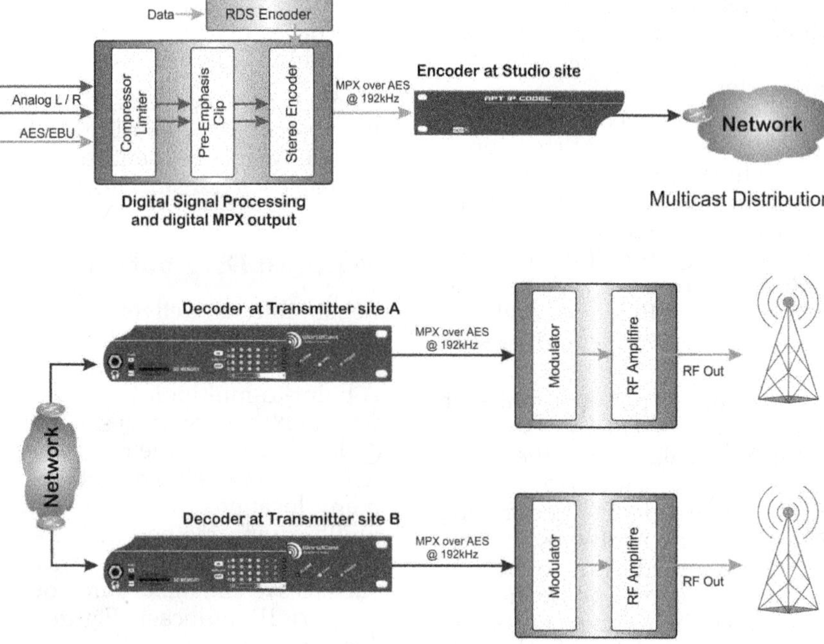

FIGURE 6.4-29 The principle of multicast distribution of IP networks

load-balancing function means that the layer-2 switch is configured to share both links for the passing back and forth of the Ethernet frames. In the event one fails, the layer-2 switch will start passing all of the traffic over the remaining link. Care must be taken not to "oversubscribe" the path, though, meaning that the normal amount of traffic passing over either link is less than 50% of its ultimate capability.

- Using a router to switch routes at layer-3. In this case, both ends of the link will terminate in a router (operating at layer-3). Normally, the router will pass all traffic over one of the two links. Configure the router to move all the traffic to the backup link in the event that the primary link fails. When the primary link comes back, the traffic will start routing over the primary once again. The primary link would be on one network segment, and the secondary link would be on another.
- Two separate networks, isolated from one another. This is a bit of a turn away from the self-healing aspect; however, because many AoIP codecs have two network ports, and can be configured to switch automatically themselves, this method should be considered. At the far end, some hosts would be on one network, and some on another. If one link were to fail, some hosts (those with only one NIC) would become unavailable to the far end, while others would remain available.

In each of the three cases, AoIP traffic will continue to flow (with perhaps a very brief gap) after one of the links fails. Using a remote control with simple network management protocol (SNMP) will allow those in charge to learn of the failure.

The first of the self-healing system configuration ideas we'll look at will make use of a licensed IP radio and a telco backup (see Figure 6.4-30).

One link providing IP connectivity in this case will be provided by a licensed IP radio operating in the 11 GHz band specified in Part 101. High-speed IP connectivity must be available between the remote site and the studio facility. If the local telco provider is AT&T, the service is called ASE (ATT Switched Ethernet); if the provider is Verizon, the service is called Ethernet Virtual Private LAN; for Frontier, it is called Ethernet Local Area Network; and for CenturyLink, it is called "Plain" Old Metro Ethernet.

Deciding which of the two links becomes the "primary" and which is the "secondary" may take some doing. In an ideal world, both will be very reliable; but in reality, one will likely be better than the other. Clearly the better one should be the primary.

The second of the self-healing configurations is seen in Figure 6.4-31. This is a variation of the first method; the second link is now provided by an unlicensed radio link (operating in one of the ISM (industrial, scientific, medicine) bands. From an economic standpoint, it might make more sense to use an unlicensed radio system, but there are a couple of aspects to consider. First (and this is not to cast aspersions on the quality of the ISM radios) many of them are very inexpensive, and meant to be "throw-aways." They are simply not as rugged as "carrier grade" radios. Second, because they work in unlicensed spectrum, they are subject to interference, and the user has no recourse, except changing channels or abandoning the link.

The third self-healing configuration (Figure 6.4-32) is simple variation on the first two; however, in this

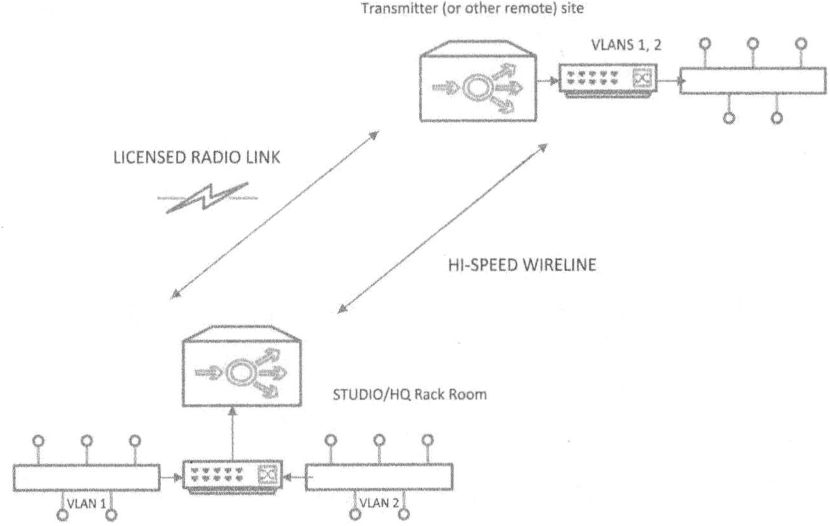

FIGURE 6.4-30 Self-healing network using both radio and wireline links

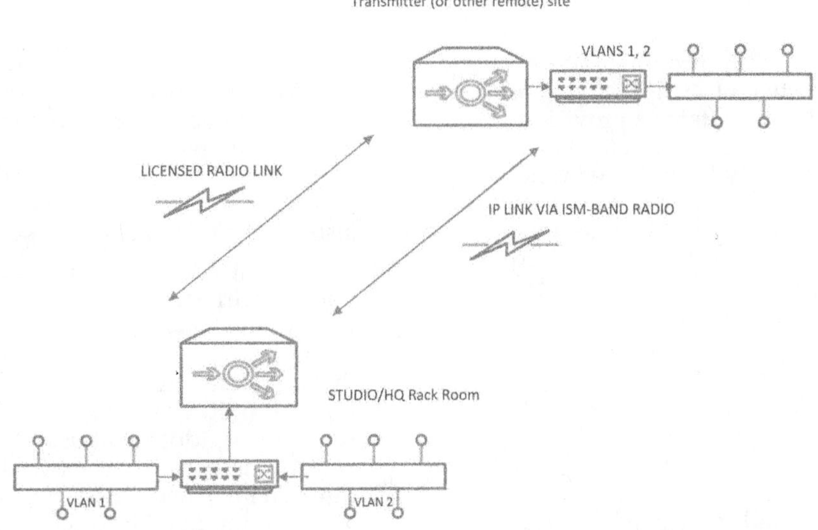

FIGURE 6.4-31 Self-healing network using licensed and unlicensed radio links

case, there are three sites, instead of two. An example of this in practice would be one studio location, a main transmitter site (location A) and an auxiliary transmitter site at location B. The studio to A link would be the normal route; in the event that failed, traffic would flow from the studio, to site B, then to site A. In any case, both sites have a redundant path.

Needed Features in an IP-Based Radio System

Many manufacturers are now making microwave radios because there is a large market for "backhaul" from cellular telephone sites. That's an advantage for broadcasters because as more radio manufacturer fight for market share, pricing gets better. Here are some of the features to consider when specifying a radio system:

- Layer-2 (Ethernet) interface. This radio link acts as if it were a simple Ethernet cable. (This goes for the metro Ethernet connections discussed earlier as well.) In addition to their Ethernet interfaces, some radios still have "auxiliary" TDM interfaces that can carry one or more T1s; if some time is needed to transition between TDM and IP, this is something to consider.
- VLAN priority support. It's quite likely that the radio link or the metro Ethernet will be used as a VLAN trunk because it's a good idea to separate AoIP traffic from all other traffic to and from the other end. It's typical that a link such as this will support nearly 100 mbps throughput (or more), so anyone downloading a manual (as an example) on the far end is not going to interrupt or even slow down AoIP traffic; however, at the very

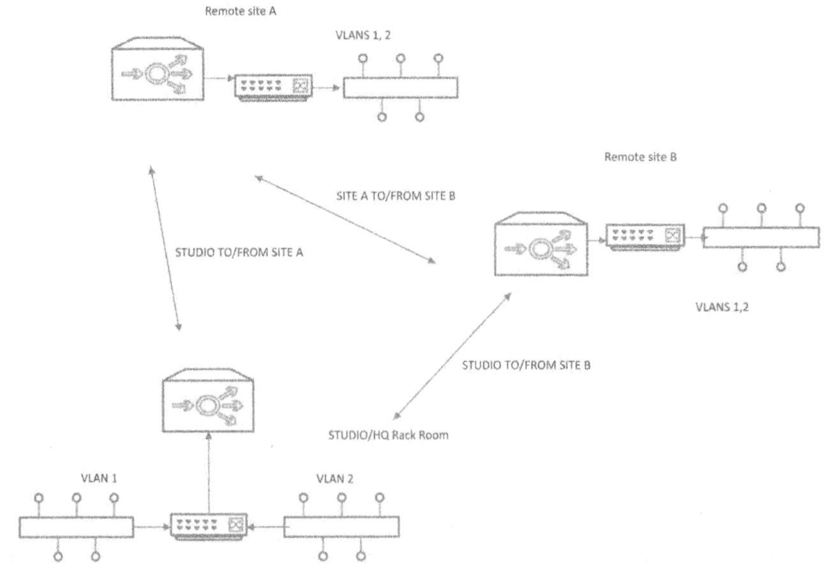

FIGURE 6.4-32 Self-healing network using round-robin connectivity among sites

least, consider it to be future-proofing. The 802.1Q standard defines a system of VLAN tagging for Ethernet frames and also contains a provision for a QoS prioritization scheme known as 802.1P, which indicates the priority level of the frame. The 802.1Q standard adds this information to the Ethernet header. The priority level values range from zero (best effort) to seven (highest). These values can be configured to prioritize different classes of traffic such as AoIP versus "everything else" you might use to and from the transmitter (or other) remote site. One more thing to keep in mind when building a system such as this: frames for the native VLAN are not tagged, and thus can't be prioritized; don't send AoIP traffic in the native VLAN.

- Adaptive modulation and coding (AMC). This is a feature of many radio systems, whereby the modulation, coding, and other signal and protocol parameters are adjusted to match the conditions on the radio link. For example, if the link fades, the modulation rate is adjusted to lower the throughput of the system. The idea is that higher-priority traffic will still pass through, even in adverse conditions.
- Remote access. Radio links such as the ones we've been talking about are designed to operate in a networked IT environment, and this remote access is a given. They will support have full http access to all configurations and parameters. SNMP support is a standard feature of this type of system. Having an additional "backup" network at the remote end is something to seriously consider, for two reasons: First, if the radio link fails, and it provides all the connectivity, then you'll be completely blind as to what is going on at the far end. Second, even if the radio link is simply being aimed by a rigging crew, it's really tough to rely on it to tell you its own signal strength at the remote end. With that backup network in place you can log in to the remote receiver and see what is going on. Nothing fancy is needed—perhaps the old DSL line you've been using up until this point. The scenarios depicted in Figures 6.4-30 and 6.4-31 will also allow you to have "backup" access, of course.
- Indoor unit/outdoor unit. Many of the older microwave systems use a very expensive waveguide between the radio and the antenna up on the roof or tower. Today it's more likely that a system will comprise an indoor unit and an outdoor unit. The indoor unit can live in the rack room; the outdoor unit usually mounts right on the back of the antenna. The two devices communicate through intermediate RF frequencies that can be accommodated through a much more readily available coax, such as LMR-400. The outdoor unit is also powered over the same cable. It's typical of ISM radios to be mounted directly on their dishes as well, and to be connected directly via a category 5 or 6 cable, which also carries the power. This greatly simplifies installation. Use direct-burial category cable for this type of connection.

DIGITAL RADIO SYSTEMS

Today, analog microwave is viewed as a "bandwidth hog." Analog STLs are, by today's standards, archaic and nonefficient users of valuable spectrum. Microwave transport using modern digital modulation techniques mitigates this problem. It is not only possible but very much in practice today to transport multiple video, audio, and data services in a single efficient manner. Perhaps the greatest benefit of IP transmission is that is agnostic to the payload; any data formatted for IP transits the network in the same way and becomes available in its original format on the opposite end of the link. Whether a digital radio system has a TDM or IP interface, its method of modulation and demodulation remains the same.

Modulation Techniques

The FM system is one in which the frequency of the carrier is caused to vary in accordance with some specified information-carrying signal. For the simple case of a sinusoidal modulating signal at frequency f_m, the corresponding frequency modulated sign is given by

$$f(t) = a \cos \omega_m t,$$

where a is the fixed amplitude of the modulated signal and $\omega_m (\omega_m = 2\varpi f_m)$ is the rate of modulation of the modulating signal.

Simply stated, there is a fixed amplitude sine wave of varying angular frequency ω_m. In the FM system all of the information is encoded as a function of the continuously varying angular frequency ω_m.

By contrast, in digital systems, the amplitude, frequency, and phase properties of the RF are quantized by the modulating signal. Digital implies a fixed set of discrete values. A digital radio waveform, then, can assume one of a discrete set of amplitude levels, frequencies, or phases as a result of the modulating signal.

Frequency Shift Keying

Frequency shift keying (FSK) consists of shifting the frequency of a sinusoidal carrier from a mark frequency (corresponding to a binary 1) to a space frequency (corresponding to a binary 0). This can be expressed as:

$$f_{c1}(t) = a \cos \omega_1 t$$

$$f_{c0}(t) = a \cos \omega_2 t,$$

where a is the fixed amplitude of the modulated signal, a binary one corresponds to frequency $f1$, and a binary zero to frequency $f2$. An alternative representation of an FSK waveform consists of letting $f1 = f_c - \Delta f$ and $f2 = f_c + \Delta f$. The two frequencies then differ by $2\Delta f$ hertz and thus we can write:

$$f_c(t) = a \cos (\omega_c \pm \Delta\omega)t,$$

where $\Delta\omega = 2\varpi\Delta f$. The frequency then deviates $\pm\Delta f$ about f_c. Δf is commonly called the frequency deviation.

Phase Shift Keying

Phase shift keying (PSK) consists of shifting the phase of a sinusoidal carrier 180 degrees with a unipolar binary signal. This can be expressed as:

$$f_c(t) = \pm a \cos \omega_c t$$

Here a binary 1 in the baseband binary stream corresponds to positive polarity and a binary 0 to negative polarity. This modulation is commonly called binary phase shift keying (BPSK).

For BPSK, the values of 0 degrees and 180 degrees are typically used. These values are used to maximize (or to separate) the decision distance. There is no reason why 45 degrees cannot be assigned as the binary 1 and 225 degrees as the binary 0, as long as the distance between the two states is maximized.

Continuing along this line, instead of using two phase states, four can be used, each separated by 90 degrees. The binary values to each of the four phase states can now be assigned as: 0° = 0,1; 90° = 0,0; 180° = 1,1; and 270° = 1,0. This type of multi-symbol signaling is commonly called quaternary phase shift keying (QPSK) modulation or 4-PSK modulation. It can be expressed as:

$$S_i(t) = a_i \cos \omega_c t + b_i \sin \omega_c t$$

A simplified block diagram of a QPSK modulator is shown in Figure 6.4-33.

It is useful to represent the signal of $S_i(t)$ in a two-dimensional diagram locating the various points (a_i, b_i). The horizontal axis corresponding to the location of a_i is called the in-phase axis; the vertical axis,

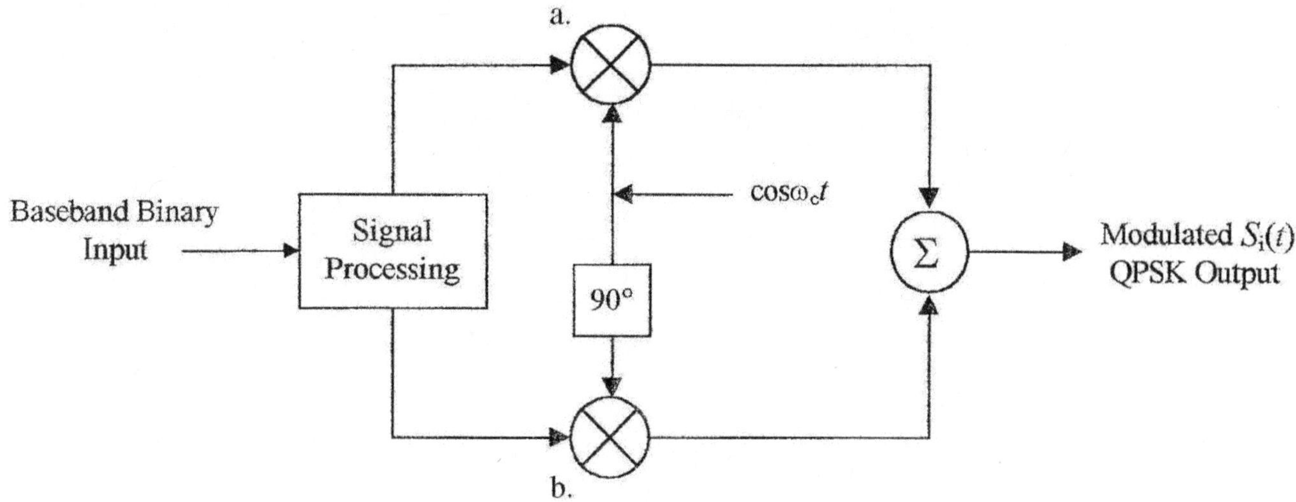

FIGURE 6.4-33 Diagram of QPSK modulator [2]

along which b_i is located, is called the quadrature axis. The four signals of $S_i(t)$ assigned earlier then appear as shown in Figure 6.4-34(a). The points are said to represent a signal constellation, or as it is sometimes called, a signal state diagram.

More general types of multi-symbol signaling schemes may be generated by letting ai and bi take on multiple values themselves. The resultant signals are called quadrature amplitude modulation (QAM) signals. These signals may be interpreted as having multilevel amplitude modulation applied independently on each of the quadrature carriers. A simplified block diagram of a QAM modulator is shown in Figure 6.4-35. The output of such a modulator can be expressed as:

$$S_i(t) = r_i \cos(\omega_c t\ \theta_i),$$

where the amplitude r_i and the phase angle θ_i are given by the appropriate combinations of (a_i, b_i). The signal constellations for 4-QAM, 16-QAM, and 64-QAM modulation are shown in Figure 6.4-34.

If the constellations shown in Figure 6.4-34 were measured with a vector signal analyzer (VSA) they would demonstrate nearly perfect vector alignment. This would be representative of maximum signal-to-noise, low-phase noise, good error vector magnitude (EVM), and modulation error ratio (MER).

Additional constellations are shown in Figure 6.4-36. The tool used for these complex measurements is a VSA. This is a tool that is rapidly replacing the swept-tuned spectrum analyzer as a design and diagnostics tool by RF engineers. The ability of the VSA to time-capture and record these complex waveforms allows for the in-depth time domain and frequency spectrum analysis. These analyzers combine high speed digital signal and analog-to-digital conversion processes, super heterodyne frequency conversion, and advanced time domain analysis.

In each part of Figure 6.4-36, the upper left quadrant is a vector display of a QAM modulated signal's I

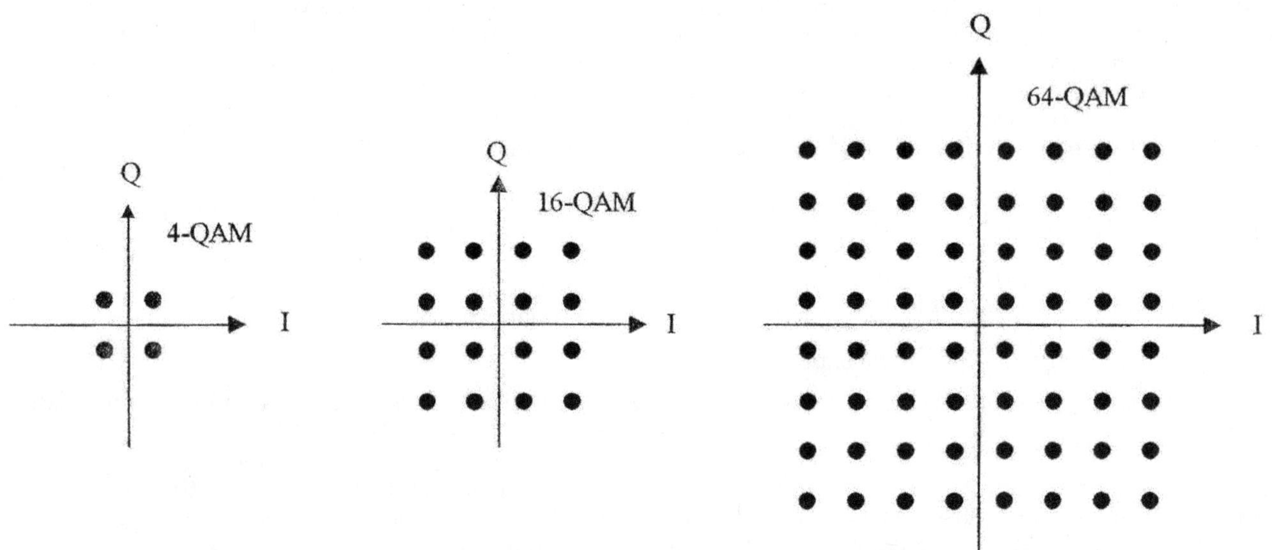

FIGURE 6.4-34 Signal constellations for (a) 4-QAM (identical to QPSK), (b) 16-QAM, and (c) 64-QAM [2]

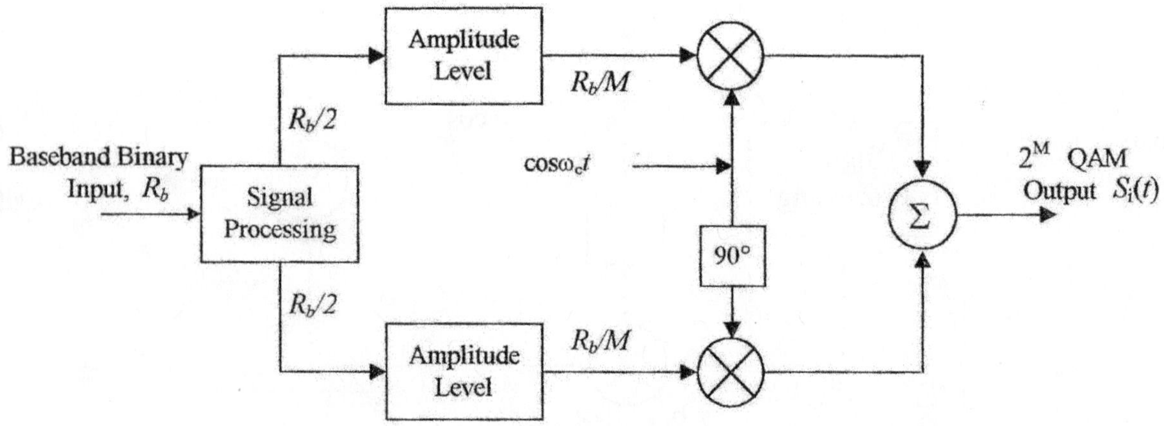

FIGURE 6.4-35 Diagram of QAM modulator [2]

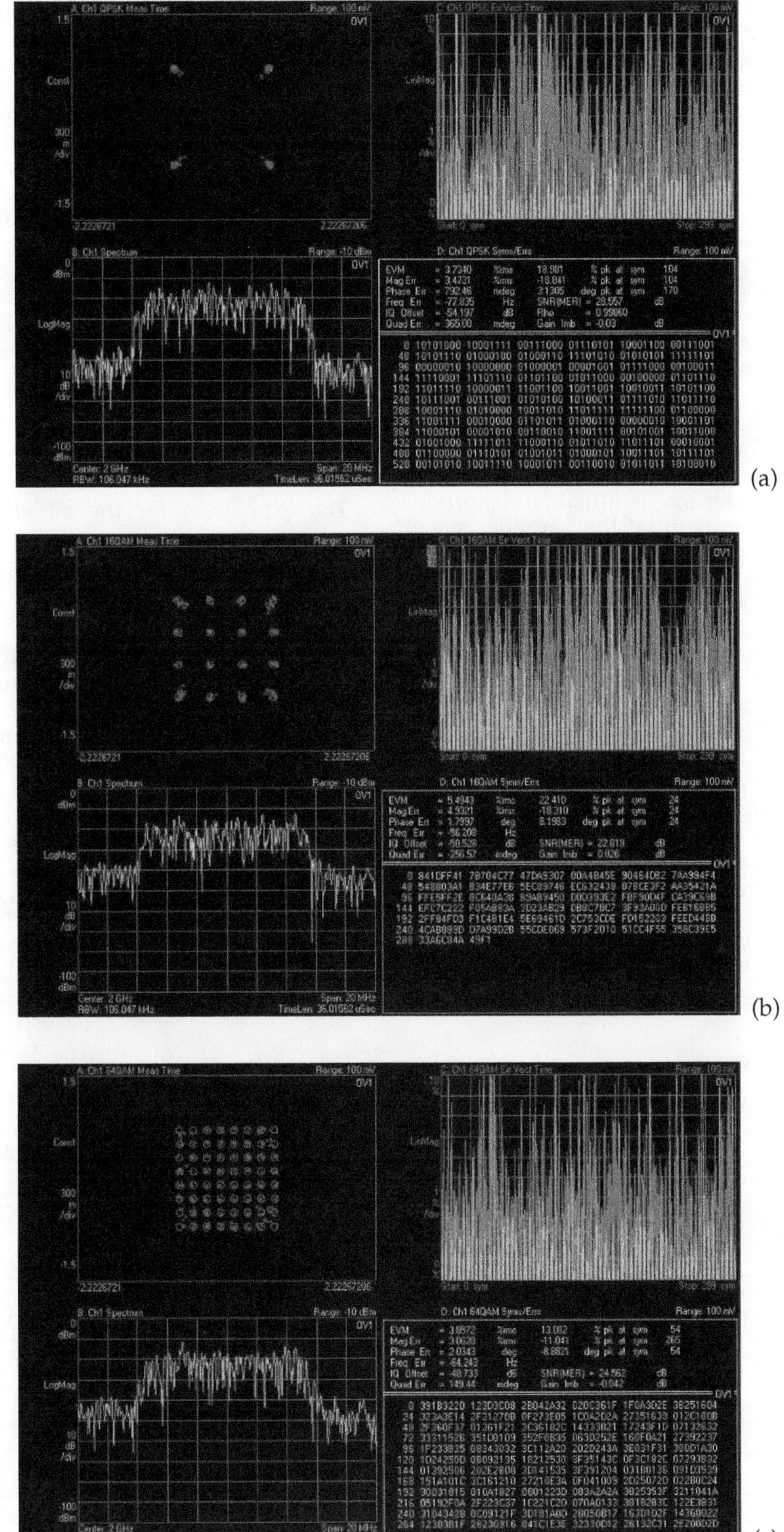

FIGURE 6.4-36 (a) Diagram of 4-QAM (QPSK) signal showing signal distortion, (b) diagram of 16-QAM signal showing signal distortion, and (c) diagram of 64-QAM signal showing signal distortion

(in-phase) and Q (quadrature) magnitude components. This is also referred to as the constellation display. The display in the lower left quadrant is representative of the swept-tuned frequency spectrum of the modulated input signal under test. The upper right quadrant represents the time component of the QAM signal. Digital modulation is three-dimensional; the IQ constellation represents two of these dimensions, the time domain being the third dimension. Last, at the lower right, is the symbol table. This represents the error summary data for all of the analyzed bits represented below the data table.

EVM analysis is a useful tool for transmitter and receiver signal chain optimization and allows prediction of dynamic system performance.

Notice that in all cases shown in Figure 6.4-36, a large percentage of the symbols are falling outside the appropriate vector alignment. This is representative of noise due to poor path performance and radio distortions (such as third-order intermodulation distortion, also called spectral regrowth) at the amplifier stage. This analysis shows poor pedestal definition in the spectrum display due to poor C/N; note the blurred constellation points, and the pedestal being only a few dB above the noise floor. Depending on where these measurements were taken this could represent poor amplifier performance due to oversaturation at the amplifier input (nonlinear operation) or, if at the receiver IF, this could represent poor path performance (low signal level or multipath distortion) or signal gain error in the down-conversion stage. Frequency errors could amount to signal distortion, however, they would not account for the system noise floor as viewed in this spectrum display.

After careful examination of the PSK and QAM signal constellations, one begins to appreciate the complexity required in a modem that can resolve the amplitude and phase information in multi-symbol signaling schemes. One has to be more impressed with the modem that can resolve this information when the digital radio system is subject to external noise factors. To get a feel of how different digital modulations compare in the presence of thermal noise we can look at the bit energy-to-noise density ratio, E_b/N_o, where E_b is the energy per bit and N_o is the single-sided noise spectral density. E_b/N_o can be expressed as the ratio of receive signal level (RSL) to the bit rate; in terms of log arithmetic, this can be written as:

$$E_b = RSL_{dBm} - 10\log(\text{bit rate}).$$

The single-sided noise spectral density of a perfect receiver in a 1 Hz bandwidth with a noise figure (NF) can be expressed as:

The expression for the bit energy-to-noise density ratio is:

$$N_0 = -174\text{dBm} + NF_{dB}$$

The expression for the bit energy-to-noise density ratio is:

$$E_b/N_0 = (RSL_{dBm} - 10\log(\text{bit rate})) - (-174\text{ dBm} + NF_{dB})$$

$$E_b/N_0 = (RSL_{dBm} - 10\log(\text{bit rate})) + 174\text{ dBm} - NF_{dB})$$

This expression shows that as RSL increases, so does the E_b/N_o. In other words, the greater that E_b/N_o needs to be, the less sensitive the receiver will be. This will be an important factor when determining the type of modulation to employ when converting existing analog links to digital.

Table 6.4-2 shows the E_b/N_o ratios based on a BER of 10^{-6} for some common digital modulation formats and their bandwidth efficiencies based on the Nyquist bandwidth (bandwidth numerically equal to the bit rate). The parameters listed in Table 6.4-2 are idealistic because the only source of errors is due to thermal noise in the receiver. No modulation implementation loss is considered.

TABLE 6.4-2
E_b/N_o at a BER of 10^{-6} for various modulation schemes (without FEC)

Modulation	E_b/N_o (dB)	S/N (dB)	Nyquist Bandwidth
2-state FSK with discriminator detection	13.4	13.4	B
3-state FSK (duo-binary)	15.9	15.9	B
4-state FSK	20.1	23.1	B/2
2-state PSK with coherent detection	10.5	10.5	B
4-state PSK with coherent detection	10.5	13.5	B/2
8-state PSK with coherent detection	14.0	18.8	B/3
16-state PSK with coherent detection	18.4	24.4	B/4
16-QAM with coherent detection	10.%	12.8	B/4
32-QAM with coherent detection	12.76	15.24	B/5
64-QAM with coherent detection	17.79	20.46	B/6
128-QAM with coherent detection	20.96	24.04	B/7
256-QAM with coherent detection	26.71	30.63	B/8

With the information given in Table 6.4-2 and from the E_b/N_o equation, the receiver threshold for a T3 (45 Mbps) 16-QAM radio system with a radio noise figure of 4 dB can be calculated.

$$RSL_{(min\ dBm)} = E_b/N_o + 10\log(\text{bit rate}) - 174\ \text{dBm} + NF_{dB}$$

$$RSL_{(min\ dBm)} = 17.0\ \text{dB} + 10\log(44.736 \times 10^6) - 174\ \text{dBm} + 4\ \text{dB}$$

$$RSL_{(min\ dBm)} = -76.5\ \text{dB}$$

For a 64-QAM system the $RSL_{(min\ dBM)} = -71.0$ dBm

For a 256-QAM system the $RSL_{(min\ dBM)} = -65.7$ dBm

The example clearly shows that for higher-order modulation schemes a greater E_b/N_o is required to enable the demodulator to extract the phase and amplitude information from the modulated signal. This inherent digital radio feature also causes the system gain to decrease as the modulation complexity increases. The previous example limits the distance between the transmitter and receiver as a function of the modulation scheme.

As a means to increase system gain, modem designers employ error-correcting codes (or forward error correction, FEC) to help improve performance. Error correcting coding usually requires redundancy (increasing the bit rate) and, therefore, poses a contradictory requirement to obtaining maximum spectral efficiency. There is no set standard for the amount of FEC a manufacturer will employ in a given modem. During the modem design phase, tradeoffs are made concerning BER performance, dispersive fade margin, spectral efficiency, and cost.

The Nyquist bandwidth, in this context, represents the minimum theoretical bandwidth required to transmit the modulated signal if a brick wall filter were used in the modulation process. As a practical matter, the actual bandwidth required is always greater than this.

For the applications being discussed in this chapter, satellite modems require more FEC than terrestrial modems to improve the system performance at threshold. Satellite links cover great distances and thermal noise becomes the dominant impairment. The composite bit rate at the output of a satellite modulator can be as much as 50% higher than the data rate into the modulator and as much as 10 dB of coding gain can be realized.

A well-designed terrestrial digital link is not fighting thermal noise as its major impairment. A modem designed for terrestrial use has to operate in a more hostile dispersive environment than the vacuum of space. The terrestrial modem designer has to balance coding gain, dispersive fade margin, spectral efficiency, and cost. Terrestrial modems typically employ 5%–15% of error-correction redundancy. Table 6.4-3 shows the effect that 6.7% of error-correction redundancy has on QAM modulation (compare these values with the corresponding values in Table 6.4-2).

Again, FEC is not a cure-all for digital radio performance. The amount of error-correction redundancy is up to the discretion of the modem designer and is just one of many factors that need to be optimized during the modem design.

The Transmitter

The transmitter in the digital radio link is the device that up-converts the modulated signal, amplifies it, and then delivers it to the antenna system for transmission into free space. A simplified block diagram of a digital radio transmitter is shown in Figure 6.4-37.

As shown in Figure 6.4-37, an IF signal (typically 70 MHz or 140 MHz) is first generated by the modulation portion of the system, and then up-converted by an image-reject mixer. This translation of the modulated IF signal up to the RF carrier frequency is made by a mixing operation or multiplication between two frequencies. The result of this mixing operation is a summation of the frequency components expressed as:

$$nF_{LO} \pm mF_{IF},$$

where both n and m = 1, 2, 3, . . .

The frequencies that are of most concern for the case when n = m = 1, or:

$$nF_{LO} \pm mF_{IF}, \text{ and}$$

$$F_{LO} \pm F_{(IF)}.$$

The desired frequency for transmission is:

$$F_c = F_{co} + F_{(IF)}.$$

$F_{LO} - F_{(IF)}$ is the difference frequency, which is commonly called the image frequency. The image frequency is typically suppressed 20–25 dB by the image rejection properties of the mixer.

TABLE 6.4-3
E_b/N_o at a BER of 10^{-6} for various modulation schemes (with FEC)

Modulation	E_b/N_o (dB)	S/N (dB>	Nyquist Bandwidth
16-QAM with coherent detection	7.75	9.62	B/4 × (1 + .067)
32-QAM with coherent detection	10.83	13.32	B/5 × (1 + .067)
64-QAM with coherent detection	16.37	19.03	B/6 × (1 + .067)
128-QAM with coherent detection	19.83	22.97	B/7 × (1 + .067)
256-QAM with coherent detection	25.90	29.83	B/8 × (1 + .067)

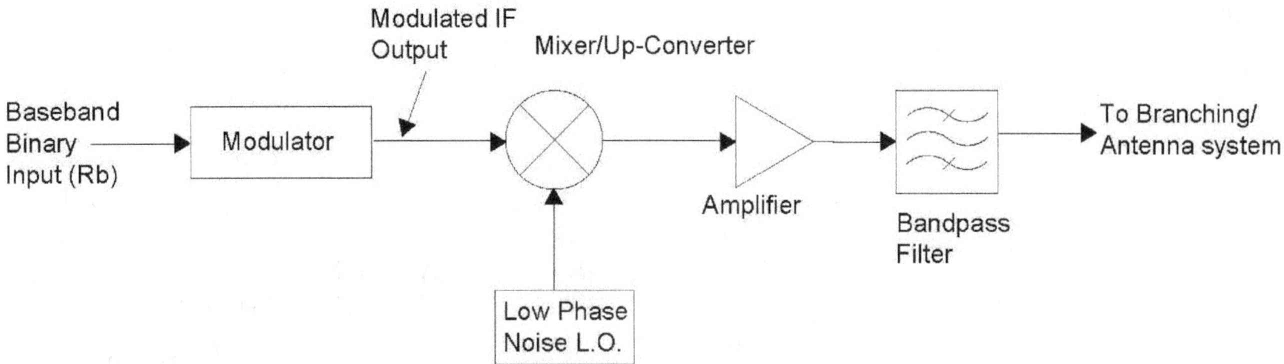

FIGURE 6.4-37 Simplified digital radio transmitter block diagram

In addition to the sum and difference frequencies, there will be some LO leakage present at the output port of the mixer. The LO leakage is attenuated approximately 20 dB by the mixer's LO to RF isolation. Some digital transmitter designs may employ a band pass filter to remove the image and LO leakage signals.

The local oscillator that drives the mixer in a digital radio is a high-stability, low-phase noise frequency source. The low-phase noise LO is required when implementing multi-symbol modulation. Phase noise (or short-term instability, or jitter) is seen near the carrier and causes degradation of BER. As can be seen in the signal constellation diagrams shown in Figure 6.4-36, LO phase noise causes the constellation points to jitter about. If the LO in the transmitter and receiver were an ideal frequency source the constellation points would not move at all. The more phase noise the LO induces onto the modulated IF signal, the more difficult signal recovery becomes. Figure 6.4-38 shows the empirically derived phase noise requirements for QPSK, 16-QAM, and 64-QAM modulation. Notice that as the modulation complexity increases the LO phase noise needs to decrease.

Continuing the journey through the transmitter, the signal out of the frequency translating mixer is input into a linear operating amplifier, which is an amplifier that is operating below its 1 dB gain compression point. The 1 dB gain compression (called G_{1dB}) is defined as the power gain where the amplifier's nonlinearity reduces the power gain by 1 dB over the small signal linear power gain. That is:

$$G_{1dB}(\text{dB}) = G_o(\text{dB}) - 1,$$

where $G_0(\text{dB})$ is the small signal linear power gain in decibels. Because the power gain is defined as:

$$G_p = P_{out} / P_{in}, \text{ or}$$

$$P_{out}(\text{dBm}) = G_p(\text{dB}) = P_{in}(\text{dBm}).$$

The output power at the 1 dB gain compression point, called P_{1dB}, is

$$P_{1dM}(\text{dBm}) = G_{1dM}(\text{dB}) = P_{in}(\text{dBm}).$$

Substituting $G_{1dB}(\text{dB}) = G_0(\text{dB}) - 1$ into this equation gives

$$P_{1dM}(\text{dBm}) = P_{in}(\text{dB}) = G_o(\text{dBm}) - 1.$$

The equation shows that the 1 dB gain compression point is that point at which the output power minus the input power in dBm is equal to the small signal power gain minus 1 dB. A typical plot of P_{out} versus P_{in} illustrating the 1 dB gain compression point is shown in Figure 6.4-39.

In the digital radio transmitter that employs linear modulation, distortion can be caused by the power amplifier operating near or beyond $G_{1dB}(\text{dB})$. This distortion is caused by intermodulation (IM) products that arise when two or more sinusoidal frequencies are applied to a nonlinear amplifier. The output of the amplifier will contain additional frequencies called intermodulation products. For example, if two sinusoidal signals:

$$v(t) = a \cos 2\varpi f_1 t + b \cos 2\varpi f_2 t$$

are applied to a nonlinear amplifier whose output voltage can be represented by the power series

$$v_0(t) = \alpha_1 v(t) + \alpha_1 v(t) + \alpha_3 v^3(t)$$

the output signal will contain frequency components at DC, f_1, f_2, $2f_1$, $2f_2$, $3f_1$, $3f_2$, $f_1 \pm f_2$, $2f_1 \pm f_2$, and $2f_2 \pm f_1$. The frequencies $2f_1$ and $2f_2$ are the second harmonics, $3f_1$ and $3f_2$ are the third harmonics, $f_1 \pm f_2$ are the second order intermodulation products (because the sum of the f_1 and f_2 coefficients is 2), $2f_1 \pm f_2$, and $2f_2 \pm f_1$ are the third-order intermodulation products (because the sum of the f_1 and f_2 coefficients is 3), and so on. The input and output power spectra for a typical solid-state amplifier are shown in Figure 6.4-40.

Figure 6.4-40 shows that the third-order intermodulation products at $2f_1 - f_2$ and $2f_2 - f_1$ are close to the fundamental frequencies f_1 and f_2 and (typically) fall within the amplifier bandwidth, producing distortion in the output.

The digital radio system that employs linear modulation and operates the power amplifier near or into saturation (beyond G_{1dB}) will see distortions that are caused by the third-order IM products. The distortions can be seen as a spreading of the spectrum that can occupy up to three times the bandwidth of the original spectrum. Third-order IM distortion will degrade BER

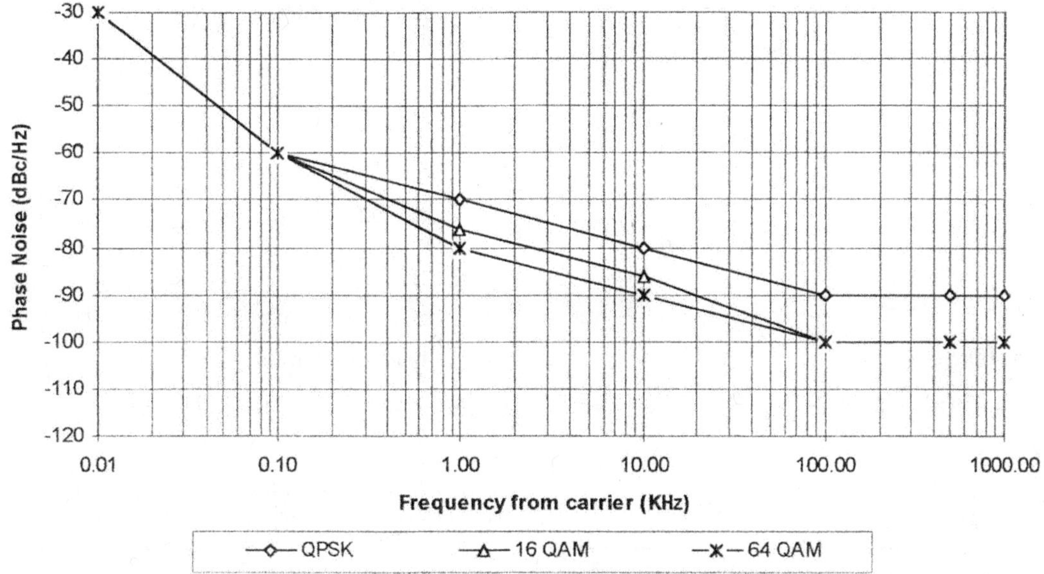

FIGURE 6.4-38 LO phase noise required for QPSK, 16-QAM, and 64-QAM [2]

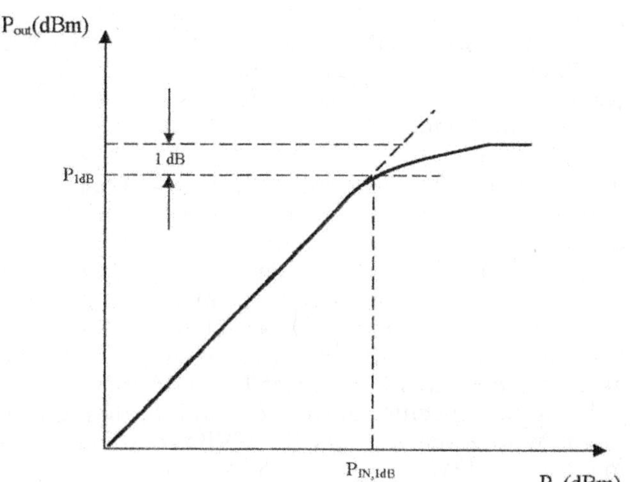

FIGURE 6.4-39 One dB amplifier gain compression [2]

performance and can cause interference to adjacent channels.

The severity of third-order IM distortion is a function of where the amplifier output level is operating in relation to its P_{1dB} point. In addition, the more complex the modulation scheme, the larger the required P_{1dB} back-off to minimize IM distortion. Table 6.4-4 shows the typical power amplifier back-offs for FSK, QPSK, and QAM modulation.

Figure 6.4-41 shows the spectrum plots of a 16-QAM 70 MHz modulated signal at the modem output and the same signal up-converted to 7037.8 MHz at the transmitter output. In this case the transmitter power amplifier was backed off by 6 dB from P1dB. It is important to note that even at this operating point, some spreading still occurs. Finally, the amplified signal is band pass filtered and directed to the antenna system for free-space transmission. At the far end of the free-space path the modulated microwave signal is captured by the receive antenna system.

The Receiver

The receiver in the digital radio link is the device that receives the modulated signal from the antenna system; filters, amplifies, and down-converts it to IF; filters and amplifies it again; and then delivers it to the demodulator for signal processing back to baseband. A simplified block diagram of a digital radio receiver is shown in Figure 6.4-42.

In the digital link, the receiver performs the lion's share of the work. In Figure 6.4-34, the 4-QAM, 16-QAM, and 64QAM signal constellations were shown. The figure shows that as the value of M (M = 4, 16, 64) increases, the space between the constellation points decreases. For the digital link to operate, the digital receiver must be capable of resolving which signal point was transmitted. For 64-QAM modulation it has to resolve 64 points; for 512-QAM modulation, it has to resolve 512 points!

All of the transmitter issues previously discussed concerning LO phase noise and amplifier linearity hold true for the digital receiver. In addition to these internal concerns, the receiver is also susceptible to outside disturbances that tend to deteriorate system performance. It is an understatement to say that the receiver must be capable of resolving which signal point was transmitted. The receiver must be capable of resolving which signal point was transmitted in the presence of transmitter distortions, receiver distortions, white Gaussian noise, co-channel and/or adjacent channel interference, and multipath distortions.

The transmitter and receiver internal distortions can be reduced by employing low-phase noise local

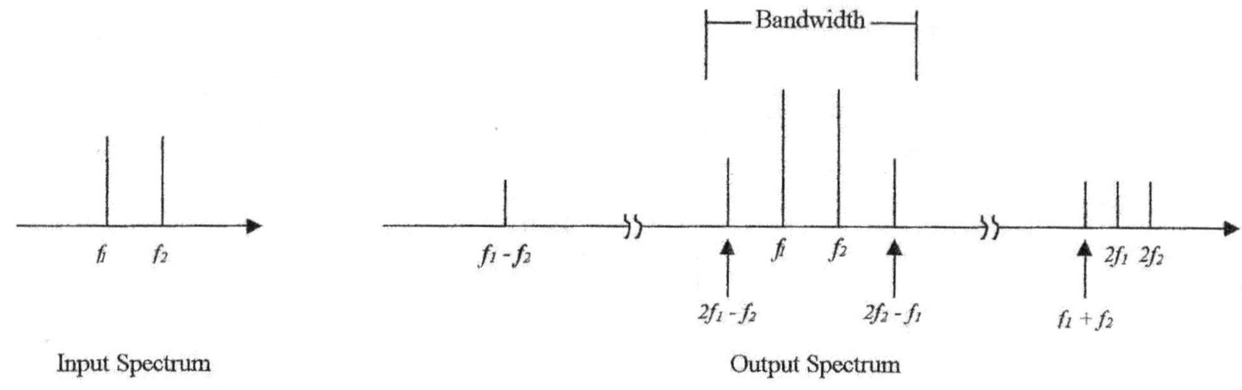

FIGURE 6.4-40 Input and output power spectrum for a typical solid-state amplifier [2]

TABLE 6.4-4
Typical power amplifier P1dB back-off

System	Variants	Typical Back-Off from P_{1dB}
FSK	2-state FSK with discriminator detection	0
	3-state FSK (duo-binary)	0
	4-state FSK	0
PSK	2-state PSK with coherent detection	I
	4-state PSK with coherent detection	2
	8-state PSK with coherent detection	4
QAM	16-QAM with coherent detection	6
	64-QAM with coherent detection	10
	256-QAM with coherent detection	12

oscillators and operating internal amplifiers sufficiently below their P1dB point.

Additive white Gaussian noise is unavoidable. The bandwidth-efficient modem requires a higher E_b/N_o for a given symbol rate as the number of bits per second per hertz is increased. Co-channel and/or adjacent channel interference in most cases can be avoided by frequency allocation management, channel filtering, and meeting spectral masks.

Multipath distortions are caused by the transmission channel itself. To minimize dispersive distortions the broadcast engineer must design the link to avoid ground reflections, implement space and/or frequency diversity, and employ adaptive equalization.

System Considerations

The implementation of digital microwave links for both terrestrial line-of-sight (LOS) and ENG mobile radio links is accelerating in the broadcast industry. This is due primarily the emergence of MPEG-2 technology. The transmission of the ATSC suite of content (over the STL) typically contains one or more MPEG-2 compressed digital video signals as primary or elementary content on the transport stream. These digital video streams take the form of standard and or high-definition content. MPEG-2 compression in the ENG environment has primarily been standard definition (SD) utilizing coded frequency division multiplex (COFDM) transmission as the de facto industry standard for modulation.

Today, the industry sees the migration to high definition (HD) field contribution and ENG. This approach to news and content delivery complicates the ENG effort all the more. The broadcast engineer must make decisions concerning the type of ENG infrastructure necessary to meet the station's requirements. Cameras, compression schemes, and interfaces all need to be considered relative to the available bandwidth. Current COFDM modulation will support specific bandwidth and bit rate requirements with restriction per the DVB modulation format for COFDM.

The current RF channel plan in the 2 GHz band specifies seven channels, each with 12 MHz bandwidth. With the convergence of varying MPEG and interface formats, the need for broader bit rates is at hand. Various types of modulation schemes are available to allow the broadcast engineer this capability.

High definition transport from ENG vehicle or airborne platform is all in operation today. There has been unheralded success in transporting respectable bit rate, approximately 20 Mbps, from aircraft with COFDM modulation in QAM operating modes. Airborne challenges have been overcome by specifically concentrating on improved MER performance of the transmitter, minimizing ground reflections by incorporating auto elevation control of the antenna element, and minimizing the use of omnidirectional antennas. The next obvious step is to focus on the delivery of higher bit rates. With the improvements in the single-carrier core QAM modulator, and the demodulator equalizers, along with focus on improvements in the areas of multipath performance and frequency errors due to Doppler

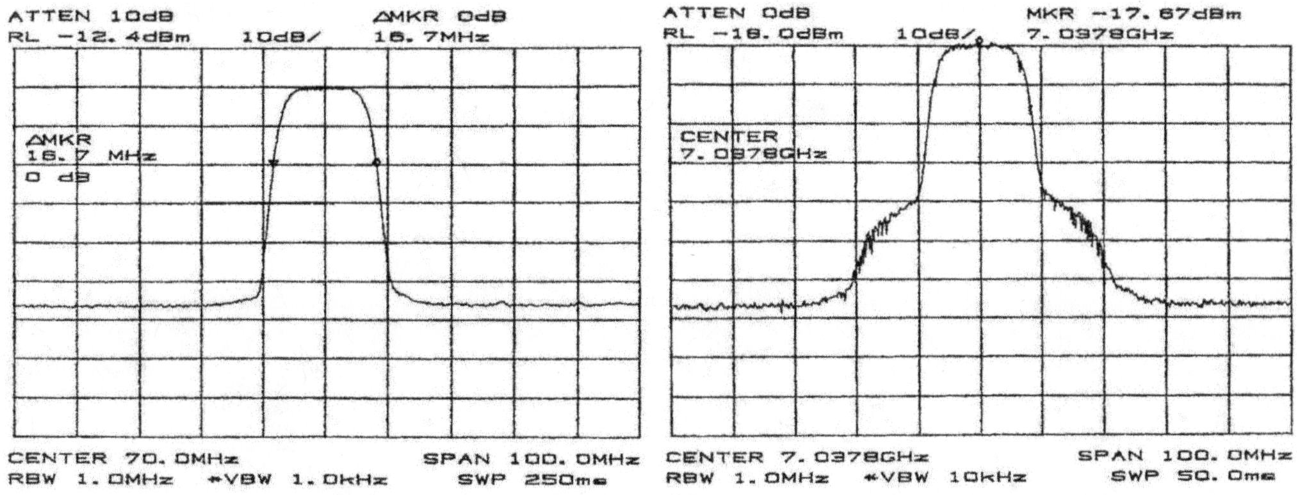

FIGURE 6.4-41 The RF spectrum for a 16-QAM signal at the modulator output and transmitter output [2]

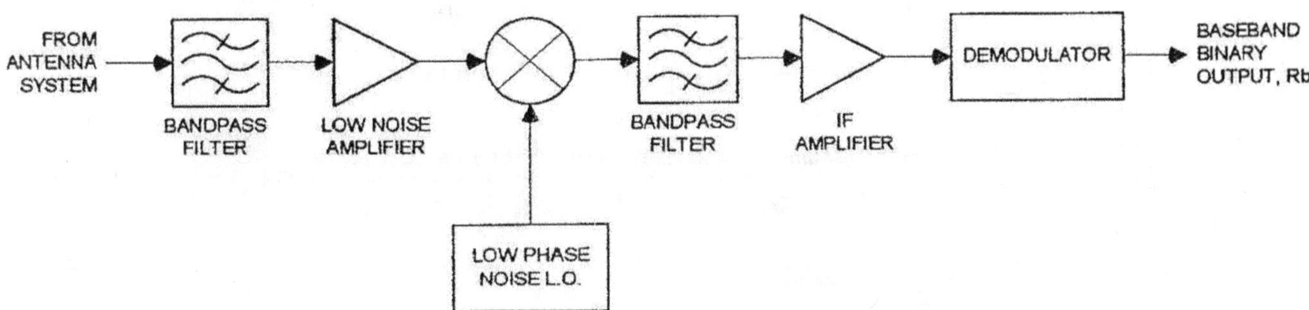

FIGURE 6.4-42 Simplified block diagram of a digital receiver from the 9th edition of this handbook [2]

shift, it has been proven that single-carrier QAM modulation is adequate to meet these demands.

There continues to be many discussions on the appropriate modulation platform to deliver this bandwidth efficiently to the broadcast environment. Single carrier modulation has been in use in Europe and Asia for many years. There, the infrastructure has been developed with this implementation in mind. To help with this discussion there are several charts given here that compare the different modulation formats. The user needs to define what the goals and challenges for the broadcast station are. The chart in Figure 6.4-43 depicts COFDM and single-carrier modulation as applied to 12 MHz and 25 MHz channel bandwidths. Recently, tests have proven that delivering 36 Mbps from aircraft utilizing single-carrier modulation operates as effectively as COFDM at half the bit rate.

Strictly from an ENG BAS perspective it is important to understand exactly what is indeed possible. With the 12 MHz channel of channel bandwidth, the broadcast environment has to operate in a much tighter spectrum. The telecom industry has been operating this way for many decades. The difference is that telecom traditionally operates in the terrestrial domain with focus on spectrum planning to minimize interference from adjacent or co-channel operators. In the mobile BAS services this is not a luxury that can be afforded; it is necessary to know what the system parameters are before deployment. Quick link analysis can be performed to allow confidence in the field deployment. Interference criteria are well published by the various equipment manufacturers. One advantage the broadcast operator has is a flexible equipment configuration. For example, channel interference by adjacent operators can be mitigated by the ability to adjust polarization of the antenna, IF filter selection, and dynamic gain control of amplifiers and LNAs.

The following charts (Figures 6.4-44 through 6.4-46) are shown as an aid to extract some of this common data on two predominant types of modulation: COFDM multicarrier modulation and single-carrier QPSK and QAM modulation.

Primary Power and Backup

With improvements in equipment and systems, engineering the reliability of the primary power source becomes a major factor in overall system availability. Where the terminals are in studios or TV transmitter sites the AC supplies may well be protected with standby generators; in this case, AC sourcing is the obvious choice. At repeaters, the AC may be supplied to a remote site by overhead lines and these may be subject to interruption under adverse weather

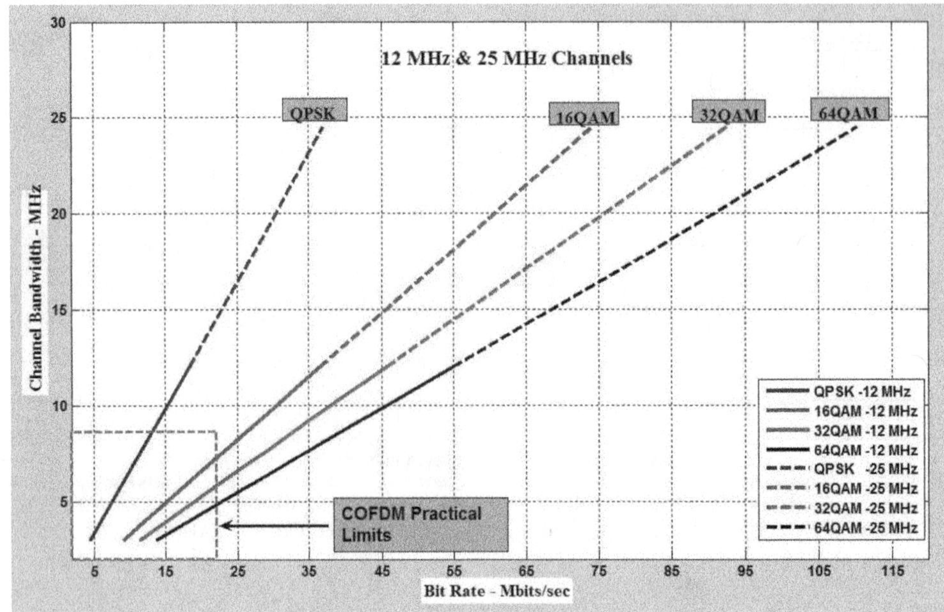

FIGURE 6.4-43 RF bandwidth versus bit rate

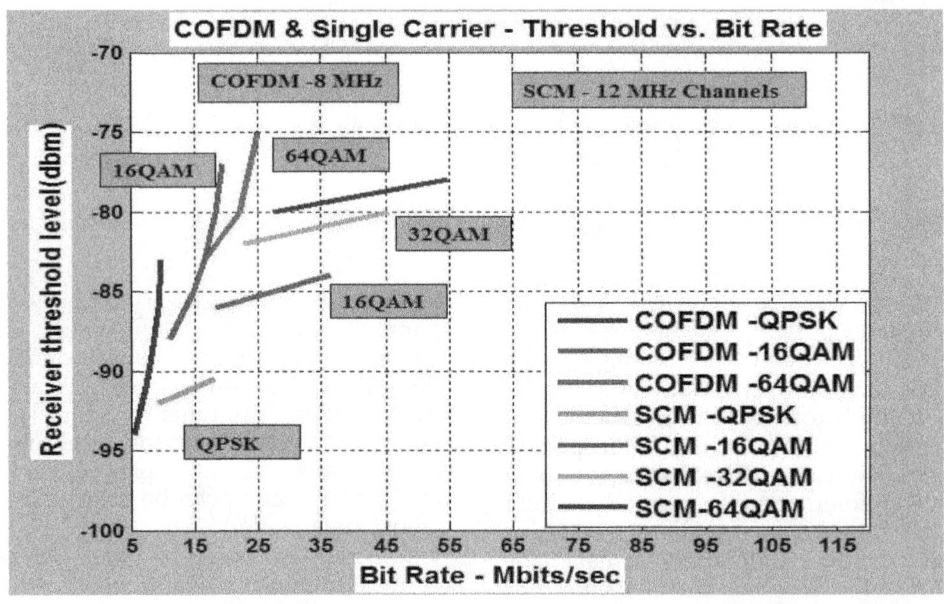

FIGURE 6.4-44 12 MHz data rate versus modulation

conditions. Because solid-state microwave equipment has essentially low power requirements, batteries are the most popular form of standby power. Where long outages are to be expected (12 hours or more) an AC generator will be needed to recharge the batteries.

Industrial lead-acid batteries are the most common type. They can be sealed to prevent evaporation of liquid, reducing maintenance and the need to ventilate the explosive gas given off when a battery is charged at too high a rate. The cells are floated across the charger, which supplies the station load until the AC fails. Because different types of lead-acid cell have different float voltages (2.15, 2.2, and 2.23 VDC for lead-antimony, lead-calcium, and lead-selenium, respectively) it is important that the charger and battery are matched—for this reason it is recommended that both are ordered from the same supplier.

It must be remembered that a battery of 12 cells with a nominal voltage of 24 VDC will float at 26.8 VDC, dropping to 24 when the charging fails and finally falling to 21 VDC at the end of the standby time. At a minimum, the electronic equipment must accept this range. To determine the capacity of each cell required to give a particular standby time (H hours) the steady current drain must be determined (A amperes). Depending on the type of regulation used in the electronic equipment

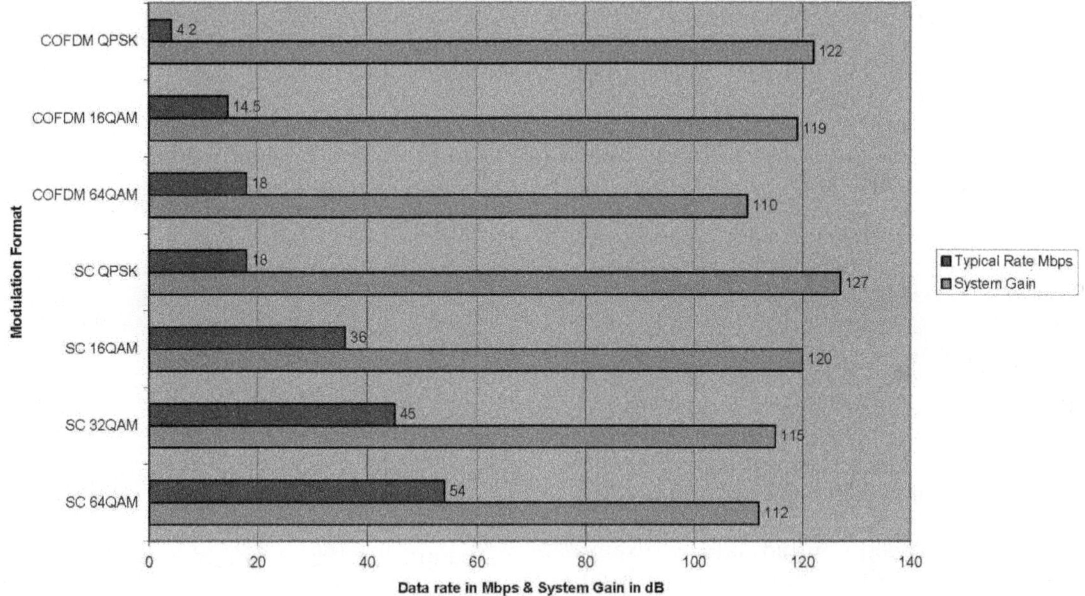

FIGURE 6.4-45 System gain comparison—12 MHz data rate versus modulation (referenced to 2 GHz BAS with 312 MHz channel)

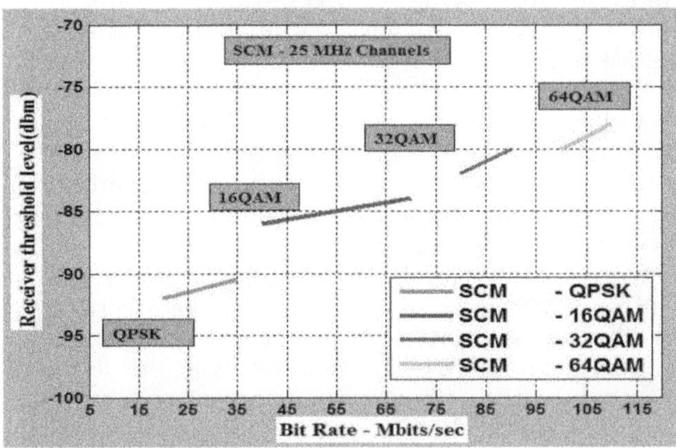

FIGURE 6.4-46 System threshold performances for single-carrier modulation

this may be higher at 26.8 VDC or 21 VDC (usually the latter)—take the higher value.

To arrive at the required cell capacity, multiply A by H to get the capacity in ampere-hours (AH). The minimum size of charger (C amperes) to meet the station load and at the same time recharge a discharged battery in R hours is given by

$$C = (A + AH) \times 1.1/R$$

Because charger failure would lead to station failure after H hours it is common practice to use a duplicated charger for full protection at key sites; in that case each charger need only have the capacity A amperes rather than C as calculated here. The chargers must be designed to share the load.

The battery acts as a large capacitor, reducing the ripple voltage generated by the charger and protecting against power line surges. A good charger will have low-voltage protection to isolate the battery when discharged and over-voltage protection and charge rate limiting to protect against surges and gassing by the cells. When diesel generators are used as the primary source of power, they should be run at 75% or more of their rating (after allowing for any derating for altitude).

Solar power using photovoltaic cells is attractive as there are no moving parts requiring maintenance. However, the batteries used to maintain the supply during hours of darkness or heavy overcast (which could exist for several days in some parts of the country) can be the most expensive part of the installation. It may pay to have a small AC generator and charger to back up the batteries if the solar input fails for more than 3 days, rather than batteries to maintain the supply for 6 days, for example. Such a generator could also power lights and test equipment during routine site visits.

FREQUENCY COORDINATION AND LICENSING

Late in 2003, the rules for BAS frequency coordination were changed by FCC Docket 01–75. Until that time, coordination had been done locally, on a case-by-case basis through the cooperation of the respective chief engineers, and under the guidance of a regional frequency coordinator appointed by the Society of Broadcast Engineers (SBE). The SBE regional coordinators consist of a voluntary group that maintained an up-to-date log of existing licenses and proposed applications.

The antenna standards in Part 74 did not change, but the newly imposed frequency coordination rules require a more careful study of the potential interference that may be caused to other licensees by a new or modified microwave link. Except for the 950 MHz aural band, the only antennas that meet existing standards are parabolic reflectors.

In the 950 MHz band, the only stated FCC requirement is found in §74.536, which states: "Aural broadcast STL and ICR stations are required to use a directional antenna with the minimum beam width necessary, consistent with good engineering practice, to establish the link." Experience has shown that the FCC considers 22 degrees to be the maximum half-power beam width that they will license in this band. Based on this de facto standard, it is reasonable to assume that actual standards may be imposed sometime in the future.

For TV STLs and intercity relay systems, the FCC has minimum rules that must be met with regard to the radiation pattern, as spelled out in §74.641. Table 6.4-5 summarizes those standards. The side-lobe suppression specifications have been omitted, and an example of the minimum acceptable antenna size has been added. Antenna sizes were based on data taken from vendor catalogs, however, they may vary plus or minus one size depending on feed structure and reflector shaping. It is good practice to review a vendor's specifications for compliance in the different categories.

While these standards apply to transmitting antennas, it is important to note that the FCC also states the following in §74.641: "The choice of receiving antennas is left to the discretion of the licensee. However, licensees will not be protected from interference which results from the use of antennas with poorer performance than identified in the table of this section." Station engineers are urged to strongly consider applying the FCC antenna standards equally at both ends of the path. For example, if a path has been coordinated using an 8 foot transmit antenna, it is good practice to use a matching, compliant antenna at the other end, even if it is only used for receiving.

For digital systems, the rules change significantly due to the profound differences in their tolerance for co-channel and adjacent channel interference, which varies as a function of the modulation type and data rate employed. Digital systems are also characterized by their threshold to interference, or T/I ratios, which may be positive or negative, depending on the circumstances.

With regard to where category A or category B antennas should be deployed, there is some ambiguity in the rules in that regard; no clear reference is given other than, without a commitment to the contrary from the FCC, one must anticipate that category A antennas will be required, and that high-performance antennas would be recommended to all concerned.

TABLE 6.4-5
Typical minimum antenna sizes per FCC §74.641

Band	Category	3 dB B/W	Minimum Gain	Minimum Size (ft)
2.02–2.11	A	5	n/a	8
	B	8	n/a	6
6.8–7.1	A	1.5	n/a	8
	B	2	n/a	6
12.7–13.2	A	1	n/a	6
	B	2	n/a	4
17.7–19.7	A	2.2	38	2
	B	2.2	38	2

FREQUENCY COORDINATION REQUIREMENTS AND PRIOR COORDINATION NOTIFICATION (PCN)

Under the new Part 74 coordination rules, frequency assignments for fixed stations above 944 MHz will apply the interference protection criteria specified in FCC Part 101.105(a), (b), and (c), and the frequency usage coordination procedures of FCC Part 101.103(d) for each frequency authorized. Exceptions are made for mobile/ENG operations in most bands, and for fixed operation in the 2 GHz band, all of which are still coordinated locally as they always have been, and the SBE continues to provide local coordinators. Table 6.4-6 shows Part 74 frequency coordination requirements before and after the rules changes in FCC Docket 01–75.

To license a new frequency, or to make a major change to an existing license, as defined by the FCC, the licensee must enlist the services of a coordinator who is capable of providing an interference analysis meeting the FCC rules.

Once the analysis is complete, a prior coordination notice is sent to all other co-channel and adjacent channel licensees within a specific set of boundaries known as the "keyhole." The keyhole refers to the shape of the area that must be analyzed for potential interference. For fixed, point-to-point service below 15 GHz, the area is essentially a circle with a diameter of 200 km (125 miles), except within ±5 degrees of the main beam, where the area extends to 400 km (248.6 miles). The size of the keyhole also varies with frequency and class of service. At higher frequencies, and when mobile stations are involved, the area is smaller.

TABLE 6.4-6
BAS coordination requirements

Band	All before 12/03	Fixed after 12/03	Mobile after 12/03
2Ghz	Local	Local	Local
2.5 GHz	Local	Prior	Local or prior
6.5 GHz	Prior	Prior	Prior
7 GHz	Local	Prior	Local or prior
13GHz	Interference criteria	Prior	Local or prior
18GHz	Local	Prior	Prior

BAS Bands and OFS Bands

Most broadcast microwave links are operated in the traditional broadcast auxiliary services bands, however, it is important to recognize that other options are available and should be employed when required. Most specifically, broadcasters may obtain licenses to operate in the private operational fixed service (OFS) bands FCC Part 101 bands between 5.9 and 6.8 GHz, and 10.7–11.7 GHz.

PCN Process

As noted earlier, Part 74 licensees must now employ the same frequency coordination procedures that Part 101 licensees use. With the exception of ENG/mobile operation and 2 GHz fixed links, which are still coordinated locally, all other fixed links must use the Part 101 procedure.

In essence, if a new path is to be constructed, or a major change is made to an existing path, coordination is required. While it is possible to manually coordinate by following the procedure outlined in the FCC rules, it would take an inordinate amount of time to research every licensee within the area of interest and calculate the potential for interference. A much easier solution is working with one of the commercial coordinators, such as those listed on the FCC website at http://wireless.fcc.gov/services/index.htm?job=licensing_1&id=microwave.

Once a coordinator has been chosen, an interference analysis will be conducted in accordance with FCC rules, and the licensee will be presented with a list of frequency and polarization options, assuming that some are available. If a frequency is not available on the band of choice, another band may be an option. If the study shows that a frequency is available to support a new or modified path, the coordinator will send out written prior coordination notices on behalf of the licensee, by email or direct mail, to all co-channel and adjacent channel licensees advising them of the proposed operation, and providing a statement that the FCC's rules have been followed in allowing 30 days for a response. If the 30-day waiting period ends without a protest, the licensee is free to file for a new or modified license.

Major Versus Minor Changes

The criteria for minor versus major changes may be found in Part 1 of FCC rules. The most current rules may be found on the FCC website, in the Wireless Telecommunications Bureau pages, at http://wireless.fcc.gov/rules.html. In summary, if a microwave station licensee makes any of the following changes, it will be considered major, and will require coordination:

1. Any change in transmit antenna location by more than 5 seconds in latitude or longitude for fixed point-to-point facilities (e.g., a 5-second change in latitude, longitude, or both would be minor).
2. Any increase in frequency tolerance.
3. Any increase in bandwidth.
4. Any change in emission type.
5. Any increase in EIRP greater than 3 dB.
6. Any increase in transmit antenna height (above mean sea level) more than 3 meters.
7. Any increase in transmit antenna beam width.
8. Any change in transmit antenna polarization.
9. Any change in transmit antenna azimuth greater than 1 degree.

Also, if a number of minor changes are made, they may add up to a major change. Refer to the FCC website for the complete rules.

FCC Data Resources and Online License Applications

The FCC website includes many informational resources as noted in previous sections, and also supports online filing of licenses and amendments. Any transaction with the Commission requires the individual or company to obtain an FCC registration number, or FRN. This may be obtained free of charge on the FCC website www.fcc.gov/help/getting-fcc-registration-number-frn-universal-licensing-system-uls.

Once an FRN is obtained, it is a relatively easy matter to apply for or modify licenses online. Provisions are made for payment of associated fees online, as well as the ability to upload required supporting documentation, such as a supplemental showing certifying that frequency coordination procedures have been completed in accordance with FCC rules.

ALTERNATIVE STL SYSTEMS

Over the years, the largest percentage of STL systems for both radio and television has been private microwave; however, alternatives have always been available. Leased telephone circuits have been the dominant alternate in radio, while leased fiber circuits have become the standard in television. The growth of digital broadcasting and IP connectivity has spawned a whole new generation of alternatives.

Carrier-Based Fiber Optic Systems

Incumbent local exchange carriers (ILECs) and competitive local exchange carriers (CLECs) have become increasingly active in offering a greater variety of broadcast-oriented services. The types of circuits that are now available include:

- DS1 for transport of AES audio in radio STL and remote pick-up applications.
- DS3 for compressed video.
- 19.39 Mbps for ATSC 1.0 STL applications.
- OC-3/ STS-1 for multiple videos or lightly compressed HDTV.

- OC-48 for multichannel video.
- Metro Ethernet for direct end-to-end layer-2 connectivity.

One thing to remember is that LECs are generally bound by tariffs, which amount to nonnegotiable prices. The only way to achieve better terms in most cases is a longer agreement. The pricing structure is normally based on an initial installation fee, plus monthly charge for the basic service and an additional charge for total mileage from point to point and any local access and transport area (LATA) crossings.

Interexchange carriers (IXCs) can provide long-distance connectivity, such as might be needed between New York and Chicago, for example. The fees are generally unregulated and negotiable except for the "last mile," which may be supplied by an LEC at one or both ends of a circuit.

In carrier circuits, the connection is virtual. The carrier provides a demarcation point at the customer premises that terminates in the protocol or connection that is being leased. For example, a carrier would provide a 75 ohm unbalanced BNC interface at the correct level for an SDI circuit, or 110 ohms balanced for a DS1.

Private IP Systems

In some cases, broadcasters have chosen to deploy a high-speed data circuit provided by their LEC. Metro Ethernet is a good example of a new type of service; pricing is based on a service level agreement (SLA) that specifies the data rate and end-to-end latency.

Radio broadcasters sometimes make use of ISPs, and thus the public Internet, for dedicated connections between their studio facility and other remote locations such as transmitter sites. Some ISPs will offer business-class services with a guaranteed QoS. The prices will be higher, but the danger of lost packets will be reduced considerably.

Unlicensed Microwave Band Systems

Unlicensed microwave spectrum in what is known as the industrial, scientific, and medical (ISM) bands has been available for over 20 years, and is popular with many communications users. The good news is that a web search on unlicensed microwaves will turn up quite a few companies that can provide radio equipment; the bad news is that an operator has absolutely no protection against interference. If another user causes interference, there is no recourse. Still, some have chosen the unlicensed route as a way to establish a low-cost link for an STL or TSL application. While this may work in suburban applications, the closer a path gets to a metro area, the more likely the chances for interference. One problem is power and EIRP limitations as shown in Table 6.4-7.

Two modulation types are allowed: frequency hopping spread spectrum (FHSS) and direct sequence spread spectrum (DSSS). FHSS tends to be more robust, is highly resistant to multipath, and has a higher immunity to interference, while DSSS can support much higher packet sizes and therefore greater throughput. The FCC allows FHSS systems a top-end emission bandwidth of 500 kHz in the 900 and 2.4 GHz bands, and 1.0 MHz in the 5.8 GHz band. Direct sequence is the opposite, where the FCC allows a minimum emission bandwidth of 500 kHz.

TABLE 6.4-7
Unlicensed ISM bands available for point-to-point applications

Band	Maximum Power Output	EIRP Limits
902–928 MHz	+ 30 dBm (1 watt)	+36 dBm
2.400–2.485 GHz	+ 30 dBm (1 watt)	+36 dBm
5.725–5.850 GHz	+ 30 dBm (1 watt)	+53 dBm

The 900 MHz and 2.4 GHz FHSS bands are mostly used for T1 and IP LAN traffic, while the 5.8 GHz radios may support a DS-3 rate, or even higher. The EIRP limits imposed on these bands do not lend themselves to long paths, and most radios automatically throttle down the throughput on longer paths to optimize path reliability versus data capacity.

SUMMARY

FM techniques have been the mainstay in microwave STL communications for decades, but are clearly on the decline as the digital transition increases the complexity of STLs and TSLs. While microwave radio still accounts for the largest proportion of STL systems in use, this chapter has examined the alternatives so as to provide information that will help an engineer decide which is best for a particular situation.

ACKNOWLEDGMENTS

Richard Miller, Microwave Radio Communications, Chelmsford, Massachusetts, and George Maier, Orion Broadcast Solutions, Sudbury, Massachusetts contributed to the material in this chapter as the update authors for the 10th Edition of the NAB Engineering Handbook. Doug Irwin would like to thank the following industry colleagues for their contributions to this chapter: Frank Foti, Telos Systems, for MPX over AES; Tony Peterle and Hartmut Foerster, Worldcast Systems, for MPX over AES; Keyur Parikh and Junius Kim, GatesAir, for 950 MHz systems with IP hardening; and Kevin Campbell, Worldcast Systems, for IP stream distribution over redundant IP networks.

References

[1] White, Robert. *Engineering Considerations for Microwave Communications Systems*, 2nd ed. GTE Lenkurt, San Carlos, CA, 1983, pp. 61–62.

[2] Whitaker, Jerry C. *A Primer: Digital Aural Studio to Transmitter Links*, TFT, Santa Clara, CA, 1994.

Bibliography

Anderson T., Detweiler, J., Gopal, M., and Iannuzzelli, R. *HD Radio™ Data Network Requirements*. Columbia, MD: iBiquity Digital Corporation, 2006

Andrew Catalog 38. Orland Park, IL: Andrew Corporation, 1997, at www.andrew.com

Cablewave Systems Catalogs 720C and 800. North Haven, CT: Radio Frequency Systems, Inc., Cablewave Systems Division, 1992

Campbell, Kevin. *IP Audio in the Real World: Pitfalls and Practical Solutions Encountered and Implemented When Rolling Out the Redundant Streaming Approach to IP Audio*. Presentation at 135th Audio Engineering Society Convention, October 17–20, 2013, New York, NY, USA

C.C.I.R. Documents of the Seventeenth Plenary Assembly, Düsseldorf, Germany, and Geneva, Switzerland: ITU, 1990

Effects of Multipath Propagation on the Design and Operation of Line-of-Sight Digital Relay Systems, ITU-R Rec., F.1093, 1994 F Series Vol., Part 1, Geneva, Switzerland: ITU, 1994

Electrical Performance Standards for Television Relay Facilities, Standard EIA\TIA 250-C. Washington, DC: Telecommunications Industries Association, 1990Feher, K. *Digital Communications Microwave Applications*. Englewood Cliffs, NJ: Prentice-Hall, 1981

Foerster, Hartmut and Peterle, Tony. *Digital MPX in the Broadcast Chain*. NAB Broadcast Engineering Conference Proceedings, April 2015

Foti, Frank. *MPX Over AES and MPX Over IP: Digital Connectivity for FM-Stereo*. Paper written and presented by Foti and Shay at the IEEE Broadcast Symposium, San Diego, CA, 2013

Freeman, Roger L. *Radio System Design for Telecommunications*, 2nd ed. New York: John Wiley & Sons, Inc., 1997

Freeman, Roger L. *Telecommunications System Engineering*, 3rd ed. New York: John Wiley & Sons, Inc., 1996

Hickin, E.M. Microwave Engineering for the Broadcaster, in NAB Science and Technology, *NAB Engineering Handbook*, 8th ed., pp. 653–669. Washington, DC: National Association of Broadcasters, 1992

Hogg, D.C. Statistics on Attenuation of Microwaves by Intense Rain, *Bell System Technical Journal*, 2949–2962, November, 1969

Interference Criteria for Microwave Systems, TIA Telecommunications Systems Bulletin TSB 10-F. Washington, DC: Telecommunications Industries Association, 1994

Irwin, Doug. System, Heal Thyself. *Radio Magazine*, May, 2016

Ivanek, Ferdo. *Terrestrial Digital Microwave Communications*. Norwood, MA: Artech House, 1989

Kim, Junius and Parikh, Keyur. *Maximizing STL Robustness Through Intelligent Interplay of RF and IP*. NAB Broadcast Engineering Conference Proceedings, April 2013

Kolberg, Erik L. *Microwave and Millimeter-Wave Mixers*. New York: IEEE Press, 1984

National Archives and Records Administration. Code of Federal Regulations, Title 47 Telecommunications, chapter 1, parts 0–101, October 1, 2005

Oster, J., and Bachner, E. *Angle Diversity: A Practical Technique for Reducing Fades, SP20–45*. Orland Park, IL: Andrew Corporation, 1989

Pathloss, Contract Telecommunication Engineering Ltd., n.d. Coquitlam, BC, Canada

Schwartz, M. *Information Transmission, Modulation, and Noise*, 3rd ed. New York: McGraw-Hill, 1984

Serafin, R. LO Phase Noise Requirements for QSPK, 16-QAM, and 64QAM Modulation, Microwave Radio Communications Memo, Microwave Communications, Chelmsford, MA, September 3, 1996

Additional information may be found at the following websites: HD Radio, at http://hdradio.com/

Marti Electronics, at http://bdcast.com/products/audio-data-links/marti

Moseley Associates, at www.moseleysb.com/

Nucomm, Inc., at www.imt-broadcast.com/

CHAPTER

6.5

Transmitter Remote Control and Monitoring Systems

PAUL SHULINS

Director of Technical Operations, Greater Media
Boston, MA

INTRODUCTION

From the early years of broadcasting, the Federal Radio Commission and later the Federal Communications Commission (FCC) required a licensed operator be physically at a broadcast transmitter at all times when radiating to observe its operation and make any required adjustments. Then in 1960, the FCC authorized class D noncommercial FM educational stations to operate via remote control. At that time, the FCC required any interruption in the remote control signal to cause the transmitter to shut off. This was called "fail safe" remote control, and it remained a requirement until 1984.

In 1953, the FCC authorized remote control of FM stations, and AM stations with a power of 10 kilowatts or less. In 1957, the FCC further relaxed the rules, permitting high power and directional stations to operate via remote control, providing that these stations submit logs demonstrating reliable and stable operation. In 1963, the FCC authorized automatic logging for AM and FM stations, and at that time chart recorders were required. Every 10 minutes an automatic alarm check was required as well. In 1969, the FCC also approved the use of sub-audible tones to transmit telemetry back to a control point, and authorized remote reading of directional parameters for AM stations. By 1971, the FCC had adopted rules that allowed VHF TV stations to operate by remote control.

In 1984, the FCC dropped the fail safe requirement, meaning that a loss of positive control no longer would require a station to immediately go off the air. However, loss of telemetry for more than 3 hours would require the transmitter to shut down automatically. These new rules opened up a whole new way to implement remote controls, because a permanent, dedicated circuit was no longer required. This was the beginning of the dial-up era of broadcast remote control. Less expensive equipment, along with the low cost of Plain Old Telephone Service (POTS, i.e., use of public switched telephone circuits), made this type of control very popular in the 1980s and beyond.

In 1995, the FCC adopted new rules that permitted radio and TV broadcast stations to operate without a person standing by on-site to monitor the transmitter operation. This was done to allow stations to take advantage of new technology that allowed reliable monitoring of broadcast equipment, as well the fact that most modern transmission gear is inherently more stable than that of several decades ago when the state of the art did not permit a high degree of stability or reliability.

RADIO FACILITY CONSOLIDATION

My experience as a director of technical operations for radio station company Greater Media in Boston has, along with the FCC Ownership rule changes in the mid-1990s, allowed me to construct one of the first radio station consolidation clusters in the country in 1998. One of the biggest challenges of this consolidation has been coping with all the transmitter sites and data. Because back in 1998, affordable commercial remote control systems were not available to give me the consolidated real-time information I wanted, I started to consider my options. This prompted me to invent systems that allowed me to "see" all the critical analog readings and status items in one consolidated area, greatly increasing awareness, and therefore allowing me to catch problems before they got out of hand.

This chapter will focus on straightforward ideas that can be implemented by any radio or TV station to allow for innovative and up-to-date methods to monitor and control a TV or radio facility.

A REVIEW OF THE RULES

The FCC now allows radio and TV stations to operate unattended. No formal notification to do so is required, but it is important to realize that unattended operation does not mean immunity from the rules and regulations regarding keeping the station within the licensed and legal parameters outlined by the FCC's technical regulations. This is specified in 47 CFR Section 73.1400. Thus, the burden is placed on the broadcaster to insure legal operation with whatever method the station owner chooses to employ.

It is important to realize that just because you choose to operate with unattended remote controls, it does not relieve you of the requirement to maintain a main studio location. Unless a waiver was granted or you are a low power TV or radio station (or a booster station), you must maintain a main studio as required by 47 CFR Section 73.1125.

The FCC also provides rules that explain the guidelines for connections to the transmitter site. Today there are many reliable ways to connect your remote controls to the transmitter site, but in the 1990s, perhaps one of the most popular was by phone lines. If you do use a phone line, it is an FCC requirement that you employ a dedicated phone line to be used exclusively for the purpose of transmitter site monitor and control while the site is unattended. This is specified in 47 CFR Section 73.1350. This rule does allow the phone line to be used for other purposes when there is a person at the transmitter site.

In general, the licensee is required to be able to correct any problem that could cause interference or turn the transmitter off within 3 hours. Some problems must be rectified in 3 minutes, like malfunctions that cause interference to other stations, or that threaten safety of life.

Another relevant topic when discussing remote controls and unattended operation is that of the Emergency Alert System (EAS). The FCC requires stations to be able to receive, evaluate, rebroadcast and log EAS tests and alerts; section 47 CFR Part 11 covers this subject in detail.

FM translators, boosters, and low power TV and radio stations are able to operate unattended and the rules are covered in 47 CFR Section 74.734. These rules state that if the transmitter is not accessible 24 hours a day during all seasons of the year, means must be provided to turn the transmitter off from a point that is accessible all the time. In addition, if the input signal is lost the transmitter must automatically shut off immediately. The FCC also needs to know who the person is responsible for turning on and off the transmitter.

The FCC requires each station to establish written monitoring procedures and maintenance schedules used to insure the station remains in compliance with the Rules. The FCC leaves it largely up to the broadcaster to create a written system that allows the licensee to provide indications that the stations are operating legally. While not specifying intervals for recording technical parameters, it is generally suggested that every several hours is a good starting point.

It is also important to designate in writing a chief operator who will be ultimately responsible for reviewing operating logs and signing off on them. The operating logs need to include tower light operation, EAS tests, and any out-of-tolerance conditions and corrective actions taken. In the case of AM directional stations, it is also a requirement that the directional system parameters be logged every 3 hours as specified in 47 CFR Section 73.68.

BROADCAST STATION REMOTE CONTROL SYSTEMS

A remote control system is an extension of sensors, meters, and control functions that brings these capabilities to an operator's position that can be a few feet away, or in some cases hundreds of miles distant. Early on, remote control systems used DC circuits from the public utility to provide connectivity between the studio and transmitter facility. Electromechanical stepper switches at the transmitter site were used to select the meter sample to be transmitted to the operator. Generally, only one parameter at a time could be read.

Traditionally, two control functions were associated with each analog parameter called "raise" and "lower." These control channels utilized a second DC circuit. By manipulating the polarity of the DC signal and the voltages, the raise and lower signals sent from the studio could be distinguished from each other.

Early remote control systems were not digital, and usually had readings on an analog meter scale without decimal points. Therefore, a transmitter reading directly off the meter face needed to calibrated by multiplying or dividing by a factor of 10, 100, or 1000. Usually, an operator needed to know the value to expect in order to choose the correct factor. After microprocessors became commonplace in the mid 1980s, digital transmitter readings became routine, eliminating the need for multiplication factors.

Another great technical advance was electronic multiplexers that were able to quickly switch different samples into a single analog-to-digital converter (ADC), saving on time, space, and cost. Suddenly, a remote control with up to 32 channels of analog samples could be fit into a one- or two-rack unit piece of equipment.

Perhaps the most innovative advancement in remote control systems was the personal computer (PC), and the advent of the Internet. A PC allowed many remote

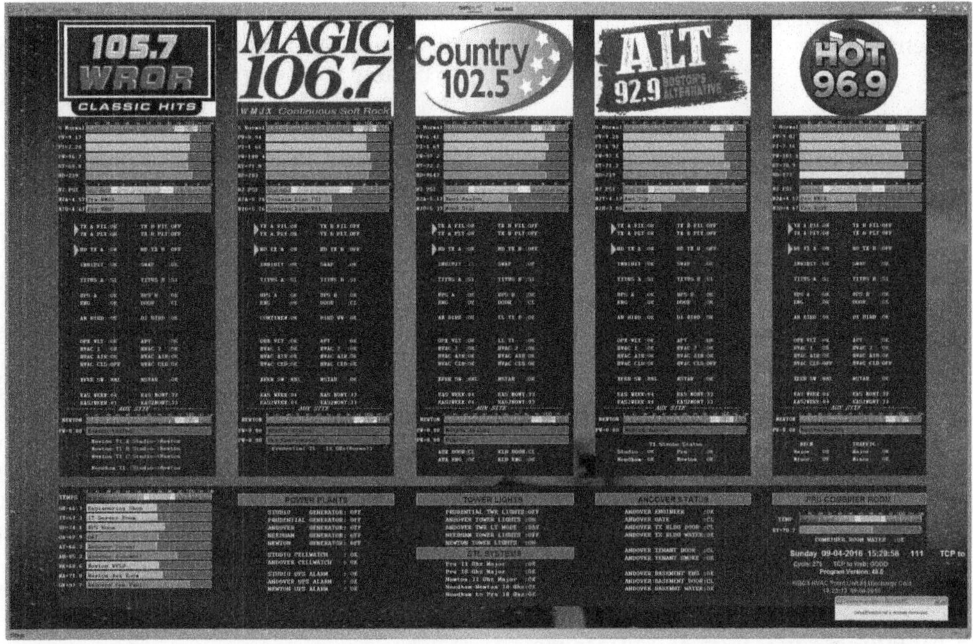

FIGURE 6.5-1 This screen shows a remote control system summary of five FM stations using both analog and digital (HD Radio) signals in Boston, Massachusetts

control parameters to be displayed simultaneously (see for example Figure 6.5-1), and in a way that was easy for the operator to interpret. When the Internet came along, this telemetry and control capability became securely available at multiple places remotely located from the transmitter site.

PLANNING A REMOTE CONTROL SYSTEM FOR YOUR FACILITY

When designing a remote control system for your facility, it is important to come up with a list of the parameters that are important to monitor both from a legal perspective and a safety perspective. For example, it is necessary to be able to tell the power level of the transmitter and the status of the tower lights (if you are responsible for them) to comply with federal regulations, however it is also good to be able to tell what the transmitter room temperature is, or the status of the transmission line air pressure, in order to be pro-active and correct any problems before they become costly.

There are three basic functions for a remote control system:

1. Monitor status items (on/off).
2. Monitor analog values, usually DC samples proportional to the analog sample you are trying to read. These should be calibrated to read accurately at the remote (studio) end.
3. Control: usually a pair of relays, commonly labeled raise/lower, to turn equipment on or off, to make power adjustments, or to change the status of a switch.

Here is a list of some of the more common priorities for monitoring:

1. Transmitter power output (analog)
2. Transmitter VSWR (reflected power)
3. Coax switch status (status)
4. Transmitter plate voltage (analog)
5. Transmitter plate current (analog)
6. Transmitter room temperature (analog)
7. Generator run status (status)
8. Tower light status (status)
9. Audio status (status)
10. Nielsen portable people meter (PPM) encoder status (status)
11. Tower phase and loop current values (for analog AM stations)
12. Building security (doors/windows status).

Of course, it is also important to be able to control certain functions at the transmitter site. These commonly used control functions include:

1. Transmitter on/off (filament if applicable) and plate control
2. Transmitter power level adjustments
3. Transmitter coaxial switch position control
4. Audio/video switching functions
5. Reset of VSWR and other types of transmission system faults.

This is just a starting point, and many more functions could be required or desired. There is no limit to the extent of the control and monitoring you can do to help you manage your remote sites.

Motorized Coaxial Switches

In many cases, main/alternate transmitters may be employed at a transmitter site for redundancy or backup. Normally these transmitters are switched into the antenna using either a patch panel or a motorized coaxial switch. Use of the motorized coaxial switch, which can be remotely controlled, allows an alternate transmitter to be put on the air by a mechanical action without having an operator physically on site to reroute the RF feed. These motorized switches (an example is shown in Figure 6.5-2) have been a great asset for broadcasters, providing for remote switching scenarios that would be otherwise impossible. However, with this new technology comes the requirement that the switches be properly interlocked to prevent the transmitter from producing power while the coax switch is either in transit, or simply sitting in an intermediate position. This is accomplished by taking advantage of contact closures that are available on most switches that provide closures when the switch is fully seated in a particular position.

Pressurized Transmission Lines

Broadcast transmission lines are critical parts of a transmission system. Many types of line employ an air dielectric to separate the inner and outer conductor of the coaxial cable, so as to establish the proper impedance. This air dielectric must be kept clean and moisture-free to avoid potential electrical arcing due to the relatively high voltages existing in the transmission line. A popular method for insuring that moisture is not present in these lines to pressurize them with an inert gas like nitrogen, or simply with dry air provided by a dehydrator.

This pressure is critical for keeping water and humidity out of the transmission lines. Therefore, it is desirable to be able to monitor the pressure using the remote control system. There are several types of sensors that may be attached to the pressurized side of a transmission line to provide a remote indication of pressure (see Figure 6.5-3). One type of sensor is simply a pressure-sensitive switch that will provide a dry contact closure when the pressure in a transmission line falls below a pre-set level. Other, more sophisticated types of pressure sensors will provide both a contact closure along with an analog DC sample that is proportional to the pressure, so that an actual reading of line pressure can be obtained via remote control. This has the added advantage of being able to alert the operator to undesirable trends before an out-of-tolerance condition is encountered.

The value of having these pressure sensors remotely monitored cannot be overstated. The types of transmission line failures generally encountered are due to internal transmission line arcing and are generally catastrophic; they are also very costly both in terms of down time and material repair costs.

Data Acquisition Modules

Data acquisition modules (see Figure 6.5-4) are a great tool for many industries including broadcasting. When used as an overly on top of existing commercially available remote control systems, these modules have the potential to bring thousands of status items to a centralized computer screen, at a fraction of the cost of a commercial remote control system. The types of data acquisition modules that are most helpful in remote control and monitoring applications for broadcasting are 16-bit ADCs and 16-channel I/O status modules. These modules are addressable and can be "daisy chained" on an RS-485 data bus. The modules can then be polled several times per second to return meaningful and timely data to a centralized computer for display and alerting.

FIGURE 6.5-2 A typical motorized coaxial switch capable of handling up to 50 kilowatts

FIGURE 6.5-3 Photo of analog and digital line gas pressure sensor installed on a transmission line in Boston, Massachusetts

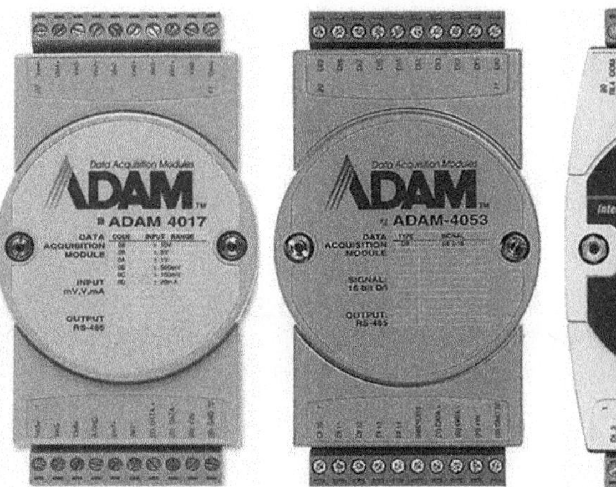

FIGURE 6.5-4 ADAM data acquisition modules, typical of the type used by the author for supplemental data acquisition on top of the commercial remote control layer

Using a computer, usually located at the studio, to poll these modules and display the results in a consolidated manner on a single screen can be very helpful in controlling and monitoring a group of facilities that are geographically diverse. In many cases the data can be acquired through Ethernet or RS-232 channels linked to the individual transmitter sites. In addition to displaying the analog and status values in a color-coded summary, the computer also can send text messages or email alerts of status changes or parameters that fall out of tolerance. It is a good idea to place the computer display in a prominent place at the studio or master control area, so that it is easy to glance at it every time you walk in the room so as to raise awareness of the general health of the systems.

EIA-485 Protocol

EIA-485 is a two-wire balanced data transmission system. It has the advantage of allowing many devices to be daisy chained along a bus and addressed individually. Although no specific data format is specified, this electrically balanced standard allows for higher data speeds than RS-232, and longer distance runs, with less noise. It is a very good system for communicating with data acquisition modules.

The diagram in Figure 6.5-5 shows a typical arrangement with a computer that generates the display with multiple serial communication ports. In most cases the RS-232 is transferred by studio-to-transmitter link (STL) to the various geographically diverse sites, where it is then converted to RS-485 signal levels. At this point it can communicate with the daisy-chained data acquisition modules throughout the facility. In addition, the local computer at the studio can drive a converter module that can display scrolling LED messages to personnel in master control as well as the studios.

Figure 6.5-6 shows a more complex arrangement where a serial port multiplier is utilized at the host computer. In this case a "rocket port" is employed that allows the computer to add 16 extra serial ports. These ports are each dedicated to a particular site, and the computer can strobe each site in sequence. It is also possible for the computer to make decisions regarding the health of the STL links, and should a segment fail, switch to a backup link for telemetry and control functions.

The computers that are running critical programs need to be monitored as well. Writing in a "heartbeat loop" of code is a good idea. One method for insuring that all the computers are running all the time is to write a short text file each minute containing the current time and date information. This can then be transferred to another PC that generates alerts, and web pages showing the confidence level that the critical machines are on-line and running the program they are assigned to run. Figure 6.5-7 shows such a web page formatted for a mobile phone; Figure 6.5-8 shows computer monitors in the master control area of Greater Media Boston displaying the real-time status of five transmitter sites.

Autonomous Local Control

While technically not considered remote control, a computer system that is located at the transmitter site, and can be accessed remotely, can offer an extra layer of reliability should STLs or other types of control links to the studio be severed for whatever reason. A computer that is overlaid on the official remote control system can operate locally and autonomously by keeping an eye on the RF levels from each transmitter, and taking corrective action if a transmitter ceases to produce the licensed level of RF. Using a set of data acquisition modules, the local computer can detect a loss of RF and safely and automatically switch the alternate or auxiliary transmitter on the air. This computer can also keep track of main and backup program audio sources as they arrive at the transmitter site, detect a loss of proper audio, and switch a backup source on the air without regard to connectivity to the studio location.

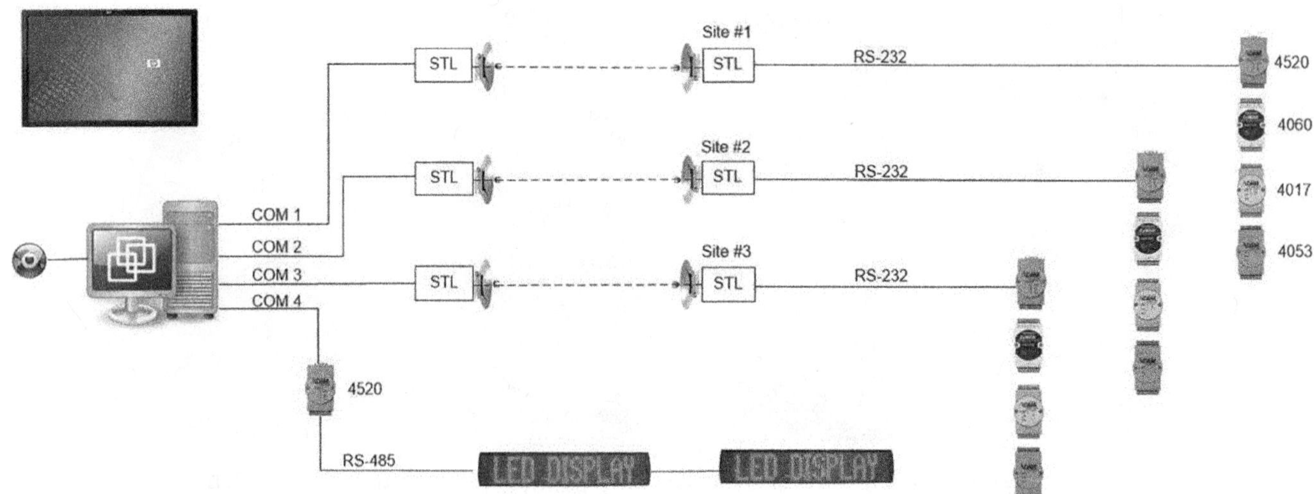

FIGURE 6.5-5 PC with multiple serial ports strobing data acquisition modules at Greater Media Boston

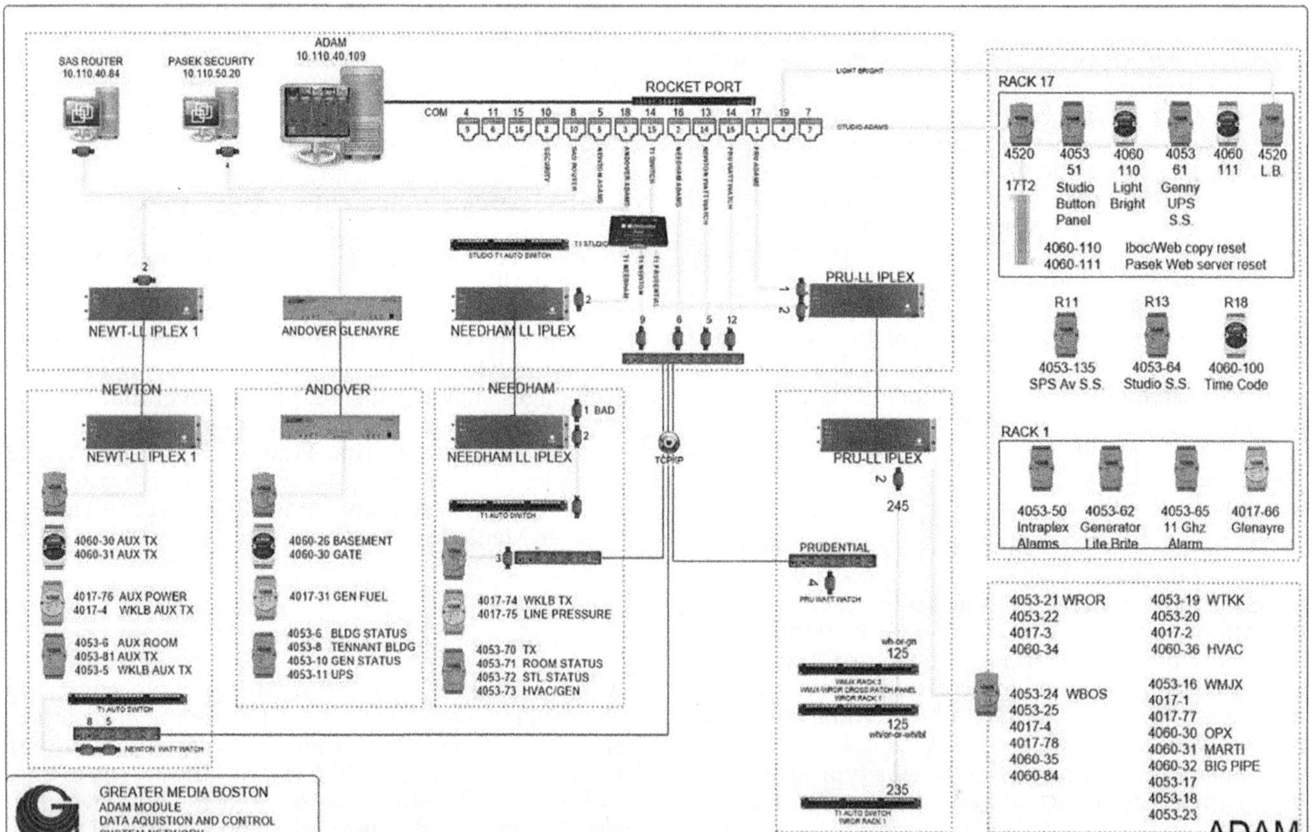

FIGURE 6.5-6 More complex arrangement showing a PC with multiple serial ports strobing data acquisition modules at Greater Media Boston

It should be noted that there should always be a hard relay interlock panel independently and permanently installed at the transmitter site to provide a level of safety over and above the computer, to insure that no switching of transmitters can occur unless the proper interlocks are satisfied. Figure 6.5-9 shows an example autonomous control computer screen.

Analog Sample Trend Analysis

Many remote control systems now offer the ability to view the short-term trends in analog values for different parameters. Usually presented in graphic form, these trends can be very useful for determine when a piece of equipment is changing its characteristics and

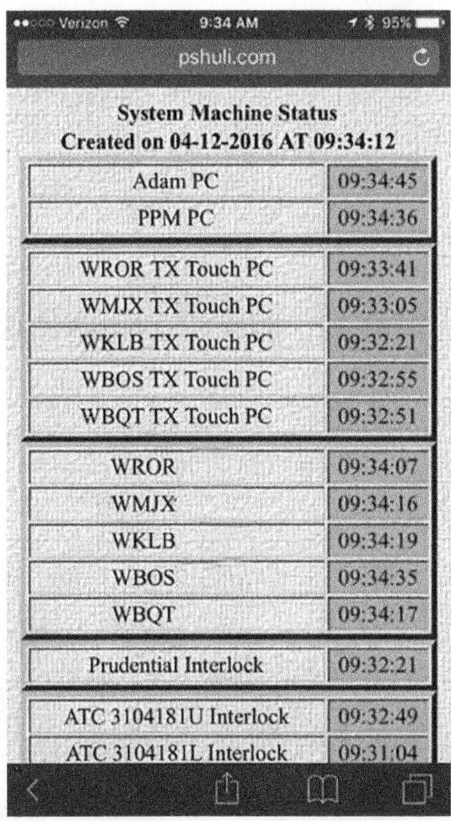

FIGURE 6.5-7 Image of web page formatted for a mobile phone that shows the times of the last refresh of the "heartbeat file"

may need attention soon. For example, monitoring the screen voltage in a tube-type transmitter over a period of time will give the operator an idea of when a tube change may be required. In areas where utility power varies substantially, the line voltage (which is directly proportional to the plate voltage in a tube transmitter), may be analyzed to see if the utility company is providing consistent power, especially during a time when your transmitter parameters are changing. Graphing room temperature may give early warnings of HVAC problems, and of course looking at the trends of transmission line gas pressure can easily save the day if leaks are detected early. Many remote control systems allow trend analysis on custom views, showing graphs for each parameter over a period of time (usually 24 hours). This is just another advantage of computer-based remote control systems.

Portable People Meter Integrity Monitoring

In many medium and major markets, radio station ratings are now measured electronically, as opposed to the old days where panelists manually filled out diaries, keeping track of what radio stations were listened to during a particular week. Today, radio station audio is often encoded with psychoacoustic watermarks that are eventually detected by devices, carried around by panelists, that acoustically detect these watermarks, and assign credit to the radio stations that the panelist is exposed to. The system, operated by Nielsen Audio, is commonly referred to as portable people meter (PPM) ratings. For a station using the PPM system,

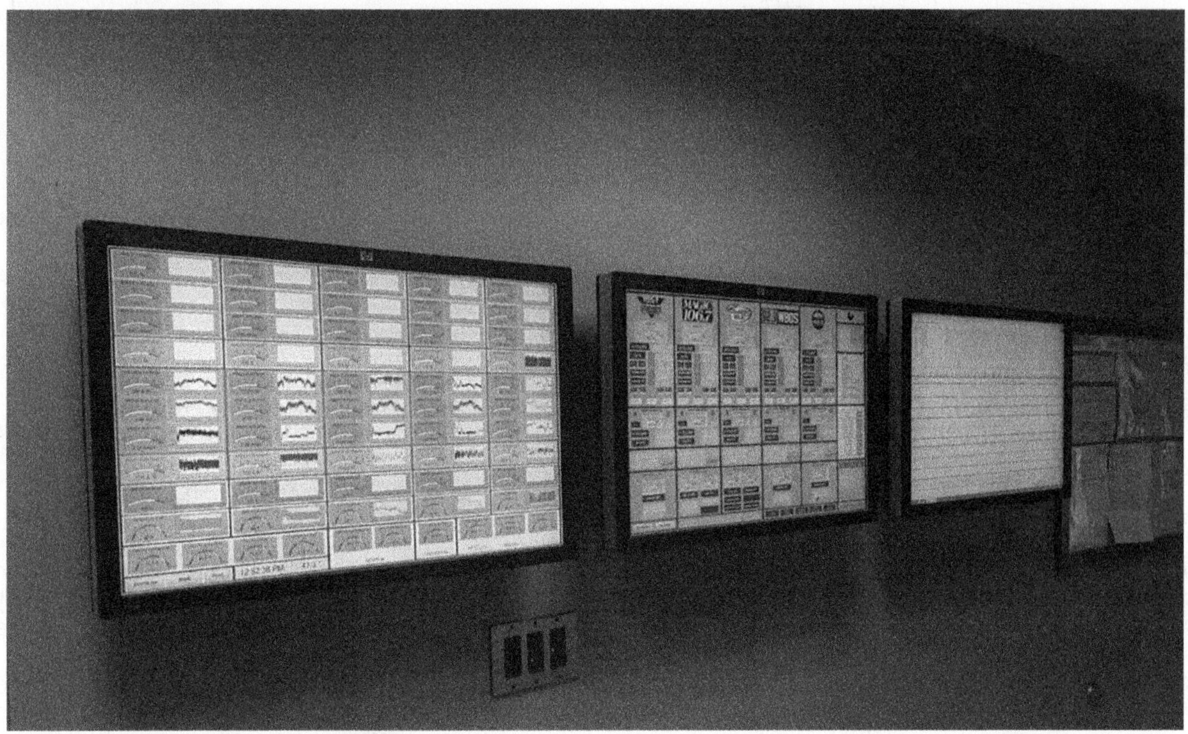

FIGURE 6.5-8 Image of three 30″ computer monitors in the master control area of Greater Media Boston showing the real-time status of five transmitter sites

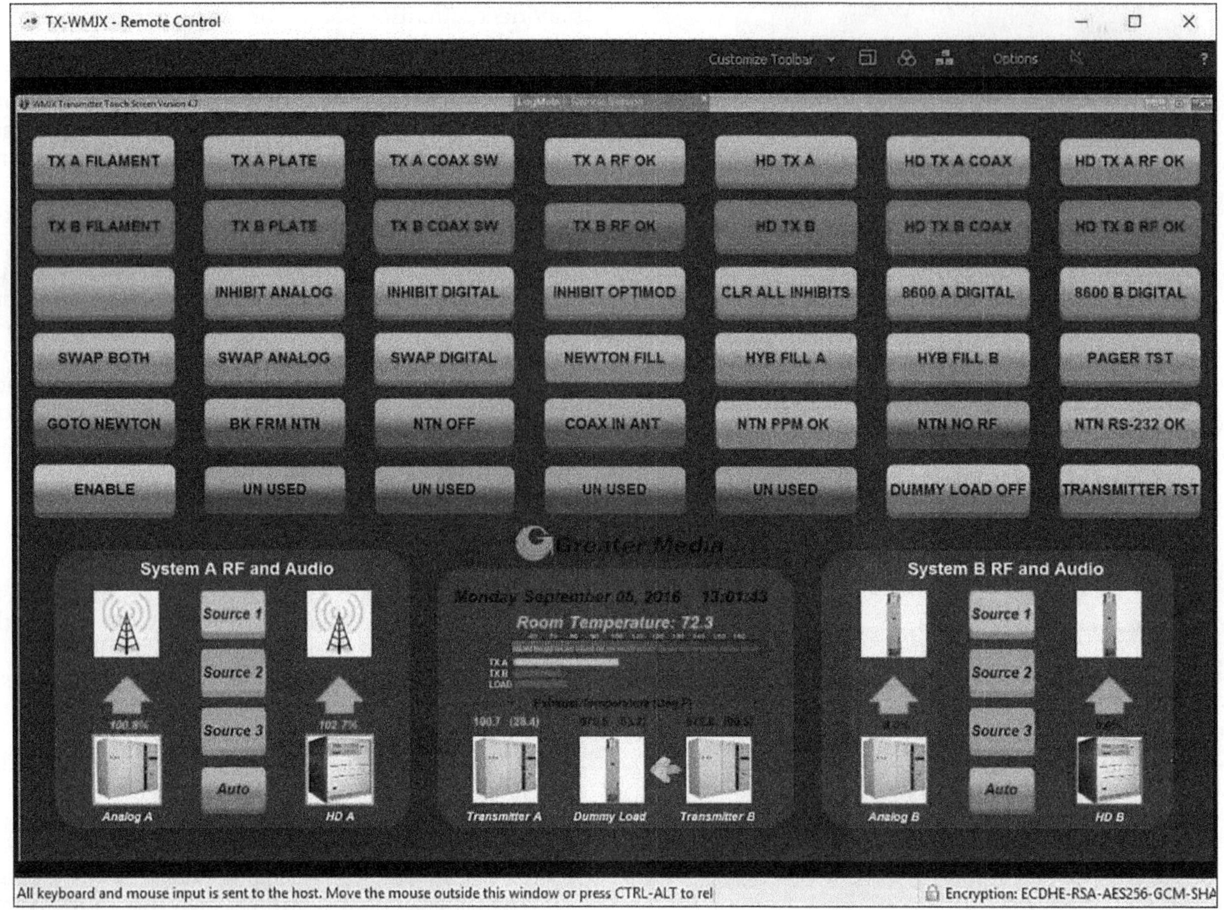

FIGURE 6.5-9 Example autonomous control computer screen showing control and status functions for an FM station operating with analog and digital (HD Radio) using separate amplification. Also shown are audio status icons for primary, secondary, and tertiary audio sources.

these watermarks that are encoded onto the audio are critical for making sure the facility gets proper credit for listening.

The challenge here is that because the watermarks are by design inaudible, it is never obvious to the station engineering staff as to whether the proper codes are making it on-air and therefore available to the panelist's PPM appliance for reporting appropriate credit for listening. Nielsen Audio normally provides PPM confidence monitors to broadcasters that are designed to listen to an on-air audio signal, and report via an LED display and a contact closure that the PPM codes are present. Remote controls are helpful for monitoring these PPM confidence monitors that are so very critical to the financial success of the radio station.

This situation can get quite complex because in a radio station cluster with five stations, each radio station may have half a dozen audio sources that require encoding. For example, there may be a primary, secondary, and tertiary source of audio, as well as companion HD Radio multicast programs, and Internet audio streams, all with distinct PPM codes. So again, a computer program that is designed to monitor dozens of PPM monitors is appropriate to bring all this information together on one screen that is easily visible from a master control area (see Figure 6.5-10). A program that is designed properly can color code boxes on a screen that indicate proper PPM codes are being broadcast and received. The computer can also be responsible for paging engineers, and closing additional relays that can actuate enunciators in studios to alert operators on duty of a possible PPM encoding problem.

Part 101 Licensed STL Links

Several years ago, the FCC relaxed its rules governing the so-called Part 101 radios for licensed microwaves. Recognizing that HD Radio signals have high data transfer rate requirements, along with the added requirements for other supported transmission equipment (now typical for today's modern radio station), the FCC decided to allow high bandwidth radios that are licensed and protected to be used for STLs. This was a great boon for remote control systems, because much more data could be reliably and securely transferred between the studio and transmitter locations. Along with Internet protocol (IP)-based security cameras, much of the equipment sold today contains a

FIGURE 6.5-10 Photo of PPM confidence monitor summary screen in master control

Courtesy Greater Media Boston

network interface card that allows the user to directly monitor and control it independently as long as an IP-based connection is possible. These new Part 101 radios typically allow for high bandwidth IP connection and an extension of the business network to the transmitter sites.

Some of the more common applications for IP-based remote control include:

1. Audio processor adjustments
2. Transmitter remote control directly via IP
3. Off-premise telephones derived from the PBX telephone system at the studio
4. Security cameras
5. HD Radio transmission (Importer to Exporter connections)
6. RDS encoder commands
7. UPS status
8. Generator status and control
9. Transfer switch status and control.

Additionally, it is possible to construct redundant microwave links and in some cases arrange them in a sonnet ring (see Figure 6.5-11), where the data normally flows clockwise, however upon failure of a segment of the ring, the data in certain segments can flow counterclockwise and preserve data connectivity between all sites.

Uninterruptible Power Supplies

At the studio or transmitter, normally uninterruptable power supplies (UPS) need to be employed, due to the sensitive nature of the microprocessor equipment utilized today. These centralized UPS systems are great, but they do come with associated costs to operate and maintain. One of the more critical functions to monitor with these UPS systems is the health of the batteries that are used to provide power in a brownout or blackout. Fortunately, most modern UPS equipment provides a status output that can indicate the health of the batteries. Because many of the UPS units will be located at transmitter sites, remotely monitoring the UPS battery health is a simple matter of tying the UPS into your remote control system. In many cases, it can be as simple as a contact closure that indicates a battery problem is detected, and in other cases, the UPS can provide much more detailed diagnostic information via a web interface or RS-232 serial data. Whatever your UPS provides, it is a great idea to take advantage of it.

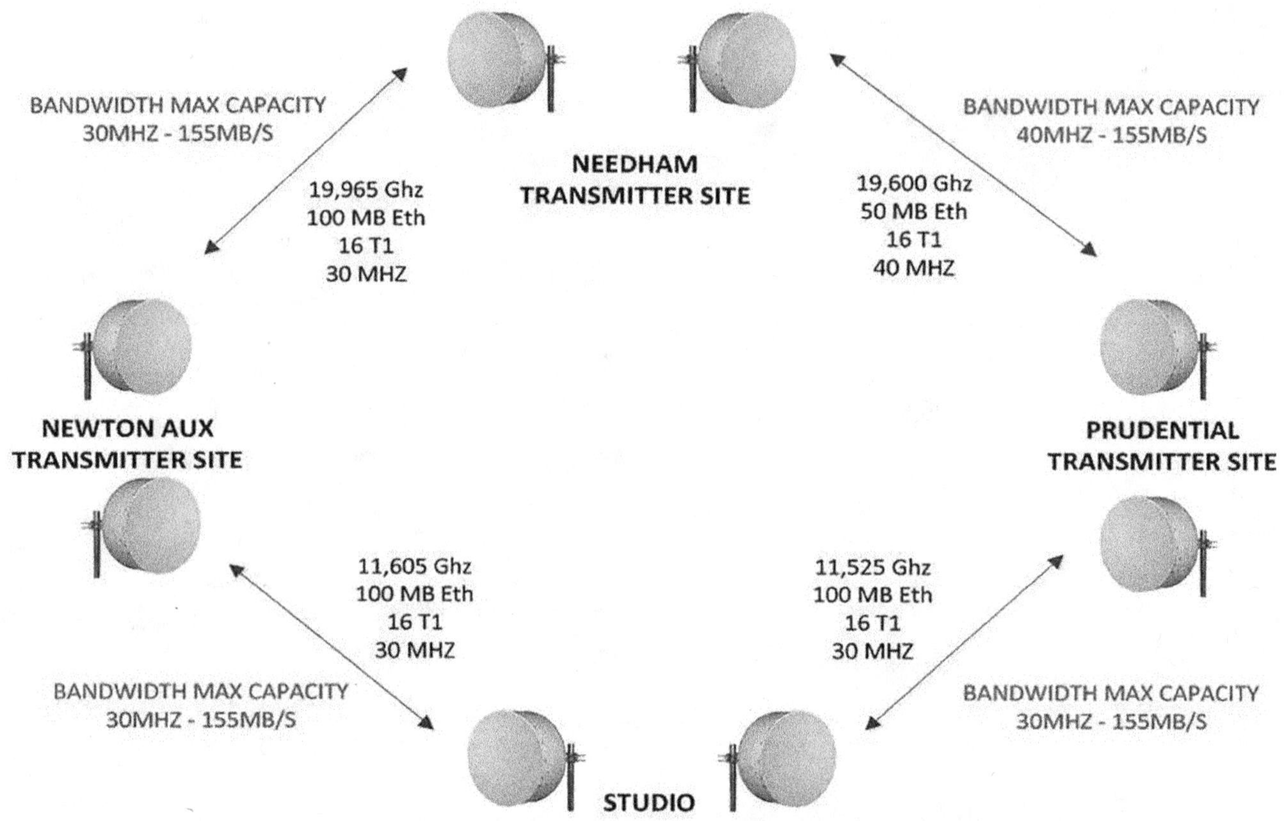

FIGURE 6.5-11 Sonnet ring STL architecture, allowing any of the four segments to fail, having the data reverse direction to maintain connectivity

Courtesy Greater Media Boston

Batteries only last a few years, and will need replacement. The UPS can give you advance warning of several weeks so ordering fresh batteries can be done before total failure occurs. In a larger UPS system, usually used at studio locations, large and expensive strings of batteries are needed to support the power requirements of many studios until the generator can start. These batteries also need periodic replacement. However, in this case it is important to be able to detect batteries going bad as soon as there are symptoms, because left to fail completely, a single battery in a chain of 40 batteries can adversely affect the entire chain. Therefore, catching a bad cell as soon as possible can save thousands of dollars in other cell replacements. In order to detect these failures, a new technology is employed, referred to as "cell watch." This computer-based monitoring system looks at each individual battery several times per minute for small changes in voltage and internal resistance (see Figure 6.5-12). Out-of-tolerance conditions are immediately reported and a contact closure can trigger external alerts and alarms to call attention to the failing cell. This, of course, can also be tied into the data acquisition modules, and the overlaid remote control.

Web Page Monitoring

A centralized computer that collects data can also be programmed to generate real time HTML web pages that, when properly formatted, can present attractive and informative real time information on mobile devices (an example is shown in Figure 6.5-13). For the engineer who is traveling between sites, this up-to-date information can be extremely helpful in diagnosing problems, and giving direction to control room operators. Access control is critical, however, in these cases where the pages are open to the public. It is important to set up user names and passwords to limit access to monitoring and control functions on the web.

Network Health Monitoring

The various sites that are connected together via a TCP/IP network are vulnerable to data disruptions, particularly if a low-level (or no level) of redundancy is used. One desirable tool to check the connectivity in real-time is the use of a "ping monitor." Dedicating a PC and a screen to this function allows engineers to see at a glance if various sites are connected together, and can ping each other with data packets continuously. A large screen running a ping test program continuously gives a real time indication that all is well. Figure 6.5-14 shows a screen shot of a typical ping monitor display. It is advisable in larger networks spanning many locations to run this kind of tool continuously and to use a large display that can easily be interpreted.

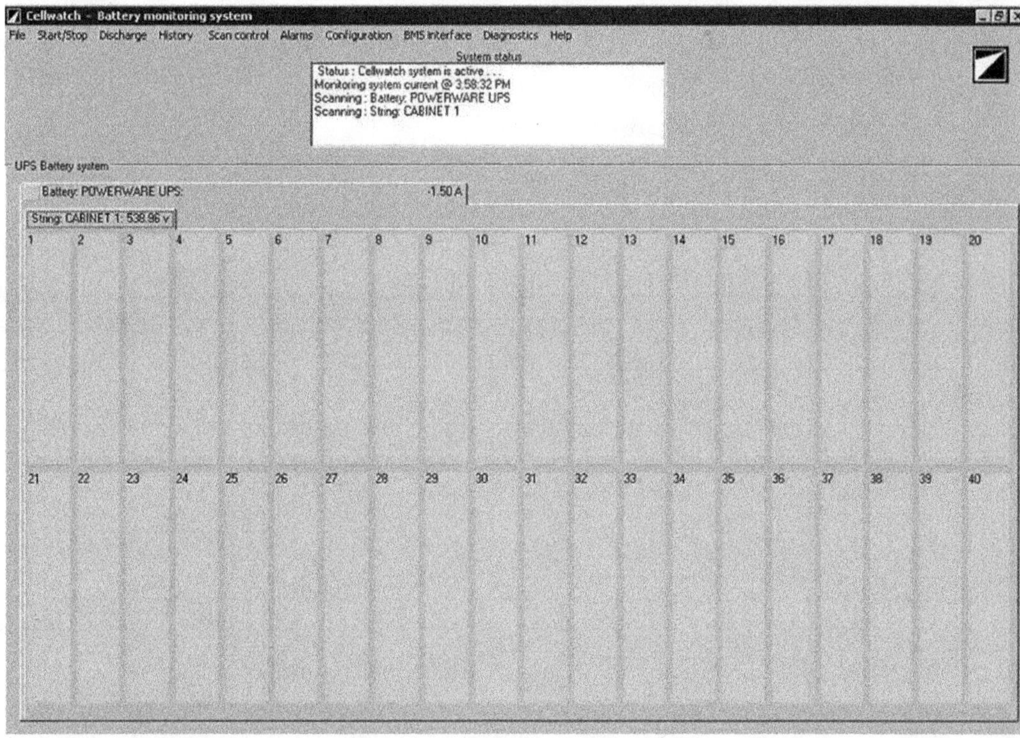

FIGURE 6.5-12 Screen shot of a typical cell watch software package indicating battery cell performance for 40 cells simultaneously

Courtesy Greater Media Boston.

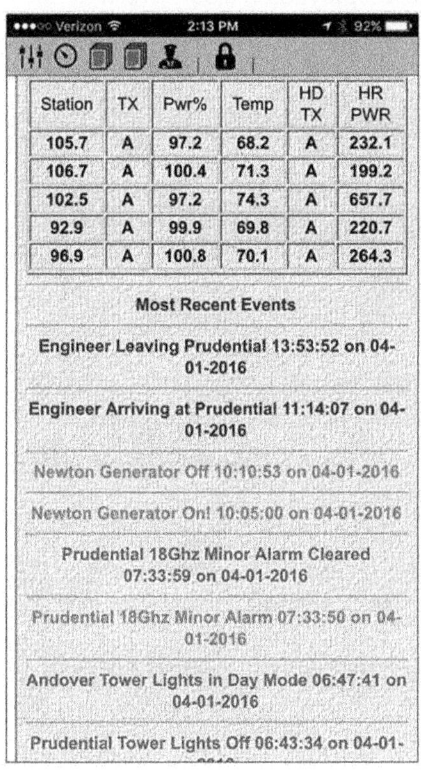

•••○○ Verizon 2:13 PM 92%

Station	TX	Pwr%	Temp	HD TX	HR PWR
105.7	A	97.2	68.2	A	232.1
106.7	A	100.4	71.3	A	199.2
102.5	A	97.2	74.3	A	657.7
92.9	A	99.9	69.8	A	220.7
96.9	A	100.8	70.1	A	264.3

Most Recent Events

Engineer Leaving Prudential 13:53:52 on 04-01-2016

Engineer Arriving at Prudential 11:14:07 on 04-01-2016

Newton Generator Off 10:10:53 on 04-01-2016

Newton Generator On! 10:05:00 on 04-01-2016

Prudential 18Ghz Minor Alarm Cleared 07:33:59 on 04-01-2016

Prudential 18Ghz Minor Alarm 07:33:50 on 04-01-2016

Andover Tower Lights in Day Mode 06:47:41 on 04-01-2016

Prudential Tower Lights Off 06:43:34 on 04-01-

FIGURE 6.5-13 Example of a real-time updated web page formatted for display on a mobile phone

Courtesy Greater Media Boston

Simple Network Management Protocol

Simple network management protocol (SNMP) is a popular protocol for network management. It is used for collecting information from, and configuring, network devices, such as servers, printers, hubs, switches, and routers on an IP network. In the broadcast world, many devices are now being sold that communicate directly via SNMP. Again, using TCP/IP communications, SNMP is becoming popular.

Developed in the late 1980s, its use became standardized in the 1990s. SNMP is a highly complex set of protocols used by manufacturers to allow control and monitoring of their specific equipment over TCP/IP networks. One of the big advantages of this standard is the ability to control not only broadcast gear, but other devices such as routers, servers, and printers. This emerging protocol shows much promise for the future.

SUMMARY

Broadcast facility remote control technology has grown tremendously over the last several decades. The technology available today is much more capable than the tools available just a decade ago. This is a good thing, because keeping pace with the ever-expanding complexities of broadcast equipment and the requirements for controlling and monitoring the quantity of equipment used today requires remote control systems to be efficient, clever, and more capable.

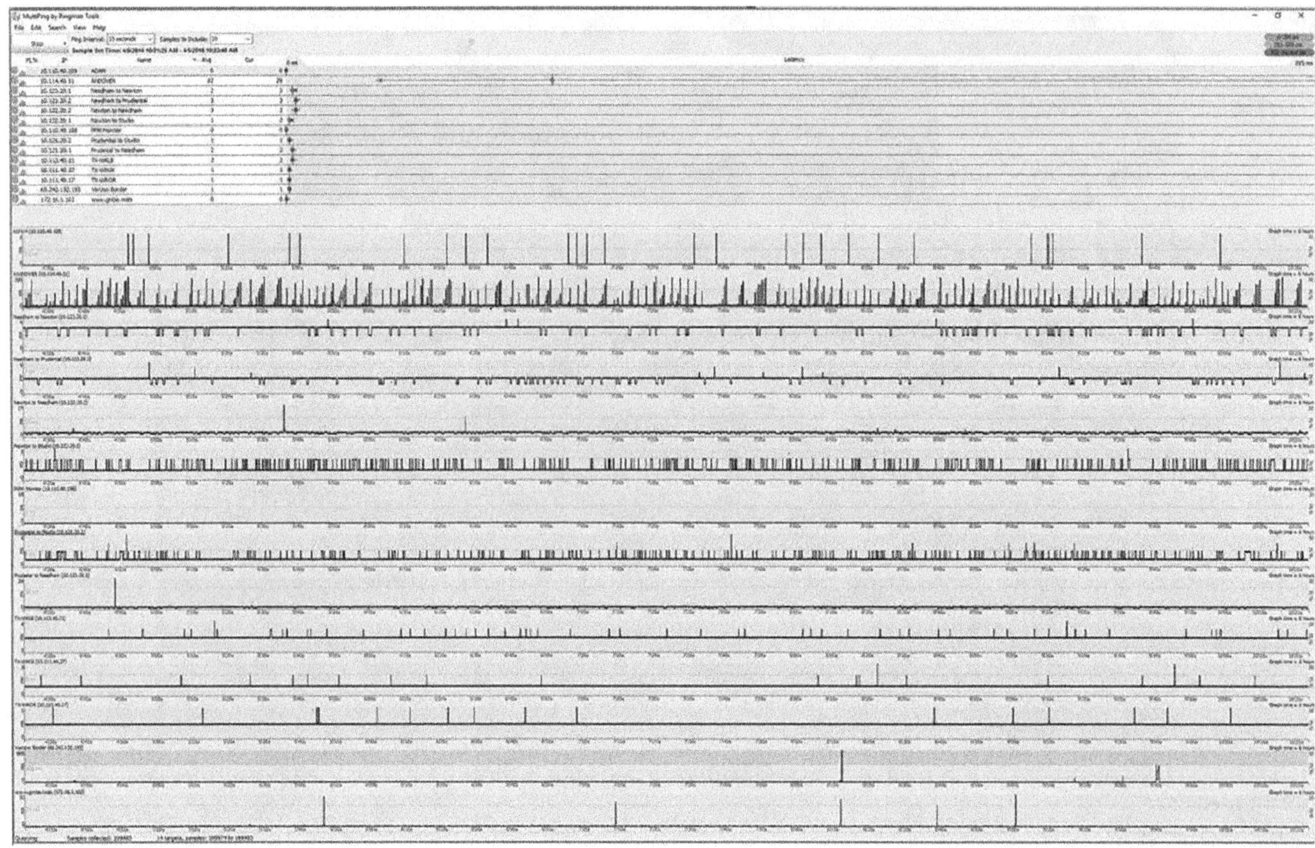

FIGURE 6.5-14 Screen shot of a typical ping monitor software package, displayed on a 30″ screen in master control

Courtesy Greater Media Boston

CHAPTER

6.6

Fiber Optic Transmission Systems

JEFF JUNIET

Director of Engineering, WFTV-DT (a Cox Media Station)
Orlando, FL

INTRODUCTION TO FIBER OPTICS

Fiber optics is a method of carrying information using optical fibers. An optical fiber is a thin strand of glass or plastic that serves as the transmission medium over which information is sent. It thus fills the same basic function as a copper cable carrying a telephone conversation, computer data, and audio or video. Unlike the copper cable, however, the optical fiber carries light instead of electrons. In so doing, it offers many distinct advantages that make it the transmission medium of choice for applications ranging from telephone calls, television, and machine control.

Development of Fiber Technology

The earliest work in optical communication came from Native North Americans and the Chinese in the form of smoke signals. The French also had the optical telegraph late in the 18th century. The 19th century saw the work in light and optics moving into controlling the direction of the light. In 1841 Jean-Daniel Colladon, a Swiss professor at the University of Geneva, was able to show the fluid flow through various holes in a tank and the breaking up of water jets by collecting and piping sunlight through a tube and focusing the light to the water tank. Passing through the water tank, the light was made to strike the edge of the jet at a glancing angle. Total internal reflection within the jet trapped the light in the liquid forcing it to follow the curved path until the water jet broke up. Instead of traveling in a straight line, the light followed the curvature of the water flow.[1]

Later, in 1870, John Tyndall was able to demonstrate that light used internal reflection to follow a specific path. As water poured out through the spout of one container, Tyndall directed a beam of sunlight at the path of the water. The light followed a zigzag path inside the curved path of the water.[2] This simple experiment marked the first research into the guided transmission of light.

Other experiments followed to look for a practical application of guided light, including Alexander Graham Bell developing an optical voice transmission system he called the *photophone*.

The 20th century saw a great jump in the development of fiber optic technology. In the 1950s development of lasers and light-emitting diodes (LED) as transmission sources for communications was realized followed by semiconductor lasers in the early 1960s. However, open-air laser signaling was far too inefficient and the fiber available at the time was only usable for a short distance. Refinements in the glass used to transmit the light came along in 1970 when Corning developed a glass fiber that exhibited attenuation at less than 20 dB/km, the threshold for making fiber optics for communication a viable technology.[3]

Early Applications

The U.S. military moved quickly to use fiber optics for improved communications and tactical systems. In the early 1970s, the U.S. Navy installed a fiber optic telephone link aboard the USS *Little Rock*. The Air Force followed suit by developing its Airborne Light Optical Fiber Technology (ALOFT) program in 1976.

Encouraged by the success of these applications, military R&D programs were funded to develop stronger fibers, tactical cables, ruggedized, high-performance components, and numerous demonstration systems ranging from aircraft to undersea applications.

Commercial applications followed soon after. In 1977, both AT&T and GTE installed fiber optic telephone systems in Chicago and Boston, respectively. These successful applications led to the increase of fiber optic telephone networks. By the early 1980s, single mode fiber operating in the 1310 nm and later the 1550 nm wavelength windows became the standard fiber installed for these networks. Initially, computer, information network, and data communications designers were slower to embrace fiber, but today they, too, find a use for a transmission system that has lighter weight cable, resists lightning strikes and electromagnetic interference (EMI), and carries more information faster and over longer distances.

The broadcast industry also embraced fiber optic transmission. In 1980, broadcasters of the Winter Olympics in Lake Placid, New York, requested a fiber optic video transmission system for use in backup video feeds. The fiber optic feed, because of its quality and reliability, soon became the primary video feed, making the 1980 Winter Olympics the first application of fiber optic television transmission. Later, at the 1994 Winter Olympics in Lillehammer, Norway, fiber optics transmitted the first ever digital video signal, an application that continues to evolve today.

In the mid-1980s the U.S. government deregulated telephone service, allowing small telephone companies to compete with the giant, AT&T. Companies like MCI and Sprint quickly went to work installing regional fiber optic telecommunications networks throughout the world. Taking advantage of railroad lines, gas pipes, and other natural rights of way, these companies laid miles of fiber optic cable, allowing the deployment of these networks to continue throughout the 1980s. However, this created the need to expand fiber's transmission capabilities.

In 1990, Bell Labs transmitted a 2.5 Gbps signal over 7500 km without regeneration. The system used a soliton laser and an erbium-doped fiber amplifier (EDFA) that allowed the light wave to maintain its shape and density. In 1998, they went one better as researchers transmitted 100 simultaneous optical signals, each at a data rate of 10 Gbps, for a distance of nearly 250 miles (400 km). In this experiment, dense wavelength division multiplexing (DWDM) technology, which allows multiple wavelengths to be combined into one optical signal, increased the total data rate on one fiber to one terabit per second (= 1000 gigabits per second).

ADVANTAGES OF FIBER

Fiber optic technology has proven to be a tremendously valuable medium for communications systems. Fiber is lighter and smaller than copper wire. A dense bundle of multiple fiber strands can be installed in the same space as only a few copper cables and can carry signals over one hundred times the distance. For example, a copper coaxial cable with a diameter of 0.274 inches will fill a 1-inch conduit to about 40% with five cables where 35 fiber optic cables each with a diameter of 0.11 inches can be placed in the same space. While care must be taken not to damage the fiber strands when installing fiber optic cable, the size and light weight make adding fiber to a facility relatively simple. Because fiber optic cables are made of glass or plastic instead of a metallic conductor, the transmitter and the receiver are electrically isolated. Along with the physical advantages, fiber provides a large bandwidth; for some applications, over a thousand times the bandwidth of a coaxial cable.[4]

There is no need for a connection to electrical ground to transmit data using fiber. This electrical isolation makes fiber ideal for high radio frequency (RF) or EMI environments such as in a broadcast transmitter or near electric motors. Electrical isolation also allows signals to travel without generating interference or EMI. Being decoupled from ground eliminates ground loops and can also save equipment from lightning strikes and sparks.

Bandwidth could be viewed as the greatest advantage of fiber optic transmission. With a capacity significantly greater than copper, fiber cabling has advantages in applications for data, audio and video networks. The major limiting factor of copper, reactive capacitance, does not apply to fiber, though this is not to say that fiber has no limitations.

The mode, or the way the light travels down the fiber, determines the bandwidth of the connection. The further down the fiber the signal travels, the more the light will disperse and the signal will ultimately become unusable. The topics of modes and dispersion will be reviewed in greater detail later. The bandwidth of an optical fiber directly relates to the data-carrying capacity and is expressed as the product of the data frequency and the distance traveled (MHz-km or GHz-km, typically). For example, a fiber with a 400-MHz-km bandwidth can transmit 400 MHz for a distance of 1 km, or it can transmit 20 MHz of data for 20 km.[5]

LIGHT

As mentioned earlier, the first experiments with controlling the movement of light were to move visible light in a medium (water). Fiber optic systems for media and data networks operate in the infrared portion of the spectrum. The visible-light portion of the electromagnetic spectrum falls between 400 and 700 nanometers (nm) and the infrared portion is just above that, between 800 and 1800 nm. Contained within that range of infrared wavelengths are the optical "windows" that fiber optic networks use to transport data. These windows are defined by the operating wavelengths that fall in between attenuation peaks.

There are advantages and disadvantages to each operating window ranging from system performance versus the cost to application and distance. The first optical window, at 850 nm, has the advantage of using low-cost lasers and detectors but has a high 3 dB/km loss. The second and third operating windows, 1310 and 1550 nm, offer better loss figures—0.5 and 0.2 dB/km, respectively, versus the first window—but the cost of installation is higher. The optical windows and their relationship to attenuation and wavelength are shown in Figure 6.6-1.

Within the second optical window, the International Telecommunication Union (ITU) has specified bands for use with single mode fiber, shown in Table 6.6-1. The ITU specification recommends the U-band for infrastructure testing and maintenance and not for signal transmission.

Index of Refraction

In a vacuum, light travels at approximately 186,284 miles per second in a straight line. However, these parameters change when light travels through a material with an index of refraction other than one. The index of refraction of a material will determine the speed of the light traveling through it and can be calculated by 186,284 / n, where n is the index of refraction. Because the fastest speed light can travel is in a vacuum, the index of any material must be greater than 1. The index of refraction for several types of materials are listed in Table 6.6-2.

In addition to the change of speed, when light travels through two materials with different indices of refraction there is a change in direction. A common example of the bending of light is viewing an object submerged in water. If you were to reach for the location where the object appeared to be, you would reach past the object. The differences in the index of refraction of the air and the water cause the change in direction of the light.

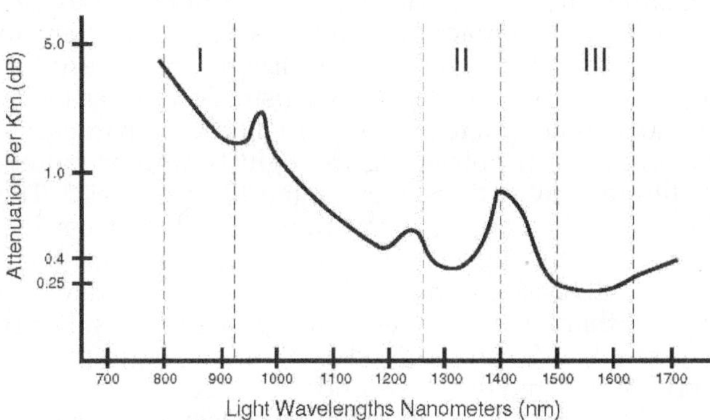

FIGURE 6.6-1 Fiber attenuation versus light wavelength characteristics

TABLE 6.6-1
ITU-specified bands for use with single mode fiber

"Original" or O-band	1260 nm to 1360 nm
"Extended" or E-band	1360 nm to 1460 nm
"Conventional" or C-band	1530 nm to 1565 nm
"Short wavelength" or S-band	1460 nm to 1530 nm
"Long wavelength" or L-band	1565 nm to 1625 nm
"Ultra-long wavelength" or U-band	1625 nm to 1675 nm

Snell's Law

Snell's Law expresses the relationship between the index of refraction and the angle that the light bends. The angle the light makes with a line perpendicular to the junction of the two materials. Snell's Law is shown mathematically as

$$n_1 \sin\theta_1 = n_2 \sin\theta_2$$

Using this information we can gain an understanding of how light travels down a fiber. If n for material 1 is greater than n for material 2, then the angle of refraction for material 2 will be greater than the angle for material 1. This principle can be illustrated in Figure 6.6-2.

TABLE 6.6-2
Index of refraction

Material	Index
Vacuum	1.00000
Air at STP	1.00029
Ice	1.31
Water at 20°C	1.33
Ethyl alcohol	1.36
Typical crown glass	1.52
Sapphire	1.77
Diamond	2.417

http://hyperphysics.phy-astr.gsu.edu/hbase/tables/indrf.html

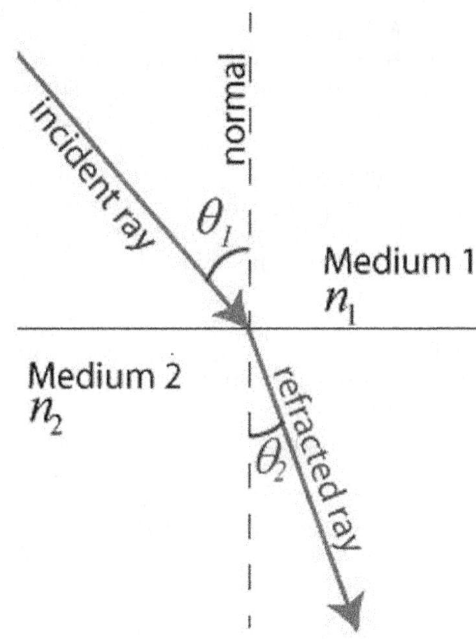

FIGURE 6.6-2 Illustration of Snell's law for two materials with different indices of refraction

As we increase θ_1, then θ_2 will also increase. θ_2 cannot increase past 90 degrees, or the light will reflect rather than refract through the junction and into the material. Once $\theta_2 = 90$ then the angle calculated for θ_1 is called the critical angle θ_c.

Optical fiber consists of a core surrounded by cladding. Both the core and the cladding are optically transparent materials, but the material used to make up the core has a greater index of refraction than the material used to make the cladding. As long as the light travels along the core at angle θ and this angle is greater than the critical angle θ_c the light will be continually reflected back into the fiber. Because the light never leaves the core, this is called total internal reflection.

Propagation

Light moving down the core of fiber optic cable can be viewed similarly to microwaves traveling through waveguide. The signal reflecting back into the fiber core from the junction of the core and the cladding defines the mode of propagation. Modes of propagation, dispersion, and scattering combine to determine the way light moves down a fiber and the strength of the signal available at the output.

Modes

Modes of propagation for the signal are determined by factors such as the wavelength of the light, the indices of refraction of the core and cladding, and the size of the core. The light waves refracting down the fiber all travel at different angles creating the mode. Fiber optic modes are ordered from high to low where the mode of order m will have m peaks. The smaller the diameter of the fiber, the fewer number of peaks that can be created by the signal and the lower the order of the mode. With single mode fiber, the lowest order mode travels down the core without reflecting off of the sides and is defined by the cutoff frequency. The smallest wavelength that will propagate just the fundamental mode is the cutoff frequency. Any longer wavelength will produce higher-order modes in the fiber and a corresponding loss of energy into the cladding. Because single mode fiber maintains the majority of the signal's energy within the core of the fiber, this is an ideal mode for long distance transmission.

In multimode fiber, multiple rays of light travel down the fiber cable, including both the low-order and higher-order modes. The multiple beams are traveling different length paths down the fiber creating dispersion at the output. As the light reflects down the length of the fiber, the waves of light interact, adding and canceling, similar to signals in a waveguide. Also, as the light travels down the fiber the beam does not completely reflect back into the core of the cable, but some of the light enters the cladding. This portion of the signal is lost to heat in the cladding and attenuates the signal as it travels the length of the fiber. All of the factors acting to attenuate the signal traveling down the fiber limit the useful distance of multimode fiber.

Dispersion

As the light pulses travel along the fiber, they broaden or stretch. As the pulses broaden they begin to overlap and deteriorate the signal, eventually making the data being transmitted in the fiber unrecoverable. This form of attenuation can have a significant impact on the signal traveling through the fiber. Dispersion can be mitigated by the use of specially designed fiber.

Chromatic dispersion is wavelength dependent and is the result of the differing wavelengths in the fiber traveling at different velocities. All of these varying wavelengths are present at the light source and spread out over time across the distance of the fiber. All data rates are susceptible to the adverse effects of chromatic dispersion, and when the distortions of the signal become too great, errors and loss of the information being transmitted can occur.

Modal dispersion occurs in any order modes but is of particular concern in multimode fiber. As the multiple modes of light travel different paths down the fiber, they reach the receiver at different times causing the dispersion of the pulses. Unlike chromatic dispersion, modal dispersion is not wavelength dependent. Instead, this type of dispersion results from the light wave expanding while traveling as expected down the fiber. Chromatic dispersion is measured in ps / (nm-km) (picoseconds per nanometer wavelength change and kilometer propagation distance. Fiber containing intentional impurities, known as dispersion-shifted fiber, can be implemented to offset chromatic dispersion.

Polarization mode dispersion (PMD) is dispersion resulting in the imperfections of the fiber cabling. In addition to manufacturing imperfections, the fiber can be stressed during installation or splicing to result in an asymmetry of the fiber. This distortion from a cylindrical geometry creates birefringence, a characteristic where each polarity of the light beam encounters a different refractive index in the fiber material. The two polarities are now traveling at different speeds, spreading the light pulses over the distance of the fiber. PMD becomes noticeable when the data rates are greater than 5 Gbps; lower data rates are not as severely impacted. PMD is described as differential group delay and is measured in picoseconds. PMD compensators or polarization-maintaining fiber are designed to induce birefringence to minimize the PMD. Compensation is typical on long-haul fiber transmission, especially at high data rates.

Scattering

Scattering of the light traveling down the fiber is a significant source of attenuation. Scattering is the change of optical power from a propagating mode to a non-propagating mode, with the light photons moving away from a straight path. Impurities in the fiber cause density variations resulting in points where the refractive index varies in the core, scattering the beam. This form of scattering is referred to as Rayleigh scattering and is wavelength dependent.

Operating at longer wavelengths lowers the amount and effect of Rayleigh scattering in a fiber. Similar to Rayleigh scattering is Mie scattering, resulting from misshapen fiber. A fiber core that is not round or has imperfections at the core cladding junction will create Mie scattering. Carefully controlled manufacturing and a high-quality fiber limit Mie scattering. Both Rayleigh and Mie scattering are linear types of scattering, whereas Brillouin and Raman scattering are nonlinear forms of light scattering. Brillouin scattering is the result of nonlinearities in the material used for transmission. Photons are converted from higher energy to lower energy and are propagated, usually backward, from the interaction with the acoustic phonons of the material via electrostriction. Raman scattering involves the nonlinear response of optical phonons. The effect on the light beam in fiber is the transmitting of the differential energy of high energy photon and the converted lower energy photon by the optical phonon.

THE FIBER MEDIUM

The fiber medium that the light is traveling on plays a significant role in the propagation and attenuation of the signal. The construction of the fiber needs to meet specific standards to minimize the adverse properties of light traveling through the medium. Fiber optic systems may be used to move the broadcaster's content around the facility, to a transmitter site, or connecting multiple sites to share or distribute content. Whatever the application, the basic building blocks of a fiber transmission system are the fiber optic cable, the transmitter, and the receiver.

The Fiber

The basic construction of a fiber cable is the core, the cladding, and the jacket. The core and the cladding are the pieces that combine to propagate the wave. The jacket, sometimes consisting of multiple materials, provides strain relief and protection. The type of fiber is determined by the operating mode of the signal, single mode or multimode. The core and the cladding of the fiber are made of glass or plastic or a combination of the two, with the core having a higher index of refraction than the cladding. The more expensive glass fiber is a lower-loss fiber than plastic fiber. However, for some applications, the lower cost of the plastic fiber makes it desirable. The difference in the indices of refraction is achieved by doping the material used for the core and the cladding. Germania, phosphorus pentoxide, and alumina are common dopants used to raise the index; fluorine or boron oxide can be used to lower the index. A combination of dopants may be used to raise or lower the index as required.

The amount of signal that can travel down the fiber is determined by how much light enters the fiber. With core diameters of 8 to 100 microns (1 micron = 1 millionth of a meter), the light needs to be focused properly in order to enter the core of the fiber. Snell's Law dictates that light travels down the fiber at angles, reflecting back and forth along the fiber. The light must also enter the fiber at the correct angle. This is the acceptance angle, or cone, of the fiber. The light that is incident on the end of the fiber at an angle greater than the acceptance angle will enter the fiber but will not be reflected back into the fiber. Instead of propagating down the fiber this light will enter into the cladding and be greatly attenuated or lost as heat. Light entering the fiber at an angle less than or equal to the acceptance angle continues down the core of the fiber. This light collecting ability of the fiber is measured by the numeric aperture (NA). This information is usually found in the fiber manufacturer's data sheet. The numeric aperture (NA) is defined as:

$$NA = \sqrt{n_1^2 - n_2^2}$$

where n_1 is the index of refraction for the core and n_2 is the index of refraction for the cladding.

Fiber Types

A multimode fiber system allows the transmission of multiple wavelengths of light to travel in differing modes along the length of the fiber. These multiple modes of light are subject to the effects of group delay and modal dispersion limiting the bandwidth. This is not to say that multimode fiber is undesirable; for lower bandwidth needs multimode fiber can be cost-effective. With a core of 50 μm or greater, multimode fiber gives a greater numerical aperture, making the coupling of light into the fiber easier and giving wider tolerances for connectors. Additionally, less expensive light sources and in some application less expensive plastic fiber can be used. Multimode fiber is used in many data centers and LANs for cost reasons.

As previously discussed, a single mode light transmission is a single wavelength traveling down the center of the fiber core. The size and composition of the fiber's core determines the mode of the signal. The diameter of single mode fiber is on the order of 5–12 microns with a cladding of 125 microns. Because modal dispersion is not an issue, single mode fiber can provide large bandwidths in the GHz range and transmission of several kilometers.

To this point, we have looked at the junction between the core and the cladding as the single point where two indices of refraction meet, and for some fiber construction this is the case. These are known as step-index fibers. In a step-index fiber, the core and the cladding have a consistent index of refraction. It is at this single junction that the light is reflected back into the core. A step-index multimode fiber allows for the maximum number of modes to develop based on the size of the core. Both multimode and single mode fiber can be step-index construction.

There is also what is known as graded-index fiber. This design has a graded change in the index of refraction from the center of the core to the junction with the cladding. The core is made of multiple layers each with a lower index of refraction. This design allows the light rays closer to the cladding, the modes that are on the longest path down the fiber, to travel faster and match the rays traveling the shorter path closer to the center axis. This design compensates for some of the modal dispersion of a multimode signal and allows for higher bandwidth and greater distance.

Because single mode fiber is only transmitting a single wavelength and modal dispersion is not an issue, graded-index single mode fiber is usually only used in special applications. Pulse broadening from polarization mode dispersion is a factor in single mode fiber, and there are fibers designed to offset this distortion of the signal. In the 1550 nm band, a single mode fiber exhibits significant dispersion. The 1310 nm band does not have the dispersion issue but has a greater loss than the 1550 nm band. By modifying the refractive index of the core of the fiber the zero dispersion properties of the 1310 nm band are pushed to the 1550 nm band. This fiber design is dispersion shifted fiber.

Fiber Material

Fiber optic cable is manufactured with two materials, glass and plastic. The all-glass, core, and cladding type of fiber is the most common and provides the best performance. This fiber is typically an ultra-pure, ultra-transparent, silicon dioxide or fused quartz glass. During the manufacturing process, impurities are added to obtain the desired index of refraction. Some of the impurities used raise the index, such as germanium or phosphorus pentoxide, and the index can be lowered using fluorine or boron oxide. There may also be other impurities that do not get removed during the fabrication process that increase the attenuation from absorption and scattering.

Fiber optic cabling is also manufactured using plastic, usually polymethyl methacrylate (PMMA), also known as acrylic. Plastic optical fiber (POF) has high loss and low bandwidth. However, for applications where losses are acceptable POF can be very cost effective. In addition to being inexpensive to manufacture, POF is rugged and has a tight bend radius allowing for installation in adverse environments such as industrial machinery. POF also has a large acceptance angle enabling these cables to be used with inexpensive transmitters and receivers. Some of the common applications of POF are home networks, digital home appliance interfaces, and car networks.

Plastic-clad silica (PCS) fiber optic cable is made with a glass core and a plastic cladding. PCS fiber has lower performance characteristics than glass fiber and can be difficult to work with when attaching connectors or splicing.

Coatings

Fiber optic cable is actually strong relative to the size of the strand, but in spite of this strength the fiber needs protection. Scratches and moisture damage the fiber and increase attenuation and loss of light. Over the cladding is a coating, usually several layers of plastic, to aid in strengthening the fiber and act as a shock absorber. This buffer coating is removed when making a cable splice or attaching a connector.

Next, packaging of the fiber in loose tube and tight buffer construction provides the needed protection from the environment. In loose tube construction, the fiber(s) are contained in a gel-filled tube along with strength members to keep the fibers free from stress. Tight buffer fiber uses a layer of plastic directly over the fiber coating. Tight buffer construction has better resistance to crushing and is more flexible than loose tube and is usually preferred for indoor installations. Loose tube better isolates the fiber from temperature fluctuations and moisture, making this construction type preferred for outdoor applications.

Additional strengthening and protection elements may be added to the fiber construction that have advantages in certain installation environments and applications. Strengthening members frequently used in fiber optic cable include Aramid yarn, fiberglass epoxy rods, and steel wire. The fiber may also be protected by an armor jacketing of corrugated steel that adds crush and expansion protection along with rodent proofing for outdoor installations.

Connectors

Optical connectors provide a pluggable interconnect for assembling a fiber network. The type of connector chosen can depend on the application and the equipment being connected. Like an electronic connection, an optical connection depends on a solid physical connection (PC), but of equal importance is proper alignment. Connectors need to be in alignment and the size of the fiber core and cladding need to be matched to maximize the transfer of the light. The addition of a connector to a fiber cable will increase the insertion loss of the cable. Insertion loss is the reduction in optical power across a junction and even in the most precision of connector, there will be some loss.

Insertion loss is calculated as:

$$IL = 10 \log(P_1 P_2)$$

where IL is the insertion loss, P_1 is power through the connectorized fiber, and P_2 is the uncut fiber.

Another consideration when selecting an optical connector is the return or reflection loss. This loss is the measurement of the power reflected back down fiber toward the source or junction expressed as a negative dB value with a higher the value being better. Reflected light can be severely detrimental especially to longer wavelength optical signals, 1500 nm and longer. There are connectors that reduce the return loss by angling the end of the connection. The angled physical contact (APC) connector angles the reflected signal back into the cladding of the fiber. APC connectors cannot be used with a nonangled connector.

Traditional methods of terminating fiber have involved cleaving, polishing, and the use of epoxy to attach the connector. This is a tedious method that usually requires practice to make a good connection.

Connectors that crimp or do not require any tools are now available and can make fiber termination easy and consistent.

TRANSMITTERS AND RECEIVERS

The other two parts of a fiber network are the transmitter and the receiver. Optical transmitters deliver the modulated optical signal to the fiber optic cable. The two most common optical sources are the light-emitting diode (LED) and the laser diode. Most LED-based transmitters operate in the 850 nm band, and some designs can operate in the 1310 nm band. LED sources emit light over an optical bandwidth and transfer several wavelengths into the fiber. These multiple wavelengths traveling down the fiber are prone to distortion of the signal from dispersion. Dispersion limits the practical distance for LED transmitters to below 10 km. LED transmitters also have bandwidth limitations but are usually less expensive than laser diode transmitters. A laser diode can operate in the 1310 and 1550 nm wavelengths and has a narrow optical bandwidth enabling transmission of a single wavelength. Laser diodes are more expensive but are suitable for long distance and high data rate transmissions. In order to maximize the transfer of optical power into the fiber, an optical transmitter must be able to transfer the light into the acceptance angle of the fiber requiring a small output aperture. A lens may be used to improve the coupling from the emitter to the fiber. The amount of power transferred into the fiber can be calculated by

$$P_F = \frac{1}{2} P_s (n+1) NA$$

where:

- P_F = the power coupled into the fiber
- P_S = the power from the source
- n = the index of refraction of the fiber's core
- NA = the numerical aperture of the source.

Fiber networks can transport both analog and digitally modulated audio, video and data signals. In addition to AM and FM modulation, digital modulation schemes are used in fiber links such as phase-shift keying (PSK), quadrature phase-shift keying (QPSK), and quadrature amplitude modulation (QAM). With possible data rates of 100 Gbps for QPSK and 400 Gbps for QAM, these are two frequently used digital modulation schemes used for video transport. QAM is a complex modulation scheme but offers a high data rate and a low bit error rate (BER), making QAM desirable for today's high bandwidth and high-resolution video formats.

Receivers

The final piece of our fiber link is the receiver. Receivers consist of three basic parts, the detector, the amplifier, and decoder. Detectors come in two varieties: the positive intrinsic negative (PIN) type or the avalanche photodiode (APD) type. The PIN (or PN) photodiode is the more common of the two, based on the cost of construction and operation. Because the PIN photodiode needs only a DC voltage of 5–20 volts and does not require a large current to operate, the power supply for an optical receiver using a PIN is less expensive. The APD type has considerably more sensitivity than the PIN type; on the order of 5–10 dB or higher. The APD design requires a highly stable voltage of about 100–200 V. Because of their higher cost and increased sensitivity, APD units are used for long-haul connections requiring the detection of a signal traveling hundreds of kilometers or more.

Because the output from a detector circuit is a small signal, an amplifier is needed to feed the decoder portion of the receiver. The quality of the amplifier is going to play a critical role in the performance of the receiver and therefore needs to be a low noise device and needs to be able to amplify the signal linearly for about 20–40 dB of gain. After the amplifier section, there may be wave shaping and filtering circuits to refine the waveform further before feeding into the decoder. The decoder circuit simply demodulates the information from the converted light wave. A decoder can be designed to demodulate either analog or digital signals, however digital signals in the video world are becoming the norm.

Because the receiver is performing the opposite task of the transmitter, these are usually sold as pairs. A matched pair of transmitter and receiver would be especially critical in an encrypted link. For most purposes, a transmitter and receiver are housed in the same package (a transceiver) and are matched with another transceiver unit for a bidirectional link.

FIBER NETWORK OR LINK CONSTRUCTION

Several factors play into the successful transmission of information across a fiber link including transmitter power, the receiver's sensitivity, losses in the fiber, and coupling losses. All of these together go into determining the link budget (also known as loss budget or power budget) of a fiber system. There are several online calculators available to determine the link budget of a system, but the calculations are not complex. Most of the data needed to determine a link budget is found in manufacturer's data sheets. Some manufacturers suggest a 2–3 dB link margin for unknown losses such as patch cables and bend losses. The link budget gives a good approximation of the loss versus the available optical power margin of a fiber link.

Types of Loss

As we have seen, there are losses associated with fiber construction. The typical attenuation of the fiber is given in the manufacturer's data sheet and is calculated as the length in km multiplied by the loss in dB/km (line attenuation = $L_{km} \times a_{dB/km}$ L). As discussed, dispersion also adds to the attenuation of a fiber link. As the length of the cable run exceeds 5 km for multimode systems and about 15 km for single mode, the effects of dispersion in ps / (nm-km) (picoseconds per nanometer-kilometer) begin to impact the distance and the bandwidth of the link. For a simple link budget calculation, dispersion is not usually added in but should be considered for long-haul and complex transport systems.

Other common losses are from the coupling loss. These losses would be from patch panel and adapter connections, and splices. The typical loss for a patch panel is between 0.5–0.75 dB of loss, and the loss of a precision adaptor is about 0.1 dB. For a rough estimate of a link budget, the individual connectors are not included. However, for precision links and links with multiple patches even these small losses should be considered.

Losses from splicing the fiber can be a little more complicated. A poor splice can create a point of high loss and high reflection in the opposite direction. When splicing fiber cable, both sides of the splice must be of the same type of fiber in both physical dimensions and refractive index. Misalignment or a gap in the fibers when making a splice will also create problems with the signal reaching the receiver. A well-done splice will have a minimal loss of about 0.1 dB. The most common type of splice is the fusion splice, but simpler mechanical splice kits are still available. A fusion splice will usually have a loss of about 0.1 dB whereas a mechanical splice will have about 0.5–1.5 dB of loss. Fusion splicing equipment has traditionally been expensive but the price has started to drop, and some large facilities have found the purchase of a fusion splicer cost-effective. For link budget purposes each splice should be counted. Once the individual losses are calculated, simple addition allows us to find the total attenuation.

Once the total loss is determined, this is compared to the optical power margin that is simply the difference in the minimum transmitter launch power and maximum receive sensitivity of the receiver. For example, if a transmitter has a −6 to −8 dBm output (dBm is the power level expressed relative to 1 mW) and a receiver has an input of −26 to −28 dBm, and the link has two patch points with a loss of 0.75 each, a splice with 0.1 dB of loss, and a 750 m length using a multimode cable at 1310 nm with a loss of 1.5 dB/km, we would have the following total attenuation (TA):

$$\text{TA} = (n \times \text{C}) + (c \times J) + (\text{L} \times a) + M$$

where:

- n = number of connectors (patch points)
- C = attenuation for one optical connector (dB)
- c = number of splices in the elementary cable section
- J = attenuation for one splice (dB)
- L = total length of the optical cable in km
- a = attenuation for optical cable (dB/km)
- M = system margin (patch cords, cable bend, unpredictable optical attenuation events, and so on, should be considered around 3dB).

So for this example,

$$\text{TA} = (2 \times 0.75) + (1 \times 0.1) + 0.75 \times 1.5) + 3 = 5.725 \text{ dB}$$

We can calculate the optical power margin as

$$(-6) - (-26) = 20 \text{ dB}$$

Consequently, with 20 dB of margin and only 5.73 dB of estimated loss, we are well within the loss parameters of this link. This calculation is very useful in planning a fiber link, but a complete power measurement will give a complete and accurate picture of the status of a link.

Testing and Certification

A visual inspection of the fiber network or link can be helpful in finding problems with the installation but measuring the performance and finding faults and errors requires test equipment. Some equipment can be as simple as a visible light source. A simple continuity test for shorter distance links can aid in determining if a link will pass a signal. However, a continuity test will really only indicate that there is a complete length of fiber, and this may be all that is required for short links, but this test will not give an indication of bandwidth or loss.

Calculating a link budget will give a good estimate of the reliability of a fiber link, but a power loss test provides an accurate measurement of loss. Most power test sets consist of a light source and a power meter. The source unit generates a stable reference signal at the operating wavelength. The power meter measures the power exiting the fiber and displays the results in dBm or mW. Optical power, just like electrical power, is expressed as dBm or mW and calculated in the same fashion:

$$\text{P}_{\text{mW}} = 10^{\left(\frac{\text{P}_{\text{dBm}}}{10}\right)}$$

$$P_{dBm} = 10\log(P_{mW})$$

The tests just discussed are part of the Tier 1 Testing and Certification per the ANSI/EIA/TIA-568-C Standard. This level of testing will certify the fiber network's insertion loss, polarity, and length of the link. The length of the fiber cable is used to determine if modal dispersion will be a problem for the link in an installation.

Tier 2 Testing and Certification requires the use of optical time domain reflectometry (OTDR) for certification. The OTDR is able to detect faults, verify length, measure loss, and gives an accurate reading of the performance of a fiber installation. Fault detection is particularly useful on new installations to detect breaks, extreme bends, or poorly installed connectors. An OTDR operates by sending a pulse of light into the fiber and as the light is scattered in the core of the fiber some is scattered in a backward direction. Additionally, there are points of reflection in the fiber where the light is directly reflected back to the OTDR receiver, in particular, at connectors, splices or bends. The OTDR measures the light being returned to the unit, calculating the distance and the loss. A skilled technician with proper training and experience can use the data from an OTDR trace to pinpoint faults and get detailed performance measurements for the fiber network. However, most units available today have an automatic measuring mode and can be operated by a technician with a minimum of training and still produce sufficient results.

APPLICATIONS

Bandwidth demands for end users are continuously growing. Whether an Internet service provider or television station, media and data infrastructures need to keep up and fiber optic technology provides the backbone.

Telecommunications

Telecommunications companies started the adoption of fiber optic transmission for phone services in the 1970s. In 1977 GTE and Bell System sent telephone traffic over fiber. Now leveraging digital encoding, telephony technology carriers can transmit hundreds of millions of voice calls every day across fiber. This combination of digital technology and fiber optic bandwidth has become the physical backbone of not just telephone traffic but the Internet as well.

Data Transmission

Networks handling 40 Gbps data and Internet protocol (IP) video streams are the next frontiers for distribution, and the requirements for these are at or exceeding the limits of what can be done using copper cabling. Fiber optic transmission is enabling large leaps in data bandwidth and currently the most high end, and some midlevel, network gear is equipped with fiber connectivity. Fiber right-to-the-desktop workstations are not unheard of for new installations. Of course, the long distance capabilities of fiber enable not just distant parts of a facility to be connected, but multiple buildings on a campus can be connected to create a seamless network.

Industrial

With fiber's immunity to outside interference from EMI, there are numerous applications in industry as well as communications for fiber optic networks. Control, telemetry, and communications data are implemented in modern industrial machines and processes. A fiber optic connection can provide the bandwidth for a machine operator to control equipment and monitor the operating parameters and surrounding conditions remotely. Video can also be sent along the fiber connection, giving a high-resolution view of the work being performed. These types of applications have been deployed in the mining and drilling industries to protect workers and create more efficient processes. Automotive and aeronautical industries use fiber for similar real-time control and status applications. Fiber optics can be used in adverse or changing conditions such as the high temperatures of an engine, monitoring multiple parameters and sending the data back to a computer that will make the necessary adjustments to optimize performance.

Video Transport

The flexibility of fiber seems almost ready-made for broadcast content. Uncompressed high definition (HD) video requires on the order of 1.5 Gbps and is transmission of this is limited to about 150 m on coaxial cable. While short runs of signals between equipment racks can easily be handled by coaxial cable, moving video and high-resolution graphs from a studio across the building, or across a campus, to a tech center can be done quickly and usually less expensively with fiber. Fiber feeds can move video and graphics to a news studio for display with on-set monitors and video walls. Images and video from a weather graphics system located in the same studio can run across a fiber connection, reducing cabling costs and eliminating the concern for path induced signal degradation.

Sporting venues use miles of fiber infrastructure to show video and graphics on multiple displays around the arena or stadium and large format displays over and around the playing field or court. Fiber connections from video editing systems to shared storage allow real-time editing of large files in a collaborative environment for both TV stations and production facilities. HD video and surround sound audio have given way to 4K and 8K video with immersive audio with metadata going along with the video and audio. As the terminology might imply, these video formats are four and eight times the resolution, and bandwidth, of HD video. While these new formats are not yet being transmitted over the air (however, 4K is supported by the ATSC 3.0 Standard currently under development), acquisition and editing in 4K is becoming more affordable and desirable.

Multichannel distribution using compressed video transport streams over fiber allows links from broadcasters to various distribution points reliably, free from interference and leveraging the efficiencies of video compression and the bandwidth of fiber. Using transport stream multiplexing to transport multiple channels over a single fiber becomes an economical way to move multiple channels over long distances. Central distribution points such as a network operations centers, cable head-ends, or satellite teleport could also use this method of channel multiplexing over fiber to move signals around the plant to encoders.

Multiplexed Signals

Moving multiple pieces of data, audio, or video across a single fiber can be done using a multiplexed signal. Multiplexed systems can be found in both multimode and single mode applications and can be as simple as two channels or as complex as 80 channels.

Wavelength division multiplexing (WDM) is the combining of multiple wavelengths into one fiber.

This could be combining the signals at the transmit point and separating them back apart at the receiving point, or having two wavelengths traveling in opposite directions in a single fiber. Coarse wavelength division multiplexing (CWDM) uses wavelengths at regular intervals from 1270 to 1610 nm, the frequency band from the second through the third transmission windows, to create 18 channels that can be transmitted down a single fiber. Each channel has an individual bandwidth of 1 to 3.125 Gbps, giving enough total bandwidth to carry multiple HD video signals on a single fiber. Highly stable lasers are used for CWDM transmission to avoid any drift because even with the relatively large tolerance of 20 nm of spacing between the channels, significant data loss could occur if one wavelength were to interfere with an adjacent one.

Dense wavelength division multiplexing (DWDM) has a denser spacing between the channels and operates in the third operating window. Forty, 80, or 160 channels can be multiplexed into a single fiber using channel spacing of 100, 50, or 25 GHz, respectively. This type of multiplexing provides a potential overall bandwidth of hundreds of GHz with each channel operating between 1 and 100 Gbps. Temperature-stabilized lasers are required to maintain precise wavelengths for the channels. Additionally, dispersion and nonlinearities of the fiber need to be planned or compensated for to maintain the signals through to the output. Hardware and fiber expenses make DWDM financially viable in long-haul and very high bandwidth applications such as Internet backbone traffic.

Time division multiplexing (TDM) is also used for fiber transmission. In TDM the data from multiple sources is converted to light and the different channels are sequentially pulsed into the fiber. These sequential pulses give a varied arrival time for each channel at the receiver. To increase the number of channels the time between the pulses is reduced, but care must be taken to balance the pulse duration and channel power to avoid interference between the channels. A combination of WDM and TDM can be used to create multiplexed transports with even larger bandwidths potentially in the THz range.

RF Transport

Fiber is also able to transport RF in addition to baseband signals. Using the satellite teleport example again, the main hub facility for the signals could be centrally located in a metropolitan area while the actual dish antenna is several miles out of town. Everything, including the modulators, could be in the main plant leaving minimal equipment out at the transmission site. Fiber's immunity to EMI and lightning effectively keep distortions and noise out of the signal.

SUMMARY

This chapter has provided background information into fiber optic technology including how this technology is useful in the broadcast facility. The reader is encouraged to use this information as a starting point in learning about how fiber optics can be applied at his or her station. Higher resolution video and expanded audio channels along with web and metadata content are all in the future of broadcast. Bandwidth needs are growing each year as broadcasters expand from over the air distributors to the web, mobile and over the top (OTT) distributors. No longer used for specialty purposes, with the need for new formats and increased content loads, fiber optic connections could become as commonplace as coaxial cable in a broadcast facility.

Notes

[1] Brief History of Optical Fiber—Olson Technology Inc., www.olson-technology.com/mr_fiber/fiber-history.htm (accessed May 10, 2016).

[2] Ibid.

[3] Ibid.

[4] The Fiber Optic Association—Tech Topics, www.thefoa.org/tech/fo-or-cu.htm (accessed June 26, 2016).

[5] Fiber Optics: Understanding the Basics Photonics, www.photonics.com/Article.aspx?AID=25151 (accessed May 10, 2016).

Bibliography

Alwayn, Vivek. *Optical Network Design and Implementation*, Cisco Press, 2004, accessed June 26, 2016, www.ciscopress.com/articles/article.asp?p=170740Belden Cables website at www.beldencables-emea.com/en/index.phtml

Brief History of Optical Fiber—Olson Technology Inc. at www.olson-technology.com/mr_fiber/fiber-history.htm

Corning website at www.corning.com/worldwide/en.html

The Encyclopedia of Laser Physics and Technology at www.rp-photonics.com/encyclopedia.htmlFOA's Guide to Fiber Optics & Premises Cabling at www.thefoa.org/tech/ref/contents.html

Fluke Networks website at www.flukenetworks.com/

Goff, David R. *Fiber Optic Video Transmission: The Complete Guide*. Boston: Focal Press, 2003.

Photonics Handbook website at www.photonics.com/EDU/

SECTION

7

RADIO TRANSMISSION

CHAPTER

7.1

Planning Radio Transmitter Facilities

PAUL SHULINS

Director of Technical Operations, Greater MediaBoston

THOMAS R. RAY III

President, Tom Ray Broadcast Consulting

INTRODUCTION

Today's modern radio station transmitting facilities require a different kind of planning than was needed only a decade ago. The changes in technology, including different cooling methods for FM transmitters, HD Radio advances, and the ubiquitous availability of remote control and monitoring for just about every piece of gear, allows broadcasters to plan transmitter sites with more monitoring and control than ever before. Today different pieces of equipment can talk to each other, as well as report back to the studio or engineer charged with maintaining the proper operation of the facility.

This chapter will be presented in two parts. The first part will deal with the considerations for planning a new FM transmitter facility, and the second part will deal with planning AM transmitter sites.

Keep in mind that decisions made today should be made with the future in mind. One example is bringing in the correct amount of HVAC and electrical power today to accommodate likely future expansion either for your company or other tenants that you may host. This type of planning can save significant time and money in the long run.

Also the choices made in the planning stages have considerable leverage over the long-term viability and operation of the site. Some factors must be considered early, such as size; layout; electrical power; heating, ventilation, and air conditioning (HVAC); and radio frequency (RF) screening. Regulatory matters such as local tower regulation, zoning, and building codes must also be researched before committing to a particular locale.

FM TRANSMITTER FACILITY PLANNING

Paul Shulins

When planning a new FM transmitter facility, thought should be given to the location first. Realizing that FM is basically a line of site service, it is generally advantageous to locate your transmitter on a hilltop, mountaintop, or a tall building that has good line of site to the community you are licensed to serve and the population that you are interested in reaching. The costs associated with the construction vary greatly and are dependent on many factors. While erecting a short tower on a tall building may seem like an easy solution to cover a community in an urban area, some of the major considerations are local zoning, Federal Aviation Administration (FAA) approval, building lease costs, availability of adequate AC power, and proximity to other broadcasters.

Normally there are regulations that restrict the type of fuel used for backup power generators in tall buildings. If backup power is important to you, and you are considering locating your FM transmitter on a tall building, researching the possibility of backup power with the city and building landlord before committing to a lease will save you some time.

Often, tall buildings are home to many broadcasters and public radio services like fire and police. It is quite common for a new broadcaster to install equipment properly; however the result of the installation may result in undesired intermodulation (IM) products. This energy, when it reaches a certain level, can be disruptive to legitimate licensed broadcast stations, and is prohibited by the Federal Communications

Commission (FCC). Generally the last station on the air from a multi-use site is required to ensure that any intermodulation products meet or exceed the FCC's specification, and if those products do not, generally the new broadcaster has the responsibility to clean up the site by remedying the problem. This is usually solved by installing specially tuned filters to trap the undesired energy and prevent it from being radiated. Not all interference is a result of intermodulation distortion. It can come from co-channel interference, atmospheric conditions, and man-made noise generated by medical, welding, and heating equipment. Most intermodulation occurs in a transmitter's nonlinear power amplifier (PA). The next most common mixing point is in the front end of a receiver. Usually it occurs in the unprotected first mixer of older model radios or in some cases an overdriven RF front-end amp. Intermodulation can also be produced in rusty or corroded tower joints, guy wires, turnbuckles and anchor rods or any nearby metallic object, which can act as a nonlinear "mixer/rectifier" device. Normally when a new broadcasters joins a site, the FCC requires an IM study to be done after the new transmitter is turned on to make sure that intermodulation products remain below a certain level to ensure there is no new undesired signal generated that could interfere with other licensed services.

HD Radio Considerations

HD Radio is becoming more popular today, and planning a new facility with the capability of this type of transmission usually makes good sense. In the past there were different methods in use to generate the HD Radio signal. One popular method involved separate amplification where a separate HD Radio transmitter was employed for the digital signal, leaving the traditional analog transmitter alone. This technique involved separate transmission lines, feeding either a common antenna or separate antennas. The method worked well but was complicated by the need for magnetic circulators in series with the digital signal to prevent the two types of transmitters from mixing signals. Today a technique called low-level combined HD Radio is widely used and is considered a more elegant solution. This low-level combined method requires only one transmitter, one transmission line, and one antenna. It makes good sense to look into transmitters that are capable of this technique, and get quotes on models that have the power capability to satisfy your licensed needs. Currently the FCC rules allow for most stations to broadcast an HD signal at a level of at least 14 dB below the analog carrier. In many cases it is possible to be authorized to broadcast at levels as high as 10 dB below the analog carrier. Before constructing a facility, look into the maximum power level you are able to legally broadcast at, and size your transmitter and transmission lines appropriately. In the future even higher power levels may be authorized, so keeping a conservative view on the future of HD Radio may be wise to plan for enough headroom to accommodate higher power levels in the future.

FIGURE 7.1-1 Example of some of the many users of a crowded rooftop in a major city. This illustrates the need for proper filters to avoid intermodulation products.

RF Radiation Considerations

Another consideration for tall buildings is the radio frequency exposure limit. Today there are state and federal radio frequency radiation (RFR) guidelines that specify the maximum radiation that a human can be exposed to in the proximity of a radiating antenna. On a typical tall building rooftop, it is very likely that there are areas that are routinely required to be accessed by contractors servicing HVAC units, window washing machinery, roofing maintenance, painting, and so forth. In most cases a careful survey of the rooftop is required to define the areas that are safe and unsafe for both public and occupational exposure. The power level, antenna type, frequency, and physical location are all factors that need to be considered when locating a new antenna on a building rooftop. If in fact you are adding to the RF environment as a new proposed tenant, the chances are likely that you will be responsible for a new RFR survey to determine the impact of the new station on the environment. This additional cost should be factored into your construction budget.

In some cases it is desirable to install an auxiliary antenna on building rooftops, such that the physical placement and power levels allow for operation from this auxiliary antenna while keeping the RF levels within acceptable standard around the master antenna so that the tower crew can work during daylight hours without the need to take the station(s) off the air or transition to a geographically diverse site. In these cases where a master antenna filter system is not employed, individual filters may be required to prevent RF products from mixing and causing spurious products to be radiated. Figure 7.1-4 illustrates one example of smaller filters for an individual auxiliary antenna.

FIGURE 7.1-2 Typical transmitter site layout showing rack equipment

FIGURE 7.1-3 Photo of the roof of the Prudential Building in Boston showing guidelines for RF safety limits

Master Antenna Systems

An elegant solution to the problems stated earlier is to employ a master antenna system that is normally a shared resource, and that can handle many FM radio stations at once. Normally an optimized broadband antenna is used that has been designed to be utilized simultaneously by radio stations across the band. This essentially means that the antenna acts as an efficient radiator for many frequencies in the FM band and can be designed to have an optimized radiation pattern that minimizes the nearfield downward radiation. In the case where this is installed on a building rooftop, it can greatly improve the chances of creating hot spots on the roof that need to be designated as problem areas requiring special RFR procedures. Another big advantage to master FM antennas is that because they need to be fed by an FM combiner system containing filters, most intermodulation products are suppressed without the need to purchase or special external filters. A master antenna system sometimes grows from a simple idea. Keep in mind that FCC rules and business models change over time, and the facility that you build today may someday be able to serve multiple stations. Keeping that possibility in mind, it is a great idea to leave room for expansion, as well as having your FM antenna modeled for other frequencies that are in use in your market to determine the directional performance for other stations that may someday utilize your antenna and pay you rent! Now would be the time to have those measurements made. Most

FIGURE 7.1-4 Greater Media Prudential Building RF filters for auxiliary antenna

antenna manufacturers offer the service of measuring the antenna performance on a "test range" before the product is delivered to you. It is never a bad idea to have this data available to you before you mount the antenna on the tower. This will give you the opportunity to orient the antenna favorably for the best coverage of the community you are trying to serve.

Tower Lighting Systems

Your radio station license may require you to install a tower lighting system. These systems are now generally LED lighting systems, which are very reliable compared to the old days when incandescent bulbs or electronic strobe systems needed constant attention. The tower lighting system needs to be automatically controlled by a photocell mounted on the north side of a building in the Northern Hemisphere. The photocell switches the tower lights on at dusk and off at dawn, or toggles the tower light mode between day and night, depending on the requirements of the license. It is an FCC requirement that the tower light operation be monitored, and outages be reported to the FAA within 30 minutes, so plan on a monitoring system if you are the owner of the tower. Even if you are not the owner of the tower, it is a good idea to monitor the operation of the tower lights. In most cases where a tower is less than 200 feet above ground level the tower will not require lighting. Whether a tower is painted or not can also have an effect on the type of tower lighting system specified in your license.

FIGURE 7.1-5 The master antenna systems at the top of the Prudential Building in Boston, serving seven FM radio stations. Unused TV antenna on top, original master FM antenna below, and the Greater Media master antenna on the bottom.

Studio to Transmitter Links

When planning a transmitter site, another major consideration is getting your studio signals to the transmitter site. Normally this is accomplished with a studio to transmitter link (STL). In the 1970s and earlier, 950-MHz unidirectional links were very popular and were the mainstay of the radio STL architecture, along with analog phone lines leased from the local utility. Today, bidirectional links are usually required due to the IP-based equipment that is universally sold. It is also required for HD Radio transmission. These modern STLs normally fall into two categories: microwave and landline.

The landline links are becoming less popular these days due to the recurring monthly charges from the utility and the somewhat limited bandwidth of the services. Also, having a landline as your only link usually means surrendering control of your reliability to the utility you are contracting with to provide the link. Having the link go down during off hours or on holidays can sometimes be a problem for some providers, so be sure to look into their policies. Most companies will provide you with a number that represents their service level agreement, stating the number of minutes per year, or the percentage of time that they will guarantee the link will be operating correctly. This number should be studied carefully, and considered when determining the maximum tolerance for downtime for your facility. Normally a number representing "five nines," that is an uptime guarantee of 99.999%, is considered a minimum reliability number. This equates to link unavailability of less than 5.26 minutes per year. An advantage to installing a landline is that it can be often installed rather quickly (often within 30 days), and the upfront cost is quite low. There are no permits that are required, and very little equipment is needed. The link availability is generally unrelated to weather events, and there is no tower or antenna required on either end for this service.

The other option is to install a microwave link. Microwave links can save you from the recurring monthly costs of leasing a line from the phone company, but have the disadvantage of higher up-front installation costs. The privately owned microwave also gives you the advantage of having the reliability of your STL link within your control. Being able to repair or replace STL components on your timetable can be an important factor in keeping your system up and running. Today, there are new rules in place that allow the use of licensed wideband links for STLs. These so-called part 101 links are authorized under Part 101 of the FCC's rules, and provide for a secure link that is protected from interference. These links are obtained by an application to the FCC that requires a prior study be done to ensure the intended RF path is clear of interference on your proposed frequency, and also that there are no objections to the proposed new service by previously licensed users. Selecting a frequency band for the proposed STL is also an important consideration. Generally the 23 GHz and 18 GHz bands are better for shorter links; as the length increases, another option that performs better is the 11 GHz band. Consulting with a design engineer who knows the availability of clear paths in your area allows for an informed decision about what band may work best for your situation.

Another microwave option is the so-called spread spectrum technology. This unlicensed service utilizes a digital technique called frequency hopping, and is an unlicensed and unprotected service. The radios for this type of service can be easily purchased and deployed without the need for frequency studies or license applications. While usually less expensive and quicker to deploy, these types of radios are not protected by the FCC and are subject to possible interference. However, in rural areas where frequency congestion is at a minimum, this may be a quick and cost-effective solution for the installation of a wideband bidirectional STL.

Remote Control Systems

Remote control systems have matured over the years, and still today offer impressive capabilities. Planning a remote control system to not only take care of turning transmitters on and off, switching coax switches, and monitoring power levels, but also to monitor the environment at the site is a smart idea. Parameters like room temperature, transmission line pressure, and building security can all be built into a modern remote control system. Installing sensors at the time of the building construction makes for a cleaner installation.

Physical Plant Considerations

A transmission plant needs to be built with a few important things in mind. Physical security is an important consideration and is in fact mandated by the FCC. Keep in mind that most of the time the site will be unattended (unless it is co-located at the studio), and this means the building may need to be fenced in, have good locks on the door, and depending on the risk factor, have an alarm system or surveillance cameras installed. This type of technology has become inexpensive and can also be tied into your transmitter remote control system or through the Internet.

FIGURE 7.1-6 An example of a licensed 11-GHz STL dish mounted on the side of a transmitter at the Prudential Building in Boston

Windows in a transmitter building are not necessary and are usually undesirable for a number of reasons. Security is an important reason; second, windows tend to compromise building insulation and this leads to higher energy costs.

The physical room size is an important consideration. Enough room to comfortably work on the equipment is crucial. This means taking into consideration the door swing radius; leaving space for a workbench to repair equipment, and storage shelves for spare parts, is also a good idea. There are electrical code issues that need to be respected as well. These mostly involve leaving the proper clearance around circuit breaker panels. Finally, leaving room for future growth is important, not only for your radio station, but for a future tenant that may want to rent tower space from you and will require equipment space in your transmitter building.

Some logistical build considerations include identifying and employing a good design team for the project. Residential building contractors are generally a poor choice, because they usually lack the experience of the structural, electrical, and mechanical requirements of a modern broadcast plant. Consultants should be used as a source of recommendations for qualified builders who may have prior experience with broadcast facility construction. Solicit multiple bids for the construction work, and encourage your contractors to bring their subcontractors to any pre-bid meetings to gain a better understanding about what the job will require. For new tower construction, soil conditions need to be evaluated by structural engineers to determine the type of foundation required for a building, tower base, and guy wire anchors. Most localities require soil tests prior to issuing building permits. Finally, planning for a staging area for building materials including tower sections is important. Keep in mind that in many cases a paved road is not yet available to new remote building sites, and wet weather may impede access by heavy construction vehicles.

Inside the building, plan carefully the locations of the transmitters and associated rack mounted equipment. In some cases it is a good idea to provide a covered trough under the equipment racks for the control cables and AC power to be routed. This allows for a clean-looking installation and also future expansion. Another option is to run a wire bridge above the racks and transmitter and to lay the cables in there. It is good practice to use a divided trough for AC power and audio and control signals, and is actually required by electrical code unless armored AC power cable is used. Planning for the locations of the transmission lines, coaxial switches, harmonic filters, and dummy loads is also a prime consideration that should be sketched out early in the building design process. The support required for these components will almost certainly require a tie-in into the superstructure of the building. Finally, feed-through panels that allow for a weather-proof seal for the egress of transmission lines should be factored into the building construction at an early stage. These panels get installed like windows and provide a weathertight seal around rigid or heliax transmission lines without damaging the line. They are available in a variety of sizes and configurations, and need to be included in your building plans at an early stage.

Prefabricated concrete buildings are often an attractive choice over "stick-built" construction. Many companies can provide prefabricated building materials manufactured to your specifications, including security, electrical pre-wiring, transmission line feed through ports, and even HVAC. Ordered months in advance, these durable shelters are literally dropped on a pre-poured cement pad at the site and are ready for immediate occupancy. If time is of the essence, the transmission equipment can even be preinstalled at the factory where the building is poured, allowing the station to be on the air in a very short amount of time.

HVAC Considerations

Another consideration is building cooling. The heat load for the building should be calculated including the sum of the British thermal unit (BTU) outputs of the transmitters, dummy load(s), and electronics that are mounted in the racks. This information is usually available from the manufacturers of the individual equipment utilized at the site. An HVAC consultant or vendor should then be able to recommend the type and size of the HVAC units to be installed. HVAC is a critical part of keeping your plant clean, reliable, and cool. It is recommended to use a closed loop system whenever possible. This technique involves exhausting the transmitter air into the open room, and using mechanical cooling to cool the room to a comfortable temperature. This technique allows for the air to be recirculated, and keeps dirty outside air from infiltrating the building. Transmitters and associated equipment are all based on computers, and these types of circuits need to be kept clean and cool to provide reliable operation. The penalty to pay for using a closed loop cooling system is higher energy costs, but in the long run, the reliability you gain will more than make up for energy usage. In winter, the transmitter exhaust will probably be enough to heat your transmitter room without costing a dime in energy.

An open loop system can be employed as well, where transmitter exhaust air is ducted directly out of the building, but make-up air is constantly being drawn in from the outside. While this method is less expensive to implement and operate, the filtered air that is drawn into the room will routinely clog filters and ultimately lead to poor airflow and dirty transmitters. Unfortunately, once a transmitter gets dirty, it is hard to bring it back to a pristine operating state. This is particularly true for transmitters that utilize tubes and operate at very high voltages. Not only do these high voltages attract dirt, but transmitters that are not kept clean eventually lead to arcing and component failure. This is another reason to keep the room clean and under positive pressure, as a closed loop system will do.

Redundancy is an important consideration that will be a topic throughout this chapter. This applies to HVAC as well. Whether you choose to use a closed loop or open loop cooling system, having two HVAC units capable of handling the load is a great idea. This way, during a failure of one unit, or just routine maintenance, the second unit can handle the heat load. Another backup method used if all else fails is to install an emergency exhaust fan system that will suck filtered outside air through the room and cool it, even if the HVAC system(s) are inoperative. This is usually considered a "doomsday" backup, because it lets in outside contaminated air and does not cool all of the room as effectively as an HVAC system. However, these simple exhaust fan systems are inexpensive to install and very reliable.

Some newer FM transmitters employ liquid cooling. This technology has been around for decades in television transmitters, but has only recently been commercialized for FM transmitters. This kind of cooling uses liquids to internally cool the transmitters, and is sustained by a heat exchanger outside the building. The advantages of a liquid cooled transmitter are dramatically reduced cooling costs, smaller HVAC units, and quieter operation. The disadvantages are slightly higher maintenance costs and the limited availability of transmitters with this type of cooling built in.

Whatever type of cooling you choose, make sure it is sized appropriately for your expected heat load, with a 25% reserve for future growth. As a rule of thumb, a typical class "B" FM transmitter plant, with tube type transmitters running at 65% overall efficiency (AC to RF), will typically require at least five tons of mechanical cooling for a closed loop system.

FIGURE 7.1-7 Greater Media Prudential Building showing HVAC condenser units secured to the building

HVAC status monitoring, temperature monitoring, and an automatic changeover from one HVAC unit to the other (made by the manufacturer) is highly recommended. A plant that loses HVAC cooling for whatever reason will be rendered useless after only a short period of time. This type of failure normally leads to stressed electrical components and shorter service life. HVAC is an important part of your plant.

Electrical Service Considerations

The transmitter plant will require sufficient electrical service. This of course needs to be calculated to determine the appropriate service to be installed. One of the first considerations to look at is the availability of three-phase power, and the requirement (if any) of the transmitter for three-phase power. If the transmitter requires three-phase power, and the power utility cannot provide it (or it is just too expensive), a mechanical rotary converter is a possible solution; however, due to the complexity of the device it is not usually recommended. The electrical power for each piece of equipment should be considered and summed. If you plan on installing a backup transmitter, make sure its power requirements are factored into the total electrical load. Of course the HVAC and the rack mounted electronics need to be considered as well. A 50% buffer should be added for future expansion as well. As a rule of thumb, a typical class B FM transmitter plant, with tube type transmitters running at 65% overall efficiency (AC to RF) will typically require a 400-amp service at 240 volts. It is generally accepted that for three-phase power, a "wye" connection is superior to a "delta" connection in terms of lightning and surge protection. However, a worthwhile investment is always an external surge protection system. This can be either an inline series system (more effective but expensive) or a shunt or parallel surge suppression system that, while still effective, it much easier to install and less expensive.

At the core of any electrical power system is the grounding system. If the transmitter is located very close to the tower, the ground can be the same for both, however if this is not the case, an array of ground rods should be provided. On the tower side of the building is the best place for the ground rods and they should be long enough to reach below the deepest frost line. The rods should be placed at distances of a least two to three times their length away from one another, and should be joined together by bare copper wire of at least 1/0 in size. Cad welding should be used as opposed to mechanical clamps, as these connections will not oxidize. Then all conductors leaving the building should be bonded to this ground rod array system. This then becomes the center of the grounding system (sometimes referred to as a star system), and the transmitter grounds and the electrical grounds should be tied to this point using copper strap. Also, due to the fact that high currents can pass through these grounding conductors—and they tend to expand when this happens—it is a good idea to provide a larger conduit for the wire to allow for expansion if needed during a lightning strike.

Today's transmission gear is universally built with computer technology. And while this is a wonderful step forward, the downside is that it is sensitive equipment and needs to be protected from power surges as well as power interruptions. An uninterruptable power supply (UPS) is now considered indispensable at transmitter sites. While the high-power transmitter may be impractical to power from a UPS (although I have done it before), all the other equipment including STL exciters, remote controls, and audio processors should all be supplied with clean UPS power. In a system with redundant transmission equipment, it only makes sense to have redundant UPS, otherwise a failure of the UPS can render the entire site useless. A smart arrangement is to have one total transmission system (exciter, audio processor, remote control, etc.) fed by UPS "A" and the other fed by UPS "B." In this case you would be covered in the event of a UPS failure, and would still have a second system to count on.

The bane of the UPS existence is the batteries. Battery life for a UPS is finite, typically 2–3 years for a good quality UPS. The batteries are heavy and can get expensive to ship and change, but need to be factored in as a cost of doing business these days. All UPS of this size and caliber have a number of diagnostics available that can remotely warn you in advance of a battery going bad or another internal UPS problem. These remote indications should always be monitored because they all will eventually present an important alarm that will require attention. Generally two larger UPS that can each power 50% of your equipment with a 30% or greater safety factor for each unit is recommended. A final consideration for the installation of UPS systems is a "make before break" bypass switch. This allows you to be able to maintain or bypass the UPS when required without interrupting the load.

Emergency backup power is not required, but can be a great asset. Consider the location of your transmitter site, and your tolerance for downtime, and then factor that into your decision as to whether an emergency power plant is a good idea for your transmitter site.

FIGURE 7.1-8 An example of an array of two UPS that are suitable for use at a transmitter site

FIGURE 7.1-9 These "make before break" bypass switches allow a UPS to be maintained without interrupting the critical load

Here are some considerations: Are you in a rural area that loses power often? Are you on a hilltop or mountaintop that is prone to power outages? Are there other utilities or broadcasters at or near your site that may already have a generator? If so, can they tell you how often it is used? What kind of fuel can you store at your transmitter site? Can you get fuel deliveries easily? Are you likely to be able to obtain necessary fuel storage permits? How much generator run time do you need, and how much fuel do you need to store?

Emergency Power Considerations

If you are located in an area that is prone to natural disasters, such as hurricanes, snowstorms, or earthquakes, this will influence your decision as to whether emergency power is required, and the amount of fuel storage you require. In climates with cold weather seasons, a water heater that circulates the coolant through the radiator is usually required to keep the engine block warm enough to allow reliable starts in extremely cold weather conditions.

Oftentimes a tall building will not allow for fuel storage, so careful research needs to be completed before committing to a building generator. However, the good news is that in the case of a tall building, the power feeds are always underground and redundant, leaving the need for a generator a minor issue in many cases.

After a generator is installed, it is important to engage a company that will perform preventative maintenance on the generator, usually on a quarterly basis, including checking belts and hoses and performing routine oil changes. After all, when you are on the generator, you really need it! It should be at peak performance or your investment will be for nothing. A generator is usually set up to run on a weekly basis either with the load or without the load. This is triggered by an exercise clock, and the successful run time can be reported by a remote control and monitored to ensure proper operation every week. This is the best way to be sure that the generator will be available when needed.

Transmission Line Types

The method for transferring RF energy from your transmitter to your antenna on the tower utilizes transmission line. There are many sizes of transmission lines; the two basic types are rigid transmission line and heliax (or flex) transmission line. Rigid transmission line is more difficult to install, but offers the advantage of allowing you to replace damaged sections easier. Flexible transmission line can be easier to install but is more difficult to repair. Most transmission lines used for moderate to high power transmission employ an air dielectric that helps form the dielectric constant property of the transmission line that determines its operating impedance. This dielectric must be kept dry to ensure no internal arcing occurs that would damage the line. This is usually accomplished by pressurizing the transmission line with either dry air from a dehydrator, or an inert gas (usually nitrogen) from either a cylinder or a nitrogen generator. You should plan on not only enough space for the pressurization equipment, but also a method of monitoring the pressure so that you can be alerted if there is a leak in the system. Using a cylinder of nitrogen to pressurize the system has a disadvantage of quickly emptying the bottle if a leak is formed anywhere in the system, and if not noticed will lead to an unpressurized system that can be dangerous for the health of the transmission line. It cannot be overstated how important it is to pressurize and monitor the transmission line to ensure the dew point remains very low, minimizing the chance of arcing. In some situations (usually very low power cases), the transmission line dielectric may be made of foam, and in this case there is no pressurization that is required.

The transmission line diameter is proportional to the amount of power the transmission line can safely handle. The calculation should be done to determine the required transmitter power output (in kilowatts), and adding a safety factor of 50%, determine the diameter of the transmission line that will be required to safely carry the energy from the transmitter to the antenna. There are many factors that go into this calculation, but the major ones are factors such as the transmission line loss (in dB per 100 feet), the length of the transmission line run, the gain of the FM antenna array (this is specified by the manufacturer and is determined by the number of elements and bays), and any losses in filters or switches.

Transmission line switches and patch bays offer the broadcaster the ability to switch RF sources. A good example is a transmitter plant with two or more transmitters. A remote-controlled motorized transfer switch offers the broadcaster the ability to swap transmitters remotely and sometimes automatically. When planning the building layout, it is important to consider the placement of these switches and the routing of the transmission lines. These switches are usually suspended with metal angle iron from the ceiling of the building. Patch bays offer similar functionality to motorized switches, but require manual manipulation of the transmission line links to reroute the RF. The final consideration with switches or patch bays is to positively interlock these devices with the sources of RF (the transmitters). The idea here is to prevent the transmitter from producing RF power when the RF is not properly routed to an appropriate load. This occurs when the motorized switch is in transit, or is stopped between positions. It can also occur when a patch panel link is not in place. This electrical interlocking is available on all transmitters and motorized switches as well as most patch panels. If power were to be applied to a switch or patch panel that was not in the correct position, major damage is almost certain to occur.

In order to be able to exercise or test a transmitter that is not on the air, a destination must be provided for the RF energy to go to. This is called a dummy load. The job of a dummy load is to provide a matched load that the transmitter will perform well into, without radiating RF into free space. These dummy loads come in many different styles and can be cooled by forced air, convection with an oil sump, or flowing water. For power levels above 5 kW, forced air cooled loads are popular. Water cooled loads are useful for making precise power measurements by carefully measuring the temperature rise from the intake to the discharge of the load. Such equipment is referred to as a calorimeter, and is useful for determining transmitter efficiency. However, flowing water (or any water service) in a transmitter facility is not an advantage for obvious reasons. If you are able to stay away from water, it is obviously a good thing at a transmitter plant. Finally, a dummy load should always be electrically interlocked to the transmitter(s) so that if the load is not in a ready state (i.e., water or air is not flowing to cool the load), then the transmitter will shut off, saving the load from damage.

FIGURE 7.1-10 Photo of mounted array of motorized coax switches

Reflected Power or VSWR (Voltage Standing Wave Ratio) Protection

One of the most important things you can do is to install equipment to protect your transmitter, transmission line, and antenna from damage due to a change in the match between the transmitter and the antenna. These changes can occur at random times due to aging transmission line or antennas, loss of pressurization in a transmission line and subsequent arcing, or other physical changes outside your control. When this happens, the RF energy from your transmitter is no longer being efficiently transferred to your antenna to be radiated. That energy is being dissipated somewhere else in your system; that is always undesired and usually leads to major damage that is expensive and time-consuming to repair. Installation of a watt meter that is calibrated and measures the forward and reflected power is a wise idea. These meters are adjustable to provide a contact closure or other signal when the reflected power reaches a preset threshold. The idea here is to use that signal to immediately turn off the transmitter to prevent heating, arcing, and damage to the antenna or transmission lines. While this will of course take your station off the air, it will do it for a very good reason, and will likely save the downstream line and antenna from major damage.

Summary

Building a new broadcast facility can be an exciting and rewarding opportunity. Starting from the ground up allows for you to construct a facility that is reliable and suits the needs of the company today, and can

FIGURE 7.1-11 A photo of an array of bird watt watchers that allow for VSWR protection for a single or combined FM antenna system

allow room for future expansion. Pre-planning is key to a successful project. Making wise decisions today will serve you well in the future, and will provide an economical and stable platform for serving the public for many decades.

CONSTRUCTING AN AM TRANSMITTING FACILITY

Thomas R. Ray III

Introduction

Constructing a new Amplitude Modulated (AM) transmitting facility is one of the most important jobs ever given to an individual or individuals at a broadcast station. The programming department can produce the finest programming known to mankind, but if the public—your consumers—cannot hear it, it does not do any good.

The process of constructing a new AM transmitting facility can be frustrating and there are many things to consider—many that perhaps you may not have thought of. This chapter will help guide you through the process so that your project will go smoothly and, hopefully, will help minimize or eliminate any mistakes.

Among the topics discussed are how and where to locate the facility, zoning and building issues you may run into, facility control, getting audio to the facility, and much more. Many of the items discussed will make you think, "Well, that is obvious." Many are very easy to overlook, as building a transmitter facility, particularly a facility with a directional antenna, is a massive undertaking and what seems obvious is not so when faced with an extremely large project.

Location, Location, Location

When starting the project to construct a new AM transmitting facility, whether this is a completely new station allocation or whether you are reconstructing an existing station, consideration must be given to where to locate the transmitter facility. This is not always completely obvious, as thought must be given to where the population of the served area is and, if possible, to predict where the population may migrate in future years. There is more than one story of an AM transmitter facility that correctly located their site to serve the area, only to have the population either move further away from the site or, in the case of a directional antenna (DA), move to the "wrong" side of the antenna into a null area.

One first must look at whether the station will be transmitting a nondirectional (ND) signal or directional (DA) signal. For an ND station, the antenna should be located as close to the center of the area to be served as is practical. For a DA station, the antenna should be located so as to direct maximum signal over the area to be served. This sometimes is more difficult than it sounds.

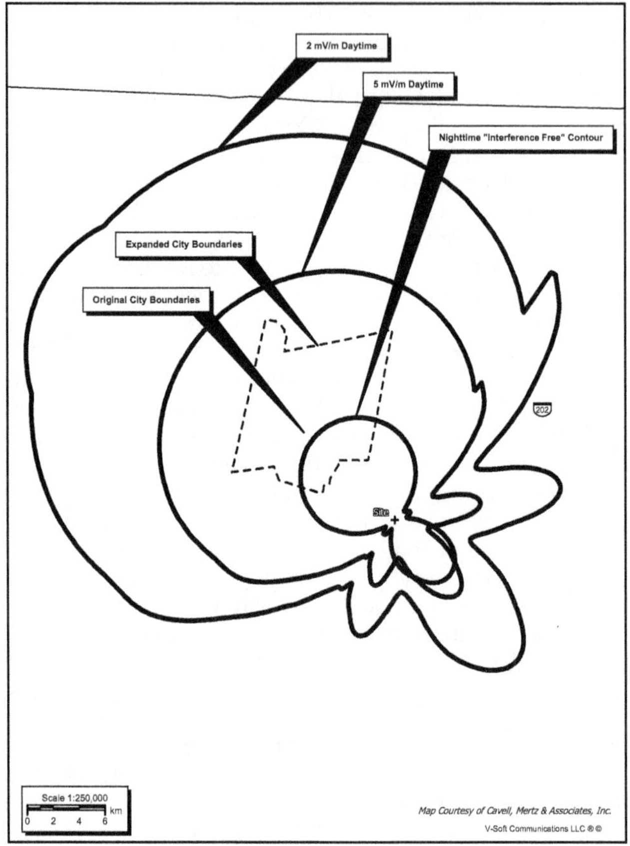

FIGURE 7.1-12 The 2-mV/M contour of WSNG in Torrington, Connecticut, shows that the signal goes right over the city of license

The station should discuss locating the facility with their consulting engineer. The consultant is extremely important in the planning of placement of the station, in addition to the design of the antenna coupling system and, if directional, the phasing and coupling equipment. The consultant can also identify a given area where the antenna may be moved, as spacing from co-channel and adjacent channel signals may affect the area where the site may be located.

Once the area is identified where the transmitter site may be located, the search for land begins. First, one must consider the ground system in identifying the size and shape of the property to locate the antenna system. A consultant can also help with this chore with a preliminary antenna system design to determine tower placement. An often overlooked part of the antenna system, the ground system is a critical component to the success of the build. The ground system will be discussed later in this chapter.

In looking for land, you need to consider several factors. The size and shape of the property to appropriately contain the towers and ground system are a start. If at all possible, the property should be located in a relatively wet area to help with ground conductivity. Try to avoid rocky land if at all possible.

Once a property is determined to work for the antenna system, a "tour of the neighborhood" should be undertaken to see what the area looks like. In particular, you are looking for possible sources of re-radiation: cellular towers, metal utility towers, and towers of other broadcast stations. Ideally, none of these will exist within 3 kilometers (approximately 2 miles) of the chosen site. If these items do exist within 3 kilometers of the site, their location should be noted and passed along to your consulting engineer to determine if these items need to be detuned to make a DA work properly.

Another item to look at would be utilities. If your transmitter will require three-phase power, is it available? What telephone facilities will you need? Are they available at the site? It can be very expensive to bring three-phase electrical power, as an example, from the main roadway to your location. You can request that the utility companies send a representative out to the location to determine what is available for facilities in that area. This will help you determine if you can actually put a facility on the air at the location.

Dealing With Various Government Agencies

Once a property has been located for the antenna system, it is time to consider talking to the various governmental agencies that may have jurisdiction over the property. Contact the local building department and zoning board. They can direct you to the agencies that may be involved. When I built the WOR transmitter facility in Rutherford, New Jersey, in 2005, the following agencies were involved:

- Rutherford building department
- Rutherford zoning board
- New Jersey Meadowlands Commission
- Meadowlands Wetlands Commission
- Army Corps of Engineers
- Bergen County Soil Conservation Commission.

Each agency had their own particular area of concern, however, the Meadowlands Commission had final say over all of them. This made for some interesting "fireworks" in getting the project completed.

You will most likely be asked several questions for zoning approval, among them:

- Why do you need as many towers as you do?
- Why do the towers need to be that high?
- Why do you need to light the towers? Can't they be dark?
- Is there any other way to light the towers besides red (or strobe) lights?
- Will the towers be painted and why aviation orange and aviation white?
- What are the health concerns regarding the emissions of the facility?
- What type of vehicular traffic can we expect?
- What type of waste do you generate?
- How many people will work at the site on a regular basis?

The answers to these questions can be pretty simple to those of us who deal with broadcast facilities on a day-to-day basis. The agencies do not; therefore, answers should be kept as straightforward and as simple as possible.

For example, the answer to the questions about the amount and the height of the towers do not need to be answered in strictly technical terms. The answer can be along the lines of: "The FCC requires we generate a certain transmission pattern for our signal. The laws of physics dictate this amount of towers will generate that pattern to meet FCC compliance." For tower height: "The FCC has determined that we require a certain efficiency in our antenna system to meet their requirements. Shorter towers do not exhibit these characteristics." But, of course, follow the lead of your legal representative and engage your consulting engineer to help answer these questions.

At this time, you also should ask what permits are required to allow surveying and testing on the property, and what other agencies will be involved. For example, a permit may be required to cut the field flat for surveying. If you are building on swampy land, you should also determine where the bedrock is located. The tower bases should rest on bedrock to prevent them from sinking, and guy anchors should be anchored to bedrock to prevent them from pulling out. This normally requires drilling to locate the level of the bedrock.

Also be prepared that the zoning regulations may require that you have working sanitary facilities at the

FIGURE 7.1-13 This is a guy anchor point. Note the steel I-beams that were placed down to bedrock, and the angled I-beams to help prevent the guy anchors from being pulled out.

site. There may be a loophole for unmanned facilities in their regulations—make sure to ask. Once many transmitter sites are tuned up, especially when they are new, time required at the site is minimal, usually consisting of a few hours every week to check the system, with occasional time for maintenance. Talk to the authorities to find out the requirements and if they can be relaxed.

Design of the Facility

Once you have the final antenna system coupling and phasing design completed, you can then start laying out your facility on the property. In many cases, your consulting engineer will help you determine where the tower bases should be located. He/she will provide you with the proper spacing and orientation between the towers based on the location of the reference tower.

A licensed surveyor should be engaged to set stakes for the location of the tower bases on site. It is important to make the surveyor understands that everything must be referenced to TRUE NORTH, not magnetic north. There is a difference between true north and magnetic north—here in the New York/New Jersey area where I am located, that difference is −13.5 degrees. This means that if the tower bases are located using magnetic north, they will be 13.5 degrees off and the pattern will not be able to be adjusted correctly. There are many ways that magnetic declination can be determined: the National Oceanic and Atmospheric Administration (NOAA) has a declination calculator on their website at www.ngdc.noaa.gov/geomag-web/ if you or your surveyor needs this information.

At this point, your consulting engineer can give you an idea of the physical size of your coupling and phasing equipment. This will help you determine the size

FIGURE 7.1-14 At the WOR transmitter facility, sanitary facilities (sewer and water) were not available—yet the authorities were insistent that the site have running water and a working restroom. The solution was to install a septic holding tank and install a 500-gallon freshwater tank. The water tank was filled by a swimming pool water provider as required.

of the tuning houses you will need (unless you will be buying custom-built weatherproof cabinets from your supplier) and the size of your transmitter building. You will need to check requirements with the building department as to what they expect for anchoring of the structures and the foundations required.

One thing to keep in mind with modern transmitting equipment: everything is a computer. As opposed to transmitters of years past, today's transmitters require relatively clean environments and proper temperature control to operate trouble-free. Keep this in mind when designing the transmitter building. It is best to air-condition the building to keep the temperature and humidity moderated. If this is not practical, air intakes on the building should be plentiful and filtered to avoid drawing dirt into the building and, consequently, into the equipment. A good amount of air should be moved through the building. It is a common mistake to undersize the air intakes. More is better. Air conditioning should be sized to handle the heat load of the size of the building, then adding the heat generation of the transmitter and associated equipment on top of that.

When discussing air conditioning with an air conditioning contractor, they generally speak about "tons" of air conditioning. For reference, one ton of air conditioning is equivalent to approximately 12,000 BTU of heat. Believe it or not, I have run into air conditioning contractors who do not know this. Most transmitter specifications state waste power in kilowatts. There are numerous online calculators to convert kW to BTU. And if your dummy load is located in the transmitter room, make sure to accommodate the heat it will generate with a transmitter under test while the other transmitter is on the air.

Avoid putting building air intakes near the transmitters where possible. Additionally, the air intakes should have a hood on the outside to pull air from a downward direction. Just putting a set of louvers in the side of the building will generally result in rain and/or snow being drawn in through the air intake. This should be avoided.

If the land you are building on is swampy or prone to flooding, the building department will usually require that the facility be built to the level of a 100-year flood. I recommend building it to the 200-year flood level. As an example, the WOR transmitter site in the New York City area was built to the 200-year flood level, or 14 feet above nominal. We were told to build to the 100-year flood level, which was 9 feet above nominal. Hurricane Sandy produced flooding that was at 11 feet. Had the facility been built to the 100-year flood level, the building would have been inundated with floodwater. At 14 feet, it was dry and remained on the air. The towers, guy anchors, and buildings were all built to 200-year flood level. It is best to be conservative when discussing flooding.

Obtain the weights of the transmitter(s), phasing equipment, and any other items to be placed in the transmitter building. The floor should be designed to handle the weight (per square inch) of the transmitter, which will most likely be the heaviest item. Do not

FIGURE 7.1-15 The floor of the transmitter building before pouring the concrete. Steel rebar helps distribute the loading of the floor, preventing the concrete from cracking and making the floor stronger.

FIGURE 7.1-16 The Harris 3DX50 transmitter (now the GatesAir Flexiva DX50) weighs approximately 3,500 pounds, is about 10 feet long by 6 feet high, and is a little over 40 inches deep. The floor must be designed for its weight and you must have room to maneuver around the transmitter.

skimp on rebar or concrete quality. Having the transmitter fall through the floor will not do the station any good.

Likewise, you should make sure the floor loading specification on a prefabricated building will accommodate the weight of the transmitter. It should be reinforced as required.

Transmission Line Runs

It is common to put the transmission line runs, the power runs, the sample line runs, and control cabling runs underground to the towers. This first and foremost will prevent theft and vandalism, and keep the lines relatively safe from the environment. Transmission and sample line is rated for direct burial. It is possible to get direct burial power cable. If this cannot be obtained, power cabling should be in conduit, as should control cabling.

If going underground, you should dig down to below frost level. The trench should have an initial bed of small crushed stone. That should be topped with a foot or more of sand. The transmission lines should be laid on top of the sand, then covered with at least a foot of sand, then crushed rock, and finally backfill dirt.

If going above ground, there are numerous methods to do this—including steel poles and wooden poles to suspend the cabling. It is recommended to run either an ice bridge from pole to pole or, at the very least, a steel cable to act as a "carrying wire" to anchor the cables so they do not droop.

It should be noted that several manufacturers make an armored direct burial control cable. This cable consists of several #12 AWG insulated wires in a flexible steel shell, surrounded by PVC insulation. This is something to consider rather than snaking numerous wires through conduit.

FIGURE 7.1-17 The freshly dug transmission line trench. It should be below frost line. This trench is ready to be lined with crushed stone and sand before the transmission lines, power lines, sample lines and control lines are laid in it.

At the tuning house end, the transmission and sample cable should be brought into the tuning house or box and properly terminated: the transmission line onto a J-plug to feed the tuning network, and the sample line into either a toroidal transformer or into a sampling system decoupling coil. Power should be terminated into a distribution panel. The control cable should be terminated onto a barrier strip for connection to the various devices such as control relays for pattern switching contactors and tower light sampling.

At the transmitter building, the transmission line will terminate either at the transmitter (nondirectional), antenna switch, or phasor. The sample lines will terminate at the antenna monitor. The power will terminate at a distribution panel. The control cable should terminate on a barrier strip at a convenient location.

The Antenna Sampling System

The sampling system is required in a directional antenna system to determine if the towers are being fed the proper power levels at the proper timing. This, along with tower placement and spacing, determines the shape of the pattern. The FCC requires that tower power ratios be within ±5% of that specified on the license. They require the phase of the towers to be within ±3 degrees of that specified on the license.

If a tower is over 120 electrical degrees in height, current sampling must be done using sample loops on the towers. These loops are placed at a height on the tower where it is predicted to have the highest RF current. They will normally be placed at the same level on each tower, providing the towers are all the same height. A coaxial cable, usually 3/8″ or 1/2″, is brought down the tower from the sample loop and into the building. Note that on the tower, the sample line is bonded to the tower. The base of the AM tower, however, is hot with RF and insulated from ground. At this point, the sample line must be insulated and brought into the building. It is important to note that the outside of the cable is at tower potential—that is, the same voltage as the tower—and is electrically "hot." It therefore must be insulated. Once in the tuning building, it is connected to a coil built with the same type of coax cable. The "top" of this coil is at tower potential. The "bottom" of the coil is connected to ground. This coil is an inductor, and will effectively isolate the RF at the top from ground at the bottom. So at the bottom of the coil, the outer conductor is now at ground potential. The center conductor of the coax will contain the signal induced into the sample loop on the tower. The connector at the bottom of the coil is connected to the sample line running back to the transmitter building.

If the tower is 120 degrees or less in height, the sample line from the building is typically terminated into a toroidal transformer at the output of the tuning house. The signal to the tower passes through this transformer and produces a voltage that is sent to the antenna monitor in the building.

The following is extremely important regarding the sampling system. The cables running down the towers and into the buildings must ALL be the same physical (and electrical) length, as well as the same brand and type of cable. The tower crew should determine the correct length and cut cable for each tower exactly equal. Ideally, a device called a vector network analyzer (VNA) should be used to determine that all cables are exactly equal. Additionally, all sample lines running to the building must be of equal length. While it is tempting, and logical, to say that if the closest tower is 400 feet from the building, while the farthest tower is 950 feet, to cut the cables to length (and therefore save costs), this cannot be done. The FCC specifies that all cables must be equal length. In this example, all the sample cables would be approximately 1,100 feet (depending on where the cables need to run: they may need to be longer). This should be verified with a VNA. All cables must also be treated to the same environmental conditions, so typically, excess sample cable is coiled up and buried. This must be taken into consideration when planning cable runs to the towers. Because all of the cables must be the same length, excess cable cannot simply be cut off—it must be either buried or coiled in a safe place.

If the facility is close to other AM broadcast stations, the antenna monitor should be equipped with filtering for the other stations. This ensures that the monitor will not be influenced by induced signal from the other stations. Signal levels from the other station will need

FIGURE 7.1-18 Part of a tuning network at a tower base. Note the J-plug connected to the transmission line toward the bottom left in the picture. TB-101, at the lower right, is the termination of the control cable.

FIGURE 7.1-19 Another portion of the tuning network. Note the J-plugs and, at the lower right of the picture, two toroidal transformers: one for the base current ammeter and one for the sample system. The sample line is dressed to the back of the network, entering the picture on the right side.

to be measured on site, and this information provided to the consultant to specify filtering requirements with the antenna monitor manufacturer.

There are circumstances in which you may use toroidal transformers on towers greater than 120 degrees. Many times, these are tuned up using sample loops, then the antenna monitor is connected to the toroids, readings logged, then the isolation coils disconnected and the system retuned to the logged readings on the toroids. This is not very common, but know that it can happen.

Grounding and the Antenna Ground System

This is where it gets interesting with the building department, because they do not typically deal with RF or with AM towers.

AM towers are typically electrically "hot," unless you are installing a folded unipole type of antenna where the tower is grounded. These sit on a ceramic insulator and cannot be grounded. This typically puzzles the electrical inspector who expects the tower to be grounded for lightning protection.

RF grounding requires the use of copper strap. The reason for this is the skin effect. In normal AC wiring, the type that, for example, connects to the electrical outlets in the wall, electrons move through every possible place in the copper wire. At radio frequencies, however, the electrons are moving much more rapidly and typically do not penetrate much below several one thousandth of an inch into the copper. A number 1 cable, typically as big around as your thumb, makes a great ground wire at the 60-Hz power line frequency, and can typically handle 130 amps. Not so with an RF signal, as the entire wire is not utilized.

We get around this by flattening out the wire into copper strap. Because the electrons flow only on the skin of the copper, this is much more efficient for RF grounding than plain old wire. And this really confuses most electrical inspectors.

For lightning protection on an AM tower, typically there are "ball gaps" located at the base of the tower. One ball connects to ground. The other ball is connected to the tower base, and the balls are spaced a given distance apart (distance is typically determined by carrier power level and modulation). If lightning strikes, it will seek the fastest path to ground and will jump that gap. The typical lightning strike is on the order of several hundred million volts—much higher than your RF voltage, and easily jumps the gap.

Additionally, the feed from the tuning house to the tower is typically a copper pipe. And typically, it has one or two turns in the pipe to form a coil. A coil is an inductor. A lightning strike creates a fast rising voltage and becomes AC. Inductors oppose AC. The inductor formed by this feed line is almost nothing at RF—but for the lightning pulse, it is very large. Therefore, little to any lightning should get past this inductor into the tuning house.

Just in case, there is typically a horn gap located in the tuning house to shunt any additional voltage that gets through the inductor to ground.

And, there is typically a static drain choke connected from the tower feed to ground in the tuning house. Because the tower is insulated, wind, snow, and rain can all statically charge the tower, much like shuffling across the floor in your stocking feet in the winter can charge you up to produce a static discharge when you touch a doorknob. The static drain choke is present to drain off this static charge and hopefully avert a lightning strike. The choke is sized to be a huge inductance at RF frequencies, but static is a DC voltage and will discharge through the choke to ground. This is your lightning protection and should make sense to the electrical inspector.

The tower ground system starts with copper ground strap brought down from the bottom of the base insulator. Silver soldered to this strap is a copper mesh screen, typically 10 feet by 10 feet to 25 feet by 25 feet, depending on power level. The edges of this mesh screen are silver soldered to copper strap, forming a square. Usually, the copper strap is nailed to landscaping timbers.

Attached to the outer copper strap are 120 #10 bare copper wires spaced 3 degrees apart. These wires are typically one-quarter wavelength long and extend out into the ground field around the towers. These wires should be buried to a depth of approximately 12 to 18 inches.

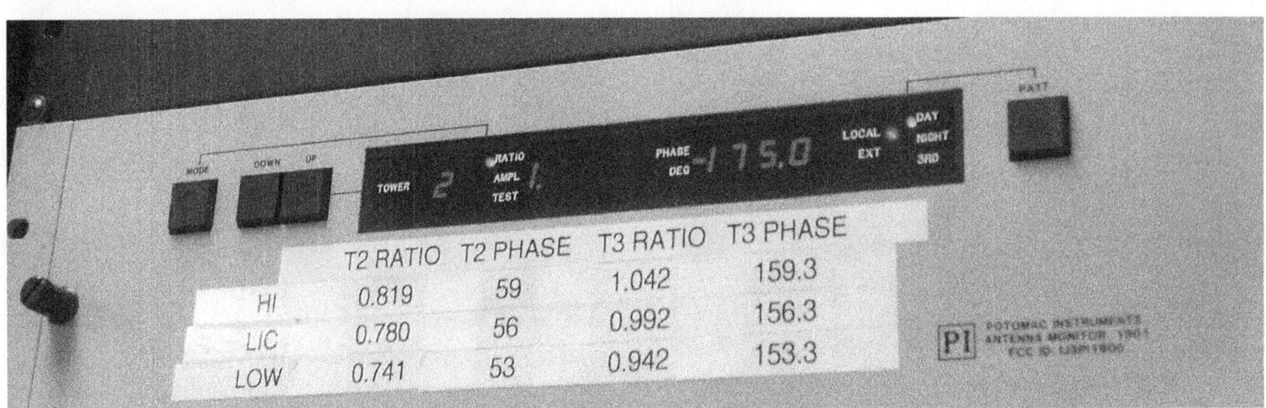

FIGURE 7.1-20 The front panel of a Potomac Instruments AM-1901-3 antenna monitor

FIGURE 7.1-21 The tower base showing a set of ball gaps

FIGURE 7.1-22 Two turns of copper pipe form an inductor that will help retard lightning egress into the tuning house

FIGURE 7.1-23 Copper strap nailed to a pressure-treated landscaping timber showing a radial attached ready for silver soldering

In the case of a multi tower directional array, where ground wires cross, a copper strap should be silver soldered to the radials—they do not need to be run all the way out past where they intersect.

Additionally, a copper strap should be run from the central strap between towers to the transmitter building. Not doing so will have the building and towers at different ground potentials, which can cause numerous issues.

At the transmitter building, I generally like to surround the building with copper strap, with the straps terminated, at a minimum, at each intersecting corner to an 8-foot ground rod. If you are building a building and have steel I-beams into the ground, you can also weld the copper to the building steel.

In the building, I like to run copper strap around the perimeter of the building floor, with strap running outside in several places to the strap surrounding the outside of the building. The strap from the tower field would connect to the outside strap. The transmitters and associated equipment are connected to the inside strap.

Of course, the utility equipment needs to be grounded according to local codes. RF ground is for RF equipment.

Make sure the copper straps outside the building and at the base of the tower(s) are buried to prevent theft and provide good contact with the ground.

It may be that a counterpoise ground system is proposed for your installation for several reasons. I know of at least one station where the Environmental Protection Agency (EPA) would not let them disturb the swamp to bury a normal ground system. With a counterpoise system, the ground system is elevated above ground and is suspended, usually on wooden poles. There are typically eight or more radials in this counterpoise system.

Work with your consultant. Make sure the counterpoise ground is up high enough, typically 10 to 12 feet, so

FIGURE 7.1-24 The copper strap is stubbed up through the fresh concrete floor of the transmitter building and, in this case, is attached to steel I-beams placed in the ground

that the average human cannot simply reach up and grab it. Provisions should also be made to protect the pole at the end of each wire from both vegetation and climbers. The ends of a counterpoise ground system are typically electrically hot and can hurt someone. And I know of at least one station that let the vegetation get a bit too close to the end of one of the wires and set the forest on fire when the end of the wire arced to a nearby vine.

Emergency Power

It is a good idea to put an emergency generator into the facility. The first consideration is that it should be sized to provide the same amperage as the incoming power feed.

Don't forget that an AM transmitter is additive with modulation: in other words, the power requirement of the transmitter changes continuously with modulation. It is not a steady power draw as with an FM transmitter. Discuss this with your generator provider. Depending on the generator manufacturer, you may need to install a prime rated generator rather than a standby rated generator.

Getting Audio to the Site

There are many ways to get audio to the transmitter site. If this is a remote site (not co-located with the studio), you will need some type of studio to transmitter link (STL).

If you are using 950-band microwave, you will need to mount the receive antenna. This can be mounted on the building if you have line of sight to the studio at this height: it can be mounted on a small utility tower adjacent to the transmitter building, or it can be mounted on one of the AM towers.

If you mount the STL antenna to an AM tower, you need to decouple the RF from the transmission line with an isocoupler at the base of the tower. The isocoupler isolates the hot AM side of the line from the line going to the STL receiver in the building. Isocouplers typically add a small amount of capacitance to the tower, which can be easily adjusted out in the tuning network.

An alternate form of isolation for the hot line would be to make the STL line on the tower a quarter wave shorted stub. A quarter wave shorted stub presents a very high impedance to the AM frequency—it would technically not be seen by the AM signal. The STL line is insulated from the tower from the base to the quarter wave point on the tower, then bonded to the tower to the STL antenna. The bottom of the line at the base of the tower is grounded. This, unfortunately, requires a great amount of maintenance to ensure that the insulators are intact and that the bonds are intact and proper. It is very easy to burn the line off the tower if an insulator fails or if a bond fails.

In many parts of the country, it is no longer possible to get "broadcast lines" from the phone company. We therefore will discuss alternatives.

One alternative is to put in a T1 from the phone company. A T1 is a data line with a guaranteed data rate of 1.544 megabits per second. Many companies make T1 audio interface devices with many different configurations. You would need to look at the different companies to determine the configuration that is right for your facility. T1 lines tend to be very reliable and are available over copper or over fiber, depending on what is available.

Another alternative is IP-based audio. Again, there are many different devices available. These devices will operate with data transport either on a private network or over the public Internet.

FIGURE 7.1-25 The generator ready to be installed in a room not yet built to house it

FIGURE 7.1-26 These 950-band STL antennas are mounted on a utility tower next to the transmitter building

And there are several ways to get IP data to the site. There are traditional Internet services through cable companies or the telephone company. There is satellite-delivered Internet. Or, you could install a 5.8 GHz unlicensed data radio between the studio and the transmitter site. Understand, however, that any IP audio link will have a delay—too much delay for monitoring live off the air.

Electrical Power

The power company will deliver power to your site. You first need to determine what you need. Unless you are running a 50,000-watt AM facility, you typically will need 208-volt three-phase power. Transmitters 5,000 watts and under can typically run on 240/120-volt single phase.

Determine the current draw of your transmitter and all other equipment at the site, including tower lights. If you have an auxiliary transmitter, make sure to include its draw in your calculation. You want to size the incoming power feed for the full load of the building if everything were turned on and operating. This ensures you have adequate power for testing. And, you may want to upsize the feed just a bit for future expansion and power sags during summer months.

You may be offered the choice of a delta or wye feed on three-phase power. I personally prefer a wye feed—it is much easier to understand and, as opposed to a delta feed, all phases measure the same voltage to ground. For devices that require 120 volts (audio processors, remote control systems, building lighting), this is derived from going between one phase of the power line and neutral, which is typically ground in many areas. In a delta system, two of the legs will measure 120 volts. One will measure around 240 volts. This becomes an issue when wiring the building and adding single-phase circuits in the future as you need to avoid the "wild leg." If you can get a wye feed, I would do so.

If you require three-phase power but it is not available, there are devices called rotary phase converters available that take single-phase power and output a three-phase supply. You would need to do the online research for these devices as the typical electrical contractor will have no knowledge of them.

Facility Control

Now that you have a brand new facility, you need a method to control it. There are many devices on the market to accomplish this feat. There are basic units that allow you to meter certain parameters, control power, change pattern, and turn the transmitter on and off. There are more sophisticated models that allow much more data and control.

Basic models will allow control over a regular telephone line by touch tone, and can be programmed to call out for help should there be issues. Call outs are set up using alarm limits: for example, if the common point current is below 8 amps, something is wrong, so call for help. The unit will then call out using an internal phone list until it finds someone who will "talk" to it and acknowledge the alarm.

More sophisticated models are IP based, but will also allow basic dial-up telephone control. The IP functions allow a full-time screen in the studio for operators to see, will provide you with access through any web browser (think cell phone web browser), and will email and text you when something is amiss. It is up to you and your budget to decide what unit you will purchase. But there is great comfort in having a world of information at your fingertips in real time as opposed to having to continually enter touch tone codes. It would be much faster to get back on the air if you can see all the data as opposed to bits and pieces of it.

At a minimum, you need to monitor antenna input power (nondirectional tower base current or directional common point current), antenna monitor phase and ratio if directional, and tower light status. It is helpful to monitor the power amplifier voltage and current on the transmitter, in addition to monitoring the incoming power feed and if the generator is on line. Additional information such as temperature in the transmitter room is helpful to determine if you have had either an air conditioning failure or a blower failure causing the equipment to overheat. Other helpful information is STL signal level if you are using a radio STL, IP STL connection, which STL is on the air, which transmitter is on the air, if the transmitter site door has been opened.

One thing to keep in mind regarding remote control systems: most require at least 0.75 volts to create a measurement. Most of the RF ammeters produce this or sometimes less. You may need to either build a precision DC amplifier to interface the ammeter to the control system, or purchase one through the ammeter manufacturer. The more voltage you present to the remote control system (without exceeding its upper limit), the more accurate the measurement will be.

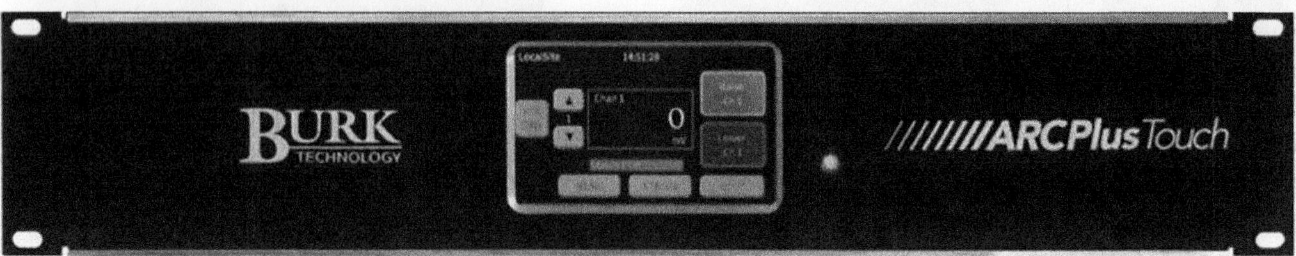

FIGURE 7.1-27 One of many IP-based remote control systems on the market

Tower Lights and Monitoring

Because the base of an AM tower is hot, you need to isolate the RF from the AC for tower lights. This is typically done in one of two ways.

If the towers are 120 degrees or less, a tower lighting choke can be placed at the base of the tower. These chokes typically have three windings—two for the AC feed, either 240 volts or separate beacon and side light circuits, and one for neutral. The "top" of the choke is at tower potential, the "bottom" at ground potential as far as RF is concerned.

If the towers are greater than 120 degrees, an Austin ring transformer is typically used. One ring of the transformer is mounted on the tower. The other ring is mounted on the tower base. The rings pass through each other but don't touch. They will pass AC frequencies but not RF.

FCC regulations require that tower lights be checked "at least once per 24 hour period." For stations on automatic control (which is most likely most stations these days), it is required that any problem with the lighting system be reported immediately to personnel. It is required to report a failure to the FAA within 30 minutes of the discovery of the failure.

There are numerous ways to determine if the lights are on. The easiest method is to purchase a tower lighting controller with alarms at the time of the purchase of the tower(s). Otherwise, you will need to put together some type of current sensing, either on your own or through a remote control system provider, to monitor current to the tower lights. This can be fairly straight forward if using a tower lighting choke. If using an Austin ring transformer, this can be accomplished, but you will most likely be monitoring all of the current into the lighting system and will need a system that can determine the extinguishment of a beacon bulb or a side light, as the FCC requires that a system be able to sense *any* bulb out on the tower.

FIGURE 7.1-28 An Austin ring transformer used to isolate the RF on the tower from the power line

Building the Site

Once the construction permit (CP) has been issued by the FCC, construction can begin on the facility. While waiting for FCC approval, it is a good idea to get all your local permits in place. Note that you should *not* put a shovel in the ground until that FCC CP has been issued. Discuss this with your legal representative, but in general, nothing should be done on the site until the FCC CP has been issued so as not to appear that construction has begun prematurely.

Unless your site is in the general area of an airport where this height may be lower, once a tower reaches 200 feet in height during construction, it must start using temporary lighting. Some tower crews have special temporary lighting systems available that are battery powered. Many will put up temporary lighting from the tower's lighting kit. Be prepared to provide power to the tower crew for this purpose. You may need to put an extended run generator in place at the base of the tower to power the tower lights until commercial power is available.

Make sure the contractors lock up building(s) at night, including job materials. Make sure you have a key or the combination to these locks.

It is also a good idea to check daily on the site: once during the day and once at the end of the day. I once had a contractor decide he did not want to continue the project and, rather than inform me, simply moved his gear out. I found this the next morning. Luckily, I had a key to the lock on the chain on the building and could get in without much trouble and discovered all this contractor's items gone.

While it is tempting to start moving transmitters and equipment into the building(s) as soon as you can, it is a good idea to wait until tower crews and contractors are out of the way. This is to prevent equipment being inadvertently damaged and to keep your frustration level as low as possible.

Fencing and Signage

FCC regulations require that the tower bases be fenced and locked to keep the public away from the tower base, as it is energized. It is a good idea to set the fencing to just beyond the general population/uncontrolled exposure limits as specified in FCC OET65. This will ensure that anyone coming up to the fence will not be exposed to levels of RF radiation in excess of the safe limit for the general population.

Additionally, signs need to be posted on the fence warning of the risk of RF exposure beyond the fence limits. A sign should be posted at the base of the tower warning of high voltage. If the towers are tall enough to require an antenna structure registration, you must have the registration number posted at the tower base. I also like to post this number on the fence so the public can easily see it.

It is also a good idea to fence and lock the guy anchors to keep people away from the items that are holding your tower up.

The Tune-Up

Tuning up a nondirectional antenna is fairly straightforward. A vector network analyzer (VNA) or RF bridge is used and the tuning network is adjusted to present an impedance of 50 ohms resistive, 0 ohms reactive (50j0) to the transmission line. At this time, the actual operating impedance of the tower should be measured, as you will need it to determine the base current for 100% power output, and to provide on the FCC paperwork.

For a directional array, if you are doing a conventional proof of performance, you should double-check the sample system to ensure the lines are equal length, then proceed.

If you are doing a moment of methods (MoM) proof, there are additional measurements that must be made with a VNA before proceeding. These items are:

- The base impedance of all towers with all towers floating;
- The base impedance of all towers with all towers not under measurement shorted;
- The impedance of all of the toroidal transformers;
- The output of all toroidal transformers to ensure they are all identical;
- The electrical length of all sample lines;
- The characteristic impedance of all sample lines.

Once these items have been measured and determined, you may proceed with the tune-up.

FIGURE 7.1-29 The inside of a phasor. You will be in this quite a bit during tune-up.

A directional array is brought up slowly, antenna parameters read, and the phasor adjusted to bring the parameters to licensed value. Once this is done, the transmission lines are "flattened." In this procedure, the input to each tuning network at each tower is measured with an operating impedance bridge and the tuning networks adjusted to present 50j0 to the transmission lines. After each adjustment, it is normally required to adjust the phasor to bring the system back to parameters, then move onto the next tower until all inputs are at 50j0. It can be long and tedious as everything interacts.

The Proof

If you are doing an MoM proof, once the system is tuned up, you need to pick three points on numerous radials from the site and measure them. These are reference points and are normally measured in the nulls, surrounding the nulls, and at least one radial in a major lobe. The data listed in the previous section, along with the reference point measurements, are tabulated and presented to the FCC for licensing. You will also need the GPS coordinates of the measurement locations, as well as the datum used by the GPS. Note that the sample system for a MoM proof must be recertified every 2 years.

If you are doing a conventional proof of performance, you first must take the station nondirectional at a reduced power level (as determined by your consultant), and measure points on certain radials determined by the shape of the pattern. The points between the base of the tower (actually, not closer than one wavelength or 5 times the height of the radiator) and 3 kilometers out will establish the nondirectional field at 1 kilometer. The points between 3 kilometers and 15 kilometers will be used to ratio against the directional field readings to determine the shape of the pattern. These readings establish the ground conductivity of the area and are used in determining if the pattern is within tolerance.

Once the nondirectional measurements have been made, the measurements are done again with the directional antenna operating to determine the inverse distance field (IDF) along each radial and therefore the pattern shape. The IDF on each radial must fall below the maximum specified by the standard pattern established by the FCC. There are circumstances where a radial can be authorized to exceed the maximum limit as an augmentation.

Monitoring points will be established in each of the null radials. Maximum limits will be set for these points, and they must be read whenever it is suspected that the antenna system may be out of adjustment. It is good practice to measure and log the monitoring points at least once every 120 days, as signal levels will change with the seasons and resultant changes in ground conductivity.

Completion of the Project

Once the new site is on the air, you need to complete the performance measurements as specified in the FCC

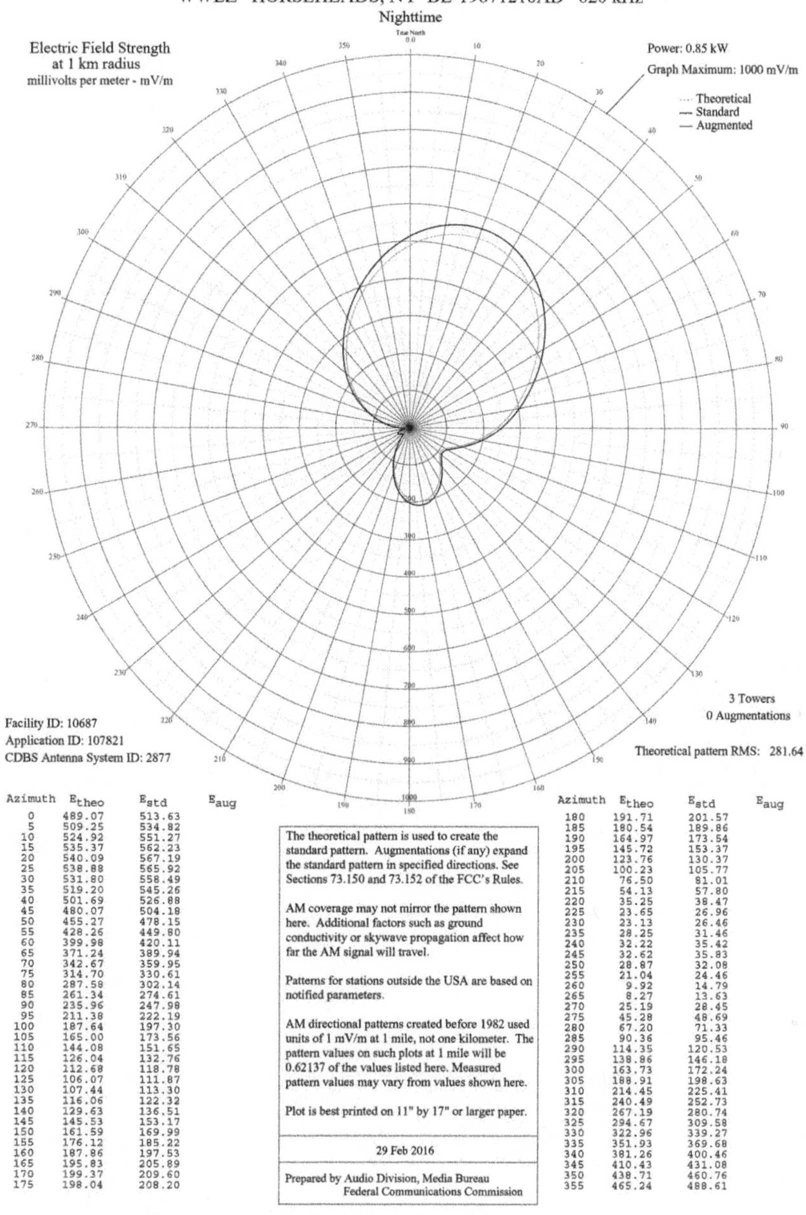

Azimuth	E_{theo}	E_{std}	E_{aug}
0	489.07	513.63	
5	509.25	534.82	
10	524.92	551.27	
15	535.37	562.23	
20	540.09	567.19	
25	538.88	565.92	
30	531.80	558.49	
35	519.20	545.26	
40	501.69	526.88	
45	480.07	504.18	
50	455.27	478.15	
55	428.26	449.80	
60	399.98	420.11	
65	371.24	389.94	
70	342.67	359.95	
75	314.70	330.61	
80	287.58	302.14	
85	261.34	274.61	
90	235.96	247.98	
95	211.38	222.19	
100	187.64	197.30	
105	165.00	173.56	
110	144.08	151.65	
115	126.04	132.76	
120	112.68	118.78	
125	106.07	111.87	
130	107.44	113.30	
135	116.06	122.32	
140	129.63	136.51	
145	145.53	153.17	
150	161.59	169.99	
155	176.12	185.22	
160	187.86	197.53	
165	195.83	205.89	
170	199.37	209.60	
175	198.04	208.20	

Azimuth	E_{theo}	E_{std}	E_{aug}
180	191.71	201.57	
185	180.54	189.86	
190	164.97	173.54	
195	145.72	153.37	
200	123.76	130.37	
205	100.23	105.77	
210	76.50	81.01	
215	54.13	57.80	
220	35.25	38.47	
225	23.65	26.96	
230	23.13	26.46	
235	28.25	31.46	
240	32.22	35.42	
245	32.62	35.83	
250	28.87	32.08	
255	21.04	24.46	
260	9.92	14.79	
265	8.27	13.63	
270	25.19	28.45	
275	45.28	48.69	
280	67.20	71.33	
285	90.36	95.46	
290	114.35	120.53	
295	138.86	146.18	
300	163.73	172.24	
305	188.91	198.63	
310	214.45	225.41	
315	240.49	252.73	
320	267.19	280.74	
325	294.67	309.58	
330	322.96	339.27	
335	351.93	369.68	
340	381.26	400.46	
345	410.43	431.08	
350	438.71	460.76	
355	465.24	488.61	

FIGURE 7.1-30 Polar plot of the WWLZ, Horseheads, New York, nighttime pattern. Measurements would need to be made in the nulls, adjacent to the nulls, and in the main lobe.

rules, 47 CFR §73.44. If this is an HD Radio facility, additional performance measurements must be made per National Radio Systems Committee (NRSC) standards to ensure compliance with emissions.

CONCLUSION

There are many things to consider in the undertaking of construction of a new AM transmitter facility. Attention to detail will pay off well in the long run. It is easier to build the facility correctly the first time rather than try to correct errors after it is on the air and operating.

CHAPTER

7.2

Transmission Audio Processing

ROBERT ORBAN

Orban Labs, Inc.
Pennsauken, NJ

INTRODUCTION

Transmission audio processing is both an engineering and artistic discipline. The engineering goal is to make most efficient use of the signal-to-noise ratio and audio bandwidth available from the transmission channel while preventing its overmodulation. The organization using audio processing sets the artistic goal. It may be to avoid audibly modifying the original program material at all, or it may be to create a distinct *sonic signature* for a broadcast by radically changing the sound of the original. Most broadcasters operate somewhere in between these two extremes.

If the transmitted signal meets regulatory requirements for modulation control and RF bandwidth, there is no well-defined right or wrong way to process audio. Like most areas requiring subjective, artistic judgment, processing is highly controversial and likely to provoke exceedingly opinionated arguments between its practitioners. Ultimately, the success of a broadcast's audio processing must be judged by its results—if the broadcast gets the desired audience, then the processing must be deemed satisfactory regardless of the opinions of audiophiles, purists, or others who consider processing an unnecessary evil.

One mark of the professionalism of broadcast engineers is their mastery of the techniques of audio processing. The canny practitioner has a bag of tricks that can be used to achieve the processing goal specified by the station's management, whether it is purist or "squashed against the wall."

ELEMENTS OF AUDIO PROCESSING

Compression reduces the dynamic range of program material by reducing the gain of material whose average or root mean square (RMS) level exceeds the *threshold of compression*. The amount by which the gain is reduced is called the *gain reduction* (G/R).

Above threshold, the slope of the input/output curve is the *compression ratio*. Low ratios provide loose control over levels, but generally sound more natural than high ratios, which provide tight control.

The *knee* of the input/output level graph can show an abrupt transition (*hard-knee*) into compression, or a gradual transition (*soft-knee*), in which the ratio becomes progressively larger as the amount of gain reduction increases. Figure 7-2.1 shows a typical

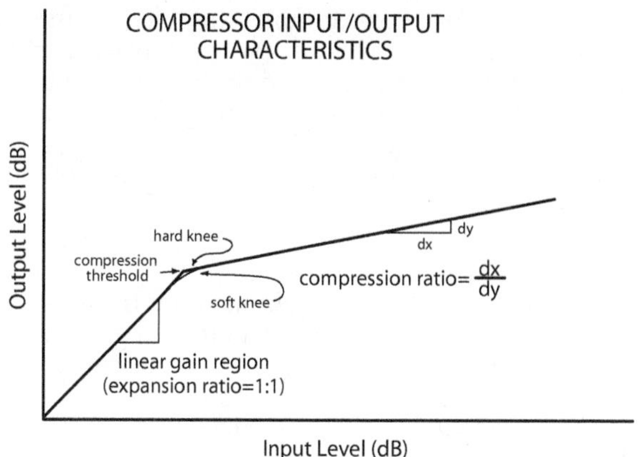

FIGURE 7.2-1 Input versus output levels for compressors

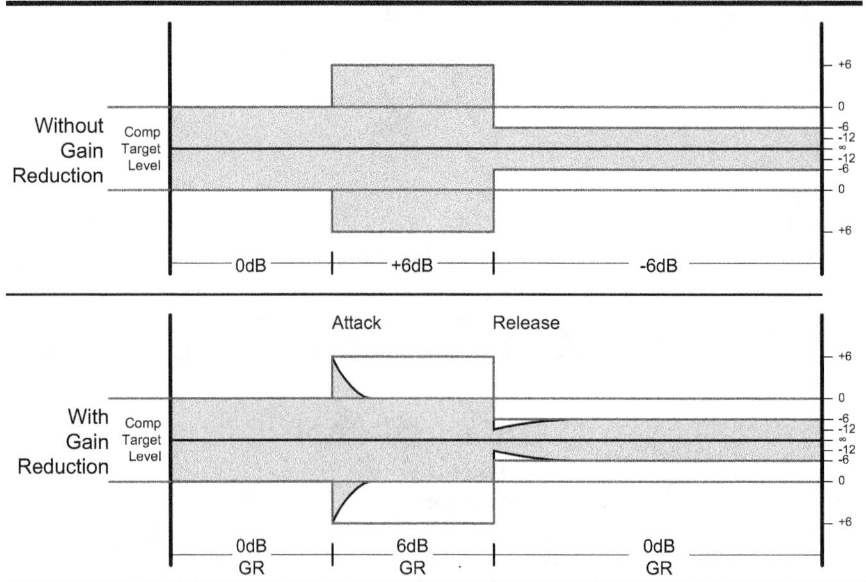

FIGURE 7.2-2 Attack and release characteristics of a compressor

Greg Ogonowski

compressor's output level as a function of its input level.

The *attack time* is, generally, the time that it takes the compressor to settle to a new gain following a step increase in level. Some engineers measure attack time as the *time constant*—the time necessary for the gain to achieve 67% of its new value, while others measure it as the time required for the gain to reach 90% of its new value for a given amplitude step (often 10 dB).

The *release time* is the time necessary for the gain to recover to within a certain percent of its final value after the level of the input signal to the compressor has been reduced below the compression threshold. It is sometimes convenient to specify the release time in dB per second if the shape of the release time is a straight line on a dB versus time graph. However, this shape often is not linear. *Multiple time constant* (sometimes called "automatic") release time circuits change the release rate (in dB/second) according to the history of the program, and according to how much gain reduction is in use. For example, the release time will temporarily speed up after an abrupt transient, to prevent a hole from being punched in the program by the gain reduction. The release time may slow down as 0 dB gain reduction is approached to make compression of wide-dynamic-range program material less obvious to the ear.

Delayed release holds the gain constant for a short time (typically less than 20 ms) after gain reduction has occurred. This prevents fast release times from causing modulation of individual cycles in the program waveform, thus reducing the tendency of the compressor to introduce harmonic or intermodulation distortion when operated with fast attack and release.

Figure 7.2-2 shows the envelope versus time of an illustrative toneburst test signal before (top) and after (bottom) it is applied to a compressor whose threshold is 0 dB and whose compression ratio is infinite. After the attack has finished, the compressor applies 6 dB of gain reduction to the +6 dB part of the signal. After the input drops to −6 dB, the compressor releases and its gain reduction moves back to 0 dB. The figure shows a straightforward single-time-constant release characteristic without delayed release.

EXPANSION

There are two types of expansion—*downward expansion* and *upward expansion*. Both increase the dynamic range of program material. Downward expansion reduces gain when the program level is lower than the *threshold of expansion* (see Figure 7.2-3), while upward expansion increases the gain when the program material is above the threshold. The primary purpose of downward expansion is to reduce noise, either electronic or acoustic, while upward expansion can reduce the effects of excessive compression or limiting on the input program by partially restoring peaks. Downward expanders are often coupled to compressors so that low-level program material is not amplified, thus reducing the noise that would otherwise be exaggerated by the compression. Expanders have attack times, release times, and expansion ratios that are analogous to those for compressors.

GATING

There are two fundamental types of gates, the *compressor gate* and the *noise gate*. The *compressor gate* prevents any change in background noise during pauses or low-level program material by freezing the compressor gain when the input level drops below the *threshold*

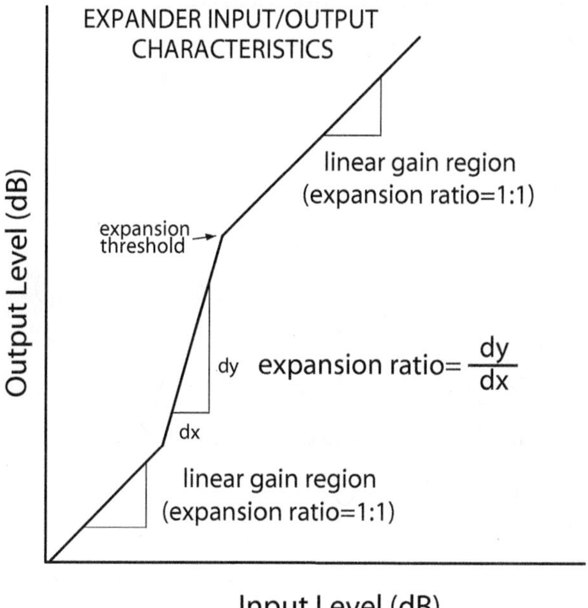

FIGURE 7.2-3 Input versus output levels for expanders

of gating. Because it produces natural sound, it is very popular in broadcasting.

Instead of freezing, many compressor gates will cause the gain to move very slowly to a nominal value (typically 10 dB of gain reduction) if the gating period is long enough. This prevents the compressor from getting stuck with an unusually high or low gain.

The *noise gate* is a downward expander with a high expansion ratio. Its purpose is to reduce audible noise. Because it causes gain reduction when the input level drops below a given threshold, the ear is likely to hear the accompanying gain reduction as a fluctuation in the noise level, sometimes called *breathing*. This can sound unnatural. Therefore, the noise gate is most useful when applied to a single microphone in a multi-microphone recording. Usually, the other microphones will mask any breathing, yet the noise reduction provided by the noise gate will still be appreciated during quiet program material.

MULTIBAND COMPRESSION AND FREQUENCY-SELECTIVE LIMITING

These techniques divide the audio spectrum into several frequency bands and compress or limit each band separately (although some interband coupling may be used to prevent excessive disparity between the gains of adjacent bands). This is the most powerful and most popular contemporary audio processing technique, because when done correctly it eliminates spectral gain intermodulation. This occurs in a wideband compressor or limiter when a voice or instrument in one frequency range dominates the spectral energy, thus determining the amount of gain reduction. If other, weaker elements are also present, their loudness may be audibly and disturbingly modulated by the dominant element. Particularly unpleasant effects may occur if the dominant energy is in the bass region, because the ear is relatively insensitive to bass energy, so the dominant bass energy pushes down the loudness of the midrange, seemingly inexplicably. Figure 7.2-4 shows the frequency responses of typical crossover filters in a five-band compressor, although there is no general agreement on the optimum number of bands.

Another type of frequency-selective limiting uses a program-controlled filter. The filter's cutoff frequency, its depth of shelving (explained under the heading "Equalization"), or a combination of these parameters, is varied to dynamically change the frequency response of the transmission channel. Such program-controlled filters are most often used as *high-frequency limiters* to control potential overload due to pre-emphasis in pre-emphasized systems like FM (VHF), and analog television audio (NTSC and PAL), and in FM-modulated transmission channels such as microwave links and satellite circuits.

Because it causes less noise modulation, frequency-selective downward expansion creates fewer audible artifacts than broadband downward expansion when used for noise reduction. With digital signal processing, it is common for frequency-selective downward expansion to use dozens (or even hundreds) of frequency bands.

EQUALIZATION

Equalization is changing the spectral balance of an audio signal, and is achieved by use of an *equalizer*. In broadest terms, an equalizer is any frequency-selective network (filter) placed in the signal path. In audio processing, an equalizer is usually a device that can apply a *shelving* or *peaking* curve to the audio. Equalizers can be static (applying a fixed curve) or dynamic (applying a curve whose shape varies as a function of the program material).

A shelving curve starts off at a specified gain. As frequency changes, the gain increases (boost) or decreases (cut) asymptotically. Finally, the gain shelves off and does not change with further changes in frequency. Figure 7.2-5 shows typical high-frequency and low-frequency (6, 12, and 18 dB/octave) shelving magnitude responses.

A peaking curve is bell-shaped on the frequency axis. As opposed to a shelving curve, it has a well-defined peak frequency. The simplest and most common peak equalizer is *second order*. The shape of a second-order curve is uniquely defined by three *parameters*: the *amount of equalization* (in dB), the *frequency of maximum equalization* (in Hz), and the *Q*, which is a dimensionless number that describes whether the curve is broad or sharp.

A *parametric equalizer* provides several peaking equalizers, in which the user has control of all three parameters. This type of equalizer is generally considered the most flexible and musical-sounding equalizer. Some parametric equalizers can also be used

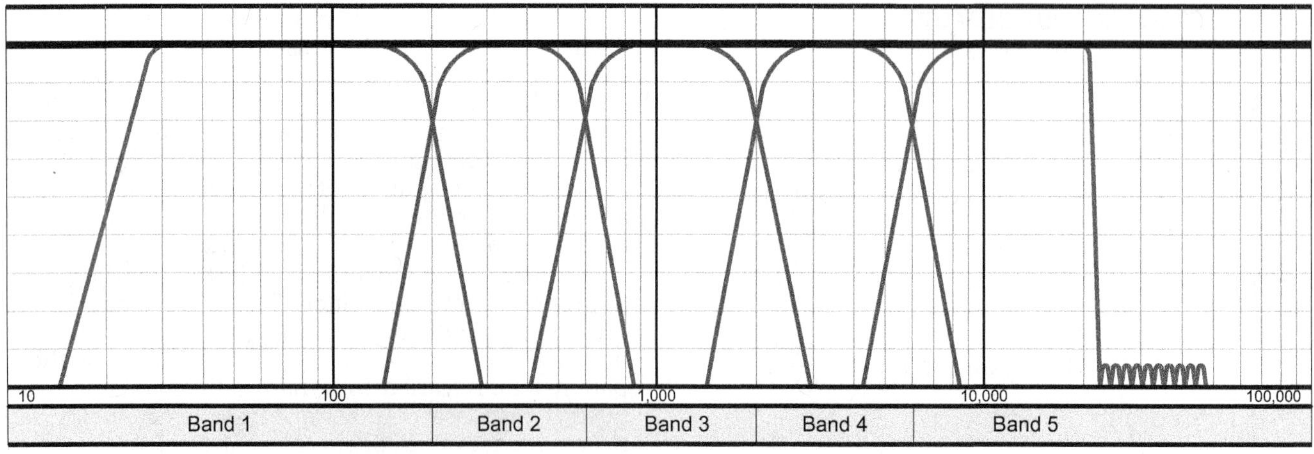

FIGURE 7.2-4 Frequency responses of typical multiband crossover filters

Greg Ogonowski

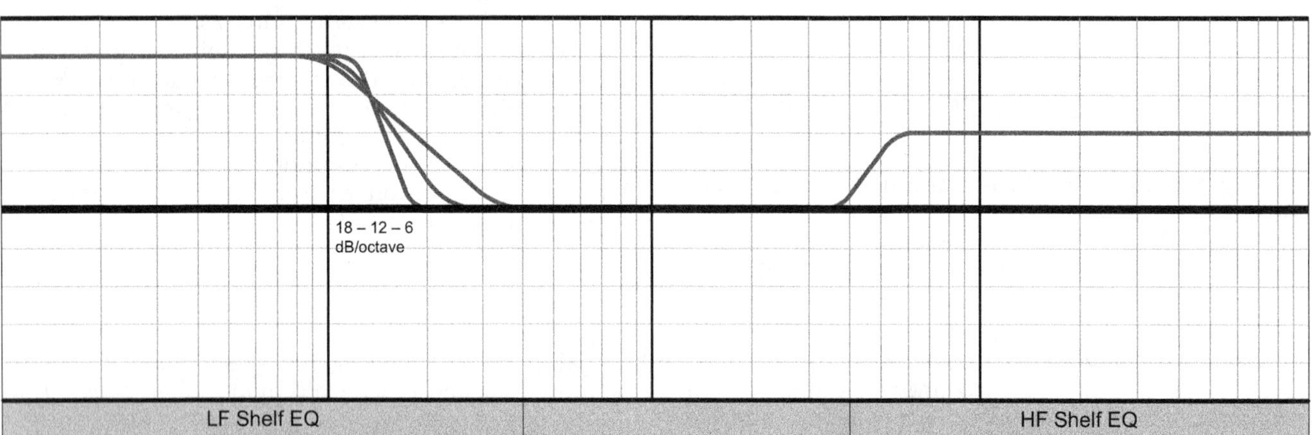

FIGURE 7.2-5 Low- and high-frequency shelving equalizer magnitude responses

Greg Ogonowski

as notch filters. Figure 7.2-6 shows typical peaking boost and cut magnitude responses for low and high bandwidth settings at each of three different center frequencies.

A *graphic equalizer* provides a number of peaking equalizers (usually 8–31) distributed on octave or fractions of octave (one-quarter or one-third) spaced frequency centers throughout the audible range. The controls for the amount of equalization are linear-throw faders, and are arranged on the panel in order of frequency. The positions of the controls, when considered together, thus provide a very rough graphic display of the amount of equalization provided by the entire equalizer. The advantage of a graphic equalizer is that it is easy to understand and quick to adjust. Its primary disadvantage is lack of flexibility. Usually, only the amount of equalization is adjustable, the Q and center frequency being fixed. However, a few manufacturers make parametric equalizers with graphic-style controls. These provide the advantages of both types.

High-frequency enhancers apply program-adaptive dynamic equalization that brightens dull-sounding input material without significantly affecting material with adequate high-frequency energy.

Lowpass and highpass filters remove spectrum at the top and bottom of the audible range, respectively. They are usually used to remove unwanted high- or low-frequency noise, and can also produce special effects (like telephone simulation). These filters usually come with their rate of cutoff fixed in multiples of 6 dB/octave; 6 dB/octave, 12 dB/octave, and 18 dB/octave are popular. In addition, the shape of the region around the cutoff frequency has a considerable effect on the listening quality of such filters. Bessel (constant delay) filters have a gentle transition into cutoff, and sound pleasant and musically neutral. Butterworth (maximally flat magnitude) filters have a sharper transition into cutoff. They are more effective at removing noise than Bessel filters, but have a more colored listening quality.

Equalizers are sometimes used on-line in transmission to create a certain sonic signature for a broadcast. Any of the types above may be used. Commercial audio processors may include equalizers for program

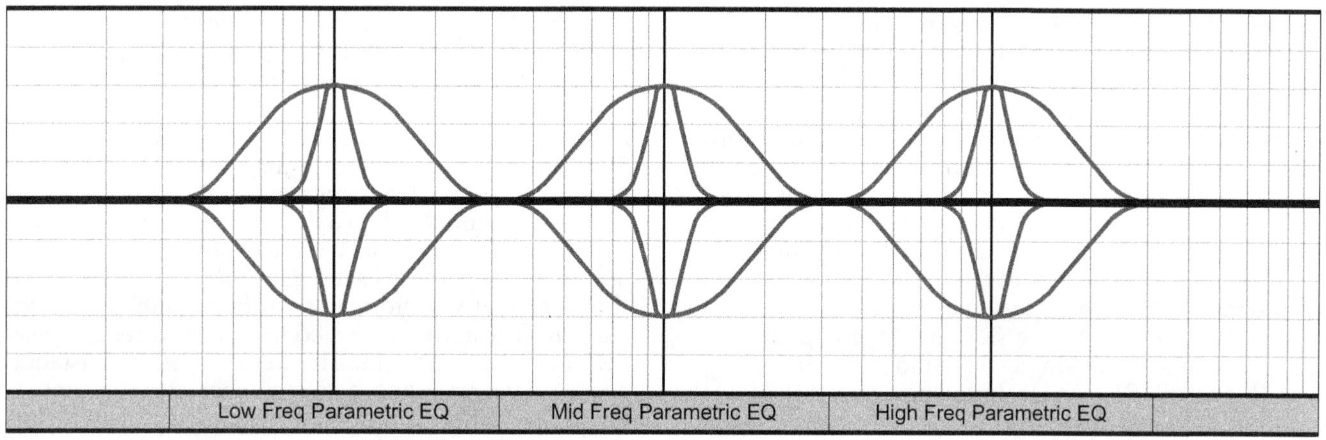

FIGURE 7.2-6 Parametric equalizer magnitude responses

Greg Ogonowski

coloration, or for correcting the frequency response of previous or subsequent transmission links. Sometimes the various bands of a multiband compressor or limiter are used as an equalizer by adjusting the gains of the various bands to achieve the desired equalized frequency response.

Filters that constrain the audio bandwidth to meet the requirements of a transmission are typically much sharper than filters used for program equalization. Elliptic-function filters are the most selective possible filters of a given filter order. Their passbands and stopbands exhibit *equiripple* behavior: the frequency response variations in the passband all have the same magnitude. The same is true of the response variations in the stopband.

Digital signal processing can readily implement *finite impulse response* (FIR) filters. Their main advantage is that they can easily be designed to produce constant time delay at all frequencies, preventing *phase distortion*. FIR filters are versatile and can be used for program equalization or bandwidth control. They are typically realized as tapped delay lines, where the taps' outputs each have a selected gain applied to them and are then summed together.

SUBHARMONIC SYNTHESIS

Subharmonic synthesizers generate bass one octave below bass present in the original program to correct bass-shy material in a way impossible with simple bass equalization. Subharmonic synthesis can help some older material with little energy below 70–80 Hz, but users must be aware of the equal-loudness curves, which show the ear's perception of the loudness of pure tones as a function of frequency and sound pressure level.[1] Material below 50 Hz takes lots of peak level to produce significant loudness. Close spacing of curves means small changes in amplitude lead to large change in subharmonic loudness. Because of the amount of peak level they use up, subharmonics will always make a broadcast sound quieter for a given amount of processing artifacts/distortion. They use up peak level that otherwise could be dedicated to audio to which the ear is more sensitive: an inevitable effect of the equal-loudness curves. Subharmonic synthesis is safest when created in the production studio, where the effect can be monitored by humans before it is broadcast.

Subharmonic synthesis should be used in conjunction with high-compression-ratio multiband compression to ensure consistent loudness of the subharmonics. Bass intermodulation distortion reduction techniques in the on-air compressor and peak limiter are essential. Subharmonics should track the level of their generating frequencies in a frequency-dependent way to keep the amount of LF enhancement subjectively constant.

Subharmonics make male voices sound weird, so when used in a transmission audio processor, a subharmonic synthesizer should be used in conjunction with automatic speech/music detection and should not be applied indiscriminately to material above 90 Hz.

PEAK LIMITING AND CLIPPING

Peak limiting is an extreme form of compression characterized by a very high compression ratio, fast attack time (typically less than 2 µsec) and fast release time (typically less than 200 ms). In audio processing for AM and FM radio, a peak limiter by itself usually limits the peaks of the *envelope* of the waveform, as opposed to individual instantaneous peaks *in the waveform*. These are usually controlled by *clipping*. As a matter of good engineering practice, peak limiters are usually adjusted to produce no more than 6 dB of gain reduction to prevent offensive audible side effects. *The main purpose of limiting is to protect a subsequent channel from overload, as opposed to compression, the main purpose of which is to reduce the dynamic range of the program.*

Peak clipping is a process that instantaneously chops off any part of the waveform that exceeds the *threshold of clipping*. This threshold can be either symmetrical or asymmetrical around 0; asymmetrical clipping is typically used in AM broadcast to control positive and negative peak modulation separately.

While peak clipping can be very effective, it causes audible distortion when overused. Modern audio processors frequently use distortion reduction techniques that remove some of the clipping-induced distortion in a frequency-selective way, which may be program adaptive and may use information about psychoacoustic masking[2] to reduce only the parts of the distortion that would otherwise be audible.

Radio frequency clipping (RF clipping) is peak clipping applied to a single-sideband RF carrier signal (a typical RF carrier frequency is 1 MHz). All clipping-induced harmonics fall around harmonics of the carrier (for a 1 MHz RF carrier the first harmonic is at 2 MHz, etc.). Upon demodulation, these harmonics remain at high frequencies and are removed by a lowpass filter. Thus RF clipping produces only intermodulation distortion and no harmonic distortion. Ordinary or audio frequency (AF) clipping produces both. RF clipping is substantially more effective than AF clipping on voice because intermodulation distortion is considered less objectionable than harmonic distortion in this application. On the other hand, RF clipping is considered much more objectionable than AF clipping on music.

The *Hilbert transform clipper* combines the features of RF and AF clippers. It acts as an RF clipper below 4 kHz (the region in which most voice energy is located), and acts as an AF clipper above 4 kHz to prevent excessive intermodulation distortion with music. It is realized at audio frequencies by using a mathematical model of an RF clipper;[3] no RF modulation actually occurs.

Unless a limiter has an attack time of less than about 10 µsec, it will exhibit *overshoots* at its output. If the goal of the processing is to precisely constrain the instantaneous values of the waveform to a given threshold, it usually sounds better to control these overshoots by a limiter with a moderate attack time (2 ms is a good starting point) followed by a clipper. Attempting to provide all peak control with the limiter does not sound as good, because the clipper affects only the offending overshoot and does not apply gain reduction to the surrounding signal.

When used in this way, clippers can cause audible distortion on certain program material. However, fast-attack limiters will cause audible clipping of the first half-cycle of certain program material, such as solo piano, harp, and nylon string acoustic guitar. Such distortion can be eliminated by a *delay line limiter*, often called a *look-ahead limiter*.[4] In its basic form this device consists of two audio paths, as seen in Figure 7.2-7, which also shows typical waveforms at various points in the signal path. The audio is applied to a delayed-release peak detector with threshold. The detector has an instantaneous attack time and a release characteristic consisting of a hold time followed by a near-instantaneous release. The output of this block is 1 below threshold and 1/(peak detector output) above threshold. If this signal were multiplied by the audio input, the product would be constrained to 1, but the multiplication process would generate considerable distortion because of the gain control signal's abrupt edges and insufficient smoothing. To remedy this, the gain control signal first applied to a lowpass filter to smooth its sharp edges. To compensate for the lowpass filter's group delay (which delays application of gain control to the audio), the audio is delayed equally prior to the multiplier's input. This time-aligns the gain control signal with the audio being controlled. The hold time of the peak detector is set equal to this delay so that release does not commence prematurely. The step response of the lowpass filter has a slight overshoot so that the limiter can reach full gain reduction in a finite amount of time.

One can model any peak limiter as a multiplier that multiplies its input signal by a gain control signal. This is a form of amplitude modulation. Amplitude modulation produces sidebands around the "carrier" signal. In a peak limiter, each Fourier component of the input signal is a separate "carrier" and the peak limiting process produces modulation sidebands around each Fourier component.

Considered this way, a hard clipper has a wideband gain control signal and thus introduces sidebands that are far removed in frequency from their associated

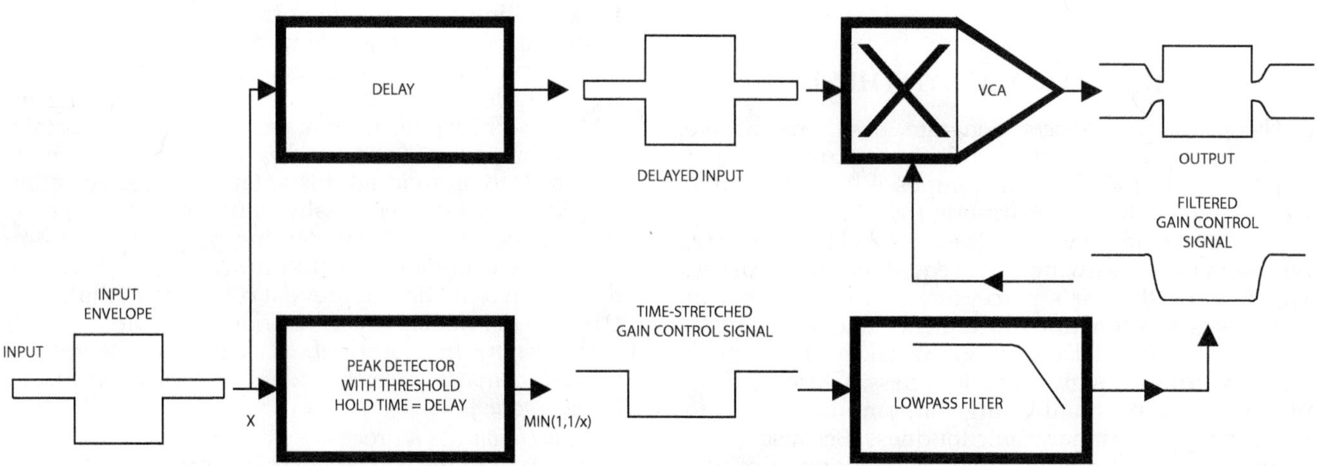

FIGURE 7.2-7 Look-ahead limiter waveforms

Fourier "carriers." Hence, the "carriers" have little ability to mask the resulting sidebands psychoacoustically. Conversely, a look-ahead limiter's gain control signal has a much lower bandwidth and produces modulation sidebands that are less likely to be audible. Accordingly, the psychoacoustic model in a transmission codec rejects these sidebands as inaudible and the codec does not attempt to encode them. This makes a look-ahead limiter a good choice to protect lossy codecs, such as those used in netcasting and almost every digital broadcasting channel.

Realizing clippers with digital signal processing (DSP) is tricky because clipping adds energy that may exceed one-half the sample frequency (the "Nyquist frequency"). Clipping products can fold around the Nyquist frequency, introducing audible aliasing components from frequencies that would be ultrasonic and inaudible in an analog system. In DSP, it is common to use extreme oversampling (or similar techniques) to prevent such aliasing from becoming audible.

PEAK LIMITING AND SPECTRAL CONTROL

The most fundamental job of a transmission audio processor is to prevent peak overload of the output channel. In streaming and digital broadcasting, the overload point is 0 dBTP.[5] In RF transmission, the audio processor must control the peak modulation of the RF carrier to the standards required by the governing authority, such as the FCC in the United States. For AM broadcasting, this usually means that negative carrier pinch-off must not occur at any time because this would cause splatter interference into adjacent channels. In FM and analog television (NTSC and PAL), the peak deviation of the carrier must be controlled so that the modulation monitor specified by the governing authority does not indicate overmodulation.[6] Because the rules often permit the modulation monitor to ignore very brief overshoots, the instantaneous peak deviation might exceed the peak modulation as indicated on the modulation monitor.

When using clipping to limit peaks, it is hard to achieve tight peak control and tight spectrum control simultaneously. Applying a clipped signal to a linear filter will cause the filter to overshoot and ring when removes spectral energy from the clipped input. This occurs because of two mechanisms: *spectral truncation* and *time dispersion*.

Peak-clipped signals resemble square waves, which because of their simplicity are useful for analyzing overshoot. One can build a square wave by summing its Fourier components together with correct amplitude and phase. Analysis shows that the fundamental of the square wave is approximately 2.1 dB higher than the amplitude of the square wave itself, and that any band-limited approximation to a square wave that simply truncates higher-order harmonics has no less than about 0.75 dB of overshoot.[7] As each harmonic is added in turn to the fundamental, a given harmonic's phase is such that the peak amplitude of the resulting waveform *decreases* by the largest possible amount. Simultaneously, the RMS value *increases* because of the addition of the power in each harmonic. This is the fundamental theoretical reason why simple clipping is such a powerful tool for improving the peak-to-average ratio of broadcast audio: clipping adds to the audio waveform spectral components whose phase and amplitude are precisely correct to minimize the waveform's peak level while simultaneously increasing the power in the waveform.

If a square wave (or clipped waveform) is applied to a lowpass filter with constant time delay at all frequencies, the higher harmonics that reduce the peak level will be removed, increasing the peak level and with it the peak-to-average ratio. Thus even a perfectly phase-linear lowpass filter will cause overshoot.[8]

If the sharp-cutoff filter is now allowed to be minimum phase, it will exhibit a sharp peak in group delay around its cutoff frequency.[9] Because the filter is no longer phase-linear, it will not only remove the higher harmonics required to minimize peak levels, but will also change the time relationship between the lower harmonics and the fundamental. They become delayed by different amounts of time, causing the shape of the waveform to change. This *time dispersion* will therefore further increase the peak level.

When a square wave is applied to a linear-phase filter, overshoot and ringing will appear symmetrically on the leading and trailing edge of the waveform. If the filter is minimum phase, the overshoot will appear on the trailing edge and will be about twice as large. In the first case, the overshoot and ringing are in fact caused by spectrum truncation that eliminates harmonics necessary to minimize the peak level of the wave at all times; in the second case, the overshoot and ringing are caused by spectrum truncation *and* by distortion of the time relationship between the remaining Fourier components in the wave.

One of the biggest challenges in transmission audio processor design is simultaneously controlling peak levels and the spectrum at the processor's output. *There is no sharp-cutoff linear lowpass filter that is overshoot-free*: reducing the overshoot of a band-limited square wave to 0 dB requires changing the relative amplitudes of the harmonics, not just truncating the higher-order ones. Nonlinear filters do this. Specific to each manufacturer, these nonlinear filters (commonly called *overshoot compensators*) typically consist of a combination of clippers and linear filters and are embedded within the processing to achieve overshoot-free spectral control to FCC or ITU-R standards.

Look-ahead limiters seldom add significant spectral energy. To control a nonclipping audio processor's output spectrum, one can put a linear lowpass filter before a look-ahead limiter. The limiter will control any peak overshoots that the filter adds without need for additional downstream lowpass filtering.

DE-CLIPPING

De-clipping is a controversial technique that attempts to reconstruct the original audio waveform from a clipped or aggressively peak-limited version of it.

There are many different de-clipping algorithms commercially available. De-clipping has attracted interest because a distressing amount of contemporary source material has been "hypercompressed"[10] in mastering or production, and record labels have then provided this material to broadcasters to use on-air, despite the fact that transmission audio processors often exaggerate the resulting audible distortion.

Information is 100% lost in flat-topped areas and cannot be recovered: a flat-topped waveform is a mathematical "singularity." Hence, de-clippers must make educated guesses about what's missing based on interpolation from material surrounding the clipped samples. To do this, the interpolation must use a model of the clipping process. However, many waveforms that look they have been hard-clipped have, in fact, been peak-limited by more complex limiting processes with sidechains and memory, and each limiter manufacturer has a proprietary way of computing the sidechain. For competitive reasons, these are seldom made public. Even if the sidechain is public knowledge, if the compression ratio is infinite, it is still impossible to deduce what the limiter's input was.

De-clippers can increase punch on transients by increasing peak levels by guessing what the missing waveform is. However, this is not the same as cancelling IM distortion. Distortion cancellation depends on having a precise, invertible model of the peak limiting process. This is usually impossible. In fact, because de-clipping is a nonlinear process, it can make its own IM distortion that adds to any IM distortion present in the original source. If simple peak clipping was used on a given track, then de-clippers can help, but sometimes they make things worse. The better the original peak limiting algorithm, the more likely it is that de-clipping will add IM distortion, not cancel it. Therefore, the safest place for a de-clipper is in the production studio, so that human ears can determine if the de-clipper is helping or adding another layer of distortion. Moreover, in the broadcast processing chain, de-clipped waveforms force the on-air processor's peak limiter to work harder. So use de-clippers with care.

STEREO ENHANCEMENT

Stereo enhancement increases the apparent width of the stereo image and exists in different manufacturer-specific forms. Some techniques include frequency-selective upward expansion of the stereo difference signal (L-R), and addition of a time-delayed version of the L-R signal to the enhancer's output. To prevent the increased L-R energy from objectionably increasing multipath distortion in FM analog transmission, most enhancers dynamically constrain the amount of enhancement to a user-adjustable maximum L-R level.

MEASURING AND CONTROLLING LOUDNESS

One of the main uses of audio processing is to increase perceived loudness within the peak modulation constraints of a transmission channel. Assessing the effectiveness of audio processing thus requires a means of measuring loudness. *Loudness is subjective: it is the intensity of sound as perceived by the ear/brain system.* No simple meter, whether peak program meter (PPM) or VU, provides a reading that correlates well to perceived loudness. A meter that purports to measure loudness must agree with a panel of human listeners.

Three important factors correlate to subjective loudness:

- The spectral distribution of the sound energy. The ear's sensitivity depends strongly on frequency. It is most sensitive to frequencies between 2 and 8 kHz. Sensitivity falls off fastest below 200 Hz.
- Whether the sound energy is concentrated in a wide or narrow bandwidth. For a given total sound power, the sound becomes louder as the power is spread over a larger number of *critical bands* (about one-third octave). This is called *loudness summation*.
- The duration of the sound. A given amount of sound power appears progressively louder until its duration exceeds about 200 milliseconds, at which point no further loudness increase will occur. This is called *temporal integration*.

THE ITU-R BS.1770 LOUDNESS MEASUREMENT ALGORITHM

In 2006, the ITU-R published Recommendation ITU-R BS.1770: "Algorithms to measure audio programme loudness and true-peak audio level." Developed by G.A. Soulodre, the BS.1770 loudness meter is a power-sensing (RMS) meter preceded by a "K-weighting" equalizer, which roughly models the equal-loudness curve of human hearing. As such, it takes into account the frequency dependence of hearing, but it does not attempt to model loudness summation or the integration time constants of human hearing. Its unit of measure is "LU" (sometimes called Lk).[11] A change in loudness of 1 LU corresponds to a change of 1 dB.

A major disadvantage of the original BS.1770 meter is that it weights silence and low-loudness material the same as high-loudness material. This causes the meter to under-read program material (like dialog) having substantial pauses that contain only low-level ambience because louder program material contributes most to a listener's perception of overall program loudness. To address this problem, the BS.1770–2 algorithm added gating to BS.1770–1, and gating has been retained through subsequent revisions. With gating, the meter ignores silence and is weighted toward louder program material, which contributes most to a listener's perception of loudness. The gated measurement indicates only sounds that fall within a floating window that extends from the loudest sounds within the preset integration period to sounds that are 10 dB quieter than the loudest sounds. There are two steps in the gating process: first, an absolute gate removes silent passages;

second, a relative gate weights louder parts of the program more heavily than quieter parts.

A more detailed explanation of the algorithm is this:

1. Using the BS.1770–1 algorithm (i.e., a K-weighting filter followed by RMS summation and averaging), calculate the RMS value in a 400-ms time window. One number is computed for every 400-ms time window. Start computing a new 400-ms window every 100 ms, so there is 75% time overlap between windows. Continue computing the RMS values of new 400-ms windows throughout the entire duration of the measurement and store all of these results—one number for each 400-ms window.
2. If any 400-ms window has a value below −70 LUFS, throw it away.
3. Compute the average of the remaining windows over the total time period of the measurement. If any window is less than 10 dB below this average, throw it away.
4. Compute the average of the remaining windows. Display this reading on the meter.

EBU Tech 3341 defines three integration time constants for the measurement: *momentary, short-term*, and *integrated*. The momentary measurement integrates the output of the RMS detector over 400 ms, the short-term measurement integrates over three seconds, and the integrated measurement integrates over an entire program segment. All samples within the specified integration time are weighted equally, and only the integrated measurement is gated. The integrated loudness of a program segment is called the *Program Loudness* ("PL").

The momentary meter is intended for live program monitoring by operators (like a modern form of VU meter). The short-term meter is also intended for program monitoring, but averages out momentary loudness fluctuations. It is also useful for detecting potentially annoying short segments of a program that may have much higher loudness than the PL. Several loudness recommendations[12] advise constraining the maximum short-term loudness of commercial announcements to 5 LU above the PL.

The integrated measurement is most useful offline, as it characterizes the average loudness of program segments that may last for several minutes. It can be used to normalize the loudness of program segments in a playout system by statically adjusting the level of each segment to achieve the target loudness. It is usually possible to play these segments back-to-back without annoying listeners with excessive loudness variations. (Research has established that typical listeners have a "comfort zone" of +2, −5 dB, within which listeners are unlikely to feel the need to adjust their volume controls.) This static loudness normalization retains the full dynamic range of each program segment, and is the simplest possible form of "audio processing."

EBU Tech 3342 defines a *loudness range* descriptor, which statistically quantifies the dynamic range of a given piece of program material, adding useful information to BS.1770's integrated loudness measurement. In general, the higher the loudness range, the less effectively the integrated PL describes the perceived loudness of program material, as material with a high loudness range will have portions that are significantly quieter and louder than the PL. A broadcaster may wish use dynamic compression to make a program segment with a large loudness range more comfortable to hear, particularly in noisy environments like moving vehicles.

The audio community has widely adopted BS.1770 for loudness normalization. Extensive testing against human listeners indicates that on average, it is approximately as accurate as more complex loudness meters. However, Figure 13 in the BS.1770–3 Recommendation shows errors as large as 5 dB on certain program material, which may be caused by the meter not taking into account many of the psychoacoustic elements affecting loudness perception. Human hearing is still the best judge of loudness, and if one's hearing disagrees with the meter, one should trust one's ears.

Using an "anchor element" for loudness normalization is controversial. ATSC A/85 recommends defining an anchor element (usually dialog),[13] and normalizing so that the anchor element is at the target loudness, while nonspeech material may be above or below the target loudness according to the artistic requirements of the program's producers. EBU R 128 recommends normalizing the loudness of the entire integrated program and discourages use of an anchor element, asserting that using it is an "active process requiring input from an experienced operator."

In addition to the recommendation in note 9, the Audio Engineering Society (AES) has published a considerable amount of work on the technology of loudness measurement. This work, mainly done in Europe and Canada, was summarized in the *AES Journal*.[14] These summaries provide references to the original papers.

THE JONES AND TORICK LOUDNESS MEASUREMENT ALGORITHM

Jones and Torick published a paper describing a meter for measuring the loudness of broadcast signals.[15] The FCC did an informal validation of the results of this meter, and concluded that it was effective in assessing whether commercials in television were noticeably louder than the surrounding entertainment programming.[16] Audio processors incorporating this algorithm in a closed-loop automatic loudness controller have been used extensively in television broadcasting.

THE PSOPHOMETER

According to the Network Performance, Reliability and Quality of Service Committee of the Alliance for Telecommunications Industry Solutions (ATIS), a *psophometer* is an instrument that provides a visual indication of the audible effects of disturbing signals of various frequencies.[17] A psophometer usually incorporates a weighting network. The characteristics of the weighting network depend on the type of circuit under investigation, such as whether the circuit is used for high-fidelity music or for normal speech. The standard for a psophometer intended to measure undesired signals on a wideband program line is ITU-R Rec. 468–2. For telephone lines, it is the ITU-T Rec. Vol.V.P.53.

SUBJECTIVE EFFECTS OF PROCESSING ELEMENTS

Reducing the peak-to-average ratio of the audio increases loudness. If peaks are reduced, the average level can be increased within the permitted modulation limits. The level with which this can be accomplished without introducing objectionable side effects (like clipping distortion) is the single best measure of audio processing effectiveness.

Density is the extent to which the amplitudes of audio signal peaks are made uniform (at the expense of dynamic range). Programs with large amounts of short-term dynamic range have low density and highly compressed programs have high density.

Compression reduces the difference in level between the soft and loud sounds to make more efficient use of permitted peak level limits, resulting in a subjective increase in the loudness of soft sounds. It cannot make loud sounds seem louder. Compression reduces dynamic range relatively slowly in a manner similar to "riding the gain." Limiting and clipping, on the other hand, reduce the short-term peak-to-average ratio of the audio.

Limiting increases audio density. Increasing density can make loud sounds seem louder, but can also result in an unattractive, busier, flatter, and denser sound. It is important to be aware of the many negative subjective side effects of excessive density when setting controls that affect the density of the processed sound.

Clipping sharp peaks does not produce any audible side effects when done moderately. Excessive clipping will be perceived as audible distortion.

Except as required to achieve very specific artistic goals (most notably in some major-market high-energy hit-music formats), the processed audio should be free from unnatural subjective side-effects, such as:

- *Pumping* (a sense that the gain is constantly and unnaturally changing—a characteristic side-effect of wideband compressors and limiters when driven heavily);
- *Breathing* (audible pulling up of background noise, cured by a compressor gate);
- *Hole-punching* (a sudden drop in loudness after a program transient, caused by the transient inducing a large amount of gain reduction that then does not decay quickly, and cured by multiple time-constant release time circuitry).

PROCESSING FOR STEREO

Processing for two-channel stereophonic transmission is similar to processing for monophonic transmission, except that two audio processing chains are used. To preserve stereo imaging completely, the gains of the left and right automatic gain control and compression circuitry must be identical, although many processors allow some gain control elements to be more loosely coupled to allow the processing to correct left/right channel imbalances in the source material. Conversely, experience has shown that fast peak limiting and high-frequency limiting circuits sound best when operated independently (without stereo coupling), because the ear does not perceive channel-imbalance-induced spatial shifts with these fast time constants. However, the ear can perceive the loudness of one channel's being modulated unnaturally by a dominant element in the other channel when the channels are coupled.

The requirements of the transmission service determine how gain-controlling elements should be coupled. In FM, the channel requiring the greatest amount of limiting determines the gain of both channels because the composite modulation is essentially the higher of the left and right channels. The processor operates by sensing the higher of the left and right channels and determining the gain of both channels such that the higher channel does not exceed a given level at the processor's output.

In matrix (also called "sum and difference") processing the left and right channels are matrixed into the L+R and L-R format. L+R and L-R pass through separate multiband automatic gain controller (AGC)/ compressor sections. This method changes the perceived width of the stereo image and can be used as a form of stereo enhancement. To prevent objectionable side effects, the maximum permitted gain difference between the sum and difference channels is usually constrained to a user-selectable value.

In FM stereo transmission, increasing the L-R level increases multipath distortion. In FM stereo processors it is common to allow matrix processing only in the AGC, while operating the multiband compressor in left/right mode so it can better control the audio level applied to the peak limiter. Digital transmission channels are analogous because the left and right channels individually overload at 0 dBFS and the peak limiters operate in the left/right domain.

In AM stereo, the gain of both channels is controlled by sensing and controlling the level of their sum (L+R), because the envelope modulation represents the sum of the channels. Matrix processing is therefore particularly useful for AM stereo.

GENERAL DESIGN REQUIREMENTS FOR HARDWARE AUDIO PROCESSORS

The audio processor hardware should conform to the following general design requirements:

- It must be packaged so that it is easy to operate and maintain, and can work in high RF fields without compromise.
- It should have setup controls with enough versatility to enable the subjective effect

to be readily tuned to the requirements of the broadcasting authority operating it. For mixed-format applications the processor may have several presets, selectable by remote control, that permit the operator to set the amount of compression, limiting, clipping, and other parameters to complement the program material being transmitted. The processor must be able to save user-modified presets for later recall.

- It should ordinarily be equipped with sufficient remote control facilities to enable it to be interfaced efficiently with modern, automated plants. Most of the required facilities are specific to the application: for example, AM (MW), shortwave (HF), FM (VHF), or television. Modern processors are capable of communicating via network (TCP/IP), modem, and serial communications. Firmware upgrades is possible on some units via a network connection to a PC.
- It should have sufficient metering to permit it to be easily set up with tones or program material. The metering should also provide operations and diagnostic capabilities. Metering usually includes input level, output level, and gain reduction (the amount of limiting or compression) occurring in each variable-gain stage.

BUILDING A SYSTEM

Combining several audio processors into a good sounding system is tricky because of headroom and time constant considerations. The device driving a given processor must be able to drive that processor into full compression or limiting. If the driving device (for example, a distribution amplifier) runs out of headroom before full limiting occurs in the driven device, then that device cannot achieve its full capability. This consideration is particularly critical in mixed digital/analog facilities when setting up the input analog-to-digital (A/D) converter of a digital audio processor. The analog drive level to the A/D converter must be set so that the A/D converter does not clip when receiving levels sufficient to cause full gain reduction in the following audio processor.

Beware of interactions between the attack times and release times when cascading several processors. It is wise to start the system with the slowest device. This is usually a compressor or automatic gain controller (AGC) with slow attack and release times and a compressor gate to prevent noise breathing. Such a processor does not significantly increase the *density* of the audio; it simply does gentle gain riding to ensure that following stages are driven at the correct level.

A multiband compressor with moderate attack and release times often follows the slow AGC. Correctly designed multiband processors have these time constants optimized for each frequency band. The low-frequency bands have slower time constants than the high-frequency bands. This multiband compressor usually does most of the work in increasing program density.

The amount of *gain reduction* determines how much the loudness of soft passages will be increased (and, therefore, how consistent overall loudness will be). Our hypothetical system reduces gain with the broadband AGC and the multiband compressor. The broadband AGC should be adjusted to control average levels and compensate for a reasonable amount of operator error. It should *not* be used to substantially increase the short-term program density; the multiband compressor and peak limiters do this.

Modern audio processing systems usually add other elements to the basic system described here. For example, it is common to incorporate an equalizer to color the audio for artistic effect. The equalizer may be any of the types described earlier and is usually found between the slow AGC and the multiband compressor. The multiband compressor itself functions as an equalizer when the outputs of its bands are summed with unequal gains.

Various lowpass filters are often included in the system to limit the bandwidth of the output signal to 15 kHz (for FM), 10 kHz (AM in NRSC countries), 4.5 kHz (AM in EBU countries, and shortwave worldwide) or other bandwidths as required by the local regulatory authority. As discussed in the previous section, "Peak Limiting and Spectral Control," clipping-based peak limiters require nonlinear final lowpass filters to prevent overshoot, while look-ahead limiters can use a linear lowpass filter before the limiter.

Highpass filters may be incorporated to protect the transmitter. This is particularly important in high-power AM and shortwave installations exceeding 100 kW carrier power.

A transmitter equalizer that corrects the pulse response of the transmitter is found on some AM processors.

In Europe, some countries are required to control their FM multiplex power according to ITU-R Recommendation BS412. The integrated power of the composite multiplex signal (including the stereo pilot tone) in any arbitrary 60-second window must be less than or equal to the integrated power of a sine wave that modulates the FM carrier ±19 kHz. Many modern FM processors include an automatic multiplex power controller. The controller measures the multiplex power and reduces the drive into the audio processor's peak limiting system to ensure that the standard is obeyed.

Figure 7.2-8 shows a simplified block diagram of a modern audio processing system.

LOCATION OF SYSTEM COMPONENTS

The best location for the processing system is as close as possible to the transmitter, so that the processing system's output can be connected to the transmitter through a circuit path that introduces the least possible change in the shape of the carefully peak-limited waveform at the processing system's output. One possible configuration is shown in Figure 7.2-9, in which a studio to transmitter link (STL) is utilized,

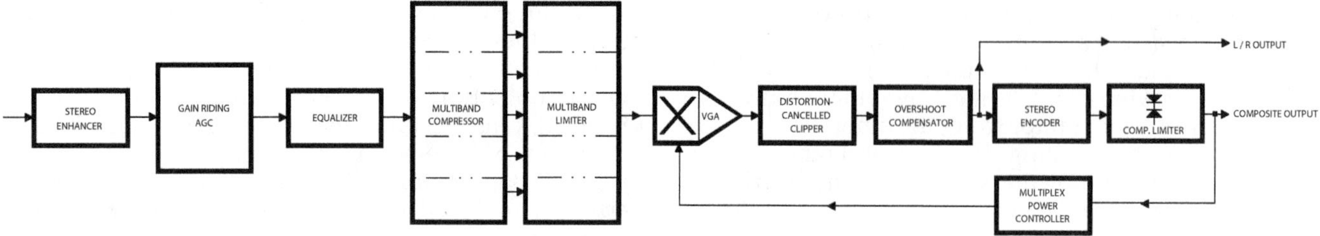

FIGURE 7.2-8 Modern audio processing system

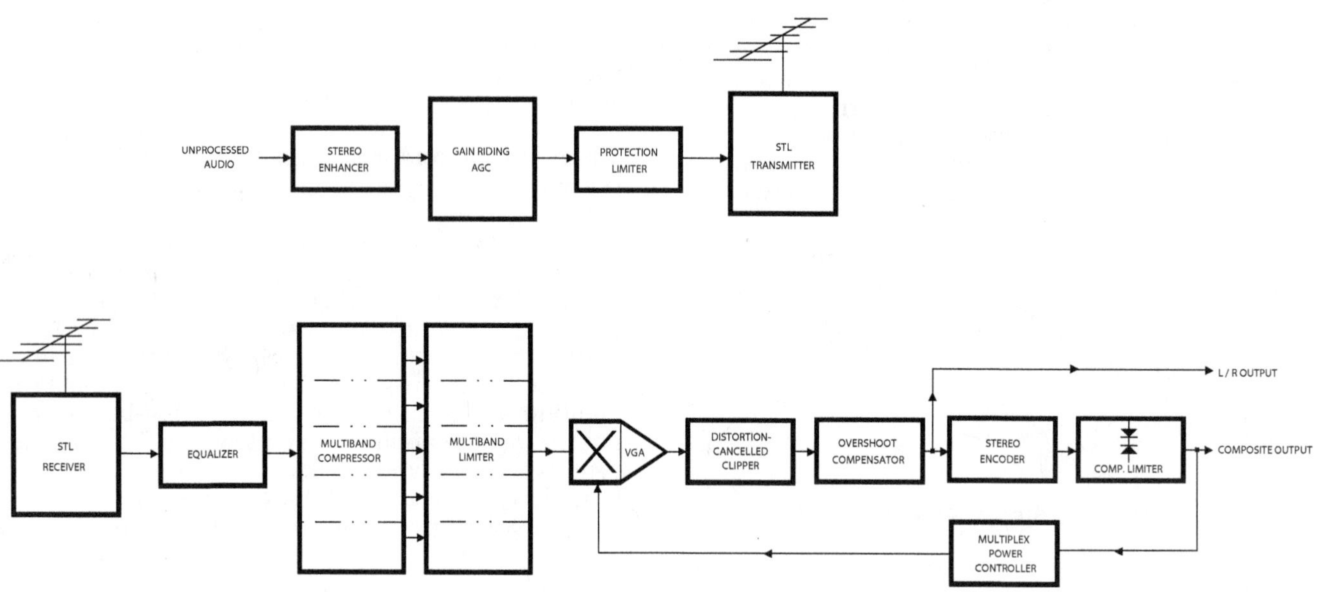

FIGURE 7.2-9 Equipment configuration with processing done after STL

and the processor is located between the STL and the transmitter. The STL might be telephone or post lines, analog microwave radio, or various types of digital paths. Sometimes, it is impractical to locate the processing system at the transmitter, and it must instead be located on the studio side of the link connecting the audio plant to the transmitter. This situation is not ideal because artifacts that cannot be controlled by the audio processor can be introduced in the link between studio and transmitter or by additional peak limiters placed at the transmitter.

In this case, the audio output of the processing system should be fed directly to the transmitter through a link that is as flat and phase-linear as possible: for example, an uncompressed digital link without sample rate conversion. Deviation from flatness and phase-linearity will cause spurious modulation peaks because the shape of the peak-limited waveform is changed. Such peaks add nothing to average modulation. Thus, the average modulation must be lowered to accommodate those peaks within the carrier deviation limits dictated by FCC modulation rules.

This implies that if the transmitter has built-in highpass or lowpass filters (as some do), these filters *must* be bypassed to achieve accurate waveform fidelity. Modern processing systems contain filters that will protect the transmitter. These filters are located in the processing system where they do not degrade control of peak modulation.

WHERE ACCESS TO THE TRANSMITTER IS AVAILABLE

The audio received at the transmitter site should be of as good quality as possible. Because the audio processor controls peaks, it is not important that the audio link (STL) feeding the processing system's input terminals be phase-linear. However, the link should have low noise, flattest possible frequency response from 30 to 15,000 Hz, and low nonlinear distortion.

If the audio link between the studio and the transmitter is noisy, performing the compression function at the studio site can minimize the audibility of this noise. Compression applied before the audio link improves the signal-to-noise ratio because the average level on the link will be greater. If the STL has limited dynamic range, it may be desirable to compress the signal at the *studio* end of the STL. To apply such compression, split the processing system, placing the AGC and multiband compressor sections at the studio, and the peak limiter at the transmitter.

WHERE ACCESS TO THE TRANSMITTER PLANT IS NOT AVAILABLE

In some situations, the organization originating the program does not have access to the transmitter, which is operated by a separate entity. In this case, all audio processing must be done at the studio, and any damage that occurs later must be tolerated. These systems usually include a peak limiter at the transmitter to provide protection against overmodulation.

If it is possible to obtain a broadband phase-linear link to the transmitter, the processing system at the studio location can feed the STL. Where only an audio link is available, feed the audio output of the processing system directly into the link. If possible, transmitter protection limiters should be adjusted for minimum possible action as the processing system does most of that work. Transmitter protection limiters should respond only to signals caused by faults or by spurious peaks introduced by imperfections in the link.

Where maximum quality is desired, it is important that all equipment in the signal path after the studio be carefully aligned and qualified to meet the appropriate standards for bandwidth, distortion, group delay, and gain stability, and that such equipment is requalified at reasonable intervals.

REQUIREMENTS FOR STLS

If the STL is ahead of the audio processor, the STL signal-to-noise ratio (SNR) must be sufficient to pass unprocessed audio. This means that the SNR of the link must be better than the sum of the desired SNR of the transmitted signal plus the maximum gain of the audio processor plus about 6 dB (a useful rule of thumb). If the STL follows the audio processor, its SNR should be 6 dB better than the desired SNR of the transmitted signal. To ensure that the STL does not distort the shape of the audio waveform (causing peak overshoot), the frequency response must be flat (±0.1 dB) throughout the operating frequency range, typically 20–15,000 Hz. The group delay must be essentially constant throughout this range. Deviation from linear phase should be less than ±10°. Phase correction can be applied to meet the requirement at high frequencies. At low frequencies, by far the best way to meet the audio specification is to extend the −3 dB frequency of the STL to 0.15 Hz or lower and to eliminate any peaking in the infrasonic frequency response prior to the roll-off frequency.

For highest quality, the nonlinear distortion of the STL system should be less than 0.1% total harmonic distortion (THD) throughout the operating frequency range.

Digital links may pass audio as straightforward Pulse Code Modulation (PCM) encoding, or they may apply data-rate-reduction processing to the signal to reduce the number of bits per second required for transmission through the digital link. Such processing will almost invariably increase peak levels, and such links must therefore be carefully qualified before used to carry the peak-controlled output of the audio processor to the transmitter or stereo encoder. For example, the MPEG-1 Layer 2 algorithm can increase peak levels up to 4 dB at 160 kB/sec by adding large amounts of quantization noise to the signal and removing those parts of the program that its psychoacoustic model deems inaudible. While the desired program material may psychoacoustically mask this noise, it is nevertheless large enough to affect peak levels severely. For any lossy compression system, the higher the data rate, the less the peak levels will be increased, so use the highest data rate practical in the system. Even with the maximum available data rate, overshoot will probably be large enough to require use of an overshoot compensator at the STL receiver (or in the equipment being driven by it). Some modern FM exciters and stand-alone stereo encoders are now equipped with such compensators.

Other links may use straightforward PCM without lossy data rate reduction. These can be very transparent and can exhibit accurate pulse response if their input anti-aliasing filters and output reconstruction filters are rigorously designed to achieve constant group delay over the frequency range that contains significant program energy. This is not particularly difficult to do with modern oversampled converter technology.

Because the output spectrum of most modern audio processing systems is already tightly band-limited, any anti-aliasing filters in digital links driven by such systems may be bypassed. This ensures the most accurate possible transient response.

USING LOSSY DATA REDUCTION IN THE STUDIO

Many stations are now using lossy data reduction systems to increase storage time of digital playback media. In addition, source material is often supplied through a lossy data reduction system, whether from satellite or over landlines. Sometimes, several encode/decode cycles will be cascaded before the material is finally presented to the audio processor's input.

All such algorithms operate by increasing the quantization noise in discrete frequency bands. If not psychoacoustically masked by the program material, this noise may be perceived as distortion, *gurgling*, comb filtering, or other interference. Psychoacoustic calculations are used to ensure that the added noise is masked by the program material and cannot be heard. In addition, there are at least two other mechanisms in broadcasting that can cause the noise to become audible at the receiver. First, a modern multiband transmission audio processor performs an *automatic equalization* function that can radically change the frequency balance of the program. This can cause noise that would otherwise be masked to become unmasked because the psychoacoustic masking conditions under which the masking thresholds were originally computed have changed. Second, the frequency response of the radio receiver (particularly in AM) can remove frequencies that were used to make the psychoacoustic masking calculations and that would otherwise have masked the added quantization noise.

Accordingly, if lossy data reduction is used in the studio then the highest data rate possible should be selected. This maximizes the headroom between the added noise and the threshold where it will be heard. Also, the number of encode/decode cycles should be minimized, because each cycle moves the added noise closer to the threshold where it will be heard.

Because the cost of hard disk storage has dropped so far, there is no longer a good argument for using lossy compression in audio playout systems. Using linear PCM minimizes the number of cascaded codecs in the transmission path.

TRANSMISSION LEVELS AND METERING

Engineers at the transmitter and the studio consider transmission levels and their measurements differently. Transmission engineers need to know the peak level of a transmission. Studio engineers need to know the lineup (or reference) level of a transmission commonly measured by a VU meter (as the approximate RMS level) or by a peak program meter (as the PPM level).[18]

The VU meter is an average-responding meter (measuring the approximate RMS level) with a 300-ms rise time and decay time; the VU indication usually lags the true peak level by 8 to 14 dB. PPM indicates a level between RMS and the actual peak. The PPM reading has an attack time of 10 ms, slow enough to cause the meter to ignore narrow peaks and lag the true peak level by 5 dB or more. For mixing and master control applications, some organizations are now providing operators with BS.1770 "momentary" loudness meters, which have ballistics similar to those of a VU meter.

TRANSMISSION LEVELS

The transmission engineer is primarily concerned with the peak overload level of a transmission to prevent overloading. This peak overload level is defined differently, system to system. In digital, it is the largest possible digital word, although the headroom of downstream D/A converters and/or sample rate converters is commonly taken into account by using a "true peak" oversampled meter per ITU-R BS.1770. With analog tape, the peak overload point is defined as the level producing the amount of harmonic distortion considered tolerable—often 3% THD at 400 Hz. In FM, microwave, or satellite links, it is the maximum permitted RF carrier deviation. In AM, it is negative carrier pinch-off. In analog telephone transmission, it is the level above which serious crosstalk into other channels occurs, or the level at which the amplifiers in the channel overload.

STUDIO LEVELS

The studio engineer is primarily concerned with what is commonly called the reference level, operating level, or line-up level. This line-up level aids studio engineers in providing adequate headroom between line-up level and the overload level of equipment to allow for the peaks that the meter does not indicate. In facilities that use VU meters, line-up level is usually at 0 VU, which corresponds to the studio standard level, typically +4 or +8 dBm. In systems that use PPM, line-up level may be at PPM 4 (for the BBC standard) or at the studio standard level (often +6 dBm).

In studios using digital links like AES3, the commonly used line-up levels are −20 dB relative to digital full-scale (dBFS) and −18 dBFS. The most commonly used is −20 dBFS and is preferable because it allows a generous amount of headroom.

For facilities that have incorporated loudness management using BS.1770, the EBU has published EBU—TECH 3343: "Practical guidelines for Production and Implementation in accordance with EBU R 128," a comprehensive guide to setting up loudness-managed signal paths and workflows.

TRANSMISSION-LINK LIMITING

Transmission-link limiting devices are sometimes used ahead of the transmission link to protect it from overload. (These links might be STLs, satellite uplinks, inter-studio digital links, etc.) These devices are usually used below-threshold (that is, with no gain reduction) as protection limiters to control peak levels. They only produce gain reduction when abnormally high levels are applied to their input due to operator error or unforeseen level variations at the source. This is useful to transmission engineers concerned with overload, and as useful to studio engineers concerned with headroom. For the needs of both engineers, such a limiter's output must be adjusted to be at or slightly below the peak overload level of the transmission channel.

To properly match the studio line-up level to the transmission protection limiter, the desired headroom must be known. For example, assume that the transmission protection limiter produces 0 dBm at its output at 100% modulation of the transmission link. Further, assume that the line-up level in a production facility is designed to allow 8 dB of headroom. The input attenuator of the transmission protection limiter would then be adjusted so that studio line-up tone produces −8 dBm at the output of the transmission protection limiter.

This assumes that the link between the studio and the input of the transmission protection limiter has enough headroom to drive the transmission protection limiter into gain reduction without clipping this link. The transmission protection limiter only protects a link connected to its output. In the previous example, if the transmission protection limiter provides 15 dB of maximum protection, the system prior to the transmission protection limiter requires 8 + 15 = 23 dB of headroom above studio line-up level.

AUDIO PROCESSING REQUIREMENTS FOR MW AND HF BROADCAST STATIONS

In amplitude-modulated services, reception is usually compromised by noise and interference and may be further compromised by acoustic noise in the listening environment (such as the automobile). The processor must therefore compensate for noise (electrical and acoustic) and interference by reducing dynamic range. This is most readily done by multiband compression and limiting to achieve lowest peak-to-average ratio without significant processing-induced side effects.

The processor must provide absolute negative peak control to prevent AM carrier pinch-off, which would otherwise cause out-of-band emissions. Additionally, the processor must incorporate overshoot-free filtering to control the audio input spectrum to the transmitter, thus preventing out-of-band emissions and interference. National (FCC) or international broadcast authorities (most notably ITU-R) usually specify the permissible occupied bandwidth to make most efficient use of available radio frequency spectrum.

The processor may also be equipped with a receiver equalizer that compensates for the poor frequency response of the typical MW or SW radio due to narrowband RF and intermediate frequency (IF) stages.

SPECIAL CONSIDERATIONS FOR INTERNATIONAL SHORTWAVE (HF) BROADCASTS

HF broadcast using normal double-sideband AM modulation with full carrier or modulation-dependent carrier control[19] requires processing similar to AM medium wave processing, except that higher density in the midrange and higher frequencies is required to get through interference and narrow receivers. Additionally, bass is usually rolled off to avoid a muddy sound due to the receiver's limited HF response, to avoid wasting transmitter power, and to minimize transmitter power supply bounce (which is further explained later). In 1984, the World Administrative Radio Conference recommended a 150-Hz first-order highpass filter for this purpose.

Audio spectral contouring is usually achieved by a combination of fixed equalization and a multiband compressor with appropriate compression thresholds. Speeding up release times in the multiband compressor can increase density in the desired frequency range. The Hilbert transform clipper (discussed earlier) has been successfully used for peak limiting (particularly with speech programming) to provide extremely low peak-to-average ratios.[20]

Single sideband transmission needs special care in peak limiting because the envelope of the modulation is usually not the same as the audio waveform due to the Hilbert transform operation required to generate the SSB signal. It is possible to use an RF clipper (discussed earlier) between the SSB exciter and RF power amplifier to control SSB modulation. However, the RF clipper must be followed by a filter to control out of band emissions, and this filter can overshoot and ring, compromising modulation control. To achieve good envelope and spectral control without RF clipping per se, it is possible to adapt the Hilbert transform clipper principle to SSB transmission by performing calculations in the I/Q (complex) domain.[21]

THE NRSC-1 AUDIO STANDARD

As the North American AM band became more crowded, interference from first and second adjacent stations became more of a problem. Receiver manufacturers responded by producing receivers with decreased audio bandwidth, so that the encroachment of an adjacent station's modulation extremes would not be audible as interference.

This truncating of the bandwidth had the effect of diminishing the receiver's high-frequency response, but it was decided that lower fidelity would be less annoying than interference. To address these problems, the National Radio Systems Committee (NRSC) in 1987 formalized NRSC-1, a standard for pre-emphasis and lowpass filtering for AM broadcast to provide brighter sound at the receiver while minimizing interference. Visit the NRSC Website for more information on NRSC work.[22]

NRSC STANDARD PRE-EMPHASIS AND LOWPASS FILTERING

The NRSC undertook the difficult task of defining a voluntary recommended pre-emphasis curve for AM radio that would be acceptable to broadcasters (who want the highest quality sound on the majority of their listeners' radios) and to receiver manufacturers (who are primarily concerned with interference from first-and second-adjacent stations).

A modified 75-microsecond pre-emphasis/de-emphasis standard (NRSC-1) was approved (see Figure 7.2-10). That provides a moderate amount of improvement for existing narrowband radios while optimizing the sound of wideband radios. Most importantly, it generates substantially less first-adjacent interference than do steeper pre-emphasis curves.

The second part of the NRSC-1 standard calls for a sharp upper limit of 10 kHz for the audio presented

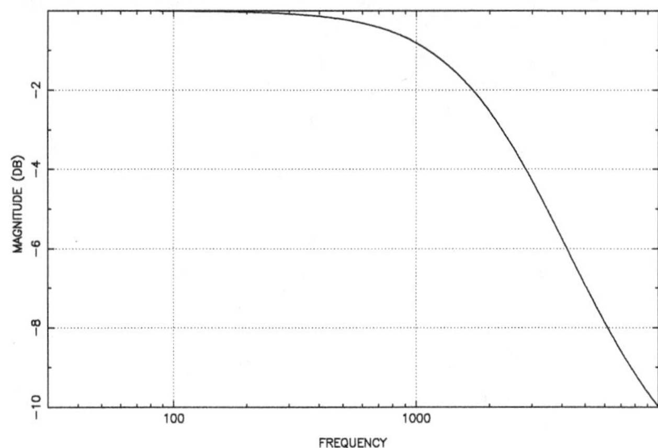

FIGURE 7.2-10 NRSC-1 de-emphasis curve

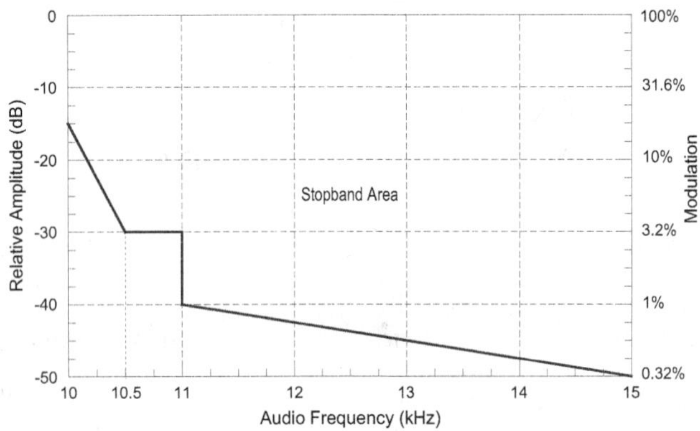

FIGURE 7.2-11 NRSC-1 lowpass filter curve

to the transmitter (see Figure 7.2-11). This essentially eliminates interference to second and higher adjacencies, making radios with 10 kHz bandwidth feasible.[23]

THE NRSC-2 STANDARD

On April 27, 1989, the FCC released a *Report and Order* that amended section 73.44 of the FCC Rules by requiring all U.S. AM stations to comply with the occupied bandwidth specifications of the NRSC-2 standard by June 30, 1990. The NRSC-2 standard is an *RF mask* that was derived from the NRSC-1 audio standard by the NRSC. The purpose of the NRSC-2 RF mask is to provide a transmitted RF occupied bandwidth standard that any station with a properly operating transmitter will meet *if NRSC-1 audio processing is used prior to the transmitter, and the station is not overmodulating*.

THE NRSC AM BANDWIDTH PREFERENCE STUDY

In 2006 the NRSC published a subjective study[24] whose goal was determining the perceptually optimum AM transmission bandwidth with then-available AM radios and audio processors. The studies started by measuring many AM radios, and concluded that a majority of then current AM radios have audio bandwidths of less than 5 kHz. Subjective tests using three representative radios were conducted with typical consumers, and the data suggested that in general, consumers prefer lower bandwidths (5–7 kHz) to higher bandwidths with those radios and well-regarded commercial AM audio processors, taking into account the fact that AM reception is often impaired by noise and interference. The study concluded: "Since consumers seem to be most critical of 'noise' and seem to tolerate more constrained bandwidth when they receive a clean signal, it is likely that lower bandwidths will satisfy consumers in most conditions."

Because of the study, North American AM broadcasters may wish to voluntarily reduce their audio bandwidth to 7 kHz during the day and 6 kHz at night. This will cause little or no audible change in frequency balance as heard on the average AM radio (whose audio bandwidth at –3 dB continues to be approximately 3 kHz), but the 6-kHz bandwidth will significantly reduce nighttime interference to first-adjacent channels compared to 10 kHz. (At night, sky-wave propagation can convey such interference to stations hundreds of miles from the interferer.) Broadcasters are also likely to find that they can process their audio harder for a given amount of processor-induced distortion, increasing on-air loudness and coverage. This is particularly important at night because interference increases.

AM TRANSMITTER EQUALIZATION

An AM audio processor may provide a transmitter equalizer to eliminate tilt, overshoot, and ringing in the transmitter and antenna. Accurate reproduction of the shape of the processed waveform requires that the transfer function between the audio input and the modulated RF envelope represent a constant delay (which may be any positive number or 0) at all frequencies contained within the audio input signal. Failure to meet this criterion can result in tilt, overshoot, and *ringing* in the modulated RF envelope. The cause of overshoot and ringing as spectrum truncation and time dispersion at the high-frequency end of the system bandpass were discussed earlier. Tilt, on the other hand, is caused by problems at low frequencies.

Figure 7.2-12 shows the response of a 10-kW plate-modulated transmitter to a 50-Hz square wave. The transmitter causes the waveform to tilt, which increases peak modulation in both positive and negative directions. The magnitude of the transmitter's frequency response is essentially flat to 50 Hz; the problem is caused by infrasonic rolloff. This rolloff is equivalent to that of a highpass filter and is minimum-phase, which introduces time dispersion, causing the shape of

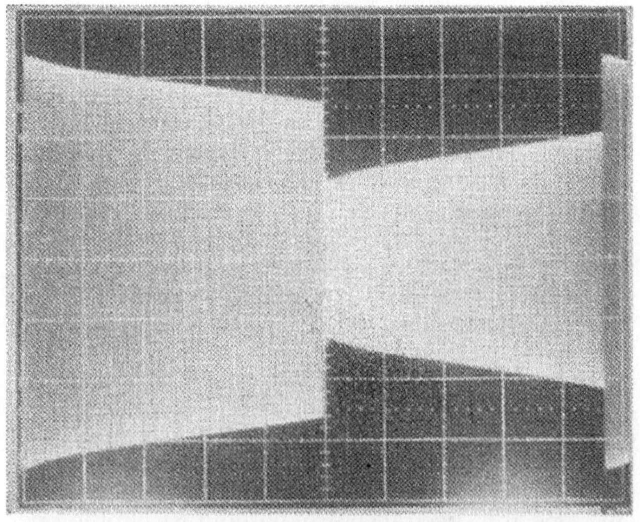

FIGURE 7.2-12 Tilt in plate modulated transmitter

the waveform to change and further increasing the peak level. This is an example of a transmitter with inadequate LF response that requires transmitter equalization to avoid introducing tilt and overshoot. The rule of thumb is that the equalized transmission system must have its −3 dB frequency at 0.15 Hz or lower in order to avoid significant tilt-induced peak modulation overshoot. Modern solid-state TXs are likely to have been designed with the required 0.15-Hz cutoff frequency so they do not require external equalization.

Some transmitters contain highpass filters at their audio inputs to protect high-power stages. *This location is absolutely inappropriate*; these filters can easily increase the peak-to-average ratio of the input audio by 3–4 dB. The correct location for a protection highpass filter is in the audio processor where measures can be taken to prevent the highpass filter from increasing the peak-to-average ratio at the audio processor's output.

BOUNCE

Pre-distorting the waveform in the audio processor (i.e., adjusting the waveform in anticipation of known errors in the transmission path) can equalize linear errors. However, one major nonlinear error, commonly called *power supply bounce*, is caused by resonances in the LC filter elements of the transmitter's high voltage power supply. These resonances superimpose a subaudible modulation onto the power supply voltage, resulting in a form of very fast carrier shift that is too quick to be seen on a conventional carrier shift meter. The net result is to compromise the control of modulation peaks, particularly on strong bass transients that cause momentarily large current demands on the power supply, and that excite the resonance.

In some older transmitters, bounce has been known to compromise achievable modulation by up to 3 dB. Because bounce is not linearly related to the modulation, small-signal equalization cannot cure it. Bounce can be minimized by preventing excessive bass energy from being applied to the transmitter.

SLEW RATE LIMITING (TRANSIENT INTERMODULATION DISTORTION)

Transmitters using pulse-duration modulation (PDM) schemes are prone to problems with slew rate limiting. Because the PDM lowpass filter is located within the audio feedback loop of the transmitter, and because this filter is typically a multi-pole elliptic function filter with a cutoff frequency below 70 kHz, it will introduce substantial delay into the feedback loop. This has two consequences: stability requires the amount of feedback applied around the transmitter to be limited, and it also requires that the open-loop gain of the modulator be rolled-off at a very low frequency. The first issue makes it difficult to design PDM transmitters with THD below 1%—2% at midrange frequencies, while the second renders *transient intermodulation distortion* (TIM, nonlinear behavior of the amplification stage prior to the frequency compensation stage) probable.[25] To minimize the probability that TIM will be bothersome, any amplification stage before the frequency compensation stage should be designed to be very linear to its clipping point, and to have sufficiently high headroom to accommodate the maximum rate of change to be expected at the transmitter's audio input.[26]

A transmitter can be qualified for TIM by one of the various difference-frequency intermodulation distortion tests. If the tests indicate that the transmitter has a low slew rate, it will not respond well to pre-emphasized audio and pre-emphasis will have to be reduced until the first derivative of the processed audio waveform seldom, if ever, exceeds the slew rate limit of the transmitter. Because of the benefits of pre-emphasis at the receiver, it is desirable to modify such transmitters to increase their slew rate, even if this means somewhat compromising harmonic distortion performance at low frequencies.

AUDIO PROCESSING REQUIREMENTS FOR FM (VHF) BROADCAST TRANSMISSION

The processor should provide a comfortably listenable dynamic range in domestic and automotive listening environments. To achieve this, it is usual necessary to apply dynamic compression to the signal. Such compression should subtle and natural sounding unless the program director requests otherwise to achieve a signature sound.

The processor must provide high-frequency limiting to complement the pre-emphasis employed (50 µsec or 75 µsec, depending upon the region in which the transmission occurs).

The processor must provide accurate peak control (as measured by a modulation monitor meeting the standards of the governing authority) in both the positive and negative directions. To ensure that absolute peak control will be retained at the system output, any system elements following the processor must have flat frequency response (±0.1 dB) and constant group delay (deviation from linear phase < ±10°). Because the pre-emphasis networks and lowpass filters ordinarily found in stereo encoders usually do not meet these requirements, they should be bypassed. Thus, the processor should provide pre-emphasis and band limiting for the transmission system. Its output must contain negligible energy above the bandwidth limit of the transmission system. In FM stereo broadcasting by the world-standard pilot-tone method, this bandwidth is limited to less than 19 kHz to prevent aliasing from the stereo subchannel into the main channel, and vice versa. To protect the pilot tone itself (ensuring correct operation of the phase-locked loop subcarrier regeneration circuitry in the receiver's stereo decoder), the bandwidth must be further limited to no greater than 17 kHz. In practice, it is customary to begin the high-frequency rolloff at slightly above 15 kHz to minimize group delay distortion in the lowpass filters used to effect the bandwidth limit. Nonlinear lowpass filters are usually used to prevent overshoot, enabling the processor to control peak deviation absolutely.

The processing system must be readily adjustable to achieve the subjective effect desired by the broadcasting authority operating it. To achieve a competitive sound in markets where many stations compete for listeners, it may be necessary to add additional multiband limiting to the basic audio processing system (which usually consists of compressor, HF limiter, and peak limiter/clipper). Adding additional multiband limiting can create greater program density than the basic processing system alone without introducing spectral gain intermodulation.

In addition to AGC, multiband compression, HF limiting, and peak limiting, modern FM audio processors may include stereo enhancement (that increases the apparent width of the stereo image), high-frequency enhancement (program-adaptive dynamic equalization that brightens dull-sounding input material without significantly affecting material with adequate high-frequency energy), subharmonic synthesis (that generates bass one octave below bass present in the original program to correct bass-shy material), and phase skew correction (to eliminate phase cancellations in mono reception and to minimize multipath distortion).

AUDIO PROCESSING REQUIREMENTS FOR DIGITAL RADIO TRANSMISSION

The United States has adopted the HD Radio system, an In-Band-On-Channel (IBOC) digital transmission system developed by iBiquity Digital Corporation (now DTS, Inc.).[27] Other popular digital radio transmission systems in use around the world include the DAB+ and the Digital Radio Mondiale (DRM) systems.[28] To encode the audio digitally, all digital radio systems use lossy data reduction with no pre-emphasis. The specific system of lossy data reduction that is used depends on the particular digital radio system. DRM launched with support of three codecs: MPEG-4 HE-AAC, MPEG-4 CELP, and MPEG-4 HXVC; the latter two were intended for use with speech program at very low bit rates. DRM has since been updated to support MPEG-4 xHE-AAC, an implementation of MPEG Unified Speech and Audio Coding. DAB+ uses MPEG-4 HE-AACv2. HD Radio uses the HDC codec, a proprietary codec developed by iBiquity Digital Corporation.

PROCESSING FOR HD RADIO, DAB+, AND DRM

While the ensuing discussion centers on HD Radio, the HD Radio digital channel has many features in common with DAB+ and DRM, so this discussion applies to the latter two systems too.

There are several differences between the audio broadcast over an FM analog channel and that broadcast using an HD Radio digital channel in the FM band. First, FM analog has a theoretical audio bandwidth limit of slightly less than 19 kHz but is most commonly limited to 15 kHz or slightly above. HD Radio digital audio for the FM band has a maximum audio bandwidth of 20 kHz.[29] Another significant difference is that analog FM uses 75µs or 50µs pre-emphasis, depending on geographic region, while HD Radio, DAB+, and DRM do not use pre-emphasis.

In HD Radio, the main channel digital audio signal and the main channel analog audio carry the same program, but the analog channel is typically peak-limited differently (using distortion-controlled clipping instead of look-ahead limiting) and is delayed by several seconds with respect to the digital channel so that transmission path disturbances will not degrade identical parts of the program. (This delay is called "diversity delay.") To prevent audio muting, HD Radio receivers automatically cross-fade between the digital and analog audio whenever the receiver cannot demodulate the digital signal (which is at a much lower power level). To achieve an unobtrusive crossfade, the two signals should have nearly the same loudness (which is proportional to a given channel's RMS level),[30] and the delay introduced at the transmitter must be matched in the radio to sample accuracy to prevent audible comb filtering during the crossfade.

Digital audio channels challenge an audio processor differently than do analog audio channels. In the analog system, the processor must control peaks precisely, must manage the pre-emphasis curve to avoid audible distortion generated by the processing, and must include a highly selective lowpass filter to protect the 19-kHz pilot signal of the multiplex stereo system.

A processor for digital audio broadcast has a different set of requirements because it must work with lossy codecs but does not need to handle pre-emphasis. By design, the gain in the digital path of the HD Radio receiver is 5 dB higher than the gain in the analog FM path. The digital path therefore has 5 dB more peak headroom than the analog path at low frequencies. This advantage increases at high frequencies because the pre-emphasis/de-emphasis process further constrains the analog channel's peak headroom. Two factors cause this. First, because pre-emphasis increases peak levels, average modulation must be reduced to prevent overmodulation or objectionable HF limiter-induced artifacts, and this reduction affects all frequencies. Second is receiver de-emphasis. For example, 75 µsec de-emphasis attenuates 15 kHz by 17 dB—at 15 kHz, 100% modulation of the RF carrier produces only 14% modulation at the receiver's output. The high-frequency limiting and clipper-based peak limiting required to deal with analog FM's pre-emphasis creates the familiar "FM radio" sound. The digital channel's headroom advantage is clearly audible; it sounds noticeably punchier than the analog FM audio, and high frequencies are clearer and less constrained. The difference is particularly audible on drums and percussion.

Particularly at the lower bit rates used in digital radio systems, their codecs can sometimes produce unique artifacts not encountered in analog transmission. Just as a processor for FM analog must manage the pre-emphasis curve, a digital broadcast processor must manage the audio spectrum to minimize codec artifacts. Given knowledge of the codec, it is possible to predict what spectral conditions will stress it. Dynamic algorithms in the processor can ameliorate these conditions and prevent unwanted artifacts, especially at higher bit rates such as 96 kbps. In essence, the audio processor can improve the efficiency of the encoder.

PROCESSING CONFIGURATIONS FOR THE HD RADIO SYSTEM

There are two basic audio processing configurations for an HD Radio system. The first uses two separate audio processors. It is mainly appropriate for broadcasters who want their analog and digital channels to have a noticeably different texture at the expense of having analog/digital crossfades sound obvious. It is usually possible to achieve this style of processing while avoiding crossfades that are so annoying that they are likely to cause tune-outs. This is the trickiest style of processing to set up and administer effectively.

The second configuration uses a processor that simultaneously provides dedicated processing for both the IBOC and FM analog signals. This configuration is appropriate for the broadcaster who wants to ensure unobtrusive crossfades while still exploiting the digital channel's ability to sound punchier. Moreover, dual-mode processors often have enough independent adjustability in the analog FM and digital channels to allow them to sound noticeably different if that is the station's goal. Dual-mode processors make it easy to ensure that the analog and digital signals remain time-synchronized at the radio's crossfade mixer, which is important to prevent audible comb filtering during crossfades.

Field measurements have shown that even with sample-accurate time synchronization within the audio processor, other factors in the transmission system can cause the delay difference between the FM analog and digital signals to drift over a range of several hundred samples. It is important to lock all sample frequencies together at the studio and transmitter by using precision frequency references like GPS-disciplined word clock. Additionally, several commercially available modulation monitors can measure the delay between the FM analog and digital signals. Some devices can send commands to a compatible audio processor to trim its diversity delay, correcting delay errors, while other devices have self-contained variable delays for correction. To prevent frequency shifts that can compromise ratings encoder watermarks, delay correction should be slower than 50 samples/second.

The following block diagrams, Figures 7.2-13 and 7.2-14, show the internal functions of stock processors for FM and AM broadcasting. The key functions required of an audio processor are illustrated.

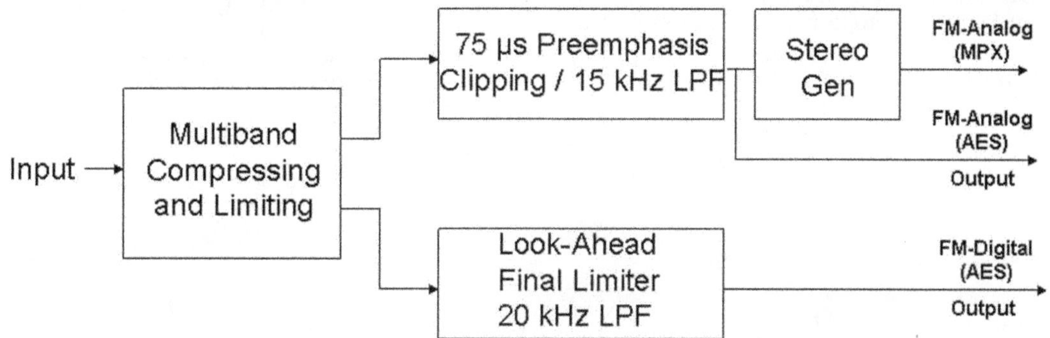

FIGURE 7.2-13 Block diagram of combined FM audio processor

Frank Foti

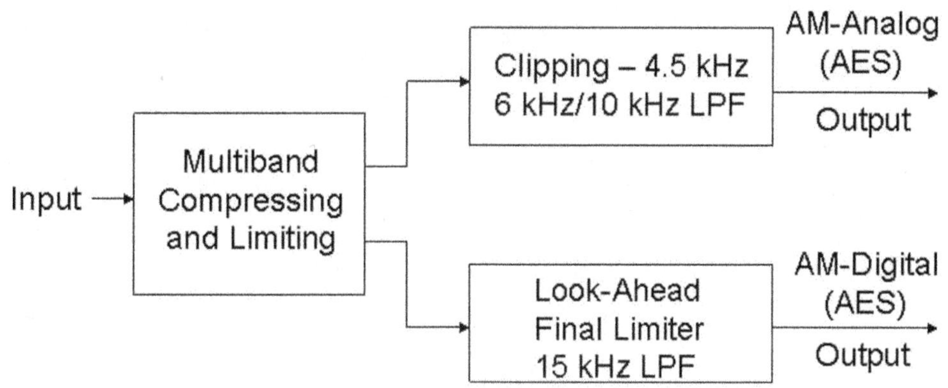

FIGURE 7.2-14 Block diagram of combined AM audio processor

Frank Foti

Audio processors in this configuration need to be inserted into the overall system shown in Figure 7.2-15, for use with HD Radio. The heavy line represents the audio path for the analog transmission.

All HD Radio exciters include a diversity delay line for the analog signal. This accepts and emits AES3 digital signals in left/right form and is transparent to the signal so it does not compromise peak control. If the audio processor does not contain a diversity delay, the processor's analog-processed output must be delayed by routing it through the HD Radio exciter. This creates an additional point of failure within the facility, as any failure of the HD Radio exciter will shut down the conventional audio path, forcing both the digital and analog systems off the air. For HD-FM, a further advantage of including the diversity delay in the processor is that the processor's internal stereo coder and composite limiter (if any) can be used, which eliminates the need to use an external stereo encoder or the exciter's built-in stereo encoder (see Figure 7.2-16).

AUDIO PROCESSING FOR MULTICAST

The HD Radio system can be used to support *multicasting*, which enables additional audio channels to be multiplexed into the HD Radio data stream. Multicast audio channels are not simulcast with the analog audio signal and cannot benefit from the system's analog backup feature. Broadcasters can choose how they wish to divide the digital channel's total available data payload: the more program channels, the lower the bitrate and audio quality. Each audio stream requires a separate processor. Processing designed for streaming audio or netcasts is appropriate.

Because of the low bit rate of a multicast channel, one important function of the processor should be to condition the audio to minimize codec artifacts. However, many broadcasters will also want to process the signal to achieve source-to-source consistency, both spectrally and dynamically. Failure to do this can make a bad impression on listeners who have come to expect this kind of smooth, professional presentation in a broadcast. To achieve consistency while minimizing codec artifacts, the audio processor must not significantly increase the density of the signal, must not aggressively peak-limit the signal, and must prevent excessive build-up of high-frequency energy. It is possible to use multiband processing to achieve these goals but the processor must be carefully set up. It is wise to use little or no stereo enhancement, slow release times, a well-chosen compressor-gating threshold, and very little look-ahead limiter gain reduction. Moreover, the thresholds of the highest frequency bands of the processing must be set low enough to tame excessive high-frequency energy.

PROCESSING FOR SURROUND AND DIGITAL TELEVISION

The most common surround formats are 5.1 and 7.1 channel, where ".1" refers to the low-frequency effects

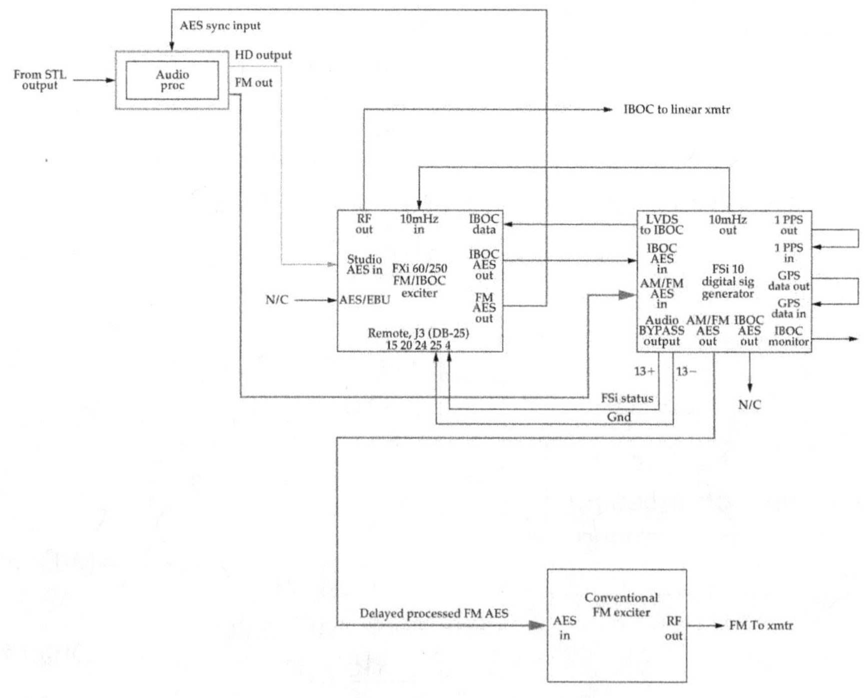

FIGURE 7.2-15 HD Radio stock setup

Frank Foti

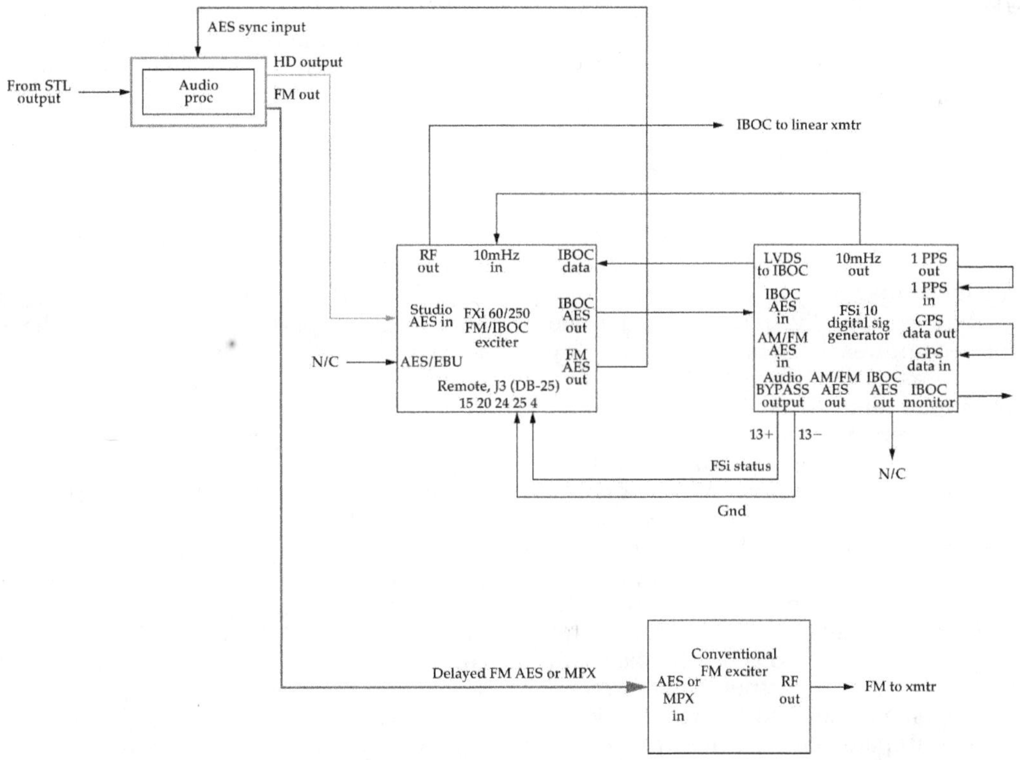

FIGURE 7.2-16 HD Radio setup showing how analog and digital portions of IBOC signal can be isolated

Frank Foti

channel, whose bandwidth is typically limited to 120 Hz.

The ATSC A/53 digital television system specified by the Advanced Television Systems Committee (ATSC) and used in the United States as of this writing supports mono, stereo, and 5.1 channel audio. The 5.1 channels are left, center, right, left surround, right surround, and a subwoofer channel (the 0.1). The audio is digitally compressed using the Dolby AC-3 codec (aka "Dolby Digital"). This codec includes several important metadata elements. *Dialog normalization* ("dialnorm") metadata provides information to the receiver about the nominal loudness of the dialog so that the receiver can hold this constant by adjusting gain inversely to the value of dialnorm. This lets broadcasters trade off headroom (for loud sound effects, for example) against noise floor without changing the loudness of dialog at the receiver.

In November 2009 the ATSC published Recommended Practice A/85: "Techniques for Establishing and Maintaining Audio Loudness for Digital Television," and this has been revised several times since. A/85 recommends setting dialnorm to the target loudness of the "anchor element" (usually dialog) in the program, and requires measuring loudness using the latest revision of the BS.1770 loudness meter (using gating). To implement the CALM[31] Act, in December 2011 the FCC released Report and Order FCC 11–182, which required broadcasters, cable providers, and direct broadcast satellite providers to constrain the loudness of commercial announcements to comply with ATSC A/85 recommendations. These forbid the integrated loudness of a commercial from exceeding the target loudness (i.e., dialnorm) by more than 2 LU.

ATSC A/85 allows "agile," "preset," or "fixed" dialnorm metadata. Agile and preset dialnorm are functionally similar: The transmitted value of dialnorm is set individually for each program element, and there must be some means for attaching the correct dialnorm value to each program element and retaining it through the program chain so that the final transmission conveys it to the receiver. Agile dialnorm applies external metadata with the correct dialnorm value to the AC3 encoder, while with preset dialnorm, AC-3 encoder "presets" are programmed, each with different dialnorm values, and are engaged via a "General Purpose Interface" (GPI) or other control interface.

In large facilities, keeping track of the appropriate dialnorm value for each program element can be very challenging. Because of this complication, fixed dialnorm is more common, using the A/85-recommended value of −24. While agile dialnorm allows each program element to be loudness-normalized by setting its associated dialnorm metadata to be the same as its BS.1770 integrated loudness, static dialnorm requires applying a fixed gain or loss to the audio data itself unless the anchor element is already at −24 LUFS.

In August 2010, the EBU published its Loudness Recommendation EBU R 128. It specifies how broadcasters and netcasters can measure and normalize audio using Loudness meters instead of Peak Meters (PPMs) or VU meters only, as has been common practice. EBU R 128 recommends normalizing the integrated loudness of each program element to −23 LUFS ±0.5 LU (±1 LU for live programs), measured using the BS.1770–2 (gated) algorithm or higher. The metering approach can be used with virtually all material. Unlike ATSC A/85, R 128 does not specify an anchor element, instead requiring the entire program to be included in the integrated (gated) loudness measurement.

Metadata in AC3 enables the receiver, under the viewer's control, to compress the dynamic range. Dolby named the two dynamic range control (DRC) metadata components "line mode" and "RF mode," but consumer electronics manufacturers typically call them something else like "night mode." Line mode is intended for gentle dynamic range reduction in high-quality consumer systems like home theater receivers, while RF mode provides more extreme control and was originally designed for use with set top boxes connected to TV sets via RF modulators. Each mode has various profiles like "Film Mode" available, which the broadcaster must choose at the encoder side. Each U.S. television network has its own technique for generating this metadata and for routing signals around the facility. Some networks insist that affiliates pass the network audio and metadata through without modification (with the assumption that the network has controlled the loudness of commercials to comply with A/85); others permit affiliates to use additional online audio processing.

In U.S. facilities, it is most common to use a static dialnorm value of −24 (equivalent to a BS.1770 target loudness of −24 LUFS for the anchor element) and to let the AC3 encoder generate the dynamic range control metadata just before transmission to the consumer. The DRC input versus output gain curves have "neutral zones" of 1:1 compression centered around the dialnorm value, so commercials whose loudness is otherwise A/85-compliant will continue to comply if dynamic range control is activated by the consumer at the receiver.

ONLINE PROCESSING FOR DIGITAL TELEVISION

The original intent of the dynamic range control metadata was to allow program material with full dynamic range to be transmitted to consumers while permitting them to set their preferred dynamic range at the receiver. In practice, many consumers either do not understand how to do this or find the amount of available dynamic range reduction insufficient. The main advantages of using online processing to control dynamic range during transmission are that it controls loudness without consumer intervention, and it is aware of the transitions between program elements so these transitions can be smoother than they would be if statically normalized elements were simply spliced together. It is also the most straightforward way to comply with the CALM Act.

Online loudness controllers often have the familiar AGC/multiband compressor/peak limiter architecture, and some include additional loudness control before the peak limiter. The loudness controller typically adjusts gain by using a loudness measurement algorithm like BS.1770 or Jones and Torick[32] as a reference. Other, more minimalist controllers may just use wideband gain adjustment based on BS.1770.

In surround processing, the gains of the AGCs and compressors for the various audio channels might be fully coupled, or the processor's designer might choose to implement partial coupling to achieve certain subjective goals. Slow compression is best achieved by using the power summation (RMS) of the channels to control their common gain because this a good model for how humans perceive the overall loudness (a single value) produced by multiple sources playing in a room. This is the way the BS.1770 meter handles channel coupling.

If the target loudness is −24 LUFS, peak limiting will almost never occur because the channel has adequate headroom. If limiting is needed, applying it to each channel individually without coupling prevents peaks in one channel from audibly modulating the loudness in another channel.

The center channel is special because it typically carries dialog. Some designers apply independent compression to the center channel while constraining the maximum gain difference between it and the remaining channels. Quasi-independent compression of the center channel can prevent energy in the remaining channels from pushing down dialog level and compromising its intelligibility.

In conventional 5.1 and 7.1 channel systems, the overload point of each channel is 0 dBFS. However, because almost all the systems use a lossy codec that introduces overshoot, it is wise to set the peak limiter threshold to constrain the maximum peak level into the codec to −1.5 dBFS, or lower if a low-bitrate codec like HE-AACv2 with MPEG Spatial Audio Coding[33] is used. Moreover, for reasons discussed above in the section on peak limiting (above), it is best to use a true-peak-aware peak limiter to prevent overloads after digital to analog conversion.

Because loudness control to government requirements is typically an important goal in surround processing, many processors have at least one gain control element that is BS.1770-aware. However, AES TD1004.1.15–10[34] states that normalizing all genres of audio to the same target loudness can result in inartistic balances between speech and music, with speech typically being too loud. Hence, this writer believes that is inappropriate to rely solely on BS 1770 as a reference for an automatic loudness controller, and that it should be used instead as a final safety element following conventional audio processing. This is also a reason not to blindly normalize all elements in a playout system to a single BS 1770 target loudness without considering genre.

On the other hand, normalizing each program segment in a playout system to a specified BS.1770 target loudness is always beneficial if an online audio processor is used for final loudness control, as static normalization minimizes the amount of gain adjustment that the processor needs to do. Without such normalization, a much larger AGC range is needed and transitions between program elements are unlikely to be as smooth.

So-called object-based systems (like MPEG-H and Dolby Atmos) are becoming more widely deployed to consumers, and the ATSC 3.0 system (still in testing as of this writing) supports such audio. In these systems, various components of the soundtrack (like dialog) can be placed in arbitrary locations in the sound field, and the system can automatically adapt to the number of loudspeakers in the listening environment while placing objects as precisely as possible for a given speaker setup. Soundtrack producers can allow consumers to adjust the loudness of selected elements or even to mute them. As of this writing, object-based systems are just starting to be deployed by satellite, cable, and Internet providers. Because mixers who are aware of modern loudness control requirements are creating soundtracks for object-based systems, and because object-based systems include dynamic range control metadata that allows consumers to choose the amount of dynamic range reduction they prefer (if any), we expect online transmission audio processing for these systems to be uncommon.

TECHNICAL EVALUATION OF AUDIO PROCESSING

Common swept frequency response, harmonic distortion, and intermodulation distortion tests are often used to evaluate audio processors. Therefore, it is useful to discuss why these tests may at times produce misleading results.

DEFINITION OF LINEARITY

A system can be tested for linearity as follows. Apply an input signal A to the system and measure its output. Let X be the output signal caused by input A. Then, remove A from the input and apply another signal B. Let Y be the output signal caused by the input B. The system is linear if the following things happen:

1. If the input waveform is multiplied by a factor k to scale it, the output waveform also becomes scaled by a factor of k, but its shape is not distorted by the process of scaling.
2. If inputs A and B are applied to the system simultaneously, the system's output is X + Y (*superposition*).

It is clear that expanders, compressors, and limiters are strongly nonlinear systems. The output of such a device is not scaled proportionally to its input; it is expanded or compressed. Similarly, when two signals are applied to such a device, its output is not the same as the sum of its response to either signal individually; superposition does not hold. Clippers are similarly nonlinear.

SINE WAVE MEASUREMENTS AND NONLINEARITY

When predicting a system's response to program material by measuring its response to individual sine waves, certain assumptions are made. The first assumption is that program material can be adequately represented as a sum of sine waves (Fourier analysis). The second assumption is that superposition holds, so that the response of the system to single sine waves also applies when several sine waves are added together at the system's input. Thus, the sine wave results can be extrapolated to program material.

Because dynamic audio processing (compression, limiting, clipping, expansion, gating) is strongly nonlinear, the usual assumptions of superposition and scaling, which permit sine wave measurements to be extrapolated to complex program material through Fourier analysis, do not hold. Conventional harmonic and intermodulation distortion measurements, historically designed to measure slight departures from linearity in weakly nonlinear systems, are of very limited usefulness. Swept or spot frequency response measurements are not useful.

When making distortion measurements with tones, their relevance must be assessed psychoacoustically. Does the system output *sound* distorted when listening to the tones? For example, when measuring harmonic distortion using fundamentals in the 50–1000 Hz region, the higher harmonics are more significant than the lower harmonics because the higher harmonics are less readily masked by the desired fundamental. However, as the fundamental frequency is increased, the harmonics become less troublesome because the ear becomes less and less sensitive to them. Eventually, their frequency exceeds the passband of the system and they become irrelevant.

Similarly, SMPTE intermodulation distortion methods measures the level of 50- or 60-Hz sidebands around a high-frequency tone induced by system nonlinearity.[35] Because these sidebands are within a single critical band (approximately one-third octave) of the high-frequency tone, they are maximally masked by it. Therefore, rather high amounts of measured SMPTE IM distortion are not necessarily cause for concern. On the other hand, ITU-R difference-frequency intermodulation distortion measurements measure the low-frequency difference tone caused by two high-frequency tones. Because the difference tone is far removed in frequency from the desired tones, it is not well masked by them, and high amounts of ITU-R IM are of some concern. See Chapter 10.1, "Audio Signal Analysis," for more information on distortion and measurements. In all cases, it is not appropriate to attempt to extrapolate the results of tone tests to program material, because superposition does not hold.

SUBJECTIVE LISTENING TESTS

There are few, if any, measurement techniques that can adequately predict whether the subjective effect of an audio processor will be satisfactory. The only effective way to evaluate nonlinear broadcast audio processing is by *subjective listening tests*. These must be done over a long period, using many different types of program material, because a processor that sounds good on a certain type of program material may sound unsatisfactory on other program material having markedly dissimilar spectral balance or dynamics.

Usually, an important subjective goal of broadcast processing is to avoid artifacts that are objectionable to the audience. In the case of processing in highly competitive major market stations, some degradation of the program (as perceived on a high-quality monitor) is often accepted for the sake of maximizing dial presence and loudness. Moderate quality compromises are usually masked on smaller and lower cost radios and are noticeable only on higher-quality radios by critical listeners.

CLARIFYING AUDIO PROCESSING OBJECTIVES

When adjusting processing or comparing processors, set clear objectives. More presence? More loudness? More consistency? More openness? Lower fatigue? Higher cumulative audience share? Longer time spent listening? Do you want to sound more like other stations in your market or stand out uniquely?

Processing can directly affect a station's finances, so engineering, programming, and management should all participate with the goals of using time most efficiently and minimizing future disagreements.

PRACTICAL SUGGESTIONS FOR ADJUSTING PROCESSING FOR AUDIO-ONLY MEDIA

Modern audio processors are complex and powerful. Many allow adjustment on at least three different levels: using factory presets as-is, adjusting a preferred factory preset using simplified control that affects many settings at once using pre-programmed optimizations created by the processor's designers, and fine-tuning the preset using all of the dozens of controls that a modern processor typically makes available. Each level takes more time and expertise and gives the user more opportunities to get into trouble.

With such a processor, adjustment should proceed step by step through the levels. Start by choosing a factory preset that is a close as possible to your goal. Then, try fine-tuning the preset with simplified control to get closer. If you still aren't where you want to be, try making further incremental changes using the full control set. If you have followed the first two steps, there should be no need to make large changes.

It is OK to stop before completing all the steps. Experienced processing designers created the factory presets, and many successful broadcasters use them unaltered. The simplified control settings also reflect the experience of the factory programmers and may be all that you need. Generally, only experienced on-air sound designers should adjust the full control set because there are many pitfalls, traps, and snares that require experience to avoid.

A common mistake is optimizing processing for just a few pieces of program material. This can be particularly tempting if a given track causes minor subjective problems in a few places. However, adjustments that improve problematic program material can compromise the sound on the vast majority of nonproblematic material. (In fact, such minor problems are almost always small compared to normal RF path degradations like FM multipath distortion.) To understand the effect of a change on the overall sound of the broadcast, it is important to listen for a long time, using a wide variety of material typical of your format.

High-frequency hearing loss is a politically sensitive problem. It occurs normally with age and accelerates if people listen to headphones too loudly for too long. It is no secret that the radio industry is skewing older. People with HF loss cannot reliably adjust and evaluate aspects of processing that affect frequencies they cannot hear, so it is important to use people with normal hearing to assess processing intended for younger demographics. That being said, older people may detect problems in the midrange and low frequencies seem unimportant to someone whose can hear all of the highs, so it is wise to sanity-check the processing using several age groups of both genders.

Experience has shown that underprocessing (making the transmission uncomfortable to hear because dynamics are wider than the listener wants) creates more complaints than overprocessing. However, overprocessing has the following potential downsides:

- Reducing the punch and dynamics that are an important part of the musical experience and "deadening the music."
- Introducing overt clipper-induced "crunchy-sounding" distortion that makes the transmission fatiguing to hear. Additionally, intermodulation can cause "gargling" or "warbling," particularly with speech over music.
- Damaging branding ("Your station sounds like a boom box with weak batteries").
- Having male engineers and program directors (possibly with HF hearing loss) adjust processing for female-skewing formats and driving off most of the audience with overcooked, possibly distorted high frequencies.

Avoid optimizing processing for music at the expense of speech. Music is more degraded by IM products while speech is more degraded by harmonic products. Clipper-induced distortion tends to be more audible with speech than with music. Some processors use automatic speech/music detection that reduces the need to use compromise processing that is adequate, not optimum, for both types of material.

Do not process for the wrong radios. According to the Consumer Technology Association's (CTA) CES (Consumer Electronics Show) data, far more clock and small table radios are sold than any other types of dedicated radio—over 10 million per year. This means that processing for the midrange frequencies is most important to the mass audience that listens through loudspeakers. Meanwhile, more and more radio listening occurs through earbuds attached to portable music players and smartphones. Before locking down their processing, PDs and engineers should listen to their stations through a portable music player because this kind of listening makes crunchy distortion overtly audible.

Netcasts are typically processed with look-ahead limiters, not clippers, and don't crunch. However, these limiters can introduce their own artifacts, including muddy-sounding IM distortion and gain pumping.

Process for the volume levels that your audience is likely to use. Except for talk radio, most loudspeaker-reproduced radio is heard quietly. Psychoacoustic equal-loudness curves tell us that bass is the first victim of this kind of listening. Once again, the midrange is crucial and you can't rely on "big bass" to mask crunchy distortion or pumping in the midrange. On the other hand, earbud-reproduced radio is often heard loudly and bass can be very apparent. Premium auto sound systems tend to have good bass response and may be played more loudly than small table radios. Bass is important, but it must never be allowed to damage the midrange with pumping or "gargling" intermodulation—many listeners don't hear bass, either because their radios are played quietly or because their radios cannot effectively reproduce bass at any volume level.

Use processing techniques, like bass pre-limiting, that ensure that bass is not allowed to damage midrange. Even upmarket table radios like those from Bose, Cambridge, Polk, Boston Acoustics, and others have little response below 70 Hz, so for the mass audience, the mid-bass is more important than the bottom two octaves. Choose whether to emphasize mid-bass or low bass performance based on target audience and the radios they are likely to be using. Speaker or earbuds?

Detecting small audible problems in the program and transmission chains requires good monitoring. Modern audio processors are remote controllable, which facilitates adjusting the processor in a quiet, acoustically treated studio with calibrated monitoring, away from machine room noise. Sanity-check your settings with radios and player devices your target audience is likely to use, but don't rely on them for primary quality control.

A transmission cannot sound good if the source material is distorted, and audio processors can exaggerate problems in the source. This has become a particular problem with the advent of "hypercompressed," overtly clipped source material. Try applying de-clipping in the production studio when material is prepared for playout, but listen carefully, as de-clippers can degrade some program material by adding additional IM distortion.

Always use the cleanest authentic source material available. Some hit music has been rereleased many times and quality can vary widely. Make sure that the version you use was sourced from the master of the hit and not from a remake, even if the nonauthentic version might have better technical quality. If authenticity demands that you use technically sketchy source material, use conservative amounts of peak limiting in the audio processor to avoid exaggerating source problems and do not set the processor to exaggerate high frequencies, which will also exaggerate distortion. If the source material is already hypercompressed, use slow release times in the multiband compressor so that it gently re-equalizes material to achieve your target spectral balance but does not significantly increase the density of the audio.

Use a microphone processor on all live microphones. This facilitates matching the density and spectral balance of the microphone channel to other source material, and will allow you to optimize the transmission processor for that material.

Normally, announcers hear their voices through a microphone processor when monitoring with headphones, and a familiar combination of microphone and processor can help them perform at their best.

Always adjust the microphone processor by listening to it through the final transmission audio processor, as this is the only way to assess how it will sound to your audience. While some processors have speech/music detectors and can change their parameters accordingly, none of these detectors is foolproof. It is safest to set "music" and "speech" parameters close to each other so that switching between them does not cause abrupt and obvious changes in sound texture if the speech/music detector makes a mistake. Tune the microphone processor so that it sounds as good as possible with the processor's "music" parameters; this will minimize the need to set speech and music parameters differently.

Make sure the input and output levels of the system are correct and that no clipping occurs anywhere in the transmission chain before and after the processor. The only clipping should be distortion-controlled clipping in the processor's peak limiter.

Use a modulation monitor to set modulation accurately to the limits established by the governing authority in your country. Do not overmodulate. Not only can this cause interference to adjacent channels, but many radios distort severely when encountering even mild overmodulation. A transmission that sounds fine through a modulation monitor may sound objectionably distorted to many listeners.

Use the modulation monitor to match modulation levels between two processors you may be comparing; otherwise, your results will be meaningless because listeners often perceive small loudness differences as differences in quality and texture instead of differences in volume.

When testing and adjusting potential changes in processing, start with your existing sound and make small, incremental changes from there. Don't be hasty; give yourself time to become accustomed to the results of a given change and listen with several pieces of source material before going further. Don't change a large number of parameters at once—doing so can make it impossible to tell which change improved the sound. Record each change as a separate user preset in the processor so that you can readily compare changes and undo later ones if you ultimately decide that they made the sound worse.

Listening fatigue sets in after about 30 minutes.[36] Fatigue causes people to become less sensitive to the effect of changes and makes it more and more difficult to judge whether a change is beneficial, so limit the duration of adjustment sessions. Wait at least another 30 minutes before further assessing the results of an adjustment session. It is often best to wait until the next day to decide.

Stop once you are unable to improve the sound further using the disciplined techniques above. Do not tinker endlessly.

It is important to measure your audience. If you know how many people listen, their demographics, and how long they listen, wait and see if the changes you made have affected listening negatively or positively. Accurate assessment takes a long time because short-term variations in ratings are often measurement "noise" that eventually averages out, and ratings can vary for many programming reasons that have nothing to do with processing.

One of the hardest lessons for processing professionals to learn is that different people have different tastes and preferences: one person's "bad" might be another person's "good," and that's OK. It's just how humans are, and psychologically projecting one's personal preferences onto an audience can lead to surprises. The best processing is processing that maximizes your overall audience share, and this requires research and discipline.

SUMMARY

Audio processing offers broadcasters an opportunity to create a signature sound and to help compensate for problems in the audio and broadcast chains, but it must be used with care and with the knowledge of what it can and can't do. The material in this chapter provides broadcasting engineers with information on how processors work, some pitfalls to avoid, and how to put processors to best use.

Notes

[1] These are popularly known as the "Fletcher-Munson" curves, named after the Bell Labs researchers who published them in the 1930s. However, later researchers revised and corrected these curves, resulting in the ISO 226:2003 Standard.

[2] Psychoacoustic masking is the ability of certain elements in a sound to render certain other elements completely inaudible by "drowning them out." Masking suppresses weaker elements, particularly those close to the frequency of the dominant element. Psychoacoustic masking is what allows lossy compression, like MP3 or AAC, to work.

[3] M.A. Gerzon, "Decomposition of Nonlinear Operators Into 'Harmonic' Components, With Applications to Audio Signal Processing." *Electronics Letters*, January 8, 1976.

[4] British Broadcasting Corporation Engineering Division, "The dynamic characteristics of limiters for sound programme circuits," Research Report No. EL-5, 1967.

[5] dBTP is a "true peak" reading as described in ITU-R BS.1770-3, Annex 2. This anticipates the peak level of output of a digital-to-analog converter by oversampling the digital signal by at least four times before measuring it. True peak levels can be significantly higher than the maximum values of their related digital samples (the more familiar dBFS reading). True peaks, if not taken into account, can cause audible distortion by clipping the output of the D/A converter if the D/A converter design does not allow extra headroom for them. Many popular consumer players do not allow this headroom.

[6] SECAM customarily uses AM sound, with the usual requirements for preventing carrier pinch-off.

[7] https://en.wikipedia.org/wiki/Gibbs_phenomenon.

[8] A phase-linear filter has constant delay with frequency.

[9] A minimum phase filter has no zeros in the right half of the s-plane. As its name implies, there is no filter with the same magnitude response that can have less phase shift. Given the magnitude response of a minimum-phase filter, its phase shift can be computed (with the Hilbert transform). This means that if a minimum-phase filter has constant group delay in its passband, this is associated with a certain type of magnitude response that rolls off gently around the filter's cutoff frequency: a minimum-phase filter with constant group delay in the passband cannot simultaneously have a highly selective magnitude response. Many textbooks provide the well-known mathematical details. See, for example, H.J. Blinchikoff and A.I. Zverev, *Filtering in the Time and Frequency Domains*, New York, Wiley, 1976, pp. 89–94.

[10] A term originally coined by Lynn Fuston.

[11] "LkFS" is the preferred unit of measure in ATSC Recommendation A/85.

[12] AES TD1004.1.15-10, "Recommendation for Loudness of Audio Streaming and Network File Playback," 2015. In addition to its recommendations, this document has a bibliography and links to useful loudness standards.

EBU R 128 s1, "Loudness Parameters for Short-Form Content (Adverts, Promos, Etc.)," 2014.

[13] Dolby Laboratories appears to have originated the concept of using dialog as an anchor element for loudness normalization. Dolby's widely-used AC3 codec (introduced for cinema sound in 1992 and later extended to consumer media including digital television broadcast) includes "dialnorm" metadata to be used to adjust the level of the playback system so that the loudness of dialog is invariant across program sources. Later, this chapter discusses dialnorm more fully.

[14] See for example: "How Loud Is My Broadcast?," *J. AES*. Vol. 52, No. 6, June 2004, pp. 662–666; "Loudness Revisited," *J. AES*. Vol. 62, No. 12, December 2014. Other loudness-related AES Journal and conference publications can be searched from the AES's website: www.aes.org/publications/.

[15] B.L. Jones and E.L. Torick, "A New Loudness Indicator for Use in Broadcasting," *J. SMPTE*, September 1981, p. 772.

[16] R.A. Haller, "An Update on the Technology of Loud Commercial Control," OST Technical Memorandum FCC/OST TM83-1, February 1983.

[17] ATIS is a U.S.-based industry standards setting organization; its Network Performance, Reliability and Quality of Service Committee (formerly designated T1A1) develops and recommends standards, requirements, and technical reports related to the performance, reliability, and associated security aspects of communications networks, as well as the processing of voice, audio, data, image, and video signals, and their multimedia integration.

[18] A good discussion of the theory and operation of PPM may be found on the Internet at http://en.wikipedia.org/wiki/Peak_programme_meter; for additional information on the VU meter see Glen Ballou (editor), "Handbook for Sound Engineers: The New Audio Cyclopedia,", Section 21.2, Indianapolis, Howard Sams, 1987.

[19] See for example, National Radio Systems Committee NRSC-G101: "AM Modulation-Dependent Carrier Level (MDCL) Usage Guideline," April 2013.

[20] R. Orban, "Increasing Coverage of International Shortwave Broadcast Through Improved Audio Processing Techniques," *Journal of the Audio Engineering Society*, June 1990, p. 419.

[21] David Hershberger, "Controlled Envelope Single Sideband," *QEX*, 2014, November/December, p. 3.

[22] NRSC Standards are available on the Internet at www.NRSC standards.org.

[23] AM broadcasters transmitting HD Radio signals must constrain the audio bandwidth of the accompanying analog AM channel to 5 kHz to avoid interfering with the digital carriers. However, the digital signal has 15 kHz audio bandwidth.

[24] National Radio Systems Committee, "Consumer Testing of AM Broadcast Transmission Bandwidth and Audio Performance Measurements of Broadcast AM Receivers," September 2006. www.nrscstandards.org/AFAB/AMSTG%20report%20summary.pdf.

[25] Simply stated, almost all feedback systems contain a filter that forces the open loop characteristic to be either lowpass (all-pole, *lag compensation*) or lowpass shelving, with poles *and* zeros (*leadlag compensation*). Feedback forces the amplifier before this filter to present a pre-emphasized signal to the filter's input such that the total response of the system is flat. If the filter rolls off at 6 dB/octave starting at 15 Hz (a typical situation in an op amp like the TL072 or the LF353), this pre-emphasis *rises* at 6 dB/octave starting at 15 Hz. High frequencies applied to this system will obviously challenge the headroom of the amplifier prior to the filter. For example, 20 kHz will be up 62.4 dB! If high frequencies drive the amplifier prior to the filter into clipping or substantially nonlinear operation, transient intermodulation distortion occurs. Because a filter with a lowpass characteristic follows the clipping process, harmonics generated by clipping will be de-emphasized, so difference-frequency IM tests are more sensitive than THD tests to this mechanism.

[26] For a maximum audio bandwidth f, the required slew rate in percent modulation per microsecond is $0.0002\varpi f$ % per µs. For 4.5 kHz, this is 2.827% per µs.

[27] See "NRSC-5-A, Digital Radio Broadcasting Standard," for additional information on the iBiquity HD Radio system (as standardized by the National Radio Systems Committee), www.nrscstandards.org.

[28] See European Telecommunications Standards Institute (ETSI) standard "ETSI TR 101 495 V1.4.1 (2012–03), "Digital Audio Broadcasting (DAB); Guide to DAB standards; Guidelines and Bibliography," for additional information on the DAB+ system, and "ETSI TS 201 980 v2.1.1 (2003–12), DRM System Specification," for additional information on the DRM system, www.etsi.org.

[29] For HD Radio digital audio in the AM band, the maximum audio bandwidth is 15 kHz.

[30] As discussed earlier, this is not true for loudness measurement with arbitrary program material. However, here we are comparing two signals that represent the same program material. They differ only in their peak-to-average ratios and possibly in their high frequency spectra if substantial HF limiting has been applied to the analog FM channel.

[31] Commercial Advertisement Loudness Mitigation Act, passed in 2010 by the U.S. Congress.

[32] See note 16.

[33] ISO/IEC 23003–1.

[34] See note 12.

[35] See RP 120–1994—SMPTE Recommended Practice—Measurement of Intermodulation Distortion in Audio Systems.

[36] Recommendation ITU-R BS.1116–1: "Methods for the Subjective Assessment of Small Impairments in Audio Systems including Multichannel Sound Systems," section 4.2. While specifically applicable to extremely critical listening tests, this document contains much generally useful information about setting up reliable listening tests.

CHAPTER

7.3

Internet Radio Interfacing and Streaming

GREG OGONOWSKI
StreamS/Modulation Index, LLC
Diamond Bar, CA

INTRODUCTION

Internet streaming has become a dominant force for delivering both static and live audio and video content. Advances in streaming and mobile technology have made it is easier than ever to reach large potential audiences via desktop computers and mobile devices.

The state of the art in Internet streaming changes quickly. The information in this chapter was up to date at the time of this writing (late 2016), but before designing and deploying a streaming infrastructure, practitioners should do online research to familiarize themselves with the latest information about these topics.

Unlike AM and FM broadcasting, several different streaming protocols and codecs are used to deliver content on Internet radio systems. Some of these are standards-based and some are proprietary. Failure to follow standards has impeded streaming player application development. Over time, many early delivery protocols have fallen out of favor, which served to simplify the development of player applications because they only have to support the standards-based survivors. Some standards-based protocols are natively supported in operating systems now, which further simplifies development.

One basic difference between traditional broadcast and Internet broadcast is that current streaming delivery is mostly unicast, requiring a discrete connection for every user, while traditional broadcast carries no such burden, and is infinitely scalable to any audience size given its unidirectional, point-to-multipoint architecture. As the number of users increases in a unicast environment, so does the content provider's aggregate bandwidth. Currently, high-capacity Internet connections and servers are used to overcome this hurdle. There are also new mobile technologies, such LTE multicast that may be able to address this problem in the near future.

REQUIREMENTS

Streaming Protocol

Not all streaming protocols provide equal performance. It is crucial to choose a streaming audio protocol that can support the features and performance you need. Many streaming protocols are dictated by your chosen content distribution network, so choose wisely. Like the protocols themselves, not all content distribution networks (CDNs) offer the same performance, and it is crucial that their developers and technical support departments understand your broadcast and performance goals. This includes choosing efficient audio codecs, and specifying network transports that are optimized for audio-only delivery instead of specifying a video protocol with an audio payload. Many video protocols are very inefficient for audio-only streams, driving up bandwidth and cost while reducing reliability. This is especially important on crowded mobile networks.

Audio streams consist of three basic components: the transport, audio codec and metadata. The transport carries the audio and metadata if the metadata is in-band and synchronous. If the metadata is out-of-band, then the transport only carries the audio, and the metadata is sent through a separate connection

NAB ENGINEERING HANDBOOK

and is considered asynchronous. This means that even though two streams may be using the same codec, if they are using different transport protocols, they may not both be compatible with the same player client.

Audio

To compete with terrestrial and satellite radio for audience share, streaming audio must provide broadcast-quality audio and performance. Achieving this requires optimizing the streaming encoding hardware; implementing audio processing to achieve loudness, consistency and peak control; and choosing an appropriate audio codec.

Most professional, high-performance streaming audio encoders run on a computer that combines features of a workstation and server. Hence, many streaming encoder suppliers have opted for custom computer solutions to prevent the not-enough or too-much syndrome.

For professional audio applications such as a live streaming audio encoder, use professional audio interface standards, which exist for analog, digital, and network audio.

Analog audio requires support for professional operating levels: typically a +4 dBu reference level with a clipping level of greater than +20 dBu. Using a sound device designed for −10 dBV consumer levels risks clipping and distorting the audio because of inadequate headroom; additionally, sound devices built into computer motherboards often have poor noise performance. Lowering the audio level with the computer's software level controls will not prevent the clipping; it will just make quieter distortion. Moreover, the onboard sound's unbalanced analog audio inputs and outputs are vulnerable to ground loops, noise and hum. Using a professional sound card with balanced I/O eliminates these problems.

For digital audio, AES3 is used for professional interfacing. Although digital audio input and output exist onboard in some computer systems, it is usually S/PDIF (IEC958). This consumer format typically uses a TOSLINK optical connection that needs format conversion to interface to a professional facility. A professional sound card with AES3 input and output resolves these issues.

Network audio protocols (or "Audio over IP" [AoIP]) such as AES67, RAVENNA, Lawo *R3LAY*, Axia *Livewire*, Audinate *Dante*, and Wheatstone *WheatNet* are becoming increasingly prevalent. A professional streaming audio encoder should be capable of supporting any of these. The network audio vendor will often make this easy by providing virtual audio driver software for the host computer; this makes the network audio channels appear to the operating system as just another set of hardware audio devices.

To attract large audiences to streaming audio, it is wise to draw on the experience of terrestrial broadcasters by using broadcast-grade audio processing. This provides the necessary automatic gain, spectral, loudness and peak control to make your streams shine with professional polish that holds your listeners and keeps them coming back for more. Both DSP and x86 native versions of professional audio processing software are available from several vendors.

The streaming audio encoder application is responsible for first encoding the audio input with an audio codec and then assembling the encoded audio bitstream and metadata into a streaming transport protocol to be sent to a streaming or web server. Different transport protocols offer different streaming features and are usually bound to a particular streaming server type. Many CDNs provide their own, often proprietary, streaming encoders. Audio quality and performance vary widely, so proceed with caution. Remember that you always have the choice of choosing a standards-based encoder and a CDN that will support it. This gives you maximum flexibility and compatibility.

To reach the vast majority of the audience today via desktop and mobile devices, the audio codecs of choice are in the AAC family. These include AAC-LC (Low Complexity) and HE-AAC (High Efficiency). These codecs sound better than MP3 and are more efficient, resulting in lower cost to stream and lower cost to consume. AAC-LC is the codec used by the most popular music download service, iTunes Music Store.

Reducing the bitrate necessary for entertainment-quality audio increases stream reliability, resulting in fewer stream dropouts and less buffering. For example, a minimum of 128 kbps is required to achieve high fidelity 15-kHz stereo audio using MP3. HE-AAC as low as 32 kbps still produces a 15-kHz stereo audio stream, compared to a 5.2-kHz bandwidth for MP3 at the same bitrate (see Table 7.3-1).

The AAC family of standardized codecs is mature and proven, dating from 1997. It is supported natively on all modern computer and mobile operating systems today. It is therefore the preferred codec set for development today.[1]

Some streaming servers offer transcoding. Transcoding allows the streaming encoder to send one stream to the streaming server using a single audio codec and stream packetization. The streaming server can then derive other bitrates and/or audio codecs to make multiple streams available. Transcoding usually decreases audio quality, so if audio quality is one of your main objectives, avoid audio transcoding. It is better to use a streaming audio encoder that has direct support for the target audio codec or codecs, such as AAC and/or MP3, instead of double-encoding from another lossy codec.

Some streaming servers offer transpacketization. This allows the streaming encoder to send one stream to the streaming server using packetization of a single stream type, from which the streaming server will derive multiple packetizations to make multiple stream types available. This can be useful for reaching streaming players that only accept certain types of streams without having to use a separate streaming server for those devices, thereby reducing cost. Because different packetizations support different metadata protocols, however, transpacketizing can cause metadata incompatibilities and synchronization problems. This feature therefore is not recommended, and it is rarely needed today because all modern operating systems

TABLE 7.3-1
Audio encoder parameters—Comparison of MP3 to AAC-LC, HE-AACv1 and HE-AACv2 codecs

Codec Parameters			MP3 Player	AAC Player	HE-AACv1 Player	HE-AACv2 Player
Channel Mode	Bit Rate (kbps)	Sample Rate (kHz)	Audio Bandwidth (kHz)			
1ch-Mono	8	8, 11.025	3.3, 3.3		8.2, 8.0	
1ch-Mono	16	11.025,12,16	4.7, 4.7		12.3, 11.7, 12.0	
1ch-Mono	18	11.025,12,16	5.1, 5.1			
1ch-Mono	20	11.025,12,16	5.5, 5.5		14.5, 14.8, 15.4	
1ch-Mono	24	16, 22.05, 24	6.2, 6.2, 6.2	7.2, 7.2, 7.2	15.3, 14.8, 15.4	
1ch-Mono	32	16, 22.05, 24	7.6, 7.6, 7.6	10.0, 10.0, 10.0	16.0, 16.2, 16.9	
1ch-Mono	40	22.05, 24, 32	9.1, 9.1, 9.1	12.4, 12.4, 12.4	16.0, 17.6, 18.4	
1ch-Mono	48	22.05, 24, 32	10.5, 10.5, 10.5	13.5, 13.5, 13.5	16.0, 20.3, 20.3	
1ch-Mono	56	32, 44.1, 48	12.0, 12.0, 12.0	15.5, 15.5, 15.5	16.0, 20.3, 20.3	
1ch-Mono	64	32, 44.1, 48	13.5, 13.5, 13.5	15.5, 15.5, 15.5	16.0, 20.3, 20.3	
1ch-Mono	80	32, 44.1, 48	16, 16.8, 16.8	16.0, 17.6, 17.6		
1ch-Mono	96	32, 44.1, 48	16, 19.3, 19.3	16.0, 22.05, 24.0		
1ch-Mono	112	32, 44.1, 48	16, 22.05, 22.2	16.0, 22.05, 24.0		
1ch-Mono	128	32, 44.1, 48	16, 22.05, 24	16.0, 22.05, 24.0		
1ch-Mono	160	32, 44.1, 48	16, 22.05, 24	16.0, 22.05, 24.0		
1ch-Mono	192	32, 44.1, 48	16, 22.05, 24			
1ch-Mono	224	32, 44.1, 48	16, 22.05, 24			
1ch-Mono	256	32, 44.1, 48	16, 22.05, 24			
1ch-Mono	320	32, 44.1, 48	16, 22.05, 24			
2ch-Stereo	16	8, 11.025, 12	3.4, 3.4, 3.4			
2ch-Stereo	18	8, 11.025, 12	3.6, 3.6, 3.6			
2ch-Stereo	20	8, 11.025, 12	3.9, 3.9, 3.9			
2ch-Stereo	24	11.025,12,16	4.3, 4.3, 4.3	6.6, 6.6, 6.6	13.8, 12.7, 13.1	15.3, 14.8, 15.4
2ch-Stereo	32	11.025,12,16	5.2, 5.2, 5.2	6.6, 6.6, 6.6	15.3, 14.8, 15.4	16.0, 16.2, 16.9
2ch-Stereo	40	16, 22.05, 24	6.0, 6.0, 6.0	8.5, 8.5, 8.5	16.0, 16.2, 16.9	16.0, 17.6, 18.4
2ch-Stereo	48	16, 22.05, 24	7.5, 7.5, 7.5	8.5, 8.5, 8.5	16.0, 16.2, 16.9	16.0, 20.3, 20.3
2ch-Stereo	56	22.05, 24	8.9, 8.9	12.5, 12.5, 12.5	16.0, 17.6, 18.4	16.0, 20.3, 20.3
2ch-Stereo	64	22.05, 24	10.3, 10.3	12.5, 12.5, 12.5	16.0, 20.3, 22.1	
2ch-Stereo	80	22.05, 24, 32	11.0, 11.0, 11.0	13.5, 13.5, 13.5	16.0, 20.3, 22.1	
2ch-Stereo	96	32, 44.1	11.9, 11.9	15.5, 15.5, 15.5	16.0, 20.3, 22.1	
2ch-Stereo	112	32, 44.1, 48	13.6, 13.6, 13.6	16.0, 16.0, 16.0	16.0, 20.3, 22.1	
2ch-Stereo	128	32, 44.1, 48	15.8, 15.8, 15.8	16.0, 16.0, 16.0	16.0, 20.3, 22.1	
2ch-Stereo	160	32, 44.1, 48	16, 20.2	16.0, 17.6, 17.6		
2ch-Stereo	192	32, 44.1, 48	16, 22.05, 24	16.0, 22.05, 24.0		
2ch-Stereo	224	32, 44.1, 48	16, 22.05, 24	16.0, 22.05, 24.0		
2ch-Stereo	256	32, 44.1, 48	16, 22.05, 24	16.0, 22.05, 24.0		
2ch-Stereo	320	32, 44.1, 48	16, 22.05, 24	16.0, 22.05, 24.0		

NOTE: Not all encoder implementations are created equal. These values are typical.

Legend:

Low Fidelity with lower than 15-kHz audio bandwidth (unacceptable for high-quality music)

High Fidelity with a minimum of 15-kHz audio bandwidth

Recommended minimums

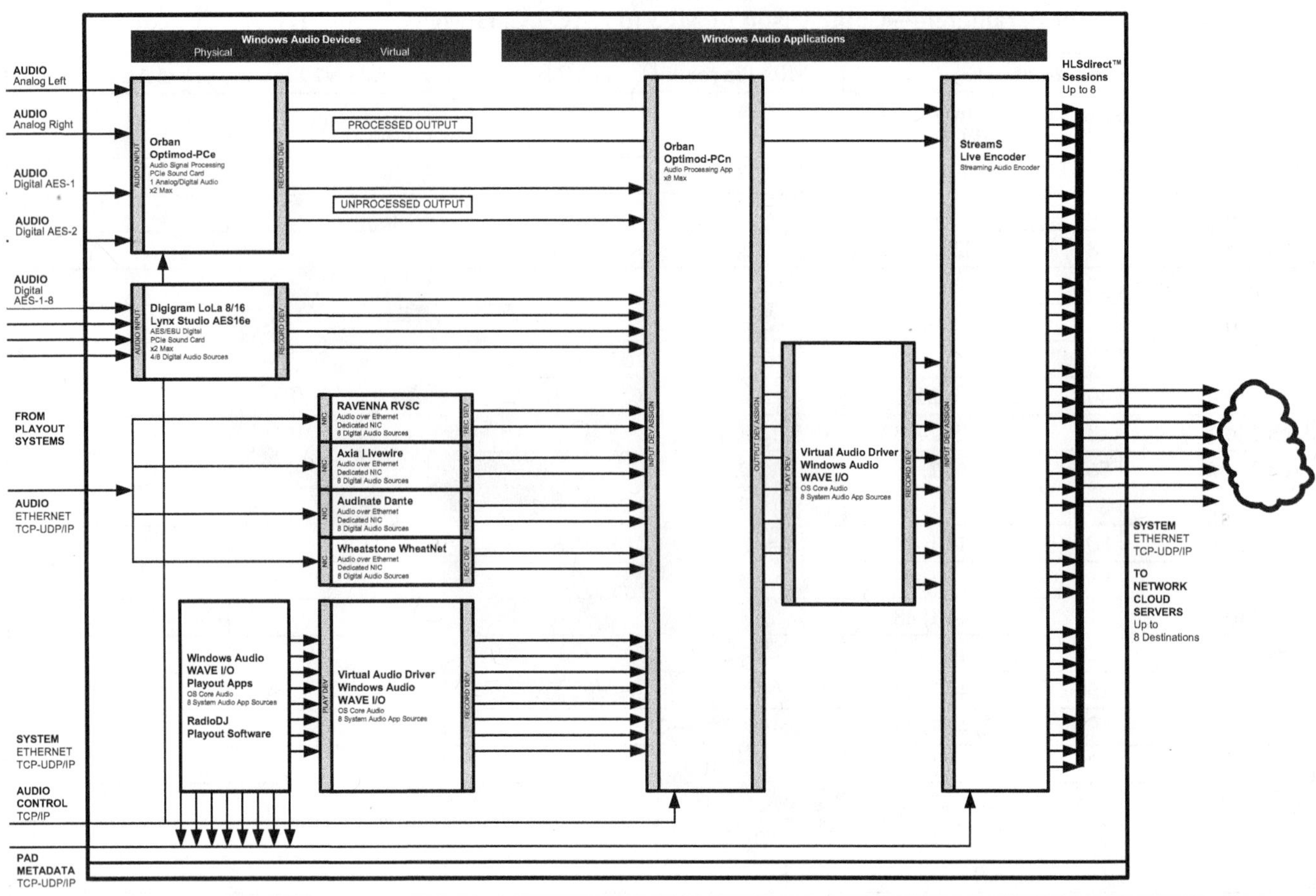

FIGURE 7.3-1 Typical multiple-source streaming audio encoder system, showing example products

and mobile devices support standards-based streaming protocols.

Metadata

Metadata contains information about the streaming content, such as song artist and title or program name, and possibly control cues. The streaming audio encoder usually transmits the metadata along with the streaming audio. Because Internet streaming occurs worldwide, it is important that the metadata be capable of handling any international character set. Metadata protocols that support UTF-8 are preferred, because all modern operating systems are now Unicode-compliant.

Metadata must perform correctly on all intended streaming players and devices, but metadata is often misunderstood by both content providers and player developers. Until recently there has been very little standardization of metadata, making player application development challenging because many metadata protocols needed to be customized. Hence, even though an audio stream may play in a particular player, its metadata may not display or control the stream correctly.

Metadata is typically sourced from an audio playout system. There are several ways to do this. The best way is to send the metadata (sometimes referred to as "now-playing" information) using XML and TCP or UDP directly to the streaming encoder. Some applications may require some sort of format conversion or bridge software to ensure data compatibility between the playout software and the streaming encoder. Usually such bridge software is also capable of outputting now-playing information in several other formats simultaneously. For example, metadata could also be sent to RDS, HD-Radio Program Service Data (PSD), and digital signage services, as well as sent in scheduled text messages. Avoid playout systems that output metadata or now-playing information using a text file. This is not a real-time solution, and may be troublesome or unreliable.

Content Insertion and Cueing

A further issue is how the metadata is transmitted with the audio. Various streaming servers handle this very differently. Some servers cache the metadata; others do not. There are in-band and out-of-band methods of delivering metadata. Some in-band metadata protocols and all out-of-band metadata protocols are asynchronous, which makes their timing insufficiently deterministic to achieve satisfactory remote cueing or content-insertion control such as server-side ad injection. This will cause sloppy transitions between commercials and programming. Accuracy to professional standards requires a streaming metadata protocol that has synchronous in-band metadata. Not only will this allow on-time delivery of the typical artist and title metadata, but it will also provide accurate cueing and control.

Content Protection—DRM, Encryption and Watermarking

Content providers may wish to protect their content by some form of digital rights management (DRM) or encryption. This protection is difficult to achieve, because any content that can be heard can be recorded. Recording the stream via a virtual audio device is always possible, easy and results in excellent quality, although the user will not get the original encoded bitstream. Although streams may be protected with various forms of DRM, SSL/TSL or encryption methods, which make the streams a bit more difficult to download, they can still be recorded in real time. Protection simply prevents direct download or stream-ripping of the original streaming bitstream.

Another potential method of protection is adding an inaudible watermark to the bitstream, where program audio psychoacoustically masks the watermark. The data in the watermark could also be used for ancillary signaling purposes. Although watermarking doesn't prevent the content from being recorded, the watermark data cannot be removed, so it is copied along with the program material and can always be traced by the rights holder. Some professional streaming encoders offer a watermarking option.

Content Delivery

There are several ways to deliver the streaming bitstream to an audience. Most content providers use a content distribution network, known as a CDN, although technically determined netcasters can readily provide these services themselves, given enough available network bandwidth. CDNs offer various delivery options and server options to complement the chosen streaming protocol and intended audience. There are several factors to consider, such as intended number of listeners, intended geographic location and, of course, cost.

For a simple, basic streaming service, a single server may suffice. Single servers can reach the entire world. Depending upon the distance, however, latency can affect results; excessive buffering and dropouts in the player might occur. Also, if a single server reaches its bandwidth capacity, the size of the audience cannot expand further. To avoid this, one can deploy multiple servers with load balancing, a method sometimes called *point-to-point streaming*.

Edge servers can improve on point-to-point streaming. Edge servers are basically multiple servers with load balancing, except the servers usually cache the streaming content and are located at multiple geographic locations to reduce latency. This results in improved stream reliability. Listeners will connect to the edge server closest to their geographic location instead of always connecting to a single server or a simple load balancer. This method is sometimes referred to as *point-to-multipoint streaming*.

There is another streaming protocol called *IP multicast*. Multicast is a connectionless protocol using UDP rather than the typical TCP. All streaming data appears at all player clients, so the bandwidth used by content provider does not change as the number of users increases. IP Multicast should not be confused with the more common unicast UDP streams. The public Internet is not multicast enabled, however, so this method of streaming is not currently used for general Internet streaming. (Multicast *can* be enabled in limited, private network environments, and multicast mode is the typical approach used in professional AoIP systems within broadcast facilities.) With advancements in LTE multicast technology, the use of IP multicast for streaming to the public eventually may become possible.

Logging, Analytics, Metrics

It is important to know how well your streams are working for you. Most media-centric CDNs offer some form of analytics reporting. This information should include not only network-related statistics about your audience, but also broadcast ratings-type information such as average quarter hour (AQH), time spent listening (TSL) and CUME. More details on this can be found at the Nielsen website.[2]

It is crucial to validate these numbers so that they are accurate and meaningful. Some algorithms are flawed and can misrepresent your streams and business. Simple issues like outdated IP databases will affect TSA (Total Survey Area) and other geographically based information. Moreover, users often leave computers and mobile devices on and walk away, logging exceedingly long (but invalid) listening times.

It can be very difficult to log and report streaming listening within large organizations. Often, the streaming server cannot accurately identify these streams, which frequently appear to be served to one IP address, because most organizations use a network routing protocol called NAT, or Network Address Translation. This challenge can be overcome by using a custom streaming player application that supports unique player IDs, but this only works accurately if the content provider is fully in charge of the player. Allowing other players or devices to play the streams will distort the data. This can be partially ameliorated by trapping the player application user-agent in the logging and analytics. (Every network software application that connects to a server has what is called a user agent. This is usually the name of the application and version number.) So using this technology to your advantage, it is possible to know what kind of "radio" your audience is using to hear your stream.

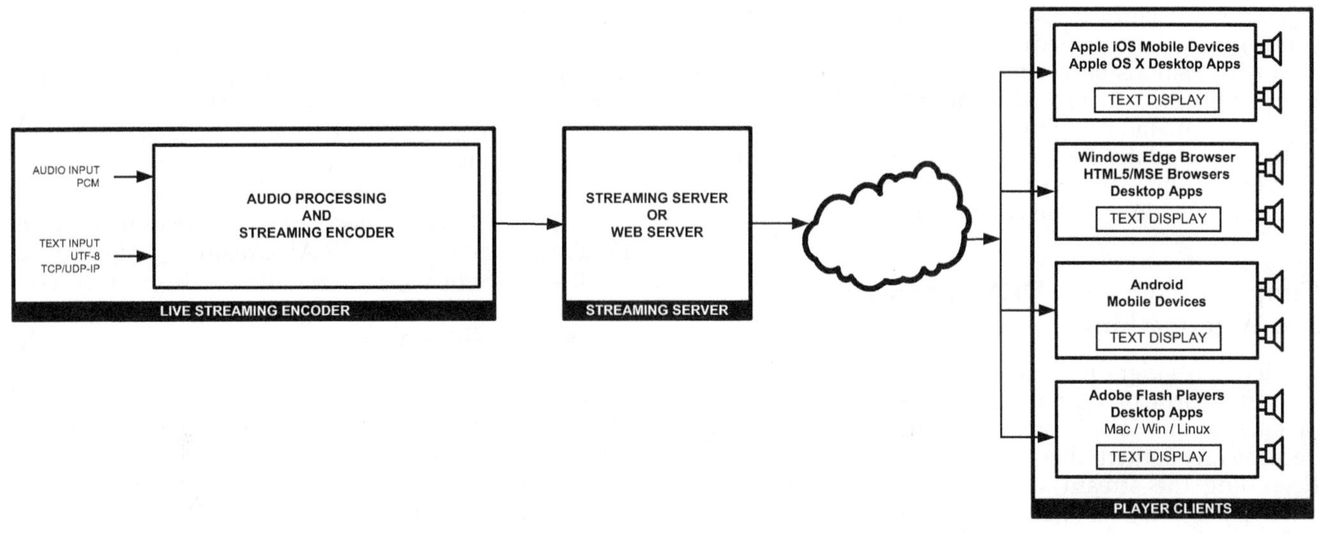

FIGURE 7.3-2 Typical basic streaming audio block diagram

SERVER PROTOCOLS

SHOUTcast

SHOUTcast is one of the original simple streaming audio protocols. It was developed in 1998 by Nullsoft, which also developed the popular streaming player Winamp. It uses the ICY (I Can Yell) protocol, which is a simple bitstream over HTTP. The server is proprietary with a limited feature set, but SHOUTcast 2 using the UVOX (Ultravox) protocol has subsequently addressed some of these limitations. It now supports UTF-8 synchronous metadata if the player also supports the UVOX protocol, which most do not. (Winamp supports the UVOX protocol.) Because the server is closed-source with few options to customize, it is not generally used for large-scale enterprise deployment. Because of its simplicity, however, it continues to be one of the most popular streaming audio protocols, especially among smaller streamers and content providers.

Icecast

Because the SHOUTcast server is proprietary, and there is no way to customize or add additional features, Icecast appeared shortly after SHOUTcast became available in 1999. Its name is an obvious play on the ICY protocol (which Icecast also uses), but Icecast has many more features. Because it is an open-source project, anything can be modified or added. In 2004, Icecast2 was released as a complete rewrite to address multi-format support and scalability. As of this writing, Icecast2 is the most popular streaming audio server in use. When the Icecast2 server is used in its normal form with a standard ICY streaming encoder, it does not support synchronous metadata, making it difficult to use with any kind of professional server-side content insertion or ad replacement system. However, at least one commercial encoder supports in-band ID3v2.4 metadata as well as the standard ICY metadata to implement synchronous metadata using Icecast2 servers.

RealAudio

RealAudio was another proprietary streaming audio protocol developed by RealNetworks in 1995. The server was based on a proprietary implementation of RTSP/RTP with both TCP and UDP modes. RealAudio was retrospectively considered ahead of its time, but its low-bitrate audio quality was poor (although it was still the best available at the time of its introduction). The subsequent development of standards such as HE-AAC were disruptive to RealAudio. With proprietary streaming servers and protocols falling out of favor and with the availability of easier ways to stream with more features, RealNetworks' RealServer and Helix Universal Server faded away and are now part of streaming audio history.

Apple QuickTime, RTSP/RTP Unicast/ Multicast and 3GPP

QuickTime Streaming Server and the open-source version Darwin Streaming Server were developed by Apple. Darwin Streaming Server was the first open-sourced RTSP/RTP streaming server, released in 1999. Apple was instrumental in the early development of mobile streaming and standards. The QuickTime file format was the basis of the MPEG-4 file format and became an ISO standard. RTSP/RTP was the original mobile streaming protocol for 3GPP. RTSP/RTP is feature-rich with many different options available. It supports both UDP and TCP transports with very capable HTTP tunneling protocols for firewall penetration. Partially due to its complexity, it has been mostly superseded by HTTP Live Streaming (HLS), which no longer requires complex streaming-specific server software to deliver live streams, and is now natively supported by Apple and Microsoft operating systems. HLS has a published IETF specification.

Microsoft Windows Media

Windows Media Services (WMS) was a streaming media server developed by Microsoft. It was a proprietary form of RTSP/RTP, supporting TCP, UDP, unicast and multicast streams. It only supported Windows Media codecs, so its popularity declined in favor of servers that support standards-based codecs, segmented streaming protocols (see the section "Segmented Streaming") and multiple operating systems. While reaching millions of Windows computers and mobile devices, WMS had metadata limitations and low-bitrate audio quality that was not competitive with HE-AAC.

Windows Media Services has now transitioned to IIS Media Services, which is the Microsoft IIS web server with media extensions. It provides support for standards-based codecs and segmented streaming using several different protocol choices, including Microsoft Smooth Streaming, Apple HLS and MPEG-DASH. Microsoft IIS web servers can be used without the media extensions to deliver live streams using a streaming encoder that supports direct segmented streaming.

Flash RTMP/FLV

Adobe Flash streaming has many facets. It is not just for streaming media alone. It is a complete system that has been used for streaming within its own ecosystem, and as separate components with other streaming encoders and servers. Flash includes a very capable player API and has been one of the most popular player platforms, although as of this writing HTML5 is replacing it. This is another example of a standards-based solution displacing a proprietary one. Flash's addition of AAC/HE-AAC support in 2007 helped move its streaming audio away from MP3. Because the Flash player component was already installed on the vast majority of computer systems, it was no longer necessary for users to download and install anything (other than a Flash player update) to play the then emerging AAC/HE-AAC audio streams.

There have been many hacks to deal with the Flash player's audio bitstream limitations. The most popular in use as of this writing is a special version of the Icecast2 Streaming Server, known as the KH Build, named after the developer, Karl Heyes. These modifications of the Icecast2 Streaming Server allow it to output a stream in the Flash FLV format, which is actually a file format, but the KH Build allows streaming over HTTP, which is what Icecast2 uses. Metadata is supported and picked up from the out-of-band metadata method that Icecast2 supports and is then placed in the FLV stream. While not fully synchronous and still suffering from a bit of latency, it is closer than the standard Icecast2 inline metadata protocol. Many high-profile streaming content providers use this method to reach Flash players.

Adobe Flash also has its own media server, formerly Flash Media Server, now transitioned to Adobe Media Server. It ingests streams from a streaming encoder using another proprietary streaming protocol, RTMP. Adobe has publicly released an incomplete version of the RTMP specification, making it difficult to implement reliably in a streaming audio encoder. Adobe Media Server can output RTMP-standard streams, as well as segmented streaming formats (see the next section) such as Adobe HDS and Apple HLS. HDS and HLS are served from a companion install of Apache web server with Adobe Media Server. HDS and HLS metadata is supported, although its performance has not been fully vetted as of this writing. Because segmented streaming protocols can use simple web servers and/or cloud storage, complex streaming servers are no longer necessary and can be avoided as a point-of-failure or expense.

Wowza Media Server is another streaming server with features very similar to Adobe Media Server. It has more stream ingest options, including RTMP, RTP and ICY. It outputs in various streaming protocols such as RTMP, RTP and several segmented streaming protocols: Apple HLS, MPEG-DASH, Adobe HDS and Microsoft Smooth Streaming.

Flash has been plagued with numerous security issues, and suffers from no support on Apple iOS, along with other technical limitations such as excessive computer resource requirements and limited audio codec bitstream support. As of this writing, Adobe has recognized that HTML5/MSE will likely replace Flash, after which Flash also will fade into streaming history.

Segmented Streaming

As of this writing, segmented streaming is ascendant, and is on track to be fully supported natively in Apple OS X and iOS, Microsoft Windows 10, and Android. There are several segmented streaming protocols available, including Apple HTTP Live Streaming (HLS), MPEG-DASH, Microsoft Smooth Streaming and Adobe HTTP Dynamic Streaming (HDS). Their mutual goal is to simplify streaming, reduce deployment and operating costs, and increase reliability to support the ever-increasing demand for streaming media. All of these protocols work similarly; the main differences are the details of the actual file formats used and client player support. As of this writing, HLS has by far the most complete support for audio-only streaming using efficient elementary streams, complete with frame-accurate timed metadata (as opposed to transport streams, which are used for video streams and which are significantly less efficient). HLS is supported natively in Apple OS X and iOS, Microsoft Windows 10, and Android, making it easy to develop player clients. MPEG-DASH is progressing, although audio-only metadata support is immature at this writing. Adobe HDS and Microsoft Smooth Streaming use proprietary file formats. This makes them less attractive to users who prefer a standards-based solution, which is one of the primary goals of segmented streaming. Hence player development is also slower there.

Segmented streaming differs fundamentally from traditional streaming. Segmented streaming can use a standard web server or cloud storage to deliver both static and live content using a direct-upload type encoder. In a sense, segmented streaming is the same as file streaming from a web server (commonly used by many music streaming services) except that segmented streaming has controlled file segments. Compared to simple file streaming, segmented streaming also has more security options available for content protection. With the appropriate segmenting tools software, it is also possible to segment static files and deliver these to the same player clients for audio-on-demand applications.

To deliver streaming content, it is no longer necessary to be bound by the requirements and expense of dedicated streaming servers. Segmented streaming does not create a constant connection to the server like traditional streaming. Instead, the streaming encoder frequently uploads segmented media files along with an updated playlist or manifest index file, and the player clients download these files and play them. Full-featured streaming encoders have a file management option to keep only the most recent segments, although some CDNs do not allow this management. Because the heavy lifting is done in the streaming encoder and player clients rather than in a streaming server, simpler, less costly web servers can be used to deliver live and static media content in the same way. This has several other benefits. Retrieving the segments creates client-server keep-alive handshaking just like web content, and offers self-buffering media delivery to the player clients. This produces the equivalent of long buffer times to increase reliability, which is especially important on crowded mobile networks and other bandwidth-challenged environments. The encoder is also fully in charge of the metadata, allowing frame-accurate, synchronously timed metadata that is not affected by server loads or transactions. This allows accurate metadata cueing and content insertion, such as server-side ad replacement with professional results.

Segmented streaming is also able to take advantage of the well-established web infrastructure using standard cache and edge-server technology already in place throughout the Internet. It is no longer necessary to use dedicated streaming server cache and edge servers, which are all protocol-dependent. It is also easier to find knowledgeable IT professionals already familiar with HTTP content delivery than it is to find specialized, streaming-server expertise. Segmented streaming's multiple benefits combine to support its ascendency. For comparison of the two approaches, see Figures 7.3-3 and 7.3-4.

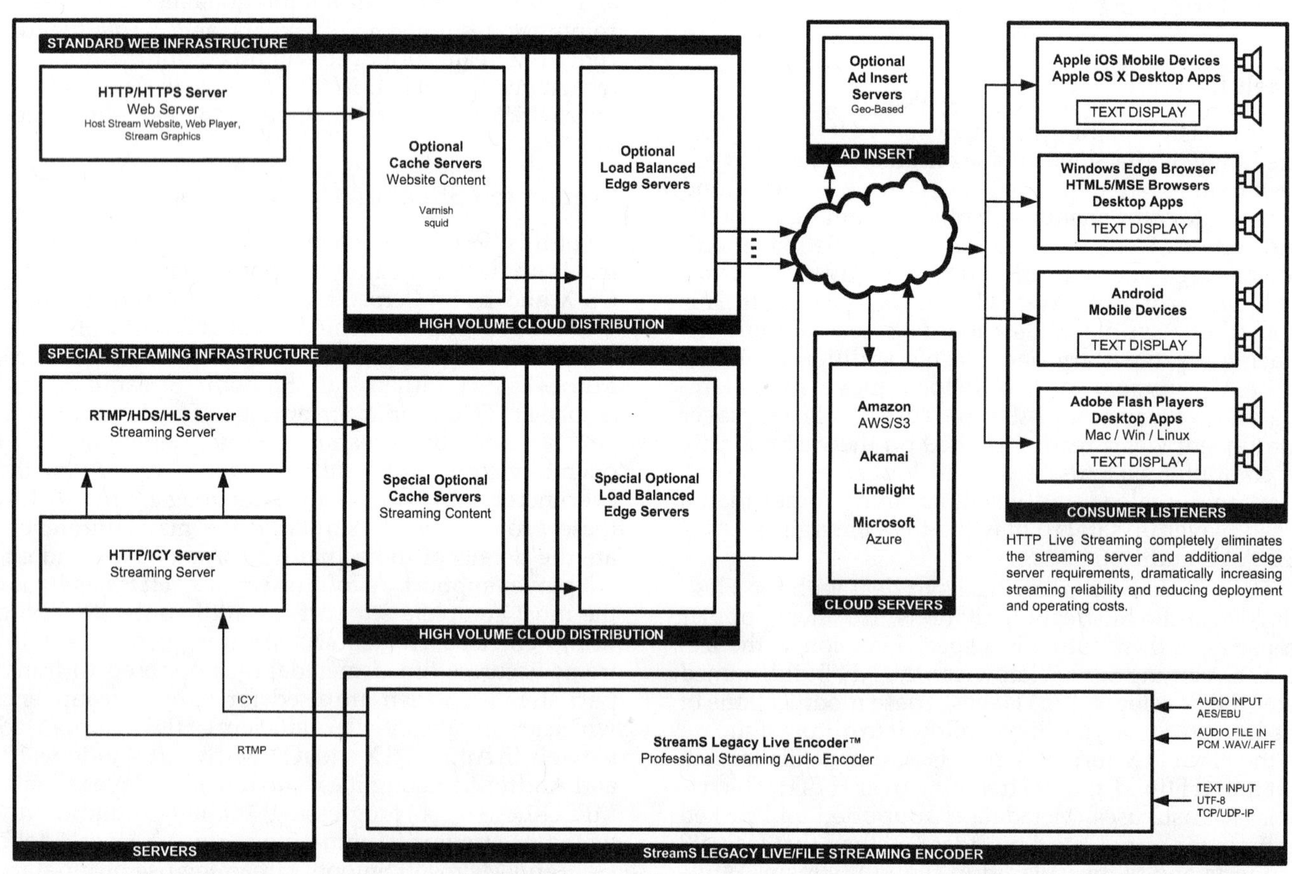

FIGURE 7.3-3 Typical legacy ICY/RTMP streaming system block diagram, showing some example products

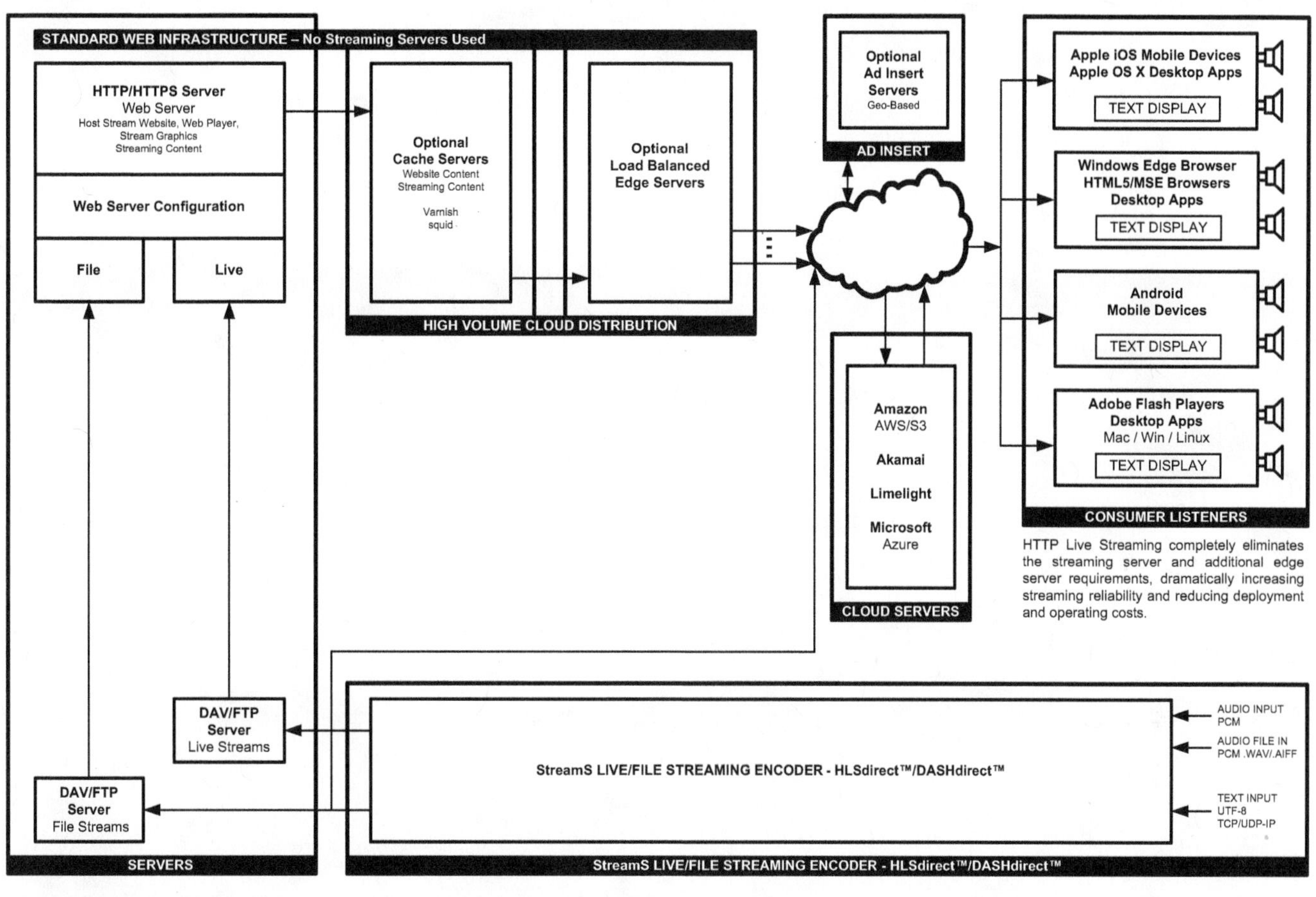

FIGURE 7.3-4 Typical segmented HTTP streaming system block diagram, showing some example products

Players

Streaming players are available in several forms, ranging from computer player applications to browser-based players, mobile apps and Internet radio appliances. It is important to identify your intended target players to plan your streaming deployment.

As of this writing, popular computer player applications include Windows Media Player, iTunes, and Winamp. There are many others, including mobile apps such as iHeartRadio and TuneIn Radio. These players usually support the standards-based metadata from each streaming protocol they support. For example, if you transmit any custom metadata that does not adhere to the streaming protocol, there is a very good chance that this metadata will not display correctly in these generic streaming player applications. If you are designing a custom player app and are completely in control of the player side of things, then you can pretty much do anything you want. But if you want your streams to be playable on generic player apps and Internet radio appliances, you need to consider the options very carefully to prevent metadata from being displayed incorrectly. No broadcaster wants an automobile dashboard player to display garbage characters when playing its audio service.

Adobe Flash players dominated the Internet for many years, but as explained above, Flash is obsolescent, so it is unnecessary to consider any further Flash development.

HTML5/MSE (Media Source Extensions) is the ascendant computer player platform. It will work natively in all modern web browsers. With segmented streaming (see previous section) and HTML5/MSE, it is now easier than ever to build custom full-featured players that work well and are Flash-free.

The streaming audio world continues to advance at a rapid pace, and radio broadcasters continue to grow their audiences reached via online platforms. Therefore this form of service delivery demands broadcast engineers' constant attention to optimally serve their services' and listeners' interests.

Notes

[1] MP3 and AAC Explained: www.iis.fraunhofer.de/content/dam/iis/de/doc/ame/conference/AES-17-Conference_mp3-and-AAC-explained_AES17.pdf.

[2] Nielsen: www.arbitron.com/downloads/terms_brochure.pdf.

CHAPTER

7.4

AM Radio Broadcasting

THOMAS G. OSENKOWSKY

Radio Engineering Consultant

The earliest form of wireless audio delivery employed amplitude modulation (AM). Transmitters used vacuum tubes to generate a carrier upon which audio was impressed by various means with high-level plate modulation being the most popular. AM broadcast signals propagate via groundwave during daylight hours and also via ionospheric reflection during nighttime hours. This reflection causes AM stations to be heard at great distances from the transmission site.

In MM Docket 87–267 FCC divided the Standard Broadcast Band (535–1705 kHz) into three categories: clear, regional and local channels. Four Classes of stations were defined. These Classes are Class A, Class B, Class C and Class D. The category and class of each station is determined by its frequency, power and licensed operating hours. The United States also has agreements with Canada and Mexico that govern the allowable coverage and operating hours on certain channels.

Some AM stations must reduce power and/or change directional pattern at sunset. Others are licensed for daytime-only operation and are required to sign off at local sunset as dictated by the terms of their FCC station license. Such stations may sign on at times specified on their station license. Many stations have pre-sunrise (PSA) and/or post-sunset (PSSA) authorizations that allow them on a permissive basis to operate at reduced power.

Antennas employed by AM stations most often take the form of vertical towers. Most are insulated from ground at the base of the structure where RF is fed from the transmitter. Typically there is an antenna tuning unit (ATU) at the tower base where the operating impedance of the tower is transformed to the characteristic impedance (Zo) of the transmission line, which is typically 50 ohms. The electrical height of the tower determines the vertical radiation characteristic. This is especially important for nighttime operation when the radiation towards the ionosphere determines the propagation of the signal at great distances. It also determines the amount of signal propagated along the ground that provides local coverage. Common tower heights range from 60° to 225°.

An ideal AM signal that is sine wave modulated 100% has two sidebands each containing 25% of the audio. It takes one-half the carrier power to modulate an AM carrier 100%. The relationship of the sidebands to the carrier is important for high-quality transmission and reception. For nondirectional stations this relationship can be affected by the phase shift of the ATU, transmission line length and transmitter switching apparatus, if used. For directional arrays the pattern bandwidth or attenuation of the carrier and sidebands can affect the sideband relationships. Every transmitter has an output matching network whose phase shift rotates the sidebands relative to carrier between the antenna terminal and the RF final amplifier. This must be considered in analyzing the overall quality of the transmission plant. Sideband symmetry is most important at the RF final amplifier.

The technology behind AM broadcasting has considerably evolved since the first broadcast around 1920. Most transmitters are solid state, replacing older tube technology. Allocation studies, array designs and phasing/coupling systems are conceived and designed on computer resulting in an efficient, broadband and high quality transmission system.

Elsewhere in this section we will examine transmitter facilities design, audio processing, AM transmitters, antenna system, coupling/phasing system design, diplexing AM stations, modern moment method computer proof-of-performance and analysis, synchronous AM operation and AM antenna system maintenance. In the Signal Measurement and Analysis section, AM radio performance measurements and AM radio antenna system and field strength measurements are examined in detail.

As of this writing testing has begun on all-digital AM broadcasting. The impact on co-channel and adjacent channel reception is under evaluation in both day and night modes of operation.

CHAPTER

7.5

Medium Wave (AM) Transmitters

JEFF WELTON
Nautel Limited

JOHN O. STANLEY

GENERATING THE RADIO FREQUENCY SIGNAL

Virtually all AM transmitters installed in the last decades have used solid state technology that will be discussed later in this chapter. However, many stations still maintain older transmitters using tubes. There are still some places where tubes are being used in currently manufactured equipment such as high-power shortwave and FM transmitters. The vacuum tube can be compared to the field effect transistor that is its closest solid state counterpart. Except for the very oldest transmitters, the low-level stages will use transistors, even if the high-power stages use tubes. Therefore, this discussion will center on the final radio frequency amplifier stage. However, many of the same principles will apply to tubes used in lower-level stages and even audio.

Tubes, unlike their solid state counterparts, have a finite operating life and as they age, the emissivity of the cathode will gradually decrease. Once there is no longer enough emission to provide a sufficient number of electrons to carry the desired currents through the vacuum to the anode, the tube must be replaced. This finite life, along with the power required to heat the cathode that contributes nothing to the output power, is an important part of why tubes have been supplanted by solid state devices. Ensuring that tubes still have sufficient emission is an important task for the technical staff in a station still using that technology. Filament voltage should be measured with a true RMS voltmeter and adjusted as needed to ensure that the emission is sufficient to provide positive modulation peaks.

A simplified typical RF amplifier stage is shown in Figure 7.5-1. The various components shown in this diagram function as follows. The choke from the HV to the tube plate provides DC to the tube while blocking RF voltage. The capacitor from the plate to the tune capacitor passes the RF while blocking the DC. The tune and load capacitors, along with the tank inductor between them form a PI lowpass filter and also match the 50-ohm output to the required plate load resistance, typically between 500 and 3000 ohms.

The design of this network uses a Q in the range of 10 for good filtering of harmonics and, if components are chosen to provide 135 degrees of delay, will also separate the tune and load functions so that the tune only sets the network to resonance at the operating frequency and the load only raises and lowers the output power with little interaction between the two functions. The load capacitor effectively changes the resistance that the tube output sees, thus changing its power out. Not shown in this simplified diagram are the filament transformer, the screen bypass and bias and the tuned grid, which guarantees that the input to the grid is a good sine wave. Forcing the grid and plate voltages to be pure sine waves with resonant circuits means that normal operation will trace out a straight line on the constant current curves of the tube because voltage is plotted on the x-axis and grid voltage on the y-axis. This makes analysis by graphical methods much simpler.

If one of the tuned circuits is detuned from resonance, the operating line will open into an ellipse, with loss of power output and an increase in plate current. Thus, proper tuning is indicated by both a dip in the plate current and a peak in the output power, grid and

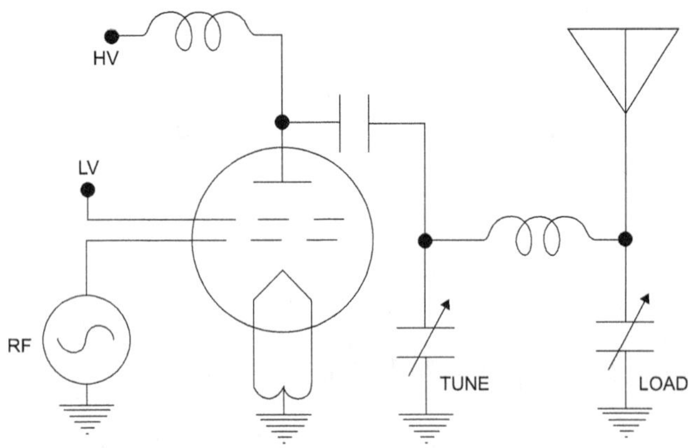

FIGURE 7.5-1 Simplified Diagram of a Class C Tuned Radio Frequency Amplifier

J. Stanley

screen currents as the plate capacitor goes through resonance. If these things do not happen at the same time, lack of neutralization is indicated, meaning that some power is feeding back from plate to grid via the grid to plate capacity or other paths. For frequencies in the standard broadcast band, neutralization is usually not an issue with tetrodes as it is with triodes or at higher frequencies.

MODULATION

Modulation is a technical term for changing. In amplitude modulation, we change the power of the transmitted wave in order to carry information. When there is no program, the standard AM transmitter produces a constant "carrier" wave. With modulation this wave is increased or decreased in keeping with the positive and negative voltage in the program source.

We usually use the term amplitude modulation to refer to amplitude changes intended to convey information, but it can also refer to extraneous rapid changes in amplitude that are not intended, such as hum modulation coming from an improperly filtered power supply. Slower changes in the amplitude, such as inadequate carrier amplitude regulation, day to night power adjustments, and turn on and off are not normally called modulation, even though they clearly represent a change in the transmitter output power.

Amplitude modulation methods can be grouped under four headings; that is to say, there are four ways to change the amplitude of the delivered signal. These are (1) changing the voltage of the power source, (2) switching something on and off, (3) changing the efficiency of the DC power to RF power transformation and (4) changing of the load. These can be visualized with the diagram in Figure 7.5-2.

The on/off switch may be for turn on or used for control of average power output via duty cycle.

The DC to RF stage consists of "active" components, either tubes or transistors.

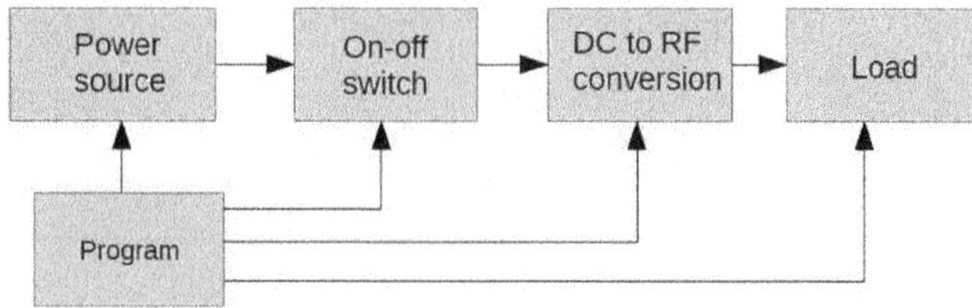

FIGURE 7.5-2 Simplified Amplitude Modulation Block Diagram

J. Stanley

The resistive load may represent an antenna, but for analysis may be considered to be a complex impedance or even just a resistor. Its value can be transformed by networks between the actual load, normally an antenna, and the active source.

The program to be broadcast can control any of these four things. Morse transmission, which is the simplest form of AM as well as the earliest form of digital communication, changes the amplitude between zero and full output, usually by the switch between the power source and the DC to RF conversion stage. In audio broadcasting, the audio program changes one or more of the four parts of the transmitter in accordance with the audio amplitude.

All four can be changed in a dynamic fashion; that is, quickly enough to follow a program of audio or video and so forth. Thus, the power to the load will be rapidly changing in amplitude based on the instantaneous value of the program input, above and below a fixed level, which we take to be the carrier power. A negative audio voltage will reduce the power to the load while a positive audio voltage will increase it. We assume for the moment that the active stage is able to produce up to four times the carrier power as is needed for 100% positive (peak) modulation. It is also capable of having its power output reduced to zero to provide 100% negative modulation.

Power supply modulation

If the DC input voltage is made to vary between zero and the value required to produce four times the carrier power level, the power to the load should also vary by that same proportion if we are to have undistorted amplitude modulation. The ability of various types of active stages to accomplish this will be of great interest. It will also be of interest to find a method of making the power supply voltage vary from zero to a high value at a rate suitable for the program audio or video. Some systems switch in additional power supplies instead of or in addition to varying a single supply. We often attempt to avoid unwanted changes in the DC input voltage that might change the transmitter output power. This would occur, for example, when hum is present or the power line voltage fluctuates due either to power grid instability or excessive loading of the power lines by the transmitter itself.

Efficiency modulation

If we leave the DC input voltage constant, but vary the efficiency of conversion of the active stage between zero and some maximum value, we can also send power to the load that will be able to reproduce 100% negative and positive modulation compared to some carrier value. Because the active stage will not be able to exceed 100% efficiency, the efficiency under carrier conditions cannot be more than 25% if we are using only changes in efficiency to reduce power.

Efficiency modulation methods typically are about 30%–40% efficient under carrier conditions, indicating that they are not using only efficiency reduction to reduce the output power but other methods as well. While legacy methods of efficiency modulation were used, modern transmitters do not use efficiency modulation, but rather attempt to maintain high efficiency at all times. Most transmitters will suffer somewhat from lowered efficiency when operating at reduced power.

Load modulation

If the power supply voltage is kept constant and the efficiency of the active stage is kept constant, the output power can still be changed by changing the resistance of the load. If we reduce the value of the load resistance to one-half its value under carrier conditions, we would expect that the output power will double, just as hooking two lamps on a 120-volt circuit will produce twice as much light as a single lamp. If we make the load resistance very high, the power transferred will approach zero. Thus, we can modulate the amplitude of the power to the load by changing the value of resistance into which the active devices feed.

A number of transmitters use this system as a part of their modulation method. Understanding how changes in the load, especially the resistive part of the load, affects delivered power is also very important where the bandwidth of the load is less than the bandwidth of the transmitted signal. On the low end of the AM band, this may be a problem even with an audio program. With the wider digital modes where bandwidth can be 30 kHz, adjusting the antenna network so that the active devices in the transmitter see a nearly constant resistance over the range occupied by the sidebands can be very important. Therefore, changes in the load presented to the active devices (tubes or transistors) can be part of a modulation scheme and also something to be avoided to prevent unwanted effects.

In methods that use load modulation, a very useful network is sometimes required. This is the so-called quarter wave network (Figure 7.5-3), so named because it emulates the behavior of a quarter wave length of coaxial cable. If we construct a network composed of three elements in either a T or a PI configuration and make all the elements with a reactance of the same value, that network will act like a quarter wave section of coax with a Zo of that value. This network has the ability to invert impedances relative to its own Zo value. For example, a quarter wave network of 1000 ohms will show 500 ohms at its input if a 2000-ohm load is connected. It makes open circuits look like a short and vice versa. It also converts inductance into capacity, and voltage sources into current sources. This very useful network is used in both vintage and modern transmitters as well as antennas and thus, should be understood by anyone seeking to understand modern transmitters.

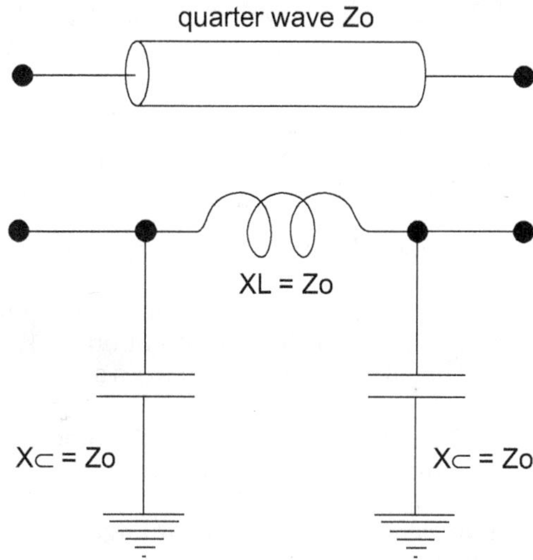

FIGURE 7.5-3 Quarter Wave Network

J. Stanley

Duty cycle modulation (PDM /PWM)

In this method power to the active stage is switched off and on with a variable duty cycle. Thus, the power supply runs at peak power at times, but is off at other times. The average power is variable from zero to peak value based on the on time to off time ratio. Filters must be used to smooth out the rapid on/off variations into a smooth continuous variation in DC power to the active stage.

Many AM transmitters use a combination of these methods; some use mainly one. Examples of transmitters using the various methods are:

Power supply modulation: Modern PDM and PSM transmitters, Push pull plate–modulated transmitters, Heising modulation, series modulation;

Efficiency modulation: Grid modulation, screen modulation;

Load modulation: Doherty, outphasing modulation, Harris DX series;

Duty cycle modulation: PDM, PWM, PSM, and other methods involving switching. Because these change the voltage to the RF stage, they can be considered as power supply modulation methods if the modulator is considered part of the power supply rather than part of the transmitter.

In some transmitters where we can separate power supply, audio and RF circuits into subparts, we can analyze the modulation process in different ways. Further, we can see the different types of modulation being active during different parts of the audio cycle. The Doherty transmitter uses efficiency (grid or screen) modulation for negative and somewhat for positive peaks, and adds load modulation for the positive peaks. The outphasing system does somewhat the same thing.

It is important to remember that some of these types of modulation may inadvertently occur when a transmitter is operated. For example, if a power supply is not sufficiently "stiff," the voltage may sag on modulation peaks. So, while some other method, for example load modulation in a Doherty transmitter, is trying to push the peaks up, the sagging power supply may be contributing to a decrease in the power. Similarly, while power supply modulation may be going positive, the transmitter may be resisting this increase due to efficiency loss in the active device due to improper loading, for example. Whenever a transmitter fails to produce good positive peaks, negative modulation from another modulation method may be contributing to that failure. Hum on one or more power supplies may be adding 120-Hz modulation. Understanding all of the ways that amplitude modulation occurs, either accidentally or on purpose, is worthwhile.

A brief review of historical methods used to produce amplitude modulation will illustrate some of these methods. The very earliest AM transmitters used a carbon microphone in series with the antenna. This absorbed some of the energy that varied as the resistance of the microphone changed in response to the sound waves falling on it. The efficiency of the system was changed based on what percentage of the RF energy went to the antenna and what percent was dissipated as heat in the microphone, which for all but the lowest powers had to be water-cooled.

Putting a carbon microphone in series with the power supply was another approach, but for a given transmitted power, even more heat was lost in the microphone as the DC to RF conversion was less than 100%. Once power levels exceeded the ability of the carbon microphone to absorb heat, a vacuum tube operating as a variable resistor was placed in series with the DC supply.

This tube operated in Class A and absorbed some of the power supply energy, which varied with the program content. With a series modulator, either the RF amplifier or the Class A modulator tube had to float above DC ground. The filament and grid power supplies must be isolated from ground.

When the modulator was connected on the ground side, this was called cathode modulation.

Note that none of these systems was pure efficiency modulation because constant power was not supplied due to the changing load. During negative peaks, the load on the supply is less, so the efficiency does not suffer as much as it would with pure efficiency modulation.

In some transmitters, modulating the grid or screen voltage was used to change the efficiency of the RF stage itself. To achieve positive modulation, it was necessary to load the tube for efficient operation at the peak power needed, then to back off the screen voltage or increase the negative grid bias until the output dropped to about 25% of the peak power. Then further variation of the screen or grid voltage by the audio signal would produce negative and positive peaks. Dropping the power by reducing grid and screen voltages puts the RF stage into a quite inefficient mode of operation with much of the input power going into heating the tube rather than producing the output signal. However, screen or grid modulation achieved some measure of modulation without the need for a second tube.

All of these modulation schemes were largely efficiency modulation methods as they began with a large amount of raw power and deliberately "wasted" some or all of it, allowing the remainder to ultimately pass to the antenna as RF output. The overall average efficiency of the system was correspondingly low, typically about 30% under carrier conditions. In addition, the maximum modulation percentage was low and the distortion was high. Clearly something better was needed.

Heising Modulation

The advent of the Heising modulator was an important advance. With this method the modulator tube, which served to absorb some of the DC power, was placed in parallel with the RF tube instead of in series so both tubes had their cathodes connected to ground. For the modulator to be able to affect the percentage of power going into the RF stage, it was necessary for the DC input to act as a current source rather than a voltage source. This was achieved by feeding the DC through a large inductor that tended to keep the current constant. The supply voltage had to be only one-half of that required with the series modulator, an advantage with higher power operation. As the modulator tube shut off, all of the current was forced into the RF tube with a corresponding rise in its plate voltage. As the modulator drew all of the available current, the voltage across both tubes went low. Thus the Heising system can be viewed as a form of power supply modulation, albeit a relatively inefficient type of it.

The power flow in this system during the modulation cycle illustrates that the inductor serves as more than a constant current device as it also stores energy during the modulation troughs and delivers it back during the peaks. Therefore, the power supply sees a constant load.

An example for a 5 kW (input to the RF stage) transmitter shows that under carrier conditions, the RF tube receives half of the total power provided by the supply. The modulator tube also receives 5 kW, all of which it dissipates in its plate. The supply delivers a constant

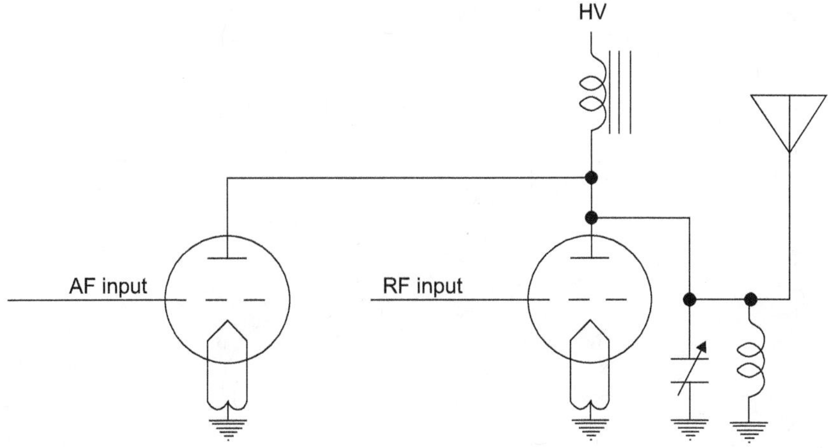

FIGURE 7.5-4 Heising Modulator

J. Stanley

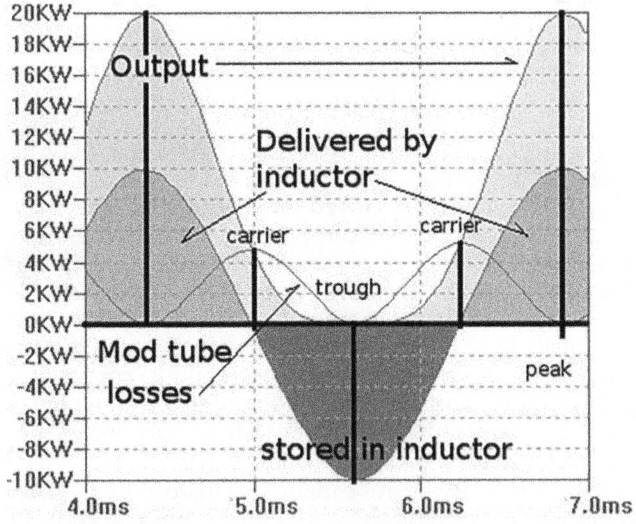

FIGURE 7.5-5 Power Flow in Heising Modulator

J. Stanley

10 kW. During modulation troughs (100% negative modulation), neither the modulator tube nor the RF tube receives any power as the voltage on their plates drops to zero. All of the 10 kW from the supply goes into the inductor. This power is released on modulation peaks so that in addition to the 10 kW from the supply, an additional 10 kW is available to the RF tube to produce the needed positive peak power of 20 kW. The modulator tube is drawing zero current so it again dissipates no power. Thus, we see that the modulator tube draws no power on either peaks or troughs as it does at the carrier level. On peaks it has zero current and during troughs, zero voltage. Averaged over time with 100% sine modulation, the power lost is 2.5 kW. This means that 7.5 kW average is supplied to the RF amplifier as is needed for 100% sine modulation. The losses in the modulator decrease with modulation, just the opposite of many other modulator arrangements.

Note that the preceding analysis assumes an ideal modulator tube and 100% efficiency in the RF amplifier. Neither is realistic, so in the real world the overall efficiency will be somewhat lower. Under carrier conditions the overall efficiency is under 35% and this only increases very marginally with normal program.

Push Pull Modulation

The development of the "push pull" class B plate modulation method was a great advance in efficiency as well as distortion. This approach allowed overall system efficiency to be more than 50% and the distortion levels to drop below 5% even at 90% positive and negative modulation levels. This became the standard AM transmitter configuration for many years so that even in the 21st century many of these transmitters are still in operation, at least as backup transmitters in stations that have installed modern solid state transmitters. Because it is possible that some stations may still be using such transmitters from time to time, some attention will be given to their operation.

The great advantage of the push pull modulator over the Heising is that it is more efficient under carrier conditions. This is because the modulator tubes operate in Class B or AB and draw very little power with no or low audio input. At full 100% modulation, they draw maximum power. The audio power required for full sine modulation is 50% of the input power to the RF amplifier. With most programing, the average modulating power is considerably less than for a sine wave, so efficiency under carrier conditions is more important than efficiency under full 100% sine wave conditions.

We can see that the push pull modulator bears some similarity to the Heising modulator. It is the use of two modulator tubes that opens the possibility of using Class B or AB instead of Class A for greater efficiency. This comes at the cost of some circuit complexity, mainly in the addition of a large modulation transformer. It also raises the possibility of crossover

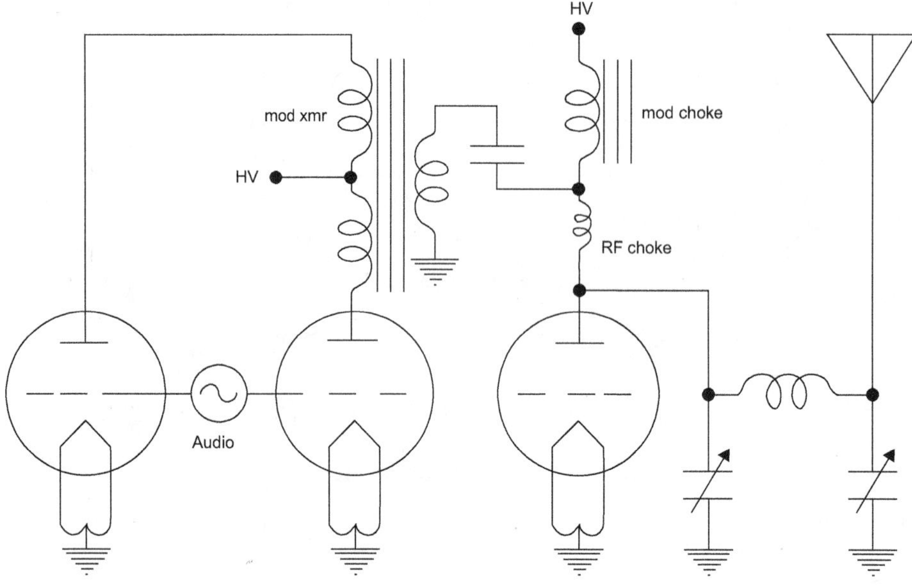

FIGURE 7.5-6 Push Pull Modulator

J. Stanley

distortion that must be controlled by proper choice of the resting bias of the modulators as well as often requiring audio feedback from the transformer primary to the low-level audio stages.

Because many of these transmitters still may be in use in standby service, even though obsolete, understanding how to make them work when parts may fail and no replacements are available may be useful, where it would not be economically justified to either replace the transmitter itself or expensive transformers, chokes or tubes. The following alternative connections are presented to cover such situations. With all of these connections, the transmitter should not be run at full power as that may lead to further damage to the transformer, choke and tubes. Running at about one-half normal power will usually keep components within acceptable limits.

Figure 7.5-7A shows how a connection can be made that does not require the modulation choke. This connection is actually used in some lower power transmitters. Allowing the DC current to pass through the secondary of the modulation transformer may compromise its performance, but if power is reduced it may give acceptable service.

Figure 7.5-7B shows how the same connection can be used with only one audio tube working. In this case the modulator tube must be rebiased to Class A with the corresponding loss of efficiency.

Figure 7.5-7C shows how the transmitter may be operated if the secondary of the modulation transformer is defective. However, this defect must not involve shorted turns. A short to ground or open turns are acceptable. One can also use a similar connection but still use the modulation choke and its associated capacitor so DC to the RF stage does not pass through the modulation transformer.

Figure 7.5-7D shows how the transmitter may be operated with a single tube in a mode that is equivalent to a Heising connection in terms of efficiency. Again, Class A is required for the modulator. The slight advantage of this connection is that the DC for the modulator and for the RF pass through different windings on the modulation transformer, thereby reducing overheating of the windings themselves and reducing core saturation. Unlike the Heising, this connection puts a highly variable load on the high voltage power supply with modulation so a stiff supply is needed with a large capacitor on its output.

Of course, if the modulation transformer is completely unusable, but the choke intact, one can connect the circuit into a Heising configuration. Using both modulator tubes in parallel may be useful to prevent a single one from overheating in the required Class A operation.

It is sometimes assumed that one of the modulator tubes "drives the positive peaks" and the other the negative peaks, meaning that one of the modulator tubes does most of the work. In fact, the storage of energy in the core of the modulation transformer tends to share the load between the two tubes. Thus, even with the normal push pull configuration, one can operate with one weak modulator tube or even one tube absent, and still modulate both positive and negative peaks. However, some distortion and loss of modulation capability should be expected. Rebiasing to Class A will reduce distortion at a loss of efficiency. Because the "one tube drives the positive peaks" understanding is not completely without merit, it may be worth swapping two unequal tubes to see which way gives the better overall modulation. Any of the preceding methods should not be considered a permanent solution, but for temporary use where the alternative is to be off air, they may be useful. With today's modern solid state transmitters as

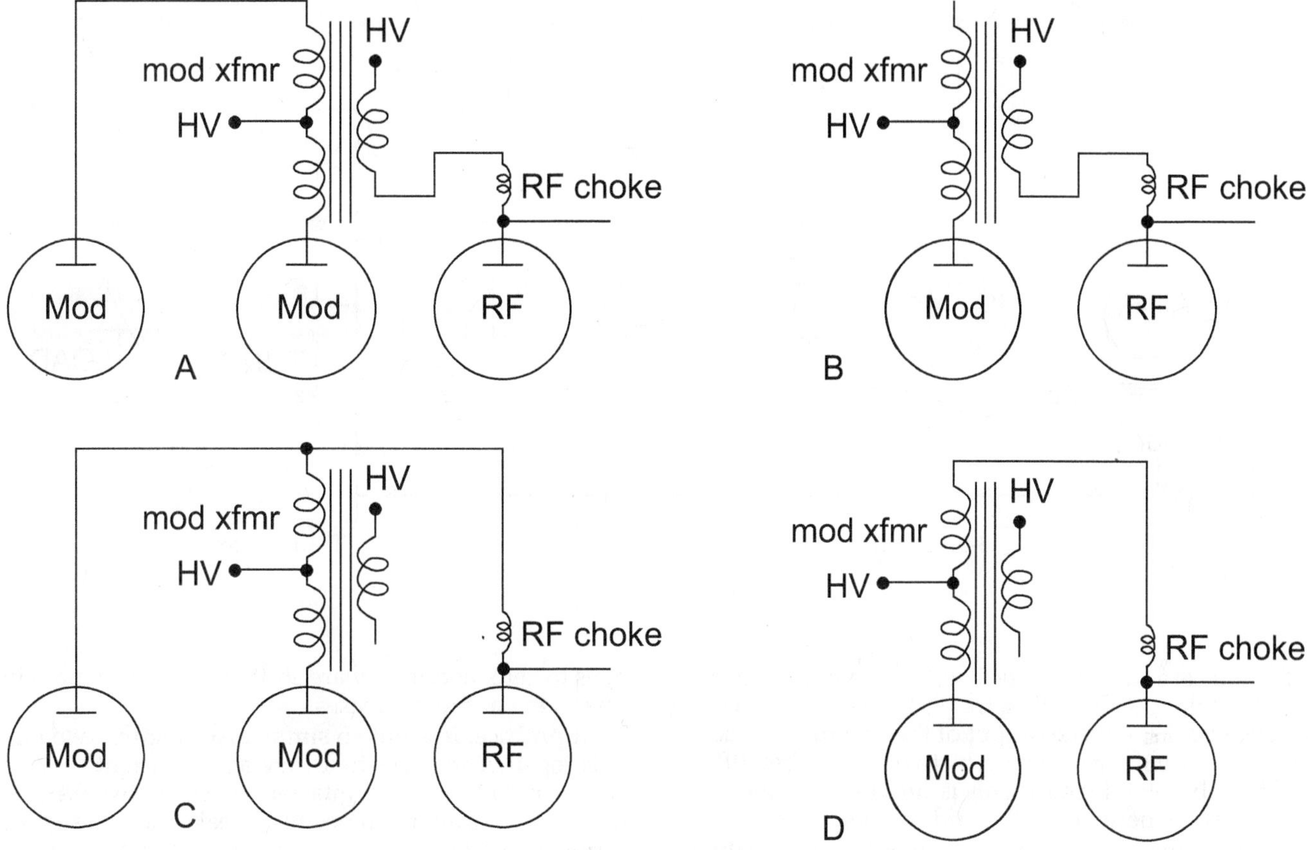

FIGURE 7.5-7 Various Push Pull Configurations

J. Stanley

the primary service, standby service should not normally be used very often or for long periods.

One disadvantage of the push pull system was the large amount of "heavy metal" in the modulation transformer and choke with their frequency limitations. Also the crossover distortion inherent with push pull audio amplifiers was a limiting factor in achieving low distortion. This and the desire to discover patentable and advantageous methods led to further developments.

Screen Modulated Doherty

The Doherty circuit was first put forth as an efficient linear amplifier that would be driven by a low-level modulated stage. These have long since passed from AM broadcast service. However, a further development using the Doherty concept was invented in 1938 using screen modulation of the peak and carrier tubes. Many of these were built and, no doubt, a few still are being used, mainly in standby service for more modern solid state transmitters. Doherty methods are also used in modern solid state linear amplifiers.

The basic Doherty concept is to use a single RF tube for operation under carrier conditions. This tube can be biased and loaded for efficient operation at that power level. When peak power is needed, a second tube is brought into operation by providing it with positive screen voltage. As this tube begins to operate, its power is added to that of the carrier tube and in addition, it changes the impedance seen by the carrier tube so as to load that tube heavier and cause its output to increase. This requires that the two tubes act as voltage sources in series. Actually putting them in series would involve floating one of the tubes above ground with serious complications, so instead they are made to act as if they were in series by separating them by a -90 degree network.

Another way to envision this is to say the quarter wave network makes the two voltage sources act as current sources and add power when connected in parallel. Perhaps the best way to illustrate this is with actual numbers. If the carrier tube is connected via a 2000-ohm quarter wave network to the load, and the load adjusted for 1000 ohms, the network will raise the impedance seen by the carrier tube to 4000 ohms and the carrier tube can be biased and driven for efficient operation at, for example, 1 kW. When the peak tube appears directly across the load, it will provide half the power to the load, making the load appear as a 2000-ohm load to each tube. This 2000 ohms will be seen by the carrier tube as 2000 ohms when viewed through the 2000-ohm quarter wave network. Thus the carrier tube will put out double its normal power and

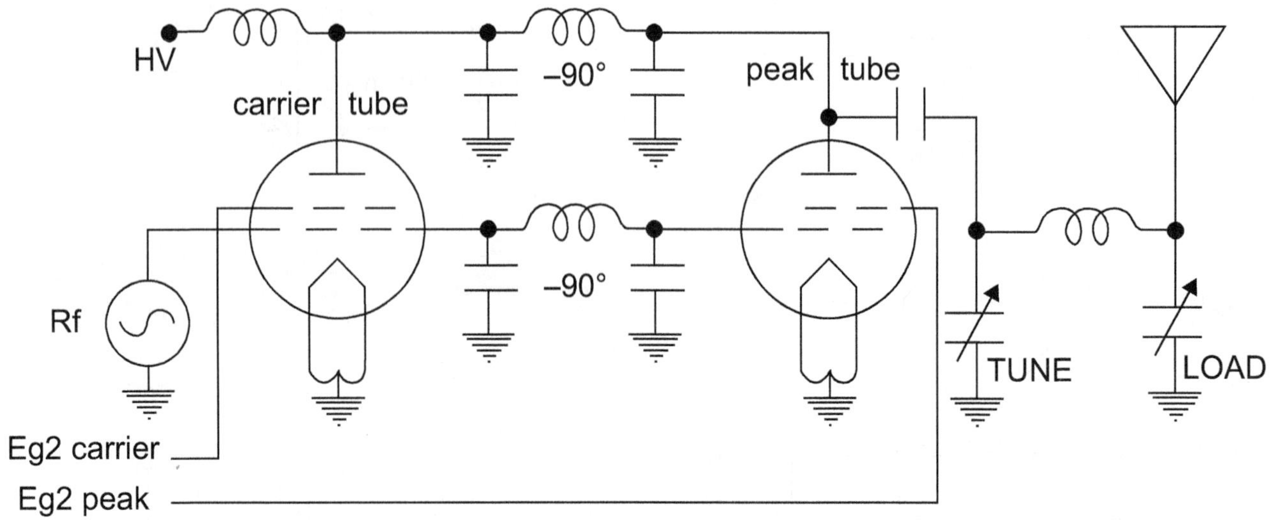

FIGURE 7.5-8 Simplified Diagram of a Screen Modulated Doherty Transmitter

J. Stanley

when this is added to the peak tube, also putting out 2 kW, the total power will be the 4 kW needed for 100% positive peaks. For the output of the two tubes to add in phase, they must be driven by two signals that differ in phase by −90 degrees. This is provided by another quarter wave network in the grid circuits.

This system uses screen modulation on negative peaks when only the carrier tube is operational, but on positive peaks it uses screen modulation as well as impedance modulation. By providing the tube(s) with an optimum load for good efficiency under both carrier and peak conditions, overall efficiency can be kept high. With pure screen modulation, one must optimize efficiency either at the carrier condition, which precluded good positive peaks, or at the peak condition, which makes the efficiency very low under carrier conditions.

Adjusting the circuits for proper phase shift between the two tubes is usually made easier by the provision of an oscilloscope that displays a Lissajous pattern that is adjusted to the proper shape.

In an emergency, a screen-modulated Doherty transmitter can be operated with only the carrier tube operational by reducing the screen voltage and carrier power to the value that allows the carrier tube alone to produce positive peaks. This, of course, makes the transmitter into a conventional screen modulated transmitter with its characteristic low efficiency.

Outphasing Modulation

Another system that also uses impedance modulation is the outphasing method marketed by RCA under the Ampliphase name. In this method, two identical RF amplifiers are driven with partially out of phase signals so that the sum of the two signals is relatively small. If the signals are differentially phase shifted so that when they go 180 degrees out of phase, the sum goes to zero, and if they are shifted so as to be more in phase, the output increases.

In practice, it is not so simple. Adding the two signals together requires that they act as current sources when paralleled, so quarter wave networks are needed. In addition, on negative peaks the two signals seriously fight with each other unless their amplitudes are reduced by reducing voltage to the screens of both tubes. RCA called this "drive regulation," but it is simply a form of screen modulation.

Like the screen-modulated Doherty, the system uses efficiency modulation for negative peaks and load modulation between two tubes for positive peaks. The system provides clean signals with good efficiency when properly tuned. However, difficulty in tuning and the tendency to produce incidental phase modulation when unbalanced were the main drawbacks.

As there are probably none of these transmitters still in service, even in standby mode, further attention will not be given to this system except to add the warning that unless you are very sure of what you are doing, do not attempt any adjustments. In emergency situations, one could be operated as a screen-modulated transmitter at reduced power, using only one tube, provided that the phase-modulation circuit was disabled to avoid serious phase modulation on the final signal.

Pulse Duration Modulation

Pulse duration modulation (PDM) transmitters, also called pulse width modulation (PWM), use what amounts to a switching power supply that can vary its output voltage at audio frequencies. In its most basic form the modulator acts as a step down switching regulator while taking its power from a DC supply equal to the maximum voltage need for positive modulation peaks. Like other step down switching power supplies, the voltage output is controlled by the duty cycle of the

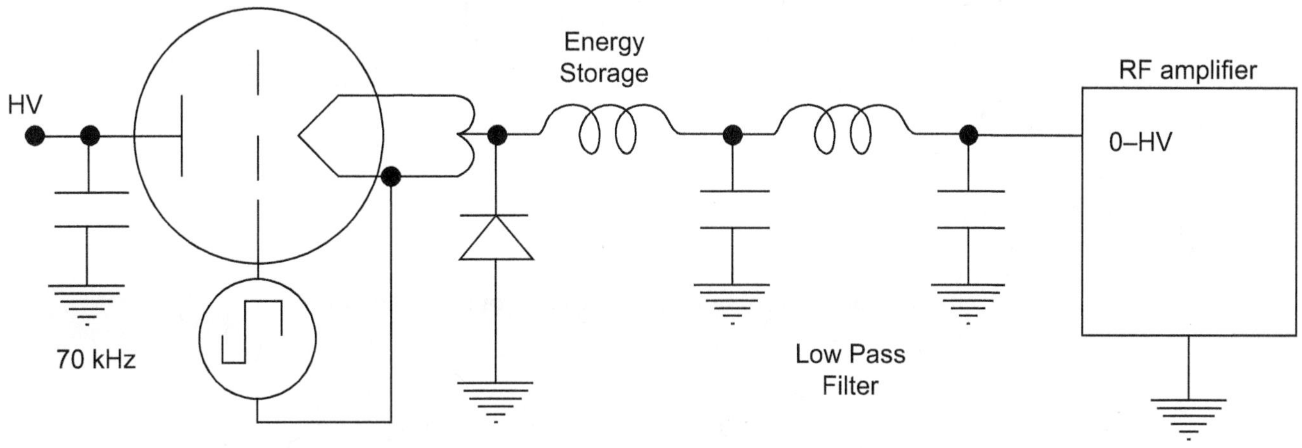

FIGURE 7.5-9 Diagram of a Pulse Duration Modulator (PDM)

J. Stanley

square wave driving the switching component, which consists of a vacuum tube or transistor.

This switch charges an energy storage inductor with current. When the switch is open, the inductor draws current from ground by virtue of its property of tending to keep the current in itself constant. This diode supplied current supplements the current through the switch so that as the voltage drops, the available current rises. Thus, the step down regulator acts as a DC transformer reducing the voltage with very little loss of power.

The output voltage can be changed at any rate from DC up to about one-half of the switching frequency so setting of the carrier level becomes very easy. This opens the possibility of modulation dependent carrier level in which the carrier is adjusted to different levels as modulation changes. This is used as a cost-saving feature.

Slowing raising the transmitter power at turn on also provides a soft start, which is easier on both the transmitter and the power system. Many transmitters using pulse modulation methods use a switching frequency of 70 kHz as this keeps some of the spurs within the 75 kHz limit for stronger spurious emissions. The storage inductor is followed by a lowpass filter that removes switching components that are above the highest audio frequency to be transmitted. Note, however, that higher switching frequencies have become common in transmitters configured to run with digital technologies such as In-Band-On-Channel (IBOC) or Digital Radio Mondiale (DRM), in order to provide accurate reproduction of the digital components (more on this later in the chapter).

Several different arrangements for PDM have been used in addition to the one shown in Figure 7.5-9. These essentially differ in where the ground connection is made and whether the switching modulator tube or the RF amplifier tube floats above ground. Other systems, sometimes called pulse step modulation (PSM), use multiple power supplies in series that switch in and out as needed to follow the instantaneous plate voltage required to produce the AM envelope. In these systems, a controller will monitor the audio signal, plus a power control signal (including adjustment for line voltage variations) and switch amplifiers on or off as needed to attain the desired level of instantaneous RF output.

Further to this, many PWM/PDM transmitters use a sample of the carrier frequency to generate the PDM frequency. This sample is then fed through a divider, which divides the sample by a factor of N, where the value of N is determined to result in a signal that is equal to the desired PDM sampling frequency for the modulator section. By using the divider circuit described, all harmonics of the PDM, when mixed with the carrier, will result in a product of either zero, or multiples of the PDM frequency, which is inaudible. This prevents audio problems caused by beating of PDM harmonics with the carrier. Note that the Nyquist theorem dictates that the PDM frequency must be at least two times the highest frequency component in the audio signal—typically PDM frequencies will vary between 60 and 80 kHz for analog transmitters and between 140 and 180 kHz for transmitters capable of transmitting using HD Radio technology.

Referring to Figure 7.5-10, we can begin to understand the basics of how this signal is generated.

The PDM square wave generator circuit is an oscillator that produces a square wave output. The square wave output of the PDM square wave generator is applied to the ramp integrator circuit and converted to a linear ramp waveform (triangle wave) at the frequency of the PDM square wave generator.

This signal is fed into a differential amplifier that compares the linear ramp waveform from the ramp integrator with the audio signal and produces a rectangular waveform, at the PDM square wave generator's frequency, as the pulse-width modulation signal. Remember that the PDM frequency is much higher than the highest audio frequency component, so the input signal shown in Figure 7.5-11a as a reference for the audio signal will appear as a variable DC voltage in comparison. The on/off ratio of the resulting pulse duration modulation signal will ultimately determine

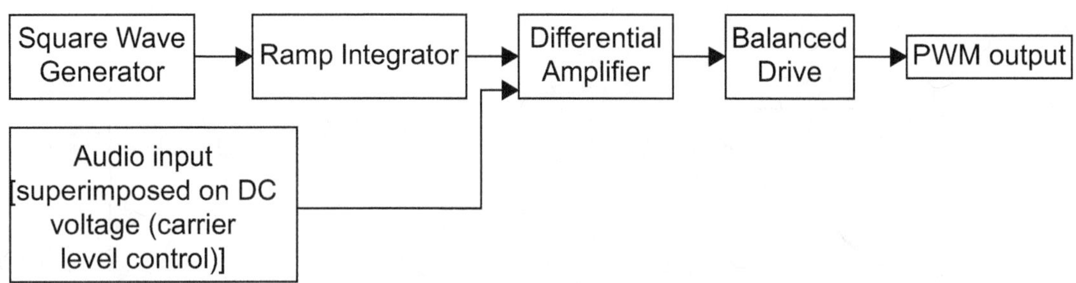

FIGURE 7.5-10 Simplified Block Diagram of PDM Circuit

J. Welton

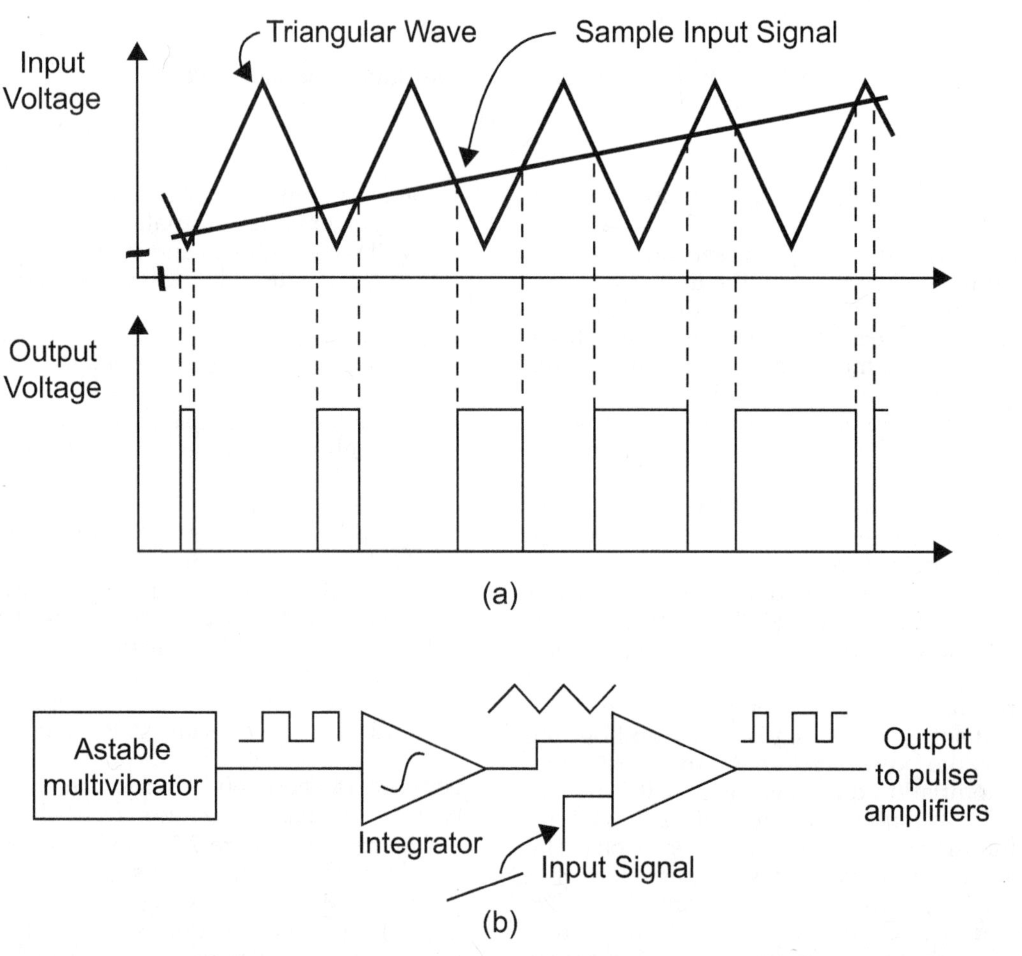

FIGURE 7.5-11 PDM Timing Diagram: (a) Waveform, (b) Sample Generator Circuit

J. Welton

the RF output level of the transmitter. It is also important to note that, in addition to providing a carrier-level control voltage to the audio input signal (for setting output power), many transmitters will add additional adjustments to compensate for changes in AC mains voltage or to stabilize output power under changing load conditions.

When there is no audio present, the output of the transmitter will be the desired unmodulated RF carrier output level, as determined by the DC level of the carrier level control voltage. When audio is present, the on/off ratio will vary in proportion to the amplitude of the audio component. In many transmitters, this signal is fed into multiple inverter/buffer circuits that take several instances of the PDM signal and phase shifts them, resulting in multiple PDM signals, which are at varying phases with reference to each other. These signals will provide the reference voltage B+Vdc for different amplifiers in the transmitter, reducing levels of PDM related spurs on the output and simplifying the amount of PDM frequency filtering required.

Typically, once the PDM signal is generated, it is fed through a Class D amplifier of some sort to provide an increase in amplitude, then the signal is fed through a lowpass filter that strips all of the PDM frequency component, leaving the original audio signal, varying between 0 and a maximum of the power supply voltage, as well as incorporating any correction voltage that was added for line voltage compensation or output stability adjustment. Some transmitters will use a Class D amplifier as a modulator to provide the B+Vdc for each RF amplifier; some will use one modulator for several amplifiers.

As FPGAs (forward programmable gate arrays) and DSPs (digital signal processors) become more powerful, it is becoming more common to see the exciter replaced with a single board that directly generates both a digital RF drive signal for the power amplifiers and multiple interphase PDM signals for the modulators, tied together by the processor's clock frequency for stability and reduced intermodulation between the PDM frequency and carrier frequency.

Pulse Step Modulation

Another type of modulation system that lends itself well to solid state, with multiple amplifiers contributing to the final combined RF output, is pulse step modulation. In this system, several amplifiers are connected in series, and then a controller turns them on as required, based on the modulating signal. So, for maximum power and a modulation peak, all amplifiers would be turned on, for minimum power, or for 100% negative modulation, all amplifiers would be turned off. This modulation system has a benefit of not having the requirement to filter a PDM frequency and less chance of PDM related spurious components, but can be subject to higher levels of thermal cycling of individual components for the top level amplifiers, which are not always on, which can potentially lead to higher failure rates if not designed well.

In addition, pulse step modulation and pulse width modulation systems can be combined—in this type of system, control algorithms are produced so that modules that are not being driven by the PSM controller are driven with higher duty cycle signals from the PWM controller. This has dual advantages—it can reduce the amount of filtering required for a PWM-only signal, and also provides much more constant on cycles for each amplifier, reducing the amount of thermal cycling. However, it does have the drawback of requiring a much more complex controller—with the advent of DSP technology, this is much less an issue than it would have been even 5–10 years ago.

CLASS D MOSFET AMPLIFIERS

In the early 1970s, solid state devices had become commonplace enough that they were starting to make inroads into broadcast transmitters, albeit mostly at lower power levels (1 kW or less). These transmitters primarily used linear amplifiers, biased for Class AB or Class B operation and due to cost and complexity were of very limited distribution.

In the late 1970s to early 1980s, it was discovered that using MOSFETs (metal-oxide semiconductor field-effect transistor) to create Class D amplifiers could result in a much more efficient amplifier and combining techniques were devised to allow multiple amplifiers to operate in parallel, producing higher output power capability with much better efficiency and stability.

Essentially, a MOSFET can be represented by a switch—as a control signal is applied to the gate, the drain and source will switch between a very low impedance and a high impedance.

In Figure 7.5-13, if the switch is opened and closed with a 50% duty cycle, a square wave at the switching

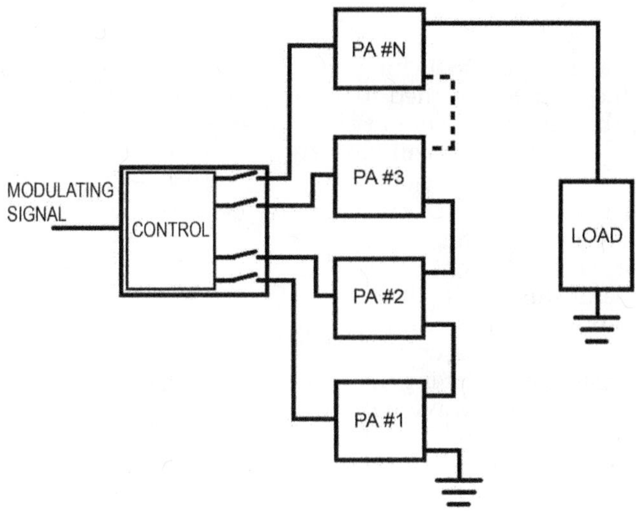

FIGURE 7.5-12 Simple Pulse Step

J. Welton

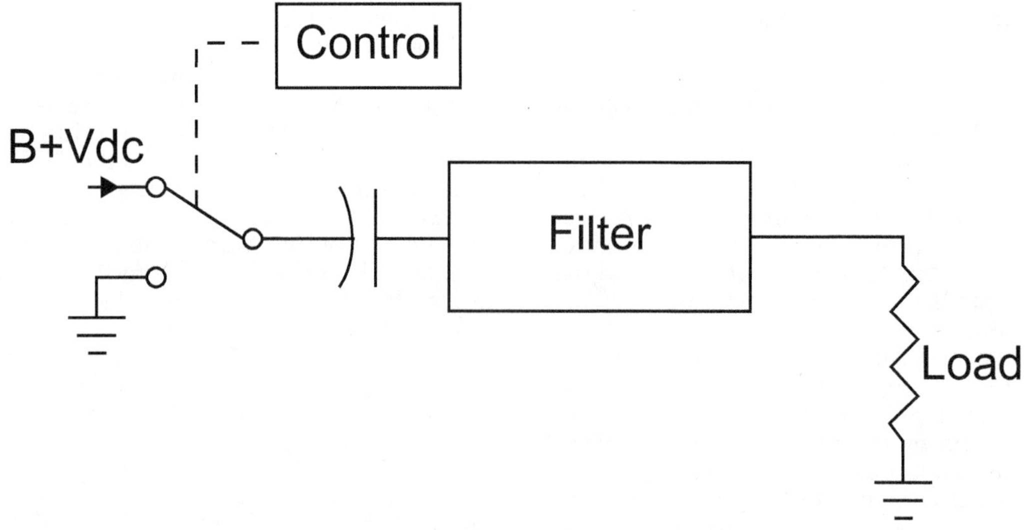

FIGURE 7.5–13 Simple Class D Operation

J. Welton

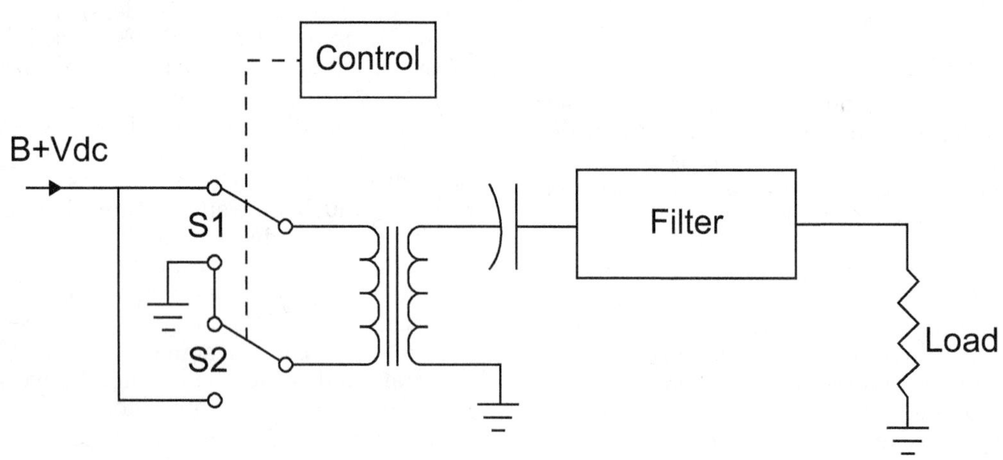

FIGURE 7.5-14 Push-Pull Class D Operation

J. Welton

frequency will result at the filter input, switching between 0Vdc and B+Vdc. If the filter is designed to pass the switching frequency, but attenuate its harmonics (such as a lowpass filter), a sine wave will be applied to the load.

Figure 7.5-14 shows a similar situation, with push pull operation. If S1 and S2 are opened and closed with a 50% duty cycle, a square wave of current, at the switching frequency and double the amplitude of the power supply voltage, will be applied to the primary of the transformer and transformed to its secondary. Again, if the filter is designed to pass the switching frequency, but attenuate its harmonics, a sine wave will be applied to the load.

Power MOSFETs can be used to replace the switches as depicted in the simple class D operation (Figure 7.5-13) and push pull class D operation (Figure 7.5-14) examples. It will be noted the switch contacts have been replaced by the phase oriented

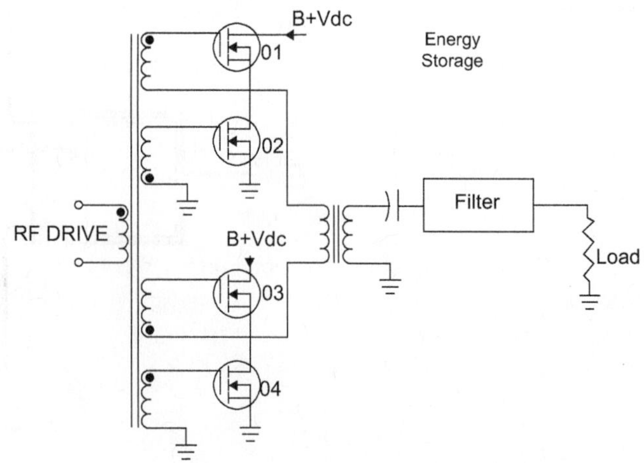

FIGURE 7.5-15 Power MOSFET

J. Welton

secondaries of an RF drive transformer. Q1 corresponds to S1-A, Q2 to S1-B, Q3 to S2-B and Q4 to S2A as depicted in the push pull class D operation example (Figure 7.5-14). Q1 and Q4 turn on/off together and Q2 and Q3 turn on/off together. In this configuration, a positive voltage at the RF drive input will produce positive voltages at the gates of Q1 and Q4, providing a path from B+Vdc to ground through the drain/source junctions of those transistors and a positive pulse rising from ground to the B+Vdc level at the secondary of the output transformer. A negative voltage at the RF drive input will turn off Q1 and Q4, as well as producing a positive voltage at the gates of Q2 and Q3, resulting in a negative pulse from ground to the B+Vdc level at the secondary of the input transformer.

This type of modified H-bridge configuration forms the basis of many of the solid state amplifiers in use for AM broadcast transmitters today. Transorbs (back-to-back Zener diodes) are frequently placed across the RF drive transformer secondary windings to protect against switching transients and hot carrier diodes across the drain/source junction of each MOSFET, to maintain current flow while the field in the primary of the output transformer is collapsing and to protect the MOSFETs when their loading is mistuned and highly reactive. With this type of circuit, the output that is produced is a sine wave, at the RF drive frequency, which is amplified to 1.414 times the level of the supply voltage (B+Vdc). By varying the supply voltage between 0V and the limit of the B+Vdc source, an amplitude modulated waveform can be created.

COMBINING

The RF combiner/filter section of a transmitter consists of two modules, the combiner and the RF output filter. The purpose of this section is to combine the outputs from the RF power modules, filter out undesired harmonic frequencies and impedance match the output of the transmitter to 50 ohms. Where many tube transmitters had loading and matching controls that could be adjusted to match the transmitter output to the load impedance over a very wide range of load impedances, the majority of solid state transmitters are factory tuned to be matched into 50 ohm, j0 loads. While many solid state transmitters do still contain user adjustable load-matching devices, they are effective over a narrower range and may not cover the variety of load impedances that tube amplifiers could be set to work into.

The combiner section takes multiple inputs (the outputs of the RF power modules) and combines them to a common point that will be the input to the RF output filter. This combining is done so that there is no phase difference between the individual signals at the common point leading to the filter, and the impedance at the output of the combiner matches the impedance at the input of the filter.

There are many ways that this combining can be achieved. Earlier solid state transmitters used simple series combining, where the amplifier output transformers were simply connected in series, forming a complex primary winding to a transformer with a single secondary that became the combined output. Transmitters of this type are still quite common and it is important to note that an amplifier module *cannot* be removed from a series-combined transmitter while the RF is on, unless the transmitter employs some sort of bypass circuitry to provide current continuity for the remaining amplifiers. Removing an amplifier from a series-combined transmitter while it is in operation will present an extremely high voltage at the open circuit that will result once the module connection is broken and will usually cause significant damage to the combiner circuit as well as to the remaining power amplifiers.

Other transmitters have used Wilkinson or other 90-degree combiners. These provide the benefit of providing a way of switching in a bypass to present a load to the remaining amplifiers in the form of the 90-degree isolation transformer. The downsides were complexity, in that they require reject loads, and inefficiency if the source amplifiers weren't well matched.

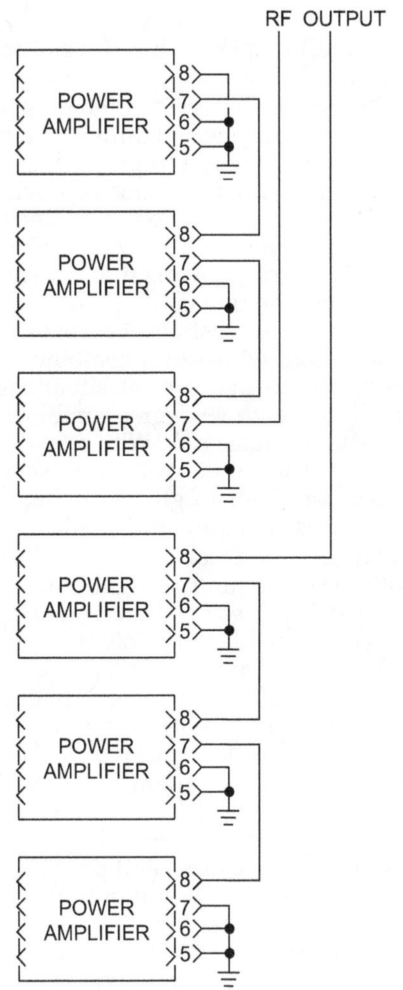

FIGURE 7.5-16 Series Combiner

Nautel Limited

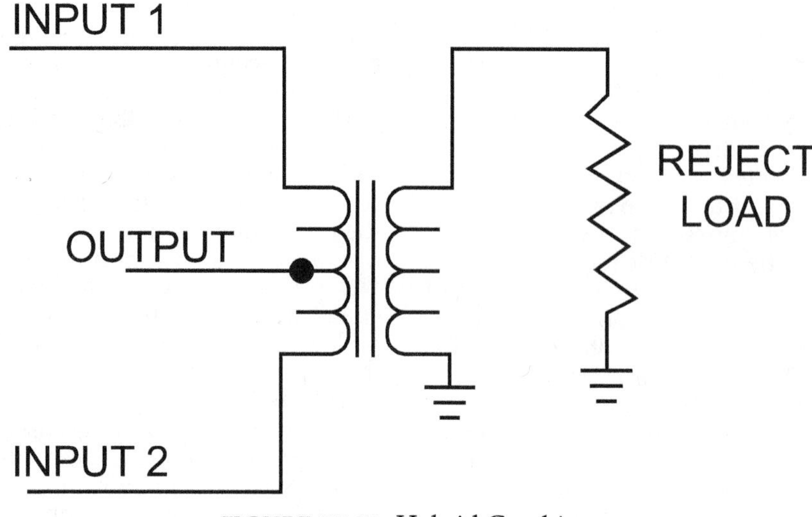

FIGURE 7.5-17 Hybrid Combiner

J. Welton

However, they do provide a degree of isolation that made it possible to work with a transmitter while it was still on air.

As well, there are 60-degree combiners, where the power amplifiers are connected through a series inductor that is nominally wound to provide 60-degrees of phase shift. In this manner, isolation is provided so that switching the output of a failed amplifier to ground provides a load to the remaining amplifiers through the 60-degree inductor, allowing the failed amplifier to be removed while the transmitter is on air. It doesn't require reject loads, so cost and complexity are minimized. In addition, with series combiners a shorted device could lead to overcurrent situations creating additional failures; with 90-degree combiners an open circuit results in a similar risk. With a 60-degree combiner, assuming that the mechanism to short the output of a failed amplifier to ground is functional and used properly, these risks are mitigated.

The point of this discussion is that it is important to be familiar with the combining configuration of any solid state transmitter being worked on, in order to ensure that the proper procedures are followed for that particular configuration. Do not ever assume that, simply because there are multiple amplifiers in an easy-to-remove configuration, an amplifier can be removed while the transmitter is producing output power, at any level.

Finally and no less important than any preceding part of this chapter, there is the RF output filter. The filter removes unwanted harmonics from its RF input (from the RF power modules), and provides the transmitter's final RF output. The RF output filter is typically a multi-pole, lowpass filter with a trap tuned to either two or three times the carrier frequency. Tube transmitters and linear amplifiers will tend to produce higher even-order harmonics, whereas Class D (switching) amplifiers will tend to produce higher odd-order harmonics, so the harmonic notch will be determined by transmitter configuration. As a rule, the RF output filter has a nominal input impedance of whatever is needed to match the final amplifier impedance, and an output impedance of 50 ohms (for solid state transmitters, the input impedance, after the combiner, can typically be estimated as total amplifiers divided by 50 ohms, although this does not account for any leakage inductance or stray capacitance due to wiring).

Most RF output filters will contain probes for monitoring voltage and current—these will be used to provide metering for forward and reflected power, as well as for various protection circuits (overcurrent and the like). Typically, maintenance of an RF output filter is limited to regular cleaning and checking integrity of connections, as well as inspecting for any damage to wire insulation or standoff insulators. One thing that is important to note, especially when changing frequency, or when performing repairs in the event of component failures in the filter, is that it is necessary to remember that the cabinet that encloses the filter is part of the circuit. Circulating currents in the filter cabinet will result in changes to tuning when access panels are removed or installed, so it is important to make an adjustment, then reinstall access panels before taking readings of filter impedance.

When interfacing with external equipment (phasors and antenna tuning units [ATUs], or antenna systems in general), two things are crucial. Even for monaural AM broadcast, but especially for proper AM stereo or HD Radio technology implementation, having symmetrical sidebands at the RF amplifier output is critical for best quality of transmission. Virtually every transmitter made in the last two decades can either be measured for phase shift through the combiner/filter network, or factory test data can be obtained from the manufacturer. Second, impedance bandwidth is extremely important for proper implementation of HD Radio technology—having a relatively flat

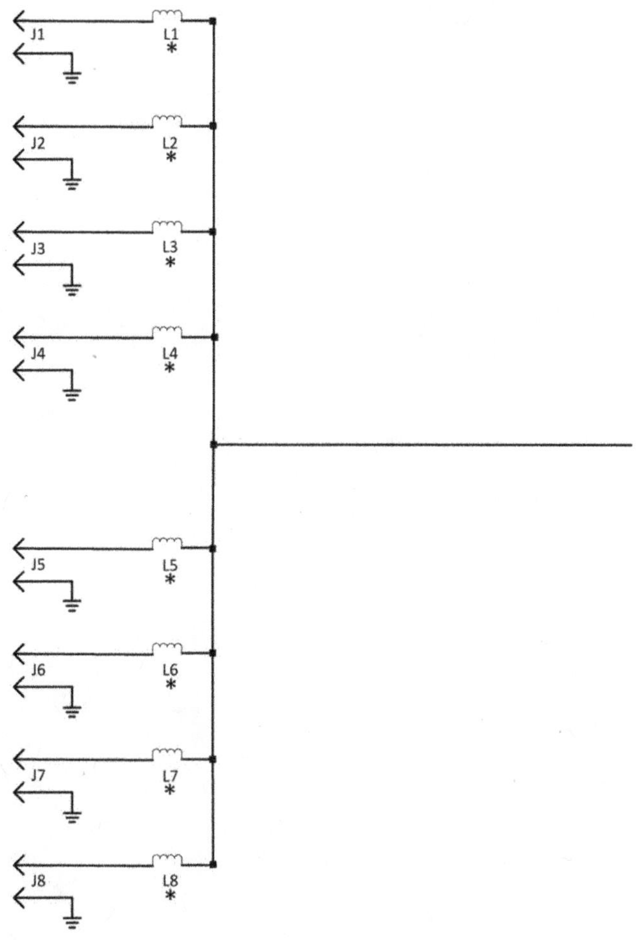

FIGURE 7.5-18 60-Degree Combiner

Nautel Limited

response out to 15 kHz is not always easily obtainable, but necessary for peak performance. That's beyond the scope of this chapter, but does bear mentioning. From the transmitter perspective, focusing on phase shift through the final filter/combiner is something that simply must be done. While this is definitely most critical for IBOC/HD Radio technology, it will also result in performance improvements for analog broadcasts, especially those in AM stereo, where transmitted bandwidth tends to be higher than for a basic monaural signal. Consulting the manufacturer for either factory data on phase rotation through the filter stage, or on best methods for measurement, is always the best starting point. With poor phase rotation, or narrow pattern bandwidth, symptoms can range from poor coverage to receivers not being able to lock on to the digital signal.

For a detailed description of IBOC refer to Chapter 7.6, "AM and FM IBOC Systems and Equipment."

For a detailed description of DRM, refer to Chapter 7.7, "Worldwide Digital Radio Systems."

CHAPTER

7.6

AM and FM IBOC Systems and Equipment

JEFF DETWEILER
DTS Inc.
Columbia, MD

INTRODUCTION

Enhanced performance, increased reliability, flexibility and cost effectiveness are the principal motivations for industries transitioning from analog to digital technology. For many countries including the United States, insufficient spectrum exists to accommodate the transition of radio from analog to digital.

HD Radio, a digital radio broadcast technology, developed by iBiquity Digital Corporation and now owned by DTS Inc. (DTS), operates within the Federal Communications Commission (FCC)–allocated amplitude modulation (AM) and frequency modulation (FM) bandwidths. DTS technology operates within the AM and FM bands (In-Band) while making use of existing broadcaster analog channel assignments (On-Channel), thus establishing the In-Band, On-Channel (IBOC) digital radio broadcasting (DRB) system.

The DTS HD Radio brand of IBOC technology allows broadcasters to add digital signals to their existing analog broadcasts, providing the capability for an eventual transition to digital-only transmission. Two modes define how audio and data are transmitted in the HD Radio system: "hybrid" and "all-digital." The hybrid mode includes both existing analog and new digital services. Broadcasters are using this mode during rollout of IBOC digital radio technology to permit the continued operation of analog-only radio receivers. In the future, when the marketplace is sufficiently saturated with HD Radio receivers, broadcasters can switch to the all-digital mode. Hybrid receivers employing HD Radio technology are backward and forward compatible, allowing them to receive current analog broadcasts in addition to all-digital broadcasts.

Conversion to Digital Technologies

HD Radio was designed to provide a unique opportunity for broadcasters and consumers to convert from analog to digital broadcasting without service disruption, while maintaining the current dial positions of existing radio stations. Consumers who purchase digital radios will receive AM and FM stations broadcasting with digital signals with digital sound quality. Additionally, consumers will have the capability to receive new multicast audio channels (for FM HD Radio) and wireless data services rendered on a radio's display screens, similar to those offered on Internet-connected devices like smartphones. Program service data (PSD) such as artist, title, program content, news, sports, local traffic and weather are available today using HD Radio technology.

National Radio Systems Committee Test Reports

DTS and its predecessor companies, iBiquity Digital (iBiquity), USA Digital Radio (USADR) and Lucent Digital Radio (LDR), have actively supported the National Radio Systems Committee (NRSC) and its evaluation of IBOC technology. The NRSC has provided a forum for all parties interested in IBOC to evaluate the HD Radio system and to provide industry input to the FCC.

In 1999, the NRSC developed its first set of IBOC test guidelines, which were designed for testing prototype IBOC systems. On December 15, 1999, USA Digital Radio filed its Report on Laboratory and Field Testing with the NRSC. On January 24, 2000, Lucent

Digital Radio filed its Test Report, which demonstrated system performance using the PAC audio CODEC technology. These reports confirmed the ability of IBOC technology to provide upgraded audio quality and robustness in both the AM and FM bands without causing harmful interference to existing analog broadcasts in those bands.

During 2000 and early 2001, the NRSC continued its analysis of IBOC technology. In December 2000, the NRSC adopted new FM laboratory and field test procedures to be used for final IBOC validation testing. In April 2001, the NRSC adopted new laboratory and field test procedures for AM IBOC testing.

During 2001, iBiquity completed an extensive test program of its FM and AM technology. These tests followed the NRSC test procedures and were conducted using independent laboratories in Alexandria, Virginia, Cincinnati, Ohio, and Austin, Texas. In addition, field tests were conducted using seven commercial and one experimental FM station and three commercial and one experimental AM station. NRSC observers actively participated in and oversaw these tests. The tests were designed to assess both the AM and FM HD Radio system performance in a variety of interference and impairment scenarios typically found in the real world. These tests also studied the potential impact of the IBOC system on analog operations of the host station as well as co- and adjacent channel analog stations. The tests involved the use of analog receivers representing OEM auto, aftermarket auto, home and portable receivers. Simultaneous tests were conducted on the digital and analog receivers to allow for direct comparison of digital and analog broadcasting. In addition to objective measurements obtained in the lab and field, these tests also involved an extensive subjective evaluation of digital and analog audio samples. In the case of FM, more than 480 listeners subjectively evaluated thousands of audio samples.

Additional information about the NRSC and its evaluation of IBOC technology can be found at the NRSC website, www.nrscstandards.org.

FCC Approval Process

USADR initiated the FCC's consideration of IBOC by filing a Petition for Rulemaking on October 7, 1998. After a series of comments supporting the USA Digital Radio Petition, on November 1, 1999, the FCC issued a Notice of Proposed Rulemaking (NPRM, Proceeding 99–325) on terrestrial digital radio broadcasting. USADR filed both Comments and Reply Comments on the NPRM. The FCC's NPRM can be viewed at https://apps.fcc.gov/edocs_public/attachmatch/FCC-99-327A1.pdf.

On July 2, 2001, the National Association of Broadcasters (NAB) filed with the FCC its Radio Board's resolution endorsing the DTS HD Radio system and encouraging the FCC to take steps to enable a fast rollout of the technology.

On November 29, 2001, the NRSC completed its evaluation of iBiquity's FM HD Radio system and recommended that the FCC authorize the technology as an enhancement to the current analog FM broadcasting system in the United States. The NRSC's evaluation was based on DTS' extensive field and laboratory tests of the FM IBOC system conducted in accordance with the NRSC's test procedures. The NRSC completed its evaluation of its AM IBOC system in April 2002.

On December 19, 2001, the FCC issued a Public Notice seeking comments on the evaluation of iBiquity's FM system by the NRSC. Supporting comments endorsing IBOC were filed by all components of the radio industry. The comments and reply comments overwhelmingly supported FCC adoption of IBOC as the digital broadcasting standard for the United States.

On April 19, 2002, the FCC issued a Public Notice seeking comment on the NRSC's endorsement of iBiquity's AM IBOC system. Comments were due June 18, 2002.

Domestic Regulatory Status

On October 10, 2002, the FCC issued a Report and Order that approved iBiquity's HD Radio technology for FM and AM daytime broadcasting. This approval allowed broadcasters to move forward with the implementation of HD Radio technology and gave receiver manufacturers enough confidence in the systems' marketability to begin development of HD Radio products.

Further Notice of Proposed Rulemaking

On April 15, 2004, the FCC adopted a Further Notice of Proposed Rulemaking and Notice of Inquiry on HD Radio technology. The Further Notice proposed to amend the existing rules for AM and FM radio to further the introduction of the HD Radio system. The Further Notice also sought comment on several new initiatives including the introduction of nighttime digital AM broadcasting and multicasting of the digital service to allow for the introduction of multicast (supplemental audio) and datacasting services.

AM Nighttime Transmission

When the FCC approved HD Radio technology, it held off on its approval of AM nighttime transmission and requested that additional tests be conducted. iBiquity subsequently carried out additional tests under the oversight of the NRSC and submitted the results for the NRSC's review. Based on the test results and the NRSC's review, on March 5, 2004, the NAB submitted an endorsement of IBOC AM nighttime transmission to the FCC, asking for approval of this service. In their letter, the NAB wrote that the dramatically improved audio quality from IBOC service is well worth the predicted and limited reductions in analog coverage. In support of the NAB's endorsement, iBiquity submitted a technical report to the FCC on March 5, 2004, consisting of an AM nighttime compatibility report and two field studies of HD Radio's AM nighttime performance. On April 14, the FCC issued a Public Notice seeking comment on the NAB's endorsement. Comments were due June 14 and reply comments due July 14. The NAB's endorsement and the AM nighttime reports are available on the FCC's website at www.fcc.gov/ecfs.

Second Report and Order, First Order on Reconsideration, and Second Further Notice of Proposed Rulemaking

On March 22, 2007, the FCC issued a Second Report and Order, First Order on Reconsideration, and Second Further Notice of Proposed Rulemaking that sought to establish the goals of adopting service rules and other requirements for terrestrial digital radio. However, they also found it necessary to ask additional questions, in particular on how to preserve free over-the-air radio broadcasting while permitting licensees to offer new services on a subscription basis. The FCC also resolved and disposed of several petitions for reconsideration that were filed in response to the First Report and Order.

In summary, the Commission, in the Second Report and Order, First Order on Reconsideration, and Second Further Notice of Proposed Rulemaking:

- Refrained from imposing a mandatory conversion schedule for radio stations to commence digital broadcast operations;
- Allowed FM radio stations to operate in the extended hybrid digital mode;
- Required that each local radio station broadcasting in digital mode provide a free over-the-air digital signal at least comparable in audio quality to its analog signal;
- Continued to require that the main digital broadcast stream simulcast the material aired on the analog signal;
- Adopted a flexible bandwidth policy permitting a radio station to transmit high-quality audio, multiple program streams, and datacasting services at its discretion;
- Allowed radio stations to time broker unused digital bandwidth to third parties, subject to certain regulatory requirements;
- Applied existing programming and operational statutory and regulatory requirements to all free digital radio programming streams, but deferred the issue of whether and how to apply any specific new public interest requirements;
- Authorized AM nighttime operations and FM dual antenna configurations;
- Considered and addressed other technical matters, such as FM translator and booster operations and TV Channel 6 interference issues;
- Deferred discussion of whether the Commission should impose content control requirements that would prevent listeners from archiving and redistributing digital musical recordings transmitted by digital broadcast stations;
- Recognized that further negotiations between the United States and the international community are taking place to resolve possible disputes about the implementation and operation of digital services by domestic radio stations;
- Dismissed several pending Petitions for Reconsideration and Petitions for Rulemaking that asked the Commission to reconsider the adoption of iBiquity's IBOC system as the technology chosen for digital radio transmission;
- Sought further comment on appropriate limits to the amount of subscription services that may be offered by radio stations.

IBOC Compatibility and AM Modulation Dependent Carrier Level

Modulation-Dependent Carrier Level (MDCL) is a technique whereby AM broadcast signals reduce the level of the analog carrier as a function of the analog modulation index in order to reduce power consumption in an AM transmission. As part of the AM Revitalization Order released on October 23, 2015, the FCC amended Section 73.1560 of the Commission's rules to "permit AM stations to commence operation using Modulation Dependent Carrier Level (MDCL) control technology without prior Commission authority, provided that within 10 days after commencement of MDCL operation, the AM station licensee submits a completed FCC Form 338, AM Station MDCL Notification.

Two MDCL methods are presently employed, dynamic carrier systems (DCS) and adaptive modulation companding (AMC). DCS reduces the analog carrier power dynamically, but leaves the IBOC sideband power unaffected. This function reduces analog carrier power at low modulation levels, and return the carrier to full power as modulation is increased. AMC systems decreases the level of both the analog carrier and IBOC sidebands together as the modulation is increased. Maximum signal compression can vary from 1 to 6 dB, with 3 dB commonly used. As part of the AM Revitalization Order, the FCC issued a general waiver for any AM stations wishing to implement MDCL. This includes stations broadcasting hybrid AM HD Radio signals, provided the signal still lies within the spectral mask and the ratio of analog and digital levels remains constant.

Concerned that dynamic shifts in both analog carrier and digital sideband power might have a deleterious effect on reception, the NRSC formed a working group to study the potential impact of MDCL on HD Radio signals. In January of 2016, NRSC-G101-A AM Modulation-Dependent Carrier Level (MDCL) Usage Guideline was released with the results of the working group study. The results of the testing and analysis may be summarized as follows:

> For static AMC MDCL as tested, the impact of compression on hybrid AM HD Radio (both Core and Enhanced modes) digital point of failure (POF) is small and may be approximated as negligible. For dynamic AMC MDCL, Core and

Enhanced mode POF are degraded slightly: less than 1 dB on average in POF level for 1 dB of AMC compression, and are degraded slightly less than 3 dB in POF level for 3 dB of AMC compression.

Use of Separate Antennas for Analog and Digital Signals

On March 17, 2004, the FCC approved the use of separate analog and digital station antennas to initiate HD Radio FM transmissions provided broadcasters apply for an STA (Special Temporary Authority). This separate antenna option will enable many FM stations to use existing equipment and reduce HD Radio implementation costs. The FCC's decision was based on the recommendation and report filed by the NAB on July 24, 2003. This report included extensive field tests conducted by iBiquity that were specified and witnessed by the NRSC. Effective with the release of the FCC's Public Notice, FM stations may file requests for STAs to begin HD Radio transmission using dual antenna systems provided they meet the following criteria:

- Digital transmission must use an auxiliary antenna;
- Auxiliary antenna must be within three seconds of latitude and longitude of the main antenna;
- Height above average terrain (HAAT) of the auxiliary antennas must be between 70% and 100% of the height above average terrain of the main antenna.

FM Digital Power Increase

On October 23, 2008, the FCC Media Bureau released a Public Notice inviting comments on a request filed on June 10, 2008, asking the Commission to modify the maximum integrated power for FM digital audio broadcasting. The request was filed by a group of 18 broadcasters and 4 manufacturers of broadcast transmission equipment, who collectively identified themselves as the "Joint Parties." The Joint Parties requested that the Commission increase the maximum permissible digital operating power of FM stations from the current level of 1% of a station's authorized analog power (–20 dBc) to a maximum of 10% of a station's authorized analog power (–10 dBc). Filed concurrently in support of the Joint Parties' request was a report prepared by iBiquity. On July 18, 2008, National Public Radio (NPR) submitted a Corporation for Public Broadcasting (CPB)–funded study on digital radio coverage and interference. The October 23 Public Notice sought comment on the Joint Parties' Request and these related technical studies.

On January 27, 2010, the FCC Media Bureau adopted an Order establishing new notification and application procedures to permit FM stations to voluntarily increase hybrid digital effective radiated power (ERP) levels to a maximum of 10% of authorized analog FM ERP. The Order also announced new interference mitigation and remediation procedures to resolve promptly allegations of digital interference to analog FM signals. The Order became effective on May 10, 2010. Subsequently, on May 17, 2010, in Public Notice DA 10–866 titled "Guidance Provided to FM Licensees Regarding Operations with Increased Digital Power," the Bureau provided additional guidance on the procedures that licensees should follow when implementing digital power increases.

FM Asymmetrical Sideband Operation

On October 4, 2011, iBiquity and NPR met with the Media Bureau to discuss the possibility of permitting FM stations to operate with unequal (asymmetrical) digital sideband power levels. Asymmetric sideband technology may be employed to independently adjust sideband levels in order to maximize sideband power for a given protection ratio. Concurrent with the submission of the meeting notice, iBiquity filed a technical report that discusses the field performance of asymmetric digital sideband operation by FM stations. Based on the submitted report, iBiquity and NPR requested that the Commission authorize voluntary asymmetric digital sideband power operation for FM stations. In a separate filing on October 24, 2011, NPR filed a Report that describes the results of field testing of asymmetric FM digital sidebands used in conjunction with testing of a new algorithm for reducing the peak-to-average power ratio in digital transmitters. The NPR Report concluded "broadcasters are best off maintaining symmetrical sideband levels, but coverage improvements are possible with an increase (over –20 dBc) of only one sideband."

On November 1, 2011, the Media Bureau solicited public comment on "FM Asymmetric Sideband Operation and Associated Technical Studies" in the digital audio broadcasting proceeding, MM Docket No. 99–325. The comment period closed on January 3, 2012, and as of the writing of this chapter (fall 2016) this proposal awaits FCC action. Until such time as the FCC acts on this matter, stations must request experimental test authority from the FCC in order to invoke asymmetric sideband operation.

International Developments

DTS is actively pursuing the adoption of its IBOC technology outside of the United States. In April 2001, the International Telecommunication Union (ITU) adopted Recommendation ITU-R BS.1514, which includes HD Radio AM IBOC technology as a recommended system for digital broadcasting in the broadcasting bands below 30 MHz. Similarly, Recommendation ITU-R BS.1114–9 includes the HD Radio FM IBOC system (referred to by the ITU as Digital System C) as a recommended system for digital audio broadcasting above 30 MHz. Further information about DTS' activities at the ITU can be found in iBiquity's AM and FM reports submitted to the ITU in March 2002, including the All-Digital Testing of Digital System C, IBOC DSB System—Ability to Meet Service Requirements, and Laboratory and Field Testing of the IBOC Digital Sound Broadcasting System Operating Below 30 MHz.

DTS also works with other international organizations to pursue international recognition of IBOC technology. DTS regularly participates in the meetings of CITEL, the Inter-American Telecommunications Commission of the Organization of American States (OAS). CITEL's Permanent Consultative Committee II (PCC II) has received regular briefings on the development of DTS' system. DTS also provides regulatory and technical briefings to broadcasting organizations in the United States and several countries abroad.

Mexico has also selected IBOC technology to implement digital radio. Adoption of HD Radio technology was formally announced by the CoFeTel (now known as El Instituto Federal de Telecomunicaciones or IFT) on June 16, 2011.

In Canada, the Canadian Radio Television and Telecommunications Commission (CRTC) released a targeted policy review of the commercial radio sector, Broadcasting Regulatory Policy CRTC 2014–554, on October 28, 2014. Included in this review was a flexible approach for introducing HD Radio technology in Canada, designed to allow for innovation and experimentation. Under this approach, licensees are required to inform the Commission in writing of any experimentation with HD Radio (or other digital radio technologies) undertaken, including the type of service they intend to provide.

OVERVIEW OF HD RADIO IBOC CONCEPTS

In the hybrid mode where both the analog and digital signals coexist, upper and lower low-level digital sidebands with multiple carriers are added to the analog spectrum as shown in Figure 7.6–1 for FM and Figure 7.6-2 for AM. These carriers are modulated with redundant information to convey digital audio and data. Because the analog AM signal is amplitude modulated, a quadrature phase component can carry digital information in the same frequency space and thus it can be placed directly beneath or in quadrature to the analog modulation as shown in Figure 7.6-2. This additional information is transmitted at a lower power level to avoid increasing noise to the analog signal.

AM and FM Hybrid and All-Digital Spectra

In the FM all-digital mode, Figure 7.6-1, the analog signal is removed and additional data carriers are added. The main channel stereo audio and its associated data information is unchanged from the hybrid mode, however the power level may be increased to provide a more robust service. In the region between +100 kHz and +130 kHz, digital carriers are added to carry the digital audio backup, tuning channel and ancillary data. In the region bounded between +100 kHz and −100 kHz, which previously carried the analog audio, additional carriers are added to carry new supplemental services like multicast audio channels, wireless data, surround sound and more.

In the AM all-digital mode, Figure 7.6-2, the analog signal is removed and the primary digital carriers are moved to occupy the center of the AM channel.

IBOC TECHNOLOGY BUILDING BLOCKS

A digital radio transmission system comprises three main building blocks:

- Audio coder/decoder;
- Modulator;
- Protocol stack.

DTS' HD Radio modem technology utilizes patented and patent-pending digital signaling techniques:

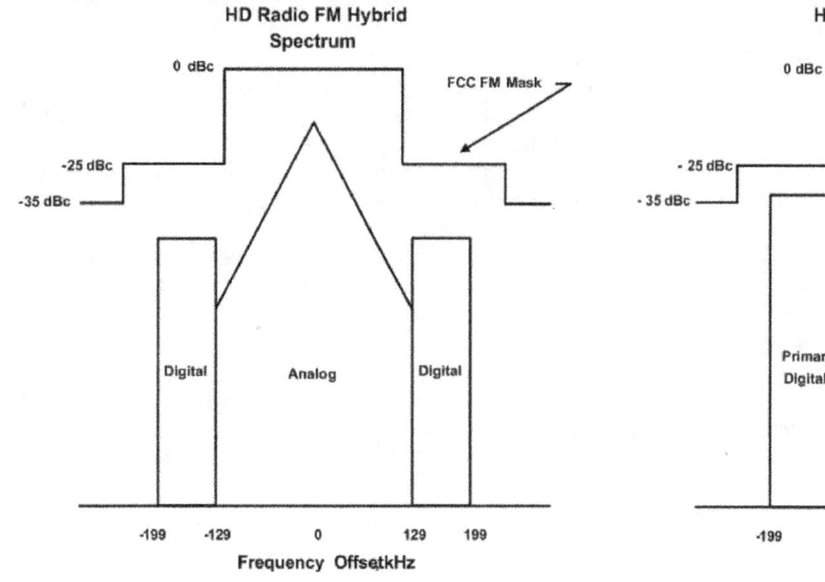

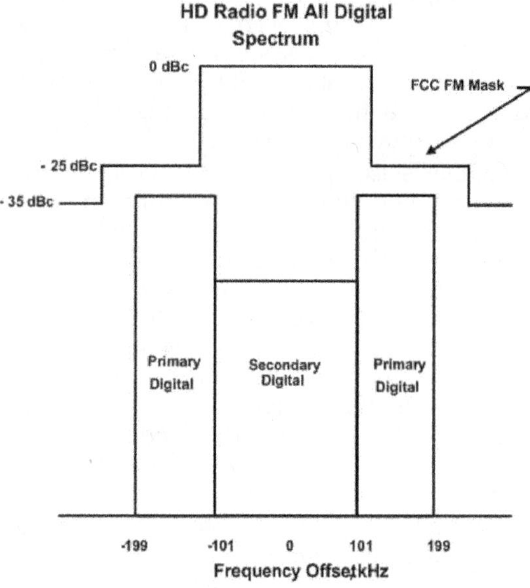

FIGURE 7.6-1 HD Radio hybrid and all-digital FM spectrum

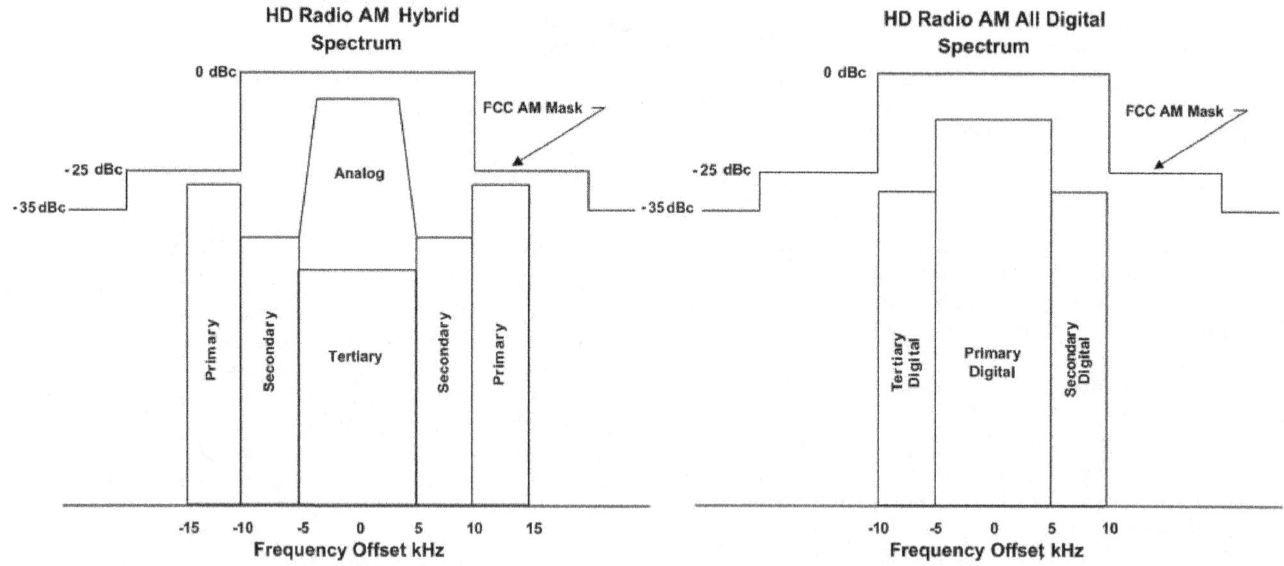

FIGURE 7.6-2 HD Radio hybrid and all-digital AM spectrum

quadrature phase ship keying (QPSK) modulation with orthogonal frequency division multiplexing (OFDM), interleaving, channel coding/error correction, time and frequency diversity and others.

Audio Coder/Decoder (CODEC)

HD Radio technology is designed to deliver compact disc (CD)–like audio quality within the FM broadcast band and FM-like stereo audio quality in the AM band. DTS' HD Radio technology, like other digital broadcast technology systems, does not have sufficient bandwidth to carry the full digital data stream delivered by a CD player, defined as 16 bits per channel sampled at 44.1 kHz (44,100 × 16 bits × 2 channels = 1,411,200 bits per second, or bps). Therefore some method of compression must be employed to reduce the audio data bit rate. The HD Radio systems uses a proprietary audio CODEC that is branded HDC and is specifically designed to reduce the number of bits required to transmit a given quality audio signal.

Audio CODECs use advanced signal processing and psycho-acoustic modeling to interpret human hearing and eliminate redundancies and irrelevancies in the audio signal. The HDC method employed in DTS' HD Radio technology for FM compresses the audio at a ratio of 15:1, enabling 96 kilobits per second (kbps) to deliver CD-like quality audio. The AM audio CODEC delivers FM-like audio quality with a compression ratio of approximately 40:1. HDC offers compatibility with broadcast multichannel audio modes, for example, 96 through 64 kbps for 5.1 channel audio configuration, built-in data channels and transport features.

DTS' HDC is uniquely optimized for use in a digital audio broadcast system. In particular, it incorporates powerful error concealment techniques for mitigating the effects of channel errors. These features allow improved broadcast system design that is better matched to the prevailing channel conditions and interference.

Modulator

A modulator is used to modulate the digital audio and data information onto an RF carrier. There are many signaling techniques used for digital modulation that vary in complexity and suitability to a given application. DTS' HD Radio technology employs OFDM to impress the data onto the RF channel as shown in Figure 7.6-3. OFDM signaling employs multiple, overlapping, orthogonal subcarriers. Trade-offs between throughput and robustness can be made with increasing number of phase states and signal levels, leading to more throughputs with less robustness in fading channels.

DTS' HD Radio technology uses OFDM techniques for robustness in the presence of multipath fading, interference and noise. Additionally, OFDM is flexible for adding optional subcarriers and allows the placement of the most sensitive digital information in more robust portions of the spectrum. DTS' HD Radio technology for FM modulates the OFDM subcarriers with phase shift keying (PSK) modulation. OFDM signals modulated with low-complexity PSK are quite robust in the presence of interference and multipath fading.

HD Radio on AM operates in very narrow channels that are free of multipath fading. The narrow bandwidth leads to an OFDM modulation technique that is optimized for higher throughput. DTS' HD Radio technology for AM uses quadrature amplitude modulation (QAM) on each subcarrier. In QAM systems, multiple phase and amplitude states are applied to each subcarrier in the OFDM waveform. Multiple phase and amplitude states allow greater data throughput in a

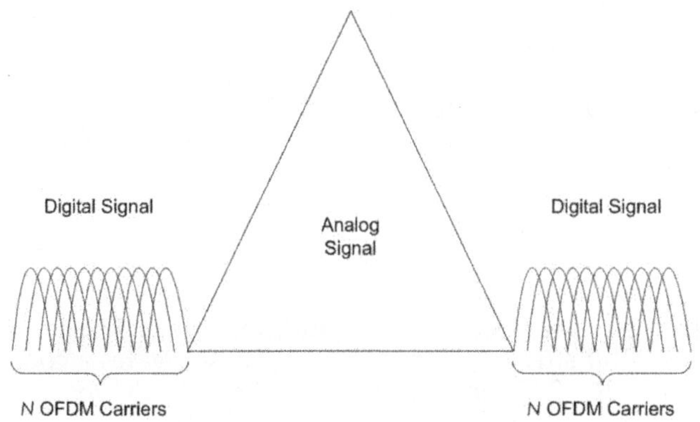

FIGURE 7.6-3 HD Radio spectrum depicting multiple overlapping OFDM carriers adjacent to FM signal

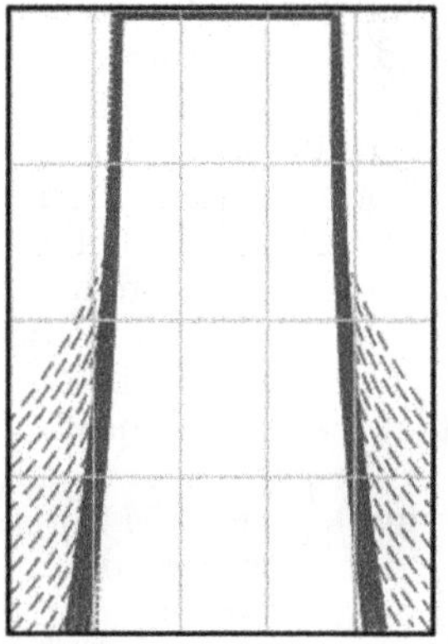

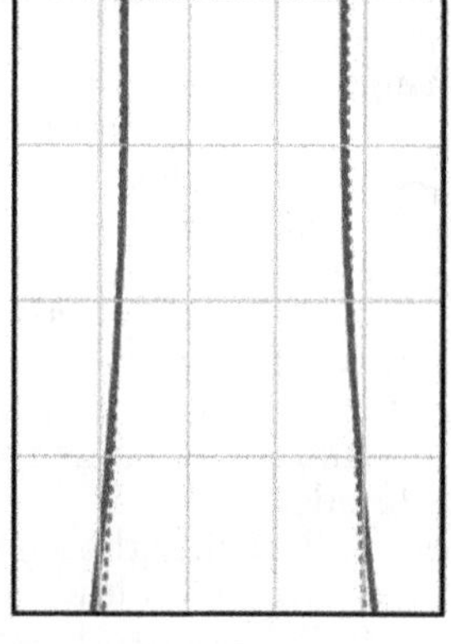

FIGURE 7.6-4 Side-lobe reduction by pulse Nyquist root-raised-cosine tapering

FIGURE 7.6-5 Example of interleaver and de-interleaver

given channel bandwidth. For AM systems, the timing of the symbols is optimized to ensure that duration of noise pulses is much shorter than the symbol duration of the QAM modulation, thus ensuring robust digital reception in the presence of static and noise that is prevalent in the AM channel.

DTS' HD Radio technology applies pulse Nyquist root-raised-cosine tapering pulse shaping to its OFDM waveform. Pulse shaping aids in acquisition and relaxes frequency-tracking requirements while reducing spectral side lobes. The root-raised-cosine function creates a guard band between the symbols that improves inter-symbol interference by reducing energy in the region where subcarriers overlap.

The benefit of side-lobe reduction and its effect on out-of-band attenuation is clearly demonstrated in Figure 7.6-4. The reduced side lobes result in less interference to adjacent channels, more robust performance in multipath and less interference to the host analog signal.

OFDM is a highly flexible modulation technique that is ideally suited for digital radio broadcasting systems. This flexibility allows DTS' HD Radio technology to provide a high quality digital audio broadcast, protect adjacent channels and provide robust performance in the AM and FM bands.

Interleaving

Digital error correction techniques are enhanced if errors in transmission are spread in a manner that minimizes data loss in successive bits. Interleaving is a technique that scrambles the bits in a predetermined manner upon transmission and reassembles them in the receiver. The length of the interleaver is normally optimized to spread errors over a longer period of time than would exist in a channel fade. Figure 7.6-5 shows an example of bit interleaving.

In the example, the middle three bits 1, 7 and 5 are successive in the transmission path (i.e., after interleaving and before de-interleaving). If these bits are corrupted in the channel, the receiver de-interleaving process will de-spread these bits and reassemble the bit stream to ensure no two consecutive bits are in error.

Digital systems typically employ interleavers to improve the performance of the system in the presence of channel impairments. Designers of digital systems must make trade-offs between interleaver length and system performance. Long interleaver lengths are desirable for providing robust reception at the expense of acquisition time. DTS' HD Radio technology for the hybrid mode employs interleavers that are optimized for performance while using a backup channel for rapid tuning. The all-digital mode adds a backup digital audio channel with a short interleaver for rapid tuning. In each mode, the main program service (MPS) has a rapid tuning channel provided as a backup signal service to ensure continuity of service in fading channel conditions.

Channel Coding/Error Correction

A digital bit stream, when passed through a transmission channel, is likely to encounter various forms of impairments including noise, multipath distortion, fading and interference. Digital systems employ various error correction techniques to correct transmission errors. These algorithms improve robustness through the introduction of error correction bits. These error correction bits are used in the receiver to verify the accuracy of the recovered bit stream, detect errors, and provide restoration of the transmitted bit stream.

Error correction codes are typically specified by their coding rate, R = information bits/total bits. For example, a coder with R = 1/2 channel coding represents an algorithm where half of the bits carry information and the other half carries the overhead of the error correction algorithm. Some error-correcting codes implement what is called forward error correction (FEC). With DTS' HD Radio technology, FEC and interleaving are closely tied together. In the presence of adjacent channel interference, the outer OFDM subcarriers are most vulnerable to corruption. The information, coding and interleaving are specially tailored to deal with this nonuniform interference such that the communication of information remains robust in the presence of such interference. DTS' HD Radio technology for FM transmits a full-quality digital audio signal on each digital sideband (upper or lower) of the analog carrier. Each of these digital sidebands is detected and decoded independently with an FEC coding gain achieved by a rate 4/5 convolutional coder on each sideband. This redundancy permits operation on one sideband while the other is corrupted. Using a patented technique known as Code Combining, the receiver combines the digital signals from both sidebands to provide additional signal power and coding gain. Similar techniques are employed in DTS' HD Radio technology for AM.

DTS' HD Radio technology makes use of a patented technology known as Convolutional Punctured Pair Codes (CPPC). CPPC techniques take advantage of the redundancies in the upper and lower sidebands by allowing the error correction in these sidebands to be combined to create a more powerful error correction algorithm. In order to effectively achieve coding gain when the pair of sidebands is combined, the code on each sideband consists of a subset of a larger (lower rate) code. Each subset has been designed through complementary puncturing of the lower rate code. A simple way of constructing a code for this application is to start with an R = 1/3 convolutional code. This code can be generated as shown in Figure 7.6-6.

The R = 1/3 convolutional encoder produces three encoded bit streams each at the same bit rate as the input. The combination of these three bit streams produces the R = 1/3 code output. To create the complementary code pair, for example, a subset of the output code bits is assigned to the lower digital sideband and a different subset is assigned to the upper digital sideband. Each subset must contain at least the same rate of bits as the information input rate, plus some additional bits to provide some coding gain. An R = 4/5 code on each sideband requires 25% additional bits. The combined bit streams of the two sidebands creates a code rate of R = 2/5. An additional feature of CPPC codes is the ability to combine the power in the two sidebands and recover a signal in a way that is 3 dB more robust than using either sideband by itself.

Error correction algorithms add to the number of bits that must be transmitted to reliably recover digital data after passing through an RF channel. This extra overhead is necessary to ensure reliable reception. DTS' HD Radio technology uses error correction algorithms to provide a robust digital broadcast while minimizing the number of additional bits needed to reliably convey the digital information.

Time Diversity Backup Channel and Blend

An effective method for dealing with channel fading in a mobile environment is to provide a second channel conveying the same information. Transmitting the information on the second channel shifted in time can enhance the total system performance when the two channels are recombined at the receiver. This technique is called "time diversity." DTS' HD Radio technology includes a time-diverse backup channel in all AM and FM modes for the main channel audio signal. If the diversity delay is sufficiently large such

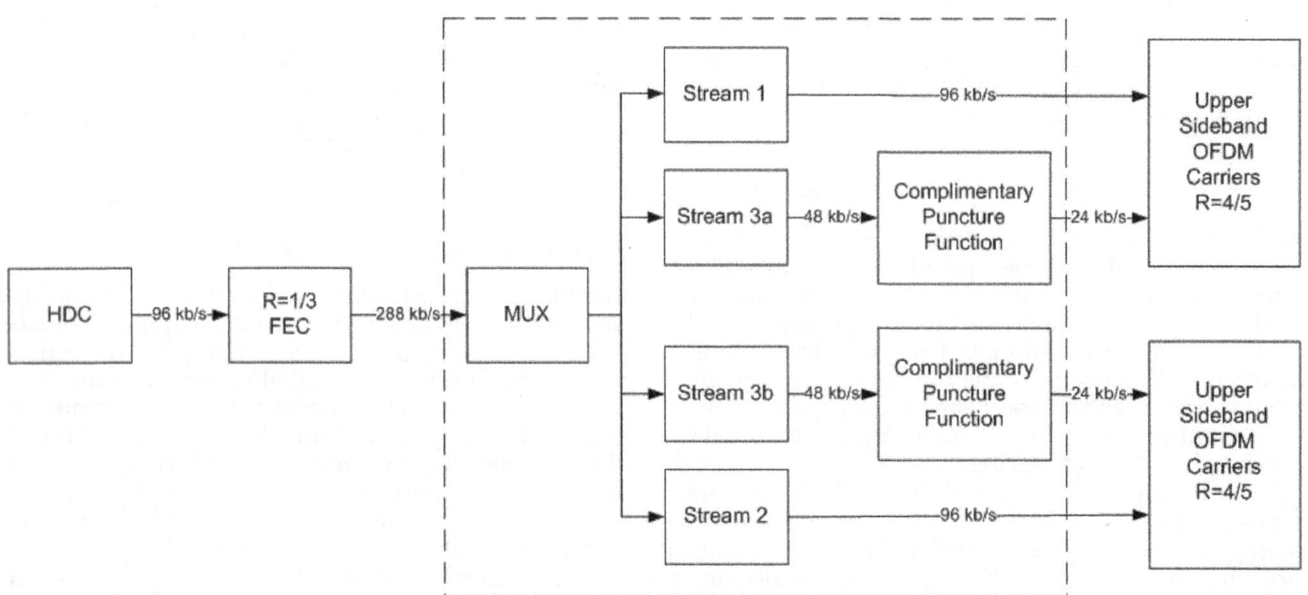

FIGURE 7.6-6 Example of an R = 1/3 convolutional encoder

that the transmission outages are independent, then the probability of an outage after diversity is the square of the probability of outage without diversity. For example, if the probability of an outage in either channel is 1.0%, then the probability of outage after diversity is 0.01%, a significant improvement. The autocorrelation function represents the probability of channel outage after diversity improvement as a function of time offset. An example autocorrelation function is shown in Figure 7.6-7, however, an actual autocorrelation function depends on distance from the station, terrain, propagation conditions, and interference.

DTS' HD Radio technology takes advantage of time diversity by delaying backup transmissions by approximately four seconds and realigning the digital and analog signals in the receiver. Figures 7.6-8 and 7.6-9 show the effect of the loss of information due to a signal blockage such as would be encountered when listening to AM in an automobile while traveling under an overpass. Figure 7.6-8 shows that the analog signal

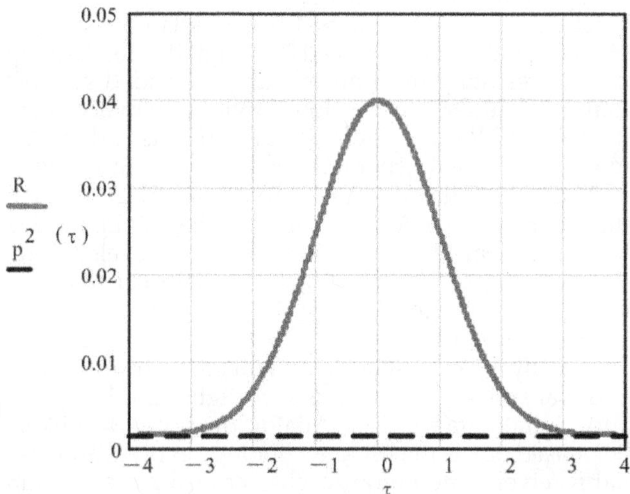

FIGURE 7.6-7 Autocorrelation function of channel loss due to blockage or severe impairment. The bell curve shows statistical distribution where p^2 = success probability, R = probability of outage, and t = time offset in seconds.

is delayed by a fixed amount from the digital signal at the transmission point. In a receiver with DTS' HD Radio technology, the digital reception delay is set to be identical to that of the analog transmission. These two time delays allow for a seamless blend to occur between the two transmissions.

The shaded blocks of Figure 7.6-9 depict what happens when a blockage (impairment) occurs. In this case, the channel outage affects segments 2 and 3 of the analog signal and 6 and 7 of the digital signal—effectively de-correlating the outages in the analog and digital paths. In this example, the receiver is decoding the digital signal in time segments 1 through 5 and then seamlessly blends to the analog signal to recover segments 6 and 7 and then blends back to the digital transmission for segments 8 and 9, thus maintaining continuity of programming.

In the future, when the analog transmission is discontinued and transition to an all-digital service occurs, DTS' HD Radio technology will continue to provide a time-diverse backup channel for main-channel audio signals. For the all-digital mode, this backup channel is a separate time-delayed digital signal. The use of DTS' patent pending time-diverse blending technique provides continuity of service during channel blockages. For this reason, time-diverse backup channels are an important part of DTS' HD Radio technology.

Frequency and Sideband Diversity

Transmission systems do not necessarily experience uniform fading across the channel bandwidth. In fact, frequency selective fading is common in the AM and FM bands. If the information is transmitted in two different parts of the spectrum and there is sufficient frequency separation between these transmissions, it is possible to mitigate the effects of a fading channel. The DTS HD Radio technology employs sideband diversity as a form of frequency diversity. The system transmits identical digital information on both upper and lower sidebands. Although the data in the upper and lower sidebands are identical, the CPPC error correction codes are different. In the receiver, the sidebands are independently detected and decoded. Additional coding and power gain is achieved when both sidebands

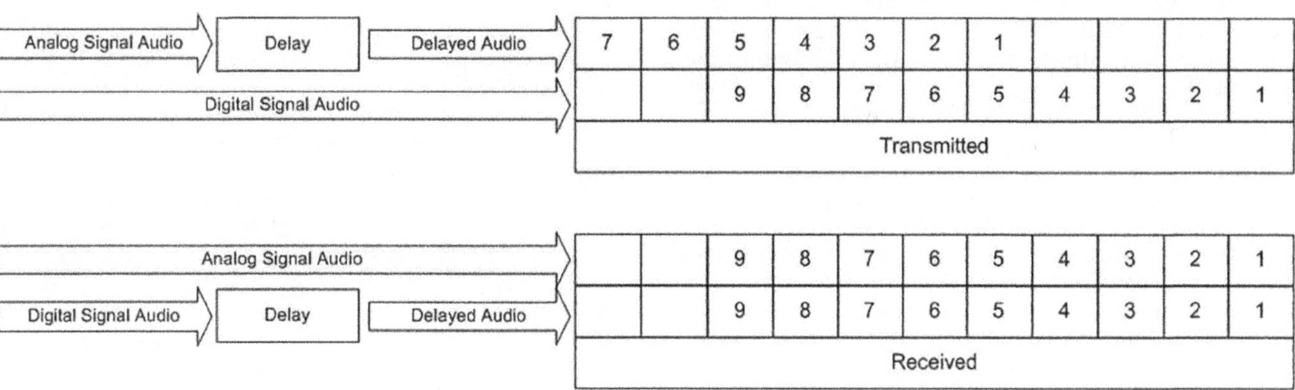

FIGURE 7.6-8 Diversity delay under normal reception

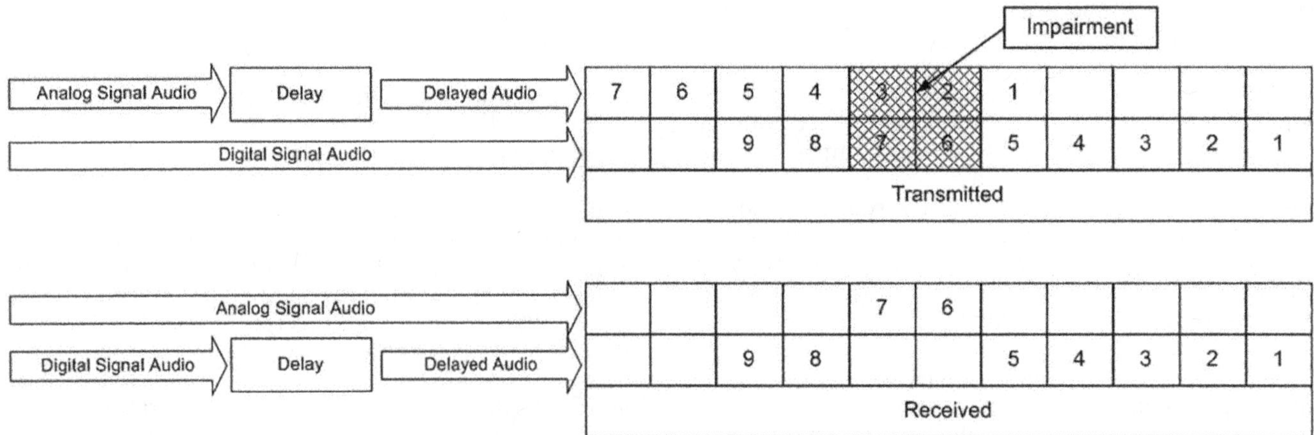

FIGURE 7.6-9 Blend transition during signal impairment

are combined. DTS' HD Radio technology essentially carries the same audio information in two different parts of the spectrum. Sideband diversity plays a key role in the system's robustness when either sideband experiences the effects of interference or multipath fading.

First Adjacent Channel Canceller

DTS' HD Radio technology signals operate at a lower power level than the analog signals in the FM band. The FCC allocations provide for a 6 dB desired-to-undesired (D/U) ratio at the protected contour for adjacent channels. Therefore, at the protected contour the analog interferer is 19 dB higher in power than the digital sideband. DTS' HD Radio technology employs a patented First Adjacent Canceller (FAC) to mitigate the effects of first adjacent channel analog interference. The FAC, which is simultaneously active in the upper and lower digital sidebands, essentially tracks the instantaneous frequency of a first adjacent channel analog interferer and nulls its effect on the digital information. Figure 7.6-10 shows two hybrid signals as they would appear at the protected contour. The arrow to the left (desired signal) depicts an analog carrier that is swinging back and forth with the analog modulation. The first adjacent channel interferer (undesired signal) is located to the right of the desired signal and is depicted as being about 6 dB lower than the desired analog signal. As shown in the left drawing, the analog interferer is higher in level than the digital subcarriers. The drawing to the right depicts the effect of FAC in eliminating the analog interference.

First adjacent channel interference from analog FM broadcasts is significantly higher in level than the digital signals. To ensure coverage of DTS' HD Radio technology near the protected contour, an effective means of interference cancellation must be employed. FAC makes possible digital reception, in a mobile environment, in the presence of analog interference that is about 19 dB higher in power level than the digital signal.

HD RADIO DEVELOPMENT CONCEPTS

DTS' HD Radio broadcasting system represents the next generation of broadcasting in the AM/FM bands. Existing analog AM and FM broadcasts may be augmented and even replaced by digital broadcasting modes, resulting in improved and expanded services to the consumer. This section provides an overview of the DTS HD Radio system design concepts and what it means to the next generation of broadcasting systems.

The HD Radio concept encompasses all of the features currently envisioned for IBOC-based digital systems for near-term and longer-range development. Features being developed are broad in concept and have a wide range of appeal to both the service provider and the consumer.

Initially, broadcast systems will support processing and transmission of analog and digital audio signals and digital data services. Analog processes will likely be phased out as the IBOC technology evolves and digital receiver penetration reaches critical mass. System design concepts have been modularized and grouped, for discussion, according to features or services to be offered. Those features are listed and described in the following sections.

HD Radio System Features

Listener flexibility in the selection of services offered is a major objective of the HD Radio system. Individual broadcasters may elect to not offer all HD Radio services or expeditiously update their capabilities as the system evolves. Therefore stations will likely offer variations of the wide variety of defined and developed HD Radio digital data services such as multicast audio, on-demand audio and traffic data, to name a few. The HD Radio broadcast system will support a wide array of receiver and broadcaster configurations to accommodate user preferences.

The HD Radio broadcasting system offers improved audio quality and services over existing analog systems. In the FM band, HD Radio can provide CD-like audio quality or multiple high-quality audio streams

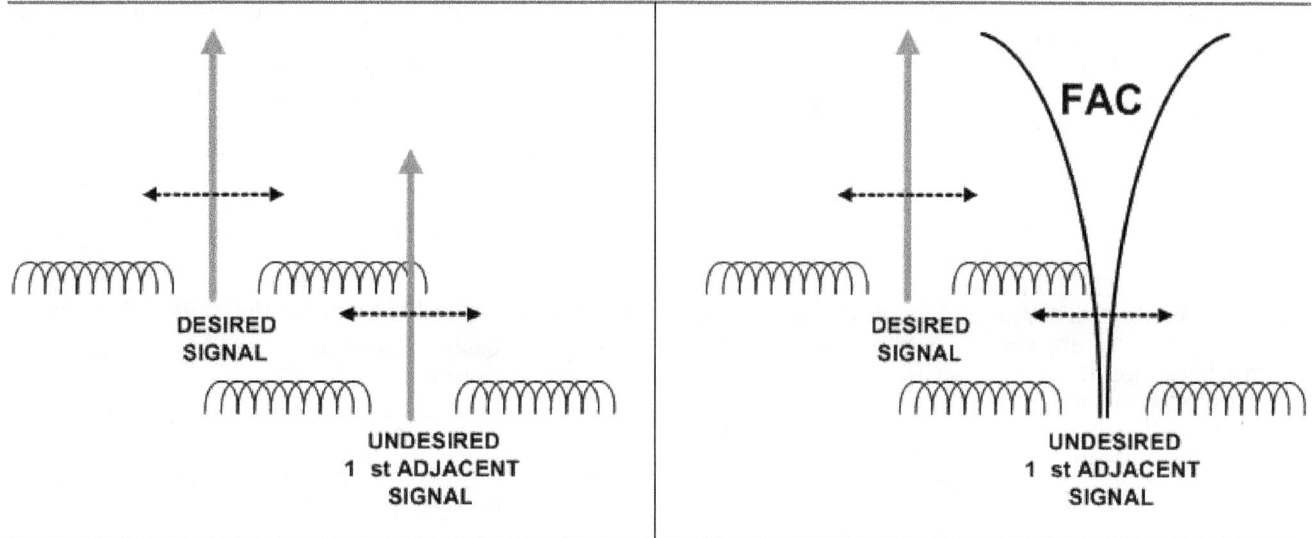

FIGURE 7.6-10 The left drawing shows digital sidebands impacted by FM carrier. The right drawing shows FAC cancellation of analog carrier interference.

surpassing FM quality. In the AM band, it provides audio quality similar to that of existing analog FM systems. HD Radio can offer new audio features, such as surround sound, not available with existing analog services, and will support multiple audio programs (multicast) on a single RF frequency assignment.

HD Radio introduces new data services that greatly expand the range of broadcaster functions and applications. Digital audio appears to the system as a source of data that may be traded off against data capacity to optimize both. Because of this, it permits synergistic cooperation between the audio and data services to:

- Permit the trade-off of digital audio fidelity (via bit rate adjustment) against data capacity;
- Permit the trade-off of analog audio fidelity (via bandwidth adjustment) against data capacity;
- Provide additional opportunistic variable data capacity based on audio activity;
- Provide various levels of data service quality (via adjustable error protection coding).

The HD Radio system offers defined levels of audio fidelity and robustness and provides resistance to multipath, Doppler shift, adjacent channel, grounded conductive structures and impulse noise interference.

Compatibility With Existing Analog AM and FM Services

HD Radio broadcasting represents a major departure in the way future broadcasting equipment and receivers must be built and function. HD Radio has been introduced to the consumer and broadcasters gradually, in a way that does not immediately obsolete the current radio broadcast infrastructure (transmitters and receivers). The current population of radio receivers continues to work without modification until such time as analog broadcasting is completely supplanted by the all-digital mode. This is made possible by the initial introduction of hybrid modes consisting of analog signals augmented with digital signals.

HD Radio features are incorporated as the technologies to perform those features are developed and refined. Features development is based on both transmitter and receiver functionality. Some broadcast systems may continue to operate entirely in the analog environment while others are converted, in stages, from analog, to hybrid, to the all-digital HD Radio system. For those reasons, flexibility in design and performance characteristics is essential as broadcasters evolve from analog to hybrid to all-digital technologies.

HD Radio broadcasting does not require new allocation of frequencies and the HD Radio over-the-air waveform coexists within existing FCC spectral emissions masks, thus minimizing the impact on existing analog services. The HD Radio system:

- Does not require a change in current FCC protected contours;
- Minimizes interference to the analog host in hybrid modes;
- Avoids harmful interference to co-channels and adjacent channels;
- Avoids harmful interference to and remain compatible with, existing Subsidiary Communications Authorization (SCA) services in hybrid modes;
- Is compatible with the existing Emergency Alert System (EAS);
- Does not require changes to existing analog radiated power in hybrid modes;
- Is compatible with existing radio frequency translators and boosters;

- Minimizes the interference of co-channels and adjacent channels on the digital signals, in both hybrid and all-digital modes;
- Does not impair acquisition time of existing analog receivers.

On-Channel Repeaters for FM Systems

The use of OFDM modulation in the FM HD Radio system allows on-channel digital repeaters to fill areas of desired coverage where signal losses due to terrain and/or shadowing are severe. A typical application occurs where mountains or other terrain obstructions within the station service areas limit analog or digital performance. On-channel repeaters may take the form of hybrid and digital-only implementations.

Seamless Transition

The introduction of the HD Radio broadcasting system is incremental and evolutionary in nature. Market and regulatory forces will necessitate evolutionary changes to the system. Such changes will, in general, conform to the following principles to make changes as seamless as possible:

- Analog sunset (mandatory elimination of analog broadcasting modes) will not be required;
- The system exhibits forward and backward audio compatibility, supporting future audio CODEC enhancements while maintaining audio compatibility with earlier receivers;
- The system exhibits forward and backward data compatibility, supporting future data service enhancements while maintaining compatibility with earlier receivers;
- The system allows each broadcaster and consumer to upgrade according to individual needs;
- Initial HD Radio receivers provide analog, hybrid and all-digital functionality, so as to remain useful throughout the entire evolution of the broadcasting industry from existing analog service to digital service;
- The HD Radio broadcasting system provides a migration path for existing SCA services when their host stations transition to an all-digital service;
- As the system evolves, it will seek to maximize use of existing electromagnetic spectrum.

Economical Implementation

The HD Radio system features a wide variety of options, not all of which is of interest to all broadcasters, all users or all user situations. Some features will interest travelers or mobile users but are of little interest in the home or place of business and vice versa. For that reason, manufacturers will likely offer specialty broadcast systems configured to the desired services for the various user situations or applications. Conversion to the HD Radio broadcasting system is designed to minimize new infrastructure requirements and maximize existing broadcaster infrastructure investment.

BROADCAST SYSTEM FUNCTIONAL TYPES

Implementation of HD Radio broadcasting systems is generally accomplished by one of the following strategies:

- Systems converted from "analog only" operation;
- Systems designed and constructed from the outset to be HD Radio–compliant.

It is anticipated that most systems will follow the first (conversion from analog) strategy. HD Radio broadcasting systems, however implemented, are of the following defined types.

Commercial Hybrid Types

Hybrid HD Radio broadcasting systems transmit an analog carrier signal that is compatible with existing analog radio receivers and is substantially identical to the analog carrier signal of existing analog-only systems, augmented with digital sidebands that carry digital audio and data, in the manner prescribed herein.

Commercial All-Digital Types

All-digital HD Radio broadcasting systems have been developed for future applications. The analog carrier portion of the broadcast signal is removed and secondary digital subcarriers utilized in the manner prescribed herein. The primary subcarriers are identical to those of hybrid systems for FM, and for AM the primary subcarriers are moved into the center of the band.

COMMON AM/FM SYSTEM SERVICE DEFINITIONS

Functional Layer Definitions

The HD Radio broadcasting system functionality has been defined in terms of functional layers. These layers correspond in definition to those of the International Standards Organization Open Systems Interconnection (ISO/OSI) model. Layers communicate at defined interfaces through structured data blocks called Protocol Data Units (PDU). A PDU is the structured data block in the HD Radio system that is produced by a specific layer (or process within a layer) of the transmitter protocol stack. The PDUs of a given layer may encapsulate PDUs from the next higher layer of the stack and/or include content data and protocol-control information originating in the layer (or process) itself. The PDUs generated by each layer (or process) in the transmitter protocol stack are inputs to a corresponding layer (or process) in the receiver protocol stack.

Layer 1, or the Waveform/Transmission Layer, generally performs the functions of a modem. It receives several logical channels over which all-digital data (including main audio and control and status data) is transferred and its output is the modulated electromagnetic waveform broadcast. Logical channels are implemented as data streams internal to the processing functions of the broadcasting system. Organizing the Layer 1 input in terms of multiple logical channels, instead of a single bit stream, affords a convenient and more effective way of utilizing the data transmission resources of the HD Radio broadcasting system.

The upper layers (Layer 2 or higher) generally deal with generating and packetizing the data to be sent, including implementing the tradeoffs between audio quality and data throughput.

Broadcasting Mode Definitions

Hybrid mode is defined as simultaneous broadcasting of analog and digital signals within the same channel, in such a way that existing analog receivers can satisfactorily receive the analog portion of the signal. HD Radio receivers will receive both the analog and digital portions of the signal. The analog portion of the signal functions as backup to the digital audio signal to mitigate outages and fades in the main channel audio portion of the signal.

In the all-digital mode, digital audio and data occupy the entire HD Radio broadcast spectrum. For the FM mode, the primary digital portion of the spectrum are those carriers that are placed, in a noninterfering manner, on either side of the existing analog channel and are present in both hybrid and all-digital modes. The primary extended portion of the spectrum is optional with additional carriers placed next to the primary carriers on either side of and adjacent to the analog channel. The secondary digital portion of the spectrum includes those carriers that are placed in the spectrum vacated when the analog carrier spectrum is removed for the all-digital mode.

For the AM mode, the digital carriers are grouped into mutually exclusive frequency sidebands designated as:

- Primary subcarriers, occupying the outer part of the spectrum;
- Secondary subcarriers, occupying the middle part of the spectrum;
- Tertiary subcarriers, occupying the inner part of the spectrum located below the analog signal.

Primary AM subcarriers are further designated as core subcarriers and collectively the secondary and tertiary subcarriers are designated as enhanced subcarriers. The tertiary subcarriers spectrally overlap the analog signal, but are phase orthogonal with it for minimal mutual interference.

Service Mode Definitions

The HD Radio broadcasting system functionality affords many degrees of freedom in the choices that can be made to optimize system end-to-end performance. These include:

- Audio robustness versus audio quality and latency;
- Data throughput versus data robustness;
- Audio quality versus data throughput tradeoffs (upper layers);
- Scrambling (Layer 1);
- Channel coding (Layer 1);
- Interleaver design (Layer 1);
- Allocation of the various subcarriers (Layer 1).

By choosing and combining these factors in appropriate ways, a wide variety of information data rates, audio/data combinations and bit error rate performance for given channel conditions can be achieved.

For the HD Radio broadcasting system, several such combinations have been defined, enumerated and designated for the hybrid FM system, the all-digital FM system, the hybrid AM system, and the all-digital AM system. These designated combinations are the defined service modes of the HD Radio broadcasting system.

Audio Service Definitions

In hybrid HD Radio broadcasting systems, the analog portion of the air interface is identical to the air interface of existing analog-only broadcasting systems. Any deviations from this requirement to implement HD Radio is compatible with existing analog-only broadcasting equipment and the transmitted signal is compatible with existing analog-only receivers. Analog audio is the means by which the broadcast system maintains compatibility with analog-only legacy receivers. It is used with digital main channel audio in hybrid mode as backup audio, to provide improved robustness at the receiver under circumstances of signal fading and blockage. Multicast or supplemental program services (SPS) are additional audio channels that may be added in addition to the main program service (MPS). The exact bandwidth of the MPS and SPS channels will be determined by service mode and the total number of audio streams transmitted. Digital audio is the primary or main audio delivery mode in both hybrid and all-digital modes. In all-digital mode it is augmented with another lower latency and bandwidth digital channel that acts as backup audio in lieu of the analog channel. While the MPS channel is backed up by the analog signal, the SPS channels are stand-alone and mute at the point of failure (POF).

Audio Classes

The following definitions are applicable throughout this document.

In the hybrid modes, Class 2 and Class 4 audio is accomplished by the analog channel. In all-digital modes, Class 2 and Class 4 audio are separate, possibly independent, encoded, low latency, and low bandwidth data streams. This relationship is defined in Table 7.6-1.

TABLE 7.6-1
Audio service classes

Class	Service	Number of Audio channels	Minimum Audio Frequency Response	Minimum Stereo Separation (dB)	Dynamic Range	Quality Level
1	FM main	stereo	20–20,000 Hz	70 dB	96 dB	CD-like
2	FM backup	monophonic	20–15,000 Hz	none	65 dB	FM mono
3	AM main	stereo	20–15,000 Hz	70 dB	72 dB	FM-like
4	AM backup	monophonic	20–10,000 Hz	none	60 dB	AM mono

Audio Quality Definitions

Audio quality is measured as produced by the end-to-end HD Radio system, from unencoded source audio at the broadcasting system to the decoded audio output of the receiver. This is quantified in terms of two methodologies to test whether a system has reached a quality target: (1) degradation category rating (DCR) and (2) absolute category rating (ACR).

Digital Data Service Definitions

The HD Radio broadcasting system will offer extensive datacasting services. These services extend to both the HD Radio AM and FM systems. Six data transport services have been defined for AM and FM as follows:

- CODEC (HDC) digital audio transport;
- Text transport;
- Control channel transport;
- Packet/message transport;
- File transport;
- Generic streaming data transport.

To support these transport services, the HD Radio broadcasting system has defined the following classes of data as shown in Table 7.6-2.

Dedicated fixed-rate services employ a fixed data rate that cannot be changed by the broadcaster. This data rate is allocated to the various data services defined by the upper Layer functionality. It includes, among others that may be defined, the Station Information Service (SIS) that offers an array of Radio Data System (RDS)-like services to the broadcaster.

Adjustable rate services operate at a rate that, unlike fixed-rate services, is selectable and changeable by the broadcaster by trading off audio quality or robustness for data throughput. To increase the data rate, the broadcaster reduces the audio bit rate (and therefore audio quality or robustness) and reallocates it to data. To decrease the data rate the reverse occurs. The adjustable rate services operate by dynamically allocating digital subcarriers among error correction, audio and data services.

Opportunistic variable-rate services offer data rates that are dynamically related to the complexity of the encoded digital audio. Simpler audio passages (e.g., simple tones, narrow bandwidth audio, and silence) require lesser bit rates, permitting the unused throughput to be used for data. The audio encoder dynamically measures audio complexity and adjusts data throughput accordingly, without compromising the quality of the encoded digital audio.

Station Information Service (SIS)

The SIS provides broadcast station identification and control information. SIS is transmitted in a series of SIS Protocol Data Units (PDUs) on the primary IBOC data service (PIDS) logical channel. The PIDS channel is a fixed rate channel that delivers basic control messages that carry service information. Service is information about the services carried in real time, other information such as schedules or service event calendars, and station-related system broadcast information similar to existing RBDS services.

Datacasting System

Datacasting is defined as delivering content from a content provider to a receiver end user via the HD Radio system. Datacasting affords expanded data functions over those provided by the IBOC data system (IDS) channels. These include, but are not limited to, the following:

- Streaming perceptual audio CODEC (HDC) applications;
- Still and streaming video applications;
- Message/packet-based applications;
- File-based applications;
- Audio storage and retrieval applications;
- Billing and management;
- Text/XML (extensible markup language) applications;
- Specialized applications with specialized receivers;
- Datacasting services with various defined levels of quality of service for each;
- Control data beyond that provided by the IDS channel.

HD Radio System Datacasting Users Are of Three Types

- Content providers who create and package content for broadcast over the HD Radio system;

TABLE 7.6-2
Classes of data service

Class	Service	System	Minimum Rate	Maximum Rate
1	Dedicated fixed rate	Hybrid, all-digital	860 bps FM 430 bps AM	N/A
2	Adjustable rate	Hybrid, all-digital	0	Equal to at least the maximum rate of audio CODEC
3	Opportunistic variable rate	Hybrid, all-digital	0	Equal to maximum rate of audio CODEC

- Operations, administrative and maintenance (OAM) users who manage the broadcast system for content delivery, billing and other administrative task support;
- Receiver end users who make use of the content broadcast by the HD Radio system.

The HD Radio broadcasting system interfaces content providers to receive content for broadcast, with the receiver systems (via the air interface) to deliver the content and with OAM users to administer and maintain the data service. For purposes of datacasting definition and specification, the HD Radio broadcasting system consists of two parts:

- Broadcast network system that receives content from content providers and delivers it to individual broadcast station systems;
- Broadcast station systems that receive content from the broadcast network system, or from local content providers for broadcast.

AM HD RADIO SERVICES

The AM IBOC system is capable of supporting the following services:

- Main Program Service (MPS)

  The MPS preserves the existing analog radio-programming formats in both the analog and digital transmissions. In addition, the MPS includes digital data, which directly correlates with the audio programming.

- Personal Data Service (PDS)

  Unlike the MPS, which broadcasts the same audio program to all listeners, the PDS enables the user to select the data services desired and when they are presented. This provides personalized, on-demand, user-valued information.

- Station Identification Service (SIS)

  The SIS provides the necessary control and identification information that indirectly accommodates user search and selection of IBOC digital radio stations and their supporting services.

- Advanced Application Service (AAS)

  This service allows a virtually unlimited number of custom and specialized IBOC digital radio applications to exist concurrently. Advanced applications can be added at any time in the future.

Support of the above services is provided via a layered protocol stack illustrated in Figure 7.6-11. This layered protocol stack is based on the International Standards Organization Open Systems Interconnection (ISO/OSI) layered model. Source material is received from the broadcaster in Layer 5; source encoded in Layer 4, multiplexed into logical channels in Layers 3 and 2 and formatted for over-the-air broadcast in Layer 1:

- Layer 5 (Application) accepts content from the user (i.e., program source).
- Layer 4 (Encoding) performs the necessary audio compression or data formatting of the various source materials.
- Layer 3 (Transport) provides one or more application specific protocols tailored to provide robust and efficient transfer of Layer 4 data.
- Layer 2 (Service MUX) provides limited error detection and addressing. Its main function is to format the data received from Layer 3 into discrete transfer frames for processing by Layer 1.
- Layer 1 (Physical Layer) provides the modulation, FEC, framing and signaling necessary to convert the digital data received from higher Layers into an AM IBOC waveform for transmission in an existing allocation in the MF band.

Layer 1 can be thought of as simply a conduit for broadcast data with a specific grade of service; the source coding, formatting and multiplexing of the program content are performed at the higher protocol Layers. However, the AM IBOC system provides a number of different configurations, called service modes, in which the number, throughput and robustness of the "data pipes," called logical channels, can vary. Therefore, after assessing the requirements of their candidate applications, higher protocol Layers select service modes that most suitably configure the logical channels. The number of logical channels and service modes reflects the inherent flexibility of the system, which supports simultaneous delivery of various classes of digital audio and data.

AM FUNCTIONAL DESCRIPTION

Several aspects of the AM HD Radio system are discussed in this section. First, the modulation and

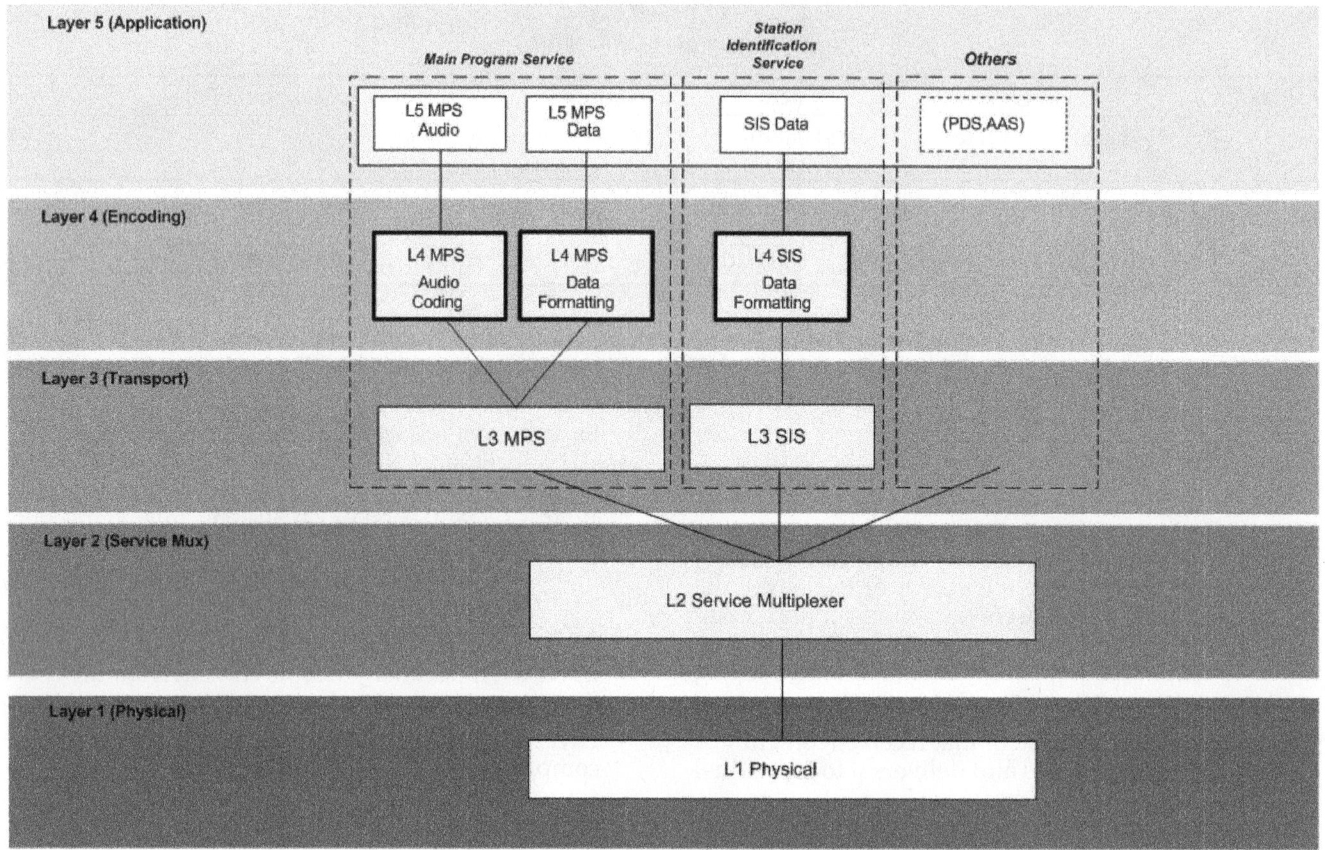

FIGURE 7.6-11 International Standards Organization Open Systems Interconnection (ISO/OSI) layered model

spectral occupancy of the waveforms are described, followed by a description of the available service modes, as well as the various broadcaster options within each service mode. Next, the logical channels in each service mode are described in terms of throughput, robustness and latency. Lastly, the functional components of AM waveform generation are presented, describing the processing necessary to convert the digital data in the active logical channels into an AM IBOC waveform.

Transmission Subsystem

The transmission subsystem formats the baseband AM HD Radio waveform for transmission through the medium frequency (MF) channel. Functions include symbol concatenation, pre-compensation and frequency up-conversion. In addition, when transmitting the hybrid waveform, this function filters and modulates the baseband analog audio signal before coherently combining it with the digital portion of the waveform.

The input to this module is a complex, baseband, time-domain OFDM symbol from OFDM signal generation. A baseband analog audio signal after application of diversity delay is also input from an analog source when transmitting the hybrid waveform. The output of this module is the MF AM HD Radio waveform.

Figure 7.6-12 and Figure 7.6-13 for functional block diagrams of the hybrid and all-digital transmission subsystems, respectively.

AM HD RADIO FUNCTIONAL COMPONENTS

Pre-Compensation

The pulse-shaping function used in the AM system guarantees that the frequency domain side lobes meet spectral limits imposed by the FCC, as outlined in FCC Rules Section 73.44.

Symbol Concatenation

The individual time-domain OFDM symbols output precompensation are summed to produce a continuum of pulses.

Diversity Delay

The first step in generating the HD Radio signal is the application of diversity delay to the baseband analog audio signal. An adjustable delay is applied to the baseband analog audio signal so that, at the output of the analog/digital combiner, it lags the audio con-

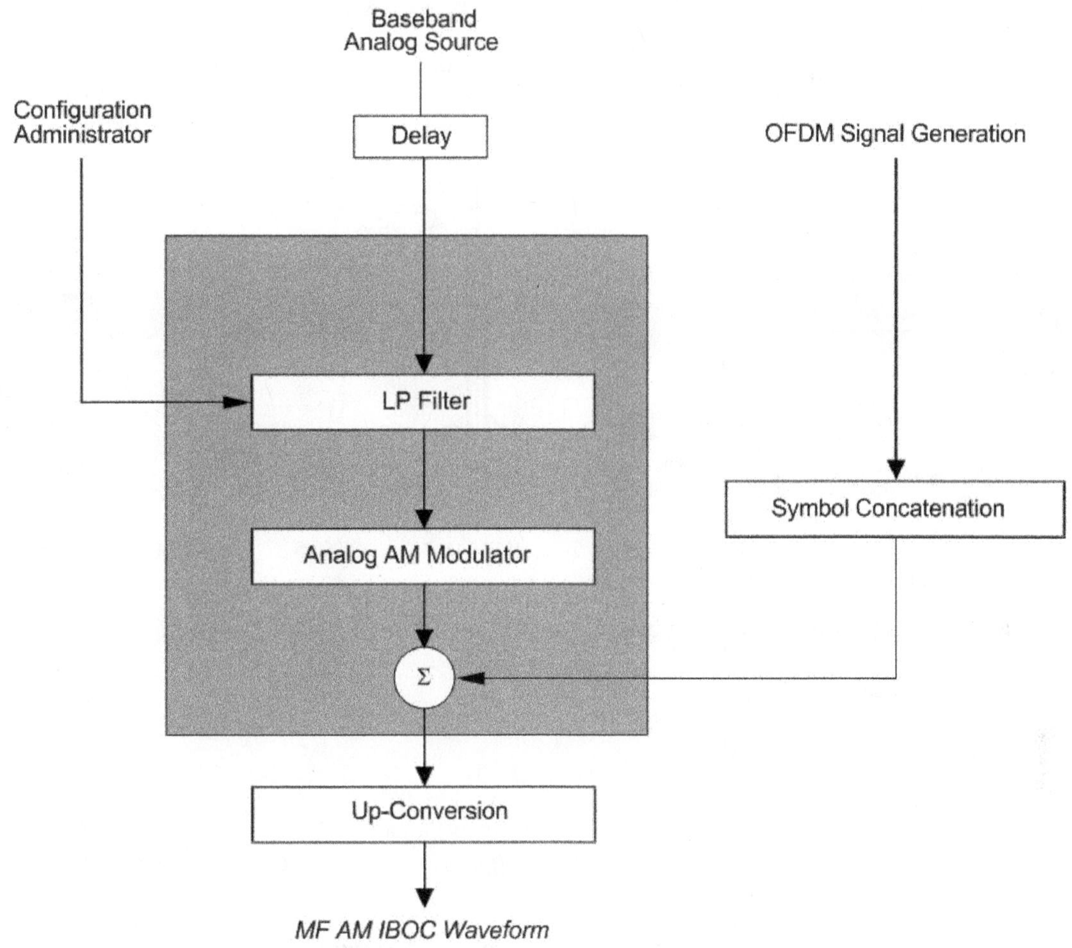

FIGURE 7.6-12 Hybrid transmission subsystem functional block diagram

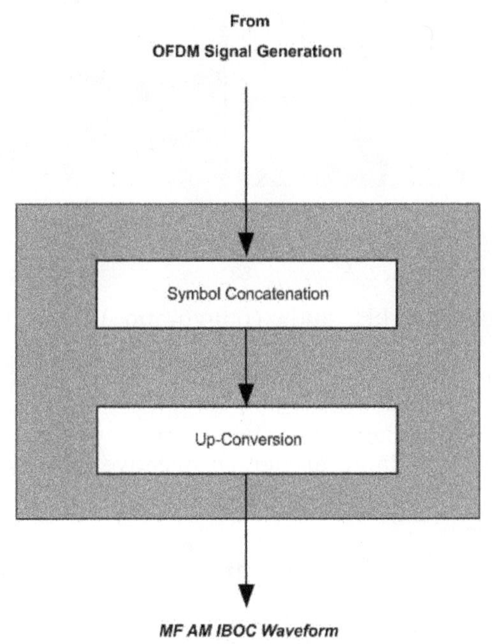

FIGURE 7.6-13 All-digital transmission subsystem functional block diagram

tent of the corresponding digital signal. For example, if both the analog and digital signals carry the same audio program, the analog audio would be delayed from the corresponding digital audio at the output of the analog/digital combiner. The delay is adjustable to account for processing delays in the analog and digital chains.

Lowpass Filtering

In hybrid mode, this process lowpass filters the analog audio data according to the state of the AAB control received from the configuration administrator. If the control bit is zero, the analog audio is filtered to a 5 kHz bandwidth. A stopband frequency of 5116 Hz with > 55 dB attenuation is used. If the control bit is one, the analog audio is filtered to an 8 kHz bandwidth. Here, a stopband frequency of 9 kHz with > 60 dB attenuation is used (see Figures 7.6-14 and 7.6-15). In both cases, the passband specifications are set to achieve the best audio possible. This lowpass filtering can also be performed in external audio processors. For hybrid transmissions configured for the reduced digital bandwidth configuration (RDB = 1), the power spectral density of the modulated AM carrier measured with

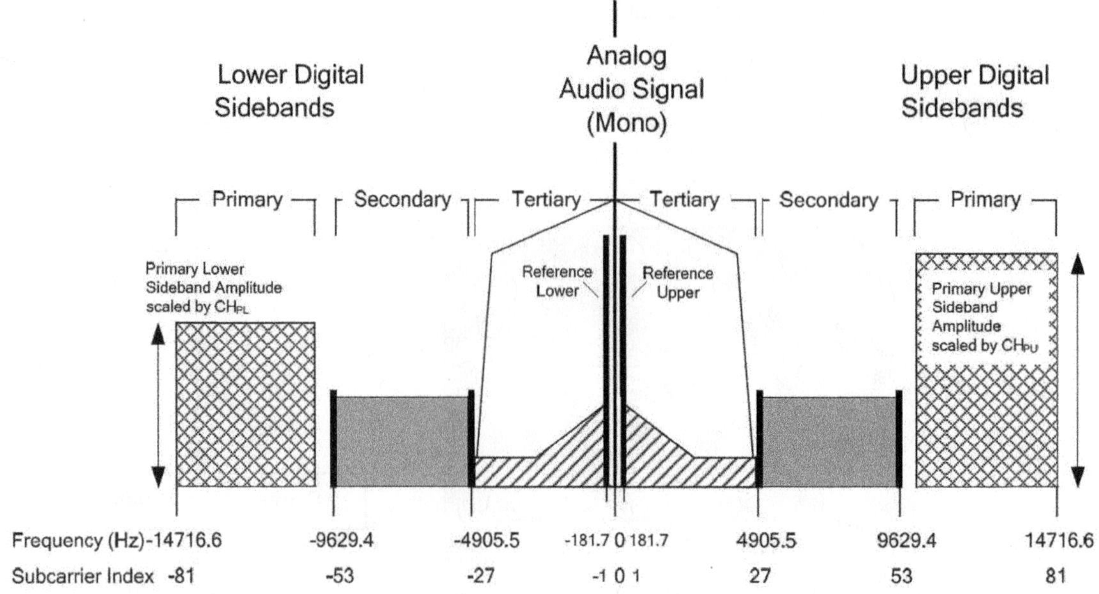

FIGURE 7.6-14 AM HD Radio hybrid waveform spectrum (5-kHz audio configuration)

DTS

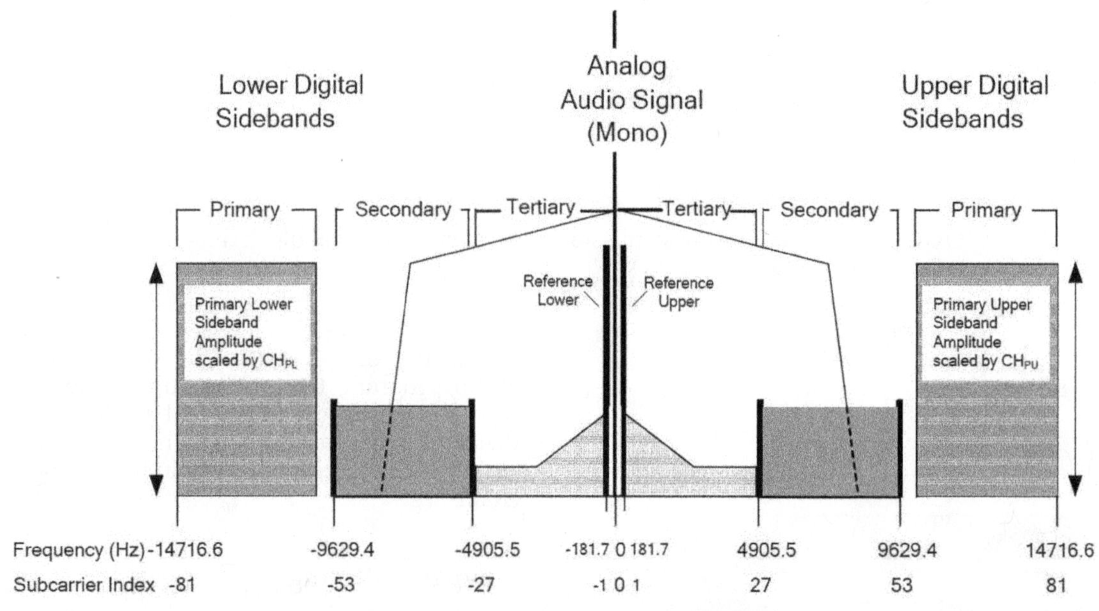

FIGURE 7.6-15 AM HD Radio hybrid waveform spectrum (8-kHz audio configuration)

DTS

the HD Radio digital component disabled, at frequencies removed from the carrier frequency by more than 9.4 kHz and up to 20 kHz shall not exceed –65 dBc/300 Hz (see Figure 7.6-16).

Analog AM Modulator

When broadcasting the hybrid waveform, the envelope of the analog AM signal is processed by applying a modulation index and adding a DC offset. The analog input signal must be pre-processed external to the AM HD Radio exciter, so negative peaks do not exceed 100%.

Analog/Digital Combiner

When broadcasting the hybrid waveform, the real analog AM baseband waveform is coherently combined with the digital baseband waveform to produce the complex baseband AM HD Radio hybrid waveform.

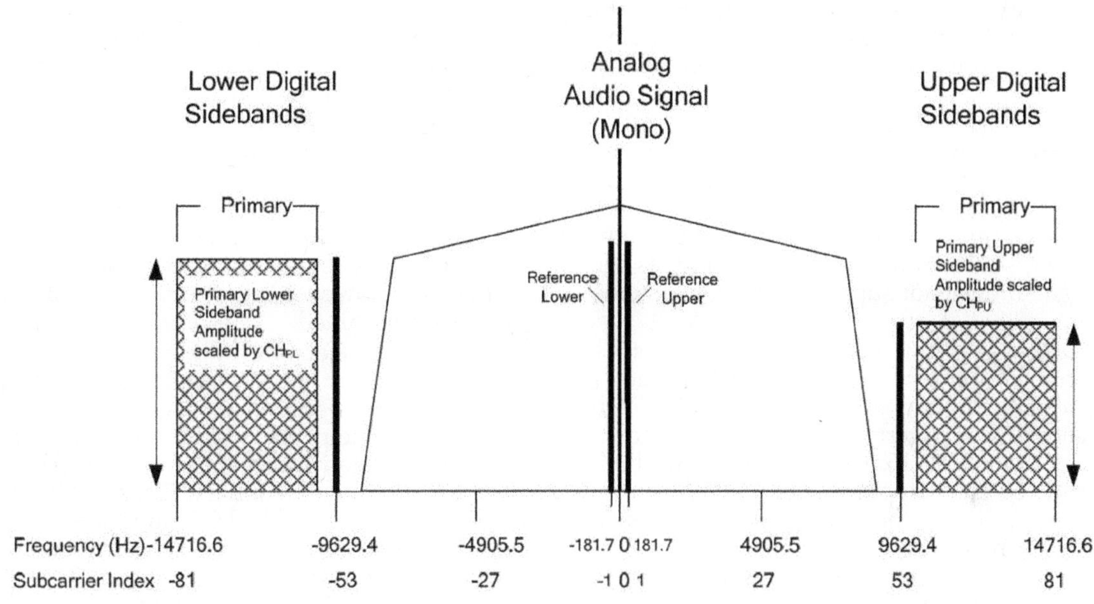

FIGURE 7.6-16 AM HD Radio hybrid waveform spectral emissions limits for RDB = 1 (RDB configuration)

DTS

OFDM subcarrier mapping appropriately scales the levels of the digital sidebands in the output spectrum.

Up-Conversion

The concatenated digital signal is translated from baseband to the RF carrier frequency. The AM HD Radio waveform is broadcast in the current AM radio band and its power levels and spectral content are limited to be within the spectral mask as defined in FCC Rules Section 73.44. The carrier frequency spacing and channel numbering schemes are compatible with FCC Rules Section 73.44. Channels are centered at 10-kHz intervals ranging from 540 to 1700 kHz. Both the analog and digital portions of the hybrid waveform are centered on the same carrier frequency.

GPS Synchronization

In order to ensure precise time synchronization the transmitted signal for each station may be synchronized in time and frequency to the Global Positioning System (GPS). In cases where transmissions are not locked to GPS, time and frequency synchronization, the accuracy requirements are relaxed and transmissions cannot be synchronized with other stations.

RF Carrier Frequency and OFDM Symbol Clock

For Level I transmission facilities all transmissions shall phase lock their L1 frame timing (and the timing of all OFDM symbols) to absolute GPS time within ±1 μs. If this specification for a Synchronization Level I transmission facility is violated, due to a GPS outage or other occurrence, it shall be classified as a Synchronization Level II transmission facility until the preceding specification is again met.

GPS Phase Lock

For Level I transmission facilities, all transmissions will maintain phase lock to absolute GPS time within ±1 μs. If this specification in a Level I transmission facility is violated due to a GPS outage or other occurrence, it will be classified as a non-GPS synchronization Level II transmission facility until the preceding specification is again met.

Waveforms and Spectrum

Digital data and audio cannot be directly propagated over RF channels; therefore a modulator is used to modulate the digital information onto an RF carrier. The AM IBOC system employs OFDM for robustness in the presence of adjacent channel interference and noise. OFDM is a parallel modulation scheme in which the data streams modulate a large number of orthogonal subcarriers that are transmitted simultaneously. OFDM can be tailored to fit an interference environment that is nonuniform across frequency.

The narrow AM bandwidth leads to an OFDM modulation technique that is optimized for higher throughput. DTS' IBOC technology for AM uses QAM on each OFDM subcarrier.

The current design calls for symbol durations of 5.8 msec and an OFDM subcarrier spacing of 181.7 Hz. In addition, the timing of the symbols is optimized to ensure that duration of noise pulses is much shorter than the QAM symbol duration, ensuring robust digital reception in the presence of static and noise prevalent in the AM channel.

The design of the AM IBOC system provides a flexible means of transitioning to a digital broadcast system by providing two new waveform types: hybrid and all-digital. The hybrid waveform retains the analog AM signal, while the all-digital waveform does not. The analog source must be monophonic, as the AM IBOC system does not support AM analog stereo broadcasts.

In the hybrid waveform, the OFDM subcarriers are located in primary and secondary sidebands on either side of the host analog signal, as well as underneath the host analog signal in tertiary sidebands as shown in Figures 7.6-14 and 7.6-15. Each sideband has both an upper and lower component. Status and control information is transmitted on reference subcarriers on either side of the main carrier.

In addition to the primary, secondary, tertiary and control subcarriers, there are two additional subcarriers between the primary and secondary and the secondary and tertiary sidebands on either side of the main carrier. These are known as IDS subcarriers and are primarily used for low latency, low data rate applications such as SIS or RDS that are currently being used in analog FM systems.

The number of OFDM subcarriers in the secondary, tertiary and IDS sidebands is twice the number needed to transmit the QAM constellation values This is because the overall-digital signal must maintain a 90° phase relationship (quadrature) to the AM carrier, thereby minimizing the interference to the analog signal when detected by an envelope detector. Placing these subcarriers in quadrature to the analog signal also permits demodulation of the tertiary and IDS subcarriers in the presence of the high level AM carrier and analog signal. The price paid for placing these subcarriers in quadrature with the AM carriers is that the information content on these subcarriers is only half of that for nonquadrature digital carriers.

During transmission, the phase relationship between the digital and analog signals must be maintained. Amplification of the analog and digital signals by a single transmitter is the most straightforward approach. Separate amplifiers can be used for the analog and digital signals and the transmitter outputs combined. However, proper phasing of the signals must be maintained.

The total power of all the digital sidebands in the hybrid waveform is significantly below the power in the AM analog signal. The power level of each OFDM subcarrier in the primary sidebands is fixed relative to the unmodulated main analog carrier. However, the power level of the secondary, IDS and tertiary subcarriers is adjustable. Table 7.6-3 summarizes the spectral characteristics of the hybrid waveform. Individual subcarriers are numbered from −81 to 81 with the center subcarrier at subcarrier number 0. Table 7.6-3 also lists the approximate frequency ranges, bandwidth, levels and modulation types for each sideband. Referring back to Figures 7.6-14 and 7.6-15, note that the subcarriers 54 to 56 and −54 to −56 are not represented because they are not transmitted in order to avoid interference with first adjacent signals.

The greatest system enhancements are realized with the all-digital waveform. In this waveform, the analog signal is replaced with higher power primary sidebands. The unmodulated AM carrier is retained and the secondary sidebands are moved to the higher frequencies above the primary upper sideband. In addition the tertiary sidebands are moved to the frequencies below the primary lower sideband. The secondary and tertiary sidebands use half the number of subcarriers, as compared to the hybrid waveform, because there is

TABLE 7.6-3
AM hybrid spectral summary

Sideband	Subcarrier Range	Subcarrier Frequencies (Hz from channel center)	Frequency Span (Hz)	Power Spectral Density, dB/Subcarrier	Modulation Type
Primary Upper	57 to 81	10,356.1 to 14,716.6	4360.5	−30	64-QAM
Primary Lower	−57 to −81	−10,356.1 to −14,716.6	4360.5	−30	64-QAM
Secondary Upper	28 to 52	5087.2 to 9447.7	4360.5	−43 or −37	16-QAM
Secondary Lower	−28 to −52	−5087.2 to −9447.7	4360.5	−43 or −37	16-QAM
Tertiary Upper	2 to 26	363.4 to 4723.8	4360.4	−44 to −50	QPSK
Tertiary Lower	−2 to −26	−363.4 to −4723.8	4360.4	−44 to −50	QPSK
Reference Upper	1	181.7	181.7	−26	BPSK
Reference Lower	−1	−181.7	181.7	−26	BPSK
Upper IDS1	27	4905.5	181.7	−43 or −37	16-QAM
Upper IDS2	53	9629.4	181.7	−43 or −37	16-QAM
Lower IDS1	−27	−4905.5	181.7	−43 or −37	16-QAM
Lower IDS2	−53	−9629.4	181.7	−43 or −37	16-QAM

no longer a need to place them in quadrature with the analog signal because it is unmodulated. As a result, the power of both the secondary and tertiary sidebands is increased.

These changes result in an overall bandwidth reduction, making the all-digital waveform less susceptible to adjacent channel interference. The reference subcarriers are located on either side of the unmodulated AM carrier, as in the hybrid waveform, but at a higher level. The spectrum of the all-digital waveform is illustrated in Figure 7.6-17. The power level of each of the OFDM subcarriers within a sideband is fixed relative to the unmodulated analog carrier. Table 7.6-4 summarizes the spectral characteristics of the all-digital waveform.

Both the hybrid and all-digital waveforms conform to the currently allocated emissions mask per FCC Rules Section 73.44 and summarized in Table 7.6-5. All measurements assume a resolution bandwidth of 300 Hz.

Service Modes and System Options

The AM IBOC system provides two service modes: MA1, MA3. Service mode MA1 is used with the hybrid waveform, while service mode MA3 is used with the all-digital waveform. Specifics of the various service modes are detailed in the next section, where the logical channels for each service mode are described. (Note that service modes MA2 and MA4 are no longer a part of the system as they were specific to the PAC codec. The PAC codec was replaced by the HDC codec at the request of the NRSC.)

In addition to the two service modes, the broadcaster has the option of configuring service mode MA1 using three additional controls: power level control, analog audio bandwidth control and enhanced carrier control.

The power level control selects one of two levels for the secondary, tertiary and PIDS subcarriers. The

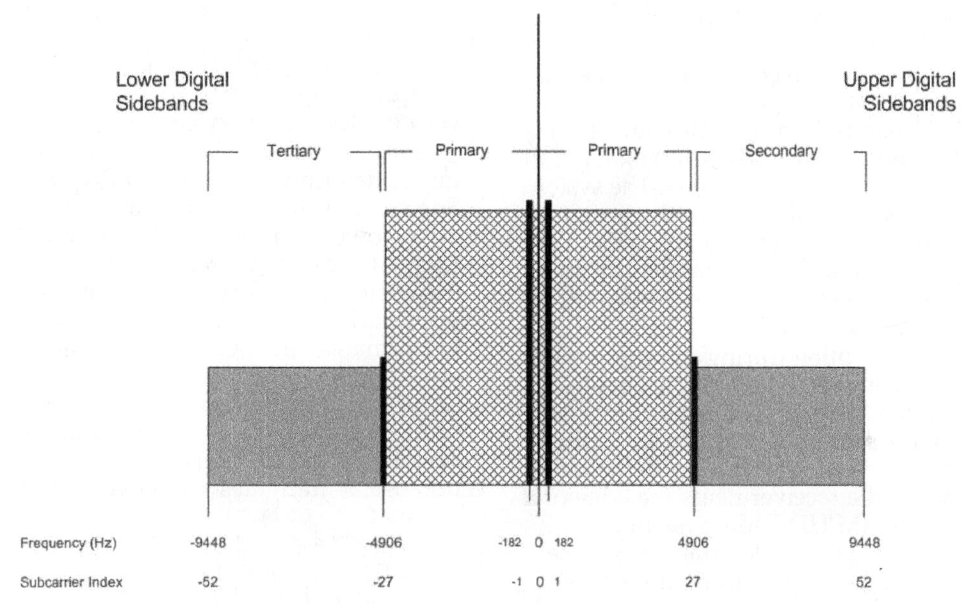

FIGURE 7.6-17 AM all-digital waveform spectrum

TABLE 7.6-4
AM all-digital waveform spectral summary

Sideband	Subcarrier Range	Subcarrier Frequencies (Hz from channel center)	Frequency Span (Hz)	Power Spectral Density, dB/Subcarrier	Modulation Type
Primary Upper	2 to 26	363.4 to 4723.8	4360.5	−15	64-QAM
Primary Lower	−2 to −26	−363.4 to −4723.8	4360.5	−15	64-QAM
Secondary	28 to 52	5087.2 to 9447.7	4360.5	−30	64-QAM
Tertiary	−28 to −52	−5087.2 to −9447.7	4360.5	−30	64-QAM
Reference Upper	1	181.7	181.7	−15	BPSK
Reference Lower	−1	−181.7	181.7	−15	BPSK
IDS1	27	4905.5	181.7	−30	16-QAM
IDS2	−27	−4905.5	181.7	−30	16-QAM

TABLE 7.6-5
AM analog (NRSC 4) spectral emissions mask

Offset From Carrier Frequency	Level Relative to Unmodulated Carrier
10.2 to 20 kHz	−25 dBc
20 to 30 kHz	−35 dBc
30–60 kHz	-5 dBc—1 dB/kHz
60–75 kHz	−65 dBc
> 75 kHz	−80 or [−43 − $\log_{10}$ (power in watts)] dBc, whichever is less

higher power levels increase the robustness of the digital signal at the expense of decreasing compatibility with certain classes of existing analog radios, namely portable radios. As receiver manufacturers produce IBOC radios or if a broadcaster is not as concerned with compatibility with these classes of existing analog radios, this option allows broadcasters to increase the robustness of the digital signal.

The analog audio bandwidth control allows the analog audio to be broadcast using either a 5 kHz bandwidth or an 8 kHz bandwidth. Broadcasting 8 kHz analog reduces the robustness of the digital signal in the presence of second adjacent interferers. The system provides for a reduced digital bandwidth configuration where the secondary and tertiary subcarriers are disabled allowing the analog audio bandwidth to be increased to 9.4 kHz. This configuration is selected by setting the control signal RDB to 1.

To provide robust reception during outages typical in a mobile environment, the AM IBOC system applies time diversity between independent analog and digital transmissions of the same audio source. In addition, a blend function allows graceful audio degradation of the digital signal as the receiver nears the edge of a station's coverage. The AM HD Radio system provides this capability by delaying the analog transmission several seconds relative to the digital audio transmission. When the digital signal is corrupted, the receiver blends to analog, which by virtue of its time diversity with the digital signal the audio does not experience the outage.

Logical Channels

A logical channel is a signal path that conducts data through Layer 1 with a specific grade of service determined by service mode. Layer 1 of the AM IBOC system provides four logical channels to higher Layer protocols: P1, P2, P3 and PIDS. P1, P2 and P3 are intended for general purpose audio and data transfer, while the PIDS logical channel is designed to carry the IDS information.

The P1 and P2 logical channels are designed to be more robust than the P3 logical channel. Therefore, P1 and P2 typically transmit the "core" audio information, while P3 transmits the enhanced audio information (such as the stereo signal). Logical channels P1 and P3 are available for all services modes.

Logical channels are defined by their characterization parameters and configured by the service mode. For a given service mode, the grade of service of a particular logical channel may be uniquely quantified using three characterization parameters: throughput, latency, and robustness. Channel code rate, interleaver depth, diversity delay and spectral mapping are the determinants of the characterization parameters.

The throughput of a logical channel is its allowable data rate and is typically defined in terms of kbps. Latency is the delay that a logical channel imposes on the data as it passes through Layer 1. The latency of a logical channel is defined as the sum of its interleaver depth and diversity delay. It does not include processing delays in Layer 1 nor does it include delays imposed in upper Layers. Robustness is the ability of a logical channel to withstand channel impairments such as noise, interference and grounded conductive structures (GCS). There are 10 relative levels of robustness designed into Layer 1 of the AM IBOC system. A robustness of 1 indicates a very high level of resistance to channel impairments, while a robustness of 10 indicates a lower tolerance for channel-induced errors. As with throughput and latency, higher layers must determine the required robustness of a logical channel before selecting a service mode.

Spectral mapping, channel code rate and interleaver depth determine the robustness of a logical channel. Spectral mapping affects robustness by setting the relative power level, spectral interference protection, and frequency diversity of a logical channel. Channel coding increases robustness by introducing redundancy into the logical channel. Interleaver depth influences performance in GCS and impulsive noise, thereby affecting the robustness of the logical channel. Finally, some logical channels in certain service modes delay transfer frames by a fixed duration to realize time diversity. This diversity delay also affects robustness because it mitigates the effects of the mobile radio channel.

Tables 7.6-6 and 7.6-7 list the logical channels P3 and PIDS and multiple robustness values for the service modes. This is because there are two power levels associated with these logical channels. The lower relative robustness number, indicating greater robustness, is associated with the higher power level settings.

For a given service mode, each logical channel is applied to a frequency sideband. Figures 7.6-18 and 7.6-19 show the spectral mapping of each logical channel for each service mode

TABLE 7.6-6
Logical channel characterizations—service mode MA1

Logical Channel	Throughput (kbps)	Latency (sec.)	Relative Robustness
P1	20.2	5.94	6
P3	16.2	1.49	7 or 10
PIDS	0.4	0.19	4 or 8

TABLE 7.6-7
Logical channel characterizations—service mode MA3

Logical Channel	Throughput (kbps)	Latency (sec.)	Relative Robustness
P1	20.2	5.94	1
P3	20.2	1.49	5
PIDS	0.4	0.19	3

Figure 7.6-18 reveals that the P1 logical channel is transmitted on both the upper and lower primary sidebands. These are redundant copies of the same information allowing the P1 logical channel to operate in the presence of a strong interferer on either the lower or upper adjacent channel. In addition to the frequency redundancy, the P1 logical channel in service mode MA1 also contains time redundancy. This is realized by transmitting redundant information that has diversity delay imposed at the transmitter. This redundant

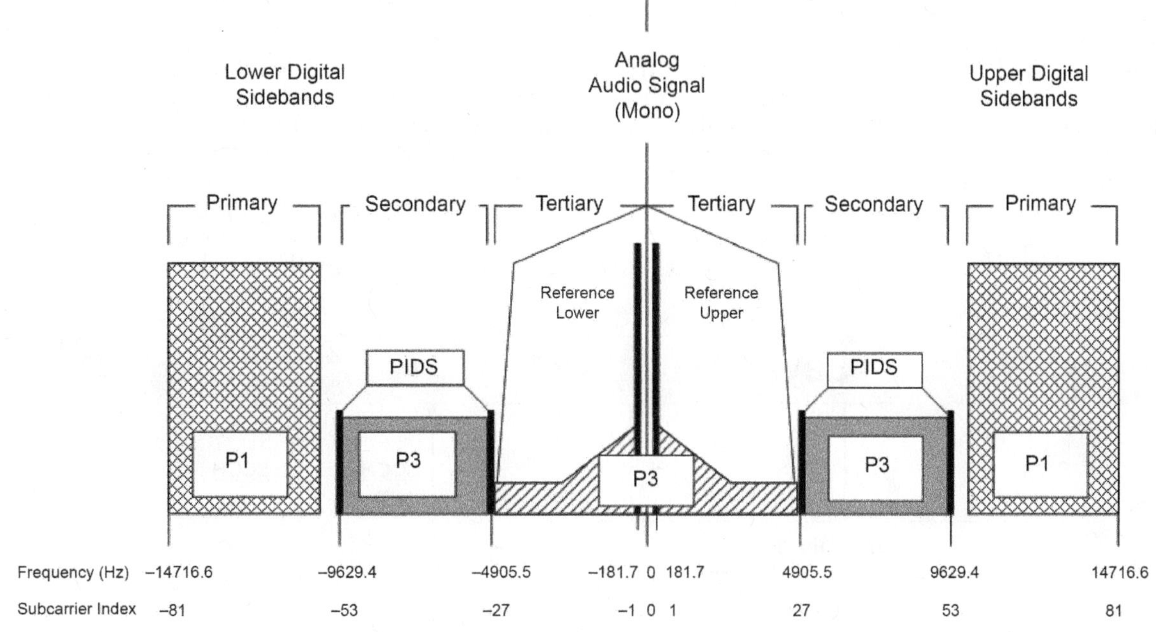

FIGURE 7.6-18 Logical channel spectral mapping—service mode MA1

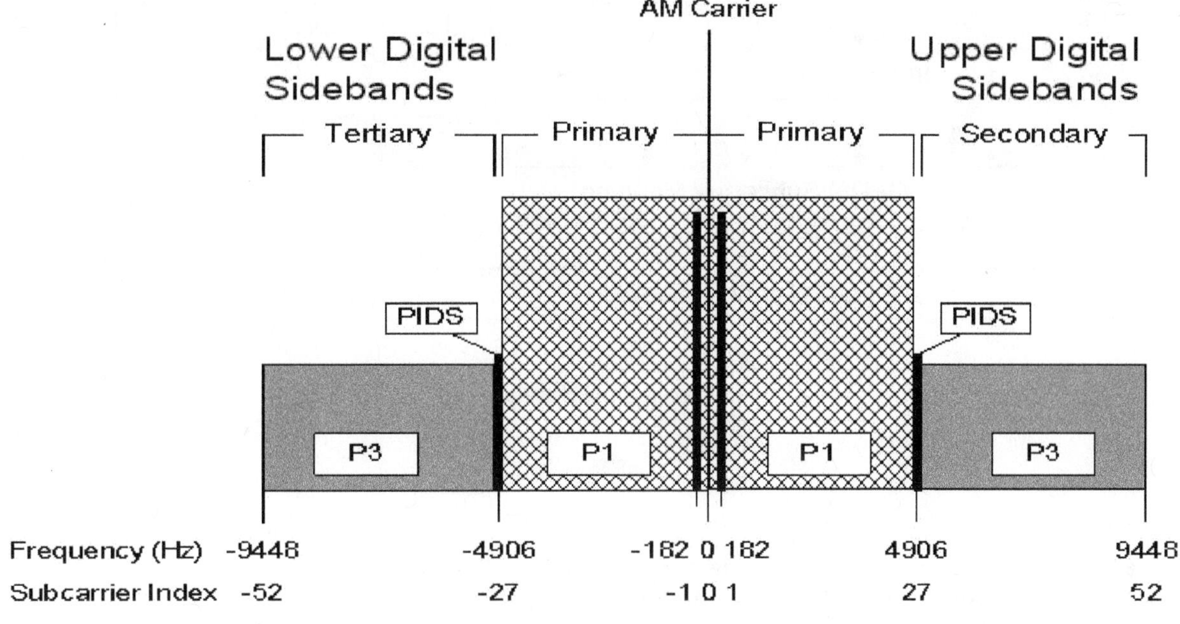

FIGURE 7.6-19 Logical channel spectral mapping—service mode MA3

information uses a short interleaver so that the digital audio may be acquired quickly. It also serves as a backup to the nondelayed information, providing robustness to short-term outages such as those caused by GCS. This is why in Table 7.6-6 and Table 7.6-7 the latency of the P1 logical channel is larger than the other logical channels. The P1 logical channel, with its high degree of robustness, was designed to transmit core audio information.

Because of the requirement to minimize interference to the host analog signal, the carriers in the secondary and tertiary sidebands are placed at low levels, and as previously mentioned they maintain a phase relationship with the host analog signal. The P3 logical channel is transmitted on these carriers and therefore is less robust than the P1 logical channel. Because of this reduced robustness, audio enhancement information such as the stereo signal is typically transmitted on this logical channel. The P3 logical channel contains no time redundancy, but contains frequency redundancy in the upper and lower secondary sideband, as long as the host analog signal is transmitted with a 5-kHz bandwidth or less. If the analog audio extends into the secondary sidebands, decoding may be impaired because both sidebands are needed to demodulate the digital signal. This is why the digital signal is less robust to second adjacent interferers when 8-kHz analog signals are transmitted.

Figure 7.6-19 shows that service mode MA3 is the all-digital equivalent of service mode MA1. Because there is no analog signal to serve as a backup channel, the time diversity inherent in the P1 logical channel serves this purpose.

Layer 1 Functional Components

Several processing steps are necessary to convert the various logical channels into an AM IBOC system waveform. Figure 7.6-20 shows a functional block diagram of the Layer 1 processing. In this diagram, the single underline notation for a logical channel name indicates that data is passed between the various functions as vectors. During the interleaving process, logical channels lose their distinctiveness as they are combined or split by the interleaving process.

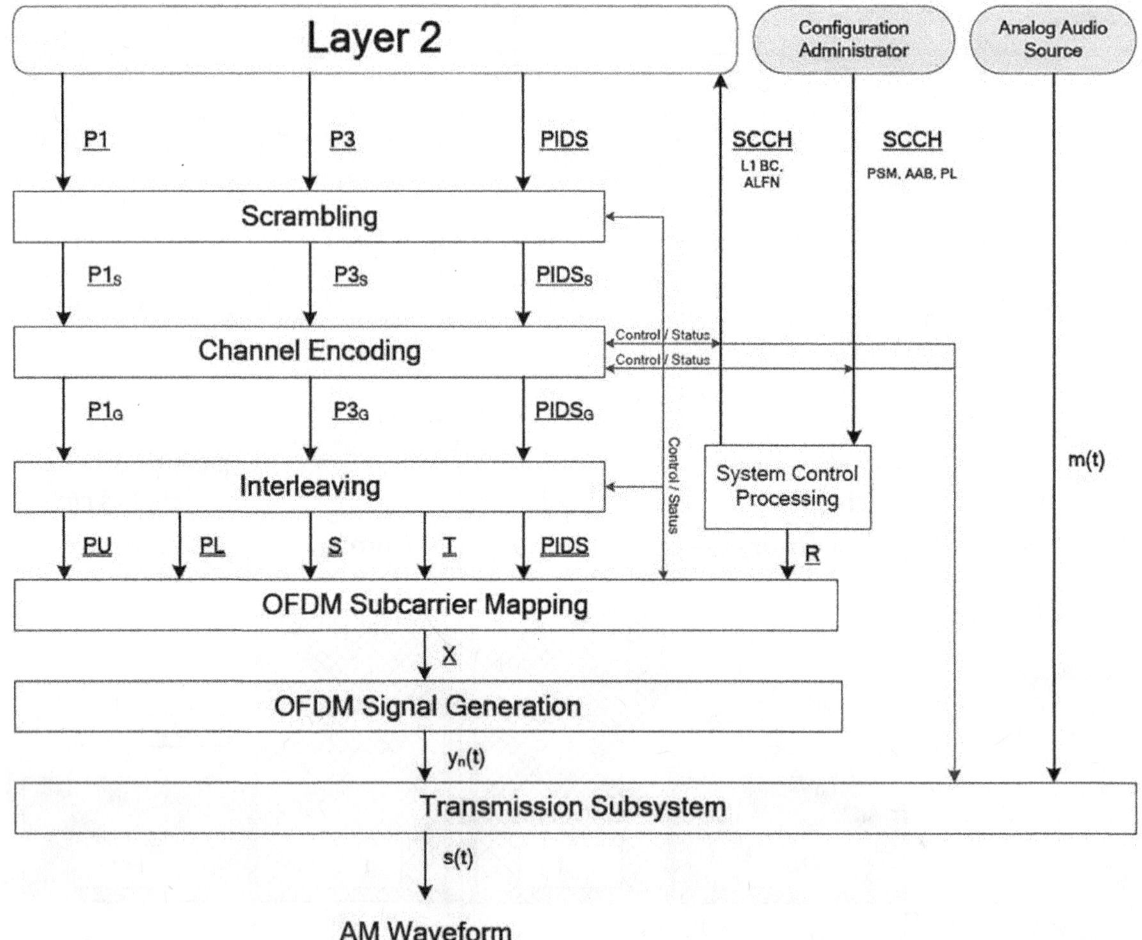

FIGURE 7.6-20 AM Layer 1 functional block diagram. The diagram illustrates the processing that takes place in Layer 1 after receiving data from Layer 2 at the top of the diagram.

DTS

Scrambling

This function randomizes the digital data carried in each logical channel to mitigate signal periodicities and aid in receiver synchronization. At the output of scrambling, the logical channel vectors retain their identity.

Channel Encoding

A digital bit stream, when passed through a transmission channel, is likely to encounter various forms of impairments including noise, distortion, fading and interference. Digital systems employ various error correction techniques to restore transmission errors. These algorithms improve robustness through the introduction of error correction bits that are used in the receiver to verify the accuracy of the recovered bit stream, detect errors and provide restoration of the transmitted bit stream.

Error correction codes are typically specified by their coding rate (R = information bits/total bits). For example, a coder with R = 1/2 channel coding represents an algorithm where half of the bits carry information and the other half carries the overhead of the error correction algorithm.

The size of the logical channel vectors is increased in inverse proportion to the code rate. The encoding techniques are configurable by service mode. At the output of the channel encoder, the logical channel vectors retain their identity.

Interleaving

Interleaving in time and frequency is employed to mitigate the effects of burst errors. Digital error correction techniques are enhanced if errors in transmission are spread in a manner that minimizes data loss in successive bits. Interleaving is a technique that jumbles the bits in a predetermined manner upon transmission and reassembles them in the receiver. The interleaving techniques used in the AM IBOC system are tailored to the AM nonuniform interference environment and are configurable by service mode. In this process, the logical channels lose their identity. The interleaver output is structured in a matrix format. Each matrix consists of information from whole or partial logical channels and is associated with a specific portion of the transmitted spectrum. Diversity delay is also imposed on selected logical channels. It is through this function that the time redundancy of logical channel P1, in service modes MA1 and MA3, is created.

System Control Processing

This function generates a vector of system control data sequences that includes system control information received from Layer 2 (such as service mode and configuration options) and status for broadcast on the reference subcarriers. This information is used at the receiver to determine how to process the AM IBOC system waveforms.

OFDM Subcarrier Mapping

This function assigns the interleaver matrices and system control vector to OFDM subcarriers. One row of each active interleaver matrix and one bit of the system control vector is processed for each OFDM symbol to produce one output vector, which is a frequency domain representation of the signal.

OFDM Signal Generation

This function generates the digital portion of the time-domain AM IBOC waveform. The input vectors are transformed into a shaped time-domain baseband pulse defining one OFDM symbol.

Transmission Subsystem

This function formats the baseband waveform for transmission through the MF channel. Major subfunctions include symbol concatenation and frequency up-conversion. When transmitting a hybrid waveform, this function modulates the AM analog audio source and coherently combines it with the digital signal to form a composite hybrid signal, ready for transmission.

FM HD RADIO SERVICES

FM Functional Description

The OFDM signal generation functional block receives complex, frequency-domain OFDM symbols from the OFDM subcarrier mapping functional block, and outputs time-domain pulses representing the digital portion of the FM HD Radio signal. A conceptual block diagram of OFDM signal generation is shown in Figure 7.6-21 The input to OFDM signal generation is a complex vector, representing the complex constellation values for each OFDM subcarrier. The output of OFDM signal generation is a complex, baseband, time-domain pulse, representing the digital portion of the FM IBOC signal.

FM HD Radio Transmission Subsystem

The transmission subsystem formats the baseband FM HD Radio waveform for transmission through the VHF channel. Functions include symbol concatenation and frequency up-conversion. In addition, when transmitting the hybrid or extended hybrid waveforms, this function modulates the baseband analog signal before combining it with the digital waveform.

The input to this module is a complex, baseband, time-domain OFDM symbol from the OFDM signal generation function. A baseband analog signal is also input from an analog source, after application of diversity delay along with optional subsidiary communications authorization (SCA) signals, when transmitting the hybrid or extended hybrid waveform. The output

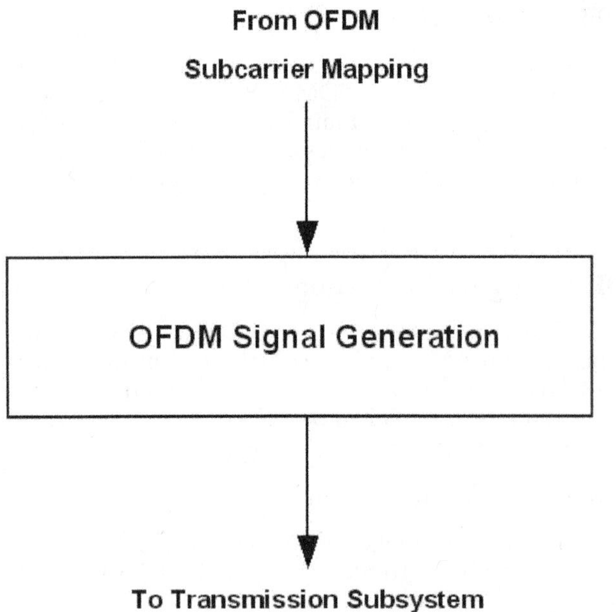

FIGURE 7.6-21 OFDM signal generation conceptual block diagram

of this module is the VHF FM HD Radio waveform. Refer to Figures 7.6-22 and 7.6-23 for functional block diagrams of the hybrid and all-digital transmission subsystems, respectively.

The FM HD Radio system affords broadcasters the ability to modify their digital audio broadcasts to meet their own specific needs. During the transition period to all-digital broadcasting, each station will have the opportunity to convert at its own pace, beginning with a hybrid analog/digital waveform and eventually turning off the analog and broadcasting an all-digital signal.

To support a wide of broadcast requirements, the FM HD Radio system was designed with a high degree of flexibility. All waveforms, hybrid or all-digital, may be configured in a number of ways by sensibly adjusting the throughput, latency and robustness of the audio and data program content as it is converted into an IBOC waveform. The following section describes the structure and generation of the FM IBOC waveforms and presents the various configurations from which a broadcaster can choose in order to transmit digital audio or data in a manner that best supports the broadcaster's needs.

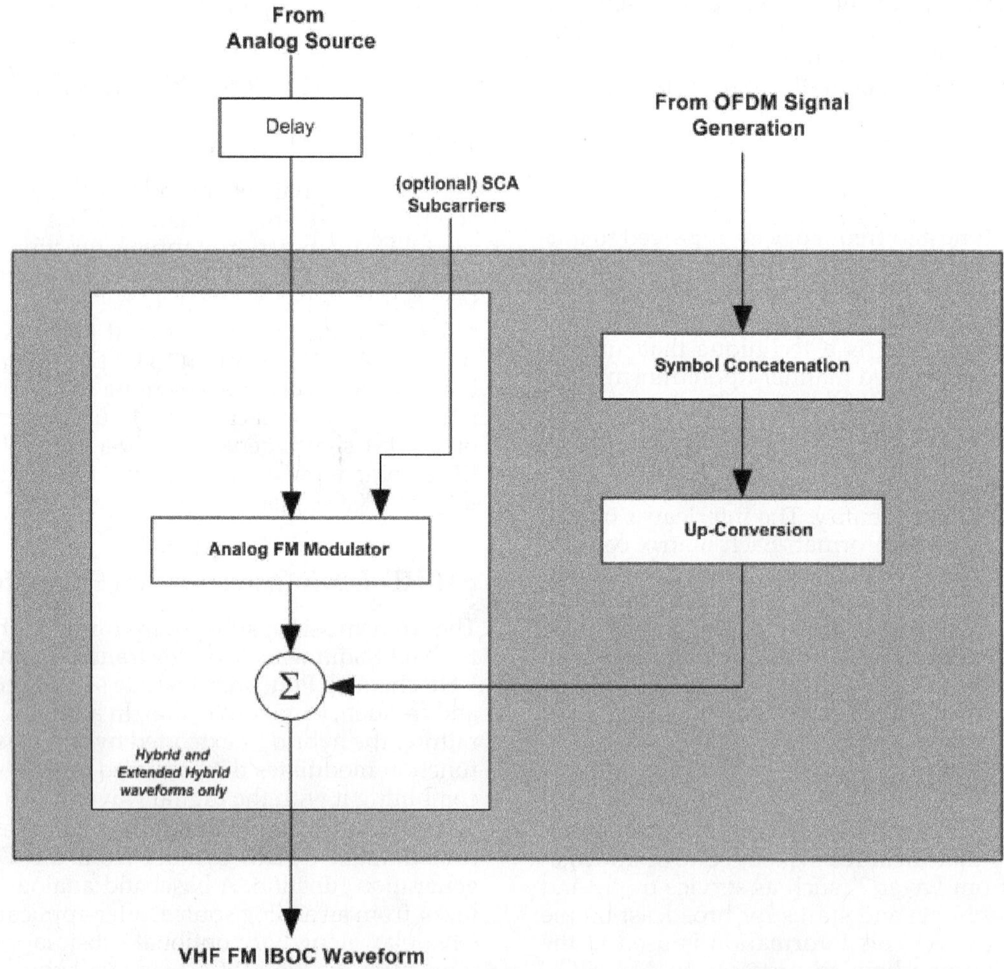

FIGURE 7.6-22 Hybrid transmission subsystem functional block diagram

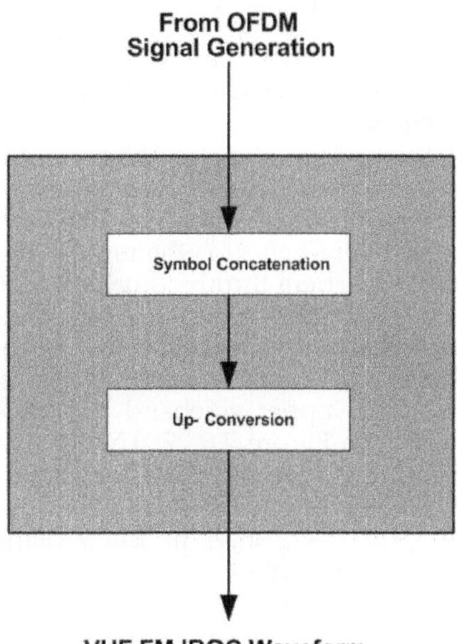

FIGURE 7.6-23 All-digital transmission subsystem functional block diagram

FM HD RADIO SERVICES AND PROTOCOLS

In order to provide broadcaster flexibility and enhance the listening experience, DTS' FM HD Radio technology supports a variety of digital program services. These include a Main Program Service (MPS), Program Service Data (PSD), Station Information Service (SIS) and Advanced Application Service (AAS).

The MPS delivers existing programming formats in digital audio along with digital data that directly correlates with the audio programming. Whereas the MPS broadcasts a traditional audio program to listeners, the PSD enables listeners to select on-demand data services, thereby providing personalized, user-valued information.

The SIS provides the control and identification information required to allow the listener to search and select IBOC digital radio stations and their supporting services. SIS is also utilized to support the HD Radio Emergency Alert broadcast protocol to convey a Common Alerting Protocol (CAP) compliant message definition adopted by the Federal Emergency Management Agency (FEMA) and other emergency organizations.

The HD Radio Emergency Alert broadcast protocol allows both AM and FM broadcasters to send multilingual announcements, text alerts, and image data to alert-enabled receivers. These receivers also have the ability to "wakeup" from a sleep state to notify the listener of local emergency events. Digital radio broadcast of alerts provides expansion beyond the analog Emergency Alert System (EAS). HD Radio Emergency Alerts will enable:

- Multilingual audio, provided simultaneously on stations supporting HD2/HD3/HD4;
- Multilingual text; provision for conveying images allows providing photos, maps and more;
- Alert parameters including alert "matter," locations, target audience and "severity";
- The primary message can support up to 374 text characters and target locations.

The AAS allows a virtually unlimited number of custom and specialized IBOC digital radio applications to exist concurrently. Auxiliary applications can be added at any time in the future. Simultaneous support of these services is provided via the layered protocol stack illustrated earlier in Figure 7.6-11. Source material (audio or data) moves down the protocol stack from Layer 5 to Layer 1 at the transmitter, is broadcast over the air and upon reception is passed back up the protocol stack from Layer 1 to Layer 5.

At the transmitter, Layer 5 receives audio or data program content from the broadcaster. Layer 4 provides content-specific source encoding (such as audio compression), as well as station identification and control capabilities. Layer 3 ensures robust and efficient transfer of Layer 4 data and Layer 2 provides limited error detection, addressing and multiplexing.

Layer 1 receives the formatted content from Layer 2 and creates an FM HD Radio waveform for over-the-air transmission in the FM band. Because most of the digital signal processing required to generate an FM IBOC waveform occurs in Layer 1, it will be explained in greater detail later.

Formatted program content is received from Layer 2 in discrete transfer frames via multiple logical channels. A transfer frame is an ordered collection of bits originating in Layer 2, grouped for processing through a logical channel. A logical channel is simply a signal path that conducts transfer frames from Layer 2 through Layer 1 with a specified grade of service. The service mode defines the active logical channels and their associated transmission characteristics.

The HD Radio system design provides a flexible means of transitioning to a digital broadcast system by providing three new waveform types: hybrid, extended hybrid and all-digital. The hybrid and extended hybrid types retain the analog FM signal while the all-digital type does not.

In all waveforms, the digital signal is modulated using OFDM. In a single-carrier digital modulation scheme, the digital symbols are transmitted serially, with the spectrum of each symbol occupying the entire channel bandwidth during its appointed signaling interval. Conversely, OFDM is a parallel modulation scheme in which the data stream simultaneously modulates a large number of orthogonal subcarriers. Instead of a single, wideband carrier at a high signaling rate, OFDM employs a large number of narrowband subcarriers that are simultaneously transmitted at a much lower composite symbol rate. The long symbol times of OFDM provide superior robustness in the presence of multipath fading and interference. OFDM is also inherently flexible, readily allowing the mapping of specific logical channels to different groups of subcarriers.

The following sections describe the transmitted spectrum for each of the three digital waveform types. Each spectrum is divided into several sidebands, which represent various OFDM subcarrier groupings. All spectra are illustrated at baseband, with an upper and lower sideband centered around 0 Hz.

Frequency Partitions and Spectral Conventions

The OFDM subcarriers are assembled into frequency partitions. Each frequency partition is comprised of 18 data subcarriers and 1 reference subcarrier, as shown in Figure 7.6-24 (ordering A) and Figure 7.6-25 (ordering B). The position of the reference subcarrier (ordering A or B) varies with the location of the frequency partition within the spectrum.

For each frequency partition, data subcarriers d1 through d18 convey digital program content, while the reference subcarrier conveys system control. OFDM subcarriers are numbered from 0 at the center frequency to ±546 at either end of the channel frequency allocation.

Besides the reference subcarriers within each frequency partition, depending on the service mode, up to five additional reference subcarriers are inserted into the spectrum at subcarrier numbers −546, −279, 0, +279 and +546. The overall effect is the regular distribution of reference subcarriers throughout the spectrum. For convenience, each reference subcarrier is assigned a unique identification number between 0 and 60. All lower sideband reference subcarriers are shown in Figure 7.6-26. All upper sideband reference subcarriers are shown in Figure 7.6-27. The figures indicate the relationship between reference subcarrier numbers and OFDM subcarrier numbers.

Each spectrum described in the remaining subsections shows the subcarrier number and center

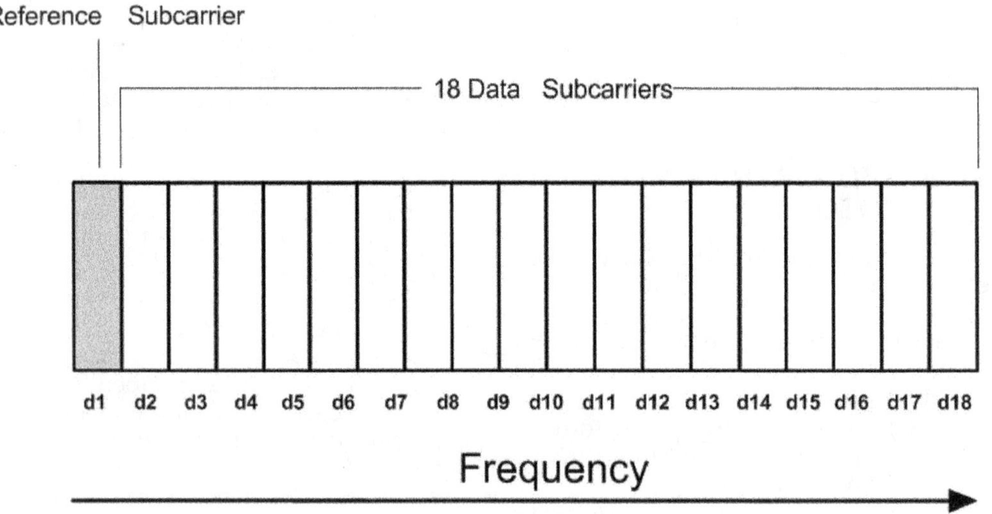

FIGURE 7.6-24 Frequency partition—ordering A

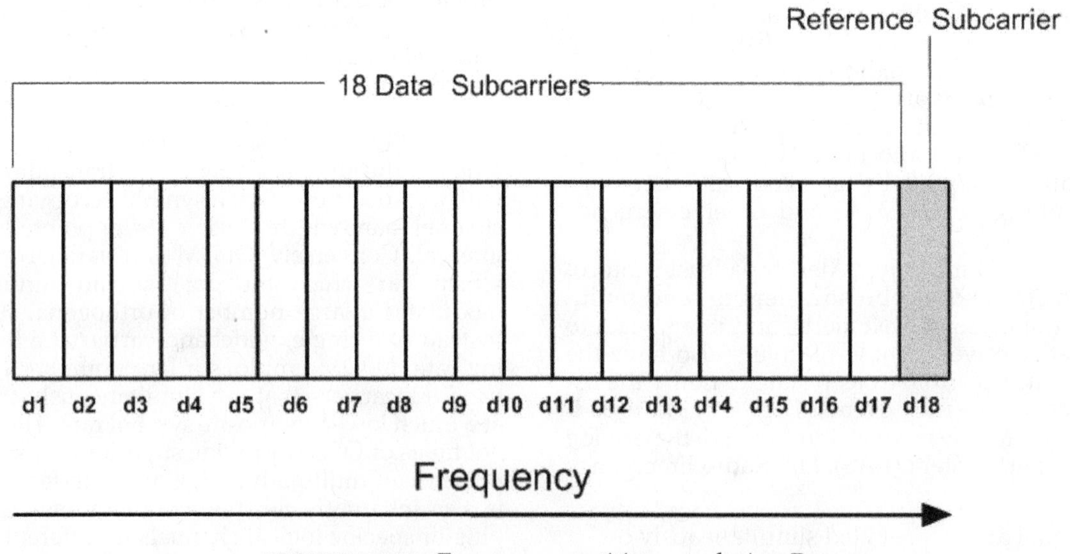

FIGURE 7.6-25 Frequency partition—ordering B

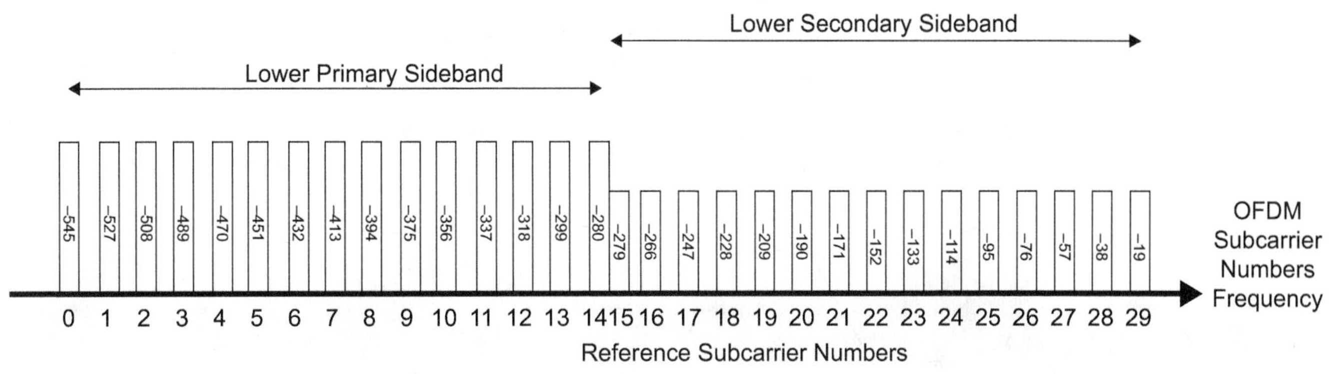

FIGURE 7.6-26 Lower sideband reference subcarrier spectral mapping

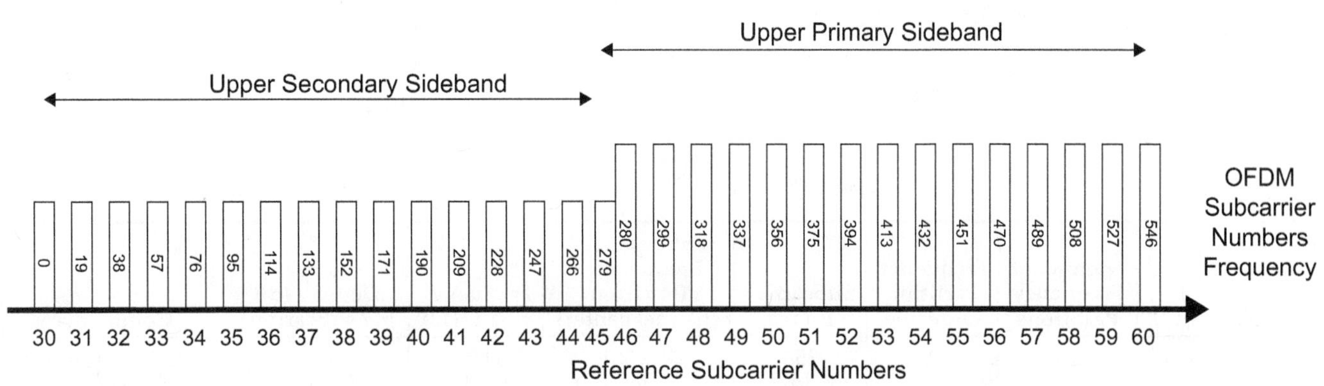

FIGURE 7.6-27 Upper sideband reference subcarrier spectral mapping

frequency of certain key OFDM subcarriers. The center frequency of a subcarrier is calculated by multiplying the subcarrier number by the OFDM subcarrier spacing $\Delta f \approx 363.373$ Hz. The center of subcarrier 0 is located at 0 Hz. In this context, center frequency is relative to the radio frequency (RF) allocated channel.

FM Hybrid Waveform

In the hybrid waveform, the digital signal is transmitted in primary main (PM) sidebands on either side of the analog FM signal, as shown in Figure 7.6-28. The analog signal may be mono or stereo and may include SCA channels. Each PM sideband comprises 10 frequency partitions, which are allocated among subcarriers +356 through +545 or −356 through −545. Subcarriers +546 and −546, also included in the PM sidebands, are additional reference subcarriers. Figure 7.6-28 summarizes the upper and lower primary main sidebands for the hybrid waveform.

The power spectral density of each OFDM subcarrier in the PM sideband, relative to the host analog power, and spectral summary is given in Table 7.6-8. A value of 0 dB would produce a digital subcarrier whose power was equal to the total power in the unmodulated analog FM carrier. The value was chosen so that the total average power in a primary main digital sideband (upper or lower) is 23 dB below the total power in the unmodulated analog FM carrier.

Extended Hybrid Waveform

The extended hybrid waveform is created by adding OFDM subcarriers to the primary main (PM) sidebands present in the hybrid waveform, as shown in Figure 7.6-29. Depending on the service mode, one, two, or four frequency partitions can be added to the inner edge of each primary main sideband. This additional spectrum is termed the Primary Extended (PX) sideband. Table 7.6-9 summarizes the upper and lower primary sidebands for the extended hybrid waveform.

The power spectral density of each OFDM subcarrier in the PM and PX sidebands, relative to the host analog power, is given in Table 7.6-9. Like the hybrid waveform, the value was chosen so that the total average power in a primary main sideband (upper or lower) is 23 dB below the total power in the unmodulated analog FM carrier. The level of the subcarriers in the PX sidebands is equal to the level of the subcarriers in the PM sidebands.

All-digital Waveform

The all-digital waveform is constructed by disabling the analog signal, fully expanding the bandwidth of the primary digital sidebands and adding lower-power secondary sidebands in the spectrum vacated by the analog signal. The spectrum of the all-digital waveform is shown in Figure 7.6-30.

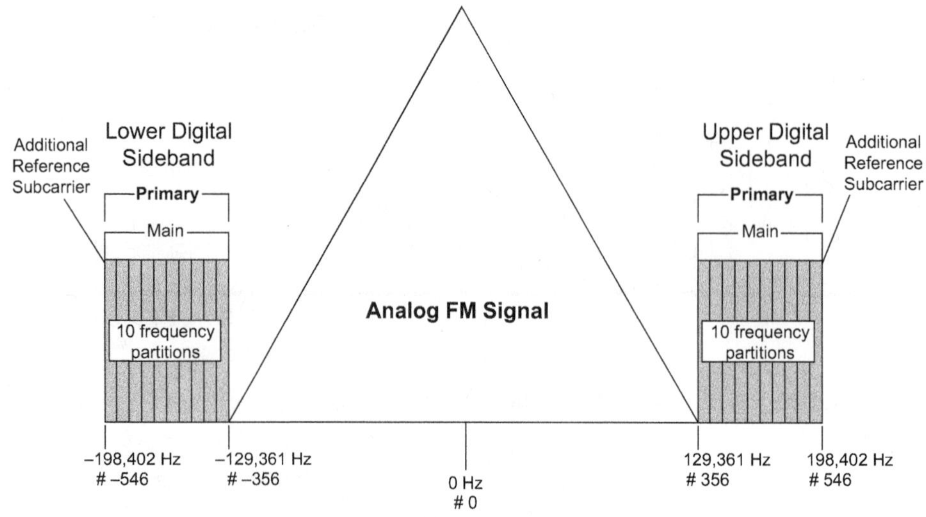

FIGURE 7.6-28 Spectrum of the hybrid waveform

TABLE 7.6-8
Hybrid waveform spectral summary

Sideband	Number of Frequency Partitions	Frequency Partition Ordering	Subcarrier Range	Subcarrier Frequencies (Hz from channel center)	Frequency Span (Hz)	Power Spectral Density (dB per subcarrier)	Comments
Upper Primary Main	10	A	356 to 546	129,361 to 198,402	69,041	–45.8	Includes additional reference subcarrier 546
Lower Primary Main	10	B	–356 to –546	–129,361 to –198,402	69,041	–45.8	Includes additional reference subcarrier –546

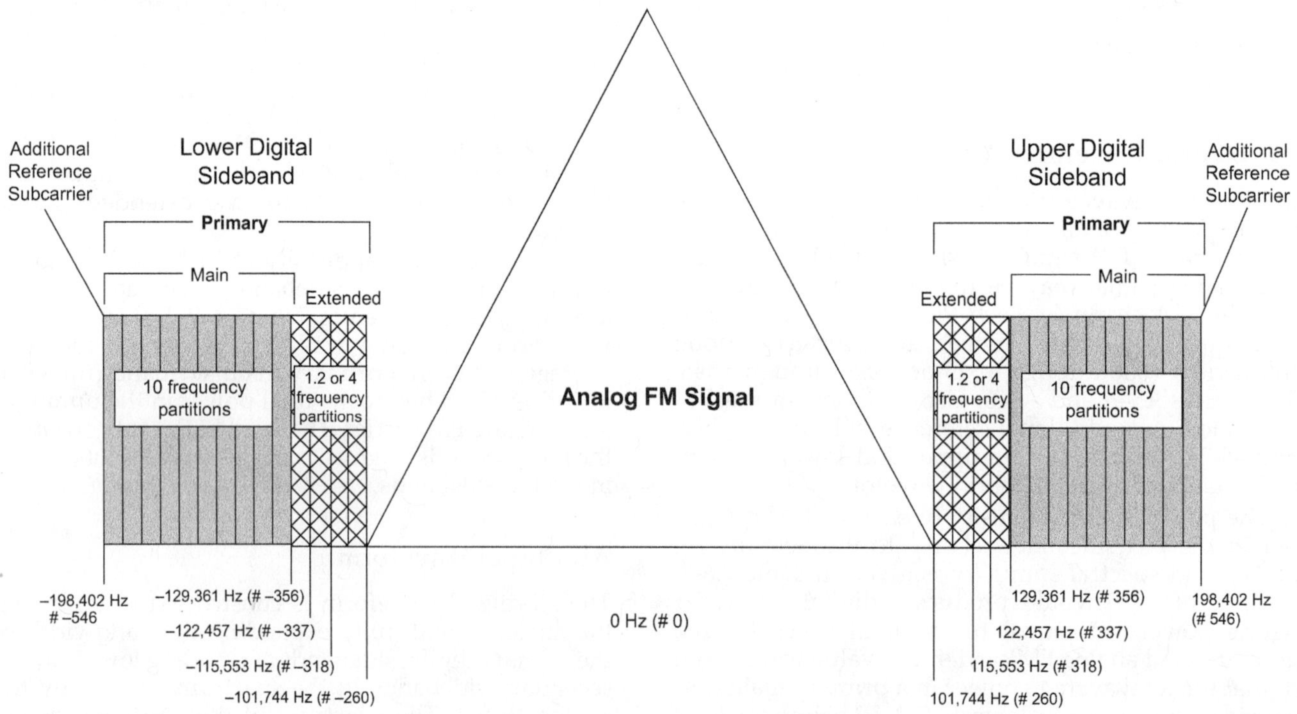

FIGURE 7.6-29 Spectrum of the extended hybrid waveform

TABLE 7.6-9
Extended hybrid waveform spectral summary

Sideband	Number of Frequency Partitions	Frequency Partition Ordering	Subcarrier Range	Subcarrier Frequencies (Hz from channel center)	Frequency Span (Hz)	Power Spectral Density (dB per subcarrier)	Comments
Upper Primary Main	10	A	356 to 546	129,361 to 198,402	69,041	–45.8	Includes additional reference subcarrier 546
Lower Primary Main	10	B	–356 to –546	–129,361 to –198,402	69,041	–45.8	Includes additional reference subcarrier –546
Upper Primary Extended (1 frequency partition)	1	A	337 to 355	122,457 to 128,997	6,540	–45.8	none
Lower Primary Extended (1 frequency partition)	1	B	–337 to –355	–122,457 to –128,997	6,540	–45.8	none
Upper Primary Extended (2 frequency partitions)	2	A	318 to 355	115,553 to 128,997	13,444	–45.8	none
Lower Primary Extended (2 frequency partitions)	2	B	–318 to –355	–115,553 to –128,997	13,444	–45.8	none
Upper Primary Extended (4 frequency partitions)	4	A	280 to 355	101,744 to 128,997	27,253	–45.8	none
Lower Primary Extended (4 frequency partitions)	4	B	–280 to –355	–101,744 to –128,997	27,253	–45.8	none

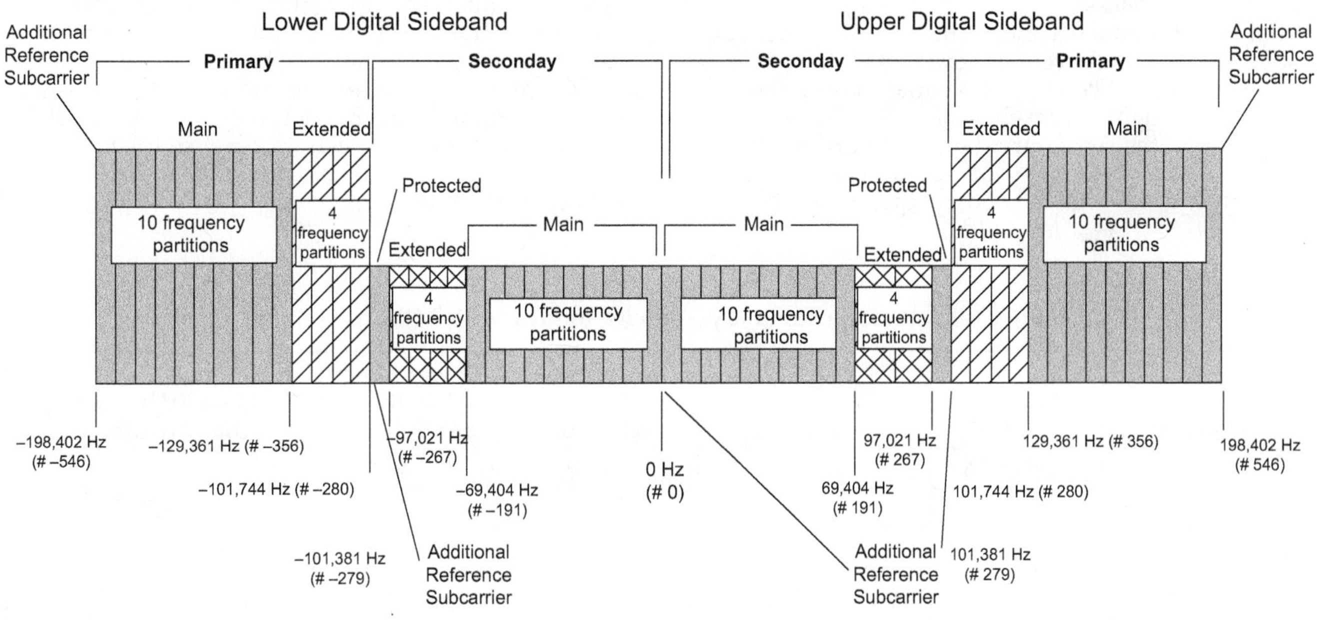

FIGURE 7.6-30 Spectrum of the all-digital waveform

In addition to the 10 main frequency partitions, all 4 extended frequency partitions are present in each primary sideband of the all-digital waveform. Each secondary sideband also has 10 secondary main (SM) and 4 secondary extended (SX) frequency partitions. Unlike the primary sidebands, the secondary main frequency partitions are mapped nearer to channel center with the extended frequency partitions farther from the center.

Each secondary sideband also supports a small secondary protected (SP) region consisting of 12 OFDM subcarriers and reference subcarrier +279 or −279. The sidebands are referred to as "protected" because they are located in the area of spectrum least likely to be affected by analog or digital interference. An additional reference subcarrier is placed at the center of the channel (0). Frequency partition ordering of the SP region does not apply because the SP region does not contain frequency partitions as defined in Figure 7.6-30. The total frequency span of the entire all-digital spectrum is 396,803 Hz. Table 7.6-10 summarizes the upper and lower, primary and secondary sidebands for the all-digital waveform. The power spectral density of each OFDM subcarrier is also provided.

For uniformity, as with the hybrid and extended hybrid waveforms, the values are referenced to the power level of the unmodulated analog FM carrier allocated by the station's FCC license (even though the analog carrier is not transmitted in the all-digital waveform). The primary sideband level sets the total average power in a primary digital subcarrier at least 10 dB above the total power in a hybrid primary digital subcarrier. Any one of four power levels may be selected for application to the secondary sidebands. The four secondary power levels set the power spectral density of the secondary digital subcarriers (upper and lower) in the range of 5–20 dB below the power spectral density of the all-digital primary subcarriers. A single secondary power level is evenly applied to all secondary sidebands.

For the all-digital waveform, the value of a_1 was chosen so that the total average power of all the primary digital subcarriers combined is equal to one. The values for a_2 through a_5 were chosen so that the total average power in the secondary digital subcarriers (upper and lower) lies in the range of 5–20 dB below the total power in the all-digital primary digital subcarriers. The selection of one of the values a_2 through a_5 is determined by the amplitude scale factor (ASF) select received from Layer 2.

Logical Channels

A logical channel is a signal path that conducts program content through Layer 1 with a specific grade of service, as determined by the service mode. There are 10 logical channels, although not all are used in every service mode. The variety of logical channels reflects the inherent flexibility of the system.

There are four primary logical channels, denoted as P1, P2, P3 and PIDS. There are six secondary logical channels that are used only with the all-digital waveform. They are denoted as S1, S2, S3, S4, S5 and SIDS. Logical channels P1 through P3 and S1 through S5 are designed to convey digital audio and data, while the PIDS and SIDS logical channels are designed to carry IBOC data service (IDS) information.

FM HD RADIO CHARACTERIZATION PARAMETERS

The performance of each logical channel is completely described through three characterization parameters: throughput, latency and robustness. The service mode sets these characterization parameters by defining the spectral mapping, interleaver depth, diversity delay and channel encoding for each active logical channel.

Throughput

Throughput defines the Layer 1 audio or data capacity of a logical channel, excluding upper Layer framing overhead. The block-oriented operations of Layer 1 (such as interleaving) require that it process data in discrete transfer frames rather than continuous streams. As a result, throughput is calculated as the product of transfer frame size and transfer frame rate. Spectral mapping and channel code rate determine the throughput of a logical channel, because spectral mapping limits capacity and coding overhead limits information throughput.

TABLE 7.6-10
All-digital waveform summary—service modes MP5, MP6, MS1–MS4

Waveform	Service Mode	Sidebands	Amplitude Scale Factor Notation	Power Spectral Density, dBc per Subcarrier	Power Spectral Density in a 1-kHz Bandwidth, dBc
All-digital	MP5–MP6	Primary	a_1	−27.3	−22.9
	MS1–MS4	Secondary	a_2	−32.3	−27.9
		Secondary	a_3	−37.3	−32.9
		Secondary	a_4	−42.3	−37.9
		Secondary	a_5	−47.3	−42.9

Latency

Latency is the delay that a logical channel imposes on a transfer frame as it traverses Layer 1. The latency of a logical channel is defined as the sum of its interleaver depth and diversity delay. It does not include processing delay or delays through higher protocol Layers.

The interleaver depth determines the amount of delay imposed on a logical channel by its interleaver. Diversity delay is also applied to some logical channels to improve robustness. For example, in some service modes, logical channel P1 presents dual processing paths; one path is delayed and the other is not.

Robustness

Robustness is the ability of a logical channel to withstand channel impairments such as noise, interference and fading. There are 11 relative levels of robustness in the FM IBOC system (as contrasted with only 10 for the AM IBOC system). A robustness of 1 indicates a very high level of resistance to channel impairments, while an 11 indicates a lower tolerance for channel-induced errors. Spectral mapping, channel code rate, interleaver depth and diversity delay determine the robustness of a logical channel. Spectral mapping affects robustness by setting the relative power level, spectral interference protection, and frequency diversity of a logical channel. Channel coding increases robustness by introducing redundancy into the logical channel. Interleaver depth influences performance in multipath fading. Finally, some logical channels in certain service modes delay transfer frames by a fixed duration to realize time diversity. This diversity delay also affects robustness, because it mitigates the effects of the mobile radio channel.

Assignment of Characterization Parameters

Tables 7.6-11 through 7.6-20 show the active logical channels and their characterization parameters: throughput, latency, and relative robustness for a given service mode. A broadcaster might use these tables as a basis of comparison when selecting a service mode.

TABLE 7.6-11
Logical channels—service mode MP1

Logical Channel	Throughput (kbps)	Latency (seconds)	Relative Robustness
P1	98.4	1.49	2
PIDS	0.9	0.09	3

TABLE 7.6-12
Logical channels—service mode MP2

Logical Channel	Throughput (kbps)	Latency (seconds)	Relative Robustness
P1	98.4	1.49	2
P3	12.4	0.19	4
PIDS	0.9	0.09	3

TABLE 7.6-13
Logical channels—service mode MP3

Logical Channel	Throughput (kbps)	Latency (seconds)	Relative Robustness
P1	98.4	1.49	2
P3	24.8	0.19	4
PIDS	0.9	0.09	3

TABLE 7.6-14
Logical channels—service mode MP11

Logical Channel	Throughput (kbps)	Latency (seconds)	Relative Robustness
P1	98.4	1.49	2
P3	49.6	0.19	4
PIDS	0.9	0.09	3

TABLE 7.6-15
Logical channels—service mode MP5

Logical Channel	Throughput (kbps)	Latency (seconds)	Relative Robustness
P1	24.8	4.64	1
P2	73.6	1.49	2
P3	24.8	0.19	4
PIDS	0.9	0.09	3

TABLE 7.6-16
Logical channels—service mode MP6

Logical Channel	Throughput (kbps)	Latency (seconds)	Relative Robustness
P1	49.6	4.64	1
P2	48.8	1.49	2
PIDS	0.9	0.09	3

TABLE 7.6-17
Logical channels—service mode MS1

Logical Channel	Throughput (kbps)	Latency (seconds)	Relative Robustness
S4	98.4	0.19	7
S5	5.5	0.09	6
SIDS	0.9	0.09	8

TABLE 7.6-18
Logical channels—service mode MS2

Logical Channel	Throughput (kbps)	Latency (seconds)	Relative Robustness
S1	24.8	4.64	5
S2	73.6	1.49	9
S3	24.8	0.19	11
S5	5.5	0.09	6
SIDS	0.9	0.09	10

TABLE 7.6-19
Logical channels—service mode MS3

Logical Channel	Throughput (kbps)	Latency (seconds)	Relative Robustness
S1	49.6	4.64	5
S2	48.8	1.49	9
S5	5.5	0.09	6
SIDS	0.9	0.09	10

TABLE 7.6-20
Logical channels—service mode MS4

Logical Channel	Throughput (kbps)	Latency (seconds)	Relative Robustness
S1	24.8	0.19	11
S2	98.4	1.49	9
S3	24.8	0.19	11
S5	5.5	0.09	6
SIDS	0.9	0.09	10

Spectral Mapping and Service Modes

For a given service mode, each logical channel is assigned to a group of OFDM subcarriers or frequency partitions. This spectral mapping contributes to the throughput and robustness of the logical channel. Hybrid modes include MP1, MP2, MP3, and MP11. (Note: MP11 replaced MP4; note that MP4 had an error in carrier mapping that failed to have an all-digital equivalent. As a result, a station implementing channels in hybrid MP4 would not be in a position to field the same services in MP6. The carrier groupings were harmonized and MP4 was replaced by mode MP11.) All-digital modes include MP5 and MP6. Secondary modes that may be added to MP5 or MP6 are MS1, MS2, MS3 and MS4.

Because this is a digital system, the various logical channels are simply conduits for the delivery of bits; the content of the bits is immaterial. However, the service modes were designed with specific services in mind for the active logical channels. As a result, although not strictly required, the recommended use of the logical channels is described along with the spectral mapping in the following sections.

Primary Spectral Mapping

The following sections describe the assignment of logical channels to the primary sidebands and describe the intended application of the logical channels for each primary service mode.

- Service mode MP1:

  The assignment of logical channels to OFDM subcarriers in service mode MP1 is shown in Figure 7.6-31. Both the P1 and PIDS logical channels are mapped to the upper and lower primary main sidebands. In service mode MP1, the P1 logical channel is designed to carry the MPS audio, while the PIDS logical channel would carry SIS data. Identical program material is carried on each sideband (upper and lower), so that the alternate sideband would be available if the other sideband were corrupted.

- Service mode MP2:

  The assignment of logical channels to OFDM subcarriers in service mode MP2 is shown in Figure 7.6-32. The transmitted spectrum for MP2 is identical to MP1, with the addition of a single extended frequency partition to each primary sideband. As in service mode MP1, the P1 and PIDS logical channels carry MPS audio and SIS data on each primary main sideband. In addition, the P3

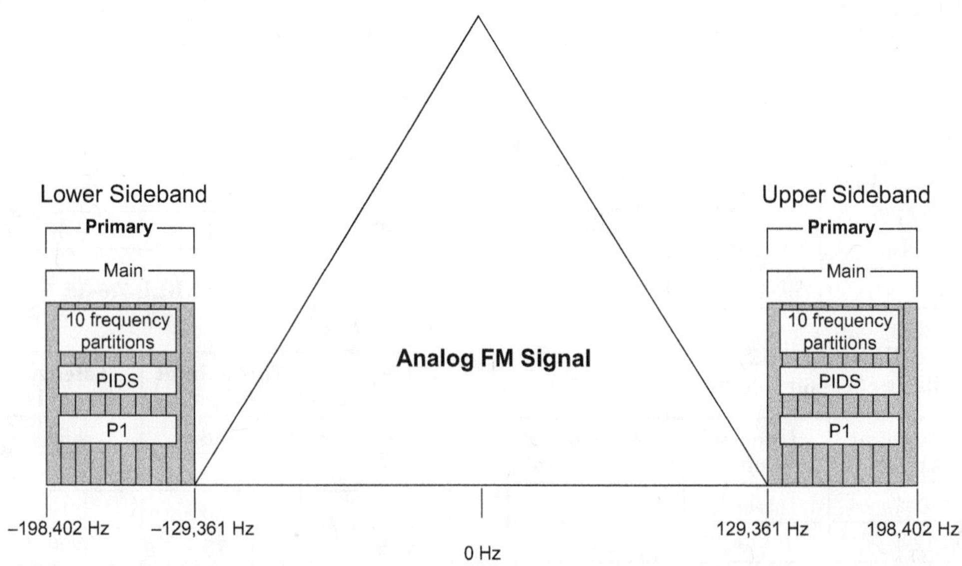

FIGURE 7.6-31 Spectral mapping—service mode MP1

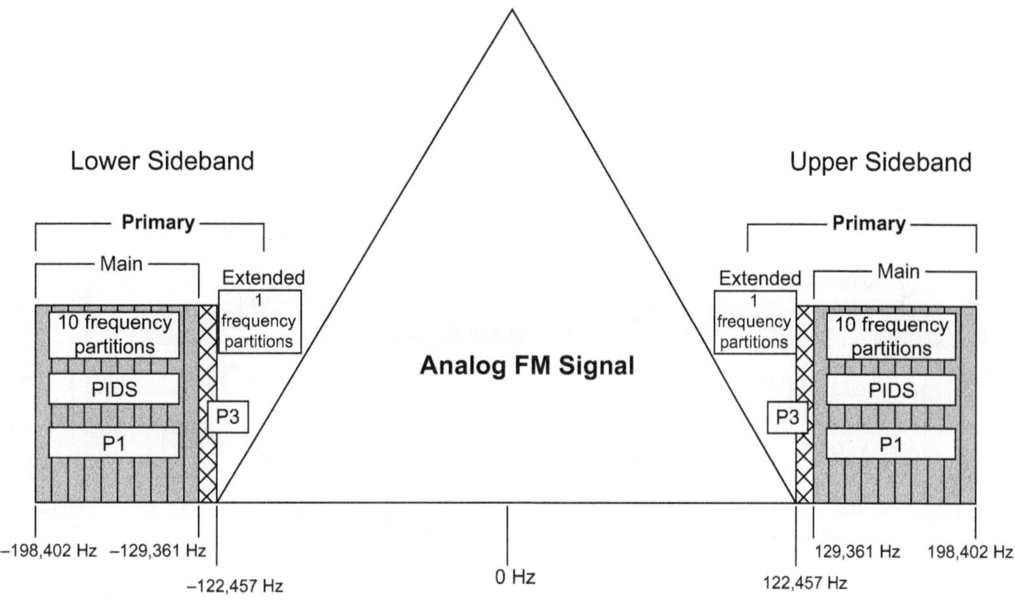

FIGURE 7.6-32 Spectral mapping—service mode MP2

logical channel is designed to carry additional data services, such as MPS, PDS or AAS data, on the primary extended sidebands. Identical program material is carried on each sideband, so that the alternate sideband would be available if the other sideband were corrupted.

- Service mode MP3:

The assignment of logical channels to OFDM subcarriers in service mode MP3 is shown in Figure 7.6-33. The transmitted spectrum for MP3 is identical to MP1, with the addition of two extended frequency partitions to each primary sideband. As in service mode MP1, the P1 and PIDS logical channels carry MPS audio and SIS data on each primary main sideband. In addition, the P3 logical channel is designed to carry additional data services, such as MPS, PDS or AAS data, on the primary extended sidebands. Identical program material is carried on each sideband, so that the alternate sideband would be available if the other sideband were corrupted.

- Service mode MP11:

The assignment of logical channels to OFDM subcarriers in service mode MP11 is shown in Figure 7.6-34. The transmitted spectrum for MP11 is identical to MP1, with the addition of all four extended frequency partitions to each primary sideband. As in service mode MP1, the P1 and PIDS logical channels carry MPS audio and SIS data on each primary main sideband. In addition, the P3 logical channel is designed to carry additional data services, such as MPS, PDS or AAS data, on the primary extended sidebands. Identical program material is carried on each sideband, so that the alternate sideband would be available if the other sideband were corrupted.

Service modes MP1 through MP11 provide essentially the same program services, with varying data capacity via the P3 logical channel on the primary extended sidebands.

- Service mode MP5:

The assignment of logical channels to OFDM subcarriers in service mode MP5 is shown in Figure 7.6-35. The transmitted spectrum is identical to MP11. However, the spectral mapping for MP5 allows operation as either an extended hybrid or all-digital waveform.

In service mode MP5, the MPS audio is divided into core and enhanced audio streams. The core audio is a stand-alone, low bitrate (~25 kbps), backup audio stream. When the core audio is combined with enhanced audio, the result is CD-like quality (~98 kbps) audio stream. The enhanced audio is not autonomous; it can only be used in combination with the core audio stream.

In service mode MP5, the core MPS audio stream is carried by the P1 logical channel and the enhanced audio is carried by the P2 logical channel. Both P1 and P2 are mapped together in the primary main sidebands. In addition, the same P1 logical channel is diversity delayed and separately mapped to the inner two extended frequency partitions of each primary sideband. At the receiver, the two P1 channels are combined to form a more robust backup core audio stream.

In hybrid and extended hybrid waveforms, the analog host provides fast tuning and a diversity-delayed backup channel for graceful degradation of audio near the edge of coverage. In the all-digital waveform, the analog host no longer exists. In this case, the robust P1 logical channel, carrying core audio, acts as the backup for graceful audio degradation and fast

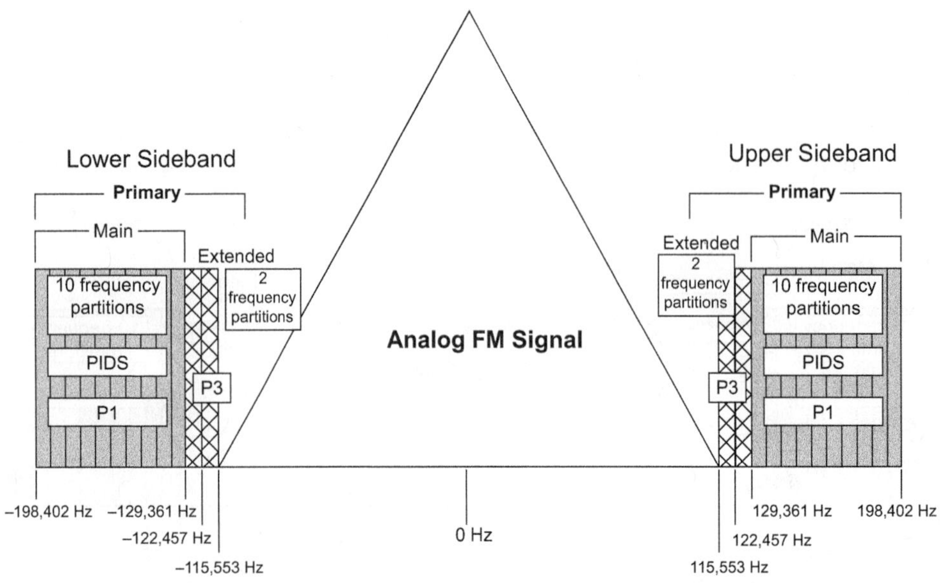

FIGURE 7.6-33 Spectral mapping—service mode MP3

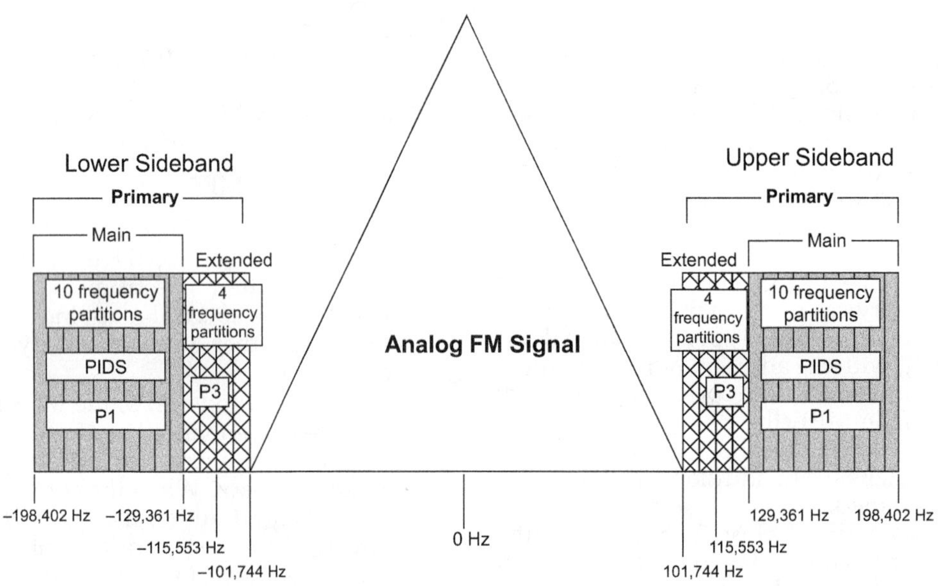

FIGURE 7.6-34 Spectral mapping—service mode MP11

tuning (because it is lightly interleaved). When the enhanced audio is not available, the receiver reverts to the backup core audio stream.

The P3 logical channel also carries additional data services, such as MPS, PDS or AAS data, on the primary extended sidebands. As in service modes MP1–MP11, the PIDS logical channel carries SIS data over the primary main sidebands. Again, identical program material is carried on each sideband, so that the alternate sideband would be available if the other sideband were corrupted.

- Service mode MP6:

The assignment of logical channels to OFDM subcarriers in service mode MP6 is shown in Figure 7.6-36. The transmitted spectrum is identical to MP5. However, for MP6, the size of the core audio stream is doubled to a higher quality ~50 kbps. As a result, all four frequency partitions in the primary extended sideband are required to carry the backup core audio and capacity is no longer available for data. Thus, the increased data capacity in service mode MP5 is traded for core audio quality in service mode MP6.

In service mode MP6, the core MPS audio stream is carried by the P1 logical channel and the enhanced audio is carried by the P2 logical channel. Both P1 and P2 are mapped together in the primary main sidebands. In addition, the same P1 logical channel is diversity delayed and separately

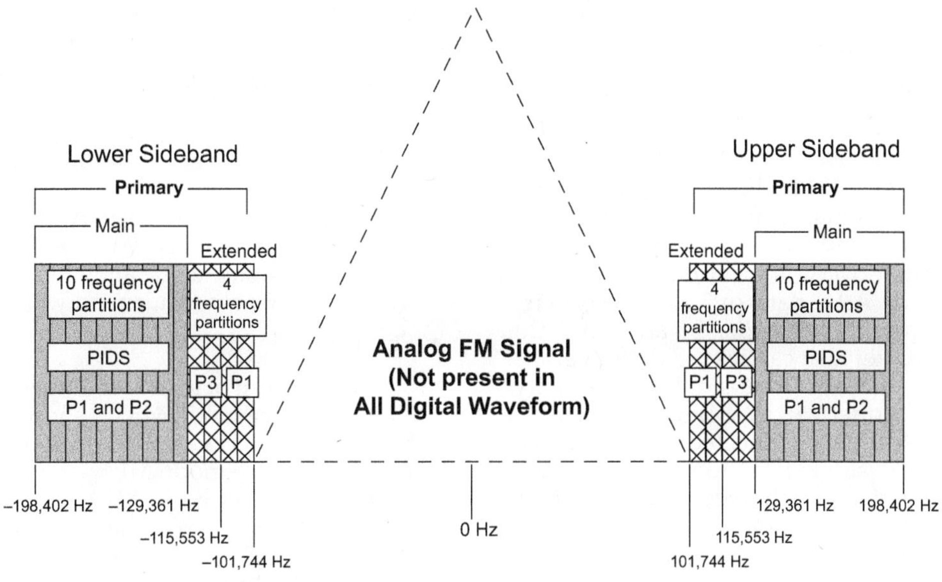

FIGURE 7.6-35 Spectral mapping—service mode MP5

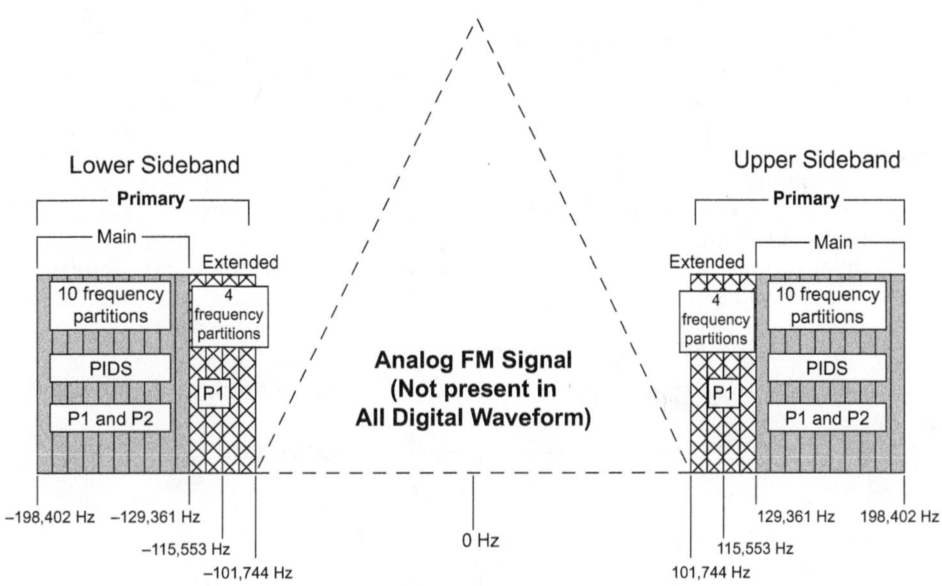

FIGURE 7.6-36 Spectral mapping—service mode MP6

mapped to all four extended frequency partitions of each primary sideband. The PIDS logical channel also carries SIS data over the primary main sidebands. Identical program material is carried on each sideband, so that the alternate sideband would be available if the other sideband were corrupted.

As in service mode MP5 the P3 logical channel carries additional data services, such as MPS, PDS or AAS data, on the primary extended sidebands. The PIDS logical channel also carries SIS data over the primary main sidebands. As always, identical program material is carried on each sideband, so that the alternate sideband would be available if the other sideband were corrupted.

Secondary Spectral Mapping

The following sections describe the assignment of logical channels to the secondary sidebands and describe the intended application of the logical channels for each secondary service mode. Note that secondary sidebands are present only in the all-digital waveform. Only the secondary sidebands are presented in the following subsections; the presence of primary digital sidebands in service modes MP5 or MP6 is implied.

"Implied" means that only the MS1–MS4 carrier groups are illustrated and they are to be used in combination with the MP5 or MP6 modes. It is possible to run MP5 or MP6 with the analog signal present.

If the analog signal is removed, there is room to add MS1 through MS4 secondary carriers in place of the analog signal (see Figure 7.6-30 earlier).

- Service mode MS1:

  The assignment of logical channels to OFDM subcarriers in service mode MS1 is shown in Figure 7.6-37. Service mode MS1 is intended for the transmission of secondary broadband data.

  In service mode MS1, logical channel S4 carries MPS, PDS or AAS data over the secondary main and extended sidebands. In addition, the SIDS logical channel also carries SIS data over the secondary main and extended sidebands. Finally, the S5 logical channel carries MPS, PDS or AAS data over the secondary protected sidebands. As with the primary sidebands, identical program material is carried on each secondary sideband (upper and lower), so that the alternate sideband would be available if the other sideband were corrupted.

- Service mode MS2:

  The assignment of logical channels to OFDM subcarriers in service mode MS2 is shown in Figure 7.6-38. Service mode MS2 is the secondary equivalent of primary service mode MP5.

  In service mode MS2, the S1 and S2 logical channels might carry core and enhanced auxiliary audio (such as surround sound), intended to enhance the MPS audio broadcast on the primary sidebands. Both S1 and S2 are mapped together in the secondary main sidebands. In addition, the same S1 logical channel is diversity delayed and separately mapped to the outer two extended frequency partitions of each secondary sideband.

  The S3 logical channel carries additional data services, such as MPS, PDS or AAS data, on the secondary extended sidebands. The SIDS logical channel also carries SIS data over the secondary main sidebands. Finally, the S5 logical channel carries MPS, PDS or AAS data over the secondary protected sidebands. Identical program material is carried on each secondary sideband, so that the alternate sideband would be available if the other sideband were corrupted.

- Service mode MS3:

  The assignment of logical channels to OFDM subcarriers in service mode MS3 is shown in Figure 7.6-39. Service mode MS3 is the secondary equivalent of primary service mode MP6.

  As in service mode MS2, the S1 and S2 logical channels might carry core and enhanced auxiliary audio (such as surround sound), intended to enhance the MPS audio broadcast on the primary sidebands. However, in service mode MS3, the size of S1 is doubled and capacity is no longer available for S3 data.

  Both S1 and S2 are mapped together in the secondary main sidebands. In addition, the same S1 logical channel is diversity delayed and separately mapped to all four extended frequency partitions of each secondary sideband. The SIDS logical channel also carries SIS data over the secondary main sidebands. Finally, the S5 logical channel carries MPS, PDS or AAS data over the secondary protected sidebands. Identical program material is carried on each sideband, so that the alternate sideband

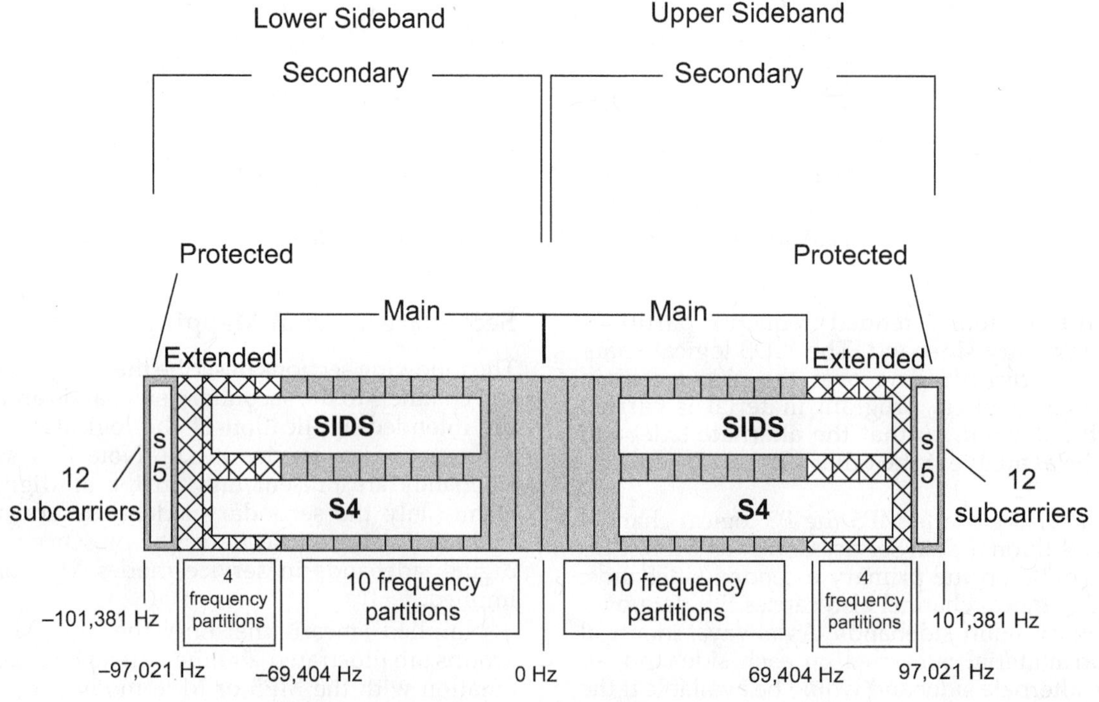

FIGURE 7.6-37 Spectral mapping—service mode MS1

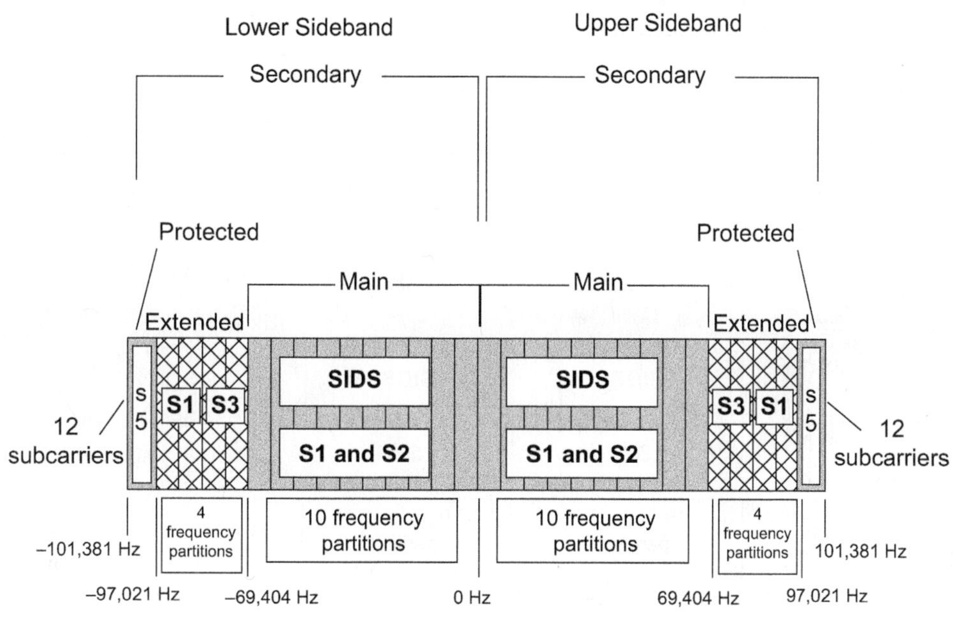

FIGURE 7.6-38 Spectral mapping—service mode MS2

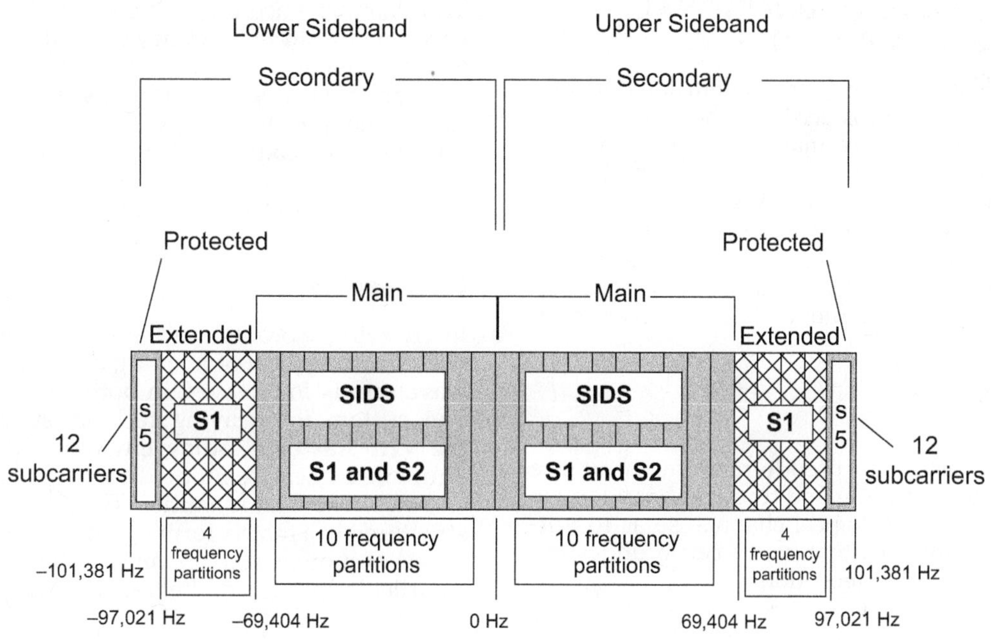

FIGURE 7.6-39 Spectral mapping—service mode MS3

would be available if the other sideband were corrupted.

- Service mode MS4:

The assignment of logical channels to OFDM subcarriers in service mode MS4 is shown in Figure 7.6-40. It is intended for broadcast of a single, low-bitrate audio stream, with the remaining capacity reserved for data services.

In service mode MS4, the low-bitrate audio is carried by the S1 logical channel, which is mapped to the outer two extended frequency partitions of each secondary sideband. Logical channel S2 carries MPS, PDS or AAS data over the secondary main sidebands. As in service mode MS2, the S3 logical channel carries additional data services, such as MPS, PDS or AAS data, on the secondary extended sidebands. The SIDS logical channel also carries SIS data over the secondary main sidebands. As always, identical program material is carried on each sideband, so that the alternate sideband would be available if the other sideband were corrupted.

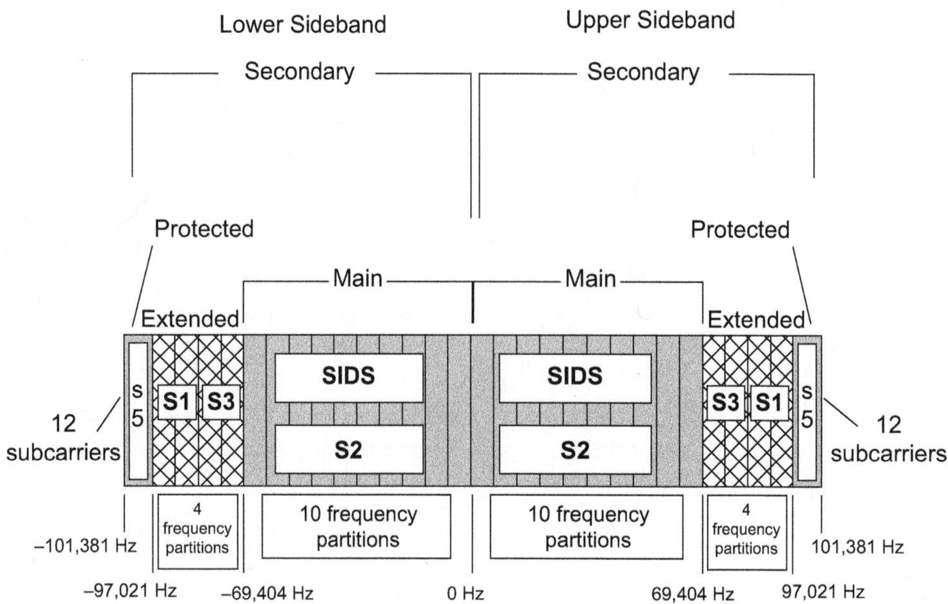

FIGURE 7.6-40 Spectral mapping—service mode MS4

FM HD RADIO FUNCTIONAL COMPONENTS

The conversion of audio program content and data into the FM IBOC waveform is accomplished by the Layered protocol stack. Source material is received from the broadcaster in Layer 5; source encoded in Layer 4, multiplexed into logical channels in Layers 3 and 2 and formatted for over-the-air broadcast in Layer 1.

This section includes a high-level description of each Layer 1 functional block and the associated signal flow. Figure 7.6-41 is a functional block diagram of Layer 1 processing.

Scrambling

This function uses a pseudo-random sequence employed in the transmitter and receiver to randomize the digital data in each logical channel, so as to mitigate signal periodicities that could cause undesired emissions and degraded reception.

Channel Encoding

A digital signal, when passed through an RF transmission channel, is likely to be impaired by noise, fading and interference. Digital systems employ error correction techniques to correct bit errors caused by these impairments. FEC algorithms improve signal robustness by adding error correction bits to the signal prior to transmission. These FEC bits are used by the receiver to correct bit errors and regenerate the transmitted bit stream. FEC codes are typically specified by their coding rate, which is simply the number of information bits divided by the total number of transmitted bits. For example, in a rate 1/2 code, half of the bits carry information and the other half carry the FEC overhead.

The channel encoding function uses convolutional encoding to add redundancy to the digital data in each logical channel, in order to improve its reliability in the presence of channel impairments. The size of the logical channel transfer frames is increased in inverse proportion to the code rate. The encoding techniques are configurable by service mode. Diversity delay is also imposed on selected logical channels.

Interleaving

Interleaving reorders the transmitted bits to disperse burst errors typical of a fading channel. The FM IBOC waveform is interleaved in both time and frequency. The custom interleaving techniques are tailored to the VHF Rayleigh fading environment and are configurable by service mode. In this process, the logical channels lose their identities. The interleaver output is structured in a matrix format; each matrix is comprised of one or more logical channels and is associated with a particular portion of the transmitted spectrum.

System Control Processing

This function generates a matrix of system control data sequences, which includes control and status (such as service mode), for broadcast on the reference subcarriers.

OFDM Subcarrier Mapping

This function assigns the interleaver matrices and the system control data matrix to the OFDM subcarriers. One row of each active interleaver matrix is processed every OFDM symbol time to produce one output vector X, which is a frequency-domain representation of

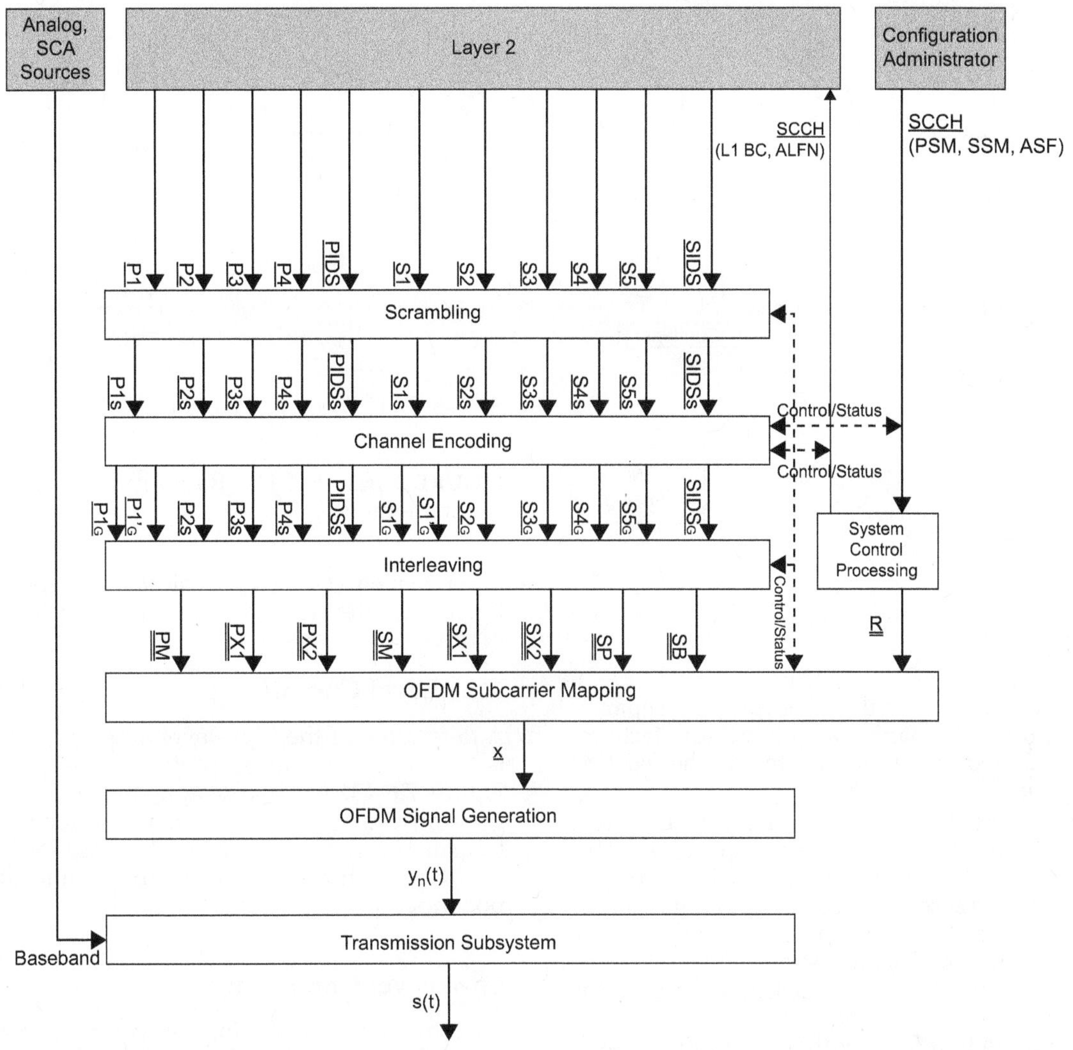

FIGURE 7.6-41 Layer 1 functional block diagram. The diagram illustrates the processes included in Layer 1 after the data is received from Layer 2 at the top of the diagram.

DTS

the signal. The mapping is specifically tailored to the nonuniform interference environment and is a function of the service mode.

OFDM Signal Generation

This function generates the digital portion of the time-domain FM IBOC waveform. The input vectors are transformed into a shaped time-domain baseband pulse defining one OFDM symbol.

Transmission Subsystem

This function formats the baseband waveform for transmission through the VHF channel. Major subfunctions include symbol concatenation and frequency upconversion. In addition, when transmitting the hybrid or extended hybrid waveforms, this function modulates the analog source and combines it with the digital signal to form a composite signal ready for transmission.

FM HYBRID IMPLEMENTATION

FM hybrid transmission modes MP1, MP2, MP3 and MP11 all require the simultaneous transmission of an analog and a digital signal. The following methods can be used by the broadcasters to transmit the two signals at the proper analog-to-digital power ratios. Four methods exist for producing the HD Radio hybrid FM signal.

High-Level Combining

Initial station conversions utilized what is known as high-level combining or separate amplification shown in Figure 7.6-42. With this method, the existing station

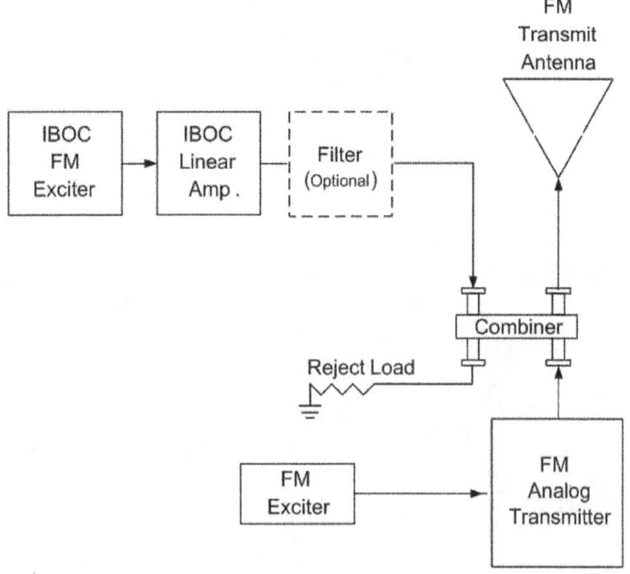

FIGURE 7.6-42 FM HD Radio high-level, separate amplification method

transmitter is combined with the output of a separate digital transmitter compatible with HD Radio technology. The combined signal is then fed to the existing station antenna.

HD Radio FM high-level combining uses two transmitters to produce the transmitted signal. This approach requires the addition of an HD Radio digital transmitter and the associated combiner, filter and digital exciter. Because both an analog and digital transmitter will be operated at the site, power demands may require the upgrade of electrical service to the facility. Heat load will also increase and may require additional cooling to remain within acceptable limits.

The high-level combining method is inefficient due to the combining technique employed. In order to achieve the requisite isolation and linearity, combiners used for HD Radio sacrifice 10% (~0.5 dB) of analog power and 90% (~10dB) of digital power to the reject load. When stations began using high-level combining, digital power was limited to –20 dBc. At this low injection level the digital signal is 1% of the analog power, so loss is more tolerable. Additionally, because the digital signal varies in amplitude as well as frequency, the peak-to-average ratio (PAR) will vary about 5.5 dB. For example, in the case of an FM station with an analog total power output (TPO) of 10 kW, the carrier power of the digital signal would be 100 W. Assuming combiner loss as listed earlier, the analog transmitter must be increased to 11.1 kW to overcome combiner insertion loss. The digital transmitter would require an average power of 1 kW to overcome the 10 dB combiner loss. The IBOC transmitter must be sized to accommodate 5.5 dB of additional overhead for PAR. This sizing for peak will amount to approximately 3.5 times the average power. When the FCC authorized digital power to increase to –10 dBc (10% of analog), this method of combining became less desirable as the losses increased. Chapter 7.18, "FM Combining and Filter System Design," explores FM combining systems in greater detail.

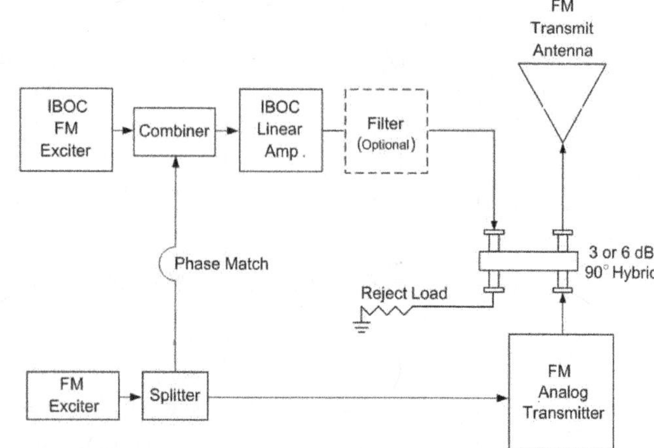

FIGURE 7.6-43 FM HD Radio mid-level, split amplification method

Mid-Level Combining

A derivative of the high-level amplification technique is mid-level or split amplification, as represented in Figure 7.6-43. In this design, the FM analog signal generation is shared between a traditional Class-C FM transmitter and a transmitter that has been linearized for amplification of both the analog and digital components.

Low-Level Combining

The low-level combining or common amplification method is depicted in Figure 7.6-44. In this implementation, the output of an analog FM exciter is combined with the output of an HD Radio exciter both at low RF levels. The combined signal is fed to a common broadband linear amplifier to raise the power to the desired TPO. This method is both power and space efficient and reduces the number of independent elements in the broadcast chain. Manufacturers are evaluating linearized versions of their transmitter design to determine optimal levels of headroom and linearity. This common amplification is both power and space efficient.

Low-level combining utilizes a shared transmitter power amplifier to boost the HD Radio digital signal and the host analog FM signal to desired output levels. This commonality reduces the demand on equipment space and may reduce power demands by increasing overall efficiency.

Separate and Dual-Input Antennas

The fourth implementation is the separate or dual input antenna method. This methodology takes one of two forms: a physically separate antenna, as depicted in Figure 7.6-45, or a dual-input antenna, as shown in Figure 7.6-46.

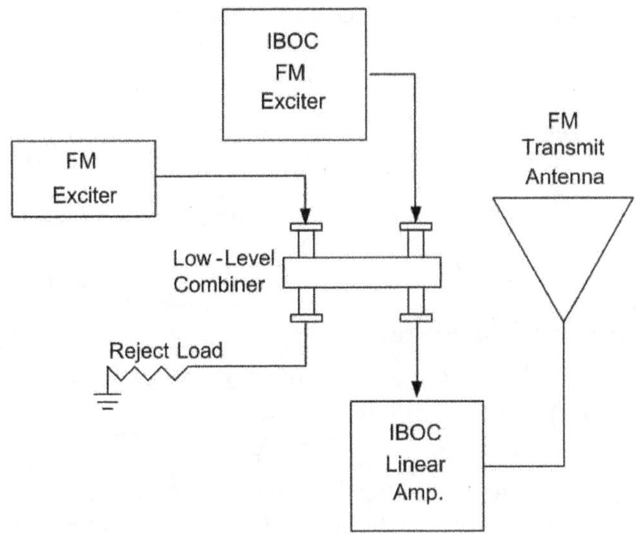

FIGURE 7.6-44 FM HD Radio low-level, common amplification method

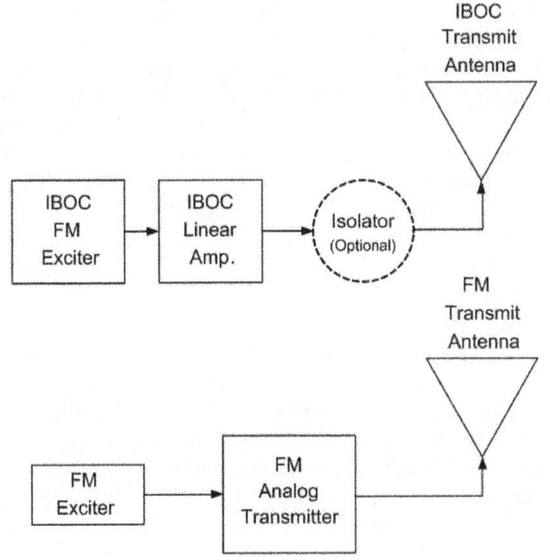

FIGURE 7.6-45 FM HD Radio separate antenna implementation

Separate antenna implementation routes the signal from independent IBOC and analog amplifiers to dedicated radiating elements for each signal. Two methods of separate antenna implementation are in use today. The basic form is an independent antenna, often previously installed as an analog backup. The second method, known as an interleaved antenna, places a digital bay at the midpoint of the analog radiating elements. In this design the digital elements phase is typically inverted (installed upside down) to provide additional isolation.

Regardless of which separate antenna method is employed, the FCC authorizes their use as follows:

- The digital transmission must use a licensed auxiliary antenna;

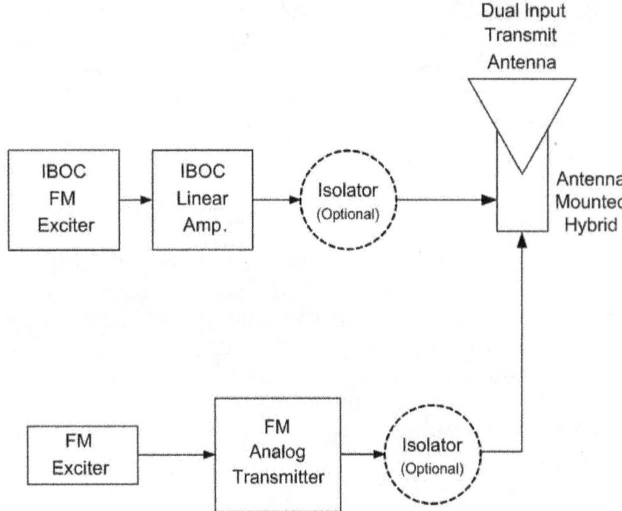

FIGURE 7.6-46 FM HD Radio dual-input antenna implementation

- The auxiliary antenna must be within 3 seconds of latitude and longitude of the main antenna;
- The height above average terrain of the auxiliary antennas must be between 70% and 100% of the height above average terrain of the main antenna.

Digital antennas require a minimum of 40 dB of isolation from the analog antenna in order to keep intermodulation products within acceptable limits. Careful placement and measurement of the antenna elements and RF isolator on one or both transmitters may be required to minimize mutual coupling.

Replicating radiating element placement with regard to tower leg and crossbar for both the analog and digital transmitting elements will ensure the patterns of the digital and analog signals are congruent. Because physically separate radiators are employed in this design, analog coverage may be superior to the digital due to the height difference between the centers of radiation. Despite this drawback, the advantage of separate antennas is the elimination of the combiner loss. This increase in system efficiency results in a significantly smaller IBOC transmitter required to develop the HD Radio carrier ratios.

Dual-input antennas utilize a hybrid to feed the two dipoles of the radiating element with independent analog and digital RF signals. In this implementation the resultant signal has right-hand circular polarization for the analog and left-hand polarization for the digital. Essentially, the dual-input antenna is a combination of free space signal combining in concert with the use of a 3-dB hybrid. Analog-to-digital coupling may be as high as 22 dB, typically necessitating the use of additional RF isolation. Dual-input antennas have the advantage that the center of radiation is identical for both the analog and digital aperture, resulting in near-identical analog and digital RF coverage.

Because system designs vary in dimension as well as configuration, the physical space and implementation constraints should be reviewed with equipment manufacturers to determine the appropriate solutions.

AM HD RADIO HYBRID TRANSMISSION

AM HD Radio requires the amplification of complex modulation. Unlike FM, AM IBOC requires strict phase coherency between the analog and digital signal. The most straightforward solution is to amplify both the analog and digital signals in a single transmitter. To accomplish this, AM HD Radio transmitters must provide ample bandwidth and minimize phase distortion.

The nonlinear transformation of complex I/Q to phase/magnitude requires infinite bandwidth to perfectly reconstruct the signal within spectral limits. Ideally, the transmission of discrete I/Q data would be desirable over a phase/magnitude transform, as it constrains frequency to the bandwidth of modulation. Transmitter bandwidth is a consideration as negative modulation peaks reach pinch-off (100%). As these values approach the baseline, infinite bandwidth products are produced. Although these emissions are infinite, sample rates and filter characteristics employed in HD Radio transmitters constrain the products to acceptable emission limits.

HD Radio transmission requires similar response and phase characteristics from the antenna as did analog AM stereo. State-of-the-art transmitter designs employ multiple solid state amplifiers that are summed to obtain the required power. For optimal transfer, it is desirable to provide the appropriate match at the summing point of the final amplifiers. Multiphase PDM transmitters with switching frequencies higher than 150 kHz and digitally modulated solid state AM transmitters are generally compatible with minor input filter modifications. Figure 7.6-47 depicts the key transmission elements of an AM HD Radio implementation.

Traditional AM tube amplifier designs show insufficient phase and frequency fidelity to pass the HD Radio waveform. However, several tube manufacturers are exploring new methods of increasing tube frequency and phase performance.

AM Antenna System

Antenna characteristics should be viewed from the driving point of the transmitter, as this is where linear distortion characteristics will be introduced. Ideally, this driving point would be situated at the output of the final amplifier ahead of any matching network. Today's solid state transmitters combine multiple amplifiers, each with different electrical lengths (delay) to the output network. It is therefore desirable to use the average of the electrical path lengths to determine the optimal point for achieving a match. This point in the circuit is known as the "amplifier summing point" of a transmitter that precedes the output matching network and sometimes has very low impedance, making it difficult to match to a higher load impedance. While no formal AM antenna specification exists, field trials indicate a sideband conjugate match to achieve optimal transfer with the fixed source impedance. The transmitter manufacturer can furnish this information as it is model dependent. Typically, the output network performs the functions of impedance matching and harmonic attenuation.

Hermitian symmetry of the transmitted signal ±5 kHz is desirable to keep quadrature information in quadrature and to minimize crosstalk from the analog into the digital and digital into analog. Figure 7.6-48 depicts the Hermitian symmetry vector relationship. The spectrum fc to +5 kHz and fc to −5 kHz is complementary QPSK. If the carrier information is transmitted

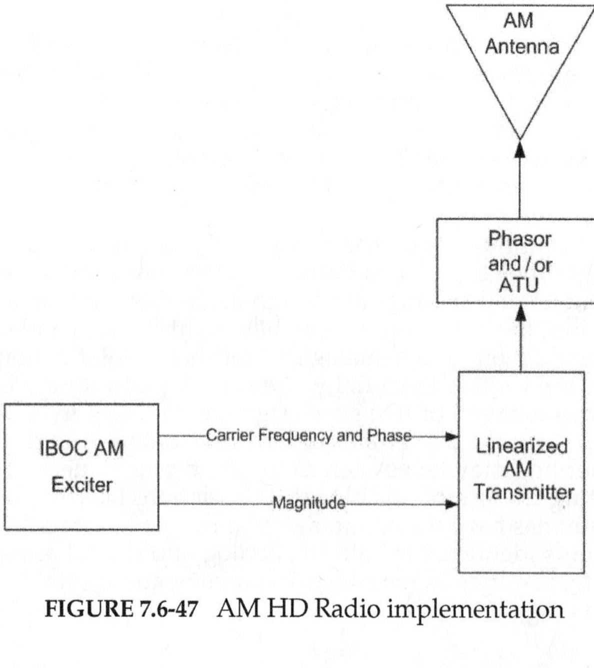

AM Antenna
Phasor and / or ATU
IBOC AM Exciter
Carrier Frequency and Phase
Magnitude
Linearized AM Transmitter

FIGURE 7.6-47 AM HD Radio implementation

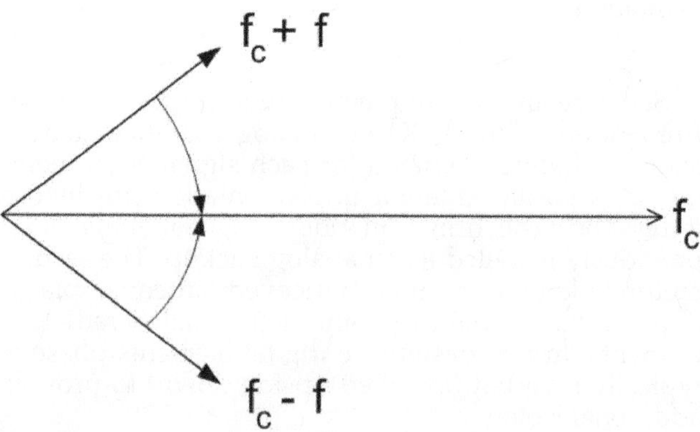

FIGURE 7.6-48 Hermetian symmetry (named after Charles Hermite, French mathematician. See www-history.mcs.st-andrews.ac.uk/Biographies/Hermite.html).

and then symmetrically filtered in the receiver, the signals will combine and null tertiary carrier interference.

DTS is exploring ways to specify the Hermitian symmetry in terms of a single quantity like voltage standing wave ratio (VSWR) or the magnitude of the reflection coefficient of one sideband normalized to the other sideband. While it is not traditional to calculate the VSWR of one sideband referenced to another, it allows more flexibility in defining a specification. A limited range of load impendence and a reactance could disqualify many designs in use today. This is the rationale for DTS suggesting a Hermitian symmetry: with a normalized sideband conjugate match within 1.035:1 VSWR ±5 kHz. This equates to an amplitude response variation of 0.5 dB across the region fc to +5 kHz and fc to −5 kHz.

Complex conjugate matching produces the maximum small signal transfer of power from a source that is not a transmission line to a load. Real-world antennas do not have sideband impedance characteristics required for absolute Hermitian symmetry, making compromise in the resistance and reactance necessary. A conjugate match is typically employed when the situation involves source impedance that is stable. Making the phase angle of the load voltage and current equal is desirable, as complex load impedances produce an elliptical load line, which generates distortion. As long as a conjugate relationship is maintained, even high VSWR relationships work satisfactorily up to the point where an amplifier depletes headroom. The departure from a conjugate relationship is tolerable while the AM carrier angle-modulation is limited to avoid disrupting the HD Radio sidebands. Unfortunately, VSWR cannot directly be related to symmetry, as it is a relationship of two complex numbers to each other. The degree they depart from a conjugate relationship without regard to the AM analog carrier must be observed. In the far field of the antenna, pattern bandwidth effects move HD Radio sidebands just as the AM analog carrier. Because the antenna system is passive and linear, the equalizer in the receiver can track the sidebands and differentiate them from what is going on with the AM analog signal.

For acceptable IBOC reception, the following amplitude response and group delay limits are suggested:

Amplitude Response:

0.5 dB for the region fc to +5 kHz and fc to −5 kHz;
4.0 dB for the region fc to +5 kHz to fc to +15 kHz and fc to −5 kHz to fc to −15 kHz.

Group Delay:

5 μsec for the region fc to +15 kHz and fc to −15 kHz.

See Figures 7.6-49 and 7.6-50 for a graphical representation of these limits.

Chapter 7.9, "Antenna Coupling and Phasing Systems," provides details of various matching networks employed in AM transmission systems.

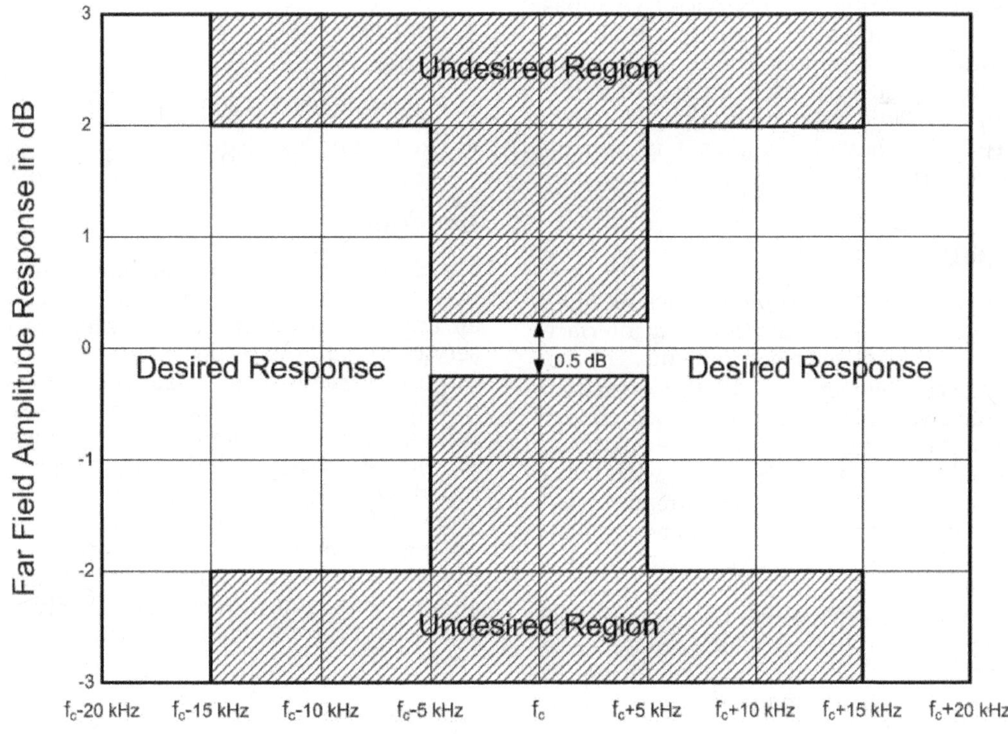

FIGURE 7.6-49 Desired AM HD Radio amplitude response limits

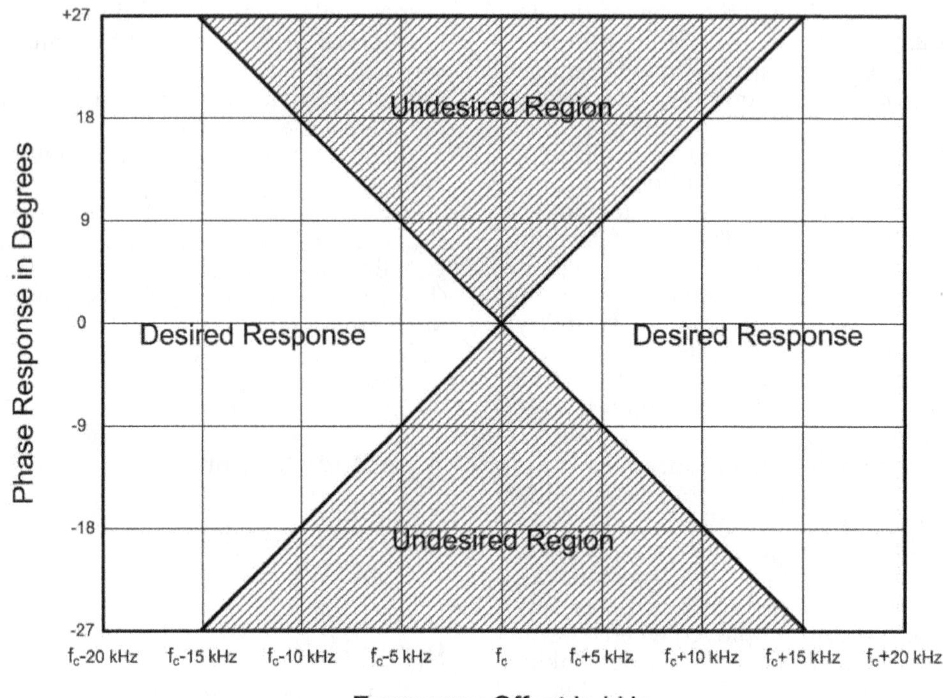

FIGURE 7.6-50 Desired AM HD Radio group delay limits

HD RADIO NETWORKING

The advent of HD Radio generation-3 hardware architecture and Advanced Application Service (AAS) has led the broadcast industry to re-evaluate the station network infrastructure. Minimizing network contention and resulting signal outages is crucial to delivering a quality HD Radio implementation. Network topology and protocol have a significant impact on overall system performance.

Hardware Topology

A discussion of HD Radio networking basics requires an understanding of the basic function and interface requirements of the transmission components.

The first- and second-generation HD Radio exciters are monolithic hardware platforms sourced with linear 44.1 kHz sampled AES-3 digital audio. This design was acceptable when only a single MPS stream was delivered to the exciter. When multiple SPS streams were added, the audio payload requirements increased several fold. For a radio station to convey linear 20-kHz audio from the studio to the transmitter requires 1.4112 Mbps throughput for each stereo channel. If broadcasters were to connect multiple stereo audio channels to the transmitter site for MPS and SPS, it is efficient to employ the systems HDC bit-reduced audio. Transport efficiency is accomplished by bit reducing MPS and SPS audio and PSD at the studio end of the system and transmitting a single multiplexed data stream to the HD Radio exciter at the transmitter site. See Figure 7.6-51 for an overview of the system topology. This implementation known as the generation-3 architecture has four major elements:

- Importer
- Exporter
- Synchronizer (EASU)
- Exciter Engine (Exgine).

Importer

The Importer contains the hardware and software necessary to deliver AAS. Data service providers use an application programming interface (API) to pass service data to the Importer over the service link. The Importer establishes session connections between multiple service providers. Once a session is established, service providers can pass service data over the Importer-to-Exporter link (I2E), which in turn will be broadcast to HD Radio digital receivers. In addition to the AAS from data service providers, the Importer also accepts SPS and PSD. The Importer multiplexes all of the service provider data, multicast audio, and data streams into a full-duplex TCP/IP or bidirectional user datagram protocol (UDP) output.

Exporter

The Exporter contains the hardware and software required to generate the MPS and the SIS. The SIS provides station information: call sign, station slo-

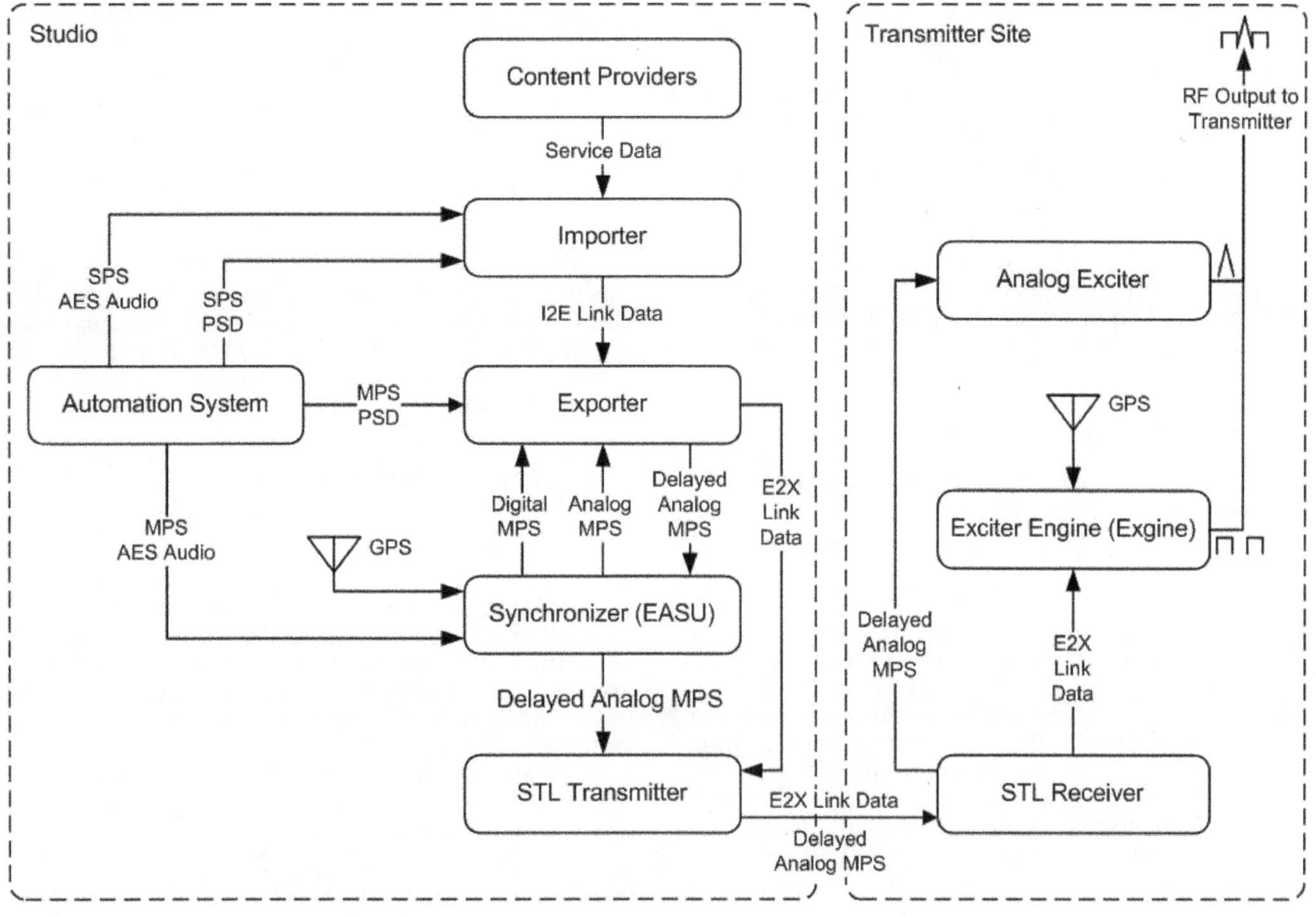

FIGURE 7.6-51 HD Radio generation-3 system architecture

gan, absolute time, and position correlated to GPS. The Exporter accepts digital MPS audio over its audio interface, bit reduces (i.e., compresses) the audio, and outputs that audio to the Exgine over the simplex Exporter to Exgine link (E2X). The Exporter applies preprogrammed delay to the analog MPS audio over its audio interface and broadcasts it as the backup channel for hybrid configuration. The delay compensates for the digital system latency, allowing receivers to seamlessly blend between the digital and analog program without a shift in time. In the FM system, the delayed analog MPS audio is returned to the synchronizer, which in turn is fed into the STL, and is stereo multiplexed and modulated by the analog exciter at the transmitter site.

The original gen-2 systems synchronizer function, also known as the exciter auxiliary service unit (EASU), would accept MPS audio and rate-convert it to the proper system clock. It output two rate-converted and reference-clock-locked copies of the MPS audio to the Exporter. One output drives both the digital and analog MPS audio. The digital MPS audio is bit reduced and modulated in the digital portion of the waveform and the analog MPS audio is modulated in the analog portion of the hybrid waveform. In most installations, audio processing is employed on the analog and digital audio paths between the synchronizer and the Exporter. In gen-3 hardware this synchronizer functionality was incorporated into the Exporter functional hardware.

The GPS receiver portion of the synchronizer function provides the master 10-MHz system clock used by the Exporter. This 10-MHz clock is also used as the reference input for the 44.1-kHz word clock used to synchronize the analog and digital AES audio streams.

Exgine

The Exgine (exciter engine) subsystem accepts the E2X data from the exciter's host processor and performs the OFDM modulation for the digital portion of the HD Radio waveform. The Exgine element is composed of a Texas Instruments C6415 processor, SDRAM, and flash memory and enables the Layer 1 modulation to be executed on the digital signal processor (DSP).

This configuration may be added to many manufacturers' digital implementations of their analog exciter offering an integrated solution for low-level analog and digital signal generation. In this distributed architecture, the HD Radio data stream is fed to the Exgine over a simplex UDP ethernet connection.

To generate the multiplex over the E2X requires a new element in the system topology. The Exporter accepts the MPS AES audio and PSD from the automation server at the studio end as well as the multiplexed SPS audio, SPS PSD, and advanced application service data from the Importer. The Exporter may be visualized as the final multiplexer of MPS and all AAS data prior to developing the simplex UDP stream to the STL. An Exporter is essentially a gen-2 exciter without the RFU (RF up-converter) and DUC (digital up-converter) subassemblies. Some manufacturers allow maximization of the original equipment investment by providing an upgrade path from existing Exciter to Exporter.

NETWORK IMPLEMENTATIONS

There are two distinct network implementations employed in AAS multicasting:

- Importer to Exporter/Exciter (I2E);
- Exporter to Exciter Engine (E2X).

The I2E network path connects an Importer to an Exporter or Exciter via either bidirectional UDP or transmission control protocol (TCP) over an ethernet connection. In implementation, an AES digital audio path is required for the digital MPS as well as a bidirectional broadband ethernet/IP connection for the supplemental programs and data between the Importer and Exporter/Exciter. Only the AAS signal components are transported by the I2E link and do not affect the main program digital service. The I2E configuration can be implemented only over a bidirectional network connection between the Importer and the Exporter/Exciter due to the inherent command-response design. Co-locating the Importer and Exciter within the same isolated local area subnet or over a dedicated WAN/WAN extension can help constrain implementation costs. This can be accomplished either via wireless (RF) or wired data links.

Importer to Exporter/Exciter

Prior to Importer version 1.2.1 and the iBiquity Reference System Software (IRSS) Exciter/Exporter software version 2.2.5, the only network protocol available for I2E was bidirectional UDP/IP. UDP protocol's lack of error correction makes it susceptible to faults. The loss of one IP packet over the link will result in the loss of an entire frame, which represents 1.48 seconds of SPS audio or PSD. Importer version 1.2.1 and the Exciter IRSS software version 2.2.5 introduced TCP/IP as the default protocol. Using TCP/IP affords up to 20 frames of receive buffering on the I2E link, allowing sufficient time for retransmission of dropped packets. The low overhead and additional robustness offered by TCP leave little reason to use the UDP configuration on the I2E data link.

With up to 1% packet loss and 100 msec latency, the I2E path continues to perform well when configured for TCP. With TCP/IP adequate bandwidth is the key to system recovery from lost packets. Configuring to a maximum of 60% of the available WAN bandwidth will produce successful results under all but the most adverse network conditions.

The average bandwidth of the I2E link ranges from 17 kbps to 156 kbps dependent on the service mode and IP protocol employed. A WAN bandwidth of at least 90 kbps and latency below 100 msec are required for FM hybrid mode MP1 operation with a single 32 kbps SPS over TCP. A 128 kbps LAN/WAN extender or two DS0 data circuits will provide sufficient bandwidth for any MP1 configuration. For MP3 mode, the highest capacity hybrid mode available commercially, a minimum bandwidth of 156 kbps is required. In practice, this would require three DS0 channels for a total of 192 kbps.

Exporter to Exciter Engine Link

Co-locating the Importer and Exporter at the studio enables bandwidth efficient communication across Exporter to Exciter Engine (E2X) link. This implementation requires a distributed system architecture consisting of the Exporter and Exgine. With this implementation, a single data stream is conveyed to the transmitter site containing the main program digital audio along with the supplemental programs and all of the associated data.

Typically, the Exporter and Importer along with the audio processing are located at the studio. The Exporter performs four primary functions:

- Input of AES MPS digital audio, MPS PSD;
- Input of Importer networked multiplex (SPS audio, SPS PSD, and AAS data);
- Diversity delay of analog AES audio path;
- Multiplexing of all services for transport over the STL.

The main program digital audio is delivered to the synchronizer (some manufacturers incorporate this functional element in the Exporter hardware) that splits the audio into two streams that feed the analog and digital processing chains. Both streams are returned to the Exporter after processing. The audio destined for the legacy analog transmitter is time aligned with the digital audio and then sent to the AES digital input of the STL for transport to the legacy analog transmitter. The main HD Radio program audio is encoded into the MPS signal.

The Importer accepts AES audio and PSD of the multicast programs. These services are encoded and sent from the Importer to the Exporter over a local network as bidirectional UDP or TCP data over Ethernet as the I2E configuration. The program service and the advanced services (SPS and data) are then combined in the Exporter into a single data stream destined for the Exciter over the STL or WAN link.

Transport of the E2X data stream is currently supported only as simplex UDP and can operate over most unidirectional STL systems of sufficient bandwidth and robustness. IRSS Version 2.3.3 has provision for E2X over TCP, requiring a bidirectional link; however,

TABLE 7.6-21
Predicted dropout at percent packet loss

	Mode	Packets Per Day × 1000	0.1% Packet Loss	0.01% Packet Loss	0.001% Packet Loss	0.0001% Packet Loss
			UDP—Expected Daily HD Dropouts			
I2E	MP1-w/ HD2 @ 24 kbps	6400	640	64	6	1
	MP1-w/HD2 @ 48 kbps	873	873	97	9	1
	MP2/3 HD2	1396	1396	140	14	1
	MP2/3 HD2 and HD3	2036	2036	204	20	2

the TCP connection is not fully supported in the IRSS Version 2.3.3 software load. Some manufacturers have independently enabled TCP service.

DTS' lab tests reveal the E2X configuration running UDP must have IP packet loss better than 10^{-5} for a successful implementation. It is not uncommon for WAN and STL systems to deliver only 10^{-3} performance (one dropped packet in every 1000), which will result in poor HD Radio system performance. Table 7.6-21 predicts the number of HD Radio dropouts that can be expected for a given network packet loss based on the number of packets sent per day when running UDP without error correction. Without error mitigation, 6 to 31 momentary dropouts per day can be expected, depending on the mode and configuration, when using UDP over a 10^{-3} link.

The TCP implementation of E2X can perform significantly better than UDP, tolerating packet losses of up to 0.03%. This level of performance is well within the capabilities of healthy WAN/STL systems.

E2X average bandwidth ranges from 120 kbps to 168 kbps, dependent on configuration and IP protocol. A WAN/STL bandwidth of at least 128 kbps with latency of less than 100 msec is necessary for MP1 mode UDP. A 128 kbps LAN/WAN extender or two DS0 data channels will provide sufficient bandwidth for any MP1 configuration. For FM mode MP3, 256 kbps or four DS0 channels should be considered for UDP and 320 kbps or five DS0 channels should be more than sufficient for TCP. With UDP, packet loss across the link becomes a critical factor and must be kept below 10^{-5} for successful operation due to a lack of error recovery.

SUMMARY OF PROVISIONING

TCP/UDP

In order for a TCP data stream to function properly under adverse conditions, the link that carries it must have reserve bandwidth above and beyond the data rate of the stream. This is necessary to accommodate the higher data rate that occurs when the stream recovers from packet loss. Additional bandwidth beyond the recommended guideline allows operation under poorer conditions, but with diminishing returns. In general, bandwidth should not be used to adjust for a poor network.

When using TCP, the WAN link must have a minimum of 40% overhead (reserve bandwidth) in order to function properly. This overhead should be calculated on the total traffic through the WAN link, which can consist of the following components:

- IBOC data stream;
- Utilities such as VNC, telnet;
- Broadcast or multicast traffic.

The aggregate of this traffic should occupy no more than 60% of the available bandwidth. If other traffic is going through the WAN, the link should have class of service, QOS, or other prioritization techniques employed to ensure that the HD Radio traffic has the necessary bandwidth. For UDP, the total traffic can be no more than 75% of the provisioned bandwidth.

Traffic Control

Through field investigations, DTS found that the only sure way to prevent extraneous traffic from traversing the link is to make the link its own IP subnet. This separates the link from the production network and spares the link from having to carry any broadcast or multicast traffic from the production and/or office network. Traffic that switches or is directed to all ports should be kept to a minimum. In addition to broadcast and multicast packets, this category also includes Unicast packets sent to media access control (MAC) addresses that are not in the switch's forwarding table. This is a situation that occurs when a device from outside the local network sends packets to a device that was recently operating in the subnet. The device is still in the router's address resolution protocol (ARP) table, but has already "aged out" of the switch forwarding tables. The switch aging timer is typically shorter than the router's ARP aging timer, resulting in a discrepancy. There are steps that the network manager can take to prevent this situation. These steps include reducing the router's ARP aging timer to match the switch forwarding table aging timer or programming the switch to block Unicast traffic sent to unknown MAC addresses.

Using TCP, audio drops occur whenever the receive buffers are depleted. In most circumstances, the audio stream will not resume until the receive buffers are all restored. Because each receive buffer corresponds to 1.48 seconds of audio, depletion of 20 receive buffers will result in an audio drop of 30 seconds or longer. UDP has the following advantages:

- Operates over a simplex STL or WAN;
- Has shorter broadcast delay due to fewer buffers;
- Operates with less overhead;
- Has shorter audio drops when they occur (although they may occur more frequently).

With UDP transmission, the loss of a single packet results in the loss of the entire audio frame of which it is a part. The resulting outage will last for the duration of that single audio frame—1.48 seconds. For any constant packet loss rate, one would expect fewer audio drops with TCP but of shorter duration with UDP.

Detailed resources on HD Radio network implementation may be found in the whitepaper section of the DTS website at http://hdradio.com/broadcasters/engineering-support/iboc-white-papers.

SUMMARY

The concept of the IBOC system is to provide broadcasters a means to deliver robust, high quality digital audio and data to receivers making optimum use of their FCC-assigned frequency without the need for additional spectrum. Additionally, data and other services may be added at the broadcaster's choice. IBOC is compatible with existing analog receivers and allows a means to an eventual all-digital broadcasting format.

ACKNOWLEDGMENTS

The author wishes to acknowledge the contributions of Tim Anderson, Jeff Baird, Kathi Cover, Harvey Chalmers, Denise Cammarata, Ashruf El-Dinary, Steven Johnson, Brian Kroeger, Stephen Mattson, Marek Milbar, Russ Mundschenk, Paul Peyla, Mike Raide, Al Shuldiner, Glynn Walden, Girish Warrier and all present and past members of the DTS team.

Bibliography

Anderson, T. B. "HD Radio™ Data Network Requirements", iBiquity Digital Corporation, Oct. 2006.

Federal Communications Commission, Code of Federal Regulations, Title 47, Part 73.

"HD Radio™ Air Interface Design Description—Advanced Application Services Transport", Doc. No. SY_IDD_1019s, Rev. G, iBiquity Digital Corporation, Aug. 2011.

"HD Radio™ Air Interface Design Description—Audio Transport", Doc. No. SY_IDD_1017s, Rev. G, iBiquity Digital Corporation, Aug. 2011.

"HD Radio™ Air Interface Design Description—Layer 1 AM", Doc. No. SY_IDD_1012s, Rev. F, iBiquity Digital Corporation, Aug. 2011.

"HD Radio™ Air Interface Design Description—Layer 1 FM", Doc. No. SY_IDD_1011s, Rev. G, iBiquity Digital Corporation, Aug. 2011.

"HD Radio™ Air Interface Design Description—Layer 2 Channel Multiplex", Doc. No. SY_IDD_1014s, Rev. I, iBiquity Digital Corporation, Aug. 2011.

"HD Radio™ Air Interface Design Description—Program Service Data", Doc. No. SY_IDD_1028s, Rev. D, iBiquity Digital Corporation, Nov. 2007.

"HD Radio™ Air Interface Design Description—Program Service Data Transport", Doc. No. SY_IDD_1085s, Rev. C, iBiquity Digital Corporation, Feb. 2005.

"HD Radio™ Air Interface Design Description—Station Information Service Transport", Doc. No. SY_IDD_1020s, Rev. I, iBiquity Digital Corporation, Aug. 2011.

"HD Radio™ AM Transmission System Specification", Doc. No. SY_IDD_1082s, Rev. F, iBiquity Digital Corporation, Aug. 2011.

"HD Radio™ FM Transmission System Specification", Doc. No. SY_IDD_1026s, Rev. F, iBiquity Digital Corporation, Aug. 2011.

Johnson, S. A. "The Structure and Generation of Robust Waveforms for AM In-Band On-Channel Digital Broadcasting", iBiquity Digital Corporation, Mar. 2000.

Kroeger, B. W., and D. M. Cammarata, "Robust Modem and Coding Techniques for FM Hybrid IBOC DAB", *IEEE Transactions on Broadcasting*, vol. 43, no. 4, pp. 412–420, Dec. 1997.

"Petition for Rulemaking to the United States Federal Communications Commission for In-Band On-Channel Digital Audio Broadcasting", Appendix C, p. 7, USA Digital Radio Corporation, Oct. 1998.

Peyla, P. J. "The Structure and Generation of Robust Waveforms for FM In-Band On-Channel Digital Broadcasting", iBiquity Digital Corporation, Mar. 2000.

CHAPTER

7.7

Worldwide Digital Radio Systems

KENNETH E. COLWELL
St. Ambrose University
Davenport, IA

INTRODUCTION

Guglielmo Marconi is quoted as asking: "Have I done the world good, or have I added a menace?" It might have occurred to those developing and promoting digital radio to ask that same question. Is their technology going to make the same kind of difference? Will it become as ubiquitous as Marconi's?

Because digital radio is relatively new, and digital TV is evolving from ATSC 1.0 to ATSC 3.0 (in the United States and elsewhere), the answer to those questions is for the future. Our goal here is to take the position of someone in North America and to explore what digital radio looks like elsewhere in the world. We will explore the technologies employed, their significant features, costs, upgrade paths, propagation potential, energy consumption, increased programming options, and spectrum implications. We will do this from the perspective of the broadcaster and the listener. Therefore, this is an overview for someone interested in understanding how digital broadcasting is developing elsewhere.

Obviously this is a very fluid situation as people, their governments, and the technology change. In light of that, we can only provide a snapshot of what is currently in front of us. Hopefully it will provide the reader with a good starting point for understanding where digital radio is currently located, and where it might be going in the not-too-distant future.

We will supply as many references as possible to help the reader explore more deeply the various aspects of our topic. An in-depth mining of all things digital is not possible at this time. We have not included spectrum masks for the same reason, given the wide variety of options available. It is hoped that the links noted can provide more depth, more clarity, or more up-to-date information. Note that Chapter 2.5, "Analog and Digital Terrestrial Radio Standards," provides additional information on the topics being covered here.

There are many advantages to digital radio broadcasting over analog. Among them are better signal-to-noise ratio (SNR), reduced interference, less multipath distortion, automatic tuning, the possibility for single frequency networks, and a wide variety of auxiliary data services.[1] Digital radio generally employs one of four different modes: HD Radio, Digital Audio Broadcasting (DAB and DAB+), Digital Radio Mondiale (DRM30 and DRM+), and China Digital Radio (CDR), although the distinction between radio and TV blurs at the transmission level as modes usually employed for digital TV can be used as well.

Each mode is fundamentally a "data-pipe" transmitting digital signals—audio and data—in a form appropriate to the medium in question. These modes have much in common. For example, they all use coded orthogonal frequency-division multiplexing (OFDM) modulation, which requires linearity (no distortion) in the power amplifier process of the transmitters. All can be employed for optimized coverage in a single frequency network (SFN) scheme. All can transmit multiple program channels from a single transmitter. All but DAB can simulcast with the analog signal.[2] It is worth noting that a side-by-side comparison of each standard for selecting the "best" one is not appropriate. The standards are fundamentally complementary rather than competitive. Each one fulfills a need, such as a particular frequency band, for example, that may not be addressed by all of them. Each one stands, therefore, as a solution for some, but not everyone's needs. The discerning engineer will be the best digital engineer.

While we will focus most of our time on the four radio modes in this chapter, we wish to note that TV modes, like the DVB-T2 Lite standard (the mobile and radio subset of the DVB-T2 standard), and the ATSC 3.0 standard, being a digital delivery system, have the capacity to deliver radio programming as well. When DVB-T2 Lite is introduced into the mix (we don't know yet all that ATSC 3.0 may have in store) things become a lot more interesting. DVB-T2 Lite can operate in a digital radio alone mode, or in a mix of TV and radio modes. We will discuss this more following our overview of the digital radio standards.

Because HD Radio is covered elsewhere (see Chapters 2.5 and 7.6 in particular), we shall only provide a brief list of the features of that system. HD Radio includes:

- AM and FM In-Band On-Channel (IBOC) operation;
- Compliance with existing spectrum masks for each service;
- System capacity of eight audio services (FM, if in all digital mode) but limited by bandwidth to four or five services, or a single audio service (AM);
- Maximum coding rate of 98 kbps with combinations of:
    - Mono at 8.1–64 kbps
    - Parametric Stereo—8.6–66 kbps
    - Stereo—16–96.8 kbps.
- Sound quality dependent on combination of audio codec and bit rate employed;
- Operation in both single-frequency and multiple-frequency networks;
- Automated receiver tuning and associated signaling;
- HDC Audio coding for bit-rates from 32 kbps to 124 kbps;
- Value-added data services.[3]

Visit HD Radio's website (http://hdradio.com/broadcasters) for a more complete exploration of the features of HD Radio.

Next we will introduce in greater detail the DRM and DAB technologies, and note how the CDR system differs from the other three systems, and then address the issues of receivers, station programming, areas of impact for each technology, and lastly the issue of "who benefits" by the transition to digital broadcasting.

DIGITAL RADIO MONDIALE (DRM)

The system specification for DRM we will use is covered in ETSI ES 201 980 v4.1.1 (January, 2014, see www.etsi.org/deliver/etsi_es/201900_201999/201980/04.01.01_60/es_201980v040101p.pdf). It is a digital radio standard that is intended to replace analog AM and FM broadcasting, although the specification describes it as useable at any frequency below 300 MHz. It will operate in the same channel configuration and frequencies that are used for analog. DRM includes operation from LF to UHF frequencies in either of its two variant forms (DRM30 and DRM+), so its impact extends beyond the usual broadcasting modes.[4] DRM30 service allows broadcasters to exploit the propagation characteristics of the MF and HF bands. It also provides for significantly improved audio quality than offered by analog services in those bands, easier tuning, and added data services. In the VHF band, DRM+ can be configured to use less spectrum, lower transmission power, and increased coverage area over conventional FM. Studies also indicate that DRM+ requires less transmitting power than DAB or DAB+ to cover the same service area.[5] By providing for an array of operating modes and techniques, DRM permits the broadcaster to adapt the system to meet the needs of their particular market or service area.[6]

DRM is also an "open system" that permits the broad and innovative development of software and hardware. This is important because it permits a fast path for receiver design that (in theory) increases the availability of receivers to the public (a key aspect of standard acceptance and use). However at the time this chapter was written (fall 2016), there were essentially no commercially available DRM receivers. The open system approach makes for the possibility of using software defined radio (SDR) listening (find more information on DRM software radio at http://drmdx.org/).

The DRM Consortium was formed in Guangzhou, China, in 1997 with the goal of digitizing AM broadcasting. The initial standard—DRM30—is for operation up to 30 MHz. DRM30 was published by ETSI in 2001.[7] (Visit www.etsi.org for the current version of the standard.)

While DRM is intended to replace analog, it can coexist (i.e., share the band), or simulcast (as IBOC), with analog broadcasting. It can be configured to comply with existing spectrum masks and frequency grids. This is especially advantageous during the transition period where analog will continue in some form, for some time. DRM also permits "suitable" transmitters to be converted from analog to digital operation, thus enabling a cost savings for certain operations. It has also been reported that these transmitters can run more efficiently in digital, thus making for a savings in energy costs.[8]

DRM allows for a wide range of transmission modes that allows the broadcaster to balance bit rate capacity, signal robustness, transmission power, and coverage as needed for a particular station or network of stations. Altering modes is possible dynamically to adjust for atmospheric variations or day/night propagation changes, for example, without the audience being aware of these adjustments.

Let's see how all this is put together. To start, we have two system considerations: signal bandwidth parameters and transmission efficiency parameters.

Generally, below 30 MHz the signal bandwidth is 9 or 10 kHz, although it could vary from 4.5 kHz to 20 kHz depending on type of service desired (e.g., simulcast with analog or large transmission capacity). Above 30 MHz, the bandwidth is 100 kHz.[9]

Transmission efficiency parameters impact the trade-offs between useful bit rates and ruggedness relating to RF noise, multipath fading, and Doppler (change of wavelength caused by motion of the receiver). With DRM there are multiple variables: coding rate and constellation parameters (determining what rate and constellation pattern is employed to convey the data) and OFDM symbol parameters (which defines the structure of the OFDM symbols to be used based on propagation conditions).[10] In order to achieve one or two levels of protection as required by the service used, these levels can be established by the code rate of the channel encoder, the constellation order, or by hierarchical modulation.[11]

In order to optimize the performance of the DRM system, the transmission COFDM parameters (carrier-spacing, guard interval, pilot density, etc.) need to be matched to the characteristics of the RF channel. Propagation in the AM bands can range from groundwave, where electrical noise is the predominant interference mechanism, to skywave with varying degrees of channel complexity and where both differential delay and Doppler effects are additional adverse factors. In some circumstances signals may reach some locations in the coverage area by means of both types of wave propagation. In all of these cases the received signal is likely to suffer from the addition of distortions or noise caused by the transmission path.[12] To address these variations in propagation and reception, the system defines five "modes," labeled A to E respectively, which are outlined in Table 7.7-1.

The selection of a transmission mode includes a choice of modulation (quadrature amplitude modulation—QAM—constellation) and coding (Viterbi) rates for the main service channel. In all DRM30 modes the option exists to choose either 64-QAM or 16-QAM for the main service channel (MSC), and this choice should be influenced by the signal-to-noise plus the interference ratio that can be achieved in the target reception area. The more robust 16-QAM option is normally chosen where the SNR is expected to be too low to support 64-QAM. For DRM+ (Mode E), it is possible to employ either 16-QAM or 4-QAM for the MSC.[13]

Therefore, given the channel bandwidth permitted and the level of robustness desired, the broadcaster can then determine from the table the bit rate available, and based on that capacity, form the number and type of services—audio and data—for the MSC accordingly.

With the signal bandwidth-related parameters and transmission efficiency-related parameters in mind, let's turn to the three channels within the DRM system.

In the DRM system, three channels of data are available: the MSC, the fast access channel (FAC), and the services description channel (SDC). These three channels feed into the OFDM cell mapper, which feeds the OFDM signal generator, which in turn feeds the modulator.[14] Refer to Figure 7.7-1.

Because each of these channels make different demands on the multilevel coding (MLC) scheme, different but appropriate coding schemes are applied to each. There are both unequal error protection (UEP) and equal error protection (EEP) schemes available and they can be combined with hierarchical modulation. FAC and SDC requires EEP. Using UEP at two code rates allows MSC data to be split and assigned to either a higher protected part, or lower protected part.[15]

The MSC consists of the audio and data services to be broadcast. The audio source(s) are encoded using one of three audio coding schemes: MPEG xHE-AAC (extended high-efficiency advanced audio coding); MPEG-4 AAC (advanced audio coding) including spectral band replication (SBR) and parametric stereo; or MPEG Surround (MPS). Typically the xHE-AAC is the preferred coding scheme.[16] An important aspect of the encoded audio requires that it be compressed in such a way that the audio superframes be of a constant length of 400 ms.[17]

The data stream for the MSC is fed into a precoder that adapts the input streams into a format suitable for transmission. Both the audio and data encoder/precoder outputs may be split into a higher and lower protected part, each of which will be assigned one of two different protection levels within the channel encoder (which comes after the multiplexer). These streams are now fed to the multiplexer that combines the protection levels of all data and audio services.[18] "The MSC multiplex may contain up to four services, any one of which can be audio or data. The gross bit rate of the MSC is dependent upon the channel bandwidth and transmission mode being used. In all cases, it is divided into 400ms frames."[19]

The FAC is also built around 400-ms frames and contains a set of core parameters required to quickly check for available services within the multiplex signal (such as when scanning quickly among the available channels) and to determine other settings—like interleaving depth—to allow the demodulation of the DRM signal. The FAC block will also contain information relating to the robustness mode selected to help ensure efficient decoding of the signal.[20]

The SDC carries advanced information like audio and data coding parameters, service labels, current time and date information, and alternative frequency signaling (AFS) tables. The data capacity will vary depending upon a variety of parameters, and must be carefully managed.[21] It has a frame periodicity of 1200 msec, which enables seamless switching between alternate frequencies.[22]

The output of the multiplexer, the FAC precoder, and the SDC precoder are then fed to their respective energy dispersal blocks. Energy dispersal provides for the randomizing of the bits that reduces the possibility of unwanted, systematic regularity in the transmitted signal.[23]

Following the energy dispersal is the channel encoder. The channel encoder adds redundant bits to

TABLE 7.7-1
DRM service modes

Standard	Mode	MSC QAM			Nominal signal bandwidth				Robustness	Typical uses(by mode)
		Options	4.5KHz	5 KHz	9 KHz	10 KHz	18 KHz	20 KHz	100 KHz	Level	
			Available bit-rates (approximate) with equal error protection and standard mapping								
DRM30	A	64	14.7	16.7	30.9	34.8	64.3	72		Min	LF and MF Ground-wave, 26MHz band line-of-sight
DRM30	A	64	9.4	10.6	19.7	22.1	40.9	45.8		Max	
DRM30	A	16	7.8	8.8	16.4	18.4	34.1	38.2		Min	
DRM30	A	16	6.3	7.1	13.1	14.8	27.3	Min		Max	
DRM30	B	64	11.3	13	24.1	27.4	49.9	56.1		Min	MF and HF Sky-wave
DRM30	B	64	7.2	8.3	15.3	17.5	31.8	35.8		Max	
DRM30	B	16	6	6.9	12.8	14.6	26.5	29.8		Min	
DRM30	B	16	4.8	5.5	10.2	11.6	21.2	23.8		Max	
DRM30	C	64				21.6		45.5		Min	Difficult HF Sky-wave
DRM30	C	64				13.8		28.9		Max	
DRM30	C	16				11.5		24.1		Min	
DRM30	C	16				9.2		19.3		Max	
DRM30	D	64				14.4		30.6		Min	Near Vertical Incidence Sky-wave
DRM30	D	64				9.1		19.5		Max	
DRM30	D	16				7.6		16.2		Min	
DRM30	D	16				6.1		13		Max	
DRM+	E	16							186.3	Min	VHF above 30MHz
DRM+	E	16							99.4	Max	
DRM+	E	4							74.5	Min	
DRM+	E	4							37.2	Max	

The selection of a transmission mode includes a choice of modulation (quadrature amplitude modulation—QAM—constellation) and coding (Viterbi) rates for the main service

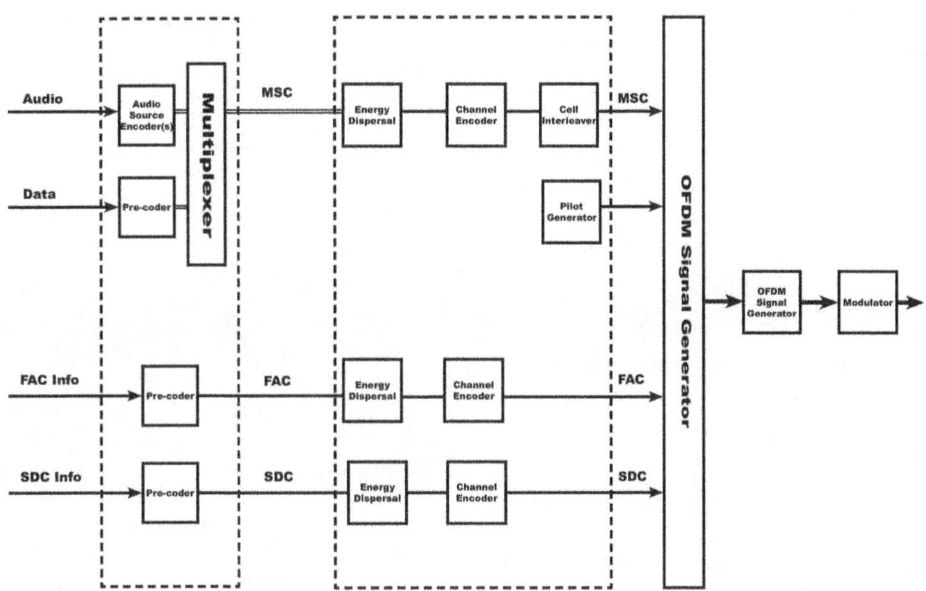

FIGURE 7.7.1 DRM block diagram illustrating the relationship between the main service channel (MSC), fast access channel (FAC), and the services description channel (SDC)

the data in a predetermined way in order to provide a means for error protection and correction, and defines the mapping of the digitally encoded information into QAM cells. These are the basic carriers of the information supplied to the transmitter for modulation.

The MSC requires QAM cell interleaving, which rearranges the time sequence of the signal bits in a systematic way as a means of "scrambling" the signal, so that the final reconstruction of the signal at a receiver will be less affected by fast fading than would be the case if the audio data were transmitted in its original continuous order.

The last element to feed into the OFDM cell mapper is the pilot generator. The pilot generator injects non-data carriers of prescribed amplitude and phase that permits a receiver to derive channel state information, thus allowing proper demodulation of the signal.

As shown in Figure 7.1-1, the MSC, FAC, SDC, and pilot generator all feed into the OFDM cell mapper. As the name suggests, "the OFDM cell-mapper collects the different classes of cells and places them on a time-frequency grid, in effect distributing the information across the sub-carriers."[24] The mapping strategy employed depends on the channel and the robustness mode selected. All data cells can be 4-QAM, 16-QAM, or 64-QAM.[25]

The output of the OFDM cell mapper feeds into the OFDM signal generator. Then the ensemble of cells (with the same time index) are transformed to a time domain representation of the signal, with a plurality of carriers. The complete time-domain OFDM symbol is then obtained from this representation after inserting a guard interval, which is a cyclic repetition of a portion of the signal.[26]

"The coding/modulation scheme used is a variety of coded orthogonal FDM (COFDM) which combines OFDM with MLC (multilevel coding) based on convolutional coding."[27] Taken with other elements in the chain, the net result is a lessening of the effects of short-term fading, creating excellent transmission and signal protection possibilities.[28] "Its main drawback has been that it requires a significant level of signal processing to extract the data from the carriers and reassemble it in the correct fashion. However, signal processing ICs are now sufficiently powerful and at a reasonable cost to make the use of this form of modulation viable."[29]

The digital representation of the OFDM signal is fed into the modulator, which converts it into an analog signal for transmission. However, we're not quite done. Certain nonlinear high-power transmitters require that the amplitude and phase components be split (ideally in the digital domain), and recombined in the transmitter. Frequency up-conversion, digital-to-analog conversion, and filtering to enforce mask/spectral requirements must also be performed.[30]

Both the digital phase and amplitude information of the DRM RF signal is corrupted to different degrees as the signal propagates. HF propagation provides a variety of challenges for signal fading, varied path delays, and Doppler effects. Error protection and error correction that has been incorporated in the DRM system lessens these issues in most cases and permits the accurate decoding of the transmitted information.[31] The FAC and SDC channel information enable the DRM receiver to detect the mode used in transmission, and decode/demodulate the signal accordingly.[32]

(An in-depth look at OFDM is beyond the scope of this paper. We have included a limited number of references at the end of this chapter for the convenience of the reader.)

The SFN capability of DRM can provide for greater spectrum efficiency because an entire area/region/country could be covered by transmitters all operating on the same frequency with some signal overlap. Provided that multiple signals arrive at the receiver's input within the period required by the guard interval, a seamless switch can occur without the listener noticing the change. If a multiple frequency network is required, the DRM signal can contain information to facilitate the seamless switching to an alternate frequency, again without a break in the audio or data.[33]

DRM "content" can consist of four elements:

1. Required metadata (FAC and SDC);
2. Nonmandatory data (e.g., text messages);
3. Audio content, including encoding and quality information;
4. Additional data services selected by the broadcaster.

All DRM receivers must be capable of processing the content from elements one through three.[34]

As one would expect, the broadcast metadata includes the service ID, service/program title, program type, program language and optionally, country of origin. The latest xHE-AAC audio coder is available with bit-rates starting at 6 kbps. Also, in the 18/20 kHz DRM30 and the DRM+ modes, the MPEG 4 stereo compatible 5.1 surround sound is possible.[35]

Besides the additional "value-added" services available in DRM, of particular note is a mandatory feature called the DRM emergency warning feature (EWF). By using the AFS feature, the EWF switches the receiver to a channel that is broadcasting emergency information. It may also turn on receivers to warn of the emergency.[36]

The reader should visit the DRM website for expanded and up-to-date information on DRM at www.drm.org/.

DIGITAL AUDIO BROADCASTING (DAB)

The acronym DAB refers both to the generic notion of digital audio broadcasting, as well as the specific broadcast standard (and in fact DAB is a registered trademark of the DAB digital radio standard). The ETSI standard from January 2006, EN 300 401 v1.4.1 (see www.etsi.org/deliver/etsi_EN/300400_300499/300401/01.04.01_40/en_300401v010401o.pdf) is the one we are using for this chapter, and we will be discussing in this section the specific, open standard (see www.worlddab.org or www.etsi.org/standards).

DAB evolved from the Eureka-147 DAB System Project, which began in 1981. The original audio with PAD data services standard was approved in 1995. DAB+ was introduced in 2007, providing enhanced audio efficiency.[37] DAB+ and DMB (Digital Multimedia Broadcasting) have their technical base in DAB. "In other words, the physical layer is still the same . . . just new applications, new transport protocols and a second error-control coding layer have been added."[38] The main difference between DAB and DAB+ is the use of the HE AAC v2 audio coder and the full use of the PAD channel in the DAB+ system. Also, DAB receivers do not decode DAB+ transmissions. DAB broadcasters generally use Band III (174–240 MHz), although the DAB scheme allows for operation above 30 MHz.

EN 300 401 describes the DAB system as being

> designed for delivery of high-quality digital audio program and data services for mobile, portable and fixed reception from terrestrial or satellite transmitters in frequency bands from 30 MHz to 3 GHz. The DAB system is designed to provide spectrum and power efficient techniques in terrestrial transmitter network planning, known as single frequency network (SFN).[39]

DAB is a digital-only transmission mode with a maximum multiplex bit rate of 2.25 Mbps. It uses three channels to transport data: the synchronization channel (SC), the fast information channel (FIC), and the main service channel (MSC). The SC provides frequency and timing reference to the receiver. The FIC's main function is to deliver the information necessary for the receiver to interpret the configuration of the MSC. This information is called the multiplex configuration information (MCI), which is basically the plan for the multiplex structure. The FIC can also carry service information (SI), conditional access (CA) management information, and fast information data channel (FIDC). "In order to allow a rapid and safe response to the MCI, the FIC is transmitted without time interleaving, but with a high level of protection against transmission errors."[40] The transmission frame that contains these three channels is 96 msec long.

The MSC consists of common interleaved frames (CIF) and service components (SC). CIFs are data fields, and the SC consists of two different transport modes: stream mode and packet mode. The reader is left to explore this in more detail, but suffice it to say that the MSC is a multiplex of subchannels.[41]

When one reviews the DAB block diagram in the ETSI EN 300 401 standard document, while it contains quite a number of elements, the basic signal flow is very similar to the DRM plan. Audio and data flow to the main service multiplexer, whose output is combined with the FIC in the transmission frame multiplexer. A simplified DAB block is shown in Figure 7.7-2.

What is particularly noticeable in the DAB and DAB+ system is the large number of audio services available. This is referred to as the "ensemble" structure of DAB. Each ensemble has a total bit rate capacity of 1152 kbps. "Each ensemble has its own ensemble label, its own unique ID code, can carry a unique identifying code of the transmitter (TII), and a signaling channel (FIC)."[42] Assuming an audio channel at 64 kbps, DQPSK, and FEC+0.5, 18 program channels could be transmitted per DAB+ ensemble, and 24 if the bit rate were lowered to 48 kbps.[43] This makes for spectrum, transmitter, and energy consumption efficiencies with fewer frequencies and transmitters needed to create the same amount of programming. (A similar ensemble technology concept for HD Radio is the digital HD Multiplex plan proposed by Philipp Schmid, research engineer at Nautel.[44] DVB-T2 Lite also employs an ensemble technology as well, with the largest number of program channels of any system.)[45]

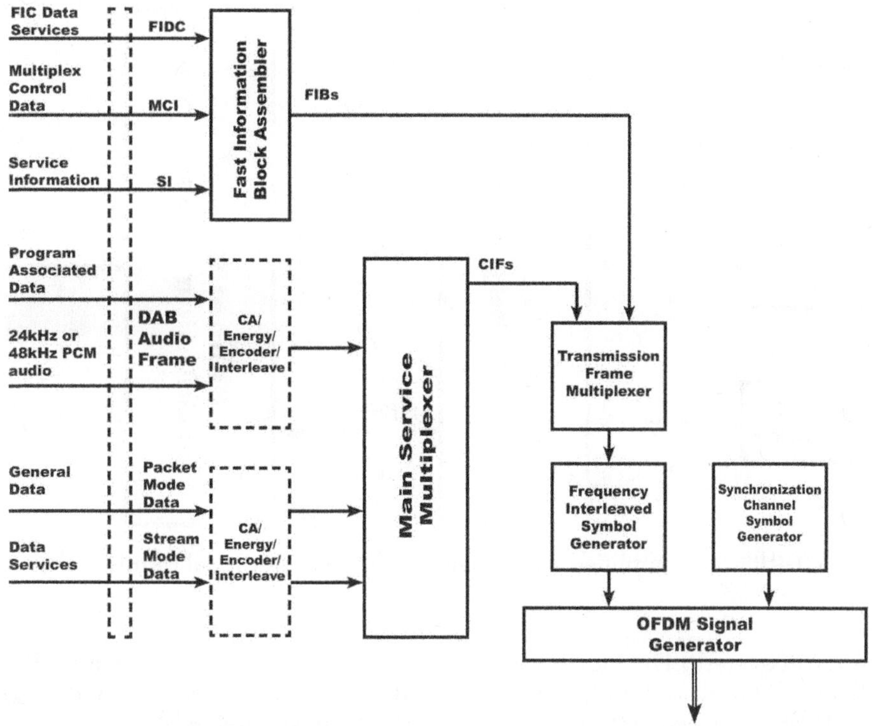

FIGURE 7.7-2 Simplified block diagram of the DAB digital radio system

Program associated data (PAD) includes the possibility for a variety of services. These include but are not limited to text delivery, slideshow images, electronic program guide, traffic info, and other custom services.

Also of note is the fact that each DAB transmitter can operate in DAB, DAB+, and DMB modes without any changes required at the transmitter. It is the playout equipment (audio encoding, video encoding (DMB), data server, and error protection) that determines the mode of operation for DAB.[46]

CHINA DIGITAL RADIO (CDR)

The CDR standard was made effective by the State Administration of Press, Publication, Radio, Film and Television of the People's Republic of China (SAPPRFT) in November 2013.[47] The CDR standard is for both AM and FM broadcasts, but with flexible spectrum modes that permit adaptation to current AM and FM band-plans, and future all-digital bands.

The audio codec used in CDR, DRA+, was "originally designed for multichannel audio processing like digital cinema or HDTV systems. Enhanced with spectral band replication (SBR) and parametric stereo (PS) technologies, the new DRA+ codec can provide near FM stereo service at 24 kbps."[48] This means the DRA+ can match the AAC+ v2 codec in terms of nominal features.[49]

CDR employs three channels: a system information channel (for transmission mode, FEC code rate, and modulation level); a service description channel (for details of the multiplex information of the services); and a main service channel (for the audio and data information). One of the constraints for the system was to use a system that would minimize the cost of developing and producing a receiver for both all-digital FM and MW in the future. To address this, CDR defines three transmission modes to be selected based on application scenarios. Mode 1 is for SFN and large areas. It is noted for its longer guard interval. Mode II is for highway or high-speed rail reception where high-speed applications are needed. In Mode II, more carrier space is needed to adapt to Doppler shift. Mode III is for future high-speed transmission applications.

The CDR system defines the bandwidth for each "sub-band" as 100 kHz in the 400-kHz wide FM channel. The subband is a digital signal that can be decoded by the CDR receiver. In the simulcast (IBOC) mode, the sub-band is placed beyond the analog FM baseband. But unlike in the HD Radio system where there is a 70-kHz digital signal above and below the analog, the CDR system employs a 50-kHz digital signal that would be located at 150 kHz above and below the analog. This lowers the interference between the analog and digital signals. The CDR standard also allows for a possible 800-kHz wide signal when the analog signal is shut off, but maintains the 100-kHz bandwidth for each subband.

> Depending on the Low Density Parity Check (LDPC) code rate and the adopted transmission mode, net bit rate ranges from 36 to 356.4 kbps per each 100 kHz subband, with a maximum spectrum efficiency of 3.56 bit/sec/hertz. A single 200 kHz channel can then carry a useful bit rate ranging from 72 to 712.8 kbps.[50]

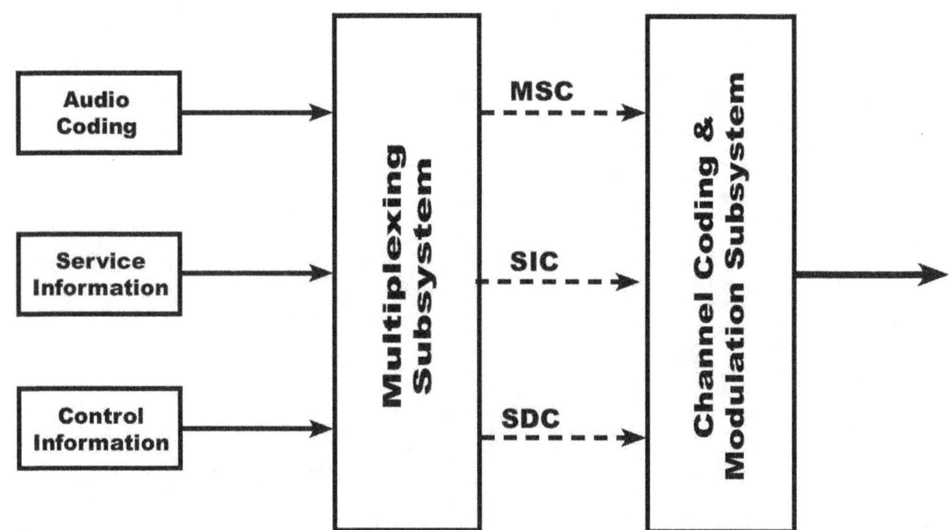

FIGURE 7.7-3 Simplified block diagram of the CDR digital radio system

By employing the rediscovered LDPC for forward error correction (FEC), CDR employs it to protect the code on the main services channel. Four different LDPC matrices were designed for different code rates. For efficiency reasons, the other two channels still use the traditional convolutional code that is also used in the DRM, DAB, and HD Radio systems.[51]

The MSC is scrambled, LDPC coded, constellation mapped, and subcarrier interleaved before OFDM modulation. The SDC is scrambled, and it along with the SIC is one-quarter convolutional coded, bit interleaved, and constellation mapped. These three channels, along with scattered pilots will be OFDM modulated.[52] Figure 7.7-3 shows a block diagram for the CDR system.

By looking ahead to all-digital service receivers that can decode both analog and digital signals, the CDR system will allow for an easy shutoff of analog when that time comes. This is significant when one considers that the Chinese market is over 1 billion listeners! An easy transition is an important requirement.

Quick Summary

As the reader stops to take a breath and before we move on to other topics, a good summary of many factors involved with digital radio was written by Lindsay Cornell, principal systems architect at the BBC in 2014. His presentation is titled "Digital Radio in All Its Variations" and may be found at http://cwbackoffice.co.uk/Presentation/DigitalSIG_06.11.14_LindsayCornell_DigitalRadioInAllItsVariations.pdf.

DVB-T2 LITE

In Sweden, Open Channel (a Copenhagen-based broadcasting service) asserted in 2010 that the DVB-T2 technology was superior to the DAB+ standard, and that the DVB-T2 Lite subset be used to provide digital radio and mobile broadcasting in an over-the-air trial in the greater Copenhagen area. The service has been on-air since June 2012.

> DVB-T2 provides a 2.5 to 4 times increase in capacity compared to the DAB/DAB+ standard under the same broadcasting conditions. It is also better for indoor reception and less sensitive to impulse noise, as well as being better suited for in-car reception.[53]

The DVB-T2 standard (www.dvb.org/standards/dvb-t2) is a digital transmission platform that can transmit a variety of digital signals, and when the T2-Lite subset is added, can create transmissions optimized for mobile reception with the added benefit of minimizing requirements for the receiver.

When operating in T2-Lite mode alone, the bandwidth is 1.7 MHz, the same as the DAB+ signal used in the VHF Band III scheme. However, by using the HE-ACC codec, T2-Lite has the capacity for 56 music programs at 64 kbps (with DAB+ having the capacity for only 16 at that rate).[54]

> DVB-T2 uses the Low Density Parity Check (LDPC) codes offering excellent performance in the presence of high noise levels and interference. Transmission quality and reliability is improved by the introduction of advanced clipping functions, named PAPR techniques. . . . Finally, DVB-T2 group has defined a method called Future Extension Frame (FEF) which can be implemented over an existing DVB-T2 network.[55]

For T2-Lite,

> the mobile services may be transmitted as a stand-alone signal, but may also be transmitted on the same radio frequency channel with other DVB-T2 services. This can be achieved thanks to the FEF feature of the standard. In this case, the mobile transmission will be ignored by normal DVB-T2 receiver and the standard DVB-T2 transmission, also called T2-base, is ignored by the mobile receiver.[56]

Bit Rates, Codecs, and Audio Quality

While it is good audio engineering practice to maintain uncompressed audio as far down the audio chain as possible, at some point it becomes necessary to feed the audio into an audio coder. The quality and efficiency of the coder assists in the balancing act of bit rate versus audio quality.

While newer coders may help lower the bit rate and thus increase the number of channels possible with the given bandwidth, the digital radio engineer who is concerned with the quality of their transmitted audio needs to understand that many listeners use their appreciation of FM audio as the standard by which they judge digital radio. Therefore, it is important to be aware of what current research says about that, and be a critical listener. For example, the NPR Tomorrow Radio project indicates that for HD Radio, a bit rate of 48 kbps is an acceptable rate,[57] where an AES paper titled "Perceived Audio Quality of Realistic FM and DAB+ Radio Broadcasting Systems" stated that not even at a 192 kbps rate can DAB+ reach a quality comparable to a modern FM system.[58] Therefore, it goes without saying that not only is the bit rate of the coder important, but also is the digital transmission standard.

Why Digital Radio? Why Now?

No doubt many a radio listener has asked those questions. The rising noise floor for MW operation makes it imperative that a fresh look at an old problem be made before AM is crippled, perhaps permanently. At the end of 2015, German and French MW transmitters were shut down. "Radio France and Deutschlandradio both said listener numbers have dwindled significantly, and the affected services will shift to FM, digital, and Internet broadcast outlets."[59] To help provide some perspective on this issue, Hal Kneller wrote a useful article, "New Life for AM With Digital Transmission (a Look at the Options)."[60] He discusses the HD Radio and DRM30 options available in this paper and notes the distinct advantages of digital AM over FM and other digital broadcast modes.

But it isn't just MW stations shutting down. Norway plans an FM switch-off in January of 2017 of the major FM government stations (commercial stations will remain on-air). This means about 200 FM stations in small towns and villages will remain on for at least five more years to avoid extra costs, and, perhaps most importantly, "to guarantee regional content."[61] In the Flanders region of Belgium they are looking to do away with FM radio within the next five years, and switch to DAB+ in order to free up spectrum space for additional channels.[62]

In a number of countries in Europe, the thinking is that an FM shutdown is inevitable. Costs (and emphasis on green technology), limited spectrum, and a call for more programming all point to digital radio as a solution. DAB+ appears to be the answer for many European countries, with the timing of the shutdown being triggered by a majority of listeners having access to the DAB+ broadcasts. In Finland, an early adopter of DAB in 1998, all DAB broadcasts were shut down in 2005 in favor of using the digital TV network (DVB-T and DVB-T2) for their terrestrial digital radio network.[63] While DAB has gained popularity in Europe, the use of DVB-T2 is gaining in popularity elsewhere; especially Asia. There appears to be the opinion that DAB (the original, not DAB+ or DMB) is an obsolete standard. Similar opinions have been expressed in Thailand, India, Sweden, and Denmark as well.[64]

Some countries, like Switzerland, allow for a voluntary shutdown of FM when the broadcaster can show that their DAB+ coverage will match that of their FM signal.[65] According to the European Broadcasting Union's Media Intelligence Service (the EBU is a strong supporter of DAB), the "leaders" in Europe in the transition to digital radio are Denmark, Norway, Switzerland, and the United Kingdom, all employing DAB and DAB+.[66] A lot of data about digital radio is available from the EBU (see www.ebu.ch/home).

To promote digital radio, the European Digital Radio Alliance was created in March 2016. The alliance is between commercial and public service broadcasters in Europe. It is headquartered at the EBU in Switzerland and lead by Helen Boarden of the BBC. It has over 300 member radio stations, operated by 12 companies, reaching over 130 million listeners. These broadcasters wanted to send the clear message that they are committed to digital radio.[67]

In April 2016, the first European workshop, "Flexible and Complete Digital Radio for Europe" was held in Erlangen, Germany, hosted by Fraunhofer IIS. At the workshop, it was emphasized that DRM is the ITU-endorsed and internationally adopted standard for international programming and broadcasting. They called on receiver manufacturers to build receivers that will embrace both DRM and DAB+, including putting them in smartphones![68] (Speaking of smartphones, the LG Stylus DAB+ Smartphone was introduced as the first smartphone for DAB, in April 2016.)[69]

But, of course, Europe isn't the only place where digital radio is growing. In China, the CDR standard is being employed by three trial networks, and by the end of 2016 it was expected that CDR broadcasts will be heard in more than 560 cities across China.[70]

Australia is also a leader in DAB+ according to WorldDAB. The DAB+ service began in metropolitan areas in late 2009. It is estimated that over 25% of the population listens to DAB+ in a week's time (see www.worlddab.org/country-information/australia).

India is an interesting example of a country expanding both analog FM and DRM30 for MW services. "A total of 266 private FM radio channels are available as part of a radio broadcasting expansion in India, announced the country's Information and Broadcasting ministry."[71] This impacts 92 cities across India. As expected, there are critics of the expansion of the analog service, citing the preference to move to a digital service mode.[72] Digital radio is moving ahead for All India Radio with the move to 27 high power, MW radio transmitters employing DRM30 (see www.nautel.com/solutions/digital-radio/high-power-drm-mw-am-transmitters-air/). According to Nautel, this "comprised the world's largest digital radio deployment to date."[73] The first of the transmitters went on the air in May of 2014. Doordshan, the Indian public service broadcaster, has decided to use DVB-T2 Lite for digital radio broadcasting instead of DAB or DAB+.[74]

THE IMPORTANCE OF RECEIVERS

If one starts with the assumptions that (1) radio is an important medium providing a significant public service, especially in emergencies; and (2) that people will listen to radio provided it is available on devices they use a lot (or find most convenient), then one can understand the push by the EBU for what is called the Euro-Chip. This is a

> reception chip which can be easily installed in mobile devices and cars. Integrating the main transmission technology standards, whether digital or analog, it will help audiences access their favorite radio broadcast services anytime, anywhere and can help develop new business models which will make radio fully part of the digital economy.[75]

What is critical is to recognize that radio as a stand-alone device is no longer as popular as it once was. Therefore, attention to radio receivers in the hands of the public (literally) is equally as important as a radio in the car. For the user of DVB-T2 Lite, their TV receiver would also be their radio receiver.[76]

In the UK, digital radios are in 80% of the new cars sold, according to WorldDAB data from early 2016. Norway and Switzerland are the other European countries with digital radios in new cars exceeding 60%. Most of the rest of Europe is at or below 25% (according to that same report). We have the usual chicken and egg conundrum. People can't listen to digital broadcasts if they don't have a digital receiver, but what good is a digital receiver if no one is broadcasting digital? Measuring the success of digital radio is impossible in that case.

But if one of the goals of digital radio is to make listening to all radio as easy as possible, then from the listener's perspective, they don't care about the technology—analog, DRM, DAB, or DVB. They only care that the broadcaster transmit excellent programming that sounds excellent! As Graham Dixon, EBU's head of radio, has said, DAB should stand for "delivering audience benefits."[77] The focus should be on the quality of the product, not how high-tech the packaging is. As is being shown in the radio expansion in India, and has been confirmed in informal interviews the author conducted with college students from India, quantity, quality, and diversity of programming is what the audience seeks.[78] It's the broadcaster's and government's responsibility to find ways to make that happen. Analog or digital, DRM or DAB or HD Radio or CDR or DVB, the public is usually oblivious to that. Just do it!

Who Benefits From Digital Radio?

With the development of the SFN and ensemble technology, often the question of "who benefits?" is placed in the arena of the "local versus national" broadcaster. The SFN may make for all kinds of economies—spectrum, transmitters, energy consumption, and so on—for the national broadcaster, but it is also a way of creating barriers to entry. Where might the local broadcaster fit in? As we saw with Norway's FM shutdown, exceptions will be made for the smaller, local broadcasters. In the UK, Ofcom (the UK's broadcast radio regulator) has been experimenting with lower power digital radio transmission schemes that would make it possible for smaller and more localized broadcasters to operate and compete with the ensemble broadcasts.[79]

However, in some places in Asia where the demand for programming is quite high[80] (or anywhere, for that matter), the use of ensemble technology at a regional or local level can create the opportunity for many programmers who have been locked out because of scarce spectrum to get on the air as one of the many ensemble channels. While the initial cost of building the digital RF transmission system is still there, the cost of adding channels (who might be thought of as "tenants" on the digital transmission platform) is not that significant, and the tenants could be a source of revenue and a means of supporting the construction and operation costs.[81]

SUMMARY

It is hoped that we have provided a useful starting point for understanding how digital radio is developing outside of North America. It is also hoped that the notes and references supplied will assist you in your understanding of this growing and changing arena. Feel free to contact the author with comments and updates to this topic. Marconi's quote is one we are asking ourselves: "Have I done the world good, or have I added a menace?" Hopefully the former, and not the latter! Stay tuned.

Notes

[1] Herbert P. Jacobson, "Advantages of Digital Broadcasting," *NASB 2007* (2007), www.shortwave.org/mediafiles.htm.

[2] Tim Anderson, "Advanced Digital Radio: HD Radio, DRM, DAB & CDR" (2015), www.gatesair.com/media-center/presentations?&item=10.

[3] See note 2.

[4] Nigel Laflin, ed., *The DRM Digital Broadcasting System: Introduction and Implementation Guide* (Geneva: DRM Consortium, 2013), 8, www.drm.org/wp-content/uploads/2013/09/DRM-guide-artwork-9-2013-1.pdf.

[5] Joachim Lehnert, "Study on the Comparison of the Transmitter Power Between DRM+ and DAB/DAB+ in VHF Band III to Cover the Same Service Area" (2015), http://drm-forum.de/images/dokumente/2015-11-09_Comparison_DRMplus-DAB.pdf.

[6] DRM.org website, "Summary", www.drm.org/?page_id=99.

[7] Ian Poole, "What Is DRM—Digital Radio Mondiale", www.radio-electronics.com/info/broadcast/drm/drm-digital-radio-mondiale.php.

[8] Laflin, *The DRM Digital Broadcasting System*, 9.

[9] ITU-R BS.1514–2, "System for Digital Sound Broadcasting in the Broadcasting Bands Below 30 MHz" (2011), 1, www.drm.org/wp-content/uploads/2012/10/Recommendation-ITU-R-BS.1514-2-201103-IPDF-E.pdf.
ETSI ES 201 980 v4.1.1, "Digital Radio Mondiale: System Specification" (2014), 17, www.etsi.org/deliver/etsi_es/201900_201999/201980/04.01.01_60/es_201980v040101p.pdf.

[10] ETSI ES 201 980, 14, 17.

[11] See note 10.

[12] Laflin, *The DRM Digital Broadcasting System*, 20.

[13] See note 13.
[14] ETSI ES 201 980, 15.
[15] ETSI ES 201 980, 86.
[16] ETSI ES 201 980, 16.
[17] ITU-R BS.1514–2, 9–10.
[18] ETSI ES 201 980, 14–16; ITU-R BS.1514–2, 8–9.
[19] ITU-R BS.1514–2, 10–11.
[20] ETSI ES 201 980, 46–52.
[21] ETSI ES 201 980, 52–78.
[22] See note 20.
[23] See note 18.
[24] Laflin, *The DRM Digital Broadcasting System*, 19.
[25] See note 18.
[26] Ibid.
[27] ITU-R BS.1514–2, 11.
[28] See note 28.
[29] See note 7.
[30] See note 19.
[31] ITU-R BS.1514–2, 12.
[32] See note 32.
[33] Laflin, *The DRM Digital Broadcasting System*, 23–24.
[34] Ibid., 28–29.
[35] Ibid., 29–30.
[36] Ibid., 30–35.
[37] Frank Herrmann, Larissa Anna Erismann, and Markus Prosch, "The Evolution of DAB", *EBU Technical Review* (2007), 2, https://tech.ebu.ch/docs/techreview/trev_311-dab_dmb.pdf
[38] See note 38.
[39] ETSI EN 300 401 v1.4.1, "Radio Broadcasting Systems: Digital Radio Broadcasting (DAB) to mobile, portable and fixed receivers" (2006), 20–21, www.etsi.org/deliver/etsi_en/300400_3004 99/300401/01.04.01_40/en_300401v010401o.pdf.
[40] See note 40.
[41] Ibid.
[42] Les Sabel, "Overview of the DAB+ System" (2013), www.worlddab.org/public_document/file/442/DAB__Overview_2013.pdf?1394188480.
[43] Jens Stockman, "Economic Advantages of DAB+" Gates Air, www.worlddab.org/public_document/file/556/economic-advantage s-of-dab-jens-stockmann-gatesair.pdf?1429721538.
[44] Philipp Schmid, "Digitizing Terrestrial Radio With Digital HD Radio Multiplex: A Transition Strategy to All Digital HD Radio Broadcasting" (2016), www.nautel.com/wp-content/uploads/2016/04/Digitizing-terrestrial-radio-philip p-schmid-NAB-2016.pdf.
[45] Allan T. Rasmussen, "Digital TV in Thailand: Think Inside the Box" (March 25, 2014), www.yozzo.com/news-and-information/media-broadcasting/digital-tv-in-thailan d-think-inside-the-box.
[46] See note 2.
[47] GY/T 268.1-2013, see https://books.google.com/books?id=lK-kCDAAAQBAJ&pg=PA7&lpg=PA7&dq=SAPPRFT+GY/T+268.1-2013&source=bl&ots=j-2bkvR0hp&sig=uD3YWQUM6Y7vektk-rghtm7fpEI&hl=en&sa=X&ved=0ahUKEwivsLzF2enNAhWFYiYKHdBCC6oQ6AEIHDAA#v=onepage&q=SAPPRFT%20GY%2FT%20268.1-2013&f=false or www.chinesestandard.net/PDF-Excerpt/ShowPDFexcerpt.aspx?ExcerptID=GY/T%20268.1-2013.
[48] Peng Gao and Feng Zou, "Design and Implementation of China Digital Radio System", *2016 NAB BEC Proceedings* (2016), 16.
[49] Davide Moro, "China Presents Its Digital Radio Standard", *Radio World* (July 5, 2016).
[50] Ibid.
[51] See note 49.
[52] Yun Wang, "Software-Defined Radio Receiver Design and Development for China Digital Radio (CDR)", (master's thesis, TUDelft, 2016), 6.
[53] See note 47.
[54] Ibid.
[55] "DVB-T2 System Implementation and DVB-T2 Lite Extension: Comprehensive Understanding of the DVB-T2 System", Thomson-Broadcast white paper (July 2014), http://thomson-broadcast.com/wp-content/uploads/2014/08/DVB-T2-T2-Lite-White-paper-v6-July-2014.pdf.
[56] Ibid.
[57] Ellen Sheffield, John Kean, Mike Starling, Jan Andrews, Kyle Evans, and Sunny Khemlani, "Results From Subjective Testing of the HD Codec a6 16–96 kbps", *IEEE Transactions on Broadcasting*, 52, No. 2 (June 2006), 245–251.
[58] Jan Berg, Christofer Bustad, Lars Jonsson, Lars Mossberg, and Dan Nyberg, "Perceived Audio Quality of Realistic FM and DAB+ Radio Broadcasting Systems", *Journal of the Audio Engineering Society*, 61, No. 10 (October 2013), 775–777.
[59] "Medium-Wave Transmitters in Germany and France Shutting Down at Year's End", ARRL Website (December 30, 2015), www.arrl.org/news/medium-wave-transmitters-in-germany-and-france-shutting-down-at-year-s-end.
[60] Hal Kneller, "New Life for AM With Digital Transmission (a Look at the Options)" (2013), www.nautel.com/resources/white-papers/new-life-for-am-with-digital-transmission -hal-kneller-nautel-ltd-nab/.
[61] Davide Moro, "FM Switch Off Plans Shift From Country to Country", *Radio World* (June 29, 2015).
[62] "FM Shutdown Eyed Within Five Years in Flanders", *Radio World* (May 18, 2016).
[63] "DAB Development in the Nordic Countries", *NORDICOM* (March 27, 2015).
[64] Allan Rasmussen, "DVB-T2 Lite vs DAB+ for Digital Radio" (December 5, 2014), www.yozzo.com/news-and-information/media-broadcasting/dvb-t2-lite-vs-dab-for-digital-radio.
Allan Rasmussen, "The 80s Called, They Want Their DAB Technology Back" (May 26, 2015), www.yozzo.com/news-and-information/media-broadcasting/the-80s-called,-they-w ant-their-dab-technology-back.
[65] See note 62.
[66] EBU Media Intelligence Service, "Digital Radio 2016: Market Report", www.ebu.ch/publications/digital-radio-2016.
[67] "The European Digital Radio Alliance", news release (March 15, 2016).
[68] "First DRM European Workshop 'Flexible and Complete Digital Radio for Europe' Asks European Stakeholders to Support and Manufacture DRM and DAB+ Receivers", news release (April 7, 2016), www.drm.org/?p=4313.
[69] Claire Reilly, "LG Stylus DAB+ Smartphone Makes Digital Radio the Star", C|NET website (April 27, 2016), www.cnet.com/au/news/lg-stylus-dab-plus-smartphone-make s-digital-radio-the-star/.
[70] See note 49.
[71] "India Accepting Applicants for Private FM Channels", *Radio World* (June 27, 2016).
[72] Frederick Noronha, "India's FM Auctions Center on Big Cities", *Radio World* (December 14, 2015).
[73] Nautel. "High Power DRM-AM-MW: Building to a Billion", website, www.nautel.com/solutions/digital-radio/high-powe r-drm-mw-am-transmitters-air/.
[74] See note 47.
[75] "The Euro-Chip", EBU website, www.worlddab.org/public_document/file/311/Euro-Chip_FAQ.pdf?1364911629.
[76] See note 47.
[77] Graham Dixon, "DAB: Delivering Audience Benefits", *Radio World* (January 11, 2016).
[78] Kenneth Colwell, interview with college students from India visiting St. Ambrose University, Davenport, Iowa (June 16, 2016).
[79] Lawrie Hallett, "Small-Scale DAB Presents Opportunity, Challenges", *Global Digital Radio Developments* (Radio World International Edition, 2016), 20–23.
[80] See note 65.
[81] "This Week in Radio Tech," #275, podcast (September 25, 2015).

References

Anderson, Tim. "Advanced Digital Radio: HD Radio, DRM, DAB & CDR", 2015. www.gatesair.com/documents/slides/2015-0 9-Anderson-Advanced-Digital-Radio-HD-Radio-DRM-DAB-CDR.pdf

ARRL. "Medium-Wave Transmitters in Germany and France Shutting Down at Year's End", website. www.arrl.org/news/medium-wave-transmitters-in-germany-and-france-shutting-down-at-year-s-end

Berg, Jan, Christopher Bustad, Lars Jonsson, Lars Mossberg, and Dan Nyberg. "Perceived Audio Quality of Realistic FM and DAB+ Radio Broadcasting Systems", *Journal of the Audio Engineering Society*, 61, No. 10, October 2013, 755–777. www.radiojackie.com/im/paq.pdf

Bower, A.J. "Digital Radio—The Eureka 147 DAB System", *Electronic Engineering*, 4, 1998, 55–56. http://downloads.bbc.co.uk/rd/pubs/reports/1998-10.pdf

Clark, Marguerite, ed. "Digital Radio Around the World". (*Radio World International* eBook, 2014.)

Clark, Marguerite. "Q&A: Ruxandra Obreja, Chairman, Digital Radio Mondiale: The Emergency Warning Function of Digital Systems Saves Lives" in "Global Digital Radio Developments", *Radio World International*, 2016.

Cornell, Lindsay. "Digital Radio in All Its Variations", *BBC*. http://cwbackoffice.co.uk/Presentation/DigitalSIG_06.11.14_LindsayCornell_DigitalRadioInAllItsVariations.pdf

Cridland, James. "AM Radio—Dead in Europe?", 2015. www.allaccess.com/radiofuturescope/archive/22763/am-radio-dead-in-europe

Cridland, James. "DAB+ and DAB—the Differences, and Its Use in the UK", website, March 8, 2016. https://media.info/radio/opinion/dab-and-dab-the-differences-and-its-use-in-the-uk

Dixon, Graham. "DAB: Delivering Audience Benefits", *Radio World*, January 11, 2016.

———. "Making Radio Modern", *tech-i*, March 8, 2016.

DRM.org. "First DRM European Workshop 'Flexible and Complete Digital Radio for Europe' Asks European Stakeholders to Support and Manufacture DRM and DAB+ Receivers", press release, April 7, 2016. www.drm.org/?p=4313

———. "Key Features", video. www.youtube.com/watch?v=2cV7W-cDD4xU&feature=youtu.be

———. "Summary", website. www.drm.org/?page_id=99

DVB.org. "On the Radio With DVB-T2 Lite", August 24, 2011. www.dvb.org/news/on-the-radio-with-dvb-t2-lite

Enders Analysis. "Radio—Set for a Digital Future?", 2015. www.endersanalysis.com/content/publication/radio-%E2%80%93-set-digital-future

European Broadcasting Union. "Digital Radio 2016: Market Insights", *Media Intelligence Service*, 2016. www.ebu.ch/files/live/sites/ebu/files/Publications/EBU-MIS%20-%20Digital%20Radio%20Report%202016.pdf

———. "The Euro-Chip", website. www.worlddab.org/public_document/file/311/Euro-Chip_FAQ.pdf?1364911629

———. R-138, "Digital Radio Distribution in Europe", 2013. https://tech.ebu.ch/docs/r/r138.pdf

———. "The Use of Band III in Europe", Technology Fact Sheet. https://tech.ebu.ch/docs/factsheets/ebu_tech_fs_band_III_use.pdf

———. "WorldDAB General Assembly Speech by EBU Head of Radio Graham Dixon", London, November 3, 2015.

European Digital Radio Alliance. News release, March 15, 2016.

European Telecommunications Standards Institute. www.etsi.org/standards

———. EN 300 401 v1.4.1. "Radio Broadcasting Systems: Digital Audio Broadcasting (DAB) to Mobile, Portable and Fixed Receivers", 2006. www.etsi.org/deliver/etsi_en/300400_300499/300401/01.04.01_40/en_300401v010401o.pdf

———. ES 201 980 v4.1.1. "Digital Radio Mondiale (DRM) System Specification", 2014. www.etsi.org/deliver/etsi_es/201900_201999/201980/04.01.01_60/es_201980v040101p.pdf

Gao, Peng and Feng Zou. "Design and Implementation of China Digital Radio System", *2016 NAB BEC Proceedings*, 16–20.

Garlinger, Douglas. "International Shortwave Broadcasting", in *SBE Broadcast Engineering Handbook: Hands-On Guide to Station Design and Maintenance*, New York: McGraw Hill, 2016, 2.101–2.114.

Hallett, Lawrie. "Small-Scale DAB Presents Opportunity, Challenges" in "Global Digital Radio Developments", *Radio World International*, 2016, 20–23.

Harnack, Kirk. "HD and AM Transmission with Gary Liebisch," TWiRT #275, September 25, 2015. http://thisweekinradiotech.com/twirt-home/2015/9/25/twirt-275-hd-and-am-transmission-with-gary-liebisch.html

Hederstrom, Christer. "Community Radio: FM and Digitalization", *Community Media Forum Europe* (green paper), May 12, 2014.

———. "Rethinking Your Digital Strategy", (guest commentary), *Radio World*, March 2014.

Herbert P. Jacobson, "Advantages of Digital Broadcasting", *NASB 2007*, 2007. www.shortwave.org/mediafiles.htm/

Hermann, Frank, Larissa Erismann, and Markus Prosch. "The Evolution of DAB", *EBU Technical Review*, 2007. https://tech.ebu.ch/docs/techreview/trev_311-dab_dmb.pdf

International Telecommunications Union. ITU-R BS.1514-2, "System for Digital Sound Broadcasting Bands Below 30 MHz", 2011. www.drm.org/wp-content/uploads/2012/10/Recommendation-ITU-R-BS.1514-2-201103-IPDF-E.pdf

———. ITU-R BS.2384-0, "Implementation Considerations for the Introduction and Transition to Digital Terrestrial Sound and Multimedia Broadcasting", 2015. www.itu.int/dms_pub/itu-r/opb/rep/R-REP-BS.2384-2015-PDF-E.pdf

Irwin, Doug. "DAB: Success or Failure? Depends on What You Read", *Radio Magazine*, July 5, 2016.

———. "GatesAir Delivers Its First China Digital Radio Systems", *Radio Magazine*, January 29, 2016.

———. "Nautel and All India Radio Partner on High-Power, MW Digital Radio Deployment", *Radio Magazine*, February 1, 2016.

Kneller, Hal. "New Life for AM With Digital Transmission (a Look at the Options)", 2013. www.nautel.com/wp-content/uploads/2013/05/New-Life-AM-Digital-TransmissionHal-Kneller-BEC-NAB-2013.pdf

Laflin, Nigel, ed. *The DRM Digital Broadcasting System Introduction and Implementation Guide*. Geneva: DRM Consortium, 2013. www.drm.org/wp-content/uploads/2013/09/DRM-guide-artwork-9-2013-1.pdf

Lehnert, Joachim. "Study on the Comparison of the Transmitter Power Between DRM+ and DAB/DAB+ in VHF Band III to Cover the same Service Area", 2015. http://drm-forum.de/images/dokumente/2015-11-09_Comparison_DRMplus-DAB.pdf

Moro, Davide. "Assessing Digital Sound Performance" in "Global Digital Radio Developments", *Radio World International*, 2016, 12–14.

———. "China Presents Its Digital Radio Standard", *Radio World*, July 5, 2016.

———. "Digital Radio Summit Offers Overview", *Radio World*, March 28, 2016.

———. "FM Switch Off Plans Shift From Country to Country", *Radio World*, June 6, 2015.

———. "Varied Pace, Approach to Digital Radio" in "Global Digital Radio Developments", *Radio World International*, 2016, 4–8.

Nautel. "High Power DRM-AM-MW: Building to a Billion", website. www.nautel.com/solutions/digital-radio/high-power-drm-mw-am-transmitters-air/

———. "Nautel HD Multiplex Technology FAQ", *2015 NAB BEC*. www.nautel.com/wp-content/uploads/2015/04/Nautel-HD-Multiplex-FAQ.pdf

NORDICOM. "DAB Development in the Nordic Countries", website, 2015. www.nordicom.gu.se/en/media-trends/news/dab-development-nordic-countries

Noronha, Frederick. "India's FM Auctions Center on Big Cities", *Radio World*, December 14, 2015.

NXP Semiconductors. "Software-Defined Radio: One Global Platform for Car Entertainment", video, 2012. www.youtube.com/watch?v=eTTF3piDLm8

Ocampo, Rogelio. "Digital Radio in Latin America Advances Slowly" in "Global Digital Radio Developments", *Radio World International*, 2016, 9.

Open Digital Radio. www.opendigitalradio.org/ & http://wiki.opendigitalradio.org/Main_Page

Park, Kyung-Won, Seong-Jun Kim, and Kyung-Taek Lee. "Transmission Rate Enhancement Schemes for Digital Radio Mondiale Systems", *International Journal of Multimedia & Ubiquitous Engineering*, 8, No. 1, 2013, 225–234. www.sersc.org/journals/IJMUE/vol8_no1_2013/21.pdf

Poole, Ian. "What Is DAB? Digital Radio Tutorial", website. www.radio-electronics.com/info/broadcast/dab/what-is-digital-audio-broadcasting-tutorial.php

———. "What Is DRM—Digital Radio Mondiale", website. www.radio-electronics.com/info/broadcast/drm/drm-digital-radio-mondiale.php

Public Service Council. "The Facts Behind the DAB Radio Failure in Sweden", March 23, 2016. http://public-service.net/docu/DABFactsSweden.pdf

Radio World. "Europe's Digital Radio Rollout, Part I", May 16, 2016.

———. "FM Shutdown Eyed Within Five Years in Flanders", May 18, 2016.

———. "India Accepting Applications for Private FM Channels", June 27, 2016.

———. "MW in Decline as Many Euro Broadcasters Shut Off Transmitters", May 9, 2016.

———. "Q&A: Graham Dixon, European Broadcasting Union", June 20, 2016.

Rasmussen, Allan. "The 80s Called, They Want Their DAB Technology Back", May 26, 2015. www.yozzo.com/news-and-information/media-broadcasting/the-80s-called,-they-want-their-dab-technology-back

———. "Digital TV in Thailand: Think Inside the Box", March 25, 2014. www.yozzo.com/news-and-information/media-broadcasting/digital-tv-in-thailand-think-inside-the-box

———. "DVB-T2 Lite vs DAB+ for Digital Radio', December 5, 2014. www.yozzo.com/news-and-information/media-broadcasting/dvb-t2-lite-vs-dab-for-digital-radio

Reilly, Claire. "LG Stylus DAB+ Smartphone Makes Radio the Star", website, 2016. www.cnet.com/au/news/lg-stylus-dab-plus-smartphone-makes-digital-radio-the-star/

Sabel, Les. "Overview of the DAB+ System" WORLD DMB, 2013. www.worlddab.org/public_document/file/442/DAB__Overview_2013.pdf?1394188480 and video www.youtube.com/watch?v=z2B48gaM2S4

Schmid, Philipp. "Digitizing Terrestrial Radio with Digital HD Radio Multiplex", 2016 NAB. www.nautel.com/wp-content/uploads/2016/04/Digitizing-terrestrial-radio-philipp-schmid-NAB-2016.pdf

Sheffield, Ellyn, John Kean, Mike Sterling, Jan Andrews, Kyle Evans, and Sunny Khemlani. "Results From Subjective Testing of the HD Codec at 16–96 kbps", *IEEE Transactions on Broadcasting*, 52, No. 2, June 2006.

State Administration of Radio, Film and Television. GY/T 268.1–2013. https://books.google.com/books?id=lKkCDAAAQBAJ&pg=PA7&lpg=PA7&dq=SAPPRFT+GY/T+268.1-2013&source=bl&ots=j-2bkvR0hp&sig=uD3YWQUM6Y7vektk-rghtm-7fpEI&hl=en&sa=X&ved=0ahUKEwivsLzF2enNAhWFY-iYKHdBCC6oQ6AEIHDAA#v=onepage&q=SAPPRFT%20GY%2FT%20268.1-2013&f=false & www.chinesestandard.net/PDF-Excerpt/ShowPDFexcerpt.aspx?ExcerptID=GY/T%20268.1-2013

Stockman, Jens. "Economic Advantages of DAB+", Gates Air. www.worlddab.org/public_document/file/556/economic-advantages-of-dab-jens-stockmann-gatesair.pdf?1429721538

Thomson Broadcast DVB-T2 Solutions. "DVB-T2 System Implementation and DVB-T2 Lite Extension: Comprehensive Understanding of the DVB-T2 System", July 2014. http://thomson-broadcast.com/wp-content/uploads/2014/08/DVB-T2-T2-Lite-White-paper-v6-July-2014.pdf

Vogg, Christian and Fernandez Quijada. "Moving Forward to Digital Radio", *Radio World*, March 6, 2015.

Wang, Yun. "Software-Defined Radio Receiver Design and Development for China Digital Radio (CDR)", TUDelft, thesis, 2015.

Wenzel, Kenneth. "DVB-T2Lite | First Deployments, First Experiences | Next Generation of Digital Radio & Mobile TV". www.slideshare.net/KennethWenzel/dvbt2-lite-next-generation-of-mobile-broadcasting

Working Group RSPG10–349. "The Future of Radio Broadcasting in Europe", *Radio Spectrum Policy Group*, 2010. http://rspg-spectrum.eu/wp-content/uploads/2013/05/rspg10_349_report_futureradio.pdf

World DAB. "Australia", website. www.worlddab.org/country-information/australia

———. "Digital Radio Europe & Asia Pacific", 2016. www.worlddab.org/public_document/file/709/WorldDAB_Infographic_2016.pdf?1458228290

———. "Digital Radio Toolkit", 2016. www.worlddab.org/public_document/file/521/EBU-MIS_Digital_Radio_Toolkit.pdf?1418724468

———. Eureka Spectrum & Regulation Newsletter. http://worlddabeureka.org/category/spectrum-and-regulation/

———. "How the DAB Family of Standards Work", website. www.worlddab.org/technology-rollout/standards/how-the-dab-family-of-standards-work

———.www.worlddab.org/

———. "Introduction to Digital Radio Spectrum". www.worlddab.org/public_document/file/630/Introduction_to_Digital_Radio_Spectrum_edited_KS_19-08-15.pdf?1440003250& www.worlddab.org/public_document/file/631/Introduction_to_Digital_Radio_Spectrum_edited_KS_19-08-15.pdf?1440084073

———. "IP for DAB Distribution Is Gaining Momentum", newsletter, 2013. http://worlddabeureka.org/2013/09/05/ip-for-dab-distribution-is-gaining-momentum/

———. "WorldDAB Global Summary", June 2016. www.worlddab.org/public_document/file/742/WorldDAB_Global_Summary_16.06.2016.pdf?1466086970

World DAB Forum. "Digital Radio Receiver Profiles", 2016. www.worlddab.org/public_document/file/711/WorldDAB_Receiver_Profiles_2016-03-16.pdf?1458300904

CHAPTER

7.8

Medium Wave AM Broadcast Antenna Systems

RONALD D. RACKLEY, P.E.
du Treil, Lundin & Rackley, Inc.
Sarasota, FL

INTRODUCTION

Medium wave broadcast (AM) antenna systems can reach a high degree of sophistication, much of which is based on advanced mathematics. The purpose of this chapter is to provide the station engineer with an understanding of some of the basic concepts of antenna design and an appreciation for the complexities of this specialty. An extensive bibliography is included at the end of the chapter for further study of the subject.

The chief purpose of a broadcast antenna system is to radiate efficiently the power supplied to it by the transmitter. A simple antenna can do this job quite well. This is often a single vertical tower that radiates its signal equally in all directions along the ground in a so-called nondirectional or omnidirectional pattern. A second purpose of an AM antenna system is often to concentrate the power in desired directions to cover populated areas and to suppress it in other directions to protect the coverage of other stations sharing the same or closely adjacent channels. This directionality may require a complicated directional antenna system with several towers if the requirements are stringent.

The signals radiated from the antenna are propagated through space to each receiving antenna. The factors affecting the strength of the received signal include the strength of the signal radiated by the broadcasting station in a particular direction, the distance to the receiving site, losses incurred by the less-than-perfect conductivity of the ground along the propagation path, terrain obstructions (large hills cast shadows even at AM frequencies), and in the case of skywave transmission, the ionospheric conditions that determine how much of the radiated signal will be reflected back to each distant receiving location. Signal strength in a particular direction can also be affected by the presence of structures such as buildings or towers near the transmitting antenna.

The polarization of the transmitted waves is also a factor; for medium wave broadcast stations, vertical polarization is used because of its superior ground-wave propagation. The Federal Communications Commission (FCC) has established maximum transmitter power limits for each of the four classes of AM channels (A, B, C, and D), so the only variables available to the design engineer attempting to maximize the coverage of a radio station involve the antenna location, the pattern design, and a limited choice of power levels. These factors go hand in hand when designing a directional antenna system. Severe constraints are usually imposed on transmitter site selection because of aeronautical, zoning, environmental, and coverage requirements. The constraints encountered in the pattern design relate to the size and shape of the transmitter site, the extent to which the necessary signal suppression can be achieved at the desired transmitter power level, and the cost to design, construct, adjust, and maintain the multi-tower system. The pattern design can also seriously affect the stability, efficiency, and bandwidth of the completed system.

TRADITIONAL AND MODERN ANTENNA ANALYSIS METHODS

For simplicity of analysis, AM antennas have been traditionally deemed to function as though the currents carried by their elements were purely sinusoidal in nature, such as would theoretically be produced in a standing-wave pattern by two sinusoidal traveling waves of identical magnitude "passing" in opposite directions. This may be visualized as the forward wave that is imposed by the current that is conducted into the base of a tower "passing" the wave that is reflected at the tower top on its way back down. The sum of the two form a sinusoidal standing wave that may be readily integrated using classical mathematics, which is very convenient for relating an antenna's current to its far-field radiation characteristics.

The problem with the sinusoidal current distribution assumption is that it simply cannot be true if radiation is taking place, as the forward and reflected waves on the tower are subject to continuous attenuation as they travel its length due to the energy that leaves the antenna to form the far-field radiation pattern. In other words, the current magnitude of the forward wave grows smaller from the bottom to the top of the tower, and the current magnitude of the reflected wave grows smaller still as it travels back down from the top of the tower to the base. Summing the two waves, therefore, does not form a perfectly sinusoidal current distribution on the tower.

It has always been recognized that antennas cannot actually have sinusoidal current distribution, but integrating true current distribution would have been exceedingly difficult in the early days of AM antenna design. Studies were done in the 1920s, and it was determined that, for determining far-field radiation, sinusoidal current distribution assumptions gave results without sufficient error to warrant more complicated analysis. It was thereafter generally agreed that antenna theory based on sinusoidal current distribution would be used for antenna analysis. The methods for calculating far-field radiation performance in textbooks and the FCC Rules have been based on sinusoidal current distribution ever since.

For directional antennas, an additional layer of error is associated with the sinusoidal current distribution assumption due to the effects of mutual coupling between the towers of an array. This may be understood by visualizing the current distribution of a tower fed at its base and operating in the nondirectional transmitting mode and again with the tower base connected to ground through an impedance while it receives the energy that induces current in it from an external field operating in the receiving mode. The transmitting mode current distribution will approximate a sinusoidal shape; the receiving mode current distribution will depend on the terminating impedance at the tower base and might not resemble the transmitting mode current distribution at all. A common method for detuning a quarter-wave tower, for instance, is to place a reactance at its base to produce a sharp minimum in tower current at approximately one-third of its height—where the current is near maximum in the transmitting mode. As each tower in a directional antenna functions in both the receiving mode and the transmitting mode at the same time, its current distribution will be determined by both. In general, the current distributions of the towers of a directional antenna will differ from one another. Nonetheless, sinusoidal current distribution has traditionally been assumed for AM directional antenna analysis because the resulting far-field radiation errors are not great when compared with other known errors in the processes used to analyze interference between stations, such as are inherent in the statistically derived nighttime skywave signal propagation models that are used for allocation studies.

Although sinusoidal current distribution is a very convenient fictional assumption for AM antenna far-field radiation analysis—where its errors have historically been held to be acceptable—much is left to be desired when it is used to deal with practical matters related to feeding energy into antennas. There can be very large errors in base impedance, particularly for towers near one-half wavelength in height, and in the ratios and phases of array element base currents necessary to produce the far-field radiation parameters of a desired directional antenna pattern. Because of these errors, the FCC always required that a proof-of-performance based on field strength measurements be made after the operating tower currents were adjusted to produce the desired directional antenna radiation pattern, generally through experimentation. The field adjustment process historically bridged the gap between the ideal design world of sinusoidal current distribution and the real world of the field strength proof-of-performance. The unknowns involved always dictated that phasing and coupling systems be designed to match a range of expected tower base impedances based on experience as well as calculation.

Great advances in antenna analysis along with the development of modern computers facilitate the numerical analysis of mathematical problems that defy solution using classical methods. It is now possible to model an AM directional antenna (through what is known as *moment method or Method of Moments modeling*) as a large number of small conductor segments and to take into account the contributions of current that are both conducted from adjacent segments and induced through mutual coupling from all of the other segments. This makes it possible to calculate tower base impedances and drive currents using their real-world current distributions, which facilitates the design of performance-optimized phasing and coupling equipment and eliminate the need for experimentation when tuning AM directional antenna systems. When careful checks of an antenna monitoring system are made, it is possible to know that operating conditions are correct for a desired pattern shape using tower currents or base voltages calculated using moment method modeling. In 2008 the FCC Rules were changed to allow directional antenna systems that met certain requirements having to do with the sampling of their operating parameters to be proofed using measurements of their internal characteristics instead of external field strength measurements, which had always been required before that.

It is known that moment method modeling can also be used to refine the process of AM antenna far-field analysis, but the traditional methods based on sinusoidal current distribution remain standard for that as they form the basis for both the FCC Rules and international regulations and treaties. For that reason, and because traditional methods are more familiar than moment method modeling for antenna analysis today, this section will employ them while acknowledging the significant advantages of moment method modeling for certain tasks.

TRADITIONAL AND MODERN DIRECTIONAL ANTENNA DESIGN METHODS

The process of determining what array geometry and field parameters are necessary to produce a desired directional antenna radiation pattern has become more automated with the advent of modern digital computers that can perform far-field calculations very rapidly. It is common today for a new pattern design that is filed with the FCC to be the last of a series of hundreds or even thousands of designs that were evaluated automatically in a search for the optimum solution to a specified set of radiation characteristics. Most directional antenna patterns in use today were designed long before such computations were possible, however, using straightforward mathematical techniques. Inline patterns having more than two towers and parallelogram patterns were generally developed from two-tower "building blocks" using the pair multiplication process.

Directional antenna patterns are explained in this section using the traditional mathematical approach, as it is applicable for understanding most existing patterns and is good for establishing the starting point for further iterative computer optimization in new pattern design. It should be remembered, however, that every directional antenna pattern is determined by how the individual tower vector contributions add in various directions and at various vertical angles, no matter what process was used to design it. Pair multiplication is just a shortcut to finding the parameters that will produce a multi-tower pattern having nulls in the desired directions, for example, instead of a method for designing patterns that are distinctly different from patterns designed by other methods. A thorough discussion of moment method modeling is contained in Chapter 7.13.

RADIATION AND FIELD STRENGTH

Two independent factors determine the signal strength at any given point within a station's service area. First is the strength of the signal radiated in that direction; second is the path attenuation between the transmitting and receiving antennas. Attenuation is determined by distance and the conductivity and dielectric constant of the ground along the propagation path. The dielectric constant influences the effective depth to which the currents associated with the vertically polarized electromagnetic wave flow beneath the surface, and the conductivity determines the loss that occurs within that effective depth. For calculating AM station signal strengths, the FCC has published families of curves of field strength *versus* distance for a number of frequency spans covering the AM broadcast band. See Figure 7.8-1. Because all of the curves were plotted with an assumed dielectric constant of 15, which was chosen to represent the average value for soil in the United States, it is customary to simply refer to the conductivity value of the curve that best fits measured data along a radial as the *conductivity* of that radial.

Field strength measurements are often graphically analyzed with conductivities that differ from the actual soil conductivity, when the dielectric constant differs significantly from 15, because it is more convenient to use the existing graphs published by the FCC than to develop alternative curves based on dielectric constants other than the assumed 15. Analyzed ground conductivity values that cascade downward in segments with increasing distance, for instance, sometimes indicate that the dielectric constant is significantly lower than 15 and that the data would plot along a single conductivity curve calculated for the actual dielectric constant. It is customary, however, to avoid consideration of differing dielectric constants by using the published FCC curves except in rare cases of extremely poor soil where it is necessary to custom develop curves based on dielectric constants lower than 15 to be able to fit the field strength data to curves. Conductivity is normally the only term that is mentioned to describe the characteristics of earth for groundwave propagation analysis.

It is customary to express the radiation in units of millivolts per meter at 1 km, unattenuated. This is the field that would exist at 1 km over perfectly conducting earth. In this case, the field strength would follow a straight line depicting values inversely proportional to the distance from the transmitting antenna; hence, the radiation is also described as the *inverse distance field*. It is convenient to present antenna radiation pattern information in terms of unattenuated field strength because it standardizes antenna analysis without regard to the soil conductivity in the vicinity of the antenna. The unattenuated radiation cannot be measured directly but can be inferred with great accuracy if sufficient field strength measurements are made to determine the ground conductivity.

The standard distance for specifying unattenuated radiation in North America was changed from 1 mile to 1 km in the early 1980s. To convert field at 1 km to field at 1 mile, the multiplication factor is 0.62137.

Field strength coverage is always dependent on radiation, distance, and ground conductivity. It is important to realize that specification of antenna radiation is an entirely different process than specification of field strength coverage.

The term *efficiency* is sometimes used to refer to unattenuated radiation. As applied for decades in FCC practice to define radiation, the word is utilized in an unconventional sense. It does not express an output/input relationship of an antenna in percent as it is used to define amplifier efficiency. It expresses the

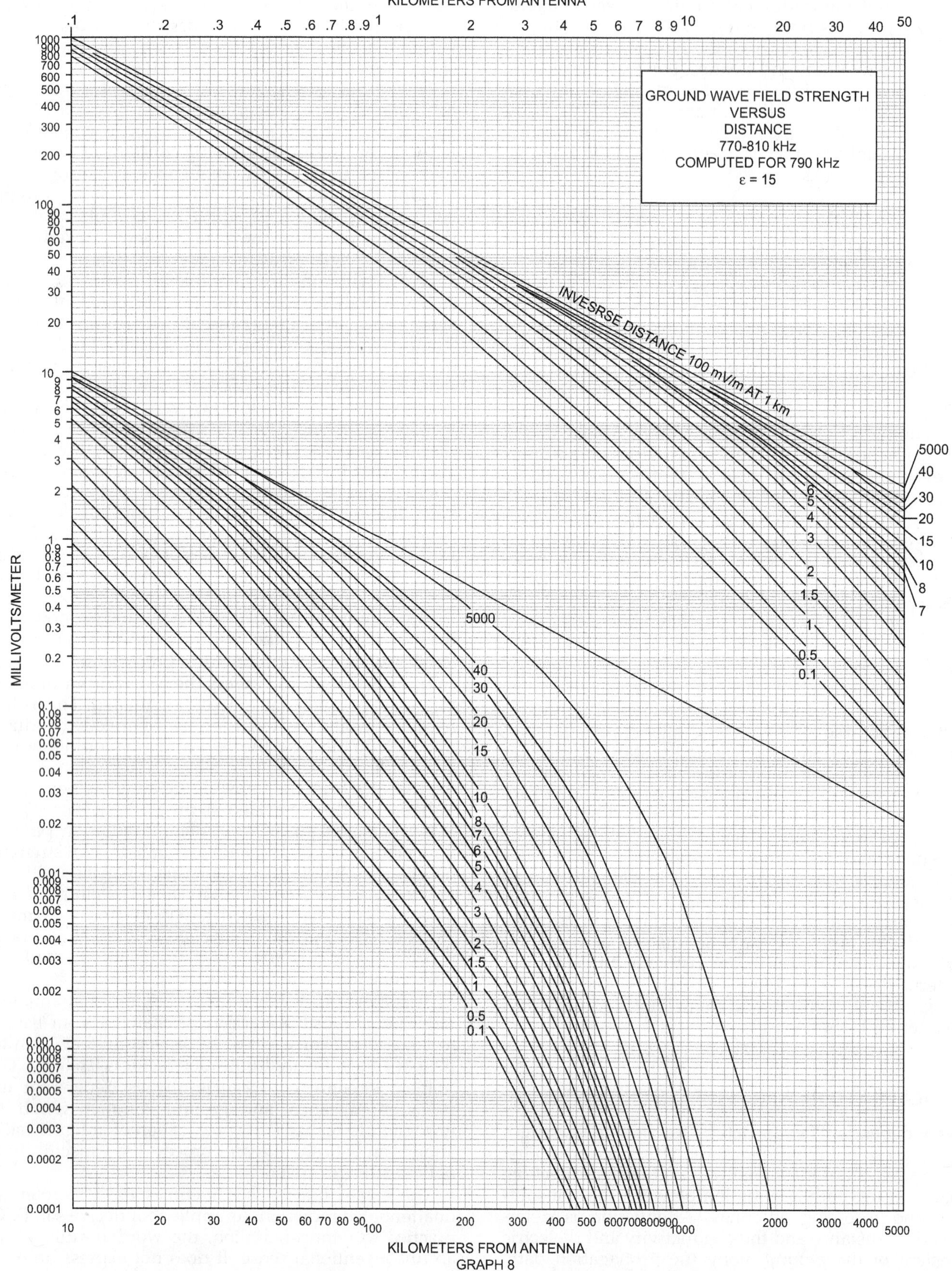

FIGURE 7.8-1 Field strength versus distance for family of conductivity curves.

unattenuated field strength in the horizontal plane of a nondirectional antenna or the *root mean square* (RMS) of the horizontal plane radiation pattern of a directional antenna pattern with a reference input power level of 1.0 kW.

THE SINGLE TOWER NONDIRECTIONAL ANTENNA

The majority of single-tower antennas is neither top loaded nor sectionalized, and most of them are insulated from ground. For such simple towers, the current is deemed to be sinusoidal and to reach a maximum 90 electrical degrees down from the top (or at the base if the tower is shorter than 90° in height). The distance along the height of a tower, measured in electrical degrees, differs from the physical distance slightly because the velocity of propagation along the tower structure is slower than the velocity of propagation in free space. This is primarily because the cumulative effect of the small currents that flow onto cross members from the legs of a tower, due to the cross members' individual capacitances to ground, makes the tower structure function in a sense as a delay line. The amount of delay depends on the cross section of the tower and the size and number of its cross members. Rather than attempting to model the intricacy of a tower's cross-member structure, it is customary to consider it to be a simple wire extended in length to account for velocity-of-propagation effects when calculating its base impedance. Past experience with towers having similar structural characteristics is often considered in selecting the velocity factor for analysis. A typical guyed tower that is 90° high physically is known to be about 95° to 98° high electrically when analyzed as a wire conductor, for instance.

Although it is beneficial to consider velocity-of-propagation effects when calculating tower impedances, antenna radiation characteristics have traditionally been calculated assuming thin wires equal in height to the towers they represent. The FCC Rules are based on these assumptions. The approximate shape of the current distribution on a thin wire of uniform cross section is given by:

$$i_a = I_a \sin(G - y)$$

where:

i_a = current (in amperes) at height y

I_a = maximum current (in amperes)

G = tower height (in degrees)

y = height (in degrees) of the current element i_a.

As an example, the general shape of the current and voltage distribution on a thin tower 210 electrical degrees high is shown in Figure 7.8-2. For shorter towers, the distribution would approximate that shown but with the lower portions cut off—there always being a current node and a voltage maximum at the top of any such tower that does not employ top loading. It is important to visualize the shape of the voltage

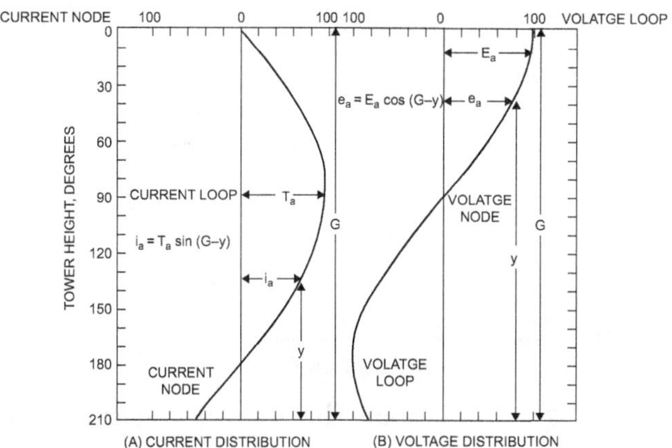

FIGURE 7.8-2 Theoretical current and voltage distribution on a vertical radiator.

distribution along the tower because of the need for good insulators at the high-voltage points; otherwise, corona or arc-overs may result and disrupt broadcasting service.

The real-world tower current and voltage are not zero at the nodes shown along the tower; rather, they reach minimum values and shift rapidly approximately 180° in phase when traversing the node region. When towers considerably taller than 180° in height are considered, the current near the base is in the opposite direction from that in the upper portion of the tower. Under these conditions, when viewed in the horizontal plane, the radiation from the lowest part of the tower is canceling a portion of the radiation from the part above the current minimum. Any increase in tower height above the optimum would actually reduce horizontal plane radiation.

VERTICAL RADIATION CHARACTERISTICS

Maximum groundwave radiation occurs for a tower 225 electrical degrees high (5/8 wavelength). The variation in tower current distribution with increasing tower height defines the shape of the radiation characteristic in the vertical plane. Figure 7.8-3 shows the size and shape of the vertical plane radiation patterns for a single tower of various heights atop a perfect ground system fed with 1 kW of power. The high-angle lobe of the vertical pattern of a 225° tower shows why that height is generally avoided by stations that are concerned about self-interference due to skywave signal within their groundwave coverage areas at night. Most Class A clear-channel stations that provide wide-area skywave coverage at night use towers no higher than approximately 190° for this reason.

INSULATED TOWER BASE IMPEDANCE

The base impedance of a single nondirectional tower is determined principally by its electrical height, its cross section, the extent of the ground system, and the elevation of the feed point above ground. For typical guyed

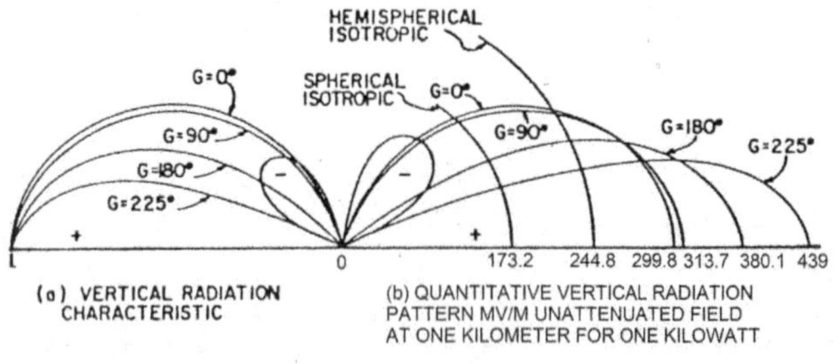

$$f(\theta°) = \frac{\cos(G° \sin\theta°) - \cos G°}{(1 - \cos G°)\cos\theta°}$$

FIGURE 7.8-3 Radiation characteristics in a vertical plane.

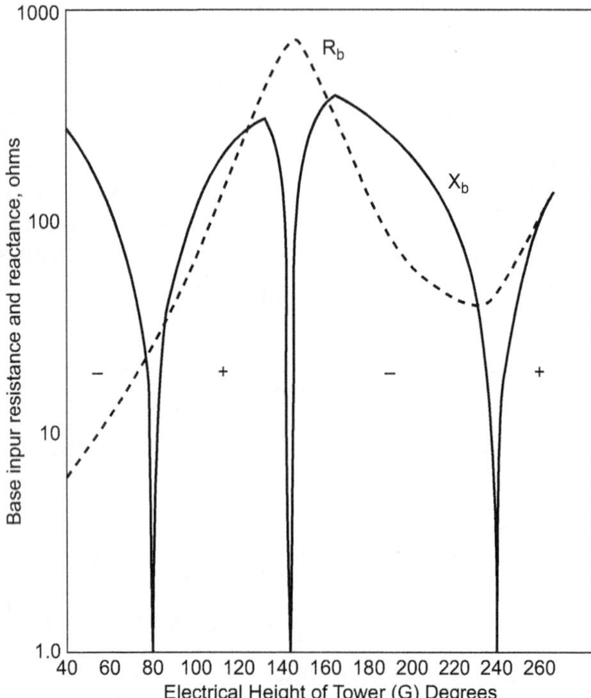

FIGURE 7.8-4 Typical base input resistance and reactance of a uniform-cross-section, base-insulated, guyed tower.

towers of uniform cross section that are base insulated and fed 4–5 feet above ground level, the resistive and reactive components of the base impedance approximate the values shown in Figure 7.8-4. The base impedance of self-supporting towers departs radically from the values shown, not only because of their large and tapering cross section, but also because of the high capacitance about the bases of such towers.

Electrically short towers are inefficient radiators, not only because of the shape of their vertical radiation characteristics, as shown in Figure 7.8-3, but also because of proportionately higher ground losses. For example, a tower 48° high with a base resistance of only 9 Ω will waste approximately 18% of the available power in the ground system resistance losses (typically 2 Ω with a full ground system).It must be remembered that the base input impedance of a tower, when measured at the output terminals of the antenna tuning unit that is used to feed it, includes the shunt effects of stray capacitance and any conductive circuits that are connected across its base insulator, as well as the series inductance of the conductor used to make the connection to the tower base. The shunt effects include those of isocouplers, isolation coils, lightning chokes, and the like. The series hookup inductance can add as much as several tens of ohms of reactance to the feedpoint impedance when a conductor having a "lightning retard coil" of one or more turns is employed as the feedline. A matching network must be designed to match the load impedance presented by the calculated tower base impedance with these other factors considered to the extent that they can be known before construction. To the extent that they cannot be known before construction, the design must allow for a sufficiently large range of adjustment.

GROUNDED TOWERS AND SHUNT-FED AND FOLDED MONOPOLES

Occasionally, towers without insulated bases must be utilized as AM radiators. Such structures include towers that are also used for land-mobile communication and FM and TV stations. Although the impedance at the base of such a tower is necessarily essentially zero, the impedance rises with increasing height of the feed point. It is a simple matter to determine experimentally the height at which a shunt-fed tower must be driven to provide a desirable input impedance. A common technique is a *slant-wire*[1] feed, where a wire is attached to the tower at a selected height above ground and brought down to near ground level at an angle approximating 45° to serve as the antenna input terminal. A slant-wire feed can distort the otherwise omnidirectional pattern of a single tower and tend to suppress radiation over the sector on the side where the slant wire is attached. This effect is much greater when towers on the order of one-half wavelength are shunt fed rather than with towers on the order of one-quarter wavelength, but it can be avoided for all height towers if, instead of the slant wire, the feed conductors are insulated at the base, brought up outside of the tower, and bonded to the tower nominally 90° above ground to form a folded monopole (sometimes called a *folded unipole*). The conductors, in this concentric arrangement, in effect form the outer conductor of a coaxial transmission line with a short to the tower at the 90° point and an open at the base insulators. This quarter-wave, open-circuit transmission line in effect puts an insulator at the tower base. The currents that flow up on the outer conductors and down on the tower essentially cancel as far as radiation is concerned. An additional component of current flows up the skirt and onto the tower, if the tower is taller than the skirt, and goes to zero at the top to produce the radiation, so the tower with this insulated skirt performs essentially as a base insulated tower. The concentric arrangement of conductors, usually six, are tied together above the

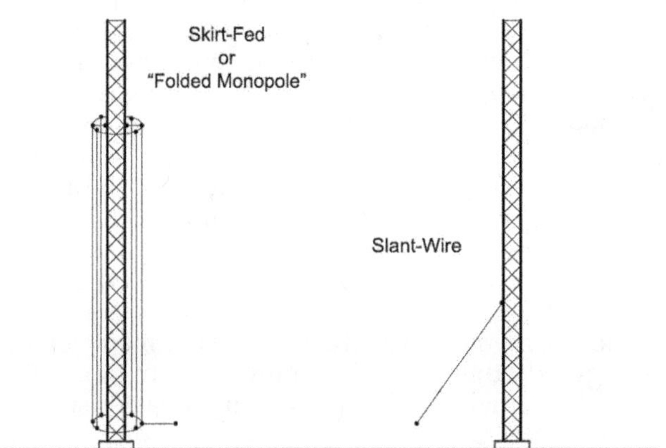

FIGURE 7.8-5 Shunt-fed grounded towers.

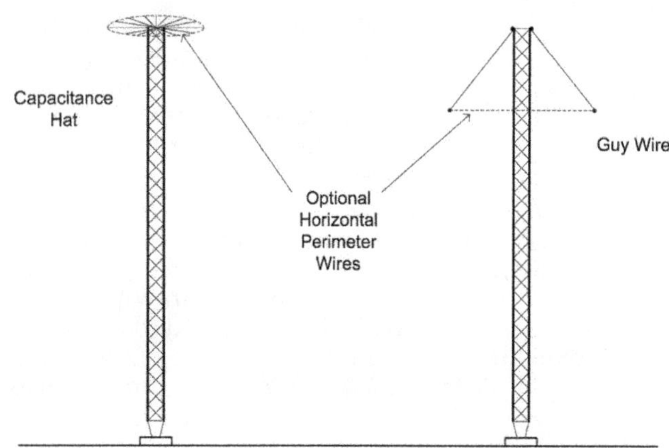

FIGURE 7.8-6 Top-loading methods.

ground level conductor insulators and fed like a base insulated tower.

Although the traditional method for matching both slant-wire and folded monopole antennas has involved experimentation with regard to the physical connection points of the feed wires, moment method modeling is sometimes used today to design optimized feed arrangements. Examples of both slant-wire and folded monopole shunt feeding are shown in Figure 7.8-5.

Slant-wire feeds are no longer accepted by the FCC for critical hours and night operation.

TOP LOADING

The performance of an electrically short tower (significantly less than 90°) can be improved, both as to radiation efficiency and bandwidth, by means of top loading. Top loading is also sometimes used to provide vertical radiation characteristics that would otherwise require construction of taller towers where tower heights are sufficient such that radiation efficiency and bandwidth are not the major concerns. Top loading is accomplished by increasing the capacitance to ground from the top of the tower. The physical realization can take the form of either a flat, more or less circular horizontal disk (usually consisting of a number of conductors fanning out from a central point) attached to the top of the tower or sections of guy wires bonded to the top of the tower and extending down a useful distance before encountering the first of the guy wire insulators (see Figure 7.8-6). The former arrangement is commonly called a *top hat*.

Moment method modeling can be used to design the physical arrangement of top-loading conductors for new towers with quite good accuracy before they are built. Most existing top-loaded towers were designed using approximate mathematical formulas based on the capacitance added by the top-loading conductors and the calculated average characteristic impedance of the tower viewed as a single wire transmission line. Because of the approximate nature of the traditional methods for top-loading design, the FCC has historically required that current distribution measurements be made on top-loaded towers before the stations using them are issued licenses. With the rule change to allow moment method modeling for proofing AM antenna systems has come acceptance of detailed moment method modeling of top-loading arrangements instead of current distribution measurements for licensing stations that use top-loaded towers. Many variations of top loading are possible. Most recent installations use sections of the three upper guy wires for top loading, although some have used 6 or even 12 nonstructural wires for top loading. By interconnecting the lower ends of the top-loading wires, the capacitive loading is increased for a given guy wire length. As the current flowing downward in each guy wire produces a far field that opposes that produced by the tower itself, which should be minimized for optimum performance, interconnecting the lower ends is recommended where relatively large amounts of top loading are required. Top loading is less desirable than increased tower height but is useful where tower height is limited due to aeronautical limitations, local regulations, or other practical considerations. Top loading increases the base resistance and lowers the capacitive base reactance, thus reducing the Q and improving the bandwidth on towers less than 90° high. When the tower height is of the order of 130°, top loading can be used to increase the electrical height of the tower to improve groundwave radiation and minimize skywave radiation at critical elevation angles.

SECTIONALIZED TOWERS

A utopian vertical radiator would have a constant current of unchanging phase throughout its height, but in real life the current must ultimately reduce to zero at the tower top or at the end of the top-loading cables. The current can be made to diminish less rapidly by inserting an inductance in series with the tower at a point partway up its height. This is the same technique as the familiar loading coil near the center of the vertical whips often used for mobile radio systems. Towers approaching one wavelength in height can be employed to provide increased horizontal plane radiation and greater suppression of high-angle radiation when they are fed at approximately half of their physical height. Such center-fed towers are commonly known as *Franklin* antennas. Examples of how both Franklin and sectionalized, loaded antennas can be constructed by placing insulators within tower structures at appropriate places are shown in Figure 7.8-7.

It is also possible to use the technique of skirt-wire feeding for sectionalizing towers where it is not feasible to use insulators, as shown in Figure 7.8-8. Although not shown because the towers are not drawn to scale, the skirt wires would typically be shorted to the tower at approximately 90° from the open ends of the skirts. It is also possible to eliminate the need for tuning boxes when skirts are used by adjusting the points at which the skirt wires are bonded to the tower to produce the required net reactances across the open skirt ends.

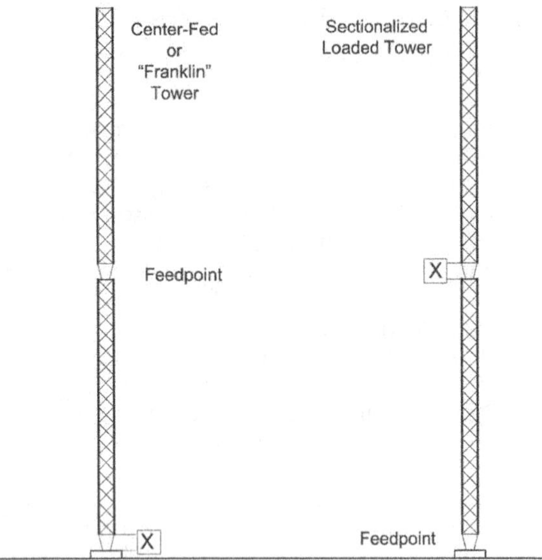

FIGURE 7.8-7 Sectionalized towers.

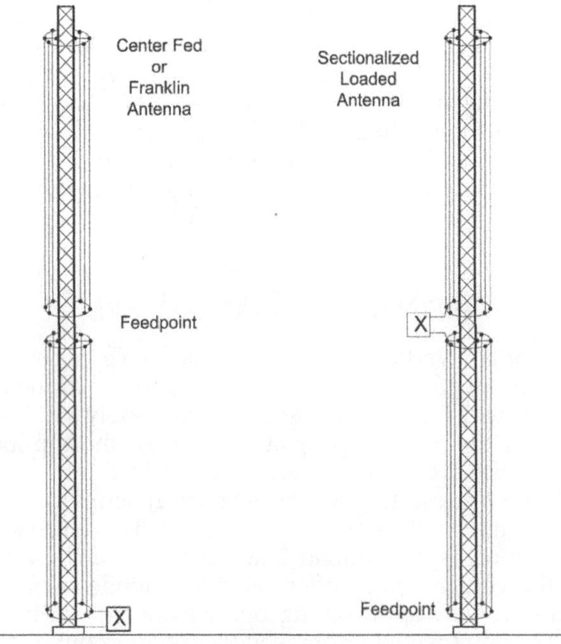

FIGURE 7.8-8 Sectionalizing with skirts.

The FCC Rules contain formulas for calculating the vertical radiation characteristics of sectionalized and Franklin antennas and specify how the parameters that describe their physical characteristics must be specified in applications. Because most existing sectionalized and Franklin antennas were licensed before the current Rules were enacted (when there were no standard methods for their analysis), many of them are grandfathered and require custom analysis to determine their vertical radiation characteristics. The information on their physical characteristics contained in the FCC's engineering database does not conform to the standards outlined in the Rules, and different mathematical formulas are used in the software employed for allocation studies on a case-by-case basis.

TOP-LOADED SECTIONALIZED TOWER

For a simple vertical radiator, the radiation characteristic can be improved by increasing the tower height up to 225° for maximum groundwave signal where skywave self-interference from the high-angle lobe that is present for tower heights greater than 180° during nighttime or transition hours is not a concern. This in effect raises the position of the current loop with respect to the ground. This principle can also be applied to the top section of a sectionalized tower. The purpose of top loading a sectionalized tower is to provide a means of further controlling the current distribution on the tower. Considering efficiency and stability, it is often possible to achieve a more favorable radiation characteristic of the whole tower by employing top loading and sectionalization together (see Figure 7.8-9). In the case of tall towers used to support FM or TV antennas, it may not be practical to employ top loading.

GROUND SYSTEMS

The current on a tower does not simply disappear. It returns to earth through the capacitance between the earth and each incremental element of the tower and the top-loading conductors, if used. For towers not exceeding 90° in height, the tower current is greatest at the base. For such towers, the radial ground current is greatest near the tower and decreases with increasing distance from the tower. For single towers, the ground currents are radial from the tower base. The ground losses are greatly reduced if the tower has a radial copper ground system, so the ground current will flow in the low-loss copper ground system rather than in the earth, which has a much higher resistance. A solid copper sheet of infinite radius would be the ultimate ground system, but experiments and experience have defined the dimensions of an adequate ground system. A system of 120 radial ground wires, each 90° long and equally spaced out from the tower base, constitutes a standard ground system (see Figure 7.8-10).

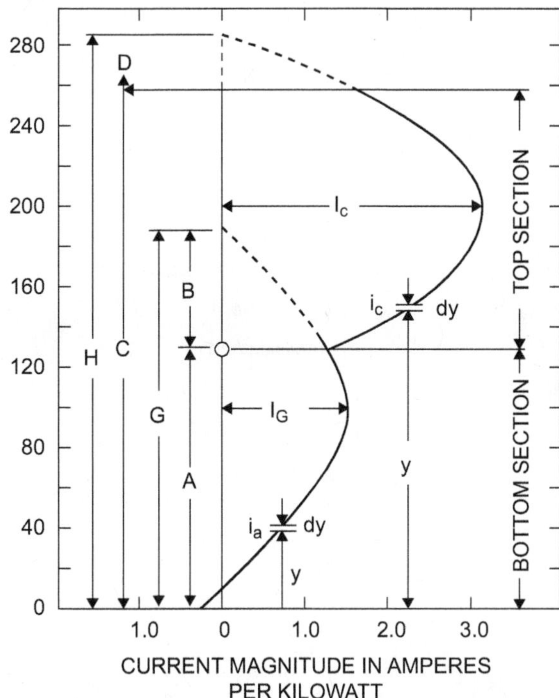

FIGURE 7.8-9 Theoretical current distribution on top-loaded sectionalized tower.

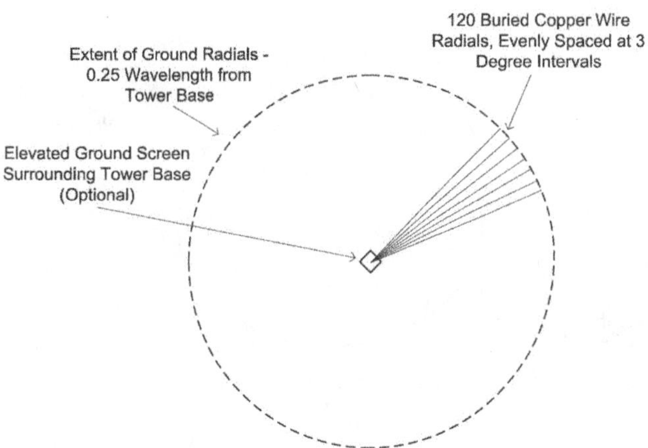

FIGURE 7.8-10 Nondirectional antenna ground system.

This arrangement is sometimes augmented with an additional 120 interspersed buried short radials (often 50 feet long) or an expanded copper-mesh ground screen, 25 to 50 foot square and centered at the tower. A superior ground screen material is the copperweld, mesh ground mat often utilized by power companies for lightning protection under electrical substations. Where wet/dry or seasonal soil variations might impact antenna stability, elevated ground screens consisting of mesh ground mats above beds of crushed rock or elevated short radials are sometimes employed.

There is no magic in a standard ground system for nondirectional towers; such a system simply represents a reasonable balance between cost and radiation efficiency. The antenna system loss including the

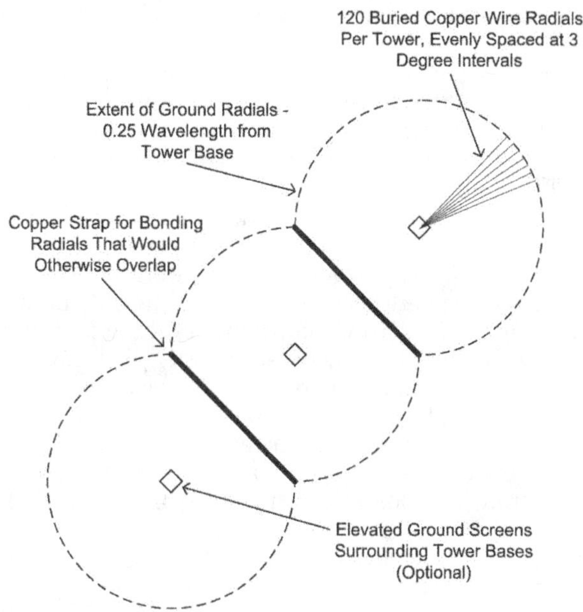

FIGURE 7.8-11 Directional antenna ground system.

tower and ground system is normally assumed to be 2 Ω and is added to the base resistance of the tower for simplified analysis. Most ground systems under directional antenna arrays consist of the usual 120 radials per tower truncated and bonded to traverse copper straps where the radials from the towers would otherwise intersect, as shown in Figure 7.8-11.

Ground system losses are minimized if the radial wires are placed above ground; thus, the E-field voltage from the tower and top-loading cables (if any) terminate on these radial conductors so the H-field current can return to the tower base without penetrating the lossy earth. Some ground systems have been installed with elevated ground radials, and it has not been found necessary to have the full 120 radials in such cases. Ground radials, however, are usually buried 6–8 inches underground for mechanical protection. Burial up to 24 inches is feasible where necessary to permit deep plowing for agricultural crops; however, the ground system should be very near the earth surface in the immediate vicinity of the tower. The earth losses are greater for the buried ground system. Changes in weather conditions can change the dielectric constant and conductivity of any unshielded earth to the detriment of base current stability, but the effects are minimized when 120 buried radials are used.

TWO-TOWER DIRECTIONAL ANTENNA

When a nondirectional antenna, with a given power, does not radiate enough field strength to serve the community of interest or fails to protect other radio stations, then it is logical to resort to a directional antenna system to achieve these objectives. The FCC Rules spell out the protection requirements to be provided to the various classes of stations, both daytime and nighttime, on the same and adjacent channels. These limits, which must be met in the directional antenna design, tend to define the shape and size of the required antenna pattern. Because the distances and directions to the other stations requiring protection are rarely the same, most directional antenna patterns are tailored to meet the specific requirements in various directions.

A directional antenna functions by carefully controlling the amplitude and phase of the radio frequency currents fed to each tower. The resulting field in any direction is the vector sum of the individual tower radiation components. To visualize the resulting pattern in the horizontal plane, one must consider the individual tower radiation components when viewed from distant points in different directions. The relative amplitudes from the individual towers remain unchanged, but the relative phases shift with azimuth because the signal from the closest tower arrives first. In a directional antenna system, one tower is defined as the reference tower, and the amplitude and phase of each other tower are measured relative to this reference. The ratio of the field from each other tower relative to the reference tower field is a fractional number sometimes expressed as a percent of the reference tower field.

The phase of the field, radiated by each tower relative to the reference tower, has two components when viewed from any distant point of observation. The phase resulting from the phase of the current fed to the tower is one component and is adjustable. The second component is the phase that appears to lead or lag the reference tower by virtue of being more distant or closer than the reference tower to the point of observation. This is termed the *space phase component* and varies continuously for each tower in a sinusoidal manner as the observation point is moved in azimuth along a distant circle around the array.

Figure 7.8-12 shows three simple directional antennas and their resulting patterns, which are easy to visualize. Figure 7.8-12a shows two towers arranged along a north-south line separated by 180° and fed with equal currents in phase. When viewed from the east or west, the fields from the two towers are in phase and the maximum field strength results. When viewed from the north or south, the field from the more distant tower is delayed by the 180° of additional distance, thus canceling the field of the closer tower so as to result in a minimum or null. The deepest minimum or null occurs only when the fields are exactly equal in amplitude and opposite in phase. Figure 7.8-12a is termed a *broadside array* because the maximum radiation is broadside to a line through the towers. Figure 7.8-12b shows a similar arrangement, but with the phase of the current in the north tower shifted by 180°. In this case, the fields from the two towers cancel each other when viewed from the east or west but would produce maximum radiation from north or south. This arrangement would be termed an *end-fire array*, because maximum radiation coincides with a line through the ends of the array. Figure 7.8-12c alters the spacing to 90° and phasing to 90° so as to produce a cardioid pattern. Other combinations of tower spacing and phasing can produce a great variety of pattern shapes.

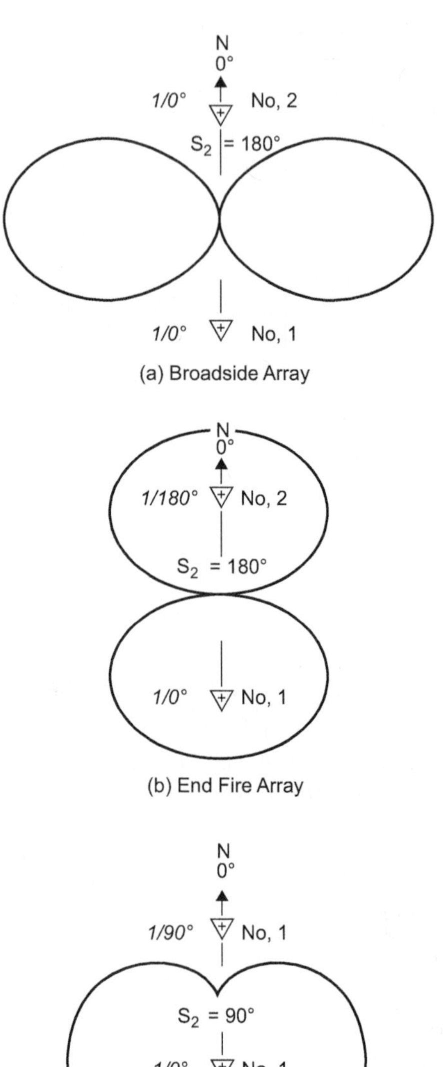

FIGURE 7.8-12 Three simple directional antenna patterns.

MULTIPLICATION OF TWO TOWER PATTERNS

Perhaps the most widely used method of controlling pattern shape involves the multiplication of two tower patterns and is known as *pair multiplication*. This is illustrated in Figure 7.8-13. When a two-tower pattern such as pattern 1 with nulls at $\pm\phi_{n1}$ is multiplied by pattern 2 with nulls at $\pm\phi_{n2,}$ the result is pattern 3 in a three-tower array. The directions of the two-tower pair nulls are maintained in the three-tower array. This is a very powerful design technique for protecting other stations and still serving a desired service area. In this special case, the spacings S_2 and S_3 are equal, resulting in an inline array with the fields of towers 2 and 3 being added in the center tower, and the end tower of the three-tower array is the multiplication of these fields, as shown in pattern 3 of Figure 7.8-13.

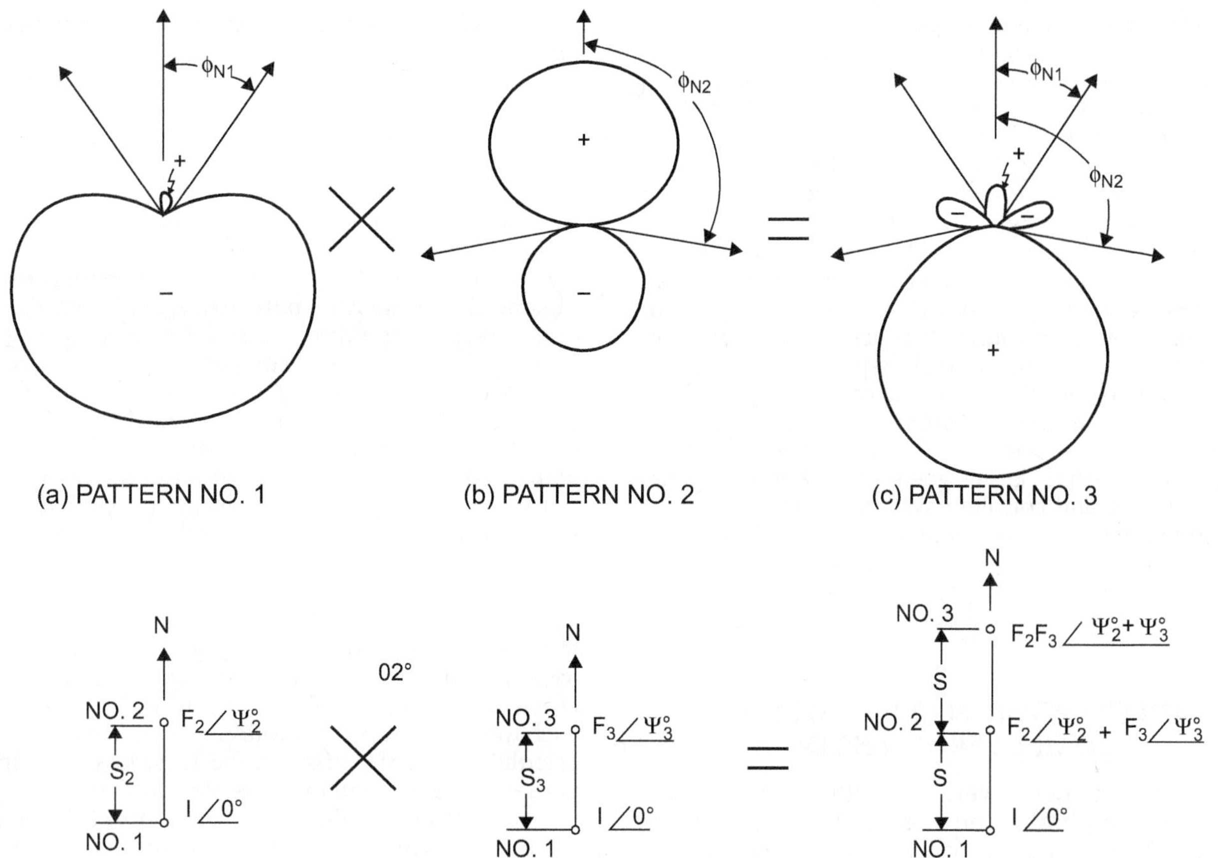

FIGURE 7.8-13 Multiplications of patterns to produce a three-tower inline array.

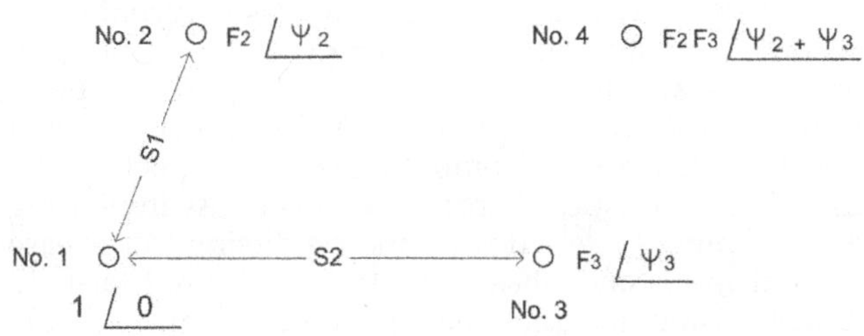

FIGURE 7.8-14 Four-tower parallelogram.

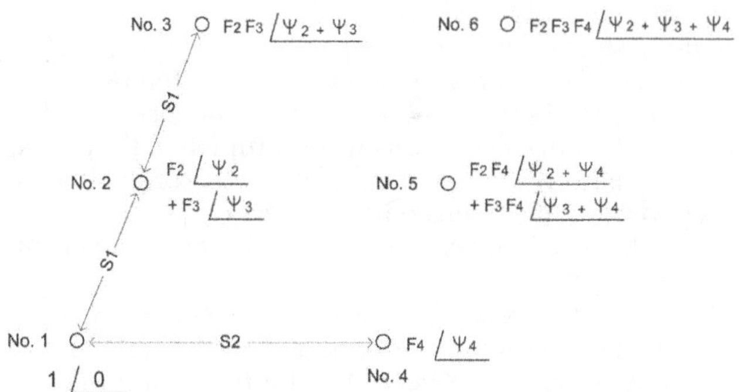

FIGURE 7.8-15 Six-tower parallelogram.

In the event that the protection directions are not symmetrically located, the two tower pairs can be placed on different azimuth angles and have different spacings to produce a four-tower parallelogram array (as shown in Figure 7.8-14). The nulls of the pair 1 and pair 2 patterns are maintained in the resulting four-tower parallelogram array. This approach can be expanded for a larger number of design pairs by using a larger number of towers. Figure 7.8-15 shows how the parameters of a six-tower parallelogram array can be developed from three design pairs. The design includes two pairs having the same spacing, S_1, and the same orientation multiplied by an additional pair with a different spacing, S_2, and a different orientation. With this approach, a wide variety of asymmetrical patterns are possible with relatively simple design calculations.

PATTERN INVERSION (OR MODING)

Directional antenna pattern designs using towers of identical height that do not have zero-field nulls often offer a choice of base impedances and power division. This occurs when a pattern has a nonunity field ratio or one or more embedded design pairs have nonunity field ratios and the same degree of null fill can be achieved by field ratio inversion. A two-tower pattern that can be produced with a field ratio of 0.8, for example, can also be produced with the same phase and a field ratio of 1.25—and each option will have a distinct set of operating base impedances and power division. The embedded pairs in a pattern can be inverted in any combination. For example, a three-tower pattern derived from two design pairs has four possible sets of parameters: no inversion, pair 1 inverted, pair 2 inverted, and pairs 1 and 2 both inverted. A pattern without any embedded design pairs can be inverted at least once by rotating all of the towers about the center point (adding 180° to each azimuth) and changing the sign of each phase angle.

It is often beneficial from the standpoint of bandwidth performance to consider all possible parameter inversions before a pattern design is considered final. In general, better performance is achieved by selecting the design with the most nearly equal operating resistances and, where there are towers having negative power flow, with the minimum total negative power. A parameter inversion study is recommended whenever a new pattern is designed and when a replacement phasing and coupling system is designed for an existing system having a geometry that makes it possible without moving towers. Software is available for reverse-engineering patterns larger than two towers to determine their embedded design pairs, if there are any.

PATTERN DESIGN USING MODERN COMPUTER METHODS

Most existing patterns were originally designed using the pairs multiplication process. Those designed prior to about 1970 may have had their parameters developed with the design engineer's slide rule and had their radiation pattern calculations done on paper with the assistance of tables of trigonometric functions and mechanical calculators. In those days, complicated pattern shapes were developed by specifying where pattern nulls would be produced, and, if the radiation in other directions was found be satisfactory after the overall pattern calculations were completed, they were considered final. Pattern shapes were often biased toward meeting interference protection requirements without, for instance, optimization of null fill on the less critical protection azimuths.

It is now possible to use computers to develop designs to fit specified radiation pattern shapes iteratively, with the major benefit being the ability to optimize radiation over all important spans of azimuth rather than simply meeting the radiation limit requirements at specific azimuths chosen to have nulls when pairs are multiplied. Software is available that varies directional antenna tower field parameters, locations, and heights while comparing calculated radiation with the specified requirements in an attempt to converge on the optimum design. In addition to radiation pattern optimization, it is sometimes possible to design patterns with fewer towers than would be necessary in a parallelogram designed using the pair multiplication technique. For a given set of pattern requirements, however, approximately the same size property is required. The pair multiplication technique remains useful for quickly estimating the property requirements for a new directional antenna system and for giving the computer software used for design optimization a good starting point to minimize the amount of random effort required of it.

RADIATION PATTERN SIZE

The pattern size is usually determined by integrating the energy flow outward through an imaginary hemispherical surface surrounding the directional antenna array. This method does not give information regarding the distribution of power radiated from the various towers of the directional antenna array, but it is very useful for making comparisons of pattern size. This computation method is available in digital computer programs and is used by the FCC. There are other methods of determining pattern size, such as the *mutual resistance method*, which employs Bessel functions, and the *driving point impedance method*, which uses mesh circuit equations with self- and mutual impedance information.

The FCC Rules require that directional antenna radiation patterns submitted in proposals for new and changed facilities be calculated based on the loss assumption of 1.0 Ω per tower. It is known that the best efficiency possible with a full ground system about each tower corresponds more closely to what would be calculated for 2.0 Ω loss, but the 1.0 Ω loss assumption is required for calculations related to new directional antenna patterns so the standard patterns that are used for interference calculations in station-to-station allocation studies might be based on worst-case (for interference) assumptions.

Moment method computer software, which is described elsewhere in this chapter, has in recent years become available for determining current distribution on towers and top-loading cables, base driving point impedances, and the patterns of directional antenna arrays. It has found common use for predicting the drive characteristics of array elements to use in phasing and coupling system design and predicting tower base current ratios and phases to produce desired far-field pattern shapes for adjustment purposes. The calculation of pattern shapes and sizes for regulatory and international notification purposes, however, is not done using moment method software because of the requirements of regulations and international treaties based on older methods in use when they were originally enacted.

RSS-TO-RMS RATIO

Each directional antenna pattern calculated to modern standards has specified for it both an RMS, which is the root mean squared of its horizontal plane radiation, and an RSS, which is the root sum squared of the individual field values radiated by the various towers to produce the pattern. The RMS corresponds to the area inside a directional pattern that is plotted to scale in millivolts per meter. It is a measure of how much radiation leaves the antenna system. The RSS, on the other hand, is a measure of how much field is required from the towers in aggregate to produce the far-field pattern. For a non-

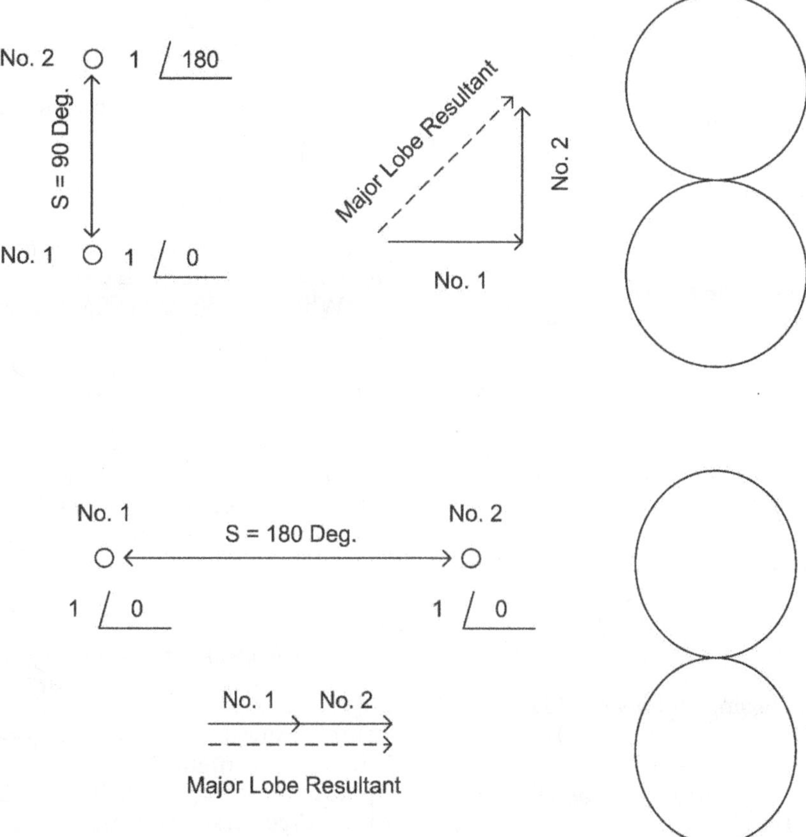

FIGURE 7.8-16 High and low RSS design approaches.

directional antenna, the RMS and RSS are both equal to the radiated field that is the same in all directions in the horizontal plane, and the RSS-to-RMS ratio is 1.0.

The RSS-to-RMS ratio for a given directional antenna pattern is the closest thing available to a quality factor for judging its characteristics relative to other patterns. A high ratio means that the combination of array geometry and pattern shape forces the individual tower fields to be high for the amount of power that is radiated into the far field. As an indicator of relatively high stored energy within a system, a high RSS-to-RMS ratio indicates that a pattern can be expected to be more difficult to adjust and maintain in adjustment because the individual tower contributions to the vector summations at null azimuths are larger than would be the case for a low RSS pattern. It is also an indicator of relatively poor expected bandwidth, as high tower fields translate to high tower currents that, in turn, requires lower base resistances to satisfy the requirement for conservation of energy. Because of their higher element currents, directional antenna patterns having high RSS values tend to be more sensitive to loss resistances associated with towers and their ground systems. In general, high RSS patterns have lower RMS efficiency than do low RSS patterns.

An RSS-to-RMS ratio value as close as possible to 1.0 is to be desired. The selection of array geometry is the biggest factor in determining what the ratio will be. Figure 7.8-16 shows two possible solutions for producing a two-tower figure-eight directional pattern with nulls to the east and west. It is obvious from the lengths of the tower field vectors relative to the major lobe resultants—where they fall most nearly in phase—that the broadside array with the towers spaced 180° apart has a significantly lower RSS than the end-fire array with towers 90° apart. In general, it is desirable to employ one or more broadside tower pairs where it is necessary to squeeze a pattern side to side. End-fire tower pairs are best used for cardioid patterns where nulls are used to reduce radiation on one side only. Parallelogram arrays often use a combination of wider spaced broadside and closer spaced end-fire pairs to make the best use of both in producing an overall required pattern shape. In order to achieve the best pattern design with a computer program that uses iterative techniques, it is a good idea to first design a pattern having low RSS characteristics to use as the starting point.

NEAR-FIELD VERSUS FAR-FIELD CONDITIONS

Theoretically, a directional antenna pattern is not fully formed except at an infinite distance, where the separate towers can be considered as one point source.

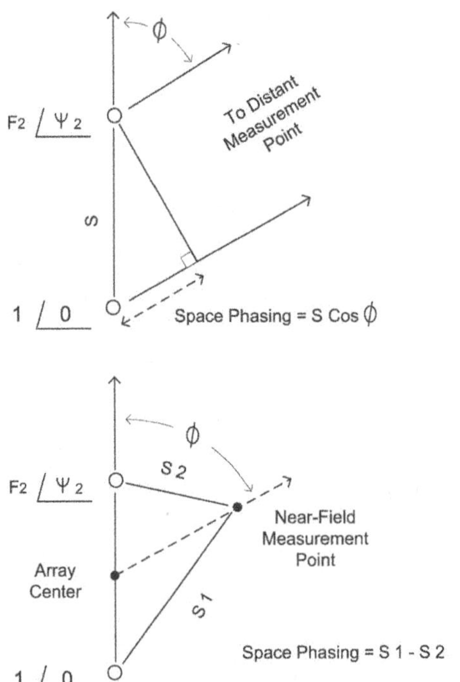

FIGURE 7.8-17 Space phasing and near-field errors.

Figure 7.8-17 shows the space phasing between two towers of an array for far-field conditions, where the rays pointing from the two towers to a very distant observation point are virtually in parallel, and for an observation point in the near field, where the right-triangle assumption for calculating the space phasing term does not apply.

Measurements made at points in the near field, therefore, will not fall on the unattenuated field strength *versus* distance line calculated for far-field conditions. As a practical matter, significant near-field effects can persist as far as 32 km (20 miles) or more from an antenna before far-field conditions prevail. This is especially true in the deep minima of wide-spaced arrays; however, misleading measurement results can often occur under apparently innocent circumstances. Near-field calculations involve consideration of the actual inverse distance attenuation and the actual phase delay from each antenna element to the observation points along a radial.

Figure 7.8-18 shows the results of such calculations on a minimum radial and the resulting analysis of field strength measurements. Line A is the inverse distance line for the theoretical unattenuated radiation at 1 km. Curve B is the result of the near-field calculations assuming only inverse distance attenuation (that is, no soil losses). It converges with the inverse distance line with increasing distance. Curve C represents a soil conductivity of 10 mmhos/m as drawn in the conventional manner from analysis of nondirectional measurements on the radial. Curve D is a composite of curves B and C. It includes the near-field calculations and is attenuated with distance in accordance with the soil conductivity previously established. This composite line converges with the near-field calculations at short distances where soil attenuation is negligible and converges with the soil conductivity line at great distances where near-field effects disappear. Because curve D accounts for both near-field effects and soil losses, it is the proper curve against which the directional field strength measurement data should be fitted. Note the good fit to the measurement data, both close to the array and at distant points, even though the first 19 measurement points fall considerably above the inverse distance line A.

When using near-field analysis on directional antenna field strength data, it must be remembered that a statistical analysis of field strength data with "corrections" applied to the individual point ratios can be very misleading if the pattern is not adjusted to the parameters for which the near-field values were calculated in the first place. When properly analyzed, the measurement data should generally conform to the shape of the calculated near-field curve as shown in Figure 7.8-18.

SEASONAL VARIATION OF FIELD STRENGTH

Field strength measurements on a previously licensed directional antenna may appear to indicate a change in pattern shape or size when the change was in fact due to changes in soil conductivity. Such changes affect distant measurements more than close-in measurements. In some areas of the United States, the conductivity is typically higher during winter and spring months when the soil is more moist than in summer and fall months, with the conductivity being the highest when the ground is frozen. Seasonal conductivity variations are not observable in some portions of the country, yet are extreme in other areas. One well-documented case showed a seasonal doubling of signal strength at 32 km (20 miles) in the main lobe of a correctly adjusted system operating on 1380 kHz. To avoid the misleading effects of seasonal conductivity changes that might appear to distort measured directional antenna patterns in size or shape, the FCC requires that all the field strength measurements in a directional antenna proof-of-performance be made under "similar environmental conditions."

STANDARD PATTERNS

Theoretical (also called calculated) patterns can have nulls wherein the radiation at specific azimuths goes completely to zero. In practice, it is not possible to prove by field strength measurements that a null exists. Reradiation and scatter from objects external to the array limit the depth to which a pattern minimum can be proven. Additionally, operational variations in phase and ratio parameters will increase radiation in any direction where the deepest possible minimum has been previously established. To accommodate these limitations, the FCC authorizes a standard pat-

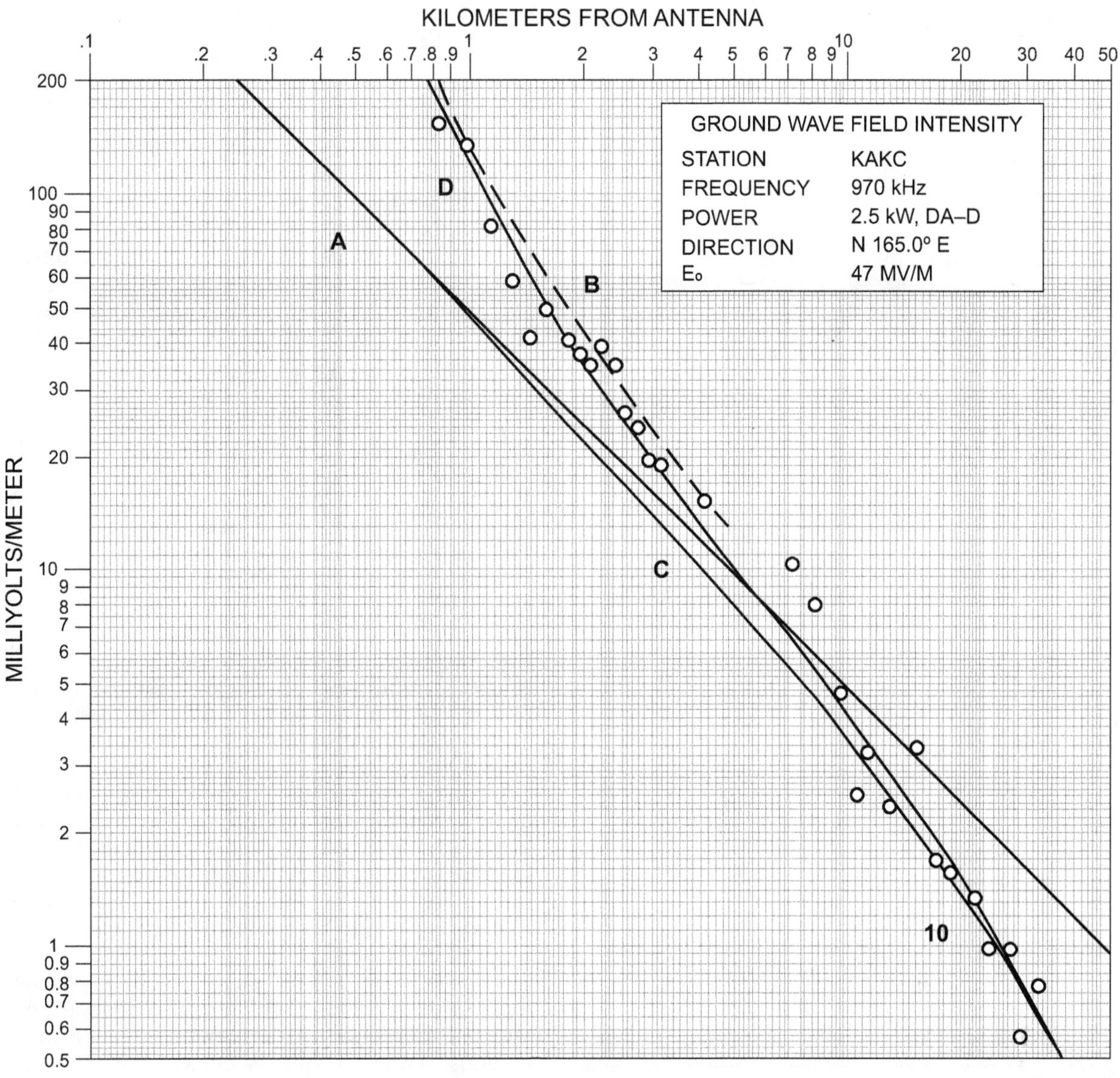

FIGURE 7.8-18 Near-field effects.

tern for each directional antenna station. Standard patterns exceed the theoretical pattern at all azimuths by specified and easily calculated amounts. It is required that the radiation from a directional station not exceed its standard pattern. All U.S. stations employing directional antennas have FCC-specified standard patterns. These supersede all earlier patterns based on theoretical calculations or on field strength measurements. The standard pattern radiation values are now used exclusively in all calculations of coverage and interference.

AUGMENTED PATTERNS

Augmentation is applied to the standard pattern when the measured field strength is exceeded in discrete directions but does not cause interference to other stations. When augmentation is required, it is achieved by applying Equation B-9 in Appendix B. Figure 7.8-19 shows the theoretical and augmented standard patterns of a simple two-tower example directional antenna.

DRIVING POINT IMPEDANCE

The input impedance of each tower in an array (called the *driving point impedance*) is not what it would be if the tower were used as a nondirectional antenna. This is because of the effects of mutual coupling with the other towers of the array. The base current in one tower in a two-tower array, depicted as I1 in Figure 7.8-20, induces a voltage across the base of the other tower, E2, and vice versa.

The ratio of E1 to I1 without the second tower present would be its self-impedance. The ratio of E1 to I2 is the mutual impedance between the two towers. When both towers are radiating to produce a directional

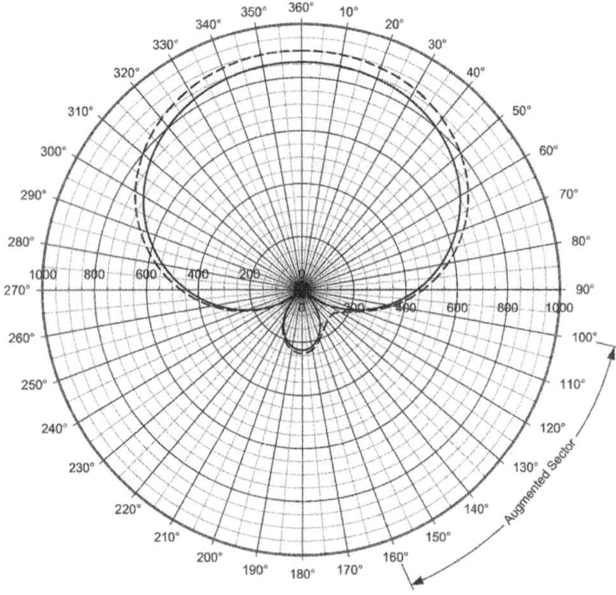

FIGURE 7.8-19 Theoretical and augmented standard directional antenna radiation patterns.

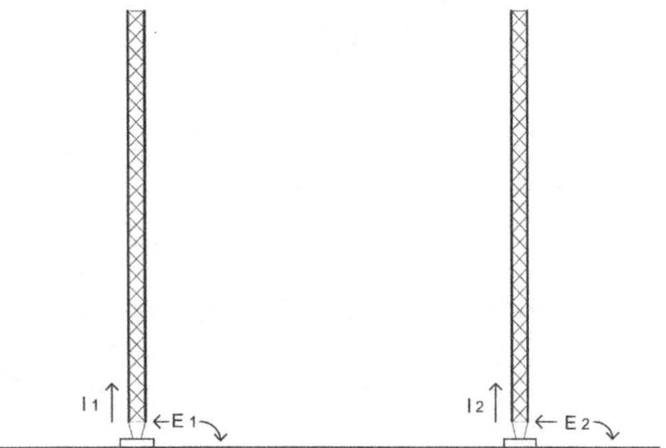

FIGURE 7.8-20 Mutual coupling.

antenna pattern, each tower base voltage is the sum of the value produced by the base current flowing into its self-impedance and the value produced by the current flowing into the other tower through the mutual impedance. The operating, or driving point, impedance of each tower in an array is the ratio of its total base voltage to its base current while functioning in the array. In other words, the driving point impedance contains the self-impedance plus the mutual impedances multiplied by the current ratios that exist with other towers in the array as driven to produce the desired pattern. The driving point impedance will depend on the array parameters and can even have a negative resistance component so the tower draws power from the other towers and delivers it back to the phasing system.

Calculated driving point impedances are used in the design of new phasing and coupling equipment that is designed before the towers are erected, but it is sometimes desirable to measure the operating impedances of towers in an existing directional antenna system. Because the operating impedance is affected by the currents in the other towers, it can only be measured by an operating impedance bridge inserted at the tower feed point while the other towers are operating with their correct current magnitudes and phases. For precise measurement, it is necessary to readjust the system with the bridge in place to produce the correct parameters, because the series inductance and stray capacitance that it adds can significantly impact tuning.

BASE CURRENTS VERSUS RADIATED FIELDS

In a directional array, the tower base current ratios will usually depart substantially from the calculated radiated field ratios when the pattern is correctly adjusted. This is caused by the nonsinusoidal current distribution of each tower as described elsewhere in this section. Thus, it has been standard practice for directional antenna patterns to be initially proven to have their correct shapes by means of a series of field strength measurements in significant radial directions from the station rather than by assuming that measurement of tower currents and phases can establish the correct pattern. It is possible, with moment method modeling, to determine base current ratios and phases that correspond very well with those necessary to produce a required directional antenna pattern shape. Using these methods, the amount of experimental adjustment work necessary before acceptable proof-of-performance field strength measurements can be made is greatly reduced and, in many cases, eliminated. Method of Moments proofing, which was approved by the FCC in a rulemaking that concluded in 2008, makes it possible in most cases to rely on observed tower currents or base voltage (which to use depending on certain requirements) to demonstrate that a directional antenna pattern shape is correct based on antenna monitor ratios and phases.

DETUNING STRUCTURES NEAR AM ANTENNAS

It is sometimes necessary to detune a tower on a directional antenna system's property, such as a tower that is used in one pattern but not another. For tower heights below approximately one-half wavelength, this can usually be done by placing a reactance from the base feedpoint to ground to cause the current distribution on the tower to have the general shape shown in Figure 7.8-21. The far-field radiation in the horizontal plane is reduced to virtually zero because the contributions from the currents that flow on the tower above and below the minimum, at approximately one-third of the tower height, are equal and 180° out of phase. The height of the minimum for optimal detuning ranges from approximately 33% of the total tower height for towers in the range of 90° up to approximately 38% for towers approaching 180° in height. It, and the reac-

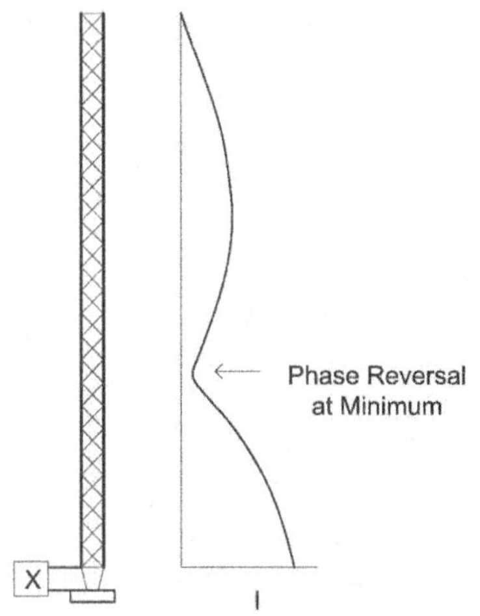

FIGURE 7.8-21 Detuning with base termination.

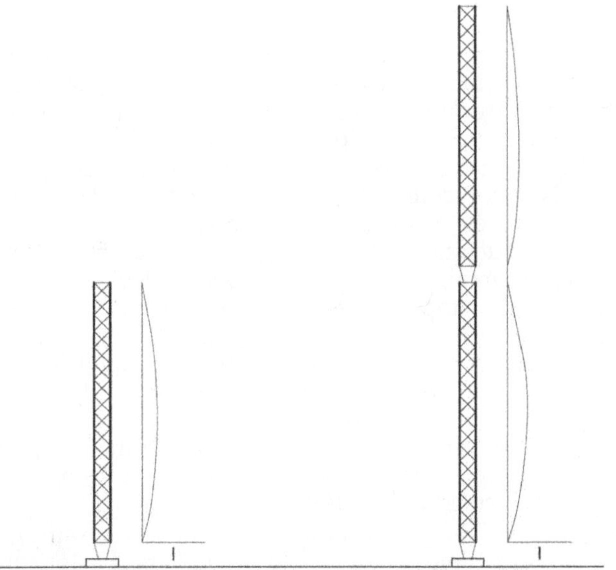

FIGURE 7.8-22 Detuning by breaking up current.

tance value necessary at the tower base, may be computed using moment method modeling.

Where the degree of detuning is not as critical, such as with communication towers and other structures that are located well away from the towers of the directional antenna system, but close enough to scatter enough field to fill out pattern nulls if they are untreated, it is usually possible to eliminate reradiation problems by breaking up current on them as shown in Figure 7.8-22. The tower on the left shows the method applied to a tower on the order of 90°, while the tower on the right shows a tower on the order of 180°. For taller towers, it may be necessary to break the current up into more than two pieces. The general principle is to limit reradiation by breaking the structure into current conducting pieces on the order of 90° in length, or shorter. This can be done by placing insulators within the structures, as shown in Figure 7.8-22, or by surrounding the structures with wire skirts, as discussed for feeding towers elsewhere in this chapter.

DETUNING POWER LINES NEAR AM ANTENNAS

Power lines near AM antennas are more complicated to analyze than simple vertical structures, as their support poles or towers are generally interconnected at their tops by one or more horizontal ground conductors. Reradiation can take place at levels far in excess of what would be expected for the vertical height involved when the circuit path between adjacent support structures, including the ground between them and the top ground wires, is near resonance at the AM station's frequency. This phenomenon defies simple theoretical analysis because of the influence of the physical characteristics of the support structures and the electrical characteristics of the soil between them. The most expedient remedy when power line reradiation is a problem is to insulate, or RF isolate, the horizontal ground wires from the tops of the support structures that are closest to the AM antenna. This can be accomplished by installing insulators with lightning gaps and/or RF chokes across them at the tops of the vertical structures. In many cases, no further measures (such as detuning the vertical structures themselves with wire skirts) will be required.

MONITORING DIRECTIONAL ANTENNA OPERATING PARAMETERS

Antenna monitoring systems are used by AM stations that employ directional antennas to monitor the ratios and phases of the currents flowing in the array elements so they can be maintained at the values that are known to produce the required pattern shapes, as shown in Figure 7.8-23. Antenna monitors are designed to meet the FCC requirements for accurate monitoring of the ratios and phases of the current samples that are fed into them, through the sampling lines, from the current sampling devices. The sampling lines are typically 3/8-inch or 1/2-inch foam dielectric transmission

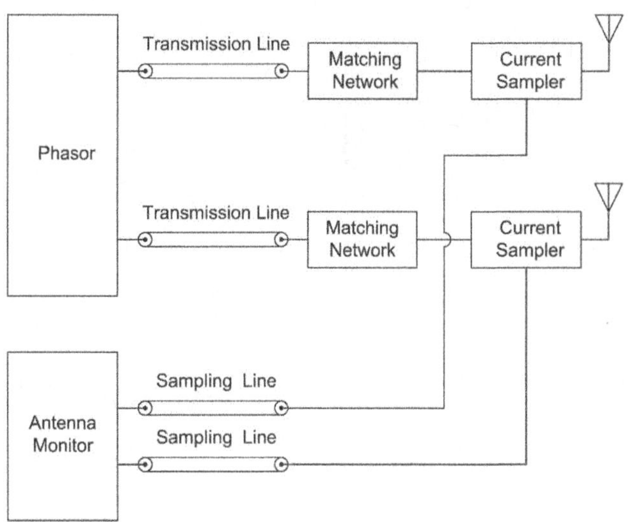

FIGURE 7.8-23 Antenna monitoring system.

lines and are semiflexible with solid outer conductors. The current sampling devices are either current transformers through which the tower base currents pass or tower-mounted inductive pickup loops, as shown in Figure 7.8-24. With the Method of Moments rules that became effective in 2008, voltage sampling became acceptable for arrays within an acceptable range of tower height. Whatever sampling device is used, they must be rated to produce voltages within the acceptable range of the antenna monitor with full power into the antenna system.

Base current transformers differ from tower-mounted sampling loops in a very important way: the current that is sampled is the sum of the actual tower current and other currents flowing to ground after the transformer due to stray capacitance in the base region and any circuits—such as lightning chokes, ring transformers, and isocouplers—that are across the base insulator. For this reason, it has been customary to use loop instead of current transformer sampling for towers much over 90° in height, where base impedances are high. For towers of any height, the relationship of the base sampled current to the actual tower current can change with changes to the circuits across the base. It should be understood that it might be necessary to operate with a different antenna monitor ratio and phase to maintain the same radiation pattern if changes are made across the base of a tower employing current transformer sampling (such as changing an isocoupler or adding a new one) and that a partial proof-of-performance might be necessary to have the FCC license modified to reflect the new parameters. Base voltage sampling is not subject to error because of shunt effects across the tower base and that is one of its advantages over base current sampling for towers within the height range where it can be used. Three methods for coupling the sampling line from a tower-mounted loop to ground are shown in Figure 7.8-25. For sampling loops that essentially monitor the base current, which are typically mounted 10 feet above the tower base, it is possible to insulate them from the tower and have them operate at ground potential. The

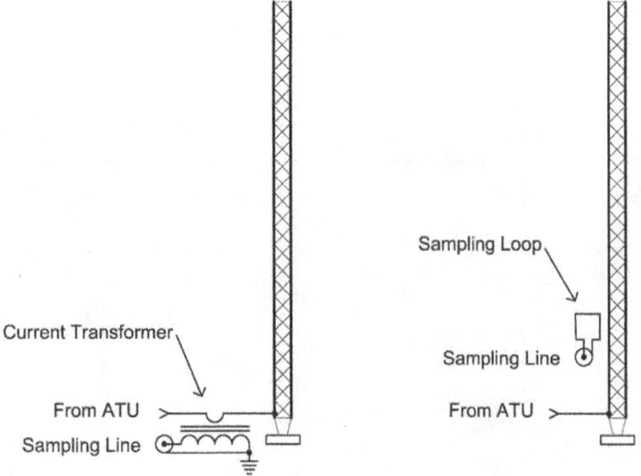

FIGURE 7.8-24 Base and loop sampling.

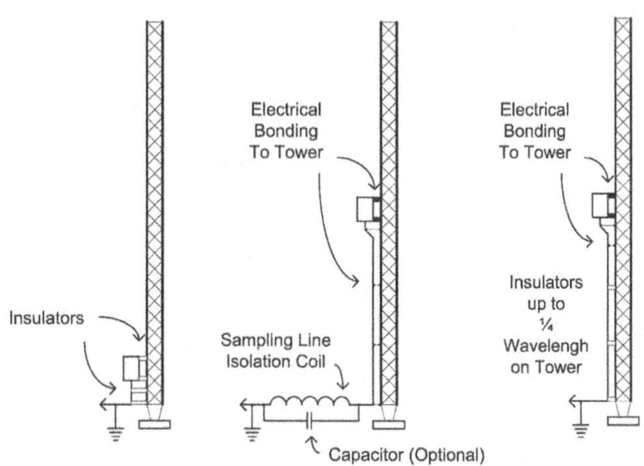

FIGURE 7.8-25 Insulated and uninsulated sampling loops.

added capacitance of the loop and sampling line on the lower portion of the tower can have a significant effect on the base impedance, but that can be compensated for at the time of the initial adjustment. Most antenna systems that employ tower-mounted sampling loops use isolation coils wound of sampling line across their tower bases. Sampling line isolation coils typically have inductances in the range of 100 to 150 μH and have minimal, and sometimes beneficial, effects on the tower base impedances. Where it is desired to compensate for the effects of the parallel inductance of an isolation coil across a tower base, a capacitor can be employed in parallel with it for that purpose. Some systems with towers on the order of one-half wavelength use quarter-wave isolation (with the sampling lines insulated from the tower for the bottom 90° of height) instead of isolation coils.

Figure 7.8-26 shows the heights at which sampling loops are typically mounted on towers. Historically, sampling loops have been mounted where the current is theoretically maximum, assuming sinusoidal current distribution. Most existing antenna systems employing sampling loops have them mounted as shown in Figure 7.8-26a (but no lower than about 10 feet above ground level to avoid stray effects from conductors near ground level) on towers 90° or less in height and as shown in Figure 7.8-26b for towers greater than 90° in height. Figure 7.8-26c shows where the sampling loops may be mounted to give indications closest to the theoretical field parameters, at the heights where the current minimum occurs for the detuned condition. This height may be determined by moment method modeling and can give measured tower currents bearing a direct 1:1 relationship to field parameters where the towers are identical. Where the towers are not identical, there is still an advantage in that the incremental changes in ratio and phase will agree between the sampled currents and tower fields even when their absolute values are not in agreement. For Method of Moments proofing, sample loops may be used if all of the towers have identical cross section characteristics and must be mounted as shown in Figure 7.8-26c.

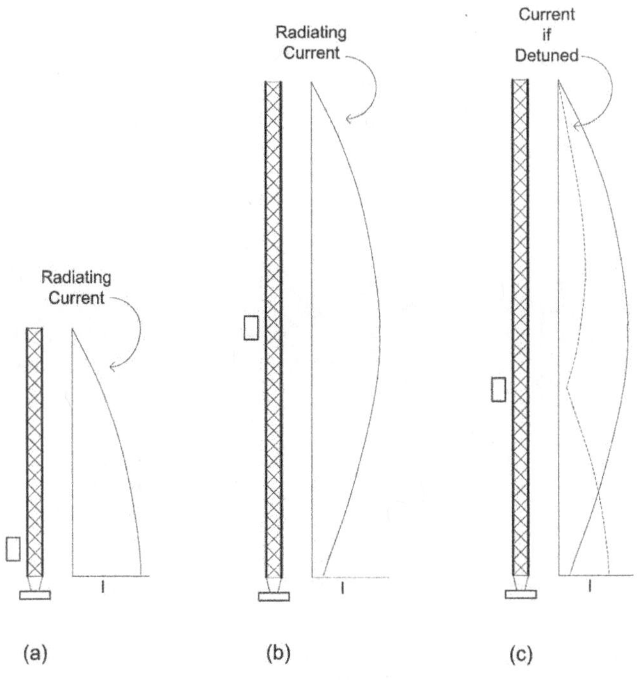

FIGURE 7.8-26 Sampling loop tower placement.

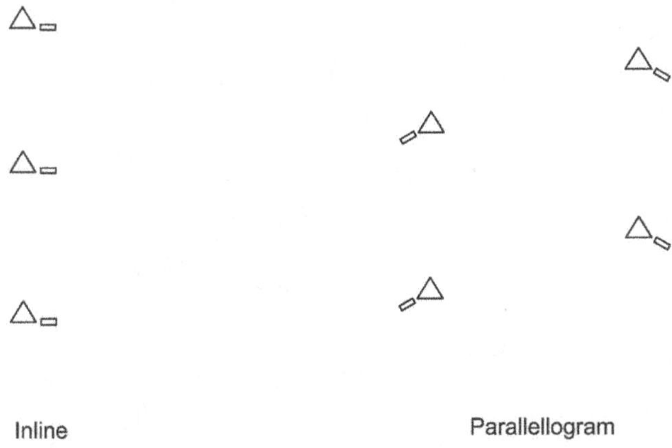

FIGURE 7.8-27 Sampling loop orientation on array elements.

When placing sampling loops on the towers of an array, care should be taken to minimize pickup from other towers. The sampling loops should be mounted with their planes including the structures of the towers on which they are located while excluding the closest nearby towers as much as possible, as shown on Figure 7.8-27.

ELECTRICALLY SHORT ANTENNAS

There is considerable interest in AM transmitting antennas that are much shorter than the typical quarter wave tower. Such antennas are useful in situations where conventional towers cannot be constructed because of environmental or aeronautical concerns, or for emergency backup antennas at stations that have conventional towers. The difficulties with such antennas center on radiation efficiency and bandwidth issues, as they typically have low base resistances and, therefore, high base currents for the power that is fed into them.

The FCC is primarily interested in the efficiency and radiation properties of antennas. Before a station can be licensed with an antenna that is shorter than would be necessary for a conventional tower antenna, the FCC must be satisfied that its efficiency meets the minimum requirements specified in the Rules and that its vertical radiation pattern is known. If an application for a construction permit is filed specifying an antenna of insufficient height under the Rules, the FCC may grant it with the added requirement that a proof-of-performance be run to demonstrate that its radiation efficiency meets its minimum requirement before a license will be issued. If there is any question about the current distribution on the antenna as it might impact the vertical radiation pattern, such as arise when top loading and center loading are employed, current distribution measurements or moment method analysis may also be required.

Antenna designs have been proposed in recent years that claim to overcome the disadvantages of short antennas by employing controversial principles that are not acknowledged by the larger peer group of antenna design engineers, such as synthesis of the electric and magnetic field components of a propagating wave separately in a small space using methods that are outside the bounds of analysis using Maxwell's equations. Despite claims of exceptional performance during informal tests and considerable discussion of a theoretical nature, several examples of such antennas have been the subject of much speculation without evidence ever being provided that they meet the FCC's minimum efficiency requirements based on objective, scientific tests.

There are well-known methods for obtaining better efficiency than is expected for short conventional tower antennas, however. Several principles for optimizing the efficiency of very short antennas, as shown in Figure 7.8-28, have been recognized for decades. Figure 7.8-28a shows how top loading can be used to maximize the value of the integral of antenna current over the vertical length of a short conductor, which is the condition for maximum radiation. Figure 7.8-28b shows how the radiating current can be divided between multiple closely coupled antenna conductors to reduce ground losses. Rather than intensifying directly under one conductor, the ground return currents are divided to flow under the multiple conductors. Figure 7.8-28c shows how what is called *center loading* (although it does not necessarily have to be placed at the exact center of the antenna to be effective) can be used to reduce the voltage across the base of a short antenna that would otherwise have a high capacitive reactance component and, thus, reduce stored energy in the region of the feedpoint. These principles may be used together to improve the performance of electrically short antennas.

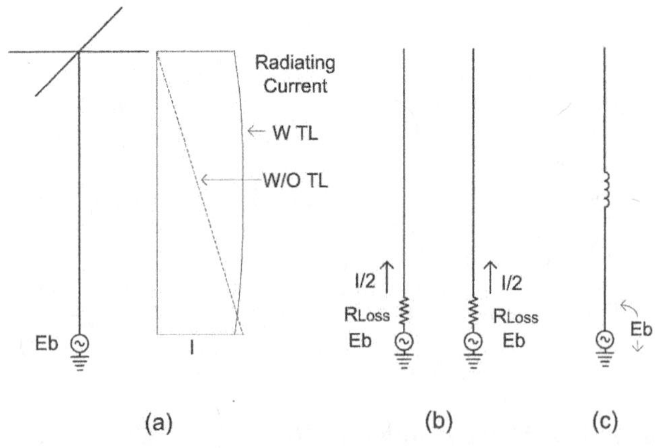

FIGURE 7.8-28 General principles for short antenna efficiency improvement.

The electrically short antennas that have been licensed by the FCC have generally used one or more of the principles illustrated by Figure 7.8-28. Two short antennas available from manufacturers have been approved for nondirectional use by stations in the United States. One consists of four top-loaded vertical wires supported by wooden poles over a quarter-wave radial ground system and has been pre-approved for both efficiency and vertical radiation, so it can be employed without the need to make proof measurements following construction. It must be constructed with its height, top-loading dimensions, and ground system scaled for frequency. Another consists of a fiberglass pole with both top and center loading, which has been pre-approved to meet FCC requirements within a specified frequency range for its standard height.

APPENDIX A DIRECTIONAL ANTENNAS FOR PATTERN SHAPE

SPACE CONFIGURATION

The plan configuration of the *k*th tower in an array is shown in Figure 7.8-A1. A space view of the *k*th tower and observation point *P* is shown in Figure 7.8-A2.

VECTOR DIAGRAM

The field strength at point *P* in space for the *k*th tower is shown in Figure 7.8-A3. The space phasing in the horizontal plane is shown in Figure 7.8-A4, and in the elevation plane the space phasing is reduced further, as shown in Figure 7.8-A5.

GENERALIZED EQUATION

The vector equation to express the vectors in Figure 7.8-A6 is the generalized equation that can be used to express the pattern shape for a directional antenna array of *n* towers. The equation in condensed form is:

$$E = \sum_{k=1}^{k=n} E_k f_k \Theta \beta_k \tag{A1}$$

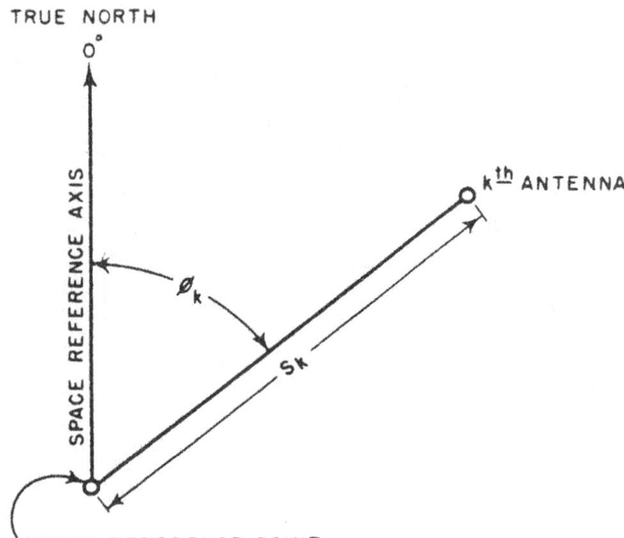

FIGURE 7.8-A.1 Plan view of space configuration of *k*th antenna.

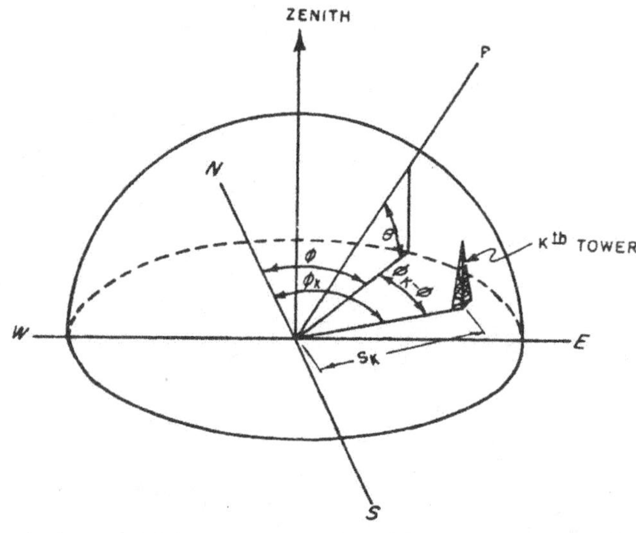

FIGURE 7.8-A.2 Space view of observation point *P* and the *k*th tower.

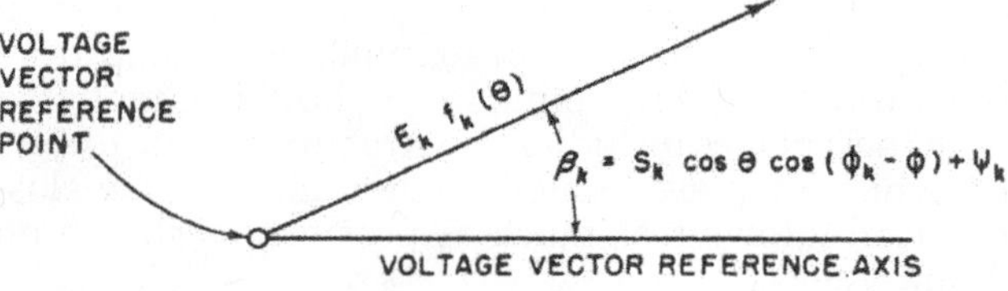

FIGURE 7.8-A.3 Voltage vector diagram for the *k*th antenna.

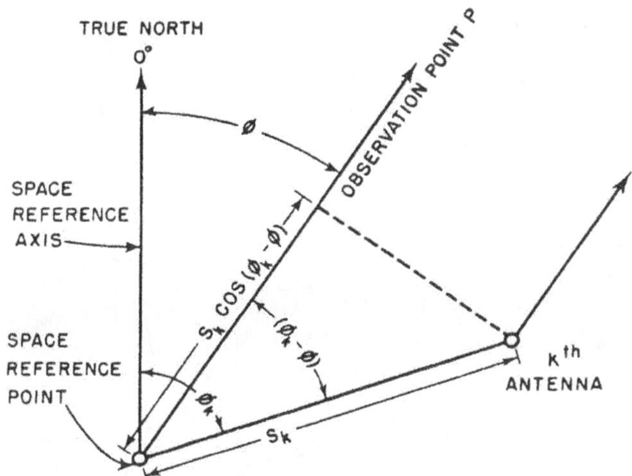

FIGURE 7.8-A.4 Plan view of *k*th antenna showing space phasing in the horizontal plane.

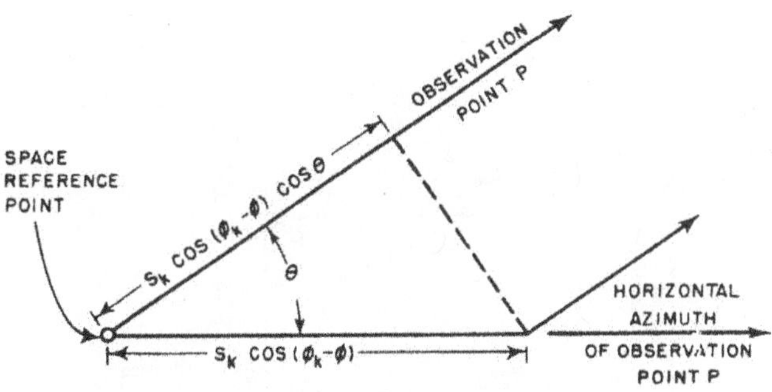

FIGURE 7.8-A.5 Elevation angle u shortens the spacing S_k to the value of $S_k$cosu.

where:

- E = the total effective field strength vector at unit distance (P) for the antenna array with respect to the voltage vector reference axis. This vector makes the angle β with respect to this axis, as shown in Figure 7.8-A6.
- k = the *k*th tower in the directional antenna system.
- n = the total number of towers in the directional antenna array.
- E_k = the magnitude of the field strength at unit distance in the horizontal plane produced by the *k*th tower acting alone.
- $f_k(\Theta)$ = vertical radiation characteristic of the *k*th antenna as given in Equation A3.
- Θ = elevation angle of the observation point P measured up from the horizon in degrees.

$$\beta_k = S_k \cos\Theta \cos(\phi_k - \phi) + \psi_k \qquad \text{(A2)}$$

is the phase relation of the field strength at observation point P for the *k*th tower taken with respect to the voltage vector reference axis.

$$S_k \cos(\phi_k - \phi)\cos\Theta \qquad \text{(A3)}$$

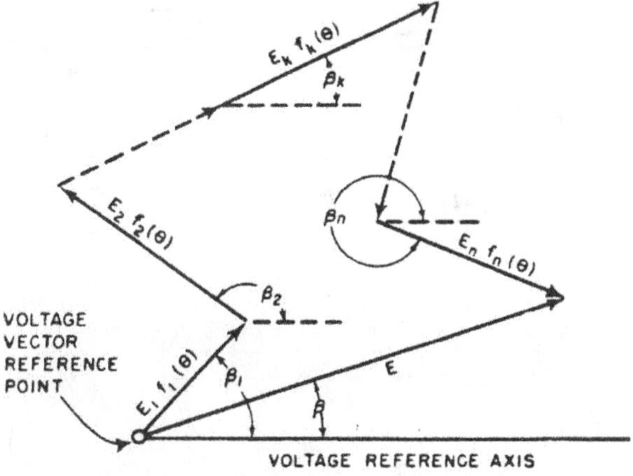

FIGURE 7.8-A.6 Summation of field strength vectors for *n* antennas in the directional antenna array.

is the space phasing portion of β_k due to the location of the *k*th tower, and ψ_k is the phasing portion of P_k.

- S_k = electrical length of spacing of the *k*th tower in the horizontal plane from the space reference point.
- ϕ_k = true horizontal azimuth orientation of the *k*th tower with respect to the space reference axis.
- ϕ = true horizontal azimuth angle of the direction to the observation point P (measured clockwise from true north).
- ψ_k = time phasing portion of β_k due to the electrical phase angle of the voltage (or current) in the *k*th tower taken with respect to the voltage vector reference axis.

The shape of any directional antenna pattern can be computed by applying the preceding equations; however, many directional antenna arrays can be designed by simplified versions of this equation. For a vertical antenna having a sinusoidal current distribution with a current node at the top, the vertical radiation characteristic takes on the form:

$$f(\Theta) = \frac{\cos(G\sin\Theta - \cos G}{(1-\cos G)\cos\Theta}$$

where:

- $f(\Theta)$ = vertical radiation characteristic.
- G = electrical height of the antenna in electrical degrees.
- Θ = elevation angle of the observation point measured up from the horizon in degrees.

The vertical radiation characteristics in Equation A3 are graphed in Figure 7.8-A7.

For a top-loaded tower the formula is:

$$f(\Theta) = \frac{\cos B\cos(A\sin\Theta) - \cos G - \sin B\sin\Theta\sin(A\sin\Theta)}{\cos\Theta(\cos B - \cos\ G)}$$

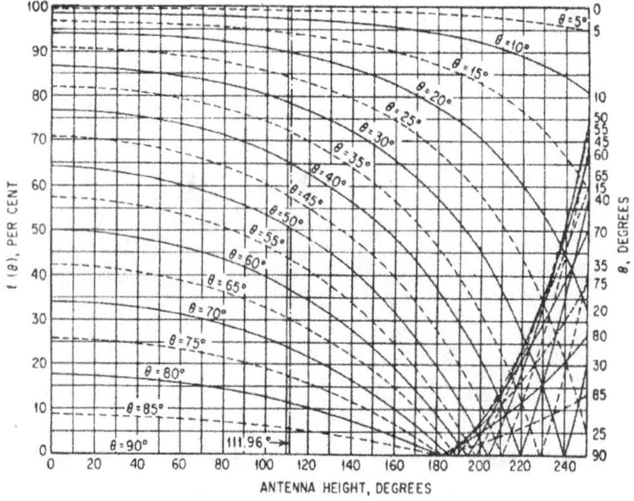

FIGURE 7.8-A.7 Vertical-radiation characteristics as a function of electrical tower height for various values of elevation angle.

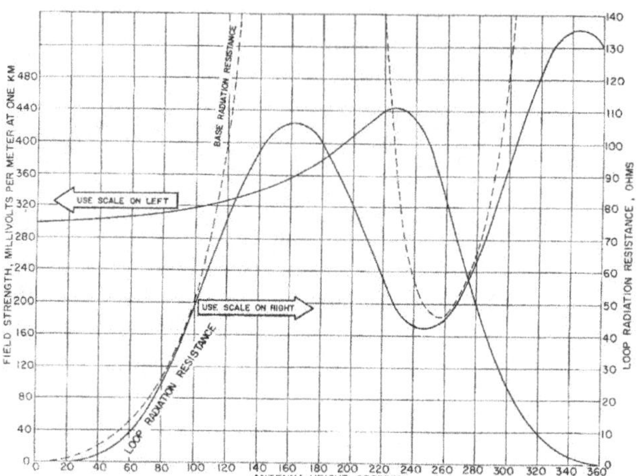

FIGURE 7.8-A.8 Inverse field strength at 1 km for 1-kW loop and base radiation resistance as a function of tower height over a perfectly conducting earth.

This is the vertical radiation characteristic for a top-loaded tower of height A and top loaded to a height of $G = A + B$. For a two-section top-loaded tower as shown in Figure 7.8-A5, the formula is:

$$f(\Theta) = \frac{\cos B \cos(A\sin\Theta - \cos G + \dfrac{\sin B\cos(H-C)\cos(C\sin\Theta)}{\sin(H-A)} - \dfrac{\sin B \sin\Theta \sin(H-C)\sin(C\sin\Theta)}{\sin(H-C)} - \dfrac{\sin B\cos(H-A)\cos(A-\Theta)}{\sin(H-A)}}{\{\cos B - \cos G + [\sin B/\sin(H-A)](\cos H - C\cos H - A)\}}$$

This is the vertical radiation characteristic equation for a two-section sectionalized tower. The same procedure can be applied if more than two sections are involved (see Figure 7.8-A6).

THEORETICAL SELF-LO6OP AND BASE-RADIATION RESISTANCE

It is useful to know the theoretical loop and base resistance of a vertical radiator. This information is presented graphically in Figure 7.8-A8, along with the theoretical inverse field strength at 1 km.

MUTUAL IMPEDANCE CURVES

The value of mutual impedance for most tower heights and spacing is given in Figure 7.8-A9. The loop mutual impedance between quarter-wave towers is shown in Figure 7.8-A10.

HORIZONTAL RMS FIELD STRENGTH

The field strength gain or loss of a two-tower array for various values of phasing and spacing is shown in Figure 7.8-A11.

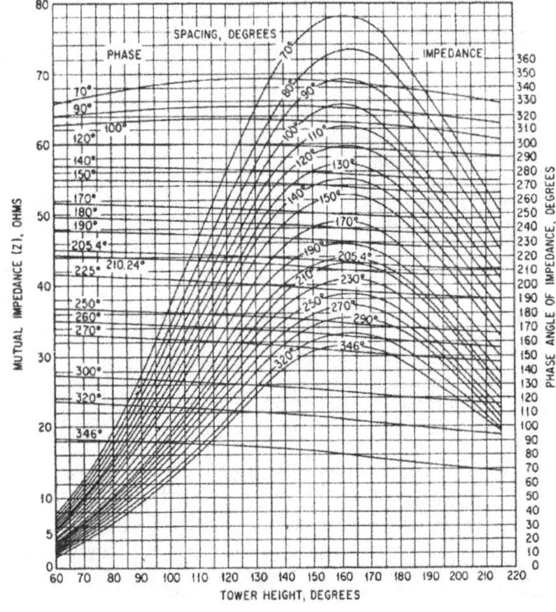

FIGURE 7.8-A.9 Loop mutual impedance and phase angle between two towers of equal height.

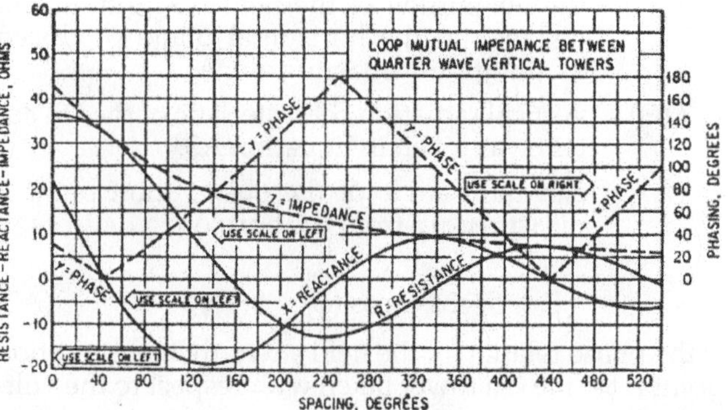

FIGURE 7.8-A.10 Loop mutual impedance between quarter-wave vertical towers.

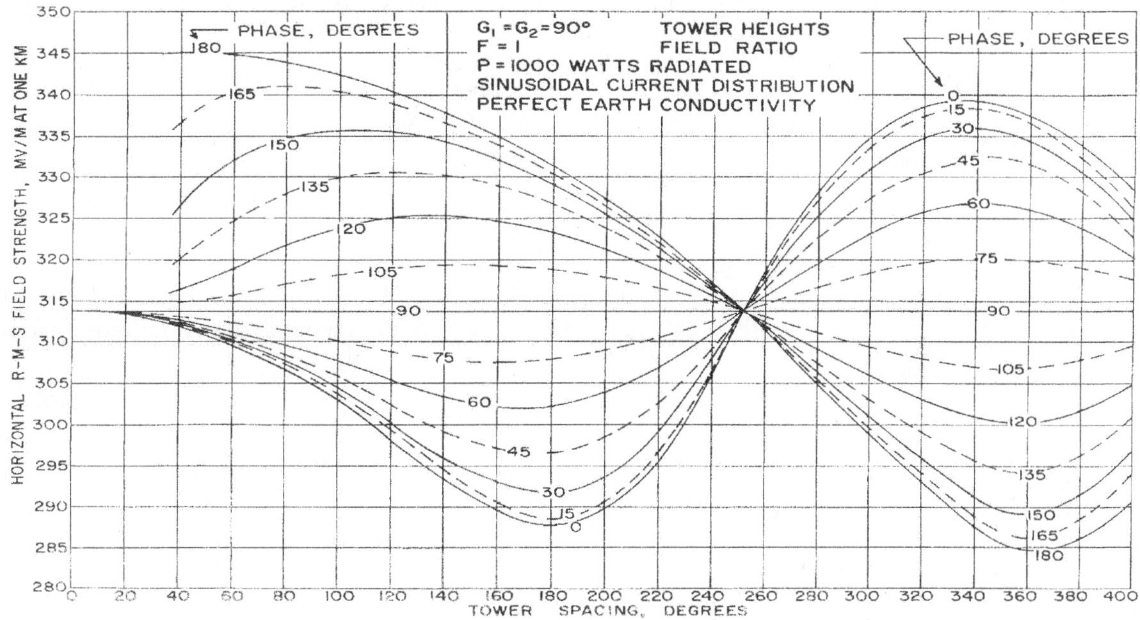

FIGURE 7.8-A.11 Horizontal RMS field strength of two-tower directional antenna.

APPENDIX B PATTERN DEVELOPMENT OF DIRECTIONAL ANTENNAS

THEORETICAL PATTERN EQUATION

The theoretical pattern equation of Appendix A can be written as follows by changing the *k*th tower to the *i*th tower to conform with the FCC practice, thus:

$$E(\phi,\Theta)th = \left|k\sum_{i=1}^{n} F_i(\Theta)\,/\,S_i\cos\Theta\cos(\phi_i-\phi)+\psi_i\right| \quad (B1)$$

where *k* is a multiplying constant that determines pattern size.

STANDARD PATTERN EQUATION

The standard pattern equation is obtained from Equation B1 by adding the quadrature term *Q* to fill minimums and increase the size by 5%, thus:

$$E(\phi,\Theta)_{std} = 1.05\sqrt{[(\phi,\Theta)th]^2+Q^2} \quad (B2)$$

where *Q* is the greater of the quantities:

$$0.025g\,(\Theta)\,E_{rss} \quad (B3)$$

or

$$10.0g(\Theta)\sqrt{p_{kw}} \quad (B4)$$

where $g(\Theta)$ is the vertical plane distribution factor, $f(\Theta)$, for the shortest element in the array (see Equation B2; also see FCC Rules Section 73.190, Figure 5). If the shortest element has an electrical height in excess of 0.5 wavelength, $g(\Theta)$ is computed as follows:

$$g(\Theta) = \frac{\sqrt{\{f(\Theta)\}+0.0625}}{1.030776} \quad (B5)$$

$$E_{rss} = \sqrt{\sum_{i=1}^{n} E_i^2} \quad (B6)$$

As an example, consider a two-tower array. The theoretical pattern equation (Equation B1) becomes:

$$E = E_1 f_1(\Theta)\ /0^0 + E^2 f_2(\Theta)\ /S_2\cos\Theta\cos(\phi_2-\phi)+\psi_2 \quad (B7)$$

Now, for 5-kW, 90° towers and the following parameters:

Tower Number	Height (G^0)	Field Ratio	Spacing (S^0)	True Bearing (ϕ^0)	Phase (ψ)
1	90	1.0	0	0	0
2	90	1.0	90	0	–90

E_{rss} (theoretical pattern) = 691.92 V/m.

Q (quadrature term) = 21.60 V/m.

E_{rms} (standard pattern) = 726.87 V/m.

A plot of the theoretical and standard patterns is shown in Figure 7.8-B1.

The minimum horizontal field strength (at 1 km) when the theoretical field strength goes to zero is given by Equation B2 for a standard pattern along the

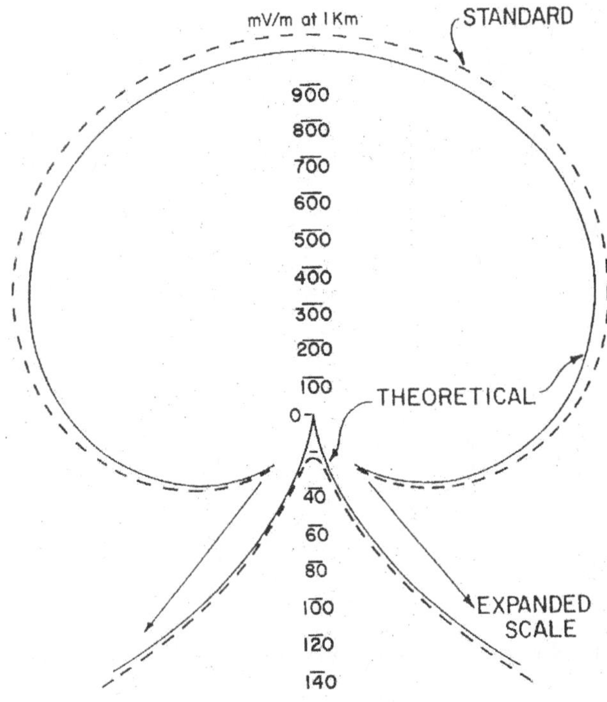

FIGURE 7.8-B.1 Theoretical and standard pattern.

ground using Equation B4 with $g(\Theta) = 1.0$. For 1 kW and under, Q is 6 according to FCC Rules. For various FCC licensed values of power, the minimum field strength values are as follows:

P_{kw}	Q	E_{min}
0.25	6.0	10.14
0.50	6.0	10.14
1.00	6.0	10.14
2.50	9.49	16.03
5.00	13.42	22.67
10.00	18.97	32.06
25.00	30.00	50.69
50.00	42.43	71.69

The minimum field strength (at 1 km) for any elevation of a standard pattern, by Equation B5, is:

$$g(\Theta) = \frac{\sqrt{0 + 0.0625}}{1.030776} = 0.2425 \qquad \text{(B8)}$$

AUGMENTED PATTERN EQUATION

The augmented pattern equation is obtained by adding an augmentation quadrature term to the standard pattern, as given here:

$$E_{(\phi,\Theta)aug} = \sqrt{\{E_{(\phi,\Theta)std}\}^2 + \{g(\Theta)\cos(180\frac{D_A}{S})\}} \qquad \text{(B9)}$$

where:

$E_{(\phi,\Theta)aug}$ = augmented pattern radiation value at azimuth, elevation.

$E_{(\phi,\Theta)aug}$ = standard pattern radiation value at azimuth, elevation.

A = augmentation constant.

$$g(\Theta) = \frac{\sqrt{\{f(\Theta)\}^2 + 0.0625}}{1.030776}$$

D_A = angular distance from center of span.

S = span of augmentation in degrees.

The principle of augmentation is illustrated in the cardioid pattern of Figure 7.8-B2.

The FCC has converted all augmented directional patterns to a table for each station, as shown in the example of Figure 7.8-B3. In this case, there were six augmentations, as tabulated in Figure 7.8-B4 and shown on the polar chart of Figure 7.8-B5. It should be noted that, where the spans overlap, Equation B9 is applied repeatedly, once for each augmentation, proceeding clockwise from true north.

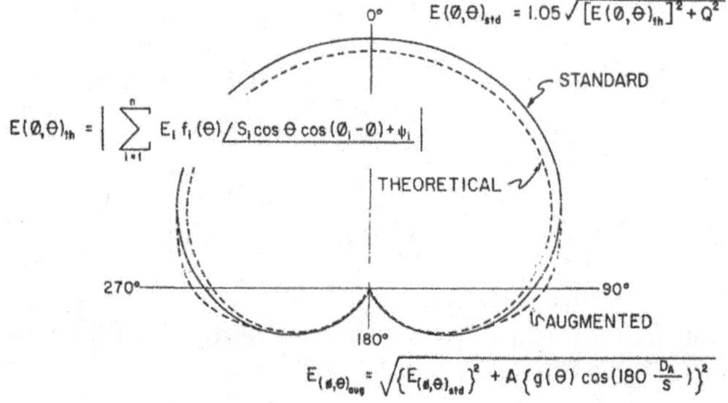

FIGURE 7.8-B.2 Theoretical, standard, and augmented patterns.

Technical Parameters Resulting From Conversion of AM Broadcast Stations To Standard Patterns

STANDARD PATTERN CONVERSION NO.: 1280–22

FREQ. KHZ	CALL LETTER	CITY	STATE	PATTERN HRS.	PATTERN STATUS	CLASS
1280	WHVR	HANOVER	PA	N	LIC.	B

POWER KW	LATITUDE	LONGITUDE	PAT-MULT MV/M	TH-RMS MV/M	STD/AUG RMS-MV/M	PAT-RSS MV/M	Q-FACTOR
.500	39-49-11	77-00-25	131.27	143.00	150.54	185.65	6.0000

TOWER NO.	PHYS-HT (A)-DEG	TL-HT (B)-DEG	TOT-HT (C)-DEG	TL-HT (D)-DEG	FIELD RATIO	PHASE DEG.	SPACING DEG.	ORIENT DEG-TR	REF FLG
1	91.0	.0	.0	.0	1.000	149.5	.0	.0	
2	91.0	.0	.0	.0	1.000	.0	90.0	178.0	

AUGMENTATION DATA

CENTRAL AZIM. DEGREES TRUE	SPAN DEGREES	FIELD AT AZIM. MV/M
64.0	12.0	17.0
260.5	55.0	103.0
288.0	14.0	7.5
288.0	10.0	21.2
295.0	14.0	30.0
295.0	10.0	43.3

HORIZONTAL PLANE STANDARD/AUGMENTED RADIATION VALUES

AZ. DEG	FIELD MV/M	AZ. DEG	FIELD MV/M	AZ. DEG	FIELD MV/M	AZ. DEG	FIELD MV/M	AZ. DEG.	FIELD MV/M	AZ. DEG	FIELD MV/M
0	136.8	60	29.1	120	174.0	180	239.4	240	163.9	300	42.2
10	132.8	70	9.0	130	196.3	190	237.0	250	136.2	310	71.0
20	123.1	80	43.5	140	213.5	200	231.5	260	104.7	320	95.4
30	107.5	90	80.0	150	225.9	210	221.4	270	69.3	330	114.4
40	86.3	100	115.0	160	234.0	220	207.2	280	31.3	340	127.7
50	59.8	110	146.7	170	238.4	230	188.0	290	20.4	350	135.1

CONSTRUCTION PERMIT LIMITS

AZIMUTH DEG. TRUE	PRESENT MV/M	NEW MV/M
64.0	17.0	17.0
231.0	179.0	185.8
288.0	33.0	21.2
352.0	131.0	135.9

PATTERN MINIMA

AZIMUTH DEG. TRUE	FIELD MV/M
68.5	6.8
284.5	17.3
290.0	20.4
299.0	40.9

PATTERN MAXIMA

AZIMUTH DEG. TRUE	FIELD MV/M
178.0	239.4
288.0	21.2
296.0	44.2
358.0	136.9

FIGURE 7.8-B.3 FCC method of specifying augmentation.

CENTER AZIMUTH OF AUGMENTATION	SPAN DEGREES	EXTENT OF SPAN	FIELD CNTER SPAN AT 1 KM
64	12 (6)	(58–70)	27.4
260.5	55 (27.5)	(233–288)	165.8
288	14 (7)	(281–295)	12.1
288	10 (5)	(283–293)	34.1
295	14 (7)	(288–302)	48.3
295	10 (5)	(290–300)	69.7

FIGURE 7.8-B.4 Table of augmentation data.

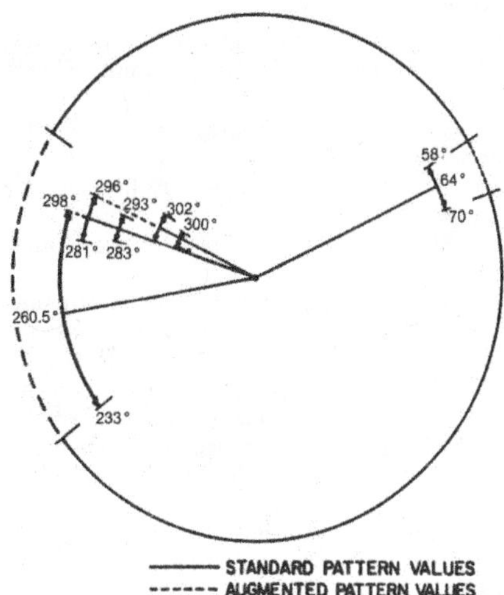

FIGURE 7.8-B.5 Augmented pattern flowchart showing overlapping spans.

Note

[1] Slant-wire feeds are no longer accepted by the FCC for critical hours and night operation.

Bibliography

Anderson, H. and Pinion, D., Short low loss antennas, *IEEE Trans. Broadcasting*, BC-32(3), 37–43, 1986.

Ballantine, S., On the optimum transmitting wavelength for a vertical antenna over perfect earth, *Proc. IRE*, 12, 833–839, 1924.

Beverage, C.M., Compact medium wave transmitting antennas, *IEEE Trans. Broadcasting*, 50(2), 142–147, 2004.

Breakall, J.K., Jacobs, M.W., King, T.F., and Resnick, A.E., Testing and results of a new, efficient low-profile AM medium frequency antenna system, in *Proc. of the 57th Annual National Association of Broadcasters Broadcast Engineering Conf.*, April, 2003, pp. 235–243.

Brown, G.H., A critical study of the characteristics of broadcast antennas as affected by current distribution, *Proc. IRE*, 24, 48–81, 1936.

Brown, G.H., Directional antennas, *Proc. IRE*, 25(1), 79–145, 1937.

Brown, G.H., Lewis, R.F., and Epstein, J., Ground systems as a factor in antenna efficiency, *Proc. IRE*, 25(6), 753–787, 1937.

Bruner, P. and Waniewski, B., Directional MF antennas using self-supporting towers with driven wire cages, *IEEE Trans. Broadcasting*, 35(2), 193–199, 1989.

Chiodini, T., Moment method predicted impedances compared to actual measured impedances of directional arrays, *IEEE Trans. Broadcasting*, 35(2), 1909–1192, 1989.

Christman, A. and Radcliff, R., Impedance stability and bandwidth considerations for elevated-radial antenna systems, *IEEE Trans. Broadcasting*, 35(2), 167–171, 1989.

Christman, A. and Radcliff, R., Using elevated radials with ground mounted towers, *IEEE Trans. Broadcasting*, 37(3), 77–82, 1991.

Christman, A. and Radcliff, R., AM broadcast antennas with elevated radials: varying the number of radials and their height above ground, *IEEE Trans. Broadcasting*, 42(1), 10–13, March, 1996.

Christman, A., Radcliff, R., Adler, D., Breakall, J., and Resnick, A., AM broadcast antennas with elevated radial ground systems, *IEEE Trans. Broadcasting*, 34(1), 75–77, 1988.

Dawson, B.F., Sharing AM transmitter sites by diplexing antenna systems, in *Proc. of the 41st Annual National Association of Broadcasters Broadcast Engineering Conf.*, March, 1987, pp. 31–36.

Dawson, B.F., Analysis of a sectionalized tower as an element in a medium wave phased array using the method of moments, *IEEE Trans. Broadcasting*, 35(2), 185–189, 1989.

Dawson, B.F., Modern analysis methods for medium wave antenna design, in *Proc. of the International Broadcasting Convention of the IEE*, Brighton, UK, September, 1990.

Design and Operation of Directional Antennas, Smith Electronics, Inc., Cleveland, OH.

Directional Antenna Patterns, Smith Electronics, Inc., Cleveland, OH.

Directional Antenna Design Example, Smith Electronics, Inc., Cleveland, OH.

Directional Antenna Pattern Shapes, Smith Electronics, Inc., Cleveland, OH.

Frese, G.M., Paran antenna for use where horizontal and vertical space is limited, *IEEE Trans. Broadcasting*, 38(3), 163–165, 1992.

Harmon, R.N., Some comments on broadcast antennas, *Proc. IRE*, 24, 36–47, 1936.

Hatfield, J.B., Analysis of AM directional arrays using method of moments, in *Proc. of the 42nd Annual National Association of Broadcasters Broadcast Engineering Conf.*, April, 1988, pp. 84–87.

Hatfield, J.B., Relative tower currents and fields in an AM directional array, *IEEE Trans. Broadcasting*, 35(2), 176–184, 1989.

Hatfield, J.B., Verifying the relationships between AM broadcast fields and tower currents, in *Proc. of the Applied Computational Electromagnetics Society Conf.*, Monterey, CA, March, 1989.

Hatfield, J.B., Computer analysis of antenna systems, in *The Electronics Handbook*, Whitaker, J.C. (Ed.), CRC Press, Boca Raton, FL, 1996, pp. 1353–1367.

Hatfield, J.B., Computer simulation of AM radio antenna systems, in *NAB Engineering Handbook*, National Association of Broadcasters, Washington, DC, 1999, chap. 4.10, 875–896.

Hatfield, J.B. and Leonard, P.W., A comparison of the fields of a medium wave directional antenna as calculated by the FCC method and the numerical electromagnetic code, in *Proc. of the Applied Computational Electromagnetics Society Conf.*, Monterey, CA, March, 1987.

Instructions for Installation of Radio Broadcast Stations Ground Systems, Smith Electronics, Inc., Cleveland, OH.

Introduction to Directional Antenna Systems, Smith Electronics, Inc., Cleveland, OH.

Jeffers, C.L., An antenna for controlling the nonfading range of broadcasting stations, *Proc. IRE*, 36, 1426–1431, 1948.

Leonard, P.W., Computer models of Kabul, Afghanistan, medium-wave antennas, *IEEE Antenna Propagation Soc. Mag.*, 48(2), 38–41, 2006.

Log Periodic Antenna Design Handbook, Smith Electronics, Inc., Cleveland, OH.

Lundin, J.A. and Rackley, R.D., Medium frequency broadcast antennas, in *Antenna Engineering Handbook*, Johnson, R.C. (Ed.), McGraw-Hill, New York, 1993, pp. 25-1–25-28.

Morrison, J.F. and Smith, P.E., The shunt excited tower, *Proc. IRE*, 25, 673–696, 1937.

Parasitic Reradiation, Smith Electronics, Inc., Cleveland, OH.

Rackley, R.D., Modern methods in mediumwave directional antenna feeder system design, in *Proc. of the 45th Annual National Association of Broadcasters Broadcast Engineering Conf.*, April, 1991, pp. 43–54.

Rackley, R.D., Engineering exhibit star-H experimental antenna, in *Proc. of the 57th Annual National Association of Broadcasters Broadcast Engineering Conf.*, April, 2003, pp. 401–439.

Rackley, R.D. and Folkert, M., Antennas for medium-frequency broadcasting, in *Wiley Encyclopedia of Electrical and Electronics Engineering*, Webster, J.G. (Ed.), John Wiley & Sons, New York, 1999, pp. 578–594.

Radiation Characteristics of Transmitting Antennae: An Introduction to Directional Antenna Pattern Design, Smith Electronics, Inc., Cleveland, OH.

Radio Broadcast Ground Systems, Smith Electronics, Inc., Cleveland, OH.

Raines, Jeremy, K., *Folded Unipole Antennas: Theory and Applications*, New York: McGraw-Hill ©2007 ISBN-13: 978-0-07-147485-6.

Roberts, W. Van, Input impedance of a folded dipole, *RCA Rev.*, 8, 289, June 1947.

Smith, C.E., *A Critical Study of Several Antennas Designed to Increase the Primary Coverage of a Radio Broadcasting Transmitter*, thesis, Ohio State University, Columbus.

Smith, C.E., A critical study of two broadcast antennas, *Proc. IRE*, 24, 1329–1341, 1936.

Smith, C.E., Short low loss antennas, *IEEE Trans. Broadcasting*, 35(2), 237–249, 1989.

Smith, C.E., Hall, J.R., and Weldon, J.O., Very high-power long-wave broadcasting antennas, *Proc. IRE*, 42(8), 1222–1235, 1954.

Smith, C.E., Hutton, D.B., and Hutton, W.G., Performance of sectionalized broadcasting towers, *IRE Trans.*, 22–34, December, 1955.

Smith, C.E. and Johnson, E.M., Performance of short antennas, *Proc. IRE*, 35, 1026–1038, 1947.

Smith, J.L., A method for modeling array elements when using NEC and MININEC, *IEEE Trans. Broadcasting*, 44(2), 186–193, 1998.

Smith, J.L., A method to determine the detuning reactance for unused elements in directional arrays, *IEEE Trans. Broadcasting*, 47(3), 259–262, 2001.

Standard Broadcast Antenna Systems, Smith Electronics, Inc., Cleveland, OH.

Theory and Design of Directional Antennas, Smith Electronics, Inc., Cleveland, OH.

Trainotti, V., Height radius effect on MF AM transmitting monopole antenna, *IEEE Trans. Broadcasting*, 36(1), 82–88, 1990.

Trainotti, V., The nocturnal service area of MF AM broadcast stations using vertical arrays, *IEEE Trans. Broadcasting*, 36(1), 74–81, 1990.

Trainotti, V., Short medium frequency AM antennas, *IEEE Trans. Broadcasting*, 47(3), 263–284, 2001.

Trainotti, V., MF AM folded monopole characteristics, *IEEE Trans. Broadcasting*, 48(4), 324–330, 2002.

Trueman, C.W., Modelling a standard broadcast directional array with the numerical electromagnetics code, *IEEE Trans. Broadcasting*, 34(1), 39–49, 1988.

Trueman, C.W. and Kubina, S.J., Power line tower models above 1000 kHz in the standard broadcast band, *IEEE Trans. Broadcasting*, 36(3), 207–218, 1990.

Trueman, C.W. and Kubina, S.J., Scattering from power lines with the skywire insulated from the towers, *IEEE Trans. Broadcasting*, 40(2), 53–62, 1994.

Trueman, C.W., Kubina, S.J., and Baltassis, C., Ground loss effects in power line reradiation at standard broadcast frequencies, *IEEE Trans. Broadcasting*, 34(1), 24–38, 1988.

Trueman, C.W., Roobroeck, T.M., and Kubina, S.J., Stub detuners for free standing towers, *IEEE Trans. Broadcasting*, 35(4), 325–338, 1989.

Westberg, J.M., Matrix method for relating base current ratios to field ratios of AM directional stations, *IEEE Trans. Broadcasting*, 35(2), 172–175, 1989.

CHAPTER

7.9

Antenna Coupling and Phasing Systems

TOM KING
Kintronic Laboratories, Inc.

BOBBY COX, PhDEE
Kintronic Laboratories, Inc.

JAMES MOSER
Kintronic Laboratories, Inc.

FOREWORD

The techniques used to design the AM passive RF networks that are required to yield maximum power transfer over a maximally flat channel bandwidth between the transmitter and the antenna have matured to a more refined degree as a result of the widespread adoption of Method of Moments (MOM) modeling for towers and their feed configurations, the development of state-of-the-art network topology computer software tools, such as SPICE,[1] and the adoption of the MOM methodology using Numerical Electromagnetics Code (NEC),[2] Mini-NEC and other similar computer programs for the design of phasing and coupling systems for AM directional antennas. As more and more antenna systems have been designed, constructed, installed and tested using these modern software tools, the empirical data derived from these station implementations has been used to refine the computer models to the point that the required commissioning time has been reduced dramatically.

The use of MOM tower modeling has served to (1) open up the possibility to explore new innovative methods for feeding towers to achieve wide audio bandwidth and maximum efficiency; (2) facilitate the modeling of the effects of nearby television towers, FM towers or cellular telephone towers on existing AM antenna patterns; and (3) enable the design of tower modifications to eliminate or minimize the effects of such towers on the AM pattern. The most complex tower structures or array of towers can be modeled in minimal time with the powerful desktop computers available today. Modern computer hardware and software tools fulfill a critical role in the design of AM HD Radio antenna systems that require a maximally flat impedance over a f_c ±15 kHz channel bandwidth.

PREFACE

This chapter of the *NAB Engineering Handbook* will provide an overview on the subject of single frequency AM analog- or digital-capable omnidirectional or directional antenna system design. Several tower feed options and the relative merits of each will be addressed followed by transmission line considerations, the basic theory of single tower matching networks, directional antenna phasing and matching system design, treatment of unused towers in a directional array, directional antenna sampling systems, RF component selection considerations and an overall system design approach for digital-capable HD Radio antenna systems.

TOWER/MAST COUPLING TECHNIQUES

The primary goal in the design of the RF feeder network is to match the drive impedance of the AM radiating structure(s) to the output impedance of the transmitter with maximally flat input audio bandwidth and minimum loss. For the purposes of this dialog the tower/mast can be a single-guyed or self-supported tower that is either base insulated or electrically grounded and shunt fed. It can also be an array of such elements where the RF feed includes the networks required to properly divide and phase the transmitter power such that the array of elements produces the desired radiation pattern. Most typically, the vertical radiating structures are modeled with a MOM program and the parameters dictating drive voltages and currents are derived from this model for both the single tower nondirectional antenna or for the multi-tower directional case. For the directional array, these parameters take the form of drive impedances, power ratios, and the relative phase of the drive voltages or currents of the towers. One tower (typically the high power tower) is the reference for the drive voltage or current for each licensed pattern. With these parameters, the feed networks can be designed. The resulting system is analyzed for both pattern and impedance bandwidth, which requires the additional step of evaluating the networks at the sideband frequencies of interest. This is complicated by the mutual coupling between the elements in the array, which is often significant. This mutual impedance, along with the frequency dependent characteristics of the feed networks, affects the individual element drive conditions at the sidebands, and as a result, the system bandwidth. A full evaluation requires a network analysis of the feed system (using a nodal circuit solving program such as LTSpice[1]). The circuit equivalent of the antenna elements over the frequencies of interest must be included in the analysis. The details of the development of the radiation pattern or the techniques to extract the drive boundary conditions or circuit equivalent of the array are beyond the scope of this section and are treated elsewhere. In this discussion it is assumed that drive impedance of the radiating element, and in the case of arrays, the power and phase budget of the elements are known.

To explicitly state what is implied here, a well-designed feed system requires that the entire system is taken into consideration. This includes the radiating element characteristics, the RF network topology and component choices, the transmission line characteristics, and especially in the case of digital broadcast systems, the characteristics of the final harmonic filter stages in the transmitter. For multiplexed installations or those with nearby stations and signals that need to be rejected, the required filters also need to be included in the analysis. Both the filters' effects on bandwidth and their level of isolation need to be evaluated. The design of filters or that of the networks to precondition drive impedances for filtering are also beyond the scope of this article and treated elsewhere in this handbook. It is key that such a system approach be used to produce the best quality in audio transmission.

Impact of Tower Feed Configurations

Various types of broadcast antennas used for the AM band are presented in Chapter 4.3. Among these are the simple base insulated tower and the folded unipole fed tower. A brief discussion of advantages and disadvantages of these two major feed methods is offered along with practical design considerations for implementing folded unipoles.

Base Insulated, Series Fed

If a tower is only to be used for AM broadcast service with no expectations of supporting other significant auxiliary antennas, then the simple series–fed, guyed tower is generally the most inexpensive feed configuration. Because the tower is conducting RF current as an AM radiator, the tower section design must exhibit solid, low-resistance electrical connections at each leg joint, and the base insulator surface must be kept clean.

A key advantage of the series-fed, guyed tower is simplicity. The configuration gives less wind loading of the tower and has fewer modes for failure than the shunt-fed configurations. A key drawback of this configuration is that adding other services on the tower requires that transformers be used to pass cables across the base insulator. Even adding simple tower lighting to the tower requires special care to convey the AC power conductors onto the tower without electrically shorting the base insulator. The transformers necessary to add tower lighting, FM or TV antennas, STL antennas, telecommunications antennas and so forth are all readily available and can allow a great many other services to share space on a series fed AM tower. Series-tuned transformers, such as isocouplers, or lighting chokes and isolation inductors, will load the tower base insulator with parallel reactance that can shift the effective tower base impedance. This can require retuning of the antenna tuning unit (ATU) to compensate for the new effective tower impedance that results. If the tower is part of a multi-tower directional array, then the effects on the tuning of the array must also be carefully considered. Each added transformer device is also another element that needs to be included in the system design and bandwidth analysis.

Folded Unipole

If a tower is to be used for services other than the AM, then a grounded tower can be a more cost-effective tower feed choice. Without the presence of a base insulator, cables used for feeding tower lights, FM antennas, studio-to-transmitter link (STL) antennas, telecom antennas and so on can be routed directly onto the tower without the expense of various isolation transformers. The tower should be structurally evaluated for the added structural and wind loading. The folded unipole is the most common method of feeding a tower with a grounded base.[3]

For a folded unipole antenna to work efficiently, several design factors must be considered, including the number of vertical wires used for the skirt, the diameter of the wires, their standoff distance from the tower, the metals used in the skirt hardware, the use of shorting stubs, and the height of the skirt on the tower. These all affect the ultimate bandwidth of the antenna system. Although each of these factors warrant lengthy discussion, they will only be very briefly covered here.

For the typical triangular cross section, guyed tower, the number of skirt wires should be at least three in order to encircle the tower symmetrically. For large cross section towers, and self-supported towers, a greater numbers of skirt wires may be used to more symmetrically encompass the tower structure. Using more skirt wires will generally enhance the bandwidth performance of the antenna as well.

The diameter of the skirt cables should be chosen to accommodate the maximum RF current with modulation with minimal resistive loss. MOM modeling of the unipole skirt can identify the peak currents in the skirt wires so that the wire (or cable) can be sized conservatively. Sizing the unipole skirt wires so as to yield RF currents with no more than 15 amps (RMS, root mean square) per circumferential inch is recommended.

The standoff distance of the folded unipole wires from the tower is a choice often made poorly. Electrically speaking, there is no practical standoff distance too great, but there are certainly standoff distances that are too small. A good rule of thumb is to have the unipole skirt wires spaced from the tower by a distance that is no less than 30 inches. This allows tower climbers to clear the wires more easily when servicing the tower. For towers wider than 24″, greater standoff distances should be used when possible and practical. Standoff distances on the order of 1.5 times the tower face width are a good, general goal for guyed towers. Using these larger standoff distances not only helps prevent damage from tower crews, but it also can help prevent excessive circulating currents between the skirt and tower that can diminish the efficiency of the antenna system. Long standoff distances also tend to allow superior bandwidth performance for the folded unipole antenna. Tower standoff insulators of up to 72″ in length can be built without much difficulty. Beyond this length, their cost and loading on the tower can become a problem. As with any additions to the tower, their effects on the tower's structural loading must always be evaluated. For self-supporting towers with greater cross sections, the 1.5*x* rule of thumb for standoff insulator length clearly doesn't apply. Keeping these insulators as long as is practical and economical without going below 30″ is advised.

The types of metals used in the RF current carrying path of a folded unipole antenna are also important. Copperweld wires (copper clad steel) are a popular choice for the skirt wires, but they introduce the risk of dissimilar metal junctions in the antenna because the tower itself is usually galvanized steel. If copper clad steel wires are used, then the hardware used to bond the skirt wires together at the bottom feed ring, and to the tower at the top of the skirt (and possibly at a shorting stub position) must be carefully chosen to minimize galvanic incompatibility. Oxidation preventive pastes can be used to help reduce corrosion problems. An alternate choice is to use galvanized steel cables, such as extra high strength (EHS) guy strand, for the skirt wires. This choice is not only economical, but also allows the use of lower cost galvanized steel hardware for bonding the wires together and for bonding them to the tower structure. The resulting zinc-to-zinc connections are less prone to long-term corrosion problems. The skirts built with EHS guy cable also tend to hold up better over time than smaller copper clad steel wire skirts.

Folded unipoles are often built with a set of tuning stubs that short the skirt back to the tower structure at a point below the upper support brackets to adjust the drive impedance. This is often used to tune the drive impedance of the antenna to 50 ohms resistive. In this case a reactance compensating component in series with the tower feed can be used to match the antenna to the transmission line. While this is a common practice, it rarely yields the best input audio bandwidth performance.

The currents flowing in the unipole skirt wires are generally highest below a tuning stub resulting in reduced currents in the portion of the skirt and tower above the tuning stubs. Placing a tuning stub a significant distance below the top support brackets can result in concentrating the currents in the lower portion of the skirt to such a degree that the efficiency of the antenna is reduced. It is advisable to use the full skirt without a tuning stub and allow the ATU to match the line when possible. If doing so results in an impractical antenna impedance for the ATU to match, then using a tuning stub may be reasonable. However, the tuning stub should be kept as high on the skirt as practical.

The height of the folded unipole skirt on the tower is another key design choice. The folded unipole skirt can span the full height of the tower, or can span only a lower portion of the tower. Folded unipoles are often put on the lower portion of tall towers in order to leave the upper, unskirted portion open for FM antennas or other antennas. When this is done, the folded unipole skirt should not stop at a point too low on the tower, yielding a skirt that is electrically short. Keeping the folded unipole skirt at least 60 degrees tall is advisable. A more ideal choice is a skirt that is roughly between 80 and 110 degrees tall. Note that the overall tower height is used by the FCC for determining the AM tower height, and it must not exceed 5/8 wavelength in order to prevent loss of ground wave signal strength. A MOM wire model of the tower and folded unipole skirt is recommended to evaluate the expected bandwidth performance for a folded unipole feed of a tower. Folded unipoles are discussed further in previously published work.[4]

Face Width (Including Guyed or Self-Supported)

The face width of the tower is generally driven by structural requirements rather than electrical ones. However it is worth noting that the tower cross section does have an influence on the bandwidth performance of an AM antenna. Wider face width results in a lower reactance component of the drive impedance resulting in a lower Q, wider bandwidth performance. This is most notable when using self-supporting towers with wide base cross sections.

Alternate Configurations (KinStar)

One more recent special case configuration of AM antenna is the KinStar,[5] which is a low-profile design intended for use in areas where traditional towers are not permitted due to height restrictions. Areas near airports or with local zoning restrictions may still be possible for consideration as an AM broadcast site when using the KinStar. The KinStar antenna is a 1/12-wave tall antenna that essentially yields the same horizontal radiation performance as a 1/4-wave tall vertical mast on the same frequency. Full details of the KinStar design are outside the scope of this chapter.

TRANSMISSION LINE IMPACT

Characteristic Impedance (Semi-flexible/Rigid and Quasi-coaxial Open Wire)

Modern AM/MW transmitters are typically designed to operate with coaxial transmission line having a characteristic impedance of 50 Ω. Semi-flexible coaxial transmission lines are available in sizes up to 2–1/4 inches using foam dielectric, which are convenient in that they don't require pressurization. These lines can comfortably operate with carrier power up to roughly 30 kW, although they have been operated at 50 kW carrier power when operated with a 50 + j0 line match and a fast reacting transmitter reflected power protection capability. For transmitter carrier powers of 50 kw and higher, semi-flexible and rigid air coaxial lines are available with an outer conductor diameter of up to 12 inches. For transmitter carrier power levels of 400 kW or higher, quasi-coaxial open wire transmission lines rather than pressurized air lines are often used due to their lower cost and ease of maintenance. The characteristic impedance of open wire transmission lines tend to be higher (typically 120 Ω to 300 Ω), which has the advantages of reducing the RF currents in the ATU components and helping enhance bandwidth performance as well. If used with a modern transmitter, open wire lines require an impedance matching "L" network to transform from the impedance of the line to the 50 Ω transmitter output port.

Implications of Operating With a Mismatched Transmission Line

It's important for the efficient operation of a modern AM broadcast station to operate with the transmission line(s) matched to a voltage standing wave ratio (VSWR) less than 1.05:1 and preferably with a 1:1 match. A mismatched line will cause reflected power that returns to the transmitter rather than being radiated by the antenna. Even if a matching network is used on the transmitter end of the line, whether a built-in fine matcher in the transmitter or a fully separate secondary matcher outside the transmitter, a mismatched line is not desirable. One reason is that the reflected power will result in standing waves in the transmission line that can stress the voltage limits of the line. Even if the line is rated adequately to handle the standing wave stresses, the line will have less voltage safety margin available to draw upon when a voltage spike occurs, such as during a lightning event. An arc in a transmission line can cause costly damage that can also take a great deal of time to locate and repair.

A common exception to maintaining the line VSWR < 1.05:1 is the case where a line is feeding a very low-power tower in a directional array. These towers can be very difficult to achieve a good match on carrier while obtaining the required pattern parameters. The VSWR may exceed 2:1 before the parameters are achieved to yield the licensed pattern.

MATCHING NETWORK CONFIGURATIONS

Although resistive elements can be used to improve an impedance match, they are inefficient and not suitable for most broadcast applications. The focus of this discussion will be the design of matching networks using passive components: inductors and capacitors. Impedances will be assumed to be complex, and it will be assumed that a complex impedance Z_A consists of the real and imaginary parts R_A and X_A, respectively. It is assumed that the reader is comfortable with operations involving complex numbers. The values for inductors or capacitors are assumed to be pure imaginary numbers for the design. In reality these elements have an associated loss that serves to temper overly optimistic estimates of filter isolation and have the greatest effect when dealing with very low drive resistances. Their impact can be assessed in a post design analysis. Of course, any network branch that has a capacitive reactance can be implemented with a combination capacitor and inductor and similarly, inductive branches can contain capacitors when used for sideband impedance shaping.

There are a couple of results from basic circuit theory that are useful for both analysis and design and are presented here. The first is the parallel equivalent impedance of a complex series impedance $Zs = Rs + jXs$, which will be denoted by $Zp = Rp + jXp$, where

$$Rp = \frac{R_S^2 + X_S^2}{Rs} \tag{1}$$

$$Xp = \frac{(R_S^2 + X_S^2)}{Xs} \tag{2}$$

The reverse transformation is

$$Rs = \frac{R_P X_P^2}{R_P^2 + X_P^2} \tag{3}$$

$$Xs = \frac{X_P R_P^2}{R_P^2 + X_P^2} \tag{4}$$

The second useful result from circuit theory is the equation for the phase shift in the current due to a shunt element jXp across a complex load Z_L. The ratio of the current across the shunt impedance is

$$\frac{I_{OUT}}{I_{IN}} = \frac{1}{1 + \dfrac{Z_L}{jX_P}} \tag{5}$$

and the phase shift in the current would be

$$Angle\left\{\frac{I_{OUT}}{I_{IN}}\right\} = \tan^{-1}\left(\frac{(R_L X_P)}{\left[X_P(X_L + X_P)\right]}\right) \tag{6}$$

Some of the displays in the following sections will show impedances on a Smith chart, which is an extremely useful tool for circuit analysis. Space does not permit discussing the implementation and use of this tool, and the reader is referred to outside sources for a discussion of Smith charts.

L Network

The L network is the simplest matching network, consisting of two branches—a series and a shunt branch—as shown in Figure 7.9-1.

The typical usage is to transform a specified load resistance into another purely real value using only reactive elements. The equations for determining the branch values[6] are:

$$X_{SERIES} = \pm\sqrt{R_{IN}(R_L - R_{IN})} \quad \text{if } R_L > R_{IN} \text{ (Shunt across Load)} \tag{7A}$$

$$\text{And } X_{SERIES} = \pm\sqrt{R_L(R_{IN} - R_L)} \quad \text{if } R_{IN} > R_L \text{ (Shunt across Input)} \tag{7B}$$

$$X_{SHUNT} = \mp\frac{R_L\sqrt{R_{IN}}}{\sqrt{R_L - R_{IN}}} \quad \text{if } R_L > R_{IN} \text{ (Shunt across Load)} \tag{8A}$$

$$\text{And } X_{SHUNT} = \mp\frac{R_{IN}\sqrt{R_L}}{\sqrt{R_{IN} - R_L}} \quad \text{if } R_{IN} > R_L \text{ (Shunt across Input)} \tag{8B}$$

with a current phase shift β across the element of

$$\beta = \mp\cos^{-1}\left(\sqrt{\frac{R_{IN}}{R_L}}\right) \quad \text{If } R_L > R_{IN} \text{ (Shunt across Load)} \tag{9A}$$

$$\text{And } \beta = \mp\cos^{-1}\left(\sqrt{\frac{R_L}{R_{IN}}}\right) \quad \text{if } R_{IN} > R_L \text{ (Shunt across Input)} \tag{9B}$$

Note that the branch reactances are of opposite signs with the sign of the phase shift corresponding to that of the shunt branch. Also, the phase shift is directly related to the resistance transformation, that is, there is not independent control of that parameter. A capacitive shunt branch yields a phase delay, and an inductive shunt branch yields a phase advance. From a practical standpoint, it should be observed that it is more intuitive to adjust the network when transforming a higher resistance to a lower one, given that the shunt reactance directly corresponds to the input resistance and the series reactance to zeroing the input reactance. For a given set of resistances, it is possible to also match a range of complex impedances using the parallel-series or series-parallel transformations presented in Equations (1) through (4) to alter the load or required input impedance into a form that matches the orientation of the L network. The range of complex impedances for which a match can be achieved is limited by the values that cause the operand in the square root operation of Equations (7) and (8) to become negative. Once designed, the phase shift of the network can be determined by Equation (6) presented earlier. The L network also is limited by becoming a degenerate case of a zero series element and infinite shunt element when the load and input resistance are the same value. This prevents its use for simple phase rotation networks in which case the purpose is to only introduce a phase rotation without transforming the impedance level. The solution is to use a network with one more branch, which provides independent control of both impedance and phase.

T Network

The most straightforward network providing independent control of the phase and impedance is the T network, as shown in Figure 7.9-2.

The topology of the network makes it trivial to incorporate the matching of a complex load impedance to a complex input impedance by simply adding the reactive component of the complex impedance to the appropriate branch of the network. The most common application is to match a complex load impedance Z_L, such as that of a tower impedance, to a real input resistance R_{IN}, such as the characteristic impedance of the transmission line. The network connections are shown in Figure 7.9-2, with the corresponding reactances a function of the input and load resistance, load reactance and desired current phase shift (β) through

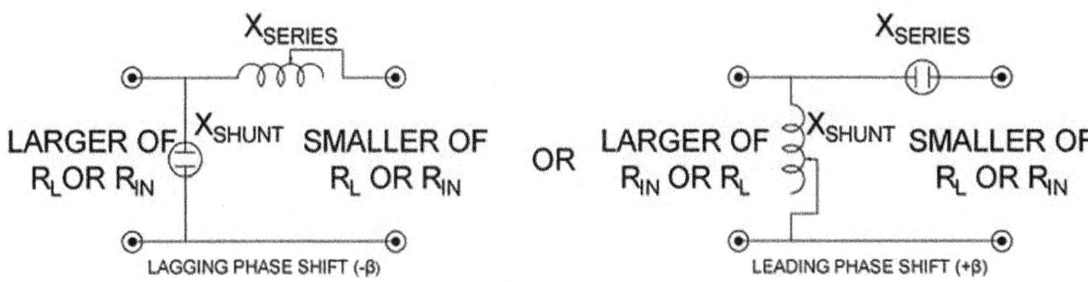

FIGURE 7.9-1 Examples of L networks

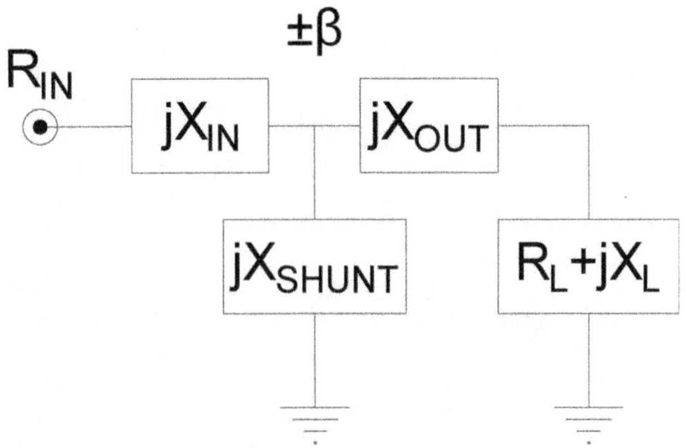

FIGURE 7.9-2 T network

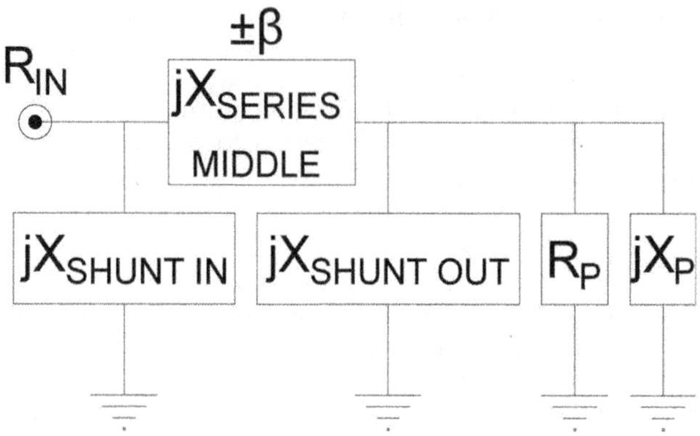

FIGURE 7.9-3 Basic pi network

the network. The branch reactances to simultaneously match these three parameters can be derived from Kirchhoff's voltage law for this two-port network, defining the ratio of input and output currents as the respective impedance values, and equating the real and imaginary components. The resulting equations[6] are:

$$X_{IN} = -\frac{\sqrt{R_{IN}R_L}}{\sin(\beta)} + \frac{R_{IN}}{\tan(\beta)} \quad (10)$$

$$X_{OUT} = -\frac{\sqrt{R_{IN}R_L}}{\sin(\beta)} + \frac{R_L}{\tan(\beta)} - X_L \quad (11)$$

$$X_{SHUNT} = \frac{\sqrt{R_{IN}R_L}}{\sin(\beta)} \quad (12)$$

Note that a negative phase shift will result in a capacitive shunt branch and typically inductive input and output branches. Similarly, a positive phase shift will result in an inductive shunt branch with typically capacitive input and output branches. In practice, each of the branches may be implemented by combinations of inductors and capacitors to achieve adjustment range, increased component ratings, or to improve the bandwidth. The combination of components in each branch will still need to produce the reactance values based on the previous equations at the carrier frequency.

Pi Network

The basic pi network as shown in Figure 7.9-3 is the dual to the T network and similarly has three parameters that can be used to simultaneously match a complex load to a real input impedance while incorporating a desired phase shift.

In dealing with a pi network, it is necessary to transform the series load impedance to its parallel equivalent, R_p and X_p as defined in Equations (1) and (2). This also makes the field adjustment of a pi network less intuitive and more dependent on careful settings guided by calculation. Adjustments to fine tune one parameter (input resistance, reactance or phase shift) will usually require changes to multiple branches. A modified phase parameter is required to compensate for the phase introduced by the impedance transformation given by the standard pi-network equations. The adjusted phase shift to be used in the equations to implement the desired network phase shift β is defined by

$$\beta' = \beta + \arg\left(\frac{1}{1 - j\frac{R_P}{X_P}}\right) \quad (13)$$

where the "arg" function yields the phase of the complex ratio shown.

If $X_L = 0$, then $\beta' = \beta$.

The resulting branch equations[6] are:

$$X_{SHUNT\ IN} = -\frac{R_{IN}R_P\sin(\beta')}{R_P\cos(\beta') - \sqrt{R_{IN}R_P}} \quad (14)$$

$$X_{SERIES\ MIDDLE} = -\sqrt{R_{IN}R_P}\sin(\beta') \quad (15)$$

If $X_L = 0$

$$X_{SHUNT\ OUT} = -\frac{R_{IN}R_P\sin(\beta')}{R_{IN}\cos(\beta') - \sqrt{R_{IN}R_P}} \quad (16A)$$

If $X_L \neq 0$

$$X_{SHUNT\ OUT} = -\frac{1}{\frac{R_{IN}\cos(\beta') - \sqrt{R_{IN}R_P}}{R_{IN}R_P\sin(\beta')} + \frac{1}{X_P}} \quad (16B)$$

As with the T network, the signs of the reactances will determine if a branch is inductive or capacitive and the signs of the shunt elements will correspond to the sign of the overall phase shift through the network.

Broadbanding

For both T and pi networks, the reactance values derived in the preceding sections provide a match on carrier. For good bandwidth, the antenna system must also provide the transmitter with an impedance characteristic that does not deviate from the carrier match excessively at the upper and lower sideband frequencies. These off-carrier sidebands can be modified for improved bandwidth by replacing the single inductive or capacitive element for a given branch reactance by an inductor + capacitor pair that are selected to produce the same combined value on carrier as the original element. A properly chosen combination will serve to "flatten the load" and provide a more uniform input resistance and reactance over a wider bandwidth. For T-networks, the most common location for this broadbanding or slope-correction is in the input branch, but there are cases where modifying the output branch or even the shunt branch produces the best results. For T networks, the broadbanding elements are typically series combinations, and for pi networks, they are applied in parallel with the input or output branch. The use of a Smith chart or the numerical perturbation of a network model can often provide useful insight as to the best location for bandwidth improvement of a network with a given load. Such insight can also indicate when the reverse is true, that an equivalent combination of elements produces poorer bandwidth than a single element in a particular branch. For example, this may arise in situations where a network design uses a fixed capacitor in series with an adjustable inductor as a cost-saving implementation for a negative reactance. There are instances where using a single variable vacuum capacitor in that location will noticeably improve the network bandwidth. Excessive strap or tubing in a branch of a physically large network can also have this effect and consideration should be given when planning the physical connection of a network as well.

A detailed presentation on techniques for enhancing bandwidth performance is beyond the scope of this chapter, but references are presented for further study.[7,8]

Cascaded Networks

Network designs often incorporate adjustment range to be able to accommodate a range of input or output impedances or phase shifts to minimize the need for any component changes during the adjustment of the networks in the field. To implement this tuning range, series inductor capacitor pairs are often used and, as in the preceding example, this can lead to a bandwidth degradation. This is especially true for more extreme phase shifts. In general, a design phase shift for a T network between the values of −60° and −120° (also between 60° and 120°) tends to have broader band performance than networks with phase shifts outside those values due to the branch reactances being required to span a larger range. A similar problem can occur when a large impedance step is required, such as when an input resistance differs from the load resistance by a multiplicative factor of 10 or more. In both of these cases, using cascaded networks can be beneficial in producing superior bandwidth compared with that for a single network. When applied to extreme phase shifts, the phase shift is broken into segments that fall within the preceding ranges. When applied to large steps in resistance transformations, intermediate values between the two resistances are used.

Analog Versus Digital (HD or DRM) Design Consideration

For nondirectional operation, the impedance bandwidth is the primary concern. For directional arrays, the pattern bandwidth is also of concern to assure optimum audio fidelity at the receiver. VSWR is the typical measure of impedance bandwidth, comparing the sideband impedances to that on carrier. At the input of a matching network, the design impedance is typically 50 Ω, which is the characteristic impedance of most transmission lines used in AM antenna systems. For analog broadcast, VSWRs of less than 1.5:1 at the ±10 KHz sidebands are desired to avoid transmitter VSWR shut-backs. For digital broadcast (HD-IBOC or DRM), the target design levels are recommended to be lower, as in Table 7.9-1, with the additional constraint that the impedance sweep yields a locus on a Smith chart with Hermitian symmetry at the transmitter finals, that is, at the input to the harmonic filter section of the transmitter. Ideal Hermitian symmetry has equal real (resistive) values, and equal but opposite signed imaginary (reactive) values at the sidebands equally distant from carrier. This results in an impedance locus either "cusp left" or "cusp right" on a Smith chart at the transmitter finals. Note that this assumes that at least the phase shift characteristics or network model for the output stages of the transmitter be known and incorporated in the design.

For directional arrays, there is the additional constraint of pattern bandwidth. Whereas the impedance bandwidth is a measure of the input audio bandwidth of the antenna system, the pattern bandwidth is indicative of the range of variation in the field parameters versus sideband frequency based on the overall network design topology. Ideally the array's radiated sideband fields will be as uniform as possible relative to the carrier radiated field, both in magnitude and phase, across the station's broadcast bandwidth. The pattern bandwidth is of primary concern for digital broadcast, and the design guidelines are summarized in Table 7.9-1.[8]

Techniques for Matching Very Low Impedance or Very High Impedance Towers

A particular application of cascaded networks is with very low or very high impedance towers. Quite often a shunt element can be used to raise a low drive resistance to a value more within the range of a standard network design. The shunt element in parallel with the reactive component of the load serves to raise the

TABLE 7.9-1
Bandwidth Design Guidelines

Quantity	Frequency (KHz)	Design Target for Analog	Design Target for Digital Operation
Impedance Magnitude	Carrier	Matched VSWR = 1.0	Matched VSWR = 1.0
	±4.5 or 5 KHz	VSWR < 1.2:1	VSWR < 1.05:1
	±9 or 10 KHz	VSWR < 1.5:1	VSWR < 1.2:1
	±13.5 or 15 KHz	NA	VSWR < 1.4:1
Impedance Rotation and Symmetry	±13.5 or 15 KHz	NA	Hermitian Symmetric, Sideband normalized to complex conjugate < 1.035:1 at ±4.5 or 5 KHz
Field Amplitude (Directional)	±13.5 or 15 KHz		±2 dB with respect to carrier
Field Phase (Directional)	±13.5 or 15 KHz		Group Delay < ±5μs ⇨ Phase Slope < ±9°/5KHz

resistance. For loads with low reactances as well, it may be required to first use a series element to provide a more desirable reactance. The sign of the shunt reactance is opposite that of the load reactance to have the desired effect. The shunt element will introduce a phase shift that needs to be considered in the overall system design, either as a contribution to the total phase shift for a tower from the reference bus in a directional array or to achieve the proper rotation for a design compatible with digital broadcast requirements.

For very high impedance towers, stray capacitance introduced across the tower base by components such as lighting chokes, isolation coils or isocouplers can dramatically affect the drive impedance being presented to a matching network. It is not practical to incorporate a wide adjustment range in the network to compensate for this effect and trying to do so can lead to the reverse slope-correction problems discussed earlier. For these cases, cascading a shunt element between the matching network and the load to transform a high impedance to a lower impedance can greatly stabilize the network design. If the shunt element has the same sign as the load reactance, but is smaller in magnitude, it can be used to lower the drive resistance to a reasonable value for the matching network. An opposite sign shunt reactance can either raise the effective load resistance or if greater in magnitude, lower the resistance and reverse the sign of the effective load reactance. There are applications where both of these techniques can be used to an advantage, and once again, it requires looking at what is required for the overall system design, as well as the effect on bandwidth to determine the best course of action. Those with experience using Method of Moments calculations will recognize that there is a rapid variation in impedance as a function of frequency for high impedance towers. The use of a shunt impedance greatly reduces the drive impedance variation required in the matching network. In addition to improving the design, the shunt elements provide an excellent method of compensating for local stray capacitance on site during tuning, or to eliminate component changes required in a matching network with the addition of future services on a tower that require auxiliary elements, such as lighting chokes or isocouplers. In directional arrays, when isolation coils are required to isolate the sampling lines feeding current sampling loops on electrically tall towers, the isolation coil can be incorporated as a shunt tuning element and serve dual purposes.

DIRECTIONAL ANTENNA PHASING AND MATCHING SYSTEM DESIGN

Power and Phase Budgets

As discussed in other chapters, directional arrays require each element to be driven to produce a particular field contribution, which is expressed as a relative field and phase. Method of Moments modeling provides a powerful tool to derive not only the drive impedances for each element in the directional mode, but also the relative power division and current or voltage phase that must be applied to each tower to produce the required pattern. The common point bus, which feeds all towers, provides a common voltage for the feed to each of the towers. Each tower presents an impedance to the common bus, some of which may have both resistive and reactive components. Given that these impedances are in parallel, the power delivered to each tower is governed by the relation

$$P_i = \frac{V_{BUS}^2}{Rp_i} \tag{17}$$

where Rp_i is given by equation (1). Because the bus voltage is common to all towers, the expression relating the power division and parallel resistances of two towers can be written as

$$P_1R_1 = P_2R_2 \tag{18}$$

There is a degree of freedom in selecting the reference resistance, which is normally taken to be that of the high power tower. A nominal 50 Ω is often chosen, but a larger value can be chosen for arrays with more towers in order to raise the resistance being fed at the common point. Towers with a complex impedance at the bus also introduce a relative phase shift. The phase contribution is normally small and can be calculated by comparing the ratio of current being fed to a particular tower with that being supplied to a reference tower with a real equivalent resistance. Most towers have a positive power flow with a positive resistive component of the drive impedance. It is possible to have a tower with a negative resistive component, which indicates power being contributed from adjacent towers to meet the boundary conditions on a particular tower's base voltage and current set by the field parameters and array geometry. Negative towers are more common for inline, closely spaced (electrical spacing < 70°) arrays resulting in high mutual coupling between tower elements. The previously discussed equations still apply to the negative resistance cases with the absolute value of the resistance used in the network design equations, and the resistance with its negative sign included used in the preceding power division equations.

The common point bus also serves as the phase reference for the towers. Each tower has a drive current phase φ_i subject to the following relations:

$$\sum_{\substack{j=1 \\ Tower\ i}}^{M\ networks} \beta_j - \sum_{\substack{j=1 \\ Reference\ Tower}}^{M\ networks} \beta_j = \varphi_i \quad Positive\ Power\ Towers \tag{19}$$

or

$$\sum_{\substack{j=1 \\ Tower\ i}}^{M\ networks} \beta_j - \sum_{\substack{j=1 \\ Reference\ Tower}}^{M\ networks} \beta_j = \varphi_i = N * 180^0 \quad Negative\ Power\ Towers \tag{20}$$

The summations add the current phase shifts introduced by all elements in the chain for a particular tower including power dividing networks, transmission line, tower impedance matching networks, prematching elements, and any parasitic elements not directly included in the method of moment model calculations. In expression (20) N is an odd integer multiplier, which ensures that the phase of a signal leaving the reference tower and returning through a negative power tower arrives at the common point with the proper phase (an odd multiple of 180°) with the contributions of the other towers.

For some arrays, the absolute phase of the reference tower has a direct impact on impedance and pattern bandwidth of the system. Not all arrays are sensitive to the relative phase of the reference tower and for the arrays with a single negative tower, the reference phase can be chosen such that the phase shift from the common point bus to the negative tower is an odd multiple of 180°. For this case, switching the ground tap position on a tank power divider can compensate the transition from a positive to negative resistance on the tower without a readjustment of the branches of other networks in the feed system to that tower.

Comparison of Various Power Divider Designs

Ohm's Law

Ohm's law or resistive power dividers are a bit of a misnomer in that they are no more lossy than the other power dividing elements. The term simply implies that the impedance presented to the common bus is designed to be real, having no reactive component. The T-network is a typical resistive power divider, where the shunt branch with its direct impact on input resistance tends to have a high degree of control over the power division to the associated tower. Adjusting the input branch affects the reactance presented to the bus, which normally correlates higher with the phase of the associated tower. However depending on the values of the impedances relative to the other impedances presented at the common point, this is not a strict dependence and adjustments of either branch can affect both parameters. The degree of mutual coupling between array elements is also a factor, making it likely that each element in a directional array responds differently when adjusting a particular branch.

Shunt and Tank

Another design approach is to use parallel shunt elements directly at the bus to divide the power to each tower. This approach is especially useful for towers with power levels considerably lower than that of other towers in the array. Although stacked capacitive elements could be used, the most straightforward approach is to use an inductive element where the entire coil is across the bus for each tower and the coil is tapped for appropriate parallel resistance, that is, power level for that tower.[9] Because the single element controls primarily the power to the tower, additional phase shifting components are required between each shunt element and the respective transmission line feeding that particular tower. The phase shifting networks can take the form of full T-networks, or can be series elements that affect the phase by changing the complex impedance presented to the bus as it is translated through the shunt element. Figure 7.9-4 presents a typical phasing system topology showing the power division, calculated equivalent input resistance values, and phase budget.

Note that T-networks are used as power dividing networks (tower 1 and 4), to implement phase shift (tower 3), and for the antenna matching units and common point match. A variable shunt element is used as the power divider for the low power tower (3). Fixed shunt elements are used as a discrete phase shift element (tower 4), and as prematching elements for the low impedance towers (3 and 4).

In applying equations (19) and (20) for the phase budget of this example, the high power tower, which is tower 2, should be taken as the reference tower. The sum of the phases from the common point bus to the tower is −156.9°. Similar sums are −152.2° and −51.7° for positive power towers 1 and 4 respectively. Applying equation (19) arrives at the tower drive current phases of −152.2° − (−156.9°) = 4.7° and

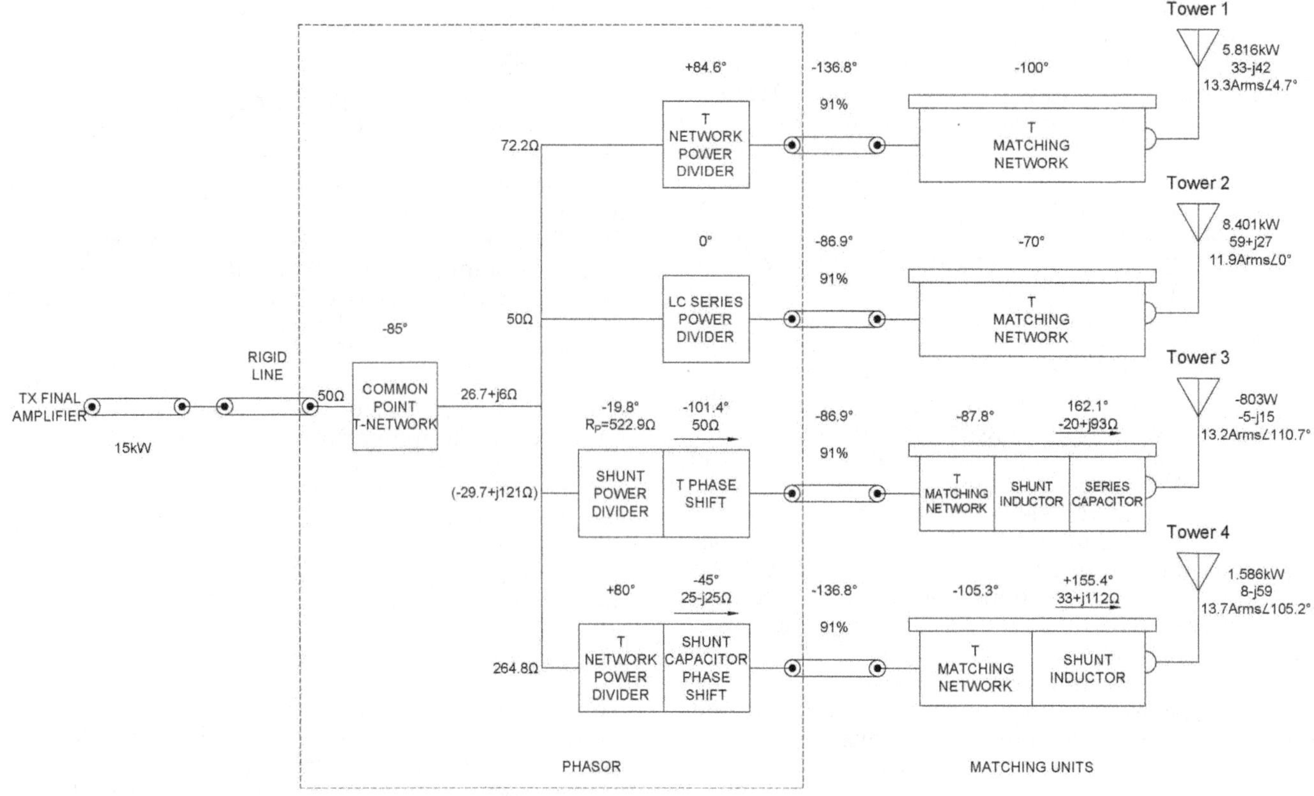

FIGURE 7.9-4 Typical phasing and coupling system for directional operation

−51.7° − (−156.9°) = 105.2° respectively. For the negative tower 3, the sum of the phase shifts from the common point bus is −133.8° and applying equation (20), the result is −133.8° + (−156.9°) + 110.7° = −180°, which is an odd multiple of 180° as required for the negative tower.

For very low power towers (which have a very high equivalent parallel resistance), the shunt inductor can be parallel resonated via a fixed capacitor to form a tank circuit in order to provide a very high impedance for the tower to the common point bus. The closer the tower is to being truly a parasitic element (having no power driving it), the greater its equivalent resistance. These towers typically have limited control from the phasor and are most affected by adjustments at the tuning unit. Towers with this characteristic can easily switch from positive to negative power flow, presenting an additional challenge to the array tuner and network designer. The tank circuit can be of benefit here, providing a relatively quick way to introduce a 180° phase change in a particular tower's branch by moving the location of the ground tap relative to the lower tap of the resonating capacitor for the tank circuit.

In cases where a tower is solidly negative, with the percentage power high enough that it is not likely to switch power flow direction, the tower can be terminated in a resistive load. Figure 7.9-5 presents an example of an array with such a negative tower.

The dashed portion indicates the functional blocks that are in place for operation when the negative tower is connected to the common point bus. They are absent when the negative tower is terminated in the resistive load. In this instance the pattern and impedance bandwidths of the system were both improved by terminating the negative tower in a resistive load and increasing the transmitter power to compensate for the power dissipated in the load. Both the current at the common point input and the current into the resistive load are typically required to be monitored during operation.

Complex Impedance and Quadrature

Other schemes of power division exist. Quadrature power dividers are a particularly elegant method that minimizes the component count, which can be beneficial for high power arrays. For this type of design, two resistive L networks are chosen for the power division such that the shunt elements form equal and opposite reactances that parallel resonate and as a result, can be removed from the circuit. The remaining series elements provide both the power division and a −90° phase shift between the two towers. Note that this arrangement provides a minimal footprint of components to accomplish this task, but at the cost of independent control of the power division and phase. Any additional phase adjustments are required to be made using other networks in the chain forming the feed for the particular tower. Given that the parallel resistances of the two towers can be designated R_1 and R_2 as in

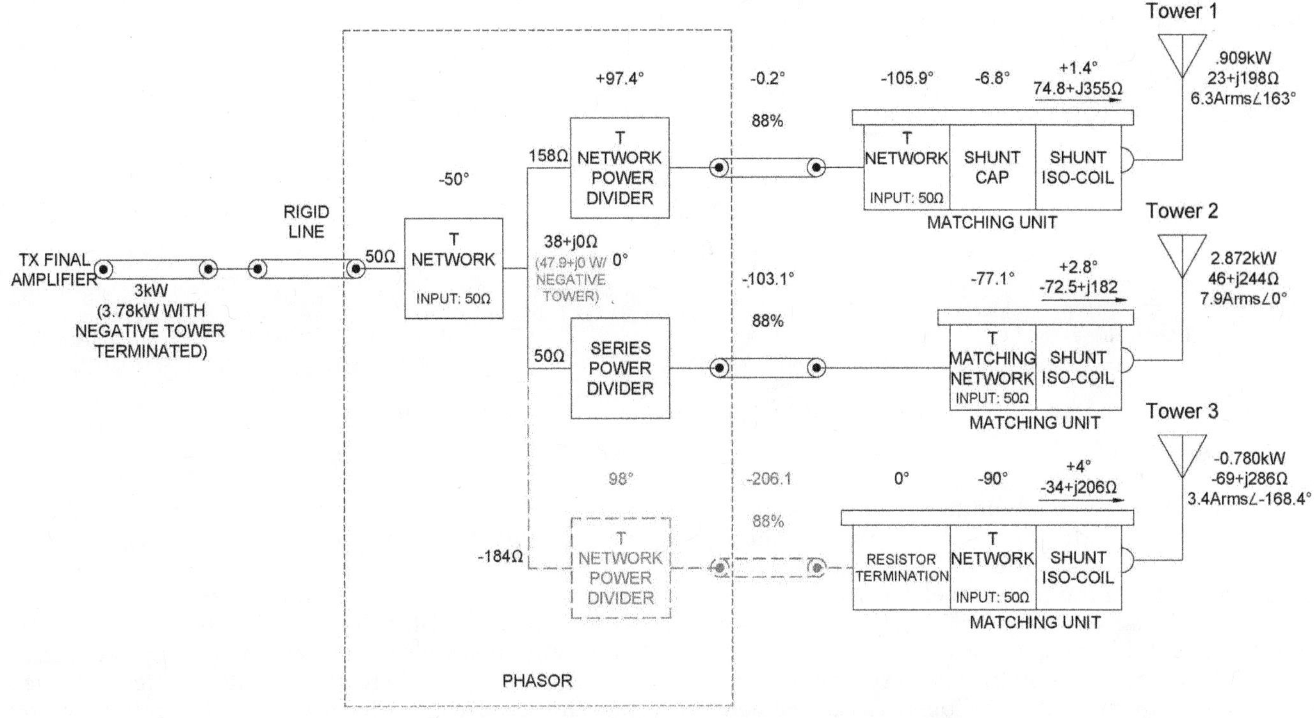

FIGURE 7.9-5 Array containing a negative tower

Figure 7.9-6, the equivalent input resistance combining these two parallel resistances is

$$R_{IN} = \frac{R_1 R_2}{R_1 + R_2} \tag{21}$$

Combining with equation (18) results in

$$R_1 = \left(\frac{P_2}{P_1} + 1\right) R_{IN} \tag{22}$$

Setting the combined input impedance R_{IN} to be the characteristic line impedance and using (21) and (22) to determine the values of R_1 and R_2 to be applied to in the L network equations (7B or 8B), the reactances for the respective series and shunt branches can be determined. The terminating R_L in equations (7 and 8) are the characteristic impedance of the transmission line and the shunt elements are across the input that is larger than the load impedance by virtue of equation (22). Opposite signs should be selected for the two series reactances. Because the combined equivalent input resistance R_{IN} is equal to the characteristic line impedance, when the substitutions for R_1, R_2, and R_L are made using equations (21) and (7 or 8B), the result is

$$1 = (\cos \beta_1)^2 + (\cos \beta_2)^2 \tag{23}$$

which is satisfied when $\cos \beta_2 \cos \beta_2 = \sin \beta_1, \sin \beta_1$, or in particular, for the quadrature condition when

$$\beta_1 = \beta_2 - 90° \tag{24}$$

With an opposite sign chosen for the two L networks, the shunt branches will be equal to one another and of opposite sign, effectively in parallel resonance and can be eliminated from the network's physical implementation.

Complex impedances can also be used for effective power division. These are most advantageous in high power designs where full T-networks and large banks of capacitors would both raise the cost of the equipment and result in high-voltage nodes that may experience undesired corona effects. Typically an L network topology is used for one or more of the towers, and it presents a complex impedance at the common point bus. In this regard it is similar to the shunt power divider except that the magnitude of the impedance is such that it has a greater effect on the overall bus impedance, and the current phase shift due to that impedance varies much more rapidly than for shunt or resistive power dividers. There also tends to be a greater coupling between the ratio and phase for a tower and the parameters of the other towers may also be affected. The loss of tuning independence and requirement for a more iterative tuning approach guided by the network model calculations are the tradeoffs for the lower component voltage stresses and reduced component count.

Designing for Impedance Bandwidth

For a directional system, once the nominal phase and power budgets are determined, the choice of network component values determines the sideband performance. The techniques discussed earlier for the broadbanding of single networks can be applied to the individual power dividing and matching networks in the system. Often there are a few network branch

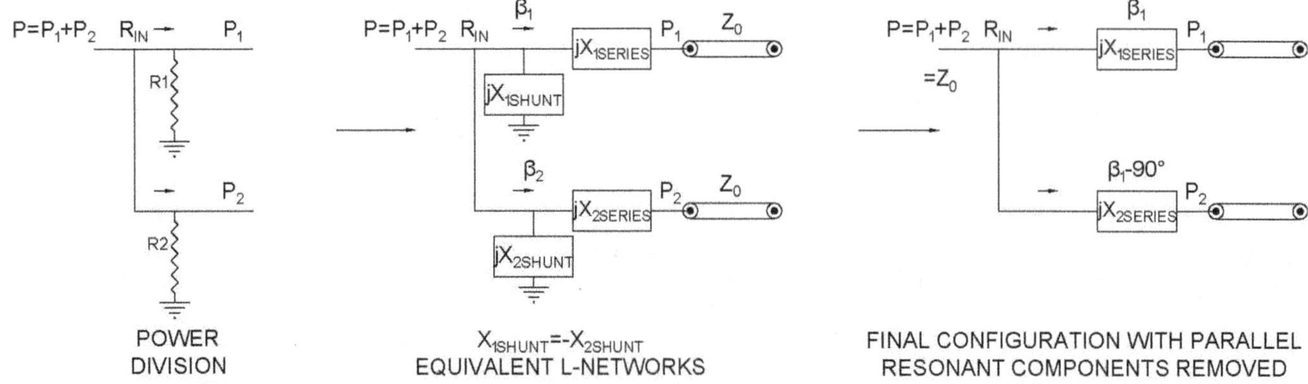

FIGURE 7.9-6 Power divider equivalent circuits

locations where the bandwidth can be improved by implementing an inductor/capacitor pair instead of a single element. Focusing on the higher power towers often proves to be the most beneficial. Once a potential broadband solution has been devised, a full nodal simulation should be used to verify the impact that the sideband changes have on the entire system. In arrays with significant mutual coupling between elements, this will often be an iterative process.

Designing for Pattern Bandwidth

For directional operation, pattern bandwidth is also of concern for analog or digital broadcast. A useful technique is to compare the normalized sideband patterns, where each sideband frequency is assumed to be transmitting the same power as the pattern on carrier. Doing so eliminates the consideration of the effects of the input impedance match on the pattern shape and allows the input match and phasing network topology to be decoupled to some degree and for them to be treated as separate problems. Of course, the impedance bandwidth of the phasing networks does affect the input match, and once the network topology is determined that provides the best pattern bandwidth, the common point load will be defined and possible broadbanding or rotation requirements for the input match can be addressed. Ideally, for a good pattern bandwidth design, the main lobe, side lobes, and nulls for the normalized sideband patterns will be very similar to those of the pattern on carrier, and will be more uniform the closer the sideband frequency is to the carrier frequency.

A key parameter affecting the pattern bandwidth is the phase shift from the common point bus to the high power tower of the array, which for design and tuning purposes should also be the reference tower of the array.

Different Parameters for the Same Pattern

One of the themes running throughout this chapter is the necessity of viewing the transmission chain as a system in order to produce the best performance. Traditionally there has been a tendency to consider the transmitter, transmission line routing, RF networks, array geometry, and individual tower structures as being independent. While each of these antenna system components have their own particular constraints, there is more interdependence than is often assumed and in some cases, the solution to a better system requires modifying the characteristics of one to benefit another and the system as a whole. For a case in point, take an inline, end-fire array that has been designed in such a way that it has a negative tower to produce the rear lobe. The radiation pattern of many inline and parallelogram arrays can be considered to be a derivative of a pattern multiplication design, whether or not they were actually designed in this manner. For these cases, the field parameters can be reverted to their constituent pairs, which can then be used to generate alternative parameter sets that produce the same pattern, but have different power division and drive impedances, which can have a substantial impact on overall performance.[10]

This is a very useful technique for patterns in which one or more towers have a drive impedance with an R component < 10 ohms or slightly negative impedance that could easily switch from positive to negative during the field adjustment. Such a tower must remain connected to the bus when it transitions to a positive resistance, but could be terminated via a dummy load if the negative power being fed back to the bus is at least 10% of the total input power. The negative resistance indicates that the tower is coupling energy from the array's near field and returning it to the bus. Removing this recycled energy from the bus can dramatically improve system performance of the array, especially where the pattern is characterized by a high RSS/RMS (Root Sum Square/Root Mean Square) ratio, at the cost of slightly more transmitter power. Table 7.9-2 presents an example of an inline array where inverting the parameters results in tower 3 becoming solidly negative in the main lobe, allowing it to be resistively terminated and improve the bandwidth. The center tower is the high power tower, with tower 3 in the main lobe and tower 1 in the back lobe. Parameter set A was the

TABLE 7.9-2
Parameter Variations

	Parameter Set A (Original)	Parameter Set B (Preferred)	Parameter Set C	Parameter Set D
Tower 1 Power (W)	−440	909	205	−356
Tower 2 Power (W)	2658	2872	2998	3775
Tower 3 Power (W)	783	−780	−203	−418
Tower 1 Drive (Ω)	−28 + j165	24 + j208	8 + j215	−16 + j170
Tower 2 Drive (Ω)	44 + j250	48 + j255	45 + j240	59 + j260
Tower 3 Drive (Ω)	24 + j268	−72 + j300	−13 + j310	−19 + j270

original set, with the rear negative tower accounting for less than 15% of the total radiated power. For parameter set B, the end tower in the primary lobe is solidly negative, with a comfortable margin at 30% of the radiated power, which provides confidence that the tower can be terminated in a resistive load.

When a set of array parameters can be reduced to its constituent pairs, each of the design pairs can be written as

$$1 + Ae^{k_A \cdot r + \alpha} \tag{25}$$

where A and α are the amplitude and phase of the complex root, $\mathbf{k}_A$ is a vector relating to the physical geometry of the pair, and $\mathbf{r}$ is the vector to the far field observation point. For a given geometry, a set of simultaneous equations can be developed between the constituent root pairs and the field parameters by multiplying out the terms of equation (25) for each pair, collecting the terms with like *k*-vectors that correspond to the tower locations, and equating the root coefficients expressions with the field parameters. The solutions to those equations are the root pairs. The alternative parameter sets can be derived by replacing the magnitude of each root pair A with 1/A. Two design pairs were used in the preceding parameter set, resulting in the four parameter set possibilities of (A, B), (A, 1/B), (1/A, B), and (1/A, 1/B). The number of alternative parameter sets increases as 2^N where N is the number of root pairs. This technique does not apply to arrays that cannot be broken down into a pair based geometry. The pair design approach is often used for inline or parallelogram arrays.

Treating Negative Power Towers

As seen earlier, when the root solutions can be found, the resulting design can have a much improved bandwidth, either by providing more suitable drive point impedances or by presenting the opportunity to terminate a negative resistance tower with a resistive load. Higher resistance towers are more amenable to diplexing, given that the sideband reactances introduced by the filters has less impact on the towers bandwidth because a given reactance is a smaller percentage of the drive resistance. For arrays with negative towers in which a good bandwidth solution cannot be found, terminating the negative tower with a resistive load can often be the solution as in the case of the example presented here. The tradeoff for better bandwidth requires that the transmitter output increase by the amount of power being terminated in the resistive load, and that the power being dissipated be monitored, usually via the remote output on a toroidal current meter at the resistive load location. The resistive load requires a reactive matching network to match the load to the required drive impedance (resistance and reactance) of the negative tower. Often the load can be placed at the input of the negative tower's tuning unit, eliminating the need for a transmission line to that tower. The pattern and impedance bandwidth improvements that result from applying this technique as described in the diagram of Figure 7.9-5 are shown in the comparisons of Figures 7.9-7 and 7.9-8, respectively.

Figures 7.9-7 and 7.9-8 show the normalized pattern bandwidth of each configuration. The towers are driven with the base conditions as determined by the network elements and mutual tower impedances for carrier and the ±5 and ± 10 KHz sidebands. In Figures 7.9-7 and 7.9-8, the radiation pattern for each frequency is normalized to the carrier power in order to remove the effects of impedance bandwidth (i.e., mismatch) from the pattern bandwidth comparison. When this technique is used for towers where it is not clear that the tower will be a negative resistance tower, the load can be placed at the phasor location, with an additional matching unit taking the place of the power dividing network for that tower. Figures 7.9-9 and 7.9-10 show the Smith chart impedance sweeps for parameter set A and B, respectively.

Detuning Versus Floating Unused Towers

Unused towers in a directional array must be considered in order to ensure that they don't contribute to the pattern as parasitic re-radiators. How to handle unused towers will depend on their feed method and on their electrical height. Series fed towers can generally be floated (left with no network across their base) if they're 90 degrees or less in electrical height.[11] However, if the tower has significant reactance across the base insulator from isolation transformers (isocouplers, isolation coils, ring transformers), then simple floating may not be effective. In these cases, when

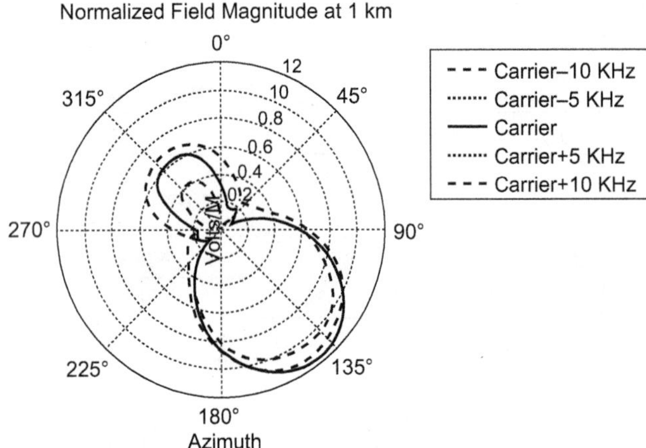

FIGURE 7.9-7 Normalized field patterns for parameter set A with negative tower

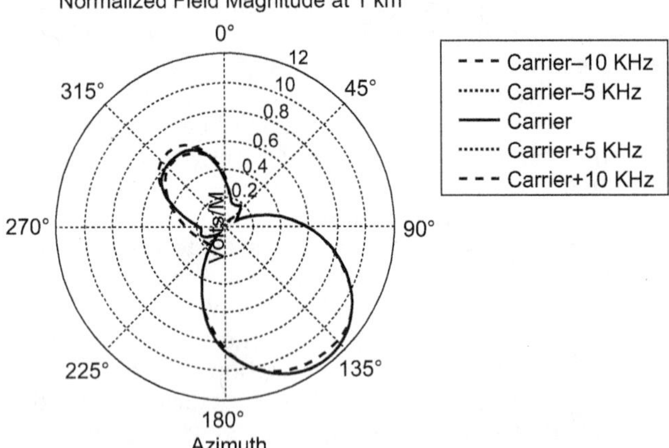

FIGURE 7.9-8 Normalized field patterns at for parameter set B with negative tower resistively terminated

the tower is taller than 90 degrees, or whenever the designer wants a more precise detuning of the tower, a MOM model should be used to estimate the reactance necessary to effectively detune the tower, i.e., reduce the re-radiation of the tower to a minimum.[12] For tower heights less than half-wave, the detune reactance that must be applied across the base insulator is generally inductive. For towers nearing half-wave and taller, the reactance becomes very low and may reverse sign to be capacitive. The detuning element (inductor for most cases) should be adjustable in order to allow for compensation for stray reactance from transformers and so on that may also contribute to the effective detune reactance across the tower base. For all unused towers that are shunt fed, a MOM model is recommended to determine the terminating reactance that will effectively detune the tower. Note that there may also be cases of highly coupled towers in an array that can't be effectively detuned by simply terminating them in a passive network. Some cases may require that the tower be driven as part of the active array.

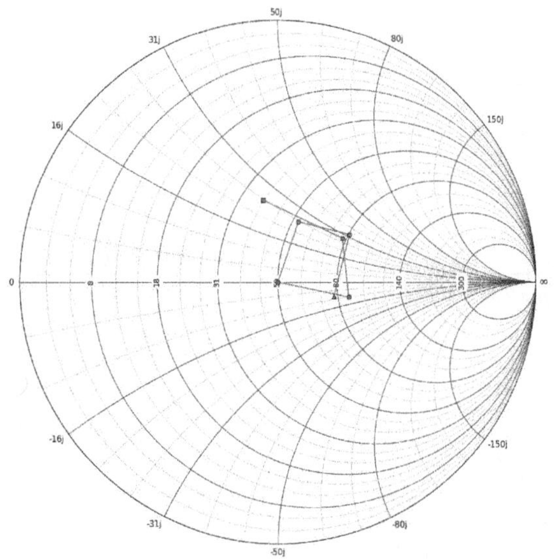

FIGURE 7.9-9 Impedance sweep for Set A

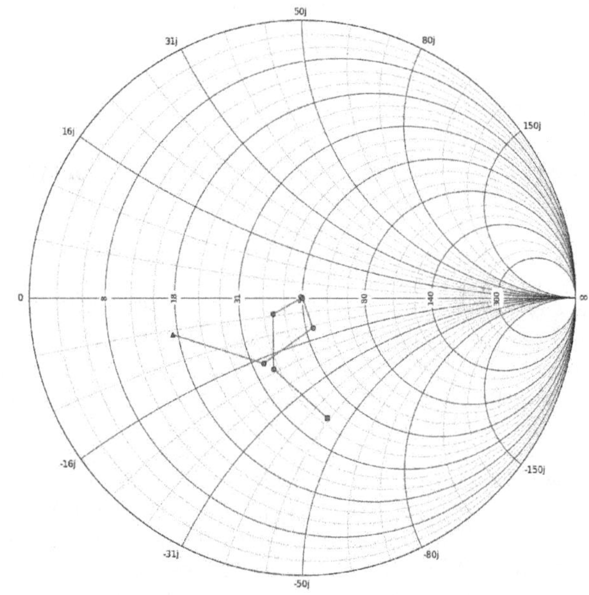

FIGURE 7.9-10 Impedance sweep for Set B

Sampling and Power Measurement Considerations

Sampling loops, base sampling toroids or base voltage sampling units can be used to monitor the array parameters based on FCC guidelines. Avoid placing base current meters at high impedance nodes where the shunt capacitance of the meter toroid can adversely affect the impedance value and where the associated peak voltage may exceed the voltage rating of the current meter sampling toroid. The meter toroid should be placed on the transmitter side of a shunt pre-matching element located in shunt across the high impedance point to reduce the magnitude of the impedance or in the case of a diplexed AM antenna system it should be placed on the transmitter side of each station's series trap filter.

RF COMPONENT CONSIDERATIONS

Adjustment Range

When selecting inductors and capacitors in matching network designs, both adequate tuning range and adequate safety margin for current and voltage must be ensured. The amount of variation to allow for in any particular circuit branch will depend on its position in the system. For a typical ATU using a T network, the input is tuned to the characteristic impedance of the input transmission line, which is typically (50 ± j0) Ω. As such, the current in this branch should be very close to the nominal current expected into 50 Ω. The output branch, on the other hand, has the uncertainty of the load resistance at play. If the design is based on a MOM model, then some degree of uncertainty can be expected for the antenna resistance depending on the on-site ground conditions and the possible addition of any other equipment across the base insulator, such as a tower lighting choke or isocoupler, and a corresponding level of variation should be allowed for in the output branch current. The shunt branch current variation will also be driven by this uncertainty in the antenna resistance. In addition to the variations being allowed for due to uncertainty in the antenna resistance, some added variation is recommended to allow some adjustment of the phase shift through the network. These variations from the nominal antenna resistance and nominal phase rotation result in a series of cases that can be quickly compiled with most network design software. The worst case currents and voltages for each branch in the network can then be reviewed when selecting the inductors and capacitors. Some knowledge of any other uncertainties that may be present for the network may warrant higher levels of variation, such as the case of an ATU at a very low power tower in an array. Similarly, knowledge of the load with greater certainty, such as when accurately measured data is available, can allow for less variation to be used in the design.

Another example case is that of a T network power divider connected to the power divider bus of a phasor. This network should be designed to accommodate a wide variation in (1) input resistance (the R_p off the bus) and (2) the corresponding power to that tower. The output current of a power dividing T network is somewhat less uncertain because it is likely to be feeding into a matched transmission line. Allowing for both phase shift variation and power variation for a power dividing T network in a phasor is good design practice.

Voltage and Current Ratings

When designing AM tuning networks, variations of the load, the phase shift and so on are considered, as mentioned in the previous section. This approach leads to branch-by-branch worst-case RF currents and voltages for the inductors and capacitors involved. When selecting components for operation in the AM band, the main driving factors are the worst case rms current for inductors and for mica capacitors, and the worst case peak voltage for vacuum capacitors using a peak modulation factor (often 125%). The following are the mathematical expressions for calculating the maximum current with modulation and the peak voltage with modulation:

$$\mathrm{I_{MOD}} = \sqrt{1+\frac{\mathrm{m}^2}{2}}\, I_{RMS} = \text{effective modulated current} \tag{26}$$

$$\mathrm{V_{MOD}} = \sqrt{2}\,(1+\mathrm{m})\,\mathrm{V_{RMS}} = \text{peak modulation voltage} \tag{27}$$

In equations (26) and (27), m is the modulation factor (1.0 for 100% modulation, 1.25 for 125% modulation, etc.). The inductors and mica capacitors should be rated for the worst case $\mathrm{I_{MOD}}$ and the capacitors for the worst case $\mathrm{V_{MOD}}$ for the range of load variation and phase shift network design goals. Care must also be taken to ensure that the interconnecting bus work is correctly sized in accordance with the same current rating as the inductor ribbon or tubing size in each branch of the network.

In the AM band, mica capacitors are driven by their current ratings rather than the voltage ratings. The worst case RMS current expected through a mica capacitor should not be greater than 80% of the current rating printed on the capacitor. Vacuum capacitors, on the other hand, are driven by their peak voltage ratings. The worst case peak voltage dictated by the design for a vacuum capacitor should generally not be any greater than 80% of the rated test voltage printed on the vacuum capacitor.

A good RF design can still be compromised if the components are not arranged in a cabinet with good spacing from other components, from the cabinet walls and with proper inductor orientations. All inductors should be spaced from cabinet walls by a minimum of one inductor radius (i.e. one diameter or more spacing from the axis of the coil to the wall, floor, ceiling, or other components). Adjacent inductors should also be arranged along axes that are orthogonal to one another. Inductors that must be arranged with their axis aligned should ideally be placed with their axis at least 5 diameters apart, based on the larger inductor diameter. Failing to mount the inductors with adequate spacing and with orientations carefully chosen can result in excess coupling between inductors, which can make network tuning more difficult. Excess stray capacitance to the cabinet can also result, leading to unexpected transformations of the impedance and further tuning difficulties.

Capacitors and bus work should be mounted with proper separation from ground to prevent arcing for the worst case modulation peak voltage. A good general practice rule of thumb is to use an insulator length based on 7.5 kV per inch of insulator length. When operating with voltages above 30 kV, metal end caps should be used over the ends of the insulators to shield the threads of the mounting screws within the insulator, preventing internal corona and cracking of ceramic insulators. Adding metal end caps also requires that the overall length of the insulator be increased to yield no more than 7.5 kV per inch between the two end caps at their closest spacing.

Altitude and Climate

Guidelines driven by voltage, such as insulator sizing, must be applied more conservatively when a system is to be operated at a high altitude, and/or in an area with high air pollution levels. Literature is available from the utility industry as well as from equipment manufacturers providing recommended derating factors for voltage based on site elevation above sea level. For example, voltage ratings up to 1000-meter elevation can be considered the same as those at sea level. At 2000 meters, they're 91% of sea level ratings, 80% at 3000 meters, 72% at 4000 meters and 56% at 6000 meters.[13] Clearly insulators should be chosen more conservatively at high altitude locations. The use of air-conditioning and/or circulating fans is also recommended to help prevent corona and arcing issues when cabinet space is limited at a high altitude site.

SUMMARY

The Need for a System Perspective

Modern tools such as the Method of Moments (NEC) and nodal circuit analysis (Spice and its derivatives) programs permit effective matching and phasing networks to be routinely designed. The network and tower characteristics such as effective electrical height and mutual coupling can be accurately characterized, especially when field measurements are incorporated into the model to calibrate and fine tune the antenna model. To achieve the best end-to-end performance of an AM transmission system from the output of the combined RF amplifiers to the wideband AM analog or digital receiver, a total RF system design approach is required.

Bibliography

[1] "LTspice IV Getting Started Guide." Linear Technology. N.p., n.d. Web. 28 Mar. 2016. www.linear.com/designtools/software/#LTspice.

[2] Burke, G.J. *Numerical Electromagnetics Code NEC-4, Method of Moments, Part 1: User's Manual*, Livermore, CA: Lawrence Livermore National Laboratory, January 1992.

[3] Laport, Edmund A. "2.3.4 The Folded Unipole." *Radio Antenna Engineering*. New York: McGraw-Hill, 1952. 109–111. Print.

[4] Rackley, R.D., Cox, B.L., Moser, J.R., and King, T.F. "An Efficiency Comparison: AM/Medium Wave Series-Fed vs. Skirt-Fed Radiators." *National Association of Broadcasters, Broadcast Engineering Conference*, Las Vegas, NV, April 1996.

[5] Breakall, J.K., et al. "A Novel Short AM Monopole Antenna With Low-Loss Matching System," *52nd Annual IEEE Broadcast Technology Symposium*, Washington, DC, October, 2002.

[6] Terman, Frederick Emmons. "Circuit Theory." *Radio Engineers' Handbook*. New York: McGraw-Hill Book, 1943. 210–215. Print.

[7] Cox, B. L., and Moser, J. R. "AM Antenna System Case Studies for DRM and IBOC DAB," *2002 Broadcast Engineering Conference Proceedings*, NAB Department of Science and Technology, Washington, DC, pp. 150–167, 2002.

[8] Rackley, Ronald D. "Evaluation and Improvement of AM Antenna Characteristics for Optimal Digital Performance." n. p, 2004. Web. 28 Mar. 2016. http://dlr.com/pdfs/AMDigital.pdf

[9] Westberg, Jerry. "4.4 AM Antenna Coupling and Phasing Systems." *National Association of Broadcasters Engineering Handbook*. 10th ed. N.p.: Focal, 2007. 746–747. Print.

[10] Rackley, R.D. "AM Directional Antenna Parameters and Pattern Inversion Techniques," *IEEE Broadcast Technology Society, 2010 Broadcast Symposium*, Alexandria, VA, October, 2010.

[11] Edison, Edward. "2.4 AM Broadcast Antenna Systems, Part II: Antenna Coupling and Phasing Systems for AM Broadcast." *National Association of Broadcasters Engineering Handbook*. 7th ed. N.P.: n.p., n.d. 2.4–103. Print.

[12] Hatfield, James B. "4.5 Computer Simulation of Broadcast Antenna Systems." *National Association of Broadcasters Engineering Handbook*. 10th ed. No.p.: Focal, 2007. 759. Print.

[13] Huamán, Dany. "MV Air-Insulated Equipment De-Rating for High Altitude Installations." *GE Industrial Solutions, Digital Specification Program*, Latin America. Web, March 24, 2016, http://apps.geindustrial.com/publibrary/checkout/LAPL0083?TNR=Articles | LAPL0083 | generic

CHAPTER

7.10

AM Diplexing, Combining, and Filter System Design

RONALD D. RACKLEY, P.E.
du Treil, Lundin & Rackley, Inc.
Sarasota, FL

INTRODUCTION

Combining transmitter sites of AM radio stations, commonly referred to as diplexing, has become an increasingly popular trend in recent times. Factors that militate toward multi-frequency operation at transmitter sites include the economy to be realized through savings in antenna tower and land costs, the difficulty with which new transmitter sites must be approved by regulatory authorities other than the FCC, and the common ownership of multiple stations serving the same markets.

The history of combining AM transmitter sites goes as far back as the 1930s. It was not uncommon to have stations share antenna tower sites before the FCC multiple ownership rule, which made it impossible for one company to own more than one AM station in a market, was enacted in 1943. With all AM stations serving a given coverage area belonging to competing licensees after that, the practice was for the most part abandoned outside of Hawaii—where the scarcity of land dictated that common tower sites be used by many of the stations on the islands. Interest in the technology saw a resurgence within the 48 states in the 1980s, as the economic picture for AM station owners changed under competition from FM stations. Owners needing to relocate stations sometimes sought more economical solutions than developing new transmitter sites, and diplexed AM transmitter sites began to be developed in significant numbers. The 1996 common ownership rule change that once again allowed a single licensee to own more than one AM station in a market resulted in increased interest in diplexing, as owners sought to combine the facilities of their stations. A recent trend has been for stations to move to combined transmitter sites so that their owners can realize income from the sale of transmitter site property while keeping them in operation.

GENERAL CONSIDERATIONS RELATED TO DIPLEXING

Successful diplexing at a single transmitter site must include consideration of many factors. First of all, antenna performance and signal coverage issues related to the available tower(s) at the transmitter site and its location relative to the desired coverage area must be suitable. If tower(s) must be added to produce the directional antenna pattern needed by the station that is moving to the site, there must be available land for that, including area for the associated ground system. The FCC's minimum antenna efficiency and city-of-license coverage requirements must be met, considering both the tower height(s) and ground system area requirements for the frequency of the new station at the transmitter site. If a directional antenna system is required, the array geometry must be acceptable for producing the required radiation pattern(s). These are threshold requirements for being able to use a diplexed transmitter site, but there are other important considerations with regard to the practicality of doing it.

Once a site is found to be acceptable on the basis of its tower(s), ground system and location, there are other factors that can significantly impact the cost of construction. The guy wire insulators must be spaced acceptably for all frequencies, which can mean adding interspaced insulators in cases where a new station at a much higher frequency is added at an existing site. The guy wire and base insulators must be acceptably rated for the peak voltages that will be present across them, which means the sum of the peak voltages under modulation for all of the frequencies. Voltage stresses at multi-frequency sites cannot be viewed in terms of a summation of the station power levels, which would be appropriate to evaluate heating effects, but rather must consider vector addition of the constituent RF waveforms at each instance where the peaks occur simultaneously under full positive modulation. It is straightforward to derive the expected voltages for each frequency from Method of Moments modeling once the antenna characteristics have been determined and apply the appropriate factors to have their absolute peak values. It must be done.

Where directional antennas are involved, the antenna monitor sampling system must be specifically designed for diplexed operation with all of the frequencies that will be used at the site. The choice of how and where sampling is accomplished and what type of antenna monitor is used will impact whether filters are required at the antenna monitor inputs and whether the built-in termination loads have acceptable power ratings for the voltages at all frequencies. Once it has been determined where the antenna currents or base voltages will be sampled, considering the requirements of the FCC rules, and whether one sampling system or separate sampling systems will be employed for the stations, the antenna monitor manufacturer must be given the expected input voltages for all stations and modes of operation so that proper diplexed operation can be assured.

Whether all stations will be capable of transmitting with acceptable bandwidth through the system of filters required for diplexing is another concern, particularly when one or more of the stations wishes to transmit in a digital mode with more stringent transmitter load bandwidth requirements than is the case for simple analog transmission. Whole system nodal analysis based on Method of Moments–derived tower base characteristics over the desired passband of each station is necessary to accurately define bandwidth performance, if that needs to be known before making a decision to move forward with a diplexing proposal.

FCC REQUIREMENTS FOR DIPLEXED STATIONS

Construction permits for diplexed stations have special conditions related to spectrum performance issues that must be met following construction. Typically they require that "sufficient data shall be submitted to show that adequate filters, traps and other equipment has been installed and adjusted to prevent interaction, intermodulation and/or generation of spurious radiation products which may be caused by common usage of the same antenna system" by the stations at the transmitter site. In most cases, measurement of all first-, second- and third-order products of the station carrier frequencies that fall within the range of field strength meters such as are commonly used for making measurements on AM stations, covering up to 5 MHz, are acceptable for meeting this requirement. There is another requirement that "there shall be filed with the license application copies of a firm agreement entered into by the stations involved clearly fixing the responsibility of each with regard to the installation and maintenance of such equipment."

FILTER DESIGN FOR DIPLEXING

The performance to be expected through the filters that must be used to isolate the transmitter outputs from each other is an important consideration. The filters serve their function to avoid the kind of transmitter-to-transmitter coupling that causes the creation of intermodulation products on undesired frequencies and avoid directional antenna pattern disturbances, where directional antennas are involved, and they must also pass the signals of the stations that are fed into them without significantly degradation.

In general, filtering for diplexing involves tower base series pass/reject filters to isolate the multiple feeds from the stations' antenna matching networks and, also, shunt filters between the transmitter and the base series filters to sufficiently block the unwanted signals and prevent intermodulation in the transmitters' final amplifiers. For nondirectional antennas, the shunt filters appear most often immediately before the series filters with both being between the antenna tuning unit (ATU) matching network and the tower base. For directional antennas, there is typically one shunt filter across the common point in the phasor cabinet to work with the series filters at the tower bases to avoid intermodulation products.

Whole system nodal analysis can be used to predict the amount of attenuation there will be for input-to-input coupling of undesired signals based on reasonable component Q assumptions in the filters. With this information, intermodulation product suppression can be calculated based on the transmitter's turn-around loss and output network response at the frequencies of interest—which may be obtained from the manufacturer(s) of the transmitters that will be used at the site. This needs to be done in order to know whether additional filtering will be necessary at the transmitter output to attenuate intermodulation products after they have been produced.

Experience has shown that serious performance related mistakes can be made if the various aspects of filter performance are not considered relative to each other in the proper perspective. Consideration of the simplest filtering requirements alone, without studying their bandwidth effects, can result in catastrophic performance deficiencies for transmitted signals. Choices can be made when designing AM antenna systems with filters for multi-frequency operation, on the other hand, to result in significant performance improvements over generic "textbook" designs—particularly with regard to bandwidth issues that impact the quality of coverage for both analog and digital transmission.

The designs of individual filter sections determine their stand-alone characteristics, but how they work together with other filter sections to have their insertion effects complement each other and result in better overall system performance must be studied with whole-system computer modeling—which considers filter, matching network and tower impedance characteristics together. Modern computer modeling techniques make it possible to predict not only how well filters can be expected to perform in terms of their pass/reject characteristics, but also in terms of their insertion effects on signal components passing through them on the required carrier and sideband frequencies.

THE ROLE OF STORED ENERGY

Filters work to filter because they have different characteristics at different frequencies. They have different characteristics at different frequencies because the reactances of the components in their network branches change with frequency, and it is possible to arrange them to produce the desired end-to-end pass and reject characteristics. The changing reactances with frequency of filter components also alter the signals that pass through them.

The rate at which the reactance of an inductor or capacitor changes with frequency is related to the amount of energy stored in it. Stored energy is defined by the current passing through a reactive component and the voltage developed across it, just as dissipated energy is defined by the current passing through a resistance and the voltage developed across it. Dissipated energy in a resistance is described in terms of power flow, normally expressed in watts. Stored energy in reactive components, being responsible for what is called in complex mathematical notation "imaginary power," is described in terms of volt-amperes reactive ("VARS") rather than watts.

SYNERGISTIC FILTER DESIGN

It is known that pass-reject filter characteristics for altering signals that pass through them have different impedance characteristics for different filter implementations. Series pass-reject filters and shunt pass-reject filters can have approximately opposite pass signal impedance characteristics under resonant (zero reactance at the pass signal's carrier frequency) conditions if they are designed to have complementary stored energy values. This principle can be used to great effect for improving bandwidth performance with synergistic filter design.

EXAMPLE OF BASIC FILTERING

Figure 7.10-1 shows the most elementary case where a single pass/reject filter is placed in series with an antenna input. Such a filter is used where a transmitter must be isolated from the signal of another nearby station. The hypothetical circuit is a standard design to pass the 1000 kHz signal and reject 1400 kHz—a rather benign frequency separation for AM station diplexing. The filter has an inductor and capacitor in series that are adjusted for series resonance at 1000 kHz with a capacitor in parallel with that branch to parallel resonate the net inductive reactance at 1400 kHz—to produce an impedance zero at 1000 kHz and an impedance pole at 1400 kHz across the filter, in other words.

For easy reference, its insertion effects have been modeled assuming a constant 50-ohm load across its passband. In other words, the calculated insertion

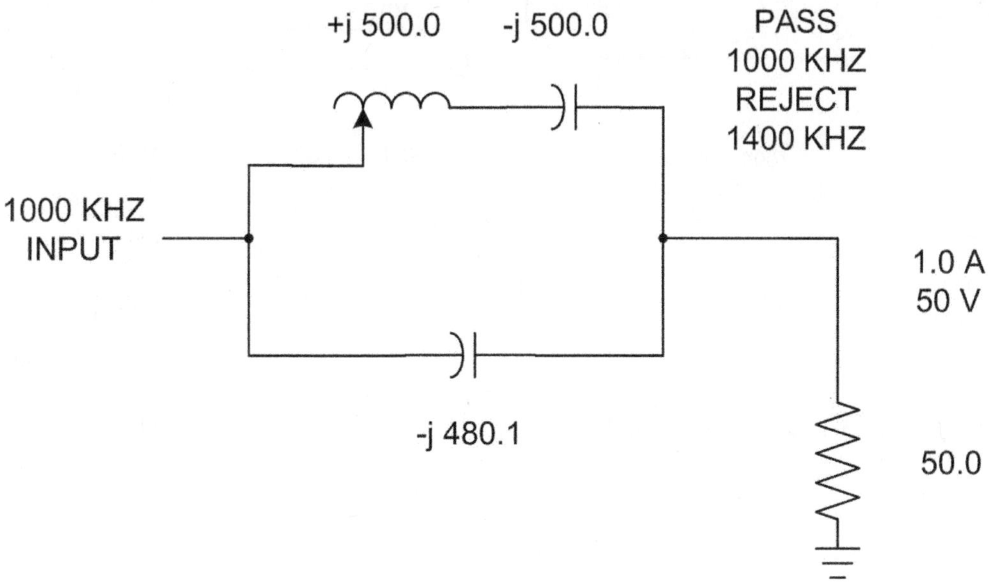

FIGURE 7.10-1 Series Pass/Reject Filter

effects are from the filter itself without any interaction involving changing antenna sideband frequency impedances.

As can be seen from Table 7.10-1, the sideband impedances predicted by the model are altered significantly from the perfectly flat characteristic of the assumed load impedance. In this case, it is due predominately to the series slope characteristics of the filter branch that is series resonant at 1000 kHz.

Example of Synergistic Two-Stage Filtering

Intuition might seem to suggest that a more complicated filter with more sections would harm the impedance sweep even more. Popular conventional wisdom relates bandwidth directly to stored energy and suggests that increased stored energy in filters can only make matters worse. Such is not the case, however. Stored energy can be distributed in reactive networks—such as filters—so as to be self-compensating to a great extent and actually improve bandwidth with synergistic filter design.

Figure 7.10-2 shows the same pass/reject filter of Figure 7.10-1 with an added filter section in shunt with its input. The shunt filter has a series branch to ground on the right in the drawing consisting of an inductor and capacitor adjusted to produce an impedance zero at 1400 kHz. The series branch on the left provides the net inductive reactance required to parallel resonate the net capacitive reactance of the right branch at 1000 kHz, but with a twist. It does not simply use an inductor to do that. It uses a series inductor and capacitor with values chosen to produce the required net inductive reactance while having voltages across both that, with the current flowing through them, add enough stored energy to that of the right branch to make the total stored energy of the shunt filter approximately the same as that of the series filter at 1000 kHz. Note that VARS values—a convenient way of expressing stored energy—are shown for each component and that the total for all components in the series and shunt filter sections are approximately equal.

A comparison of Table 7.10-2 with Table 7.10-1 shows the improvement in passband characteristics resulting from doubling the total stored energy by the addition of the "VARS—matched" shunt filter section. At the same time, the attenuation of the undesired signal is significantly increased.

TABLE 7.10-1
Insertion Effects of Series Pass/Reject Filter

Frequency kHz	Input Resistance	Input Reactance	Normalized VSWR
985	50.0	–j 14.7	1.34
990	50.0	–j 9.8	1.22
995	50.0	–i 5.0	1.11
1000	50.0	0.0	1.00
1005	50.0	+i 5.0	1.11
1010	50.0	+i 10.2	1.22
1015	50.0	+i 15.4	1.36

TABLE 7.10-2
Insertion Effects of Series and Shunt Pass/Reject Filters With Matching Stored Energy

Frequency kHz	Input Resistance	Input Reactance	Normalized VSWR
985	54.1	+j 3.0	1.10
990	51.9	+i 1.0	1.04
995	50.5	+i 0.2	1.01
1000	50.0	0.0	1.00
1005	50.5	+i 0.1	1.01
1010	52.1	+i 0.2	1.04
1015	54.7	–j o.1	1.09

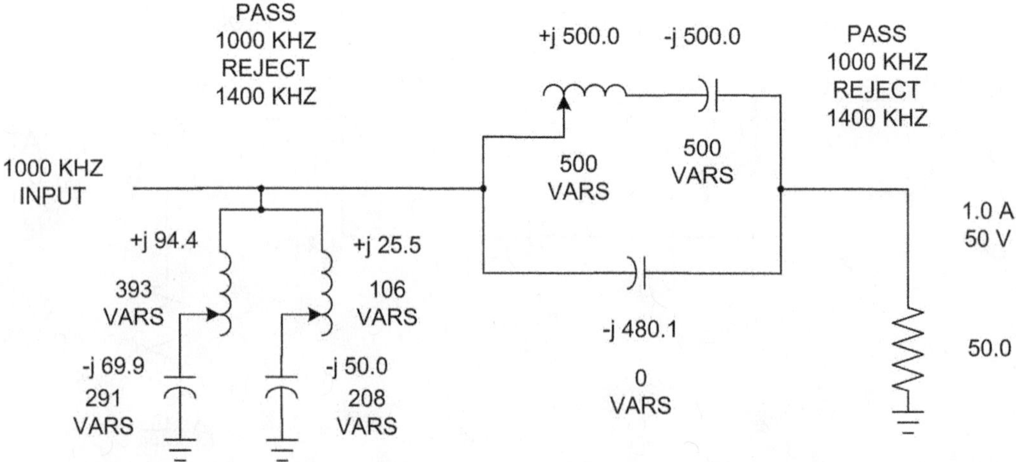

FIGURE 7.10-2 Series and Shunt Pass/Reject Filters With Matching Stored Energy

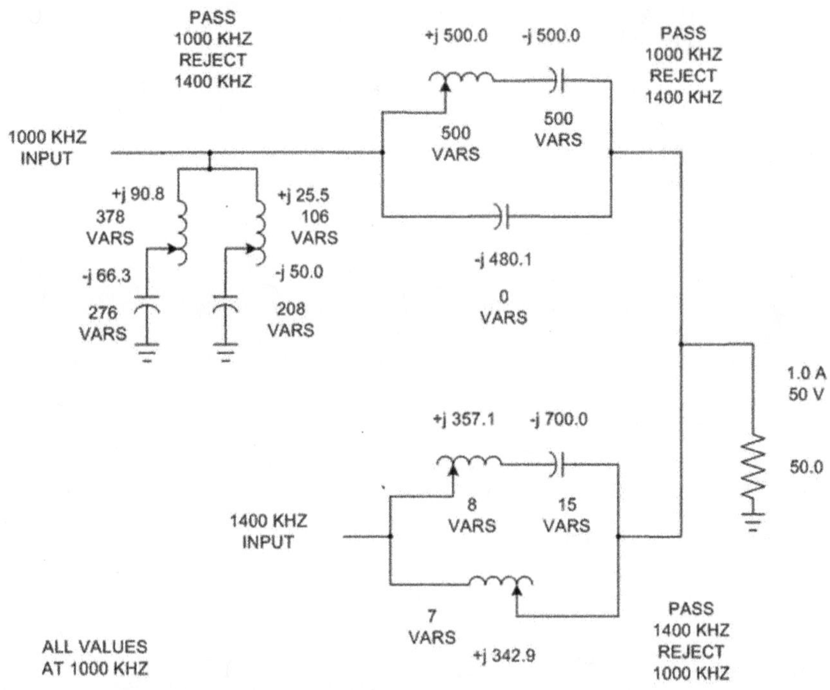

FIGURE 7.10-3 Series and Shunt Pass/Reject Filters for Diplexing With Matching Stored Energy

Example of Synergistic Diplexer Filtering

Figure 7.10-3 shows a design for a diplexed system with both 1000-kHz and 1400-kHz inputs. Because the pass 1400-kHz filter appears in shunt across the load at 1000 kHz, the design of the shunt filter across the 1000-kHz input was altered so the stored energy in it combined with the stored energy in the pass 1400-kHz filter at 1000 kHz add together to provide "VARS—matched" conditions for both shunt filters with the series pass 1000-kHz filter section. A comparison of Table 7.10-3 with Table 7.10-2 shows virtually identical bandwidth results for both synergistic filter designs.

TABLE 7.10-3
Insertion Effects of Series and Shunt Pass/Reject Filters for Diplexing With Matching Stored Energy

Frequency kHz	Input Resistance	Input Reactance	Normalized VSWR
985	53.9	+j 2.7	1.10
990	51.8	+i 0.9	1.04
995	50.5	+i 0.2	1.01
1000	50.0	0.0	1.00
1005	50.5	+i 0.1	1.01
1010	51.9	+i 0.2	1.04
1015	54.4	–i 0.1	1.09

Nonresonance at Pass Carrier Frequency

Figure 7.10-4 shows the same circuit with one alteration: the insertion of a reactance in series with the load.

A comparison of Table 7.10-4 and Table 7.10-3 shows that resonance in this case is necessary for synergistic filter performance as far as the input bandwidth is concerned.

COMPROMISES IN FILTER DESIGN

Unless optimization of one frequency at the expense of the other into a diplexed antenna is desired, it is generally not possible to make both load impedances at the filter output resonant with the simple addition of a series inductor or capacitor. (It is possible sometimes to use a full matching network with stored energy chosen for its series and shunt branches to resonate both frequencies of a diplexed antenna at the filter output, but the details of doing that are beyond the scope of this discussion.)

Sometimes decent results can be found with a compromise in matching for the two frequencies of diplexed stations. To illustrate, Figures 7.10-5 and 7.10-6 show filter designs calculated for two stations that are separated in frequency by only 6% with the filters connected directly to the tower, while Figures 7.10-7 and 7.10-8 show a compromise design with a series capacitance chosen to produce the same ratio of carrier frequency resistance to reactance at the output of each filter.

Tables 7.10-5 and 7.10-6 show that the insertion of filters can have very detrimental effects on input impedance bandwidth. Tables 7.10-7 and 7.10-8 show that a compromise involving the addition of few

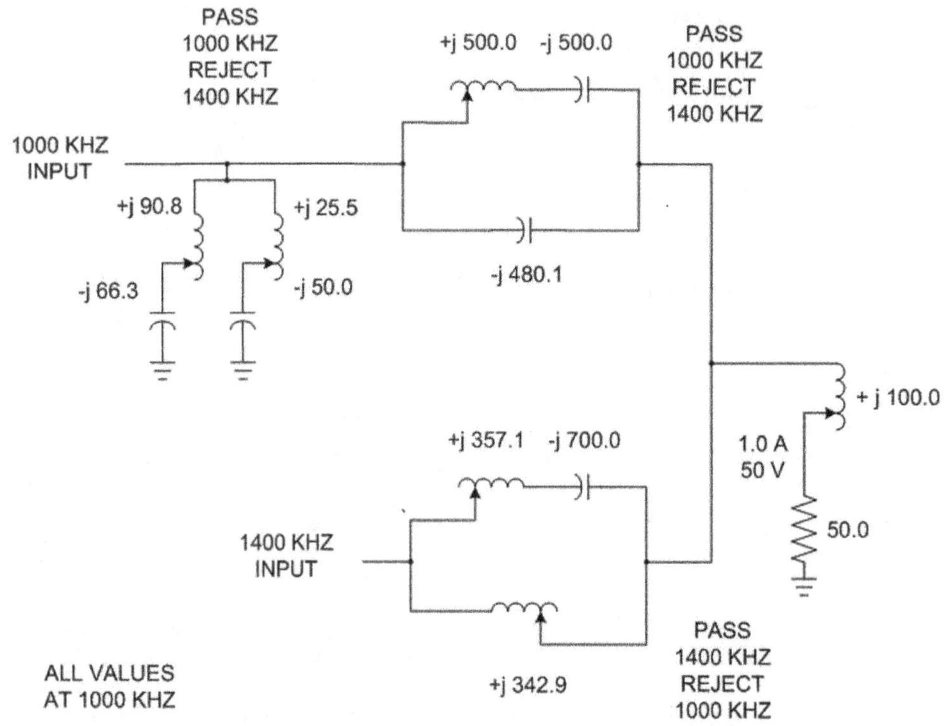

FIGURE 7.10-4 Series and Shunt Pass/Reject Filters for Diplexing With Reactive Load

TABLE 7.10-4
Insertion Effects of Series and Shunt Pass/Reject Filters for Diplexing With Reactive Load

Frequency kHz	Input Resistance	Input Reactance	Normalized VSWR
985	20.1	59.2	4.31
990	25.8	69.4	2.82
995	34.8	82.6	1.73
1000	50.0	100.0	1.00
1005	78.1	122.6	1.77
1010	135.3	145.6	3.06
1015	248.9	127.9	5.04

components—in the case shown only one capacitor—can provide a very worthwhile improvement in performance.

DIPLEXED DIRECTIONAL ANTENNA SYSTEMS

A recent example that illustrates some of the issues that may be faced with multi-frequency directional antenna operation involves the addition of a station on 930 kHz to a six-tower site of stations on 1230 kHz and 1580 kHz. The 930-kHz station uses either tower 4 or tower 5 for nondirectional operation in the daytime and towers 3, 4 and 5 for directional operation at night. The 1230-kHz station uses towers 2 and 6 fulltime for directional operation. The 1580-kHz station uses four towers in the daytime and all six at night. The towers used for 930-kHz transmission were chosen so that no tower of the six would carry the signals of more than two stations at once.

Figure 7.10-9 shows the directional antenna feed system for the 930-kHz station. The night-DA common point matching system employs cascaded T networks to rotate the sideband impedances of the power divider input so that the shunt network for rejecting 1230 kHz and 1580 kHz serves for broadbanding, using the principles of synergistic filter design. Whole-system nodal modeling based on an admittance matrix representation of the tower bases derived from Method of Moments modeling was employed to predict the power divider sideband impedance characteristics and manage the stored energy in the input shunt filter appropriately. The phase shifts of the nondirectional matching networks were chosen to work synergistically with the input shunt filter also.

Figure 7.10-10 shows all of the tower base filters and Figure 7.10-11 shows the common point filters that were added at the inputs to the 1230-kHz and 1580-kHz phasor cabinets when the 930-kHz station moved to the site. To reduce the number of components required for filtering, the towers that are not used on 1230 kHz are detuned by tapping the sampling line isolation coils for the required reactance so that isolated filter input ports are only necessary for towers 2 and 6. All six towers have isolated ports for

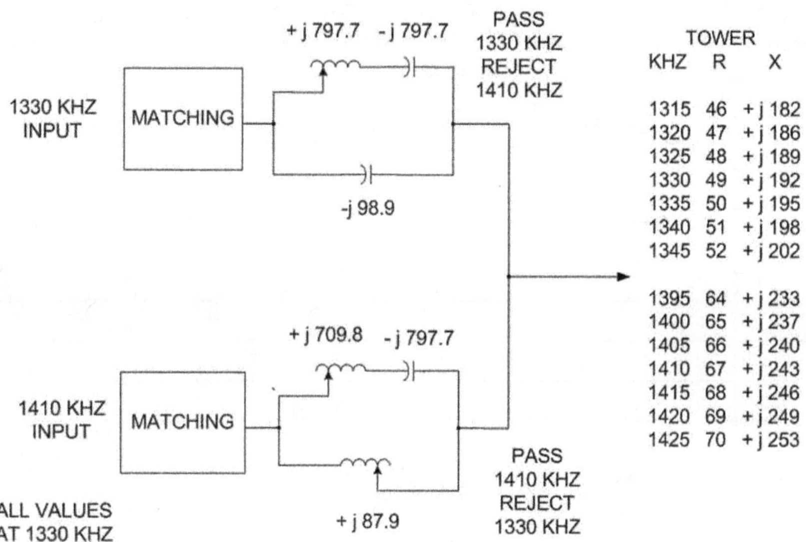

FIGURE 7.10-5 Series Pass/Reject Filters for Diplexing With 6% Frequency Difference Directly Into Tower Base at 1330 kHz

TABLE 7.10-5
Insertion Effects of Series Pass/Reject Filters for Diplexing With 6% Frequency Difference Directly Into Tower Base at 1330 kHz

Frequency kHz	Input Resistance	Input Reactance	Normalized VSWR
1315	10.2	+j 29.3	6.67
1320	18.9	+i 34.4	4.02
1325	38.9	+i 32.5	2.14
1330	50.0	0.0	1.00
1335	24.8	–i 15.0	2.25
1340	10.9	–i 9.5	4.77
1345	5.5	–i 3.5	9.22

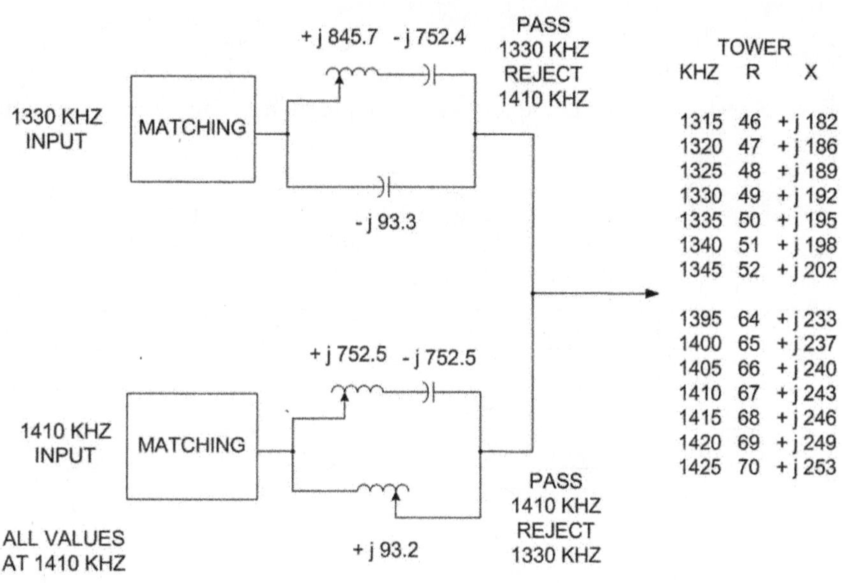

FIGURE 7.10-6 Series Pass/Reject Filters for Diplexing With 6% Frequency Difference Directly Into Tower Base at 1410 kHz

TABLE 7.10-6
Insertion Effects of Series Pass/Reject Filters for Diplexing With 6% Frequency Difference Directly Into Tower Base at 1410 kHz

Frequency kHz	Input Resistance	Input Reactance	Normalized VSWR
1395	6.1	+j 27.0	10.53
1400	15.0	+i 34.6	5.03
1405	38.7	+i 34.9	2.25
1410	50.0	0.0	1.00
1415	27.1	–i 13.9	2.04
1420	14.5	–i 10.5	3.61
1425	8.9	–j 6.2	5.70

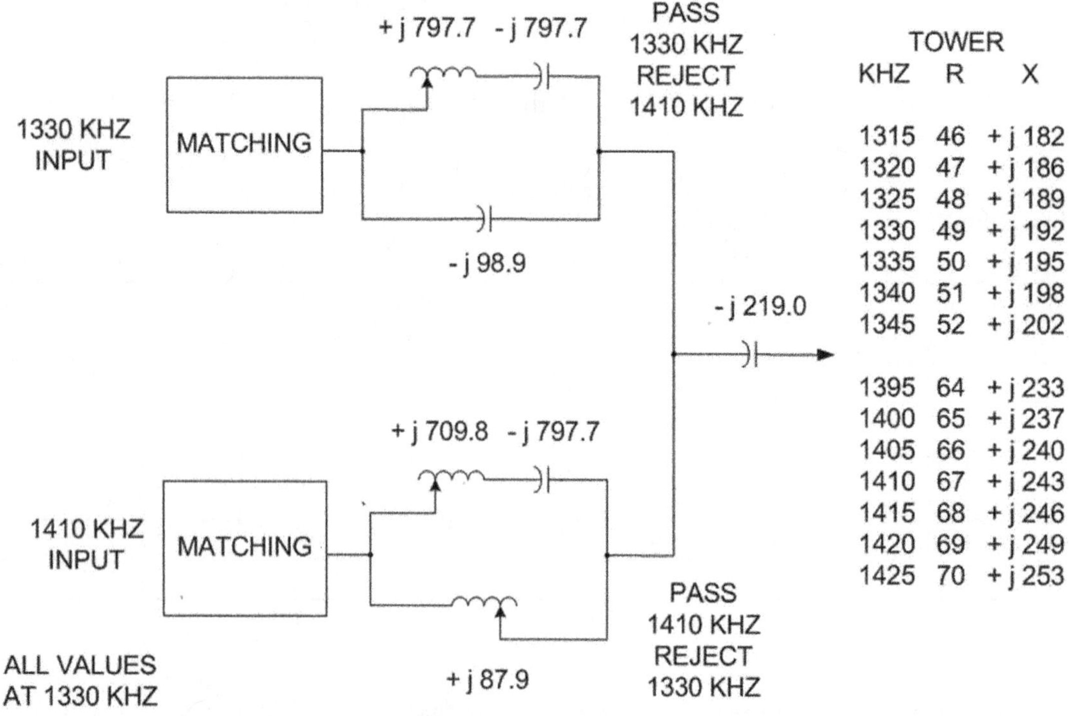

FIGURE 7.10-7 Series Pass/Reject Filters for Diplexing With 6% Frequency Difference Through Reactance Balancing Capacitor at 1330 kHz

TABLE 7.10-7
Insertion Effects of Series Pass/Reject Filters for Diplexing With 6% Frequency Difference Through Reactance Balancing Capacitor at 1330 kHz

Frequency kHz	Input Resistance	Input Reactance	Normalized VSWR
1315	37.1	+j 17.8	1.66
1320	43.2	+i 13.9	1.39
1325	47.7	+j 8.1	1.19
1330	50.0	0.0	1.00
1335	48.8	–i 8.9	1.20
1340	44.1	–i 16.7	1.45
1345	36.4	–i 21.9	1.81

TABLE 7.10-8
Insertion Effects of Series Pass/Reject Filters for Diplexing With 6% Frequency Difference Through Reactance Balancing Capacitor at 1410 kHz

Frequency kHz	Input Resistance	Input Reactance	Normalized VSWR
1395	50.9	+j 22.1	1.54
1400	53.9	+i 13.5	1.31
1405	53.0	+i 5.9	1.14
1410	50.0	0.0	1.00
1415	46.1	–j 4.0	1.12
1420	42.0	–i 6.4	1.25
1425	37.7	–i 8.0	1.40

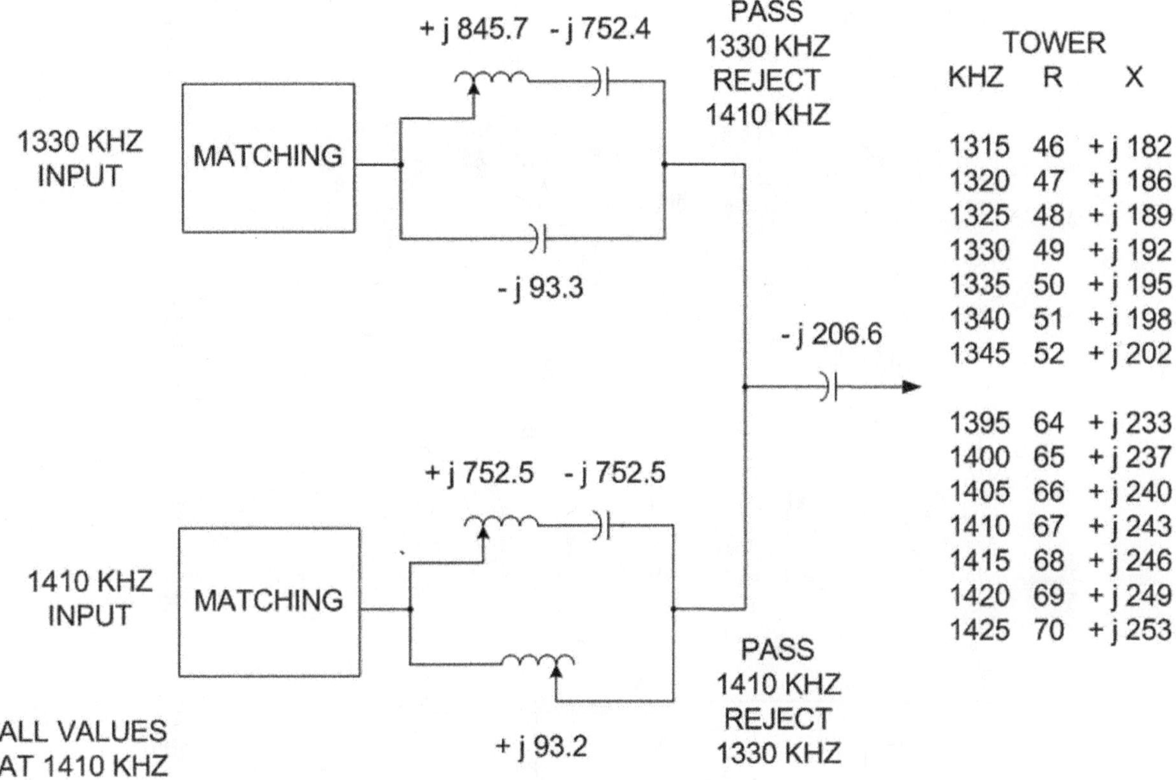

FIGURE 7.10-8 Series Pass/Reject Filters for Diplexing With 6% Frequency Difference Through Reactance Balancing Capacitor at 1410 kHz

930 kHz and 1580 kHz to facilitate either feeding them for directional operation or detuning them. The sampling loops are on the towers at the height required for a Method of Moments proof-of-performance, so it is possible to confirm their detuned condition by nulling the loop current at the desired frequency. The shunt coils that are across the input and output ends of the series filter are to cancel the stray capacitance effects within the filter boxes to reduce the impedance transformation between the filter inputs and outputs.

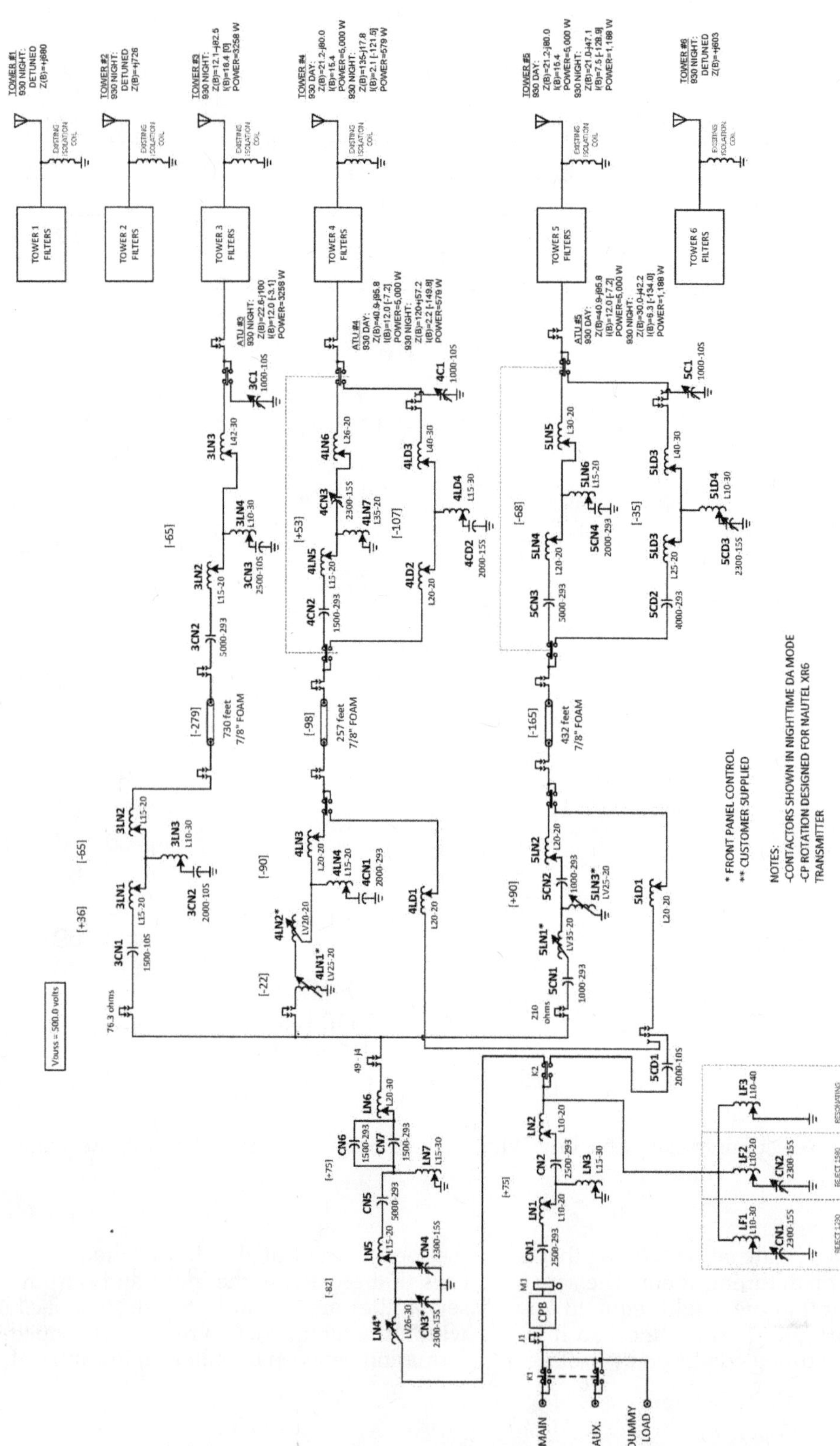

FIGURE 7.10-9 Phasing and Coupling System for 930-kHz Station Added to Diplexed 1230-kHz and 1580-kHz Six-Tower Array

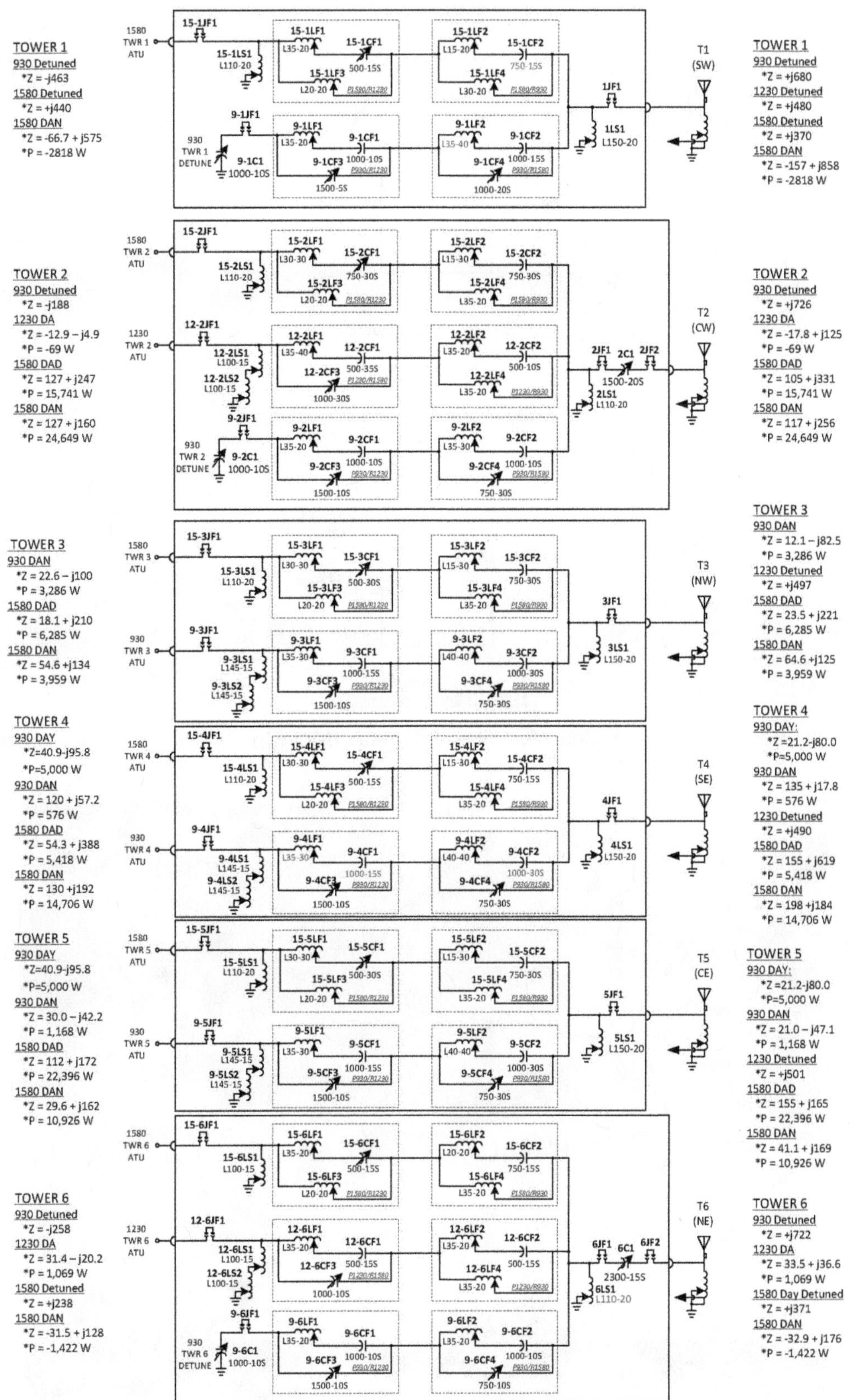

FIGURE 7.10-10 Tower Base Filters and Detuning Circuits for 930-kHz, 1230-kHz and 1580-kHz Six-Tower Array

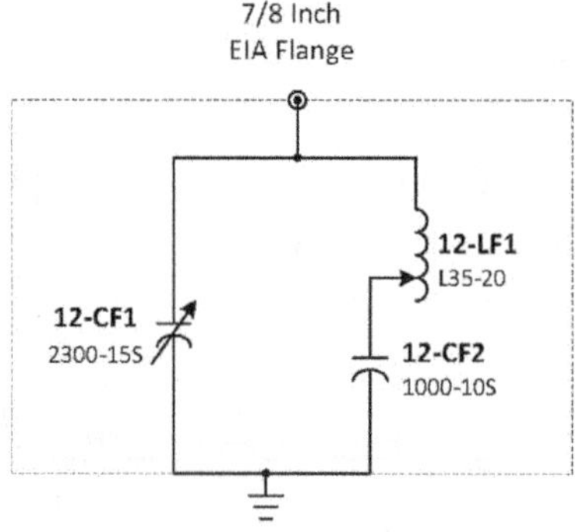

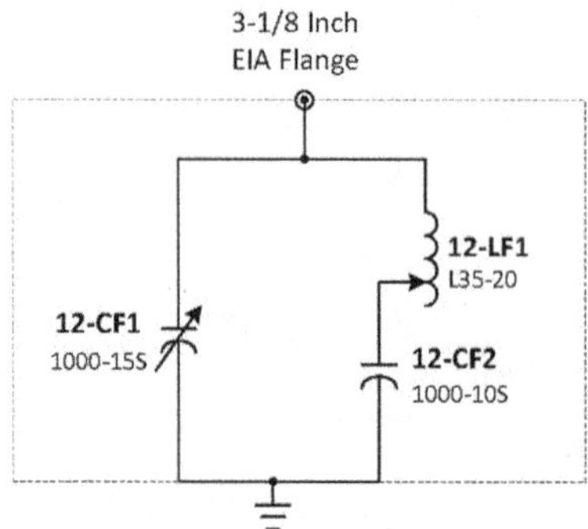

FIGURE 7.10-11 Common Point Filters Added With 930-kHz Station's Move to 1230-kHz and 1580-kHz Six-Tower Array

FIGURE 7.10-12 Base Filter Units in Weatherproof Enclosures and the Original 1580-kHz Tuning Houses at the Tower Bases

CHAPTER

7.11

Synchronous AM Broadcasting

STEPHEN F. SMITH
Consultant

AM transmitter carrier synchronization using low-cost, off-the-shelf GPS timing-reference hardware has the potential to increase the effective coverage of co-channel AM stations with overlapping contours by eliminating beat frequencies and the associated noise artifacts that serve to make the reception in the fringe areas unlistenable, both in daytime and nighttime scenarios. This chapter will address the basic carrier synchronization system design and will present results, including field measurements, that will serve to demonstrate the improvements in reception quality, both static and mobile, in the region of overlapping fringe-area co-channel contours that can be realized with transmitter carrier synchronization techniques. In addition, the long-term successful implementation and operation of two local-area synchronous AM station networks utilizing multiple co-channel boosters will be discussed. These systems have been operated over the last 20 years by the WAPA Radio Network, covering large portions of the island of Puerto Rico. Included are numerous practical insights on setup and operation of the multi-site systems, as well as the basic conclusion that these systems, once properly configured, are both stable and very low maintenance, resulting in increased coverage and revenue generation.

BACKGROUND

The initial idea of carrier-synchronizing AM stations to improve coverage in overlapping areas actually dates back to the mid-1920s [1]. Over subsequent years, multiple technical groups in the U.S., the U.K., and in continental Europe ran numerous extensive field tests, usually focused on synchronizing AM transmitters in both wide and more localized areas, all of which clearly demonstrated the reception benefits of synchronization. Generally the actual synchronization-control signals were carried over leased telephone lines to highly customized oscillator control hardware integrated into both master and slaved transmitter units; extensive field listening tests were conducted to study the nature and extent of the interference-induced distortion in the signal-overlap zones [1, 2].

FCC Rule Section 73.1545 requires that AM broadcast stations operate within ±20 Hz of their assigned carrier frequencies, which theoretically permits difference-frequency beats between co-channel stations of up to 40 Hz. Typically, co-channel stations presently operate at much smaller frequency errors; co-channel beat frequencies are usually now below 5 Hz, and are sometimes even below 1 Hz. Unfortunately, in areas that are located far from the local transmitter (i.e., the desired station), distant-station co-channel carrier-beat components generally cause fast fluttering-type modulations of, and/or large-amplitude "swishing" sounds in, the desired station's audio at the receiver and concurrent distortion of the audio modulation by the more distant station(s), even if they are sub-hertz in nature, because the received desired-station's carrier's amplitude and phase are noticeably modulated by the distant station's carrier signals. These effects are very evident to the listener in nighttime scenarios, even on many "clear" channels. Most AM radio receivers are provided with automatic gain-control (AGC) circuitry. However, the typical receiver's automatic gain-control, also called automatic volume control (AVC), to avoid bass distortion, usually responds far too slowly to "average out" or suppress these inter-carrier beat modulations; thus, these highly annoying modulation effects are largely passed on intact to the listener.

These beat-related effects (excepting external noise) are a (if not *the*) principal factor in the degradation of evening and nighttime AM fringe-area reception quality and the resulting loss of listeners for virtually *all* AM stations. Perhaps the most deleterious aspect of these beats is the listener-annoyance factor, in that the high-level artifacts (volume modulation, cyclic distortion, and pronounced swishing sounds) often quickly induce listener tune-outs. This situation is not only progressively worse further into the fringe areas of the desired stations (usually in the outer suburbs of the city of license), but also occurs much closer in, in the deep nighttime nulls of directional stations. The current poor state of repair of many AM directional arrays, plus the low-power pre-sunrise/post-sunset (PSRA/PSSA) operations at many nondirectional Class-D stations, only exacerbates these problems. The consequence of these minor differences in operating frequencies of such co-channel stations is that broadcasts in the overlapping areas of ground-wave (as well as sky-wave) coverage will produce carrier-beat frequencies that will generate the loud "swishing noise," rendering both stations much less listenable. The plot in Figure 7.11-1 [3] reveals the relative audibility of these fringe-area co-channel beats in both synchronized and unsynchronized settings with two interferers at different levels; the overall result is that synchronization reduces the beat perception by some 6–10 dB, depending partially on the program material of both sources.

Obviously, in this scenario the effective co-channel interference-limited coverage area is increased, both day and night, for all stations involved. Aside from the usually larger carrier difference-frequency beats (assumed now to be zero) and much smaller transient sideband beats [5], there are also Doppler beats induced by the relative velocity of a vehicular receiver with respect to the various stations. In the diagram, "M" indicates music; "V" is varied voice; "FV" is fast-tempo voice; and "SV" is slow voice. On the abscissa are the two relative interference levels; the ordinate conveys the International Telecommunications Union (ITU) audio quality designations. The *mobile* reception scenario for overlapping co-channel ground-wave signals is depicted in Figure 7.11-2; the equations (1) and (2) for the received signals are thus:

$$f_{beat(total)} = \Sigma n f_{beat(n)} \quad (1)$$

$$f_{beat(n)} = (\mathbf{v}_{Rn} \cos \theta_n)(f_0/c) \quad (2)$$

where $f_{beat(n)}$ is the *n*th beat frequency in Hz, $\mathbf{v}_{Rn}$ is the receiver velocity in m/s relative to station *n*, θ_n is the angle of the trajectory from the radial from station *n*, f_0 is the original carrier frequency in Hz, *n* is the number of received co-channel stations, and c is the speed of light in m/s. Thus the combined Doppler-beat signal is merely the sum of the Doppler frequency components due to the relative radial velocities with respect to each station, multiplied by the inverse of the nominal RF wavelength [3].

In static conditions, where cancellations due to carrier out-phasing are deeper, the regions of higher distortion are shown in light shading (darker if audio-synchronized). As will be shortly explained, however, these effects are generally no worse than in FM multipath and are well tolerated by most all listeners. For moving receivers, the consequence of the Doppler effect is the *very* low-frequency beat-modulation of the audio envelope in mobile receivers, although several factors ameliorate the situation in real vehicular listening environments. First, the apparent modulation from near 0 to 0.3 Hz (typically less than 0.2 Hz) is largely suppressed by the action of the radio's internal feedback AGC circuitry, which rapidly and effectively levels these slow-amplitude variations to maintain a fairly constant detected carrier magnitude.

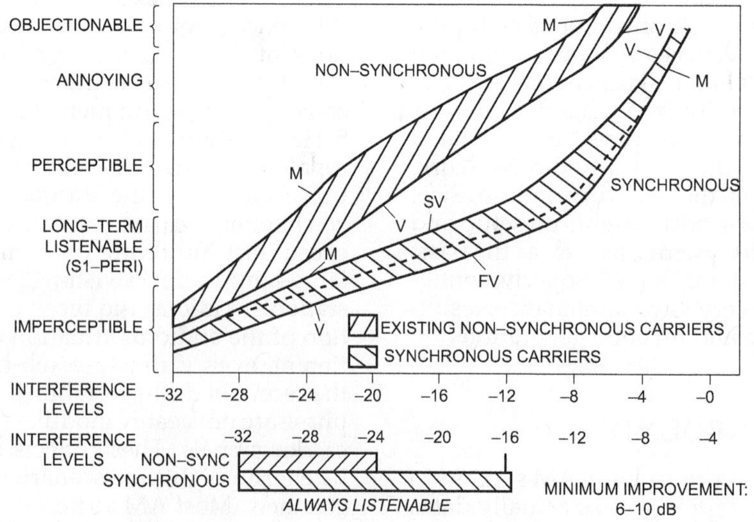

FIGURE 7.11-1 Audibility of co-channel fringe beats

IEEE

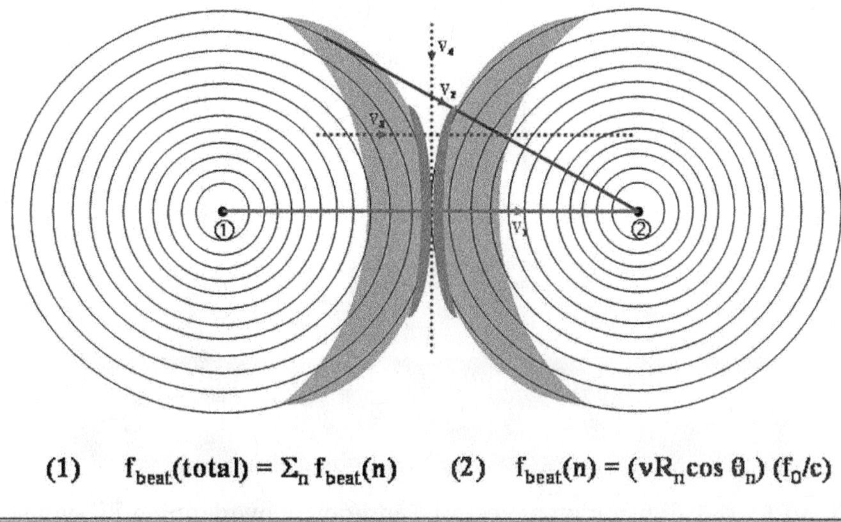

FIGURE 7.11-2 An illustration of the AM co-channel ground wave contour overlap area where the carrier beats occur, including Doppler effects for mobile receivers

NAB

Second, the presence of relatively high levels of ambient road noise in the vehicle at higher speeds, particularly in the low-frequency region of the audible spectrum, serves to mask these cyclic but low-level variations. Third, local RF field irregularities, including receiver antenna pattern nonuniformities, also cause overall level variations that dither (randomly modulate) these cyclic field variations; these variations also tend to mask the beats. When the vehicle slows and thus produces less road noise to mask the beats, their frequencies drop to negligible values and generally fall below audibility. Finally, the dynamic nature of most types of music and voice broadcast programming also inherently tends to aurally mask these very low-frequency components.

Obviously, the magnitudes of the beats will be dependent on the relative amplitudes of the two co-channel signals being received; for most areas, where the signals are at least 10 dB different in level, the resultant beats are very weak. The bottom line is that these Doppler effects are overall very minor. Compared with the standard static-receiver synchronous AM reception case, the presence of these sub-hertz Doppler beats in *mobile* listening environments typically causes a degradation (i.e., increase) in the overall beat audibility of only about 1–1.5 dB compared with the curves in Figure 7.11-1.

HOW DOES AM SYNCHRONIZATION WORK IN THE FIELD?

It is useful to examine how the phases/delays of the audio and RF components of the AM radio signals can affect reception quality in the field, particularly in signal-overlap regions. For instance, the RF signal delay is very roughly 1 millisecond for 186 miles (corresponding to the speed of light in air). At a point equidistant from two omnidirectional, co-phased (synchronous) transmitters with equal power and propagating via groundwave mode over land paths of identical RF conductivity, the two RF signals will arrive with equal amplitudes and delays (phases). Now if we assume that the RF carriers and the sideband audio signals are precisely in phase (matched in time) as they leave the two antennas, at the exact midpoint between the two transmitters the RF signals and the detected audio will also be in phase; the signals can be added algebraically to calculate the resultant. Now for points *not* equidistant from the two transmitters, the RF signals will add vectorially, as illustrated in Figure 7.11-3.

In general, there will be augmentations and cancellations of the two waves occurring at spatial intervals of one-half wavelength, essentially the same as is the case for standing waves on a mismatched transmission line. Modulation distortion will be minimal near the 0°-additive points and rise somewhat at quadrature-phase contours, and peak as the summed signal approaches null at the 180° points. Obviously, near the equal-signal points, the standing wave patterns will exhibit maximum variations; in fact, FCC Rule Section 73.182(t) [5] defines the region of "satisfactory service" for synchronous stations as areas where the ratio of field strengths is ≥ 6 dB (≥ 2:1). The Rules as quoted did not assume the accurate time-synchronization of both audio components; as cited by Reply Comments of Blanco-Pi and duTreil, Lundin & Rackley in the initial (2014) FCC AM Revitalization action [3], and confirmed by previous lab testing at Oak Ridge National

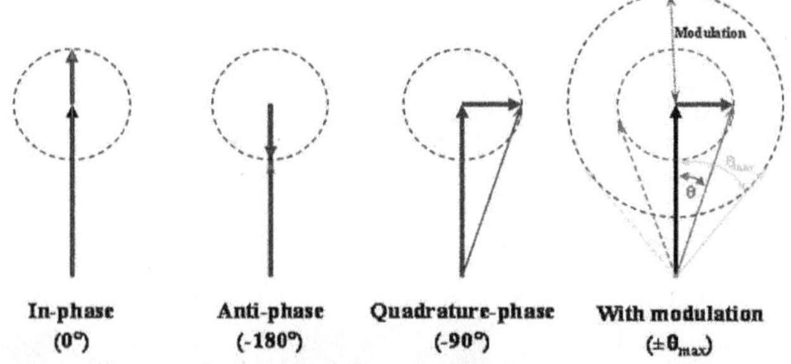

Effective interference-induced modulation & distortion levels can be calculated for stationary signals but the audibility effects are best studied via listening tests.

FIGURE 7.11-3 Illustration of vectorial addition of overlapping RF signals

NAB

Class	Channel	Contour (Day) μV/m	Contour (Night) μV/m	Interfer. (Day) μV/m	Interfer. (Night) μV/m
A	Clear	100	500 (50%SW)	5	25
B	Clear Regional	500	2000 (GW)	25	25
C	Local	500	—	25	—
D	Clear Regional	500	—	25	—

Class A stations are protected to 0.1 mV/m (0.5 at night); interferers are ≥ 26 dB down.
Class B, C, D stations are protected to 0.5 mV/m (2.0 for B night); interferers ≥ 26 dB down.

FIGURE 7.11-4 Current FCC Section 73.182(q) co-channel rules

Laboratory [3], precise audio time-matching among the transmitters significantly mitigates the apparent distortion and reduces the area of discernible distortion (darker-shaded crescent zones in Fig. 7.11-2). The current FCC Section 73.182(q) co-channel interference limits are shown in Figure 7.11-4.

The interference patterning in the synchronous overlap zone can be further reduced by phase-dithering of the booster signal(s), either in a cyclic or random-phase fashion. Terrain variations, vehicle antenna reception pattern asymmetries, buildings, and other groundwave scatterers or diffractors (i.e., multipath sources) will also reduce the magnitude of these overlap-zone disturbances via the inherent dithering of carrier phase. In moving vehicles, the audible effects will be even less, especially on speech programming. It has also been long known [2] that the distortion zones can be designed to fall over less-populated areas and major arteries. Numerous further theoretical details in the implementation of optimal synchronization of AM stations, beyond the scope of this chapter, are available [2, 3, 4, 8, 9].

AM SYNCHRONOUS STATION FIELD FRINGE-AREA TEST

In January 2015 Kintronic Laboratories conducted fringe-area AM synchronization field tests in conjunction with two 1-kW Class-C AM stations on 1400-kHz, WKPT in Kingsport, Tennessee, and WGAP in Maryville, Tennessee, which demonstrated the audio artifacts that result from the frequency difference in the overlapping regions of multiple co-channel stations.

These audio files were compared for the same geographic location with those resulting from the Global Positioning System (GPS) time-based, phase-locked carriers of the two stations' transmitters to illustrate the advantages of frequency-locking of the carrier frequencies to eliminate carrier beats in the overlapping region

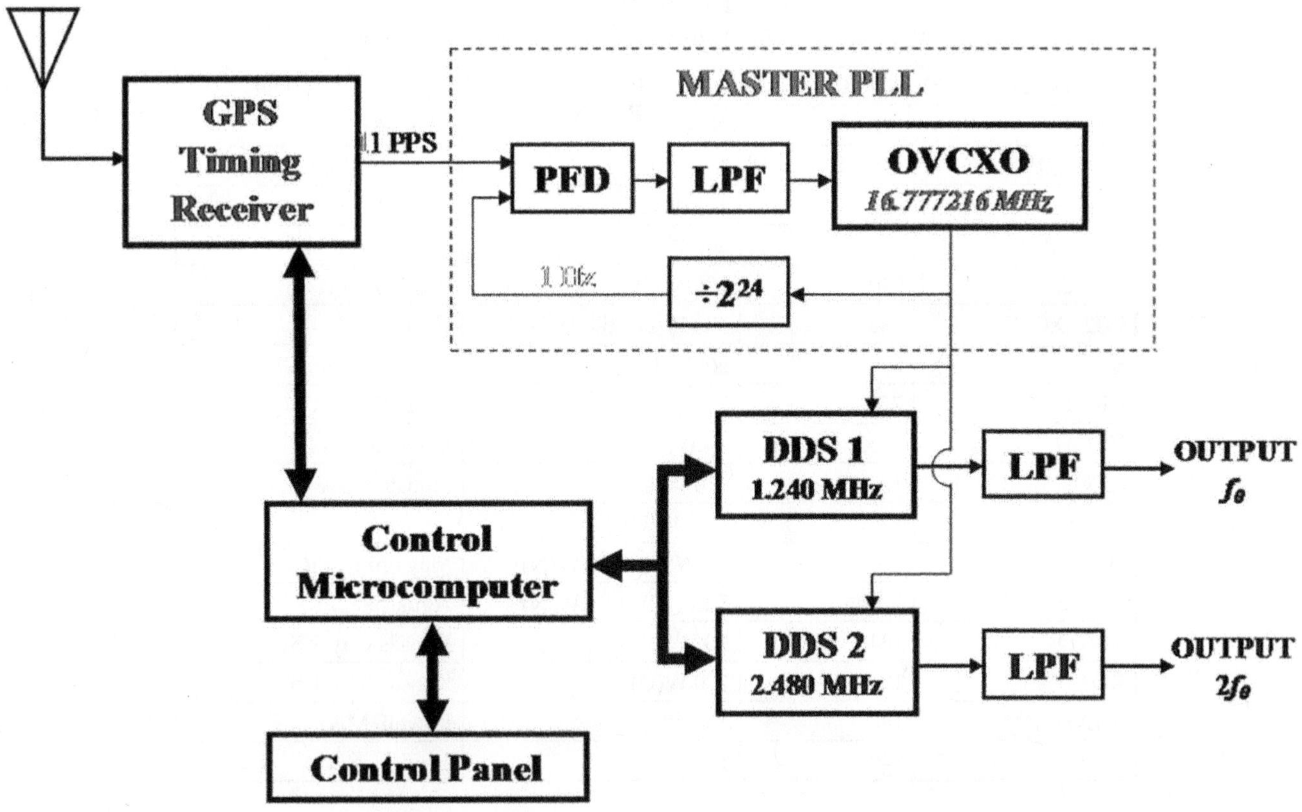

FIGURE 7.11-5 Typical AM transmitter synchronizer unit

of the two co-channel stations. The basic result was confirmed that synchronous operation results in a roughly 6–10 dB reduction in carrier beats; during mixed music and newscasts on WGAP and sports-talk programming on WKPT, the effect was to largely eliminate the background audio of the weaker station, whereas the carrier beats were quite noticeable in the same location while the two stations were offset by 1–2 Hz [8].

A representative architecture of a current-generation GPS-based transmitter synchronizer unit for 1240 kHz, using direct-digital synthesis (DDS) techniques to generate both the main carrier and a second-harmonic, to interface both newer and older types of transmitters, is shown in Figure 7.11-5. Final lowpass filters convert the stepwise DDS waveform outputs to sinusoidal signals for interface to the respective transmitters. Obviously other, more traditional types of synthesizers (i.e., integer-ratio dividers with a conventional voltage-controlled oscillator, or VCO) will also work satisfactorily for the synchronizer unit.

CURRENT AM SYNCHRONOUS BOOSTERS

The concept of utilizing AM co-channel GPS-based synchronous boosters to augment the coverage of existing AM stations is not new in the US broadcast market. There are currently eight AM stations in the United States and Puerto Rico as listed in Table 7.11-1 (from the FCC website's AM search page) that are utilizing AM synchronous boosters (denoted with an asterisk). The WAPA Radio Network in San Juan, Puerto Rico (WAPA and WISO), has been successfully operating full-power AM synchronous boosters without listener complaints for more than 20 years. All of these boosters are currently being operated under FCC experimental licenses. Very importantly, in contrast to many earlier synchronous AM implementations, both these networks have employed precise delay-matching of the RF signals *and* the respective audio modulation signals from each transmitter; this greatly reduces the magnitude and extent of the signal-distortion "mush" zones to a level comparable to the localized multipath-induced cancellation effects experienced with FM stations in hilly terrain.

Based on the lower costs of current GPS receiver technology and the accumulated experience of these AM stations currently operating AM synchronous boosters, it is our opinion that this technology should be immediately licensed by the FCC as a flexible, low-cost means to extend the coverage of existing nondirectional or directional stations, especially in populated but poorly covered areas.

Consider the example of the synchronous boosters being operated by the WAPA Radio Network in Puerto Rico, a small but densely populated island (3.7 million people in a 100-mile by 35-mile area). The unpopulated areas are relatively scarce and small. Regarding the two synchronized WAPA transmitters operating on 680

TABLE 7.11-1
Current U.S. synchronous AM & booster* stations

STATION	FREQ. (kHz)	POWER	LOCATION
WRJR	670	20 kW/3 W(N)	Claremont, VA
WR2XJR*	670	700 W(N)	Portsmouth, VA
WAPA	680	10kW(U)	San Juan, PR
WA2XPA*	680	400 W(D), 570 W(N)	Arecibo, PR
WIAC	740	10kW(U)	San Juan, PR
WI2AXC*	740	500 W(D), 100 W(N)	Ponce, PR
KKOB	770	50kW(D/N)	Albuquerque, NM
KKOB*	770	230 W(U)	Santa Fe, NM
KCOH	1230	1 kW(U)	Houston, TX
KCOH*	1230	410 W(U)	Houston, TX
WISO	1260	2.5 kW(U)	Ponce, PR
WI2XSO*	1260	5 kW[D), 1.8 kW(N)	Mayaguez, PR
WI3XSO*	1260	5 kW[D), 4.8 kW(N)	Aguadilla, PR
KDTD	1340	1 kW(U)	Kansas City, KS
KDTD*	1340	200 W(U)	Kansas City, KS
WLLH	1400	1 kW(U)	Lowell, MA
WLLH*	1400	1 kW(U)	Lawrence, MA

kHz, a typical listener can drive his vehicle from the northeast over to the central and northwestern areas of the island without detecting a problem in the station's signal quality due to the overlap of the synchronized signals; the 10-kW main station and the 400/570-watt booster have overlapping coverage areas as shown in the coverage plot of Figure 7.11-6.

The shaded land area is the 2 V/m contour overlap of the WAPA main transmitter and the WA2XPA booster. It is also important to note the small 5-mV/m overlap area along the northern coast. WA2XPA could be considered as a "fill-in" synchronous booster for WAPA, because the WA2XPA transmitter site is *inside* the 2 mV/m coverage contour of WAPA, but could be also considered a contour-expanding booster for WAPA, because the booster improves coverage to the west, without causing co-channel interference to other stations on the frequency. To fully exploit these benefits of synchronous AM coverage, the FCC will need to adopt new allocation and technical rules defining new AM synchronous boosters and their implementation; such rules are anticipated to be straightforward and consistent with existing standards for AM stations.

Although not currently permissible, the use of synchronous AM boosters to explicitly increase the coverage area of existing stations in populated areas should not be impaired or restricted, as long as no significant increase in interference to either co-channel or adjacent-channel stations would be so generated, again in full compliance with existing FCC allocation rules. In addition, boosters may be nondirectional or directional, as exemplified by the existing licensed synchronous stations listed in Table 7.11-1. The extended operational experience of the WAPA/WISO networks in Puerto Rico has confirmed the practical long-term efficacy of these techniques.

OPERATING EXPERIENCES WITH LOCAL SYNCHRONOUS AM BOOSTERS

Based on reports from a local radio engineer in Puerto Rico, who is a more qualified listener than most in the public, some *slight* cancellations are evident in the respective WAPA and WISO signal-overlap area audio, although these are probably less severe than those produced by a typical directional antenna pattern null while driving. In fact, no complaints about perceived signal distortion are ever received from the general listening public. Several other pertinent observations: first, when driving over a highway the cancellations are practically unnoticeable, as evidenced by received audio recorded in the field in these locations; second, while driving if one makes a stop at the specific site the cancellation occurs (or if a home receiver is located at a cancellation site), the spatial cancellation would be noted as a slight-to-moderate distortion in the audio, but not enough to impede listening to the station.

Moving a few feet eliminates or practically eliminates the most critical cancellation distortion. In the case of home receivers, the listener won't likely notice the cancellation because of the radio's directional ferrite antenna; in vehicles, the nonuniformity of the antenna pickup pattern will also serve to mask the cancellations. In long-term operations in Puerto Rico, Eng. Wifredo Blanco-Pi and his son Jorge G. Blanco-Galdo,

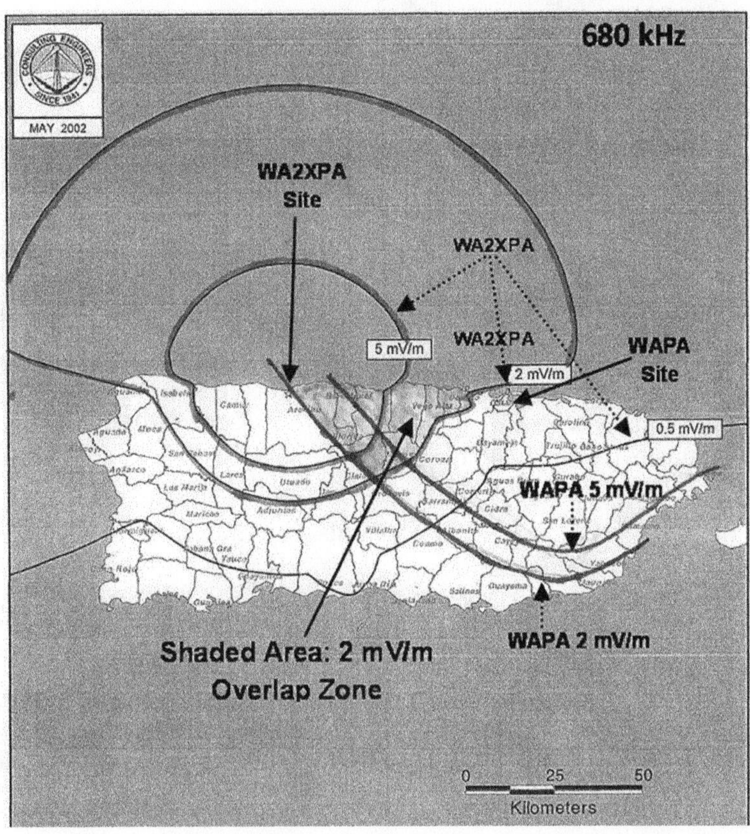

FIGURE 7.11-6 Predicted coverage contours of WAPA and its WA2XPA synchronized booster

NAB

in the implementation of their AM synchronous booster systems associated with WAPA and WISO, note that there are several guidelines that should be observed when optimally synchronizing AM stations:

1. When utilizing a transmitter without an output network at one synchronous site and a transmitter with an output network at the other site, it is difficult to accurately synchronize the signals. Both transmitters must operate either with or without an output network (this equalizes the RF delay).
2. All audio processors at both transmitters both should be the same type and set up identically in their operational parameters.
3. Don't attempt to use an audio path (analog *or* digital) between synchronized stations where there is no control over all the equipment, the path being used, or (stated more simply) over the total delay in the paths. If, for example, a fiber-optic cable from an external provider is used as opposed to a microwave link, it will not be possible to properly synchronize the audio.
4. It is essential that the net audio delay path difference be accurately compensated on the order of microseconds, not milliseconds.
5. If the main and alternate/auxiliary transmitters are not of the same type, another delay unit will be necessary to compensate between the difference in the transmitters' audio delays. It is thus straightforward to synchronize the RF frequency, the audio phase and/or delay of the synchronized stations, once all the equipment RF and audio delays are systematically analyzed. Broadbanding of the antenna systems is also highly desirable to minimize RF phase errors, both at carrier and sideband frequencies. The synchronous AM system, once set up, is very stable and generally won't need readjustment unless there is an equipment change or failure. In general, such synchronous boosters are in every respect technically identical to a standard AM transmitting installation, except for the addition of the RF and audio synchronization gear.

Several of the WAPA/WISO boosters have directional antenna arrays and are able to provide strong-signal coverage to population areas not adequately served by the main stations. Other practical uses of low-power, low-cost synchronous AM boosters would include useful fill-ins for covering populated areas lying in existing directional pattern nulls. Boosters can further be used daytime and/or nighttime as needed to provide improved coverage of outlying suburbs (especially true for Class-C stations). Obviously, the utilization of additional boosters can be easily handled under existing AM allocation rules [6].

CONCLUSIONS

A synchronization scheme to minimize beat-related interference among co-channel AM stations, both local and distant, can be implemented at relatively low cost. Although the general technique of *local* synchronization has been known and studied in the past for AM stations (both local and distant), only recently has the feasibility of assembling an *economical* wide-area (continental to even worldwide) synchronization system for AM broadcasting emerged, largely due to the availability of very low-cost (< $50) GPS timing receivers and inexpensive electronic devices such as microprocessors and logic chips. A GPS-referenced frequency-synchronizer unit capable of holding both modern and older transmitters to well within 0.5–1.0 ppb of the assigned frequency, can eliminate carrier-beat interference among co-channel AM stations. According to both our lab and field tests, the net result of the ≥ 6-dB beat reduction would be a near-doubling of the existing co-channel interference-limited coverage radius of stations (and nearly a quadrupling of the serviceable reception area), as depicted in Figure 7.11-7. Universal AM synchronization would also be extremely useful in mitigating the potentially much-increased beat levels resulting from the relaxed AM overlap rules recently (2016) proposed by the FCC [7].

For a hypothetical station in a high-conductivity area with co-channel interference-limited contours (i.e., on Class A or B channels), a 6-dB reduction in interference due to beats can provide a corresponding almost 2:1 gain in listenable range. Obviously, other local factors such as power-line noise may well intervene, but both daytime and nighttime listenable areas will improve significantly. Although essentially all stations will benefit during early-morning and near-sunset (critical) hours, this is of paramount importance to local Class-C stations, which often can no longer cover their population centers at night due to suburban sprawl, as well as Class-D stations with very limited nighttime power levels. The broad use of wide-area AM synchronization technology would also significantly improve long-distance sky-wave reception by minimizing the carrier beat effects from co-channel interfering signals, and the same synchronization techniques have been successfully proven for over 20 years in overlapping full-power booster implementations in Puerto Rico. Practical low-power boosters to fill in directional nulls (equivalent to FM translators) are also now feasible. Figure 7.11-8 provides a typical implementation scenario for a fill-in synchronous booster.

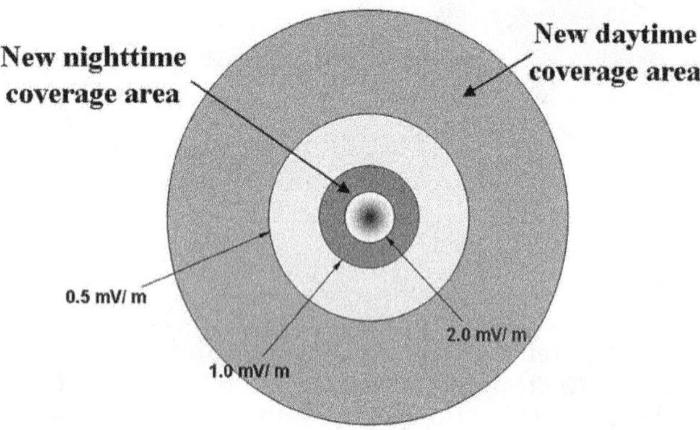

FIGURE 7.11-7 Effective synchronous day/night interference-limited coverage improvements

IEEE

ACKNOWLEDGMENTS

The author gratefully acknowledges the assistance, encouragement, and valuable technical insights offered by Eng. Wifredo G. Blanco-Pi and Jorge G. Blanco-Galdo, WAPA Radio Network, San Juan, Puerto Rico; Tom King and Jim Moser, Kintronic Labs; Jim Sexton, WGAP; George DeVault and Lyle Musser, Holston Valley Broadcasting; Jim Glogowski, Multi-Cultural Broadcasting; Brad Lynch, Trimble Navigation, Ltd.; Tony Moore, Oak Ridge National Laboratory (Retired); and Trevor Swoyer, Consultant.

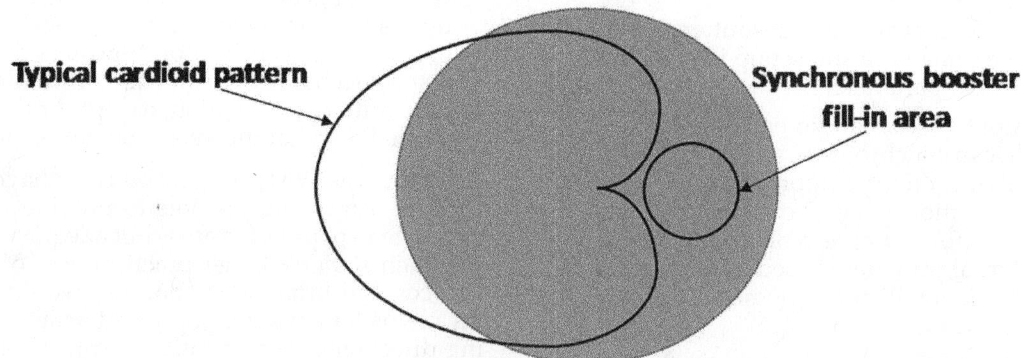

FIGURE 7.11-8 Synchronous AM booster for null fill-in

NAB

References

[1] L. McC. Young, *Present Practice in the Synchronous Operation of Broadcast Stations as Exemplified by WBBM and KFAB*, Proceedings of the IRE, Vol. 24, No. 3, March 1936, pp. 433–446.

[2] Whitaker, George, Sr., *Case History: Synchronous Broadcasting*, SBE Proceedings, 1991, pp. 101–107.

[3] Smith, S.F. and Moore, J.A., *Carrier-Frequency Synchronization System for Improved Amplitude Modulation and Television Broadcast Reception*, U.S. Patent 6,563,893, May 13, 2003 (now in public domain).

[4] Part 73 FCC Rules for AM, FM, and TV, §73.182, www.fcc.gov/mb/audio/bickel/amfmrule.html

[5] Kahn, Leonard R., *POWER-Side*, IEEE Transactions on Broadcasting, Vol. 34, No. 3, September 1988, pp. 407–420.

[6] Kintronic Laboratories, *Reply Comments in FCC Docket 13–249, "AM Revitalization"*, filed March 20, 2014, on fcc.gov Website.

[7] Kintronic Laboratories, *Reply Comments in FCC Docket 13–249, FCC 15–142, "AM Revitalization"*, First Report and Order, Notice of Proposed Rulemaking (FNPRM), and Notice of Inquiry, filed March 18, 2016, on fcc.gov Website.

[8] King, Thomas F.; Smith, Stephen F.; Blanco-Pi, Wifredo G.; and Blanco-Galdo, Jorge G., *Field Trial Results of AM Transmitter Carrier Synchronization*, Proceedings of the NAB Engineering Conference, Las Vegas, NV, April 2015.

[9] U.S. Patents 7,881,416; 7,587,017; and 7,218,696* to S.F. Smith and J.A. Moore, assigned to Oak Ridge National Laboratory (*now in public domain).

CHAPTER

7.12

AM Antenna System Maintenance

JOHN F. WARNER

Vice President AM Engineering, iHeartMedia

OVERVIEW

It is important to understand that antenna systems do not operate without considerable supervision and maintenance. Difficulties can arise either within the antenna system itself, or the environment in which it operates.

Maintenance of AM antenna systems can be broadly divided into issues of

- Electrical changes or failures
- Mechanical changes or failures.

Problems typical of AM antenna systems can be generally defined as

- Catastrophic, such as caused by lightning
- Gradual degradation or "drift."

PREVENTIVE MAINTENANCE

A schedule of routine preventative maintenance is essential to reliable operation and should be established and rigidly adhered to by appropriate technical staff. In the past, many transmitter sites were manned and rigid preventative maintenance plans were in place. This is not to say they cannot be continued today. While deregulation and advances in equipment reliability have removed operators from most transmitter sites, preventive maintenance remains key to reliable and FCC compliant operation. Record keeping is an important part of any maintenance program. Although computerized databases are normally thought of for maintenance record keeping, a paper logbook or notebook is satisfactory. Historical documentation of previous problems and their resolution is essential to solving problems. Records dating back to the original tune-up, including the consultant's field notes, are often found to be useful. Copies should be made of all documents and stored at another location other than the transmitter site. The loss of a transmitter building to fire or natural disaster not only destroys the facility, but likely destroys the information needed to replicate it as well. Appropriate lock/tag-out practices should be in force at all transmitter sites. Working alone can be dangerous and should be avoided. Work practices are governed by company policies as well as by government agencies such as OSHA.

MECHANICAL CONSIDERATIONS

Reliability of mechanical components such as contactors on relays and switches is dependent on cleanliness, good mechanical condition, and proper lubrication. Tuning unit enclosures should be sealed against the elements and the entry of insects, vermin and snakes. If the enclosure is properly sealed and in good repair, routine cleaning of the area every few months should be sufficient. A good "spring cleaning" should be scheduled to remove anything that has taken up residence over the winter. The use of a shop-type vacuum cleaner is preferred to blowing dirt around with an air compressor. Standard cleaning techniques can be used on the building or enclosure. Avoid using soap-based cleaning products on electrical components as they may leave a residue. Denatured alcohol or ammonia is often used for this purpose, but the area should be well ventilated and proper personal protective equipment such as gloves should be used. Follow label cautions as to mixing various cleaning agents. Lubricants that are oil-based are not appropriate in areas where dust can collect. Pivoting parts of contactors should be cleaned as necessary with a solvent and lubricated with graphite. Solvent label instructions and warnings should be followed. If possible, the contactor should be removed to a well ventilated area. Many stations stock a spare for each type of contactor that allows for removal of a unit from service for maintenance.

Repeated use of abrasives on silver-plated surfaces is not a good practice. Spring finger contacts should be replaced if they have become burned or distorted. Even though overheated contacts can often be bent back into shape, their spring properties will be lost. Spring contacts from various vendors may look the same, but it's always best to buy the correct ones. Close examination of contact surfaces is needed to differentiate between contacts that have burned black and the natural oxidation process. Silver oxide is black and shiny and should not be disturbed as it is still a good conductor.

Improper operation of micro switches is often a source of intermittent problems with directional antenna systems. Micro switch mounting hardware should be tight and the position of switch in relation to the moving arm of the contactor should be as specified by the manufacturer. In general, when the movable arm of the contactor is fully seated, the associated micro switch should be fully depressed and a slight amount of pressure exerted on the movable actuator of the switch. A switch that is working properly will usually emit an audible "click" when its actuator is fully depressed. An ohmmeter can be used to check the electrical functioning of a micro switch once the associated control circuit has been de-energized. Most contactors have two pairs of micro switches. One pair is a tally of contactor position for the related pattern change logic and the other is to ensure that only one contactor solenoid can be energized at a time in the event of logic failure. In the case of traditional pattern control logic systems, the failure of one of the micro switches in series with the solenoid coils will prevent that coil from being energized when a pattern change command is applied. The failure of the tally micro switch will make the controller think the contactor hasn't moved and will hold the transmitter off.

Many approaches have been taken to pattern change logic design. They often contain a large number of traditional relays that can be sources of failure or intermittent operation.

Systems that use highly reliable programmable logic controllers (PLCs) are now available. PLCs are used by the millions in industrial applications and they perform well in systems that are exposed to voltage transients and other anomalies. The timing and sequencing functions are programmed using a ladder logic program and then loaded into the PLC. External relays are sometimes used to buffer the controller from high AC switching voltages used on older systems. Modern systems use low-voltage DC for control and tally indicators. Figures 7.12-1 and 7.12-2 show a well-designed and laid out control and tally panel for a multi-tower dual AM transmitter facility. Note that the control buttons and tally indicators are clearly marked. A key switch disables remote control functions to eliminate accidental changes while conducting local maintenance. All control and status indicators can be sent to a remote control and metering point.

Mechanical drive mechanisms and turns counter in phasors should be checked for freedom of movement and setscrews should be checked for tightness each time the system is shut down for inspection and maintenance. As with tuning units, the phasor should be cleaned on a regular basis. If knife switches are used for meter switching, they should be kept clean and

FIGURE 7.12-1 Control panel of PLC transmitter and antenna system controller

Tunwall Radio

FIGURE 7.12-2 PLC controller of current design. Logic is contained within the PLC and can be reprogrammed to change timing functions as needed.

Tunwall Radio

their contact pressure should be firm enough to ensure proper operation.

Signs of heating or discoloration at connections points are usually a sign of loose hardware. Bolted connections should be checked for tightness when the system is de-energized. Once the system is locked/tagged out, carefully touch the components and connections in the system to determine that excessive heating is not occurring. Infrared temperature sensors are excellent for finding hot spots. Records should be kept of all turn counter positions if counters are present in the system. If line or base current meters are present, their indications should be noted before the system is shut down. Taking time to provide thorough documentation of the antenna system will pay off in the long run.

ELECTRICAL MAINTENANCE

The amount of electrical maintenance needed by an antenna system is directly related to its age, complexity, and maintenance history. Two of the biggest issues seen in phasing and coupling systems are the deterioration and failure of mica capacitors and intermittent operation of variable inductors.

A mica capacitor is essentially a stack of metallic foil sheets interleaved with layers of mica insulating material. In effect, these layers can be viewed as a group of capacitors in series. As depicted in Figure 7.12-3, the total capacitance of the device follows the formula for capacitors in series.

Should one section of the capacitor short, the total capacitance of the device will increase. This potentially begins a domino effect. As the capacitance increases, the capacitive reactance decreases. A decrease in reactance allows more current to flow, increasing the heating of the capacitor. Additional heating can cause an additional section or sections to fail and the process repeats itself, often to the point of short circuit of the device. Heating can be detected by touching the component once the system has been de-energized, or through the use of an infrared temperature sensor. Infrared sensors are important tools for the broadcast engineer. The cost of these sensors had dropped to the point where they are affordable by the smallest station or contract engineer. Another sign of impending failure of a capacitor, inductor, or transformer is the appearance of potting material oozing from the component.

The gradual failure of a mica capacitor is usually accompanied by the gradual drift of a parameter or parameters that can't be attributed to environmental effects. This process begins as the change or drift of a system parameter to the point where corrective steps

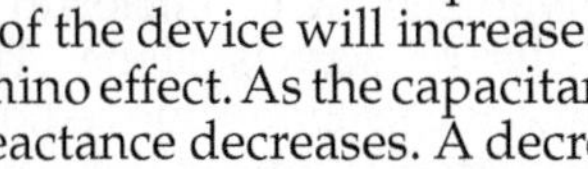

$$1/Ct = 1/C1 + 1/C2 + 1/C3 + \ldots$$

FIGURE 7.12-3 Mica capacitors are actually a group of capacitors in series

are necessary. To compensate, an adjustment or adjustments are made to the phasing system. The drift continues in the same direction until another adjustment is needed. This process continues until the adjustable element comes to the end of its range. This is a critical point in the failure process. At this point, further adjustments should be avoided and an attempt should be made to locate the failing component. Anytime a parameter continues to drift in the same direction, non-user adjustable components should be evaluated. An inexpensive capacitance meter can be purchased for less than the cost of an hour of a broadcast consultant's time. Any capacitor that has increased in capacitance by a factor or 10% or more is suspect and should be replaced.

Unlike mica capacitors, vacuum capacitors are not known to change in value. Either they short or are physically destroyed by lightning.

Inductors with a roller contact are often sources of trouble if they have to be moved after years of sitting in the same position. The protective circuits in modern transmitters are so fast and so sensitive that visible arcing usually doesn't develop. Bad contacts in roller coils can often be heard to be "microphonic," in concert with modulation. The area where movable rollers contact the inside of coils can be cleaned by using silver polish sparingly, or they can be removed, disassembled, and cleaned ultrasonically. If any chemical cleaner is used it should be removed thoroughly by flushing with clear water and drying before reassembling the coil.

LIGHTNING AND SUDDEN COMPONENT FAILURE

Sudden component failures are often related to lightning activity or the inadvertent application of power exceeding design values. Most lightning discharges can be effectively dealt with on insulated towers by the proper installation and adjustment of spark gaps. Gaps should always be installed across tower base insulators and can be installed across vacuum caps and at the ends of transmission lines. A preliminary spark gap setting can be calculated based on the voltage present using Ohm's Law referenced to the complex base impedance and the base current.

$$E = I_b Z_b$$

The voltage at carrier multiplied by a factor of four (X4) will yield an approximation of the voltage to be expected with full modulation. Under dry conditions, and with large conducting surfaces, the breakdown voltage of air is considered to be 30 kV/inch. This spacing can be used to calculate the initial gap setting when ball gaps are used, and the balls are large in relation to the gap dimension. As a practical matter, begin with this spacing and decrease the gap till either arcing or transmitter protective reaction is noted and then open it slightly till the arcing stops. In most cases gaps are found to be excessively large although they are often left larger to prevent premature firing during rain, snow or icing events.

If lightning damage is a chronic problem the grounding and bonding around the tower base should be investigated. Ground straps at the tower base can be supplemented with ground rods driven adjacent to the tower base. All grounding and bonding connections should be made with silver solder or by brazing. For an in-depth discussion of lightning protection see Chapter 9.2, "Lightning Protection for Towers."

The buildup of static charges on the tower proper can be dissipated by a properly grounded static drain choke or by bonding the neutral wire of the tower lighting choke to the tower on one side and ground on the other side. Local interpretations of the National Electrical Code may forbid this practice in some jurisdictions, so a local electrical contractor or the regulating body should be consulted.

Tall towers at the low end of the AM band are often plagued with static discharges across guy insulators. The electromagnetic fields generated by these discharges cause the operation of the protective circuits of modern transmitters and nuisance carrier interruptions. Inductive chokes or high value resistors across guy insulators may be helpful in these situations, but the assistance of an experienced consulting engineer will probably be necessary.

FIGURE 7.12-4 Multiple ground straps connect the antenna tuning unit to the ground radial system. The straps have been bolted to the tuning unit legs for mechanical stability and to reduce vibration due to wind.

John F. Warner

GROUND SYSTEMS AND BONDING

Grounding and bonding of antenna tuning units and other hardware in proximity of tower bases is of utmost importance to maintain stable antenna system operation. All ground radials should terminate at a common ring of copper tubing surrounding the antenna base pier. Ground straps that cross under the base insulator should be bonded to the ground ring as well. These straps provide a point of attachment for the tower spark gap. The ground ring should be supplemented with a minimum of four (4) driven ground rods for lightning protection purposes. A minimum of one 4-inch ground strap should bond the antenna tuning unit ground surface to the tower ground system described earlier. The outer conductor of the transmission line as well as the sample line should be bonded to the antenna tuning unit ground surface. The use of multiple ground straps to the antenna tuning unit is encouraged as they provide a low impedance connection between the tuning unit and ground. An example of base grounding is shown in Figure 7.12-4. Multiple ground straps from the antenna tuning unit cabinet are shown as well as the ground straps that cross the tower base, under the base insulator. All straps have been brazed to each other and to the copper ring that terminates the radials at the tower. In order to keep vegetation under control, the area surrounding the tower base has been leveled and covered with several layers of "viscolene" sheeting or landscape fabric. The sheeting is then covered with washed gravel to hold the sheeting in place and to keep the ultraviolet rays of the sun from causing deterioration. This method prevents weed seeds from reaching the soil and sprouting.

When mechanical connections are not silver soldered or brazed, be careful to not mix dissimilar metals. Discussions of the connection of dissimilar metals and the galvanic series can be found in engineering texts. In general, copper should not be mechanically connected to galvanized steel enclosures or tower members with bolts or screws alone. The point of attachment should have its zinc removed down to bare steel. Using a flux-coated brass rod the point of attachment is "tinned" with a thin layer of brass. The copper feed conductor can then be brazed or silver soldered to the tinned area. Usually a tower will have a small nonstructural tab for attachment purposes. The structural elements of the tower should not be excessively heated.

If antenna tuning units are to be housed in metal cabinets or buildings the seams of the cabinets should be bonded and the structure then bonded to the antenna ground system. The same procedure should be applied to metal doors on frame or masonry buildings, especially when dealing with high power levels.

If a new transmitter building is being built or an existing building refurbished, it is important that attention be given to grounding and bonding. In a new installation, a ground strap buried outside the perimeter will serve to keep lightning and other transients out of the building. The strap should be buried below grade and supplemented with driven ground rods at intervals equal to the length of the rods—that is 8-foot rods at 8-foot intervals. All equipment that is installed in the building should be bonded to ground straps that are attached to the floor. These straps should leave the building at grade and attach to the buried perimeter strap. The electrical service grounding conductor as well as telephone company grounds should be bonded at a common point outside the building. The outer conductors of the transmission and sample lines should be

FIGURE 7.12-5 Interior grounding straps through slot in side of building bonded to building perimeter strap and electrical service grounding conductor

John F. Warner

bonded to this same common strap where they enter the building using grounding kits supplied by the cable manufacturer.

If a prefabricated transmitter building is to be used, slots can be provided at floor level for the passage of ground straps to the outside. Slots can be cut into existing structures as well. An example of this concept is shown in Figure 7.12-5.

The purpose of this important practice is to have all potentials rise and fall together. A lightning strike that enters a transmitter building on a power line and exits via the RF transmission line can cause considerable damage. For more information on facility grounding see Chapter 11.2, "Facility Grounding Practice and Lightning Protection."

GROUND SYSTEMS, RF PERFORMANCE

Array instability and a gradual deterioration of coverage are indicative of ground system defects or deterioration. The 1937 article on ground systems by George H. Brown remains the definitive work on the subject and should be read and reread by serious students of AM antenna systems. A copy of this and other Brown articles can be obtained from IEEE for a nominal fee at http://ieee.org/publications.

The antenna should be considered as a series circuit consisting of the radiation resistance of the antenna, the reactance of the antenna, and the ground resistance. In the case of antenna efficiency, the reactive power does no work and can be ignored. The amount of energy radiated is controlled by the ratio of the radiation resistance divided by the sum of the ground and radiation resistances and is expressed as

$$Rr / (Rr + Rg)$$

where Rr is the radiation resistance and Rg is the ground or loss resistance. Note that as the ground resistance increases, the efficiency of the system decreases. In determining efficiency the FCC assumes a ground resistance of 1 ohm. A standard AM ground system consists of 120 radials at least one-quarter (1/4) wavelength long. Radials around towers taller than ninety (90) degrees are often as long as the tower is tall. In addition, a ground screen or mesh may have been in the area around the base of the tower. The size of the screen varies from installation to installation but usually extends 30–40 feet from the tower base. Radial wires should be bonded to the perimeter of the screen as well as to the tower base ring. Instead of using ground screens, the practice of installing an additional 120 shorter radials between the full length radials is a popular option. These shorter radials are usually 50 feet in length.

In the case of the nondirectional antenna, the base resistance can be measured and compared to historical measurements of base resistance as an indicator of ground system condition. If the ground system has deteriorated the measured resistance will rise as the sum of the radiation resistance and the ground resistance increases. In the case of the directional antenna, deterioration of the ground system is often indicated by a decrease of all the tower base currents compared to historical values. The deterioration can be indicated by parameter changes with changes in soil moisture as well.

A physical inspection of radials in the area of the tower base can be performed by careful excavation of the area with a shovel. Unfortunately, soft or lead solder was occasionally used in the past and radials have often become disconnected from the rest of the system. Radials can be evaluated by using a shielded loop connected to the external input of a field strength meter, or by using a locator such as those used by utility companies. The latter method is used when the antenna system is de-energized. If adjacent radials can be found at their outer ends, an ohmmeter can be used to determine if they are continuous to the tower base and back. The resistance of the entire circuit to and back from the tower base should be comparable to that calculated from wire resistance values found in textbooks. The period immediately after World War Two saw a surge in AM antenna construction. At that time copper clad wire was readily available and was often used for ground radials. Over the years the copper has deteriorated and left the steel to rust. After 70 years there is little evidence to be found of a ground system.

Another, although time-consuming, method of evaluating ground systems is to make field strength measurements to determine the unattenuated field strength at 1 kilometer. The unattenuated field strength is determined by analysis of field strength measurements made between 0.5 and 3.0 kilometers. Measurements should be made at intervals as close as possible to every 0.2 km and are made on six or more radials. The data is plotted on log-log paper and compared to a family of curves found in the FCC Rules. Field strength data may also be provided to a qualified consulting engineer for analysis and recommendations.

In general, it is important to pay particular attention to that portion of a ground system that is installed in a marshy area or in areas of clay, peat, or other expansive soils. These areas move more during freeze and thaw cycles and the periodic movement of the conductors can cause work hardening and breakage. Stations in the northeast part of the United States appear to suffer more ground system deterioration than other areas due to acidic soil caused in part by acid rain. Acid soil conditions can be readily identified by soil testing and the presence of acid-loving plants. A local agricultural extension agent can be consulted for direction in remedying acid soil conditions. A few tons of lime spread over a tower field is much less expensive than the replacement of a ground system.

TRANSMISSION AND SAMPLE LINES

Transmission lines are often buried and therefore "out of sight, out of mind." Little can be done to inspect buried transmission lines visually. The polyethylene-like material used for modern coax jackets are suited for direct burial if not damaged in the burial process. Cables should be buried below the frost line in soil that is free of sharp rocks or rubble. If stone layers such as shale are present the trench should first be backfilled with a 6-inch layer of clean sand, followed by the cables and another cushioning layer of sand. Sharp stones should be removed from the backfill material to the extent possible. Where the budget allows, installation of large diameter conduits is preferred. The additional material expense will likely be a small percentage of the cost of the excavation.

CABLE FAULT LOCATION

A broken outer conductor of a transmission line is very difficult to locate. If this is suspected, simple DC continuity measurements may be of limited value in locating the fault due to the existence of numerous parallel ground paths. A time domain reflector (TDR) can be used for fault location but the operator should be experienced in its use. If an incorrect velocity factor is used with a TDR, the distance to the fault (in feet or meters) will be wrong. It is important to measure the distance to the fault from both ends of the cable and use that information to confirm the fault location proportionally. If the cable is determined to be open or shorted by an ohmmeter an RF bridge can be used to locate the fault as described later. If available a vector network analyzer (VNA) can be used in place of a bridge. Current leakage through the dielectric of a cable can be quantified with a high-voltage leakage detector, known as a "megger." This device applies several hundred volts across an open line and measures the current. A good cable will have a leakage in the range of a few microamperes or tens of microamperes.

The reactance values presented by a resonant transmission line are important in antenna array analysis, design, and troubleshooting. These principles are explained in depth in various books on antenna and transmission line theory as well as in the *ARRL Radio Amateurs Handbook*. Some of those basic principles include the following.

If a transmission line is terminated in a short, the short will repeat itself every half wavelength. When using an impedance bridge for these measurements the reactance at resonance will be zero (0), but there may be a few ohms of resistance due to the resistance of the copper conductors. A short length of copper strap should be used to short the line at the far end. A clip lead is not suitable for this application. Look for the lowest frequency at which your impedance bridge indicates zero (0) reactance. The length of the line under test, in degrees, can be calculated as a simple ratio.

Null frequency (kHz) / 180 = Operating frequency (kHz) / line length in degrees

Likewise, an open line that is 1/4-wavelength long will appear as a short at the far end. The ratio for this calculation is:

Null frequency (kHz) / 90 = Operating frequency (kHz) / Line length in degrees

When converting these distances from electrical degrees to feet, remember to use the correct velocity factor in your calculations.

Pressurized lines will help to keep moisture out and provide more reliable operation. While foam dielectric lines are generally less troublesome they can create problems if the foam becomes waterlogged. In addition to a decrease in insulation resistance, water in the foam will change the dielectric constant and velocity factor of the line, therefore the electrical length of the line. Because dielectric constant can't always be reliably determined with a TDR, impedance bridge measurements as described earlier are usually more accurate. A bridge or network analyzer can be used to determine the actual characteristic impedance of a line as well. The line is measured both shorted and open at any convenient frequency. The frequency selected should not be near the resonant frequency of the cable under test. The open and shorted resistance and reactance values are first converted to polar form. An inexpensive hand calculator can be used for the conversion. An example follows.

A length of very old RG11 cable was measured both open and shorted at 1.5 MHz. The reactance readings were normalized for frequency with the bridge in use and the results were as follows:

Open	Resistance	4.7 ohms
	Reactance	–j105 ohms
Shorted	Resistance	2.1 ohms
	Reactance	+j105 ohms

Using the handheld calculator the rectangular values were converted to polar values as follows:

Open	105 @ –87 degrees
Shorted	56 @ +87 degrees

The polar values are multiplied using standard polar convention by multiplying the magnitudes (105 × 56 = 5880) and adding the angles (–87 + 87 = 0). Because the angles cancel they can be ignored and the characteristic impedance of the cable found by taking the square root of the magnitude (5880), which yields a characteristic impedance of 76.68 ohms. This is close to the characteristic impedance (75 ohms) published for this cable by the manufacturer.

When making this or any other RF bridge measurement it is important to use a very short conductor when short circuiting the cable. A small piece of copper strap or braid is preferred.

SAMPLE SYSTEM ISSUES

As often as not, perceived array problems may actually be sampling system problems. If the monitor point values are within limits and the common point is presenting the proper load to the transmitter, chances are the improper antenna monitor values are due to a failure of some part of the sample system.

Antenna monitor values are meant to be an indication of the fields in the individual elements of an array. Samples from each tower are extracted either by loops mounted on the tower or by toroids that sample the current at the base of the tower. Toroids, while popular, due to their simplicity are inappropriate in the case of electrically tall towers and towers that support other systems such as FM antennas. This is because the toroid cannot distinguish between the current flowing in the tower and the current flowing in an isocoupler or other circuit element that is shunted across the tower base. These effects and therefore the current magnitudes and phases can be calculated using nodal analysis. This process is explained in Chapter 7.13, "Antenna System Moment Method Modeling."

Samples that are extracted from the towers are sent to the antenna monitor via lengths of small diameter coaxial cable. Specifications for antenna sample systems can be found in the FCC Rules as well as FCC policy statements on good engineering practice. Sample lines should be of equal length. This preserves the phase relationships of the towers with those indicated on the monitor. In an array where the sample system has been properly designed, installed, and is operating properly, there will be close agreement between the monitor values and the theoretical array parameters. An excellent paper, "Review and Analysis of Medium Wave Directional Antenna Sample Systems" by Lockwood, Jones, and Folkert, is available at nominal cost from http://ieee.org/publications.

Difficulties with sample systems employing loops are usually mechanical in nature. Broken leads on the sample loops or poor connections to the loops are often the cause of incorrect or intermittent measurements. The hardware used to mount the coax connector to the loop often comes loose due to vibration caused by wind. Circular breaks of the outer conductor may also occur. These breaks are often hidden under the polyethylene jacket of the cable. An open sample line outer conductor in the vicinity of the loop will cause a higher than usual voltage to be produced because the effective area of the loop is increased. Because unterminated sample lines can develop voltages high enough to cause painful burns, be careful when handling them.

If the reference tower sample is missing or very low, the antenna monitor indications of the other towers may be random and variable. If the reference sample is high due to a poor cable outer connection, the higher voltage present will make *all* the other ratio indications low in amplitude, but the phase indications will be correct or nearly so.

Many of the modern antenna monitors have an amplitude mode that allows for the measurement of the actual RF voltages applied to the monitor. These voltages and their phases should be recorded for future reference.

Because the antenna monitor is essentially an RF voltmeter with phase-measuring capabilities, a dual-trace oscilloscope can be substituted for troubleshooting. The voltages indicated on the scope can be compared to find the voltage ratio values, and the time difference between zero crossings can be used to mathematically determine phase. A tee connector can be installed where the sample line attaches to the antenna monitor and the oscilloscope lead attached to the other side of the tee. This will terminate the sample line. This is most important with loop sampling as high voltages may be present if the line is not terminated.

A length of flexible coax and an RF oscillator can also be used to determine proper antenna monitor operation. The oscillator is fed to the flexible cable and then to the reference input of the monitor through a tee connector. The other end of the coax is attached to the other monitor inputs in sequence. The loop ratio indication of the reference and tower under test should be the same within a few percent. The tower under test will have a negative phase indication that is related to the length of the flexible cable. A 100-foot length of polyethylene dielectric coax such as RG141 (velocity factor = 0.66) is about 55 degrees long at 1 MHz or 83 degrees long at 1.5 MHz. In the case of those antenna monitors that are made to operate on only one frequency, the numbers given here can be scaled to find lag at the frequency of interest.

Aging of power supply filter capacitors and the failure of mercury wetted relays are common problems in antenna monitors, along with lightning damage. Although power supply problems can usually be repaired locally, returning the unit to its manufacturer for repair and calibration is usually the best choice. If measurements from the monitor are to be used in an FCC filing, the calibration date of the monitor should be provided. The Commission is looking for a calibration in the last 3 years.

While a simple continuity check can be used to test the integrity of a sample loop and line, a traditional ohmmeter is often affected by RF energy in the vicinity. A simple alternative involves the use of a transformer with a low-voltage secondary, such as a

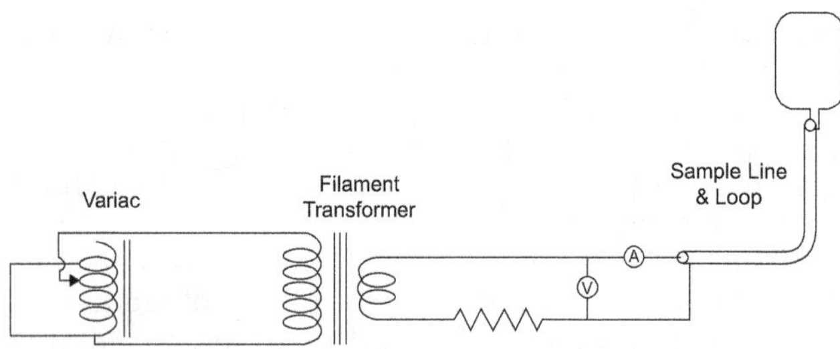

FIGURE 7.12-6 Configuration of low-voltage transformer for determining sample system continuity

filament transformer, an AC ammeter, and a "variac." The primary of the transformer is connected to the AC line through the variac and the secondary is connected across the sample line at the transmitter building. A current limiting resistor of 1 or 2 ohms can be connected in series with the secondary as well. The variac is adjusted to produce 2 or 3 amps of current flow in the secondary. The secondary voltage of the transformer is then read and the resistance of the loop and line calculated using Ohm's law. A schematic of this configuration is shown in Figure 7.12-6.

The resistance of the loop and line should be in the low single ohms range. More importantly, if all sample lines are the same length, as is usually the case, the resistance readings should be very close to the same value. A significantly higher resistance on one line indicates a problem with a connector, a broken outer conductor, or a poor mechanical connection on the loop.

Toroids can be checked by moving them around in the system to determine if an abnormal reading follows one particular toroid. Most toroids used for sampling employ an internal termination resistor that can be damaged by lightning. Toroids are not field repairable and should be sent back to their manufacturer for repair and calibration. When removing and replacing toroids, note the arrow stamped on top of the case. While reversing the toroid won't make a difference in the sample amplitude, the phase will be reversed by changing the current sense relationship between its winding and the conductor passing through it.

BASE CURRENT METERS

Although FCC Rules no longer require base current meters, they are still an effective troubleshooting tool. Provision should be made for measuring base currents in newly designed systems. Meter jacks should be present at the output of each antenna tuning unit and appropriate meters should be purchased and kept safely in the transmitter building. Meters stored this way will be safe from lightning and comparisons can be made with currents measured when the array in built and tuned. If the budget allows, toroidal meters can be installed in each antenna tuning unit.

MONITOR POINTS

Monitor points are selected on each protected radial in a directional array when its initial proof-of-performance is conducted. Points are selected to be indicative of the signal strength on the radial in question. Points that were selected many years ago may no longer be suitable due to changes in the electromagnetic environment. If the field strength at a monitor point is above the licensed limit, further measurements should be made before any adjustments are made. Eight to ten additional points should be measured on the radial and the measurements compared to previous measurements. If the entire radial is found to be above its construction permit limit, measurements should be made on the other monitor point radials to establish a baseline for readjustment. While this may seem time-consuming, it is important to note that parameter adjustments affect radiation in all directions, not just the direction with the abnormal measurement. The concept can be visualized by putting a rubber band on a table. If you take your forefinger and thumb and squeeze on the rubber band, it will get smaller where the pressure is applied, but bulge out in other directions.

If any ratio parameter has moved more than 5% from its licensed value or any phase moved more than 3 degrees, a partial proof-of-performance will have to be conducted in order to license the new parameters. The partial proof-of-performance process is discussed in detail in Chapter 10.3, "AM Radio Antenna System and Field Strength Measurements."

RE-RADIATORS AND THEIR IDENTIFICATION

Before any attempt is made at readjustment the area around the array should be scouted for new tower or power line construction. These re-radiators are usually close to the array if they have enough influence on the array to raise the level of a whole radial.

Re-radiators are usually found in the main lobe of the array where they are illuminated with a high signal level as opposed to in a minima area where levels are much lower. If the re-radiator is within a few wavelengths of the antenna it is possible but not probable that the antenna monitor values have changed. Re-radiators that are located in pattern minima may affect the readings obtained at a monitor point, but usually won't move an entire radial above its limit.

The increase in the number of cell and other communications towers built in the last 30 years have had significant impact on AM directional arrays and the engineers who maintain them. FCC Rules require owners of these structures to notify broadcasters when they are built within 1 km of nondirectional antennas and 3 km of directional facilities. Measurements must be made both before and after construction of the new tower to assure that the antenna pattern of the station has not been affected. The directional station will have to cooperate in any measurement process by operating with their night pattern during daytime hours. Before and after measurements should be conducted within as short a time span as possible so that changes in environmental conditions don't affect the results. The process can be expedited by supplying the person making the measurements with copies of relevant measurement locations found in proofs as well as copies of maps used previously. In lieu of field measurements, moment method models of the directional array with the suspect tower included in the model can determine the effects of the re-radiator. The Rules governing disturbance of antenna operation by nearby structures can be found in FCC Rules Part 1.

IDENTIFYING RE-RADIATORS

AM antennas are often located in places where extensive development may be taking place. Their environment is made up of the ground they are built on, the surrounding terrain, the ground radial system that is installed, and other nearby man-made structures that may be new or modified. The likelihood of significant re-radiation is related to the height and effective radius of any new metallic structure and its proximity to the array. A method of determining if an object is reradiating significantly is the use of a field strength meter in the vicinity of the suspected structure. This method takes advantage of the highly directional characteristics of the loop antenna of the instrument.

A point is located a distance away from the suspected structure approximately equal to the height of the structure. The point should be located such that when the field strength meter antenna is pointed directly at the array, it is oriented orthogonally, or ninety (90) degrees off axis of the reradiating object.

TEST EQUIPMENT

Measurements of AM radio frequency energy require the use of specialized test equipment. The most commonly used devices are the RF impedance bridge and an RF generator/detector. All bridges operate on the principle of a Wheatstone bridge in which unknown quantities of resistance and reactance are compared to closely controlled and known values of resistance and reactance in opposing arms of the bridge. Each type requires a signal source and a means to detect the balanced condition of the bridge.

There are two common types of bridges described by the means in which they operate. The bridges such as the General Radio model 916 or model 1606 are known as cold bridges as they are used with the circuit under test de-energized. The cold bridge is calibrated against standards of resistance and reactance before use. The unknown is then connected to the bridge, along with an RF generator and detector, and the resistance and reactance controls adjusted until there is a minimum signal indicated at the detector output. The dial indications are then read and corrected for the operating frequency in use by dividing the dial numbers by the operating frequency in megahertz.

Many of these fine old bridges are found lying around transmitter sites but are missing their operating instructions. Copies of these manuals can sometimes be found on the Internet.

The other type of bridge commonly found is the Delta operating impedance bridge (OIB). This bridge is known as a hot bridge because it can be used in an energized circuit as long as the power applied is below the rating of the bridge. This mode of operation is shown in Figure 7.12-7

In this example the bridge is measuring the impedance at the input of an antenna tuning unit. Note that the J plug has been removed to allow the bridge to be inserted into the circuit. The input on this type of bridge is on the right side of the instrument and the output is on the left. Most phasing system manufactures strive to design their products so that these bridges can be inserted without crossing the leads. In the case of a circuit where there is negative power flow, the leads are connected in the opposite direction. In that case, the tuning unit becomes the input and the transmission line becomes the output or load. The dial readings on this type bridge are multiplied by the operating frequency in megahertz to obtain the correct values. This is the opposite of the General Radio bridge. Recently the vector network analyzer has taken the place of the traditional bridge for many applications. The paper "Swept Frequency Techniques for Evaluating AM Antenna System Bandwidth" by Ron Rackley is an excellent tutorial on the technique. It can be downloaded from http://dlr.com.

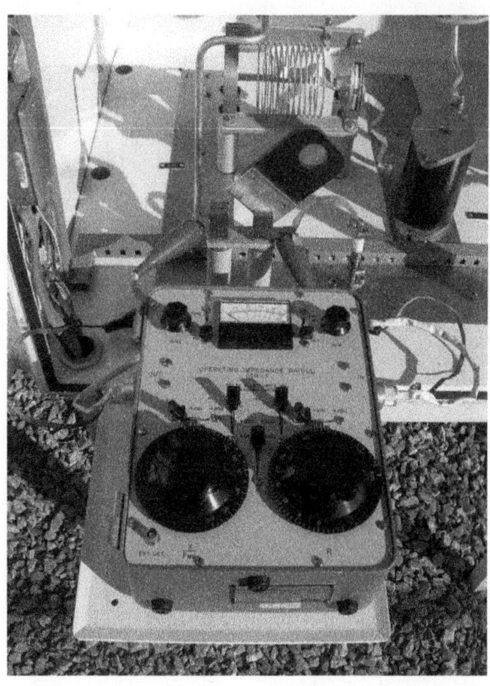

FIGURE 7.12-7 Operating impedance bridge at input to antenna tuning unit

SUMMARY

The maintenance of an AM antenna system includes not only transmitter and tower but all components related to the system as a whole such as antenna tuning units, sample lines and monitoring. The work must be meticulous and well documented. Testing takes patience and the proper test equipment. The result of conducting AM antenna system maintenance in a methodical and careful manner will be reliable and stable operation for the long run.

CHAPTER

7.13

Antenna System Moment Method Modeling

W. CRIS ALEXANDER
Crawford Broadcasting Company
Denver, Colorado

OVERVIEW

Sinusoidal Current Distribution

Since the earliest days of radio, analysis of the vertical radiators and directional arrays of AM radio stations has made use of assumed current distribution for each tower. For the sake of simplicity, uniformity and convenience, sinusoidal current distribution was assumed. This simplified the mathematical integration of current along the length of each radiator for calculating its radiation characteristics. The reality is that this wasn't a bad assumption—it was the best we had to work with. But out in the real world, and especially when dealing with a multi-tower directional array, top loading and towers in the presence of nearby conductive objects, that validity of that assumption went out the window.

When radiation occurs, energy leaves the antenna and the forward and reflected tower currents are attenuated as they propagate on the tower. As a result, the current vectors get progressively smaller as they go up and then down the tower. When you sum the forward and reflected current vectors together, understandably they cannot trace a sinusoidal curve.

Conventional methods of directional antenna analysis not only assumed sinusoidal current distribution for each tower, but they also required uniform current distribution for elements in an array. Also, the tower currents used in the calculations were not directly responsible for the far fields for each tower.

In the directional mode, there are two or more currents flowing in any element of the array. One is the current in the transmitting mode; the others are the currents in the receiving mode, that is, currents induced into the tower from the incident fields from all the other array elements. The actual directional antenna (DA) tower current is the sum of the transmitting mode current and all the receive mode currents. Because of this, the current distributions of the towers differ, and the relationships of the base currents to the tower far-field contributions are not uniform.

As a result of all this, the tower currents required to produce a required directional pattern could not be accurately calculated using traditional approaches. Theoretical parameters provided a good starting point, but the only way to adjust the array to produce the proper pattern was through cut and try. This was a lengthy process that started with a set of nondirectional (ND) field strength measurements on all the null and lobe radials followed by a few directional measurements at some of the same points. Once one or more points on each radial could be determined to be "representative" of the overall radial, the array was adjusted through trial and error to achieve the proper ratios at each of these "tune points." Again, until Method of Moments modeling came along, this was the best way we had to adjust an array to produce the proper pattern shape and size.

A Better Way

In moment method modeling of an antenna, the radiator is divided up into segments. Tower segment currents are solved numerically, taking into account the field coupling between segments on the radiator and the currents conducted from adjacent segments. If the

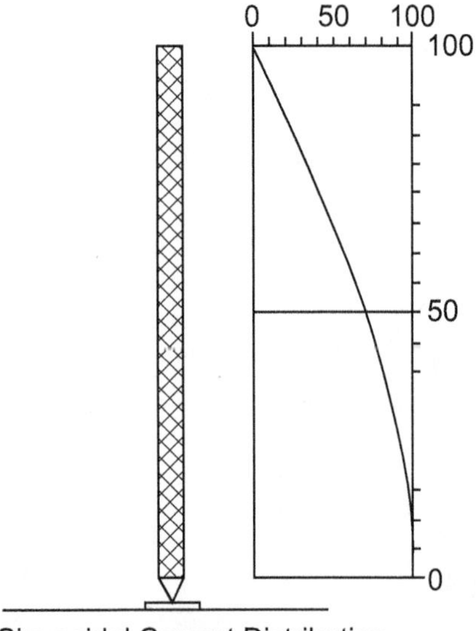

FIGURE 7.13-1 Sinusoidal Current Distribution

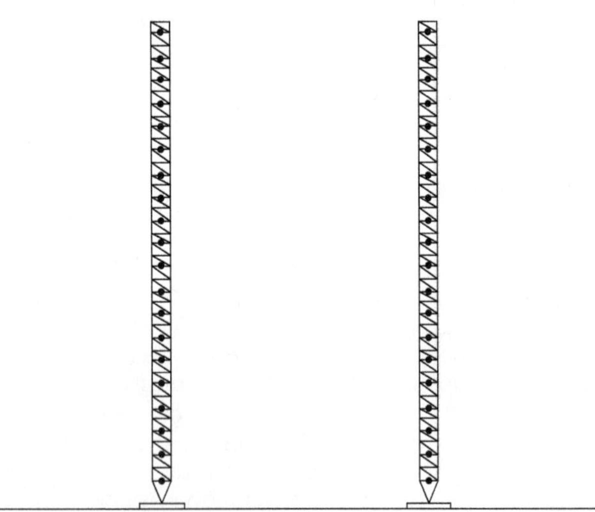

FIGURE 7.13-2 Moment Method Modeling Segments

modeler is careful in his selection of segment length, a close approximation of actual current distribution can be obtained from the model. This has the benefit of predicting base drive impedances much more accurately than any other method. It's not hard to see how knowing the actual current distribution on the tower and the driving point impedances would benefit the engineer designing a phasing and coupling system or tuning up an array.

Advantages of Modeling

Moment method modeling has several big advantages over the traditional ways of directional antenna analysis. First, as mentioned earlier, modeling solves for very close approximations of tower current distributions within a directional antenna. This is probably the most important advantage, and the one from which all the other advantages stem.

Modeling predicts tower currents and voltages that are directly related to directional antenna parameters. This makes it possible to adjust the array to the desired parameters using antenna monitor parameters, provided that they accurately represent the tower currents or voltages. This is the big time saver. It eliminates all the trial and error that is so much a part of the traditional array tune-up.

Finally, modeling predicts base driving point impedances very accurately, allowing close design of the phasing and coupling system. Using traditional analysis, the designing engineer had to allow for large variations in power distribution, phase and load impedance because it was unknown exactly what operating parameters (and the resulting driving point impedances) would be required to "make" the pattern. Modeling eliminates that guesswork, allowing the phasing and coupling system to be designed without all the "slack" for the unknowns.

These advantages convert directly to savings—time savings by eliminating all the field work to "talk" a pattern in through trial and error and to proof the pattern thereafter; money savings by eliminating all the man-hours to get the pattern right by trial and error and the labor required to produce a traditional directional proof-of-performance based on field measurements; and money savings by allowing proper component selection at design time.

Another advantage is the fixed cost represented by the modeling option. A traditional directional antenna tune-up and proof is an open-ended process. At the outset, no one knows how long it will take or how much it will cost because there is no way to know how many iterations it will take to get the pattern right. If there are re-radiators or other factors that distort the measured field, months or even years can be added to the process. All this is eliminated through modeling.

We all have to do more with less these days. The time and cost savings represented by the modeling option makes this a very attractive choice indeed.

LIMITATIONS OF TRADITIONAL FIELD STRENGTH MEASUREMENTS

Since the very beginning, broadcast engineers have employed field strength measurements as a means of verifying the performance of AM antenna systems, nondirectional and directional. Indeed, this has been the only means we had to gain some assurance that an antenna was performing as it should.

But also since the very beginning, we have known that there are big problems with field strength measurements used for this purpose. Anyone who has made radial field strength measurements on an AM station knows that there is often no rhyme nor reason to their values. In a perfect world, we would see an E × D relationship from the tower to the end of the radial, or at the very worst an E × D relationship modified for changes in ground conductivity. But we don't live in a perfect world. Measurements plotted on a piece of log-log graph paper often better resemble a shotgun blast than any kind of inverse distance line or conductivity curve. There are many reasons for this, some understandable and others that are tough to quantify.

The field intensity measurements that have been traditionally used to verify AM antenna performance are, in reality, magnetic field measurements that are presumed to relate in a certain way to the real parameter of interest, the electric field.

Maxwell's Equations

Maxwell's field equations, derived from Faraday's induction law, Ampere's rule and the Biot-Savart law, show the basic relationship between electric currents and magnetic fields. These are the same equations that give us the speed of light as 3×10^3 meters per second. These equations describe the relationship between currents and fields. Remember that the currents flowing in an antenna create fields. At a distance from the antenna, those E (electric) and H (magnetic) fields are described by the relationship $E = 120\varpi H$. Note that Ohm's law states that $Z = E / H$, so the impedance of free space is given as 120ϖ or 377 ohms.

Once we get some distance from the antenna, this starts to fall apart, however. Radial field intensity measurements simply do not work well in any situation other than over uniform, smooth, high-conductivity terrain. Indeed, in Norton's 1942 paper "The Calculation of Ground-Wave Field Intensity Over a Finitely Conducting Spherical Earth," certain simplifying assumptions are made. Real-world effects such as changes in conductivity, most diffraction effects and discontinuities in dielectric constant are not considered at all. These simplifying assumptions are, however, necessary to the process. Ben Dawson, P.E. wrote in his paper "The Inadequacy of Magnetic Field Measurements for Antenna Performance Verification" that "Even the simple case of mixed conductivity with no diffraction and uniform dielectric constant is difficult."

As was mentioned earlier, one of the assumptions is a perfectly conducting plane surface. The far field in such a case would diminish in a 1 / R relationship. But there are a couple of problems with that analysis. First, the earth is a sphere, not a plane, and second, the earth is not a perfect conductor and it is not homogenous.

Maxwell's $120\varpi H$ relationship comes from an assumption of a permittivity of $4\varpi \times 10^7$ henrys/meter = μ, and a dielectric constant of $1 / (36\varpi \times 10^9)$ farads/meter = e. Because neither the permittivity nor the dielectric constant is in fact constant (they are anything but), it is apparent that the $E = 120\varpi H$ relationship is not a good fit beyond a certain distance from the antenna.

Magnetic Field Disturbances

There are a number of factors that can and do disturb the magnetic field at any given location. Here are just a few:

- Surface layer impedance dielectric discontinuities due to vegetation;
- Conductivity changes due to soil or other surface geology changes;
- Dielectric constant changes due to soil or surface geology changes;
- Conductivity and dielectric constant changes due to bodies of water;
- Diffraction effects due to topography;
- Diffraction effects due to rugose topography;
- Diffraction due to abrupt changes in conductivity/dielectric constant changes;
- Electric field distortion due to electrically small vertical scatterers;
- Magnetic field distortion due to finite sized loops of conducting material;
- Absorption by poor conductor structures (e.g., wet concrete);
- Quasi-transmission line or ducting effects by urban streets with parallel rows of structures;
- Reflection or multipath effects from slopes of good conductivity soil;
- Quasi-free space propagation in curving sloped terrain;
- Near-field effects from arrays of radiators;
- Localized near-field effects from re-radiators;
- Layered conductivity effects resulting from 1 / R propagation.

There are many other possible sources of magnetic field variations that will distort the indication on a field intensity meter in any given location.

Analysis of Measured Data

To anyone who has done any analysis of measured radial field strength data, it would come as no surprise at all that there is an enormous range of variability—in some cases more than 20 dB—of the relationship between the magnetic and electric fields. That variability is one of the primary reasons that we analyze such data graphically. It simply won't fit any equation on a consistent basis.

Distribution of measured MW radial field intensity data is not Gaussian. It tends, instead, to be Rayleigh distributed, as do most all other propagation situations, with the bulk of the data to one side of the inverse distance line as opposed to even distribution on both sides.

As such, arithmetic analysis of measured data will not in many cases provide a correct answer. Keep in mind that such analysis is trying to solve for two variables—the apparent conductivity and the inverse distance field, and the conductivity variable itself is actually more than one unknown.

So the best we can do is plot the measured data on a piece of log-log paper and graphically analyze it to see what curves best fit the data. Clearly this is a "best guess" methodology at best.

METHOD OF MOMENTS BASICS

This chapter is not intended to be a comprehensive tutorial on antenna modeling. There are other venues for that, including the excellent *ARRL Antenna Modeling Course* (www.arrl.org). That course as well as others can give the student a much more in-depth look "under the hood" of the Method of Moments modeling process, including the features of and differences between the many programs and cores available (NEC-2, NEC-4, MININEC in its several versions, etc.). J.L. Smith's book *Basic NEC With Broadcast Application* (Focal Press) also provides an excellent treatment of AM broadcast antenna modeling in the NEC environment.

So what does a moment method program do? The short answer is that it solves for the current distribution on a wire based on a numerical solution of an integral equation representation of the electric fields. In other words, at its heart, a moment method program provides the user with an accurate prediction of the current distribution on an antenna.

Moment method modeling is based on the premise that an antenna is characterized by a collection of arbitrary thin, straight wires in free space or over a ground plane. The process starts with several assumptions:

- The radius of the wires is very small with respect to the wavelength and the wire length.
- The wire must be subdivided into short segments so the radius is assumed small with respect to segment lengths.
- The currents in the wires are axially directed (no circumferential currents on the wires).

In the AM broadcast antenna world, all our work is over a ground plane. Ground planes are accommodated in moment method programs by a process called the "method of images." Where a wire attaches to the ground plane (i.e. one end of a "wire" has a Z coordinate of zero), a current basis function is automatically added to the wire end point connection to ground.

Segmentation and Wire Radius

Current distribution is solved by dividing the wire into a number of segments. There are some rules established by the limitations of the NEC and MININEC cores related to segmentation and wire radius:

- Segment length, Δ, should be less than about 0.05 wavelengths and longer than 10^{-3} wavelengths at the desired frequency.
- Extremely short segments (less than 10^{-3}y) should be avoided.
- The wire radius, α, should be such that λ / α is greater than 30 and $2\varpi(\alpha / \lambda)$ is much less than 1.
- The ratio Δ / α must be greater than about 8.
- Segments may not overlap.

The FCC rules for modeling of AM broadcast antennas have some further limitations on segmentation and radius. Except for some large aperture free-standing towers, each tower is represented as a wire "cylinder."

- The radius of each cylinder must be between 80% and 150% of the radius of a circle with a circumference equal to the sum of the widths (S) of the tower sides ($3S / 2\varpi$).
- There must be no less than one segment per 10 electrical degrees of the tower's physical height.

There is a practical limit to the maximum number of segments, established by the preceding rules. Generally speaking, the more segments that are used, the better the "resolution" of the model. However, there is a point of "convergence" beyond which increasing the number of segments will produce no improvement in model accuracy. Most models of broadcast towers of reasonable height will produce good results using between 10 and 20 segments.

One quick word about a fundamental difference between NEC (NEC-2, NEC-4, etc.) and MININEC cores. The NEC core puts voltage sources in the middle of the specified segment whereas the MININEC core puts it at the end. This has a very practical effect on AM broadcast antenna models. A NEC model of a broadcast tower with the voltage source placed on segment 1 will, in actuality, have its source placed some distance up the tower whereas a MININEC model will have the source placed at the actual base of the tower. To make a NEC model of a broadcast tower work, it is thus necessary to use a greater number of segments, which will result in smaller segments. The FCC rules require that base calculations must be made for a reference point at ground level or within one electrical degree elevation of the actual feed point. With the 1-degree restriction, a segment length of no more than 2 degrees must thus be used for NEC models.

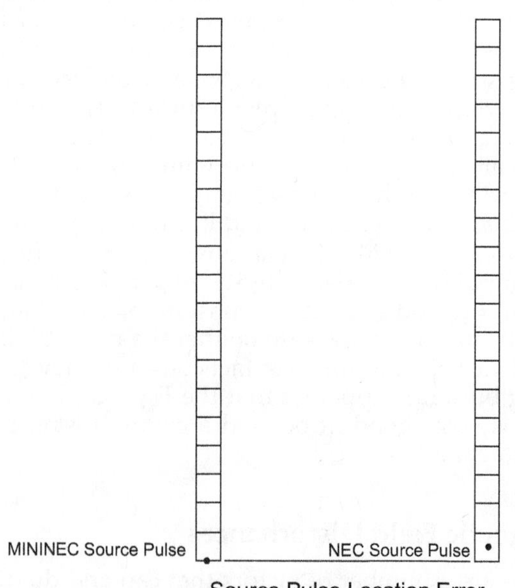

FIGURE 7.13-3 Source Pulse Location Error

Another means of overcoming this NEC limitation without resorting to very short segments is to employ a separate short (less than one electrical degree) wire between ground and the bottom of the tower wire for the base connection.

One more word about NEC and MININEC. Do not be deceived by the names. MININEC is not a scaled-down version of the NEC core. The MININEC core was originally written so that moment method computations could be made on the minicomputers of that day. Even the slowest and smallest of today's PCs far exceeds the capabilities of the fastest and biggest computers of the early 1980s. The MININEC core has continued to evolve and represents a computing engine every bit as powerful as its NEC counterpart.

Once the program solves for the current distribution on the towers, it can then solve for the impedance at each source (tower base).

Directional Antenna Sources

Current distribution and impedance are not all we need, however. What we need to properly model a directional array are the voltages and phases of the sources at each tower base. A separate source is required for each element in the array. In the broadcast world, we work with current ratios and phases, not voltages and phases. Ratios and phases of the source voltages and currents are generally not the same as the ratios and phases of the desired field parameters. So what we need is a way to translate the tower current moments to source voltages and phases.

Source parameter determination requires summing the current moments of the individual towers with a constant source driving each tower base with the others shorted to ground. The summed current moments are normalized to the reference tower to determine the field parameters. Because the relationships of tower currents to base voltages are linear, linear algebra can be used to determine the source parameters as follows:

For a two-tower directional antenna:

$$F_1 = V_{11}T_{11} + V_{12}T_{12}$$
$$F_2 = V_{21}T_{21} + V_{22}T_{22}$$

With tower 1 driven and tower 2 shorted:

$$F_1 = T_{11}$$
$$F_2 = T_{21}$$

With tower 2 driven and tower 1 shorted:

$$F_1 = T_{21}$$
$$F_2 = T_{22}$$

For a directional array with n towers:

$$F_1 = V_{11}T_{11} + V_{12}T_{12} \ldots + V_{1n}T_{1n}$$
$$F_2 = V_{21}T_{21} + V_{22}T_{22} \ldots + V_{2n}T_{2n}$$

$$\vdots$$

$$F_n = V_{n1}T_{n1} + V_{n2}T_{n2} \ldots + V_{nn}T_{nn}$$
$$[F] = [T] \times [V]$$
$$[S] = [T]^{-1}$$
$$[V] = [F] \times [S]$$

In the early days of moment method modeling of AM antennas, the consulting engineers who pioneered the use of modeling used iteration to determine the proper source voltages and phases. While this worked, it was very time-consuming and not particularly accurate. Today, a series of matrices is used to determine the source parameters to produce the requisite element fields.

There are external matrix inversion programs available for this purpose (such as Westberg Consulting's DRIVE), and at least two commercial MININEC applications (Expert MININEC Broadcast Professional, which is no longer available, and ACSModel) have an internal pattern synthesis routine.

Coordinate Systems

The NEC and MININEC cores use the spherical coordinate system consisting of X, Y, and Z axes with units in meters. Azimuth displacement in the X-Y plane is measured from the +X axis and increases in the counter-clockwise direction. The elevation angle is measured from the +Z axis with the zenith being 0 degrees.

In the broadcast world, however, we work in the geographic coordinate system based on distance and bearing referenced to true north. The reference tower in an array usually has a spacing and orientation of zero and zero, respectively, and the other towers are specified using distance in electrical degrees and bearing referenced to true north from the reference tower.

While some moment method wrappers, such as ACS-Model and Expert MININEC Broadcast Professional,

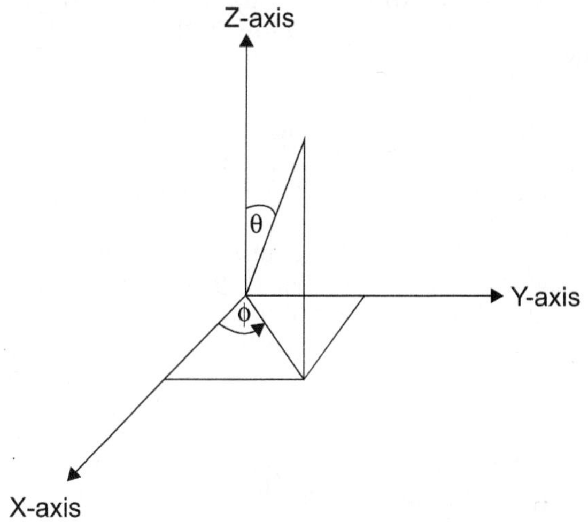

FIGURE 7.13-4 Conventional Spherical Coordinate System

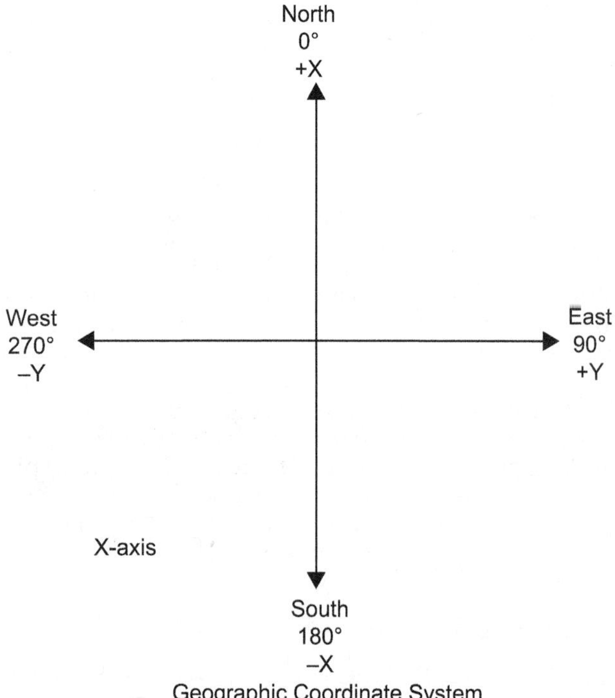

FIGURE 7.13-5 Geographic Coordinate System

will allow entry in the geographic coordinate system, programs that do not will require conversion to the metric X-Y-Z coordinate system. Conversion is not difficult. The reference tower is specified at X-Y coordinates 0,0. The distance to other array elements is then converted from degrees to meters, and the X-Y coordinates are determined using the following equations:

$$X = \text{Spacing} \cos(\theta)$$

$$Y = \text{Spacing} \sin(\theta)$$

Z axis coordinates are straightforward. Each tower base has a Z coordinate of zero. The Z coordinate of the tower top is simply the tower electrical height in meters.

Calibrating the Model

To use a moment method model for proofing with internal current or voltage measurements, the model must be "calibrated" by comparing measured base impedances to those predicted by the model. This is done by first making a set of matrix measurements.

The base impedance of each tower in the array is measured with all the other towers either floated or shorted, depending on tower height. Shorter towers are generally floated while taller towers are shorted—whichever condition is closest to the "detuned" condition for that height of radiator.

The model is then run and changed to account for environmental factors, such as velocity factor, loading of guy wire insulators and so on. The modeled tower height will generally be different (usually greater) than the physical height because of this velocity factor. Generally speaking, larger aperture towers will have lower velocity factors due to the greater number and length of structural cross-members. The modeled tower height and wire radius will be based on the observed impedance matrix.

Radiation Pattern

Amateur radio experimenters and others who use moment method modeling are primarily interested in two things by way of program output: input voltage standing wave ratio (VSWR)/bandwidth and radiation pattern. In the broadcast engineering world, we don't really care about VSWR because we employ matching networks at each array element, although the driving point impedances are important for a number of reasons. And while we do care about radiation pattern, the fact is that if we can produce the correct field from each tower in the array, the pattern will be right.

Most if not all NEC and MININEC packages have a radiation pattern provision, allowing the user to output the horizontal plane radiation pattern at any elevation. This output is not really of interest to the broadcast array modeler other than as a confirmation that the model has been properly constructed. If, for example, a pattern should have nulls at 45 and 315 degrees and a lobe at 180, the moment method package radiation pattern output should show that and not something else.

FCC MODELING RULES

The FCC's modeling rules are contained in 47 C.F.R. Sections 73.151 and 73.155. The actual text of the rules is available on the FCC's website at www.fcc.gov, so they are not included herein. Rather, a digest of the rules as they exist at the time of this writing is provided next. In essence:

- A Method of Moments program must be used.
- The model must be constructed in such a manner that it does not violate any of the internal constraints of the program used.
- Only arrays consisting of series-fed elements are eligible for the modeling option.
- Matrix impedance measurements must be made at the base and/or feed point of each array element.
- The physical characteristics of array elements (height and radius) may be varied to calibrate the model.
- Model impedances must agree with the measured impedance matrix within ±2 ohms and ±4%.
- Actual spacings and orientations must be used.
- Towers may be modeled using vertical wires or with multiple wires representing legs and cross-members.
- Drive point impedances must be determined from the model output.

- The radius of each array element must be between 80% and 150% of the radius of a circle with a circumference equal to the sum of the widths of the sides.
- No less than one segment per 10 electrical degrees of the tower's physical height must be used.
- Base calculations must be made at ground level or within one electrical degree of the actual feed point elevation.
- For nontapered towers, the modeled height of each element must be between 75% and 125% of the physical height.
- For tapered towers, stepped-radius wire sections may be used to simulate a tower's taper, or the tower may be modeled using wires for the legs and cross-members.
- The series feed impedance between the ATU output and tower base must be less than 10 uH unless measured higher.
- The shunt feed capacitance to model the base region effects must be less than 250 pF unless measured or specified by the manufacturer to be higher.
- If the shunt feed capacitive reactance is less than five times the magnitude of the base operating impedance, it must be considered in the model.
- The tower positioning must be confirmed post-construction by a surveyor or registered professional engineer.
- Operating parameters must be determined from the output of the computer model.
- Samples may be current transformers or voltage sampling devices at the base of each element, or tower-mounted loops.
- Loops must be located at the elevation where the current in the tower would be at a minimum if the tower were detuned.
- Loops may be used only on towers of identical cross-sectional structure, including leg and cross-members.
- Loops on unequal height towers must be mounted with identical orientations at the proper elevations.
- If tower heights other than the physical heights are used in the model, loops must be mounted at the same percentage height as indicated in the model.
- Sample lines must be equal in length within one electrical degree and characteristic impedance within two ohms, as confirmed by measurement.
- Base current sample transformers may be used for towers of 120 degrees or less or 190 degrees or greater in height.
- Base voltage sample devices may be used for towers of greater than 105 electrical degrees.
- Tower-mounted sample loops may be used on towers of any height.
- Base current sample transformers or voltage sampling devices must be calibrated against one another within the manufacturer's specifications.
- Antenna monitor sample indications must agree with the model-determined ratios within ±5% and with the model-determined phases within ±3 degrees.
- Three reference field strength measurements must be made on each pattern minima and maxima radial.

More detail on the FCC rules regarding calibration and certification of the sample system and the required biennial recertification thereof are presented in the section dealing with field procedures for AM moment method modeling.

USING MOMENT METHOD MODELS FOR DIRECTIONAL ANTENNA PROOFS

Benefits of Modeling

There are many reasons why using moment method modeling for tune-up and performance verification of AM directional antennas is superior to traditional methods. We have covered most of these in prior sections of this course. Among them are:

- Accurate locations for tower-mounted sample loops
- Accurate driving point impedance data for every array element
- Accurate pre-construction bandwidth analysis
- Eliminating dependence on field measurements for accurate adjustment.

The bottom line, however, is cost savings. With more accurate data going in, without the open-ended prospect of trial-and-error adjustment and without the requirement for hundreds of nondirectional and directional field measurements, the DA tune-up/proof project employing moment method modeling will have much lower expenses and a more or less fixed cost.

A direct benefit of lower, fixed costs is better compliance. It stands to reason that station licensees that would otherwise opt to roll the dice on an inspection, notice of violation and forfeiture for having an array out of tolerance because of the often huge costs of bringing it into compliance, would be more inclined to spend a much lower and fixed amount to eliminate that risk. That, in turn, results in reduced interference, which benefits everyone on co-channels and adjacent channels.

Facility and Model Eligibility

The FCC rules now permit moment method modeling as a means of performance verification of AM directional arrays in certain cases. A digest of these rules is contained in another section of this course. To be eligible for the modeling proof option, a facility and model must meet the following criteria:

- Only series-fed antenna elements;
- Only accurate current/voltage samples;
- The moment method model must match the measured base Z matrix;
- For cylinder models, the effective radius assumption is limited to 80%–150% of the physical radius;
- For cylinder models, the effective height assumption is limited to 75%–125% of the physical height;
- At least one segment for each 10 degrees of physical height;
- The model feedpoint must be at ground level or within 1 degree of physical elevation;
- Series inductance must be 10 uH or less unless measured higher;
- Parallel stray capacitance must be 250 pF or lower unless measured higher.

Modeling Software

While the FCC rules do not specify a modeling platform, and while good results can be obtained from any platform employing a NEC or MININEC core, the FCC Media Bureau has opted to use *Expert MININEC Broadcast Professional* (reportedly no longer in production) and ACSModel (available from the SBE Store at www.sbe.org/sections/store_books_listings.php or directly from Au Contraire Software at www.aucont.com). Employing the same platforms as the FCC will ensure comparable results, presumably reducing processing delays.

Examples in this chapter were taken from ACSModel.

MEASUREMENTS FOR MOMENT METHOD MODELING OF AM ARRAYS

Base Impedance Matrix Measurements

The one thing that differentiates a typical moment method antenna model and an AM broadcast directional array moment method model is that the latter is calibrated against real-world conditions. This calibration process removes some of the assumptions, moving them from the realm of the unknown to the known. We can assume a velocity factor for a tower, for example, because the modeled tower height at a certain effective height behaves the same as the prototype tower in the real world.

Precedent to construction of a method moment model, then, is a set of base impedance matrix measurements. The procedure for making these measurements is presented in detail here with the assumption that the one making the measurement knows how to operate his test equipment, presumably a General Radio 1606B or equivalent impedance bridge or Delta Operating Impedance Bridge (OIB) in one of its forms, and either with a suitable signal generator and detector.

Sample System Measurements

Once the model has been constructed, calibrated and run, the sample system must be calibrated—proofed—before the directional array can be adjusted to the model parameters. The FCC has set certain guidelines for the lengths and impedance of sample lines, sample current transformer accuracy and loop dimensions and orientation. All these are covered in the chapter *FCC Modeling Rules* in the Modeling tutorial within this course. If you have not yet read that chapter, you are encouraged to do so now because the material therein is foundational to much of what will be presented in this tutorial.

The base impedance matrix measurements are requisite to model construction and are essentially a one-time affair. Unless something changes on one of the towers—the addition of a studio-to-transmitter link (STL) antenna or replacement of an isocoupler with a different type, for example—there is no reason to repeat them.

The sample system measurements, on the other hand, are a biennial event. The FCC requires that the station licensee "recertify the performance of that directional pattern at least once within every 24 month period" including measurements "made to verify the continuing integrity of the antenna monitor sampling system." This requirement has proven to be a bit off-putting to some engineers, but this is more because of a lack of understanding of the requirements than any particularly difficult aspect of the recertification.

The reality is that the recertification measurements need consist of nothing more than spot-checks of the information that was found in the full certification when the array was tuned up. For example, the sample lines must be measured to determine their resonant frequency and characteristic impedance. Much of the work in this regard is in finding the unknowns. Once certified, those variables have been solved. Recertification can be done by simply measuring the impedance of the open-circuited sample lines at the frequencies on which they were found to be resonant and confirm that they are still resonant. Characteristic impedance measurements can likewise be quick measurements on two frequencies, one above and one below the resonant frequency, which were determined in the initial system proof.

The entire sample system of even a large array can be measured in a few hours. While there is certainly nothing wrong with hiring a consulting engineer to do the recertification, any competent RF engineer who can operate a bridge can do it; the biennial recertification is well within the capabilities of most AM station chief engineers.

Reference Field Strength Measurements

The FCC's AM modeling rules require that "reference field strength measurement locations shall be established in directions of pattern minima and maxima." In essence, once the pattern has been adjusted to the model parameters, three radial field strength measurements must be made on each of the null and lobe radials. These measurements must be repeated during the biennial recertification.

It should be stressed that these reference field strength measurements are *not* "monitor points"; there are no licensed maxima applied to them and there is no need to measure the field strength at any of the points between biennial recertifications. Further, if, during the recertification measurements, one or more of the measurements is found to exceed the original values, it means nothing. The measurements were included in the rules simply to give the licensee a means of verifying the gross pattern shape and size. If measurement of the reference field strength points were to show, for example, an overall shift in the pattern (some fields up and others down), that *might* be an indication of another problem—but then again, it might *not*. There is no statutory duty to act on a changed reference field strength.

We will not cover in this chapter the reference field strength measurements beyond this brief discussion. It is assumed that the engineer responsible for the care and feeding of an AM directional array is well acquainted with the proper operation of a field intensity meter and the selection of good radial measurement point locations.

The measurements described earlier will be fully discussed later in this chapter.

MATRIX IMPEDANCE MEASUREMENTS

We will deal with field measurements of impedance in more detail in the section on field work for moment method modeling for performance verification, but a brief discussion of the requirements is needed here.

The term "base impedance matrix" sounds complicated and perhaps a bit off-putting to the engineer new to modeling, but the opposite is actually true. Measurement of the base impedance matrix consists of nothing more frightening than bridging the base of each tower with all the other towers either floated or shorted. Measuring at the antenna tuning unit (ATU) output or the tower base itself is optional, and many modelers measure both, but the general rule is that if you are base sampling (from a current or voltage sample device at the ATU output), you must measure the matrix at the ATU output. Loop sampled towers should be measured at the base itself.

In the simple case of a two-tower array employing 90-degree towers, start with tower 1. Assuming that measurements will be made at the base of each tower, disconnect all appurtenances from both towers, including isocouplers, lighting chokes, sample lines, feed tubing and the like, watching out for unwanted capacitive coupling (such as a disconnected sample line still lying in proximity to a tower member).

With the base of tower 2 now open, back at tower 1, bridge the base of the tower right at the top of the base insulator, at the point where the feed tubing would normally connect, and note the resistance and reactance (remember to correct the reactance for frequency), marking the notation as taken with tower 2 open.

If desired, now short tower 2 with at least two low-inductance straps made from copper strap or 1-inch or wider braid. Short directly across the base insulator to the RF ground under the insulator. Repeat the base impedance measurement at tower 1 and note the R and X, correcting the X for frequency. Be sure to note that tower 2 was shorted.

Then reverse the process. Open tower 1 and bridge the base of tower 2. Then short tower 1 and bridge the base of tower 2. That's it, a complete set of base impedance matrix measurements. If there are more towers in the array, continue with this process, measuring the base impedance of each tower in turn with all the other towers either floated or shorted.

What is the criterion for measuring with the unused towers open or floating? A good rule of thumb is to employ the condition that would best approximate a "detuned" condition for the tower. Electrically short towers, say less than 120 electrical degrees, are better floated while electrically tall towers, say more than 150 electrical degrees, are better shorted. Towers of intermediate heights should be both open and shorted in the matrix. As the model is run, the condition that the model best matches can be used to calibrate the model.

What is the criterion for measuring at the tower base versus at the ATU output? If it is possible to measure at the tower base with the feed tubing completely disconnected, it is better to do so because it eliminates the several unknown variables of series and shunt stray reactance. Those reactances can often be measured separately and dealt with outside the model structure. In many cases, however, the feed tubing has been welded or otherwise electrically bonded to the tower structure, and removing it would be difficult. In such cases, measure at the ATU output. The strays can still be measured with some degree of accuracy and treated outside the model.

A STEP-BY-STEP MODELING EXAMPLE (ACSMODEL)

Constructing the Model

With all the details of the directional array in hand, including element spacing, orientation, height, ratio, phase, tower face width, tower configuration and impedance matrix, the next step is to construct the model. Run ACSModel and click File | New.

You will be prompted for a callsign, a description and a comment. Enter any pertinent informational data. The information entered here will be printed on each page of the output. For our example, enter:

Callsign: KLVZ

Description: 3-Tower Daytime Array

Comment: Tower 2 driven, all others floated

This information is printed on the output, so including pertinent information in the description and comment fields will help you keep track of what each model run represents. This is important when you find stacks of paper being produced, as is common in a modeling project. If you want to change the callsign, description or comment, click Problem Definition | Description (or type Ctrl-D).

ACSModel, which was written specifically for AM broadcast, assumes perfect ground and a ground model environment.

Next, click Problem Definition | Frequency (or click Ctrl-F). Set the units to kHz, set the initial frequency to 810, then click OK.

Wires

With the preliminaries out of the way, now it's time to construct the actual model. Click Problem Definition | Wires (or click Ctrl-W).

Our tower geometry is as follows:

Twr.	Spacing	Orientation	Height
1	75.9	220	77
2	0	0	77
3	75.9	40	77

First select the Geographic coordinate system and 3 wires.

In the wire 1 End One Spacing box, enter 75.9; in the orientation box, enter 220; in the Z box, enter zero. Now in the wire 1 End Two Spacing box, enter 75.9; in the orientation box, enter 220; in the Z box, enter zero. In the Radius box, enter 0.2911 (in this example, the tower has a 24-inch face, or 0.6097 meters—$3 \times 0.6097 / 2 \times \varpi = 0.2911$ meters effective radius). Enter 20 as the number of segments. This is a convenient number that exceeds the FCC minimum and will provide for good model performance. This establishes the wire representing tower 1.

Continue by adding the bottom and top geographic coordinates for towers 2 and 3. Your screen should look like this:

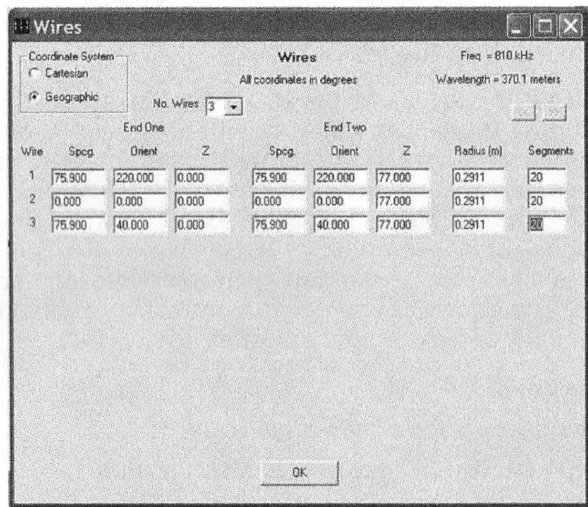

FIGURE 7.13-6 Antenna Geometry Entry

Click OK to save the wires. It's a good idea to save your work, so click File | Save (or click Ctrl-S), enter KLVZ-2 as the filename, navigate to the desired folder and click Save.

Calibrate the Model

The next step involves either floating or shorting two of the towers and driving the other to get a look at the impedance compared with the base impedance matrix for that same condition. In this case, because the towers are electrically short, the matrix was measured with the towers floated (open), so we will need to insert an open circuit at the base of two of the towers at a time. In MININEC, wires with one end at a Z value of 0 are presumed to be connected to ground (shorted), so we have to do something to lift that ground. We do this with a lumped load.

Click Problem Definition | Loads (or type Ctrl-L). Remember that we are using 20 segments per tower, so pulse 1 (with a Z axis value of 0) is the base of tower 1, pulse 21 is the base of tower 2 and pulse 41 is the base of tower 3. We want to place an open circuit at the bases of towers 1 and 3, the end towers. The easiest way to do this is to enter a very high capacitive reactance at the nodes for the bases of those towers.

In the Load box, select 2 for the number of loads. For the Load 1 Pulse, enter 1 for the base of tower 1. In the Resistance box, enter 0. In the Reactance box, enter −10000. This value is not critical. It needs to be a high enough reactance that no significant amount of current will flow through it at the operating frequency. Repeat this for Load 2 at node 41 for the base of tower 3. Your screen should look like this:

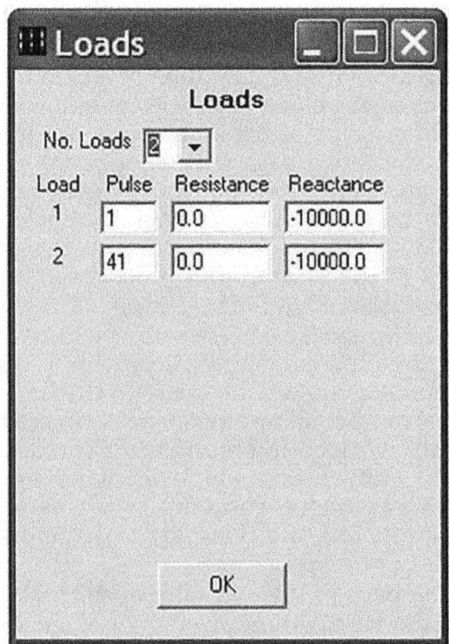

FIGURE 7.13-7 Load Entry Screen

Click OK to close the screen. Save your work.

Now we need to drive tower 2 with a voltage source. Click Problem Definition | Sources (or type Ctrl-V). The program defaults with a source at node 1. Type 21 in the Source current node box. The Sector box should default to 1. In the Magnitude box, type 1. This is not critical; any convenient value will work. The Phase box should default to zero. Click Modify. Your screen should look like this:

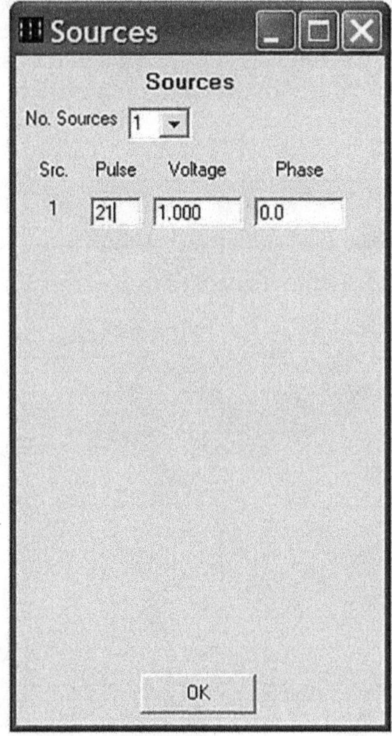

FIGURE 7.13-8 Source Entry Screen

Click OK to close the screen.

Now we're ready for our first model run. Click Run | Run Model (or press F5). The model will run and a WordPad window will open displaying the output file. The numbers we care about are the resistance and reactance displayed under Source Data. Scroll down to find this data:

```
********************     SOURCE DATA     ********************
Pulse  21       Voltage = (1.0, 0.0j)
                Current = (0.0082, 0.0164j)
                Impedance = (24.316, -48.793j)
                Power = 0.004091 Watts
```

FIGURE 7.13-9 Tower #2 Initial Impedance Prediction

The measured tower 2 base impedance with towers 1 and 3 floated measured 28.5 + j8.5. Our first try model run, using the physical height of the tower and nominal radius, produced an impedance of 24.316 – j48.793. This is outside the FCC-specified allowance of ±2 ohms and ±4%. We need to adjust our model.

RF traveling in a steel tower moves slower than the speed of light because of all the cross-members and because of the loading effect of the guy insulators. So we can with some assurance assume the tower is electrically longer than its 77-degree physical length. Let's assume a velocity factor of 0.9448 and rerun the model. (This value was arrived at after several iterations. Finding the right number is a sometimes lengthy trial and error process. The intermediate steps are omitted here in the interest of brevity.)

Close the WordPad window, then click Ctrl-W to look at the wires again. Modify each tower's top coordinates to have a Z of 81.5 degrees (77 / 0.9448), then click OK. 81.5 degrees is within the 75%–125% physical height limit. Press F5 to run the model.

When the WordPad window comes up, again scroll down to Source Data to see the calculated impedance at the base of tower 2 with towers 1 and 3 floated.

```
********************     SOURCE DATA     ********************
Pulse  21       Voltage = (1.0, 0.0j)
                Current = (0.0195, 0.0172j)
                Impedance = (28.764, -25.376j)
                Power = 0.009775 Watts
```

FIGURE 7.13-10 Tower #2 Try #2 Impedance Prediction

Note that the impedance is now shown as 28.8 – j25.4. The resistance is within the ±2 ohm/±4% window. In fact, it's very close to the matrix value of 28.5. But the reactance is off. Remember that the FCC rules allow a series inductance of up to 10 uH. If we assume a series reactance of the opposite of +j8.5 + j25.4 (+j33.9), the model output impedance for tower 2 driven with towers 1 and 3 floated becomes 28.8 + j8.5, very close to the measured value, 33.9 / 2ϖF = 6.7 uH, within the allowable 10 uH limit.

Next, we need to drive tower 1 and float towers 2 and 3. Type Ctrl-L to bring up the Loads window. Click in the Pulse box for load 1, change it to 21 and click OK. That action got rid of the open at the base of tower 1 and moved it to the base of tower 2. Now we need to move the voltage source to the base of tower 1. Click Ctrl-V to bring up the Sources window. In the Pulse window for the source at pulse 21, type 1 and click OK. This moved the source to the base of tower 1, so now we have towers 2 and 3 floated with tower 1 driven. Click Ctrl-D and change the description and comment to "Tower 1 driven, all others floated." Click Ctrl-S to save and change the filename to KLVZ-1.

Now we'll run the model again. Press F5. When the WordPad window comes up, scroll down to the Source Data.

```
********************     SOURCE DATA     ********************
Pulse  1        Voltage = (1.0, 0.0j)
                Current = (0.0201, 0.0166j)
                Impedance = (29.523, -24.434j)
                Power = 0.010051 Watts
```

FIGURE 7.13-11 Tower #1 Impedance Prediction

Note that the impedance from the model run is shown to be 29.5 – j24.4. The matrix impedance for tower 1 with tower 2 and 3 open measured 29.0 + j7.0, so this value of R is within the ±2 ohm/±4% window. If we add a series reactance of +j31.4, the reactance component becomes +j7.0, right on the nose. Again +j31.4 corresponds to an inductance value of less than the 10 uH FCC maximum, so we're good.

Close the WordPad window.

Now we repeat the exercise by moving the pulses of the loads at the bases of towers 2 (pulse 21) and 3 (pulse 41) to the bases of towers 1 (pulse 1) and 2 (pulse 21). Then move the voltage source from pulse 1 to pulse 41, change the comment to "Tower 3 driven, all others floated" and save as KLVZ-3. Run the model (F5). When the WordPad window opens, scroll down to Source Data:

```
********************     SOURCE DATA     ********************
Pulse  41      Voltage = (1.0, 0.0j)
               Current = (0.0201, 0.0166j)
               Impedance = (29.523, -24.434j)
               Power = 0.010051 Watts
```

FIGURE 7.13-12 Tower #3 Impedance Prediction

The impedance is shown to be 29.5 – j24.4. The matrix Z of tower 3 driven with towers 1 and 2 open measured 29.0 + j7.0. The predicted R is within the ±2 ohm/±4% window, so it's fine. If we use a series impedance of +j31.4, we get an X of +j7.0, right on the money.

We can now consider our model to be calibrated.

In some cases, it could be that once we get the first tower calibrated to the model and run the second tower with the others floated or shorted, the R is outside the ± 2 ohm/±4% window. It may then be necessary to adjust the height of the driven tower a bit to bring the R within the window.

Once that is done, the bad news is that you get to rerun the previous tower again with the changed height on the second tower. That tower may now need a tweak as well. Then, if the first tower is changed, the second tower must be rerun and perhaps adjusted again.

The process repeats with all the towers in the array until the full matrix is matched. This can be a time-consuming process, so be patient. An experienced modeler quickly gains a "feel" for what effect each parameter has, speeding up the iteration process considerably.

Modeling the Directional Pattern

Once the model has been calibrated, we can move on to the next step: determining the source voltages and phases that will produce the proper fields in the individual radiators. In ACSModel, this is done internally when you enter the theoretical directional antenna parameters.

Start by removing the loads producing the open circuits at the tower bases. Type Ctrl-L, set the number of loads to zero and click OK. Next, go to the description (Ctrl-D) and change the comment to "Directional Day" and click OK. Save the file as KLVZ-D.

Now click Problem Definition I Directional Parameters (or type Ctrl-B). Set the number of towers to 3 and enter the power as 2200 watts. Next, enter the field ratios and phases for the array. Our pattern's theoretical parameters look like this:

Twr.	Ratio	Phase
1	0.525	−127.5
2	1.000	0
3	0.510	+124.5

Note that the pulse numbers have already been set for the bases of the three towers, but make sure they are correct (1, 21 and 41). Now type the proper ratio and phase in the appropriate boxes for the three towers. The DA parameters should look like this:

Click OK to save the parameters and close the window.

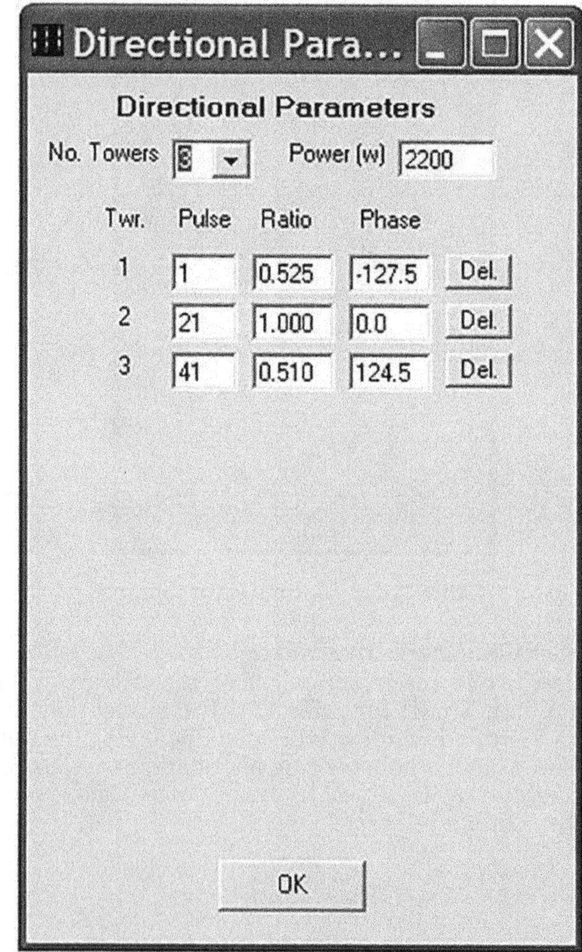

FIGURE 7.13-13 DA Parameter Entry

Now press F5 to run the model. When the WordPad window appears, scroll down to the Source Data section. This displays the source voltages (in complex rectangular format) that were computed from the theoretical directional parameters. These source voltages are important only internally; you do not need them for any other purpose but should know how they were derived.

```
********************     SOURCE DATA     ********************
Pulse  1        Voltage = (116.6441, -151.5995j)
                Current = (-4.153, -5.5248j)
                Impedance = (7.392, 26.67j)
                Power = 176.56 Watts

Pulse  21       Voltage = (266.8355, -239.4331j)
                Current = (14.1669, 0.4586j)
                Impedance = (18.269, -17.492j)
                Power = 1835.21 Watts

Pulse  41       Voltage = (212.9386, 212.4528j)
                Current = (-4.3595, 6.1414j)
                Impedance = (6.637, -39.384j)
                Power = 188.23 Watts

Total Power = 2200.000 Watts
```

FIGURE 7.13-14 Directional Mode Sources

There is, however, information in this section that you will need. Print this page. The impedances shown, once compensated to include the series inductance values that we determined for each tower in the calibration process, will become the driving point impedances for the towers. Keep in mind here that these numbers are for the tower bases. If there are significant amounts of stray capacitance, that must be considered and factored into the driving point impedance seen at the output of the antenna tuning units. More on that later.

Close the WordPad window. Now click Run | Moments (or type Ctrl-M). A WordPad window will open displaying the current moments for the array. If desired, divide the current magnitude value for the two end towers by that for the reference tower to confirm that the field ratios are correct. Check the phases to make sure they are correct as well.

```
Current Moments (amp-meters) Peak

Frequency: 810 kHz
Input Power: 2,200 Watts

                                        Vert. Current Moment
Wire        Real          Imag          Magnitude     Phase
 1       -243.2334     -316.9879        399.5545    -127.50
 2        761.0563        0.0000        761.0563       0.00
 3       -219.8442      319.8752        388.1386     124.50

Medium wave array vertical current moment (amps-meters) peak
(Calculation assumes tower wires are grouped together.
The first wire of each group must contain the source.)

Tower       Real          Imag          Magnitude     Phase
 1       -243.2334     -316.9879        399.5545    -127.50
 2        761.0563        0.0000        761.0563       0.00
 3       -219.8442      319.8752        388.1386     124.50
```

FIGURE 7.13-15 Current Moments

Close the WordPad window.

Next, press F5 again to run the model. When the WordPad window opens, scroll down to the on the Current Data section. You should see something like the following listing. Note the magnitude and phase at the base node of each tower (1, 21 and 41). These, normalized to the values for tower 2 (reference tower), become the tower base operating parameters for the array if you are employing base sampling. ACSModel calculates and prints the base operating parameters for you, so there's no need to do the math.

```
********************     CURRENT DATA     ********************

Wire No.  1 :
Pulse        Real         Imaginary     Magnitude     Phase
 No.         (Amps)       (Amps)        (Amps)        (Degrees)
 1           -4.153       -5.5248       6.9116        -126.9326
 2           -4.2134      -5.565        6.9801        -127.1298
 3           -4.2226      -5.5537       6.9767        -127.2465
 4           -4.2002      -5.5058       6.925         -127.3389
 5           -4.1487      -5.4234       6.8282        -127.4146
 6           -4.0693      -5.3076       6.688         -127.4776
 7           -3.9631      -5.1592       6.5057        -127.5301
 8           -3.8309      -4.9792       6.2824        -127.5736
 9           -3.6735      -4.7685       6.0194        -127.6092
 10          -3.4918      -4.528        5.7179        -127.6378
 11          -3.2867      -4.2586       5.3794        -127.6601
 12          -3.0591      -3.9613       5.005         -127.6767
 13          -2.81        -3.6373       4.5963        -127.6882
 14          -2.5403      -3.2874       4.1545        -127.6951
 15          -2.2508      -2.9124       3.6808        -127.6978
 16          -1.9421      -2.5131       3.1761        -127.6968
 17          -1.6144      -2.0894       2.6405        -127.6925
 18          -1.2671      -1.6404       2.0728        -127.6854
 19          -0.8976      -1.1624       1.4687        -127.6758
 20          -0.4984      -0.6457       0.8156        -127.664
 E           0.0          0.0           0.0           0.0
```

FIGURE 7.13-16 Tower #1 Currents

```
Wire No.  2 :
Pulse        Real         Imaginary     Magnitude     Phase
 No.         (Amps)       (Amps)        (Amps)        (Degrees)
 21          14.1669      0.4586        14.1743       1.8543
 22          14.0145      0.3308        14.0184       1.3521
 23          13.8288      0.2496        13.8311       1.034
 24          13.5799      0.1812        13.5811       0.7644
 25          13.2651      0.1217        13.2657       0.5256
 26          12.8842      0.0695        12.8843       0.309
 27          12.4377      0.0238        12.4377       0.1098
 28          11.9271      -0.0157       11.9271       -0.0754
 29          11.354       -0.0493       11.3541       -0.249
 30          10.7205      -0.0772       10.7208       -0.4125
 31          10.0289      -0.0994       10.0294       -0.5677
 32          9.2815       -0.1159       9.2822        -0.7154
 33          8.4808       -0.1268       8.4818        -0.8569
 34          7.6292       -0.1322       7.6303        -0.9928
 35          6.7286       -0.132        6.7299        -1.1239
 36          5.7807       -0.1262       5.7821        -1.2509
 37          4.7858       -0.1148       4.7872        -1.3744
 38          3.7417       -0.0977       3.743         -1.4951
 39          2.6407       -0.0744       2.6418        -1.6138
 40          1.4609       -0.0442       1.4615        -1.7325
 E           0.0          0.0           0.0           0.0
```

FIGURE 7.13-17 Tower #2 Currents

```
Wire No.  3 :
Pulse        Real         Imaginary     Magnitude     Phase
 No.         (Amps)       (Amps)        (Amps)        (Degrees)
 41          -4.3595      6.1414        7.5314        125.369
 42          -4.2465      6.0237        7.37          125.1826
 43          -4.1483      5.9117        7.222         125.0574
 44          -4.038       5.7788        7.0498        124.9446
 45          -3.913       5.6221        6.8498        124.838
 46          -3.7722      5.4408        6.6205        124.7346
 47          -3.6156      5.2348        6.3621        124.6327
 48          -3.4435      5.0045        6.0748        124.5311
 49          -3.2563      4.7505        5.7594        124.4291
 50          -3.0547      4.4736        5.417         124.3261
 51          -2.8394      4.1746        5.0487        124.2217
 52          -2.6112      3.8545        4.6557        124.1155
 53          -2.371       3.5142        4.2393        124.0075
 54          -2.1197      3.1547        3.8007        123.8973
 55          -1.8579      2.7768        3.341         123.7849
 56          -1.5862      2.3811        2.8611        123.6703
 57          -1.305       1.9677        2.3611        123.5533
 58          -1.0139      1.5357        1.8402        123.4338
 59          -0.711       1.082         1.2947        123.3115
 60          -0.3908      0.5975        0.714         123.1846
 E           0.0          0.0           0.0           0.0
```

FIGURE 7.13-18 Tower #3 Currents

The base magnitudes and phases are listed as:

Twr.	Mag.	Phase
1	6.91	−126.9
2	14.17	1.9
3	7.53	125.4

Normalized to the reference tower, the base operating parameters become:

Twr.	Mag.	Phase
1	0.488	−128.8
2	1.000	0.0
3	0.531	+123.5

As with the aforementioned driving point impedances, remember that these numbers are at the tower base. If there is a significant amount of stray capacitance, it must be considered outside the model to determine the effect on the impedance, current and phase at the ATU output.

The base driving point impedance is modified by a shunt capacitive and series inductive reactance as follows:

$$R_A = R_B X_S^2 / (R_B^2 + (X_B + X_S)^2)$$

$$X_A = +jX_S(R_B^2 + X_B^2 + X_B X_S) / (R_B^2 + (X_B + X_S)^2) + jX_L$$

Where:

Z_{BASE} = $R_B + jX_B$

Z_{ATU} = $R_A + jX_A$

X_S = Shunt reactance

X_L = Inductive series reactance

The ATU output current and phase are modified by a shunt capacitive reactance as follows:

$$I_{ATU}\ \text{Magnitude} = ((1 + X_B / X_S)^2 + (R_B / X_S)^2)^{1/2}$$

$$I_{ATU}\ \text{Angle} = \arctan(-R_B / X_S) / (1 + X_B / X_S)$$

Where:

I_{ATU} = ATU output current for unity base current with no phase shift

Z_{BASE} = $R_B + jX_B$

X_S = Shunt reactance

While you can calculate the offsets in the ATU output currents and phases using these formulas, it is much more elegant to do so using a nodal analysis program such as WCAP, available from Westberg Consulting (www.westbergconsulting.com).

Radiation Pattern

As mentioned previously, the radiation pattern calculated by the moment method program really has very little value to the broadcast antenna modeler. The directional pattern is produced by the combination of field current ratios and phases in the elements. If the current moments in the model are correct, the pattern will be correct. However, it's not a bad idea to check the output radiation pattern of the model to be sure it closely resembles the desired pattern. If it does not, chances are that you have a problem in your model.

Close the WordPad window if it is still open. Then in ACSModel, click Solution | Radiation Pattern (or click Ctrl-P). A window will open presenting you with two sets of variables: elevation and azimuth. The defaults are set for a 360-degree sweep at 1 km with RMS values, which is what the way an AM broadcast theoretical pattern is normally displayed. Click OK to close.

Press F5 to run the model. After the model runs, a WordPad window will open showing the tabular output. A GUI window will also open showing the model-determined radiation pattern. You should see this display:

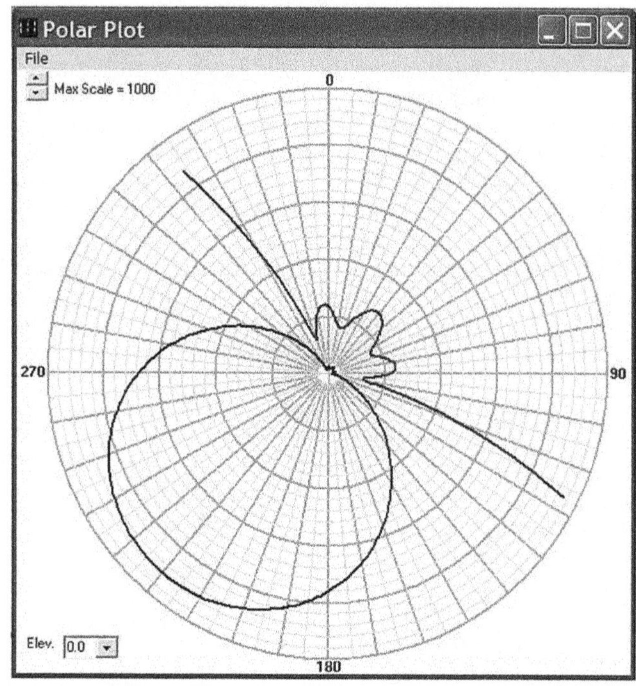

FIGURE 7.13-19 Modeled Directional Pattern

TOWER MOUNTED LOOP SAMPLING

The example thus far has dealt with base sampling, but what if tower-mounted loop sampling is desired? The procedure is a little different.

To determine the sample loop location, we must find the elevation where the current in the tower would be at a minimum if the tower were detuned. The procedure to do this is to drive one tower only. We'll show you here how to do this in ACSModel.

First, set up a new model (Ctrl-N) with a single towers of the same height (81.5 degrees), radius (0.2911 m) and number of segments (20) as used in the calibrated model.

Now click Run | Detune. The model will run several times, iterating internally to find the reactance necessary to detune the tower. A window will open showing that value of reactance.

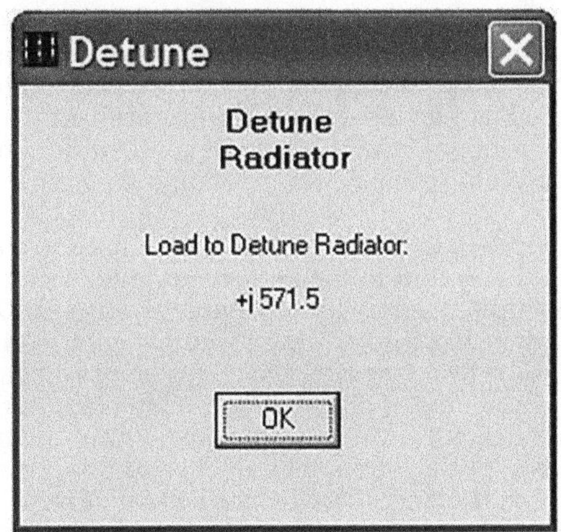

FIGURE 7.13-20 Modeled Detuning Reactance

******************** CURRENT DATA ********************

Wire No. 1 :

Pulse No.	Real (Amps)	Imaginary (Amps)	Magnitude (Amps)	Phase (Degrees)
1	0.1008	-0.089	0.1345	-41.45
2	0.0732	-0.0646	0.0976	-41.445
3	0.0556	-0.0491	0.0741	-41.4238
4	0.0407	-0.0359	0.0543	-41.3703
5	0.0277	-0.0243	0.0369	-41.245
6	0.0163	-0.0141	0.0215	-40.9138
7	0.0062	-0.0051	0.0081	-39.4416
8	-0.0025	0.0027	0.0037	132.8466
9	-0.01	0.0095	0.0138	136.6199
10	-0.0163	0.0152	0.0223	137.1058
11	-0.0213	0.0197	0.0291	137.2605
12	-0.0252	0.0232	0.0342	137.3114
13	-0.0277	0.0256	0.0377	137.3154
14	-0.0291	0.0268	0.0396	137.2932
15	-0.0292	0.027	0.0398	137.2544
16	-0.0281	0.026	0.0382	137.2037
17	-0.0256	0.0238	0.035	137.1437
18	-0.0219	0.0204	0.0299	137.0761
19	-0.0168	0.0156	0.0229	137.0014
20	-0.01	0.0093	0.0137	136.9188
E	0.0	0.0	0.0	0.0

FIGURE 7.13-21 Detuned Radiator Currents

Make a note of that value.

Now create another tower located 1000 m away. Because the segment length should be the same for both towers, it's best to simply make the second tower identical to the first but at the 1000-m location. Press Ctrl-W, specify two wires and the Cartesian coordinate system. Click in the spacing box for end one of wire 2. Before typing anything in, select the Cartesian coordinate system. Note that the coordinate field labels change from Spacing and Orientation to X and Y, respectively. Note that the 81.5-degree tower height (wire 1 end two Z) changed to 83.792 m. For the second wire, enter a X coordinate of zero, Y coordinate of 1000, an end one bottom Z coordinate of zero and an end two Z coordinate of 83.792 m. This puts the second tower at 1000 m and due north from the first tower.

Next, create a +j571.5 ohm load at the base of tower 1 (pulse 1). Press Ctrl-L, specify one load, enter pulse 1, zero ohms resistance, 571.5 ohms reactance and click OK.

Now we need a source at the base of tower 2. Because tower 2, which will produce a plane wave for us, is located 1000 m away, we need to hit it with a considerably voltage to induce a meaningful current into tower 1. Create a source (Ctrl-V) at pulse 21 with a magnitude of 1000 volts and a phase of zero. Click OK.

Press F5 to run the model. When the WordPad window opens, scroll down to the Current Data section and find the currents for wire 1 (tower 1).

Note that at pulse 8, the current magnitude is at a minimum (and the phase of the current flips positive). Pulse 8, then, is the location at which the sample loop must be positioned on this particular tower. If we scroll up to the Antenna Geometry section for wire 1, we can see that pulse 8 is 29.3272 meters above the base. In this example, all three towers calibrate at the same height, so we need do this only once. If, however, we had towers of different calibrated heights, we would have to repeat this entire process for each tower, thus determining the detuning reactance and sample loop elevation for each tower. If the differences in tower heights are less than half a segment, you can get away with going through the process just once, for the tower closest to the median height of the array.

The FCC rules require that if the effective height of the array element has been adjusted to match the Z matrix, the loops must be mounted at the same percentage height as indicated in the model. In this example, our towers are 77 degrees physical height and the model effective height is 81.5 degrees. Segment 8 on the 81.5-degree radiator is centered at 28.525 degrees above the base, so the loop height on the actual tower is computed as:

$$28.525 / 81.5 \times 77 = 26.95 \text{ degrees, or } 90.84 \text{ feet above the base of the tower.}$$

What parameters would we look for, then, on our antenna monitor with the loops mounted at this height? Look at the currents again on pages 1267 through 1269.

Note that in the directional model the pulse 8 tower 2 current (node 28) is 11.9271 amps at 0 degrees. That corresponds to a ratio of 1.000 and a phase of 0 for the reference tower. Note the segment 8 tower 1 current (node 8) is 6.2824 amps at −127.5 degrees. That represents a ratio of 0.527 and a phase of −127.5 degrees. Note the segment 8 tower 3 current (node 48) is 6.0748 amps at 124.5 degrees. That represents a ratio of 0.509 and a phase of +124.5 degrees. The loop-sampled array parameters would then be:

Twr.	Mag.	Phase
1	0.527	−127.6
2	1.000	0.0
3	0.509	+124.5

SAMPLING SYSTEMS FOR MODELED DA PROOFS

The Traditional Sample System

A traditional directional sample system consists of either tower-mounted pickup loops or current transformers installed at the output of the antenna tuning units (ATUs), an antenna monitor, and a transmission line connecting each pickup device to the antenna monitor. In most cases, the sample transmission lines are of equal length, and in many cases, the sample lines are phase-stabilized, which means that they have been temperature cycled over a specified range in an oven of sorts at the factory.

Over the years, those constructing and adjusting AM directional arrays have made a good effort to insure the accuracy of the sample system, at least to the degree possible with test equipment on hand. Sample loops are constructed as close to identically as possible, current transformers are certified by the manufacturer to be accurate to within 2% for magnitude and to within 3 degrees for phase, and sample lines are carefully cut to have equal electrical length.

The Modeled Array Sample System

While the sample system performs the same job in the moment method modeling world, its accuracy and stability take on new import. In a traditional measurement-verified directional antenna system, the sample system's primary job is to monitor the stability of the operating parameters. In a model-proofed array, it is used to monitor stability but also serves as the means by which the array is set up. As such, its accuracy, stability and linearity are critical.

A sample system must be able to measure voltage or current as it is modeled for each directional antenna element. Current samples can be, as noted earlier, tower-mounted loops or current transformers at the ATU output. A voltage sampler that provides an accurate means of voltage sampling is available from Kintronic Laboratories.

The sample produced by these devices must exclude influences that cannot be modeled accurately. Some examples are towers with nonradiating sampled currents that cannot be modeled, such as skirt-fed towers (folded unipoles) and slant-wire fed towers.

Base Sampling

A tower that is base sampled should have its current transformers located at the same point where the base impedance matrix measurements were made. As such, it is important in base-sampled towers to measure the matrix at the ATU outputs. A tower-mounted sample loop cannot be used to base sample a tower.

Sampling devices in model-proofed arrays should be mounted in such a way that they can be easily removed in order to calibrate them against one another. Most manufacturers of phasing and coupling equipment make provision for straightforward removal of current transformers.

The FCC modeling rules require that sample current transformers (TCTs) be calibrated against one another in the model-proof process and once every 24 months thereafter.

The easiest way to do this is to construct a test jig on a piece of plywood or OSB that is wide enough to mount all the TCTs side by side. Place a piece of 4-inch strap along the length of the board, and drill through the strap for the TCT mounting holes at each mounting location. Place an insulator of the correct height at each end of the board, and run a single piece of rigid tubing from insulator to insulator, passing it through each of the TCTs.

A short length of sample line is then used to connect each TCT's output to the antenna monitor input; each short sample line must be of equal length. An RF current is then fed through the assembly to a load. This can be done by connecting the test jig across an input or output J-plug in the phasor. The antenna monitor must indicate that the TCT outputs agree within the manufacturer's specifications, usually 2% and 3 degrees. Any samples that are outside this range must be returned to the manufacturer for repair or replacement.

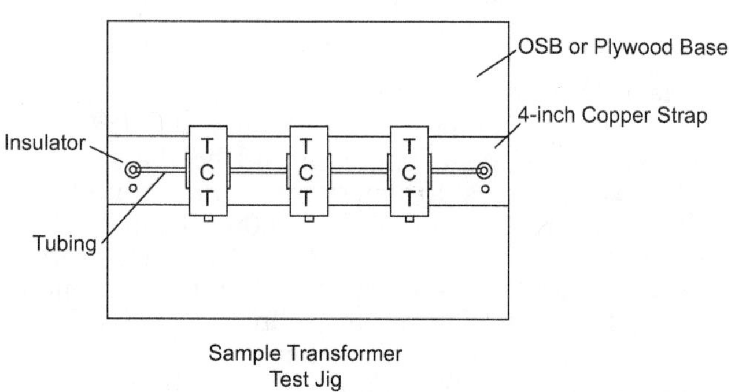

FIGURE 7.13-22 Sample Transformer Test Jig

Loop Sampling

Sampling loops are permissible in model-proofed directional arrays. They must be identical in construction and mounting, and they may be used only on towers with identical legs and cross-sectional structure to insure identical pickup sensitivity. Loops could not be used, for example, on an array where one tower had been replaced with a similar tower that had legs of a different diameter. Loops should generally be leg-mounted and oriented perpendicular to the opposite face. A good test of loop orientation is to stand under the tower and look up at the loop. It should line up with the guy wires on that leg.

Loops must be electrically bonded to the tower, but the sample line can be either at tower potential or insulated.

The conventional location for a sample loop is near the base of a short tower, high enough above the base insulator that it is no longer influenced by the base effects but still in the assumed maximum current area of the tower. For taller towers, the conventional location is roughly one-quarter wavelength down from the top of the tower, at the current loop (assuming sinusoidal current distribution).

On model-proofed towers, the loop must be located at a height where tower current closely corresponds to the tower's far-field contribution for the directional pattern in ratio and phase. The tower current at this

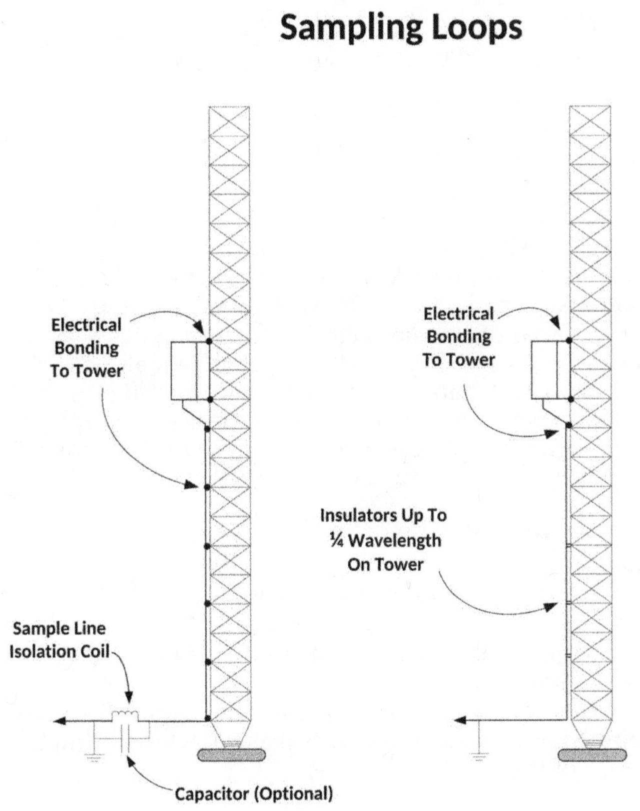

Sampling Loops

Electrical Bonding To Tower

Electrical Bonding To Tower

Insulators Up To ¼ Wavelength On Tower

Sample Line Isolation Coil

Capacitor (Optional)

FIGURE 7.13-23 Sample Loop Installation

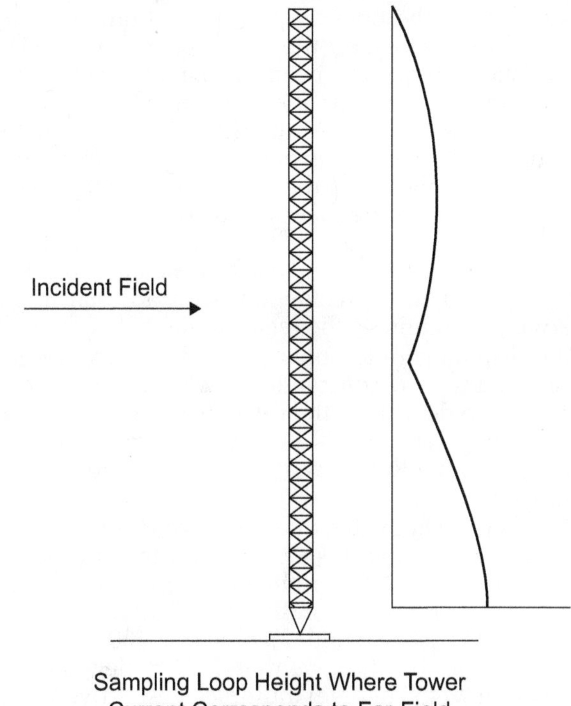

FIGURE 7.13-24 Sample Loop Location

height corresponds to the tower's far-field contribution to the radiation pattern as delta changes in ratio and phase are made. This location is the elevation where the current in the tower would be at a minimum if the tower were detuned. The previous section details how to determine this location through modeling.

Sample Line Measurements

There have been a number of approaches to sample line calibration over the years. Some engineers simply match the physical lengths of the sample lines and assume that they maintain a uniform phase shift and characteristic impedance. Others employ a bridge to carefully resonate the lines to the exact same frequency, insuring that each will have the same phase shift. Yet another method is to employ a vector impedance meter to resonate the lines, watching for the phase to cross through zero as the frequency is slowly increased.

As we mentioned earlier, because a model-proof relies on the sample system to match the model-indicated operating parameters to create the proper directional pattern, it is critical that the sample system be carefully calibrated. Sample line length (within 1 electrical degree) and characteristic impedance (within 2 ohms) must be the same from line to line. The FCC rules spell all this out, along with a procedure for measuring the lines.

The rules call for determining the open-circuit resonant frequency closest to the carrier frequency to establish length, and for measuring at frequencies corresponding to odd multiples of one-eighth wavelength immediately above and below that resonant frequency and at carrier frequency to establish characteristic impedance. By far the fastest and easiest way to do this is with a network analyzer, although traditional bridge measurements can certainly be used. Sample line measurements should be made with the base sampling devices, isocoils or lines to the tower-mounted loops disconnected and the sample lines open at the far end.

In an unterminated transmission line, impedance zeros will occur at odd multiples of 90 degrees electrical length (i.e. 90 degrees, 270 degrees, 450 degrees, etc.) as the electrical length is varied by changing frequency. An impedance zero is where the reactance passed through zero and, for lines with loss, resistance is low. The procedure for this is to determine the multiple of 90 degrees for the zero closest to the carrier frequency, then ratio that length to the length at carrier frequency.

Again, a network analyzer presents the fastest and easiest means of making all the required measurements. Start with the known physical length of the line (or if unknown, use a TDR to find the physical length). Use the equation $F = 4 \times 984 / L / VF$ to find the approximate one-quarter λ frequency for that length of line, where L is the physical length of the line and VF is the assumed velocity factor as a decimal.

Now disconnect the tower end of the sample line and leave it open. Set up the analyzer for a one-port (S11) measurement, Smith chart R +jX.

Go through the open and short calibration procedure for your analyzer, then set up the sweep. The lower (start) frequency needs to be well below the one-quarter λ frequency and the upper (stop) needs to be well above the three-quarter λ frequency. If the one-quarter λ frequency is below the lower frequency range of the analyzer, set the sweep to start well below the three-quarter λ frequency and to end well above the five-quarter λ frequency. Connect the monitor end of the sample line to the S1 port of the analyzer. Use enough power to overcome any stray RF that might otherwise contaminate the measurement.

The display should show a spiral that crosses the left side of the zero X line at least twice. Use the markers feature to determine the frequency of each left side zero crossing. The X will be zero or very close to it at this point; the R represents the resistive loss of the sample line.

Because you know the quarter-wavelength multiple at each zero crossing, it's easy to find the one closest to the carrier frequency. In the following example, the carrier frequency of the station is 670 kHz. The zero crossings (markers 1 and 2) occur at 600 kHz and 1006 kHz for the three-quarter λ (270 degree) and five-quarter λ (450 degree) multiples, respectively. The 600 kHz crossing is closest to carrier frequency. From that we can determine the electrical length of the sample line in degrees:

$$670 / 600 \times 270 = 301.5 \text{ degrees}$$

Next, we must determine the characteristic impedance of the line. Again, the FCC rules call for determining the open-circuit resonant frequency closest to the carrier frequency to establish length, and for measuring at frequencies corresponding to odd multiples of one-eighth wavelength immediately above and below that resonant frequency. So we're looking for the frequency that is one-eighth λ (45 degrees) above and below the resonant frequency. We find those as follows:

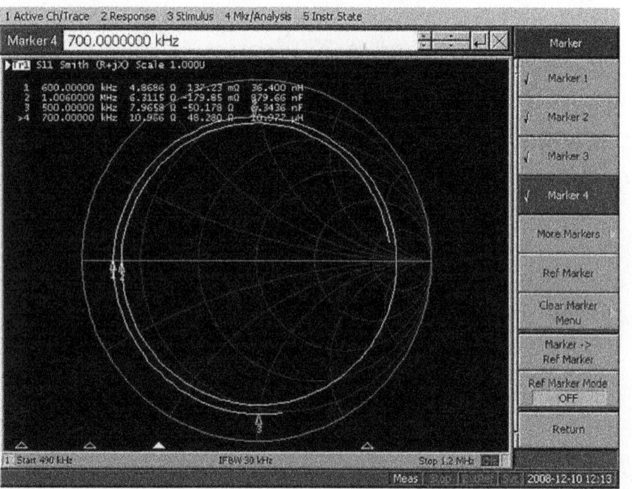

FIGURE 7.13-25 Sample Line Electrical Length and ZO Measurement

$$270 - 45 = 225 \text{ degrees}$$

$$225 / 270 \times 600 = 500 \text{ kHz}$$

$$270 + 45 = 315 \text{ degrees}$$

$$315 / 270 \times 600 = 700 \text{ kHz}$$

So we measure the impedance at each of those frequencies, which should be 90 degrees on either side of the zero crossing on the Smith chart (markers 3 and 4 in the preceding figure). Note that the 500 kHz impedance is 8 –j50.1 and the 700 kHz impedance is 11 +j48.3.

A piece of transmission line theory states that the measured impedance magnitude will equal the transmission line characteristic impedance at odd multiples of one-eighth wavelength (±45 degrees of multiples of 90 degrees) when open circuited. The formula for calculating the magnitude of the impedance is:

$$Z_O = (R_{IN}^{\ 2} + X_{IN}^{\ 2})^{1/2}$$

So . . .

For the –45 degree frequency, $Z_O = (8^2 + 50.1^2)^{½} =$ 50.7 ohms

For the +45 degree frequency, $Z_O = (11^2 + 48.3^2)^{½} =$ 49.5 ohms

The geometric mean of the –45 and +45 degree impedance values represents the characteristic impedance of the line:

$$Z_0 = (Z_{-45} Z_{+45})^{\frac{1}{2}}$$

In this case, $Z_O = 50.1$ ohms.

While from an engineering perspective it would be desirable to have a measured characteristic impedance that matches the output impedance of the sample device and the input impedance of the antenna monitor, the important factor from a model-proof standpoint is that the characteristic impedances of the lines are all within ±2 ohms of one another.

In many sample systems, there are several segments of sample lines per tower. Systems with loops and isolation coils have separate horizontal runs, isolation coils and tower runs. The entire length can be measured as a single line or the sections may be measured separately and their lengths added together to determine the total length.

Measuring Sample Lines Into Sample Loops

Tower-mounted sample loops present a somewhat more difficult situation than base sample current transformers. Tower-mounted loops are difficult to access, and generally speaking, once a loop connection has been made and weatherproofed, it's best to leave it alone (we have all likely had bad experiences with tower crews opening and re-sealing a weatherproofed connection).

When an array is model-proofed for the first time, it is likely that the loops will have to be relocated on the tower to place them at the proper elevation. At that time, there is a unique, one-time opportunity to measure the entire length of the open sample line for each tower, and you should take advantage of that. But once the loop is connected, it would be expensive and unwise to send a tower worker up each tower to open each connector, especially because the loop impedance (actually the impedance of the sample line terminated in the loop) must be measured anyway.

Sample line measurements can be made with the loops connected. Obviously, all excitation must be removed from the array before this is done. Loops, as they are required to be identical, may be assumed to be identical terminating loads for the unmatched sample lines. As the sample lines are equal in length, measurements on different lines will confirm equal lengths and equal loop characteristics by examining the observed reflection coefficients as follows:

$$K = \frac{Z_{IN} - Z_o}{Z_{IN} + Z_o}$$

Reflection coefficients are calculated for measured impedances looking into the antenna monitor ends of the sample lines while connected to the loops. The differences in sample line length are one-half the calculated reflection coefficient angle differences.

The procedure, then, for measuring in such a situation would be to first measure Z_{IN} on the carrier frequency at the antenna monitor end of the sample line with the line connected all the way through to the loop. Then break the sample line at a convenient place, say at the input to the isocoil, and measure Z_{IN} on the carrier frequency at that point looking towards the loop. While the line is broken at that point, measure the resonant frequency of the line back to the antenna monitor and calculate the electrical length on the carrier frequency. Then run each value of Z_{IN} through the previous equation to find K_1 and K_2, then find the difference between K_1 and K_2 and halve it to find the length difference between the lines. Apply that factor to determine the total electrical length of the sample line, from antenna monitor to loop.

Network Analyzer Versus Bridge

All of the preceding measurements are a snap with a vector network analyzer (VNA), and that represents by far the best means of making the measurements.

Using Reflection Coefficients
To Determine Lenght Differences
With Sample Loops Connected

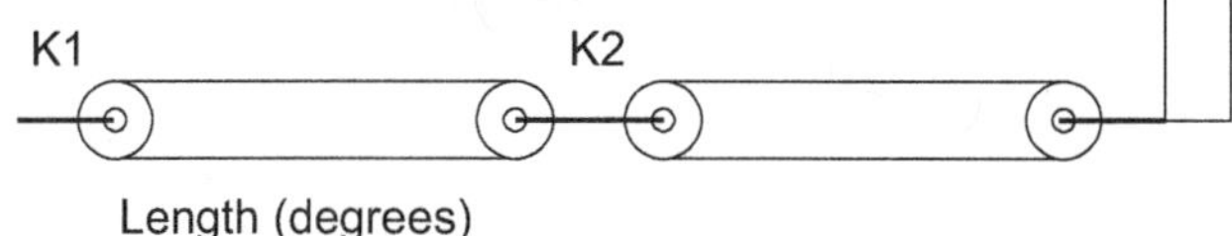

Length (degrees)

Angle difference between
K1 and K2 is 2 X Length
(degrees)

FIGURE 7.13-26 Using Reflection Coefficients to Determine Differences With Sample Loops Connected

However, not everyone owns or has access to a VNA. Much more common is the impedance bridge, either the General Radio 1606B type that has been the mainstay of the broadcast consulting engineer for static impedance measurements for decades, or the operating impedance bridge (OIB) in its various forms.

The good news is that most everything required by way of sample line and sample system calibration measurements can be done with a bridge. It just takes more time and patience. A bridge, particularly the GR-type bridge, has some advantages over the VNA. It is likely more accurate for spot measurements. The frequency of zero reactance and low resistance is unmistakable with such an instrument. The bad news is that finding that frequency means constant initial balance adjustments, repeated at every new frequency. It takes a lot of iterations to nail down the resonant frequency of the line.

Still, it can be done with a bridge, and that fact should encourage those who perhaps might otherwise avoid the model-proof option because they do not have access to a VNA.

Another advantage of a bridge over a VNA is its ability to work in the presence of (relatively) high RF fields. Consider the case of a directional array in a large city where there are many AM signals, some of their sites quite close by (within 5 miles or so) and with high incident fields. Most VNA inputs have a maximum level of +20 dBm, which can easily be exceeded in such a situation—particularly when coupling to a tower-mounted sample loop. An impedance bridge, which with its accompanying detector is highly frequency selective, has no such limitation.

CONCLUSION

Moment method modeling represents an excellent option for many directional antennas. Eligible arrays can be licensed using this technique and do away with often troublesome monitor points forever. Further, a good, calibrated model will provide an excellent prediction of driving point impedances that can be used to accurately design a phasing and coupling system. Once licensed pursuant to a moment method model, the biennial recertification measurements can, in most cases, be done by the station engineer.

The keys to a good model are good measurements and attention to detail in creating and running the model.

CHAPTER

7.14

FM Radio Broadcasting

GEOFFREY N. MENDENHALL, P.E.

GatesAir
Mason, OH

TIM ANDERSON

GatesAir
Mason, OH

INTRODUCTION

Although the mathematical principles explaining frequency modulation (FM) have been known for many years, the advantages and practical application to radio broadcasting were not realized until the 1930s, when Major Edwin H. Armstrong conducted extensive developmental work proving that FM radio transmissions were possible and practical. Many theoreticians claimed to have proof that Armstrong's experiments were impossible based on mathematical models claiming that an infinite transmission bandwidth would be required. He never received proper credit for his many contributions to the radio communications industry during his lifetime [1].

The advantages of FM include freedom from static, wide audio bandwidth, and the ability of an FM receiver to capture the stronger of two signals transmitted on the same carrier frequency. More recently, an in-band, on-channel (IBOC) digital radio system, also known as HD Radio, has been added to many FM broadcast stations [26, 33, 34]. This chapter will cover conventional analog FM technology as well as digital HD Radio transmission technology.

FCC TRANSMISSION STANDARDS

The Federal Communications Commission (FCC) regulates and enforces the technical standards that apply to radio broadcasting in the United States. In theory, this ensures that the public is provided with a consistently high standard of transmission quality from station to station. The rules and regulations covering radio broadcast services, including those for FM broadcast transmitters, are set forth in Part 73 of Title 47 of the Code of Federal Regulations (CFR), available from the U.S. Government Printing Office in Washington, D.C. and online at www.gpo.gov/fdsys/browse/collectionCfr.action?collectionCode=CFR. The rules and regulations are periodically changed to keep pace with new technology and changes within the broadcast industry. Every broadcast engineer should have access to a current copy of these rules and regulations so a radio station's technical performance is maintained within the prescribed limits.

FREQUENCY MODULATION THEORY

Angular Modulation

Frequency modulation (FM) and phase modulation (PM) are both special cases of angular modulation. In any angular modulation system, both the frequency and phase of the carrier vary with time as a function of the modulating signal. The relationship between the frequency deviation of the carrier, the phase deviation of the carrier, and the sinusoidal modulating frequency is defined as the modulation index (m), where:

$$m = \frac{\textit{frequency deviation}\,(\textit{peak} - \textit{to} - \textit{peak hertz})}{\textit{modulating frequency}\,(\textit{hertz})}$$

Because FM and PM are both subsets of angular modulation, they are virtually indistinguishable from one another except in the modulator characteristics.

In a PM system, the modulating signal causes the phase of the carrier wave to vary according to the instantaneous amplitude of the modulating signal. A phase modulator generates a constant amount of phase deviation of the carrier with a constant amplitude modulating signal, independent of the frequency of the modulating signal. The frequency deviation of the carrier produced by a phase modulator does increase as the modulating frequency is increased even though the level of the modulating voltage is held constant. The net effect is that the phase modulator behaves as if it were a frequency modulator with a 6 dB/octave rising slope on the modulating signal input.

An FM modulator generates a constant frequency deviation of the carrier with a constant amplitude modulating signal, independent of the frequency of the modulating signal. The phase deviation of the carrier produced by a frequency modulator decreases as the modulating frequency is increased even though the level of the modulating voltage is held constant. The net effect is that the frequency modulator behaves as if it were a phase modulator with a 6 dB/octave falling slope on the modulating signal input. In FM broadcasting, the RF carrier should have frequency deviation that is proportional to the amplitude of the modulating signal but independent of the frequency of the modulating waveform.

The instantaneous frequency (rate of change of phase) of the RF output wave differs from the carrier frequency by an amount proportional to the instantaneous amplitude of the modulating waveform. For example, consider a 100 MHz carrier wave frequency modulated by a 1000-Hz audio tone and assume that a 1-volt input to the modulator causes ±20 kHz of frequency deviation on the positive and negative peaks of this tone. If the audio input amplitude is increased to 2 volts, the peak deviation will become ±40 kHz varying in sine-wave fashion from one peak of deviation to the other and back again at the 1000-Hz rate. In FM broadcasting, 100% modulation results in a peak frequency deviation of ±75 kHz of the RF carrier.

When pre-emphasis is used ahead of the frequency modulator, the system becomes a phase modulator at audio frequencies above the turnover point of the pre-emphasis network. This is because the frequency response of the pre-emphasis network rises at the rate of 6 dB/octave above this point. FM broadcasting with pre-emphasis really becomes a mixture of FM at low modulating frequencies and PM at high modulating frequencies.

PRE-EMPHASIS

The standards adopted for FM broadcasting in the United States allow the use of pre-emphasis. The standard pre-emphasis curve is defined in FCC Rules Section 73.317(e) as an ideal resistance capacitance network with a time constant equal to 75 microseconds. The 3-dB point for 75-microsecond preemphasis is at a frequency of:

$$f = \frac{1}{2\pi RC} = \frac{1}{2\pi\left(75\times10^{-6}\right)} = 2,122\,Hz$$

The 75-microsecond curve and the tolerance allowed by the FCC are shown in Figure 7.14-1. The frequency response characteristics in decibels for several popular pre-emphasis time constants are given in Table 7.14-1.

The noise voltage in a narrow bandwidth (for example, 1 Hz) increases directly with frequency; therefore, the power spectral density increases as the square of frequency. When de-emphasis is used in the receiver, the noise voltage is attenuated above 2.1 kHz so it remains constant with frequency. The power spectral density is also constant above 2.1 kHz. The high-frequency noise at the receiver would be much greater without de-emphasis.

Pre-emphasis is practical because program content energy tends to peak at several kilohertz and then falls off rapidly at the higher frequencies. For this reason, the higher frequencies may be boosted in amplitude without causing an excessive increase in modulation level. Modern audio processing equipment takes the pre-emphasis curve into account when controlling peak modulation levels.

The location of the pre-emphasis network in the system depends on the operating mode. Stereo transmission requires that the FM modulator have a flat response to the composite baseband signal from the stereo generator, so the individual pre-emphasis

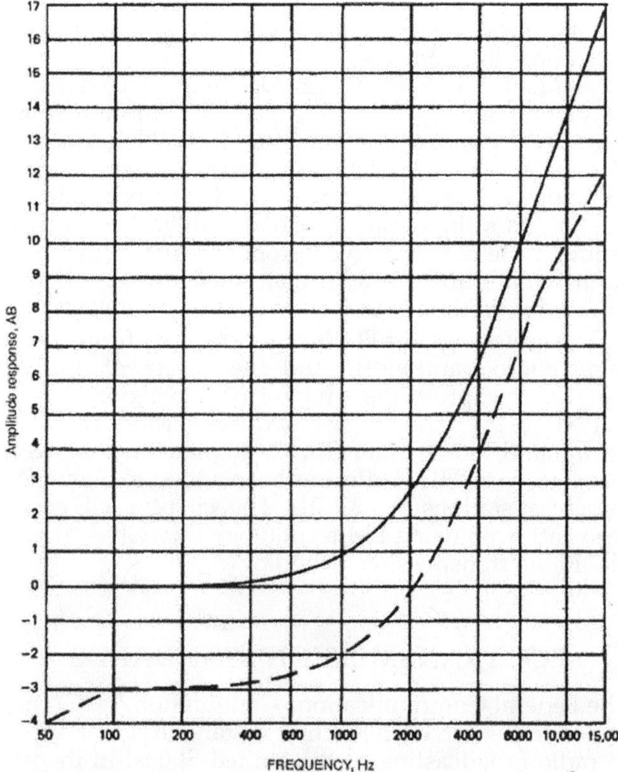

FIGURE 7.14-1 Typical 75-μs pre-emphasis curve (solid line) and tolerance limits between solid and dashed lines.

TABLE 7.14-1
Precise Amplitude Response Values for Various Amounts of Pre-emphasis

Freq (Hz)	Pre-emphasis amplitude response *versus* time constant (dB)			
	25 μs	50 μs	75 μs	150 μs
50	0.000	0.001	0.002	0.010
100	0.001	0.004	0.010	0.038
400	0.017	0.068	0.152	0.577
1000	0.106	0.409	0.871	2.761
2000	0.409	1.445	2.761	6.583
3000	0.871	2.761	7.1469	9.540
4000	1.445	4.115	6.583	11.822
5000	2.087	5.400	8.164	13.656
6000	2.761	6.583	9.540	15.182
7000	3.442	7.661	10.749	16.486
8000	4.115	8.643	11.822	17.623
9000	7.1469	9.540	12.785	18.630
10,000	5.400	10.362	13.656	19.534
11,000	6.005	11.120	14.451	20.353
12,000	6.583	11.822	15.182	21.103
13,000	7.145	12.475	15.858	21.793
14,000	7.661	13.084	16.486	22.433
15,000	8.164	13.656	17.073	23.029

Notes: Values in shaded area are subject to roll-off by audio lowpass filter section of subcarrier generator. Values shown in decibels are relative to $f_m = 0$ Hz.

networks for the left and right channels are located in the stereo generator before the left and right audio channels are multiplexed into the composite baseband signal. In the case of a digital stereo generator, the Audio Engineering Society (AES) [22] serial audio data contains both the left and right channel information in alternating data frames. The digital stereo generator or the upstream digital audio processor applies the pre-emphasis required and 15 kHz audio lowpass filtering using digital signal processing (DSP) techniques.

FM Sideband Structure

The frequency-modulated RF output spectrum contains many sideband frequency components, theoretically an infinite number. Consider, as an example, a radio frequency (RF) carrier of frequency *fc* at a frequency much greater than the modulated sinusoidal signal *fm* with frequency (*fm* ≪ *fc*). The spectrum consists of pairs of sideband components spaced from the carrier frequency by multiples of the modulating frequency. When the modulation index is small (m = 0.5), the amplitude of the second- and higher-order sidebands is small so the output consists mainly of the carrier and the pair of first-order sidebands, as illustrated in Figure 7.14-2(a). The total transmitter RF output power remains constant for a given modulation waveform, but the distribution of that power into the sidebands varies with the modulation index such that power at the carrier frequency (and lower order sidebands) is reduced by the amount of power added to the higher order sidebands.

As the modulation index is increased (as in wide deviation FM broadcasting), the higher order sidebands become more prominent. The amplitude and phase of the carrier as well as the sidebands can be expressed mathematically by making the modulation index (m) the argument of a simplified Bessel function, as shown in Figure 7.14-3.

In a monophonic FM broadcast transmitter, the modulation index can become very high at low modulating frequencies. With a 50-Hz audio input signal of sufficient amplitude to produce 75-kHz deviation (100% modulation), the modulation index is:

$$m = \frac{75,000}{50} = 1,500$$

With a 15,000-Hz input at the same deviation (also 100% modulation), the modulation index is only:

$$m = \frac{75,000}{15,000} = 5$$

Figure 7.14-2 illustrates the frequency components present for modulation indices of 0.5 (a), 5 (b) and 15 (c). Note that the number of significant sideband components becomes very large with a high modulation index. For a given modulation index, the total bandwidth occupied can extend beyond ±75 kHz from the carrier depending on the modulating frequency. This

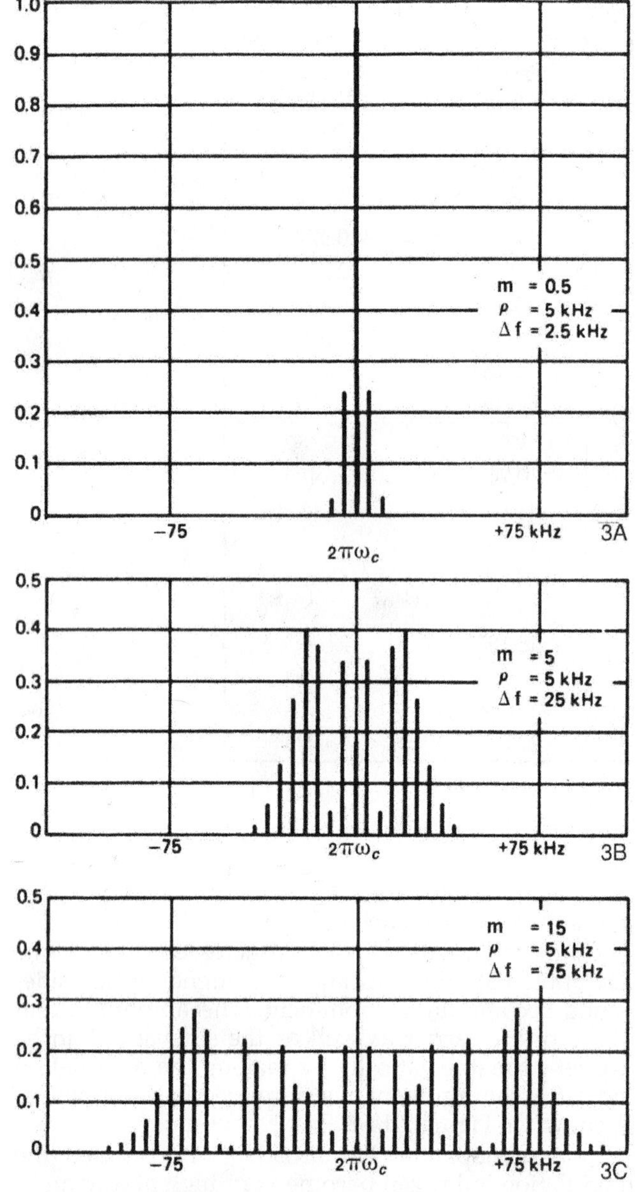

FIGURE 7.14-2 RF spectrum with modulation indexes of (a) 0.5, (b) 5.0, and (c) 15.

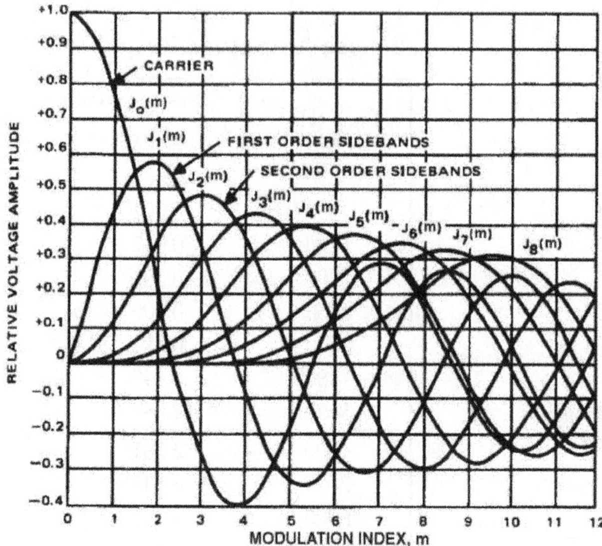

FIGURE 7.14-3 Relationship of carrier and sideband amplitudes to modulation index.

single tone modulating frequency analysis is useful in understanding the general nature of FM and for making tests and measurements. When audio program modulation is applied, there are many more sideband components present. They vary so much that sideband energy becomes distributed over the entire occupied bandwidth rather than appearing at discrete frequencies.

Bessel Nulls

At certain modulation indices, the amplitude of the carrier component of the signal goes to zero, with all the transmitted power being distributed at frequencies other than the carrier frequency [2]. This carrier null phenomenon is useful as an extremely accurate method for measuring the frequency deviation and to check the calibration of modulation monitors. Referring again to Figure 7.14-3, note that the carrier amplitude goes to zero and reverses sign at several values of modulation index, including 2.405, 5.520, and 8.654. Figure 7.14-4 is a photograph taken from an RF spectrum analyzer showing the first Bessel null (M = 2.405) of a carrier at a frequency of 100 MHz.

To determine the audio input level required to achieve 75-kHz deviation, apply an audio tone of exactly 8667 Hz (75,000 divided by 8.654). Starting from zero amplitude, increase the audio level until the carrier disappears for the third time (as 8.654 corresponds to the third Bessel null, as shown in Table 7.14-2). At this audio level, the deviation is exactly ±75kHz. The carrier amplitude (null) detector must have sufficient selectivity to separate the carrier from the sidebands and could be a spectrum analyzer or a receiver with a narrow IF bandwidth. The FM signal can be heterodyned down to a convenient frequency for measurement. Heterodyning does not alter the modulation index; however, when a frequency (or phase) modulated wave is multiplied or divided, this also multiplies or divides the frequency deviation and the modulation index by the same amount. A listing of useful carrier and first-order sideband nulls as function of the modulation index (M) and the modulating frequency (F_m) is given in Table 7.14-2.

Occupied Bandwidth

After examining the Bessel functions and the resulting spectra, it becomes clear that the occupied bandwidth of an FM signal can be far greater than the frequency deviation of the signal. In fact, the occupied bandwidth is infinite (if all sidebands are taken into account), so it is apparent that a frequency modulation system would require the transmission of an infinite number of

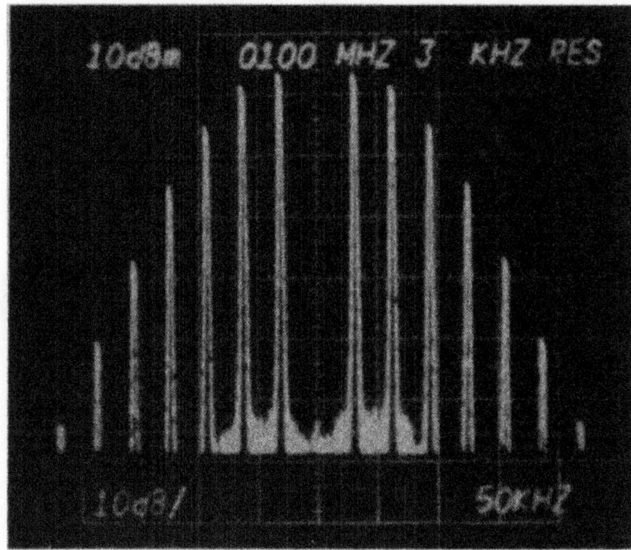

FIGURE 7.14-4 Photograph taken from an RF spectrum analyzer showing the first Bessel null (M = 2.405) of a carrier at a frequency of 100 MHz @ ±75 kHz.

TABLE 7.14-2
Sideband Nulls as a Function of Modulation Index and Modulating Frequency

	Modulation Index (M)		F_m for 75 kHz Deviation	
Null	Carrier	First sidebands	Carrier	First sidebands
1st	2.405	3.832	31,187	19,574
2nd	5.520	7.016	13,587*	10,690
3rd	8.654	10.174	8667	7372
4th	11.792	13.324	6361	5629
5th	14.943	16.471	5023	4554
6th	18.071	19.616	4150	3823
7th	21.212	22.759	3536	3295

* This tone is recorded on track 38 of the NAB Broadcast Audio System Test CD.

sidebands for perfect demodulation of information. In practice, a signal of acceptable quality can be transmitted in the limited bandwidth assigned to an FM channel.

Effects of Bandwidth Limitation

Practical considerations in the transmitter RF circuitry and available spectrum make it necessary to restrict the RF bandwidth to less than infinity. As a result, the higher-order sidebands will be altered in amplitude and group delay (time). Bandwidth limitation will cause distortion in any FM system. Consider the block diagram shown in Figure 7.14-5(a), where a perfect FM modulator is connected to a perfect demodulator via an RF path of infinite bandwidth. The demodulated audio shown in Figure 7.14-5(b) contains no distortion components. In Figure 7.14-6(a), a bandpass filter is inserted between the modulator and demodulator to restrict the bandwidth. Audio distortion products now appear at the output of our perfect demodulator, as shown in Figure 7.14-6(b). These distortion products are due solely to the bandwidth restriction (300 kHz = 3 dB BW) imposed by the bandpass filter.

Figure 7.14-7 show the effects of a narrowband RF bandpass filter on the RF spectrum of a composite signal consisting of a stereophonic subcarrier modulated only on the left channel with 4.5 kHz plus a 67-kHz unmodulated Subsidiary Communications Authorization (SCA) subcarrier. In Figure 7.14-7a, through a wideband RF path, there are no baseband distortion products in the demodulated spectrum. Figure 7.14-7b shows the corresponding effects observed on the demodulated baseband spectrum for the same signal through a narrowband filter. The only distortion evident on the RF spectrogram is the loss of some sidebands greater than 150 kHz from the center frequency and some amplitude differences between the upper and lower sideband pairs. Note the creation of many undesired intermodulation terms in the demodulated baseband spectrum that cause crosstalk into both the stereophonic and SCA subcarrier bands. The change in the RF spectrum is subtle, but the resulting spectrum after demodulation is clearly affected. As a result, the distortion in any practical FM system will depend on the amount of bandwidth available as well as the transmitted modulation index.

Group Delay Symmetry Versus Amplitude Response Symmetry

Although both the amplitude response and time response (group delay) across the FM channel have an effect on the amount of distortion added to the FM signal, the symmetry of the group delay response is more important than the total group delay variation or the amplitude response. Best FM performance is always obtained when the system is tuned for symmetrical group delay (time) response. Depending on the circuit topology, the tuning conditions for symmetrical group delay response may not coincide with the symmetrical amplitude response.

Limiting Factors Within an FM Transmitter

Relating the specific quantitative effect of the bandwidth limitations imposed by a particular transmitter to the actual distortion of the demodulated composite baseband is a complicated problem. Some of the factors involved are:

- Total number of tuned circuits involved;
- Amplitude and group delay response of the total combination of tuned circuits in the RF path;
- Amount of drive (saturation effects) to each Class C stage;
- Nonlinear transfer functions (AM-AM and AM-PM for HD Radio) within each amplifier stage.

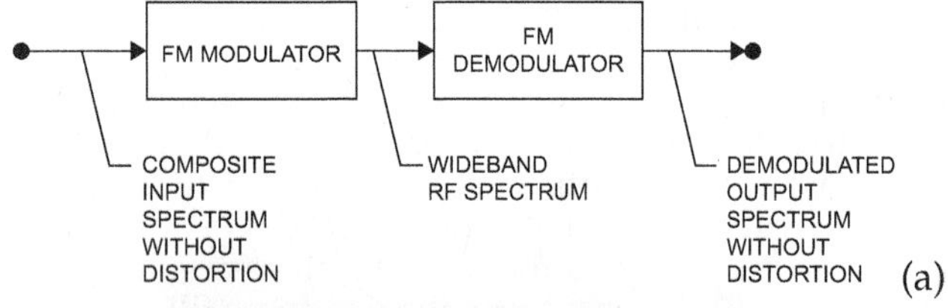

(a)

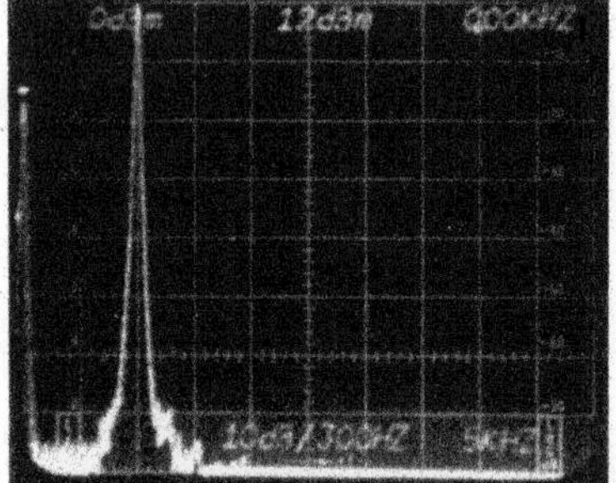

BASEBAND SPECTRUM TO FM MODULATOR

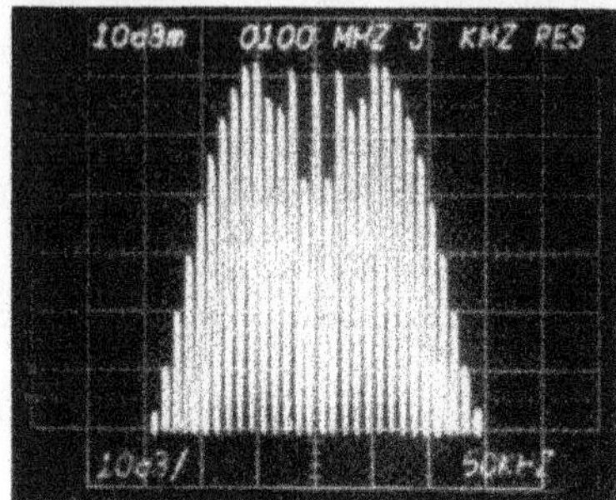

RF SPECTRUM TO DEMODULATOR

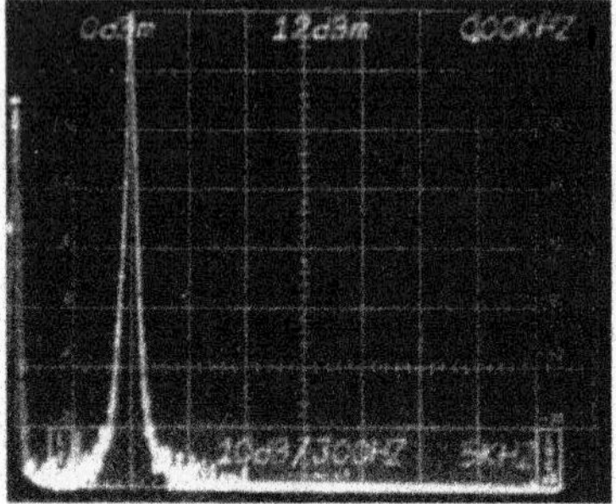

DEMODULATED BASEBAND SPECTRUM (b)

FIGURE 7.14-5 (A) Bandwidth-limited RF path. (b) Single-tone (10-kHz) modulation through narrowband RF path.

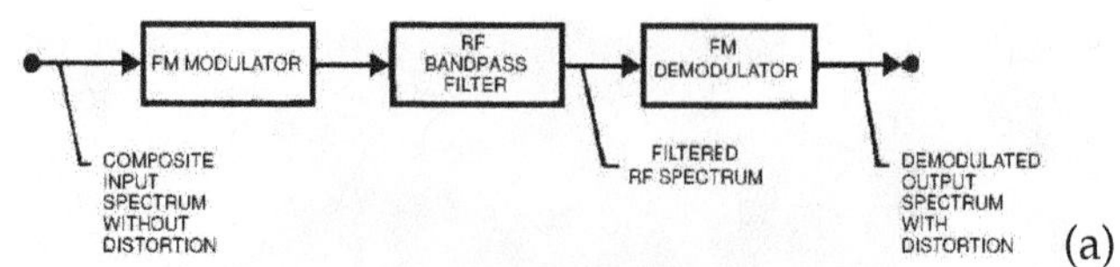

(a)

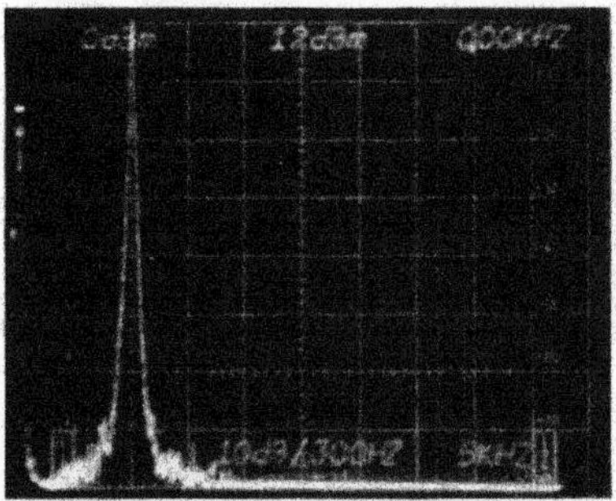

BASEBAND SPECTRUM TO FM MODULATOR

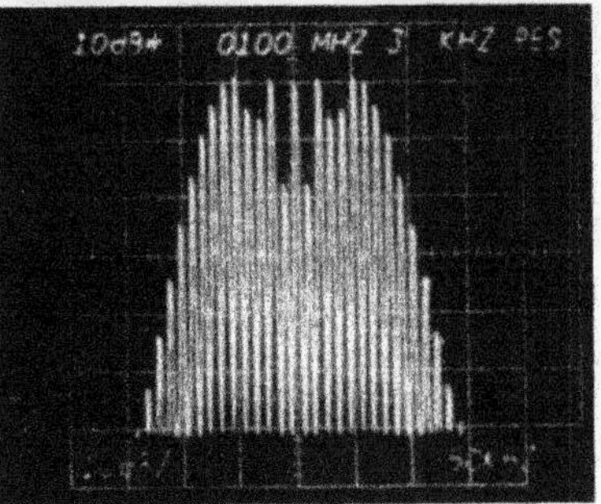

BANDWIDTH LIMITED RF SPECTRUM TO DEMODULATOR

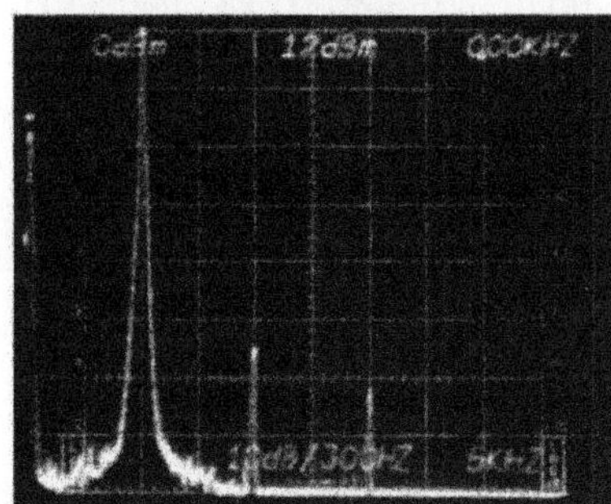

DEMODULATED BASEBAND SPECTRUM (b)

FIGURE 7.14-6 (A) Wideband RF path. (b) Single-tone (10-kHz) modulation through wideband RF path.

Improvement of the RF Path

The following design techniques can help improve the transmitter's bandwidth:

- Maximize bandwidth by using a broadband exciter and a broadband intermediate power amplifier (IPA) stage;
- Use a single-tube design or a broadband, completely solid-state design where feasible;
- Optimize both grid circuit and plate circuit of the tuned stage for the best possible bandwidth and symmetrical group delay response;
- Minimize the number of interactive tuned networks;
- Use a broadband antenna system that provides a low standing wave ratio on the transmission line.

For more information about FM and digital modulation theory, see references [2, 3, 4, 21, 24].

FM ANALOG AND DIGITAL RADIO TRANSMITTERS

The purpose of the FM transmitter is to convert a main channel audio signal and its associated audio or data subcarriers, or an AES3 serial digital audio data bit stream, into a frequency-modulated, radio frequency signal at the desired power output level to feed into the radiating antenna system. In its simplest form, the FM transmitter can be considered an FM modulator and an RF power amplifier packaged into one unit as shown in Figure 7.14-8. In fact, an FM transmitter consists of a series of individual subsystems, each having specific functions:

- The FM exciter converts the analog audio baseband or serial, AES3, digital audio data into frequency-modulated RF and establishes the key qualities of the signal;
- An intermediate power amplifier (IPA) is required in some transmitters to boost the RF power level up to a level sufficient to drive the final RF power amplifier stage;
- The final power amplifier further increases the signal level to the value required to drive the antenna system;
- A transmitter control system monitors, protects, and provides commands to each of these subsystems so they work together as an integrated system;
- The RF lowpass filter removes undesired harmonic frequencies from the transmitter's output, leaving only the fundamental output frequency;
- A directional coupler provides an indication of the power being delivered to and reflected from the antenna system;
- The power supplies convert the input power from the AC line into the various DC or AC voltages and currents required by each of these subsystems.

BASEBAND SPECTRUM TO FM MODULATOR

BASEBAND SPECTRUM TO FM MODULATOR

RF SPECTRUM TO DEMODULATOR

BANDWIDTH LIMITED RF SPECTRUM TO SPECTRUM

DEMODULATED BASEBAND SPECTRUM

DEMODULATED BASEBAND SPECTRUM

FIGURE 7.14-7 (A) Left group of images: stereo (L or R = 4.5 kHz) plus SCA (unmodulated) modulation through wideband RF path. (b) Right group of images: stereo (L or R = 4.5 kHz) plus SCA (unmodulated) modulation through narrowband RF path.

Analog FM Exciter

The function of the exciter is to generate and modulate the carrier wave with one or more inputs (mono or stereo audio, SCA, data are the most common) in accordance with appropriate standards. Stereo audio transmission places the most stringent performance requirements upon the exciter. Because the exciter is the origin of the transmitter's signal, it establishes most of the signal's technical characteristics including

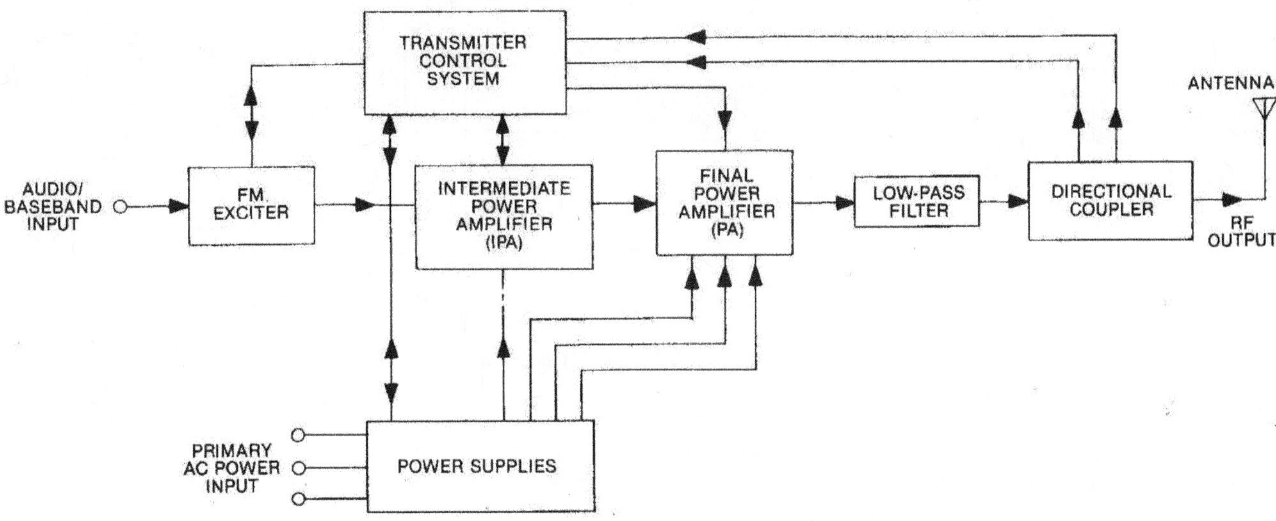

FIGURE 7.14-8 Simplified block diagram of an FM broadcast transmitter.

signal-to-noise ratio (SNR), distortion, amplitude response, phase response, and frequency stability.

Waveform linearity, amplitude bandwidth, and phase linearity must be maintained within acceptable limits throughout the analog baseband chain from the stereo and subcarrier generators to the analog FM exciter's modulated oscillator. The introduction of AES3 [22] digital audio transport and digital FM modulation techniques such as direct digital synthesis (DDS), direct to carrier, and digital-to-analog (D/A) conversion eliminate the distortions introduced by analog circuits.

In a digital FM exciter, the left and right audio data is converted into a digital representation of stereo baseband by digital signal processing (DSP). This data is then further converted into a frequency-modulated carrier by a DDS numerically controlled oscillator (NCO). From here, the FM carrier is usually amplified in a series of Class C nonlinear power amplifiers, where any amplitude variation is removed. The amplitude and phase responses of all the RF networks that follow the exciter must also be controlled to minimize degradation of the signal quality.

Direct FM

Direct FM is an analog modulation technique where the frequency of an oscillator changes in direct proportion to an applied voltage. Such an oscillator, called a voltage-tuned oscillator (VTO), was made possible by the development of varactor tuning diodes that change capacitance as their reverse bias voltage is varied (also known as a voltage-controlled oscillator, or VCO). If the composite baseband signal is applied to the tuning terminal of a VTO, the result is a direct frequency-modulated oscillator. Figure 7.14-9 is a block diagram representative of most direct FM exciters on the market. The SNR of an FM exciter is dependent on the short-term stability of the modulated oscillator by factors, such as:

- Operating level;
- Noise figure of the oscillator transistor;
- Circuit configuration;
- Method of amplitude limiting;
- Loaded "Q" of the oscillator tank circuit;
- Mechanical stability of components.

Optimization of these factors has resulted in an SNR of better than 90 dB below 100% modulation in the current generation of analog FM exciters.

Analog FM Modulator Linearity

Nonlinearities in the FM oscillator alter the waveform of the baseband signal and create distortion in the

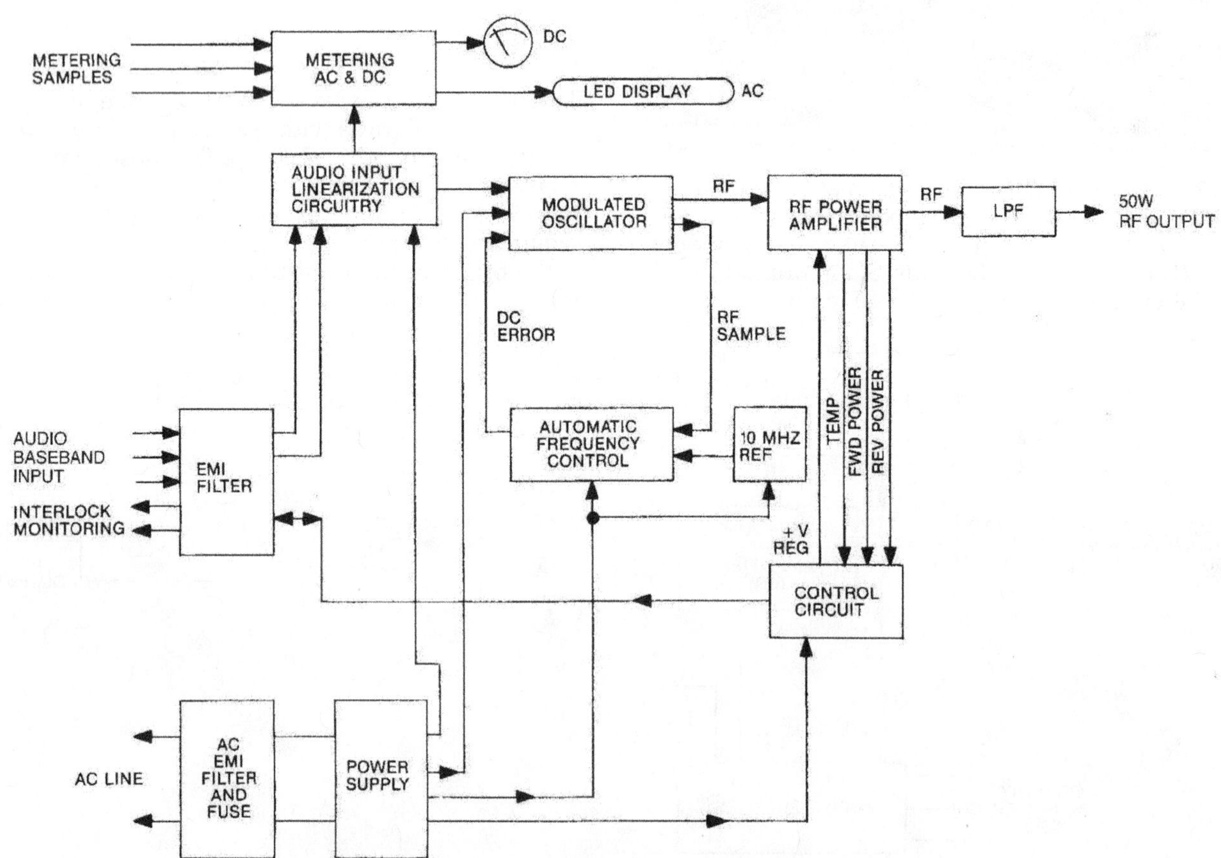

FIGURE 7.14-9 FM exciter block diagram using direct FM.

demodulated output at the receiver. A secondary effect of this distortion may include stereo crosstalk into the SCA subcarrier signals [12]. The composite baseband signal is frequency modulated onto an RF carrier by the modulated oscillator. Frequency modulation is achieved by applying the composite baseband signal to a voltage tuned RF oscillator. The modulated oscillator usually operates at the carrier frequency and is voltage tuned by varactor diodes operating in a parallel LC circuit. To have perfect modulation linearity, the RF output frequency must change in direct proportion to the composite modulating voltage applied to the varactor diodes. This requirement implies that the capacitance of the varactor diodes must change as the square of the modulating voltage.

Unfortunately, the voltage versus capacitance characteristic of practical varactor diodes is not the desired square-law relationship. All varactor-tuned oscillators have an inherently nonlinear modulating characteristic. This nonlinearity is predictable and repeatable for a given circuit configuration, making correction by complementary pre-distortion of the modulating signal feasible. Suitable pre-distortion can be applied to the composite baseband signal by using a piecewise linear approximation to produce the desired complementary transfer function. Figure 7.14-10 shows how the pre-distortion network is cascaded with a nonlinear voltage-tuned oscillator to produce a linearized frequency modulator.

It is also possible to improve both the linearity and signal to noise ratio (SNR) of the modulated oscillator by demodulating its RF output to baseband and then feeding some of this baseband with the proper phase relationship back into the composite input of the modulator. This configuration places the entire modulated oscillator within a negative feedback loop and transfers the responsibility for maintaining linearity to the demodulator. Digital demodulation schemes can be made very linear, but the additional complexity and the potential problems with loop stability have limited the applications of this approach to linearization.

Analog modulator linearization has reduced harmonic and intermodulation distortion to less than 0.01% in the current generation of equipment. Any distortion of the baseband signal caused by the modulated oscillator will have secondary effects on stereo and SCA crosstalk, which are quite noticeable at the receiver in spite of the rather small amounts of distortion to the baseband. For example, if the harmonic distortion to the baseband is increased from 0.05 to 1.0%,

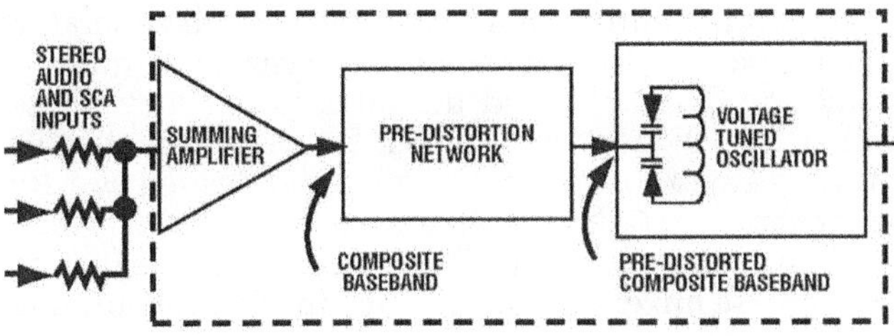

FIGURE 7.14-10 Linearized FM modulator block diagram.

as much as 26 dB of additional crosstalk into the SCA can be expected.

Transient intermodulation (TIM) distortion is usually not a factor in varactor-tuned modulated oscillators. The modulation bandwidth capability is generally more than ten times the composite signal bandwidth and no negative feedback is used to maintain linearity.

Ensuring that the composite baseband signal undergoes minimal distortion in the modulation process will reduce undesired harmonic and intermodulation products in the baseband, making the FM exciter transparent to the signals coupled into it. All exciter stages after the modulated oscillator operate as broadband amplifiers with minimal bandwidth limitations. Analog FM exciter technology is currently capable of transmitting near compact disc quality with less than 0.01% distortion and an SNR of better than 90 dB. Digital FM exciter technology utilizing direct digital synthesis can exceed the limitations of analog modulators to provide the full 16-bit digital performance of a compact disc.

Phase-Locked Loop Automatic Frequency Control

The frequency stability of direct FM oscillators is not good enough to meet the FCC frequency tolerance of ±2,000 Hz. This requires an automatic frequency control system (AFC) that uses a stable crystal oscillator as the reference frequency. The modulated oscillator need not have good long-term stability because the AFC feedback loop will correct for long-term drift to keep the average carrier frequency within limits. The modulated oscillator does need excellent short-term (less than 1 sec) stability because the control loop time constant must be long enough so the AFC circuit does not try to remove desired low-frequency audio modulation. This means that the oscillator is essentially running open-loop at frequencies between 5 Hz and 100 kHz, so the noise performance of the modulator will also be determined by the short-term stability characteristics of the oscillator.

Phase-locked loop (PLL) technology has provided a means of precisely controlling the carrier's average frequency while permitting wide deviation of the carrier frequency at baseband modulating frequencies. This implies that a PLL system behaves like an audio high-pass filter with higher modulating frequencies being ignored by the control loop while lower frequencies are considered to be errors in the average frequency and are tracked out by the loop. An added advantage of PLL is the ability to synthesize the desired frequency from a single reference oscillator, thereby eliminating the need to change crystals when changing the frequency of the exciter.

The block diagram shown in Figure 7.14-11 includes the key elements in the PLL. The output of the modulated oscillator operating at the carrier frequency is digitally divided down to a frequency of a few kilohertz or less and is called the comparison

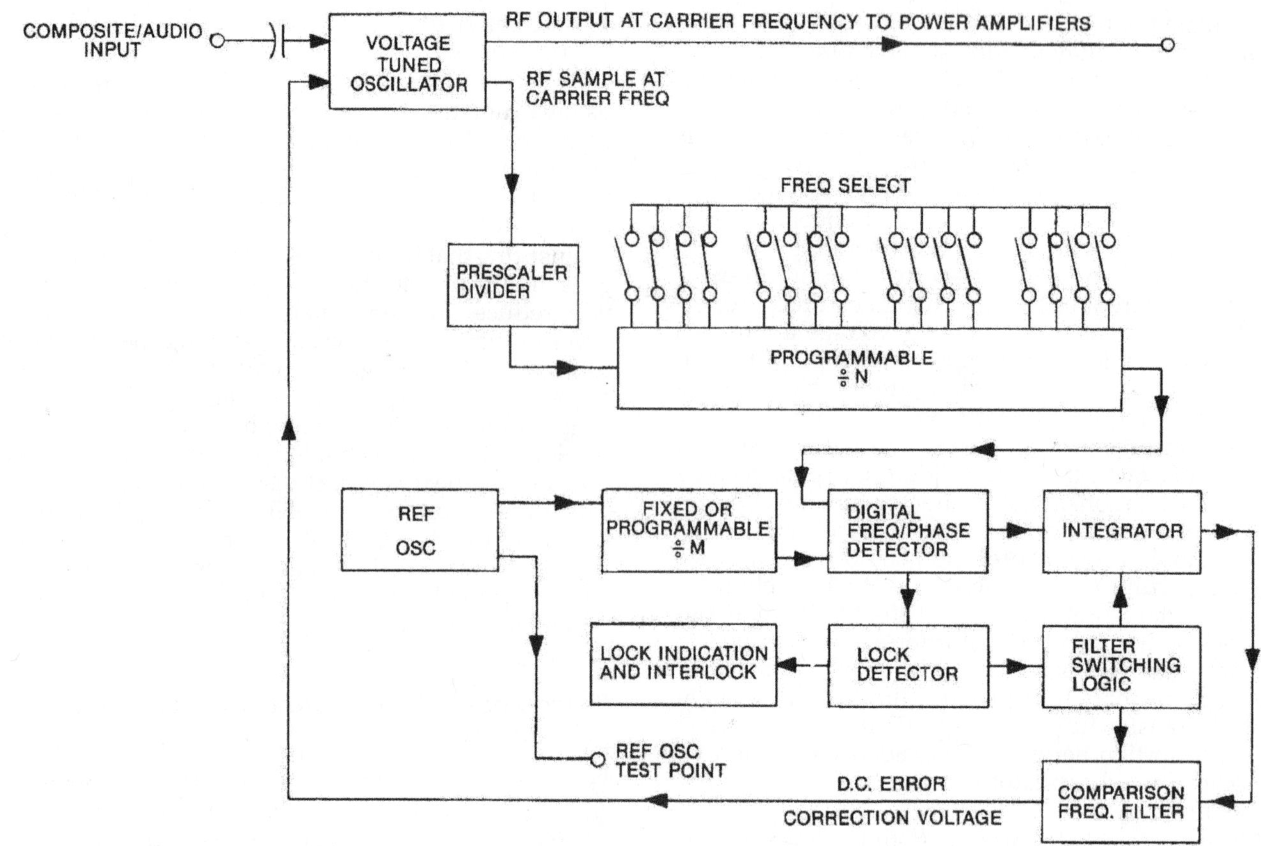

FIGURE 7.14-11 Phase-locked loop frequency synthesizer.

frequency. Likewise, the reference crystal oscillator is also digitally divided down to the comparison frequency. The two frequencies are compared in a digital phase/frequency detector to develop an error voltage that corrects the carrier frequency of the modulated oscillator. The reason for dividing the modulated oscillator frequency so many times is to reduce the modulation index enough to limit the peak phase deviation at the comparison frequency to a value that will not exceed the linear range of the phase/frequency detector. If the linear range is exceeded, the loop will lose lock. This is why some exciters may lose AFC lock in the presence of low-frequency modulation components.

The phase detector output is integrated and low-pass filtered to remove the comparison frequency and all other frequency components above about 5 Hz so the AFC circuit does not try to track low-frequency modulation. Some FM exciters use a dual-speed PLL to keep the loop turnover frequency low enough to maintain good amplitude and phase response at 30 Hz while also providing quick lock-up time. The PLL error correction circuitry must respond quickly during the initial frequency scan of the FM band to achieve lock-up to the precision reference oscillator in a few seconds. The loop bandwidth is wide during acquisition and lock-up. After lock is achieved, the bandwidth is reduced to provide the optimum modulation characteristic.

The reference oscillator is usually temperature compensated and requires no warm-up to maintain ±3 parts per million (PPM) or better accuracy over the operating temperature range. A 10 MHz source may be used as the reference frequency for convenient comparison to international or global positioning system (GPS) frequency standards. For more information about PLL frequency synthesizers, see reference [5].

FM by I/Q Modulation Techniques

Another method used to generate FM is the quasi-digital, in-phase/quadrature (I/Q) RF technique also known as an RF modem. Two identical analog RF mixers are fed in-phase (I) and quadrature (Q) analog baseband signals derived from two separate D/A converters at the output of a DSP digital stereo generator. The I mixer is also fed the in-phase carrier frequency, while the Q mixer is fed the carrier frequency phase shifted by 90° from the I mixer. When the RF outputs of these two mixers are summed, the desired FM RF spectrum is produced if the system is perfectly balanced and matched. Practical limitations in the matching of these mixers require that feedback incorporating sophisticated nonlinear DSP adaptive correction be placed around these mixers and the I and Q D/A converters driving them. This technique has the advantage of producing FM at the carrier frequency instead of at an intermediate frequency.

Digital FM Exciter Using Direct Digital Synthesis

DDS eliminates the need for a PLL in the FM modulation process. It does so by directly synthesizing the carrier frequency, including FM modulation, from a sine wave look-up table in a programmable read-only memory (PROM) device operating in conjunction with a digital phase accumulator and a fast D/A converter. When this technique is combined with DSP technology, the entire process of generating stereo baseband with an Radio Data System (RDS) subcarrier and SCAs, then frequency modulating this digital baseband information onto the RF carrier is accomplished entirely in the digital domain. The cost-to-performance ratio of DDS/DSP technology has made it competitive with the analog technology. The full benefit of DDS/DSP technology requires digital transmission of audio information as an uncompressed, digital bit stream all the way from the digital audio source through a digital console, digital audio processing, and an uncompressed, digital studio-to-transmitter link (STL) to the AES3 digital input port of the DSP/DDS exciter [22, 23]. This same technology is used in the fully digital radio services including the HD Radio system, Digital Radio Mondiale (DRM), and the Eureka 147 (EU-147) DAB transmission standards, which are implemented worldwide.

With DDS, the complete FM waveform is generated entirely in the digital domain. As digital modulation is an inherently linear process, no pre-distortion is required. The FM signal generated by a DDS device has low noise and distortion for true 16-bit digital audio quality (−96 dB FM SNR and 0.0016% harmonic distortion for ±75 kHz deviation and 75 µs pre-emphasis/de-emphasis).

The current generation of DDS exciters uses a 32-bit NCO. The basic resting frequency of the NCO is set by a 32-bit tuning word. Frequency modulation occurs when modulation data varies the structure of the tuning word within the phase accumulator section of the NCO. The modulated output of the NCO is converted to analog FM, up-converted, filtered, and amplified to become the RF signal for a conventional FM broadcast transmitter RF amplifier chain. The recent generation of combined FM and HD Radio exciters uses direct-to-carrier, D/A conversion that directly converts the digital output of the NCO to the carrier frequency without the need for up-conversion or I/Q mixing at carrier frequency. A block diagram of a DDS digital FM exciter is shown in Figure 7.14-12.

DDS FM exciters also eliminate several basic limitations found in analog exciters using direct FM via the modulation of VCOs. Very low audio frequencies must be filtered from program signals feeding a VCO/PLL to avoid affecting the circuits of the analog exciter, which see very low modulating frequencies as an off frequency condition that requires correction. A DDS-based FM exciter has no such limitation and the modulation frequency response extends virtually to DC (zero hertz). These lower octaves of program material are important to accommodate digital audio source material and to preserve the phase correlation existing in the original program.

Digital IBOC HD Radio Exciter Functionality

Most digital HD Radio exciters today incorporate the 3rd or 4th generation Exgine-based HD Radio architecture that separates the codec and data multiplexing functions ("Exporter") from the RF signal-generation function of the exciter subsystem ("Exgine" is a contraction of exciter and engine). The Exporter can be located at the transmitter site or the studio with only a single, unidirectional User Datagram Protocol (UDP) Ethernet connection between the two units. The unidirectional UDP stream carries the primary and supplementary audio and data services that have been multiplexed by the Exporter to the Exgine that creates the digital orthogonal frequency division multiplex (OFDM) signal for conversion to RF and amplification by the transmitter. Today's Exgine-based exciters use DSP and field programmable gate array (FPGA) hardware platforms that run on embedded, real-time, operating systems for greater reliability and much faster restart. Figures 7.14-13 and 7.14-14 illustrate the Exgine and direct to carrier exciter architecture.

Fourth Generation Exgine

The generation and amplification of the hybrid HD Radio waveform presents unique challenges due to the inherently high peak-to-average ratio (PAR) of the signals and complex system of intermodulation products any system nonlinearity produces. In order to create and amplify these signals cleanly and efficiently, it is necessary to employ crest factor reduction (CFR) and distortion pre-correction techniques. The biggest challenge with amplifying OFDM waveforms used for HD Radio and all other digital radio formats is their high crest factor. OFDM modulates vectors of information symbols in parallel over a large number of individual orthogonally-spaced subcarriers. Because of these peaks, the power amplifiers used in the transmitters need to operate within their linear range with appropriate power back-off to minimize peak distortion. Peak distortion caused by nonlinearity introduces intermodulation noise interference to the subcarrier modulation, and causes out-of-band emissions.

Direct to Carrier Digital-to-Analog Conversion

The latest generation of HD + FM exciters incorporate high-speed D/A converters that can convert the digital representation of the HD plus FM signals directly to the carrier frequency in the 87–108 MHz FM broadcast band. This eliminates the need for modulation at a lower intermediate frequency and then up-converts the signal to the FM band. Direct to carrier digital A/D conversion provides a more accurate creation of the RF signal sideband structure than analog techniques.

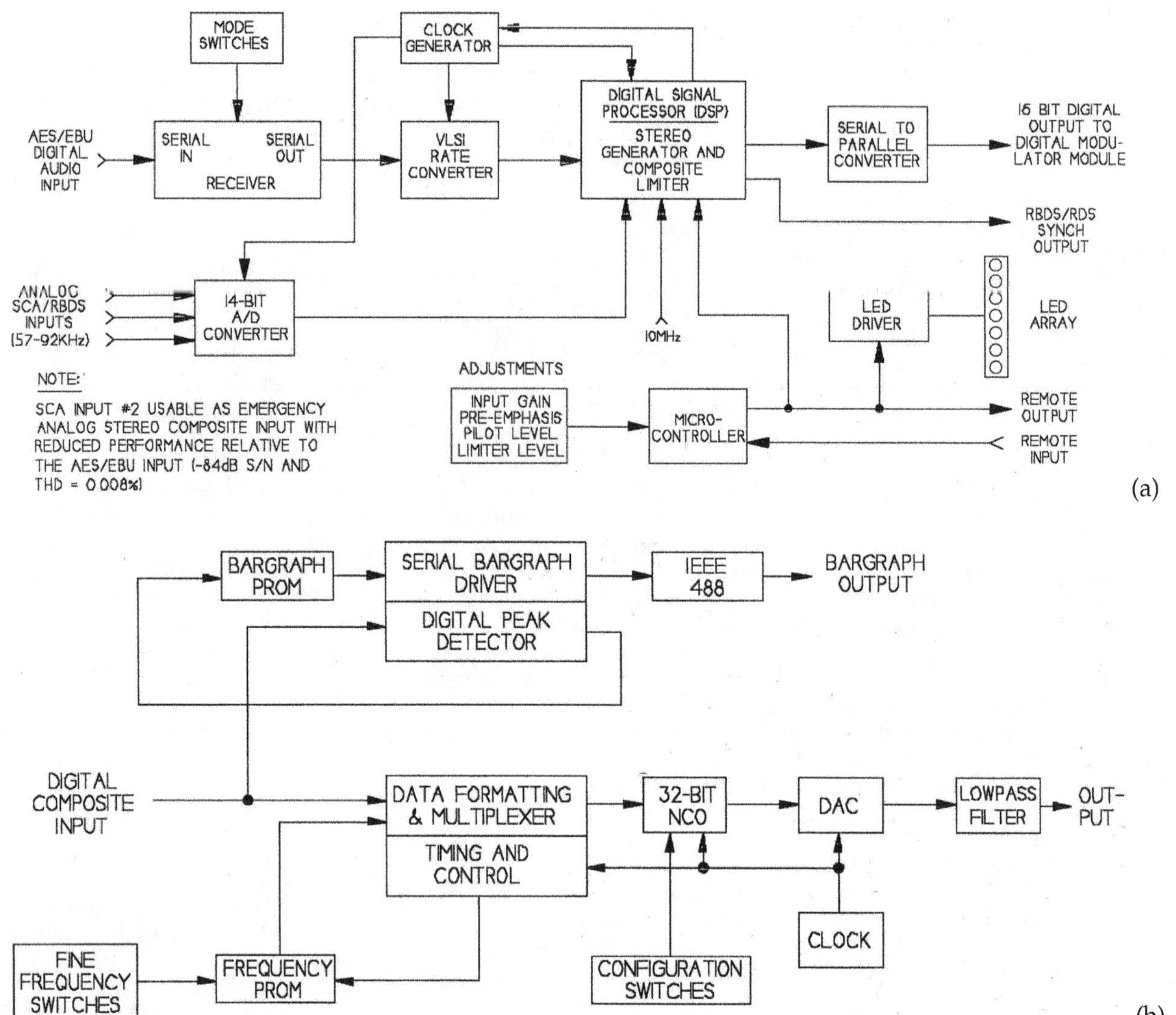

FIGURE 7.14-12 (a) Harris DIGIT-CD digital input module and DSP stereo generator. (b) Digital modulator in the Harris DIGIT-CD FM exciter.

HD Radio System Architecture

Figure 7.14-15(a) illustrates the studio system block diagram for the HD Radio studio audio and program associated data encoding and multiplexing. Figure 7.14-15(b) illustrates the HD Radio transmission plant system block diagram.

Exciter RF Output Stage

The broadband RF amplifier in the exciter raises the output of the modulated oscillator from a power level of a few milliwatts up to the range of 5 to 50 W. The output stage is normally protected against damage that could be caused by an infinite voltage standing wave ratio (VSWR) on the output at any phase angle. The typical exciter RF amplifier is designed to have a bandwidth of at least 20 MHz, using successive broadband impedance matching sections for each stage. Each group of matching sections consists of micro-strip or lumped elements. The broadband performance of the RF amplifier eliminates the need for adjustments to any particular frequency within the FM band. The exciter output is transparent to the signal generated by the modulated oscillator, and the amplifier stability accommodates varying load conditions. A micro-strip directional coupler in the RF amplifier output network provides automatic control of power output level and protection against operation under high VSWR conditions. All current-generation FM exciters typically produce at least 50 W of RF output; thus, the exciter may be used as a complete transmitter for educational or low power stations with a harmonic lowpass filter at the output. For higher power level requirements, the exciter drives an external power amplifier.

Exciter and Final Linear Amplification Requirements for HD Radio

The analog FM signal has a constant amplitude and does not require linear amplification. Typically, RF

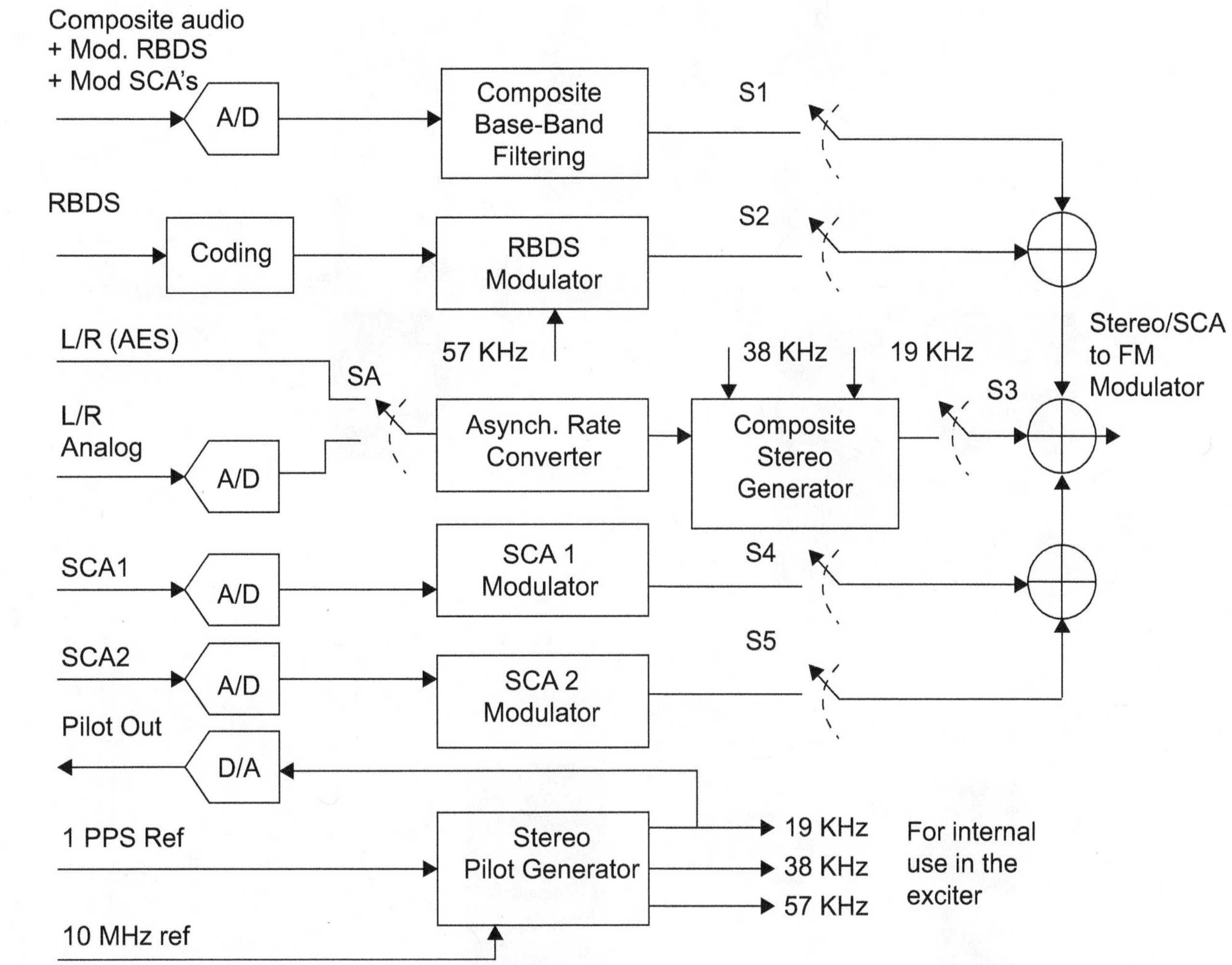

Composite audio + Mod. RBDS + Mod SCA's
A/D
Composite Base-Band Filtering
S1
RBDS
Coding
RBDS Modulator
S2
57 KHz
L/R (AES)
SA
L/R Analog
A/D
Asynch. Rate Converter
38 KHz
19 KHz
Composite Stereo Generator
S3
Stereo/SCA to FM Modulator
SCA1
A/D
SCA 1 Modulator
S4
SCA2
A/D
SCA 2 Modulator
S5
Pilot Out
D/A
1 PPS Ref
Stereo Pilot Generator
19 KHz
38 KHz
57 KHz
For internal use in the exciter
10 MHz ref

FIGURE 7.14-13 Flexstar HDx FM digital signal processing stereo generation and SCA modulation.

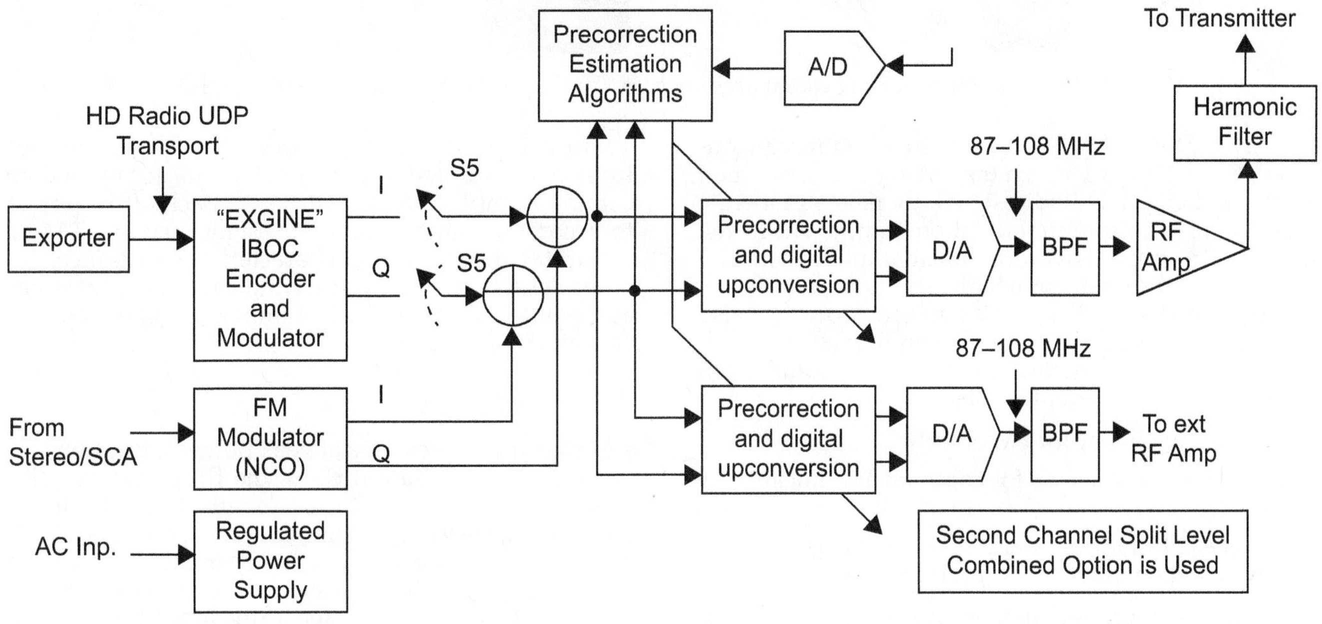

FIGURE 7.14-14 Flexstar HDx digital FM and IBOC and direct to carrier frequency D/A conversion.

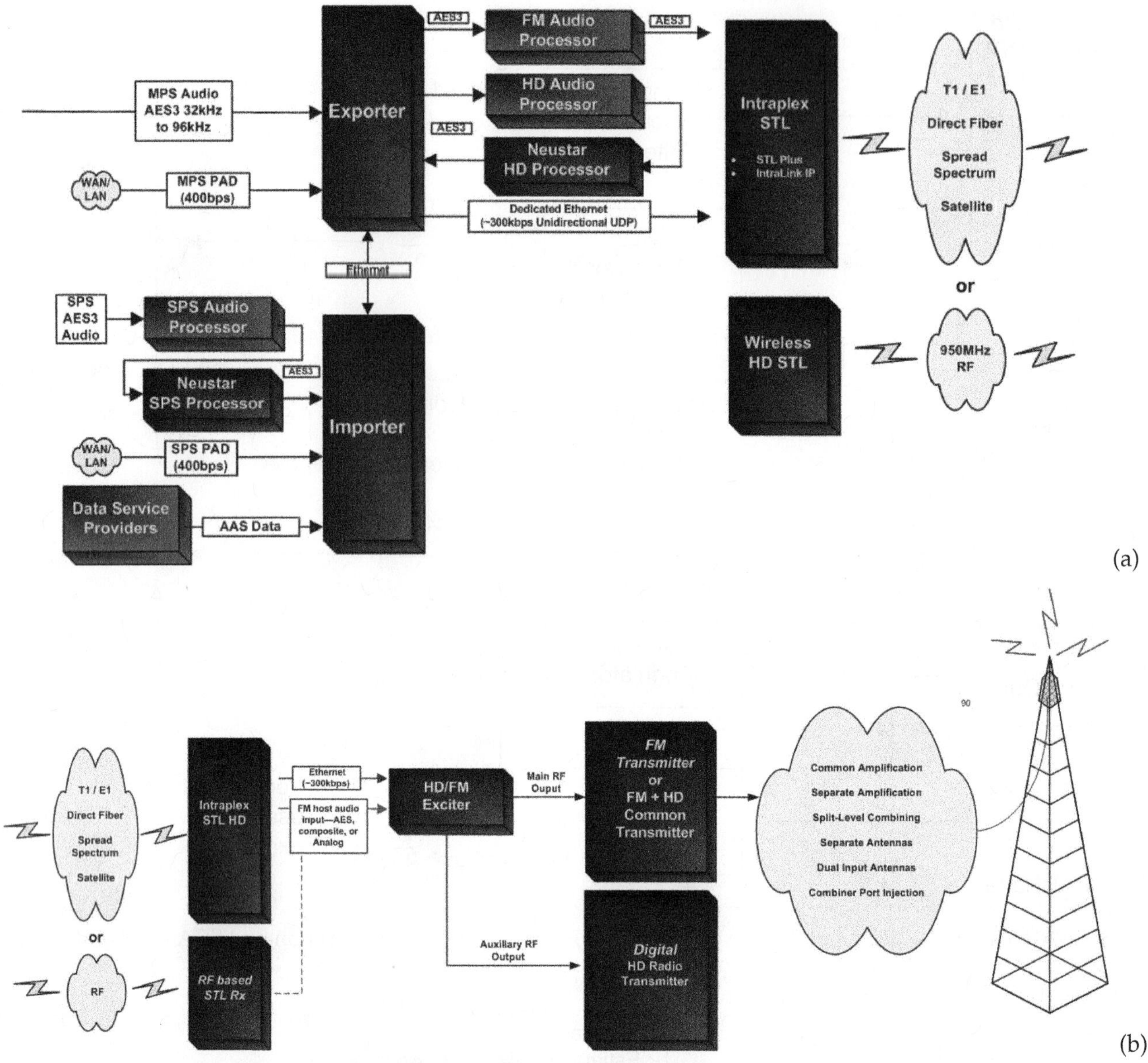

FIGURE 7.14-15 (a) Exgine studio system architecture. (b) Exgine transmission system architecture.

power amplifiers for FM operate in a nonlinear, saturated, Class C mode. When the FM signal is combined with the digital HD Radio signal, there is a moderate crest factor amplitude modulation component to the combined signal that requires linear amplification. The digital signal by itself has a high crest factor amplitude component due to the OFDM modulation that also requires linear amplification. Depending on the modulation mix requirements, the exciter RF amplifier can operate in three different modes:

- Class C, nonlinear for FM only;
- Class AB, quasi-linear, for common amplification of FM + HD (moderate crest factor);
- Class AB, full linear operation for separate amplification of HD only (high crest factor).

Changes in operating modes are accomplished by changing the bias and drive level to the RF amplifier. There is typically a 1.6 dB back-off in power output from the saturated Class C power to accommodate common amplification and a 5 dB back-off in power to accommodate a digital-only signal. Figure 7.14-16 illustrates the saturation characteristics of typical RF amplifier where the power output eventually saturates as the power input drives the amplifier into the nonlinear region of operation.

Adaptive Pre-correction

AM to AM nonlinearities at peak power saturation and near-zero power output cause the RF power amplifier's output amplitude to not exactly track the input amplitude, creating intermodulation products. Typically, these nonlinearities occur near the peak output power of the amplifier where saturation effects cause the output response to flatten as the input continues to increase, as can be seen in Figure 7.14-17.

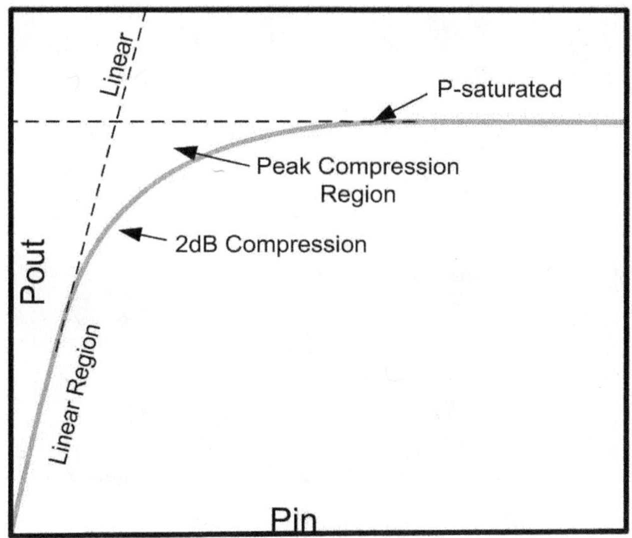

FIGURE 7.14-16 Nonlinear operation of a typical RF amplifier.

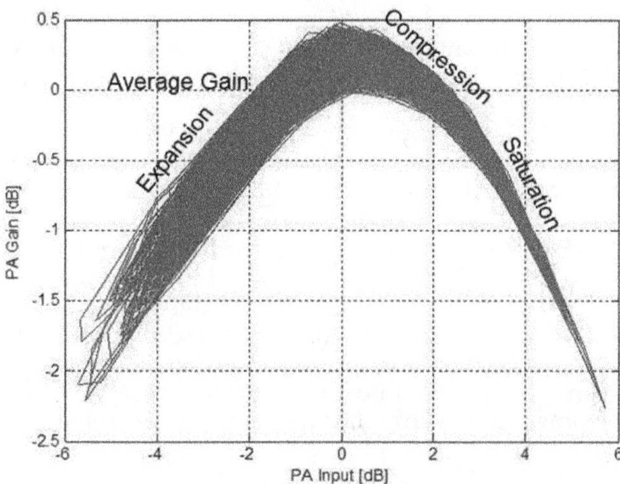

FIGURE 7.14-17 The power amplifier's nonlinear gain as observed with the IBOC+FM waveform.

AM to PM nonlinearities cause the RF power amplifier's output RF phase to not track the input phase. The amplifier acts like a phase modulator as the power output varies to follow the input signal. This is caused by changes in the RF power transistor's output capacitance over the dynamic range of the signal as shown in Figure 7.14-18.

All RF amplifiers also have some temporal memory effects that make the shape of the nonlinearities change over time with changing modulation data states. The memory effects can also be seen in Figures 7.14-17 and 7.14-18 as a blurring of the gain and phase traces. This is known as "memoryful" behavior. Nonlinear pre-correction including memoryful correction is applied to reduce undesired RF gain and phase intermodulation products to meet National Radio Systems Committee (NRSC) RF mask compliance.

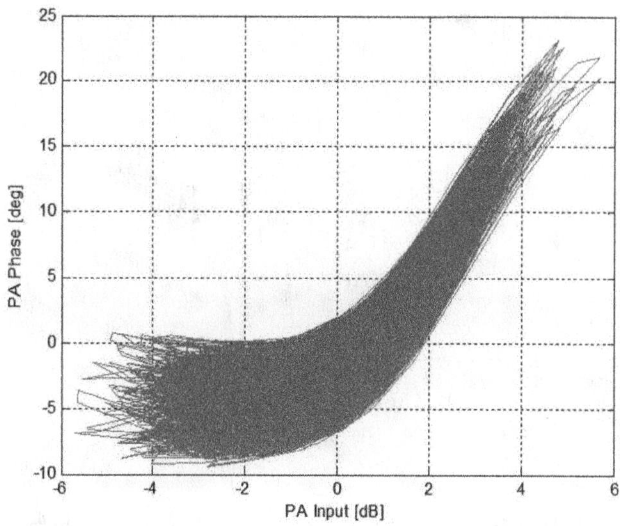

FIGURE 7.14-18 PA introduces phase distortion in the saturation region.

Most modern FM + HD exciters have some form of digital adaptive pre-correction for both linear and nonlinear distortions in the high power portions of the transmission system including a multi-station "filterplexer." Linear distortions include amplitude and group delay variations in the transmitter and any filters following the transmitter. The exciter typically accepts signals from RF sampling points at the output of the power amplifier and after any RF filters at the feed point to the antenna. These RF samples are demodulated to digital baseband and compared with the original digital signal. The difference between these signals is used to drive the linear and nonlinear pre-correctors in the exciter to minimize undesired RF spectral components that would fall outside the RF mask limits.

Real time, adaptive pre-correction automatically maintains the optimum spectrum and enables greater power utilization of the RF amplifiers. Figures 7.14-19 and 7.14-20 show the RF spectrum before and after the application of adaptive pre-correction [32]. Adaptive nonlinear pre-correction takes an RF output sample from the transmitter output or antenna system sample point and analyzes the waveform to characterize the gain and phase nonlinearities of power amplifier (shown in Figures 7.14-17 and 7.14-18) compared to the original input signal. Based on that analysis, a set of algorithms is applied to the signal generation process in order to create a gain and phase pre-distorted waveform. The resulting corrected output of the power amplifier is shown by the lighter (and lower) trace in Figure 7.14-20. This pre-correction process is repeated continually at a regular interval. The corrected output is displayed on the transmitter's spectrum analyzer in Figure 7.14-21.

Exciter Spectrum Analysis

The RF to digital baseband circuitry used for adaptive pre-correction allows the demodulated digital signal

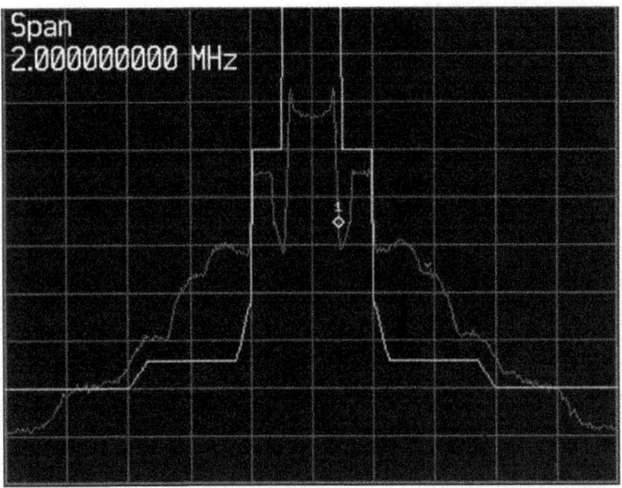

FIGURE 7.14-19 Uncorrected spectral plot of FM+HD signal at −10 dBc through an amplifier approaching saturation.

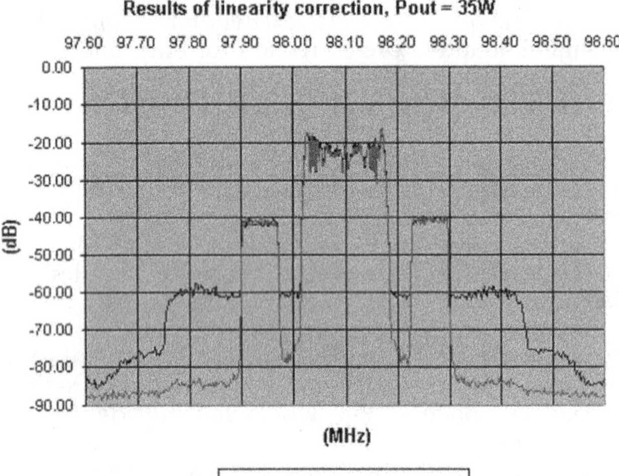

FIGURE 7.14-20 Adaptive pre-correction results. The light trace (on bottom) shows the corrected waveform.

to be displayed as an RF spectrum on the front panel display of the exciter, as shown in Figure 7.14-21.

Crest Factor Reduction (CFR)

CFR techniques in the 4th Generation HD Radio architecture uses a predictive summation of the peak magnitude of both the analog FM and digital waveform to calculate a resulting peak vector and apply multiple iterations of demodulation, clipping, restoration, and re-modulation of the input signal vectors for improved power amplifier utilization. In order to get reasonable RF power amplifier utilization, the peak-to-average ratio (PAR) of the digital signal must be reduced. This process introduces some degree of distortion, but intelligent techniques can be employed to minimize this distortion.

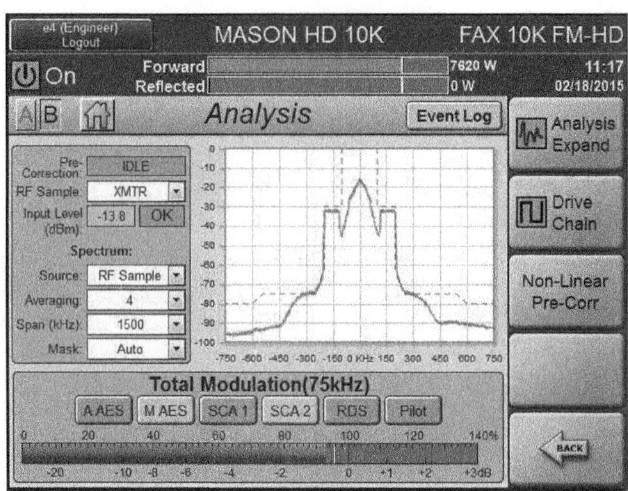

FIGURE 7.14-21 IBOC HD Radio exciter corrected spectrum display.

The amplitude of the digital signal has a statistical distribution that can be measured by estimating the complementary cumulative distribution function (CCDF). Instantaneous peaks at all levels are averaged over a large number of samples. Generally, the PAR determines the RMS power back-off necessary to meet the required digital SNR (EVM or Modulation Error Ratio, MER) and the RF mask based on peaks that occur 0.01% of the time as shown in Figure 7.14-22.

MER is a measurement standard used to quantify the digital signal quality of a digital radio transmitter and is the standard measurement of the digital SNR in most digital transmission systems including HD Radio. Modulation error ratio is equal to the ratio of the root mean square (RMS) power of the reference vector to the power of the error. It is defined in dB as: [4]

$$MER(dB) = 10log\left(\frac{P_{signal}}{P_{error}}\right)$$

Where P_{error} is the RMS power of the error vector, and P_{signal} is the RMS power of the ideal transmitted signal. The higher the MER, the better the SNR and thus the ability of the receiver to decode the data stream.

Received MER is degraded by the intermodulation noise products caused by peak clipping of the CFR as well as by propagation-induced channel noise. With 4th Generation CFR, consideration is given to maintaining the average MER at or above the NRSC-5 FM hybrid HD Radio transmission specification minimum of 14 dB. Within each iteration, the distortion is constrained to a maximum threshold value whereby the out-of-band spectral components created by the clipping process, lie within a predetermined spectral mask that is below the receive threshold signal-to-noise decision point, as can be seen in Figure 7.14-23.

A typical exciter user interface to adjust and optimize hybrid CFR versus MER and signal quality monitoring is shown in Figure 7.14-24.

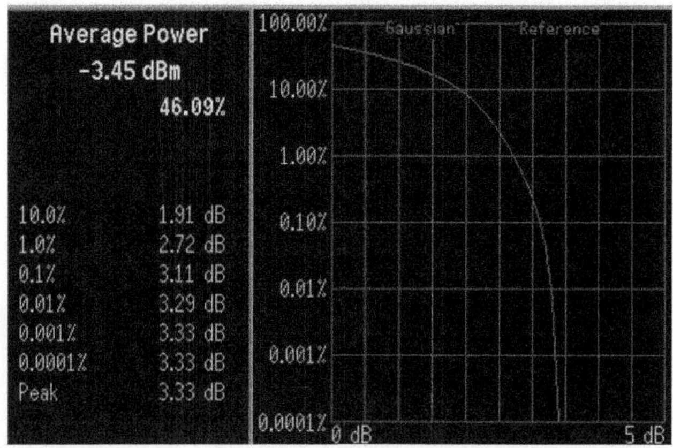

FIGURE 7.14-22 0.01% CCDF of 3.11 dB resulting from eight iterations of PAR2 CFR.

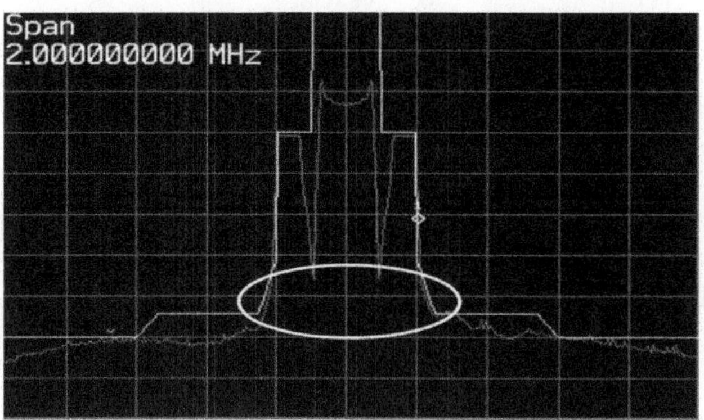

FIGURE 7.14-23 The noise from the clipping distortion is constrained to lie within the predetermined spectral mask and below the useful MER threshold.

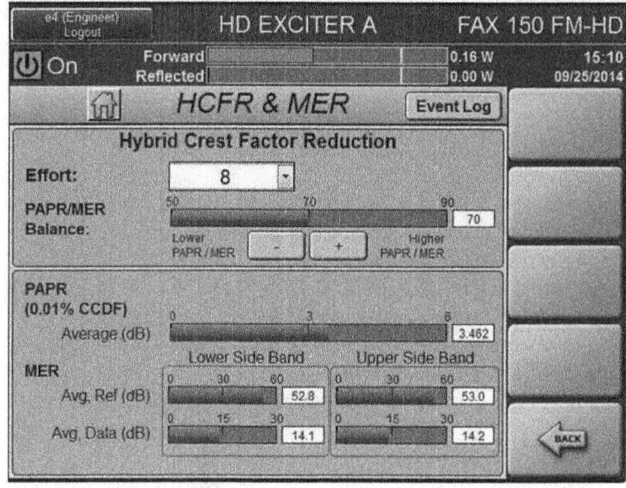

FIGURE 7.14-24 HCFR controls, PAPR and MER displays allow quality monitoring and "fine-tuning" of the CFR versus MER.

Asymmetrical HD Radio Sidebands

Asymmetrical HD Radio digital sideband generation is an option within the Exgine OFDM modulation process and has been implemented in the 4th-generation HD Radio architecture. Operating with asymmetrical digital sidebands can allow stations to increase digital sideband power on one side of the station's channel while operating the digital sideband on the other side at a lower power so as to better protect adjacent channel stations with a closer-spaced protection contour on the other side. A power increase in only one sideband does not bring the full improvement in digital coverage that raising both sidebands would bring, but increasing the HD Radio power in one sideband above the other sideband will provide some coverage improvement compared to operating both at the lower power.

The left side of Figure 7.14-25 shows symmetrical IBOC sidebands with a total injection power that is 14dBc below the 10 kW unmodulated FM carrier level, which places each of the individual sidebands at −17 dBc. The right side of Figure 7.14-25 shows asymmetrical IBOC sidebands with the lower sideband at −17 dBc and the upper sideband at −13 dBc.

The benefit to coverage will be dependent on multipath and fading, but a +4 dB increase in one sideband could provide as much benefit as if both sidebands had been increased by approximately +2 dB. Table 7.14-3 gives a summary of common amplification, FM+HD transmitter power ratings for typical asymmetrical sideband configurations. The sideband-1 and sideband-2 levels are scaled to agree with the NPR Labs IBOC Power Allowance Calculator presentation. The actual individual sideband RMS powers are 3dB less as shown in parentheses, but together they add up to the combined PAR power shown in Table 7.14-3. The PAR values are for an average between operating modes MP-1 and MP-3 with standard, iBiquity (now DTS), CFR. improvements will also be required to the digital only, CFR to accommodate asymmetrical sideband operation.

A web-based NRSC FM IBOC total power calculator (screen shot shown in Figure 7.14-26) makes accessible the information from Tables 1-5 of NRSC-G202 recommendation at this link: www.nrscstandards.org/nrsc_iboc_power.asp.

Exciter Control and Monitoring

The exciter control and monitoring system provides automatic regulation of the power output and protection against impedance mismatch of the RF output stage. The exciter's frequency control system is normally interlocked with the transmitter control system to prevent off-frequency operation. Displays of important operating parameters are typically provided by a combination of analog metering, LED displays, or a graphical user interface (GUI). These parameters usually include supply voltages, RF power amplifier operating parameters, forward output power, reflected power, modulation level, and operating mode.

A color-coded peak reading display may be provided to constantly monitor the peak FM deviation. A high-speed peak detector enables accurate peak modulation readings. A peak hold function is often included to provide a clear indication of short transient

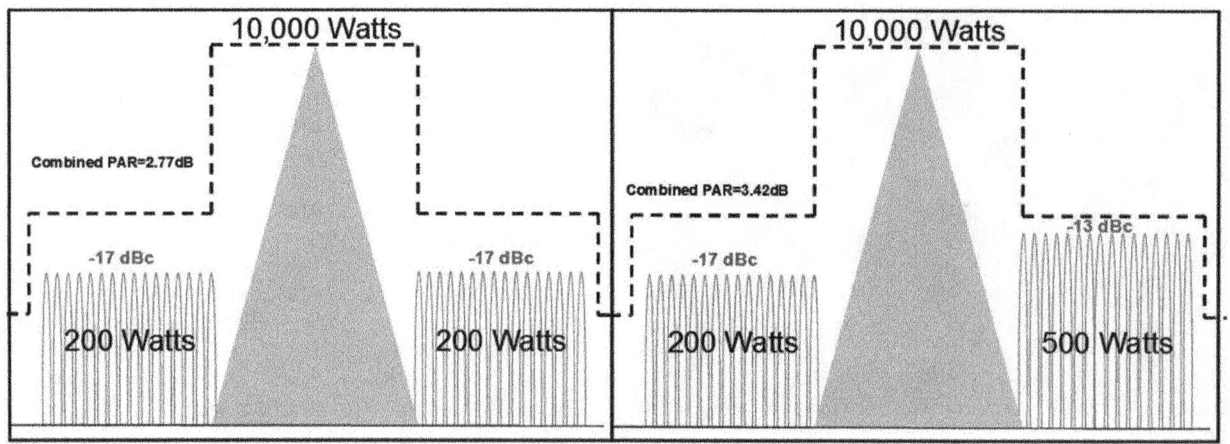

FIGURE 7.14-25 (a) Left side: −14 dBc lower and −14 dBc upper sideband power, total integrated power of −14 dBc or 4%. (b) Right side: −14 dBc lower and −10 dBc upper sideband power, total integrated power of −11.55 dBc or 7%.

TABLE 7.14-3
Asymmetrical IBOC Sideband Power Ratios

HD Sideband-1 dBc	HD Sideband-2 dBc	Combined dBc	HD Power %	Combined PAR dB	% of −14dBc Power Rating
−14.0 (−17)	−14.0 (−17)	−14.0	4.0	2.77	100
−13.0 (−16)	−14.0 (−17)	−13.5	4.5	2.90	97
−12.0 (−15)	−14.0 (−17)	−12.9	5.2	3.06	94
−11.0 (−14)	−14.0 (−17)	−12.2	6.0	3.23	90
−10.0 (−13)	−14.0 (−17)	−11.5	7.0	3.42	86

NRSC FM IBOC
Total Power Calculator

STEP 1 - select service mode

MP 1
MP 2
MP 3
MP 11

STEP 2 - select individual sideband operating points

Nominal Digital/Analog Power

-13 dBc (-10 dBc*) Upper Sideband Power

-17 dBc (-14 dBc*) Lower Sideband Power

*Note: the numbers above in parentheses are the corresponding total sideband power in the symmetrical case

Calculate

Total Digital Sideband Power = -10.8 dBc or 8.38 % of Analog

Website programming by:

FIGURE 7.14-26 A web-based NRSC FM IBOC total power calculator makes accessible the information from Tables 1-5 of NRSC-G202.

peaks exceeding 100% modulation. Local control is provided through either front panel switches or a GUI. Remote control of the exciter is normally provided by a parallel hardware interface, serial port, or Ethernet connection. Most exciters provide both parallel and serial interfacing for flexible remote control.

Exciter Packaging

Protection of sensitive circuits within an FM exciter from external electromagnetic interference (EMI) is important because the exciter is often located in the near field of multiple broadcast antennas operating over a broad range of frequencies. The exciter should be protected from conducted EMI by use of RC and/or LC filters on all leads entering the cabinet, including the AC line. The power supply transformer may have an electrostatic shield between the primary and secondary windings.

The modulated oscillator or frequency synthesizer is usually very sensitive to EMI, magnetic fields, and vibrations. It must be well RF shielded and is often shock mounted to prevent the transmission of mechanical vibrations from the transmitter's blower. This avoids microphonic pick-up that would degrade the FM SNR and HD Radio bit error rate (BER). Magnetic shielding of the oscillator or a hum-bucking circuit may be used to prevent hum pick-up from nearby transformers. The mechanical construction of most present day exciters incorporates a modular approach, which allows easy removal of subassemblies for repair or replacement.

FM Transmitter RF Power Amplifiers

The remainder of the FM transmitter consists of one or more power amplifiers, each having from 8 to 20 dB of power gain. Ideally, the transmitter bandwidth should be as wide as practical with a minimum of tuned stages. Broadband solid state amplifiers usually eliminate the need for tuned networks in the RF path. High-power transmitters in the multi-kilowatt range may use multiple tube stages each with relatively low gain such as a grounded-grid configuration or a single grid-driven power amplifier (PA) stage with high gain and efficiency. The cost, redundancy, and wide bandwidth benefits of solid-state transmitters make them attractive at power levels up to 30 kW. At higher power levels, the lower cost per watt of high-power, single-tube transmitters is still attractive even though the modulation performance and reliability are less than that of a solid-state transmitter. Design improvements in tube-type power amplifiers have concentrated on improving bandwidth, reliability, and cost effectiveness while design improvements in solid state amplifiers have focused on cost reduction to make them competitive with tube technology at high power levels.

RF Power Amplifier Performance Requirements

The basic function of the power amplifier is to bring the power of the exciter output up to the desired transmitter power output level. Most of the overall transmitter performance characteristics are determined by the exciter, but a few are established or affected by the following power amplifier characteristics:

- The RF output level at harmonics of the carrier frequency is almost completely a function of the attenuation provided by the power amplifier output matching circuit and output lowpass/notch filters. The FCC limit in decibels is (43 dB + 10 log [power in watts] dB) or 80 dB, whichever is lower. The specification is 73 dB for transmitters with 1 kW output increasing to 80 dB above 5 kW output power;
- The major source of asynchronous AM noise usually originates in the last power amplifier stage;
- The RF power output control system must keep the output within +5% and −10% of authorized power;
- Inadequate power amplifier RF bandwidth that affects phase linearity (constant time delay) across the signal bandwidth can reduce stereo separation and cause crosstalk to and from the SCA subcarrier;
- The presence of standing waves on the transmission line between the power amplifier and the antenna may also interact with the power amplifier to cause degraded stereo separation and SCA crosstalk;
- If the transmitter is amplifying an HD Radio signal, linear operation is required to prevent the generation of excess RF intermodulation products.

The power amplifier should provide trouble-free service and be easy to maintain and repair. Good overall efficiency is also desirable to reduce the primary power consumption and heat load released into the transmitter room.

Power Amplifier Bandwidth Considerations

As discussed earlier, the FM signal theoretically occupies infinite bandwidth. In practice, however, truncation of the insignificant sidebands (typically less than 1% of the carrier power) makes the system practical by accepting a certain degree of signal degradation; therefore, the transmitter power amplifier bandwidth affects the modulation performance. Available bandwidth determines the amplitude response and group delay response. There is a trade-off between the bandwidth, gain, and efficiency in the design of a power amplifier [9, 18]. The bandwidth of an amplifier is determined by the load resistance across the tuned circuit and the output or input capacitance of the amplifier. For a single tuned circuit, the bandwidth is proportional to the ratio of capacitive reactance to resistance:

$$BW \propto \frac{K}{2\pi f_c R_L (C)} = \frac{K(X_C)}{R_L}$$

Where:

- BW = bandwidth between half-power points (BW3 dB)
- K = proportionality constant
- R_L = load resistance (appearing across tuned circuit)
- C = total capacitance of tuned circuit (includes stray capacitances plus output or input capacitances of the tube)
- X_C = capacitive reactance of C
- fc = carrier frequency

The load resistance is directly related to the RF voltage swing on the tube element. For the same power and efficiency, the bandwidth can be increased if the capacitance is reduced.

Effects of Circuit Topology and Tuning on FM Performance

Analog FM broadcast transmitter RF power amplifiers are typically adjusted for minimum synchronous AM (incidental amplitude modulation), which results in symmetrical amplitude response. This ensures that the transmitter's amplitude passband is properly centered on the FM channel. The upper and lower sidebands will be attenuated equally or symmetrically, which is assumed to result in optimum FM modulation performance. This is true only if the RF power amplifier circuit topology results in simultaneous symmetry of both amplitude and group delay responses [16].

The tuning points for symmetrical amplitude response and symmetrical group delay response usually do not coincide, depending on the circuit topology; therefore, simply tuning for minimum synchronous AM (symmetrical amplitude response) does not necessarily result in best FM performance. In fact, symmetry of the group delay response has a much greater effect on FM modulation distortion than does the amplitude response. Tuning for symmetrical group delay will cause the phase/time delay errors to affect the upper and lower sidebands equally or symmetrically. The group delay response is constant if the phase shift versus frequency is linear. In this case, all components of the signal are delayed in time, but no phase distortion occurs.

Measurements taken on a typical FM transmitter as well as computer simulations show that tuning the RF power amplifier for symmetrical group delay response resulted in minimum distortion and crosstalk and confirmed that group delay response asymmetry causes higher FM distortion and crosstalk than amplitude response asymmetry [17]. Therefore, the transmitter should be tuned for the symmetrical group delay response that results in best FM performance rather than the symmetrical amplitude response that results in minimum synchronous AM.

Intermediate Power Amplifiers

The IPA is located between the exciter and the final amplifier in transmitters that require more than about 50 W of drive to the final amplifier. The IPA may consist of one or more tubes or solid-state amplifier modules.

Interstage Coupling Circuits

The IPA output circuit and the input circuit to the final amplifier are often coupled together by a coaxial transmission line. Impedance matching is usually implemented by the input circuit to the tube. The interconnecting transmission line between the coupling circuits should be matched to avoid a high VSWR. Directional wattmeters may be placed in the line to measure forward and reflected power from which the standing wave ratio can be determined. The VSWR is established by the match at the load end of the transmission line. Solid-state RF power devices present a low load impedance at the device output terminal, so an impedance transformation is required to couple these devices into the relatively high impedance of the final amplifier grid circuit.

Therefore, virtually all solid-state IPA systems have a 50 Ω impedance point within the system that can be used to feed the antenna should it be necessary to bypass the power amplifier. The tube in the final amplifier stage of most high-power transmitters requires between 150 and 600 watts of drive. This permits the use of solid state, wideband power amplifier modules to boost the exciter output power to the level required to drive the grid of the final amplifier tube.

Linear Amplification Requirements for IBOC HD Radio

The PAR for the IBOC HD Radio signal in separate amplification (without FM) is about 7 dB after CFR and causes the amplifier output level to go to zero when the vector sum of all the OFDM carriers is zero. This requires that the RF amplifier be linear throughout its entire dynamic range. Linear amplification is needed to meet the RF emission mask and to minimize in-band RF intermodulation distortion (IMD) products that degrade the digital SNR. Typically, a back-off of at least 5 dB from Class C operation is required to maintain enough linearity. The peak-to-average ratio for the combined analog FM + HD Radio signal does not require the RF amplifier output to go to zero due to the presence of the larger analog FM signal in the mix, which reduces the overall crest factor. Typically, a back-off of 1.6 dB from Class C is enough to obtain the required linearity at the upper end of the RF amplifier's power output range. Table 7.14-4 shows the average power (average of all the coded orthogonal frequency-division multiplexing [COFDM] carriers plus analog FM) available as a percentage of the rated analog-only transmitter power and the power back-off from rated analog power required (in dB) for each mode of operation.

High-efficiency amplifiers operate with less than 180° of a full RF cycle. This creates a nonlinear response of the envelope, specifically at low transition levels where the power output approaches zero. To correct for this, linear amplifiers are biased to conduct at least half the RF cycle, which is achieved by placing them in Class AB mode. Class A mode, the most linear operating mode, is not used in high-power transmitters due to its inherent low efficiency. The operating efficiency of the amplifier drops as much as 50% when changing from Class C to Class AB operation [27].

Efficiency Penalty of Linear Amplification

Efficiency is lost when the amplifier is operating in the region between saturation and cutoff, where the loading on the amplifier is not optimum for the RF waveform at that instant and when the static bias current through the amplifying device is a significant portion of the total current at that point in the waveform envelope.

TABLE 7.14-4
Average Power for Different Modes as a Percentage of Class C (FM Only) Operation

Modes	Average Power (%)	Back-Off (dB)
FM only	100	0
FM + digital	70	−1.6
Digital only	32	−5.0

Common Amplification in a High-Power, Vacuum-Tube Amplifier

High-power vacuum tube amplifiers can also be operated in Class AB linear mode to provide common amplification of the analog FM signal and the HD Radio signal. When operated in linear mode, the plate efficiency of the amplifier tube typically drops from approximately 80% to about 63%. The reduced operating efficiency of linear operation needs to be taken into consideration when sizing the AC power and building cooling requirements.

HD Radio Power Increase

On January 29, 2010, the FCC Media Bureau adopted an order to modify the FM digital audio broadcasting rules to expand digital coverage. This change to the rules allows virtually all US FM stations to increase their HD Radio digital power by +6dB from the previously authorized −20 dBc level (1% digital injection) to −14dBc (4% digital injection). This amounts to a digital power increase of four times the current HD Radio power. The rule changes also allowed licensees to submit an application for an increase in FM digital ERP beyond −14 dBc up to −10 dBc (10% digital injection). Increasing injection above −14 dBc requires a special temporary authority (STA) application with the FCC and must meet protection requirements. Limits are placed on power increases for stations currently licensed in excess of class the maximums, that is, "super-powered" stations, to protect analog radio service from interference.

Implementing an HD Radio power increase requires careful consideration. The first difficulty that we encounter in increasing the IBOC carriers is the fact that we must de-rate the transmitter's available output to accommodate them. The more we increase the power in the IBOC carriers, the more we must de-rate the transmitter. Due to the relatively high 5- to 6-dB effective peak-to-average AM component of the IBOC carriers, HD Radio transmitters must be de-rated from their saturated FM (CW)-only maximum total power output (TPO) when operating FM+HD or HD only. As the level of the digital carriers are increased, the RF inter-modulation products must be suppressed proportionally further to maintain mask compliance.

Transmitter efficiency is reduced significantly as digital injection levels are increased. Peak voltages on antenna and transmission line components are increased dramatically as digital injection is increased due to higher peak-to-average power. High-level combining methods becomes exponentially more difficult to implement as the IBOC to analog power ratio increases. Typically, high-level combining is accomplished using separate analog and digital amplification and a 10-dB hybrid combiner, in which 10% of the analog power and 90% of the digital power is sent to the reject load and wasted as heat.

Split-level combining that worked well at −20 dBc is not practical for higher digital sideband power. Split-level combining becomes more difficult, if not impossible, to implement because as sideband levels increase, the maximum digital to analog ratio that can be obtained from the transmitter and still maintain NRSC mask compliance is not sufficient to provide efficient split-level combining.

Common amplification is generally considered the best solution for increased HD Radio injection. As the injection is increased, more of the available power is needed for the peak-average power, requiring greater headroom to maintain NRSC mask compliance due to the nonlinearity of the amplifier. Increasing HD Radio sideband power requires further back-off of the transmitter power.

Space combining is also a practical and efficient solution but requires increased digital amplifier power, proportional improvement in RF IMD suppression and proportional improvement in analog/digital antenna isolation. For the transmitter operating in digital-only, the PAR is increased even further, but without the constant envelope analog FM signal being present, the maximum available TPO is reduced even further. At −20 dBc injection, the IMD products need to be suppressed approximately 35 dB as shown in Figure 7.14-27.

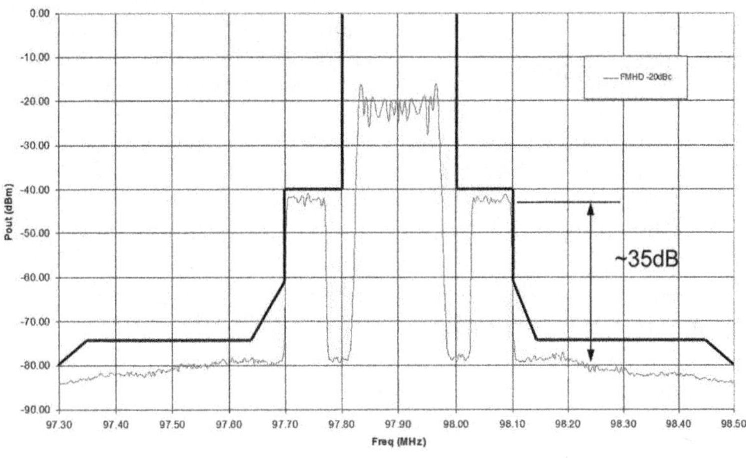

FIGURE 7.14-27 35 -dB IMD suppression for −20-dBc injection.

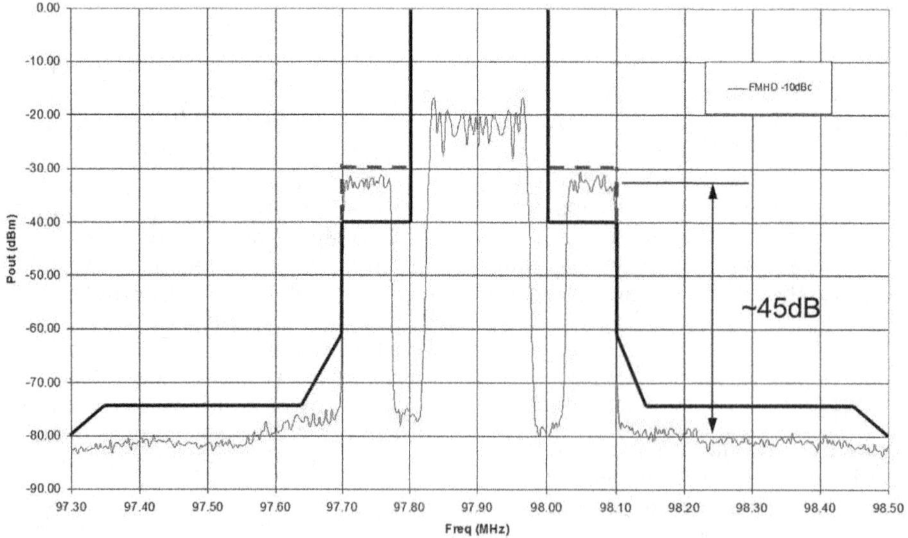

FIGURE 7.14-28 45-dB IMD suppression for −10-dBc injection.

At −10 dBc injection, the inter-modulation products need to be suppressed approximately 45 dB, an additional 10 dB (100×) from that needed at −20 dBc to maintain mask compliance as shown in Figure 7.14-28.

AMPLIFICATION OF HD RADIO SIGNALS

Separate Amplification

Figure 7.14-29 shows a block diagram of separate amplification of the FM analog and IBOC digital signals by the addition of new components from exciter to antenna.

Common Amplification

Figure 7.14-30 shows a block diagram of the FM analog and IBOC digital signals combined into a common power amplifier and using the same transmission line and antenna. For more information on HD Radio amplification, see references [33] and [34].

Solid-State RF Power Amplifier Systems

A solid-state RF power amplifier usually consists of multiple individual amplifier modules that are combined to provide the desired power output. The advantages of using several lower power modules instead of a single high-power amplifier include the following:

- Redundancy is provided by isolating the input and output of each module to permit uninterrupted operation at reduced power if one or more of the modules fail;
- Failed modules can be repaired without having to go off the air;
- More effective cooling of each power device is achieved by splitting the concentration of heat to be dissipated into several areas instead of one small area;
- Better isolation between the amplifier modules and the input circuit of the final power amplifier or antenna is provided by the combiner/isolator;
- Redundant power supplies and air-cooling systems for each module improve overall reliability.

Each RF power amplifier module consists of one or more solid-state devices with broadband impedance transformation networks for input and output matching. Vertical metal oxide semiconductor field-effect transistors (VMOSFETs) or the newer lateral diffused metal oxide (LD-MOS) field-effect transistors permit the design of broadband amplifier stages that exhibit both high efficiency and the wide bandwidth necessary to cover the FM broadcast band.

The input impedance to the solid-state device is always lower than the desired 50 Ω input impedance, so a broadband impedance transformation scheme is required. This is usually accomplished by using a combination of coaxial BALUNs and push-pull coaxial line sections that are cross-coupled to provide 4:1 or higher transformation ratios over the FM band.

By operating two devices in push-pull, the input impedance is double that of a single ended circuit, and the suppression of even order harmonics is obtained. Two devices fed in this manner also provide some degree of redundancy within the module itself, as partial RF output can be obtained with one failed device. In a similar manner, the low output impedance of these solid-state devices can be transformed up to the desired 50 Ω module output impedance where combining occurs. Figure 7.14-31 illustrates a simplified schematic of a broadband, 425-watt, VMOSFET RF amplifier module utilizing the push-pull configuration. Figure 7.14-32(a) is a photograph of this RF amplifier module.

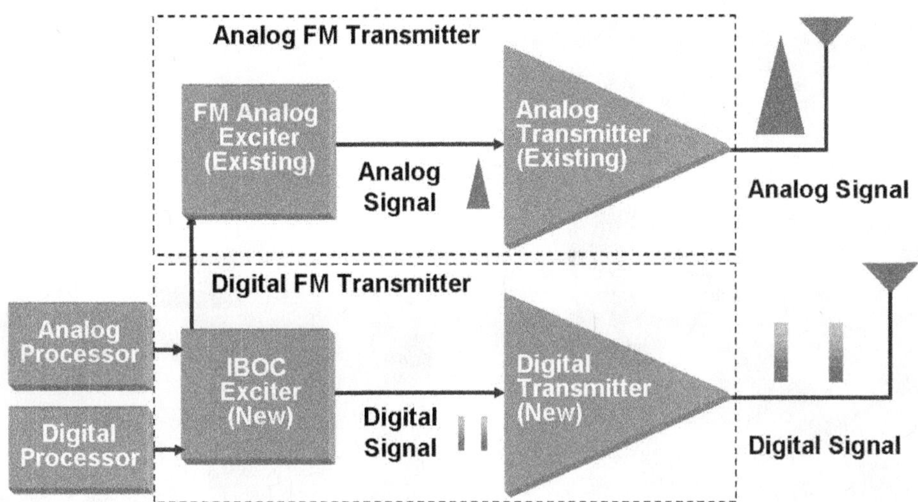

FIGURE 7.14-29 45-dB IMD suppression for −10-dBc injection.

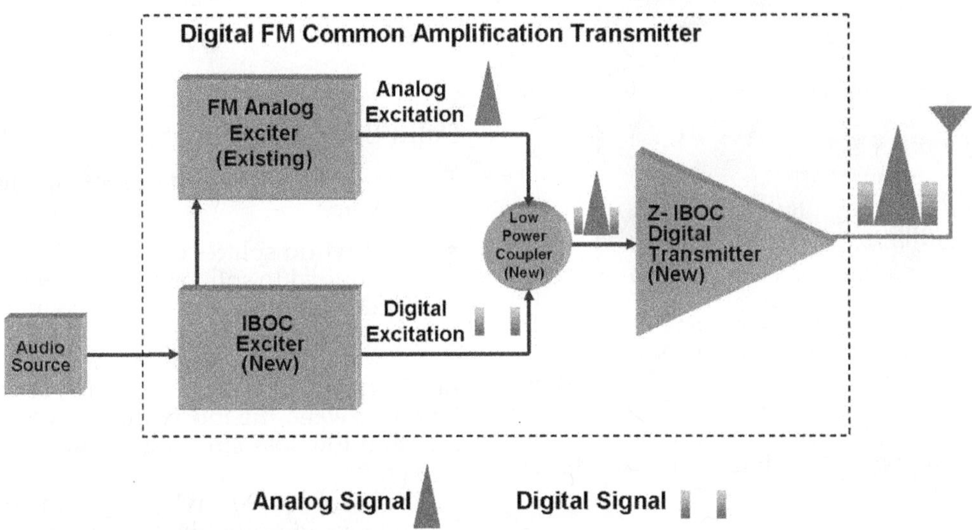

FIGURE 7.14-30 Simplified block diagram of common amplification transmitters.

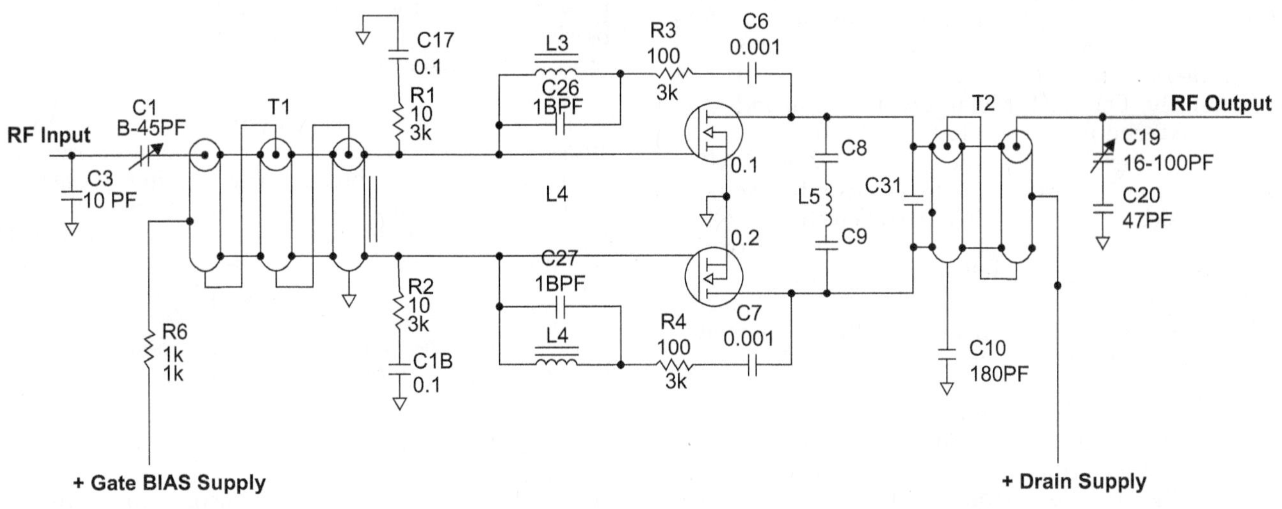

FIGURE 7.14-31 Typical push-pull configuration of LD-MOS or V-MOS RF power amplifier.

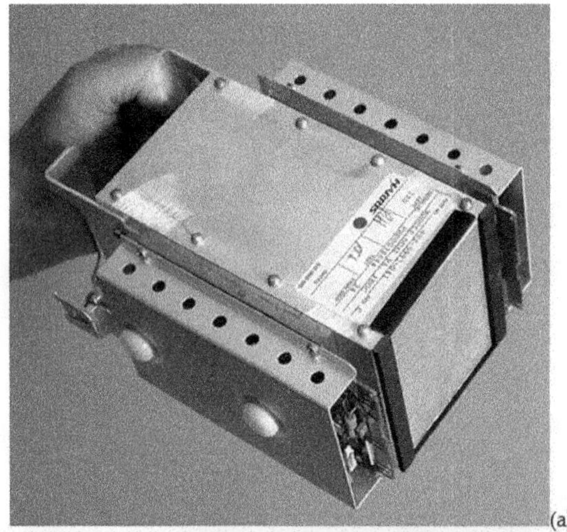

(a)

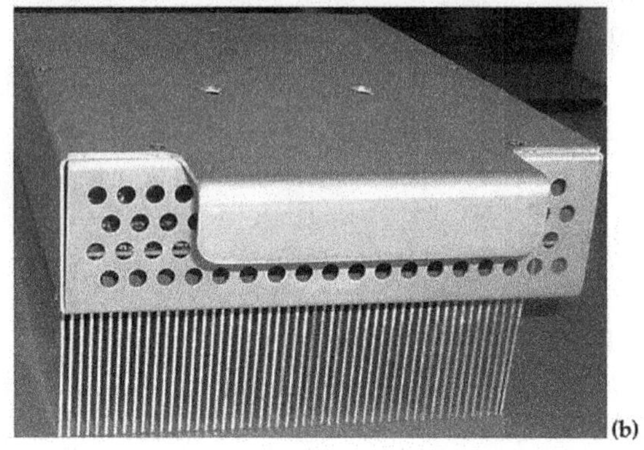

(b)

FIGURE 7.14-32 (a) Example of a broadband, 850-watt, VMOSFET RF power amplifier module. Note the relative size of the module compared to the hand holding it. (b) Example of a broadband, 1-kW LD-MOS RF power amplifier of similar power level. Both types of modules are "hot replaceable" while the transmitter is on the air.

New Solid-State RF Power Devices

Several new solid-state RF power device technologies have recently replaced previous generation V-MOS and T-MOS devices. The most popular new solid-state power devices for VHF FM applications are based on Laterally Diffused Metal Oxide Semiconductor (LD-MOS) technology (see Figure 7.14-33).

LD-MOS devices, like their V-MOS and T-MOS ancestors, are also metal oxide semiconductor field effect transistors (MOS-FET). LD-MOS technology has the advantage of placing the source terminal of the FET at ground potential, with a geometry layout that improves thermal performance, lower capacitances, and source inductance. All this combined translates into a device with higher gain, higher power density, and higher device efficiency. The 50-volt LD-MOS device is well matched to the use of low-cost, 48-volt telecom power supplies. The 50-volt LD-MOS technology was first introduced to broadcast transmitters in October 2008 and the reliability of this technology has now been field proven.

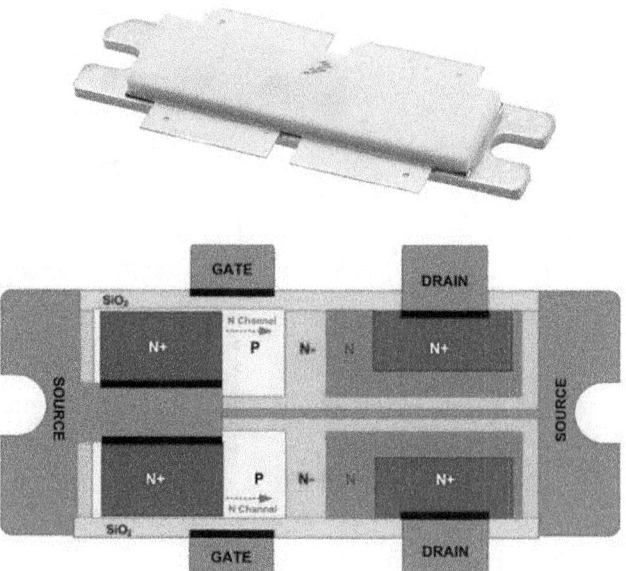

FIGURE 7.14-33 50V LD-MOS device structure.

Solid-State Amplifier Splitting and Combining

Two frequently used types of splitting/combining schemes are:

- A 90° hybrid splitter or combiner; N – 1 hybrids are required to split or combine N inputs (see the section on transmitter output combining);
- A Wilkinson N-way in-phase splitter or combiner.

Either type of splitter/combiner must provide isolation between the individual power amplifier modules and low loss splitting or combining of the total power.

The cascaded 90° hybrid system shown in Figure 7.14-34 provides double isolation between the power amplifiers and the load by first combining the two pairs of amplifiers and then combining the outputs of the first two combiners. A portion of the reflected power, caused by a mismatch at the output, will be dissipated in reject loads so the power amplifier modules will see a lower VSWR than exists at the output.

The Wilkinson system shown in Figure 7.14-35 is a simple and effective way to split and combine modules operating in phase, but usually requires a balanced reject load that makes reject power measurements more difficult. By adding additional coaxial BALUN sections to the Wilkinson combiner (Wilkinson/Gysel), it is possible to use unbalanced reject loads. The unbalanced 50 Ω reject loads are accessible for monitoring of reject load power, which is useful in determining the balance of the system [6, 7].

RF Module Combiner Configuration

Both the 90° hybrid and Wilkinson combining systems require resistive RF power reject loads to provide

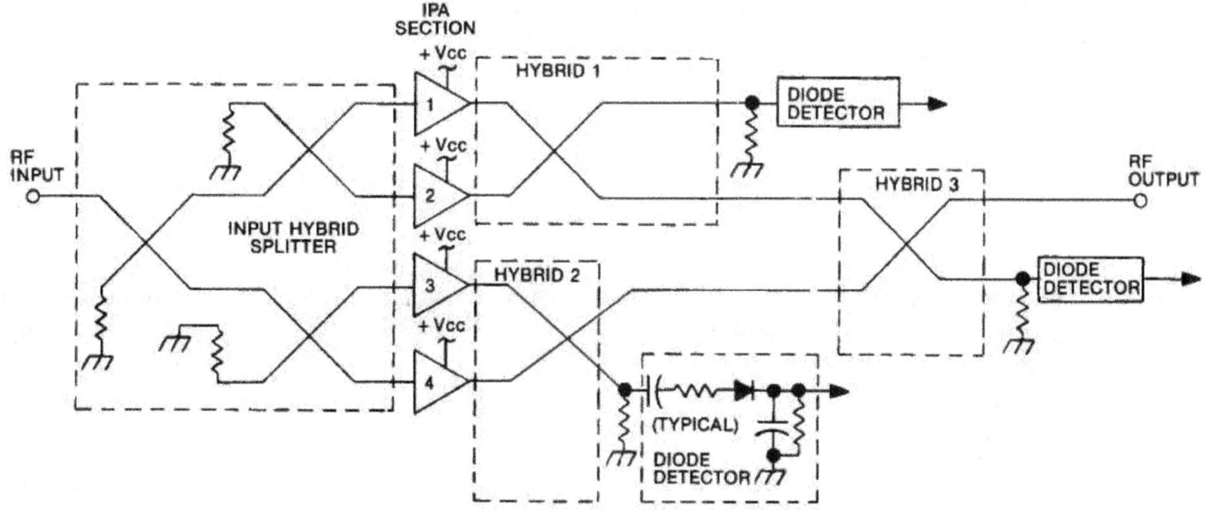

FIGURE 7.14-34 Cascaded 90° hybrid splitting/combining system.

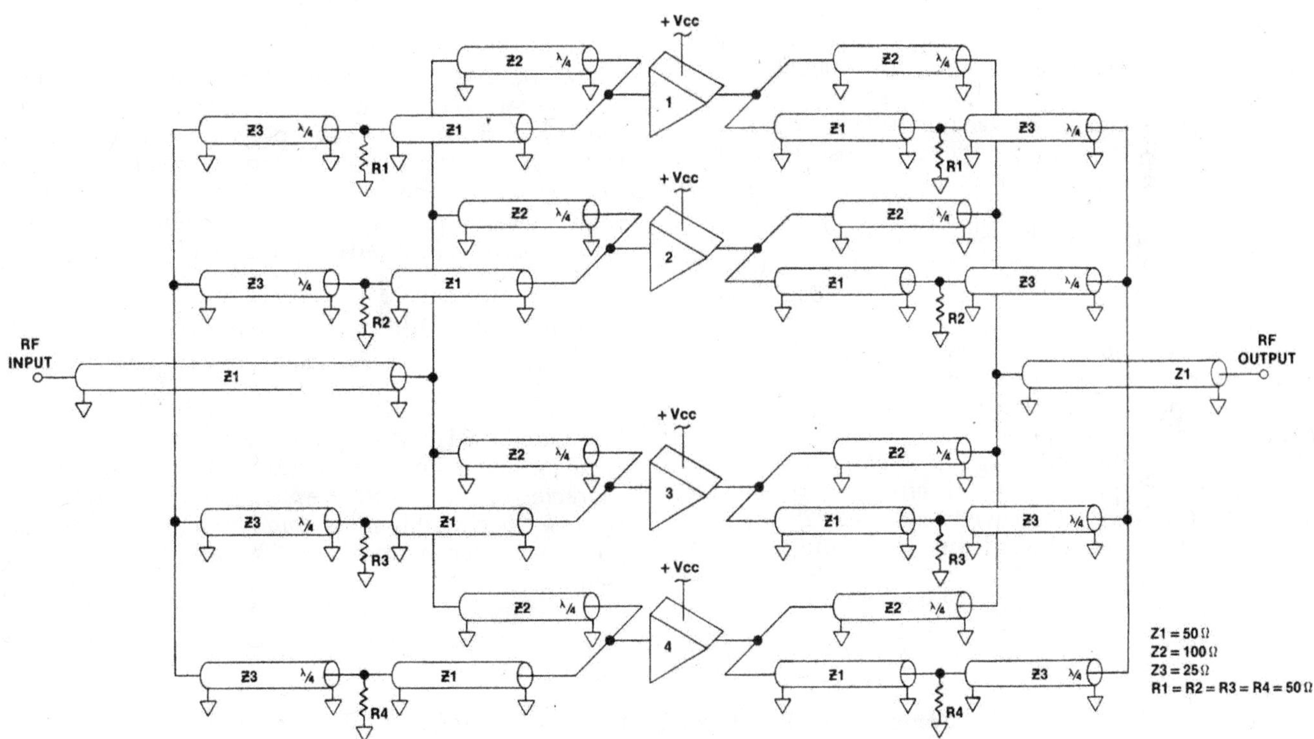

FIGURE 7.14-35 Wilkinson/Gysel in-phase splitting/combining system with unbalanced reject loads.

isolation between the amplifier modules in the event that one or more of the modules fail. A portion of the RF power from the remaining modules is wasted in the reject loads instead of being delivered to the output. Because most splitter/combiner systems are designed with 50 Ω input and output impedances, these systems can be easily used as low-power stand-by transmitters by routing the output to the antenna system. An RF low-pass filter (LPF) is required only when directly feeding the antenna system. The harmonic suppression of the IPA is not as critical when driving a nonlinear power amplifier that also generates harmonics, because the final amplifier stage will have its own LPF.

SOLID-STATE FM BROADCAST TRANSMITTERS

The combining techniques just described can be used to construct a completely solid-state transmitter using arrays of combined modules for the final output stage.

An additional RF lowpass filter is usually required to meet FCC emission requirements.

Advantages of Solid-State Transmitters

The advantages of a solid-state transmitter when compared to a single-tube transmitter, include the following:

- Built-in amplifier and power supply redundancy;
- Improved FM performance;
- The ability to cover the entire FM band without the need for retuning;
- Elimination of tube replacement costs;
- Less maintenance than tube-type transmitters.

Solid-State Transmitter Design Considerations

Several manufacturers offer solid-state FM broadcast transmitters with power outputs ranging from 100 W up to 80 kW, but current economic factors still favor the single-tube FM transmitter for power levels above 30 kW. Figure 7.14-36 illustrates the cost crossover point for solid-state versus vacuum tube technology. In order for a solid-state transmitter to be cost and power consumption competitive with a single-tube transmitter, the efficiency of the solid-state RF power amplifiers and combining system should approach the 80% efficiency obtainable from tube-type RF amplifiers, a level that has been achieved with LD-MOS solid-state devices at VHF frequencies.

The introduction of 50V LDMOS RF transistors has made a significant improvement in transmitter power amplifier operating efficiency. The 50V LD-MOS technology offers higher amplifier gains of 21 dB in FM operation, thereby reducing the inefficiency of the drive stages. Typical DC to RF efficiency of 84% in FM operation can be achieved and over 72% overall efficiency, from AC power input to RF output. This is somewhat better AC to RF efficiency than a typical single-tube transmitter, which is normally less than 65%.

Trends in the newest solid-state FM transmitters are to supply redundant RF amplifiers, power supply, and control circuits to keep the transmitter on the air at reduced power in the event that one or more components should fail. Identical and interchangeable IPA and PA modules offer additional redundancy. RF modules that can be removed and inserted in an operating transmitter also provide the advantage of not requiring an off-air period for some maintenance services.

Solid-state transmitter layouts using direct, cable-free connection of the RF modules to the RF combiner further enhance transmitter reliability and stability. Another enhancement provided in some current solid-state FM transmitters is a microprocessor-based control system that can monitor parameters within the transmitter and provide intelligent control of the transmitter system. Figure 7.14-37 shows a block diagram of a solid-state transmitter.

VACUUM-TUBE POWER AMPLIFIER CIRCUITS

The amplitude of an FM signal remains constant with modulation so efficient, nonlinear, Class C amplifiers can be used. Vacuum-tube power amplifier circuits for FM service have evolved into two basic types. One type uses a tetrode or pentode tube in a grid driven circuit while the other uses a high-mu triode in a cathode-driven (grounded-grid) circuit.

Grounded-Grid (Cathode-Driven) Triode Amplifiers

The high-mu, zero-bias triodes used in cathode-driven (grounded-grid) FM amplifiers were originally developed for linear single sideband (SSB) amplifiers. Their characteristics are well adapted to FM broadcast use because the circuit is simple, and no screen or grid bias power supplies are required. Figure 7.14-38 shows the basic circuit configuration. In this case, the grid is connected directly to chassis ground. In this case, the grid is connected directly to a chassis ground. The output

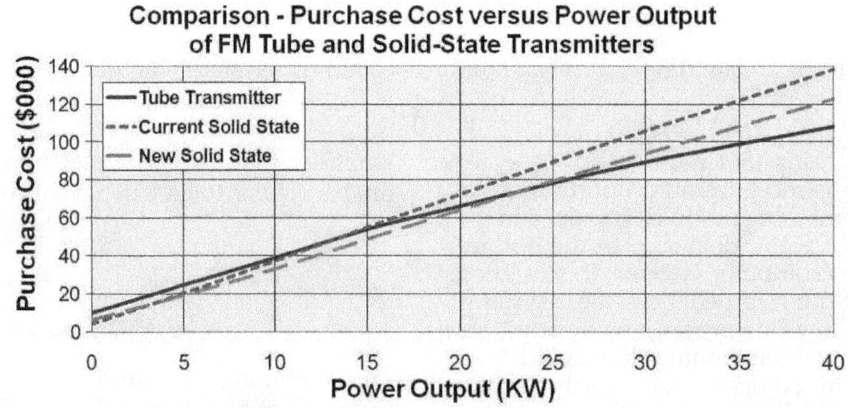

FIGURE 7.14-36 Solid-state vs. vacuum tube cost comparison.

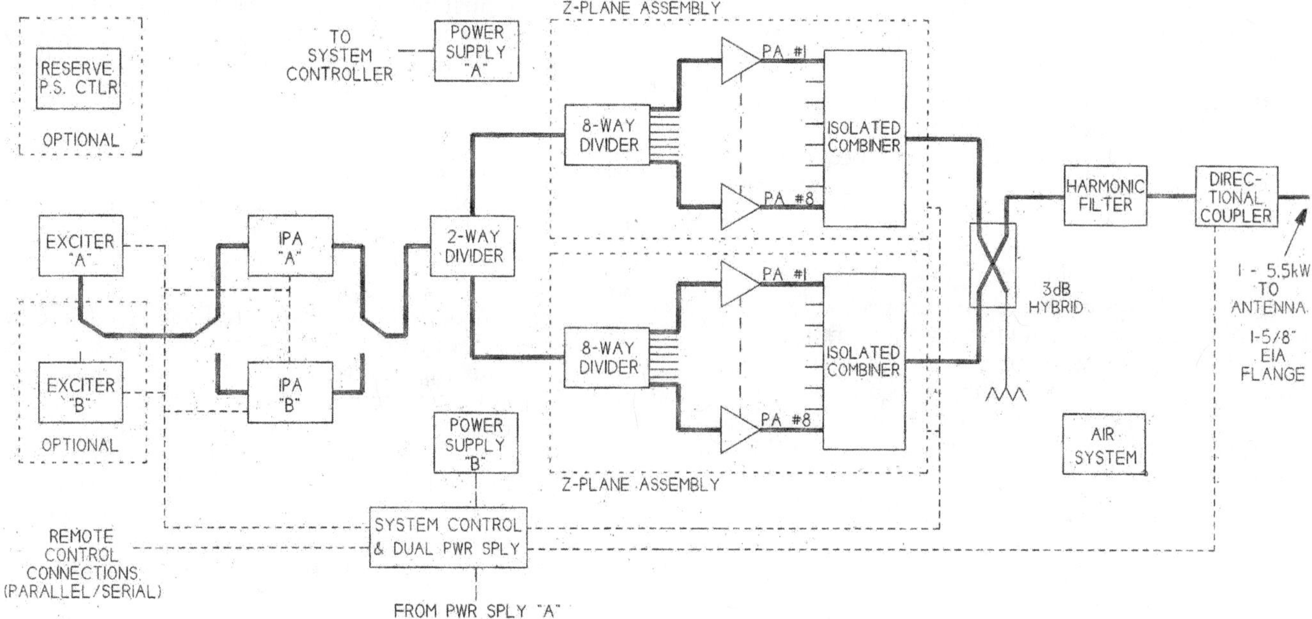

FIGURE 7.14-37 Solid-state transmitter block diagram.

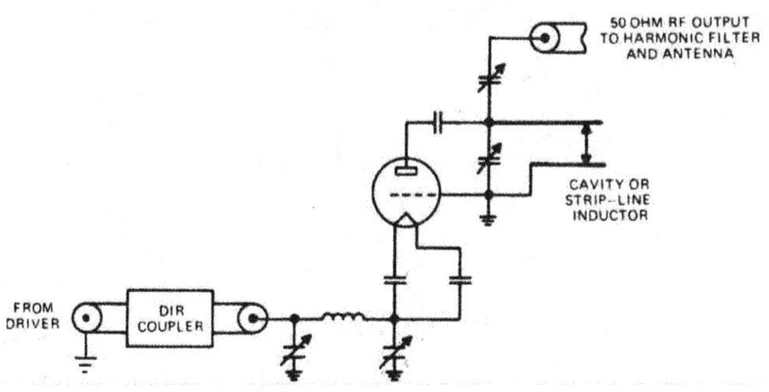

FIGURE 7.14-38 Cathode-driven, grounded-grid triode power amplifier.

tank circuit is a shorted coaxial cavity that is loaded by the tube output and stray circuit capacitance. A small capacitor is used for trimming the tuning, and another small variable capacitor is used for adjusting the loading. A pi-network matches the 50 Ω input to the tube cathode.

The triodes are usually operated in the less efficient, Class B mode to achieve maximum power gain, which is on the order of 20 (13 dB). They can be driven into high-efficiency Class C operation by providing negative grid bias. This increases the plate efficiency but also requires increased drive power and a bias power supply.

Most of the drive power into a grounded-grid amplifier transfers through the tube and appears in the stage's output. This increases the apparent efficiency so the efficiency factor given by the manufacturer may be higher than the actual plate efficiency of the tube. The true plate efficiency is determined by dividing the output power by the total input power, which includes both the DC plate input power (plate current times plate voltage) and the RF drive power. Because most of the drive power transfers through the tube, any changes in loading of the output circuit will also affect the input tuning and driver stage.

With RF drive voltage on the cathode (filament) of the tube, some means of decoupling is required to block the RF from the filament transformer. One method employs high-current RF chokes, as the inductance can be very low at this frequency range. The other commonly used method feeds the filament power through the input tank circuit inductor. Cathode-driven stages are normally used only for the higher power stages. The first stage in a multi-tube transmitter is nearly always a tetrode because of its higher power gain.

Grounded-Grid Versus Grid-Driven Tetrode Operation

There are several trade-offs between the performance of grounded-grid and grid-driven configurations of a tetrode PA with respect to gain, efficiency, amplitude bandwidth, phase bandwidth, and synchronous AM under equivalent operating conditions:

- When driving a grounded-grid PA into saturation, the bandwidth is limited by the output cavity. The PA bandwidth in the grid-driven amplifier is limited by the input circuit Q, which is determined by the input capacitance and the amount of swamping resistance.
- Output bandwidth under saturation can be improved in either configuration by reducing the plate voltage. This involves a trade-off in efficiency with a smaller voltage swing. The bandwidth improvement can be obtained with a loss of PA gain and efficiency.

- A grounded-grid saturated PA improves bandwidth over a grid-driven saturated PA at the expense of amplifier gain.
- The best performance for FM operation is obtained when the amplifier is driven into saturation where little change in output power occurs with increasing drive power. Maximum efficiency also occurs at this point.
- The phase linearity in the 0.5 dB bandwidth is better in a grid-driven configuration. The grounded grid PA exhibits a more nonlinear phase slope within the passband yet has a wider amplitude bandwidth. This phenomenon is due to interaction of the input and output circuits because they are effectively connected in series in the grounded-grid configuration. The neutralized, grid-driven PA provides more isolation between these networks, so they behave more like independent filters.

Grid-Driven Tetrode and Pentode Amplifiers

Transmitters using tetrode amplifiers usually have one less amplifying stage than transmitters using triodes. Because tetrodes have higher power gain, they are driven into Class C operation for high plate efficiency. Against these advantages is the requirement for neutralization, along with screen and bias power supplies. Pentode amplifiers have higher gain than their tetrode counterparts. The circuit configuration and bias supply requirements for the pentode are similar to the tetrode because the third (suppressor) grid is tied directly to ground. The additional isolating effect of the (suppressor) grid eliminates the need for neutralization in the pentode amplifier.

Impedance Matching Into the Grid

The grid circuit is usually loaded (swamped) with added resistance. The purpose of this resistance is to broaden the bandwidth of the circuit by lowering the circuit Q and to provide a more constant load to the driver. It also makes neutralizing less critical so the amplifier is less likely to become unstable with varying output circuit loading. Cathode or filament lead inductance from inside the tube, through the socket and filament capacitors to ground can heavily load the input circuit. This is caused by RF current flowing from grid to filament through the tube capacitance and then through the filament lead inductance to ground. An RF voltage is developed on the filament that in effect causes the tube to be partly cathode driven. This undesirable extra drive power requirement can be minimized by series resonating the cathode return path with the filament bypass capacitors or by minimizing the cathode to ground inductance by using a specially designed tube socket using thin-film dielectric "sandwich" capacitors for coupling and bypassing.

High-power, grid-driven Class C amplifiers require a swing of several hundred RF volts on the grid. To develop this high-voltage swing, the input impedance of the grid must be increased by the grid input matching circuit. Because the capacitance between the grid and the other tube elements may be 150 picofarads (pF) or more, the capacitive reactance at 100 MHz will be very low unless the input capacitance is parallel resonated with an inductor. Figure 7.14-39 shows two popular methods of resonating and matching into the grid of a high-power tube. Both methods can be analyzed by recognizing that the desired impedance transformation is produced by an equivalent L network.

In Figure 7.14-39(a), a variable inductor (L_{in}) is used to raise the input reactance of the tube by bringing the tube input capacitance (C_{in}) almost to parallel resonance. Parallel resonance is not reached because a small amount of parallel capacitance (C_p) is required by the equivalent L network to transform the high impedance (Z_{in}) of the tube down to a lower value through the series matching inductor (L_s). This configuration provides a lowpass filter by using part of the tube's input capacitance to form (C_p). Figure 7.14-39(b) uses

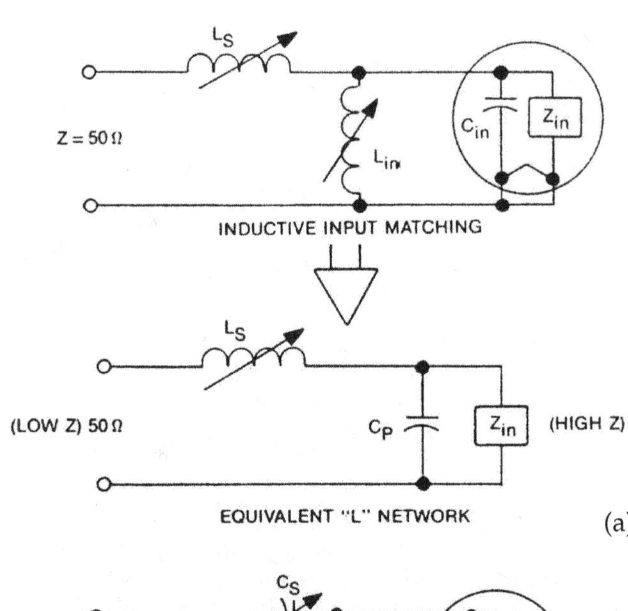

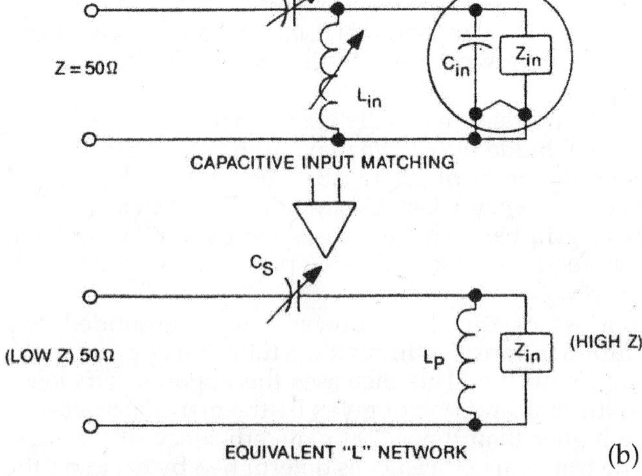

FIGURE 7.14-39 (a) Inductive input matching. (b) Capacitive input matching.

variable inductor (L_{in}) to take the input capacitance (C_{in}) past parallel resonance so the tube's input impedance becomes slightly inductive. The variable series matching capacitor (C_s) forms the rest of the equivalent L network. This configuration is a highpass filter.

Neutralization

Cathode-driven, grounded-grid amplifiers utilizing triodes do not require neutralization. The grid-to-ground inductance, both internal and external to the tube, should be kept low to maintain this advantage. Omission of neutralization will allow a small amount of interaction between the output circuit and the input circuit through the plate-to-filament capacitance. This effect is not very noticeable because of the large coupling between the input and output circuits through the electron beam of the tube.

Grid-driven, high-gain tetrodes require accurate neutralization for best stability and performance. This is particularly true if the amplifier is to be operated in Class AB linear mode for common amplification of FM + HD Radio signals. Self-neutralization can be accomplished by placing a small amount of inductance between the tube screen grid and ground that is usually in the form of several short, adjustable-length straps. The RF current flowing from plate to screen in the tube also flows through this screen lead inductance. This develops a small RF voltage on the screen, of the opposite phase, which cancels the voltage fed back through the plate-to-grid capacitance. This method of lowering the self-neutralizing frequency of the tube works only if the self-neutralizing frequency of the tube/socket combination is above the desired operating frequency before the inductance is added. Feedback neutralization utilizes a small coupling capacitor, usually in the form of a small plate located near the anode of the tube. The sample of the RF voltage from the anode intercepted by this plate is coupled through a 180° phase-shift network into the grid circuit. This technique has the advantage of providing neutralization over a very broad range of frequencies, if implemented correctly and stray reactances are minimized.

Special attention must also be given to minimizing the inductances in the tube socket by integrating distributed bypass capacitors into the socket and cavity deck assembly. Pentodes normally do not require neutralization because the suppressor grid effectively isolates the plate from the grid [8].

Power Amplifier Output Circuits

The output circuit usually consists of a high-Q (low loss) transmission line cavity, strip line, or lumped inductor that resonates with the tube output capacitance. A means of trimming the tuning and a means of adjusting the coupling to the output transmission line must also be provided by the output circuit. The tank circuit loaded Q is kept as low as practical to minimize circuit loss and to maintain as wide an RF bandwidth as possible.

Quarter-Wavelength Cavity

The quarter-wavelength coaxial cavity is the compact and popular PA output circuit illustrated in Figure 7.14-40. The tube anode is coupled through a DC blocking capacitor to a shortened quarter-wavelength transmission line. The tube's output capacitance is brought to resonance by the inductive component of the transmission line that is physically less than a quarter-wavelength long. Plate tuning can be accomplished either by adding end-loading capacitance at the high impedance end of the line with a variable capacitor or by changing the position of the ground plane at the low impedance end of the line. The plate tuning capacitor may be a sliding or rotating plate near the anode of the tube.

The center conductor of the transmission line (air exhaust chimney) is at DC ground while the anode of the tube operates at a high RF and DC potential. High voltage is fed through an isolated quarter-wavelength decoupling network inside the chimney to the anode of the tube, while the plate blocking capacitor prevents DC current flow from the anode into the chimney. RF power may be coupled from a quarter-wavelength cavity to the transmission line by a capacitive probe placed at the high RF voltage point located at the anode end of the quarter-wave line. The loaded Q of this circuit varies with the degree capacitive coupling. Another method of coupling power from the quarter-wavelength cavity uses a tuned loop located near the grounded (high current) end of the line. In this case, the tuned loop operates as both an inductive and a capacitive pick-up device. The quarter-wavelength cavity has approximately twice the operating bandwidth of the half-wavelength cavity.

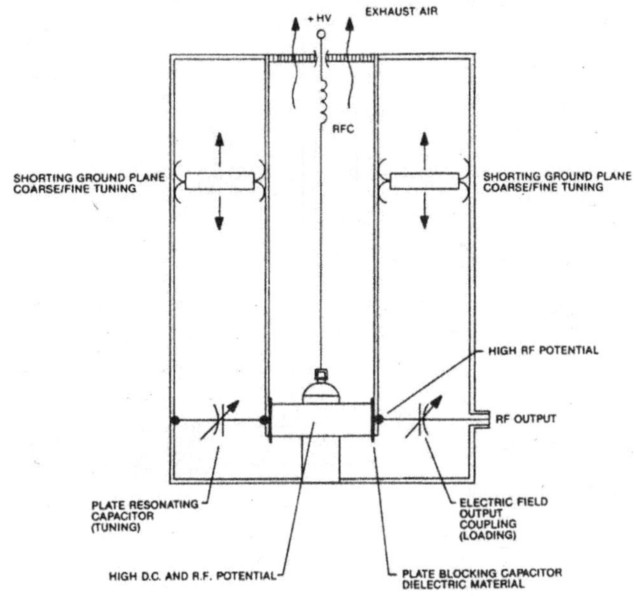

FIGURE 7.14-40 The quarter-wavelength cavity.

Folded-Half-Wavelength Cavity

Another approach to VHF power amplification uses the folded-half-wave cavity design illustrated in Figure 7.14-41. The DC anode voltage is applied to the lower portion of the plate line through a choke at the RF voltage null point. The half-wave line is tuned by mechanically expanding or contracting the physical length of a flexible extension (bellows) on the end of the secondary transmission line stub, which is located concentrically within the primary transmission line (air exhaust chimney). Coarse frequency adjustment is accomplished by presetting the depth of the top secondary section of plate line into the tank cavity. Power may be coupled from the half-wave line by an inductive loop located in the strong fundamental magnetic field near the center of the cavity.

Other power amplifier configurations may use lumped components or hybrid combinations with distributed transmission line elements to achieve similar results. The discrete circuit elements are chosen for their individual inductance or capacitance instead of being operated in a purely quarter-wave or half-wave mode. Stray inductance and capacitance add to the component values resulting in the hybrid nature of these circuits. Regardless of the specific configuration, the output circuit must transform the high resonant plate impedance down to the output transmission line impedance of 50 Ω. References at the end of this chapter give detailed information about the design of tube-type RF power amplifiers [8, 13].

Power Supplies

Power supplies provide the appropriate DC or AC voltages to the various subsystems with the transmitter. In a typical FM transmitter, the voltages and currents can range from less than 5 volts at a few milliamperes to over 10,000 volts at several amperes. Safety must therefore be a prime consideration when working around potentially lethal power supplies [19]. Power supplies must be designed with adequate bleeder resistors and interlocks to discharge high voltages before an operator can come in contact with these circuits. The degree to which the AC ripple components are filtered out of the DC outputs of the power supplies will, in large part, determine the asynchronous (without FM modulation) AM noise of the FM transmitter. FM transmitters usually contain multiple power supplies for each of the functional blocks within the system. These power supplies fall into two general categories:

- Single-phase supplies (single input winding on the transformer);
- Polyphase supplies (three or more input windings on the transformer).

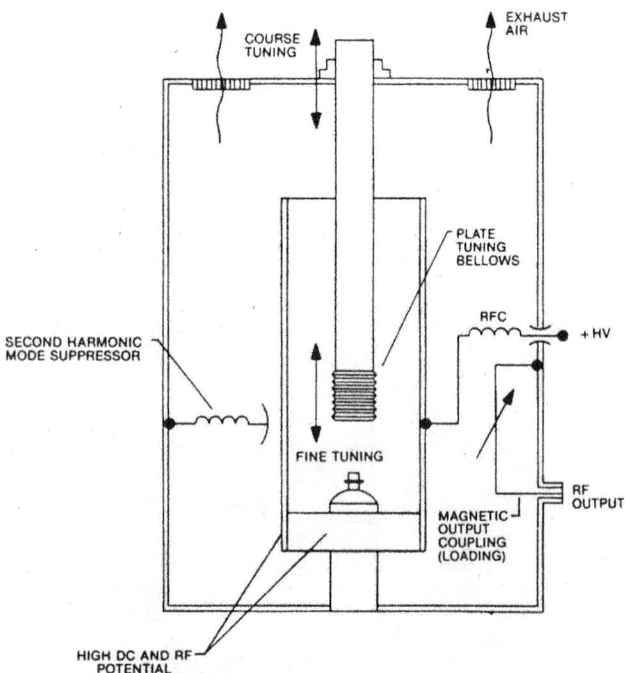

FIGURE 7.14-41 The folded half-wavelength cavity.

Single-Phase Power Supplies

Single-phase power supplies with conventional full-wave rectification and filtering are used most often for the FM exciter, the control system, bias supplies, and the IPA. A single-phase supply requires a larger filter choke to achieve the critical inductance requirement and a greater value of filter capacitance to maintain acceptably low ripple content compared to a polyphase supply. Large-value filter components also mean that the greater stored energy in these components can have a more destructive effect if an arc-over occurs. Choke-input filter sections are normally used to help limit the in-rush current while the shunt capacitor is charging during turn-on. This reduces stress on the transformer and rectifiers by keeping the charging current nearly constant, producing the best filtering action. Choke-input filters have the undesirable characteristic of poor voltage regulation over a wide range of loads. The output voltage will rise well above the nominal value with no load unless there is enough current through the bleeder resistor to keep the choke in the constant current range. Fortunately, in an FM transmitter application, the load on the power supply is relatively constant because the power output of the transmitter does not vary significantly with FM modulation. In higher power transmitters with a three-phase main power source, it is important to balance each of the individual single-phase loads among the three phases so the total load on each of the individual phases is equal.

Polyphase Power Supplies

Polyphase power supplies are used for the final power amplifier, high-voltage supply in most high-power transmitters. Sometimes they are used for tube or solid-state IPA supplies. Large blowers, used to cool transmitters, are usually operated from a three-phase power source. It is important to make the three-line connections to the blower motor in the proper sequence so the motor will turn in the proper direction. The most common type of polyphase supply is three-phase with full-wave rectification and LC filtering. Other polyphase systems encountered in broadcast equipment are usually multiples of the three phases, with 12 phase rectifiers becoming more popular. The main advantages of a polyphase power supply are:

- Division of the load current between three or more lines to reduce line losses and the size of each of the lines;
- Greatly reduced filtering requirements after rectification due to the low ripple at the output of a polyphase full-wave rectifier;
- Better voltage regulation with a choke input filter, with typically 6% or less variation from no load to full load;
- Greater choice of output voltages from a given transformer by selection of either a delta or wye configuration.

The main disadvantage of polyphase systems is their susceptibility to phase imbalance, which causes degraded performance of the power supply. If significant imbalance exists in a polyphase system, ripple rejection will be reduced in the polyphase rectifier with a resulting increase in AM noise.

The broadcast engineer should determine that the local utility does, in fact, provide true three-phase power to the transmitter site. This can usually be verified by making sure that there are three transformers on the utility pole feeding the transmitter site. In many rural areas, the utilities are still synthesizing pseudo three-phase service by providing the so-called open delta (V–V) or Scott (T–T) connection with two transformers instead of true three-phase service. Operation on an open-delta service will degrade the transmitter's performance and increase the susceptibility of the transmitter to damage from transients on the line. Most transmitter manufacturers state that their warranty is void if the transmitter is connected to an open delta system.

Regulated Power Supplies

In some cases, phase control switching regulation is applied to the high-voltage power supply feeding the final output tube. The regulation is accomplished by switching "thyristors" in the AC mains ahead of the primary winding of the transformer. As the switching duty cycle is reduced, the plate voltage is also reduced. It is important to protect solid-state devices connected to the main power line from transients. Heavy-duty transient suppressors are available for this purpose. Completely solid-state power amplifiers require lower voltages at much higher currents than tube amplifiers. Voltage regulation of these high-current supplies is necessary to suppress ripple, but the design of these specialized regulators is different from the typical high-voltage power supply. Linear regulators are used at the lower power levels although they are low in efficiency, because they are simple and provide excellent ripple rejection without the need for suppression of switching transients. The linear regulators use series or shunt devices that change resistance dynamically to provide regulation with changes in load and therefore dissipate some of the power within the dynamic resistance.

Switched-Mode Power Supplies

Switched-mode power supplies are used at higher power levels because they are high in efficiency, but they are more complicated and require additional suppression of the switching transients. The high efficiency comes from the digital "on" or "off" nature of the switching regulator that reduces resistive losses by using low-loss reactive components to store energy during switching.

The latest generation switching power supplies have significantly improved the conversion efficiency from the AC mains input power to the DC power required by the LD-MOS RF transistors. Typical power supply conversion efficiencies have improved from 85% to 95%. Switching power supplies have also eliminated the need for outboard voltage regulators with their associated operating, cooling, and maintenance costs. Switching power supplies make it possible to use either polyphase or single phase AC power sources. Remote transmission facility locations where polyphase power is not available can eliminate the cost of bringing polyphase power to the facility or the costs of converting single phase power to polyphase power.

Switched-tap power supplies can provide good voltage regulation over a limited range at higher efficiency than a switched-mode regulated supply. A solid-state controller switches banks of silicon controlled rectifiers (SCRs) between several taps on the secondary of the power transformer to maintain a constant output voltage with changes in load. This type of power supply is simple and efficient and has an excellent power factor (PF).

Low-voltage, high-current power supplies contain extremely large amounts of stored energy. This can be dangerous due to the high peak currents that can occur during a short circuit across a component with high stored energy. For this reason, pay special attention to methods of safely discharging these circuits without damaging components or injuring the operator. The voltage regulator should provide short-circuit protection with some type of current limiting. The main danger to the operator from this type of power supply is burns due to the nearly instantaneous heating of metallic tools and other conductors (such as a metal watchband or ring) that accidentally get into a high-current path (such as a short across the filter capacitor or, in the case of tube-type amplifiers, the filament transformer secondary) [10].

Step-Start

Step-start is often used in large power supplies where peak in-rush currents become excessively high when the power supply is initially turned on at full power level. These peak currents are caused by the need to overcome the hysteresis effect (to initially magnetize the core of the transformer) when AC power is applied and to charge the filter capacitor. Step-start systems temporarily insert a resistance or reactance in series with the power lines to limit the current to a reasonable value until initial magnetization of the core and filter charge is completed.

Transmitter Control Systems

Transmitter control systems are often overlooked or given little priority in the selection and maintenance of a broadcast transmitter. The transmitter control system serves several important purposes including:

- Basic on/off control of the transmitter;
- Overload protection to protect the transmitter from damage;
- Safety interlock protection to prevent injury to people and accessory equipment such as RF switching equipment or RF loads;
- A means of controlling the transmitter output power;
- Remote control capability and interfacing at installations where the transmitter is not at the same location as the control operator;
- Warm-up and cool-down timing sequences of filaments or other time-sensitive operations.

Additionally, a transmitter control system may provide for:

- Status indications of overloads or other critical parameters;
- Automatic regulation of the transmitter output power;
- Local and remote diagnostic indications to aid in adjustment and maintenance;
- Totally automatic operation of the transmitter plant;
- Integrated remote control capability;
- Control and monitoring via Internet connection.

The transmitter's ability to stay on the air will only be as good as the reliability of the control system, so the selection and correct operation of the transmitter control system is important. Solid-state logic and microprocessor control systems are well protected against damage from high voltages by optical isolation, shunt protection techniques, and radio frequency interference (RFI) filtering. Operating experience with the current generation of transmitters has proven that a properly designed solid-state control system is far more reliable than older relay based designs.

Automatic Power Control

Many transmitters also provide automatic power control (APC) circuitry to maintain the transmitter's power output within preset limits by correcting for changes in line voltage, component aging, or small amounts of drift in operating parameters. The APC circuitry compares a sample of the transmitter output power to a reference and then adjusts the RF drive or other voltages within the transmitter to bring the output power within tolerance. Some of the more sophisticated APC circuits also utilize proportional VSWR fold-back of the transmitter output power. If a sample of the reflected power on the transmission line exceeds a safe limit, the transmitter output power is proportionally reduced to a safe level until the problem is resolved. This feature prevents lost air time during antenna icing or other limited VSWR situations. Of course, APC circuits should provide fast-acting shutdown of the transmitter during a catastrophic failure of the antenna system such as a short or open circuit.

Computer Control Systems

Most transmitters are now equipped with microprocessor-based control systems. Microprocessor technology lends itself well to industrial control applications like broadcast transmitters. The hardware can be made just as reliable as hard-wired digital logic. Changes and growth in the operational features are made by simple modifications in software instructions rather than a complete redesign of the hardware. Some of the features that distinguish these control systems from nonmicroprocessor systems include:

- Built-in trouble tree with fault location and diagnostic read-outs and user-friendly messages on GUI;
- Simultaneous read-outs of all operating parameters;
- Real-time calculation of efficiency, dissipation, VSWR, and other parameters requiring calculations;
- Adjustment of parameters to maximize efficiency and minimize reject load dissipation;
- Built-in clock/calendar for logging changes in operating status, power failures, and overloads;
- Tolerance flagging on key operating parameters as warnings for logging and for preventive maintenance;
- Ability to communicate with the outside world for remote control or logging purposes through a standard serial interface or Transmission Control Protocol (TCP)/Internet Protocol (IP);
- Integrated remote control capability without external remote control equipment and interfacing;
- Provision to customize the system features to the station's individual requirements through the use of software menus;
- Tuning aids that allow the operator to adjust the system for peak efficiency, minimum dissipation, and minimum VSWR by means of a real-time display of these calculated parameters.

The method of communicating information to the operator varies among systems, but most use LED or LCD read-outs with codes or alphanumeric messages. Microprocessor controllers can also be equipped with GUIs so a large amount of information can be displayed in an easily read and understood format.

Controller Back-Up Systems

Some degree of redundancy is desirable in the transmitter control system so the transmitter can stay on the air if a portion of the system fails. There are several approaches currently in use to provide back-up systems. A multilevel hierarchy can be used that automatically hands over basic control functions to a primary life-support controller in the event of microprocessor hardware or software problems. Good system design separates diagnostic and supervisory functions from basic control functions so a failure in a higher level function will not affect the ability of the system to remain on the air without interruption. Watchdog circuits and software are embedded within the control system to detect failures and initiate corrective action before an interruption in service occurs. It is also possible to have distributed microprocessor systems with multiple processors that can automatically pick up the tasks of a failed processor without affecting the ability of the transmitter to remain on the air. The ability to quickly replace a controller subsystem while remaining on the air is a feature of the recent generations of transmitter control systems.

Remote Control Interfacing

Regardless of the type of control system used, the ability to interface easily with standard remote control systems is very important. Most transmitters have a parallel control interface with control lines for each individual function requiring a momentary contact closure of 24 volts DC or less at a current of 50 milliamperes or less. These levels are compatible with relay logic or optically isolated solid-state logic. Analog levels output from the transmitter for remote meter readings generally are fully buffered and fall into the range of 0 to 5 volts DC for a full-scale reading of a particular parameter at an impedance of less than 10 kΩ. The advent of microprocessor based control systems has been accompanied by a trend toward using a standard computer asynchronous serial interface in addition to the parallel interface. Serial interfacing reduces the number of connections to the transmitter and can carry both control functions and digitized meter readings through the same interface. By converting analog information into digital information before transmission of data to the remote control point, the need for calibration and recalibration of the remote metering point is reduced. The current trend is toward open, nonproprietary, serial data protocols and interface standards such as the Simple Network Management Protocol (SNMP) and TCP/IP. This gives the user maximum flexibility in choosing dedicated remote control equipment or using software based network management tools. Microprocessor-based control systems also allow the remote control system to have access to more in-depth information about the transmitter than is practical by parallel interfacing with an external remote control system. A personal computer or laptop computer can be used to control the transmitter through an ordinary dial-up phone line, Internet, or radio link.

Internet Protocol Remote Control and Web Access

Broadcast radio transmitters increasingly include IP for remote control and monitoring. The serial interfaces are usually either RS232 or Ethernet. When RS232 is used, an extra device is needed to convert the information to IP. Internet-based devices are the most extensive, and some generate GUIs on remote PCs via a web browser such as Internet Explorer or Firefox, as illustrated in Figure 7.14-42.

TCP/IP Protocol Suite

Internet-based services are built upon the TCP/IP protocol suite. TCP and UDP are both in this suite. UDP is known as a connectionless protocol. The connection between a server and a client is not guaranteed because there is no handshake. The protocol presumes one-way communication. A server knows where to send a message but its receipt is not acknowledged. UDP is therefore considered an unreliable protocol, but it is efficient when the network link is reliable.

TCP is a connection-based protocol. Computers on each end of the communication link know about the other. As long as the physical link exists, TCP will ensure that messages are received reliably. It does this by acknowledging and requesting information multiple times if necessary.

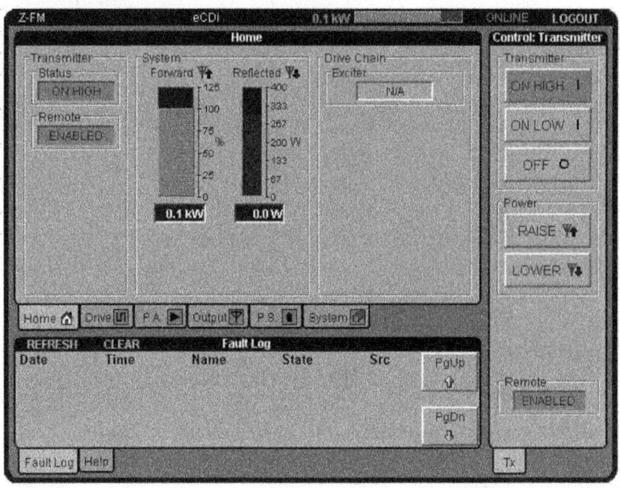

FIGURE 7.14-42 FM transmitter remote graphical user interface (GUI).

HTTP and HTML

Hypertext Transfer Protocol (HTTP) is the basis of Internet-based web pages. Hypertext Markup Language (HTML) is a means of describing the layout and content of web pages and is the content of the files served to web browsers via HTTP. Web browsers decode HTML sent in these files to generate the proper display. HTTP and HTML work well for static graphical and textual images. A more powerful mechanism exists using Java applets. Java applets can be embedded within HTML and executed client-side, thereby relieving the server-side processing burden. This approach can be applied to radio transmitter systems because their embedded control systems are generally limited in order to be as cost effective as possible.

SNMP

The Simple Network Management Protocol (SNMP) was created to manage routers on a large network, but it has been applied to a multitude of other equipment, including radio transmitters. Data in the transmitter is identified with an object identifier (OID) that is unique; no two OIDs are the same. A management information base (MIB) describes the OIDs. An SNMP Manager is required on the remote computer, whereas the transmitter must be an SNMP Agent. An issue raised with the use of any of these protocols is that of setting up the gateway between the Internet and the control equipment. It is necessary to explicitly allow access through all the necessary ports; however, in general, the more ports that are open, the more vulnerable a network will be to hackers and viruses. For more information, see Chapter 6.5, "Transmitter Remote Control and Monitoring Systems."

RF Output and System Filtering

The high efficiency, nonlinear RF power amplifiers used in FM broadcast transmitters generate significant amounts of energy on frequencies that are integer multiples (harmonics) of the desired fundamental frequency. The output circuit alone does not provide enough harmonic attenuation to meet FCC regulations. To comply with Section 73.317 of the FCC Rules and to prevent interference to other services, a lowpass filter must be installed at the output of the transmitter. The FM band is narrow enough that one lowpass filter design can be used for any FM channel carrier frequency. These filters usually consist of multiple LC sections arranged so frequencies within the FM band are passed with typically 0.1 dB or less attenuation while frequencies above the FM band are attenuated 60 dB or more.

The most common type of filter used in this application is a reflective filter in which the frequency components outside the passband are reflected out of the filter and back toward the source because the filter exhibits an impedance mismatch at these undesired frequencies. The filter can be constructed using either lumped inductors and capacitors or by using a section of nonconstant impedance transmission line to form distributed inductors and capacitors. The filters designed for low-power transmitters often employ lumped elements (coils and capacitors) because these elements are compact and can be integrated into the transmitter cabinet. The distributed type of filter is most often used with high-power FM broadcast transmitters because of its simplicity, extreme ruggedness, and ability to handle higher power levels. The distributed filter has the disadvantage of having larger physical dimensions than a similar lumped filter, which may require mounting the filter external to the transmitter cabinet. Figure 7.14-43 shows a cut-away view of a typical distributed lowpass filter. Note that the areas where the center conductor of the transmission line is smaller than that required for input Z_0 are inductive, while the areas where the center conductor is larger in diameter are capacitive.

When two filters (such as the output cavity and the harmonic filter) are connected together by a transmission line, the total harmonic attenuation will vary with interconnecting line length. The attenuation characteristics of the harmonic filter are specified for the condition where both the source and load impedances are equal to the desired transmission line impedance. In practice, the source impedance at the output of the tank circuit is much less than the 50 Ω load impedance presented by a properly terminated filter. If an incorrect length of line is selected, the harmonic attenuation may be insufficient and the transmitter tuning may be affected. This undesirable condition can be corrected by changing the line length by approximately one-quarter wavelength. The line length between the transmitter output circuit and harmonic filter is usually supplied precut to the appropriate length by the transmitter manufacturer.

Harmonic Notch Filters

In some cases, a second-harmonic notch filter is required in addition to the lowpass filter if the second-harmonic component from the amplifier is high in amplitude and the cut-off slope of the lowpass filter is not steep enough to provide sufficient second harmonic attenuation. The additional attenuation required (typically 30 dB) can be provided by a notch filter that places a short circuit across the transmission line at the second-harmonic frequency while exhibiting a high impedance at the fundamental frequency. A one-quarter wavelength (at the fundamental frequency) shorted coaxial stub is often used for this function. The second-harmonic energy is partially reflected back into the power amplifier and partially dissipated in the equivalent series resistance of the series-tuned circuit formed by the stub. This shorted stub provides a very low inductance and a DC path from the center

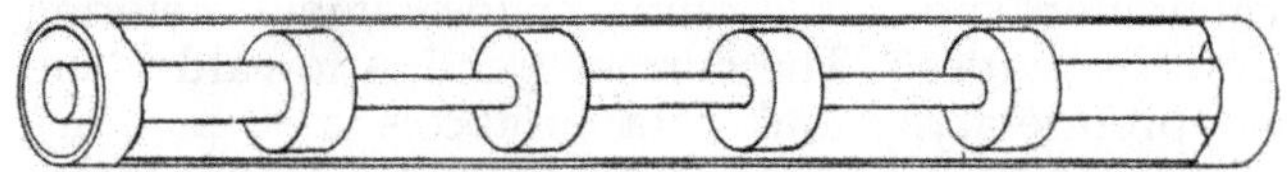

FIGURE 7.14-43 Distributed lowpass filter.

conductor of the transmission line to ground, providing a separate, protective advantage by shunting static discharges, such as lightning, to ground. Some transmitters have internal second-harmonic suppressors that eliminate the need for an external notch filter.

Transmission Line Power and VSWR Measurements

Directional wattmeters are instruments that measure the forward power and reflected power in a transmission line. The net power delivered to the load (antenna) is the difference between the forward power and the reflected power. If the transmission line is perfectly matched, all the forward power will be absorbed by the load (antenna) and there will be no reflected power. The peak voltage at each point along the line will be the same value; similarly, the current at each point along the line will also have a uniform value. If the transmission line is mismatched, there will be reflected power with a resulting standing wave on the line. This means that the voltage and current distributions along the line will no longer be uniform, with high values at certain points on the line and low values at points one-quarter wavelength away. The ratio of the high-voltage value to the low-voltage value is the voltage standing wave ratio (VSWR).

VSWR Measurement

Although some FM transmitters can operate into a VSWR of greater than 1.8:1, the VSWR on an FM antenna transmission line should normally be kept to a value of 1.1:1 or better for good stereo performance. It takes very little reflected power to produce substantial VSWR. For this reason, the reflected power is usually read on a more sensitive meter position. Problems in the antenna system, such as loose connections or icing, may cause excessive VSWR. Instruments external to the transmitter are available that monitor reflected power and activate an alarm if it becomes excessive. As long as the transmitter power output is relatively constant, the use of reflected power to indicate excessive VSWR is simple and adequate [14].

Combined Transmitters

It is possible and practical to combine the output of two RF power amplifiers to obtain higher power output levels. The important advantage is that the broadcast signal is not interrupted should one amplifier fail. The radiated signal strength merely drops 6 dB until the failed amplifier is repaired and put back on the air. A dual amplifier system costs more than a single amplifier for a given total power output, but it offers the economic advantages of reducing lost air time and eliminating the need for a separate standby transmitter. Automatic or manual output switching can be used to route the full power of the remaining amplifier directly to the antenna.

Two methods may be used to bypass the output combining hybrid to allow 100% of the power of the remaining transmitter to be sent to the antenna if one transmitter of a combined pair should fail. The first method uses three motorized switches (or patch panels) to bypass the 3 dB hybrid combiner while connecting the operating transmitter directly to the antenna and the failed transmitter directly to the test load. This allows recovery of the 50% power lost in the reject load when one transmitter is off the air. A disadvantage is that the system must be taken off the air for several seconds for operation of the coax switches. A second method provided by some transmitter suppliers uses a pair of 3 dB hybrids interconnected with one fixed and one variable RF phasing section. The phasing section is constructed to operate while under RF power and can redirect the full output of either transmitter directly to the antenna and place the other transmitter into the test load without taking the system off the air. A dedicated system controller allows automatic or manual control. This so-called switchless combiner offers the highest possible on-air availability for combined FM transmitters. With complete redundancy in the RF power amplifier chain, some stations go one step further and also install dual exciters with automatic switching so if one exciter fails the other unit is quickly switched into service.

Hybrid Couplers

Hybrid couplers are reciprocal four-port devices that can be used either for splitting or combining RF sources over a wide frequency range. Figure 7.14-44 shows an exploded view of a typical 3-dB, 90° hybrid coupler.

The coupler consists of two identical parallel transmission lines that are coupled over a distance of approximately one-quarter wavelength and are enclosed within a single outer conductor. Ports at the same end of the coupler are in-phase, while ports at opposite ends of the coupler are in quadrature (shifted by 90° in phase) with respect to each other. The phase shift between the two inputs or outputs is always 90° and is somewhat

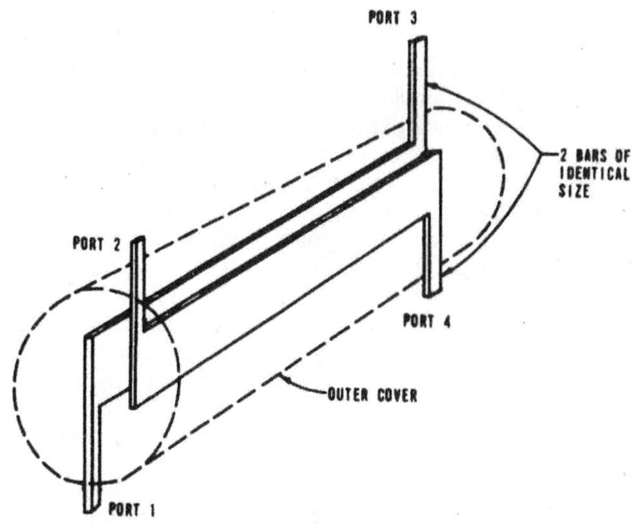

FIGURE 7.14-44 Model of 90° hybrid coupler.

independent of frequency. If the coupler is being used to combine two signals into one output, these two signals must be fed to the hybrid coupler in phase quadrature. The reason this type of coupler is also called a 3-dB coupler is that when used as a power splitter, the split is equal or half-power (3 dB) between the two outputs. Hybrid couplers can also be made with other coupling ratios. For example, a 6-dB coupler is often used for split-level combining of HD Radio and analog FM signals. A 10-dB coupler is used for high-level combining of HD Radio and analog FM signals.

Hybrid Combiners

The output hybrid combiner effectively isolates the two amplifiers from each other. Tuning adjustments can be made on one amplifier, including turning it on and off, without appreciably affecting the operation of the other amplifier. Good isolation is necessary so if one transmitter fails the other will continue to operate normally instead of in a mistuned condition. Two of the ports on the hybrid coupler are the inputs from the power amplifiers. The sum port is the antenna output terminal and the difference port goes to a resistive load called the reject load, as only the rejected power due to imbalance appears here.

When the power fed to each of the two inputs is equal in amplitude with a phase difference of 90° the total power of both inputs is delivered to the sum port (antenna). Very little of the power appears at the reject load if the phase relationship and power balance are correct. If the phase relationship is reversed between the two amplifiers, all the power is delivered to the reject load, so it is important that the proper one of the two possible 90° phase relationships is used. When all the ports on the hybrid combiner are terminated, an isolation of 30 dB or more can be achieved between the power amplifiers. For perfect isolation between the amplifiers, the load impedance on the sum and difference ports must be exactly the same. This is approached in practice by providing a 1.0:1 VSWR with a resistive 50 Ω load for the termination (reject load) on the difference port and then reducing the VSWR on the antenna transmission line as low as possible by trimming the antenna match. This will prevent the input port impedances from changing significantly when one amplifier is not operating.

The input ports will present a load to each transmitter with a VSWR that is lower than the VSWR on the output transmission line. This is because part of the reflected power coming into the output port will be directed to the reject load and only a portion will be fed back into the transmitters. The straight line in Figure 7.14-45(a) shows the effect of the output port VSWR3 on the input port VSWR1. The curved line in Figure 7.14-45(a) shows the change in isolation (dB) between the input port and output port as a function of the output port VSWR3 for a 3-dB hybrid combiner. Figure 7.14-45(b) shows the reduction in isolation between coupler ports caused by antenna and/or reject load VSWR.

If the two inputs from the separate amplifiers are not equal in amplitude or not exactly in phase quadrature, some of the power will be dissipated in the difference port reject load. The match in input power and phase is not critical. Figure 7.14-46 shows how the percent of useful power available at the combiner output port is related to the square root of the power input imbalance between transmitter A (Pa) and transmitter B (Pb). Figure 7.14-47 shows how the percent of useful power available at the combiner output port is related to the phase error from the desired 90° between transmitter A and transmitter B.

The power lost in the difference port reject load can be reduced to a negligible level by touching up the amplifier tuning and by adjusting the phase shift. For example, if one amplifier is delivering only half the power of the other amplifier, only about 3% of the total available power remaining will be dissipated in the reject load; 97% of the remaining power, or 48.5%

($0.97 \times 0.50 = 0.485$), is still fed to the output transmission line. If one transmitter fails completely, half of the working amplifier's output goes to the antenna, and the other half is dissipated in the difference port reject load. This is why the radiated output drops by 6 dB or to one-fourth of the original combined power. The reject load must be rated to handle a minimum of one-fourth of the total combined power, but often the reject load is rated to handle one-half the total power so it can also be used as a test load for one of the transmitters.

Hybrid Splitting of Exciter Power

Figure 7.14-48 shows a block diagram of a pair of combined amplifiers with dual exciters. The exciters cannot be operated in parallel like the amplifiers because their RF outputs would have to be on exactly the same carrier frequency and almost exactly in phase under all modulation conditions. An exception would be to use a dual RF channel, digital exciter that can provide matched modulation on each of its RF outputs feeding each of the combined transmitters. An automatic or manual exciter switcher is used to direct the output of one exciter to the combined transmitter while the other exciter is routed to a dummy load. The one exciter in use feeds a hybrid splitter/phase shifter that transforms one 50 Ω input into two isolated 50 Ω outputs that have a 90° phase shift between them, with half the power going to each output. The operation of this hybrid splitter is the reciprocal of the hybrid combiner described earlier.

The exciter must have enough power output capability to drive both power amplifiers. In some cases, IPAs are inserted between the splitter and power amplifier to boost the drive level. The length of coax cable from the power splitter to each amplifier input must be cut to a precise length so the amplifiers will be fed in the proper phase relationship.

Each of the power amplifiers is assumed to have equal gain and phase shift. In practice, it may be difficult to get the amplifiers tuned so their gains and phase shifts are equal at the same time. For this reason, a line stretcher or variable phase shift network is usually included with the exciter splitter to adjust phasing independent of amplifier tuning. Recent generations of digital FM and HD Radio exciters have dual RF channels that digitally provide the necessary phase shift between the two outputs without external components.

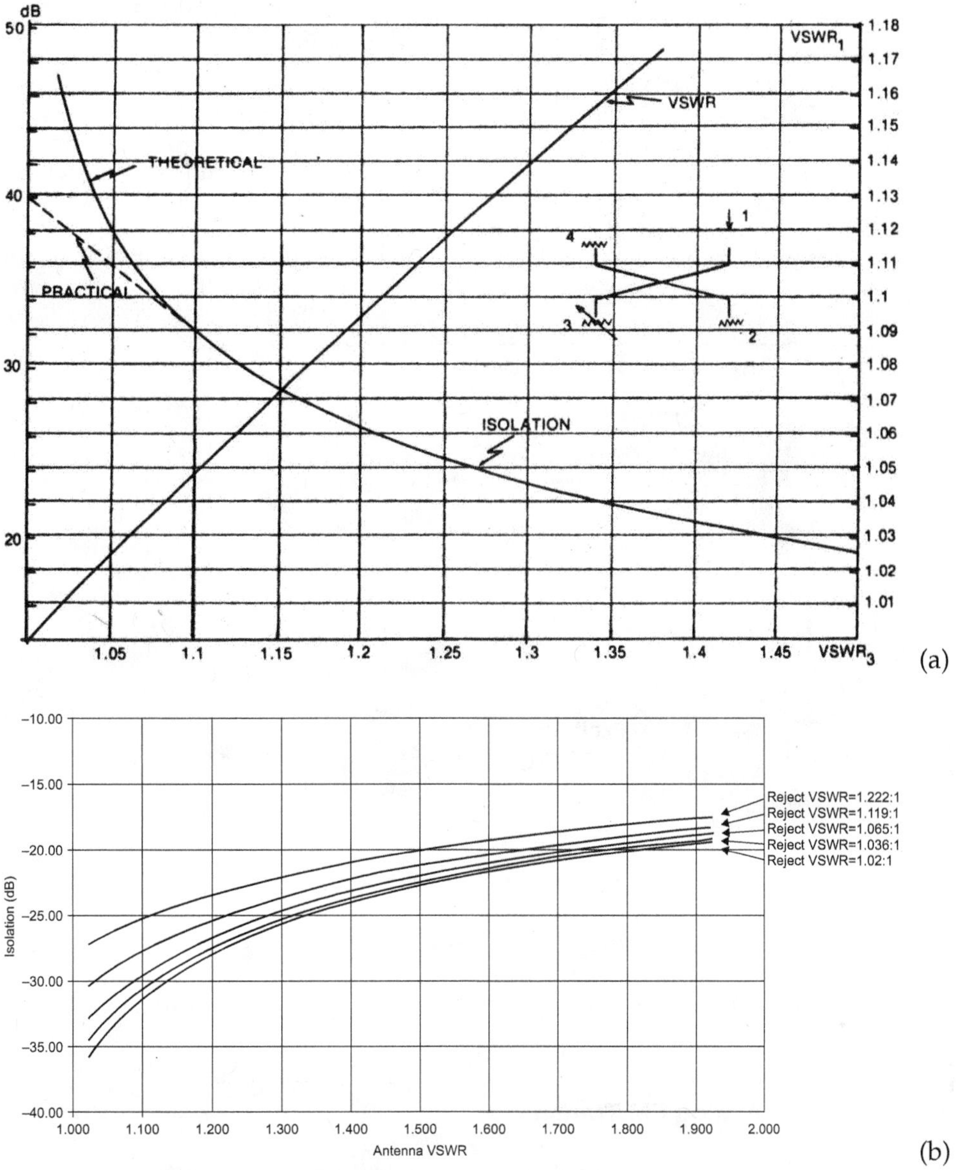

(a)

(b)

FIGURE 7.14-45 (a) Input port VSWR1 as a function of output port VSWR3 (straight line) and port-to-port isolation (in dB) as a function of output port VSWR3 (curved line). (b) Coupler isolation (in dB) as a function of output port VSWR. (Traces are in same order as legend.)

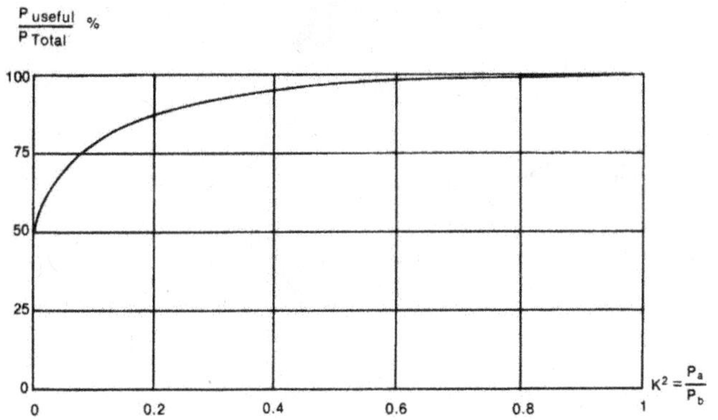

FIGURE 7.14-46 Power output as a function of imbalance in hybrid coupler.

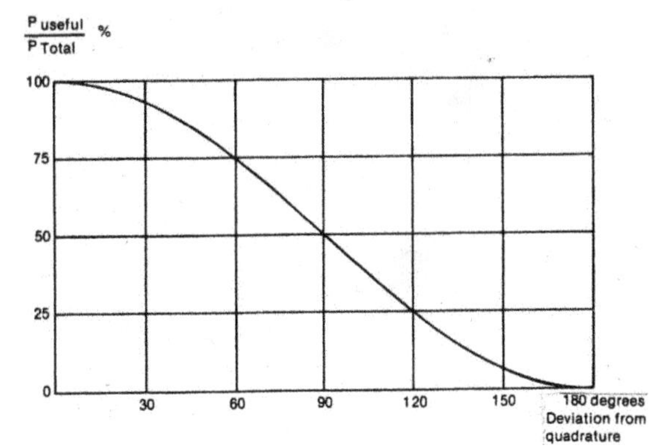

FIGURE 7.14-47 Power output as a function of phase error in hybrid coupler.

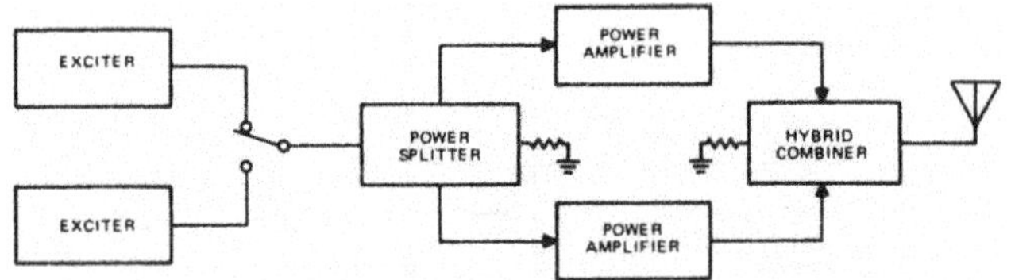

FIGURE 7.14-48 Block diagram of transmitter with two power amplifiers, a hybrid splitter, a hybrid combiner, and dual exciters with automatic switching.

High-Level Combining HD Radio With Analog FM

High-level IBOC combining has been widely used since the beginning of HD Radio conversions. This method requires a new digital transmitter, about one-third the size of the main transmitter, whose IBOC output is combined at a high level with the FM signal, using a 10-dB hybrid coupler as shown in Figures 7.14-49 and 7.14-50. Two limitations of this method are the

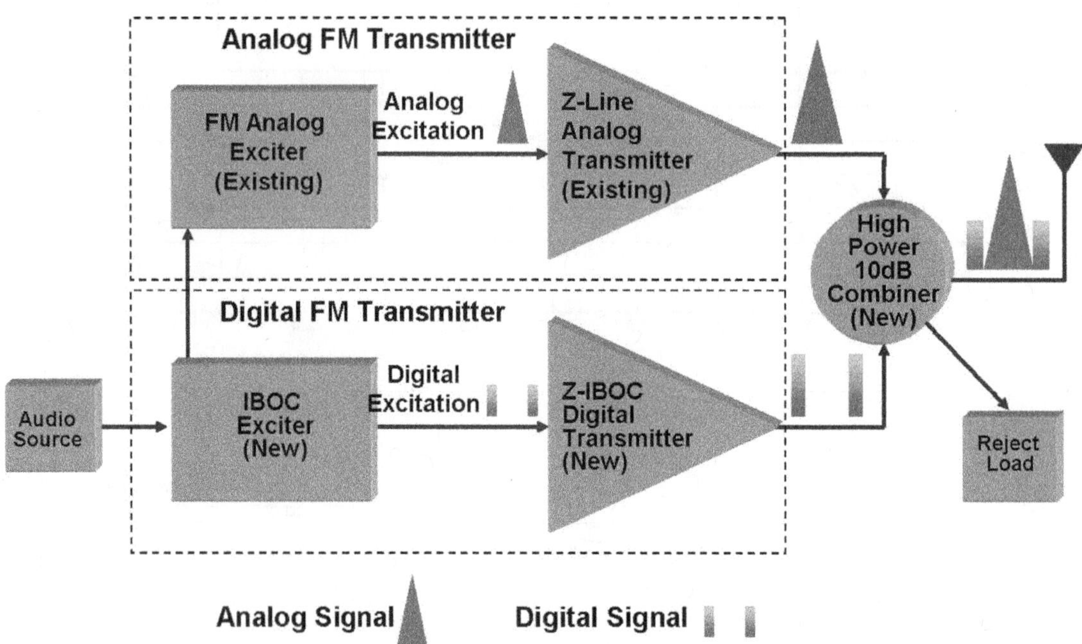

FIGURE 7.14-49 High-level combining with separate exciters.

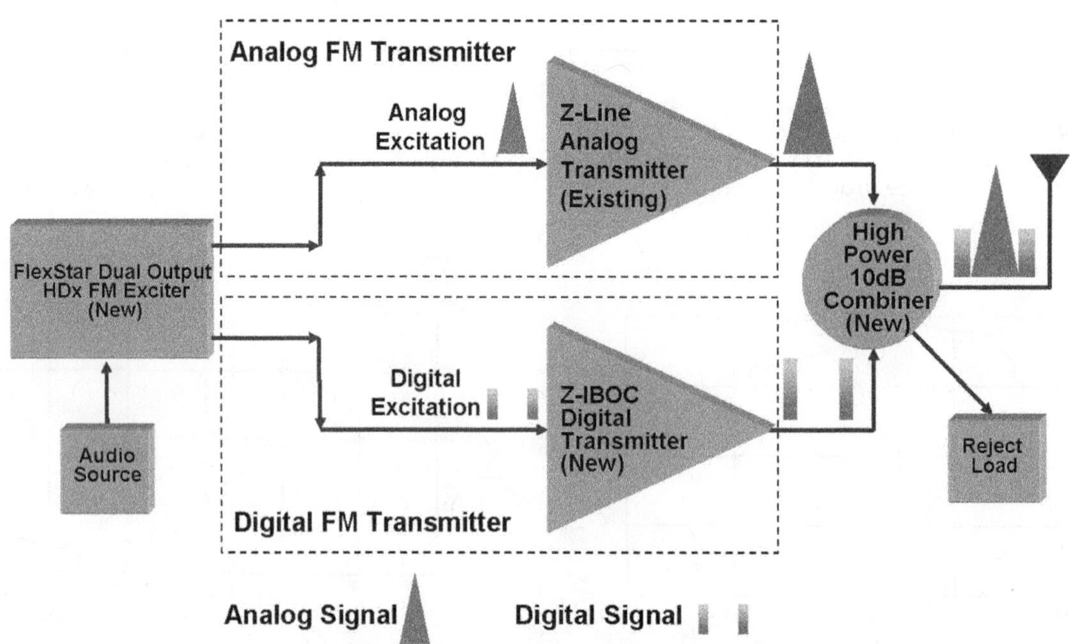

FIGURE 7.14-50 High-level combining with dual output exciter.

power lost in the reject load and the +10% additional power headroom demanded of the main analog FM transmitter to compensate for the loss of analog power into the reject load.

Split-Level Combining HD Radio With Analog FM

Split-level combining is an improved HD Radio combining scheme similar to high-level combining and is shown in Figures 7.14-51 and 7.14-52. The digital transmitter carries not only the IBOC signal, but also a portion of the FM signal that has been phased to add at the output combiner with the signal generated by the main FM transmitter. Split-level combining allows upgrading to HD Radio without increasing the power output requirement of the main FM transmitter. It also reduces the power lost in the reject load by about 50%.

The coupling coefficient for this configuration is about 6 dB, instead of the 10-dB coupling used in high-level separate amplification. The coupling value is chosen to minimize the size required for the digital transmitter and to reduce the dissipation in the reject load while boosting the overall system efficiency by 3% to 4%. The total reject power is reduced by 50%. The digital transmitter power should be 35% of the main transmitter's power output to provide the secondary FM injection level that is required to make the main FM path lossless. Note that this value (35%) is comparable to the size required for separate amplification. Figure 7.14-53 is an example of a split-level system calculation [28–31]. A redundant FM + HD Radio combining system block diagram is shown in Figure 7.14-54.

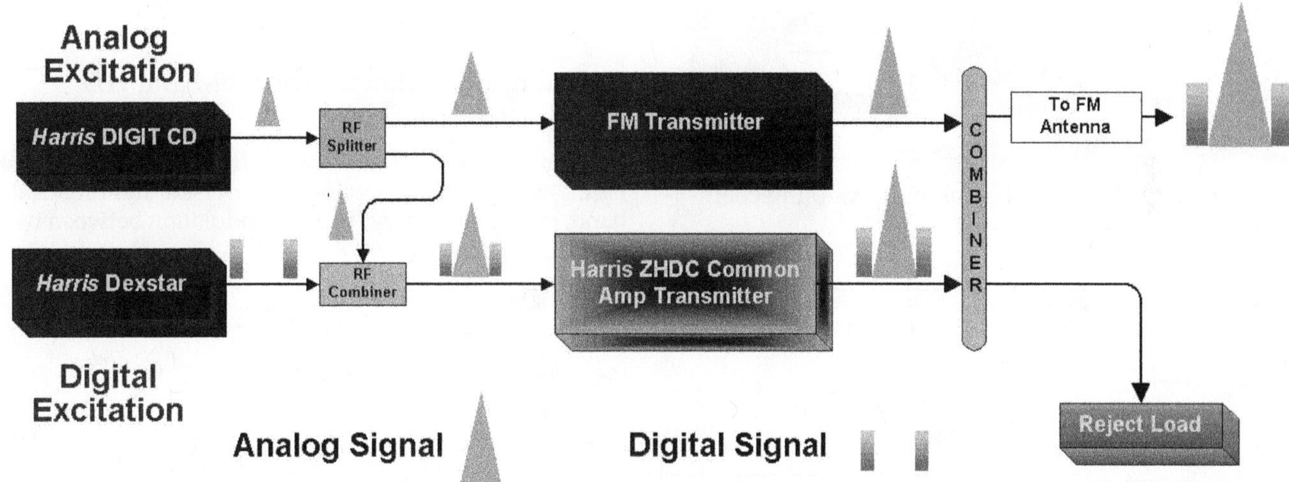

FIGURE 7.14-51 Split-level combining with separate exciters.

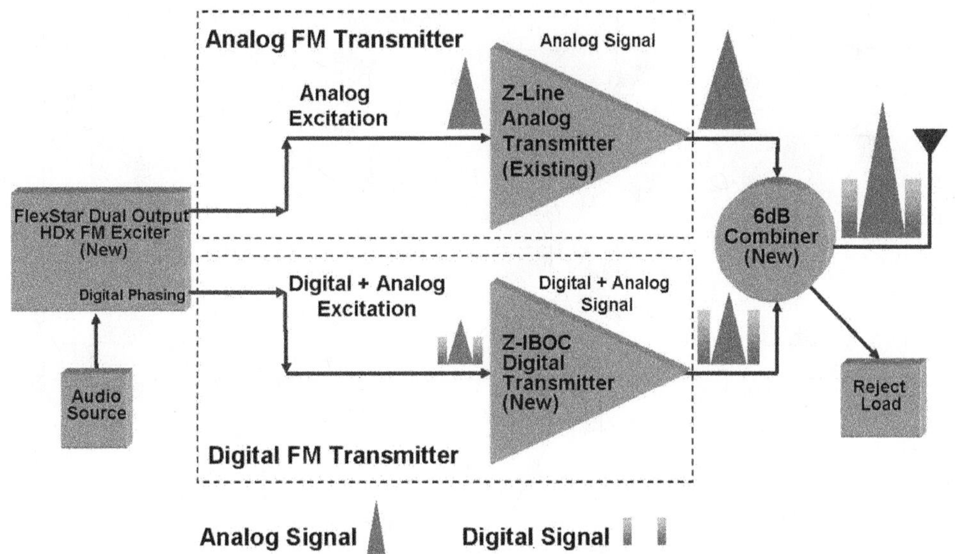

FIGURE 7.14-52 Split-level combining with dual output exciter.

COMBINING MULTIPLE FM STATIONS ON A SINGLE ANTENNA

It is common practice for several FM stations to share a single broadband antenna system by combining the multiple signals into a single, broadband antenna system.

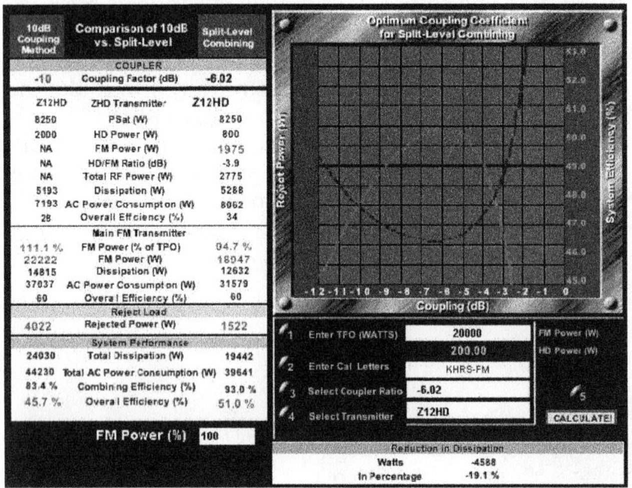

FIGURE 7.14-53 Split-level combining coupling coefficient calculator.

Filterplexing

A device called a filterplexer (also known as an RF multiplexer) is used to connect several transmitters on different frequencies together onto one antenna system. The filterplexer provides isolation between the multiple transmitters while efficiently combining their power into a single transmission line. This is usually accomplished by a system of bandpass filters, band reject filters, and hybrid combiners. The isolation is required to prevent power from one transmitter from entering another transmitter that can result in spurious emissions and to keep the rest of the system running in the event of the failure of one or more transmitters. The wideband port on a constant impedance filterplexer can also be used to sum multiple HD Radio signals into a dual-feed antenna system. An important consideration in the design of a filterplexing system is the effect on the phase response (group delay characteristic in the passband) of each of the signals passing through the system due to individual bandwidth limitations on each of the filterplexer inputs.

RF Intermodulation Between FM and HD Broadcast Transmitters

Interference to other stations within the FM broadcast band, as well as to other services outside the broadcast band, can be caused by RF intermodulation between two

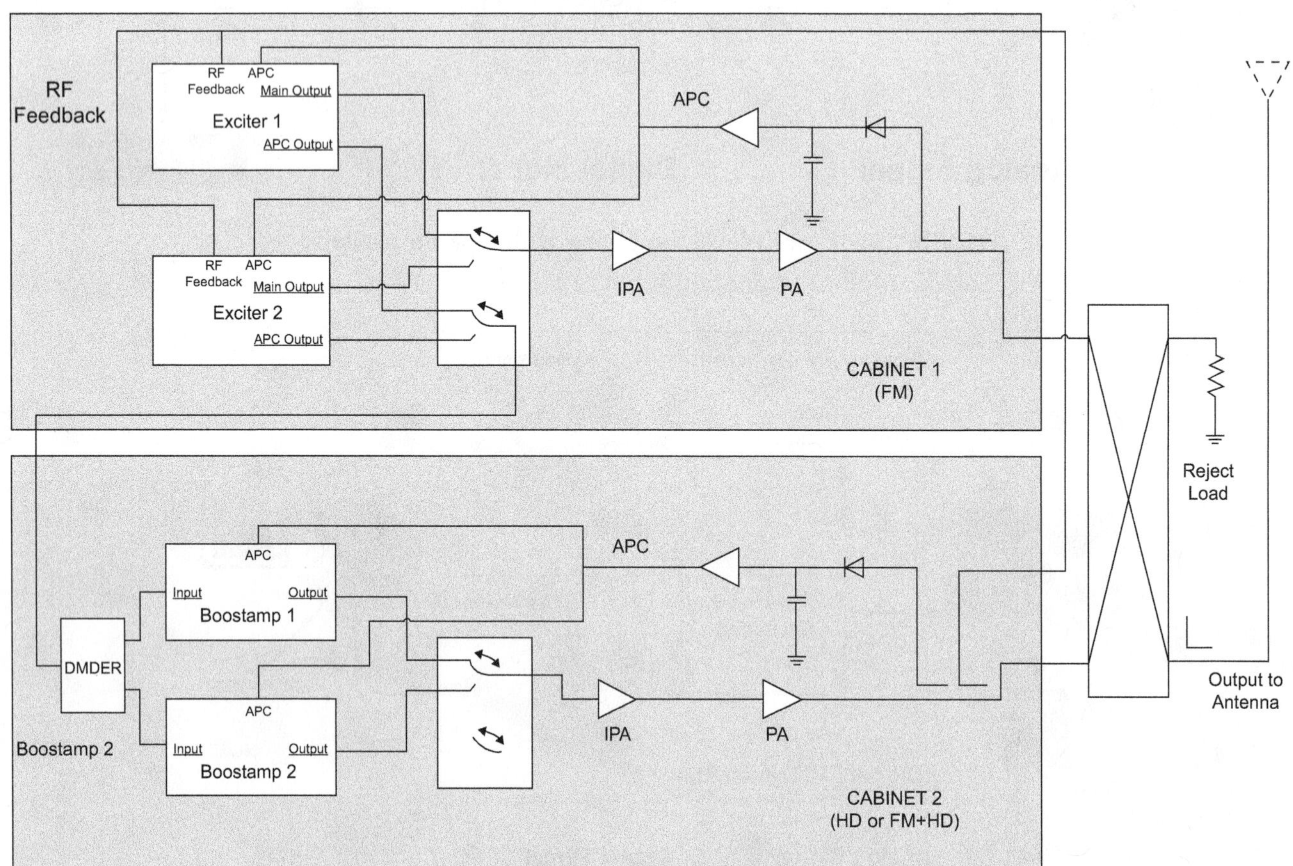

FIGURE 7.14-54 Redundant FM + HD Radio combining system.

or more FM or HD Radio broadcast transmitters. Transmitter manufacturers have begun to characterize the susceptibility of their equipment to RF intermodulation so this information is becoming available to the designers of filterplexing equipment. The degree of intermodulation interference generated within a given system can be accurately predicted before the system is built if the actual mixing loss of the transmitters is available when the system is designed. Accurate data on mixing loss or turnaround loss not only speeds the design of filterplexing equipment but also results in higher performance and more cost-effective designs because the exact degree of isolation required is known before the system is designed. Filterplexer characteristics, as well as antenna isolation requirements, may be tailored to the specific requirements of the transmitters being used. The end user is assured, in advance of construction, that the system will perform to specification without fear of overdesign or under design of the components within the system.

Mechanisms That Generate RF Intermodulation Products

When two or more transmitters are coupled to each other, new spectral components are produced by mixing the fundamental and harmonic terms of each of the multiple output frequencies; for example, if two transmitters are involved, the third-order intermodulation (IM_3) terms could be generated in the following way. The output of the first transmitter (f_1) is coupled into the nonlinear output stage of the second transmitter (f_2). If there is not complete isolation between the two output stages, f_1 will mix with the second harmonic of f_2, producing an in-band third-order term with a frequency of ($2f_2 - f_1$). In a similar fashion, the other third-order term will be produced at a frequency of ($2f_1 - f_2$). This implies that the second-harmonic content within each transmitter's output stage, along with the specific nonlinear characteristics of the output stage, will have an effect on the value of the mixing loss. It is possible, however, to generate these same third-order terms in another way. If the difference frequency between the two transmitters ($f_2 - f_1$), which is an out-of-band frequency, remixes with either f_1 or f_2, then the same third-order intermodulation frequencies are produced. Empirical measurements indicate that the ($2f_2 - f_1$) type of mechanism is the dominant mode generating IM_3 products in modern transmitters using a tuned cavity for the output network. Figure 7.14-55 shows an example of how the intermodulation product frequencies may be calculated. Figures 7.14-56 and 7.14-57 show the resulting frequency spectra.

Intermodulation as a Function of Turnaround Loss

Turnaround loss or mixing loss describes the phenomenon whereby the interfering signal mixes with the fundamental and its harmonics within the nonlinear output device. This interfering signal can be another FM signal or the HD Radio OFDM sidebands on the host analog FM signal. The mixing of these signals occurs with a net conversion loss. Thus, the term turnaround loss has become widely used to quantify the ratio of the interfering level to the resulting IM3 level.

3rd ORDER INTERMODULATION PRODUCTS

f_1=100.3 MHz. f_2=101.1 MHz.

$2f_1-f_2=[2(100.3)-(101.1)]=[200.6-101.1]=99.5$ MHz.

$2f_2-f_1=[2(101.1)-(100.3)]=[202.2-100.3]=101.9$ MHz.

OR

$[f_1-(f_2-f_1)]=[100.3-(101.1-100.3)]=[100.3-0.8]=99.5$ MHz.

$[f_2+(f_2-f_1)]=[101.1+(101.1-100.3)]=[101.1+0.8]=101.9$ MHz.

FIGURE 7.14-55 Calculation of third-order RF intermodulation product frequencies.

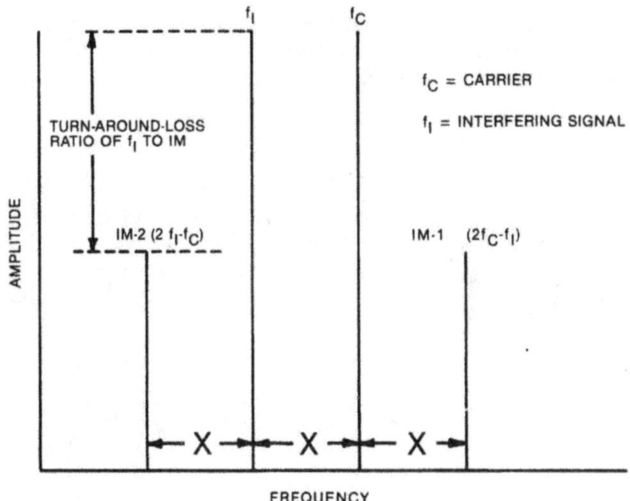

FIGURE 7.14-56 Frequency spectrum and turnaround loss ratio of third-order IM products with the interfering signal level equal to the carrier level.

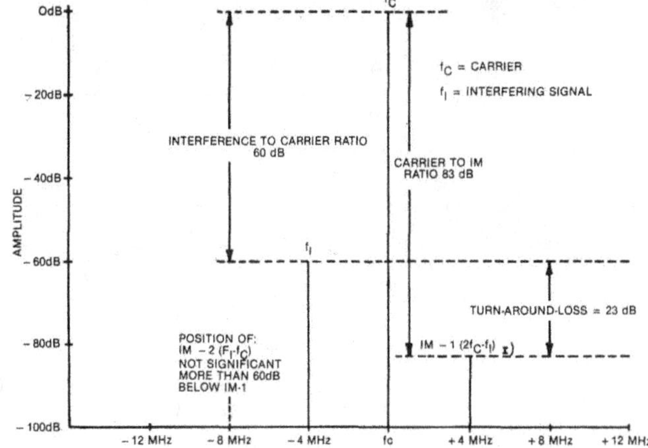

FIGURE 7.14-57 Typical frequency spectrum of third-order IM of a broadcast FM transmitter when interfering signal level is 60 dB below carrier level.

A turnaround loss of 10 dB means that the IM_3 product fed back to the antenna system will be 10 dB below the interfering signal fed into the transmitter's output stage.

Turnaround loss will increase if the interfering signal falls outside the passband of the transmitter's output circuit, varying with the frequency separation of the desired signal and the interfering signal. This is because the interfering signal is first attenuated by the selectivity going into the nonlinear device; the IM_3 product is then further attenuated as it comes back out through the frequency selective circuit.

Turnaround loss consists of three components:

- Basic in-band conversion loss of the nonlinear device;
- Attenuation of the out-of-band interfering signal due to the selectivity of the output stage (HD Radio interference is in-band and therefore not attenuated);
- Attenuation of the resulting out-of-band IM3 products due to the selectivity of the output stage.

As the turnaround loss increases, the level of undesirable intermodulation products is reduced, and the amount of isolation required between transmitters is also reduced.

The transmitter output circuit loading control directly affects the source impedance. This therefore affects the efficiency of coupling the interfering signal into the output circuit, where it mixes with the other frequencies present to produce IM_3 products. Light loading reduces the amount of interference that enters the output circuit with a resulting increase in turnaround loss. In addition, the output loading control setting will change the output circuit bandwidth (loaded Q) and therefore will also affect the amount of attenuation that out-of-band signals encounter passing into and out of the output circuit.

Second-harmonic traps or lowpass filters in the transmission line of either transmitter have little effect on the generation of intermodulation products. This is because the harmonic content of the interfering signal entering the output circuit of the transmitter has much less effect on IM_3 generation than the harmonic content within the nonlinear device itself. The resulting IM_3 products fall within the passband of the lowpass filters and outside the reject band of the second harmonic traps, so these devices offer no attenuation to RF intermodulation products. Figure 7.14-58 gives an overview of the various filtering options to prevent excessive IM_3 products.

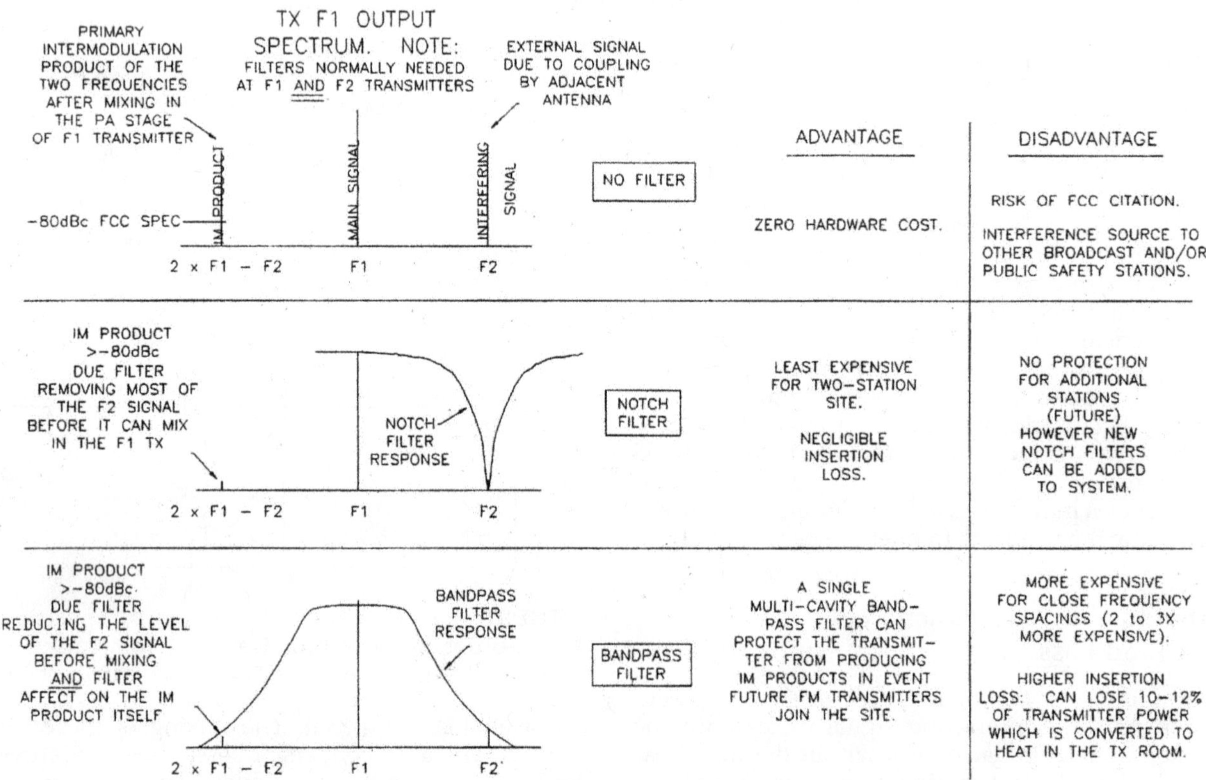

FIGURE 7.14-58 Three RF filtering options with advantages and disadvantages.

RF Intermodulation Between the FM and HD Radio Transmitters

Insufficient isolation between the analog FM transmitter and the digital HD Radio transmitter will cause undesirable RF intermodulation products. The required isolation can be difficult to achieve and depends on frequency in systems where space combining of the FM and HD Radio signals occurs in the antenna system. The desired isolation in a high-level or split-level system can be achieved by adjusting the impedances presented to the coupler ports.

Reduction of RF Intermodulation in Solid-State Transmitters

Depending on the topology of the combining system, broadband, solid-state transmitters may be more or less likely to generate RF intermodulation products than single-tube transmitters, which have a narrowband selective cavity in the output stage; however, some solid-state designs use balanced N-way module combiners together with a conventional 30-dB hybrid as the final output combining stage. This topology tends to provide a uniform 15-dB minimum turnaround loss to incoming RF interference, regardless of the frequency separation of the solid-state transmitter's FM carrier from an interfering carrier.

For frequency spacings closer than about 5 MHz from the interfering carrier, a solid-state transmitter using this combining technique actually produces fewer RF intermodulation IM_3 products than a single-tube FM transmitter in which close-in turnaround loss can be less than 6 dB. In some cases, a solid-state FM transmitter of this design may not require as much (or any) external RF filtering. In other cases, it may require more RF filtering than a tube/cavity transmitter. The turnaround loss of a narrowband tube/cavity transmitter is usually better than a broadband solid-state transmitter when there is a large frequency separation (> 5 MHz) between the carrier frequency and the interfering frequency. This is due to the skirt selectivity of the narrowband final amplifier cavity. Co-sited FM transmitters are likely to require external RF filtering to prevent the generation or transmission of unacceptable RF intermodulation products, regardless of whether they are of tube/cavity or solid-state design. The turnaround loss of the transmitter may also be improved by 20 dB or more with the addition of an external RF circulator to the output of the transmitter. The circulator is a unidirectional device that allows the transmitter output power to pass through it outbound with little attenuation to the antenna system while diverting incoming interfering RF signals to a reject load [11, 25].

FM TRANSMITTER OPERATIONAL MEASUREMENTS

The FCC considers certain parameters important enough to justify almost continuous observation of modulation level, carrier frequency, and output power level.

NRSC-5 FM + HD Radio RF Mask Compliance

With the introduction of HD Radio, it is important to keep the transmission system operation within the RF emission mask prescribed by the NRSC-5 Standard. This is due to the potential to generate RF intermodulation products between the host FM signal and the HD Radio sidebands. The NRSC-5 RF emission mask limits are shown in Figure 7.14-59. The spectrum analyzer setup parameters required to accurately measure the proposed NRSC-5B emission limits are listed in Table 7.14-5. The block diagram for the NRSC-5 test setup is shown in Figure 7.14-60. Note: Measurements are made by averaging the power spectral density (PSD) of the signal in a 1 kHz bandwidth over a 30-second period of time. The total FM power is defined as the reference at 0 dBc.

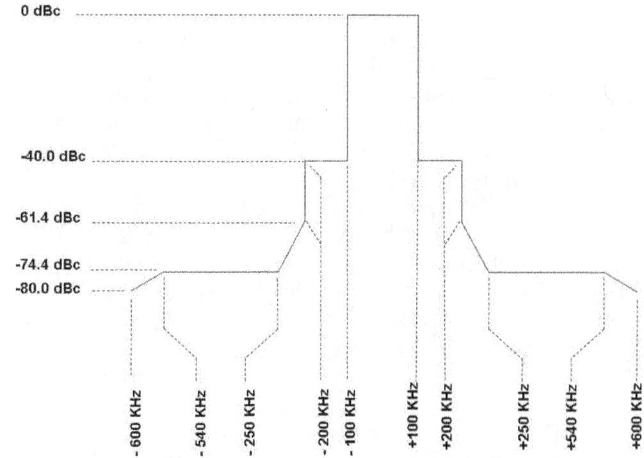

FIGURE 7.14-59 NRSC-5 RF emission limits.

TABLE 7.14-5
Nominal Settings of Controls on Spectrum Analyzer to Measure Emission Limits of Digital Radio Signal

Resolution bandwidth	1 MHz to set unmodulated FM carrier reference, then switch to 1 kHz
Span	2 MHz
Detect	Sample (not peak)
Video BW	1 kHz
Average	ON, at 100
Sweep	Leave in AUTO, sweep speed dependent on analyzer model
Sample points	400 minimum, use maximum points allowed
Marker	Set to peak
Marker	Reference level
Marker	Zero offset
Average type	Video

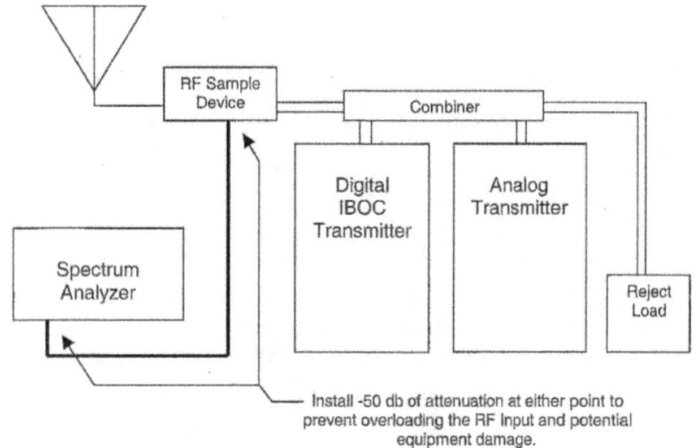

FIGURE 7.14-60 NRSC-5 RF emission measurement test setup.

FM Modulation Measurement

Measuring modulation levels can be accomplished with a broadcast-type modulation monitor or with a modulation analyzer test instrument. Most FM exciters have accurate built-in peak modulation displays for convenient setup and adjustment. Once the initial levels are correctly set, modern audio processing equipment will usually hold the modulation levels to the desired levels. Modulation monitors have a peak indicating device that can be preset to flash at the particular level of interest. This device should be used instead of the meter to determine peak modulation conditions of the transmitter.

The reason for setting a peak deviation limit is so the related occupied bandwidth does not increase to the point of interfering with stations on adjacent channels and causing excessive audio distortion in the FM receiver by exceeding the receiver's IF bandwidth. The FCC currently enforces the modulation limit by monitoring the instantaneous peak deviation of the station as displayed on an oscilloscope. This method of measurement does not exactly correlate with the station's occupied bandwidth because the duty-cycle of the modulation peaks is not taken into account. As a result, many sophisticated peak-limiting and overshoot control devices have appeared on the market to maximize loudness without exceeding the peak deviation limit. These devices remove the low-energy peaks that would extend beyond 100% modulation. The use of these devices can cause some degradation of the audio quality, and they might not be used if the method of modulation measurement were changed to one based on occupied bandwidth. Recently introduced modulation measurement devices ignore short duration overshoots and provide modulation level indications that more accurately reflect the resulting occupied bandwidth. The FCC now enforces occupied bandwidth requirements for the FM band by measuring compliance with the FM emissions mask defined in FCC Rules Section 73.317.

Carrier Frequency Measurement

The average carrier frequency must be measured with an accurate frequency monitor and maintained to within ±2,000 Hz of the assigned channel. These monitors fall into two categories: analog display of the frequency error from the nominal carrier frequency and digital display of the absolute carrier frequency. The trend is toward the digital counter because of its accuracy and ease of use. Current generation exciters utilize a high-stability crystal oscillator or an external GPS frequency reference that can maintain the carrier frequency accuracy an order of magnitude better than the FCC requirement.

Historically, the FCC required FM stations to utilize type-approved modulation and frequency monitors, but this requirement has been eliminated. Each station is still required to maintain its frequency, modulation, and audio performance within the FCC limits defined in Part 73 of the Rules, but the responsibility for selecting the method of measurement and type of measuring equipment is now up to the licensee. Good operating practice is to have on hand the equipment necessary to accurately measure the broadcast signal; for example, modern modulation analyzers provide frequency agility as well as greater measurement capability than the more specialized modulation monitors. General purpose frequency counters are now available with sufficient accuracy to measure the carrier, subcarrier, and stereo pilot frequencies directly. Spectrum analyzers provide a wide range of capability, including the measurement of harmonic and spurious frequencies at the carrier frequency, composite baseband, Bessel nulls, occupied bandwidth, stereophonic and SCA crosstalk, and synchronous AM.

Measurement of RF Power Output

The methods for determining RF output power are specified in the FCC Rules. An accurately calibrated directional wattmeter provides an acceptable way of making a direct measurement of RF output power. Until recently, the directional wattmeter was seldom used as the primary RF power determining method because of the requirement for recalibration to a traceable standard at regular intervals. Use of the indirect method of power measurement avoided this requirement. The FCC permits the use of the transmitter power output meter directly, if it is periodically calibrated by comparison with the indirect method, instead of with a dummy load and standard wattmeter.

Using the indirect method, the output power is calculated from a measurement of the DC input power multiplied by the efficiency factor of the final amplifier stage. The efficiency factor is provided by the transmitter manufacturer on the final test data sheet or in the instruction manual and must be applicable to the particular frequency and power level in use. The power input to the final amplifier stage is normally defined as the product of plate voltage and plate current to this stage. Multiple output stages that are combined for the total power must have their individual DC power inputs arithmetically summed to obtain the total power input. The directional wattmeter can be used as a check when compared to the power output calculated by the indirect method to determine if the efficiency factor has changed due to incorrect tuning, changing antenna conditions, or a weak output device.

Measurement of AM Signal to Noise

The perfect FM transmitter will have a constant output, regardless of FM modulation or power supply variations. In practice, there will be some residual amplitude modulation of the FM transmitter. Two types of AM signal-to-noise ratio (SNR) are of interest to the FM broadcast engineer:

- Asynchronous AM SNR measured without FM modulation is required by the FCC Rules and is primarily related to power supply ripple;
- Synchronous AM SNR or incidental carrier AM (ICAM) measured with FM modulation is not required by the FCC Rules and is related to the tuning and overall bandwidth of the system.

Asynchronous AM

Residual amplitude modulation of the transmitter output without FM modulation, due primarily to power supply ripple, is measured with an AM envelope detector. Most FM modulation monitors include an AM detector for this purpose. The detector must include 75-µs de-emphasis of its output. AM noise measurements must be made directly at the transmitter output (or an accurate sample of its output). No amplifying or limiting equipment may be used between the transmitter output and the AM detector, as this equipment would modify the residual AM noise level present.

The FCC Rules require residual AM noise to be 50 dB below the level that would represent 100% amplitude modulation of the carrier. Because the transmitter cannot be amplitude modulated, this reference must be established indirectly by a measurement of the RF carrier voltage. (Refer to the instructions of the detector manufacturer to determine the reference level.) If the transmitter is unable to meet the 50 dB requirement, the problem can usually be traced to a power supply component or to line imbalance in a three-phase system.

Synchronous AM

Synchronous AM is a measure of the amount of incidental amplitude modulation introduced onto the carrier by the presence of FM modulation. Although this measurement is not required by the FCC Rules, it provides information about the amplitude response and tuning of the transmitter. Measurement of synchronous AM also gives the station engineer an idea of the overall system bandwidth and whether the passband is positioned correctly. Because all transmitters have limited bandwidth, there will be a slight drop-off in power output as the carrier frequency is swept to either side of the center frequency. This slight change in RF output level follows the waveform of the signal being applied to the FM modulator causing AM modulation in synchronization with the FM modulation. The concept is similar to the slope detection of FM by an AM detector used in conjunction with a tuned circuit.

Synchronous AM measurements are made directly at the transmitter output (or an accurate sample of its output). No amplifying or limiting equipment may be used between the transmitter output and the AM detector, as nonlinearities in this equipment could modify the synchronous AM level present. Because the transmitter cannot be fully amplitude modulated, an equivalent reference level must be established indirectly by a measurement of the RF carrier voltage. (Refer to the instructions of the detector manufacturer to determine this reference level.) Generally, the reference level is determined by setting a carrier level meter to obtain a specific DC voltage level at the output of the detector diode without modulation. It is important, when making these measurements, that the test setup does not introduce synchronous AM and give erroneous readings that would cause the operator to mistune the transmitter to compensate for errors in the measuring equipment.

The input impedance of the envelope detector must provide a nearly perfect match so there is a very low VSWR on the sampling line. A significant VSWR on the sampling line will produce synchronous AM at the detector because the position of the voltage peak caused by the standing wave moves along this line with FM modulation. A thru-line type of directional coupler normally used to drive the wattmeter movement has the envelope detector diode built into the sampling element. This provides a DC component that the meter movement responds to plus the demodulated synchronous AM component to which the meter movement does not respond. If the thru-line element output is fed to an oscilloscope instead of the wattmeter movement, the synchronous AM waveform can be accurately measured. This approach eliminates the errors due to VSWR on the sampling line, because the detector is located at the sampling point. The manufacturer of the thru-line coupler can supply the special connectors and/or cables to connect its output to the oscilloscope. It is important to avoid hum pick-up from AC ground loops while making these low-level measurements. Both the thru-line element detector and the precision envelope detectors have some residual RF on their DC output, so an RF filter network may be required between the detector and the input the oscilloscope.

Most FM demodulators cannot be relied upon to make accurate synchronous AM noise measurements, so it is necessary to cross-check the demodulator reading directly against the demodulated output of a precision envelope detector. This can be done by first measuring the DC component of the waveform with a voltmeter or by DC coupling the scope input. The scope is then AC coupled, and the input sensitivity is increased until an accurate peak-to-peak measurement of the AC modulation component can be made. The peak-to-peak AC voltage is then divided by twice the DC component to obtain the voltage ratio; 20 $\log_{10}$ of the voltage ratio is the actual synchronous AM noise level in dB below equivalent 100% AM modulation. Multiplying the voltage ratio by 100 yields the percent of AM modulation. Note that the peak detected value of the carrier must be doubled to convert it to the peak-to-peak value of the carrier. The ratio of the peak-to-peak modulation component to the peak-to-peak carrier is then used to calculate the percentage of synchronous AM modulation [15].

Acceptable Level of Synchronous AM

Synchronous AM of 35 dB or more below equivalent 100% AM is considered to be acceptable because the limited bandwidth of the IF filter in the receiver will reintroduce higher levels of synchronous AM to the FM signal before demodulation. Higher levels of synchronous AM can cause increased chopping of the signal at the receiver near the limiting threshold under weak signal fringe area conditions and can exacerbate multipath problems. Excessive synchronous AM is also an indirect indication of passband-induced distortion problems that degrade stereo performance and increase SCA crosstalk.

Many older multi-tube transmitter designs currently in use will have as much as 6% (−30 dB) synchronous AM when simply tuned for best power output and efficiency even though the asynchronous AM (without modulation) may be better than −50 dB. Some of the newer single-tube transmitters can be adjusted for 50 dB or more suppression of synchronous AM. The synchronous AM level of virtually any FM transmitter can be improved by proper tuning techniques. An approximation to the overall system bandwidth can be related to the synchronous AM as shown in Table 7.14-6.

Limitations of Synchronous AM Measurements

Synchronous AM measurements are an indirect way of evaluating and optimizing FM performance. Even though synchronous AM measurements are a helpful aid to correctly tune an FM transmitter, these measurements tell only the amplitude response half of the total story. Transmitter tuning also affects the group delay (time) response that in turn affects the relative time delays of the higher order FM sidebands. Even though the amplitude response appears flat when the grid is heavily driven, the group delay (time) response still has a serious effect on the higher order FM sidebands.

Synchronous AM Versus Symmetrical Group Delay Response

Computer simulations as well as empirical measurements made on FM transmitters showed that group delay asymmetry results in much more distortion than asymmetrical amplitude response [17]. As long as the group delay response is symmetrical, the amount of synchronous AM will have little effect on the FM modulation performance and distortion. Most FM transmitters will exhibit a significant increase in synchronous AM when tuned for symmetrical group delay response even though this results in the best FM modulation performance. Tuning for minimum synchronous AM is a good starting point, but it is more desirable to finish tuning at the symmetrical group delay point. Fine tuning the input and output for minimum even-order harmonic distortion will optimize the group delay (time) response. Transmitters that utilize wideband solid-state IPAs will add less distortion to the FM signal because both the amplitude and group delay (time) response will be better than systems utilizing several tuned stages.

Tuning the Transmitter for Best Performance

When properly adjusted as discussed in the preceding sections, modern power amplifiers can operate with high reliability and high-power efficiency without compromising subcarrier performance. All optimization should be made with the APC system disabled so the APC will not chase the adjustment to keep the output power constant. The transmitter should be connected to the normal antenna system rather than to a dummy load. This is because the resistance and reactance of the antenna will be different from the dummy load, and the optimum tuning point of the transmitter will shift between the two different loads. The tuning sequence is described next.

TABLE 7.14-6
Approximate System Bandwidth as Related to Synchronous AM

Synchronous AM Below 100% AM (+75 kHz at 1 kHz FM)	Approximate Bandwidth of Transmitters (−3 dB point)	RF Level Variation at Receiver Limiter	
		(%)	(dB)
−30 dB	410 kHz	6.32	0.57
−35 dB	550 kHz	3.54	0.31
−40 dB	730 kHz	2.00	0.18
−45 dB	1.00 MHz	1.12	0.10
−50 dB	1.34 MHz	0.64	0.06
−55 dB	1.82 MHz	0.36	0.03
−60 dB	2.46 MHz	0.20	0.02

Initial Tuning and Loading

The transmitter is first tuned for normal output power and proper efficiency according to the manufacturer's instructions. The meter readings should closely agree with those listed on the manufacturer's final test data sheet if the transmitter is being operated at the same frequency and power level into an acceptable load.

Input Tuning and Matching

The input tuning control should first be adjusted for maximum grid current and then fine-tuned interactively with the input matching control for minimum reflected power to the driver stage. Note that the point of maximum grid current may not coincide with the minimum reflected power to a solid-state driver. This is because a solid-state driver may actually produce more power at certain complex load impedances than into a 50 Ω resistive load. The main objective during input tuning is to obtain adequate grid current while providing a good match (minimum reflected power) to the coaxial transmission line from the driver. In the case of an older transmitter with a tube driver integrated into the grid circuit of the final amplifier, the driver plate tuning and the final grid tuning will be combined into one control that is adjusted for maximum grid current.

Output Tuning

The output tuning control adjusts the resonant frequency of the output circuit to match the carrier frequency. As resonance is reached, the plate current will drop while both the output power and screen current rise together. Under heavily loaded conditions this dip in plate current is not very pronounced, so tuning for a peak in the screen current is often a more sensitive indicator of resonance.

Output Loading

There is a delicate balance between screen voltage and output loading for amplifiers utilizing a tetrode tube. Generally, there is one combination of screen voltage and output loading where peak efficiency occurs. At a given screen voltage, increasing the amplifier loading will result in a decrease in screen current, while a decrease in loading will result in an increase in screen current. As the screen voltage is increased to get more output power, the loading must also be increased to prevent the screen current from reaching excessive levels. Further increases in screen voltage without increased loading will result in a screen overload without an increase in output power.

Automatic Power Control Headroom

APC feedback systems are utilized in many transmitters to regulate the power output around a predetermined set-point with variations in AC line voltage or changes in other operating parameters. Modern FM broadcast transmitters may utilize a high gain tetrode as the final amplifier stage with adjustment of the screen voltage providing fine adjustment of the output power. For each power output level there is one unique combination of screen voltage and output loading that will provide peak operating efficiency. If the screen voltage is raised above this point without a corresponding increase in loading, there will be no further increase in power output with rising screen voltage and screen current. If the screen voltage is raised without sufficient loading, a screen current overload will occur before the upward adjustment in power output is obtained.

To avoid this problem, tune the transmitter with slightly heavier loading than necessary to achieve the desired power output level and allow for about 5% headroom in adjustment range. The output loading can be adjusted for a peak in output power of 5% over the desired level, and then the screen voltage can be reduced enough to return to the desired level. This procedure will allow headroom for an APC system controlling screen voltage and will result in about a 1% compromise in efficiency, but it will ensure the ability to increase power output up to 5% without encountering a screen overload.

Centering the Passband

A simple method for centering the transmitter passband on the carrier frequency involves adjustment for minimum synchronous AM. If the bandpass is narrow or skewed, increased synchronous amplitude modulation of the carrier will result. A typical adjustment procedure is to FM modulate 100% at 1 kHz and fine-adjust the transmitter's grid tuning and output tuning controls for minimum 1-kHz AM modulation as detected by a wideband envelope detector (diode and line probe). The FM modulating frequency of 1 KHz is used instead of the 400 Hz so the audio highpass filter in the audio analyzer can be used to eliminate the AC line frequency related asynchronous component from the synchronous AM component. It is helpful to display the demodulated output from the AM detector on an oscilloscope while making this adjustment.

Note that, as the minimum point of synchronous AM is reached, the demodulated output from the AM detector will double in frequency to 2 kHz. This is because the fall-off in output power is symmetrical about the center frequency, causing the amplitude variations to go through two complete cycles for every one FM sweep cycle as shown in Figure 7.14-61. It should be possible to minimize synchronous AM while maintaining output power and efficiency in a properly designed power amplifier. If an oscilloscope is not available for direct observation of the demodulated AM waveform, the 19 kHz bandpass filter and metering circuit used to measure pilot injection level in a stereo modulation monitor may be used as a tuning aid to center the passband. In this case, the main carrier is FM modulated with a 9.5-kHz tone (without pilot or any other modulation). If the transmitter is tuned

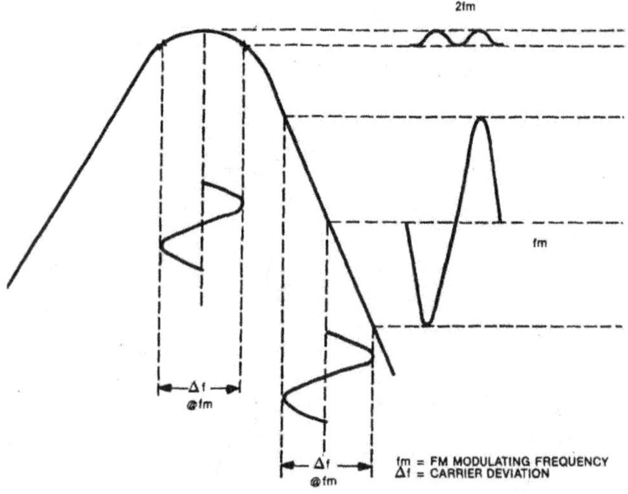

FIGURE 7.14-61 Synchronous AM waveforms.

for symmetrical AM response, the demodulated AM signal will have a strong second harmonic component at 19 kHz that falls within the passband of the pilot metering circuit. The output of the AM detector is fed into the composite baseband input of the stereo modulation monitor. The transmitter is then tuned for a maximum reading on the pilot injection metering position.

Effect of Transmitter Tuning on the FM Sidebands

The higher-order FM sidebands will be slightly attenuated in amplitude and shifted in time (group delay) as they pass through the final amplifier stage. The alterations in the sideband structure that are introduced by the amplifier passband result in distortion after FM demodulation at the receiver. The amount of distortion depends on the available bandwidth versus the modulation index being transmitted. For a given bandwidth limitation, the distortion can be minimized by centering the passband of the amplifier around the signal being transmitted. This will cause the amplitude and group delay errors to affect both the upper and lower sidebands equally (symmetrically).

Tuning an amplifier for minimum plate current or for best efficiency does not necessarily result in a centered passband. One way to center the amplitude passband is to tune the amplifier for minimum synchronous AM modulation while applying FM modulation to the transmitter. Because the circuit topology of most transmitters exhibits a difference in tuning between the symmetrical amplitude response and the symmetrical group delay response, FM modulation performance can be further improved by tuning for symmetrical group delay rather than for minimum synchronous AM. The symmetrical group delay tuning point usually does not coincide exactly with the symmetrical amplitude tuning point but rather falls between the point of minimum synchronous AM and the point of maximum efficiency.

The transmitter may be tuned for minimum intermodulation distortion in left-only or right-only stereo transmissions. Stereo separation will also vary with tuning. For stations employing a 67 kHz SCA, transmitter tuning is important for minimizing crosstalk into the SCA. Modulate one channel only on the stereo generator to 100% with a 4.5 kHz tone. This will place the lower second harmonic (L–R) stereo sideband on top of the 67 kHz SCA. Activate the SCA at normal injection level without modulation on the SCA. Tune the transmitter for minimum output from the SCA demodulator. This adjustment can also be made by listening to the residual SCA audio while normal stereo programming is being broadcast.

A more sensitive test is to tune for minimum even order harmonic distortion that will result in a symmetrical group delay response and will optimize distortion, separation, and crosstalk.

Modern power amplifiers have been designed to operate without compromising subcarrier performance. By providing broadband matching circuits, adjustment of these transmitters for optimum FM modulation performance (such as minimum distortion, minimum crosstalk, maximum separation) is repeatable and stable.

The following field adjustment techniques are listed in ascending order of sensitivity:

- Tune for minimum synchronous AM noise;
- Tune for minimum IMD3 in the left or right channel only;
- Tune for minimum crosstalk into the unmodulated SCA subcarrier;
- Tune for minimum even-order harmonic distortion (symmetrical group delay).

In any of these tests, the grid tuning is frequently more critical than the plate tuning. This is because the impedance match into the input capacitance of the grid becomes the bandwidth limiting factor. Even though the amplitude response appears flat when the grid is heavily driven, the group delay (time) response has a serious effect on the higher order FM sidebands [11, 12, 15, 16].

OPTIMUM TUNING VERSUS EFFICIENCY

VHF amplifiers often exhibit a somewhat unusual characteristic when tuning for maximum efficiency. The highest efficiency operating point does not exactly coincide with the lowest plate current because the power output continues to rise on the inductive side of resonance coming out of the dip in plate current. If the amplifier is tuned exactly to resonance, the plate load impedance will be purely resistive, and the load line will be linear. As the output circuit is tuned to the inductive side of resonance, the plate load impedance becomes complex, and the load line becomes elliptic instead of linear because the plate current and plate voltage are no longer in phase. Apparently, best efficiency occurs when the phase of the instantaneous plate voltage slightly leads the plate current. This effect is believed to be caused by the nonlinear gain characteristics of the power amplifier tube operating on an elliptic load line.

Care of Power Tubes

The operating life of high-power vacuum tubes can be extended by proper care. Most high-power tubes utilize a directly heated cathode composed of a thoriated tungsten filament structure. The key points to extending the life of RF power tubes include:

- Store tubes upright, along the axis of symmetry, not on their side, to keep the internal elements concentrically aligned.
- Use care when handling tubes to prevent mechanical shocks to the delicate internal structure. Do not set a tube on a hard surface without padding.
- Keep the tube seals and anode cooler free of dust and dirt by weekly cleaning even in a clean environment.
- Keep a spare tube on hand and rotate the tubes every few months to help keep the chemical "gas-getter" active so the tubes remain gas-free.
- Keep a regular record of all tube operating parameters so any trend of changes will be noticeable. If a tube fails during the warranty period, this data will be necessary to receive credit on a replacement tube.
- Monitor the filament voltage on a true RMS-responding instrument and log any changes for future reference. The sampling point for this voltage measurement should be located as close to the tube's filament contacts as possible to minimize errors due to voltage drops in the filament wiring.

A properly operated tube will gradually lose emission from the cathode until it is no longer useful because the emissive material has been consumed. The carcass of the tube can then be rebuilt with a new cathode and recycled back into service. Tube life is not directly related to plate dissipation (within the ratings) but is related to the filament operating temperature (filament voltage) and the current density (milliamperes per heater watt) emitted by a given size filament. This means that operating a given tube type at a lower filament voltage and plate current will proportionately increase the life of the tube. For directly heated thoriated tungsten filaments, the plate current should be less than 4 milliamperes per watt of heater power for extended life.

Normally a new tube will deliver full output at a reduced filament voltage. By operating the tube at the optimum filament voltage, the filament life can be significantly extended. The optimum value may be found by slowly reducing the filament voltage from the manufacturer's rated value until the RF power output drops about 2% and then increasing the filament voltage until the RF power increases back up 1%. Informational bulletins from tube manufacturers on extending tube life permit the filament voltage to be reduced more than 5% below the manufacturer's rating as long as operation is closely monitored to stay above the point where there is a 1% drop in power output [20]. A brand new tube should be operated at the full rated filament voltage for the first 200 hours before the voltage is reduced to the optimum value for long life. This will ensure that the "gas-getter" is properly activated. As the tube ages, the filament voltage will have to be increased to stay at the optimum value until RF output power cannot be maintained at or above the rated value of filament voltage. At this point, the tube's useful life comes to an end. Check the manufacturer's data sheets and application notes that are often enclosed with the tube. Guides to the proper care of power tubes and forced air cooling are listed in references [8] and [20].

INSTALLATION CONSIDERATIONS

Adequate planning and care in the installation of an FM + HD Radio broadcast transmitter and associated equipment will help avoid many problems that may be difficult and expensive to correct later; for example, poor grounds and ground loops may cause high electrical noise levels.

FM + HD Radio Transmitter Plant

The HD Radio transmission facility is usually an extension of the existing analog FM transmitter site. Consideration must be given for the additional floor space, AC power, and cooling required by the digital transmitter, RF power combiner, and RF reject load. Figure 7.14-62 shows a simplified block diagram of a typical FM + HD Radio transmission facility. For more information about converting a transmission facility to HD Radio, see references [26], [33], and [34].

Wiring the Transmitter Plant

Separate metallic shielded conduits or troughs should be provided for the audio and the AC wiring. A third conduit should be used if computer logic levels are employed for equipment control. High-voltage and high-current wiring should be isolated from low-level audio, control logic, and metering signals to prevent the coupling of transients that could produce fault conditions. Conduits or wiring troughs may be either overhead or below the cabinets. The AC wiring should be well separated from the audio pairs to prevent the induction of unwanted hum and noise into the audio circuits.

Audio shields should be grounded at only one point to prevent ground loops in the shields. This point may have to be found experimentally to give the lowest noise pick-up. The equipment racks and transmitter should be connected together by copper straps at least 2 inches wide tied to a good earth ground at one point.

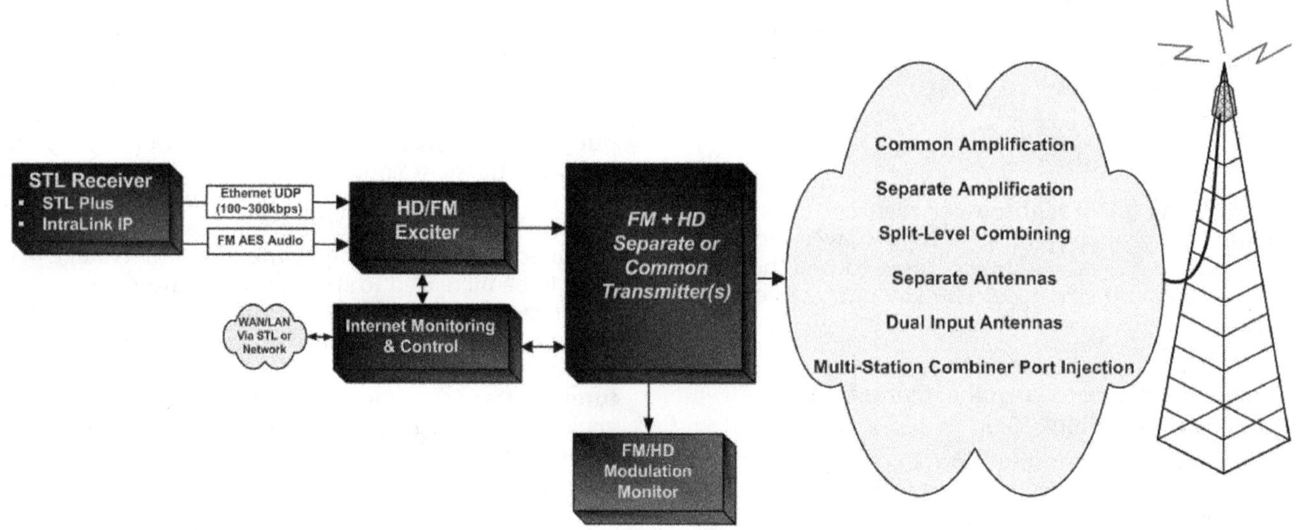

FIGURE 7.14-62 Simplified block diagram of a typical FM + HD Radio transmission facility.

If a good ground screen is not available, a satisfactory ground can be provided by driving four or more copper ground rods 8–10 feet long into the ground spaced about 3 feet apart. These ground rods should be tied together with a wide copper strap. The straps connecting the equipment to the earth ground should be as short and direct as practical. See Chapter 11.2, "Facility Grounding Practice and Lightning Protection," for more information on grounding techniques.

It is often difficult to remove VHF RF from the equipment by grounding because at FM carrier frequencies, nearly any connection to an earth ground has an appreciable impedance. The best way to keep RF out of sensitive low-level circuits is by keeping them enclosed within an RF shield and by filtering leads that enter the shielded unit. Filters in the audio lines may be appropriately sized shunt capacitors and small bi-filar RF chokes that add common mode inductance without adding differential inductance that would affect audio performance.

For stereo transmission, it is necessary to keep the L and R audio lines phased properly. To ensure proper monaural compatibility, correct audio polarity must be maintained throughout the station from the microphones and other audio sources through all of the audio equipment to the stereo generator audio input terminals. Stereo phone line pairs or separate RF studio-to-transmitter links should also be checked for correct polarity and equal phase delay. The transmitter equipment should be located and arranged to provide sufficient clearance around the front, sides, and rear for easy access during servicing and maintenance.

Transmitter Cooling

Almost all FM broadcast transmitters require forced air cooling to remove heat from the output stage and other assemblies within the cabinet. An important consideration in locating the transmitter is the provision for adequate cooling air. If the overall efficiency of the analog FM transmitter is about 50%, the transmitter will generate about the same number of kilowatts of heat as it does RF power output. HD Radio transmitter systems that require linear amplification will operate at lower efficiency than a comparable power FM transmitter. The extra heat load from the HD Radio transmitter and related signal combining reject load must be considered when planning the air-conditioning system.

Figure 7.14-63(a) shows a transmitter located in an air-conditioned room. This type of closed-loop system requires no special ducting and has the advantage that the transmitter intake air is usually much cleaner than outside air. The transmitter exhaust air places a substantial heat load on the air conditioner during the summer, but it becomes a source of heat in the winter. The transmitter manufacturer can usually supply data on the number of cooling BTUs required, so the proper size of air conditioner may be selected. This method is frequently used with the lower power transmitters. A protective system should be provided to prevent overheating of the transmitter if the air conditioner fails.

Figure 7.14-63(b) shows a transmitter located in a wall separating an air-conditioned room and a ventilated but not air-conditioned room. A large exhaust fan is provided in the ceiling to remove the rising hot air, and an adequate cool air intake is provided in the lower portion of an outside wall. A filtering system is required to keep the transmitter interior clean.

Figure 7.14-63(c) shows a transmitter located in an air-conditioned room with intake and exhaust air ducts to the outside. An auxiliary blower or fan may be required to overcome pressure drop in the ducting. This type of system requires careful design to make sure the air flow through the transmitter is not impeded by the duct work. Additional air interlocks may be required to protect the transmitter from a failure of the

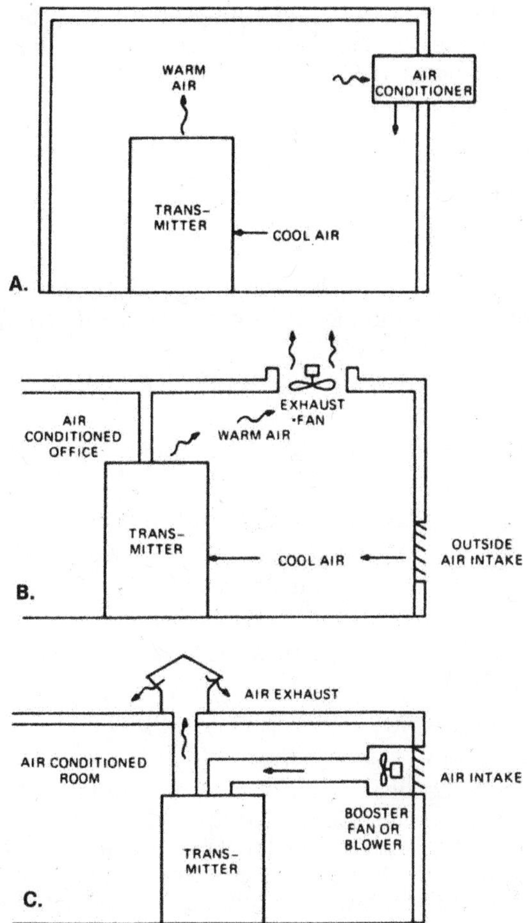

FIGURE 7.14-63 Three methods of providing cooling air for the transmitter (see text).

auxiliary blower. The air intake and exhaust openings to the outside should be provided with rain shields, insect screens, and dust filters as dictated by the environment. The location of the air intake and exhaust openings should be arranged so wind pressure will not impede the air flow.

Liquid Cooling

There is renewed interest in liquid-cooling systems for high-power FM + HD Radio common amplification transmitters. Liquid-cooling systems offer the ability to remove large amounts of heat from the transmitter room without significant changes to the existing HVAC system, and they are already widely used in UHF digital television transmitter systems.

The higher efficiency of liquid media heat removal from the power amplifier devices greatly reduces the liquid-cooled heat sink temperatures and allows higher peak and average power output from the same RF transistors that would be otherwise limited by the higher operating temperature of an air-cooled heat sink.

The energy consumption of a liquid-cooling system is a small fraction of the energy consumption of an air-cooling system that moves a similar amount of heat from the building. If a new facility or an existing facility is converted to all liquid-cooled transmitters, the amount of air-conditioning can be dramatically reduced to remove only the small amount of latent heat from the transmitters and auxiliary equipment racks.

Liquid-cooled transmitters typically have twice the power density, per unit volume, of comparable air-cooled technology. This is because the volume needed by the open area of the air-cooled heat sinks is compressed down to the thickness of the liquid-cooled chiller plates. A 2:1 reduction in floor space and vertical height (volume) can significantly reduce the cost of facility space especially in high rise buildings.

Within the transmitter power block, coolant is circulated through chiller plates attached directly to the power FETs in the seven hot-swappable, hot-pluggable power amplifier (PA) modules (Figure 7.14-64) by means of a distribution manifold/chiller plate attached to the combiner isolation loads (Figures 7.14-65 and 7.14-66). This provides evacuation of up to 86% of the waste heat generated by the transmitter. Figure 7.14-67 is a block diagram of a typical, redundant liquid-cooling system used in a solid-state FM or HD/FM transmitter.

Figure 7.14-68 illustrates the size and system components of a liquid-cooling pump module. Figure 7.14-69 illustrated a typical liquid-to-air heat exchanger for a liquid-cooled transmitter.

Sharing Cooling Systems

There is a significant benefit to sharing a common liquid-cooling system across multiple transmitters. In some cases, it may even be possible to share the building's liquid-to-air heat exchanger, thereby eliminating the need for separate liquid-to-air heat exchangers. This would be particularly attractive for transmission facilities atop high buildings. Another cooling option is the use of a geothermal cooling loop that is trenched into the ground adjacent to the transmitter building.

Variable Speed Cooling Systems

Regardless of whether an air-cooling system is used or a liquid-cooling system is used, the energy consumption of the cooling system can be significantly reduced by carefully matching the system capacity to the ambient conditions. Variable speed cooling systems adaptively utilize thermal feedback to consume only as much energy as required for the current temperature and heat load.

Transmission System Total Cost of Ownership

There is high interest among broadcasters and RF network operators to reduce their overall transmission facility cost of ownership. The total cost of ownership (TCO) includes much more than the initial cost, energy consumption and maintenance of the transmission equipment. The acquisition, maintenance, operating,

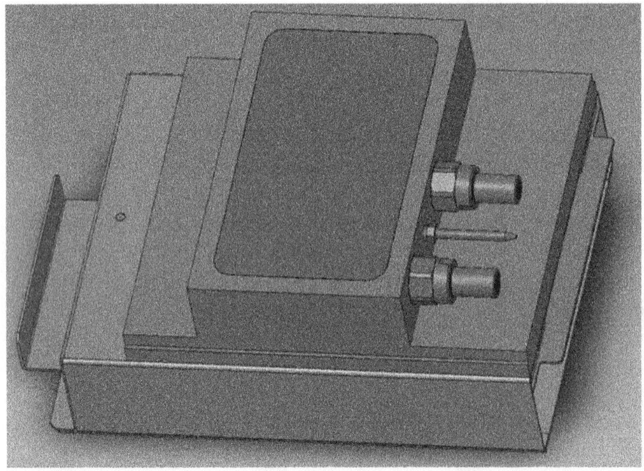

FIGURE 7.14-64 GatesAir FLX FM hot-pluggable, liquid-cooled PA Module with chiller plate.

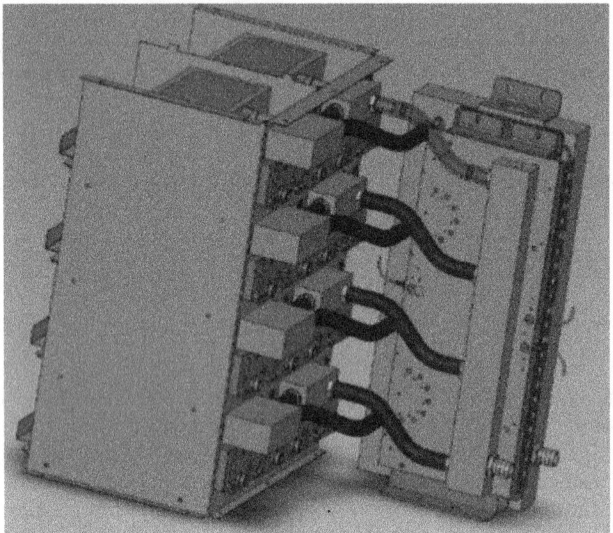

FIGURE 7.14-65 GatesAir FLX liquid coolant distribution system.

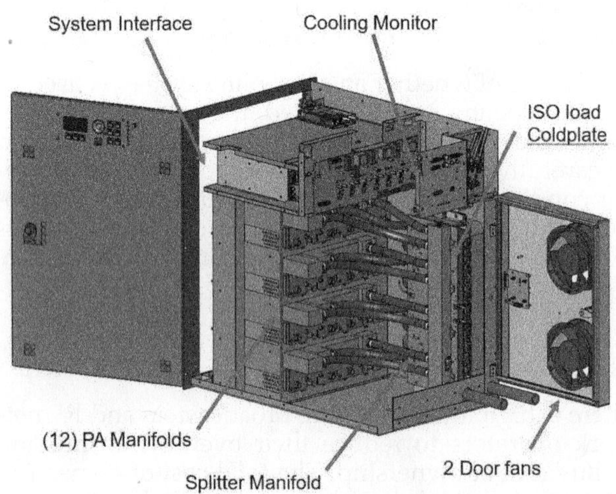

FIGURE 7.14-66 GatesAir FLX liquid coolant distribution system.

and training costs of all the other required systems in the transmission facility must also be considered to arrive at lowest TCO.

Contributors to TCO include the following:

- Acquisition and installation cost of transmission equipment;
- Operating cost of transmission equipment power consumption;
- Maintenance cost of transmission equipment;
- Acquisition and installation cost of facility cooling system (HVAC, etc.);
- Operating cost of facility cooling system power consumption (HVAC, etc.);
- Maintenance cost of facility cooling system (HVAC, etc.);
- Periodic replacement cost for facility cooling system (HVAC, etc.);
- Acquisition and maintenance cost of power conditioning or other special equipment;
- Acquisition or recurring lease cost of floor space required;
- Training costs for all required equipment.

As outlined earlier, the total cost of ownership must take into account far more than the initial purchase cost of the transmission equipment. Many facility design and purchase decisions are still being made on the basis of the transmitter and antenna costs in isolation from the many additional costs associated with the rest of the supporting infrastructure required to make the facility work. To understand the true cost of ownership, all of the preceding facility cost items must be taken into account.

Transmitter selection can have the following impact on TCO:

- Physical size and foot print (ft^2);
- Power density of transmitter (watts/ft^3);
- Type of cooling—air or liquid;
- Operating efficiency (AC to RF) of the transmitter;
- AC power requirements—(1) phase or (3) phase (polyphase power may be costly to bring to some sites);
- AC input voltage and voltage regulation requirements;
- Maintainability—hot-swappable modules, air or liquid filter replacement, all other maintenance tasks;
- Commonality of transmitter types—spares sharing, common training, and operational requirements;
- N + 1 transmitter system architecture can reduce the number of backup transmitters required.

The choice between air or liquid cooling for the transmission equipment can make a significant difference in the overall cost of ownership. Although the initial purchase cost of a liquid-cooled transmitter is

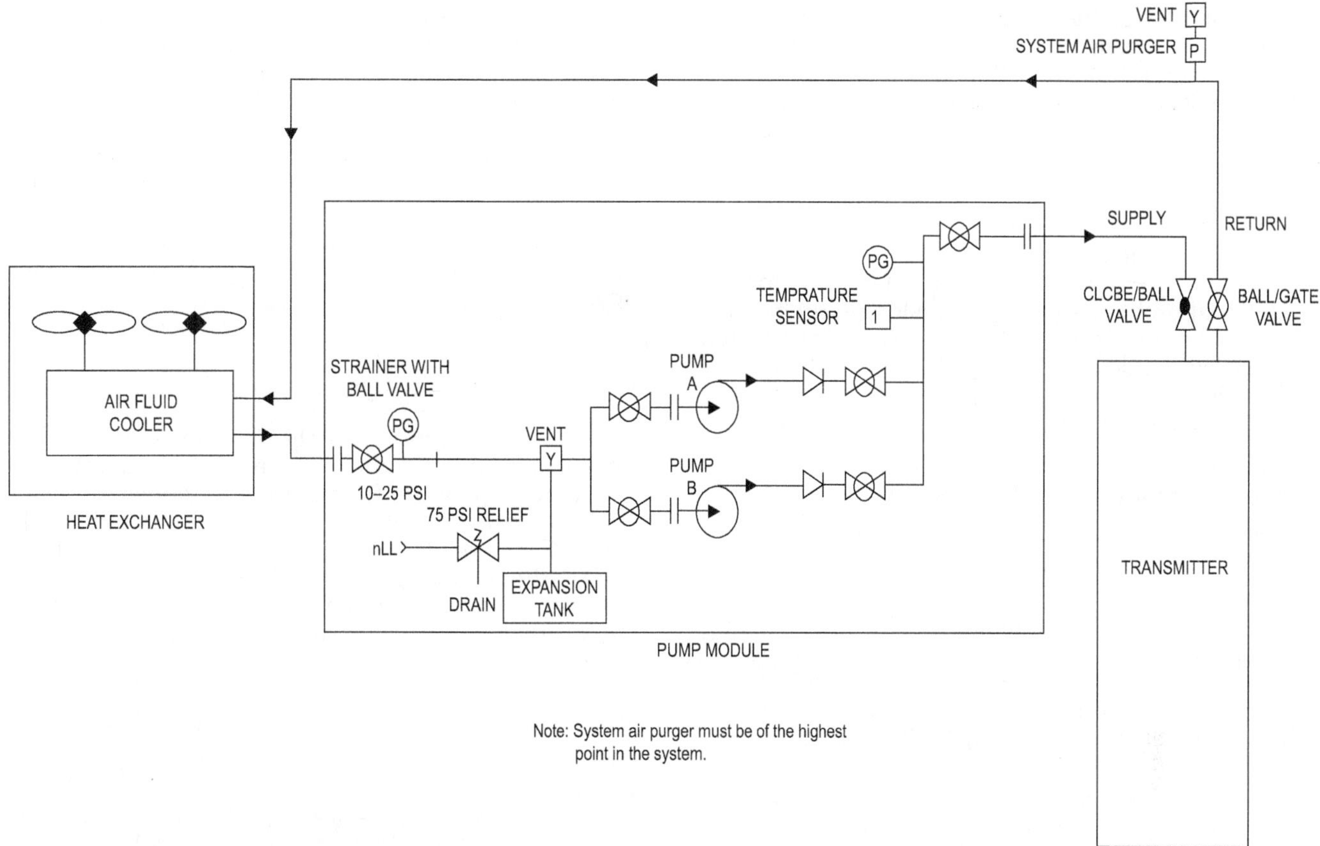

FIGURE 7.14-67 Typical liquid-cooling system block diagram.

FIGURE 7.14-68 High-efficiency liquid-cooling pump module.

FIGURE 7.14-69 Vertical air flow—liquid-to-air heat exchanger.

typically higher than an equivalent air-cooled transmitter, the elimination of most of the costs associated with air-conditioning can make the break-even period short on the cost difference of the transmitter and provide significant long term operational savings.

Many broadcast facility operators are finding that the cost of purchasing and operating air-conditioning systems is increasing. The typical air-conditioning unit requires ongoing maintenance and eventual replacement on a 5- to 8-year cycle. The combined operating, maintenance, and replacement costs become a significant part of the overall facility operating cost. Liquid-cooled transmitters have significantly lower energy consumption than the combined energy consumption of an air-cooled transmitter and the required air-conditioning.

Preventive Maintenance

Preventive maintenance is equipment inspection and maintenance performed at regular intervals before an operational problem develops. The long-term benefits are great because potential problems are discovered and solved while they are still easily manageable. A checklist of a few typical preventive maintenance items for an FM transmitter plant might include:

- Weekly overall internal and external cleaning and inspecting for damage or excessive wear;
- Lubricating motors, tuning gears, and other moving parts at intervals recommended by the manufacturer;
- Checking and logging all meter readings, including daily checks of filament voltage and comparing these readings with the previous set of readings as an aid to diagnosing a developing problem (this can be done automatically with digital transmitter control systems and remote control systems);
- Regularly exercising the automatic power control and any other mechanical servo systems;
- Checking the antenna lighting and deicer systems;
- Checking the transmission line pressurization and VSWR;
- Checking all air filters in the transmitter plant and cleaning or replacing as required;
- Checking the proper operation of all monitoring and remote control equipment.

Air filters should be periodically cleaned or replaced according to the transmitter manufacturer's instructions. This is important because dust or insect clogged air filters may reduce the cooling air flow enough to cause overheating of some components. The probability of component failure increases rapidly when cooling is insufficient. Particular attention should be paid to removing dirt and dust from high-voltage components during regular maintenance after all power is removed and all components are discharged. Dust should be cleaned from the transmitter with a suitable brush and vacuum cleaner or as otherwise recommended by the transmitter manufacturer. Weekly cleaning is usually sufficient. In addition, good overall housekeeping will pay big dividends in the long run by keeping equipment clean and free of problems that would otherwise be caused by dirt build-up.

Maintenance Systems

The key to making any maintenance program work is to set up formal systems for checklists, logging, parts inventory management, and repair scheduling. These systems provide the discipline required to keep the maintenance routine accurate and complete. Each station should develop a system suited for the particular physical plant involved. When there is more maintenance and repair work requiring attention than there is time to do it all, set priorities for completing each item so no item is forgotten. Accurate notebooks, in either hard copy or software database format, describing all installation and maintenance work are a very helpful part of any maintenance system especially when work spans many years.

SUMMARY

Because of the importance of FM transmission, whether analog or digital, to radio broadcasting, this chapter has emphasized the need for a thorough understanding of the technical principles of frequency modulation theory, the methodology for implementing digital IBOC (HD Radio) systems, and how digital signal processing techniques have made significant improvements in the signal quality, efficiency, and reliability of broadcast transmitters.

References

[1] Lessing, L., *Man of High Fidelity: Edwin Howard Armstrong*, Bantam Publishing, New York, 1959.

[2] Westman, H. P. (Ed.), *Reference Data for Radio Engineers*, Fifth ed., Howard W. Sams and Co., Indianapolis, IN, 1970.

[3] Terman, F. E., *Electronic and Radio Engineering*, Fourth ed., McGraw-Hill, New York, 1955.

[4] Clarke, K. K. and Hess, D. T., *Communications Circuits: Analysis and Design*, Addison-Wesley, Reading, MA, 1978.

[5] Rohde, U. L., *Digital PLL Synthesizers: Theory and Design*, Prentice Hall, Englewood Cliffs, NJ, 1983.

[6] Krauss, H. L., Bostian, C. W., and Raab, F. H., *Solid-State Radio Engineering*, John Wiley & Sons, New York, 1980.

[7] Howe, H., Jr., Simplified design of high power, N-way, inphase power divider/combiners, *Microwave J.*, 51–53, December, 1979.

[8] *Care and Feeding of Power Grid Tubes*, Eimac Division of CPI, Inc., Palo Alto CA, 2001 (www.cpii.com/eimac/).

[9] Hershberger, D. and Weirather, R., *Amplitude Bandwidth, Phase Bandwidth, Incidental AM, and Saturation Characteristics of Power Tube Cavity Amplifiers for FM*, Harris Corp. Broadcast Division, Quincy, IL, 1982.

[10] Hnatek, E. R., *Design of Solid-State Power Supplies*, Second ed., Van Nostrand Reinhold, New York, 1981.

[11] Mendenhall, G. N., *A Study of RF Intermodulation Between FM Broadcast Transmitters Sharing Filterplexed or Co-Located Antenna Systems*, Broadcast Electronics, Inc., Quincy, IL, 1983.

[12] Mendenhall, G. N., *The Composite Signal: Key to Quality FM Broadcasting*, Broadcast Electronics, Inc., Quincy, IL, 1981.

[13] Lyles, J.T.M. and Shrestha, M. B., *Transmitter Performance Requirements for Sub-Carrier Operation*, Broadcast Electronics, Inc., Quincy, IL, 1984.

[14] Bruene, W. B., An inside picture of directional wattmeters, *QST Mag.*, April, 1959.

[15] Mendenhall, G. N., *Techniques for Measuring Synchronous AM Noise in FM Transmitters*, Broadcast Electronics, Inc., Quincy IL, 1988.

[16] Shrestha, M. B., *The Significance of RF Power Amplifier Circuit Topology on FM Modulation Performance*, Broadcast Electronics, Inc., Quincy, IL, 1990.

[17] FMSIM, *FM Stereo Simulation and Analysis Program*, Quantics, Nevada City, CA, 1990.

[18] Anthony, E. J., *Optimum Bandwidth for FM Transmission*, Broadcast Electronics, Inc., Quincy, IL, 1989.

[19] Shrestha, M. B., *Personal Safety Considerations With Broadcast Transmitters*, Broadcast Electronics, Inc., Quincy, IL, 1989.

[20] Artigo, R., Extending transmitter tube life, *Broadcast Manage. Eng. Mag.*, March, 1982 (revised March 1990 and reprinted as Eimac Application Bulletin AB-18, Eimac Division of Varian Corp., San Carlos, CA).

[21] Twitchell, E. R., A digital approach to an FM exciter, *IEEE Trans. Broadcast.*, 38, 106–110, 1992.

[22] AES3–2003, AES Recommended Practice for Digital Audio Engineering: Serial Transmission Format for Two-Channel Linearly Represented Digital Audio Data (revision of AES3–1992, including subsequent amendments).

[23] Mendenhall, G. N., Implementing an Uncompressed Digital Path From the Studio to the On Air Signal, Harris Corp. Broadcast Division, Quincy, IL, 1997.

[24] Dittmer, T. W., *Advances in Digitally Modulated RF Systems*, Harris Corp. Broadcast Division, Quincy, IL, 1997.

[25] Agnew, D., *FM Harmonic Measurement Correction Factors*, Harris Corp. Broadcast Division, Mason, OH, 2003.

[26] Mullin, K., *A Planning Guide: Determining the Best IBOC Migration Path for Your AM or FM Radio Station*, Harris Corp. Broadcast Division, Quincy, IL, 2003.

[27] Cabrera, G., *Reducing FM IBOC Transmission Costs With the Proper Configuration and Linearization Techniques*, Harris Corp. Broadcast Division, Mason, OH, 2003.

[28] Cabrera, G., *Dual FM Injection Improves IBOC High-Level Combining*, Harris Corp. Broadcast Division, Mason, OH, 2004.

[29] Cabrera, G., *Improving Efficiency with Split-Level Combining*, Harris Corp. Broadcast Division, Mason, OH, 2004.

[30] Cabrera, G., *Understanding Split-Level Combining*, Harris Corp. Broadcast Division, Mason, OH, 2005.

[31] Fluker, S., *Split Level Combining Explained*, Cox Radio Corp., Orlando, FL, 2004.

[32] Mattsson, A., *Linearizing HD Radio Transmitters: A Technology Survey*, Harris Corp., Broadcast Division, Mason, OH, 2005.

[33] Xperi Corporation, BOC White Papers, http://hdradio.com/broadcasters/engineering-support/iboc-white-papers.

[34] GatesAir, Media Center White Papers, www.gatesair.com/media-center/papers.

CHAPTER

7.15

VHF (FM) Radio Transmitters

JEFF WELTON
Nautel Limited

OVERVIEW

When Major Edwin H. Armstrong first patented frequency modulation (FM) in 1933 with the first experimental station, W2XMN going on air in 1937, there was significant resistance to the concept—the government resisted it, World War II delayed development and it was moved to a new band in 1945 (the current 88–108 MHz band used in North America and much of the world). Since then, however, there have been many developments, greatly improved efficiency of operation in tube amplifiers, development and improvement of solid state amplifiers and movement from signal generation using Voltage Controlled Oscillator circuitry, and discrete components to Direct to Channel signal generation with Digital Signal Processors. In this chapter, the theory of all stages of the FM broadcast transmitter will be discussed, as well as the evolution of each.

An FM modulator generates a constant frequency deviation of the carrier with a constant amplitude modulating signal, independent of the frequency of the modulating signal. The phase deviation of the carrier produced by a frequency modulator decreases as the modulating frequency is increased even though the level of the modulating voltage is held constant. The net effect is that the frequency modulator behaves as if it were a phase modulator with a 6 dB/octave falling slope on the modulating signal input. In FM broadcasting, the RF carrier should have frequency deviation that is proportional to the amplitude of the modulating signal but independent of the frequency of the modulating waveform. The instantaneous frequency (rate of change of phase) of the RF output wave differs from the carrier frequency by an amount proportional to the instantaneous amplitude of the modulating waveform. For example, consider a 100-MHz carrier wave frequency modulated by a 1000-Hz audio tone and assume that a 1-volt input to the modulator causes ±20 kHz of frequency deviation on the positive and negative peaks of this tone. If the audio input amplitude is increased to 2 volts, the peak deviation will become ±40 kHz varying in sine-wave fashion from one peak of deviation to the other and back again at the 1000-Hz rate. In FM broadcasting, 100% modulation results in a peak frequency deviation of ±75 kHz of the RF carrier. [1]

TRANSMITTER STAGES

As seen in Figure 7.15-1, any radio transmitter can be broken down into five primary stages. The stages that are unique from one frequency range to the next are the Exciter, RF Amplifier stage and the Combiner/Filter stage. These are the items that will be discussed in this chapter, although some reference may be provided to Control and Power Supply stages, as well.

EXCITER STAGE

The exciter stage is one of the sections of the FM transmitter that has seen some of the greatest changes over the years. Early FM exciters used discrete components and Voltage Controlled Oscillators (VCO), where the audio signal modulated a reference signal, which was then mixed with a control signal to generate the FM carrier. In later exciters, the control signal also would incorporate a feedback signal from the exciter output, in order to provide Automatic Frequency Control (AFC). A simplified circuit is shown in Figure 7.15-2.

In this circuit, the audio is pre-amplified then fed into a varactor diode that makes up part of an oscillator circuit, in conjunction with frequency multipliers to approach the desired carrier frequency. This oscillator circuit will typically employ VCO circuitry to allow connection of an external correction input. When that is employed, an additional oscillator, with variable switching to allow setting to any frequency in the FM band, makes up a Phase Lock Loop (PLL) circuit that stabilizes the primary oscillator. A feedback loop goes into the PLL to provide frequency correction as needed. The signal then goes through a buffer amplifier and can then be fed either directly to a transmitter or through a lowpass filter in the event the exciter is being connected directly to an antenna.

As integrated circuit technology advanced, FM exciters evolved accordingly. The next generation used Direct Digital Synthesis (DDS) to create an extremely stable frequency reference that was easily programmable for the desired carrier frequency. Many of these still used VCO circuitry for Automatic Frequency Control, but the end result was an exciter that was much more stable with respect to frequency, had better signal to noise specifications for both AM and FM noise (typically 55–60 dB for AM synchronous noise and up to or beyond 90 dB for FM noise) and were much more transparent to the modulating audio, with distortion specifications as low as 0.1% THD+N. In addition, advanced IC technology allowed for more comprehensive control and monitoring, as well as increased flexibility with respect to the ability to change frequency, utilize alternate audio inputs and monitor all operational aspects of the unit.

Further improvements in the areas of Digital Signal Processors permitted the development of the first Direct to Channel FM exciters, where there was no need for VCOs, or the associated tracking filters and occasionally complex alignment procedures. With these systems, the audio is converted to a digital signal and fed into the DSP, where it is digitally added to the carrier to create the modulating waveform. These systems have the advantage of even better audio specifications, improved noise specs and typically a lower parts count with respect to critical components, helping to reduce the Mean Time Between Failure (MTBF). There are still power supply and amplifier stages, however, compared to the maintenance required on earlier generation exciters, reliability was typically increased significantly.

The next logical step was to minimize parts count even further by packaging the entire exciter section onto a microprocessor-based circuit board—this allowed for even greater monitoring and control, full integration into transmitter control circuitry, and the

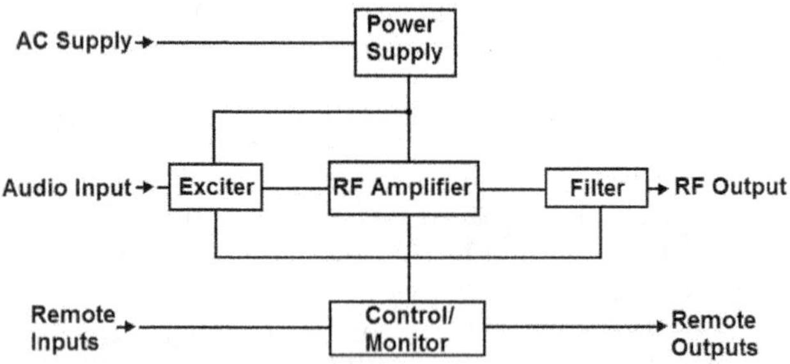

FIGURE 7.15-1 Simplified Block Diagram of a Radio Transmitter

J. Welton

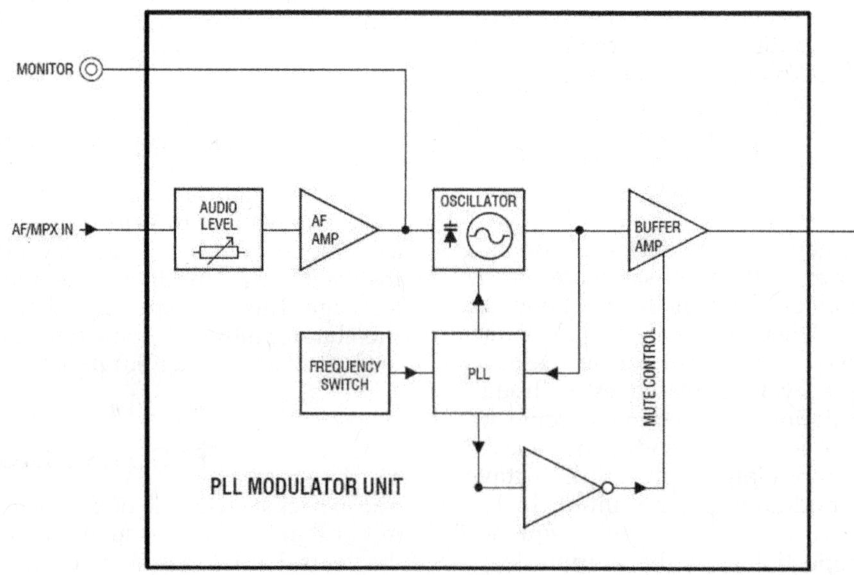

FIGURE 7.15-2 Simplified Block Diagram of an Early FM Exciter

Nautel Limited

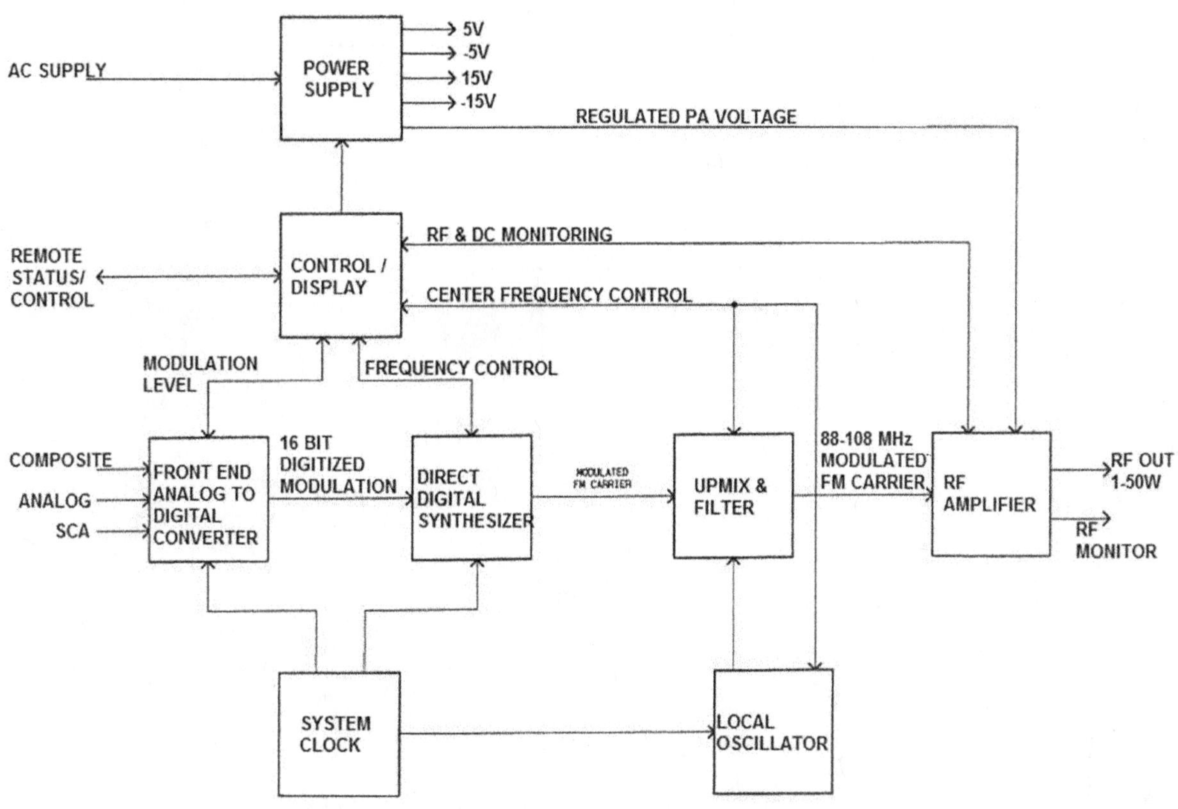

FIGURE 7.15-3 Simplified Block Diagram of DDS-type FM Exciter

Nautel Limited

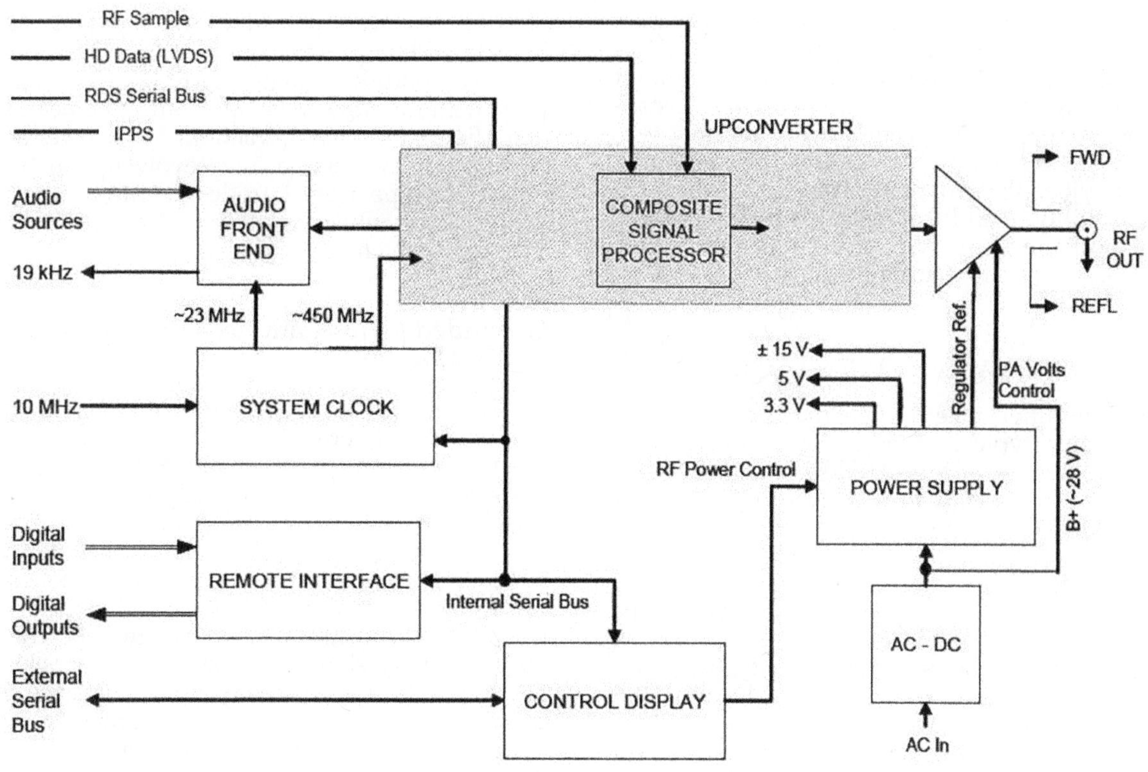

FIGURE 7.15-4 Simplified Block Diagram of Early DSP-type FM Exciter

Nautel Limited

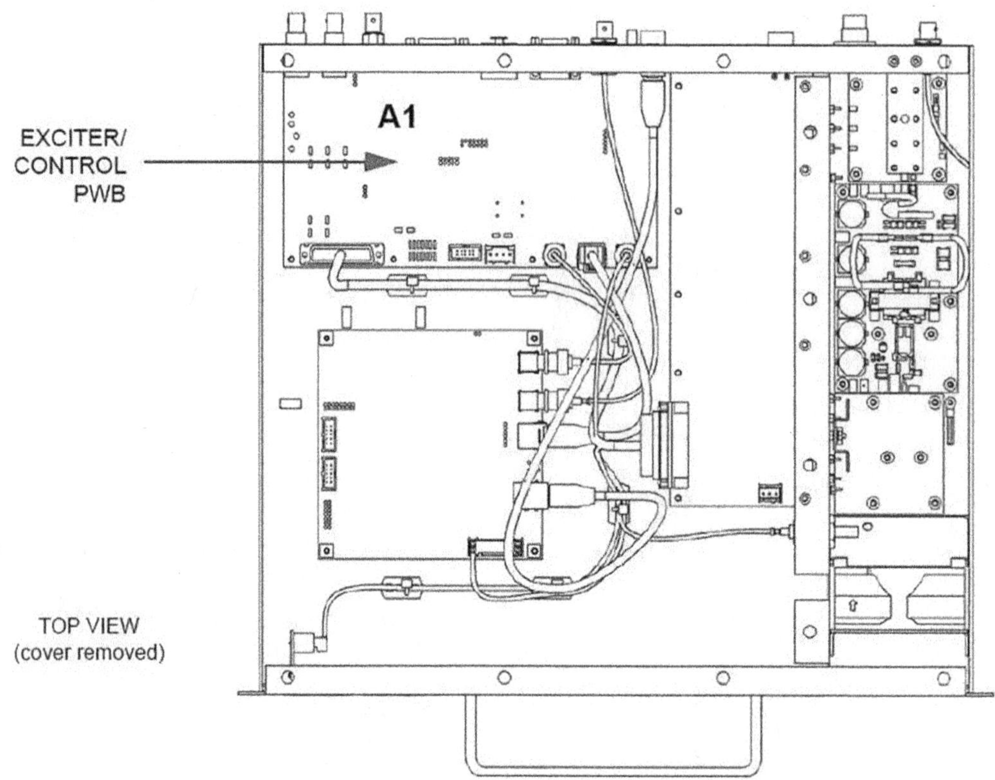

FIGURE 7.15-5 Mechanical Diagram of Direct to Channel FM Exciter

Nautel Limited

ability to switch between traditional analog operation or various combinations of analog and DRM+ or HD Radio technology. With this implementation, upgrades that were traditionally done by cutting track or installing jumpers and changing components are now typically handled by simply upgrading the software/firmware payload that drives the exciter. In addition, it allows the ability to add features such as automation, playlists, audio backup with real time monitoring, RDS and SCA generators, to name a few.

POWER AMPLIFIER STAGE

The power amplifier stage is where the low level signal from the exciter (typically 5–50 W for analog systems and in the milliwatts for digital (DRM+ or HD Radio™ technology)) is amplified to the desired Transmitter Power Output (TPO) needed to achieve the station's licensed Effective Radiated Power (ERP). ERP is a function of TPO, less coaxial cable loss, multiplied by the gain of the associated antenna and is beyond the scope of this chapter.

There are several power amplifier configurations in common use today, comprising both tube and solid state configurations, with and without Intermediate Power Amplifier (IPA) sections. When used, the IPA is simply a pre-amplifier that amplifies the exciter output to a level sufficient to drive the final power amplifier stage. For the most part, IPA stages in use today consist of solid state amplifiers, although there are still a few tube based IPAs in use.

The amplitude of an analog FM signal remains constant with modulation so efficient, nonlinear, Class C amplifiers can be used. Vacuum-tube power amplifier circuits for FM service have evolved into two basic types. One type uses a tetrode or pentode tube in a grid driven circuit while the other uses a high-mu triode in a cathode driven (grounded-grid) circuit.

Grounded-Grid (Cathode-Driven) Triode Amplifiers

The high-mu, zero-bias triodes used in cathode-driven (grounded-grid) FM amplifiers were originally developed for linear single sideband (SSB) amplifiers. Their characteristics are well adapted to FM broadcast use because the circuit is simple, and no screen or grid bias power supplies are required. Figure 7.15-6 shows the basic circuit configuration. In this case, the grid is connected directly to chassis ground. In this case, the grid is connected directly to a chassis ground. The output tank circuit is a shorted coaxial cavity that is loaded by the tube output and stray circuit capacitance. A small capacitor is used for trimming the tuning, and another small variable capacitor is used for adjusting the loading. A pi-network matches the 50 Ω input to the tube cathode.

The triodes are usually operated in the less efficient, Class B mode to achieve maximum power gain, which

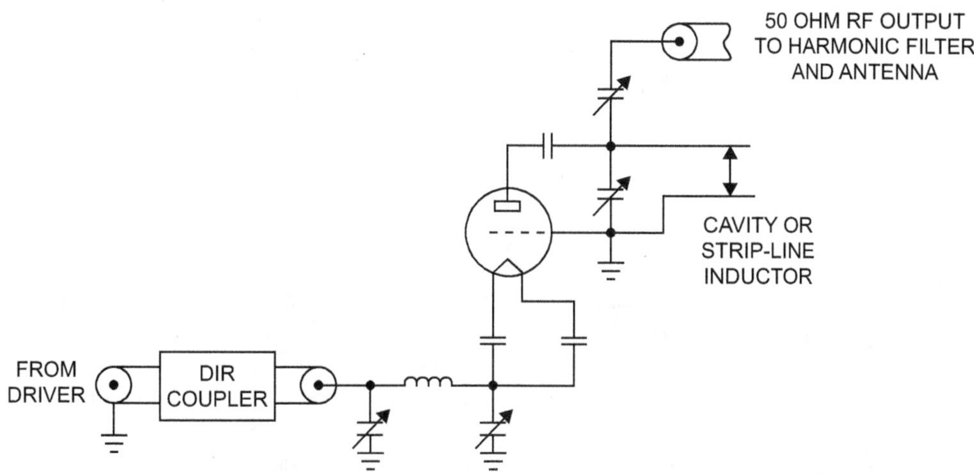

FIGURE 7.15-6 Cathode-Driven, Grounded-Grid, Triode Power Amplifier

NAB

is on the order of 20 (13 dB). They can be driven into high-efficiency Class C operation by providing negative grid bias. This increases the plate efficiency but also requires increased drive power and a bias power supply.

Most of the drive power into a grounded-grid amplifier transfers through the tube and appears in the stage's output. This increases the apparent efficiency so the efficiency factor given by the manufacturer may be higher than the actual plate efficiency of the tube. The true plate efficiency is determined by dividing the output power by the total input power, which includes both the DC plate input power (plate current times plate voltage) and the RF drive power.

Because most of the drive power transfers through the tube, any changes in loading of the output circuit will also affect the input tuning and driver stage. With RF drive voltage on the cathode (filament) of the tube, some means of decoupling is required to block the RF from the filament transformer. One method employs high-current RF chokes, as the inductance can be very low at this frequency range. The other commonly used method feeds the filament power through the input tank circuit inductor. Cathode-driven stages are normally used only for the higher power stages. The first stage in a multi-tube transmitter is nearly always a tetrode because of its higher power gain.

Grounded-Grid Versus Grid-Driven Tetrode Operation

There are several trade-offs between the performance of grounded-grid and grid-driven configurations of a tetrode PA with respect to gain, efficiency, amplitude bandwidth, phase bandwidth, and synchronous AM under equivalent operating conditions:

- When driving a grounded-grid PA into saturation, the bandwidth is limited by the output cavity. The PA bandwidth in the grid-driven amplifier is limited by the input circuit Q, which is determined by the input capacitance and the amount of swamping resistance.
- Output bandwidth under saturation can be improved in either configuration by reducing the plate voltage. This involves a trade-off in efficiency with a smaller voltage swing. The bandwidth improvement can be obtained with a loss of PA gain and efficiency.
- A grounded-grid saturated PA improves bandwidth over a grid-driven saturated PA at the expense of amplifier gain.
- The best performance for FM operation is obtained when the amplifier is driven into saturation where little change in output power occurs with increasing drive power. Maximum efficiency also occurs at this point.
- The phase linearity in the 0.5 dB bandwidth is better in a grid-driven configuration. The grounded grid PA exhibits a more nonlinear phase slope within the passband yet has a wider amplitude bandwidth. This phenomenon is due to interaction of the input and output circuits because they are effectively connected in series in the grounded-grid configuration. The neutralized, grid-driven PA provides more isolation between these networks, so they behave more like independent filters.

Grid-Driven Tetrode and Pentode Amplifiers

Transmitters using tetrode amplifiers usually have one less amplifying stage than transmitters using triodes. Because tetrodes have higher power gain, they are driven into Class C operation for high plate efficiency. Against these advantages is the requirement for neutralization, along with screen and bias power supplies. Pentode amplifiers have higher gain than their tetrode counterparts. The circuit configuration and bias supply requirements for the pentode are similar to the tetrode because the third (suppressor) grid is tied directly to

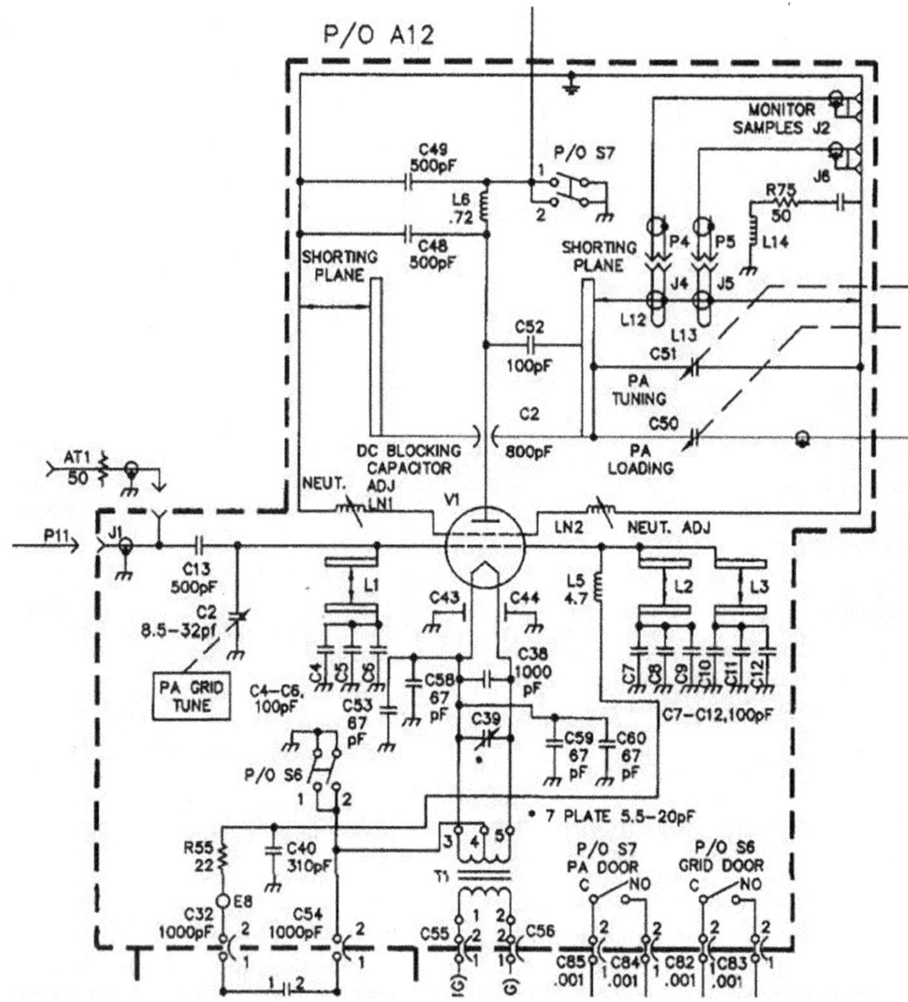

FIGURE 7.15-7 Final RF Amplifier Section, Grid-Driven Tetrode Amplifier

Continental Electronics 816HD-20 manual

ground. The additional isolating effect of the suppressor grid eliminates the need for neutralization in the pentode amplifier.

Impedance Matching Into the Grid

The grid circuit is usually loaded (swamped) with added resistance. The purpose of this resistance is to broaden the bandwidth of the circuit by lowering the circuit Q and to provide a more constant load to the driver. It also makes neutralizing less critical so the amplifier is less likely to become unstable with varying output circuit loading. Cathode or filament lead inductance from inside the tube, through the socket and filament capacitors to ground can heavily load the input circuit. This is caused by RF current flowing from grid to filament through the tube capacitance and then through the filament lead inductance to ground. An RF voltage is developed on the filament that in effect causes the tube to be partly cathode driven. This undesirable extra drive power requirement can be minimized by series resonating the cathode return path with the filament bypass capacitors or by minimizing the cathode to ground inductance by using a specially designed tube socket using thin-film dielectric "sandwich" capacitors for coupling and bypassing.

High-power, grid-driven Class C amplifiers require a swing of several hundred RF volts on the grid. To develop this high-voltage swing, the input impedance of the grid must be increased by the grid input matching circuit. Because the capacitance between the grid and the other tube elements may be 150 picofarads (pF) or more, the capacitive reactance at 100 MHz will be very low unless the input capacitance is parallel resonated with an inductor. Figure 7.15-8 shows two popular methods of resonating and matching into the grid of a high-power tube. Both methods can be analyzed by recognizing that the desired impedance transformation is produced by an equivalent L network. In Figure 7.15-8a, a variable inductor (L_{in}) is used to raise the input reactance of the tube by bringing the tube input capacitance (C_{in}) almost to parallel resonance. Parallel

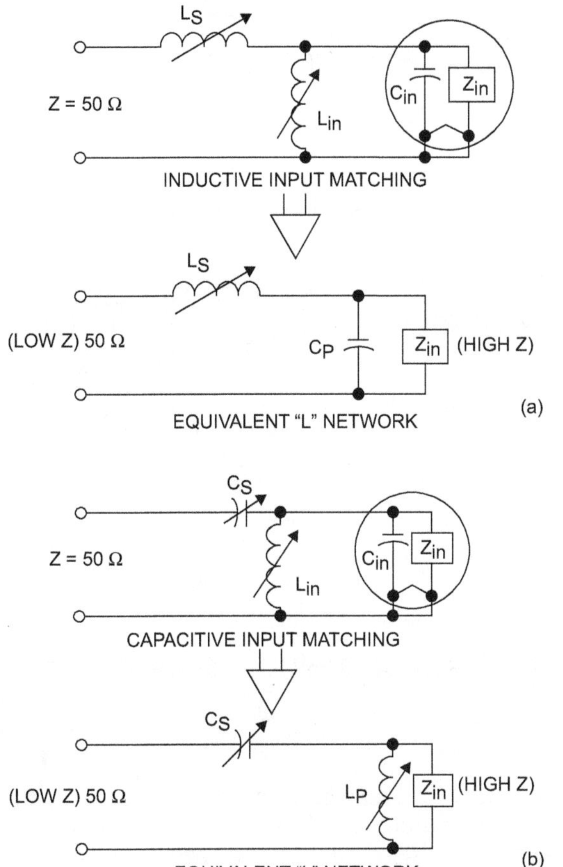

FIGURE 7.15-8 (a) Inductive Matching (b) Capacitive Matching

NAB

resonance is not reached because a small amount of parallel capacitance (C_p) is required by the equivalent L network to transform the high impedance (Z_{in}) of the tube down to a lower value through the series matching inductor (L_s). This configuration provides a lowpass filter by using part of the tube's input capacitance to form (C_p). Figure 7.15-8b uses variable inductor (L_{in}) to take the input capacitance (C_{in}) past parallel resonance so the tube's input impedance becomes slightly inductive. The variable series matching capacitor (C_s) forms the rest of the equivalent L network. This configuration is a highpass filter.

Neutralization

Cathode-driven, grounded-grid amplifiers utilizing triodes do not require neutralization. The grid-to-ground inductance, both internal and external to the tube, should be kept low to maintain this advantage. Omission of neutralization will allow a small amount of interaction between the output circuit and the input circuit through the plate-to-filament capacitance. This effect is not very noticeable because of the large coupling between the input and output circuits through the electron beam of the tube.

Grid-driven, high-gain tetrodes require accurate neutralization for best stability and performance. This is particularly true if the amplifier is to be operated in Class AB linear mode for common amplification of FM + HD Radio signals. Self-neutralization can be accomplished by placing a small amount of inductance between the tube screen grid and ground that is usually in the form of several short, adjustable-length straps. The RF current flowing from plate to screen in the tube also flows through this screen lead inductance. This develops a small RF voltage on the screen, of the opposite phase, which cancels the voltage fed back through the plate-to-grid capacitance.

This method of lowering the self-neutralizing frequency of the tube works only if the self-neutralizing frequency of the tube/socket combination is above the desired operating frequency before the inductance is added. Feedback neutralization utilizes a small coupling capacitor, usually in the form of a small plate located near the anode of the tube. The sample of the RF voltage from the anode intercepted by this plate is coupled through a 180° phase-shift network into the grid circuit. This technique has the advantage of providing neutralization over a very broad range of frequencies, if implemented correctly and stray reactances are minimized. Special attention must also be given to minimizing the inductances in the tube socket by integrating distributed bypass capacitors into the socket and cavity deck assembly. Pentodes normally do not require neutralization because the suppressor grid effectively isolates the plate from the grid. [2]

Power Amplifier Output Circuits

The output circuit usually consists of a high-Q (low loss) transmission line cavity, strip line, or lumped inductor that resonates with the tube output capacitance. A means of trimming the tuning and a means of adjusting the coupling to the output transmission line must also be provided by the output circuit. The tank circuit loaded Q is kept as low as practical to minimize circuit loss and to maintain as wide an RF bandwidth as possible.

Quarter-Wavelength Cavity

The quarter-wavelength coaxial cavity is the compact and popular PA output circuit illustrated in Figure 7.15-9. The tube anode is coupled through a DC blocking capacitor to a shortened quarter-wavelength transmission line. The tube's output capacitance is brought to resonance by the inductive component of the transmission line that is physically less than a quarter-wavelength long. Plate tuning can be accomplished either by adding end-loading capacitance at the high impedance end of the line with a variable capacitor or by changing the position of the ground plane at the low impedance end of the line. The plate tuning capacitor may be a sliding or rotating plate near the anode of the tube. The center conductor of the transmission line (air exhaust chimney) is at DC ground while the anode of the tube operates at a high RF and DC potential. High

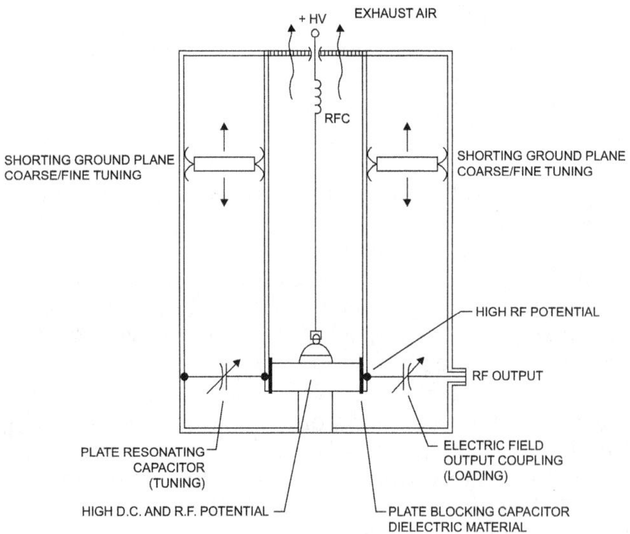

FIGURE 7.15-9 The Quarter-Wavelength Cavity

NAB

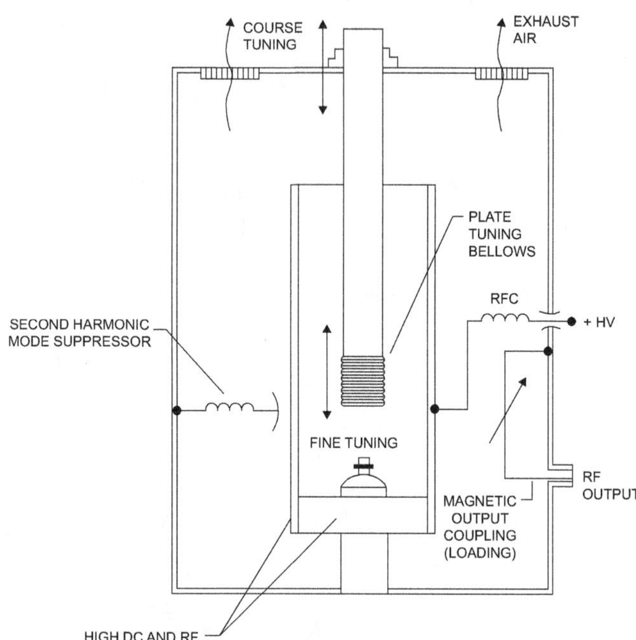

FIGURE 7.15-10 The Half-Wavelength Cavity

NAB

voltage is fed through an isolated quarter-wavelength decoupling network inside the chimney to the anode of the tube, while the plate-blocking capacitor prevents DC current flow from the anode into the chimney. RF power may be coupled from a quarter-wavelength cavity to the transmission line by a capacitive probe placed at the high RF voltage point located at the anode end of the quarter-wave line. The loaded Q of this circuit varies with the degree capacitive coupling.

Another method of coupling power from the quarter-wavelength cavity uses a tuned loop located near the grounded (high current) end of the line. In this case, the tuned loop operates as both an inductive and a capacitive pick-up device. The quarter-wavelength cavity has approximately twice the operating bandwidth of the half-wavelength cavity.

Folded-Half-Wavelength Cavity

Another approach to VHF power amplification uses the folded-half-wave cavity design illustrated in Figure 7.15-10. The DC anode voltage is applied to the lower portion of the plate line through a choke at the RF voltage null point. The half-wave line is tuned by mechanically expanding or contracting the physical length of a flexible extension (bellows) on the end of the secondary transmission line stub, which is located concentrically within the primary transmission line (air exhaust chimney). Coarse frequency adjustment is accomplished by presetting the depth of the top secondary section of plate line into the tank cavity. Power may be coupled from the half-wave line by an inductive loop located in the strong fundamental magnetic field near the center of the cavity.

Other power amplifier configurations may use lumped components or hybrid combinations with distributed transmission line elements to achieve similar results. The discrete circuit elements are chosen for their individual inductance or capacitance instead of being operated in a purely quarter-wave or half-wave mode. Stray inductance and capacitance add to the component values resulting in the hybrid nature of these circuits. Regardless of the specific configuration, the output circuit must transform the high resonant plate impedance down to the output transmission line impedance of 50 Ω. References at the end of this chapter give detailed information about the design of tube-type RF power amplifiers [1, 2, 3].

Solid State Technology

Historically, the poor efficiency that resulted from traditional approaches to solid state FM (in the vicinity of 50%) was a deterrent to solid state VHF amplifier development. This was resolved through changes in amplifier biasing and as semiconductor technology developed in the early 1990s. In addition, the high levels of intermodulation experienced with broadband solid state FM technology have also been a concern. Earlier transmitters resolved this by using tuned circuitry in the amplifier stages, as seen in Figure 7.15-11.

One of the benefits of most solid state FM designs is the reduction in AM noise, both synchronous and asynchronous, over a tube-type transmitter, meaning that broadcasters can finally realize the full potential of their exciter. In this figure, the amplifier uses a 50-volt dual device in a push pull configuration to produce 250 watts. The device is biased Class C/E for increased efficiency. The input and output circuits each use a transformer and an inductor/capacitor to match the FET inputs and outputs to 50 ohms. The input and output circuits require tuning for the frequency used, making the amplifiers fairly wideband with respect to their frequency capability.

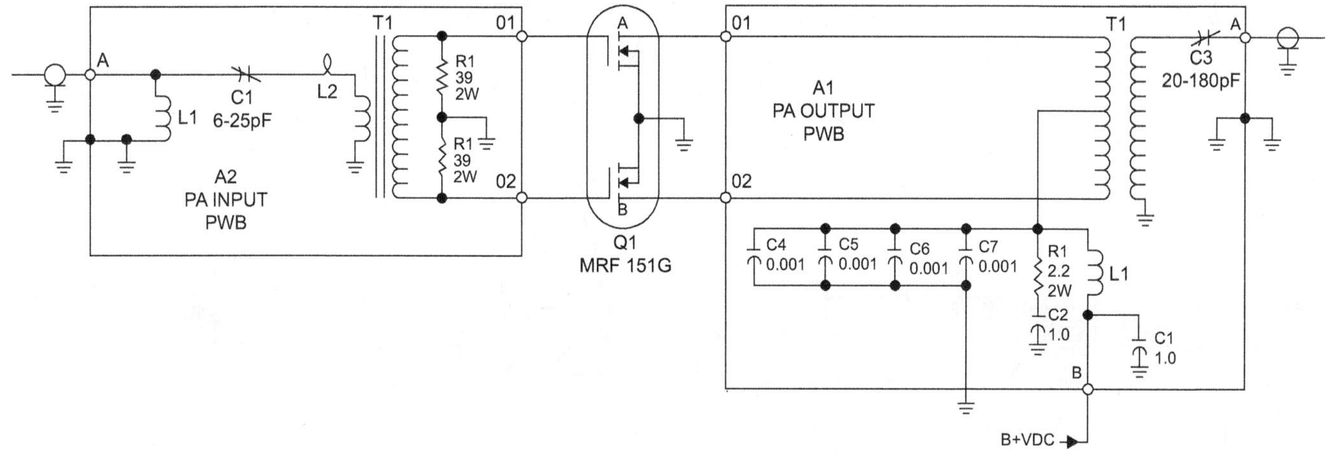

FIGURE 7.15-11 Schematic Diagram, Solid State Class C/E Amplifier

Nautel Limited

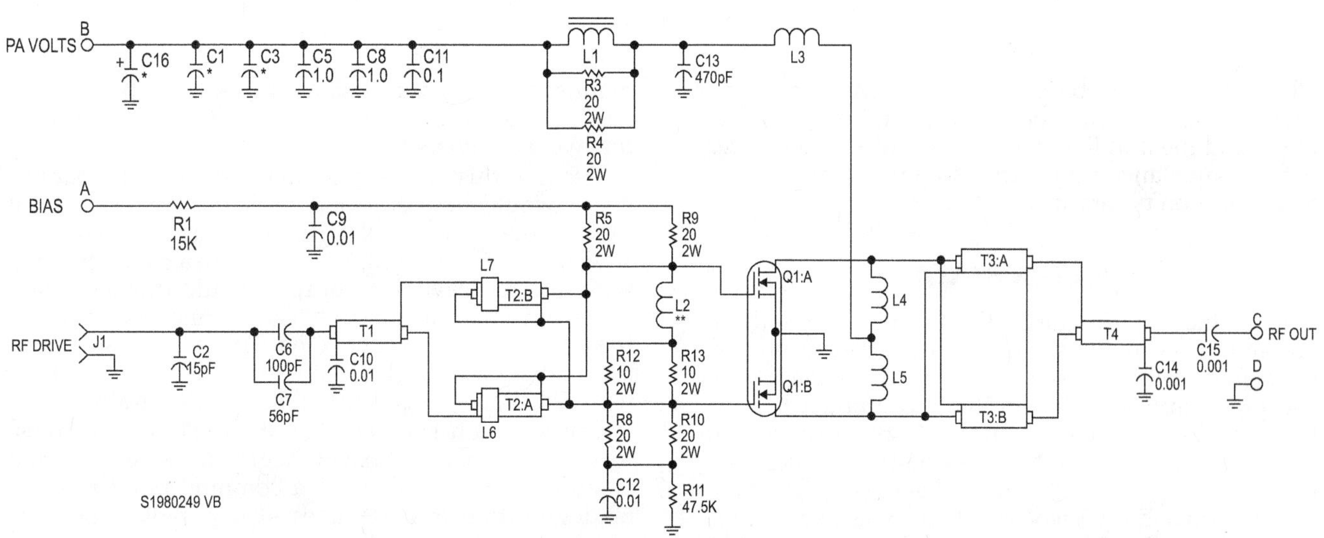

FIGURE 7.15-12 Schematic Diagram, VMOS Amplifier

Nautel Limited

Over the past couple of decades, the bulk of the improvements in amplifier technology have been related to the advent of more efficient devices, as VMOS then LDMOS transistors were developed. VMOS (Vertical, or V-groove, Metal Oxide Semiconductor) devices were so named because of the V-shaped groove cut in the substrate material, creating a vertical direction of electron travel and allowing higher current flow between source and drain. This allowed power output levels per device sufficient to reduce device count to the point that a solid state transmitter could be built economically. LDMOS (Laterally Diffused Metal Oxide Semiconductor) offered even higher power output per device, with lower drive levels and much improved efficiency, resulting in transmitter AC to RF efficiency breaking the 70% mark, finally putting solid state transmitters ahead of tube transmitters in this regard.

Combining multiple solid state amplifiers offers several benefits. Frequently, in transmitters with an Intermediate Power Amplifier (IPA) stage, a power amplifier module can be used as a temporary IPA, in the event of an IPA failure. Multiple amplifiers also results in reduced chance of being off-air relative to a single device amplifier stage. In addition, depending on design power headroom and controller configuration, some transmitters provide dynamic biasing of amplifier stages, to allow the controller to automatically maintain output power in the event of an amplifier failure.

Typically, solid state amplifiers are built with the transistor directly mounted on a heat sink, over which cooling air is driven by one or more fans. This results in airflow directly over the heat sink and not over the components, reducing the risk of component failure due to dust or airborne contaminants. Most amplifiers are designed to have 50-ohm input and output,

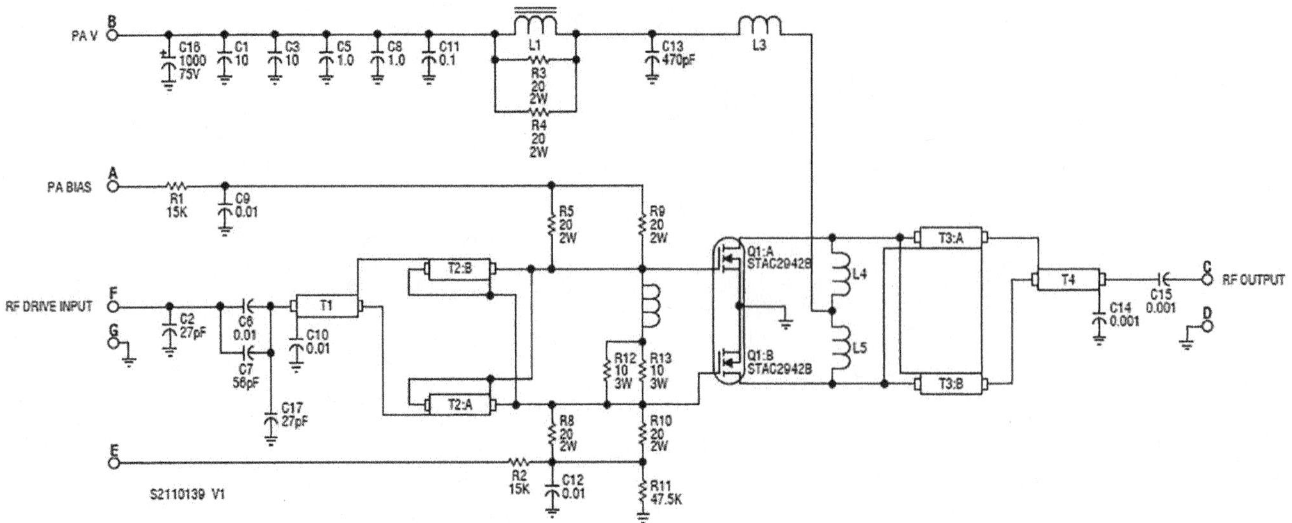

FIGURE 7.15-13 Schematic Diagram, LDMOS Amplifier

Nautel Limited

although they may be packaged in a module configuration such that the module input and output are 50 ohms and the amplifiers at a lower impedance. Alternately, impedance transformers may be used to keep all connection points at 50 ohms.

FILTER STAGE

One of the earliest methods of combining individual amplifiers in solid state FM transmitters was the 60-degree combiner, used to combine multiple 50-ohm PA outputs into one coherent 50-ohm output. It also contained a filter stage to filter the harmonics to an acceptable level. At the output of the Combiner/Filter there was a power probe to outputs voltages proportional to the forward and reflected power levels, as well as a –50 dB monitor point. Many earlier solid state transmitters used tuned amplifiers and filters, as the narrow banding provided better attenuation of undesired signals. However, these did not lend themselves well to the type of broadband operation required by later developments, such as HD Radio technology and DRM+.

COMBINER

A 60-degree combiner is named for the frequency selected cables connecting the amplifier outputs to the combiner common point—these are cut to be a nominal 60 degrees at the assigned carrier frequency. In earlier Medium Wave transmitters, where MOSFET devices had lower gains, it was determined that combining with hybrids would have proved costly, due to the need for dissipative resistors and the associated wiring. It was noted that Class D designs, using power FETs, would almost certainly fail in a short-circuit mode and it became possible to effectively series connect all of the outputs to produce the composite output. With this method, a failure would only slightly change the operating load impedance of the nonfailed outputs. In this manner, theoretical isolation was replaced by "acceptable isolation," and the dissipative dummy load resistors were eliminated.

Because this combining was done with transformers, it would be costly, and difficult to implement at higher power levels and higher frequencies, such as the VHF bands. Connecting the outputs to a common point with quarter wave transformers would transpose parallel connections to effective series connections as far as the PA outputs were concerned. This method would be acceptable for a short at the PA output, but would not tolerate an open circuit situation at the PA output.

Investigation showed that if the 90-degree transformer was changed to a 60-degree transformer—that is, all outputs combined to a common point through 60-degree transmission lines—"acceptable isolation" would be realized.

As with other combining techniques, if all PAs are operating and delivering identical, coherent outputs, each PA is loaded with 50 ohms. If one now fails and becomes a virtual short circuit, then, in general, the remainder become equally loaded with an impedance of:

$$\alpha + j\beta\Omega$$

where:

$$\alpha = R\left(\frac{n(n-1)}{n^2 - 2n + 2}\right)$$

$$\beta = \frac{a}{(n-1)}$$

R = total load impedance

n = the total number of PAs in the transmitter divided by the number of failed PAs

In the case of a transmitter with 28 PAs, one PA shorting would present an impedance to the other outputs of 51.8 + j1.9 ohms. It will be noted that for large "n,"

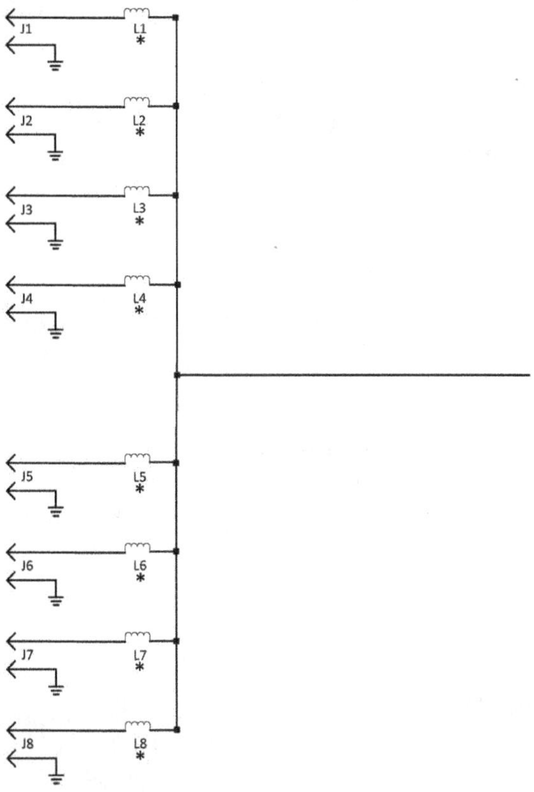

*L1 through L8 are frequency determined lengths of coaxial cable.

FIGURE 7.15-14 Schematic Diagram, 60-Degree Combiner

Nautel Limited

the modules remain almost perfectly matched. Even with n = 8, the load impedance only rises by 13%, at a phase angle of 8°.

It follows that for a failed open-circuit—that is, one PA removed,

$$\alpha = R\left(\frac{n-1}{n}\right)$$

$$\beta = \frac{R}{n}$$

Therefore, in a transmitter, with 28 PAs, if one PA fails open-circuit, the other, nonfailed PAs see an impedance of 48.2 + j1.78 ohms. Again, for high quantities of "n," the nonfailed PAs remain almost perfectly matched.

For low values of "n," however, the load will decrease and may cause overheating of PAs. This is a problem that can easily be remedied with an allowance for additional cooling in the design process.

This provided a very cost-effective and simple method of combining multiple outputs to one coherent output with "acceptable isolation." It also allowed the removal of a power module for maintenance, without switching the transmitter off the air. This is known as soft failure with on-air maintenance. Note that removal for maintenance was achieved by disconnecting the power module at the input to the combiner. The resultant impedance seen by the non-failed PAs was still well within acceptable limits and would not significantly affect transmitter operation.

While quite innovative in its time, and allowing for maximum isolation, serviceability and performance, the 60 degree combiner, as previously mentioned, was inherently narrowband. Thus, it was limited to fixed frequency, non-linear amplification and was not suitable for applications requiring frequency agility, or for the linear amplification and broader bandwidth needed for HD Radio™ technology or DRM+.

HYBRID (WILKINSON) COMBINER

One of the more common ways of combining multiple amplifiers at VHF frequencies is the Wilkinson combiner (Figure 7.15-15), or variations thereof. With this combiner, multiple signals, either in phase, or separated by a specific phase orientation, depending on configuration, are fed into the inputs, with the output providing a phase oriented sum of the input signals. A reject load dissipates any power lost due to residual imbalances in either magnitude or phase. The inherent advantage to this type of combiner is that it can be quite broadband, easily covering the entire FM frequency range, negating the need for tuning to a specific frequency. This leads to ease of use for N + 1 and backup applications, as well as providing the bandwidth required for operating with HD Radio technology or DRM+. The disadvantage is that an imbalance in power, such as in the case of an amplifier failure, will cause power to be dissipated in the reject load and subsequent need for increased cooling.

Newer technology in the areas of DSPs and processor based controllers allows for individual amplifiers to be biased independently and dynamically, in order to minimize power dissipation under normal operation, as well as to allow for the use of variable speed fans, to increase cooling of the reject loads in the event of an amplifier failure and the resulting increase in reject load temperature, as more power is dissipated.

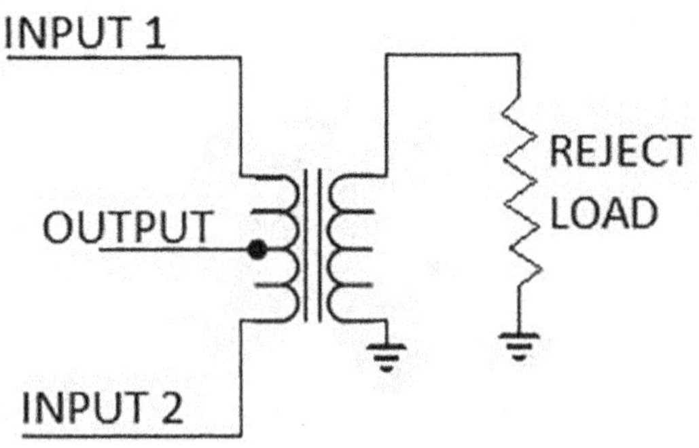

FIGURE 7.15-15 Simplified Diagram of a Hybrid Combiner

Nautel Limited

A typical transmitter will have multiple amplifiers packaged in a power module, with a hybrid combiner for each pair of amplifiers, and additional combiners for combining multiple pairs in some cases. This provides a single output from the module, which can then go through additional hybrid combiner stages as needed, in order to combine with additional modules. Referring to Figure 7.15-16, we see a power module containing four amplifiers. PA#1 and #2 as well as PA#3 and #4 are combined with in phase combiners, then the outputs are fed into the A and B inputs of a 90-degree hybrid combiner. Stripline coupling is used to provide a DC sample of combiner reject load power dissipation, to increase fan speed for additional cooling if needed and to provide metering of reject load power from the module.

FILTER

The purpose of the filter section of any transmitter is twofold—to match the impedance of the RF Drive stage to the load impedance and to attenuate any out

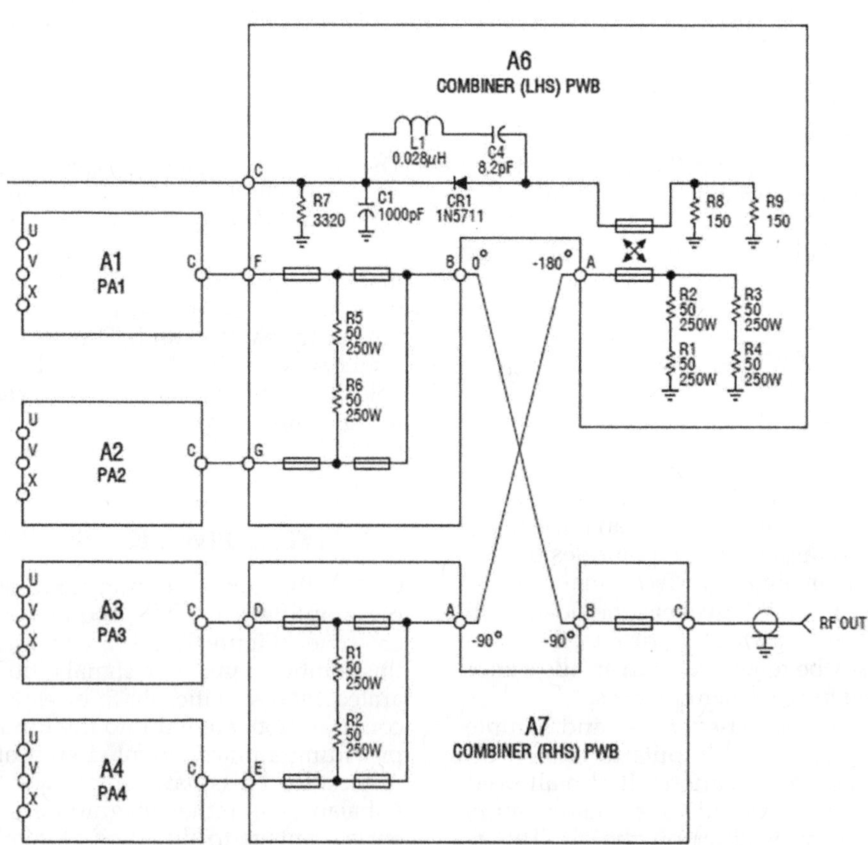

FIGURE 7.15-16 Schematic Diagram, Hybrid Combiner

Nautel Limited

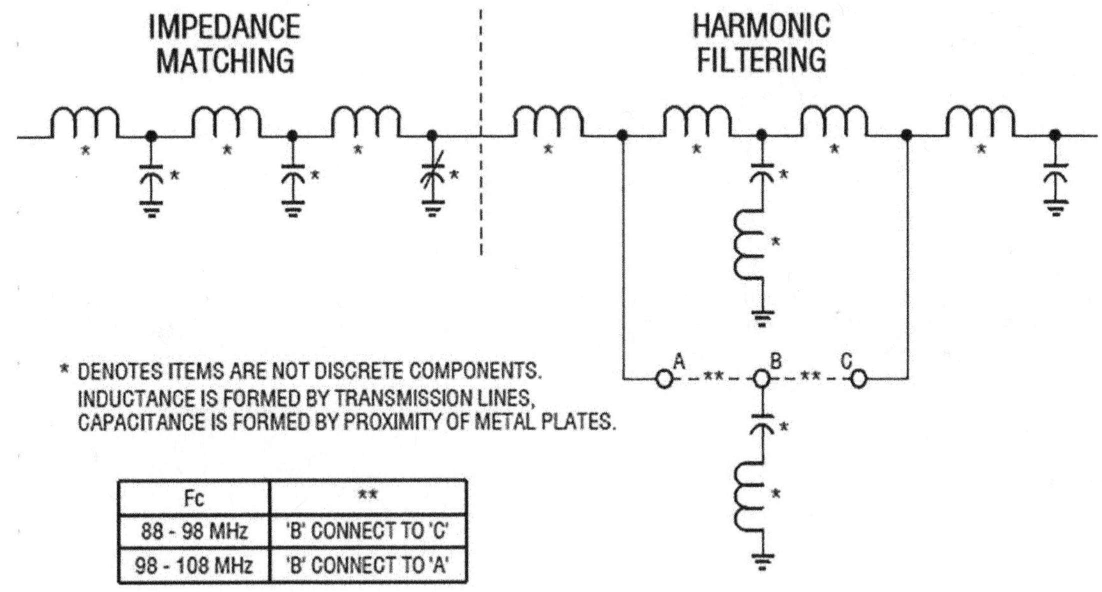

FIGURE 7.15-17 Schematic Diagram, RF Output Filter

Nautel Limited

of band signals created by nonlinearities in the amplifier stages. As previously discussed, filters will typically be either quarter wave or half wave, similar to with tube transmitters. Given the multiple combining stages in current solid stage transmitters, it can be assumed that some attenuation of out of band signals occurs in the combining process, which frequently leads to a lesser requirement for attenuation in the final filter. In addition, because most transmitters currently built are designed to have a nominal 50-ohm impedance at the output of each stage, impedance matching is also less critical. This leads to the ability to provide wider bandwidth filters, using line sections for inductance and insulated plates for capacitance, in order to provide the desired schematic layout.

A schematic diagram of an earlier tuned filter network is shown in Figure 7.15-16. This schematic shows the electrical equivalent of the device, however, all components shown are either lengths of transmission line (inductors) or insulated plates (capacitors). For a variable capacitor, all that is required is to provide the ability to mechanically move the plate for tuning. Troubleshooting is rarely needed and is usually limited to locating and replacing insulators that may have arced due to dust ingress or transient energy.

Typically, the final stage in all output filters will be a stripline coupler (or several couplers) to provide an in-phase sample for forward power, an out of phase sample for reflected power—both of which are rectified to provide a DC sample—and an additional coupler to provide an attenuated sample of the RF output, for connection to station monitoring equipment (modulation monitor or spectrum analyzer, for example). The forward and reflected samples may be used for internal and remote metering, as well as for protection circuitry within the transmitter, depending on the specific design.

CONTROLLER

More often, the controller of a modern transmitter will very closely resemble a computer motherboard. In fact, this may actually be the case in some systems. Advances in processor technology have made it possible to monitor every aspect of the transmitter, to the individual amplifier transistor current draw, as well as allowing for advanced monitoring in the form of MER (modulation error ratio) or even spectral analysis of the transmitter output. In addition, it allows for much more "behind the scenes" control, including the previously mentioned dynamic amplifier biasing, allowing a system running HD Radio technology to automatically set bias of the amplifiers to a slightly less linear level, improving overall efficiency, while monitoring for sufficient NRSC mask clearance and providing improved user interface, with things like email alarms, or in some cases, the ability to connect the transmitter (through a firewall) to the Internet and permitting it to directly contact the manufacturer should there be a situation requiring support or parts service.

POWER SUPPLIES

Power supplies will vary significantly from one make and model of transmitter to another. Older tube transmitters will usually have high-voltage, rectified DC supplies, whereas current solid state transmitters will more often have several switching power supplies that connect directly to the AC mains, without need of power transformers. Many transmitters, especially solid state, will offer redundancy in the form of multiple low-voltage power supplies for control circuitry (or exciter(s), if internal to the transmitter). In between these two extremes, earlier solid state transmitters also frequently contained rectified DC supplies, although these were typically at much lower voltage, depending on the devices used in the power amplifiers (for example, a typical VMOS amplifier would require a 48V DC PA power supply).

ACKNOWLEDGMENTS

The author would like to acknowledge his reliance on Geoff Mendenhall's contributions on tube amplifiers that were presented in the 10th Edition of the NAB Engineering Handbook. This earlier work, as well as his excellent overview on radio transmitters, was extensively employed in this chapter.

Bibliography

[1] Lessing, L., *Man of High Fidelity: Edwin Howard Armstrong*, Bantam Publishing, New York, 1959.

[2] *Care and Feeding of Power Grid Tubes*, Eimac Division of CPI, Inc., Palo Alto CA, 2001 (www.cpii.com/eimac/).

[3] Lyles, J.T.M. and Shrestha, M.B., *Transmitter Performance Requirements for Sub-Carrier Operation*, Broadcast Electronics, Inc., Quincy, IL, 1984.

CHAPTER

7.16

Radio Data System

SCOTT A. WRIGHT

Updated for the 11th Edition by

ALAN W. JURISON

iHeartMedia
Syracuse, NY

INTRODUCTION

The Radio Data System (RDS) provides for a method of transmitting a low-bit rate digital signal using analog FM radio. It is intended for application to VHF/FM sound broadcasting transmitters that carry stereophonic (pilot-tone system) or monophonic sound broadcasts specified in ITU-R Recommendation BS.450–3. Radio Broadcast Data System (RBDS) was the official name used for the U.S. version of the RDS Standard, when it was first adopted in 1993 by the National Radio Systems Committee (NRSC) with revisions in 1998, 2005 and 2011.[1] The RDS Standard was originally adopted in 1982 by the European Broadcasting Union (EBU) for transmitting low bit rate (on the order of 1 kilobit per second) digital information to appropriate receivers using conventional FM radio broadcasts. The RDS Standard is now an International Electrotechnical Commission (IEC) standard; the current IEC version is 62106 Ed.3 (2015).

The RDS system supports the transmission of numerous types of information, including time and station identification. The two standards (RDS and NRSC-4) are nearly identical, with only a few differences, including the manner in which numbers are assigned to each of 31 musical and other program formats, this field is called Program Type (PTY). The RDS signal is a component of FM broadcasting that uses an encoder to create a signal that is combined with other components of the FM baseband including the mono (L+R) and stereo multiplex (L–R) program audio-derived signals, and the 19 kHz stereo pilot tone. The RDS signal, also called the RDS subcarrier, is a 1,187.5 bits per second (bps) data stream (with approximately 674 bps of usable data) encoded into a 4-kHz-wide suppressed-carrier amplitude modulated subcarrier centered at 57 kHz. The 57-kHz frequency was chosen for being the third harmonic (3×) of the 19-kHz pilot tone, so it would not cause interference or intermodulation with it, or with the stereo difference signal at 38 kHz (2×). The data format utilizes error correction. RDS defines many features, including how private or other undefined features can be "packaged" in unused program groups.

At the time this chapter was written, an enhancement to the RDS Standard, called "RDS2," was under development by the RDS Forum, the European standards-setting body that maintains the RDS Standard. With RDS2, up to three additional subcarrier signals (each essentially identical to the "main" RDS subcarrier but centered at different frequencies in the FM baseband) may be included, greatly expanding the data capacity of the system. Additional information about RDS2 can be obtained from the RDS Forum at www.rds.org.uk/2010/Overview.htm.

THE RDS AND NRSC-4 STANDARDS

The Radio Broadcast Data System (RBDS) Standard (now NRSC-4) was first adopted in January 1993. It was developed by the National Radio Systems Committee (NRSC), a technical standards-setting body that is jointly sponsored by the National Association of Broadcasters (NAB, www.nab.org) and the Consumer Technology Association (CTA, www.cta.tech). Thus, NRSC-4 is an agreed upon standard between broadcaster and receiver manufacturer interests.

Revised editions of both the European RDS Standard and the U.S. RBDS Standard were adopted in 1998. The RBDS Standard was revised again, in April 2005, and later in April 2011. The April 2011 revision of the RBDS standard, NRSC-4-B was significant as it streamlined the document, including only those sections that differ between the U.S. and European versions of the Standard.

In the U.S. all references to RDS technology generally use the term "RDS" rather than "RBDS." RBDS is generally used only to refer to the previous versions of the NRSC Standards document itself (before 2005) or to the name of the NRSC Subcommittee given responsibility for maintaining the Standard. At the time of initial adoption, the NRSC felt that it was important to distinguish the U.S. Standard from its European counterpart by giving it a slightly different name. In the U.S., the NRSC-4 Standard is a voluntary standard. Because its success is largely dependent on the radio listener's ability to use the same radio data system receiver in the same manner in any location, it is hoped that broadcasters and equipment manufacturers will comply with the spirit and the letter of the Standard.

NRSC-4 Related Guideline Documents and Reports

The NRSC has issued a series of informative RDS guidelines and reports in recent years. These documents were designed with practical RDS usage in the U.S. in mind, but have global implications as well. The intended audience for this work includes broadcasters, broadcast equipment manufacturers, receiver manufacturers and data service providers.

As of this writing, here are the current versions of these documents:

- NRSC-G300-B: RDS Usage Guideline (September 2014);
- NRSC-G301: Creation and Distribution Practices for Audio Program Metadata Guideline (April 2013);
- NRSC-G200-B: Harmonization of RDS and IBOC Program Service Data (PSD) Guideline (January 2017);
- NRSC-R300: Program Associated Data (PAD) Field Length Study (November, 2011).

To obtain a copies of the current U.S. NRSC Standards, Guidelines and Reports, visit the NRSC web page (www.nrscstandards.org). The European RDS Standard was initially issued by the European Committee for Electrotechnical Standardization (CENELEC) under Document Number EN 50067 but is now an International Electrotechnical Commission (IEC) standard, IEC 62106 Ed.3 (2015), and may be obtained from the IEC website at www.iec.ch/.

The RDS Subcarrier

In the U.S., the FCC allows almost any technology to be used for FM subcarriers as long as certain criteria are met. This means that analog or digital or a mixture of both may exist within the subcarrier frequency range as long as the total modulation of the FM signal does not exceed 110% (82.5 kHz).[2] RDS has proven to be compatible with existing services due to its low injection requirements and narrow bandwidth. In addition, by phase locking the AM double-sideband suppressed-carrier RDS signal with the FM stereo pilot signal, there is little risk of interference to the main audio program, even under multipath conditions. RDS is carried on many classical music networks throughout Europe and the U.S. with no degradation to the audio quality. Figure 7.16-1 depicts the baseband audio spectrum of an FM broadcast station.

RDS nests into the 57-kHz position between the stereo multiplex and the commonly used analog 67- and 92-kHz subcarrier channels. Injection levels as low as 1.3 kHz allow RDS to be easily implemented without giving up program audio power. Because RDS lies at 57 kHz, it is compatible with nearly every subcarrier in use or development today. RDS was shown, in the NRSC's evaluation of the FM in-band/on-channel (IBOC) digital radio system, to be compatible with the hybrid mode of operation of the HD Radio system.[3] Due to the low implementation cost of RDS and high adoption rate among receivers, it is not likely to be displaced by another system for many years.

RDS Data Structure

The RDS signal is modulated and demodulated as a synchronous bit stream. The 57-kHz RDS subcarrier is amplitude-modulated by a shaped and biphase-coded data signal. The spectrum of this modulated signal is shown in Figure 7.16-2.

The RDS data rate is 1187.5 bits per second. The data transmission is fully synchronous, and there are no gaps between the groups or blocks. The RDS group structure or baseband coding is depicted in Figure 7.16-3. This baseband structure is described as follows:

- The largest element in the structure is called a group (104 bits);
- Each group comprises 4 blocks of 26 bits each;
- Each block comprises an information word and a checkword;

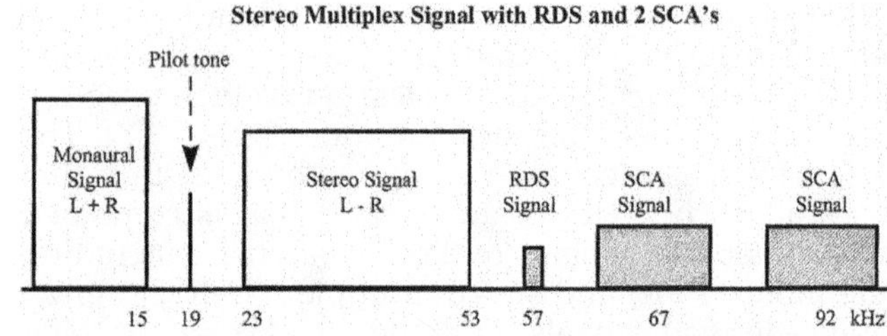

FIGURE 7.16-1 FM baseband spectrum showing main, pilot, stereo, RDS and two SCA channels.

- Each information word comprises 16 bits;
- Each checkword comprises 10 bits;
- All information words, checkwords, binary numbers, or binary address values have their most significant bit (MSB) transmitted first.

Information Word

The information word contains the actual data for each block. This information word is composed of 16 bits. The most significant bit (m_{15}) is transmitted first.

Checkword and Offset Word

The checkword and offset word consist of 10 bits. The most significant bit (c_9) is transmitted first. The checkword allows the receiver to detect and correct errors that occur during data reception. The error-protecting code has the following error-checking capabilities:

- It detects all single and double bit errors in a block;
- It detects any single error burst spanning 10 bits or less;
- It detects about 99.8% of bursts spanning 11 bits and about 99.9% of all longer bursts;
- The code is also an optimal burst error-correcting code and is capable of correcting any single burst of span 5 bits or less.

Table 7.16-1 shows the possible offset words. The offset word allows detection of the block number within the data group. This allows critical data such as the PI, PTY and TP codes (described later) to be decoded without reference to any block outside the one that contains the information. This is essential to minimize acquisition time for these kinds of messages and to retain the advantages of the short (26-bit) block length. To permit this to be done for the PI codes in block 3 of version B groups, a special offset word (C') is used in block 3 of version B groups. The occurrence of offset C' in block 3 of any group can then be used to indicate directly that block 3 is a PI code, without any reference to the value of B0 in block 2.

Synchronization of Blocks and Groups

The blocks within each group are identified by their offset words A, B, C or C' and D. The beginning and

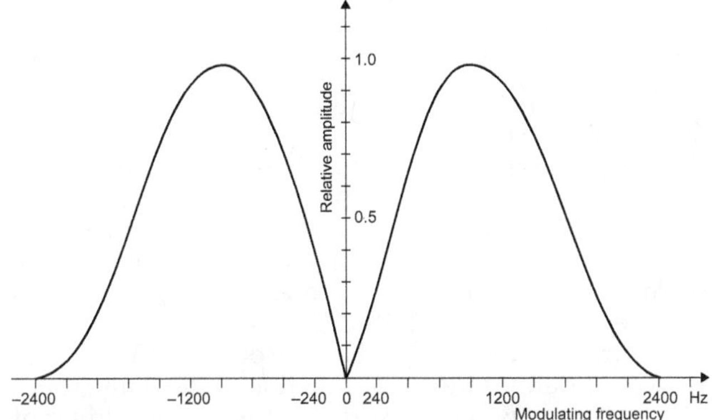

FIGURE 7.16-2 Frequency spectrum of RDS signal: this spectrum is centered at 57 kHz in the FM baseband.

TABLE 7.16-1
Offset Words*

Offset Word	Binary Value									
	d_9	d_8	d_7	d_6	d_5	d_4	d_3	d_2	d_1	d_0
A	0	0	1	1	1	1	1	1	0	0
B	0	1	1	0	0	1	1	0	0	0
C	0	1	0	1	1	0	1	0	0	0
C'	1	1	0	1	0	1	0	0	0	0
D	0	1	1	0	1	1	0	1	0	0
E	0	0	0	0	0	0	0	0	0	0

*Eight bits (i.e., d_9 to d_2) are used for identifying the offset words. The remaining two bits (i.e., d_1 and d_0) are set to logical level zero.

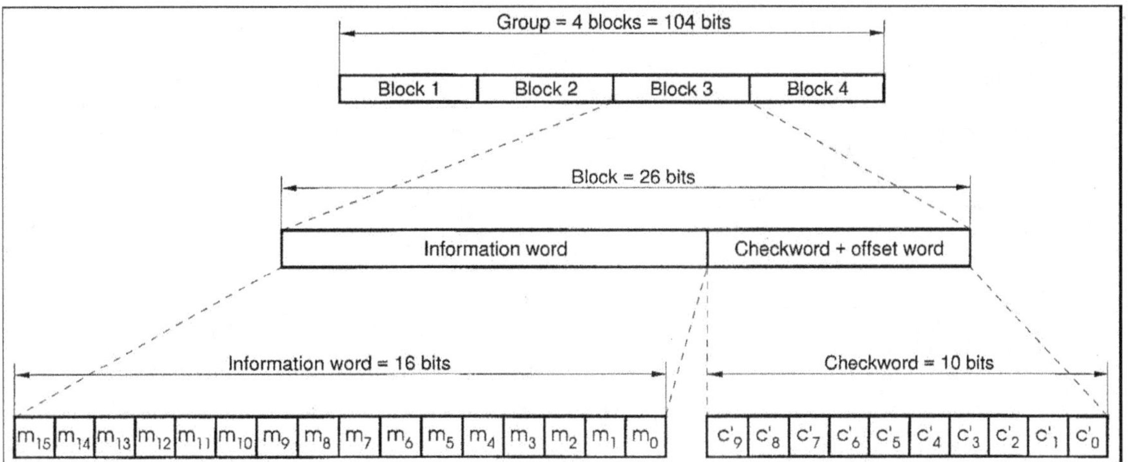

FIGURE 7.16-3 Structure of the RDS baseband coding.

end of the data blocks may be recognized in the receiver decoder by using the fact that the error-checking decoder will, with a high level of confidence, detect block synchronization slip as well as additive errors. This system of block synchronization is made reliable by the addition of the offset words (which also serve to identify the blocks within the group). These offset words destroy the cyclic property of the basic code so that in the modified code, cyclic shifts of codewords do not give rise to other codewords. A detailed explanation of a technique for extracting the block synchronization information at the receiver is given in Annex C of IEC 62106.

Group Structure

The information coded into each group has a common fixed structure, as depicted in Figure 7.16-4.

All data groups share a common fixed structure of information coding to allow critical data to be transmitted in a fixed, highly repetitive pattern to ensure the best reception even under adverse reception conditions. This fixed structure is described as follows:

- Block 1: Block 1 of every RDS data group contains only the program identification or PI code. Block 1 is identified by offset word A.
- Block 2: Block 2 contains the group type code, version code, traffic program code and program type (PTY) code. All codes are binary coded with the MSB transmitted first:
    - Group type code: The data groups are identified through the 4-bit group type code identified as A3–A0, along with the 1-bit version code, B0. The bits A3–A0 yield decimal values 0 through 15;
    - Version code: Each group then has an "A" version and a "B" version based on the state of B0 where
        - B0 = 0 is identified as group type A.
        - B0 = 1 is identified as group type B.
    - All B type groups repeat the program identification (PI) code in blocks 1 and 3.
    - Traffic Program (TP) code: This single-bit code is utilized with the traffic announcement (TA) code to provide the traffic feature.
    - Program type (PTY) code: This five-bit code describes the audio program type being aired. There are 31 possible PTYs available.
    - Block 2 is identified by offset word B.
- Block 3: Dependent on the group version code (A or B) or offset word (C or C′), block 3 may carry two distinct types of data.
    - Version code A: This block carries data defined by the group type code.
    - Block 3 is identified by offset word C for version code A groups.
    - Version B type groups carry only the PI code.
    - Block 3 is identified by offset word C′ for version code B groups.
- Block 4: This block carries data defined by the group type code.
    - Block 4 is identified by offset word D.

RDS Group Types

RDS data groups are referred to by both group type and version (examples: 0A, 14B, where the number is the group type and the letter is the version). Taking into account both the group type and version codes, there are 32 defined data groups. Some of the group features are defined in their entirety and cannot be defined in any other way, whereas other data groups are openly defined such that the actual data contained in the group can be defined by the operator. The group type code defines the basic function of the data group. The "A" and "B" versions differ only in the coding of

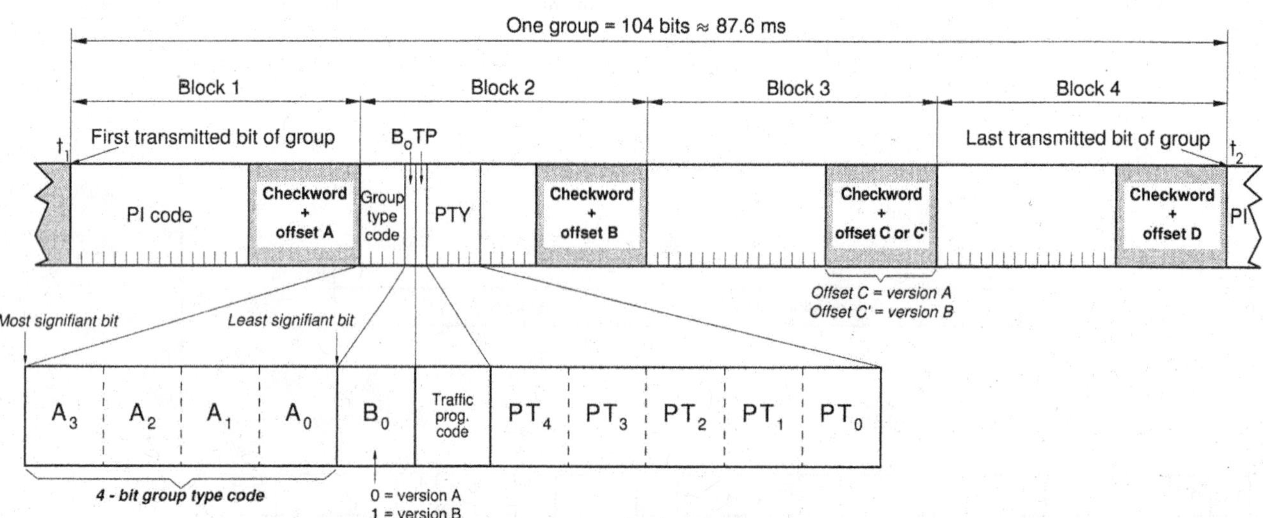

FIGURE 7.16-4 RDS group coding structure.

the data for the group. Because RDS data is transmitted in functional groups that share a common structure, the operator can select from these group types and transmit all or only a portion of them depending on the features in use. In this manner, there is no need to transmit data that is not in use, thereby maximizing data throughput. Table 7.16-2 summarizes the possible applications for all possible group types. Only group type 0A or 0B must be included in the transmission sequence.

RDS FEATURES

RDS provides primary and secondary features. Primary features are either contained in every information

TABLE 7.16-2
RDS Group Types (Taken From NRSC-4-B)

Group Type	Group Type Code/Version					Flagged in Type 1A Groups	Description
	A_3	A_2	A_1	A_0	B_0		
0A	0	0	0	0	0		Basic tuning and switching information only
0B	0	0	0	0	1		Basic tuning and switching information only
1A	0	0	0	1	0		Program Item Number and slow labeling codes only
1B	0	0	0	1	1		Program Item Number
2A	0	0	1	0	0		RadioText only
2B	0	0	1	0	1		RadioText only
3A	0	0	1	1	0		Application Identification for ODA only
3B	0	0	1	1	1		Open Data Applications
4A	0	1	0	0	0		Clock-time and date only
4B	0	1	0	0	1		Open Data Applications
5A	0	1	0	1	0		Transparent Data Channels or ODA
5B	0	1	0	1	1		Transparent Data Channels or ODA
6A	0	1	1	0	0		In-House applications or ODA
6B	0	1	1	0	1		In-House applications or ODA
7A	0	1	1	1	0	Y	Radio Paging or ODA
7B	0	1	1	1	1		Open Data Applications
8A	1	0	0	0	0		Traffic Message Channel or ODA
8B	1	0	0	0	1		Open Data Applications
9A	1	0	0	1	0	Y	Emergency Warning System or ODA
9B	1	0	0	1	1		Open Data Applications
10A	1	0	1	0	0		Program Type Name
10B	1	0	1	0	1		Open Data Applications
11A	1	0	1	1	0		Open Data Applications
11B	1	0	1	1	1		Open Data Applications
12A	1	1	0	0	0		Open Data Applications
12B	1	1	0	0	1		Open Data Applications
13A	1	1	0	1	0	Y	Enhanced Radio Paging or ODA
13B	1	1	0	1	1		Open Data Applications
14A	1	1	1	0	0		Enhanced Other Networks information only
14B	1	1	1	0	1		Enhanced Other Networks information only
15A	1	1	1	1	0		Open Data Applications
15B	1	1	1	1	1		Fast switching information only

group or are included in type OA or OB groups, which are always transmitted. Secondary features are transmitted only when necessary or desired. This prevents data capacity from being wasted on unused features.

Primary Features

Program Identification (PI) Code

The PI code is a four-digit hexadecimal code that is unique for each station. In Europe, the PI code is assigned. In the U.S., the PI code is in most cases calculated from an FM station's call letters; however, this calculation method is not absolute. NRSC-4-B and NRSC-G300-B outline several cases where the traditional call-sign calculation method cannot apply.[4] The guiding principle is that each FM station with a unique program audio stream has a unique PI code. Some receivers can use the PI code to identify a station rather than using the reception frequency. RDS supports frequency diversity such that if a station simulcasts on another frequency and if both of these signals include the same PI code then a suitably equipped receiver can automatically tune to the strongest of the two stations. Duplicate PI codes on stations with dissimilar audio programming in the same geographic area can cause listener disruptions, as some radio receivers may automatically switch between these two stations. If a station simulcasts, then it must pick a single PI code and use it on all of the simulcasting stations. While the PI code is transparent to the user, its proper use is vital to proper receiver operation. Specialized PI codes are available for use for networked programming as well.

The addition of RDS TMC networks and significant increases in the number of FM translators that broadcast AM radio stations and HD multicast channels have challenged the original call sign conversion method of PI code calculation, which did not consider these new transmission scenarios. As of this writing, the NRSC is considering refinement of the existing PI code structure within the U.S. to better accommodate FM translators. While NRSC-4-B and NRSC-G300-B currently offer guidance to broadcasters in this area, this guidance is considered not complete. The NRSC is evaluating several proposals, including one in which PI codes for FM translators in the U.S. would be assigned through a registration process. It is best to check the NRSC website for the latest guidance for PI code assignment in the U.S.

Program Service (PS) Name

The Program Service (PS) Name was initially designed to be whatever name the station chooses to present to its listeners. It could be call letters like "WXZH-FM," or a slogan like "X-100." The PS Name is often displayed prominently on an RDS receiver. When the listener tunes to an RDS station, the station's name will be shown on the radio display. The PS Name value itself cannot be longer than eight characters. Any character referenced in IEC 62106 Table E.1 may be used, but special characters might not be displayable on certain low-cost displays. It should be noted that in practice, use of the PS Name has changed since the initial RDS specifications were created. Use of a "dynamic," "framing" or "scrolling" PS is a common practice within the U.S. that changes the PS Name value frequently, for example, to display song title and artist information. NRSC-G300-B Section 6.9 offers best practice information for stations that elect to employ this enhanced feature.

Traffic Program (TP)

Stations offering listeners traffic bulletins should consider using this feature. The TP identifies the transmitting station to the listener as a station that offers traffic programs.

Traffic Announcement (TA)

Usage of the TA feature is discouraged by the NRSC, see NRSC-G300-B Section 5.4. When an actual traffic bulletin is broadcast, this information can be set to a logic "1." Some RDS receivers detect this and will automatically stop any playback device that may be in use and will return to the FM tuner mode.

Feedback from listeners is that this feature can be confusing and annoying. Since the original development of the TP/TA feature set in RDS, more efficient and advanced traffic reporting features have been made available to the general public. Instances of these include Global Position Systems (GPS) navigation devices with embedded FM or HD Radio receivers that can decode traffic incident data and display it on a map and provide rerouting assistance. Also, out-of-band data provided on the Internet and viewed on smartphones provides another way in which to convey this information to the public.

Program Type (PTY) Codes

The program type code is used to identify the type of program material being broadcast. There are predefined codes for Country, Rock, Top 40 and so on. RDS receivers can automatically search for stations by PTY, thus allowing listeners to find their favorite type of programming without tuning to all available stations. Advanced receivers can even interrupt listeners when, for example, news is broadcast. A definition of the terms used to denote North American Program Types from NRSC-4-B is given in Table 7.16-3.

Alternate Frequencies (AF)

The alternate frequencies feature allows an RDS receiver to automatically tune to the best signal when multiple transmitters or translators are broadcasting the same program. Regional or national programs that are broadcast over large areas can even be linked together, providing the listener with the illusion of one very powerful transmitter. It is also possible to link stations only during specific times when the program material is common (perhaps during a sporting event) using the linkage actuation feature. Competitors cannot steal listeners away using this feature because an RDS receiver will look only for alternative frequencies

TABLE 7.16-3
North American Program Types (Taken From NRSC-4-B)

Program Type	Name	Description
1	News	News reports, either local or network in origin
2	Information	Programming that is intended to impart advice
3	Sports	Sports reporting, commentary and/or live event coverage, either local or network in origin
4	Talk	Call-in and/or interview talk shows either local or national in origin
5	Rock	Album cuts
6	Classic Rock	Rock-oriented oldies, often mixed with hit oldies, from a decade or more ago
7	Adult Hits	An up-tempo contemporary hits format with no hard rock and no rap
8	Soft Rock	Album cuts with a generally soft tempo
9	Top 40	Current hits, often encompassing a variety of rock styles
10	Country	Country music, including contemporary and traditional styles
11	Oldies	Popular music, usually rock, with 80% or greater noncurrent music
12	Soft	A cross between adult hits and classical, primarily noncurrent soft rock originals
13	Nostalgia	Big band music
14	Jazz	Mostly instrumental, includes both traditional jazz and more modern smooth jazz
15	Classical	Mostly instrumentals, usually orchestral or symphonic music
16	Rhythm and Blues	A wide range of musical styles, often called "urban contemporary"
17	Soft Rhythm	Rhythm and blues with a generally soft tempo
18	Foreign Language	Any programming format in a language other than English
19	Religious Music	Music programming with religious lyrics
20	Religious Talk	Call-in shows, interview programs, etc., with a religious theme
21	Personality	A radio show where the on-air personality is the main attraction
22	Public	Programming that is supported by listeners and/or corporate sponsors instead of advertising
23	College	Programming produced by a college or university radio station
24	Spanish Talk	Call-in shows, interview programs, etc. in the Spanish language
25	Spanish Music	Music programming in the Spanish language
26	Hip-Hop	Popular music incorporating elements of rap, rhythm and blues, funk and soul
27–28		Unassigned
29	Weather	Weather forecasts or bulletins that are nonemergency in nature
30	Emergency Test	Broadcast when testing emergency broadcast equipment or receivers. Not intended for searching or dynamic switching for consumer receivers
31	Emergency	Emergency announcement made under exceptional circumstances to give warning of events causing danger of a general nature. Not to be used for searching—only used in a receiver for dynamic switching.

that are specified by the station that is currently tuned in. An example of an AF network is shown in Figure 7.16-5.

Secondary Features

Program Type Name (PTYN)

The program type name feature allows additional flexibility for the broadcaster that wants to stand out from competitors. While the PTY codes are predefined, the PTYN can be any eight characters the broadcaster desires to further describe the current program. For example, a broadcaster that is currently using the PTY "Personality" may set the PTYN to "Limbaugh" to further describe the current program. A "Rock" station may set the PTYN to "BobTom" to describe their morning team. An RDS receiver cannot search by PTYN but will display the PTYN in place of the PTY once tuned to a particular station.

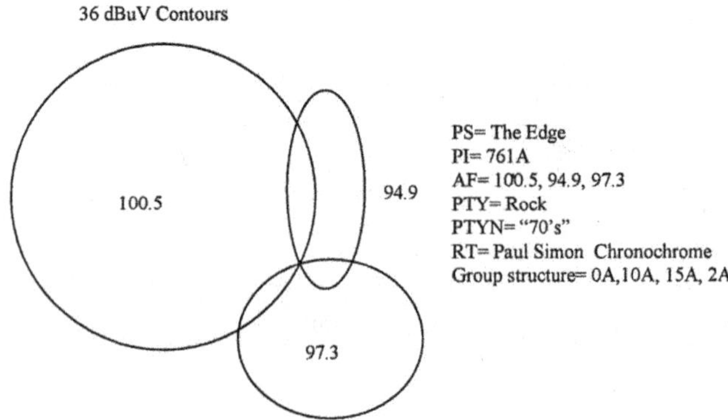

FIGURE 7.16-5 Example of an RDS alternate frequency network.

RadioText (RT)

RadioText allows a station to transmit up to 64 characters of information to the listener. Information such as the current artist and song title, station promotional information, local events and even additional information about the advertiser whose commercial is currently playing can be sent. NRSC-G300-B Section 6.8 offers broadcasters significant information on how to optimize their RT transmissions.

Open Data Application (ODA)

The open data application allows the definition of an unused RDS data group based solely on the Application Identification (AID) code. Thousands of AID codes are available, meaning that the remaining data groups could be defined thousands of different ways. An up-to-date listing of ODA AID codes may be found on the NRSC website at www.nrscstandards.org/sg/ODAlist.pdf. The NRSC manages ODA AID registration within the U.S. while the RDS Forum does so for all other locations.

Group 3A is for ODA application identification. Other, previously unused groups are assigned for ODAs as well. The use of an "A" type group allows low data capacity applications to exist entirely within Group 3A. The ODA allows multiple reuse of unused data groups rather than fixed definitions, ensuring a long future for RDS. The actual use of data can be either public or private. The ODA allows encryption of data for fee-based services.

The ODA is much more flexible than original RDS data-only groups such as the transparent data channel because it can be located by the group 3A packets and then automatically tuned (by use of the AID) and tracked regardless of any other data being transmitted. This is a dynamic and potentially very profitable data group. Most of the newer, extended features of RDS such as RadioText Plus (RT+) function as an ODA.

RadioText Plus (RT+)

RT+ is an ODA data stream defined in IEC 62106 Annex P that can be added to an RDS signal, making it possible to identify each field of text being encoded in a RT field. Until the RT+ Standard was developed, there was no way to know what the specific parts of the RT data were, from a hardware or software standpoint, and this is important for song tagging. RT+ offers the ability to classify different types of information (such as title) and their location within the RT field.

NRSC-G300-B Section 6.10 offers additional guidance regarding RT+ and recommends that all broadcasters deploy RT+ tagging for Item.Title, Item.Artist, Item.Album, as well as StationName.Short and StationName.Long. These are the most commonly used RT+ fields in receivers today.

Clock Time and Date

The clock time and date feature enables broadcasters to transmit the current date and time once every minute, thus enabling RDS-equipped receivers (clock radios, car radios, etc.) to be automatically set after a power failure or time zone change. Broadcasters using this feature should ensure the accuracy of the transmitted time to ±2 sec.

Emergency Alerting and RDS

The original Emergency Alert System (EAS) Open Data Application for RBDS (Annex Q) was removed from the NRSC-4-B standard in April 2011. As of this writing, all of the RDS-based alerting systems in the U.S. are closed systems (not open), only accessible to receivers specifically designed for use with these systems. The most commonly deployed private system is the GSS/Alert FM System. NPR Labs has been involved in a modernization of the (now retired) RBDS Annex Q methodology in a research project in conjunction with the Department of Homeland Security (DHS) and the Federal Emergency Management Agency (FEMA). However, this is presently a closed system and is only operational in about 25 pilot stations in the Gulf Coast region of the U.S. More information about emergency alerting systems in use in the U.S. can be found in the NRSC-G300-B guideline document in Section 8.

In-House Application

Data contained in the in-house application group is to be used only by the broadcaster. Remote control applications, station telemetry, or paging applications can be supported using this data group.

Radio Paging (RP)

Paging services including numeric or alphanumeric can be supported by RDS.

Traffic Message Channel (TMC)

The traffic message channel feature allows traffic information to be coded and displayed on specialized consumer receivers (i.e., navigation systems) so that normal audio bulletins are not required. More information about these systems can be located in Annex 1 of the NRSC-G300-B Guideline.

DIFFERENCES BETWEEN RDS AND NRSC-4

While many of the changes to the RDS and NRSC-4-B Standards over the years have made these two Standards more similar, there still remain differences. These differences are mainly market driven. From a listener point of view, most of these differences are transparent. In exploring these differences, it will become evident that the receiver manufacturer must be well versed in the differences to ensure proper receiver operation.

With the advent of the NRSC-4-B version of the NRSC-4 Standard, the Standard document itself serves as a reference highlighting the differences between the RDS and NRSC-4 Standards. The principal content of NRSC-4-B consists of sections that replace in their entirety the corresponding sections in the IEC 62106 (RDS) Standard document. Only sections where there are differences between the two Standards are included in NRSC-4-B.

Summary of Major Differences Between RDS and NRSC-4

- Program Type Definitions (PTY)

To accommodate differing broadcast styles, the PTY code definitions are different. These differences may be accounted for through the use of a look-up table within the receiver.

- Program Identification Coding (PI)

Due to the high penetration of network-based programming throughout Europe compared to the largely independent single station structure of North America, the derivation of PI codes is different. In NRSC-4-B, PI codes are primarily based on a station's call letters from 0000–AFFF and are network-based from B000–FFFF. This means that PI codes above B000 in North America are treated by the receiver the same way that all PI codes are treated in Europe. Below B000, PI codes in North America do not employ the regionalization feature. While alternate frequency switching is still employed below B000, variants based on changes in the second nibble of the PI code do not exist and should be ignored. PI code definitions can be found in Annex D of the Standard.

RDS DATA TRANSMISSION CAPACITY LIMITS

An analysis of the limitations of RDS transmission capacity limits is given here, and is based on the use of RDS in North America. Due to differences in broadcasting requirements between North America and Europe, the importance of certain features is not the same. In Europe the networking of broadcasts is facilitated by the use of Enhanced Other Networks (EON). The use of EON consumes approximately 8% of the available RDS bandwidth.

The use of features to achieve automated tuning, essentially for mobile reception, requires a considerable portion of the available capacity. RDS offers a choice of some 20 well-defined features, but the limited data capacity restricts the number that can be implemented in a single channel.

Calculation of RDS Capacity

With a bit rate of 1187.5 bps, the RDS channel capacity is a rather limited resource. Because the four checkwords in each group of 104 bits occupy a total of 40 bits, and each group address needs 5 bits, the useful bit rate is $1187.5 - [(1187.5 / 104) \times (40 + 5)] = 673.7$ bps.

Analysis of RDS Capacity

In order to better understand the usage of RDS capacity for each of the features, one can group them into categories according to their impact on the channel capacity. By examining the features required for RDS receiver tuning, then, one can identify the remaining capacity for non-program-related features.

As shown in Table 7.16-4, the program-related features can be divided into three categories:

- The primary features—AF, PI, PS and TP/TA—mainly required for the automated tuning process;
- A group of features—CT, DI, MS, PIN, PTY and PTYN—require relatively little RDS capacity to be implemented. Note that PIN requires a repetition rate of only one group type 1A per minute; however, in connection with RP it must be increased to one per second;
- RadioText.

Furthermore, the non-program-related features—IH, RP, TDC and TMC—are identified. The following can be seen from Table 7.16-4:

- AF, PI, PS, TP/TA: These primary functions of RDS, essentially supporting the automated tuning process, already require 48.35% of the available channel capacity.
- PTY, MS, DI, PIN, CT: All of these features require relatively little RDS capacity. Whereas the first three require 9.65% whether implemented or not, the latter two add an additional 0.21% (5.6% if RP is implemented), bringing the total required to 58.2% (or 63.59% with RP).
- RT: RadioText requires an additional 17.58%, bringing the total required for program-related features up to 75.78% (or 81.17% with RP).

This then leaves 24% (18% with RP), for the implementation of the non-program-related features (IH, ODA, TDC and TMC); that is 162 bps (or 121 bps with RP). Each of them will require, if implemented, a significant proportion of the remaining capacity.

However, due to the time-multiplexing possibilities of many of these features, the average capacity available for other features will be greater than the indicated "peak demand." Note that the NRSC has

TABLE 7.16-4
RDS Program-Related Features

Application	Feature	Group Types Containing This Information	Appropriate Minimum Group Repetition Rate per Second	Number of Occupied Bits per Group	Number of Occupied Bits per Second	Percentage of 673.7 Bits per Second	Accumulated RDS Capacity
Automated tuning	PI	All	11.4	16	182.4	27.07	48.35
	PS	0A	4.0	16	64.0	9.50	
	AF	0A	4.0	16	64.0	9.50	
	TP	All	11.4	1	11.4	1.69	
	TA	0A	4.0	1	4.0	0.59	
Various other program-related features	PTY	All	11.4	5	57.0	8.46	58.20/63.59*
	MS	0A	4.0	1	4.0	0.59	
	DI	0A	4.0	1	4.0	0.59	
	PIN	1A(B)	0.02/1.0* 37†	0.74/37*	0.11/5.5*	0.10	
	CT	4A	0.02	34	0.68		
RadioText	RT	2A(B)	3.2‡	37	118.4	17.58	75.78/81.17*
Various non-program-related features	RP	1A/4A/7A/13A					
	TDC	5A(B) 6A(B)					
	IH	3A/8A					
	TMC						

*If Radio Paging is used, an increased repetition rate of 1 per second is necessary.

†Although 16 bits are actually used for PIN, the associated 16 undefined and 5 spare bits must also be taken into account because this represents used capacity.

‡A total of 16 type 2A groups are required to transmit a 64-character RadioText message, and therefore, to transmit this message in 5 seconds, 3.2 type 2A groups will be required per second.

researched RDS uses in the U.S. and offered recommendations on how to optimize a station's encoding equipment to best cater to the wide variety of receivers out in the marketplace. This research is encapsulated in NRSC-G300-B Section 6.8.2 and Table 1.

In conclusion, RDS has a finite capacity, but with careful planning, it can be optimized to provide many beneficial services to the broadcaster and the consumer.

SUMMARY

The NRSC-4-B Standard and its European counterpart, IEC62106, the RDS Standard, provide FM broadcasters and receiver manufacturers with the tools necessary to provide a meaningful supplement to FM audio programs. These standards continue to evolve with the goal of staying relevant as broadcasters transition to digital radio broadcast systems.

Notes

[1] In 2005 the NRSC changed the name of the RBDS Standard to NRSC-4. NRSC-4-B is now current edition of the U.S. version of the Standard as of this writing.

[2] See 47 CFR 73.1570(b)(2). Other countries have modulation different regulations.

[3] See "DAB Subcommittee—Evaluation of the iBiquity Digital Corporation IBOC System, Part I—FM IBOC," November 29, 2001 (available on the NRSC web page).

[4] See NRSC-4-B Section D.7 and NRSC-G300-B Section 5.1.

Bibliography

European Committee for Electrotechnical Standardization (CENELEC), *Specification of the Radio Data System*, EN 50067, Brussels, Belgium, www.cenelec.org

European Committee for Electrotechnical Standardization (CENELEC), *Specification of the Radio Data System (RDS) for VHF/FM Sound Broadcasting in the Frequency Range 87.5 to 108.0 MHz*, EN 50067:1998, Brussels, Belgium, www.cenelec.org

International Electrotechnical Commission (IEC), *Specification of the Radio Data System (RDS) for VHF/FM Sound Broadcasting in the Frequency Range from 87.5 to 108.0 MHz*, IEC 62106, Geneva, Switzerland, www.iec.ch

International Telecommunication Union-Radiocommunication Sector (ITU_R), *Transmission Standards for FM Sound Broadcasting at VHF*, ITU-R Recommendation BS.450–3, 2001, ITU-R, Geneva, Switzerland, www.itu.int/dms_pubrec/itu-r/rec/bs/R-REC-BS.450-3-200111-I!!PDF-E.pdf

Kopitz, Dietmar, and Bev Marks, *RDS: The Radio Data System*, Norwood, MA, Artech House, 1998

National Radio Systems Committee, *NRSC-4-B, United States RBDS Standard*, Washington, DC, www.nrscstandards.org

National Radio Systems Committee, *NRSC-G200-A, Harmonization of RDS and IBOC Program Service Data (PSD) Guideline*, Washington, DC, www.nrscstandards.org

National Radio Systems Committee, *NRSC-G300-B, RDS Usage Guideline*, Washington, DC, www.nrscstandards.org

National Radio Systems Committee, *NRSC-G301, Creation and Distribution Practices for Audio Program Metadata Guideline*, Washington, DC, www.nrscstandards.org

National Radio Systems Committee, *NRSC-R300, Program Associated Data (PAD) Field Length Study*, Washington, DC, www.nrscstandards.org

Wright, Scott, *The Broadcasters Guide to RDS*, Waltham, MA, Focal Press, 1997

CHAPTER

7.17

FM Broadcast Antennas

THOMAS B. SILLIMAN, P.E.
Electronics Research, Inc.
Chandler, IN

ERIC R. WANDEL, P.E.
Wavepoint Research, Inc.
Crane, IN

INTRODUCTION

This chapter covers antenna systems for FM radio broadcasting and attempts to provide a comprehensive review of the many factors that must be taken into consideration by broadcast engineers, technicians, and station managers when making decisions regarding these systems. To achieve the best coverage in the station's market in addition to enjoying long-term, reliable performance, many factors must be considered. The site location and antenna height perhaps has the most significant impact on the market coverage, with an ideal site location providing line-of-sight coverage to the stations desired coverage area. However, many other factors can help optimize performance in terms of not only pattern and coverage, but also reliability when subjected to high input power conditions and mechanical stresses due to tower deflections as well as harsh environmental conditions.

There are many different antennas available today for FM broadcast stations. There are side-mount antennas that can be mounted on the side of a tower below the top guy strand of a guyed tower, and there are top-mounted antennas that can be mounted on a pole, a tower, or a custom lattice tower with structural members spaced proportionally to the bay spacing of the antenna so that each antenna element is mounted in an identical reflective environment. There are also top-mounted custom panel antennas that are broadband for multi-station use that can achieve pattern circularity typically better than ±2 dB.

The horizontal relative field patterns of FM antennas that are mounted on the side of a pole or a tower can have significant pattern distortion resulting from the current that is induced onto the steel support structure. Pole-mounted FM antennas will have a significant vertical polarization null behind the side-mounted antenna. Tower-mounted antennas with a lattice tower in the antenna aperture will have pattern distortion in both the horizontal and the vertical polarized horizontal plane relative field patterns. The relative field pattern distortion will depend on where on the lattice tower section the antenna is mounted, so with multi-bay FM antennas mounted on lattice tower sections, each bay of the array will have a unique pattern distortion unless the tower lattice is custom spaced to the frequency of operation.

Another consideration is selection of polarization. Antennas for FM broadcasting use horizontal polarization (H-pol), vertical polarization (V-pol), or circular polarization (CP). Cross-polarization, that is alternate use of H and V, is used as a means to prevent co-channel interference in some European countries but not in the Western Hemisphere. CP, together with its special form, elliptical polarization (E-pol), was introduced in the United States in the early 1960s as a means to provide greater signal penetration into the many different forms of FM receiving antennas, which are now found in the service area. H-pol is the standard in the United States, but CP or E-pol may be used if desired. V-pol only is permitted for noncommercial FM stations seeking to limit interference to TV channel 6.

FM radio receivers use a variety of antennas including extendable monopoles (whips), dipoles, and capacitive coupling to power cords and headphone leads, as well as antennas that are embedded in the glass windows of automobiles. Receiver antennas differ from their transmission counterparts, which have a fixed polarization.

For economic and technical reasons, the desired effective radiated power (ERP) should be produced with a balance between antenna gain and transmitter power. The height of the antenna over the service area, distances to areas of population, ERP, and economics are items that must be considered.

In the past decade, the addition of a digital signal in the form of digital sidebands adjacent to the existing analog signal (popularly referred to as In-Band On-Channel, or IBOC) has become more prevalent. These digital sidebands have expanded the bandwidth of operation for stations employing IBOC, and there are several solutions for adding (simulcasting) the IBOC digital carriers to the FM signal, including common mode transmitters as well as solutions involving multiplex filters, dual-input FM antennas, and interleaved FM antenna arrays. With interleaved FM antenna arrays, the two FM antennas are mounted so that the analog FM antenna array is interleaved with the FM digital antenna array. The isolation between the two arrays can be increased by having one of them right-hand circular polarized and the other left-hand circular polarized. Typical achievable isolation between digital antenna input to analog antenna input is 45 dB, and for this reason, a circulator is typically not necessary to maintain the 40 dB isolation needed. If circulators are used to isolate FM analog signals from FM digital signals, there is a likelihood of intermodulation products occurring due to nonlinear characteristics of the circulator, so a passband filter following the circulator to the antenna output may be required. Dual input side-mount FM antennas are also available with better than the necessary 40 dB isolation between analog and digital inputs. Circulators can be added for additional isolation, but once again, whenever a circulator is used in an FM transmitter output, a pass band filter may need to be added to eliminate the intermodulation products that circulators can create due to the fact that circulators are nonlinear devices.

PROPAGATION

Before discussing the details concerning the design, selection and placement of FM broadcast antennas, we first review some propagation phenomena that have an impact on some of the design choices and tradeoffs. Here we only raise an awareness of some propagation issues that motivate choices in antenna mounting height, elevation, azimuth patterns, and polarization, to name a few.

FM broadcasting, operating in the VHF frequency band of 88–108 MHz, enjoys relatively consistent propagation conditions between day and night, and therefore, FM stations have relatively uniform day and night service areas. FM propagation loss generally includes everything that impacts the radiated signal in the propagation channel between the transmitting antenna and the receiving antennas. The channel impacts include the free space path attenuation of the wave and such factors as refraction, reflection, depolarization, diffraction, absorption, scattering, ducting phenomenon influenced by changes in atmospheric pressure and humidity, Fresnel zone clearances, grazing, and Brewster angle problems.

Propagation is dependent on all these properties out to approximately 40 miles (65 km). Some additional factors enter the picture at greater distances. Radio wave propagation is further complicated because some of these propagation variables are functions of frequency, polarization, or both, and many have location and time variations.

It is the intent of the FM broadcast transmission system to put a signal into FM receivers of sufficient strength to overcome noise and interference. For analog FM it is generally desired to provide at least 20 dB carrier-to-noise ratio, which will provide at least 30 dB of stereo separation. The required RF signal level varies from about 2 μV/m (microvolt per meter) for high-sensitivity FM stereo tuners in the suburbs to about 500 μV/m for less-sensitive portables. Automobile receivers typically have wide-ranged sensitivity values. With the addition of digital radio in the so-called IBOC implementation, there can be additional sources of interference from the digital sidebands adjacent to the FM analog signal.

FM antenna manufacturers typically do not guarantee coverage, rather they supply antennas that provide certain radiation pattern requirements and gain. Many antennas are assumed to have an omnidirectional (omni) pattern, which can be approximately achieved for an antenna element in free space, but an omni pattern is rarely achieved in practice due to sources of distortion, such as support structures as well as feed lines, brackets and other appurtenances on the support structure. FM antennas are usually designed to provide a horizontal plane pattern circularity of about ±3 dB when mounted on the side of a specific tower or pole, but higher variations are possible. Pattern measurements and optimization are discussed in more detail later in this chapter.

Radiation pattern and propagation are two distinctly separate conditions that together are used to determine coverage. The pattern is the radiation that is transmitted by a given antenna in any given direction, without any propagation limitations, and can be measured on a good antenna test range, in an anechoic chamber or with near-field measurement techniques. Propagation depends on the path of intervening terrain and environmental conditions existing between the transmitting antenna and the receivers.

The actual service area signal strength contours are based on two probability factors—location variability and time variability. The FCC signal coverage charts referred to as the F(50,50) curves are used as a measure of analog FM coverage and are based on a probability of occurrence of certain field strength levels for at least 50% of the locations, at least 50% of the time at a receiver height of 9 meters. These FCC signal propagation contour charts (FCC Rules Section 73.333) are based on the assumption that average propagation conditions exist. One or more of the propagation conditions mentioned earlier may reduce the measured signal strength from the predicted values substantially.

Propagation Loss

The power radiated from an FM transmitting station incurs loss as it propagates away from the antenna. In general, the signal power varies inversely to the square of the distance from the source. The Friis transmission formula for computing power received from power transmitted includes factors for free-space loss [1], and the free-space loss (FSL) in decibels (dB) can be represented as:

$$\text{FSL(dB)} = 36.6 + 20 \log D(\text{miles}) + 20 \log F(\text{MHz})$$

An easy reference to remember is that at 100 MHz and a distance of 15 miles, the free-space loss is 100 dB. Doubling the distance increases the free-space loss by a factor of four, or 6 dB, so for example at 30 miles the free-space path loss is 106 dB. The free-space loss is not the only factor acting to attenuate the signal, however. A typical FM station signal travels through air with weather variables, over the imperfect earth's surface (ground), and through vegetation and around buildings, all of which cause additional attenuation and other propagation issues.

Refraction, diffraction, and reflection from hills, other terrain, buildings, structures, and other objects may occur in the propagation path between the transmitting and the receiving antennas. These, along with absorption, scattering, lack of Fresnel zone clearances, and other factors all can reduce the signal strengths and also cause changes in signal polarization.

Finally, care should be taken to minimize the incidental AM noise on the FM transmitter output signal. This can be done using an incidental AM noise detector and adjusting the FM transmitter loading and tuning circuits to minimize the AM noise in the transmitter output. This will reduce the noise in the receivers that are located in the fringe coverage area of the station.

Multipath Considerations

For analog FM broadcasting, a strong direct signal source is desired. When energy from two or more paths reaches the receiver (due to reflections), a condition called multipath reception occurs. Poor reception is experienced when there is insufficient strength difference between the direct and the reflected signals, because they can cancel each other where the geometry places them out of phase. Because the path lengths change relative to each other from the transmitted signal and from the reflected signal as a vehicle moves through the multipath coverage area, the two signals will go from adding to cancelling as a function of driving distance.

An important factor in achieving the goal of delivering an optimal signal to the listener is the location of the transmitting antenna. Great care must be exercised to find a suitable tower site because poor selection of the site can result in unfavorable signal propagation and poor signal quality. One very serious result of poor site selection is multipath propagation in some directions.

As an example, the transmitter should not be located so that strong reflections take place from nearby hills or mountains. This can happen when the transmitter is placed on one side of a large city and the other side of the city has a high mountain range, or in general if the coverage area is along or surrounded by hilly terrain. This is illustrated in Figure 7.17-1 where a mountain range causes reflections back into the desired coverage area. Radiation into the city directly from the transmitting antenna, as well as reflections from the nearby hills and mountains, will create two or more signal paths toward many receivers. These reflected signals can be strong enough compared to the direct signal strength to cause severe multipath problems in receivers. Beam tilt and directional azimuth patterns can be used to mitigate the impact of multipath by minimizing the signal level directed toward these natural reflectors. Ideally, it is best to locate the transmitter site on the side of the mountain with the antenna oriented toward the coverage area and away from the source of the multipath, in this case the mountain itself. It should be noted that some digital radio modulation formats, such as Orthogonal Frequency Division Multiplexing (OFDM), offer techniques for mitigating multipath interference within certain performance bounds.

In coverage areas where multipath is a problem, such as close-in coverage areas near mountains with the antenna site on the top of the mountain, careful consideration of the antenna gain is important. Too much gain will result in antenna minor lobe coverage of the desired coverage area, adding to multipath problems. Further, consideration of the amount of vertical polarization radiation from the antenna can be considered. Reducing the vertical ERP to 15% to 20% of the horizontal ERP has proven effective in reducing multipath problems in some sites with difficult terrain issues.

Multipath reflections are easy to identify. On an automobile radio, the signal will drop out, sometimes abruptly, as the vehicle moves. This effect may be rhythmic with distance while traveling slowly. It is sometimes called the picket fence effect as it acts like a picket fence alternately blocking and letting the signal pass. A field strength meter will usually reveal great variations of signal when moving, say, 100 feet (30 m) in a line with the transmitter.

Ground Reflections

Reflections from the surface of the earth and other structures play an important part in the strength and

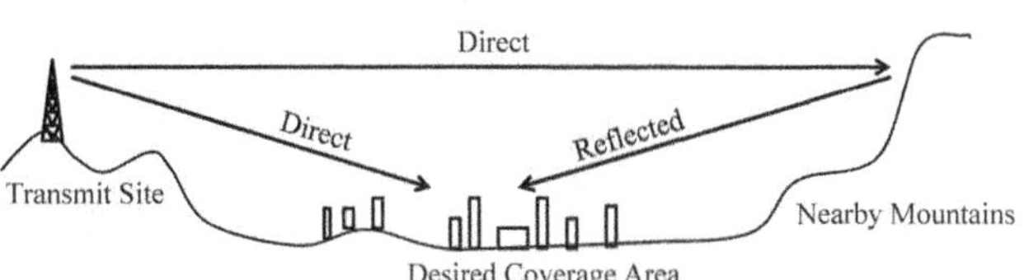

FIGURE 7.17-1 Example of multipath in mountainous environments

the quality of the signal in FM broadcasting. The geometry of the path from a typical FM broadcast antenna to receivers in the coverage area involves the angles of depression, incidence, and grazing, which are defined here. As shown in Figure 7.17-2, the depression angle is defined with respect to the local horizontal at the transmit antenna, the grazing angle is defined with respect to the local tangent to the surface, and the incidence angle is defined with respect to the normal to the surface. In general the depression and grazing angles are not equal as would be the case for a flat earth, although close to the antenna a flat earth assumption may be valid.

It is quite difficult to predict accurately the reflection coefficient (efficiency), which varies considerably as a function of polarization, frequency, grazing angle, surface roughness, soil type, moisture content, vegetation growth, weather, and the season. There are complex formulas for predicting the ground conductivity at the frequency of interest. For 100 MHz, a value of 10 millisiemens per meter (mS/m) ground conductivity is often used, with a permittivity of 25, as being about the average for the continental United States [8].

The ground reflection coefficient versus grazing angle is quite different between H-pol and V-pol [3,4]. As an example using a permittivity value of 25, the V-pol is attenuated considerably more than the H-pol, as shown in Figure 7.17-3. Recall that the grazing angle decreases toward zero as the distance from the

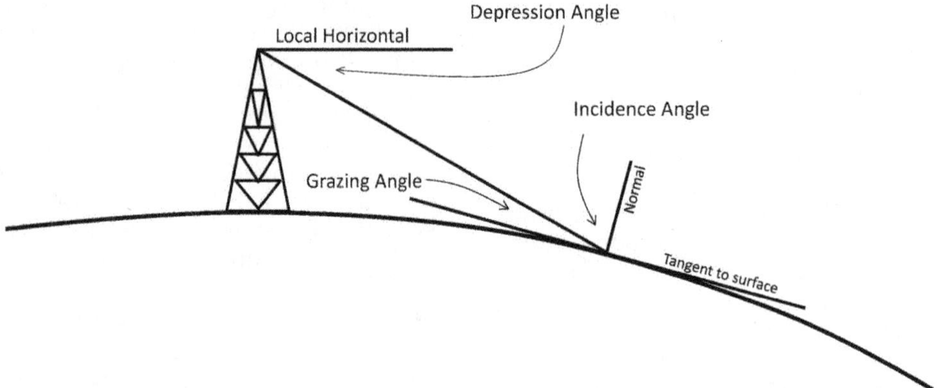

FIGURE 7.17-2 Angles used in describing the geometry of the transmit site and ground reflections

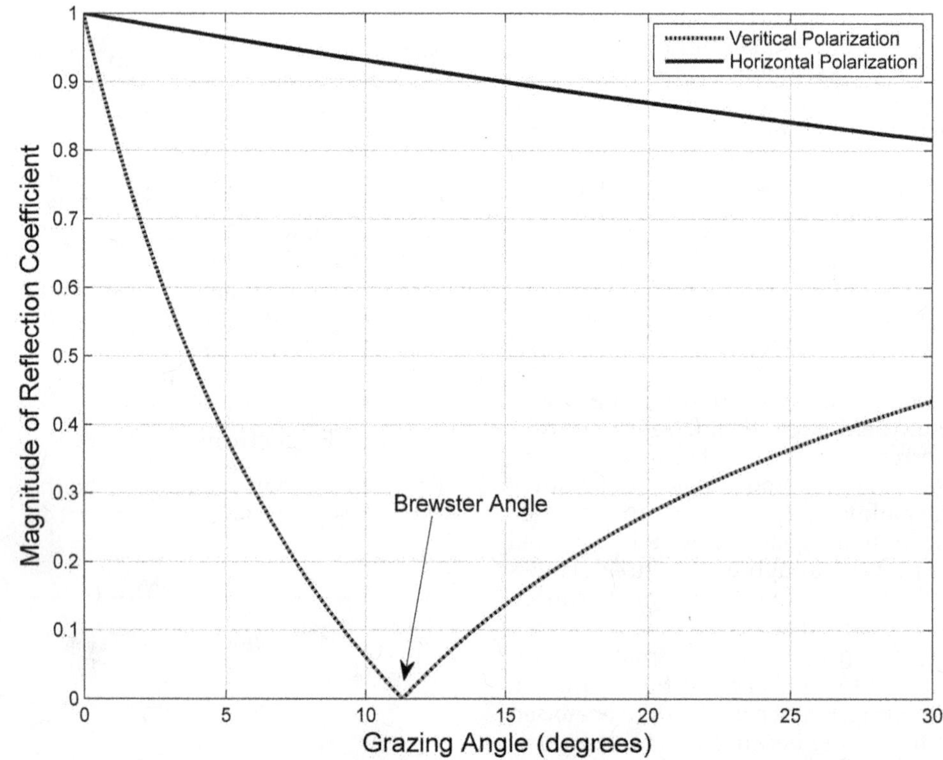

FIGURE 7.17-3 Magnitude of reflection coefficient versus grazing angle for horizontal and vertical polarization

tower increases. The phase of the V-pol reflection also changes substantially with angle, while H-pol remains nearly the same. At these useful low-propagation angles, there is considerably less V-pol signal reflected than H-pol. Field measurements confirm this fact [7]. For this reason, it is very difficult to accurately measure the axial ratios in the service area. To be meaningful, the H-pol and V-pol ratios must be measured on a good antenna test range.

Notice that in Figure 7.17-3 the minimum reflection coefficient for V-pol occurs at a grazing angle of about 11 degrees. Below this angle, the reflection coefficient rapidly increases to unity. The angle at which the minimum reflection coefficient occurs is called the *Brewster* or *polarizing angle*, after the English mathematician who first discovered this phenomenon.

For ground reflections occurring near the Brewster angle, the reflection coefficient is much smaller for V-pol than the H-pol. Therefore, the reflected V-pol signal component of CP is attenuated considerably. The greatest attenuation for V-pol from ground reflection occurs at this angle.

Field measurement of V-pol signals will usually show a significant variability of H-pol to V-pol ratios due to this Brewster angle phenomenon. The Brewster angle is also a function of soil conductivity and may change from place to place, as well as from season to season [5].

Using the results from Figure 7.17-3 and using a permittivity value of 25 where the Brewster angle is shown to be approximately 11 degrees, the distance to the Brewster grazing angle from the tower as a function of tower height can be plotted as shown in Figure 7.17-4.

Fresnel Zone Clearance

An often neglected consideration in FM transmitting antenna location and height is *Fresnel zone* radius clearance in the path to the service area. Microwave engineers always make certain that their signal paths have this important clearance.

The effect of clearance above ground or other obstacles was studied by August Jean Fresnel, a French scientist who first discovered this phenomenon in optics. Fresnel zones are circular areas surrounding the direct line-of-sight path of a radius such that the difference between the direct and indirect path length to the zone perimeter is a multiple of a half-wavelength longer than the direct path. This is illustrated in Figure 7.17-5. The zone diameter varies with frequency and path length. The greater the path length, the larger the required mid-path clearance required for full signal.

Fresnel also discovered that the entire first zone radius is not required for full signal strength. Six-tenths of the first zone would suffice, which is fortunate because the radius is quite large at FM frequencies. The equation for determining the first Fresnel zone radius for 4/3 earth curvature is

$$d = 1140\sqrt{\frac{f}{R}}$$

where d is the path length in miles, f is in MHz, and R is in feet for the first radius.

In Table 7.17-1, the required 0.6 first Fresnel zone radii clearances at the middle of the path are shown for 98 MHz and service areas up to 52 miles (92 km) from the transmitter. The idea is to raise the height of the transmitting antenna so that the mid-path height is as high as or higher than shown in the table. Due to the geometry of the Fresnel zone, if the terrain is relatively flat, the mid-path radius will control and be larger than that required elsewhere along the path. If the mid-path clearance is less than the values shown, the FM signal will be attenuated in accordance with the curve shown in Figure 7.17-6, where it is shown that an attenuation of 6 dB occurs when there is zero clearance and zero attenuation occurs just as the Fresnel zone clearance exceeds 0.6. [6]

The center-of-radiation heights of the antennas in Table 7.17-1 are actual and not height above average terrain (HAAT). Some of these recommended heights will reduce the allowable ERP in accordance with FCC Rules Section 73.211(b), depending on the class of station and the zone. However, it is better to have the Fresnel clearance than the maximum low height ERP

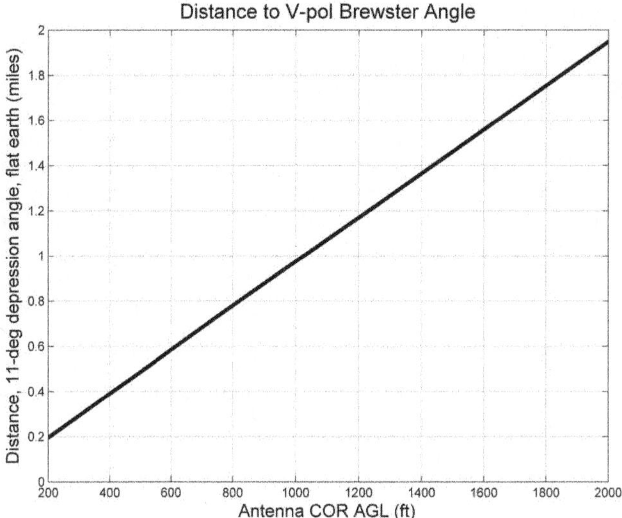

FIGURE 7.17-4 Distance to V-pol Brewster angle versus tower height

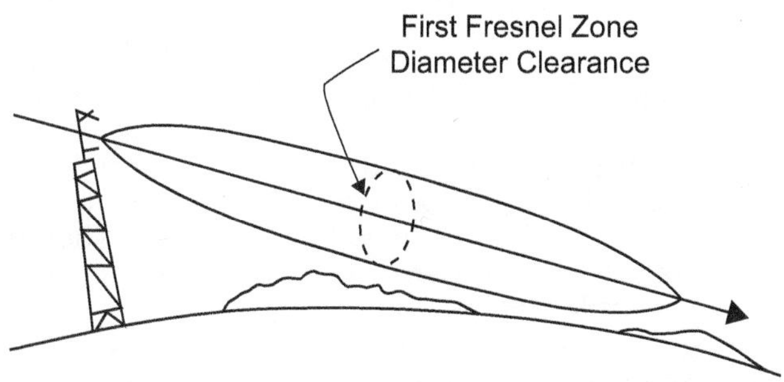

FIGURE 7.17-5 Illustration of Fresnel zone diameter at path midpoint

TABLE 7.17-1
Recommended Minimum Antenna Heights (for Flat Terrain and 98 MHz)

Service Area Radius Required (km)	Fresnel Zone Six-tenths Clearance (m)	Recommended Min. Antenna Height (m)	Probable FCC 80–90 Class
8	47	95	A
12	58	115	A
16	66	130	A
24	81	167	A, B, C-2
32	94	188	B-1, C-2
40	105	213	B-1
48	115	230	B
56	125	250	B
64	133	267	C-1
72	141	282	C-1
80	149	297	C-1
92	159	318	C

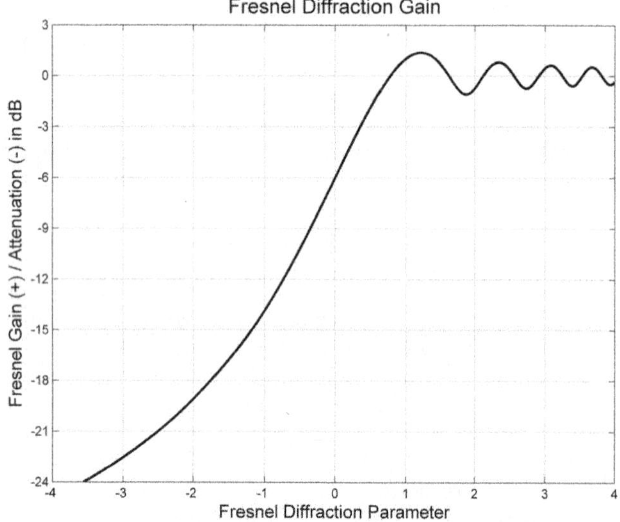

FIGURE 7.17-6 Attenuation of FM propagation when the path between transmitter and receiver lacks Fresnel zone clearance in the ratios shown

values, as the higher heights will produce stronger signals.

It is a well-known propagation axiom that greater heights are more useful in producing higher signal strengths far from the antenna than ERP levels, everything else being equal. Without the first Fresnel clearance of 60%, the signal level at the distant point may suffer. This reduction will follow the curve shown in Figure 7.17-6 for different values of clearance and worst-case reflection conditions.

In order for the FCC prediction curves to be valid, the recommended minimum antenna heights should be employed. These heights not only provide line-of-sight conditions to the service limits but also proper Fresnel clearances. Both conditions are required for the FCC F(50,50) curves to be valid.

The values in Table 7.17-1 are for relatively flat terrain, but take into consideration the FCC suggested roughness factor of up to ±150 feet (50 m). Where the tower height is limited by HAAT values or other limitations, the signal strength will suffer due to those factors.

Linear Height Gain Effect

By raising the receiving antenna above the immediate effects of the soil, the signal level will be increased. Actual field measurements have shown a 9 dB increase in signal when the dipole was raised from 3.28 feet (1 m) to a level of 30 feet (9.1 m). This is due to reflection phenomena in the foreground of the receiver, not ground conductivity.

FCC Service Contours

From the FCC coverage prediction charts, it is possible to draw contours of the various grades of service for a given ERP and antenna height above average terrain. These predictions, for at least 50% of the locations at least 50% of the time, constitute the basis for the service contours. The city grade contour is 70 dBµV/m (3.16 millivolts per meter) and primary service contour is 60 dBµV/m (1.0 millivolts per meter). The FCC Rules Section 73.333 charts for these predictions have a built-in terrain roughness factor, as explained earlier.

GENERAL COVERAGE STANDARDS

There are certain height and power levels established by the FCC for various classes of stations. The United States has been divided into three geographical zones based on population density as well as propagation refractive index levels. For each zone ERP and height values have been set to prevent co-channel and adjacent channel interference.

Zone I, generally speaking, is the northeastern part of the United States. Zone I-A includes Puerto Rico, the Virgin Islands, and that portion of California lying below the 40th parallel. Zone II includes Alaska, Hawaii, and the remainder of the United States not in the above two zones. The zones are more fully described in FCC Rules Section 73.205 and Figure 1 of Section 73.699.

Under the FCC Rules that resulted from Docket 80–90, *Modification of FM Broadcast Station Rules to Increase the Availability of Commercial FM Broadcast Assignments*, in 1983, new ERP levels and additional classes of stations were created. The distance to the 60 dBu (1 V/ m) signal contour is the controlling factor so that the ERP based on the HAAT is adjusted to produce that level and no more at a specific distance for a particular class station.

Table 7.17-2 shows for each FM class station, the zone, the maximum ERP, the maximum HAAT, and the distance to the 60 dBu contour calculated by using the maximum ERP and HAAT, and then rounding to the nearest kilometer and mile.

Stations may be upgraded using the easiest method, which is to increase existing location tower height. Such factors as local zoning laws and aircraft flight patterns may preclude this approach, however.

FM Signal Measurements

The signal strength received at 5 feet (1.5 m) above ground, which is about average for auto whip antennas, is 10–15 dB lower in level than at the standard FCC measurement height of 30 feet (9.1 m). This fact should be taken into consideration when comparing low height measurements with the FCC Rules Section 73.333 prediction charts, which are based on a 9-meter (29.53 feet) receiving height.

TABLE 7.17-2
FM Station Classes, Zones, and ERP (Section 73.211)

Class	Zone	Max. ERP (kW)*	Max. HAAT (Meters)	Distance to 60 dBuV/m (km)
A	I, I-A, II	6	100	28
B1	I, I-A	25	100	39
B	I, I-A	50	150	52
C3	II	25	100	39
C2	II	50	150	52
C1	II	100	299	72
C0	II	100	450	83
C	II	100	600	92

*Note that the max ERP shown here corresponds to the max HAAT. For antennas mounted above the max HAAT, the ERP must be reduced.

Signal levels inside houses, apartments, offices, and other structures vary widely. Levels depend on the type of building construction, but in nearly all cases will be lower than those outdoors. Reflections inside the building reduce stereo separation, and cause crosstalk problems with Subsidiary Communications Authorization (SCA) channels. Outside FM receiving antennas generally provide good reception.

Field strength measurements should not be used to determine the transmitting antenna radiation pattern or efficiency except under controlled conditions. The propagation factors discussed previously camouflage the true antenna performance. The only technically acceptable way to determine the antenna's characteristics is on an antenna test range or more accurately in an anechoic chamber.

This information may be used to determine the actual quality of service and the areas where usable signal levels in fact exist. Predicted contours may be considerably different from actual measured values. Chapter 10.5, "FM Radio Field Strength Measurements," details the measurements process.

Required Signal Strength

The history of FCC proceedings provides the rationale for the following levels for minimum satisfactory signal strength and maximum for the listed coverage areas:

- 34 dBu = 0.05 V/m rural areas
- 60 dBu = 1.00 V/m suburban areas
- 70 dBu = 3.16 V/m principal community
- 82 dBu = 12.64 V/m highest useful level.

The first three levels were established by the FCC in the early 1950s when tube receivers and H-pol antennas were popular. Modern receivers have much better sensitivity. CP has added greater signal penetrating power than H-pol when the levels were first established.

The FCC defines two grades of signal contours on applications. The first is based on the 70 dBu contour (3.16 V/m) required to cover the principal community of license. The second is the 60 dBu contour (1 V/m), which defines the primary service area.

The FCC also stated that, in rural areas, levels as low as 50 μV/m were useful. Indeed current home stereo tuners and FM auto radios operate very well with only 25 μV/m. In practice 50 μV/m (0.05 V/m) provides good quieting in nearly all automobile and portable radios receiving a stereo signal from a CP station antenna. Therefore, 50 μV/m should be considered the minimum useful signal level.

If the highest level of 3.16 V/m is quadrupled, it will be 12.64 V/m. This is a 12 dB increase, equal to increasing the FCC power level by more than 15 times. It can be safely said that this level of 12.64 V/m is considerably more signal than necessary by any present-day working FM radio. Any signal level higher than this at the receiving antenna has not proven to be of significant value.

Blanketing

Excessive RF signals can overload the front end of receivers and make satisfactory reception impossible. The FCC Rules Section 73.318 defines the 115-dBu (562-mV/m) level as the *blanketing contour*, and adopted the free-space prediction method to predict how far this contour extends.

New or modified FM stations have the responsibility to satisfy all complaints at no cost to the complainant, of blanketing-related interference inside this contour within 1 year of commencement of operations.

The distance to the 115-dBu contour is determined using the following equation:

$$d(\text{dBu}) = 0.394\sqrt{P}$$

or

$$d(\text{dBu}) = 0.245\sqrt{P}$$

where P is the ERP, in kilowatts of the maximum radiated lobe, irrespective of vertical directivity. For directional antennas, the horizontal directivity shall be used.

ANTENNA CHARACTERISTICS

Antenna gain can be increased by adding additional radiating elements (bays) to the antenna at the cost of narrowing the radiated beam. High-gain antennas concentrate the energy into such a narrow beam that often null fill must often be employed to achieve the desired signal strength within the first few miles to the tower. For mountain top antenna farms, second null fill as well as first null fill is usually specified.

Directional antennas achieve increased gain over nondirectional antennas by limiting the radiated energy to specified azimuth directions. Directional antennas are useful when the tower is located near a large body of water, mountain range, or other areas where energy radiated in those directions is otherwise wasted. They are also employed to avoid interference where stations are not far enough apart.

Antenna gain is expressed in power ratio or in dB. For example, an antenna with a power gain of 2 is also said to have a gain of 3.0 dB.

FCC Rules Section 73.310(a) defines antenna gain as the inverse of the square of the root mean square value of the free-space field strength produced at 1 mile in the horizontal plane, in millivolts per meter for 1 kW antenna input power to 137.6 V/m (in metric units, 1 km and 221.4 V/m). Note that this gain is in reference to a horizontally polarized half-wave dipole. For a CP antenna, the gain is half for the same input power.

A two-bay H-pol antenna has a power gain of approximately 2. But a two-bay CP antenna in FCC terminology has a gain of about 1 because the other half of the power is V-pol and is not considered in the gain calculations. The vertically polarized energy must not exceed the H-pol (except for noncommercial, educational FM facilities attempting to minimize interference to TV channel 6 reception).

The power gain of an antenna is used with the transmitter power and transmission line and other losses when determining the ERP. Consider for example a 10 kW transmitter and an antenna power gain of 5. Neglecting transmission line loss, the ERP is 10 kW × 5 = 50 kW ERP. If the antenna gain were 10 and the transmitter power were 5 kW, the same ERP of 50 kW (5 kW × 10 = 50 kW ERP) would be obtained.

The FCC defines ERP to mean the product of the antenna input power (transmitter output power less transmission line and other loss) times the antenna power gain. Where circular polarization is used, the term ERP is applied separately to the H-pol and V-pol of radiation. For allocation purposes, the ERP is the H-pol component of radiation only. The V-pol component power normally must not exceed the H-pol power.

Beam Tilt

FM broadcasting antennas are normally mounted on towers such that the peak power (main beam) of the elevation pattern is perpendicular to the tower axis and directed horizontally. For a typically symmetric elevation pattern in a standard FM antenna without any beam tilt, more than one-half of the total radiated power is above the horizon when earth curvature is taken into account. The higher the antenna above the terrain, the greater the elevation angle down to the earth's horizon.

In order to strike the farthest service area from a high HAAT, the beam may need to be tilted down toward the earth. Electrical beam tilt lowers the beam angle equally in all azimuth headings and is chosen more frequently than mechanical tilting, which exhibits different effects in different directions. Choose enough tilt to position the center of the main beam on the farthest edge of the desired coverage area or just below the horizon, whichever is closer. Refer back to Figure 7.17-2 for an illustration of depression angle, which defines beam tilt, and how it relates to the grazing angle.

For low gain antennas (two to four bays), the main beam is very broad, and if the antenna HAAT is less than 500 feet, there is little to be gained with beam tilt. On the other hand, beam tilt makes a large difference on high-gain antennas mounted on towers with a high HAAT.

Beam tilt is usually accomplished electrically, by delaying the currents to the lower bays, and advancing the phase of the upper bay currents during the design and construction of the antenna at the factory. This can be achieved by offsetting the center feed input to the antenna toward the upper half of the array. Typical values for beam tilt range from 0.5 to 1.0 degree of tilt, depending on the antenna height, distance to the far service area, and the antenna elevation pattern. An example elevation pattern for an 12-bay full-wave spaced antenna with 0.75 degree beam tilt is shown in Figure 7.17-7.

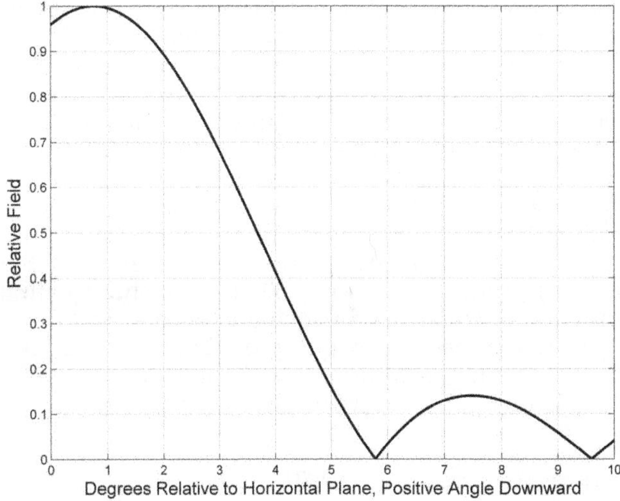

FIGURE 7.17-7 Example of an elevation pattern for an 12-bay full-wave spaced antenna with 0.75-degree beam tilt

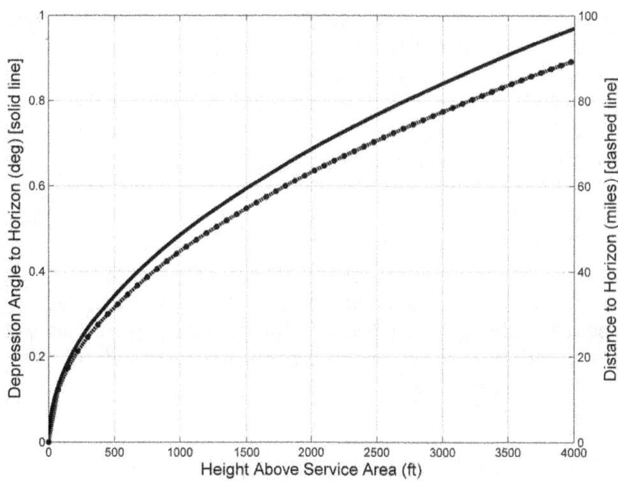

FIGURE 7.17-8 Depression angle and distance to the horizon as a function of height above service area

The station class, antenna elevation pattern gain, the distance to horizon as compared to desired service area, and the depression angle to the horizon are all factors to consider when making decisions about beam tilt. Figure 7.17-8 shows the depression angle to the horizon as a function of height of antenna above the service area and also shows the distance to horizon—both values based on the 4/3 earth curvature. As an example, at a height of 1000 feet, the distance to the horizon is 44.7 miles and the depression angle is 0.49 degrees.

Null Fill

While the beam tilt puts more signal into the far reaches of the service area, it does not solve the problem sometimes caused by high-gain antennas within several miles of the transmitter. Elevation angle nulls common to all antennas with two or more bays appear farther and farther away from the antenna as its gain is increased with more bays.

When multiple bay arrays are employed, lobes and nulls occur in the elevation pattern. As the number of bays increases, the main beam narrows and the first null radius increases. The advantage of beam tilt and null fill varies depending on factors such as tower height, site elevation, number of bays, and relative locations of communities to be served. A simple rule of thumb is that null fill is beneficial when there is desired service area within the radius of the first null.

In most FM applications, the null is relatively close to the antenna, thus a small amount of null fill (5%–10%) takes care of the problem. Larger amounts of null fill are unnecessary and reduce the gain of the antenna. Note that null fill has no effect on distant coverage.

Null fill can be employed in one or more of the pattern nulls, where the first null is referenced as the first null next to the main beam and farthest away from the transmitter site. Null fill can be achieved with variations in either or both of the power and phase distributions in the antenna array factor.

VSWR Bandwidth

According to theory, the bandwidth of an FM signal is infinite if all the sidebands are taken into account. Also, at certain modulation indices, the carrier amplitude goes to zero and all the transmitted power is on frequencies (sidebands) other than the carrier frequency. Practical considerations in the transmitter and receiver circuitry make it necessary to restrict the RF bandwidth to less than infinity.

Prior to 1984 the maximum deviation for FM stations was 75 kHz, representing 100% modulation. In that year the FCC changed the maximum deviation to 82.5 kHz (110%) for those stations with 10% injection of subcarrier channels. This additional deviation requires greater antenna system bandwidth than previously needed.

System bandwidth is measured at the point in the antenna system where the transmitter is connected. This usually includes the harmonic filter, the main coaxial transmission line, and the antenna.

The significant sidebands are usually considered to be those whose amplitude exceeds 1% of the unmodulated carrier. With 110% modulation (82.5-kHz deviation) these sidebands produce a bandwidth of 260 kHz.

The voltage standing wave ratio (VSWR) bandwidth is the range over which the system under consideration has a reflection coefficient of less than 5%—a VSWR of 1.1:1. Digital radio transmission systems using IBOC technology place greater emphasis on VSWR bandwidth as the sideband relationship to carrier is essential for optimum operation.

Checking System VSWR

The VSWR of the narrowband antenna system should be checked regularly and adjusted as necessary. One of several methods for checking VSWR in coaxial line systems using test equipment may be used. These include a signal generator test setup, an impedance test set, or a network analyzer.

The VSWR should be measured to ensure that the reflection response is balanced to 200 kHz on each side of the carrier frequency, which accommodates digital radio sidebands. With transmission lines longer than 300 feet (100 m), it is suggested that the VSWR bandwidth should be less than 1.08:1 for a bandwidth of 260 kHz. The additional delay due to increasing line length becomes more of a problem for digital signals, so the amplitude of the reflection must be reduced for best operational results.

Antenna Impedance Matching Techniques

The impedance bandwidth of an antenna is usually defined as an operating frequency band over which the VSWR is below a certain value. For FM antennas mounted on tower and building rooftops, stability of the VSWR during variations in environmental conditions such as temperature, precipitation, icing, and wind conditions is also important to consider.

Ideally the desired impedance bandwidth is achieved at the feed point of the antenna and then transformed to the antenna system input through a series of transformers and the feed hardness of the antenna system. Despite achieving a desired bandwidth in the antenna system at the factory, the VSWR of the antenna can be degraded due to the reflections from the mounting structure when installed. For this reason, "fine matching" the antenna after installation is commonly performed to optimize the system. The two most common methods are to use either adjustable fine matching sections or "slug" tuning to add reactive elements at specific locations in the transmission line below the antenna input.

Fine matching sections typically employ a series of plungers or tuning screws spaced along a section of rigid transmission line such that the reactance of the antenna impedance can be zeroed while matching the antenna impedance to the characteristic impedance of the transmission line. Fine matching designs exist that allow the addition of capacitance or inductance. These fine matching sections are typically built to allow pressurization and can be adjusted externally without disconnecting line sections for access to the inner conductor. In general, one fine matching section is capable of matching the antenna impedance at a single frequency of operation. The bandwidth is still somewhat dependent on the inherent bandwidth of the antenna. Multiple fine matching sections can be used to match an antenna to multiple frequencies.

Slug tuning can be accomplished by fixing dielectric rings, metal collars, or metal sleeves to the inner conductor of rigid transmission line, typically using a stable, aggressive adhesive vinyl tape, set screws, or even soldering the sleeves to the inner conductor. These slugs function in a similar way as the fine matching sections by adding capacitance to the line at specific locations in order to zero the reactance of the antenna impedance while matching the antenna impedance to the characteristic impedance of the transmission line. Note that because these add capacitance, they generally must be placed at an inductive location on the line. Slug placement requires access to the inner conductor, so the transmission line must be disassembled when a slug is placed or moved. The length of the slug determines the shunt capacitance the slug adds to the antenna impedance. The correct amount of slug is added to the antenna at a location where the antenna input impedance is 50 ohms in shunt with a known inductance, and the slug is selected to have a capacitive impedance equal to the impedance of that shunt inductance. Multiple slugs can be used to tune the antenna for use at multiple frequencies.

Importance of Low VSWR

The VSWR shown by the transmitter reflectometer does not have a significant impact on the range of the signal, but increased VSWR can impact coverage by reducing the quality of the signal and specifically the stereo signal. Degraded reception can be observed in fringe coverage areas. VSWR values above 1.1:1 may also decrease the final amplifier efficiency in older transmitters or cause transmitter fold back (reduced output power). Other possible negative effects of high VSWR are increased intermodulation products and increased AM synchronous noise [9].

Intermodulation and SAM Distortion

Intermodulation distortion and synchronous AM (SAM) noise can be caused by narrow VSWR bandwidth in the antenna system, as well as by final amplifier circuitry [10]. SAM is important in FM transmitter facilities employing subcarriers. SAM is AM modulation of the carrier caused by frequency modulation of the carrier frequency in the VSWR notch. At the notch the reflected energy is the lowest. As the deviation takes place, the greater the frequency swing, the greater will be the reflections due to the VSWR notch. With a flat VSWR curve, SAM does not take place. If the VSWR curve is skewed, SAM will occur and intermodulation and stereo crosstalk will increase. Where fringe coverage is important, synchronous AM noise should be minimized by making transmitter adjustments while measuring SAM.

Directional Antennas

The FCC sometimes requires that the azimuth radiation pattern be made directional to reduce the distance spacing of normally allocated ERP toward a given short-spaced station, or for other reasons. [See the FCC Rules Sections 73.213, 73.215, and 73.316(b)–(c).] To conform to these specifications, most broadcasters order antennas that are pattern adjusted, measured, and certified to the FCC's requirements, and the antenna proof is filed with the FCC.

Directional antennas are licensed for peak ERP values based on the azimuth pattern. The V-pol ERP may not exceed the H-pol ERP. The amplitude away from the null for short-spaced stations cannot climb more than 2 dB per 10 degrees of azimuth, and the depth of the null cannot exceed −15 dB [reference Section 73.316(b), Directional Antennas of the FCC Rules]. The final RMS (root mean square) of the measured antenna pattern must be at least 85% of the authorized composite directional antenna pattern in relative field.

Directional antennas may be top mounted on its own structure, side mounted on the face or leg of a tower, or side mounted on a standoff pole that is then mounted to a tower. Because the support affects the pattern, the antenna is specified and measured with the pole or tower on which the antenna is mounted. Exacting details of the planned mounting structure are needed in order to accurately conduct a pattern study, as all conducting tower members, conduits, transmission lines, and other appurtenances will impact the pattern shape.

Creating a directional pattern is a combination of the natural pattern resulting from side mounting as well as the use of parasitic elements. Using these two factors, a wide variety of directional patterns can be achieved.

For directional antennas, the gain is a function of the azimuth gain in addition to the elevation gain and an efficiency factor. The elevation gain relative to a half-wave dipole is a function of the array factor and individual element patterns. A horizontal dipole has a nominally omnidirectional element pattern in the elevation plane normal to the length of the dipole, for example, but a vertical dipole exhibits an element pattern with a null directly up and down. These differences in element patterns have an impact on the overall elevation gain.

It is worth noting here that the array factor and elevation patterns of the H-pol and V-pol components may be different for certain types of antennas, such as the four-dipole flat-panel antenna shown in Figure 7.17-18, so it is possible that different elevation gains will be applied to each polarization.

For dual polarized antenna systems, the power split between polarizations must also be taken into account. For instance, in a circularly polarized antenna, the power is split equally between the H- and V-polarizations and therefore reduces the overall gain per polarization by a power factor of 2. This factor along with the efficiency factor is typically accounted for in the overall elevation gain that is reported for an antenna.

The azimuth gain of a pattern to be used as a directional is calculated by using the relative field values taken from at least 36 evenly spaced radials for the entire 360 degrees of azimuth, where the maximum relative field value is normalized to a value of 1.0. The RMS value is calculated as:

$$RMS = \sqrt{\frac{1}{N}\sum_{N}(Rel\,Field_n)^2}$$

and the power gain, G, is given by:

$$G = \frac{1}{(RMS)^2}$$

The total antenna gain in each the H-pol and V-pol are calculated by the formulas:

$$\text{Antenna gain (H-pol)} = G_{AZ,\,H\text{-}pol} \times G_{EL,\,H\text{-}pol}$$

$$\text{Antenna gain (V-pol)} = G_{AZ,\,V\text{-}pol} \times G_{EL,\,V\text{-}pol}$$

where the elevation gains include the contributions due to the array factor, the element elevation pattern, and the array efficiency. A typical efficiency factor of 0.95 may be used to account for insertion and other losses in the feed harness, although more exact calculations or measurements can be made.

In order to limit the V-pol ERP to be less than the H-pol ERP in a CP antenna where there is an equal power division between polarizations and when the elevation gains are identical, it follows that the V-pol azimuth gain must be less than or equal to the H-pol azimuth gain: $G_{AZ,V\text{-}pol} \le G_{AZ,H\text{-}pol}$. Because the power input to the antenna must be based on a maximum allowable ERP in the horizontal polarization, the ERP of the V-pol signal may in general be less than the H-pol ERP, although with skilled personnel at antenna test ranges, the directional pattern requirements can generally be met while achieving essentially equal gain in each polarization.

ANTENNA POLARIZATION

Radio waves are composed of electric and magnetic fields at right angles to each other and to the direction of propagation. When the electric component *E* is horizontal, the wave is said to be horizontally polarized. Such a wave is radiated from a horizontal dipole. References are with respect to the earth plane. If the desired electric component is vertical, a vertical dipole could be used to produce the vertically polarized wave.

Circular Polarization

When the two plane waves are equal in magnitude, and if one plane wave lags or leads the other by 90 electrical degrees, the field will rotate as shown in Figure 7.17-9, at the speed of the carrier frequency, and will be polarized circularly.

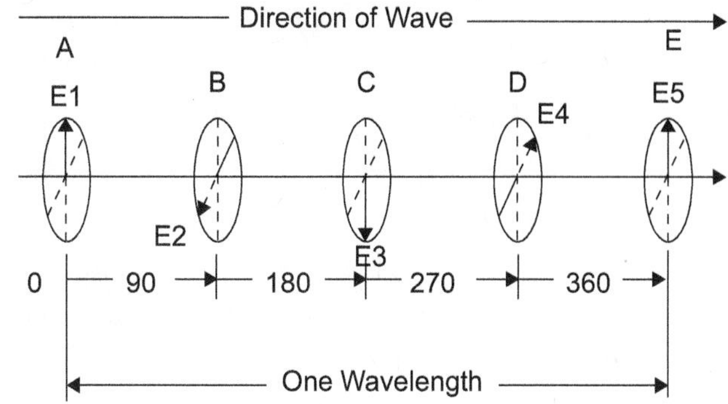

FIGURE 7.17-9 Circularly polarized wave propagation in one wavelength of travel, showing right-hand rotation. Note vector rotation with wave travel.

Only in the special case where the horizontal and vertical components are equal in strength with a 90-degree phase difference is the radiation said to be CP.

The direction of rotation shown by the vector arrows in Figure 7.17-9 depends on the relative phase of the two components. Thus, the polarization of the wave will appear to have either clockwise or counterclockwise rotation, as shown. The FCC has set clockwise rotation as the technical standard in order that similar sense of rotation antennas may be used for reception in the future.

Notice that in Figure 7.17-9 the polarization rotates as the field propagates in time and space. Importantly, vertical and horizontal components are in quadrature phase. It is this rotation that enhances the signal penetrating qualities of CP, tending to help avoid a total dropout of signal in the received polarization.

The axial ratio as shown in Figure 7.17-10 is that between the maximum and minimum voltage component at any orientation of the reference measuring test dipole that is placed perpendicular to the direction of propagation. An axial ratio of 1:1 (0 dB) is perfect. In practice, axial ratios of 2 dB or better are considered to be excellent and commercially available. Axial ratios over 4.9 dB (1.75 to 1 voltage ratio) are considered to be elliptically polarized, a hybrid form, and not as good in signal-penetrating qualities as CP.

Because most receiving antennas are linearly polarized, the introduction of CP does not increase the net power received because the vertical and horizontal components never occur during the same instant. Even though CP requires twice the transmitter output power, the signal does not propagate additional distance from the transmit antenna. Thus, using CP does not necessarily mean an increase in coverage. However, the introduction of CP eliminates the requirement of the receiving antenna to have a specific polarization (H or V). Thus, CP allows more consistent coverage within the contour. Its rotating vector can penetrate areas where linear polarization is stopped, shadowed, or cancelled due to out-of-phase reflections.

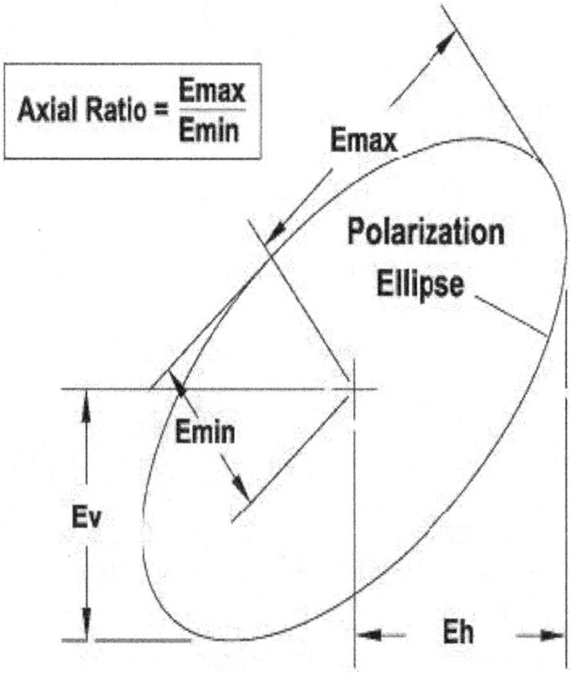

$$\text{Axial Ratio} = \frac{E_{max}}{E_{min}}$$

FIGURE 7.17-10 Axial ratio is the ratio of the larger polarized component divided by the smallest. This may be expressed in dB as $20\log(E_{max}/E_{min})$.

MATCHING COVERAGE AND ANTENNAS

Table 7.17-3 shows the FCC-predicted signal strengths for a typical Class A facility on a relatively flat plane, with the antenna center at 328 feet (100 m) HAAT. A power of 3 kW is used. The first two columns show the distances, with the farthest being the horizon from this height. The third column indicates the true earth angle from the antenna to the distances shown. From the elevation information the ERP from each antenna was determined at each vertical angle. This ERP value was used to find the signal strength from the FCC F(50,50) FM prediction chart, FCC Rules Section 73.333, Figure 1. Under the signal level in millivolts-per-meter (mV/m) column, the predicted field strengths shown in Table 7.17-3 are based on the above procedure. From 5 miles (8 km) to the horizon, the signal strengths are identical. This is due to the shape of the antenna elevation pattern near the maximum.

Departure occurs as the depression angle to the receiver becomes larger. Beyond 4 miles (6.4 km), the one-bay antenna and the six-bay antenna produce nearly the same signal level.

Going toward the transmitter from 4 miles (6.4 km), the field increases in favor of the one-bay antenna. In this example, the table clearly indicates that the high-power transmitter, low-gain combination antenna does not improve the signal strength available to the receivers beyond about 4.5 miles (7.25 km). The signal level starts to increase between 4 and 5 miles (6.4 and 8 km). Any increase above this level is useless because full limiting has certainly taken place in even the poorest FM receiver.

In Table 7.17-3 the same signal strength of (16 V/m) at 5 miles (8 km) and beyond comes from either transmitter-antenna combination. This is due to the fact that the ERP power at the vertical angle of −0.74 degrees is about the same from both antennas. The ERP at 0.0 degree elevation pattern will of course be exactly the same for both combinations. The field does not change measurably until observation is made beyond 1.5 degrees from the peak value in a six-bay antenna.

The signal strengths in Table 7.17-3 were based on relatively flat terrain for an antenna 328 feet (100 m) HAAT. The true earth curvature distance to the horizon is 25.56 miles (41.23 km). Therefore, the useful signal drops off very rapidly beyond this point for the typical Class A station.

There are no nulls in a one-bay antenna pattern. In a six-bay antenna the first null occurs at about −10 degrees, which is approximately 0.37 miles (0.6 km) from the tower. Antenna arrays are never perfect, so

TABLE 7.17-3
Transmitter Power Versus Antenna Gain for a Class A 3kW ERP Station*

Service Distance		Vertical Angle	Signal Level (mV/m)	
(miles)	(km)		7.5 kW TX 1-bay Antenna	1 kW TX 6-bay Antenna
1	1.6	3.58	275	210
2	3.2	1.80	88	81
3	4.8	1.21	42	40
4	6.4	0.92	24	22
5	8.0	0.74	16	16
6	9.6	0.63	11	11
7	11.3	0.55	8.5	8.5
8	12.9	0.49	6.2	6.2
9	14.5	0.44	5.0	5.0
10	16.1	0.41	3.7	3.7
12	19.3	0.36	2.5	2.5
14	22.5	0.33	1.8	1.8
16	25.7	0.31	1.4	1.4
18	28.9	0.29	1.1	1.1
20	32.2	0.28	0.85	0.85
22	35.4	0.28	0.70	0.70
24	38.6	0.27	0.55	0.55
26	41.8	0.27	0.40	0.40

*Showing the signal level at various distances for two transmitter/antenna combinations—zones 1 and 1-A maximum HAAT 328 feet (100 m)

the null is never zero power. With a minimum radiation of 5 W in the first null, the predicted signal would be 31 V/m. The second null is closer to the tower and with the same 5 W ERP would be even stronger in this example. So in practice there may be no need to fill in the nulls of the six-bay antenna.

Another consideration is that the nulls may fall close to the tower and the number of people within the null areas may be small. Thus, problems resulting from these close-in nulls would be minor.

Class B and C-2 Station Coverage

The same comparisons of transmitter-antenna combinations can be made for Class B and the new Class C2 stations, operating with a HAAT of 492 feet (150 m) with 50 kW ERP. This is shown in Table 7.17-4. A 55 kW transmitter with high-efficiency coaxial lines and a two-bay CP antenna would provide the 50 kW ERP. It is compared with a 10 kW transmitter feeding a 10-bay CP antenna. The terrain flatness is assumed not to exceed ±150 feet (50 m).

Table 7.17-4 also shows that the signal levels are the same from 4.6 miles (7.4 km) to 35 miles (56 km) under similar columns as for the Class A station comparisons. The FCC uses a receiving height of 30 feet (10 m) so the horizon is a bit farther away, at 31.3 miles (50.5 km).

TABLE 7.17-4
Transmitter Power Versus Antenna Gain, Class B, C-2 50 kW EFP Zone 1, 1-A, and C-2

Service Distance		Vertical Angle (deg)	Signal Level (mV/m)	
(miles)	(km)		55 kW TX 2-bay Antenna	10 kW TX 10-bay Antenna
1	1.6	4.97°	900	140
1.5	2.5	3.25°	562	165
2	3.2	2.49°	310	230
3	4.8	1.67°	153	135
4	6.4	1.26°	92	88
4.6	7.4	1.15°	71	71
5	8	1.02°	57	57
7.5	12	0.70°	22	22
10	16	0.55°	13	13
15	24	0.41°	6.5	6.5
20	32	0.36°	3.1	3.1
25	40	0.33°	1.9	1.9
30	48	0.328°	1.1	1.1
35	56	0.332°	0.7	0.7

From the transmitter out to about 2 miles (3.2 km), the signal rises much more rapidly in the two-bay antenna than in the 10-bay, the latter being somewhat similar to a cosecant curve. There is surplus signal close in and more than is needed or can be tolerated. This is one of several problems with high transmitter power and low-gain antenna combinations. With the two-bay antenna there is 900 V/m at 1 mile (1.6 km) and 562 V/m at 1.55 miles (2.5 km). This is above, or at, the blanketing level of 562 V/m discussed earlier. The high-gain antenna does not cause this type of problem under identical conditions.

The signals from both combinations are much more than necessary for present-day FM receivers out to about 10 miles (16 km). There is no practical difference technically in usable signal strengths presented to receivers in the entire market area from either antenna. There is, however, a great deal of savings in capital costs as well as operating expenses between the two combinations.

One antenna factor is not clearly indicated in Table 7.17-4. The two antennas have elevation pattern nulls. The two-bay antenna null at −30 degrees falls 852 feet (260 m) from the base of the tower and can be disregarded. The 10-bay antenna null can be filled to as little as 2.5% field, which will not affect its gain. This would represent a minimum ERP at the nulls of 31 W. Although seemingly very small, it is effective, as shown in Table 7.17-5.

TABLE 7.17-5
2.5% Null-Fill, 10-Bay Antenna

Null	Angle	ERP	Distance	Field
First	−5.75°	31 w	4,800 ft	31 V/m
Second	−11.50°	31 w	2,240 ft	70 V/m
Third	−7.25°	31 w	1,512 ft	109 V/m

Note that the 10-bay antenna nulls can easily be filled to produce signal levels in excess of those required. If the transmitter is located in a populated area, these high levels prevent the loss of stereo separation and noise in the SCA (if there are reflections from high-level lobes in the built-up areas). This problem is common to TV transmitters that produce ghosts from high signal level lobe areas reflecting into null areas. This problem is greatly and satisfactorily reduced with null fill, as shown in Table 7.17-5.

MATCHING TRANSMITTER POWER AND ANTENNAS

Several available combinations of antenna gain and transmitter power will provide the necessary ERP. The choice is complicated by the nature of the terrain in the service area such as flat land, some rolling hills, mountains, or perhaps a large valley. It now becomes necessary to understand the regulatory limitations on the tower height.

Important considerations when choosing the transmitter power and the gain combination to produce a given ERP are as follows:

- Transmitter
- Feed system
- Antenna
- Final amplifier
- Tower
- AC power consumption.

The transmitter, antenna, tower, and coaxial feed line are one-time capital costs for the station. Tubes and commercial power, however, are continuing hour-by-hour cost factors. A low-power transmitter with higher gain antenna is often a much more economical solution than a high-power transmitter.

The ERP is the product of the antenna power gain and the antenna input power. Many different combinations of power gain and input power will yield the same ERP. The azimuth pattern will be quite similar for many different antenna power gains. The only difference in various combinations is the elevation pattern. As discussed previously, there is no significant or important difference in serving listeners from very different transmitter/antenna ratios.

The signal strength at any given location is a direct function of the ERP from the antenna elevation pattern angle to that location, the height of the antenna, and the propagation path. The ERP at the pertinent angle is the product of the elevation pattern relative amplitude at that angle squared, times the maximum ERP.

In practice there is no significant difference between a 3-kW ERP Class A station using a 7.5-kW transmitter and a one-bay CP antenna, or, one using a 1-kW transmitter and a six-bay CP antenna, all other factors being equal.

Normally, all the power radiated above the antenna elevation pattern to the horizon is wasted. It is the radiated power below the angle to the horizon that reaches FM receivers. Therefore, only the radiated power toward the earth should be considered useful.

The ideal antenna system would put the same signal level from the base of the tower all the way out to the horizon. This requires an antenna whose elevation pattern is a cosecant curve, the normalized reciprocal of sine. It would be the most efficient antenna elevation pattern. Although this curve is impossible to achieve, it is approached as the antenna gain becomes greater.

ANTENNA SITE SELECTION

The transmitter location must be carefully chosen and can be an iterative process between satisfying FCC requirements—separation, city of license coverage, and other interference protections—and achieving desired performance. Achieving desired coverage objectives is a primary objective, but site economics also plays a role. Fresnel zone clearances and other factors outlined in this chapter should be considered.

If a candidate site has an existing operating FM or VHF TV station, it makes an excellent source of signals to check propagation for a new station, although care must be taken to understand the operating parameters of the existing facility to make sure you are comparing apples to apples, including whether or not the existing antenna pattern is directional or otherwise distorted by the mounting structure.

In the past, a good field strength meter would be used to measure the actual signal from the existing station where relative rather than absolute readings are important. Checks would be made for reflections as well as level changes within a short walking area of about 100 feet (30 m); also checks for stereo separation. Using this information, the operation of a new station near the one being checked can be directly compared before moving or submitting the FCC application.

There now also exist a great number of computer-based tools for assessing and analyzing potential site performance. It is easy to search for potential existing tower sites, either on the FCC website or on the websites of the so-called vertical real estate tower owners. Terrain profiles between a candidate site and locations in desired coverage areas can be easily checked with free online 3D mapping programs, for instance, to check Fresnel zone clearance past terrain features and to assess required tower heights. Further, there are a number of propagation modeling programs that provide several terrain-based coverage analyses to very accurately estimate field strength and contours while taking into account terrain, ground clutter, urban clutter, and other factors. With these tools, the initial vetting of a potential site can be done from behind a computer.

High-Gain Antenna Contradictions

The advantages of high-gain, low-power transmitter combinations to produce the required ERP have been discussed. Their superiority in relatively flat land applications cannot be disputed. There is, however, the matter of unusual height over average terrain to be considered. As examples, if the transmitter is located on Mount Wilson, in California, or on a very tall building in Chicago or New York, the elevation pattern issue can become a serious problem. This is true particularly when a significant portion of the audience is near the sites, as is the case for these three locations.

Mount Wilson, which serves the greater Los Angeles metropolitan area, is more than 1 mile (1.6 km) above most of its listeners. Coverage is required from 11 miles (17.75 km) out to the horizon, which is −0.57 degrees at 105 miles (168 km). Pasadena, the nearest city, is 13 degrees below the horizon. A high-gain antenna tilted down 0.5 degrees would serve the far reaches well, but would not lay down a moderate signal at −13 degrees.

FCC Rules Section 73.211 limits the ERP for over-height antennas such as those on Mount Wilson with 2,900 feet (884 m) HAAT. New stations using that height must reduce ERP in accordance with the equivalence calculation, so that the predicted signal at the 1-mV/m contour does not extend beyond 32 miles (52 km) for Class B stations.

In these situations a moderate (rather than high) gain antenna should be considered. From Mount Wilson several existing four- and five-bay antennas now provide excellent service.

TV CHANNEL 6 PROBLEM

Television channel 6 occupies the band 82–88 MHz with the sound carrier at 87.75 MHz. The FM broadcast band is in the range 88–108 MHz. Noncommercial educational FM stations are assigned 88–92 MHz. Interference can exist between the two, with the TV station viewers receiving sound and picture interference from the FM stations and channel 6 signals interfering with low-powered FM stations in the lower part of the FM band. The FM receiver is relatively selective with a response to about 200 kHz, but the TV receiver has a bandwidth of at least 6 MHz. (See FCC Rules Section 73.525.)

Three techniques can be employed to minimize channel 6 interference from FM stations: collocation, locating the FM station in an area of low population density, and antenna cross-polarization.

Collocation

The purpose of collocation (that is, placing the FM transmitter at the channel 6 transmitter site) is to achieve the same propagation path for both TV and FM stations, thus maintaining a nearly constant desired-to-undesired signal ratio in the service area. If possible, both antennas should be mounted on the same tower. If not, a maximum separation of 0.25 miles (400 m) between the two is still considered as collocation.

The horizontal and vertical plane radiation patterns of both antennas should be similar because the objective is to maintain a near constant desired-to-undesired signal ratio. The HAAT should be similar, thus the desirability of collocating on the same tower. The maximum ERP of the FM stations operating on this basis is specified in FCC Rules Section 73.525(d), table B.

Alternate Locations

The FM station may not be intended to serve the same community as the TV station, or collocation may not be possible. In this event, the FM broadcaster should be located in an area of relatively low population density by imposing a limit on the population that may be included within that area where a particular undesired-to-desired protection ratio is exceeded.

Two ratios were proposed by a committee that studied this problem in 1983 [8]. Their recommendation varied according to the separation between the FM station from the channel 6 aural frequency of 87.75 MHz. In any event, the interference area should not have more than 3,000 people living in it. [See FCC Rules Section 73.525(c) and (e).]

Cross-Polarization

Several organizations have made discrimination tests in the United States and in Europe with cross-polarized antennas from which it has been established that a discrimination of 16 dB can be expected in rural areas and 10 dB in urban areas between two stations with one using V-pol and the other using H-pol, and the receiving antenna being similarly polarized. This is sufficient in most cases to resolve the FM channel 6 problem.

While technically cross polarization will help solve the problem, the FCC rules do not require it. This is left as an option for the FM applicant to use. Most TV channel 6 receiving antennas will remain H-pol, while automobile FM antennas will stay V-pol. So if the TV station remains H-pol, this interference problem may be cleared up if the FM station switches to V-pol. [See FCC Rules Section 73.525(e)(4).]

COMMERCIALLY AVAILABLE ANTENNAS

There are several basic classes of antennas available for FM broadcasting. These and variations of them are made by several manufacturers in different models, gains, and input power ratings. They may be broken down into the following classes:

- Ring stub and twisted ring
- Shunt- and series-fed slanted dipole

- Multi-arm short helix
- Panel with crossed dipoles.

These antennas have many things in common. For example, nonsymmetrical antennas are designed for side mounting to a steel tower or pole, as shown in Figure 7.17-11. Radiating elements are shunted across a common rigid coax line that eliminates the problems associated with the older corporate feed system using semiflexible solid dielectric low-power cables. Shunting elements every one wavelength across a transmission line makes impedance matching simple. Bandwidth is limited by the VSWR of the individual elements and the use of an internal transformer.

With more than about seven bays, the first three of the above antennas are more difficult to match and there is undesirable "beam squint," because the elevation beam angle changes with frequency deviation by the transmitter. Antennas with more than seven bays are fed from or near the center, thus dividing the phase change in half and effectively eliminating the beam squint. Center feeding the antenna also simplifies the VSWR matching.

A means for tuning out reactance after the antennas have been installed on the tower is also common with all the antennas. Located at the input to the antenna, the VSWR tuner consists of adjustable location dielectric or metal slugs on the inner conductor of the main coax line. Several fixed-position variable capacitors, spaced one-eighth wavelength along the main feeder near the antenna input, are also used on some side-mounted antennas to adjust the VSWR to low levels.

Another variety of antenna has curved radiating elements around a circumference whose diameter is determined by the number of element arms. Each radiator consists of two, three, or four such circular arms, depending on the model. Each element is fed through a shunt arrangement and then shunted across the vertical rigid feed coaxial line.

FIGURE 7.17-11 Example of ring-stub FM antenna element

Wideband panel antennas are becoming popular where high buildings, favorable mountain sites, or high towers are available. Several firms make wideband panel antennas. Some have very wideband VSWR features in each radiator. Others with not so broad VSWR use phase impedance compensation similar to the European scheme, which uses 90-degree phase quadrature impedance compensation.

Phase quadrature compensation makes it possible to cover the entire 88–108 MHz band with a VSWR under 1.1:1 while maintaining excellent elevation and azimuth patterns, together with very good axial ratios. This is especially important in IBOC operation where a single antenna is employed for both analog and digital signals. Power ratings up to several hundred kilowatts are offered so that many FM stations can be diplexed into one such antenna.

Only the wideband community FM antenna design now uses a corporate feed system, while the others are shunt fed from a common rigid coax line. This corporate feed system, using air dielectric semiflexible line at the lower power levels, is very successful. It splits the input power to many different dipoles at the correct amplitude and phase.

Standard Side-Mount Antennas

Most of the FM antennas that are in use today are side-mounted antenna arrays. The advantage of using a side-mounted FM antenna array is cost, ease of installation, and ease of maintenance. They are mostly circularly polarized, but many are horizontally polarized and some are vertically polarized to protect channel 6 TV channels. Some are elliptically polarized.

The antennas that are fed with a rigid inter bay coaxial harness are usually spaced so that the phase rotation in the harness is 360°. Arrays that are spaced 360° have an array factor that has a grating lobe up and down, and the array works by multiplying this array factor times the antenna element radiated field pattern, which theoretically has a null up and down. Because of scattering and other issues with antenna design, side-mounted antenna elements always have some radiated field up and down resulting in some radiation up and down the tower. If the antenna array is a half wave spaced array, the phase rotation in the harness will be 180°, so in order to make the array work, every other element in the array must be installed upside down to make the elements radiate in phase. The advantage of half wave arrays is that the array factor of a half-wave array has a zero up and down rather than a grating lobe that you have with a full-wave array. Another efficient array design is the (N – 1) / (N) array where N = number of elements in the linear array. For example, the standard two bay half-wave array where N = 2 is a (N – 1) / (N) array. Array factors for linear arrays of this design offer a more efficient use of the antenna aperture on the tower because the gain will be greater than a half-wave array of more than two elements while maintaining an array factor that has a null up and down the tower.

The actual horizontal plane relative field pattern of a side-mounted antenna array can be measured or calculated. With the development of Numerical Electromagnetic Code (NEC), side-mounted antenna elements or arrays can be modeled in computers to get fairly good results of the predicted horizontal radiation pattern of the side-mounted antenna array. The antenna can be measured to get very accurate results. Measurement can be done on a full scale range, a scale model range, and in an indoor anechoic chamber. Full scale ground reflection range measurements have the advantage of being able to measure the actual antenna on the tower it will be mounted on. The disadvantage of full scale ground reflection ranges is that the ground reflection path on the range changes phase as the antenna model is rotated on the turntable. This results in a small error on the measured range pattern. This error can be eliminated or at least minimized by modifying the measurement by placing the phase center of the antenna over the center of the turntable. By doing this, the ground reflection becomes mostly a constant rather than a variable. With model range measurements, you have scaling errors as well as reflections, and the electrical properties of insulating material can change with frequency. Because of this, special care must be used in measurements of this type. The most accurate way to measure full scale FM antennas is with anechoic chambers, which use absorption materials and have a measurable zone where the antenna under test is in the far field. The antenna is rotated in the chamber to measure the antenna's actual radiation pattern. Because these ranges are indoor ranges, measurement is not affected by weather. Using antenna measurement techniques, side-mounted arrays that are licensed as directional arrays as well as antennas that are licensed as nondirectional are available. Parasitic elements are used to create custom directional patterns for side-mounted FM antennas. These added elements have currents that are induced on them that result in reradiated fields that shape the antenna's relative field pattern to achieve a legal directional pattern. Parasitic elements can also be used to improve the pattern circularity of nondirectional side-mounted antennas.

Side-Mounted Antenna Array Elements Available Today

Side-mounted antenna arrays can be used by single stations or by many stations. Using careful array design with broadbanded antennas in the array, side-mounted arrays can be made to work for use as multiplex antenna arrays with usable bandwidths of up to 20 MHz with input VSWR less than 1.15. These arrays require custom feed harnesses in order to make sure that the individual elements in the array are phased properly.

Standard side-mount antennas come in a variety of shapes and forms and are currently used in the majority of applications. Their chief advantages are low cost, easy installation, relatively high gain, and low tower constraints. They are available in linear polarized configurations (H-pol or V-pol) or circularly polarized (CP).

Most side-mount antennas comprised a series of radiating elements, or bays, which are fed via a rigid inner-bay feed line. The most typical feed lines used may be either rigid or flex lines and range from 1/2-inch foam flex line for low power and translator facilities up to 6-1/8-inch line for high power (typically multiplexed) facilities up to 120 kW. Even higher powers can be achieved by using dual inputs feeding different halves of the antenna or by using larger rigid line (less common). Most antenna elements come in high- and low-power versions. These antennas are mounted directly to the side of a tower or pole, as shown in Figure 7.17-12. Leg and face mounts are typical on tower structures.

Some manufacturers with test ranges offer side-mount antennas with custom directional patterns. The pattern shaping is accomplished by optimizing the mounting and adding parasitic reflectors, which are on the order of one-half wavelength. Repeated range tests have shown that side-mount antennas have largely distorted patterns due to feed lines, mounting structures, and other conductive items in or near the antenna aperture.

Series-Fed V-Dipole Antennas

This antenna has similar bandwidth to its shunt-fed counterpart, but the array is typically intentionally tuned high in frequency. The combination of this tuning technique and the internal protection of its feed allows this antenna to be somewhat resistant to light icing. With heavy ice conditions with two inches of radial ice on these antennas, the input VSWR will

FIGURE 7.17-12 Shunt-fed slanted dipole antenna element

reach a value of VSWR = 2:1. These antennas are available with de-icing heaters and with radomes.

This model is larger in size and heavier than other types of side-mount antennas so that tower constraints may become an issue. The antenna is typically field tuned for an optimized match. Careful placement of ceramic slugs can produce a good VSWR over the stations useful bandwidth.

Ring Radiators

There are several antennas that are simple adaptations of ring radiators and were designed and manufactured in the 1950s and 1960s for horizontal polarization. By adding vertical stubs to the ends of the radiator or twisting the ring, elliptical polarization (of sorts) is achieved. The ring stub and the twisted ring are not really circularly polarized because the axial ratio varies considerably with azimuth. At best they may be said to be elliptically polarized.

The design has been improved by adding a second horizontal ring and improving the feed. Reducing bay spacing reduces high axial side lobes. The antenna has good circularity in free space, but like other types of side-mounted antennas, it is strongly affected by its support structure and feed line.

The radiation patterns are strongly affected by the tower mounting environment. Being of relatively high Q design, they are more susceptible to detuning because of icing. Radomes and electrical de-icers are available to overcome this problem.

Ring-Stub Antennas

The H-pol radiation from these antennas comes from the ring portion whose plane is parallel with the earth. There is a minor lobe from each radiator, which is strengthened with vertical stacking for additional power gain. This nadir-zenith lobe is the result of 360-degree stacking on the rigid coax feed line. It reduces the gain and presents a lobe at the tower base that is detrimental to low-level audio equipment and personnel located in a building at the base of the tower.

In order to keep the cost down, like the twisted-ring antenna, the ring-stub is manufactured in several radiator-to-radiator spacings across the FM band. This results in some minor beam tilt up or down depending on the frequency. Most higher-priced slanted dipole and helix antennas are spaced exactly 360 degrees and are usually tested to assure this spacing during production.

Shunt-Fed Slanted Dipole Antennas

The slanted dipole antenna in its present configuration was developed and patented in 1970 [11]. It consists of two half-wave dipoles bent 90 degrees, slanted and fed in-phase. The slant angle is critical as it is the factor that determines the ratio of vertically and horizontally polarized radiated power. When fed through a vertical support pole on which the antenna was mounted during initial development tests, the axial ratio varied less than 1 dB.

In the internal fed design of this style, at the feed point the element impedance is ~80 ohms, is transformed to 100 ohms at the tee where the two dipole impedances add in shunt to 50 ohms. While different impedance transforms are used from that point based on the number of bays, mutual coupling and other factors, this 50-ohm impedance at the tee has the benefit that a single element can be quickly converted to a one-bay auxiliary antenna by using a 50-ohm inner conductor in the stem.

The commercial adaptation uses a horizontal boom containing a step transformer. This boom supports two half-wave dipoles in which the included angle is 90 degrees, as shown in Figure 7.17-13. The two sets of dipoles are rotated at 22.5 degrees from the horizontal plane. Two opposite arms of the dipoles are delta matched to provide a 50-ohm impedance at the radiator input angle. All four dipole arm lengths may be adjusted to resonance by mechanical adjustment of the end fittings. Shunt feeding, when properly adjusted, provides equal currents in all four arms resulting in excellent azimuth circularity.

Short Helix Antennas

An asymmetrical radiator is the four-arm shunt-fed helix. By using four dipoles, curved so that their circumference is about one wavelength, a CP antenna is produced, as illustrated in Figure 7.17-14 [12]. Each dipole is about one-half wavelength and is shunt fed. These are supported on a four-arm structure, one end of which is tied to the supporting structure. The dipoles overlap so that the current flow around the circumference is circular. The four feed arms are

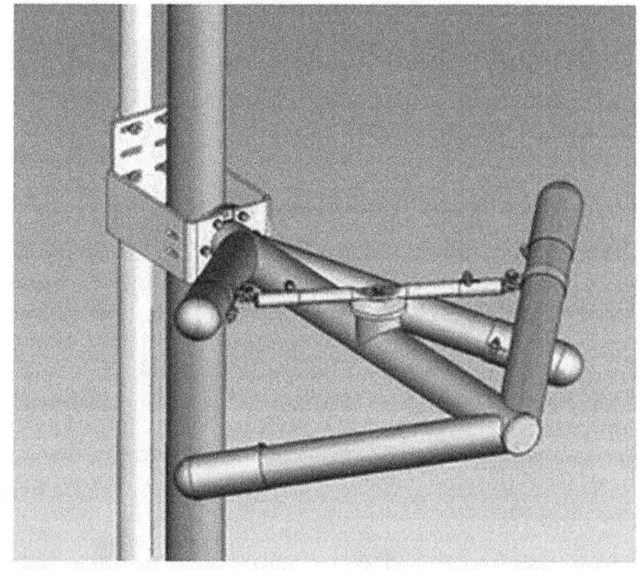

FIGURE 7.17-13 High-power shunt-fed slanted dipole antenna element

FIGURE 7.17-14 Omnidirectional circularly polarized 3-arm normal mode helical antenna, Stonehenge Tower, Portland, Oregon

Ernie Hopseker

connected in shunt and the feed impedance is quite low, but may be improved with an internal step transformer. The CP quality of the four-arm, side-fire, short helix is good. Three- and two-arm models are also available, but their axial ratio is not as good as the four-arm model. Pattern circularity is ±1 dB for the four arm, together with an axial ratio of about 3 dB. These radiators are stacked about one wavelength apart on a rigid coax feed line to obtain the necessary power gain. Like other asymmetrical FM antennas its patterns are strongly affected by the supporting structure. See the "Pattern Optimization" section in this chapter for the need and methods to circularize the azimuth pattern.

Electrical de-icers using the stainless-steel dipole arms as one-half of the heating circuit are available. Heat is created by passing a large current at low voltage through each arm from voltage-dropping transformers placed at each bay level. Plastic radomes are also available to keep snow and ice off the sensitive VSWR parts of the antenna.

Twisted-Ring Antennas

This type consists of one or more rings, which have been partially twisted so that the open ends of the ring are about 10 inches (25 cm) apart. One semicircular arm of the ring is fed with a small loop or by a direct tap on that arm, as shown in Figure 7.17-15. A number of these rings are fed in the same manner as the ring stubs, and have the same zenith-nadir lobe problem. The mechanical twist is not the same when viewed in all the azimuth directions. Therefore, the current is not the same, with the end result that in some directions there is much more elliptical radiation than in others.

These antennas are very simple and relatively inexpensive for single frequency use, but have some serious operational limitations for CP operation. They do not have the same signal penetrating effect as the slant dipole, short helix, or the flat panel antenna type of CP antennas.

Short Helix—Multi-arm Antennas

The number of arms may be increased to four instead of the two in the slanted dipole variety. To provide CP, the arms are curved to form a one-wavelength circumference. These short multi-arm helices are also stacked in the conventional manner, like the others in this series for power gains as desired. This design uses two-wavelength feed straps to feed all the elements in phase. This antenna is shunt fed, and is arrayed and mounted similarly to the slanted V-dipole antenna.

The azimuth pattern of all these nonsymmetrical antennas is affected by the supporting steel structure. With pattern optimization, the pattern can be made quite omnidirectional (see "Pattern Optimization" section in this chapter).

Series-Fed Antennas

A similar arrangement of arms supported by a tee arrangement may be series fed. In this style of antenna, part of the outer end is insulated from the rest of the dipole and fed across the insulated break as shown in Figure 7.17-16. To allow for adequate power-handling capacity and to increase the VSWR bandwidth, 3-inch (75-mm) diameter tubing is used. The antenna has a VSWR bandwidth of about 1%, so it makes an excellent single channel FM antenna. The antennas are usually mounted on the side of a supporting tower or pole and stacked vertically to achieve required power gain.

The large diameter was used in the design of this antenna to achieve an improved VSWR bandwidth and to avoid corona damage. Electrical de-icing as well as plastic radomes are available for this antenna.

Flat-Panel Antennas

The panel antenna was developed in Europe to provide a wide bandwidth for several collocated FM stations without the need to change antennas when a new channel was added or if the operating frequencies were changed.

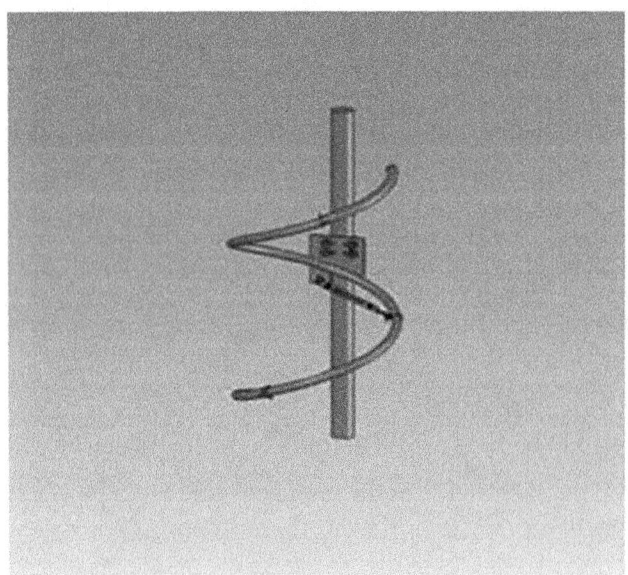

FIGURE 7.17-15 Twisted ring radiator

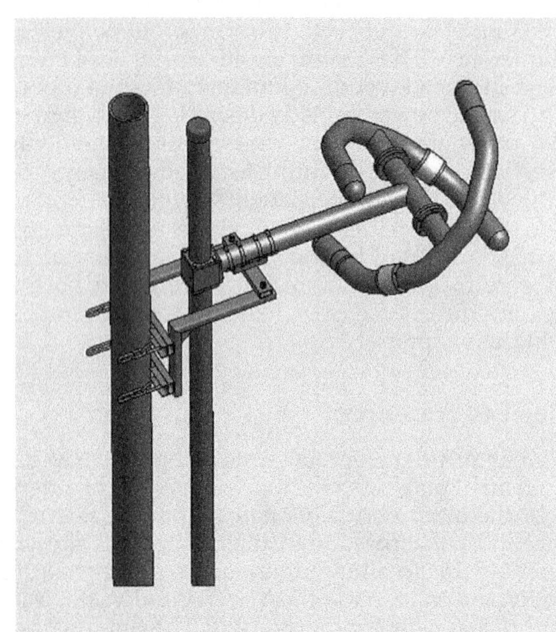

FIGURE 7.17-16 Side-mount dual-dipole FM antenna element fed across insulated breaks

Panels typically cover the face of the tower and are rectangular in the flat configuration. Additionally, wings on the corners of the flat panels are added to improve the horizontal plane relative field pattern circularity. In the cavity style they are about 8 feet (2,450 mm) in diameter and about 3 feet (1,000 mm) deep. A heavy metal frame is often used over which a large diameter wire mesh has been welded. The wire mesh screen openings vary from 4 to 12 inches (100–300 mm). Electrically they are considered nearly solid metal. These openings produce relatively low wind loads. The entire flat frame or cavity is strong enough to support a man on its mesh openings. Some manufacturers hot-dip galvanize their steel after fabrication; others use stainless steel construction.

For FM, two crossed dipoles are used as the illuminating source for each panel or cavity, as shown in Figure 7.17-17. Each dipole is fed in phase quadrature. That is, one dipole receives its peak current 90 degrees after the other, to produce CP. A typical set of electrical and mechanical specifications for a CP eight-bay cavity community antenna are shown in Table 7.17-6.

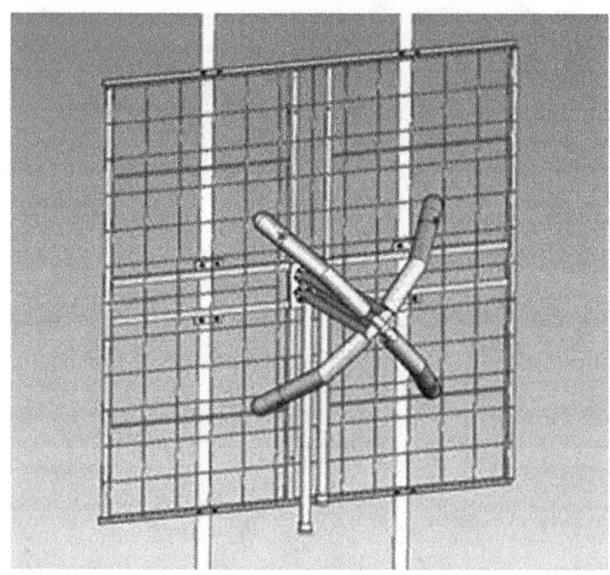

FIGURE 7.17-17 Dual-arrowhead dipole flat-panel antenna

TABLE 7.17-6
Typical Measured Community Antenna Performance

Parameter Description	Parameter Value
Operational frequency range	88 to 108 MHz
Safe RMS input power rating	200 kW
Power gain ratio, each polarization	4.4 (6.43 dB)
Maximum VSWR any channel between 88 and 108 MHz	1.15:1
Elevation pattern beam tilt	–0.5°
Polarization	Right hand circular
Axial ratio	Better than 2 dB
Azimuth circularity V-pol or H-pol	Better than ±2dB
Antenna dead weight, less than	7,000 lbs (3,183 kgs)
Active wind load, RS-222-C 50/53 PSF	8,000 lbs (3,636 kgs)
Antenna input flanges, two, size	6-1/8 inch
Number of bays (stacks)	8-bay typical
Radiator type	Circularly polarized, panel

Flat-panel antennas are typically side-mounted on large face-size towers. The screen panels greatly reduce interaction and distortion between the antenna and tower. The panels are directional, thereby requiring three or four panels to be mounted around the tower to achieve acceptable azimuthal circularity. These antennas are usually branch fed, and often the array's top and bottom halves are fed separately. This allows operation of either half of the antenna separately when it is necessary for temporary emergency operation. Circular polarization is achieved on each panel by feeding two perpendicular dipoles 90 degrees out-of-phase. This phase offset helps this antenna achieve usable bandwidths on the order of 10 MHz. Wings on the corners of the screens are often added to achieve excellent pattern circularity.

By pulling the dipole back on its feed support arms, the arrowhead-shaped dipoles control both V-pol and H-pol azimuth patterns. Rotating the dipoles 45 degrees with the earth-ground reference further improves the polarization ratios.

Round dipoles made of tubing as large as 6-1/8 inches (155 mm) in diameter are used along with a single line quadrature feed. This combined arrangement makes an excellent wideband CP panel to cover the entire FM band. Power splitters, dividers, and cables, along with a number of these panels, complete the antenna design.

On large face towers, circularity in the H-pol can be quite good, on the order of ±2 dB. On standard configurations, the V-pol pattern is quite different. As a result, the axial ratio of this antenna ranges from good at some azimuth headings to rather poor at others. This is because the azimuth pattern of an H-pol dipole is like a figure eight, or cosine function, while the pattern for a V-pol dipole is not directional in its azimuth plane. Therefore, each polarization will react quite differently when mounted in front of a panel. This problem is corrected by adding wings to the panel at the corners of the flat panels. This does add wind load to the array.

Dipoles on these panels are often mounted at 45 degrees referenced to the ground. This has no effect on the axial ratio or pattern performance; it instead is done for tuning considerations to compensate for mutual coupling.

Flat panel arrays can be easily designed to have separate inputs for right-hand circular polarization and left-hand circular polarization. Often, the analog signals of the users are fed into the right-hand circular polarization input while the digital IBOC signals are fed into the left hand circular polarized input. Typical isolation between inputs is −25 dB.

Improvements to this design have reduced the differences between the patterns of the polarizations.

These techniques are effective for applications requiring only a few MHz of bandwidth. One method optimizes the angle of the dipole bend as well as its distance to the panel. This design requires three panels to be mounted around a tower. Axial ratio and pattern circularity are improved at the cost of system bandwidth.

Another method uses four dipoles forming a square shape in front of the panel, as shown in Figure 7.17-18. By adjusting the spacing between the dipoles, the beam width of a panel can be controlled. Over a small bandwidth, the pattern performance is greatly improved. It is necessary to mount four panels around a tower for a circular pattern. A large amount of panel interaction and leakage are severe design limitations.

For projects that require wider bandwidth, skew mounting is often used. This physical configuration allows the panels to be fed in mode 1 (0°, 90°, 180°, 270° phase for four around), which can increase the bandwidth of the system at the input. Although skew mounting deteriorates pattern performance, the increase in bandwidth extends its applications.

These panel arrays have one significant disadvantage. Because each bay level of this panel array consists of one vertically polarized dipole and a two-bay horizontally polarized array, the array factor design for the vertically polarized component of the array is not the same as the array factor for the horizontally polarized component of the array. Because the stacking gains of the vertical component and the vertical component are different, a special feed harness is required to split the input power to the array into unequal amounts for the vertical dipole array and the horizontal dipole array. Because of this, if the antenna is phased for right-hand circular polarization, it can't be fed as a dual-input antenna using right-hand and left-hand circular polarization inputs.

Cavity-Backed Panel Antennas

The use of a cavity screen instead of a flat panel has greatly improved axial ratios. The cavity acts as a resonator with little leakage toward the tower. The shape of the azimuth pattern in each plane becomes both controllable and symmetrical. System bandwidth is

FIGURE 7.17-18 Broadband four-dipole flat-panel FM antenna

improved over the flat-screen design. By adjusting the diameter of the cavity structure, beam widths can be altered to meet specific requirements. Mounting three cavities around a tower gives good pattern circularity. Axial ratios usually range from good to excellent.

The cavity antenna uses the reflective properties of the flat-screen panel. In the cavity however, the illuminating dipoles are flat instead of round and all four arms are parallel to the plane of the cavity, as shown in Figure 7.17-19. Like the flat panel with its round dipole supporting balun, the cavity also holds its flat dipoles with a double coaxial balun.

The dipoles in the cavity get their wide VSWR bandwidth through the sleeve dipole principle [11]. Capacity is provided by a metallic ring close to all four dipole arms placed between them and the back of the cavity. Circulating surface currents flow on the dipole arms, which results in evenly radiated patterns in all polarization planes. The bandwidth of a single cavity can cover the full 20 MHz band with a VSWR better than 1.1:1. Therefore, it is not necessary to skew mount these antennas for bandwidth considerations.

This antenna has the advantage over some other designs of greater VSWR bandwidth, however flat panel and Cogwheel antennas both can be designed to cover the entire FM frequency band. Antenna arrays of cavity back radiators on the same face tower as a flat panel (reflector screen on face of tower with broadband FM panel antenna in the center of the face mounted to the reflector screen) will produce a horizontal relative field pattern with deeper nulls than the horizontal relative field pattern of the flat panel antenna. This is due to the fact that the phase center for the element radiation from a panel antenna mounted on a flat screen is at the edge of the screen, but the phase center for the same panel element mounted inside a basket is at the opening of the basket. For this reason, to achieve an optimum pattern from a basket antenna design, one needs a small face tower.

Cavities and flat panels can be modeled using a computer. Factors, such as tower size and orientation, as well as the phase and skew of the elements, can be modeled to determine optimum mounting and feeding. This is useful in projects that require a directional pattern. A station should take the pattern constraints and gain requirements to an antenna manufacturer to determine what is feasible.

FIGURE 7.17-19 Dual-dipole cavity-backed panel antenna

Cogwheel Top-Mounted FM Master Antenna Arrays

In addition to flat panel and basket FM master antennas, three pole-mounted FM broadband panel elements can be mounted on a pole with screen partitions to isolate the elements of each bay level of the array as shown in Figure 7.17-20. In order to meet the structural requirements of a top-mounted antenna, the outer edge of the three screens is made with structural components to carry the wind load on the antenna to the top of the structure below the antenna. Cogwheel antennas typically have a VSWR of less than 1.15 over the entire FM band as well as a pattern circularity of better than ±2 dB. These antennas like both flat panel arrays and cavity-backed antenna arrays can be fed as dual fed arrays with one input producing right-hand circularization and the other input producing left-hand circularization. Typical input isolations between polarizations is 25 dB.

FIGURE 7.17-20 Three-sided panel style around a pole with structural screens

Crossed Dipole Theory

Common to the flat panel and the cavity is the operation of the dipoles that generate CP. The dipoles are fed currents in phase quadrature through a coaxially balanced balun that provides equal currents to all four arms of the two dipoles. They excite the entire cavity or flat panel with a rotating radio frequency (RF) field in a plane parallel to the dipoles. The RF field is thus CP and may be ideally represented by a rotating vector of constant magnitude revolving one revolution per wavelength of propagation distance. It is right-hand polarized as the field rotation is clockwise as viewed from behind the screen, looking toward the direction of propagation, if the phasing between the two crossed dipoles is properly made.

Radiation patterns, associated beamwidth, and directivity are determined to a large extent by the size of the cavity or flat panel. The geometry of the dipole has less effect than the reflector size. The size and shape of the dipole control the antenna impedance and the VSWR. The screen panel, be it flat or a cavity, fulfills the following five important electrical functions:

- Isolates the radiating elements from the tower or the mounting structure, and reduces mutual coupling.
- Provides sharper beam width and more gain than achievable with the dipoles alone.
- Furnishes pattern control so that the beam width is nearly equal for both horizontal and vertical plane polarization.
- With an effective balun feed system, the crossed dipole radiated pattern phase is very uniform as the amplitude changes normally with azimuth.
- Computer-aided designs are easily achieved in production for various width towers because the pattern is simply pure electrical geometry.

Patterns achieved from flat screens versus cavities vary greatly due to the relative phase of each style—phase front is the screen on flat panel versus front of basket for baskets.

Antenna Element Spacing and Downward Radiation

Most FM antennas have elements that are spaced one wavelength apart (9–11 feet) for reasons such as gain considerations, mutual coupling effects, and ease of feed design. There are cases, however, that require different element spacing. High levels of downward RF radiation are the most common reason, although considerations such as aperture constraints and beam shaping also factor into decisions to utilize other than one-wavelength spacing.

Radio frequency radiation (RFR) safety levels must be considered in nearly all site locations. Power radiated in the lowest sidelobe at or near the base of the antenna mounting structure can cause a variety of problems, including human exposure hazards and electromagnetic interference (EMI) to other electronic. See Chapter 3.5, "Human Exposure to Radio Frequency Energy," for further information on this subject.

When antenna elements are arrayed, the resulting elevation pattern contains lobes and nulls. The farthest sidelobe from the horizon typically peaks between 70 degrees and 90 degrees below the horizon for full-wave spacing. This lobe occurs because the physical path results in no phase cancellation in that direction, and thus the downward radiation of each element is additive. Shortening the spacing changes the difference in the elements physical path length, and results in some phase cancellation.

Antenna array pattern in elevation plane is a sum of multiple stacked sources as follows:

$$E(\theta) = \sum A_n(\theta) e^{-j\frac{2\pi}{\lambda}\bullet z(n)\cos(\theta)}$$

where $0° < \theta < 180°$ is the elevation angle from zenith, $A_n(\theta)$ is a complex elevation pattern of a single radiating element, and $z(n)$ is a vertical location of radiating element ($n = 1, \dots, N$).

For uniformly excited and linearly phased array of equally spaced elements, the elevation pattern takes the form of

$$E(\theta) = A(\theta)\sum e^{-j\frac{2\pi d}{\lambda}\bullet n\bullet\left(\cos(\theta)-\frac{\Delta\phi\lambda}{2\pi d}\right)}$$

where Δ is a phase shift between adjacent elements to tilt the beam, if required.

When spacing $d = \lambda$ and $\Delta = 0$ the pattern of an array of horizontal dipoles (for which $A(\theta)$ = const) has three equal peaks: main beam at $\theta = 90°$ (horizon) and grating lobes at $\theta = 0°$ (zenith) and 180° (downward). For an array of vertical dipoles the upward and downward radiation is suppressed by element factor A_{max} / A_{min}. Most suppression on the order of −20 dB or better is for straight vertical dipoles, less suppression is provided by bent vertical dipoles, and no suppression due to element alone for horizontal dipoles.

The expression above for uniformly excited and linearly phased array of equally spaced elements is a finite geometric series and can be reduced to a single term:

$$|E(\theta)| = \left| A(\theta)\bullet\frac{\sin\left(\frac{\pi N d}{\lambda}\bullet\cos(\theta)\right)}{N\bullet\sin\left(\frac{\pi d}{\lambda}\right)}\right|$$

At $\theta = 0°$ and 180° the normalized pattern represents additional suppression of upward and downward radiation by an array factor, and is a function of spacing d / λ and the number elements N, as follows:

$$|E(0°,180°)| = \frac{\sin\left(\frac{\pi N d}{\lambda}\right)}{N\bullet\sin\left(\frac{\pi d}{\lambda}\right)}$$

Full suppression takes place at $d / \lambda = (N - 1) / N$. Full suppression also takes place when $d / \lambda = 0.5$ for any even number N of vertically stacked radiating elements. The $(N - 1) / N$ array for $N > 2$ results in a more efficient array design.

Half-wave spacing, for example, greatly suppresses the levels of the side lobes, while increasing the width of the main beam. Despite the lack of power in the side lobes, the extra width of the main beam causes the pattern to be less directive. Spacing the elements $(N - 1) / N$ wavelengths apart improves sidelobe suppression and the gain of the antenna is not greatly affected. As an example, for a 5-bay antenna, the spacing would be 0.8 wavelengths. The resulting downward radiation due to the array factor (not taking into account element radiation kernels) as a function of relative bay spacing is shown in Figure 7.17-21 for arrays of two, four, six, and eight bays. In Figure 7.17-22 the elevation pattern for a 4-bay $(N - 1) / N$ spaced array is shown.

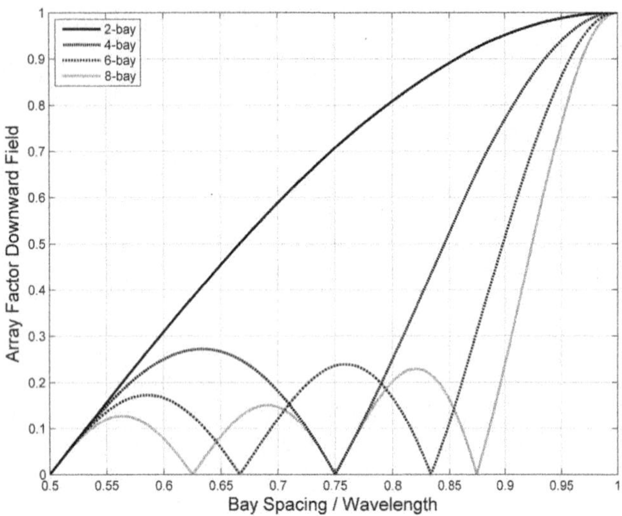

FIGURE 7.17-21 Array factor downward field as function of bay spacing

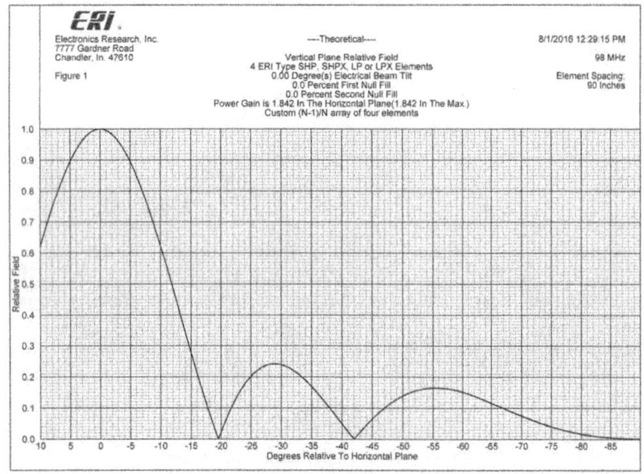

FIGURE 7.17-22 Elevation pattern for $(N - 1) / N$ array for N = 4-bay

Short-spaced antennas are fed either by a shunt or branch feed system. For half-wave spaced antennas, a shunt line delivers each element 180 degrees out-of-phase with the next element. This phase distribution problem can be overcome by flipping every other element upside down, thus inducing a 180-degree phase shift in the feed. Other spacings, such as 0.8 wavelengths, can be accomplished with a branch feed, which can deliver equally phased signals to each element regardless of spacing. Also, in the shunt feed case, an extra elbow complex in the form of U-link in between each pair of elements provides an additional path to deliver in-phase signals to elements spaced shorter than one wavelength. $(N - 1) / N$ spaced arrays provide the best aperture efficiency for a given number of bays for a design providing the total downward radiation suppression. The half-wave array of two bays is the first order of the $(N - 1) / N$ spaced array.

Typical circularly polarized FM antennas are side-mounted dual bent dipoles that are series fed. The series feed design has many advantages. First, the antenna feed is internal and pressurized and thus less affected by ice in the winter. Second, because the antenna feed is internal, it does not produce any additional radiation that can produce excessive downward radiation. The series fed design achieves circularization by spacing the two bent dipoles electrically 90°. The two bent dipoles each radiate both horizontal and vertical polarization, and together due to the element spacing produce a circular polarized radiated signal.

The panel antenna is another common type of FM antenna. Its dual dipole is phased in quadrature and backed by cavity or screen. With the size of screen and depth and size of a larger cavity, suppression (A_{max} / A_{min}) of downward radiation by the elements alone improves.

Calculation of power density at the base of a broadcasting facility can be approximated as

$$P = 1.6 \times ERP \times \frac{\left(|E(180°)| \times \frac{A_{max}}{A_{min}} \right)^2}{4\pi R^2}, \ \frac{W}{m^2}$$

where factor 1.6 is used to account for ground reflection, and R is the height of the antenna above ground.

Field strength at the base is $F = 377\ P$ in V/m, and should be less than the FCC-mandated threshold for human exposure, which currently stands at 27.5 V/m (FCC OET Bulletin 65, Edition 97–10, August 1997, table 1).

Problems With Side-Mounted Antennas

Single-station FM antennas are typically side-mounted on a pole or tower. Unlike panel antennas, the support structure greatly affects or distorts the radiation pattern. The resultant pattern may have large peaks and nulls that can result in coverage and reception problems.

In addition, the V-pol and H-pol patterns react quite differently to these distortions. Due to the geometric complexities of the CP radiating elements and tower structure, it takes computer modeling to accurately predict pattern effects. Therefore, the use of a test range is required to determine how an antenna behaves when mounted on a tower section similar to the one on which the antenna will eventually be used. For nondirectional stations, a test range can determine the proper mounting of an antenna to achieve an acceptable circularity. Depending on the tower size, the depths of nulls can be greater than 10 dB. An optimized mounting configuration can make the nulls less significant and oriented in areas where service to the primary coverage area is not hurt. Parasitic reflectors are often used to improve the circularity of nondirectional antennas.

The tower or pole movement and deflection has a direct impact on the life of the antenna as well as potential impact on beam steering fluctuations. Excessive pole movement can cause damage to the antenna—one parameter of interest is to ensure that the pole does not have a deflection radius of curvature (under load) of less than 1500 feet.

Custom towers are available today such as the Lambda Tower, which have controlled deflection under load as well as structural member designs that are customized to be spaced at the user's frequency wavelength. By spacing the tower members at a half wave, each antenna element in the side-mounted array mounted to the tower will see the same geometry. This is true for full-wave spaced arrays and for half-wave spaced arrays. For this reason, each element will produce tower scattering that is identical in magnitude and phase. With coherent scattering, the array will produce an elevation relative field pattern that is as close as possible to the calculated elevation pattern of the array.

When the top spot on a tower or structure is available, pole mounting is often preferred. A pole provides a stable and symmetrical support structure that has low interaction with the horizontally polarized component. In combination with the feed line, the pole typically induces a null in the V-pol pattern, directly opposite of the elements. Proper orientation of the element can reduce the effects of pattern distortion.

Mounting an antenna on the side of a tower can produce unpredictable results. The positions of the peaks and nulls vary greatly from the orientation of the elements. As the face size of the tower increases, the distorting effects magnify. To compensate, many stations use smaller sections of tower at the top, where they plan to install the FM antenna. Eighteen-inch and smaller face towers tend to produce good results. With careful planning, the use of a 24-inch or larger face tower can also be successful. Note that the patterns of each polarization react differently, and thus axial ratios can be quite poor.

PATTERN OPTIMIZATION

Single-station FM antennas are usually side mounted on a pole or tower. This is economical and it frees the tower top for other possible uses. Unfortunately, the pole or tower tends to distort the radiation pattern, seriously affecting station coverage in some directions.

This problem can be serious if the FM antenna has been randomly attached to a support tower. Most FM antenna makers do not manufacture and sell towers. A few have made supporting poles on which the FM antenna has been mounted, adjusted, and pattern tested. TV antenna makers, on the other hand, usually build the antenna as a complete self-supporting structure to be mounted on top of a support and are usually not faced with this side-mounting problem. The logical but more expensive solution would be to make the FM antenna a self-supporting structure just like TV antennas.

Improper FM antenna side-mount installation on a tower can cause serious pattern problems. Measured patterns have indicated that, in some cases, the maximum radiation can actually be in the opposite direction from the desired direction.

When Optimization Is Necessary

Side mounting an FM antenna on a tower may have serious consequences with an FM station's coverage. Nulls may be toward important service areas. Nulls as low as 1% of the RMS power have been measured with towers varying in width from 18 to 120 inches (0.5–3 m). Another problem is that with nulls come lobes. Lobes as great as 9.8 dB over RMS power have been found. When used without pattern optimization, this lobe would produce an ERP in a given direction nearly 10 times the FCC-licensed value. Translating this to a 50 kW ERP station there would be radiation in some directions of only 0.5 kW and others with 477 kW. This is a maximum to minimum ratio of 29.8 dB, and clearly not acceptable.

CP creates other problems, such as the H-pol and the V-pol ratios are not always the same and vary moderately in any given azimuth. This ratio can be as great as 15 dB and must also be addressed in order to resolve the horizontal plane circularity problem. The axial ratio could be degraded causing the V-pol radiation in certain directions to be much stronger than the H-pol. This violates the FCC's requirement that with CP, the V-pol must not be stronger than the H-pol component.

Section 73.316 of the FCC Rules covers FM antennas but does not specifically address the problem of azimuth circularity. In fact the FCC assumes that FM nondirectional broadcast antennas have perfectly circular horizontal radiation patterns [13]. In actual practice, they seldom do.

In order to produce a horizontal plane pattern that even approaches a circle requires considerable work by the firm making the antenna. Because it is nearly impossible to produce a circular pattern with a nonsymmetrical side-mounted antenna, the term technique *optimization* (to do the best possible) is in common usage now.

Theory of Optimization

When optimizing an antenna as it will be mounted on the structure, an attempt is made to either find an optimal standard mounting scheme or to control the reradiation of H-pol and V-pol fields with the use of parasitic elements, typically only for an antenna that will be filed as a directional.

When considering the impact to the pattern for a pole-mounted antenna, for instance, the pole diameter is small in wavelength in the direction of the H-pol electric field, thus scattering is minimal and not much H-pol reradiation can be expected. However, when the V-pol dipole excites the pole, a large amount of energy is intercepted and reradiated because the large dimension of the pole (length) is parallel to the electric field. A similar effect is produced by the vertical transmission line that is common to the antenna itself. The result is appreciable distortion of the vertically polarized azimuth pattern.

Figure 7.17-23 shows typical H-pol and V-pol patterns for a pole-mounted side-mount antenna. The pole and/or vertical coaxial line have transformed the V-pol pattern from circular to a cardioid, while the H-pol pattern remains essentially omnidirectional. The null of the cardioid can be more than 7 dB down from the RMS value. This phenomenon is well known, and as a compromise, broadcasters generally install the antenna on the side of the tower support structure facing the main service area. There are many exceptions to this, as some measured patterns on triangular towers of standard construction have shown.

In Figure 7.17-23 the V-pol is stronger than the H-pol in direction that the antenna is oriented but is suppressed in the opposite direction. This makes for a very poor CP antenna in some directions. (Note that for directional antennas, horizontal parasitic elements can be very effective in aligning the H-pol pattern with the V-pol pattern.) Towers and poles under about 2 feet (0.6 m) in cross section will exhibit the same effects on the antenna patterns. Towers greater than this size will increase the complexity of scattering effects. Three or four tower legs, the horizontal and diagonal cross members, transmission lines, ladders, tower lighting, and de-icer conduits all will be excited by the vertical and horizontal currents from the radiators, which in turn will reradiate and affect the horizontal plane patterns.

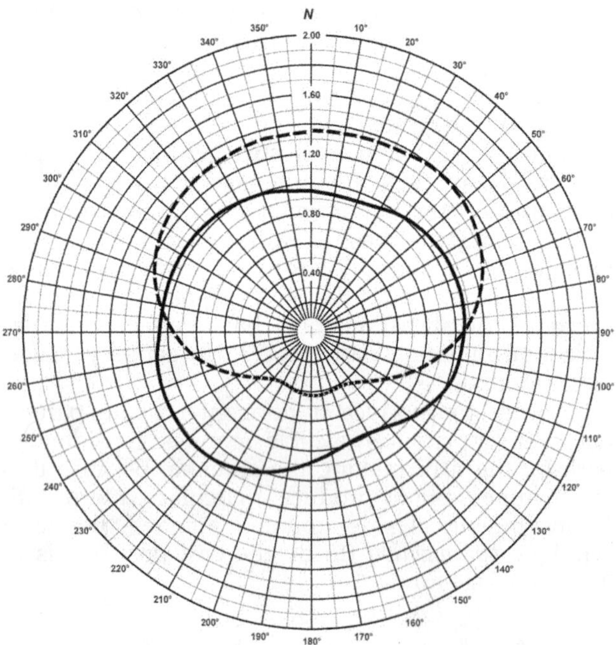

FIGURE 7.17-23 Typical H-pol (solid) and V-pol (dashed) azimuth patterns for pole-mounted FM antenna

In contrast to the simplicity of the antenna on the side of a pole, the tower-supported antenna may be mounted on the face or on a corner, at or between horizontal cross members, or tilted at various angles compared to the tower—all multiplying the complex factors affecting the patterns.

Finally, the side-mounted antenna array could be mounted on a wavelength matched tower. In this case, each bay of the antenna array will have identical tower member reflections reducing the complex factors affecting patterns on towers that do not have tower members spaced at the half wavelength of the user's frequency.

Optimization Methods

The most popular technique to achieve the desired pattern is through the use of Yagi antenna principles, wherein parasitic elements are placed in the field of the radiator to modify its radiation pattern. For example, a shortened dipole with length slightly less than a half-wave dipole (director) placed in proximity to a radiator reinforces radiation in the forward direction and suppresses the signal in the opposite direction. If the parasitic element is longer than a half-wave dipole (reflector), the effect is reversed. The signal is suppressed on the side of the parasitic element and reinforced in the direction of the radiator. Similarly, parasitic elements can be used with FM antennas mounted on the sides of towers or poles to produce pattern changes. As discussed here, both directors and reflectors may be used. Both are frequency sensitive. The effects of the supporting structure are also frequency sensitive.

Therefore, an arrangement of parasitic elements for a given FM frequency will not necessarily be the same for another, nor will the pattern be the same for a given arrangement, if it is moved up or down the tower by as much as 1.5 feet (0.5 m). For the wavelength-matched tower, however, the effect on each element in the array is the same.

The resulting patterns cannot be easily predicted. There are many factors that affect the horizontal plane pattern. Precise methods of modeling are needed so that the patterns can be adjusted and properly measured. Range testing, anechoic chamber facilities, and computer modeling can all be effectively utilized to determine resulting patterns for complex mounting arrangements. Therefore, the cost for doing this is high, because it is time-consuming and requires qualified antenna technicians, especially for full-scale antenna range pattern studies. In addition, the final parasitic arrangement must be permanently fabricated and installed. However, the results are generally well worth the expense and time.

Pattern Service

There are three basic types of pattern service furnished by antenna manufacturers in the United States. FM antennas may be adjusted for the best omnidirectional pattern possible or they may be adjusted to proven minimum ERP values in particular azimuth directions. The minimum required values, plotted on a polar chart by the broadcaster, may be combined with the tower orientation. Using the customer's make and model tower, two or more bays of the antenna are fabricated, installed on a section of the tower, and put on the test range. Adjustments are made such as leg chosen, distance from the leg, and the orientation of the antenna with respect to that leg. Parasitic elements are then used to further improve and shape the pattern.

These measurements can be made on a ground reflection range, or these measurements can be made in an anechoic chamber. The results achieved from the use of the anechoic chamber will be more accurate than those results achieved using a ground reflection range.

Antenna measurements can also be made in anechoic chambers, although at the VHF frequencies of FM radio, full-scale models are difficult to use in a chamber due to size and ability to have a sufficiently large quiet zone in the chamber. Scale models are typically used for chamber modeling, which of course requires precisely scaled components—including towers, poles, transmission line, conduit, antennas, and so on—and also requires that the scale model antennas are driven exactly like the full-scale versions.

As an example, Figure 7.17-24 illustrates a typical optimized directional antenna pattern where the H- and V-pol patterns have been brought into alignment using parasitic elements. This example shows a face-mounted side-mount antenna on a 24-inch face width tower. Notice that only vertical parasitic elements were used in this example to bring the V-pol pattern into alignment with the H-pol pattern, which essentially did not change.

Various methods have been used to optimize FM antenna azimuth patterns. Some firms use models at twice the operating frequency. Others use theoretical methods, backed up by experimental proof. The final optimized antenna is match marked on the tower sections so that it will be assembled exactly as it was made and tested at the fabricator's plant, and tested on the antenna range.

Optimization of nondirectional FM antennas should only be done to improve the pattern circularity of the side-mounted antenna array. Optimization resulting in increased directivity is not allowed by the FCC.

A complete set of installation prints must be provided so that the antenna is assembled exactly as

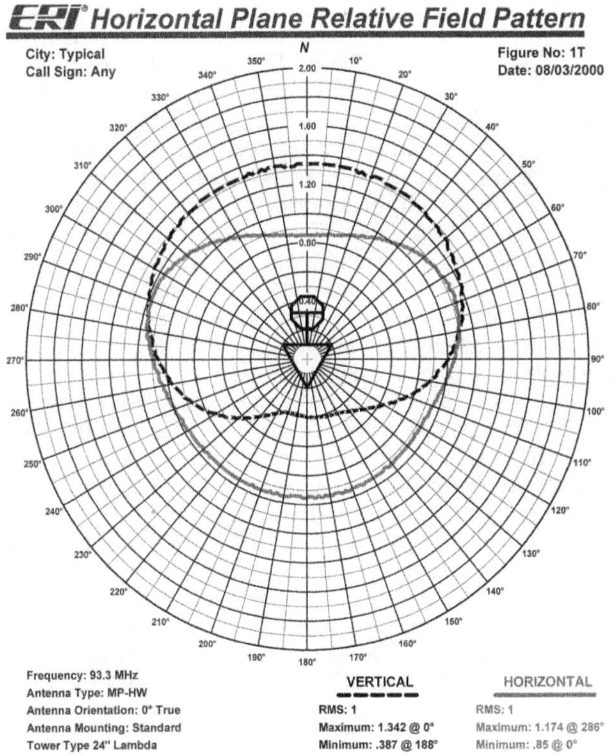

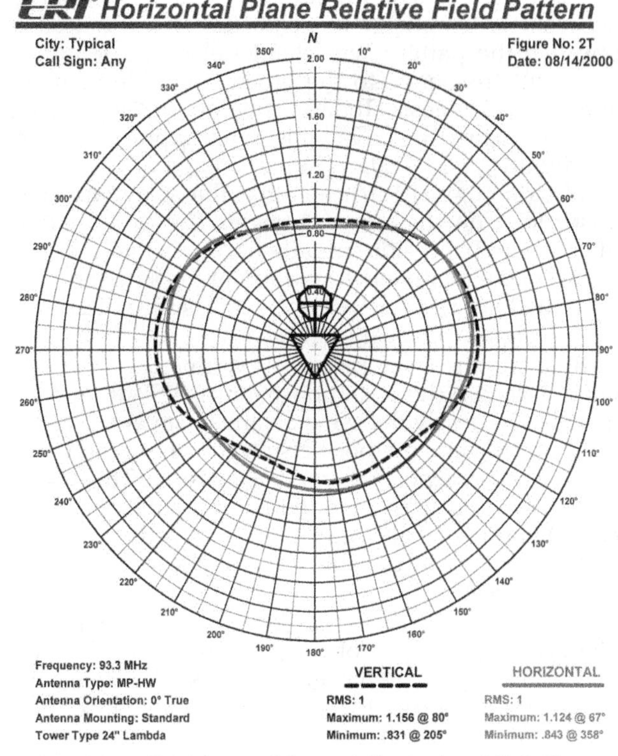

FIGURE 7.17-24 Typical H-pol and V-pol azimuth patterns for tower-mounted FM antenna, with directional pattern optimization

tested, with all the correct locations, and angles of all the parasitic elements.

WIDEBAND COMMUNITY AND MULTI-USER ANTENNAS

The most significant developments in FM antennas have come in broadband designs. Flat panels, cavities, helicals, and a side-mounted design have given broadcasters a range of choices to fit their application. Factors such as pattern requirements, tower constraints, and system budget can narrow the options for selecting the FM broadcast antenna.

In order for an antenna to be useful throughout the 20 MHz FM band, its operation must be the same on any frequency. The VSWR at 88 MHz, for example, must be just as good at 108 MHz. The CP azimuth pattern should remain the same on one end of the band as the other, as must the axial ratio. This is a much more rigorous requirement than those placed on the single-channel slanted dipole or the ring stub.

In the wideband panel antennas, several factors go together in order to meet these important requirements:

- Basic wideband dipole radiators
- Screen-panel pattern control
- Quadrature phase distribution.

Wide band side-mounted antennas are also available today. Pattern optimization can't be used to improve the pattern of side-mounted broadband antennas because the length of the parasitic element used to improve the pattern will change electrical length as the input frequency to the broadband side-mounted antenna changes. Today, sidemount broadband antenna arrays are available with input frequency bandwidth of 18 MHz. The advantage of the broadband side-mounted array is that the array will have lesser load on the support tower. The disadvantage of the broadband side-mounted array is that the horizontal plane relative field pattern of the side-mounted array will not be as omnidirectional as the pattern of a well-designed panel antenna array.

By using the three principal parameters in a wideband antenna, the radiation pattern, VSWR, and the gain can be nearly the same on any channel within the FM band. Several methods are used to make the VSWR of the crossed dipoles as good as possible. The dipoles are usually fed with a folded balun or the split-tube-type balun [14]. This improves the impedance match, phase, and amplitude linearity of the resulting azimuth pattern.

The length to diameter (or width) ratio is usually about five. This not only reduces the Q but also increases the voltage flash-over levels. The low Q also increases the bandwidth by decreasing the rate of reactance change with frequency. A natural factor aiding the VSWR problem is the fact that in order to obtain CP from two crossed dipoles, there must be a phase quadrature of the two currents feeding the crossed dipole. The two reflections, as a result of VSWR, return back to the phasing device 180 degrees out-of-phase with each other. Being the same amplitude, they cancel. The phase quadrature of the two currents feeding the crossed dipole antenna element is usually achieved by using a 90 degree hybrid. Often, the hybrid is designed with four ports, and the fourth port can be used to add a second input to the crossed dipole that is isolated from the other port by polarization. Using this method, for example, a station's digital IBOC signal can be radiated left-hand circularization while the station's analog signal can be radiated right-hand circular polarization. Typical right-hand circular polarization to left-hand circular polarization is 25 dB.

All of these factors, plus two or three more levels of quadrature reflection cancellations, bring the overall system input VSWR at the antenna to as good as 1.08:1 across the band. This cancellation technique eliminates the need for electrical de-icers or plastic radomes, as the VSWR is not affected by moderate ice coatings. However, radomes may be necessary with flat panels to physically protect some radiators from falling ice.

These and other factors all contribute to make the panel antenna the best possible for either single channel or community antenna use. Cavity-backed antennas also make excellent broadband multi-user antennas, however the horizontal plane patterns for cavity-backed radiator arrays will typically have more scalloping in the horizontal plane relative field pattern.

Top-Mount Antennas

There are a few broadband designs that require top mounting on a pole. The first incorporates a series of dipoles that are mounted on a pole. Each dipole is branch fed individually. By isolating crossed dipoles using screens, circular patterns with pattern circularity of better than ±2 dB can be achieved. The use of screens between dipoles not only keeps pattern circularity at an acceptable level, but also provides isolation between the antenna elements resulting in improved impedance bandwidth for the array. Large bandwidths and good patterns have been measured on this system. The considerable amount of feed cables that run through the aperture is controlled by mounting the antenna elements on a pipe. The feed cables can be run inside the pipe protecting them from the radiated field from the array. In order to maintain structural integrity, the screens can be designed as structural components. The second, normal mode helical, antenna design, is also being used for FM applications. This traveling wave–type antenna has bandwidth on the order of 15%–20%. Fed from the top, a series of wires are wrapped around the conductive pole over an aperture of two to three wavelengths. The pitch angle of the wire wrap controls the pattern characteristics. Pattern circularities of ±1 dB over a 20-MHz band have been measured. Axial ratios better than 2 dB are considered typical. The feed system is enclosed by stainless-steel feed cans that eliminate pattern distortion by the feed system, a common fault of other top mounts. In regions where icing is a problem, heating elements inside the radiating wires will stabilize the antenna's performance.

Multichannel Side-Mount Antennas

Another broadband FM broadcasting antenna is the side mount. This design incorporates individual groups of four series internal fed dual dipole elements configured in half-wave modules. The individual modules are then branch fed to maintain the required input voltage phase for each of the four modules. Up to four modules can be used to achieve the desired array input power gain. Excellent circularity and axial ratios have been measured in free space for these series internal fed antenna elements. Interaction with feed lines and support structures still distort the pattern.

A single FM element has been measured to have a bandwidth of 5.5 MHz with a VSWR of 1.1:1. This bandwidth does not limit the spread of two combined stations. By careful design of the four bay modules of this FM element, the antenna can tune two stations over 15 MHz apart with a VSWR of 1.15:1 over each channel. This comes at no cost of poor performance other frequencies, thus many stations over 15 MHz are achievable. As a side mount, this antenna has substantially reduced costs and tower constraints over a panel type. The number of users for this side-mount design is limited by the input power rating of the array. Individual four-element modules are rated at up to 100 kilowatts input power.

Power-Handling Capacity

A multichannel antenna must be designed to not only accommodate the bandwidth requirement for all users, but also to handle the input power due to multiple users. Considerations for both peak power and average power must be made.

The average power capacity of an antenna refers to its ability to operate below some critical temperature. The antenna (including its feed system) will convert a certain portion of its input power into heat, and that heat must dissipate into the environment without raising the temperature of any critical components, such as insulators, to extreme levels. The generation of this heat is due to RF currents and the attendant resistive losses in the component materials. The rating of coax is based on the amount of power that will raise the temperature of the inner conductor of the coax to 100°C (212°F). The average power of an analog FM transmitter may be considered constant with time and the average power due to multiple (*N*) FM transmitters each delivering P_o watts of average power to the antenna is simply $N \times P_o$.

The peak power capacity of an antenna refers to its ability to operate without producing excessive electric fields of such magnitude that arcing or corona discharge occurs. While average power capacity is about amperes and cooling, peak power is about volts and conductor geometries. Corona or arcing can occur where electric fields are concentrated by sharp points. On a dipole structure, the dipole ends are the location of the highest voltages, so larger radius parts are often used there to prevent discharge.

In multiple channel systems the voltage at a point in the system is equal to the sum of the voltages for each channel. In the absence of subcarriers or digital modulation, the peak power (as with the average power) of an analog FM transmitter may be considered constant with time. The peak power due to *N* nonsynchronous FM transmitters each delivering *P* watts of average power to the antenna is $N^2 \times P$.

Pressurization and altitude have an effect on both the peak and average power-handling capacity of an antenna. Generally at lower pressure the voltage at which corona discharge or arcing may occur will be lower, which bears on the peak power capacity. Also, because lower pressure reduces the cooling efficiency, the average power-handling capacity is impacted. Pressurization of feed system components may be used to increase the power-handling capacity of that portion of the antenna system. High relative humidity and fog can apparently precipitate corona discharge from the ends of dipoles, so these factors must also be considered in multiple-frequency antennas.

Transmission line is rated for average power and peak power, and the safety factor for most published specifications for transmission line is 2:1. In calculating the safe operating input power to a common antenna with both IBOC and FM Analog, the IBOC peak voltage calculation is different from the FM Analog peak voltage calculation. The Peak Power Rating can be reduced to a Peak Voltage Rating as shown next.

Peak Power and Voltage Considerations

Mismatches between the characteristic impedance of the line and the load or source impedances lead to a voltage standing wave ratio (VSWR) that results in voltage maxima and minima, for example along the length of interbay transmission line. It can be shown [4] that the ratio of maximum voltage to input voltage due to a certain VSWR is given by

$$\frac{V_{max}}{V_{in}} = \left(\frac{2 \bullet VSWR}{VSWR+1}\right)$$

This factor should be applied to peak voltage calculations when considering system ratings. As can be seen in Figure 7.17-25, the VSWR derating factor approaches a multiple of two as VSWR increases. It should be noted that while VSWR at the input of most broadcast antenna systems would not normally see a VSWR greater than 1.5:1, higher VSWR values are possible within the feed harness of an antenna where individual element impedances may be transformed to higher or lower impedances as needed for power division and impedance matching.

A power derating factor (DF) based on VSWR is given by:

$$DF = \frac{VSWR^2+1}{2 \bullet VSWR} + \frac{F'\left(VSWR^2-1\right)}{2 \bullet VSWR}$$

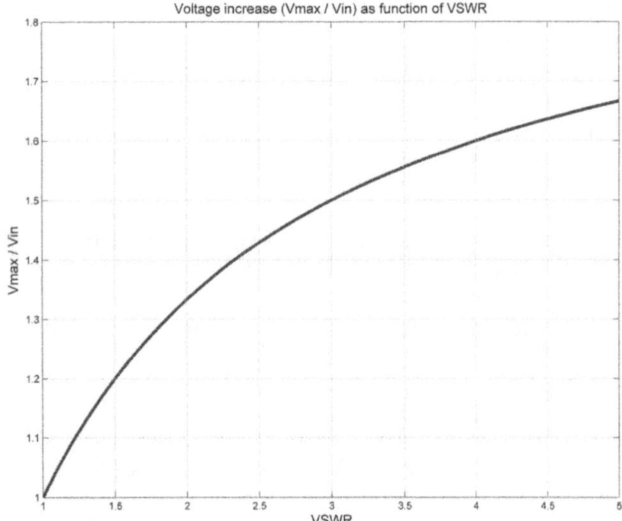

FIGURE 7.17-25 VSWR derating factor for voltage

where F′ is a function of both frequency and line size. Table 7.17-7 provides this factor for frequencies across the FM band and for common transmission line sizes.

Once the derating factor has been determined by calculation, it can be used to determine a safe operating power for the various coaxial components of an antenna harness or any coaxial feed line in an RF system. The derating factor at 98 MHz is plotted in Figure 7.17-26.

For example, using this derating calculation, safe operating powers for a system at 98 MHz are shown for various coaxial rigid line sizes in Table 7.17-8.

For a 3-1/8″ rigid 50-ohm coaxial component in the antenna harness, for example, the sum of all the analog power and all the IBOC power must be less than 43.3 kW.

Peak voltage must also be considered. The allowable peak voltage values are listed in Table 7.17-9 for various rigid line sizes.

The RMS voltage V_0 for FM analog in 50-ohm line is given by:

$$V_0 = \sqrt{50 \bullet P_0}$$

where P_0 is the power in watts.

The peak voltage for an analog signal is given by

$$V_{peak} = V_0\sqrt{2}$$

For a digital signal using multicarrier modulation formats, such as the orthogonal frequency division multiplex (OFDM) used for HD Radio (TM), the signal envelope varies based on a statistical variation of the multiple carriers adding and subtracting. The ratio of the maximum power to the average power is referred to as the peak to average power ratio (PAPR). For the IBOC implementation currently in use, efforts have been made to reduce the PAPR, and it has varied from around 6 dB to lower values in recent years. Using a PAPR value of 6 dB equates to a voltage ratio of 2-to-1,

TABLE 7.17-7
Values of F' for US in VSWR Derating, 50-Ohm Line

Line Size	F' @ 88 MHz	F' @ 98 MHz	F' @ 108 MHz
7/8″	0.55	0.51	0.24
1-5/8″	0.40	0.35	0.16
3-1/8″	0.41	0.33	0.16
4-1/16″	0.35	0.30	0.12
6-1/8″	0.29	0.22	0.09

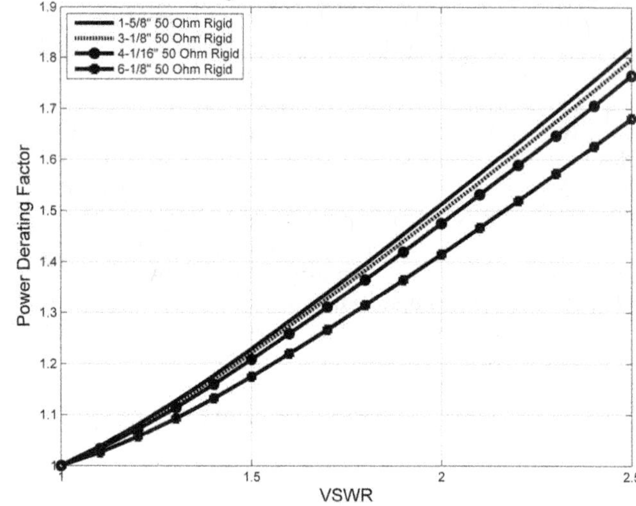

FIGURE 7.17-26 Power derating factor for various line sizes at 98 MHz

TABLE 7.17-8
Safe Operating Power, 50-Ohm Line at 98 MHz

Line Size	Rating at VSWR = 1.0:1	Rating at VSWR = 1.5:1
1-5/8″	15.1 kW	12.3 kW
3-1/8″	52.9 kW	43.3 kW
4-1/16″	84.0 kW	69.5 kW
6-1/8″	180.5 kW	153.6 kW

TABLE 7.17-9
Peak Voltage Ratings for Rigid Coaxial Line, 50 Ohm

Line Size	RMS Voltage (V)	Maximum allowable peak voltage (V)
1-5/8″	2576	3643
3-1/8″	4695	6640
4-1/16″	5959	8427
6′1/8″	8764	12,394

and in terms of the average digital voltage, $V_{IBOC,0}$, the peak voltage in the digital waveform is given by:

$$V_{IBOC,peak} = V_{IBOC,0}\ \sqrt{2} \cdot 2$$

As an example, peak levels for a system with analog power input at 98 MHz of 20 kW with digital power at −14 dB relative to analog are calculated next.

Analog peak voltage:

$$V_{peak} = \left(\sqrt{50 \bullet 20000}\right)\sqrt{2} = 1414 Volt.$$

IBOC power level (at −14 dB):

$$P_{IBOC} = 20kW \bullet 10^{\left(\frac{-14}{10}\right)} = 796W$$

For a peak to average ratio (PAPR) of 6 dB, the IBOC peak voltage is

$$V_{IBOC,peak} = \left(\sqrt{50 \bullet 796}\right)\sqrt{2} \bullet 2 = 564 Volts$$

Additionally, transmission line rating will be affected by ambient temperature, altitude, and internal dry air or nitrogen pressure. Refer to Chapter 9.4 for more detailed discussion of transmission line ratings.

Community Antenna Economics

A community (shared) antenna system can provide many benefits for multiple station facilities when there is a limit to available tower aperture, when a common filter/combiner system is the best approach to controlling intermodulation issues, or simply to allow costs to be shared so all parties benefit from a superior antenna that each station independently could not economically justify.

If enough planning is done in advance it may be possible to install all the FM stations in one community on one tower at considerable savings to all users. Some exclusions include lack of adequate mileage separations, the existence of excellent facilities, and FAA tower height limitations.

One of the first large-scale community antenna projects in the United States was the Alford multiplex facility in New York city located on the Empire State Building. That system was installed and put into service in 1966. In 1970, a group of nine Houston broadcasters went together to utilize the 12-bay multiplex antenna that was installed in 1971. That system was purchased by the One Shell Plaza Building. Later, when that system became impractical due to shadowing from new buildings in downtown Houston, that same group of nine stations formed the Senior Road Tower Group and installed a 2,049-foot (625 m) tower supporting a 12-bay community FM antenna system, with an HAAT of 2,000 feet (610 m) [15]. This height permitted the maximum service allowed. This system replaced the ERI master antenna system on the Shell Building, and it utilized a Harris FM antenna and a Dielectric combiner. Then, in 2006, the Harris/Dielectric multiplex antenna system was replaced by an ERI Master Antenna and a pass band constant impedance combiner for the nine stations.

The original two runs of 8-3/16 inches (208 mm) diameter coaxial lines are now used to feed the ERI antenna so that the analog power in both the lines causes right-hand CP from the antenna for nine FM stations. The new ERI antenna is a dual-input antenna, and the digital output from the nine stations is fed to the antenna using two flex lines feeding the left-hand CP antenna input. The nine stations use one diplexer each, all nine of which are housed in one 2,400 square foot (223 m^2) room. The nine constant impedance filters are back fed with digital signals, and circulators are utilized to offer additional isolation to the digital transmitters. The 9-port modular multiplexer has a total power-handling capability of 270 kW. The insertion loss for each station varies. The isolation between the various transmitters meets FCC spurious emission (intermodulation) requirements. The constant impedance multiplex filters are monitored at a central operating rack that displays each multiplexer's forward, reflected, and rejected power. This permits troubleshooting in an orderly and rapid manner. Electrically operated coaxial switches permit each station to be connected to the dummy load for individual testing. Air conditioning and chilled water are used to remove heat produced during operation.

Technical Advantages

Besides the financial advantages cited under the economics heading, there may be the competitive advantage of protecting the channel classification and using the same height antenna as the competitor. Other advantages include its emergency upper-lower half feature for transmission line or antenna half backup, however separate auxiliary antennas are recommended in case work needs to be done on the main antenna. The flat VSWR curve is highly useful for SCA and IBOC operation. Stations sharing this type of antenna will all experience less intermodulation interference than if they had separate but closely placed antennas.

STATION MULTIPLEXING

Multiplexers are passive devices used to combine the power of two or more stations and feed the combined power to a common transmission line and/or a common transmitting antenna. This system of utilizing one well-sited, high-quality antenna has become popular, convenient, and economical.

Wideband panel antennas, although expensive for use by one station, are cost-effective for two or more stations. These antennas maintain their omnidirectional horizontal plane patterns and VSWR throughout the FM broadcasting band from 88–108 MHz. Thus, they make the ideal antenna for multi-station diplexing. As mentioned earlier, side-mounted antennas can also be used for multiplexed operations and are commonly used as such.

FM ANTENNA INSTALLATION ON AM TOWERS

The current trend is to locate FM transmitters in places where the best service may be rendered to the most listeners. This usually permits the maximum possible height to be used. Sometimes, however, it may be economical and convenient to install the FM antenna on a tower used for AM broadcasting. If the steel AM tower is not base insulated but is grounded and shunt fed, the FM coaxial line may be connected to the tower, without any further problems.

TRANSFORMER ISOLATION

However, if the AM tower is insulated at the base, an isolation transformer may be used to couple the FM power across the base insulator without introducing objectionable mismatch into the FM antenna feed line. Isolation transformers do add capacitance to the AM tower base impedance. An isolation transformer is especially desirable for feeding high-impedance AM radiators or AM radiators that are part of an AM-directional antenna system that might be adversely affected by a quarter-wave isolation system.

Some of these transformers have two tightly coupled RF coils that are resonant at the FM operating frequency. An air gap is provided for the AM power to pass through the two resonant loops. The capacity is quite low, resulting in a very high capacitive reactance placed across the tower base insulator. Figure 7.17-27 shows the internal basic construction of a typical isolation transformer. The insulation for AM under the top of the box may be high-density polyethylene, Teflon, or fiberglass. The metal top provides a rain shield as well as protection from dust, mud, or snow. The isolation transformer is adjusted for minimum return loss at the FM frequency and minimum FM voltage

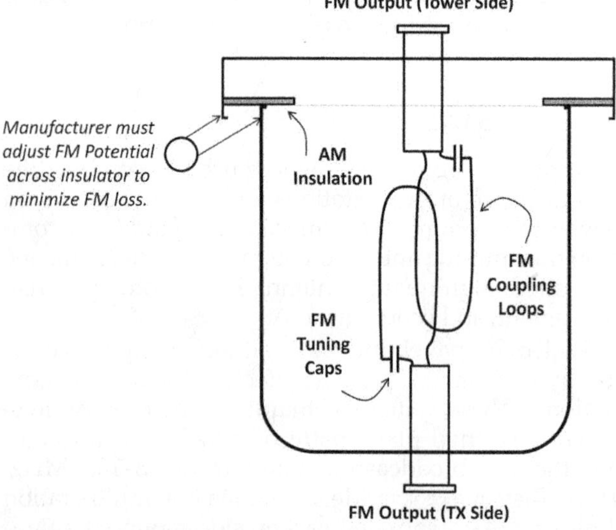

FIGURE 7.17-27 FM isolation transformer (isocoupler) for AM tower mounting

potential difference across the gap on the output that isolates the AM from ground. This minimizes the FM loss in the isolation transformer. When adjustments are completed, the AM capacitance to ground is measured.

The use of these isolation transformers permits the AM tower to operate undisturbed by the presence of the FM antenna. It also allows the FM coaxial line to be connected in the usual manner, except for the placement of the isolation transformer. Some have internal gas blocks and permit the passage of dry air pressure through the transformer via a plastic tube, and others are pressurized. Pressurizing the isolation transformer protects the components in the transformer from moisture, insects, and any weather-related problems.

In addition to lower cost, the isolation transformer method has another advantage in directional AM tower use. It does not distort the AM radiator current distribution that may adversely affect the AM radiation pattern. Typically, the isolation transformer is mounted next to the AM tower base insulator with the insulated end of the transformer next to the tower. The feed line from the antenna is then mounted to the insulated end of the transformer. There is a rain shield that protects the insulator from moisture that could cause arc tracks during rain storms. Ideally, isolation transformers should be internally pressurized.

Quarter-Wavelength Isolation

A less popular and older method is to use the technique of quarter-wavelength transmission lines. Simply stated, the opposite end of a shorted quarter-wavelength line has high impedance. This high impedance is placed across the AM tower base and may be successfully used to provide the necessary isolation. It is more difficult to physically accomplish this as the tower should be at least one quarter-wavelength high, and the FM antenna coax line must be insulated all the way down the tower. In practice the insulated part may be as short as 75 degrees of line, as the line hangers and distributed capacity of the line tend to electrically increase the physically shorter line. For best results, the FM line should be placed within the tower body. The location of the shorting point of the outer copper sheath of the coaxial cable to the tower is experimentally found by first measuring the base impedance prior to cable installation and then choosing a point that produces the same impedance (the quarter-wave short up the tower produces an open circuit at the base).

Guy Cable Considerations

The presence of continuous steel guy cables going through the FM antenna level on a steel supporting tower have a small effect on the horizontal polarization pattern, but can have a more significant effect on the vertical polarization pattern. This impact can be seen in both the azimuth and elevation patterns. As a historical reference, guy cables in the vicinity of FM antennas didn't surface as a significant problem until V-pol was authorized for use by FM stations.

In addition to pattern anomalies, the steel cables reradiate near the ground. This may cause RF problems in some high-power installations with low-level audio equipment located in a building near the base of the tower.

Strong currents may be induced when the steel guys are in the immediate vicinity of the radiators. If that guy passes close to the side of the radiator, the field on that particular side of the antenna element will induce currents in the guy wire. The radiator currents can become unbalanced and the impedance of the element is disturbed, changing its VSWR and radiation pattern. Because of these effects, it is common practice to break up the guys using insulators, fiberglass rods, or plastic guy cable, within 10 feet (3 m) of the antenna radiators.

When a CP antenna is side mounted on a guyed tower, the vertically polarized field will have an appreciable component parallel to the guy wire in its aperture and will induce currents in the wire. If the guys are continuous, a progressive wave traveling toward the ground will result, and will radiate most of its energy before reaching the ground. The energy will be radiated in cones concentric with the wire. A small amount of the V-pol power will thus bleed off.

On tall towers FM antennas are typically mounted between guy wire attachment point. The bottom bay of an FM sidemount antenna should be mounted just above the guy wire attachment point. The top bay of the array should be at least 30 feet below the guy wire attachment immediately above the array, otherwise consideration of the impact on the pattern should be given.

Porcelain Insulators

If the FM antenna is side mounted on an AM tower, which usually will have metal guys and porcelain break-up insulators, the insulators will probably be spaced several FM wavelengths apart. The induced currents will form standing waves on the sections between insulators and radiate multilobed patterns into space at many angles from the wire axis. If the sections between insulators happen to be of a resonant FM wavelength however, currents in the guy wires and their radiated fields will be considerable. A single isolated piece of guy wire with its ends insulated can only resonate in multiples of one-half wavelength, so this spacing of insulators must be avoided. In fact, with the capacitive end loading of the insulators, the resonant length of wire will be somewhat less than one-half wave, so three-eighths wavelengths should also be avoided. A one-quarter wavelength is much better, but this would be quite expensive as it requires insulators every 30 inches (762 mm).

Alternatively, and a better approach, the guys through the FM antenna aperture on the tower could be replaced with plastic cable or fiberglass rod insulators, which is transparent to RF energy.

Nonconductive Guys

In order to eliminate FM pattern distortion, any guy cable going through the antenna level should be of nonconducting material. Plastic fiberglass (GRP) insulating rods as well as flexible plastic rope covered with a PVC plastic jacket can be used. The black jacket prevents deterioration due to ultraviolet sunlight radiation, which may be injurious to the plastic strands of the rope. Plastic rope has been used successfully for more than 25 years.

TABLE 7.17-10
Typical Data for Nonconductive Tower Guy Ropes

Outside Diameter (inches)	Break Strength (lb)	Weight (lb/1000 ft)
0.16	1,200	8
0.25	4,000	21
0.42	11,200	55
0.53	20,800	95
0.66	35,000	147
0.92	58,300	276
1.38	130,000	595
1.95	252,000	1055
Typical data; taken from Phillystran Tech Bulletin 301–09/13, Phillystran (R) HPTG Tower Guy Ropes.		

The idea is to remove metallic RF-conducting steel guy cables from within the antenna aperture. The rest of the guy may be of steel construction. The length of the steel guy from its attachment point near the antenna to a point well below the antenna is simply replaced with an equal length of fiberglass rods or plastic rope. Plastic rope is available in continuous lengths of up to 10,000 feet for smaller diameters and kits are available for installing the end fittings.

The cable may be purchased in strengths exceeding similar diameter steel guy cable. These strengths for a sample of diameters are shown in Table 7.17-10. This data can be used to match to the corresponding size of commonly used extra high strength (EHS) steel guy wire.

INSTALLATION PROCEDURES

If the installation is not properly planned and carried out, there may be unwarranted delay and cost associated with putting the FM antenna on its support tower. The following suggestions are offered to avoid unnecessary delays and expenditures.

Planning the Installation

The first step is to ensure the structural integrity of the tower. This should be performed by a qualified, licensed Professional Engineer. Some tower manufacturers offer this service. Because of the high cost of rigging services, it is essential to carefully plan the installation. Make sure that all parts are on hand. The installation of the antenna should be planned by a technically qualified person who must supply accurate tower construction information to the antenna manufacturer. If this information contains errors, these will be carried through the design and fabrication of mounting hardware, and finally show up in the field to frustrate and confuse the installation crew, wasting time and money at every stage of the process. Before the installation is scheduled, the installer should provide a rigging plan for the project.

The station should consider hiring a tower rigging company that is financially qualified and mechanically well equipped to do the work. A written contract should exist between the station and rigging firm, with a fixed price. The rigging contractor should be licensed as a contractor in the state. The contractor should also supply an insurance policy holding the station harmless and making the station and its personnel coinsured. Only in this manner will the broadcaster be protected in the event of injury or property damage.

Two qualified climbers with current training in fall protection and rescue with knowledge about antennas and coax line should inspect the tower and check out the mounting design of the brackets before the full rigging crew arrives. If errors are found, contact the factory immediately. Particular attention should be paid to the following:

- Fit of mounting brackets to tower member.
- Freedom from interference of the mounts with gussets, leg flanges, guys, and their attachment points, tower face members, and obstruction lights.
- Compatibility of transmission line and antenna input coax terminals.
- Location of transmission line run relative to antenna input terminals.
- Availability of proper voltage, current, and cable size for de-icers if required.

Receiving and Unpacking

The shipping boxes are usually numbered and the total number is indicated on each box; contact the shipper if not all boxes are delivered, or if equipment is received damaged. If possible, do not store the material outdoors, boxed or otherwise.

As soon as the antenna is received, open and examine it for shipping damages so that any necessary claims may be filed with the shipping company immediately. Check the material against the parts list and installation drawing.

The box with the installation drawing and instructions is usually marked. Open it first, so that the balance of the items may be easily identified and counted. Contact the factory immediately if any material appears to be missing or is damaged during transportation.

Do not arrange for the installation crew to be on site until all antenna and coaxial line are at the site. Otherwise, unnecessary delays and costs may result.

Installation Tips

Broadcast antenna manufacturers furnish detailed installation instructions that should be closely followed by the rigger. Together they will ensure a perfect installation saving time and money.

The following items are specifically called to the attention of the broadcast engineer (in addition to all those stated before) to permit proper installation and good long-term performance:

- Follow manufacturer's instructions.
- Do not leave antenna parts where rain or moisture can enter. Store indoors and keep units capped as received.
- Do not allow dirt or other foreign matter to enter any coaxial part.
- Protect all antenna parts from physical damage and abuse.
- Hoist antenna members carefully, with a tag line to prevent damage by striking against the tower.
- Install on the tower as indicated by the manufacturer's instructions, remembering that the antenna bays are installed in the correct order.
- Riggers should lubricate O-rings with a small amount of silicone grease before mating flanges.
- The full complement of flange bolts must be used and they should be as tight as instructed.
- Tuners or individual element devices, if used, should be adjusted only after the entire antenna and tower installation has been completed.
- Rigid transmission lines should be properly installed with two hangers per 20-foot (6 m) length, and with the inner conductor retaining pin on the top of each section. If rigid transmission line is used that has bellows for inner conductor movement with temperature, these line sections will have captive bullets at the top of each section.
- If semi-flexible cable, such as Heliax or Wellflex, is used, it should be firmly tied down at least every 5 feet (1.5 m) for 3-1/8-inch (76 mm) line, and every 3 feet (1 m) for 1-5/8-inch (43 mm) coax line. The line manufacturer's hangers should be used. The line should not be attached to the tower using plastic tie wrap straps. This line must be installed by using the manufacturer's recommended installation technique.
- After physical installation has been completed in accordance with the manufacturer's recommendations, the main transmission line should be pressurized with dry air through a dehydrator, air pump, or by using dry nitrogen gas. (See the "Air Pressurization" section in this chapter for more information.)
- Dry air or gas pressure should be maintained at all times. Most antenna warranties are not valid unless this is done. It is the installer's responsibility to make certain that the entire coax and antenna hold air pressure.

Installation of FM Antennas

Care must be taken when FM antennas are installed on towers. For side-mounted FM antennas mounted on guyed towers, care must be taken to keep the top bay of the antenna as far below the tower metal guy support wires. If there is not enough space to keep the FM side-mount antenna away from the wire, replacement of the metal guy wire with an insulated wire should be considered. Side-mounted FM antennas should be partially pre-assembled before being installed. If the FM antenna array has a rigid coaxial feed harness, one FM antenna with bracket attached and with one rigid inter bay coax section attached should be assembled on the ground. The rigid line section should be attached above the one bay of the array. A rope can then be attached to the antenna bay and to the rigid line section so that the bay can be raised up to the tower in a vertical position. The bottom bay of the side-mounted array should be installed first, and then the additional elements are installed one bay at time above the bottom bay. Once the antenna array has been completely installed, dry air or nitrogen should be used to pressurize the array. Typical pressure is 5 psi. When looking for leaks, however, up to 15 psi can be applied to the system. Because dry air and nitrogen are heavier than water vapor (water vapor has one oxygen and two hydrogen atoms and is thus lighter than both oxygen and nitrogen, which have two atoms of oxygen or nitrogen), the installed antenna should be purged at the highest point in the array to eliminate any moisture in the array. Once the antenna is installed, the antenna return loss can be adjusted. Antenna match adjustments are always done as close to the antenna input as possible and will require a physical adjustment. There are three types of antenna match adjustment available for side-mounted circular polarized FM antennas. These methods are custom adjustable fixed capacitor tuning sections, custom adjustable probe tuning sections where the probe can be capacitive or inductive, and the ceramic slug method. Only the ceramic slug method requires opening up the antenna input coaxial section to make the match adjustment. Both the position of the slug and the amount of slug are critical to achieve the best input impedance match of the antenna.

Face-mounted FM antennas are very stable on towers during high winds, but leg-mounted FM antennas can depending on the outer diameter of the tower leg rotate around the leg during high winds. In order to stabilize the antenna, anti-rotation brackets can be added. These brackets brace the leg-mounted antenna to an adjacent leg, thus resisting antenna rotation due to exposure to high winds.

All FM antennas are affected by the weather. Ice forming on an FM antenna will cause the FM resonate frequency to drift down. With very broadband FM antennas, this change can be tolerable, but with narrow-band FM antennas the change will result in an excessive voltage standing wave ration causing the station transmitter to trip off. In areas of the country where ice can be expected during winter months, heaters and/or radomes are often added for protection against ice. The heaters are installed at the factory and are usually inside the actual antenna elements. These heaters are generally controlled by a circuit on the tower that turns the heaters on when the outside temperature is inside a range of, for example, 20°F to 40°F. As an alternative, radome covers are often added to the individual FM antenna bays. These radomes are made of nonconducting material and are designed to keep the ice forming on the antenna bay several inches away from the actual radiating portion of the antenna bays.

Often, FM side-mount antennas are installed on active AM towers. Installing an FM coaxial feed line on an AM tower that has a base insulator can be done by installing an FM Isolation Transformer at the base of the tower or by using a quarter wavelength at the AM frequency of the FM feed line on insulators so that it is isolated from the first quarter wave of tower height. FM Isolation Transformers are designed to pass the FM carrier while creating an insulator at the AM frequency. These devices do add capacitive loading to the tower AM base impedance. During the FM antenna installation, the tower should be isolated from the AM array, and the base insulator should be bypassed with a ground to protect the tower workers. Careful care should be used to make sure that the rigging does not shock the ground crew. Often, a ground rod is driven into the ground where the load line picks the antenna bays up. While handling the rigging, a ground should be attached to the load line grounding it to the ground rod.

Installations on building tops should be carefully reviewed by an engineer to make sure that workers on the building are not exposed to excessive radiation once the antenna has been installed and put into service.

When installing custom wrap around panel FM antennas, each face of the tower will have an array of FM antennas with reflector screens. These antenna elements should be partially assembled on the ground with one element, that element's reflector screen, and the associated mounting bracket for that sub-assembly assembled prior to installing on the tower. Then, each tower face array should be installed on the tower along with the associated power distribution harness. After installation, the array should be purged.

Installation of directional FM antenna arrays requires care to make sure that the array is orientated correctly. When the antenna has been installed, a licensed surveyor must be retained to work with the installation crew to make any final orientation adjustments of the array to guarantee that the antenna orientation is correct.

Once the antenna has been installed, the pressurization of the antenna must be maintained. Both the antenna-reflected power and pressurization should be monitored daily. Monitor and control systems are readily available that will identify problems and provide a record of any fluctuations in the antenna/feed line system parameters.

Antenna Maintenance After Installation

The antenna system should be checked by no less than two qualified climbers with current certification in fall protection and rescue every time the obstruction lights are replaced, or if lights are not used, at least once a year. The qualified climber should look for vibration and storm damage, loose or broken coax hangers, and signs of arcing across exposed insulators. A dry rag soaked in 91% isopropyl or other solvent alcohol or equal should be used to wipe clean all exposed insulators in each antenna element. (Do not use carbon tetrachloride!) All safety and OSHA rules must be followed during antenna installation and maintenance.

STRUCTURAL CONSIDERATIONS

Most FM antennas in the Western Hemisphere are installed on the sides of a steel tower, between 18 and 60 inches (45–152 cm) wide. The antenna and its transmission line together with all mounting brackets introduce wind loading, in addition to their dead weight. The live wind loading is a result of the amount of physical surface presented to the wind. It is sometimes called the *wind catch area*. This consists of either flat or round antenna members, coaxial lines, mounting brackets, and hardware, all represented as surfaces that are exposed to the wind.

The dead weight of the antenna system is fixed and is always present on the tower. The live load is a variable, depending on the wind velocity, and is added to the dead load for the total amount present.

Various building codes determine the rated winds to be considered in the design of the tower system. Typically, these building codes specify a recent tower design standard that should be used to calculate the wind load of the antenna with and without ice.

ANTENNA POLE MOUNTING

Nonsymmetrical antennas may also be installed on a round pole, made of various diameters of steel pipe. Several antenna manufacturers supply these as a complete system and will optimize the horizontal plane pattern. The advantage of pole mounting on top of a tower or building is that the pattern may be more easily contoured. This provides more signal in the service area because the antenna orientation is not limited by a fixed triangle formed by a guyed tower.

TRANSMISSION LINE SYSTEMS

Two types of coaxial transmission lines may be used to feed FM antennas. One is rigid coaxial line sections, each approximately 17.5 feet long, and requires elbows, flanges, spring hangers, and other devices to attach the line sections to the tower. Sections of rigid coax 17.5 feet long are used for broadband FM antenna feeds because the repetition of supporting insulators for these sections don't resonate in the FM band, whereas 20-foot sections resonate at mid-band.

The other has a semi-flexible coaxial line that is available with either air or foam dielectric and uses fixed hangers. Semi-flexible cable is available on a spool whose diameter depends on the line size. EIA end flanges mate to the antenna flanges as well as other RF equipment. Detailed information on FM RF transmission lines can be found in Chapter 9.4.

AIR PRESSURIZATION

If the antenna is operated without positive pressure of dry air or nitrogen, the manufacturer will not assume responsibility for failure under power. Moisture or the accumulation of water within the coaxial transmission line is a very serious matter. Its presence causes the VSWR to rise and corrosion to occur. When a sufficient amount of moisture is present, arcing will take place burning the line or antenna radiating elements. High humidity or moisture will cause the inside of the coaxial transmission line to corrode over time, thereby increasing the line loss. For this and other reasons, the entire antenna system must be dry-air pressurized.

After the antenna is installed and the transmission line connected, the system is purged with dry gas or dry air to remove trapped moisture before RF power is applied. The complete system purge requires a considerable volume of dry gas.

Before expending this amount, it is good practice to perform a quick check for major leaks. The system pressure is typically raised to 10–15 PSI and the source of pressure supply shutoff. A pressure gauge should be installed on the antenna side of the shutoff valve. The pressure, when corrected for temperature, should not fall at all. If the pressure should fall to half its initial value in a 24-hour period, the leaks should be found and repaired as soon as possible. The system should be checked with a leak detector or soap suds to locate the leak. A pinched or missing O-ring, a gap in the weld on a flange connection, a crack in a the outer of a coax section, or a damaged insulator can cause serious leaks.

Once the system is known to hold pressure, it should be purged with dry air or nitrogen gas. Either must be dry enough to have a dew point well below the coldest temperature expected to be encountered. When using nitrogen, it should be of the oil-dried type, to remove nearly all the moisture from the gas.

Pressure of 5–8 psig (0.34–0.48 atmospheres) should be maintained in the system at all times to ensure that no moisture will be able to enter. Very small leaks will pull in moisture if there is a drastic change in temperature, such as when a rainstorm occurs. This is due to the pumping action due to expanded dry air/gas pressure cooling down, and contracting below the outside air pressure during cold ambient temperature changes.

PROTECTION FROM ICING

High-Q antennas are subject to increased VSWR ratios as well as pattern distortion, with light to moderate coatings of ice. Low-Q antennas, such as the panel type, are usually not affected in this manner. Where climatic conditions cause sufficient ice or in some cases snow to affect the antenna's performance, there are two remedies. The radiating element may be covered with a radome, or it may be electrically heated to melt or prevent the formation of ice on its sensitive surfaces.

Electrical Heaters

By far the more popular method of de-icing is to order electrical heaters at the time the antenna is ordered.

Electrical de-icing equipment is supplied as an option and is factory installed. Kits are furnished for interbay connections, but the broadcaster must supply power from the building to the center of large arrays, or the bottom element on smaller antennas. Local electrical codes of course must be followed.

While a thermostat may be used with smaller de-icer wattages, a power relay operated by the thermostat is required. Due to high power costs, a sophisticated de-icer control, which operates when temperature and humidity conditions produce sleet or icing, is desirable.

Most de-icers use a resistance heating element that is inserted inside the antenna radiator arms. One manufacturer, however, did use a different method, dropping the 230/240 volts to a few volts with a transformer located at each bay level. The low voltage is passed through the ice-sensitive arms of the radiator and connected to the far ends by a heavy Teflon-coated wire. The current return was by the stainless-steel antenna element, whose resistance is sufficient to produce enough heat to melt or keep the ice off. This method is now obsolete as the transformers are expensive and heating costs are rising, as hourly electrical rates go up. Further, the voltage dropping transformer technique is not as efficient as direct heaters.

A word of caution when selecting an FM antenna with electrical heaters. Some de-icers use 1 kW of power for each bay as previously described and will slightly increase the wind loading. Some have a switchable power option feature using 125/500 W per element, with only about a 15% increase in wind loading, when compared to an antenna without electrical de-icers. The continuing cost of electrical de-icers is a consideration of the operational cost and should not be overlooked.

Automatic de-icers are those with a thermostat for mounting near the antenna for accurate temperature sensing of the actual ambient temperature. The temperature zone of 20°F to 35°F (−7°C to 2°C) is the most likely icing range, depending on humidity conditions. De-icers should be turned on at 35°F, prior to ice formation, because it is better to prevent icing than to remove it once it has formed. Power should be turned off when the temperature goes below 20°F because ice does not usually form below this temperature.

Radomes

A radome is a protective dielectric housing for an antenna-radiating element. Its function is to protect the antenna not only from ice, but snow and physical damage due to ice dropping from above. Radomes also help protect the radiating element from environmental corrosive atmospheres.

The primary purpose of using radomes on FM antennas is to prevent the VSWR from rising with the formation of ice, if the site and height cause icing to occur during the winter months. Ice formation detunes high-Q radiators, increases the VSWR, and causes vertical plane pattern changes. Figure 7.17-28 shows typical radomes enclosing side-mount antenna elements.

Ice may form on the radome but does not particularly affect the operation of the radiator if that ice is kept at least 0.05 wavelengths (approximately 6 inches) from the sensitive portions of the antenna element.

Radomes are particularly desirable in heavy icing environments where de-icers are not adequate even with very high heat density. They are also useful in protecting antenna elements from falling ice when they are so exposed.

Radomes are cost effective with single channel high-Q antennas where electrical de-icer heating power costs are expensive. The de-icer power cost is a continuing one, while radomes are a one-time capital investment, which may be depreciated over time.

Radomes are generally composed of low-loss dielectric material with low values of dielectric constants and loss tangents. Laminated fiberglass, using glass cloth reinforcement, has a constant of about 4.1 and a loss tangent of about 0.15. Water absorption by the radome increases its dielectric constant and loss tangent. Materials that do not easily absorb water or those treated with a protective gel coat are often used to shed water and prevent the adhesion of ice.

FIGURE 7.17-28 Typical radomes installed on side-mount antenna

Good radome designs take into consideration operating temperature, a relative humidity of 100%, safe wind pressures, ice, hail, snow loads, rain adhesion, wind, and supporting tower vibration. They are made from fire-retardant plastic, and must safely withstand air contaminants over the useful life of the antenna. ABS plastic is a common material used. All these factors increase the cost, but are necessary for a long, useful life. Radome shapes are dictated by the form of the radiating element in most instances. Careful care must be taken to make sure that there are no flat surfaces inside the radome where water can accumulate. Water accumulating inside the radome during rain storms will cause the antenna VSWR to increase, and that VSWR will slowly decrease as that water evaporates.

In all cases radomes are supplied by the antenna manufacturer, and usually supplied in two pieces that are held together with stainless-steel fasteners. Radomes are usually not pressurized. When not pressurized, it is important to pay attention to how any moisture can escape, typically by putting drain holes in the radome and, as mentioned earlier, making sure there are no areas where water can pool.

LIGHTNING

Because FM towers are usually located on high ground, hilltops, or high buildings, they require lightning protection because they are likely recipients of lightning strikes. The type of damage that can be caused by lightning to a FM tower varies. Smaller coaxial lines will usually melt; larger coax (such as 1-5/8 to 3-1/8 inches) will also melt in some cases, and others will conduct the heavy current into the transmitter building to do damage there.

If an antenna with a radome is hit by lightning, the temperature of the volume of air inside the radome instantly increases causing the radome to explode. If the antenna experiences corona, the radome will implode as it melts due to the electric discharge from the enclosed dipole element.

When hit by lightning, the FM antenna itself may heat, arc, melt, or otherwise be damaged. Holes in the outer conductor, burns, and melting at flanges are common. Teflon or polyethylene insulation will burn, depositing a film of carbon, causing further damage if RF from the transmitter continues after the strike.

Protection of the FM antenna system may be provided to some degree by taking several precautions. The top of the tower should have a lightning rod, about 1 foot (0.3 m) higher than the uppermost obstruction light part. The FM antenna itself should be firmly grounded to the tower. If the coaxial cable is buried between the tower and the transmitter building, it must be at least 6 feet away from any tower base grounding system copper wire or strip [16].

A ground system should be located immediately around the base of the tower. This should have a direct current loss of less than around 10 ohms to earth ground, although a level of 5 ohms is commonly used. This low resistance may be obtained by using ground wires buried in the soil. Six radials spaced at 60-degree intervals, buried as deep in the soil as possible and running out up to 150 feet (46 m) each, should provide a suitable ground of less than 10 ohm, even if the soil is shallow or rocky.

For soil with a soil resistivity of ρ, a single driven rod with radius of *a* and a depth of *L* will have resistance *R* given by:

$$R = \left(\frac{\rho}{2\pi L}\right)\left(ln\frac{4L}{a} - 1\right), \Omega$$

When multiple grounding rods are used, spacing of at least *2L* should be used to get the full benefit of multiple rods [2].

Guyed tower anchors should also be grounded. This is covered in Chapter 9.1. It is important to install the proper number of ground rods and/or copper wire radials in order to obtain a connection to earth ground of less than 5 ohms. In any event these ground rods or radial wires must be tied together with AWG 4 or larger copper wire, or 2-inch (5 cm) copper strap. This is to provide for thousands of amperes of current flow for less than one second in the event of a direct lightning hit.

If the FM antenna is located on an AM-insulated base tower, then the spark gap should be set at the lowest point that still provides protection for the highest AM modulation peak voltage.

Another way to protect the FM transmission line isolator (if one is used), as well as the tower and FM antenna, is to use an RF choke across the insulated tower base. This tends to reduce the static build-up voltages due to passing thunderstorm clouds, snow, hail, or dust storms. Arc-overs due to these less severe sources usually do not cause damage, but may trip the FM transmitter reflectometer because they will create a current flow through the reflectometer circuitry.

If the base-insulated tower supporting the FM antenna is located in an area of regular thunderstorms, a way to protect both antennas from lightning is to ground the AM tower at the base, and shunt feed it.

Several excellent methods exist. The folded unipole method not only grounds the tower for lightning purposes, but may improve the VSWR bandwidth. See Chapter 9.2 for more information on lightning protection for towers.

FM SCA MULTIPLEXING

In order to prevent crosstalk and interference from in band modulation, the input antenna bandwidth should be such that the antenna input VSWR is less than 1.1:1 for carrier frequency ±200 KHz.

Spurious Emissions

Interference to other stations within the FM broadcast band as well as to other services outside the band can be caused by RF intermodulation product energy developed between two or more FM broadcast transmitters.

It may be due to coupling through a diplexer or coupling between two antennas. This phenomenon has been well documented [17].

When RF energy from two or more transmitters is combined, new spectral components are produced by mixing the fundamental and the second harmonic of each of them. The dominant intermod product generated by each transmitter is at twice the transmitter's frequency minus the interfering frequency. For example, 101.1 MHz and 102.7 MHz transmitters would produce two intermod signals. The 101.1 MHz transmitter would generate a product appearing on 99.5 MHz, and the 102.7 MHz transmitter would generate a product appearing on 104.3 MHz. This dominant intermodulation product always occurs on a frequency that is two times the carrier frequency minus the interfering frequency. The bandwidth of the intermodulation product is twice that of the FM transmitter frequency, and for this reason, intermodulation products are easy to recognize using a spectrum analyzer. This product is typically referred to as a (2A-B) product, where A is the frequency of the transmitter in which the product is generated, and B is the frequency of the interfering signal.

Second harmonic traps or lowpass harmonic filters in the transmission line of either transmitter prior to the diplexer have little effect on the generation of intermod products because both the carrier frequency and the interfering signal are above the lowpass filter rejection band and below the harmonic filter reject band. The resulting intermod falls within the passband of the lowpass filters and below the reject band of the second harmonic traps, so these devices offer no attenuation to intermod products. For separate antennas mounted on a common tower, notch filters can be used to provide the added isolation between transmitters to eliminate intermodulation products.

Diplexers, however good, will couple some of the undesired energy back to each transmitter, generating intermod products. The key to this problem is to keep that undesired power level as low as possible using proper transmitter output circuitry and tight diplexer specifications. Diplexed transmitter installations should be checked for the presence of excessive intermod products after they are installed.

HARMONIC FILTERS

FCC Rules Section 73.317(d) calls for the harmonics of FM transmitters to be as much as 80 dB or more below the transmitter output. This requirement is usually met by using a lowpass filter that passes the station carrier frequency power but attenuates its harmonics.

The transmitter provides some harmonic attenuation of course, and is usually 25–38 dB for single-ended amplifiers. The worst-case harmonic is the third. Harmonic filters by several firms provide a minimum of 50 dB for harmonics from the second through the tenth. Adding the transmitter attenuation to that of the filter normally provides more than the required level.

The high level of rejection is made possible by using high-impedance (inductance) and low-impedance (capacity) coaxial sections for m-derived three to five section filters, with half-pi end sections. Harmonic filters are commonly made in three production schedules. Some are adjusted to the customer's operating frequency, so that there are no attenuation gaps in the higher harmonics. They are not tunable outside the factory as the insertion loss, and attenuation along with passband VSWR are closely related.

The insertion loss in the passband varies from 0.05 to 0.08 dB while the rejection for the second through the tenth harmonic is 50–60 dB, depending on the number of internal filter midsections. The VSWR in the passband varies from 1.05 to 1.1:1. Harmonic rejection is due to the very high VSWR on the harmonic frequencies, which may be as high as 15:1. This rejected power is passed back to the transmitter amplifier where it is absorbed.

Harmonic filters are available in straight rigid coaxial line sections and may sometimes be pressurized. Power capacity varies from 10 kW for 1-5/8 inch EIA line size to 50 kW for the 6-1/8 inch size.

IBOC ANTENNA OPERATION

There are three methods of transmitting FM IBOC (in-band on-channel). The first employs a single transmitter that generates both analog and digital signals and feeds them to a common antenna. The second method employs a separate analog and digital transmitter combined in the transmitter building and fed to a common antenna. The third employs separate analog and digital transmitters, each feeding signals through separate transmission lines and antennas. The IBOC digital signals are lower in power than the analog signal so antenna and transmission line power ratings used exclusively for the IBOC signal may be smaller in size.

ANTENNAS FOR DIGITAL RADIO APPLICATIONS

There can be different types of antenna arrangements. The most important issue is isolation between analog and digital signals when combining them using shared or interleaved antennas. The analog signal (typically +20 dB above digital) will appear at the digital transmitter port as reflected or coupled signal, respectively.

Dual input side-mount antennas are available for combining separate analog and digital transmitters. These antennas use an input hybrid to feed a dual-input side-mount antenna. In order to eliminate downward radiation, the bay spacing for the dual-input arrays is designed to produce a null in the elevation pattern. The dual-input side-mount antennas use polarization to isolate the analog and digital input power to the antenna, and typical isolation is 45 dB between the two inputs.

Interleaved Antennas

Two antennas shown in Figure 7.17-29 are separate FM side-mount antennas sharing the same aperture and fed by separate transmission lines. Each antenna

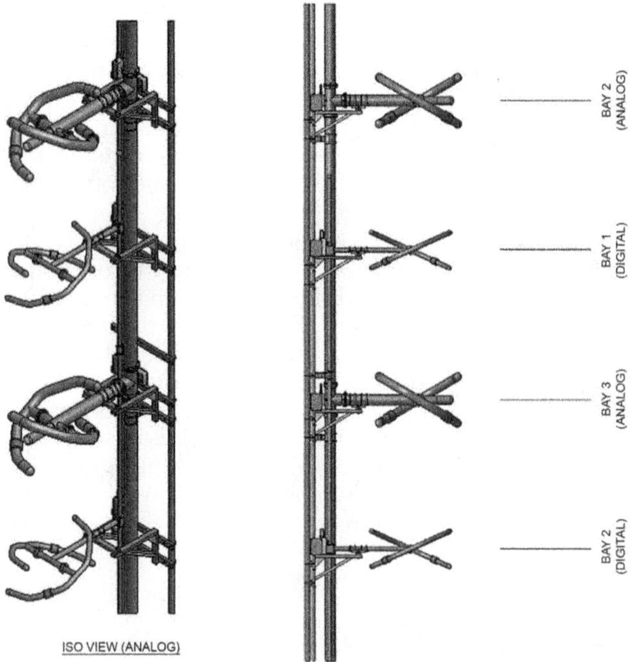

FIGURE 7.17-29 Interleaved FM antennas for analog and digital transmission

consists of wavelength-spaced elements sharing the same aperture at half-wavelength spacing of each other. Because of half-wavelength spatial and small frequency separation between the two antennas the coupled signal is large (typically −15 dB for H-pol signals). Therefore, the isolation at the digital input is poor. Coupling can be suppressed and isolation therefore improved by making one of the antennas left-hand circularly polarized (LHCP) and the other right-hand circularly polarized (RHCP). Then the signals are effectively cross-polarized and the coupling between them is suppressed and isolation is improved to −40 dB or better. With proper adjustment, typically isolation of 45 dB can be achieved.

Low-Level Injection

This technique, low-level injection, was used early on to combine analog and digital signals into a shared antenna and shared transmission line. To isolate the antenna-reflected analog signal from the digital transmitter a four-port 10 dB hybrid combiner is used for the two transmitters, as shown in Figure 7.17-30. The hybrid combiner dumps 90% of digital transmitted power and 10% of analog signal into a dummy load and passes 90% of the analog and 10% of the digital signals to the antenna. As the reflected analog signal travels back from the antenna (VSWR of 1.1:1) to the hybrid combiner, the 10 dB hybrid sends 10% of analog-reflected power toward the digital input and transmitter. If the antenna is tuned to have a VSWR of 1.1:1 (−26.5 dB) over an FM channel, then the reflected analog signal at the digital input (transmitter) is −36.5 dB below rated FM power.

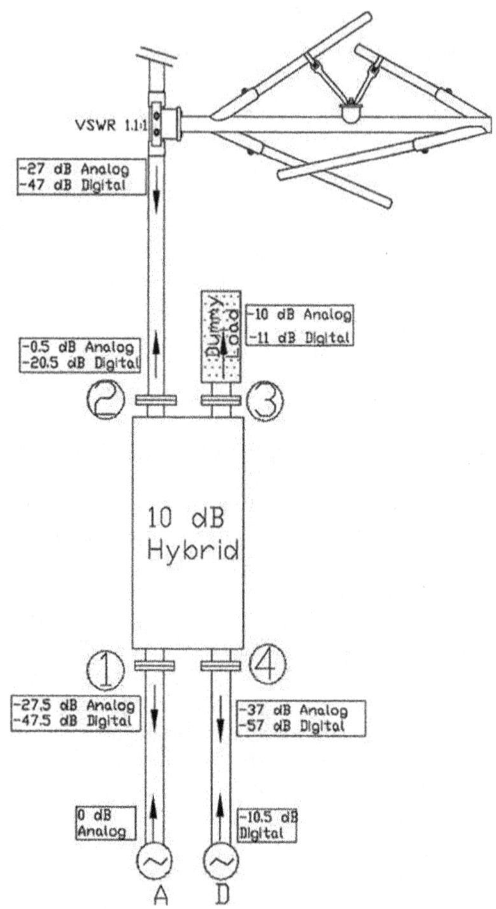

FIGURE 7.17-30 Example of 10 dB hybrid combiner used to couple analog and digital transmitters into a common transmission line and antenna. Note relative values for forward and reflected power.

Now, there is a far more efficient method used to combine analog and digital transmitters that uses a low loss combiner. This combiner, the All Pass Combiner, shown in Figure 7.17-31, uses phase to combine analog and digital transmitters in a constant impedance combiner that uses two hybrids connected by a circuit that selectively shifts the phase of the two digital carriers by 180 degrees. Thus, the analog and digital signals both exit the combiner at the antenna output port.

High-Level Injection

A shared antenna and two separate transmission lines are practical for an array of panel antennas, such as by using panel elements as shown in Figure 7.17-32. The typical FM panel antenna has dual inputs (two polarization ports) fed in quadrature phase (i.e., 0° versus ±90° for RHCP and LHCP, respectively). Polarization ports can be injected at high level with the digital signal via 3-dB hybrid combiners (at shared antenna) making analog and digital signals LHCP and RHCP, respectively. Two different arrangements of digital radio injectors are considered here (narrowband single FM channel and broadband).

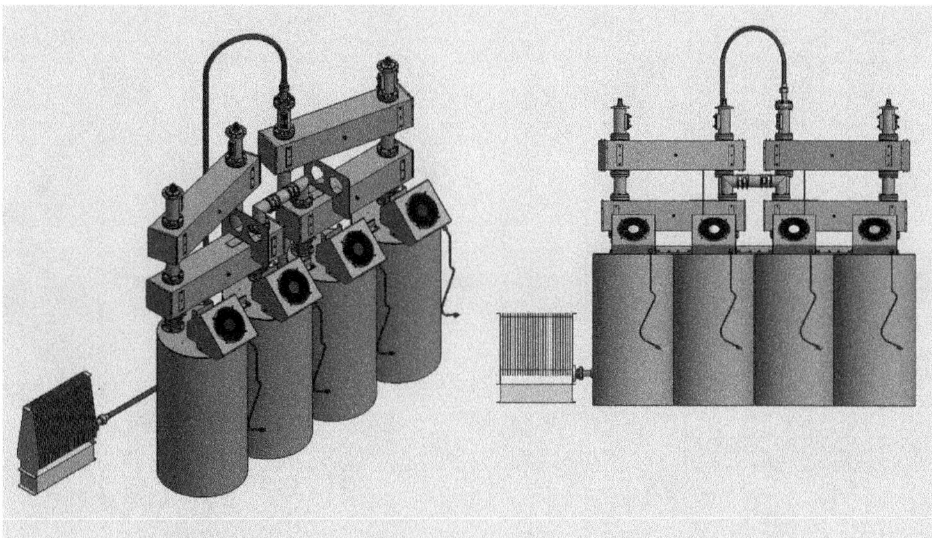

FIGURE 7.17-31 Schematic Diagram of All Pass Combiner

FIGURE 7.17-32 FM dual dipole circularly polarized panel antenna array

Broadband Digital Radio Injector

A schematic with hybrid combiners at every panel is shown in Figure 7.17-33. Broadband isolation of the analog signal at the digital input and vice versa is achieved by combining both analog and digital with feeds phased in a turnstile manner, that is, with incrementally increased phase-feeding panels of one bay.

Superturnstile Feed

Analog-reflected signals seen at port 4 are first suppressed by the close match of individual dipoles S_{22} and S_{33}. Then as reflections travel down toward the digital input they follow the turnstile route and therefore are greatly suppressed over a broadband by phase compensation at the digital input. With four-around panels per bay, each is fed with –90 degrees phase increments to produce an omnidirectional azimuth coverage. The panel lateral offsets are used to optimize circularity for both analog and digital azimuth pattern. Typical turnstile omnidirectional pattern circularity is within ±1.5 dB.

Back-Fed Multiplex Filters

Where dual-input antennas are used by several different FM stations and where constant impedance pass band combiners are used to combine the station frequencies, the combiners can be used to combine the digital transmitters for the various stations by modifying the combiner input hybrid dummy load port to a digital transmitter input port with a circulator. The reverse flow in a single combiner module is illustrated in Figure 7.17-34. The final combiner module dump load port then can be connected to the left-hand circular polarized input port of the antenna. In this way, the combiner is used to feed the analog transmitter to the right hand CP input to the antenna and the digital transmitter to the left hand CP input to the antenna.

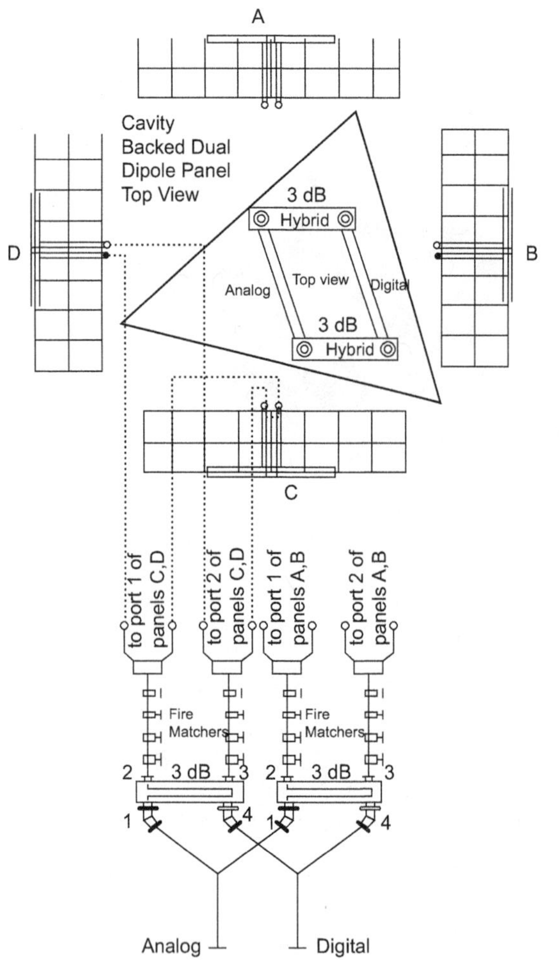

FIGURE 7.17-33 Diagram of high-level injection for shared antenna and separate transmission lines

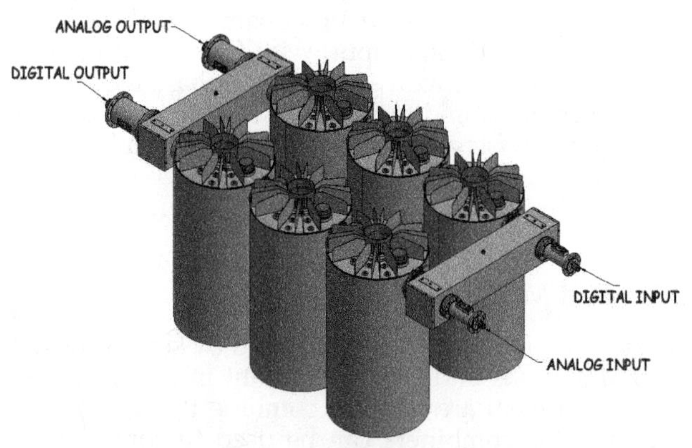

FIGURE 7.17-34 Diagram of back-fed combiner system

With three-around panels the bays produce a wide cardioid azimuth coverage either when fed in-phase or in 60-degree phase increments. Slight asymmetry of the latter can be compensated with lateral panel offsets of about 10 inches. Figure 7.17-35 shows the top view and a schematic of three-around panels with both analog and digital signal feeds to the panels in 60-degree phase increments. Reflected analog signals that show up at the digital input are effectively cancelled because the reflected signals from the three panels come back to the digital port in 120-degree phase increments.

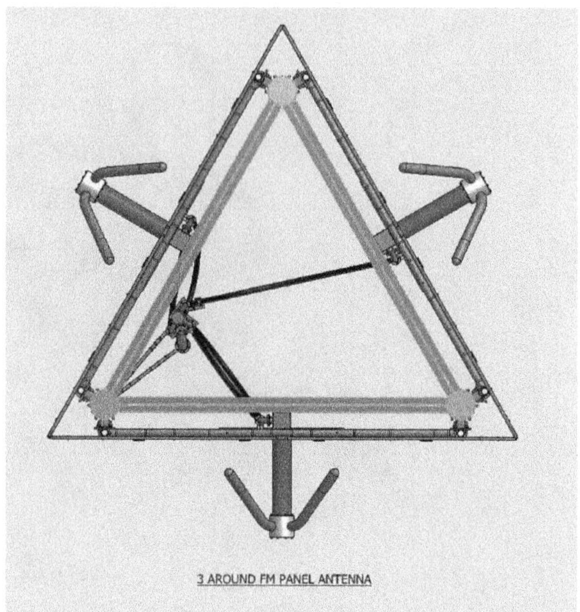

FIGURE 7.17-35 FM dual dipole antenna—3-around configuration

Circulators with bandpass filters add isolation. The bandpass are needed because circulators can be "dirty," creating spurious emissions, especially as they are driven with higher power relative to their size and power rating.

ACCESSORY ANTENNA SYSTEM EQUIPMENT

There are multiple devices and systems that support the operation of a broadcast antenna. The dry air pressurization of coaxial transmission line was discussed earlier in this chapter.

Reflectometers

The reflectometer is a device for detecting the ratio of power into the antenna (forward) and the rejected power from the antenna (reverse). It consists of a short coaxial line section about 12 inches (305 mm) containing diode detectors, coupling loops, and terminations that produce the DC current that drives a suitably calibrated VSWR meter.

Reflectometers are wideband devices and therefore must be placed *after* the harmonic filter. Putting them between the transmitter and the filter causes them to read the rejected harmonic power along with the reflected, thus giving an erroneous reading.

Dummy Loads

A very useful test device in an antenna system is the terminating (dummy) load. One is sometimes used when a number of diplexers are used in a community antenna arrangement to terminate the broad input port to the multiplex filter. Dummy loads are available in several power levels up to 50 kW or more and are cooled by air or water.

RF Switches

Often used with a dummy load, coaxial line switches are available to provide electrical or manual switching of transmitter power to diplexers, antennas, standby transmitters, and so on. They typically are not pressurized.

SUMMARY

With the introduction of digital radio broadcasting using in-band, on-channel techniques it is essential that the FM broadcast engineer fully understand the concepts and practices of good antenna design, installation, and maintenance. Antennas deliver the signal to the audience and therefore are the single most important part of the transmission system. By understanding the interrelationships between antenna site, antenna design variables, propagation, and local terrain, the broadcast engineer can develop the best configuration for a given FM station facility.

Selection of the proper antenna system for an FM station is essential to delivering optimum signal to the receiver. The antenna type, number of bays, mounting structure, height above ground, and antenna location relative to the desired audience are all essential factors in achieving that goal.

ACKNOWLEDGMENT

The authors of this chapter wish to acknowledge the work of Peter K. Onnigian, who passed away in 2015. He provided an excellent baseline for this chapter that was used in many prior editions of this handbook. He founded Jampro Antennas in 1959.

References

[1] Johnson, Richard C., *Antenna Engineering Handbook, Third Edition,* New York: McGraw-Hill, Inc., 1993, pp. 1-12–1-13.

[2] Silliman, Thomas B., "Tower Site Grounding Design and Measurement", IEEE Broadcast Symposium, October 14, 2015.

[3] Skolnik, Merrill I., *Introduction to Radar Systems, Second Edition,* New York: McGraw-Hill Publishing Company, 1980.

[4] Pozar, David M., *Microwave Engineering,* New York: Addison-Wesley Publishing Company, 1990.

[5] Stutzman, Warren L., and Thiele, Gary A., *Antenna Theory and Design, Second Edition,* New York: John Wiley & Sons, Inc., 1998.

[6] Elmore, William C., and Heald, Mark A., *Physics of Waves,* New York: Dover Publications, 1969.

[7] Moeller, Adolph. Effects of Ground Reflections on Antenna Test Range Measurements, *Microwave Journal,* March 1966, pp. 47–54.

[8] Armstrong, A. Study of Electromagnetic Wave Propagation at 112 MHz, *Proceedings of the IREE Australia,* April 1969, pp. 105–110.

[9] Onnigian, Peter. "Stereo Degradation as a Function of Antenna System VSWR", *Audio Engineering Society Annual Meeting,* Audio Engineering Society, New York, 1976.

[10] Mendenhall, Geoff. The Composite Signal—Key to Quality FM Broadcasts, *Technical Monograph,* Quincy, IL: Broadcast Electronics, Inc., 1984.

[11] Onnigian, Peter. Circularly Polarized Antenna, U.S. Patent 3,541,470.

[12] DuHamel, Ray. TV and FM Transmitting Antennas, in *Antenna Engineering Handbook,* 2nd ed., eds. R.C. Johnson and H. Jasik. New York: McGraw-Hill, 1984, chapter 28, pp. 8–9.

[13] FCC Public Notice. Criteria for Licensing of FM Broadcast Antenna Systems, Notice 84–437 25004. Washington, DC: Federal Communications Commission, September 14, 1984.

[14] Rudge, A. *Handbook of Antenna Design,* vol. 2. London: Peter Peregrinus Ltd., 1983, pp. 917–922.

[15] Fisk, Ronald. Design and Application of a Multiplexed Nine-Station FM Antenna, Senior Road Tower Group, *Technical Monograph,* Quincy, IL: Harris Corporation, 1983.

[16] Marshall, J.L. Lightning Protection, in *Canadian Broadcasting System.* New York: John Wiley & Sons, 1973, pp. 53–54.

[17] Mendenhall, Geoff. Study of RF Intermodulation Between FM Broadcast Transmitters Sharing Diplexed Antenna Systems, *Technical Monograph,* Quincy, IL: Broadcast Electronics, Inc., 1983.

CHAPTER

7.18

FM Combining and Filter System Design

ROBERT A. SURETTE

Harrison, ME

SHIVELY LABS

Bridgton, ME

There have been many changes in the applications for filters, in today's transmission facilities, in order to keep up with the ever-increasingly crowded FM spectrum. An ideal filter response curve looks like a perfect square wave as shown in Figure 7.18-10. Figure 7.18-47 shows a typical response when cross-coupling loops are added to a bandpass filter. The good news is with today's computer programs and experienced technicians, while that elusive square wave has not been achieved, it is getting closer and closer every year. In general, these larger and faster computers are still using the same basic filter concepts in use since the early days of FM transmission. But the application of these basic concepts has improved over the years.

Eight graphical plots show how close modern designs are getting to the aforementioned square wave response. The first set of four plots were measured using two sets of bandpass filters with cross-coupling loops configured into a balanced combiner module (see the "Balanced Combiners" section of this chapter). For years, combining stations at 800 kHz spacing was considered the state of the art. Figures 7.18-1 to 7.18-4 show the frequency response, insertion loss, group delay and VSWR of a 10 kW bandpass balanced combiner module tuned for stations at 600 kHz spacing. Other than a small increase in insertion loss, the only parameter that has changed is the group delay, which has more or less doubled.

In the second set of plots (Figures 7.18-5 through 7.18-8), a bandpass filter with cross-coupling was used in a branched combiner configuration to combine stations at 700 kHz spacing on a crowded broadcast tower. Please note in the frequency response curve that there is a notch at 400 kHz above the tuned frequency. This is because there is another antenna on the same tower physically mounted close enough to this combined antenna making it necessary to provide filtering in order to eliminate intermodulation products (spurs).

The preceding examples are just a quick look at some of the changes in tuning of four-cavity bandpass filters. In 1985 a group delay equalizer was added to a combiner module to reduce the effects of group delay on a transmitter. Recently two group delay equalizers were placed in series to make one leg of a runout combiner. Again, we are seeing basic concepts being applied to new applications.

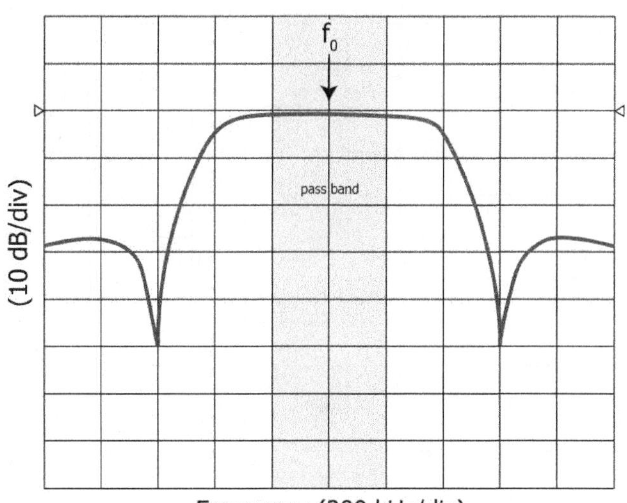

FIGURE 7.18-1 Frequency response, balanced combiner, 600 kHz spacing

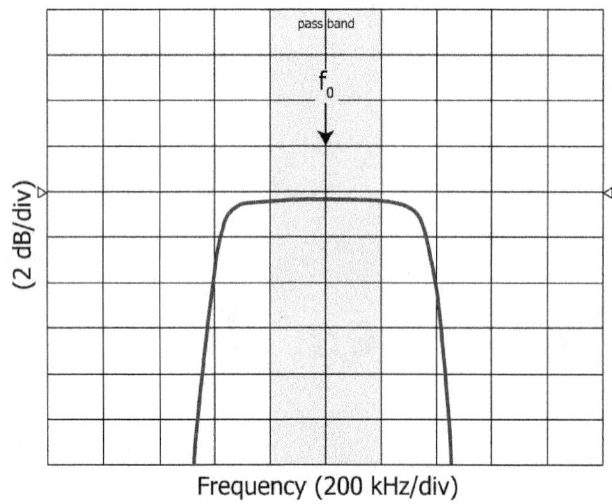

FIGURE 7.18-2 Insertion loss, balanced combiner, 600 kHz spacing

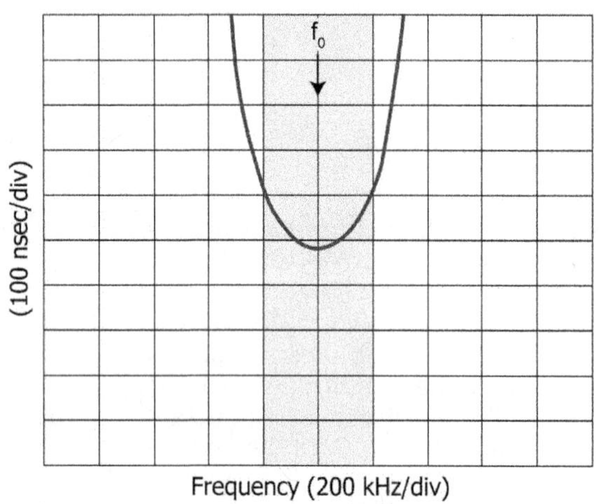

FIGURE 7.18-3 Group delay, balanced combiner, 600 kHz spacing

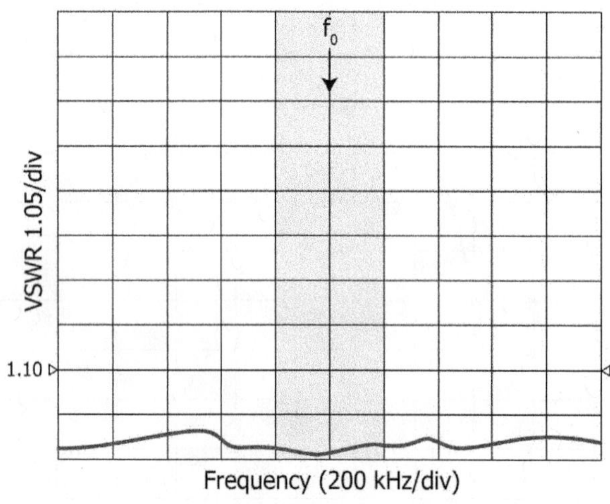

FIGURE 7.18-4 VSWR, balanced combiner, 600 kHz spacing

VSWR bandwidth is due to the input hybrid

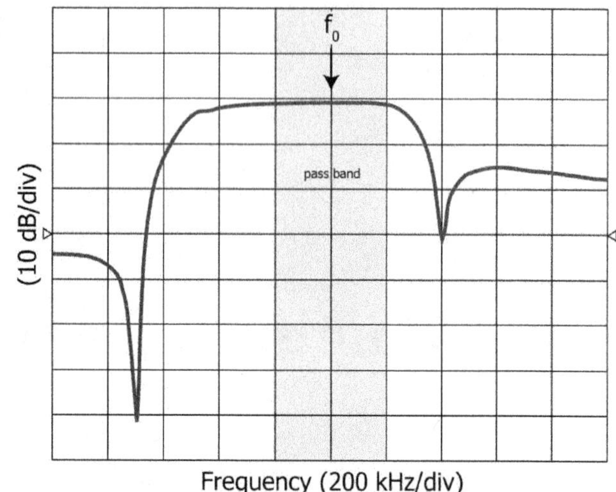

FIGURE 7.18-5 Frequency response, branched combiner, 700 kHz and 400 kHz notches

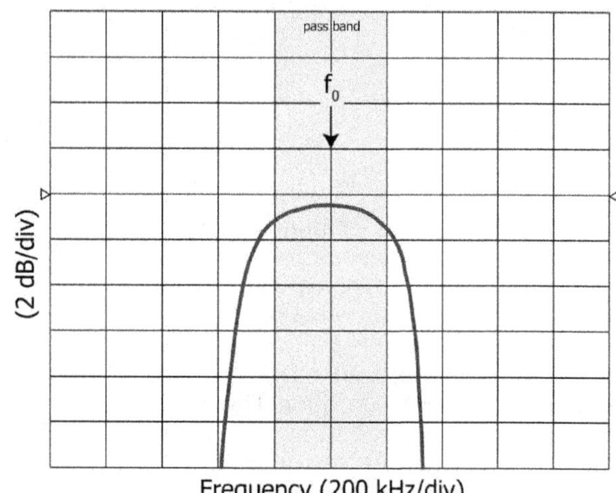

FIGURE 7.18-6 Insertion loss, branched combiner, 700 kHz and 400 kHz notches

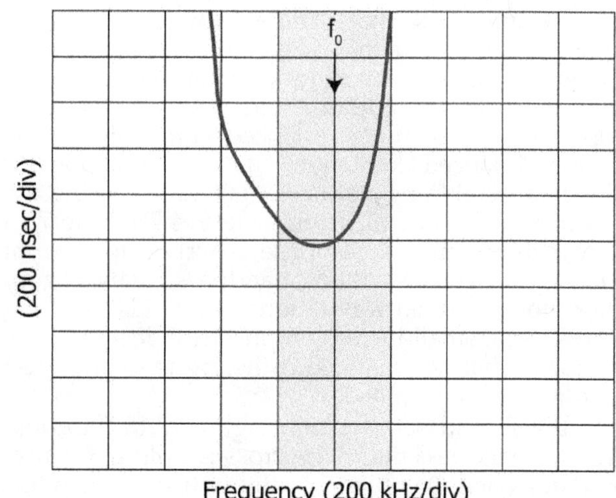

FIGURE 7.18-7 Group delay, branched combiner, 700 kHz and 400 kHz notches

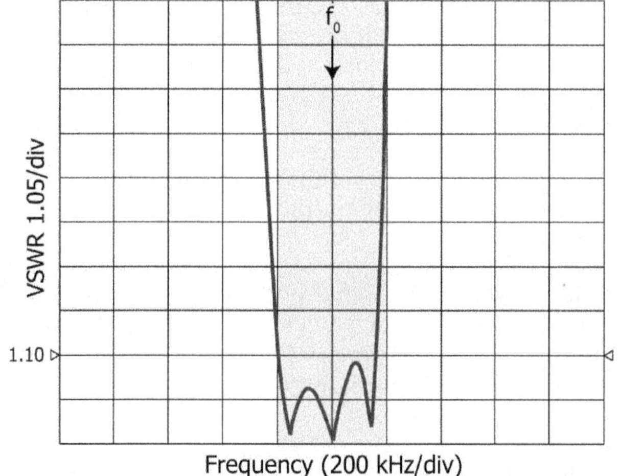

FIGURE 7.18-8 VSWR, branched combiner, 700 kHz and 400 kHz notches

INTRODUCTION

Transmitting several frequencies from a single broadband antenna system requires the use of a combining system, or combiner, composed of radiofrequency (RF) filters and interconnecting transmission line. In general, a combiner can be categorized as one of two types: branched (star point) or balanced (constant impedance). Any of these types may employ band-reject (notch) or bandpass filters. This chapter discusses the use of filters, other components in FM combiners, and the hardware used to combine an in-band-on-channel (IBOC) digital signal into an analog signal.

APPLICATIONS

For years, both the FM spectrum and the FM channel were straightforward and uncomplicated. Until the early 1980s, the number of stations on the air in all but the largest metropolitan areas was low by today's standards. In most areas, the frequency spacing between stations exceeded the 0.8 MHz minimum that is common in all parts of the country today. These wider frequency spacings, the relative ease of developing new tower sites, and the limited station ownership in any market worked against the economics of combining stations into common antennas; therefore, most stations operated on single-frequency antennas, and large, multi-station antennas were generally only found in a few of the largest markets.

In the late 1980s, with the arrival of Docket 80–90, the FM spectrum in the United States became increasingly crowded. Ensuing changes in ownership regulations that tightened zoning regulations changed the economics of combining. It has become increasingly common to combine stations in even the smallest markets, including stations with very low power levels. Further complicating the spectrum is the dramatic increase in auxiliary antennas that began as a result of the need to accommodate digital television (DTV) construction. Expansion of combined systems has not been limited to small and medium markets. Large metropolitan stations that rarely exceeded 10 combined stations in the 1990s are now routinely being replaced by systems with room for 20 stations or more.

At the same time that changes in the FM spectrum made combining attractive and increased filtration a necessity, the FM channel itself became increasingly complicated. In the 1980s, the 67 kHz Subsidiary Communications Authorization (SCA) became more widely used. This was quickly followed by the 92 kHz SCA, pushing critical information to the ±100 kHz fringe of the FM channel and closer to potentially interfering signals. With the introduction of IBOC in the early 2000s, the channel has increased in size to ±200 kHz from the center frequency, and its full channel width is being utilized. Even this enhanced channel is becoming more crowded as digital multicasting becomes commonplace.

The net result is that, as the FM channel becomes larger and more complex, filters and combiners have had to evolve to provide the necessary isolation between closely spaced signals at the same time that their own pass bands must be more tightly controlled to pass the desired channel. Today's combiners are even being used to isolate separate signals on the same channel to facilitate the combining of analog and digital signals.

WHY COMBINERS ARE USED

As populations migrate to suburban areas, it has become more desirable to construct large broadcasting facilities that can reach these heavily populated areas from more central locations. Of course, these prime locations have become more valuable, so it important to use each location to its fullest potential. This can best be done by sharing a transmitter site and a common antenna among several users. To accomplish this, the broadcast industry uses combiners of various types and sizes. For example, in San Francisco (Mount Sutro), Toronto (CN Tower), Montreal (Mount Royal), New York City (Empire State Building and 4 Times Square) and Chicago (John Hancock Building and Willis Tower), tall towers or towers on skyscrapers have been used to consolidate as many broadcasting facilities as possible, including VHF-TV, UHF-TV, FM and land mobile communications services. This approach has proven very effective, not only by using real estate economically but also by spreading the tower costs over many users.

Shortage of Prime Locations

Group ownership of FM stations in a market has led to proliferation of combined stations and, with the implementation of DTV systems, FM stations are being forced off existing towers, making it even more imperative that they share tower space, which increases the demand for combined systems.

FCC ISOLATION REQUIREMENTS

When more than one signal is broadcast over a single antenna, the signals must be combined in such a way that signals from one transmitter are not permitted to feed back into another transmitter. Failure to do so would cause intermodulation products to be generated within the final amplifier stages of the transmitters and subsequently radiated from the antenna. These intermodulation products are generally referred to as *spurs*. Spurs created between FM stations can occur not only in the FM band but also within the low-band VHF channels and above the FM band, causing interference to the aviation band. FCC Rule Section 73.317(d) specifies that spurs more than 600 kHz removed from the carrier must be attenuated below the carrier frequency by 80 dB or by 43 + 10log10 (power in watts) dB, whichever is less. In practice, stations operating transmitter output powers of 5 kW or greater must usually meet the 80 dB requirement, while stations running lower transmitter power output (TPO) fall under the computational method.

Experience provides ways to prevent spurs. Each transmitter must be isolated from all others in the system by a minimum of 40 dB, with 46 to 50 dB ensuring regulatory compliance. Spur attenuation is accomplished by a combination of transmitter turnaround loss and filtering. Turnaround losses are inherent to the way spurs are created in the transmitter. These losses typically run in the 6 to 13 dB range for tube-type transmitters, while 15 to 25 dB is typical for solid-state units. An off-frequency signal is attenuated 40 dB as it passes through the bandpass filters of the combiner module toward the transmitter, with the spur it creates exiting the transmitter an additional 6 to 25 dB below the level where the signal entered. This spur is then attenuated 40 dB as it passes back through the bandpass filters. The result is spur attenuation of at least 80 dB, with 100 dB or more possible.

IN-BAND-ON-CHANNEL COMBINING

The IBOC signal is transmitted above and below the standard FM analog signal (Figure 7.18-9) and is discussed in Chapter 7.6 of this handbook. An IBOC signal can be combined in a modified analog transmitter. This is referred to as *low level combining*. The IBOC signal can also be combined by using a dual-input antenna or separate antennas. The "Combining Digital

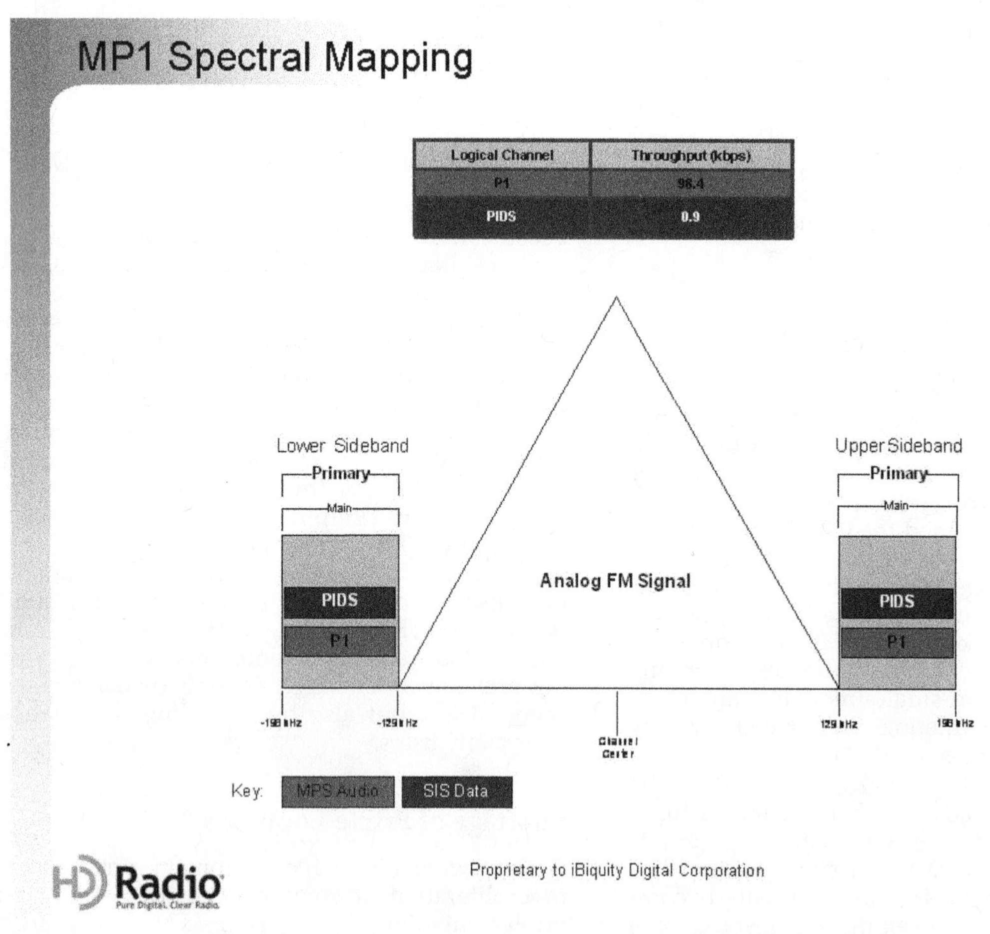

Logical Channel	Throughput (kbps)
P1	98.4
PIDS	0.9

FIGURE 7.18-9 Spectral mapping of an FM channel

DTS, Inc.

and Analog Signals" section of this chapter discusses combining the digital and analog signals from separate transmitters into a common transmission line before sending them to the antenna.

COMBINER CHARACTERISTICS

Important characteristics of combiners are frequency response, insertion loss, group delay, impedance, physical size and tuning compromises. Improving one parameter may result in a reduction in another.

Frequency Response

Energy transfer through the bandpass filter is highest, or least attenuated, at the resonant frequency and drops off at frequencies above and below that frequency. This frequency response is the fundamental property that enables a filter cavity to sort frequencies. If it were possible to design an ideal filter, its frequency response plot would be as shown in Figure 7.18-10. Response would be flat within the pass band, with a vertical roll-off at the edges of that band. Figure 7.18-11 shows the frequency response of a real world single-cavity bandpass filter. Note that the energy transfer is highest at the resonant frequency (f_0) and drops off gradually away from f_0.

Insertion Loss

Even at the resonant frequency f_0, energy transfer is not perfect; some energy is lost along the way, which is expressed as insertion loss—that is, the loss of energy at the resonant frequency. The lost energy is converted to heat and dissipated in the metal surfaces of the cavity. A cavity that is larger in size is more efficient than a smaller sized cavity in that it will provide a lower insertion loss at the resonant frequency with comparable frequency response. Coupling efficiency also affects insertion loss; curve B of Figure 7.18-12 shows the effects of coupling adjustment. The theoretical ideal filter would show no insertion loss in the pass band.

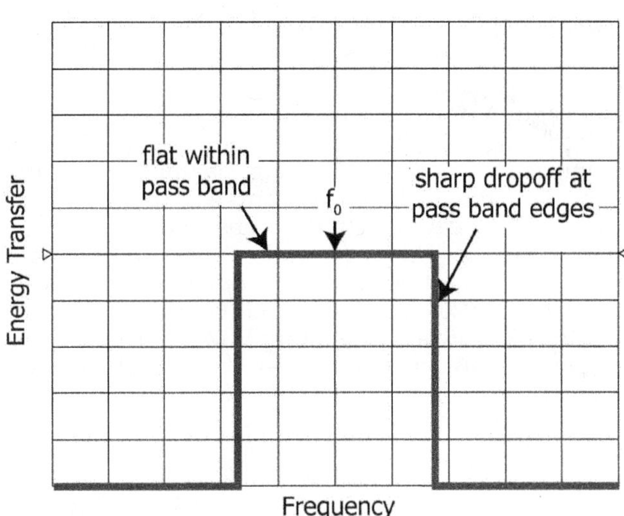

FIGURE 7.18-10 Ideal filter frequency response

Group Delay

An RF signal takes a finite amount of time to pass through the cavity, and just as more energy is lost, more time is taken at nonresonant frequencies. Figure 7.18-13 shows a plot of time *versus* frequency and shows that, as the frequency changes further away from f_0, the signal takes more time to pass through the cavity. This is termed *group delay difference*, or *group delay* for short. Excessive group delay within the pass band can result in signal distortion. An ideal filter would have no group delay difference; that is, the curve would be a horizontal line, at least across the pass band. IBOC requires the full channel bandwidth, so it is important to limit group delay across the full channel.

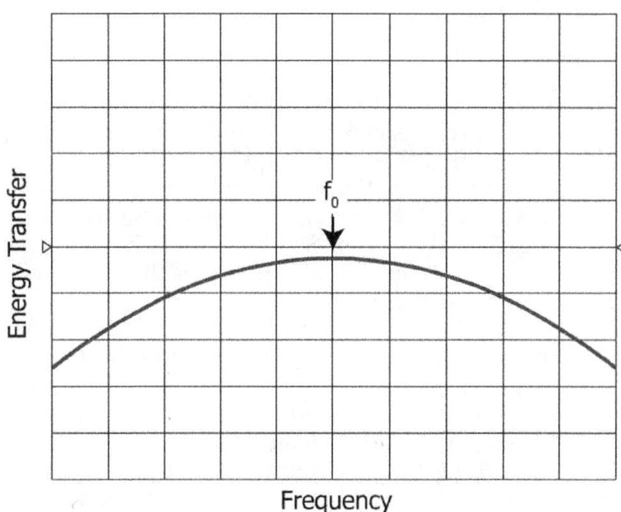

FIGURE 7.18-11 Frequency response for a single-cavity filter

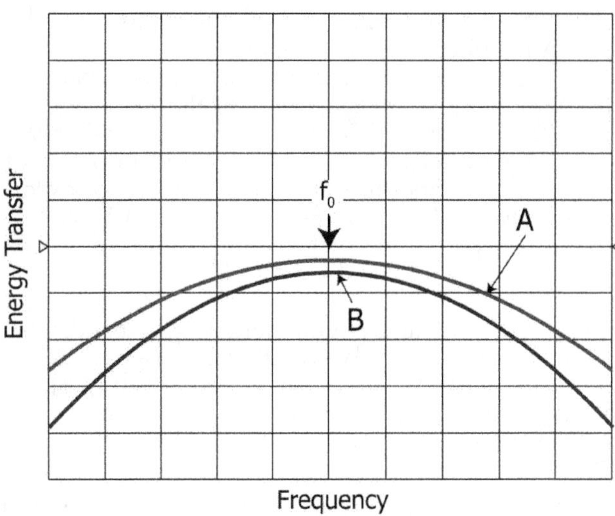

FIGURE 7.18-12 Insertion loss for a single-cavity filter

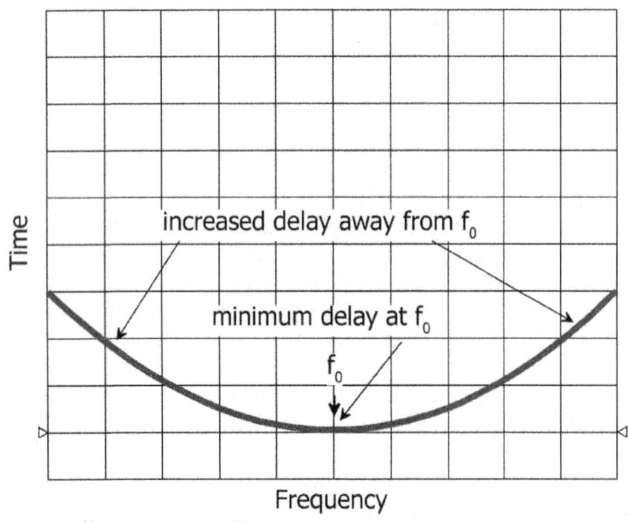

FIGURE 7.18-13 Group delay for a single-cavity filter

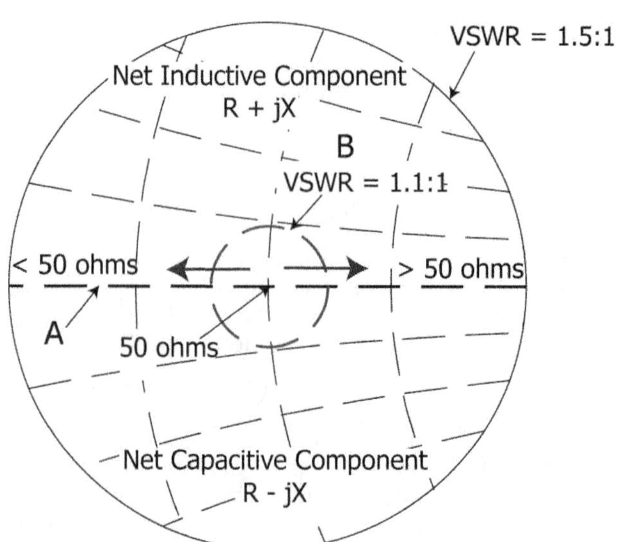

FIGURE 7.18-14 Smith chart components

Impedance

Current flow in any RF circuit must overcome resistance, capacitive reactance and inductive reactance. The vector sum of these is termed impedance. Because this is a complex function, it may only be fully represented on a complex diagram known as a Smith chart. While a full discussion of Smith charts is beyond the scope of this chapter, a few features of the chart will aid in the understanding of filter performance and tuning. Figure 7.18-14 shows an expanded Smith chart. The center horizontal axis (A) represents a state of pure resistance. In a properly tuned system, this state exists at f_0, the resonant frequency, where the inductive and capacitive components cancel each other out. The center point on line A represents a resistive value of 50 ohms. To the left, the resistive value decreases, approaching a short circuit (0 ohms); to the right, it increases, approaching an open circuit (infinite ohms).

The region above the horizontal axis represents a state when the vector sum of the circuit is inductive in nature; conversely, below the axis the circuit is capacitive. Any point on the chart may be expressed as R ± jX, where R is the resistive component, j is a constant and X represents the magnitude of the net inductive or capacitive component of the circuit. A circle drawn around the center point would be a locus of points of equal voltage standing wave ratio (VSWR); for example, circle B in Figure 7.18-14 represents a VSWR of 1.1:1. Points within the circle then represent conditions of VSWR less than 1.1:1. Our ideal filter would be plotted as a dot at the center of the chart, representing a pure 50-ohm resistance throughout the pass band, with no capacitive or inductive components. Figure 7.18-15 shows the Smith chart of a single-cavity bandpass filter. At the resonant frequency f_0, the impedance is pure resistance and 50 ohms at the chart center. As the frequency changes away from f_0, the inductive and capacitive components grow, forming a vertical arc. The slight offset to the right of chart center represents insertion loss. The small circles (beads) on the curve indicate the

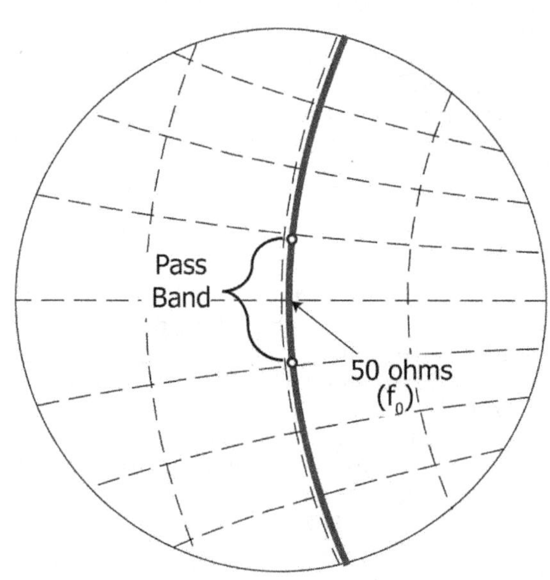

FIGURE 7.18-15 Smith chart for a single-cavity filter

pass band. A pass band of ±200 kHz is generally considered acceptable for a filter system. Figure 7.18-16 is an impedance diagram showing manipulation of the coupling through the cavity. Curve A (truncated for emphasis) is a cavity with the loops adjusted for maximum coupling. This curve almost passes through the center of the chart (R = 50 ohms) due to insertion loss, and the entire 200 kHz pass band (between the beads) is within the circle representing VSWR = 1.1:1. As the coupling is adjusted to achieve increased isolation (curve B) and extended for still more isolation (curve C), the center of the curve moves into the R > 50 ohm area to the right of chart center, an indication of greater insertion loss. In addition, the beads representing ±200 kHz move outward, well outside the 1.1:1 VSWR area. Again, this illustrates the tradeoff between increased isolation and increased insertion loss.

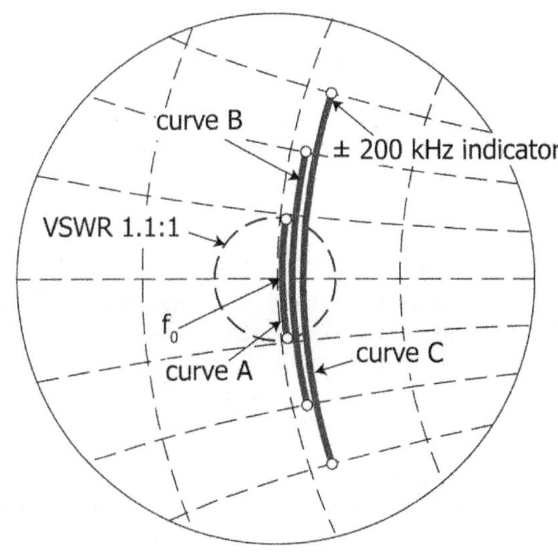

FIGURE 7.18-16 Tuning of a single bandpass cavity

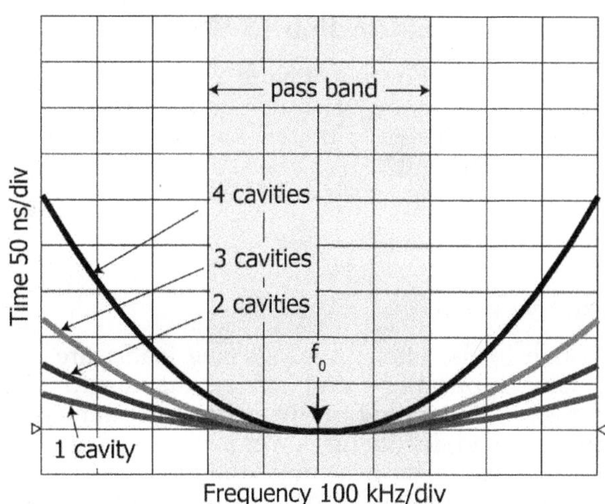

FIGURE 7.18-17 Group delay for one-, two-, three- and four-cavity filters

Physical Size

The physical size of the cavity is established for the purpose of power capacity and electrical performance. The cavity is then tuned to optimize the performance for a given application.

Tuning Compromises

Note that an ideal filter would have a 50-ohm impedance (unity VSWR), no insertion loss, no group delay and flat frequency response within the pass band. Figures 7.18-11, 7.18-12 and 7.18-13 show that actual cavity type filters do not meet these ideal parameters. It is important to remember that filters are always designed for best real world overall performance and that, at times, a little performance must be sacrificed in each parameter to improve overall performance. To obtain increased isolation to meet today's standards, the number of cavities in a filter system must be increased, but this occurs at the cost of increasing group delay and insertion loss. In a four cavity system, the group delay curve becomes so steep as to be unacceptable (Figure 7.18-17); therefore, the tuning is modified to decrease group delay to an acceptable level, as shown in Figure 7.18-18. This adds some minor distortion to the frequency response (Figure 7.18-19). Although none of the individual parameters is optimized by itself, the overall performance of the filter is optimized and acceptable. A four-cavity bandpass filter is as large a filter system as is needed for most high isolation applications.

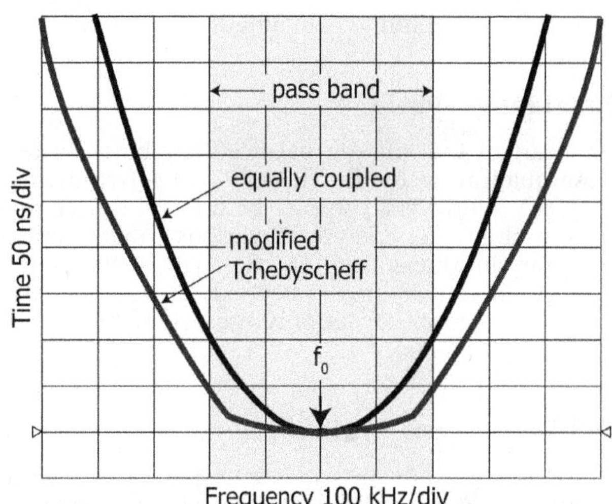

FIGURE 7.18-18 Group delay for a four-cavity filter tuned for group delay

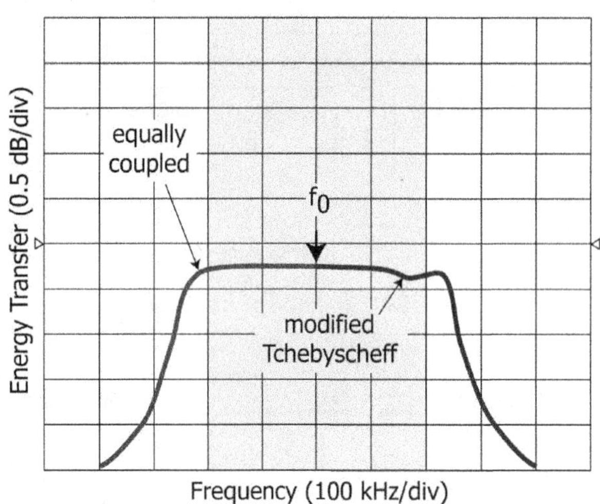

FIGURE 7.18-19 Frequency response for a four-cavity filter tuned for group delay

COMPONENTS OF COMBINERS

Combiners are often made up of multiple elements that as a system provide the desired results. Separating the elements can lead to mismatches or other incompatibilities.

Tee or Star-Point Junction

A tee junction, shown in Figure 7.18-20, is a coaxial component that allows two RF signals to flow into a common path; a star-point junction is a tee with more than two input paths. This basic coaxial component is one of the building blocks of a branched combiner.

Resistive Load

Resistive loads, often called *dummy* loads, are used in many applications and can be manufactured in many sizes depending on the power requirement. In a dummy load, incoming power is absorbed and converted to heat. The heat must then be dissipated to the surrounding air, so the power rating of a dummy load is determined by the size of the resistor and the amount of heat that can be dissipated before the resistor overheats and fails. If enough resistors can be chained together with enough cooling, they can dissipate almost an unlimited amount of RF energy.

Quadrature Hybrid

The heart of the modern balanced combiner system is the quadrature hybrid (usually just referred to as *hybrid*). A hybrid is a complex broadband device that has the ability to operate in various modes, either singly or simultaneously. The detailed mathematical explanation of a hybrid is beyond the scope of this work; this section covers only the use of hybrids in combining systems.

Hybrid as Signal Splitter

In Figure 7.18-21, the hybrid is acting as a power splitter and phase shifter. When an RF signal is applied to port 1 (TX1), the hybrid splits the signal in half, and the phase of port 4's output is delayed with respect to port 3's output by 90°. Port 2 is called the isolated port, because the isolation between ports 1 and 2 is approximately 35 dB and is usually terminated with a 50-ohm resistive load. If two inputs are required, port 2 can be used as an additional transmitter input (TX2). In this configuration, the output power levels are the same as earlier, but the phases are reversed (Figure 7.18-22).

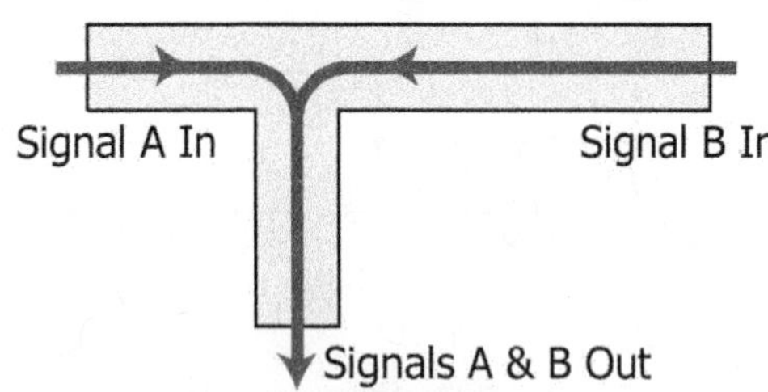

FIGURE 7.18-20 Tee junction

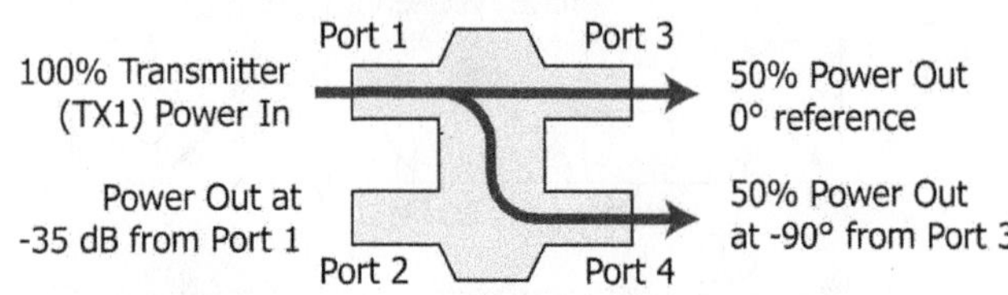

FIGURE 7.18-21 Hybrid as a signal splitter

Hybrid as Signal Combiner

A second mode uses a hybrid in reverse, for combining transmitters, as shown in Figure 7.18-23. If two equal RF signals, with the proper phasing, are introduced at ports 1 and 2, the combined signal exits the hybrid through port 4. If the phase of the two input signals is reversed, the signal will exit the hybrid through port 3. Again, the isolated port is usually terminated with a resistive load. The hybrid can be used to combine two incoming signals in the exact reverse of Figure 7.18-22. If two incoming signals with the correct phasing are present at port 1 and two at port 2, as shown in Figure 7.18-24, then port 4 is an output for one combined signal TX1 and port 3 is the output for the other combined signal TX2.

Hybrid as Signal Reflector

The third hybrid mode of operation is the reflected mode (Figure 7.18-25). When two identical devices with high impedance, such as bandpass filters tuned to another frequency or band-reject filters tuned to the incoming frequency, are attached to ports 3 and 4

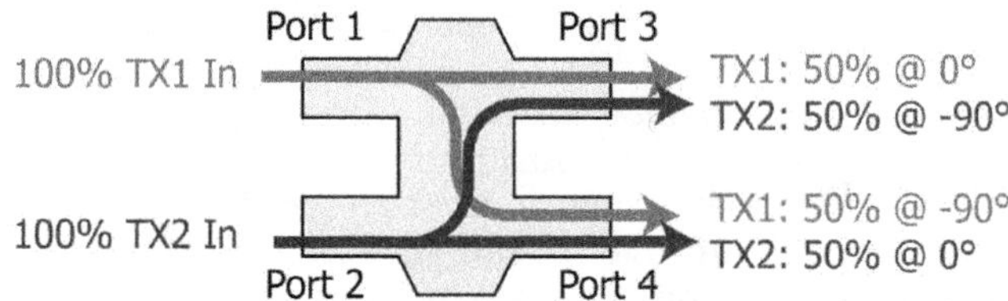

FIGURE 7.18-22 Hybrid splitting two input signals

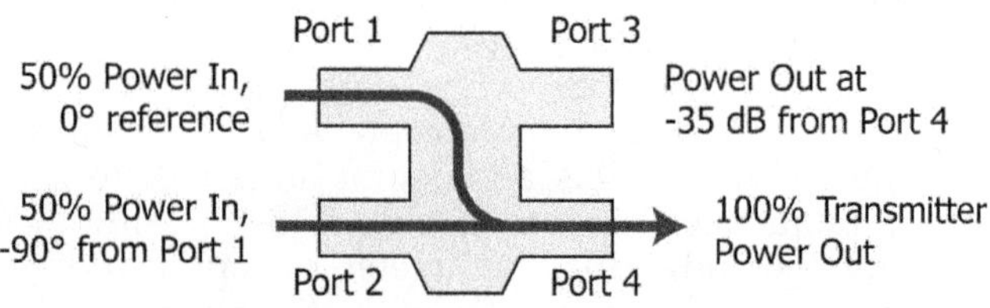

FIGURE 7.18-23 Hybrid as a signal combiner

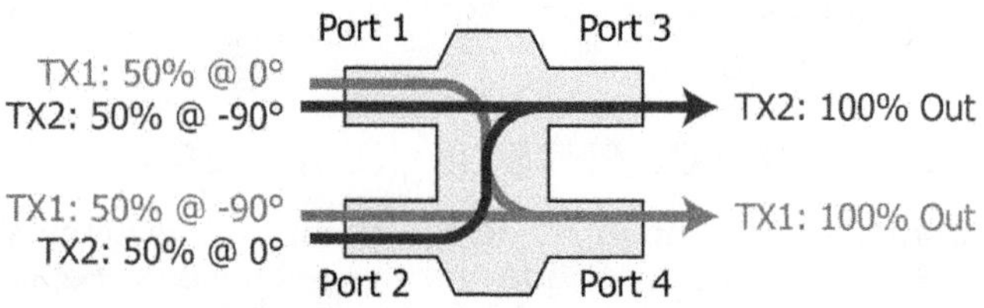

FIGURE 7.18-24 Hybrid combining two signals

of the hybrid, the signal entering at port 1 is reflected and exits the hybrid through port 2. Again, the hybrid is symmetrical; if a second signal enters port 2 it will be reflected and exit port 1. The characteristics of this mode make the hybrid useful in conjunction with other hybrids and cavities in combining systems. A hybrid can operate in all three modes simultaneously. With power moving in many different directions at once, it is imperative that the hybrid have good electrical characteristics and that it be as balanced and symmetrical as possible, both mechanically and electrically. Balanced and symmetrical hybrids show the same electrical characteristics through each port. The more identical the electrical paths through these ports are, the greater the isolation that can be achieved and the lower the VSWR at each port. Figure 7.18-26 shows the performance curve of a well-balanced and symmetrical hybrid.

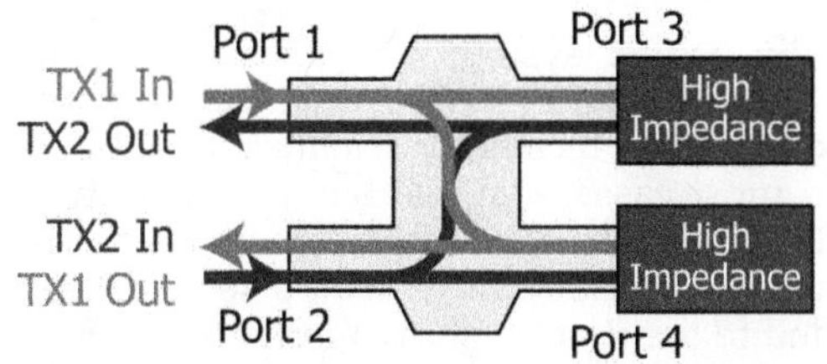

FIGURE 7.18-25 Hybrid as a signal reflector

Hybrid Ring

When two hybrids are used in a ring configuration (Figure 7.18-27) to both split and combine a single input signal, virtually 100% of the signal exits the ring through the hybrid leg opposite the input. In a balanced and symmetrical hybrid ring, if the signal is introduced at port 1, the outgoing signal will be at port 8, with isolation at ports 2 and 7. Likewise, if the signal is introduced at port 7, it will emerge at port 2, with isolation at ports 1 and 8, and if it is introduced at port 8, it will emerge at port 1, with isolation at ports 2 and 7. Energy can flow in all four directions at the same time without the signals mixing (Figure 7.18-28). The multiple flow paths of the hybrid ring make it the backbone of the balanced combiner.

FILTERS

Filters sort RF frequencies, attenuating some while allowing others to pass readily. Depending on the design, a filter may either attenuate (band-reject type) or pass (bandpass type) a relatively narrow bandwidth.

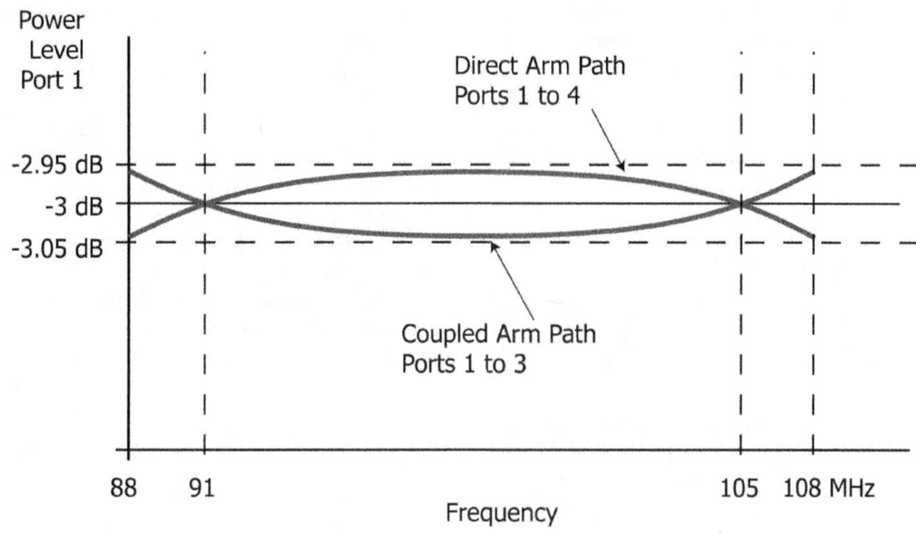

FIGURE 7.18-26 Hybrid frequency response

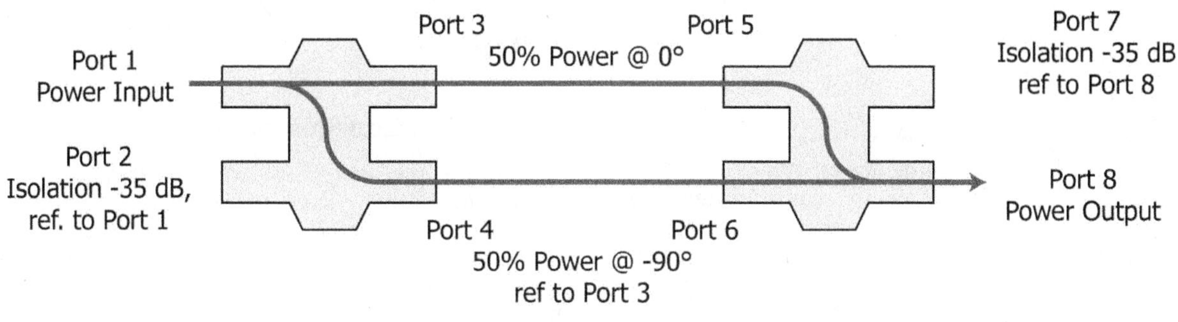

FIGURE 7.18-27 Hybrid ring

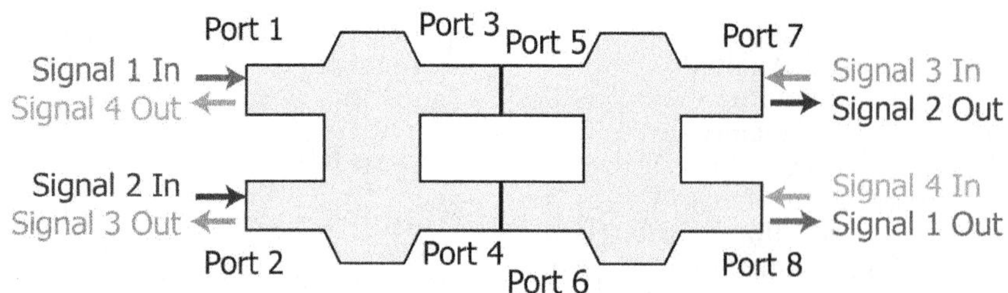

FIGURE 7.18-28 Hybrid ring multiple flow paths

Band-Reject or Notch Filter

There are several ways to design a band-reject or notch filter (Figure 7.18-29), but they all accomplish the same purpose. In one form, a cavity with only an input coupling loop is mounted off the transmission line by means of a matched tee. This provides a path that removes the tuned frequency from the system, allowing other frequencies to pass with minimum loss. Other designs employ some form of capacitive coupling into the cavity.

Multiple Notch Cavities

The frequency response of a typical notch cavity is shown in Figure 7.18-30. When more isolation is required, two notch cavities are coupled in sequence. The resonant frequencies of the cavities may be identical, yielding a response curve with a very deep narrow band notch, as shown in Figure 7.18-31, or they may intentionally be staggered to give a broader notch response, as shown in Figure 7.18-32.

Performance and Limitations

The impedance plot of a typical notch cavity is shown in Figure 7.18-33. When a single notch cavity is used, an impedance matching network is added to the filter to improve the impedance bandwidth. The group delay plot of a notch cavity (Figure 7.18-34) distorts signal quality. No practical device has been marketed to equalize the group delay of a notch-cavity system; however, this has not been a major issue, because about the same time (mid-1980s) that group delay was recognized as an issue, the industry was turning toward bandpass filtering anyway.

Bandpass Filter

Figure 7.18-35 shows the basic mechanical configuration of a bandpass filter cavity. When RF energy is applied to the input coupling loop, the loop inductively couples the energy into the cavity. Energy is transferred through the cavity and inductively coupled to the output coupling loop. The resonant frequency of the cavity is tuned by adjusting the tuning probe. The transfer of energy is maximized at the resonant frequency; therefore, a filter of one or more identical cavities can be used to attenuate frequencies other than the resonant frequency.

Multiple Bandpass Cavities

Generally, a filter system is considered adequate if it provides a VSWR of 1.1:1 over a frequency range of ±200 kHz. This is termed the *bandwidth* of the filter system. In most cases, a single bandpass cavity will not yield this much bandwidth. To increase the isolation and increase VSWR bandwidth, a second cavity may be added to the first, as shown in Figure 7.18-36.

When two identical cavities are coupled a quarter wave apart, the impedances superimpose themselves, as shown in Figure 7.18-37. Note that Figure 7.18-29 shows two Smith charts superimposed 180° apart. The small circles (beads) representing the ±200 kHz bandwidth fall on a VSWR circle of about 1.3:1. When their impedances are added together mathematically, due to phase cancellation, the VSWR bandwidth improves to about 1.1:1 (Figure 7.18-38). Curve A of Figure 7.18-39 shows the frequency response of the two-cavity filter.

When still more isolation is required, more cavities can be added. Figure 7.18-39 shows the frequency responses of two-, three-, four- and five-cavity systems. As more cavities are added, the curve becomes more square and flatter across the pass band, with a sharper roll-off; that is, it begins to approach our ideal filter shown in Figure 7.18-10. Consider, however, that the five-cavity filter does not show a great improvement over the four-cavity filter, and, in fact, the four-cavity filter represents the best compromise among isolation, insertion loss and physical size for close-spaced stations transmitted through a combining system. Figure 7.18-40 shows Smith charts for a three-cavity system and a four-cavity system. Note that the beads indicating the ±200 kHz points are within the 1.1:1 VSWR circle.

Mechanical Constraints

To obtain the optimum mathematical cancellation shown in Figure 7.18-38, the cavities must be spaced at one-quarter electrical wavelength. As the frequency increases, the electrical wavelength decreases; therefore, the physical length of the inter-cavity coax must be shortened. At the higher frequencies of the FM band, the large cavities used for high power applications are

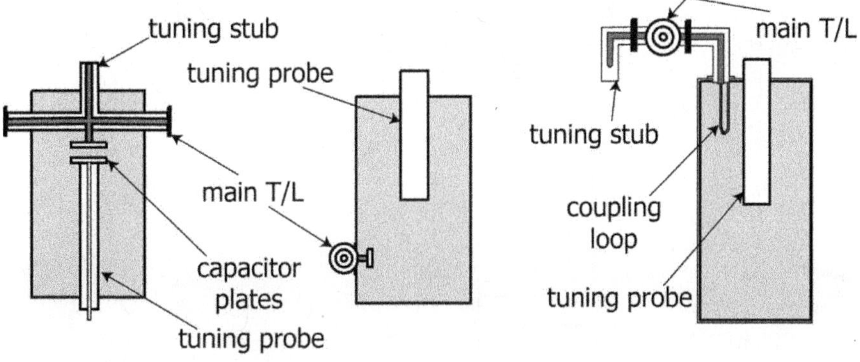

FIGURE 7.18-29 Notch filter configurations

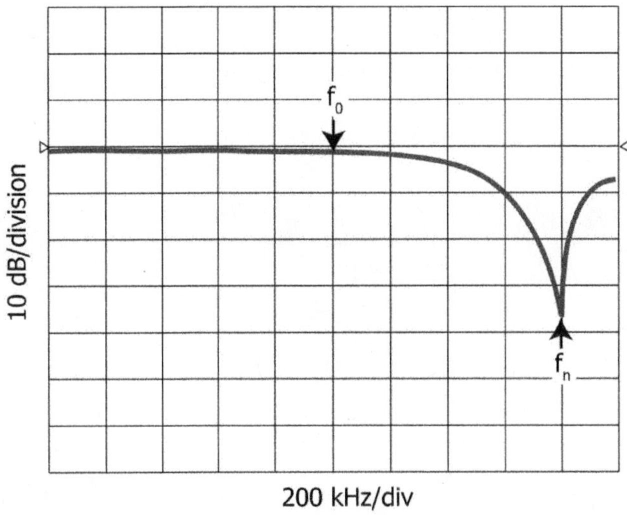

FIGURE 7.18-30 Frequency response for a single notch cavity

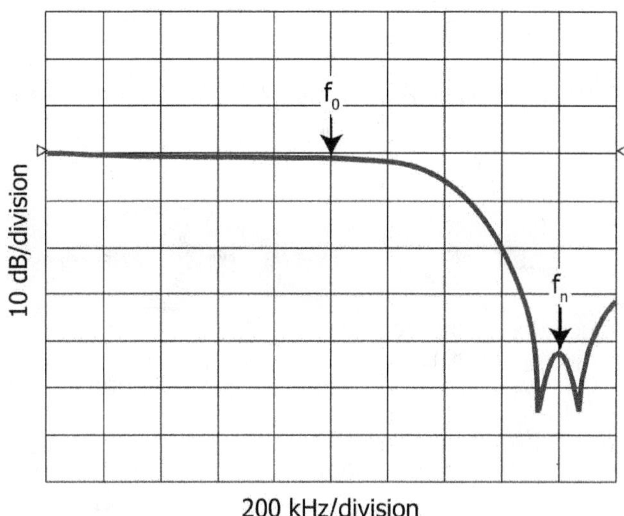

FIGURE 7.18-32 Frequency response for staggered dual notch cavities

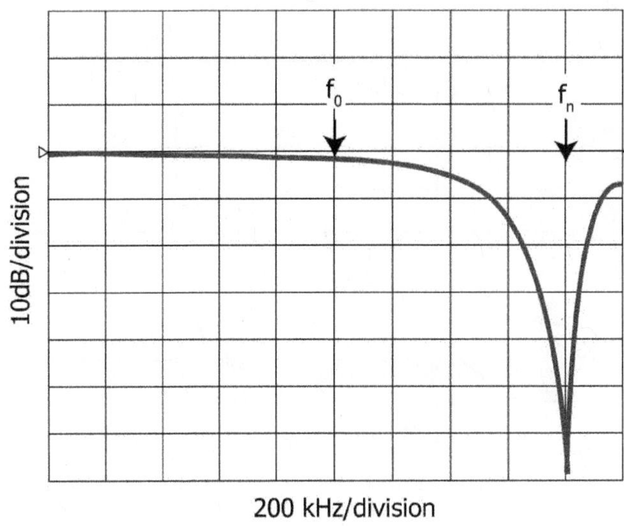

FIGURE 7.18-31 Frequency response for dual notch cavities

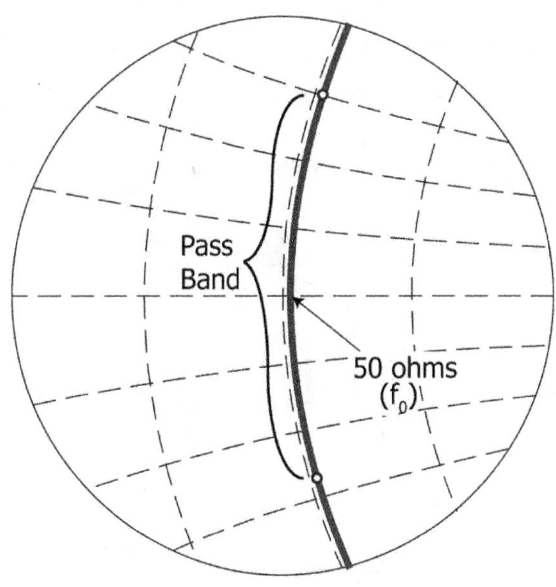

FIGURE 7.18-33 Impedance plot for a single notch cavity

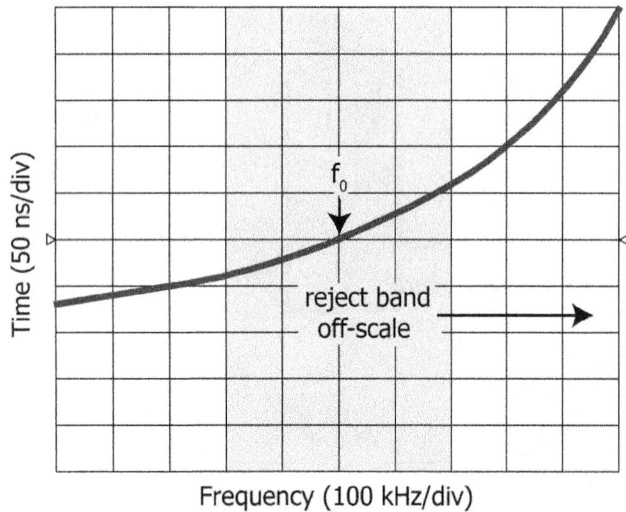

FIGURE 7.18-34 Group delay for a single notch cavity

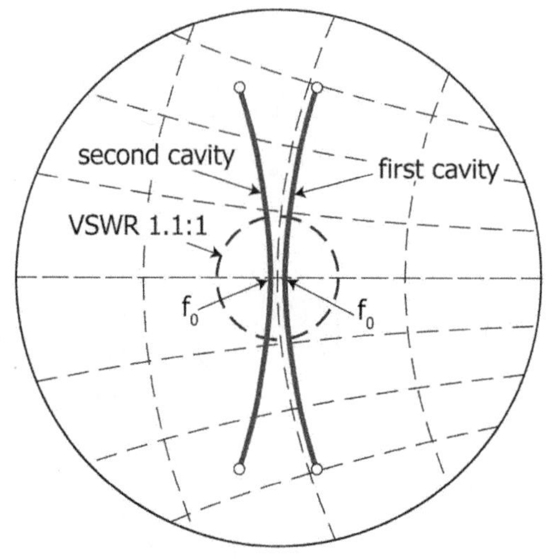

FIGURE 7.18-37 Superimposed impedance curves of a two-cavity bandpass filter

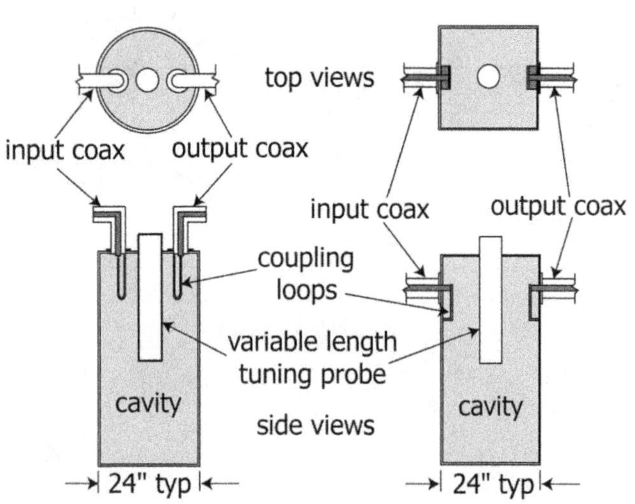

FIGURE 7.18-35 Bandpass cavity configurations

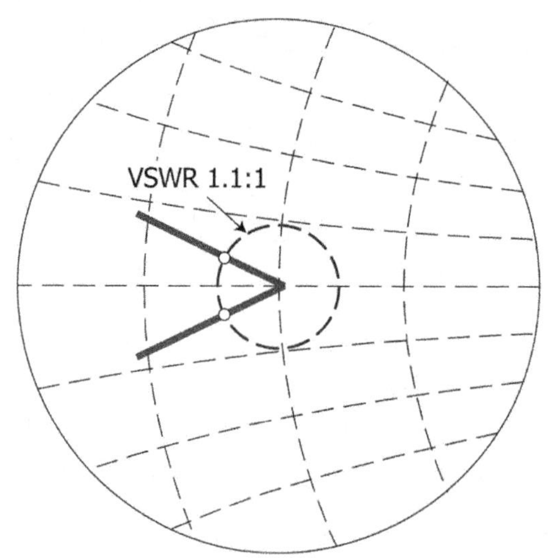

FIGURE 7.18-38 Impedance plot for a two-cavity bandpass filter

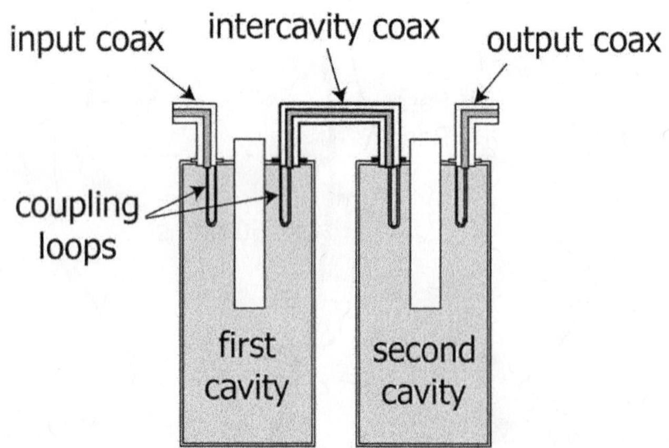

FIGURE 7.18-36 Two-cavity bandpass filter

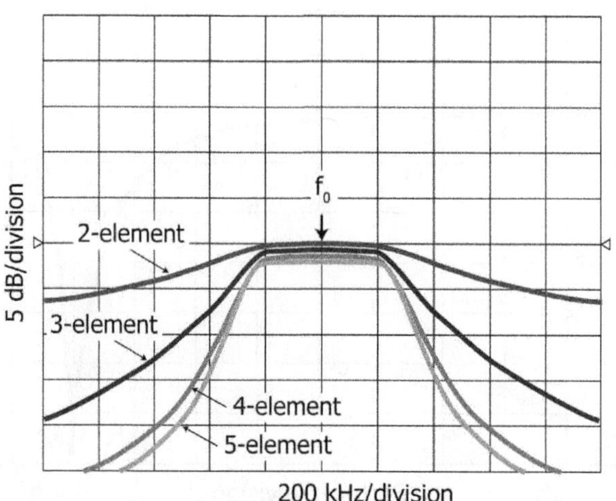

FIGURE 7.18-39 Frequency response for two-, three-, four- and five-cavity bandpass filters

difficult to link together, because the cavities themselves approach one-quarter electrical wavelength. As a result, when the inter-cavity coax is added, the electrical spacing is longer than one-quarter wavelength. In this case, the coupling loops must be manipulated to compensate for the extra length so the impedance bandwidth of the cavities is maintained.

Common-Wall Coupling

The spacing problem can be prevented by building the cavities contiguous to each other and coupling them through a tuned opening in the wall between them, as shown in Figures 7.18-41 and 7.18-42. The one-quarter electrical wavelength spacing is maintained by the coupled fields between the cavities.

Coupling Options

Although common-wall coupling can be accomplished using a true iris placed away from the top of the cavity (Figure 7.18-42), the size of the iris is difficult to control and adjust. Another method, shown in Figure 7.18-43, of coupling energy from one cavity to the next is neither an iris nor a slot but a trapezoidal opening designed so no adjustments are needed to couple the energy from one filter to the next across the FM band. Like most broadband-tuned networks, however, it is difficult to optimize a filter set at any one particular frequency. Perhaps the best configuration, shown in Figures 7.18-41 and 7.18-44, is a slot at the very top end of the cavities where the magnetic fields are at their strongest point and the size and shape of the slot can be manipulated externally for ease of adjustment in tuning the filters.

Interdigital Filters

Interdigital filters have only recently been introduced as an alternative to loop- and iris-coupled filters at FM frequencies. Interdigital filters do not employ individual cavities that must be coupled together. As shown in Figure 7.18-45, the energy is directly coupled to the input and output tuning probes. Parts counts are minimized and interdigital filters are significantly smaller than iris-coupled filters. Because of their smaller size,

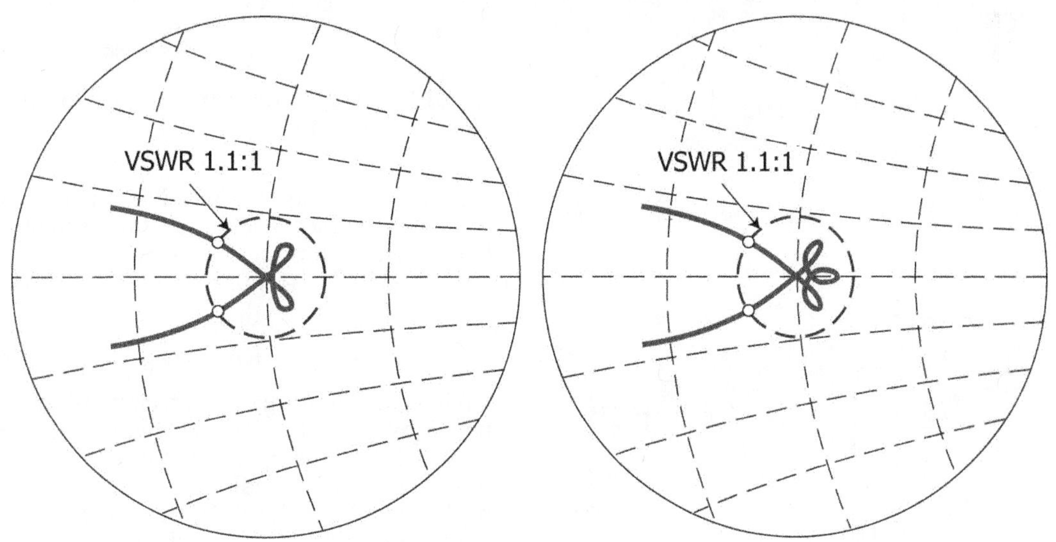

FIGURE 7.18-40 Impedance plots for three- and four-cavity equally coupled bandpass filters

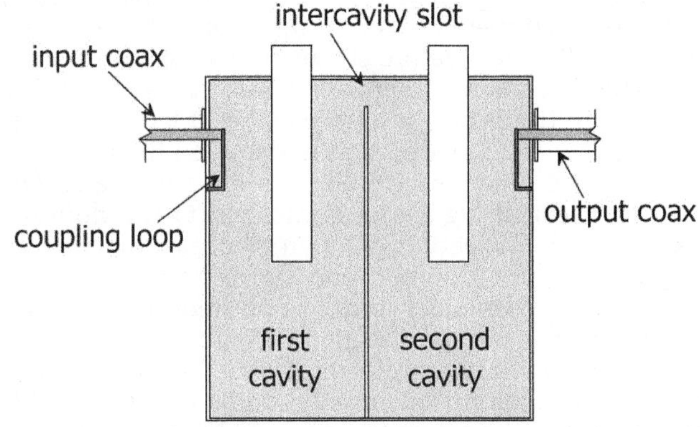

FIGURE 7.18-41 Slot-coupled cavities

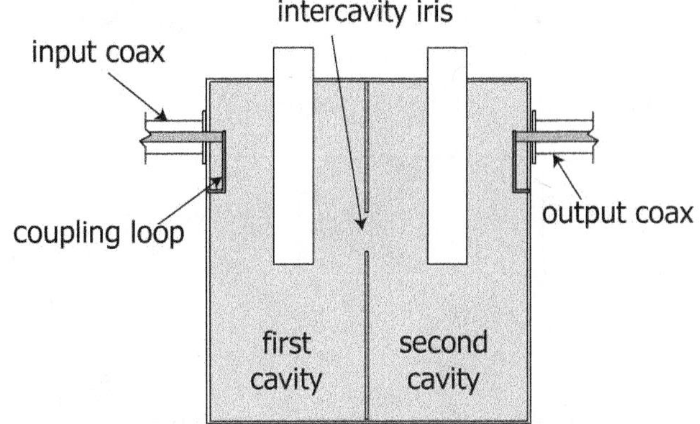

FIGURE 7.18-42 Iris-coupled cavities

interdigital filters have higher insertion losses than either loop- or iris-coupled filters of the same power rating, and careful attention must be paid to the thermal properties of the filter. Interdigital filters have better out-of-band isolation than cavity-style systems and are ideal for balanced combiners because of the ease of maintaining identical tuning across the channel.

Cross-Coupling

If a transmission line segment is added between the first and last bandpass sections (Figure 7.18-46), a parallel transmission channel is created. This line segment is then tuned to achieve specific phase and amplitude characteristics, so unwanted frequencies at both ends of the filter cancel each other out. It therefore acts as a band reject component, creating notches at the edges of the pass band (Figure 7.18-47).

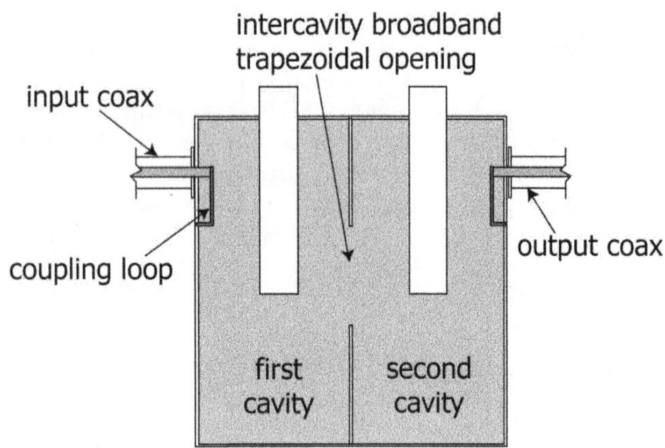

FIGURE 7.18-43 Cavities coupled by broadband trapezoidal opening

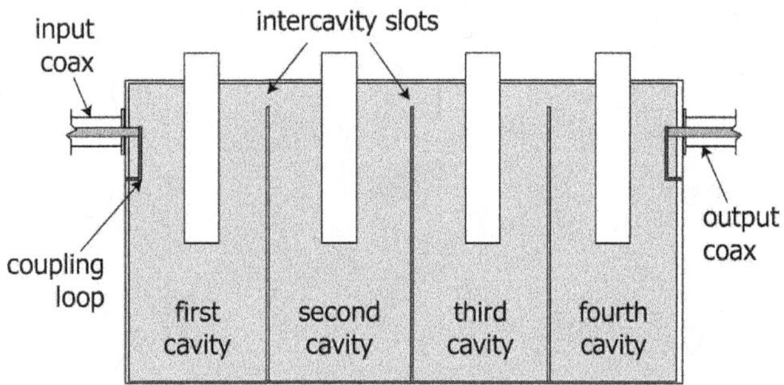

FIGURE 7.18-44 Four-cavity slot-coupled filter

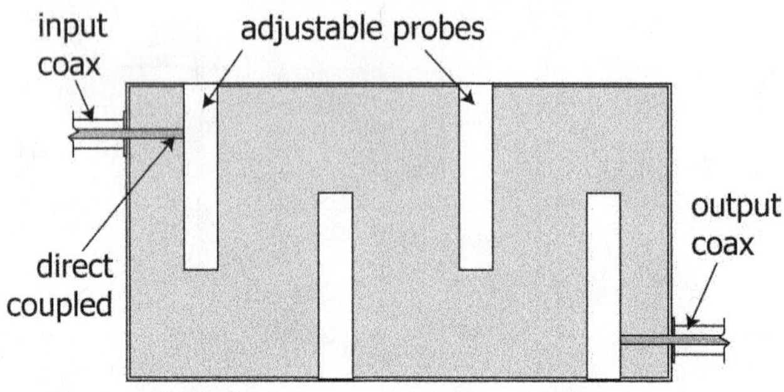

FIGURE 7.18-45 Four-pole interdigital filter

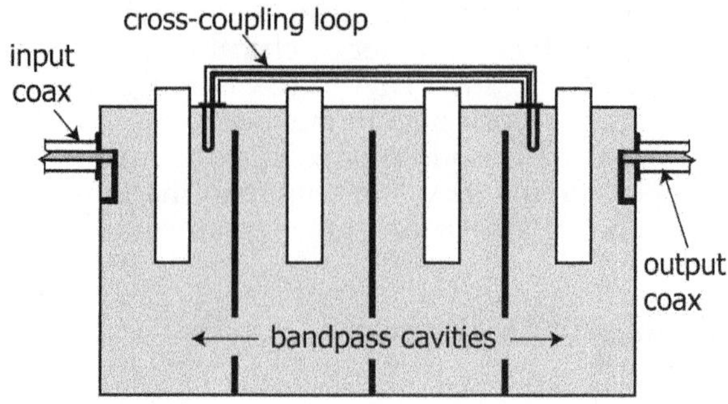

FIGURE 7.18-46 Filter with cross-coupling

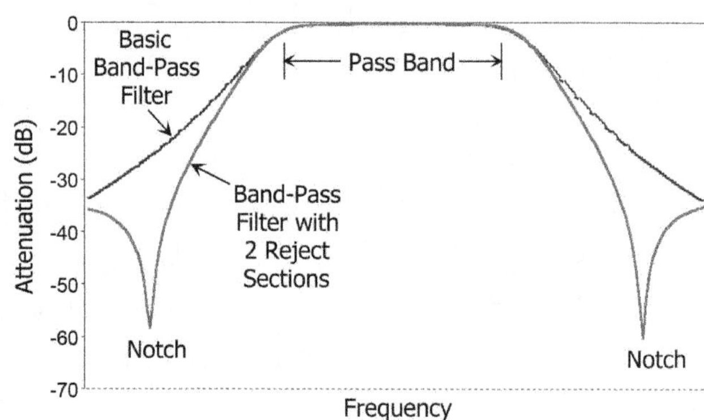

FIGURE 7.18-47 Frequency response for a filter with cross-coupling loops

ISOLATORS

An isolator is comprised of a circulator and a load. Signals move between legs in only one circular direction, giving the device its name. Although it is theoretically possible for the signal originating at any given leg to reach any other leg, this is prevented by the existence of one high-impedance leg, which traps energy trying to move across it and shunts it off to a dummy load. Thus, it is possible to configure the circulator to allow the signal from the transmitter to flow freely out the adjacent antenna leg, but energy returning through the antenna leg is interrupted before it can reach the transmitter leg. This is shown in Figure 7.18-48. The signal from the transmitter is fed into the isolator at leg 1. It flows out leg 2 on the transmission line toward the antenna. At the same time, any signal from the antenna enters the circulator at leg 2 and is directed to the dummy load at leg 3. The actual isolation value is a function of the match of the dummy load and is typically 26 dB. This ability of isolators to divert on-frequency signals headed in the wrong direction is key to a number of modem combining strategies that employ separate digital and analog transmission paths and where the combining method does not afford at least 35 dB of isolation between the digital and analog transmitters.

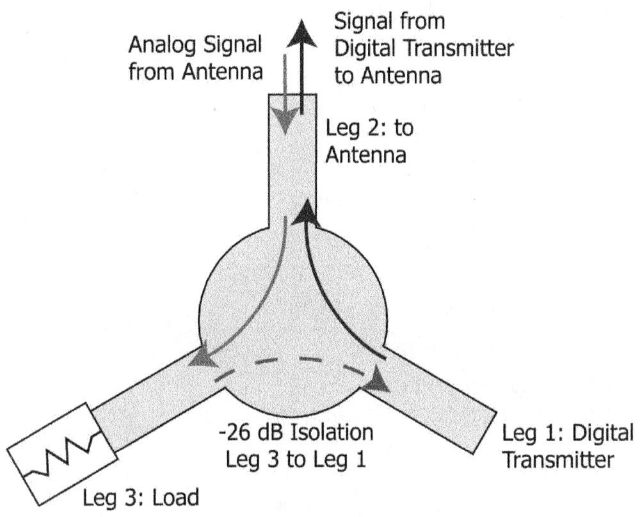

FIGURE 7.18-48 Isolator

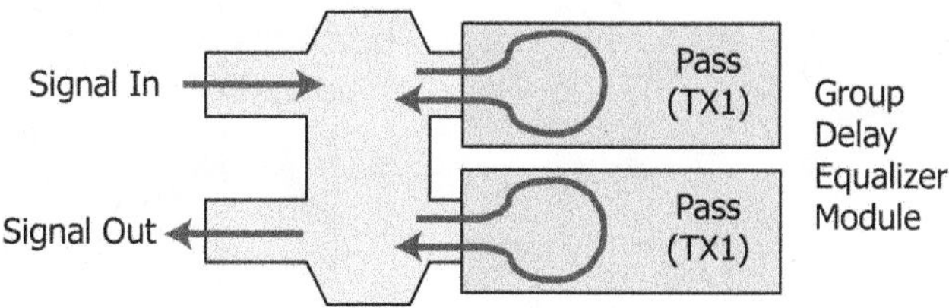

FIGURE 7.18-49 Group delay equalizer

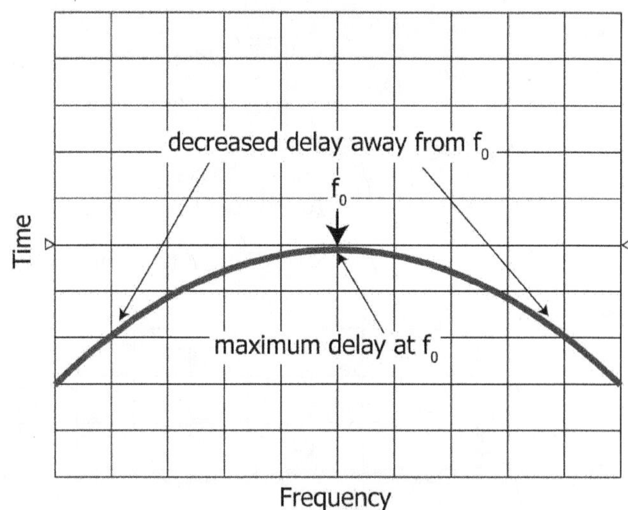

FIGURE 7.18-50 Group delay of group delay equalizer

DIRECTIONAL COUPLERS

Precision directional couplers are commonly found on each broadband output of a combiner system and are convenient ports for taking FCC-required test measurements, enabling diagnostics and serving as a port for any protection and monitoring system the combiner may employ. Directional couplers located on the inputs to each module further enhance the versatility of the system.

GROUP DELAY EQUALIZER

A group delay equalizer consists of a quadrature hybrid and two identical bandpass filters that have only one coupling loop, so the energy is coupled in and out of the cavity by the same loop (Figure 7.18-49). The tuned frequency is delayed for longer than the off-resonant frequencies (Figure 7.18-50).

TYPES OF COMBINERS

In addition to using hybrids as combiners, a series of specialty devices are employed for combining multiple RF signals into a common path. These include the branched or star-point, balanced, notch filter and bandpass filter combiners.

Branched or Star-Point Combiners

A branched combiner is a simple combination of a tee junction and the required number of filters to ensure a sufficient amount of isolation to prevent spurs. For example, an FM branched combiner consisting of a three cavity bandpass filter in series with two band-reject cavities (Figure 7.18-51) may be used to provide the isolation required for two close-spaced frequencies 0.8 MHz apart. TX1 and TX2 are the signals from transmitters 1 and 2 as they enter the combiner. The signals pass through the notch and bandpass filters and arrive at the tee junction. The length of the coaxial line between each set of filters and the tee junction is adjusted to provide a very high impedance (approaching an open circuit) to the other frequency, so the power flow of each signal is through its own filter, out of the tee junction and up to the antenna.

Performance

Refer again to Figure 7.18-38, the frequency response curve for a three-cavity bandpass filter, and Figure 7.18-31, the frequency response curve for a two-cavity, staggered-frequency band-reject filter. When these filters are used in combination, the resulting curve is as shown in Figure 7.18-52. Note that the insertion loss for the pass frequency f_1 is only about 0.25 dB and the isolation at the reject frequency f_2 is greater than 50 dB across the channel. The impedance plot, Figure 7.18-53, is likewise the combination of impedance plots for the same filter combination.

Branched Combiners With Feedback Loops

Although many branched combiners still in operation use notch cavities for enhanced isolation, most modern branched combiners have gone to feedback loop technology (Figure 7.18-54) for this purpose. Figure 7.18-55

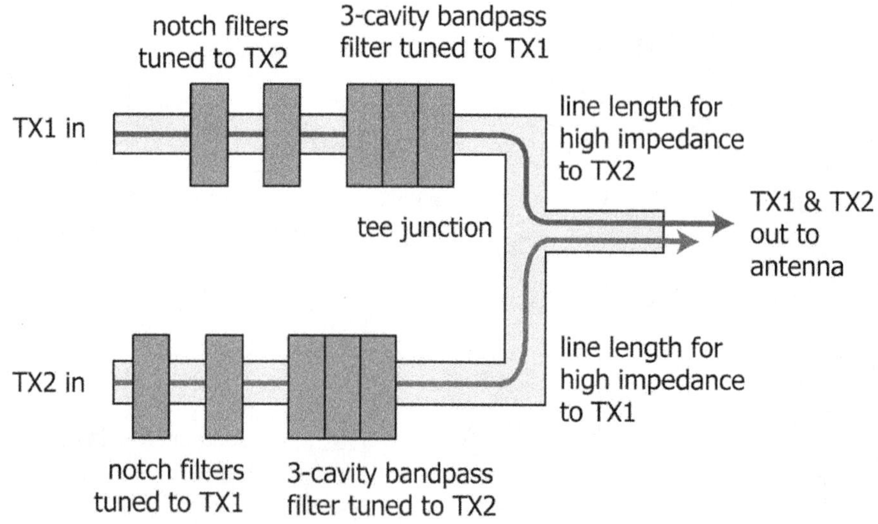

FIGURE 7.18-51 Branched combiner with notch cavities

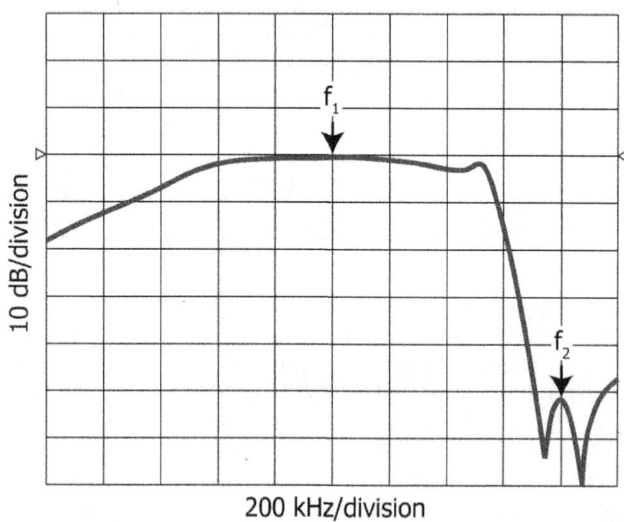

FIGURE 7.18-52 Frequency response of a three-cavity bandpass filter in series with two notch cavities

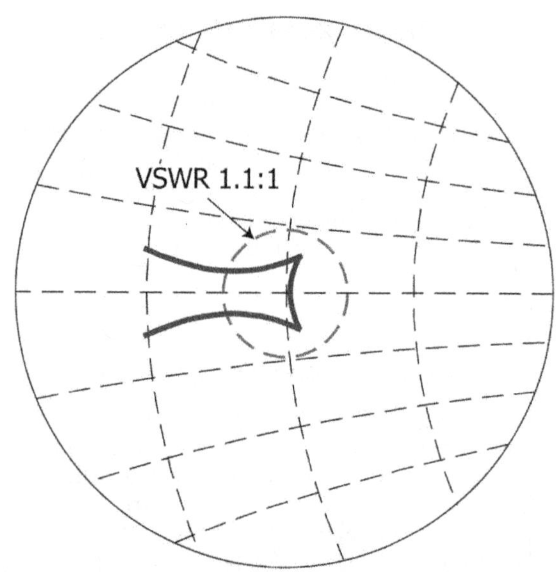

FIGURE 7.18-53 Impedance of a three-cavity bandpass filter in series with two notch cavities

is the frequency response curve of the three-cavity bandpass filter with feedback loops. Notice that the curve is smoother through the pass band and, even though it only has one notch, the isolation at f_2 still exceeds 50 dB. The impedance plot of a branched combiner with feedback loops is almost identical to that of the combiner with notch filters (Figure 7.18-53).

Limitations

A branched combiner is very efficient for a two station installation and has been used for as many as four stations, but a tee junction for more than four stations becomes impractically large, and adjusting the lengths of interconnecting coax becomes prohibitively complex as well. Also, a branched combiner cannot easily be expanded later to include more stations, although it can be expanded by integrating it with balanced

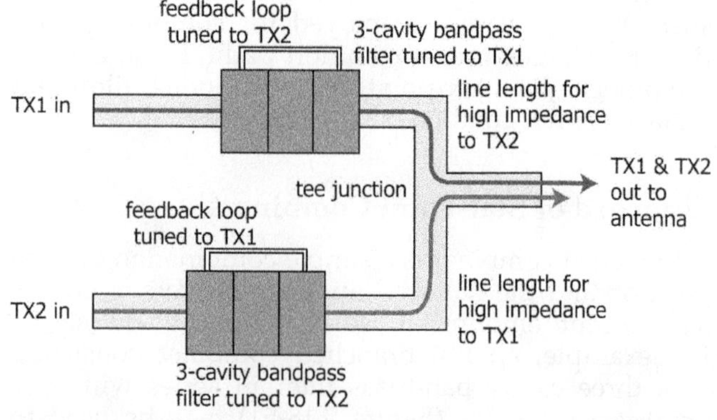

FIGURE 7.18-54 Branched combiner with feedback loops

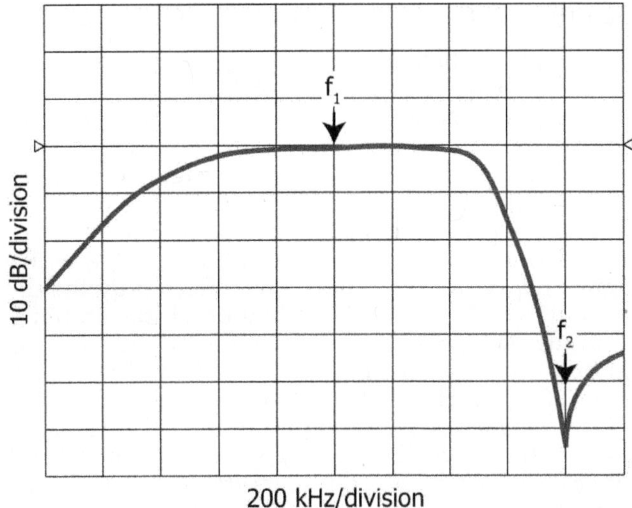

FIGURE 7.18-55 Frequency response of branched combiner with feedback loops

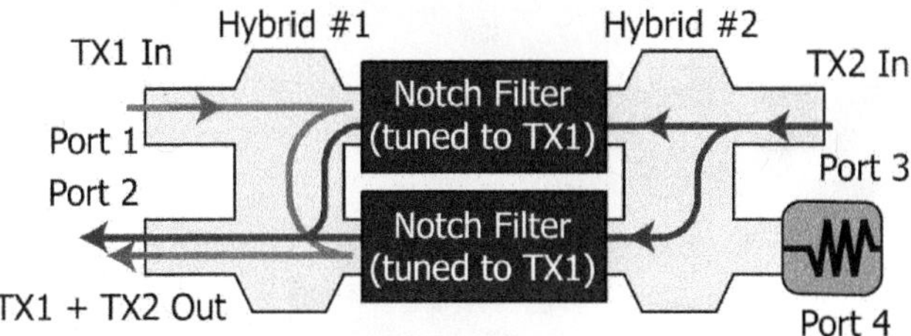

FIGURE 7.18-56 Two-station, one-module notch-filter balanced combiner

combiner modules. To combine more than four stations, a balanced combiner becomes more practical and cost-effective.

Balanced Combiners

The balanced combiner is based on a hybrid ring. Each leg of the ring contains an identical set of either bandpass or band-reject filters, hence the term *balanced*. It is imperative that the filters of all modules be tuned to have as close to the same response characteristics as possible. The goal is to have the hybrids react identically to the filters. Small differences in electrical length through the hybrids quickly add up to an increased VSWR. For example, a phase difference of ±2° in the legs of a hybrid produces a VSWR of 1.07:1 (or a return loss of 29 dB). If that phase difference degrades to ±4°, the VSWR deteriorates to 1.15:1 (23 dB). Most early balanced combiners used notch filters.

Notch-Filter Balanced Combiners

In the notch-filter balanced combiner (Figure 7.18-56), both notch filters within the hybrid ring are tuned to reject TX1's frequency, which enters the combiner at port 1. That signal is reflected by the filters and exits at port 2. TX2 enters port 3, the broadband input port of the module, and passes through in the diagonal mode (shown in Figure 7.18-27) with minimal loss in the reject cavities.

Performance

The isolation of transmitter 2 from frequency TX1 is the sum of the hybrid ring isolation of 35 dB and the isolation of the notch cavities and can approach 35–40 dB. However, the isolation of transmitter 1 from frequency TX2 is only that of the hybrid ring—about 35 dB; therefore, additional filtering, either bandpass or band-reject, is required to ensure that no spurs are generated within transmitter 1. This added filter is shown in Figure 7.18-57.

External Bandpass Filtering

A better way to reject multiple unwanted frequencies, of course, is to use a bandpass filter tuned to the desired frequency. For example, Figure 7.18-58 shows a five station, four module combiner. In this example, each input filter is a bandpass filter tuned to the frequency of that input. If reject filters were to be used at the various inputs, each input would have to filter all the frequencies previously introduced; therefore, port 3 of module 2 would have to contain two notch filters; port 1 of module 3, three filters; and port 3 of module 4, four filters. This proliferation is avoided by the use of input bandpass filters.

Emergency Input Port

In some cases, instead of having a station located at port 3 of module 1, that port is terminated in a 50-ohm load and can be used as an emergency input for any station in the system. Providing an extra port in this way allows a damaged module to be bypassed. Because of the nature of that particular port, as long as the input filter at port 1 of module 1 is a bandpass filter, no further input filtering is necessary.

Limitations of Notch-Filter Balanced Combiners

A problem with using notch filters within the hybrid rings is that if the two filters in any module are not identically tuned, an imbalance occurs within the hybrid ring, thus reducing the isolation to a point where a spur can be generated within a transmitter. Once a spur has been generated, there are no filters within the system to reject that spur because the filters are tuned only to the expected frequencies; therefore, the spur is broadcast. A second disadvantage of using internal notch filters is that each module in turn has to conduct the accumulated power of all the previous modules. For a multiple high power system each module must be larger than the previous one, and the power rating of the system is limited

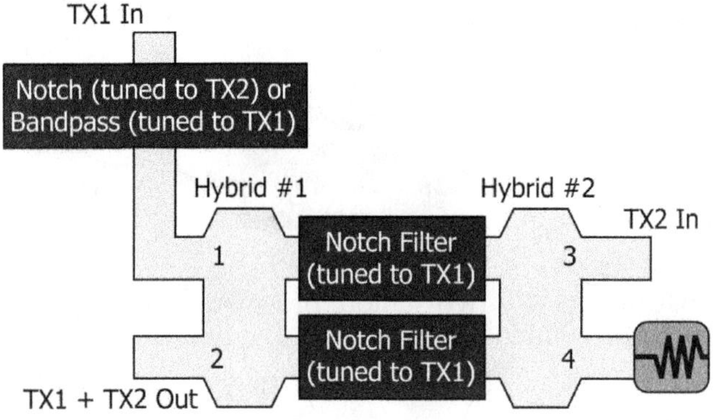

FIGURE 7.18-57 Two-station, one-module notch-filter balanced combiner with input filter at TX1

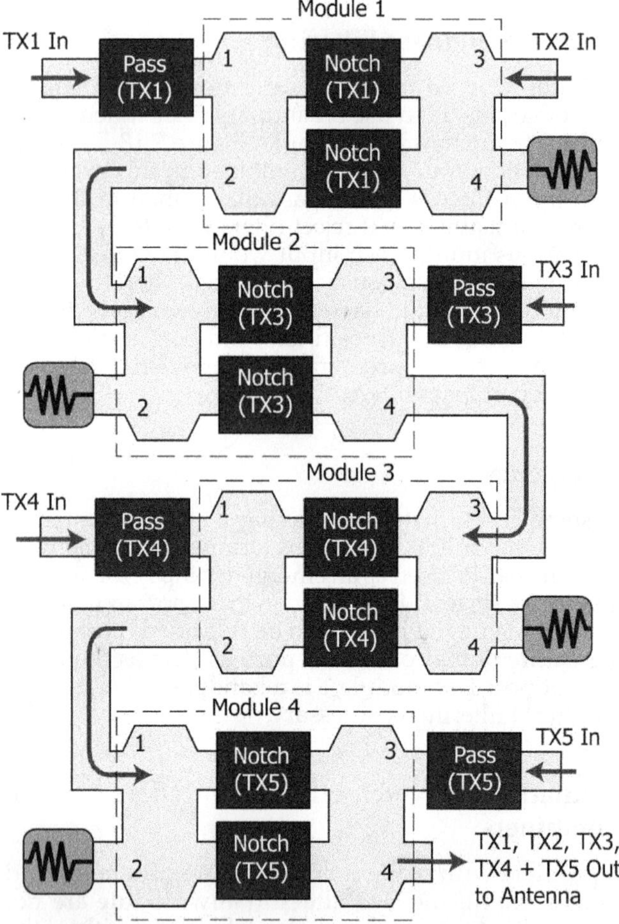

FIGURE 7.18-58 Five-station notch balanced combiner with input bandpass filters

by the size of the final module. Third, notch-filter combiners are impractically narrowband in nature for today's wideband IBOC channels, especially when the frequencies combined are closely spaced. Because of these limitations, notch-filter systems are no longer used. Modern FM combiners use bandpass filters.

Bandpass Filter Balanced Combiners

In a bandpass balanced combiner system, bandpass filters are used within the hybrid ring. The basic system layout is similar to that of a notch combiner. The power flow is shown in Figure 7.18-59 (compare to Figure 7.18-56). In the notch system, the filters rejected signal TX1 entering port 1. In the bandpass system TX1 also enters port 1 but passes through the hybrid ring's bandpass filters and out port 4, while signal TX2, entering at port 3, is reflected by the filters and exits at port 4. The isolation of transmitter 1 from frequency TX2 is the sum of the hybrid ring isolation (35 dB) and the isolation of the bandpass filter (about 25 dB). However, the isolation of transmitter 2 from frequency TX1 is only the hybrid ring isolation of about 35 dB; therefore, an additional filter must be added between transmitter 2 and its input port (Figure 7.18-60), similarly to the single module notch filter balanced combiner shown in Figure 7.18-57.

Alternatively, a second module may be added to port 4 of module 1, and port 3 terminated in 50 ohms (and available as an emergency input port). Signal TX2 is then introduced at port 1 of module 2, as shown in Figure 7.18-61. No input filter is necessary now for TX2, because it is isolated by the bandpass filters in module 2. The emergency input port now sees both frequencies TX1 and TX2, reduced 35 dB below each transmitter's power level. A multiple station bandpass balanced combiner (Figure 7.18-62) is an extension of the latter configuration, where each frequency has its own module. In a bandpass system, the accumulated power entering each module flows only through the output hybrid, so the power handling capacity of the system is limited only by the size of the output hybrids and interconnecting transmission line, not the entire module.

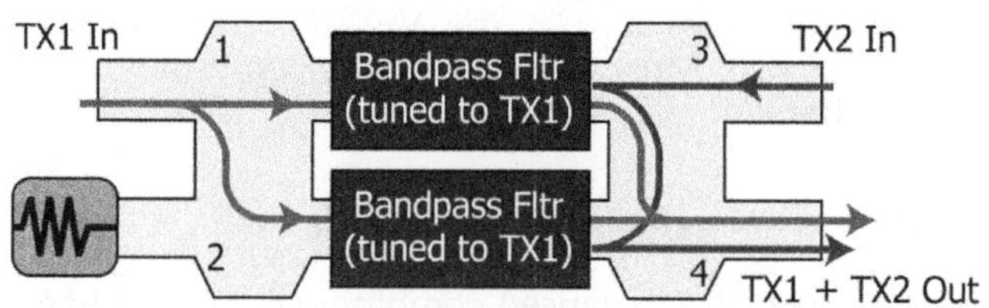

FIGURE 7.18-59 Two-station, one-module bandpass filter balanced combiner

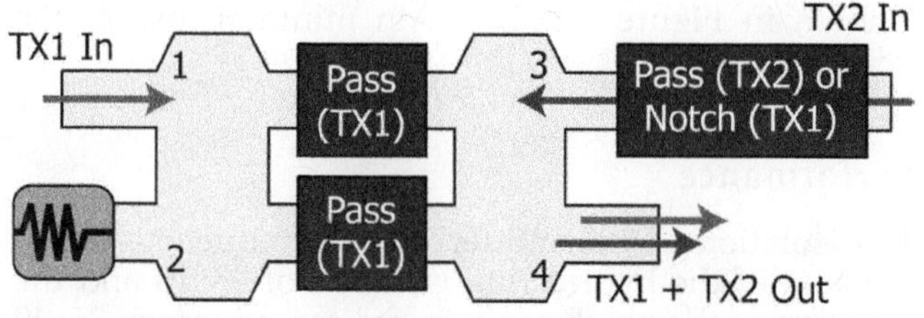

FIGURE 7.18-60 Two-station, one-module bandpass filter balanced combiner with input filter at TX2

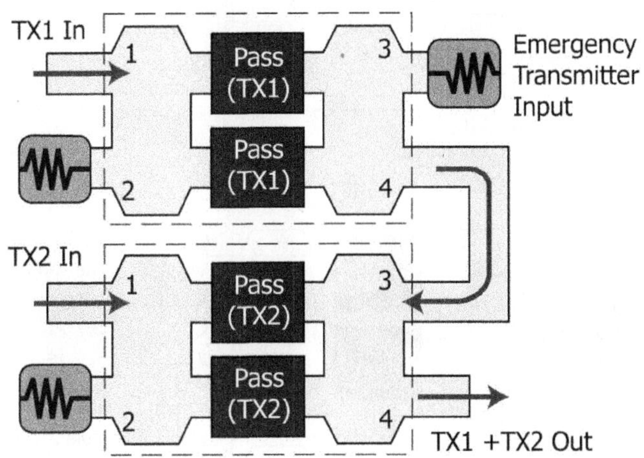

FIGURE 7.18-61 Four-station bandpass filter balanced combiner

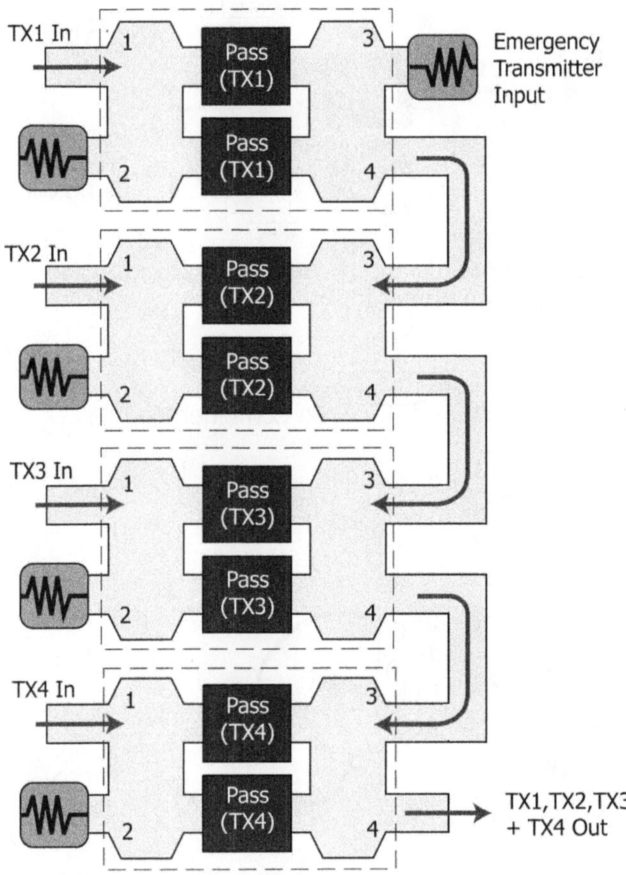

FIGURE 7.18-62 Two-station, two-module bandpass filter balanced combiner

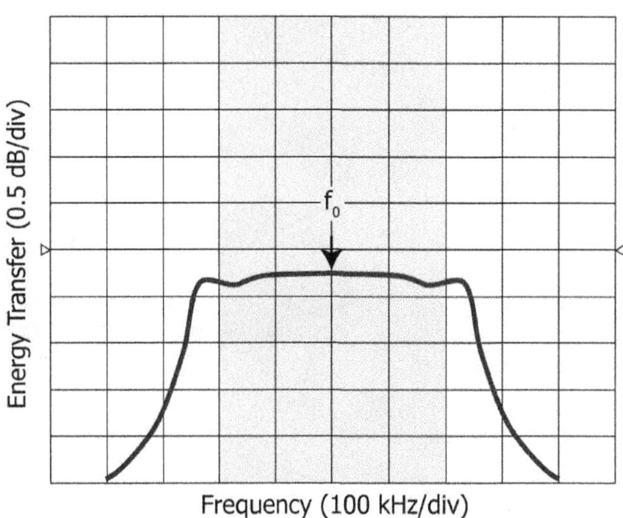

FIGURE 7.18-63 Frequency response for a bandpass filter balanced combiner

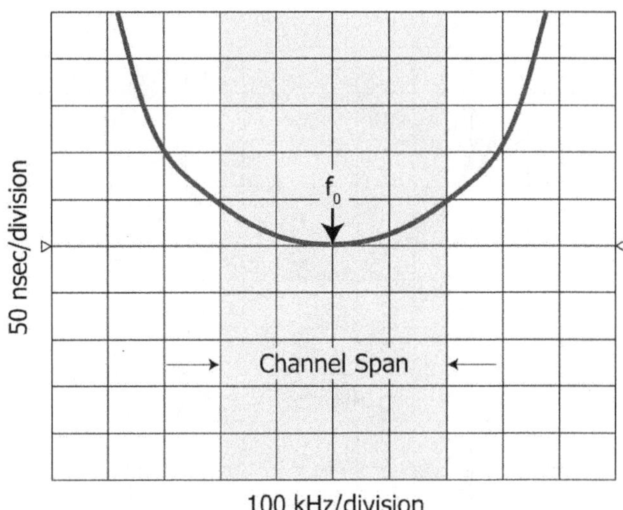

FIGURE 7.18-64 Group delay for a bandpass filter balanced combiner

Performance

The frequency response, the group delay, and the impedance diagram for this combiner are shown in Figures 7-18.63, 7-18.64 and 7-18.65, respectively.

Group Delay Effects

When two stations are 1.2 MHz apart or closer, the bandpass filter will not provide quite enough isolation and will allow a small amount of signal interaction. This affects the group delay curve of the module that is farthest from the antenna, as shown in Figure 7.18-66. A group delay equalizer can be installed either at the combiner input, using high power components (Figure 7.18-67), or between the transmitter's exciter and the IPA, using similar low power components (Figure 7.18-68).

COMBINING DIGITAL AND ANALOG SIGNALS

Of particular interest is the combining of traditional analog FM and new digital IBOC signals on the

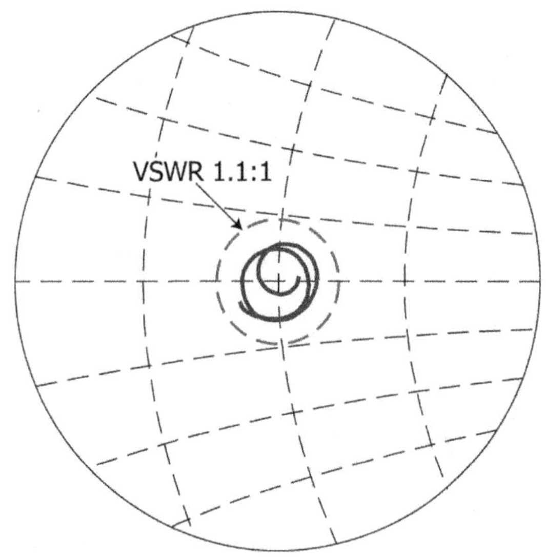

FIGURE 7.18-65 Impedance for a bandpass filter balanced combiner

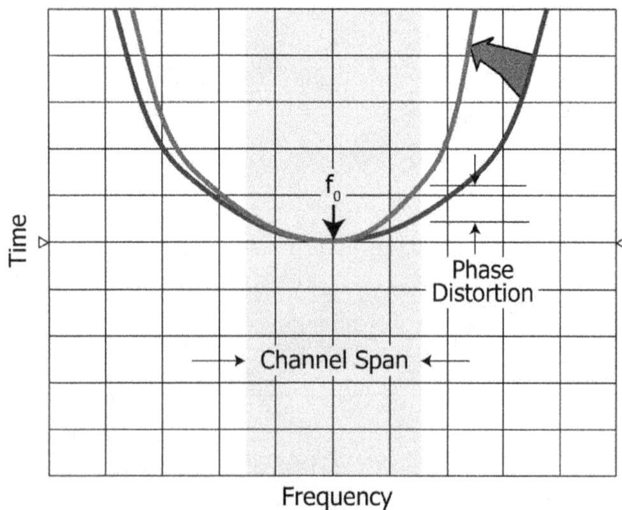

FIGURE 7.18-66 Distorted group delay

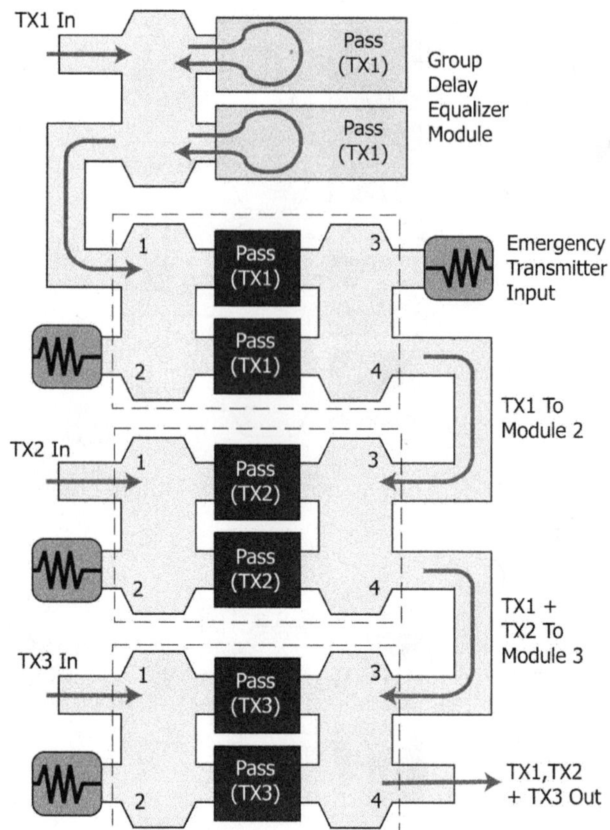

FIGURE 7.18-67 Balanced combiner with group delay equalizer at combiner input

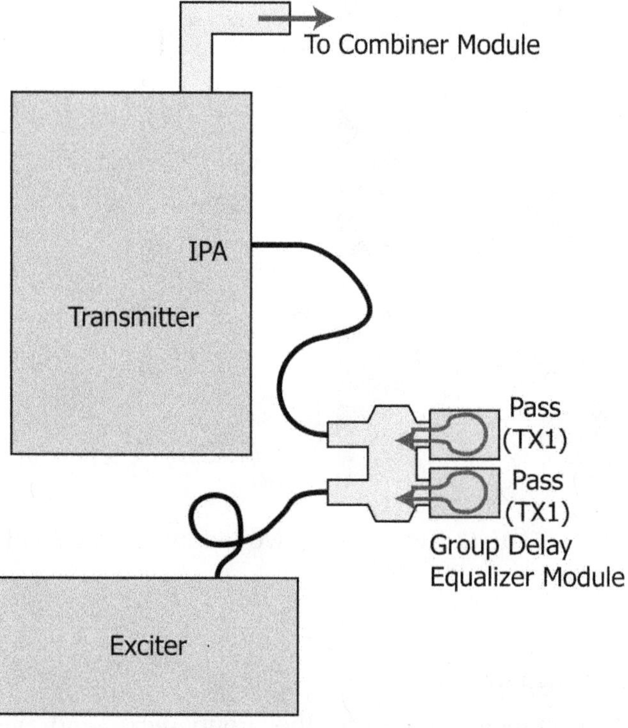

FIGURE 7.18-68 Group delay equalizer between exciter and transmitter IPA

same FM channel. Several methods to do this at high and medium levels and with separate antennas are described.

High-Level Combining

Small probes or small strip lines forming a precision directional coupler have been used historically to couple a small amount of energy out of the transmission line with coupling factors of anywhere from −40 dB to −60 dB from the RF power level being transmitted within that line. Typical uses of this level signal are to monitor or measure the high power signal with low signal level devices. Losses to the main path of the signal due to this sampling system are insignificant to the analog signal.

High-level combining uses a directional coupler (Figure 7.18-69) that has been mechanically enlarged to handle power levels in the kilowatt range, with a nominal coupling factor of −10 dB, and can be used in reverse to inject the digital signal into the analog RF stream. One strip carries the RF energy from the analog transmitter, which is considered the main line of the transmission system. The other strip is considered the coupling strip. The spacing of the strips determines the amount of coupling between the two signals. Increasing the coupling to −10 dB introduces a loss of 10% of the analog power, which is dissipated in a dummy load.

The digital signal enters the directional coupler at the reject port of the coupler, referenced to the analog input. Because it is a −10 dB coupler, only 10% of the digital signal is coupled to the main line. The remaining 90% flows to the dummy load. Several iterations of a high power combiner/injector have been tried. The −10 dB value is a good compromise for minimizing the loss to the analog transmitter while keeping the size of the digital transmitter to a reasonable level. An injector with a coupling factor smaller than −10 dB will increase analog losses, while a larger coupling factor will require a substantially larger digital transmitter. Note that the digital transmitter operates at only a fraction of the power of the analog transmitter. This method of combining analog and digital is normally used for stations that only have one single-input antenna and an analog transmitter with the reserve capacity to make up for the 10% loss in power. Depending on the reserve capacity of the analog transmitter and the size of the digital transmitter, the coupling factor can be adjusted to optimize almost any installation.

Mid-Level Combining

Mid-level combining (Figure 7.18-70) was developed by incorporating a standard −3 dB quadrature hybrid and using two analog transmitters: one standard analog transmitter and one linearized transmitter equipped to transmit digital along with the analog. It has been a well-established practice to combine two analog transmitters into a quadrature hybrid so most of the power goes to the antenna with minimal loss to the hybrid's dummy load. When the digital component of the linearized transmitter is turned on, the signal enters port A with its associated analog signal. Because no digital signal is entering port B, the digital signal is split in half. The benefit of this method over high-level combining/injection is that there is no significant loss to the analog signal and only a 50% loss to the digital signal rather than a 90% loss. As is the

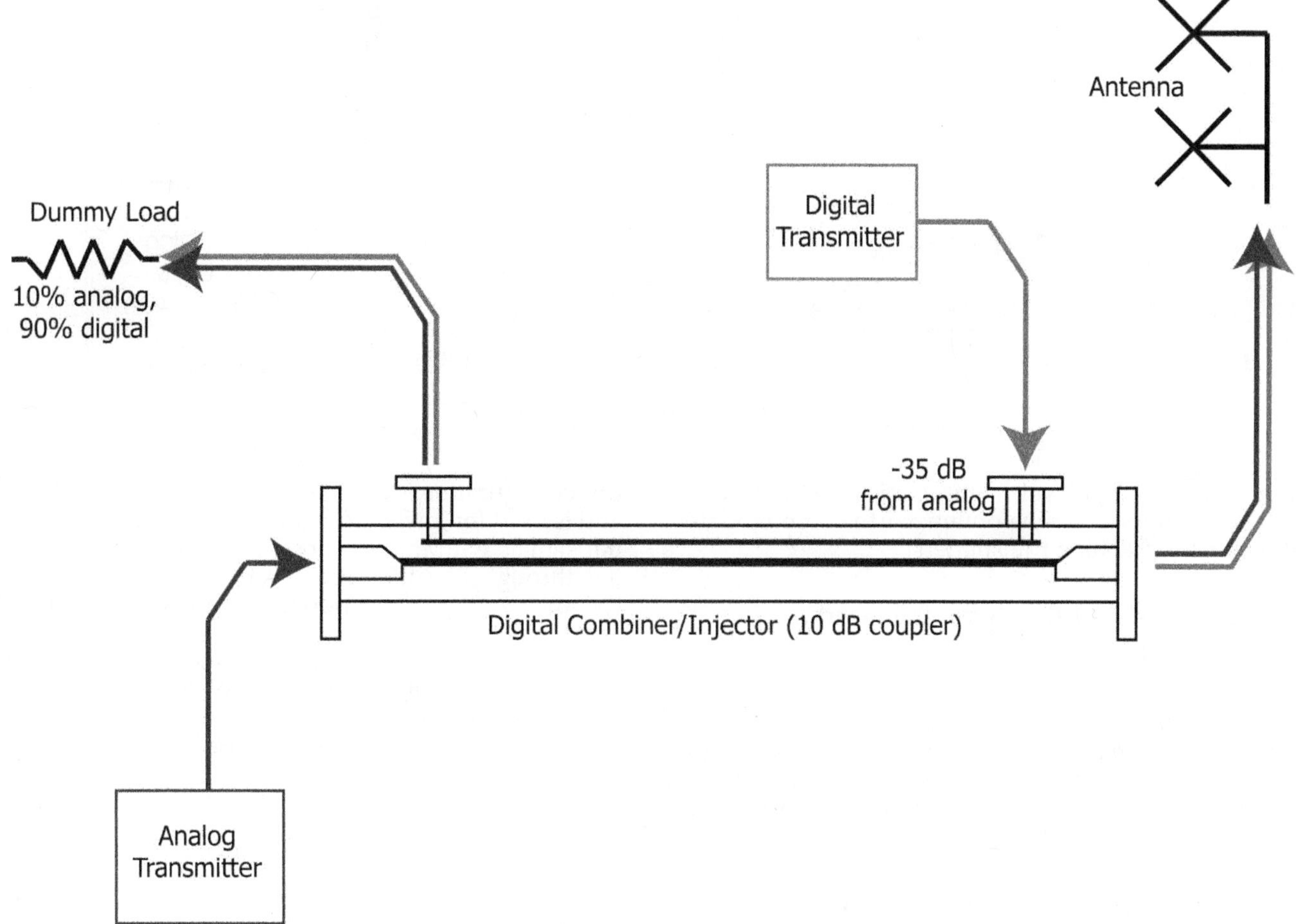

FIGURE 7.18-69 High-level combiner/injector

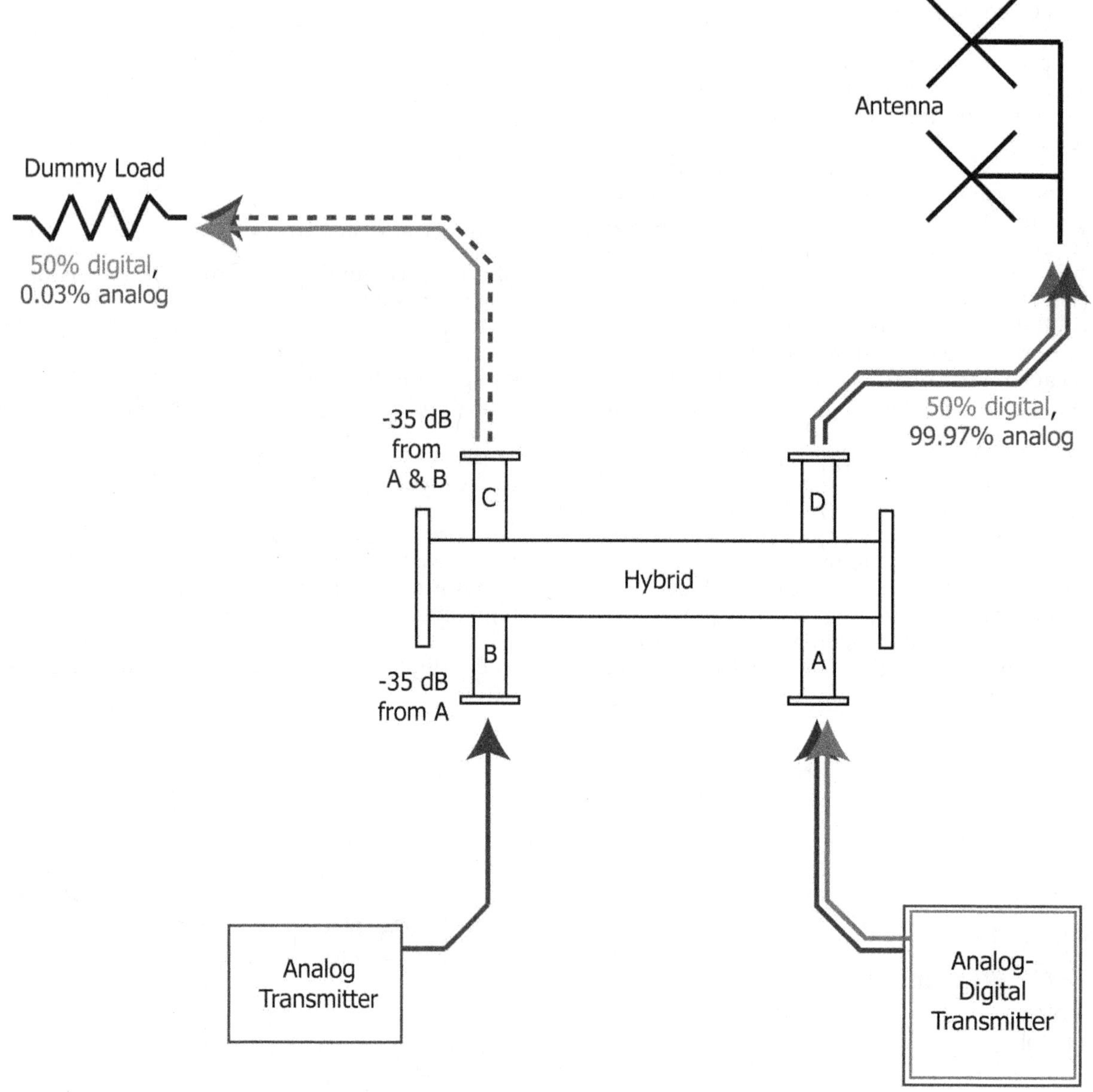

FIGURE 7.18-70 Mid-level combining

case with the high-level coupler/injector, the power split of the hybrid can be optimized to accommodate different-sized analog transmitters.

Combining Using Bandpass Balanced Combiners

Strategies that combine analog and digital signals in antenna radiators or use separate analog and digital radiators in close proximity are among the most popular IBOC implementation strategies because they minimize the size and cost of the digital transmitter and reduce the energy wasted.

Back-Feeding IBOC Into a Balanced Module

The simple use of balanced combiner modules shown in Figure 7.18-62 is termed *single-feeding*. A variation on this configuration is called *back-feeding* (Figure 7.18-71), and is used for low level combining of analog and digital signals with minimal loss. Digital transmitters are fed through isolators into the hybrid ports opposite the analog transmitter ports, and a combined digital signal exits the wideband port (top left) normally occupied by the system reject load in a single-fed combiner. The only added hardware is the isolator in place of the dummy load of the single-feed combiner, to prevent analog on-channel signals from feeding back into the digital transmitter.

Purpose of the Isolator

Transmission systems that do not have enough isolation between the analog and digital components require isolators. When combiners are configured for backfeed operation, on-channel power is coupled from

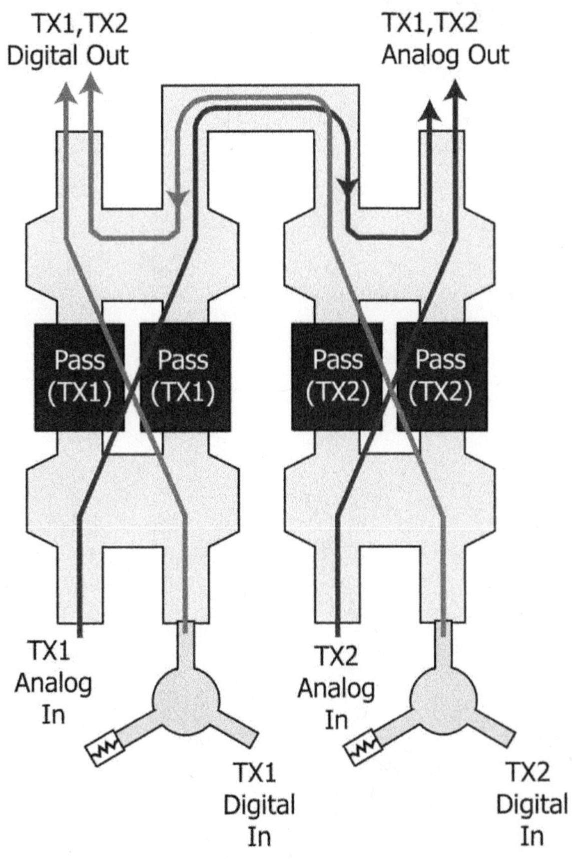

FIGURE 7.18-71 Back-feed configuration

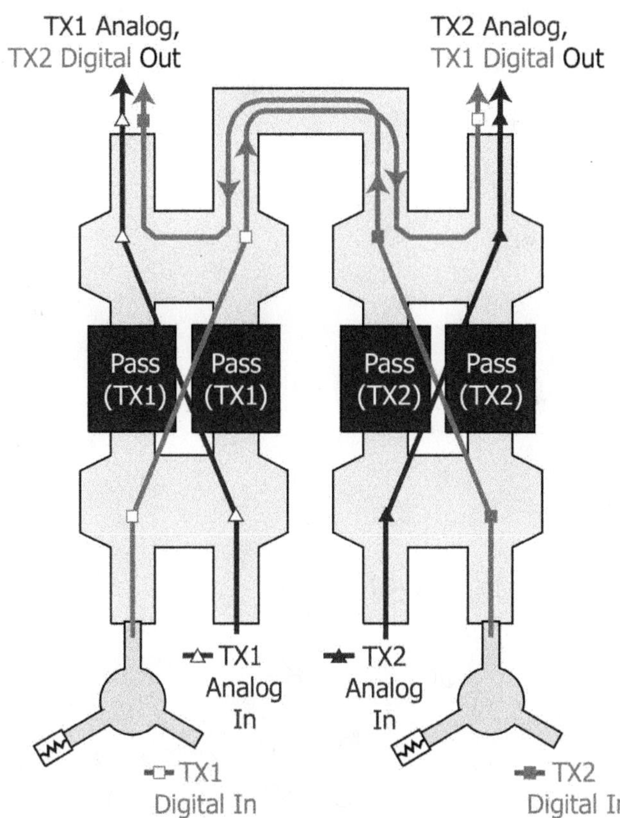

FIGURE 7.18-72 Cross-feed configuration

one transmission path into the other via the antenna elements and feeds back into the module through the opposite leg from which it exited. Although an efficiently operating antenna will minimize the energy coupled between paths, there still will be sufficient energy returned to require a dummy load for the port opposite the analog transmitter input. If a station runs an analog-only or high-level combined analog/digital signal, a stand-alone dummy load is used on this port. When the port is occupied by a digital transmitter, the dummy load becomes part of an isolator assembly. Isolators are not used where the analog and digital signals are already combined in the transmitter (low level), combined through a hybrid providing at least 35 dB of isolation (mid-level), or combined using a coupler/injector providing at least 35 dB of isolation (high-level).

Cross-Feeding IBOC Into a Balanced Module

The cross-feed, or split-feed, configuration (Figure 7.18-72) is a further extension of back-feeding. Rather than segregating digital and analog signals into separate transmission lines, it combines the analog signals of some stations with the digital signals of others. Again, an isolator is used to provide additional isolation between the analog and digital transmitters. Usually, the analog power is split as evenly as possible, thus minimizing both the average and peak power any broadband line component carries. Thus, 9-inch components are eliminated in all but the largest systems. Using equal-sized transmission lines also provides redundancy. A failure in a transmission line or portions of the antenna feed system can be overcome by directing a station's primary transmitter (either analog or digital) over the remaining transmission line.

SUMMARY

Combiners are required when it is necessary to transmit multiple signals from a single antenna. Without proper combining, signals will interact in each other's transmitters, producing intermodulation products. This chapter has discussed the fundamentals of combining and the use of combiners in FM broadcasting. Several designs and many different components and configurations were described. The various types of combiners have their own advantages and disadvantages. The system designer must be aware of each so the appropriate filter system or systems can be selected for the specific combining application.

Books on Related Topics

Matthaei, G.L., Young, L., and Jones, E.M.T., *Microwave Filters, Impedance-Matching Networks, and Coupling Structures*, Artech House Books, Dedham, MA, 1980.

Smith, P.H., *Electronic Applications of the Smith Chart in Waveguide, Circuit, and Component Analysis*, McGraw-Hill, New York, 1969.

CHAPTER

7.19

FM and AM Translators and Boosters

RON CASTRO

WHAT ARE THEY AND WHY ARE THEY USED?

In the early days of FM, communities that lacked adequate FM coverage due to terrain blockage, or simply because they were too far away from existing FM stations, could be served by applying to the FCC for a license to operate a low-powered station located such that it could receive signals from an existing full-service station and retransmit or "translate" them onto another FM channel that covered the community. The FCC sanctioned these stations provided there was no financial connection between the translator station licensees and the full-service stations they carried. Today, in certain situations, the licensees of full-service stations, both FM and AM are permitted by the FCC to use translators to fill in areas where terrain blocks predicted service contours of FM stations or to provide an FM signal to augment AM coverage. Recently, FM translators have been used to carry the HD-2 programming of a primary FM station within that station's coverage contour.

FM APPLICATIONS

The method prescribed by the FCC to predict the service area of an FM station is based on a 1947 paper by Kenneth Bullington.[1] In those days, this was the basis of a practical method of estimating the usable coverage, as well as the potential interference of an FM station. Even though newer propagation prediction methods, aided by today's powerful and affordable computers, have revolutionized our ability to predict signal coverage and interference, the FCC still uses the Bullington methodology as its primary standard, although it is generally acknowledged that its main purpose today is more of an allocation tool than to accurately predict FM signal coverage.

The Bullington method still holds up pretty well in predicting signals against the newer technology in relatively flat terrain situations, but its inability to account for the specific location of terrain obstructions in the signal path causes anomalies such as the one illustrated in Figure 7.19-1. This shows a comparison of the 54 dBu coverage contour, which is the thick black line, of a typical Class B station serving San Francisco as predicted by the FCC's method, and one using a more modern prediction method over the very diverse topography of the Bay Area. In some directions, the coverage, shown in gray, extends beyond the FCC contour, but in others it falls short. The white areas in the illustration are where the actual signal is predicted to be below 54 dBu. Often, as in this situation, a broadcaster finds that the FCC service contour includes many communities and highways where the actual signal is poor or nonexistent.

FM licensees can remedy this anomaly by operating a "fill-in" translator or an on-channel booster to provide coverage within its licensed FCC-calculated service contour. For commercial stations, the service contour of the translator or booster itself cannot extend beyond that of the primary station. Leeway is granted to noncommercial broadcasters to use "non-fill-in" translators to extend their coverage beyond that of their main stations' service contours, but in no case may any translator cause interference to other full-service stations or other existing translators.

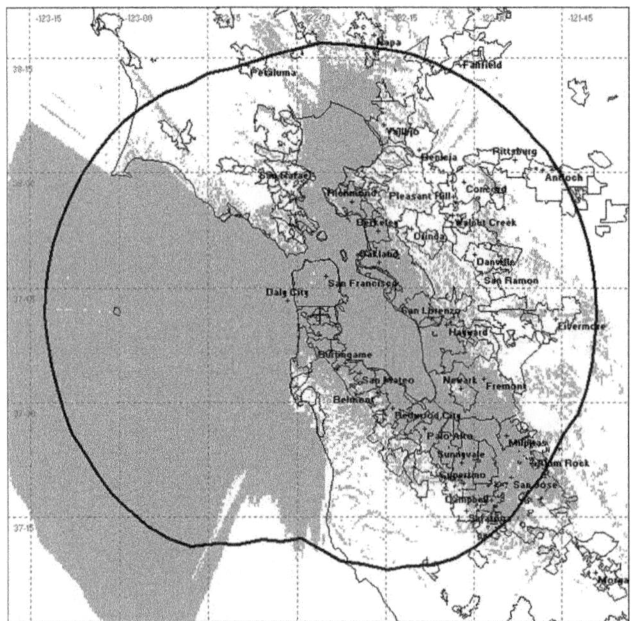

FIGURE 7.19-1 The 54 dBu FCC F(50,50) contour of a typical San Francisco Class B FM station. The gray areas show 54 dBu coverage using an alternative, "point-to-point" terrain-sensitive methodology.

THE TRANSLATOR

Simply put, a translator is an FM transmitter that rebroadcasts program material from another FM station on a different FM channel. This differentiates it from a station that gets its programming from an originating source other than a licensed, full-service FM station, or from a "booster" that retransmits the signal of a licensed, full-service FM on the same channel. Figure 7.19-2 shows the most basic translator, which consists of an FM receiver tuned to channel "A," connected to a directional antenna pointed at the station to be rebroadcast. The audio output of the receiver is used to modulate an FM transmitter that rebroadcasts the audio from the receiver on channel "B" and is connected to an antenna that provides coverage to the intended community. Such stations are usually located on mountain tops that have line-of-sight paths to both the primary station and the community to be served.

Noncommercial stations operating boosters in the nonreserved part of the FM band must use receivers to feed their translators, but in all other situations, other feed methods can be used.

THE BOOSTER

Boosters differ from translators in that they retransmit the received station on the *same* FM channel. The advantages of this arrangement are that by using the same channel as the main transmitter in an area where the FCC already presumes coverage, there is less likelihood of interference to other full-service stations, and also that listeners don't have to tune their radios to a different channel as they move from one area to another in order to continue receiving the programming. One disadvantage is that the booster transmitter can cause interference to the main transmitter (and vice versa) in wide areas where the signals of both transmitters are similar in strength. Another disadvantage is that the simple feed method used by the translator shown in Figure 7.19-2 quite often does not work due to feedback from the output to the input, which in this case are on the same frequency. Despite these challenges, boosters are often the preferred and sometimes the only way to provide a usable signal to a blocked area in the main station's coverage contour.

DIFFERENT TYPES OF BOOSTERS

There are three types of boosters that are generally in use today, each with its own unique advantages and disadvantages. The first, the amplified bandpass booster, which is considered to be the "original" type because it was the first to be authorized by the FCC, consists of a directional antenna connected to a bandpass filter tuned to the primary station's frequency, followed by a high-gain amplifier that provides output power to a direction transmit antenna sufficient to provide fill-in service to the desired area. This is shown in Figure 7.19-3. The high-gain amplifier has an automatic level control that keeps the output power at a fixed level, and a device that shuts down the amplifier if the input signal drops below a certain threshold.

Besides being a simple design, the advantage of this type of booster is that the main and booster transmitters are always on the exact same frequency, preventing a destructive heterodyne that could be caused if the two they were on slightly different frequencies.

There are two major drawbacks to this system. One is that the input to the amplifier can often be as little as −60 dBm, and the desired output can be in excess of 40 dBm, giving a typical system a gain of more than 100 dB. With that much gain, it is easy to see that a disaster can result if the input does not have enough isolation from the output to prevent regenerative feedback that would result in wide areas of interference and no beneficial coverage. The other is that the input can be plagued by co-channel or adjacent channel interference or noise that will appear in the transmit signal if it gets through the bandpass filter.

An additional drawback that can be common to all boosters is that signals from the primary transmitter and the booster transmitter can arrive at a given location with similar signal strengths but at different times of arrival, resulting in severe multipath that can cover significant areas, including some that had no interference before the booster was established.

While many of these types of boosters have been successfully constructed and operated over the years, many have failed due to these problems. The FCC recognized this early on and soon allowed broadcasters to mitigate the feedback problem by choosing alternate feed methods, leading to the second type of booster, one that is fed by the same audio as the main transmitters, but not directly from the main transmitter's

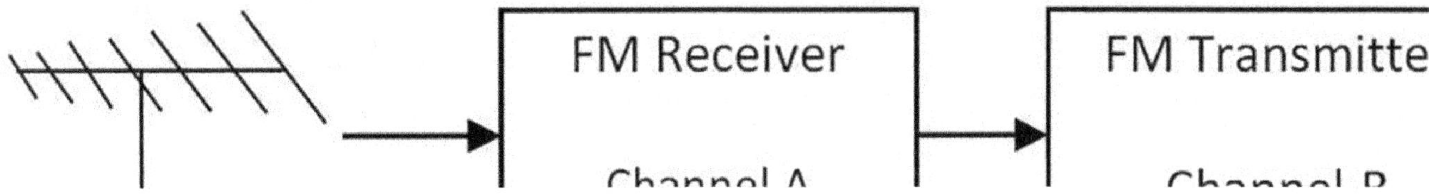

FIGURE 7.19-2 Block diagram of a basic translator showing the receive antenna on the left and the transmit antenna on the right

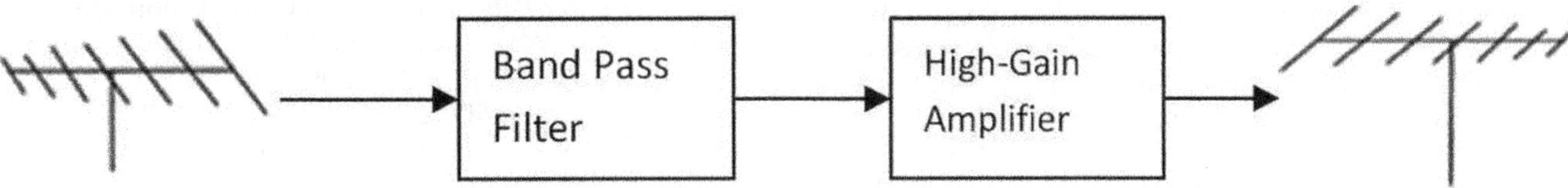

FIGURE 7.19-3 The amplified bandpass booster

over-the-air signal. There are many alternate feed methods that can be used with boosters as well as with translators.

The third and perhaps most interesting type of booster, the synchronous booster, is a more recent concept that was described in a paper by Stanley L. Salek, P.E.[2] that addresses the problem of mutual interference created by the main and booster transmitters in "overlap" areas where the signals are similar in strength. Such areas of interference in conventional boosters were often so pervasive that many boosters were ultimately shut down because they caused more loss of coverage to main transmitter's signal than they added.

Salek posited that these areas of interference were a result of the two transmitters' signals arriving at slightly different times where they were similar in strength, and that by delaying the signal of one of the transmitters so that both signals arrived at a targeted area at approximately the same time, the interference could be mitigated. His research further demonstrated that not only did the signals have to arrive at nearly the same time, they also had to have exactly the same carrier frequency, preferable phase-locked to a common source, and identical modulation characteristics. Although the first type of booster discussed, the bandpass amplifier, inherently has identical frequency and modulation because it is nothing more than a straight-through amplifier, there is no practical way to control the important time-of-arrival issue. Conventional alternately fed boosters can lend themselves to time-of-arrival control with the addition of equipment for that purpose, but the typical design does not allow for phase-locking the carrier to a common source or for perfectly replicating the modulation of the main transmitter. In the properly designed synchronous booster, all three factors are addressed.

Power, Coverage Permitted and Ownership

The FCC has very specific standards for effective radiated power and areas of coverage permitted for boosters and translators, as well as for their location and operation, which are covered under Part 74 of the Rules and Regulations.[3] The rules are nearly as complex as those for primary stations, and anyone considering an application for a booster or translator would be well served by reading the rules carefully and seeking advice from a qualified consulting engineer and attorney.

The FCC recognizes two distinct types of translators: fill-in and non-fill-in. Fill-ins can be used and owned by the primary station licensee if they do not extend the FCC-protected coverage contour of the station.[4] Non-fill-ins are used to extend the coverage of a primary FM station beyond its protected coverage contour into communities that they were not predicted to cover. For commercial stations, this type of translator may only be owned and operated by an entity with no financial connection to or support from the licensee of the primary station. There is an exception to that rule for the few areas of the country where there is no coverage from any FM or AM stations. Noncommercial stations may operate their own fill-in and non-fill-in translators. For all commercial licensees, only fill-in translators are permitted. Boosters are permitted for fill-in use only and may only be owned and operated by the licensee of the primary station. People or entities with no connections to primary stations may operate fill-in or non-fill-in translators for those stations, provided they have the express permission of primary station.

Fill-in translators may operate with a power of up to 250 watts, with no restriction on the height above average terrain (HAAT),[5] however they must meet certain requirements regarding contour overlap that are covered in the FCC Rules.[6] There are additional power restrictions for translator stations that are located within 320 km (199 miles) of the Canadian or Mexican border. Boosters, which may be used only by FM stations, are permitted to operate with an effective radiated power of up to 20% of that permitted for the class of the primary station. As an example, a Class B station is allowed a maximum of 50 kW, therefore a booster for a Class B would be allowed an ERP of up to 10 kW.

As with fill-in translators, there is no restriction on the HAAT of a booster as long as the predicted protected contour of the main station (57 dBu for Class B1 stations, 54 dBu for Class B stations and 60 dBu for all other FM stations) fully surrounds the same contour of the booster. There are no minimum amounts of power for boosters or translators.

Translators are not permitted to be used *solely* for the purpose of relaying a signal to another translator.

The FCC has provided an easy-to-read overview of the regulatory issues of boosters and translators complete with hypertext links to the relevant regulations on its website at www.fcc.gov/mb/audio/translator.html. This is an excellent resource for potential applicants as well as current licensees.

Cost Analysis

Determining if a translator or booster is economically feasible is an important first step, and several factors must be carefully assessed before moving forward.

What Is the Population to Be Served and Can You Legally Serve It?

Referring to the FCC Rules and Regulations and previous paragraphs, you can determine if you are eligible for a fill-in or non-fill-in station. With that in mind, you need to look at the areas that you want to serve and define them in terms of geographic area and population.

Potential for Revenue

For commercial stations, this can be estimated with help from the sales and programming departments who can help determine how the added signal will enhance the ability to compete. Will the added coverage be in your Arbitron market and add to your existing ratings, hence adding to your share of market revenue? Are there potential advertisers in the area? For non-coms, the mission must be considered and balanced against the cost. For those who are not licensees but who are looking to import programming into a community, how will you defray the cost?

Choosing the Right Method of Coverage

If you have identified the area you want to serve, how will you best cover it? Will you choose a translator or a booster? If you want a booster, will it be synchronous or nonsynchronous? High or low elevation site? High power or low? Directional or nondirectional antenna? Will you be on an existing site or building a new one? Will you be purchasing an existing translator or licensing a new one? If you purchase an existing translator, will you need to move it to a new location? Your plans may change as you move forward, but it's good to have a starting point to work from for the next steps.

Costs to Be Considered

At this stage, you should be planning the initial construction costs of the project and the ongoing expenses. The best advice is to use a spreadsheet program and get more specific with each item as you progress with the project. Here are some of the costs you should research:

1. *Engineering and legal costs*: These would include consulting engineers who will prepare the FCC application and its various exhibits, work by your FCC attorney, fees charged by the FCC for the CP and the license, and the cost of an engineer to do the installation and testing of the equipment.
2. *Cost of purchasing an existing facility*: If you are doing this, you'll need to talk to a broker or go direct to a current owner and determine if there is an interest in selling. Broker's fees as well as moving the facility if needed will have to be determined.
3. *Equipment*: This will be hard to estimate at this stage of the project because you may not know for certain the power level you will run, how you will feed the system with program material, what kind of antenna you will use and so on. Equipment costs can range widely based on many factors, and the best way to do an estimate is to look at a high range, a low range and a reasonable range in between. Update as your plans change.
4. *Permitting and land use*: Every local jurisdiction will be different, and costs will range widely based on certain land use factors even within a particular jurisdiction. The local zoning board might allow a large tower in an area that's zoned for industrial use with just a simple application, but they might require some very expensive environmental reports, special remediation work and perhaps a public hearing in an environmentally or esthetically sensitive area. It's a good idea to contact a local land use consultant and learn what you might be dealing with. This may heavily influence where you will locate the station, or even if the project is feasible.
5. *Construction:* Anyone who has had home improvement work done knows the old adage about construction costs: "make a budget, then double it!" You will need to consider the cost of bringing in utilities and connecting to them, work that may be needed for access and security such as roads, fencing and gates, the building or enclosure you intend to use, HVAC, electrical, tower costs including soils reports, foundation and erection of a new tower or improvements that may be needed for an existing tower, land improvement such as grading, trenching, drainage, tree removal and so on.
6. *Ongoing expenses*: You should examine recurring costs such as rent, utilities, insurance, equipment maintenance, land and tower maintenance, local and state taxes and FCC annual regulatory fees.

Justifying the initial cost and on-going expense: Now that you have defined an area that you want to serve, and you have a working idea about construction and ongoing costs, the next step is to justify the costs to management. For noncommercial operations, this may be a simple as showing the cost per additional potential listener that will be added to the service area.

For commercial operations, you should collect as much data as you can from sales, programming and management people about how better coverage can be used to generate revenue. The departments should work collaboratively with the goal of estimating increased cash flow. There is certainly going to be a lot of guesswork here, but the final decision-maker will need some well-thought-out projections on which to base a decision.

To help each of the departments in making their projections, you should prepare coverage maps and population data that depict the coverage before and after the addition of the translator or booster. If possible, these should be based on "real-world" signal projections[7] rather than on FCC contours. Put the maps and other information you have gathered into a presentation to help you explain the benefits and drawbacks of adding a translator or a booster.

The final decision to go forward will most likely be made on the basis of a "capital budgeting project," which is a way of comparing the costs to the benefits in an objective manner. Some of these evaluations can be made on a simple determination of how long it will take to amortize the initial cost with the new cash flow generated. Another quick calculation is to determine the net annual cash flow increase, then multiply that by the "cash flow multiples" that are used in determining the valuation of a station. If that number is significantly greater than the initial costs, plus any negative cash flow in the early months or years of operation, the project will have a better chance of gaining approval.

A more formal method of capital budgeting used by some managers is "risk adjusted net present value," which is based on the proposition that a dollar in your hand today is worth more than a dollar you hope to get in the future. A manager using this method will look at money inflows and outflows, including such items as depreciation and taxes, for each of several years with the assumption that the project will end and the equipment will be liquidated after the final year. The net cash flow for each year is *discounted*, along with the liquidation value of the equipment, to make it comparable to value of money today. This discount is determined by adding both the projected cost of borrowing money and a factor to adjust for the risk that the investment might not reach its goal. These "adjusted cash flows" are then added to find the risk adjusted net present value of the project. If that value is greater than the initial investment, the project is considered to be viable. You should understand that managers are presented with many proposals aimed at increasing the value of the station, but money is limited, and the manager will choose the ones that will bring the greater return on investment for the available capital.

Placement Guidelines for Translators and Boosters

Boosters operate on the same channel and within the existing coverage of the primary station, but finding a channel for a translator station will be a challenge in most populated areas of the country.[8] The charts shown in Section 74.1204 of the FCC Rules and Regulations show the various contours that must be protected by a translator station. There are several computer programs that can plot these contours and help you determine if there is a channel available to you to serve a particular area, and if so, where it can be located and what the boundaries of your coverage contour will be. Boosters have few restrictions with regard to other stations, except that their signal contours must be 6 dB below the protected contours of any existing first-adjacent channel station.[9]

Figures 7.19-4, 7.19-5 and 7.19-6 show how a fill-in translator can be located to serve a community within the FCC protected contour of a primary commercial FM station. In this example, the gray areas in Figure 7.19-4 shows the actual 60 dBu Longley-Rice projected coverage of the primary station, while the black line shows the FCC F(50,50) 60 dBu coverage contour. A highway running through the community is also shown.

A carefully placed translator can be constructed by the licensee to serve this community if it does not extend the 60 dBu FCC coverage contour, and does not cause a prohibited contour overlap with any existing station. In Figure 7.19-5, the 60 dBu F(50,50) FCC contour of a proposed translator on a nearby hill is shown, the contour that could interfere with a first-adjacent station, the first-adjacent station's protected contour and the actual area of 60 dBu or better signal as determined using a terrain-sensitive, Longley-Rice-based mapping program. The FCC contour plots show that the proposed translator would meet regulatory requirements, and that the actual coverage, shown by the gray area, would serve the community, as well as its surrounding areas and portions of the highway.

In Figure 7.19-6, the combined coverage of the primary station and the translator are shown in the gray areas.

The selected site should provide good line of sight and a good signal intensity to the population to be served, but must not have prohibited contour overlaps. The parameters that need to be evaluated in reaching these three goals are:

1. Topography
2. Elevation
3. ERP (effective radiated power)
4. Directional pattern
5. Co-channel and adjacent channel interference to be overcome.

In most cases, translators have a maximum power of 250 watts ERP[10] and boosters can use as much as 20% of the ERP allowed for the class of the primary station, but if there is a contour overlap, there will be a trade-off between elevation and ERP that may play a significant role in the signal intensity in the targeted area. Is some situations, a lower elevation site, closer to the targeted area with a higher ERP may produce a more usable signal than a very high site with lower ERP even though the FCC coverage contour of the

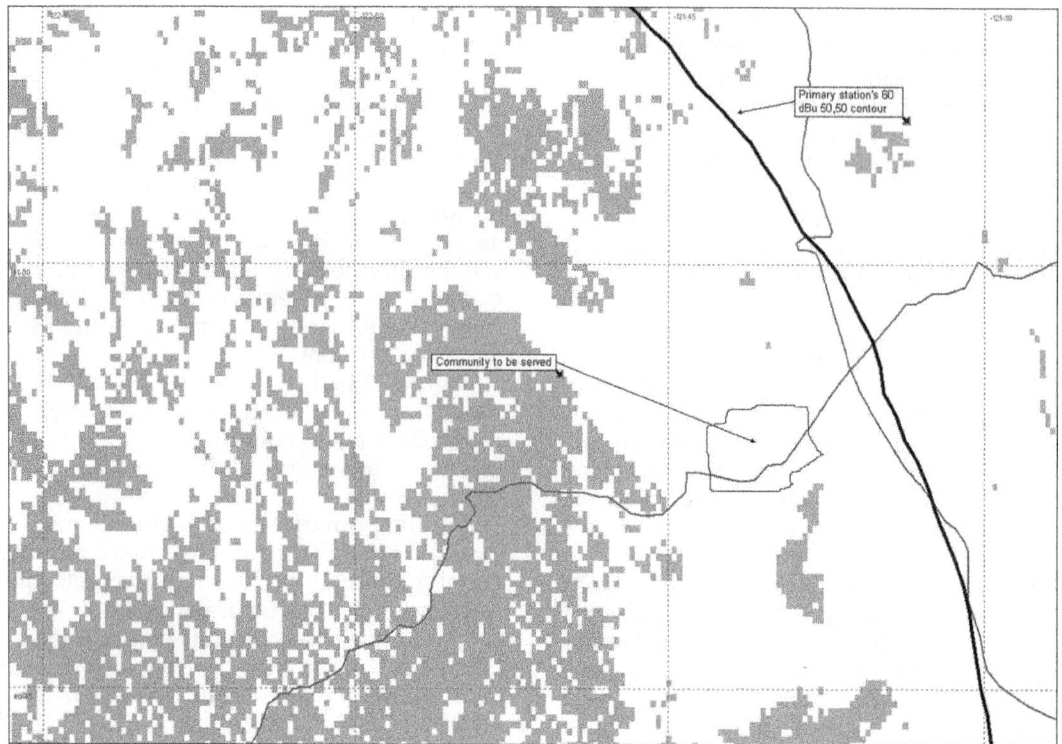

FIGURE 7.19-4 Map showing "primary station" actual 60 dBμ Longley-Rice projected coverage (gray shaded areas) versus the conventionally predicted 60 dBμ contour for the station

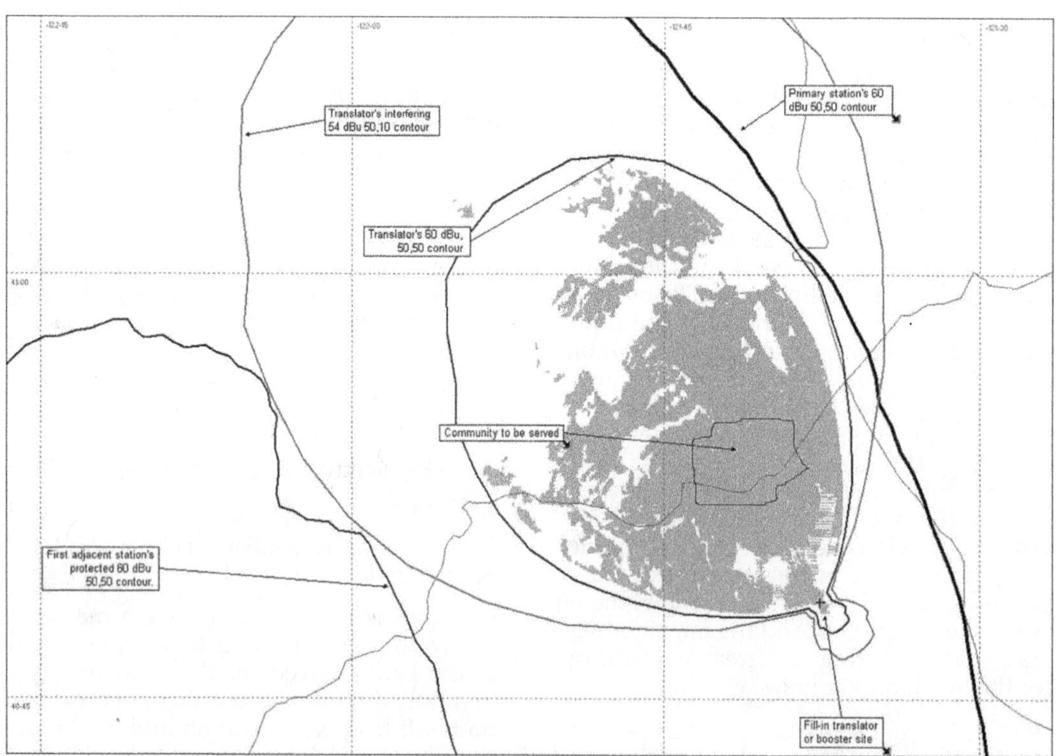

FIGURE 7.19-5 Map showing the 60 dBμ F(50,50) FCC coverage contour of a proposed translator, along with the contour from the translator that could interfere with a first-adjacent station, the first adjacent station's protected 60 dBμ contour and the actual area of 60 dBμ or better signal as determined using a terrain-sensitive, Longley-Rice based prediction program

higher site may appear to extend the signal farther. In other situations, the exact opposite may be true. Terrain sensitive propagation mapping programs are very valuable in comparing sites that may be candidates, and it is a good idea to model as many diverse sites as possible if the software is available before making a final choice. Such programs, which are described in Chapter 1.8, can also be used to determine if a line of sight exists for any signal path that may be needed to feed the program material to the site.

In the case of boosters, an important area of concern is signal overlap between the primary and booster signals. While this overlap is generally legal (unless it causes interference in the city of license), it is undesirable and unfortunately unavoidable in most cases. Due to the "capture effect" of FM receivers, only the stronger of the two signals should be heard by a radio in the overlap area. In theory, this is true, but in practice, one signal generally has to be 15–20 dB greater than the other for it to be "captured" to the point where the weaker signal causes no noticeable interference. The more similar in strength the signals are, the greater the potential for interference.

The best general advice is that the booster signal should be "cut off" from the primary signal by terrain, directional pattern and/or ERP to result in as little overlap as possible, and that as much of the overlap areas as possible should be over sparsely populated or traveled areas. Interference in some small areas of overlap that fall over highly populated or traveled areas can, in many cases, be mitigated by use of a synchronous booster system. Plotting the critical time of arrival versus signal intensity difference issue is best done using terrain-sensitive propagation mapping techniques described in Chapter 1.2 and by actual signal measurement techniques that are covered in more detail in Chapter 10.6.

SITE SELECTION

Selecting a site for a translator or booster can be as challenging as selecting one for a primary station. As you get to the stage of choosing the site where you will finally locate the station, there are several important factors that should be considered.

Obstructions: Unless you are planning to use a satellite or land-line feed, you will need a good line-of-sight path from the primary station transmitter, or from some alternate feed source such as a 950 MHz or 5 GHz link. Make sure there are no terrain features, trees or buildings that will cause blockage or multipath of the transmitted signal or of the received feed signal.

Accessibility and security: Will you be able to get to the site when you want and will you be able to keep unauthorized people out?

Nearby RF sources: Are there any nearby RF sources that could prevent reception of the feed signal or mix with the transmitter signal to produce intermodulation distortion? If you are locating at an existing communications site, this could be a serious issue. FCC-mandated coordination is required for 950 MHz links, but a transmitter operating in 850 MHz cellular band at the same site requires no coordination and could require filtering to prevent receiver overload.

Power and telco: Is there existing power at the site? If not, you will need to investigate the costs and problems of getting it. The same is true of telephone, T-1, cable or fiber lines if your system will require them.

Land use issues: If you are proposing to build a new site in an environmentally protected area or in a residential area, you may run into problems getting the permits needed from local authorities. Many elevated areas that make excellent transmitter sites have been designated as "scenic resource areas" and may be off limits or require significant visual mitigation and/or expensive environmental impact reports (EIRs). Check the applicable zoning regulations, or better yet, hire a zoning consultant who has experience with the local authorities. *Know Thy Neighbors* . . . they may consider your project to be an eyesore! You may need to get variances, which may require hearings and notification of nearby property owners. In situations like this, a little public relations goes a long way, and both neighbors and community leaders are more likely to respond positively to your project if they have been included in the planning from the early stages.

THE RIGHT ENVIRONMENT FOR YOUR EQUIPMENT

A translator or booster generally occupies less space, gives off less heat and requires fewer visits than a main transmitter, therefore the shelter needed can be smaller and significantly less elaborate than a main transmitter site. Many creative solutions have been used to house translators and boosters that represent a good compromise between utility and cost. If you plan on locating at an existing communications site, most of the issues relating to shelter will already be addressed, but if you are building a site from scratch, you will need to consider the following:

Type and size: Buildings can be of cinder-block, wood, fiberglass or metal. They can be prefabricated housings such cargo containers or sheds. Don't overlook any existing structures such as barns, garages or other buildings. Depending on the local environment, you may only need a small weatherproofed rack, such as the one shown in Figure 7.19-7.

Temperature control: If the climate might reach extremes, heating or air conditioning may be needed in addition to reasonable ventilation to stay within the equipment manufacturers recommended limits. If there will be more equipment added later, plan ahead for the additional heat.

Dust, dirt and moisture: These culprits can cause failures and shorten equipment life. Make sure you have adequate filtering and change or clean filters as needed.

Power issues: Even if there is reliable power at the site, you may want to plan for short-term outages

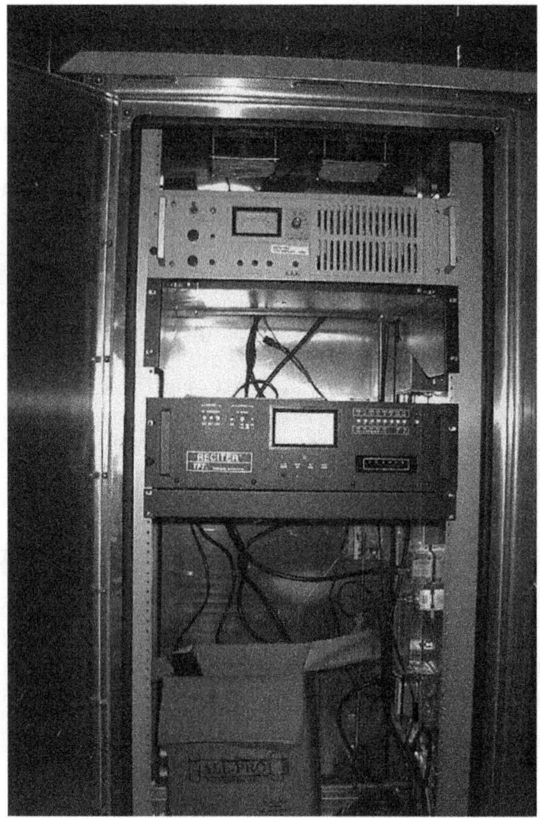

FIGURE 7.19-7 A weatherproofed rack, surrounded by an 8-foot redwood fence, provides shelter for this synchronous booster and 250-watt amplifier. Because the station was located in an area with a mild climate, only the two small fans on the ceiling were needed to control the temperature.

Ron Castro

or occasional "brownouts" with a UPS system, or longer-term outages with a generator.

Grounding: Consider the three major types of grounding: safety, RF and lightning. Safety grounding in AC power circuits is usually spelled out in local building codes, which should be followed. RF grounding can be important if the equipment is located in a high RF field and especially if is located at or near an AM broadcast site. Lightning grounding is an extremely important and often overlooked area that should be carefully researched, designed and implemented. Surge suppression equipment should be installed on incoming AC power lines and telco lines and transmission lines should be grounded at a common ground "window" at the building utility entrance that is connected to the racks and equipment.

Critters: Crawling, flying, walking, biting, stinging and generally making a mess! They can be avoided by using weathertight, sealed entrances for transmission lines, power and telco, as well as for doors and windows. All air ventilation should be filtered. Rats can get in through large conduits, so the conduit openings should be stuffed with stainless steel or copper wool.

Feed Methods for Translators and Boosters

Once the site for the translator or booster has been chosen, the next challenge is to determine how the program signal will get from the studio to the site. There are a variety of methods to choose from.

Direct Feed: If the chosen site is in line of sight from the main transmitter, a feed system consisting only of a high-performance receiver with its composite output connected to the translator transmitter's composite input is a good candidate, and there are manufacturers who make transmitters with a high-performance receiver built into a single chassis for just such a purpose. A form of direct feed can be used for FM+HD translators or boosters, however a sophisticated digital up-conversion system is needed.[11] Nonsynchronous analog boosters can simply use a selective set of filters to feed a booster transmitter if there is adequate isolation between the receive and transmit antennas. This is the most economical method, but the main transmitter's signal at the site must be of adequate signal strength and free of multipath as well as co-channel and adjacent channel interference and noise. While AM stations using FM boosters can use an AM receiver feeding audio to a monaural FM transmitter, a better way is to feed stereo audio from the board to a processor and stereo generator using one of the following methods.

A 950-MHz "Single Hop" Analog or Digital Feed: If the studio has a clear line of sight to the translator or booster site, a feed consisting of a 950-MHz STL transmitter at the studio and a matching receiver at the site is a good choice. A composite analog system can be used, or if a stereo generator is installed ahead of the translator or booster transmitter, a discrete channel analog or digital system can be used.

"Multi-Hop" Feeds: Often terrain or other obstacles will block a line-of-sight signal from the studio to the site, but if an intermediate site is available that has line of sight to both locations, the 950-MHz signal can be "translated" to another 950-MHz channel and sent on to its destination. These STL translators work well provided the transmitted signal does not overload its receiver. Another creative method is to locate a site with line of sight to the translator or booster site that also receives a good signal from the main transmitter or the studio. The main station's signal, or a studio STL signal can be received and retransmitted on 950 MHz to the translator or booster site using only a single hop.

IF Translators: A typical 950-MHz translator uses a receiver with a composite output feeding a 950-MHz transmitter on another frequency. This system works well, however it can make subtle changes in the modulation that can be detrimental in the case of a synchronous booster system. This problem can be mitigated with the use of an STL translator that works by heterodyning the received signal down to an IF frequency, which is then heterodyned back up the output frequency in the transmitter, thus producing an exact replica of the received signal. At present, no known manufacturers are supplying such equipment, however older systems using this method may still be encountered on the field.

Land-Line and Spread Spectrum RF Feeds: Terrain, distance or overcrowding may make use of the 950-MHz band impractical. In that case, a land-line solution may be a viable alternative. Because equalized and phased pairs of 15-kHz analog phone lines are rarely available at a reasonable cost anymore (if at all), digital transmission using cable or fiber IP, ISDN, Fractional T-1 or T-1 might be the right choice. These methods will cost more initially for equipment and will incur an ongoing cost, but both of these costs have been trending downward in recent years. Less expensive alternatives such as DSL and wireless broadband have in the past been considered too unreliable for full-time delivery, however reliability has been improving, and with adequate buffering to mitigate drop-outs, they may be worthy of consideration.

Point-to-point spread spectrum equipment operating in the unlicensed 2.4 GHz and 5.8 GHz UNII (Unlicensed National Information Infrastructure) and ISM (Industrial, Scientific and Medical) bands are being successfully employed in many feed systems, however, because they are unlicensed, anyone can turn on equipment using the same frequencies and cause interference. Many consumer products, such as cordless phones and wireless LANs, share these frequencies as well as many Internet service providers who cover wide areas with service on the same channels. The advantage of using this method is that a very wideband digital link, capable of delivering much more than just two or four uncompressed, high-quality digital audio channels can be established with no need to obtain a license. Several manufacturers are offering reliable, easy-to-implement systems that can cost as little as a few hundred dollars for hardware and no ongoing costs. These are well worth the time to investigate.

Satellite Feed: In cases where a single program may feed multiple translators that are separated from the main transmitter by long distances or challenging terrain, feeding by way of a satellite link may be an option. The cost of satellite transmission is high and varies widely based on many factors such as bandwidth and signal strength. It is rarely a cost-effective solution except where many translators are being fed from a common source.

RECEIVING THE FEED SIGNAL

Assuming that the translator or booster will be fed by an RF method, several important factors must be considered during the design phase of the project.

Line of Sight: This can be verified with visual observation, but often distances require that computer mapping programs be used to verify that there are no obstructions.

Predicting the Signal Strength: Whether picking up the signal on the FM band or at 950 MHz or above, a study should be conducted to determine the received signal strength. For the FM band, the existing signal at the proposed site can be measured with a field intensity meter, or it can be predicted by computer propagation analysis software. For feeds at 950 MHz and above, there are programs, some available free on the Internet, that can predict received signal strength based on variables such as transmitter power, type and length of transmission lines, antenna gains, path distance and Earth curvature.

Antenna Types: For FM signal feeds being received "off-air" on the main transmitter's frequency, directional Yagi or log periodic dipole arrays are most often employed. Choose one with adequate gain, side and back rejection at the operating frequency to mitigate interference and front-end overload. Select one that is sturdy enough to handle the environment at the site and plan to place it at an elevation high enough to receive a solid signal, but not so close to the transmitting antenna that front-end overload becomes a problem.

Interference From Near and Far: A properly coordinated 950-MHz link should be free of interference from co-channel and adjacent channel signals if well constructed, however nearby transmitters far removed from the receiving frequency can cause front-end overload to the receiver. This is most prevalent when sharing a site with multiple users or when operating an STL translator, but can also be caused by paging transmitters or cell sites in the area. The potential for interference from these services can be reduced by using bandpass filters on the receiving frequency alone or in combination with band-reject filters on the frequencies of the offending transmitters.

Off-air FM signals are mainly affected by the transmitted signal from the translator output, and by strong signals from other nearby FM transmitters. Bandpass and band-reject filters on the receiver can usually cure this problem. Often the received signal from the main transmitter is weak enough to require a low-noise preamplifier. If a preamplifier is used in a high RF environment, it should always be placed *after* a set of filters, just ahead of the receiver.

Isolation: Wherever interference from nearby RF sources might be problematic, increased isolation of the receive antenna from the source of radiation can help reduce those problems. Consider both vertical and horizontal separation, positioning the antennas in each other's null, cross-polarization or placing them on opposite sides of obstructions such as buildings or hills if the situation allows. Avoid bundling receiving transmission lines with those used by transmitters on the same tower.

Transmitting Antennas for Booster and Translators

Once you have determined the proposed area of coverage and have chosen either a nondirectional (ND) or directional antenna (DA), determining the specifics of the antenna is the next job. Here are some areas to consider:

The tower: How much "vertical real estate" will you have, and what is the wind-load capacity of the tower? If you are building a new tower, you'll have more choices although economics may dictate the ultimate design of the antenna. If you are using space on an existing tower, you are likely to have some limitations. It's wise to seek the advice of a competent, state-licensed structural engineer and to have the final design analyzed and wet-stamped. Municipal jurisdictions, tower and property owners as well as insurance companies may require such documentation.

Polarization: Boosters and translator antennas may be vertically, horizontally, dual or circularly polarized. There is no true consensus as to which type performs better, although the majority of systems are vertical. While circular polarization is now nearly universal for primary station antennas, choosing either vertical-only or horizontal-only will require less transmitter power, less antenna hardware and less wind load than circular or dual polarization.

Stacking antennas: Both DA and ND antennas can be "stacked" vertically to narrow the beam pattern in the vertical plane, which increases gain and requires less transmitter power for a given ERP. Stacking also mitigates radiation toward the ground, which may be an issue if the elevation above ground is low or a multi-user site is already close to its radiation limit.

ND antennas are not much different than primary station ND antennas in how they operate, although they can most often be designed to cost less by reducing their power-handling capability. Most antenna manufacturers sell inexpensive, low-power "educational" antennas that may suffice.

Directional antennas, which are most commonly used in boosters and translators are of two basic varieties:

The parasitic array: Referred to as a "Yagi" or "Yagi-Uda" array, these usually consist of a horizontal boom with a dipole element to which RF is fed, a passive "reflector" element behind the dipole and usually one or more "director" elements in front of the dipole, which when all combined cause the antenna to radiate the signal in a unidirectional beam. All of the elements are mounted on the boom. The elements can all be vertical or horizontal, or both can be combined on the same boom for circular or dual polarization. The advantage of a Yagi is that a lot of gain, consistent with a narrow beam pattern, can be achieved in a relatively small amount of space. The disadvantages of Yagi arrays are (1) that they must be designed and tuned for one single FM channel, (2) the front-to-back ratio of the patterns are not as high as LPDAs and (3) that nearby conductors, such as other antennas sharing the tower, can cause the patterns to distort by detuning the parasitic elements.

The LPDA: "Log periodic directional array" antennas look similar to their Yagi cousins, but they consist of a series of dipole elements on a boom, giving it a distinctive "fishbone" look. RF is fed to the front dipole in the array, and each dipole following dipole is tuned to a successively lower frequency and fed out of phase with the dipoles before and after it. The combination of parasitic reradiation and phase relationship of the dipoles that are nearest to resonance at the operating frequency cause the array to have a unidirectional beam pattern. Although the LPDA has less gain for a given boom length than a Yagi, the design is inherently broadband, and a typical design can cover the entire FM band with no tuning, low VSWR and little change in gain or pattern. It is not as susceptible to pattern distortion from nearby antennas as the Yagi is. It also has a very high front-to-back ratio, which is desirable in certain situations that require protection of another station. LPDAs are made for vertical or horizontal polarization only, however vertical and horizontal arrays can be arranged on a tower to deliver circular or dual polarization.

It is possible for multiple boosters or translators using the same site and covering substantially the same area to share a single broadband antenna with the use of a system of tuned cavities called a combiner. Conversely, a single translator or booster transmitter can be used to serve more than one geographic area from a single site by splitting the signal into multiple beam antennas pointed in different directions.

Propagation and Coverage Modeling for Boosters and Translators

Longley-Rice-based, terrain-sensitive mapping computer programs have taken much of the guesswork out of planning a successful translator or booster.

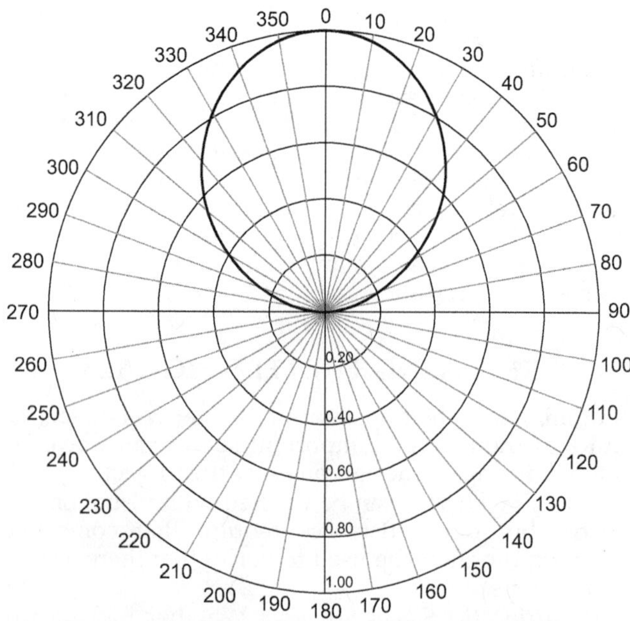

FIGURE 7.19-8 The azimuth pattern of a commercially available 8-element vertically polarized LPDA. On the right, a circularly polarized 5-element Yagi.

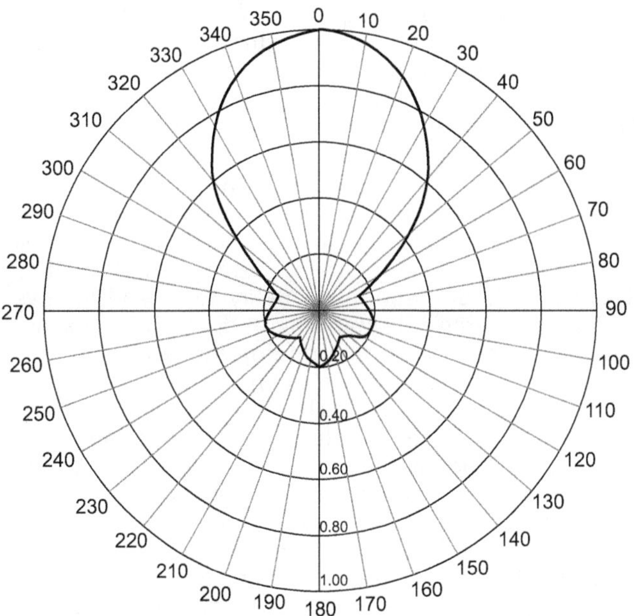

FIGURE 7.19-9 The azimuth pattern of a commercially available, circularly polarized 5-element Yagi

A detailed discussion of these programs can be found in Chapter 1.2. Here are some specific applications of these programs useful to anyone planning a translator or booster project:

Analysis of FCC contours: In planning a booster or translator, it will be necessary to determine if there will be any prohibited overlaps of coverage contours using the FCC's contour prediction method.

Analysis of directional antennas: By applying manufacturer's specifications of directional antennas being considered for a project, along with ERP and elevation at a particular set of coordinates, a pattern can viewed both in terms of FCC predicted contours and "real-world" signal coverage.

Comparative analysis of sites: Often several sites may be considered. Using a propagation analysis program, the best suited site can be chosen.

Preparation of FCC license application exhibits: If you prepare your own application, you will have to demonstrate compliance with certain requirements, especially where there is potential for interference to another station. The FCC uses such programs to check your application, and if you have done your homework, you can avoid having your application returned, delayed or dismissed.

Evaluation of interference zones and timing issues for boosters: This is essential to get the best performance and minimize interference in a booster system. Most programs will allow you to see graphically what will happen as you adjust the directional pattern and relative timing and signal between the primary transmitter and the booster.

Checking for line of sight for the input feed: A path free of obstructions is necessary for any system using an RF feed for the program input.

Are such programs accurate? There are many caveats to the use of Longley-Rice-based propagation programs, particularly in regard to the predicted signal strength once an obstruction has been encountered by the signal. Because the purpose of a translator or booster is to bring coverage to areas where coverage is less than desirable, determining just how bad existing coverage is may in some cases stretch the reliability limits of such programs. In the case of boosters, it is important to know where the signal strengths of the primary and booster stations overlap, and in cases where one or both signals traverse urban areas, hilly or mountainous terrain, it may be necessary to verify the signal conditions by field measurement, using a mobile, GPS-based data collection system as discussed in Chapter 10.6. Propagation analysis programs can help in displaying the collected signal data and comparing that to the predicted signals. Most consulting engineering firms can arrange to have this type of analysis done for you.

Booster Interference Zones

A common problem that can accompany the use of an on-channel booster station is interference it causes to the primary station. If all receivers had perfect capture ratios, this would not be a problem, because the receiver would demodulate only the strongest of the two signals on the channel and completely reject the other, but in the real world, capture ratios on common consumer radios are far less than perfect. Any area where the two signals are near to each other in strength is likely to be an interference zone. Interference zones can be mitigated to some degree if we can control the three parameters of the signal that cause interference: carrier frequency difference, modulation difference and time-of-arrival difference. If any of these three are not addressed in the design of the booster system, interference between the two transmitters is almost certain to occur at some location, particularly if the booster is desired to have seamless coverage between its service area and that of the primary transmitter.

The first job in mitigating interference zones in booster station design is to design the system from the beginning in such a way as to minimize them, or to cause them to fall over unpopulated areas. Careful use of Longley-Rice-based, terrain-sensitive propagation mapping programs can help in this job, as well as a thorough knowledge of the local terrain. Here are areas that should be investigated during the design phase:

Terrain shielding: The most effective way to separate the two signals is to use the terrain to block one signal from interfering with the other. This may mean locating the booster at a lower elevation to limit its coverage back into the coverage area of the main.

Directional antennas: You may need to try several antenna designs and test them in the propagation software at various azimuth angles to get the null in a position that protects the primary station.

Power levels: The tendency is to run the maximum amount of power that can be authorized, but in many situations, less is better. The converse can also be true: running a higher ERP can move an interference zone over a less important area.

Synchronization: If the interference zones have been minimized, precise synchronization of carrier frequencies can abate the undesirable heterodyne that results from two transmitters on slightly different frequencies. Synchronizing modulation such that each station's frequency swings precisely the same amount and in the same direction as the other station's at all modulation percentages will reduce the multipath-like noise that results from random differences in unsynchronized systems. The third parameter of synchronization involves controlling the time-of-arrival of the two signals such that they arrive simultaneously at a target area where the signals are nearly the same strength and where a good quality signal is desired.

Synchronizing carrier frequency and modulation alone will reduce mutual interference problems; however, combining that with time-of-arrival control for full synchronization is the gold standard of booster systems. It's easy to see why interference can result from two FM stereo signals of the same strength as they arrive at slightly different times and combine in a receiver. The top of the upper sideband of the stereo subcarrier is about 53 kHz. The time it takes for one cycle at 50 kHz is 20 µS, so a signal 90° out of phase requires only a 5 µS delay, and is enough to cause slightly audible distortion in the stereo signal due to the effect of phase cancellation. At 10 µS delay, the 50 kHz signal is 180° out of phase, and now 25 kHz is 90° out of phase. The distortion gets progressively worse as the difference in time of arrival of the signals at a certain location increases.

If we set a design goal of keeping the delay below 5 µS where the signals are equal or within ± 5 dB, we can keep the distortion to minimal levels. As the signals depart in level, phase cancellation is reduced and the delay can increase on a sliding scale without causing significant interference until the signals are different by approximately 15 dB. In the range of ± 5–15 dB, a reasonable design goal is to keep the delay to less than 10 µS. When the difference in signals is greater than 15 dB, phase cancellation is minimal and the capture effect of the receiver will reduce the interference to the point where time of arrival is no longer an issue. Note that every receiver will deal with the two signals differently due to differing capture ratios, but the preceding rule of thumb has worked successfully in the field.

How much of an interference zone area can time-of-arrival synchronization mitigate? Considering that a radio wave travels just under 1 mile in 5 µS, and that if you are in the middle of two signals each traveling toward you at the speed of light, the synchronization area for two signals that are within ±5 dB can be as small as one-half mile in either direction, or 1 mile total. On the other hand, if the primary transmitter and the booster transmitter both line up exactly in the direction of the interference zone to be synchronized and that zone is not located between the two transmitters, the synchronization zone can extend indefinitely. If the vector differential of the distances from the primary and the booster transmitter to a target are less than 1 mile, time-of-arrival difference can be synchronized to the point of insignificance.

Time-of-arrival synchronization is accomplished with digital delay devices that can be added onto analog or digital booster systems. The process can also be performed in software in some digital exciters. Determining the amount of delay for a specific system is an exercise in mapping and requires choosing a target location and determining the total difference in time-of-arrival delay *including delay caused in the STL paths* in each signal. The delay is inserted at the transmitter with the shortest total path to the target, allowing the other transmitter to "catch up," and the amount of delay is generally set to 5.37 µS for each mile difference in the distances to the target. Fine tuning the delay time is done by tweaking it for minimum distortion on the stereo pilot at the target.

Synchronization of the other two parameters, the carrier frequency and modulation, is inherent in an off-air fed system that simply amplifies the primary stations' signal, making no significant changes to it. This can also be inherent in certain digital designs in which identical, pre-processed digital audio streams are fed to identical digital exciters that are both phase-locked to a common source, such as GPS. With analog exciters, the most common method is to phase-lock the master oscillators of both the primary and the booster transmitters to the 19 kHz stereo pilot. If this method is used, care should be taken to ensure that the pilot frequency is always within ±2 PPM.

Modulation synchronization in the analogue domain requires a complex system to ensure that both transmitters are fed with and transmit identical modulation. This can best be achieved by feeding both exciters with 950-MHz signals from a single STL transmitter that are then heterodyned down to the operating frequency. Alternately, a signal fed to either the primary or booster exciter by any feed method is extracted from the IF stage and heterodyned to 950 MHz for transmission to the other site, where it is then heterodyned back down to the operating frequency.

CONCLUSION

Translators and boosters can bring important benefits to many FM broadcast stations but are often overlooked due to their complexity and their low priority in relation to overall station operation, but as communities grow, populations move and competition heightens, station managers are increasingly considering their value. Now that AM stations are permitted to use FM translators within their coverage areas, an entire new avenue of support for AM operators is available.

The engineering department is usually charged with the responsibility of investigating, designing and constructing a system, but often lacks the time, the tools and the practical experience to maximize the benefits a booster or translator can bring. That is why it is important to seek advice from experienced technical advisors who can help organize the project and work with the local engineer and management to see the project through to a successful completion. As with any project, planning is the key to success, particularly in these areas:

- Regulatory
- Financial
- Environmental
- Structural design
- Electrical design
- Coverage prediction and maximization
- Interference prediction and mitigation
- Construction
- Final evaluation.

The future of radio is as bright as it has ever been, and translators and boosters will be an important factor in helping us bring new and exciting programming and technologies to millions of listeners and to better serve our communities.

Notes

[1] K. Bullington, "Radio Propagation at Frequencies Above 30 Mc/s", *Proc. IRE* 35, 1947.

[2] P.E.S. Salek, "Analysis of FM Booster System Configurations" 1992 NAB Engineering Conference. www.h-e.com/pdfs/ss_nab92.pdf.

[3] A summary of FCC rules can be found on the FCC's website at www.fcc.gov/mb/audio/bickel/part74rule.html.

[4] AM stations may use FM fill-in translators provided the 60 dBu contour of the translator does not extend the 2.0 V/m contour of the AM primary station, *and* the FM translator is located within 25 miles (40 km) of the AM station's transmitter site.

[5] Non-fill-in translators may be required to operate at lower power as described in Section 74.1235(b).

[6] See Section 74.1204 of the FCC Rules and Regulations.

[7] For a detailed look at real-world coverage maps, see Chapter 1.2.

[8] The FCC generally only accepts applications for translators during "windows" of time that happen infrequently. As of this writing, two windows are tentatively planned for AM station licensees to apply for FM translators. One window would allow an AM licensee to purchase an existing FM translator or Construction Permit and move it to any channel and up to 250 miles. The second window would allow applications to be submitted by AM licensees for new FM translators. Any licensee or private citizen may purchase an existing translator or CP at any time, although channel and geographic moves are limited.

[9] An FM licensee may apply for a booster construction permit at any time, and such applications are routinely granted.

[10] Non-fill-in translators are limited in power based on HAAT as described in Section 74.1235(b).

[11] A full discussion of FM+HD or HD-only translators and boosters is beyond the scope of this chapter, however, an excellent article "Extending Your HD Radio Footprint" by Geoffrey N. Mendenhall, P.E. delves into this topic. The article can be found at www.gatesair.com/products/transmit-radio/fm-transmitters/flexstar-hdx under Product Literature.

CHAPTER

7.20

HF Shortwave Radio

J. FRED RILEY, JD, MBA
W8OY

JOHN O. STANLEY
K4ERO

INTRODUCTION

HF Broadcasting Definition

HF (High Frequency) broadcasting uses the 2–26 MHz range of frequencies—from just above the MW (Medium Wave) AM (Amplitude Modulation) broadcast band to near VHF; three octaves of frequencies coverage.[1] The 2–5 MHz section is generally denoted Tropical Band. The transmissions are typically conventional AM.

General Introduction to HF Broadcasting

HF broadcasting is different from conventional AM on the MW (Medium Wave) broadcast bands. Everything about HF broadcasting is dependent on propagation via skywave. Domestic MW in the days of the clear channels was dependent on skywave and ground wave at night but now is primarily a ground-wave service. HF skywave propagation is dependent on frequency, antennas, sunspot cycles, seasons, and other changing conditions. These propagation changes require that the transmitters for HF broadcasting be frequency agile, the antennas be able to cover many frequencies with predictable performance and that the antennas may need to change configuration as conditions change.

HF antennas are generally horizontally polarized and are balanced. (All modern HF transmitters use single-ended PA stages requiring a BALanced-to-UNbalanced—BALUN—transformer somewhere in the system.) Antennas may have beams that are aimed straight up so they bounce off the ionosphere to cover a fairly local region. They may have take-off angles that vary from a few degrees above the horizon to moderate angles depending on frequency. The antennas for international broadcasting for long distances are typically broadband curtain array composed of many broadband folded dipoles fed in parallel that cover a full octave of frequencies. Most of these curtain antennas also divide the columns of the dipole arrays. By inserting time delays into the RF paths to the columns the direction of the beam can be bent (slewed) left or right. Slews of ±30-degrees are typical on larger arrays, often in 10-degree steps.

International broadcasting at medium distances may use fixed or rotatable Log-Periodic arrays that have broader horizontal beams and have take-off angles significantly above the horizon, 30–60 degrees.

Transmitter power for government-owned international HF broadcasting was almost universally 250 kW for many years until 1985. An MIT study commissioned by the VOA in the early 1980s looked at the practicality of increasing transmitter power from 250 kW to 500 kW. This study concluded that 500 kW—despite the expense of operation, transmission line and antenna modifications—was a reasonable upgrade given the presence of jamming. Although some countries persist in its use, jamming operations have steadily decreased. But man-made noise has increased over the intervening years—particularly in urban areas—so much that the increase in power from 250 kW to 500 kW is necessary on some circuits just to deliver the same signal-to-noise.

The typical HF station does not exist; each is a unique version of a combination of one or more of transmitters, antennas, program sources, transmission lines, a matrix for audio switching, a matrix for

RF output switching, baluns, slew switches, configuration switches and so on. The antennas, support towers, slew switches, configuration switches and phasing lines are all located outside the transmitter building in the elements with only as much protection as is absolutely necessary. Even the baluns are often located in external locations. Regular maintenance is normally required for all these outside elements due to the wear and tear of weather.

Transmitters and RF matrix switches are normally mounted in the main building but some locations have external matrices in shelters. Frequency agility of the transmitters necessarily involves internal configuration changes that slide, twist, rotate, and bend. The rotating contacts of the matrices also involve various forces that require periodic maintenance.

HF Propagation

Propagation limitations are everything to an HF broadcaster. Propagation dictates the time of day, the frequency of operation, type of antenna, probability of coverage and transmitter power. Everything about the path between the location of the transmitting station and the location of the listener in the target area is a function of the condition of the ionosphere: the ionized region of the sky that can reflect the signals so that they arrive at the target.

Adding to the HF broadcaster's concerns, the ionosphere changes with time of day, season, sunspot cycle (long and short) and solar disturbances. It is typical that an HF broadcaster has four schedules for the four seasons that change from year to year. This requires that the HF broadcaster must make a guess as to what the ionosphere will be like months into the future when the operating schedules are prepared. Propagation tools have been devised over the years to help make sense of all these factors of which VOACAP[2] is the most easily accessible.

The first decision an HF broadcaster must make is where the target audience is located. The second decision is what times of day the target audience is to receive the signal. That information can be entered into a propagation prediction program that will give the optimum frequency for that time or times of day. These prediction programs also estimate the height of the reflecting portion of the ionosphere and predict the optimum take-off angle of the antenna to be used. The third decision is the desired signal-to-noise ratio (SNR) at the receiver location. Path loss and receiver-site noise levels set the minimum power necessary for an adequate SNR. Typical transmitter power levels available are 100, 250 and 500 kW. Sometimes 50-kW transmitters are used, but they are typically just 100-kW transmitters with smaller power supply capabilities.

It must be noted that the HF broadcaster has an even more difficult problem with noise sources in population centers. Everything from computers to light dimmers generates noise in the HF spectrum. A recent survey found that over the past 20 years or so the RSS (root-sum-square) value of HF noise has increased at least 3–6 dB in most urban areas. Obviously, delivering the same signal-to-noise ratio over a path requires ever higher antenna gain, transmitter power or proximity of the transmitter to the target.

The most simplistic example is that of an HF broadcaster wishing to cover areas inside their own country. Australia is an example of a country with a desire to cover a wide area of the country where coverage with regional AM or FM stations would be cost prohibitive. NVIS (near-vertical-incident skywave) antennas are used to beam the transmitted signal over very high take-off angles, that is, straight up. The choice of the right frequency allows the signal from the transmitter site to bounce off the ionosphere and return to earth in a pattern that is nearly circular and several hundred to a thousand miles wide. Receivers located in this pattern can easily receive the transmitted signal despite poor ground conductivity, intervening obstacles or other phenomena that may make MW or FM signals difficult or impossible to receive. Experiences in Australia and other countries have shown that 100 kW is adequate for this type of broadcasting. Frequencies for this type of broadcasting vary from a low of 2.2 MHz to a high of about 9.5 MHz.

HF broadcasting for medium distances, 1000–2000 miles, typically use broadband log-periodic type antennas. In the case of medium-distance broadcasting the target audience is typically spread over a wider geographical area relative to the transmitter site. This requires two things: a modest take-off angle and a fairly large horizontal beamwidth—features that the log-periodic antenna has as part of its practical design. Because the distances to be covered are moderate, a typical power level would be 100 kW. Single-frequency transmission over medium distances using a single frequency and a narrowband antenna are known but are not generally the rule.

International HF broadcasting over long distances typically uses large curtain arrays, high transmitter powers and flexibility in operating frequency. Schedules of time and frequency must be prepared and distributed months in advance. It is not unusual to broadcast the same program over several transmitters and antennas operating on different frequencies at the same time to ensure that at least one of the possible paths delivers a usable signal to the target area.

Propagation predictions deal in probabilities, not certainties. Propagation prediction can almost certainly tell you when certain things won't happen. For instance, Marconi's first transatlantic telegraph transmission: an analysis of the propagation between the UK and the US shows that Marconi could not have made the receptions he reported making on his nominal transmitting frequency using a 25-kW spark transmitter.[3] Luckily, he was using a transmitter with tremendous harmonic content. And his receiver was a very sensitive type of diode with a largely untuned input. An analysis of the sunspot cycle, time of day and other factors indicates a wonderful opening of HF on that day and at that time and it is believed by most who have reviewed the data that he did hear his transmitter from England in America via the HF circuit that was open at that time—not the LF or MW frequency he thought he was using.

Selective Fading

The VOACAP plot shown in Figure 7.20-1 is for a path between Greenville, North Carolina, and Barcelona, Spain. It shows which bands and times are useful during the day in a visual format. The percent reliability is shown by the various colors. Dots are used to indicate the optimum times and bands.

These data are for a sunspot number of 50 and springtime conditions; 500 kW to a 4×4 curtain are used. An evening transmission, Barcelona time, should be possible on the 31-, 25-, 22- and 19-meter bands, or a morning transmission on the 49-meter band. A somewhat less reliable alternative is 31 meters. In Region 2, 41 meters cannot be used, and 60 meters is not an international band. This very quick and easy analysis does not tell the whole story. Tabular data in Figure 7.20-1 is taken from a VOACAP printout that shows additional details about each meter band. These data are for 21 UTC (Universal Coordinated Time) for the same path and conditions. We see that for 19 meters (15.4 MHz) the required take-off angle is 9.0 degrees, well within the pattern of typical HF antennas. The greatest reliability is on either the 22- or 19-meter bands, but looking at the bottom line, we find that 22 meters has an 84% chance of multipath probability. This is in keeping with a general rule that lower frequencies are more likely to have multipath.

Multipath occurs when two or more signals arrive at the receiver via different paths. Refer to Figure 7.20-2.

We can guess that in this case, both the 3F2 and the 4F2 signals are arriving, although 2F2 may also be present. A more complete analysis using the "all modes" printout feature of VOACAP shown in Figure 7.20-3 confirms this. As these two signals mix they will either add or cancel depending on the relative phase, which will constantly change as the exact path lengths vary due to perturbations in the F layer.

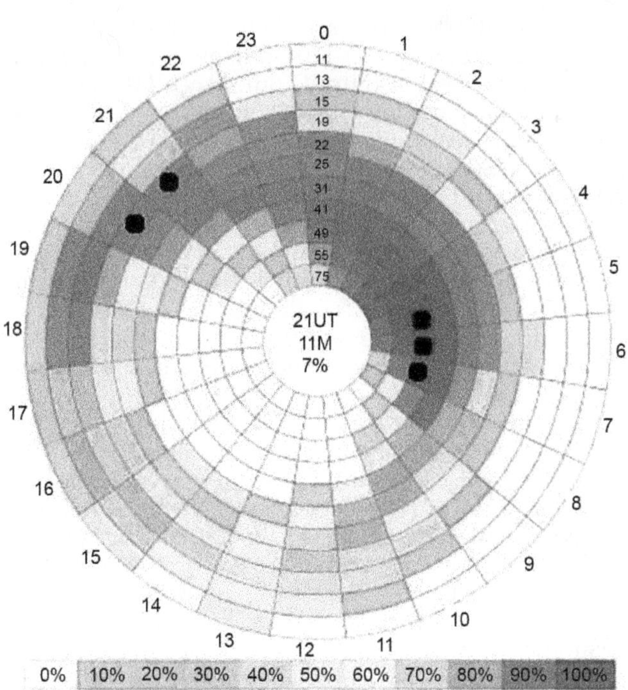

FIGURE 7.20-1 VOACAP Plot of Path from Greenville, North Carolina, to Barcelona, Spain

This constructive and destructive interference will produce nulls in the channel frequency response as shown in Figure 7.20-4. When these nulls fall on the sideband frequencies of the AM signal they produce only slight distortion. When they fall on the carrier frequency, as shown here, the carrier is insufficient to properly demodulate the sidebands and severe distortion results. As the notch frequency moves up and down, the distortion will come and go.

Selective Fading is also important in digital modes such as DRM[4] (Digital Radio Mondiale). There is a one-way digital path from the transmitter to the receiver. There is no receiver-to-transmitter path (as with the Internet) to ask for repeats when an excessive error rate

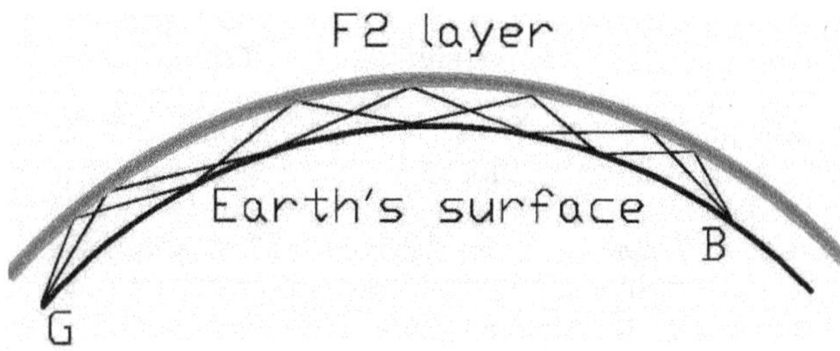

FIGURE 7.20-2 Skywave Propagation and Resultant Multipath

21UT	6.1	7.2	9.7	11.9	13.7	15.4	17.7	21.6	25.9	FREQUENCY
	3F2	3F2	3F2	3F2	3F2	3F2	2F2	2F2	2F2	MODE (Hops and layers)
	7.7	7.5	7.6	7.9	8.3	9.0	2.0	5.4	5.4	TAKE OFF ANGLE
	1.00	1.00	0.99	0.96	0.87	0.74	0.82	0.47	0.15	MUF PERCENT OF DAYS
	62	71	82	88	89	89	73	73	41	SIGNAL to NOISE RATIO dB
	0.56	0.78	0.92	0.93	0.92	0.92	0.74	0.73	0.17	RELIABILITY
	0.41	0.69	0.87	0.87	0.84	0.00	0.00	0.00	0.00	MULTIPATH PROBABILITY

FIGURE 7.20-3 VOACAP All Modes Printout

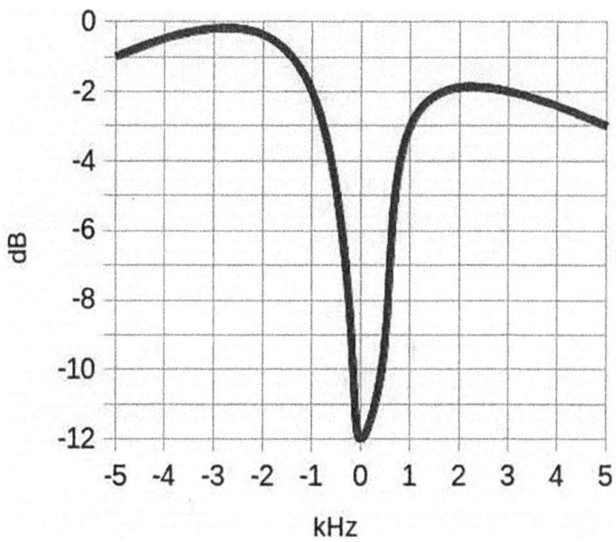

FIGURE 7.20-4 Illustration of Notch Frequency Caused by Selective Fading

occurs. DRM addresses this by actually transmitting the information twice, the second time at lower fidelity. Different modes allow more redundancy and more time delay in the repeat channel. The fidelity may drop momentarily but the signal is never completely lost.

Selective fading is familiar to nighttime MW listeners. This is due to the addition of ground wave and sky wave signals and causes distortion to come and go rather slowly. MW 170- to 195-degree anti-fading antennas help reduce medium-angle sky waves to reduce selective fading distortion.

Selective fading on HF varies from less than a second to a few seconds. The use of an antenna with a narrow vertical beam to excite fewer modes can be useful as is selecting a frequency where multipath is less likely. MUF (Maximum Usable Frequency) is the highest frequency that supports propagation to the target. Using the highest possible frequency to minimize selective fading may mean "riding the MUF," using a frequency that is as high as possible without having too many days where the signal is lost due to low MUF values. In Figure 7.20-3 it will be noted that the reliability for the next three lower bands is similar to that for 19 meters. Where possible, for this service one or more of these frequencies would be used in addition to the 19-meter band. This gives the listener two options. On most days when 19 meters is propagating, it will provide the less distorted signal. On the few days where low MUF is below 15 MHz on the 19-meter broadcast band, the lower bands will still allow reception.

A similar situation exists for the morning transmission noted above at 5 to 7 UTC on the circular chart. The 31-meter band may be somewhat less reliable, but on days when it does propagate well, the signals will show less selective fading distortion than the 49-meter band. A tabular VOACAP printout for those hours predicts that 49 meters will be 96% reliable with an 84% chance of selective fading whereas the 31-meter band will be 86% reliable—but with no selective fading. Again, a two-frequency transmission arrangement would be desirable.

VOACAP is available online and only requires the user to input the required information to get propagation predictions. In addition there are several other online alternative propagation engines that can be used for propagation predictions, some specifically aimed at Amateur Radio Operators but generally useful for HF broadcasters who want a second opinion. In addition NTIA keeps sunspot and other indices available for the general public for HF Propagation prediction use.

HF Transmitter Overview

Transmitters for HF broadcasting are generally manufactured for two power levels: 100 kW and 500 kW. A 500-kW transmitter with a smaller power supply and perhaps- a smaller PA tube are used for 250/300 kW service. A 100-kW transmitter with a smaller power supply is used for 50-kW HF service. All modern tube transmitters use plate-modulated Class C output stages[5] using high-gain tetrode tubes. All are modulated with some form of PDM power supplies to supply the audio + DC required that provides power to amplify the carrier and generate the modulation sidebands. Transmitters start to diverge widely from manufacturer to manufacturer in the type of output network used to provide output matching and harmonic suppression for the Class C PA stage. Tetrode tubes also require modulation of the screen and this requirement will be discussed in the modulator section.

Historically, transmitters were designed for 300-ohm balanced output to match the standard HF transmission line to feed a balanced antenna. Push-pull PAs were standard up until the early 1970s. Hughes, General Electric, RCA, Gates, Collins and other used push-pull triodes or tetrodes (usually the EIMAC 4CV100,000C) for 250 kW, the old standard transmitter power for international broadcasting; 100 kW and 50 kW were used for less-demanding paths. Starting in the 1980s VOA (now IBB) settled on a standard power of 500 kW based on the MIT study mentioned earlier and a whole new generation of transmitters was developed. Most of these had 50- or 75-ohm unbalanced outputs. The main driver for unbalanced output was the smaller size of the transmitter-to-antenna-switching matrix transmission line; coax instead of balanced box line.

It is a general rule that the higher the anode voltage the higher the PA and overall efficiency. But physics imposes an upper limit on the voltage that can be applied to the PA output stages. Whether 100 kW or 500 kW the voltage applied to the PA is normally in the 14–15 kV range. The applied voltage for 100% positive modulation is twice the value for carrier so the peak voltage required is 28–30 kV. The major impact is that the anode impedance of a 100-kW transmitter's PA stage is about 5 times higher than for a 500-kW transmitter.

PA output networks for 500-kW transmitters to match the approximate 170-ohm anode impedance to the 50- or 75-ohm output impedance fall into two design types. One type uses a foreshortened cavity arrangement with link coupling between the PA primary and PA secondary network. The second group uses a hairpin-type anode inductor with sliding contacts. Each type of network has advantages and disadvantages and both require careful arrangement to prevent spurious resonances that could result in arcing or component failure.

Pi output networks for 100 kW transmitters match the approximate 900-ohm anode impedance to the 50- or 75-ohm output network and are more conventional with rotary or band-switched plate tank inductors.

Almost all single-ended and push-pull transmitters use the Pi network as a fundamental part of the output network. The Pi network can be easily modified to a Pi-L network as shown in Figure 7.20-5. For example the Pi can be designed to match 150-ohms and the L to match from 150 to 75-ohms. By combining the two Cs of the Pi and the L network an extra 12 dB of second harmonic suppression can be achieved.

An interesting part of the Pi Network design is when the phase shift is an odd multiple of 45 degrees. The tuning control changes only the reactance at the

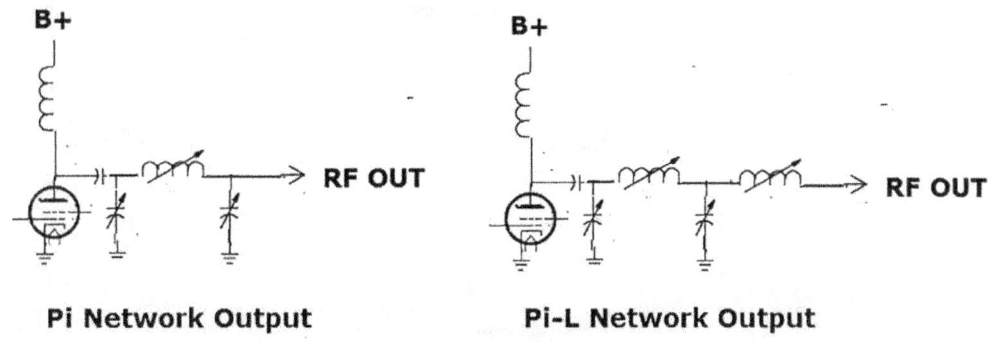

FIGURE 7.20-5 Pi and Pi-L Output Network Design

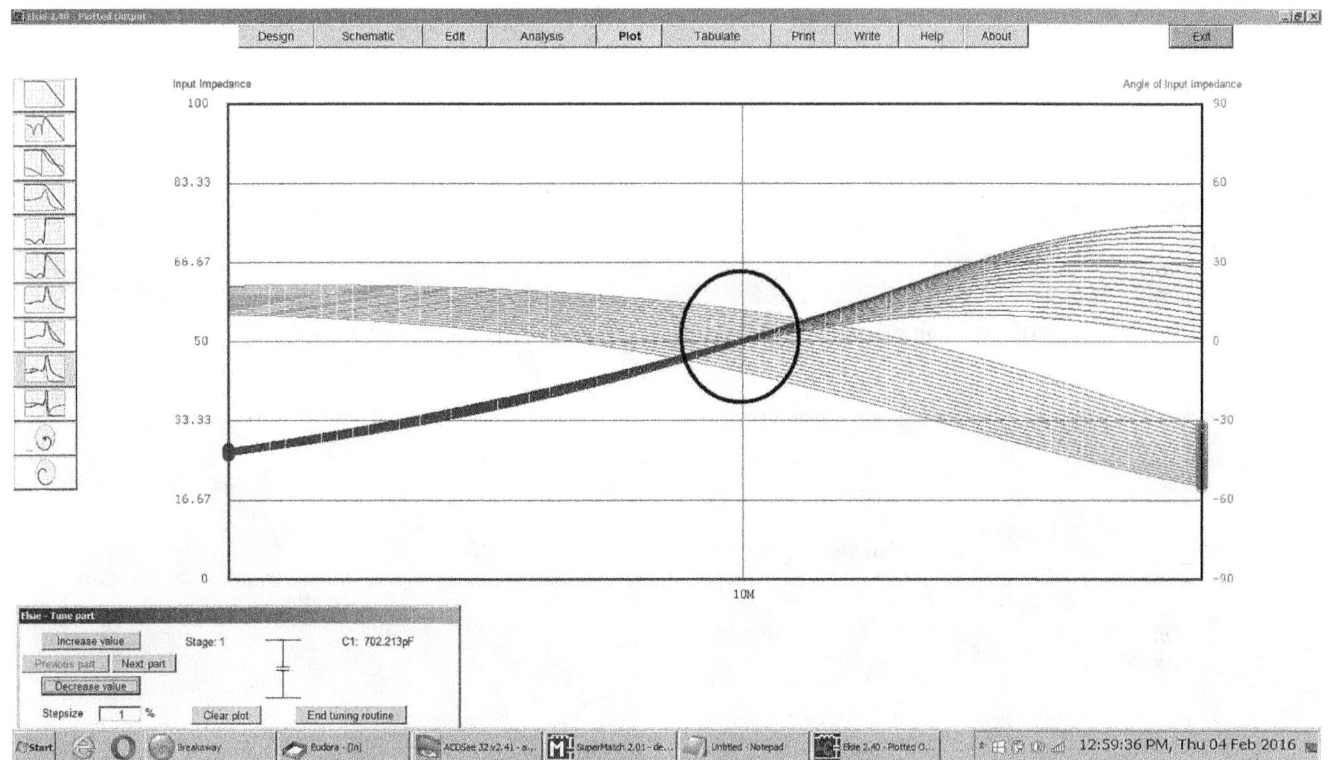

FIGURE 7.20-6 Effect on Input Impedance Versus Tuning Change

anode and the loading control changes only the resistance change at the anode—that is, the controls do not interact. Unfortunately, lower anode impedances result in more departure from 135 degrees for a given anode Q. The greater the departure, the more control interaction is experienced. Thus, for an anode impedance of 900 ohms the phase shift might be close enough to 135 degrees (±15 degrees) so that the tuning and loading control do not interact. At 250 or 190 ohms there would be significant interaction. Figure 7.20-6 shows the effect on the input impedance as the tuning is changed; only the reactance at the input changes materially. Figure 7.20-7 shows the effect on the input impedance when the loading is changed; only the resistance at the input changes materially. The low interaction between the two controls can be seen in a way that makes the benefit of the 135 degree network clear. The graphs were prepared using the excellent ELSIE and PI-EL freeware from Jim Tonne.[6]

The plate tank coils for many HF transmitters trace their origin to an old BBC (Brown-Boveri Company) transmitter. To conserve space the tank coil was wound in a circular hairpin configuration, shown in Figure 7.20-8 with a rolling slider shorting out the unnecessary inductance. The back resonances of the unused section had to be tuned away from spurious resonances by a capacitor as shown in Figure 7.20-9. This design, with a modified contact arm, was later used in the 500-kW AEG-Telefunken S4005. Its final form (when the solid-state modulator freed up cabinet

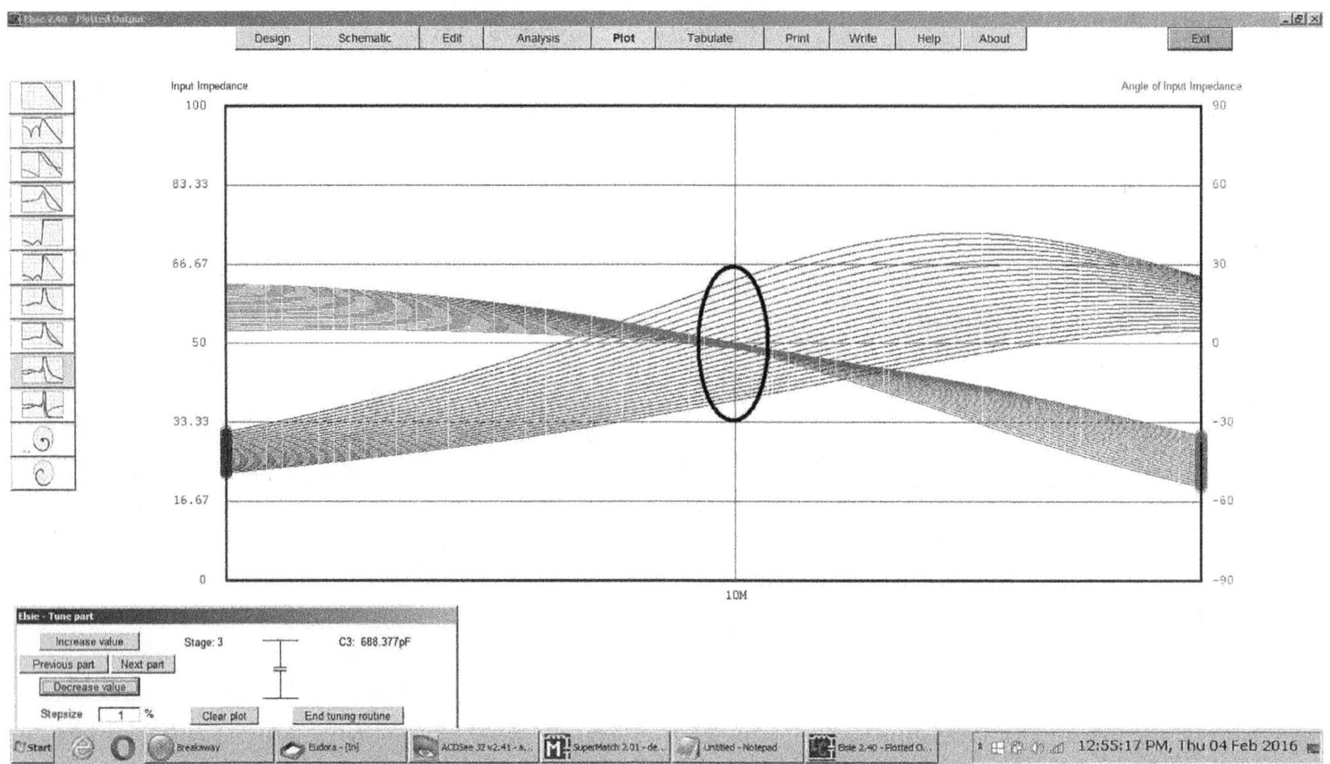

FIGURE 7.20-7 Effect on Input Impedance Versus Loading Change

FIGURE 7.20-8 Tank Coil Wound in Circular Hairpin Configuration

FIGURE 7.20-9 Spurious Resonance Elimination Using Capacitor

space) resulted in straightening the tuning inductor out from a circle form to a flat "hairpin" inductor and using a sliding contact, rather like a strip line inductor. Spurious back resonances in this tuning inductor were controlled with a single capacitor switched in and out as necessary.

One of the problems with this initial BBC design and in the AEG S4005 was that the contact pressure of the roller, shown in Figure 7.20-10, was high enough to eventually deform the tubing that made up the inductor deforming it leading to increased maintenance on the contact pressure over time. The later sliding point contact eliminated that problem.

It should be noted that the shorting plunger type band switched coil designed by George Woodard (formerly VP, Eng., RFE/RL and IBB) used in the CEC 100-kW and older 250-kW transmitters with their 1/3 turn shorts eliminated spurious back-coil resonances amazingly well and is used in more 100-kW transmitters than any other design.

FIGURE 7.20-10 Sliding Point Roller Contact

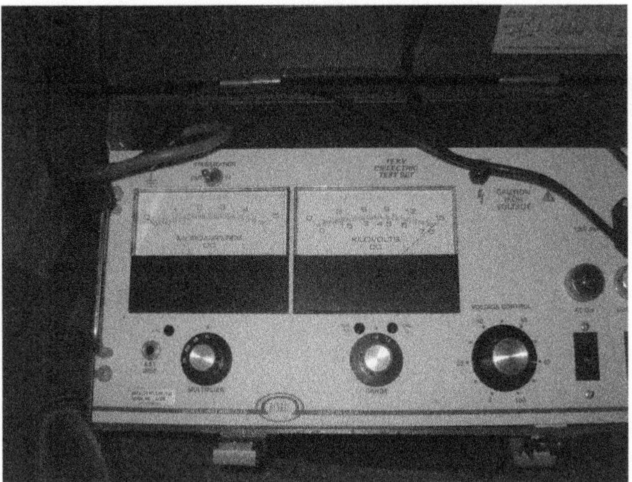

FIGURE 7.20-11 30 kV DC Hi-Pot Tester

FIGURE 7.20-12 Vacuum Capacitor Under High Voltage Test

Additional output network components are necessary after the first PA output network components to provide additional harmonic suppression and impedance matching. But at the 50- or 100-kW level the Pi-L network makes the required harmonic suppression (spurious) easier to accomplish; these are discussed in detail later.

Vacuum variable capacitors are an essential ingredient in the design and construction of most HF transmitters. The combination of high-voltage and high-current capability combined with small size and low series inductance make them essential output network components. At the same time, vacuum variable capacitors are generally agreed to be the single component that fails most often in HF transmitters. A vacuum variable capacitor has an internal bellows that allows the capacitor plates to be pulled apart to vary capacitance. Unfortunately this bellows is also a current-carrying element resulting in cycles of heating and cooling and work hardening of the bellows material. Alternative designs have been proposed to minimize work hardening.[7]

At higher transmitting frequencies the bellows is pulled tighter and the folds of the bellows become more compressed. A vacuum variable capacitor may tolerate a million adjustments between 90% and maximum capacitance—but only a few thousand cycles when adjusted between 10% and minimum capacitance. Despite all the limitations and failure rate they are an essential part of HF broadcasting. A vacuum variable capacitor is shown under HVDC test in Figure 7.20-12.

Figure 7.20-11 shows one of the second or third most important pieces of test equipment in any HF Station. DC hi-pot testers, sometimes called dielectric testers, can be used to clean out the gas inside a capacitor and restore it to usable standoff voltage.

AC high potential units are smaller, lighter and more portable. If an HF broadcaster has the funds, then both DC and AC should be available. If funds are limited, a DC unit is much to be preferred for diagnostic testing but an AC unit is still useful because, as noted earlier, a DC unit can be used to clean out the capacitor gas restoring it to a usable standoff voltage. Many arc points in a capacitor can be smoothed out using a DC unit.

It should be noted emphatically that no capacitor or vacuum tube should ever be installed in a transmitter without subjecting it to a DC leakage test. A full 60-kVDC unit is to be desired so allow tube and capacitor clean up. But a 15- or 30-kVDC unit will diagnose and identify most problems.

Digital and AM/PM Modulation Modes

HF Broadcasters and equipment manufacturers (including receiver manufacturers) joined in a consortium to bring higher fidelity and more reliable coverage to HF broadcasts. The system that resulted was DRM (Digital Radio Mondiale). This DRM AM/PM Modulation system could deliver a full fidelity signal, in stereo or mono, using existing HF transmitters that were coupled to DC. It could also transmit a high-fidelity

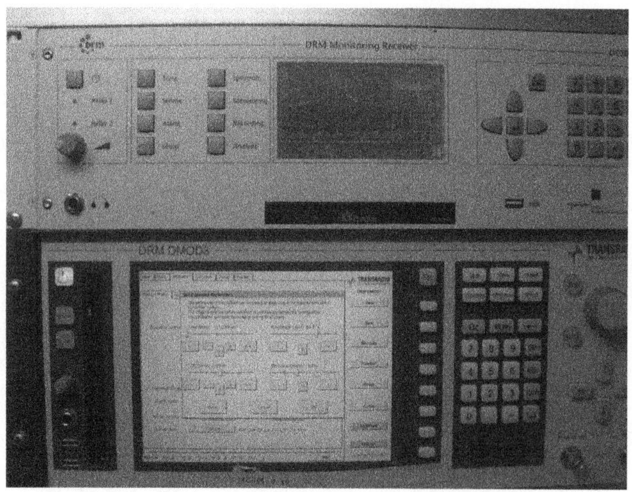

FIGURE 7.20-13 DRM Exciter and Monitor

digital signal and a conventional AM signal simultaneously. All modern transmitters have been or can be modified to allow DRM transmissions with the appropriate exciter. Figure 7.20-13 shows the process of mating exciter and transmitter with a DRM exciter and DRM monitor receiver. Refer to Chapter 7.7, "Worldwide Digital Radio Systems," for a detailed discussion of DRM technology.

A DRM transmission has 100 individual carriers spaced 100 Hz apart, a 10-kHz bandwidth signal. The phase and amplitude components are separated. The PM components are amplified by the high-efficiency PA stage while the AM components are amplified by the high-efficiency PDM modulator stage. The delay through the audio path is always much longer than the delay through RF path. Delay is introduced into the PM (RF) path so that the AM and PM components arrive at the PA stage at exactly the same time. Active and passive pre-distortion is used to ensure the linearity of the AM path. PM pre-distortion is also introduced if necessary to correct for any residual Incidental Phase Modulation (IPM) in the RF chain. This results in a transmitted bandwidth without spurious sidebands that could cause interference to any nearby signals.

AM/PM modes such as SSB/ISB/DSB (with or without carrier) and DRM are all possible with modern transmitters with PDM modulators. All use EER (Envelope Elimination and Restoration) where the amplitude (AM) and phase (PM) elements are separated. The phase signal replaces the normal, unmodulated RF drive to the PA while the amplitude signal is handled by the wideband PDM modulator. Amplification of these AM/PM signals puts a requirement on the PA stage that its IPM be kept low so that the PM sidebands will be properly combined with the amplitude signal from the modulator in the PA stage. The methods and uses of these techniques are discussed at length in Chapter 7.7.

Single Sideband is a very effective HF transmission mode. Amateur Radio operators—the first developers of HF communications—use SSB almost exclusively. The SSB signal can be amplified with low-efficiency linear amplifiers. As noted earlier, however, the AM and PM components of the SSB signal can be separated and amplified separately. Time delays and linearity corrections as described for DRM are used to ensure the transmitted bandwidth is tightly controlled. It goes without saying that ISB (independent sideband) and DSB (double sideband), as well as SSB—with varying or no pilot carrier—can be transmitted with the appropriate exciter.

Tetrode-Tube Transmitters

All modern tube transmitters for HF Broadcasting use anode (plate) modulation and tetrode tubes. This necessitates screen modulation as well to keep the screen current from rising and falling as the anode voltage swings up and down.[8] The voltage applied to the screen to keep the screen current constant over the modulation cycle is highly nonlinear and varies from one tube type to another. The most reliable way to accomplish the required task is to place a large audio reactor in series with the screen supply. The reactor stores energy in a magnetic field that increases as the screen current rises. The reactor then releases the energy as extra voltage when the screen current falls. This self modulation of the screen voltage is not coupled to DC as the anode modulation but a relatively slow-speed supply voltage variation can be provided by using a PDM screen-voltage supply. In some cases resistors are inserted in line from a fixed screen supply such that the maximum screen dissipation is not exceeded even if the anode voltage is zero.

Modulation of the RF drive is not required by most high-gain tetrode tubes, although it is fairly easy to achieve. Drive sufficient to support the crest of 100% positive modulation is applied at all times and, because of the very high gain of the tube, the required drive level is typically about 20 dB below carrier level. One favored tube type can benefit from drive modulation because of a limit on the maximum constant bias voltage that can be applied. The perceived difficulty in providing this modulation usually results in it being omitted even though the linearity is much improved with drive modulation. Again, this is peculiar to one widely used tube type.

The worst of all possible worlds occurs when the screen modulation reactor has a large magnetic field built up from several amps of screen-grid current and RF drive is suddenly lost to the PA tube. The magnetic field collapses instantaneously and can over-voltage the screen or screen bypass capacitor. Some designers use an arc ball gap across the reactor terminals. The problem with that is that the time constant for a reactor is $TC = \omega L/R$, where ω is 2p. The lower the R—as in ball gaps—the longer it takes for the energy to be dissipated in the resistance of the winding of the reactor. It is far better to use a small resistor in series with the arc balls. Or, better yet, resistance in series with MOVs designed to clamp at less than twice the applied screen voltage as shown in Figure 7.20-14.

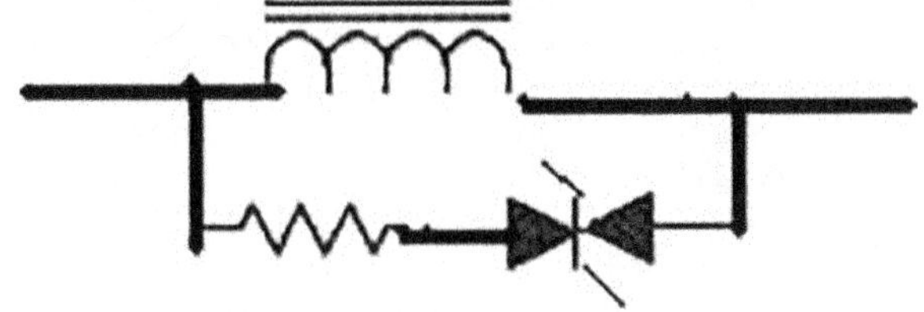

FIGURE 7.20-14 MOV Used to Clamp Voltage

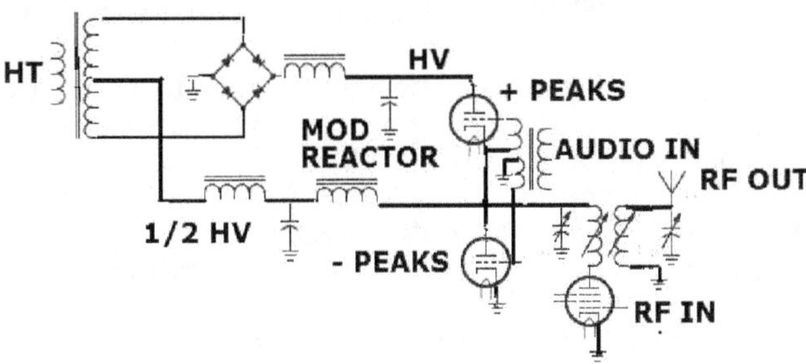

FIGURE 7.20-15 HCJB 500 kW Class AB2 Modulator Scheme

FIGURE 7.20-16 Hughes 250 kW HC-114 Transmitter

FIGURE 7.20-17 GE 4BT250A 250 kW Transmitter

Older Transmitter Development

A notable exception to the use of PDM modulators coupled to DC on large HF transmitters or the use of old Class B (AB2) Modulators was developed by the HCJB engineering team for their 500-kW transmitter. The DC anode voltage was applied to the PA through a large modulation reactor as shown in Figure 7.20-15. At the output of the modulation reactor, two modulator tubes were used: one to pull the anode voltage up to twice its carrier value for +100% modulation and another to pull the voltage down to near zero for −100% modulation. This unusual arrangement was, of course, not suitable for AM/PM modes but could provide an excellent AM signal.

No HF Transmitter discussion would be complete without mentioning the 250-kW HF Doherty transmitter. Only twelve were ever built and ten of those went to the VOA Relay station in Kavala, Greece, on the sea near Xanthi. The Screen-Impedance-Modulated Doherty[9] was the first transmitter to break the 60% overall efficiency barrier; previously 50% had been considered normal. GE, Hughes, RCA, Gates and Collins had stuck with high-level push-pull modulators until that point and had accepted the lower efficiency.

The Doherty amplifier relies on 90-degree phase shift networks between the grids and the anodes and differing anode impedances in the absence of modulation. A full discussion is beyond the scope of this section. At the crest of positive modulation, both PA tubes, the Peak and the Carrier each supply half of the peak power, that is, 500 kW for a 250-kW transmitter. Screen Modulation of both tubes with a DC offset between screens of the Peak and Carrier Tubes was patented in 1963 by Joe Sainton of Continental Electronics. (The drive was not modulated as in a true Doherty, a linear amplifier.) The Doherty uses two PA tubes instead of one. But you can use a higher-than-normal anode voltage (for greater PA efficiency) and the DC Anode voltage does not change during modulation, a tremendous benefit. As an aside, the humble Doherty is making resurgence in the VHF, UHF and higher frequencies.

An interesting side story is that the Kavala Relay Station was so good for reaching the targets in the Soviet Union that the USSR paid for and built an installation of 500-kW Jammer transmitters at a site at Plovdiv, Bulgaria, just 90 miles over the mountains from the Kavala Relay Station in an attempt to duplicate its propagation. But 90 miles closer couldn't overcome the fantastic ocean-side Kavala site and antennas of the Kavala Relay Station. Steve Lockwood of Hatfield and Dawson has provided Figure 7.20-16 of the famous (or infamous) 250-kW Hughes HC-114 used at the Tinang Philippines Relay Station. Ten were supplied.

The GE 4BT250A 250-kW HF transmitter shown in Figure 7.20-17 is renowned for its service in Monrovia, Liberia. Jim Hawkins has a picture on his Greenville, North Carolina, Relay Station Tour web page.

Another famous transmitter was the Collins 821–1A shown in Figure 7.20-18. One is on display at the old Bethany, Ohio, VOA/Crosley Relay Station location and another was rescued from Delano, California, by the Collins Collectors Association. The 821-A1 was designed to have a tuning time of 20 seconds whereas most HF transmitters usually require up to 30 seconds to tune. All the old standard-power 250-kW transmitters except the previously mentioned 250-kW screen-impedance-modulated Doherty transmitters in Greece all used conventional high-level plate modulation. The VOA's first 500-kW transmitters made by Continental Electronics were located at Greenville, North Carolina.

A large number of CEC 419F 250-kW HF transmitters, shown in Figure 7.20-19 were also used (and continue to be used) in Tinian, Kuwait and other sites.

HF Transmitter Efficiency

The arrivals of the Solid-State Modulator, ever-better tetrode final tubes and higher anode voltages some 15 years later finally brought the overall HF transmitter efficiency from 50% or less to 75% at 300 and 500 kW and over 70% at 100 kW. The solid-state modulator also eliminated all the big-iron used in the modulation transformer, modulation reactor, filter reactors, etc. With coupling to DC, tilt and overshoot were eliminated so clipped audio that contains square waves could be passed through the modulator stages.

Shortwave listeners may wonder about the absence of specifics about Soviet, Russian or Czechoslovakian-Tesla transmitters. Those designs were brute force in approach using grounded-grid triodes in the high-power final stage. They worked—even up to one 2000 kW—but the overall efficiency was abysmal.[10]

Some recent non-U.S. designs appear to be direct copies of American, French and Swiss transmitters with a single known exception: Radio Tirana in Albania. The transmitter there seems to be a copy of an American 500-kW transmitter scaled down to 100 kW. A few pictures of the Radio Tirana transmitter might remind any reader of the 419/420-series transmitters.[11]

The control system for the servos and the modulation controller are exact copies of the 419G/420C components. See Figure 7.20-20. They also use the Advantech computer and PTS synthesizer used by CEC and shown in Figure 7.20-21 to simplify this copy.

The PA combines the best of the 418E and the 419G/420C circuitry. Notice the CEC 418-style neutralization paddle and decoupling inductor in Figure 7.20-22 plus the reduced size of the CEC 420's cavity lines to match the higher anode impedance of this 100 kW transmitter.

A fantastic database of HF transmitters and HF transmitter locations can be found at www.shortwave.be/.

FIGURE 7.20-18 Collins 821-1A Transmitter

FIGURE 7.20-19 Continental 419F 250 kW HF Transmitter

HF Baluns

Every modern HF antenna is balanced with respect to ground while every modern RF transmitter has an unbalanced output. There are two widely used methods for achieving the BALanced-to-UNbalanced output. One is the tunable 4:1 balun, discussed in more

FIGURE 7.20-20 Transmitter Control and Monitoring

FIGURE 7.20-21 Advantech Computer and PTS Synthesizer

FIGURE 7.20-22 CEC 418 Final Amplifier Stage

detail later, that uses two shortened cross-connected 150-ohm coaxial line sections, one insulated from ground at its characteristic impedance inside a box with the other outside the box. Open-wire transmission lines have very low transmission losses compared to coaxial line. If the transmitter output impedance is set to 75 ohms a 4:1 balun results in a perfect 300-ohm match. TCI (Technology for Communications International, a division of SPX) makes a 75:300 or 50:300 ohm Ferrite Balun that has been widely used with great success at 100 kW.

Coaxial cable can be used for the transmission line to the antenna. The transmission losses are much higher but they may be acceptable for shorter transmission line runs because of the protection they provide from the weather and environmental factors. In such cases a Telefunken-style balun is used.[12] A moderate length of coax is insulated from ground after its input point. At the output terminal a compensation line is run from the center conductor (parallel to the outer conductor) back to the input ground point. This is a 1:1 balun whose output impedance is equal to the input impedance.

The output of the Telefunken-type balun can be connected to the antenna or an open-wire transmission line by using a tapered transmission-line transformer. The balun uses a length of rigid coax and a short from the output center conductor of the coax back to the outer conductor of the coax section. The 50-ohm output has to be transformed up to the impedance of the transmission line, typically 300 ohms, balanced. Starting with two large plates that form a 50-ohm open-wire line the size of the plates is gradually decreased and the spacing between the plates is gradually increased at an approximate logarithmic rate over distance until the desired 300-ohm impedance is achieved. Such a tapered-line impedance transformer can also be used at the antenna input. There the 300-ohm impedance is transformed down to the antenna input impedance. If coax is used the balun is normally located near the antenna input and then the 50-ohm balun coaxial impedance can be transformed to the 60, 120 or other impedance of the antenna.

The balun is a critical component. The single most important characteristics of any balun is the perfection of output balance and the rejection of unbalanced current on the input feed line—typically coaxial line if the balun is located near the transmitter. If we denote the current in one output line of a balanced line as I1 and the other output line as I2, the two currents should be identical and exactly 180 degrees out of phase. Any departure from equal currents or 180-degree phase relationship gives rise to an unwanted unbalance current, I3, flowing in the input feed line. I3, the unbalance, causes the incoming line to radiate from the incoming line and the equipment to which it is attached. The result is high RF fields from the coaxial line and transmitter. This results in what is termed a "hot" situation. RF burns can occur from the touching the coax or transmitter and the local RF fields can exceed safety limits, discussed in greater detail in the Personnel Safety section. And unbalance on the transmission line to the antenna can also result in excess feedline radiation, reduction of power delivered to the antenna and poor antenna performance. Hence, the quality of a balun is given by the suppression of this unbalance current, I3, and is generally given and stated as decibels of common-mode rejection.

Antenna Overview

The Rhombic antenna was the antenna of choice for many years in the early years of HF Broadcasting.[13] The long-wire nature of this antenna makes it much less susceptible to fading because radiation is occurring all along the length of this long antenna. The

Yagi-Uda antenna had the highest gain for its size of any antenna. Unfortunately, the end voltages of the elements of the Yagi go to very high values as the power increases. This led to the development of the cubical quad antenna with is equivalent to two two-element Yagi antennas but with manageable element voltages.

The basic problem with the rhombic is that four supports are required. Being a long-wire antenna it is also bidirectional unless terminated resulting in about half the transmitter power being dissipated as heat. Engineers at the old Crosley HF station in Bethany, Ohio—later owned by the VOA and IBB—worked out a series of phasing lines to form what they called a re-entrant rhombic where a unidirectional pattern without power loss was achieved.[14] Low take-off angles on lower frequencies for long-distance transmissions would have resulted in extremely tall support towers because the vertical take-off antenna for HF is related to the antenna's height above ground. It can be said truthfully that the rhombic antenna occupied more space per dB of gain than any other antenna. Other smaller, higher-gain antennas slowly made the rhombic obsolete on a space versus gain comparison. A Sterba Curtain, for instance, was a single-band antenna with a very small footprint that was capable of moderate, bidirectional gain. Used with a reflector it could provide a unidirectional pattern with only two supports.

The log-periodic antenna is a wideband antenna with multiple elements whose length tapers at a logarithmic rate. The design criteria is beyond the scope of this article, but the log-periodic is a rather large antenna with modest gain and modest front-to-back ratio. Its wideband nature makes it a favorite for frequency agile operation. Like all HF antennas its take-off angle is related to its height above ground. At practical heights a rotatable log-periodic has a take-off angle that varies with frequency but can be generalized as in the 30- to 45-degree range. This is ideal for the mid-distance coverage described earlier as 600–1000 miles.

A fixed log-periodic can be made where the take-off angle is almost constant with frequency. The antenna elements are high for the low-frequency elements and lower for the high-frequency elements. The height of the front and rear support towers can be selected to give the most desired take-off angle for a particular target. The fairly broad vertical and horizontal beam of this type of antenna can be of benefit when a particular target is known—even with the varying height of the ionosphere. The ionosphere sets the optimum take-off angle.

The broadband curtain antenna was developed by Weldon and Carr. It consists of a four-high, two-wide array of broadband full-wave multi-wire dipoles all paralleled together, fed in phase, and is described in more detail later. A four-high, three-wide curtain with slewing (electrical steering) capability was later proposed and installed for Radio Free Europe by A.D. Ring & Associates. The first versions covered two bands. Today, over a full octave is common. Typical ranges are 5.8–12.3 MHz, 11–22 MHz, 13–26.1 MHz and 9–18 MHz.

The most common arrangement in use is a sixteen-dipole array consisting of four rows of four dipoles in each row and column. This is also normally arranged into independent column feeds. At the base the time/phase relationship of each column can be altered and the antenna can be steered. This is called slewing, and the base switch configuration can steer—or slew—the main beam up to 30 degrees from the perpendicular—that is, 60 degrees of coverage. A slew switch example is shown later. The reflecting screen can be put to double use by putting another dipole array behind it resulting in 120 degrees of coverage. Only three antennas, oriented at 120 degrees to each other, can cover the full 360 degrees. (There is some deterioration in the beam characteristics when slew is used. Because sidelobes arise their presence is not felt to be material because the main beam suffers only slightly.) The only limit to the number of rows and columns of dipoles is set by the height and width of the support towers and the maximum wind speed hitting the array. A counterweight system is used to allow the antenna to deform without breaking.

The HRS system is used to designate the characteristics of a curtain antenna where the bottom row is 0.5 wavelength off the ground is a 4/4/0.5. If the bottom row is 1 wavelength off the ground it would be a 4/4/1.0. Adding more rows lowers the take-off angle. A very long-throw antenna with a low take-off angle might be a 4/6/1.0.

TCI[15] in the US and Ampegon's[16] German factory are two well-known antenna suppliers. Every antenna manufacturer has a proprietary design for its dipoles, the side-to-side spacing of the dipoles, the dipole-to-reflector spacing and other design elements. A third manufacturer, Sino-Sky Hi-Tech[17] of Beijing, China, is new to the international antenna market but recently completed two pair (5.8–12.5/13–26.1) of front/back curtain antennas in the Middle East.

RF Matrix

Sites having multiple transmitters and/or multiple antennas normally use an RF matrix to route the selected transmitter's RF output to the best antenna for that target and frequency. Logic in the matrix controller is normally used to prevent a transmitter operating on a low frequency from being routed to an antenna designed for higher frequencies.

Two types of matrices have been used at 500 kW: unbalanced 50 or 75 ohms and balanced 300 ohms. The matrix impedance is determined by the location of the balun(s) in the system—modern transmitters being single-tube units—and the desired feedline impedance. RF matrix switches for 50 or 75 ohms can be quite compact, a desirable feature. The larger size and greater spacing of the 300-ohm, balanced RF matrix results in it often being located above the transmitter room or in an adjacent building or shelter. The balun can be made an integral part of the transmitter and co-located with the transmitter.

One of the most interesting RF Matrix installations is the distributed RF matrix at the IBB (International Broadcasting Bureau) relay station at Tinang in the Philippines. Transmission lines from ten transmitters exit the transmitter hall on one side and the transmission lines are routed through four separate sections of the distributed matrix. Any transmitter may be routed to antennas at the first, second, third or fourth RF matrix houses. This arrangement was necessitated by the long, narrow piece of property where the station was built during the mid-1960s.

Broadcast Targets

HF broadcasters select their coverage targets based upon the mission of the broadcasting entity that can vary over the years. Governments typically target geographical regions with differing political beliefs for influence purposes (propaganda), to provide their interpretation of news, to reassure allies, or to offer educational or cultural programing. Religious organizations target distant regions where they believe their message is important to be received. The use of HF for strictly commercial purposes, while once more common, is increasingly rare.

The IBB is the technical arm of the United States Broadcasting Board of Governors and provides the transmission facilities for different program services including VOA. The VOA charter is quite restrictive in limiting the amount of propaganda that can be incorporated into its programming.

Religious broadcasting has remained a force even in the United States. Whether renting otherwise unused air time or building stations, religious broadcasters of all manner have maintained a presence on HF. EWTN (Eternal Word Television Network), for instance, operates an HF broadcasting facility in Alabama, bringing Roman Catholic programs in different languages to listeners in the Americas, Africa and other target areas. EWTN is one of a few stations in the United States that has 500-kW transmitters available. Most other religious broadcasters in the United States use 50–100 kW of power with less sophisticated antenna systems.

Many religious broadcasters have or operate facilities whose target areas are the Muslim countries of South Asia and the Middle East. Saudi Arabia maintains a significant HF presence. They broadcast Wahhabi Islam programming to countries across North Africa, China, the Balkans and former Soviet republics where Islam is the dominant religion. Kuwait and Iran are also prominent religious broadcasters although they also incorporate programs about the culture of their countries.

LESEA (Lester Sumrall Evangelistic Association) operates a facility on Palau broadcasting religious and secular programming. Some of the secular programming is anti-Malaysian government directed at Borneo's West Coast, much to the chagrin of the Malaysian government.

The Vatican operates a relay station outside Rome at Santa Maria Galleria to broadcast programming of interest to Catholic listeners. One of the more interesting antennas at Radio Vatican is a full-size 4/4/0.5 High/Low curtain antenna. The end support towers are on railroad tracks so that the entire curtain can be rotated to cover targets throughout the world from one site.

Family Radio operated a 10-transmitter site North of Lake Okeechobee (and in Maseru, Lesotho) until recently when the Okeechobee facility was sold to Radio Miami International. The Maseru facility was the old BBC MW and HF relay station. The Florida facility has antennas for Europe, Asia, Africa and the Americas.

The *World Radio TV Handbook* (WRTH),[18] among other sources, has more information on current operations of shortwave stations around the world.

Frequency Coordination

The High Frequency Coordination Conference (HFCC), a sector member of the International Telecommunication Union (ITU), is responsible for coordinating the transmitting frequencies (and modes of transmission) of international (and regional) HF Broadcasting. Although the coordination process is beyond the scope of this section, the Asian Broadcasting Union (ABU) has prepared an excellent primer on the subject of coordination of broadcasts. United States HF Broadcasters receive frequency authorization from the Federal Communications Commission (FCC). The United States is a signatory nation to the ITU and any frequency authorization is subject to frequency coordination by the HFCC.[19] The purpose of frequency coordination is to prevent unwanted interference caused by different HF Broadcasters around the world operating on the same frequency at the same time to the same (or adjacent) target audiences.

Jamming

The existence of jamming (intentional interference) is a type of censorship and typically exists where a free press does not function, typically totalitarian regimes that are afraid of criticism from internal or external sources. Jamming—intentional interference to international HF broadcasting—has been a problem since the earliest days of international broadcasting. Not every target country desires that the listening audience in the target country receive the international broadcasts beamed towards them. Jamming is a very expensive business that requires a high degree of coordination between the multiple jamming transmitter sites, very high power and often involves interference to other international broadcasters not "beaming" to the country in question.

HF BROADCAST SYSTEM ELEMENTS

Frequency Agility and Transmitter Autotuning

Single-frequency or manually tuned transmitters are of limited utility to International HF broadcasters that operate with many antennas and on many frequencies throughout their broadcast schedule. Thus most frequency-agile transmitters have some type of an autotuning system. These transmitters all follow a general formula of using servo motors in the transmitter to coarsely position the tuning elements of the transmitter. Tuning is accomplished by either using a position potentiometer on the servo assembly and a reference potentiometer (or voltage) for that frequency. Newer models may use stepper motor steps to move between positions. Many transmitters in the 50/100 kW range stop any coarse tuning after fine tuning or IPA autotuning is completed.

Modern high-power transmitters supplement the coarse tuning with a system of fine tuning for all tunable stages in the transmitter. A circuit is resonant when the voltage and current in that circuit are in phase. Samples of voltage and current are taken and compared in a phase discriminator. A phase discriminator has zero output when its two inputs are 90 degrees out of phase.[20] This works because a current sample gives an output 90 degrees out of phase from a capacitive sample.

A grounded-grid amplifier may have an inductive (current) sample taken from the combined driver anode/PA Grid tuning network of the PA stage. The driver anode/PA grid current sample is compared with a voltage sample from the driver cathode. An inductive sample is 90 degrees out of phase from the actual current. When the current and voltage samples are 90 degrees out of phase with each other, the anode voltage and current are in phase with each other—that is, the driver is in resonance. A phase discriminator gives a zero output voltage when input samples are 90 degrees out of phase. The discriminator output can be used to drive a motor through an amplifier until the two signals to zero output, that is, resonance. Driver tuning is normally the most sensitive tuned element and benefits the most from fine tuning. The tuning process in many older channelized transmitters stop with only driver anode tuning. The assumption was that the PA/Output tuning has been preset to the correct antenna characteristics. That may be acceptable at 100 kW. But 250, 300 and 500 kW, transmitters all take autotuning considerably further because of the higher voltages and currents involved. For instance, the driver anode/PA grid current sample can be compared to a voltage sample of the PA Anode Voltage to bring the PA anode into resonance by motors driven by that discriminator output signal.

Figure 7.20-23 shows a simple autotuning system. A reference voltage from a positioning preset pot or a computer are compared against a signal from a positioning potentiometer geared to the tuned component. The simple system is capable of quite close positioning of the tuned component. A discriminator increases the accuracy by bringing the stages to resonance. Errors in the discriminator performance can be entered into the control computer so that they are automatically compensated out.

Modern transmitters step through the tuning process rapidly: (1) tune the driver plate/PA grid; (2) adjust the antenna matching; and (3) adjust everything else simultaneously. PA tuning is done by the process described earlier. PA loading is accomplished by adjusting the PA DC anode voltage and DC anode current to the correct ratio. If the output network uses cavity tuning, a discriminator ensures that the phase shift across the network is 90 degrees by simply comparing the primary and secondary voltages of the inductively coupled (link coupled is another term for the same thing) output network.

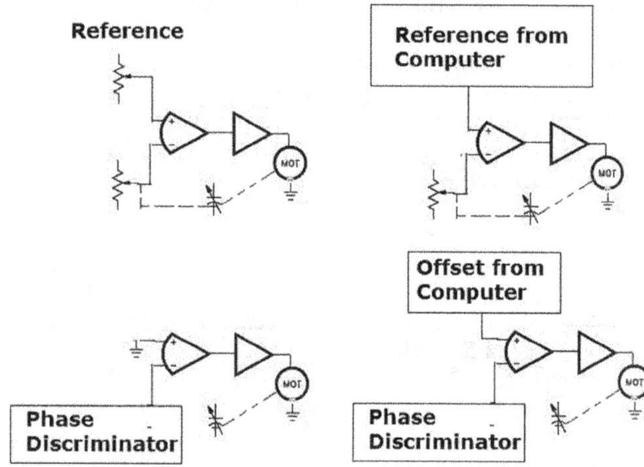

FIGURE 7.20-23 Simple Autotuning System Elements

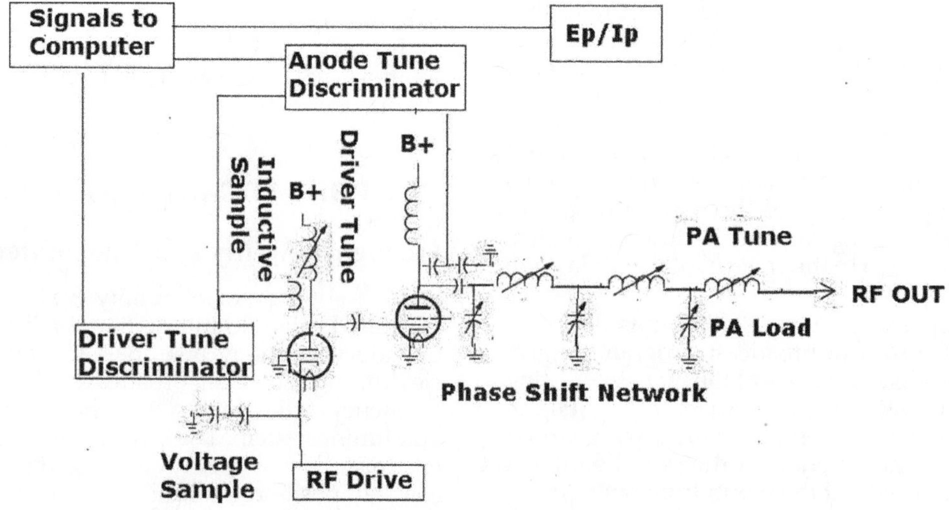

FIGURE 7.20-24 Ampegon Transmitter Autotuning System

The Ampegon transmitter in Figure 7.20-24 uses an interesting variation of this theme: the coarse tuning sets the PA output network to a specific phase shift for 100 ohms. See Figure 7.20-25. The last stage in the transmitter output network is an L network that is able to compensate for both Resistance and/or Reactance errors in the load. PA tuning and loading is accomplished by the L and C components of the output L network. The proper operation requires the phase shift total before the output L network be either an odd or even multiple of 45 degrees. If an even multiple, the inductor of the L Network becomes the tuning element while the capacitor of the L Network is for loading. For an odd multiple of 45 degrees the output inductor of the L Network becomes the loading and the capacitor of the L Network becomes the tuning. (By setting the phase-shift network to a point where spurious resonances are avoided parasitic resonance issues at the anode are avoided.)

FIGURE 7.20-25 Autotuning System in Ampegon Transmitter

CEC high-power HF transmitters, shown schematically in Figure 7.20-26, use a tunable balun or external antenna matching unit to set the actual impedance of the load to zero reactance. (This occurs after, first, bringing the driver anode/PA grid to resonance. Note that zero reactance is not the same as minimum VSWR.) Then the PA anode is adjusted for proper tuning and loading with vacuum variable capacitors whose servo motors are driven from discriminator signals at the tube anode. The inductive/link coupling is adjusted by a variable capacitor on the output inductor for 90 degrees of phase shift. The cavity/link-coupling inductors, shown in the schematic of Figure 7.20-26 and pictorially in Figure 7.20-27, use large diameter copper lines to serve as the inductors of the link coupling circuit. Their inductance is coarse adjusted by a movable shorting plane shown at the back of picture.

Modern transmitters generally break the HF frequencies they cover into 100 kHz segments. Tuning information for a perfect load situation is stored in the logic of the tuning system and all tuning components are moved to their coarse position. After switching to an imperfect load—like an antenna, the driver stage is first fine-tuned to deliver maximum drive and then output components compensate for the antenna/load imperfection and allow minor touch-ups of the main PA tuning and loading components. The output T network capacitor is shown in Figure 7.20-28.

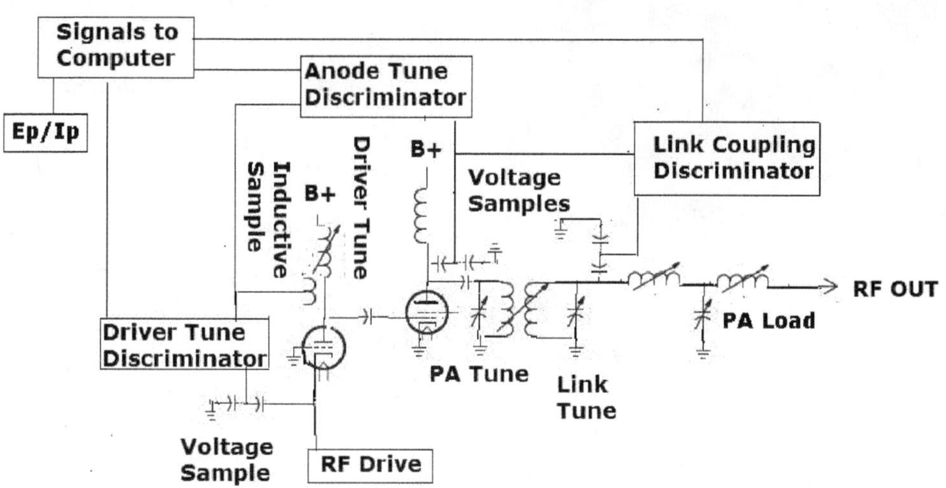

FIGURE 7.20-26 Continental HF Transmitter Output Matching Network

FIGURE 7.20-27 Continental HF Transmitter Link Coupling Inductors

FIGURE 7.20-28 Output T Network Components

A nonobvious benefit of the link coupling in the CEC design is the rejection of other HF signals from other adjacent HF transmitters and reduced intermodulation products.

Many channelized transmitters were built in the 1970s and 1980s for the 50- or 100-kW power level and typically had about ten channels. On this type of transmitter, the tuning conditions and frequency could be set for each of the ten channels. That was generally enough for that time. As microprocessor-based controllers became more robust and reliable they were introduced into servo control systems. These microprocessor-based tuning controllers allow an almost infinite number of channels to be memorized, far more than the original ten-channel autotune systems.

A 500-kW transmitter like the CEC 420C has 223 coarse tuning presets for every 100-kHz frequency segment between 3.9 and 26.1 MHz. These presets are for an ideal load impedance. Operating into an actual antenna with resistance and reactance errors requires that the transmitter adjust its antenna matching components to eliminate load reactance so the rest of the output network need only match a pure resistive load.

Tuning time can be further improved by storing the final tuning information in channels. Rather than call on a frequency of 17,775 kHz into an perfect antenna, the tuning on 17,775 kHz into Antenna 5 with +30 degrees of slew can be memorized in, for example, Channel 001 while 17,750 kHz into Antenna 6 with 10 degrees of slew can be memorized into Channel 002. They are both in the same 100-kHz segment. But they may have quite different tunings. By starting at preset channels, the tuning time if reduced to a minor touch up rather than a complete tuning sequence.

Large curtain antennas are sensitive to the spacing between the driven elements and the reflector curtain. These two parts of the antenna have different wind resistances—TCI's curtain being a notable exception. Counterweights on the array try to keep the antenna straight, but increasing wind speed can cause the antenna to physically deform resulting in impedance changes. Most high-power HF transmitters offer a tuning system that is always active (or that can be activated) to keep the PA always properly tuned and loaded despite wind conditions.

The real world is a lot more difficult than the picture painted in the previous paragraphs. Phase detectors are often sensitive to amplitude; ideally they shouldn't be. Voltage pickup capacitors, as shown for anode voltage in Figure 7.20-29 and secondary voltage in Figure 7.20-30 and Figure 7.20-31, are often imperfect in their phase response and the RF voltage at the anodes of the driver and PA are full of harmonic energy. Current samples are also frequency sensitive and introduce phase shifts. All cables to the phase discriminators must be matched in length to eliminate undesired phase errors. Introducing filters into the sample lines to eliminate harmonic energy also introduces unwanted phase shifts that require compensation. Mechanisms must be found to minimize the effect of signal levels on the phase discriminators and to correct any errors in the voltage and current samples.

Inductive current pickups like the driver anode toroid in Figure 7.20-32 are heavily swamped with resistance to ensure they are both broadband and give proper current phase samples. Voltage divider capacitors are made with plates of metal mounted solidly to a stiff support with dielectric between the pickup

FIGURE 7.20-29 Anode Voltage Sample

FIGURE 7.20-30 Output Matching Network Components

FIGURE 7.20-31 Secondary Voltage Sample

FIGURE 7.20-32 Driver Anode Current Pickup

plate and the support—a coaxial capacitor. Limiters can be used to stabilize inputs to the phase detectors. The increased logic capability of the microprocessor-based tuning systems allows errors to be compensated by using offsets stored in the tuning system. For example, the phase discriminator on a stage should equal zero at resonance but instead gives 30 millivolts when tuning is correct. This offset is stored in the computer memory allowing the tuning system to adjust the tuning to the correct 30 millivolts rather than zero. Tuning with modulation is even more trying. Low-pass filtering and zero-crossing detectors are used to strip off audio components on the phase detector signals. (Phase detectors are, unfortunately, sensitive to input voltage.) The result of all this care is transmitters that can tune themselves whatever the conditions automatically.

It is important to mention again that a change in frequency of a transmitter involves turning, twisting, bending, sliding and so forth. All can result in slipping connections, broken couplings and a significant mechanical maintenance requirement. Add to that the problem with vacuum capacitors mentioned before and it is obvious that the tuning system must be quick to recognize a tuning problem and stop the process before component damage can occur. Timers within the tuning programs monitor progress and shut down the process and alert the station operator/operation system if a problem occurs. Normally, programs are prioritized so that a high-priority transmitter failure (or failure to tune) is immediately remedied by pulling a lower-priority transmitter off schedule to allow the station to keep the high-priority schedule.

HF Transmitter Component Peculiarities

The object of the PA tuning network in HF transmitters is to match the anode impedance of the PA to the antenna/feedline and to suppress harmonics. The necessity of being able to tune over two or three octaves has consequences. Unwanted, parasitic resonances will occur in the output networks components. There are finite lengths of connections between the PA and the components of the output network. These finite lengths can become resonant with other components at a harmonic. When a parasitic resonance occurs at the wrong harmonic of the operating frequency, excessive voltage and currents may occur in or around the components.

The grid of the PA tube acts like a waveform clipper and multiple harmonics are generated of significant level at least up through the sixth or seventh harmonic. Class C operation of the PA results in the PA current waveform looks like a clipped waveform—again with significant harmonic energy through at least the sixth or seventh harmonic. If one of the associated components in the PA grid or anode circuit is excited by one of these harmonics the result can be serious component damage.

Every component—especially solenoid inductors used for decoupling—can have multiple resonances throughout the HF and VHF range.[21] Solenoid inductors are modified to eliminate unwanted resonances. One way is by altering the winding pitch over the harmonic frequency range or changing diameter as in Figure 7.20-33. If that change in pitch is insufficient, shorting relays or contact arms must also be used to detune the components away from the spurious harmonic frequency resonance. Odd shapes may also be necessary to avoid spurious resonances.

Harmonic energy can also excite parasitic resonance in the plate tuning/blocking capacitor circuit or in the output inductor itself. Mitigation of these problem required early designers to sweep the PA area with a grid-dip meter to find these unwanted, parasitic resonances. Later, the vector impedance meter was widely used to find and correct unwanted resonances.

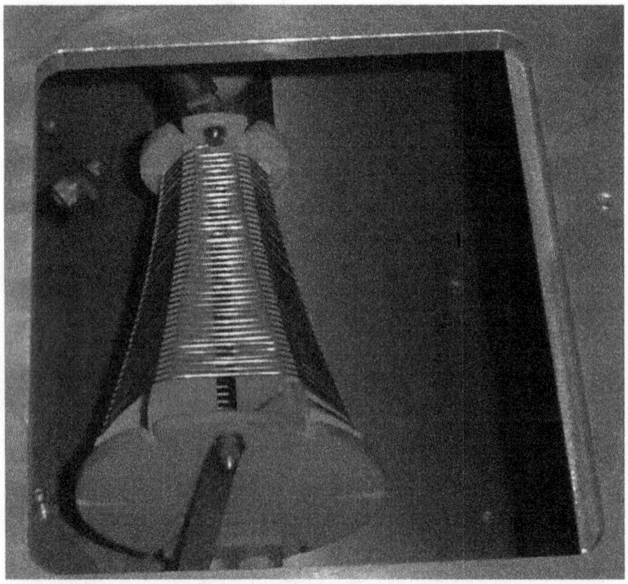

FIGURE 7.20-33 Decoupling Inductor to Eliminate Unwanted Resonances

Network analyzers are used now to identify frequencies where problems may occur and to find solutions. Whether the output network is a hairpin or cavity style, solving the problem of parasitic resonances in the actual output network may require that components not be operated at their theoretical position. This work of finding and eliminating spurious resonances is pre-formed when the transmitter is designed. Deviations from the theoretical values and the detuning of resonant inductors are programmed into the coarse tuning presets of the transmitter.

Other techniques used are absorbers that may be tuned to a harmonic frequency that is not particularly high in amplitude but might slip through the output network and show up as excess harmonic energy in the transmitter output. Absorbers, as in Figure 7.20-34, are seldom found in the stages of a transmitter after the PA grid due to the large power that may be dissipated. There are very limited exceptions. An absorber is normally recognized as a noninductive resistor that is not connected to anything, or a noninductive resistor tied to a circuit element on one end and nothing on the other end relying on distributed, stray capacitance to the chassis ground.

It is often necessary or desirable to parallel two capacitors to get an adequate variable capacitance range and/or current-handling capacity in some transmitter designs. It is found that such arrangements can exhibit resonances at VHF frequencies. A common way to minimize the effect of these VHF resonances is to place noninductive resistors across the interconnect plate to lower the Q of the VHF resonance and to absorb any voltage developed across the interconnect plate. The interconnect plate is the inductor of the

FIGURE 7.20-34 Harmonic Absorbers

FIGURE 7.20-35 100 kW Band-Switched Output Network

FIGURE 7.20-36 Solid State MW Transmitter Series PA Combiner

VHF resonance circuit and the resistors shown in Figure 7.20-35 prevent voltage buildup across that interconnect plate. Absorbing the VHF resonance energy at the source prevents it from appearing in the output of the transmitter.

Solid-State HF Transmitters

Low power MW or FM transmitters typically use Wilkinson-type combiners to get 5 or 10 kW operating power. The Wilkinson-type combiner is frequency dependent using quarter-wave line sections. By loading the coax with ferrite it is possible to form a star combiner that is good for combining a large number of modules although at a low output impedance. The lower the impedance, the higher the resistive losses and the lower the efficiency.

The star combiner is a method of using coaxial cable. Each cable from each separate amplifier is loaded with ferrite to form an isolated output. The center conductor from the first amplifier's coax is connected to the shield of the second coax's jacket, the second coax's center conductor is connected to the shield of the third coax's shield, and the Nth coax's center conductor is connected to the first coax's shield. Obviously this is a series combiner just as much as using ferrite coupling transformers to couple to a combiner rod. The star is good for about one octave of frequency coverage with ferrite losses climbing at band ends.

Higher power MW transmitters use a series combiner with each module inductively coupled to a combiner pipe so that each module can add its contribution as shown in Figure 7.20-36.

Solid-state transmitters can have quite high efficiency. Spurious-free, high-efficiency RF combiners where three-octave coverage is required can result in reduced overall DC-to-RF conversion efficiency at HF. Contrast this with tube-type RF amplifiers that have 85–90% efficiency coupled with 95%–97% modulator efficiency over the same three-octave range.

VHF and UHF Harmonics

ITU Recommendation ITU-R SM.329–10 requires that spurious emissions be limited to 50 milliwatts while the FCC requires that spurious emission be limited to −80 dB measured up through 1 GHz. Class C PA amplifiers are prodigious generators of RF harmonics. Thankfully, suppression of these harmonics by the PA matching and subsequent networks is almost complete. There are always "holes" in the rejection characteristics of any two- or three-octave output network as the tuning components always have parasitic inductance, capacitance and resistance that can allow spurious emissions to appear at the output.

Harmonic attenuation at the 100-kW level can be done by absorption and discrete-component networks of proper design. A low-Q absorber may be all that is needed for a single troublesome spurious range. Outside that single range, other rejection/suppression devices are required.

A typical rejection lowpass filter design would be two 90-degree lagging Pi networks back-to-back combining the output capacitor of the first network with the input capacitor of the second network making a single center capacitor. By turning the shaft of the center capacitor twice as fast as the end capacitors the center capacitor always has twice the capacitance of the output and input capacitors. (There is a straight-line relationship between turns and capacitance in standard vacuum variable capacitors.) The Pi-network inductors are spiral wound and can also have a straight-line (linear) inductance versus turns relationship. These twin-Pi filters, shown in Figure 7.20-37, have sufficient rejection into the VHF and UHF spectrum to attenuate all harmonics to the required specification. This is primarily because the voltages and currents are low enough at the 100-kW power level that the components can be small enough and close enough so that the components approximate perfect components. Spiral inductors,[22] like that shown before and in Figure 7.20-38, are to be found in almost every HF transmitter.

FIGURE 7.20-37 Spiral Wound Twin-Pi Filters

FIGURE 7.20-38 Example of Spiral Wound Inductor

The situation for VHF and UHF harmonic control at the 250/300/500 kW powers is not simple. The higher voltages and currents require larger components with less-than-ideal performance because of their physical size and spacing. The solution is to use filters made of microwave analogs of inductors and capacitors as shown in Figure 7.20-39. Designs for the VHF/UHF filters are beyond the scope of this chapter but, typically, sections of 6-ohm line are used as capacitors and interconnecting rods are used as inductors.

COTS (commercial, off the shelf) coaxial VHF/UHF lowpass filters are available. They are typically water-cooled as are designs constructed by transmitter

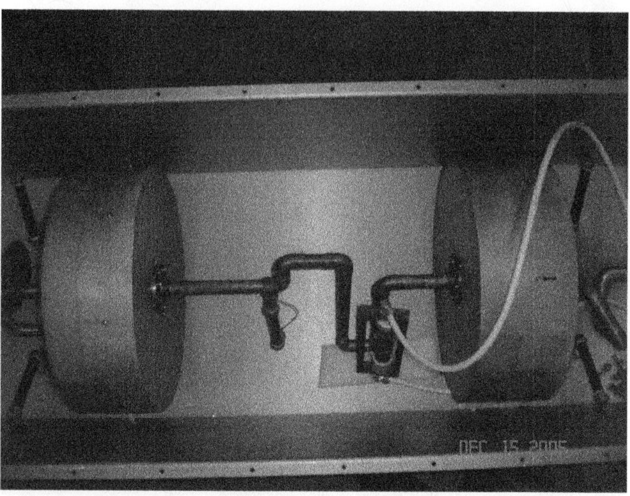

FIGURE 7.20-39 VHF-UHF Lowpass Filter

manufacturers. Balanced-line VHF/UHF lowpass of 50, 75 and 300 ohms have been developed by transmitter manufacturers.

Other Spurious Emissions Control

All high-power PA tubes with circular geometry emit Barkhausen Radiation. Typical tube-element spacing results in radiation at about 500 MHz in the tube area. This seldom escapes from the transmitter because the panels and doors of the transmitter PA area are designed and tested to ensure that this radiation does not escape the cabinet and irradiate transmitter plant personnel. This radiation results when the electrons from the filament are forced back to the filament by negative screen and/or anode voltage.

The operating voltages of high-power PA tubes result in X-rays and soft X-rays being emitted from the PA tube. The transmitter's shielding panels and doors are responsible for X-ray control and shielding. The exit point of X-rays from the PA tubes is known to be a cone around the ceramic insulation between the anode and screen grid. Soft X-rays are the most biologically active but are also the easiest to shield.

Measurement of Spurious (Harmonic) Emissions

No section on harmonics would be complete with a brief discussion of techniques that must be used to correctly measure harmonics. The first requirement is a good sample using something like a coaxial capacitor that forms a voltage divider that is measured and proven to be accurate over a wide frequency range. Unless the sampling system is frequency compensated the measurement results will be corrupted. The best is a frequency compensate sample taken from the dummy load. Calibration of the sample can be done with a tracking generator, if available.

Figure 7.20-40 shows a perfectly fine driver cathode voltage sample for use in driver anode tuning. It

FIGURE 7.20-40 Driver Cathode Voltage Sampling

would be a poor model for a voltage sampler for harmonic measurements. At some frequency the inductance of the capacitor leads will go resonant with the C of the capacitor resulting in an exaggerated response. The design of the coaxial bypass capacitor to ground is perfectly acceptable. A proximity sample similar to that used to sample the PA anode voltage is almost frequency independent. One factor the must be considered is that the coaxial capacitor to ground be relatively high capacitance with low capacitive reactance relative to the input impedance of the spectrum analyzer, typically 50 ohms. If the capacitive reactance is not low enough the pickup capacitance will form a voltage divider with the input R of the spectrum analyzer resulting in a 6 dB per octave rising frequency response.

The second requirement for accurate measurements is a fundamental-reject filter, a simple, well-shielded and relatively high-Q parallel resonant circuit in series with the signal applied to the spectrum analyzer. If a fundamental-reject filter is not used you will most likely measure the dynamic range of the instrument rather than the actual harmonic level.

The third requirement is that the spectrum analyzer (or other instrument) be operated at the point where its own spurious responses are lowest, that is, where its dynamic range is greatest. The general rule is that this highest dynamic range point occurs at 0 dBm. A step attenuator or fixed attenuator to reduce the signal to the optimum level assures that the instrument is operated at its most accurate point. Because 0 dBm is 50–57 dB below the transmitter power output level, interconnecting coaxial lines must be of the highest quality with the greatest possible amount of braid cover and external signal rejection. And all elements (spectrum analyzer case, step attenuator case, fundamental-reject filter case, etc.) must be solidly grounded together with short pieces of wide braid or strap.

To make a proper measurement—after all the preceding precautions have been taken, the transmitter is first operated with into a dummy load with the carrier level input to the spectrum analyzer set to 0 dBm and the fundamental-reject filter out of the circuit. The carrier is then nulled for minimum value and the harmonic level noted. As shown in Figure 7.20-41, a transmitter operating on 13.65 MHz has a harmonic

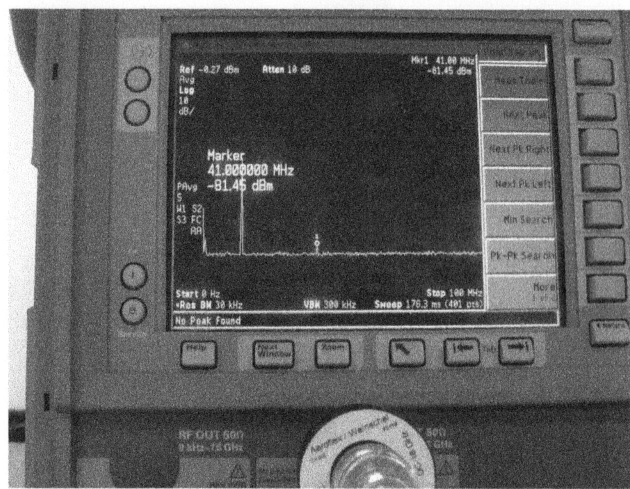

FIGURE 7.20-41 Spectrum Analyzer Display

output level of −81.45 dB, meeting both the 50 V international requirement and the US −80 dB requirement. If a fundamental-reject filter were not used, if the analyzer's front end were overdriven and/or if sufficient care in the signal sampling and equipment grounding is not taken, the displayed result might be far different than that measured when the proper precautions are followed.

Measurements made into an antenna are generally not accurate because the harmonic frequencies are normally well outside the normal operating range of the antenna components. This results in VSWR, which can cause the harmonic level to constructively or destructively interfere at the sample point.

Field measurements of harmonics can be made but a calibrated antenna must be used to ensure that the performance of the sample antenna is known and all other precautions are followed.

Great care and careful procedures will give accurate results. Poor attention to detail will give poor results at substantial variance with the actual harmonic level.

Modern Modulators

HF Broadcast transmitters almost always use a high efficiency PA that is supplied with DC + Audio to achieve amplitude modulation. All use PDM modulators of one form or another.[23] A short review of PDM modulator basics is in order. Refer to the schematic in Figure 7.20-42 in this discussion. There are five essential elements of any PDM modulator: a power supply, a switch, a commutating circuit consisting of a commutating diode, a commutating inductor and a commutating capacitor. The switch may be a single tube or multiple switches in series that use IGBTs (Insulated Gate Bipolar Transistors) that require low drive but have low turn-on voltage drop. When the switch is closed, a voltage appears across the inductor that represents the difference between the output voltage and the supply voltage. Energy is stored in the magnetic field of the commutating inductor. When the switch is

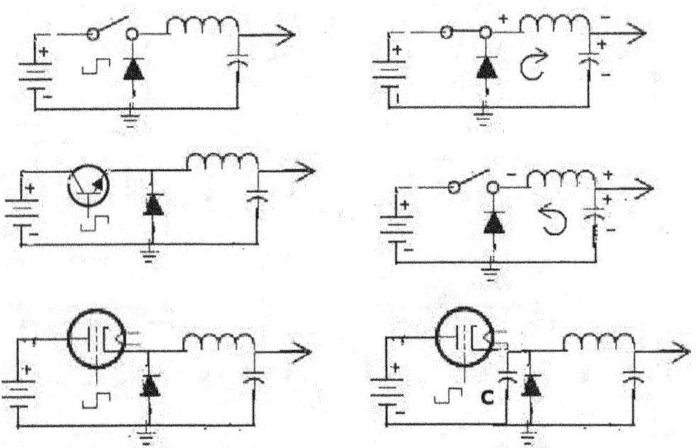

FIGURE 7.20-42 Schematic Diagrams of PDM Circuits

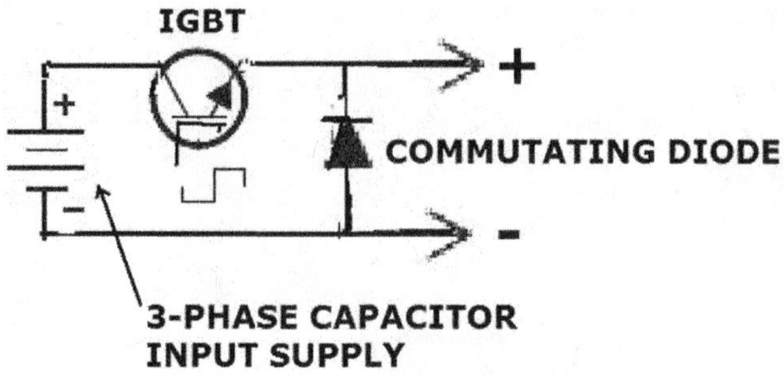

FIGURE 7.20-43 Power Supply Switch

opened, the magnetic field starts to collapse and the voltage across the inductor reverses. The commutating diode conducts the stored energy to the commutating capacitor. A 50% duty cycle on the switch results in a voltage output of the supply that is 50% of the supply voltage.

The earliest PDM HF Broadcast transmitters[24] used what is often called a "high deck" PDM Modulator. A tetrode tube whose anode voltage is the supply voltage (more than twice the RF anode DC voltage) is mounted on an insulated platform and is powered by an isolation transformer. The tube is either fully cut off or fully conducting to minimize loss. Unfortunately, the large amount of stray capacitance between the tube cathode and ground results in large charging and discharging currents in that stray capacitance.[25] The switching frequency of this type of modulator was typically about 50 kHz and the upper modulating frequency range was about 7 or 8 kHz, about one-quarter the switching frequency. The stored energy in the stray capacitance leads to undesirable effects especially in the negative trough of the modulation waveform. An almost universal fix was to underdrive the grid of the switching tube as pulses get shorter and shorter. This degraded efficiency but, as the power was low, the lost efficiency was immaterial. In the late 1960s both Gates Radio and Collins Radio developed PDM systems to considerably reduce the amount of and effect of the capacitance at the modulator anode. In the Gates Radio system the PA section was in an isolated box that was bypassed to ground with the input inductor of the PDM commutating filter inductor connected to the anode. In the Collins Radio system the PA anode was grounded and negative anode voltage was applied to the modulator cathode with the modulator anode also connected to the input inductor of the commutating filter.

The commutating diode has a different problem. When the diode is conducting energy from the commutating inductor to the commutating capacitor the diode will not stop conducting instantaneously when the switch tube turns ON. The switch tube turns on into a momentary short circuit. The duration of that short depends on the recovery time of the commutating diode or diode assembly. Gold doped diodes recover from the conducting to nonconducting condition in about 300 nanoseconds. Platinum doping allowed recovery times in the 100-nanosecond range. Power was lost in the voltage drop across the diode and the short circuit energy when the switch tube first came on. Faster diodes give higher efficiency and are almost universally used.

The C in the diagram in Figure 7.20-42 is the stray capacitance that must be charged and discharged—the performance limiting problem. The diagram shows the magnetic field of the inductor being generated and collapsing as the switch closes and opens.

Resistance in the inductor and the capacitor also resulted in additional minor losses but, as a system, the modulator efficiency could easily reach 90%—far better than the old push-pull Class B modulators used before. For 500 kW service a typical modulator supply delivered 30–33 kVDC.

A major series of design changes resulted in the use of 32 to 48 individual 700- or 900-volt supplies (depending on the manufacturer and generation) connected in series. Rather than have all the switches all switch on and off at the same time an ingenious system followed. Consider a group of 32 individual 900-volt switches connected in series. The commutating inductor and commutating capacitor summed into single units as shown in Figure 7.20-43. For a desired anode voltage of 14 kV the power supply duty cycle is just under 50%. Thus 16 switches had to be on at about 95% at any time to give the necessary 14 kV. Using GTO (Gate Turn-Off) Thyristors on each switch, half the switches would be on and half would be off. A 70 kHz signal was injected into the DC component controlling the switches such that switches were turned on and off sequentially such that only half the switches were on at any time—but the average number was always just fewer than sixteen.

One hundred percent Positive Modulation resulted in all 32 switches being turned almost fully on so that the applied voltage was 28,000 volts. At 100% negative modulation no modules were on. Because the 32 switches were incremented and decremented at 70 kHz the average switching speed for each gate turn off transistor was only 70 kHz/32 or ~2.2 kHz.

This design was improved enormously by using 48 independent 700 VDC switches and using IGBTs (insulated gate bipolar transistors) as the switch elements.

The switching frequency was lowered to 70 kHz/48 or ~1.45 kHz. The overall efficiency of such a modulator design could exceed 97% in 500 kW service. Because 33,600 volts was available with 48 individual 700 VDC switches a number of switch modules could be defective without any effect on positive modulation capability. Typical 500-kW anode voltage is 15 kV.

The 4CV100,000C tube, for example, operates best at 100 kW with an anode voltage of 14 kV—a perfect fit for the 33.6-kV modulator. Other transmitter manufacturers cut the number of switches at the 100-kW level to only 32 individual 700-volt switches so that the maximum anode voltage available for the PA tube is only 11 kV with no reserve modules. Other tube types designed for the lower anode voltage and higher anode current had to be used while sacrificing some anode efficiency.

Transformers to supply the switches are wound with 16, 24 or 32 windings. When 16 or 24 winding transformers were used, they were configured with their primaries wound in extended deltas to give a +15 and −15 degree phase shift; one transformer would be power ABC and the second ACB, reversing the phase shift. The result was a twelve-pulse ripple on the whole modulator: 600 Hz at 50-Hz line frequency and 720 Hz at 60-Hz line frequency. In a large operation transformers with +15 and −15 degree phase shifts plus others with +7.5, −22.5 and −7.5, +22.5 degree phase shifts were used and resulted in 24-pulse service with very high power factor and low harmonic distortion.

It is also possible to get twelve-pulse ripple using a single transformer if the secondary windings are wound half wyes and half deltas; the primary can be a standard (not extended) delta. Refer to the block diagram in Figure 7.20-44. Transformer manufacturers are skilled at providing delta and wye windings that have, within a few tenths of a volt, the same under-load output voltage from the switches. All the windings are ~490 VAC phase-to-phase to give 700 VDC output of each switch and ~420 VAC for the 600 VDC switches. (The primary is always in a delta to keep down harmonics reflected back into the supply line.)

It is not possible is to get the same exact voltage from all windings, although a designer can get very close. Even with exactly the same number of secondary turns, the windings near the center of the transformer will have slightly more voltage that those at the top and bottom of the transformer. PDM modulator designers compensate for this by firing switches in a low-high-middle sequence to keep modulator noise well under the desired value. (If the switches were fired sequentially the voltage differences would show up as noise.) When two transformers are used the same technique is used but the next switch incremented will be one from the second transformer while the decremented switch will be from the first transformer. The actual sequence is stored in logic. The FPGA (Field Programmable Gate Array) is used in modern transmitters to ensure that the switches are fired in the correct order, to exclude switches from the firing order when their status signal is lost, to ensure that all modules are on an equal amount of time and so forth. The logic could be hard-wired components but the FPGA-based logic allows

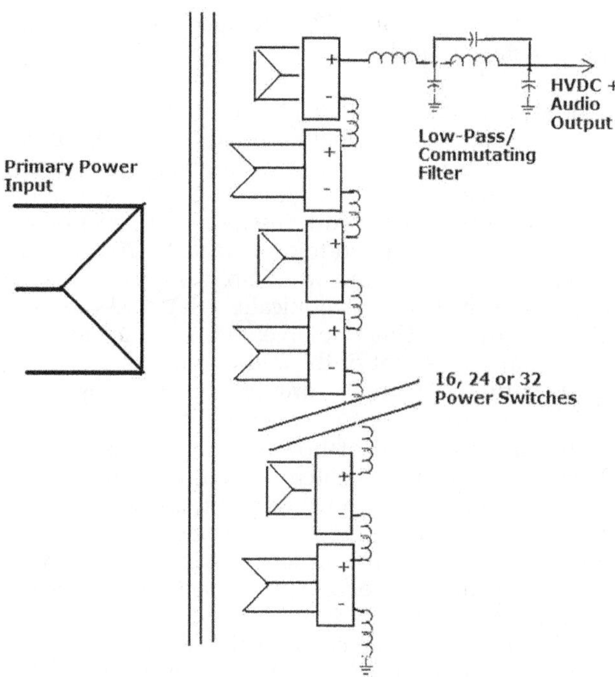

FIGURE 7.20-44 PDM Block Diagram

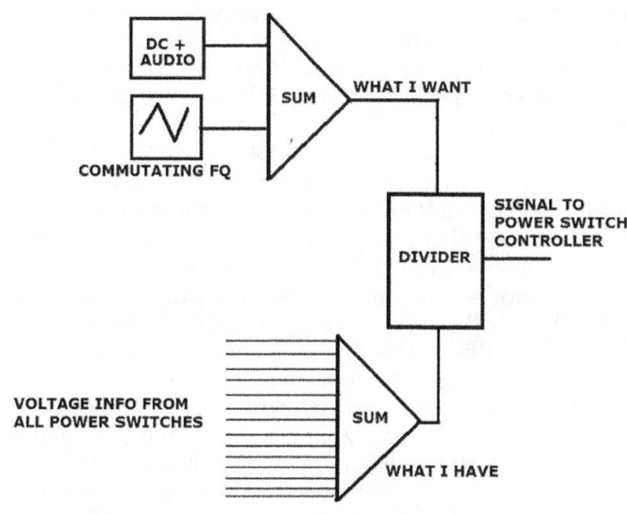

FIGURE 7.20-45 Low Level PDM Modulator Stage

alterations in unique cases to be compensated for by a simple software modification, not rewiring.

Note that when each switch turns on it must charge the stray capacity in the transformer of every switch above it. The input (commutating) inductor is broken into smaller sections and placed between each switch to limit the peak current on the IGBT.

DRM has audio and RF components from DC up to the 40–50 kHz range. Obviously, 50 or 70 kHz was no longer an adequate switching frequency. The switching frequency was changed to the 150–200 kHz range. This increased the switching speed of each IGBT to ~4 kHz just as faster IGBTs with reduced losses were becoming commercially available. The result has been a threefold increase in switching speed with the same high (97%) overall efficiency.

Two further subtle improvements were made to high-power, solid-state PDM modulators. First, the switching frequency (whether 50 kHz, 70 kHz or 200 kHz) was dithered so that a fixed audio modulating frequency was always sampled at a different point, that is, uncertainty. Second, the increment and decrement clock was designed to operate at 10–20 times the switching frequency and was also designed so that it could never become harmonically related to the dithering frequency. The implementation of uncertainty in the sampling point of the audio and uncertainty in the exact time of incrementing or decrementing of the switches improved performance significantly.

The determination of how many switches should be on and how many switches should be off is determined in the lowest-level stages of the PDM modulator shown in Figure 7.20-45. Status signals from each switch are fed back to a summing circuit that provide a sum voltage that is an exact replica (including hum and noise) of the total amount of voltage available. A DC signal representing the carrier value desired is mixed with a bandwidth-limited audio signal to form a DC + Audio signal that is the exact representation of the desired modulated waveform plus the dithered sampling frequency. The desired signal is divided by the sum signal to determine how many switches should be on at any instant. As the sum signal contains any residual hum or noise the resultant division contains an equal and opposite signal to cancel any modulator ripple noise.

Status signals from the 700 VDC power switches are also used to test each switch for proper operation. If the correct signal is not received from the power switch status fiber-optic lines, the switch is simply taken out of the available pool and normal operation continues.

It is not immediately obvious but the system described for power control is self-regulating and noise cancelling.[26] The incoming AC line voltage may rise or fall. That results in a change in the present voltage: the status sum signal. The division by the desired (audio + DC) voltage changes the fractional value passed on to the subsequent stages. The output voltage from the modulator remains absolutely constant while the incoming voltage may vary widely. Further, any hum/ripple coming from the supply shows up the denominator. Thus the fractional value is changing with any hum component. The ripple (or hum) out of the total supply voltage, status sum signal, is completely cancelled.

Self regulation is important in eliminating any carrier shift due to line voltage sags. And the status signals operate in the microsecond rate while any ripple is at a millisecond rate. The output signal from the modulator is a pure representation of the input DC + Audio signal with zero linearity error. (Nonlinearity may be introduced by the PA, but not the modulator.)

DRM requires much wider amplitude bandwidth (40–50 kHz) than a conventional AM signal. (SSB requires about 20 kHz.) The amplitude component of the DRM composite signal is sent on to next stage without bandwidth limitation to generate the higher frequencies necessary. For conventional AM the bandwidth is limited. The next state (after division) is a group of FPGA (Field Programmable Gate Arrays) that determine how many switches are to be ON and when to turn others on and off based on the result of the present versus desired division and ensure that all switches are on for the same average amount of time to distribute the load equally. It should be noted that all DRM exciters include circuits to linearize the PA modulation and to cancel and IPM (Incidental Phase Modulation) due to poor neutralization or tuning errors. Most DRM Exciter pre-distortion circuits are proprietary but a technique called "Pure Signal" is equivalent and perhaps easier to understand.[27]

AM on HF has a recommended 4.5-kHz cutoff frequency and some manufacturers put this frequency limit in their low-level stages rather than allow the audio processor to do the job. Care should be taken that the only one or the other cutoff filter is used to prevent excessive high-frequency roll-off.

The post-modulator lowpass filter that passes the desired audio and eliminates the switching remnants from the modulator has a time delay. In addition, the IGBTs do not turn on for about 200 nanoseconds after the ON signal is received. The DRM exciter must delay the Phase component of the DRM signal fed to the RF input of the transmitter so that the amplitude and phase modulation components arrive at the PA simultaneously. Typical RF delays required range from 12–18 microseconds depending on the post-modulator lowpass filter design; higher impedance filters (as at the 100-kW level) have more time delay due to the higher impedance of the filter's termination (about 1800 ohms) and the large values of C in the RF bypass capacitors.

The design requirements of the post-modulator lowpass filter (the commutating inductor and capacitor) are beyond the scope of this chapter. The inductor and capacitor normally become a series of inductors and capacitors in a complex configuration. The frequency response of the final design must be virtually flat to 40–50 kHz, rejection of the commutation frequency or frequencies must be high (> 80 dB) and the phase shift of the filter must be linear or constant.

A final note about solid-state modulators concerns the IGBT turn-off time. It is roughly 150%–200% of the IGBT's turn-on time. This precludes the type of tail-biter circuit often seen in lower-power PDM (or polyphase PDM) transmitters without elaborate/expensive time constant circuits.[28] The power lost in the tail-biter dissipation resistor would also result in a decrease in the overall efficiency of the modulator. Low-power transmitters use transistors without the delay of IGBTs.

Tetrode Screen Modulation

The screen voltage of a tetrode tube must be modulated.[29] Screen voltage mμ is very much higher than anode mμ. We say that anode current is a function of screen voltage. The screen voltage and screen current must be kept high enough to keep the tube in saturation. Depending on the tube the modulating voltage required for proper screen modulation can be very nonlinear (for linear modulation) and is normally full of second harmonic distortion. The easiest way to achieve the screen modulating voltage is with self-modulation. A large (10 Henry or more) audio reactor is placed in series between the screen power supply and the screen of the tube. Energy is stored in the reactor. As the anode voltage rises during positive modulation the screen current starts to fall. But the energy in the reactor's magnetic field collapses and attempts to keep the screen current constant. As the anode voltage decreases during negative modulation more energy is stored in the magnetic field as the reactor attempts to keep the screen current constant. This screen voltage self-modulation allows the tube to get the screen voltage it needs regardless of the tube's characteristics.

A variation on this theme is a much higher screen voltage supply that uses both a modulation reactor and screen-voltage-dropping resistors in series. In some cases the anode voltage must rise above a certain value before the screen voltage comes on.

CEC's SSM (solid-state modulator) units include a PDM screen-voltage supply that operates to provide a screen voltage that is variable and proportional to the anode voltage. This removes the necessity of screen dropping resistors or having a minimum anode voltage before the screen power supply is turned on. Protection of the screen Grid and control Grid from internal arc damage is covered later.

Primary Power Control

There are two ways of applying power to the Solid-State Modulators used in modern transmitters. One is simply to apply the primary AC voltage at full voltage to the primary(ies) of the power transformer(s.) If this happens at a zero-axis crossing on one phase the core becomes magnetized and the primary on that core section tries to become a linear motor and throw itself off the core. The transformer bracing and reinforcement are strained each time this happens.

Cast-resin windings are often used in Europe and some of these have failed because the stress on the winding from high turn-on current cracked the cast-resin cover. The 250-kW Doherty transmitters at Kavala had multiple transformer failures from both turn-on shocks and operation of the "crowbar" circuits. (Recall that crowbar circuits short the HVDC directly to ground, and the short on the primary of the transformer persists until the circuit breaker finally opens.) The solution at Kavala was to put large air-core starting inductors in series with the primaries of the HV transformers.

The shock to the transformers from crowbar operation has been eliminated by the high-speed over-current sensors in modern SSM designs. But the turn on current problem remains and the surge current is limited only by the impedance of the transformer and the resistance of the primary windings—and the resistance is very low while impedance is about 6%.

A much better way to turn on a high energy supply is to use contactors to charge the capacitors of the SSM through a series resistance and then short out the starting resistance, that is, a "step-start" circuit as shown in Figure 7.20-46. One major caution: the contactors are not normally rated for opening or closing under fault currents. The contactors typically use DC coils, so it is no problem to just introduce a reverse-biased diode across the coils so the contactor opening is delayed about 250 milliseconds. This allows the primary-power circuit breaker that *is* rated for opening under fault currents) to handle the fault.

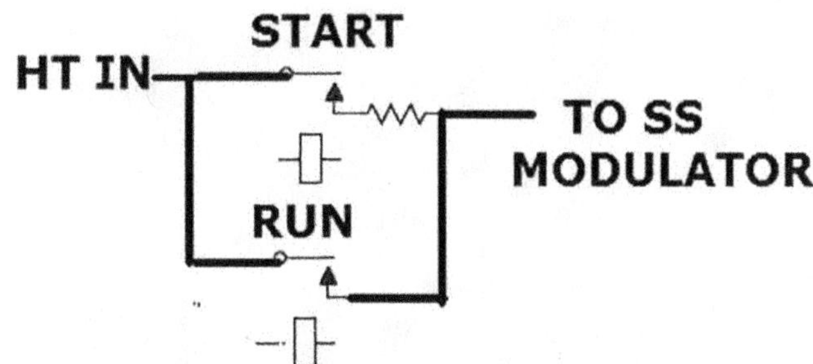

FIGURE 7.20-46 Step-Start Circuit Configuration

A second caution when using step-start: the start contactor should open after a time period that allows the run contactor to close. Almost every manufacturer that has used a step-start circuit has experienced a failure of a run contactor to close or a defective section of a run contactor. Either circumstance can result in the failure of the resistors due to overheating.

Selection of the amount of starting resistance is critical as is the time the resistance is in the circuit. The resistors for this type of service are rated in joules (watt-seconds). Typical primary voltages vary from 380 V to 20 kV. Most SSM capacitor charging times (to 90% full voltage) are about 500–700 milliseconds. The IBB standard HT (high tension) AC is 4160 volts, 2400 volts to ground. Assume there is a short to ground so that 2400 volts is applied across the starting resistor. Also assume that the run contactor closes in one second. For a charging resistor of 67 ohms selected for a peak turn-on current of 36 Amps, $E^2/R = 2400^2/67$ or 86,000 joules. A check of the Kanthal catalog says the 18" long resistor, an 1038AS-67, is rated for 119,000 joules and 120,000 volts, peak.

Primary Power Components

The installation engineer should ensure the ability and experience of each and every individual associated with installation of all primary power components. In the failure incident shown in Figure 7.20-47, the technician made a shortcut on the center conductor of the HV cable. In an effort to prevent having to run a new cable he patched the center conductor with a splice and electricians tape and hid his work with a stress cone. The result was, of course, a difficult repair under heavy time pressure.

Cooling

Almost all HF transmitters use a combination of forced air and water cooling. Water is used for cooling

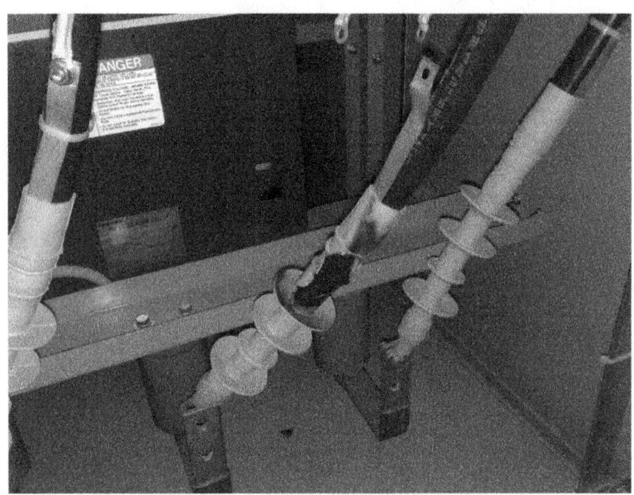

FIGURE 7.20-47 Stress Cone Installation Failure

FIGURE 7.20-48 Modified Marconi Water-Cooled Output Stage

medium and large vacuum tubes and vacuum capacitors especially in the PA tuning area. See Figure 7.20-48. The high voltages at the PA anode result in high currents in the PA tuning capacitors. At a Q of 10 the capacitive reactance is only one-tenth the RF impedance. Skin effect and resistance loses result in heat that must be taken away by water. The PA tubes are also kept smaller by using water cooling because of the high heat capacity of water. Other high-current components in HF transmitters also cooled by water can include the plate tank coil and, in 500 kW transmitters, any inductors in the transmitter as in Figure 7.20-49. Even ceramic blocking capacitors can be water cooled and shown in Figure 7.20-50.

Forced air cooling is only used for hot air removal from the transmitter and for spot cooling of hot spots. Most transmitter designs try to minimize the amount of heat removed by forced air cooling by maximizing the amount of heat removed by water. This philosophy minimizes the amount of outside air necessary for transmitter room cooling. Heavy insulation is used in the doors and cabinets of well-designed transmitters to prevent heat radiation into the transmitter room.

The water used to cool the transmitter is plumbed outside so the heat can be transferred to the outside air in nonfreezing climates. Two 250-kW transmitter heat exchangers for 40°C maximum outside temperature are shown in Figure 7.20-51. The sealed water system totally isolates the transmitter from contaminants in the outside air except for (hopefully filtered) flushing air. The outside heat exchanger must be rated for both the amount of heat to be dissipated under worst-case conditions of outside temperature and transmitter heat dissipation. The heat exchanger is subject to all the elements that the transmitter is shielded from and special precautions must be taken in the selection process. Corrosive atmospheres can attack the aluminum fins on the outside of the copper tubes of the water circuits. In exchange for slightly reduced cooling efficiency, a thin vinyl or hirsute coating can be applied to the cooling fins to prevent corrosion and eventual heat exchanger failure.

FIGUR 7.20-49 Water-Cooled Output Stage Inductor

FIGURE 7.20-50 Water-Cooled Output Stage Ceramic Capacitor

FIGURE 7.20-51 Two 250-kW External Heat Exchangers

Selection of the water-to-air heat exchanger depends on multiple factors including water flow rate. Turbulent water flow increases the efficiency of the transfer of heat from the water to the air. If the water flow in each circuit of the heat exchanger goes below 3–4 gallons per minute (GPM), laminar flow may result. Laminar flow is nonturbulent water flow where the outer water flow in the cooling tube slows and acts as an insulator between the inner water flow in the heat exchanger pipes and the efficiency of the heat exchanger drops dramatically. The circuits wrap back and forth in the heat exchanger as many times as necessary to get the rated heat transfer rating. Because the minimum water flow per heat exchanger circuit can vary with cooling pipe diameter, it is best to work closely with the heat exchanger manufacturer to make the optimum selection for each application.

Note that a heat exchanger that is out of service for even a short period of time cannot be left open to the air. Any water remaining inside quickly becomes acidic and the heat exchanger oxidizes from the inside out. When received new, a heat exchanger will have all water removed and will be pressurized with dry nitrogen to prevent damage. A heat exchanger taken out of service—whether water or steam—must be treated to protect it from damage. The best way is to use a high pH liquid to prevent the copper from oxidizing, similar to the high pH antifreeze used in modern automotive engines. Most heat exchanger manufacturers recommend a solution of NaMBT (sodium mercaptobenzothiazole) to protect an out-of-service heat exchanger. If not available, automobile antifreeze can be used, although it may be more difficult to remove all the additives found in antifreeze from the heat exchanger when placed back into service.

Steam cooling of the anode(s), also called vapor-phase cooling, was widely used for a 10- to 20-year period starting in the late 1960s to cool high power transmitters of all frequency ranges. A 250-kW PA with the glass steam exit pipe is shown in Figure 7.20-53. There are two problems with the first implementations of steam/vapor-phase cooling. The first, most obvious problem is that the water temperature must be raised to 100°C to boil and the steam must be routed to the steam-to-water condenser outside. The higher temperatures inside the transmitter and the steam plumbing heat loss results in much higher heat loss to the room.

FIGURE 7.20-52 250-kW PA With Glass Steam Exit Pipe

The second problem with steam cooling was much more subtle. Many implementations that used no water pump worked when used in LW, MW, FM and TV service where the transmitters were always on. In HF the transmitter RF output normally must shut down for frequency changes so there are continual on/off cycles of steam generation. When transmission resumed again and the water had risen to boiling temperature and steam generation started, steam pressure had to first force the air out of the steam lines and condenser before the cooling could start. This results in high back pressure on the anode. Back pressure can force water and debris from the anode back into the water-supply line and, even worse, lower the water level on the anode. The top of the anode becomes uncovered and overheats. Eventually the back-pressure problem resolves but damage may occur because of this repeated over-temperature at the top of the PA anode or failure of the water-supply line.

The solution to this second problem was a transmitter design that had a water pump that always supplied the anode with a surplus of water that was returned to the water cooling circuit over the top of the PA anode boiler. This did not resolve the room heat radiation but did slow the rate of PA tube damage or loss due to anode-top exposure. The pump, of course, could be much smaller than required for a totally water-cooled pump. It made surplus pressurized water available for cooling other circuit areas without a large air-cooling blower. It should be noted that all vapor-phase tube designs either use a large number of sharp-edged fins or a "pineapple" surface design of raised sharp squares to promote boiling at the lowest possible dissipation.

Most PA and driver tubes in the 250–500 kW HF power range use water-in/water-out cooling. PA tubes at 250/300/500 kW typically use hypervapotron tubes. A 250 kW PA stage using a hypervapotron tube is show in Figure 7.20-52. These water-in/water-out tubes do boil water at the anode but also have built-in condensers to convert the steam back to water before leaving the tube. Hypervapotron tubes have been used at the 100 kW power level but in very limited numbers.[30]

Continental Electronics developed a now widely used modification for older vapor-phase transmitters that uses a shell and tube heat exchanger, shown in Figure 7.20-54, on top of the transmitter. This condenses the steam back to the water phase immediately.

FIGURE 7.20-53 Hypervapotron PA Tube and Tuning Coil

The steam heat exchanger can be eliminated. Use of extra insulation on the hot-water outlet line keeps excess heat from the room. The efficiency of the water-to-air heat exchanger increases as the water/air temperature differential increases. A shell and tube steam condenser is shown. An alternative to the shell and tube heat exchanger could be a plate heat exchanger, as is shown in Figure 7.20-55.

Glass is almost universally used for steam lines. CPVC (chlorinated polyvinyl chloride) tubing cannot be used for steam lines despite its electrical suitability. CPVC strength is down to 50% at 60°C and is almost nonexistent at 100°C, the steam temperature. Water-cooling, where freezing temperatures are encountered, requires a primary pure-water loop, a water-to-water heat exchanger and a second pump and water/ethylene glycol cooling loop. Pure ethylene glycol is not required; automobile antifreeze may be used where it is more readily available. (Pure glycol has no corrosion inhibitors and should be buffered to a high pH to prevent oxidation of the copper components.)

This second loop introduces a second temperature differential between the temperature of the outside cooling air and the return water to the transmitter. Flow in this secondary loop should be kept as high as possible to increase the efficiency of the water-to-water and water-to-air heat exchangers.

A well-designed transmitter also has the maximum anode water flow consistent with system pressure limits. The TH558/4CM500,000A is a popular tube used in 250/300/500 kW transmitters. Some older design transmitters used only 30 GPM of anode water flow in 500 kW service and tube life was markedly reduced. A better flow rate is 55 GPM. Mistuning with 30 GPM water can result in steam not being fully condensed in the anode's internal condenser. Steam escapes into the return water line often resulting in failure of the flexible anode-water return line. Tube life and system reliability is markedly improved by the additional cooling water flow.

FIGURE 7.20-54 Shell and Tube Heat Exchanger

FIGURE 7.20-55 Plate Steam Condenser

Pure Water Systems

Pure water has a pH of 7.0 and has a resistance of about 20 meg-ohms/cm^2. It has been assumed historically that this was the optimum situation. If at any point the water is allowed to mix with air the water pH drops and the system starts to corrode inside. All transmitters include a cooling-water resistivity meter. The meter does not identify what may be decreasing water resistivity, whether it is air contamination, metal contamination or dirt. The water conductivity/resistivity meter is a critical indicator of when it is time to change water. If air contamination is the problem then air contamination would show up as both a drop in resistivity and pH. A pH meter or pH strips are a highly recommended instrument at any water cooled transmitter plant.

Empirical analysis has suggested that a small increase in the cooling water pH to approximately 7.5 provides better long-term corrosion resistance than pure water—contrary to long-held assumptions. A small amount of NaOH (sodium hydroxide) can be used with a pH meter or test strips to keep the pure-water system slightly basic. In all cases the water should never be allowed to drop below a pH of 7.0. At a pH of 7.0 water can have a resistance of 20 meg-ohms/cm^2 and it is highly corrosive. The cooled parts have both DC and/or RF on them and a high resistance helps minimize losses in the water feed and return lines. Pure water of 20 meg-ohms tends to leach out silver and copper metal and other impurities that contaminate the cooling water. By maintaining the water pH at 7.5 the resistance of the water drops to about 1 meg-ohms/cm^2. There is a very slight increase in the electrical power lost in the water columns but it is insignificant and is more than offset by the noncorrosive characteristic of the slightly buffered water. Resistivity should be kept between 500 K and 1 meg-ohms/cm^2. A summary is given at the end of this section.

The pump used for the water system is important. For most applications a brass impeller and coated iron housing provide the best results. Seldom is a multi-stage pump or stainless-steel construction required when a centrifugal pump is properly selected.

An important element in any pure-water system is continuous water demineralization, often called water polishing. A typical system uses a 50/50 resin mix of anion/cation mixture. Cooler water from the water-water or water-air heat exchanger flows at a slow rate—about 1 gallon per minute—and reactive elements in the water are passivated. The output of the demineralizer is routed to the water reservoir tank. A standard 2 CF (cubic foot) demineralizer tank is shown in Figure 7.20-56.

Two common mistakes found in demineralizer installation are feeding the demineralizer tank from the hot water side rather than the cooled return water from the heat exchanger. Hot water causes the demineralizer resin to be damaged. A second mistake is reversal of the input and output of the demineralizer tank. This particular mistake can result in rupture of the tank from overpressure and/or transfer of the demineralizer resin to the water reservoir tank where water is not flowing through the resin and not being polished. Properly installed, the demineralizer resin has a very long service life usually exceeding 5 years.

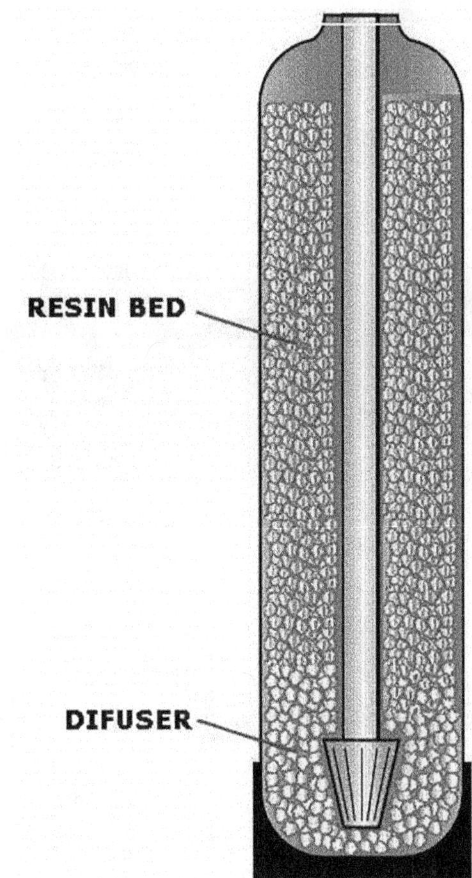

FIGURE 7.20-56 Demineralizer Tank Construction

Demineralizers are not used in vapor-phase water systems without a pump as the return water is too hot and there is no system pressure to force water through the resin bed. Without a pump, the accumulated contaminants in the water are deposited in the tube anode structure.

A final, critical element in any pure-water system is a fine-mesh water filter on the pump output to catch dirt or particles in the water system. Fine-mesh filter have significant pressure drop as they capture material. The normal indication of a mesh clogged by impurities is high pump output pressure and reduced system flow rate. The filter should be cleaned any time this is observed. Fine mesh filters are not used in vapor-phase systems without a pump because return water lacks adequate pressure to overcome the back pressure introduced by the filter.

Figure 7.20-57 shows a complete cooling subsystem with a stainless-steel reservoir. The output of the pump is first routed through a strainer and a temperature switch before being routed to the transmitter. The cooled return-water line (below) also had a thermal protection switch and the tap-off for the water demineralizer (blue 2 cubic-foot Culligan tank) filled with 50/50 resin. Flow sensors and regulators inside the

FIGURE 7.20-57 Complete Stainless Steel Reservoir Cooling Subsystem

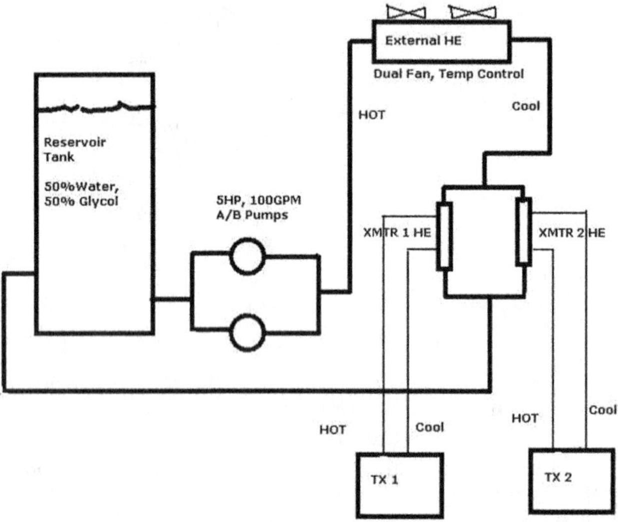

FIGURE 7.20-58 Dual Transmitter Cooling Subsystem

transmitter ensure that all systems are cooled. The large size of the reservoir tank serves a second, less-obvious purpose; it eliminates turbulence and allows any air in the system to percolate out of the system water to prevent acidification.

A dual-transmitter, single-outside-loop system for freezing temperature zones is shown in Figure 7.20-58. Each transmitter has a pure-water system and a plate heat exchanger to give up the transmitter heat to the water/glycol mix. Note that the mix of glycol and water may have to be adjusted depending on the minimum external ambient temperature at the specific site.

There is only one modern frequency-agile HF Broadcast transmitter at the 50–100 kW level that uses air cooling only. Until recently the last such transmitter used an air-cooled 4CX35000C in 50 kW HF service. Ampegon, however, is now using a TH581 in air-cooled service for their 100 kW transmitter. Transmitter sites in cooler climates can benefit from using the transmitter cooling water to heat the transmitter building. The glycol water loop is routed from the water-to-water heat exchangers to water-to-air heat exchangers to warm the building. The use of economical variable speed drives can be used to reduce the pump speed to keep the water temperature at a desirable temperature, say 100°F/28°C. It is important to use temperature-controlled pressure bypasses to keep the pump outlet pressure from going too high for the internal water-to-air heat exchangers. Over-pressure on the pump itself in such a system is not a problem because the only effect is to lower motor current. What must not happen is a reduction in the cooling water flow to the transmitter. The bypasses serve to reduce flow to the external, cold water-to-air heat exchangers.

Pure Water Cooling Summary

Keep the system closed to air. If there is some oxygen, it reacts with hydrogen in the water and removes the gaseous isolation from the piping surface, so the corrosion process accelerates.

Carbon dioxide reacts with water forming carbonic acid, which, in turn, increases hydrogen ion concentration (and so lowers pH) in the water. This significantly accelerates corrosion (assuming all other conditions are equal) because hydrogen in contact with metal walls (piping surface) accepts electrons from the metal, thus leaving this metal in ionic form, which easily goes into solution.

Very low resistivity of cooling water is not desirable because it is well known that pitting corrosion is a serious problem when water has high mineral content. Having cooling water of very high resistivity is not desirable. In this case water also becomes very corrosive and it starts to dissolve any material that it contacts.

The acceptable pH range of water is 7.5–8.5. If the pH drops below 7, pH adjustment (with a NaOH solution or other recommended chemical) is required. Raising pH will protect copper in the water system from corrosion.

Control Circuits

Control circuits are common causes of transmitter failure. It often happens after a maintenance period: a D connector is dislodged, an interlock wire pulled loose, a corroded fuse holder—the list is endless. It could happen in operation: a relay coil failure, a sand accumulation in a flow switch and so forth. Modern transmitters offer multiple visual and computer diagnostics to aid in quick resolution. A few fundamentals are in order and will be reviewed.

Control logic for transmitter turn-on is almost always arranged in a two-step sequence: low level and high level. Refer to Figure 7.20-59. Starting first with the application of power, the next step is normally the cooling system's water, air or both. The second step in the first group of systems is the filaments and driver power. Assuming all interlocks are closed the last element of the first step is application of PA bias. One

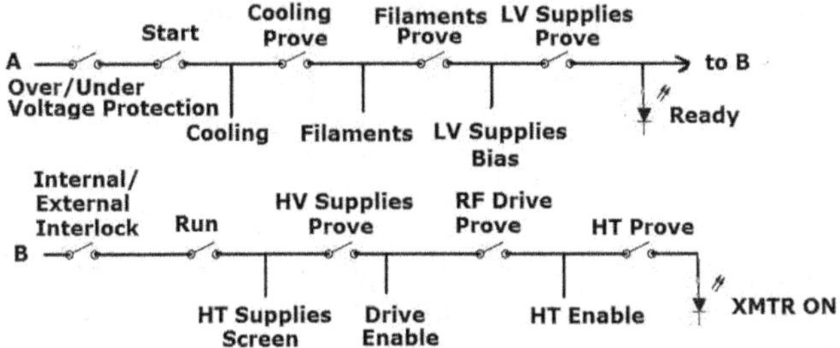

FIGURE 7.20-59 Typical Transmitter Control Circuits

transmitter design also applies high voltage (HV) to the solid-state modulator at the end of the start procedure although, of course, the modulator is muted with zero output.

The second step in turn on is normally application of HT power to the solid-state modulator, HV power to the RF Driver, RF Drive to the low-level states, driver and PA Grid and, after that, application of PA anode and screen voltages resulting in RF power output. One older design mentioned earlier held off the screen voltage supply until the anode voltage had increased to ~5 kV. (The screen supply was a fixed voltage dropped by resistors to limit screen current to a tolerable current value.)

High-power HF transmitters have a multitude of protection circuits that are interwoven throughout the control circuit and they are addressed first. There are reflected power sensors to prevent operation into bad loads, arc sensors to shut down HV if corona or an arc is detected, under-drive sensors to prevent low-efficiency or unstable operation, overload relays to protect the various stages from over-current and/or over- or under-voltage, and various other modes of protection in individual modules or supplied.

HV power supplies can store enormous amounts of energy that can cause substantial damage until primary circuit breakers or overloads can activate and remove power. Still, the stored energy must go somewhere in an arc event—and hopefully not into the arc. Series tube-type modulators could quickly detect overloads and shut down the HVDC to the PA but they could not protect the modulator stage itself. A solution in older designs was a mercury-tube crowbar circuit that grounded the output of the HVDC supply when an overload was detected. All the energy in the HVDC supply and the continuing energy until the primary circuit breaker opened was absorbed in the crowbar tube itself. A secondary circuit/switch was required to keep all voltage off the crowbar tube to reset the system. It was complicated but it allowed the old supplies to meet the 50-joule test.

Most HF transmitting tubes specify a 50-joule limit on applied power to any tube element. (VHF and UHF tubes often have a limit of 20 joules). A common 50-joule test is to use a 32-gauge wire cut to a length in inches equal to the test voltage. The wire is grounded at one end and attached to the HVDC supply by a switch. The wire should survive the test.

Solid-state modulator designers like Bryan Weaver and Gil Hulsey of CEC saw the possibility of additional protection. By mounting optically coupled over-current sensors in the HV output and ground return of the solid-state modulators any ground fault inside the modulator or the HV output could be detected and the fault energy removed at the opening speed of the IGBTs in the supply, in about 800 nanoseconds under fault current. Depending on the energy stored in the bypass and blocking capacitors the fault energy can easily meet the 50-joule test and even the 20-joule test. An innovative grid and screen protection circuit to keep excess energy out of the tube's internal structure has been developed by CEC and is discussed further in the "Vacuum Tubes" section of this chapter.

The sensor in the ground return of a solid-state modulator circuit is also useful in another way: any fault to ground anywhere in the solid-state modulator itself is immediately detected and, again, the IGBTs of every switch in the modulator immediately open to relieve the fault in the solid-state modulator.

Almost all fast-acting overload sensing circuits (VSWR, HVDC arc, arc sensor, etc.) can be set for repeat tries. In all cases a short delay—about 300 ms—is also inserted to allow corona or plasma gases to dissipate from the area where the fault was detected.

Arc sensors come in two major groups: visible light and UV. Visible light sensors can respond in as little as 1 microsecond but external light must be excluded from the cabinet. UV sensors are also sensitive to cosmic rays and that requires a counter circuit be integrated into the detection circuit to ensure the UV signal detected is real. This can extend the detection time to 10–50 milliseconds. The benefit of UV sensors is they are "blind" to visible light. A warning: all photocells deteriorate over time. A typical cell will deteriorate to 50% sensitivity at 7 years. Replacement should be scheduled to keep sensitivity high.

As an aside, another feature that may or may not be included in the control circuit is a bypass of the filament run-up circuit. High power tubes typically use a large mesh filament structure that has a specified run-up time that varies by manufacturer. The slow, steady application of filament voltage from zero up to the recommended value occurs over the manufacturer-specified time period, 5–15 minutes. This allows the tube's internal elements to slowly come up to temperature without stress. Some transmitter manufacturers include a time-delay bypass to instantly reapply full filament voltage despite the danger to the tube. This allows the transmitter to return to the air quickly if a high-priority program is being transmitted.

Vacuum Tubes

Every high-power tetrode tube used is a larger version of the original RCA 6L6 beam-power tetrode. The screen grid and the control grid wires are in direct lines between the cathode/filament and the anode.

Most driver tubes are relatively small and use thoriated-tungsten filaments that require little or no warm up time. A 100-kW tube, the 4CV100,000C also has a bar filament structure that allows operation with very little warm-up time. A picture of the internal structure is shown in Figure 7.20-60. Larger mesh filaments like that shown in Figure 7.20-61 are tightly sprung and must be warmed up slowly to prevent damage because of their physical mass. If the tube voltage is repeatedly cycled fully on and completely off, kinks can develop in the filament mesh and bulge outward enough to result in a filament-to-grid short. Most transmitters use a grounded-grid triode stage (or a triode-connected tetrode in one case) for the driver stage. All these tubes are a direct descendant of the Eimac 3–500Z high-mµ triode; typical mµ is 200. A grid spaced close to the filament (or cathode) allows zero-bias operation and power gains of well over 10 dB. A solid-state driver stage typically provides the RF Drive to the grounded-grid triode driver stage.

Although mesh filaments are normally used in very large tubes, one 500-kW tube, the CQK-650–1, developed by ABB (now Ampegon) used a 96-bar

FIGURE 7.20-60 Internal Structure of 4CV100,000 Ceramic Tube

FIGURE 7.20-61 Large Filament Mesh of Power Grid Tube

thoriated-tungsten filament that required very little warm-up time. Unfortunately this was a proprietary, handmade tube that was very expensive to manufacture and was only used by ABB and never sold on the open market. Another oddity was the tungsten screen and grid structures with an astonishing 270 pF of capacitance between the anode and screen grid; no tuning C used above 10 MHz. The high cost has driven it out of new transmitter designs in favor of the Thales TH576, even with its mesh filament.

One final aspect is tube filament voltage management. Readers are referred to Application Bulletin 18 available from CPII/Eimac.[31] This application bulletin recommends lowering the filament voltage on a vacuum tube slowly until something changes. For modulated stages, distortion may be the best measure and any change is material—even lower distortion. The tube filament voltage should then be operated a few tenths of a volt above the point where a change is detected. For example, the nominal (in name only, not normal) filament voltage of a 4CX3000A is 9.0 volts. Longer life and full performance can be achieved by operating the tube at 8.5 volts. The nominal voltage of a 3CW20,000A7 is 7.5 volts, but 7.1 volts is normally adequate. Even on high power tube, 23 volts is the nominal voltage for a TH558 but operation at 21 volts is perfectly adequate for normal operation at 300 kW. Again, the real test is to lower the voltage until some change is noticed and then increase the voltage above that value.

Do not lower the filament voltage more than ~7% of the nominal voltage. The getter assembly inside the tube is indirectly heated by the filament structure. If the voltage is too low, the getter assembly will not be hot enough to keep gas out of the tube. Reduced heater voltage is *not* recommended for tube with oxide-coated cathodes.

Loss of filament emission in large transmitting tubes is not normally the tube failure mode in HF Broadcasting operation. In almost every case the end of tube life is signaled by a grid-to-filament or a grid-to-screen short. CEC has incorporated a special high-speed grid current and screen current sensor that uses SCRs to instantly ground the screen and grid voltages to prevent the stored energy in the supplies from being dissipated in the internal arc. It is shown schematically in Figure 7.20-62. The very close grid-to-filament and grid-to-screen spacing is shown in Figure 7.20-63.

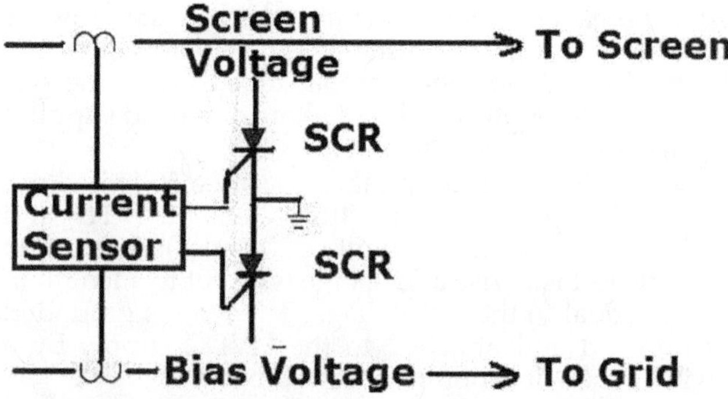

FIGURE 7.20-62 SCR Crowbar Protection Circuit Schematic

FIGURE 7.20-63 Close Element Spacing in Ceramic Power Tube

FIGURE 7.20-64 Gas-Filled Variable Neutralization Capacitor in 500 kW Transmitter

As tube filaments age, methane gas (CH_4) is released from the filament structure. If the getter does not bind this gas quickly enough internal tube arcs can (and do) occur. Internal tube arcs are a fact of life for large tubes.

Tube rebuilders have gotten better and better over the years and in many cases can repair tubes with internal shorts. The extreme long life of the filament structure—especially when tube filament voltage management techniques are used—allows rebuilders to provide rebuilt tube with performance as good as a new tube as half the cost of a new tube. Lower power filaments simply do not have enough filament surface area to allow for such long tube lives. Small tube may have lives as short as 2 years while medium-size tubes used as driver have a 4- to 5-year tube life. If those tubes are operated at full, nominal voltage, tube life may be cut in half.

Neutralization

Every neutralization circuit is the same yet different in implementation.[32] A sample of the anode RF voltage is fed back to the grid of the PA tube through a phase inverter circuit. Some type of variable C is used to adjust the amount of feedback so as just to cancel the amount of positive feedback from the PA anode to the PA grid to cancel (neutralize) this positive feedback signal. Although the amount of C between the PA anode and grid remains constant, the feedback circuit will have to be adjusted over the frequency range to keep neutralization perfect. DRM has caused this requirement for perfect neutralization because a wrong value of feedback at different frequencies—a value that changes slightly due to parasitic inductances and capacitances—results in the generation of IPM sidebands that have no equivalent amplitude sideband to cancel them. Every band will have a different, optimum position. Figure 7.20-64 shows a gas-filled (SF6) variable neutralization capacitor and the drive-rod connection to its servo motor.

Neutralization itself is fairly simple using a spectrum analyzer and an audio oscillator: operate the transmitter at full power and modulate to 90%–95% with a 2-kHz signal. Use a spectrum analyzer (with a properly attenuated input) to see the sidebands appearing 4 kHz, 6 kHz, 8 kHz and so on away from the carrier. Adjust the neutralization so that the 4-kHz sidebands are minimized and balanced—balanced even if not minimized; the two points will be very close. Watch the other sidebands at 6 kHz and 8 kHz away from the carrier to ensure that they are also minimized and balanced as close to the 4-kHz balanced sideband point as possible. The IPM correction circuits in the DRM exciter (if used) will take care of the rest. The transmitter distortion is often improved when the transmitter is perfectly neutralized due to PM sensitivities in most AM detectors—an unexpected finding.

Black Heat

Black Heat is not a term familiar to most engineers but it is a common feature and common requirement for large HF tubes manufactured by Thales. The filament voltage on the final tube is never reduced to zero except in unusual circumstances. In the transmitter's normal off condition the filament of the PA tube is reduced to 4–5 volts and either a small circulating pump or a convection heat exchanger allows water to circulate in the anode. The filament blower (if the tube is socketed) is never turned off unless the filament voltage goes to zero in unusual circumstances—such as tube replacement or maintenance in the socket area. The temperature of the filament is reduced to a value that keeps the filament structure warm enough to stay expanded but not enough to decarburize the filament.

Some American experts believe that black heat is not necessary and the filament voltage should be kept at full voltage at all times whether the transmitter is on or off. This is based on the fact that the filament structure undergoes phase changes at about 85% of nominal filament voltage and any reduction of voltage below that point will result in filament material phase changes and stress. European tube manufacturers promote black heat while other manufacturers believe the voltage should be turned off or left at full voltage.

Transmitter Rundown Cooling

Smaller bar filament tubes require no such black heat treatment. One surprise to most engineers is the unwanted effects of the usual water and blower run-down circuits after the bar filament voltage is removed. Rundown after the filament voltage is removed while full cooling remains on causes the internal tube structures to be cooled far more rapidly than the internal springing can accommodate.

The first reaction to the idea of no cool-down cycle is usually astonishment. But ask a couple of quick questions: Has the transmitter site every lost power unexpectedly? Did the tube fail when that happened? It can be readily established that it did not. If you use a bar filament tube and you cannot defeat the run-down cycle, set the timer for the shortest possible run-down. An improvement in tube life will result. Experience shows that filaments rarely fail. When a failure does occur the usual failure mode is a seal crack at an anode, grid or socket seal that allows air (oxygen) inside the tube. Thoriated tungsten filaments are typically consumed as the oxygen causes their temperature to rise above the melting point. The structure typically falls into the grid. A mesh filament can become (what is called) "unzippered" due to a circular failure of the filament wires as shown in Figure 7.20-65. The spring on the filament mesh then pulls the two segments apart, with an air leak most likely the cause.

The compound that joins the ceramic and metal of the tube is kovar and kovar contains iron. Kovar matches the thermal expansion characteristics of the copper and ceramic. The sealing material is heavily overlaid with silver to protect the kovar. If the silver plating is damaged during installation or improperly done at the factory, water molecules in the air can create a hole in the kovar seal by rusting the iron in the kovar. (This also applies to vacuum capacitors that use a similar seal.) Both vacuum tubes and vacuum capacitors can be temporarily restored to full voltage operation by slowly raising the high-potential applied to the tube or capacitor. Although an occasional arc may occur the voltage standoff of the tube or capacitor will slowly come back to rated value. This is not a permanent fix but may allow several more weeks or months of capacitor or tube operation. Repeat as necessary until the replacement arrives.

High-Voltage Testing of Transmitting Tubes

Almost every high-power tube except the 4CV100,000C is designed for anode UP mounting. It is important that any tube be tested only in the vertical orientation and not on its side. The internal supports are not designed for large shear forces and a sudden bump could break an internal structure. The screen and grid (in wire-gridded) tubes could sag giving false, low breakdown readings. When checking between anode and screen grid the control grid and the filament structure should be well connected to the screen terminal. It is typical to support the tube on a tube hoist, connect the anode to the ground terminal of the high-potential generator and the HV negative terminal to the screen. All PA tubes in the category should easily support 60 kV of standoff voltage. Grid-to-screen and grid-to-filament voltage tests values are taken from the tube specification sheet. One way not to test a tube is shown in Figure 7.20-66. The tube survived, regardless.

Figure 7.20-67 from Tom Yingst (who was at Eimac when many high power tubes were developed) is useful. It shows how the graphite grid and screen structures are formed so precisely given the difficulty of working with graphite. As mentioned before, the grid and screen "wires" are in a straight line between the filament and the anode.

Simultaneous Transmission of the Same Program

Many times two or more transmitters may be carrying the same program at the same time. Some broadcasters use time delays to prevent excessive current modulation

FIGURE 7.20-65 Ionized "Zippered" Power Grid Tube Filament

FIGURE 7.20-66 Improper Method of Testing Power Grid Tube

FIGURE 7.20-67 Laser Cutting of Grid Structure

of the power lines. But one other solution shouldn't be overlooked. The audio feed of one transmitter can simply be inverted relative to the other; while one is going positive, the other is going negative. It may not be as sophisticated as time delays but the results are satisfactory.

Simultaneous Transmitter Turn-Off

HF transmitter sites are often in rural locations away from the main power generation plant. Others operate—as in Tinian—as the largest single customer of the local power plant. No matter the case it is not a good idea to turn off all transmitters at the same time during a frequency change or at the end of a schedule. A 1- or 2-second pause should be inserted between each turn off to allow the power gensets (motor-generator sets) and power distribution network to stabilize.

Safety

HF Broadcasters deal with an equipment system costing millions of dollars. Providing safety for the equipment and for the many personnel who maintain these expensive systems is paramount.

Equipment Safety

It is desirable in any HF system that no movable contacts in the matrix or antenna (or transmitter) operate while the transmission is on. VOA adopted a system many years ago that has been time tested and proven to work effectively and it is presented schematically in Figure 7.20-68. Dual interlocks, Forward and Reverse, are used to prevent damage to the transmitter itself or related equipment. The Forward interlock is present only when the transmitter is off or muted. The Forward interlock allows the associated matrix switch(es) to operate, the

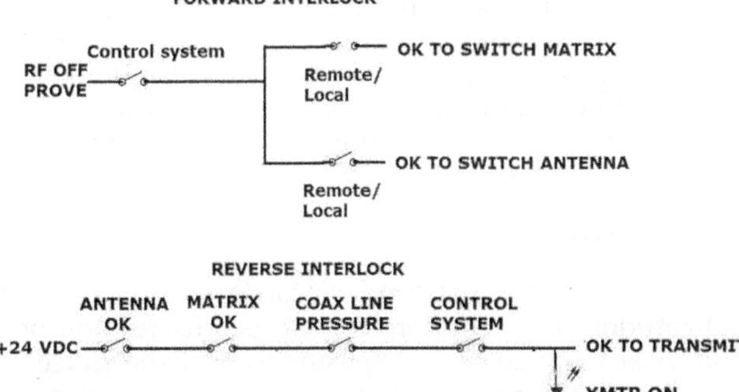

FIGURE 7.20-68 Forward-Reverse Safety Interlock Schematic

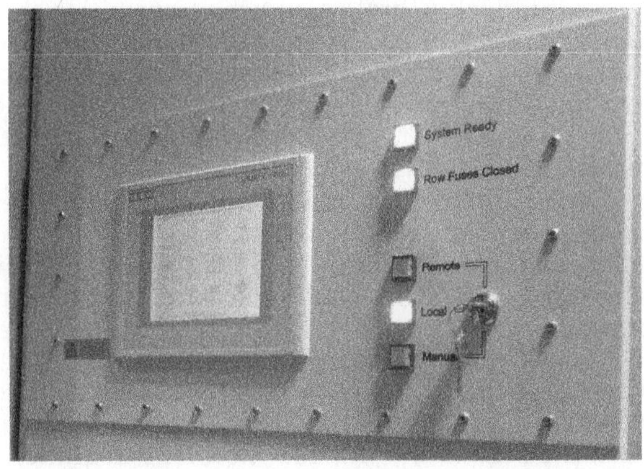

FIGURE 7.20-69 Keyed Transmitter Matrix Control Panel

associated antenna configuration switch(es) to operate and to allow the system to finish its final configuration and provide a Reverse interlock. Nothing moves until the Forward interlock is present and sustained.

The Reverse interlock is used to keep the transmitter off or muted when matrix or antenna configuration switch(es) are moving into position. The Reverse interlock can also be used to ensure that the transmit frequency and the transmit antenna are in the same range. In addition, arc sensing circuits mounted in the matrix or antenna base equipment can be used to shut down the transmitter when and if an arc occurs during transmission. The Forward and Reverse Interlock signals pass through auxiliary switches on the matrix to ensure that the Forward Interlock is routed to the antenna selected by the matrix. The Reverse Interlock signal insuring the antenna is in a valid position is routed back through the matrix and any pressurized line and so on that is important can be included in the signal that comprises the total Reverse interlock.

The Control System is normally responsible to ensure that the operating frequency range of the antenna and the transmitting frequency are in agreement. It also ensures the reported antenna/matrix situation is as desired. Remote/Local Switches prevent movement of the switches when no change is desired as in Figure 7.20-69 of the Matrix control.

Station control systems monitor the Forward and Reverse interlock lines at every functional point in the entire system and report their status and status changes to the station operators and logs. All of the subsystems should also have a key lockout system as shown.

ARC Sensors

A good UV (ultraviolet) detector can ignore visible light but can detect arcs or coronas and is a useful adjunct. The disadvantage of UV sensing is that the detector tube can also be triggered by cosmic rays. Counter circuits inside the UV detector ensure that the event is a real arc and not a burst of cosmic rays. The counter circuits mean a typical activation time of about 10–50 milliseconds from the time of the arc event until the Reverse interlock is broken and the transmitter shuts down. CEC developed UV arc sensors that are pulsed faster and have reduced the reliable arc-detection time down to 10 milliseconds. Remote UV sensors can be included in the Reverse Interlock circuit to allow them to protect the system from remote arcs.

Note that inside the transmitter visible-light sensors are used for arc and corona detection. They are not sensitive to cosmic rays and no counting circuits are required. A good visible-light sensor can detect and report an arc or corona even in as little as a microsecond, a 10,000 times improvement. This requires that the inside of the transmitter be in total darkness for such circuits to work effectively. Also note, as mentioned earlier, the sensitivity of most photocells degrades to 50% sensitivity at 7 years.

FIGURE 7.20-70 Grounding Switch Assembly

FIGURE 7.20-71 Safety Ground Key Switch Control Panel

Personnel Safety

Many older transmitter designs relied solely on electrical interlocks for personnel safety. This can still be found in FM and MW transmitters designed for the domestic market. Transmitters designed for the international market, primarily HF transmitters, must be designed and built using mechanical interlocks so that all exposed voltages are grounded before safety keys are released to allow access to the transmitter. IEC 60215 (sometime called IEC 215) governs the transmitter safety system design. Transmitter manufacturers repeatedly test equipment to ensure that it is safe for personnel. Quality Control personnel conduct safety tests. An interlocked safety system meeting the IEC215 requirement with a view sight to observe the position of the grounding switch is shown in Figure 7.20-70.

Every door that would permit access to dangerous voltages must have a keyed interlock. The key for the doors are not released from the key-exchange block until all safety grounds are in place. A typical installation is shown in Figure 7.20-71. Notice the view port where the operator can observe that the grounding switch is actually in place as an addendum to the front-panel grounding switch labels.

A system with a series of heavy-duty, long-throw door latches is often used in conjunction with extensive electrical interlocking. The electrical interlocks are not the primary safety system but can provide additional operator protection. Another function of the electrical interlocks is to provide a front-panel display of any open or improperly closed doors.

Open-Wire Transmission Line RF Hazard

Open-Wire Transmission Line is a potential radiation hazard to personnel. In the perfect site the antenna load at the end of the transmission line (usually 300 ohms) is balanced, 150 ohms to ground measured on either side. The current flowing in the open wire line has equal and opposite current flowing in the two wires. The wires have space between them; they do not and cannot occupy the same physical space resulting in some radiation from the transmission line. The voltage on the line is sensitive to the height of the line above ground; there's charging and discharging of the stray capacitance to ground. The field between a perfectly balanced line can be quite high and personnel should be restricted from the area near the open-wire lines.

Another problem occurs when the load is not perfectly balanced. Just as with a poor balun, if the line is not terminated in a perfectly balanced antenna there is a difference current, I3, the unbalance current, that causes power to radiate from the transmission line in an uncontrolled manner. This radiated power increases the loss of power by the transmission line and also increases the field in the area of the transmission line at the antenna.

The maximum allowable exposure for personnel inside the antenna area is approximately (610 V/M)/F where F is the frequency in megahertz. Exposure limits are found in ICNIRP 1998 and ANSI/IEEE C95.1–1992. These standards diverge slightly but present guidance suggests that a limit of 61 V/M should be observed. Present exposure limits for people outside the antenna area are lower, 27.5 V/M. A good policy is to have a 2 m fence (well grounded) 20–30 feet (6–9 m) away from any open-wire lines to keep all personnel out of the area of high RF fields. Any access doors to the fenced area should be interlocked with a safety key and, for monitoring purposes, an electrical interlock with a reporting line back to the control point. The actual standards presently are:

ICNIRP 1998

Occupational 1–10 MHz E = 610/f (MHz)
10–400 MHz E = 61 V/M

General public 1–10 MHz E = 87/sqrt(f- MHz)
10–400 MHz E = 28 V/M

ANSI C95.1–1999

Controlled environment: 0.1–3 MHz E = 614 V/M
3–30 MHz E = 1842 V/M/f(MHz)

Uncontrolled environment: 1.34–30 MHz
E = 823.8 V/M/f(MHz)

Grounding Versus Earthing

Earthing is a term used to describe a system to return power-line AC currents to earth. A system of ground rods is used throughout the HF system for both lightning and transient protection using ground rods and cable. Ground rod and cable are invisible at RF frequencies used at HF. A 4/O cable has over 4 ohm/MHz/foot of inductive reactance at 1 MHz and ten times that at 10 MHz.

Grounding is a term used to describe a system to return stray RF currents to ground, ground being some point below the surface of the earth. The only way to properly return stray RF to ground is by using a combination of both a substantial number of 4″ wide internal ground strap lattices as shown in Figure 7.20-72 and

FIGURE 7.20-72 1-Meter Grounding on Transmitter Room Floor

FIGURE 7.20-73 Exterior Ground Rod Radial Anchor

an extensive radial network outside the actual plant. Ground radials are often terminated in a ground rod as shown in Figure 7.20-73 but the ground rod's function is to hold the radial in place rather than to increase the grounding ability. Each ground radial should be spaced every few meters, one-quarter wavelength long at the lowest transmitting frequency.

What is actually happening in a good grounding system is the formation of a large distributed capacitor between the transmitting equipment and RF ground—somewhere below ground level.

Every structure whether building, antenna mast support, feed line support and so forth should have any reinforcing steel tack welded at every crossing as shown in Figure 7.20-74 to ensure no floating reinforcement bars inside. Feedline supports have been known to crumble when internal supports were not tack welded and grounded properly.

After all of the cross matrix of 4″ wide straps are installed, care should be taken to ensure that stub-ups are located properly to attach to all the major system

FIGURE 7.20-74 Tack Welding Rebar at Intersections

FIGURE 7.20-75 Welding of Dissimilar Metals on Exterior Support Structure

components before the final floor is laid. See Figure 7.20-74. Silver solder or brazing should be used at all stub-up points and at each crossing of the 4″ strap to ensure the whole network is electrically complete.

Any external structures near the transmitter building should also be connected to the transmitter hall grounding system to ensure that they have the best possible RF ground.

Where copper and steel or copper and aluminum meet, as in Figure 7.20-75 Cad Welding (also called Thermite Welding) should be used. The high temperature of the "bomb" fuses the dissimilar metals into single compound that will not corrode due to cathodic action in the soil.

Properly installed, a grounding system complements the earthing system and brings the entire transmitter building as close to RF ground as possible.

It goes without saying that (1) every structural element in the entire station should be connected to the ground system, (2) every conduit or other metal conductor should and must be connected to the grounding system, and (3) where nonmetallic building members are used, ground straps should be imbedded into the walls and support columns of the building. The goal is to form a Faraday Cage to keep outside RF and transients outside of the transmitter plant itself.

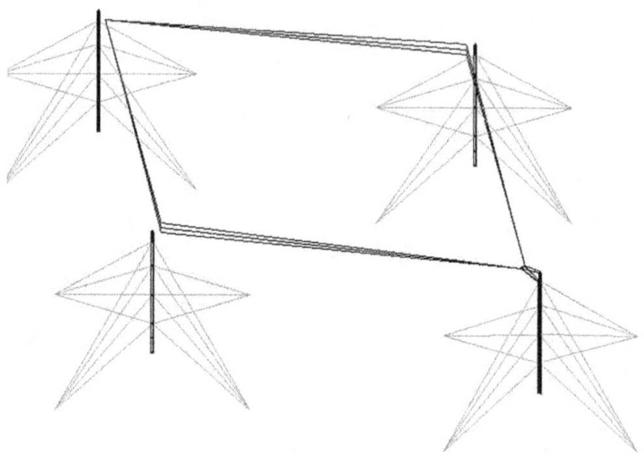

FIGURE 7.20-76 Rhombic Antenna Design

Specific Antenna Systems

The Rhombic antenna shown in Figure 7.20-76, was the antenna of choice for many years. The long-wire nature of this antenna makes it much less susceptible to fading because radiation is occurring all along the length of this long antenna.

The basic problem with the rhombic is that four tall supports are required, the taller the better. Being a long-wire antenna it is also bidirectional unless terminated resulting in about half the transmitter power being dissipated as heat. Engineers at the old Crosley HF station in Bethany, Ohio—later owned by the VOA and IBB—worked out a series of phasing lines to form what they called a re-entrant rhombic where a unidirectional pattern without power loss was achieved. Unfortunately, documentation on the actual performance of this arrangement is not readily available except in *Radio-Electronic Transmission Fundamentals* by Benjamin Whitfield Griffith (p. 476). For any Rhombic, low take-off angles on lower frequencies for long-distance transmissions would have resulted in extremely tall support towers.

It can be shown mathematically that the rhombic antenna occupies more space per dB of gain than any other antenna. Other smaller, higher-gain antennas slowly, over time, made the rhombic obsolete. A Sterba Curtain, for instance, was a single band antenna with a very small footprint that was capable of very high gain. Used with a reflector it could provide a unidirectional pattern with only two supports.

The Yagi-Uda antenna had the highest gain for its size of any antenna. Unfortunately, the end voltages of the elements of the Yagi go to very high peak values as the power increases. This lead to the development of the cubical quad antenna with is equivalent to two two-element Yagi antennas but with manageable element voltages. The bandwidth of the Yagi-Uda and cubical quad's parasitic elements function over only 100–200 kHz but the band allocation may be 500–600 kHz wide.

The log-periodic antenna is a wideband antenna with multiple elements whose length tapers at a logarithmic rate. The design is beyond the scope of this article, but it is fair to say that the log-periodic is a rather large antenna with modest gain and front-to-back ratio.

A log-periodic antenna can also be aimed straight down at a ground-mounted reflector screen. The signal bounces off the ground straight up skyward. This is appropriate for NVIS (Near-Vertical -Incident-Skywave) coverage such as at Alice Springs in Australia. NVIS antennas are used for local/regional coverage using the HF bands.[33] An example is shown in Figure 7.20-77.

The wideband nature of the log-periodic makes it a favorite for frequency agile operation. Like all horizontal antennas its take-off angle is sensitive to its height above ground. The fairly wide horizontal beam width allows a wide area to be covered horizontally ±30 degrees. The higher take-off angle beam gives reliable medium-distance signal coverage without the low-angle radiation that would go beyond the target area.

At practical heights a rotatable log-periodic has a take-off angle that varies with frequency but can be generalized as in the 30- to 45-degree range making it ideal for mid-distance coverage described earlier as 600–1500 miles.[34]

Shown in Figure 7.20-78 are a 50-kW Antenna Products log-periodic (foreground) and a new TCI 521–4 100-kW rotatable log-periodic. The comparison between the construction styles is obvious. A lower-power version of the Antenna Products LP was and is still used throughout the US Air Force's communications circuits for HF communications over a wide frequency range.

Shown next in Figure 7.20-79 is a 250-kW Rohde and Schwarz rotatable log-periodic. The size of the elements is large to handle the power rating but the number of elements is actually too few to have really acceptable VSWR over the different broadcast frequencies. This large antenna has a limited number of elements simply because the weight and wind load. The engineering compromise is to have an antenna that could be rotated and, in an ideal situation, would have an acceptable VSWR in the HF Broadcast bands.

FIGURE 7.20-78 50-kW Log-Periodic (Foreground) and TCI 521–4 100-kW Rotatable Log-Periodic Antennas

FIGURE 7.20-77 Australia Log-Periodic Scatter Beam Antenna

A fixed log-periodic can be made where the take-off angle is almost constant with frequency. The antenna elements are high for the low-frequency elements and lower for the high-frequency elements. Such an antenna is shown in Figure 7.20-80. The height of the front and rear support towers can be selected to give the most desired take-off angle for a particular target. The fairly broad vertical and horizontal beam of this type of antenna can be of benefit when a particular target is known—even with the varying height of the ionosphere, the thing that sets the optimum take-off angle.[35]

A simple two-dipole example is shown in Figure 7.20-81. Even without a reflector screen this antenna could be used for NVIS, high-angle take-off, local coverage use.

Most curtain arrays use two-wire folded dipoles. A folded dipole does *not* have a wider bandwidth than a two-wire, center-fed dipole—contrary to popular belief. Figure 7.20-82 shows the two types of dipoles that have the same VSWR bandwidth. The impedance of a folded dipole is, however, ~300 ohms while a two-wire dipole is ~72 ohms—in free space and with no other objects nearby. When paralleling 2, 4, 8, 16 or 24 dipoles the resultant impedance of all the dipoles rises, at least doubling or tripling depending on the folded dipole design. The first curtain array at IBB was done by Weldon and Carr in the 1940s to give a two-band antenna. It used three-wire dipoles and the mutual impedance brought each dipole up to 450 ohms

FIGURE 7.20-79 250-kW Rotatable Rohde and Schwarz Log Periodic Antenna

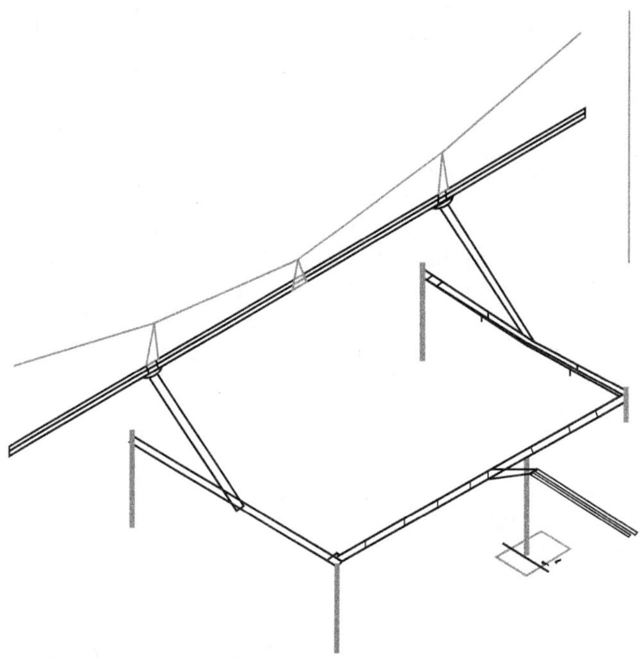

FIGURE 7.20-81 Two Dipole Antenna Layout

FIGURE 7.20-80 Log-Periodic Antenna Array in Malaysia

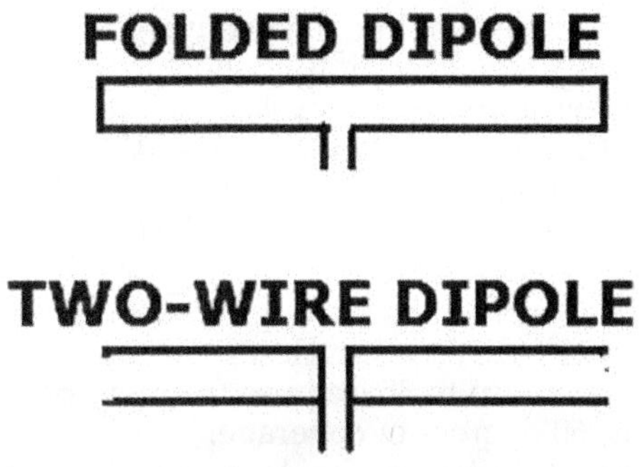

FOLDED DIPOLE

TWO-WIRE DIPOLE

FIGURE 7.20-82 Two Different Dipole Implementations

FIGURE 7.20-83 Russian Dipole Antenna Consisting of Multiple Wires and Matching Transformer

impedance. It was only eight dipoles paralleled and is shown in Figure 7.20-83. The array had a 62-ohm feed impedance.

A dipole consisting of multiple wires used by Russian antenna engineers is shown in Figure 7.20-83. Two things to note are that the mutual coupling to the dipoles above and below in an array raise the dipole antenna impedance to about 140 ohms. On each dipole of this array there is a 140-to-300 transmission-line transformer so that the input impedance of the array is four times that of a collinear dipole array, shown earlier, in a side-by-side arrangement. Look closely at the center feed to see the transmission-line transformer.

A unique antenna solution by Dr. Ron Wilensky of the American company TCI is shown in Figure 7.20-84. This dipole configuration is employed in the widely used TCI Model 611 folded-dipole curtain array; it is still available but not advertised. A line drawing of the TCI dipole shows in Figure 7.20-84 its unusual construction. It can be seen that each folded dipole is a six-wire flat cage that is a version of the two-wire folded dipole shown earlier. The six conductors are held in tension by means of the two side catenaries. The upper three and lower three wires form, respectively, upper and lower conductors that have a larger electrical diameter than the single conductors used in the two-wire folded dipole. In addition, it is possible to insert a shorting wire, as shown, that connects the

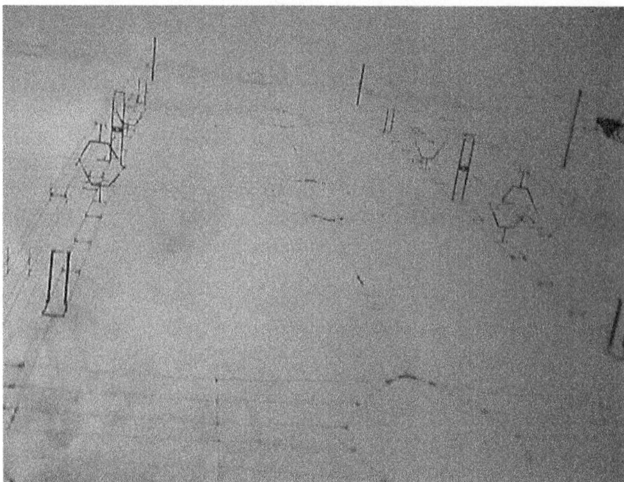

FIGURE 7.20-84 TCI Dipole Design Layout

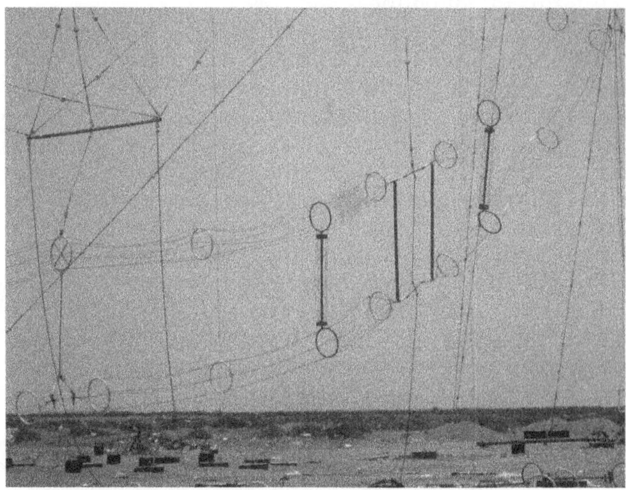

FIGURE 7.20-85 Four-Wire Box Dipole Antenna

upper and lower conductors from the end to a position closer to the center of the dipole. The larger electrical diameters and optimization of the position of the shorting wire increase the impedance bandwidth of the dipole and also helps increase the feed impedance.

When used in a two-, four- or six-high array the driving point impedance of the individual dipole rises to 600 ohms or more because of mutual coupling to the other dipoles in the array and the dipole design. This is a convenient impedance level that enables the transmission lines that interconnect the dipoles to have impedance levels in the 300-ohm range that can be implemented using two-conductor transmission lines having practical dimensions without any transmission-line transformer necessary.

One approach that is very much less intuitive is shown in Figure 7.20-85 and was used by Telefunken in constructing the first arrays for the IBB transmitting station on Tinian. This is a four-wire box dipole, and despite its appearance has a very flat frequency response over an entire octave. Although Telefunken, now TransRadio Sendersystem, Berlin, no longer produces antennas the design is instructive as to what can be done with a "fat" dipole design. This design is credited to Dr. Peter Brueger. The foundational element is a "fat" folded dipole.

Another take on the "fat" folded dipole design uses large-effective-diameter elements for broadbanding the antenna's frequency response. Sino-Sky, Beijing, has used this arrangement widely in China and has begun exporting this design.

All of the dipoles in all these arrays are fed in phase. Power must be delivered to all the dipoles at the same instant in time. Refer to Figure 7.20-86. This requires an in-phase feed system. As shown all the dipoles are fed in phase. On the ground all the dipoles are paralleled at the center of the antenna base works at the slewing (directional) switching. The beam towards the left (rear of the antenna) bounces off the reflector screen, flips phase and is bounced back in phase to constructively interfere in the beam direction creating a unidirectional pattern.

The most common arrangement is use is a 16-dipole array consisting of four columns with four dipoles in each column. A front view is shown in Figure 7.20-87. The feed systems of the four columns are paralleled through a series of time-delay feedlines at the back (or front) center of the antenna base works. As the time/phase relationship of each column is altered the antenna can be slewed, that is, steered. The schematic, below, shows the 300-ohm feed lines as a single line for simplicity. The slewing switch has two levels. The lower level steers the outer two columns, while the upper level steers the inner two columns. The system shown provides slew depending on the dimensions of the eight delay lines. Three-, five- and seven-position slewing systems having, respectively, 6, 10 and 14 delay lines have been implemented where smaller angular slew steps are desirable.

Dr. Wilensky has provided the basis for a helpful diagram in Figure 7.20-88 showing how slew is accomplished in a typical left slew, no slew and right slew system.

A similar system is used in the phased aperture radar is used to steer—or slew—the main beam. HF Antennas slew up to 30 degrees from the perpendicular, that is, 60 degrees of coverage.

The reflecting screen can be put to double use by putting another dipole array behind it resulting in 120 degrees of coverage. Such a front/reflector/back arrangement is shown in Figure 7.20-89. This allows three antennas oriented at 120 degrees to each other to cover the full 360 degrees. There is some deterioration in the beam characteristics when slew is used; sidelobes arise. But their presence is not material because

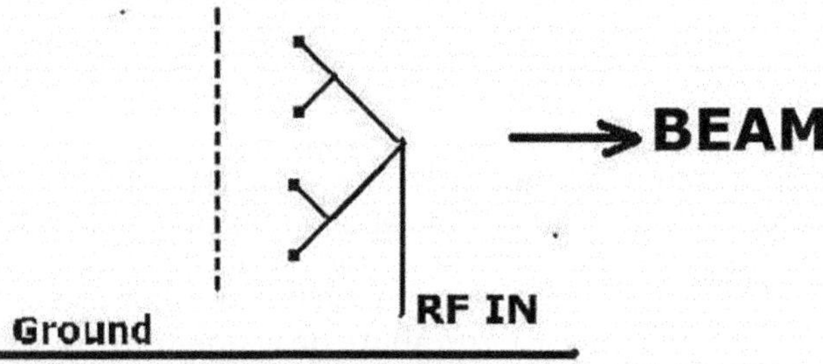

FIGURE 7.20-86 Schematic Representation of Reflector Screen Implementation

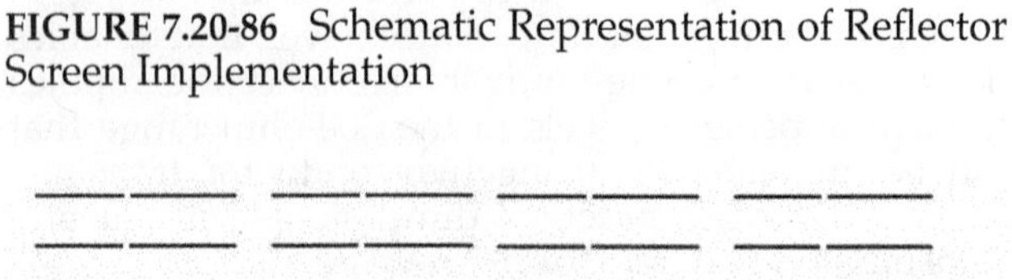

Ground

4 X 4 Array Front View

FIGURE 7.20-87 Front View of 4 × 4 Dipole Array

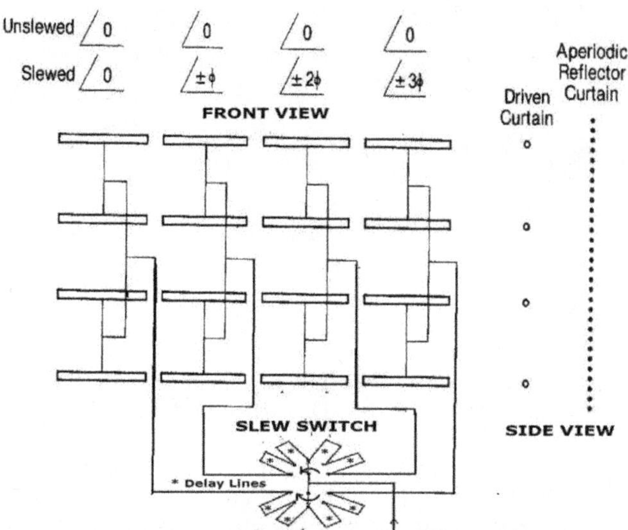

FIGURE 7.20-88 Slew Switch Operation

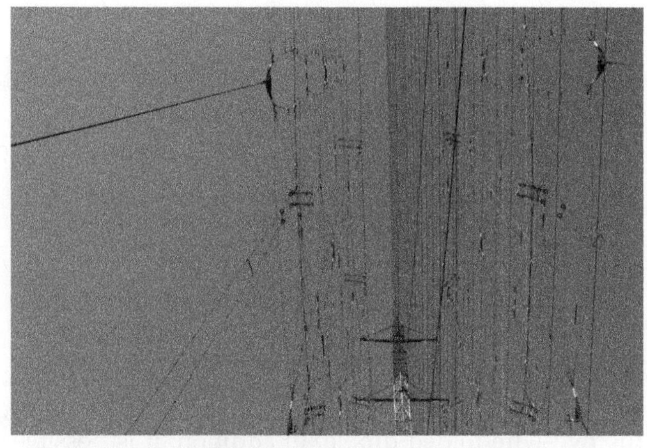

FIGURE 7.20-89 Russian Front Back Array

FIGURE 7.20-90 Slew Switch for Front/Back Operation in Saudi Arabia

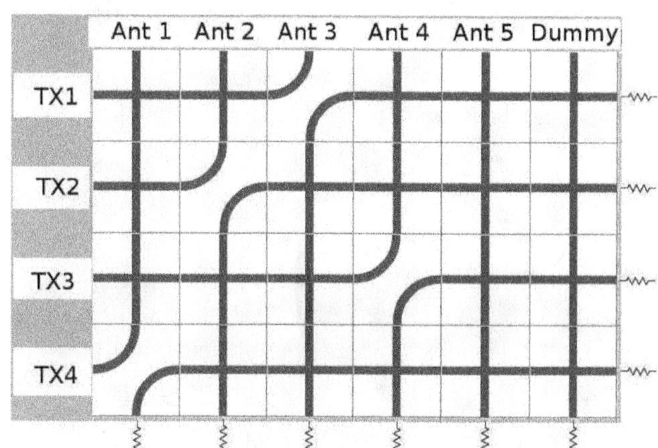

FIGURE 7.20-91 4 × 6 Matrix Switching Schematic

the main beam suffers only slightly. The only limit to the number of rows and columns of dipoles is set by the height and separation of the support towers—and the cost.

An effective reflector curtain can approximate a solid surface and keep the two arrays sufficiently isolated to allow simultaneous operation of the front and back antenna arrays using only a single reflector screen.

A Front/Back Transfer Switch like the one shown in Figure 7.20-90 is also used if the antenna is only to be used Front *or* Back and not Front *and* Back.

All of the elements (dipoles) in the array are coupled together not just by their RF drive but their electromagnetic fields. The design of the dipoles, the side-to-side spacing and the vertical spacing all impact the final antenna product to give it a one-octave bandwidth with the lowest VSWR occurring in the broadcast bands.

RF Matrices

Sites having multiple antenna feedlines and multiple transmitters normally use a matrix to send the transmitter's RF output to the best antenna for that target and frequency—everything from 1 × 2 to 8 × 13 and larger. Logic in the matrix controller is normally used to prevent a transmitter operating on a low frequency from being routed to an antenna designed for high frequencies.

Figure 7.20-91 shows a 4 × 6 matrix schematically. Note that all unused ports are terminated in resistors to terminate any small amount of RF coupled into the unused lines.

a. Transmitter 4 connected to Antenna 1
b. Transmitter 3 connected to Antenna 4
c. Transmitter 2 connected to Antenna 2
d. Transmitter 1 connected to Antenna 3
e. Each unused Row and column terminated in (not shown) resistors to the right and below the matrix. The resistors eliminate any voltage induced into an unused line.

Two types of matrices have been used at 500 kW: 50 or 75 ohms, unbalanced, and 300 ohms, balanced. The matrix impedance is determined by the location of the balun(s) in the system and the desired feedline impedance. RF switches for 50- or 75-ohm coax can be quite compact, a desirable feature even for 12″ line rather than the normal 9″ line. The larger size and greater spacing of the 300-ohm balanced, open-wire matrix usually results in it being located above the transmitter room or in an adjacent building or shelter.

The modified TCI matrix shown in Figure 7.20-92 with the inverted termination resistors is a TCI 50-ohm, 8 × 13 with two of the input rows folded back on the bottom and inverted. The building height was not sufficient when two additional transmitters were added. Two rows were inverted, mounted and then connected to the bottom of the original 6 × 13 matrix. This was possible by moving the 500-ohm termination resistors (used to keep isolation high) from the bottom of the original matrix to the top of the addition.

A portion of a larger 300-ohm matrix is shown showing the Straight or Rotate positions that the switches can take. A 300-ohm balanced switch is substantially larger than a 50- or 75-ohm unbalanced switch. A typical DASTO[36] 300-ohm Matrix switch is shown in Figure 7.20-93. Notice the anti-corona rings where attachments are made to the switch where other switches or feedlines are connected.

FIGURE 7.20-92 Modified 8 × 13 TCI Matrix Switch

FIGURE 7.20-93 DATSO 300-Ohm Matrix Switch

The larger size of the 300-ohm matrix has generally resulted in far more 50-ohm and 75-ohm matrices. It should be noted, however, that the total cost of Z cross points, when all factors are considered, is about the same whether 300-ohms balanced or 75- or 50-ohms balanced. A two-transmitter 300-ohm matrix is one of the simplest forms, as shown in Figure 7.20-95. The outputs of the matrix feeding the antennas on the South side of the matrix are shown in Figure 7.20-94.

A system designed for six transmitter inputs and six antennas plus a dummy load outputs, that is, a 4 × 7 matrix, for 50-ohms shown in Figure 7.20-91 is smaller but, of course, requires externally mounted baluns.

FIGURE 7.20-94 Building Feeder Lines Exit and Anchoring

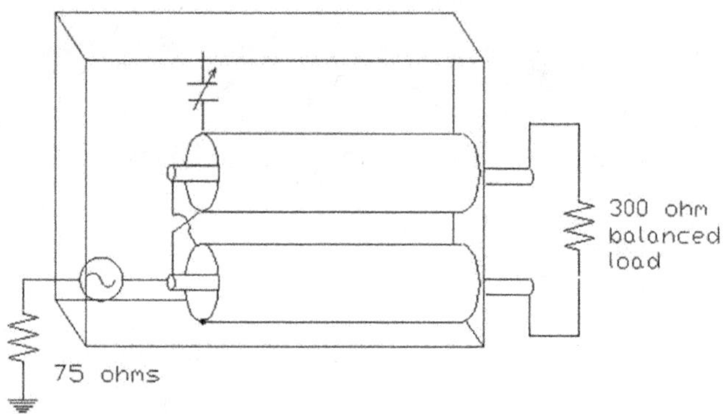

FIGURE 7.20-95 75- to 300-Ohm Balun Schematic

Baluns

A tunable 4:1 balun designed by George Woodard when at Continental Electronics is shown in Figure 7.20-95 that uses two shortened cross-connected 150-ohm coaxial line sections, one insulated from ground. The one isolated from ground inside a box of such size that its outer conductor has a 150-ohm impedance. With 180 degrees of phase inversion the output is 300 ohms balanced. The tunable nature of the balun also allows it to serves as a low-Q bandpass filter, a useful adjunct.

Coaxial cable can also be used for the transmission line as noted in the introduction. The losses may be acceptable for shorter transmission line runs because of their isolation from the weather. In such cases a Telefunken-style balun is used as shown in Figure 7.20-96. Described in less detail earlier, a moderate length of coax is insulated from ground after its input point. At the output center-conductor terminal a compensation line is run from the center conductor (parallel to the outer conductor) back to the input ground point. This is a 1:1 balun whose output

FIGURE 7.20-96 50- to 300-Ohm Matching Line and Telefunken-Style Balun

FIGURE 7.20-97 75- to 300-Ohm Tapered Line Transformer in Spain

impedance is equal to the input impedance, 50 or 75 ohms. The reactance of the outer jacket of the coaxial line, aided by the current in the compensation line, suppresses the unbalance current, denoted I3, at the input.

The reverse is true at the antenna; the low impedance of the paralleled dipoles of the antenna must be stepped up to the 300-ohm impedance over the frequency range when 300-ohm line is used. On the other hand, the approximate 60 ohms, 120 ohms, 300 ohms or other impedance of the antenna slew switch input must be stepped down to the coax output impedance where coax is used.[37]

Figure 7.20-97 shows the 75:300 ohm transmission line transformers of Radio Nacional Espana in Noblejas, Spain. Each transmission-line transformer covers the 5.8–26 MHz frequency range.

HF Transmission Line

Transferring the power out of the transmitter to the radiating antenna is the function of the transmission lines. A transmission line may use high impedance and higher voltages just as power lines use higher voltage to minimize resistive losses. Or lower impedance and lower voltage may be used for local transmission line(s). Transmission lines of 300 ohms balanced, and 50 or 75 ohms unbalanced, are standard values for open-wire and coaxial feed lines in HF broadcasting. (As an aside, 75-ohm coax line has the lowest loss and 30-ohms has the highest power-handling capability.) Whether 50 ohms is best or not is irrelevant because it has become the de facto standard for coaxial lines. Construction of coaxial line (nonflexible) is fairly simple and (at HF frequencies) it is not prone to impedance variations over its length. Another potential benefit of coax is that—assuming a good balun at the end—it is nonradiating and the RF hazard to people is nonexistent. Open-wire line does radiate, sometimes significantly, and can be an RF hazard. This is discussed later in more detail.

Open-wire line is exposed to the weather and/or vegetation that can result in additional maintenance requirements. It should be noted that the design of a 300-ohm line is also not a trivial matter. Four-wire 300-ohm line is more often used than two-wire 300-ohm line. While more expensive to fabricate and install, four-wire line has lower loss and better impedance characteristics. If one inspects a four-wire line installation carefully, spacer plates of varying length are used for matching the characteristic impedance of the line. The droop of the line between support poles and the metallic supports causes impedance bumps that must be equalized out. Alumiweld wire is normally used because the steel core adds mechanical strength that allows the line to be pulled tighter reducing the number of support poles required. Doubling the number of conductors decreases the resistance loss.

The choice of the transmission line/feeder type is a balance between the competing factors of loss and maintenance. There is no single best solution. When all factors are taken into account there is little if any cost difference between a 9″ 50-ohm coaxial feeder and a 300-ohm open-wire feeder. Coax can be mounted close to the ground whereas open-wire line requires 4 or 5M support (and support guying). More coax supports are required than open-wire supports. Coax requires more precautions for expansion and contraction during temperature changes. Coax is manufactured in 38.5-foot lengths to allow shipment in a 40-foot container. It is not a trivial matter to construct 300-ohm line and have it work properly. Where open-wire lines are used with different transmitters, and proximity is necessary, dual runs of 150-ohm Quasi-Coaxial line (300 ohms total) as shown in Figure 7.20-98 may be used to prevent crosstalk between the lines and the transmitters. Where many lines must be routed side by side the isolation of coaxial line may make the additional loss of coaxial line tolerable.

FIGURE 7.20-98 Dual Runs of 150-Ohm Quasi-Coaxial Line

FIGURE 7.20-99 Transmission-Line Transformers and Balun in Saudi Arabia

Coaxial-line sizing is dependent on custom. In Europe 4″ coaxial line is used for 100-kW service whereas in the US 6 1/8″ line is common. In Saudi Arabia 12″ line is used for 500-kW service. In most of the rest of the world 9 1/16″ line is considered adequate. Larger coax has lower loss and may be used where a longer coax run to the antenna is necessary. But there is always the tradeoff between loss and convenience. In Tinian, IBB installed new antennas well South of the original antennas at a distance where too much power would be lost in a coaxial line, even in 9″ line. Reducing the loss in the long run to the new antennas necessitated a hybrid solution. Nine-inch coaxial cable was used from the matrix to an isolated area across the antenna access road then two 50-ohm baluns and the two 50:300 ohm transmission-line transformers were mounted at the coaxial outputs, and the long runs to the new South antennas were completed with two 300-ohm open-wire lines.

Two 12″ lines from the matrix to the baluns at the Korais Road transmitting station in Riyadh, Saudi Arabia, are shown in Figure 7.20-99. Note the compensations windings under the coax sections that form the 50-ohm balanced to 50-ohm unbalanced baluns. The

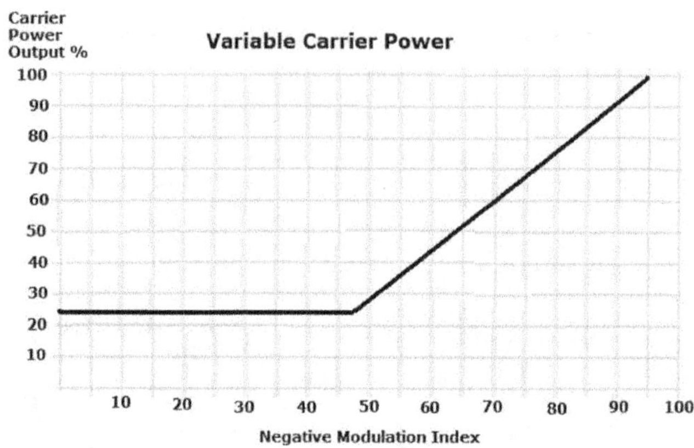

FIGURE 7.20-100 Graph Illustrating Dynamic Amplitude Modulation

50:300 ohm transmission-line transformer lines are made from multiple wire runs rather than solid plates. The number of wires steps down as the spacing steps apart, giving the impedance transformation necessary. The proximity to ground almost guarantees that there will have to be substantial adjustments to length and spacing to give the lowest VSWR over the broadcast bands.

As in the IBB situation in Tinian described previously, coaxial line—even 12″ line—would have losses that were too high on the long runs from the transmitter/matrix area to the antennas. Nine-inch line did, however, allow the size of the matrix to be kept relatively small even with the large number of transmitters and antennas at that site.

Variable Carrier Power

Reducing the average power consumption of HF Broadcast transmitters is highly desirable. One way to achieve this is to reduce the carrier power when possible. The current use of modulators coupled to DC allows a low-level DC signal to control carrier power.

There are two general schemes used.[38] One is to reduce the carrier power when no modulating signal is present. The second alternative is to reduce the carrier power when audio is present. In either case, when the carrier power is reduced, the AVC (Automatic Volume Control) voltage at the receiver is also reduced and the signal-to-noise ratio is reduced by the amount of carrier reduction. The best empirically derived value of carrier reduction for a system that reduces carrier power without modulation is 6 dB, one-quarter power.

The most widely used system in international broadcasting is shown in Figure 7.20-100. A variation on this carrier-reduction-with-no-modulation theme is DAM, or Dynamic Amplitude Modulation. The carrier power rises from 25% to 50% of the normal carrier power during extended no-modulation periods. The resulting signal-to-noise degradation is 3 dB rather than 6 dB. Of course, prolonged no-modulation periods are usually not encountered. Many variable-carrier-power systems adjust the carrier amplitude rapidly as modulation level rises and falls. Others moderate the carrier drop time. But the effect without prolonged no-modulation periods is exactly as shown in Figure 7.20-100.

A major innovation was developed by Continental Electronics in its CCM (Controlled Carrier-level Modulation) system.[39] First, the comparator that determines the carrier level looks only at the negative modulation level. This allows positive modulation in the presence of asymmetry to rise to as much as 200%. Second, after the correct carrier level (for negative modulation) is determined and adjusted there is a delay before the carrier level drops. Slower speech rates are used by most HF Broadcasters during presentations, approximately 150–160 words per minute. The delay between words is therefore about 300–400 milliseconds. By delaying any carrier level decrease by 400 milliseconds there is no carrier level (or receiver AGC level) change between words or sentences during speech. This delay in carrier-level drop minimizes the ability of the listener to detect the operation of this variable carrier-power system in operation. It should probably be noted here that asymmetrical modulation is not used in HF broadcasting by both convention and transmitter design.

Whether CCM or DAM the transmitters never over-modulates negatively because of lack of carrier. The DC value representing the carrier level rises as fast as necessary to adequately demodulate the negative modulation. Even with a PDM rate as slow as 50 kHz the rise time of the carrier could be as fast as 20 microseconds—a value ten times as fast as necessary with a 4.5 kHz (standard) audio cutoff frequency. The carrier level rises to 100% at audio input level equal to 95% negative modulation in both DAM and CCM to prevent negative over-modulation.

A second system of carrier reduction is denoted AMC (Amplitude Modulation Companding) and is shown in Figure 7.20-101. During 100% peak modulation the carrier is reduced to one-fourth of the normal carrier power. A 500-kW transmitter, during full modulation, would become a fully modulated 125-kW transmitter. This system obviously lowers PEP (Peak Envelope Power) to the normal full carrier power, that is, 500 kW while CCM and DAM use the full 2000-kW PEP. The PEP capability of the transmitter is wasted because the PEP remains constant.

The goal of all these system is power-consumption reduction. The systems that reduce carrier power during low or no modulation, DAM, CCM and so forth reduce power consumption from 17% to 25%, the amount of power consumption reduction depends on the type of program modulation. The most savings occurs during voice-only modulation where, during breath pauses, the carrier falls to 25% power. The system that reduces carrier power during modulation, AMC, can show up to 50% power-consumption reduction. AMC is the most power conserving IFF the 6 dB signal-to-noise reduction during modulation can be tolerated.

AMC is used in the United States on MW in a modified form. The rate of carrier-level change is slowed significantly. This is to allow compatibility with the proprietary HD (High Definition) Radio system demodulator algorithm used in the United States. It is not for any inherent benefit to the AMC system. The reduced speed of the carrier drop does, however, allow the full PEP of the transmitter to be used after a long input signal pause. Conversely, the carrier level is held down for longer period when a modulating signal is removed. With few gaps in MW programming the transmitter is kept at 25% power most of the time.

Variable Carrier Power Modulation Monitors

Variable carrier power and the accurate determination of modulation depth has been a continuing problem. This problem has been solved by the American company Belar Electronics. The Model AMMA-2[40] shown in Figure 7.20-102 has a special function MDCL for modulation dependent carrier level. In this mode the carrier level reference is automatically adjusted so that accurate modulation level measurements are possible even when the carrier power is changing dynamically. Note in Figure 7.20-102 that the carrier level setting may be anywhere between 25% and 150% and the readings remain accurate.

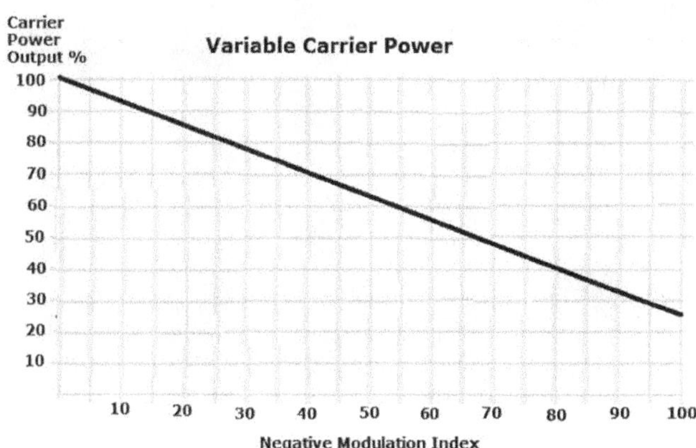

FIGURE 7.20-101 Graph Illustrating Companding Principle

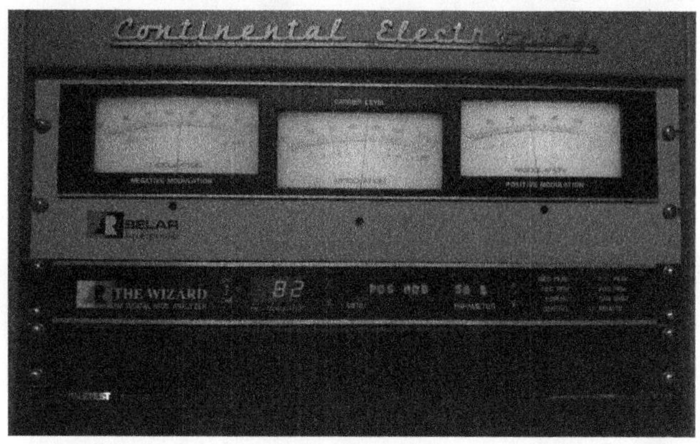

FIGURE 7.20-102 Belar MDCL Compatible Belar AMMA-2 AM Modulation Monitor

Low-Power HF Broadcasting

In addition to its use for international broadcasting, the shortwave bands are also used for domestic broadcasting in many other parts of the world.

Non-U.S. in-country broadcasters use the bands below 8 MHz and use antennas that direct the energy at high angles to avoid a skip zone in the coverage. This type of broadcast is often referred to as NVIS, for Near Vertical Incidence Skywave. Some bands are allocated internationally as tropical bands, and include the 120-, 90- and 60-meter bands. While mainly intended for domestic use in the tropics, some users of these bands seek to cover both in-country and regional audiences and are well outside the tropics. Other bands, 75, 49 and 41 meters, are within the usable NVIS frequency range but are allocated as international bands. However they are often used for in-country (outside the US) broadcasting especially in daytime when the tropical bands may give weak signals and no international interference is present.

The relative number of stations using the various NVIS bands for in-country broadcasts is shown in Figure 7.20-103. This clearly indicates that the most popular band for in-country shortwave is the 60-meter band that lies near 5 MHz. Second is the use of 90 meters, with 75 and 49 meters being a close second. The 120-meter band is almost unused, and there are a significant number of "out of band" frequencies that are in use. These have the advantage of very much reduced interference but may not be available on some receivers.

Frequencies for the tropical bands are usually allocated for in-country use with little or no international coordination and stations tend to stay on the same frequencies without seasonal changes, as is common in international broadcasting. Transmitter power used by in-country broadcasters vary widely. Figure 7.20-104 shows the percentage of stations using various transmitter powers on the 60-meter band. More than half of the stations use 10 kW or less with 23% using 1 kW or less. A considerable number of US and in-country stations use 50- or 100-kW transmitters. These are mainly government stations in larger countries, whereas the lower power stations tend to be private, often commercial stations and religious or cultural broadcasters.

One can legitimately ask how a station using 1 kW or less can provide a viable service, but it should be remembered that some of these service areas where there is no electricity grid and noise levels will be low except when local thunderstorms are present. Some of the lowest power stations are intended to serve only one small town and thus are equivalent to lower power stations on the standard AM band, but which have chosen to use the shortwave frequency either to escape local noise, to have relaxed antenna requirements or because the AM band had no open allocations. The moderately powered stations are often commercial stations simulcasting an AM or FM frequency for local coverage but who wish to serve a wider in-country audience by using 5 or 10 kW to a nondirectional antenna that will extend their daytime coverage into rural areas that have few or no daytime AM or FM stations.

Low-Power Transmitting Equipment

Transmitters for 50- and 100-kW in-country broadcasting are often identical to the transmitters used for international broadcasts and may in fact be the same transmitters that are operated during hours when the propagation to international targets is not favorable. Because many domestic stations use only a single frequency, however, the multiband, rapid band switching features of the typical transmitter designed for international use are expensive and offer no advantage. Therefore, some of the 50 kW transmitters may be built for a single-frequency.

Lower power transmitters are almost always designed for only one or two frequencies and bear more resemblance to a typical AM transmitter than a high-powered HF transmitter. Many of them are either

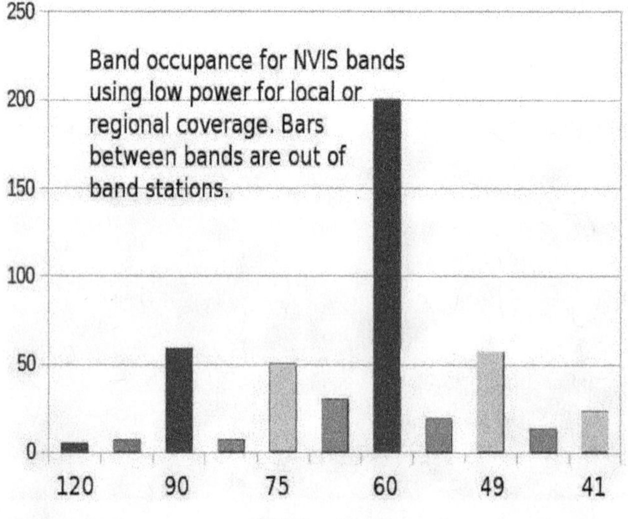

FIGURE 7.20-103 NVIS Bands for In-Country Broadcasts

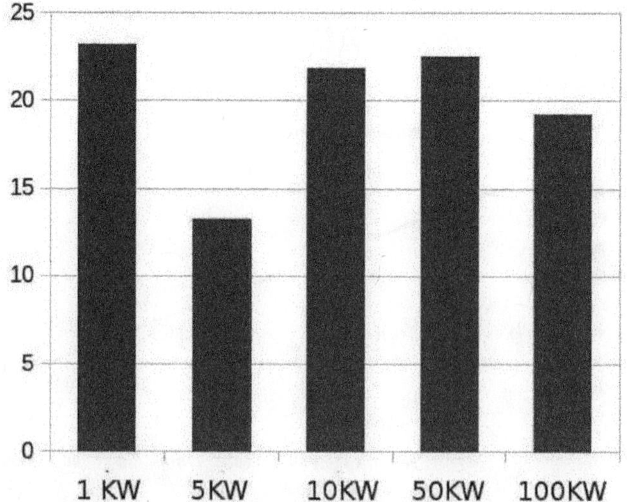

FIGURE 7.20-104 Graph Illustrating 60-Meter Band Transmitter Power Versus Number of Stations

converted tube-type MW transmitters or are made by the same companies that make AM transmitters with only the frequency determining components changed. This opens the possibility of using an AM transmitter that had been made redundant by a modern solid state transmitter and converting it for use on the shortwave bands. Quite a few of the 1- to 50-kW transmitters presently used in domestic service are the result of this type of conversion.

Push-pull modulated types are probably easiest to convert, along with those with a PDM modulator. Doherty types are somewhat more complicated to convert, and Chireix outphasing types virtually impossible. Often replacing the stages up to the final amplifier with broadband solid-state RF generation and amplification circuits is an advantage. Equipment manufactured for the amateur service (ham radio) can be used for these low level stages.

The equipment used is not always crude. The CIA is said to use a large number of frequency-agile HF transmitters in shelters that were designed and built by several companies. Single-frequency or dual-frequency receivers tuned to the operating frequency or frequencies are widely distributed to the local populace and are used to receive the programs broadcast. HCJB has been following this model over the past few years. Their SonSet program uses FM transmitters (or HF) along with fixed-tuned, solar-powered receivers.

Completely solid state transmitters for the NVIS bands are gradually becoming more available. The limiting factor is that the efficiency of Class D amplifiers drops off rather rapidly as the frequency goes up due to limitations in the transistors themselves. In addition, the number of modules and the complexity of the combining networks increases rapidly as the power rises, making about 25 KW the point where a tube design becomes competitive for HF transmitters.

Antennas for NVIS service are quite different from those used in international service. The gain is almost always lower and the main beam is directed upwards rather that at a low angle towards the horizon. Many stations use a simple dipole. Arrays of dipoles are also used. Some designs are slewable, meaning the coverage footprint can be varied somewhat from the unslewed pattern that is centered on the antenna.

US Domestic Shortwave Broadcasting

Even in the U.S. shortwave broadcasters often incidentally serve a domestic audience even though their listed target audience is international—as required by U.S. law. FCC rules do not currently permit shortwave broadcasting to U.S. territories. These rules are found in the U.S. Code of Federal Regulations. Some key regulations are addressed next.

Section 73.701 Definitions

The following definitions apply to terminology employed in this subpart:

(a) International broadcast stations. A broadcasting station employing frequencies allocated to the broadcasting service between 5900 and 26100 kHz, the transmissions of which are intended to be received directly by the general public in foreign countries. There are both Federal and non-Federal Government international broadcast stations; only the latter are licensed by the Commission and are subject to the rules of this subpart.

Section 73.751 Operating Power

No international broadcast station shall be authorized to install, or be licensed for operation of, transmitter equipment with:

(a) A rated carrier power of less than 50 kilowatts (kW) if double-sideband (DSB) modulation is used.

Section 73.753 Antenna Systems

All international broadcasting stations shall operate with directional antennas. Such antennas shall be designed and operated so that the radiated power in the maximum lobe toward the specific zone or area of reception intended to be served shall be at least 10 times the average power from the antenna in the horizontal plane. Radiation in all other directions shall be suppressed to the maximum extent technically feasible. . .

US based HF stations often enjoy incidental domestic reception as a consequence of operating antenna systems (such as a Rhombic) below their natural cutoff frequency.

HF Diplexers

The use of diplexers for operating multiple transmitters on a single antenna is well known in both MW and VHF FM broadcasting. It may also be useful in unusual circumstances in HF broadcasting where, for example, space does not permit construction of multiple antennas for each direction to which transmission is desired.

A short review of the properties of one-quarter wavelength and one-half wavelength lines is in order. A one-quarter wavelength line is impedance inverting. If you take a quarter-wave section of 50-ohm transmission line and terminate it in 25 ohms you will measure 100 ohms at the other end. If you short one end of a quarter-wave transmission line it will appear as an open circuit at the other end. A short becomes an open and an open becomes a short. A half-wavelength line is impedance repeating. If you take a one-half wavelength of coax and terminate it into 25 ohm you will measure 25 ohms at the other end. A short remains a short; an open remains an open.

Diplexers can be very useful. The IBB station on Saipan has only one antenna capable of coverage of North Korea. A TCI wideband diplexer allows the operation on both the 9 and 11 MHz bands simultaneously. Advanced design allows wideband performance. When Far East Broadcasting was on Saipan they also used a diplexer for their favored direction where only one antenna was available.

It is possible to make a diplexer for HF using sections of transmission line, as is done for VHF diplexers. An example of such a diplexer made with 300-ohm balanced lines is shown in Figure 7.20-105.

The operation can be understood as follows: the 15.3 MHz signal feeds point A where it sees a 15-meter line terminated in a short. The 15-meter shorted line is nearly three-quarters of a wavelength away, electrically, so will look like an open circuit to the 15.3-MHz signal. Thus all of the energy feeds to the antenna via the 7.5-meter section of line to point B. Also at this point the 15.3-MHz signal sees an open circuit towards point C. This open circuit is produced by a shorted half-wave (10 meter) section of line that is a half wave and produces a short at point C. This short reflects through the quarter-wave (5 m) section back to point B, as an open circuit. Thus all of the energy feeds into the antenna.

The 9.9-MHz signal arrives at point C and sees a 10-meter section that looks like a capacitance in parallel with the 7.5-meter section that looks like an inductance. Currents in these two sections will cancel leaving only the power traveling towards B. The 9.9-MHz signal sees an open circuit looking from point B to A due to the shorted half-wave, 15-meter section, that is then transformed into an open at point B.

The frequency response of this circuit is shown in Figure 7.20-106. Note that the rejection notches are quite narrow. The diplexer is adjusted by moving shorts on the transmission line where indicated so as to null out the unwanted signals when higher rejection is required. After that the matching stubs, where needed, are adjusted for best VSWR at the desired frequency. Because the 15-meter stub already is open at 15.3 MHz, no additional matching stub is needed. This design uses only shorted stubs that are preferred because they are easily adjusted with a movable short that can be grounded to prevent voltage buildup. For some frequency combinations a short open stub may be used provided that corona treatment is used to prevent large voltages from building up on the stub ends.

Caution must be observed with regard to excessive voltage going to the antenna. The entire antenna system must be rated for the sum of the peak voltages of the two transmitter's RF signals not the simple sum of the transmitter powers. When two 100-kW transmitters are combined an antenna rated for 400 kW is needed; 500 kW is the nearest standard size. Failure to observe this power limitation can cause antenna arcs that reflect power back into the diplexer, that is, high VSWR. The high reflected power could do damage to the antenna, diplexer and/or transmitter matching components.

Wideband designs that cover the entire 9-MHz and 11-MHz Broadcast Bands with minor adjustment are possible are possible even with this narrow frequency spacing between transmitting frequencies. Shown in Figure 7.20-107 is a design that covers 9.5–9.9 MHz and 11.65–12.05 MHz. Matt Folkert formerly of RFE/RL (Radio Free Europe/Radio Liberty), now of duTreil, Lundin & Rackley actually built this design for RFE/RL and was kind enough to share it. In turn, each pair of stubs compensate each other with their respective 0.2 λ separation, line lengths 3 and 9. Isolation between ports is then ensured when zeros are turned into poles with 0.25 λ length lines, 5 and 7. Note that the 0.2 λ sections are for 9 MHz at 3 and 11 MHz at 9.

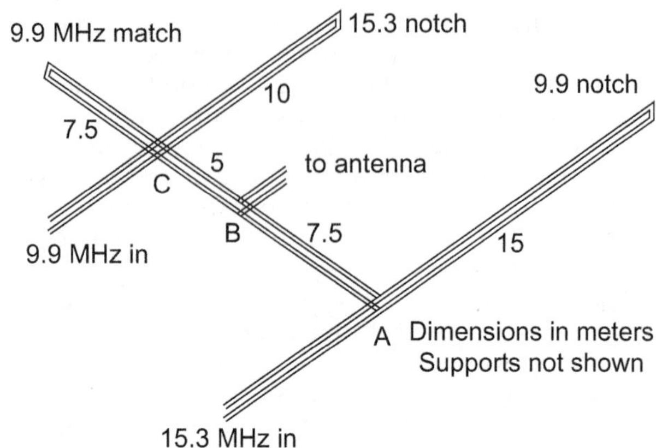

FIGURE 7.20-105 9- and 15-MHz Balanced Line Diplexer Schematic

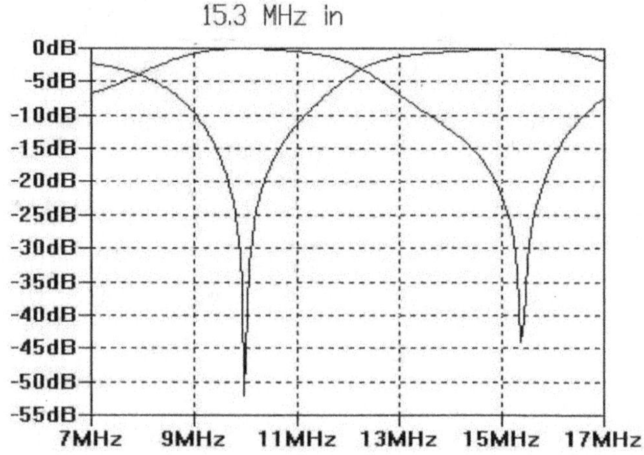

FIGURE 7.20-106 Diplexer Frequency Response

Matt also has some triplexer designs that were also field proven at RFE/RL but they are beyond the scope of this section. Assuming all transmitter powers are equal for a triplexer, the peak voltage rating of the feedlines and antennas would be nine times that for a single transmitter.

HF Audio Processing

Audio processing (equalization, compression and limiting) is employed to reduce the peak-to-average modulation ratio and increase loudness at the receiver. This also helps overcome noise at the receiver. Audio processing is covered in detail in other chapters in this handbook.

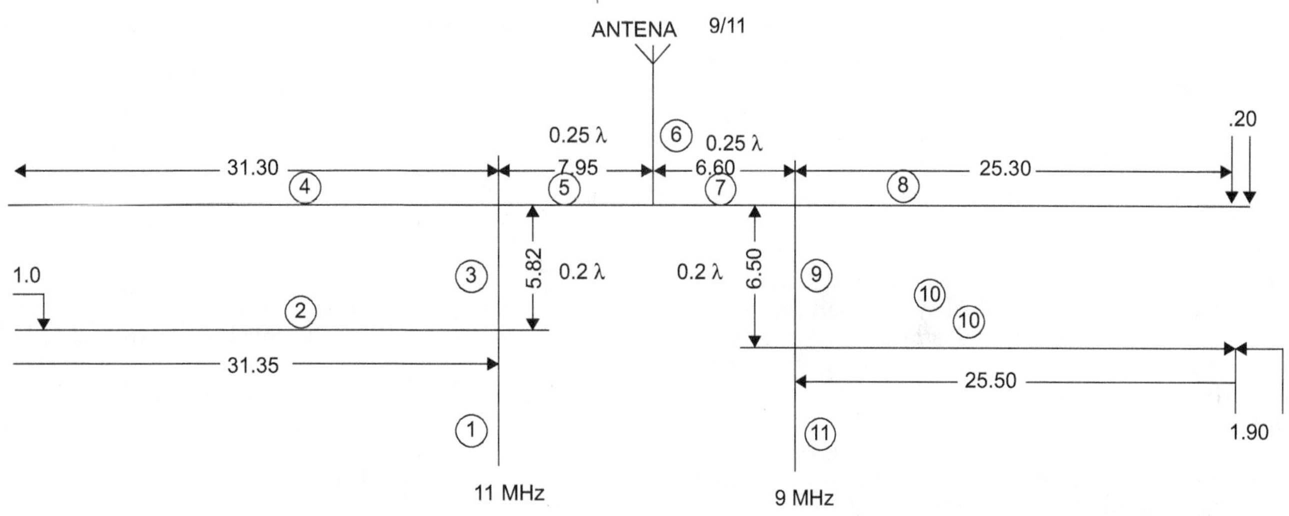

FIGURE 7.20-107 Diplexer Covering 9 and 11 MHz bands

Pie Racks

PIE (Program Input Equipment) is normally located in the control room although in some installations much is located adjacent to the transmitter. Co-location is typical where no central control room is used. A typical PIE Rack is shown for four transmitters in Figure 7.20-108.

Satellite links, microwave links, test equipment, local sources and so on are grouped together, audio processing and routing are grouped together and the modulation monitoring equipment and off-air speaker monitoring is grouped together. Patch panels are provided for test purposes. Most patch panels are configured to allow headphone monitoring of each point along the audio path without disturbing the program content. There is no standard arrangement for PIE racks but the example shown contains and demonstrates the major elements required. A UPS for the PIE rack and control console is shown to the left of the PIE racks.

Many engineers include a small audio speaker at the local control point of each transmitter so engineers in the transmitter hall—right at the transmitter—can monitor each transmitter's audio. The audio for these near-transmitter speakers need not come from the control room or modulation monitors. An RF sample and a simple diode detector are more than adequate for monitoring conventional AM signals.

FIGURE 7.20-108 PIE Racks

Control Systems for Automatic/ Remote Station Operation

There is no single best control system because each HF site is unique in the number of transmitters, antennas, programs and so on vary. Each station control typically builds on a standard platform. One expandable system to control every aspect of the broadcast day is shown in Figure 7.20-109.

Each transmitter and system component is monitored. Perhaps more important, the stations operating

FIGURE 7.20-109 Transmitter Control Console

schedule with its frequency, program, antenna direction (front/back), slew and so on is set into the system computer. Once the programming information is entered, the system automatically controls each aspect of the station operation. Control is also coordinated. The system, for instance, ensures that the transmitters are not all switched off simultaneously. Where program priority is used, the system will automatically switch configurations to keep the highest-priority program on the air. The requirements on the control system can be complex or simple. The control system can provide as much system automation as desired. The more extensive and complete the control required, the greater the system cost. The reduction in operator involvement (and cost) can ensure a return on investment.

All modern control systems use a series of checks to ensure that two programs are not scheduled simultaneously on the same transmitter or antenna. This prevents conflicts during operation.

Another system control function is to keep the station's logs recording every parameter of the stations operation in a nonvolatile form so that maintenance engineers have a complete picture of what happened before, during and after any fault occurs.

As noted in the introduction to HF Broadcasting the entire system may change many elements of its configuration several times a day. The required maintenance level of an HF station cannot be compared to a single-frequency MW or FM station. Station control can be a useful and important adjunct but completely unattended operation should be employed with caution.

Notes

[1] Jerome S. Berg, *Broadcasting on the Short Waves, 1945 to Today*, Jefferson, NC: McFarland & Company, p. 41, 2008.

[2] VOACAP, "Propagation Modeling Program," download at: www.greg-hand.com/versions/, itshfbc_151021.exe, multiple tutorials at: www.voacap.com/ The 32-bit version is at: http://elbert.its.bldrdoc.gov/pc_hf/hfwin32.html

[3] Joe Craig, "Marconi's First Transatlantic Wireless Experiment," archived at: www.ucs.mun.ca/~jcraig/marconi.html

[4] DRM Consortium, "The DRM Digital Broadcasting System Introduction and Implementation Guide," September 2013, archived at: www.drm.org/wp-content/uploads/2013/09/DRM-guide-artwork-9-2013-1.pdf

[5] R. A. Heising, "Transmission System," U.S. Patent No. 1,655,543, January 1928.

[6] James Tonne, "PI-EL," Tonne Software, 2015, archived at www.tonnesoftware.com; James Tonne, "Elsie," Tonne Software, 2015, archived at www.tonnesoftware.com.

[7] Douglas M. Beuerman and Mark H. Burleson, "Double-Bellows Vacuum Variable Capacitor," U.S. Patent 6268995, July, 2001.

[8] Bernard T. Ellis Jr., "Combined Plate and Screen Grid Modulated Amplifier Circuit," U.S. Patent 2939090, May, 1960.

[9] Sainton, J. B., "High Efficiency Amplifier and Push-Pull Modulator," U.S. Patent No. 3,314,024, April, 1967.

[10] Theodore Khvilivitzky, "Forty Years of Soviet Short Wave Broadcast Transmitters: From 50 to 2,000 kW in A Unit," Unpublished, 1998 [your consultant or frequency manager may have a copy].

[11] Continental Electronics, "419H-DRM 300 kW, 420D-DRM 500 kW, Shortwave Broadcast Transmitters," archived at: www.contelec.com/sw419420drmfeatures.htm.

[12] Von Baeyer Hans Jakob Ritter and Werner Buschbeck, "Antenna transformer," U.S. Patent 2187014, January, 1940.

[13] Charles T. Rauch, "Rhombic Antennas, V-Beam, and Inverted V," 2008, archived at: www.w8ji.com/rhombic_antennas.htm.

[14] Benjamin Whitfield Griffith, *Radio-Electronic Transmission Fundamentals*, Atlanta, GA: Noble Publishing Corporation, p. 476, September, 2000.

[15] TCI, www.spx.com/en/tci/products-services/.

[16] Ampegon AG, www.ampegon.com/products/antenna-systems/.

[17] Sino-Sky, Beijing, www.sino-sky.com.cn/en/html/Product Center/SWAntenna/.

[18] WRTH Editors, *World Radio TV Handbook 2016: The Directory of Global Broadcasting*, Oxford, UK: WRTH Publications Ltd., 2016.

[19] WBU-TC Task Group on Coordination Procedures for HF Broadcasting, "Shortwave Broadcasting: A Primer on Coordination of Seasonal Schedules," pp. 1–5, archived at: www.abu.org.my/upload/PrimerHF.pdf.

[20] Mini-Circuits, "Coaxial Phase Detector," Mini-Circuits, 2013, archived at: www.minicircuits.com/pdfs/ZRPD-1+.pdf.

[21] Eimac Staff, *Care and Feeding of Power Grid Tubes*, "Stabilizing the Amplifier," CPI, Eimac Division, 2003, archived at: www.cpii.com/library.cfm/9, pp. 77–81 (Section 3).

[22] Frederick W. Grover, "Tables for the Calculation of the Inductance of Circular Coils of Rectangular Cross Section," National Bureau of Standards, 1922, pp. 467–469, archived at: https://archive.org/stream/tablesforcalcula18451grov#page/n3/mode/2up.

[23] Bohumil Kyrian and Wilhelm Tschol, "Amplitude Modulated Broadcast Transmitter," U.S. Patent 4,864,635, September, 1989; Bryan Weaver and Daniel Dickey, "Power Amplifier Having Multiple Switched Stages and Method of Operating Same," U.S. Patent 5,099,203, March, 1992.

[24] Gil Hulsey and Bryan Weaver, "RF-AM Transmitter With Pulse Width Modulator," U.S. Patent 4,776,036, October, 1988.

[25] Ernest R. Kretzmer, "Distortion in Pulse Duration Modulation," *Proceedings of the Institute of Radio Engineers*, Vol. 35, No. 11, pp. 1230, November, 1947.

[26] Gil Hulsey and Bryan Weaver, "AM-RF Transmitter With Compensation for Power Supply Variations," U.S. Patent 4,747,161, May, 1998.

[27] Warren C. Pratt, "Digital Predistortion: Linearizing Our Amplifiers," Ham Radio Friedrichshafen, 2014, pp. 15–21, archived at: http://n3sh.org/Tech%20Section/PureSignal_final_selection.pdf.

[28] Ky T. Luu, Wayne Duello and Anders Mattsson, "Advances in AM Modulation Techniques to Improve Digital Transmission of HD Radio and DRM," Harris Broadcast Communications, 2004, p. 6.

[29] Eimac Staff, *Care and Feeding of Power Grid Tubes*, "Amplitude Modulation," CPI, Eimac Division, 2003, archived at: www.cpii.com/library.cfm/9, pp. 73–74 (Section 3.11).

[30] Joseph Milnes, "Computational Modelling of the HyperVapotron Cooling Technique for Nuclear Fusion Applications," PhD Thesis, 2010, archived at: https://dspace.lib.cranfield.ac.uk/bitstream/1826/5555/1/Joseph_Milnes_Thesis_2010.pdf, p. 5.

[31] Eimac Staff, "Extendingtransmitter Tube Life," archived at: www.cpii.com/docs/related/22/AB18.pdf.

[32] Eimac Staff, *Care and Feeding of Power Grid Tubes*, "Neutralization Below VHF," CPI, Eimac Division, 2003, archived at: www.cpii.com/library.cfm/9, pp. 109–113 (Section 5.)

[33] TCI, "Model 615 Short-Range Broadcast Antenna," archived at: www.spx.com/en/tci/pd-615-short-range-broadcast-antenna/.

[34] TCI, "Model 521 Rotatable Log-Periodic Antenna," archived at: www.spx.com/en/tci/pd-521-rotatable-log-periodic-antenna/.

[35] TCI, "Model 527 and 527B Super High Gain Log-Periodic Antennas," archived at: www.spx.com/en/tci/pd-527-527b-super-high-gain-log-periodic-antennas/.

[36] DASTO, "RF Switchs," archived at: www.dasto.com/rf_components/default.htm#coaxial switches.

[37] Edmund Laporte, Radio Antenna Engineering, "Tapered Transmission-Line," Epina eBook Team, 2007, archived at: www.vias.org/radioanteng/radio_antenna_engineering_04_04_09.html.

[38] National Radio Systems Committee, "NRSC-G101, AM Modulation-Dependent Carrier Level (MDCL) Usage Guideline," April, 2013, archived at: www.nrscstandards.org/SG/nrsc-G101.pdf.

[39] J. Fred Riley, "Comparison of CCM Techniques," Continental Electronics Corporation, IEEE Broadcast Symposium, September 22–23, 1994.

[40] Belar, "MODEL AMMA-2, DSP-Based Microprocessor-Controlled Digital AM Modulation Monitor And Analyzer," archived at: www.belar.com/AM/amma2.htm.

CHAPTER

7.21

Hybrid Radio and the RadioDNS Open Standards

NICK PIGGOTT
Project Director, RadioDNS
London

INTRODUCTION

Broadcasting in the AM, FM and shortwave frequency bands has been the primary distribution technology for radio since the inception of the medium, and the capabilities of the radiofrequency (RF) technology have defined the listening experience. Access to broadcast spectrum has generally been regulated, creating a stable market and relatively high barriers to entry for new entrants.

Internet protocol (IP) creates an alternative distribution model for incumbents and new entrants, who are using the flexibility of connected device environments to create innovative media experiences. This is increasing competition for audiences and advertising spend.

Broadcast and IP each have unique strengths. The capabilities of broadcast are well understood and use standards that have proved to be reliable and durable for decades, albeit creating a certain inflexibility. In contrast, while TCP/IP (the most commonly used Internet protocol) has been stable since the introduction of v4 in 1995 (www.rfc-editor.org/rfc/rfc1812.txt), the applications deployed on top of it change frequently and many become obsolete relatively quickly.

Concerns exist, to varying degrees, about the ability of IP, and specifically wireless IP, to cost-effectively scale to match the reliability, coverage, accessibility and ubiquity of broadcast. Alongside the technical risks of replicating a broadcast system in an architecture designed for short-range two-way communication from battery powered devices (specifically, using smartphones, the most ubiquitous mobile connected devices), there are risks from regulation, or lack of it, commonly referred to as "net neutrality." Depending on how these complex discussions develop, the cost of implementing IP networks, both fixed and wireless, will divide in different ways between the consumer and the content provider, with the potential to change assumed costs and business plans for using IP as a primary distribution platform for radio.

Conversely, broadcast is failing to provide functionality now considered to be critical for attractive media experiences. Consumption cannot be directly measured, individual listeners can't be identified or targeted to, and there's no capability to personalize the experience.

Incumbent radio broadcasters have continued to support broadcast technology, as it delivers the overwhelming majority of the current audience. They have also embraced IP as a new distribution technology, but entirely in parallel to their broadcast distribution. Originally the two distribution methods targeted two nonoverlapping groups of devices; broadcast reached radio devices, particularly in automotive; IP reached devices such as personal computers, and more recently, mobile phones and connected home audio devices.

The growing pervasiveness of IP connectivity is causing these two previously distinct classes of device to overlap, as more devices are incorporating IP alongside broadcast radio functionality, albeit functioning entirely separately (see Figure 7.21-1). Examples include domestic radios that combine both broadcast radio and IP streaming functionality, and cars that are connected to wireless IP networks.

NAB ENGINEERING HANDBOOK

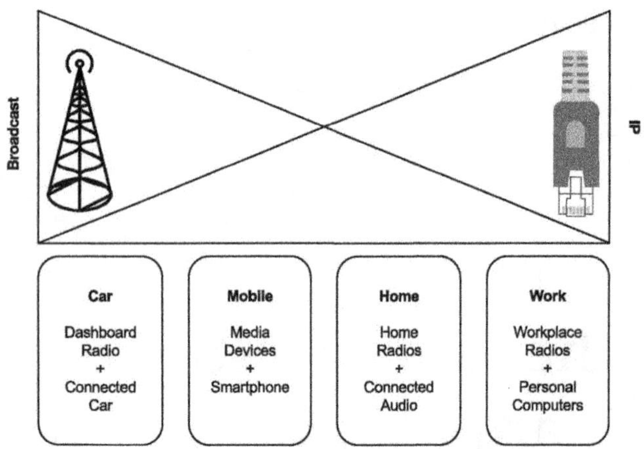

FIGURE 7.21-1 The growing overlap of broadcast technology and IP connectivity in more devices and locations.

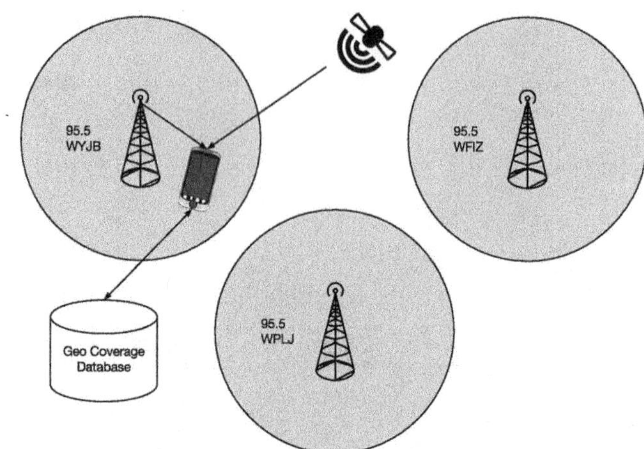

FIGURE 7.21-2 A receiver can use strategies like GPS (satellite) or cellular network localization information to determine its location.

HYBRID RADIO

The principle of hybrid radio is to bridge broadcast and IP technologies to leverage the strengths of each individual technology and create a radio experience that is better than could be delivered by using either technology in isolation. This approach could enable a smooth transition to all-IP distribution (i.e., streaming audio), or using the two technologies together could be a permanent solution. The technicalities of combining the two technologies should be largely invisible to the end user, and the process should be entirely automatic.

Fundamental to the hybrid radio process is the ability to discover and identify radio stations in both the broadcast and IP domains. This allows a receiver to make an IP connection to the correct address on the Internet, and transact with servers to retrieve additional content and services.

Discovering and Identifying Radio Services and the Corresponding IP Services

The pre-eminent method of discovering broadcast radio services is to scan the relevant frequency range, looking for signals of suitable quality, where suitability is based on a set of factors determined by the receiver manufacturer. This can be an imprecise method, particularly with AM modulation where it can be difficult to discriminate between a genuine signal and noise. An alternative approach is to use an authoritative database of radio services, created by a trusted and neutral provider such as a national regulator, although this approach may only be available in certain territories.

STRATEGIES FOR IDENTIFYING RADIO SERVICES

Geolocation

Simply knowing the frequency of a radio service is insufficient to identify it. As frequencies are reused across regions, one needs knowledge of both the expected transmission coverage and the location of the receiver to allow the radio service to be identified. This requires a database that is kept regularly updated with any changes to planned coverage areas, and also relies on an assumption that coverage does not significantly deviate from the intended coverage area. By cross referencing the physical location of a receiver to geographic information in a database, and matching frequencies received to frequencies in the database, each radio service can be identified.

Creating and maintaining such a database is non-trivial and it requires a very high availability, creating a cost of operation that would grow as the system coverage became more expansive and requests to it became more frequent. In addition, it requires that there are sufficient guard areas between frequency reuse, and particularly (but not exclusively) with AM signals, actual coverage can vary considerably from planned coverage, leading to an increased risk of misattribution of a signal to a station.

A receiver must accurately identify its location to correctly look up radio services available to it. A receiver can use strategies like GPS location or cellular network localization information (https://en.wikipedia.org/wiki/Mobile_phone_tracking) to locate itself to varying degrees of accuracy (depicted in Figure 7.21-2), but without those technologies it is reliant on positioning information based on its IP address and the network it is connected to, and this "geo-IP" approach can create some very inaccurate results (www.theguardian.com/technology/2016/aug/09/maxmind-mapping-lawsuit-kansas-farm-ip-address). Collectively, these requirements and inaccuracies make assumptions of received radio signals based on geo-information undesirable as a universal solution.

Embedded Identifiers

An alternative approach to identifying radio stations is to use identifiers embedded in the radio transmission.

This definitively identifies a radio service without having to make assumptions of actual coverage, maintenance of a complex geo-database or a requirement for receivers to know their physical location. The receiver needs to be capable of retrieving the identifiers from the transmission. Such identifier embedding exists now as part of radio transmission standards.

FM Radio Data System (RDS, IEC 62106 Edition 2.0, 2009–07; NRSC-4-B)

RDS specifies the program information (PI) code, a 16-bit service identifier transmitted every 57 ms. An optional 8-bit extended country code (ECC) can also be transmitted. The first nibble (4 bits) of the PI code is a country code for the country of origin of the radio service, but this only allows for a maximum of 15 different values (as 0 is a disallowed value for the first nibble). In order to create sufficient values to represent globally unique radio services, the ECC code should also be transmitted. The ECC code is a two nibble (8-bit) code, assigned to countries according to Table X of TS 101 756 [1]. When combined with the country code, this creates a globally unique service identifier. An RDS encoder can be added to an existing FM transmitter for approximately USD160.

HD Radio (FM/AM)

The HD Radio in-band/on-channel digital radio system transmits a globally unique facility code identifier. No separate equipment is required to encode this identifier.

DAB Radio (EN 300 401)

The DAB digital radio system includes identifiers for country (8 bits ECC), ensemble (16 bits EId) and individual services (16 bits SId) that collectively provide 40 bits of globally unique identifier information, transmitted every 100ms. No separate equipment is required to encode these identifiers.

AM Radio/AMSS (ETSI TS 102 386: Digital Radio Mondiale, DRM); AM Signaling System (AMSS)

AMSS allows transmission of a globally unique 24-bit service identifier (SId) every 1 second. No encoder equipment is currently commercially available, but its complexity is not significantly different from an RDS encoder.

Limitations

Real world implementations display three problems:

- The FM RDS ECC code is often not transmitted;
- Due to poor coordination, two or more different radio services in a country may transmit the same RDS PI code;
- Service identifier codes are not coordinated between all systems (AM, FM, DAB, HD), so it is necessary to distinguish which system service identifiers relate to.

These issues must be considered when designing systems that uniquely identify a radio station.

RESOLVING IP ADDRESSES FOR RADIO SERVICES

A receiver can tune into a transmission and acquire identifiers from a radio service. Having identified the radio service, the receiver needs to know how to reach the radio service over the Internet, which requires translation of the numeric identifier codes into an IP address. One approach might be to operate a database, similar to that described in the earlier section on geo-location, but it would suffer from the same problems of scale, speed, reliability and cost.

DNS—Domain Name System

All addressing on the Internet uses either IPv4, notated as 4 decimal numbers separated by dots (e.g., 192.168.123.12), or increasingly IPv6, which uses an altogether more verbose addressing structure of 8 groups of 4 hexadecimal digits (e.g., 2001:0db8:85a3:00 00:0000:8a2e:0370:7334).

In order to make addressing more usable, the domain name system (DNS) was created (RFC 822, November 1983). This allows users to enter an address as a name, and DNS can return the relevant destination IP address. DNS uses a hierarchical addressing system, using periods to separate levels of the hierarchy, creating the familiar pattern of names used today (e.g., www.google.com).

As DNS is critical to the functionality of the Internet, it is deployed on a massive scale. Typically every local area network will have at least one DNS server, meaning that queries from devices are answered very quickly. DNS records are cached and replicated between all machines, and there is high resilience designed into the system. The costs of operating DNS are borne by each network operator, not by a central organization.

DNS is an ideal candidate to improve the lookup process between a service identifier and the corresponding IP address of the radio station. DNS queries require a text domain name as an input, so one approach might be to format the numeric service identifier into a text string. Thus, the RDS PI code 0xCAFE could be presented to DNS as "Domain Name: café." However, a domain name without a hierarchy is assumed to be a host name, referencing a machine on the local network. Such queries are only resolved by the local DNS server, and so cannot be globally referenced. In order to create a globally accessible DNS record, the domain name needs to be a Fully Qualified Domain Name (FQDN), which requires a domain hierarchy to be appended to it.

The RadioDNS Hybrid Lookup for Radio Services

The RadioDNS Hybrid Lookup for Radio Services specification (ETSI TS 103 270 v1.2.1) standardizes the construction of a FQDN using identifiers from radio transmissions. It is an open specification, which can be implemented freely without any contractual arrangements or royalty payments.

All RadioDNS FQDN constructs follow the same basic format:

<system specific identifiers>.<system name>.radiodns.org

The suffix "radiodns.org" is a domain managed by RadioDNS, according to a published domain management policy (https://radiodns.org/organisation/). RadioDNS operates all the root name servers that act as authorities for the radiodns.org domain.

Extended Country Codes (ECC) and Global Country Codes (GCC)

As noted earlier, the ECC code is used in combination with the country codes to allow PI codes to be globally unique. In order to identify a specific country, the ECC and country codes must be concatenated to make a 24-bit value. For the purposes of constructing FQDNs that can be mapped to specific countries, the concept of a global country code (GCC) is defined as

<Country Code><Extended Country Code>

For example, a PI code of "c479" and an ECC of "e1" represents a radio service in the United Kingdom. Therefore the GCC code for the United Kingdom is "ce1". Some countries have multiple country codes assigned to them, and the U.S. is such a country. In this case, the GCC codes from "1a0" to "ba0", "da0" and "ea0" all refer to the U.S.

Deriving ECC When It Is Not Transmitted

Often the ECC code is not transmitted. In this case, the RadioDNS specification provides guidance on to calculate the correct ECC for the radio service being received, taking into account the case where signals may be crossing a border from another country. This process requires the very coarse geolocation information of the country where the receiver appears to be located. This can be approximated using many techniques, including geo-IP lookup and cellular network identifiers, and does not require GPS information.

An Example of a RadioDNS FQDN for FM

A combination of PI code and ECC code should create a globally unique identifier for every FM radio service, but as noted earlier there are inconsistencies in the allocation of PI codes and transmission of ECC codes. To overcome the risk of clashing PI codes, the construction of the FQDN includes frequency, although this can be replaced with a wildcard value ("*") if the broadcaster can be certain that no other radio service is using the same PI code in the same region. The frequency is notated as an integer number to an accuracy of 10kHz with no decimal places, and with leading 0s to create a fixed length field of 5 characters. The full construct is

<frequency | *>.<PI code>.<GCC code>.fm.radiodns.org

For example, Z100 in New York transmits on 100.3MHz, with the PI code of 192B, so their RadioDNS FQDN would be:

10030.192B.xa0.fm.radiodns.org

Construct information and examples for other systems (DAB, HD, AM) are shown in the RadioDNS Hybrid Radio Lookup specification.

Using a RadioDNS FQDN to Locate the Broadcaster's Own Domain

The DNS specification allows a query to return a CNAME record, which specifies a domain name as a result, rather than an IP address. This could be considered analogous to a "telephone call diversion" or "web page redirect." RadioDNS uses this technique to translate a RadioDNS FQDN to a broadcaster domain rather than a specific IP address. This allows a broadcaster to register multiple radio services in DNS and point them all to a single FQDN that is under their exclusive control. For example:

Query FQDN for CNAME: 09580.c479.ce1.fm.radiodns

Response FQDN of CNAME: rdns.musicradio.com

The FQDN rdns.musicradio.com is controlled by the radio group and is used by all radio services owned by that group.

To summarize the process of discovering and identifying radio services and their IP services (see Figure 7.21-3):

- The radio service includes service identifiers in their broadcasts;
- These service identifiers are received by the receiver;
- The service identifiers are used to construct a RadioDNS FQDN;
- This RadioDNS FQDN is sent as a query for a CNAME to the nearest DNS service, which responds with a CNAME response containing the FQDN for that radio service.

The time between tuning into the transmission and resolving to the broadcaster's FQDN is typically less than 500 ms.

Requests to register records in the radiodns.org domain are handled by RadioDNS in accordance with the published trust model. This method of using DNS to resolve service identifiers to a broadcaster FQDN means that no information passes through servers operated by RadioDNS, and all connections from device to broadcaster are made directly.

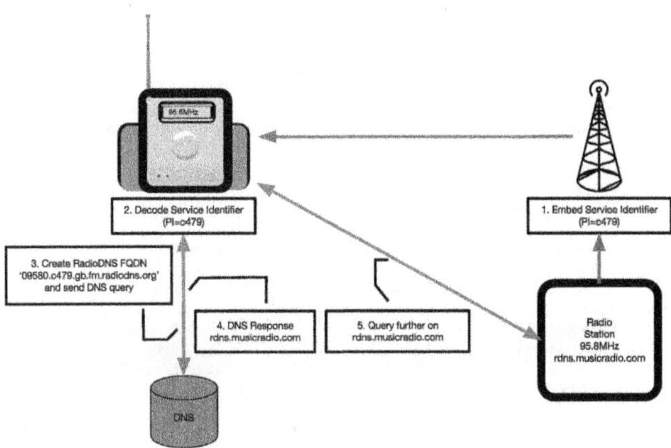

FIGURE 7.21-3 Illustration of the process of discovering and identifying radio services and their IP services using the RadioDNS method.

HYBRID RADIO APPLICATIONS

Having identified a radio service and located its presence on the Internet, it is possible to open an IP connection to the specified host and start exchanging data over IP in parallel to receiving audio over broadcast. In some cases, applications of hybrid radio are closed and proprietary, and can only be implemented on and accessed by receivers that have entered into the relevant agreements with the application provider. RadioDNS defines applications as open standards, building on the framework of FQDN discovery. These applications can be freely implemented and guarantee interoperability between providers and devices, allowing for an open and competitive market in technical solutions.

Service and Program Information

As the variety and volume of media experiences have increased, the mechanisms for discovery and presentation of media have become similarly more sophisticated. Almost all of this functionality requires accurate metadata about the various media sources in order to function correctly, and various systems have been devised to carry metadata alongside the media to which it relates. For example, most file formats for media include the capability to include metadata to describe the media, such as ID3 Tags for MP3 files. Similar functionality is enabled by broadcast systems such as RDS, HD Radio and DAB, which allow definition of station names, textual information about the current event, visuals and program guides. Some of this functionality has evolved in ways not originally foreseen, which now hinders some of the originally intended uses for transporting metadata.

The growing number of media sources requires more volume of metadata to be transmitted to organize navigation for the user. Using only over-the-air methods means that the metadata arrives relatively slowly and less reliably compared to the same metadata delivered over an IP connection. An example of this are station logos delivered as image files. As receiver display sizes and resolutions have increased and manufacturers have developed more complex user interface designs, the range of logo sizes that can be delivered over-the-air is falling short of receiver manufacturer requirements. The receiver could request a logo of specific size and resolution over IP and receive a response almost instantly.

RadioDNS has defined Service and Program Information (TS 102 818 v3.1.1) as a standard for defining metadata about Radio Services, the programs that each service transmits, and individual events within those programs [2]. It uses an XML (eXtensible Markup Language) format, which allows for the amount of detail in the definition to be customized according to the metadata available, and for new elements to be added as new requirements arise. The standard uses the RadioDNS hybrid lookup for radio services process to identify the FQDN of the radio service. It then uses DNS SRV records of the format _radioepg._tcp to locate the host(s) that will provide service and program information metadata for that service. For example:

- RadioDNS FQDN Query for CNAME: 09580.c479.ce1.fm.radiodns.org
- DNS Response of CNAME: rdns.musicradio.com
- Service & Program Information Query for SRV: _radioepg._tcp.rdns.musicradio.com
- DNS Response of SRV record:epg.musicradio.com

This two-step process, of a first step to identify the broadcaster's FQDN, then a second step to identify the host serving the specification application, is typical of the RadioDNS process for connecting to applications. The receiver can now connect and query the host "epg.musicradio.com" to retrieve detailed metadata about that radio service. The transport protocol is HTTP or HTTPS, and the URL structure is:

- http(s)://<hostname:port>/radiodns/spi/3.1/SI.xml to retrieve service information
- http(s)://<hostname:port>/radiodns/spi/3.1/<service identifiers>/<YYMMDD>_PI.xml to retrieve program information for the specified service and day (YYMMDD).

The detailed explanation of how to construct the URLs is given in TS 103 270 [3]. The Service Information (SI) file can contain information on one or more radio services, which might represent all the services provided by one broadcaster or technology provider (example file provided in Figure 7.21-4). The services do not have to be geographically co-located. The benefit of this approach is that a single HTTP request can return a single XML document with information on multiple radio services received by the receiver, thus removing the requirement for the receiver to make individual queries for every single station. If the broadcaster allows for the document to be cached and the receiver has sufficient memory, the SI file can also provide information on radio services that have not yet been received but may do so in the near future, for example during a drive to a bordering area.

The SI file can contain a very wide range of information, but important elements are:

- <mediumName>—The name of the radio service, as used on-air (e.g. "KISS FM"), not the legal call letters or regulatory identifier;
- <description>—The radio service description, as used on-air (e.g. "More music variety");
- <mediaDescription>—URLs to a range of logos, which can be of various standard dimensions plus any free-form dimensions chosen by the broadcaster;
- <genre>—Information to describe the radio service using the structured data defined by TV Anytime;
- <bearer>—Information to describe the broadcast frequencies and identifiers and the streaming URLs used by that radio service, and the relative offset times of each bearer.

Switching Between Broadcast Radio and Streaming Audio

If the broadcaster provides bearer definitions for a radio service that are a mix of broadcast and streaming URLs, it is possible for the receiver to use this information to switch between receiving the radio service on broadcast or using streaming audio. Bearer definitions can also specify the relative time-offsets between each bearer, to account for different transmission delays.

The receiver manufacturer can make a decision on what constitutes "poor" quality broadcast radio reception, and connect to the most suitable streaming URL for their device. They can use the time-offset information in the bearer information to pre-delay the various sources to approximate time-alignment, so that the transition is not as noticeable to the user. The receiver can continue to monitor the broadcast signal, and when it is considered to be "good," switch back to broadcast reception and end the stream.

This practice, also called "service following," allows the user to continue listening to the radio service while using minimal IP data, tactically filling in the gaps of broadcast radio. The broadcaster can help improve the experience of this switching by ensuring the streams specified in the SI document are as close to simulcast with broadcast as possible, including an absence of "pre-roll" advertising, or content that "stretches" the stream time offset relative to the original broadcast ("ad insertion"). Using a similar audio processing and encoding quality will also help.

Program and Program Event Information

The program information (PI) files hold information on daily program schedules and events within each program. That information can include titles and descriptions, and visuals, as well as links to related resources. This information can be used when a user searches for

```
<?xml version="1.0" encoding="UTF-8"?>
<serviceInformation creationTime="2016-03-27T23:01:16+01:00"
originator="Broadcaster Group" xml:lang="en"
xmlns="http://www.worlddab.org/schemas/spi/31"
xmlns:xsi="http://www.w3.org/2001/XMLSchema-instance"
xsi:schemaLocation="http://www.worlddab.org/schemas/spi/31 spi_
31.xsd">
  <services>
    <service>
      <shortName>Radio</shortName>
      <mediumName>Radio Station</mediumName>
      <longName>Radio Station New York</longName>
      <mediaDescription>
        <shortDescription>New York City's No.1 Station
</shortDescription>
      </mediaDescription>
      <mediaDescription>
        <multimedia height="32" mimeValue="image/png"
url="http://cdn.radiostation.com/2.0/id/25/logo/32x32.png"
width="32"/>
      </mediaDescription>
      <mediaDescription>
        <multimedia height="32" mimeValue="image/png"
url="http://cdn.radiostation.com/2.0/id/25/logo/112x32.png"
width="112"/>
      </mediaDescription>
      <mediaDescription>
        <multimedia height="128" mimeValue="image/png"
url="http://cdn.radiostation.com/2.0/id/25/logo/128x128.png"
width="128"/>
      </mediaDescription>
      <mediaDescription>
        <multimedia height="240" mimeValue="image/jpg"
url="http://cdn.radiostation.com/2.0/id/25/logo/320x240.jpg"
width="320"/>
      </mediaDescription>
      <mediaDescription>
        <multimedia height="600" mimeValue="image/jpg"
url="http://cdn.radiostation.com/2.0/id/25/logo/600x600.jpg"
width="600"/>
      </mediaDescription>
      <mediaDescription>
        <multimedia height="800" mimeValue="image/jpg"
url="http://cdn.radiostation.com/2.0/id/25/logo/800x800.jpg"
width="800"/>
      </mediaDescription>
      <genre href="urn:tva:metadata:cs:ContentCS:2004:3.6.10">
Hit-Chart/Song Requests</genre>
      <link mimeValue="text/html" uri="http://www.radio.com/"/>
      <link mimeValue="text/html"
uri="http://twitter.com/radiostation"/>
      <link mimeValue="text/html"
uri="http://www.facebook.com/radiostation"/>
      <bearer bitrate="48" cost="70"
id="http://streaming.com/radiostation" mimeValue="audio/aacp"
offset="16000"/>
      <bearer bitrate="128" cost="73"
id="http://streaming.com/radiostationMP3" mimeValue="audio/mpeg"
offset="10000"/>
      <bearer cost="30" id="fm:ce1.c479.09580"/>
      <radiodns fqdn="www.radiostation.com"
serviceIdentifier="radiostation"/>
      <serviceGroupMember id="radio"/>
    </service>
</serviceInformation>
```

FIGURE 7.21-4 Example of service information (SI) XML file.

content they are interested in, or used in a recommendation algorithm.

The timing information in the program can be used to trigger an automatic recording process, similar to a TV DVR. The program can also have streaming URLs associated with it, which allow the program to be listened to on-demand, rather than in real-time. Program information can be provided into the future, where it is useful for setting timed alerts or recording, but can also be provided historically to allow easy access to on-demand content. Additionally, a PI file can be provided wholly without any timing or broadcast information, which represents a catalogue of audio that can be searched and accessed on-demand.

PI for Syndicated Program Distribution

The PI structure can also be used as a data interchange format between broadcasters who produce, distribute or ingest syndicated programming. Elements allow for the definition of acquisition methods and times, such as timed transmissions or re-transmissions on satellite distribution, or retrieval of files using HTTP or FTP during specified time windows.

Enhancements to the SPI Specification

It is expected that the following enhancements will be made to TS 103 270 in January 2017:

- Addition of phoneme information, which will aid voice navigation of radio services in the automotive environment;
- Addition of client identification keys, which will allow information (such as "clean" streaming URLs) to be provided only to trusted manufacturers and devices;
- Addition of detailed information on the role of individuals in radio programs, to allow differentiation between presenters, contributors and guests.

Visuals

An increasing number of media devices and media experiences include the capability to show visual content to the user. People who are used to seeing album art when playing back media files or using streaming music services should be presented with a similar experience when listening to radio.

Digital radio platforms have the functionality to transmit images, but the bandwidth typically available is low (16 kbit/s or less), which means acquisition time is higher even for relatively modestly sized images. To get the image to display coincident with the start of the audio, the broadcaster often has to start transmitting the image some minutes ahead of its required time, and transmit it multiple times to ensure a successful reception of the entire image in the presence of errors in the broadcast channel. The growing size and resolution of displays also demands higher resolution visuals, which in turn creates much larger file sizes and acquisition times over broadcast become impracticably long.

Text information can also be transmitted over both analog and digital platforms, and the restrictions on bandwidth are less noticeable to the end user, however systems like FM RDS only allow for 64 character strings to be sent using a limited character set.

```
<?xml version="1.0" encoding="UTF-8"?>
<epg xml:lang="en" xmlns="http://www.worlddab.org/schemas/spi/31"
xmlns:xsi="http://www.w3.org/2001/XMLSchema-instance"
xsi:schemaLocation="http://www.worlddab.org/schemas/spi/31 spi_
31.xsd">
 <schedule creationTime="2016-08-31T13:37:09"
originator="Broadcaster Group">
  <scope startTime="2016-09-02T22:00:00+01:00" stopTime="2016-09-
04T06:00:00+01:00"/>
  <programme
id="crid://epg.radiostation.com/programmes/5838/3343785"
shortId="3343785">
   <shortName>Wkend</shortName>
   <mediumName>Weekender</mediumName>
   <longName>The Weekender With Studio 54</longName>
   <location>
    <time actualDuration="PT8H0M" actualTime="2016-09-02T22:00:00+
01:00" duration="PT8H0M" time="2016-09-02T22:00:00+01:00"/>
   </location>
   <mediaDescription>
    <longDescription>8 hours, non-stop in the mix, ad free. The
weekend lives on!</longDescription>
   </mediaDescription>
   <mediaDescription>
    <multimedia height="320"
url="http://cdn.radiostation.com/ArtWork/139545082c77.png"
width="600"/>
   </mediaDescription>
   <mediaDescription>
    <multimedia height="2048"
url="http://cdn.radiostation.com/ArtWork/a5f02ecab058.png"
width="2048"/>
   </mediaDescription>
   <memberOf id="crid://epg.radiostation.com/programmes/5838"
shortId="5838"/>
   <link uri="http://www.radiostation.com/weekender"/>
  </programme>
  <!-- More programmes would be defined here -->
 </schedule>
</epg>
```

FIGURE 7.21-5 Example of program information (PI) XML file.

The RadioDNS specification Hybrid Digital Radio Slideshow (TS 101 499 v3.1.1) translates the existing functionality of broadcast visuals and text into the IP domain, allowing access to the higher burst transfer speeds available in IP [4]. This means that images and text can be prepared and transferred to the receiver at higher resolutions and faster than using broadcast, often not requiring pre-transmission.

The broadcaster advertises availability of visual content by providing DNS SRV records of the format _radiovis._tcp and/or radiovis-http._tcp to locate the host(s) that will provide visual content for that service. For example:

- RadioDNS FQDN query for CNAME: 09580.c479.ce1.fm.radiodns.org
- DNS response of CNAME: rdns.musicradio.com
- Visuals query for SRV: _radiovis._tcp.rdns.musicradio.com _radiovis-http._tcp.rdns.musicradio.com
- DNS response of SRV record: vis.musicradio.com 61613 vis.musicradio.com 80

The Visuals specification provides for two transport protocols:

- STOMP—the Simple Text Orientated Message Protocol. A very simple protocol for passing messages over an IP network, which is relatively easy to implement and has a low transport overhead. Some networks or proxies may not allow the STOMP connections;
- HTTP COMET—a technique for using HTTP to mimic push messaging. This protocol is more complex than STOMP and requires that the receiver supports HTTP and JSON technologies. However, the generic use of HTTP means it will generally traverse all networks and proxies without being blocked.

Broadcasters are recommended to support both protocols in tandem. Receiver manufacturers are recommended to support both protocols, and use whichever is appropriate for the IP network the device is connected to. For example, a receiver may attempt a STOMP connection but have it blocked by a proxy or firewall, so it can attempt an HTTP connection as a fall back. The message content and functionality is identical in both transport protocols. The process for displaying visuals is (see Figure 7.21-6):

1. The receiver connects to the Visuals host (using STOMP or HTTP) and subscribes to messages relating to that radio service. One server may handle visual for multiple services. The receiver waits for a message from the server.
2. The receiver receives a message with a unique Message ID from the server, instructing it to SHOW an image at a unique URL, with the Trigger-time at which the image should be shown (denoted in UTC time). If the Trigger-time is absent, the receiver is just being asked to cache the image. If the Trigger-Time is "NOW," the image should be shown immediately.
3. The receiver connects to the URL provided in the message, which may be on server different to the visuals server, such as a Content Delivery Network (CDN). The receiver provides information on its preferred resolution for visuals, giving x and y resolution and x and y dot density information.
4. The server responds with a binary image, in JPG or PNG format, which will fit entirely within the rectangular bounding box specified by x and y. In an ideal circumstance, the broadcaster will use techniques like responsive designed HTML to render the image to the exact resolution requested by the receiver. The receiver caches the image, using the URL as the index.
5. The receiver will display the image at the Trigger-time associated with the image.

The receiver continues listening for more messages. In the HTTP COMET protocol, it must reopen a new connection to the visual server, and provide the Message ID of the last received message, in order to synchronize with the server message queue. The image formats supported for visuals are JPEG and PNG. The PNG support extends to animated PNG (A-PNG) images, allowing short animations in the image. The receiver is not obliged to show the animation, and can instead show a single static image defined by the image creator.

Categorization and Linking

The message providing the URL, Trigger-Time and Message ID parameters may also provide two additional parameters to support enhanced functionality:

- Category—the slide can be categorized using a free-form text format, such as "News" or "Sports." This allows the receiver to enable the user to browse through cached images under categorized headings.

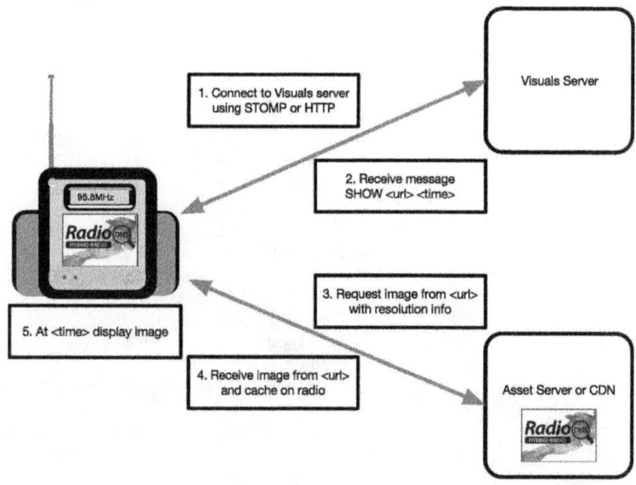

FIGURE 7.21-6 Illustration of the process of displaying visuals using the RadioDNS method.

- Link—the visual can have a URL associated with it. The receiver can provide a way for the user to "click" on the link, and handle it as it would a link in a webpage. This function is usually only relevant on receivers that have web browsing functionality.

Using Visuals as an Audience Measurement Proxy

One consequence of the visuals messaging protocol is that it creates a proxy for audience measurement. Each receiver will create a connection when it tunes into the radio service, and break that connection when it tunes away. It's possible to monitor the changing number of connections as an indicator to listening patterns. If the number of connections is consistently high, it may be considered a valid sample size.

Interactivity

Much radio listening happens as a secondary activity in ambient conditions. For example, while driving a car, or in the kitchen. Radio is capable of attracting people's attention, but that peak of interest quickly declines and often listeners forget interesting information. Radio also suffers in comparison with connected media service of being unable to accurately measure and report on the engagement people have with the medium.

RadioDNS has created a standard for interactivity that is designed to capture interest in radio without creating a distraction. This standard, RadioTAG, helps listeners quickly and simply make a tag on events in programs, and then be able to review them at a later stage. A key use case is the driver in a car, who can push a physical button to register their interest, and then review their list of tags once they're not driving.

A key element of this framework is the ability to capture interest on one device, such as the receiver in a car, but show the results on another device, such as a smartphone or tablet. This requires that the same user can be identified on multiple devices that they might use. To achieve this, RadioTAG enables use of the Cross Platform Authentication standard. Cross Platform Authentication is an authentication framework designed to span devices with embedded operating systems, such as connected car radios, connected TVs, and devices with operating systems such as smartphones and browsers. The user identifies themselves and is authenticated with an agreed central registry. Once authenticated, the user's device is able to aggregate their tag information from various devices to show a composite view of their activity.

A broadcaster advertises availability of a RadioTAG service by creating a SRV record _radiotag._tcp on their FQDN:

- RadioDNS FQDN query for CNAME: 09580.c479.ce1.fm.radiodns.org
- DNS response of CNAME: rdns.musicradio.com
- RadioTAG query for SRV: _radiotag._tcp.rdns.musicradio.com
- DNS response of SRV record: tag.musicradio.com

The RadioTAG protocol uses HTTPS as a transport layer, and the process is as follows:

- A tag request consists of the service identifiers and the date and time in UTC format. The broadcaster is required to refer to their own logs of broadcast events to resolve this back to a response for the listener;
- As the response is to an individual, the content of it can be customized. For example, if the request happens the beginning of an item, such as the beginning of an advertisement, the broadcaster can return information on that event plus prior events, to account for the listener not pressing the tag button quickly enough. Similarly, the response can be altered based on existing knowledge of the user's behaviors, potentially to offer different content, different offers.

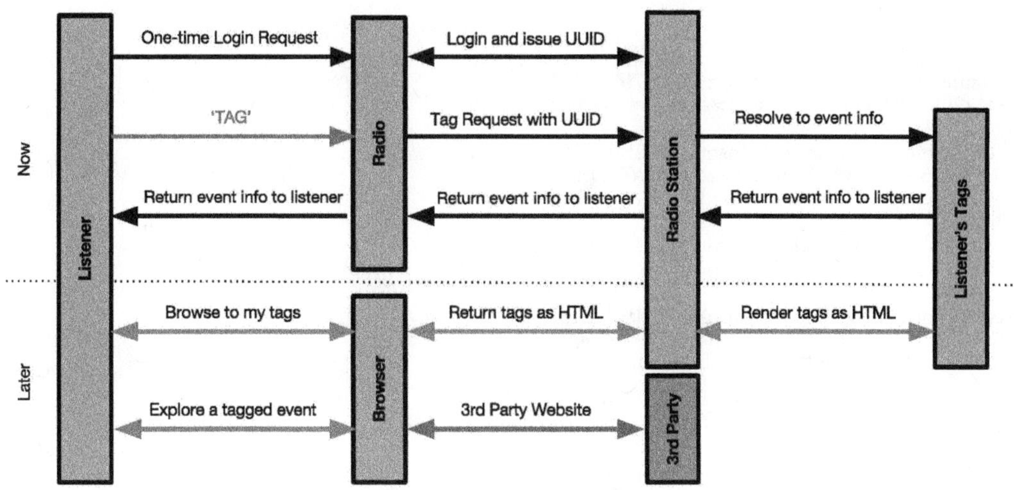

FIGURE 7.21-7 Diagram illustrating the RadioTAG protocol.

RadioTAG has been designed to capture two levels of action:

- Interest—the listener has heard something of interest;
- Engagement—the listener wants to actively engage.

In the development of the Standard, all use cases have been considered, not just those around commerce activity. By presenting the button to the listener as a tag or bookmark, rather than a "buy" option, it is assumed the listener will use the function more, allowing value to be derived from analyzing interest and not just engagement.

Future Applications

RadioDNS encourages the development of applications based around the service discovery framework standardized in TS 103 270. Members of RadioDNS can call together a group to investigate and develop new applications, and present them for standardization. Currently in development are applications to present web content alongside radio, and to allow the insertion of targeted audio into radio broadcasts.

SUMMARY

Broadcast is an appropriate technology for broadcasting content to large audiences in large areas, and the addition of IP connectivity alongside it allows for some of its shortfalls to be addressed. IP is ideally suited to handle these transactional activities, working well within the current design of IP networks in terms of reliability and availability.

RadioDNS leads development of an open and interoperable approach to combing broadcast and IP, which emulates the freedom and control that broadcasters have over their existing radio transmissions, and is being adopted as a global standard for hybrid radio.

References

[1] European Telecommunications Standards Institute (ETSI), ETSI TS 101 756 V1.6.1 (2014–05), Digital Audio Broadcasting (DAB); Registered Tables, 2014.

[2] European Telecommunications Standards Institute (ETSI), ETSI TS 102 818 V3.1.1 (2015–01), Hybrid Digital Radio (DAB, DRM, RadioDNS); XML Specification for Service and Program Information (SPI), 2015.

[3] European Telecommunications Standards Institute (ETSI), ETSI TS 103 270 V1.1.1 (2015–01), RadioDNS Hybrid Radio; Hybrid lookup for radio services, 2015.

[4] European Telecommunications Standards Institute (ETSI), ETSI TS 101 499 V3.1.1 (2015–01, Hybrid Digital Radio (DAB, DRM, RadioDNS); SlideShow; User Application Specification, 2015.

SECTION

8

TELEVISION TRANSMISSION

CHAPTER

8.1

Planning a Television Transmitter Facility

DAVID SANDERFORD AND MATTHEW A. SANDERFORD, P.E., CPBE

Marsand, Inc.

INTRODUCTION

The planning of a television transmitter facility, as discussed in this chapter, applies to any work performed at the transmitter site. This includes main and standby equipment, replacing of existing equipment, and new installations. Chapter 6.1 in the *NAB Engineering Handbook, Tenth Edition* is an excellent treatise authored by John M. Lyons, The Durst Organization. He has successfully addressed the necessary areas relating to the planning and installation of a television transmitter facility. With his permission, this chapter includes much of what he has written, as it accurately brings out the elements in a clear and concise manner. The purpose of this chapter is to expand on the existing groundwork Mr. Lyons provided to include practical detail for the many engineers in the stations who need to know the steps necessary for implementing a successful installation. Many, if not most, engineers do not have the budget or means to contract with consultants and project managers who do this on a daily basis.

The following areas are addressed:

- Station Responsibilities—meeting FCC requirements and project timetable
- RF Budget—Formulas for determining power requirements
- Antenna RF System Selection—antenna and transmission line
- Tower Work—studies, inspections, modifications, and tower crews
- Transmitter Selection—reuse, buy used, or buy new
- Transmitter RF System—mask filter, RF switches/patch panels, and combiners
- Transmitter Installation—space, utility, and other site requirements
- Safety and Code Compliance—RFR, OSHA, NEC, and local codes
- Ancillary Equipment—microwave, fiber, ENG, monitoring, remote control, and IT considerations

This chapter not only deals with the various complex aspects of a new facility but also addresses modifying existing facilities to accommodate replacement equipment, to add auxiliary facilities, or to meet requirements for a channel change.

STATION RESPONSIBILITIES

Once a station has a construction permit (CP) from the FCC, this CP will have an expiration date for the completion of the facility construction. In order to meet this deadline, a plan needs to be developed to establish realistic steps to meet this goal. The following list provides a starting point:

Meeting ADA Requirements

It is important for the proposed installation to be in conformance with Americans with Disabilities Act (ADA) legislation. Enacted in 1992, the intent of the ADA is to protect disabled persons from discrimination in public accommodations and commercial facilities. New construction and existing facility renovations must adhere to the ADA provision in every regard. Conformance to the act includes having doorways and paths with sufficient room for wheelchair passage and replacing doorknobs with lever-type handles. Restrooms must also comply with the act; fixtures must meet stringent requirements for wheelchair access, even if no disabled or wheelchair-bound persons are currently working at the facility. Always confirm with the applicable codes pertaining to your site for compliance.

Radio Frequency Exposure Issues

Radio frequency radiation (RFR) is a concern wherever an RF facility is planned. For projects involving a renovation or change, an RF profile of the current site should be performed, and computer modeling of a proposed site should be completed prior to the start of the construction. At a multiuser site, the lease or license agreement will usually spell out the site owner or manager's responsibilities with regard to making the RF exposure from the site conform to, for example, FCC rules and Occupational Safety and Health Administration (OSHA) regulations (see Figures 8.1-1 and 8.1-2). Typically, the site owner or manager will either require engineering statements from individual tenants regarding RF exposure or will have such studies done themselves at the tenants' expense. In some instances, during the actual construction, RF power will need to be reduced to protect tower workers. This power reduction will be spread across all the stations using one master antenna if that is the section of spectrum that is over the limit. This procedure allows the other users on other antennas (whether they are single antennas or another master antenna) to remain at full

FIGURE 8.1-2 Multiuse Tower Example Located in Cedar Hill, TX

Photo courtesy of Electronics Research, Inc.

FIGURE 8.1-1 Tower Farm Example Located on South Mountain, Phoenix, AZ

Photo courtesy of Electronics Research, Inc.

power. There are also ways to monitor an active site, such as positioning area RF monitors at the hot spots of the site where maintenance personnel are likely to be located. Warning lights and audible alarms can warn personnel when they are entering an area on a rooftop where the RF exposure may exceed or exceeds the safe human exposure levels. See Chapter 3.5, "Human Exposure to Radio Frequency Energy," for more information on this important issue.

Interference Resolution

Resolution of interference between multiple users at a common site is generally the responsibility of the most recent user or modifier at the site. Most lease/license agreement language specifies this fact, and the tower owner may be able to force the assumed offender to lower power to a point where there is no interference or, if this cannot be accomplished, to shut down until the problem can be resolved. In an effort to keep a site interference free, leases/license agreements usually specify the frequencies of all existing users at the site and the owning or managing company will usually require an interference analysis with the addition of each new tenant so there are no surprises when the newest user, or one making modifications, turns on the transmitter. Included in the frequency list are the new user's services, main television or radio frequencies, and all ancillary frequencies, antenna gains and beamwidth and radiated power for STL, TSL, ENG, RPU, spread spectrum, or any other services at the common site. Further, the proximity of a new antenna and/or transmission line to other radiators can adversely affect the associated facilities by altering antenna patterns or causing VSWR problems. The placement of the new equipment must be studied not only for structural considerations but also for detrimental effects to existing radiators on the tower.

Lockout/Tagout

Lockout/tagout procedures must be in place as part of RF control and monitoring. Lockout is secured by the turning of a switch (e.g., the shorted position) and a lock being placed through the control to prevent unauthorized tampering. Tagout is a procedure whereby the person locking out the switch places a tag through the lock shackle with information as to who performed the lockout and who has the key. Lockout/tagout monitoring should also be a part of site management controls.

Tower Light Monitoring

Tower light monitoring techniques vary from site to site. Usually the site owner or manager is responsible for monitoring, logging, and reporting outages to the FAA and having repairs performed. In some cases, the user of the top antenna takes responsibility for all monitoring obligations. The FCC, however, holds all the tenants equally responsible for notifying or taking corrective actions for any obstruction light violations on the tower and can issue fines accordingly. If the construction proposal includes increasing or changing the height of the tower, the FAA should be consulted to make sure that marking and/or lighting do not also need to be amended.

Union Issues

In many markets, trade union members may be required to perform all the tasks associated with transmitter/antenna installation. Their costs are often higher than nonunion work and their "standard" workday shorter. For electricians performing work on the tower, or places on the site that are 16 ft. or more above the floor, a premium known as "high time" may be charged above the basic hourly rate. Overtime/overnight hours and weekends carry heavy premiums but may be necessary if the project is behind schedule or if the other tenants on the tower insist that a portion of work be done overnight.

Hazardous Material Removal

Replacement of older transmitters and other RF system components may include the removal of the old components. Engineers should check to determine if any of the old equipment contains hazardous materials before it is removed from the site. If hazardous materials are involved, the proper contractors must be hired to document and remove them.

Tower Work

If the station leases space on the tower or will need to lease space on a tower, the station will need to contact the tower owner and/or management company as early as possible to include them in the planning process. They will have to run any proposed changes through their structural engineering firm. Often there will be a project manager representing the owner with whom the station will need to coordinate throughout the project. Any tower crew used for any aspect of the project will have to be approved by the tower owner before it is allowed to work on the tower. It is critical to secure and schedule tower crews well in advance of the work to be done.

If the station owns the tower, the process to identify and select a structural engineering firm and a tower crew must begin as soon as possible. Ideally, there is already a structural engineering firm of record that has the current tower specifications and inventory. If not, a detailed tower survey will be in order to provide a "map" of all of the equipment and transmission line on the tower, whether it is in use or not, so that the chosen structural engineer has accurate and well-developed knowledge of everything on the tower from which to create a "baseline" structural assessment. Once the initial tower assessment is complete, the proposed equipment is submitted to the structural design firm to determine if the existing structure is capable of supporting the added weight and moment.

If the station cannot find a recent structural study, an attempt should be made to find, at minimum, the original design drawings that the original tower erector used to stack the tower when it was new. Many tower manufacturers keep detailed records for every tower they have ever designed. Some tower companies have acquired the assets of defunct tower companies, including the original drawings. In cases where there are multiple tenants on the tower, a dialog with these tenants should begin early in the planning stages as well. They will need time to understand how any proposed changes will affect their operations, particularly during the installation process. If the work will affect a tenant, and the tenant does not have auxiliary facilities, then steps must be taken to minimize the impact to the tenant. These could include working within certain time windows and providing temporary antennas and line for the duration of the work.

If the tower site falls within a jurisdiction that requires permitting for work on the tower, application for permit must be completed at the earliest date possible because the permit must be granted before any work on the tower begins.

In cases where new tower construction is required, extensive studies and outside consultants will be required including, but not limited to, the following:

- FAA consultant
- Indian Affairs consultant
- Environmental consultant

Equipment Lead Times

Whereas some smaller solid-state transmitters and low-power antennas with "off the shelf" patterns can be delivered in 30–60 days, most full-power transmitters require 60–90 days, and antennas range from 90–120 days. These dates are ARO—after receipt of order. The process leading up to the final quotation on which to place the order can take several weeks if not months. These lead times can stretch out considerably depending on backlog. Further, most RF system equipment manufacturers not only require a down payment with the order but also may require most of the rest of the payment, up to 90%, before the equipment ships. Make sure that enough time is allotted for the corporate financial offices to prepare and release all the funds necessary.

Medical Notifications

The FCC requires notifying hospitals and nursing/residential care facilities providing inpatient and residential care services that lie within the station's service contour at least 30 days in advance of commencing over the air broadcast for a new television transmitter facility. This would include a channel change, power increase or antenna pattern change. This requirement can be met by using an outside service to identify the medical facilities and contact them with a letter stating what the new operating parameters will be. A report detailing the list of medical facilities contacted, a copy of the letter, and a description of the methods used should be included in the station's Public File.

FCC Filings

The details necessary to successfully file an application with the FCC are beyond the scope of this chapter. Here the services of a consultant are helpful and are typically used by most stations; however, the various forms required to build out a television transmitter facility will be briefly mentioned.

In order to obtain a construction permit (CP), the station must prepare and file the appropriate form in the FCC's LMS system:

- Form 2100 Schedule A (replaces Form 301) for a full power television station
- Form 2100 Schedule E (replaces Form 301-CA) for a Class A television station
- Form 2100 Schedule C (replaces Form 346) for a low power television station, TV translator or TV booster station

The purpose of the CP is to establish a television transmitter facility's location, antenna height, antenna pattern, and effective radiated power (ERP). The service area contour is calculated and overlaid on a map. An interference study must be completed and the results included in the application to indicate that the proposal does not add more than an additional 0.5% interference to another station. A statement must be included regarding radio frequency exposure to humans (RFE Statement) and steps to be taken to meet the guidelines established per OET65. Often, fencing or other access restrictions are required to keep the general population away from RF exposure, so if this is a new tower or modification of an existing tower that does not have such devices, they need to be included in the schedule and the budget.

Immediately upon completion of the television transmitter facility, an application for license to cover (also filed in the FCC's LMS system) is required:

- Form 2100 Schedule B (replaces Form 302-DTV) for a full power television station
- Form 2100 Schedule F (replaces Form 302-CA) for a Class A television station
- Form 2100 Schedule D (replaces Form 347) for a lower power television station, TV translator, or TV booster station

This form can be filled out by the consulting engineer that prepared the original application for CP, but it may also be filled out by station personnel as long as the final RF budget is understood and the form is filled out properly.

RF BUDGET

The CP issued by the FCC contains the authorized effective radiated power (ERP) for the station. To achieve this ERP, the following formula in Table 8.1-1, consisting of the gains and losses of the various components, is used to help select the correct antenna, transmission line, and transmitter power:

TABLE 8.1-1
RF Budget Calculation

ERP = Effective Radiated Power in kW
(The power specified in the station's construction permit or license.)

ERP_{dBk} = *Effective Radiated Power in dBk*
(Decibels referenced to 1kW)

TPO = Transmitter Power Output in kW
(average power at input to transmission line, after any filter attached to the transmitter, if used)

TPO_{dBk} = *Transmitter Power Output in dBk*
(Decibels referenced to 1kW)

$Line\ Loss_{dB}$ = *Transmission Line Loss in dB*
(Total power loss for the entire length of transmission line used from the gas barrier to the antenna.)

Determine TPO from ERP:

$$ERP_{dBk} = 10 \times \log_{10}(ERP)$$

$$TPO_{dBk} = ERP_{dBk} - AntennaGain_{dB} + Line\ Loss_{dB}$$

$$TPO = 10^{\left(\frac{TPO_{dBk}}{10}\right)} \text{kW}$$

The antenna is selected according to the necessary gain and required pattern to achieve the ERP in the CP. While the $LineLoss_{dB}$ can vary a bit depending on transmission line selection, the transmitter size must be chosen to complement the antenna gain. A lower antenna gain requires a larger transmitter power, while increasing the antenna gain needs less transmitter power in order to meet the same ERP.

ANTENNA RF SYSTEM SELECTION

The antenna RF system includes the antenna and transmission line after the mask filter and any combiners. Selecting the antenna and transmission line must take into account the transmitter size required to meet the ERP and the tower loading that may require structural modifications. If it is possible to reuse existing transmission line and replace the antenna with one of similar dimensions and weight, then the need for structural modifications may be avoided. However, if transmitter size is the limiting factor, then the antenna and line must be chosen to match. Often this involves balancing the costs of tower modifications against antenna gain and costs and transmitter costs.

Antenna Selection

The antenna can be designed for a single channel or broadband to handle a wide range of channels. In some instances, antennas typically designed for single channel can be designed to accommodate two or three channels if they are adjacent to each other. The antenna must be able to handle the RF power introduced at its input, which is calculated from the far end of the transmission line run.

FIGURE 8.1-3 Antenna Pattern Test Range Located in Raymond, ME

Photo courtesy of Dielectric, Inc.

An antenna can be top or side mounted on the tower. Ideally, the top-mount antenna is the most desirable as it has the least amount of interference from any nearby structure, such as the tower itself. The top-mount antenna is structurally heavier, and there must be space available on top of the tower to accommodate it. The side-mount antenna is lighter in weight, but the pattern is affected by the tower structure to which it is attached. The antenna manufacturer can run "scatter studies" that predict the effects of the tower structure on the requested antenna pattern and simulate the resulting, effective pattern. The antenna location, distance from the tower, and orientation can be optimized using the results of the studies (see Figures 8.1-3 and 8.1-4). Side-mounted antennas perform better when directional patterns are required.

Some multi-tenant sites employ master antennas that are broadband, typically panel style, antennas that accommodate a wide range of channels either in the UHF or VHF bands. Coordination with the master antenna owner or site manager is required in order to add an additional channel to the system. Typically, these antennas are omnidirectional patterns. It is important to check the beam tilt of an existing broadband panel if one is being considered, as different stations may been in need of different beam tilt setting for individual null fill requirements.

FIGURE 8.1-4 Main/Standby Antenna Example Located in Henderson, KY

Photo courtesy of Electronics Research, Inc.

The decision for choosing a top mount vs. a side mount may also be dependent not only on space availability but on the capacity of the tower itself to handle the additional transmission line and antenna load. A tower study must always be performed to determine the capacity and safety factor of the proposed installation. Any modification of an existing tower requires that the tower be brought up to the current ANSI-TIA standard, which is 222-Rev G at this writing. The costs associated with the tower modification could very well determine the antenna selection as well.

Most high-power antennas are provided with radomes. Radomes are necessary to protect the antenna radiation elements and minimize ice buildup on the antenna elements that lead to increased SWR reflected power.

Transmission Line Selection

Selection of the transmission line depends on power-handling capability, channel frequency, and tower capacity (see Figure 8.1-5). Transmission line can be either coaxial or waveguide. Coaxial line can be in rigid line sections or continuous flexible in structure.

Coaxial rigid line is channel dependent due to flange reflections associated with the channel average wavelength. Supplier catalogs list the channels with their associated lengths in their catalogs. For quick reference, the typical lengths are shown in Table 8.1-2:

FIGURE 8.1-5 Transmission Line Routing Example Located on Lookout Mountain, Denver, CO

Photo courtesy of Electronics Research, Inc.

TABLE 8.1-2
Channels Available per Rigid Transmission Line Length

Channels available in the 19.5 ft. line sections: 9, 10, 13, 17, 21, 25, 26, 29, 30, 33, 34, 37, 38, 42, 46, 50
Channels available in the 19.75 ft. line sections: 16, 20, 24, 28, 32, 36, 40, 41, 45, 49
Channels available in the 20 ft. line sections: 2, 3, 4, 5, 6, 7, 8, 11, 12, 14, 15, 18, 19, 22, 23, 27, 31, 35, 39, 43, 44, 47, 48, 51

If the transmission line will be used for more than one channel, and the other channels do not share the same line length, then the manufacturer can calculate and provide a custom length to minimize the flange reflections for each channel.

An alternative to the standard cut lengths in Table 8.1-2 is a progressive, incremental increase in line lengths over the entire transmission line run. The incremental difference between each line length reduces the overall SWR flange reflections to an almost insignificant value. This also allows the line to be used for any channel.

Flexible coaxial line is also broadband and can be used on any channel. The cable loss will vary according the channel frequency. The higher the channel, the more loss will be incurred in the coaxial line.

The coaxial transmission line, whether rigid or flexible, is sized not only to handle the output power of the transmitter but also to minimize the overall attenuation loss distributed over the length of the line run. Some manufacturers provide a coating treatment internal to the coax that allows for increased power handling through better heat transfer. Another thing to consider is an external coating that minimizes the heating effects and temperature changes caused by the sun. The coating is provided by certain manufacturers, applied during production, and has proven to be reliable and effective.

Typical power handling for coaxial cable for digital broadcasting is shown in Table 8.1-3:

TABLE 8.1-3
Coaxial Transmission Line Sizes and Power Capacity

Transmission Line Type	Lower Channel/ Average Power	Upper Channel/ Average Power
3-1/8″ 50 Ω	CH 2 71 kW	CH 51 20 kW
4-1/16″ 50 Ω	CH 2 113 kW	CH 51 31 kW
6-1/8″ 50 Ω	CH 2 243 kW	CH 51 68 kW
6-1/8″ 75 Ω	CH 2 211 kW	CH 51 59 kW
7-3/16″ 75 Ω	CH 2 285 kW	CH 51 79 kW
8-3/16″ 50 Ω	CH 2 410 kW	CH 38 121 kW
8-3/16″ 75 Ω	CH 2 362 kW	CH 51 101 kW
9-3/16″ 50 Ω	CH 2 508 kW	CH 26 160 kW
9-3/16″ 75 Ω	CH 2 428 kW	CH 37 127 kW

Waveguide comes in rectangular, circular, or elliptical shapes. Since waveguide does not use a center conductor like coaxial transmission lines, the power handling capability is much greater. However, each line section is shorter—from 11.5 to 12 ft.—and the cross-sectional length of the waveguide is restricted to a group of finite channels. The larger the waveguide, the more wind load on the tower. Table 8.1-4 shows typical channel capabilities of waveguide:

TABLE 8.1-4
Rectangular Waveguide Channel Bands

WR1800—Channels 14 through 41
WR1500—Channels 20 through 51
WR1400—Channels 25 through 51
WR1150—Channels 43 through 51

TOWER WORK

The tower is another critical element in the installation project. The following items must be considered from the output of the transmitter to the antenna:

- Tower Study
- Up-To-Date Tower Inspection
- Tower Crew Selection
- Tower Work Management

Tower Study

Once the antenna type, model, and transmission line size are chosen, a tower structural study must be performed to examine the effect of the changes on the capacity of the tower. With rare exceptions, the tower study will be made according to the current ANSI-TIA standard. A qualified structural engineering firm licensed in the state of the associated tower site must be used for the study. The structural engineering firm must have a proven track record and experience with similar projects. The study should take into consideration any items that the station wants to *remove* from the tower as well as add and provide the recommended structural changes along with detailed drawings of the components to be used, assuming that the tower can be modified. There will be cases where the tower in question cannot be modified and therefore cannot accommodate the additional wind load proposed. In this case, a new tower or alternate tower site must be considered. In the case of a leased tower or master antenna, the tower study would be coordinated through the site manager and/or owner.

Up-To-Date Tower Inspection

An up-to-date tower inspection is helpful for the tower planning stage, although not always required. Accurate record keeping and regularly scheduled maintenance reports can also be used. This includes the tower inventory, or mapping, as well as the structural condition of the tower elements, antennas, transmissions lines, etc. This information is applied in the latest tower EIA standard.

Selection of Tower Crew

Selection of a tower company is one of the most critical aspects of the construction phase. Prior to selection, the station engineer or project manager must prepare a scope of work for the tower company to bid on. The scope of work must include the modifications to the tower, parts to be provided, the removal of any existing equipment on the tower such as unused antennas, lines, dishes, etc., installation of the new antenna and transmission line, and time allowance for any adjustments at the end of the project. The drawings produced by the structural engineer for tower modification should be included with the scope of work. The scope of work must be detailed enough so that it specifies the level of responsibility for cranes, forklifts, storage containers, removal of debris from the site, etc. Anything not included in the scope of work or final contract that the tower crew performs adds up to additional costs and time delays. The tower company must have experience and qualifications to work on the size of the tower in the project. There are many tower companies with crews available to work on towers shorter than 1,000 ft. tall. Very few tower companies are equipped and qualified to install full power television station antennas. Even fewer are qualified and equipped to work on towers 1,000 to 2,000 ft. tall. The tower crew must provide references for similar work performed, showing its capability to handle the antenna size, tooling, and experience. Liability insurance and, in some cases, bonding is necessary to protect the project. Corporate lawyers and insurance agents should be asked about the amount of insurance that they feel most comfortable with requiring. In the case of leased or group-owned sites, preapproved tower companies who have complied with the site's guidelines will be the only ones authorized to work at that site.

Following the selection of a tower company, company staff should immediately be involved in the project planning from that time forward until the completion of their part. This is essential, as it is their responsibility to coordinate delivery and off-loading of antenna, transmission line, and components to modify the tower. Bad weather days need to be factored into the schedule and allowance made for work around existing antennas that may need to reduce power or go off air for the crew's safety.

A staging area should be available at the site for setting up the crane, winch, or gin pole assembly and storing of parts such as transmission line, hangers, etc. (see Figure 8.1-6). It is highly advisable to use lockable storage containers to store these items at the site, even if a fence already surrounds the site. Storage containers can be rented or leased only as long as necessary and provide safety from the weather as well as additional security from theft.

In some cases, the staging area may include a landing location for a helicopter if one is needed in the assembly of a tower or lifting of antennas. Typically, the tower company will recommend or subcontract any helicopter service needed.

FIGURE 8.1-6 Top Mount Antenna Lift with Gin Pole for WFYI

Photo courtesy of Electronics Research, Inc.

Tower companies must ensure that their workers are in full compliance with OSHA safety standards. They must require daily, documented safety meetings prior to any activity on the tower and must be in open and constant communication with the designated engineer or project manager.

Management of the Tower Work Process

The management of the tower work process is primarily a timeline issue. The following sequence is typical of the procedural checklist to follow:

1. Upon issuance of the CP, determine specifications of equipment to go on the tower such as antenna, transmission line, microwave dishes, etc.
2. Submit items with specifications and locations on the tower to the tower structural engineering firm for the tower loading study.
3. Evaluate the results of the tower study to determine if any changes need to be made to the proposed equipment and if the tower can be modified or another location chosen.
4. Prepare a scope of work for the bidding process to submit to qualified tower companies for the tower work and equipment installation on the tower.
5. Make the final choice of the tower company to use and involve company staff immediately in the remaining planning and implementation process.
6. Prepare a realistic timeline of the work on the tower while keeping in mind the time of year and inclement weather days that may be expected.
7. Ensure that the transmission line(s) is(are) pressurized when the crew leaves each day during installation and the line purged before and after final connection is made into the antenna system (see Figure 8.1-7).

FIGURE 8.1-7 Multiple Transmission Line Pressurization Manifold Used With Nitrogen

Photo courtesy of Electronics Research, Inc.

8. Always have a station representative or site manager at the site at all times that work is being performed on the tower. Unexpected situations will arise that require time-sensitive decisions by the station or site manager that could cause delays and add costs to the project.
9. At the conclusion of the tower work, however extensive it may be, have a qualified third party tower company check and certify that the work on the tower has been properly completed.

TRANSMITTER SELECTION

Transmitter selection is integral to determining the RF budget. Since the antenna selection affects power gain and the transmission line determines power loss, both affect the power level required of the transmitter. Selecting the appropriate antenna and transmission line can greatly affect transmitter costs.

The RF budget specifies the transmitter power output (TPO) required at the input of the antenna RF system in order to produce the specified effective radiated power (ERP). TPO is typically measured after the transmitter's mask filter, output switch/patch panel, and any channel combiners if used. In practical terms, this would be the power measured at or near the gas barrier that separates the transmitter RF system from the antenna RF system.

It is often helpful to know the "prefilter" transmitter power output ($TPO_{PreFilter}$). This is the power required out of the transmitter before the loss due to any channel combiners, RF switches/patch panels, and mask filter. Many transmitter models are specified by their manufacturers in terms of prefilter TPO. Table 8.1-5 explains how to calculate the $TPO_{PreFilter}$:

TABLE 8.1-5
Transmitter RF System Losses and Power Calculation

$TPO_{PreFilter} = PreFilter\ Transmitter\ Power\ Output\ in\ kW$

(The average power output of the transmitter before the channel mask filter, channel combiners, switches, etc.)

$TPO_{(dBk)} = Transmitter\ Power\ Output\ in\ dBk$

(The transmitter power output in dB referenced to 1kW that enters the antenna RF system. See "RF Budget.")

$FilterLoss_{dB} = Output\ Filter\ Loss\ in\ dB$

(Includes harmonic and mask, or channel, filter)

$CombinerLoss_{dB} = Combiner\ Loss\ in\ dB$

(Used if station is combined with another channel into a common RF system)

$MiscLoss_{dB} = Miscellaneous\ Loss\ in\ dB$

(Loss due to miscellaneous components such as RF switches, patch panels, elbows, and transmission line sections inside the building leading up to the input to the transmission line.)

$$TPO_{PreFilter} = 10^{\left(\frac{TPO_{dBk} + FilterLoss_{dB} + CombinerLoss_{dB} + MiscLoss_{dB}}{10}\right)}$$

There are three choices in selecting a transmitter: reuse the existing transmitter, purchase a used transmitter, or purchase a new transmitter. The following outlines the requirements of each as well as some advantages and disadvantages.

Reusing the Existing Transmitter

Requirements:

- The transmitter must be able to operate at the assigned frequency. In general, recently manufactured solid-state transmitters are frequency agile within their operating band (i.e., VHF or UHF) (see examples in Figures 8.1-10 and 8.1-11). Tube-type transmitters are agile across the UHF band (see examples in Figures 8.1-8 and 8.1-9). The manufacturer can best answer whether or not it is possible to retune the transmitter and, if so, which components need to be replaced. Table 8.1-6 lists the frequency dependent items within a transmitter:

TABLE 8.1-6
Frequency Dependent Items Within a Transmitter

Solid State	Tube Type
Exciters	Exciters
Pre-amplifiers/Drivers	Pre-amplifiers/Drivers
Amplifiers	Linearizer circuits
Combiners/Circulators	Circulators

- The transmitter will have to meet the power requirement to make the effective radiated power (ERP) specified in the construction permit (CP) at the new channel frequency.
- If the channel mask filter is not designed to be tuned to the new channel, a new mask filter is needed.
- Besides the frequency dependent components inside the transmitter, there will be other external RF components to consider as to whether or not they will operate properly at the new channel. These include combiners, phase shifters/magic tees, water column loads, and RF switches.

Advantages of Reusing the Existing Transmitter:

- Reusing the existing transmitter can save on costs over purchasing new.
- The majority of the equipment needed is already installed. Transmitter installation costs are reduced to transmitter conversion costs. The station engineer should consider this balance carefully. While some transmitters are capable of being upgraded or retuned, there is a point at which

the retune becomes more expensive than a new purchase.

- In most cases, no additional electrical installation is required.
- In most cases, no additional floor space/building capacity is required.
- In most cases, there is no change to the heat load in the room, so there are no changes to the HVAC required.

Disadvantages of Reusing the Existing Transmitter:

- In the absence of a standby transmitter or auxiliary site, converting the existing transmitter will require some down time with loss of service over the air while the work is completed and the transmitter is "proofed" at the new channel.
- The transmitter may be out of warranty or near end of life. Some existing transmitters are no longer supported by their manufacturer. Depending on the age of the transmitter, there may not be a source for certain replacement parts to continue to service the transmitter.
- Older transmitters are not as efficient as new ones, and installing a new transmitter may save significantly on the power bills. One more scenario worth consideration is if the existing transmitter will not meet the increased power requirement, whether due to introducing cross-polarization or other demands that may accompany any changes to the antenna RF system, it may be possible to add to the existing transmitter the capacity to meet the new power requirement. In order to accommodate these changes, electrical power, HVAC, and building floor space availability must be considered.

In summary, if the station's existing transmitter is capable of operating at the new channel and power, then reusing the transmitter can save on costs, but without proper planning and, in most cases, a standby or auxiliary facility, the down time to effect the transition may be untenable.

Purchasing a Used Transmitter

Requirements:

- The same considerations must be taken into account as when reusing an existing transmitter. If the used transmitter is not already set up on the new channel, it will have to be converted. If the used transmitter comes with a mask filter and any other RF components, those will also need to operate at the assigned channel or be converted or replaced.
- The used transmitter, cooling system, electrical, and associated RF must fit in the transmitter room. If it includes outdoor heat exchangers and/or power supplies, there will need to be a place for these as well on an appropriate equipment pad.

Advantages:

- Used transmitters are usually very inexpensive, especially when compared to a new transmitter.
- Having an additional transmitter provides a separate transmitter for the transition and allows the existing transmitter to remain on the air during the transition.

Disadvantages:

- It may be necessary to uninstall, pack, and ship the used transmitter from the original site.
- A second transmitter requires additional floor space, building modification, or an addition.
- The addition of a used transmitter will require the installation of a new, separate electrical power system (conduits, breakers, disconnect switches); sufficient electrical power capacity must also be available.
- An additional transmitter could require additional HVAC.
- An additional transmitter could also require an outside equipment pad for beam supplies and/or heat exchangers.
- There may not be an accurate history of operation available; the station acquires a used transmitter with hidden problems.
- There may be little to no warranty; equipment may be approaching end of life.

One way to make use of a used transmitter would be on a temporary basis to allow for converting an existing transmitter, installing a new transmitter, or replacing an existing antenna RF system. An inexpensive used transmitter could be placed on the air at a temporary facility. There may be a broadband antenna with space for lease in the market or tower space for a lightweight antenna and flexible line to help facilitate an economical transition. The used transmitter could be retained as an auxiliary site backup after the main transmitter is operational.

Purchasing a New Transmitter

Requirements:

- The new transmitter must, of course, operate on the assigned channel at the required power output level.
- There must be space available to install the transmitter with all of its components.
- Electrical power capacity must be available for the new transmitter.
- Cooling capacity must be available for the new transmitter.

FIGURE 8.1-8 Comark DCX Paragon L3 MSDC IOT High Efficiency Tube Type UHF Transmitter at WGBH in Boston, MA

Photo courtesy of Hitachi Kokusai Electric Comark LLC

FIGURE 8.1-9 GatesAir PowerCD E2V ESCIOT High Efficiency Tube Type UHF Transmitter Located at Lookout Mountain, Denver, CO

Photo courtesy of Electronics Research, Inc.

Advantages:

- The existing transmitter can remain on the air while the new transmitter is installed.
- New transmitters come with a warranty, typically with a 10–15 year serviceable lifetime.
- This will lower operating costs compared to older transmitters due to higher efficiencies.

Disadvantages:

- The cost of the new transmitter plus installation can be higher than converting an existing transmitter or purchasing a used transmitter.
- A new transmitter will require additional floor space and may require building modification or addition.
- The new transmitter will also need electrical installation; there must be electrical power capacity to add the new transmitter to the room.
- Additional HVAC capacity may also be needed.
- An additional (or new) outside equipment pad could be necessary for outside components.

Purchasing a new transmitter can be the most expensive option as it involves the cost of the transmitter, new electrical service to support the transmitter, possible building modifications including an equipment pad or even a new building, and possible HVAC addition. However, with the new transmitter come the advantages of the latest technologies, which include improved power efficiency and easier maintenance duties—both of which reduce operating costs.

Table 8.1-7 lists the advantages and disadvantages of the two types of transmitters available for purchase today:

TRANSMITTER RF SYSTEM SELECTION

The transmitter RF system consists of the RF components between the transmitter and the antenna RF system. This is usually delineated by the output of the transmitter's final combiner up to the gas barrier; however, it can involve more complex variations. At its simplest, the purpose for the transmitter RF system is to filter the RF coming out of the transmitter in order to meet the FCC specifications for out of band emissions. Therefore, the simplest and sometimes only transmitter RF system component is the mask filter. Sometimes the transmitter RF system is more complicated, and can include:

- A coaxial switch between main and standby transmitters
- A coaxial switch between main and standby antennas
- A coaxial route between the top and bottom halves of an antenna or a "split feed" routing to individually feed both halves of the antenna
- A coaxial switch between multiple mask filters for frequency agile transmitters to be used as backup for multiple stations
- A combiner to combine multiple transmitter feeds together to feed a broadband antenna

The possible combinations are too numerous to cover every one in detail in this chapter. Therefore, this section will list some of the things to consider when choosing a transmitter RF system.

Mask Filter Selections

The mask filter is a critical component of the overall system. Without it, the FCC requirements for out of band emissions would not be met. Most high-power filters

TABLE 8.1-7
Solid-State vs. Tube-Type Transmitters

Solid State	Tube Type
Available in the high VHF and UHF bands, and low VHF in some cases	Available in the UHF band only
Air cooled available up to 7 kW Liquid cooled available from 1.5 kW—100 kW	Up to 33 kW per cabinet; can be combined with other cabinets to achieve the required power
43–50% Efficient—Measured at the panel Doherty Amplification System efficiency including pumps, fans, and heat exchangers is 40–48%.	*52–55% Efficient*—Measured at the tube ESCIOT/MSDC IOT System efficiency including pumps, fans, drivers, and heat exchangers for the ESCIOT and MSDC IOT transmitters is 33–35%.
Operates using standard line voltages to power 50VDC power supplies.	Operates using voltages up to 36kVDC. Requires specialized skills to maintain and troubleshoot.
The loss of an amplifier results in a small drop in overall power.	The loss of a tube results in: • an off-air condition for a single tube transmitter • half-power for a dual tube transmitter • two-thirds power for a three-tube transmitter (assumes functioning switchless/"magic tee" combiners)
Liquid cooling system can be plumbed with hoses, allowing more flexibility in placing the transmitter components, shortens installation time, and lowers installation cost.	Liquid cooling system is typically plumbed with rigid copper pipe due to the high flow rates and pressures involved, which lengthens installation time and increases installation cost.
The cost to replace a failed amplifier typically involves paying a fixed service fee, and the replacement unit can usually ship same day.	The cost to replace a failed tube is over $50K. Tubes typically take 4–8 weeks to acquire if not in stock.
There is very little technical skill required to change out a failed amplifier. Following the proper procedure, the amplifier can be changed out while the transmitter remains on the air.	A high level of expertise is involved to change out a tube, requiring specialized test equipment and experienced personnel.
Coolant can be used indefinitely unless contaminated.	Coolant recommended to be changed out every 5 years.
Amplifiers are rated to last the lifetime of the transmitter: 15–20 years.	Tubes, if run at or near their rated power, have an expected lifespan of 80,000–100,000 hours. However, as the tube ages, it requires more frequent maintenance.
Air cooling is noisy due to airflow. Liquid cooling is quieter; some models are almost silent.	Noisy due to blowers used for the tubes and high rate of water flow through the cooling system.
Higher initial purchase cost	Lower initial purchase cost

in use today use a channel-specific, cut cavity-style filter. Low power filters are usually tunable, but may not be tunable across the entire band. In planning a television transmitter facility, it is beneficial to include a fully tunable mask filter in the transmitter RF system (see Figures 8.1-12, 8.1-13, and 8.1-14). These filters are tunable across their respective bands and match the frequency agility of most modern transmitters. A tunable mask filter and frequency agile transmitter along with a properly designed transmission line system and broadband antenna make up the solution for an adaptable transmitter facility. If budget is a consideration, a single-channel filter can certainly be used.

Mask filter selection criteria:

- Power Rating—it must handle the required transmitter power output; some very high-power models use liquid cooling to increase the power rating.
- Frequency—it must operate at the correct frequency.
- Meets the Transmitter Manufacturer's Specifications—the mask filter will often be provided through the purchase of a transmitter in order for the manufacturer to certify the system performance will meet the FCC requirements.
- Reflective vs. Constant Impedance Filter (CIF):
  - Tube-type transmitters require a CIF: this configuration involves using two matched, reflective filters with a hybrid splitter on the input and a hybrid combiner on the output to minimize power reflected to the transmitter.
  - A CIF configuration doubles the power-handling ability of a standalone reflective filter and can be used with both types of transmitters.
  - Solid-state transmitters normally operate well into a reflective filter.

FIGURE 8.1-10 GatesAir ULXT-50AT High Efficiency Solid State Liquid Cooled Transmitter Installed by Marsand, Inc. at KOKI in Broken Arrow, OK

Photo courtesy of Marsand, Inc.

FIGURE 8.1-11 Comark E-Compact High Efficiency Solid State Air-Cooled Transmitter

Photo courtesy of Hitachi Kokusai Electric Comark LLC

- Tunable vs. Cut to Channel:
  - Planning for the future favors a filter that is tunable across the band.
  - Tunable reflective filters are available up to 25 kW air cooled and 50 kW liquid cooled.
  - Tunable filters in a CIF configuration are available up to 50 kW air cooled and 100 kW liquid cooled.

Depending on the size of the filter vs. space available, with cost also a factor, a higher powered filter

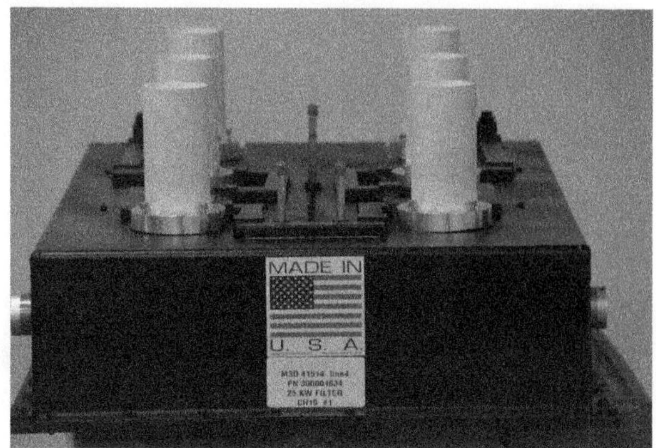

FIGURE 8.1-12 Dielectric 25-kW UHF Liquid Cooled Six-Pole Tunable Channel Mask Filter

Photo courtesy of Dielectric, Inc.

FIGURE 8.1-13 Dielectric VHF 6-Pole Tunable Channel Mask Filter

Photo courtesy of Dielectric, Inc.

for a multiple-cabinet tube-type transmitter could be replaced with several smaller filters located at the output of each transmitter cabinet, the outputs of which would be combined. While this configuration provides some redundancy in that there is not just one filter in the system, it does make it cumbersome to match the different filter responses.

A note concerning filter design: Current FCC requirements are met with a six-pole filter design in most cases. Channel 14 may require up to 12 poles to achieve the level of protection required to the Land Mobile band just below it. With the capabilities of today's television exciters, eight-pole filters are seldom needed, however, and with the advent of ATSC 3.0 on the horizon, the need for eight-pole filters should be investigated on a case-by-case basis.

RF Switches and Patch Panel Selection

RF switches come in two styles: rectangular waveguide and coaxial. Patch panels are coaxial only. The advantage of a switch over the patch panel is the

FIGURE 8.1-14 RFS 50 kW Liquid Cooled 6-Pole Tunable Channel Mask Filter Overhead Installation

Photo courtesy of Marsand, Inc.

ability to switch quickly and remotely. A summary of each follows:

Rectangular Waveguide RF Switch

Limitations:

- Rectangular waveguide is "banded"—certain sizes will pass only certain ranges of frequencies. Please see the "Rectangular Waveguide Channel Bands" table (Table 8.1-4) for a list.
- The switch must be exercised periodically to prevent it from seizing.

Advantages:

- Rectangular waveguide switches can handle higher power than coax switches.
- Switching time is short.

Since waveguide is not broadband in nature, it is not a good match for a system that is frequency agile in all other aspects. However, if reusing an existing transmitter RF system is an option because the new channel lies within the operating band of the existing waveguide, then reusing the waveguide switch will work and save on costs.

Coaxial RF Switch

Limitations:

- Coaxial RF switches are limited in their power-handling capabilities according to the coaxial line size. Please see the "Coaxial Transmission Line Sizes and Power Capacity" table (Table 8.1-3) for a breakdown.
- The practical limit for a coaxial switch is 6-1/8"in size.
- The switch must be exercised periodically to prevent it from seizing.

Advantages:

- Coaxial RF switches are compact.
- Switching time is short.
- They are broadband.

Because coaxial RF switches can be broadband, they are a good fit in systems that have been designed to be frequency agile. Their compact form allows them to be easily integrated into a system, and the ability to switch quickly and remotely works well for systems that have main and standby antennas or main and standby transmitters feeding one antenna.

Patch Panel

Limitations:

- Like coaxial RF switches, patch panels are limited in power handling according to line size.
- Switching time is long, as the patch is manually pulled out of the panel and reinserted in the new position.
- Patch panels must be located where they are accessible to manually switch them.

Advantages:

- Patch panels are available in any coaxial line size, allowing them to handle power up to the limit of the coaxial line at the channel in question.
- They are broadband.
- There are some models available that are designed to be ergonomic, allowing one person to switch patches relatively easily on the larger line sizes.
- They are less expensive than switches.

Because patch panels are constructed of coax, they are broadband and therefore integrate well in systems designed to be frequency agile. They are able to handle higher power than switches.

RF Combiner Selection

RF combiners are used to join two or more sources together into one output. With that said, combiners used in the reverse direction can split a single input into two or more outputs. Solid-state transmitters use combiners with multiple inputs to sum the amplifier drawer outputs. Transmitters with two or more cabinets use high-power combiners to sum the outputs of the cabinets. Channel combiners allow two or more transmitters on different channels to feed a common antenna RF system.

Waveguide Combiner

Waveguide combiners used to combine transmitter cabinets inherit the band limitations shown in the "Rectangular Waveguide Channel Bands" table (Table 8.1-4) and must be chosen accordingly. The standard 3 dB hybrid combiner is a fixed RF device in which if one cabinet fails, the output power drops to 25%. Other combinations of hybrid combiners are also available. A switchless combiner/magic tee uses phase shifters to dynamically feed one or the other input directly to the output in the case of a cabinet failure.

A channel combiner uses designs that use custom transitions to adapt various waveguide sizes to allow channels across the band to be combined. These are not frequency agile and may require extensive modification to accommodate a channel change on one or more of the inputs.

Coaxial Combiner

Coaxial combiners are smaller in size than their waveguide counterparts. Their power handling is limited by the coaxial line size. Although not as common as the 3 dB hybrid combiner, there are phase-shifting coaxial combiners as well.

Coaxial channel combiners make use of the CIF design using channel filters and 3 dB hybrid combiners to add transmitters on different channels together. These can be cascaded to add additional channels as long as the line, 3 dB combiners, and CIF also increase in size to handle the increasing power. Coaxial channel combiners are frequency agile if the CIFs are tunable across the band.

Reject/System Load Selection

For each combiner in a system, one or more reject loads are required to absorb power imbalances (or "waste" power) in the combiner. A system load is located on one of the ports of the final RF switch or patch panel and is used for testing purposes. It replaces the antenna RF system and absorbs the RF power allowing the transmitter to operate at full power without broadcasting over the air.

The load types are summarized in Table 8.1-8.

TRANSMITTER INSTALLATION

Once the transmitter is selected, planning for the installation cannot begin soon enough. The transmitter manufacturer should be able to provide most of the information required for successful installation of its particular model of transmitter including general layout, electrical and cooling requirements, and RF connections. Some manufacturers offer various levels of installation and planning services all the way up to a "turnkey" package. If not, then the services of a broadcast engineering consultant or independent transmitter installation expert can be used to plan and implement the transmitter installation.

The following sections will cover general transmitter installation planning guidelines followed by some practical considerations for three types of transmitters: air cooled solid state, liquid cooled solid state, and tube type.

TABLE 8.1-8
RF Load Types

Oil cooled resistive	• Compact; can be suspended • Broadband • Coaxial input • Models available that handle up to 10 kW (UHF)
Air cooled resistive	• Large; usually floor mounted • Broadband • Coaxial input • Models available that handle up to 40 kW (UHF)
Liquid cooled resistive	• Very compact and light; can be suspended • Coaxial input • Requires a water-cooling system to remove the heat from the resistor • Is susceptible to voltage spikes in the RF (DTV "Crest" factor)

Liquid-cooled column loads	• Coaxial
	  • Compact; can be suspended
	  • Broadband (older models may be cut to channel)
	• Waveguide
	  • Long and heavy; can be suspended
	  • Channel banded
	• Requires a water-cooling system to remove the heat from the load
	• Robust

Layout

Some type of CAD or drawing software is very helpful for visualizing the proposed transmitter site layout. The transmitter manufacturer can provide typical layouts that show recommended minimum clearances and maximum allowable run lengths for cooling and electrical where applicable.

Besides planning for the required electrical clearances and transmitter spacing, the most important aspect to consider is load bearing. The floor under the various pieces of equipment must be able to support the equipment. The roof structure, if it is to be used to hang the RF system, must be able to support the weight of the RF system. Supporting the equipment means more than "not collapsing" under the weight; the structure must provide enough rigidity that the transmitter cabinet does not shift or sag and the RF system does not sway. The outside equipment pad, if needed, should be large enough to fit the equipment with allowance for airflow and provide clearance for maintenance access as well as support the weight of the equipment.

Any outside equipment should have overhead ice protection. This usually involves erecting an ice shield using metal grating. Any transmission line runs between the building and the tower should hang under an "ice bridge" (see Figure 8.1-15) The transmitter building itself should have a roof designed to withstand ice falling off the tower.

The following is a list of practical site considerations when planning for a transmitter installation:

- Large door opening for bringing in the transmitter: 6 ft. wide × 8 ft. tall
- Forklift on site for off-loading, with qualified operator
- Good road access and room to turn around for large trucks

FIGURE 8.1-15 Ice Bridge Example Located at WEIU, Charleston, IL

Photo courtesy of Electronics Research, Inc.

- A clean site with all trash, undergrowth, debris, and old or unused equipment removed
- Sealed or finished floor and walls
- Good lighting inside and outside for security and maintenance
- Fencing or some method of securing the outside equipment pad

Electrical

Electrical planning should include an electrical engineering consultant. A reputable master electrician licensed to work at the transmitter site will need to oversee the electrical work to make sure national and local codes are met and provide the proper electrical permits.

Power Requirements

In general, the power requirements are determined by dividing the transmitter's rated power output by its worst-case system efficiency. Note that the rated power output is the maximum power for which the transmitter is designed. The actual transmitter power output (TPO) is determined in the station's RF budget calculations, and it should be less than or equal to the transmitter's rated power output.

Using the transmitter system efficiency, the calculated power requirements include all of the electrical capacity that is needed for running the transmitter. The only additional items that would need to be considered are HVAC, lighting, station equipment racks, and utility outlets.

Now for an example as shown in Table 8.1-9:

Some solid-state transmitter models have the ability to operate in standard efficiency mode if for some reason the high efficiency Doherty mode fails. For those models, the worst case efficiency would be 35% for the purpose of sizing the electrical power requirements.

To complete the power requirements for the site, the following must be considered in the total heat load for the HVAC system so that electrical needs for the HVAC system can be added to the total current load in the building:

- Transmitter heat load requiring additional cooling in the room: This figure should be provided by the HVAC system manufacturer on a per model basis.
- Environmental heat load requiring additional cooling in the room: This figure should be provided by your HVAC contractor or mechanical engineering consultant. It depends on the climate in your region, size of the room, and insulation characteristics of the building walls and roof.
- It should be noted here that with modern liquid-cooled transmitter systems, most of the waste heat is expelled outside of the building so that, if the installation is in a cold climate, provisions for and electrical specifications for

TABLE 8.1-9
50 kW Transmitter Power Requirements Example

Transmitter Rated Power Output: Current = 386A *Transmitter System Efficiency:* $\eta_{eff} = 40\%$ *Power Factor:* $pf = 0.9$	$TPO_{Rated} \div \eta_{eff} = Power_{In}$ $50\text{kW} \div 40\% = 125\text{kW}$ $Power_{3phase} = \sqrt{3} \times Voltage \times Current \times pf$ $@480\text{V:}125\text{kW} = \sqrt{3} \times (480\text{V}) \times \text{Current} \times 0.9$ $\text{Current} = 167\text{A}$ $@400\text{V:}125\text{kW} = \sqrt{3} \times (480\text{V}) \times \text{Current} \times 0.9$ $\text{Current} = 200\text{A}$ $@400\text{V:}125\text{kW} = \sqrt{3} \times (480\text{V}) \times \text{Current} \times 0.9$ $\text{Current} = 334\text{A}$ $@208\text{V:}125\text{kW} = \sqrt{3} \times (208\text{V}) \times \text{Current} \times 0.9$ $\text{Current} = 386\text{A}$

building heat should be included unless the waste heat is recycled in some manner to heat the transmitter building.

- Station equipment such as microwave gear, switch gear, monitoring and test equipment, air dryers, etc.
- Room lighting.
- Utility outlets.
- Tower lighting.

Note that in planning for electrical service, the maximum operating load should not exceed 80% of the available service. If additional incoming utility service is needed, engage the local power company as soon as possible in the project.

Uninterruptible Power Supply (UPS)

Modern transmitter power amplifiers are designed to withstand the usual power bumps and power loss at a transmitter site. However, there is a "boot up" time required to bring the exciters back online after a power loss. As exciters are software driven, often a temporary loss of power will cause them to "blue screen." An uninterruptible power supply (UPS) can "hold up" critical functions for a short period of time while a generator starts up and switches on line or utility power is restored. Another benefit of the UPS is double power conditioning: The utility power is "filtered" by the UPS inverter and clean power is output. The UPS must be sized properly for its purpose, and the power storage technology must be selected.

There are two approaches to sizing a UPS at the transmitter site: Either to back up the whole transmitter's power needs, or just back up the remote control system and exciter(s) in the transmitter. Backing up the entire transmitter requires a large, specialized UPS and is generally designed to maintain electrical power for just a few minutes after loss of utility power. This provides enough time for the station to switch to generator power if the utility power has not been restored in time. Backing up the transmitter control circuits and exciter requires a much smaller UPS. It may also involve rewiring some of the power distribution circuits within the transmitter. Directions from the manufacturer and/or help from a qualified transmitter installer may help with this. The UPS systems for a TV transmitter installation can range from a standard, "off the shelf" unit available at an office supply store to a three-phase unit requiring an electrician to install, with price tags to match.

A UPS relies on power stored within batteries or on a system using a flywheel design in the event of utility power loss. Batteries require maintenance and periodic (3–5 years) replacement. Typically, the smaller UPS units will utilize battery storage; replacement of batteries in this size UPS is simple and relatively inexpensive. For a UPS large enough to hold up a 50 kW transmitter, however, a flywheel design is almost always preferable; the flywheel design will usually not require much maintenance and does not require planned replacement, especially if it is using magnet bearings in a vacuum. These costs should be included during planning.

If a UPS is used, its size and ventilation requirements must be known early on to include in the room layout. Battery cabinets are usually vented at the top and should be located close to the UPS inverter cabinet. The floor must support the weight of the UPS with batteries. An electrical path and transfer switch locations must be determined for the electrical, as the feed for the protected load must pass through the UPS.

Generator

A generator provides means of staying on the air if utility power fails for a longer period than just a few seconds. It can be sized to provide power to the entire transmitter building or, at the minimum, to power the transmitter and its support systems (i.e. room cooling, switch gear, etc.). An automatic transfer switch (ATS) is used to sense utility power loss, start the generator, switch the load over to the generator, and switch back at some point after utility power is restored.

In choosing a generator, the larger units are going to use diesel fuel, whereas the smaller ones can be chosen to use propane or liquid natural gas. Large diesel tanks require spill containment structures, environmental assessments and, in some jurisdictions, a yearly inspection. Propane tanks may require inspections as well. Designs using lower RPM and liquid-cooled engines will usually last longer and require fewer repairs. Generators should be started and run monthly. Tests under load are highly recommended.

A concrete pad or adequate space on an existing pad is required for a generator. Sometimes a surrounding structure is necessary for noise dampening and to secure the generator. The pad must be able to support the weight of the generator and fuel if there is a tank included under the generator. If a new pad is required, it may be beneficial to run electrical underground to stub up in the pad to feed the generator rather than overhead or on the ground after the generator is in place. A 3-1/2 to 5 ft. clearance all around the generator is usually sufficient; however, some generators come with enclosures that have doors that must swing clear of any obstruction.

Automatic Voltage Regulator (AVR)

Depending on the power quality at the transmitter site, an automatic voltage regulator (AVR) may be in order. These use a mechanical means to keep the voltage feeding the transmitter within a preset tolerance. For larger units, the floor load-bearing strength must be considered, as they are quite heavy.

The AVR will be floor mounted and should have 3-1/2 ft. of clearance in front of the unit for maintenance.

Transient Voltage Surge Suppressor (TVSS)

Most transmitter manufacturers will include or specify a surge suppressor to protect their equipment. An inline suppressor requires the electrical circuit that feeds the transmitter panel board to pass through it. In the case of a large electrical spike, the inline style suppressor may self-sacrifice and open up, causing the transmitter to power down until power can be routed around the failed unit. A parallel suppressor requires a dedicated breaker in the transmitter panel board to feed it.

Planning for the transient voltage surge suppresser (TVSS) requires having floor space for the larger units, wall space for the smaller units, and a dedicated breaker for the parallel style. Planning also should include 3-1/2 ft. clearance in front of the TVSS.

Grounding

Proper grounding will alleviate future problems. Lightning strikes will wreak havoc at transmitter sites. Whether through a strike on the tower or via the utility electric, the induced currents and voltages due to lightning will find a way to ground. Many excellent articles go into the detail of designing a grounding system. The following are some guidelines for planning the grounding at the transmitter site.

There must be a copper ground system installed in the ground. For sites with towers, multiple ground rings around the tower base are used with radials attached to the tower base and joining the rings. A ground loop can be extended from this to encircle the transmitter building. If the soil is arid or rocky, chemical ground rods can be used by locating them periodically along the ground loop. Wherever possible, exothermic or cad welding should be used to insure good bonding.

Ground strap, 2 to 4 inches wide, should be installed around the transmitter room and along any outside equipment pad. Bond the outside and inside straps together and to the in-ground system. Tie all electrical panels, transformers, suppressors, etc., to the ground strap. Make sure that there are ground bars installed and tied to the equipment in the equipment racks and tie the ground bars to the ground strap in the transmitter room (see Figures 8.1-16 and 8.1-17). The transmitter cabinets, pump assemblies, and heat exchangers should be grounded as well.

Transmission line should be bonded periodically to the tower using grounding kits provided by the manufacturer. At the very least, any transmission line, waveguide, or coax should be bonded in three locations; near its antenna, at the base of the tower, and where it enters the building. Placing ground bars at the base of the tower and at the building entry help organize the grounding jumpers where multiple lines are involved.

FIGURE 8.1-16 Grounding Bar Example with Cadweld® Connections

Photo courtesy of Marsand, Inc.

FIGURE 8.1-17 Voltage Transformers, Grounding, and Hose Installation for a Rohde & Schwarz Solid-State LiquidCooled Transmitter at KERA, Cedar Hill, TX

Photo courtesy of Marsand, Inc.

HEATING, VENTILATING, AND AIR CONDITIONING (HVAC)

A mechanical engineering consultant should be used to design a heating, ventilating, and air conditioning (HVAC) system for the site. A reputable, licensed HVAC contractor can in many cases provide an adequate design as well as install the system.

Most transmitter sites will require air conditioning. As transmitters only convert somewhere between 40 and 50 percent of the AC energy into RF energy, transmitters can exhaust a large amount of waste heat into the building. To know the amount of A/C needed, the heat load will have to be calculated, as mentioned earlier. The manufacturers of all the various electronic and RF components can provide the heat load figure for the equipment. In addition, the environmental heat load must be considered: climate, roof composition, building insulation, amount of exposure to direct sunlight, etc. The system design should include redundant air conditioning units; each should be able to handle the full heat load on its own (see Figure 8.1-18). An important item to consider is to make sure that the air conditioning vents always point away from the mask filter, as the swings in temperature when the unit cycles can affect tuning.

AIR-COOLED SOLID-STATE TRANSMITTER CONSIDERATIONS

Layout

The air-cooled solid-state transmitter is the least complex to install. Physical sizes of the transmitter itself range from a rack-mounted unit taking up three to five RUs to a couple of amplifier cabinets slightly larger than a deep equipment rack. Amplifier cabinet heights vary from just under 6 ft. to over 7 ft.

FIGURE 8.1-18 Redundant Air Conditioners for the Transmitter Room

Photo courtesy of Marsand, Inc.

There should be 3-1/2 to 5 ft. clearance in front and behind the transmitter to allow for access for operation and maintenance. Care should be taken to ensure nothing obstructs the path of the airflow.

The mask filter may be mounted within the amplifier cabinet or is rack mountable for lower power units. Otherwise, space is required to suspend or floor mount the mask filter. The mask filter should be mounted in such a way that it is secure and protected.

Electrical

The air-cooled solid-state transmitters will usually accept several different voltages. When placing the order for the transmitter, the operating voltage must be specified. If the electrical configuration available at the site does not match any of the transmitter options, a transformer will be needed.

Three-phase power may be required for transmitters that use three-phase blower motors; however, most manufacturers are moving away from three-phase blowers. If three-phase power is available at the site, most transmitters will use line-to-line voltages for 208–240 V and line-to-neutral for 350–480 V systems. For very low-power transmitters, 120 V may be an option. Note that the amplifier power supplies are single phase, and usually so are the blowers.

HVAC

Air-cooled solid-state transmitters can exhaust either into the transmitter room or outside duct. If exhausted into the room, the HVAC must be able to handle the additional heat load. Keep in mind that the size of the room can come into play. In a small room or a room with a low ceiling, care must be taken to route the exhaust in such a way that it is not recirculated to the transmitter air intake before the HVAC system has a chance to cool it. The position of HVAC ducts for proper air-cooled transmitter cooling is critical. The best position for cooling ducts is near the floor of the building (since cold air falls) for ingest into the transmitter, with exhaust ducts above the transmitter cabinets to carry away the heat.

In some climates, outside air can be used to cool the transmitter. This air will need to be filtered well, and the moisture content in the air must meet the manufacturer's specification. Otherwise, the cooling fins will eventually clog with debris, and the electronics may corrode.

LIQUID-COOLED SOLID-STATE TRANSMITTERS

Layout

Liquid-cooled solid-state transmitters at the very minimum require plumbing and an external heat exchanger (see Figures 8.1-19, 8.1-20, 8.1-21, and 8.1-22 for examples of different layouts). Higher powered systems usually have a pump assembly and a heat exchanger per amplifier cabinet, although this can vary. The plumbing can be industrial-style hoses or copper pipe or either, depending on the transmitter manufacturer's requirements.

The amplifier cabinet heights vary between just under 6 ft. to just over 7 ft.; 3-1/2 to 5 ft. of clearance must be maintained in front and behind each cabinet for maintenance. With the weight of the amplifiers and power supplies installed, a transmitter cabinet can be extremely heavy. The floor must be able to support the final weight of the transmitter as well as personnel and test equipment without shifting or giving. Often, if the frame of the cabinet is not square, the amplifiers will not seat properly into the combiner. This can be caused by a floor that gives too much under load.

Thought must be given to routing the plumbing. For copper pipe, it is important to route the pipe to minimize joints, provide adequate support, and avoid crossing over electrical components such as distribution panels. For multiple cabinets, routing the copper plumbing can become quite intricate. Hoses are usually laid in a wire tray. Use of sweep elbows instead of straight elbows helps minimize flow restrictions and aid in the noise reduction in the room. It is usually faster to install hoses than copper pipe. The tendency for larger systems is to bundle the hoses where they

FIGURE 8.1-20 Rohde & Schwarz THU Style Pump Assembly

Photo courtesy of Marsand, Inc.

FIGURE 8.1-19 GatesAir ULXT-40AT Transmitter Pump Assemblies Using Copper Pipe at WJYS & WJAX, Jacksonville, FL

Photo courtesy of Marsand, Inc.

FIGURE 8.1-21 GatesAir Dual ULXT-40AT Solid State Liquid Cooled Transmitters With Combination Hose and Copper Pipe Plumbing at WJYS & WJAX, Jacksonville, FL

Photo courtesy of Marsand, Inc.

FIGURE 8.1-22 Rohde & Schwarz THU9–4 Solid State Liquid Cooled Transmitter With Hose Plumbing Style Pump Assembly at KERA, Cedar Hill, TX

Photo courtesy of Marsand, Inc.

FIGURE 8.1-23 Rohde & Schwarz THU9-4 Heat Exchangers Installed Horizontally at KERA, Cedar Hill, TX

Photo courtesy of Marsand, Inc.

overlap each other. This may not affect normal operation, but when a hose must be replaced, it is very difficult to remove. It may be worth the time and extra expense to route the hoses in such a way that they lie flat and do not overlap.

The pump assemblies may be located inside the amplifier cabinet for low-power transmitters where there is room available. Usually, the pump assembly is a freestanding unit that is installed indoors. Some pump assemblies can be installed outside. The assembly often comes with a standby pump installed for redundancy. An area 3-1/2 to 5 ft. in front of the pump assembly must be kept clear for maintenance.

The heat exchangers are installed outside in most cases. There are sites located in colder climates that can make use of the heat from the heat exchangers to maintain a comfortable temperature in the building; however, these installations would need to include a well-thought-out ventilation system to duct the heat outdoors during the warmer seasons as well as to regulate the temperature indoors. The heat exchangers come in either a vertical or horizontal orientation (see examples, Figures 8.1-23 and 8.1-24). Some are convertible between the two. In the vertical configuration, the area on the exhaust side of the fans must be kept free of obstruction for 6 ft. For the horizontal unit, the area above the fans must be free of obstruction for 6–8 ft. If multiple heat exchangers are used in a system, care must be taken to avoid any areas where heat may build up and find a path to recirculate. This can happen when a solid roof is used over horizontal units even if the roof is 10–12 ft. above the units. Place the heat exchangers where they are shaded from the afternoon sun. The site altitude above mean sea level may derate the heat exchanger; this should be brought up during specification before purchase.

Electrical

The liquid-cooled solid-state transmitters will usually accept several different voltages. When placing the order for the transmitter, the operating voltage must be specified. If the electrical configuration available at the site does not match any of the transmitter options, a transformer will be needed.

FIGURE 8.1-24 GatesAir ULXT-40AT Heat Exchangers Installed Vertically at WJYS & WJAX, Jacksonville, FL

Photo courtesy of Marsand, Inc.

Three-phase power may be required for transmitters that use three-phase pump motors and fan motors on the heat exchanger; however, some models use single-phase power throughout. If three-phase power is available at the site, most transmitters will use line-to-line voltages for 208–240 V and line-to-neutral for 350–480 V systems for their amplifier power supplies.

Some transmitter models provide power from the amplifier cabinet to feed the pump assembly and heat exchanger directly, whereas others require a branch circuit or circuits from the station's electrical panel to feed the pump assembly and heat exchanger.

HVAC

Liquid-cooled solid-state transmitters do not require external cabinet cooling for the amplifiers, as the liquid cooling removes the heat and dissipates it outside the building. However, the transmitter still presents a heat load to the room due to radiant heat from the RF components—notably the harmonic filters and combiners—as well as power supplies. HVAC planning must take into account the additional heat load to the room.

TUBE-TYPE TRANSMITTERS

Layout

Tube-type transmitters are the most complex to install. They require liquid cooling, so there are pump assemblies and heat exchangers besides the amplifier cabinets. Also, each amplifier cabinet will have an associated high voltage power supply (HVPS). The transmitter placement requires careful planning:

- Each cabinet should be placed as close to its HVPS as possible to minimize the high voltage cable runs.
- The floor under as well as in front of each amplifier cabinet must be able to support the weight of the IOT carriage assembly. These assemblies roll on narrow casters that are known to crumble tile laid on week subfloor.
- Allow room in front of the cabinet, 5 to 6 ft., for the IOT carriage assembly to roll out for maintenance.
- Copper plumbing is used in the cooling system, so there must be a plan to route the copper pipe, standard electrical conduit, high voltage conduit, and RF line to the top of each amplifier cabinet.
- If the pump assembly is located indoors, it may require a separate room due to the excessive noise of the pumps.

Electrical

Tube-type transmitters require 480 V three-phase voltage. Separate branch circuits are required for each amplifier cabinet, pump, and heat exchanger. Some models require a second branch circuit for each amplifier cabinet. A service disconnect is required on the feed to each HVPS, and it must be located where it is visible to anyone servicing the associated HVPS.

HVAC

Although tube-type transmitters dissipate the majority of the heat the IOT generates outside via the heat exchanger, there is still a large heat load placed on the room HVAC system. The manufacturer can provide the heat load numbers, and the HVAC system should be sized accordingly.

Remote Control and Monitoring

Remote control and monitoring are typically items that are ordered independently of the transmitter, but they can be included in the purchase package. Several remote control systems are available that offer control, status, and alarm functions for the transmitter. They can also be expanded to include various functions in the building such as door and fire alarms, temperature sensing, etc. The newer transmitters are remote accessible via IP. Most of their functions are available over SNMP.

SUMMARY

The intent of this article was to cover many of the practical aspects of planning a television broadcast facility. In planning a project of this complexity, this article may help populate a checklist and highlight how important it is to use the resources available in the industry, make no assumptions, and communicate early and often. Along with planning a television transmitter facility, additional considerations include documentation and implementing a maintenance program in order to improve the prospects of having a dependable and serviceable broadcast facility.

CHAPTER

8.2

Analog Television Transmitters

FRED BAUMGARTNER
TV Product Manager, Nautel

INTRODUCTION

As this is written, analog television transmission is waning worldwide, with approximately 50 countries left to complete the transition to digital by 2020. Analog TV is still found in those corners of the world that have yet to complete their digital TV transition as well as in some low-power, translator and closed-circuit situations in the United States. Until recently, virtually all digital television receivers would recognize and display an analog television signal. Analog reception quality, sensitivity, selectivity and tolerance of unstable video have been declining as television receiver manufacturers reduce or eliminate support for analog TV. Some receiver manufacturers no longer publish analog specifications, and some models don't support analog at all.

It can be argued that analog TV has one advantage—analog degrades "gracefully" as the signal fades into noise, whereas digital signals reach a "digital cliff" beyond which the digital signal is suddenly no longer recoverable. Actually, excellent digital transmission using roughly one-fifth the power output covers more area than an analog transmitter with a fair to good quality picture.

The story of broadcast television standards is one of technological advances, politics, governments, corporations and standard-setting organizations. There is considerable history, and there are competing firsts around the world (see Figure 8.2-1). To keep perspective, consider that NTSC analog television is based on EIA RS-170, where the RS means *recommended standard.* While millions upon millions of pieces of equipment comply with RS-170 or RS-170a, it remains a *recommended standard*, not an *adopted standard*.

As far as an analog transmitter's video is concerned, the differences in the world's analog TV standards occur in the bandwidth allocated for the video and how one takes advantage of the structure of the video waveform to increase efficiency (pulsing and other tricks used to handle the peaks of sync). For the most part, the transmitter isn't affected by the number of lines the picture has, whether the active video is inverted, whether it is progressive or interlaced, the frame rate, or the method of adding color information. The aural portions of analog TV transmitters are FM, with various pre-emphasis, deviations, and means of accommodating stereo (if they do at all). AM audio was used in some early analog transmission systems, but those are now museum items. The test equipment, test signals and monitors for a given analog video format have to support that standard, even if the TV transmitter itself will support a wide range of video formats. As far as TV receivers are concerned, there is no compatibility between the formats or, for that matter, between variations of each standard, although some receivers support more than one format.

PAL was developed in the early 1960's as an improvement over NTSC and stands for *phase alternating line* format (see Figure 8.2-2). PAL served most of Western Europe and Australia. PAL and derivative PAL-N have 50 Hz field rates with 625 lines, while derivative PAL-M emulates NTSC's 525 lines and 60 Hz nominal field rate. The wider 8-MHz channels permit more detail than NTSC. By reversing the color phase with each line, some of the color-shifting

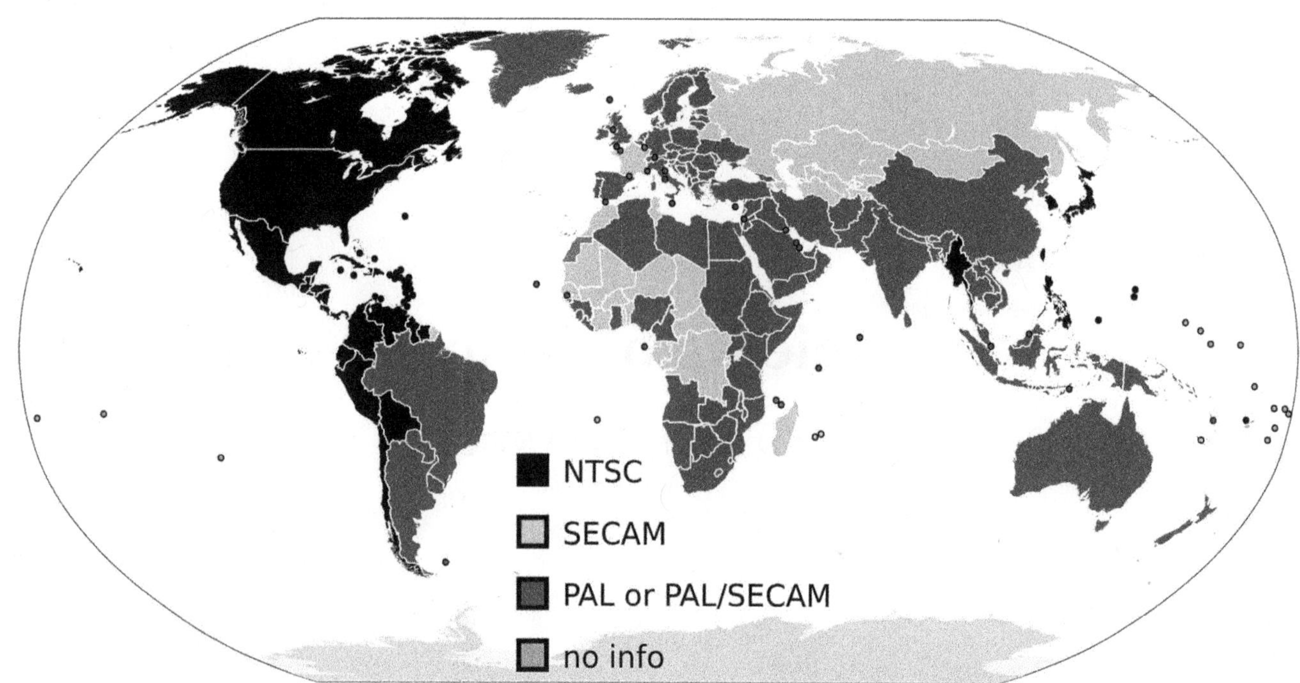

FIGURE 8.2-1 There are three common analog TV standards throughout the world.

Courtesy Wikipedia Commons

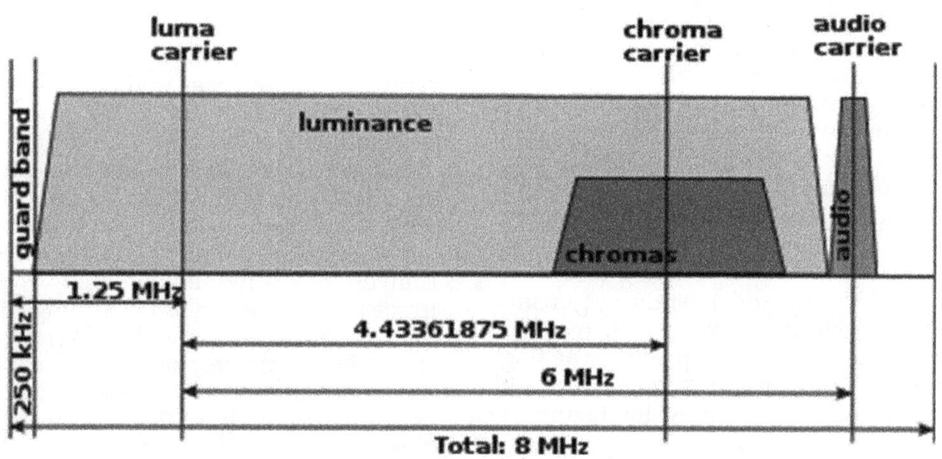

FIGURE 8.2-2 PAL occupied bandwidth
The PAL bandwidth is wider than NTSC.

Courtesy JM Gonzalez (Own work) [Public domain], via Wikimedia Commons

distortion inherent in NTSC is canceled out, but at the cost of a delay line and additional circuitry in each color TV receiver. The wider PAL channel also allows a higher color subcarrier frequency of 4.43361875 MHz as compared to 3.579545 MHz for NTSC. Both NTSC and PAL use quadrature-amplitude-modulated subcarriers to carry the chrominance information, added to the luminance video signal to form a composite video baseband signal.

SECAM differs from the other color systems by the way the R-Y and B-Y color signals are carried.

SECAM uses frequency modulation to encode chrominance information on the subcarrier instead of quadrature amplitude modulation, like PAL and NTSC (see Figure 8.2-3). Also, instead of transmitting the red and blue information together, it only sends one of them at a time, and uses the information about the other color from the preceding line. This requires a delay line in the receiver for storing one line of color information. Thus the informal "Sequential, With Memory" name. Because SECAM transmits only one color at a time, it is free of the color artifacts present in

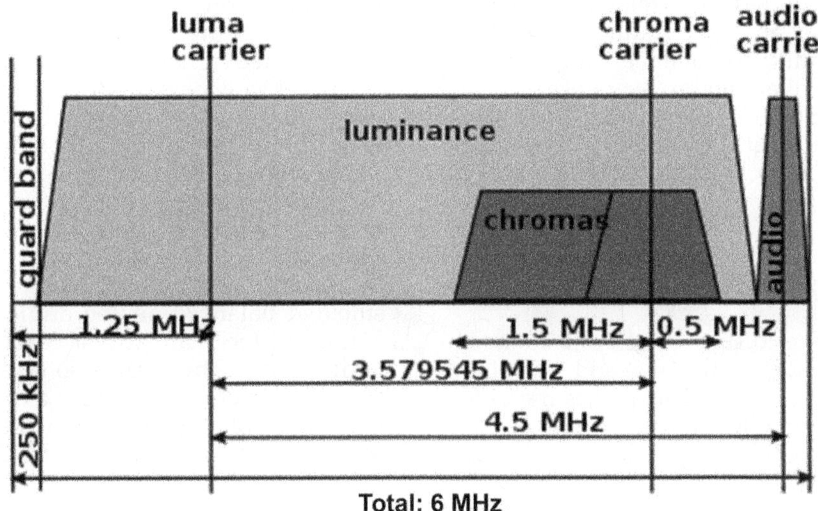

FIGURE 8.2-3 NTSC occupied bandwidth

NTSC was the domestic U.S. standard prior to DTV and is still used in some areas of the world and some systems and subsystems.

Courtesy JM Gonzalez (Own work) [Public domain], via Wikimedia Commons

NTSC and PAL that result from the combined transmission of both signals.

In North America, NTSC (National Television Systems Committee, the predecessor of ATSC) defines a picture with 525 lines of scanned video that is repeated 30 times (frames) per second. NTSC *interlaces* odd with even lines at 60 *fields* per second. Each frame of video is composed of two fields. NTSC has many other small compromises designed to save spectrum and improve picture quality.

NTSC will be the focus of this chapter. For the most part, tube transmitters will be the focus of this chapter, and solid-state transmitters in the digital TV transmitter chapter (Chapter 8.3).

FIGURE 8.2-4 RCA UHF transmitters were popular klystron transmitters. This belongs to Milwaukee Area Technical College and uses three cabinets to achieve 110 kW visual on Channel and 11 kW aural. The RF and air plumbing is visible above.

Author's collection.

COMMERCIAL TRANSMITTERS

A traditional analog transmitter is really two transmitters, an audio (aural) transmitter and a video (visual) transmitter that are combined together at *high level* (full transmitter power output levels) and filtered ahead of the transmission line and antenna. A *combined mode* analog transmitter mates the video and aural signals at *low level* before the combined signal is fed to a single amplifier system (see Figure 8.2-4).

Most of the analog transmitters remaining in the field are either older equipment, being kept functioning in anticipation of being replaced in a digital transition, or *dual-cast* transmitters that can be switched between analog and digital as needed. Dual-cast transmitters are popular because they offer an option for the transition to digital without the need to retain and maintain old equipment or having to retire a recently purchased analog-only transmitter with remaining value. The transition plans for some countries call for transmissions to regularly switch between analog and digital. While this makes transmission easier—it is hard on digital TV receivers—the channel scanning process doesn't make provisions for alternating transmission modes. Many transition plans replace legacy analog transmitters over time with dual-cast transmitters. On a given date, when enough new transmitters and receivers are in place, a nation will switch to digital.

Dual-cast transmitters are in almost every way a digital transmitter. They utilize an exciter than can produce an analog signal as well as a digital one, or two exciters—one for analog and a separate one for digital. The analog signal to be transmitted must be a single NTSC signal. The digital signal must be a multiplex of all of the signals required for a television receiver to decode the multiplex and may contain several services. A digital transmitter is a linear amplification system that can amplify an analog waveform as well as any number of digital TV waveforms. The analog waveform is less demanding to amplify, so a given transmitter is capable of delivering an analog output power twice that of digital.

ANALOG VIDEO AND AUDIO

All analog video has very distinct synchronization signals that take up about 25% of the time and 50% of the overall energy of the TV signal. This makes the video easy for a cathode ray tube television monitor to display. There are many horizontal synchronization "pulses" (15,750 per second in NTSC) that mark the end of one scan line and the start of the next. There are far fewer vertical synchronization pulses (60 per second in NTSC) marking the top and bottom of the screen. Many televisions are *multi-scan* and are capable of receiving multiple standards and often many variants.

At any given instant, the analog video signal level represents the electrical value of a given "spot" (the analog equivalent to a pixel) in the active video, or a value that represents some point of a synchronization pulse (see Figure 8.2-5). With very rare exceptions, the analog video signal is transported throughout the facility as a 1 volt peak-to-peak signal with a bandwidth covering the spectrum from DC to about 8 MHz. This is normally carried on a 75 ohm coaxial cable to the transmitter. Technically however, no video above 4.25 MHz makes it through an NTSC transmitter, because to do so would interfere with the aural transmitter's signal of 4.5 MHz. Audio is usually delivered to the transmitter as a monophonic or stereophonic pair of shielded balanced analog audio signals at line level. Some exciters can also accept an AES-EBU digital audio signal. There can also be a secondary language (SAP) channel. In the case of a dual-cast transmitter or later series of analog exciters, the digital signal is presented to the transmitter as an SDI signal.

Vestigial Sideband in Television Transmission

Analog transmitters modulate the video signal onto an amplitude modulated (AM) carrier. This is the same modulation that is used in AM radio; however, the bandwidth of video is a thousand times the width of audio (when a 5-kHz wide audio signal is AM modulated, it fits within a 10-kHz wide channel. If a 5-MHz

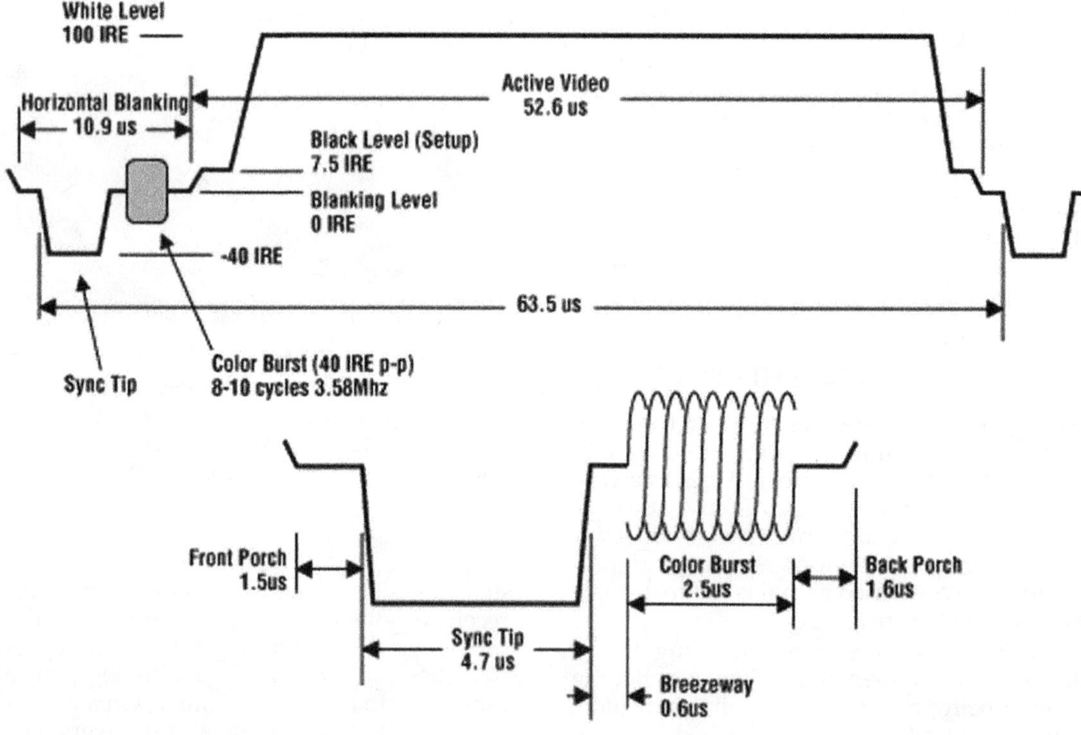

FIGURE 8.2-5 NTSC baseband video waveform.

Viewing a single line of video, one can see that it begins with a horizontal synchronization pulse and ends with a horizontal synchronization pulse. The active video space in this case is white and, if repeated for the full frame, the picture would be a white screen. More complex video would vary the value of the active video with the scene. A burst of color reference is seen on the back porch of the horizontal sync pulse.

Courtesy Wikipedia Commons

wide video signal is modulated, it will fill a 10-MHz wide channel).

Video "Tweaks" Specific to Analog Transmission

There are a series of small modifications to the video signal itself that improve the quality and or efficiency of the transmission system. NTSC and most analog TV transmission systems invert the video signal for transmission so that a "negative" of the desired picture is transmitted. The purpose of this inversion is to disguise noise acquired in the transmission path. Since black video elements require more transmitter power than white video elements, during the period in time when the synchronization pulses are "blacker than black," have no luminance at all and actually extend below the level at which there is no luminance displayed on a monitor, the transmitter is at maximum power. Small spikes of noise generally appear as white sparkles, which blend in better with a white picture than a black picture. Against a black background a "sparkle" would be most objectionable. By inverting the picture, a dark picture's higher energy overpowers the noise. A bright picture element, with lower power, is more susceptible to noise but hides the noise in the bright content.

Gamma encoding of images is used to optimize the bandwidth used to transport an image by taking advantage of the nonlinear manner in which humans perceive light and color. Additionally, in CRTs and photo pick up devices the video voltage produced and reproduced is not a linear representation of the light levels. With proper gamma compression and correction, a larger dynamic range of light can be compressed into the limited useful dynamic range of the video signal. The nonlinear gamma transfer curve is fixed and not dynamic, as is the similar companding often used with audio. Gamma (gamma correction, gamma nonlinearity, gamma encoding or usually simply gamma) also serves to reduce the apparent noise in the received video.

Practical analog television also employs a number of techniques to make the signal more spectrum-efficient. NTSC uses a 6-MHz channel to transmit video, audio and color information. In order to fit a 4.25-MHz video signal into that channel, the lower sideband of the AM signal is suppressed in the same manner that amateur radio operators use single sideband (SSB) to save bandwidth. Since the original sidebands are identical, only one is needed to carry the information. Unlike the suppressed carrier SSB ham radio operators favor, the TV transmitter carrier remains at full strength and the lower sideband is not completely eliminated. The NTSC carrier is placed 1.25 MHz up from the lower channel band edge. For example, U.S. channel 3 covers 60 MHz to 66 MHz. The video carrier is at 61.25 MHz. Signals below 60 MHz and above 66 MHz are removed. The vestigial sideband (VSB) transmits the first 1.25 MHz as an AM signal with no sideband suppression and the remaining higher frequency signal as an upper sideband (USB) SSB signal. Since greater picture detail requires wider bandwidth, limiting the transmission bandwidth limits the video resolution.

Audio

Television audio is sent on a separate carrier located 5.75 MHz above the lower channel edge, and thus 4.5 MHz above the visual carrier and only 250 kHz below the upper channel edge. For NTSC U.S. channel 3, the audio carrier is located at 65.75 MHz. Audio is frequency modulated (FM) in the same manner that FM radio is, although the deviation is narrower (25 kHz as opposed to broadcast FM's 75 kHz). The 75 μs pre-emphasis curve is the same as broadcast FM in the Americas. Stereo for NTSC TV systems follows the Multichannel Television Sound (MTS) standard that the Broadcast Television Systems Committee (BTSC) adapted from the broadcast FM stereo standard. MTS sound borrows the horizontal 15,750 Hz timing frequency for the stereo reference as opposed to FM's 19-kHz *pilot* that must be inserted in the FM radio transmission for an FM stereo radio to be able to decode the stereo signal. MTS stereo also employs *companding* (processing that employs a nonlinear dynamic range function) to reduce noise, which broadcast FM does not.

Typically, the aural carrier is transmitted at a level 10 dB below the visual carrier level. Even so, the narrower bandwidth and more noise-immune FM signal is likely to be received without noticeable impairment even when noise makes it impossible to lock the video (early TV systems used the same power for the aural as the visual). Likewise, the choice of a narrow FM deviation was driven by the desire to allow the few early TV receivers with AM audio to *slope detect* the new FM audio.

Color

Color in analog systems is transmitted as differential color information, where the green-minus-luminance (G-Y) and blue-minus-luminance (B-Y) information is transmitted in quadrature on a subcarrier with suppressed carrier at 3.579545 MHz. A few cycles of this 3.58-MHz timing reference is transmitted as part of each horizontal sync pulse, synchronizing an oscillator in the TV so that the color subcarriers can be decoded—3.58 MHz was chosen to place the sidebands of the color information and the sidebands of the black and white video in such a way as to minimize their interaction. The red, green and blue needed to display a color picture are mathematically derived from the luminance (Y), G-Y and B-Y information. The color information is sent with low resolution in order to conserve spectrum, taking advantage of human eyesight's similarly poorer resolution of colors. Color was added to the NTSC standard in this backward-compatible manner so that black and white television receivers would continue to operate with a minimum of degradation from the additional color information being transmitted.

Frequency Offset

Analog TV channel's carrier frequencies may be offset slightly from one another to reduce the effects of co-channel interference in fringe reception areas. In the United States, a table of allocations specified the either positive or negative offset of the visual and aural carriers.

Differences Between Analog and Digital TV Transmission

Unlike digital transmission, where each of the OFDM or 8VSB carriers are simultaneously transmitting individual pieces of the complete stream of data, analog transmission is entirely serial; there is only one analog value for video, colors and audio at any given instant. There is only one carrier for video and one for audio, and the bandwidth occupied is a function of the modulation and resultant sidebands of these two independent carriers. In analog transmission, there are no buffers, error correction or any other signal manipulation typical of digital systems; hence the only delay in the transmission system is the propagation of the signal through the transmission medium—basically the speed of light. Watching the image of an analog signal leaving the station adjacent to the picture received off-air rarely shows a perceptible difference in time. What you do see on the off-air monitor that you don't see prior to the transmitter in the studio is the combined effect of all impairments in the transmission system, especially noise and multipath (reflections of the transmitted signal). In analog transmission there is no opportunity or mechanism to repair the image or audio information.

TRANSMITTER ARCHITECTURE

Because analog TV transmission employs separate visual and aural carriers, the architecture can either be separate visual and aural transmitter systems combined together at high power in an RF diplexer, or it can be co-amplified, where the visual and aural are combined at a low power level and the combined signal is amplified for transmission.

Traditional Independent Aural and Visual Systems

Legacy tube-based transmitters almost always employed separate transmitters for efficiency and cleaner output. The aural transmitter in an analog system is usually FM, transmitted at 10% of the visual carrier level, which can be amplified by a very efficient, nonlinear, Class C amplifier. The visual amplification is AM and requires a much more linear Class A or Class AB amplifier. Unfortunately, the separate aural and visual components in all but the most linear of amplifiers will cross-modulate each other, placing undesirable *intermod* products both in the video and outside the TV channel, requiring filtering.

In a traditional diplexed analog transmitter (Figure 8.2-6), separate aural and visual transmitters are combined in a filter diplexer, which is a special form of combiner that is frequency specific in that it provides isolated inputs for the two transmitters. If the aural fails, the visual remains at full power, and vice versa. An aural notch filter removes visual sidebands that would interfere with the aural carrier.

Diplexers

The key element in this traditional design is the diplexer. Even at UHF frequencies, diplexers are large, often the size of the transmitters themselves. At low VHF channels, they can be massive. These are either mounted on the floor, often some distance from the transmitter, or above the transmitter if sufficient vertical clearance is available (why TV transmitter buildings usually have high ceilings). Access must be provided to allow the diplexer to be tuned, something that needs to be done rarely after the initial installation.

The general construction of a diplexer is that of any cavity-style RF filter system. Its job is to take the visual and aural transmitter outputs and combine them into the desired 6-MHz (in the United States) bandwidth. Additionally, the diplexer has to shape the visual signal to fit the allocated vestigial sideband spectral shape. It also has to remove any visual sidebands that would land on top of the aural carrier, with the use of a "notch diplexer." In the case of UHF channel 14, additional filters may be needed to reduce interference to land

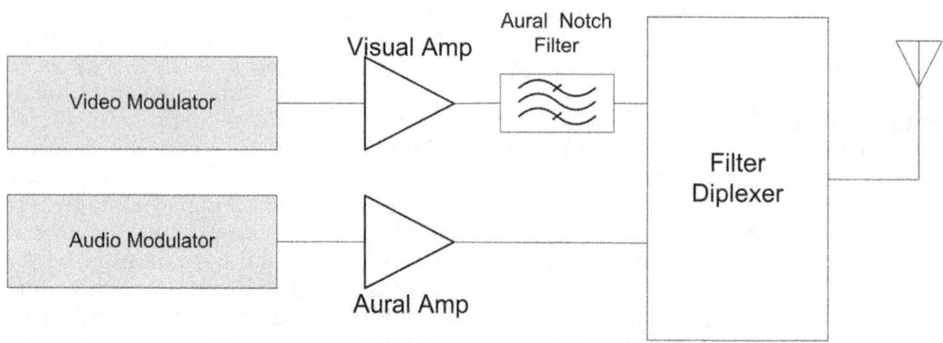

FIGURE 8.2-6 Diplexed analog transmitter

mobile and public safety services located just below the channel. The tuning of cavity filters changes with the temperature of the unit. Therefore, adjustments are only made after the filter is at operating temperature. The antenna, the diplexer and related filters have a big impact on the frequency response and phase linearity of the system.

Combined Amplifiers

Increased antenna gain disadvantages the coverage. Too much gain and the elevation pattern might not be broad enough to cover the desired population. A lower gain antenna has better coverage characteristics because there would be fewer nulls in the radiation pattern and the main lobe would be broader. With lower antenna gain, higher transmitter power output (TPO) is needed to get to the same ERP (effective radiated power). As these factors are considered, more transmitter power may be needed to produce the required antenna input power. When powers are required beyond what a single-transmitter/amplifier configuration can provide, amplifiers can be combined (see Figure 8.2-7). A typical amplifier combiner is a four-port device with two inputs, a combined output and a reject port. The amplifiers must have the correct phase and amplitude at the combiner to add together at the output port and cancel as much energy as possible at the reject port. The reject port requires a load capable of handling at least half the power of one transmitter and provisions to shut down the system if this is exceeded. Each exciter signal is split to feed the required amplifiers. Through the use of variable attenuators to balance the power and adjustable delay lines to adjust the phase, the individual amplifier cabinets are kept in balance. Imbalance in either the phase or amplitude results in power appearing at the reject port that should go to the output port and thus is turned to waste heat by the "absorber," or reject load.

FIGURE 8.2-7 Combining amplifiers to reach higher power levels is common. In this case, two cabinets are combined at WGBH.

Courtesy Comark.

Combining multiple amplifiers has the added benefit of creating redundancy. If one amplifier fails, the transmitter will still deliver 25% of the original TPO to the antenna and 25% to the reject load. By reconfiguring the RF plumbing, the surviving amplifier can bypass the combiner, which allows the station to operate at half-power, keeping the combiner's waste energy out of the reject load. Chapter 8.3, "Digital Television Transmitters," has more detailed information on combining RF amplifiers.

Combining schemes can be quite complex. At one time, there were UHF stations operating with four 60-kW visual cabinets for 240-kW TPO and a pair of aural amplifiers operating one at a time (a main and a cold or warm spare) for 24-kW aural TPO. Switching schemes and control logic can also be complex. Some transmitter systems can be a maze of RF switches, loads, filters and logical controllers that allow the system to be rapidly reconfigured to compensate for a failure or allow maintenance of an amplifier cabinet while the rest of the transmitter is on line.

Co-amplification Systems

Low power transmitters, typically less than 1 kW used for translators and gap fillers, usually co-amplify the aural and visual signals using a single power tetrode tube design or a single solid-state module (or array of modules) at both VHF and UHF. Dual-cast transmitters and most modern solid-state transmitters employ the co-amplification technique (Figure 8.2-8).

Co-amplification has been used in analog TV transmitters as means of recovering from an amplifier failure. In the instance of a transmitter with both an aural

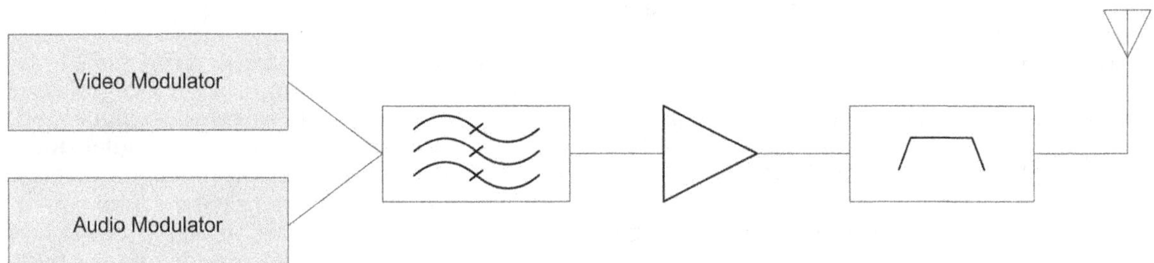

FIGURE 8.2-8 Co-amplification for analog TV is popular with modern solid-state transmitters. The aural and visual signals are combined at a low level, and the combination is then amplified for transmission.

and a visual amplifier where one fails, the surviving amplifier can be used in a co-amplification mode with some effort and temporary RF plumbing. The surviving RF amplifier cabinet is excited with the combined signal developed by an auxiliary exciter used for this purpose or is assembled by special circuity and filters either in the exciter or a low level combiner external to it. When running a transmitter in this configuration, the power output of the amplifier has to be reduced to a point where the amplifier is linear enough that intermodulation products and cross-modulation of the visual and aural signals are not objectionable. Any pulser circuitry needs to be disabled to avoid amplitude modulating the FM aural carrier. The RF path from the transmitter cabinet output to the antenna needs to be clear of any aural notch filters; however, the vestigial sideband filter should remain in the path.

Consider, as an example, a two-cabinet transmitter with a 60-kW visual amplifier and one 30-kW aural operating at 6 kW. If necessary, it can be operated with just the 60-kW amplifier in co-amplification mode; however, the TPO will need to be reduced to 20–30 kW in order to operate at a linear enough point to deliver a reasonably clean signal. One reason to have a 30-kW capable aural cabinet when only 6 kW is needed is to make it suitable for a co-amplification mode backup with a TPO of 10–15 kW as opposed to the 2–3 kW that a smaller amplifier would allow. Another reason to run a 30-kW amplifier instead of a 6-kW is that the tube can be operated at a very efficient lower current and will thus have an extraordinary life expectancy.

THE EXCITER

An example of a basic modulator is the tiny analog channel 3 or 4 modulator one can still buy for a few dollars or find in some set top boxes, games and other household devices that still support analog TV. Add a linear amplifier and antenna and you have a basic transmitter. It's not going to be a very good transmitter; it will have undesired radiation outside the channel's bandwidth and marginal video quality, and while its sync pulses might make a TV receiver work, they will likely be out of specifications.

A professional exciter has additional circuitry to compensate for limitations in the amplifiers, filters, antenna and other pieces of the transmission system. Traditional exciters are built-in drawers that can be pulled out, and a (usually hinged) cover lifted to expose dozens or hundreds of adjustments. In analog systems, everything, including the wiring used in the cabinet, affects the output signal in the form of impairments or degradations. No practical amplifier is completely linear. After everything else is adjusted properly, the exciter, and sometimes an external video processing ("proc") amp, will be used to "pre-" distort the signal in such a way that the end product is as close to the original video as is practical. Many hours can be spent with test signals such as color bars, ramps and special test signals designed to expose the limitations of the system. Most analog transmitter sites have an analog video test signal generator.

A practical exciter modulates the visual carrier and filters the visual sidebands to achieve the vestigial sideband waveform suitable for transmission. Typically, the modulated oscillators operate at some standard intermediate frequency and are mixed or up-converted to the desired transmitter output frequency. This allows the use of mass-produced surface acoustic wave (SAW) filters. SAW filters are compact electromechanical devices that perform even better than a large array of tuned circuits in a complex filter. Electrical signals are converted to a mechanical wave in a device constructed of a piezoelectric crystal or ceramic. The wave is delayed as it propagates across the device before being converted back to an electrical signal. The delayed outputs are recombined to produce a direct analog implementation of a finite impulse response filter. Virtually all traditional TV exciters use SAW filters to shape the visual waveform.

In a traditional separated system, the aural and visual modulators are separate, sometimes in two separate enclosures. There is little reason for the visual modulator to communicate with the aural modulator outside of basic control and level functions. If the modulator is designed for co-amplification, a low level diplexer is included in the design so that the output conforms to the channel bandwidth as much as possible. This type of exciter can be used in cable systems or to feed a co-amplification transmitter. It is not uncommon to use a cable-grade modulator as a backup exciter, backup transmitter, or low-power translator or community station.

Analog exciters usually experience a certain degree of "microphonics." Vibration from fans and pumps in particular can physically move some components, especially tuned inductors, enough to cause changes in their values that fluctuate in step with the vibration. The exciter becomes an unintentional microphone that modulates frequency, phase and gain of the video and audio. In some situations, such microphonics are visible in the picture and audible in the sound. To suppress microphonics, exciters are often mounted in separate racks that are not attached to the transmitter itself.

Digital Exciters for Analog TV

Exciters have evolved over the lifetime of analog TV from tube-based devices filled with variable resistors, capacitors, coils and tubes to transistors, op-amps and eventually all-digital platforms using digital-to-analog converters and RF mixers to operate on the desired frequencies. In particular, dual-mode exciters often implement the exciter and correction functions in firmware. With enough processing power, a digital exciter can support any number of waveforms, digital and analog. The precorrection is done in the digital domain. The chapter on digital television covers this in more detail.

Silicon-based analog exciters implement all of the filtering and analog signal processing in software and firmware. The exciter may be built in any number of forms including programmable arrays, application-specific integrated circuits and, of course, microprocessors. In effect, these are software-defined exciters. These exciters are often smaller than a traditional exciter, and their adjustments are made via a user interface. Most of these units are self-adjusting. If the correction process is activated with software or other user interface, the exciter will send test signals, look at the video returned from the station demodulator and build and execute a correction profile. It may also use a set of test signals embedded in one line of the vertical interval to constantly adjust the transmitter. An advanced exciter can set up a transmitter in a few minutes or seconds, sometimes without interrupting the program.

Impairments

A television transmitter must compensate for a series of impairments. These impairments are common to all transmission systems; however, they are manifested differently in digital and analog TV. The language used to describe impairments in analog video is somewhat unique, and the approach to deal with these impairments is built on analog descriptions and the related functions to compensate for them.

Clamping and DC Restoration

Analog video relies on absolute values for the key pieces of the waveform. The tip of sync has a defined voltage value and makes an excellent reference for restoring the DC voltage values of the analog video waveform. Video circuits and links often put the video through a capacitor or equivalent device in which the AC signal values are retained but the DC values are not. "Hum" induced into the video cause the DC values to wander with the power system's induced AC sine wave. A processing amplifier installed ahead of the modulator will "clamp" the tip of sync to a desired defined value. The AC video waveform will then be referenced to this value, effectively removing induced hum and DC variations. The transmitter needs to have the video sync at the value that causes maximum power and the white levels restricted to values that represent the minimum power the visual carrier should reach. Without this DC restoration, the transmitter power would swing wildly, exceeding the amplitude modulation limits imposed by both physics and regulation.

TV receivers often rely on the existence of some visual carrier (not pinched off at 100% negative modulation), even when the picture is at the highest white levels, to mix with the aural carrier so that a receiver's aural demodulator can demodulate the resultant 4.5-MHz FM signal. Tube-type "intercarrier" receivers often tapped the 4.5-MHz aural signal off of the video applied to the picture tube, taking advantage of the amplification and filtering of the video path. In intercarrrier receivers, a filter in the video path, often right at the base of the cathode ray tube, removes the FM signal so that it does not appear as "sound bars" in the video. Even digital receivers receiving analog signals may rely on the video carrier never reaching zero. The processing amplifier that handles these tasks is often external to the modulator and performs functions additional to clamping the DC levels and clipping the white levels. The processing amplifier permits adjustments of color gain and phase and adjusting color burst phase and level, as well as some equalization functions ahead of the modulator ("proc" functions can also be built into the transmitter's modulator).

Gain Linearity

A practical RF power amplifier needs to be operated well into RF compression. A perfectly linear amplifier's output will increase exactly in step with the input level. As an amplifier reaches the limits of its power output, increases in input level produce less increase in output until the amplifier reaches its limit (saturation), where increasing input level produces no further increase in output level. Further, as amplifiers reach their limits they produce more undesirable distortion products including RF energy outside the channel bandwidth, which includes harmonics and intermodulation products. Amplifiers are more efficient when operated closer to their limits. In analog TV, the peak power required is the tip of sync. A practical exciter will include a "sync stretcher" that permits predistorting the video waveform dramatically so that sync pulses going to the amplifier are much higher in level than what is seen at the transmitter output as the amplifier compresses the sync portion of the video signal.

A less severe form of gain nonlinearity is seen in the active video. Gain varies slightly over the bandwidth of the video and as a function of the bandpass filtering required for the vestigial sideband and filtering that reduces the video signal to near the 4.5-MHz aural carrier. This *differential gain* is compensated for in the exciter with a series of controls that allow for adjustments much like an audio equalizer. The gain must be adjusted at various different frequencies through the video passband.

Phase Linearity and Group Delay

The nonlinearity of electronic devices often results in level dependent phase shift. This *differential phase* distortion results in shifts in colors because the color information in an analog TV transmission system is communicated as phase information. Differential phase will result in inaccuracies in the video reproduction and is most noticeable in shifts in skin colors (what humans are most sensitive to) between the shadows on a face and the highlights. The exciter allows for adjustments in differential phase across the bandwidth of the luminance as well as both color signals.

All components of a signal are delayed as they pass through the exciter and transmitter. This delay is different at different frequencies in practical analog transmitters. This results in *group delay*. Group delay issues are observed as a loss of sharpness, color shifts and interference between video elements. Group delay is also adjustable over the bandwidth of the transmission system.

RF POWER AMPLIFIERS

TV transmitters were first used in the VHF band. Band I VHF (channels 2–6) covers 47 to 88 MHz. At channel 2, the 6-MHz wide TV signal represents a bandwidth that is 12% of the operating frequency (47 MHz). The only broadcast service with a similar challenge is the bottom of the AM broadcast band, with a bandwidth-to-operating frequency ratio of 3.5%. Maintaining the integrity of any signal over that much relative bandwidth is an engineering challenge. While most amplification devices will perform well enough, the matching networks will be a compromise.

For VHF, the use of power tetrode tubes was popular until solid-state amplifiers became more cost effective. As this is written, the state-of-the-art solid-state UHF transmitter has just reached the point where it is competitive with klystrons and IOTs.

Intermediate Power Amplifier

Exciters typically output only a few watts, typically too low to drive the final power amplifiers. An intermediate power amplifier (IPA) may be incorporated into the exciter if only tens of watts are required, or there may be one or two separate amplifier stages if the drive needed is in the hundred-watt or higher range.

The IPA has to have sufficient headroom to accommodate the peaks that result from precorrection. It is not uncommon for the tip of synch to be stretched to three times its normal peaks. Likewise, the IPAs need to be very linear and add a minimum of noise and undesired artifacts. IPAs are a key piece of the transmission chain and are usually expensive and very conservatively designed to operate without trouble for the life of the transmitter.

IPAs have been solid state for some time. Even transmitters that once had tube-type IPAs often have had them removed in favor of replacement with broadband solid-state "amplifier bricks." Solid-state replacements for aural exciters in particular often include a solid-state IPA to replace the entire chain up to the aural final. Co-amplification and visual replacement exciters often are purchased with replacement IPAs.

Tetrodes

Most analog VHF transmitters were built around tetrodes. A tetrode is a four-element vacuum tube that for this application is nearly entirely constructed of ceramic and has a plate assembly composed of air cooling fins. Air under pressure is forced through the cooling fins to remove the waste heat. The typical configuration is grounded grid with neutralization designed into the amplifier assembly. It is rare to require that a TV transmitter tube amplifier be neutralized. The filament and grid connections are usually concentric rings on the base of the tube, although on very early transmitters, copper straps might be used for the high current filament connections.

Tetrodes typically need to be mounted and maintained in a vertical orientation so that gravity doesn't pull the hot internal grids into a shorted condition. They should be stored vertically as well.

With age, tetrodes, like all vacuum tubes, lose emission capacity from the cathode. In a television transmitter, the filament voltage can sometimes be increased in the last stages of tube life to increase emission for a short period of time. Eventually, the tube will most likely simply reach a point where there is not enough amplification to create the required ERP. Tubes running with higher filament current to extend their useful life may simply burn out like a light bulb. Occasionally a tetrode fails if the control grid, which is very close to the cathode, shorts to the cathode either from thermal stress, vibration (from airflow through the tube) or defect. Most tubes in current use are rebuilt, and the supply of tubes is diminishing. On rare occasion, the screen grid will short to the plate (anode), which can cause damage in the screen grid supply as the full plate supply voltage is applied to the circuit. Tetrodes have typical plate voltages in the 5,000-volt range, though this can vary considerably. Even though it is not as much voltage as required for an IOT, it is equally lethal.

The TV transmitter tetrode design is very similar to that used in broadcasting's other VHF service, FM. The use of broad-banded cavities, a push-pull pair of 250 watt range intermediate power amplifiers (IPA) and air cooling is very similar (if a solid-state IPA is not used). FM transmitters and TV aural amplifiers are Class C amplifiers for efficiency. The video is essentially AM modulation, and Class A-B amplification is best to keep the linearity within the range for which the exciter can correct while still maintaining efficiency. Tetrodes for TV tend to deliver around 30-kW output at VHF, and both single-tube and double-tube push-pull designs exist. High-power tetrodes can also be water jacketed and employ water cooling systems. Those that are air cooled are fairly large and heavy in order to achieve the necessary surface area for proper air cooling.

Tetrodes were also popular in UHF low-power TV transmitters, typically in the 1-kW output range. There were probably more of this type of low-power TV translator built than any other type. Air cooling is convenient for remote locations and design for operation at altitude is well within the range of these tubes. They are almost always a single-tube design and coamplify the aural, making them a simple device to maintain. Tetrodes are very rugged, especially at this power level, a good trait where power is unreliable and dirty and in locations where air conditioning may fail for hours or days on end. Direct translation (receiving the primary station, mixing it with a local oscillator to the frequency desired, filtering this and providing the resulting signal to the transmitter) is not common, but not unheard of. Most use a receiver to deliver audio and video to an exciter with limited but affordable control capabilities.

Klystrons

Beyond a few kilowatts, tetrodes are not practical at UHF. Klystrons were used in UHF transmitters at the beginning of UHF broadcast television. UHF stations were allowed much more power than VHF stations to compensate for the propagation of spectrum that was considered undesirable at the time for NTSC transmission and poor TV receiver sensitivity. Klystrons are large, heavy, delicate and expensive. Special winches and handling equipment are required to work with them. The tube is composed of a high-power heater on one end and a liquid-cooled collector on the other, with a series of ceramic windows (RF gaps) that couple to either three or four external cavities. Between each cavity a circular DC-powered magnet focuses the electron beam created between the heater/cathode and the collector/anode. The electron beam is supported with a DC voltage generated by a beam power supply, generally placed outside of the transmitter building. A special high voltage insulated single wire carries the relatively low current but very high voltage from the beam supply to the amplifier cabinet. This wire runs in grounded metal conduit until it enters the transmitter. A nick in this wire's insulation can be extremely hazardous. The beam supply is a large power transformer with a rectifier array that is built in to the transformer's container, and usually covered by transformer oil that creates roughly 25,000 volts DC at 5 amps for a typical 60-kW klystron amplifier. Often, the collector is operated at DC ground and the heater/cathode assembly is operated at very high negative potential. Most transmitters place the collector at the top, but there is some argument that having the collector at the bottom makes leaks less of an issue.

Internally, there is a set of drift tubes (essentially short copper pipes) that assist the magnets in keeping the electron beam focused. Electrons that collide with the drift tubes eventually deform the drift tubes, reducing their effectiveness and create "body current." Body current is typically metered and represents the limits in the ability of the electron beam to be focused. The current in the magnetics is adjusted to minimize the body current.

Each of the four cavities is resonated by a mechanism that allows the physical size of the cavity to be changed. Usually these mechanisms employ chains, gears and moving panels, with conductive finger stock securing them electrically to the sides of the square cavity. Cavities are stagger tuned according to the manufacturer's instructions to achieve the desired bandwidth.

The first cavity, or input cavity, is closest to the heater/cathode. The input cavity is excited by a coupling loop with just a few watts of power. Typically, the input loop is fed via a high loss (often 10–20 dB) attenuator pad. This isolates the mismatch of the coupling loop from the driver amplifier. The input level to the input loop, the positioning—and thus coupling—and frequency response are all adjustable.

IOT and klystron transmitters used in digital UHF service, like solid-state transmitters, have solid-state preamps and IPAs. IOT and klystron devices might require low levels of drive, but the input circuit is often a very poor match for the standard 50 ohm IPA output. The solution is to put a large attenuator pad between the driver amplifier and the tube input coupling. Thus, if the attenuator is 20 dB and one watt is needed to drive the tube, then 100 watts of drive is necessary, which may require two or more stages of IPAs. If all of the power for the tube is reflected back to the IPA (the worst case) with the 20 dB pad, there will be only 0.01 watts (10 mW) of reflected power reaching back to the IPA. It's not efficient, but effective. If a transmitter has a failed IPA or the output tube is reaching end of life, the pad can often be temporarily reduced with adequate results.

The next two cavities are for amplification. Klystrons are velocity modulation tubes. The input RF bunches up the electron beam, and the bunching becomes more pronounced as it passes each cavity. The final cavity is the output cavity. There is a much heftier (than the input) coupling loop followed by a matching network to output a signal level from a few watts to 60 kW of RF power from the tube and send it to the antenna via the transmission line.

Replacing a klystron takes about four hours longer if significant cleaning or repairs to the finger stock are required. Retuning the tube typically requires another four hours. Readjusting the exciter after a tube change takes yet more time. The old or failed tube is disconnected after all voltages are safely removed and secured, and the approximately 400-pound carriage is rolled out of the transmitter cabinet, where work can be done on it. A set of specialized tools are required. The cavities must be removed, the hardware organized and saved and the heater connections and any supports disconnected to remove the tube. The klystron must be lifted directly upward from its carriage using the harness and hardware provided by the manufacturer and some type of block and tackle, typically attached to the ceiling supports or a special cart. Taking a klystron in and out of a special, gimballed shipping container requires that the container be upright. This work is best done with at least two people. The tube can then be shipped back to the factory to be rebuilt.

Installation of the new tube requires lightly abrasive, soap-free cleaning supplies including alcohol (or other approved solvent) to clean any finger stock and the tube's ceramic windows. If any part of the assembly is taken apart for maintenance, retuning or inspection, it should be clean and free of any residue and contaminants before it is replaced. This is not a good place to be stingy with time, effort or quality cleaning supplies. Nothing should be loose or likely to come loose with ordinary vibration.

Before the carriage is returned, the cavities can be "cold tuned" close to their operating frequency, with tools including a tracking spectrum analyzer and a return loss bridge. Once the carriage is reinstalled into the transmitter cabinet, the tube can be tuned under power. Reduce the operating voltage if at all possible, and start with low levels of drive.

If a spare klystron is kept for a period of time, most manufacturers recommend that an ion pump be attached. This reduces the contamination to what is a very large area of vacuum compared to other electron tubes.

Cavity arcs can be common and problematic. Usually the arc occurs in the output cavity or the previous (usually third) amplifier cavity. It has to be quenched very quickly to minimize damage to the cavity and tube. Photo sensors are placed in the last two cavities and sometimes in critical locations in the output network. The cavities are normally dark, so any bit of light is either an arc forming or a light leak. Most sensors have a tiny light bulb or LED attached that allows them to be tested. Some can be tested with a flashlight shining into a small opening in the cavity. While the transmitter can often be reset after an arc is detected, arcs will almost certainly reoccur with increasing frequency and increasing damage. Following an arc, the cavity should be removed and cleaned, any finger stock that has been stressed should be replaced, and the tube's ceramics and contact rings cleaned and any rough spots buffed out.

Pulsing

The klystron's beam has to accommodate the highest power required by the analog video waveform, the tip of sync. The rest of the video waveform is not as demanding. One means of increasing the efficiency of the transmitter is to add a pulser circuit, which increases the beam current during the sync intervals. While pulser circuits may save on the utility bill, they have complications of their own. As the beam current increases, so does the incidental phase modulation (ICPM). During the pulsed periods, the exciter has to change its phase correction to accommodate the new and temporary operating parameters of the tube. Pulsers are high stress components installed adjacent to the heater assembly. Pulsers operate at the same negative 25,000 volts DC as the heater assembly but need to get the sync pulse information from the video waveform operating in the 1-volt peak-to-peak range at ground potential, so optical coupling is typical of later transmitters. Pulsers are notorious for wear-and-tear failures, and it often proves less expensive to simply not use them and accept the higher average beam current and its higher power bill.

MSDC Klystrons

Multistage depressed collector (MSDC) klystrons offer additional efficiency. The energy that isn't converted to RF energy leaves the tube as heat through the collector. The basis of the collector design is usually a series of five elements, positioned and biased in such a way as to slow the electron beam before it hits the final collector. The elements each have their own cooling channels, and each one is biased about 6,500 volts below the previous (0, −6,250, −12,500, −18,750 and −25,000 volts, respectively). The cooling fluid has to be nonconductive to prevent currents between the collector elements. There is usually an additional focus magnet as part of the assembly to compensate for the beam spreading the MSDC creates.

This generation and variety of transmitter usually generates any bias voltages using resistor divider networks composed of high voltage wire-wound ceramic resistors (see Figure 8.2-9). These can also be high wear and tear components that require periodic replacement.

FIGURE 8.2-9 A Comark IOX transmitter with front door open revealing the IOT final amplifier assembly. Courtesy Comark

Courtesy Comark

IOTs

Inductive output tubes (IOTs) are a type of klystron combined with a tetrode that in many ways are both simpler and easier to handle than klystrons and offer higher efficiency (see Figure 8.2-10). The major difference between the operation of a klystron and an IOT is how the electron beam is bunched. Klystrons velocity modulate the beam of electrons at an RF input gap that is the first cavity. This is followed by a drift space in which electrons have been sped up at the input gap to "catch up" with electrons that passed the first cavity earlier but have subsequently slowed down. This causes bunches of electrons to form.

In an IOT, the RF input signal is applied between the cathode and a grid positioned close to and in front of the cathode. This causes the electron beam to be "density modulated" within the electron gun region itself. A DC bias voltage (V_G) of −80 volts relative to the cathode biases the grid, so that with no RF drive a quiescent current of about 500 mA flows. The cathode is held at the negative beam potential of −30 kV to −36 kV and the density-modulated beam is accelerated through an aperture in the grounded anode to the output cavity. A conventional klystron output system is then used to couple power from the IOT to the transmission line. In some countries, where an 8-MHz channel bandwidth is required or better performance is desired, a double-cavity coupling system

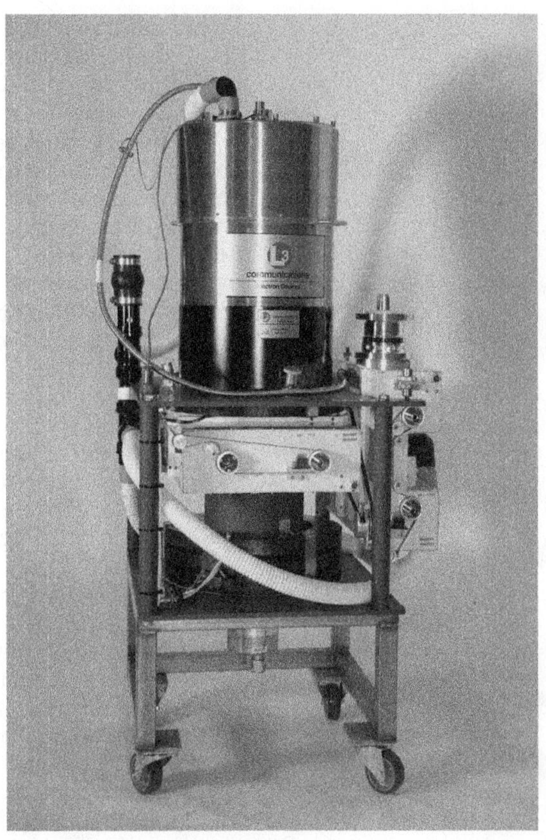

FIGURE 8.2-10 An IOT in a carriage that is rolled into a transmitter, where water, power, and RF connections are made to the appropriate supporting subsystems.

Courtesy Comark

can be used to achieve the required bandwidth and efficiency.

Since 1991, IOTs have been the tube of preference for UHF high power. After 25 years, they have functionally replaced all but the oldest klystrons and are themselves being supplanted by solid-state devices. IOTs and a few klystrons will remain in service for some time, in particular as backup equipment and in parts of the world with longer required equipment life-cycles. IOTs are fairly easily converted to digital service, and many have been.

MSDC IOT

IOTs can also benefit from multistage depressed collectors. The efficiency gain can be as high as 1.5 times that of a single collector IOT (see examples of IOTs in Figures 8.2-11 and 8.2-12). The theory and practicality is essentially identical to that of the MSDC klystron.

Cooling

A key component to any high-power transmitter is the cooling system. As high-power transmitters are notoriously inefficient, more energy leaves the transmitter building in the form of hot water or air than actually exits into the transmission line to the antenna in the form of RF energy. Sometimes, the waste energy can be practically used to heat buildings or water or for some other purposes. The efficiency of the cooling system is a component of the overall transmitter system efficiency.

The RF system must shut down or reduce power when the cooling system's limits are reached or when it fails. Tubes and other devices can quickly be destroyed or harmed if they overheat.

As in all other transmission equipment, redundancy of cooling system components is desirable and can be achieved in several ways; parallel cooling systems, dual pumps and multiple fans or heat exchangers. Cooling systems monitor temperature, air- and water flow. "Delta-T," or the difference in temperature between the input and output of the cooling system, remains remarkably stable over a wide range of ambient temperatures if the transmitter's operating point and air- or water flow remain constant, as well as the transmitter's operating point. A steady delta-T is a good indicator of the cooling system's health.

Air cooling is possible with some low-power tetrode designs and in some solid-state designs. Ducted outside air can be used in some air-cooled designs but comes with the necessity of keeping the air clean, filtered and within the desired temperature range.

FIGURE 8.2-11 An IOT in the box before installation.

Courtesy EEV

Dumping the heated air into the transmitter room and circulating and removing it with an air conditioning system is convenient and clean but adds the air conditioning to the potential sources of failure.

Typically, only the output tube's collector is water cooled, with all of the other components of the transmitter being air cooled. This generation of transmitter can have large fans, using three-phase motors in some designs, which are quite loud. Fan capacity must often be increased with altitude.

The most interesting and efficient cooling system (at least in quantity of water used) is the vapor-phase system. Here the plate or collector heat boils water that is conducted out of the building and condensed back to liquid water, typically in an exterior, fan-cooled condenser unit. There are many complications in a vapor-cooled system. The water can freeze, evaporate or become contaminated. Maintaining the proper fluid levels is burdensome.

Liquid and water cooling systems are considerably more popular and much easier to maintain and operate. The system can be closed or have an open reservoir. The system may be two stage if the first stage requires nonconductive water and the second stage requires antifreeze or connects to a building's cooling system. A transmitter located at the top of a building may require cooling via the building's environmental system. Exterior heat exchangers can clog with plant material and other debris or can freeze without adequate antifreeze (see Figure 8.2-13).

For cooling, a typical 60-kW klystron tube requires a flow of about 30 gallons per minute of clean water or water and glycol. As transmitter efficiency increases and better cooling fluids are used, the same 60-kW RF output may only require a conservative 20 gallons of cooling fluid per minute in a single collector IOT.

Liquid systems can also cool power supplies, filters, absorbers and station loads. As liquid systems should

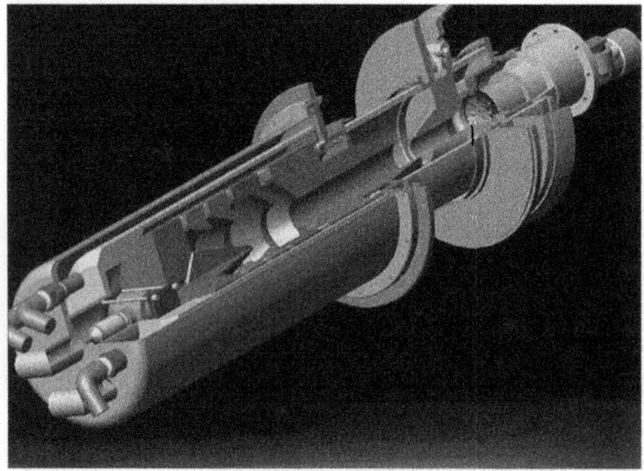

FIGURE 8.2-12 Breakaway view of an MSDC IOT showing the multiple stages of depressed collectors.

Courtesy EEV

FIGURE 8.2-13 Heat exchangers are placed outside and protected from falling ice and vandalism.

Courtesy Comark

not be allowed to go to steam, on very hot days in very hot climates, there may not be adequate outside cooling of the liquid to enable a transmitter to stay on the air at full power.

Liquid cooling systems can leak. They may experience electrolysis issues with dissimilar metals. They may harbor organic organisms. Any dirt or small particle contamination becomes abrasive, wearing down the system components. Antifreeze shifts the pH slightly to the alkaline. Some corrosion inhibitors add to the alkalinity. Liquid cooling systems are complicated, and transmitter manufacturers specify and/or supply compatible components, coolants and occasionally other chemicals. Ideally, a cooling system will operate as the main system for at least 15 years, then as a backup for another 15 years. When assembling or repairing a cooling system, stick to the manufacturer-supplied or recommended parts. Not just any replacement part will work.

The transmitter site should have replacement water and antifreeze as well as basic equipment to clean up a leak. A leak should not cause damage. Drains and drip trays separating cooling pipes from equipment that should remain dry and floor drains are critical in the facility design.

Pumps have a defined life expectancy. Having a system with redundant pumps and a means of easily replacing one while on the air is desirable. Systems also need to have any air bubbles removed by some "bleeding" process or procedure. Air that accumulates in a critical place in the system can cause damage.

Good water flow is critical for the operation of dummy loads—water can boil off quickly and destroy the load—but the loads are fragile and vulnerable to abrasion. Leaving the water flowing for any length of time will abrade the thin resistive coating, changing the resistance from the desired 50 or 75 ohms. Ideally a set of flow sensors and logic will prevent this from happening.

Solid-state transmitters present different cooling issues covered in the digital TV transmitter chapter (Chapter 8.3) of this book.

POWER REGULATION AND SURGE PROTECTION

In the United States, 120/240 or 277/480 60 Hz three-phase AC is common. Northeastern Europe utilizes 230/400 VAC 50 Hz more often. Klystrons and IOTs are particularly sensitive to voltage fluctuations, so a voltage regulator is typically installed with the transmitter. Tetrodes too, under more extreme circumstances, may require an external voltage regulator. Tetrode transmitters usually have some internal voltage regulation on all of the critical voltages including the plate B+ and screen grid. Filament and lower control voltages are also tightly regulated. Surge protection for any transmitter is highly desirable. Generators are common for backup power. In the current generation of transmitters, uninterruptable power supplies (UPS) are not common for the amplifier stages—the UPSs would have to be too large and prohibitively expensive. However, exciters, video, audio processors, control systems, microwave studio-to-transmitter links, transmitter-to-studio links, and monitoring and control (M&C) systems are well served by a reasonably sized UPS. Exciters and M&C systems don't recover quickly, and sometimes this sensitive equipment doesn't return to its "last-known" state after even a small power interruption. A UPS is valuable for either "holding up" the delicate electronics while waiting for the generator to start and get up to speed and the transfer switch to engage or for a short-duration outage if there is no generator. Without a UPS, in this instance, the transmitter site becomes invisible from the control point, which is especially troublesome if the generator or transfer switch fails and could have been reset remotely.

The filament voltage for all tubes is a critical factor in determining the life expectancy of the tube. Too high and it burns out too soon. Too low and the cathode can become contaminated. Transmitting tubes employ a trace (1–2%) of thorium in their tungsten filaments to assist with emission. Operated properly, the filament will have thorium-emitting electrons though out its entire life. A life expectancy of 35,000 hours (7 years) is typical under normal conditions. The record is closer to 150,000 hours for some klystrons. Current IOTs are rated for 80,000 to 100,000 hours—more than the lifetime of the transmitter. Tubes cost tens of thousands of U.S. dollars, and their exact end of life unpredictable. Typically, one budgets for the replacement of an expensive tube every few years.

In some cases the operating parameters of a transmitter can be temporarily changed to extend the life of a tube or assist in troubleshooting when other problems exist. In the case of klystrons, in particular, the beam power transformer configuration could often be changed with a switch (do not change the switch state while under power) built into the beam supply, to reduce beam voltage and thus current and power. Others require rewiring (changing taps) of the beam transformer. This mode might allow continued operation when diagnostics are in progress, an arcing problem exists that can't be remedied immediately or the cooling system can't cool the tube in normal operation, and especially when tuning the cavities.

Tetrodes may also employ a means of reducing power under certain circumstances. Leaving a transmitter in black is pretty easy on a klystron or IOT but places a lot of stress on a tetrode because more current is used to make power, which creates more heat. For the most part, a tube will run for some time transmitting a black signal, but after a while, transmitting black most likely means that there are no viewers, and the transmitter can be temporarily throttled back to save the stress on the tube. Some tetrode transmitters have a circuit to do this.

High power transmitters have points at which an RF or DC (in tube transmitters) arc might form. A method of watching for and mitigating such an arc would be to use an optical sensor that "sees" the light as the arc forms and immediately *crowbars* the power supplied to that section of the transmitter. A crowbar circuit (the equivalent of throwing a literal crowbar across a power supply's output terminals) fires a device that very quickly shorts the power supply. Relays and contactors are far too slow to be useful, so active crowbars can use SCRs, transistors, gate turn off (GTO) thyristors or a forced commutated thyristor to short the circuit. Optical sensors can be used in resonant cavities, transmission lines and cabinets that are normally dark.

In IOT transmitters, a crowbar circuit is a necessity. The device of choice is generally a thyratron. One popular thyratron uses low pressure deuterium (heavy hydrogen) as the operating gas. These circuits and tubes are often supplied by the same manufacturer that supplies the IOT. There is a common test wherein a very thin wire is used to short the power supply. If the crowbar circuit works, the wire remains intact. If it does not, the wire is vaporized. A special pneumatically operated switch and protective enclosure allow the test jig to be used safely.

Ideally, the power supply will be shutting down at the same instant as the crowbar is activated, as the crowbar is absorbing a great deal of energy and the power supply will likely itself be damaged or destroyed in a very brief period of time if operation is continued. Crowbar circuits can be reset after the transient issue passes.

Linear DC power supplies (transformer, rectifiers and filter capacitors, as opposed to switching) start with their filter capacitors discharged, so they represent a near short when first turned on. To reduce the *inrush current*, a *step-start* circuit might be used. The step-start is usually a resistor bank or a silicon controlled rectifier assembly that, when engaged, can limit the initial inrush current to a less stressful level.

PROOF OF PERFORMANCE

Transmitters, when new, generally are purchased with commissioning and a full proof-of-performance as part of the package. With any luck, the results of that POP are retained, so one can see how close the transmitter is to the day it was commissioned. The POP includes measurements of everything from power and efficiency to video and audio fidelity. After significant work, like replacing the output tube, good engineering practice suggests that some portion of a POP be performed to verify that the transmitter is operating properly.

Transmitter Logs

Good engineering practice also suggests that analog transmitter performance is periodically logged. At one time in the United States, and still in some nations, transmitter logs are required. As a practical matter, the operating log records forward power, reflected power, frequency, cooling system performance (temperatures or "delta-T") and any other important parameters for both the aural and visual transmitter at regular time intervals, varying from every 30 minutes to daily. Additionally, a maintenance log is kept in which every maintenance activity is recorded. Typically, as part of that maintenance log, every measurable parameter is recorded. Many of these readings may now be automatically recorded, but originally, analog transmitters often had many readings that could only be gathered manually. The daily records were often instrumental in tracking changes in the transmitter performance, providing early warning of trouble and making troubleshooting easier by having a set of reference readings from when all was healthy. Tracking tube and cooling system performance is necessary to know when to replace tubes, clean filters, etc.

Video Performance

Analog video is surprisingly resilient. The video signal can be significantly degraded and impaired and still produce a picture that people will watch. A consumer television located at the transmitter site seldom looks very good because the signal overloads the receiver's input and bounces around inside the transmitter building, creating a large amount of objectionable ghosting (multipath). The transmitter signal obtained in the transmitter building is a warped "end fire" signal off the bottom of the antenna, with reflections of unintended minor lobes of RF bouncing back from nearby terrain and structures. Consumer televisions have limited shielding, so even if a coax with a clean signal is attached to the antenna input, it is likely to be overpowered by ingress from the transmission systems, including in building leaks from the system's components. Still, the television receiver is likely to make a picture, even if the transmitted signal has significant issues. In most cases, a consumer TV at the site can only give you a false sense of security.

Ideally for good reception at the transmitter site (analog or digital), a professional demodulator is part of the transmitter system's equipment complement. This can be fed from sample ports in the transmission line or from sample ports that are part of the exciter, intermediate amplifiers, etc. If a switch is used to select between sample points, it must have very high isolation, so much so that it is a lot more affordable and probably better to have a series of cables that can be plugged into the demodulator as desired and terminated when not in use. The demodulator can feed video monitors, waveform monitors and vectorscopes. Some exciters can take advantage of the precision decoding to automatically adjust their parameters.

Measuring Power

There are three ways of measuring power. The first is to use calibrated couplers in the RF path with calibrated power sensors. For digital transmitters, this is typically the method used. Internal "working" meters are often calibrated against a standard high quality calibrated probe with appropriate calibrated attenuators and couplers.

Digital power and FM power, both peak and particularly RMS power, remain constant regardless of the content transmitted. Analog amplitude modulated video power varies with the video content. Transmitter manufacturers provide a method of measuring power and often formulas to convert their metering to RMS power. In general, the analog TPO is measured with a specific reference video signal that places the transmitter in black. There are other details to consider, like what state any vertical interval signals should be.

Because the waveform is complex, power is measured as the root mean square (abbreviated RMS), also known as the quadratic mean, which is defined as the square root of the arithmetic mean of the squares of a set of numbers. RMS is equal to the value of the direct current that would produce the same power dissipation in a resistive load. Typically, the transmitter power is measured using one of these two methods, and the transmitter's power output metering is adjusted so that 100% represents TPO at full desired (usually licensed) power. Regulations might specify that the power metering be calibrated at intervals (usually yearly) and that TPO be maintained within a tolerance (typically 90%–105% in the Unites States).

The traditional method of measuring power and calibration is to place a *station load* in place of the antenna. The station load is water cooled and part of the cooling system. One at a time, the visual and aural transmitters are used to heat the station load. Manufacturers provide a methodology to calculate the power from knowing the change in temperature across the load (two thermometers placed in special water-filled *thermometer pots* located at the input and output of the station load), the flow rate of the cooling water and the specific heat capacity of the cooling fluid. While water has a specific heat of 1.0, a specific gravity of 1.0 and a dynamic viscosity of 1.2 at 120°F, a 50/50 solution of ethylene glycol needed to reach a freezing point of −34.2°F/−36.8°C has a specific heat capacity of 0.832, a specific gravity of 1.064 and a dynamic viscosity of 1.5 at 120° F. The specific gravity of the solution is measured to determine the ratio of the ethylene glycol-based water solution. Compensation for the specific heat of the solution can be calculated to determine how much power is required to raise the temperature a given amount. The flow meter, too, depending on design, requires compensation for the viscosity of the fluid. All of this changes with the temperature of cooling fluid, so a simple procedure is not easy to design and the potential for error is high. The advantage of this process is that it can be run continuously as the transmitter power is adjusted, requiring only that one read the identical (or corrected) thermometers and do the math to determine the RMS power.

The alternative method is to use a fixed quantity of pure water, typically 50 gallons in a drum, with a pump that is connected only to the station load. Start with the station load, pump and hoses dry. Turn on the pump. Measure the temperature of the water. Turn the transmitter on for a fixed and precise period of time, generally a minute to a few minutes, and use the formula: Watts = 3.1 × Gallons × ΔT (in °F) / Heat-Up Time (in hours) to calculate the RMS power. The disadvantage of this method is that it requires some setup and can only be done up to the point where the water gets too hot to accommodate the test. An interval can be chosen that allows two or three tests before the water needs to be replaced with cool water. If the interval is long, or the water is significantly above ambient temperature, heat loss from an uninsulated water drum becomes an issue.

Digital transmitters can use the same methods as analog transmitters, although modern digital transmitters often do not have the precision flow meters or thermometer wet pots, nor are they supplied with laboratory-grade matched thermometers. The simple, closed-loop method always works.

SAFETY AND LIABILITY

Safety is a big concern with tube-type transmitters. The reality is that transmitters must be opened and exposed for some troubleshooting and maintenance while energized. A mistake or misunderstanding of what is "hot" and what is not has resulted in any number of fatalities. It is exceedingly easy to be killed or badly injured working on a transmitter. Additionally, the cost of liability for stations with personnel working on high-power transmitters can be a serious consideration, particularly since the number of high voltage transmitters and thus the number of transmitter engineers experienced with high voltage is diminishing. This is a powerful argument in favor of low voltage solid-state transmitters.

Solid-state transmitters, with their 50-volt amplifier operating voltages, are *considerably* safer than tube transmitters. Tetrode tube transmitters with their nominally 5,000-volt plate voltages are *somewhat* safer than klystrons and IOTs, with their nominally 20,000-volt collector voltages. Lives are saved with solid-state transmitters, and owners and insurers' liability is reduced.

High-power transmitters require high-power AC feeds and generators, generally operating between 200 and 600 volts AC, delivering hundreds of amps of current. The primary power supplies have safety issues that any high-power electrical system presents. They also demand the same safety procedures including "lockout/tagout" procedures, ground fault interrupters, fusing and the use of proper clothing, eye protection, gloves and tools.

Working with high voltage requires special tools and skills. For example, a voltage probe for 35 kV is available for CRTs, but if someone was holding the probe and the ground lead became disconnected or there is even a tiny crack or hole in the lead when measuring voltage on a CRT, the circuit will deliver a shock that will hurt, but on an IOT or klystron, the shock will almost certainly be fatal. High voltage "linesman gloves" are good, but they need to be continuously tested and replaced. When old, they may develop cracks and fail. In any case, it is unlikely that they are very useful in this kind of work.

This is where training and skill come into play. Simply put, you should be nowhere near anything that can arc, arc flash or otherwise explode. Eye protection is important because of the unpredictability of events. For example, one can take a high voltage probe, connect it to a meter and turn on the transmitter with the intention of being able to see the meter well enough from a distance. A failure of some component at this voltage might cause the meter or probe to explode. It is very hard to work on these transmitters for any length of time without experiencing a potentially dangerous "event." The safety part is being safely located and equipped when the unpredictable happens. One needs to be hypervigilant, have a strong understanding of the principles of working with high voltage and high energy . . . and be very cautious.

All transmitters present some danger of *arc flash*. While seemingly safer than high voltage and the danger of electrocution, a high current arc can blast molten metal over a considerable area, causing burns and in some cases penetrating a body and causing injuries, some severe, or even death.

High power RF can present a high voltage danger (at RF frequencies), an arc flash hazard (arcs at RF are as dangerous as arcs at AC or DC) and a radiation hazard. RF is non-ionizing radiation, as is sunlight, as opposed to X-rays, which have energy levels high enough to cause cellular and genetic damage with relatively low levels of exposure. Still, high-power transmitter RF will reach power levels where exposed flesh can be cooked. Particularly hazardous is peering into an opening in a cavity or transmission line; this can cause instant and irreversible eye damage.

Some cooling systems, especially when failing, may clog or stop circulating and go to steam. The steam may cause couplings and hoses to fail; being nearby presents a scalding hazard. Typically, these scalding accidents happen when one is trying to repair a failing cooling system. The energy density on a collector or anode is considerably higher than with a solid-state device, and water can become steam in a matter of seconds.

When a water-cooled solid-state amplifier or other component needs to be removed and the cooling fluid removed for shipping or storage, keep the cooling solution away from your mouth and know what to do if it is ingested. Proper disposal is important. Glycol is poisonous to many life forms.

In legacy transmitters with ceramic tubes, an often understated safety hazard is that the ceramics contain beryllium oxide, which is safe in normal handling but a carcinogen if released into the air by mechanical abrasion. Legacy transmitters may also contain PCB high voltage capacitors, which should not be ruptured and need to be disposed of properly.

X-rays are created by IOTs and klystrons, as the voltage is high enough that electrons colliding with the collector will create a stream of X-rays. It might be possible to remove enough of what serves as shielding to present a hazard, but the collector assemblies of IOTs and klystrons are dense and absorb these X-rays. While in typical transmitter use, X-rays should pose no issues; they can present a hazard in some transmitters if they have been modified from their original design.

Additional hazards include poorly protected fan blades or working inside the equipment on the unprotected side of a fan, moving heavy equipment, chemicals, fire and the usual hazards of an industrial site.

Working at night is necessary at television stations for revenue reasons. Most mistakes happen when people are distracted or tired. Ideally, transmitter technicians shouldn't work alone. Having a coworker with a piece of lumber to knock someone free of high voltage has been the difference between life and death. Likewise, transmitter sites are often dangerous to travel to and from, especially at night and in bad weather. The transmitter site should have basic provisions and first aid and have what is needed for survival and rescue. The bottom line is that TV transmitter work, especially with high-power tube transmitters, is hazardous work. Lack of focus, lack of knowledge, seemingly innocuous shortcuts and risk taking often will literally be the death of a transmitter engineer.

ACKNOWLEDGMENTS

For research and editorial assistance:

Mark Durenberger

Peter Douglas

Paul Grzebik

Martyn Horspool

Joseph Turbolski

References/Resources

NAB Broadcast Engineering Handbook 10th Edition

Chapter 6.2 "Analog Television Transmitters"

Revised by JAY C. ADRICK, Harris Broadcast Communications Division

EBU Technical Review Autumn 1997

The inductive output tube, the latest generation of amplifier for digital terrestrial television transmission.

R. Heppinstall (EEV Ltd) G.T. Clayworth (EEV Ltd)

CHAPTER

8.3

Digital Television Transmitters

FRED BAUMGARTNER

TV Product Manager, Nautel

When TV first came into popular use, the United States and others had huge areas to cover, with significant rural populations. The United States follows a "presatellite" model where the objective was to cover both urban and rural populations simultaneously with high-power, wide area coverage transmitters supplemented with lower power translators or boosters, usually receiving the main transmitter off air and translating its programming to another channel in order to cover communities outside of the convenient reach of the main transmitter.

FIGURE 8.3-1

FIGURE 8.3-2 A typical low-power, in this case 1-kW, digital transmitter. The upper unit contains the modulator, control systems, power splitter to feed the two 500-watt amplifiers below, and combiner that takes the output of the two amplifiers and, through a coupler to monitor the VSWR and power, feeds the output connector, which in turns feeds the filter and antenna. The lower two amplifier units are identical and contain the power supplies for the amplifiers.

Photos Courtesy of Nautel

In the 1950s, most private homes needed a rooftop antenna. Multiple dwelling units (MDUs) supplied a master antenna system (MATV), where a single antenna is amplified and distributed to each living unit. Those antennas were connected with either the more efficient (lower loss), less expensive, 300-ohm flat twin-lead or the easier to install and convenient 75-ohm coax to the family television set, usually located in a living room or family room. An RF splitter might be employed to feed a second television in another room. In this early era, the lower frequency VHF channels were most desirable because of the large coverage range that often extended beyond the visible horizon. In 1962 the U.S. Congress issued the "All-Channel Receiver Act," sometimes called the "UHF Mandate," requiring new TV receivers to also receive the UHF band (channels 14–69 at the time). These early UHF receivers were not as sensitive on UHF, and to compensate for the smaller coverage at the upper UHF channels, the hefty power level of 5 million watts was authorized (50 times what was permitted on Channel 2) for UHF stations.

As this is written, television receivers come in many forms, from large ultra-high definition televisions (UHDTV) to handheld and mobile devices. An over the air (OTA) TV station's programming is often supplied to televisions via multichannel video providers (MVPDs) using cable (CATV), satellite (DBS), and IPTV. Some, if not most, television receivers are not connected to OTA antennas at all but to the MVPD, which has taken the OTA broadcast signal and translated or transcoded it onto the MVPD platforms, which typically use different modulation (QAM) and often different compression (often higher efficiency, more recent codecs).

While in the beginning, VHF Band I (54–88 MHz) and VHF Band III (174–216 MHz) were preferred, today, UHF channels (Band IV, 470 MHz to somewhere near 608 MHz, or Band V, which extends to nearly 700 MHz depending on regulation) are preferred because of the better building penetration and the ability to reach mobile devices and small indoor antennas. There is a general trend throughout the world to repurpose UHF band V for wireless data use, which is in turn is changing the design of the OTA distribution network. Single-frequency networks (SFN) composed of many synchronized transmitters on the same frequency do a better job of putting stronger signal levels into high population density areas and buildings than a single big transmitter, usually outside of town on a tall tower, running as much power as permitted. ATSC 3.0 and DVB-T2 are particularly useful when an SFN design is desired.

For these reasons, and the various legacies of TV throughout the world, as this is written, arguably the most extensive changes are being made to the OTA TV distribution network in its history: packing stations closer together, increasing payloads, moving to more agile modulation and content standards and methods—thus changing TV transmitters themselves and how they are deployed (see Figure 8.3-3).

VHF AND UHF TRANSMITTERS/TUBES AND SOLID STATE

As this is written, the UHF spectrum is being "repacked" in the United States, and some number of stations will move from UHF to VHF. Currently, there are 1,300

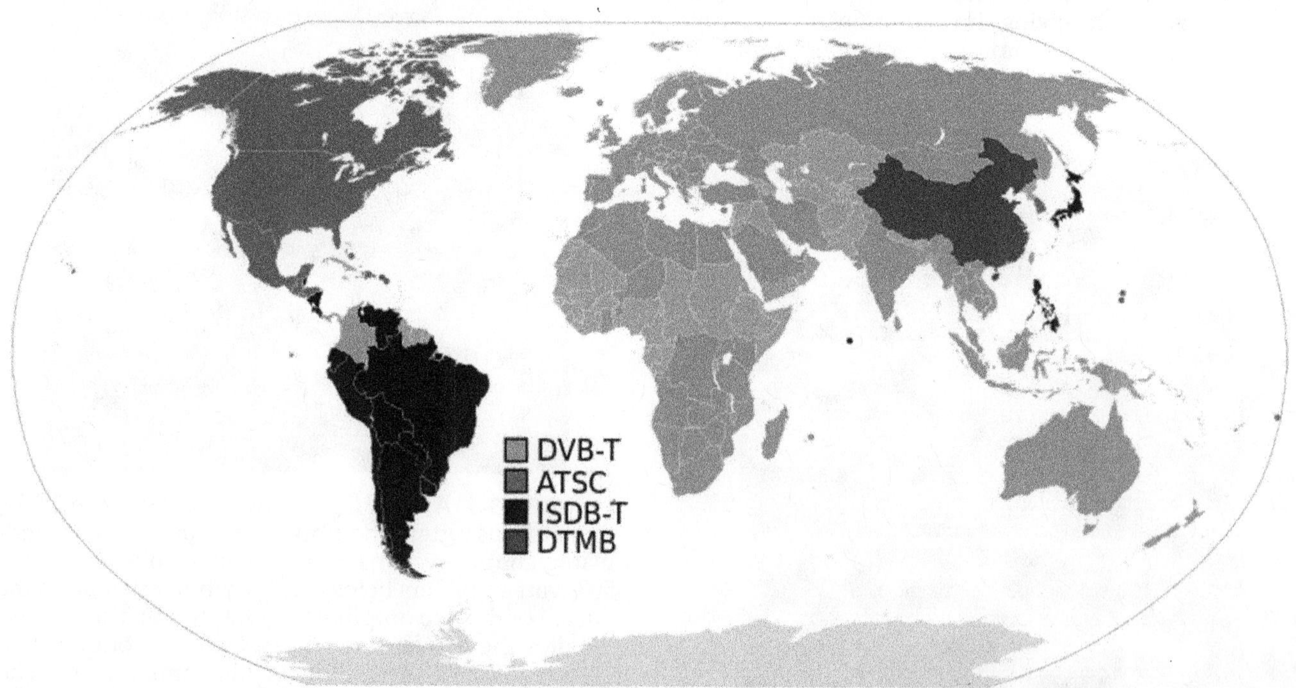

FIGURE 8.3-3 Waveforms and digital standards for broadcasting vary around the world.

Courtesy Wikipedia Commons

full-power UHF stations on the air and 450 VHF stations. Solid-state transmitters for UHF and VHF are highly similar. One difference is that the typical VHF transmitter is lower power than the typical UHF transmitter. Further, the combiners and filters for VHF are physically larger given the lower frequencies and longer wavelengths. Waveguide and waveguide filters are impractical at VHF. The architecture, construction, exciters, amplifiers, power supplies, cooling, and almost everything else is nearly or completely identical.

This chapter focuses on solid-state transmitters, while the analog TV transmitter chapter (Chapter 8.2) focuses on tubes. Readers can refer to that chapter for material on tube operation.

DIGITAL WAVEFORMS USED FOR TELEVISION

Any tube transmitter can be converted from analog to digital use. In many ways, the digital waveforms are easier to accommodate than analog waveforms. From the transmitter's point of view, the waveforms for digital TV, no matter what the standard, are surprisingly similar. Not all exciters support all formats, but all amplifiers do, with some limitations, support all formats, with the understanding that various waveforms have different *peak-to-average power ratios* or *crest factors*. The transmitter's amplifiers have to have enough *headroom* to accommodate these peaks.

TRANSMITTER ARCHITECTURE

The heart of a transmitter is the RF power amplifier. All transmitters also require a control system, power supply, and cooling system. The differences are driven mostly by scale. The higher the transmitter power output (TPO), the more complex and robust the support systems must be.

Physical Construction

The vast majority of new transmitters are solid state. Almost universally, each of a solid-state transmitter's components is mounted in extra-deep 19-inch racks. A few designs use the less common 21-inch wide rack. Amplifier assemblies are designed so that they are ideally light enough for a single technician to remove and replace. The rack will certainly be specially built or modified to integrate power combiners and dividers, as well as power and control distribution, into the assembly. The transmitter might be shipped with the amplifiers and other components installed, but it is more common for them to be integrated on site.

The alternative to rack assemblies is *box construction*, where the transmitter is generally one or more large custom metal cabinet(s) that must fit through the doors and elevators on the way to being installed. Tube transmitters are constructed in custom metal cabinets. Making the physical box for a transmitter is a major component of what a transmitter manufacturer does.

In either case, EMI (electromagnetic interference) or RFI (radio frequency interference) is a major concern for the transmitter manufacturer. While the mask filter and antenna system will reduce unwanted radiation via the antenna path, one still needs to inhibit RF radiation to and from the transmitter and transmission components. Cabinet shielding, including RF seals at access doors and removable panels, is the primary means of protection. TV transmitters, especially SFN boosters, tend to be collocated with other spectrum users. Undesired radiation from a TV transmitter might be well below the levels that present a safety issue but can easily be an issue to sensitive, low level receivers and other equipment located either in the transmitter building or nearby. Interference from channel 14 transmitters into the public safety communications band located just below the television allocation and spurs and harmonics from TV transmitters that land just above the television spectrum are experienced at many sites. Transmitter manufacturers in the United States must certify that their equipment meets EMI requirements. Operating transmitter components with covers removed, worn or missing EMI seals, missing or loose screws, or loose or improperly assembled interconnecting cables allows undesired RF to egress from the transmitter. In some sensitive locations, shielded transmitter rooms are called for to prevent leakage from the transmitter reaching other tenants of the site and beyond.

Ingress is also a possible problem. Without adequate shielding, strong enough RF may enter sensitive circuity or mix with RF in the transmitter to produce unwanted mix products or impaired transmitter operation.

In tube transmitters, there is also a concern regarding X-ray radiation. The tube assemblies and cabinetry are both designed to reduce X-ray radiation to near nothing.

Transmitter Power Output (TPO)

The maximum UHF power allocation in the United States is 1 MW (1,000,000 watts) effective radiated power (ERP), with approximately half of stations operating with less power due to license restrictions, which are necessary to reduce interference to other stations. Average gain for a TV transmitting antenna is approximately 21 dB, but antenna gain between 16 dB and 32 dB have been licensed. Thus, an average full-power UHF TV station in the United States emits about 50-kW TPO, with the practical range being between 28-kW and 80-kW.

The United States Low Power Television (LPTV) ERP maximum allocation is 15-kW. The average antenna gain used by LPTV stations is 14 dB. Translators, boosters, and gap fillers run the entire range from a few watts to full power. Thus an average full-power LPTV station may operate at 1.1-kW TPO, and the practical range is 500 watts to 4-kW TPO. Typical UHF translators in the United States are also likely to operate between 50 watts and 4-kW TPO but often are as small as 2 watts elsewhere in the world.

The power allocation for Band III VHF TV (174–240 MHz, channel 7 to 13 in the United States covers 174–216 MHz) is 160-kW for full power and 3-kW for full-power LPTV. The average Band III antenna gain is 6 dB for full-power stations and 5 dB for LPTV stations. The typical full-power TPO is thus 27-kW, and the LPTV TPO may be 600 watts. Fewer stations operate on Band I VHF (54–88 MHz, channel 2 to 6), where the maximum power allocated for full power stations is an ERP of 45-kW. Band I is unpopular because it is more susceptible to impulse and man-made noise interference, has poorer in-building penetration, and requires bigger transmitting and receiving antennas (thus ruling out practical mobile use).

Mask filters and transmitter combiners typically operate at about 90% of maximum for low-power and 95% for full-power applications. Transmission line losses are almost negligible for a full-power station with a short transmission line run operating on channel 2. A 2,000-foot length of transmission line at channel 49, however, can consume half the power the transmitter puts out. In all cases, transmitter and antenna combination selection is the result of a number of choices between size, efficiency, capital cost, operating cost, and performance. Waveguide and larger transmission line sizes are more efficient at higher frequency but are more expensive and can require stronger towers to handle the additional wind and weight loads, which in turn are more expensive. Higher gain antennas are more practical at higher frequencies, but increasing antenna gain and height make it difficult to reach an ideal coverage pattern, so additional design and testing costs may be necessary.

HOW TRANSMITTERS WORK

A block diagram of a transmitter is shown in Figure 8.3-4. A program signal is supplied to the transmitter, in which it is modulated, amplified, and filtered in order to feed the antenna system to be transmitted over the air to viewers. All of these components produce waste heat. RF power amplifiers are the least efficient of the system components in turning AC power into RF energy to feed the antenna system, so control systems are employed to turn the transmitter on and off as desired, adjust power, cooling, and other systems, and protect the transmitter from antenna system and electrical faults.

THE DTV EXCITER

The transmission process begins with a *program stream*. Eventually, the TV receiver will recover this stream of data and use it to supply content to the audio speakers, video display, electronic program guides, alternate language audio, closed captioning, video descriptive service, and any other service that is destined for the viewer. In ATSC 1.0, DVB-T, and a host of other transmission standards, there can be any number of videos, audios, data, and control services all multiplexed into the *transport stream* (TS) that feeds the transmitter. In first-generation digital standards, the supported encoding schemes are limited (usually MPEG-2) and a finite (but extensive) number of aspect ratios, resolutions, and ancillary functions supported.

Unlike analog transmission, the digital exciter is completely unaware of what is encoded into the TS.

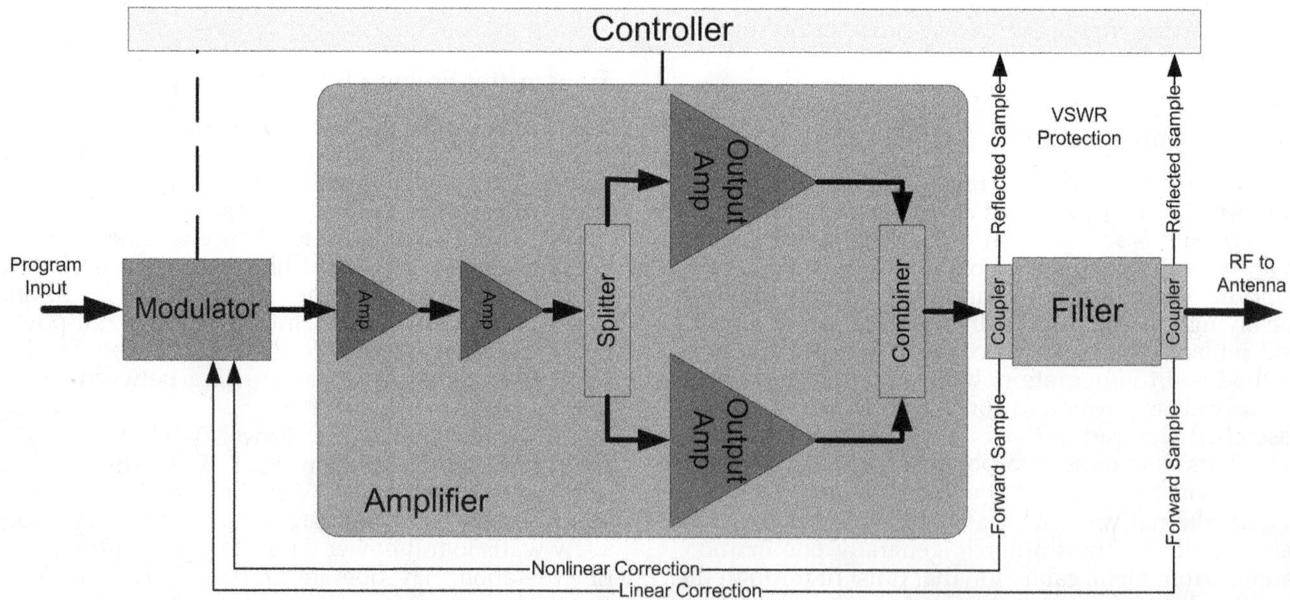

FIGURE 8.3-4 Digital TV transmitters, and all transmitters for that matter, have the same components; an exciter and amplifier(s) and the control, power supplies, and cooling to produce a high-level RF signal suitable to feed the antenna and supply the audience with enough decodable signal to watch TV.

ATSC 1.0 modulation supports a fixed bit rate and one uniform modulation method (although ATSC 1.0 mobile/handheld, known as ATSC M/H, sometimes informally but inaccurately referred to as ATSC 2.0, has a means of trading payload capacity for an increased mobile robustness). In the case of ATSC 1.0, the modulation is 8-VSB with a 19.39-Mb/s payload carried in a 6-MHz bandwidth. In DVB-T, the modulation is a bit more flexible; OFDM supports QPSK, 16QAM or 64 QAM, and several bandwidths, typically 5 (introduced as part of the DVB-H annex), 6, 7, or 8 MHz wide. DVB-T2 offers additional options of 1.7 and 10 MHz bandwidths.

In the second generation of digital TV (DTV), ATSC 3.0 (being standardized and adopted as this is written) and DVB-T2, among others, offer more complex and varied modulation options, allowing separating parts of the waveform to support both more robust modulation for mobile devices and at the same time higher bit rates for higher resolution video, but with the attendant increase in signal fragility. In addition, more efficient video and audio codecs such as *high efficiency video coding* (HEVC) are supported, as well as connection to the Internet to support interactive and personalization features. IP (Internet protocol) is favored over the packetized data of preceding methods to allow Ethernet connectivity.

The *exciter* or *modulator* both excites and modulates (see Figure 8.3-5). It *modulates* whatever TS information is desired onto an RF carrier in the desired modulation scheme. This modulated signal, in turn, is used to *excite* the transmitter's amplifiers.

All but the simplest modulators are built on a hardware platform that can accommodate a wide range of modulation schemes. Many current modulators are built with sufficient capability to handle the newer standards with the appropriate firmware. As this is written, several of the world's standards are being updated. Virtually all modulator manufacturers have announced support for everything and anything on the foreseeable horizon. Moreover, futureproofing is important, especially in the United States, where the reallocation of television spectrum is causing the replacement and upgrading of current transmitters, which will be followed by the release of ATSC 3.0.

FIGURE 8.3-5 This is a GatesAir XTE Exciter, which is ready for ATSC 3.0, etc. It provides sufficient RF drive to excite a transmitter's power amplifiers. Most of what we think of as a transmitter's capabilities, features, and functions are accomplished in the exciter. The silicon implementation for this exciter is in a field-programmable gate array (FPGA) used to give it a boot time of about 10 seconds.

Courtesy GatesAir

In Figure 8.3-6, program content can enter the modulator via transport stream over IP (TSoIP), which is typically Ethernet, or the TS can be carried as ASI or a SMPTE-310 stream over 75-ohm coax. The reason to have multiple physical inputs on an exciter is to provide for a fail-over option (backup TS on the second input) or to allow multiplexing. Many exciters allow one to *packet ID (PID) filter*, or select the desired packets based on their PIDs, out of an ASI or TSoIP that might carry other traffic, including other TSs, in order to transmit the selected program content. Most exciters do not have a built-in multiplexer that allows mixing of sources or have much in the way of TS monitoring or diagnostics built in. The expectation is that any multiplexers and TS tools exist prior to the transmitter.

Without a TS input, a transmitter has nothing to transmit. The standard response is to shut down and indicate a TS failure. Some modulators allow an internally generated TS composed of a PRBS (pseudorandom binary sequence) that contains no information to be transmitted either for the purpose of commissioning the transmitter or to keep the RF on air while the TS is missing. Automatic switching to PRBS upon loss of TS is generally an optional setting of the exciter. The PRBS signal will not make a picture, nor is it useful with anything but the basic RF transmission and instrumentation (you can set power, check for VSWR, etc., but a stream analyzer or TV set will see nothing in it).

It is very easy to exceed the payload capacity of a DTV transmitter exciter, especially with TSoIP and ASI streams with variable bit rate components that might add up to exceed the acceptable bit rate. Most exciters will simply truncate the excess bits without regard to their importance or which service is being impacted. In the mild case, this appears as an intermittent video or audio glitch that wanders between services. In more extreme cases, breakup and loss of audio on all services results from this truncation. Most exciters have some basic TS tools such as the ability to suppress null packets to make more room for useful data. A transmitter engineer's first thought when seeing these kinds of impairments is that the transmitter is failing, but it

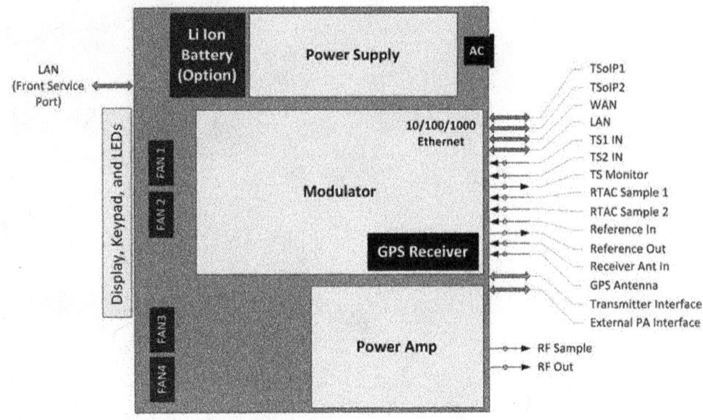

FIGURE 8.3-6 A simplified block diagram of the GatesAir XTE, multiple standard, DTV exciter.

Courtesy GatesAir

is more commonplace for the TS or its configuration to be the problem.

As mentioned the in analog TV transmitter chapter (Chapter 8.2), monitoring an OTA signal at the transmitter may be a challenge given the high RF levels, highly distorted RF environment, and poor ingress shielding of consumer TVs. A sample port, attenuators, as needed, and a reference TV, even if it is off the shelf, are useful. A professional demodulator to feed a stream analyzer is very useful, but may be economically impractical at smaller stations, translators, or boosters.

SFNs and Frequency Accuracy

Around the world, regulations as to the required frequency accuracy of a DTV signal vary from practically nearly none to accuracy standards that push the limits of affordable technology. ProTelevision™ (see Figure 8.3-7) specializes in OEM (as well as standalone) DTV modulators used by a number of manufacturers. They offer a standard 2 ppm (part-per-million) version as well as a 0.25 ppm and 0.01 ppm option for more stringent requirements.

Digital modulators almost all use a 10-MHz *reference* oscillator, generated internally or externally or locked to an external (usually GPS) source, to derive all of the critical frequencies and clocks required to regulate the modulator. The actual modulator stage might operate at 100 MHz, for example, and be mixed with a local oscillator that is phase locked to the reference oscillator to up-convert or down-convert the modulator's output to reach the desired transmit frequency. The reference oscillator might be used elsewhere in the modulator for anything from clocking processors and time keeping to stabilizing data IO interfaces.

Higher accuracy comes at higher cost. GPS is relatively convenient and inexpensive for similar frequency stability. GPS disciplined oscillators come in all price ranges, with matching performance characteristics. In the case of a single-frequency network (SFN), the one-pulse-per second output of the GPS is also required. GPS lock can be lost to interference (even jamming) or antenna failure (electrical or physical, including such annoyances as bird nests). TV channel 23's harmonics can interfere with GPS. Most GPS receivers are programmed to go into *holdover* mode where the oscillator uses its last and best training information to continue operating. A very good GPS may allow operation in holdover mode for a very long time. A not-so-good frequency and timing reference might drift enough in an hour to render the transmitter an interferer to other boosters in an SFN. Typically, it is better to shut down a rogue transmitter rather than allowing it to interfere with other transmitters in the distribution network, making it harder to receive the station in some areas than if the rogue transmitter didn't exist. In an SFN, it is critical to know the state of the GPS and what is a tolerable holdover time period. Ideally, the GPS can be fixed in time. Most telecommunications systems that require GPS stabilization utilize dual (as far apart as practical) antennas, receivers, automatic failover switching, reporting, and distribution amplifiers to feed the plethora of equipment, including SFN TV transmitters, at the shared

FIGURE 8.3-7 The ProTelevision™ OEM modulator is the core of many transmitters.

Courtesy of ProTelevision

site. Some TV transmitters designed for SFN use have dual GPS inputs, high grade oscillators to support long holdover times, notification of GPS failure functions, and automatic autonomous shutdown or power reduction at the end of the holdover period. Others for convenience may integrate a GPS receiver into the transmitter cabinet.

SFNs require that all transmitters have identical waveforms transmitted at the matching frequency. In order for this to work, a buffer stores the TS content until the exact moment it is to be released based on the GPS 1-PPS mark. SFNs are designed so that signals from one or more transmitters overlap and the signals add constructively to make a more receivable signal and thus penetrate places a single transmitter cannot reach.

Impairments

The exciter also supports the extremely important precorrection requirements of the transmitter. Later in this section, we'll look at the limitations on the RF amplifier and the distortions that are inherent in RF amplification. Further, the more efficient an RF amplifier is the more precorrection and higher RF drive levels it demands. The exciter then becomes a major part of the overall efficiency and *fidelity* of the transmitter.

The measurement of the fidelity of a digital transmitter is an expression of the variation of the waveform from the ideal. Digital transmission is deterministic—the RF energy has precise and well-defined values at defined periods of time. Imagine any digital waveform, whether it is OFDM, 8-VSB, or QAM, frozen in an instant of time. Each component of the waveform (usually viewed as a constellation) has specific desired values of amplitude and phase. Because we can measure the values and we know what they should be, we can compare the ideal and actual values against each other and achieve a value for the error. If we imagine moving forward in time just enough to arrive at the next symbol, the values have now changed, but their ability to reach the desired value is affected by how much they have had to change in the very brief interval between symbols as well as by other characteristics of the amplifiers. Because each symbol is different, we have to look at the average errors of all the components of the modulated waveform.

Lower fidelity results in reduced coverage area because the receiver has a more difficult job decoding the signal. Below the recommended maximum level of impairment, improvement in fidelity results in noticeable improvement in the link performance of the system. As the system reaches maximum fidelity, incremental improvement reaches the point of diminishing returns.

The modulation error ratio, MER, is a measure used to quantify the performance of a digital TV transmitter or receiver in a communications system. MER is expressed in dB, and modern transmitters are expected to have an MER of 30 dB or better.

Error vector magnitude, EVM, (on occasion also called *receive constellation error* or RCE), is another measure used to quantify the performance of a digital TV transmitter or receiver. EVM can be expressed as a logarithmic value in dB; however, it is more popular to express it as a percentage. A modern digital TV should have an EVM of less than 2%.

Signal to noise, SNR or S/N, is the classic expression of the value of the signal to the noise in the channel. It is defined as the ratio of signal power to the noise power, often expressed in dB. In TV transmitters, SNR is sometimes used to describe something different than MER or EVM. SNR tends to be used to describe the entire link between transmitter and receiver. Sometimes, SNR is used to differentiate a calculated and/or weighted MER from a measured value. There is no real convention for the use of SNR; however the choice between the use of EVM and MER as the desired measurement is often regional. Domestically, MER is used most often. The means of measuring and calculating MER, EVM, and SNR might vary, but in the end, all three express the fidelity of the transmitter and for our purpose are interchangeable.

Maintaining an acceptable MER is as important as maintaining the transmitter's power output. Poor MER performance (ATSC quantifies this as less than 27 dB) has a number of origins; transmitter phase noise, nonlinear distortions, linear distortions such as micro reflections, amplitude ripple and group delay, poor amplifier alignment, poor interstage impedance matching, interstage signal clipping, amplifier compression, etc.

The obvious considerations for digital decoding at the receiver are path impairments and power. The signal needs to be strong enough and the path needs to be clean enough for the signal to reach the receiver intact enough to be decoded. Noise, attenuation, and reflections impede reception. Practical receivers correct for linear distortions fairly well using adaptive equalizers, but the other component of MER is nonlinear distortion, which cannot be compensated for by the receiver. Arguably, the most important requirement for an effective transmitter is also the most difficult—nonlinear correction of the amplifiers.

Clipping

Many exciters offer an option to clip the highest peaks. The crest factor or PAPR is high for the COFDM modulation used in ATSC 3.0, DVB-T, DVB-T2, etc. It is not economically viable to dimension the headroom of the amplifier design high enough to allow for the highest naturally occurring peaks. It is preferable for both longevity (a very high crest factor is very hard on amplifiers) and performance to clip these peaks before they reach the amplifiers. Further, a peak affects the performance of the amplifier for some time after the peak has been experienced. If the clipping is done in the modulator, this limits the distortion to the exact period of the undesirable peak. It also limits the effects of the peak on the rest of the signal, reducing the overall noise.

Controlled clipping also makes it possible to shift the frequency distribution of the clipper noise to optimize the performance even further. While it is possible to manually set the clipper for best performance, it can also be done dynamically by some exciters. There are a number of trade names for this function that results in a higher operating level and thus higher power level in the transmitter.

Precorrection

All the power in the world is of no use if the signal is distorted to the point that it cannot be decoded. The amplifiers need to be linear and the filter and antenna system need to cleanly pass the signal. Practical amplifiers and filters will degrade the signal. The exciter can *predistort* the signal that excites the amplifiers in a manner that is opposite of the known detrimental effects of the amplifiers and filters, resulting in a clean output.

It is entirely possible to have an exciter that performs no precorrections. If the exciter is used to feed an amplifier operating in a very linear manner (and thus not very efficiently) and there is no filter, there is no need for precorrection functions. In some very low-power applications, where regulation permits, a very economical and feature-free DTV exciter is possible.

Most exciters perform the precorrections in a two-step process. The first correction is termed *nonlinear* because the correction is for the nonlinear effects of the amplifier(s). In nonlinear correction, a sample of the transmitter's output is returned to the exciter's *nonlinear sense input*.

The second correction compensates for the mask filter or any other filter such as a multi-station combiner. No filter is perfect, and in order to suppress undesired emissions outside of the desired channel, as a typical DTV mask filter does, considerable attenuation and group delay artifacts are seen, primarily at the edges of the passband. These *linear* distortions are compensated for by again either a fixed algorithm or a dynamic process. For linear correction, a sample of the filter's output is returned to the exciter's *linear sense input* to enable this function.

In certain applications, these corrections are performed manually. For example, if the transmitter is small and has few variables, a set predistortion algorithm might be applied at the factory that will work for that particular model of transmitter. This is a rare occurrence. Most precorrection is dynamic and automatic. Here a sample of both the amplifier's output and filter's output is compared to the desired modulated signal. A series of measurements are made and correction information is gathered to inform an algorithm that results in the desired fidelity at the transmitter's and filter's output.

Linear compensation in a *multi-station combiner* must be set one station at a time while only that transmitter is operating. When using a combiner, the sample of the output of the combiner is used for linear compensation by all of the transmitters, so during setup, the sample must be analyzed one station at a time because the objective is to reduce any out-of-band emission, and the other signals that are not under the control of the exciter will swamp out any useful information that the sample has. Once each transmitter's precorrection information is captured, the exciter linear compensation needs to be placed in *fixed* mode so that the exciter does not attempt to correct for linear distortions based on the sample information that is contaminated by the neighboring transmitters. Nonlinear compensation can be set to be dynamically *adaptive* if the sample at the output of the transmitter and the input of the filter is clean enough (where no significant signals from the other transmitters come back through the filter). It is possible to filter the nonlinear return sample so that residual energy from the other transmitters is suppressed to the degree necessary for the exciter to correct only for the transmitter it is driving. It is more difficult to filter the linear errors in such a manner. In both cases, the effect of the additional filter in the sense lines must be taken into account.

Digital TV exciters are usually placed on uninterruptible power, usually with the controller, even if nothing else in the facility is. Some exciters can take a long time to reboot and can become "scrambled" as a result of interruptions in power.

A basic function of the exciter is to measure the distortions of the system and display or report them. It is necessary that the nonlinear sample returned from the amplifier output coupler and the linear sample return are both within the window of acceptable RF levels. If the transmitter is operating at reduced power, there may not be enough sample signal to operate the precorrection. Typically a manufacturer will recommend that the samples be a few dB below the acceptable maximum when operating at full power. This allows the precorrection to operate over the widest range practical. Most current exciters provide a spectrum analyzer kind of view as well as the basic measurements of the shoulder levels and MER (modulation error ratio).

Both linear and nonlinear correction can be operated in one of three modes:

1. *Off/Monitoring*: Here there is no pre-correction applied; however, the measurement (monitoring) functions of the exciter are still active. This is basically a test function because it allows you to see how well the transmitter is performing before any precorrection is applied. It also lets you measure the effect of the linear and nonlinear precorrection one at a time or when both are active. Operation in this mode does not always produce an acceptable MER.
2. *Static/Fixed*: Precorrection is active and running on an unchanging file of parameters either captured in a setup procedure, set manually (rare), or provided by the factory. While the exciter is still monitoring the performance of the exciter, it is not dynamically adjusting the precorrection to respond to linear and nonlinear errors. As mentioned above, this mode is necessary for the linear correction when sharing an antenna. Experience with the system may indicate that this mode is best. It may be necessary in instances where adaptive correction proves to be unstable over some time. It can also be wise in some cases where the system may be more reliable when not depending on the hardware and software that allow for adaptive precorrection.

Generally, one captures both a linear and nonlinear precorrection data file for anticipated number of different power levels. These may be associated with a *transmitter profile*. If the transmitter is required to operate at half-power because of antenna icing or SFN issues, the exciter will need to load the precorrection data associated with that power level. Generally, this is automatic because the controller is either networked to the exciter or built into the same unit.

3. *Adaptive/Dynamic*: Most systems operate in this mode continuously. The mode also has a subset of options. For these reasons, a separate section will be devoted to adaptive mode.

There is a great deal of intellectual property involved with precorrection circuits, in particular those systems that reach better than typical MER and efficiency through adaptive precorrection algorithms. These are very proprietary and protected and appear under a number of trade names.

Adaptive Precompensation

In the real world, a transmitter, when operated in a temperature-controlled environment, will operate with no significant variations in performance over a long period of time. On the other hand, if an amplifier fails, the temperature of the filter changes significantly, or the antenna characteristics shift due to temperature or ice damage, the characteristics of the transmission system change. If a transmitter employs some form of smooth power foldback, the amplifier characteristics will change with the operating power of the transmitter. In these cases, adaptive precorrection dynamically adjusts the transmitter for the best fidelity (MER, EVM, or SNR). Precorrection circuits are not a savior, however, for an RF system that is out of tolerance enough that the transmitter folds back severely.

Further, small variations in characteristics over short or long periods of time can be compensated for.

For these reasons most transmitters are operated with dynamic, adaptive precorrection.

Depending on manufacturer, there are four basic adaptive modes. As above, the nonlinear and linear correction can be operated in each mode independent from the other. Two modes require that the transmitter operator enters a target or threshold value (there is usually a factory-entered default value that appears on startup) for the shoulder performance. Exciters often keep a log of this operation. The log is important because it will provide a history of the exciter's activities. The options include:

1. *Run-to-Target:* The current adaptive precorrector characteristic is applied (even if the transmitter is brand new or any historical data has been reset and the data file is essentially blank). The precorrector algorithm will run until the preset threshold value is reached for the spectrum shoulders. It may take most of an hour for the exciter to complete the corrections and reach the target. When the target is reached, the process is halted and the transmitter continues to use this correction data. The exciter will not rerun the correction process again until commanded to do so by the operator. The *run-to-target* mode is good for gathering the required information to create the data file that the precorrector can use in fixed mode. This process may be run for each operating power profile for a transmitter. The precorrection data can be saved to a file.
2. *Idle/halt:* In this mode, the adaptive precorrection has already been run and the adjustments have been made. The operator can reactivate the dynamic process at any time, and the transmitter will start with this correction data and "tweak" it as necessary. If there has been experience with an exciter "running away" (which might be caused by external factors such as intermittent contaminated samples caused by some interfering source, or the software has some condition under which runaway or some other undesirable state occurs), this mode is probably desirable. It's good practice to capture and save precorrection data files periodically.
3. *Auto-Run:* The exciter dynamic precorrection is idle until the monitor determines that the performance thresholds are not met. Recognizing the condition, the exciter executes a run-to-target operation and, when it is completed successfully, returns to auto-run. If the unit fails to make the corrections, the exciter generally halts the adaptive process and notifies the operator with an alarm. Under certain circumstances better results are obtained with this mode than any other. This is a mode that will also accommodate smooth changes in power levels.
4. *Continuous Adaptation:* Here the exciter continues to adjust the precorrection in real time on either the nonlinear, linear, or both correctors. This is also a mode that will accommodate smooth changes in power levels.

Arguably, the majority of transmitters are operated in adaptive–continuous adaptive mode for both nonlinear and linear corrections. Different manufacturers implement different precorrection features and algorithms, with distinctively different characteristics and effectiveness. If the exciter is provided by the manufacturer of the transmitter, there will generally be a recommendation as to the most valuable precorrection settings for a given environment and situation.

The logs for the exciter and experience over time will give you the information necessary to make the right exciter configuration decisions. Combined antenna operation, different foldback schemes, the stability and options for the TS, any SFN functions, the availability of a second exciter—all are part of proper exciter setup.

RF AMPLIFIERS

The RF amplifiers are the functional heart of the transmitter. What comes out of the exciter (without any precorrection/predistortion) is the desired RF waveform for the TV receiver, but the amplifiers take a very low-power signal from an exciter and raise it to the power levels that an RF system can broadcast to the desired audience. Practical amplifiers consume appreciable amounts of power and cooling, produce undesired RF products outside of the assigned bandwidth, and add noise and nonlinearities to the signal. As mentioned previously, the rest of the transmitter functionality is to provide the ability to clean up the signal irregularities of the amplifiers and keep them from harm.

In the analog transmitter section of this book, tube transmitters were discussed. Tube transmitters can certainly be used in DTV service. Many exciters will drive a tube system just as well as a solid state, though there are differences between the characteristics of the amplifiers.

For safety, convenience and total cost of ownership, most transmitters built as of this writing are solid state.

Figure 8.3-8 is the block diagram of a typical low-power transmitter. Each of the amplifiers is solid state, and they are combined in order to reach the desired power. There are multiple stages because these amplifiers are good for about 20 dB (or a 100-fold gain). Gain increases of more than 100X produce issues with stability (self-oscillation) and solid-state devices used in this application are largely limited to gains in the 100-times range.

Power Control and Driver Amplification

In the diagram, you can see that the exciter's low signal level is fed to a *pre-amp* and *variable attenuator*. This is the circuit that controls the gain of the amplifier section of the transmitter and buffers the output of the exciter with a small preamplifier that presents a constant load back to the exciter. The attenuator is controlled either by an analog control voltage or digitally in the case of a programmable attenuator. If the transmitter is in active power control mode (many are all of the time), the output power of the transmitter is monitored and a servo circuit compares the measured TPO with the *set point*. If the power level drifts, the servo circuit adjusts the attenuator to maintain the desired constant power. Additional circuitry may override the set point and add attenuation to reduce power in the event of high reflected power or excessive heat in the amplifier. In all cases, this is the point where the entire transmitter's output power is being controlled.

There is an alternative design where each final amplifier module has its own power control attenuator on board. In this case, a common control bus or individual commands set the gain of the individual amplifier modules.

A TV transmitter typically has a minimum operating power in order to stay within the dynamic range of programmable or variable attenuators and the dynamic range of couplers and analog-to-digital converters.

From this point forward, the gain of the transmitter's amplifiers is essentially static. Only small individual adjustments (if any) can be made. These adjustments are generally for the purpose of balancing amplifiers. These adjustments are made at the factory or in the field when replacing a power amplifier module or subassembly.

The next stage is the *intermediate power amplifier* (IPA). While there may be multiple stages of intermediate amplifiers, smaller solid-state transmitters usually have only one. Some high-power transmitters

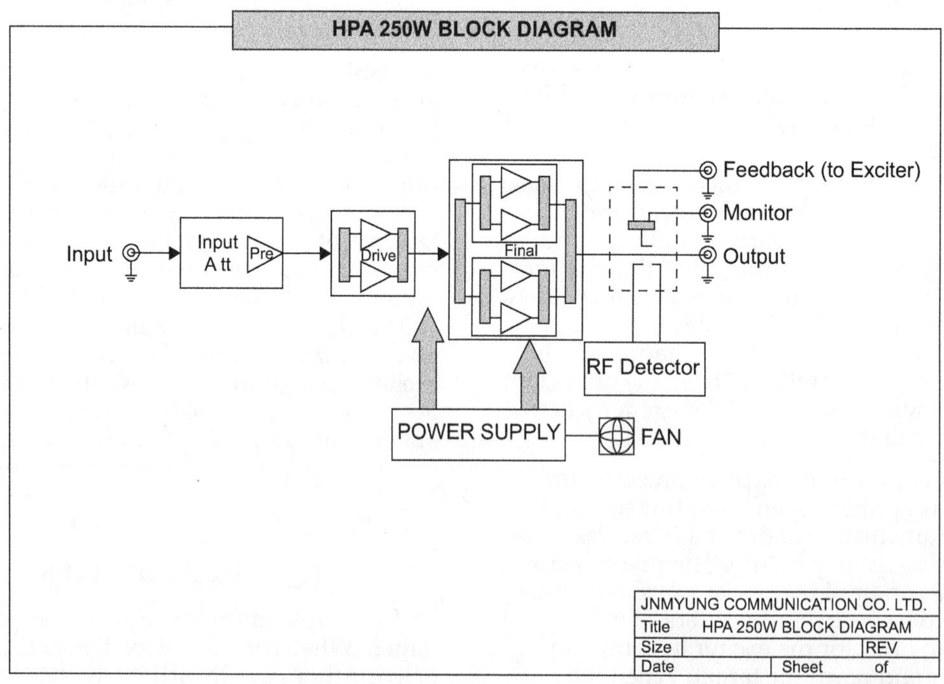

FIGURE 8.3-8 Typical 250 watt digital TV RF amplifier section block diagram.

Courtesy Nautel

need an additional stage to feed the larger number of individual power amplifier modules. The IPA is a good place to monitor not only the input from the preamp/control stage but also the power being sent to the *final amplifiers*. Troubleshooting solid-state transmitters is less complicated than troubleshooting tube amplifiers. While a tube might have any number of issues ranging from bias supplies and cavities to filament supplies and control, a solid-state amplifier is pretty straightforward. Troubleshooting solid-state amplifier modules usually consists of checking the RF power in and out of each stage, followed by checking the DC voltage and current of the devices. It is a reasonable and ordinary procedure to insert an appropriate power coupler temporarily between stages to check for power. To properly maintain a transmitter, one should have the right set of cables, connectors, and adaptors and suitable power meters, pads, and spectrum analyzer. It is also good to have permanent monitoring and logging of the interstage RF levels integrated into the transmitter, but it is unlikely that all the points necessary to troubleshoot will be monitored continuously.

An extremely useful tool is a *USB power meter* (see Figure 8.3-9). A TV transmitter has a large number of (mostly calibrated) couplers at various locations and *monitoring ports* that are placed throughout the system, intended for power meters and spectrum analyzers. USB power meters that are useful at VHF and/or UHF frequencies cost about $1,000. Having a single device that can be moved from point to point can validate the functionality of the installed sensors. A good power meter can be used to calibrate metering integral to the transmitter when used with calibrated couplers. Couplers are calibrated for the frequency of operation at the factory, can be either unidirectional or bidirectional, and are labeled with the coupling factor and directionality. Power meters typically have a maximum input power of a watt or less, so often need to be used with calibrated attenuators to avoid damage to the sensor.

One approach that is fairly common in most contemporary designs is the use of an integrated circuit RMS/peak RF detector and buffer. These are very affordable, and the voltage output from the device can drive an analog-to-digital converter, often built into an embedded controller or microprocessor. These have a fairly broad dynamic range, age well, and are so affordable that they can be used widely throughout a transmitter for RF level monitoring, troubleshooting, and protection.

FIGURE 8.3-9 A typical USB power meter is powered by the computer it is attached to and accepts power to be measured via typically an N-connector.

Photo courtesy of Fred Baumgartner

The IPA and any preamp have one additional critical requirement—they must pass the precorrected signal from the exciter to the final amplifiers with a minimum of distortions. This precorrected signal from the exciter has a much higher peak-to-average power ratio than the transmitter output. The preamps and IPA are designed to maximize the purity and headroom of the signal while adding a minimum of noise and distortion. Conversely, the output power amps are designed to maximize power and efficiency.

Splitters and Combiners

Solid-state power amplification requires a series of amplifiers operating in parallel. In order to do this, the drive power from the IPAs needs to be distributed to each amplifier module, and the resultant power needs to be combined to make up the final TPO. Splitters and combiners can be of any number of designs, but most are a version of a *90-degree splitter/combiner*. This is also known as a *hybrid combiner*. When each output port of a splitter represents one-half of the input power, it is called a *3 dB coupler*. Generally, a splitter is a combiner used in reverse, and vice versa. Electrically, the splitters and combiners that make up a transmitter are the same, though, and each is sized for the power intended. The first set of combiners in Figure 8.3-10 must handle a thousand times the power that the last set of splitters that drive the amplifiers are required to handle.

In the drawing, the 90-degree splitters and combiners are *four-port* devices. When used as a combiner, two ports are inputs, one port is the sum of the two inputs, and the fourth port is the difference of the two input signals that will be passed along to the *reject load* or *absorber*. When the two inputs are equal in power and in-phase, all of the power reaches the output port and none goes to the reject port. Note that "in-phase" in this case means that the amplifiers are operating 90-degrees apart in in-phase to accommodate the 90-degree splitters and combiners.

When used as a splitter, half of the power is directed to one output port and the other half to the other output port. Any imbalance in the load results in some power being reflected into the reject load and back to the input port. For example, if 8 watts is applied to the "signal in" of Figure 8.3-8, two 4-watt signals should be produced for the next level of splitters, where it is again divided into four 2-watt feeds, which are then divided once more into eight 1-watt feeds to the individual amplifiers. In this drawing, each of the amplifier assemblies creates 40 dB of power gain. Each 1-kW amplifier is combined with its neighbor to produce 2-kW at the first step, then 4-kW, and finally 8-kW. This is a typical transmitter design. Clearly, this is a hypothetical model, as there are small losses in the splitters and combiners.

If the top amplifier in the drawing were to fail, the combiner it feeds would be out of balance, and 500 watts would appear at both the absorber and 500 watts at the output. Likewise, the next combiner would attempt to sum 2-kW with 500 watts, and the result will be 2,250 watts out and 250 watts to the absorber. The final combiner will see 4,000 watts and 2,250 watts;

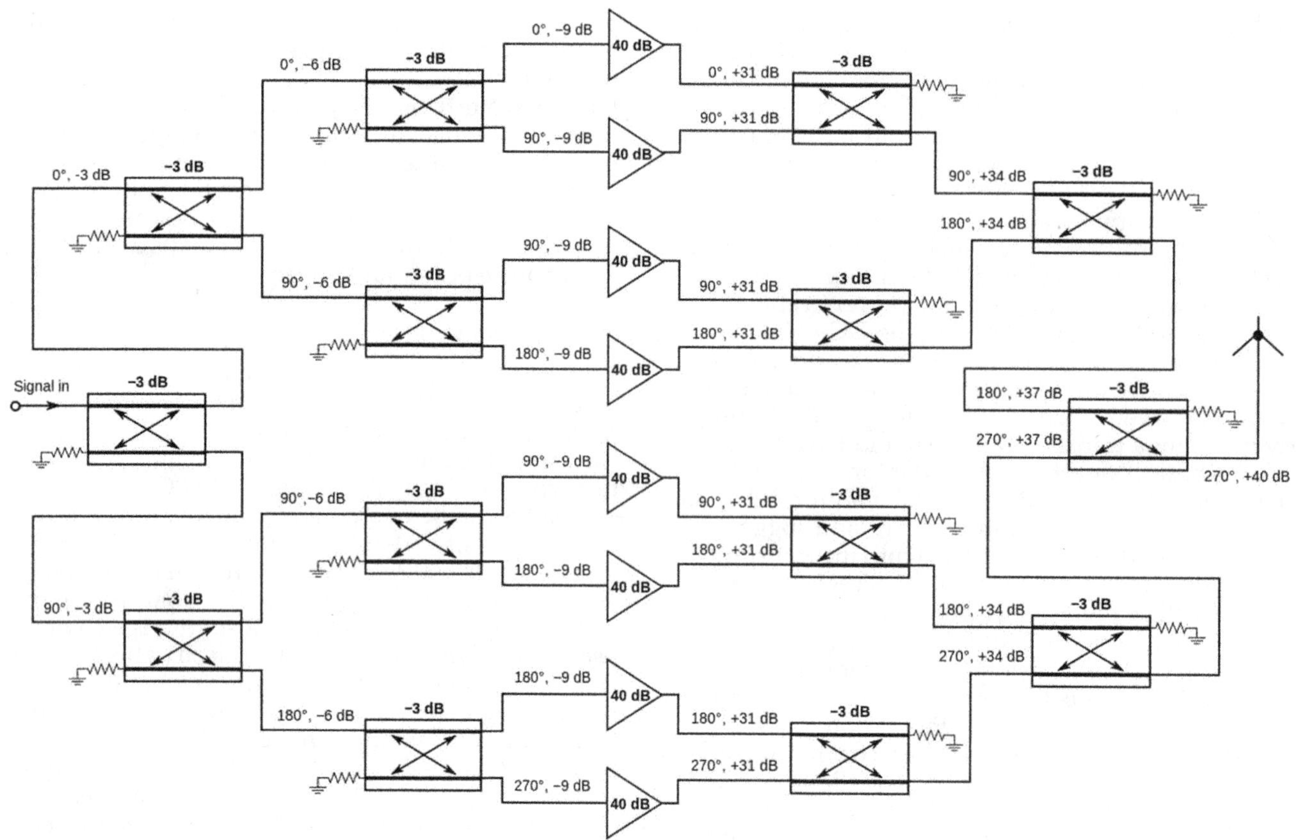

FIGURE 8.3-10 The basic architecture of a solid-state transmitter is to split the drive power to feed a series of amplifier modules, which are then recombined to produce a high-power output.

Diagram courtesy Spinningspark at Wikipedia

thus, the final output absorber will see 125 watts, and 6,125 watts will go to the output.

In this configuration, it should be noted that the loss of one 1,000-watt amplifier out of eight results not only in the loss of 1,000 watts from the failed amplifier but an additional 875 watts in combiner losses as they become unbalanced. Again, the additional real-world losses of the combiners themselves are ignored. As a practical matter, the output will be somewhat less than the theoretical 6,125 watts.

In our example, the 8-kW of amplification may not be enough to meet the desired TPO. In theory, we can combine any number of 8-kW amplifiers to reach the desired value, although, as the combiners become more numerous, the loss they introduce also accumulates. In typical construction, a rack of amplifier drawers is combined within the rack or very near to it (see Figure 8.3-11). The absorbers can be mounted anywhere, including outside the transmitter building, or in vertical space above the transmitter. Each rack then feeds the switching/combiner system. Switching is often used to allow reconfiguration of the combining network so that a group of amplifiers can be taken off-line for maintenance without the penalty of reject power losses.

In the amplifier architecture shown in Figure 8.3-10, the amplifiers are operating at various phase shifts from the input and output signal. Like the spark plugs in a car, each amplifier must be connected to the proper phase ports in and out, or serious problems can result. If an amplifier is connected to an input that is 180 degrees from what is desired, the combiner will put all of its power into the absorber and none to the antenna. Here, an absorber that, if properly connected, would see a maximum of 500 watts (as in this example) will see 2-kW. The splitters insert a series of phase shifts, and the combiners, being a mirror of the input splitters, remove the phase shifts. This is a function of the splitter design chosen, but in this case it has the added benefit of allowing some of the degradations caused by a given amplifier to be canceled out by another amplifier operating 90, 180, or 270 degrees out. These canceled products appear in the reject loads along with any results of imbalance in power and phase. The objective is to have as little rejected power and as much output power with as little distortion as practical.

Absorber heat can be significant when the transmitter is in a partially failed state. Thermostatically controlled fans and water cooling are usually employed to keep the system from overheating and being damaged. Often, the sound of the transmitter room is decidedly different when the absorber fans are running, so a transmitter engineer might know upon entering the room that a module has failed.

FIGURE 8.3-11 Multiple amplifier groups are combined to reach the desired level. There are three racks of amplifiers, where the lower and upper groups of each are brought together into three hybrid combiners immediately above each rack; the absorber for each is the gray loads behind the transmitter. The three outputs are combined in a three-way combiner, which in turn outputs to the large black mask filter on the floor near the first three absorbers. This is a factory setup, so a water-cooled load and an air-cooled large load are used for test.

Photo courtesy GatesAir

Replacing a module in the field generally means that the engineer may need to fine-adjust not only the gain of the replaced module but also the phase. Somewhere in the transmitter circuitry, often in the amplifier drawers, there is some means of making small changes to gain and phase of the amplifier, although some equipment will automatically self-align. The manufacturer will likely have a procedure to do this in the field when a component is replaced or periodically as performance drifts if required.

In a practical transmitter, the line length of the larger, less flexible, transmission lines from the output of the amplifier modules is cut so that the output ports are made as convenient as possible for installation. The resulting compensating phase shifting is done by cutting each of the drive lines to the desired length. This results in various drive-line lengths with various delays and phase shifts, so the input lines, which are much thinner and more flexible, are cut to compensate for the varied drive-line lengths. They are often specific to the amplifier they serve.

There are many splitter/combiner designs. Transmitter manufactures might use a five-port (one in, four out) 6-dB splitter or combiner when appropriate. The *Wilkinson power divider* is popular, as well as waveguide couplers of various designs. On circuit boards, the *strip line* family of *parallel transmission line* couplers is used for most sensing, combining, and splitting functions.

Splitters and couplers all introduce some *insertion loss*, which results from the resistance of the materials they are constructed of. A good practical 3-dB splitter is closer to a 2.9-dB device. Combiners are passive devices. Design and geometry limit the useful bandwidth of the splitter or combiner. Extending the bandwidth generally comes with a cost in size and efficiency.

Amplifier Design

Practical solid-state amplifiers are constructed as a printed circuit board (PCB, or more accurately printed wiring board [PWB], because there are no components embedded in the *substrate* in most designs) mounted on a heat sink or cold plate. The substrate material is often polytetrafluoroethylene (PTFE), which is low dielectric loss circuit board material for VHF and UHF applications that has very low moisture absorption (0.01%) properties and is mechanically workable. Teflon™ is probably the best known PTFE, although substrates suitable for UHF transmitter amplifiers often include a ceramic content for better thermal and dielectric characteristics.

Figure 8.3-12 is a *design reference board* from Ampleon™, a major supplier of the active devices used in TV broadcast transmitters. The same and similar transistors find use in telecommunications, military, medical, and scientific applications. Transmitter manufacturers use the design reference board as a base for their, often much enhanced, designs. The photo shows the strip-line power splitter driving the FET's gates. A phase shifter and the power combiner at the output of the device's drain provide the amplifier output.

The actual amplifying device is a laterally diffused metal oxide semiconductor (LDMOS) field effect transistor (FET). The high breakdown voltage of LDMOS

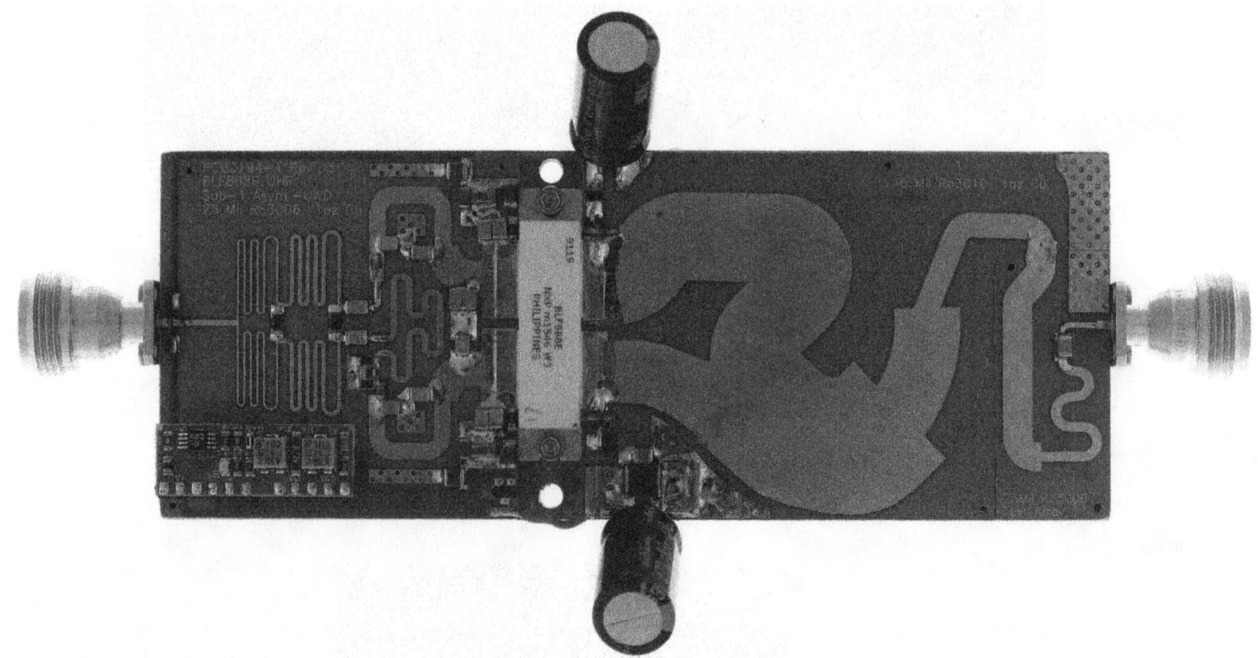

FIGURE 8.3-12 The basic component of the solid-state RF power amplifier is the device. In this case, an Ampleon™ LDMOS power transistor (bearing the NXP™ label) is shown on a design reference board.

Photo courtesy of Ampleon

is one of its most important advantages and why LDMOS dominates both VHF and UHF broadcast transmitter designs. For a given output impedance, the power output is the square of the voltage swing; therefore, there is a 12-dB power advantage going from 12 to 50 volts V_{dd}, the supply voltage employed in most broadcast transmitters. LDMOS utilizes epitaxial silicon, low-doped P-type layers grown on low-resistivity (i.e., highly doped) silicon wafers. Either a diffused sinker or a trench etched through the epitaxial layer is used to ground the source to the substrate. The source is grounded to the baseplate and typically operates at 0 volts potential. Most designs bundle two FETs together in one package, often a SOT-979 (ceramic) or SOT-1223 for lower-power devices. The catalog price for common FET devices used in broadcasting runs about $225 for small quantities. As this is written, at least two manufacturers have announced the availability of 700-watt CW devices for 600-MHz operation.

Very noticeable on LDMOS amplifier boards are the two large electrolytic capacitors, one for each device, which provide a low impedance current reservoir. The capacitors are chosen to handle the heat produced by the LDMOS devices and will be more than adequate over the lifetime of the amplifier.

Also notable is the number of components dedicated to the input and output matching networks and *broadbanding* the amplifier. LDMOS device operating impedances range from 50 ohms to much lower, while the entire amplifier pallet will have a 50-ohm input and output, courtesy of the matching networks.

Practical amplifiers can include means of measuring and monitoring operating parameters as well as control. This may be located on a daughter board or built directly on the amplifier substrate.

RF pallets are circuit boards mounted securely with thermal conducting compounds, glues, and/or screws to a machined, typically aluminum bottom plate that mechanically stabilizes the unit and carries heat to the heat sink or cold plate. The pallet can be separated from the cold plate or heat sink should it need to be replaced. Most pallets are designed with two devices and operate either as a push-pull Class AB amplifier or a Doherty configuration. The classic design takes advantage of the higher efficiency of AB operation into the nonlinear upper ranges of the device. Push-pull cancels some of the undesirable characteristics of Class AB operation and results in a more linear combined operation.

There is a circumstance where a Class A, highly linear but rather inefficient design is useful as a final amplifier. In a very low-power application, it may be more economical to use a simple exciter lacking the precorrection capability to accommodate Class AB or Doherty amplifiers. Adequate performance is achieved at low cost and complexity.

As this is written, Doherty designs are becoming more available as new LDMOS devices better support this scheme. Doherty designs employ a Class AB amplifier combined with a less linear but higher efficiency Class C *peaking* amplifier operated at 90 degrees out of phase from the primary *"carrier"* amplifier. This is a convenience, because 90-degree splitters and combiners are less complex, broader band, and easier to implement, and it also serves to conjoin the amplifiers in such a way that minimizes distortions.

The input network sets the phase shift and drives both amplifiers. The output network combines the outputs compensating for the phase shift. The Doherty design first was used for high-power AM transmitters in the 1930s, where a *carrier amplifier* did most of the work and a *peaking amplifier* was utilized in addition for handling modulation peaks. Digital transmission revived the Doherty design because the power required for digital modulation peaks is needed only a portion of the time. When not in use, the peaking amplifier is biased to idle. Additional efficiency can be achieved by actively controlling V_{dd} for the device fast enough to be available when peak power is required and to drop voltage to the device when it is not needed. While the efficiency of a Doherty design is higher than a classic AB design, the power capacity of a given pair of transistors is less so—after all, the peaking amplifier has a low duty cycle and thus contributes less to the overall output. For this reason, some manufacturers allow the user to select either AB or Doherty operation at their option.

TV transmitters are frequently built to handle either the OFDM modulation found in ATSC 3.0 and most other systems, including the DVB-T family, or the 8-VSB of ATSC 1.0. Power ratings for the same amplifier will be different depending on modulation. Generally, the RMS power rating of an amplifier is higher for 8-VSB, because the peak-to-average power ratio is lower than OFDM. Keep in mind that if a transmitter is intended for ATSC 3.0 utilization in its lifetime, the amplifiers need to be sized accordingly or expandable to reach the desired ATSC 3.0 peak power levels.

As this is written, Doherty designs are producing a DC-to-RF efficiency approaching or just into the 50% range (this is not AC-to-RF efficiency, which is discussed later). Device and transmitter manufacturer design roadmaps call for improvements in efficiency and power density, not only for devices but also for overall transmitters. In much of the world, the cost of power is high enough that increases in efficiency even in low-power transmitters is a much-desired requirement by purchasers.

Figure 8.3-13 shows an amplifier drawer with three pallets in parallel, each using two 50-volt ultra-wideband LDMOS devices in a GatesAir customized broadband Doherty implementation. The useful frequency range is 470–750 MHz.

Power supplies are mounted either below the amplifiers or externally (the module pictured uses external supplies). Maintaining transmitter efficiency over a wide bandwidth is difficult. Some manufactures offer lower-efficiency, very broadband units as well as "banded" higher efficiency units that work only on a narrower portion of the UHF TV spectrum. VHF transmitters can be even more frequency specific. Amplifiers that are useful for both VHF Band I and VHF Band III are rare if they exist at all.

Repair of a solid-state transmitter is very different from replacing a tube or repairing a tube transmitter, because solid-state transmitters are module oriented as opposed to component oriented. Replacing modules is considerably less of a task than troubleshooting a high-voltage tube-based amplifier. Today, the standard model, where practical, is *depot* repair. One sends a bad amplifier in to the repair depot to be rebuilt and sometimes retuned. The more versions of an amplifier (narrowband versus broadband) that must be supported, the more expensive that practice is.

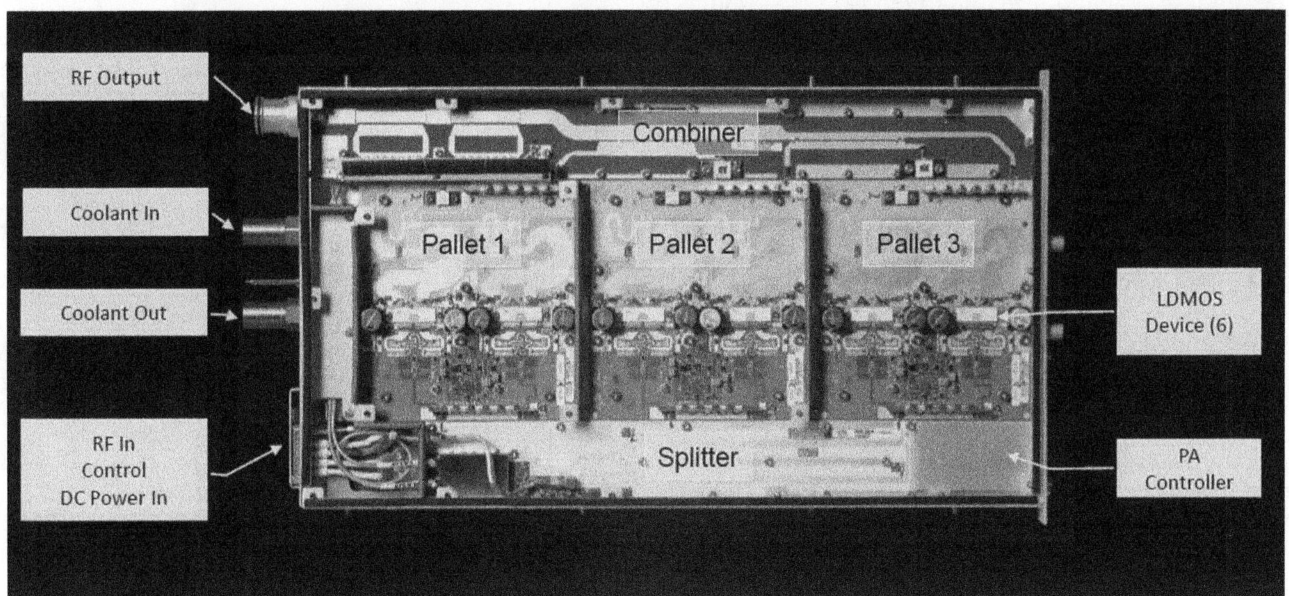

FIGURE 8.3-13 This amplifier module fits into a standard 19-inch rack with cooling, power, RF, and control connections on the back apron. This unit has three pallets with two amplifier devices each to achieve 900 watts of ATSC 1.0 power or 600 watts of OFDM power.

Photo courtesy GatesAir

As the spectrum allocated to UHF TV worldwide is reduced, the decision on how to support the bandwidth is made easier in favor of broadband units only. One reason to broadband as much as is possible in a digital transmitter is that in its approximately 15-year useful life as a main transmitter and potential additional 15 years of backup service, the frequency allocation or station at which the transmitter is used is likely to change.

In parts of the world where depot-level repair is not practical—the cost of shipping long distances is high or the import/export tax structure makes shipping a faulty unit for repair too costly—component level repair is required. Unsoldering and resoldering RF power transistors requires some skill, patience, care, and proper equipment. Identifying a bad transistor is fairly straightforward. If there is V_{dd} voltage and notably lower-than-expected current, one or both FETs in the pallet have likely failed. If the amplifier has no built-in instrumentation, an AC/DC clamp Amp. meter placed over the 50-volt supply line to the device is appropriate. It is important to check the bias on the FET's gates.

Some transmitter amplifier designs are *hot swappable*—an amplifier drawer can be removed while the transmitter continues to operate. Some modules include an indicator that identifies an amplifier fault. This makes maintenance fairly straightforward; remove and replace the faulty module per the manufacturer's instructions. Some modules use aligned connectors for everything including cooling, AC power, RF in/out, and control. In this case, simply removing the front mounting hardware allows the module to be pulled straight out. Weight is an issue and can be a safety issue for a person working alone. Cold plates and especially power supplies can be heavy. A typical module of this sort is in the 70-pound (30-kg) range. Other modules require that RF connections and or water connections are first disconnected and the drawer *isolated* (depowered and functionally removed from the RF circuit). Cooling water systems normally employ *dripless connectors* that should mate and disconnect without dripping. Still, good practice might dictate that water valves for the amplifier being removed are turned off first and a cloth to absorb any drips is strategically placed. Some manufacturers ask that the cooling fluid be drained from an amplifier before shipping, which requires a jig consisting of a pair of connectors and some tubing and a bucket to collect the fluid.

CONTROL SYSTEMS

In many ways, the transmitter control systems are the most important and most difficult-to-design piece of a transmitter. Since RF amplification devices have higher power density, a fault in the system has the potential to destroy or severely damage the amplification devices, combiners, and other RF system hardware in fractions of a second.

Transmitters apply power to antenna systems. Antenna systems may be trouble free for decades, but eventually lightning; vandalism; leaks in transmission line and antenna components caused by wind vibration, rain, ice, and thermal cycling; decaying fiberglass radomes; loosening RF bullets that connect the transmission line internal conductors; or wear and tear from constant RF voltage stress and heating will eventually cause an antenna fault. If not the antenna, the array of high-power filters, RF switching, and general plumbing between the transmitter's output and the antenna's transmission line input may also fail. The most destructive scenario is when some part of the system fails and the transmitter continues to provide power. In microseconds to minutes, components from transmission line to transistors and tubes burn up or melt. In high-power systems, these can be quite spectacular and dangerous.

There are advanced systems that quickly and even preemptively detect faults in the transmission system. These can be very elegant and ingenious. Still, their ultimate job is to tell the transmitter's control system to shut down. Most transmission systems largely rely upon the transmitter to detect the fault and respond.

The control system has to protect the transmitter against faults and perform the basic functions of output power regulation and adjustment—turn the transmitter on and off, etc. These basic functions can occur with some delay, and many functions require gathering data by polling operating parameters and calculations, which takes time. Conversely, fault protection does not have the luxury of time. For that reason the two systems are typically separated.

Fault protection is usually accomplished by relatively simple, fast-acting, single purpose circuits with sensors that almost instantly trigger protection. Typically, a transmitter will have VSWR protection consisting of an RF coupler that detects high VSWR and immediately and unconditionally shuts down the associated amplifier. In solid-state transmitters, there may be VSWR protection at each power transistor, each module, and each combiner as well as the transmitter's last output coupler. These couplers are often directional couplers used for monitoring and measuring the RF at that point. It is typical to have a coupler with several calibrated monitoring or sensing outputs in the −25 dB to −50 dB range.

Recovery from a fault should be done in such a way that further damage to the antenna, transmitter, or other systems is avoided. Some damage can be cumulative. An arcing transmission line component will become more damaged each time sufficient power to initiate the breakdown is applied. This can in turn cause future problems or other problems if metal or insulating fragments fall into places where they can cause harm or seals are stressed that will eventually leak.

Fuses are employed liberally throughout a transmitter. Transmitters are designed with the concept that any given fuse will never open in the ordinary life of the transmitter. Fuses are for safety and limiting damage, in particular fire, after a component or subsystem has failed catastrophically. An RF transistor can fail in a shorted condition, opening its related fuse to avoid damage to the power supply system and isolate the short. "Self-resetting" fuses are rarely found in a transmitter. Fuses are especially useful for troubleshooting and almost never fail without reason, even if the reason is intermittent. Transmitter fuses are often specialized with fast or slow blow times and features such as sand-filled cartridges that self-quench an arc upon opening.

Human safety is also a major concern. Transmitters of any reasonable size and design support an "interlock loop," so called because it is usually a string of switches in a series loop, forming a large "OR" circuit. If any switch is open, the transmitter interlock is "opened" and the transmitter will not transmit. Sensing switches on doors that protect people from the high AC and RF voltages inside a transmitter, position-sensing switches that will keep the transmitter from operating while an RF switch or patch is out of position, and environmental (smoke, water, temperature) sensors are typically included in the interlock loop. There may be dozens of safety and status switches in the interlock loop, any one of which will keep the transmitter off-line. The better interlock systems will indicate which switches have opened to speed up troubleshooting. The worst systems are undocumented strings of switches, which are often installed in external (to the transmitter) RF switching. These interlock systems can take some time to unravel before fixing, especially when the only indication is the transmitter's "interlock open" status indicator, sometimes buried deep in a menu.

The basic control functions are typically accomplished by a microprocessor-based controller or controllers that poll the transmitter's operating parameters and external commands, then issue commands to the various subsystems. Basically, these are a combination of servo functions and sequences. Turning a transmitter on executes a routine that typically brings up the cooling system, checks its operation, brings up exciters and other subsystems power, loads parameters into digitally controlled attenuators and programmable oscillators, and ramps up the power supplies, all the while monitoring the transmitter's performance. Most control systems also slowly ramp up the transmitter output power controlling the attenuator that is associated with the final amplifier drivers.

Transmitter manufacturers go to great lengths to test transmitter fault protection systems, emulating shorts and opens in the antenna systems, lightning strikes, RF arcing, and major faults in the supply of power. Many manufacturers encourage buyers to attend their new transmitter's acceptance testing, done on the factory floor or at the station as part of the installation and commissioning. Typically, the testing requires two to three days. Ideally a fault at any point in a transmission line, resulting in any vector of return power, will fault the transmitter. Transmitter manufacturers typically have sections of transmission line with holes at close intervals that a shorting bar can be placed in to accomplish this test. The method emulates both open and shorts at any distance and vector. A portable fault emulator is necessary in the field.

Foldback

When an antenna ices or experiences extreme heat or cold, or there is some marginal damage or irregularity in the antenna system, a practical antenna will reflect some of the power that was to be radiated back to the transmitter. A perfect antenna system would have zero watts of reflected power regardless of the transmitter forward power. Most transmitters have a dual (forward and reverse) directional coupler at the output of the transmitter and the filter. Usually these are the same calibrated high-quality couplers that provide the sense samples to the exciter's precorrectors. Consider a typical set of readings:

Transmitter Power Out:	1,050 Watts Forward Power	15 Watts Reflected Power
Filter output Power	1,000 Watts Forward Power	5 Watts Reflected Power

In this example, the antenna is reflecting back 5 watts of the 1,000 watts it is being fed by the filter. This can also be expressed as a voltage standing wave ratio, which in this case is a VSWR of 1.01:1 (spoken as "one point oh one to one"). For interest, return loss (in this case 46.02 dB) is another way of expressing the ratio of forward to reflected power used with TV transmitters in a few instances.

VSWR should remain constant as the power is increased or decreased. The reflected power marches in step with the forward power. Double the forward power results in double the reflected power as long as the ratio remains unchanged.

In the example, the transmitter is putting 1,050 watts into the filter, of which 50 watts is lost. Some of this lost power is undesirable out-of-band products the filter is absorbing or reflecting back to the transmitter (as it was intended to do), along with the additional power that the antenna is reflecting back. This makes up the 15 watts of reflected power we see at the transmitter. The rest of the loss is the inherent losses due to the resistance of the transmission lines, switching, impedance transformation, and filter components, which becomes heat. The transmitter is seeing a VSWR of 1.03:1.

Let's say that the transmitter can safely handle 50 watts of reflected power, or a VSWR 1.11:1. In order to keep the reflected power absorbed below this 50-watt maximum, the transmitter has to be reduced in power.

There are basically three methods of dealing with this. The simplest method simply shuts off the transmitter and waits for an operator to restart it, preferably at a reduced power.

The second is to *fold back* the transmitter to some fixed level. Typically this is half-power. If 50% power fails to get the transmitter below the acceptable level of reflected power, the control system may be designed to drop the power one more step to 25%. If that power level is also not acceptable, the transmitter shuts off, showing a VSWR alarm. Controllers can have automatic routines that attempt to restore full power over time. The use of fixed levels of operation works well with precorrection that is set up at discrete operating levels or *profiles*. The number of steps and the means of recovery options vary from transmitter manufacturer to manufacturer.

A more elegant foldback would be to continuously adjust the transmitter forward power to whatever point will result in either the limit set for reflected power (in this case, 50 watts), the limit set for TPO (in this case, 1,050 watts), or some other limit such as device operating temperature. In this manner, the transmitter will operate at the highest allowable power up to its desired set point, whether this is a minor power reduction or a major reduction to just a few watts. This can be a challenge for the exciter's precorrection because it needs to track the changing transmitter power and thus amplifier characteristics.

Foldback activity is not desirable when there is a fatal fault. If the tower collapses or an arc develops somewhere in the system, a foldback system that constantly attempts to drive power may cause additional damage. For this reason, as desirable as it is, foldback circuitry should be intelligently implemented.

Internal Communication

Communications within a transmitter offer a special challenge. There can be high RF fields that can interfere with the low-level signals used for communication. For this reason, almost all internal communication throughout the RF sections is via low data rate serial connections that are heavily filtered and shielded from RF. Expect to see a CAN (controller area network) bus, a two-wire structure first used in automobiles . . . in a similarly high EMI (electromagnetic interference) environment. Usually, CAN bus voltages are high (12 volts) compared to most communications circuits. RS-485, some proprietary methods (in some a DC voltage is distributed to control power output from modules), and even optical links are in use. Don't expect to see too much Ethernet, I²C or RS-232. For simplicity, RF modules in particular are often connected via a control *bus*. Placing all units on a bus means that each module or device must have a unique identifier or address if they are to be looked at or controlled or adjusted individually. The downside of using such a system is that a short on the bus halts all communications to all modules on that bus, but being able to "daisy chain" each module, rather than a *home run* from each module to the central controller, simplifies wiring immensely.

In the non-RF sections, expect all of the ordinary interfaces. Likewise, external control of pumps, RF switching, resettable breakers, etc., is often *Modbus*, a protocol that is almost universal in programmable logic controllers and process controlled devices.

Automatic Transmitter Control

Control systems range from autonomous to operator dependent or "conditional." Most modern transmitters exercise some degree of autonomy, adjusting power and the gains of subsystems to compensate for changes in temperature, antennas, and power supplied. Most operators prefer to manually intervene after an "event" to minimize damage and maximize understanding.

A fully autonomous transmitter is desired in applications where an operator cannot communicate with the transmitter or there is no staff to do so and the transmitter is so isolated that getting a technician on site requires considerable time, expense, or danger. In this application, a transmitter control system might be set up to do everything possible to stay on the air at whatever safe power level is available. It probably is also "trained" to attempt to restart and reset subsystems in an attempt to keep it on line at full power. These systems can range from quite elegant and redundant to a very simple single translator.

Many TV stations want an operator who can check weather, primary power, and fuel levels and look at logs in order to make the important decisions to raise power, switch antennas, switch to a backup transmitter, or switch to a backup program feed. Often, if a transmitter engineer can be reached, it is the transmitter engineer that makes any adjustments or decisions, sometimes via a home computer or mobile device. An operator might know that antenna icing is occurring, or the outside temperature is high enough that reduced power operation is desired. Simpler autonomous systems are often unaware of their environment. More complicated systems might be good for unattended operations, but that complexity may create unanticipated additional problems.

Transmitter control systems are usually configurable, whether they are integral to the transmitter or a more involved external monitoring and control system that can look at the larger system to accommodate a range of automatic and manual commands and controls.

External Communication

Beyond the confines of the transmitter, there has to be some means by which the transmitter communicates with a monitoring and control (M&C) system and/or directly with an operator. Normally this is done via SNMP over Ethernet. Some commands and status are communicated via GPIO (*general purpose input/output*), a simple closed contact or open collector, where a given wire might be the "on" or "off" command and another might indicate that the transmitter is on or off, attached via a multiconductor cable and connector. While either can be connected to a remote control system, SNMP is generally favored for M&C systems and the GPIO for local control, such as on-off buttons located near the transmitter.

Many transmitters also support a browser-based user interface that can allow an operator not only to control the transmitter but to look deeper into configurations and error logs. In some cases the Internet-accessible Web GUI is the only means of controlling the transmitter.

Transmitters have a 15-year life expectancy, but browsers and plug-ins are upgraded frequently and often take advantage of caching. One of the challenges for transmitter interfaces is that while they may do some ingenious tasks when they are purchased, keeping these supported over more than a decade is a real challenge. Likewise, the fact that the number of external devices that one would wish to support—computers, cell phones, tablets—changes over time can be difficult to keep up with.

All of the usual Internet-accessible device considerations apply. Most transmitters support username/password security. It should be put behind a VPN and/or a firewall to reduce the concern of external hacking. Support for tiered access and rights management might also be desirable while maintaining factory access, which can be necessary for troubleshooting. The ability for the control system to text and email alarms and notifications is largely standard on current transmitters. Having the transmitter reach out to the manufacturer to order replacement parts and modules is an interesting feature and service. Being able to record who logs into the transmitter and when along with a description of what they did is also a good feature.

It is not uncommon for a browser to be upgraded with a regular automatic update that causes issues with a control system. It's not uncommon to be looking at cached data or pages rather than real time data. It seems likely that manufacturers will have to continuously push updates to user interfaces to keep pace with changes in the Internet world, or accept a simpler, harder-to-break interface.

Security is always an issue. In truth, a TV transmitter is not likely a very attractive hacking target. It's difficult to remotely do physical damage to a properly designed and implemented transmitter system, though one might be able shut it off or reconfigure inputs.

POWER SUPPLY

While there are many similarities between tube and solid-state power supplies, there are also many differences. Tube-type transmitter power systems are covered in detail in the analog TV transmitter chapter (Chapter 8.2), and solid-state systems will be discussed here.

Most solid-state transmitter amplifier devices operate on a rather benign 50 volts DC, with currents that are consistent with the power level of the transmitter. A 1-kW transmitter will require on the order of 50 amps. While a tube's beam power supply can easily be located at some distance, the lower voltage and higher currents for solid state require heavier gauge cable, and there is far more voltage drop across that cable for any distance, which reduces the overall AC to RF efficiency. For this reason, locating the power supplies closer to solid-state amplifiers is optimum.

Most solid-state transmitters employ switching-mode power supplies rather than linear power supplies. Switching mode supplies are lighter and more efficient and offer wide ranges of operating voltages, negating the need for external voltage regulation. One common configuration places the power supplies in the same chassis as the RF amplifier they feed. Another configuration places the power supplies in the same racks as the amplifiers. Still others place the supplies outside of the same cabinet but within a reasonable distance of the amplifiers.

While a tube transmitter requires a large single power supply, solid-state transmitters typically use multiple smaller power supplies, usually paired with individual amplifiers. This configuration serves to increase the overall reliability of the transmitter, because configured in this manner, failure of a single power supply can only affect one or two amplifier assemblies.

Using *universal power supplies* allows the use of a wide range of supply voltages, for example a 90–240 VAC, 50/60 Hz per supply. Using distributed power supplies allows the use of either single-phase or three-phase power. Single-phase power is not practical or requires a phase converter for tube-type transmitters of any significant size. While three-phase power is desirable for large solid-state systems, it is not necessary for many lower-power models.

Switch mode power supplies are also useful in the depot-level repair scenario. When the power supply fails, it is simply replaced, and the failed unit is sent back for rebuilding or disposal. In some cases, the power supply is field replaceable. In others the amplifier drawer with its power supplies are sent back to be refurbished.

Switch mode power supplies can fail in a manner where the voltage regulation is lost and a higher-than-desired output voltage is delivered. To prevent damage from this failure state, overvoltage protection is employed. A common approach is to use a solid-state overvoltage protection device that shorts the power supply, causing the line fuses to open when necessary.

Because solid-state transmitters operate on 50 volts DC, which is power that can be easy to supply with sealed lead-acid cell batteries. Some smaller transmitters are configured to run off of a battery bus. Solar- and wind-powered translator sites can be built without voltage inverters. While a battery supply is inherently an integral uninterruptible power supply (UPS), it also serves to protect the transmitter from the effects of dirty power.

With a ready supply of 50 volt DC, many transmitters utilize DC-to-DC converters internally to supply +/− 12 volts, 5 volts, and any other lower voltages needed for exciters, controllers, and displays. This is often a better option than a separate AC-to-DC power supply. While distributed power supplies that paired with individual amplifiers provide a high degree of redundancy, a single point of failure is often created when a single DC-to-DC converter is used to supply all critical functions.

With the high currents in solid-state transmitters, especially those that use battery buses or singular large supplies, *arc flash* can be a danger. A short across the power supply can vaporize metal or turn it into shrapnel. Both the transmitter and personnel need to be protected from an accidental arc flash. The use of appropriately sized fuses is critical.

COOLING SYSTEMS

TV transmitters tend to be air cooled at lower power levels for the convenience of being able to utilize rack-mountable amplifiers that can be easily removed without having to disconnect coolant systems or draining amplifier system when they are not in use.

However, installing and repairing the air cooling units that remove the room heat that an air-cooled transmitter generates generally requires special tools and skills. By comparison, water cooling systems can be fairly straightforward and simple.

Air Cooling

It is possible to vent transmitter-cooling air outside with a system of air filters and dampers and some means of dealing with very hot or very cold return outside air. On very hot days, the outside air may be too hot to cool the transmitter adequately. On very cold days, the outside air might cause condensation and circuit instability. The use of dampers and air conditioning for peak heat days can solve these issues.

Still, most air-cooled transmitters don't have an easy way to collect the hot exhaust air and provide fresh cooling air. Usually the chassis involved are simply too small to have such duct connections. For this reason, the hot air is usually dumped into the room. Most transmitter facilities are not constructed with the careful attention to airflow that a computer or equipment room is. Most small transmitters follow the computer convention of pulling air in from the front panel (that often has the air filter built into the front panel hinged door or removable cover) and exhausts it through the rear panel or sides of the chassis. In any case, the airflow should not be impeded, and a minimum of the exhausted air should return uncooled as supply air.

Air cooling requires a high volume of air, which requires space, which makes compact designs impractical. Forced air also carries dust and contaminants that require periodic cleaning. Careful consideration should be given to the placement of air conditioning ducts. Ideally, cooling ducts should be placed at floor level to allow the cool air to enter the racks at the base and heat as it is drawn up and through the transmitter cabinet. Even if a transmitter is too small to be directly ducted, a properly constructed vent hood can be placed above the transmitter with an appropriate exhaust fan in or near the ceiling of the building to exhaust as much of the heated air as possible, much as a whole-house fan draws hot air out of a house. It is also possible to use the heated air to keep a transmitter building warm in the winter in colder climates. Since air cooling systems require a certain amount of outside air to be returned into the building, special attention needs to be paid to the quality of air that is returned. Air conditioning systems serve not only to cool the incoming air but also to dry it, as humid air isn't suitable for transmitter systems.

It's also important to pay attention to air filtration with an air-cooled transmitter. At minimum, the filters covering the air blowers in an air-cooled transmitter should be checked weekly and changed or cleaned as soon as dirt appears on the filter. Cooling systems that pull outside air in and cool it should also be adequately filtered, and the filters changed or cleaned often.

Liquid Cooling

The simplest water cooling systems use a mixture of distilled water and a *specified* antifreeze at a given ratio. Transmitter manufacturers specify the cooling fluid and either provide the outside heat exchangers, plumbing, and pumps, or tightly specify what specific models can be used in order to prevent electrolysis and corrosion. Manufacturers may use many different metals, alloys, composites, and natural materials in cooling a complete transmitter system. Introducing a foreign metal or other material or using an unapproved cooling fluid will frequently lead to failure. Failures in water-cooled systems result, at minimum, in messy leaks and sometimes in extended off-air time. Typical failures include the disintegration of pump impellers, fittings, and connectors and nonmetallic materials that clog small passages and filters. Maintenance is also specified by the manufacturer and generally involves periodic replacement of the cooling fluid, flexible hoses, some fittings, and pumps (see Figure 8.3-15). However, water cooling failures are rare and usually can be traced back to some defect or operator error. Maintenance intervals are often measured in years.

Outside heat exchangers are composed of thin copper or aluminum pipes welded or soldered to cooling fins (see Figure 8.3-14). This is a liquid cooling system's most vulnerable aspect. Since the fans must always be operable, it must be protected from vandals and falling ice, it cannot be allowed to clog up with leaves, dirt, or anything else. It is the most likely piece to leak, and if it ever freezes, it is very unlikely to be worth repairing. Still, these tend to operate with few issues for the life of a transmitter.

For transmitters in tall buildings, it is likely that a liquid-to-liquid heat exchanger will dump the waste transmitter heat into the building's chilled water circulating system.

Split Systems

Some transmitters use distilled water or a specialized oil that typically serves as an electrical insulator for the cooling circuit. Generally, this fluid in turn transfers its heat to the fluid used in the outside heat exchanger, typically with the addition of a *shell and tube* heat exchanger (one pipe inside another, where the inner pipe's heat is transferred through the metal pipe wall to water surrounding it inside the outer pipe) and an additional pump or redundant pair of pumps.

In other installations, building-supplied chilled water might be the destination for the heat. This is typical in tall buildings where transmitters are located on the top floors. Occasionally a building might recover transmitter heat for use in heating office or other space. Distributed heating and cooling (DHC) systems can make efficient use of waste heat in HVAC *district networks*.

REDUNDANCY AND RESERVE

High-Power Transmitters

Larger, high-power transmitters, those with two or more separate cabinets (or racks) of amplifiers that can be operated independently along with redundant

FIGURE 8.3-14 External water cooling heat exchangers.

Courtesy GatesAir

FIGURE 8.3-15 A pumping system for a water-cooled transmitter.

Courtesy GatesAir

exciters and control, have an almost built-in level of redundancy. Should a cabinet fail (whether an IOT or array of solid-state amplifiers), one can continue to operate with the survivor. With proper RF switching, the surviving cabinet can operate without taking the combiner loses outlined previously in the combiner section of this chapter.

While the amplifier portion of a high-power transmitter might be inherently redundant, the exciter and some part of the intermediate power amplifiers or drive amplifiers will not be unless there are redundant exciters and drive chain. Domestically, this is simply referred to as redundant exciters, and internationally as *dual drive.*

Small Transmitters

Small transmitters present a higher exposure to a single point of failure. Throughout the world, *passive reserve* is the most common form of protection. Domestically, this is often called 1+1 (one-for-one) protection. Here, two usually identical and complete transmitters are available to be either manually or automatically switched to the antenna, with the transmitter not coupled to the antenna connected to a dummy load. "Passive" refers to the off-line transmitter usually being in an idle mode. One can also operate the standby transmitter in *active reserve* by leaving both transmitters on continuously. The off-line transmitter can be operated into the dummy load for testing, maintenance, or setup. Because the load and the antenna present different characteristics to the transmitter, the exciters need to be operated in a mode where, when online, the correction desired is in place. In some cases, the second (backup) transmitter is smaller than the main, when a reduction in power is tolerable.

Small transmitters are often frequency agile (excluding the mask filter). One popular low-power configuration is to have several transmitters operating on different RF channels from one location and typically sharing one antenna. Here, an N+1 ("N-for-one") configuration can be economical. A single spare *protection* transmitter serves as the backup for all the other transmitters. In a simple backup configuration, the spare can be patched in manually or even taken off a shelf and installed in place of a failed transmitter. This becomes increasingly complex if one needs to do this quickly or via remote control. Pushing configuration data to the protection transmitter and routing it to the correct combiner or antenna port and mask filter involves a number of RF switches and an appropriate control system. Even more complex is a system with two spares that are available as backup for a number of transmitters (N+2).

Backup (Auxiliary) Sites and Systems

There is a point where the protection equipment costs brings one in range of simply having a second backup system, ideally in a different location, should a tower fail or a fire or other disaster take place at the main transmission facility. A completely independent backup system has its own antenna, studio transmitter link, power source, etc. A lower-power, shorter tower, smaller wind load transmission line and antenna can serve to keep a large portion of the desired population viewing when the main site cannot.

MASK FILTERS

Digital transmitters all produce out-of-band emissions. If these emissions are not severely attenuated, the transmitter likely will cause interference to other television stations or services. As mentioned earlier, out-of-band emissions tend to be more severe as power levels increase and amplifiers are driven harder to achieve higher efficiencies. The solution to these increasingly severe out-of-band emissions is a *mask filter*. The term refers to a regulatory set of charts that plot frequency against the permissible levels of out-of-band energy. Graphically, it appears as a mask under which the transmission system's emissions must fit and not exceed. Many spectrum analyzers, especially those built into the exciter, will display the mask as an overlay of the spectrum measurement.

Another way to describe the out-of-band emissions is to measure and display the *shoulder level*. The shoulder level should be anywhere from the −36 dB the FCC requires domestically to a challenging to achieve and measure −60 dB. Simply put, the shoulder levels compare the signal level in the first 500 kHz below the channel (*lower shoulder level*) and the first 500 kHz above the channel (*upper shoulder level*) to the center of the channel (*the head*), which is the desired and maximum signal level. Making these measurements in the United States is described in the IEEE document, *Recommended Practice for Measurement of 8-VSB Digital Television Transmission Mask Compliance for the USA*.

Many exciters will take these measurements for you and log them over time. Drifting or decreasing shoulder measurements is a sign of decaying transmitter fidelity and/or changes in the antenna and filter systems. Realistically, there should be no more than a 1-dB change of these values over any time period.

Regulations that restrict out-of-band emissions vary throughout the world but are generally similar to the U.S. domestic rules. One means of identifying a mask filter is to call out the number of cavities employed (see Figure 8.3-16). Four-, six-, and eight-cavity designs are popular, with six-cavity filters probably being the most common in use for low-power and translator service.

Domestically, LPTV (low-power) TV transmitter filters are governed by 47 CFR part 74.794 of the FCC regulations. The following is taken directly from the FCC rules and regulations.

> The LPTV station application must specify one of three filter specifications to be used.
>
> A *simple mask* attenuates the channel edge emissions by no less than 46 dB. More than 6 MHz from the channel edges, emissions must be attenuated no less than 71 dB. At any

FIGURE 8.3-16 A 5-kW, eight-cavity mask filter.

Courtesy GatesAir

frequency between 0 and 6 MHz from the channel edges, emissions must be attenuated no less than the value determined by the following formula: A (dB) = 46 ($\Delta f2$ / 1.44)

For a *stringent mask,* emissions must be attenuated no less than 47 dB for the first 500 kHz from the channel edges. More than 3 MHz from the channel edges, emissions must be attenuated no less than 76 dB. At any frequency between 0.5 and 3 MHz from the channel edges, emissions must be attenuated no less than the value determined by the following formula: A(dB) = 47 + 11.5 (Δf – 0.5)

For a *full-service mask,* the power level of emissions on frequencies outside the authorized channel of operation must be attenuated no less than the following amounts below the average transmitted power within the authorized channel. In the first 500 kHz from the channel edge the emissions must be attenuated no less than 47 dB. More than 6 MHz from the channel edge, emissions must be attenuated no less than 110 dB. At any frequency between 0.5 and 6 MHz from the channel edge, emissions must be attenuated no less than the value determined by the following formula: Attenuation in dB = –11.5 ([Delta] f 3.6). [Delta] f = frequency difference in MHz from the edge of the channel.

This attenuation measurement is based on a bandwidth of 500 kHz.

U.S. full-power TV stations operate under 47 CFR part 73.622 h (see Figure 8.3-17). The following is taken from those rules.

The power level of emissions on frequencies outside the authorized channel of operation must be attenuated no less than the following amounts below the average transmitted power within the authorized channel. In the first 500 kHz from the channel edge the emissions must be attenuated no less than 47 dB. More than 6 MHz from the channel edge, emissions must be attenuated no less than 110 dB. At any frequency between 0.5 and 6 MHz from the channel edge, emissions must be attenuated no less than the value determined by the following formula: Attenuation in dB = –11.5 (Δf 3.6);

Where: Δf = frequency difference in MHz from the edge of the channel.

This attenuation is also based on a measurement bandwidth of 500 kHz. Other measurement bandwidths may be used as long as appropriate correction factors are applied. Measurements need not be made any closer to the band edge than one half of the resolution bandwidth of the measuring instrument. Emissions include sidebands, spurious emissions and radio frequency harmonics. Attenuation is to be measured at the output terminals of the transmitter (including any filters that may be employed). In the event of interference caused to any service, greater attenuation may be required.

Mask filters are passive devices with very long life expectancies and very few reasons to fail. As this is written, some manufacturers are offering *field tunable filters.* This is a major change from past practice, where filters were designed and built for a specific frequency and required major rebuilding if they could be reused at all on a new frequency.

Filters are tuned at the factory, and occasionally, if a filter is moved or a major change in the transmitter facility is made, they are retuned in the field. If a filter is small enough, it makes sense to simply send it back to the factory to retune and avoid the costs of field engineering. While a filter may be field tunable,

FIGURE 8.3-17 Mask filters, RF switching, and reject loads for a full-power UHF station.

Courtesy GatesAir

the equipment and skill set required is not commonly available at the station or station group level.

Filters come in many designs, though four-section *cylindrical waveguide filters* are probably the most common for full-power UHF stations. VHF stations generally employ designs that save space.

Filters' performance and characteristics are entirely dependent on their physical dimension, which change with temperature. Aluminum is desirable due to its lighter weight, low cost and other desirable characteristics. It is a good choice for filters that will not have heating issues operating in a room with stable temperature. High-power UHF systems can have significant filter heat issues, sometimes addressed with water cooling and even regulated temperature systems.

Reducing the effects of temperature is another approach. *Invar*, whose name comes from "invariable" (known generically as FeNi36 and 64FeNi in the United States), is a nickel–iron alloy that has an exceptionally low coefficient of thermal expansion. These filters can operate at as high as 140 degrees Fahrenheit.

Filters are tuned and designed to operate at some given temperature that is reached after some period of operating time. One reason for using adaptive precorrection is to compensate for drifting filter temperature and characteristics. If a fixed or static mode of precorrection is to be used, or any element of the RF transmission system is to be adjusted, it should be done when the filter and other components are at their stable operating temperature, which can take several hours.

Tubes in digital service have one major different filter requirement. A tube does not tolerate reflected power from a filter or antenna as well as a solid-state transmitter does. For this reason, a tube needs a larger and more complicated mask filter, usually a *constant impedance filter* (CIF). A solid-state transmitter has no issues with a *reflective* filter where the RF that is to be suppressed by the filter is reflected back into the amplifier.

ALTITUDE

At increasing altitude, air becomes less able to carry away heat. Generators, HVAC equipment, transformers, and fans, and thus transmitters, all must be *de-rated* for altitude. If a device relies on air to cool it, and that means motors, power conductors, thermal breakers, computers, and even light dimmers, they are generally designed to operate at less than 3,300 feet above sea level. While this varies with the design of each device, a general derating factor of 15% at 5,000 feet and 25% at 10,000 feet is useful as a sanity check. Good engineering practice is generally to operate devices at a maximum of 80% of rated capacity. At 10,000 feet, following this rule of thumb, devices would be operating at 55% of sea level-rated capacity. Instead, we might replace standard fans with higher capacity fans in transmitters and servers and operate building space at lower temperatures, which in turn requires larger cooling units. High voltage and RF arcs occur more easily at altitude, so additional insulation (with even poorer cooling characteristics) and spacing might be required. The National Electrical Code (NEC) doesn't address the topic directly with guidelines, but power cable, thermal breakers and other devices rely on ambient air to remove heat, so the NEC does note that electrical wiring at altitude requires additional capacity.

Working above 10,000 feet presents its own dangers and limitations. Transmitter manufacturers all make provisions for operation at altitude, even if it is only to indicate that the transmitter should not be used above a given altitude.

EFFICIENCY

Transmitter efficiency becomes very important in locations where power is expensive and in stations where the power bill for the transmitter is a major portion of the station's operating budget. In all cases, efficiency is desired for "green" operation. In some locations, power is so expensive that a few additional percentage points in efficiency gain in even small transmitters have a fast payback.

The Filter's Effect on Efficiency

Transmitter efficiency is usually measured from AC to RF TPO output. This includes any cooling fans and pumps, exciters, intermediate and preamplifiers, power needed for control systems and lost in power supply efficiency. If the output filter is included, AC to transmission line efficiency is measured. The filter is not always supplied as part of the transmitter, however, so the only specification that can be supplied is AC to TPO. More complicated is that the filter's job is to remove undesired RF from the transmitter, and some transmitters have a larger amount of undesired, out-of-band RF to be dissipated by the filter. Because the AC to TPO measurements include even the undesired RF, a transmitter may appear to have a substantially better AC to TPO efficiency than AC to transmission line efficiency. If the filter can be included when comparing transmitters, it does provide a more real-world operating figure.

Pallet Efficiency

Occasionally, in solid-state transmitters, the RF amplifier pallet's DC to RF out efficiency is specified. As the transmitter increases in size, the power combiner losses increase. Some combiners may be fairly efficient and others may be fairly lossy. Likewise, RF pallets seldom have the same efficiency across any large bandwidth. In the case of UHF pallets, a manufacturer may produce a pallet that has good efficiency across 50–100 MHz of bandwidth and be useable for as much as 300 MHz. Maintaining spares with good efficiency across any large bandwidth is difficult.

Cooling Efficiency

As operators, we are interested in the *total system AC to transmission line efficiency*. A water-cooled system with outside heat exchanger radiates relatively little heat into the transmitter building, requiring less supporting air conditioning. An air-cooled system either utilizes outside air directly into the transmitter, which is difficult to do and usually dirty and difficult to maintain (and thus rare) or vents into the transmitter room, where the bulk of the heat needs to be cooled by air cooling units. High-power, high-efficiency transmitters tend to operate with 5–10% of the TPO being converted to heat. Water-cooled filters help keep this heat out of the transmitter room. Nonetheless, all transmitters radiate heat into the transmitter room and likely have some components, including power supplies and exciters, that are not water cooled.

SAFETY

The chapter about analog transmission (Chapter 8.2) concludes with a section on safety and liability that covers both modern solid-state transmitters and older tube-based designs.

While solid-state transmitters, with their lower operating voltages, are considerably safer than the tubes transmitters they replace, the RF levels are just as high as before, and we introduce the arc flash hazards from transmitters with power supplies that can provide considerable current. One still needs to practice lockout-tagout practices; work with a companion, and employees still need to be trained in first aid.

THE CHANGING DISTRIBUTION NETWORK ARCHITECTURE

Throughout the world, a growing number of distribution networks use a combination of high-power digital transmitters and lower-power booster transmitters, tightly synchronized in time and frequency to deliver building penetrating coverage in urban areas. These single-frequency networks (SFNs) will almost certainly become more common, if not the default architecture, in the United States as TV evolves from being a big screen viewing appliance to supporting the ever-increasing numbers of smaller and more mobile devices, "second screens," and more interactive IP-based multimedia systems. TV distribution will also utilize distributed antenna systems (DAS) to cover venues, transportation, offices, health clubs, and restaurants—all the places we expect a cell phone to work.

Already much of the world uses small over-the-air transmitters organized into a multiple frequency network (MFN), combined with satellite and sometimes cable to reach the audience. Outside of the United States, a good deal of the world's geography and population density, combined with the lower cost of building the low-power systems, and the fact that these systems were built in a "post-satellite" era so that there was already in place a platform for reaching the hard-to-reach, low-population density areas has enabled this type of methodology. Additionally, analog television continues to exist outside of the United States, largely because the cost of a fast digital conversion is deemed unaffordable. The digital transition is happening everywhere, but not all at the same rate or time.

ACKNOWLEDGMENTS

For research and editorial assistance:

Mark Durenberger

Peter Douglas

Paul Grzebik

Martyn Horspool

Joseph Turbolski

Doug Garlinger

Dirk Freeman

References/Resources

The major television transmitter manufacturers have a variety of papers and materials on transmitters. Some manufacturers have excellent manuals on their equipment and their use. Additionally, there is a good deal of information available on Wikipedia.

[1] Brett Jenkins and John Freberg, 2007 *NAB Engineering Handbook*, Digital Television Transmitters, Chapter 6.4

[2] LDMOS, Microwaves 101.com www.microwaves101.com/encyclopedias/ldmos on the use of LDMOS

[3] The Basics of the Doherty Amplifier, Bill SladeOrban Microwave Products, 900 Loma Verde, Suite A, El Paso, TX 79936

[4] IEEE Recommended Practice for Measurement of 8-VSB Digital Television Transmission Mask Compliance for the USA, 9 August 2008, RF Standards Committee G-2.2, IEEE 3 Park Avenue, New York, NY 10016–5997, US

[5] Liquid cooling fundamentals for high-power electronics co-written by GatesAir and Colder Products: www.electronicproducts.com/Thermal_Management/Thermoelectric_Coolers/Liquid_cooling_fundamentals_for_high_power_electronics.aspx

[6] Doug Garlinger, *DTV Handbook*, Society of Broadcast Engineers Publication, 2009

CHAPTER

8.4

Next-Generation DTV Systems

WALTER FISCHER
Rohde & Schwarz

JOHANNES SINNHUBER
Rohde & Schwarz

FROM ATSC 1.0 TO ATSC 3.0—INTRODUCTION

ATSC, which stands for Advanced Television System Committee, is a group of U.S.-based digital terrestrial television broadcast standards. The first ATSC standard was developed in the mid-1990s parallel to the first DVB standards. It describes the physical transportation layer, including the multiplex input format and the content format of the multiplex input signal, and has quite strict limiting rules for the source coding.

There are now three ATSC versions:

- ATSC 1.0, published in 1995
- ATSC 2.0, published in 2015
- ATSC 3.0, under development as of this writing

In addition, a single-frequency network (SFN) and a mobile handheld extension were standardized in 2009. The mobile handheld extension is called ATSC-MH. The ATSC 3.0 standard is expected to be finalized by the beginning of 2017 (see Table 8.4-1).

ATSC 1.0 was very much restricted to MPEG-2 video content that mainly focuses on HDTV. The audio codec is Dolby Digital AC-3 audio. The physical layer data rate of ATSC 1.0 is 19.39 Mbit/s and is fixed. The multiplex signal is the MPEG-2 transport stream, typically provided by an SMPTE310 interface. PSI/PSIP tables are used to signal the transport stream program structure, program labels, electronic program guide (EPG), time and date, etc.

ATSC 2.0 allows also advanced MPEG-4 video (H.264) and MPEG-4 AAC content.

The physical layer of ATSC 1.0 and 2.0 is the same and is an eight-level vestigial sideband (8VSB) single-carrier modulation. The RF channel bandwidth is 6 MHz, the spectrum is roll-off-filtered and has a pilot signal at the original modulation carrier position. Because the single-carrier modulation has a symbol rate of 10.76 Msps, the capability to implement SFNs is very limited and typically not used.

ATSC 3.0 uses orthogonal frequency division multiplex (OFDM) with constellations up to 4096QAM

TABLE 8.4-1
ATSC versions

Standard	Video	Audio	Multiplex signal	Modulation
ATSC 1.0	MPEG-2, HD and SD	Dolby Digital AC-3	MPEG-2 transport stream	8VSB single carrier
ATSC 2.0	MPEG-2 and MPEG-4/AVC/H.264	Dolby Digital AC-3 and HE AAC	MPEG-2 transport stream	8VSB single carrier
ATSC 3.0	tbd (HEVC)	tbd	STL, tbd	OFDM

supporting nonuniform constellations. When this chapter was written, the ATSC 3.0 standard had not yet been finalized. The chapter describes some of the main aspects of the ATSC 3.0 standard that are expected to be close to the final version of the standard.

ATSC 3.0 DESIGN GOALS AND OVERVIEW

ATSC 3.0 has the primary design goals:

- To provide efficient and flexible use of the spectrum
- To ensure robust transmission for stationary and mobile reception
- To enable UHD and immersive audio on content level
- To allow hybrid services aggregating other sources based on IP transmission

The first two points include the focus on a higher spectral efficiency that is closer to the Shannon limit compared with ATSC A/53. For background information on the Shannon limit, refer to the section on DVB-T2 presented later in this chapter. As covered in (1). ATSC 3.0, advances in terms of spectral efficiency are mainly based on the application of the most recent codes for error correction and the usage of nonuniform constellations for modulation. Forward error correction (FEC) and modulation patterns are described in sections 4 and 5, respectively. One of the ATSC 3.0 features aiming at flexible use of the spectrum in terms of stationary and mobile reception is layer division multiplexing (LDM), which is covered in subsection 5.3.

A strict focus on IP technology in the baseband and in the protocol layers enables ATSC 3.0 to be used for hybrid services. The broadcast service can be augmented by other IP-based broadband services. A broadcast video stream can be synchronized with an audio stream coming from an IP-based unicast source, for example. ATSC 3.0 studio transmitter link (STL) and ATSC 3.0 protocol layer had not yet been finalized when this chapter was written. For a more detailed view on the design goals mentioned and further ATSC 3.0 goals, refer to (2).

This chapter focuses on physical layer aspects, which are expected to be close to the final standard. Figure 8.4-1 shows a simplified block diagram of the physical layer system in ATSC 3.0 (LDM and MIMO are not illustrated in the block diagram). The main functional blocks are input formatting, bit-interleaved coding and modulation (BICM), framing and interleaving, and waveform generation. Input formatting aspects are briefly described in section 3. In ATSC 3.0, the input formatting block is followed by the BICM block, which consists of FEC, the bit interleaver and the modulation mapper. BICM generates the I/Q samples from the baseband data. There is one BICM block for each physical layer pipe (PLP). The first step of the BICM is the FEC, which is described in section 4. Bit interleaving after the FEC accounts for burst errors. Modulation mapping is covered in section 5. Section 6 focuses on framing, interleaving and waveform generation.

INPUT FORMATTING AND MULTIPLEX SIGNAL

We previously noted that when this chapter was written, the studio transmitter link (STL) and multiplex signal were still under discussion in the ATSC 3.0 standardization process and had not yet been finalized. Aspects affecting the input are therefore subject to ongoing change and extension. ATSC 3.0 systems use the ATSC link protocol (ALP) for input formatting. This section only gives basic information about ALP covered in the (3). For detailed information, refer to the most recent version of ATSC 3.0 STL specification, which specifies ATSC 3.0 input and multiplex signal in detail. The ATSC 3.0 STL specification is expected to reach candidate status in the second quarter of 2016.

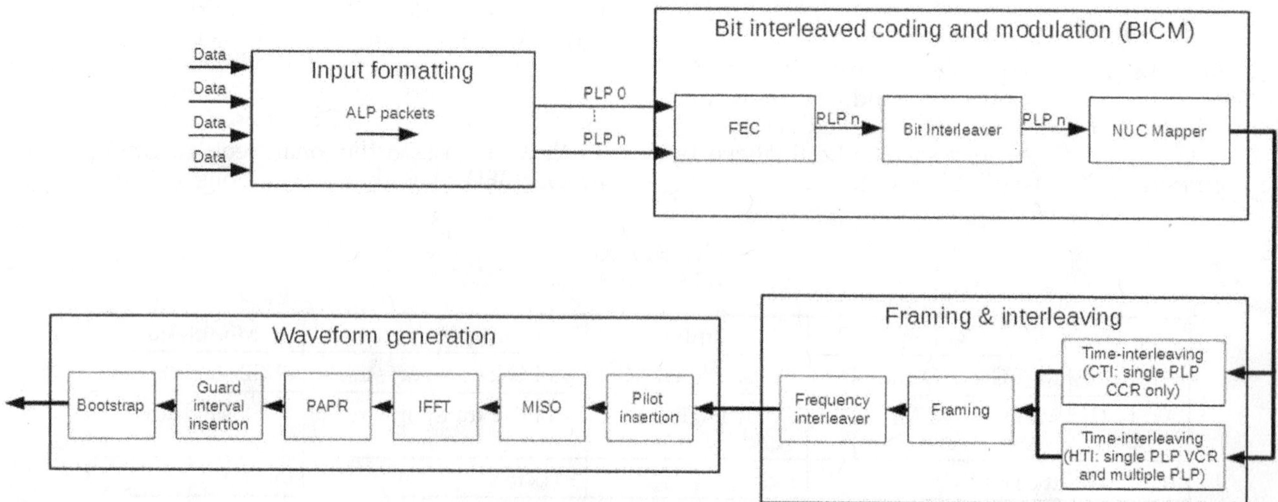

FIGURE 8.4-1 ATSC 3.0 physical layer block diagram.

ALP packets are flexible in their payload and can carry IP data as well as traditional MPEG transport streams (MPEG TS). ALP applies overhead reduction for IP data as well as MPEG transport streams. The length of an ALP packet is not fixed and is specified in the ALP packet header. There can be several data streams, each feeding a separate PLP (see Figure 8.4-1). Every PLP can have its own error coding and modulation parameters. This gives broadcasters the flexibility to adjust the data capacity and robustness of each PLP to match the needs of the associated service. ATSC 3.0 allows for 1 to 64 PLPs.

FORWARD ERROR CORRECTION (FEC)

In general, new transmission standards benefit from technical progress in the field of channel coding. As described in section 2 of this chapter, using the most recent codes in the FEC is one aspect of the ATSC 3.0 performance improvements toward the Shannon limit. Like DVB-T2, the ATSC 3.0 FEC is constructed as a concatenation of an outer and an inner code. For the optional outer code, a Bose, Ray-Chaudhuri and Hocquenghem (BCH) code or a cyclic redundancy check (CRC) code is used, while for the mandatory inner code a low-density parity check (LDPC) code is applied.

Outer Code

In order to maximize broadcasters' ability to adapt their ATSC 3.0 implementation to their specific application, the standard provides flexibility for the outer coding. Possible codes at this stage are BCH, CRC or no outer parity coding at all. The BCH code provides additional error detection and correction capabilities, while the CRC code only provides error detection capabilities. When BCH is applied as an outer code, up to 12 bit errors can be corrected and the LDPC error floor is lowered. Using no outer code at all maximizes the possible payload in the FEC frames, when the inner coding is seen to be sufficient. However, the application of an outer code should have an advantage on the overall performance in typical broadcasting environments. Figure 8.4-2 shows the structure of the FEC frame with regard to the payload and parity portions for the different scenarios.

The standard defines two possible values for the overall frame length N_{inner}. For the larger frame length (N_{inner} = 64800 bit), the BCH code adds 192 parity bits, while for the smaller frame setting (N_{inner} = 16200 bit) there are 168 BCH parity bits. When using CRC, there are 32 parity bits. The calculation of BCH codes and CRC codes relies on generator polynomials. For detailed information on how the BCH and CRC parities are calculated, refer to (3 p. 30).

Inner Code

The use of an inner code is mandatory in order to account for bit errors in the real-world transmission channel and to provide an appropriate portion of information redundancy for proper error correction capabilities at the receiver end. LDPC codes are state of the art and are also used for DVB-T2. ATSC 3.0 is flexible in this regard, since it supports a wide range of possible code rates from 2/15 to 13/15. Since the overall frame length is fixed for a certain setting (N_{inner} = 64800 bit or N_{inner} = 16200 bit), the code rates determine the relation of the inner parity portion to the baseband packet payload, as depicted in the frame structure in Figure 8.4-2. The larger frame length results in optimized performance and the closest gap to the Shannon limit. Therefore, it is expected to be first choice for most scenarios. The smaller frame length of N_{inner} = 16200 bit offers a latency advantage but does not reach the performance of the other setting. It is suitable for applications where latency is critical and/or simpler encoder/decoder structure is preferred due to smaller blocks for processing.

As a result of the wide range of code rates supported by ATSC 3.0, there are two different types of LDPC code structures. The structure of one type optimizes performance for medium and high code rates and is similar to the code structure used in DVB-T2. The other type of code structure focuses on low code rates (in general <

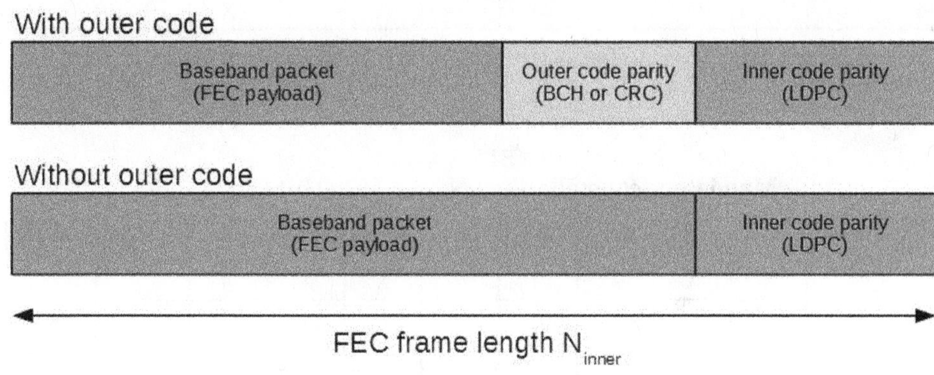

FIGURE 8.4-2 Structure of FEC frame.

ATSC Candidate Standard Physical Layer Protocol. s.l.: Doc. S32–230r21, 28 September 2015, p. 29.

6/15). ATSC 3.0 clearly defines which code structure is to be used for a specific code rate and frame length. For more details on the selection of type and calculation of parity bits in each case, refer to (3 p. 32).

MODULATION PATTERNS

Constellation Overview

Leaving the 8VSB single-carrier modulation of ATSC 1.0 behind, ATSC 3.0 considers quaternary phase shift keying (QPSK) and a range of quadrature amplitude modulation (QAM) patterns of different order for multicarrier COFDM waveforms. In contrast to DVB-T2 modulation patterns, ATSC 3.0 QAM patterns are nonuniform constellations (NUC). Subsection 5.2 gives a more detailed description of the NUC structure.

With the uniform structure, the QPSK modulation uses the same constellation mapping for all code rates. It is the most robust ATSC 3.0 modulation scheme. Higher-order QAMs offer a better spectral efficiency but at the cost of robustness. The ATSC 3.0 standard includes 16QAM, 64QAM, 256QAM, 1024QAM and even 4096QAM nonuniform constellations. In order to make use of the nonuniform performance benefit, every QAM order has a specific constellation mapping for every supported code rate from 2/15 to 13/15. To reduce implementation complexity, not all possible combinations are mandatory. Table 8.4-2 and Table 8.4-3 show all mandatory combinations of QAM order and code rate for N_{inner} = 64800 bit and N_{inner} = 16200 bit.

We noted earlier that for both possible frame lengths N_{inner}, the same constellation mapping is used in each case; 1024QAM and 4096QAM are not considered for N_{inner} = 16200 bit. The slightly diagonal structure of the table accounts for the general trade-off between robustness and spectral efficiency. In typical environments, higher-order QAM signals reasonably come with a higher FEC code rate, while a lower QAM order needs less redundant FEC information. The following subsection describes their nonuniform structures.

Nonuniform Constellation (NUC) Structures

A nonuniform constellation means that the different points in a constellation diagram are not equidistant in in-phase and quadrature components. According to (1), uniform constellation patterns result in a gap toward the theoretical Shannon limit, especially for high constellation orders. Nonuniform constellations can lower this gap. Therefore, they can be applied to get the spectral efficiency closer to the Shannon limit, resulting in a more significant performance gain for the high-order QAM patterns.

There are two-dimensional NUCs (2D NUCs) and one-dimensional NUCs (1D NUC). ATSC 3.0 considers 2D NUCs for 16QAM, 64QAM and 256QAM. Figure 8.4-3 shows a 16QAM 2D NUC for the code rate 6/15 as an example and depicts the placement of the constellation points without uniform restrictions. Via symmetry the quadrants of the constellation diagram can be derived from one quadrant only. At the receiver end, the 2D NUC structure requires a full-size 2D demapper.

TABLE 8.4-2
Mandatory constellations N_{inner} = 64800 bit (3 p. 41)

Code rate/constellation	2/15	3/15	4/15	5/15	6/15	7/15	8/15	9/15	10/15	11/15	12/15	13/15
QPSK	X	X	X	X	X	X	X	X		X		
16QAM			X	Xx		X	X	X		X		
64QAM		X	X	X	X	X	X	X	X	X		
256QAM			X	X		X	X	X	X	X	X	X
1024QAM				X		X	X	X	X	X	X	X
4096QAM						X		X		X	X	X

ATSC Candidate Standard Physical Layer Protocol. s.l.: Doc. S32–230r21, 28 September 2015, p. 41.

TABLE 8.4-3
Mandatory constellations N_{inner} = 16200 bit (3 p. 42)

Code rate/constellation	2/15	3/15	4/15	5/15	6/15	7/15	8/15	9/15	10/15	11/15	12/15	13/15
QPSK	X	X	x	x	x	X	X	X				
16QAM				x	x	X	X			x		
64QAM				x	x	X	X	X	x	x		
256QAM				x		X	X	X	x	x	x	x

ATSC Candidate Standard Physical Layer Protocol. s.l.: Doc. S32–230r21, 28 September 2015, p. 42.

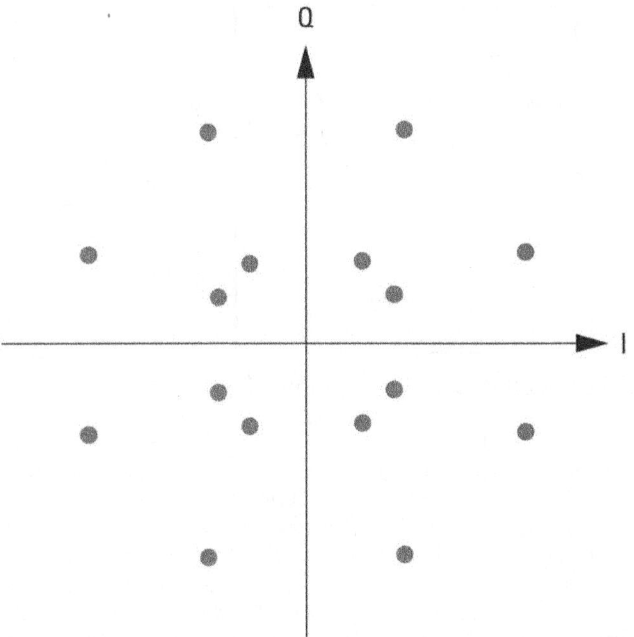

FIGURE 8.4-3 Example of 16QAM 2D NUC for code rate 6/15.

ATSC Candidate Standard Physical Layer Protocol. s.l.: Doc. S32–230r21, 28 September 2015, p. 43.

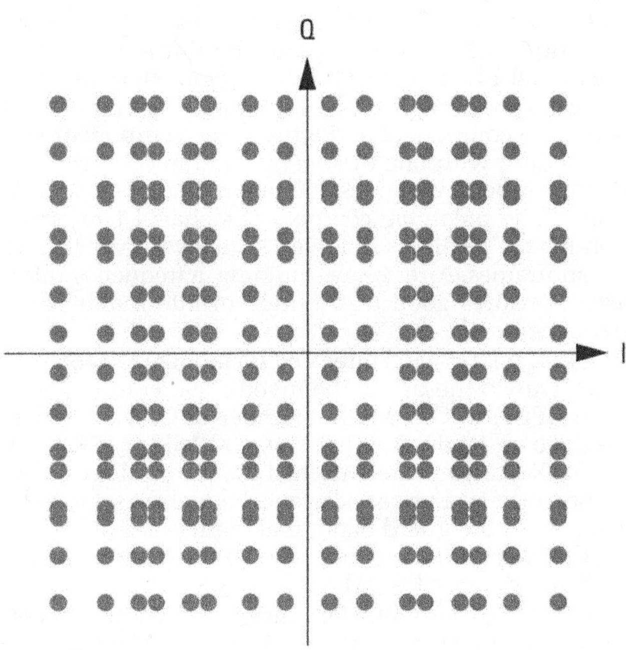

FIGURE 8.4-4 Example of 1024QAM 1D NUC for code rate 6/15.

ATSC Candidate Standard Physical Layer Protocol. s.l.: Doc. S32–230r21, 28 September 2015, p. 44.

In order to reduce receiver complexity for 1024QAM and 4096QAM, these QAM orders have a 1D NUC structure. For 1D NUCs the two-dimensional QAM constellation can be separated into two independent phase amplitude modulation (PAM) constellations. Therefore, only two separate 1D demappers are needed at the receiver end, reducing complexity.

Figure 8.4-4 shows an example of a 1024QAM 1D NUC for code rate 6/15. The square constraint structure visualizes how the constellation can be separated into independent in-phase and quadrature components.

Layer Division Multiplexing (LDM)

LDM describes the constellation superposition of multiple data streams at different power levels. The individual streams are separate PLPs and are also called layers in the context of LDM. They can have different FEC and modulation parameters. This gives broadcasters the flexibility to design the individual layers for different reception conditions having diverse robustness and payload capacities. For example, one robust layer could have a medium data rate that provides a medium resolution video stream designed for mobile reception on handheld devices having a small display. This layer can be superimposed by a lower robust layer for stationary reception, providing higher capacity for a UHD video stream delivered to large screens in living rooms. According to (3), only two-layer systems are defined.

Figure 8.4-5 shows a block diagram of two-layer LDM and visualizes how the I/Q sample vectors are superimposed in the constellation. The robust base layer is called *core layer,* while the second layer is named *enhanced layer*. The injection level controller for the enhanced layer determines the distribution of transmission power between the two layers. Changing the injection level makes it possible to adjust the robustness of the individual layers in relation to each other in addition to the FEC and modulation parameters. After the two layers have been superimposed, the power is normalized to unity. Figure 8.4-6 shows an example of the superposition of I/Q vectors in a constellation diagram. In this example, the core layer uses a QPSK modulation and is superimposed by an enhanced layer with a 16QAM constellation for a code rate of 6/15. For the sake of clarity, this superposition is only shown in the first quadrant. The I/Q sample vector of the core layer plus the enhanced layer vector yield the sum vector of the LDM system, depicted by the thicker arrow. In this example, the injection level makes it possible to adjust the length of the enhanced layer I/Q vector.

In contrast to the frequency division multiplexing (FDM) and the time division multiplexing (TDM) approaches, the entire bandwidth is used the entire time. Superposition of layered symbol energy results in a more effective use of the spectrum as described in detail in (4). To give an intuitive explanation in simplified terms, the core layer usually has higher coverage area than the enhanced layer. The SNR for the core layer is designed for the edges of the coverage area. Inside the coverage area, the additional signal power does not benefit the core layer but improves the SNR for the enhanced layer. In FDM/TDM systems this additional signal power would be "wasted." Therefore, (4) also

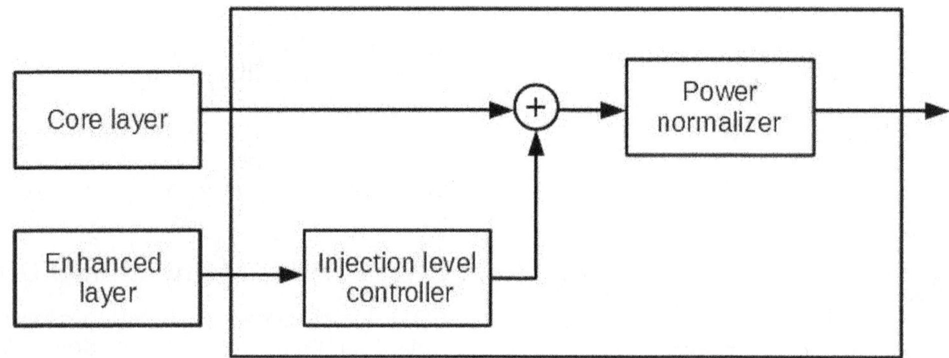

FIGURE 8.4-5 Block diagram for LDM constellation superposition.

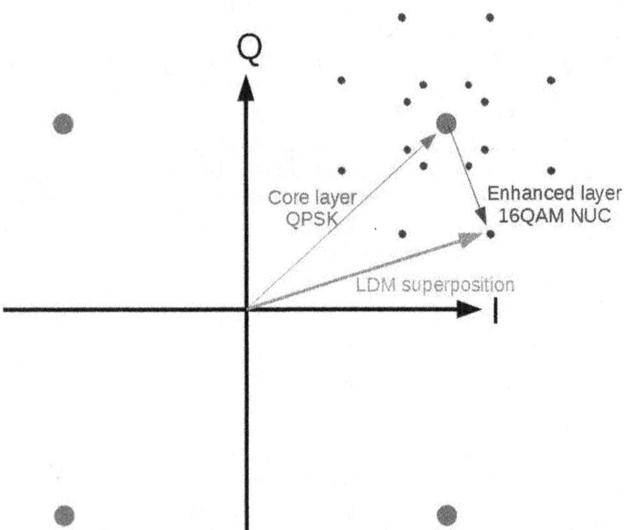

FIGURE 8.4-6 Example of LDM constellation superposition.

Johannes Sinnhuber

shows in a simulation of a typical scenario that overall channel capacity for the LDM system is higher than for a similarly configured TDM/FDM system. In TDM systems, however, receivers can be turned on and off so that only the portion of the signal dedicated to mobile reception is picked up. This means, as stated in (5), battery drain of mobile receivers is higher in LDM systems than in TDM systems.

COFDM WAVEFORM

At the end of each physical layer pipe, BICM data is mapped into cells. Cells are nothing more than pairs of I/Q values, i.e., real and imaginary values. The data volume (i.e., the number of bits) needed to build an I/Q cell depends on the QAM order, following the rule: 2^n bits = N states. Each BICM PLP output is connected to a configurable time interleaver block that interleaves the mapped cells, i.e., the pairs of I/Q values. The time interleaver runs in convolutional time interleaver (CTI) mode if there is only one PLP with constant cell rate (CCR). Hybrid time interleaver (HTI) mode is used if there is only one PLP with variable cell rate (VCR) or multiple PLPs. In CTI mode the time interleaver is a combination of two rotating data switches and several rows of shifting registers. In HTI mode, the time interleaver consists of three blocks (a cell interleaver, a twisted block interleaver and a convolutional delay line). The type of time interleaver in use depends on whether single or multiple physical layer pipes are being used. In the case of a single PLP with CCR, the CTI time interleaver is selected. In all other cases, HTI mode is chosen.

The outputs of all time interleavers end up in the framing block, which prepares the time- interleaved cells of all PLPs for OFDM generation. Here the cells are grouped into a frame structure, starting with a bootstrap component and a preamble component and proceeding with one or several subframes. The bootstrap component signals the beginning of an OFDM frame. The preamble component signals L1 information for the receivers followed by the payload cells in the subframes. After frame-building, a frequency interleaver ensures good distribution of information over frequency.

ATSC 3.0 supports three OFDM modes (8K, 16K and 32K). Due to the use of the inverse fast Fourier transform (IFFT) in OFDM, the number of OFDM carriers needs to be a power of two (i.e., 8K, 16K or 32K). Not all OFDM carriers are switched on; the border carriers to the lower and upper adjacent channel are set to zero. They build the guard band, making it possible to stop the OFDM shoulders caused by the sine$(x)/x$ functions of the modulated OFDM carriers.

The ATSC 3.0 OFDM channel bandwidth is typically 6 MHz; bandwidths of 7 MHz and 8 MHz are also supported (see Table 8.4-4). A frequency guard band is at the beginning and end of the channel, where the OFDM carriers are switched off.

Not all OFDM carriers are used for payload. They can also be used for the following:

- Building the guard band (these carriers are switched off)
- Continual pilots for frequency synchronization purposes

TABLE 8.4-4
Number of OFDM carriers and occupied OFDM bandwidth

C_{red_coeff}	Number of OFDM carriers	Number of OFDM carriers	Number of OFDM carriers	Occupied bandwidth	Occupied bandwidth	Occupied bandwidth
	8K FFT	16K FFT	32K FFT	6 MHz channel	7 MHz channel	8 MHz channel
0	6913	13825	27649	5.832844	6.804984	7.777125
1	6817	13633	27265	5.751844	6.710484	7.669125
2	6721	13441	26881	5.670844	6.615984	7.561125
3	6625	13249	26497	5.589844	6.521484	7.453125
4	6529	13057	26113	5.508844	6.426984	7.345125

TABLE 8.4-5
Guard interval (GI) sizes in ATSC 3.0.

FFT size	GI length (samples)	GI (relative)
8K	2048	25 %
8K	1536	18.75 %
8K	1024	12.5 %
8K	768	9.4 %
16K	4096	25 %
16K	3648	22.2 %
16K	2432	14.8 %
16K	1536	9.4 %
16K	1024	6.25 %
16K	768	4.7 %
32K	4864	14.8 %
32K	3648	11.1 %
32K	2432	7.4 %
32K	1536	4.7 %
32K	1024	3.1 %
32K	768	2.3 %

- Scattered pilots for channel estimation and correction
- Subframe boundary pilots
- Edge pilots

Pilots are boosted in comparison with the payload symbols, i.e., their amplitude is higher.

The pilots are inserted just before the IFFT block. The IFFT block transforms the OFDM carrier into the time domain. The result is an OFDM symbol. Its length depends on the OFDM mode and is based on the following formula known as "orthogonality condition":

$$\Delta t_{\text{symbol}} = 1/\Delta f_{\text{carrier spacing}}$$

In the time domain, a guard interval is inserted cyclically in front of each OFDM symbol. This cyclical prefix (CP) is nothing more than a repetition of the end of the following OFDM symbol. The guard interval corresponds to the maximum length of an echo path that can be handled by the OFDM system. The guard interval is for protecting the OFDM signal against intersymbol interferences in the presence of multipath conditions.

ATSC 3.0 supports four different guard interval sizes in the 8K mode and six guard interval sizes in the 16K mode and in the 32K mode. The higher the OFDM mode is, the more OFDM carriers are in use and the closer the carrier spacing (see Table 8.4-5). The more carriers are in use, the longer the OFDM symbols will be because of the orthogonality condition. Longer symbols result in longer cyclical prefixes. To allow long transmitter station distances in single-frequency networks, long guard intervals are required (see Table 8.4-6).

The peak-to-average power reduction (PAPR) block in the OFDM section is for limiting the crest factor. For PAPR, reserved carriers (TR = tone reservation) can be switched to I/Q states, which should result in a lower crest factor. Alternatively, active constellation extension (ACE) can be used for PAPR. In ACE some modification on the outer constellation points is allowed in order to reduce the crest factor. The PAPR algorithm has to be developed and implemented by the ATSC 3.0 modulator manufacturer. The PAPR is carried out in the frequency domain before IFFT, but the result has to be measured in the time domain; therefore, it is actually a back-coupled iterative process.

The traditional SISO (single input–single output) transmission principle has been in use in broadcasting applications for many years. MISO (multiple input–single output) supports multi-antenna systems, while MIMO (multiple input–multiple output) supports multiple receiver inputs. ATSC 3.0 supports SISO, MISO and MIMO mode.

All physical parameters of ATSC 3.0 can be derived from the ATSC modes' elementary periods:

- 0.1447 µs (for 6 MHz bandwidth)
- 0.1240 µs (for 7 MHz bandwidth)
- 0.1085 µs (for 8 MHz bandwidth)

TABLE 8.4-6
Carrier spacing and symbol duration in ATSC 3.0.

Mode	Carrier spacing in 6 MHz	Symbol duration in 6 MHz	Carrier spacing in 7 MHz	Symbol duration in 7 MHz	Carrier spacing in 8 MHz	Symbol duration in 8 MHz
8K	843.75 Hz	1.185 ms	984.37 Hz	1.015 ms	1125 Hz	0.888 ms
16K	421.9 Hz	2.370 ms	492.22 Hz	2.032 ms	562.5 Hz	1.777 ms
32K	210.94 Hz	4.740 ms	281.28 Hz	4.063 ms	281.25 Hz	3.555 ms

Based on these elementary periods, the symbol durations are as follows:

- For the 8K mode, the elementary period multiplied by 8192
- For the 16K mode, the elementary period multiplied by 16384
- For the 32K mode, the elementary period multiplied by 32766

SINGLE-FREQUENCY NETWORK (SFN) IN ATSC 3.0

One of the ATSC 3.0 design goals is to consider SFN capabilities. As also mentioned in the chapter on SFNs, standardization on STL and SFN distribution had not yet been finalized when this chapter was written. ATSC 3.0 SFN capabilities are therefore not covered at this stage.

Current Status of Standardization

When this chapter was written at the beginning of 2016, the ATSC 3.0 physical layer standard had Candidate status, which it had reached in September 2015. Candidate status means that the document state is expected to be close to final standard. In December 2015 the "System Discovery and Signaling" part (covering the structure of the bootstrap) also reached Candidate status. As a further core component, the SFN/STL standardization document is in the working draft status and still under discussion. Major progress toward achieving Candidate status is expected for the first half of 2016. The same is true for the working draft status of the ATSC 3.0 management and protocols standard. The ATSC 3.0 standard was expected to be finalized by the beginning of 2017.

New-Generation DTV Standards

DVB-T2

Digital Video Broadcasting (DVB) is a group of standards dealing with different ways of distributing digital video signals and associated audio and data. The first generation of DVB standards includes DVB-C, DVB-S and DVB-T and was published around 1995. At that time the main topic was to switch over from analog television to standard definition television (SDTV). The state-of-the-art multiplex input signal at that time was the MPEG-2 transport stream using MPEG-2 video and MPEG-1 layer II audio. In 2003 there was an increasing demand for higher data rates, because HDTV was visible on the horizon. As a result, DVB-S2—the new transmission standard via satellite—was developed and standardized. In DVB-S2 a new, powerful forward error correction (FEC) was introduced, pushing the net data rate closer to the physical limit. The second part of this chapter discusses the second generation of digital terrestrial video broadcasting—DVB-T2, which is based on DVB-S2.

FIRST- AND SECOND-GENERATION DVB STANDARDS: OVERVIEW AND COMPARISON

The first generation of DVB standards (DVB-x1) includes DVB-C, DVB-S, DVB-T, DVB-H and DVB-SH. The modulation process steps of these standards are coupled to the physical structure of the MPEG-2 transport stream protocol. The MPEG-2 transport stream consists of 188-byte packets having a 4-byte header and 184-byte payload. The first-generation DVB modulator interface expects such a data signal and synchronizes to it. The energy dispersal unit, which is nothing more than the data scrambler, is directly adapted to this data structure. The Reed-Solomon coder is the first DVB FEC coder and is referred to as an RS (188, 204) block coder. In other words, a 188-byte transport stream packet is taken and a 16-byte FEC overhead is added to it. It is then followed by a short Forney interleaver; in DVB-S and DVB-T, the FEC ends with a convolutional coder with puncturing.

The second generation of DVB standards (DVB-x2) needed to have much higher performance that got closer to the theoretical physical limit, known as the Shannon limit. DVB-S2, DVB-T2 and DVB-C2 are part of this state-of-the-art transmission standard group published between 2003 and 2009. All these new standards are no longer coupled to a fixed-input data stream format. Generic and multiple data stream formats and streams are allowed and therefore possible. But the MPEG-2 transport stream still contains and supports all demands made on a broadcast multiplex signal and, up to now, it is still the preferred input format for second-generation DVB standards. However, it is no longer mandatory. DVB-x2 also comes with a high-performance FEC that is built around the low-density parity check (LDPC) coder. LDPC is an

TABLE 8.4-7
Comparison and Overview of First- and Second-Generation DVB

	First-Generation DVB DVB-x1	Second-Generation DVB DVB-x2
Standards	DVB-C DVB-S DVB-T DVB-H DVB-SH	DVB-S2 DVB-T2 DVB-C2
Multiplex input signal	MPEG-2 transport stream	T2-MI (SFN, MPLP) MPEG-2 transport stream or generic stream
Forward error correction	Reed-Solomon and convolutional coder	BCH and LDPC coder
Synchronization of modulation process to input stream format	Modulation process coupled to MPEG-2 transport stream input format	No coupling of modulation process to input stream format
Number of input streams	Typically one input stream, DVB-T allows two streams (hierarchical modulation)	Single and multiple input streams, physical layer pipe (PLP) concept, optional variable coding and modulation

FEC algorithm developed in 1963 by Robert Gallager. At that time, however, there was no high-speed silicon technology developed to be able to use it. LDPC brings a 30% higher data rate in comparison with the former FEC algorithm (see Table 8.4-7).

DVB-T2: GETTING CLOSER TO THE SHANNON LIMIT

The main goal of DVB-T2 is to get more net data rate out of the broadcast transmission channel. The physical limit of the transmission channel is named after Claude Elwood Shannon, who published the "Shannon Law" in 1948. This Shannon limit is nothing more than the following formula:

$$C = B \times \log_2 (1 + S / N);\ [\text{bits/s}]\ (= \text{exact formula})$$

$$C = 1/3 \times SNR \times B;\ [\text{bits/s}]\ (= \text{coarse formula for an } SNR \text{ of 10 dB or better})$$

Where:

C = channel capacity in bits/s

B = channel bandwidth in Hz

S = signal power in watts

N = power in watts

SNR = signal-to-noise ratio in dB

Therefore, this mathematical equation represents more or less the product of bandwidth and signal quality. The wider the channel, the higher the theoretical maximum data rate in that channel; the better the SNR, the more data can be transported. Noise or any kind of interferer limits the data rate because bit errors are produced. The Shannon limit cannot be exceeded, not even with the best forward error correction. A good FEC brings the data rate closer to the Shannon limit. Depending on the signal quality, which is mainly described by the SNR, the modulation pattern and the FEC for a current application is selected. And depending on the channel bandwidth, the signal bandwidth needs to be chosen. In single-carrier modulation the following rule needs to be followed:

symbol rate = signal bandwidth < channel bandwidth

In multicarrier modulation, the rule is:

$$signal\ bandwidth = m \bullet carrier\ spacing = m \bullet \frac{1}{symbol\ duration\ OFDM} < channel\ bandwidth;$$

where m = No. of OFDM carriers

DVB-T(1) was developed and published around 1995. And the FEC used in DVB-T(1) represents the possibilities at that time, which was in principle Reed-Solomon block coding, time interleaving and convolutional coding with puncturing. Orthogonal frequency division multiplex (OFDM) was already in successful technical use in digital audio broadcasting (DAB). Due to the estimation of the SNRs in a digital terrestrial broadcast channel, the final DVB-T parameter set was selected. The result was OFDM with QPSK, 16QAM or 64QAM modulation and a Reed-Solomon RS (188, 204) FEC that was coupled to the transport stream protocol having 188-byte packets. After the Reed-Solomon coder, there is a short Forney time interleaver followed by a convolutional coder with a code rate of 1/2. The DVB-T(1) FEC ends with a puncturing unit, allowing six different final code rates to be selected. The net data rate of DVB-T(1) in an 8-MHz channel is between about 5 Mbit/s and 31 Mbit/s.

Before starting with the development of DVB-T2, all the requirements for the new standard were selected; the biggest requirement was to get as close as possible to the Shannon limit. Using a new forward error correction built around the LDPC coder, the result was 30% higher data rate. The LDPC algorithm was developed in the early 1960s and published in 1963 by Robert Gallager. But until 2000 there was not enough processing power available in the modulator and demodulators to use LDPC. In DVB-x2, the LDPC coder was combined with a BCH coder, a technology that has been in use for a long time, even in DVB-T(1) for the TPS (transmission parameter signaling) bits.

The DVB-x2 FEC consists of the following:

- Baseband scrambler
- BCH coder

- LDPC coder supporting a long 64-kbit and a short 16-kbit FEC frame
- Short bit interleaver

The LDPC/BCH FEC overhead is described by the code rate factor and can be selected according to six different robustness degrees.

The FEC is not the only screw for increasing the DVB-T2 data rate. The net data rate of DVB-T2 is higher than that of DVB-T(1), due to the following:

- The better relation of the OFDM symbol sizes relative to the guard interval sizes (e.g. 32K and 16K mode)
- The extended carrier mode
- Selectable overhead in the pilot pattern
- Higher modulation pattern up to 256QAM

Further goals of DVB-T2:

- Greater flexibility
- Support for fixed, portable and some mobile reception capability
- Support for different and multiple input signals and input signal formats
- Support for a channel bandwidth of 1.7 MHz, 5 MHz, 6 MHz, 7 MHz, 8 MHz and 10 MHz
- Support for 1K, 2K, 4K, 8K, 16K and 32K OFDM mode
- SISO and MISO mode

DVB-T2 uses only coherent modulation; differential mapping is not implemented (see Figure 8.4-7).

DVB-T2 INPUT SIGNAL FORMATS AND INPUT SIGNAL PROCESSING

DVB-T2 supports several input signal formats. Single and multiple input signals with theoretically up to 255 inputs are possible. The following input formats are used in DVB-T2:

- MPEG-2 transport streams (TS)
- Generic fixed packetized streams (GFPS)
- Generic continuous streams (GCS)
- Generic encapsulated streams (GSE)

The MPEG-2 transport stream format is well known and is used in all of the DVB-x1 standards. The MPEG-2 transport stream consists of a data structure with short packets of a constant length of 188 bytes that is divided into a 4-byte header and a 184-byte payload part. The header starts with a sync byte that has a value of 0x47. The content of the payload part is used in a flexible way, carrying, e.g., video, audio, data, etc., signaled by the packet identifier (PID). The MPEG-2 transport stream format is ideal for broadcast purposes and has all the tools needed for that application.

A generic fixed packetized stream is a generic data signal from which the constant length of the packets and the value of the sync byte is well known. The packet length is limited.

A generic continuous stream is a generic data stream with unknown or variable data-packet length and/or unknown sync byte.

A generic encapsulated stream (DVB-GSE) is a data format that was defined by the DVB group in 2007 to create a format especially for transporting IP packets. Together with the transport stream format, it is the preferred DVB-T2 data input format. But typically it is still the MPEG-2 transport stream format that is in use in most T2 applications.

A signal path from the physical input, its signal format, the forward error correction and the modulation is called a physical layer pipe (PLP). The term "PLP" is defined in the DVB-T2 standard. DVB-T2 allows the use of a single PLP (SPLP) and multiple PLPs (MPLP). The idea behind the PLP is to transport different applications with different data rates, robustness and therefore coverage. DVB-T2 transports the PLPs using variable coding and modulation (VCM) over time. Most of the time, only one PLP is used; the reason behind this is good, but the right applications are missing. This is similar to the principle of hierarchical modulation in DVB-T(1), but this was also not used except for some field trials.

If multiple PLPs or single-frequency networks are used in DVB-T2, the input data streams are combined into a special multiplex signal called a T2 modulator interface (T2-MI).

T2-MI SIGNAL

The T2-MI signal is the multiplex signal format of DVB-T2. It is built in the DVB-T2 headend (multiplex

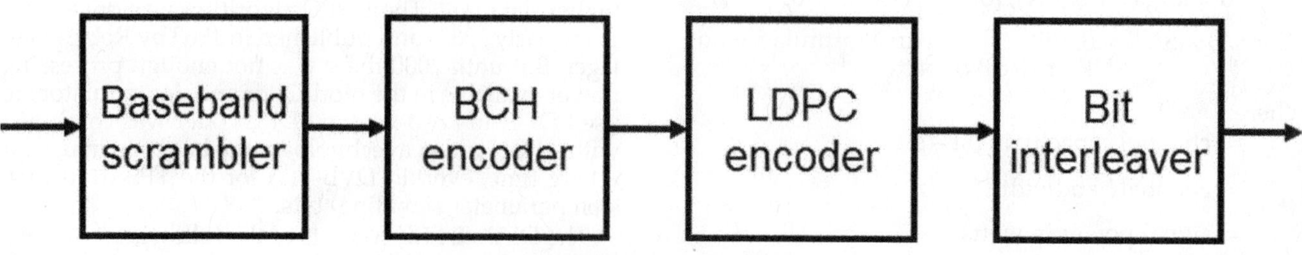

FIGURE 8.4-7 DVB-T2 forward error correction.

Fischer 2

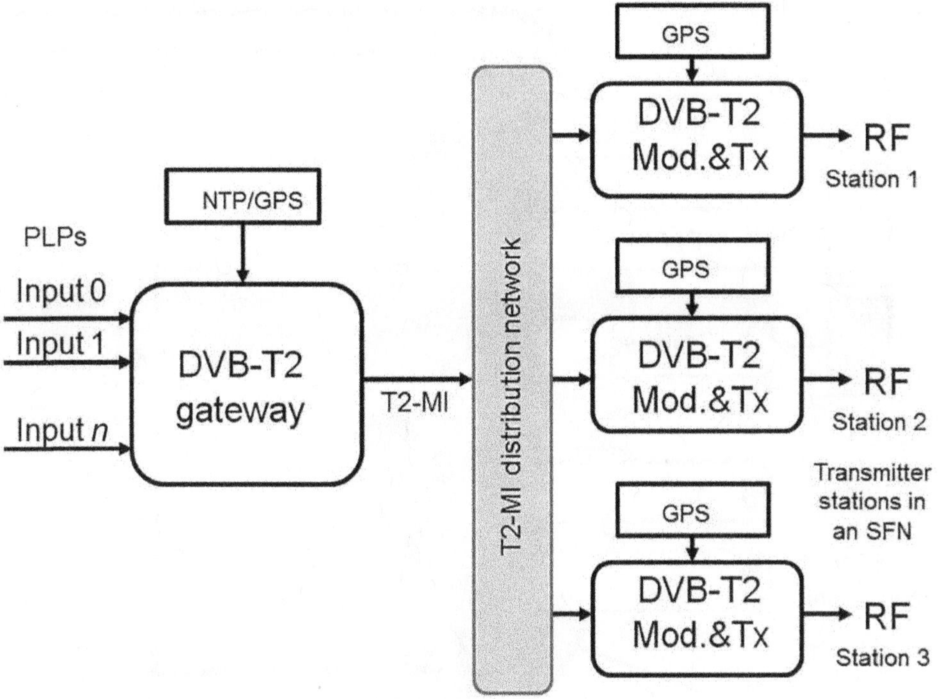

FIGURE 8.4-8 Structure of a DVB-T2 network.

center) in the T2 gateway. The T2 gateway is the final part of the T2 headend. It is fed by the different input streams (up to 255) and by GPS or network time protocol (NTP) (see Figure 8.4-8).

The T2 gateway combines all input streams, inserts SFN synchronization time stamps and layer 1 control information for setting up all T2 modulators in the transmitter stations. The result is a T2-MI signal with a packet structure consisting of a T2-MI header and a payload. The header includes the length of the T2-MI packet and information about which payload is currently transported. This payload can be one of the following:

- A baseband frame
- Auxiliary I/Q data
- L1 current information
- L1 future information
- A DVB-T2 time stamp
- Individual addressing information
- A future extension frame (FEF)

The T2-MI packets are transported on a well-defined PID via an MPEG-2 transport stream using data piping. Physically the transport stream can be distributed using TS-ASI or DVB-IP. This is done for compatibility reasons so that either existing TS-ASI distribution networks or dedicated IP networks can be used (see Figures 8.4-9 and 8.4-10).

If an MPEG analyzer were connected to an MPEG-2 transport stream carrying T2-MI packets, no PSI/SI would be detected. It is only a pure transport stream, typically using only a PID for T2-MI data piping and PID = 0x1FFF for the null packets. Alternatively, the DVB-IP protocol could be used to transport that transport stream via dedicated IP networks. DVB-IP means that a transport stream is transported using UDP, RTP and IP protocol. Advanced MPEG or TV analyzers support the analysis of T2-MI streams (e.g. the R&S DVMS1, R&S DVMS4 or R&S ETL from Rohde & Schwarz [DVMS1], [DVMS4], [ETL]).

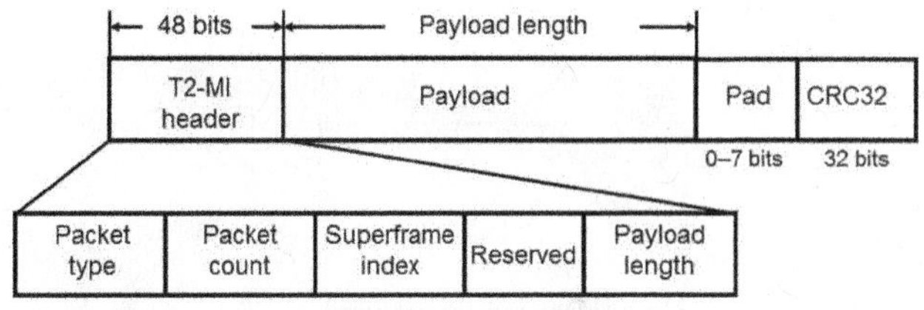

FIGURE 8.4-9 T2-MI packet.

Fischer 1

DVB-T2 FORWARD ERROR CORRECTION (FEC)

The channel coding or the FEC in DVB-T2 consists of an outer and inner coder. The outer coder uses Bose-Chaudhuri-Hoquenghem (BCH) block coding. The inner coder is a state-of-the-art LDPC block coding algorithm. The LDPC principle has been known since the 1960s but could not be used before the 2000s

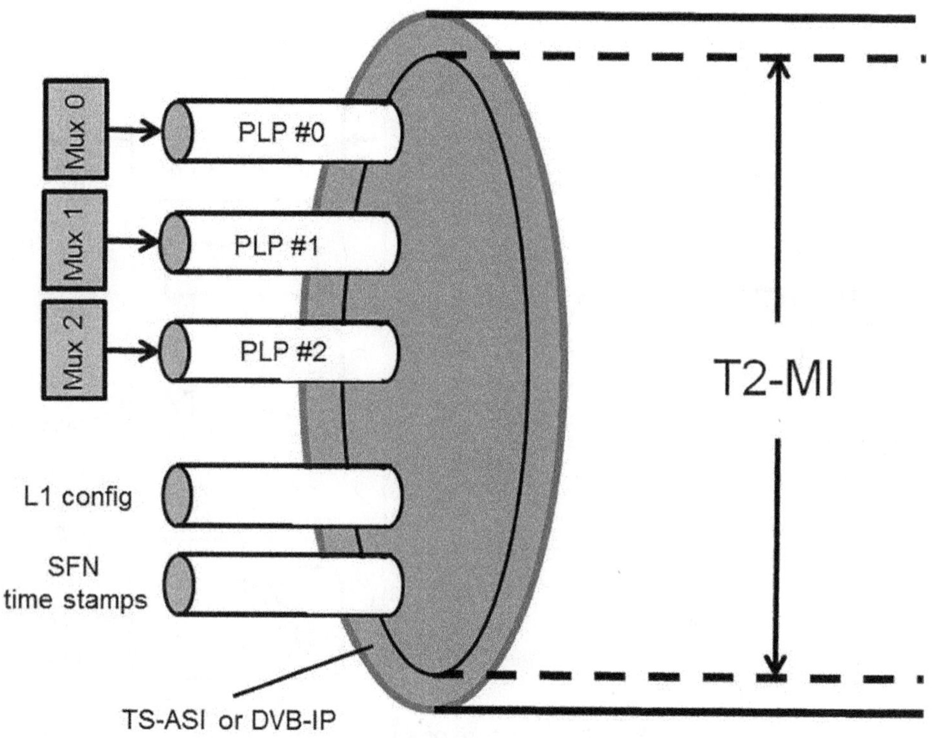

FIGURE 8.4-10 T2-MI packet signal structure.

Fischer 2

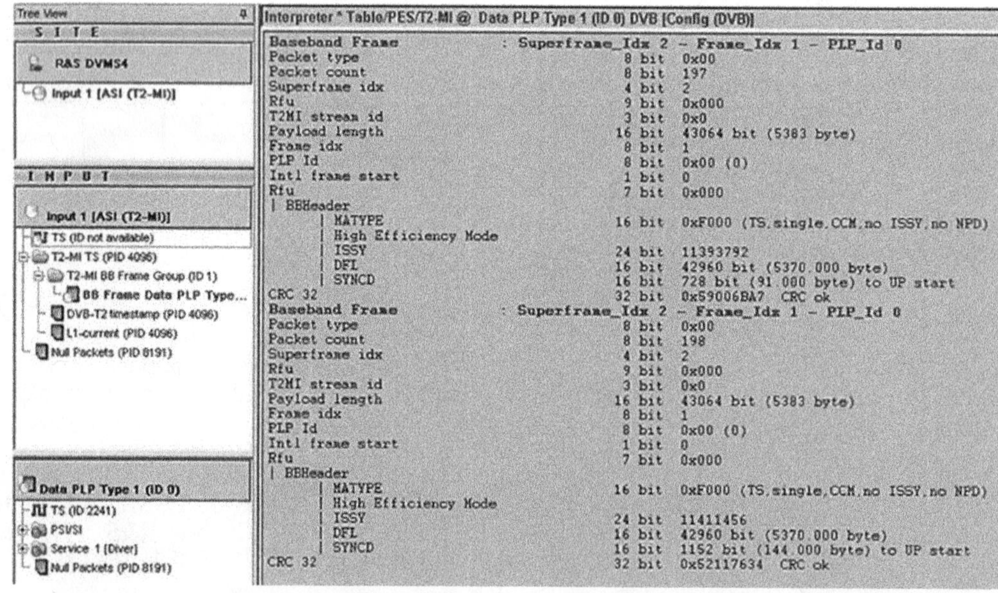

FIGURE 8.4-11 T2-MI signal structure analyzed with an MPEG analyzer [DVMS].

because of the high calculation power needed in the modulators and receivers. DVB-T2 supports six code rates:

- 1/2 (maximum robustness, minimum net data rate)
- 3/5
- 2/3
- 3/4
- 4/5
- 5/6 (minimum robustness, maximum net data rate)

The T2-FEC starts with a scrambler that mixes the data stream with a pseudorandom binary sequence followed by the BCH and LDPC coder. The last part is the bit interleaver, which is the first of at most four interleavers in DVB-T2. The scrambler is responsible for the "whitening" of the data stream. This means that possible zero or ones sequences are broken up into a more or less random data stream. In case of burst errors, the bit interleaver helps to break them up into single errors.

If there are multiple PLPs, a full FEC path is used in each PLP path, allowing different FEC code rates to be selected.

MODULATION PATTERN

DVB-T2 only supports coherent modulation. That means all the bits are coded in one symbol. The advantage, in comparison with differential modulation, is the better bit error ratio in case of stationary reception or, in other words, an SNR requirement that is 3 dB better. DVB-T2 supports QPSK, 16QAM, 64QAM and 256QAM modulation. The mapping rule follows Gray coding, and nonrotated and rotated constellation diagrams are used.

First of all, the term "cell" has to be defined. In DVB-T2, a cell is nothing more than a mapped bit combination represented by an I/Q pair or, in other words, a real and an imaginary value. This definition could also be used in any other kind of modulation or broadcast standard, such as DVB-T(1), but this was not necessary. In DVB-T2 we cannot see the result of the mapping process directly. A lot of signal processing steps—such as rotation, Q delay, several interleaving processes, etc.—will occur before the I and Q values can be placed on an OFDM carrier. Building a cell in DVB-T2 therefore involves the following:

- The mapping of n bits
- on a (re, im) I/Q pair
- in an I/Q domain with N states
- following the rule
- $2^n = N$ states

After the cell-building, the constellation diagrams can be processed in a nonrotated or rotated manner. Rotation means that the constellation diagrams are turned counterclockwise by a few degrees to get well-distributed projection points (see Figure 8.4-12) from all constellation points to the horizontal and vertical axes (I-axis, Q-axis). In the receivers, the information about the position of the projection points on the I-axis and Q-axis can be used for making soft decisions in the FEC processing block. Furthermore, the Q value is not transported on the same OFDM carrier; a Q delay process in the modulator separates the Q values from the I values after rotation (Figure 8.4-12). The reason for this is, in case of fading notches, to prevent the I and Q values from falling into the same notches. The idea behind rotation and Q delay is to make the DVB-T2 signal more robust. At the moment, however, this is pure theory; practical tests in the lab and in the field have shown that there are no advantages to using rotated constellations (see Figure 8.4-13).

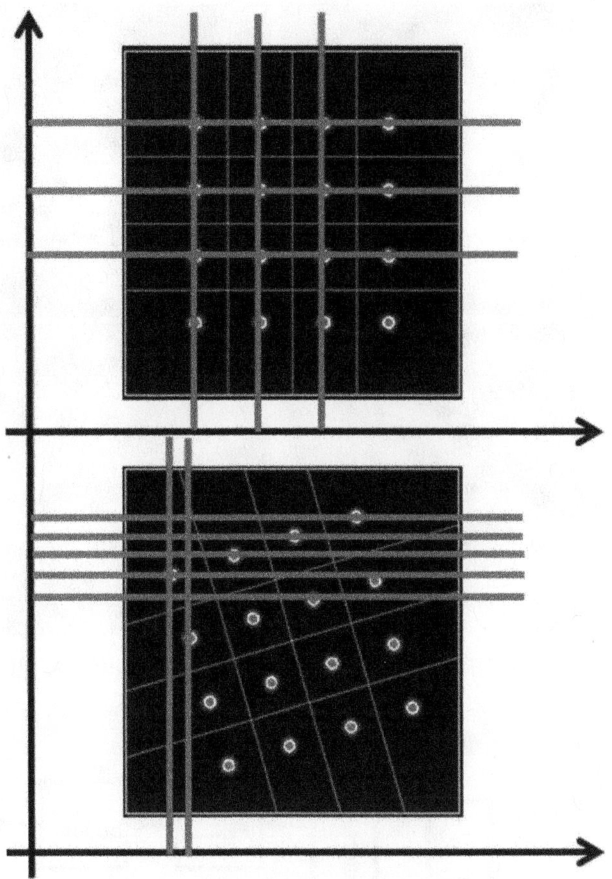

FIGURE 8.4-12 Nonrotated and rotated constellation diagram with projection points.

Fischer 1

After the optional rotation and Q-delay process, the cell interleaver distributes the cells before they pass through further signal processing stages.

INTERLEAVING

DVB-T2 uses four interleavers:

- The bit interleaver immediately after the FEC
- The cell interleaver after rotation and Q delay
- The optional and configurable time interleaver
- The frequency interleaver immediately before the main OFDM block

The interleaving process distributes data over time in order to be prepared for burst errors or similar time domain effects. Deinterleaving in the receiver returns the interleaved information to the original order and breaks up burst errors into single errors that can be more easily repaired by the receiver FEC.

DVB-T(1) has no time interleaver but only a short interleaver similar to that of DVB-T2. The time interleaver makes T2 more powerful (see Figure 8.4-14). Time interleavers are also used in DAB, DAB+ and ISDB-T systems.

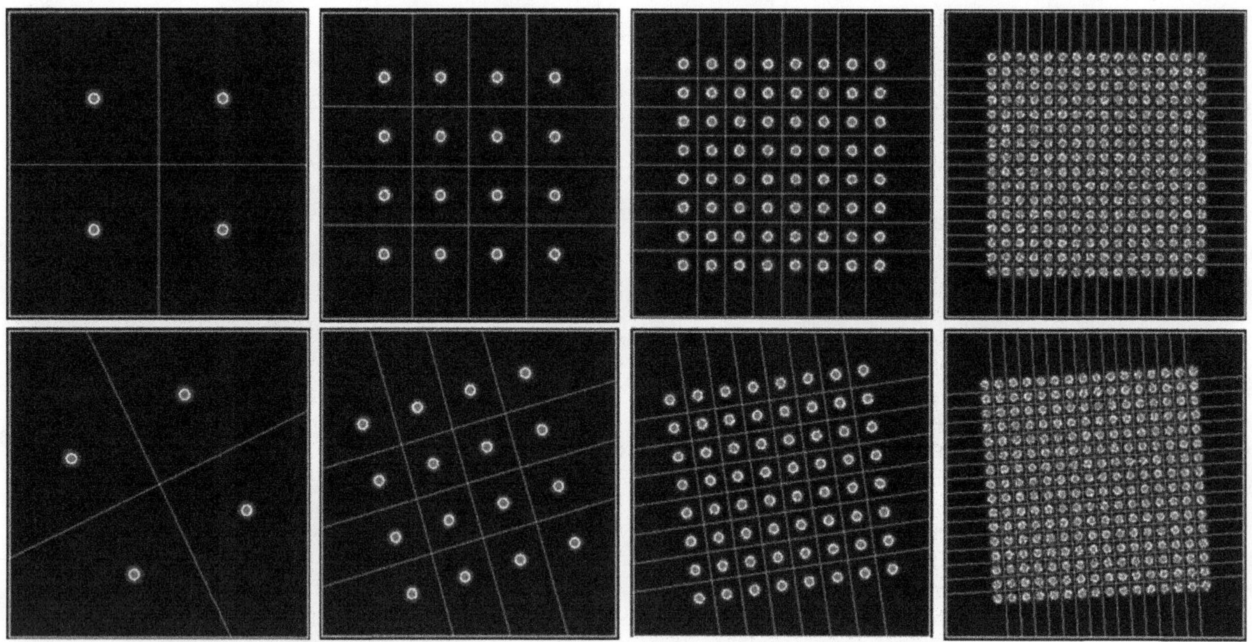

FIGURE 8.4-13 DVB-T2 constellation pattern, QPSK, 16QAM, 64QAM, 256QAM nonrotated and rotated.

Fischer 1, ETL

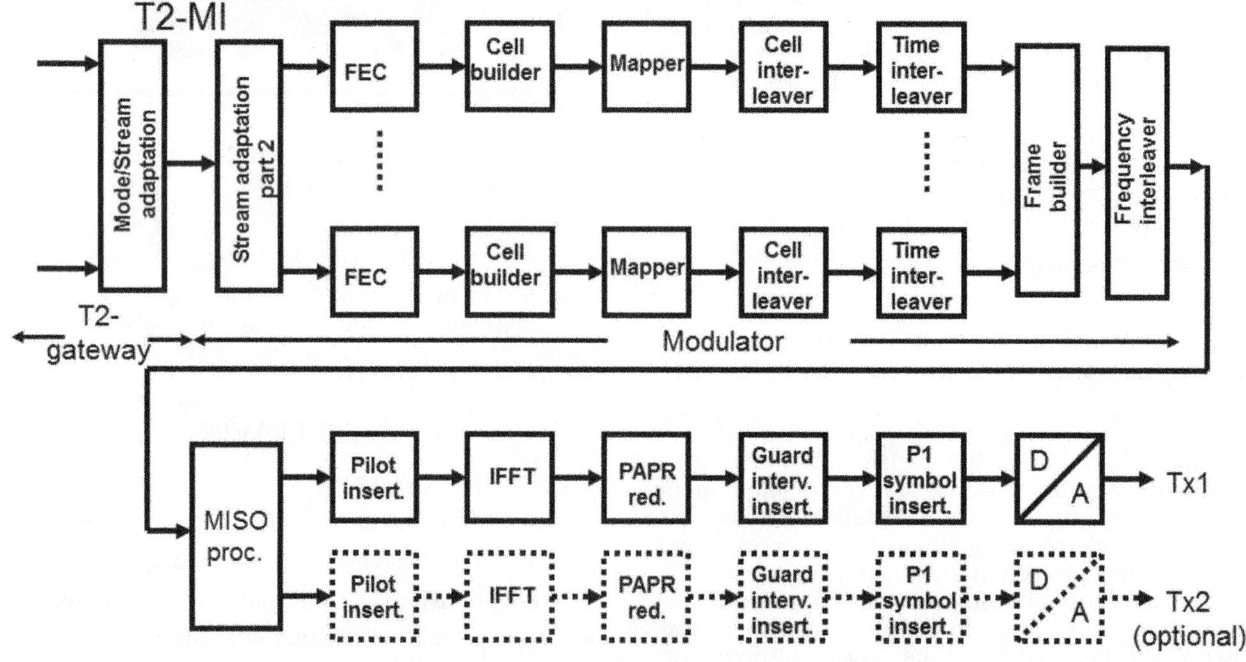

FIGURE 8.4-14 Block diagram of a DVB-T2 modulator.

Fischer 1

OFDM IMPLEMENTATION DVB-T2

DVB-T2 supports channels of different bandwidths (see Figure 8.4-15):

- 8-MHz, 7-MHz and 6-MHz terrestrial TV broadcast channels
- 10-MHz and 5-MHz L-band channels
- A 1.75-MHz DAB/DAB+ channel

The OFDM signal bandwidths, always somewhat smaller than the channel bandwidths, are approximately:

- 7.6 MHz, 6.6 MHz and 5.7 MHz
- 9.6 MHz and 4.7 MHz
- 1.536 MHz

Orthogonal frequency division multiplex (OFDM) means that instead of one I/Q-modulated carrier, many carriers are used. This is called multicarrier modulation. The reason for using OFDM is the multipath situation in the terrestrial broadcast channel. OFDM is designed to cope with that multi-echo reception situation. The main processing part in the OFDM modulator is the inverse fast Fourier transform (IFFT) block. The IFFT runs on $2^n = N$ carriers. In DVB-T2, N can have the following values

- 32K (32768)
- 16K (16384)

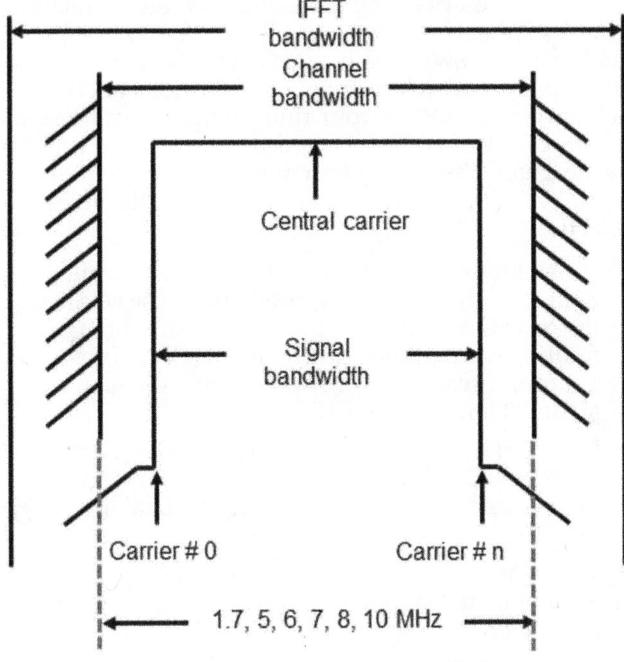

FIGURE 8.4-15 DVB-T2 channel.

Fischer 1

- 8K (8192)
- 4K (4096)
- 2K (2048)
- 1K (1024)

The more OFDM carriers in use, the closer the carrier distance $\Delta f_{\text{carrier}}$ is. The reciprocal of the carrier spacing is the symbol duration:

$$\Delta t_{\text{symbol}} = 1/\Delta f_{\text{carrier}}$$

This formula is also called "orthogonality condition." If this condition is fulfilled, then all OFDM carriers are positioned in a way that all carrier spectra have a zero where all the other OFDM carriers have their maximum. This results in minimum carrier cross talk.

All OFDM carriers have a $\sin(x)/x$ spectrum (see Figure 8.4-16). This results in "shoulders" on both sides of the useful OFDM signal. But these shoulders are lower if more OFDM carriers are in use. Consequently, a 32K-mode OFDM spectrum has a lower shoulder than a 1K-mode shoulder. Therefore, in the 32K, 16K and 8K modes an extended carrier mode is allowed, which means that the spectrum is about 100 kHz wider but the data rate is also a little higher in comparison with the "normal" OFDM mode.

The IFFT, which comes after all interleavers, is the step that creates the OFDM signal. After the frequency interleaver, there are still some gaps in the spectrum; they are the reserved positions for the pilots. DVB-T2 uses the following pilots:

- Continual pilots for the automatic frequency control (AFC) process in the receiver
- Scattered pilots for channel estimation and correction

The continual pilots are on a fixed position in the spectrum and in the constellation diagram; their position is on the horizontal axis, either in the negative

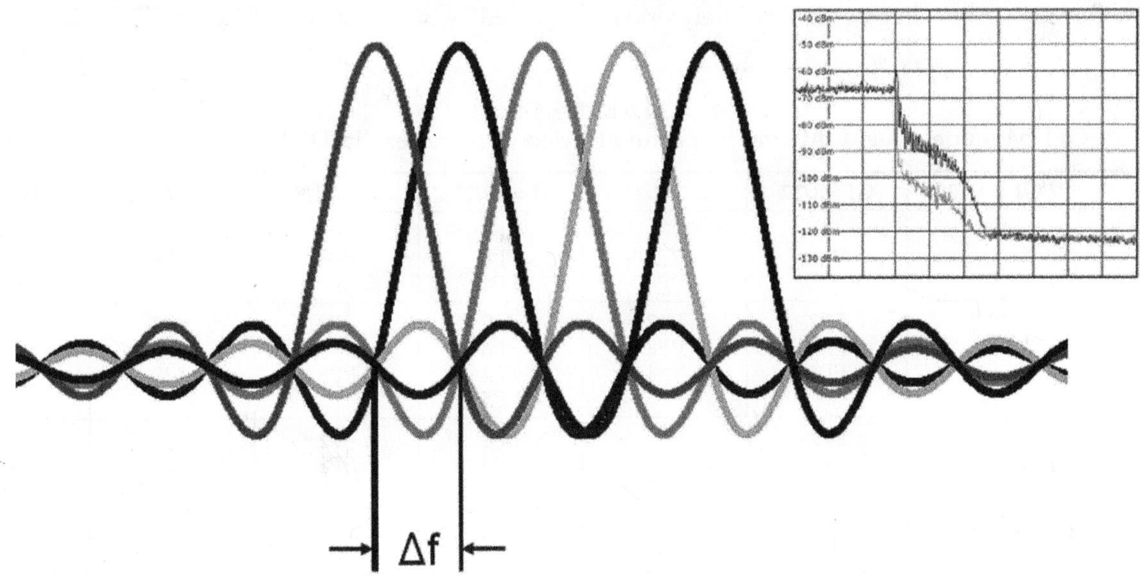

FIGURE 8.4-16 OFDM carrier spectrum ($\sin(x)/x$).

Fischer 1

or the positive direction. The amplitude of all pilots is boosted in comparison with the data carriers. The scattered pilots jump over the spectrum starting on the lower frequency end, moving toward higher frequencies. In DVB-T2, the carrier distance of the scattered pilots can be selected by the pilot pattern parameter (PP1, PP2, PP3, PP4, PP5, PP6, PP7 or PP8). PP*x* stands for pilot pattern *x*. The nearer the pilot pattern is, the better the channel estimation will predict echoes. The wider the pilot pattern position, the higher the net data rate. Not all pilot patterns can be selected in all OFDM modes and guard interval sizes.

After the IFFT, the signal is in the time domain. The OFDM symbols now have a dedicated symbol duration, which is the reciprocal of the OFDM carrier spacing. DVB-T2 uses the following approximate carrier spacings:

- 250 Hz (32K)
- 500 Hz
- 1 kHz
- 2 kHz
- 4 kHz
- 1 kHz (1K)

Consequently, the OFDM symbol duration in DVB-T2 is about:

- 4 ms (32K)
- 2 ms
- 1 ms
- 500 µs
- 250 µs
- 125 µs (1K)

The exact values depend on the channel bandwidth. The DVB-T2 OFDM mode most selected is the 32K mode. This is because the longest symbol durations are the best mode for building single-frequency networks. On the other hand, the 1K mode would be the best mode for mobile reception capability because of the longest carrier spacing.

Before the OFDM symbols, there is a break that is filled up with the end of the following symbol. It is called guard interval (GI) or cyclical prefix (CP). The guard interval is the time that protects the OFDM symbol against cross talk resulting from echoes. Consequently, the longest possible echo delay time corresponds to the length of the guard interval. In single-frequency networks (SFN), long guard intervals need to be selected. In DVB-T2 the length of the guard interval is described by a fractional value in relation to the symbol duration, e.g., GI = 1/4, 1/8, 1/16, etc. Table 8.4-8 shows which guard interval sizes can be chosen. The big advantage of the 32K mode in DVB-T2 is that its symbols are four times longer than those in the 8K mode of DVB-T(1), and therefore the overhead in the guard interval is four times lower. This results in higher net data rates. Listed below are the rules for selecting the DVB-T2 parameters:

- If a single-frequency network needs to be built, then the 32K (or 8K, 16K) mode has to be selected; the guard interval size chosen must fulfill the multipath conditions caused by the SFN. Typical guard interval sizes for SFNs are in a range of about 70 km (e.g. GI = 1/16 in 32K mode).
- In case of multifrequency network (MFN), quite short guard intervals in the range of about 15 µs can be used, which results in higher data rates than in SFNs.
- OFDM modes toward 4K, 2K and 1K are better for mobile reception but not the right choice for building SFNs.

Several OFDM symbols including the guard interval form a DVB-T2 OFDM frame. The frame starts with a very special preamble symbol (P1) that marks the beginning of the OFDM frame. The P1 symbol is followed by one or up to 16 P2 preamble symbols. The P1

TABLE 8.4-8
OFDM modes, guard interval sizes and allowed pilot pattern in DVB-T2 (SISO mode)

FFT size	GI = 1/128	GI = 1/32	GI =1/16	GI = 19/256	GI = 1/8	GI = 19/128	GI = 1/4
32K	PP7	PP4 PP6	PP2 PP8 PP4	PP2 PP8 PP4	PP2 PP8	PP2 PP8	NA
16K	PP7	PP7 PP4 PP6	PP2 PP8 PP4 PP5	PP2 PP8 PP4 PP5	PP2 PP3 PP8	PP2 PP3 PP8	PP1 PP8
8K	PP7	PP7 PP4	PP8 PP4 PP5	PP8 PP4 PP5	PP2 PP3 PP8	PP2 PP3 PP8	PP1 PP8
4K, 2K	NA	PP7 PP4	PP4 PP5	NA	PP2 PP3	NA	PP1
1K	NA	NA	PP4 PP5	NA	PP2 PP3	NA	PP1

symbol only signals some major basic parameters for the DVB-T2 receiver such as the OFDM mode, guard interval size, etc. The P2 symbols are responsible for informing the receivers about all physical PLP parameters. These parameters include the following:

- Number of PLPs
- PLP modulation
- PLP code rate

The content of the P2 symbol is called layer 1 (L1) signaling. The number of P2 symbols depends on the OFDM mode. The OFDM modes that have the longest symbols need few or only one P2 symbol. The P2 symbols consist of a pre- and postsignaling part. The presignaling part has a fixed modulation pattern, BPSK. In the postsignaling part, either BSKP, QPSK, 16QAM or 64QAM can be selected as the modulation pattern. The L1 signaling FEC code rate is fixed.

The length of a DVB-T2 frame can be adjusted between certain limits. A DVB-T2 superframe consists of a number of frames; its length can also be selected between certain limits. In DVB-T2 SFNs, all the frames and superframes in all transmitter stations have to be fully synchronized (see Figure 8.4-17).

PAPR REDUCTION

OFDM systems have a rather high relation between the maximum peak power and average power. This relation is called crest factor or peak-to-average power ratio (PAPR). DVB-T2 provides methods that reduce this PAPR. These PAPR reduction methods are called:

- Active constellation extension (ACE)
- Tone reservation (TR)

They can be optionally implemented by DVB-T2 modulator manufacturers. ACE can only be used if rotation is switched off. If TR is switched on, then some carriers are reserved for PAPR reduction, resulting in a somewhat lower net data rate. The idea of PAPR reduction is to enable the modulator to slightly manipulate some parts of the signal to counteract the high peak amplitudes without changing the signal quality. As mentioned earlier, PAPR reduction in DVB-T2 is an option; not each modulator supports this feature, and there is no standardized way to implement the PAPR algorithm. It is therefore in the hands of the modulator/transmitter manufacturer.

SFNS IN DVB-T2

Single-frequency networks are used to save frequencies. The same frequency can be used in a larger coverage area having multiple transmitters running on the same frequencies. These transmitters must meet the following requirements:

- They must be fully synchronous in frequency (better than $1 \cdot 10^{-9}$).
- They must be fully synchronous in time (all OFDM symbols, frames and superframes in all SFN transmitter stations need to start at the same time).
- They must be fully synchronous in data (all symbols, frames and superframes in all SFN transmitter stations need to carry exactly the same data).
- They must not violate the guard interval (transmitter distance within the cyclical prefix size).

SFNs are synchronized and controlled via the T2 gateway and GPS time/frequency standard receivers in the broadcast headend (multiplex center) and in the SFN

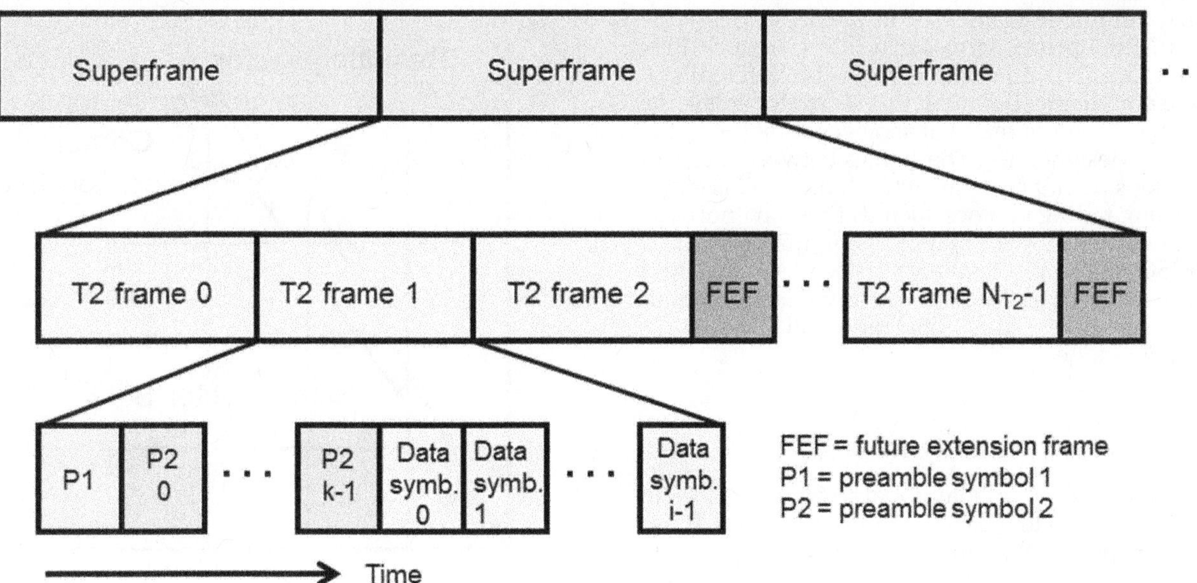

FIGURE 8.4-17 Structure of a DVB-T2 OFDM superframe and frame with P1, P2 and data symbols.

Fischer 1

transmitter stations. Use of the T2-MI signal is mandatory in T2-SFN networks. The T2-MI signal contains the baseband frame content for all PLPs, the L1 control information describing the physical parameters of the PLPs and the SFN synchronization time stamps. The time stamps are derived from a high-accuracy time standard that comes directly from GPS or via a network time protocol (NTP) server. The synchronization time stamps indicate either a relative or an absolute time. With this time information, the transmission points in all modulators in an SFN are synchronized in time. For this purpose, a DVB-T2 OFDM signal consists of symbols, frames and superframes. The beginning of each symbol, frame and superframe is controlled by the incoming T2-MI signal in each T2 modulator in an SFN. The transmission time in each station can be shifted slightly by a local adjustment parameter called static delay.

SISO/MISO IN DVB-T2

SISO (single input–single output) and MISO (multiple input–single output) are well-known terms in the modern mobile communications world. SISO is nothing more than standard transmission and reception conditions with one transmitter and one receiver antenna. Diversity receivers for mobile reception in automotive applications are already known from DVB-T(1). Such receivers may have several reception antennas, and the receiver always selects the antenna signal that has the best reception quality. According to the Alamouti principle, there can also be several transmission and reception antennas resulting in space-time diversity. In DVB-T2, a modified Alamouti principle is implemented as an option called T2-MISO mode. T2 does not support space-time diversity but space-frequency diversity. Building co-located or distributed MISO networks is described in DVB-T2. Some field trials included a number of tests for implementing the distributed MISO principle in T2 in SFNs. In such SFN MISO applications, the signals from the different stations are not identical. At first, this seems to violate the SFN principle, but in reality it achieves what the MISO principle is designed for. The signals between adjacent SFN stations do not correlate to each other. This eliminates strong fading notches such as those in "normal" SFNs. The tests showed that by using the distributed SFN MISO, the single-frequency networks in T2 could be a somewhat optimized. At the moment, no operating DVB-T2 network uses MISO mode; all T2 networks run in SISO mode.

DVB-T2 MEASUREMENTS

DVB-T2 uses the same quality parameters as those in ATSC or DVB-T(1) to measure the signal quality. These parameters include the following:

- RF level
- Bit error ratio (BER)
- Modulation error ratio (MER)
- Shoulder attenuation
- Channel impulse response function (echo pattern)

The RF level is measured by sampling and calculating the RMS voltage of the DVB-T2 signal. A typically "good" RF level for a receiver is about 55 dBμV to 70 dBμV. The minimum requested receiver input level depends on the robustness of the signal described by the PLP modulation and FEC parameter. The minimum receiver input level is thus in the range of 20 dBμV to 30 dBμV. The BER is the relation between the errored bits and the received bits, defined as:

- BER before LDPC (channel BER)
- BER before BCH
- BER after BCH

The MER is the overall signal quality parameter, describing any kind of influence on signal quality caused by, e.g.:

- Gaussian noise
- Signal jitter
- Phase noise
- Multipath propagation
- Amplitude response
- Group delay
- Any kind of additive interferer

During the MER measurement, the error vector length in the constellation diagram is measured and the relation of signal amplitude to error vector amplitude is presented as a MER value in dB. The shoulder attenuation describes the out-of-band products in the adjacent channels (see Figure 8.4-18). It is measured using a spectrum analyzer with the markers in the center of the channel and on both—the lower and upper—shoulders of the OFDM signal. Typically the

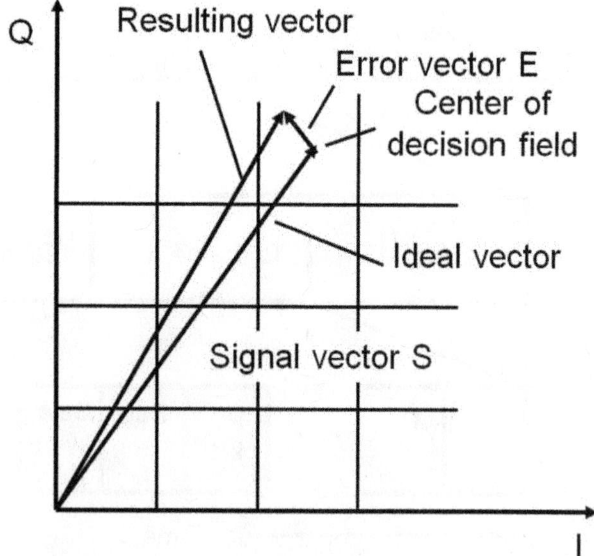

FIGURE 8.4-18 MER measurement principle.

Fischer 1

shoulders are measured about 200 kHz below and above the useful spectrum. The shoulder attenuation is about 40 dB. This measurement is a typical transmitter adjustment and acceptance test measurement. This measurement is used to set up the precorrection for the DVB-T2 power amplifiers (se Figure 8.4-19).

The echo pattern or channel impulse response is used for the coverage measurement in single-frequency networks to check whether the SFN condition is fulfilled and to make sure that no signal is outside the guard interval (see Figure 8.4-20).

DVB-T2—STATUS

DVB-T2 was published in 2008, and the first DVB-T2 networks were on air in 2010. From that year on, most of the digital terrestrial networks worldwide were switched on in T2. Most of the DVB-T2 networks use the following parameters:

- SFN mode
- SISO mode
- 8 MHz channel
- 32K extended carrier mode
- 256QAM or 64QAM
- Rotated or nonrotated (currently, there is no advantage in rotated)
- CR = 2/3 or 3/4
- Time interleaver on
- Single PLP

In most of the applications, T2-MI contains one PLP carrying an MPEG-2 transport stream with MPEG-4/AVC/H.264 video and MPEG-1 layer 1 audio, Dolby AC-3 audio and some data broadcast applications such as HbbTV® and teletext. Germany will change from DVB-T(1) to DVB-T2 in 2017 by using the new HEVC/H.265 video codec.

DVB-T2 increases the net data rate of DVB-T(1) by a factor of 1.7 to 2 depending on the application and the selected physical transmission parameters. Possible DVB-T2 data rates are within a range of about 7 Mbit/s to 1 Mbit/s. Typical practical DVB-T2 net data rates for coverage planning are approximately:

- 22 Mbit/s to 26 Mbit/s for SFNs and portable indoor device
- 30 Mbit/s to 40 Mbit/s for SFNs and fixed (outdoor roof) antenna
- Even >40 Mbit/s for MFNs and fixed (outdoor roof) antenna

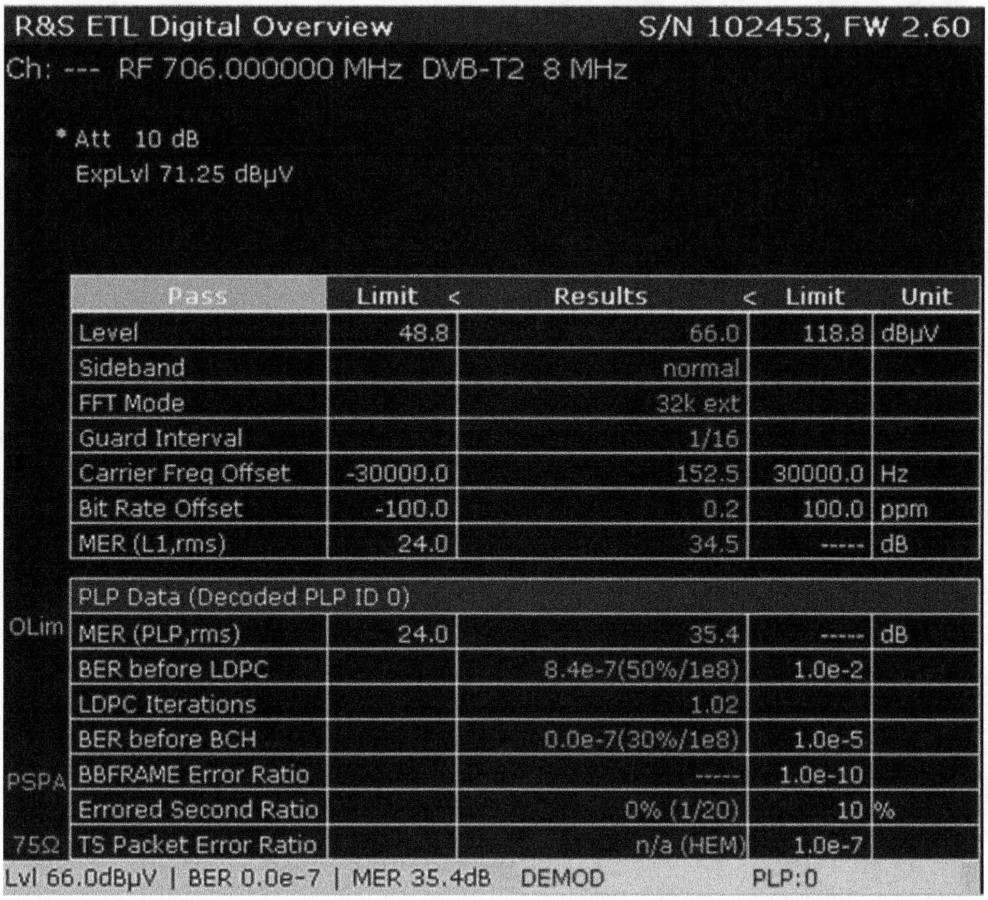

R&S ETL Digital Overview S/N 102453, FW 2.60

Ch: --- RF 706.000000 MHz DVB-T2 8 MHz

* Att 10 dB
ExpLvl 71.25 dBµV

Pass	Limit <	Results	< Limit	Unit
Level	48.8	66.0	118.8	dBµV
Sideband		normal		
FFT Mode		32k ext		
Guard Interval		1/16		
Carrier Freq Offset	-30000.0	152.5	30000.0	Hz
Bit Rate Offset	-100.0	0.2	100.0	ppm
MER (L1,rms)	24.0	34.5	-----	dB
PLP Data (Decoded PLP ID 0)				
MER (PLP,rms)	24.0	35.4	-----	dB
BER before LDPC		8.4e-7(50%/1e8)	1.0e-2	
LDPC Iterations		1.02		
BER before BCH		0.0e-7(30%/1e8)	1.0e-5	
BBFRAME Error Ratio		-----	1.0e-10	
Errored Second Ratio		0% (1/20)	10	%
TS Packet Error Ratio		n/a (HEM)	1.0e-7	

OLim PSPA 75Ω

Lvl 66.0dBµV | BER 0.0e-7 | MER 35.4dB DEMOD PLP:0

FIGURE 8.4-19 DVB-T2 measurement parameters [ETL].

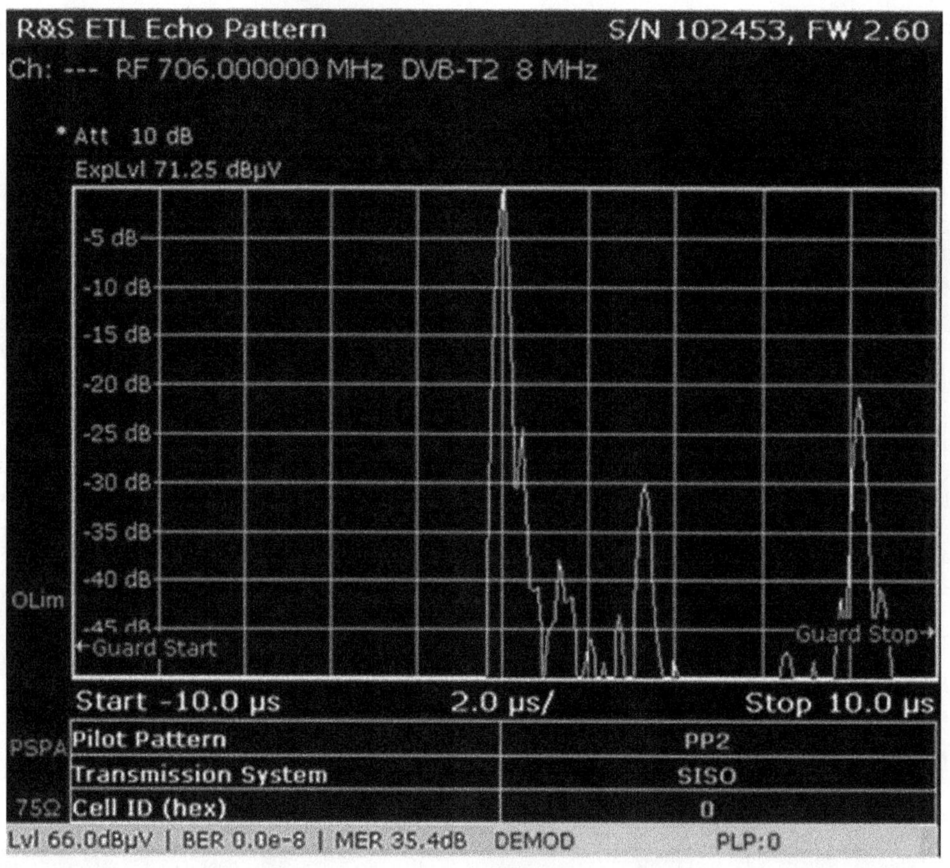

FIGURE 8.4-20 DVB-T2 echo pattern [ETL].

Some standards developed and published later are based on principles and ideas coming from DVB-T2, which are, e.g., DVB-C2 and ATSC 3.0.

Some figures in this chapter are taken from [Fischer1] and [Fischer2], Springer.

HbbTV is a registered trademark.

Additional information for ATSC may be found at ATSC.org.

References

[1] Lachlan, Michael and Gomez-Barquero, David. *Modulation and Coding for ATSC 3.0.* 2015 IEEE International Symposium on Broadband Multimedia System and Broadcasting (BMSB), Ghent, Belgium, June 17–19, 2015.

[2] ATSC. *System Requirements for ATSC 3.0.* s.l. Doc. TG3-S31-087r12, 21 October 2015.

[3] ATSC. *ATSC Candidate Standard Physical Layer Protocol.* s.l. Doc. S32–230r21, 28 September 2015.

[4] Liang Zhang, Yiyan Wu, Wei Li, Heung Mook Kim, Sung-Ik Park, Pablo Angueira. *Channel Capacity Distribution of Layer-Division-Multiplexing System for Next Generation Digital Broadcasting Transmission.* 2014 IEEE International Symposium on Broadband Multimedia Systems and Broadcasting (BMSB), Beijing, China, June 25–27, 2014.

[5] Fay, Luke. *ATSC 3.0 Physical Layer Overview.* 2015 IEEE International Symposium on Broadband Multimedia Systems and Broadcasting (BMSB), Ghent, Belgium, June 17–19, 2015.

Bibliography

[Fischer1] Fischer, W., *Digital Video and Audio Broadcasting Technology—A Practical Engineering Guide*, Springer, Heidelberg, 2010, 3rd Edition.

[Fischer2] Fischer, W., *Digitale Fernseh- und Hörfunktechnik in Theorie und Praxis*, Springer, Heidelberg, 2015, 4th Edition. Auflage.

[Alamouti] Alamouti, S., A Simple Transmit Diversity Technique for Wireless Communications, *IEEE Journal*, October 1998, pp. 1451–1458.

[DVB-T2] Digital Video Broadcasting (DVB); Frame Structure Channel Coding and Modulation for a Second Generation Digital Terrestrial Television Broadcasting System (DVB-T2), January 2008.

[T2-MI] ETSI TS 102773, Digital Video Broadcasting (DVB); Modulator Interface (T2-MI) for a Second Generation Digital Terrestrial Television Broadcasting System (DVB-T2), 2009.

[DVB-T2 Implementation Guidelines] Implementation Guidelines for a Second Generation Digital Terrestrial Television Broadcasting System (DVB-T2), DVB Document A133, February 2009.

[Reimers] Reimers, U., DVB, *The Family of International Standards for Digital Video Broadcasting*, Springer, Berlin, Heidelberg, New York, 2004.

CHAPTER

8.5

Single-Frequency Networks for DTV Systems

WALTER FISCHER

Rohde & Schwarz

In the past, typical broadcast networks transmitted different local content on each transmitter site. That was true for FM transmitter stations and analog TV transmitter stations as well as most ATSC stations. Today such a principle is called multifrequency networks. Analog TV signals in particular would create ghost pictures if we transported and received those signals via different signal paths. That was one of the reasons for using directional antennas in analog TV. Back then, the availability of free frequencies was not as big a problem as nowadays. Today frequencies are an expensive and rare resource that has to be carefully saved and handled. When digital audio broadcasting (DAB) [ETS300401] was launched in the early 1990s, a new term was introduced: SFN, or single-frequency networks. SFNs use the same frequency over a larger coverage area, encompassing several transmitter stations on the same frequency to transport the same data content fully synchronously in frequency, time and data. In particular, digital broadcast standards using the orthogonal frequency division multiplex (OFDM) principle are ideal for SFN applications. This chapter discusses the basics of SFN.

BROADCAST NETWORK STRUCTURES, MFN AND SFN

A multifrequency network (MFN, Figure 8.5-1) is nothing more than a broadcast network using the historical principle of transporting different local content on different transmitter stations using different frequencies. MFNs are in use in the following:

- Analog terrestrial television
- FM sound broadcast
- Digital terrestrial broadcasting using single-carrier modulation
- Digital terrestrial broadcasting using multicarrier modulation

Analog terrestrial broadcast and digital terrestrial broadcast using single-carrier modulation can only cope with more or less Gaussian channels, which means single-path reception. Additional signal paths are handled like error signals for the receiver.

MFNs are also used in digital terrestrial broadcast standards if the application is only to distribute local content for a city or a small region.

The development of digital terrestrial broadcast standards began in the mid-1990s. These include the following standards:

- DAB/DAB+
- DVB-T
- ATSC 1.0 and 2.0
- ISDB-T (Japan)
- DRM/DRM+
- DTMB (China)
- DVB-T2
- ATSC 3.0

The digital terrestrial broadcast standards are based on either of the following modulation types:

- Single-carrier modulation
- Multicarrier modulation (OFDM, orthogonal frequency division multiplex).

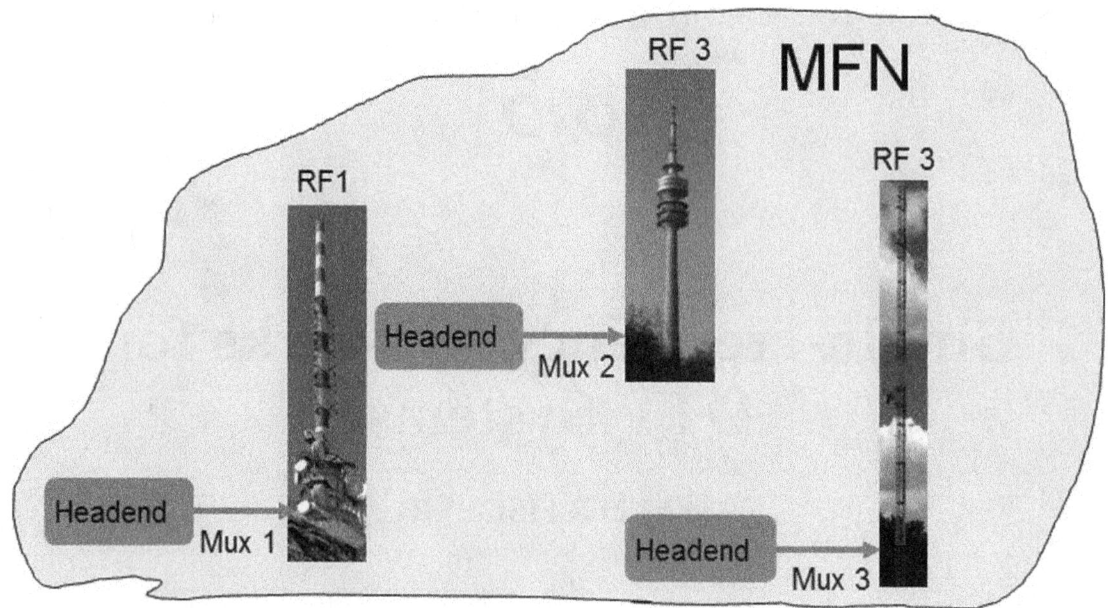

FIGURE 8.5-1 Structure of a multifrequency network (MFN).

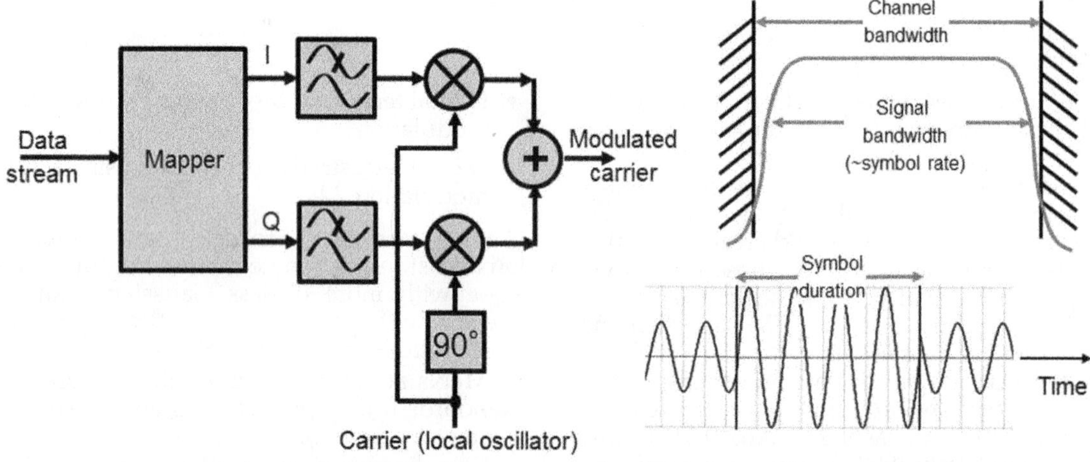

FIGURE 8.5-2 Symbols and channel bandwidth in single-carrier modulation.

To achieve a high data rate, the symbol duration in single-carrier-modulated signals is very short, e.g., about 150 ns. The reason behind this is the typical terrestrial broadcast channel bandwidth (6 MHz, 7 MHz or 8 MHz) and the fact that the symbol rate needs to be less than the channel bandwidth; the symbol duration is the reciprocal of the symbol rate (see Figure 8.5-2). Consequently, the symbol duration in double-sideband digital single-carrier modulation needs to be longer than:

- $> \frac{1}{6MHz} = 167ns$
- $> \frac{1}{7MHz} = 143ns$
- $> \frac{1}{8MHz} = 125ns$

The goal, however, is to reach the highest possible symbol rate and thus data rate in such a channel.

On the other hand, there are echoes in the terrestrial ambient. Typical echo length is up to about 15 µs, which means that the expected signal spread is much greater than the symbol duration in single-carrier-modulated digital terrestrial broadcast systems. In analog TV, such multipath reception ends in visible ghost pictures. In digital terrestrial systems, the limit for that multipath situation depends on the standard. It is a fact that single-frequency networks look like multipath reception for the receivers. ATSC 1.0 and ATSC 2.0 use the following symbol duration:

$$Symbol\ duration_{ATSC} = \frac{1}{10.76223776\,Msymbols/s} = 93ns.$$

ATSC is vestigial sideband filtered. The condition in ATSC 1.0 and 2.0 to be fulfilled is:

$$2 \bullet 93 \text{ ns} = 186 \text{ ns} > 167 \text{ ns}$$

However, 93-ns ATSC symbol duration is very short in relation to the expected terrestrial echoes. This is why there were originally no SFN solutions in ATSC 1.0 and 2.0. Later on, trials were conducted to implement some SFN functionality in ATSC 1.0 and 2.0 as well. This was done by using a defined modulator frame initialization and reset controlled by the ATSC headend and modern equalizers in the ATSC receiver. In ATSC 1.0 and 2.0, SFN capability is still highly limited, and ATSC SFN applications are not common.

DAB, DAB+, DVB-T, DVB-T2, ISDB-T and DTMB are OFDM-based transmission standards and can thus easily cope with multipath reception situations. OFDM symbols are typically much longer than single-carrier symbols. The OFDM symbol duration depends on the OFDM carrier distance. It is the reciprocal of the carrier distance described by the orthogonality condition of OFDM, namely:

$$\Delta t_{symbol} = \frac{1}{\Delta f_{carrier}};$$

The guard interval (GI), also called cyclical prefix (CP), lies between the OFDM symbols. It is a cyclic break or rather a cyclic repetition of the end of the following OFDM symbol (see Figure 8.5-3). The purpose of the guard interval is to avoid intersymbol interference or, in other words, intersymbol cross talk caused by echoes. The longest allowed echo corresponds to the length of the guard interval. In the case of OFDM, a broadcast channel is divided into N subcarriers. The more OFDM carriers there are, the closer the carrier distance and the longer the symbols. The length of the guard interval is an overhead in the signal that reduces the net data rate of the transmission system. As a result, in terrestrial broadcast systems there is a demand for long OFDM systems with long guard intervals. Typical single-frequency networks use OFDM modulation with as many OFDM carriers as possible. This results in symbol durations in the range of 1 ms or longer having a guard interval of about 250 µs. This guard interval size corresponds to an optical horizon of a transmission tower with a height of 300 meters and results in an SFN transmitter station distance of about 70 km (see Figure 8.5-4).

Both in multifrequency networks (MFN) and single-frequency networks (SFN), the broadcast data signal is generated by the broadcast headend. The broadcast data signal, which is called a multiplex

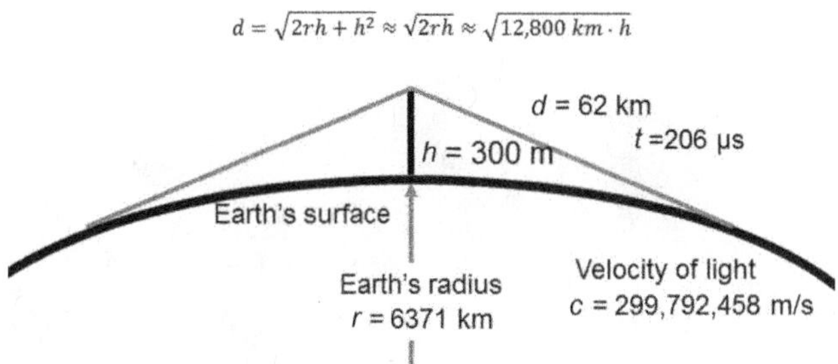

FIGURE 8.5-4 Optical horizon of a 300-meter transmitter station tower [Fischer2].

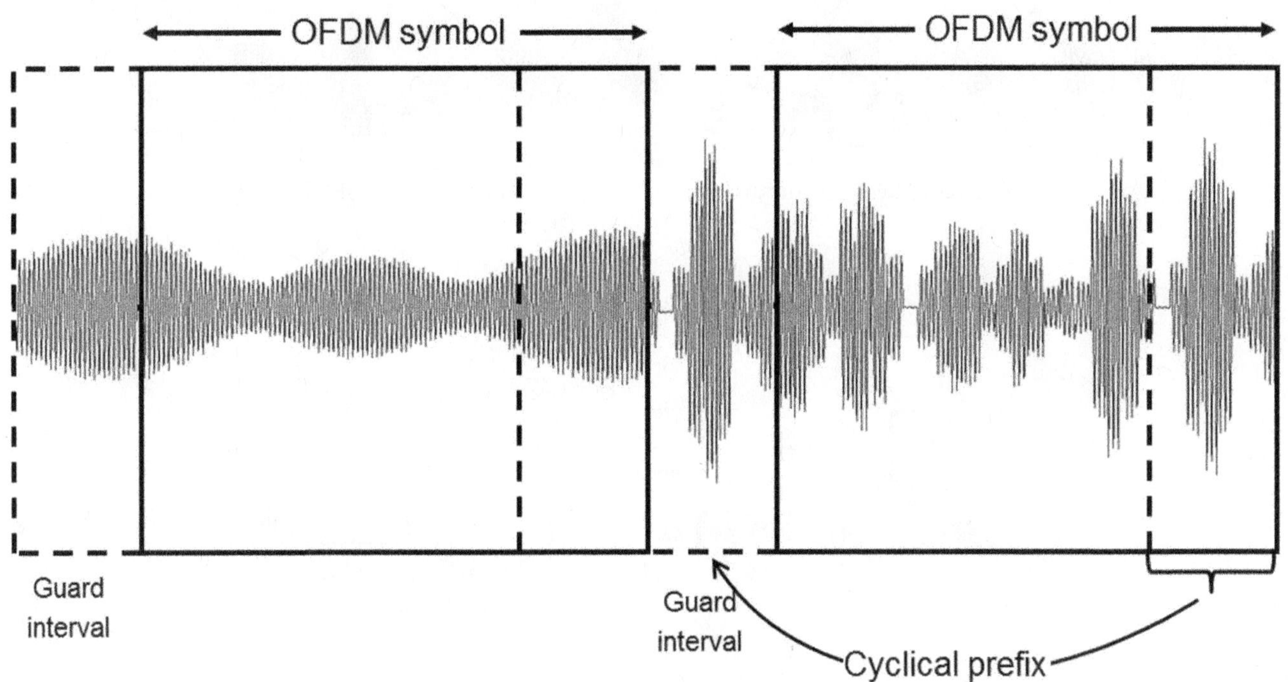

FIGURE 8.5-3 OFDM symbols with cyclical prefix/guard interval [Fischer1].

signal, and its format depend on the broadcast standard. Multiplex signal formats include the following:

- MPEG-2 transport stream with PSI/SI in DVB-T
- MPEG-2 transport streams with PSI/SI and megaframe initializing packets (MIP) in the case of DVB-T SFNs
- ETI signals in the case of DAB/DAB+
- BTS signals in ISDB-T
- T2-MI signal in DVB-T2
- MPEG-2 transport streams containing PSI/PSIP in ATSC 1.0 and 2.0

In an MFN, the broadcast headend only feeds one transmitter station. In SFNs, one broadcast headend is the source of the multiplex signal for several transmitter stations (see Figure 8.5-5).

All the transmitter stations get the same multiplex signal from the broadcast headend. The broadcast headend inserts SFN time stamps typically derived from GPS; these time stamps are used to compensate the delay between the headend and the transmitter stations. The transmitter stations also receive GPS signals with which the incoming SFN time stamps are compared and with which the transmitters are synchronized in time and in frequency (see Figure 8.5-6).

SFN BASICS

In a single-frequency network there are several conditions to be fulfilled. First of all, the main purpose of a single-frequency network is to reuse the same frequency over a bigger coverage area. Earlier documents about SFNs mentioned terms such as SFN gain and thus better coverage due to the use of SFN. But the practical reality is different. Every receiver mainly likes Gaussian channels and not multipath reception conditions caused by SFNs. Consequently, there are fading notches caused by multipath propagation between the SFN stations. In principle, each SFN station creates

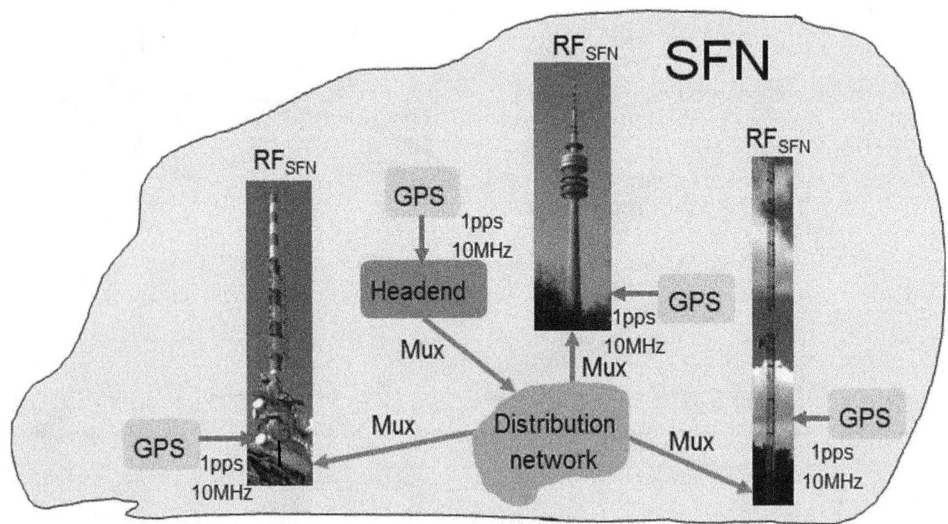

FIGURE 8.5-5 Structure of a single-frequency network (SFN).

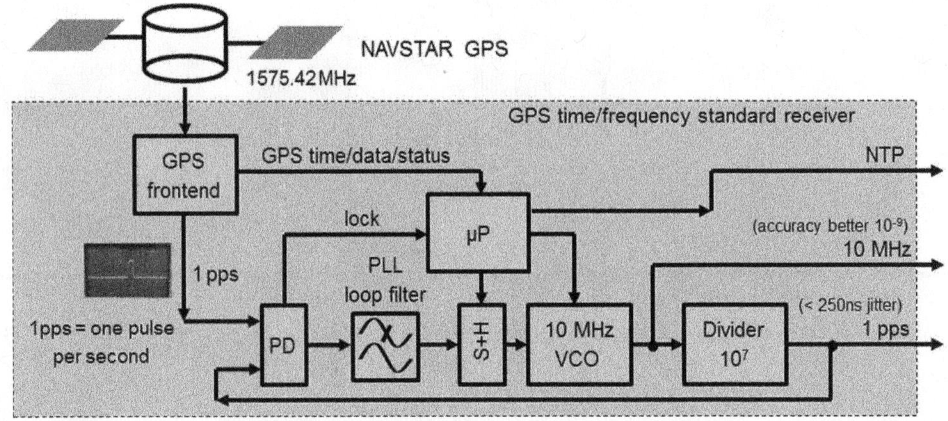

FIGURE 8.5-6 GPS time/frequency standard.

"artificial echoes," resulting in a multipath situation for each receiver, depending on where the receiver is located in the coverage area. But because of OFDM and OFDM symbols with a long guard interval, the receivers can cope with that multipath situation. The receiver picks out an overlay of all multipath signals and feeds that mix into the signal processing. The receiver signal processing cuts out the sampling window exactly in the right length, which is the symbol duration. The time position of this FFT sampling window can be shifted forward and backward to find the optimal FFT window position.

Now let us discuss the SFN conditions that have to be fulfilled. The first condition is that all transmitters must be fully synchronous in frequency. The frequency is synchronized via a GPS time/frequency standard receiver. This receiver gets the GPS signal from the GPS antenna and couples a 10-MHz voltage-controlled oscillator (VCO) to the one pulse per second (1-pps) signal coming from the GPS frontend. Both the 10-MHz signal and the 1-pps signal are used to synchronize the SFN transmitter. The second condition to be fulfilled is time synchronization. This is also done via GPS, and for this purpose the 1-pps signal is used (see Figure 8.5-6). Furthermore, all broadcast data has to be exactly the same on all transmitter stations, which must be synchronous in time and data. To achieve this, SFN time stamps are inserted in the broadcast multiplex signal by the broadcast headend equipment. The name and technical format of these time stamps depend on the broadcast standard. Examples of SFN time stamps:

- In DVB-T: megaframe initializing packets (MIP)
- In DAB/DAB+: time stamp information (TIST)
- In DVB-T2: time stamps in the T2-MI packets

Such time stamps contain either relative or absolute time information.

All the processes in the modulators need to be fully synchronized. For that reason, OFDM symbols have to be concatenated to OFDM frames and superframes (see Figure 8.5-7). The timing of all OFDM symbols, frames and superframes has to be synchronized to the incoming broadcast multiplex signal. The transmission time of the beginning of the OFDM symbols, frames and superframes is well defined in the broadcast standards and must be the same on all stations in an SFN.

SFN CONDITIONS

As mentioned earlier, in single-frequency networks we have to fulfill the following conditions:

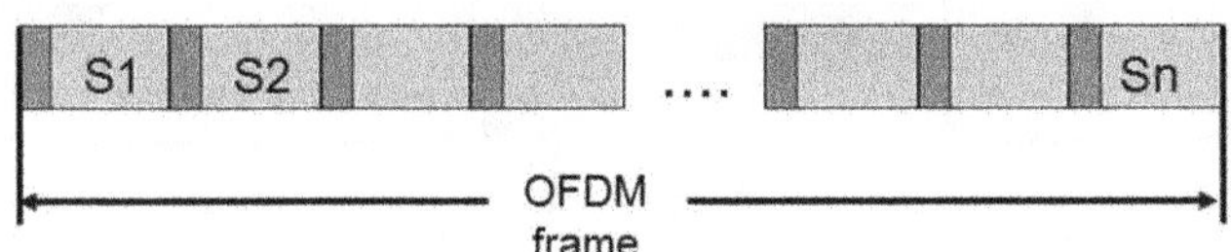

FIGURE 8.5-7 Building OFDM frames.

Fischer 1

All transmitter stations must be fully synchronous in:

- Frequency
- Time
- Data
- And must not exceed or violate the guard interval (transmitter distance).

Synchronization in frequency, time and data is achieved by having SFN time stamps in the feeding multiplex signal coming from the broadcast headend in combination with a GPS time/frequency standard receiver located in the headend and in the individual transmitter stations.

The delay between the broadcast headend and the individual SFN transmitter stations is, of course, different and has to be compensated so that the same data is combined in the same OFDM symbols in all modulators. This is done by having relative time stamps or orientation points in the feeding broadcast multiplex signal.

The broadcast transmission time in the RF domain, which is the beginning of an OFDM symbol, frame or superframe, can be shifted slightly forward and backward in the transmitters to optimize the single-frequency network. Typically this parameter is called "static delay" and can be set in the broadcast modulator. There may also be a "dynamic delay" and its automatic adjustment. The dynamic delay is caused by some dynamic changes in the feeding network. A reason for a dynamic delay could be a satellite feed with a varying orbit position of the satellite or some switching processes in the feeding network.

To fulfill the guard interval condition, the distance between adjacent transmitters needs to be within the guard interval distance. The "static delay" parameter can compensate somewhat by slightly shifting the symbols' transmission time in the individual SFN stations. It is important to plan the SFN network correctly to show whether there are any guard interval violations in the coverage area. The SFN network planning software must have the right geographic and topographic information material (maps) for the coverage area. Information about the antenna diagram, transmitter power, broadcast standard parameter and a lot of practical SFN experience is also required. The results of network planning must be verified by taking coverage measurements. During the coverage measurements, the field strength, signal quality, bit error ratio, modulation error ratio and the impulse response is measured. Investigating the impulse response makes it possible to detect guard interval violations, and these problems can be rectified by changing the static delay and antenna parameters. Planning and verifying SFNs is described later on in more detail.

SFN IMPLEMENTATION IN DIFFERENT DTV STANDARDS

All broadcast transmission standards based on OFDM offer the building of single-frequency networks. The

SFN implementation principles in the different transmission standards are similar but not the same. This section describes how SFNs are built in the following standards:

- DAB/DAB+
- DVB-T
- DVB-T2
- ISDB-T

The U.S. standards ATSC 1.0 and 2.0 are based on single-carrier modulation. Due to the physics of the very short symbols, this type of modulation is not ideal for building single-frequency networks.

But there is also a standardized way to implement limited SFNs, which is discussed in this section as well.

Any OFDM system, first of all, processes OFDM symbols with a cyclical prefix or guard interval before the symbol. The symbol is typically 1 ms to 4 ms long, and the guard interval size is in the range of about 250 µs. An OFDM frame is built from an integer number of symbols, and a number of frames are combined into a superframe. In some systems, another frame called a megaframe is defined. The superframe or megaframe size is selected to achieve a time of about half a second or one second so that the superframe or megaframe can be adjusted and identified by a one pulse per second (1-pps) signal. The trick in SFNs is to synchronize all transmitters in the network so that all OFDM symbols, frames, superframes and megaframes are fully synchronous in time. The same data from the multiplex input signal (MPEG-2 TS, ETI, T2-MI, BTS, etc.) must be placed into the same symbols, frames, superframes, megaframes, etc. For this reason, SFN time stamps are inserted into the multiplex input signal and all transmitters are fed by GPS, using its time and date information and its 1-pps output signal.

SFNS IN DAB/DAB+

Digital audio broadcasting (DAB) was a European project developed in the late 1980s. It was the first commercial standard using OFDM and SFN. The DAB standard included SFN right from the start. It is relatively easy to build SFNs in DAB because everything in the standard is synchronous. Even the multiplex signal (the ensemble transport interface or ETI) is fully synchronous, containing a TIST signal, the time stamp for the SFNs. The TIST signal contains an absolute time format and is transmitted at the end of an ETI frame. At the beginning of an ETI frame, the multiplex configuration information that controls the modulator in the transmitter station is transported. In the ETI signal, the fast information channel (FIC) and the main service channel (MSC) are delivered to the DAB transmitters. All DAB modulators are automatically synchronized to each other because of the synchronous ETI signal; this is due to the overall synchronous definition of the DAB system, which was originally built around a MUSICAM (MPEG-1 layer II) audio frame. In the modulator, a forward error correction overhead is added to the FIC and MSC data, and subsequently the data is differentially mapped to Π/4-DQPSK modulated OFDM symbols having a length of 1 ms in DAB mode I. A 250-µs guard interval is then added at the beginning of the OFDM symbol. The channel spacing in DAB mode I is exactly 1 kHz. The 1536 OFDM carriers form a spectrum with a bandwidth of 1.536 MHz. A DAB OFDM frame starts with a null symbol where all carriers are set to zero, followed by a time-frequency-phase-reference symbol (TFPR) and the data symbols containing the error-protected FIC data and the MSC data (see Figure 8.5-8). The TIST signal defines the transmission point of the beginning of a DAB frame. This transmission time can be shifted by a static delay, a parameter that can be set on each SFN transmitter station.

The physical layer of DAB+ is absolutely identical to that of DAB. The only difference between the two systems is that DAB+ has a new audio codec, called MPEG-4 high efficiency audio coding, AAC+. The AAC audio data in DAB+ is additionally protected by a Reed-Solomon block code.

DAB/DAB+ uses the following physical layer parameters:

- Convolutional coding with equal or unequal forward error correction described by the protection level
- OFDM with mode I, II, III or IV (OFDM carrier spacing 1 kHz, 2 kHz, 4 kHz, 8 kHz)
- Π/4 shift DQPSK (mobile reception in automotive receivers)

Typically mode I is used. DAB runs in VHF Band III.

SFNS IN DVB-T

Most DVB-T [ETS300744] networks since 1998 are single-frequency networks, typically running in the 8K mode with a guard interval of 1/4. Networks for portable indoor coverage use 16QAM modulation, typically with a code rate of 2/3, which gives a fall-off-the-cliff condition of about 12 dB in a Gaussian channel. Other networks are planned for fixed antenna (outdoor roof antenna) reception having a fall-off-the-cliff condition at about 18 dB. This is done by using 64QAM and a code rate of 3/4.

In DVB-T the multiplex signal is the MPEG-2 transport stream. In DVB-T SFNs, this MPEG-2 transport stream contains special megaframe initializing packets (MIP) that run on a well-defined PID of 0x15. These MIPs are inserted by the DVB-T network adapter, which is the final part of a DVB-T headend. The MIPs

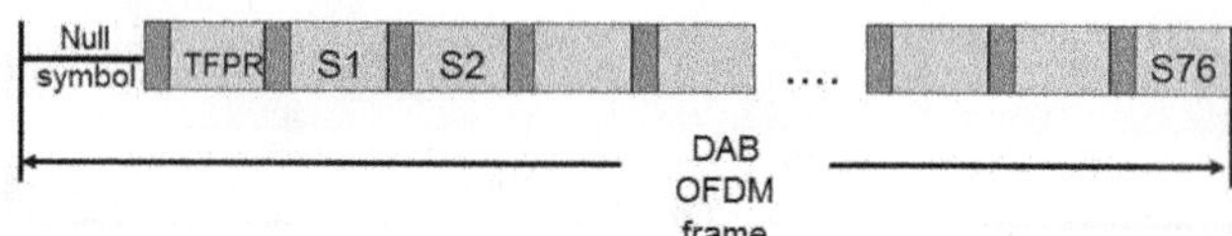

FIGURE 8.5-8 DAB OFDM frame.

Fischer 1

contain a time back to the last reception of the 1-pps signal from the GPS receiver, which is one of the input signals of the DVB-T network adapter. Sometimes this network adapter is also called MIP inserter. The following parameters are also transported in the MIP:

- Maximum network delay
- OFDM mode
- Guard interval
- Type of modulation
- Code rate
- Addressed transmitter power (optional)
- Addressed static delay of transmitter stations (optional)

Each DVB-T modulator in the SFN receives the MIP, reads out the content and compensates the delay of the distribution network, which means that all modulators delay the data signal by the same maximum delay value. The DVB-T modulator also receives the 1-pps and the 10-MHz signal from GPS. The incoming MIP signals contain a time relative to the last reception of 1 pps, and the modulator also gets the 1-pps time directly from GPS. This dual time information ensures that the right timing of the DVB-T system is achieved on all stations in an SFN.

Now we will discuss the physical frame structure of DVB-T. In DVB-T, an OFDM symbol in a channel with an 8-MHz bandwidth is 896 µs long in the 8k mode. Selecting a guard interval size of 1/4 results in an absolute guard interval length of 224 µs, which corresponds to a maximum transmitter station distance of about 67 km. An OFDM frame is built from 68 DVB-T OFDM symbols, and the 67 transmission parameter signaling (TPS) bits are transported during this OFDM frame (see Figure 8.5-9). Four DVB-T frames are combined into a superframe. DVB-T SFNs can also have a megaframe, comprising a number of superframes. The length of a megaframe depends on the DVB-T modulation parameters. To be part of a 1-pps period, this time must be in the range of 0.5 s to 1 s. The time stamp transported in the MIP describes the time of the start of the next megaframe in the MPEG-2 transport stream back to the last change of the rising slope of the 1-pps signal received by GPS. To synchronize a DVB-T SFN, all megaframes, superframes, frames and OFDM symbols must contain and transport the same data, and they must start at the same time, making them fully synchronous with each other in all modulators in the DVB-T SFN. But the real transmission time of

- the DVB-T OFDM symbol
- DVB-T frame (= 68 symbols)
- superframe (= 4 frames)
- and megaframe (= *n* superframes)

can be shifted and controlled slightly by the operator by changing the static delay parameter. This parameter can be different in all the DVB-T transmitter sites and allows an SFN to be optimized regarding the time position of the stations relative to each other.

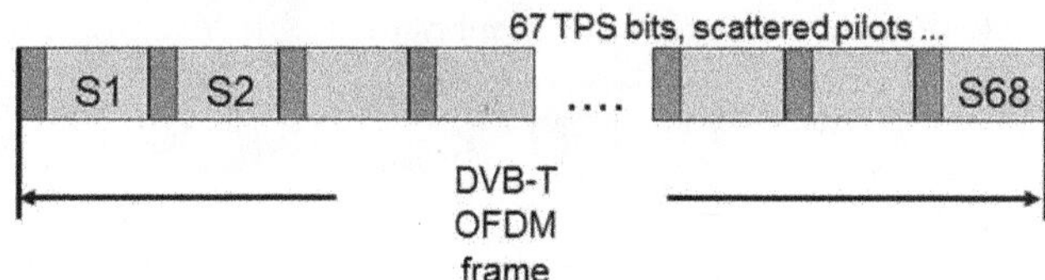

FIGURE 8.5-9 OFDM frame in DVB-T.

Fischer 2

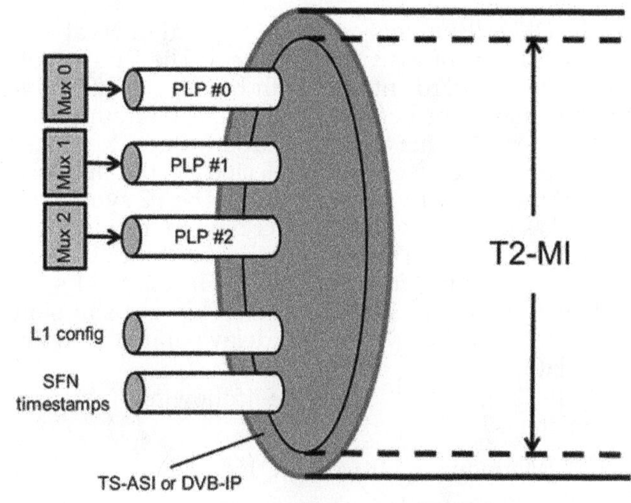

FIGURE 8.5-10 T2-MI signal.

Fischer 2

SFNS IN DVB-T2

DVB-T2 also uses OFDM. The system is very complex and can optionally transport several multiplexes with different levels of robustness in different physical layer pipes (PLP). The final part of a DVB-T2 headend is the T2-gateway, which combines

- the PLP data signals
- the layer 1 control signal
- and the SFN sync signal

to form the T2 modulator interface signal (T2-MI).

The T2 gateway gets all the different PLP input signals and a time reference signal that is 1 pps or network time protocol (NTP), derived from GPS (see Figure 8.5-10). The T2-MI signal is then distributed to the different T2 SFN transmitter stations. The T2 system is more or less a synchronous system, and the SFN control signal inside T2-MI makes it possible to synchronize all modulators in order to build a single-frequency network.

In DVB-T2 an OFDM frame consists of the following:

- A preamble symbol *P1*
- One or more preamble symbols *P2*
- Several *m* data symbols

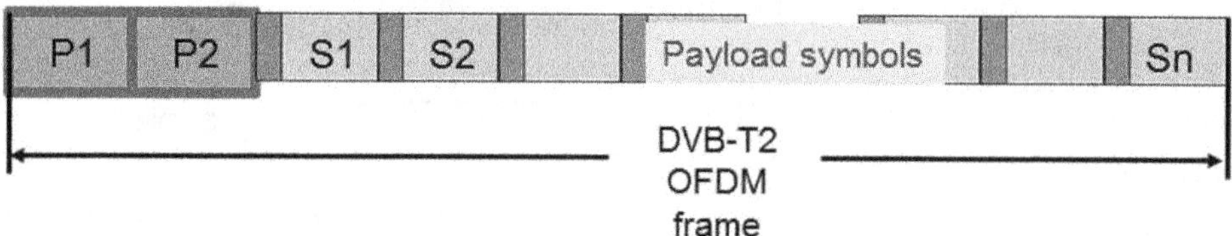

FIGURE 8.5-11 OFDM frame in DVB-T2.

Fischer1

All T2-OFDM symbols consist of a guard interval and the useful part of the OFDM symbol. The P1 symbol, which has a guard interval both before and after the symbol, marks the beginning of an OFDM frame. Several frames combine to form a superframe. The frame and superframe size can be selected by the network operator and is set via the gateway. The P2 symbol or symbols signal the physical layer data to the receivers (see Figure 8.5-11). The transmission time of the beginning of a T2 superframe is defined in the SFN sync information in the T2-MI signal. This transmission time can be shifted by the static delay parameter in the modulators.

DVB-T2 systems use the following physical parameters:

- OFDM mode (1K, 2K, 4K, 8K, 16K, 32K)
- Normal or extended carrier mode
- Guard interval size
- QPSK, 16QAM, 64QAM and 256QAM (different for the different PLPs)
- Code rate (different for the different PLPs)

Typical T2 parameters: 32kext; guard = 1/16; 64QAM or 256QAM; code rate = 2/3 or 3/4. The typical OFDM symbol length is about 4 ms, with a guard interval of about 250 µs.

SFNS IN ISDB-T

ISDB-T [ITU205] was published in 1999. Its principle is based on DVB-T and DAB. It is therefore also a digital terrestrial broadcast standard that uses OFDM. ISDB-T is the Japanese system for digital terrestrial television. The modulation is QPSK, DQPSK, 16QAM or 64QAM. The forward error correction (FEC) is Reed-Solomon block coding and convolutional coding followed by a puncturing unit, which means it is nearly identical to that of DVB-T. An ISDB-T channel is divided into 13 segments of 428.57 kHz each. The channel bandwidth is 6 MHz. The k, l and m segments can be combined into three layers, where $k + l + m = 13$. In the different layers, different FECs and different modulation patterns can be selected. ISDB-T is used in Japan, certain South American countries and the Philippines.

The ISDB-T multiplex signal is called broadcast transport stream (BTS), which has a constant bit rate of 32.5079 Mbit/s. It is a 16-byte extension to the 188-byte transport stream protocol, i.e., the BTS packet size is 204 bytes. In the 16-byte overhead, the multiplex information about the ISDB-T layer is signaled. The BTS signal is transported via the studio transmitter link (STL). The BTS signal also contains the layer 1 control information for the modulators and the SFN sync information. This information is transported in the ISDB-T information packets (IIP), which are similar to the DVB-T MIP packets. The IIP packets consist of the modulation control configuration information (MCCI) and the network synchronization information (NSI). The MCCI describes the physical ISDB-T parameters.

ISDB-T uses the following physical parameters:

- OFDM mode (mode 1 = 2K, mode 2 = 4K, mode 3 = 8K)
- Guard interval size (1/4, 1/8, 1/16, 1/32)
- Type of modulation (QPSK, DQPSK, 16QAM, 64QAM, different for the different layers)
- Code rate (1/2, 2/3, 3/4, 5/6, 7/8, different for the different layers)
- Number of segments per layer (1 to 13, different for the different layers)

In the NSI part, the following information (and other data) is transported to the SFN transmitters:

- Synchronization time stamp (STS), derived from 1 pps
- Maximum delay

SFNS IN ATSC 1.0/2.0

SFNs in ATSC [A53] [ATSC-MH] are discussed in detail in Chapter 8.6 of this handbook. The following may be regarded as an overview in context with this general discussion. SFNs were originally not planned, but around 2009, some SFN capability was standardized to allow the frame synchronization of ATSC 1.0-based systems. The ATSC multiplex signal is the MPEG-2 transport stream; it has a data rate of 19.39 Mbit/s and is typically delivered on an SMPTE310 interface. ATSC 1.0 and 2.0 use the same physical layer, which is 8VSB single-carrier modulation. As already mentioned, single-carrier-modulated systems are typically not designed for SFN use. The main reason is that they use very short symbols without a guard interval. This is also true for ATSC 1.0/2.0. However, the ATSC

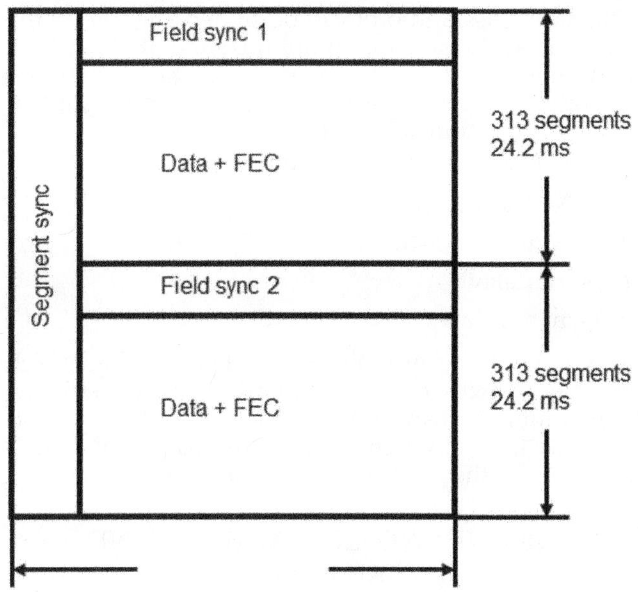

FIGURE 8.5-12 ATSC 1.0 frame.

Fischer1

1.0/2.0 physical layer frame can be synchronized by the incoming MPEG-2 transport stream signal if the headend inserts reset signals for the ATSC frame—which is exactly what was done in a later ATSC 1.0 extension for SFN systems (see Figure 8.5-12). But despite this, ATSC 1.0/2.0 provides very limited SFN capability. Together with the ATSC 1.0/2.0 SFN definition, a mobile TV extension in ATSC (ATSC-MH) was also standardized. To provide the synchronization of the modulators, even more signal processing is required in the ATSC headend.

SFNS IN ATSC 3.0

Because ATSC 3.0 is based on OFDM and experience from DVB-T/T2 networks, SFNs similar to those in DVB-T/T2 can be built. However, the terms are different. The main ATSC 3.0 technical parameters are:

- OFDM with 8K, 16K, 32K FFT size
- LDPC/BCH FEC
- QPSK, 16QAM, 64QAM, 256QAM, 1024QAM, 4096QAM modulation
- Uniform and nonuniform constellations
- Time division multiplexing (TDM) and layered division multiplexing (LDM) similar to the T2-PLP concept
- Time interleaver
- MISO, MIMO

The baseband signal is called studio transmitter link (STL), which is also used for the SFN distribution. The SFN concept is expected to be similar to DVB-T2. The ATSC 3.0 standard is not finished yet.

SFN ADAPTERS AND SFN TIME/ FREQUENCY STANDARDS (GPS)

The final part of a terrestrial broadcast headend for single-frequency networks is the SFN network adapter. It inserts all the necessary control and synchronization information for the SFN into the multiplex signal. The main SFN sync signal contains the following:

- The SFN time stamp, typically derived from Global Positioning System (GPS) or network time protocol (NTP)
- The maximum delay of the SFN distribution network
- Physical layer control information for the transmitters

The SFN time stamp indicates either an absolute or a relative time. An absolute time is a time related to UTC, Julian date or another well-defined time, e.g., the first second of the year 2000 or something else defined in the corresponding broadcast standards. Relative time typically means nanoseconds back relative to the last change of the second (1-pps change). An absolute time (GPS time) and a 1-pps signal coming from a GPS receiver drive the SFN network adapter. Alternatively, the network adapter could also be fed by an NTP server. An NTP server is a time standard for IP networks and is typically also synchronized via GPS. GPS is available in all parts of the world. It is typically used for location-finding in navigation systems in cars, aircraft, military equipment and mobile phones. The GPS system is based on atomic clocks of very high accuracy, and each GPS satellite knows the following:

- The exact time on board the satellite
- Its exact position in outer space

If a GPS receiver unit is locked to a minimum of four GPS satellites, then the receiver location's three-dimensional position can be calculated from the incoming time signals from the satellites by means of triangulation. In the SFN, only the time information is used. A professional GPS time/frequency standard receiver (Figure 8.5-6.) includes the following:

- Redundant GPS frontends, which are connected to . . .
- Redundant GPS antennas
- A high-end temperature-stabilized 10-MHz voltage-controlled oscillator, which is locked to . . .
- The GPS 1-pps signal via a high-end phase locked loop (PLL)

The GPS time/frequency standard delivers the following:

- A 1-pps signal
- A 10-MHz reference signal
- The GPS time
- The location

The functionality of a professional GPS receiver of this type is based on the following principle:

The GPS receiver frontend delivers a 1-pps signal that has its relevant time information in the rising slope of a very short pulse of only a few milliseconds. Also included is a high-end 10-MHz oscillator that can be voltage-controlled by a driving signal with a voltage range of typically 0 volts to 5 volts. This control voltage makes it possible to control the oscillator frequency in a range of several parts per million (ppm). The oscillator is temperature-stabilized and can even be coupled with a Rubidium standard. The 10-MHz signal is then divided by $1 \bullet 10^7$ to get another 1-pps signal. Both the 1-pps signal from the GPS frontend and the 10-MHz oscillator and divider are then compared in a phase comparator. The output signal from the phase comparator is then low-pass-filtered in the loop filter, which then feeds and controls the 10-MHz voltage-controlled oscillator. The control signal is also sampled by an analog-to-digital converter, making it possible to hold this signal and thus the oscillator for a certain time period in order to hold the state for the 10-MHz and 1-pps signal in case of a missing GPS signal. Depending on the quality of the oscillator, this can range from a number of minutes to several hours. Consequently, the output signal used from the GPS time/frequency standard is not directly the GPS frontend signal; it is rather the averaged 10-MHz oscillator signal and the 1-pps signal after the 10-MHz divider.

The SFN adapter in the broadcast headend and all transmitter stations in an SFN contain redundant GPS receivers. The SFN adapter takes the 10-MHz signal and locks the multiplex clock to it. It also uses either the averaged absolute GPS time information or the 1-pps pulse information and inserts high-accuracy SFN time stamps into the broadcast multiplex signal. In DVB-T(1), the packets containing such SFN time information are called megaframe initializing packets (MIP). In DAB and DAB+, this time information is referred to as time stamps (TIST).

Another parameter that comes with this SFN information inserted by the SFN network adapter is the maximum delay, or rather the expected or estimated delay of the SFN distribution network. Here, typically several hundred milliseconds are estimated and inserted. Typically, no time is spent to really measure the maximum delay. Network planners estimate the network delay and also add some more "safety time."

The modulator in each SFN transmitter station receives the incoming broadcast multiplex signal (e.g. the MPEG-2 transport stream with DVB-T MIPs) and is also connected with the station GPS receiver via 1 pps and 10 MHz. The 10 MHz signal is used to synchronize the transmitter frequency, with an accuracy in a range of 10^{-9} to 10^{-12}, i.e., typically better than 0.5 Hz (example: 650-MHz transmission frequency and 10^{-9} yields 0.65 Hz, 10^{-10} results in 0.065 Hz). The 1-pps signal and the SFN time sync information in the broadcast multiplex signal are compared, and then the multiplex signal will be delayed and adjusted to the maximum delay value of the network. After that delay, all the modulators are fully synchronous in frequency, time and data.

The RF transmission time is defined in the corresponding broadcast transmission standard. It is typically the beginning of the first OFDM symbol in an OFDM frame, superframe or any similar definition. This transmission point can be shifted slightly by the static delay parameter. To optimize the SFN network and thus avoid guard interval violations, network operators can enter this parameter using different values for every transmitter station.

We have the following definitions:

- SFN time stamp
- Maximum SFN delay
- Static delay
- Dynamic delay

The dynamic delay is the SFN network delay, which steadily changes due to dynamic delay processes in the SFN multiplex distribution signal network. This could be caused, e.g., by a changing orbit position of a satellite used for the multiplex signal distribution.

SFN NETWORK PLANNING AND VERIFICATION

A single-frequency network needs to be planned and verified. To plan a network, the following items are required:

- High-accuracy maps with geographic, topographic, morphologic information
- The transmitter locations
- The transmitter antenna diagram and gain
- The transmitter output power
- The loss in the antenna feeding network
- The physical modulation and FEC parameters
- The network planning parameters
- Network planning software

Some companies offer network planning software and network planning service. High-accuracy maps are typically supported by governmental departments. It is necessary to have maps that provide information about the following:

- The structure of the earth's surface (height profile, hills, valleys, mountains)
- Physical and chemical structure of the soil (sea, water, ground water, desert, forest, electrical characteristics, etc.)
- Buildings, vegetation, etc.
- Population, e.g., knowledge about whether the area is urban or rural

The accuracy of this geographic information should be in the range of 50 meters to 100 meters or even better depending on the requested accuracy of the coverage planning results.

The transmitter station is described by the following:

- The transmission frequencies
- The physical parameters of the transmission standard (signal bandwidth, modulation, forward error corrections, guard interval size, etc.)

- The transmitter output power
- The feeder loss (mask filter, antenna combiner, antenna cable)
- The antenna gain
- The antenna diagram (horizontal and vertical)
- The construction data of the antenna (manufacturer, number of antenna fields, coupler, etc.)
- The GPS location of the antenna
- The antenna height
- The static delay of the transmitters

If more details are known about the antenna and its construction, the accuracy of the antenna parameters increases and a greater number of details can be entered into the planning base of the network planning program. Sometimes a database for all transmitter stations in a country is available from the regulation authority.

It is also important to know the exact location (GPS data) and height of the antenna mast. Each SFN station can be shifted in time by means of the static delay parameter, which is a modulator setup parameter. It is entered in ns or µs.

Basic parameters for calculating or estimating the robustness or fall-off-the-cliff signal-to-noise ratio are the physical modulation parameters that are used in the digital terrestrial broadcast standard. These parameters include the following:

- The standard itself
- The signal bandwidth
- The OFDM mode
- The modulation pattern (e.g., QAM order)
- The forward error correction (e.g., the code rate parameter)
- The guard interval size

Another network planning parameter is the planned reception condition:

- Outdoor fixed antenna (roof antenna with gain)
- Outdoor antenna (e.g., outside a window)
- Indoor reception
- Or deep indoor reception

Furthermore, the quality of the receivers must be described, indicating the estimated equivalent noise degradation (END) or noise figure of the receiver.

Fall-off-the-cliff signal-to-noise ratios (SNR_{min}) are typically in the range of:

- 18 dB (outdoor roof antenna, fixed antenna with gain)
- 12 dB (portable indoor)
- 6 dB or less (deep indoor)

A minimum receiver input level RF_{min} can be calculated from the SNR_{min} parameter, the typical noise figure of the receiver and the signal bandwidth, by using the following equation:

$$N = -228.6dBW/(K \cdot Hz) + 10 \cdot \log(B) + 10 \cdot \log(T) + F$$

where

N = noise power in dBW

B = bandwidth in Hz

T = temperature in K

F = noise figure of receiver in dB

$RFmin = N + SNRmin$

This is automatically done by the network planning software tool.

After all the network planning parameters and all geographic map databases have been entered, the network planning process is started, and after a while the results will be shown by the software. The processing time depends on the following:

- The computer's processing power
- The size of coverage area
- The required accuracy of the planning results
- The map database
- The number of SFN stations
- The network planning model (2D model, 3D model, etc.)

The network planning software calculates the electrical field strength in all requested locations in the coverage area by using the following:

- The free-space loss in the used frequency range
- Reflections from hills, buildings, mountains, etc.
- The defraction in the geographic surface in the coverage area

The network planning process produces a coverage map showing the following:

- The field strength of each SFN station
- The sum field strength of all SFN stations in an overlay
- The good and bad interference zones
- The SFN echo pattern

Network planning can be optimized by "playing with" the modulation parameters, static delay and antenna diagram (e.g., vertical down tilt, directivity, ERP) (see Figures 8.5-13, 8.5-14 and 8.5-15).

After network planning has been concluded and the single-frequency network switched on, the verification process can be started.

FIRST TESTS AFTER INSTALLATION

After network planning has been concluded and the transmitters installed, the following initial tests are performed:

- Checking of all important technical parameters of the transmitter using a dummy load
- Antenna load test

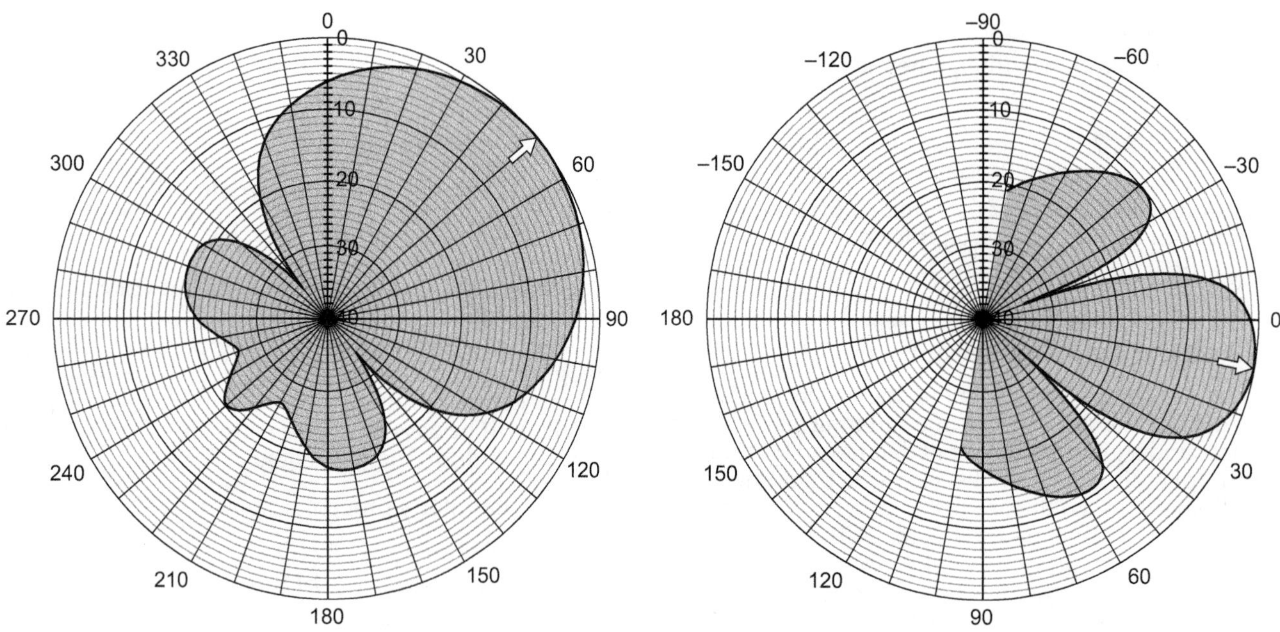

FIGURE 8.5-13 Example of antenna diagram (LStelcom ©, Chirplus ©).

LStelcom

FIGURE 8.5-14 Example of network planning result (LStelcom ©, Chirplus ©).

LStelcom

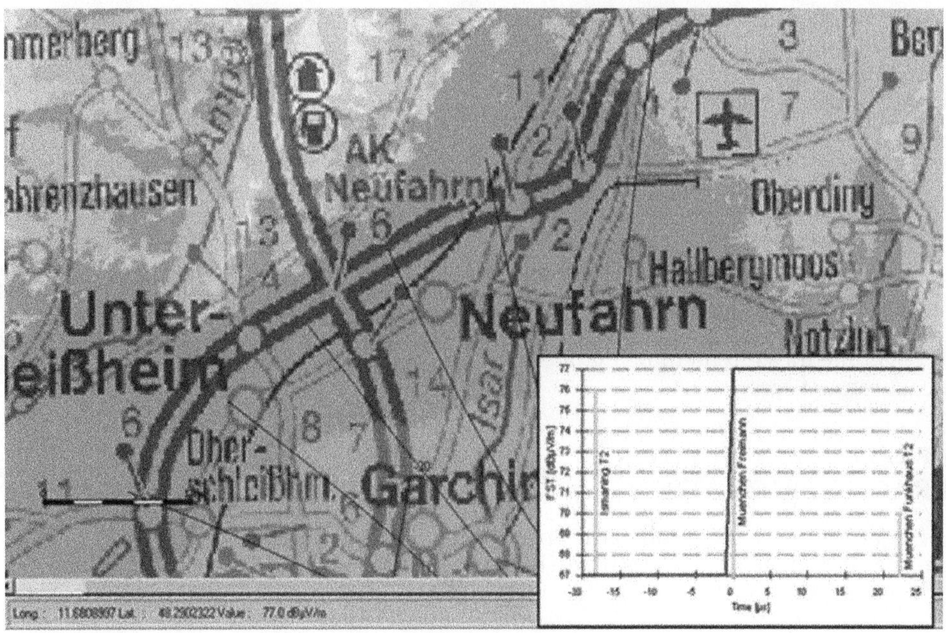

FIGURE 8.5-15 Example of a calculated SFN echo pattern (LStelcom ©, Chirplus ©) at a selected location in the SFN.

LStelcom

- Measurement of all important technical parameters behind the antenna combiner
- Initial reception tests outside the transmitter station

The main parameters of a digital terrestrial broadcast transmitter need to be tested shortly before the network is to go on air. This is done using a measurement coupler and a dummy load. The coupler should have a coupling attenuation in the range of 50 dB to 60 dB, a value that depends on the output power of the transmitter; the power of the measurement signal should be in the range of about +5 dBm to +20 dBm. There is also a measurement coupler for the reflected power; this coupler needs to have a similar coupling range. And there should be another measurement coupler behind the mask filter and behind the antenna combiner. Sometimes the mask filter is included in the antenna combiner.

The following measurement parameters are used at the transmitter output:

- (Forward) RF power, measured in dBm, watts or kW
- Reflected power, measured in dBm or watts
- Shoulder attenuation, measured in dB
- Harmonics, out-of-band products, measured in dB
- Modulation error ratio (MER), measured in dB
- Bit error ratios (BER) before the first FEC, after the first FEC and after the second FEC
- Constellation diagram
- Spectrum
- Amplitude response, measured in dB
- Group delay, measured in ns
- RF frequency
- Crest factor, measured in dB

The measurements must be done directly after the transmitter, after the mask filter and after the antenna combiner. These measurements require the following test equipment:

- A number of measurement couplers that are part of the transmitter itself (after the transmitter, after the mask filter, after the antenna combiner)
- A dummy load that is part of the transmitter station
- A broadcast test receiver (e.g., R&S ETL)
- A high-accuracy 10-MHz reference signal
- A power meter
- A high-end spectrum analyzer
- A network analyzer

Most of the parameters can be measured using the test receiver (e.g., R&S ETL). For high-accuracy frequency measurements, the test receiver needs to be fed a high-accuracy 10-MHz reference signal. To measure shoulder attenuation and harmonics, a high-end spectrum analyzer must be used (e.g., R&S FSW). If it is necessary to check the filters, a vector network analyzer is also necessary. For measuring the power, typically a high-end power meter is needed. To avoid killing the test equipment, attenuators have to be inserted between the measurement coupler and the test equipment. The measurement signal must be in a range of typically 0 dBm to +10 dBm. The maximum input level of spectrum analyzers and test receivers is typically +30 dBm.

In addition, the path to the antenna must be checked using a vector network analyzer. The main parameter is the return loss of the antenna feeder and the antenna system. The network analyzer can also be used to check all filters.

After the quality check of the transmitter and the antenna, a load test of the antenna should be performed, starting with 50% of the maximum power and then 100% power. During this test, it is important to monitor the reflected power and to visually inspect the antenna system to determine the possible occurrence of smoke.

The next test is coverage measurement using one or more coverage measurement cars.

SINGLE-FREQUENCY NETWORK COVERAGE VERIFICATION

To verify a single-frequency network, a fully equipped coverage measurement vehicle is necessary. This vehicle is typically a small bus that includes the following:

- A power supply for 110-volt AC or 240-volt AC, which can comprise a 12-volt or 24-volt battery system and a power inverter or Diesel generator
- An antenna mast (10 m)
- Multiple omnidirectional measurement antennas on its roof
- A broadcast test receiver (e.g., R&S ETL)
- A spectrum analyzer
- A GPS receiver
- A number of TV receiver examples
- A TV monitor

The main test equipment is the broadcast test receiver. It can be connected to the different antennas on top of the vehicle and to the antenna mast system. This test receiver is used to measure the following parameters:

- RF level
- MER
- BERs,
- Constellation diagram
- Spectrum
- Channel impulse response or echo pattern

If the antenna and antenna feeder cable parameters are known, the *k*-factor of the antenna system can be calculated by means of the following equation:

$$k = (-29.8 + 20 \lg (f/\text{MHz}) - g) \text{ [dB]}$$

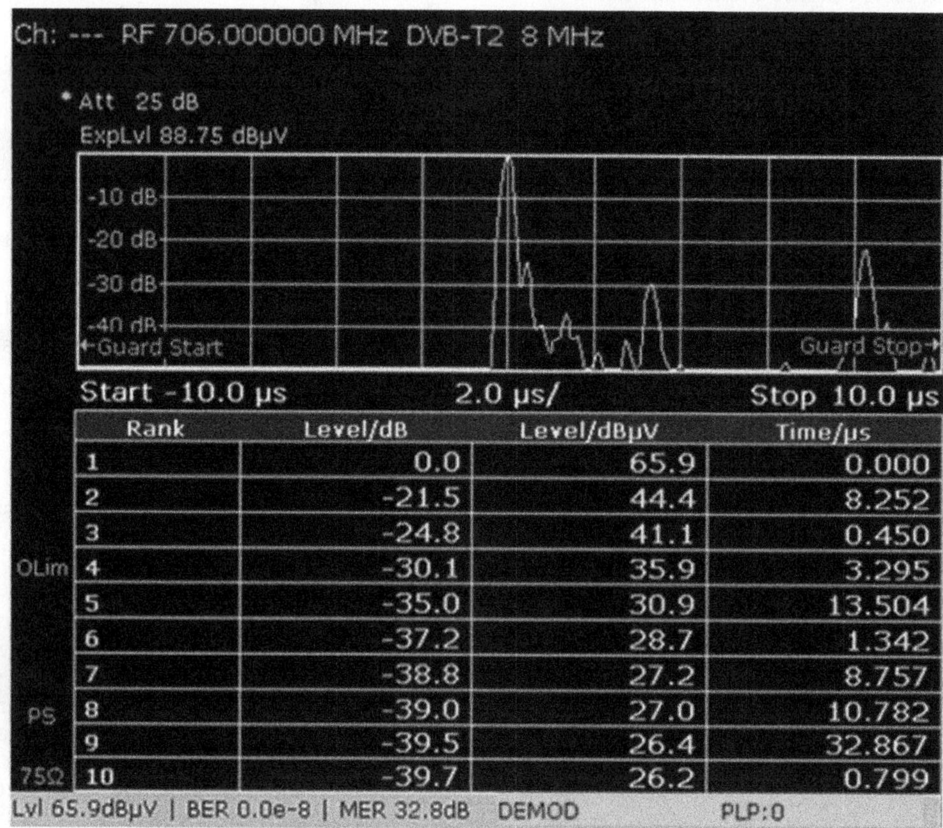

Rank	Level/dB	Level/dBµV	Time/µs
1	0.0	65.9	0.000
2	-21.5	44.4	8.252
3	-24.8	41.1	0.450
4	-30.1	35.9	3.295
5	-35.0	30.9	13.504
6	-37.2	28.7	1.342
7	-38.8	27.2	8.757
8	-39.0	27.0	10.782
9	-39.5	26.4	32.867
10	-39.7	26.2	0.799

FIGURE 8.5-16 Channel impulse response in an SFN.

ETL

where
k = antenna k-factor in dB

f = reception frequency in MHz

g = antenna gain in dB

From the k-factor and the RF level, the electrical field strength can be processed by using the following formula:

$$E\,[\mathrm{dV\mu V/m}] = k + U\,[\mathrm{dB\mu V}]$$

where
E = field strength in dBμV/m

U = antenna output voltage in dBμV

The signal quality can be checked by analyzing the MER and BER. But the main test of the single-frequency network is the analysis of the channel impulse response or echo pattern diagram. It shows the situation in the SFN and identifies possible guard interval violations. The investigation and protocol of the spectrum and constellation diagram view can be very helpful. In addition, the GPS position, time and data and the weather conditions have to be added to the measurement protocol.

When switching on an SFN, not all transmitters should be switched on at the same time. It is better to place the measurement vehicle in a central measurement position in the coverage area and then to switch on and check the transmitters, first, individually, one transmitter at a time, and then all transmitters successively until the entire SFN has been switched on. This procedure, in combination with the coverage map processed by the network planning software, helps to identify the transmitters at that measurement location.

DAB also features an optional transmitter identification information (TII) signal that helps to identify the SFN transmitter stations in the channel impulse response diagram (see Figure 8.5-16). Such an option is also defined in DVB-T2 systems but has not yet been implemented in practice.

We have now explained the basic principle and the details of single-frequency networks in all the different digital terrestrial broadcast standards. SFNs have been in use for many years. Nevertheless, the required coverage quality can only be achieved through an accurate planning and verification process.

Some figures in this chapter are taken from [Fischer1] and [Fischer2], Springer.

Bibliography

[Fischer 1] Fischer, W., *Digital Video and Audio Broadcasting Technology—A Practical Engineering Guide*, Springer, Heidelberg, 2010, 3rd Edition

[Fischer 2] Fischer, W., *Digitale Fernseh- und Hörfunktechnik in Theorie und Praxis*, Springer, Heidelberg, 2015, 4th Edition. Auflage

[ETL] R&S ETL TV Analyzer

[FSW] R&S FSW Spectrum Analyzer

[LStelcom] LStelcom, CHIRplus BC, Lichtenau, Germany

[ETS300401] Radio Broadcasting Systems; Digital Audio Broadcasting (DAB) to mobile, portable and fixed receivers, ETSI, May 2001

[ETS300744] Digital Video Broadcasting (DAB); Framing structure, channel coding and modulation for digital terrestrial television (DVB-T), ETSI, 1997

[ETS302307] ETSI EN 301 307; Digital Video Broadcasting (DVB); Second Generation of Framing Structure, Channel Coding and Modulation for Broadcasting, Interactive Services, News Gathering and other Broadcast Satellite Operations, ETSI, 2006

[ITU205] ITU-R 205/11, Channel Coding, Frame Structure and Modulation Scheme for Terrestrial Integrated Services Digital Broadcasting (ISDB-T), ITU, March 1999

[TR101190], Implementation Guidelines for DVB Terrestrial Services, ETSI, 1997

[ATSC-MH] ATSC-M/H Standard, Part 1, A/153, 2009

[A53] ATSC Doc. A53, ATSC Digital Television Standard, September 1995

CHAPTER

8.6

Single-Frequency Networks (SFNs) for Television—Digital On-Channel Repeaters, Distributed Transmitters, and Distributed Translators

S. MERRILL WEISS
Merrill Weiss Group LLC
Metuchen, NJ

Author's note: The maps in this chapter, as well as certain diagrams containing elements in color, are best viewed using online or file-based sources so that the full information content communicated by the color characteristics can be comprehended.

SFNS DEFINED

A single-frequency network (SFN) consists of several transmitters, operating on a single channel, covering a service area to deliver a unitary program service or a complement of such services. The signals from the various network transmitters overlap each other and have the potential to interfere with one another. Steps are taken in the design of the network to maximize the service and to minimize the interference within the network.

In some ways, SFNs can be thought of as being akin to cellular telephone networks. They use multiple transmitters spread throughout a service area to place those transmitters closer to receivers. They are described in terms of "cells" when discussing the areas served by the various transmitters and are sometimes called cellular television systems. But they are radically different from cellular telephone systems in that cell phone systems use different frequencies for transmission and reception in each direction within each of the sectors associated with each cell in the network. The frequencies are reused in the network at locations sufficiently separated from one another that there will be little or no interference between transmitters on the same frequencies. In SFNs, all the transmitters are on the same frequency; interference between transmitters will exist and must be treated.

The reasons for applying SFNs range from service improvement, to economic, to regulatory, to interference considerations, to spectrum efficiency. Although theoretically possible with any form of modulation, SFNs only become really practical when digital techniques are used for transmission. SFNs depend, in particular, either on certain modulation methods or on techniques routinely applied in receivers to enable operation of digital transmission systems. SFNs come in two basic configurations, having different philosophical approaches and different design objectives, involving tradeoffs between improvements in service and interference caused within their service areas.

Benefits of SFNs

SFNs enable service to be provided in places that cannot be reached using single, high-power transmitters alone. If there are areas within a station's service contour that would not receive adequate signal levels due to terrain considerations, additional transmitters in an SFN configuration can be used in conjunction with a high-power transmitter to fill in the gaps in coverage. The high-power transmitter in such a case becomes part of the SFN and serves as a main facility that is supplemented with smaller extension transmitters, often called *gap fillers*. In cases in which it is desirable not to have a high-power main transmitter, the service areas of the multiple transmitters in an SFN can be tailored to achieve the coverage and signal levels desired, while doing so with lower overall power emitted.

Because they do not necessarily require the tall towers and large antennas that single-transmitter operations do, transmitters in SFNs often can be installed on existing towers used for other purposes, such as cellular telephone and other wireless communication services. They sometimes can be installed on the roofs of buildings. If there is not already an existing tower on which to install the antenna for a single-transmitter operation, these alternatives can avoid the need for zoning and planning approvals that sometimes delay or preclude progress on the otherwise necessary tower construction. In such circumstances requiring construction or reinforcement of a large tower, depending on the particular situation, using alternative locations may be less expensive and may result in a lower overall cost of implementation.

Since they permit transmitters to be located closer to receivers, SFNs reduce the amount of fade margin that must be provided to achieve reliable communications. As a consequence, they can be operated at lower power levels to obtain a given signal level over a particular service area. With lower-power operation comes less interference caused to neighboring stations, both in frequency and geography. At the same time, the shorter distances from transmitters to receivers, combined with the increased number of transmitters, can result in a more uniform signal level being delivered to the area to be served by the station. The more uniform signal levels can enable the use of indoor and low-performance receiving antennas in larger portions of the service area.

When a transmission system is used to deliver signals to pedestrian and mobile receivers, the use of SFNs permits signals to be delivered to those receivers from multiple directions. Doing so helps to minimize the impact of buildings and terrain on reception as a receiver moves. When the signal field from one transmitter is cut off by virtue of the receiver moving behind an obstruction, it is more likely that a signal field will be available from another transmitter when a sufficient number of transmitters are in use.

Digital Differences

Digital transmission involves the communication of data through naturally occurring channels and using forms of modulation that are inherently analog in nature.[1] To recover the data at a receiver, a variety of steps must be taken to retrieve the transmitted bits and to overcome the impairments to the signal that occur along the path of the transmission channel. The types of impairments caused by the path from transmitter to receiver include the addition of noise, variations in the amplitude response across the spectrum of the signal, variations in the time delay across the spectrum, and dynamic variations in those characteristics with time. The impairments lead to energy from one information-carrying symbol being delayed in time so that it interferes with deciphering the adjacent symbols, in what is termed *intersymbol interference* (ISI), and significant parts of the technologies of the various transmission systems are devoted to overcoming the effects of ISI on reliable recovery of data from received signals.

There are two fundamentally different forms of modulation that are predominantly used to deliver the data rates and levels of robustness required for broadcasting of digital television signals, and they lead to different methods of mitigating the effects of channel impairments when the signals reach receivers. These two forms of modulation involve either the use of a single carrier to carry all of the information in a channel or of multiple carriers to carry the information.

In a single-carrier system, a very high symbol rate is matched to the spectrum width of the transmission channel, and the modulation characteristics are optimized to deliver as much data as possible, as reliably as possible, through the channel on each data-carrying symbol. A consequence of the high symbol rate is that impairments caused by propagation through a channel from transmitter to receiver typically require use of digital *adaptive equalizers* that generally operate in the time domain to compensate for what are effectively filters of various and varying sorts caused by propagation channel characteristics and resulting in ISI. A single-carrier method was used for the first digital television system worldwide—the ATSC Digital Television System defined in ATSC A/53 [1] and now called ATSC 1.0. Two types of single-carrier modulation were considered for ATSC 1.0–8-level vestigial sideband (8-VSB) and 32-point quadrature amplitude modulation (32-QAM), with 8-VSB having been selected for ATSC 1.0 and 16-QAM and 64-QAM having been selected for the first generation of digital television over cable plants.

In a multicarrier system, the data to be carried first is demultiplexed into a large number of parallel data streams. Each of the data streams is modulated onto one of a corresponding number of carriers that occupy the channel bandwidth. The carriers use any type of modulation that would be appropriate for modulating a single carrier that is centered in a channel, such as QPSK or the various flavors of QAM (e.g., 16-QAM up through 4096-QAM). By virtue of the demultiplexing process, the data rate of the data stream applied to each carrier is greatly reduced, and the symbol periods thereby are lengthened correspondingly. At the same time, the multiple carriers are modulated using a technique (called *orthogonal frequency division multiplexing*—OFDM) that permits many carriers to be tightly packed in a channel without causing interference to one another. OFDM also permits generation and recovery of the multicarrier spectrum and the data that it carries, respectively, using a *fast Fourier transform* (FFT) technique that creates and modulates the data onto all of the carriers in one processing step and that similarly recovers the data from the many carriers in one step. An improved form of OFDM involves coding the data across the carriers so that, if a particular carrier is lost due to selective fading in the propagation channel, its data is duplicated elsewhere in the channel and still can be recovered. The improved form is denoted as coded OFDM (COFDM) and has been used for quite some time in systems standardized by the Digital Video Broadcasting (DVB) project based in Europe, in the Integrated Services Digital Broadcasting (ISDB) system developed in Japan, and in other systems developed elsewhere, most recently in the ATSC 3.0 system nearing completion as this is being written.

Adaptive equalizers are applied to single-carrier (e.g., 8-VSB) receivers in recognition of the fact that channel impairments behave very much like the application of linear filters (albeit complex ones) to the signals that traverse the channels from transmitters to receivers. The filters represented by the channels cause signal distortions that could make it impossible to recover the data carried in the modulation because of the interference from one symbol to another. By characterizing the channels (often through determining their channel impulse responses), it becomes possible to fashion filters the characteristics of which are the converse of those of the channels. Applying such filters to the received signals can mitigate channel impairments other than noise and allow the data carried in the modulation to be accurately retrieved. (Noise can only be overcome through improvements in the link budget between transmitter and receiver, which is one of the reasons for using SFN techniques, as described later.) The filters used can operate either in the frequency domain or in the time domain, although time domain filters are most prevalent in consumer receiver designs as of this writing.

The channel impairments treated by adaptive equalizers largely derive from multipath transmission of the signals, i.e., the signals propagate over more than one path from a given transmitter to a given receiver. Multipath causes the receiver to experience the effect of echoes in the received signals. With analog transmission, these echoes were called "ghosts" because of the ghost images they produced in the displayed pictures that were recovered from signals experiencing multipath transmission. With digital transmission through multipath channels, there are no ghosts—only a complete failure of reception when the ISI caused by the echoes crosses a threshold beyond which the error correction system in the receiver becomes incapable of recovering the data. With adaptive equalizers, the threshold is much higher than it would be without them.

In multicarrier, particularly COFDM, systems, the issues caused by ISI can be addressed in a different way. As described earlier, the symbols used in COFDM systems are much longer than those in single-carrier systems. When a symbol transitions from one state to another, the delayed energy in the propagation channel will last for a length of time determined by the characteristics of the particular channel, and the duration of the energy delay will be the same for either type of modulation—single-carrier or multicarrier. In the multicarrier case, however, given the length of time that the symbols remain in any specific state, the period of energy delay will represent a much smaller proportion of the symbol period than in the single-carrier case. (Indeed, in the single-carrier case, the ISI can last longer than the length of a single symbol.) In fact, with enough carriers in use, the period during which the delayed energy occurs can be ignored by the receiver, thereby effectively "cutting out" the ISI and allowing the receiver to decipher the data carried by the symbol. The period during which the incoming signal is ignored is called a *guard interval*, and the length of the guard interval is a parameter that can be set at the transmitter to account for the types of propagation channels that exist in the geographic region of a broadcast station's service area.

SFNs take advantage of the techniques built into receivers for treating multipath by making the signals from the multiple transmitters in a network appear to receivers as echoes of one another. The result is that receivers then act on the multiple received signals just as they would on signals from a single transmitter received through multipath channels. Because different single-carrier receiver designs vary in their ability to deal with multipath, the extent to which SFN designs for single-carrier systems can depend on receivers to treat their multiple signals as echoes of one another also varies. In multicarrier systems, the guard interval period to be used is sent to receivers as a control parameter, thereby enabling all receivers to perform in the same way with respect to their treatment of both naturally occurring and synthetic echoes. As described in detail in the next section, the characteristics of receivers of both types with respect to their handling of multipath put limits on the designs of SFNs.

SFN Prerequisites

To enable receivers to treat the signals from multiple SFN transmitters as echoes of one another, those signals must have the characteristics of echoes when they arrive at receiver inputs. In other words, the signals must appear to receivers as though they emanated from a single transmitter, modified only by the effects of the environmental transmission channel along the path from transmitter to receiver, including multipath.

Achieving the effect of single emitted signals even though originating from multiple transmitters, SFN signals must be on essentially the same frequency. To the extent that they are on slightly different frequencies, they will have the characteristics of echoes that have been reflected by moving objects, i.e., they will seem to have had a Doppler effect superimposed on one or more of them. Receiver demodulators are designed to deal with signals having Doppler-impacted multipath, but the extent of any Doppler shift reduces their ability to handle other forms of multipath effects. Thus, the frequency difference between transmitters must be kept to the minimum practical level.

For receiver demodulators to determine the channel characteristics and for single-carrier adaptive equalizers to compute a filter that can correct for the channel, the multipath signals must contain the same data symbols, which must be separated in time from one another by no more than an amount determined by the receiver design. Thus, the transmitters in an SFN must emit the same symbols at approximately the same time. This means that the transmitter outputs must be synchronized with one another in terms of what they emit and when. As described later in this chapter, it may be desirable to offset the output timing of some transmitters to aid in achieving the required time relationships at receiver inputs in certain geographic areas.

Types of SFNs

There are two fundamental approaches to the design of SFNs: a small-cell scheme and a large-cell scheme. The primary difference between them is in the way the overlap of signals from the several transmitters in the network is handled. There also are differences related to the use of single-carrier or multicarrier modulation that result in greater or lesser ease in dealing with naturally occurring or man-made (SFN-related) multipath.

The small-cell method is implemented using lower-power transmitters that cover smaller areas and are in closer proximity to one another than in the large-cell method. In the small-cell scheme, there is intentional signal overlap to a very great extent. As discussed later, the close physical transmitter spacing has the benefit of keeping the time spacing of signals from nearby transmitters within the delay spread windows of single-carrier receiver adaptive equalizers or has the benefit of requiring less system overhead and the consequent reduced payload associated with longer guard interval duration in multicarrier systems.

The large-cell method is implemented using higher power transmitters that cover larger areas and have greater separation from one another than in the small-cell method. In the large-cell scheme, the signals from the several transmitters are separated to the extent possible using terrain shielding, more than normal beam tilt and other elevation pattern techniques, azimuth pattern shaping, and transmitter power-level choices. As discussed later, in single-carrier systems, the wide separation of transmitters tends to stress receiver adaptive equalizers and leads to portions of service areas where signals are likely to fall outside the delay-spread windows of receiver adaptive equalizers. In multicarrier systems, the wide separation of transmitters leads to longer guard intervals and consequent reduction of system payload capacity.

As of this writing, all of the SFNs implemented in North America have been large-cell designs. Small-cell designs currently are too expensive to implement in terms of the capital costs of the transmitters and antennas and the infrastructure to interconnect them. With the declining cost of electronics and the plummeting cost of terrestrial data distribution bandwidth, the economic situation for the small-cell scheme may change. If it does, the small-cell method will be the more preferable of the two approaches for reasons that will become apparent in later sections of this chapter. Because of the current greater practicality of implementation of the large-cell method, it will be the principal focus of the remainder of this discussion.

Large-cell SFN Tradeoffs

Implementing large-cell SFNs implies design tradeoffs or compromises with respect to the coverage and performance of the networks. Because of the larger separation of the transmitters in the large-cell scheme, for networks using single-carrier modulation, there are likely to be areas in which the combined signals from two or more transmitters fall outside the range in which receiver adaptive equalizers can compensate for the effective channel and accurately recover the data. It becomes part of the design process to minimize such areas and to place them where they will have the least effect on service to the public. For networks using multicarrier modulation, there will be a need to use a larger guard interval than would be required to mitigate only naturally occurring echoes, thereby reducing the payload capacity of the network.

For networks using single-carrier modulation, the design tradeoffs implicit in large-cell SFN designs, then, are between increased signal levels and service in areas where there should be no difficulty caused to receivers in recovering the transmitted data and internal network interference (INI) having characteristics such that it will create difficulties for receivers in data recovery. The tradeoffs are made so as to maximize the areas where signal levels and service are increased without INI causing problems, while minimizing those areas where INI is predicted to make reception more difficult. The maximization and minimization described generally are determined based on counting populations affected rather than geographic areas. The use of population counts for making decisions tends to place any deleterious INI in regions with low population densities. The methods applied in the design process for making such tradeoffs are discussed later in this chapter. For networks using multicarrier modulation, the tradeoffs are much simpler; they involve the separation distances between transmitters, the guard intervals necessary to accommodate such longer effective echoes, and the loss of system capacity that arises with use of longer guard intervals.

Resources, Abbreviations, and Terminology

- **A/110** is the document number of the ATSC Synchronization Standard for Distributed Transmission, which is available in a 2011 version as of this writing. It documents the distributed transmission method for ATSC 1.0 that uses a centralized data-processing model in a transmission adapter to embed timing and control information contained in transmitter control packets in the data stream to distributed transmitters. [1]
- **A/111** is the document number of the ATSC Recommended Practice: Design of Synchronized Multiple Transmitter Networks applicable to ATSC 1.0, although much of its content will remain applicable to ATSC 3.0. It describes in detail methods for design of networks using digital on-channel repeaters, distributed transmitters and distributed translators, the RF watermark, and related concepts. [2]
- **A/324** is the document that specifies the scheduler, studio-to-transmitter link (STL), and single-frequency network functionality of the ATSC 3.0 system. It includes a number of protocols to tie

together the transport and physical layers of the ATSC 3.0 system, as well as error correction coding and other technologies to be used in constructing ATSC 3.0 transmission systems. [4]

- **ATSC** is the Advanced Television Systems Committee, which documented the 8-VSB Digital Television Standard and has adopted several documents on the subject of single-frequency networks.
- **ATSC 1.0** is the original ATSC Digital Television System documented initially in 1995 in ATSC A/53 and related standards and recommended practices.
- **ATSC 3.0** is an Internet protocol (IP)-based replacement for ATSC 1.0 documented primarily in 2016 in ATSC A/300 and related standards and recommended practices.
- **DOCR** is a digital on-channel repeater, which receives signals over-the-air from a main transmitter or from another type of transmitter closer to the signal source in a network and retransmits them on the same channel as the one on which they were received.
- **DTS** is the nomenclature for distributed transmission systems technology used by the U.S. Federal Communications Commission (FCC), which it applies to SFN transmitters involving synchronized transmission.
- **DTx** is the abbreviation for distributed transmission used in the ATSC documents.
- **DTxN** is the abbreviation for distributed transmission network used in ATSC documentation, which is a network of transmitters fed through studio-to-transmitter links (STLs) and synchronized with one another using transmitter control packets inserted into the transport stream by a transmission adapter.
- **DTxR** is the abbreviation for distributed translator used in ATSC documents. DTxRs receive signals over the air and translate them to another channel, where their outputs are synchronized with one another in similar fashion to the synchronization of DTxTs.
- **DTxT** is the abbreviation for distributed transmitters used in ATSC documents. DTxTs produce outputs that are synchronized with one another from transport stream inputs that contain transmitter control packets carrying the necessary synchronization information.
- **E-DOCR** is the abbreviation for equalization DOCR, which adds an adaptive equalizer to a DOCR to overcome many of the limitations of conventional DOCRs. In particular, the signal distortion resulting from signal coupling from the transmitting antenna to the receiving antenna can be significantly reduced.
- **Gap filler** is a transmitter located at such a position that it fills in an area within a station's noise-limited contour that receives little or no predicted service due to lack of adequate signal level.
- **INI** is internal network interference, which is the interference that occurs between transmitters operating on the same channel in an SFN.
- **ISI** is intersymbol interference, which is the interference that occurs in a transmitted data stream between symbols nearby in time because of the effects of transmission through a propagation channel and the resulting echoes. ISI occurs in the signals from single transmitters due to the natural effects of signal propagation, and it occurs in SFNs as a result of reception of signals from multiple transmitters.
- **SFN** is the abbreviation for single-frequency network, a network of transmitters collectively serving an area by transmitting on the same channel.
- **TA** is the abbreviation for transmission adapter, as used in ATSC documents. TAs insert transmitter control packets into transport streams to carry the information necessary to synchronize DTxTs and/or DTxRs.
- **TCP** is the abbreviation for transmitter control packet used in ATSC documents, which are inserted into transport streams by TAs to carry the information necessary to synchronize transmitters in a DTxN.

Example Designs

To help in visualizing and understanding many of the concepts involved in the design of SFNs, two examples are included in this chapter. One is of a system that has been on the air for over 14 years as of the time of publication. The network was the first digital television (DTV) SFN to be constructed in the United States, first being authorized on an experimental basis. It was a system the design for which included a high-power main transmitter and several lower-power transmitters. There was severe terrain blockage from the main transmitter to the principal population centers on the southern extreme of the service contour of the main transmitter. Without the additional transmitters acting as gap fillers, there would be little or no service to the bulk of the public the station was intended to serve (although it would more than adequately serve its community of license, which is located near the main transmitter). The SFN was designed for a total of four transmitters—one main and three lower-power gap fillers.

The second SFN example design comprises eight transmitters. It is a design developed to overcome the problem of a station, due to local governmental restrictions, not being able, within a reasonable period, to construct a tower where it needed to do so. The station had a construction permit for maximized facilities, but the site intended for its tower was withdrawn from consideration by the same community leaders who had first suggested it. It also happened that the station was on a channel adjacent to an in-market neighbor, the reference points of the two stations being separated by about 50 km. To complicate the design further, there were co-channel neighbors in two adjacent markets on either side of the station's allotted service area. The SFN generally was designed to use low-power transmitters, with one moderate-power transmitter required to take advantage of the well-known benefits of collocation to the extent possible.

TYPES OF SFN TRANSMITTERS

Three basic transmitter types can be used in designing an SFN. They differ primarily in their spectrum utilization, in the way signals are delivered to them, and in their configurations. Each has advantages and disadvantages that make it most applicable for solving particular issues in network design. They are digital on-channel repeaters (DOCRs), distributed transmitters (DTxTs), and distributed translators (DTxRs). Their basic characteristics and designs are considered next, followed by a comparison of their respective advantages and disadvantages, which can help in determining where they are best applied. Details of their designs are presented in later sections of this chapter. It should be noted that, while the transmitter types are considered separately here and later in the chapter, they can be combined into SFNs having composite characteristics of the various methods.

DOCRs

DOCRs are the digital equivalent of the "boosters" that have been used in analog services for many years, but they can take advantage of several digital processing techniques that are not applicable to analog signals. They range from very simple combinations of a receiving antenna, a channel filter, an amplifier, and a transmitting antenna to quite sophisticated combinations of receiving antenna, receiver, digital signal processing, transmitter, and transmitting antenna. To operate within SFNs, DOCRs must be designed to meet the SFN prerequisites discussed earlier. The basic configuration of a DOCR is shown in Figure 8.6-1. Designs appropriate for SFN use are described in detail in the major section on DOCRs.

DOCRs provide the simplest method for establishing an SFN. They receive signals over the air from a main transmitter and retransmit them on the same channel. Because they are receiving and transmitting on the same channel, isolation between the transmitting and receiving antennas becomes a critical factor in the installation of a DOCR. To the extent that high levels of isolation are not maintained, the performance of a DOCR can be degraded. In the extreme, with a simple DOCR design, the feedback from transmitting to receiving antenna can make a DOCR oscillate if the amplifier gain is too high. Because of the antenna isolation factor, the power output of a DOCR is limited, and there are limitations on the locations where DOCRs can be used. DOCRs of a particular design also have a fixed time delay that depends upon the design, which further limits the locations where such a DOCR can be used. These limitations are discussed in the sections on network designs. There are certain techniques (e.g., equalization DOCRs—E-DOCRs) that have been applied in relatively sophisticated DOCR designs, which can mitigate some of the limitations in DOCR application. These later developments also are discussed below in the major section on DOCRs.

DTxTs

Distributed transmitters (DTxTs) are used in distributed transmission (DTx) networks (DTxNs). The

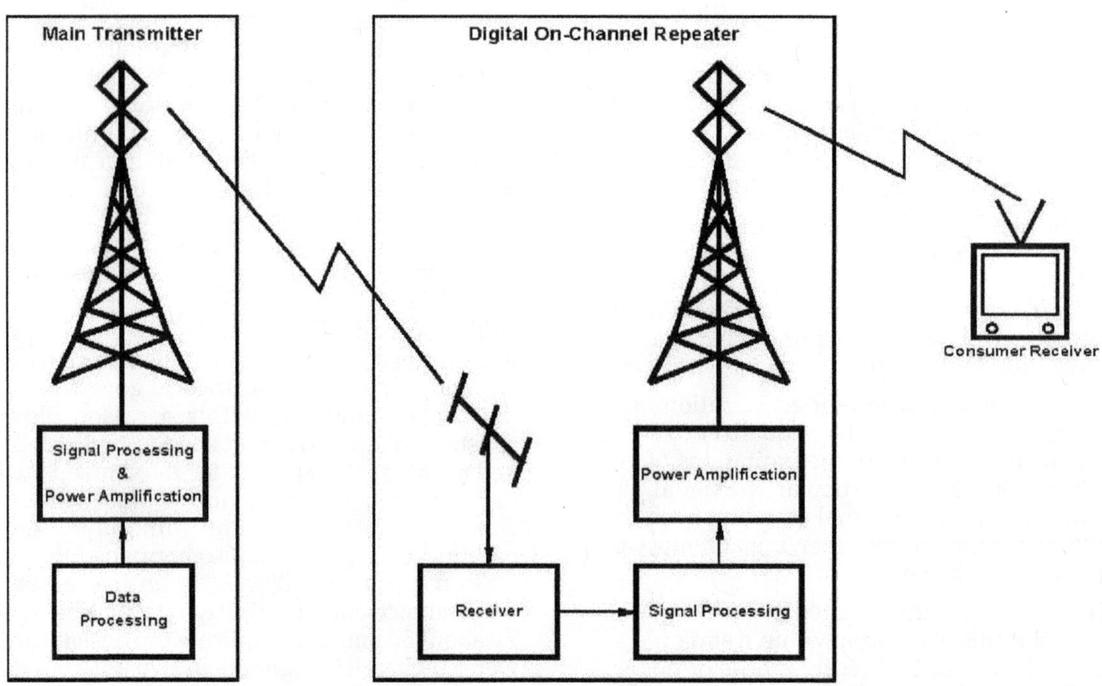

FIGURE 8.6-1 Basic digital on-channel repeater (DOCR) configuration.

primary characteristic of a DTx system is that it uses a studio-to-transmitter link (STL) to deliver signals to each of the transmitters in a network. The inclusion of the STL in the design avoids the issue of antenna isolation that is so limiting in the DOCR approach. As a result, DTxTs are not inherently limited in the power levels they can transmit or in the relative emission timing of their outputs. Because of the use of an STL to deliver signals to the transmitters in a DTxN, however, each transmitter must include its own exciter, of one form or another, depending on the DTx method used. The exciters of the several transmitters must be properly synchronized with one another in order to comply with the SFN prerequisites discussed earlier. Several schemes for synchronizing transmitters in DTxNs are described in detail later in the major section on DTxTs.

DTxTs represent the most complex method for establishing an SFN. They require both an STL and a means of sending synchronization information through the STL to all of the DTx transmitters in the network. Since they are not limited in power or timing, however, they provide the greatest flexibility in achieving any particular network design goals. In fact, they permit solutions not possible with DOCRs, in some instances; they similarly can provide solutions not possible with DTxRs. A basic DTx system is shown at the conceptual level in Figure 8.6-2.

DTxRs

DTxRs are a hybrid of DOCRs, DTxTs, and translators. They receive their signals over the air from a broadcast transmitter, the output of which also can be received by the public. Thus, they do not require STLs to feed signals to them, just like DOCRs. Under the right conditions, they are capable of virtually unlimited power and timing flexibility, similar to DTxTs. But they require additional RF channels, just like translators, which may not be available in many locations. A conceptual DTxR network involving a single tier of translators and two channels is shown in Figure 8.6-3.

DTxRs essentially receive signals over the air and repeat those signals on different RF channels. Because of the different input and output frequencies, many of the issues related to coupling between transmitting and receiving antennas can be avoided if the channels are separated widely enough in frequency to permit avoiding desensitization of the receiver by the transmitter. DTxRs do have the issue of requiring an adequate received signal level from the preceding transmitter in the network, thereby, to some extent, limiting their placement geographically, but they are not limited in the way that DOCRs are.

Advantages and Disadvantages

The relative advantages and disadvantages of the respective types of SFN transmitters are summarized in Table 8.6-1. This comparison considers limitations in emitted power, emission timing, and geographic placement, plus the need for an STL to each transmitter and the need for an additional RF channel. Also shown are the relative costs of the several types of solutions. The evaluations of the various characteristics are

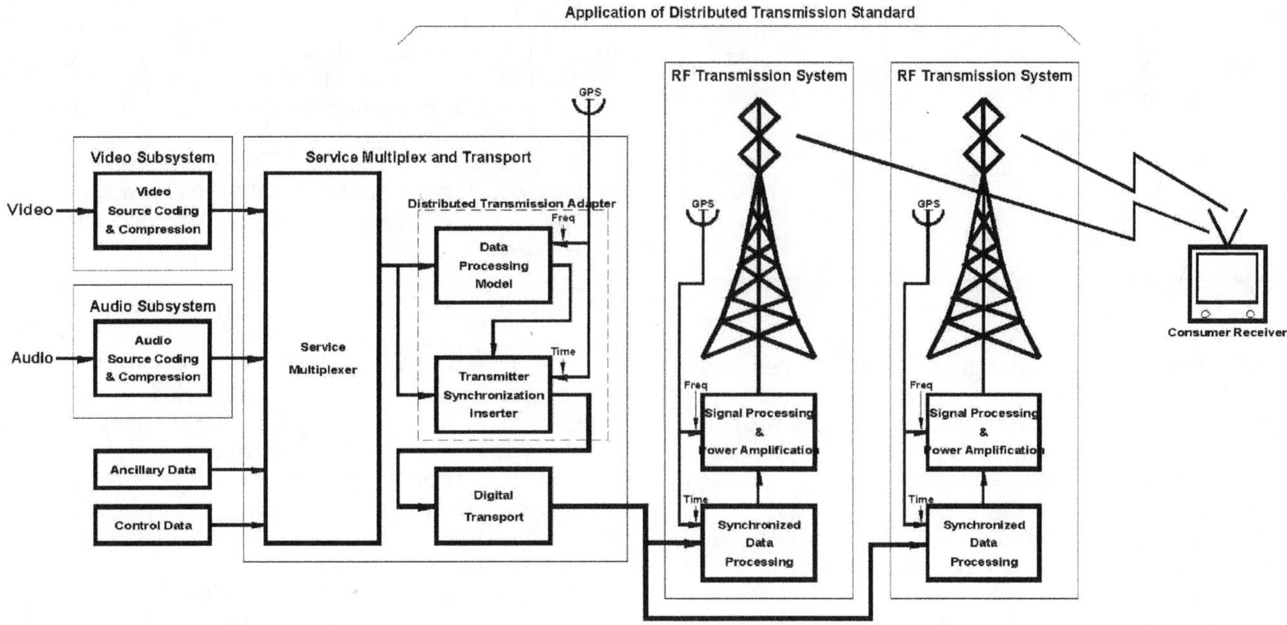

FIGURE 8.6-2 Conceptual view of a distributed transmission (DTx) system, showing a transmission adapter (TA) and multiple distributed transmitters (DTxTs).

S. Merrill Weiss

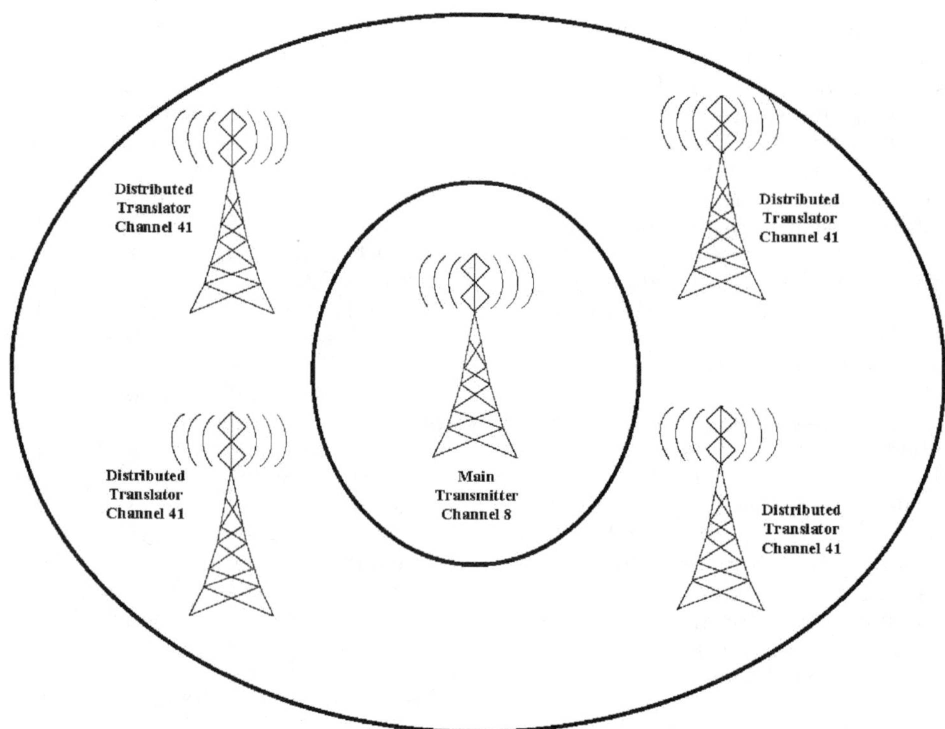

FIGURE 8.6-3 Distributed translator (DTxR) network with a main transmitter on one channel and a tier of DTxRs on a second channel.

TABLE 8.6-1
Relative Advantages and Disadvantages of SFN Transmitter Types

Characteristic	DOCRs	DTxTs	DTxRs
No power limits	–	++	+
No timing limits	–	++	++
No inherent geographic limits	–	++	+
No STL needed	++	–	++
No additional spectrum needed	++	++	–
Cost	$	$$$	$$

arranged so that a minus sign is used for a beneficial property that *does not apply* to a particular transmitter type, a plus sign indicates that the beneficial property *does apply* to that transmitter type, and two plus signs indicate that the property *strongly applies*. The number of dollar signs indicates relative cost.

As can be seen in Table 8.6-1, DOCRs have several design limitations, do not require an STL or more spectrum, and are lowest in cost. DTxTs have the greatest design flexibility, do require an STL but no additional spectrum, and are highest in cost. DTxRs have medium design flexibility, do not require an STL but do require additional spectrum, and are intermediate in cost.

NETWORK DESIGN OBJECTIVES

Designing an SFN requires careful balancing between competing objectives. On one hand, it is desirable to cover the largest area and to provide the highest reasonable signal level over the coverage area. On the other hand, higher signal levels may result in more interference to other stations, both within and outside the station's market, and they may result in more internal network interference for the station for which the network is being designed, especially for stations using single-carrier modulation. To optimize the coverage and service, it helps to begin with the characteristics of receivers and to design the network to accommodate their capabilities.

Coverage and Service

Before attempting to design a network to cover a particular area, it helps to determine what received signal levels will be considered as providing service under various conditions. The classic regulatory threshold for service, while appropriate for application and licensing purposes, may not have much meaning with respect to practical service.

Using the values for UHF operations based on transmission of first-generation, single-carrier signals, i.e., ATSC 1.0 (the values for VHF can be substituted into this discussion with the same results), the reception threshold field strengths specified in, for instance, the rules of the FCC in the United States are often held to be about 10 dB below what is the minimum necessary for semi-reliable service when an outdoor antenna is used. Thus, a target threshold value for outdoor antenna reception (at 9.1 meters, or 30 feet, above ground level) realistically might be a field strength of 51 dBu (dB microvolt/meter) rather than the 41 dBu included in the FCC rules.

When indoor service on a set top or similar antenna is considered in a suburban or rural setting, several adjustment factors must be applied to the threshold considered necessary for reliable outdoor reception. First, the antenna generally will be at a lower elevation above ground level, resulting in lower field strength. Next, the antenna gain (10 dB) assumed in arriving at the outdoor threshold likely will not be available. Then, there will be additional attenuation of the signal from passing through the roof and floors or the walls and windows of the structure. All of this leads to the need for a signal at the standard measurement elevation above ground level about 30–40 dB stronger than required for reliable outdoor reception, that is, according to a consensus among experts on the subject, having a field strength of 80–90 dBu at UHF.

When the area to be served is in a downtown canyon or a high-rise area, the signal may need to be another 10 dB stronger under standard measurement conditions in order to achieve reliable set-top reception inside apartments and similar abodes in such areas. Thus, according to the expert consensus, the signal level required for reliable reception may be as high as 100 dBu in such areas.

The use of new transmission technologies can have considerable impact on the actual reception thresholds needed under the sorts of conditions described. Comparing ATSC 1.0 systems with ATSC 3.0 systems, for example, the former provided one choice of emission characteristics, while the latter provides a plethora of such choices. With ATSC 1.0, there was one configuration of the premodulation data processing functions, resulting in one combination of operating characteristics, namely: 19.392658 Mb/s and 14.9 dB C/N threshold in an additive white Gaussian noise (AWGN) channel. With ATSC 3.0, there are choices of modulation constellations, error correcting codes and ratios, interleaving schemes, and operating modes, resulting in C/N thresholds from approximately −5 dB to +32 dB at bit rates from a few Mb/s to several tens of Mb/s if the entire channel were devoted to a single combination of parameters. The ATSC 3.0 system is capable of intermixing numerous combinations of parameters by allocating portions of the channel resources to each of them, thus offering tremendous flexibility. Analyzing all of the choices can be a complex undertaking, yet the system allows almost any practical set of requirements with respect to the trade-off between robustness and bit rate to be met. When applied to SFN operation, the range of choices permits addressing almost any application requirement within the range of parameter settings included in the standard.

Using Receiver Performance Characteristics

After providing adequate received signal level, the next most important factor in designing an SFN is either the expected capability of receiver adaptive equalizers for systems using single-carrier modulation or the setting of the guard interval for systems using multicarrier modulation. Typically, adaptive equalizers are characterized in terms of the length of the echoes (i.e., the time delay) they can correct and the amplitudes of those echoes with respect to the strongest signal arriving at the receiver input. From an SFN design standpoint, the most desirable adaptive equalizer would have the longest delay spread (i.e., time from the earliest leading echo to the latest trailing echo to which the equalizer can adapt) and the smallest echo amplitude separation at any particular delay value. In COFDM systems like ATSC 3.0, the appropriate value of guard interval must be determined to allow receivers to accept the delay spreads caused by distances between transmitters while not wasting channel capacity by setting the guard interval value to be too long. While the following discussion focuses on adaptive equalizers, the concepts are applicable to selection of guard intervals for networks using multicarrier modulation.

Since an ideal equalizer cannot be expected to be found in consumer-grade receivers, it helps to have some concept of what level of performance practical equipment may be capable of achieving. To that end, the ATSC has produced a Recommended Practice: Receiver Performance Guidelines (A/74) document that contains sufficient information to guide SFN design in the near-term. In particular, an early version of A/74 contained a chart (Figure 4.3) that showed the relationship between time delay and amplitude that could be expected in consumer receivers in the years following its publication. That chart is reproduced here as Figure 8.6-4.

The chart contains a mask in which conditions falling above the line drawn on the chart can be expected to be in the range that receiver adaptive equalizers will be able to treat, while conditions below the line may not be adequately handled by some receivers. The conditions shown are the separation in amplitude (on the vertical scale) and in time (on the horizontal scale) between a pair of arriving echoes of the same signal. One echo is considered to be the main signal path, as determined by its being the strongest received echo, and the other either leads or lags the main echo. A DTx network design that keeps signals from the several transmitters in the space above the line should reasonably assure reception on relatively modern receivers. In those parts of a DTx network service area in which the conditions fall below the line, it is far more likely that reception will be impaired.

A few observations can be made about some significant points in the chart of Figure 8.6-4. The best adaptive equalizer performance is represented by values near the bottom of the chart and near the sides. These are the values with the lowest amplitude separation between echoes and with the longest time delays. The closer an adaptive equalizer comes to the bottom corners, the better it is from the standpoint of flexibility of SFN design.

Above the level where the line in the chart flattens out to the sides, adaptive equalizer performance does not matter; the adaptive equalizer essentially is not needed above that level of amplitude separation

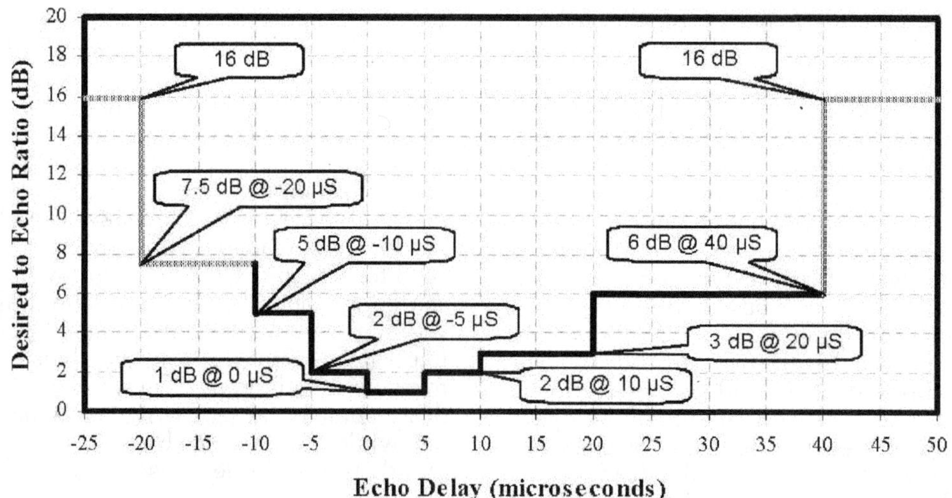

FIGURE 8.6-4 Minimum adaptive equalizer performance recommendations from early version of ATSC A/74.

Figure courtesy of ATSC, Washington, D.C. [3]

between echoes. The flat top level represents the desired-to-undesired (D/U) ratio between signals from different transmitters at which one of them becomes "dominant" and network timing adjustments become irrelevant. This relationship is discussed in detail later in the section on network design techniques.

Note that, for a given amplitude separation between echoes, there is more time delay capability for echoes trailing the main signal (i.e., arriving at the receiver after it) as opposed to the time advance capability for echoes leading the main signal (i.e., arriving at the receiver before it). This relationship results from the tailoring of adaptive equalizer designs to approximate what occurs most often in nature and from the structure used in adaptive equalizers, with infinite impulse response (IIR) filters used to compensate leading echoes and finite impulse response (FIR) filters used to compensate trailing echoes. It means that there is more design flexibility in SFNs with apparent trailing echo relationships between received signals from multiple transmitters than there is with apparent leading echoes.

The same sort of chart could be prepared to examine the application of guard intervals to multicarrier modulation. In such a case, the equivalent mask would have a height equal to the AWGN C/N ratio in use for a particular portion of the spectrum resource, plus some margin—perhaps a dB or so. The opening in the middle of the chart would have edges extending from the bottom to the top of the amplitude space and angling slightly outwards. The opening would match the value of guard interval selected for a particular operating condition. The opening effectively would cover the time period from the earliest arriving echo to the latest arriving echo that fell within the amplitude range from 0 dB to the C/N value of the applicable operating conditions. The characteristics of the mask could be applied to the network design in essentially the same way as could the mask for a single-carrier system, with the exception that the end result of the analysis could be used to select the value of guard interval for the network.

Minimizing Leading Echoes

In networks using single-carrier modulation, to put the least stress on receiver adaptive equalizers, thereby leaving the maximum margin for treating path impairments, it is important to design SFNs to minimize the leading echoes they create. Determining where leading echoes are likely to occur and their expected extent is readily done when the relationship between relative levels and timing of the signals arriving at a receiver input is understood with respect to geography. This relationship can be seen in Figure 8.6-5.

Figure 8.6-5 shows a generalized map with two SFN transmitters (Tx-1 and Tx-2). Each transmitter has a contour surrounding the region in which its signal exceeds that of the other transmitter by a ratio of field strength of 10-to-1, which corresponds to 20 dB or close enough to the flat top of the mask in Figure 8.6-4 to be suitable for purposes of discussion. (In a real design effort, the value of 16 dB from Figure 8.6-4 would be used, and the ratio of the signal levels from the transmitters would be predicted at each study point on a grid surrounding all the transmitters in the network.)

There are two other features of significance to the evaluation of a network design shown in Figure 8.6-5, both represented by dashed lines. One is the line representing the locations where the field strengths from the two transmitters are equal; the other is the line representing the locations at which the arrival times of the signals from the two transmitters are equal (i.e., there is zero delay between the two signals). The positions of the two lines can be controlled by adjusting the relative power levels of the two transmitters (for the position of the equal signal level line) and the relative emission times of the two transmitters (for the position of the equal arrival timeline).

Yet another feature of Figure 8.6-5 is the gray area between the equal field strength line and the equal arrival timeline. In that gray area, the first-arriving signal (from Tx-2) will be weaker than the second-arriving

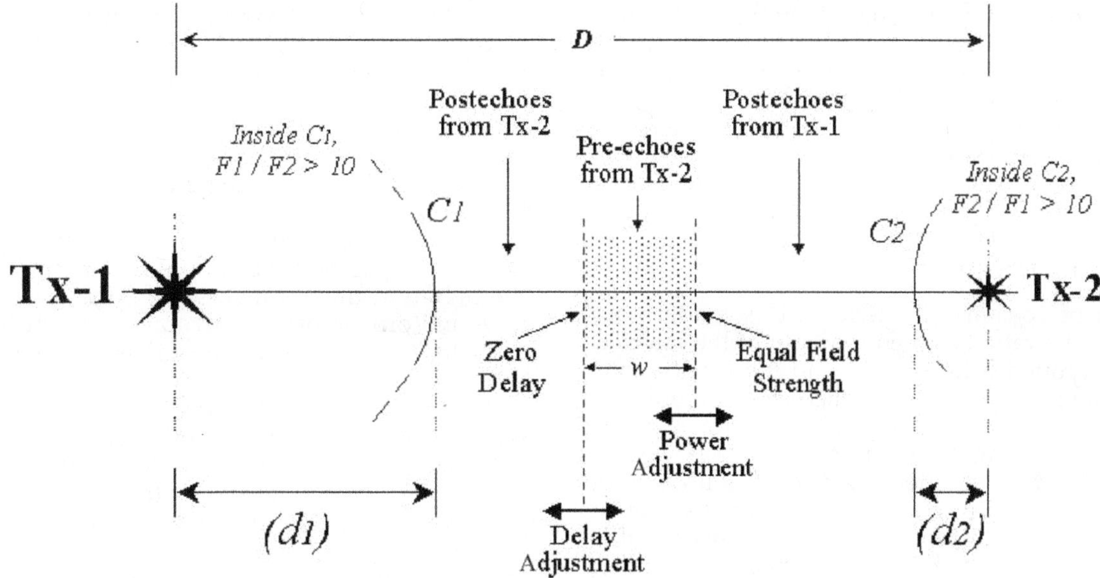

FIGURE 8.6-5 Relationship between signal levels and arrival times from a pair of transmitters at a receiver.

Figure courtesy of Y. Wu and K. Salehian, Communications Research Centre Canada, Ottawa, Ontario

signal (from Tx-1). In other words, the space between the two lines represents the area with leading echoes. As previously discussed, leading echoes put the greatest stress on adaptive equalizers. Thus, the optimum adjustment of the network parameters generally will be one that minimizes or eliminates the gray area and, with it, the leading echoes. Depending upon the choices of basic transmitter types, such an adjustment may or may not be possible.

Note that minimization of leading echoes is unnecessary in multicarrier systems that do not apply time domain adaptive equalizers.

Avoiding Internal Network Interference (INI)

In the small-cell scheme described previously, there is intentional overlap of the signals and consequent creation of INI. The effect of the INI can be minimized by placing the transmitters close enough to one another that their signals will fall within the delay spread windows of receiver adaptive equalizers in networks using single-carrier modulation and within the guard intervals selected for use in networks using multicarrier modulation. Controlling the areas in which the signals from the transmitters can be received, so that there is overlap of signals from only a small number of transmitters at any one location, helps further to reduce the stress on adaptive equalizers but may be unnecessary in systems using guard intervals.

For large-cell systems, the avoidance of INI is a much more challenging design factor. Combinations of power level selection and antenna pattern control, along with terrain shielding, can be used to minimize the overlapping of signals that produces INI in systems using single-carrier modulation but may be unnecessary in networks using multicarrier modulation. In laying out the service areas for each of the transmitters in a network, it is desirable to minimize the width of the bands between the transmitters where the signals will overlap without one of the transmitters being dominant. Reduction of the width of the overlap zones can be achieved through selection of antenna elevation patterns with characteristics that are rather different from the norm used for stations covering large areas from a single transmitter. In particular, greater-than-conventional beam tilt combined with reduction of the energy radiated from the top of the main beam can minimize the width of the overlap region.

Use of elevation pattern control of the overlap regions between cells, as well as for control of adjacent-channel interference, as discussed later in this chapter, implies the use of larger antenna vertical apertures than often would be required for the power levels and service areas involved. The increased antenna size tends to drive up the cost of using elevation pattern control in two ways: the cost of the antennas themselves and the cost of the tower space on which to install the antennas. Nevertheless, when the lower elevations of the antennas above ground level often possible are taken into account, substantial savings frequently can be achieved with respect to the cost of building a single tall tower with a high-power antenna when one does not already exist.

Minimizing Interference to & from Other Stations

There are two types of interference to neighboring stations that are of primary concern when designing an SFN: co-channel interference to stations in neighboring markets and adjacent-channel interference to stations in the same market. Co-channel interference actually can be easier to control through the use of SFNs than with a high-power, single-transmitter design, while in-market, adjacent-channel interference can become more problematic when using SFNs. Both are manageable if attention is paid to the differing characteristics of each type of interference.

Co-Channel Interference to Neighboring Markets

Treatment of co-channel interference is made manageable by the ratio between the radii of the interference zone around a transmitter and the service zone around the same transmitter, which is on the order of three to one at UHF, for example. This relationship has the effect that a smaller interference zone is created when a smaller transmitter service area is used, and, consequently, the transmitter can be placed closer to the co-channel neighbor without causing unacceptable interference. The relationship is shown in Figure 8.6-6.

Figure 8.6-6 shows a very simplified example of the same coverage as that of a single central transmitter being provided by a group of four SFN transmitters. In the diagram, the large star represents a high-power transmitter at the center of a circular coverage area, having a radius R, and the four smaller stars represent four lower-power SFN transmitters, each having a coverage area of radius $r = R/2$. Point "A" is at a distance of $3 \times r$ from the closest SFN transmitter and point "B" is at a distance of $3 \times R$ from the single, central transmitter. Substituting $R/2$ for r, the distance of point "A" can be shown to be $2 \times R$ from the single central transmitter.

Disregarding the aggregation of signals from the more distant SFN transmitters, these computations result in point "A," located a distance of $2R$ from the center of the larger circle, being the limit of the interference zone of the SFN in the particular direction shown. Since the limit of the interference zone of the single, central transmitter is point "B," located a distance of $3R$ from the transmitter, it is apparent that, in the case described, using an SFN results in reducing the radius of the interference zone to a distance of approximately 2/3 that of the single central transmitter.

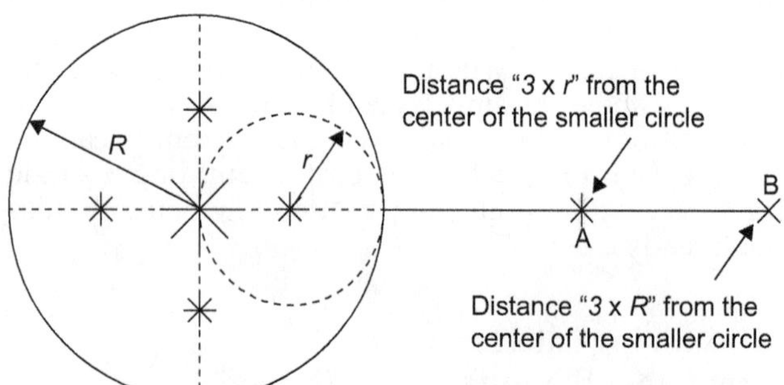

FIGURE 8.6-6 Relationship of service and interference zones of a single, central transmitter and of a group of SFN transmitters.

Figure courtesy of Y. Wu and K. Salehian, Communications Research Centre Canada, Ottawa, Ontario

In this simplified example, if an SFN transmitter were moved toward the edge of the coverage area and its effective radiated power (ERP) were reduced in the direction in which it was moved, then its interference to a station in a neighboring market in that direction would be further reduced. It is worth noting that substituting an SFN for a single central transmitter also has the effect of making reception in most parts of the service area more robust against interference from neighboring stations. Distribution of power among the SFN transmitters makes the SFN signal strength more uniform across the coverage area and more tolerant to stronger unwanted signals from neighboring stations.

The example given demonstrates that, when an SFN replaces a single central transmitter, its design can have the effect of reducing interference to stations in neighboring markets. It is important to recognize, however, that pre-existing DTV allotment plans are based essentially on single central transmitters that are assumed to replicate the coverage of previously existing analog stations. In such a situation, design and implementation of an SFN actually constitutes a replacement for an already planned single central DTV transmitter and may have to respect its protected contour and coverage area. In the planning for such a single central transmitter, adequate separation distances from neighboring market stations already would have been provided. When the single central transmitter is replaced by an SFN, the separation distances already provided likely will be more than those required by the SFN alone.

It is likely that there will be cases in which an SFN is designed independent of a planned single, central DTV transmitter. In such circumstances, the impact of the SFN on neighboring market stations can be predicted from the required protection ratios, the ERP of the transmitters, and the configuration and topology of the SFN. For a more precise determination, the cumulative effect of all the network transmitters should be taken into account. Based upon the results of the calculations, SFN design parameters can be adjusted to avoid causing unacceptable effects on neighboring market stations. It is to be expected, however, that network transmitters in an SFN service area replicating the coverage of a single central transmitter can be closer to neighboring-market stations than could the single central transmitter itself.

In-Market Adjacent-Channel Interference

Providing protection to and obtaining protection from adjacent-channel neighboring stations can be markedly different from the case with respect to co-channel neighbors. Adjacent-channel stations may be in geographically neighboring markets, they may be in the same market as the one in which an SFN is to be built, or there may be partial overlap of the coverage areas of the respective stations. When adjacent-channel neighbors are in separate markets, the analysis and design process become essentially the same as with co-channel stations in the same places; only the required D/U protection ratios are different. When there is overlap of the service areas, whether partial or complete, a number of additional considerations come into play.

The primary difference when there is overlap of service areas between stations on adjacent channels (whether first-adjacent or other channel relationships) is that transmitters usually have significantly stronger signals in the immediate areas surrounding them than they do in the far distance from their locations. This phenomenon is often described as there being "hot spots" around transmitters in which D/U ratios to other stations can become quite low (i.e., the undesired signals can be quite a bit stronger than the desired signals). It has long been understood that the best way to reduce or eliminate interference between adjacent-channel stations is to collocate them with one another, taking advantage of similar radiation patterns from nearby locations to keep the D/U ratios between the stations relatively constant. Placing transmitters on adjacent channels at substantial distances from one another seems to contravene the optimum solution.

Two ways are possible to provide protection to neighboring stations from noncollocated transmitters: reduce power of the separated transmitters to the point at which the hot spot signal levels are predicted not to cause unacceptable interference or adopt techniques that avoid creating hot spots in the first place. The power reduction method is straightforward and relatively inexpensive to implement, but it results in very significant reduction in the service that can be obtained from a transmitter so constrained. Avoidance of creating hot spots can be accomplished using specialized, but relatively expensive, antenna elevation pattern control techniques, which are described later in the major section called "Network Design Methods," in the subsection "Making Tradeoffs." Fundamentally, the method for avoiding creating hot spots requires use of a pattern that creates uniform field strength from the base of the tower on which the antenna is installed out to some distance from the tower at which the field strength begins to fall off. With such a pattern in use, it becomes possible to set the D/U ratio between the stations over the service area of the transmitter so as not to exceed a value selected in the design process.

Obtaining protection from adjacent-channel neighbors, especially those operating at high power levels on tall towers, often requires more of a brute-force approach. Ideally, adjacent-channel stations both would build SFNs together, collocating their transmitters at all sites and thereby minimizing interference to one another (and obtaining economic efficiencies, too). When such a coordinated design is not achievable, then the next best approach is to place a moderate-power transmitter in the SFN near to the high-power adjacent-channel operation, to serve the area where its signal is strongest. Doing so will allow setting the D/U ratio from the other station to the SFN signal at a selected value in the strong-signal region of the adjacent-channel transmitter. The moderate-power SFN transmitter presumably would operate at lower power than its neighbor, but the D/U ratio would be selected to maintain a workable value over the service area of the SFN transmitter. Then smaller transmitters could be situated throughout the remainder of the service area allotted to the station using an SFN, and those smaller transmitters could actually deliver stronger signals in their regions than would the high-power adjacent-channel neighbor in those same areas. Of course, protection also would have to be provided in the opposite direction (from the SFN to the other station).

Regulatory Considerations

Any SFN will be designed and built in an environment constrained by what the spectrum regulatory body having local jurisdiction will permit and license. In the United States, the regulator is the FCC. Regulatory requirements determine technical matters such as the power levels and antenna heights permitted to be used for each of the transmitters, service matters such as any requirements to serve particular areas or to provide certain minimum signal strengths in specific areas, and the methods to be used in calculating the accumulation of interference from multiple transmitters to neighboring stations.

In the United States, for example, the FCC permits use of the same power levels and antenna heights for SFN transmitters as are allowed for single-transmitter facilities. Additional limitations occur through operation of the interference protection requirements of the rules in combination with limitations on where transmitters are permitted to be located and the areas they are permitted to serve. The FCC requires that stations using DTS technology provide service to essentially all of the areas they have served from their single-transmitter operations (known as "replication") and allows them to serve all of the areas that they may have authorized in construction permits for their digital facilities (known as *maximization*) or within other defined areas. The FCC also requires that stations using DTS technology deliver signal levels to their communities of license equal to or greater than would be required from a single-transmitter facility. The FCC requires that interference to neighboring stations be calculated using an aggregation of the predicted signal levels from the several transmitters in an SFN before interference calculations are performed.

NETWORK DESIGN METHODS

Designing a single-frequency network requires careful consideration of all the factors that will impact upon or be impacted by its operation, selection of the optimum set of techniques for each application, and frequently making tradeoffs with respect to optimization of various system characteristics. It is important to recognize that the technologies of SFNs are effectively toolkits that provide a variety of solutions in most cases; the design task becomes one of finding the optimum combination.

Choosing Among Techniques

The first step in designing an SFN involves making choices of the types of transmitters to be used and the cell design approach to be applied. These choices will depend upon an analysis of the environment in which the network will be built and the objectives of the network design. Among the characteristics requiring analysis are the presence of neighboring stations that must be protected from interference, whether there already is a single-transmitter facility the service of which is to be enhanced or extended by the SFN, the presence of terrain features either that must be overcome or of which advantage can be taken in the network design, the availability of tower space and its geographic relationship to the areas to be served, and the like.

While this discussion treats each of the various characteristics as independent of the others, in fact, they interact with one another, and systems can be designed with combinations of the several techniques. Thus, a service area could have portions that use large-cell techniques and other portions that use the small-cell approach. Similarly, a single, central transmitter could be converted to one of the transmitters in a DTxN, with several, most likely smaller, transmitters added to form the network. It also is possible to use a DOCR to repeat the signals from one of the transmitters in a DTxN or to use a DTxT to feed one or more DTxRs.

Transmitter Types

Each of the transmitter types has situations in which it can be optimally applied. The detailed technical characteristics of each type of transmitter are treated later in the chapter in sections devoted to each type. For purposes of network design, it suffices to be aware of certain application implications of the differences between the transmitter types. These differences have to do with the ways in which signals are delivered to the several transmitter types, the spectrum requirements of the various types, and the geographic limitations on placement of the transmitters.

Assuming the existence of a single-transmitter facility that does not provide adequate field strength throughout its allotted service area, for example, any of the three types of SFN transmitters might be added to provide service in the underserved areas. The possibility of using DOCRs would depend on the power levels that are required of the added transmitters, the geometry of the places available to locate those transmitters relative to the locations of the areas to be served and of the original single transmitter, and the signal levels from the original transmitter that can be received at those potential DOCR locations. The potential for use of DTxRs would depend on the availability of an additional channel in the areas in which service is to be improved. If higher power is needed than is available from DOCRs, if additional spectrum is not available for the use of DTxRs, or if timing adjustments are required to minimize INI, then DTxTs, fed through some form of STL would be necessary.

If the design is not based on inclusion of an existing, central transmitter, then either a single transmitter can be used to feed a group of DOCRs and/or DTxRs, or a network of DTxTs can be established to cover all or part of the desired region. A network of DTxTs, in turn, can be extended by DOCRs and/or DTxRs. The DTxTs would provide complete flexibility in the placement of the transmitters, the power levels at which they can operate, and the relative timing at which they can emit their signals. DTxRs, of course would require a separate channel on which to emit their signals, but they provide the same flexibility on that second channel as DTxTs have on the primary channel. DOCRs would operate on the same channel as the DTxTs, but, when compared to the DTxTs, they would be limited in power output, placement, and relative timing.

Cell Arrangement

Whether to use the small-cell or large-cell approach in a given area depends upon a number of factors. Principal among these is cost. The small-cell scheme provides the optimum technical solution, but it generally costs more than the large-cell method. It involves many more transmitters, requires a means to deliver signals to each of those transmitters, and may involve lease payments at all of the transmitter sites for space to mount antennas and in which to locate equipment. Large-cell systems have similar types of costs, but the multiplier for the number of transmitters will be much lower. It also is possible to combine large-cell and small-cell regions within the service area of a station; for instance, the large-cell method can be used in rural and suburban areas while the small-cell method can be used in urban areas and high-rise canyons.

In choosing between the two approaches, it is important to start with the locations of the population centers and an understanding of their spread across the service area. Ideally, large-cell transmitters should be placed adjacent to or within isolated population centers, with the regions between population centers served from the transmitters associated with the population centers. Depending on interference considerations, this layout may allow sufficient signal levels in the population centers for reception with indoor antennas, while areas away from the population centers might require outdoor antennas. When extended urban areas exist within the service area, especially when there are high-rise canyons, use of the small-cell scheme likely is a better choice and can be justified economically by the higher population density to be served.

Important factors in deciding which type of cell to use are the level of service intended to be provided (e.g., to outdoor or indoor receiving antennas) and the interference protection that must be provided to neighboring stations. Indoor reception typically requires considerably higher field strengths (typically measured at 9.1 meters—30 feet—above ground) than does reception using outdoor antennas. The difference is on the order of 40 dB higher field strength required for indoor reception, which includes the effects of reduced receiving antenna height, reduced receiving antenna gain, and losses from penetrating into a building to reach the receiving antenna. Placing transmitters closer to the receivers reduces the necessary fade margin needed to maintain reliable service and provides more uniform signal levels throughout the service area. At the same time, if there is an adjacent-channel station with a service area that overlaps the service area of the network being designed, then special efforts must be undertaken in the design of individual transmitters to avoid creating signal "hot spots" around those transmitters that will cause interference to the adjacent-channel neighbor. The special efforts usually involve use of well-controlled elevation patterns with uniform field strengths surrounding the transmitters.

Interference Aggregation

When multiple transmitters are used, it becomes necessary to aggregate the impacts of the signals from the several transmitters in a network on the signals of neighboring stations. There are two fundamental ways in which to carry out the aggregation. In either case, it is assumed that interference is determined by computing the population receiving interference when signals from the SFN are treated as the undesired signals and those from the neighboring station being studied are treated as the desired signals. Interference is studied in each of an array of cells, called *study cells*, within a geographic grid, and all of the population within each study cell is treated as receiving interference if there is determined to be interference predicted at the reference point for the respective study cell. In the first case, the interference is determined in the normal way for the particular channel relationship being studied, and the impacted populations receiving interference from each network transmitter are totaled. In the second case, the received signal levels (RSLs) from the several transmitters in the network are aggregated, and the total power obtained is used in calculation of the interference caused to the neighboring station.

Aggregating the population affected by each transmitter in an SFN is a relatively straightforward exercise, termed *cell aggregation*. It requires only avoidance of double counting of the population receiving interference, as described in the next subsection, "Population Counting." All of the normal factors included in an interference analysis, such as pointing of the receiving antenna toward the desired station and applying the off-axis antenna discrimination to the signal from each SFN transmitter, are computed individually for each SFN transmitter with respect to each desired station studied. The process is repeated for each SFN transmitter, and the results are accumulated. To make this method work, it is necessary to collect data on the locations and populations of the study cells receiving interference from each of the SFN transmitters.

Aggregating the power of RSLs prior to determining whether interference occurs in each study cell is somewhat more complicated than the cell aggregation method and is termed *signal aggregation*. In the signal aggregation method, the predicted RSLs from each of the SFN transmitters, after consideration of antenna discrimination in the direction of each SFN transmitter and the predicted RSLs from the desired station studied, are calculated at the reference points of each of the study cells. For each study cell, the RSLs from the SFN transmitters are combined into a single value, which is used as the undesired signal level in a calculation with the RSL of the desired station studied to determine whether interference is predicted to occur in the respective study cell. If there is interference predicted in a cell, its population is included in the total population predicted to receive interference from the SFN signals. In the case of signal aggregation, there is no need to avoid double counting, because each study cell of a desired station is analyzed only once.

There are several methods by which the power of RSLs could be added together in the signal aggregation method. These include simple addition of the RSLs of the signals, a root-mean-square (RMS) calculation of the combined RSL power, and a root-sum-square (RSS) calculation of the combined RSL power. For several reasons, too mathematically complex for treatment here but having to do with the fact that the signals from the several transmitters cannot be coherent with one another by the time they reach a receiver, the RSS power summation method is the appropriate one for use in all cases in which it is necessary to combine the RSLs from multiple SFN transmitters into a single value.

Population Counting

A significant aspect of all determinations of interference, at least as practiced in the United States, is the counting of the population predicted to receive interference above a threshold D/U RSL ratio, the value of which threshold depends upon the frequency relationship between the channels used by the desired and undesired stations. The D/U ratio thresholds for various channel combinations are published in the rules of the FCC and have the force of law. Interference is considered to be permissible when the population affected by that interference falls below a specified percentage of the population potentially served by the station receiving the interference. Populations receiving interference below the specified percentage are considered to be *de minimis*.

When counting populations in evaluating interference involving an SFN, there are several possible cases that may require consideration. Earlier sections of this chapter focused on interference from SFNs to single-transmitter operations. Interference in the reverse direction, from a single transmitter to an SFN, also must be evaluated, as must interference from one SFN to another. When the interference to be evaluated is from a single transmitter to an SFN, study cells in the service area of the SFN may be served by more than one desired SFN transmitter. In such instances, the appropriate evaluation method is to consider service to the study cell to be provided by the strongest of the SFN transmitters, as seen from the study cell and evaluated by field strength over the cell, and to aim any directional antenna used for modeling toward that strongest transmitter. Without such an approach, it would be possible for interference to a particular cell to be counted as being caused with respect to the signals from more than one desired SFN transmitter, resulting in double- or multiple-counting of interference to the study cell and an incorrect total population being predicted to receive interference.

Evaluating interference between SFNs is a composite of the methods for evaluation of the interference in both directions between an SFN and a single-transmitter operation. For each cell to be studied, the desired-signal RSL should be the one from the transmitter producing the highest field strength at the reference point for the study cell, and the receiving antenna should be assumed to be aimed toward the corresponding SFN transmitter. The undesired signal RSL used for determining interference to each study cell should be the signal aggregation of the RSLs from all the transmitters in the undesired SFN, after inclusion of the effects of antenna discrimination. The result of the analysis described will be a determination of which cells are and are not predicted to receive interference. The total population served and the population predicted to receive interference then can be calculated for the pair of SFNs, and a determination can be made whether the amount of interference is acceptable based upon the *de minimis* criterion.

Software Design Tools

Designing SFNs in the modern spectrum environment would not be possible without software design tools running on reasonably fast computers. A number of different software tools can be used in designing SFNs, but certain features and characteristics of those tools are required. A tool used must be able to evaluate interference both to and from other stations when multiple transmitters are involved as the desired and/or the undesired signals. A tool is needed to compute the coverage that can be achieved with repeated iteration of the network design. A tool is needed to examine internal network interference, based on the characteristics of receiver adaptive equalizers and including consideration of the timing offsets of the transmitter emissions.

All of the enumerated software tools may be included in one software package, or they may involve separate programs. In any event, they have certain features in common. They all require a terrain-sensitive propagation model. Typically, in the United States, because of its adoption by the FCC for interference prediction, the Longley-Rice terrain-based propagation model is used. Other models (e.g., the Terrain-Integrated Rough-Earth Model—TIREM) also could be applied. The tool generally used for interference analyses between stations has been the TV_Process program, written for and published by the FCC, but that is transitioning to a new TV Study program at the time of this writing. This new program includes the Longley-Rice methodology, as described in FCC Office of Engineering and Technology Bulletin Number 69, and can be relatively straightforwardly modified to treat multiple transmitters in a network on the desired, the undesired, or both sides of the D/U ratio calculation.

The TV_Process program was not conducive to interactive network design modifications with immediate display of the results, although the TV Study program may eventually provide such capabilities. Consequently, at the time of this writing, at least one other tool is needed to deal with coverage, internal network interference, transmitter timing adjustments, and stresses placed on adaptive equalizers in systems using single-carrier modulation or determination of optimum guard intervals in systems using multicarrier modulation, ideally producing a graphic display of the results of system parameter changes. Many such tools exist in the marketplace. At least for systems using single-carrier modulation, unless and until they are modified to meet the specific needs of SFN design as described herein, it usually will be required to extract data produced by propagation modeling tools into a standard mathematical processing and data manipulation program for completion of the necessary algorithms. The manipulated data often then can be returned to one of the original software tools for display. It should be noted that integrated tools for SFN design in networks using multicarrier modulation already do exist and are available in the marketplace as of the time of this writing.

Making Tradeoffs

The art of SFN design is the art of making tradeoffs. It is highly unlikely, in any given case, that a design will be achieved that can perfectly meet all of the goals set out for the network. Consequently, it is important at the outset to establish goals for the SFN design that are given relative priorities. Then, when it becomes impossible during the design process to meet all of the goals perfectly, guidance will be available on how to balance the potential choices against one another. This, of course, is the process of making tradeoffs.

Tradeoffs in the design of an SFN typically involve matters of coverage versus internal network interference, service versus interference to neighboring stations, and the like. There also likely will be aspects of cost included in the necessary tradeoffs. For example, the number of transmitters included in the network, the use of one type of transmitter or another, the type of STL, and the power levels of the individual transmitters all affect both network performance and cost.

Part of the process of making tradeoffs is determining where to locate the INI that is inherent in the use of multiple transmitters. Typically, the choice will be made to create regions near population centers where INI will be minimized. This result can be achieved either through use of small cells, so as to avoid the effects of INI, or by placing larger cells so that one transmitter is dominant in the population center and any INI is located in low-population regions. Depending on the specific terrain, sometimes advantage can be taken of terrain features such as ridges and mountains as dividing lines between the service areas from different transmitters, thereby avoiding the effect of INI.

When considering treatment of interference to geographically neighboring stations, tradeoffs often will be between protection of the other stations, service within the service areas of the SFNs, and cost. Frequently, the best method for managing interference to other stations is through control of antenna elevation patterns. For example, it is possible to use greater than normal beam tilt in the direction of a distant station, thereby maintaining the power level in that direction while providing a reduction toward the distant service area. It also is possible to create an elevation pattern that has a notch on top of the main beam in order to further reduce the signal toward the distant station. The trade-off for achieving such elevation directivity is that the antenna becomes longer than would otherwise be required, especially at low transmitter power levels, and the added length contributes both to the cost of the antenna itself and to the cost of tower space on which to install it.

In situations in which there is an adjacent-channel station with a service area that overlaps that of an SFN, it similarly is possible to mitigate interference to the neighboring station. The interference reduction can be achieved again by use of elevation pattern control—in this case, through use of a pattern having relatively uniform field strength from the transmitter site to some substantial distance from the transmitter, at which point the field strength begins to fall off. Such a pattern is shown in Figure 8.6-7 and is termed an *inverted cosecant squared* pattern. Placing SFN transmitters that use the inverted cosecant squared pattern within the service area of the adjacent-channel station allows setting the D/U ratios to the neighboring station in such a way that the values selected are not likely to be exceeded within the coverage areas of the respective SFN transmitters. The trade-off to achieve the elevation pattern necessary for protection of an adjacent-channel station with a service area overlapping that of an SFN is, once again, the length of the antenna and its attendant costs of acquisition and of tower space.

Design Examples

As discussed earlier, two design examples are presented herein. The examples given are of a four-transmitter network in which a high-power, single-transmitter facility is extended with three added transmitters to overcome terrain obstacles, and of an eight-transmitter network designed from the beginning for SFN operation in an environment having substantial interference protection requirements. In both instances, the technique used is that of DTxTs, requiring STLs to tie the studios to the various transmitters, but other geographic configurations could have allowed for use of DOCRs, and greater spectrum availability could have allowed for use of DTxRs.

Example 1—Main Transmitter and 3 Gap Fillers

The first example is of an ATSC 1.0 system designed to overcome the terrain blockage of a major geographic feature within the service area: a ridge that blocks service from the main transmitter—a megawatt-class facility—to the three principal population centers along the southern tier of the main transmitter service contour. The effect of the obstruction can be seen in Figure 8.6-8, which shows the noise-limited service

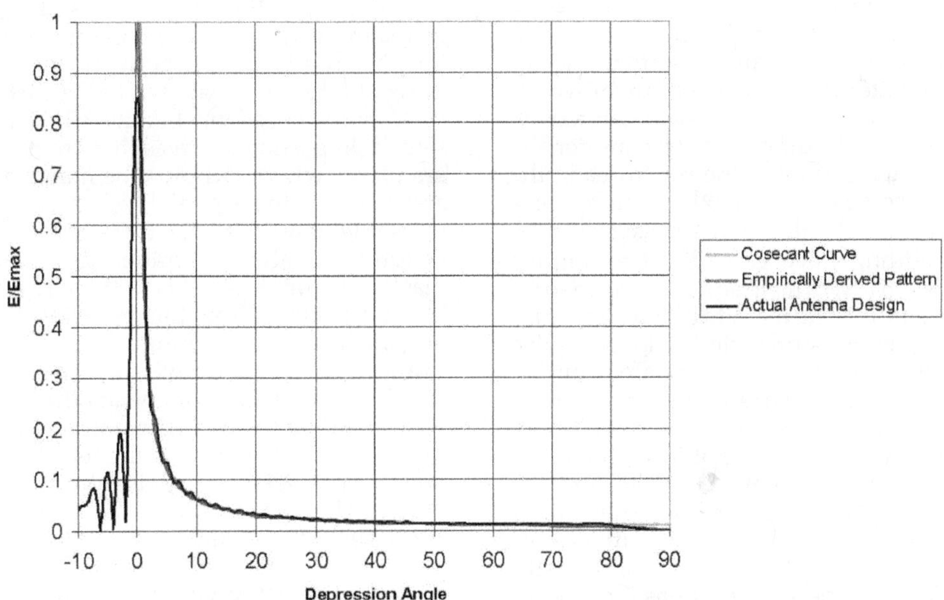

FIGURE 8.6-7 Actual inverted cosecant squared antenna design compared to empirically derived ideal elevation pattern and cosecant curve (cosecant curve applies to field values, cosecant squared to power values). The three curves essentially overlap each other. To observe what small difference there is, see the color version of the image available in the publisher's companion website for this edition. (Information regarding this link is provided in the front matter.)

S. Merrill Weiss

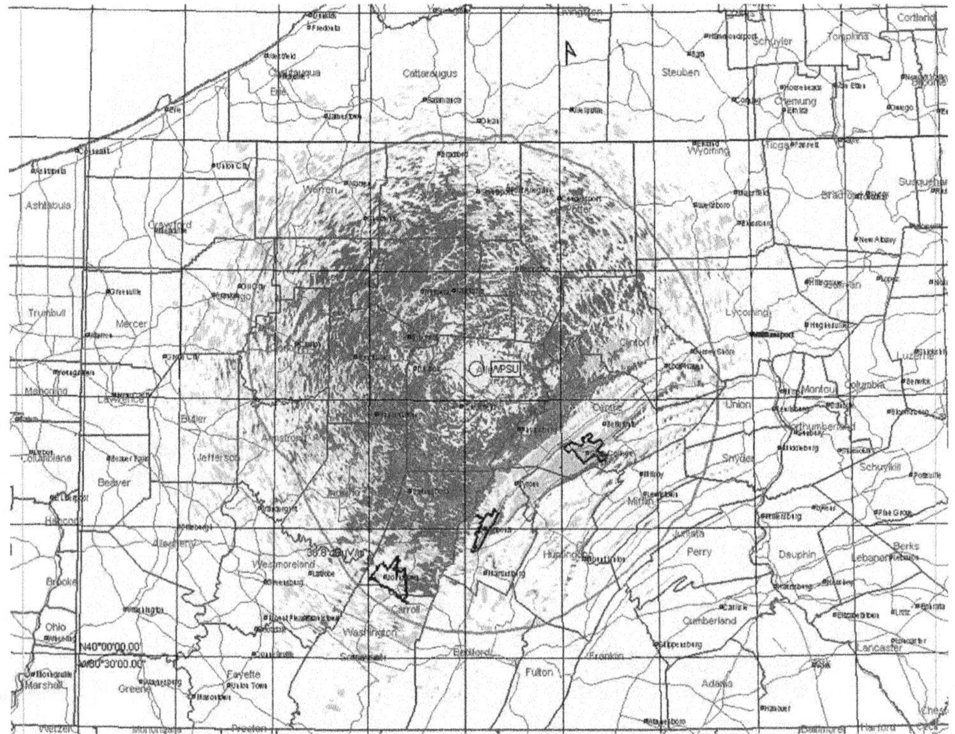

FIGURE 8.6-8 A 810-kW main transmitter at 413 meters' height above average terrain (HAAT), with major terrain obstructions to southeast and south-southwest cutting off services to all three major population centers.

S. Merrill Weiss

contour of the station together with the Longley-Rice predicted field strength at locations throughout the allotted service area of the station. The population centers all are located in isolated valleys, so that relocating the main transmitter near any one of them would not have provided service to all of them. Moreover, such relocation from the existing, main transmitter site would have eliminated service to the rural area in the northern half of the service contour, where it often provides the only over-the-air television service.

The result of adding three gap filler transmitters is shown in Figure 8.6-9, which adds to the map of Figure 8.6-8 the contours of the additional transmitters and their Longley-Rice predicted field strengths. As indicated by the yellow and orange areas on the maps, the regions receiving field strengths sufficient to permit indoor antenna reception encompass the population centers in Figure 8.6-9 but not in Figure 8.6-8. Because of the size of the valleys involved and the relative containment of service from each transmitter caused by the terrain, the added DTx transmitters have power levels of 25 kW ERP to the south and southwest, while the DTxT to the southeast has a power of 50 kW ERP. In each case, the gap-filler antennas have higher-than-normal beam tilt values, ranging from 3 to 5 degrees depression, down into their respective valleys, to provide signal coverage as widespread as possible within the valleys. The antennas used are ordinary low-power types, with transmitter power output levels on the order of 2 kW each.

Even though the terrain provides a degree of isolation between the service areas of the several DTxTs in Example 1, it still is necessary to pay attention to the INI that is caused between the transmitters. The areas of concern are shown in Figure 8.6-10, where the D/U ratios based on Longley-Rice analyses are presented. In particular, areas that are not yellow or white are places where INI must be managed. These are the areas where the signal level is above the reception threshold and where the D/U ratio is below the 16 dB adaptive equalizer operation threshold shown in the mask of Figure 8.6-4. It is in these regions that the timing of the signals from the respective transmitters must be controlled so as to minimize the effect of the INI and to allow receiver adaptive equalizers to treat the multiple signals as echoes of one another. The timing of the signals, in the regions where timing matters, is shown in Figure 8.6-11. To obtain the results shown, the transmitters were set with offsets from equal emission times of 0 μs, 115 μs, 135 μs, and 150 μs, respectively, for the main, southeast, south, and southwest transmitters.

Example 2—SFN With Numerous Interference Constraints

The second example is an ATSC 1.0 system so constrained by interference that it takes advantage of many of the possible techniques for interference mitigation described previously. The primary constraint is an in-market full power station on a first-adjacent

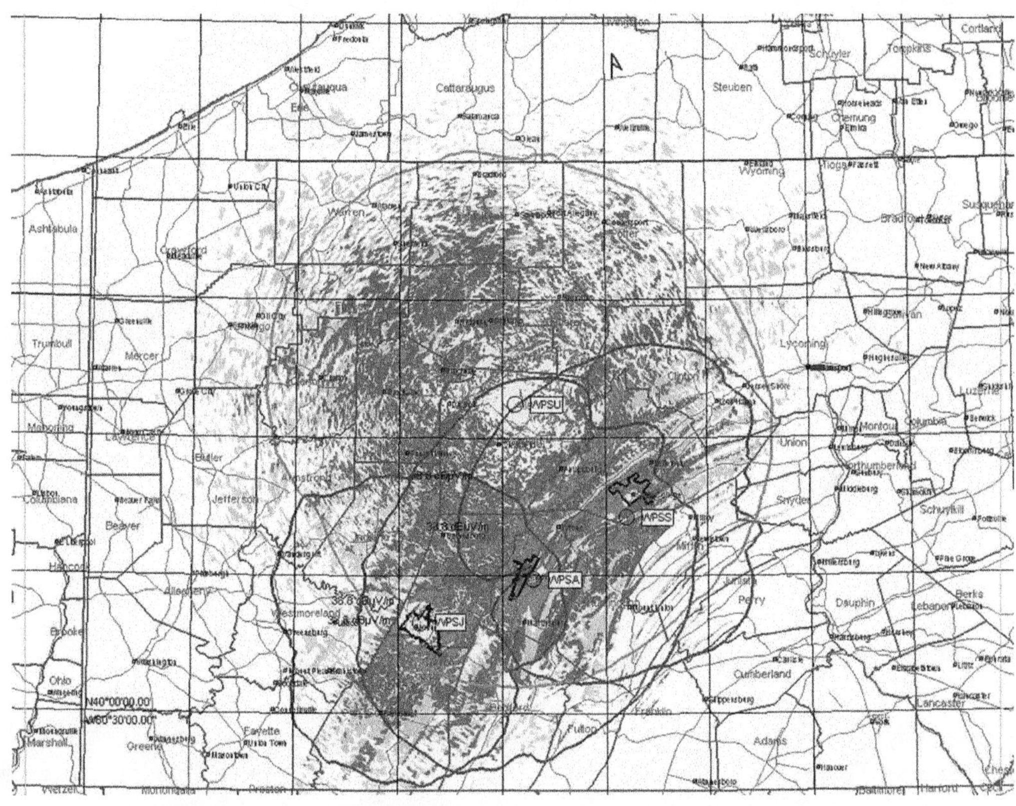

FIGURE 8.6-9 DTxN comprising main transmitter of Figure 8.6-8 and three DTxT gap fillers, operating at power levels of 25 kW, 25 kW, and 50 kW ERP from left to right (west to east).

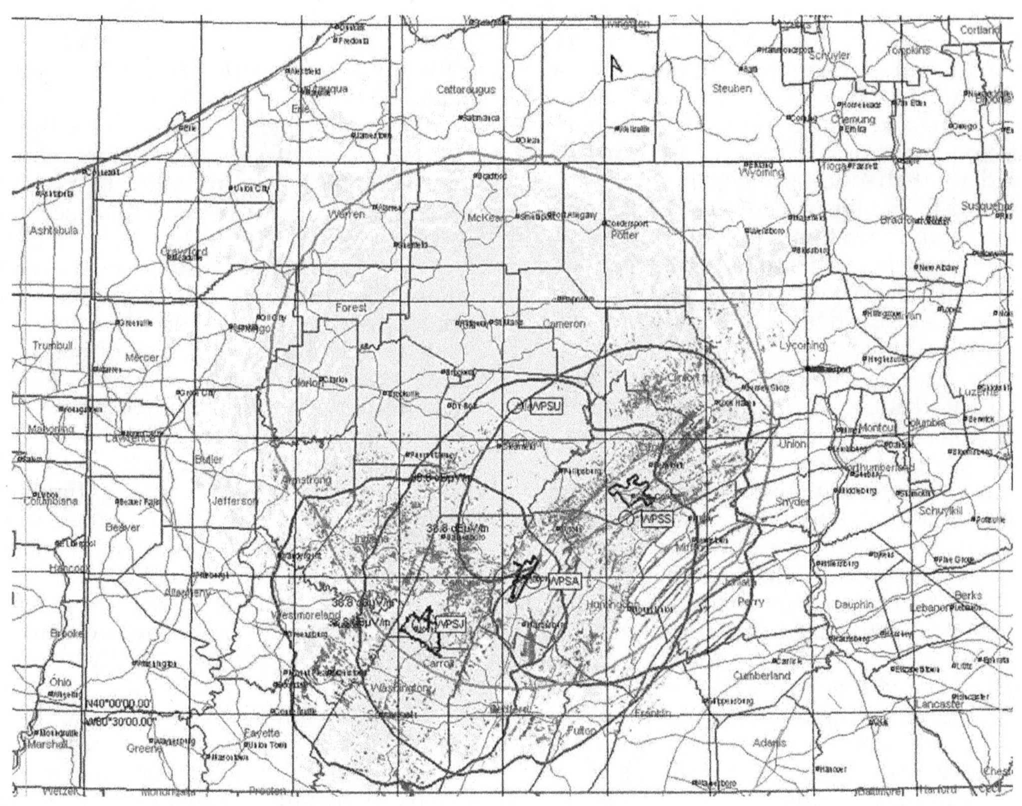

FIGURE 8.6-10 DTxN of Figure 8.6-9, showing INI in areas of concern.

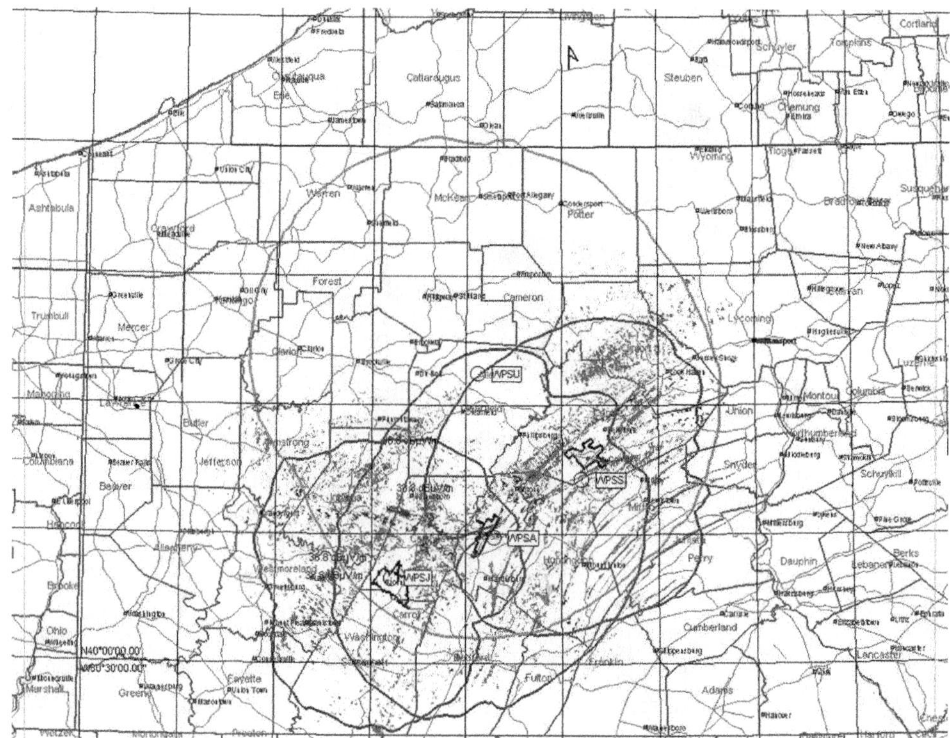

FIGURE 8.6-11 DTxN of Figure 8.6-9, showing relative timing of signals in areas having INI exceeding 16 dB threshold.

channel. There also are co-channel full service stations in adjoining markets to the northeast and southwest and a low-power station entitled to protection (i.e., a Class A station) also to the northeast. The protected contours of the several neighboring stations and the service contour authorized to the subject station all are shown on the map in Figure 8.6-12. On that map, the subject station's authorized contour is shown in orange, the adjacent-channel protected contour is shown in red, and the co-channel protected contours are shown in maroon. Protection was required in both directions, both to and from the neighboring stations, but since the other stations already were built, control of interference in both directions had to be part of the SFN design considerations.

The best way to achieve interference protection with respect to an adjacent-channel in-market neighbor is through collocation of the transmitters. That is the approach taken for the highest-power transmitter in the SFN, with its location in the same antenna farm as the adjacent-channel neighbor. The adjacent channel is a megawatt-class station, and the power of the collocated transmitter was set about 8 dB lower, at about the same antenna height, to obtain a measure of protection from the neighboring station. At the same time, the collocated transmitter must provide protection to the two stations (Class A and full service) to the northeast, and its contour must be constrained so as not to project excessively beyond the authorized service contour of the station. These goals were achieved using a panel antenna with a carefully sculpted elevation pattern that varies in the different directions. In particular, it puts a null toward the radio horizon in the direction of the stations to the northeast, resulting in a reduction of interference to both stations, as measured by population predicted to receive interference with respect to the full service station and by contour overlap with respect to the Class A station.

All of the other transmitters in the SFN use the inverted cosecant squared antenna elevation pattern described previously, in conjunction with azimuth patterns chosen to match the authorized contour to the extent possible. The ERP of the smaller transmitters in the network ranges from 110 watts to 4.25 kW, and antenna heights are between 30 and 76 meters above ground level at the center of radiation. The overall result of the design in terms of service provided is shown on the map in Figure 8.6-13, which includes the noise-limited contours of the SFN transmitters in purple, the authorized contour of the station in orange, and the Longley-Rice predicted field strengths, as shown in the legend. The areas in which INI must be mitigated are shown in Figure 8.6-14 as darker shaded areas(or in colors other than yellow or white in the companion website images), where signal levels are predicted to be too low for service. Please also see the colorversion of this map in the publisher's online companion website. The relative timing of arrival of the signals in the areas where their timing matters is shown in Figure 8.6-15. The emission timing of the various transmitters was adjusted to achieve the results shown on the map. With the combination of design parameters used, interference to other stations generally was reduced by a substantial amount, and the population served within the

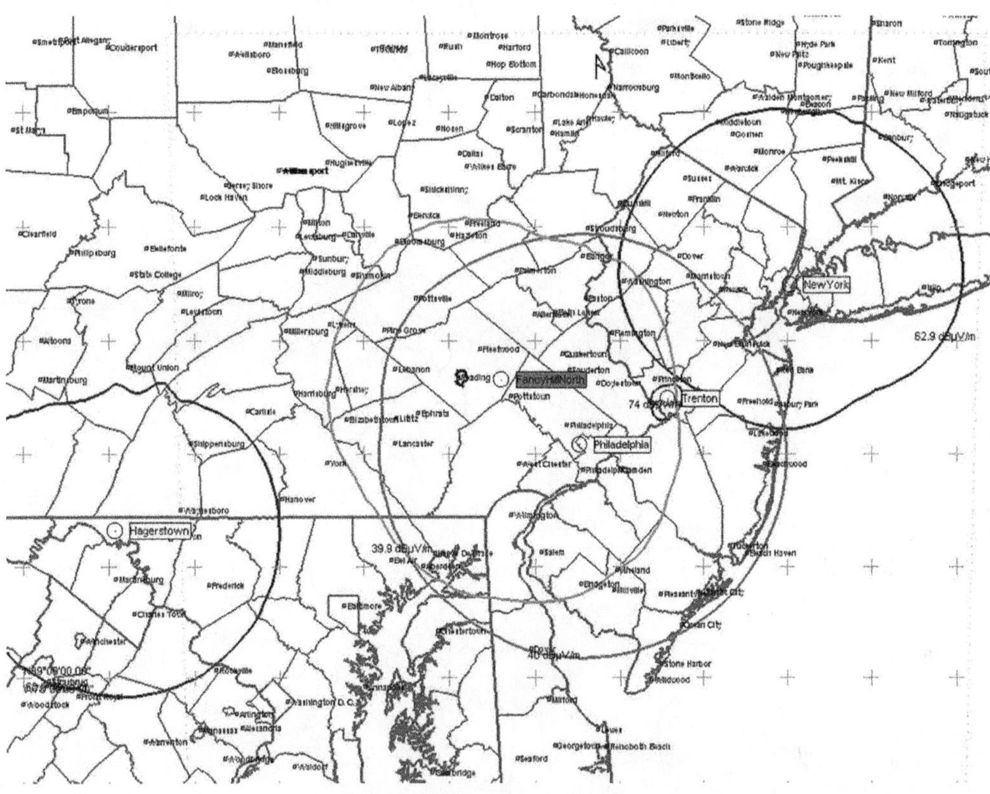

FIGURE 8.6-12 Authorized contours of station in Example 2: one in-market adjacent-channel station; two co-channel full-service stations in adjoining markets; and one low-power (Class A), co-channel, in-market station.

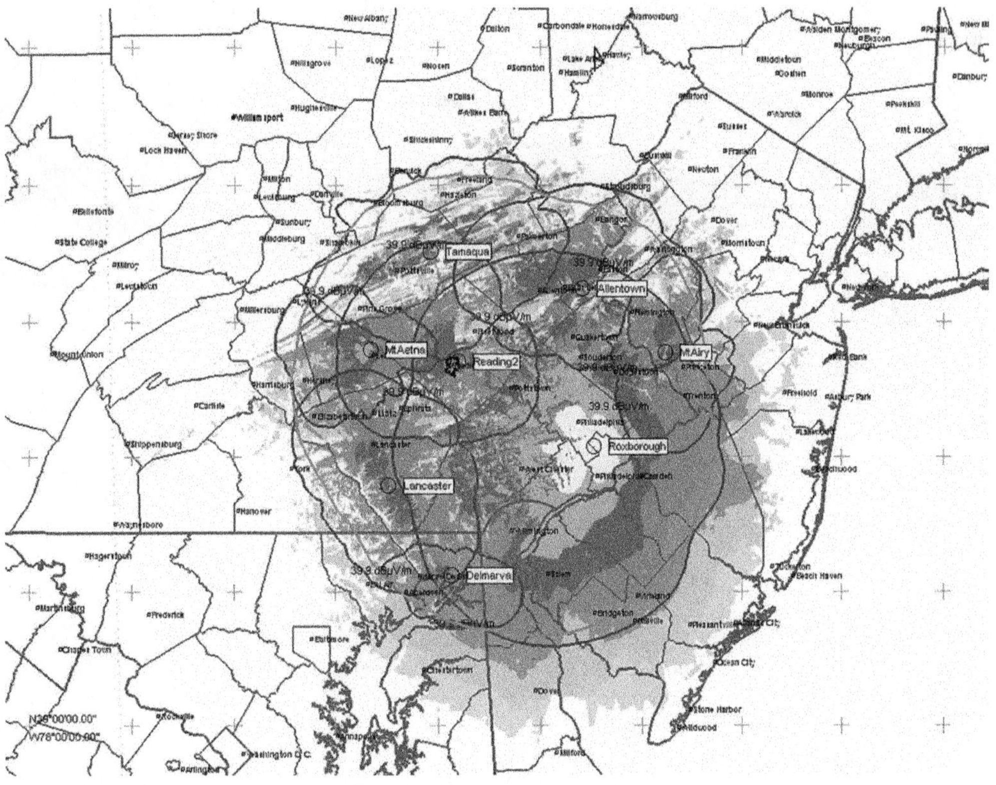

FIGURE 8.6-13 Authorized contour of station in Example 2, contours of eight DTxTs, and Longley-Rice predicted field strengths from DTxN.

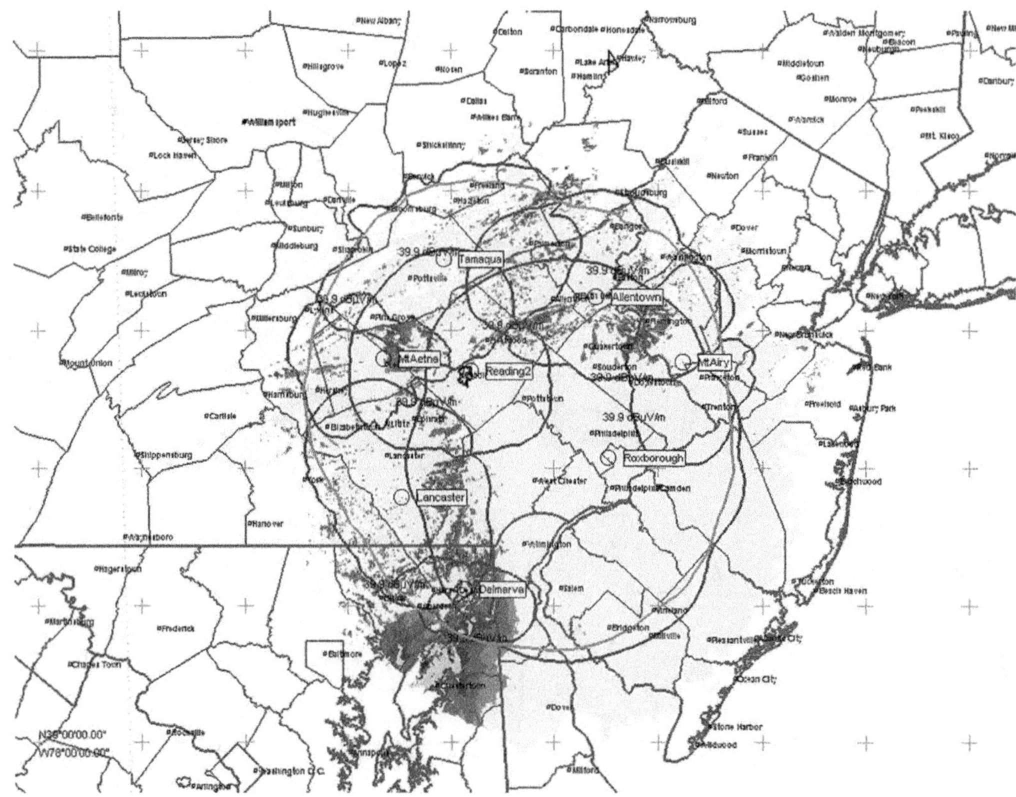

FIGURE 8.6-14 DTxN of Figure 8.6-13, showing INI in areas of concern.

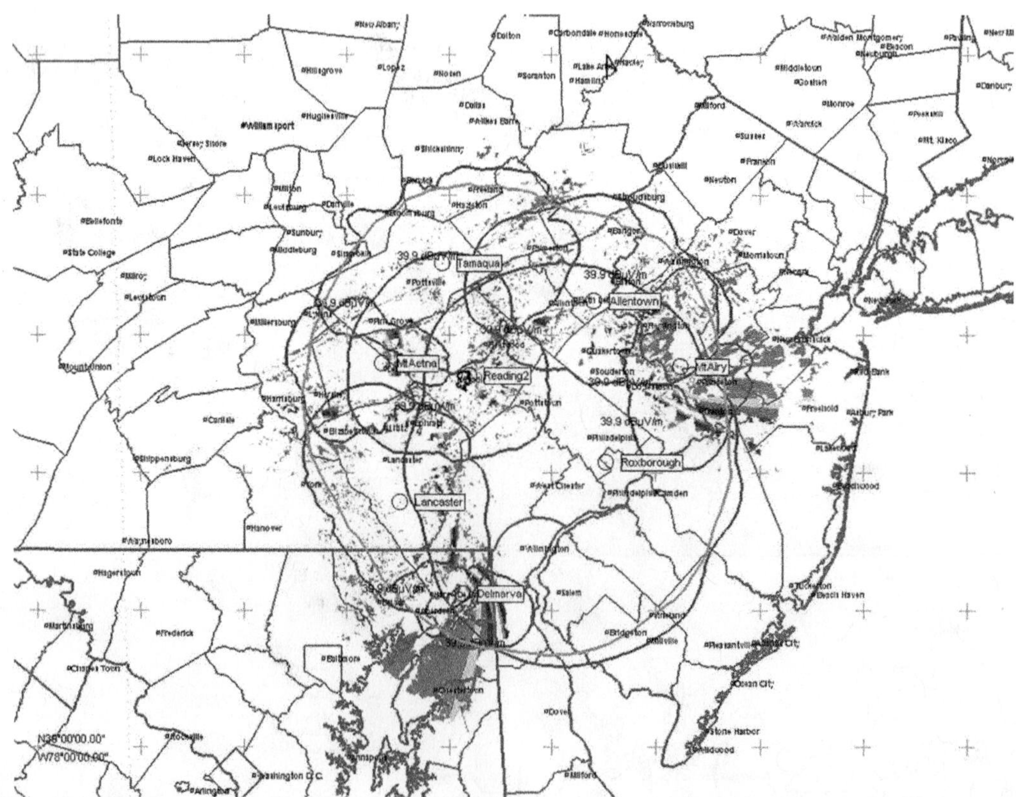

FIGURE 8.6-15 DTxN of Figure 8.6-13, showing relative timing of signals in areas having INI exceeding 16-dB threshold.

authorized contour was increased by about 3 million people over what could have been achieved with a single transmitter at the authorized location, largely as a result of overcoming adjacent-channel interference from the in-market, adjacent-channel operation.

DIGITAL ON-CHANNEL REPEATERS (DOCRS)

From an implementation standpoint, DOCRs may be the simplest of the SFN transmitters to install, but they are not the most flexible in their application and may not be the easiest to design into a network. Fundamentally, DOCRs receive signals over the air from an earlier transmitter in the network, process those signals, and retransmit them on the same channels on which they were received. Because they receive and transmit on the same channels, an important aspect of their designs, with respect to both the equipment and the installation, is the coupling that occurs from the transmitting antenna back into the receiving antenna. Installations should be designed to minimize the antenna coupling, while the equipment may be designed to minimize the effect on signal quality of any coupling that does occur.

DOCRs are characterized by a fixed time delay through them, with the actual time delay dependent on the equipment design. Generally, the more the processing that is done in a DOCR to clean up the signal, the longer the time delay through the DOCR becomes. Some clever designs have been developed to minimize the time delay while producing the highest signal quality possible from a DOCR. Coincidentally, the same designs are those that permit the highest power outputs to be obtained from DOCRs.

Because the laws of physics dictate that DOCRs always have a positive time delay from arrival of the signal at the receiving antenna to emission of the signal from the transmitting antenna, there are certain limitations that should be applied to the geometric arrangement of the transmitter that is the source of signals to the DOCR, the location of the DOCR itself, and the location of the area to be served by the DOCR. The possible arrangements are shown in Figure 8.6-16. The limitations derive from an objective of avoiding causing—or at least minimizing—leading echoes in the area to be served by the DOCR.

As can be seen in Figure 8.6-16a, if the source of signals to the DOCR (labeled as the "Main Transmitter" in the figure), the DOCR itself, and the area to be served are all in roughly a straight line, with the area to be served beyond the DOCR when looking from the signal source, then the arrival time of signals in the DOCR service area from the DOCR will be approximately the arrival time of the signals from the signal source plus the delay of the DOCR. Thus, the DOCR signal will arrive in the service area later than the signal from the signal source. Assuming the DOCR signal is stronger in the DOCR service area than that of the signal source, the DOCR signal will provide the timing reference to receivers in the DOCR service area, and the signal from the signal source will appear to receivers as a leading echo. Because the two transmitters and the DOCR service area are in a straight line and in the order shown, the leading echo from the signal source will have the minimum possible time offset from the reference signal provided by the DOCR.

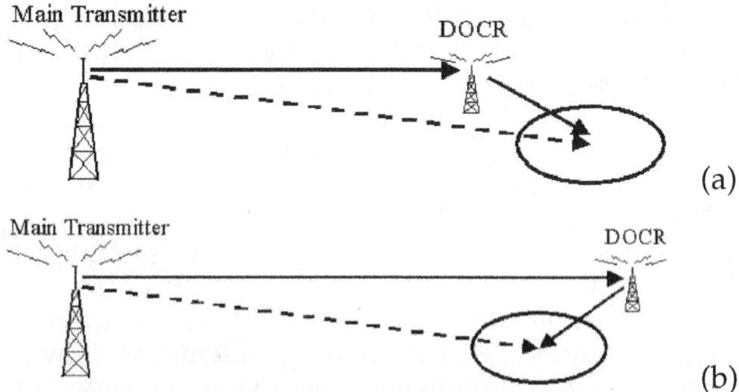

FIGURE 8.6-16 Geometric relationship between main transmitter, DOCR, and DOCR service area and its effect on leading echo time displacement. (a) Main transmitter, DOCR, and DOCR service area in approximate straight line, yielding minimum leading echo time displacement. (b) DOCR service area between main transmitter and DOCR, yielding increased leading echo time displacement.

Figure courtesy of Y. Wu and K. Salehian, Communications Research Centre Canada, Ottawa, Ontario

If the geometric relationship between the signal source (i.e., "Main Transmitter"), the DOCR, and the DOCR service area places the service area to the side of the line between the two transmitters or, worse, between the two transmitters, the time delay to receivers of signals from the signal source will become relatively shorter while the time delay of signals from the DOCR will become relatively longer, as can be seen in Figure 8.6-16b. The result will be a leading echo displaced much further in time from the reference signal from the DOCR than in the case shown in Figure 8.6-16a and described earlier. The longer leading echo will place more stress on receiver adaptive equalizers, leaving less margin for them to correct echoes occurring in the natural environment.

As was discussed earlier with respect to Figure 8.6-4, receiver adaptive equalizers in networks using single-carrier modulation generally can be expected to have less margin for handling leading echoes than for trailing echoes. Since DOCRs always convert source signals into leading echoes when receivers in the DOCR service area are able to receive both the source signal and the DOCR output, whenever possible in such networks, steps should be taken in designing the DOCR installation to minimize the leading echoes. These steps include using DOCRs primarily to serve areas that have significant terrain blockage of the signal from the signal source, so that the D/U ratio between the DOCR signal and the source signal will be as great as possible; placing the DOCR between the signal source and the intended DOCR service area, to the extent possible; and applying advanced DOCR signal processing techniques in those cases in which high DOCR power output is required, isolation cannot be achieved between the signals from the source and those from the DOCR, or the geometry of the transmitters and service area must be other than the desirable arrangement. For networks using multicarrier modulation, taking the steps outlined immediately above may not be necessary.

Turning to the technology of DOCRs themselves, several configurations are shown in conceptual block diagram form in Figure 8.6-17. In Figure 8.6-17a, the simplest arrangement of a receiving antenna, channel filter, amplifier, and transmitting antenna is shown. Since there is no frequency conversion, the prerequisite that the output frequency must match the input frequency will be met inherently. Since there is no demodulation, the prerequisite that the output data symbols must match the input symbols also inherently will be met. Limitations of the simple DOCR configuration are in adjacent-channel selectivity, since only on-channel filters are used; in the signal-to-noise ratio that can be achieved, since there is no signal processing to recover the original signal; and in the power level that can be transmitted, as a result of the other two limitations. An advantage of the simple design is that typically it has the shortest time delay of any of the designs—generally, measured in nanoseconds or low numbers of microseconds. Typically, the very simple design of Figure 8.6-17a only would be used to serve a small, isolated area and where good separation can be achieved between receiving and transmitting antennas of the DOCR.

Moving up the complexity scale, Figure 8.6-17b shows a DOCR design that uses intermediate frequency (IF) signal processing. The input signal is converted to an IF, filtered and amplified, then converted back to the same frequency for more amplification and delivery to the transmitting antenna. The IF filtering can be done with conventional tuned circuits or with a surface acoustic wave (SAW) filter, which can achieve greatly improved filtering characteristics as compared to ordinary tuned circuits. With IF signal processing, a DOCR can achieve better performance than the simple design with respect to adjacent-channel emissions and increased power output because of the narrower bandwidth of the signals that it must amplify. The trade-off for these improvements is a longer delay through the DOCR, which can be up to several microseconds, possibly low tens of microseconds, when a SAW filter is used. Note that, for DOCR designs including IF processing, use of the same local oscillator (LO) frequency for both downconversion and upconversion helps to restore the output signal to the same frequency as the input signal, to meet the prerequisite for SFN operation.

Further up the complexity scale is a design that demodulates the IF signal to baseband symbols, applies adaptive equalization to the baseband, slices the baseband signal to reduce the effects of noise on the received signal, and remodulates the signal for further amplification and transmission. This design concept is shown in Figure 8.6-17c and is termed an equalization DOCR or E-DOCR. The advantages of this design are that the effects of signal impairments in the natural transmission channel from the signal source to the DOCR and the impairments that result from coupling of the transmitted signal into the receiving antenna both can be minimized by the combination of adaptive equalization and symbol-level slicing. The result is that significantly higher gain can be applied in the DOCR, and consequently higher power can be

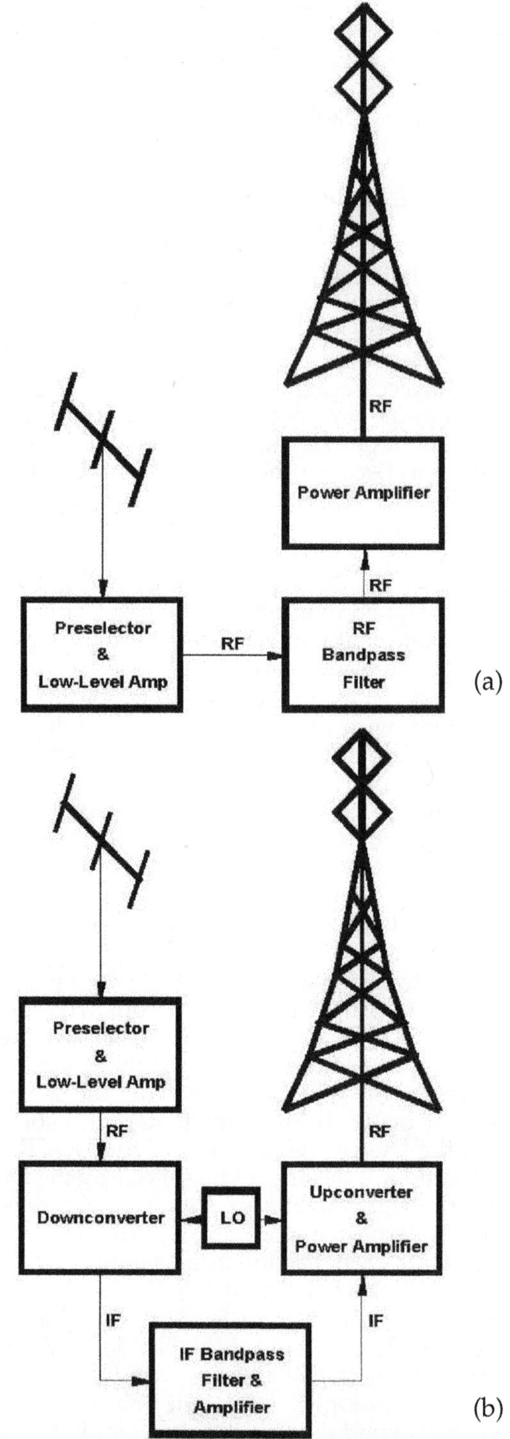

FIGURE 8.6-17 DOCR configurations: (a) RF processing DOCR, (b) IF processing DOCR, (c) baseband equalization DOCR, (d) baseband decoding DOCR.

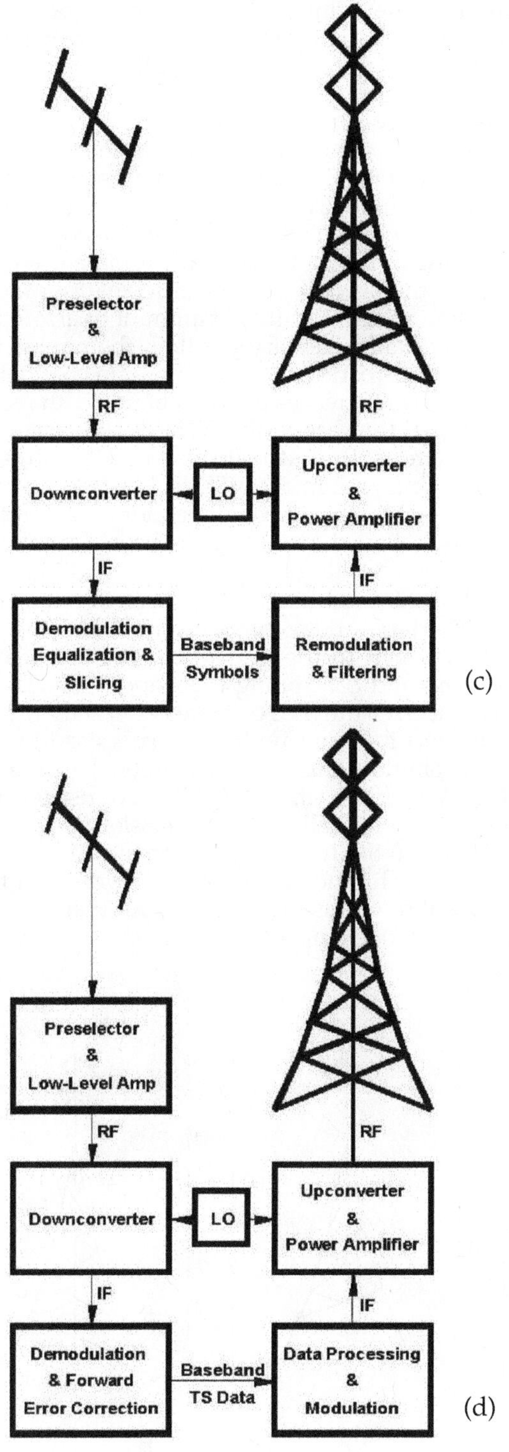

FIGURE 8.6-17 *(continued)*

transmitted relative to the received signal level than is possible with the simpler designs. The tradeoffs are that the delay through the DOCR becomes even longer than in the simpler designs, and any errors that occur in the slicing of the received signal to restore the levels of the symbols will be built into the transmitted signal, causing potentially unrecoverable errors in consumer receivers. To partially overcome these tradeoffs, careful equipment design can hold the E-DOCR delay to something on the order of 5 μs, and a network design producing a reasonably high signal-to-noise (S/N) ratio on the receiver input can help reduce retransmitted errors. If a high enough S/N cannot be obtained, then use of one of the simpler designs may be advisable.

The most complex DOCR design involves completely decoding the received signal to transport stream data, as shown in Figure 8.6-17d. In this case, both adaptive equalization and forward error correction can be applied to the received signal, resulting in a fully reconstructed, noise- and error-free signal for retransmission. To make the transmitted symbols match the received symbols, the location of the data frame sync and the states of the trellis coding in the received signal must be recovered and used in the remodulation process. While theoretically possible to build, the design of Figure 8.6-17d is likely to be impractical to implement because of the very long delay that will result from all of the signal and data processing. Such a design could only be used where there is nearly complete isolation of the DOCR service area from the source signal, and, in such a case, a simpler solution would be to use retransmission of the signal, on channel, without implicating SFN techniques.

DISTRIBUTED TRANSMITTERS (DTXTS)

DTx is a technique that uses multiple DTxTs that operate in a more-or-less conventional manner to provide service to a region. Typically, the DTxTs are fed the data stream for transmission over STLs; they individually modulate the data onto carriers at an IF frequency; the carriers are upconverted to the broadcast channel, amplified, and fed to an antenna. To operate in a DTxN, as dictated by the prerequisites for SFN operation discussed earlier, the DTxTs must be synchronized with one another in their modulation processes, and they must produce output signals on the same frequency. A basic DTx system has the configuration shown earlier in Figure 8.6-2.

While there is one basic concept for distributed transmission, there are a number of ways in which it can be implemented. The differences between methods largely involve the form in which the data stream is delivered to the several transmitters in a network and the resulting techniques required to synchronize transmitter operation and adjust the relative timing of emissions from the DTxTs. In all cases to be covered, the data is delivered to the DTxTs in digital form for local modulation and upconversion to the output channel. Not discussed in detail herein is the alternative of modulating the signal in one place and delivering an analog IF signal to all of the transmitters for upconversion. Such a system would be totally dependent on the stability of the time delays of the STLs to the several transmitters for the stability of the entire network, and experience has shown that adequate time delay stability cannot be achieved economically in the STLs for such a technique to provide the required network performance.

Synchronization Methods

There are three basic ways in which to deliver signals to the transmitters in a DTxN that uses single-carrier modulation. Networks using multicarrier modulation are considered later in this subsection. The first involves modulating the data onto a carrier at a central location and distributing the resulting RF signal to the various DTxTs. At each DTxT, a receiver must recover from the RF signal both the data and the various states of the modulation system that carried the data. The data then can be remodulated and upconverted onto a carrier at the assigned frequency of the DTx network. In the course of processing the data prior to remodulation, each DTxT can extract information from it to control adjustment of the delay through that DTxT, thereby attaining the network timing objectives. A system using the RF distribution method for delivery of the data to the DTxTs is shown conceptually in Figure 8.6-18. While it has the ability to control the transmitter emission timing, like the completely analog distribution method mentioned at the start of this section, the RF distribution method requires an analog STL with a reasonable noise margin to reach each of the DTxTs. Such a requirement can be very limiting in the types of STLs that can be used and in their availability. Moreover, the data processing at each transmitter would be much more complex than with other methods, so this approach is believed not to have been implemented as of this writing.

A second scheme for delivering the data to the several DTxTs in a network is to perform at a central location all of the data processing that normally occurs in a transmitter, delivering to all of the network transmitters the symbol data that must be input to the modulator. This approach is shown conceptually in Figure 8.6-19 and uses a digital STL to each transmitter. Since the data delivered to the DTxTs is already formatted for transmission, if any data for control of the transmitters is to be embedded in the data stream, at the transmitters, the symbol data must be converted back to its original form so that the needed information can be extracted from the data stream, or some form of multiplexing of the entire stream could be used. More significant, the normal process of formatting the data for emission adds a significant amount of error correction coding data necessary to allow the payload data to survive the broadcast RF transmission channel. This additional data increases the data bandwidth required in the STL. In the case of the ATSC 8-VSB system, the increase is from 19.39 Mb/s to over 32.25 Mb/s. This bandwidth increase in the STL makes the centralized data-processing approach uneconomical, and this method is believed not to have been commercially implemented as of this writing.

The third method uses data processing at a central location to derive data that can be embedded in the data stream sent to the network DTxTs for synchronizing them and controlling their operation. The data stream sent over the STL operates at exactly the same data rate and in the same format as is used in a conventional, single-transmitter system (i.e., for the ATSC 8-VSB system, 19.39 Mb/s of MPEG-2 transport stream packets). A device called a transmission adapter (TA) is inserted into the transport stream signal path at the input to the STL. The TA develops information that is sent to all of the transmitters to permit them to tightly control the processing of the data they receive and

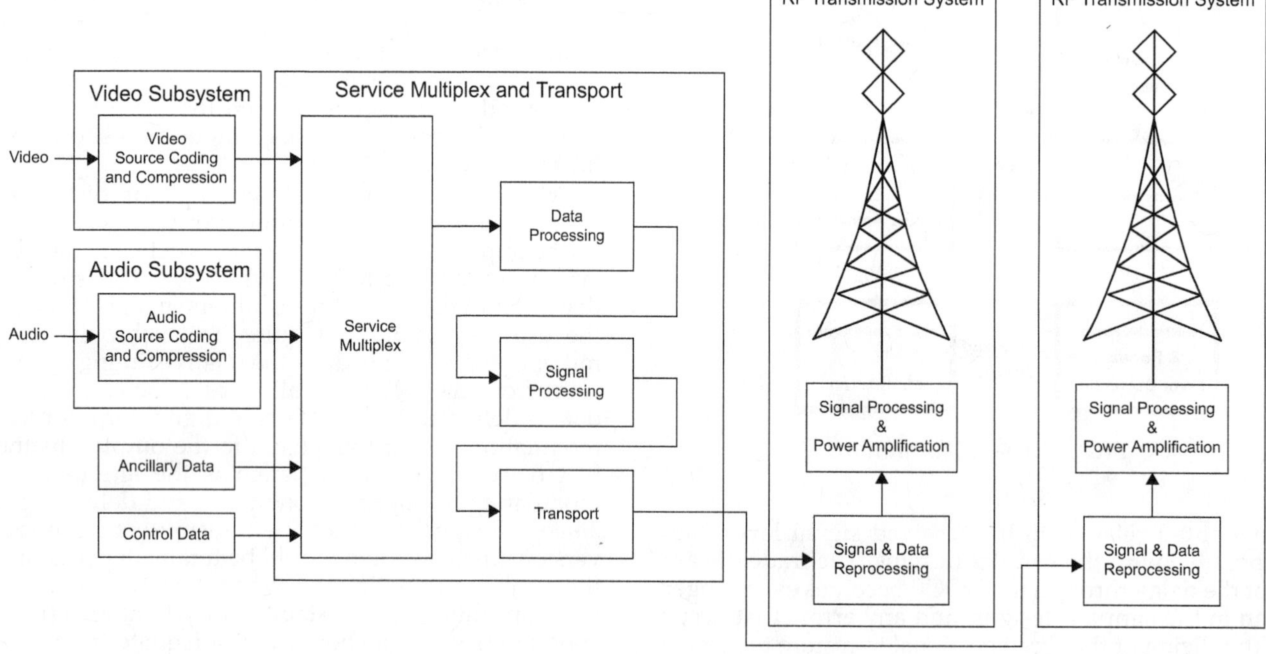

FIGURE 8.6-18 DTx system using RF distribution.

S. Merrill Weiss

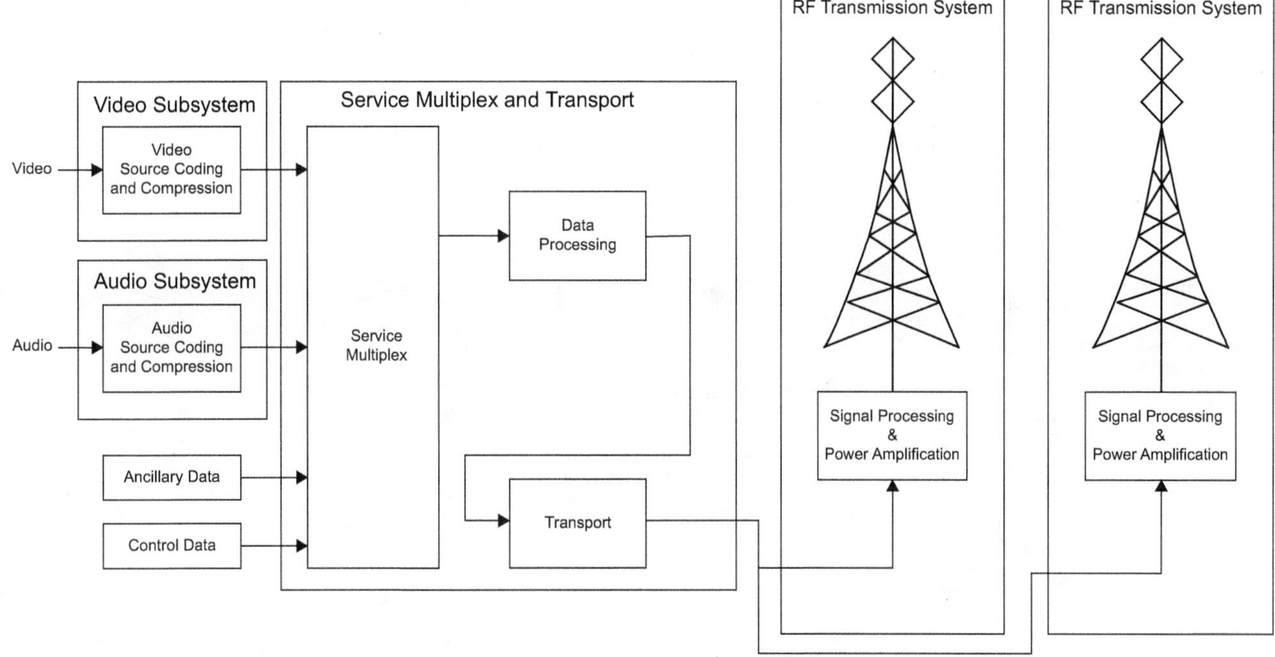

FIGURE 8.6-19 DTx system using distribution of symbol data.

S. Merrill Weiss

emit, thereby synchronizing them with one another. The data sent from the TA also permits adjustment of the network timing by individually controlling the relative emission time of each of the DTxTs. To achieve this result, a small amount of data capacity from the transport stream (typically, 1 packet per second) is required to carry the synchronization and control information from the TA to the DTxTs. This method has been adopted by the ATSC for use with the 8-VSB transmission system and is documented in the ATSC Synchronization Standard for Distributed Transmission (A/110). [1] A conceptual block diagram of the scheme is shown in Figure 8.6-20. All of the remaining discussion of DTx for networks using single-carrier (i.e., ATSC 1.0) modulation will focus on this method, in the form adopted by the ATSC. Parallel concepts for networks using multicarrier (i.e., ATSC 3.0) modulation are treated later in this section.

In the ATSC 1.0 8-VSB transmission system, there are a number of data-processing functions that operate either completely asynchronously or with a loosely defined relationship to the data in the stream that is to be transmitted. For several of these processes (i.e., data randomization, Reed-Solomon error correction coding, and convolutional interleaving), operation is synchronous between them and with the packet structure of the MPEG-2 transport stream, but there is no association of their starting points with particular packets in the stream. Other processes (i.e., precoding and trellis encoding) are completely asynchronous with one another and with the data in the stream. The result of this lack of defined relationships between the data in the transport stream and the operation of the various transmitter data-processing functions is that the data processing in the 8-VSB system can take any of 42,880,953,483,264 (almost 43 trillion) states for a given data input, and no two transmitters in a network are likely to produce the same output symbols at the same time, let alone a larger number of transmitters doing so.

To facilitate the TA approach to transmitter synchronization, the TA incorporates a model of the data processing that conventionally takes place in the transmitters. The model includes all portions of the data-processing functionality through the formation of symbols. The model processes the incoming transport stream data after randomly starting its operation with respect to that data. In doing so, it establishes a relationship between the transport stream data and the output symbols that are produced. Then it is necessary only to communicate from the TA to the DTxTs information about the relationship between the transport stream data and the symbols that are to be produced from them. The communication takes two forms: information about the relative timing of the processes that are synchronous with one another and with the transport steam packet structure (called "cadence synchronization") and information about the states of the precoders and trellis encoders at particular times (used to "jam sync" the equivalent functions in the transmitters). The information is inserted by the TA into the transport stream so that it can be extracted and acted upon identically by each transmitter in the network.

The information sent from the TA to the DTxTs is carried in a special packet type designated as an operations and maintenance packet (OMP), which has a dedicated packet identifier (PID) assigned. Each OMP begins with a further identifier for the particular

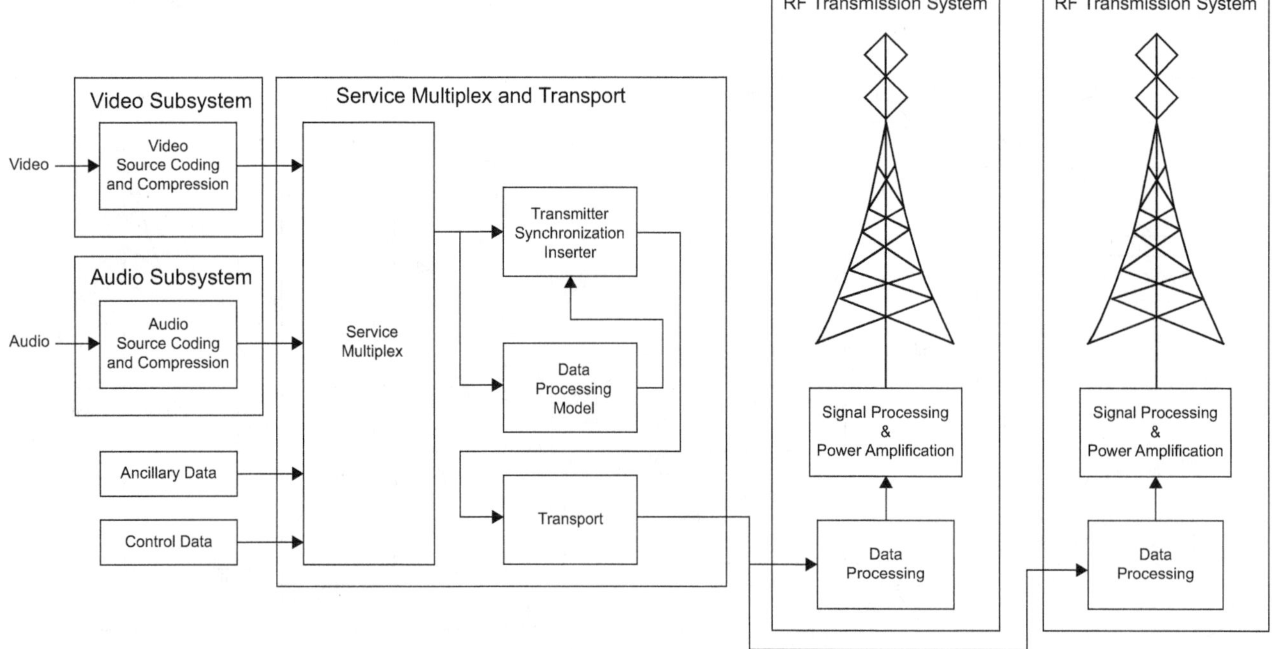

FIGURE 8.6-20 DTx system using transport stream distribution.

S. Merrill Weiss

application, in this case a transmitter control packet (TCP). In fact, 16 values of OMP identifier are assigned for TCP use to support multiple layers of DTxRs, as described later. In addition to the synchronization information for the DTxTs, the TCP carries information used to set the emission times and to control the power levels of the several transmitters. To facilitate the inclusion of TCPs in the data stream, packets termed "precursor packets," which can be converted to TCPs by the TA, are inserted into the transport stream by the service multiplexer or a remultiplexer upstream of the TA, or by the TA itself, at its input, in place of incoming null packets. The general scheme of the emission time control system is depicted in Figure 8.6-21, which shows a timeline of the various elements of the system and their relationships. It should be noted that a similar system timing scheme was used by SFN systems applicable to first-generation multicarrier modulation (e.g., DVB-T).

The time relationships illustrated in Figure 8.6-21 are all based upon an underlying reference time clock, available in common at all sites in the network and shown as 1-second clock ticks. The reference time can be derived from GPS or equivalent sources, for example. Certain information is carried in the TCP, and other information is calculated at each transmitter from the data received in the TCP plus certain locally derived values. Starting on the left side of the figure, a synchronization time stamp (STS) is sent in the TCP to identify the time at which a particular reference bit within the TCP is released from the TA. Also sent in the TCP is a value for maximum delay (MD) that is used by all of the transmitters in the network to determine when the reference emission time is for the reference bit within the TCP. MD is set as an input parameter in the TA and depends for its value on the length of time delay through the STL system to the farthest point (in time) in the network. (MD can be up to one second, so the system has sufficient capacity to treat STL time delays equal to almost four satellite hops.) All of the DTxTs in the network use the combination of the STS plus MD values to calculate the same reference emission time at each transmitter, as identified toward the right side of the timeline.

In addition to calculating the reference emission time for the network, each DTxT also calculates a number of other values to allow it to emit the signal at the intended time with respect to the other transmitters in the network. Each transmitter is sent individually, in the TCP, a value of offset delay (OD), which instructs it when to emit its signal relative to the reference emission time for the network. It also has an input parameter that sets the length of time between a specific point in the transmitter data processing and the antenna output—the transmitter and antenna delay (TAD). By adding the offset delay (which may have a positive or negative value) to the reference emission time and subtracting the TAD value, the transmitter can determine when it is supposed to release the reference point in the data stream into the modulator. By determining the TCP arrival time relative to the 1-second clock ticks, it can calculate the difference between that TCP arrival time and the intended TCP modulation time. That time difference is the time delay required in the particular transmitter (Tx delay) for proper operation of the network. The Tx delay is calculated whenever a TCP is received, and its value is maintained at least until the next TCP

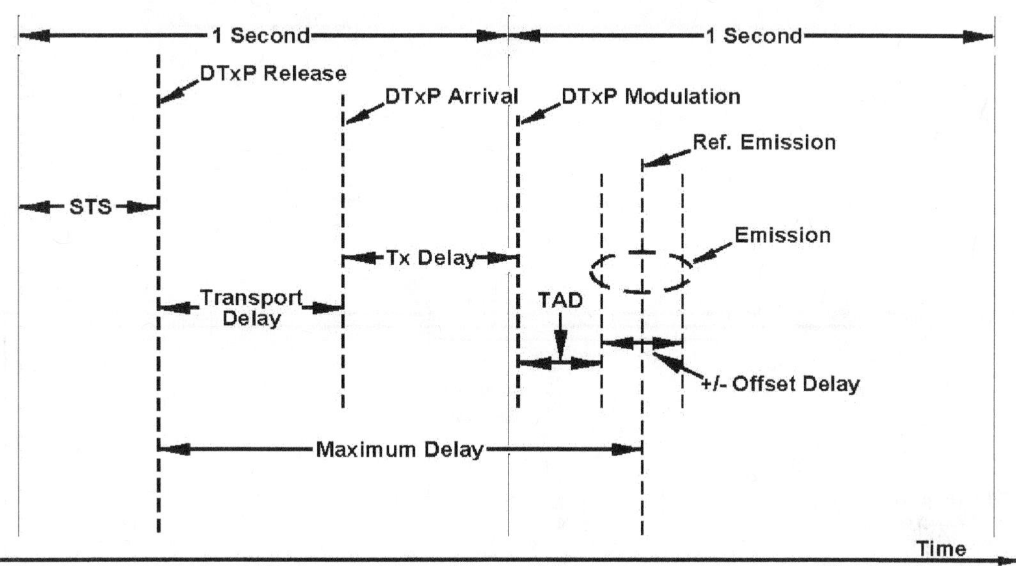

FIGURE 8.6-21 Transmitter emission timing control information.

S. Merrill Weiss

arrives. Because shifting the transmitter time delay can cause interruptions in the emitted signal, unless relatively complex steps are taken in the transmitter design to overcome them, some amount of hysteresis generally is applied to the Tx delay value, so that it will not be changed unless a significant error has accumulated.

More advanced systems intended for emission time control of transmitter outputs in SFNs use direct indication of the time of day when particular points in the signal waveform are to be emitted, instead of using just 1-second clock ticks for timing. This is true of DVB-T2, which applies MPEG-2 transport streams for carriage of data across the STL, and of ATSC 3.0, which is the first DTV system to use IP and its subsidiary protocols to format data streams for carriage across STLs. An overview block diagram of the ATSC 3.0 STL scheme is shown in Figure 8.6-22.

As shown in Figure 8.6-22, ATSC link layer protocol (ALP) packets are delivered to a broadcast gateway that includes a configuration manager and scheduler, an STL preprocessor, and an STL transport protocol (STLTP) formatter and error correction coding (ECC) encoder. The output of the broadcast gateway is a set of IP streams carrying baseband packets (BBPs) that are used to populate the payload space of a number of physical layer pipes (PLPs) provided by the physical layer of the ATSC 3.0 standard. The ALP packets can be formatted external to the broadcast gateway, as shown in the figure, in which case they would be delivered to the broadcast gateway using the ALP transport protocol (ALPTP), or they could be formatted within the broadcast gateway to encapsulate incoming IP data streams carrying the content to be transmitted, in which case use of the ALPTP *per se* would be unnecessary.

The scheduler functions as the controller for the entire emission operation, determining the exact composition of physical layer frames and their precise emission times. It communicates that information to the transmitters in an SFN by causing preparation of a preamble packet stream, the information in which is used by transmitters to configure themselves, and then is sent to receivers to instruct them precisely how the waveform of each emitted physical layer frame is constructed. The scheduler also causes preparation of a timing and management packet stream that instructs transmitters with respect to all of the information that they need that is not included in the preamble information, such as the scheduled emission time of each physical layer frame. Note that time information from a common, highly accurate source is provided to both the broadcast gateway and the transmitter(s) so that the correct time will be known by all devices in the STL network. All of the information and data sent to the transmitters is distributed using multicast/IP/UDP methods that include RTP headers to serve as a means of keeping packets in the correct sequence over the STL and also to provide a pathway for communicating information between the ATSC 3.0 transport and physical layers without creating layer violations.

At the exciter(s), IP/UDP port numbers are used to steer the various IP data streams to the correct inputs within the exciters. These include the inputs for up to 64 PLP streams, a preamble packet stream, and a timing and management data stream. A number of buffers are present in the broadcast gateway and also in the exciter(s) to manage the data flow through the system and to provide time for various functions to happen. Just as importantly, the buffers take up the slack when delays through the STL and other subsystems vary, thereby enabling an uninterrupted signal flow from the broadcast gateway to and through the exciters/transmitters.

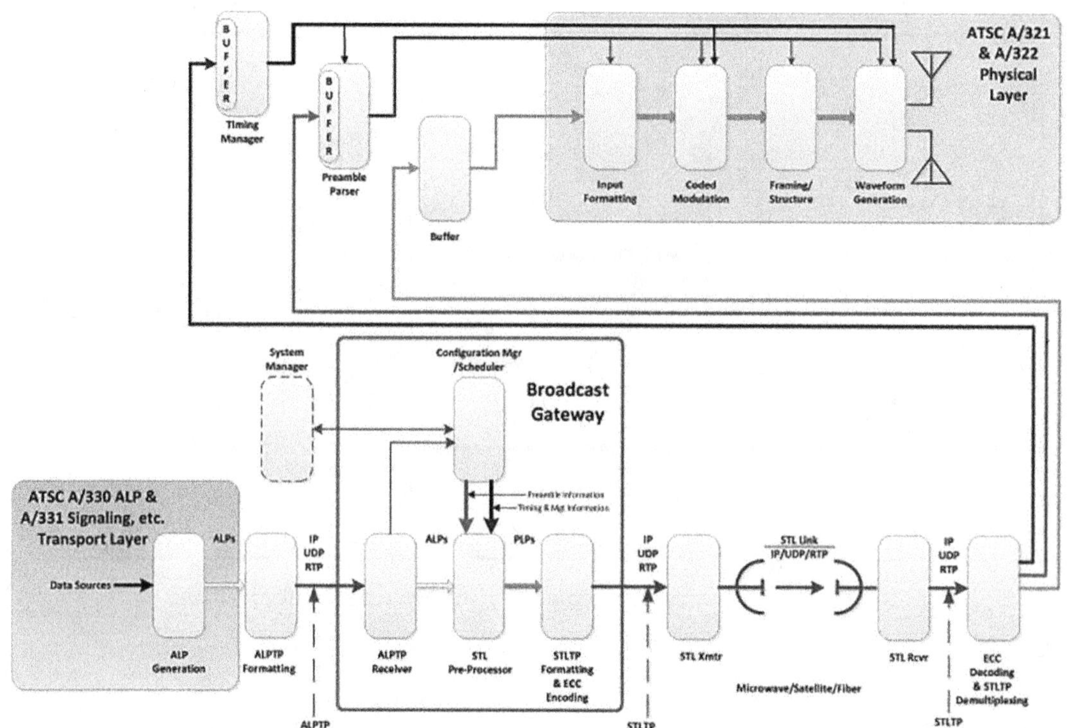

FIGURE 8.6-22 ATSC 3.0 broadcast gateway, STL, and physical layer overview block diagram.

Courtesy ATSC

STL Considerations

It is a fact that the time delay of signals transiting an STL typically varies. For conventional, single-transmitter facilities, this variation is immaterial. When an STL is embedded in a DTx network, the variation, which most likely will be different on each STL path, may become quite significant. Time delay change accumulates from slight errors in the output frequency of the link, following a buffer typically used for de-jittering the signal data. The accumulation of time delay variation can occur despite (indeed, because of) the use of long-time-constant frequency control loops to recover accurately the input frequency at the output of the link.

Consider a link carrying an MPEG-2 transport stream with a data rate appropriate for an ATSC 1.0 8-VSB transmission. The transport stream will operate with a data rate of 19.392658 Mb/s. If the error in the STL between input and output is 2.8 ppm, half the error band permitted by ATSC standards, over a period of an hour, the error will accumulate to about half a microsecond. Over a day, it will become an error of about 12 microseconds. Depending on the amount of time that the frequency control loop runs high or low in frequency before shifting in the opposite direction to maintain the average output frequency to match the input, even greater time delay changes can accumulate in the STL. Such delay changes are likely to cause the transmitter to reset its internal delay, thereby interrupting the signal while it does. Below the hysteresis threshold at which a delay reset occurs, the delay changes will cause the relative timing of the transmitters to wander back and forth, causing the locations of equal signal arrival times in the service area from the several transmitters to wander geographically. The hysteresis threshold generally is set to a moderately low value (e.g., 0.5 microsecond) to prevent too much geographic wander, thereby reducing the time between delay resets when there is much time delay variation.

To overcome the potential for STL time delay variations, a method is provided in the ATSC A/110 standard to lock the data stream frequency to an external frequency reference at the TA and at the DTxTs. Since an external time reference, from a source such as GPS or LORAN-C, already is required at each node in the network, using the frequency reference available from the same external sources involves little increase in complexity of the system. There is a change required, however, in the data processing of the TA, which must adjust the data rate of its input to be precisely the specified value on its output. To accomplish that potential frequency change, it is required periodically to insert packets into or delete packets from the data stream. It makes the insertions and deletions using null packets whenever possible. If there are no null packets in the stream for too long a time when a deletion is needed, the TA can delete the precursor packet that it would have converted into a TCP when the next one appears in the data stream.

While the operation of an ATSC 3.0 STL subsystem is based on use of IP packets instead of MPEG-2 transport stream packets, many of the considerations are the same. In both cases, the STL can consist of a microwave link, a satellite link, a terrestrial fiber link, or a combination of any of these.

DISTRIBUTED TRANSLATORS (DTXRS)

Combining the concept of a distributed transmitter with that of a conventional translator results in a DTxR. A DTxR has its input on one channel and its output on another. The benefit of DTxRs is that they can receive their input signals over the air from earlier transmitters in the network and they can share an output channel while mitigating interference between themselves in the same way as do DTxTs. Because of the different input and output channels, DTxRs can operate with more power than can DOCRs, and they do not require STLs. The trade-off for the additional flexibility of DTxRs is that they require at least one additional broadcast channel on which to operate, which additional spectrum may not be available in many places in which it would be desirable to utilize DTxRs.

Networks using DTxRs can be designed in "layers," in which transmitters sharing the same channel in the same layer are treated as part of a separate subnetwork for purposes of interference mitigation. Indeed, it is possible to design a very extensive network using DTxRs in which only two channels are required—the channels alternating with one another from layer to layer, moving away from the initial transmitter in the network. To support the synchronization and timing requirements of the transmitters in each layer, the TA must produce a separate TCP for each layer, for which purpose a number of OMP identifiers are available to indicate to which layer a particular TCP applies. Since the TCP carries information that depends on the data processing of the signal in the layer prior to the one to which it applies, there is implied in the TA data-processing model a cascading of corresponding layers of data processing to develop the symbols to which the transmitters in the respective layer are to be synchronized. The basic layout of a DTxR network with one central transmitter and a single layer of translators and using two channels appears in Figure 8.6-3.

In the DTxR itself, there is a receiver that demodulates and decodes the received signal to a baseband MPEG-2 transport stream. The receiver is followed by a modulator that is virtually identical to the one in a DTxT, as described previously. The only differences are that the data processing in the modulator examines the TCP for the OMP address associated with the layer in which the DTxR resides, and provision must be made in the modulator for identification of that layer as an input parameter. The overall configuration of a DTxR nearly matches that of a baseband decoding DOCR, which decodes the received signal to a baseband transport stream, as shown in Figure 8.6-17d. The primary difference is that the local oscillators used to downconvert and upconvert the received and transmitted signals, respectively, will be on different frequencies, since the input and output channels of the DTxR will be different.

When setting up a DTxR network, it is important to set the maximum delay (MD) for each layer, which will be an independent parameter, to a somewhat higher value for each successive layer in the network. The MD value must be high enough to allow for the signal passing through all of the preceding layers, with the attendant decoding and remodulation at each DTxR in the chain. At the same time, the MD value should be kept only a little higher than necessary to account for all of the preceding layers so as to allow for the addition of following layers. The total time available is just under one second, so there is adequate capacity in the system to accommodate as many layers as might be needed in the most complex of network designs.

NETWORK ADJUSTMENTS

As noted previously, one of the prerequisites for SFN operation is that the symbols emitted by the transmitters in a network must be identical to one another and that the transmitters must operate on essentially the same frequency. The purpose, of course, is to make the signals from the several transmitters appear to receivers as if they all were emitted by one transmitter and followed multiple paths to the receivers, thereby creating multipath that can be treated by receiver adaptive equalizers. Thus, for purposes of reception by consumer receivers, the transmitted signals must appear to be identical to one another.

When it is necessary to adjust the SFN, it is desirable to be able to differentiate one transmitter from another. The differentiation is needed to permit, at any given receiving location, the relative field strength and arrival time of the signal from each transmitter to be determined with respect to those of the other transmitters. This determination would enable adjustment of the emission characteristics of each transmitter. Unfortunately, with identical signals being transmitted and received, such a differentiation is not possible. It is possible, however, to create a hidden difference between the transmitted signals that is not detectable by consumer receivers but that can be detected by a special receiver built for the purpose. Using such a technique allows network characteristics to be measured without interfering with the operation of the network from the standpoint of delivery of signals to consumer receivers. The technique developed for the purpose of differentiating transmitters is termed an "RF watermark."

RF Watermark

An RF watermark is a signal that is added to a host signal (e.g., an 8-VSB broadcast signal) to permit the identification of that host signal and to obtain information from and about the combination of the host signal and the RF watermark without interfering with the normal operation of the host signal for broadcasting purposes. Specific RF watermark technology was developed for use with 8-VSB transmissions as part of the process of standardizing the synchronization methods for distributed transmission using ATSC 1.0 technology. The RF watermark is defined in ATSC A/110 and explained in ATSC A/111. There are several purposes for the RF watermark, which are met with very high performance potential by the particular technology adopted. There also are several additional applications for the technology that have become apparent because of its high performance under difficult conditions.

The same basic RF watermark technology developed for ATSC 1.0 also has been applied in ATSC 3.0, but in two different ways. In ATSC 3.0, the RF watermark can be used for its original purposes, as described next, but it is also useful as a part of a new method for increasing the efficiency of RF spectrum utilization caller layered division multiplexing (LDM). LDM provides a mechanism for applying different proportions of the power output of a transmitter to different parts of the transmitted waveform and also of enabling very robust operation at very low (even negative) C/N values while simultaneously supporting high data rate but less robust transmission from the same transmitter.

Purposes

As applied in an SFN environment, the RF watermark serves three main purposes: identification of transmitters, measurement of the channel impulse response (CIR) from each transmitter, and carriage of information with great robustness when compared to that of the host signal. As already discussed, when the prerequisites for SFN operation are met with respect to the signals delivered to consumer receivers, it becomes impossible to tell the signals from the several transmitters apart, if nothing additional were done. Each of the transmitters would carry, as part of its program content, the station identification of the station operating the network, but individual transmitters would be indistinguishable from one another. Thus, one of the principal purposes of the RF watermark is to carry identification information that allows differentiating one transmitter in an SFN from another, useful both when mitigating INI and when evaluating interference to other stations. For this objective, the RF watermark emitted by each network transmitter includes a code that can be recovered by a specialized receiver. The code comprises two portions: one part associated with the network and one part associated with the individual transmitter. The transmitter identification (TxID) codes are maintained in a publicly accessible database. The specialized receiver thereby is enabled to completely determine the source of any signals that it is receiving carrying an RF watermark.

The second goal of including an RF watermark in the emissions of each transmitter in an SFN is to support measurements that are necessary to adjust the network for minimum INI. Ideally, both the relative strengths and the relative arrival times of the signals from various transmitters could be ascertained at any receiving location. The CIR is designed just for the purpose of indicating the strength of a signal and its echoes over time, i.e., it characterizes the propagation channel traversed by a signal. If the CIRs could be taken independently on the signals from each of the transmitters in an SFN, then presenting them overlaid on one another (e.g., in different colors—one for each transmitter) on a common display and using a common time base would show exactly the information needed to adjust the transmitter amplitudes and timing in an SFN. The RF watermark completely supports such functionality.

Technology

The RF watermark consists of a spread spectrum signal that is added at very low level to the host signal carried by each transmitter. Because it operates in the amplitude region that normally would be considered to be noise in a transmitter output, the RF watermark is termed a *buried spread spectrum* (BSS) signal. In the case of an 8-VSB host, the BSS signal is configured with 2-VSB modulation that is synchronized to the modulation of the host signal and has the same spectrum occupancy as the host. It generally operates at a level 30 dB or more below that of the host, although higher levels can be used for out-of-service network testing and adjustment. The relationship between the signals, within the spectrum of the operating channel, is portrayed in Figure 8.6-23, where the host signal is shown in as the darker trace (or in blue in the companion color image website) and the RF watermark is shown as the lighter gray trace (or in green on the companion color image website). Given the synchronization between host and watermark signals, the RF watermark has the effect of opening up an additional set of "eyes" in the amplitude versus time representation of the combined signal. The set of small additional eyes is apparent in Figure 8.6-24.

The 2-VSB signal is modulated with a code called a Kasami sequence, which is a code that has a number

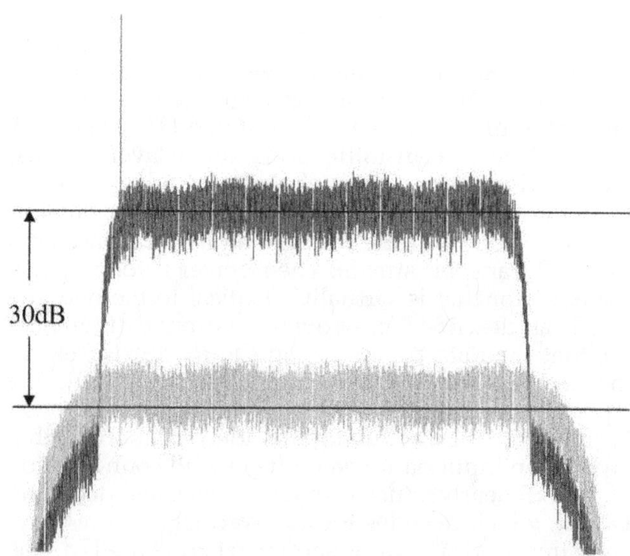

FIGURE 8.6-23 Superimposition in spectrum of host 8-VSB signal and RF watermark.

Figure courtesy of X. Wang, Y. Wu, and J.-Y. Chouinard, Communications Research Centre Canada, Ottawa, Ontario [5]

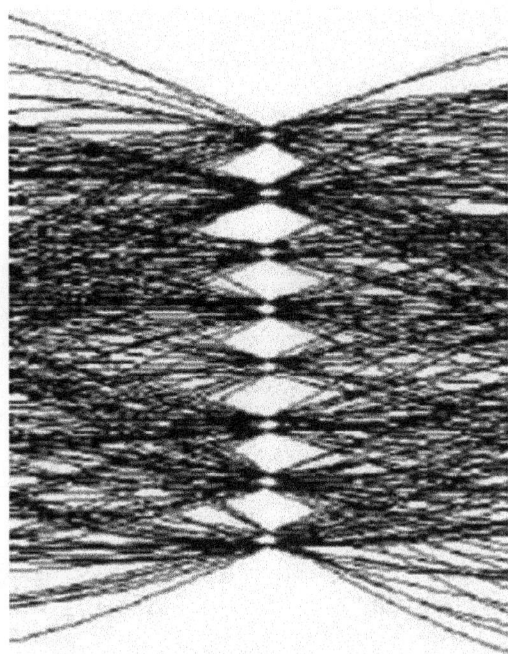

FIGURE 8.6-24 Eye openings of 8-VSB host signal with added eyes of RF watermark in between.

Figure courtesy of ATSC, Washington, D.C. [2]

of properties that make it particularly useful for the RF watermark application. The important properties relate to the autocorrelation and cross-correlation of different values of the code. The particular coding method selected permits over 16 million different patterns to be generated, thus carrying 24 bits of information. The sequences generated are sufficiently short that they can be repeated almost four times during one 8-VSB data field. The code sequences are synchronized with the 8-VSB data framing to make finding the start of a code sequence easier and, hence, faster. The particular Kasami sequence transmitted by each transmitter serves as its TxID and is unique to that transmitter, with part of the data represented by the code value serving as an identifier for the network in which the transmitter operates and the other part of the data serving as an identifier for the individual transmitter within the network. With 12 bits allocated to identifying networks and 12 bits allocated to identifying transmitters within networks, up to 4096 networks and up to 4096 transmitters within each network can be identified.

Carrying additional information on the RF watermark is accomplished by inverting the phase of the data in the Kasami sequence from one repetition of the code to the next. In a receiver for the RF watermark, finding a code or its inverse is a relatively efficient operation using standard correlation techniques. Thus, the phase of the Kasami sequence data relative to the assigned code value (i.e., noninverted or inverted) can indicate binary ones and zeroes. This capability allows the transmission of a one or a zero for each repetition of the code sequence. Since there are four usable repetitions of the code sequence in each 8-VSB data field and since there are approximately 40 data fields per second, a data rate of about 160 b/s results. While this data rate is quite slow when compared with the rates provided by modern data communications technology, it is carried on a highly robust channel, making it quite useful.

Performance

There is a tremendous challenge that must be met by an RF watermark if it is to be effective. One of its main purposes is to allow measurement of the relative levels of signals from multiple transmitters that can affect the operation of adaptive equalizers. As shown previously, adaptive equalizers become operational when the amplitude difference between multipath signals is on the order of 20 dB. Thus, the RF watermark must enable measurement of the level of an echo that is 20 dB below the level of the signal serving as the reference to an adaptive equalizer. But the RF watermark itself must be inserted on the order of 30 dB below the level of the host signal that carries it, so as not to interfere with its host. This combination of factors means that the RF watermark must be recoverable when it is 50 dB below the strongest signal in an ensemble of multipath signals. Not only must it be recoverable, it must be retrieved with a sufficient signal-to-noise ratio (i.e., about 10 dB) to permit the necessary amplitude and timing relationships to be determined. It has been demonstrated that the RF watermark is capable of just such performance.

Since the RF watermark can be received with a usable margin when it is 50 dB below the amplitude of any host signal and since it is transmitted about 30 dB below the level of its own host signal, it stands to reason that the RF watermark can be recovered at a received signal level about 20 dB below that at which an 8-VSB signal can be correctly demodulated. The 20 dB advantage permits reception in places where the host signal cannot be received. Moreover, when combined with the fact that the frequencies on which digital television signals operate can penetrate buildings much better than can higher frequencies, the RF watermark can provide reliable reception in places not reachable by many other techniques. Furthermore, because of its 2-VSB modulation, the RF watermark can be received readily by receivers that are in motion, such as those carried by pedestrians or those in vehicles. This combination of characteristics makes possible a range of potential applications for the RF watermark beyond those originally intended for it.

Field Measurements

When setting up and maintaining an SFN, it is desirable to confirm through field measurements the actual operation of the network in comparison with its design parameters. Two basic characteristics need to be measured, and there are two ways in which they can be measured. The two characteristics are the field strengths of the signals from the several network transmitters that reach any particular location at which testing is performed and the relative arrival times of the signals at that location. The two ways in which to make the measurements are through short-term visits to the test locations, such as with a mobile measuring system, and with fixed installations of monitoring equipment at critical locations within the network.

Making measurements with a mobile measurement system, if done prior to system operation or during out-of-service periods, can be facilitated by turning transmitters on and off sequentially to permit field strength data to be obtained. Determining signal arrival timing, however, requires some form of instrumentation that can respond to the presence of multiple signals and indicate their timing relationship. One way to acquire this data is with a receiver equipped to display adaptive equalizer tap weight information that can be interpreted by a human observer. Tap weights can show the presence of echoes from multiple transmitters but can make determination of the source of the various echoes very difficult, especially in natural environments having a great deal of multipath and, consequently, many active taps in an equalizer. Frequently, determining the sources of echoes also requires turning transmitters on and off. The difficulty of interpreting tap weights becomes even greater when the levels of the signals being received are radically different from one another, making one signal appear at a very low level on the tap weight display relative to the stronger signal that serves as a reference.

Routine monitoring of SFN operations while they are in service cannot take advantage of the turning on and off of transmitters to determine signal levels and for the same reason cannot benefit from the interpretation of tap weights by inspection of a human observer. Instead, equipment is needed at monitoring sites that autonomously can collect and analyze the necessary data from the signals of multiple, continuously operating transmitters. Multiple monitoring sites may be necessary in order to collect information about the operation of all of the transmitters in a given network. The results of the analyses at the monitoring sites then must be forwarded to the station for collection and aggregation with the data from the other sites to create a complete picture of network operation.

Virtually all of the needs for measurement and monitoring in SFNs are addressed by the RF watermark. It provides the mechanism through which to determine both the amplitude and the timing of each signal received at a location. With specialized receivers to recover the channel impulse responses of the signals from multiple transmitters, the data can be collected while the transmitters are in regular operation, the absolute amplitudes of the signals can be determined using information on the RF watermark insertion levels, and the timing of arrival of the signals can be measured both relative to one another and with respect to an external time reference. Such receivers can provide fully automatic analysis of the various signals even when those signals are received at significantly different levels from one another, thereby facilitating aggregation of the information and analysis of the overall operation of the network.

Note

[1] The term "channel" can have several different meanings. As used in this chapter, it can refer either to the path through the environment taken by a signal or to the portion of the electromagnetic spectrum used to transmit a signal. Efforts have been made in the text to clarify which meaning applies to each use.

References

[1] *Synchronization Standard for Distributed Transmission*, 2011 Revision (A/110:2011), Advanced Television Systems Committee, Washington, DC, April 8, 2011.

[2] *Recommended Practice: Design of Synchronized Multiple Transmitter Networks* (A/111), Advanced Television Systems Committee, Washington, DC, September 18, 2009.

[3] *Recommended Practice: Receiver Performance Guidelines* (A/74:2010), Advanced Television Systems Committee, Washington, DC, April 7, 2010.

[4] *ATSC Candidate Standard: Scheduler/Studio to Transmitter Link*, Advanced Television Systems Committee, Washington, DC, September 30, 2016.

[5] Wang, X., Y. Wu, and J.-Y. Chouinard. "Robust Data Transmission Using the Transmitter Identification Sequences in ATSC DTV Signals." *IEEE Transactions on Consumer Electronics*, vol.51, no.1, February 2005, pp.41–47.

Related Reading

Angueira, P., M.M. Velez, D. De La Vega, A. Arrinda, I. Landa, J. L. Ordiales, and G. Prieto. "DTV (COFDM) SFN Signal Variation Field Tests in Urban Environments for Portable Outdoor Reception." *IEEE Transactions on Broadcasting*, vol. 49, no. 1, March 2003, pp. 81–86.

Angueira, P., M.M. Velez, D. De la Vega, G. Prieto, D. Guerra, J.M. Matias, and J.L. Ordiales. "DTV Reception Quality Field Tests for Portable Outdoor Reception in a Single Frequency Network." *IEEE Transactions on Broadcasting*, vol. 50, no. 1, March 2004, pp. 42–48.

Bank, M. "On Increasing OFDM Method Frequency Efficiency Opportunity." *IEEE Transactions on Broadcasting*, vol. 50, no. 2, June 2004, pp. 165–171.

Bretl, W., W. R. Meintel, G. Sgrignoli, X. Wang, S. M. Weiss, and K. Salehian. "ATSC RF, Modulation, and Transmission." *Proceedings of the IEEE*, vol. 94, no. 1, January 2006, pp. 44–59.

Guerra, G., P. Angueira, M.M. Velez, D. Guerra, G. Prieto, J.L. Ordiales, and A. Arrinda. "Field Measurement Based Characterization of the Wideband Urban Multipath Channel for Portable DTV Reception in Single Frequency Networks." *IEEE Transactions on Broadcasting*, vol. 51, no. 2, June 2005, pp. 171–179.

Hershberger, D. L. "Implementation of the ATSC Distributed Transmission System." *2003 NAB Broadcast Engineering Conference Proceedings*, pp. 280–289. Washington: National Association of Broadcasters, 2003.

Hershberger, D. L. "Lessons Learned from DTx Implementation—and Applications," *2005 NAB Broadcast Engineering Conference Proceedings*, pp. 493–500. Washington: National Association of Broadcasters, 2005.

Kim, S. W., Y.-T. Lee, S. I. Park, H. M. Eum, J. H. Seo, and H. M. Kim. "Equalization Digital On-Channel Repeater in the Single Frequency Networks." *IEEE Transactions on Broadcasting*, vol. 52, no. 2, June 2006, pp. 137–146.

Lee, Y.-T., S. I. Park, H.-M. Eum, H. M. Kim, J.-H. Seo, S. W. Kim, B. Ledoux, S. Lafleche, and Y. Wu. "Laboratory and Field Test Results of Equalization Digital On-Channel Repeater (EDOCR)." *2005 NAB Broadcast Engineering Conference Proceedings*, pp. 485–492. Washington: National Association of Broadcasters, 2005.

Ligeti, A., and J. Zander. "Minimal Cost Coverage Planning for Single Frequency Networks." *IEEE Transactions on Broadcasting*, vol. 45, no. 1, March 1999, pp. 78–87.

Linfoot, S. L. "A Comparison of 64-QAM and 16-QAM DVB-T Under Long Echo Delay Multipath Conditions." *IEEE Transactions on Consumer Electronics*, vol. 49, no. 4, November 2003, pp. 978–982.

Linfoot, S. L. and L.-P. Gao. "A Soft Decision 16-QAM Demodulation Algorithm for Multipath Affected DVB-T Systems." *IEEE Transactions on Consumer Electronics*, vol. 51, no. 4, November 2005, pp. 1121–1128.

Malmgren, G. "On the Performance of Single Frequency Networks in Correlated Shadow Fading." *IEEE Transactions on Broadcasting*, vol. 43, no. 2, June 1997, pp. 155–165.

Mattsson, A. "Single Frequency Networks in DTV." *IEEE Transactions on Broadcasting*, vol. 51, no. 4, December 2005, pp. 413–422.

Nakahara, S., S. Moriyama, T. Kuroda, M. Sasaki, S. Yamazaki, and O. Yamada. "Efficient Use of Frequencies in Terrestrial ISDB System." *IEEE Transactions on Broadcasting*, vol. 42, no. 3, September 1996, pp. 173–178.

O'Leary, S. "Field Trials of an MPEG2 Distributed Single Frequency Network." *IEEE Transactions on Broadcasting*, vol. 44, no. 2, June 1998, pp. 194–205.

O'Leary, S. "Digital/Analogue Co-Channel Protection Ratio Field Measurements." *IEEE Transactions on Broadcasting*, vol. 44, no. 4, December 1998, pp. 540–546.

Oziewicz, M. "The Phasor Representation of the OFDM Signal in the SFN Networks." *IEEE Transactions on Broadcasting*, vol. 50, no. 1, March 2004, pp. 63–70.

Rebhan, R. and J. Zander. "On the Outage Probability in Single Frequency Networks for Digital Broadcasting." *IEEE Transactions on Broadcasting*, vol. 39, no. 4, December 1993, pp. 395–401.

Rinne, J., A. Hazmi, and M. Renfors. "Impulse Burst Position Detection and Channel Estimation Schemes for OFDM Systems." *IEEE Transactions on Consumer Electronics*, vol. 49, no. 3, August 2003, pp. 539–545.

Salehian, K., M. Guillet, B. Caron, and A. Kennedy. "On-Channel Repeater for Digital Television Broadcasting Service." *IEEE Transactions on Broadcasting*, vol. 8, no. 2, June 2002, pp.97–102.

Salehian, K., B. Caron, and M. Guillet. "Using On-Channel Repeater to Improve Reception in DTV Broadcasting Service Area." *IEEE Transactions on Broadcasting*, vol. 49, no. 3, September 2003, pp. 309–313.

Salehian, K., Y. Wu, and B. Caron. "Design Procedures and Field Test Results of a Distributed-Translator Network, and a Case Study for an Application of Distributed-Transmission." *2005 NAB Broadcast Engineering Conference Proceedings*, pp. 501–510. Washington: National Association of Broadcasters, 2005.

Salehian, K., Y. Wu, and B. Caron. "Design Procedures and Field Test Results of a Distributed-Translator Network, and a Case Study for an Application of Distributed-Transmission." *IEEE Transactions on Broadcasting*, vol. 52, no. 3, September 2006, pp. 281–289.

Sgrignoli, G. "DTV Repeater Emission Mask Analysis." *IEEE Transactions on Broadcasting*, vol. 49, no. 1, March 2003, pp. 32–80.

Tanyer, S. G., T. Yucel, and S. Seker. "Topography Based Design of the T-DAB SFN for a Mountainous Area." *IEEE Transactions on Broadcasting*, vol. 43, no. 3, September 1997, pp. 309–319.

Wang, J., Z.-X. Yang, C.-Y. Pan, J. Song, and L. Yang. "Iterative Padding Subtraction of the PN Sequence for the TDS-OFDM Over Broadcast Channels." *IEEE Transactions on Consumer Electronics*, vol. 51, no. 4, November 2005, pp. 1148–1152.

Wang, J.-T., J. Song, J. Wang, C.-Y. Pan, Z.-X. Yang, and L. Yang. "A General SFN Structure With Transmit Diversity for TDS-OFDM System." *IEEE Transactions on Broadcasting*, vol. 52, no. 2, June 2006, pp. 245–251.

Wang, J.-T., J. Song, J. Wang, C.-Y. Pan, Z.-X. Yang, and L. Yang. "Corrections to 'A General SFN Structure With Transmit Diversity for TDS-OFDM System.'" *IEEE Transactions on Broadcasting*, vol. 52, no. 3, September 2006, p.412.

Wang, X., Y. Wu, and B. Caron. "Transmitter Identification Using Embedded Pseudo Random Sequences." *IEEE Transactions on Broadcasting*, vol. 50, no.3, September 2004, pp. 244–252.

Wang, X., Y. Wu, and J.-Y. Chouinard. "Transmitter Identification in Distributed Transmission Network and Its Application in Position Location and a New Data Transmission Scheme." *2005 NAB Broadcast Engineering Conference Proceedings*, pp. 511–520. Washington: National Association of Broadcasters, 2005.

Weiss, S. M. "Distributed Transmission Systems—Overcoming the Limitations of DTV Transmission." *2003 NAB Broadcast Engineering Conference Proceedings*, pp. 263–279. Washington: National Association of Broadcasters, 2003.

Weiss, S. M. "Designing Distributed Transmission Systems to Meet FCC Requirements." *2006 NAB Broadcast Engineering Conference Proceedings*, pp. 152–161. Washington: National Association of Broadcasters, 2006.

Wu, Y., X. Wang, R. Citta, B. Ledoux, S. Laflèche, and B. Caron. "An ATSC DTV Receiver with Improved Robustness to Multipath and Distributed Transmission Environments." *IEEE Transactions on Broadcasting*, vol. 50, no. 1, March 2004, pp. 32–41.

Zhang, G. L., L. Yantao Qiao, and W. Zhang. "Obtaining Diversity Gain for DTV by Using MIMO Structure in SFN." *IEEE Transactions on Broadcasting*, vol. 50, no. 1, March 2004, pp. 83–90.

CHAPTER

8.7

Television Antenna Systems

ANDRE J. SKALINA
VP Technology, Dielectric, LLC

WHAT IS AN ANTENNA?

An antenna is an essential component of any wireless communication system. Antennas are not as trendy as cameras or television sets, they do not have any attractive controls or colorful touch screens, but they are the most important part of the RF broadcast system.

The antenna provides three very important attributes to a broadcast system:

1. The antenna acts like a matching transformer to maximize the power transmitted. It does this by matching the RF transmission line impedance with the intrinsic impedance of the air. The effectiveness of the impedance match is measured by the antenna VSWR.
2. The antenna distributes the power from the transmitter to the assigned coverage area, based on the station requirements. This feature provides the broadcasted signal to locations that require signal and suppresses signal to areas that are protected or do not require strong or any signal. It distributes the energy based on the distance; more energy radiates toward the horizon where it is needed and less energy close in to the transmitting site. This is usually shown as antenna radiation patterns.
3. The antenna passively amplifies the power level delivered by the transmission line by effectively focusing the energy into the desired downward radiation. Since the antenna amplification is passive and requires no power to amplify the signals, energy consumption, and thus operating cost, is significantly reduced. This amplification is defined by the antenna gain.

The antenna is usually located on top of a tower or a building high above the ground, so it has to provide long-term service in very harsh environmental conditions. If the antenna is properly designed, it delivers all three features listed, consistently without any issues, for a very long time. For many people working at the broadcast stations the antenna is "out of sight and out of mind"; only when something happens to it does it get noticed.

ANTENNA BASICS

The antenna features mentioned above can be further defined by some basic parameters of the antenna. Since there are many antenna books that define these parameters in details [e.g., 1], they will be described only briefly here.

The equation

$$P_{in} = \frac{ERP}{G_H} P_{in} = \frac{ERP}{G_H} \qquad (1)$$

defines the relation between the antenna gain (w.r.t. half-wave dipole) of the horizontally polarized component (G_H), the ERP (effective radiated power) and the power into the antenna (P_{in}). Since the ERP is assigned to stations by the regulating body (FCC in the United States) there are two inversely proportional variables, G_H and P_{in}, that are under the station's control. Therefore this simple equation is very significant to the broadcasting station since it directly affects the coverage, tower loading, size of the transmitter building, sizing of the cooling system in the transmitter building, the power capacity of the transmitter building, operating cost and cost of the equipment.

Antennas with high gain (G_H) can affect the stability of the signal in high winds, or high temperature variations. They provide high wind load on the tower, might not provide stable signal close to the tower and are more expensive. However, the lower power (P_{in}) needed to obtain the FCC-approved ERP offers lower operation cost, smaller transmitter footprint, lower cooling cost and lower transmitter cost.

Antennas with low gain provide a more stable and uniform signal, both far and close in, and are usually less expensive. Conversely, because more input power must be delivered to the antenna, they can require a higher capital investment in the transmitter and support plant as well as higher long-term operating cost.

Thus, choosing the right antenna is a compromise that lies somewhere between the very high-gain antenna and low gain. The choice of the antenna is obviously made to provide the best coverage possible, taking into account the above-mentioned economic aspects.

One should also remember that adding vertical polarization reduces the gain of the horizontal component that is used for calculating the ERP, thus increasing the output power requirement from the transmitter.

ANTENNA IMPEDANCE

Antenna impedance (Z_a) is defined as the impedance presented by the antenna at its input. The location of the input is specified by the antenna manufacturer. Most antennas used for TV broadcast have one input. However, there is a significant number of antennas with two inputs, feeding the upper half and lower half of a panel antenna or feeding different polarizations, also in a panel antenna. The antenna impedance defines how well it matches the characteristic impedance (Z_0) of the transmission line. Since the impedance is a complex number, it is difficult to determine how well the antenna is matched; a better way to show the match of the antenna is to measure VSWR (voltage standing wave ratio). The VSWR is calculated from the impedance by calculating the voltage reflection coefficient (Γ)(Γ) given by:

$$\Gamma = \frac{Z_a - Z_0}{Z_a + Z_0}\ \Gamma = \frac{Z_a - Z_0}{Z_a + Z_0} \tag{2}$$

Which is related to VSWR, thus:

$$VSWR = \frac{1+|\Gamma|}{1-|\Gamma|} VSWR = \frac{1+|\Gamma|}{1-|\Gamma|} \tag{3}$$

VSWR is a simple number usually displayed as a ratio relative to 1, a perfect match. For example, a VSWR of 2.56 is displayed as 2.56:1.

The VSWR varies from 1.0:1 when the antenna is perfectly matched to the transmission line to infinity when there is either a dead short or open. Since the VSWR varies as a function of frequency, the broadcast antenna specification is usually set as less than or equal to a maximum VSWR (e.g. ≤ 1.10 : 1) for any frequency within the operating range of the antenna.

As mentioned before, the antenna impedance, and thus the VSWR, is specified at the antenna input, so anywhere else along the transmission line the VSWR might not be the same. If there is a perfect line separating the new measurement location from the input, the VSWR will be the same; however, there is no such thing as perfect line. The nonperfect line includes components like flanges on individual sections of transmission lines, filters, elbows and reducers, all of which have their own VSWR. Therefore, the maximum VSWR of the system, including RF filter, transmission line and elbow complex, can be quite different and always worse than the maximum VSWR at the antenna input. This is due to in-phase addition of the impedances of the individual components in the system at some frequencies.

For example, for a typical broadcast RF system, the antenna specification is ≤ 1.08:1, the transmission line specification is ≤ 1.05:1, the elbow complex at the tower top is usually ≤ 1.02:1, the gas stop ≤ 1.02:1 and five elbows ≤ 1.01 each. The maximum VSWR of the system can be approximated by:

$$VSWR_{max} = 1.08 \times 1.05 \times 1.02 \times 1.02 \times 1.01^5 \tag{4}$$

Thus the maximum VSWR of a real antenna system that includes transmission line, elbow complex, gas stop and five elbows and the antenna can reach 1.22:1 at some frequencies within the operating band. This also illustrates the importance of keeping the VSWR of each component as low as possible.

The maximum VSWR is measured using a network analyzer before the transmitter is connected. Typically, however, the transmitter telemetry will measure average reflected power. This is a good indication of the average VSWR over the operating frequency range, which will typically be less than the maximum VSWR measured with the network analyzer.

ANTENNA RADIATION PATTERNS

The antenna radiation pattern is a three-dimensional representation of the power radiated from an antenna and the resulting field distribution. Since there are no receivers in the near field of the antenna, all patterns are shown in the far field, where the radiation can be considered a plane wave.

A three-dimensional pattern does not provide the information the broadcasters really need in an easy-to-interpret way; therefore, in the broadcast industry, two-dimensional patterns are used—azimuth pattern in a plane parallel to ground and elevation pattern perpendicular to ground. The orientation of the azimuth pattern is usually specified with respect to true north.

The patterns are usually shown in normalized field strength. The azimuth pattern requirements are either omnidirectional or directional, determined by the station to direct the signal toward populated areas and

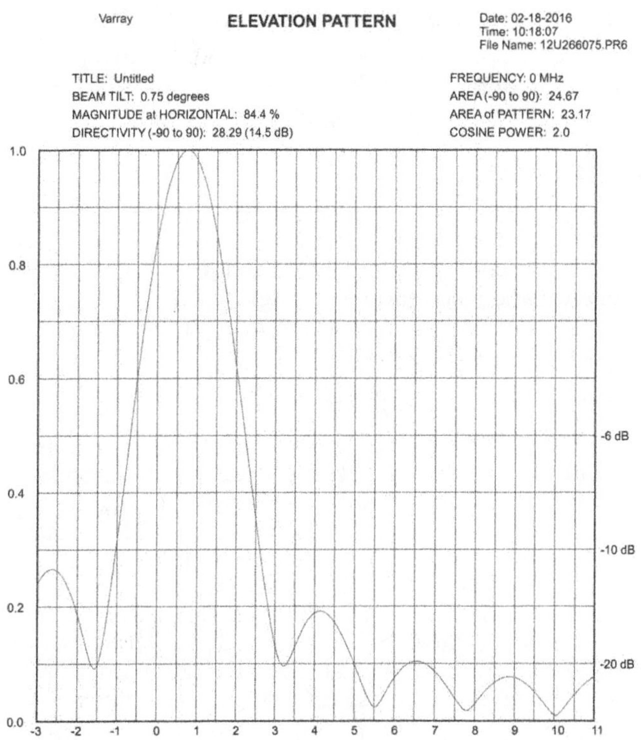

FIGURE 8.7-1 Typical elevation pattern with 0.75 deg beam tilt and 10% first null fill.

A. J. Skalina, Dielectric, LLC

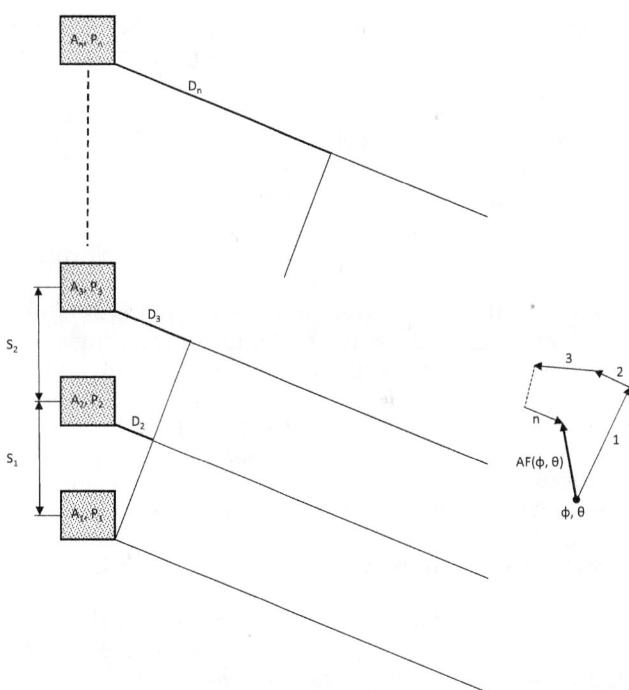

FIGURE 8.7-2 Vectorial summation of far-field pattern; the boxes represent antenna radiating elements.

A. J. Skalina, Dielectric, LLC

minimize the radiation toward unpopulated areas such as the ocean or mountains, then reviewed and approved by the regulating agencies (the FCC in the United States) to minimize interference between stations. The elevation pattern is typically determined by the antenna HAAT (height above average terrain), which defines the antenna beam tilt. Since the elevation pattern is typically scalloped, nulls will exist at a number of points along a radial from the antenna—see Figure 8.7-1.

If there is population in a null, null fill is added to the antenna design. If there is no population close to the antenna, null fill might not be required. Note that adding null fill reduces the gain of the antenna.

ANTENNA ARRAYS

Most broadcast antennas comprise many radiating elements creating an antenna array. These radiating elements can be slots in a pylon-style antenna or dipoles in a panel antenna. The larger the number of elements in a certain direction results in higher gain in that direction. Thus, since high elevation gain and low azimuth gain are desired for broadcast antennas, they are tall and have small cross section.

The formula that calculates the far-field pattern (AP) of the entire array of elements in simplified form is:

$$AP(\varphi, \theta) = EP(\varphi, \theta) \times AF(A\downarrow i, P\downarrow i, S\downarrow i, \varphi, \theta) \quad (5)$$

Here EP is the pattern of an element of the array, and AF, the array factor, is a function of the geometrical arrangement of the elements, their radiated amplitude and phase, the number of elements and their spacing from each other. The array factor (AF) is a vectorial summation of contribution from each and every antenna element of the array, based on its radiated amplitude (A_i) and phase (P_i), and relative phase delay (D_i) at the spacing (S_i) and location of the summation point (φ, θ) in space with respect to the antenna array as shown in Figure 8.7-2.

The far field elevation or azimuth patterns are obtained by geometric arrangement of the elements and by adjusting the amplitude (A_i), phase (P_i) and spacing (S_i) of each element to obtain a desired pattern.

When all amplitudes (A_i) and phases (P_i) of a linear vertical array are the same, the far-field elevation pattern has no beam tilt and no null fill. By adjusting the amplitudes and phases of each element, the required beam tilt and null fill can be obtained. The spacing between the elements is usually uniform, $S_1 = S_2 = S_3$. The number of elements defines the gain of the antenna, more elements resulting in higher gain. Keep in mind that increasing the null fill reduces the gain.

This formula for AP assumes that the azimuth radiation pattern for each antenna radiating element is the same for all elements in the array. This is true for all top-mounted antennas. However, antenna arrays side mounted on a supporting structure, where the structure cross section varies throughout the antenna aperture, might produce different individual element patterns. The resulting scatter pattern, produced by reflections from the support structure, will be different for different elements, since the supporting structure behind them is different.

Also any mutual coupling between the antenna array elements is ignored in (5). The mutual coupling disturbs the radiated amplitude (A_i) and phase (P_i) of the individual elements, and thus the elevation and/or azimuth pattern are modified. Since the up–down coupling is usually higher than side to side, the elevation pattern is affected more.

Historically, the different element patterns from different antenna elements and mutual coupling were ignored, since there was no way to accurately calculate them. The far-field pattern measurement range measures the true elevation pattern, if mounted on the supporting structure, which includes all components that exist in the antenna aperture such as transmission lines and tower platforms. The range to measure the far-field elevation pattern of full-size TV antenna arrays is usually miles long, and the turntable has to be capable of supporting heavy antennas weighing many tons. To maintain a far-field range that is several miles long is expensive; thus, near-field measurement techniques were developed that provided good results. The near-field testing does not require large real estate and is easy to operate; therefore, the far-field tests' ranges have been slowly decommissioned.

Lately, with the availability of RF modeling software and fast computers, a 3D pattern for the entire array, even with a supporting structure, can be calculated. This new technique should be used on antenna arrays with nonuniform supporting structures within the antenna aperture and antennas with heavy coupling between antenna elements. The accuracy of this technique depends on the accuracy of the modeling of the actual antenna and the supporting structure. From the 3D calculation the pattern at any elevation or azimuth cut is available, thus it fully describes the radiation properties of the antenna. The RF modeling technique is becoming more and more popular, and eventually it will be used exclusively to provide the radiation patterns.

Antennas for TV Broadcast

Since every TV station's antenna requirements are different, there is a need for many antenna types. These can be broadly categorized in several mutually exclusive groups:

Based on Power Rating

For a given ERP, the required power rating of the antenna depends on the size of the transmitter and the length and size of the transmission line. There are three large groups based on the power handling.

High power with TPO 30 kW and above.

This power is usually required for stations with high ERPs, omnidirectional coverage or stations with vertically polarized component.

Medium power with TPO 5 kW to 30 kW.

These powers are mainly for directional coverage and high-gain antennas or for low to medium ERP levels. Standby antennas may fall into this category as well.

Low-power TPO below 5 kW.

These power levels are usually for Class A, LPTV or translators. They are side mounted on a tower and typically made out of aluminum to reduce weight, cost of material and also manufacturing.

Based on Mounting Location

Top Mounted, on a Building or a Tower

This is a prime location that provides azimuth patterns not affected by the supporting structure, and usually is a single location on the tower unless a T-bar or a candelabra configuration is used. In the case of a T-bar or candelabra, the azimuth pattern is affected, but not to the same extent as side mounting since the spacing between the antennas is larger. More than one station can occupy the top of the tower when using stacked antennas or broadband antennas. In a stacked configuration, two or more antennas can be placed on top of each other.

Side Mounted, on a Building, Tower or a Pole

Side mounting is usually done on a tower, as shown in Figure 8.7-3.

FIGURE 8.7-3 Example of a side-mounted antenna.

Dielectric, LLC

This is not an ideal location, since the azimuth pattern from the antenna is distorted by the signal scattered from the tower. It is, however, suitable for directional antennas, since the antenna can be located on the tower such that the null of the azimuth pattern points to the tower. This antenna orientation minimizes the impact of the scattered signal on the antenna pattern.

Based on Operating Frequency Range

Antennas operate over a certain frequency range; usually the frequency range of operation is specified based on the VSWR, but the limitation can also be due to intolerable pattern change. How broadband the antenna is depends on the antenna element design and how the antenna elements are fed.

Broadband

Operating in full UHF or full Hi VHF band.

These are usually panel type antennas with branch feed systems. Panel-type antenna elements, coupled with branch feeding of the entire array, usually provide the best broadband performance. An antenna with 2 to 1 frequency range, a VSWR < 1.1:1 and with stable patterns can be obtained. As the frequency increases, the vertical beamwidth decreases, resulting in an increase in gain. This is due to the fixed spacing between the antenna elements becoming relatively greater when measured in wavelength. There is no compensation for this effect.

Medium Bandwidth

Operating in portion of UHF or Hi VHF band.

There could be many antenna types in this group—panel and pylon, for example. Depending on the design, the pattern stability can vary. They usually work over 12- to 60-MHz range.

Single Channels

These are usually pylon or traveling wave-style antenna. The advantage is that they have very low wind load, are very reliable and are usually less expensive because there are very few components and therefore very few RF connections.

Based on Antenna Gain

The elevation gain (G_{EL})is specified at the peak of the elevation pattern with respect to a half-wave dipole and for directional patterns also at the peak of the azimuth pattern (G_{peak})—see equation (10). The elevation gain (G_{EL}) is directly proportional to the electrical length of the antenna in wavelengths. For elliptically polarized antennas, the gain of the horizontal component is always specified, since the FCC has specified that horizontal polarization is used in equation (1) to calculate the power into the antenna.

High Gain

High gain antennas are defined as: UHF antennas with gains $G_{EL} > 25$ (14 dBd) and Hi VHF antennas with gains $G_{EL} > 12$ (10.8 dBd).

This type of antenna allows the broadcaster to use less power and thus have a lower operating cost. However, since the gain is proportional to the antenna length, the broadcasters have to make sure that there is aperture available on the tower to side mount the antenna without interference, or that the overall height of the structure allowed by the FAA is not exceeded if the antenna is mounted on the top of the tower. If the tower is located on the top of a mountain, access has to be examined to make sure there are no restrictions for a long tractor trailer.

Medium Gain

UHF antennas with gains between 12 (10.8 dBd) $< G_{EL} <$ 25 (14 dBd) and Hi VHF antennas with gains between 6 (7.8 dBd) $< G_{EL} <$ 12 (10.8 dBd) can be called medium gain antennas.

Medium gain antennas require more power than high gain antennas to get the necessary ERP but have a broader main beam. Thus the signal level is higher and more stable over larger areas around the tower.

Low Gain

Low gain antennas are UHF antennas with gains $G_{EL} < 12$ (10.8 dBd) and Hi VHF antennas with gains $G_{EL} < 6$ (7.8 dBd).

Low gain antennas require even higher power to reach the required ERP than medium gain. The operating cost is therefore higher; however, this type of antenna provides the strongest and most stable signal over large areas around the tower. These attributes make this antenna an excellent choice for the ATSC 3.0 broadcast standard. These antennas are also the shortest and easiest to transport and install.

Based on Antenna Azimuth Pattern

The azimuth pattern is defined in a plane that is parallel to the ground defined in the spherical coordinate system in a plane $\theta = 0$, for $-90° < \theta < 90°$.

Omnidirectional

An omnidirectional antenna is supposed to radiate equally in all azimuth directions, but this is difficult to achieve on practical antennas. The closeness of the actual antenna pattern to perfectly omnidirectional antenna is defined as circularity. The circularity can be calculated by:

$$Circularity = 10 \times \log\ Circularity = 10 \times \log_{10} \frac{Field\ Pat.Max}{Field\ Pat.Min} \frac{Field\ Pat.Max}{Field\ Pat.Min} \tag{6}$$

The FCC required ERP is fixed on the RMS circle that is calculated from the azimuth pattern gain—see equation (7)—and it follows that the lower the azimuth gain, the closer the circularity is to unity (0 dB). For real antennas with nonunity circularity, the required ERP is then in between the pattern maximum and minimum; thus, the ERP exceeds the nominal on the peak and is lower in the nulls. Equations (8) and (9), respectively, determine the increase of ERP over RMS (ERP_+) and decrease below the RMS (ERP_-) in the azimuth pattern nulls in dBs:

$$RMS = \sqrt{\frac{1}{AZ\ Pattern\ Gain}} RMS = \sqrt{\frac{1}{AZ\ Pattern\ Gain}} \quad (7)$$

$$\begin{aligned} ERP_+ &= 20 \times \log_{10} \frac{Field\ Pat.\ Max}{RMS} ERP_+ \\ &= 20 \times \log_{10} \frac{Field\ Pat.Max}{RMS} \end{aligned} \quad (8)$$

$$\begin{aligned} ERP_- &= 20 \times \log_{10} \frac{Field\ Pat.Min}{RMS} ERP_- \\ &= 20 \times \log_{10} \frac{Field\ Pat.Min}{RMS} \end{aligned} \quad (9)$$

Directional

If the shape of the azimuth pattern is modified to avoid interference, meet protection requirements or prevent sending signal into an area where there is no population and coverage is not required, it becomes directional. Since the protection levels are set with respect to the max ERP, the peak gain is used for these antennas. This ascertains that the signal in the protected areas does not exceed the prescribed level. In a simple pattern calculation, where the patterns for all array elements are the same, the peak gain is given by:

$$G_{peak} = G_{AZ} \times G_{EL} G_{peak} = G_{AZ} \times G_{EL} \quad (10)$$

Based on Antenna Polarization

The polarization defines the orientation of the electromagnetic waves with respect to the earth's surface. For maximum power transfer between transmitting and receiving antennas, their polarization has to be the same.

a. Horizontally Polarized

The antenna is horizontally polarized when the waves generated by the antenna are propagating parallel to the earth surface. This is the FCC standard for TV antennas.

Elliptically Polarized

The antenna is elliptically polarized when the radiated electromagnetic wave has both vertically and horizontally polarized components. The elliptical polarization is created when the amplitudes of the two components are nonzero and the phase between them is between 0 and 180. The power fed into the elliptically polarized antenna splits between horizontal and vertical polarization; therefore, in order to get the required ERP, more power into the antenna is required. The advantage of the elliptical polarization is that it provides higher receiving margin for mobile receivers.

Circularly Polarized

Circular polarization is a special case of the elliptical polarization. When the horizontal and vertical components have the same magnitude and there is a 90-degree phase difference between them, the electromagnetic waves propagate in a circular fashion. When the vertical component lags behind the horizontal, a right-hand polarization is achieved, and that is the standard in the United States. When employing circular polarization, twice as much power is required to feed the antenna to achieve the required ERP.

Variable Polarization

The antennas of this type have two inputs and can provide variable polarization by simply changing the power division between the inputs or by changing the phase of the signal between the two inputs. The phase modification is preferred, since the horizontally and vertically polarized patterns track closer and the power into the antenna can be distributed evenly between the two inputs. This type of antenna can be beneficial to stations that share an antenna and is invariably realized in a panel antenna, each panel having two inputs. In this case the stations do not have to have the same amount of vertical polarization, but one station can be horizontally polarized, another circularly polarized and a third elliptically polarized. The magnitude of the vertical polarization depends on how much transmitter power they have available and, obviously, the overall power rating of the antenna. The increased versatility of this type of antenna does make it suitable for certain applications in the United States—those in which two or more stations share a combined antenna facility and wish to operate with different amounts of vertical polarization. This has to be weighed against the fact that all stations must have the same radiation pattern, two transmission lines are required and two sets of channel combiners are required in the transmitter building.

Antenna Type

Traveling Wave

The traveling wave antenna can be built as coaxial type or a waveguide type. These types of antennas are fed from the bottom. In both cases the antenna radiating elements are longitudinal slots. These slots are all the same and have to be of certain length and width, and in order for a slot to radiate they have to be excited by a coupling bar on one side of the slot. The size and the penetration depth of the coupling bar determine the amount of power coupled into the slot. There can be one to eight longitudinal slots around the pipe circumference to obtain the desired azimuth pattern. See Figure 8.7-4.

The benefit of the traveling wave design is that null fill is inherent to the design. The beam tilt is obtained by adjusting the spacing between the longitudinal slots. Theoretically, this antenna would have to be very long in order to radiate most of the energy that is delivered to it. The length would depend on the percentage of power radiated from each slot (attenuation). The antenna has to be designed with an attenuation that provides a reasonable gain; if the attenuation is too high, the gain is low with high null fill, and, conversely, low attenuation increases the gain and provides lower null fill. Practical attenuation has to be found that provides reasonable antenna length. In any case there will be energy

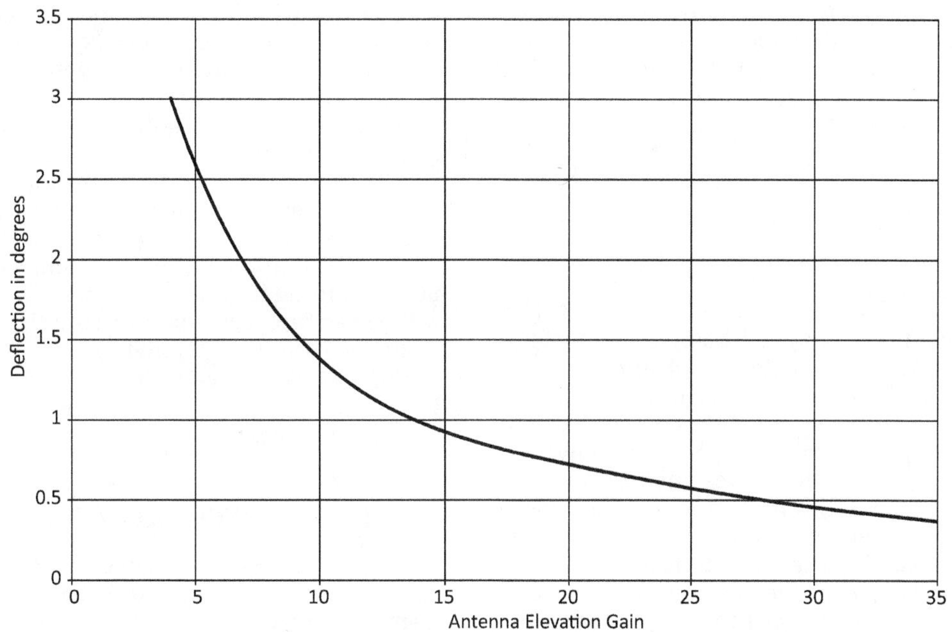

FIGURE 8.7-4 Suggested deflection vs. gain for 10% of the time.

remaining at the end of the antenna that has to be either absorbed or reradiated. Absorbing the energy reduces the antenna gain; reradiating has to be done in phase with the main slots in order not to distort the azimuth pattern and reduce the antenna gain.

The traveling type antenna can be for a single VHF or UHF channel and is mainly top mounted on a tower. This type of antenna can be horizontally or elliptically polarized. Elliptical polarization can be obtained by placing a parasitic element in front of the slot. This parasitic element is fed by the energy radiated from the slot and is part of the antenna design. The antenna cannot be retrofitted later with these dipoles.

Pylon Type

These antennas are coaxial type operating in the TEM mode. The antenna elements are slots and, as for the traveling type antenna, are longitudinal and have to be excited by a coupling bar. There can be one to eight slots around the pipe. Pylon-type antennas operate with standing waves inside the pipe; therefore, the length of each slot and the size of coupling bar can be different, as shown in Figure 8.7-5.

The pylon antenna does not have inherent null fill; thus the antenna slots have to be carefully designed to obtain a correct beam tilt and null fill while maintaining good input impedance. This antenna type is typically used for one or two adjacent UHF channels and can be side or top mounted. As with the traveling wave antenna, this antenna can also be horizontally or elliptically polarized by adding a passive dipole.

Panel Type

The antenna radiating element for this type of antenna is a broadband panel. The panel can be horizontally polarized as shown in Figure 8.7-6 or elliptically polarized.

FIGURE 8.7-5 UHF slot on a pole with a coupling bar and passive dipole in front.

FIGURE 8.7-6 Example of a broadband horizontally polarized panel antenna.

Multiple panels are used as building blocks for an antenna that is supported by a mast or assembled around a tower section.

Each panel can have one or two inputs fed by a feed system comprising power dividers and transmission line sections. The feed system can be internal to the supporting mast or positioned below the mast with long feedlines to the panels. The number of panels is determined by the azimuth pattern, the elevation gain needed and the power rating of the antenna.

Structural Considerations for Designing Antennas

All antennas have to be designed based on the latest structural codes that govern design of antennas on tall structures or buildings. The structural codes prescribe the maximum wind speed and other conditions that the antenna and its supporting structure have to withstand. They also have to be designed to operate dependably in high wind, rain and icing conditions and in seismic areas. The top-mounted antennas sway in windy conditions, and it can affect the antenna performance, especially toward the horizon. To minimize or eliminate the sway of a top mounted antenna or stack can be expensive. However, using the antenna elevation gain, one can specify the antenna deflection that can be tolerated and expensive stiff structures might not be required. The higher the gain, less sway can be tolerated due to the narrow vertical beamwidth. On the high slope of the elevation patterns, a small deflection can cause the signal level to drop many decibels. Conversely, low-gain antennas have wide beam, thus the antenna can be allowed to deflect more. The choice of wind speed at which to calculate the antenna sway is important; a good choice can be wind speed that occurs about 10% of the time at that location (see Figure 8.7-3). There is no need to use maximum wind speed, since it occurs statistically very rarely.

The deflection is calculated as an angle of the tangent at the center of the antenna to the vertical. For stacked antennas, the deflection has to be calculated for the top and bottom antenna separately.

FIGURE 8.7-7 Deltastar (broadband panel) antenna with strakes in preparation for shipment.

Also, tall cylindrical antennas and stacks potentially can sway in low wind speeds. This phenomenon is called vortex shedding and the sway is actually perpendicular to the wind direction [2]. The vortex shedding can overstress the antenna and lead to a structural failure. There are some computer simulations that can predict the vortex shedding on simple structures. For more complex structures, a wind tunnel analysis on a scale model has to be performed. If the simulation determines that the vortex shedding can occur, then it has to be mitigated using passive strakes near the top of the structure, as shown in Figure 8.7-7, or liquid or mechanical damping systems.

These strakes have to be factored into the antenna design since they can affect its mechanical and electrical performance.

ANTENNA REQUIREMENT FOR ATSC 3.0

The new ATSC 3.0 "next generation" broadcast television standard in the United States will give broadcasters many advantages over the existing ATSC 1.0. ATSC 3.0 is based on OFDM (orthogonal frequency division multiplexing) modulation that uses many carriers (8K to 32K) versus the existing ATSC 1.0, which uses a single-carrier 8VSB modulation scheme. The ATSC 3.0 standard will include multiple PLP (physical layer pipes) that can, for example, deliver a high-quality video (4K Ultra HD) signal to mainly stationary receivers and a lower-quality, but more robust, video signal to mobile devices on another PLP layer. Delivering high-quality video requires a higher signal level to deliver the required higher SNR (signal to noise ratio). That will require the antenna to provide a higher signal level within the contour while not exceeding the required 41dBμV/m level at the contour edge, as prescribed by the FCC. Increasing the input power would not work, since the max ERP cannot be exceeded.

There are three methods to increase the signal level in the coverage area; one is to increase the beam tilt, thus point the peak of the main beam to areas that are closer to the tower. This method, however, decreases the signal level at the coverage contour, thus reducing the coverage area. The second method is to decrease the antenna gain; this also would decrease the coverage contour. The third way is to introduce a higher-gain antenna with very heavy null fill. It has been shown that the gain reduction of this third method is much less severe than in the first two methods but results in a large increase of the signal level in the closer coverage area [3].

Another benefit of the OFDM delivery is the opportunity to use a SFN (single-frequency network). The SFN uses transmitting stations of various powers that work on the same frequency, in conjunction with the main transmitter site, to enhance the signal in the locations where the signal level is insufficient. Since all three methods described decrease the signal in the periphery of the coverage area, the SFN can be used to increase signal strength in those affected areas. The SFN antenna systems can be low power or medium power; each can have directional patterns and can be side mounted on the tower. The planning of the SFN has to be done carefully to achieve the required SNR within the contour while not exceeding the 41dBμV/m signal level at the contour.

To summarize, the best antenna for the ATSC 3.0 will have a very heavy null fill, will not reach 41dBμV/m at the contour and thus will need SFN to fully cover the serviced area.

RF System Troubleshooting

RF system troubleshooting usually starts when some unusual indication occurs. The indication of a problem can be higher VSWR registered by the transmitter or other monitoring device, increasing VSWR over time or calls received by the station reporting a loss of picture or a nonviewable picture. The symptoms have to be taken seriously; ignoring them might lead to an issue or issues that are costlier to repair and/or would take much longer to repair, thus longer off-air time. Being off the air or on an auxiliary antenna that is not equivalent in coverage to the main antenna can be a major problem to the station, especially when an important event is going to be broadcasted, the ratings are near or the issue is during the ratings.

VSWR ISSUES

Most of the time the stations use the transmitter-reflected power meters as an indicator of RF system condition. These meters are calibrated to indicate either the actual reflected power in watts or the percentage of forward power reflected. Either way, they provide an indirect method of monitoring VSWR (4% reflected power is equivalent to a system VSWR of 1.5:1). In addition there are several separate monitors available that can indicate VSWR directly, along with other building parameters. Regardless, monitoring reflected power or VSWR over time is a good practice to identify potential issues, but any changes have to be carefully interpreted. VSWR is a scalar number that indicates only magnitude but not the location of the potential issue. One should know that exposure to the elements and environmental conditions like rain and ice can change the antenna VSWR significantly, but these changes can be tolerated since they are transient in nature. This is in contrast with enclosed systems, like transmission line, where even a small change in VSWR can indicate, or become, a large problem if not treated. If there is a problem and the transmitter trips, it is important not to override the transmitter protection circuit, since the longer the power is applied to the system the more parts can get damaged. The transmitter power output should be reduced in consultation with the manufacturer as the root cause is investigated.

There are two distinct initial RF component failures, one due to overheating, thus average power, another due to voltage breakdown/arcing and thus peak power. If no immediate action is taken, it can be difficult to determine the root cause since overheating can easily lead to arcing and vice versa. This is another reason why it is important to investigate immediately after remedial measures have been taken.

Different symptoms mean different things and require different questions to be asked. Good log keeping can be very helpful in identifying the root cause of problems.

High VSWR—Transmitter Trips

This can be caused by issues anywhere from the transmitter up to and including the antenna. Since the reflected power or VSWR measurement does not locate the source of the high VSWR, the next step should be to call a RF technician and sweep the system using a network analyzer with a time-domain option, which should locate the issue within the system. Once located it can be investigated and corrected. In order to zero in on the fault, the line or the component can be opened by disconnecting the line at flanges near the fault location for reference. If the issue is at the antenna end, a similar technique should be employed to determine if the problem is the antenna or the elbow complex. Once the location is pinpointed, the line or the antenna have to be carefully inspected and any damaged parts replaced. The damage usually comes with large amount of soot that is generated by burning the PTFE (polytetrafluoroethylene, Teflon©) insulators. The soot propagates up and down the transmission line; therefore, it has to be inspected and cleaned. No soot should remain in the system for full power operation.

Please note that the fumes generated by burned PTFE are toxic, so any work has to be performed in well-ventilated areas, and one should not inhale the toxic air escaping from a pressurized transmission when it is being disconnected.

If the insulators cannot be cleaned, they should be replaced. Most of the time the PTFE insulators have to be replaced, since any trace of carbon left on the insulators can cause issues later.

After the repair, the system should be checked with the network analyzer to make sure the measurement returned back to normal.

VSWR Is Slowly Increasing

Small increases in reflected power or VSWR below the threshold of a fault condition are often ignored, but if they persist, or change in magnitude and frequency, they should be investigated. There could be a poor connection at a flange or a connector bullet which, if left alone, will likely overheat and burn and could cause significant damage. Such a problem can cause a high VSWR, reflecting power back to the transmitter and resulting in periodic hot spots in the RF system at current nodes.

Antennas are susceptible to environmental conditions. Rain, sleet or icing conditions can increase the VSWR from nominal, e.g., 1.08:1, to 1.2:1 or even higher. Such changes could be normal if the reflected power or VSWR return to baseline values after the environmental condition returns to normal. Again, the importance of long-term monitoring and logging of parameters to identify what is "normal" cannot be overstated. The VSWR can vary with large temperature variations as well, which is due to the thermal expansion and contraction of a long transmission line feeding the antenna. Again, if the VSWR returns to baseline, there should be no issue.

Operating at lower power until the issue is found might be prudent. The system should be measured with a network analyzer and compared to baseline measurements, and if there is any indication of a higher VSWR at a certain location in the transmission line, inspecting that location might be prudent, since a small issue, if not corrected, can escalate into a large issue, potentially damaging long sections of lines and components. The issue in the rigid transmission line usually starts at the flanged connections between line sections or a line insulator. A wideband time-domain sweep of the transmission line will locate the flanges and insulators. Then, by comparing the measurement to the baseline measurement, one can determine where to start inspecting. As noted earlier, the issue might be a bad connector and flange connection or small debris on the insulator.

Catching this issue early can eliminate major damage to the transmission line later.

Intermittent VSWR Trips

Intermittent issues are usually very difficult to find, since after the high VSWR indication the system can go back to normal and low-power network analyzer measurements cannot detect the issue. Operating at lower power levels until the root cause is found might be prudent. The root cause does not have to be in the RF system; therefore, the monitoring system and transmitter issues should be eliminated first. The line and antenna could be eliminated by running the transmitter into a dummy load. Some transmitters may allow operation directly into a load before the filter and combiner if it appears that the problem is inside the building.

It would help if there is a way to correlate the trips to some environmental conditions, like wind, rain, sleet and icing. If it is correlated with moisture, rain and ice, for example, the pressurizing system should be evaluated. It could be an air leak in the system whereby, during the rain, moisture can ingress into the line and cause a transmitter to trip. If the system is airtight, then the unpressurized portions of the RF system should be visually inspected for cracks that might let the moisture into the system. The antenna radomes should be inspected, since the radomes are designed to protect the antenna parts from the environmental conditions.

If related to windy conditions, one should look for loose connections in the RF system or insufficient support for line and/or antenna. Areas that are the most affected by windy conditions are antennas, elbow complexes and transmission lines. One should look for missing or insufficient hangers that would allow a larger-than-normal deflection of the RF components in the wind. Antenna sway over some time can also loosen some important components inside that might cause a VSWR trip.

Looking for the intermittent trips can be difficult and very frustrating, but a methodical approach, coupled with careful observation, will usually identify the source.

COVERAGE ISSUES

Loss of signal or other signal-quality problems in large areas around the antenna site might be due to coverage issues. The troubleshooting of a coverage issue is usually initiated by a large number of calls to the station from consumers or even an MVPD taking an off-air feed, complaining about signal quality. When the calls are received, the station has to ask not only what the issue is but where the receiver is located, since the location helps to determine the cause.

The coverage issue can come from these sources: antenna, multipath or environment. The transmitter can be a cause, but the assumption is that it was checked out and is working properly into the dummy load. The antenna can malfunction and provide lower-than-expected RF signal levels. Multipath is generated by a signal bouncing off man-made structures or mountains and arriving at the receiver delayed relative to the main signal and potentially causing destructive interference, which may vary with time. Environmental issues are related to attenuation of the RF signal when propagating in different weather conditions or different terrains, such as urban, suburban, forest, over water surfaces, etc. To distinguish where the coverage issue is generated, one has to know the antenna design. Knowing the antenna design helps to separate the antenna issues from multipath or environmental issues.

Before one starts the troubleshooting, one should know the location of the coverage issues with respect to the transmitting site, the type of antenna, antenna mounting and orientation.

The location of the callers should be plotted on a map. By looking at the distribution of the calls, one can prioritize the troubleshooting sequence. If the calls are located radially in a line or a narrow arc from the antenna site to the horizon, one should start by looking at the azimuth pattern, since an unexpected null in the azimuth pattern can cause this problem.

If the calls are concentrated around the antenna or in a ring around the antenna, concentration should be on the elevation pattern or input power. In both cases the antenna design should be carefully examined to determine if the observed problem could even be caused by the antenna if something is incorrect with it.

Let's look at some more specific cases.

No Coverage or Poor Coverage

This might be due to insufficient power into the antenna, or maybe an antenna design or assembly issue. The transmitter power-output level should be checked and confirmed. The antenna could cause this if an elevation pattern has an unusually large beam tilt below or high above the horizontal. If the beam tilt is high, high above horizon, the signal toward the main coverage area might be insufficiently low. If the antenna is directional covering, maybe the antenna was mounted pointing in the wrong direction. Knowledge of the antenna type and its design helps to determine if the antenna is even capable of creating a situation that would cause this issue.

No Coverage or Poor Coverage in a Small Arc

This can be due to an antenna issue or antenna-mounting issue. If a directional antenna, the antenna orientation should be checked.

If an antenna is side mounted on the tower, the mounting location should be verified. Most manufacturers can produce an azimuth pattern that includes the effects of the tower.

If the antenna is a top-mounted panel-type antenna, the orientation of the antenna with respect to the arc with an issue is important to know. Improper phasing of the RF signal between antenna faces can lead to signal cancelation on a radial. This can be caused by an improper antenna design or assembly issue.

The Coverage Issue at a Cluster Far From the Site

In this case a loss of signal occurs in a cluster in the middle of the coverage area. The coverage is good closer in and also further from the cluster. The question in this case is can an antenna generate an issue? Placing a typical antenna 1400 feet above ground with the elevation pattern shown in Figure 8.7-1 generates ground level signal strength, as shown in Figure 8.7-4.

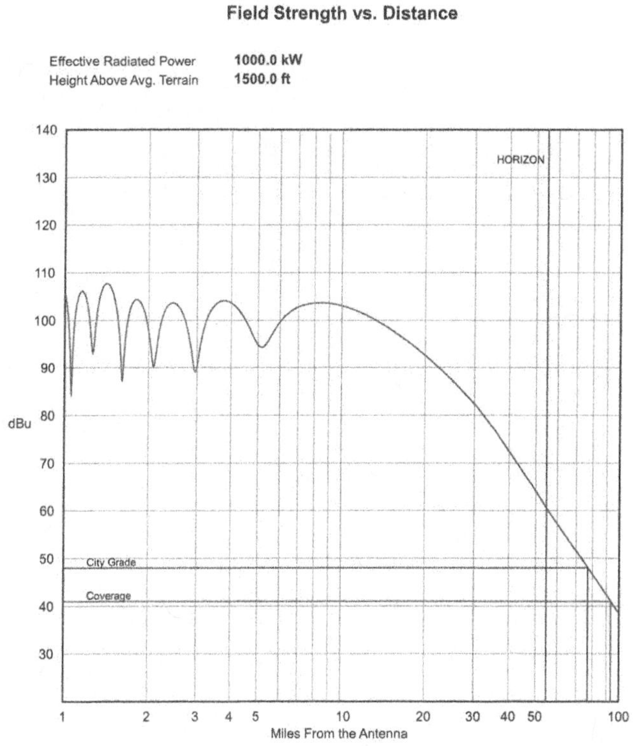

The horizon is at 0.58 degrees depression angle on the elevation pattern, and it is 53 miles away.

The important characteristic to point out is that the main beam of the elevation pattern that is between 0.58 to 2.5 degrees on the elevation pattern corresponds to 53 to 6.1 miles respectively on the signal strength chart (Figure 8.7-4.). So if there is a low signal localized about 8 to 16 miles away, one can work backwards and determine that it would fall between 1.9 and 1.0 degrees, respectively, on the elevation pattern. Can the broadcast-type antenna generate this dip in a signal in one direction between 8 and 16 miles away? Most probably not. This type of issue is usually generated by multipath in that particular region generated by the terrain or man-made structures.

Troubleshooting can be a difficult process where one has to take into the account the cost of the process and what the result will be. Most antenna manufacturers can be very helpful in resolving the issues. Hiring a field RF engineer with lots of experience in troubleshooting is advised.

FIGURE 8.7-8 Signal strength from 1400 ft. vs distance and depression angle.

ACKNOWLEDGMENTS

I would like to thank Andy Whiteside, President of Dielectric, for his many helpful comments and suggestions and invaluable support in writing this chapter. His effort is greatly appreciated.

I would like to thank my daughter Karin for her help.

References

[1] Balanis, *Constantine A: Modern Antenna Handbook*, 2008, John Wiley and Sons, Inc.

[2] Giosan, Ioan, *Vortex Shedding Induced Loads on Free Standing Structures*, 2005, Vancouver, BC, Canada.

[3] Schadler, John, *Broadcast Antenna Design to Support Future Transmission Technologies*, NAB, Las Vegas, NV, 2014.

CHAPTER

8.8

Television Filters, Combiners, and RF Components

DANIEL S. FALLON

Dielectric

This chapter is divided into two sections; the first section allows the reader to identify and understand the function of various components that comprise a passive RF system found in a television transmission facility. Individual components are presented with their basic function and physical description. Each component is characterized in terms of power handling and bandwidth.

The second section examines the processes and procedures of installation, maintenance and management of TV combiner systems at shared tower sites. The material explores the role of broadcast engineers in managing the transmission facilities from the business and relationship aspect as applied to design, construction, operations and maintenance of TV combiner systems, whether a new installation, system upgrade or existing system retrofit.

TELEVISION FILTERS, COMBINERS AND RF COMPONENTS

INTRODUCTION

The first section of the chapter briefly describes what is known at the time of writing about ATSC 3.0 and its impact on RF components. The next section describes various components including hybrids, directional couplers switches and patch panels. The following section is devoted to filters, both lowpass and bandpass. The last three sections deal with the combination of components into power combiners, channel combiners and full RF systems.

ATSC 3.0 Considerations

At the time of this writing, final approval of the A/322 Physical Layer Standard has been granted. The two most significant changes for RF system design from ATSC 1.0 to ATSC 3.0 are the use of orthogonal frequency division multiplexing (OFDM) modulation and an increase in the occupied bandwidth.

The use of OFDM is well known in broadcast; both DVB-T and the ISDB-T modulations use this type of signal. The most important characteristic of OFDM for passive RF systems is the peak-to-average power ratio (PAPR). This is the relation of the power of the short-duration peaks in a signal to its long-term average power. The PAPR of ATSC 1.0 signal is on the order of 6 dB. The PAPR of OFDM signals is on the order of 10 dB. This increase in PAPR means that components need to have additional voltage capacity to reliably pass the ATSC 3.0 signal. For most coaxial components the peak voltage rating is directly related to its DC hi-pot test value. For filters and waveguide components the peak voltage rating is based on conservative design or power testing. The increase in PAPR is more pronounced for combined channel sites where the potential additions of voltages from each channel in the common transmission line and antenna need to be accounted for.

For ATSC 1.0 most existing RF components are limited by average power rather than peak, using a 6-dB PAPR. An increase in the PAPR may reduce a component's power rating due to it being peak power limited instead. This is most likely for coaxial components such as hybrids, switches and filters. Waveguide components will typically have enough peak power headroom to absorb the PAPR increase without a change in power rating. There are a couple of user configurable PAPR reduction schemes that can be used in ATSC 3.0. The downside of using them is a negative impact on the maximum achievable data rate.

The proposed occupied bandwidth of ATSC 3.0 is 5.832 MHz. This is an increase of 452 kHz over the 5.38 MHz of ATSC 1.0. The increased bandwidth impacts the mask filter more than any other component. The ATSC mask requirements are not changing, so the out-of-band emissions must be attenuated by the same amount as ATSC 1.0. The first concern is to assure any legacy mask filters to be used with ATSC 3.0 are wide enough to handle the 5.832 MHz. Most six-pole filters should be wide enough, but eight-pole filters may not be. In either case the existing mask filter should be measured to assure a good match and minimal roll-off and group delay in the 5.832-MHz bandwidth. The other concern is the thermal stability of the mask filter. The increased occupied bandwidth decreases what is known as the guard band of the mask filter—this is the distance between the passband and the stop band of the filter. When a filter heats up with applied power, some amount of passband drift is inevitable. Even thermally compensated or invar constructions can have measurable drift on the order of 50 kHz to 100 kHz. The decreased guard band means that the thermal stability needs to be 150-kHz overall shift or less. Again, for most currently deployed filters this should not be a problem, but they will have to be verified on a case-by-case basis. ATSC 3.0 again provides a user setting that can adjust bandwidth. It is known as the "carrier reduction coefficient" and has five distinct levels that will adjust the bandwidth as low as 5.508 MHz. Obviously, reducing the occupied bandwidth will impact the achievable payload data rate.

The use of dual polarization broadcast is another commonly considered option for ATSC 3.0. Adding power to the vertical polarization of a signal does not count against the station ERP but effectively puts more power at the receiver, increasing the signal-to-noise ratio. The use of vertical polarization is a strategy to overcome both indoor antennas and physically small receive antennas. The impact on RF systems is to increase transmitter total power output (TPO) and the required power rating of the components in the RF system. The possible use of two feed lines, required for variable or adaptive polarization, complicates the required RF system by adding power dividers and more complex switching requirements to the output of the RF system.

RF COMPONENTS

Hybrids

The 90-degree hybrid is probably the most common component used in high-power broadcast RF systems. Figure 8.8-1a illustrates the function of the hybrid when used as a power divider and Figure 8.8-1b illustrates the same component used as a power combiner. In the power division application the applied power is split in half to the two output ports. The two outputs are 90 degrees out of phase. The fourth port is isolated, and very little power arrives there from the input. In the power-combining application, two ports are fed with half the desired power. The two applied signals need to be 90 degrees out of phase. The hybrid will add 90 degrees of additional phase shift to one-half of each signal and no additional phase shift to the other half. The result is a sum at one port where both signals have the same phase and a cancelation at the other port where the phase is opposite. The two inputs are isolated from one another.

The physical appearance of a 90-degree hybrid will vary depending on the band it is used for, the port size, and whether it is coaxial or waveguide. Low-band and high-band VHF exclusively use coaxial hybrids, with the port size depending on the power level required. These are typically in a "cross-over" configuration where the two ports that are physically on the same side of the hybrid are isolated from one another and power applied will be split between the ports on the opposite side. Figure 8.8-2 shows a typical highband VHF hybrid with the ports identified. The length of a coaxial hybrid is on the order of a quarter-wavelength—high-band VHF hybrids will have approximately 15-inch (38-cm) port spacing, and low-band hybrids will be longer depending on channel.

UHF hybrids are a challenge to realize the coaxial cross-over configuration. This is due to the fact that 1/4 wavelength at UHF (4" to 6") is the same as the port size required to handle the applied power (3-1/8", 4-1/16" or 6 1/8"). To overcome this limitation one solution has been to use a noncrossover configuration. Figure 8.8-3 shows a noncrossover UHF hybrid with the ports identified. This configuration allows a single section UHF hybrid to be realized with 3-1/8" ports, 4-1/16" ports or 6-1/8" ports. This style of hybrid is sometimes referred to as a "Y-brid."

Another hybrid used in high-power UHF RF systems is the waveguide hybrid. The most common of these is the "Riblet short slot" hybrid. Figure 8.8-4a shows this hybrid with the ports identified. The size and frequency range of this hybrid will vary with the size of the waveguide used at the ports. WR1800 is the largest and typically used from 470 MHz to 494

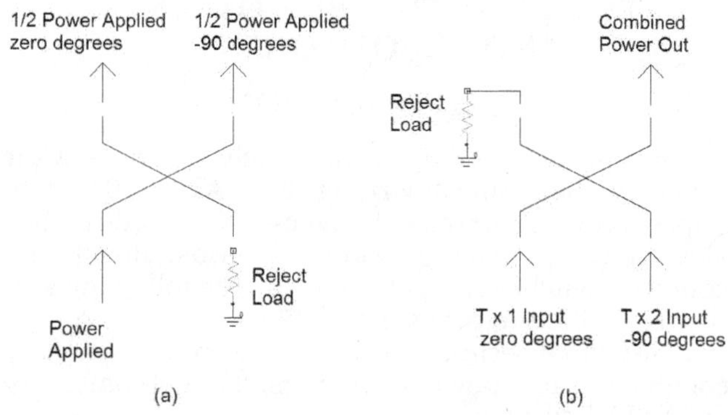

FIGURE 8.8-1 Hybrid Functionality—Combiner and Divider.

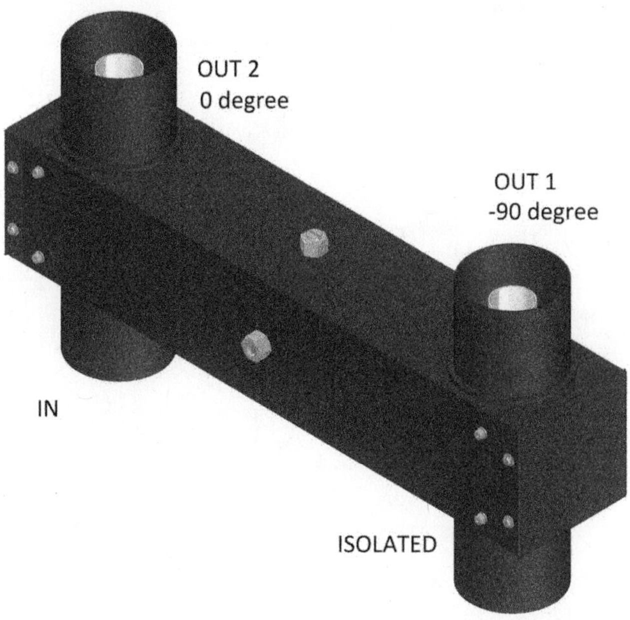

FIGURE 8.8-2 High-Band VHF Coaxial Hybrid.

Courtesy Dielectric, Inc.

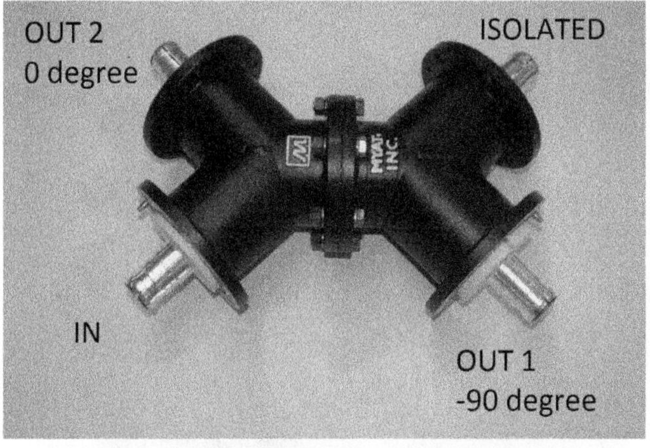

FIGURE 8.8-3 Noncrossover UHF Hybrid.

Courtesy Myat, Inc.

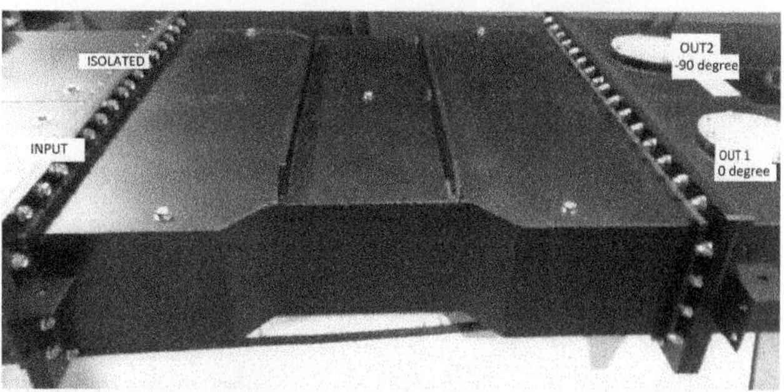

FIGURE 8.8-4A Riblet Waveguide Hybrid.

Courtesy Dielectric, Inc.

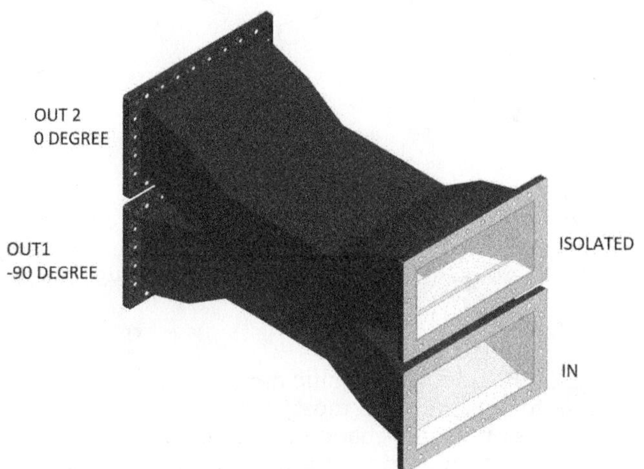

FIGURE 8.8-4B E-Plane Waveguide Hybrid.

Courtesy Dielectric, Inc.

MHz. WR1500 is the most common UHF waveguide size and is typically used from 494 MHz to 630 MHz. WR1150 is the smallest waveguide size and is used above about 630 MHz. The Riblet short slot hybrid can be tuned to very good performance over a bandwidth of up to three UHF channels. For wider band performance, the E-plane hybrid is preferred. This is shown in Figure 8.8-4b. The E-plane is also employed in cases where the port orientation is more convenient than the Riblet hybrid.

The performance figures of merit that apply to all hybrids are the port match, the isolation, the power-split balance and the phase balance. The match (return loss) and isolation are similar and typically on the order of 30 dB to 35 dB for good high-power performance. An isolation of 30 dB for 30 kW applied to an input means that 30 watts will be directed from the input to the isolated port. The power-split balance is the difference, in dB, between the input and the two outputs. A hybrid optimized over a narrow bandwidth (3% or one channel at high-band VHF, or three channels at UHF) will have an unbalance of less than 0.1 dB. A hybrid used for all of high-band VHF (21%) will have an unbalance of about 0.4 dB. A hybrid used for all of the UHF band (60%) bandwidth will have an unbalance on the order of 1 dB. In terms of percentage, an unbalance of 0.1 dB means that 50.6 % of the input power is applied to one output and 49.4% is applied to the other output. An unbalance of 1 dB, on the other hand, means that 56% of the input power is applied to one output port and 44% is applied to the other output. Figure 8.8-5 shows the power balance of a typical UHF hybrid and its variation over the band. The phase balance of almost all symmetric 90-degree hybrids is −90°+/−1°.

The power rating of all coaxial hybrids is less than the line rating for the coaxial line size used at the ports. A general rule of thumb is that the average power rating is 80% of the line rating for the frequency of use and the peak power rating is 70% of the peak power

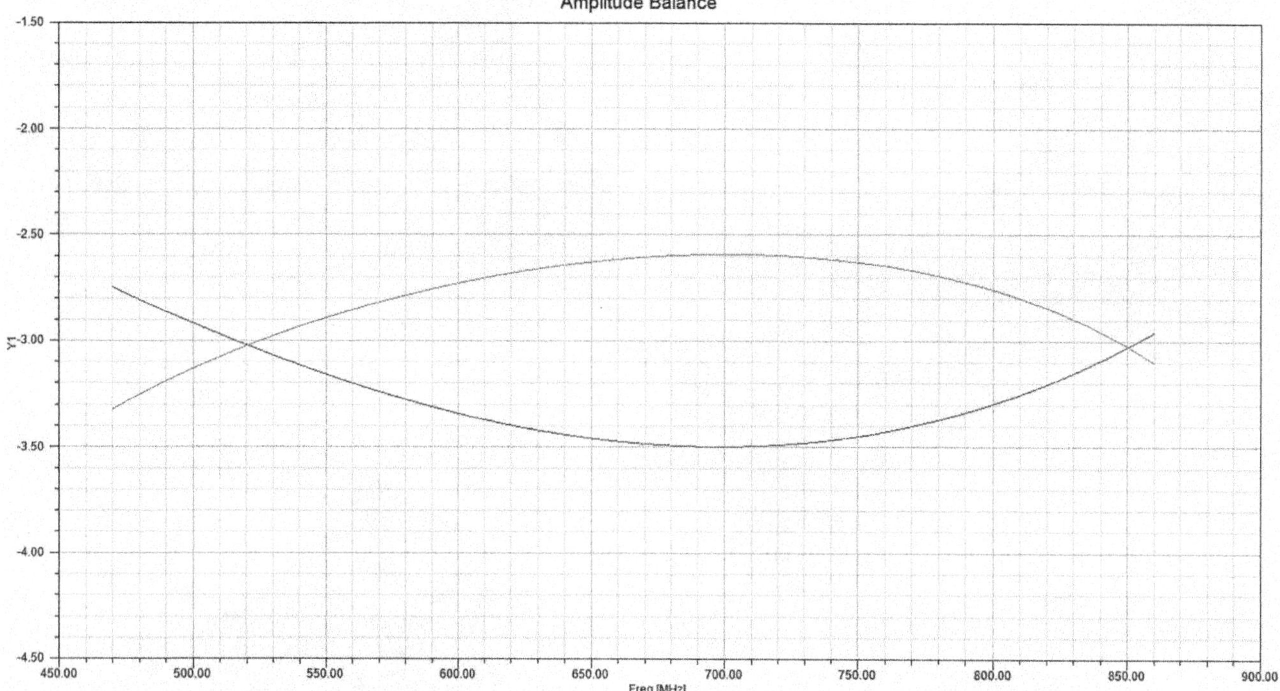

FIGURE 8.8-5 Power Balance of UHF Hybrid.

rating of the line. The specific numbers will vary from design to design. The most important concern for ATSC 3.0 or multiple-channel use is to assure the manufacturer has a high-potential ("hi-pot") test value and the applied RF peak power rating is properly derated from this value.

The average power rating of waveguide hybrids is on the order of a few hundred kW. Typically all aluminum construction is used for waveguide hybrids due to the very low absolute loss values (0.02–0.03 dB). It will be rare that a waveguide hybrid displays thermal hot spots due to power dissipation. Empty waveguide has peak power ratings on the order of several mega-watts. It is more likely that a waveguide hybrid will be peak power limited due to the tuning structures (buttons and posts) used inside. Unfortunately, waveguide structures cannot be hi-pot tested. Proper attention needs to be paid to button spacing and radii to assure good peak power handling.

Another category of 90-degree hybrids used in broadcast power combiners are unequal split hybrids. Chief among these is the –4.77 dB hybrid. This hybrid has a coupled path loss of –4.77 dB (or 33%) from the input to the first output and through line path loss of –1.77 dB (or 66%) to the other output. A common application is to combine power from three transmitter cabinets. These hybrids can be realized with either coaxial or waveguide ports.

The last hybrid device that needs to be mentioned in this section is the 180-degree hybrid. Coaxial implementation of this device is rare due to its large size, but the waveguide form is the well-known magic tee. Figure 8.8-6 shows the magic tee layout and port definitions. The magic tee has what is known as a sum and difference port. When used as a combiner, power applied to the two inputs will appear at the sum port if they are in phase. If the power applied is out of phase it will appear at the difference port. Similar to the Riblet short slot hybrid, the magic tee comes in the three basic waveguide sizes and can be tuned for good performance over approximately a 3% (three UHF channel) bandwidth. There is little need for broader band magic tee designs. Similar to other waveguide components, the magic tee will have large power-handling margin for most applications.

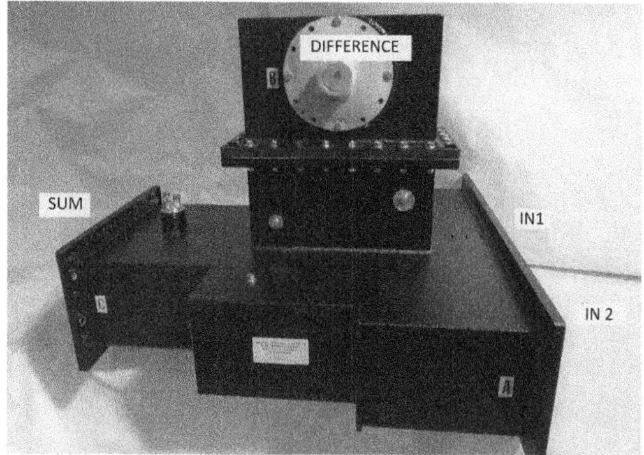

FIGURE 8.8-6 Waveguide Magic Tee.

Courtesy Dielectric, Inc.

Patch Panels

Patch panels are high-power manual switches. Even though there are some waveguide patch panels, they are typically coaxial line devices. The most common sizes are 1-5/8″, 3-1/8″, 4-1/16″ and 6-1/8″ (50 ohm). There are occasional requirements for larger (7-3/16″, 8-3/16″ or 9″—50 ohm) patch panels, and these are manufactured as custom devices. Patch panels are used for switching to a rarely used configuration (e.g., from antenna to test load) or at sites that are easily accessed to perform emergency switching (e.g., from main to alternate transmitter). They are simpler and therefore more economical than motorized switches.

The most common patching configurations are three port, four port, five port and seven port. Figure 8.8-7a shows a three-port patch panel with the function of each port identified. Figure 8.8-7b shows a seven-port patch panel with typical functionality identified. The alternate paths are connected using a U-link that connects the RF path between the coaxial ports. This U-link either consists of two coaxial elbow sections or of a short length of rectangular coax/strip line connecting the two ports.

The size of a patch panel has a great impact on its ease of use. Generally 1-5/8″ and 3-1/8″ patches are easy for one operator to remove and re-engage. A 6-1/8″ patch panel can be both difficult to disengage and heavy depending on its design. The larger patch panels can be designed for ease of use, but the features required for this are more expensive than the simple "sleeve coupling" approach.

The bandwidth of patch panels is generally very good. A VSWR of < 1.04:1 can be maintained from 60 MHz to 860 MHz for most sizes. The power rating of patch panels will vary depending on the attachment mechanism, but 90% to 100% of full line rating at the frequency of use is typical. The peak power rating is usually 100% of the line-size rating, as coaxial patch panel structures will hi-pot test to full line value.

Most patch panels are supplied with micro-switches that will change state if a U-link is moved. These micro-switches should be wired to the interlock circuit of the transmitter(s) feeding the patch panel. The intent is that the micro-switch will be actuated before RF contact is broken so the transmitter will shut off and prevent an electric shock hazard to the personnel moving the patch. This, of course, is a backup measure—the transmitter should be switched off before moving the patch. It is important to understand if the micro-switches are normally open (no) or normally closed (nc) and assure that they will function properly with the transmitter.

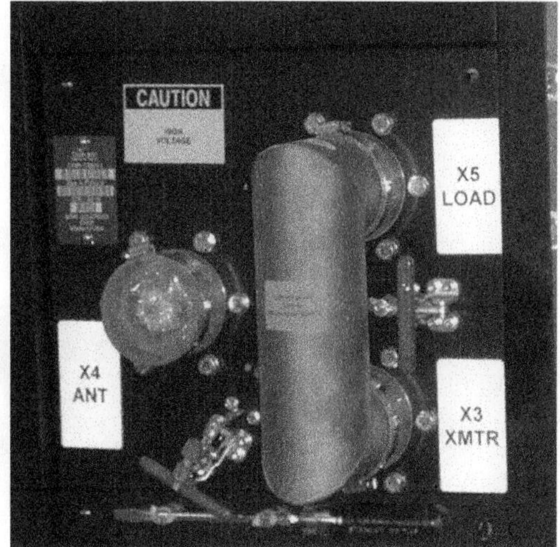

FIGURE 8.8-7A Three-Port Patch Panel.

Courtesy Dielectric, Inc.

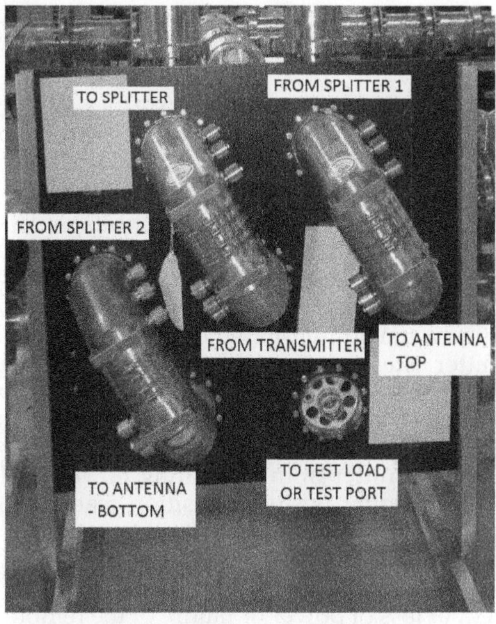

FIGURE 8.8-7B Seven-Port Patch Panel.

Courtesy Dielectric, Inc.

Switches

Motorized switches include both coaxial and waveguide configurations. They are typically used for automated operation from a remote location. This may be either for emergency switching or for an inaccessible transmitter site. Both coaxial and waveguide switches are used, and the configuration for both is double pole, double throw, also commonly known as a transfer switch. Figure 8.8-8 shows the internal paths for the transfer switch. When the switch is moved to the next position, all of the internal pieces (except coaxial inners) rotate 90°. For applications that only require a three-port switch, one of the ports of a four-port switch is capped off, creating a single-pole double-throw switch. Two or three switches can be used in cascade to accomplish more complex switching solutions (e.g., combiner bypass or main/alt transmitters with main/alt antennas).

Almost all switches use a motor-driven linear or rotary actuator to move the mechanical contacts that are in the RF portion of the switch. The motor will run

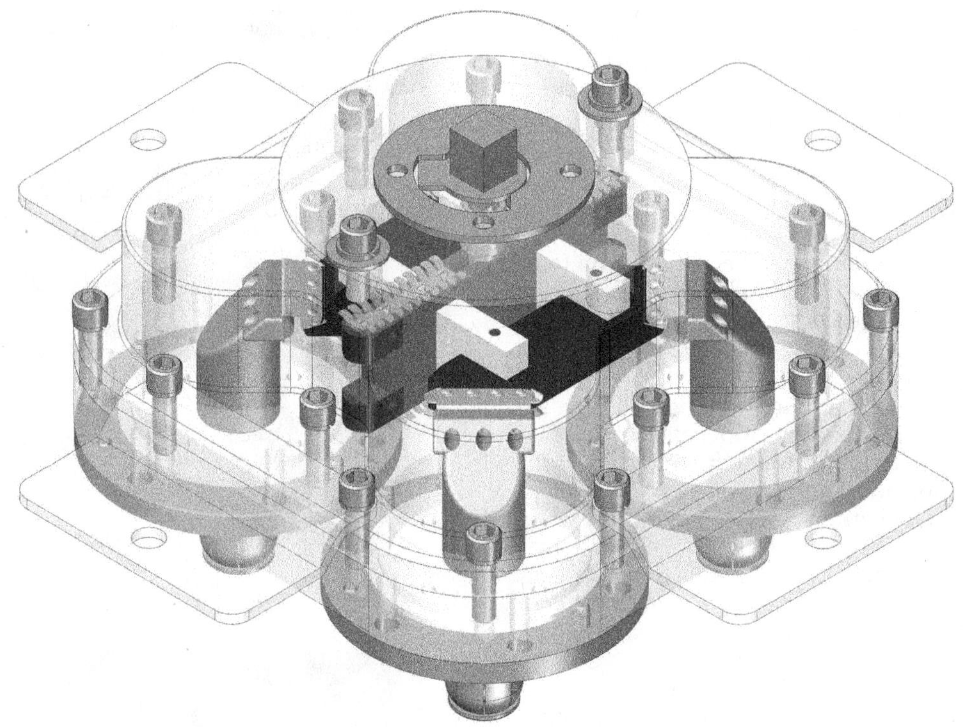

FIGURE 8.8-8 Transfer-Switch Internal Connections.

Courtesy Dielectric, Inc.

on either 115-volt AC or 230-volt AC voltage. The control or actuation voltages that give the switch the command to move position can be 115 volt/230 volt AC or 12 volt or 24 volt DC. Typically, switches can be wired into either a transmitter-based control system or a full transmitter site remote control system. The switching time is from 1 second to 6 seconds depending on switch type. There are interlock switches that should be wired to the transmitters' interlock circuit to prevent the transmitter from generating power while the switch is changing position. Position read-back indicators are also typical, so the remote system can have confirmation of the switch position. Most switches also have a manual actuation mechanism to serve in the event of loss of power or failure of the remote actuation system. The details of a switch's remote control wiring will vary from manufacturer to manufacturer. The common concern for all switch installations is providing AC power to the switch and cabling from the switch location to the remote control interface.

The physical layout of coaxial switches is typically one of two configurations. One configuration is all ports are on the same face, as shown in Figure 8.8-8. The other configuration is all ports are at right angles to each other, as shown in Figure 8.8-9.

The average power handling of coaxial switches varies between 70% and 90% of line power (at the frequency of use) depending on the manufacturer. The peak voltage handling of a switch is generally 70% that of coaxial transmission line. Again, this will vary from manufacturer to manufacturer. Typically coaxial switches are very broadband, covering all broadcast frequencies (54 MHz–860 MHz) with very good match (< 1.05:1 VSWR) and low loss. Loss is dependent on switch size.

FIGURE 8.8-9 Coaxial Transfer Switch.

Myat, Inc.

Waveguide switches are UHF only. They come in the three major UHF waveguide sizes (WR1800, WR1500 and WR1150). There are two basic types of switches depending on the port orientation. They are E-plane and H-plane. Figures 8.8-10a and 8.8-10b show each type of waveguide switch. The bandwidth of waveguide switches will vary from manufacturer to manufacturer. It is possible to make a single switch that has a good match over a wide bandwidth (e.g., WR1500 from 494 MHz to 680 MHz). The average power handling of waveguide switches is specified between 200 kW and 300 kW depending on manufacturer. The peak power range is between 80 and 120 MW. Due to the high power handling, waveguide switches are a natural fit for multichannel high-power sites.

Directional Couplers

Directional couplers have three common uses in broadcast RF systems. The first is to provide an accurate monitor of output power level so the station's effective radiated power (ERP) can be monitored for compliance with its FCC license. The second common use is to monitor reverse power for protection for the transmitter, line and RF system. There are several VSWR monitoring systems that will provide interlock capability to switch a transmitter into either foldback (reduced power) or to power down in the case of elevated system VSWR. The third typical use of directional couplers it to provide a sample of the output spectrum to be used by adaptive precorrection algorithms. This sample is generally fed back into the transmitters' exciter to adjust the applied power and assure the best signal is broadcast.

FIGURE 8.8-10A H-Plane Waveguide Switch.

Courtesy Dielectric, Inc.

FIGURE 8.8-10B E-Plane Waveguide Switch.

Courtesy Electronics Research, Inc.

Directional couplers are mounted on any coaxial or waveguide size. They do not significantly affect the main line match or insertion loss, they simply sample a small amount of power. The coupled line ports are typically type N or SMA. A detector (to convert power to a voltage value) is mounted on the sample port, and cabling is used to feed the signal back to a location where it is used. One thing to consider when positioning a coupler is the ease of attaching a cable or detector and the ease of reading the coupled port values of the coupler.

The couplers typically used in broadcast facilities have coupling values that vary between −40 dB and −60 dB. The power can then be sampled using a simple low-power detector. As an example, a 60-kW forward power will produce a 190-mW sample with a −55-dB coupler. Similarly a 200-watt reverse power will produce a 20-mW sample using a −40-dB coupler. These low coupling values are easily achieved using what is known as a vestigial loop coupler. These are electrically short (20–40 degrees electrical length) coupling loops that are positioned a suitable distance away from the inner conductor (or outside the broad-wall of a waveguide).

Vestigial loop couplers are electrically short, so they have greater coupling values at higher frequencies and lower coupling values at lower frequencies. The coupling slope is on the order of 6 dB/octave (i.e., −50 dB at 88 MHz and −44 dB at 176 MHz). When reusing a coupler at a new frequency it is best to have a precise measurement of its coupling value. There are some couplers available that are slope compensated. This means they have a coupling value that is constant (within a small margin) over frequency. This is done by adding a filtering circuit and attenuator to the sampled port.

The directivity of a coupler determines the accuracy of the power reading taken from the coupler. The directivity is a measure of the ability of the coupler to separate the forward power on a line from the reverse power on a line. For example, if a line has 20-kW forward power and the reflected power is measured at 50 watts, the measured return loss is −26 dB. If the directional coupler used to measure the reverse power has a directivity of 40 dB, then the actual return loss is between −24.5 dB and −28 dB. If the directivity is 30 dB, then the actual return loss is between −22 dB and −32.5 dB. If the directivity is as poor as −26 dB, then the actual return loss could be anywhere between −20 dB and −50 dB. Good directivity is important, and typically broadcast couplers are specified at 30 dB or 35 dB minimum.

Vestigial loop couplers are typically adjustable. The coupling is increased by moving the coupler body into the main transmission line, and the coupling is reduced by moving it away. The directivity is optimized by rotating the coupler body relative to the direction of propagation in the main line. It is important to point out that directivity should not be set with a coupler in a broadcast system. The directivity is highly dependent on the main line load when it is set. For this reason, if a coupler needs adjustment it should be pulled out of a system and set using precision coaxial line or waveguide adaptors.

Usually, loosely coupled broadcast couplers do not affect the power handling of the main coaxial or waveguide transmission line. A small hole in the outer conductor (or waveguide broad-wall) is used to couple power to the coupling loop. This small hole has no practical impact on power handling.

Phase Shifters

Phase shifters are a basic building block for what is known as a switchless combiner. A switchless combiner allows power from two or more transmitters be added together with good isolation between them. The advantage of a switchless combiner over a simple hybrid or switched combiner is that the switchless combiner can change position while both transmitters are on. This is possible because the phase shifter can change the phase added to a signal with low reflection back to the transmitter while it is moving.

Phase shifters can be constructed in any coaxial line size or in waveguide (for UHF only). Even though the construction and phase shift mechanism is different between coaxial and waveguide phase shifters, they are both typically driven by a linear actuator. When the control signal is applied, the linear actuator will move the phase shift mechanism until it reaches a preset mechanical micro-switch. The micro-switch position is factory preset for the required phase shift at the channel of operation. The linear actuator drive is typically 110 volt or 220 volt AC, and the control signals required will vary from manufacturer to manufacturer.

A coaxial phase shifter is usually of "trombone" construction. Figure 8.8-11 shows a cross section of a coaxial phase shifter. The "trombone" inner conductor is supported inside a rectangular outer conductor by insulators. Both the inner conductor and insulators move to create the phase shift. The inner conductor for the rectangular portion will slide inside of the coaxial line inner conductor. The sliding RF contact is accomplished by spring fingers between the inners. With proper design this type of phase shifter will have good VSWR at any phase shift position and can be quite broadband. The overall phase shift available is limited by length. The power handling is limited to 70–80% of line rating due to the smaller inner and slightly smaller height required in the rectangular outer.

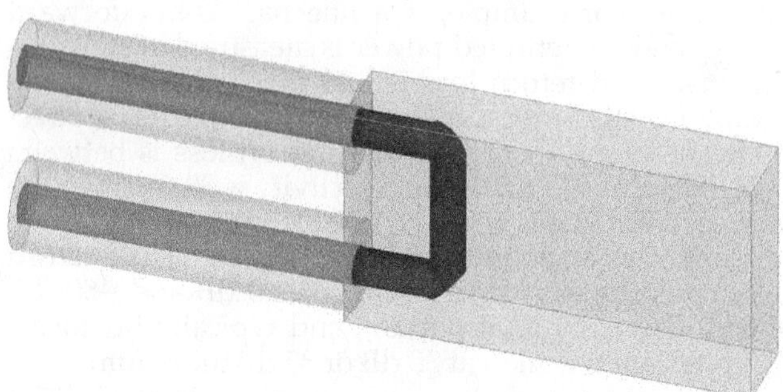

FIGURE 8.8-11 Trombone Coaxial Phase Shifter.

A waveguide phase shifter consists of an empty waveguide tube with a dielectric slab in it. The slab is stepped to create good match at the frequency of use and is typically Teflon, but other plastics can be used. The slab can move laterally across the broad dimension of the waveguide to create more phase shift. When the slab is against the narrow wall of the waveguide, the phase shift is at minimum. Moving the slab to the center of the guide provides maximum phase shift. With low-loss dielectric material used for the dielectric slab, excellent power handling can be realized. Typically the match of the slab is narrowband—a new slab would be required to move more than one channel.

Coaxial/Waveguide Tees

Coaxial and waveguide power dividers are commonly used to split power feeding panel antennas. For antennas with dual-feed transmission lines to an upper half and lower half there are commonly power dividers in the transmitter room fed by switches or patch panels. These power dividers split the power to each antenna half, and the switches/patch panels allow full power to one-half of the antenna for redundancy. Coaxial power dividers are used in panel antennas to split power to the antenna elements, both radially and in elevation.

Unlike hybrids, or the Gysel power divider, tee power dividers do not have isolation between the outputs. This means that power reflected from one output port will appear at both the input and the other output ports. Since they are generally used in antenna feed networks, this typically is not a problem.

All power dividers used for panel antennas have impedance transformers built in to assure a good input match over the required bandwidth. In contrast, unmatched tees are three port devices that do not have transformers. The unmatched tee is used in branch and manifold channel combiners. Waveguide tees are used for high-power UHF combiners; coaxial tees are used for VHF or low-power UHF combiners.

When used as a power divider, coaxial tees can be very broadband. The number of transformer sections determines the match and bandwidth. Often the power divider will have a coaxial reducer built in. For example, a 6-1/8" EIA input can be split to two 3-1/8" outputs. The power split can be two ways, three ways, four ways or more. The average power handling is generally limited by the larger coaxial size. Hi-pot voltage and therefore reliable peak power handling will be limited by the smaller coaxial size.

Waveguide power dividers can also be very broadband, up to 80% of the guide bandwidth. Again, the bandwidth and match is a function of the number of transformer sections, and therefore the better the bandwidth the larger the splitter. Waveguide splitters are going to have exceptional peak and average power handling. Generally the power limit in a waveguide splitting system will be the coaxial lines they are attached to.

Loads

Loads are used in broadcast RF systems either as a test load or to terminate an unused port. When used to test a transmitter they are referred to as test loads. In this case they must be rated for full RF power. When used to terminate an unused port they are referred to as "ballast" loads, as they help keep the system balanced. Ballast loads are used on hybrid power combiners or in constant impedance filters (CIF). In these uses they must be rated for the failure of one component—e.g., 1/4 output power for a hybrid combiner.

Before the widespread use of solid-state power detectors, water loads were commonly used to determine the output power of a transmitter. By a calorimetric calculation, the flow rate and temperature rise of a coolant can be used to determine the power dissipated in the coolant. Using solid-state power detectors, either water loads or oil loads can be used as test loads. Water loads are still preferred for high-power systems. The transmitter is run into the test load to perform the FCC proof and for transmitter maintenance and troubleshooting.

The smallest loads used in broadcast RF systems are type N or SMA terminations used on directional couplers. These are typically rated at 1/2 watt or 1 watt. In some circumstances a large fin-cooled termination can be used that is rated at 5 or 10 watts. These types of loads are known as dry loads—they are simply resistive elements in a case. Another type of dry load, used as a reject or ballast load for low-power CIF systems, are conduction-cooled dry loads. These are resistive elements mounted on a heat spreader that must be attached to a large heat sink. These are typically rated from 10 to 50 watts.

Other types of loads that can be used from 5 kW to 40 kW are forced air-cooled loads. These are composed of power dividers, resistive elements and fans. These allow high-power installations to install a test load without the complications of plumbing it to a heat exchanger or oversizing the heat exchanger to accommodate the dissipated power. The only downside of these loads are their relatively large size and, if they are to be used for long periods of time, the transmitter room HVAC must be sized to carry away the additional heat load.

Oil loads are frequently used for reject and ballast loads in channel combiners and filter systems. These are typically round rod resistive elements that are embedded in synthetic heat transfer oil. The heat dissipated in the resistive element establishes a convection current in the oil that carries the heat away to the walls of the load case. The case of the load is enclosed in fins to enhance convection to the air. For higher power ratings, fans are used to force air through the fins. Oil load ratings vary from 100 watts to 10 kW. In all cases, oil loads must be located some distance from the coaxial port they are attached to, so either type N cables, 1-5/8" coaxial line or 3-1/8" coaxial line must be used to plumb to the load. The load will come with the coaxial port size appropriate to its power rating.

Due to the aforementioned calorimetric power calibration, water loads have a well-established place in broadcast installations. Water loads make sense in installations that use water-cooled transmitters because the heat exchanger is already present. The cost of a water load is significantly less than a forced-air load of equivalent power rating. The power rating of a water load is a function of flow rate and temperature rise of the coolant. Temperature rises can be as high as 30° C, so with sufficient flow rates, power ratings in excess of 120 kW can be achieved.

There are two basic types of water loads—those that use a round rod resistive element and those that use the coolant as the resistive element. The round rod resistors have a thick film of resistive material printed on the rod. The rod is bathed in moving coolant, which carries the heat away. Care must be exercised when specifying these loads with digital modulation. The load must be rated not only for the average power but also the peak power expected from the digital modulation. The advantage of these loads is that they can generally be made shorter than loads that use coolant as the resistive element. To use the coolant as a resistive element, an impedance transformer must be included that adapts the 50-ohm input impedance to the relatively low (~10 ohm) impedance of the coolant-filled coaxial section. These loads do not have the same limitation on peak power that the resistive loads do. Their VSWR is more sensitive to coolant mix ratio and coolant temperature. They are typically optimized to a 50/50 mix of water and ethylene glycol.

The term "water" load is a slight misnomer. Most installations have closed-loop heat exchanger systems where the coolant is a 50/50 mixture of ethylene glycol and water. It is becoming more common to substitute propylene glycol for ethylene glycol for environmental reasons. This has little impact on the performance of a water load. The only factor to consider here is that the increased viscosity of propylene glycol will have reduced flow rates for the same pumping power. Open-loop systems are those in which water is applied to a load and then disposed of using a drain. For loads with low flow requirements and temporary use, this setup is economical. The caveat with open-loop systems is that most loads that use the coolant as the resistive element are optimized with a mix of ethylene glycol and water. When run with water, only the VSWR of the load will increase. The use of an open-loop water-cooled system is best with a load with a resistive rod element inside.

For very high-power (60 kW+) UHF installations, waveguide water loads are available. These loads use the coolant as a resistive element. The coolant is contained in dielectric tubes that are positioned in the waveguide to create a good broadband match.

The match of UHF water loads is generally about 1.1:1 over all of UHF (470 MHz to 860 MHz). They are typically optimized for use at a single channel by the use of a fine matcher or a coaxial or waveguide tuning slug.

FILTERS

Filters used in most broadcast installations are lowpass or bandpass filters. High pass filters are used occasionally in channel combiners and, with the exception of older analog diplexers, band stop filters are not used for high-power broadcast applications. Almost all low-power and high-power television transmitters must employ filtering to meet the FCC emissions requirements. These filters include lowpass filters to attenuate the harmonic emissions from transmitter and bandpass filters to attenuate the third- and fifth-order intermodulation products generated in the power amplifiers of the transmitter.

Lowpass Filters

Lowpass filters are also known as harmonic filters. They attenuate the harmonic power generated in the output power amplifier of a transmitter. They can be designed with very wide passbands and in most cases spurious free stopbands to the third harmonic or more. A coaxial bandpass filter can be used for all VHF transmitters and UHF transmitters with a transmitter power output (TPO) less than approximately 30 kW. For larger UHF transmitters, waveguide lowpass filters are available.

A lowpass filter is characterized by a passband with a good match and low loss and a stop band with high attenuation. Figure 8.8-12 shows the frequency response of a typical lowpass filter. For both highband VHF and UHF, the passband can be full band. The distance between passband and stop band (of a given attenuation, typically 30 to 40 dB) is controlled by the number of sections in the filter. The more sections a lowpass filter has, the steeper its attenuation and the closer the passband and stop band can be. Typical coaxial lowpass filters used for broadcast transmitters have between nine and 15 sections.

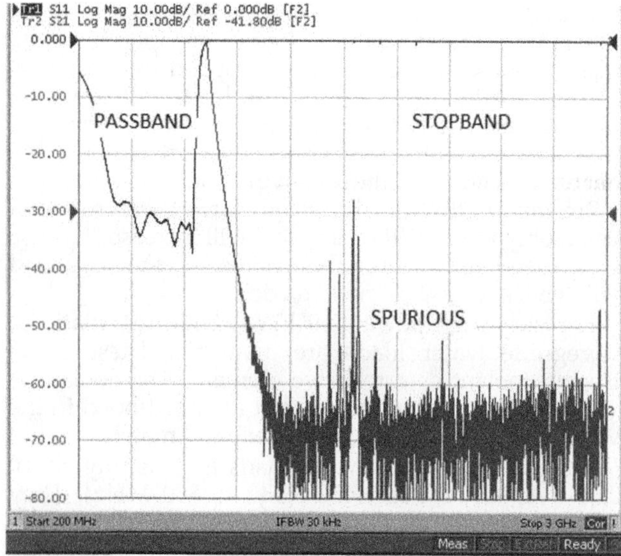

FIGURE 8.8-12 Lowpass Filter Frequency Response.

Figure 8.8-13 shows part of the inner conductor of a coaxial lowpass filter. It is known as a "stepped impedance" design. The inner conductor has large diameter (low-impedance) and small diameter (high-impedance) steps. The low-impedance sections correspond to the shunt capacitors of a lumped element lowpass filter. The high impedance sections correspond to the series inductors.

Since the passband is so wide, lowpass filters typically have very low insertion loss. In spite of this, the power rating of coaxial lowpass filters is typically only 60%–75% of line rating. Both fins and forced air cooling can increase the power handling, but a lowpass filter inner conductor will run hotter than a transmission line inner conductor due to a higher power-loss density. The peak power handling of a coaxial lowpass filter is determined by the gap between inner conductor and the outer conductor. This breakdown voltage of this gap can be measured with a hi-pot test. Good design practices will assure a gap adequate for the high peak to average power ratios resulting from OFDM modulation.

The high frequency spurious response of a lowpass filter will limit the stop-bandwidth of a lowpass filter. The spurious response is driven by the length of the inductive sections and the size of the outer in a coaxial lowpass filter. A VHF filter can have a spurious free response to the fifth harmonic due to the relatively low power required and resulting small line sizes. For UHF filters, the spurious responses are above to the third harmonic. Luckily the second and third harmonic outputs of a high-power amplifier are the strongest, with lower power output at the higher harmonics.

Coaxial lowpass filters are practical for UHF in only the smaller line sizes (1-5/8", 3-1/8" and 4-1/16"). For higher power levels (> 25–30 kW), waveguide lowpass filters are recommended. With forced air cooling, waveguide lowpass filters can be used up to 70 kW. Figure 8.8-14 is a photo of a waffle iron filter. The waffle iron is the most common type of waveguide lowpass; its name comes from the center portion of

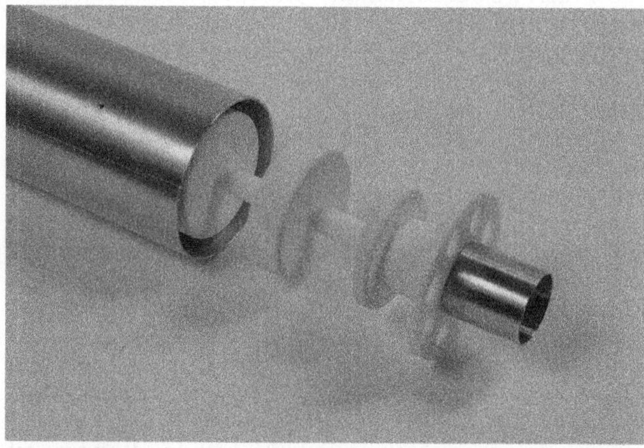

FIGURE 8.8-13 Coaxial Lowpass Filter Inner Conductor.

Courtesy Dielectric, Inc.

FIGURE 8.8-14 Waffle Iron Filter.

Courtesy Dielectric, Inc.

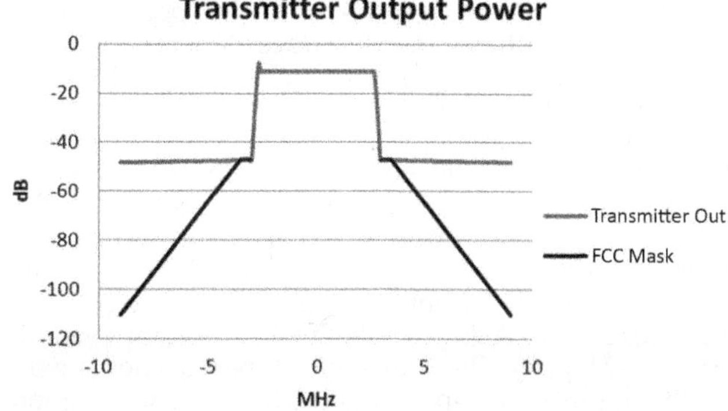

FIGURE 8.8-15 Transmitter Output Spectrum and FCC Mask.

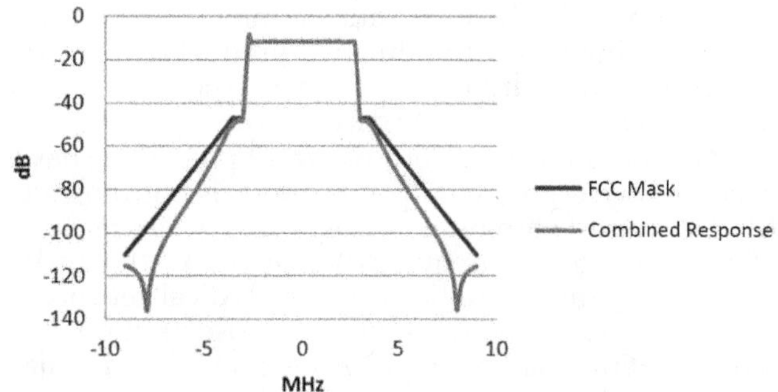

FIGURE 8.8-16 Transmitter Output Spectrum and FCC Mask.

the filter, which has a series of teeth similar to a waffle iron. A design known as the corrugated waveguide lowpass is also common; it lacks the waffle teeth but has steps in both the broad-wall and sidewall. Both types can have wide passbands and spurious free stop bands to the third harmonic.

It is important to understand that coaxial lowpass filters provide rejection for transverse electromagnetic (TEM) propagation only. This is the fundamental mode of propagation in coaxial line and is how the line is intended to be used. Higher order or waveguide modes can propagate at higher frequencies in coaxial line. The cutoff frequency for higher-order mode propagation depends on line size—it is 3.43 GHz for 1-5/8″ line, 1.73 GHz for 3-1/8″ line and 875 MHz for 6-1/8″ line. If harmonic power is present in a line at a frequency above this cutoff it may propagate in a higher order mode. In this case the lowpass filter may not attenuate the harmonic power properly. Coaxial elbows are the most common way of transferring power from TEM propagation to waveguide mode propagation at frequencies above cutoff. It is best to locate a harmonic filter as close to the transmitter as possible and avoid coaxial elbows between the HPA output and the lowpass filter input.

Similarly, waveguide lowpass filters are designed to operate in the TE11 fundamental waveguide mode. At higher frequencies several different waveguide modes can propagate. The lowpass filter will not properly attenuate power propagating in higher-order modes. Again, the component most likely to transfer power from the fundamental to the higher-order mode is a waveguide miter. Positioning a lowpass filter in an RF system prior to waveguide miters will help properly attenuate harmonic power.

In cases where it is not practical to place the harmonic filter directly at the HPA output, the approach of using two harmonic filters has been used. There are several RF systems in existence where there is both a coaxial harmonic filter on a tube output and a waveguide filter in the RF system. The chances are that the spurious response and higher-order mode response of the two filter types will not overlap—providing good composite harmonic rejection.

There is one more consideration in placement of harmonic filters that is relevant to tube transmitters. The harmonic power is reflected back to a tube. The phase of the reflection determines whether or not the power constructively remixes in the tube (generating more harmonic power). The distance between the harmonic input and the tube output needs to be correct to minimize harmonic and spurious generation.

Bandpass Filters

Bandpass filters are found with almost all television transmitters. This is because high-power amplifiers generate unwanted noise outside the channel of operation. This noise is generated by intermodulation products in the transmitter. For this reason another name for the filters paired with a transmitter is the "intermod filter." Bandpass filters are also building blocks for channel combiners, which allow more than one channel to share transmission line and antennas.

Figure 8.8-15 shows the frequency response of a typical transmitter and the FCC spectral mask. Figure 8.8-16 shows the superposition of a six-pole intermodulation filter with the transmitter output. Clearly the transmitter alone does not meet the spectral mask, and the transmitter with filter does. The mask shown is the FCC full-service mask. The FCC also has two additional masks that apply to low-power broadcasters; these are the "simple" and "stringent" masks. Both the full-service and stringent masks generally need six-pole filters for most transmitters to meet the requirements. The simple mask usually only needs a four-pole filter. There are masks known as "critical" and "noncritical" for both the DVB-T standard (used in Europe/Africa/Asia) and the ISDB-T standard (South America/Japan). The critical mask requirements generally require eight-pole filters (to maintain a good match) and the noncritical masks require six-pole filters.

Another use of eight-pole filters is in adjacent-channel combiners. An eight-pole filter has enough rejection in an adjacent channel to provide excellent isolation when used as part of a balanced filter module. Eight-pole filters were widely deployed during the U.S. DTV transition to mitigate intermodulation products caused by tube transmitters driven to saturation. The term "sharp tuned" was widely used to describe these filters.

There are a few occasions where 12-pole filters have been deployed on the output of a television transmitter. The most common is on channel 14, where there may be use of the frequencies just below the channel by land mobile radio or older medical telemetry equipment. Higher-order filters have also been used to reduce interference into wireless telephone bands adjacent to a broadcast channel.

There are two classifications of filters that broadcasters should be familiar with. The first is a "reflective" filter and the second is a "balanced" or "constant impedance" filter (CIF). The reflective filter is a two-port device with an input and an output. In simple terms it has a passband and a stop band. Power applied to the input in the passband has a good match and will appear at the output with low loss. Power applied in the stop band sees a large mismatch at the filter input port and consequently is reflected back to the transmitter. A reflective filter is simple, relatively low cost and takes up the least amount of space. Reflective filters can generally be used as mask filters with solid-state transmitters. Most solid-state amplifiers can handle reflected intermodulation power from the channel sidebands without any issue.

A constant impedance filter consists of two bandpass filters, two hybrids and two loads. Figure 8.8-17 show the schematic of a constant impedance filter. Even though this configuration is larger and more expensive than a reflective filter, it has practical application. The first is for very high-power transmitters. It is in most cases impractical or very expensive to build a reflective filter for very high-power transmitters. The input hybrid in the constant impedance filter splits the input power in half and applies half to each filter in the module. This means a CIF for a transmitter of 60 kW can be built with two filters of rating 30 kW. A tube transmitter typically needs a constant impedance filter to function properly. Intermodulation power reflected back into a tube tends to remix in the tube and generate even more out-of-band power. The reject load in a CIF performs the function of absorbing the intermodulation products so they are not reflected back to the transmitter. The input port of the CIF has a good wideband match. This is the reason it is called "constant impedance." The ballast load will absorb off-channel power that comes down the transmission line from the antenna. For this reason a CIF is a good choice for a wideband antenna deployed at a site with several other antennas. If the ballast load is removed, then the port on the output hybrid it was attached to becomes the "wideband" port. This port can be used to combine power from one or several more transmitters with the power from the CIF input or "narrowband" port. The CIF can easily be used as a channel combiner.

For broadcast use, the most significant factor of a bandpass filter is its insertion loss. The insertion loss determines a filter's efficiency; for example, what percentage of input power makes it to the output port? Except for very low-power filters (600 watts or less), filter efficiencies are 90% or better. For very high-power filters, loss must be minimized. The reason is two-fold. The first is the simple economics of converting AC power to RF power with an efficiency between 40% and 60%—once a transmitter creates the RF power, wasting the power in a low-efficiency filter does not make sense. The second is the fact that the power lost in a filter is converted to heat. If the heat is not efficiently removed from a filter, then the filter will heat up, and if it is not made from thermally stable material, this heating can impact the filter's frequency response. Most broadcast filters are designed to operate with some self-heating, and they are designed to maintain specified performance, with the case temperature resulting from heat dissipation at their rated input power. The third impact of low-efficiency filters is the added load to a transmitter room's HVAC system or the required external heat exchanger load, if the filter is liquid cooled.

The following equation determines the loss of a filter based on the electrical quantities that describe it. [1]

$$L = \frac{4.343 f_0}{\Delta f\, Q_u} \sum_{i=1}^{n} g_i$$

Where L is the mid-band insertion loss in dB, f_0 is the channel center frequency, Δf is the filter bandwidth, n is the number of sections, and Q_u is the unloaded Q of the cavity. The sum of the g_i values depends on the match of the filter and the number of sections. A few generalizations can be made by studying this equation; the sum of the g values will increase with the number of sections, so filters with eight or 12 sections will be less efficient than a six-section filter. The loss is inversely proportional to filter bandwidth, so a filter for 8 MHz will have less loss than one for 6 MHz. The loss is directly proportional to the center frequency. Lowband VHF filters generally have very little loss, highband VHF can have a moderate amount of loss, and at UHF the type and size of the resonator determines the loss. The unloaded Q of the cavity depends on the resonator type and size. Cylindrical waveguide cavities have the best Q_u, rectangular waveguide has the next highest, and coaxial cavities generally have the lowest Q_u for

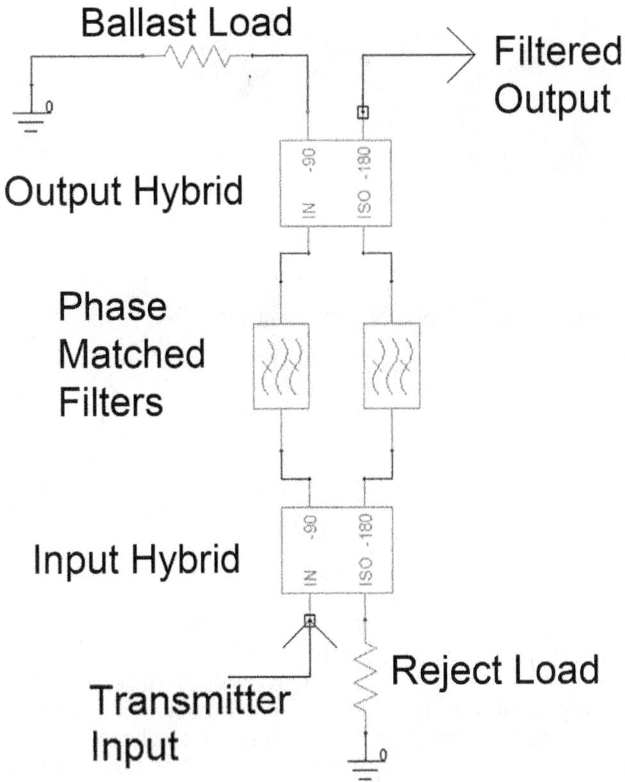

FIGURE 8.8-17 Constant Impedance Filter Schematic.

the types of resonator used in broadcast filters. The size of a coaxial cavity resonator determines its Q_u.

Distributed lowband VHF filters are large simply due to the long wavelength. As the loss equation shows, the loss will be small due to the frequency. The percent bandwidth (Δf / f_0) is large; however, this means that the best type of filter to address lowband requirements is the interdigital. The interdigital is a configuration where adjacent resonators are grounded to the case at opposite ends. This arrangement supports large-percent bandwidth filters well. Since the filters are large and the TPO requirements for lowband are relatively low, these filters normally run cool. Free convection cooling is adequate in all cases. There is a large wavelength difference between channel 2 and channel 6—for this reason, filters are generally tunable between channels 2/3 only and channels 4/5/6 only.

Highband VHF filters are typically full band tunable (channels 7–13). The style of filter of choice for this band is typically the comb-line filter. The comb-line is a rectangular box with several round resonators inside; the resonators are spaced along one axis and are grounded to the case at the same end. Unlike lowband, the TPO required for highband VHF can range from a couple of kW to tens of kW. The filter size will change depending on the TPO to keep the thermal rise to a minimum. Occasionally highband VHF requirements will be met with an interdigital filter—this style spreads the heat dissipated in the resonators more evenly and so is suited to higher power filters.

The UHF band presents several more options and challenges compared to VHF. The TPOs required for UHF are generally larger, making cavity Q_u more critical. The wavelength is reduced—making waveguide resonators practical. The relative bandwidth of the full band (470 MHz to 860 MHz) is larger, making band tunable filters more difficult. There are a larger variety of filter designs available to address these specific challenges.

There are full band tunable coaxial cavity filters available for low power levels (< 10 kW), and when liquid cooled they are available to 20 kW. The size of these filters determines the power rating. There are band tunable filters available for low-power television transmitters down to 50 watts and 100 watts and then up to 20 kW for full power television transmitters.

Rectangular waveguide filters are the next step up from large coaxial resonators in terms of Q_0. This style of filter has been widely deployed in channel combiners. The disadvantage of rectangular waveguide is that it is difficult to implement cross-coupling and thermal stability in them, making them more expensive to use as mask filters. There has recently been introduced a tunable rectangular waveguide filter with low loss and cross-coupling built in [2]. This filter is acceptable for use as a 30-kW free convection and has a higher power rating with forced convection and water cooling.

By far the workhorse of the UHF high-power mask filter technology is the dual-mode cylindrical waveguide (DMCW). This topology has the advantage of very high Q_u and relatively short length. Figure 8.8-18 shows a CIF built with two DMCW eight-pole filters. Note that there are two filters and each has four physical filter cavities. The "dual mode" means that a single filter cavity contains two filter sections. This is an advantage over rectangular waveguide filters since thermal stability is accomplished by the use of Invar material. Invar is a steel nickel alloy with a very low coefficient of thermal expansion. Invar is an expensive material. Since a DMCW filter contains two cavities per resonator, it can be built with less Invar material than an equivalent rectangular waveguide filter.

An important characteristic of bandpass filters, particularly at UHF, is thermal stability. Coaxial cavity filters are typically made with an Invar rod in the resonator. The resonator will be hotter than the filter housing and will grow in length more than the case. This will tend to move the resonant frequency down. A thermally compensated resonator will have an end cap inside the filter that is flexibly attached to the rest of the resonator (using spring fingers or a bellows connection). The end cap is attached to an Invar rod, and the Invar rod is attached to a point outside the filter housing. This assembly holds the end cap stable under heating, and consequently the frequency is thermally stable. Waveguide filter cavities are typically made with Invar walls. These are thermally stable to < 2 kHz/°C. For practical purposes, all thermally stable filters should be held to a total shift of 50 kHz—80 kHz from the room temperature tuning state to the full-power operating temperature.

FIGURE 8.8-18 Dual-Mode Cylindrical Waveguide Eight-Pole Constant Impedance Filter.

Courtesy Dielectric, Inc.

Bandpass Filter Tuning

The uncertainty caused by the 600-MHz spectrum auction and its associated channel repack has generated a desire among both low-power and high-power broadcasters to be able to change the channel of their transmission system. In response to this desire is the development and deployment of band tunable filters for power levels ranging from 100 watts to 60 kW by several manufacturers.

Unlike transmitters and exciters, which can change channel with the touch of a few buttons, high-power filters are fundamentally mechanical devices with several moving parts. To retune a six-pole ATSC full mask filter, six resonators and eight coupling adjustments need to be precisely set. The setting of a resonator will be affected by the setting of its adjacent couplings, so often several iterations are required to get the adjustments correct. The bottom line is that changing a mask filter to a new channel can be done in less than an hour by skilled technicians who perform the work on a regular basis. For those with no prior experience, it is an exercise in patience, and the task will likely take more than a day.

The options available to retune an existing band tunable filter are:

- hire a field engineer to come to your site and retune the filter
- send the filter back to the manufacturer for retuning
- purchase a new filter tuned to the correct channel
- attempt to perform the task yourself or have it done by someone in your station group

The field engineer is certainly the quickest and most convenient. The unfortunate fact is that there is a limited supply of qualified talent. If a few hundred stations want to change channel on the same night and there are 20 field engineers capable of doing the job, then this is not a viable approach. Sending the filter back to the manufacturer for retuning is practical for the smaller filters. This can be done if there is a backup transmitter or if the filter is part of a two (or more) cabinet system. Factory technicians can complete the job in less than an hour, so the labor cost is small. The cost will primarily be freight. Again, for lower-power filters, the cost for a new one is relatively low. If the timing and logistics allow, a station group can purchase a filter tuned to the correct channel and then have the old one removed. The old one could then be retuned, either by a field engineer, at the factory or by someone at the station group, and then redeployed to another site.

For the stations or station groups that have the time and personnel to dedicate to filter tuning, here is a brief synopsis of the procedure. The equipment required to retune a filter will include a vector network analyzer, associated cables and calibration kit, and test transitions from type N to appropriate line size. Some filters require special tools and wrenches as well—a complete list of tools required should be available from the manufacturer. Tuning procedures should also be available from the manufacturer. The filter should be removed from the rest of the transmission system to a location where all adjustments are easily accessible and the network analyzer can be attached and viewed during the tuning process.

All tuning procedures begin with setting the input coupling and first resonance, then move into the filter to the second resonance and coupling between first and second resonators, and then proceed to the next resonators in. Typically the very center of the filter is the last "presetting" operation required. After all resonators and couplings are set to their values, there is typically some additional fine tuning required. The transmission response at the channel is usually pretty easy to achieve; a proper equiripple passband return loss takes much more work and patience to tune in. The last step is to lock down all adjustments so the filter can be moved back to the required location and not detune. It is best to use the memory function on the network analyzer to store the tuned response and then, one by one, lock down all adjustments. Resonators in band tunable filters are typically very sensitive, and the simple process of locking down the nuts that hold them secure will cause the resonance to move. The locking down procedure typically requires overcompensating the position of the resonator and using the torque on the nut to bring the filter back into tune. Filters supplied with fine matching screws make this process much easier because the fine matchers are less sensitive than the primary resonators.

Agilent published an application note in 2005 [3] that describes a time domain method for filter tuning. (Agilent test and measurement is now known as Key-sight Technologies.) This method requires a vector network analyzer with a time domain option. The advantage of the method is that each response peak and null in the time domain has a correlation with a physical adjustment in the filter. If a reference filter response is available for comparison, it is easy to see what adjustments need to be made. This approach is recommended for those completely new to filter tuning because it makes the fine tuning process much more intuitive.

POWER COMBINERS

Power combiners are used to combine the power of two or more transmitter cabinets and feed the sum of the power to an antenna. There are two primary types in use, hybrid combiners and magic tee combiners.

The hybrid combiner is shown in Figure 8.8-1. A 3-dB hybrid is used to combine two cabinets at half-power each, and a 3-dB hybrid in cascade with a 4.77-dB hybrid is used to combine three cabinets at 1/3 power each. Hybrid combiners are typically used with solid-state transmitters. The failure mode in solid-state transmitters is typically one (of several) amplifier module at a time. It is rare that a whole cabinet will fail. The resulting unbalanced power feeding the hybrid combiner will cause a percentage of the remaining power to be shunted to the reject load. Since only one or two modules would typically be down at the same time, this is not a large consequence.

Figure 8.8-19 is a schematic of a magic tee combiner. This combiner is typically implemented in waveguide and used with tube transmitters. Unlike solid-state modules, the failure of a tube will cause a substantial reduction in power. In order to not waste any more power in a reject load, the magic tee combiner allows full efficiency with remaining power. It also allows switching modes while both transmitters are under power; for this reason, another name for the combiner is a "switchless combiner." Table 8.8-1 [4] lists the operating modes and the phase shifter positions. A waveguide phase shifter is in the zero phase position when the dielectric drive rod is all the way out. This means the dielectric slab is against the narrow wall of the waveguide. If the drive rods are in the guide, the phase shifter is in a nonzero position. For combining three cabinets, there is a cascade configuration where the first switchless combiner acts as just described and the second has phase shifters with three phase positions. This allows any combination of single, dual or all three tubes to be on air at once.

CHANNEL COMBINERS

Channel combiners are used to combine the power from multiple transmitters, each on its own channel, onto a common transmission line. These are used to either feed a single broadband antenna or to feed two separate antennas using a channel splitter in the tower. The number of channels can vary from two to up to 10 or more. The total power range can vary from a few hundred watts for type N cabling to a few hundred kW for waveguide combiners. There are two broad classes of channel combiners—junction combiners and directional combiners.

Junction combiners are the simplest type of combiner, involving the fewest parts. They consist of a bandpass filter for each channel and either a single tee combining all of the channels or a series of tees combining them sequentially. A junction combiner using a single tee is also known as a branch combiner or a star-point combiner. A two-channel branch combiner

TABLE 8.8-1
Switchless Combiner Operating Modes and Phase Shifter Positions

MODE	PHASE 1	PHASE 2
A+B to Antenna	0°	0°
A to Antenna, B to Load	–90°	0°
A to Load, B to Antenna	0°	–90°

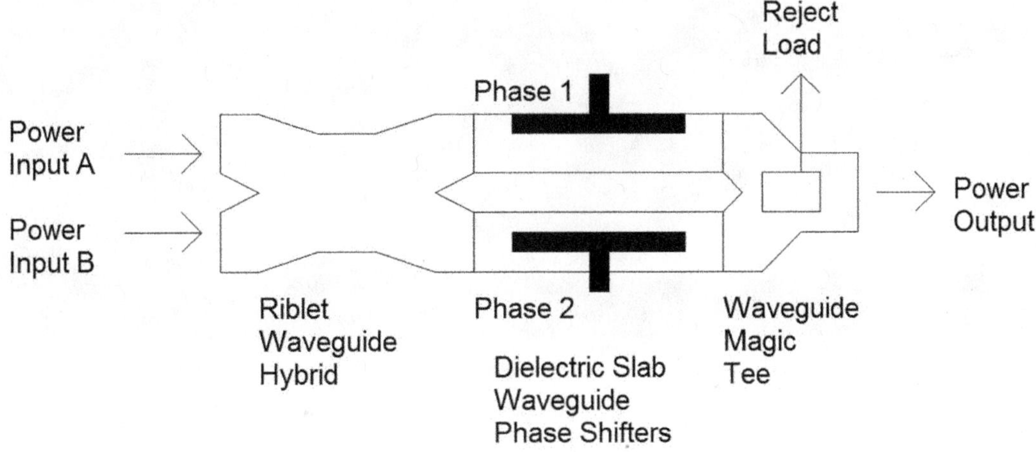

FIGURE 8.8-19 Magic Tee Combiner.

is shown in Figure 8.8-20. This is for two UHF channels with a combined power less than 5 kW. Star-point combiners for more than two channels are possible but generally not practical since manufacturing a multiport junction in small quantities is not economical. Three port junctions can be cascaded with line lengths between them to form what is known as a manifold combiner. Figure 8.8-21 shows a waveguide manifold combiner under construction.

Both branch combiners and manifold combiners involve the least amount of hardware when compared to directional combiners. This means they cost less and take up less space than the directional alternative. The disadvantage is that they are harder to add channels to after they are installed. If a multichannel site is going to have a fixed set of channels, this is not a problem. If the multichannel site will start with two or three known channels but would like to add or change

FIGURE 8.8-20 Two-Channel Coaxial Branch Combiner.

Courtesy Dielectric, Inc.

FIGURE 8.8-21 Waveguide Manifold Combiner Under Construction.

Courtesy Dielectric, Inc.

channels in the future, then a directional combiner may be a better choice. There is also a power limitation on branch combiners—all of the power for each station has to go through a single filter, and the transmission line on the output side (the "spine") needs to handle the combined power of all channels. Very high-power stations (> 60 kW) may require a directional combiner to spread the power between two filters.

The number of sections required for each filter in a directional combiner depends on the channel spacing. A spacing of two or three channels will require a six-pole filter for each channel. A spacing of four or more channels will require four-pole filters to achieve the required isolation between transmitters (typically > 35 dB). Adjacent channels are not practical with a junction combiner.

There are sites that incorporate the mask filtering function into the channel combiner. The combiner filters are then six- or eight-pole filters. Note that these sites require a method of switching each channel to a test load to run the transmitter proof of performance. If mask filtering is not required, then the combiner filter bandwidth is made as wide as possible to minimize insertion loss and group delay variation over each channel. Nonmasking combiner filters do not need to be thermally stable since the slight shift in frequency with the heating due to power or room ambient fluctuations will not significantly impact performance. For this reason, a rectangular waveguide with aluminum construction is a common way to realize UHF manifold combiners.

There are two primary implementations for directional combiners used in broadcast. The first is the constant impedance combiner (CIF) and the second is the directional filter combiner. Figure 8.8-17 shows the schematic for a constant impedance filter. If the ballast load is removed, then there is an open port on the output hybrid—this is known as the wideband port. The wideband port is isolated from the input on the narrowband hybrid, and the filter response protects the reject load from power applied. If power is applied to the wideband port in the stop band of the two bandpass filters, it is split by the output hybrid and reflected by each of the bandpass filters. Due to the phase characteristics of the hybrid, the reflected power then sums at the output port of the CIF. The output port now has the sum of the power applied at the narrowband input (on the input hybrid) and the broadband input. Figure 8.8-18 shows an eight-pole CIF implemented with dual-mode cylindrical waveguide filters and waveguide hybrids. This unit is excellent for combining two high-power adjacent channels. It can also combine channels with wider channel separation—the limiting factor being the bandwidth of the output hybrid.

Figure 8.8-22 shows a picture of a directional filter combiner. The function is similar to a CIF, but the physical layout is smaller. This design is implemented in waveguide only, using cylindrical waveguide resonators. Similar to the CIF, the directional filter has a "narrowband" end, located at the top in Figure 8.8-22, where the power for each channel is applied. It also has a "wideband" end, located at the bottom in the figure, where power from upstream channels is applied at the wideband port and combined with the narrowband power. Similar to a CIF, directional filter modules can be cascaded in a chain to combine several channels. Unlike a CIF, the directional filter is not constrained by the bandwidth of a waveguide hybrid but rather the wideband match of the downstream modules. Similar to the CIF, the directional filter is amenable to future channel additions. Higher-order filtering functions (e.g., eight-pole with dual cross-couplings) are difficult to implement in directional filters. They are well suited to three- and four-pole filtering functions, which are appropriate for channel spacing of three or more.

A generalization can be made that CIF combiners are better suited to ceiling mounting because they are low profile and directional filters are better for floor mounting because they are narrow but tall.

A set of general guidelines for selecting channel combiners follows:

- A two-channel combiner is most economical with a branch combiner—coaxial or waveguide, depending on power levels.
- If the channels are adjacent, a CIF with eight-pole filters is recommended.
- For several channels with no expectation of adding future channels, the manifold combiner is most economical.
- If there is a plan to add future unknown channels, then a CIF or directional filter chain will allow easy addition in the future. Choose the approach based on the space available.

FIGURE 8.8-22 Directional Filter Combiner.

Courtesy Dielectric, Inc.

COMPLETE RF SYSTEMS

Between a transmitter output and the wall feed through to the pressurized transmission line, there are several functions that are performed by high-power RF components. These functions typically are power combining, filtering, power monitoring and switching. For multichannel sites, channel combining and power splitting for dual-feed antennas are also required. Since all of these components are provided by the same transmission vendor, they are typically integrated into an RF system.

Figure 8.8-23 shows a typical two-tube system. This includes two inputs for tube transmitters, a magic tee combiner with associated reject load, a waffle iron filter, a constant impedance intermod filter, an output directional coupler and a waveguide switch with a water load. There is a fine matcher at the output that will feed the transmission line to the antenna. Figure 8.8-24 shows a coaxial RF system, which would be paired with a single-cabinet transmitter operating at 20 kW or below. The components contained in this system are a cylindrical waveguide bandpass filter, a patch panel to switch between the antenna and test load, the test load with a fine matcher, a directional coupler and a coaxial switch to allow switching between the main and aux antenna. There are fine matchers on each antenna line.

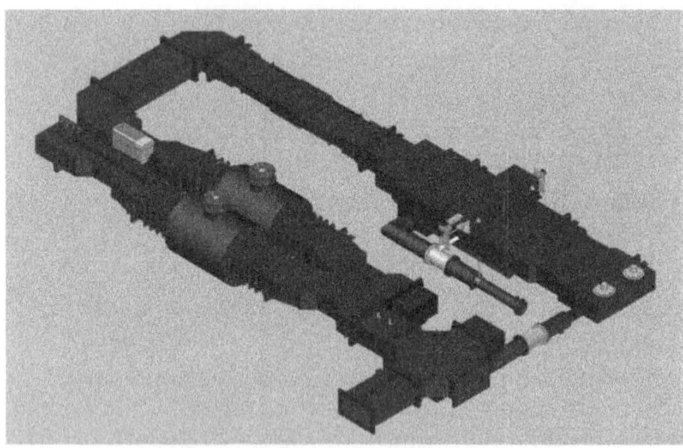

FIGURE 8.8-23 Two-Tube RF System.

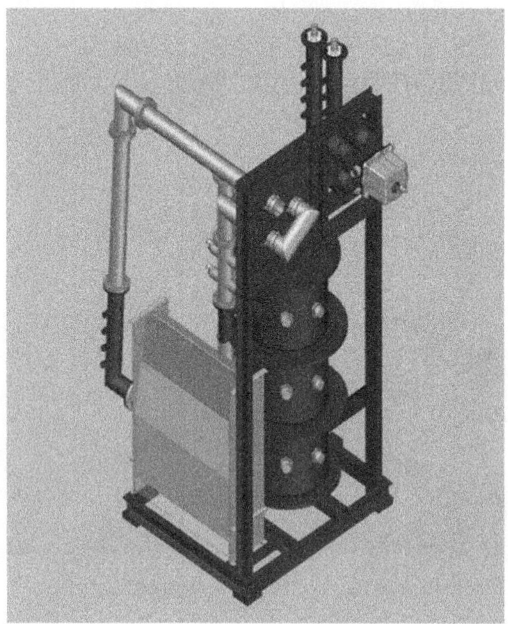

FIGURE 8.8-24 Coaxial I/O RF System.

Courtesy Dielectric, Inc.

The advantage of integration is that the RF performance can be optimized to the channel of use. As an example, a constant impedance filter, RF switch and combining hybrid are all specified to a 1.06:1 VSWR. When they are combined together, the net VSWR could conceivably be as bad as 1.18:1. When the components are integrated together in an RF system, any phase summations of the components' match can be corrected for, so the RF full system can have a composite VSWR of 1.06:1 or better over the channel of use.

The fine matchers provided with RF systems are intended to be used to optimize the match of the RF system when it is attached to the transmission line and antenna. All of the tuners are typically fully out when the system is shipped from the factory. Once all of the transmission components are in place, the installer can sweep the entire RF system and optimize the channel match into the transmitter. The fine matcher at the RF system output along with fine matchers or tunable elbows in the transmission line and antenna feed can be used to optimize the match of the system without having to open any flanged joints in the installation.

References

[1] Ian Hunter, *Theory and Design of Microwave Filters*, IEE Electromagnetic Wave Series, No. 48, Institution of Engineering and Technology (March 31, 2001), MPG Books Limited, Bodmin, Cornwall, UK.

[2] Derek J. Small, *Efficient UHF Tunable Waveguide TE10 Mode Filter*, NAB Engineering Conference Proceedings, April 2016.

[3] Agilent Application Note AN 1287–8, *Simplified Filter Tuning Using Time Domain*, Santa Clara, CA.

[4] Scott B. Durgan, *Ghost Cancellation in TV Switchless Combiner Systems*, Passive Power Products Technical Publication, 27 October 1997.

CHAPTER

8.9

Broadcast Television Co-Location Site Management

SHANE O'DONOGHUE
Empire State Building

PEGGY MILES
InterVox Communications

INTRODUCTION AND OVERVIEW

Tower Sharing Using Combiners—Multiple TV Stations

Shared tower sites offer broadcasters significant cost savings by sharing a facility and a reduction in personnel (labor expenses), location expenses (leasing) and shared equipment expenses. The broadcasting benefit for sharing tower space using a combiner system is to utilize the best broadcast location (for audience coverage) available in the stations' market.

Stations benefit by the joint effort among broadcast engineers through collaboration on site selection, position on the tower, technical design parameters and cost savings. Cost savings extend to sharing of installation, retrofit, leasing and management costs.

Tower Sharing Considerations; Cost Savings and Coverage

The location for a tower is chosen based on its signal attributes that result in the best quality and coverage for a particular station. Co-locating and sharing a combiner offer cost savings, but a technically superior location may prove to be the best value for a TV station in the long run. Tower site location is based first on the technical benefits offered by a superior geographical position and greater height above average terrain (HAAT). Then co-location and infrastructure sharing are considered for the business economics, provided the technical requirements are satisfied.

If co-location is the best option for a station, the tower owner leases (licenses) space on a combined transmission system and houses the combining system: the combiner equipment, monitoring and system controls, transmission lines and the transmitter space. The location chosen provides the best broadcast coverage to meet the FCC-licensed television market area (TMA) while minimizing interference from adjacent markets.

In large TV markets, optimal tower space is often hard to find, and many stations commonly cooperate and co-locate at one tower facility (ex: The Empire State Building in New York, NY, and Willis Tower in Chicago, IL) for many technical reasons, with the ultimate goal to attain the best audience coverage by location and the advantage of greatest height above average terrain (HAAT), while best managing adjacent-channel interference (Empire State Building Broadcasting—Empire State Realty Trust 2016) (Willis Tower 2016). (See examples of shared tower space in Figures 8.9-1 and 8.9-2.)

The economics of the tower site lease are considered alongside the technical requirements to determine if a station should join with other stations in a multiplexed combiner system and shared tower space. The shared knowledge from the TV station engineers is gained by all TV stations in the shared tower facility. This shared knowledge is not available in a tower facility with a single station antenna. Operations, maintenance and troubleshooting knowledge is shared in addition to the more quantifiable shared labor, maintenance and parts replacement costs. The TV station engineers and the tower owner will typically collaborate on writing and updating an operations and maintenance procedures manual that will be reviewed periodically to assure

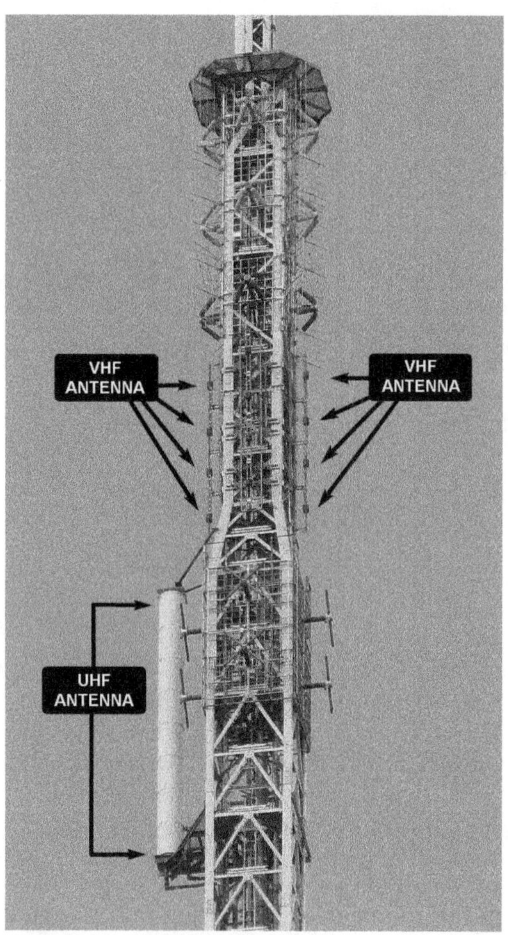

FIGURE 8.9-1 In New York, NY, Empire State Building Main VHF Antenna for WABC-TV 7, WPIX-TV 11, and WNET-TV 13. (Dielectric THA-04–2H/8UD2SP-2-HM, 4-sided). The Main UHF Antenna is below on the northwest leg of the tower for WCBS-TV Channel 33, WWOR-TV Channel 44, and WNBC-TV Channel 28. (Dielectric TUF-C3-8/24U-2-S.)

Courtesy: Tony Fonseca P.E., Turris Engineering

FIGURE 8.9-2 Main and standby antennas each accepting six channels combined, installed at Mt. Wilson California, 2003.

Courtesy: Radio Frequency Systems

that the system is operated and maintained in a standard operating procedure (SOP).

Efficiency/Shared Resources/Risk Reduction

Close collaboration between the tower owner and the broadcast TV stations are necessary for proper design, retrofit, operation and maintenance of combining systems. After each broadcast TV station establishes the shared antenna's position on the tower, the technical specifications of HAAT, total power output (TPO) and effective radiated power (ERP) are determined, and antenna patterns are developed to indicate the antenna patterns for each TV station. The TPO for each TV station's transmitter will be established to manage the power efficiency. The TV combiner selection will determine power efficiency for each TV broadcast transmitter's TPO.

Ownership & Business Structure

In many cases, the TV combiner system is owned by the broadcaster, and the majority of tower sites that house combiners serving multiple stations are owned and managed by a third party. In some cases, the tower owner owns the combiner and the antenna and licenses the right to use the system to each broadcaster. More and more stations are selling their towers to tower facility management firms (Stine 2016).

Combiner facility and tower site leases (including modification and term extension agreements) are structured to consider the future operation and maintenance of the TV combiner system. A detailed review of the existing or proposed combiner facility and tower site assures that the site is capable of handling the system and all contingencies. Thorough system design and facility selection assure the best possible coverage for all the stations deciding to co-locate and use channel combining systems.

FACTORS AND ISSUES WHEN STATIONS SHARE TOWER SPACE

Tower Site Review, Inventory, Analysis

A review of the proposed combiner facility and tower site determines if the location is capable of handling the proposed design. TV broadcasters work with the tower site to optimize possibilities with consideration to minimize interference from adjacent channels.

Combiner Specifications

The design of the combiner is customized to fit the facility space and needs of all the broadcasters. Combiner specifications are determined after review of each participating station's authorized channel frequency, the resulting patterns and necessary power levels. The possibility of expanding combiner channel capacity for new stations may figure into the specifications and facility space allocation.

Independent Broadcast Design Engineering Input and Evaluation

The use of an independent broadcast design engineer /engineering team avoids the situation where broadcasters have to rely solely on the system manufacturers to provide all design details.

Broadcast engineers for each TV station in concert with the tower facility management team select a broadcasting engineering company to design the TV and antenna system. The design process is performed by the broadcast engineering team working with antenna and combiner systems manufacturers.

Plans, Specifications, Documents for the RFP Process

A complete set of design plans, specifications and construction documents is provided by the design team. A detailed request for proposal (RFP) based on the final design is developed as a key deliverable from the design team and sent to the systems manufacturers for bids. Operations of the combiner for RF shutdown and switching between the main and auxiliary antennas are addressed and planned for during the design phase.

Lease Terms, Extensions, Negotiations

Combiner facility and tower site lease negotiations, modifications and term extension agreements factor in the future operation and maintenance of the combiner system and are considered in light of technology changes and changes in ownership or management of the broadcast station or tower facility. There may be a lease and a separate license fee along with different arrangements where a broadcaster may own the system and maintain it, or the tower location may provide part or all services. Broadcasters should review all factors and determine adequacy of terms for operations, compliance, insurance, liability and all issues. The agreements, as appropriate, should detail responsiveness to requests and maintenance issues (resolution time), facility access, security, on-site and remote monitoring capabilities and safety and training issues.

If the tower owner is providing a turnkey combiner and antenna system, determine the investment, return to the tower owner and yearly escalation charges over the lease term. That cost is understood as it relates to the lease and license fee being charged along with the yearly escalation charge. Lease agreements generally include an annual rent escalation charge or cost of living increase expressed as a compounding annual percentage increase in rent to cover the tower owner (landlord) for increases in costs associated with increases in maintaining the broadcasting facilities. These increases can be determined by the federal government's published Consumer Price Index (CPI). Alternatively, the rent escalation charge can be determined as a fixed percentage mutually agreed to by the TV broadcaster and the tower owner (landlord).

The lease and license agreement with the tower owner includes responsibilities, scope and frequency for the system maintenance, training and emergency procedures. If the broadcasters are jointly responsible for the combiner and antenna system, then a separate agreement is needed between and among the TV broadcasters sharing the combiner and antenna system for maintenance. The tower owner may be responsible for the system maintenance and charge the broadcasters for common area maintenance (CAM). This is defined in the tower lease and license agreements with each TV broadcaster on the combiner system.

The lease agreement will specify the maintenance requirement, the equipment covered, and who is responsible for maintenance. If new equipment is being installed, the lease will specify the product warranty period and what is covered. A 3- to 5-year warranty period is normal, but lately longer warranty periods at 10 years are being offered.

Training

Factory training is recommended for all operations personnel. Qualified RF engineers with broadcast engineering training and qualifications attend factory training. Their training continues on location by observing the installation and learning how to use the various controls and switching procedures. All maintenance and operations personnel are trained on the software interface controls involved in operating and maintaining the system.

Combiner system emergency and safety procedures are provided by the equipment manufacturer and are further expanded in the facility operations procedures manual (SOP). It is generally created by the tower owner and the TV broadcast engineers or the party responsible for maintenance and operation of the combiner system. The combiner owner is ultimately responsible for the manual creation, updating, documentation and compliance.

The manual details broadcast operations and system procedures and includes RF training, inspections and maintenance schedules. The recommended manufacturer maintenance recommendations (with instruction and service manuals) are incorporated in the SOP manual. Specific building procedures and tenant service information are included in these materials. The SOP materials are presented to all engineers, employees and contractors to confirm understanding and agreement. The document is signed before they can start any work or access secure areas.

(See Appendix A for checklist of documents commonly generated for co-location site management.)

USES AND APPLICATIONS

Reasons Combiners Are Utilized by Broadcasters

TV combiner systems are used when TV broadcast stations in one market decide to join together and share an existing broadcast tower site that has antenna aperture capacity to accommodate many broadcast signals. Use of a TV combiner system is predicated on a situation where the economics or the practicality of providing each broadcast station with a standalone antenna is not possible, warranted or economically feasible. Alternatively, broadcast stations in a marketplace may choose to custom build a tower transmission site to best suit all their transmission and business needs. In densely populated areas such as New York and Chicago, combiner systems are used at multi-tenanted sites such as the Empire State Building, the World Trade Center and 4 Times Square in New York and Willis Tower in Chicago. Combiner systems are used to allow a large number of stations to broadcast their signals from combined antenna systems. TV and radio combiner systems are used extensively throughout the world. The most well-known manufacturers of TV combining systems used in the United States are Dielectric LLC, Electronic Research, Inc., Jampro Antennas, Inc., and Radio Frequency Systems.

BROADCAST STATIONS—COLLABORATION AND PARTNERSHIP

Business Arrangements

Broadcast TV stations generally compete with each other in the TV marketplace but cooperate with each other when the choice is made to use shared RF channel combiners and antenna systems. The broadcast stations in the market may join in a partnership agreement to own or lease the combiner and antenna system. A common process for this type of business arrangement is that the TV station broadcast engineers select a site and, in concert with RF system designers and engineers, design a combiner and antenna system to meet the broadcast transmission design criteria developed for each TV station's frequency and the combined system.

The business details of the agreement are developed to assure the success of the overall continued operation and maintenance of the system for optimal performance over its life expectancy. At the outset, the business terms are developed to assure technical performance and system operations. Specifications are developed so that the TV stations meet their FCC license requirements. Before work starts (i.e., design requirements and parameters), the business terms are established between the TV stations and the tower owner.

Business Parameters

Stations often develop a joint venture agreement (JVA) for all broadcast participants. An example of this is the Metropolitan Television Alliance, LLC (MTVA) formed in New York City following the destruction of the World Trade Center on September 11, 2001 (Kurz 2015) (see Figures 8.9-3, 8.9-4 and 8.9-5).

Number of Stations on a Combiner

The maximum number of TV stations that will utilize the system determines the overall size of the system.

Select a Tower Site

A tower site is selected that will meet all the TV stations' specifications and FCC requirements for coverage into the designated market area.

Negotiate Terms, Develop Agreements

The TV stations create an agreement among those who agree to be on the combiner system, and each station creates a lease agreement with the tower owner. The lease agreements with the tower will specify the terms and condition for their use of the combiner. There are many different ways the TV station agreements and the tower lease agreements can be structured. Those business terms will be worked out between the parties and their legal counsels.

Begin the TV Combiner System Design Process

After establishing the maximum number of TV stations on the system for the foreseeable future, the design process begins for the combiner and antenna system.

FIGURE 8.9-3 The UHF combiner at the Empire State Building installed by CBS to accommodate all the UHF TV Stations following the 9/11 destruction of the World Trade Center serving WCBS-TV Channel 33, WWOR-TV Channel 44 and WNBC-TV Channel 28 in New York City, NY.

Peggy Miles —Intervox Communications

FIGURE 8.9-4 Another view of the UHF combiner at the Empire State Building.

Peggy Miles—Intervox Communications

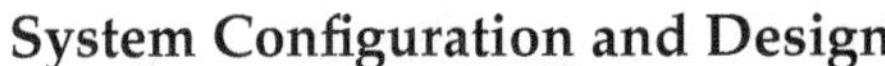

System Configuration and Design

Channel combiners or multiplexers combine multiple RF frequencies for transmission into a multiple-array antenna system. Each tower site has its own unique design requirements (see Figures 8.9-6 and 8.9-7). The design parameters depend on variables such as height of the antenna above average terrain (HAAT), the FCC license requirements, and the allowable effective radiated power (ERP) for each frequency on the combiner system.

Antenna arrays and combiner systems are usually bought together from the same manufacturer. Broadcasters working with different manufacturers have experienced delays and technical problems. Using a combiner and antenna from the same manufacturer assures consistency and compatibility of design components, such as waveguide couplers, transmission line switches, phase shifters, and power dividers.

INFRASTRUCTURE

Location/Logistical Considerations

TV broadcasters, after establishing the antenna aperture on the tower mast, then move forward to review and select the best location and infrastructure to house the combiner system.

Locate space within the tower facility as close as possible to the antenna and in close proximity to the TV stations' transmitters. For a tall tower structure, the

FIGURE 8.9-5 The VHF combiner at the Empire State Building currently accommodates TV Channels WABC-TV Channel 7, WPIX-TV Channel 11 and WNET-TV Channel 13.

Peggy Miles—Intervox Communications

combiner will be in its own facility adjacent to or in close proximity to the TV transmitters and close to the base of the tower. In facilities atop buildings such as the Empire State Building in New York and Willis Tower in Chicago, the transmission facilities are on the highest floors in utility spaces immediately below the tower mast. These spaces are designated to assure the shortest possible transmission line runs to accommodate transmission performance characteristics (see Figure 8.9-8).

Sufficient electrical power will be provided by the tower owner. This may involve running new electrical risers from the power vaults or the electric utility to assure adequate power. The largest power consumption in the TV combiner room is the air conditioning system. The air conditioning system is generally a dedicated system specifically designed to cool the heating load of the combiner. The system is usually protected from an electrical power outage by a backup electrical generator.

Consider the building environment as a whole and whether other infrastructure poses a risk to the combiner, transmission lines and antenna operations (see Figures 8.9-9 and 8.9-10). An example of this would be water damage to the combiner or its control elements caused from an adjacent suite, floor or building plumbing leak.

Space size is defined by combiner design, maintenance access, airflow ducts, cooling units and transmission line access from the individual TV station transmitters.

Combiner cooling and ventilation factors assure the adequate cooling of the combiner. Facility review includes confirmation of access to chiller or condenser water 24 hours a day, 365 days a year, and adequate ventilation for air-cooling-only applications.

Other site infrastructure considerations include height above mean sea level and the risk if in a flood zone.

Floor-loading capacity for new combiner construction is determined by a structural engineer, who determines if reinforcements are necessary to accommodate the combiner.

MANAGEMENT AND OPERATIONS

Management and Operations

Shared site coordination is addressed as early as possible in the design process and detailed in the TV stations' lease agreements with the tower owner.

FIGURE 8.9-6 One of the two chains of RFS combiners that have been earthquake proofed at Mt. Wilson overlooking Los Angeles, installed in 2004.

Courtesy of Radio Frequency Systems

Engineers need to carefully consider every aspect of the agreements (between the stations and the tower owner) and the management of the system due to the complexity and possible liabilities working with different stations, tower facilities and contractors. The TV broadcast engineers will likely be the liaisons in communicating between the tower owners, legal counsel and executives at their TV stations, to explain the new agreements and technical systems details as described in the TV station lease agreement.

(While different stations have different business structures and unique situations, the example provided in this chapter is illustrative of management structures commonly used by stations today.)

Business Structures, Individual Station and Master Agreements

Typically, the majority of tower sites that house combiners serving multiple stations are owned and managed by a third-party real estate management entity.

The combiner and the antenna are often owned and operated by the tower owner and paid for by the TV broadcasters or paid for by the tower owner but selected and approved for purchase by the TV broadcasters.

Each TV station has lease and license agreements with the tower site owner and a master agreement among each TV station and the tower site owner for the combiner, antenna and the tower aperture (including the RF shields, ice shields, deicing equipment, lightning arrestors, FAA warning lights, work lights and other appurtenances). Those agreements include the details of purchase, ownership and operation and maintenance of the combiner and its associated transmission infrastructure (HVAC, fire suppression, facility backup electrical power generator and uninterruptible power supply (UPS).

Ownership and management of the combiner and antenna infrastructure is a decision made by the TV stations leading up to and during the tower site lease agreement negotiations. Considerations during negotiations include the following:

- Infrastructure design and engineering costs
- Bidding for the combiner facility construction
- Purchase of the combiner and antenna system

FIGURE 8.9-7 One of the two chains of RFS combiners which have been earthquake proofed at Mt. Wilson overlooking Los Angeles, installed in 2004.

Courtesy: Radio Frequency Systems

FIGURE 8.9-8 Multiple transmission lines from each transmitter converge in the combiner room and then make their way up to the antenna.

Peggy Miles—Intervox Communications

FIGURE 8.9-9 Planning and design takes into consideration the complicated, varied needs of the broadcasters and possible future needs for technological upgrades. The positioning of different equipment presents unique challenges from the transmitters to the combiner and to the tower.

Peggy Miles—Intervox Communications

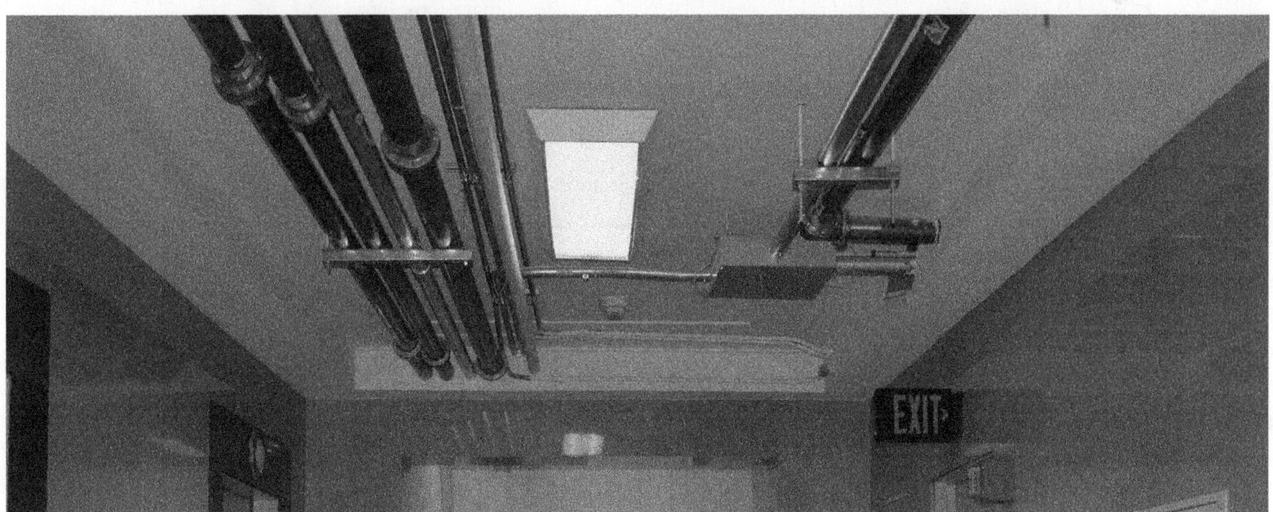

FIGURE 8.9-10 Hallway transmission infrastructure: Many different transmission lines, wiring, and the access to maintain them are considered in the planning stages.

Peggy Miles—Intervox Communications

- Ownership and financing of the combiner and antenna system
- Responsibility for operations and maintenance
- Addition and termination of TV stations
- Insurance

Who Pays for What, When?

All business terms are negotiated among the TV stations' executives and chief engineers (tenants) and with the tower site owner (landlord). The system design and engineering costs are typically divided among the TV stations on the combiner. The design and engineering costs may be initially paid for by the site owner and amortized into a master lease agreement to be paid as an additional monthly fee over the term of the lease with a known return on investment for the tower site owner (landlord). The return on investment varies and may range between 5% and 7% annually.

Third Party Facilitation of Station and Tower Bids

The construction bidding/bid leveling process for the combiner and antenna system could result in delays and conflicts due to each station and tower site having preferences for a particular design, installation, construction or equipment manufacturer. In order to assure the least biased results, the design process is facilitated by a third-party entity, a qualified engineering or construction management company on behalf of the TV stations, and the tower site owner.

The request for proposals for the purchase and installation of the combiner and antenna system are submitted to qualified antenna and combiner systems manufacturers (see a comparison of vendors in Figure 8.9-11). The request for proposals includes the manufacture, testing, certification and installation. Specifications as detailed in the proposal require that the manufacturer is fully responsible for certifying the installation. With this process, the manufacturer will

CIF COMBINER SPECIFICATION COMPARISON

ITEM	VENDOR A	VENDOR B	COMMENTS
	9 or 10 CHANNEL COMBINER		
Combiners:	**9 & 10 channel**	**9 & 10 channel**	-
VSWR for any channel	<1.1:1 (>26dB)	<1.06:1 (>30dB)	VENDOR B <1.06:1 is better than <1.1:1
Minimum Group delay for channel	<75nsec	<100nsec	Both acceptably low
Maximum combiner loss for any channel	0.85dB for 10 channel	0.75dB for 10 channel	VENDOR A combiner requires 2.3% more power for all stations
Narrow band to wideband port isolation	>30dB	>30dB	-
Input port to input port isolation	>55dB	>55dB	-
Voltage Safety Margin with -10dB	>2:1 Voltage	>2:1 Voltage	Minimum voltage safety margin required is 2:1, Both vendors meet it without pressurization.
	COMBINER MODULE		
Type:	CIF	CIF	
Cavity shape	Cylindrical	Square	Either design can be beneficial and adaptable to the installation.
Number of cavities	4	3	VENDOR B has less cavities that results in less loss, more compact, less volume and less weight.
Temperature compensated	Yes	Yes	-
Cooling	Free convection	Free convection	No forced air cooling needed for either vendor combiner.

FIGURE 8.9-11 Comparing Vendors for CIF Combiner (Summary Sample); an engineering consulting company on behalf of a tower site, and the broadcaster tenants reviewed specifications from two competing vendors for a new combining system.

Empire State Reality Trust/Empire State Building Broadcasting

certify the installation is adequate to meet its warranty. Antenna and combiner systems serving multiple broadcasters must perform reliably and, as such, multiyear warranties on maintenance and repairs include shelf spares throughout the warranty period and the expected life of the combiner and antenna system.

Tower Site Lease Agreement

Purchasing and ownership of the combiner and antenna system is specified in the tower master agreement (an agreement between the TV stations) and in the TV station lease agreement between the TV stations and the tower site owner. TV transmitters, at this point in time, are owned by the broadcasters; whether that changes in the future is unknown.

Operations and maintenance requirements are specified in the master lease agreement between the TV stations and the tower owner. The tower owner specifies in the lease terms an amortized fee for the cost of the combiner and antenna system in addition to the lease rent, or alternatively each TV station agrees to pay for the design and installation of the combiner and the antenna system up front as a lump sum payment.

Agreements can include a common area maintenance (CAM) fee to compensate the tower site owner for maintaining the combiner, antenna and the tower facility including ventilation and cooling (HVAC), electrical systems including backup power, fire suppression systems and general maintenance of the facility. CAM fees are equally divided between TV stations on the combiner and are included in the rent.

The tower site and combiner/antenna owner may develop or facilitate maintenance service agreements with qualified service vendors. In some cases, these services are provided by the equipment manufacturers or mutually acceptable vendors. Operations and staffing requirements can also be detailed in the master lease agreement.

The transmission lines from the TV transmitters to the combiner and the transmission lines connecting the combiner to the antenna are specified in the master lease agreement.

Operations and Maintenance

The allocation of resources to operate and maintain common broadcast systems such as the combiner, combiner room, transmission lines and power dividers can be managed by a third party such as the tower owner, engineering firm or contract maintenance vendor. Stations have a choice of providing a maintenance technician on call or on site and may install remote monitoring system and diagnostic systems to the extent possible and economical. Monitoring of conditions such as temperature and the status of the cooling, air condition and ventilation systems may be provided by the tower sites.

Broadcast transmitter sites are not manned and are remotely controlled from the TV studio. In many cases the tower site owner provides the monitoring of facility conditions such as temperature and the status of the combiner cooling, air conditioning and ventilation systems and dispatches repair and maintenance technicians when needed. In large transmission sites such as the Empire State Building in New York and Willis Tower in Chicago, maintenance technicians are on site 24 hours per day 365 days a year.

Qualified/Approved Maintenance Contractors and Personnel

The list of approved and qualified maintenance contractors and personnel are included in the master lease agreement as an exhibit, subject to update on a regular basis by the tower owner and the TV broadcasters. The primary vendors for maintenance and repair are the equipment manufacturers contracted as part of the equipment warranty and purchase agreement to offer maintenance and repairs throughout the expected life of the combiner and antenna equipment.

Insurance and Liability

Insurance obligations are specified in the master lease agreement between the TV stations and the tower owner. Each party is responsible for maintaining adequate insurance for repairs and maintenance for all foreseeable situations, including complete systems replacement if necessary.

Broadcast Staff Availability Requirement

Each TV station is responsible for having trained broadcast technicians available for the operation and switching of stations from their main antenna to their auxiliary when required for maintenance, repairs or other reasons.

Shutdown Schedule

A shutdown schedule is developed between the TV stations and the tower site owner that specifies exclusionary time periods for antenna shutdowns. Certain time periods, such as TV sweeps months, may be excluded on a yearly basis. Other dates prohibited for shutdown may include those of major TV events with expected large TV audiences, such as the NFL's Super Bowl, the Olympics and MLB's World Series.

Regulatory Compliance Responsibility

The tower site owner is responsible for building code compliance to the local municipality and to FCC and FAA codes. The tower structure will comply with EIA/TIA 222G for tower structural loads. Any adding or removing of antennas will require analysis, review and certification from a licensed, qualified structural professional engineer (PE).

FCC compliance is the responsibility of the TV stations. This compliance includes all RF aspects of the

TV station's assigned frequency including effective radiated power (ERP) and radio frequency radiation power levels for human exposure. As necessary, the tower owner cooperates with the TV stations by RF monitoring common areas where members of the public or other personnel may visit or perform work.

IEEE provides a free copy of the document *IEEE C95, Standards: Safety Levels with Respect to Human Exposure to Radio Frequency Electromagnetic Fields* on the IEEE website under the IEEE Get program (IEEE Standards Association 2016). The tower site owner is responsible to assure all common areas of the tower site are in compliance for RF exposure and to correct any issues to adhere to FCC and OSHA guidelines. As different states have adopted OSHA standards and enforcement policies, review of your states' standards and policies is recommended (Williams 2007).

Using a qualified RF broadcast engineer, the tower site owner performs monthly or quarterly RF inspections to monitor compliance (see diagrams and images of RF monitoring systems, Figures 8.9-13, 8.9-14, 8.9-15, 8.9-16 and 8.9-17). Records are kept by the tower site owner and are subject to audit by the FCC and OSHA.

Tower Site Security

The tower site security is the responsibility of the tower site owner. All RF broadcast areas are secured with an access control system throughout the transmission facility. This includes the transmitter and combiner rooms along with any area designated as controlled access by FCC rules and regulations.

RF Safety Signage

RF signage is installed in all areas in compliance with the latest FCC guidelines. In addition, many sites have card reader access control systems that assign site access passes to personnel who have the required FCC Radio Frequency Radiation (RFR) awareness training (see Figures 8.9-18, 8.9-19 and 8.9-20 for examples of safety precautions for these areas). This assures that only those personnel with the proper training can access RFR controlled areas. The responsible party for FCC compliance to RFR exposure limits is the TV station as the FCC broadcast license holder. The federal government's Occupational Safety and Health Administration (OSHA) issues safety standards for radio frequency radiation (RFR) exposure limits that directly reference the FCC guidelines.

TOWER WORK PLANNING CALENDAR

Emergency Work Only
Sweeps Period - Urgent Work Only
Urgent Work OK
All work OK

	Mon	Tue	Wed	Thu	Fri	Sat	Sun
					1	2	3
Jan	4	5	6	7	8	9	10
	11	12	13	14	15	16	17
	18	19	20	21	22	23	24
	25	26	27	28	29	30	31
Feb	1	2	3	4	5	6	7
	8	9	10	11	12	13	14
	15	16	17	18	19	20	21
	22	23	24	25	26	27	28
Mar	1	2	3	4	5	6	7
	8	9	10	11	12	13	14
	15	16	17	18	19	20	21
	22	23	24	25	26	27	28
April	29	30	31	1	2	3	4
	5	6	7	8	9	10	11
	12	13	14	15	16	17	18
	19	20	21	22	23	24	25
	26	27	28	29	30	31	1
May	2	3	4	5	6	7	8
	9	10	11	12	13	14	15
	16	17	18	19	20	21	22
	23	24	25	26	27	28	29
June	30	1	2	3	4	5	6
	7	8	9	10	11	12	13
	14	15	16	17	18	19	20
	21	22	23	24	25	26	27

FIGURE 8.9-12 RF Shutdown Work Planning Calendar.
Empire State Reality Trust/Empire State Building Broadcasting

PERFORMANCE AND QUALITY OF SERVICE (QOS)

Monitoring and Control

Combiner components combine, switch, route and filter the RF power from the TV transmitters. The system design as set by the manufacturer determines the overall performance of the system. The combiner system is a passive system with precisely tuned components that allows a number of TV channels to operate within a specified spectrum. (Passive RF components of a broadcast system include the antenna, cable, rigid line and patch panels, combiners and filters.) Once the system is properly tuned to combine a specified number of channels into a single antenna feed, there are several variables monitored and controlled to assure the continued stability of the system. The components controlled are:

Transmitter Power to the Master Combiner

The TV broadcast stations are responsible for their station's input power to the combiner. Fluctuations are monitored, and the impact to the combiner should be understood. The combiner design can handle a TV station or a number of TV stations going off line at any one time because the remaining input to the combiner is normally matched in enhanced combining systems, because inputs are isolated from each other.

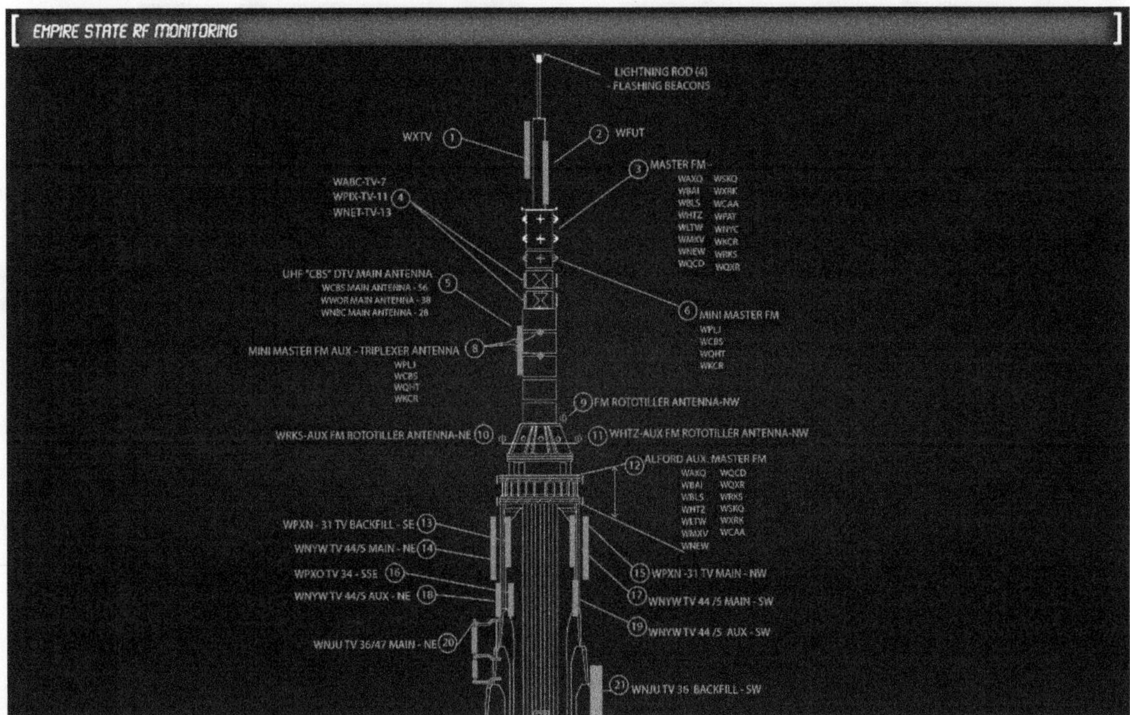

FIGURE 8.9-13 RF Monitoring System for real-time RF monitoring and alarming to assure safety and compliance to FCC regulations. Monitoring equipment installed on main tower and infrastructure for all radio, TV, main, auxiliary and other telecommunication tenants.

Courtesy of SuiteLife Systems

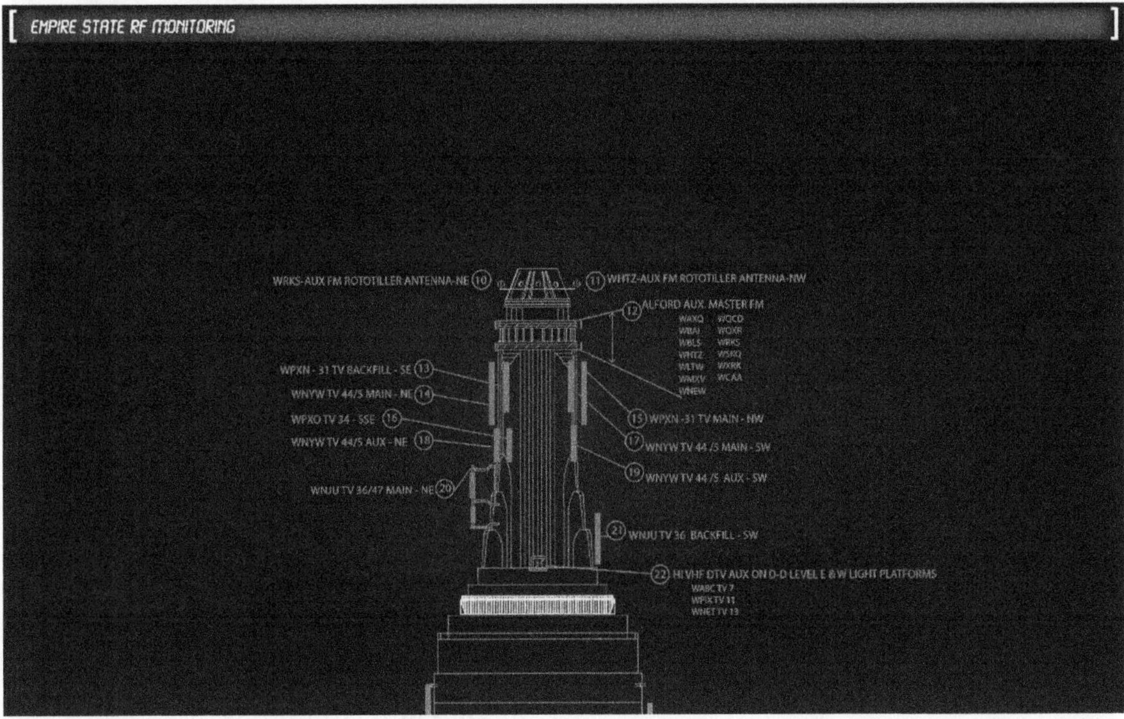

FIGURE 8.9-14 RF Monitoring System for broadcasters and telecommunication tenants for broadband, telecommunications, auxiliary transmitters and other systems.

Courtesy of SuiteLife Systems

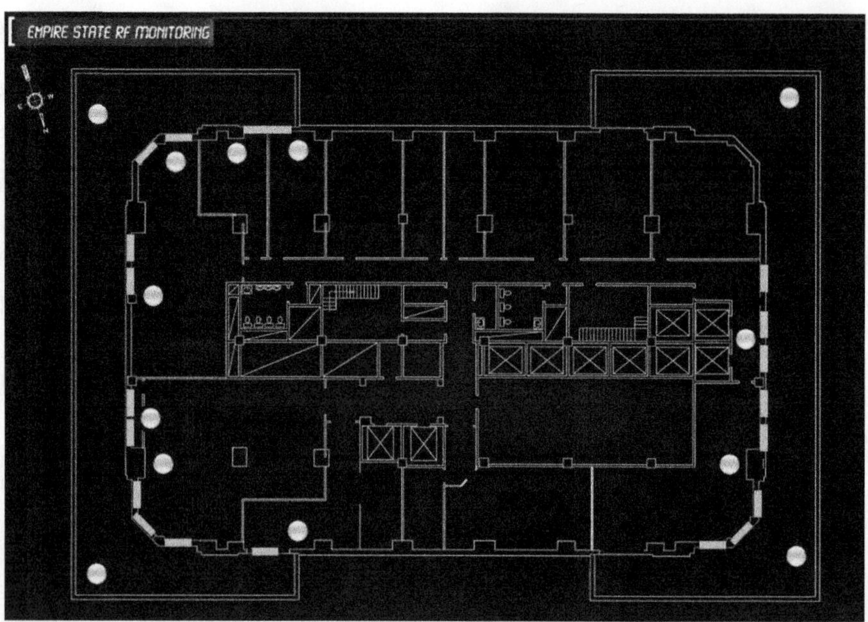

FIGURE 8.9-15 RF Monitoring System for real-time RF monitoring and alarming to assure safety and compliance to FCC regulations. Each floor that has transmitting equipment and transmission lines is installed with window and door alarms and area sensors that trigger immediate building security or emergency procedures. The round shape on the diagram signifies a sensor that triggers a nearby green/safe and red/danger indicator light. The small rectangles on the floor plan indicate access controlled sensor alarms on windows and doors.

Courtesy of SuiteLife Systems

FIGURE 8.9-16 RF Monitoring system associated RF sensor light on a wall.

Peggy Miles—Intervox Communications

Combiner Cavities and Room Temperature

Temperature of the combiner cavities are maintained through cooling and ventilation of the combiner cavities and the entire combiner room. The combiner cavities and combiner room temperature are continually remotely monitored, and any temperature increases above the specified limits are investigated and corrected in a timely manner to assure the transmission lines are cooled to specifications.

RF Power Measurement

RF power measurement devices installed in various locations at the combiner to assure proper functionality and quick diagnosis of any issues (see Figure 8.9-21). Automated alarming systems are installed to alert designated maintenance personnel of any out of specification conditions that require attention.

Voltage Standing Wave Ratios

Voltage standing wave ratios (VSWR) are continually monitored, and fluctuations outside the specification are remotely alerted to a qualified maintenance technician for immediate investigation.

Single-Point Grounding

Grounding systems designed in the combiner, transmission lines and the antenna comply with manufacturers standards focused primarily on single-point

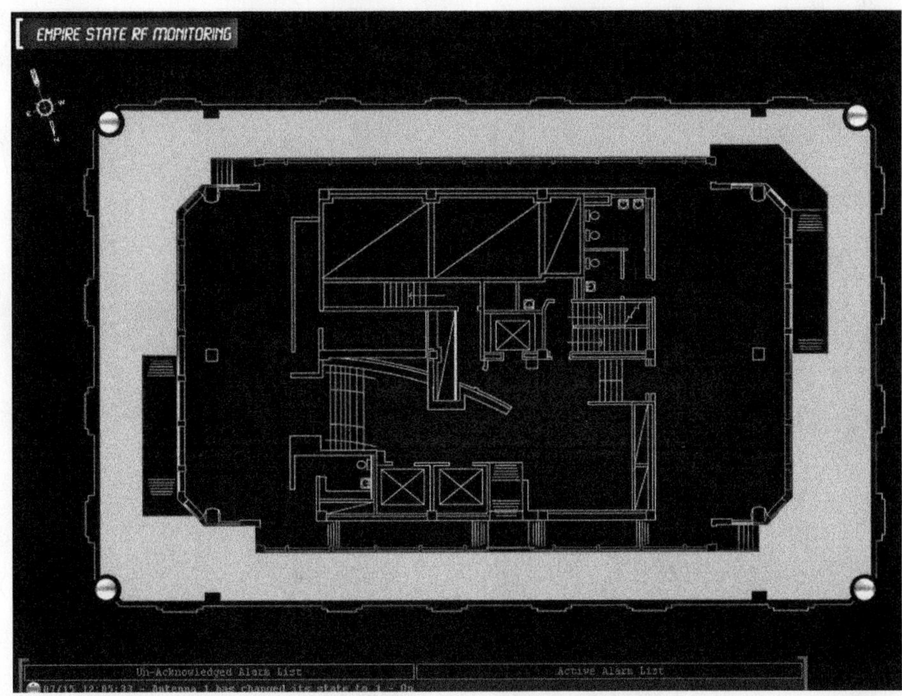

FIGURE 8.9-17 RF Monitoring System for public space. The current state is shown on this floor of the building as a light gray perimeter indicating the area is safe. (This perimeter is shown as a "green" region in the companion *Handbook* color figures website.) This public space is outside the building on an upper floor directly below microwave transmission equipment and the broadcast tower.

Courtesy of SuiteLife Systems

FIGURE 8.9-18 Complex tower sites and multi-transmission facilities have different levels of RF exposure. RF signs indicate by color the exposure environment: blue—RF exposure may exceed FCC general public exposure limits; yellow may exceed FCC rules for human exposure and exceed FCC rules for human exposure; and red indicates RF exposure exceeds FCC guidelines for human exposure.

Peggy Miles—Intervox Communications

FIGURE 8.9-19 Different areas of the building may warrant further RF instructional signs and barriers to prevent human exposure to RF radiation or protective gear.

Peggy Miles—Intervox Communications

FIGURE 8.9-20 Stairwells and emergency access corridors are clearly marked with hazard warnings and directives on alternative exits.

Peggy Miles—Intervox Communications

grounding (SPG) and follow standards such as Motorola's R-56, TIA/EIA 222G and IEEE. Scheduled inspection routines review all grounding systems from the combiner and transmission lines to the tower mast and antenna. Any loose connections or corroded points require immediate repair.

Pressure Readings

Pressure readings in pressurized lines require continual monitoring to assure proper pressures are maintained to specification throughout the transmission lines, from the combiner to the antenna elements. The gas pressure specified by the manufacturer in maintained and gas cylinders are replaced on a regular schedule to assure compliance with specifications. Remote monitoring and alarming is recommended to assure that the gas pressure is maintained to specification at specified points in the transmission lines between the combiner and the antenna.

RF Shutdowns

RF shutdowns are coordinated between the TV stations and the tower site owner. TV stations assign a broadcast technician to switch the stations to their auxiliary antenna on the tower from the switching capability in the combiner system. Stations with auxiliary sites in a different location often have the ability to shut down their transmitters remotely and power on at their auxiliary antenna locations. In systems with on-site auxiliary antennas such as the Empire State Building, the switching from the main antenna to the auxiliary antenna is performed in the combiner by trained TV broadcast engineers. This allows each station's transmitter to continue to transmit at full power without interruption while the switching takes place in the combiner switching system (Figure 8.9-22 shows a tower shutdown notice).

Due to the rise in injuries on communication towers, OSHA is focusing more on the communications sector. All facilities are subject to unannounced inspections.

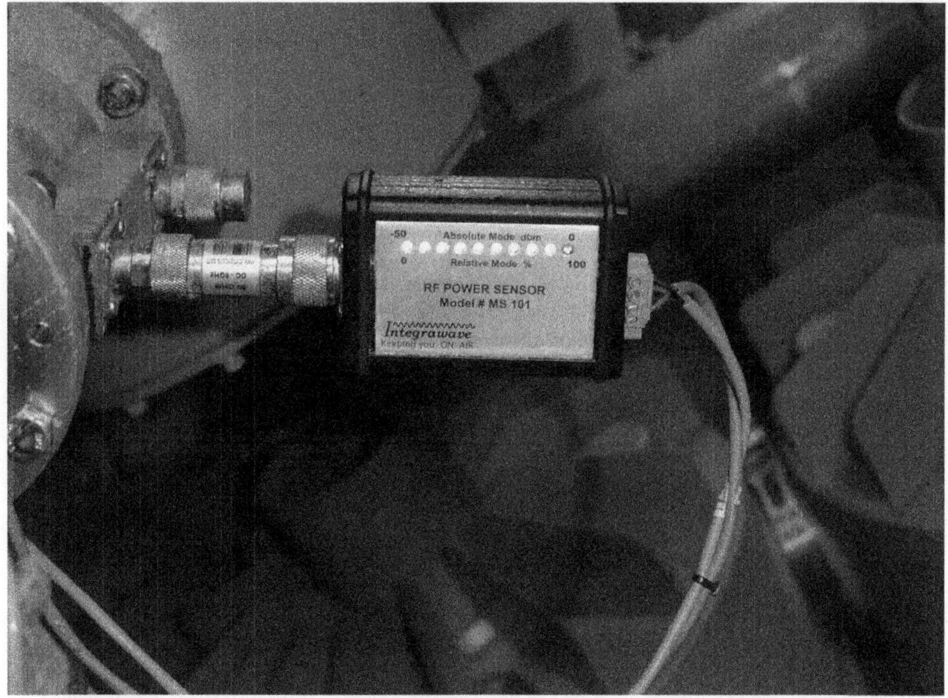

FIGURE 8.9-21 RF power measurement devices connected to transmission line.

Peggy Miles—Intervox Communications

OSHA has communicated that it will obtain "contracts in order to identify not only the company performing work on the tower, but the tower owner, carrier, and other responsible parties in the contracting chain," and further, that contractor selection should include safety criteria and close oversight of subcontracting, stating that a "simple 'check the box' contract language may not provide enough information to evaluate a contractor's ability to perform the work safely." Engineers are advised to carefully review all safety qualifications and training of all contractors that perform work on the combiner and antenna even if contracted by a firm on their behalf. To assist in this process, The FCC and OSHA issued a draft best practices document for communication tower work due to the rise in tower accidents. They continue to study the situation carefully (Occupational Safety and Health Administration and Federal Communications Commission 2016), (Occupational Safety and Health Administration 2014).

RF Safety Officer Inspection for Work Commencement

RF shutdowns of the antenna require an RF safety officer to inspect and verify that the main antenna is shut down and is not radiating RF energy (Figure 8.9-23 shows antenna RF status indicators). The safety officer gives the all-clear signal for maintenance engineers to go onto the tower to perform maintenance or repairs. The RF safety office confirms that the RF energy lockout tagout (LOTO) process is performed to assure that all engineers and technicians are aware that the RF energy has been locked out during the shutdown (Occupational Safety and Health Administration 2015).

Radio Communication

Radio communication is maintained between engineers on the tower and within the transmission facility to assure all safety rules and regulations are being maintained and that any needed information or assistance can be given to the engineers on the tower.

Inspection Routines

Inspection routines are performed as specified by the manufacturer and documented in the preventative maintenance procedures. All inspection and measurements are documented and reports provided to the TV stations and the tower site owner. All noncompliant conditions are repaired as soon as possible following detection.

Training and Emergency Procedures

Training is provided to all personnel working on the combiner and antenna systems on a regularly scheduled basis to assure that all procedures are followed and emergency procedures are practiced on a regular basis. Simulations of emergency conditions assure that personnel are aware and capable of responding to emergency conditions.

Tower Shutdown Notice (Sample)

For the tower maintenance and safety program, we will be performing electrical repairs, removal of unused wave guide and performing selective cleaning & painting per the following schedule:

Tower Shutdown – November 13-18, 2016
Tower Shutdown – November 20-24, 2016
Tower Shutdown – November 29- December 3, 2016

During this shutdown, the stations listed below **must have an engineer/technician on site and will assume the following broadcasting posture.**
Stations not listed are not affected for this shutdown.

WFUT TV	Off or Auxiliary Site
WXTV TV	Off or Auxiliary Site
Master FM	Switch to Alford or Auxiliary Site
Mini Master FM	Switch to Auxiliary Site
WNBC TV	Switch to TUA Antenna
WCBS TV	Switch to TUA Antenna
WWOR TV	Switch to TUA Antenna
WABC TV	Switch to Auxiliary
WPIX TV	Switch to Auxiliary
WNET TV	Switch to Auxiliary

No station will reactivate an antenna without the express direction from RF Safety coordinator.

At 04:45 or a time previously designated by the RF safety coordinator, all shutdown personnel will meet on the 85th floor and then the all clear will be given for each station to return to their normal operation posture. Failure to meet on the 85th Floor at the designated time may result in delay returning to normal broadcasting. Please acknowledge receipt of this notice by responding via e-mail to name@xyz.com

FIGURE 8.9-22 Shutdown notice notifying which stations will have to move to auxiliary antennas and that each station impacted must have a qualified engineer present during all work, will follow lockout/tagout procedure and will meet at the combiner room staging floor location with the RF safety engineer supervisor for all-clear before the stations are allowed to restore full power or transfer from their auxiliary antennas.

Empire State Reality Trust/Empire State Building Broadcasting

Emergency procedures documented in the manufacturers' specifications and TV stations' procedures are followed by all personnel.

Infrastructure Labeling

Infrastructure labeling is employed in the maintenance system to assure all components and documents are included in the inventory system. A bar-coding inventory labeling system implementation eases identification of components, tracking of their age and expected timing for replacement (see Figure 8.9-24).

Incident Tracking and Discrepancy Reporting

A specific-incident tracking and broadcasting discrepancy report program assures timely follow-up, corrective action and root cause failure analysis. This system provides data to the manufacturers on the performance of their equipment in the field to improve their designs to eliminate future failures and thus improve overall system reliability.

CONCLUSION

TV combiner systems are highly efficient at putting a large number of TV stations on air from a single antenna array. This results in considerable cost savings as opposed to each TV station installing its own antenna. The process of implementing a broadcasting combining system requires extensive cooperation and compromise among competitors in the market. Although broadcast stations compete on content and quality of programming, they successfully cooperate together in this case, utilizing the latest technology for their mutual benefit. The result is a significant reduction in expenses for every participating station on the combiner over the life of the system.

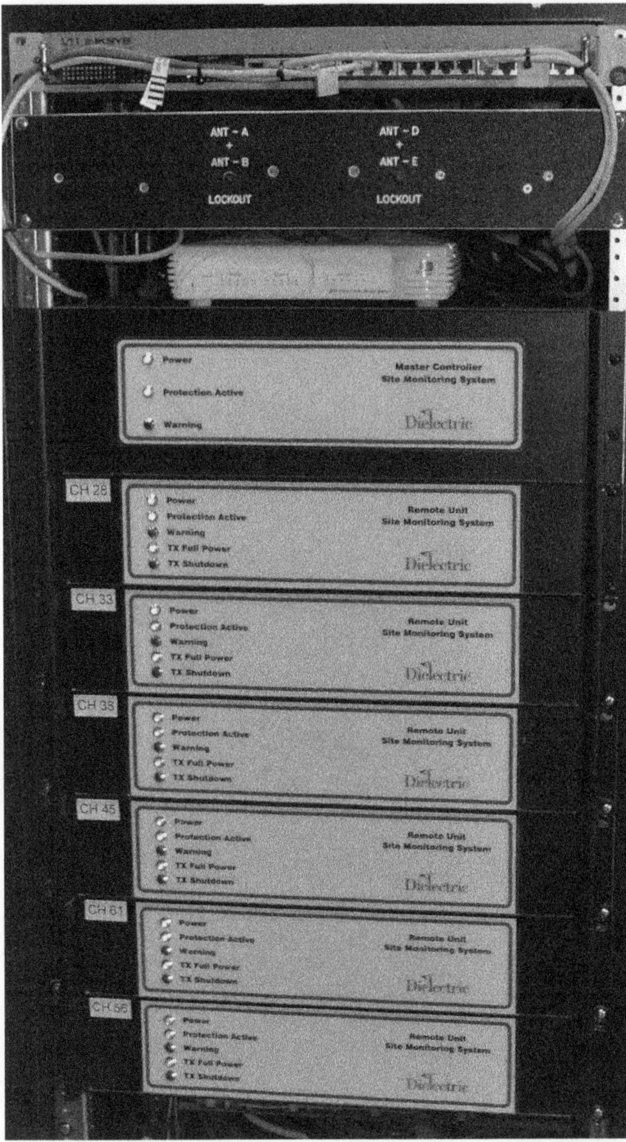

FIGURE 8.9-23 In addition to building-wide RF monitoring and transmission line monitors, lighted controls indicate which antennas are shut down and locked out. The status of each station is located on this control panel in the UHF and VHF combiner room. The FM combiner, also on the same floor, has equivalent monitors for the 19 radio stations.

Peggy Miles—Intervox Communications

ACKNOWLEDGMENTS

Shane O'Donoghue wishes to thank the following:

Tony Fonseca of P.E. Turris Engineering conducts regular inspections on the Empire State Building.

Ted Collora, Hanson Professional Services Inc.

Empire State Building Broadcasting—Empire State Realty Trust

Peggy Miles, InterVox Communications

APPENDIX A

Commonly Generated Documents

Joint Venture Agreement, Joint Partnership Agreement, Master Agreement

License Agreement

Leasing Agreement

FCC Documentation From Each Station

Operations Manual—Facility

Operations Manual—Manufacturer Components

Repository—Manuals

Historical Minutes

Broadcast Member Stations Meetings

Broadcast Members Stations Meeting With Tower Site

Structure of Broadcast Partnership—Responsibilities of Members, Bylaws

RFP Request for Proposal. Proposals, Issues, Manufacturer Agreements, Warranties

Letter of Intent, Letter of Agreement (Binding or Non-Binding)

Compliance List, FCC, OSHA, Historical Preservation, Environmental, and other regulatory issues such as local, state, and/or federal

Emergency Contact: Phone Numbers, Emergency Phone Email/Text, 1st, 2nd, 3rd Contact

Contact Database, Contractors, Manufacturers, Subcontractors, Parts Suppliers, Other

Emergency Response Activation Process

Safety Training Requirements, Safety Training Schedule

Confidentiality/Non-Disclosure Agreement (Press Contact, Who Represents the Group of Broadcasters)

Dissolution Process (Sale of Station, or Tower Ownership Transfer)

Permitted Staff by Location, Training Verification or Reason, update every 6 months or if employee issue warrants, notify and update immediately.

Calendar of Scheduled Maintenance, Schedule of Blackout Periods for Maintenance, Schedule of Upcoming Work

Incident Reporting System and Distribution Process, and Response

Building Access Procedures, Badge & Identification, Security Precautions, Safety Awareness Memo

Equipment Installation and Removal List

Shutdown Procedure, Lock-Out Procedure

Inventory Control, Spare Parts, Bar-Coded Linked Equipment/Transmission Line Database

FIGURE 8.9-24 Transmission lines, equipment and other items that travel from the transmitters to the combiner and to the antenna are identified by bar-code. Hand-held scanners can identify from a distance the source, path, ownership, use and other pertinent data.

References

Empire State Building Broadcasting—Empire State Realty Trust. 2016. *Empire State Building Broadcasting*. Accessed July 15, 2016. www.esbnyc.com/business-esb/broadcasting.

IEEE Standards Association. 2016. *IEEE Get Program*. Accessed July 27, 2016. http://standards.ieee.org/about/get/index.html.

Kurz, Phil. 2015. *TVNewsCheck*. January 8. Accessed July 15, 2016. www.tvnewscheck.com/article/82062/ny-stations-testing-signals-from-world-trade.

Occupational Safety and Health Administration. 2014. *Communication Towers, Letters to Communication Tower Industry Employees*. February 10. Accessed July 27, 2016. www.osha.gov/doc/topics/communicationtower/Letter_CT_Employers.html.

Occupational Safety and Health Administration. 2015. *Control of Hazardous Energy (Lockout/Tagout)*. Accessed July 28, 2016. www.osha.gov/SLTC/controlhazardousenergy/.

Occupational Safety and Health Administration and Federal Communications Commission. 2016. "FCC and DOL Workshop on Tower Climber Safety and Apprenticeship Program." Federal Communications Commission. OSHA's Directorate of Construction. February 11. Accessed July 20, 2016. www.fcc.gov/news-events/events/2016/02/fcc-and-dol-workshop-tower-climber-safety-and-apprenticeship-program, www.fcc.gov/file/3470/download.

Stine, Randy. 2016. "Tower Assets Bring Broadcasters Quick Cash." *RadioWorld*. June 8. www.radioworld.com/article/tower-assets-bring-broadcasters-quick-cash/278994.

Williams, Edmund, ed. 2007. *NAB Engineering Handbook*, 10th Edition. Burlington, MA: Focal Press.

Willis Tower. 2016. *Willis Tower—We're Sending Signals*. Accessed July 15, 2016. www.willistower.com/building-information/broadcast/.

SECTION

9

BROADCAST TOWERS AND TRANSMISSION LINES

CHAPTER

9.1

Tower Design, Erection, and Maintenance

JOHN WAHBA, PhD, P Eng, P.E.
Turris Corp.

INTRODUCTION

The purpose of this paper is to provide some useful and practical information regarding the design, construction and ongoing maintenance for a broadcast tower. It provides some insight on what is involved in the design, installation and maintenance of a broadcast tower

SELECTION OF TOWERS

When we think of broadcast facilities, we often think of the costly antennas and transmission methods required to provide the coverage demands. However, one key component that is often overlooked is the tower structure required to support those antenna systems at the requisite height for proper coverage. In fact, the tower structure is just as important as the antenna and other transmission equipment. One cannot be without the other in most environments.

Communication towers are a complex feat of structural engineering that must take into account everything from soil and environmental conditions, steel types, location and land area to proposed and future antenna loading. All of these factors must be taken into account to produce a structurally sound tower that will fulfill the needs of the customer both today and in the future, as requirements change. Poor tower design can often lead to catastrophic structural failure and the loss of thousands of dollars in potential revenue. When designing a tower, it is important to balance the need for the lowest-cost solution and the best investment. Most often, that balance can be satisfied with the same solution.

TOWER APPLICATIONS

In wireless technologies, which use higher-frequency bands to transmit data, the wavelengths are considerably shorter than in a TV or radio broadcast environment. As such, the towers for wireless applications tend to be much shorter than in broadcast towers. Shorter towers can be self-supporting, monopole, tripole, or guyed structures, typically no higher than 400 ft.

Broadcast radio and television broadcasters are required to transmit their signal over a very large coverage area. Because they operate at frequencies much lower than wireless technologies, the wavelengths are much longer, allowing the signal to travel greater distances. In order to maximize signal transmissions, broadcast towers are usually very high or placed in locations that have a very high elevation. Broadcast towers can be either self-supporting up to 1,000 ft. (300 meters) or guyed structures up to 2,000 ft. (600 meters) in height.

SELF-SUPPORTING TOWERS

A self-supporting tower is a freestanding structure that requires less land than a guyed tower and can accommodate a variety of antennas. Latticed self-supporting towers are often used because of restrictive real estate locations, where a self-supporting tower is the only alternative since the footprint is typically 10% of its actual height (e.g., a 400-ft. tower would require only 40 ft. at its base). However, it is important to note that all tower sites are transmission facilities and require on-site equipment in order to power, generate and transmit the data signals. This is typically housed in a shelter at the base of the tower. Therefore, real estate considerations need to accommodate the combined footprint of the tower and shelter, plus a perimeter security fence. Latticed self-supporting towers are used in broadcast applications (up to 1,000 ft.). Figure 9.1-1 shows a layout of a latticed broadcast self-supporting tower

Monopoles are also called self-supporting tubular structures and are commonly used where real estate restrictions or aesthetic considerations normally dictate the use of monopole or latticed poles (tripoles). These structures have a relatively small footprint and face width in comparison to standard self-supporting and guyed towers. For example, a 140-ft. monopole structure may require only 3 to 4 square feet of land area (plus equipment shelter). Depending on the loading and design requirements (wind and ice) and the governing municipal laws, designers may opt for one of these towers. These towers are typically used for shorter applications on rooftop and elevated terrains. Monopoles can be used in applications up to 200 ft. high in a typical environment.

Tripole designs are simply three-sided towers of equal face width supported by the three legs that make up the corners of the tower. There are no guys to support the structure; rather, the three legs are anchored into the ground and support the entire structure. Unlike most self-supporting latticed towers, the width of a tripole tower remains constant along its entire length, versus a standard self-support structure, which is wider at the bottom. It utilizes the same footprint as a monopole, while its open structure allows it to blend more easily with the surroundings. A tripole can be used in applications up to 300 ft. in height, in cases where there is little real estate available at the base.

GUYED TOWERS

A guyed tower is composed of a relatively slender mast supported by steel cables (guys) at the necessary levels. This composition makes the guyed towers versatile and will accommodate a variety of applications. Guyed towers are made in all-welded (the tower sections are shipped to the sites as single weldment assembly for each section) and knock-down configurations (where each section is assembled on the ground at the site). The towers are commonly three-sided equilateral triangular masts. However, four-sided masts are utilized for some applications. In some cases, for broadcast antennas, the face width of the mast will be

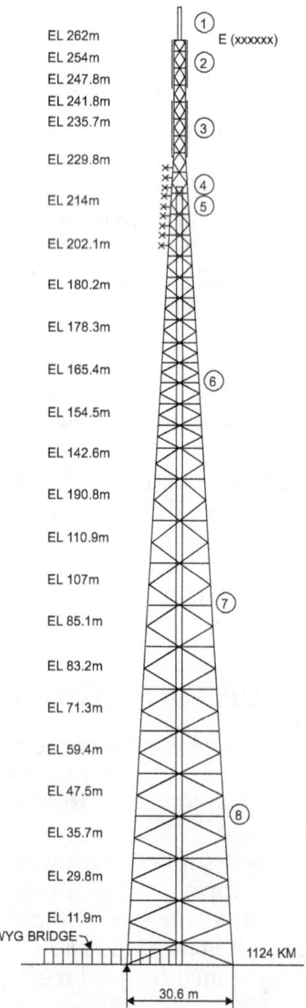

FIGURE 9.1-1 Layout of a latticed broadcast self-supporting tower.

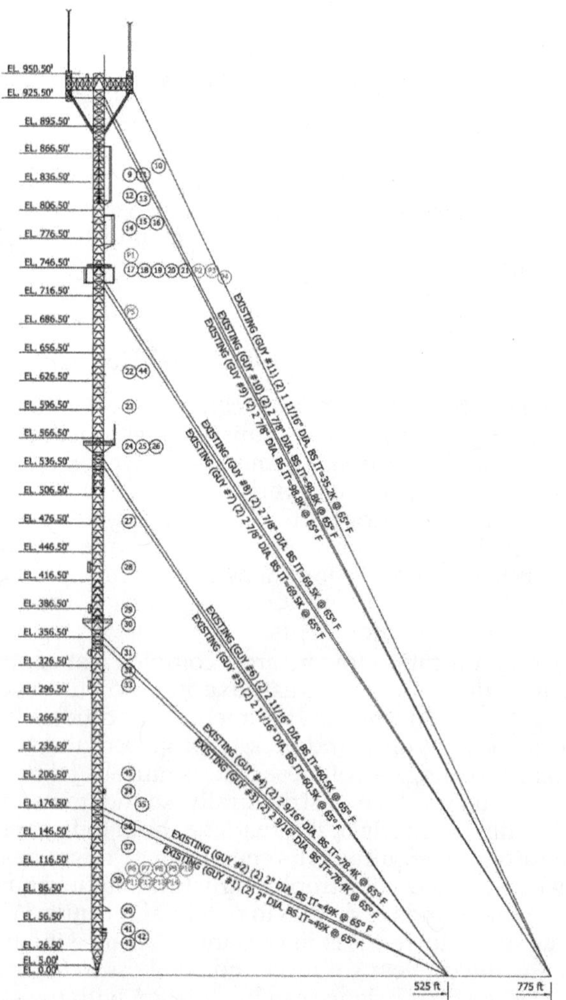

FIGURE 9.1-2 Layout of a typical guyed tower.

constrained by the need to optimize radiation patterns from panel or side-mounted antennas. These factors may dictate the size of the tower mast face width. Like self-supporting towers, guyed towers are designed and built to withstand wind and ice loads. The combined effects of wind and antenna weight will dictate the overall design and structural requirements of the tower. Because of the necessary guy wires to support the tower, a large amount of real estate is required, typically 60–80% of the structure's height, but can go as low as 35% of the structure height for cases of limited real estate. In such cases, this smaller guy radius would result in a premium to the cost of the structure. The guy radius required to support the tower loading is determined by the structural analysis and takes into account the environmental and structural factors such as wind, ice, twist and sway. Guyed towers can be used in applications varying from 250-ft. (75-meter) to 2,000-ft. (600-meter) applications. Figure 9.1-2 shows a layout of a 1,000-ft. (300-meter) guyed tower.

For a unique application, where a tall structure was required with only a small area of available land for its footprint, a unique solution such as a "hybrid" tower was developed. Figure 9.1-3 shows a layout and picture of a 1,300-ft. (400-meter) hybrid tower constructed in Jakarta, Indonesia.

Selection of Tower site

As with any tower, the ideal location would be a site on level ground. However, uneven site terrain is not uncommon. In cases where the anchor elevation is higher or lower than the base elevation, the amount of land required will be reduced or increased proportionately to the grade slope, and the uneven terrain must be accounted for in the design of the tower.

DESIGN CONSIDERATIONS

When selecting a tower, it is of primary importance to choose a site location that will satisfy the needs of the propagation study conducted by the RF engineering team. This will determine the required height of the tower needed to provide optimum coverage in that area. Additionally, antenna loads will also influence the design of the tower.

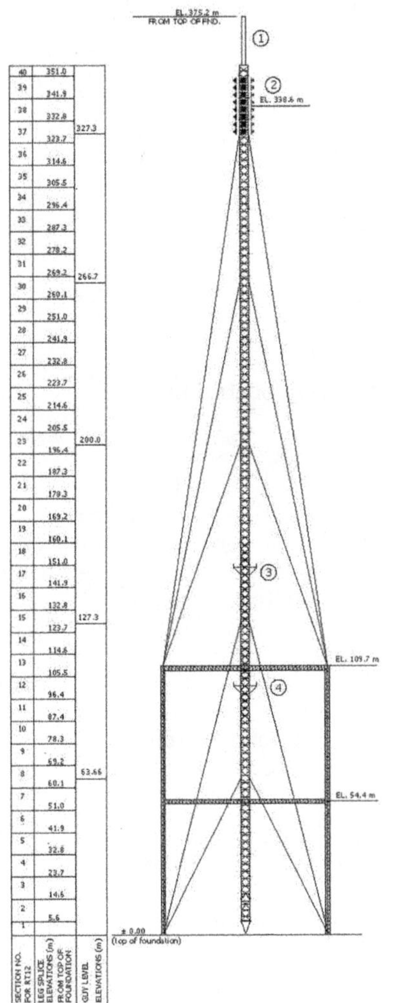

FIGURES 9.1-3A AND 9.1-3B Layout and picture of a 1,300-ft. (400-meter) hybrid tower constructed in Jakarta, Indonesia.

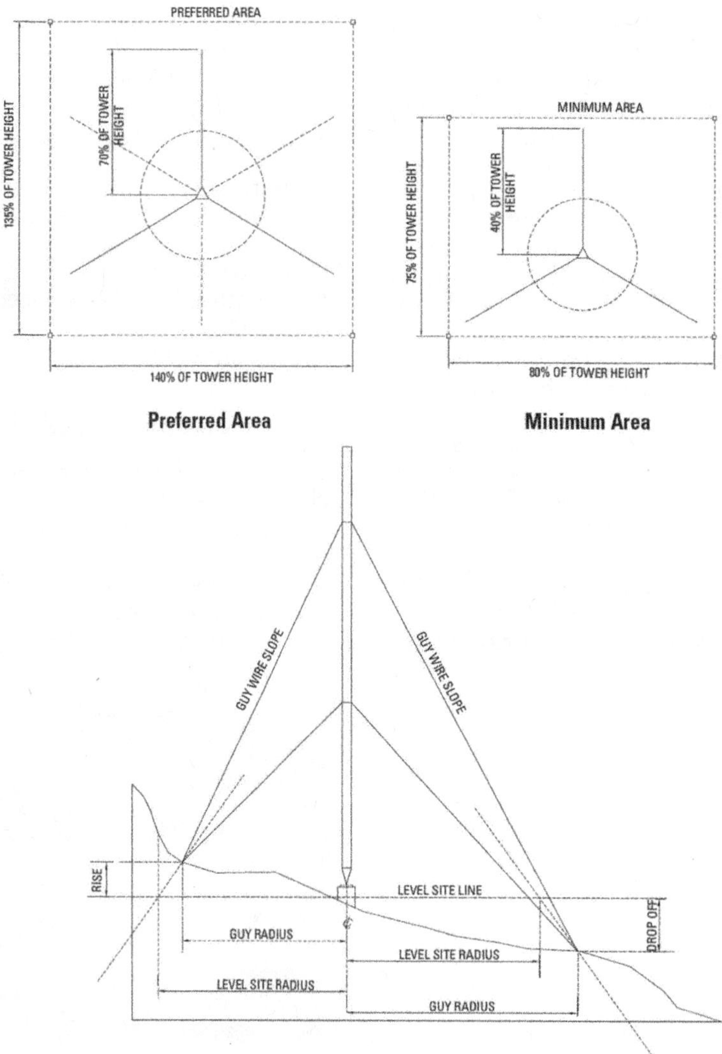

FIGURE 9.1-4 Area required for guyed tower construction.

Knowledge of the types of antennas and antenna placement on the structure is also a key consideration when designing the tower. The weight of a particular antenna combined with winds from a particular angle may cause the antenna to act as a sail, causing the tower to twist and sway with the force of the wind. This is known as deflection. Depending on the type of antenna to be mounted, designers will engineer the tower to meet maximum allowable deflection requirements in order to maintain its RF design needs. Depending on the usage requirements of the tower, performance requirements may necessitate greater strength than is required by the design standards.

The design-specific considerations required in any tower design and construction are not "optional" or "nice to have." They are critical to the structural integrity of the tower and to public safety and are crucial to allow designers to build the structure to the design conditions (wind, ice and/or seismic loads) at the particular location of the tower.

Communications towers are required to be designed and installed not only to meet the customer's requirements, such as height antennas, tilt/sway limitations, etc., but also to comply as well with local and/or national standards and building codes. In the United States, tower designs usually conform to the latest ANSI/TIA/EIA-222, "Structural Standards for Steel Antenna Towers and Antenna Supporting Structures." Design parameters include reference wind speed, topographical conditions, surrounding terrain (urban, rural or body of water), radial ice thickness, and twist and tilt limits, which are required to achieve optimum tower design.

Environmental Factors

Many environmental factors contribute to the design of a tower structure. Wind, ice and earthquake loads are some of these factors. However, the two main influences on tower design are wind and ice. Both values are dependent upon site location (e.g., ice is not a consideration in warm climates). National standards have wind and ice maps as a guideline; however, more accurate site-specific values can be requested and may be required for exposed locations.

Perhaps the largest of these design considerations is wind. Note that the higher the elevation, the stronger the wind. As the force of the wind hits the face of the tower or its attachments, it places a stress on the structure. This force causes tension in the supporting legs or guys on the side affected by the wind, effectively stretching them taut, while creating compression on the legs on the leeward side (or reducing the tension of the guys on a guyed tower) as the tower deflects (bends or gives) to the wind. A simplified illustration of the effect is shown in Figure 9.1-7. Properly

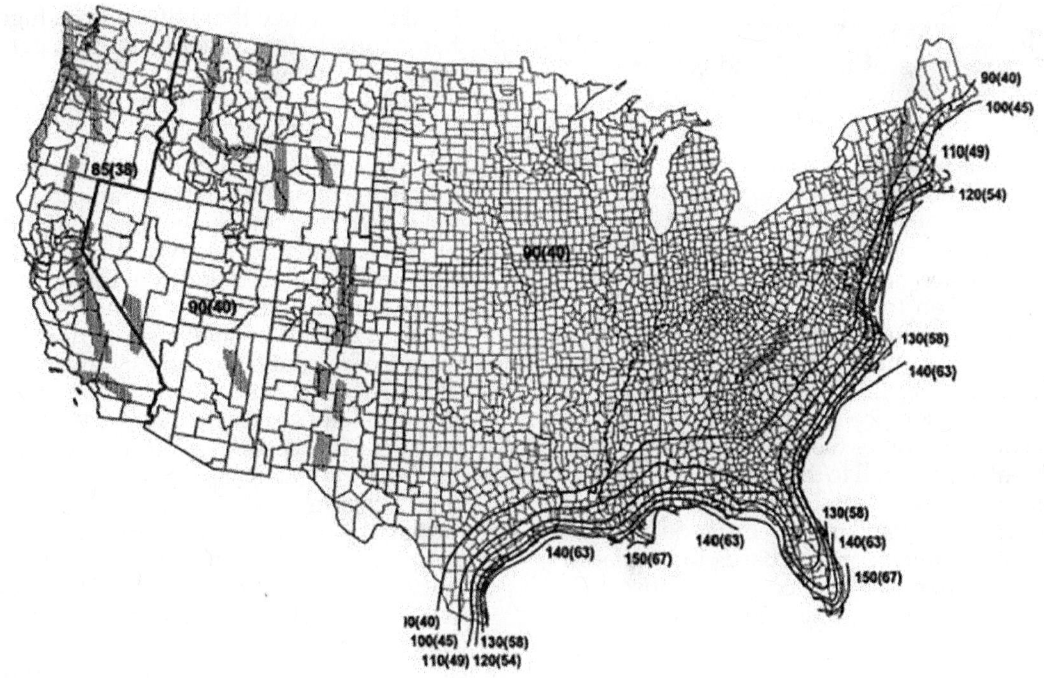

FIGURE 9.1-5 Nominal design wind speed map for the United States per ANSI/TIA-222-G.

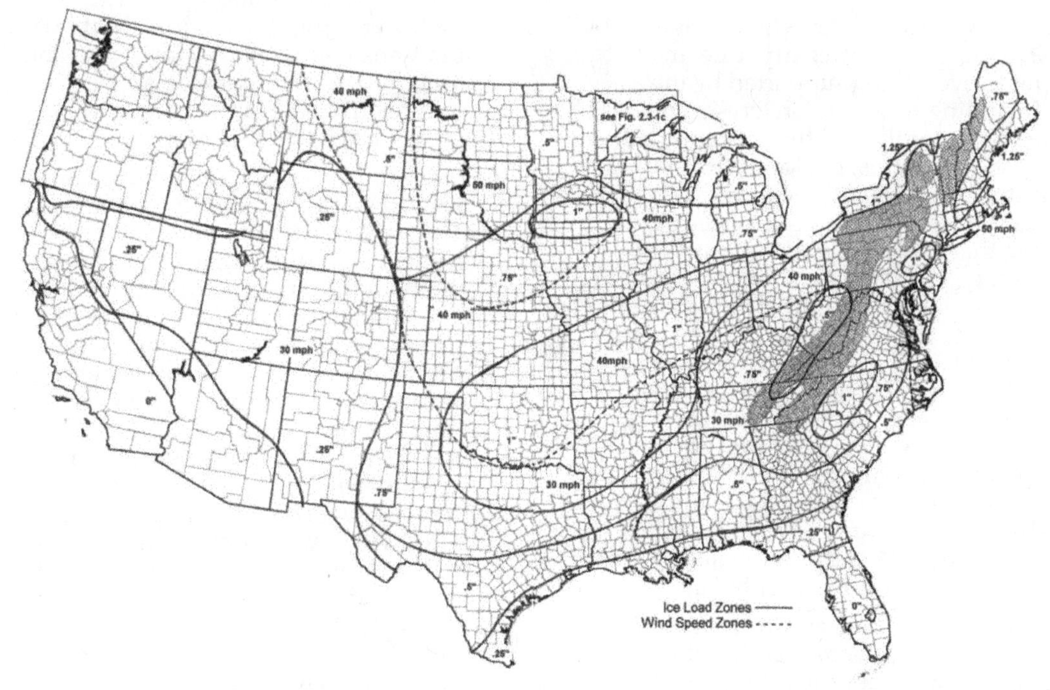

FIGURE 9.1-6 Nominal design ice thickness map for the United States per ANSI/TIA-222-G.

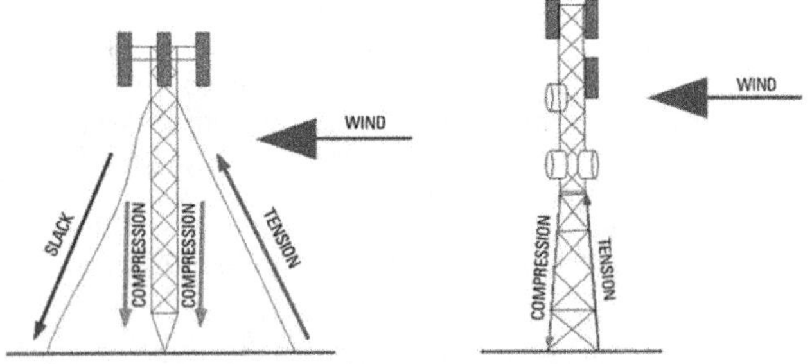

FIGURE 9.1-7 Examples of tension and compression areas of a tower as a result of wind force.

designed structures will take these factors into account and allow for the structure to return to its original position when the wind subsides.

Ice is also a primary consideration for tower design. The weight of the ice accumulated on the attached antennas, radial ice on the tower structure itself and the additional area resulting from the accumulated ice on the structure can result in overloading and structural failure of the tower or antenna mounts. Another issue that is a result from ice buildup is falling ice. As the air warms or ice becomes too heavy on a particular area, it may break off and fall to the ground, often striking antennas or waveguide cabling, causing serious damage or outright destruction. In order to prevent ice damage, it is common for ice guards to be placed just above lower antennas. The ice guard causes the ice chunk to break, thus protecting the critical device below.

DESIGN PROCESS

a) Calculation of Applied Loads: When designers refer to the loading on a tower structure, they are referring to everything supported by the structure, including transmission lines, antennas, antenna-support mounts, lighting systems, ladders and fall-arresting systems, platforms, ice guards, torsion resistors and other items, in addition to the initial tower members and components. It is worth noting that a tower purchaser must specify the quantities, type and levels of all present and future antennas that the tower will be expected to support. The wind load of an antenna or effective projected area (EPA) of it can be determined by calculating the drag coefficient of the antenna appropriate for the antenna shape and the projected surface area of the proposed antenna. These drag coefficients are usually determined by standards or by wind-tunnel test data supplied by the antenna manufacturers. Designers must also account for the EPA of the associated antenna mount and include this in the calculations.

b) Design and Layout of Transmission Lines: Transmission lines can place a significant load on a structure because they must run from each antenna on the mast down to the equipment shelter on the ground. Depending on the type of transmission line, this can result in a significant amount of loading applied to the structure. Proper placement or arrangement of the transmission lines can reduce the effect of the loading. The transmission line support mounts determine positioning of transmission lines on the tower. In addition, when ice forms around lines that are closely spaced, it can form a block that fills the gap between the lines. Designers consider this situation equal to a flat area that will attract a higher loading than a round area of a single line. Regardless of placement, transmission line shielding will be considered for every wind direction considered in the design calculations.

c) Structural Analysis and Member Sizing: The designer will use the design parameters discussed earlier to determine the general configuration of the tower structure, i.e., type of tower (guyed vs self-supporting) and width of tower, number of guy levels, etc. Based on the previously determined layout, a structural analysis is performed in order to determine the forces in the members and compare it to the member capacity in order to appropriately size each and every tower structural member. As an example of the output results of the structural analysis, Figure 9.1-8 shows a "Leg Load Diagram," which presents, in a graphical format, the force at each of the tower legs and compares it with the capacity of the tower leg at that specific elevation throughout the tower height. In terms of tower design, it is a wise choice for the designer to make an extra investment in tower leg capacity. Since the tower legs support the entire structure, adding extra capacity from the outset will eliminate costly tower modifications in the future, should additional loading be imposed upon the tower through the addition of extra antennas and transmission lines. Horizontal and diagonal structural members can be replaced or modified with relative ease; a tower leg, however, can seldom be changed once in place.

d) Twist and Sway Considerations: Twist and sway, or twist and tilt limits, refers to the amount of structural movement due to wind or seismic events that is allowed to occur yet maintain the ability of the structure to provide its service, with some signal degradation, during such events. These requirements may govern the design of the tower. Depending on the antenna size and frequency, the maximum twist and sway limits should be chosen. In a guyed tower, the positioning and size of the guys and torsion resistors will affect the amount of twist and sway in the tower. The size and position of the guy wires will have a direct effect on the amount of tilt in the tower. The torsion resistors will dictate the amount of twist. Generally, a tower with a wider face width is "stiffer" and therefore requires fewer guy levels than one with a narrower cross section. However, there is a trade off in the initial tensions and guy sizes required in a larger tower versus the

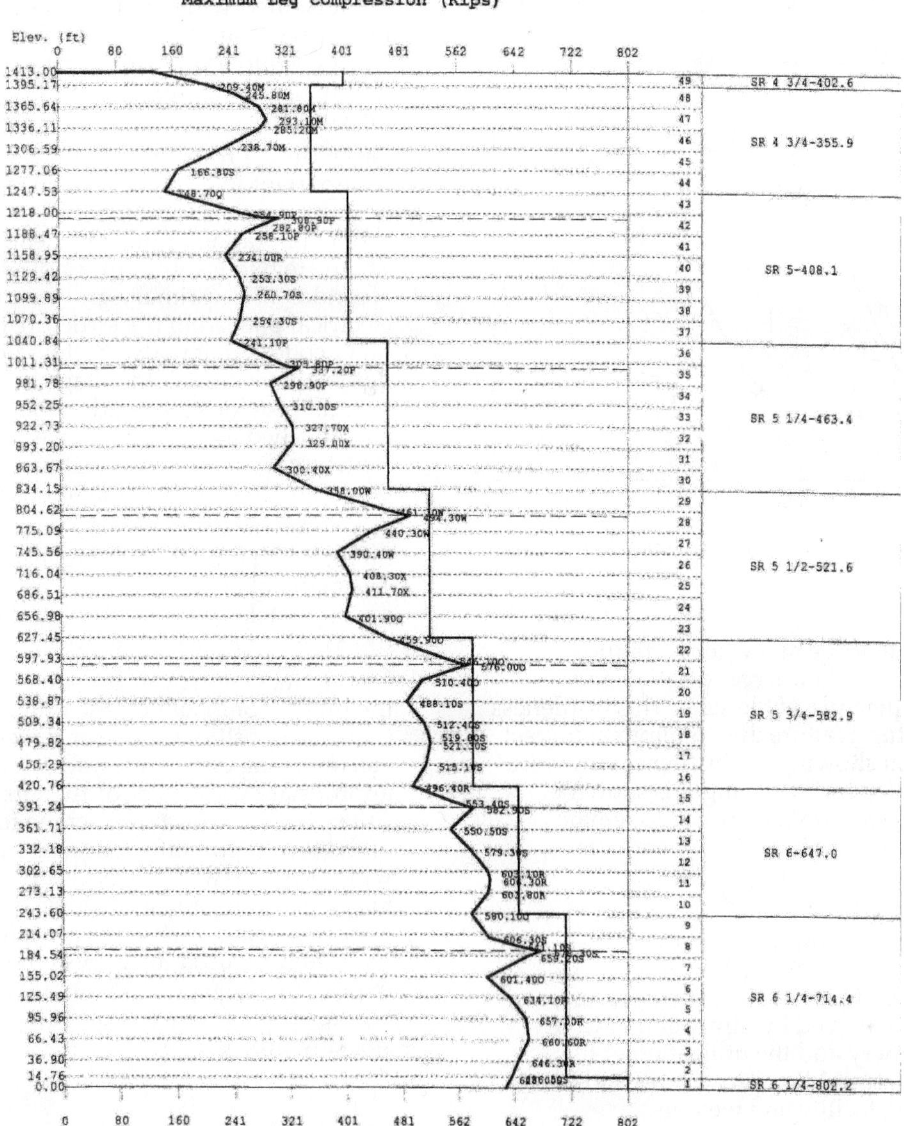

FIGURE 9.1-8 Leg load diagram.

number of guy levels and the face width required to provide an adequate structure. A good design will optimize the trade-offs so the appropriate degree of stiffness can be achieved.

e) Foundations Design: Obviously, even the best tower will fail if the foundations on which it stands have not been designed properly to meet the specific soil or rock conditions at the tower site. In order to design the most suitable foundation system for each intended tower site, a professional geo-technical engineer should perform a soil investigation and provide a soil or rock condition report. The soil condition report will include information such as allowable bearing pressure, compression strength, water table level, soil density, depth of frost penetration, corrosive nature of the soil and tension/compression capacities for pilings. Rock site condition reports should include: type and condition of rock, depth to sound rock, density, allowable bearing pressure, shear strength, water table level, recommended anchoring procedure and rock quality designation (RQD).

TOWER CONSTRUCTION

Manufacturing

a) Structural Steel: Of critical importance is the use of the correct grade(s) of structural steel for specific applications within the structure. ASTM standards list many types and grades of structural steel; however, not all are suitable for tower use. In fact, geographic location may play a determining factor as to the type of steel that is used in manufacturing.

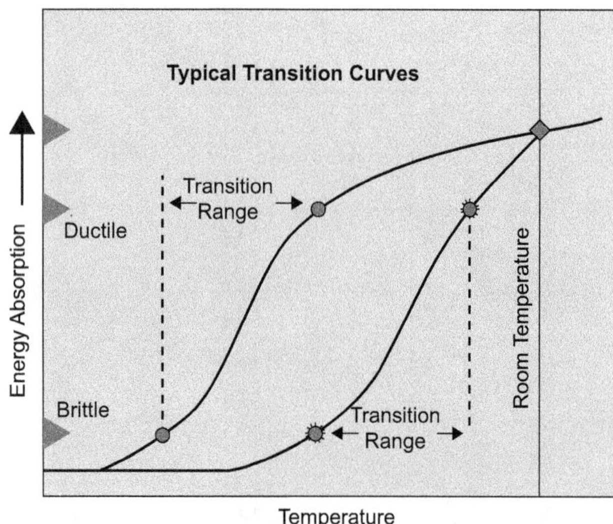

FIGURE 9.1-9 Temperature impact on steel brittleness.

Steel can become brittle at low temperatures, depending on the chemical composition and method of manufacture of the steel. This brittleness can cause structural failure due to minimal impact loads. A diagram showing the brittleness measure of steel as the temperature changes is shown in Figure 9.1-9; the steel utilized needs to remain ductile considering the lowest temperatures occurring at the site.

b) Manufacturing Quality: In order to ensure the quality of the end product is achieved, manufacturing tolerances need to be set and adhered to during the manufacturing process; this is typically achieved by utilization of CNC operated machinery and the utilization of the properly designed welding procedures and certified welding facility and welders.

c) Tower Finish: The finish of the tower structure is an important consideration in itself, with specific environmental and aviation compliance regulations. Towers are typically protected against corrosion by a hot-dip galvanizing process. This process is also regulated in accordance with ASTM A123/A153 to ensure quality standards. Galvanizing tower members and components can protect them against corrosion for up to 25 years in coastal locations and 50 years in urban locations. Painting of the towers, if required, must comply with Federal Aviation Administration (FAA) regulations in the United States or the appropriate country standards.

Installation

a) Foundation Installation: The first part of the tower to be installed is the tower foundations. An important consideration for the foundation installation is that the installation is surveyed and installed within tolerances as specified by the designer and the appropriate standards. Also, the supervision of the installation and the required inspections and material testing are important.

b) Grounding: Proper grounding of a tower structure is important in order to eliminate damage or destruction from lightning strikes and to ensure proper operation of the attached antennas. There are two levels of grounds: primary and secondary. The primary ground provides a direct electrical path from the structure to the earth or other conducting body for the safe discharging of lightning strikes, high voltage or static. Secondary ground provides the conducting connection for any attached apparatus, such as antennas to the structure. For safety's sake, the structure must also be grounded during installation.

c) Tower Assembly: Sections are typically assembled on the ground prior to being installed (stacked). The assembly bed used needs to be leveled and appropriately performed using the installation drawings and the bolt torque procedures.

d) Tower Stacking: Tower stacking or erection is the process of installing the different sections in their final positions. Shorter towers are typically erected using only cranes, which need to be appropriately selected for the lifts to be used. Taller towers would use a crane for the lower part of the tower (stub lift) followed by a ginpole/winch system, which is typically used for the installation process of sections of tall guyed towers. During the tower erection, temporary guy wires may be used to support the structure in the interim until the permanent guys are installed. Figure 9.1-10 shows a general layout of a tower stacking process.

INSPECTION AND MAINTENANCE

To ensure proper operation of the tower, regular inspection and condition assessments must be performed periodically. The ANSI/TIA-222-G recommends that the towers be inspected every 3 to 5 years and also if the tower was subjected to a major design event such as a major wind storm or a significant earthquake.

In order to perform the inspections properly, the tower is typically climbed; therefore, the RF levels of the different antennas may be reduced to allow for the safety of the climber.

The standard provides the following checklist as guidelines to the items that require inspection:

a) Structure Condition

This is checked for items such as:

1. Bent members (legs and bracing)
2. Loose members
3. Missing members

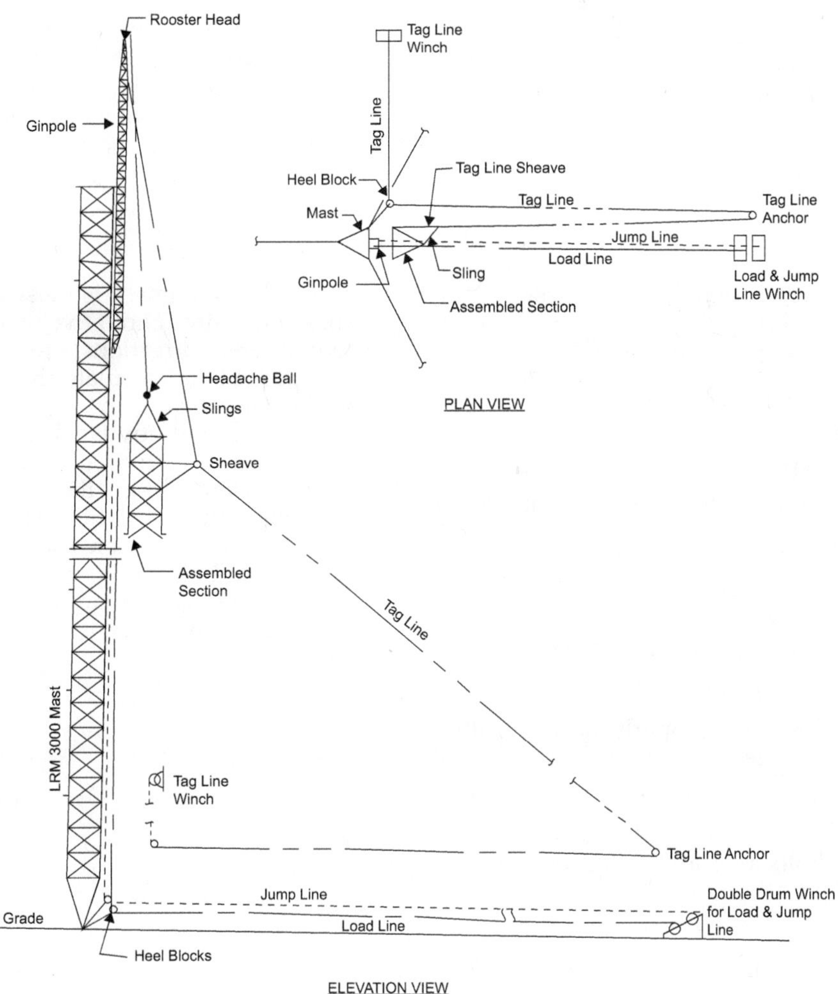

FIGURE 9.1-10 General layout of tower stacking.

4. Climbing facilities, platforms, catwalks (all secure)
5. Loose and/or missing bolts and/or nut locking devices

b) Finish

1. Paint and/or galvanizing condition
2. Rust and/or corrosion condition including mounts and accessories
3. FAA or ICAO color marking conditions
4. Water collection in members (to be remedied, e.g., unplug drain holes, etc.)

c) Lighting

1. Conduit, junction boxes and fasteners (weather tight and secure)
2. Drain and vent openings (unobstructed)
3. Wiring condition
4. Light lenses
5. Bulb condition
6. Controllers (functioning), flasher and photo control, alarms

d) Grounding

1. Connections
2. Corrosion
3. Lighting protection (secured to structure)

e) Antennas and Lines

1. Antenna condition
2. Mount and/or ice shield condition (bent, loose and/or missing members)
3. Feed line condition (flanges, seals, dents, jacket damage, grounding, etc.)
4. Hanger condition (snap-ins, bolt on, grips, etc.)
5. Secured to structure

f) Other appurtenances (walkways, platforms, sensors, floodlights, etc.)

1. Condition
2. Secured to structure

g) Insulator Condition
    1. Cracking and chipping
    2. Cleanliness of insulators
    3. Spark gaps set properly
    4. Isolation transformer condition
    5. Bolts and connection secure
h) Guys
    1. Strand condition (corrosion, breaks, nicks, kinks, etc.)
    2. Guy Hardware Conditions
        a. Turnbuckles or equivalent (secure and safety properly applied)
        b. Cable thimbles properly in place (if required)
        c. Service sleeves properly in place (if required)
        d. Cable connectors (end fittings)
            i. Cable clamps applied properly and bolts tight
            ii. Wire serving properly applied
            iii. Strand devices secure
            iv. Performed wraps—properly applied, fully wrapped and sleeve in place
            v. Poured sockets secure and showing no separation
            vi. Shackles, bolts, pins and cotter pins secure and in good condition
    3. Guy Tensions
        a. Tension within design tolerance
        b. Record temperature, wind speed and wind direction
i) Concrete Foundations
    1. Ground condition (settlement, movement or earth cracks, erosion)

       Site condition (standing water, drainage, trees, etc.)
    2. Anchorage condition
        a. Nuts and/or nut locking device (tightened)
        b. Grout condition
        c. Anchorages and/or anchor-bolt condition
    3. Concrete condition
        a. Cracking, spalling or splitting
        b. Chipped or broken concrete
        c. Honeycombing
        d. Low spots to collect moisture
j) Guyed Mast Anchors
    1. Settlement, movement or earth cracks
    2. Backfill heaped over concrete for water shedding
    3. Anchor rod condition below earth (maintain required structural capacity of anchor during exploration; attachment to temporary anchorage may be required)
    4. Corrosion control measures (galvanizing, coating, concrete encasement, cathodic protection systems, etc.)
    5. Anchor heads clear of earth
k) Tower Alignment

The tower alignment is measured from direct measurements throughout the tower height from at least two different directions. From these measurements, it is possible to calculate the degree of tilt or twist in the tower. Also, the guy tensions are measured for the different guy levels to determine their deviation from the required design values. These measured tensions are further corrected to account for the temperature variation. The standards TIA-222-G provide guidelines on the allowed tolerances for out of plumb. In a guyed tower, proper straightness may be achieved by adjusting the guy tensions at the various guy elevations.

CHAPTER

9.2

Lightning Protection for Towers

EDWARD A. LOBNITZ
TLC Engineering for Architecture
Orlando, Florida

INTRODUCTION

Understanding how lightning works as well as its effects on tower structures, grounding, and antenna systems is very helpful when trying to apply protection techniques that are practical. In fact, lightning protection is never 100% perfect, and understanding the limitations or cost-effective application of protection is an important part of the design process. Also, since no two installations are ever exactly alike, repetitive siting of towers and locating antennas on existing towers will always require a review of standard protection to assure that expected protection has not been compromised.

Lightning is potentially one of the most hazardous natural occurrences on our planet. It can kill, maim, start fires, cause explosions, damage equipment, interrupt critical data transmissions, stop a shuttle from flying, and cause many other disasters. It is also beautiful, awe inspiring, exciting, mysterious, and challenging to all who are enchanted by its technical fascination. It is both predictable and unpredictable. Statistics such as the following are both surprising and sobering when attempting to protect against such a power:

- From 1959–2003, the total injuries in the United States were 3,696.
- Fatalities by state from 1959–2003:
  - Florida: 425
  - Texas: 195
  - North Carolina: 181
  - New York: 134
  - Tennessee: 133
- A typical Florida thunderstorm can unleash 10,000 lightning strikes.
- In Florida, lightning kills approximately 10 people and injures 35 people every year. It is more deadly, in recent years, than hurricanes, tornadoes, or floods.
- Nationally, about 60 people a year die from lightning and 300 are injured.
- Lightning strikes the earth 100 times each second. Over the continental 48 states, an average of 20,000,000 cloud-to-ground flashes have been detected every year since 1989. About half of all flashes have more than one strike point, so at least 30 million points on the ground are struck each year.
- There are about 45,000 thunderstorms annually across the United States.
- Property and crop losses due to lightning totaled $25.7 million in 2003.

Methods of protection from lightning have been changing ever since Benjamin Franklin's original studies of lightning and the creation of his lightning rod system, today called the Franklin Rod System. Although the exact physics of structure lightning rod protection originally was not fully understood, extensive research has greatly enhanced our understanding of lightning physics while also creating great controversies regarding theory and protection techniques. Also, electronic equipment has entered the picture as one of the fastest-growing technologies of our century, but with inherent weaknesses to the effects of lighting. This has created yet another huge industry known as surge protection, with its own confusing mix of technologies, theories, manufacturing methods, and jargon to help further the protection methodologies and challenge protection designers with almost unlimited technical choices.

Fortunately, there are some standards that have been and are being developed to help protection designers through the maze of technologies available in order to evaluate effectiveness, technology comparisons, test results, performance claims, and other criteria. Many of these standards are listed in the references at the end of this chapter.

These standards and publications are considered *consensus* standards since their committee members or testing agencies are made up of members from across the industry or third-party testing agencies. Only equipment that is verified to be in compliance with these standards and testing agencies should be used by protection designers. Claims by other equipment manufacturers regarding performance or construction should not be considered unless full verification and proof of claims by third-party testing are provided. Lightning can cause great losses to life and property, and assumed protection must be equal to the task without question. More cannot be expected from lightning protection than it can be proven to provide, regardless of guarantee or guarantee claims.

The purpose of this section, therefore, is not only to examine protection techniques for the present generation of towers, antenna system equipment, and protection methods and equipment but to aid in preparing the reader to analyze new and ever-changing environments.

LIGHTNING AND STORM DEVELOPMENT

The intensity of lightning storm activity throughout the world varies by location. Until recently, thunderstorm activity was measured by the number of *thunderstorm days* observed and reported. In the United States, data is reported through the National Weather Service and displayed on an isokeraunic map (see Figure 9.2-1). A thunderstorm day is defined as any day during which thunder is heard at a specific observation point. Since these observations merely confirm the presence of lightning and do not provide information regarding severity, number of ground strokes, and stroke location, more complete data was needed to determine exposure risk. Atmospheric measurements throughout the United States [1] and other countries have led to new maps depicting ground stroke intensity by location (see Figure 9.2-2). Data is now available in the United States and, increasingly, throughout the world to determine not only whether lightning occurred but also stroke coordinates, discharge characteristics, flash density, direction of storm movement, and other data that can be manipulated into very accurate charts, graphs, trend analyses, histographs, etc. Access to this extensive database can be secured through private and public sources via the internet and may be useful in selection of broadcast antenna sites and level of protection.

Thunderstorms are of two general types: convection storms and frontal storms.

Convection Storms

Convection storms are usually local in extent, of relatively short duration, and account for the majority of storms in the United States. They are caused by local heating of the air near the earth and, therefore, occur mainly during summer months and in warmer climates where moisture is present, although many convective storms are generated over mountainous areas. These storms can generate to great heights of over 40,000 ft., where vast amounts of moisture and high temperatures are present, producing intense lightning activity. However, they dissipate quickly because accompanying cloud shade and rain cool the earth and dissipate the clouds' source of energy.

Frontal Storms

Frontal storms extend over greater areas, may continue for several hours, and are usually more dangerous, causing greater damage than convection storms. These storms develop from the meeting of a warm, moist weather front with a cold front that may at times extend for several hundred miles, exposing large areas to severe lightning discharges.

Figures 9.2-3 and 9.2-4 show typical convection (summer) discharges and frontal (winter) discharges, respectively.

The formation of lightning in all thunderstorms is generally believed [1] to be caused by ice/dust particles colliding with moisture particles in vertically generated air currents. The collisions create positive and negative charges within the cloud that eventually separate out so that negative charge concentration occurs near the bottom of the cloud and positive charges concentrate at the top. The negative charge at the bottom of the cloud causes a corresponding positive charge on the earth below the cloud. This charge difference between cloud and earth will commonly develop electrostatic field strengths of 100 KV per meter and higher. The field strength above the top of a 50-meter, well-grounded broadcast tower, therefore, could approach several million volts and result in upward charge streamers and corona discharge. A lightning strike to the tower is then a very real possibility.

The lightning process actually begins as a faintly visible stepped leader originating normally from the base of the storm cell. This leader carries the negative cloud charge toward the positive ground charge, forking and branching due to wind and random charge pockets in the air. As the leader nears the earth, one or more upward streamers will be initiated, usually from the tallest grounded bodies in near proximity to the downward leader. At some point, the downward leader will connect to an upward leader, and return stroke currents flow in an effort to equalize the charge difference between the storm cell and the earth. The distance between the connection of the downward leader and the object producing the upward leader is

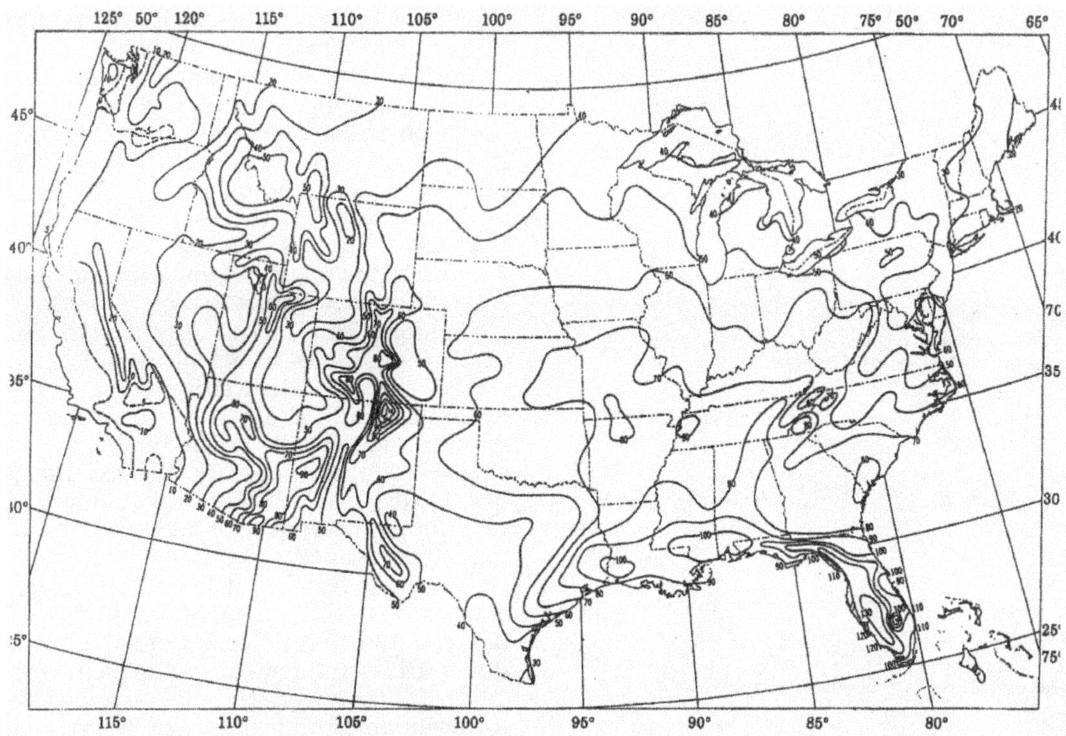

FIGURE 9.2-1 Average number of thunderstorm days annually.

Reprinted with permission from *NFPA 780, Installation of Lightning Protection Systems*, © 1997, National Fire Protection Association, Quincy, MA 02269. This reprinted material is not the complete and official position of the National Fire Protection Association; on this reference subject, the NFPA's position is represented only by the standard in its entirety.

FIGURE 9.2-2 Composite of measured lightning flash density per year, 1989–1992; flashes per square kilometer.

Data supplied by the National Lightning Detection Network

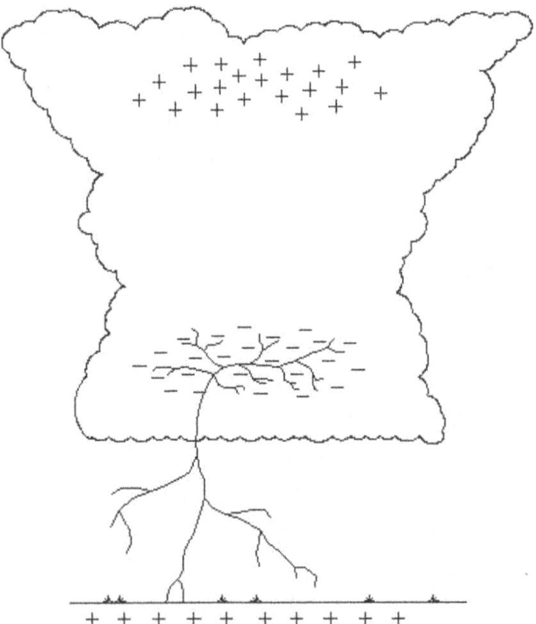

FIGURE 9.2-3 Typical mode of discharge in summer storms (negative stroke predominance).

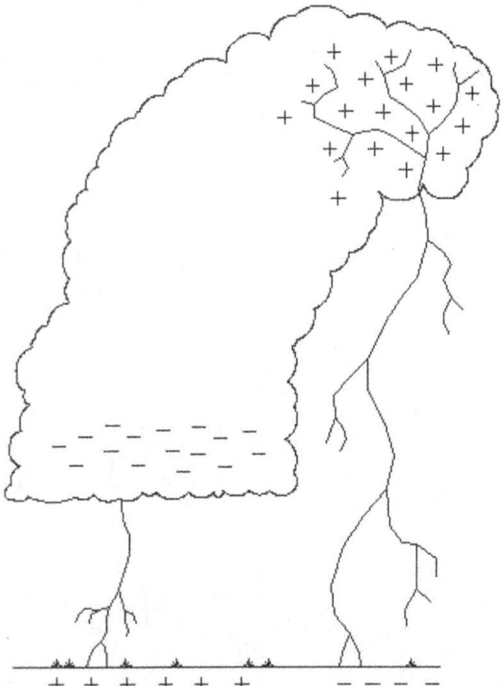

FIGURE 9.2-4 Typical discharge in winter storms (positive stroke predominance).

called the *striking distance* and is the basis for the rolling ball theory, to be discussed later. The striking distance is most greatly influenced by the amount of charge in the downward leader such that the greater the charge, the greater the striking distance. Basically, the amount of leader charge and the surge impedance of the object being struck determine the value of lightning current. Lightning, however, does not always perform as would be expected from this discussion.

In addition to bottom-of-cloud strikes, severe, long-traveling lightning strokes can emanate from the positively charged top-of-storm cells to negative earth charges. Also, grounded objects above 150 ft. tall can initiate upward-leading charges, reversing the previously mentioned process. Grounded objects taller than 150 ft., such as broadcast towers, can be struck on the side of the tower in lieu of the top and can subject side-mounted antennas to direct strikes.

Lightning currents range from an average value of 20,000 amperes to a high recorded value of over 400,000 amperes. Many strikes contain up to 30 strokes just milliseconds apart. These strokes appear to flicker or strobe; are usually wider, brighter, and hotter than single strikes; and can cause serious damage if not adequately dissipated. Figure 9.2-5 shows typical lightning propagation development. Figure 9.2-6 provides probability values of lightning stroke peak values.

Lightning can affect broadcast tower structures and associated equipment or studio buildings in basically two ways. These generally take the form of direct strikes and remote strikes. Direct strikes usually attach to the tower structure, building air terminals, tower-mounted antennas, and tower warning lighting. Remote lightning strikes can inductively couple the surge field into the tower structure, coaxial cable shields, warning lighting conduit systems, building rod systems, and antennas. Both types of strikes can cause considerable damage if proper protection is not provided in the design of the tower/building system and in applied surge/lightning protection equipment. Design techniques include proper grounding and bonding, selection of tower/building location, coaxial cable routing, and bulkhead design. Protection techniques include proper selection and placement of surge protection devices for coaxial cables, lighting warning circuit, equipment building power service, tower-mounted preamp equipment, and emergency power equipment, as well as air terminal protection for side-mounted antennas mounted above 150 ft. and for the equipment building located outside the tower-protected zone. Locating new antennas on existing towers requires a special analysis of existing conditions, as will be discussed later.

As mentioned earlier, the radiated impulse field from a remote lightning strike can pose a significant hazard to equipment and systems, particularly those that are interconnected by long lengths of cable. Figure 9.2-7 is a composite made by a number of researchers of lightning electromagnetic field measurements that have been normalized to a distance of 10 km. The figure shows a frequency domain distribution that peaks at about 10 kHz at an intensity of slightly more than 1 volt per meter. It is important to realize that nearer strikes can create field strengths many orders of magnitude higher than those shown. The predominant low-frequency component is also very effective in coupling energy into systems of wiring, producing continuous frequency, and ringing waveforms due to inductance, capacitance, and resonant conditions, even if wiring is buried in the ground.

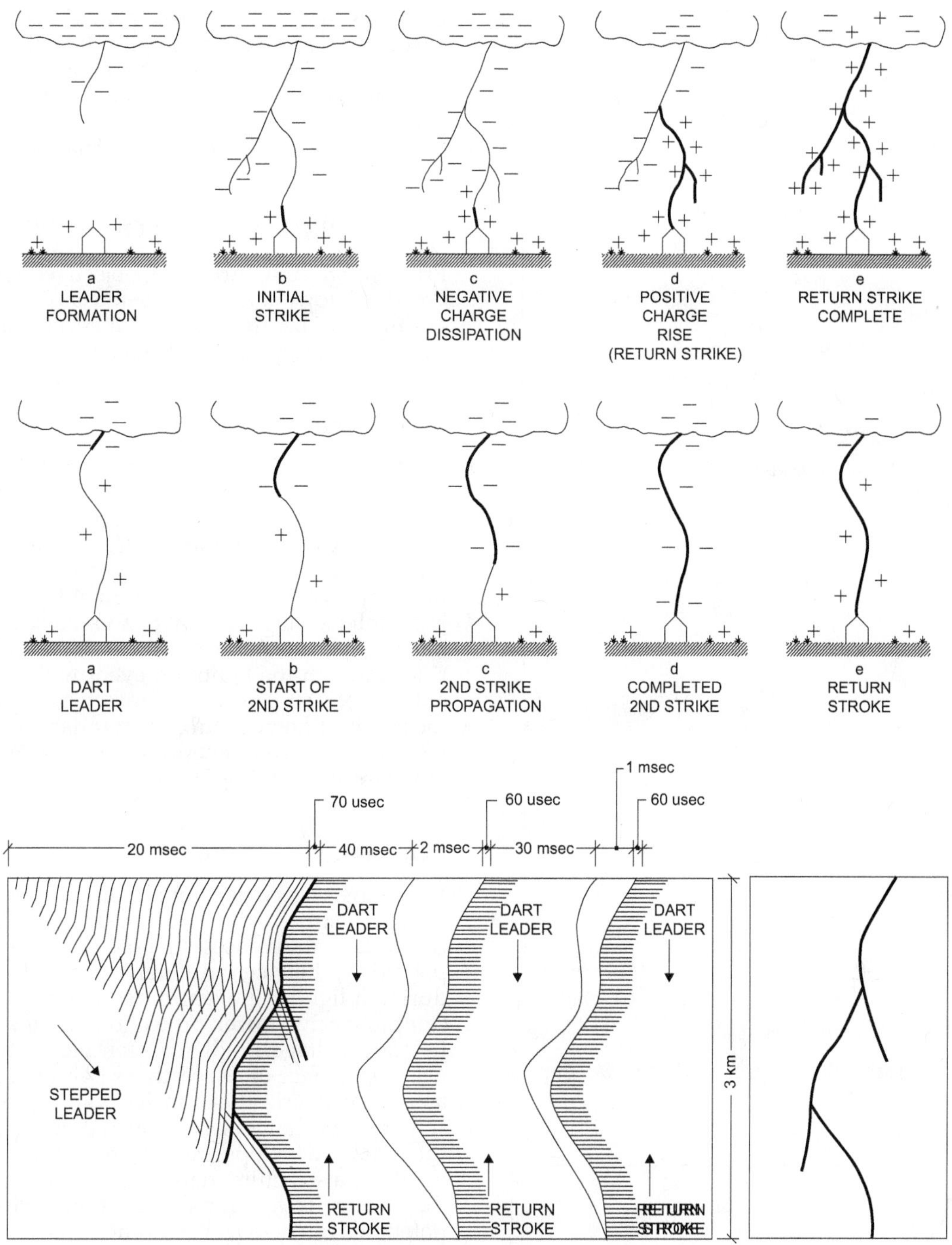

FIGURE 9.2-5 Typical lightning propagation development.

The time domain current waveform associated with a typical lightning strike is characterized by a very fast leading edge or rise time, followed by a more gradual decay. Technically, rise time is the period of time required for the wave to increase from 10% to 90% of its crest value. Decay time is normally expressed as the time measured between the wave crest and 50% of the crest value. A description of a waveform such as 1.5 × 50 μs for a 10–20 KA strike would indicate a single impulse waveform with a rise time of 1.5 μs and a decay time of 50 μs, to half peak value.

There are many variations on actual lightning waveforms seen in real-world circuits. Waveforms such as those shown in Figure 9.2-8 may be found with rise times measured in a fraction of a microsecond near the point of lightning entry to a circuit. Normally, however, the higher the lightning current, the slower the rise time and decay period, such that a 250,000 amp strike will have a much slower rise time (30 μs) and 250 μs decay compared to a 10–20 KA strike. As a wave propagates through a wiring system, the rise time and the decay time will lengthen. The polarity of the impulse may be either positive or negative. As

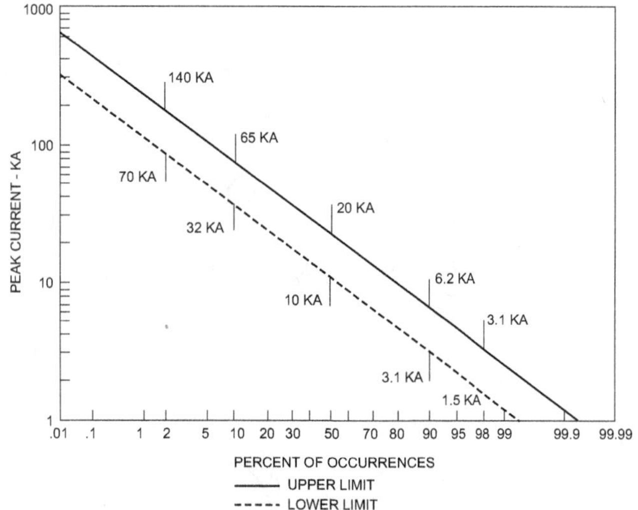

FIGURE 9.2-6 Lightning stroke intensity.

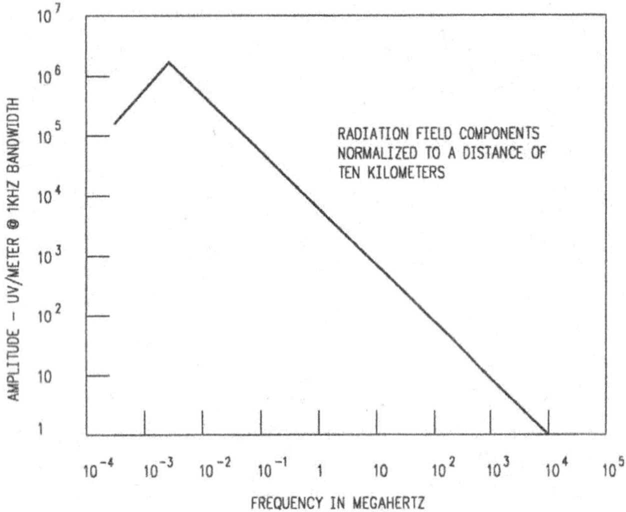

FIGURE 9.2-7 Lightning signal amplitude versus frequency.

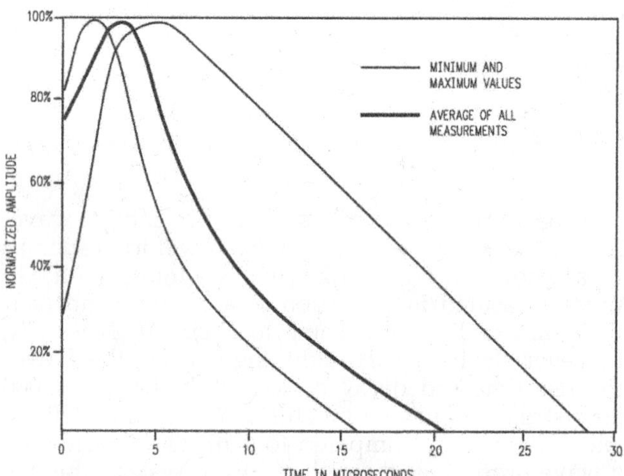

FIGURE 9.2-8 Typical lightning waveforms.

in remote strokes, inductive and capacitive properties of the wiring system may cause the circuit to act as a resonant tuned circuit, producing a ringing wave that alternates in polarity. References that categorize the waveforms and current levels for several types of circuits are listed at the end of this chapter.

PROTECTION OBJECTIVES

Primary lightning protection objectives, specifically for broadcast towers and associated building equipment wiring, may be grouped into the following two basic categories: equipment and personnel protection.

Equipment Protection

Control of small potential differences that are deadly to electronic equipment and antenna cable, as measured between active circuitry and grounded media, is key to protecting equipment. Controlling the potential differences to a value below the equipment damage threshold will ensure the equipment survives. Providing tighter voltage control to a value below the equipment upset threshold will help to ensure the system rides through the lightning event without any noticeable effect. All protection, however, is predicated on the assumption of average or standard lightning strike parameters, such as energy waveform, and can, therefore, never be 100% effective.

Personnel Protection

Protecting service personnel from the threat of a direct lightning strike, secondary flashing (side flash), and controlling differences in potential (step and touch potentials) between different parts of their bodies during a lightning event must be integrated into any comprehensive tower protection scheme. Step potentials are voltage gradients seen along the surface or near the surface of the earth, as lightning current radiates hemispherically from its point of entry into the soil or grounding system. Touch potentials are voltage differences developed in horizontal and vertical elements of a structure, natural object, or system during the passage of lightning current. Both step and touch potentials can be hazardous and must be minimized.

Secondarily, the broadcast tower and associated building should also be reviewed for protection against lightning's effects. In most cases, however, towers and equipment buildings will be self-protected, as will be discussed later, and provide a means to intercept lightning strikes, conduct the lightning current safely through or around the structure, and dissipate the current into the earth. These characteristics, however, are critical to personnel and equipment protection. Lightning current passing uncontrolled through a structure may result in deterioration of tower joints, ignition of combustible materials, generation of explosive forces in masonry and other moisture-bearing materials, and burning or tearing of roofing systems. Secondary flashing between the primary current path and nearby unbonded grounded objects may also pose a threat to persons in or near elements of the structure.

PROTECTION STRATEGIES

A typical broadcast tower installation will consist of an antenna tower in association with a studio building or an equipment hut for repeater locations. The preferred building type would consist of structural steel framing to maximize equipotential grounding and bonding characteristics. Where the buildings are not fully within the zone of protection provided by the tower or other adjacent structure, they must be provided with a Franklin Rod-type system. Design requirements for this system are adequately documented and described in NFPA 780, *Standard for the Installation of Lighting Protection Systems* [2] and as supported by UL 96A [3].

Broadcast towers are normally three- or four-legged structures constructed of galvanized tubular or structural steel with sectional vertical truss elements. They are either self-supporting or guyed and can be of various heights, from a satellite dish near the ground to a 1-600-ft. or higher antenna-support structure. Lately, however, some architecturally creative-type towers of both steel and concrete have been introduced that create some interesting challenges to lightning protection methodologies. Also, some towers are ground mounted, some are mounted on top of concrete or steel buildings of various heights, and some are large enough to span multilegged over the broadcast building. It is, therefore, impossible to describe detailed protection techniques for all type of towers. Basic requirements, however, will be presented so that application to any type tower can be easily extrapolated. Where new antennas are mounted on existing leased towers, an analysis of existing conditions and protective techniques must be made and improvements or corrections made as necessary to maximize protection. Lessors may, however, object to making changes.

The tower structure itself is the basic lightning protection element that helps dissipate direct lighting strikes; minimize voltage rise on associated antennas, coaxial cable, and equipment; and protect adjacent buildings. Where new towers are located in populated areas, increased lightning activity and dissipation of ground currents can cause local controversy, increase risks, and influence tower design complexity beyond the basic requirements presented herein.

Tower protection elements consist of the following:

- Grounding
- Coaxial cable shield bonding
- Enhanced strike attachment
- Warning light bonding
- Maintenance of joints
- Protection for side- and top-mounted antennas
- Grounded versus isolated guys

Proper tower grounding is essential for the protection of all other systems and facilities interconnected with the tower. It should consist of exothermic weld connections between the tower base or legs and foundation-reinforcing steel, anchor bolts, grounding radials, counterpoise rings, guy anchors, and adjacent building counterpoise and fences.

Coaxial cable shields from antennas must be bonded to the tower at the top, near the antenna, at each guy attachment point, at the midpoint of any 150- to 200-ft. interval between established bonding points, and at the bottom where the cable leaves the tower. The cable should exit the tower as low as possible to minimize voltage gradient and current dissipation at the bulkhead panel on the building.

Air terminals mounted on top of the tower will enhance the probability that the rod will be struck in lieu of the tower structure or any side-mounted antennas, thereby minimizing maintenance from damaged antennas and pitting of tower steel. Any antenna or other tower-mounted components should not be mounted within 12 inches of the top of the air terminal or beyond a 45-degree cone of protection from the top of the air terminals. If the tower is over 150-ft. high, side-mounted antennas above this level are vulnerable to direct hits and should be protected by mounting horizontal air terminals above and below the antennas, protruding at least 6 inches beyond the antenna. Since they are horizontal and located in the end nulls of the antenna pattern, the air terminals should not affect system performance. See Figure 9.2-9.

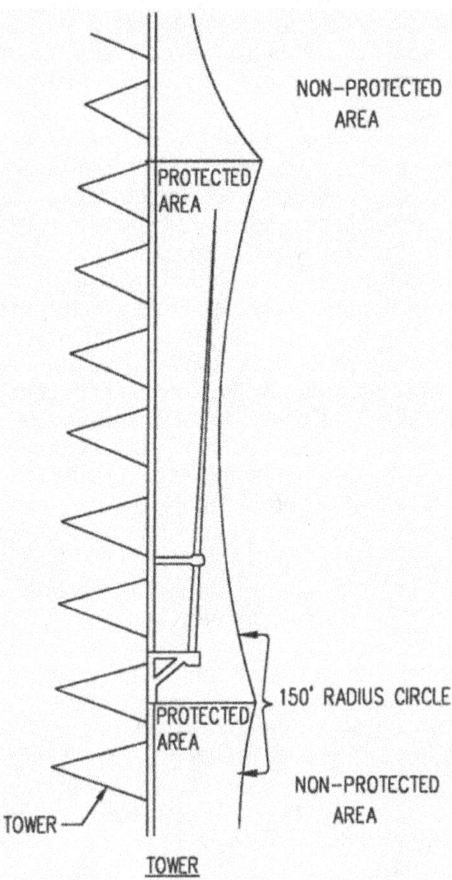

FIGURE 9.2-9 Side antenna protection method.

The warning light(s) on the tower must be circuited in rigid metal galvanized conduit, and the fixture itself should be constructed of metal to minimize surge currents on the wiring that would be difficult and expensive to eliminate from the source panel board in the adjacent building. Surge suppression on the lighting circuit would still be required at the panel board, but it could be selected to handle minimal surge current. Also, any low-voltage power or control wiring serving tower-mounted amplifiers, heaters, and other equipment must also be installed in rigid metal conduit.

A separate lightning protection down conductor is not necessary for adequate dissipation of lightning currents. The tower itself will always conduct the bulk of any lightning strike, although a portion will travel on the coax shield and warning light conduit. An additional down conductor that is attached to the tower at intervals will have a negligible effect on the division of current paths due to the inductance distribution of all components.

Lightning currents traveling down towers can, over time, cause arcing at tower joints, stripping away galvanizing and resulting in rusting. Maintenance, therefore, of the tower system is important to long-lasting, well-performing components. Air terminals subject to heavy lightning currents of long duration can cause pitting and melting of points as well as severe mechanical stresses in mounting hardware. Bonding straps, cable fasteners, and other miscellaneous hardware can loosen and allow movement or "whipping" of conductors from electromagnetic forces associated with unequal down-currents in various components of the tower system. Therefore, at least yearly maintenance is recommended.

For guyed towers, guy wires can be very effective in dividing the lightning currents into additional paths to ground, therefore lowering the current and voltage stress to tower components and the service building bulkhead. Where guy wires contain insulators, special precautions must be observed. The insulators will be subject to voltage differences during lightning strikes occurring between guy-to-tower attachment points and the ground potential at the anchors due to ground currents and induced guy wire voltages. These voltage differences can reach hundreds of thousands of volts and can create arcing or "tracking" over insulators, resulting in ineffective isolation over time. Therefore, maintenance inspection of the insulators and selection of maximum insulative values for the insulators are very important.

The rolling sphere concept is a protection strategy that is fully described in NFPA 780 and effectively applies to towers up to 150-ft. high above level ground. Figure 9.2-10 shows traditional lightning protection for a studio facility with an adjacent microwave relay tower. Protective zones are based on an imaginary "rolling sphere" that is 150 ft. in radius being passed over the structure. The sphere is also rolled around the structure tangent to earth. Air terminals are placed in such a way that the ball never contacts the structure or other objects requiring protection. Properly protected masts, light poles, and adjacent buildings may be taken into account when establishing zones of protection.

The protective zones created by the 150-ft. radius sphere have proven statistically adequate for most facilities and are the basis behind most codes and standards [4,5,6]. An interesting way to think about structural lightning protection design is to imagine rolling an inked sphere of 150-ft. radius around and over a structure. The ball should roll only on the earth, air terminals, and other suitable metallic components that have been connected to the lightning protection system. Any other area that receives ink should be considered unprotected. Where grounded guy wires are present, the sphere rolls over the wire in similar fashion to protect items under the projected zone. This is similar to overhead ground wire protection, described fully in NFPA 780.

Most broadcast towers, and much of the equipment they support, have sufficient mass and conductivity to resist damage from direct lightning contact. Localized burning, pitting, and mechanical forces may occur in the immediate area of contact with the lightning channel. However, the brief duration of a lightning stroke creates little heating in areas away from the point of lightning contact. Air terminals, shown in both Figures 9.2-10 and 9.2-11, when properly bonded to the tower, may be used to protect more sensitive objects. Whenever practical, coaxial transmission lines and other wiring should be extended up the inside face of the tower to minimize the possibility of direct lightning contact.

Personnel Protection

The threat to personnel during a lightning strike ranges from the obvious danger of direct contact with a lightning strike to the more obscure effects of step and touch voltages. Protection from a direct strike, when near or within structures, is accomplished with traditional rolling sphere concept methods. However, to ensure that an adequate protective zone is provided in areas frequented by personnel, the rolling sphere should be reduced to a 100-ft. radius.

Step and touch potentials are created as a lightning current passes through resistive soil and other available paths as it dissipates into the earth. A person in contact with only one point of the gradient will simply rise and fall in potential with the gradient without injury. A person in contact with multiple points on the earth or objects at different potentials along the gradient will become part of the current path and may sustain injury or death.

Figure 9.2-11 [7,8] indicates a number of methods for protecting personnel from the direct and secondary effects of lightning. A typical tower/transmitter site is used as an example. A technician responding to a service problem during a thunderstorm would likely exit his or her vehicle outside the gate, unlock and open the gate, and move his or her vehicle into the inside yard. The technician would then leave the vehicle and enter the building.

The threat of a direct lightning strike to the technician has been minimized by establishing a protective zone over the areas to be traversed. This zone is

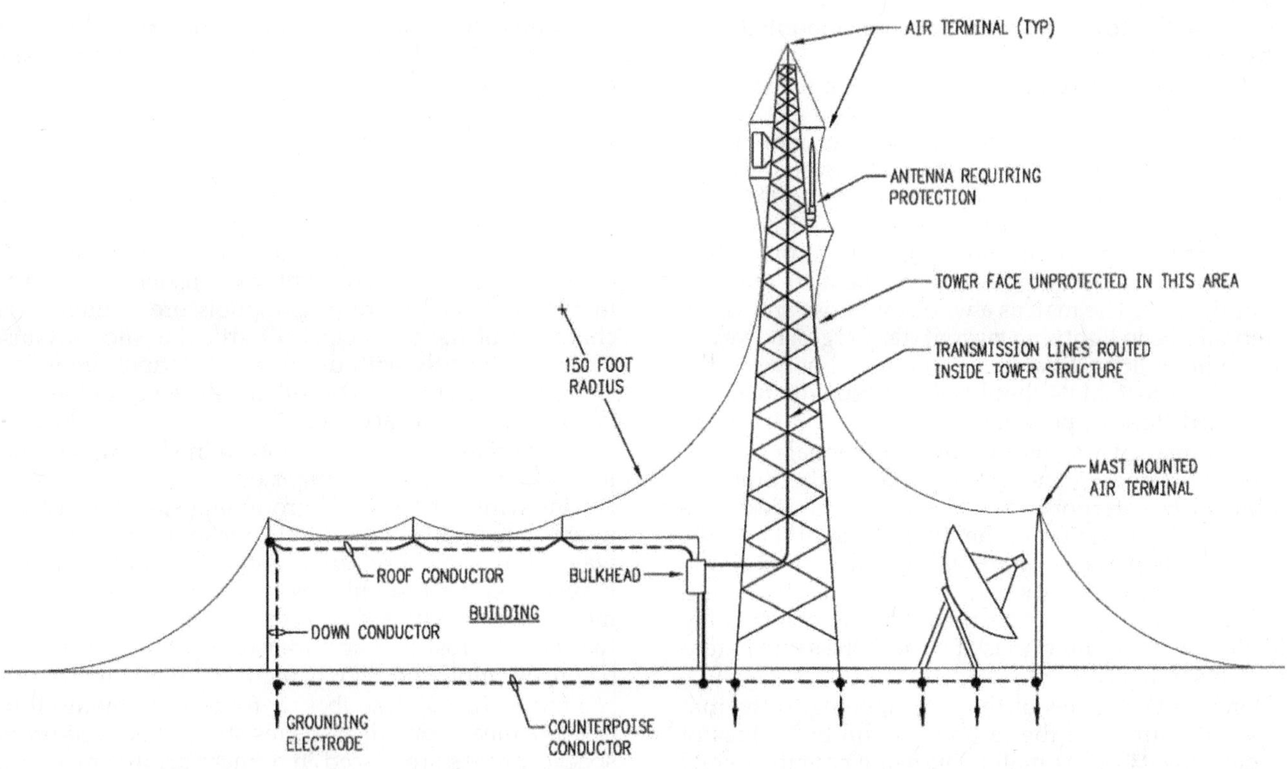

FIGURE 9.2-10 Example of rolling ball theory application.

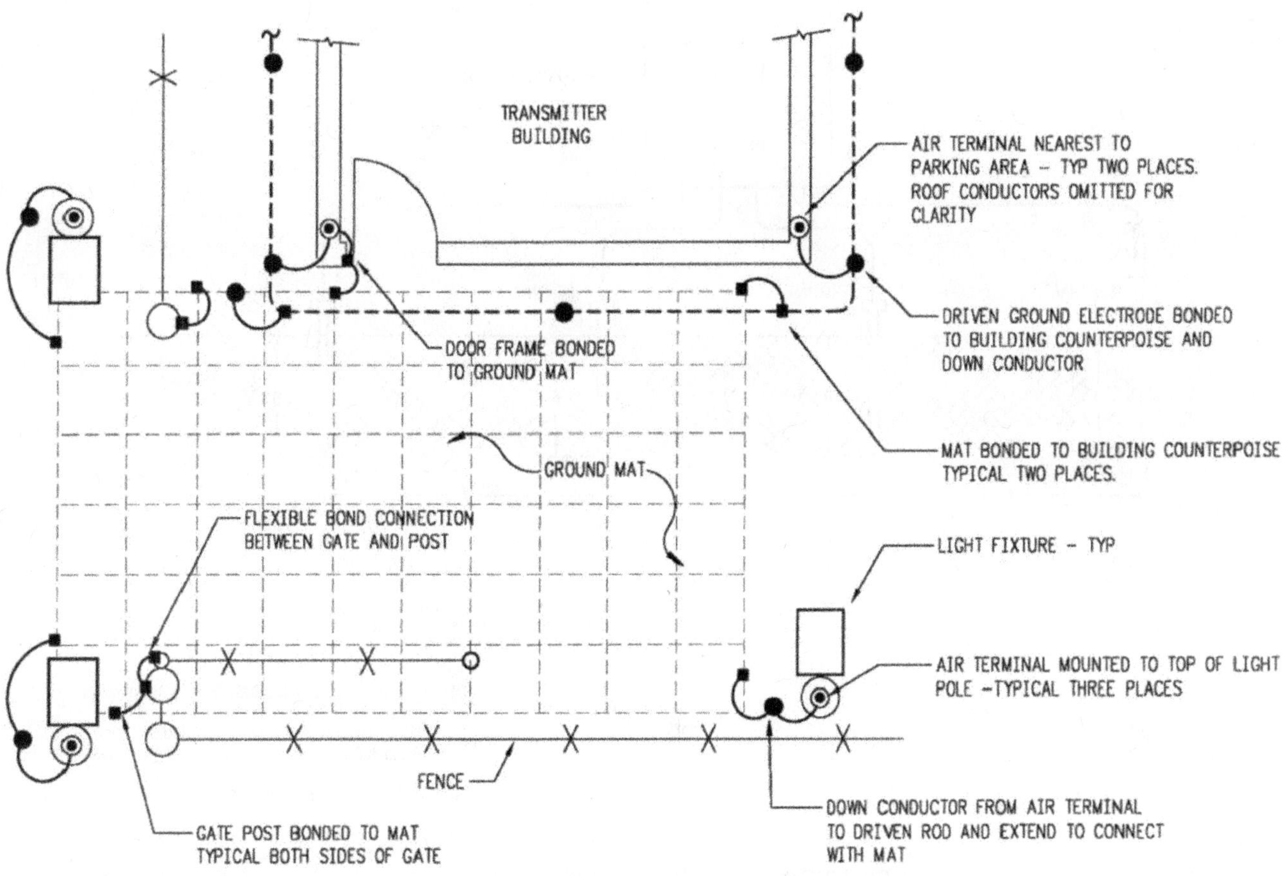

FIGURE 9.2-11 Personnel protection methods.

created by the tower and air terminals mounted atop light poles.

Step potentials are minimized through the use of a ground mat buried just below the surface of the area where the technician is expected to be outside the vehicle. Ground mats are commercially available, fabricated in a 6 × 6 inch square pattern using #8 AWG bare copper wire. Each intersection is welded, creating, for all practical purposes, an equipotential plane that short-circuits the step potential gradient in the area above the mat. The mat, as a whole, will rise and fall in potential due to lightning current discharges; however, there will be little difference in potential between the technician's feet. Mats should be covered with 6 inches of crushed stone or pavement.

The threat of dangerous touch potentials is minimized by bonding the ground mat to the building perimeter counterpoise, the fence at each side of the gate opening, bonding to the door frame of the transmitter building door, and providing a flexible bonding connection between the swing gate and its terminal post. Such bonding ensures that the object being touched by the technician is at or near the same potential as his or her feet.

Bonding both sides of the gate opening to the mat helps to ensure that the technician and both sides of the gate are at approximately the same potential while the gate is being handled. The flexible bond between the gate and its support post may be accomplished using a commercially available kit or by exothermically welding a short length of flexible 2/0 AWG welding cable between the two elements.

External Ground System

The effectiveness of a grounding system is a function of the type and extent of the electrode system used and the resistivity of the surrounding soil. Soil resistivity is dependent on the quantity of free ions (chemical salts) in the soil, temperature, and moisture content. The character of the soil below a particular site may also vary significantly with depth and location due to layering of different types of soil, the presence of hardpan layers, and subsurface rock.

Temperature is a major concern in shallow grounding systems as it has a major effect on soil resistivity. During winter months, the grounding system resistance may rise to unacceptable levels due to freezing of liquid water in the soil. The same shallow grounding system may also suffer from high resistance in the summer as moisture is evaporated from soil. It is wise to determine the natural frost line and moisture profile for an area before attempting design of a grounding system.

Figure 9.2-12 describes a four-point method for in-place measurement of soil resistivity. Four uniformly spaced probes are placed in a linear arrangement and connected to a ground resistance test meter. An alternating current (at a frequency other than 60 hertz) is

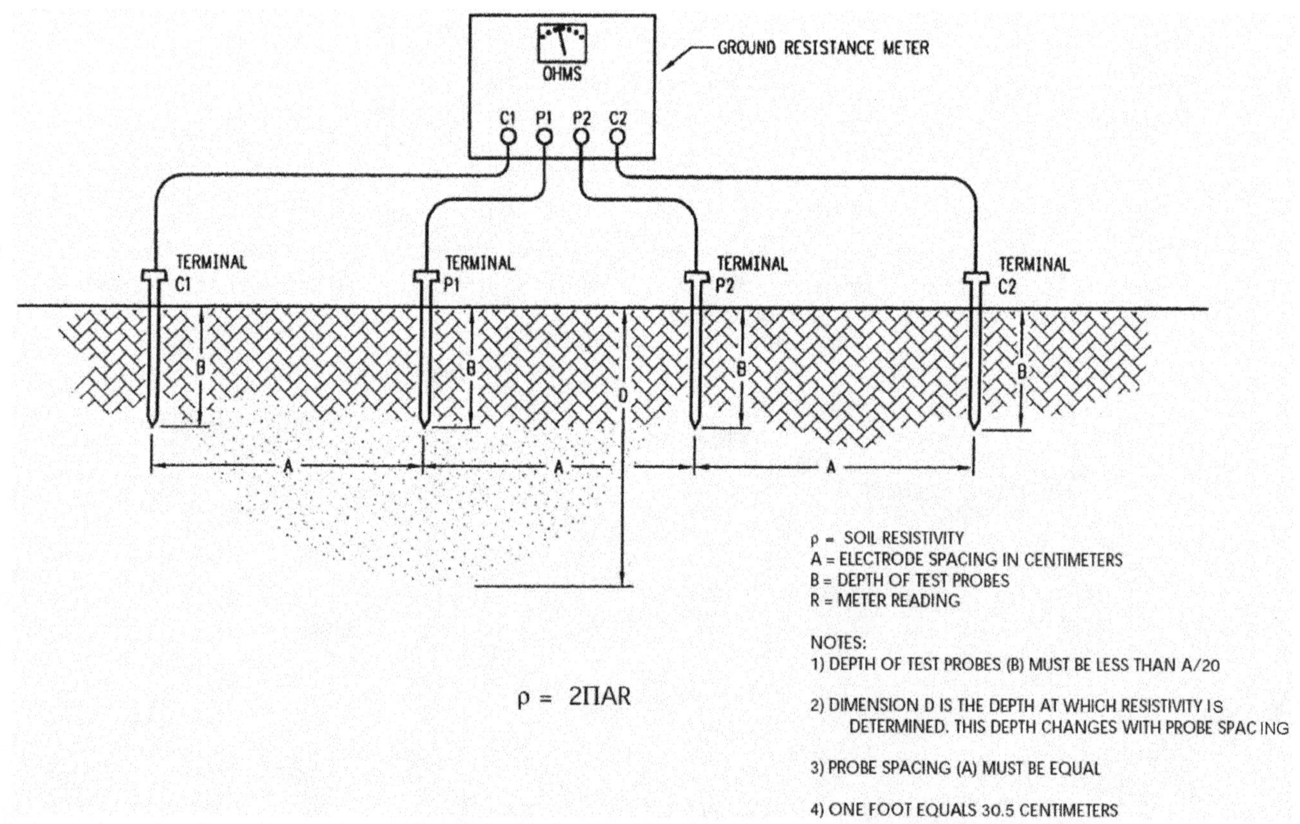

FIGURE 9.2-12 Soil resistivity test method.

passed between the two most distant probes, resulting in a potential difference between the center potential probes. The meter display in ohms of resistance may then be applied to the formula to determine the average soil resistivity in ohm-centimeters for the hemispherical area between the C1 and P2 probes.

Soil resistivity measurements should be repeated at a number of locations to establish a resistivity profile for the site. The depth of measurement may be controlled by varying the spacing between the probes. In no case should the probe length exceed 20% of the spacing between probes.

Once the soil resistivity for a site is known, calculations can be made to determine the effectiveness of a variety of grounding system configurations. Figure 9.2-13 presents equations for several driven rod and radial cable configurations that, after the soil resistivity is known, may be used for the purpose of estimating total system resistance. Generally, driven rod systems are appropriate where soil resistivity continues to improve with depth or where temperature extremes indicate seasonal frozen or dry soil conditions. See Figure 9.2-14 for a typical U.S. soil resistivity map.

Radials are also quite effective if placed below the frost line. They are often the only practical solution in areas with shallow subsurface rock. There have been instances at bald rock mountain-top sites where radials were either grouted into saw cuts in the rock or simply pinned against the face of the rock.

The performance of a grounding system in high-resistivity soil can often be improved through the addition of chemical salts or conductive concrete. Salts leach into the soil, increasing the number of free ions, with a proportionate decrease of soil resistivity in the area of the rod. Magnesium sulfate (Epsom salts), copper sulfate (blue vitriol), calcium chloride, sodium chloride (table salt), and potassium nitrate have been used for this purpose. Conductive concrete increases the volumetric area for surge dissipation and is very effective in poor soil conditions or rocky soil.

Figure 9.2-15 describes the trench and well methods for applying chemical treatment. A typical precharged chemical ground rod installation is also shown. In a precharged rod, moisture from the air enters the rod through breather holes at the top of the rod and leaches through chemicals inside, gradually exiting the rod through weep holes. As one might expect, chemically enriched grounds require recharging after a number of years to maintain their effectiveness. It is also wise to check with governing environmental agencies before introducing any foreign chemical into the soil.

The bentonite method of grounding is also shown in Figure 9.2-15. Instead of driving the rod, it is placed in the center of a 6- to 12-inch augered hole. A slurry consisting of bentonite clay and water (well drillers mud) is then poured around the rod. As the water settles out, the resulting clay remains moist through absorption of moisture from the surrounding soil. Popular additives to the slurry include up to 75% powdered gypsum (calcium sulfate) and up to 5% sodium sulfate (galvanic anode backfill). Conductive concrete is installed in a similar manner and can be utilized in horizontal trenches or mounded on top of rocky soil to enclose grounding conductors.

Ground Electrode Testing

Testing of all grounding electrodes before they are connected to form a complex network is a fairly simple process that is well described in the documentation included with all ground electrode meters and therefore will not be described here. Also, the system as a whole should be tested after all interconnections are made, providing a benchmark for future tests.

On a new site, it is often possible to perform ground system tests before the power company ground/neutral conductor is attached to the system. It is worthwhile to conduct a before-and-after test with probes in the same position to determine the influence of the power company attachment during future tests. It is also worthwhile to install permanent electrodes and marker monuments at the original P2 and C2 probe positions to ensure the repeatability of future tests. At existing sites, ground testing may be impossible without shutting down existing power systems, so strict adherence to recommended grounding practices must be followed. Also, clamp-on ground test meters are now available to test existing ground terminals for comparative analysis.

Tower-Building System

In typical tower-building arrangements, the tower is normally subject to more frequent and larger lightning currents than the station building. It is therefore reasonable to place emphasis on the tower grounding system, with less emphasis on that for the station building. Improved grounding at the tower will result in less current flowing between the tower and station building grounding systems, reducing potential differences between the two systems. The potential rise on the tower due to a lightning strike is a result of fast-rising lightning current, tower inductance, and ground resistance such that $E = IR + L(di/dt)$, where E is the potential rise from top of tower to ground, I is the magnitude of lightning current, R is the grounding resistance, L is the tower inductance, and di/dt is the lightning current rise with respect to time. Tower structures have a certain amount of inductance per foot. The amount of this inductance is dependent on the geometric configuration as well as the width of the tower. This width-to-height ratio will determine the total inductance of a tower. A 150-ft. tower, for instance, with 35 inch side widths can have an inductance of 40 µH. This value of inductance can be approximated (W/H 1%) by treating the tower as a quarter-wave antenna using:

$$\mathrm{f} = \frac{468 \times 10^6}{2(\mathrm{H\ in\ feet})} \quad \text{then inductance } \mathrm{L} = \frac{377}{2\pi f} \qquad (1)$$

Therefore, there is not much that can be done about the tower inductance once the tower structure is known. Grounding, however, can be minimized with proper grounding techniques.

Hemisphere radius a	$R = \frac{\rho}{2\pi a}$
One ground rod length L, radius a	$R = \frac{\rho}{2\pi L}\left(\ln\frac{4L}{a} - 1\right)$
Two ground rods $s > L$; spacing s	$R = \frac{\rho}{4\pi L}\left(\ln\frac{4L}{a} - 1\right) + \frac{\rho}{4\pi s}\left(1 - \frac{L^2}{3s^2} + \frac{2L^4}{5s^4}\right)$
Two ground rods $s < L$; spacing s	$R = \frac{\rho}{4\pi L}\left(\ln\frac{4L}{a} + \ln\frac{4L}{s} - 2 + \frac{s}{2L} - \frac{s^2}{16L^2} + \frac{s^4}{512L^4}\cdots\right)$
Buried horizontal wire length $2L$, depth $s/2$	$R = \frac{\rho}{4\pi L}\left(\ln\frac{4L}{a} + \ln\frac{4L}{s} - 2 + \frac{s}{2L} - \frac{s^2}{16L^2} + \frac{s^4}{512L^4}\cdots\right)$
Right-angle turn of wire length of arm L, depth $s/2$	$R = \frac{\rho}{4\pi L}\left(\ln\frac{2L}{s} + \ln\frac{2L}{s} - 0.2373 + 0.2146\frac{s}{L} + 0.1035\frac{s^2}{L^2} - 0.0424\frac{s^4}{L^4}\cdots\right)$
Three-point star length of arm L, depth $s/2$	$R = \frac{\rho}{16\pi L}\left(\ln\frac{2L}{s} + \ln\frac{2L}{s} + 1.071 - 0.209\frac{s}{L} + 0.238\frac{s^2}{L^2} - 0.054\frac{2}{L^4}\cdots\right)$
Four-point star length of arm L, depth $s/2$	$R = \frac{\rho}{8\pi L}\left(\ln\frac{2L}{a} + \ln\frac{2L}{s} + 2.912 - 1.071\frac{s}{L} + 0.645\frac{s^2}{L^2} - 0.145\frac{s^4}{L^4}\cdots\right)$
Six-point star length of arm L, depth $s/2$	$R = \frac{\rho}{12\pi L}\left(\ln\frac{2L}{a} + \ln\frac{2L}{s} + 6.851 - 3.128\frac{s}{L} + 1.758\frac{s^2}{L^2} - 0.490\frac{s^4}{L^4}\cdots\right)$
Eight-point star length of arm L, depth $s/2$	$R = \frac{\rho}{16\pi L}\left(\ln\frac{2L}{a} + \ln\frac{2L}{s} + 10.98 - 5.51\frac{s}{L} + 3.26\frac{s^2}{L^2} - 1.17\frac{s^4}{L^4}\cdots\right)$
Ring of Wire-diameter of ring D, diameter of wire d, depth $s/2$	$R = \frac{\rho}{2\pi^2 D}\left(\ln\frac{8D}{d} + \ln\frac{4D}{s}\right)$
Buried horizontal strip length $2/L$, section $\propto$ by b, depth $s/2$, $b < a/8$	$R = \frac{\rho}{4\pi L}\left(\ln\frac{4L}{a} + \frac{a^2 - \pi ab}{2(a-b)^2} + \ln\frac{4L}{S} - 1 + \frac{s}{2L} - \frac{s^2}{16L^2} + \frac{s^4}{512L^4}\cdots\right)$
Buried horizontal round plate, radius a, depth $s/2$	$R = \frac{\rho}{8a} + \frac{\rho}{4\pi s}\left(1 - \frac{7}{12}\frac{a^2}{s^2} + \frac{33}{40}\frac{a^4}{s^4}\cdots\right)$
Buried vertical round plate, radius a, depth $s/2$	$R = \frac{\rho}{8a} + \frac{\rho}{4\pi s}\left(1 + \frac{7}{24}\frac{a^2}{s^2} + \frac{99}{320}\frac{99}{320}\frac{a^4}{s^4}\cdots\right)$

Notes:

1. Approximate formulas, including the effect of images
2. Dimensions must be in centimeters to return result in ohms
3. ρ = resistivity of earth in ohm-centimeters
4. For 10 ft. (3 m) rods of $\frac{1}{2}''$ (12.7 mm), $\frac{5}{8}''$ (15.88 mm) and $\frac{3}{4}''$ (19.05 mm) diameters, the grounding resistance may be quickly determined by dividing the soil resistivity ρ in ohm-centimeters by 292, 302 and 311 respectively.
5. Data source IEE Green Book (Std. 142-1982)

FIGURE 9.2-13 Formulas for calculation of resistances to ground.

Figure 9.2-16 describes a typical grounding configuration for a guyed tower and associated transmitter building. Tower grounding may be accomplished either by a system of interconnected driven electrodes between the tower base and guy anchor points or by radial counterpoise conductors without rods. The chosen method should be determined from previously described electrode calculations and earth resistivity testing. In the driven electrode configuration, one radial counterpoise conductor is extended from the base of the tower to each guy anchor point. This conductor interconnects driven or ring electrodes near the tower base, at guy anchor locations and at intermediate points to assist in dissipating wire currents.

The rings shown feeding the system of radial conductors should stop within a few feet of the tower base. The complex mesh created by multiple bonds between the rings and radials will help to feed

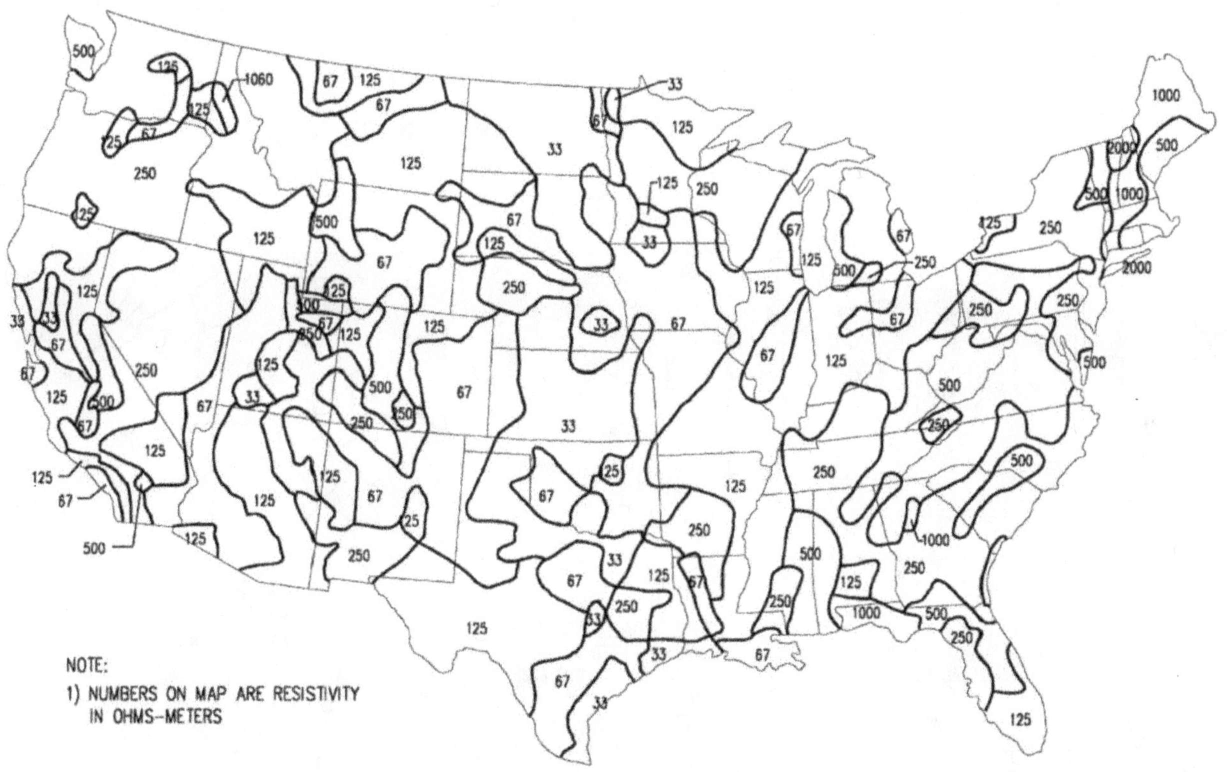

FIGURE 9.2-14 Estimated average earth resistivity in the United States.

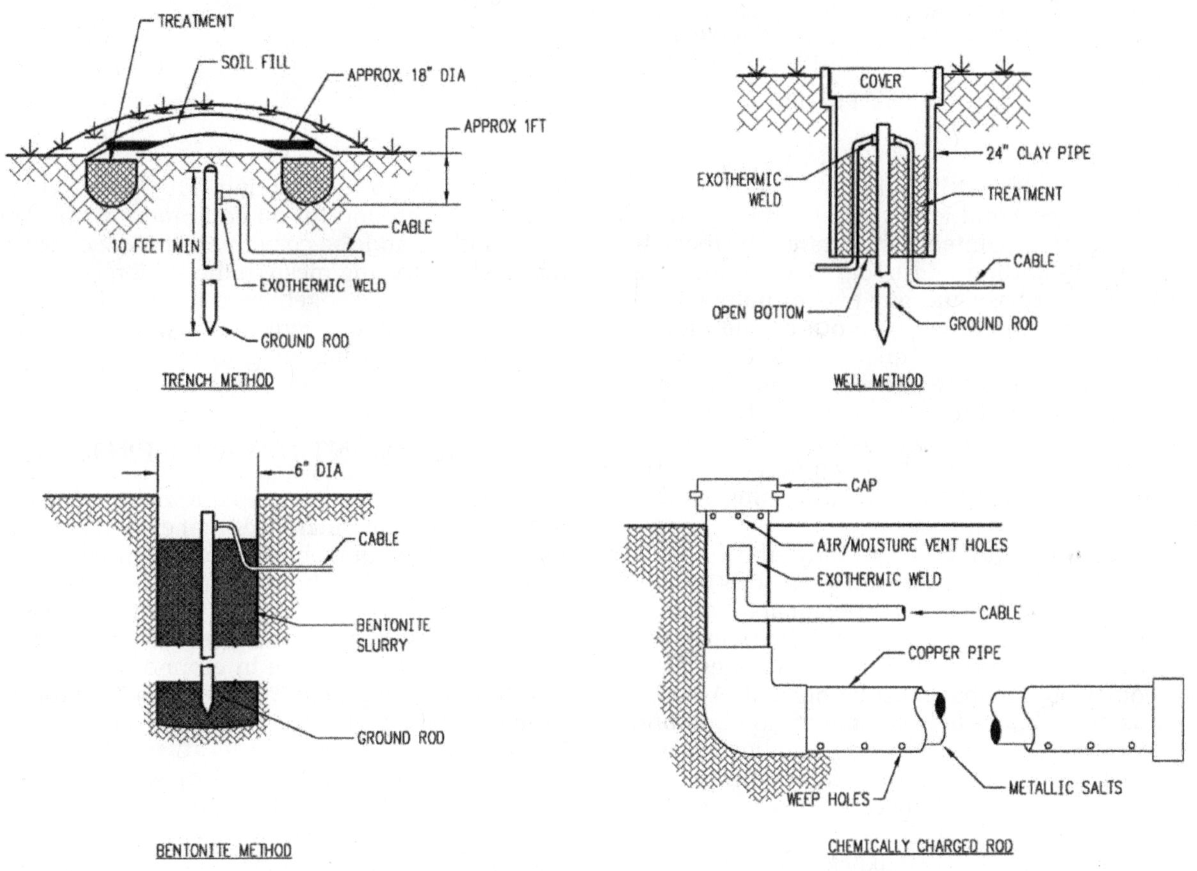

FIGURE 9.2-15 Common chemically treated ground systems.

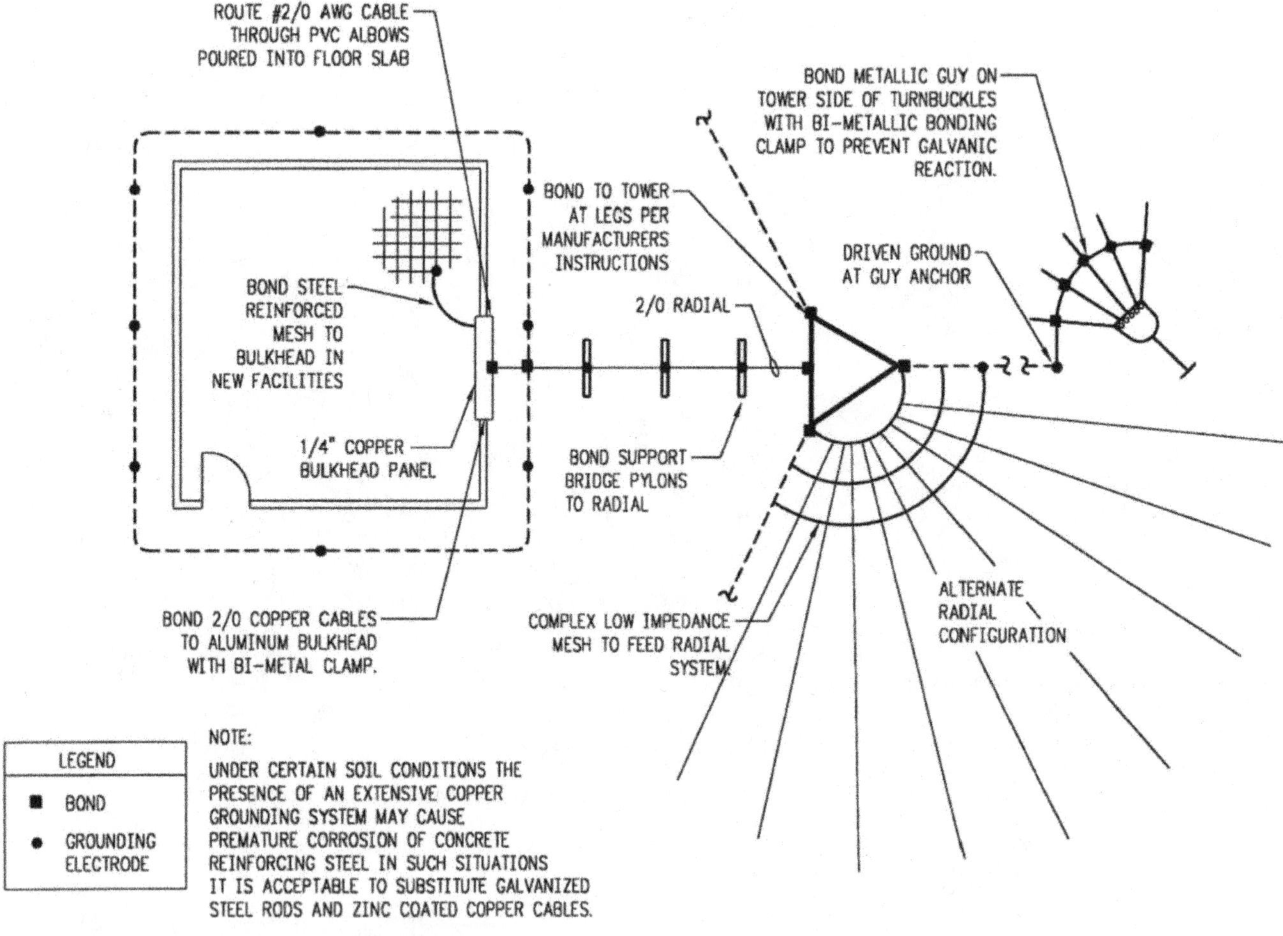

FIGURE 9.2-16 Typical tower ground system.

lightning current efficiently from the tower legs into each of the radial conductors. Apart from providing more copper in contact with the earth, there is no advantage in adding additional rings in the area between the tower base and guy anchor points. Current flowing from the tower base out on the radials will produce approximately equal potentials between adjacent radials. With nearly equal potentials at both ends, additional bonding conductors between the radials will carry little or no current. There is only a slight advantage in bonding between radials for guy wire currents introduced into the grounding system at guy anchor points.

A perimeter ground ring (actually shown as a square) or counterpoise is shown that encircles the station building. helping to equalize potential differences within the building. The station building ring also serves as a connection point for driven electrodes, fencing, and other objects that must be bonded. A bonding conductor is shown between the station building ground ring and the tower base or grounding system. This conductor will equalize potentials and minimize the level of current carried between the tower and station building by the coaxial lines.

A commercially available bulkhead plate is shown on the side wall of the station building and is bonded to the station grounding ring. This plate serves as a single-point ground for all equipment within the station building and the coax lines. In new construction, the steel reinforcing mesh in the station building floor should be bonded together to the bulkhead panel to minimize potential differences between the equipment and floor during a lightning strike.

EQUIPMENT DAMAGE PROCESS

Most lightning damage to equipment occurs as the result of potential differences that exceed the tolerance level of the equipment. These potential differences may be presented to the equipment or system through external metallic circuits as a conducted current or induced transient voltage surge. They may also occur as the result of differences in ground potential at various items of equipment that are connected together to form a system. Figure 9.2-17 describes these situations in greater detail.

Common mode surges, which are also referred to as "longitudinal modes" in many documents, arrive at the equipment with approximately equal potential on both sides of a balanced pair or on a number of circuits simultaneously. These surges may be induced into a

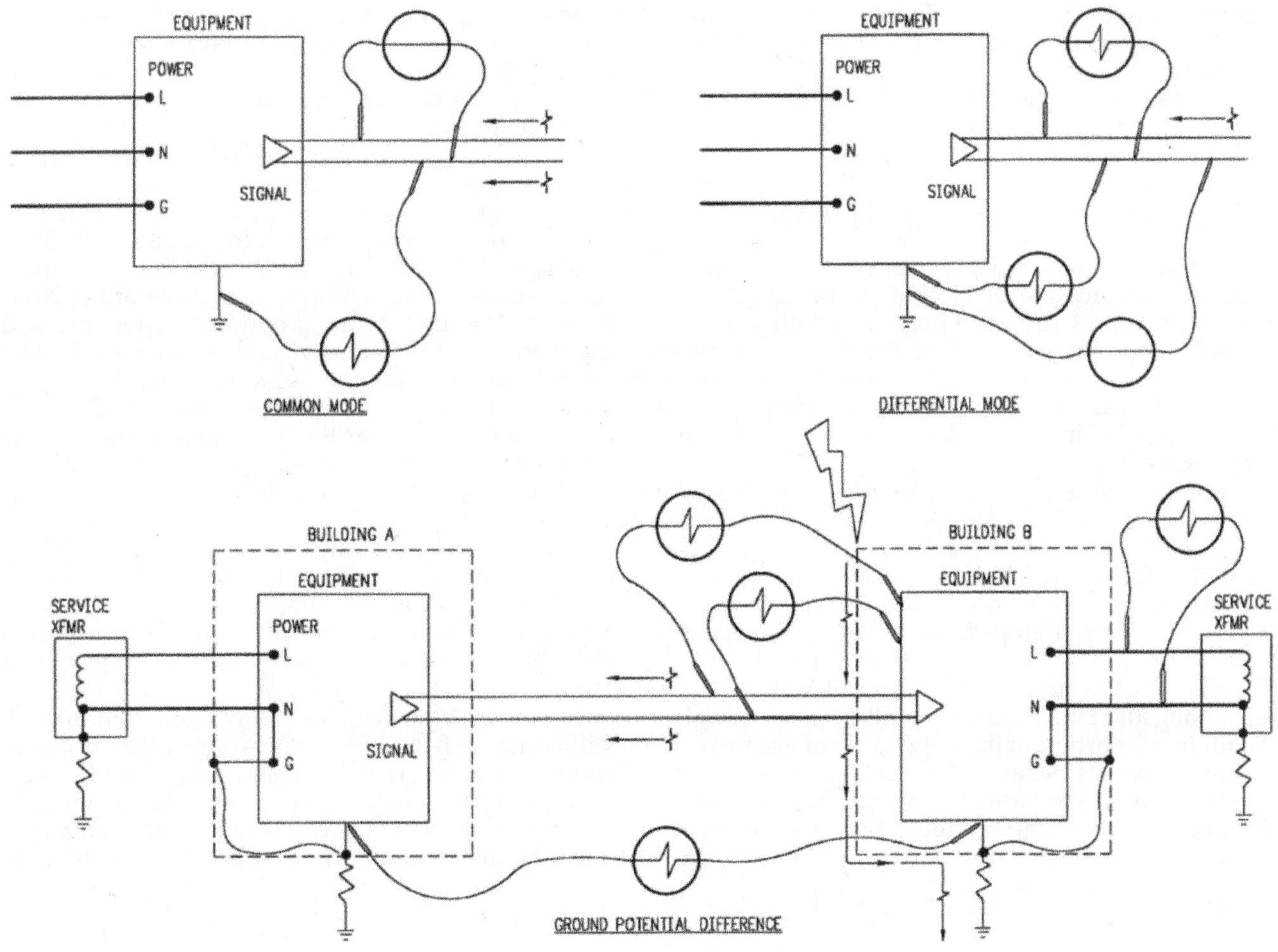

FIGURE 9.2-17 Typical equipment damage process.

wiring system by nearby lightning, directly coupled into the circuit or even created by the action of an upstream suppression device as it clamps two or more conductors together. Common mode surges may enter equipment on power phases, signal, and other circuits. Damage is normally sustained due to potential differences between the affected circuit(s) and equipment chassis or other uninvolved circuits.

Differential or normal mode surges are often more damaging than their common mode counterpart because most equipment is designed to operate in a differential fashion. In power circuits, a differential mode surge may appear on one or more phases relative to the neutral and ground conductors. In signal circuits, especially those operating on a balanced differential basis, the tolerance to differential mode surges is lower than for common mode. Common mode surges on electrical power systems are routinely converted to differential mode at electrical services where one side of the service is referenced to ground. The same conversion process can occur on a balanced circuit when an upstream surge protective device (SPD) clamps one side of a pair before the other.

Ground differential damage is a bit more obscure than either of the other mechanisms. However, it is responsible for a great deal of damage to systems with equipment in multiple locations. Equipment in different buildings, or even equipment within different areas of the same building, can be damaged through ground differentials.

In the simple example of Figure 9.2-17, assume that building B receives a lightning strike of 20,000 amperes. Also assume the grounding system resistance is 2 ohms. As the lightning current flows into the earth through the 2-ohm grounding resistance, a 40,000 volt potential rise will be produced in building B's grounding system. Since the equipment in building B references the local building through its power cord and bonding conductor, its chassis will rise to about the same potential.

Circuitry within building B's equipment will attempt to track the building ground potential rise, except for the components that attach to wiring from building A. These components see a large difference in potential between the balance of their circuitry and the wiring to building A. Building A has not been involved in the ground potential rise, so these circuits are still near ground potential.

Component breakdown occurs within the equipment in building B, and a small fraction of the total lightning current attempts to find a path to ground through the wiring leaving for building A. Upon reaching the equipment at the remote building A, this current presents itself as a common mode surge, causing damage to the equipment.

EQUIPMENT PROTECTION STRATEGIES

Protecting equipment from the effects of lightning involves a combination of grounding, bonding, and surge suppression. Grounding provides a path to introduce lightning currents into the earth. Bonding serves to equalize lightning potential differences between various elements of equipment. Surge suppression limits differences in potential on active circuits that cannot be directly bonded.

Bonding is a means of equalizing potentials during a lightning strike. Figure 9.2-18 repeats the example in Figure 9.2-17 regarding damage due to ground potential differences. This time, however, a bonding conductor is provided between the two grounding systems in an attempt to keep both at the same potential.

Kirchhoff's law tells us that current will divide itself among all of the available parallel paths through a circuit in proportion to the impedance of each path. Lightning currents behave in the same way, flowing through all available paths to ground. In direct current circuits, the voltage produced across any circuit component is the product of current and resistance. When dealing with rapidly changing lightning current, inductance of the circuit plays a far larger role than simple resistance. Recall that an inductor tends to oppose any change in current until it has stabilized its magnetic field.

The bonding conductor shown in Figure 9.2-18, by virtue of its connection to the ground at remote building A, serves as one of several paths for lightning current to follow on its way to ground. In this case, only 5% or 1,000 amperes of the lightning current flows through the 75-ft. conductor, with the remainder of the 20,000 amperes flowing into the grounding system of building B and its electrical service. Assuming a 1.5 microsecond rise time for the lightning current, the peak end-to-end voltage on the conductor is 15,030 volts, 15,000 volts of which is the result of inductance in the conductor. The remaining 30 volts are the result of the conductor's resistance.

The normal reaction to lowering the potential difference between buildings in this example is to suggest a larger cable. After all, larger cables have less resistance and lower voltage drop. Changing to a larger cable, however, has little effect on the circuit inductance, affecting primarily the 30-volt portion of the total.

Figure 9.2-19 provides a comparison of inductance values for a 1-ft. length of various sizes of round conductors, strip materials, and coaxial cables. Strip materials are considerably more effective for the same cross-sectional areas as a round conductor, making them more attractive as a bonding medium for

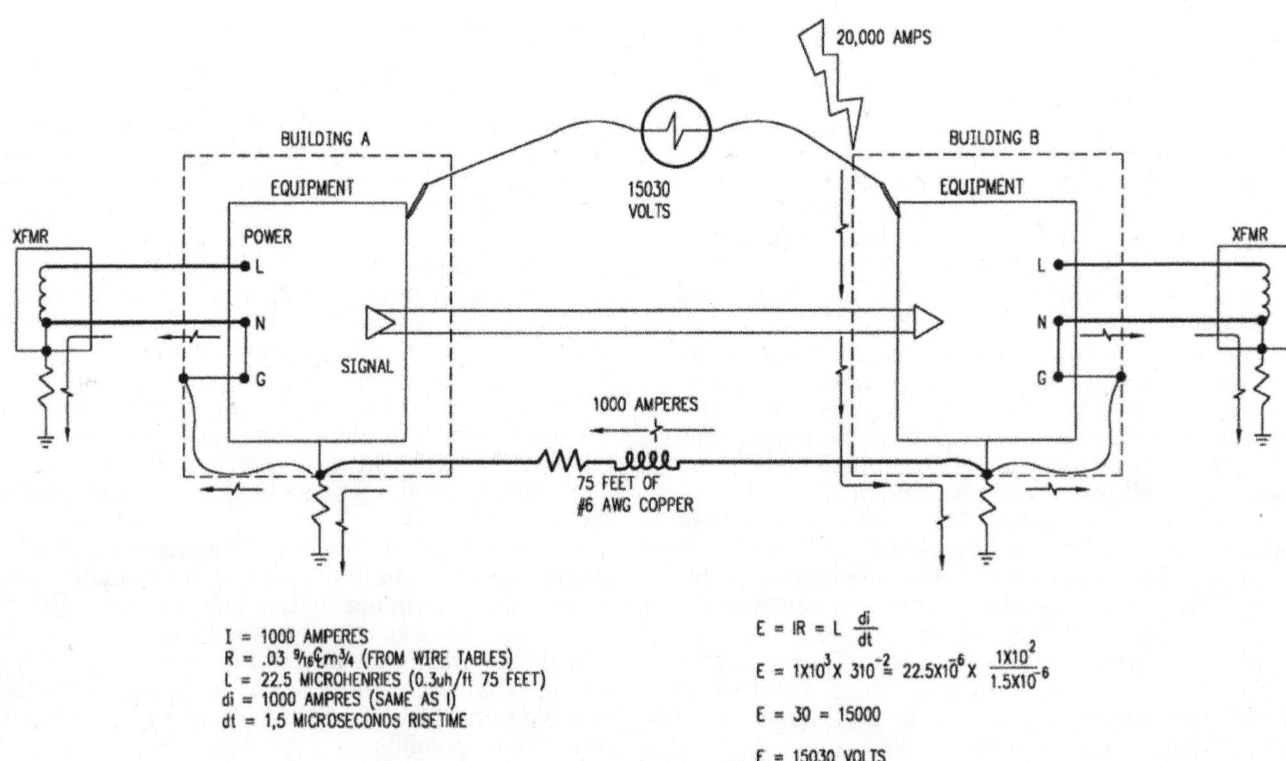

FIGURE 9.2-18 Example of ground potential difference.

lightning protection purposes and bulkhead grounding. Another interesting property of strip material is that it is not proportionally more effective once a width of about 4 inches is reached. The major reason for thickness is mechanical strength and mounting convenience.

As a final note on the inductance of bonding materials: *never expect an insulated conductor in steel conduit to carry lightning current effectively*. Ferrite beads make a reasonable effective low pass filter when placed around an insulated conductor. Steel conduit around an insulated cable creates the same effect, increasing the inductance of the cable within the conduit at least an order of magnitude similar to a choke coil. Where this condition exists (and there are many locations), a marginal compromise is to bond both ends of the cable to the conduit, permitting the conduit to serve as part of the circuit.

Single-Point Grounding

If, after the exercise in bonding, it is concluded there is no way of preventing potential differences in a conductor carrying lightning current, that is correct. There is, however, a method of preventing lightning current flow through a bonding circuit. With no current flow there can be no potential difference between the bonded items. This method is called *single-point grounding*.

In Figure 9.2-20, the equipment chassis and all metallic circuits leaving the equipment for the outside world have been bonded together and to a ground conductor at a single point. There is no possibility of a difference in potential between the circuits entering the equipment or between these circuits and the chassis, as they are all bonded together. There is also no possibility of current flow from the single-point ground into

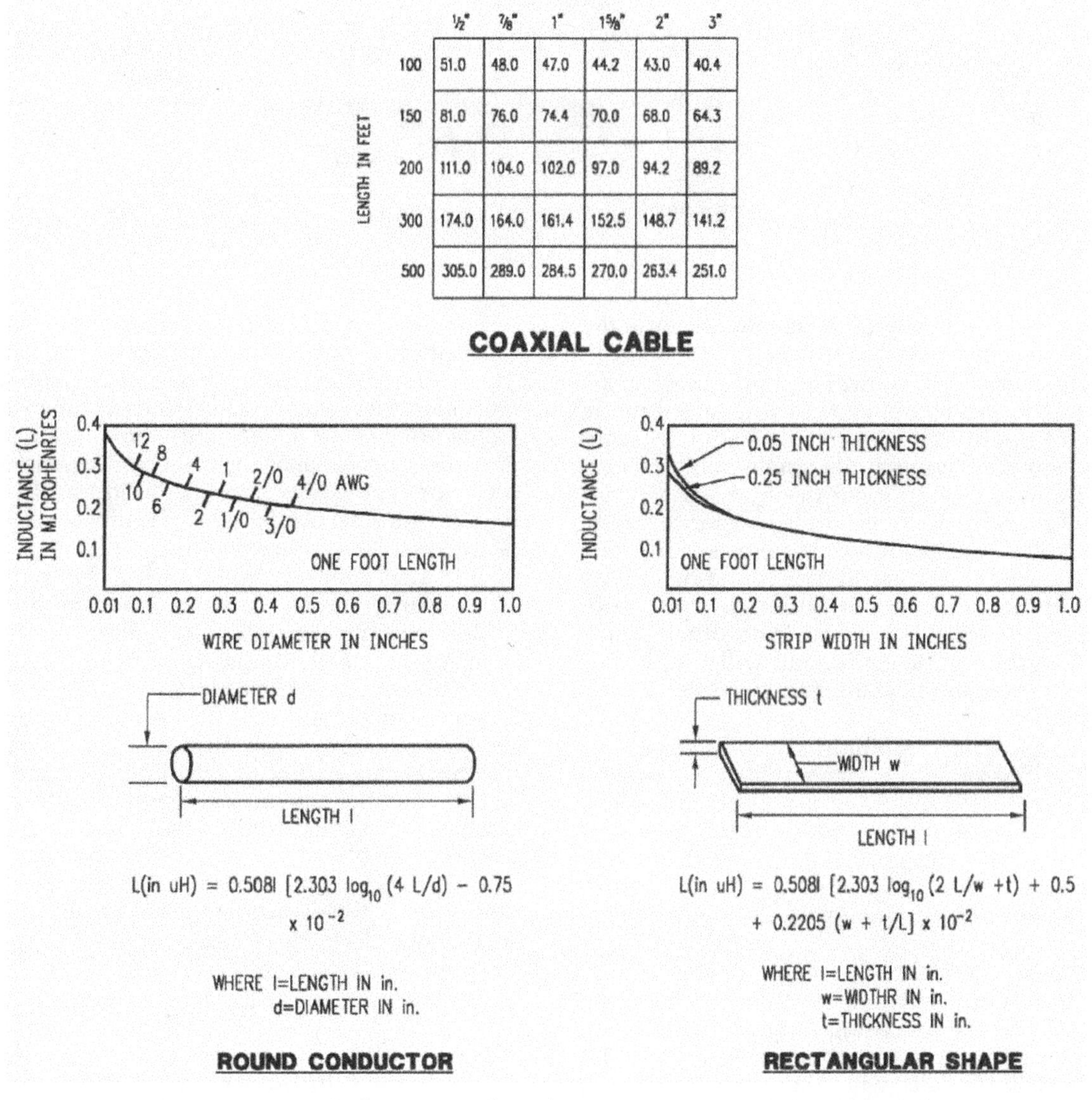

COAX DIAMETER

LENGTH IN FEET	½"	⅞"	1"	1⅝"	2"	3"
100	51.0	48.0	47.0	44.2	43.0	40.4
150	81.0	76.0	74.4	70.0	68.0	64.3
200	111.0	104.0	102.0	97.0	94.2	89.2
300	174.0	164.0	161.4	152.5	148.7	141.2
500	305.0	289.0	284.5	270.0	263.4	251.0

COAXIAL CABLE

FIGURE 9.2-19 Inductance values for common engineering materials.

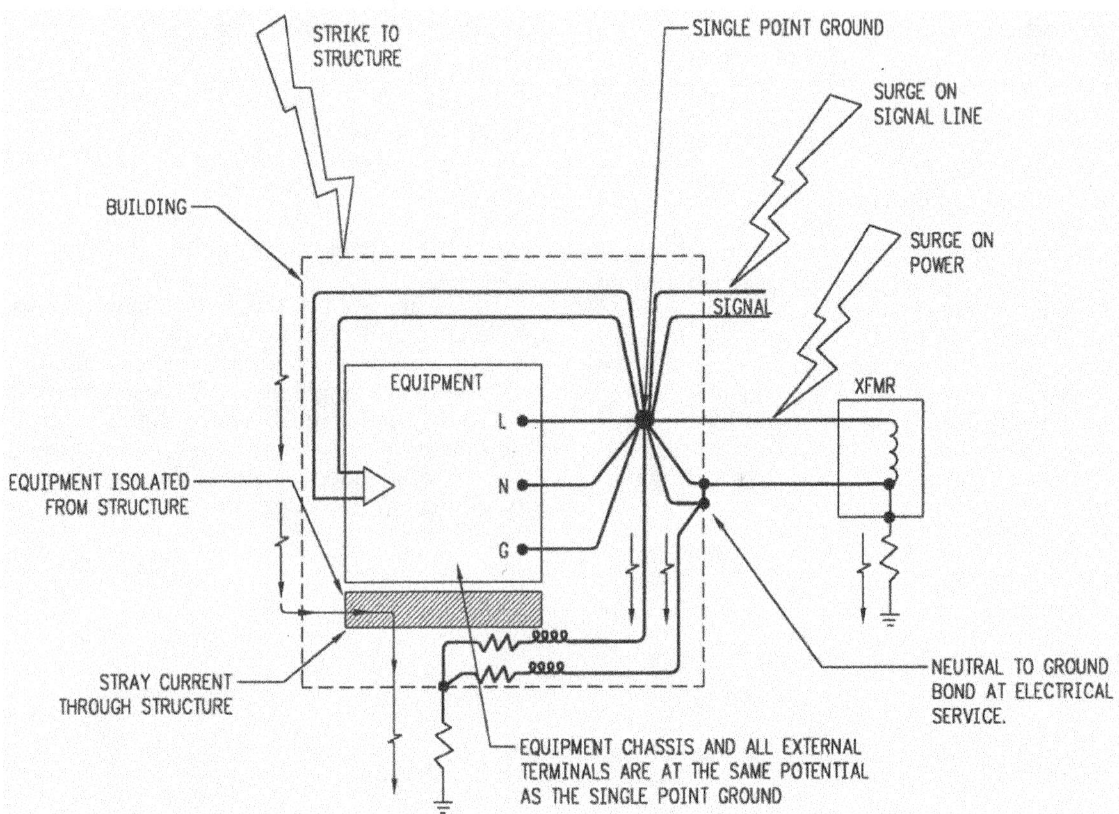

FIGURE 9.2-20 Single-point grounding concept.

the equipment through any of the circuits because the equipment is isolated from the structure.

A surge entering on the power or signal lines cannot present itself to the equipment in differential mode because the lines are all connected together. A common mode surge arriving at the single-point ground will pass harmlessly to ground through the grounding conductor and ground electrode resistance. There will be potential rise at the single-point ground due to inductance and resistance in the grounding circuit; however, no current can flow through the equipment because it remains isolated from other points of ground reference. The equipment will simply rise and fall in potential, tracking the potential of the single-point ground.

A lightning strike to the building or other structures connected to its grounding system will also cause the single-point ground to rise and fall in potential. The equipment, however, sees no potential difference, as its chassis and all external metallic circuits are tied together at the single-point ground. The only difficulty with the example in Figure 9.2-20 is that *nothing works*!

Figure 9.2-21 provides a more realistic approach to single-point grounding. The only difference between Figure 9.2-21 and the preceding example is that transient voltage surge-suppression devices (TVSS) are used on the active circuits, which, for obvious reasons, cannot be directly bonded to the single-point ground.

For the purpose of this discussion, it is helpful to think of surge protectors as a conditional bond, clamping or limiting the excursion of voltage on active circuits to a safe level relative to each other and to the single-point ground. The single-point ground may rise and fall in potential as the protectors discharge current into it or during a strike to the building, but the difference in potential presented to the protected equipment is always held within safe limits.

It is worth noting that SPDs will clamp in response to a rise in potential on their ground terminal as well as for legitimate transients on their active conductors. A strike to the building or nearby structure will cause a significant elevation in ground potential. The single-point ground will rise in potential by virtue of its connection(s) to the building grounding system. The SPD units, seeing their building ground terminals rise in potential above their remotely connected active circuits, will clamp, forcing the active circuits to track the potential of the single-point ground and the chassis of the protected equipment. Again, the voltage excursion seen by the equipment is held to a safe level, and no damage is sustained.

Applications of Single-Point Grounding

The application of single-point grounding is normally limited to equipment within a room or a group of rooms. While it is possible to design larger

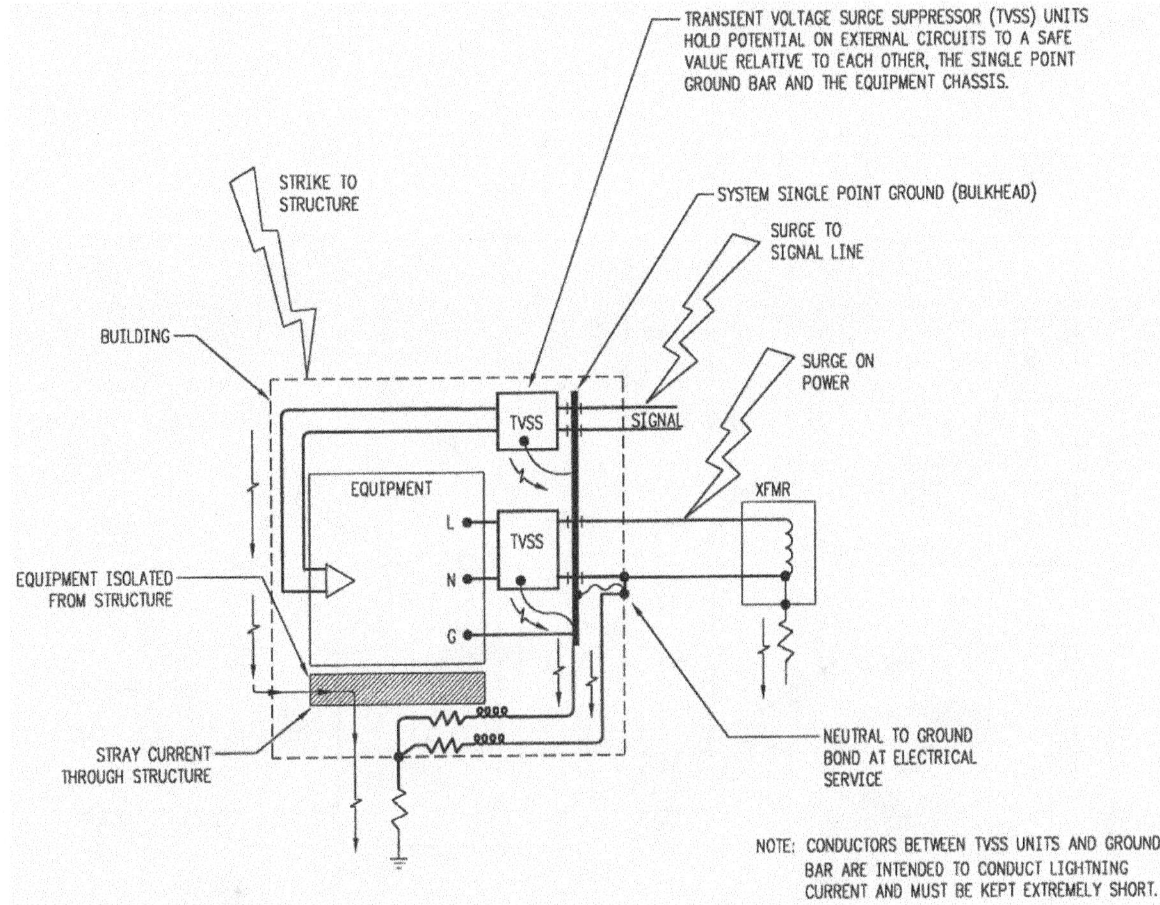

FIGURE 9.2-21 Practical application of single-point grounding with TVSS.

configurations, the need to bring circuits in at different locations soon dictates the need for multiple locations, each treated as an island of equipment with its own protection devices and single point of ground reference. Larger single-point grounding systems are also more susceptible to induced voltages from nearby lightning by virtue of their increased cable lengths.

As an example of two extremes in scale, a computer room that serves terminals throughout a station complex may be engineered with a single-point grounding system and proper surge protection on its external circuits. The terminals and their printers, however, are scattered throughout the building, referencing ground at each location through their power cords. It is possible to designate the ground pin on the receptacle for each terminal/printer combination as the single-point ground for the equipment at that location. A combination power and data protector may be provided for each location that ensures that these conductors are held within safe limits of the receptacle ground pin and chassis of the equipment. The equipment is isolated from stray grounds by placement on a desktop.

Figures 9.2-22 and 9.2-23 are examples of how single-point grounding applies to a typical broadcast tower-building installation. A bulkhead panel in the wall of the building serves as a single-point ground reference for all equipment within the small facility. All coaxial cables, waveguides, and raceways from the tower are bonded to the bulkhead as they pass into the building.

The physical size of the electrical equipment dictates that it cannot be located directly at the bulkhead panel. To minimize the effect of bonding system inductance, a 6-inch wide bonding bus is extended to each side of the bulkhead, or for large installations, an overhead ground ring is provided. Width of the strip provides the necessary low inductance. Its 1/4-inch thickness, while not necessary for electrical reasons, provides the installer with a bus that may be drilled and tapped to accept short bonding pigtails to the equipment.

Support hangers for cables and raceways serving the protected equipment are isolated from the roof structure to prevent inadvertent current flow through the raceways. Isolation is provided between the equipment feet and floor slab by a high-dielectric polypropylene pad. Such isolation may not be necessary if adequate isolation is provided by the equipment feet. Nylon bolts may be used with conventional expansion anchors to secure equipment to the floor without violating the integrity of the single-point grounding system.

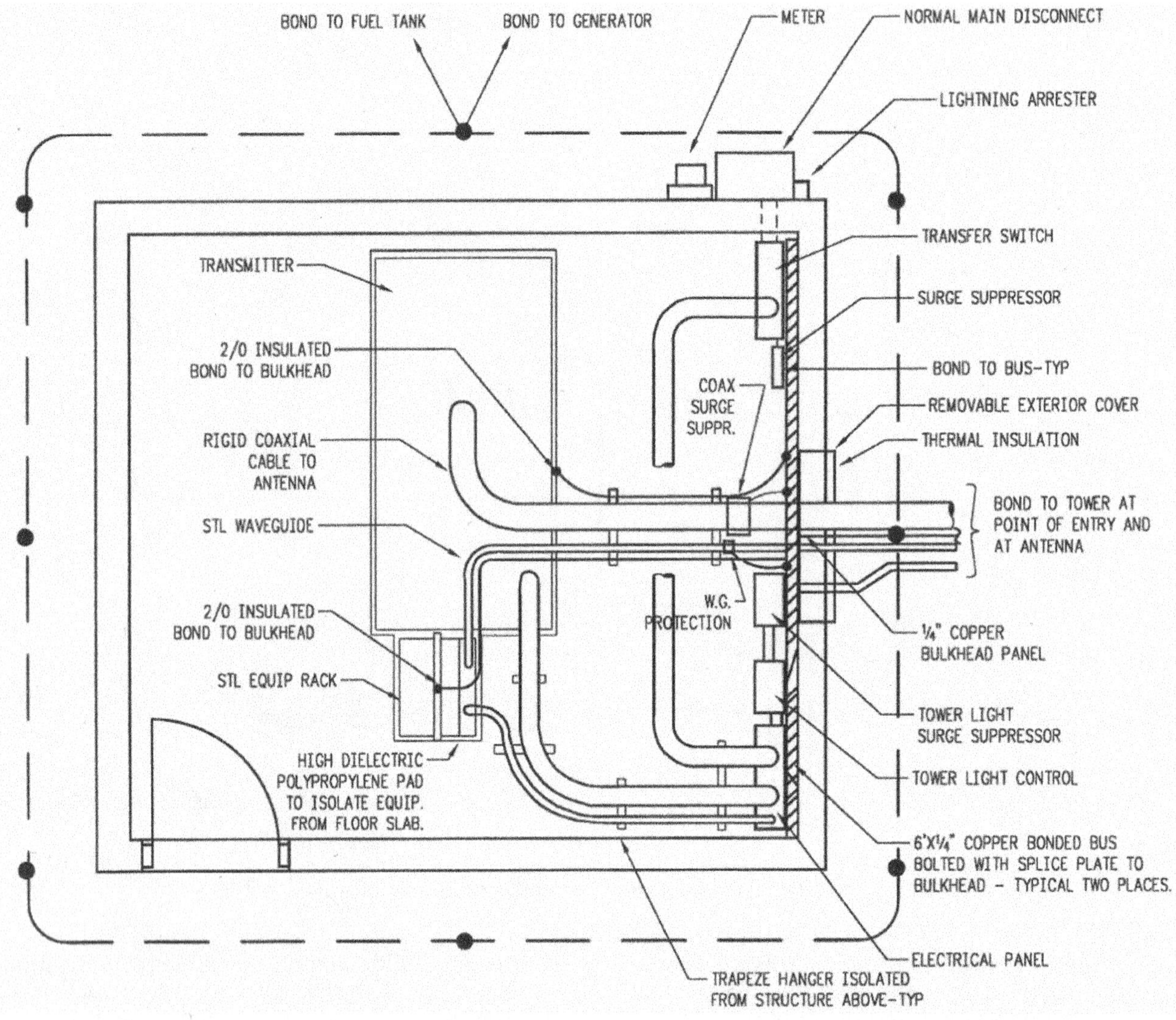

FIGURE 9.2-22 Single-point transmitter building grounding.

SURGE PROTECTION

In Figure 9.2-24, surge protectors are shown at the transfer switch emergency feed to protect the transfer switch from surges on the emergency power feeder whenever the feeder and generator starting controls are not run in rigid metal conduit. Where metal conduit is used, it must be bonded to the perimeter ground ring. Main service surge protectors should be connected through individual or integral fused switches or breakers with short-circuit interrupting ratings equal to or higher than the available service fault current. Also, a protector is installed on the circuit for tower lights. Both protectors should be bonded to the grounding bus, keeping the length of their bonding lead as short as possible. Surge protection should also be provided on all metallic, control, monitoring, and communications lines and circuits serving lighting, winches, and other electrical items outside the building. Again, locate these suppressors at the point of entry for the circuits and bond their ground leads to the bus with the shortest possible lead length.

The lightning arrester shown at the main disconnect outside the building serves an important purpose. Under normal operation, the main disconnect is closed and the transfer switch is connected to utility power. During operation on emergency power the normal transfer switch position is open. Potentially, a lightning strike to the utility line will propagate along the line as a traveling wave in both directions from the point of lightning contact. Upon reaching the open circuit input of the transfer switch, the wave will reflect back on itself, potentially doubling its initial crest value. The same condition can occur at the main utility company disconnect if it should be open during servicing, and each condition can cause equipment flashover.

Flashover within electrical equipment is serious; however, the problem is compounded when operating voltage is also present. The flashover arc provides a low-impedance path for 60-Hz follow current, and

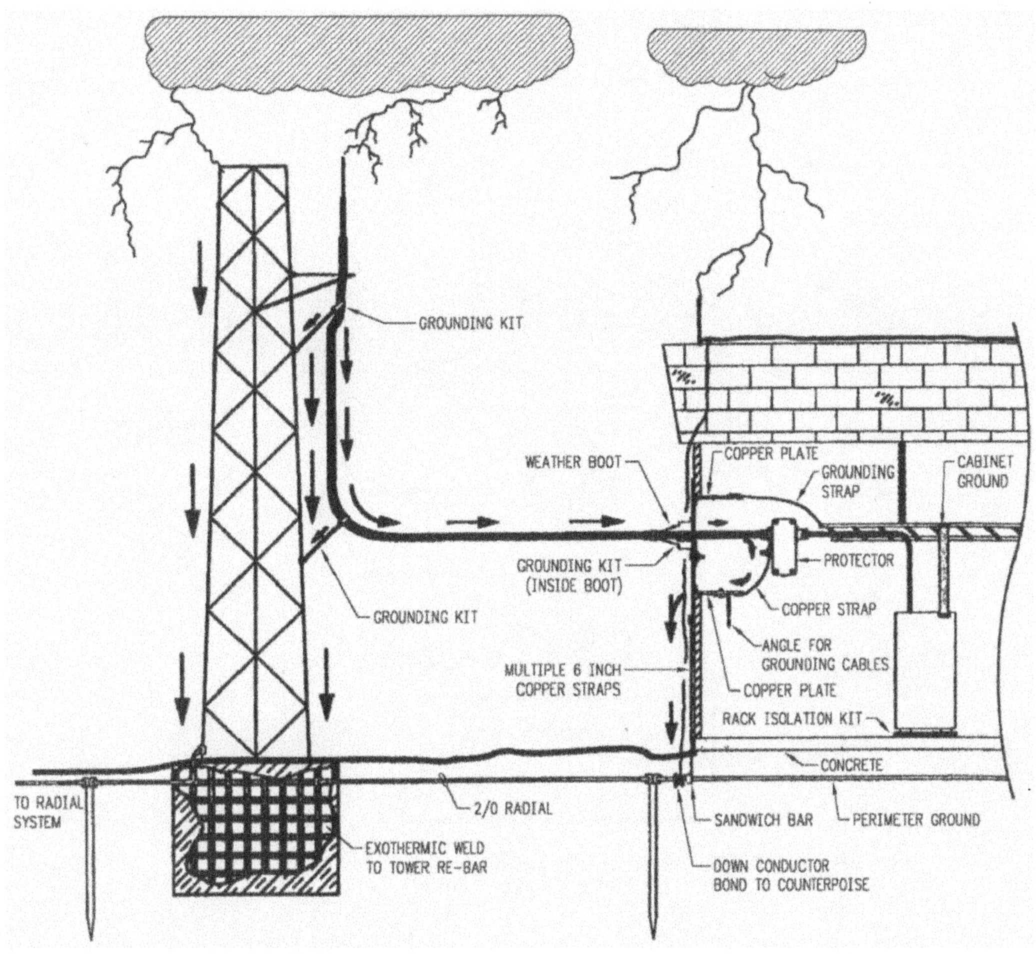

FIGURE 9.2-23 Typical bulkhead installation with single-point grounding.

significant damage to the equipment may occur. The arrester will prevent this condition from occurring by limiting the traveling wave voltage.

It is of great importance to minimize the length of the suppressor leads and that of the grounding conductor because their inductive voltage drop is additive with the initial clamping voltage of the suppressor.

Signal Line Suppressors

Figure 9.2-25 shows a common surge suppressor configuration for signal line and coax applications. Most suppressors of this type are both shunt type and multistage hybrid devices utilizing a high-energy first stage, a fast-acting second stage, and impedance in series between the two stages to coordinate their clamping behavior. Because of the multistage design, these devices must be installed in series with the protected circuits.

The treatment of shields is often an issue when dealing with signal line surge suppressors. Figure 9.2-25 also shows shields being bonded to the suppressor ground bus to force them to track the single-point ground. If ground loop or other technical restrictions prevent direct bonding of the shields, they should be protected with suppressors, like any other active circuit.

The bonding lead distance for signal line suppressors is often more critical than it is for power devices. Tolerance levels of signal circuits are normally lower than for power-supply inputs, and the relatively small voltages developed in the suppressor ground leads can become significant.

Signal line and coaxial surge suppressors are sold in a variety of shapes and sizes for different applications. The suppressors should generally clamp transient voltage on a circuit to within 150% of normal peak-operating voltages, and even lower in some applications. Since most signal line suppressors are inserted in series with the circuit being protected, it is wise to evaluate the effect of their series impedance and capacitance on the insertion loss of a circuit.

The effect of suppressor capacitance can be important in many high-speed data, RF, and video applications. One simple way of evaluating the effect of this capacitance is to equate it to equivalent cable feet. For example, if the desired suppressor exhibits capacitance of 100 pF, and the cable used in the circuit is rated at 10 pF per foot, will the circuit tolerate an additional 10 ft. of cable? If so, the suppressor capacitance should produce no noticeable effect on the circuit.

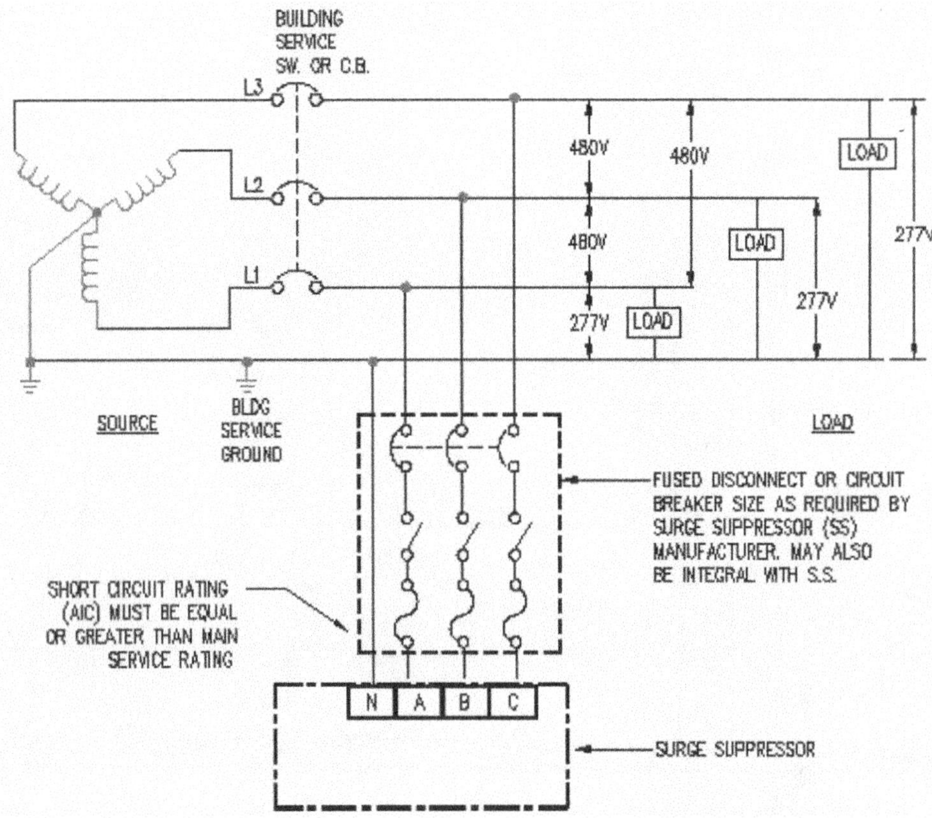

FIGURE 9.2-24 Main service surge-suppression installation.

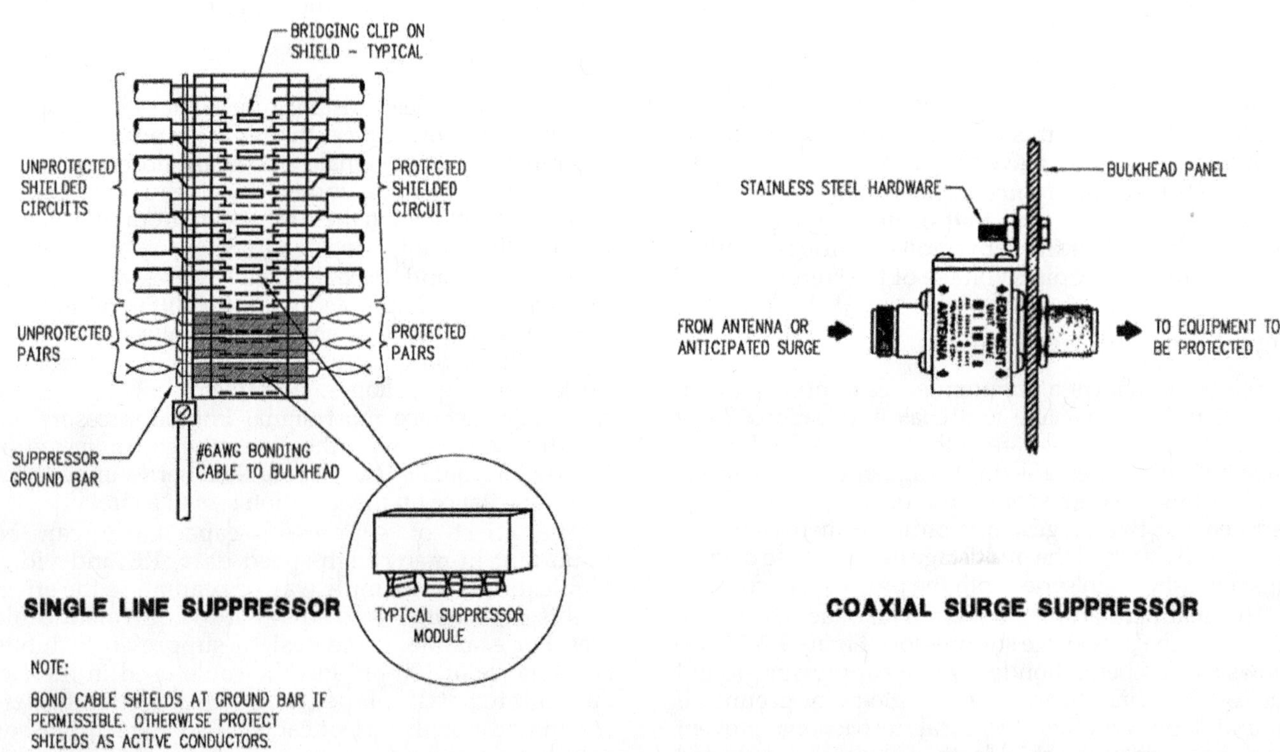

FIGURE 9.2-25 Common signal and coaxial suppressors.

Surge Protection Selection

Surge protectors and lightning arresters come in all shapes, sizes, types, and accessories, and a complete discussion of application methods and evaluation is not possible in this document. Contacting surge protector manufacturers, however, may be extremely confusing due to the myriad of choices available, but eventually a choice must be made. Contacting the chairmen of the referenced IEEE, UL and IEC standard committees may also be helpful in understanding application issues. The following guidelines, therefore, may be helpful:

- Select protectors and arresters based on IEEE C62.41.1 (2002).
- All surge protectors should be UL 1449, 2nd edition (or future 3rd edition) tested, approved, and labeled.
- Ask for published proof of testing to back up claims, especially clamping levels, energy handling capabilities (joules, watts, or current), and speed of response.
- Look for a minimum of 3- to 5-year unconditional warranties. Many companies will offer 10-year warranties, but only *unconditional* warranties are of any value.
- Review the six major performance characteristics: response time, voltage protection level, power dissipation, disturbance-free operation, reliability, and operating life. Make sure maximum power dissipation level and voltage protection level using specific waveforms are stated at the same point to avoid misinterpretation.
- Talk to other similar users for advice and recommendations.
- Consult industry standards for performance requirements. At present, there is no testing or application standard for coaxial surge suppressors, so a comparative review of reputable manufacturer data will be necessary. Compare: impedance (50 or 75 ohm); speed; frequency range; number of transmit, transceive or receive-only signals; transmit power; presence of AC or DC power with the RF signals; mounting; and connector type and sex.

ISOLATED GROUND RECEPTACLE

Figure 9.2-26 shows an isolated ground receptacle circuit commonly used in computer room grounding applications. The receptacles used in this type of circuit differ from the norm in that their ground pins are

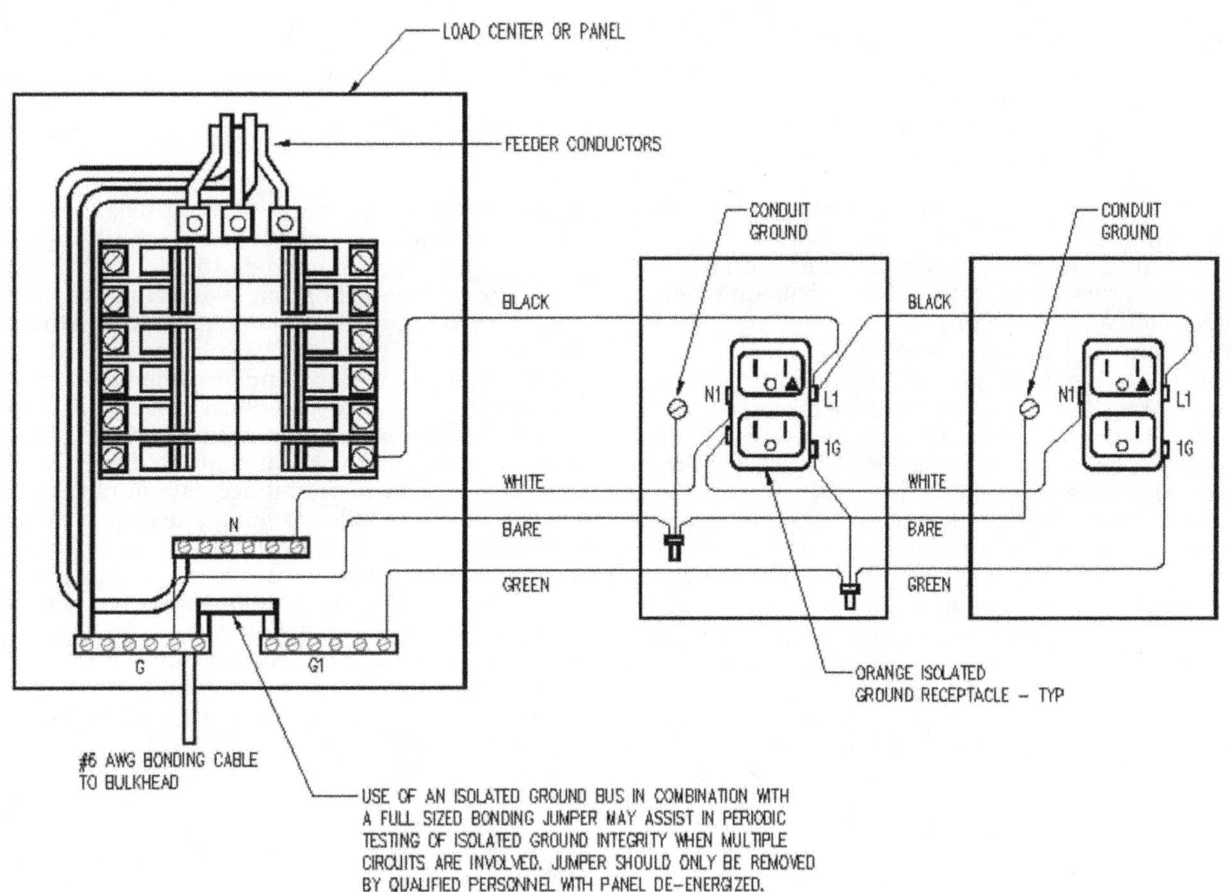

FIGURE 9.2-26 Isolated ground receptacle.

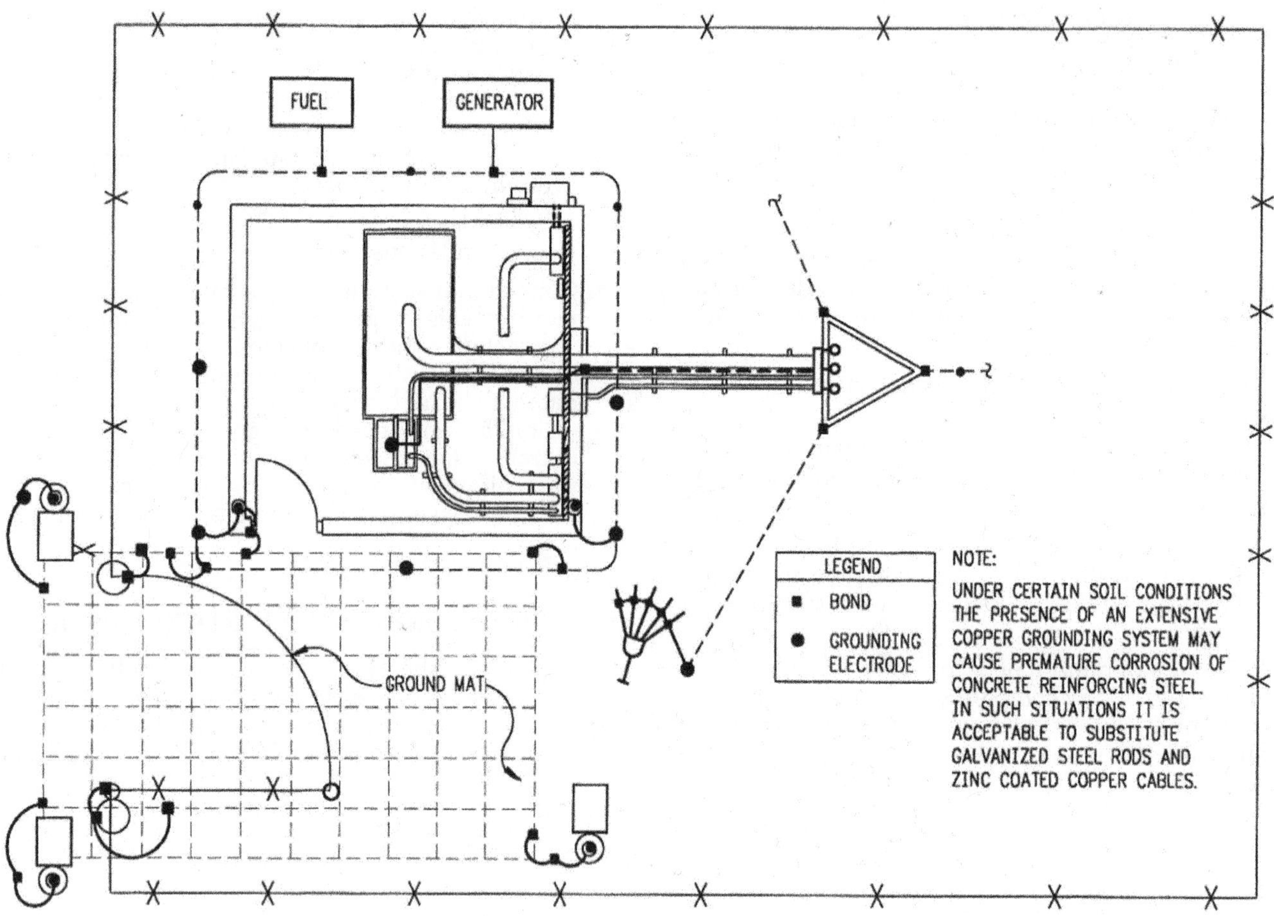

FIGURE 9.2-27 Typical composite bonding and grounding installation.

electrically isolated from their mounting tabs. They are, therefore, isolated from their outlet box and structural ground at each receptacle location. A dual system of grounding conductors ensures that equipment plugged into an isolated ground (IG) receptacle references ground first at the single-point ground.

The use of isolated ground receptacles helps to ensure that plug-in terminals, printers, diagnostic, and other ancillary equipment are properly referenced to the single-point ground and not the local structure. It only takes one item of equipment connected between the protected equipment and a remotely grounded receptacle to compromise the integrity of the grounding system.

Figure 9.2-27 is a composite of the bonding and grounding recommendations for the typical broadcast site. While complicated in appearance, each component has its purpose as part of a simple-to-understand subsystem. Finally, Figure 9.2-28 illustrates a comprehensive single-point grounding system that ties all elements of the grounding system together.

NEW VERSUS EXISTING SITES

The basic suggestions enumerated in this article should apply equally to both new and existing installations; however, sites having existing towers, buildings, grounding systems, and site appurtenances offer a more challenging design effort to establish truly complying grounding and bonding solutions. For instance, a new tower installation can take advantage of tower base reinforcing-steel grounding, whereas grounding for an existing tower must be enhanced to account for the loss of such a massive and low-impedance grounding component. Also of critical concern in existing sites is that the existing grounding systems cannot be isolated sufficiently without shutting down the entire broadcast station to permit accurate grounding measurements. The challenge for the designer, therefore, is to approximate the ideal conditions with creative technical modifications that do not violate the basic tenets of grounding and bonding presented herein and in associated publications. In most cases, calculating exact grounding and impedance values to compare to some ideal value is not as important as *relative* values between various portions of the grounding and bonding components.

Site Audit Report

Performing a site audit report for an existing site requires more time and attention to detail than at a new site, obviously, but the audit report will form the

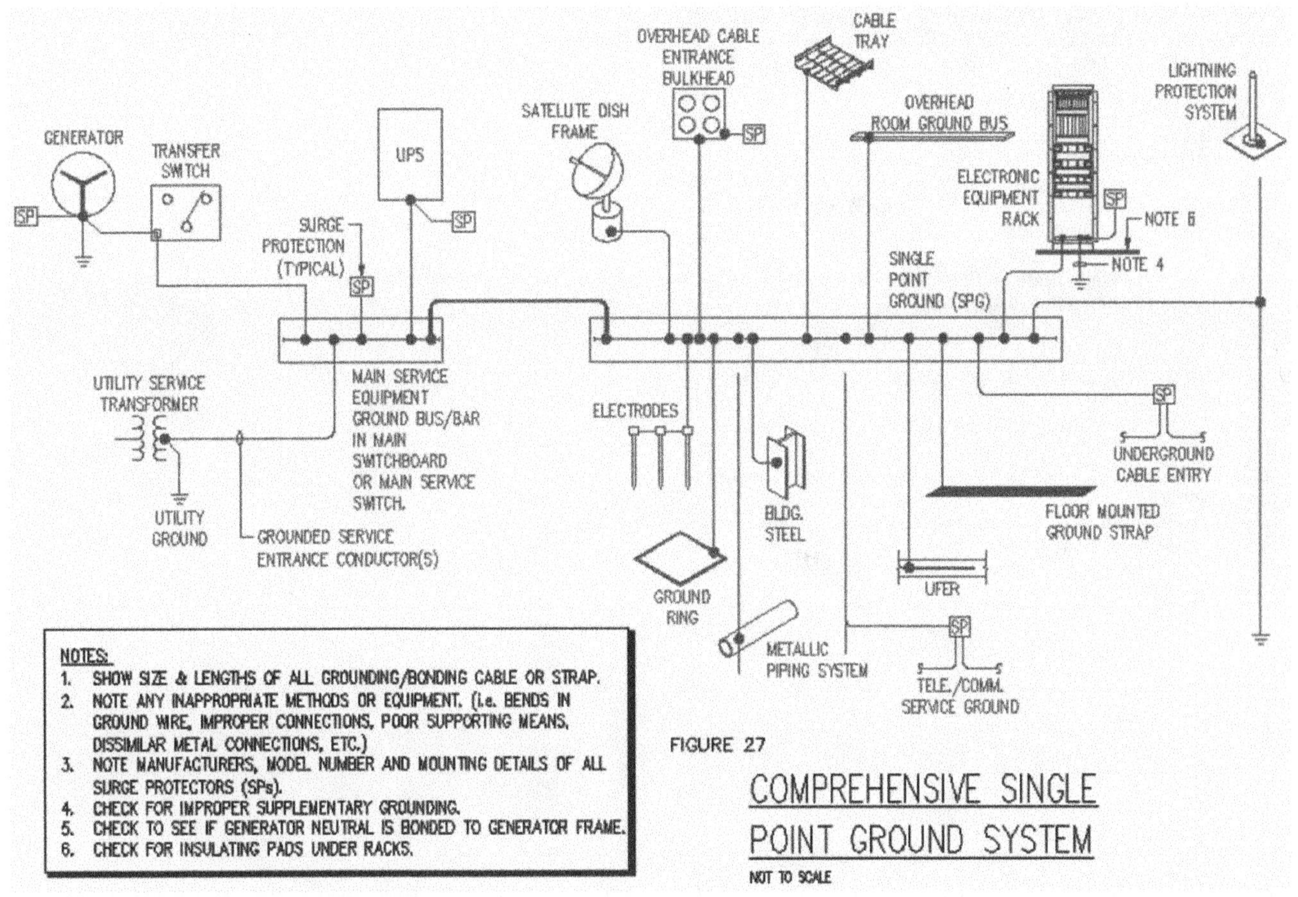

FIGURE 9.2-28 Comprehensive single-point ground system.

basis of the grounding and bonding desired. Detailed sketches of all grounding systems, pictures of all conditions, and uncovering buried systems to minimize guesswork are critical to establishing a basic understanding of the existing systems on which to build or improve.

Designing and investigating an existing site with maintenance in mind will enhance the ability of those who will be asked to maintain a secure, quality system to provide the tools necessary to do their job. Documentation of existing equipment types and model numbers, wire/cable types and sizes, conductor routings, connector and splice types and locations, sketches, pictures, plans, etc., should all be cataloged and properly tabbed for ease of future reference.

SUMMARY

This chapter has been written in tutorial form, as every site is different and no single set of recommendations will apply to every situation. The principles set forth, while tailored to a broadcast environment, apply equally to other systems. There are still a few mysteries to be solved in completely understanding lightning, but once it enters a wiring system, it becomes an electrical current that is both predictable and understandable. A list of publications that deal with the subject material in greater depth is included at the end of this chapter.

References

[1] Earle R. Williams. The Electrification of Thunderstorms, *Scientific American*, November 1988.

[2] NFPA 780, *Lightning Protection Code*, National Fire Protection Association, Quincy, MA.

[3] UL-96A, *Standard for Lightning Protection Systems*, Underwriters Laboratories, Northbrook, IL.

[4] See [2].

[5] See [3].

[6] LPI-175, *Standard of Practice*, Lightning Protection Institute, Woodstock, IL.

[7] Military Handbook 419A, *Grounding, Bonding, and Shielding for Electronic Systems*, December, 1987. Available from the U.S. Government Printing Office, Philadelphia, PA.

[8] IEEE 142, *Grounding Practices for Electrical Systems*.

Additional Resources

ANSI/IEEE C62.33 (1989), Standard Test Specifications for Varistor Surge Protective Devices.

ANSI/IEEE C62.34 (R2001) (1996), IEEE Guide for Performance Low-Voltage Surge-Protective Devices (Secondary Arresters).

ANSI/IEEE C62.35 (1987), Standard Test Specification for Avalanche Junction Semiconductor Surge Protective Devices.

ANSI/IEEE C62.36 (2000), Standard Test Methods for Surge Protectors Used in Low-Voltage AC Power Circuits.

ANSI/IEEE C62.41.1 (2002), IEEE Guide on the Surge Environment in Low-Voltage (1000 V and less) AC Power Circuits.

ANSI/IEEE C62.41.2 (2002), IEEE Recommended Practice on Characterization of Surges in Low-Voltage (1000 V and less) AC Power Circuits.

ANSI/IEEE C62.45 (1992), Guide on Surge Testing for Equipment Connected to Low-Voltage AC Power Circuits.

Federal Information Processing Standards (FIPS) Publication 94, CCITT Rec. K-17, Waveform Specification for Electronic Systems.

IEEE 142 (1991), Grounding Practices for Electrical Systems ("green book").

IEEE 518, Recommended Guide on Electrical Noise.

Military Standard (MIL-STD-220A) (1952).

NEMA LS-1 (1992), Low-Voltage Surge Protective Devices. NFPA 70 (2005), National Electrical Code.

NFPA 75 (2003), Standard for Protection of Electronic Computer Systems.

NFPA 780 (2004), Standard for the Installation of Lightning Protection Systems.

UL-1283 (2nd edition), Standard for Safety—Electromagnetic Interference Filters.

UL-1449 (1998), Standard for Safety—Transient Voltage Surge Suppressors.

CHAPTER

9.3

Tower Lighting and Monitoring

RICHARD G. HICKEY
Vertical Bridge Holdings, LLC.

INTRODUCTION

One critical component of delivering broadcast content to the public is tower ownership. Of the 118,000 plus structures listed in the Federal Communications Commission (FCC) Antenna Structure Registration (ASR) database, about 12,000 are owned by broadcasters. The vast majority of these structures are guyed or self-support towers. Also included are rooftop, pole, or other constructions (see Figure 9.3-1 for an example of a tower candelabra). A common theme with most broadcast structures is the intrusion into trafficked airspace.

The FCC and FAA (Federal Aviation Administration, an agency of the Department of Transportation) share a unique relationship in the governance of marking and lighting vertical structures for the purpose of conspicuity and avoidance by aircraft in the United States and U.S. properties. The driving theme of this allegiance is safety to low-flying air traffic. Both agencies take compliance to these regulations very seriously. Many tower owners who have taken these regulations lightly have encountered costly fines and sanctions, often well publicized and causing much embarrassment as well as financial distress.

This chapter will address the relationship of the FCC and FAA, the regulatory requirements for marking structures, acceptable methods of marking and monitoring and proper maintenance and offer insights on how to limit exposure to liabilities beyond simple compliance to the published requirements.

FIGURE 9.3-1 Broadcast Tower Candelabra

Hilights, Inc.

THE FCC AND FAA—HOW IT WORKS

The FCC was created by passage of the Communications Act of 1934, replacing the Federal Radio Commission created in 1926. Official origins of the FAA began with the passage of the Air Commerce Act, also in 1926. Unofficial regulatory origins began well before, some with the advent of mail transport in the earliest days of motorized flight.

Through the years, upper height limits of various vertical structures were extended through evolutionary improvements in architectural engineering. Buildings, bridges, chimneys, and towers intruded into trafficked airspace. The need to mark these structures for added conspicuity to pilots of aircraft was evident, and both the FCC and FAA established guidelines to make the structures more visible. However, there were some inherent problems. The regulations and determinations often varied between the two agencies. While the FCC was only focused on structures primarily intended to support broadcast and communication hardware, the FAA was responsible for any vertical construction for any purpose.

The two agencies worked together loosely to establish guidelines for marking vertical structures that enter trafficked airspace. There were several discrepancies between the guidelines of each agency. The FCC issued ASR (Antenna Structure Registration) "determinations" based on the guidelines found in Form 715. FAA ASN (Aeronautical Study Number) determinations based on Advisory Circular 70/7460-1 for towers were often ignored by tower owners since, unlike the FCC, the FAA had no enforcement division or ability to levy fines and other penalties.

The adoption of CFR (Code of Federal Regulations) 47 § 17.23 brought the two agencies into agreement on a unified set of regulations and vehicle of enforcement, taking effect on January 1, 1996. The FCC recognized FAA Advisory Circulars 70/7460-1 and 150/5345-43. A more recent statement by the FCC clarifies the relationship:

> The FCC has been given the authority by Congress to require the painting and/or illumination of antenna towers when it determines that such towers may otherwise constitute a menace to air navigation. 47 U.S.C. § 303(q). The FCC's rules governing antenna tower lighting and painting requirements are based upon the advisory recommendations of the FAA, which are set forth in two FAA Advisory Circulars. 47 CFR §§ 17.21–17.58. *Although the FAA's lighting and painting standards are advisory in nature, the FCC's rules make the standards mandatory. The standards and specifications set forth in these FAA documents are incorporated by reference into the FCC's rules, making these advisory standards mandatory for antenna towers"* [emphasis added].

As mentioned, the FAA has no enforcement mechanism. The FCC actively enforces the FAA regulations on broadcast towers with fines and related legal actions.

MAJOR CHANGE: MIGRATORY BIRD CONFIGURATIONS

The FAA made many major changes in the newly published AC 70/7460-1L (12/4/2015) predicated on the reduction of deaths of migratory birds from collisions with towers. Studies have shown that many species of birds indigenous to North America are attracted to red lighting, particularly steady-burn or nonflashing red lighting. One study cited the loss of an estimated 250,000 birds in the 1989 Exxon Valdez oil spill. By comparison, the study estimates an annual loss of 6.6 million birds due to collisions with broadcast and communication towers. Towers under 150′ AGL in height are normally below the common flight altitude of migratory birds and are therefore unaffected by the new regulations. Configurations for towers at taller heights are drastically changed. The information shared in this chapter is based on the recently published AC 70/7460-1L and AC 150/5345-43H. (NOTE: At this writing, AC 150/5345-43H was updated by the FAA on September 28, 2016. See the link in the list of important documents.

REGULATIONS FOR MARKING, LIGHTING, AND MONITORING

Owners and/or sponsors must apply to the FCC and FAA prior to beginning construction of an antenna support structure. The FCC requires Form 854 and FAA requires Form 7460-1. Specific information required includes precise latitude/longitude location and AGL (above ground level)/AMSL (above mean sea level) heights of the structure, along with other data. An official ASN letter ("determination") is issued upon processing the related form. The determination is in effect until the owner or sponsor of the structure refiles for any reason. Reasons may include correcting the recorded information contained in a previous filing including ownership, any change to the tower height or size, change of ownership, change of RF output, and change of marking method or lighting configuration.

The FCC and FAA work closely to continually review published regulations and issue updated revisions as warranted. Evolutionary changes to the environment, tower and other vertical structures, and aircraft technology are a few considered factors.

Owners of broadcast structures should review the latest determinations for their towers and other vertical structures to ensure the proper marking is displayed. Note the FCC may have issued Form 715 guidelines for marking and lighting for older structures. The FAA normally recognizes these instructions. It is also important to note that updates to FAA ACs 70/7460-1 and 150/5345-43 are not retroactive and do not affect previously issued determinations that name an earlier version of the AC. This is commonly known as "grandfathering." This should not be confused, however, with FCC requirements for monitoring tower lighting systems, where updates to the requirements for monitoring and maintaining aviation obstruction lighting apply to all structures licensed by the commission (FCC).

A list of important documents:

- FAA AC 70/7460-1L outlines marking, lighting, and monitoring regulations including configurations and other requirements: www.faa.gov/documentLibrary/media/Advisory_Circular/AC_70_7460-1L_.pdf
- FAA AC 150/5345-43H frameworks the parameters within which aviation obstruction lighting must operate and the color parameters for structure paint bands: www.faa.gov/documentLibrary/media/Advisory_Circular/draft-5345-43h.pdf
- FAA AC 150/5345-53D, published September 2012, outlines the Airport Lighting Equipment Certification Program (aviation obstruction lighting

included): www.faa.gov/documentLibrary/media/Advisory_Circular/150_5345_53d.pdf

- FAA AC 150/5345-53D Appendix 1 Addendum lists all Third-Party Certifiers along with all lighting products that have been successfully tested and certified. This addendum is updated on a monthly basis: www.faa.gov/documentLibrary/media/Advisory_Circular/150_5345_53d_addendum.pdf
- FAA Form 7460-1 Notice of Proposed Construction or Alteration: www.faa.gov/documentLibrary/media/Form/FAA_Form_7460-1_2017.pdf
- FAA Form 7460-2 Supplemental Notice: www.faa.gov/documentLibrary/media/Form/FAA_Form_7460-2_2017.pdf
- FCC Form 854 Application for Antenna Structure Registration: https://transition.fcc.gov/Forms/Form854/854.pdf (NOTE: Also see www.fcc.gov/help/antenna-structure-registration-asr-forms)
- FCC Part 17, specifically §§ 17.47 through 17.56, addresses marking and lighting, with additional requirements for monitoring: www.ecfr.gov/cgi-bin/text-idx?tpl=/ecfrbrowse/Title47/47cfr17_main_02.tpl

BROADCAST STRUCTURE MARKING BASICS

With few exceptions, the FAA regulations for marking and lighting towers or other structures are fairly straightforward. The reader is advised to thoroughly study the latest revisions of the previously named FAA Advisory Circulars and FCC Part 17.47–56 and understand the requirements therein. Key elements to tower marking and lighting are day and night marking, AGL and AMSL height of the structure, and the "four C's": color, candela, coverage, and configuration.

Acceptable lighting colors are red, which is only approved as a night marking, and white, which may be used for day, night, or both.

Candela is a measurement of directed light output from a particular point: the light fixture, in this case. One candela has a directional radiant intensity of 1/683 watt per steradian, equal to one "directed" candlepower. The candela is abbreviated cd and its standard symbol is I_v.

Coverage describes the visibility, path, and projection of the light. Each required light level must present a complete 360 degree unobstructed outward beam path, meaning multiple lights are necessary at each light level below the apex of the structure. Exception: a lightning rod, cable, bracket support, or other necessary component smaller than 7/8" is allowed with all lighting save the L-810 "side marker" light. The configuration of the lighting system in terms of required number of light levels depends on the height of the structure.

Color: Daytime Marking

Acceptable daytime markings are aviation orange and white paint bands or white flashing lighting. Red lighting of any type is not an acceptable method of daytime marking.

Paint bands are to be equally spaced with full coverage, no gaps of unpainted structure between bands. Everything on the exterior surface of the tower must be painted, with these exceptions: "Ladders, decks, and walkways of steel towers and similar structures do not need to be painted if a smooth surface presents a potential hazard to maintenance personnel. Painting may also be omitted from precision or critical surfaces if the paint would have an adverse effect on the transmission or radiation characteristics of a signal. However, the structure's overall marking effect should not be reduced." Top and bottom bands are orange, with the remaining bands alternating between orange and white. The number of paint bands required is dependent on the AGL height of the tower and is always an odd number. Structures greater than 10.5′ up to and including 700′ in AGL height require seven bands. Structures greater than 700′ up to and including 900′ in AGL height require nine bands. Structures greater than 900′ up to and including 1,100′ in AGL height require 11 bands. Greater heights in 200′ increments are marked accordingly, with two paint bands added at 1,300′, 1,500′, 1,700′, and 1,900′ each.

White flashing lights may normally be used instead of paint bands for day marking. White flashing lights are normally not allowed below 200′ AGL. Per the latest revision AC 70/7460-1L, the use of L-865 medium intensity white flashing lighting is approved on structures 700′ AGL or less in overall height. The L-865 produces a daytime candela output of 20,000 cd ±25%. high intensity L-856 white flashing lighting may be used in lieu of paint bands on structures taller than 700′ AGL. The L-865 produces 270,000 cd ±25% in the day mode and 20,000 cd in the twilight mode for dusk and dawn (see Figure 9.3-2).

Color: Nighttime Marking

Paint is unacceptable as a means of nighttime marking for obvious reasons. Aviation obstruction lighting is normally required for nighttime marking on structures over 200′ AGL and is often required on shorter structures due to various circumstances including proximity to airports/airfields/helipads, coastal regions, low-altitude flight lanes, AMSL height, and many more influences.

Nighttime marking (lighting) may be red or white in color, though white lighting is not normally allowed below 200′ AGL. Shorter structures may be marked with L-810 low intensity lighting. Most structures over 150′ AGL in height are required to have medium or high intensity lighting. However, all medium and high intensity lighting, regardless of color, has an intensity of 2,000 cd ±25% in the night mode.

Candela: FAA Nomenclature—Types of Aviation Obstruction Lighting

"Tower lighting" is broken down into different descriptive types. Note the FAA makes no differentiation in

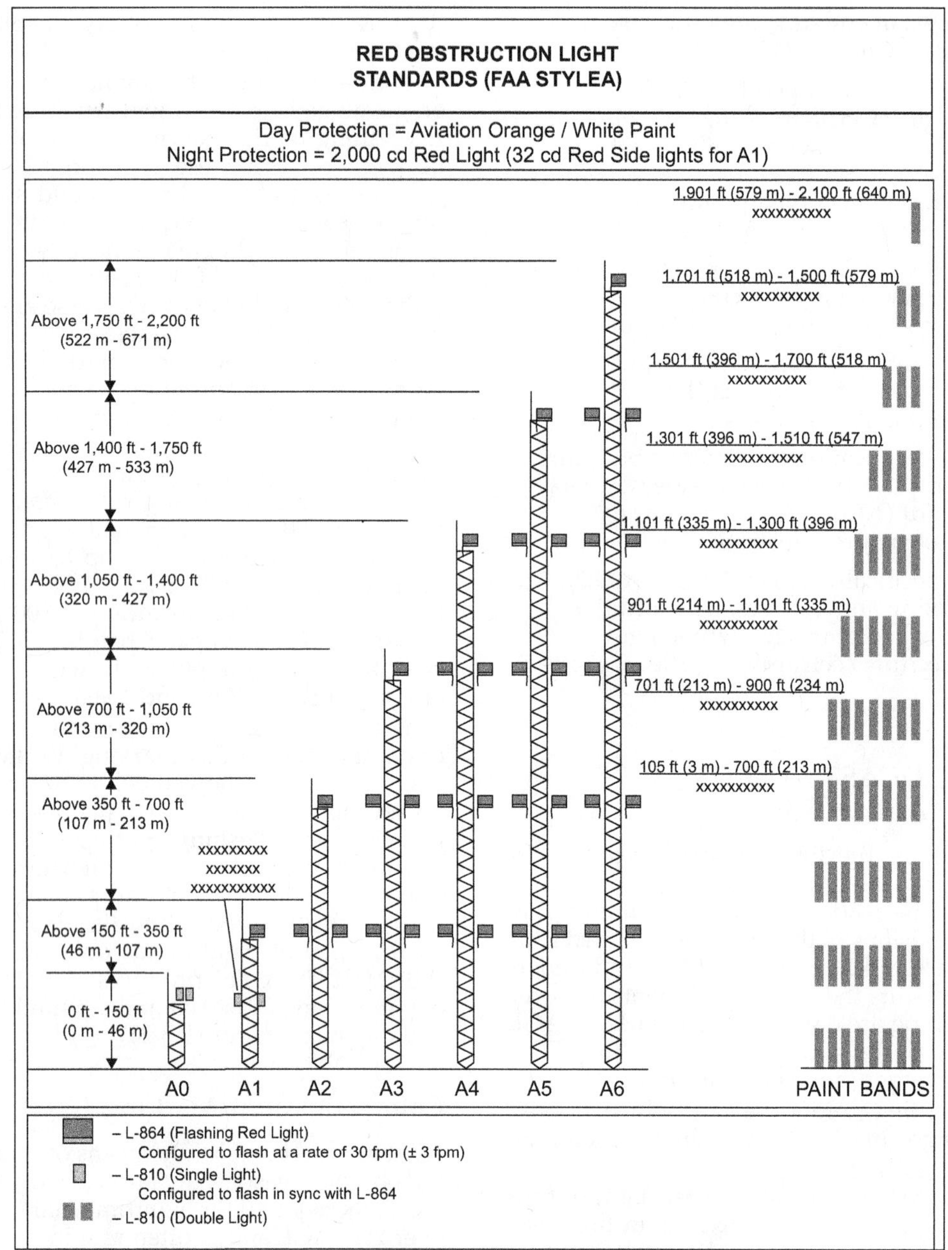

FIGURE 9.3-2 FAA "A" Lighting/Painting Configurations

FAA/Public Document

technologies (meaning, for example, an L-864 using incandescent, xenon, or LED technology is still regarded as a medium intensity L-864 red flashing beacon):

- L-810: Commonly known as "side marker lights" (see Figure 9.3-3) and used in intermediate levels on taller towers, these smaller red lights may also be used as steady-burn primary lighting on A-0 (explained later in this chapter) structures ≤ 150′ AGL. On structures with an A-1 determination citing AC 70/7460-1L, the L-810s are required to flash in synchronized form with the medium intensity red lighting also required. The required candela output is 32.5 cd at a minimum. On structures taller than 350′ AGL, the L-810 side marker lights are not required. Technologies currently available: incandescent and LED.

NOTE: Before the owner of an older lighting system can disconnect the L-810 lights or make other changes to the current lighting configuration, FCC Form 854 and FAA Form 7460-1 must be filed for the structure and permission to change received. The current red-night lighting system must be capable of a flash rate in the 27- to 33 flashes per minute (fpm) range, and the overall lighting configuration must be brought to current AC 70/7460-1L specifications.

FIGURE 9.3-3 L-810 Side Marker Light

Photo courtesy of International Tower Lighting, LLC

FIGURE 9.3-5 L-865 Medium Intensity White Flashing Light

Photo courtesy of SPX Flash Technology

FIGURE 9.3-4 L-864 Medium Intensity Red Flashing Light

Photo courtesy of International Tower Lighting, LLC

FIGURE 9.3-6 L-856 High Intensity Xenon White Flashing Light

Photo courtesy of TWR Lighting, Inc.

- L-864: Medium intensity red flashing light for night marking only (see Figure 9.3-4). Candela output is 2,000 cd ±25%. May normally be used on structures > 150′ to 2,200′ AGL. Technologies currently available: incandescent, xenon, and LED.
- L-865: Medium intensity white flashing light (see Figure 9.3-5). May normally be used for day marking on structures ≥ 200′ to 700′ AGL in lieu of orange and white paint bands. Candela output is 2,000 cd ±25% in night mode and 20,000 cd ±25% in day mode. The L-865 may be combined with the L-864 for a dual day-white/night-red combination, where the opposing color is completely extinguished while the other is in operation in the respective mode. In no case is it acceptable to display both red and white lighting at the same time. Technologies currently available: xenon and LED.
- L-856: High intensity white flashing lights for use on towers over 700′ AGL in lieu of orange and white paint bands for day and twilight marking. Candela output is 2,000 cd ±25% in night mode, 20,000 cd ±25% in twilight mode (dusk and dawn), and 270,000 cd ±25% in day mode. The L-856 may also be combined with the L-864 for a dual day-white/night-red combination, where the opposing color is completely extinguished while the other is in operation in the respective mode. In no case is it acceptable to display both red and white lighting at the same time. L-856 technologies currently available: xenon and LED (see Figures 9.3-6 and 9.3-7).

FIGURE 9.3-7 L-856/L-864 LED White High Intensity Day/Red Medium Intensity Night Flashing Light

Photo courtesy of Dialight

Coverage: Number of Lights Required at Each Level

This appears to be a simple subject, but often requires careful planning. The key is to provide a 360° unobstructed view of the lighting to pilots of approaching aircraft. Beam path is the term used to describe the light projected from the lighting fixture. A full description of beam path can be found in AC 150/5345-43H.

Any obstruction of any size in the beam path of an L-810 is prohibited. Any obstruction wider than 7/8″ in the beam path of an L-864, L-865, or L-856 is prohibited. The tower, including legs and cross members, is an obstruction. Other obstructions often include cables, light brackets, and antennas. More light units may need to be added to ensure a 360° unobstructed display is achieved.

L-856 high intensity lighting is, at this writing, only available in a unidirectional design. Since each beacon provides 120° output, a minimum of three light units is needed at each lighting tier to achieve 360° unobstructed view. More may be necessary in extreme cases where side-mount antennas or other obstructions are present.

L-864 and L-865 lights are normally omnidirectional, projecting light in a 360° radius. A single L-865 may often be used at the top of a tower provided there are no obstructions over 7/8″ wide, but larger whip, panel, or other antennas may create the necessity for more light units. Lower levels will be required to have two units at a minimum, since the tower is an obstruction. Again, more may be necessary depending on obstruction factors.

On shorter towers requiring an A-0 system (described later in this chapter), a "double" L-810 may be required. This system consists of two L-810 units mounted at the same height in a single yolk-type base. Neither L-810 is regarded as an obstruction to the other.

Configuration

Along with the number of light units per level, configuration normally describes the number of required levels and the type of lighting included in the entire tower lighting system. When more than one light level is required, the levels should be evenly distributed from the top down on the structure. If two levels are required, the levels would be at the top and midpoint. Three levels required would mean one level at the top and the other two levels evenly spaced at the 1/3 and 2/3 elevations on the structure.

NOTE: Light levels may be adjusted within 10′ when necessary to accommodate guy wires and personnel who replace or repair light fixtures. Moving a tier of lighting up or down 10′ to accommodate a service platform is allowed. Moving a tier of lighting 30′ to get the lights out of the RF zone of a side-mount antenna is not allowed.

On towers ≤ 700′ AGL, or those of any height that have orange and white paint bands for day marking, the number of light levels required is dependent on the overall height of the structure.

For towers over 700′ AGL that have L-856 high intensity white lighting for day marking, the number of light levels required depends on the height of the main skeletal structure, excluding any appurtenances. (NOTE: AC 70/7460-1L has typographical errors in 6.3.1 and 7.8. "Including" should read "Excluding," as stated in Chapter 4.)

Many towers have antenna masts or reduced-face, un-guyed "appurtenances" atop the main skeletal structure that are incapable of supporting lighting units.

Appurtenances on structures > 700′ AGL are often incapable of safely supporting a level of L-856 or L-864 lighting, especially considering wind and weight loading. When the appurtenance extends no more than 40′ above the apex of the tower, no additional lighting is required above the top of the main skeletal structure, though a 360° is required. If the appurtenance extends more than 40′ above the tower apex, the top of the appurtenance must be marked with an L-864, L-865, or dual L-864/L-865 "AOL" beacon.

On structures 700′ AGL or less that have top appurtenances of 40′ or less in height, the top light should be installed at the apex of the main structure provided the appurtenance is no more than 7/8″ in width. If the appurtenance is over 7/8″ in width, an additional top light should be installed in a diametrically opposed fashion to ensure a 360° view is attained.

If any appurtenance exceeds 40′ in height and is incapable of supporting the required top light, one or more top lights should be mounted on a pole adjacent to the appurtenance to achieve a 360° unobstructed view within 40′ of the overall top of the structure (see Figure 9.3-8).

The FAA breaks configuration descriptions for lighting and marking into alpha-numeric designations. The alpha indicator denotes the type of system and numeric denotes the number of levels required. To clarify this, the AC 70/7460-1 has chapters devoted to descriptions of the different types of lighting/marking and illustrations, or "Figures" are included in the Appendix.

AC 70/7460-1L Chapter Quick References. Chapters commonly seen referenced in determinations for broadcast towers marked with an asterisk (*):

1 = ADMINISTRATIVE AND GENERAL PROCEDURES

2 = GENERAL

*3 = MARKING GUIDELINES (normally cited in some FAA Regions)

*4 = LIGHTING GUIDELINE (normally cited in every determination)

*5 = RED LIGHTING

*6 = MEDIUM INTENSITY WHITE LIGHTING

*7 = HIGH INTENSITY WHITE LIGHTING

*8 = MEDIUM INTENSITY DUAL LIGHTING

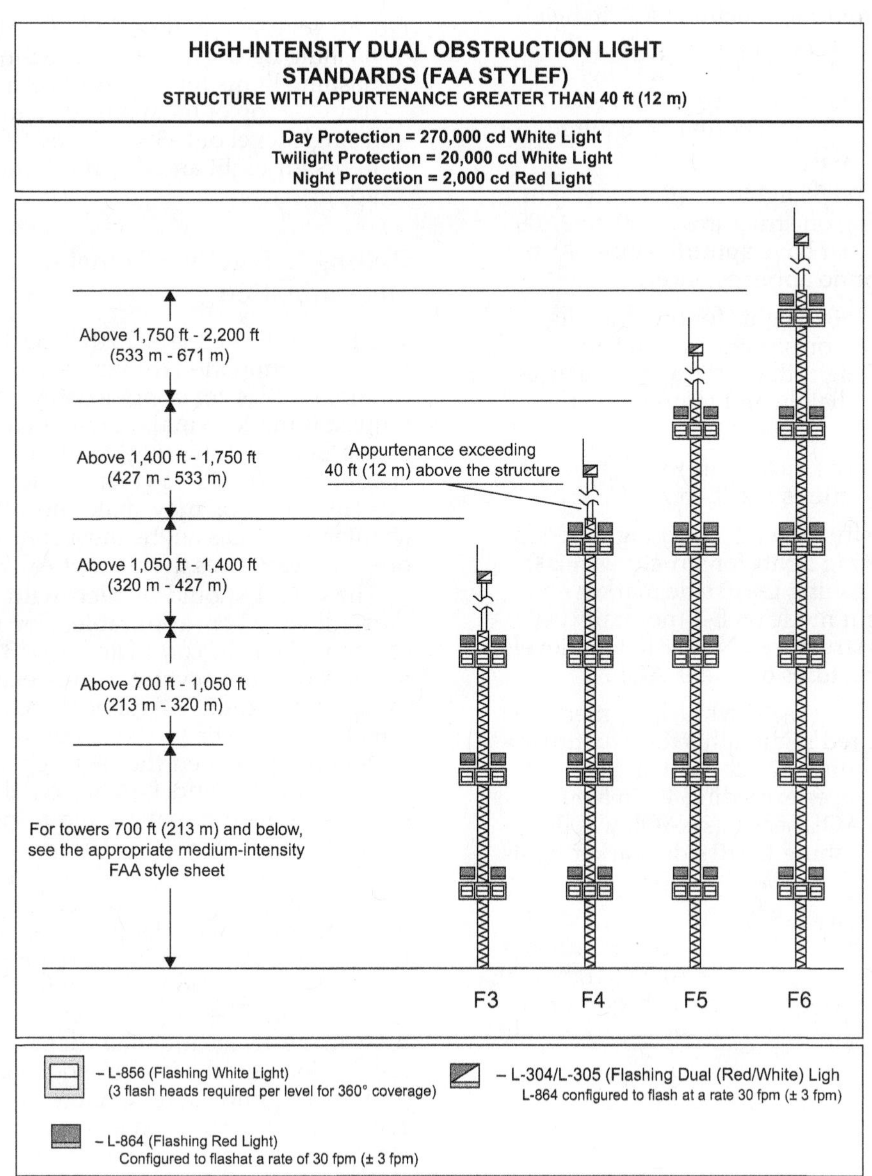

FIGURE 9.3-8 FAA "F" Lighting Configurations

FAA/Public Document

*9 = HIGH INTENSITY WHITE/MEDIUM INTENSITY RED DUAL LIGHTING

10 = CATENARY

11 = MOORED BALLOONS AND KITES

*12 = MARKING AND LIGHTING (normally cited in every determination; when cited with Chapter 5, normally indicates a requirement for paint bands)

13 = WIND TURBINES

14 = AIRCRAFT DETECTION LIGHTING SYSTEMS (Radar)

Alpha indicators normally associated with broadcast towers:

A = Red lighting for night, orange and white paint bands for day. Steady-burn L-810 red lights normally required for structures 150′ in height and shorter. L-864 flashing red lights with intermediate levels of L-810 flashing lights normally required from > 150′ to 350′ AGL. L-864 flashing red lights normally required from > 350′ to 2,200′ AGL.

B = High intensity L-856 lights for night, twilight, and day marking on structures > 700′ to 2,200′ AGL in height that have appurtenances 40′ or less in height or no appurtenance.

C = High intensity L-856 lights for night, twilight, and day marking on structures > 700′ to 2,200′ AGL in height that have appurtenances exceeding 40′ in height and require an AOL beacon.

D = Medium intensity L-865 lights for night and day marking on structures > 200′ to 700′ AGL.

E = Medium intensity dual L-864 red night/L-865 white day flashing lights for structures > 200′ to 700′ AGL. Flashing L-810 side marker lights required in night mode on E-1 (normally 200′ to ≤ 350′ AGL) structures. No L-810 light levels required on structures over 350′ AGL.

F = Dual high intensity L-856 white day/medium intensity L-864 red night lights for structures > 700′ to a maximum of 2,200′ AGL in height that have appurtenances exceeding 40′ in height and require an AOL beacon(s). AOL will be L-864/L-865. Flashing L-810 side marker lights not required.

G = Dual high intensity L-856 white day/medium intensity L-864 red night lights for structures > 700′ to a maximum of 2,200′ AGL in height that have appurtenances 40′ or less in height or no appurtenance. Flashing L-810 side marker lights not required.

Numerical indicators with AGL heights:

0 = 0–150′

1 = "A" designation: > 150′ up to and including 350′, "D" and "E" designation: > 200′ up to and including 350′

2 = > 350′ up to and including 700′

3 = > 700′ up to and including 1,050′

4 = > 1,050′ up to and including 1,400′

5 = > 1,400′ up to and including 1,750′

6 = > 1,750′ up to and including 2,200′ (2,200′ AGL is commonly regarded as the maximum allowed height for a fabricated structure)

Examples:

A-3: Three light levels at 1/3, 2/3, and top of structure, all consisting of L-864 medium intensity flashing red lights.

E-1: One level of medium intensity flashing dual lighting (red L-864 for night, white L-865 for day) at the top of the tower and one intermediate level of flashing L-810 red lights at the midpoint.

F-6: Six levels of dual L-856 high intensity white flashing lights for day and twilight with L-864 red flashing lights for night at 1/6^{t} level delineations from the top of the main structure down, plus a single level of L-864 / L-865 flashing AOL lights for night and day marking at the top of the appurtenance.

Putting It Together: Translating a Determination

The FAA will cite the AC 70/7460-1 current to the date of the determination for new constructions and refiles. In some refiles for a pre-existing structure where no request is made to make a major change or any change to the lighting/marking configuration, the FAA may cite the AC 70/7460-1 version referenced in a previous determination or may make no reference at all. This chapter will focus on the most common determination: one that references the current AC 70/7460-1L.

The 7460-1 should be filed with pertinent information added when applicable. For example, a broadcaster in Florida contracted a 483′ AGL tower on a tower farm with several towers exceeding 1,000′ AGL in close proximity. Another broadcaster contracted a 600′ AGL tower in the same area. The first broadcaster only provided the required information including coordinates and heights on the application and received a determination requiring a dual medium intensity lighting system. The second broadcaster with the taller construction included a description of the surrounding towers with heights, lighting, and the distance from each, along with a request for no lighting (types of lighting may be requested) and received a determination of No Hazard with no lighting or marking required. The FAA simply does not have enough time to perform added in-depth research for every application. It is the owner/sponsor's responsibility to provide this information. A few extra minutes or hours spent on the application can save thousands of dollars in equipment, maintenance, and paperwork through the years.

The determination letter will cite the specific location, coordinates, and heights as provided to the FAA in the Form 7460-1, the date of issue, ownership, and other information. For lighting and marking purposes, the section just underneath the location information is crucial. Typical language:

> This aeronautical study revealed that the structure does exceed obstruction standards but would not be a hazard to air navigation provided the following condition(s), if any, is(are) met:
>
> > As a condition to this Determination, the structure is marked/lighted in accordance with FAA Advisory Circular 70/7460-1L, Obstruction Marking and Lighting, [type of lighting may or may not be mentioned here]—Chapters 4 [specified chapter number] and 12.

The Appendix A of AC 70/7460-1L is rarely mentioned but is a critical source of guidance because it includes illustrations for the required configurations on all structure heights.

Examples using AC 70/7460-1L:

1. 1,040′ main structure (tower) with a top-mounted 60′ mast antenna. Overall height of 1,100′ AGL. Determination cites "Chapters 4, 9, and 12." The chapters cited translate to an "F" dual system using red medium intensity lighting for night marking and white high intensity lighting for day and twilight marking. Referencing the height of the main structure and the appurtenance to Figure A-11 in the Appendix, the system is an F-3: three levels of dual lighting, one at the top of the main structure, one at 1/3 height, and one at 2/3 height. An L-864/L-865 combination (may be two separate or one combined beacon) AOL is required at the top of the mast (appurtenance).
2. 360′ overall tower. Determination cites "Chapters 4, 6, and 12." The chapters cited translate to a "D" 24-hour white system using medium intensity lighting for night and day marking. Referencing the overall height of the structure (appurtenances are not a consideration when the overall height of the structure is ≤ 700′ AGL) to Figure A-7 in the Appendix, the system is a D-2: two levels of white lighting, one at the top (apex) of the overall structure and one at 200′ (the lowest level of white lighting must not be below 200′ AGL). Intermediate levels of red L-810 side marker lights are not required on structures over 350′ AGL under the AC 70/7460-1L.
3. 346′ overall structure. Determination cites "Chapters 4, 5, and 12." The chapters cited translate to an "A" system using medium intensity red lighting for night and orange and white paint bands for day marking. Referencing the overall height of the structure (appurtenances are not a consideration when the overall height of the structure is ≤ 700′ AGL) to Figure A-6 in the Appendix, the system is an A-1: one level of red medium intensity L-864 lighting at the top (apex) of the overall structure and one level of flashing L-810 red side markers at the midpoint of the tower.

Typical FAA ASN Determination with identifying information blurred (Figure 9.3-9). Note the HIWOL reference to "High Intensity White Obstruction Lighting":

TOWERS BETWEEN 50′ AND 200′ AGL

The FAA and FCC have not historically required broadcast and other towers in this height range to exhibit marking and/or lighting, but Act H.R. 636 "FAA Extension, Safety, and Security Act of 2016" was passed by Congress and signed by the president in July 2016. Section 2110 of the act may require owners of these towers to mark and light structures in rural areas.

Under Section 2110, the FAA is required to compose and issue new regulations within 1 year of passage requiring "covered towers" to be "clearly marked." The definition of a "covered tower" under the act:

Is self-support or guyed tower.

Is 10 feet or less in diameter at the above-ground base, excluding concrete footing.

Is between 50 feet and 200 feet in height AGL.

Has accessory facilities on which an antenna, sensor, camera, meteorological instrument, or other equipment is mounted and is located (a) outside the boundaries of an incorporated city or town; or (b) on land that is (i) undeveloped; or (ii) used for agricultural purposes.

Congress further defines "undeveloped" land as area over which the "[FAA] Administrator determines low-flying aircraft are operated on a routine basis, such as forested areas with predominant tree cover under 200 feet and pasture and range land."

Excluded under the law are any structures that:

Is adjacent to a house, barn, electrical utility station or other building, are within "the curtilage of a farmstead."

Supports electric utility transmission or distribution lines.

Is a wind-powered electrical generator with a rotor blade radius that exceeds six feet or

Is a street light erected or maintained by a federal, state, local, or tribal entity.

The law also authorizes the FAA to define and clarify 2110 "as necessary to carry out the section." This will include definable descriptions for the aforementioned "buildings" and what is considered to be "adjacent." A database will be developed and maintained, but the big question is on the enforcement of these new requirements.

As opposed to the requirements of newly issued versions of the AC 70/7460-1, these requirements will be retroactive. From Section 2110:

> (c) Application.—The regulations issued under subsection (a) shall ensure that—
>
> (1) all covered towers constructed on or after the date on which such regulations take effect are marked in accordance with subsection (b); and
>
> (2) NOTE: Deadline; a covered tower constructed before the date on which such regulations take effect is marked in accordance with subsection (b) not later than 1 year after such effective date.

Mail Processing Center
Federal Aviation Administration
Southwest Regional Office
Obstruction Evaluation Group
10101 Hillwood Parkway
Fort Worth, TX 76177

Aeronautical Study No.

Prior Study No.

Issued Date:

** DETERMINATION OF NO HAZARD TO AIR NAVIGATION **

The Federal Aviation Administration has conducted an aeronautical study under the provisions of 49 U.S.C., Section 44718 and if applicable Title 14 of the Code of Federal Regulations, part 77, concerning:

Structure:	Antenna Tower
Location:	
Latitude:	
Longitude:	
Heights:	7 feet site elevation (SE) 1536 feet above ground level (AGL) 1543 feet above mean sea level (AMSL)

This aeronautical study revealed that the structure does exceed obstruction standards but would not be a hazard to air navigation provided the following condition(s), if any, is(are) met:

As a condition to this Determination, the structure is marked/lighted in accordance with FAA Advisory circular 70/7460-1 K Change 2, Obstruction Marking and Lighting, 24-hr hi-strobes - Chapters 4,7(HIWOL),&12.

This determination is based, in part, on the foregoing description which includes specific coordinates , heights, frequency(ies) and power . Any changes in coordinates , heights, and frequencies or use of greater power will void this determination. Any future construction or alteration , including increase to heights, power, or the addition of other transmitters, requires separate notice to the FAA.

This determination does include temporary construction equipment such as cranes, derricks, etc., which may be used during actual construction of the structure. However, this equipment shall not exceed the overall heights as indicated above. Equipment which has a height greater than the studied structure requires separate notice to the FAA.

This determination concerns the effect of this structure on the safe and efficient use of navigable airspace by aircraft and does not relieve the sponsor of compliance responsibilities relating to any law, ordinance, or regulation of any Federal, State, or local government body.

FIGURE 9.3-9 Example FAA Determination Letter

FAA/Public Document

These new regulations take effect within 1 year of passage, or by July 7, 2017. Pre-existing "covered" towers will be required to exhibit the specified marking and lighting by July 7, 2018.

The previously described relationship between the FAA as a rule maker and the FCC as an enforcement body may change in light of these new regulations. The governing branch for the FAA, the Department of Transportation (DOT), has an enforcement arm capable of levying fines and other actions. Broadcast tower owners should watch for any Notice of Proposed Rule Making (NPRM) on this subject and participate in the process.

Monitoring

Proper monitoring is the single most important factor of lighting system ownership. Improper monitoring can lead to embarrassing actions and expensive fines from the FCC and can be financially devastating in the event of an accident involving an aircraft collision on the structure. Proper documentation of monitoring is critical.

Monitoring of aviation obstruction lighting is addressed by both the FAA and FCC and may be accomplished by either visually inspecting the tower lighting or through the use of automatic remote sensor technology.

Visual monitoring: The FAA requires a visual inspection of the obstruction lighting system in all intensities at least once every 24 hours. If the system is an "A" type consisting of red lighting for night marking and paint bands for day, one daily check when the system is in the night mode should be sufficient. If the system is a 24-hour white or dual (red night, white day) configuration, at least two intensities (day and night) are exhibited and must be inspected and logged each day. High intensity lighting adds a third intensity, twilight, and another level of inspection. Since no more than one of the modes is exhibited at any one time, multiple inspections must be performed and recorded each day.

In each mode, every light on the structure must be inspected. Logged entries should state the ASR #, date, time, and point from which the inspection was performed, mode, and specific observations/actions. Examples:

Date: 11/19/15

Time: 9:18 PM

ASR: 1111111

Observation point: 189 Somewhere Drive, City, State, Zip (or coordinates)

Mode: Night Red (only mode at this site)

Observation: Single top beacon and two beacons at each lower level. All lights functioning properly.

Signed: John Doe

Date: 11/20/15

Time: 9:24 PM

ASR: 1111111

Observation point: 189 Somewhere Drive, City, State, Zip (or coordinates)

Mode: Night Red (only mode at this site)

Observation: Single top beacon and two beacons at each lower level. One lamp in northwest light appears to be not working. Called station at 9:27 PM and relayed findings to Station Engineer Bill Anonymous.

Signed: John Doe

In the event of a complaint or FCC inspection, monitoring logs for the site may be requested and be reviewed. Substitute visual monitors are normally necessary due to vacations, unexpected illness, etc.

When wired or wireless means and proper equipment are available, automated remote monitoring is normally preferred to visual monitoring. Since alarm relay information is communicated expeditiously in most cases, the lighting system at each structure is only required to be called or "polled" by automated means once every 24 hours. The results of the poll, along with any communication of alarms, must be documented just as visual monitoring is logged. The documentation may be performed physically or may be recorded in a SCADA or other computer format. Critical information for inclusion is time and date of the call, results, alarms, or any other information accessed by the monitoring unit. Many third-party suppliers are also available to competently provide remote monitoring services.

The FCC requires a visual inspection of any lighting systems monitored by automated means at intervals not exceeding 90 days. This is normally referred to as a "QLI" or Quarterly Lighting Inspection. However, the FCC will issue a waiver for this requirement to structure owners who can demonstrate strict and robust procedures through advanced automated means to monitor their lighting systems either directly or through a third-party service. This "QLI Waiver" is only available through an application and rigorous investigation process by the FCC.

Failure of any flashing obstruction light (which now includes flashing L-810 intermediate lights) that lasts for more than 30 minutes, improper mode display (night cd output in day or reverse, for example), or failure of one lamp in a two-lamp fixture requires a NOTAM (NOtice To AirMen) to be filed, alerting all pilots who check for open NOTAMs to be informed of the potential hazard. The NOTAM is normally open for 15 days and closed automatically in the absence of a request to remain open, though the owner may request and receive longer-span NOTAMs if unusual circumstances warrant the lengthened time span. The NOTAM must be closed and notification issued upon restoration of normal operation. See more on NOTAMs, along with reporting information, in Chapter 2 Section 4 (2.4) in AC 70/7460-1L.

The list of recent enforcement actions by the FCC is considered "public information" and is available to view at http://transition.fcc.gov/eb/broadcast/asml.html. A brief examination will reveal an extensive list of actions and fines, most of which could have been avoided through adoption of a detailed monitoring process by the tower owner.

Maintenance and Technology

Proper lighting and marking maintenance ranks very near monitoring in importance to a tower owner. Lighting systems are often expensive and improper maintenance may cause early failure and costly repairs. Different technologies have unique applications and require varying degrees of attention. Selecting the right service company is important from both cost and liability standpoints.

Paint bands can be expensive to maintain, but may be necessary in some cases where additional marking is required, where daytime lighting is regarded as a nuisance by nearby businesses or residents, or the tower structure needs additional protection from the elements of nature. Addressing orange and white paint bands, AC 70/7460-1L states in 3.3.1:

> The FAA recommends that quality paint and materials be selected to maximize years of service. The paint should be appropriate for the surfaces to be painted, including any previous coatings, and suitable for the environmental conditions. Surface preparation and paint application should follow the manufacturer's recommendations.
>
> **Note:** In-Service Aviation Orange Color Tolerance Charts are available from private suppliers for determining when repainting is required. The color should be sampled on the upper half of the structure, since weathering is greater there.

Likewise, lighting technologies are important when formulating a future tower-ownership plan. AC 70/7460-1L addresses light fixture inspection and maintenance in 4.5. Supply voltage to the different fixtures should be measured and documented. Voltages outside the nominal range (±10% the stated voltage on

FIGURE 9.3-10 Technician Servicing Lighting System Controller

Photo courtesy of Hilights, Inc.

the manufacturer's product nomenclature) may result in non-compliant light output from xenon or some LED systems. More than 3% voltage drop to the lamp in an incandescent or some LED systems will result in noncompliant light output. Lamps should be replaced at 75% of their rated life or, of course, upon failure. Also stated in 4.5:

> Flashtubes in a light unit should be replaced immediately upon failure, when the peak effective intensity falls below specification limits or when the fixture begins skipping flashes, or at the manufacturer's recommended intervals. Due to the effects of harsh environments, light fixture lenses should be visually inspected every 24 months, or when the light fixture fails, for ultraviolet (UV) damage, cracks, crazing, dirt buildup, etc., to ensure the certified light output has not deteriorated. (See Chapter 2 paragraph 2.4 for reporting requirements in case of failure.) Lenses that have cracks, UV damage, crazing, or excessive dirt buildup should be cleaned or replaced.

TECHNOLOGY PROS AND CONS

Incandescent: Only available for L-810 or L-864 red lighting. Lowest initial cost per fixture, lowest component cost to maintain, very simple with minimal componentry. At 120 volts, takes 10.33 amps per L-864 and nearly an amp per L-810, making it the least electrically efficient option. Requires largest cable or conductor sizes (requiring larger conduit) due to electrical consumption. Lowest rated lamp life as well. Possibly a good choice for those towers that require frequent climbs for other types of maintenance, where wind and weight load are of little consideration, and where the cost of electricity is of little consequence.

Xenon: Available for L-864, L-865, and L-856 lighting. No L-810 option, meaning configurations that require L-810 fixtures result in hybrid or two-technology systems. Mid-range price for initial purchase. Capacitance-discharge technology. Quite complex with multiple components, many of which are relatively expensive. Much more efficient in terms of electrical consumption than incandescent. Smaller conductors required, though more may be necessary. Rated lamp life is normally twice that of an incandescent, requiring fewer tower climbs for unexpected maintenance.

LED: Available for all tower lighting—L-810, L-864, L-865, and L-856. Normally most expensive to purchase but least expensive to maintain. Not as simple as incandescent, but not as complex as xenon. Least expensive in terms of power consumption, requires smallest conductors. Rated life of LEDs is usually over twice that of xenon and roughly five times incandescent, meaning fewer unscheduled tower climbs required for maintenance. Warranties are normally over twice as long in duration as those for xenon products.

Maintenance companies are a varied lot. Crews who are highly trained and proficient with antennas and other assets on a tower structure may lack the knowledge and ability to properly install and/or maintain an aviation obstruction lighting system. Anytime a tower owner authorizes a third-party to climb a tower, said owner has accepted an enormous degree of liability regardless of contract language.

SUGGESTIONS FOR SELECTION

- Consult peers. Most will have trusted relationships, but that trust should not be shared until a thorough investigation has been performed.
- Consult the lighting manufacturer. Many manufacturers are closely associated with quality installers or may offer a list of technicians who have attended equipment training classes.
- Ask the service company for references, then take the time to contact each reference. A quick online search of the company and principles may reveal open legal actions or past claims.
- Ask for proof. Certifications, including OSHA or other recognized industry safety training, are normally the first stage of qualification.
- Is the company or a key person within the company a qualified, licensed electrician? The electrical systems that supply these lighting systems can be the single determining factor on how well the system functions and for how long. A study performed in 2011 revealed the majority of premature failures or an unusually frequency of issues concerning lighting systems is directly attributable to a poor installation of the electrical supply system. Some states now require fully licensed electrical contractors for installation of sensitive componentry.
- Ask for lead times, but be aware that adverse weather conditions and other safety factors may affect promised completion dates.
- Ask to see any parts replaced by the service company either in person or through photographs, but do not demand to keep the parts. The service company may be required to return the parts to the manufacturer or supplier for warranty or research reasons.
- Keep copies of the wiring schematic or design available for the crews. Extra copies are often a cheap investment.
- Do not wait until service is needed prior to performing this research and selecting a tower lighting maintenance company. Perhaps the biggest mistake in selecting a service company is often driven by the urgency of need. Take the time and make the decision in advance of a crisis.

Exposure to Liability: Compliance Does Not Close the Door

The first step to limiting liability with regards to structure lighting systems is to attain full compliance. Just as it is important to select a good service supplier for maintenance, it is also important to build a team and put an action plan in place to help confirm the systems and procedures are compliant. Are the monitoring logs up to date? Have all maintenance events been properly recorded? If the FCC Enforcement Bureau appeared on site unannounced, would everything be in place for a full inspection? Could the station defend against a NOV? Has the Compliance Plan been reviewed by a

FIGURE 9.3-11 Technician Servicing Strobe Light

Photo courtesy of Hilights, Inc.

third-party professional? Has the tower been inspected by a qualified company or individual?

Concerning certified aviation obstruction lighting, the FAA states "Other manufacturers' equipment may be used provided the equipment meets the specification requirements of AC 150/5345-43." This means lighting products that are not certified to have met or exceeded the standards published in the respective FAA AC 150/5345-43 may be used. Any accident or incident involving an impact or near-miss of a tower structure with an aircraft is normally investigated by the FAA and NTSB (National Transportation and Safety Board). The results are "public knowledge" and are published for anyone to see, including any attorneys who may be involved with the inevitable lawsuits commonly associated with this type of event. These reports often include the type of obstruction lighting employed on the tower. The sum of this information is easy; the plaintiff's attorney builds a case for a settlement, constructing the potential from each individual component in the investigation. Although the uncertified equipment is legal to use, the burden of proof that the output, direction, color, operation, and other factors are within compliance falls to the owner of the tower. The manufacturer of the lighting may or may not be able to produce documentation that the equipment was tested by a qualified third-party laboratory and meets or exceeds the criteria contained in the respective AC 150/5345-43.

This is only one of many considerations, including monitoring and methods of reporting alarms, unobstructed view, proper placement, and many more.

As a footnote: For tower lighting, marking, and monitoring compliance and liability, the FAA names the tower owner as the responsible party. The FCC secondarily names licensees or permittees as responsible parties in 47 CFR Part 17.6. The responsibility is outlined by the FCC, as seen here:

§17.6 Responsibility for Painting and Lighting Compliance

(a) The antenna structure owner is responsible for maintaining the painting and lighting in accordance with this part. However, if a licensee or permittee authorized on an antenna structure is aware that the structure is not being maintained in accordance with the specifications set forth on the Antenna Structure Registration (FCC Form 854R) or the requirements of this part, or otherwise has reason to question whether the antenna structure owner is carrying out its responsibility under this part, the licensee or permittee must take immediate steps to ensure that the antenna structure is brought into compliance and remains in compliance. The licensee must:

(1) Immediately notify the structure owner;
(2) Immediately notify the site management company (if applicable);
(3) Immediately notify the Commission; and,
(4) Make a diligent effort to immediately bring the structure into compliance.

(b) In the event of non-compliance by the antenna structure owner, the Commission may require each licensee and permittee authorized on an antenna structure to maintain the structure, for an indefinite period, in accordance with the Antenna Structure Registration (FCC Form 854R) and the requirements of this part.

(c) If the owner of the antenna structure cannot file FCC Form 854 because it is subject to a denial of Federal benefits under the Anti-Drug Abuse Act of 1988, 21 U.S.C. 862, the first tenant licensee authorized to locate on the structure (excluding tenants that no longer occupy the structure) must register the structure using FCC Form 854, and provide a copy of the Antenna Structure Registration (FCC Form 854R) to the owner. The owner remains responsible for providing to all tenant licensees and permittees notification that the structure has been registered, consistent with §17.4(f), and for posting the registration number as required by §17.4(g). [61 FR 4363, Feb. 6, 1996, as amended at 79 FR 56985, Sept. 24, 2014]

IN SUMMARY

Tower lighting and marking is often an undesirable responsibility. There is no profit to be derived, the expense is considerable, the bureaucracy is time consuming and tedious, and the liabilities can be astronomical. Devote time for education, exploration, and documentation to minimize the burden of this facet of tower management.

CHAPTER

9.4

Transmission Lines for AM, FM, HF, and TV Stations and Associated Systems

PHILLIP YOUNG
Director of Quality Assurance, Electronic Research, Inc.

NICHOLAS PAULIN
Project Engineering Manager, Electronic Research, Inc.

BILL HARLAND
Vice President of Marketing, Electronic Research, Inc.

INTRODUCTION

There are three fundamental components to creating a high-power broadcast facility. Those three items are a transmitter, an antenna, and the transmission line that connects the transmitter to the antenna (Figure 9.4-1 and 9.4-2 show examples of transmission lines).

A transmitter's purpose is to amplify a mixed RF signal from the exciter. The antenna radiates the amplified RF and produces a signal out to its viewers and listeners. This chapter is intended to provide individuals planning a terrestrial broadcast transmission system with the information needed to select and install the appropriate transmission lines to connect transmitters and receivers to their associated antennas. This chapter will cover product selection and installation of the system from the transmitter output to the antenna input. This chapter also includes sections on system maintenance, inspection protocols, and general information on repairing system damage.

There are many variables that need attention when designing a transmission line system. This chapter will explain how to answer all of the following questions.

- How much RF power handling is needed?
  - Is this system being designed for multiple frequencies?
  - What are the theoretical peak voltage requirements?

FIGURE 9.4-1 Semi-flexible air dielectric coaxial cable, rigid coaxial, and circular waveguide transmission lines.

Courtesy of Electronics Research, Inc.

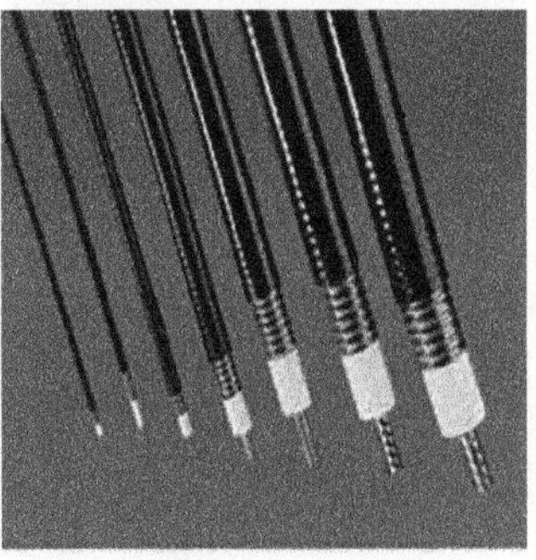

FIGURE 9.4-2 Semi-flexible foam dielectric coaxial cable.

Courtesy of CommScope

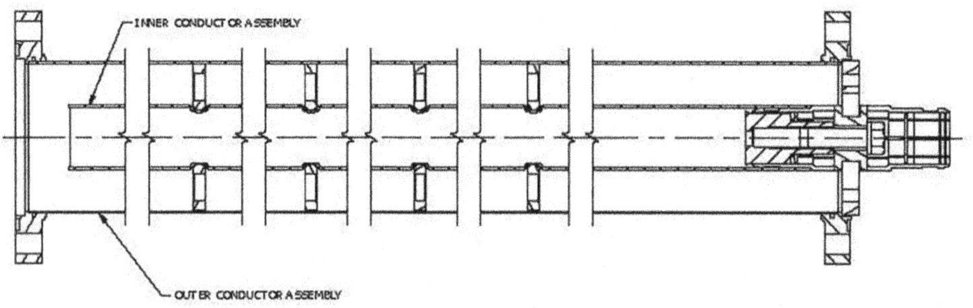

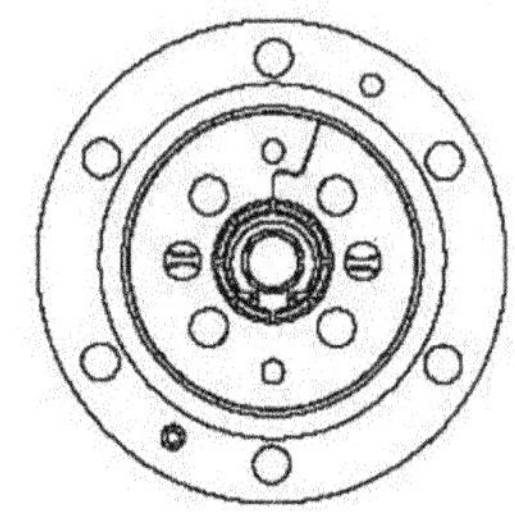

FIGURE 9.4-3 Rigid coaxial transmission line section.

Courtesy of Electronics Research, Inc.

- How much power is needed at the antenna to produce the desired ERP?
- How much transmitter power is available?
- How long is the transmission line length (vertical and horizontal runs)?
- Is 50-ohm or 75-ohm coaxial transmission line best for this application, or should waveguide be used?
  - Should rigid (see Figure 9.4-3) or flexible transmission line be used?

Transmission Line Types

There are broadly two types of transmission line in general use: These are coaxial transmission lines and waveguide transmission lines. Coaxial transmission lines are made up of two concentric conductors on the same axis (coaxial). The electric (E) and the magnetic (H) fields have no component in direction of transmission; both the E field and the H field are "transverse" to direction of transmission, and therefore the mode of transmission is called transverse electromagnetic mode (TEM). The TEM mode of propagation is illustrated in Figure 9.4-4. Waveguide transmission lines are E field is "transverse" to the direction of transmission and the H field has both transverse and longitudinal components and is therefore called transverse electric mode (TE). The TE mode of propagation is illustrated in Figure 9.4-5.

Within each of these broad categories of transmission types are many variations. In broadcast transmission applications they are grouped as semi-flexible and rigid transmission lines.

Coaxial Transmission Line Characteristics

Semi-flexible cable known as coaxial cable has unique features, including a copper corrugated outer conductor, which gives it flexibility, durability, and complete shielding (see Figures 9.4-6 and 9.4-7).

There are two different types of flexible cable: closed-cell and air dielectric. The closed-cell flex cable is one in which the inner conductor is separated from the outer by closed-cell foam—polyethylene dielectric. The air dielectric flex cable is characterized by a

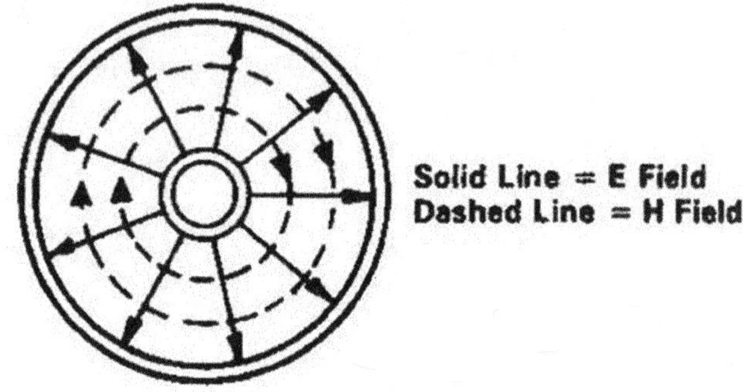

FIGURE 9.4-4 Coaxial transmission line transverse electromagnetic mode (TEM) propagation.

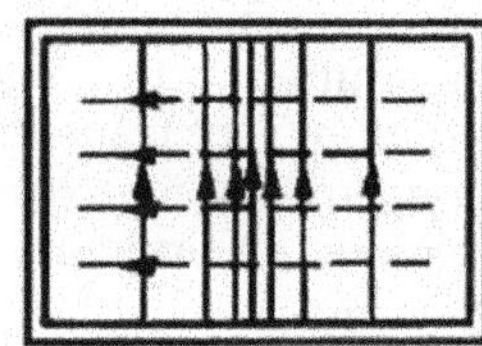

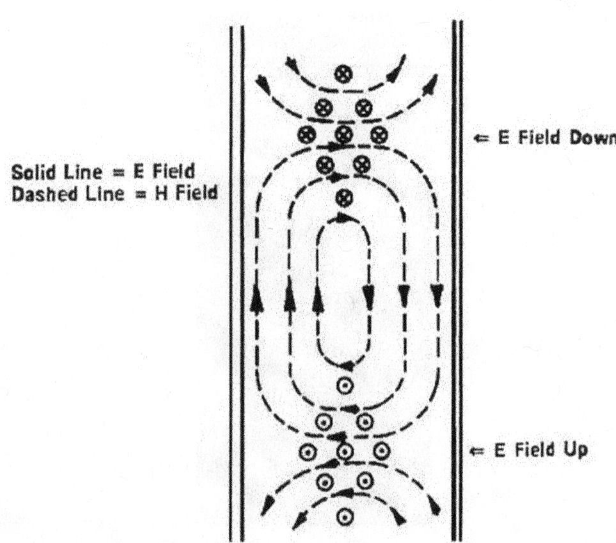

FIGURE 9.4-5 Waveguide transmission line transverse electric mode (TE) propagation.

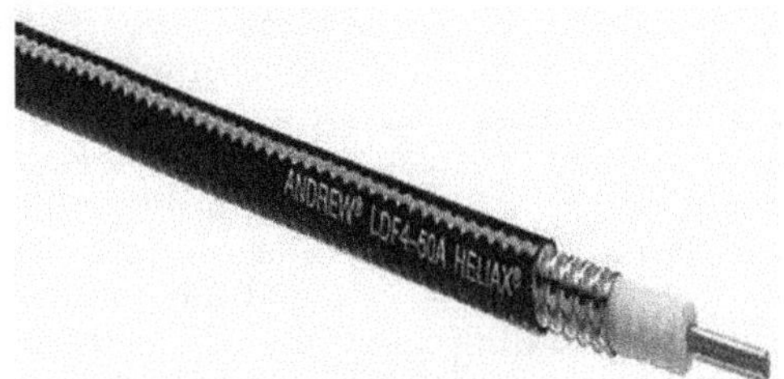

FIGURE 9.4-6 CommScope (Andrew) LDF4-50A 1/2-inch semi-flexible foam dielectric coaxial cable.

Courtesy of CommScope

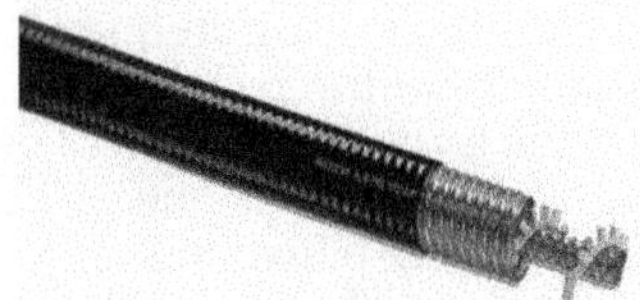

FIGURE 9.4-7 CommScope (Andrew) HJ8-50B 3-inch semi-flexible air dielectric coaxial cable.

Courtesy of CommScope

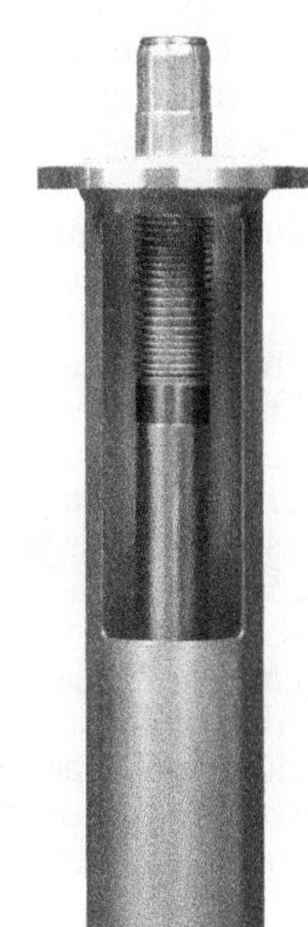

FIGURE 9.4-8 ERI 1329Line 3-1/8-inch rigid coaxial transmission line.

Courtesy of Electronics Research, Inc.

polyethylene or polypropylene spacer between the inner and outer conductor. The cables are designed to provide optimum electrical performance for a wide range of RF applications. It has low attenuation, shielding, low VSWR, and intermodulation performance. These cables range in size from 1/2 inch up to 9 inches, and the size selected depends on the power being transmitted. Using flexible cable eliminates multiple flange reflections that would show up in rigid line. Advantages are cost, long continuous lengths, installation, and reliability.

Rigid transmission line is constructed with two flanges, outer conductor, inner conductor, and multiple insulators (see Figure 9.4-8).

The flanges are designed to Electronic Industries Association (EIA) standards for the purpose of interchangeability of products by different manufacturers. The flanges and outer and inner conductor are usually made of copper material, but with the commodity's rise in price, manufactures have been using other nonferrous materials such as aluminum to manufacture rigid coaxial transmission line.

Insulators are used to hold the inner conductor in place to keep it from shorting out against the outer conductor. Insulators are made of dielectric material such as polyethylene or polypropylene, known as virgin Teflon. The insulators are designed to create no or very small impedance bumps in the signal due to the dielectric material and the spacing of them. Each piece of rigid line has an insulator or insulators, depending on the length, and they are spaced out a certain distance to cancel out its reflection.

The transmission line system is to transfer the power/signal from the transmitter to the antenna. Impedance matching is important when components of an electric circuit are connected from transmission line to antenna. The impedance ratio determines how much of the wave is transmitted forward and how much is reflected. Electrical impedance is the measure of the opposition that a circuit presents to a current when a voltage is applied. The impedance in transmission line is the difference of inner conductor outer dimension (*OD*) and the space between the outer conductor inner dimension (*ID*) and the dielectric material between them. The impedance (Z_0) equation:

$$Z_o = \frac{60}{\sqrt{\varepsilon_r}} \times \log_{10}\left(\frac{ID}{OD}\right) \tag{1}$$

where:

Z_0 = characteristic impedance

ε_r = dielectric constant of the dielectric material (1.00059 for air at 1 ATM)

ID = inside diameter of the outer conductor

OD = outside diameter of the inner conductor

For coaxial transmission lines, an impedance configuration of 30 ohms provides the maximum peak power-handling capability. For maximum average power handling, coaxial impedance should be about 60 ohms, and for minimum attenuation, coaxial impedance should be 77 ohms. To be practical from a manufacturing and system design perspective, the standard transmission line characteristic impedances for commercially available coaxial transmission line products is 50 ohms or 75 ohms. In general, AM, FM, and VHF television generally use 50-ohm transmission lines. At UHF television frequencies, 75-ohm transmission line is generally used to take advantage of its much lower attenuation at the operating frequencies.

Coaxial transmission lines have a maximum Cutoff frequency that is important to understand before choosing the size of line. The principal mode of propagation of coaxial transmission line is TEM, transverse electromagnetic. Higher order modes, also called waveguide modes, interfere with the TEM mode to a degree where efficient transfer energy is not possible. The cutoff frequency, F_{cutoff}, is the highest frequency that can operate efficiently in the TEM mode. When selecting a transmission line size, cutoff frequency should be one of the first considerations taken into account. As coaxial transmission line size increases, the cutoff frequency decreases. The calculated cutoff frequency is adjusted downward to account for other components within a system, such as elbows and elbow complexes. Using a figure of 90% of the theoretical will generally be sufficient for the useful cutoff frequency. Check the manufacturer's specifications when selecting, as each manufacturer may have slightly different figures. The cutoff frequency equation is:

$$F_{cutoff} = \frac{7510}{\sqrt{\varepsilon_r} \times (OD + ID)}$$

Where:

F_{cutoff} = cutoff frequency in MHz

ε_r = dielectric constant of the dielectric material (1.00059 for air at 1 ATM)

ID = inside diameter of the outer conductor

OD = outside diameter of the inner conductor (2)

Voltage Standing Wave Ratio (VSWR)

The goal in the design of any transmission line system is to transfer the energy from the transmitter to the antenna with a minimum of power loss. The attenuation constant of the transmission line is one factor in determining system efficiency. The other significant factor is the impedance mismatch between the transmission line and antenna. The greater this mismatch, the greater the power loss from energy reflected back down the transmission line. The incident and reflected waves combine in the transmission line and create an uneven voltage distribution. The maximum and minimum voltage and current create a standing wave along the transmission line. The VSWR is the ratio of maximum to minimum voltage and is expressed as:

$$VSWR = V_{max} / V_{min}$$

As VSWR increases, the maximum voltages in the line increase, and in severe cases this can lead to voltage breakdown. This may also cause excessive heating along the transmission line that can lead to the softening or even melting of insulators, which could cause inner conductors to shift physically.

Transmission Line Power Ratings

Transmission lines are rated for both their peak and average power-handling capabilities, and both are required to fully describe the capabilities of a given transmission line size. When evaluating a transmission line for applications where the signal to be transmitted is amplitude modulated, such as medium wave broadcast signals, or the application involves combining multiple analog and/or digital FM or television signals, then peak power is usually the limiting factor for a transmission line. In the case of a system being used for a single FM broadcast or television channel, then average power-handling capability will be the primary consideration in evaluating the suitability of one transmission line size over another.

Peak Power and Peak Voltage Limitations

In addition to knowing the impedance of the coaxial transmission line and the cutoff frequency, it is equally important to understand the power handling and loss characteristics of the transmission line. The peak power rating of a given transmission line is limited by the voltage breakdown value between the inner conductor and the outer conductor. The voltage breakdown value is independent of the RF frequency, but it does vary in air dielectric transmission lines with transmission line pressure and the density of the pressurizing gas. The formula for calculating the DC voltage breakdown of two cylinders placed coaxially is shown here:

$$\mathrm{E_p} = 3.17(10)^4 OD\delta\left(\log_{10}\frac{OD}{ID}\right)\left(1+\frac{0.273}{\sqrt{OD\delta}}\right)$$

E_p = production test voltage DC or peak AC

ID = inside diameter of the outer conductor

OD = outside diameter of the inner conductor

δ = air density factor = $3.92B/T$

Where B = absolute pressure in cm of mercury

T = temperature in degrees K

$\delta = 1.0168$ for $B = 67$ cm (29.92 inches) at $T = 293$ degrees K (20 degrees C) (68 degrees F)

The RF equivalent rms voltage is then calculated as:

$$\mathrm{E}_{rf-rms} = \frac{0.7\mathrm{E_p}}{SF\sqrt{2}}$$

where:

$E_{rf\text{-}rms}$ = RF operating voltage, maximum operating voltage

E_p = production test voltage DC or peak AC

SF = safety factor, industry standard = 2.0 for rigid transmission lines and 1.4 for semi-flexible transmission lines

The peak voltage handling is then simply

$$\mathrm{E}_{rf-peak} = \sqrt{2} \times \mathrm{E}_{rf-rms}$$

where:

$E_{rf\text{-}peak}$ = peak RF voltage handling

$E_{rf\text{-}rms}$ = RF operating voltage, maximum operating voltage

And the peak power handling is

$$P_{peak} = \frac{E_{rf-rms}^{\;2}}{Z_0}$$

where:

P_{peak} = peak RF power handling

$E_{rf\text{-}rms}$ = RF operating voltage, maximum operating voltage

Z_0 = characteristic impedance

An important item to remember is that the specification term RF peak power is misleading because it is a power term in watts based on a root mean square voltage, not a peak voltage. This is particularly important in planning transmission line systems for multiplexed systems that have more than one television or FM station combined into a single transmission line.

Derating Factors for Peak Power and Peak Voltage

The standard conditions for specifying peak RF power and voltage are usually stated as VSWR + 1.0, zero modulation, and one atmosphere absolute dry air (0 lb/in² or 0 kPa) at sea level. Therefore, the peak power rating must be derated for modulation type and VSWR. (3) The derating for modulation varies by type of service.

Modulation Peak Power Derating Formula

AM $$P_{\max} = \frac{P_{pk}}{(1+M)^2 \times VSWR}$$

$P_{\max}$ = Derated peak power

P_{pk} = Peak power rating of the transmission line

M = Amplitude modulation index (100% = 1.0)

$VSWR$ = Voltage standing wave ratio

The formula demonstrates that 100% amplitude modulation increases the peak power in the transmission line by a factor of 4.

FM $$P_{\max} = \frac{P_{pk}}{VSWR}$$

$P_{\max}$ = Derated peak power

P_{pk} = Peak power rating of the transmission line

$VSWR$ = Voltage standing wave ratio

Digital TV $$P_{\max} = \frac{P_{pk}}{VSWR}$$

$P_{\max}$ = Derated peak power

P_{pk} = Peak power rating of the transmission line

$VSWR$ = Voltage standing wave ratio

NOTE: For DTV $P_{\max}$ compared to the +7 dB peak power levels for 8VSB and not the average power level of the signal.

Analog TV

$$P_{\max} = \frac{P_{pk}}{(1+AU+2\sqrt{AU})\times \text{VSWR}} = \frac{P_{pk}}{(2.09)\times \text{VSWR}}$$

$P_{\max}$ = Derated peak power

P_{pk} = Peak power rating of the transmission line

AU = Aural to visual ration (20% Aural = 0.2)

$VSWR$ = Voltage standing wave ratio

2.09 = Modulation derating factor for TV for $AU = 2.0$

Increasing pressure and/or using high density gases increase peak and average power rating of a transmission line. Figure 9.4-9 and Table 9.4-1 show the effects of raising transmission line pressure on peak power rating.

This method is not recommended for long-term operation because a loss of pressure could have a catastrophic effect on the system. (4)

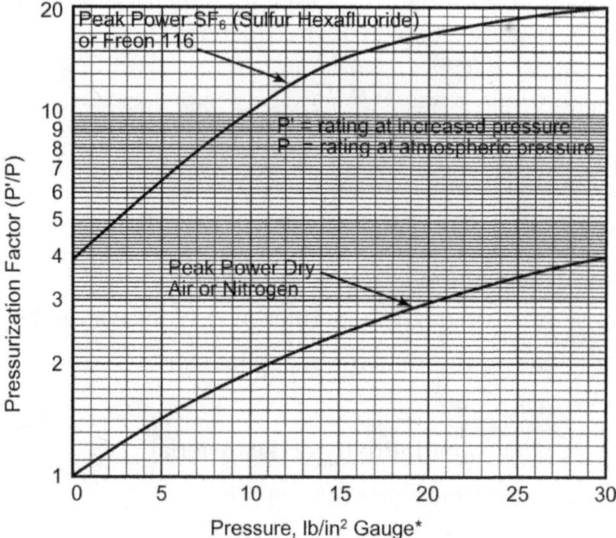

FIGURE 9.4-9 Uprating factors for peak power with increased line pressure (for kPa multiply by 6.895).

Courtesy of CommScope

TABLE 9.4-1
Uprating Factors for Average Power With Increased Line Pressure

Pressure	Factor, 50 ohm	Factor, 75 ohm
0 psig	1.00	1.00
5 psig	1.09	1.08
10 psig	1.16	1.15
15 psig	1.21	1.22
20 psig	1.26	1.28
25 psig	1.31	1.33

Attenuation and Average Power

Transmission line is made up of an outer conductor, inner conductor, and spacers. The power/signal being transferred through the line is current, and it flows on the outer conductor inner dimension (I.D.) and on the outer dimension (O.D.) of the inner conductor. This "skin effect" is the tendency of an alternating electric current (AC) to become distributed within a conductor such that the current density is largest near the surface of the conductor and decreases with greater depths in the conductor. The electric current flows mainly at the "skin" of the conductor, between the outer surface down to a level called the "skin depth." The skin effect causes the effective resistance of the conductor to increase at higher frequencies where the skin depth is smaller, thus reducing the effective cross section of the conductor. The skin effect is due to opposing eddy currents induced by the changing magnetic field resulting from the alternating current. At 88 MHz in copper, the skin depth is about 6.95 μm. At higher frequencies the skin depth becomes much smaller. Because the interior of a large conductor carries so little of the current, tubular conductors such as pipe can be used to save weight and cost. It can be calculated to see how deep the current flows on the skin of the material. (5) Skin depth equation:

$$\delta = \sqrt{\frac{2\rho}{\omega\mu}}\sqrt{\sqrt{\rho\omega\varepsilon}^2 + \rho\omega\varepsilon}$$

δ = skin depth

ρ = resistivity of the conductor

ω = angular frequency of current

μ_r = relative magnetic permeability of the conductor

μ_0 = the permeability of free space

μ = $\mu_r \mu_0$

ε_r = relative permittivity of the material

ε_0 = the permittivity of free space

ε = $\varepsilon_r \varepsilon_0$

Attenuation is reduction in signal strength (power) as it passes through transmission line and is due to the dielectric losses (coaxial line only) and the conductor losses. The dielectric loss is directly proportional to frequency, and the conductor loss varies with square root of frequency. Both factors are higher at higher frequencies. And the total attenuation is directly proportional to transmission line length. Attenuation of a particular transmission line is specified in decibels (dB) per unit length, either feet or meters. Total attenuation of a length of transmission line is called its insertion loss and is expressed in dB. Attenuation can be derived from the formula:

$$\alpha_c = \frac{13.6}{\lambda}\left(\frac{\delta_a\mu_a}{a} + \frac{\delta_b\mu_b}{b}\right)\frac{\sqrt{\epsilon}}{\log_{10}\frac{b}{a}} dB / unitlength$$

where:

αc = attenuation due to conductor losses, dB per unit length

λ = free space wavelength

δa = skin depth at frequency and rating temperature of inner conductor

μa = permeability of the inner conductor (unity of copper and aluminum, nonmagnetic)

δb = skin depth at frequency and rating temperature of outer conductor

μb = permeability of the outer conductor (unity of copper and aluminum, nonmagnetic)

ϵ = dielectric constant of air (1.00059)

a = radius of the outer of the inner conductor

b = radius of the inner of the outer conductor

Once the attenuation value is known, the efficiency of the transmission line system can be stated as the percentage of input power which reaches the antenna. The balance of the power is dissipated as heat and is lost. The efficiency of a run of transmission line can be found using the formula:

$$Efficiency = \frac{100}{10^{\frac{dB}{10}}}\%$$

dB = total transmission line attenuation loss, either calculated or measured

Designing transmission line system with higher efficiency reduces transmitter power output required to reach a given effective radiated power (ERP), and so it cuts power consumption but requires costlier (larger) transmission line. A good starting point in system design is to aim for a transmission line efficiency of 70% or better:

71% efficiency equates to an insertion loss of 1.5 dB

80% efficiency equates to an insertion loss of 1dB

50% efficiency equates to an insertion loss of 3dB

The average power-handling capability of a given transmission line type and size is expressed in watts or kilowatts of average power; the maximum average power is determined by allowable temperature rise of inner conductor and the oxidation of the dielectric material supporting the inner conductor. It is important to understand that the operating conditions that manufacturers use to determine power handling and attenuation varies. The most significant variation among competing suppliers is inner conductor operating temperature. Specifying a higher inner conductor operating temperature increases average power-handling capability, but this also reduces the useful life of the transmission line.

Derating Factors for Average Power and Attenuation

To convert rated transmitter power to average power for FM and digital television, the conversion factor is 1.0. To convert an analog television transmitter's output power to equivalent average power, multiply the peak visual output power by a factor of 0.8 (totally black picture + aural). AM broadcast in the 530- to 1,700-kHz band is limited by the transmission line's peak power capability. At the higher frequencies used in shortwave and HF communications the limitation is average power and the derating factor derived from:

$$DF = 1 + \frac{M^2}{2}$$

DF = Derating factor

M = Amplitude modulation index (100% = 1.0)

The average power rating specified for a given transmission is based on the manufacturer's standard operating conditions. In most cases the average power handling is based on an ambient temperature of 40 degrees C (104 degrees F). Figure 9.4-10 provides an adjustment factor that can be used to establish the average power rating of the line in its actual operating condition.

To derate the average power for VSWR at a given frequency, the derating factor is derived from the formula:

$$DF = \frac{2 \times (VSWR)}{VSWR^2 \times (1 + F^1) + 1 - F^1}$$

DF = Derating factor

F^1 = Factor that varies with line size and frequency and is found in Figure 9.4-11.

The average power-handling capability of a transmission line will be reduced by direct solar radiation. In most moderate climates the radiation intensity is 200 W/m^2 or less. On the other hand not dry climates at midday can have solar radiation levels of 1,000 W/m^2 or more. In these locations the derating impact of the high ambient temperatures must also be considered, as mentioned previously (Figure 9.4-10). The mean value over the course of a 24-hour day that is applicable to average power derating calculations, provided absolute maximum ratings are not exceeded, is 400 W/m^2 (see Figure 9.4-12).

Attenuation characteristics for transmission lines is established by most manufacturers at an ambient temperature of 20 degrees C (68 degrees F), and the data for attenuation variation over the useful frequency range for a given transmission line is available from each manufacturer. Attenuation increases as ambient temperatures rise. Figure 9.4-13 provides a correction factor that can be applied to derive attenuation under

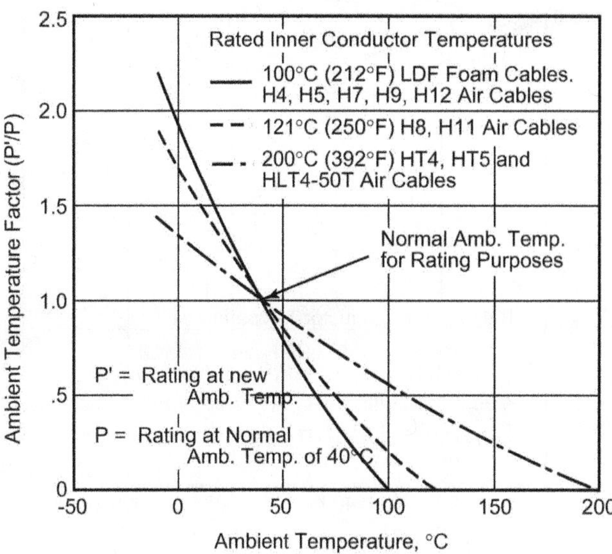

FIGURE 9.4-10 Average power-handling capability with variation in ambient temperature.

Courtesy of CommScope

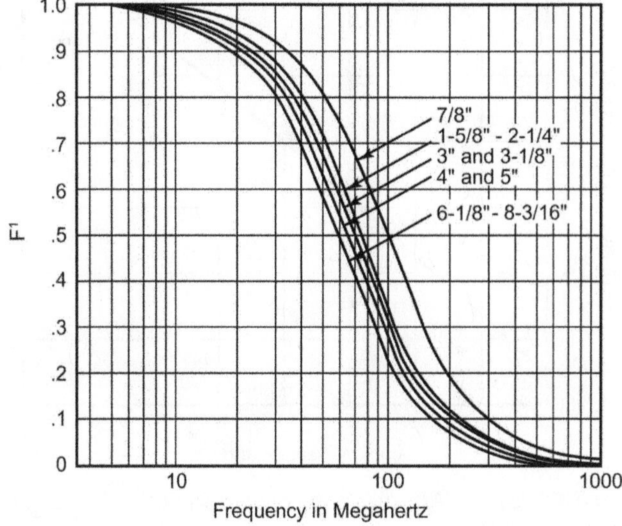

FIGURE 9.4-11 Average power derating factor with VSWR.

Courtesy of CommScope

the actual ambient temperature the transmission line is operating.

As with ambient temperature transmission line, manufacturers provide attenuation constants assuming a VSWR of 1.0. When a transmission line is attached to an antenna or other load, the VSWR increases and this increases the total loss for the system. Figure 9.4-14 provides the minimum additional loss from the effects of load VSWR, assuming a VSWR of 1.0 at the input to the transmission line.

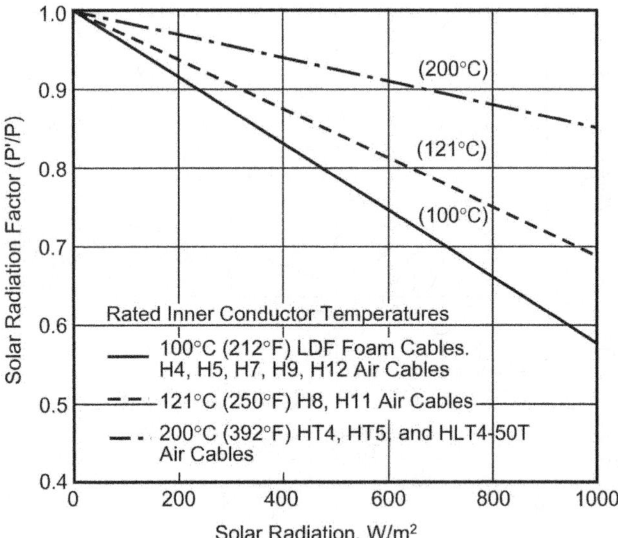

FIGURE 9.4-12 Derating factor for average power with intensity of solar radiation.

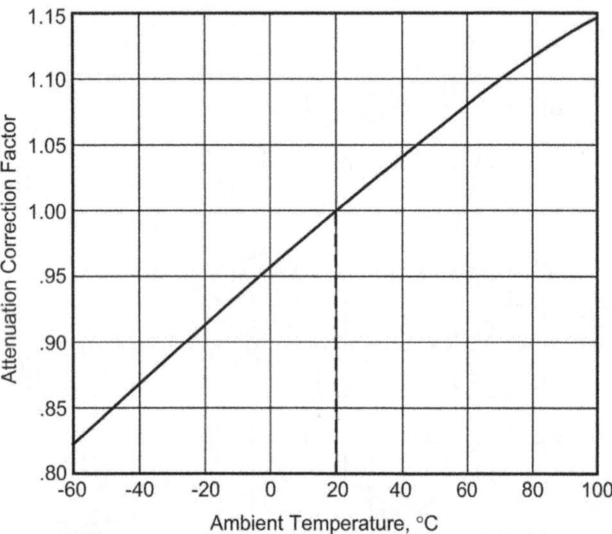

FIGURE 9.4-13 Attenuation correction factor for ambient temperature.

Courtesy of CommScope

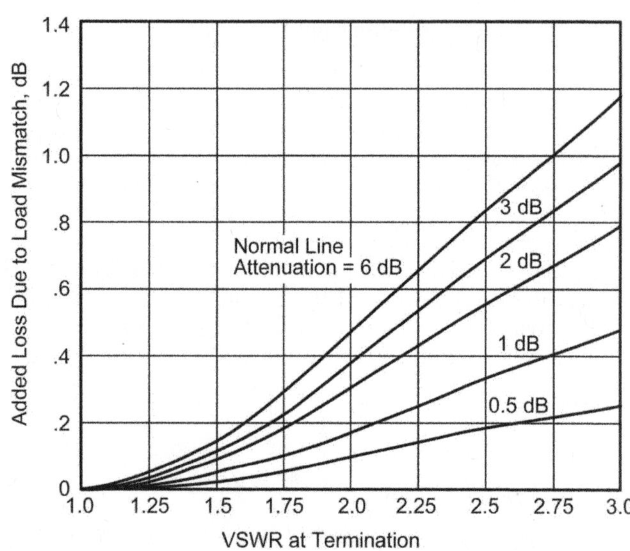

FIGURE 9.4-14 Effect of load VSWR on transmission line loss.

Courtesy of CommScope

Other Derating Factors to Consider

Transmission line average and peak power ratings are established at sea level by all transmission line manufacturers. At higher elevations, power-handling capacity for the line is reduced as elevation increases. Table 9.4-2 provides recommended derating factors for transmission lines at various elevations above sea level.

TABLE 9.4-2
Derating Factors for Transmission Line Average and Peak Power Handling With Increased Elevation

Altitude Above Sea Level (in feet)	Average Power Derating Factor	Peak Power Derating Factor
0	1.00	1.00
5,000	0.92	0.83
8,000	0.87	0.73
10,500	0.84	0.66
15,000	0.78	0.55

Special Considerations in Multichannel Combined Transmission Systems

Systems that combine multiple FM or television signals into a common transmission require special consideration to not only total power but also must take into account the type of modulation used for each carrier. Determining the total average power in the system is very straightforward, since the average power rating is based on the temperature rise of coaxial components, the sum of the transmitter power outputs is summed and used to check average power safety factor. But these systems require careful consideration of the total peak voltage in the system because the peak voltages, just like the average powers, add together. In systems that include multiple digital FM services, HD Radio™[1] or multiple digital television signals, the peak-to-average power of the digital signals must be taken into account. These digital services consist of multiple carriers, and there is a real probability that these carriers can add up at some point in time to produce a voltage that is the sum of all the carrier voltages. In evaluating these systems it is best to convert peak powers to voltage. For an analog FM signal, this is done with the following formula:

$$E_{Peak} = \sqrt{2} \times \sqrt{Z_0 \times TPO}$$

E_{Peak} = Peak RF voltage

Z_0 = Characteristic impedance of the transmission line

TPO = Transmitter power output

For digital television and digital FM signals, the peak power envelope must first be calculated from the peak to average power level of the service. For FM HD Radio and 8VSB DTV, the generally accepted factors used are +6 dB and +7 dB, respectively, which translate to a 4:1 factor for FM and a 5:1 factor for DTV. With this information we can calculate the peak RF voltage for a digital television or digital FM signal with the formula

$$E_{Peak} = \sqrt{2} \times \sqrt{PARP \times Z_0 \times TPO}$$

E_{Peak} = Peak RF voltage

PAPR = Peak to average power ratio (4 for digital FM and 5 for 8VSB DTV)

Z_0 = Characteristic impedance of the transmission line

TPO = Transmitter power output

The sum of all of the peak voltages of each station's analog and digital signals summed is used to evaluate the transmission line system by comparing the total peak RF voltage to the peak voltage ratings of the components in the system. Table lists recommended maximum peak voltage rating for most sizes of rigid and semi-flexible coaxial transmission lines.

Velocity of Propagation

Adding the dielectric material, other than air, to support the inner conductor of the transmission line causes the TEM wave to slow down its velocity. The velocity of propagation is expressed as a percentage of the speed of light in a vacuum and calculated from the dielectric constant of the dielectric material:

$$V_p = \frac{c}{\sqrt{\varepsilon_r}}$$

where:

V_p = velocity of propagation

c = speed of light

ε_r = dielectric constant of the dielectric material (1.00059 for air at 1 ATM)

This specification is provided by the transmission line manufacturer and is important for transmission lines that must be matched to be equal in their electric lengths. Any two lengths of transmission line will have variations in the amount of dielectric material between them, and so their relative electrical lengths

TABLE 9.4-3
Rigid and Semi-Flexible Coaxial Cable Maximum Recommended Peak Voltage Ratings

Rigid Line Ratings				
Size	**Impedance**	**Peak Power Rating**	**DC Product Test Voltage**	**Max Peak Voltage**
7/8-inch	50 ohms	41 kW	6 kV	2.10 kV
1–5/8-inch	50 ohms	132 kW	11 kV	3.85 kV
3–1/8-inch	50 ohms	440 kW	19 kV	6.65 kV
4–1/16-inch	50 ohms	710 kW	24 kV	8.40 kV
6–1/8-inch	50 ohms	1500 kW	35 kV	12.25 kV
6–1/8-inch	75 ohms	1069 kW	36 kV	12.60 kV
7–3/16-inch	75 ohms	1426 kW	42 kV	14.70 kV
8–3/16-inch	50 ohms	2592 kW	46 kV	16.10 kV
8–3/16-inch	75 ohms	1825 kW	47 kV	16.45 kV
9–3/16-inch	50 ohms	3296 kW	52 kV	18.20 kV
9–3/16-inch	75 ohms	2294 kW	53 kV	18.55 kV
Semi-Flexible Line Ratings				
Size	**Impedance**	**Peak Power Rating**	**DC Product Test Voltage**	**Max Peak Voltage**
7/8-inch	50 ohms	90 kW	9 kV	3.15 kV
1–5/8-inch	50 ohms	305 kW	11 kV	3.81 kV
3-inch	50 ohms	640 kW	16 kV	5.66 kV
4-inch	50 ohms	1890 kW	21 kV	7.33 kV
5-inch	50 ohms	1690 kW	28 kV	9.11 kV

will be different even if they are cut to exactly the same physical lengths. This has application for transmission lines used in AM directional array sampling systems and for dual transmission line systems for master FM and television antennas where two separate transmission line are used to feed the upper and lower halves of a broadband antenna system.

Table 9.4-4 lists the common characteristics of the most popular rigid and semi-flexible transmission lines commonly used. From the data provided, transmission line attenuation and average power handling of these transmission lines may be calculated using the following formulas:

To find attenuation for a given frequency:

$$\alpha_c = \left(F \times \sqrt{A}\right) + \left(F \times B\right)$$

ε_c = attenuation due to conductor losses, dB per 100 feet

F = Frequency of operation

A = Factor A from Table 9.4-4

B = Factor B from Table 9.4-4

To calculate average power-handling capability at a given frequency:

$$p_{AVG} = PA / \alpha_c$$

p_{AVG} = Average power-handling capability in kilowatts (NOTE: Average power rating cannot exceed peak power rating)

PA = Factor PA from table

α_c = attenuation due to conductor losses, dB per 100 feet

Waveguide Transmission Line Systems

Waveguide is a structure that guides waves, such as electromagnetic waves or sound waves. It enables a signal to propagate with minimal loss of energy by restricting expansion to one dimension, or two. There are different types of waveguides for each type of wave. The original and most common meaning is a hollow conductive metal pipe used to carry high frequency radio waves, particularly microwaves (see Figure 9.4-15).

Depending on your RF needs, waveguide comes flexible in elliptical form and rigid, which can be rectangular or circular (see Figure 9.4-16). Rigid waveguide is constructed of aluminum due to the weight of a waveguide system. A waveguide in circuit theory is described by a transmission line having a length and self-impedance. In other words, the impedance is the resistance of the circuit component (in this case a waveguide) to the propagation of the wave. This

TABLE 9.4-4
Typical Performance Characteristics of Coaxial Transmission Lines[1]

Size	Impedance	A	B	PA	Peak Power	Velocity	Cutoff Frequency
7/8 inch 50 Ohm, Rigid Line	50 ohms	0.0369181	0.0000002	1.63055	41 kW	99.8%	6000 MHz
1–5/8 inch 50 Ohm, Rigid Line	50 ohms	0.0188523	0.0000056	2.78603	132 kW	99.8%	3000 MHz
3–1/8 inch 50 Ohm, Rigid Line	50 ohms	0.0096010	0.0000053	4.77903	440 kW	99.8%	1600 MHz
4–1/16 inch 50 Ohm, Rigid Line	50 ohms	0.0073227	0.0000086	5.79190	710 kW	99.8%	1262 MHz
6–1/8 inch 50 Ohm, Rigid Line	50 ohms	0.0048998	0.0000045	8.44162	1500 kW	99.8%	806 MHz
6–1/8 inch 75 Ohm, Rigid Line	75 ohms	0.0044527	0.0000045	6.70456	1069 kW	99.8%	830 MHz
7–3/16 inch 75 Ohm, Rigid Line	75 ohms	0.0036989	0.0000058	7.99729	1426 kW	99.8%	752 MHz
8–3/16 inch 75 Ohm, Rigid Line	50 ohms	0.0032955	0.0000055	8.55601	2592 kW	99.8%	698 MHz
8 3/16 inch 50 Ohm, Rigid Line	75 ohms	0.0035750	0.0000130	10.78770	1825 kW	99.8%	580 MHz
9 3/16 inch 50 Ohm, Rigid Line	50 ohms	0.0032159	0.0000004	12.00681	3296 kW	99.8%	552 MHz
9 3/16 inch 75 Ohm, Rigid Line	75 ohms	0.0029173	0.0000001	9.18730	2294 kW	99.8%	632 MHz
1/2 inch 50 Ohm, LDF4–50 HELIAX®	50 ohms	0.0642625	0.0001890	2.30798	40 kW	88.0%	8800 MHz
7/8 inch 50 Ohm, HJ5–50 HELIAX®	50 ohms	0.0352549	0.0001652	2.77605	90 kW	91.6%	5200 MHz
1–5/8 inch 50 Ohm, HJ7–50 HELIAX®	50 ohms	0.0194719	0.0000849	3.36045	305 kW	92.1%	2700 MHz
3 inch 50 Ohm, HJ8–50B HELIAX®	50 ohms	0.0124425	0.0001697	5.99842	640 kW	93.3%	1640 MHz
4 inch 50 Ohm, HJ11–50 HELIAX®	50 ohms	0.0103863	0.0001056	7.39361	1100 kW	92.0%	1220 MHz
5 inch 50 Ohm, HJ9–50 HELIAX®	50 ohms	0.0073679	0.0000521	6.66308	1890 kW	93.1%	960 MHz
5 inch 50 Ohm, HJ9HP-50 HELIAX®	50 ohms	0.0063130	0.0001170	13.00681	1690 kW	96.4%	960 MHz

[1]HELIAX is a registered trademark of CommScope

FIGURE 9.4-15 Rectangular waveguide transmission line system.

Courtesy of Electronics Research, Inc.

FIGURE 9.4-16 Circular waveguide system; rectangular waveguide is used for the horizontal run.

Courtesy of Electronics Research, Inc.

description of the waveguide was originally intended for alternating current but is also suitable for electromagnetic and sound waves, once the wave and material properties are properly converted into current and impedance. In connecting a waveguide to an antenna, a complete transmission is usually required, so that their impedances are matched. (1) Energy cannot propagate in waveguide unless it is above the lower cutoff frequency and below the upper cutoff frequency. The advantage of waveguide over coaxial transmission line is that it has much higher power-handling capability than coaxial transmission line and it has much lower loss. The disadvantage is that waveguide is larger than coaxial transmission lines, and so it presents a significantly greater mechanical load for the tower. (6)

In broad terms, waveguide is used only at UHF and higher frequencies and is manufactured rigid line sections that are circular, elliptical, or rectangular. The lower digital television power levels have limited the need to use circular or rectangular waveguide as transmission lines for UHF DTV applications, and its use in television broadcasting is more common as transmission line for broadcast auxiliary services such as studio-to-transmitter links and intercity relays. In these applications, semi-flexible elliptical waveguide is the most common type employed.

Semi-Flexible Transmission Line System Design and Installation

In general, the selection of foam dielectric versus air dielectric semi-flexible coaxial transmission line is driven by power handling and whether the antenna connected to the transmission line requires pressurization.

In general, low power—transmitter power levels of less than 1 kW—use foam dielectric cables if the antenna does not require pressurization. Once the semi-flexible transmission line type and size has been selected, based on the acceptable transmission line loss and the required peak and/or average power-handling capability required, then the connectors and installation accessories must be added to the system bill of materials. The typical items required are shown in Figure 9.4-17.

Semi-flexible transmission lines are delivered on reels in one continuous length. If the total length of transmission line exceeds the capacity of the reels available, then more than one reel is provided, and the transmission line sections are raised and connected together with intermediate connectors. Unlike rigid transmission lines, the corrugations in semi-flexible transmission lines compensate for the differential expansion that occurs between the transmission line inner conductor and outer conductor due to the heat rise of the inner when power is applied. The corrugation also compensates for the differential expansion between the tower steel and the copper or aluminum transmission line material as the ambient temperature changes.

The focus has been on the electrical-performance aspects of the transmission line system. Now there needs to be consideration given to the mechanical specifications of the transmission line selected. From the transmission line, outer dimensions of the transmission line and its shape factor are used to derive its effective projected area (EPA), which is used to calculate the mechanical load the transmission line will exert on the tower supporting it. This information needs to be provided to the tower manufacturer if the station is building a new tower or be supplied to the structural engineer performing the structural analysis being done to confirm that the existing tower can support the added antenna and transmission line load. The minimum bending radius of the cable represents the tightest turn radius the cable can make, which is usually at the base of the tower. Do not bend the cable tighter than the radius specified. If there is insufficient space to accommodate bend radius required, terminate the

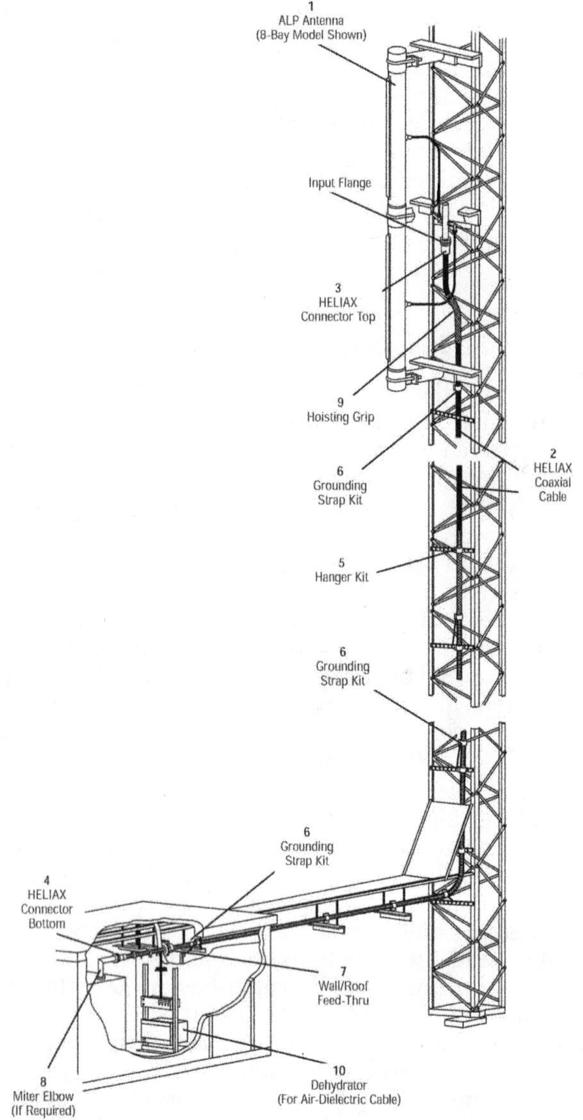

FIGURE 9.4-17 Typical transmission line installation for semi-flexible coaxial transmission lines.

Courtesy of CommScope

vertical run with a cable connector and use a 90-degree rigid transmission elbow to connect to a separate transmission line piece for the horizontal run. The minimum bend radius should also be taken into consideration if the transmission line is to be installed in conduit, as all changes in direction of the conduit run must be greater than the minimum bend radius for the transmission line being installed. The tensile strength rating of the cable also factors into conduit installation, as this represents the maximum pulling force that can be applied to the cable during installation. The cable weight is required by the transmission line installer so that the amount of weight that needs to lifted and supported during installation can be calculated. The flat plate crush strength of the cable is provided for installations where the cable is buried (see Table 9.4-5).

The installation of antennas and transmission lines should only be performed by qualified, experienced, and certified personnel (see Figure 9.4-18). The installation instruction provided for these products is written for individuals with these qualifications. When the transmission line and accessories arrive on site, carefully inspect the materials received for shipping damage. In the case of air dielectric transmission line to confirm that the transmission line is under pressure. A tire gauge can be used to check pressure.

If the cable has a pressure loss, check all joint connections, especially at the pipe threads on the connector gas inlet connections. If the connector at the antenna end of the cable is not installed, it should be installed on the cable before hoisting.

The transmission line should be lifted by a hoisting line with sufficient strength to support the weight of the entire transmission line length being installed. The hoist line should be run through a pulley installed high enough on the tower to permit the cable to be raised sufficiently to reach the antenna connection. The transmission line reel should be supported on an axle that allows free rotation of the reel as the cable is hoisted. It is generally safest, with heavily loaded reels, to have the cable payout from the top of the reel. In cases where the cable and reel can be braked by hand, then the cable reel can pay out from the bottom of the reel. If the transmission line length is short enough to have

TABLE 9.4-5
Mechanical Characteristics of Semi-Flexible Coaxial Transmission Lines

Size	Diameter Over Jacket	Minimum Bending Radius	Number of Bends, minimum	Bending Moment	Cable Weight	Tensile Strength	Flat Plate Crush Strength
1/2 inch 50 Ohm, LDF4–50 HELIAX®	0.63 inches	5.00 inches	15	3 lb-ft	0.15 lb/ft	250 lb	110 lb/in
7/8 inch 50 Ohm, HJ5–50 HELIAX®	1.11 inches	10.00 inches	15	25 lb-ft	0.54 lb/ft	800 lb	250 lb/in
1–5/8 inch 50 Ohm, HJ7–50 HELIAX®	1.98 inches	20.00 inches	15	30 lb-ft	1.04 lb/ft	750 lb	175 lb/in
3 inch 50 Ohm, HJ8–50B HELIAX®	3.01 inches	30.00 inches	15	30 lb-ft	1.78 lb/ft	750 lb	195 lb/in
4 inch 50 Ohm, HJ11–50 HELIAX®	4.00 inches	40.00 inches	15	191 lb-ft	2.50 lb/ft	900 lb	280 lb/in
5 inch 50 Ohm, HJ9–50 HELIAX®	5.20 inches	50.00 inches	15	200 lb-ft	3.30 lb/ft	1000 lb	275 lb/in
5 inch 50 Ohm, HJ9HP-50 HELIAX®	5.20 inches	50.00 inches	15	200 lb-ft	3.40 lb/ft	1000 lb	240 lb/in

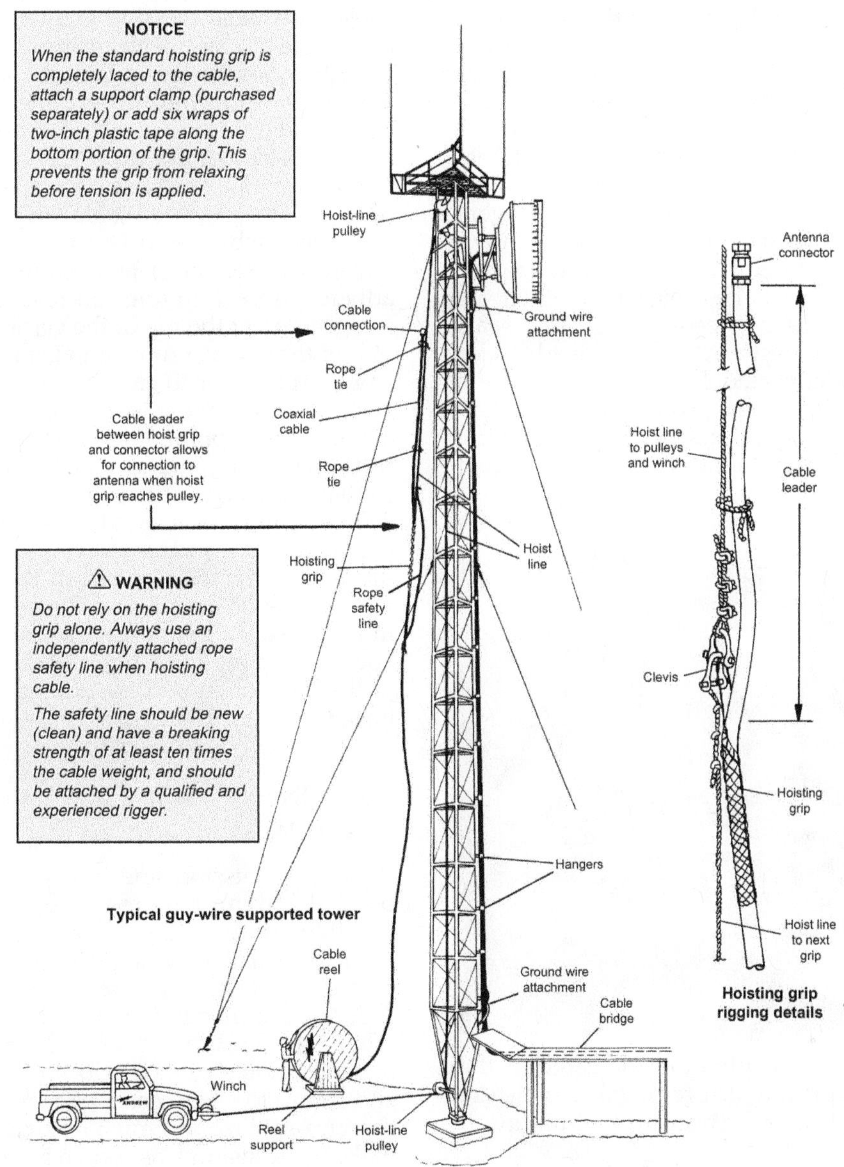

FIGURE 9.4-18 General guide for rigging and hoisting semi-flexible coaxial transmission lines. Provided as a reference guide only. Adhere to manufacturer's instructions supplied with the product purchased.

Courtesy of CommScope.

not required reel, it should be uncoiled on the ground and then hoisted up the tower. Be sure the connector is covered and protected during hoisting. The weight of the cable during hoisting is supported by cable grips that are installed far enough down the cable to allow the connector to be attached to the antenna when the cable is hoisted. Additional hoisting grips need to be installed every 150 to 200 feet along the run to support the cable weight. In addition the transmission line should be tied to the hoist line every 50 feet with fiber-reinforced tape. The objective is to be certain there is slack in the cable relative to the hoist line, so the weight of the cable is being supported by the hoist line, not by the cable. The cable should be hoisted slowly in order to prevent kinking and other damage.

Once the cable is in place, maintain tension on the hoist line until the transmission line is firmly anchored. The hoisting grips used to support the line during hoisting should be attached to the tower using a turnbuckle, shackles, and chain. They will remain in place and continue to support the weight of the transmission line. In addition to the hoisting grips, transmission line hangers must be installed along the full vertical run. These hangers should be installed at the top of the transmission line run within 1 to 2 feet of the antenna connector. The connector spacing should be based on the manufacturer's installation instructions, which generally call for the top three hangers to be installed 2 feet apart and the hanger spacing for the balance of the vertical is usually 3 feet or 5 feet, depending on cable size. Environmental conditions in some areas may require different configurations. At the tower base, leave a "drip loop" at the base of the vertical run (see Figure 9.4-19). There is a potential for moisture to accumulate between the transmission line jacket and the outer conductor; the drip loop will prevent any accumulated moisture from entering the transmitter building and causing equipment damage. The horizontal run also needs to be supported with hangers every 3 to 5 feet, as specified by the cable manufacturer. The horizontal run should be protected from falling ice and debris by an ice bridge. The cable should enter the building through a wall/roof feed-through that seals the cable entrance (see Figure 9.4-20). Jacketed cables can also be buried underground. It is important that the cable entrance and exit into the ground be adequately protected, and proper underground electrical warning tape should be installed in the trench above all buried

FIGURE 9.4-19 A drip loop should be created at the start of the horizontal run to insure that any moisture accumulation will not enter the transmitter building.

FIGURE 9.4-20 A typical multiple entrance feed-through plate with a properly installed ground bar.

cable. The cable should be buried below the frost line and at least 3 feet deep for protection against damage from heavy vehicles. A layer of sand at least 4 inches above and below the buried cable will provide adequate protection from stones and other objects that could damage the cable. Under no circumstances is it acceptable to use wrap lock or hose clamps to attach transmission line to the tower, as the cable can easily be damaged and deformed, which will negatively impact performance and useful life. The transmission line grounding should adhere to local building codes. In general, the cable run is grounded at the top of the vertical run and at the bottom of the vertical run, and again at the end of the horizontal run, prior to entering the building. In addition, vertical runs of more than 200 feet should be grounded at the halfway point. If the vertical run is more than 400 feet, then additional grounding kits should be added so that there no portions of the vertical run longer than 200 feet without a grounding kit at each end. The same applies to long horizontal runs. Some installers ground the cable every 50 feet and in these cases the grounding kits can be installed on the cable while it is hoisted. In addition, if the transmission line passes through the aperture of a high FM or television antenna, that portion of the vertical should include grounding transmission line every 12 to 18 inches.

Rigid Transmission Line System Design and Installation

Rigid transmission line is usually selected when the power-handling required exceeds the power-handling capability of semi-flexible transmission line. Since rigid transmission line is not corrugated, it has lower loss than equivalently sized semi-flexible transmission line. Also, rigid transmission line is supplied in sections; it is often easier to ship and off load at site versus a large reel of semi-flexible cable. Rigid line typically uses flanged end connectors between sections of line. Where turns or changes in direction must occur, "elbows" are used to make the transition, as shown in Figure 9.4-21.

The flanged connections between the individual rigid transmission line sections cause a small reflection, as there is a small impedance mismatch at each connection. These reflections add up in a long transmission line run and cause a high VSWR spike in the system. These critical frequencies can be determined with the following formula: (6)

$$F_c = \frac{492.15 \times V_p \times n}{L_{ft}}$$

F_c = Critical frequency in MHz

V_p = Relative velocity

L_{ft} = Transmission section line length in feet

n = Any integer

Table 9.4-6 lists the recommended standard rigid transmission line section lengths for Region II television channels and the FM broadcast band. In cases where

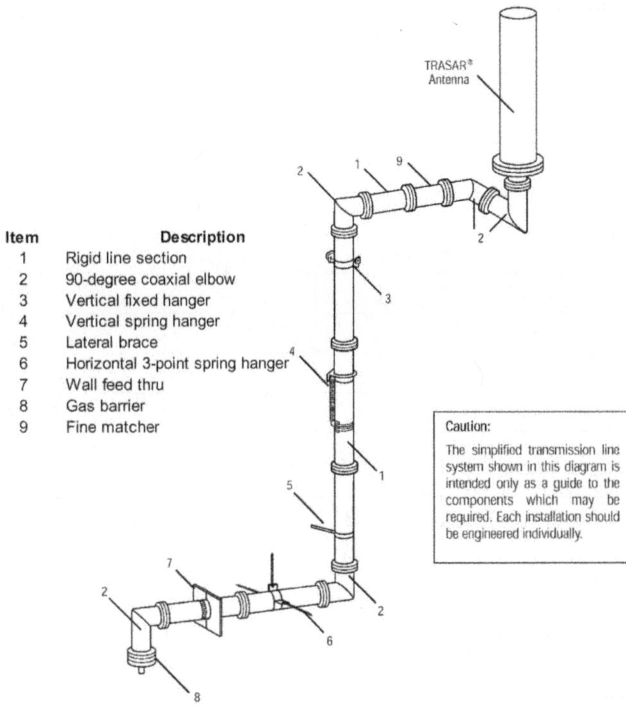

FIGURE 9.4-21 A simplified block diagram outlining the components in a typical rigid transmission line system.

Courtesy of Electronics Research, Inc.

FIGURE 9.4-22 Inner connector with watch band spring for differential expansion compensation.

Courtesy of Electronics Research, Inc.

multiple television channels are combined in a single run of transmission, a broadband rigid transmission line must be used. These systems are available from multiple manufacturers and use proprietary techniques to randomize individual transmission line section lengths to minimize the VSWR spikes caused by the flange reflections (see Figure 9.4-22).

TABLE 9.4-6
Recommended rigid transmission line section lengths for Television and FM frequencies.

Recommended Line Section Lengths		
U.S. Television Channels		
20.00 foot (6.096 m) Section Length		
	Channels: 2, 3, 5, 6, 7, 8, 9, 11, 12, 14, 15, 18, 19, 22, 23, 27, 31, 32, 35, 36, 39, 40, 43, 44, 47, 48, 51	
19.75 foot (6.020 m) Section Length		
	Channels: 16, 20, 24, 28, 33, 37, 41, 45, 49	
19.5 foot (5.944 m) Section Length		
	Channels: 4, 10, 13, 17, 21, 25, 26, 29, 30, 34, 38, 42, 46, 50	
FM Radio Frequencies		
	Foot (m)	**MHz**
	20.00 (6.096) Sections	88.1–95.9
		100.3–107.9
	19.50 (5.944) Sections	96.1–98.3
	19.00 (5.791) Sections	98.5–100.1
	17.50 (5.342) Sections	88.1–107.9

TABLE 9.4-7
Coefficient of Thermal Expansion

Coefficient of Thermal Expansion	
Steel	0.0000065 in/in/Degrees F
Copper	0.000009 in/in/Degrees F
Aluminum	0.000013 in/in/Degrees F

Rigid transmission line systems also must accommodate differential thermal expansion as a part of their system design (see Table 9.4-7). In operation the inner conductor temperature rises substantially higher, 60 degrees C, than the outer conductor, and this causes the inner conductor to expand more than the outer conductor. To accommodate this expansion the inner conductor is cutback from the outer conductor to accommodate this expansion (see Table 9.4-8). In addition, rigid transmission line sections are available that incorporate thermal expansion compensation devices in the form of a watchband spring or a spring bellows compensator (see Figures 9.4-23, 9.4-24, 9.4-25).

The copper and aluminum outer conductors employed in rigid waveguide and coaxial transmission lines expand and contract with changes in ambient temperature at a different rate than the steel used to construct the tower. The rigid transmission line expands and contracts at a greater rate than the tower steel. With a 100-degree change in temperature, the differential expansion between the tower and a copper transmission line is 3″. So this requires some method of supporting

TABLE 9.4-8
Rigid Transmission Line Flange Information and Cutback Dimensions.

Rigid Line and Flange Information						
Line Diameter, in.		1-5/8	3-1/8	4-1/16	6-1/8	6 1/8
Characteristic Impedance (ohms)		50	50	50	50	75
Flange Diameter, in. (mm)		3-1/2 (89)	5-3/16 (132)	6-3/16 (157)	8-1/8 (207)	8-1/8 (207)
Bolt Circle Diameter, in. (mm)		2-13/16 (71)	4-3/8 (111)	5-3/8 (137)	7-3/8 (188)	7-3/8 (188)
Number of Holes		4	6	8	12	12
Hardware Size, in.		5/16	3/8	3/8	3/8	3/8
Special Cut Flanged Lengths						
Cut outer conductor shorter than measured length to allow for flange(s), in. (mm)	One flange (C)	1/4 (6)	17/64 (7)	3/8 (9.5)	19/64 (8)	9/32 (7)
	Total both flanges	1/2 (13)	17/32 (13)	3/4 (18)	19/32 (15)	19/32 (15)
Cut inner conductor shorter than outer conductor to allow for inner connector, in. (mm)	One end (A)	5/8 (16)	1-3/32 (28)	1-5/16 (33)	1-9/32 (32)	1-3/8 (35)
	Total both ends	1-1/4 (32)	1-15/16 (49)	2-5/8 (66)	2-17/32 (64)	2-19/32 (66)
Unflanged Lengths						
Cut inner conductors (each side of inner connector) shorter than outer conductors, in. (mm) (D)		9/16 (14)	27/32 (21)	1/4 (6)	9/32 (7)	NA
Cutback dimensions for inner conductor operating temperatures up to 120 degrees C. (248 degrees F.), maximum.						

FIGURE 9.4-23 Copper bellows for differential expansion compensation.

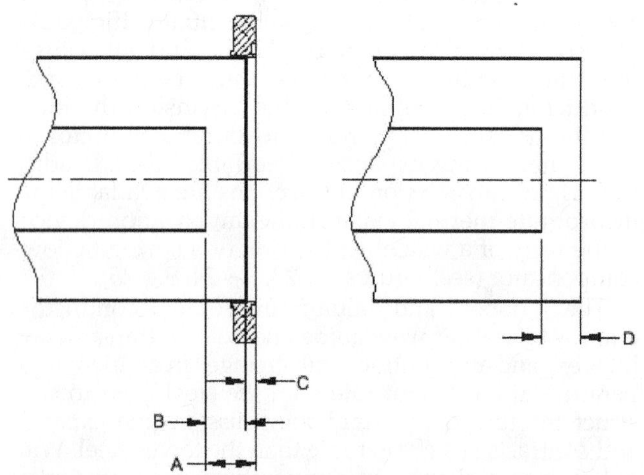

FIGURE 9.4-24 Cutback dimensions for flanged and unflanged rigid transmission lines.

FIGURE 9.4-25 Typical elbow complex shown at the input of an FM antenna.

Courtesy of Electronics Research, Inc.

the transmission line while allowing these different expansion rates as the temperature varies between day and night and through the changes in seasons.

A rigid transmission line system employs one or more fixed hangers at the top of the vertical run (see Figure 9.4-26).

These fixed hangers are installed near the antenna input. In most installations, an elbow complex made up of four 90-degree elbows is installed at the antenna input. This elbow complex provides expansion compensation between the fixed hanger(s) and the antenna input. The elbow complex can be disassembled without disturbing any part of the transmission line system or the antenna input section. The balance of the vertical run is supported by vertical spring hangers spaced an average of every 10 feet along the entire vertical run (see Figure 9.4-27).

All manufacturers provide spring tension settings as a part of their installation instructions. The spring tension settings are based on the transmission line length from the fixed vertical hanger and the ambient temperature. In most cases, the spring hangers should all be installed, and then the spring set to length and secured to the transmission line as a second step. In order to avoid large variations in tension between the spring hangers, they should be all set within hours of each other. When setting the vertical spring hangers, the tension springs should not cross over flange connections, and also be sure that the tension springs are not in contact with tower members or other obstructions. At the base of the vertical run, a single 90-degree elbow connects to the horizontal run to the transmitter building. All of the vertical expansion and contraction as well as the expansion and contraction of the horizontal run occurs at this location. To accommodate this movement requires that the horizontal transmission line run has sufficient length to allow for the expansion and contraction of the vertical run. Table 9.4-9 provides recommended minimum horizontal transmission line runs for most rigid transmission line sizes.

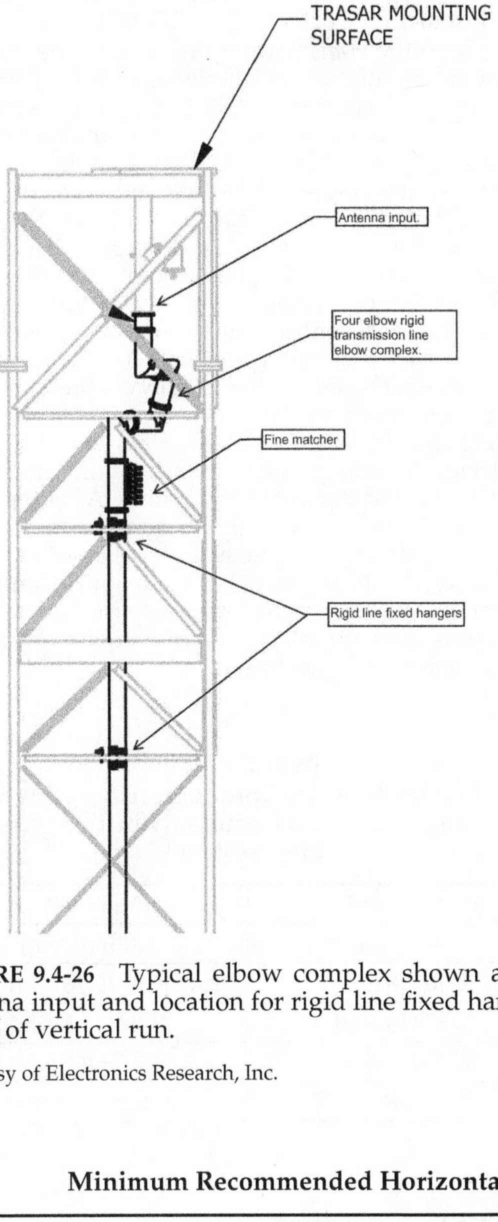

FIGURE 9.4-26 Typical elbow complex shown at the antenna input and location for rigid line fixed hangers at top of vertical run.

Courtesy of Electronics Research, Inc.

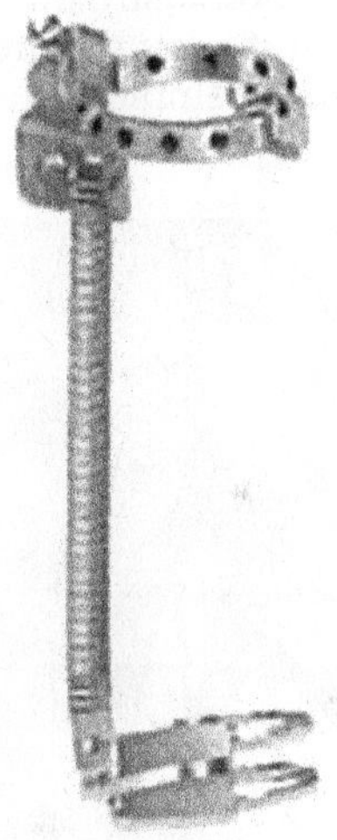

FIGURE 9.4-27 Rigid transmission line vertical spring hanger.

Courtesy of Electronics Research, Inc.

TABLE 9.4-9
Minimum Recommended Horizontal Run Lengths for Rigid Transmission Line Systems

		Minimum Recommended Horizontal Run Length									
Vertical Run Length		3-1/8 inch		4-1/16 inch		6-1/8 inch		7-3/16 inch		8-3/16 inch	
100 ft	(30.5 m)	15 ft	(4.6 m)	15 ft	(4.6 m)	15 ft	(4.6 m)	20 ft	(6.1 m)	20 ft	(6.1 m)
500 ft	(152.4 m)	25 ft	(7.6 m)	30 ft	(9.1 m)	25 ft	(7.6 m)	40 ft	(12.2 m)	40 ft	(12.2 m)
1000 ft	(304.8 m)	35 ft	(10.7 m)	40 ft	(12.2 m)	50 ft	(15.2 m)	60 ft	(18.3 m)	60 ft	(18.3 m)
1500 ft	(457.2 m)	40 ft	(12.2 m)	50 ft	(15.2 m)	60 ft	(18.3 m)	70 ft	(21.3 m)	70 ft	(21.3 m)
2000 ft	(609.6 m)	45 ft	(13.7 m)	60 ft	(18.3 m)	70 ft	(21.3 m)	80 ft	(24.4 m)	80 ft	(24.4 m)

The horizontal rigid line run is anchored at the building by its building entry flange, and the expansion and contraction is also accommodated at the elbow at the base of the vertical run. The horizontal run is supported at 10-foot intervals by horizontal spring hangers (see Figure 9.4-28) that accommodate the vertical movement of the horizontal, caused by the expansion and contraction of the tower and the vertical transmission line run and the horizontal run's expansion and contraction.

In order to accommodate that expansion and contraction, the vertical spring hangers are not used for a portion of the vertical run above the elbow. Table 9.4-10 provides information regarding placement of the last vertical spring hanger above the horizontal run.

The final item included in the rigid transmission line hanger system is the lateral brace. The lateral brace is installed near the base of the vertical run, and its function is to eliminate lateral movement of the horizontal and vertical transmission line runs. All expansion and contraction should occur along the axis of each, and no side-to-side movement should occur.

When installing a rigid transmission line, it is important that care be taken to insure the flanges and hangers do not contact tower members or other obstructions in the course of their normal movement. It is also important that all bolts are tightened in accordance with the manufacturer installation instructions (see Table 9.4-11) and that all bolts include any required flat washers and lock washers. When assembling the transmission line sections, be certain that all O rings are properly seated and not pinched into any electrical contact surfaces.

FIGURE 9.4-28 Rigid transmission line horizontal spring hanger.

Courtesy of Electronics Research, Inc.

Pressurization

When the installation of the any air dielectric transmission line is complete, the gas barrier is installed inside the transmitter building. The transmission line during installation was exposed to the atmosphere, and so it must be purged of moisture prior to the application of power.

To purge the transmission line, determine the total system volume of the transmission line and the antenna, if it is pressurized, and divide this by the dehydrator flow rate. This figure will provide the number of hours the dehydrator must run to displace the total untreated air in the system. Open the farthest end of the transmission line/antenna system as possible. Run the dehydrator (see Figure 9.4-29) long enough to replace the complete volume of air in the complete system at least three times. If it is not possible to open the far end of the transmission line, connect the dehydrator to the system and pressurize to at least 5 psig, unless it is a waveguide system, in which case it should only be pressurized to its maximum capacity. Wait 15 minutes for the dry air added to absorb moisture in the system, and disconnect the dehydrator and allow the transmission line to vent. Repeat these two steps at least 12 times. After the transmission line has been initially purged, maintain a positive 3 to 5 psig dry gas pressure. An alternate to an air dehydrator, dry nitrogen, either from a nitrogen generator or delivered in compressed gas cylinders and fitted with a pressure regulator, can also be used to purge and pressurize transmission lines.

TABLE 9.4-11
Generally Accepted Torque Specifications for Hardware Used Most Commonly in Transmission Line Systems

Hardware Size	Torque Value
1/4″ (6 mm)	7 lb-ft (9 Nm)
5/16″ (8 mm)	12 lb-ft (16 Nm)
3/8″ (10 mm)	21 lb-ft (28 Nm)
1/2″ (13 mm)	46 lb-ft (62 Nm)
5/8″ (16 mm)	76 lb-ft (103 Nm)

TABLE 9.4-10
Recommended Minimum Distance to the Last Vertical Spring Hanger

Horizontal Run Length		3-1/8 inch		4-1/16inch		6-1/8 inch		7-3/16 inch		8-3/16 inch	
20 ft	(6.1 m)	5 ft	(1.5 m)	6 ft	(1.8 m)	9 ft	(2.7 m)	10 ft	(3.0 m)	12 ft	(3.7 m)
40 ft	(12.2 m)	6 ft	(1.8 m)	7 ft	(2.1 m)	11 ft	(3.4 m)	12 ft	(3.7 m)	15 ft	(4.6 m)
60 ft	(18.3 m)	7 ft	(2.1 m)	8 ft	(2.4 m)	13 ft	(4.0 m)	15 ft	(4.6 m)	17 ft	(5.2 m)
80 ft	(24.4 m)	8 ft	(2.4 m)	9 ft	(2.7 m)	14 ft	(4.3 m)	17 ft	(5.2 m)	20 ft	(6.1 m)
100 ft	(30.5 m)	9 ft	(2.7 m)	10 ft	(3.0 m)	15 ft	(4.6 m)	18 ft	(5.5 m)	22 ft	(6.7 m)

Field Measurements/Troubleshooting

When the transmission line is installed and ready for power, a technician will take electrical measurements on the transmission line system. First, an insulation resistance test should be performed measuring the resistance between the inner conductor and the outer conductor. The insulation resistance between the inner and outer conductors should be at least 100,000 megohms, which will indicate the system is sufficiently free of interior moisture. This is particularly important if the transmission line has been stored outdoors with no assurance that positive pressure was maintained in the line.

FIGURE 9.4-29 Large capacity membrane dehydrator.

Courtesy of CommScope

Additional testing equipment used includes a vector network analyzer (VNA), coax test cables, and a calibration kit. In the RF industry, a VNA is used due to the fact it has a signal generator, receiver(s), display, and has two test ports having the capability measurement of four S-parameters: S_{11}, S_{21}, S_{12}, S_{22}. There are many different functions on the analyzer that help the technician see the electrical performance of the transmission line. One method used by technicians is TDR (time domain reflectometer), which measures DC voltage over time. TDR is expressed in units of time; it is expressed in nanoseconds. The network analyzer generates a pulse of energy and times the interval to reflections. If the velocity of the energy through the transmission line is known, distances to faults in the transmission line can be computed. For transmission line, the VNA generates a pulse onto the inner conductor and looks for any impedance mismatches on the inner conductor. If the inner conductor is uniform in impedance and is installed correctly, then there will be no major reflections. If there are impedance variations, then some of the pulse will be reflected back to the network analyzer. Using the TDR method is to characterize and locate faults in the transmission line system, for example discontinuities in a connector or insulator problems. This is to make sure the transmission line system is transmitting the correct signal to the antenna and looking for any suspicious impedance bumps that could cause a failure. Figure 9.4-30 is an example of a TDR measurement.

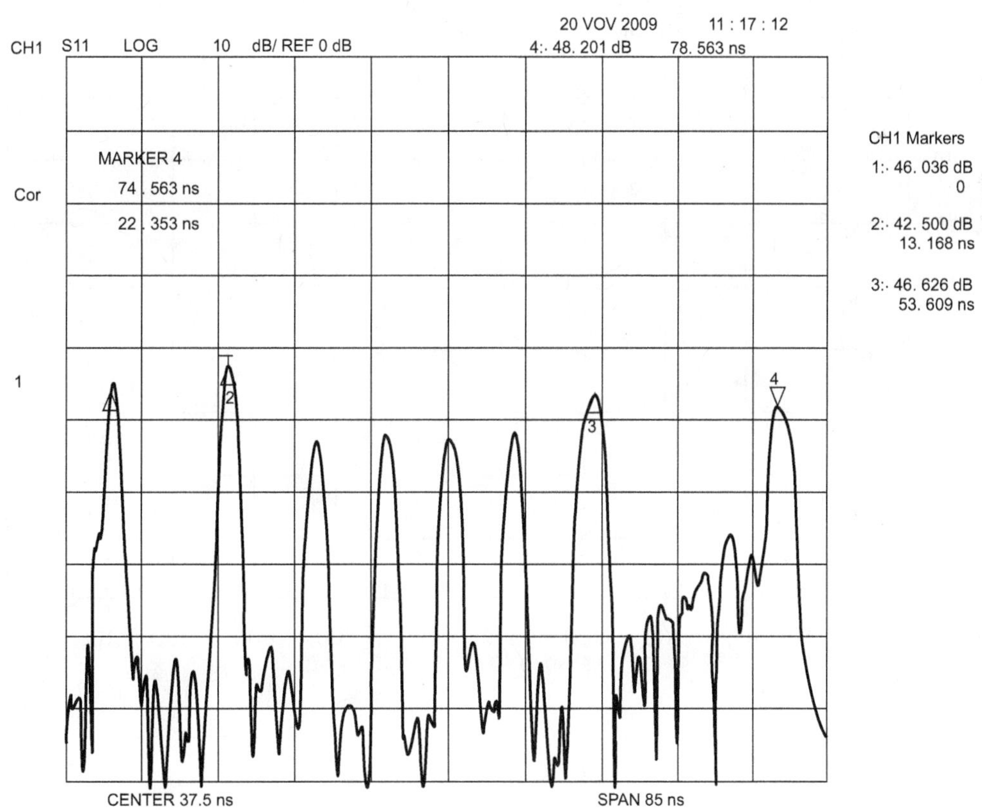

FIGURE 9.4-30 TDR measurement of a transmission line system.

Figure 9.4-30 is a screenshot (display on the network analyzer) of time domain reflectometer (TDR) measurement. On the display of the network analyzer, note that S_{11} measurement is being used for this exercise. On the display there are eight large spikes (impedance bumps); the four larger ones are connections and the four smaller ones are insulators inside the transmission line that helps support the inner conductor. The reason for the four larger spikes is because of the impedance difference on the bullet connections of the inner conductor. Marker 1 is the input from a Type-N connector to 3-1/8″ flange (reducer) connecting to 6.5′ of 3″ flex line; marker 2 is the reducer mating to the 3-1/8″ line; marker 3 is the end of the 20′ stick of 3-1/8″ 50-ohm transmission line mating to the output 3-1/8″ to a Type-N reducer; and marker 4 is Type-n into the analyzer output. When you have 2,000 feet of transmission line and the signal does not look like you expect, as explained earlier, TDR measurements can pinpoint your problem in the transmission line; for example every 1 ns equals 11.8 inches, roughly 1 foot, and if you look at the example of the screenshot (display), you notice marker 2 and marker 3 are 40 ns difference; that equals the 20-foot stick of transmission line that is being tested in this exercise.

Another method is distance to fault (DTF) is a performance verification and failure analysis tool used for antenna and transmission line service and maintenance (see Figure 9.4-31). It uses the frequency domain reflectometry (FDR) measurement technique. FDR is a transmission line fault isolation method that identifies signal path degradation for transmission lines by using a swept RF signal. DTF requires no calculating of the distance of the transmission line. This dual role of predicting future failure conditions and isolating existing problems makes DTF an important part of service and maintenance on transmission lines.

System Inspections and Maintenance

An annual inspection of the antenna and transmission line system is always a good practice. These inspections should include the electrical performance tests discussed in the previous section of this chapter. A climbing inspection should examine the transmission line and its supporting hardware. Any missing hardware should be replaced, and any indications of wear, corrosion, binding, or other mechanical stresses should be addressed. The transmission line should be touch tested for hot spots.

A complete inspection of the transmission line system would include:

1. Transmission Line System:
    a. Verify proper operation of system check valves, if equipped
    b. Missing or loose hardware

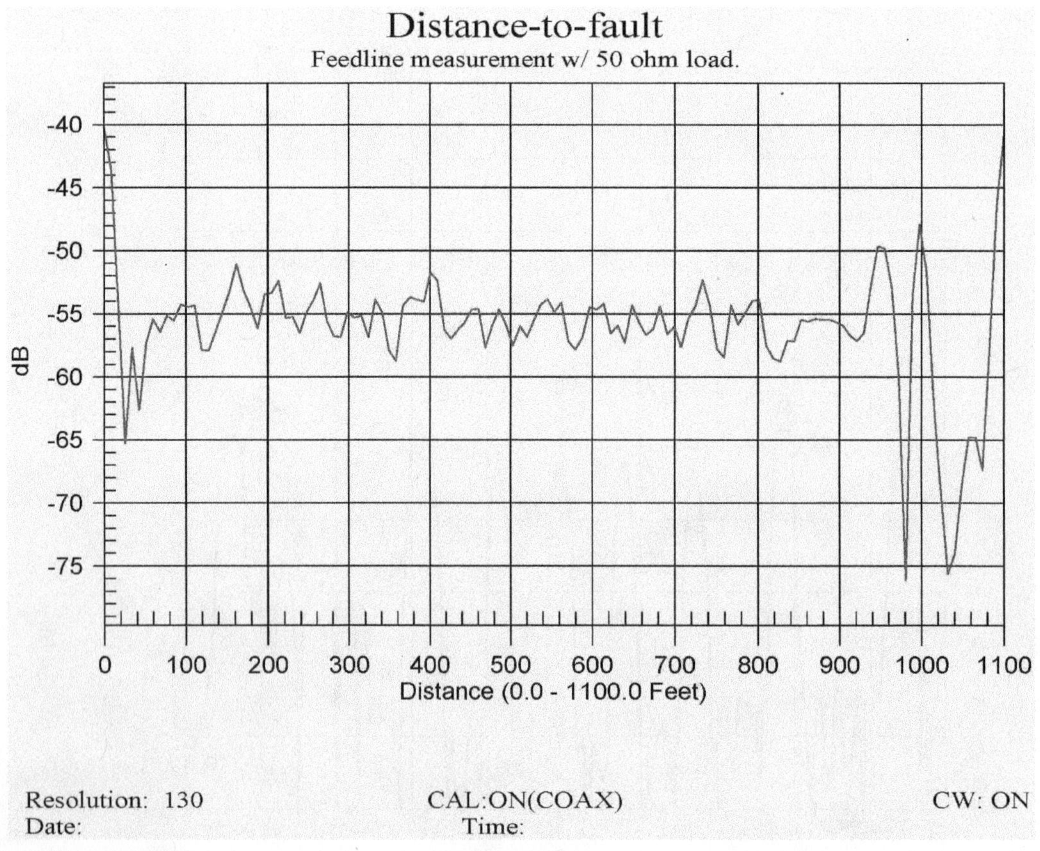

FIGURE 9.4-31 Distance to fault (DTF) measurement of a transmission line system.

   c. Fixed hanger inspection
   d. Transmission line damage dents, other voids, and corrosion
   e. Proper spring hanger support settings, corrosion, and proper clearance with tower members
   f. Pressure leaks
   g. Proper grounding kit installation
   h. Connections properly sealed
   i. Transmission line bends smaller than minimum radius
   j. Infrared heat signature (hot spots)
2. Pressurization Equipment:
   a. Missing or loose hardware
   b. Physical damage
   c. Water filters
   d. Air intake filters
   e. General operation (noise, etc.)

It is good practice to assemble the results of these annual tests into a report that is kept on file and can be readily referenced. As a matter of practice, when operating any transmission system it is best to establish a baseline system VSWR at the time the system is first installed. The system should be equipped with VSWR overload detection, and if this system triggers a transmitter shutdown, it is important that the system not be energized and operated until after the source of the fault is identified and repaired. The system should always be maintained with a positive air or gas pressure to prevent moisture from entering the transmission line. Following these steps will ensure that a transmission line system will continue to provide a long and useful life.

Note

[1] HD Radio is a trademark of DTS, Inc.

Bibliography

Andrew Corporation. *Andrew Catalog 37*. Orland Park: Andrew Corporation, 1997. p. 589.

Cozad, K.W. *NAB Engineering Handbook* 9th Edition, Coaxial Transmission Lines. Washington, DC: National Association of Broadcasters, 1999.

Fitch, C.S. At RF It's Only Skin Deep. *Radio World*. 2012, 6.

Howard W. Sams & Co. Inc. *Reference Data for Radio Engineers*. Indianapolis: Howard W. Sams & Co. Inc., 1975. 0-672-21218-8.

Leonard, B. Transmission Line Trade-Offs for DTV. *Broadcast Engineering*. July, 1998.

Ulaby, F.T. *Fundamentals of Applied Electromagnetics*. Upper Saddle: Pearson Prentice Hall, 2004.

SECTION

10

SIGNAL MEASUREMENT AND ANALYSIS

C H A P T E R

10.1

Audio Signal Analysis

STANLEY SALEK, P.E.
Hammett & Edison, Inc., Consulting Engineers
San Francisco, California

THOMAS KITE and DAVID MATHEW
Audio Precision, Inc.
Beaverton, Oregon

Updated for the 11th Edition by

JOHN KEAN
Cavell Mertz and Associates
Manassas, VA

INTRODUCTION

Audio signal analysis remains one of the fundamental chapters in the *NAB Engineering Handbook* because audio is an essential part of all broadcasting. With the 11th Edition, this chapter reflects the almost universal use of digital audio systems. Analog measurement definitions and techniques are maintained to explain fundamentals, and a section on audio loudness measurement has been added.

The ability to quantify audio signals in terms of characteristics and qualities is paramount in audio engineering. Reference is often made to established standards and requirements, including measurements relating to amplitude, frequency distribution, distortion, noise, and phase. The observation of such attributes allows a virtually complete characterization of an analog electrical audio system.

In the broadcast environment, audio measurements are used to gauge the overall quality of equipment such as amplifiers, recording systems, mixing consoles, digital audio devices, and other networks throughout the broadcast signal path. Digital versions of these audio systems may use nonlinear audio encoding that employs various types of bit-rate compression in addition to linear encoded systems. Because nonlinear encoding relies on complex psychoacoustic principles, conventional measurement can be challenging or virtually impossible. For this reason, measurement of some aspects of nonlinear systems is beyond the scope of this chapter.

AMPLITUDE ANALYSIS

The most basic of needs in audio measurement is to determine a value relating to the size, or amplitude, of an analog audio signal. Since an audio waveform is rapidly changing, methods have been developed to convert peak, root mean square (RMS), and average values of the changing waveform into corresponding proportional DC voltages that can be more easily observed.

There are specific cases in which the peak value is the most direct measure of magnitude. It gives an indication of the largest excursions (either positive or negative) of an audio waveform. As shown in Figure 10.1-1, the audio signal is applied to an absolute value circuit, which rectifies the waveform such that the output is always positive. A diode is then used to couple the signal into *C* and *R*. These serve as memory and decay time elements, respectively, that can be adjusted in value to provide the desired visual and reference ballistics. Although the output is still changing with time, following the input, the excursions corresponding to the peak values of the original waveform are much slower and more easily observed on metering devices.

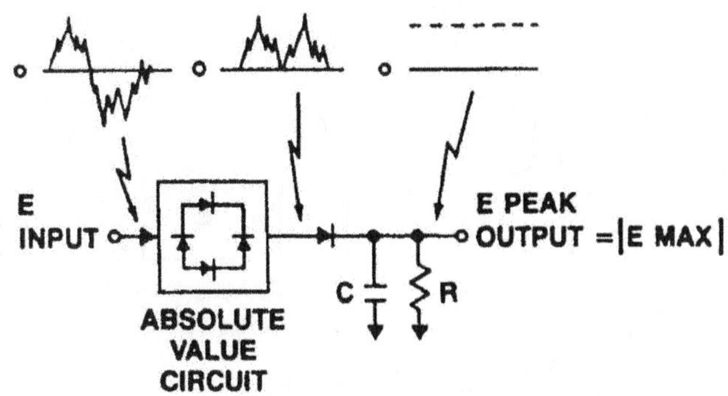

FIGURE 10.1-1 Peak value detection.

As the value of resistor R is increased, the decay time of the output is proportionally increased as well. If the resistor is completely removed, a peak hold circuit results.

Peak (actually peak-to-peak) waveforms can also be observed on an oscilloscope, although this technique may be impractical because of the difficulty in reading the rapidly changing waveforms typical in most audio material. (The pursuit of accurate, easily readable signal measurement has been a goal since the beginning of audio.)

While the peak value is important, the RMS value of an audio signal is equally important because it indicates the energy content of the signal without regard to its waveform. For a given measurement period of a waveform, the RMS level can be defined as follows:

$$E_{rms} = \sqrt{\frac{\left(E_1^{\,2} + E_2^{\,2} + \ldots + E_n^{\,2}\right)}{n}}$$

where E_1 through E_n are successive amplitudes over a total of n samples. As can be seen from its name, the value is computed by taking the average of n samples of E squared. Performing the square root function completes the calculation. This function is also commonly referred to as "true RMS" [1]. Figure 10.1-2 shows how this technique is accomplished electrically.

Through the use of the absolute value circuit of Figure 10.1-1 and the R-C configuration found in the RMS detector of Figure 10.1-2, an average detector can be made, as shown electrically in Figure 10.1-3 and mathematically:

$$E_{average} = \frac{E_1 + E_2 + \ldots + E_n}{n}$$

In terms of audio perception, the average value of an audio signal is related to program material density, where the peak value described earlier relates to a maximum. Since the peak value defines the upper limit of allowable modulation in a transmission system, it is often technically desirable that the peak-to-average ratio be as low as possible to attain highest perceived loudness and signal-to-noise ratio (SNR). Achieving this effect may require compromising aesthetic goals and may not always be appropriate, depending on the type of program material.

The decibel (dB) is a unit for comparing relative levels of voltage or power in transmission systems. In broadcast audio systems, the most common representation of decibels is dBm. A dBm is the value of a signal with reference to 1 mW into a 600-ohm load. The level in dBm of a signal can be found using the following relation:

$$\text{dBm} = 20\,log\left(\frac{E}{0.775}\right)$$

where E is in volts. The number 0.775 represents the voltage level reference of 0 dBm. Note that, strictly speaking, this formula is true only when the circuit impedance is 600 ohms. In practice, the formula is used typically without regard to the impedance level, although such application can lead to significant error. Voltage levels obtained from the peak, RMS, and average circuits described previously can be used for possible values of E. When this is done, some common types of metering can be synthesized to observe the activity of audio material.

Legacy Metering

For simplicity, the industry has long preferred to detect the average amplitude of the given waveform, referenced to the RMS value of a sine wave. The "standard volume indicator," popularly called a "VU meter," was introduced in 1939 to provide an electrically simple standard program level indicating device—see Figures 10.1-4(a) and 10.1-4(b).

The original, electromechanical VU meter is defined as a combination of an AC bridge rectifier, a resistive attenuator, and a direct-current ammeter with an approximately linear voltage scale to display average power of electrical program signals. The VU meter was calibrated to read 0 VU (100% in Figure 10.1-4(a)) when connected to a source of sinusoidal voltage of 1.228 across the 7500-ohm device. In a terminated 600-ohm system, this voltage corresponds to a power of 2.5 mW, or a power of 4 dB above 1 mW. Because the meter is powered directly by a 600-ohm (or other impedance) program line, the VU meter became a standard for telephone companies and popular with broadcast program producers [2]. Today, the great majority of VU meters are driven by amplifiers or are completely electronic,

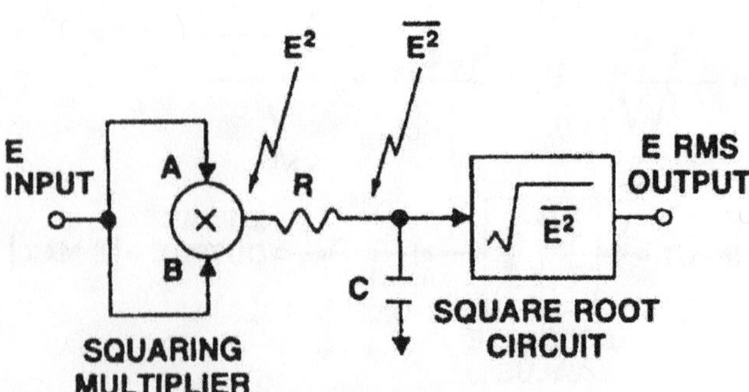

FIGURE 10.1-2 True RMS detection.

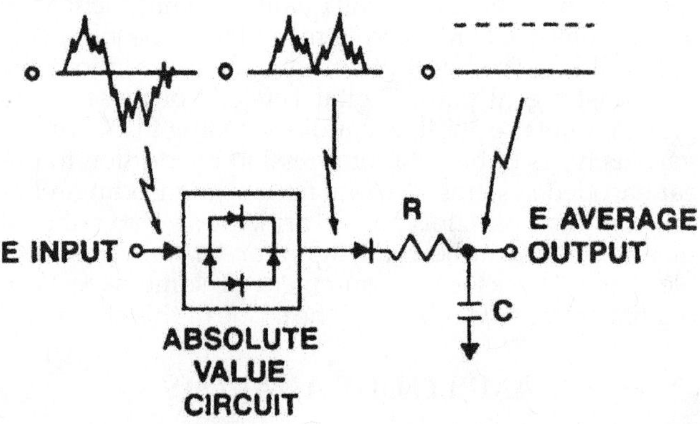

FIGURE 10.1-3 Average value detection.

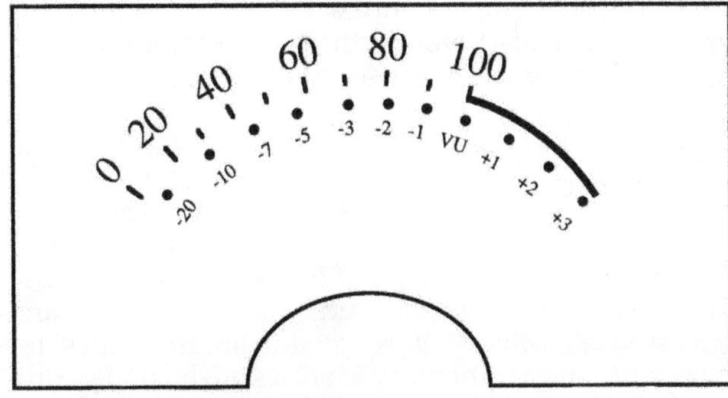

FIGURE 10.1-4(A) VU meter.

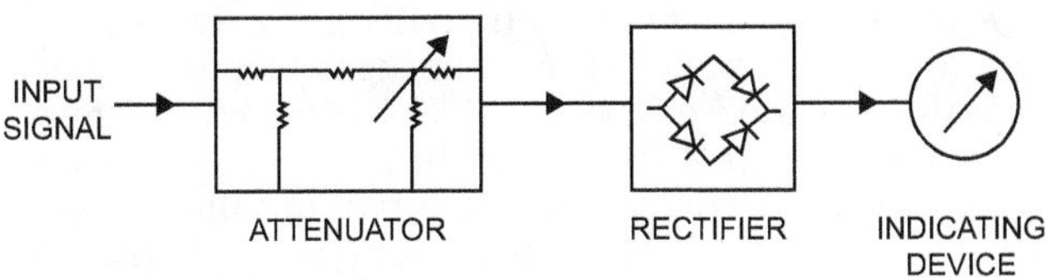

FIGURE 10.1-4(B) Block diagram of the stages of a typical VU indicator.

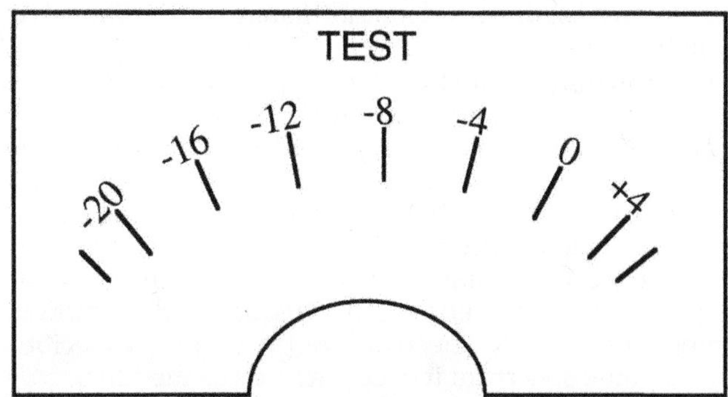

FIGURE 10.1-4(C) Arrangement of a typical PPM scale.

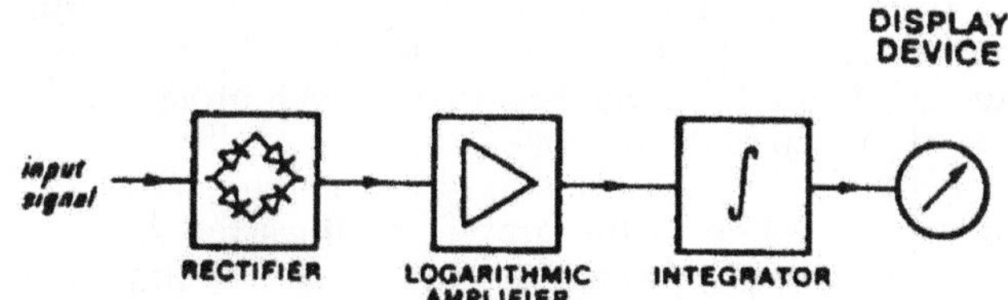

FIGURE 10.1-4(D) Block diagram of the stages of a typical PPM.

and the power volume calibration is unimportant: their reference indication is related to a desired system level in voltage, tape flux level, etc.

The VU meter's dynamic characteristics are set to display 99% of its steady-state reading on a sine wave tone burst 300-ms long, with a fall to 5% of the reading in 300 ms. This is essentially an average responding device with a characteristic that is too slow to respond to short-duration program peaks. Therefore, levels must be set with a distortion-free margin ("headroom") below the point of clipping (i.e., system overload) [3]. However, the difference between signal peaks and the instantaneous average level (the peak-to-average ratio) varies widely across different program material. This lack of correlation requires a greater headroom margin above 0 VU and results in an inefficient use of dynamic range.

To address the signal peak issues and find an efficient level below overload, peak level meters or peak program meters (PPMs) were developed to read the near-instantaneous value of the audio signal. Examples of mechanical and electronic peak meters are shown in Figures 10.1-4(c) and 10.4(d). Both types employ a signal rectifier and an integrator, producing a fast rise and slow fall effect on the display device. Typical standards require the PPM to read −2 ± 0.5 dB of the steady-state value for a tone burst of 10 ms and take 2.8 seconds for the pointer to fall 20 dB [4]. Peaks shorter than the response time of a PPM are not indicated fully. However, this integration time was determined in the late 1930s from listening tests of signal overload in vacuum tube amplifiers, which found that signal distortion for briefer intervals was generally inaudible. Compared to VU meters, the much faster response time of PPMs, such as British Broadcasting Corporation (BBC) and European Broadcasting Union (EBU) types, allows operation with smaller headroom margins, which results in better use of the available channel SNRs.

By the 1990s, a new generation of all-electronic peak metering devices were developed using multicolor LEDs, usually arranged in bar graph displays that are bright and easy to read. Freed of the electromechanical limitations of moving-pointer meters, these indicators adopted *instantaneous* signal measurement. It was a natural pairing with digital audio systems that were also coming into use, which have abrupt overload characteristics that can produce audible distortion with overloads lasting only milliseconds of time. As discussed in the following section, the latest generation of peak meters has advanced to "true peak" measurement, which estimates the amplitude of the waveforms between digital samples. Normally sampled meters may overlook these overshoots, and while they are brief, they may occur repeatedly and with duration to become audible.

Loudness Measurement

By the 2000s, instantaneous peak level displays dominated the professional and consumer market. They became common in mixing consoles, meter bridges, recorders, and other audio equipment. It was natural for producers and engineers to rely on these popular meters to control and monitor their content as it was distributed and transmitted to consumers. Unfortunately, reliance on peak-reading meters for program level led to problems with consistent program levels. This is because peak-reading meters are designed to indicate the potential for signal peak overload, but they are not good indicators of optimal audio level. To compensate for that, engineers would use their ears to mix and set program levels, but this is subjective. As explained later, loudness metering is a sensible alternative for measuring and managing the range of content in today's audio.

Fortunately, a great deal of work has been done on audio signal measurement by members of the International Telecommunications Union (ITU) and the EBU. Their research over many years led to the development of an algorithm for a better meter—one that measures program *loudness* similar to the intensity of sound as determined by human hearing. The sense of loudness depends primarily on sound's frequency distribution across the audible spectrum, or, to be more specific, *where* in the spectrum it occurs as well as the *bandwidth* of the sound and its duration. The implementation for this algorithm is currently defined by ITU Broadcast Systems Recommendation BS.1770–4 [5].

The simplified design of the ITU loudness meter is illustrated in Figure 10.1-5. The left and right channel audio is passed through separate "K-weighting" filters having a gain plateau above 2 kHz and a roll-off below 100 Hz. The frequency response of the filter, as shown by the gray line in the graph, modifies the spectral distribution to represent the ear's sense of loudness with mixed program content at moderate listening levels. Next, the signal from each filter is converted to means-square amplitude before being summed. The rectified and summed measurement is then logarithmically converted to a dB-linear scaling. Multichannel versions of ITU loudness meters with center and surround channels use identical stages before summing, with 1.5-dB lower gain than the left and right channel stages; the low-frequency effects (LFE) channel is not included, as it contributes little to the sense of program loudness. Extensive tests performed by ITU members verified that with appropriate time constants, the readings correlated well with listener impressions of program loudness for a variety of music and speech material.

Integrating filters with 3-second and 400-millisecond time windows are applied to the signal to provide "short-term" and "momentary" displays. These provide the real-time indication for immediate leveling and mixing of audio signals. For very short content, such as commercials, the audio may be mixed and processed to achieve specific maximum values for short-term and momentary loudness, which may provide better matching to the surrounding full-length program.

ITU loudness meter displays are marked in "loudness units" that are equivalent to dB and are referenced to a digital full scale. Thus, readings start at "0 loudness units related to full scale" (0 LUFS) and decrease from there. Figure 10.1-6 shows a typical bar graph scale, where the contiguous section of illuminated segments (from –20 LUFS to the bottom) is displaying the instantaneous loudness. (The actual display is colored in red, yellow, and green zones to make reading easier.)

Another feature of the ITU-R BS.1770 standard is that it provides measurement of *true peak* levels, which are well-suited to the fast overload rate of digital audio. Moreover, the signals from digital processing or lossy coding (e.g., perceptual audio coding) may increase signal peaks above those of the normally sampled signal values. True peak measurement estimates positive and negative signal swings in the continuous time domain by oversampling. These intersample peaks may be up to 1 dB higher than normally sampled

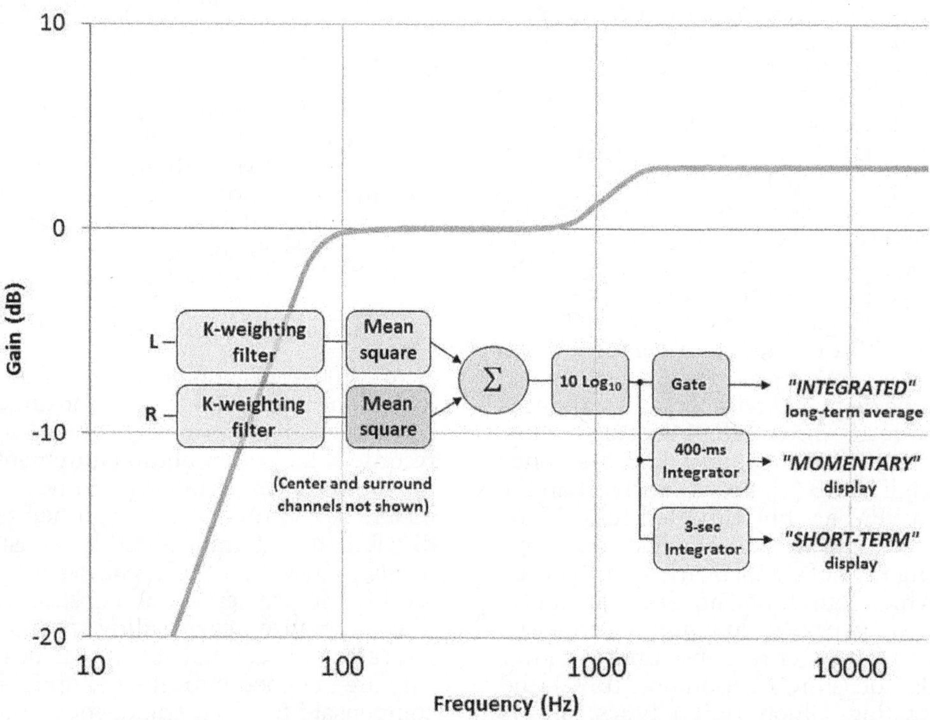

FIGURE 10.1-5 The ITU loudness meter—block diagram illustrates design, curve illustrates gain vs. frequency response.

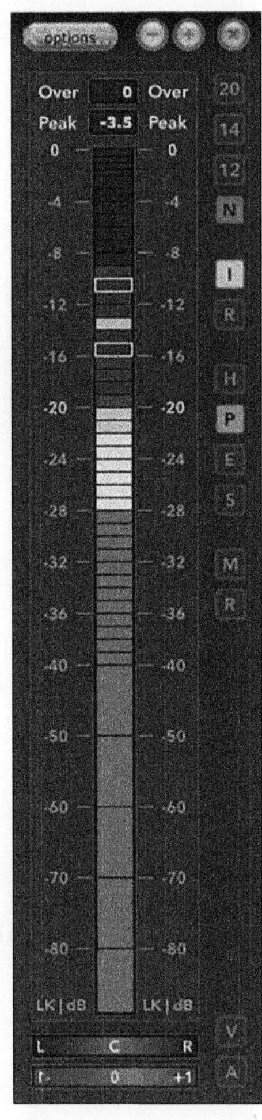

FIGURE 10.1-6 Typical ITU loudness meter bar graph display.

peaks, which if undetected may cause distortion effects in digital-to-analog converters, sample-rate converters, and audio codecs. True peak level is designated as dBTP, for decibels referenced to digital full scale. The true peak value is indicated in the bar graph meter in Figure 10.1-6 as the isolated solid bar at approximately −14 dBTP. Some meters, such as this one, also track the highest recent peak, shown with the rectangle at −3.5 dBTP, and the number of peaks exceeding full scale (zero over, in this case).

Another key loudness measurement is a long-term average, called "integrated loudness," which may be (and preferably is) measured for an entire program, regardless of the length. The integrated loudness is intended to provide one number, in LUFS, which indicates "how loud the program is on average." This is measured in the same way as short-term and momentary loudness, including a gating function. The integrated loudness measurement is divided into 400-millisecond frames that are summed with 75% time overlap. The gate causes a frame whose integrated power is significantly lower than the others to be ignored in the measurement because these quieter frames do not noticeably contribute to the perceived program loudness. Without this gating function, programs with longer periods of silence, low-level background sounds or fade outs will get too low of an integrated loudness level value.

The integrated loudness is also used to guide the mixing and production of a program to achieve a desired value, called the target level. Studies carried out by EBU members showed that a long-term average loudness of −23 LUFS is an optimal balance between using as much of the audio channel's peak headroom as possible (for minimal noise) and offering enough system headroom to provide a compelling sonic range. This target level is being adopted nearly worldwide by audio and video producers for broadcast and network distribution. It is generally not applied to radio transmission, which is still driven by competitive loudness over sonic range.

In 2011, FCC Report and Order FCC 11–182 incorporated ATSC Recommendation A/85 into the FCC Rules that apply to broadcast television, cable, and satellite. A/85 originally specified that the BS.1770–1 integrated loudness of commercial and promotional announcements (an ungated measurement) must be equal to the *dialnorm* metadata transmitted to consumers' receivers (*dialnorm* is used to control playback gain with an audio compression system). While any value of *dialnorm* is acceptable under the rule, the target loudness must always match this value. In practice, most broadcasting organizations found it more practical to use a fixed *dialnorm* of −24 instead of dynamically changing the transmitted value of *dialnorm* depending on program content. A/85 was later updated to incorporate gating, but for simplicity retained the 1:1 relationship between target loudness and *dialnorm*. While the EBU had determined that a target loudness of −24 LUFS without gating was roughly equivalent to −23 LUFS with gating, this difference is well within the +2/−5 LU "comfort zone" over which most consumers do not feel the need to adjust their volume controls.

Figures 10.1-7(a) and 10.1-7(b) show the measurements of four level indicators with two very different types of program audio. The first audio sample is the 5-minute jazz track "One More Once," directly from the CD album of the same title by Michel Camilo. As is common with commercially released music, the sound has been tightly compressed and limited, with the peak signal envelope running near (or at) the full-scale clipping point of 0 dBTP. The standard volume indicator was calibrated to indicate 0 VU with a sinusoidal signal of 0 dBTP. (This is for demonstration only. See later in this chapter for a discussion of meter calibration.) Under these conditions, the VU meter indicates the program material approximately 10 to 15 dB below the peak limits.

Also shown in the chart are the variations in the short-term loudness and the integrated loudness. While the short-term indications generally follow the VU meter, the VU meter's indications "swing" to a greater extent. This is because the short-term indicator has a 3-second integration time, compared to the 0.3-second rise time of the VU meter. The slower rate of change is generally easier to watch and make judgments of program level. (It should be noted that these charts show the maximum indication within 1-second

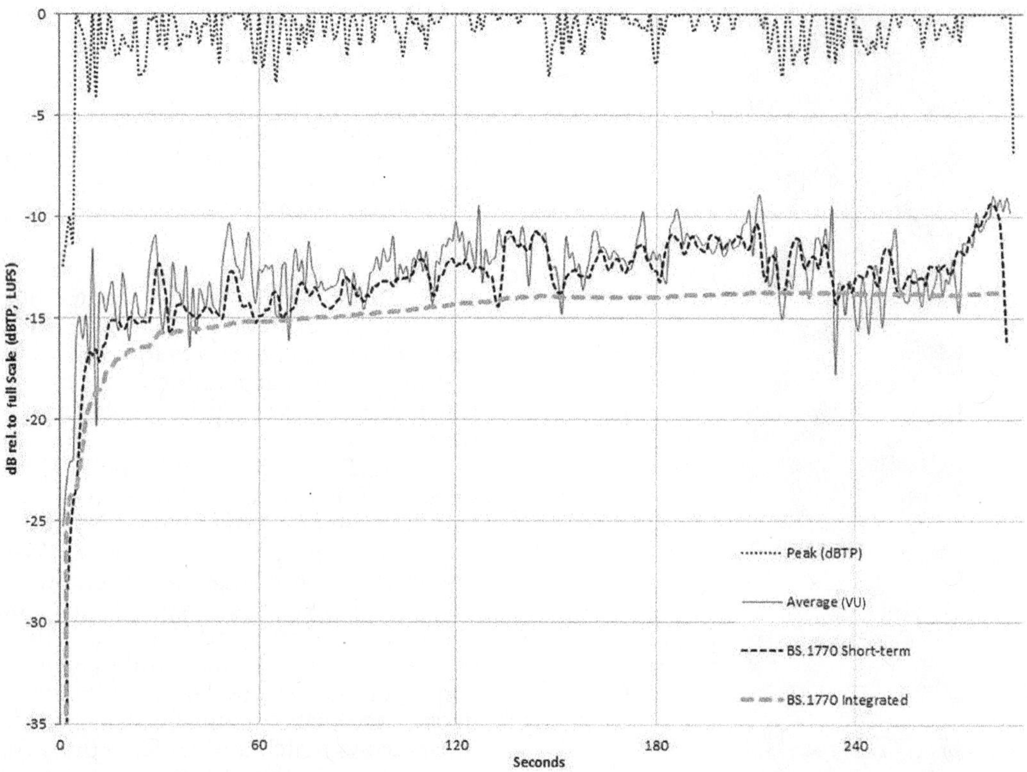

FIGURE 10.1-7(A) Measurements of 5-minute jazz track "One More Once," directly from the CD album of the same title by Michel Camilo, using four audio level indicators.

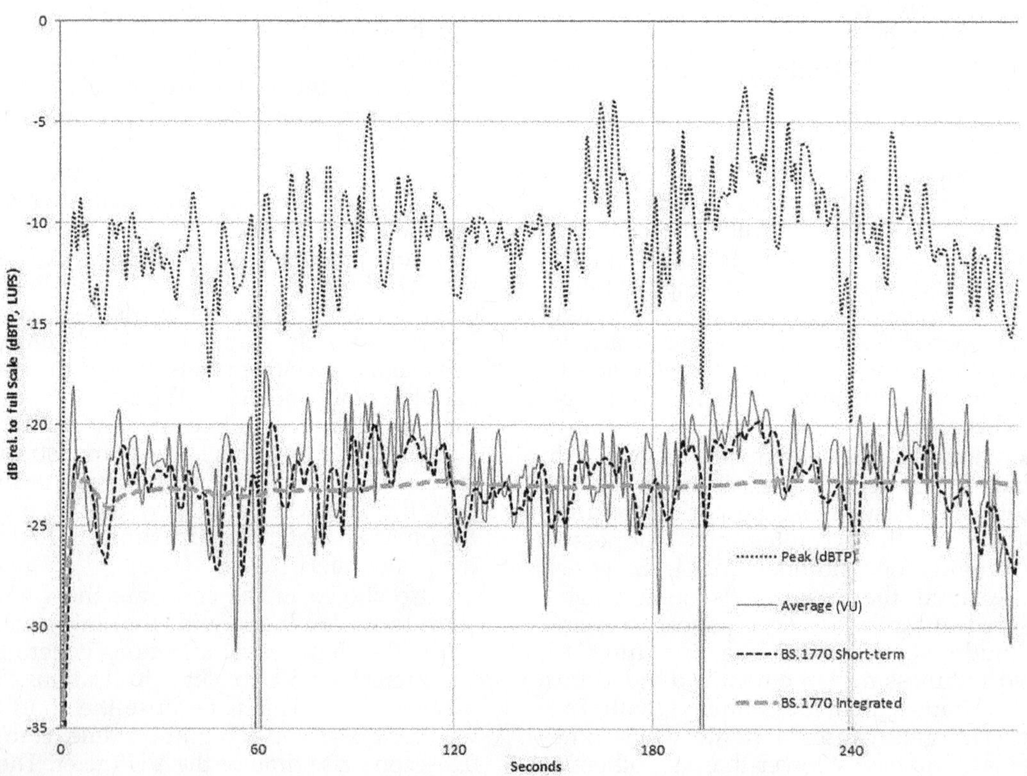

FIGURE 10.1-7(B) Measurements of the opening minute of NPR's *Weekend Edition Saturday* and a 4-minute newscast, using four audio level indicators.

intervals for each measurement. Thus, the peak and VU meters would be varying more than is indicated.) The integrated loudness measurement averages all values from the starting point and only catches up with the short-term loudness after half a minute, but it yields a very slow-changing indication that settles to approximately −13.5 LUFS.

The program material in Figure 10.1-7(b) is very different. It is the opening minute of NPR's *Weekend Edition Saturday* and a 4-minute newscast. Both are voice content from condenser microphones with no processing. The signal peaks are clearly free to move upward with no limit, reaching −3 dB below full scale. The VU and short-term loudness range between −20 and −25 dB below full scale, and the integrated loudness quickly converges on −23 LUFS. This content is said to have a *peak to loudness ratio* of approximately 20 dB [−3 − (−23)]. Comparing the ratio of 13.5 dB for the first audio sample, it is apparent how peak measurements can lead to very different program levels and inconsistent loudness.

The following provides guidance on the alignment of scales across various meters using a 1-kHz tone [6]. (These are suggestions, and it should be noted that organizations may have different program level objectives, including different requirements for headroom.) Figure 10.1-8 shows an ITU loudness meter scale at the top with a target of −23 LUFS, which is specified by EBU R128. Instantaneous and true peak meters are not shown because they are not used for setting program levels. Operation at the specified targets is expected to provide sufficient headroom to avoid full-scale clipping with most content.

The BBC and IEC peak program meters scales are legacy devices that are superseded by ITU loudness meters in most organizations. These meters involved a short integration time and specific instructions for operators on optimum levels for various types of content.

The VU meter scale is aligned to provide 20 dB of headroom above 0 VU. For example, a system operating at +4 dBm for 0 VU would require an overload level of at least +24 dBm. Again, this is a starting point that depends on how one "reads" an active VU meter, the program material, etc.

Target Loudness

In the process of live production it is helpful to know how close one is to the −23 LUFS target. In addition to a continuous display of the integrated loudness, meter manufacturers have come up with a variety of graphical methods for looking back at short-term loudness since the start of a program. This gives the mixing engineer some perspective on how appropriate the present levels are relative to the past. One such meter is shown in Figure 10.1-9, which provides a circular chart of levels winding counterclockwise around the display (in this case approximately 20 minutes of loudness

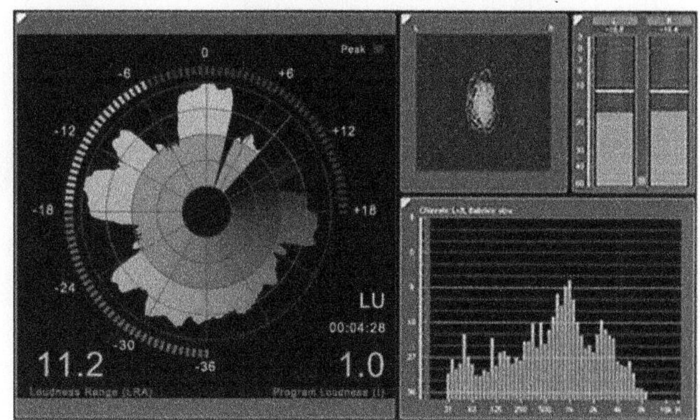

FIGURE 10.1-9 Example loudness meter showing enhancements for tracking short-term loudness.

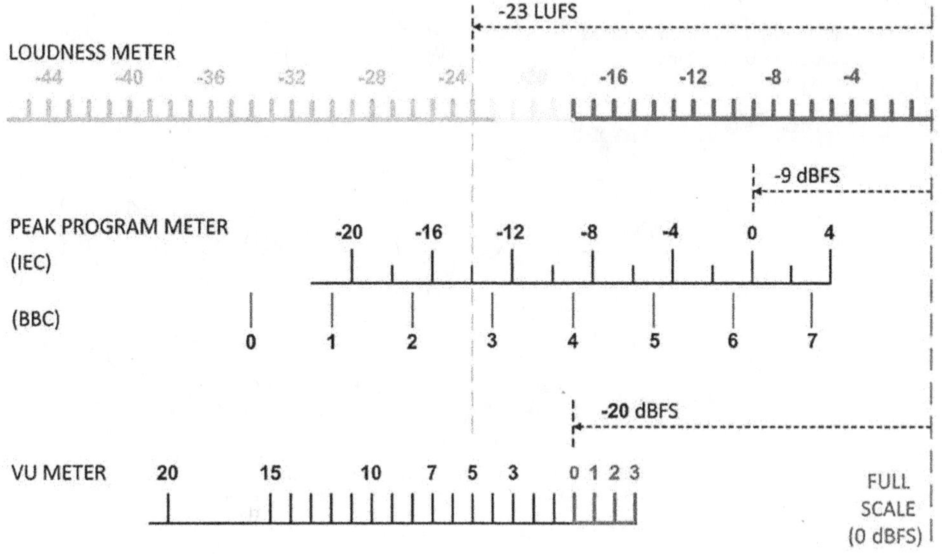

FIGURE 10.1-8 Comparison of different audio level scales and illustration of −23 LUFS level across these scales.

have been tracked). Due to this complexity, most loudness meters are software-based, either as a window on a computer screen or on a dedicated screen device. Software-generated meters allow for the display of other features: in Figure 10.1-9, three other panels are included, showing a Lissajous channel phase pattern, a stereo bar graph of loudness and peak level, and a real-time spectrum display.

Normalization

Normalization is a technique for adjusting overall program levels to a desired standard value. As discussed earlier, peak-reading meters are concerned with only the maximum permitted signal, to avoid channel overload, but shifting to a loudness target overcomes problems with volume or loudness consistency. Figure 10.1-10 shows a simplified chart with two programs having different peak levels and loudness characteristics: the program to the left is louder than the program to the right, as shown by their short-term and integrated loudness traces. Normalization simply measures integrated loudness of each program:

- The louder program, on the left, measures −14 LUFS overall or 9 LU above the target loudness. For compliance, *the entire program* is lowered in gain by 9 dB, putting the integrated loudness at −23 LUFS;
- The softer program, on the right, measures −27 LUFS overall or 3 LU below the target loudness. To correct this program, the gain is increased uniformly by 3 dB, raising the integrated loudness to −23 LUFS.

Although the peak levels may differ between the two programs, this has little effect on the ear's sense of loudness, which is governed by the frequency response and timing characteristics embodied in the ITU loudness algorithm. As a result of normalization, the programs will sound more compatible, and no sonic changes will be heard to either program.

Normalization is easy to perform on recordings, usually stored in digital audio files. There are numerous companies making software that can measure and normalize a file's program level to the target loudness. Some "watch" a computer folder for new files and proceed automatically with the normalization. Others can "batch process" a library of files.

There are two ways to achieve loudness normalization: one actually changes the gain of the recording and rewrites the normalized file to storage; the other adds metadata to the digital file's header, which is used by playback equipment to adjust the gain. The first method requires decoding, gain offset and re-encoding to normalize, a process of *cascaded* coding. This may introduce audible artifact with audio files that use lossy compression, such as MP3. This should be avoided because most lossy coders are not intended to be cascaded with themselves or any other lossy compression. This is not a significant issue with linearly coded audio, such as WAV-format files, as these are lossless encoding-decoding processes. If lossy-compressed digital audio files are involved, the most transparent method of normalization is to only embed metadata in the file as a playback gain instruction.

Normalization is not strictly applicable to live audio content, as the integrated loudness of later content cannot be known. However, collecting integrated loudness over a long period of time, perhaps for hours, provides clues as the expected loudness, which may provide an approximate gain adjustment. Ideally, loudness should be monitored during live production and mixed to achieve the desired target level. This will result in programs that are so close to the target that automatic normalization is usually unnecessary.

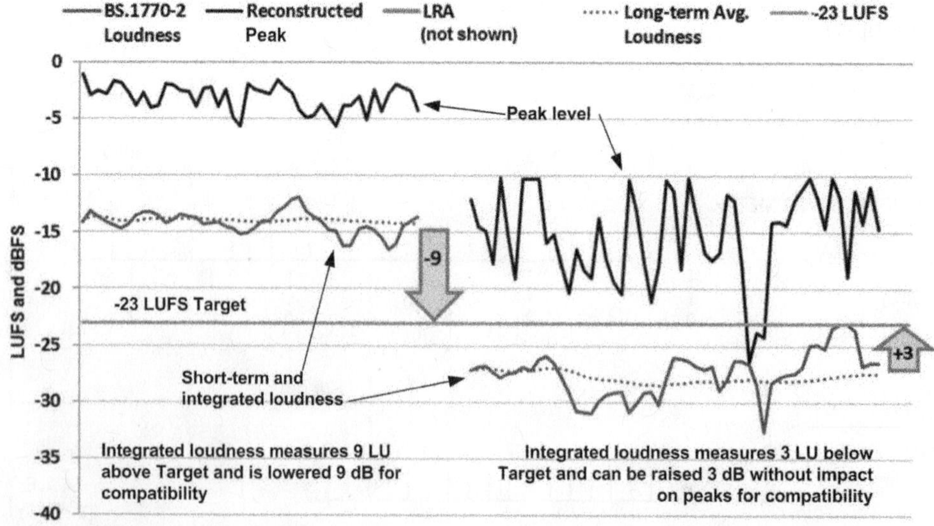

FIGURE 10.1-10 Simplified chart with two programs having different peak levels and loudness characteristics.

FREQUENCY ANALYSIS

Amplitude analysis methods, as described in the previous section, are generally used to provide an indication of signal levels simultaneously over the entire audio range. To characterize audio system performance, it is important to measure discrete frequencies, allowing frequency response measurement in addition to the dynamic measurement discussed earlier.

Simply stated, frequency response is the capability of a device or system to pass or amplify equally all frequencies within a specified range. As far as audio in the broadcast environment is concerned, the range of interest is generally 20 Hz to 20 kHz. Although few musical instruments produce fundamental frequencies greater than 4 kHz and the human voice does not contain fundamental frequencies much above 1 kHz, the reproducing device or system must be able to pass the harmonics that accompany the fundamental frequencies. Some forms of music and low-frequency sound effects may extend below 30 Hz. Without adequate bandwidth or with uneven frequency response, an unnatural coloration of the perceived sound becomes evident. To overcome this potential problem, amplifiers with flat frequency response over the specified range are employed. Since the responses of series-connected amplifiers are additive, it is important to verify the flatness of each device in a system.

Several methods are available to measure audio frequency response. They include discrete (tone) measurement and swept frequency methods. Parallel analysis and Fast Fourier Transform (FFT) techniques also can be used.

Discrete Frequency Analysis

The discrete frequency measurement method is uncomplicated and inexpensive. A simple measurement system consists of a low-distortion audio frequency oscillator and a wideband AC voltmeter. The oscillator output is connected to the input of the device or system to be characterized. The voltmeter is used to observe the level at the output of the device or at a desired intermediate point in a system.

The measurement is made by first setting the generator output level to the nominal input operating level of the device. Generally, a 400-Hz or 1-kHz frequency is chosen initially for most audio systems. The device output level is read on the AC voltmeter, and this quantity is noted as a zero dB relative reference. Provided the generator itself has a flat frequency response, measurements at frequencies through the audio band can be made while recording the corresponding dB output levels with respect to the reference. A convenient and commonly used technique is to increment the frequency in a 1, 2, 5 sequence (100 Hz, 200 Hz, 500 Hz, 1 kHz, 2 kHz, 5 kHz, etc.). This method permits plotting the final response data on four-cycle "LOG/LIN" graph paper, providing regularly spaced frequency increments horizontally. The logarithmic amplitude data are plotted along the horizontal axis, with the zero dB relative reference placed in a convenient position on the linear vertical axis.

Although the discrete frequency measurement technique is straightforward, it is also often tedious and time-consuming. Numerous frequency measurements must be made to ensure adequate testing. This method is most usable in the response measurement of single-ended devices lacking a suitable input port for connection to an audio generator, such as compact disc players. Prerecorded test signals are used to provide the tones necessary for discrete frequency response characterization [6, 7].

Swept Frequency Analysis

A faster and more efficient means of measuring frequency response is the swept frequency method. This process employs a frequency-agile generator as a signal source and measures response over the entire range of interest in one sweep. The signal detector may be a tracking type that follows the generator and measures a narrow band of frequencies centered on the source frequency. Use of a tracking detector helps ensure that the amplitude measured is that of the tone generator and is not influenced by spurious noise or harmonics.

Devices specifically designed to conduct swept frequency measurements include spectrum analyzers with tracking oscillators and dedicated automated audio test systems. Spectrum analyzers directly produce response images on a built-in display, while the dedicated systems generally use a connected personal computer (with appropriate software and hardware) for data display and storage.

Network analyzers are also useful in audio frequency domain measurements. They are used to characterize two-port networks (devices having real-time input and output ports) as to frequency, phase, and delay responses. They are employed where substantial accuracy in the measurement of these parameters is required. RF subsystem and semiconductor device design have been the major application for network analyzers, although some newer generation equipment includes audio frequency coverage. In the case of audio systems, network analyzers allow precise response measurement of amplifier and filter designs.

Figure 10.1-11 shows a typical setup for measuring amplifier frequency response using the swept frequency method. The signal source used to drive the test device is the tracking oscillator output of the analyzer. The device output is terminated with an appropriate characteristic load impedance and connected to the analyzer input. Measurement of the frequency response is made by manually or automatically sweeping the analyzer across the frequency range of interest. A plotter or personal computer (with appropriate software and hardware) can be used to record a permanent record of the test device response characteristics.

Real-Time Analysis

A real-time audio analyzer (RTA) consists of a sequential collection of one octave or one-third octave filters having individual detectors and indicators at each output. The program audio is simultaneously fed to the

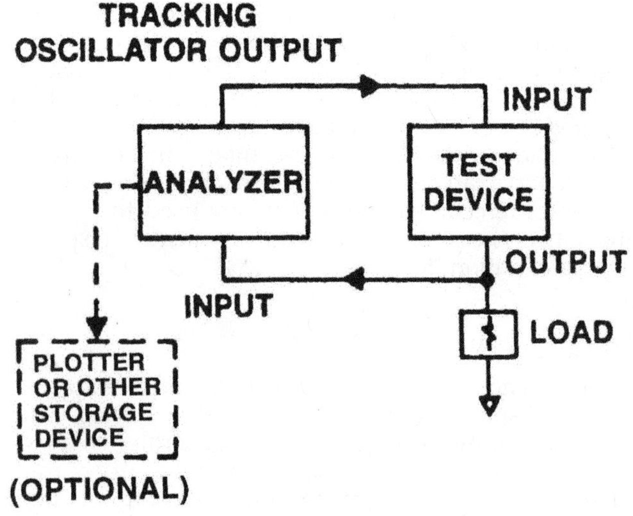

FIGURE 10.1-11 Frequency response measurement using swept method.

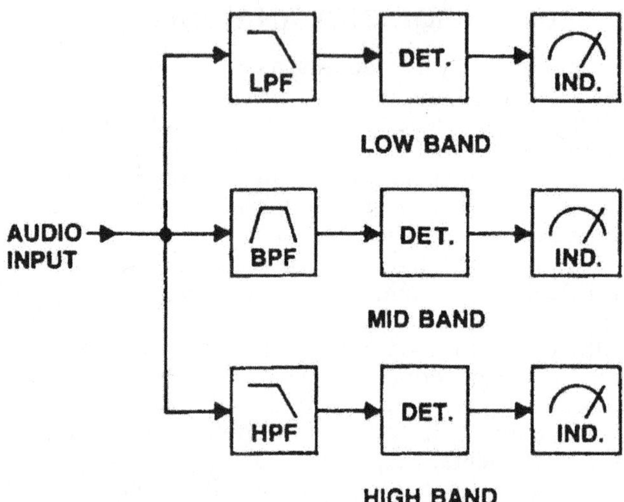

FIGURE 10.1-12 Basic real-time analyzer.

inputs of all the filters. The output signal of each filter is proportional to the amount of energy occurring in that particular frequency band. This technique is also referred to as parallel analysis.

A simplified version of an RTA is presented in Figure 10.1-12. As shown, it is intended to break the audio band into three sections using low-pass, band-pass, and high-pass filtering. Signal detectors are then used to condition the audio for display on a suitable indicator, one set for each of the three bands. The detectors can be (and often are) the same peak, RMS, or average circuits described earlier. Typical readout indicators are bar graph displays with dB-calibrated scales. When arranged side by side, the readouts provide a graphical presentation of amplitude versus frequency, similar to a spectrum analyzer display. Unlike the sweeping spectrum analyzer, however, an RTA relies on a continuous broadband signal rather than a swept frequency tone.

The RTA may be used for dynamic program material and for room acoustics analysis, as well as measuring frequency response of audio devices in conjunction with a "pink" noise source. Pink noise has a constant mean-squared voltage per octave of frequency. Because it carries a uniform amount of energy in each octave, pink noise displays uniformly from octave band to octave band with an RTA. This characteristic makes it popular in audio work. By connecting the pink noise source to the input of a device to be characterized, and the RTA to its output, a response curve can be displayed almost instantly. The characteristic response of pink noise is shown in Figure 10.1-13, as compared with white noise. White noise has a constant power spectral density per Hz (bandwidth) at any frequency, except that it is attenuated for frequencies above the audio spectrum. In the figure, the noise response lines are captured with a spectrum analyzer, which measures with constant bandwidth across the spectrum. Thus, white noise produces a flat response while pink noise has a downward slope that reduces 10 dB per decade of frequency. An RTA registers the pink noise equally for each octave band.

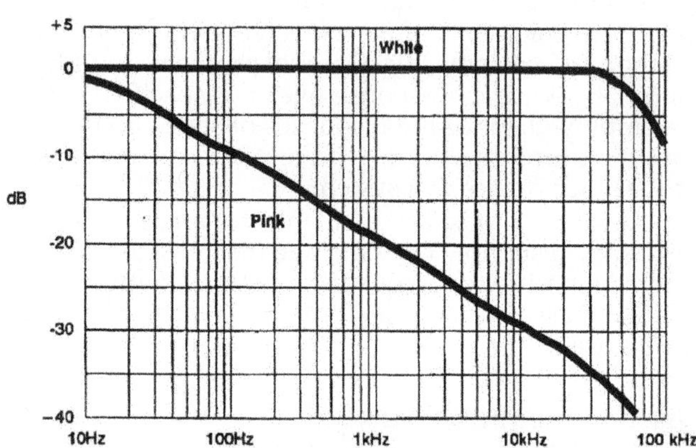

FIGURE 10.1-13 Response versus frequency for white noise and pink noise.

Modern audio analyzers perform an FFT on an audio signal and mathematically transform the result into an amplitude display of the frequency components present. The filtering and detection process is done by a specialized digital signal processing (DSP) microcomputer, which is more consistent, more accurate, and faster than previous analog techniques. A particularly fast and efficient form of swept frequency analysis uses a so-called "chirp" test signal that is synchronized to an FFT-based measurement.

DISTORTION MEASUREMENT

When a two-port device is driven beyond its range of linear operation or through areas of discontinuity, signal distortion occurs. As a result, additional frequencies appear at the device output that were not present at its input. In cases in which distortion becomes extreme, it can be identified through listening. Odd-order harmonic distortion (such as produced by signal clipping) becomes audible at approximately 1.25% (expressed as a percentage of the amplitude of the undistorted waveform). Even-order distortion, characterized by a coloration of the program material, becomes audible at approximately 5%. Generally, systems with a wider frequency response capability need to maintain lower distortion levels to be acceptable. Since distortion is a subjective phenomenon, techniques are available to measure its various types.

Simple audio distortion can be characterized in two ways: total harmonic distortion (THD) and intermodulation distortion (IMD). The two effects produce measurements that are often expressed by single, unrelated numbers. While THD content is determined by only one method, IMD has several accepted measurement practices, most notably the Society of Motion Picture and Television Engineers (SMPTE) and International Telecommunication Union–Radiocommunication Sector (ITU-R, formerly CCIR) methods. As discussed below, audio signals can be characterized using THD and IMD methods, but they also are subject to other types of degradation.

Total Harmonic Distortion

Total harmonic distortion is a measure of individual harmonic amplitudes with respect to the amplitude of the fundamental frequency. In practice, harmonics greater than third order often add little to the resultant value because of their negligible amplitude. THD is defined as:

$$\text{THD\%} = 100 \times \frac{\sqrt{A_2^{\,2} + A_3^{\,2} + A_4^{\,2} + \ldots + A_n^{\,2}}}{A_1}$$

where A_2 through A_n are the amplitudes of the individual harmonics and A_1 is the amplitude of the fundamental.

As shown in Figure 10.1-14, a 1-kHz sine wave with harmonic distortion shows only minor differences when it is overlaid with an undistorted signal (as viewed on an oscilloscope). The amplitude and slope errors do not lead directly to a numeric result. But when a spectrum photo of the same waveform is observed (see Figure 10.1-15), the preceding relation can be applied. Assuming a system with the fundamental (A_1) at 0 dBm (0.775V), the second harmonic (A_2) at −26 dBm (38.8 V), and the third harmonic (A_3) at −50 dBm (2.5 V), the harmonic distortion can be calculated as follows:

$$\text{THD\%} = 100 \times \frac{\sqrt{(0.0388)^2 + (0.0025)^2}}{0.775} = 5.0\%$$

Although spectrum analysis can produce accurate THD measurement results, a simpler and more cost-effective procedure that produces a direct numeric quantity is more popular. Figure 10.1-16 shows the block diagram of a typical THD analyzer. An oscillator (with much less harmonic distortion than the device or system to be measured) is connected to the test device input. The distorted output signal of the device is filtered to remove A_1, the fundamental component. This produces a signal that, when RMS-detected, determines the THD produced by the device being tested, relative to the amplitude of the fundamental.

THD measurement is often conducted using the same 1, 2, 5 sequence of frequencies mentioned for discrete response measurement. The THD results can be plotted on the same graph to characterize the device under test on a single figure. THD measurements may be taken over various input levels, but as the level is reduced, noise characteristics may affect the readings. In such cases, the spectrum analyzer method could produce more meaningful results.

Intermodulation Distortion

The intermodulation method of measuring distortion uses a test signal composed of two (generally

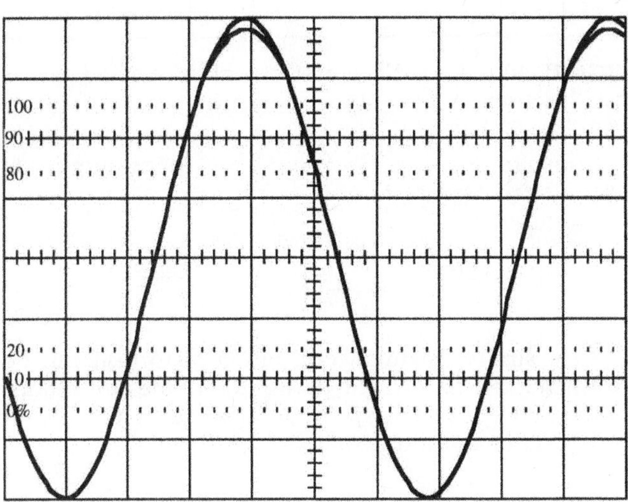

FIGURE 10.1-14 Comparing the distorted output of an amplifier with its undistorted 1-kHz input component.

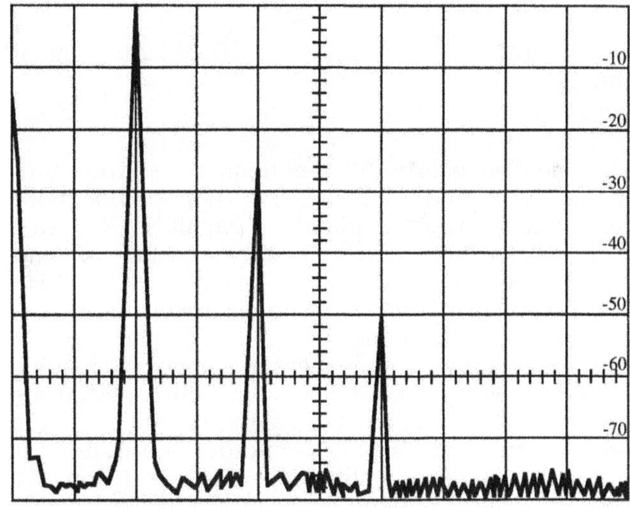

FIGURE 10.1-15 Measured THD using a spectrum analyzer (V: 10 dB/div.; H: 500 Hz/div.).

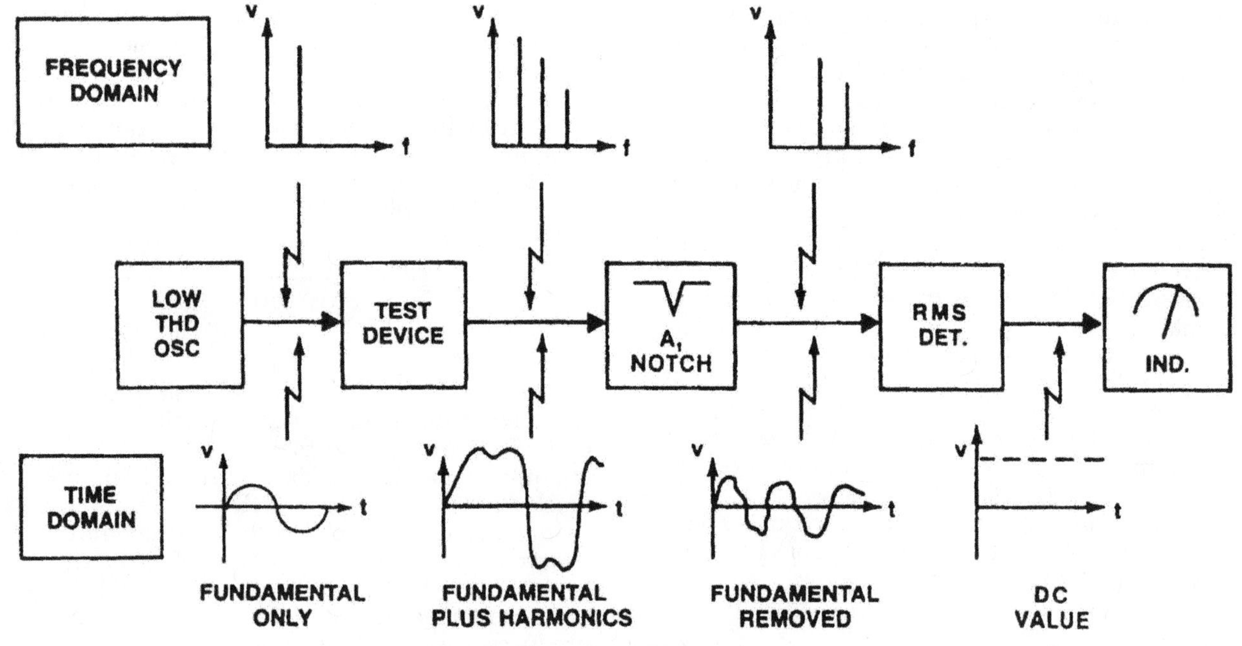

FIGURE 10.1-16 THD analyzer.

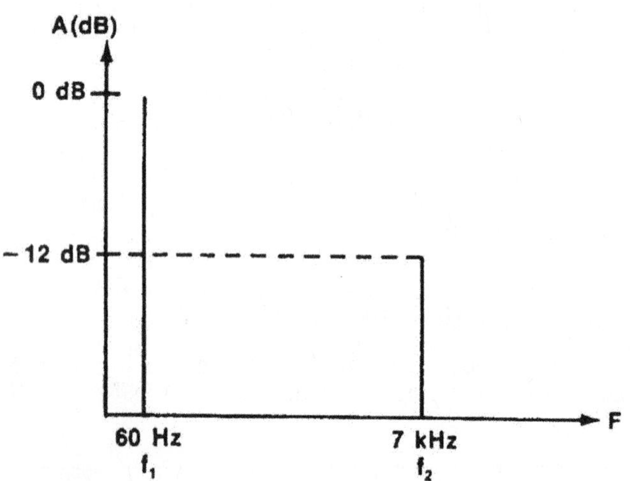

FIGURE 10.1-17 Spectrum of SMPTE IMD test signal ratios.

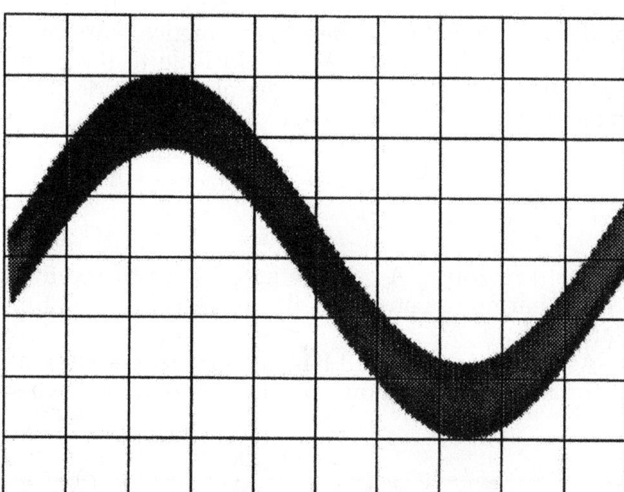

FIGURE 10.1-18 SMPTE test signal as viewed on an oscilloscope.

sinusoidal) signals of different frequencies. After summation, they produce the effect of an amplitude modulated carrier when applied to a circuit having IMD. The SMPTE method uses a low-frequency (f_1) and a relatively high-frequency (f_2) signal (usually 60 Hz and 7 kHz, respectively) that are mixed at a four-to-one amplitude ratio (see Figures 10.1-17 and 10.1-18). This method involves the measurement of the relative amplitude of the modulation sidebands added to the higher frequency signal. For diagnostic purposes, it is often useful to determine even-order and odd-order distortions separately, although this is best done by spectrum measurement techniques. Even-order distortion usually can be characterized by the ratio of the sum of the amplitudes of only the two second-order spurious frequencies, $f_2 - f_1$ and $f_1 + f_2$, to the amplitude of the carrier signal, f_2:

$$\text{SMPTE IMD\%}_{(\text{2nd order})} = \left[\frac{A_{(f_2-f_1)} + A_{(f_1+f_2)}}{A_{f_2}}\right] \times 100$$

In a similar manner, odd-order distortion can be characterized by the ratio of the sum of the amplitudes of the two third-order spurious frequencies, $f_2 - 2f_1$ and $2f_1 + f_2$ to the amplitude of f_2:

$$\text{SMPTE IMD\%}_{(\text{3rd order})} = \left[\frac{A_{(f_2-2f_1)} + A_{(2f_1+f_2)}}{A_{f_2}}\right] \times 100$$

Figure 10.1-19 shows the output signal of an amplifier with IMD, as viewed on an oscilloscope. Note the elongated trough as compared to Figure 10.1-18. As with THD, spectrum analysis can be used to determine the numerical amount of distortion present.

Intermodulation sidebands can be seen around f_2 in the spectrum photo of Figure 10.1-20. Second- and third-order distortion percentages for this example are calculated as follows:

$$A_{f_2} = -12\text{dBm} = 195\text{mV}$$

$$A_{(f_2-f_1)} = A_{(f_1+f_2)} = -38\text{dBm} = 9.75\text{mV}$$

$$\text{SMPTE IMD\%}_{(\text{2nd order})} = \left[\frac{9.75+9.75}{195}\right]\times 100 = 10.0\%$$

$$A_{(f_2-2f_1)} = A_{(2f_1+f_2)} = -58\text{dBm} = 0.98\text{mV}$$

$$\text{SMPTE IMD\%}_{(\text{3rd order})} = \left[\frac{0.98+0.98}{195}\right]\times 100 = 1.0\%$$

As shown, the contribution of even-order distortion products is usually greater than that of the odd-order. To express the result as a single quantity, the vector sum of the two quantities is taken:

$$\text{SMPTE IMD\%}_{(\text{total})} = \sqrt{(\text{IMD\%Even})^2 + (\text{IMD\%Odd})^2}$$
$$= \sqrt{10^2 + 1^2} = 10.05\%$$

As with THD, SMPTE IMD has a direct method of numeric solution, as shown in the block diagram of Figure 10.1-21. The two test frequency oscillators are summed to produce the $f_1 + f_2$ signal, which is then applied to the input of the device to be tested. The distorted output signal is high-pass filtered to remove the f_1 fundamental component, leaving only the amplitude modulated f_2 component. Using a standard AM

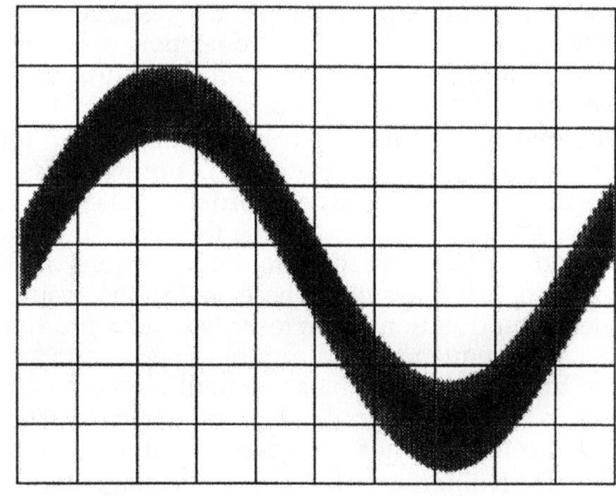

FIGURE 10.1-19 Output of an amplifier exhibiting significant SMPTE IMD.

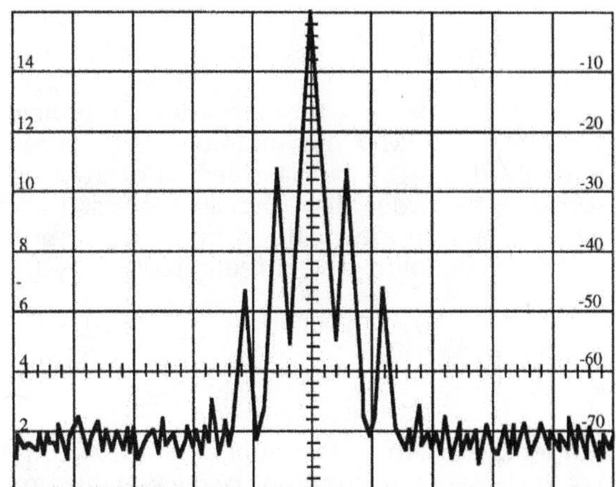

FIGURE 10.1-20 SMPTE IMD measurement (V: 10 dB/div., top of screen −12 dBm; H: 100 Hz/div., center frequency 7.0 kHz).

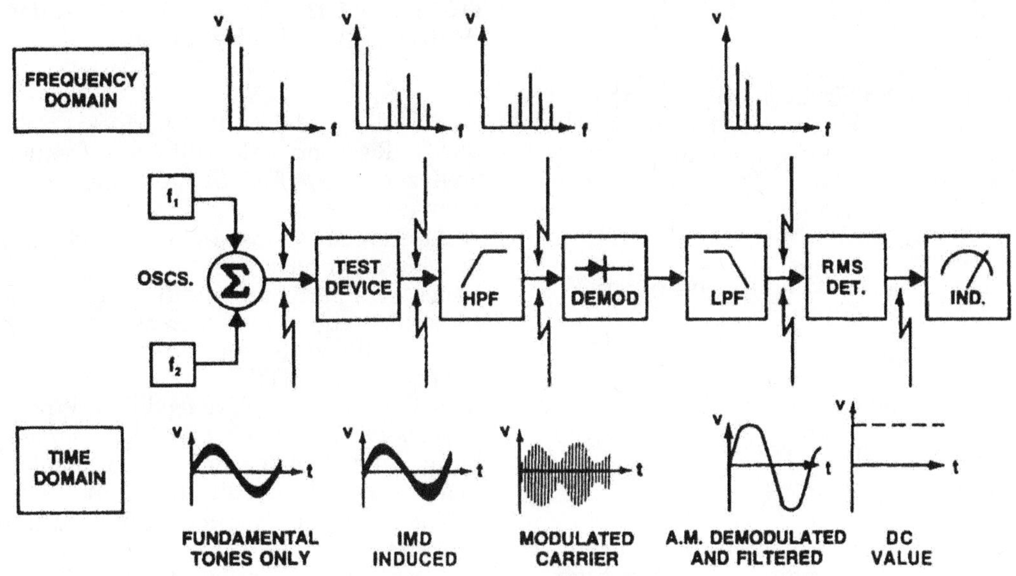

FIGURE 10.1-21 SMPTE IMD analyzer.

demodulator and low-pass filter, the residual f_1 component is obtained. After RMS detection, a DC level proportional to the distortion is produced that can be viewed on a direct-reading indicator.

Of some historical note, *wow and flutter* is a term that describes a special case of IMD normally associated with analog audio tape recorders. It is caused by variations in tape velocity across the recording and/or reproducing heads, due to imperfections in the mechanical drive system. These variations result in frequency modulation of the recorded and reproduced signal. The frequency spectrum obtained is similar to that of the SMPTE IMD measurement method, except the f_1 low-frequency signal is generated by fluctuations in tape speed and is not of any set amplitude.

The ITU-R intermodulation method uses a combination of two higher frequency sinusoidal signals (f_3, f_4) of equal amplitude. They are typically 1 kHz apart and found at 5/6 kHz, 14/15 kHz, or 19/20 kHz in many applications. One of the spurious frequencies generated is low in frequency, whereas others are gathered around the two driving frequencies. Figures 10.1-22 and 10.1-23 spectrally show the driving frequencies before and after passing through a test amplifier. As with the SMPTE IMD measurement, the generated spurious products can be classified as even-order or odd-order. Even-order distortion is expressed as the ratio of the amplitude of the difference component ($f_4 - f_3$) to the sum of the two driving frequencies (f_3, f_4):

$$\text{ITU-R IMD\%}_{(\text{2nd order})} = \left[\frac{A_{(f_4-f_3)}}{A_{f_3} + A_{f_4}}\right] \times 100$$

Odd-order distortion is determined by calculating the ratio of the sum of the amplitude of the two third-order products, $2f_3 - f_4$ and $2f_4 - f_3$, to the sum of the amplitudes of the two driving frequencies, f_3 and f_4:

$$\text{ITU-R IMD\%}_{(\text{3rd order})} = \left[\frac{A_{(2f_3-f_4)} + A_{(2f_4-f_3)}}{A_{f_3} + A_{f_4}}\right] \times 100$$

In the case of Figure 10.1-23, the driving frequencies f_3 and f_4 are at 5 kHz and 6 kHz, even-order products at 1 kHz, and third-order products at 4 kHz and 7 kHz, respectively. Distortion percentages for this example are calculated as follows:

$$A_{f_3} = A_{f_4} = 0\text{dBm} = 775\text{mV}$$

$$\text{ITU-R IMD\%}_{(\text{2nd order})} = \left[\frac{0.49}{775+775}\right] \times 100 = 0.032\%$$

$$A_{(2f_3-f_4)} = -62\text{dBm} = 0.62\text{mV}$$

$$A_{(2f_4-3)} = -61\text{dBm} = 0.69\text{mV}$$

$$\text{ITU-R IMD\%}_{(\text{3rd order})} = \left[\frac{0.62+0.69}{775+775}\right] \times 100 = 0.085\%$$

It is a common practice for direct-reading metered analyzers to measure only the amplitude of the difference product ($f_4 - f_3$) with respect to the driving signal amplitudes. A device that performs this task is called an ITU-R second-order difference frequency distortion analyzer. Figure 10.1-24 illustrates how the measurement is made.

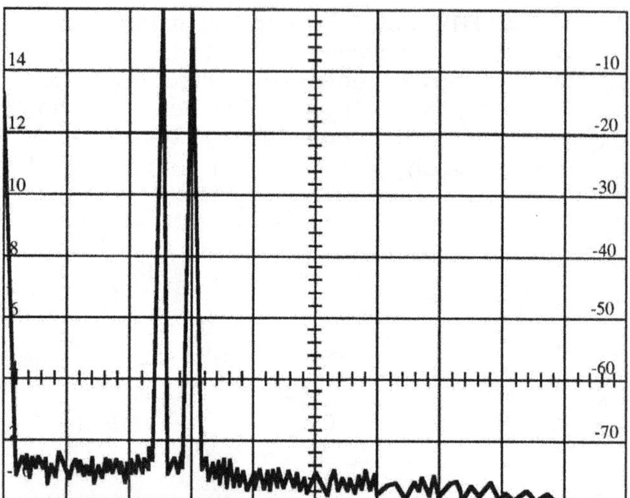

FIGURE 10.1-22 Spectrum of ITU-R IMD test signal (V:10 dB/div.; H: 2 kHz/div.).

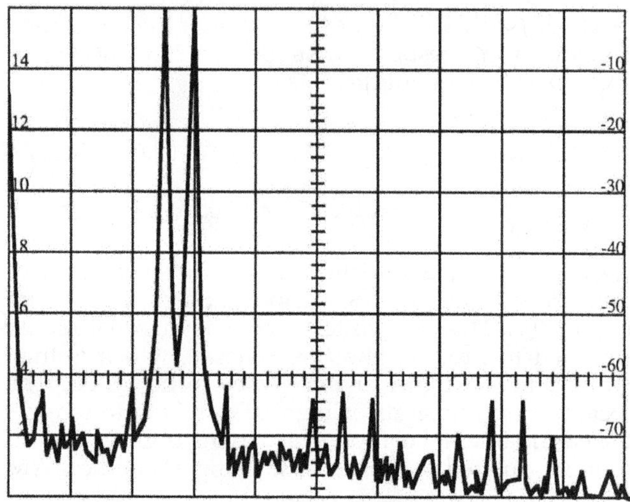

FIGURE 10.1-23 ITU-R IMD measurement example (V:10 dB/div.; H: 2 kHz/div.)

Additional forms of intermodulation distortion measurement have been devised to uncover potential issues with particular audio systems. Some use multiple audio frequencies, whereas others employ nonsinusoidal audio waveforms. One such example is transient intermodulation (TIM) distortion, found only in amplifiers that utilize negative feedback. When this feedback is excessive, a fast-rising transient signal applied to the input of the amplifier can produce an internal overshoot that saturates the circuits in the amplifier.

The most popular procedure used to measure TIM distortion is called the sine-square wave method. The test signal employed uses a square wave to induce nonlinearity in the test device by saturating

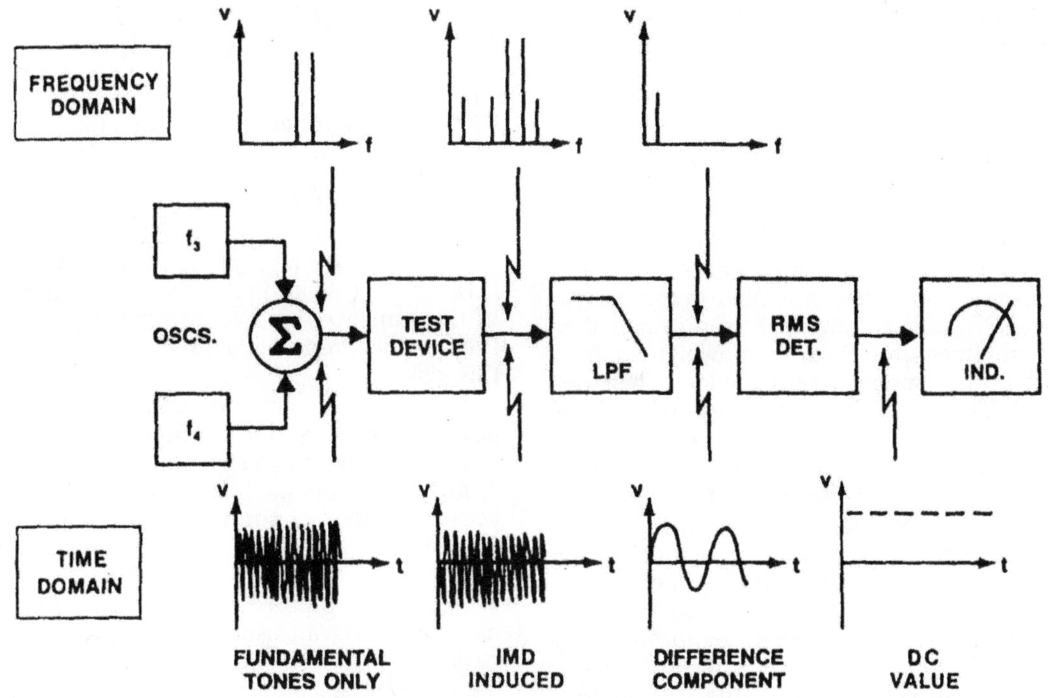

FIGURE 10.1-24 ITU-R second-order IMD analyzer.

the amplifier's internal current, caused by its alternate rises and falls. Mixed with this square wave is a low-level, high-frequency sine wave, which is unrelated harmonically. In one definition, the frequency of the square wave is 3.18 kHz and that of the sine wave 15 kHz, where the peak-to-peak amplitude ratio of the former to the latter is four to one [7]. Similar to the other IMD spectrum analyzer evaluation methods, the components present at the output can be combined to express a TIM distortion figure as a single percentage.

Other TIM measurement methods include a sawtooth wave method that takes amplifier slew rate into account and a noise-square wave method, where the sine wave of the sine-square wave method is replaced by a narrow-band noise spectrum [8].

Added Filtering in Distortion Measurement

Filtering as part of distortion measurement is often useful to remove components that are of little interest and as a diagnostic aid, especially with THD measurement. A 20-kHz or 30-kHz high-pass filter placed in series with the output of the device being measured is useful for testing broadcast equipment. This practice is often acceptable, since the harmonics produced outside the transmission bandwidth can be eliminated, producing a more realistic result.

A high-pass filter also can serve as an important analytical aid. With a cutoff frequency in the 400-Hz range, for example, it can be placed in series with the test device output and used to determine the contribution of AC line frequency hum (50 or 60 Hz) and related harmonics (100/120 and 150/180 Hz) to a THD measurement utilizing a fundamental frequency of 1 kHz or greater. Verification of adequate grounding used in the test setup also can be evaluated using this filter.

When filtering is used in IMD measurement, it is important to verify that in-band distortion components are not inadvertently removed. Also, sharp cutoff analog filter designs may produce overshoot components that could affect measurement results.

NOISE MEASUREMENT

In audio engineering, noise is a random energy distribution in which individual spectral components are not clearly resolved. Primary sources of noise in analog circuits and amplifiers are in the resistive circuit elements [9]. It is important to control noise in amplifiers as their gain increases to preserve a high signal-to-noise ratio, which is the ratio of the operating signal level to the noise level inherent in the amplifier itself.

To understand the origin of the noise, we can model a passive resistive element as a noiseless resistor in series with a noise voltage generator, E_r:

$$E_r = \sqrt{4KTBR}\,(\text{volts})$$

Where:
K = Boltzmann's constant (1.38×10^{-23} W-sec/°K)
T = Temperature in degrees Kelvin
B = Noise bandwidth (Hz)
R = Resistance in ohms

In the equation, noise voltage is a physical phenomenon that can be worsened by an increase in any of the variable factors. Therefore, noise cannot be eliminated, but it can be reduced. This optimization process is often accomplished by proper selection of the resistive components because of an additional factor known as excess noise, which is proportional to the voltage drop across the resistor and related to the material from which it is made. Of the different available types, carbon composition resistors are prone to the most excess noise contribution, whereas metal-film devices show the least.

At times, the actual spectral distribution of noise is of less importance than the noise voltage within a given bandwidth. For audio frequencies, a 15- or 20-kHz bandwidth is of interest. With a low-pass filter in this range connected in series with an amplifier output, and the input of the amplifier grounded, an unweighted but band-limited noise measurement can be made. When the noise output level is obtained, it can be expressed as a ratio with a standard operating level and reference frequency. This produces an indication of the amplifier's SNR.

When the gain of the amplifier is known, this same technique can be used to determine equivalent input noise voltage, that being the voltage of the noise that would be found at the input of the amplifier if the amplifier were completely noiseless [10].

A practical goal in the measurement of audio noise is to obtain data that correlates well with the subjective perception of noise [11]. For example, two amplifiers with identical SNRs can sound very different because one may have a uniform noise spectrum and the other may have the noise distributed unevenly over the audio frequency range. This is because the ear perceives the loudness of a signal according to the frequencies involved. To make comparative noise measurements more representative of human hearing, several weighting filters have been developed that alter noise spectra over the frequency band of interest.

"A" weighting is based on the inverse of early measurements by Fletcher and Munson of the ear's sensitivity at low sound pressure levels [12]. An improvement on A-weighting is the CCIR/ARM method, which emphasizes frequencies more in the 2-kHz to 10-kHz range and uses a meter with faster ballistics [13]. It is believed that this method, which is based on the obtrusiveness as well as the levels of different kinds of noise, provides a more commercially acceptable result when used to characterize modern, wide-range audio equipment. Figure 10.1-25 compares the two curves.

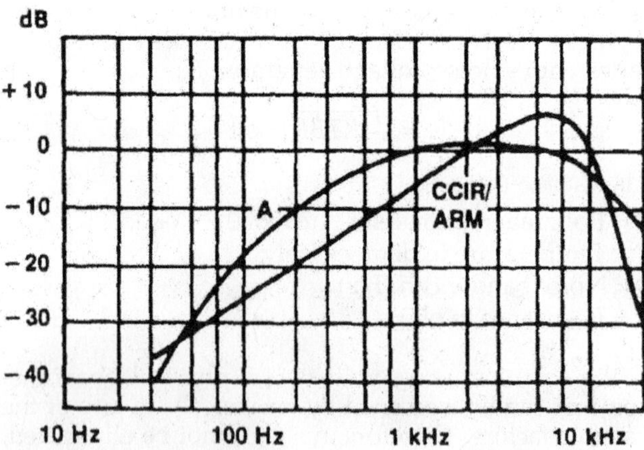

FIGURE 10.1-25 Noise measurement weighting curves.

PHASE MONITORING AND MEASUREMENT

An (L + R) summation is the monophonic-compatible signal stereo broadcasting. In analog FM systems, separation information is transmitted via an (L – R) signal. Since these two signals are created through a summation and difference process of the original left and right channel stereophonic source, it is important that they can be recombined properly at the receiver [14]. Amplitude and phase errors must be minimized in the transmission system to accomplish this goal. Phase measurement is important in accomplishing this task.

In a stereo program system, if left and right audio information is correlated but delayed in phase, the error would not be evident on a stereophonic receiver. A monophonic signal, however, would be degraded because of inexact summation. To monitor audio systems for the presence of this problem, a phase meter can be used.

A popular phase indicator is the Lissajous figure method, using an oscilloscope in the XY mode. The patterns produced are shown in Figure 10.1-26. An oscilloscope is connected such that the left channel audio causes an X-axis deflection and right channel audio produces a Y-axis deflection, as shown in Figures 10.1-26A and 10.1-26B, respectively. When each channel contains the same program material, the pattern of Figure 10.1-26C is produced along the L + R axis. If one of the channels is inverted, the pattern of Figure 10.1-26D becomes evident. This is often called the L – R axis. Monophonic program material that follows this axis is said to be inverted in polarity because no sum or L + R information is present.

The patterns of Figures 10.1-26E, 10.1-26F, and 10.1-26G are commonly seen when phase errors exist between two channels carrying the same discrete audio tone signal. Stereo program material, in unprocessed form, generally modulates the L + R axis while simultaneously deviating in the L – R direction to a lesser amount, as shown by the spread pattern in Figure 10.1-26H.

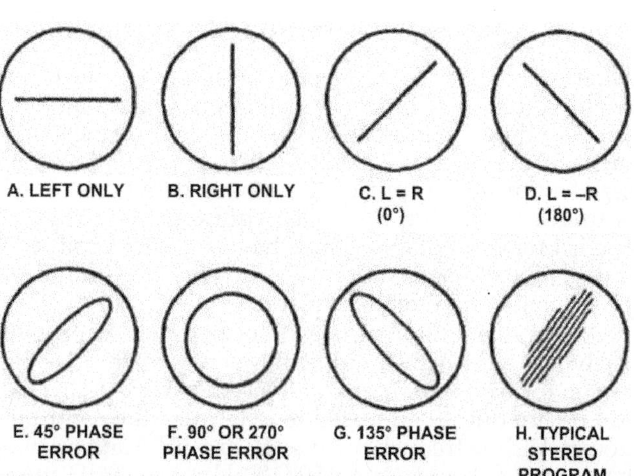

FIGURE 10.1-26 Interpretation of Lissajous patterns.

DIGITAL AUDIO

Much (if not most) of audio today is recorded, mixed, edited, stored, and transmitted digitally. Digital audio refers to the parts of the audio chain in which the audio signal is represented by a sequence of discrete numerical *samples* rather than by a continuous variation in a physical parameter, such as electrical voltage. The primary benefit of a digital representation is that it is robust—a digital signal can in principle be copied and transmitted without loss with much greater immunity to the degrading effects (noise, interference, crosstalk, etc.) commonly associated with analog systems. In addition, a numerical representation enables signal processing that would be more expensive or impractical using analog techniques.

Sampling

Analog audio can be transformed into a digital audio signal using specialized quantization hardware to convert the continuous time-varying amplitude information into a discrete series of numbers, with each related to the instantaneous value of the signal. A digital audio sample represents the instantaneous level of an audio signal at a specified time as a discrete numerical value. A digital audio stream comprises a sequence of such samples taken at regular intervals, a representation known as *discrete time*.

The *sampling theorem* states that a bandwidth-limited signal can be perfectly represented by a discrete time sequence of samples, provided that the rate at which the samples are taken is at least twice the signal bandwidth. An audio signal with a bandwidth of DC to 20 kHz therefore requires a sampling rate of at least 40 kHz to be perfectly represented by a sequence of samples. In practice, a somewhat higher sampling rate is used, to allow spectral room for physically realizable filters to operate. Figure 10.1-27 shows a 3-kHz sine wave with 48-kHz sampling instants indicated by the vertical lines.

Real-world signals are not truly band-limited, and therefore filtering is needed before sampling takes place to remove signal energy above one-half the sampling frequency. If filtering is not performed or is inadequate, tones with frequencies above half the sampling rate will *alias* into the passband, leading to potentially objectionable products in the sampled signal that are not tonally related to the original signal—an objectionable effect, even at low amplitude relative to signals in the passband.

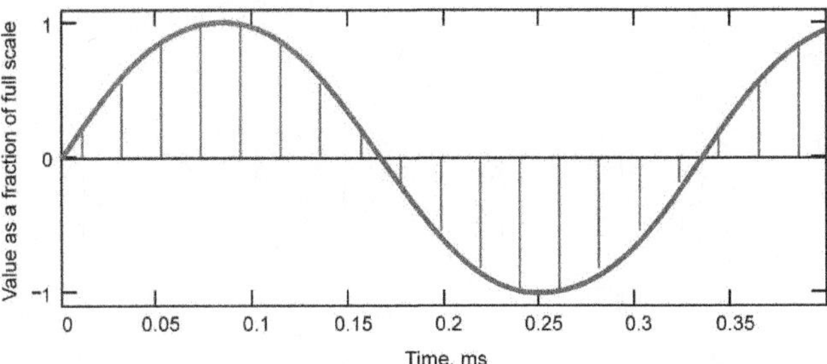

FIGURE 10.1-27 A 3-kHz sine wave with 48 kHz sampling instants shown.

For a true digital representation of audio, it is also necessary to ascribe a discrete amplitude to the instantaneous level of the signal and to represent this amplitude as a numeric value. This process is known as *quantization*. The combined processes of sampling and quantization are performed by an analog-to-digital converter (ADC).

Sampling Rate and Dynamic Range

Professional and broadcast audio typically uses a sampling rate of 48 kHz. This covers the standard 20-Hz to 20-kHz audio bandwidth, with some margin above 20 kHz for filtering (48/2 = 24 kHz). Other sampling rates commonly found in broadcast systems are 32 kHz and 44.1 kHz. Master recordings are sometimes made using 96-kHz and even 192-kHz sampling rates. The 32-kHz sampling rate is adequate for 15-kHz audio often used in broadcast systems, while 44.1 kHz is the consumer sampling rate used for compact discs and other consumer audio media. In a typical broadcast environment, sample-rate converters (SRCs) are sometimes needed to interface digital equipment operating at different sampling rates.

A digital audio system has a theoretical dynamic range that is determined by the number of discrete levels used to quantize the audio samples. In a 16-bit system (the minimum generally accepted as providing high-quality audio), each audio sample is represented as a 16-bit binary word, giving a maximum dynamic range of approximately 98 dB below a full-scale waveform. In practice, the dynamic range is somewhat less because of the combined effects of noise in the analog front end and inherent converter noise. This range does not consider the backoff in signal needed to avoid clipping, nor does it consider the noise contribution of *dither*, a minute amount of noise added deliberately to linearize the digital system with low-level input signals. Professional systems routinely employ 20-bit or 24-bit sampling depth to lower the effects of *quantization distortion* well below the weakest sounds and to allow signal mixing and processing to be added without increasing the quantization distortion to noticeable levels.

Analog-to-Digital Conversion

An ADC must perform the following functions, most or all of which are performed by modern ADCs in a single, integrated circuit:

- Bandwidth limit the input signal to satisfy the sampling theorem;
- Sample the signal at regular intervals;
- Quantize each sample to the nearest of a predefined set of discrete levels;
- Represent the discrete level as a digital word and make it available externally;
- Many ADCs include complex firmware for digital signal processing, such as filtering, equalization, and waveform compression and limiting, which frees a host computer's central processor to handle other tasks.

Typically, the digital words representing the audio signal are made available in real time on the data output of the ADC, for consumption by subsequent devices. These words are represented as pulse code modulation (PCM), a binary format with a fixed word length such as 16 bits or 24 bits. Stand-alone ADCs incorporate a digital transmitter to transmit the digital output words over a standard transport (see "Transporting Digital Audio" later in the chapter).

A sample clock is provided to the converter by an external clock circuit. In broadcast and professional audio, the sampling rate is typically 48 kHz, but other rates may be used, such as 96 kHz or 44.1 kHz. In professional systems, it is often required to lock the sampling rate of the ADC to an external reference, or house sync, so that all converters in the system run at the same rate. Internally, clock synchronization circuitry locks the local clock to the external clock. This process can lead to audible distortion or even total signal loss if there is a problem locking to the house sync. See "Interface Jitter" later in the chapter.

The block diagram of a typical stand-alone stereo ADC is shown in Figure 10.1-28. Most ADCs employ oversampling, which uses a high sample rate at the front end, allowing the use of a simple passive analog low-pass filter for band limiting. The ADC samples and digitizes the analog signal at the oversampled rate and then reduces the sample rate to the desired rate using an internal digital filter and decimator. The output of this block is typically 16- to 24-bits wide. Typically the two channels are multiplexed and serialized onto a single data line, with an accompanying bit clock and frame clock on separate lines. The I^2S standard is often used.

A stand-alone converter must make the audio data available externally for consumption by subsequent devices. Figure 10.1-28 shows one method: the I^2S data and clock signals are combined with auxiliary data and biphase mark encoded for transmission over a cable.

Transporting Digital Audio

A digital audio signal can be transported as a serial bit stream, referred to as the digital interface signal. The AES3 standard specifies the physical, electrical, and data characteristics of a transport for stereo digital audio using balanced connectors and coaxial cable (see Figure 10.1-29). The AES3id information document and SMPTE276M extend AES3 to unbalanced connectors. The consumer S/PDIF standard is similar to AES3id but uses a reduced data voltage and allows the use of optical interconnection (known as "TOSLINK" connectors). These standards share an interface format, with only minor differences in the status data carried alongside the audio information, and therefore are largely compatible with one another at the data layer level.

As shown in the figure, the digital transport signal, or *carrier*, consists of a sequence of *frames*. Each frame consists of two *subframes*, one for each channel. Each subframe consists of the following:

- A unique *preamble* signifying the start of a sample;
- Up to 24 bits of embedded audio data;
- 4 bits of *metadata*: parity, validity, channel status, and user bits.

The audio data and metadata are biphase-mark encoded. This makes the polarity of the signal unimportant, ensures that there is no DC, and allows the sample clock to be recovered at the receiver by maintaining a steady stream of signal transitions on the line. The shortest time between these transitions is known as a *unit interval*, or UI. For the AES3 interface format, there are 128 UIs per sample, so for a sample frequency of 48 kHz, 1 UI is about 163 ns.

The AES/EBU interface (discussed later) is intended to drive cables of up to 100 (AES3) or 1000 (AES3id) meters in length, whereas the S/PDIF wire interface is intended for equipment interconnection within 10 meters.

The channel status bits for each channel form a 192-bit block, delimited by a special *block preamble* that occurs every 192 samples. The first bit of the channel status block specifies whether the format is professional or consumer; the meaning of each bit in the block is different in the professional and consumer standards. It is important that the receiver of the digital bit stream react appropriately to the information in the channel status bits. For example, the audio/nonaudio bit is set when compressed audio is being carried; a DAC should recognize this and mute its output. Further information on the meaning of these bits and the expected response to them can be found in the references.

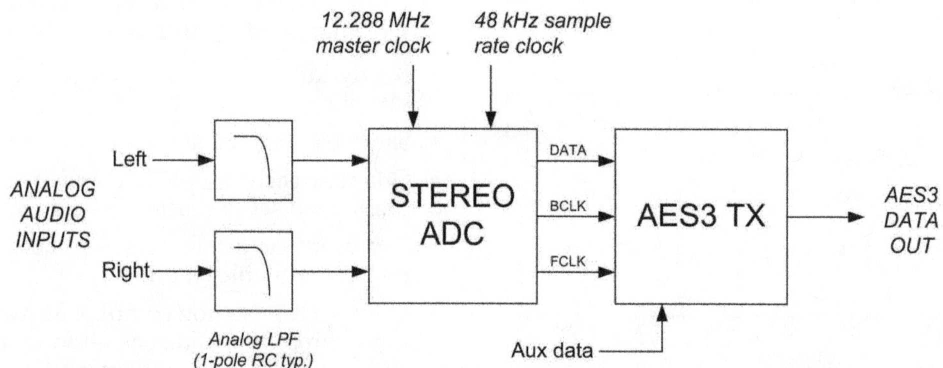

FIGURE 10.1-28 Stereo digital audio encoder (AES/EBU or S/PDIF).

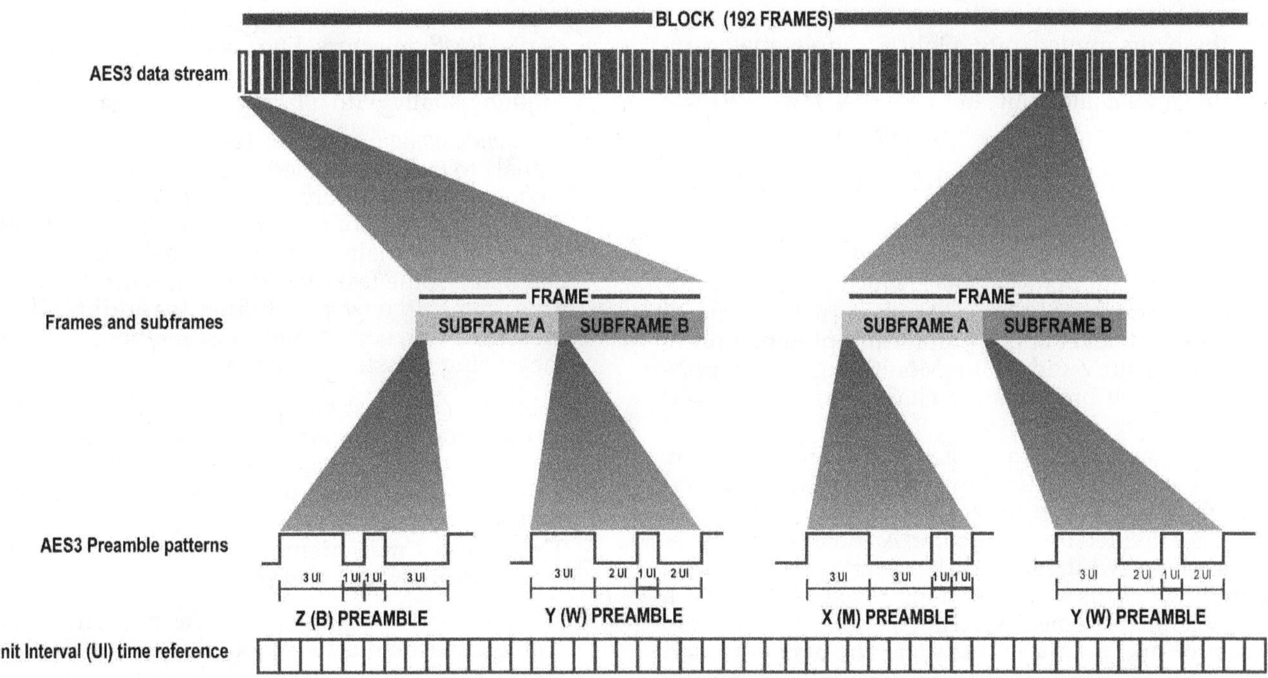

FIGURE 10.1-29 AES3 bit stream.

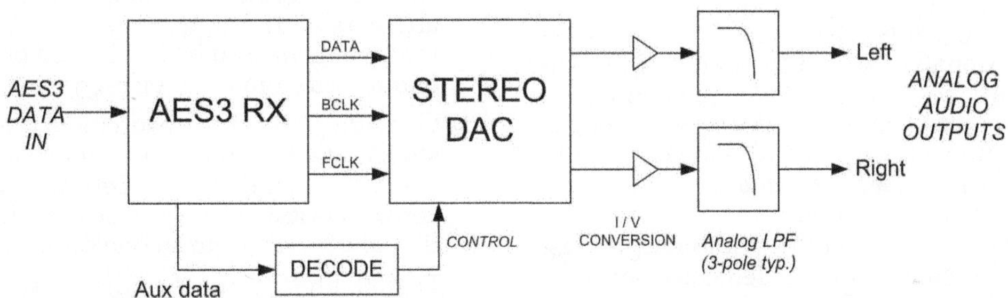

FIGURE 10.1-30 Stereo digital audio decoder (AES/EBU or S/PDIF).

Digital-to-Analog Conversion

A digital audio stream must be converted back to analog form for presentation to the user. This reconstruction is performed by a digital-to-analog converter (DAC). A DAC must perform the following functions, most or all of which are performed by modern DACs in a single integrated circuit:

- Accept a sequence of digital words from an external source;
- Convert each word to an instantaneous analog level (typically an electrical current);
- Bandwidth limit the sequence of analog levels to half the sampling rate.

The last step is necessary because the sampling process creates replicas, or *images*, of the original signal at multiples of the sampling frequency. These images must be removed during reconstruction so that high energy levels above the system bandwidth do not interfere with the operation of subsequent devices in the audio chain.

As with ADCs, the clock is provided to the DAC externally. Stand-alone DACs incorporate a digital receiver to receive the digital input words over a standard transport (see "Transporting Digital Audio," earlier). Standard transports transmit a clock signal with the audio itself; the digital receiver extracts this clock signal from the transmitted signal and feeds it to the DAC. Clock extraction can be a problem if the signal at the DAC is poor. See "Interface Jitter," later in the chapter.

The block diagram of a typical stand-alone stereo DAC is shown in Figure 10.1-30. A standard interface, in this case AES3, is used to convey the digital audio signal. An AES3 receiver conditions the signal to remove any noise incurred during transport and extracts the word clock, sampling clock, and stereo audio data. These are typically made available on three lines in an I2S format. The auxiliary data are also extracted and made available on a separate interface.

A DAC chip receives the stereo I^2S data and converts it to analog form. Most DACs employ *oversampling*, in which a digital interpolator first multiplies the sample rate, and then a digital filter removes images below half the oversampled rate. Only a simple analog filter is then needed to remove the oversampled images from the final analog output.

MONITORING VERSUS MEASUREMENT

Monitoring and metering refer to measurements made on the audio chain when it is carrying a live signal, such as program material, over which the test instrument has no control. Typically, these measurements are limited to level and a graphical representation of the stereo or surround image. Monitoring is an important part of the broadcasting chain, and is discussed elsewhere.

For proper measurement of an audio device or chain, it is necessary to insert a known signal at one point and measure a signal at a downstream point. The known signal is provided by a generator. An analyzer acquires the downstream signal and characterizes it. The term audio analyzer is also commonly used to describe a complete measurement system consisting of an audio generator and an audio analyzer. The generator and analyzer work together to make audio measurements.

DIGITAL AUDIO MEASUREMENT

The measurement of audio characteristics in the analog domain traditionally makes use of an analog generator, such as a sine wave oscillator; analog signal conditioning, such as ranging and low-pass filtering; and analog measurement, such as RMS detection and metering. A dedicated piece of circuitry is typically required for each analog function.

In contrast, the measurement of audio characteristics in the digital domain requires a dedicated software algorithm for each function. Algorithms designed specifically for generating, manipulating, and analyzing signals in the digital domain are known as digital signal processing (DSP) algorithms. The continual decline in the cost of computing power and memory has led to DSP implementations of audio analyzers that are cheaper than their analog counterparts. In addition, DSP makes possible signal generation and analysis methods that cannot be performed using analog techniques, such as the Fast Fourier Transform (FFT), and multitone analysis.

DACs and ADCs extend the reach of DSP into the analog domain. A DSP-based audio analyzer can therefore be used to measure both analog and digital audio systems. The analyzer interfaces to analog devices through DACs and ADCs and to digital devices through digital interface transmitters and receivers. The analyzer can also measure cross-domain devices by using the analog and digital interfaces in combination.

When measuring digital devices, or when using DSP and converters to measure analog or mixed-domain devices, there are two broadly defined approaches:

- *Time domain methods*: Typically, these are DSP implementations of analog techniques, using digital sine wave generators, digital filters, and digital RMS detectors. The speed and accuracy of these methods are similar to the analog versions, though usually with a larger dynamic range;
- *Frequency domain methods*: The FFT allows signals to be transformed efficiently into the frequency domain. Frequency domain processing facilitates certain operations that are impractical in the time domain, such as deconvolution, and can make some tasks easier, such as filtering. This opens up new possibilities for audio measurement, such as very fast methods of measuring transfer functions.

DSP-based audio test equipment (see Figure 10.1-31) may use time or frequency domain methods internally when making traditional measurements. Usually, the user does not need to be aware of the method being used.

The types of measurements performed by a DSP-based audio analyzer can be categorized as follows:

- Traditional measurements, such as level, frequency, and total harmonic distortion plus noise (THD+N) ratio. These measurements can be performed on analog, digital, and mixed-domain devices;
- Nontraditional measurements, such as power spectrum, continuous exponential sweep, and maximum length sequence (MLS). These measurements can be performed on analog, digital, and mixed-domain devices;
- Converter-specific measurements, such as stopband attenuation and jitter sensitivity. These are performed on ADCs, DACs, and sample-rate converters (SRCs, devices designed to convert digital bit streams from one sample rate to another);
- Digital interface measurements, such as bit exactness, jitter, and eye patterns. Although these characterize the digital interface itself rather than the audio signal carried on it, interface problems can be a major cause of audio signal degradation.

FIGURE 10.1-31 Dual-domain analog and digital audio analyzer.

Photo courtesy of Audio Precision

Traditional Measurements Using DSP

The following traditional measurements provide an overview of the performance of an audio device under test (DUT):

- *Level and gain*: The basic level measurement shows the signal level at the output of an audio device. Typically, the level is measured with a root mean square (RMS) detector. Depending on the device, the result might be displayed in volts (analog), decibels full scale (dBFS, digital), or related units. Gain is the ratio of the DUT output level to its input level when the DUT is driven with a sine wave;
- *Total harmonic distortion plus noise (THD+N)*: THD+N measurements have been used for many years as a comprehensive single-value statement of an audio device's performance. The DUT is driven with a sine wave. The output of the DUT contains the sine wave fundamental (possibly changed in level because of DUT gain), together with distortion and noise generated inside the DUT. The fundamental is removed and the level of the residual is measured. The result is usually expressed as a ratio in dB of the residual level to the total level;
- *Frequency response*: This test shows how the gain of a device varies across the audible spectrum when driven with a sine wave. For most devices, a "flat" response curve, in which all frequencies over a specified bandwidth are passed with the same gain, is desirable;
- *Signal-to-noise ratio (SNR)*: A single figure that indicates how "noisy" an audio device is. Two measurements are required: The first measures the DUT output level when driven by a sine wave (usually at either maximum level or at a nominal operating level); the second measures the residual noise in the device. The result is expressed as a ratio in dB of the signal level to the noise level;
- *Crosstalk*: In multichannel DUTs, crosstalk is a measure of signal leakage between channels. All practical devices have some crosstalk. Crosstalk is measured by applying a high-frequency sine wave to at least one channel and measuring the level of the tone in a nondriven channel. The result is expressed as a ratio in dB of the level in the nondriven channel to the level in the driven channel (or channels). Crosstalk typically increases with increasing frequency because of capacitive coupling, so high-frequency tones provide a more sensitive test;
- *Phase*: Phase is a measure of lag or lead of a single tone with regard to a reference tone. Interchannel phase measures the phase relationships between channels in a multichannel device; one channel is chosen as the reference. Absolute phase measures the phase at the output of a device relative to its input. Phase is expressed in degrees.

Nontraditional Measurements Using DSP

DSP makes available measurement methods that cannot be performed using analog circuitry. These methods have gained acceptance in the audio industry because of their accuracy, speed, and resolution. They include:

- *Power spectrum*: Also known as the FFT spectrum, this is a display of the spectral content of a signal. When the DUT is driven with a sine wave, for example, the power spectrum of the DUT output shows the harmonic distortion, noise floor, and hum level of the DUT on the same graph (see Figure 10.1-32);
- Real-time analyzer (RTA): An RTA can be implemented with analog circuitry, but DSP-based systems are much more common. The RTA uses a bank of bandpass filters of constant *Q* (fractional width) to divide the input signal into bands and then measures the level in each band with an RMS detector. The result is typically displayed as a vertical bar chart, with one bar for each band. Third-octave bands are often used, giving 31 bands over the full audio spectrum (DC to 15 kHz). In conjunction with a pink noise generator, an RTA can be used to quickly determine the frequency response of a DUT;
- *Multitone*: A periodic collection of sine tones is applied to the DUT by the generator (see Figure 10.1-33). Simultaneously, the DUT output is acquired by the analyzer. An FFT is performed on the acquired signal. The FFT is post-processed in multiple ways to compute the frequency and phase responses (using the FFT bins that contain the generator tones), the distortion performance (using the FFT bins that do not contain tones), and the noise performance (by using an acquisition of twice the generator length, leading to FFT bins that contain only noise generated in the DUT). A multitone has a dissonant but not unpleasant quality and typically lasts for less than 250 ms.

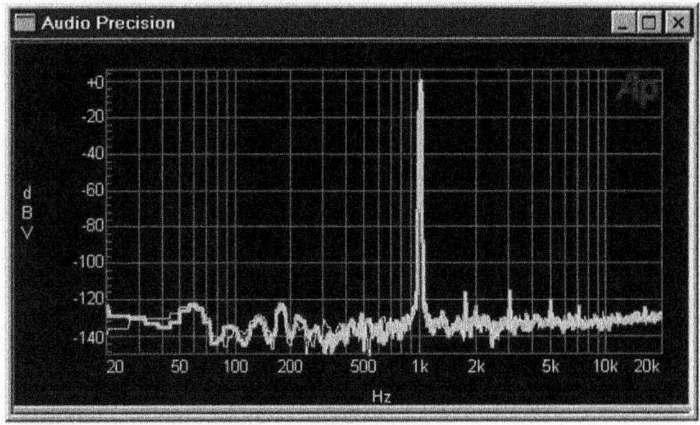

FIGURE 10.1-32 Display of power spectrum for two channels. Stimulus is a 1-kHz sine wave—note visibility of noise (at 120 dB below the stimulus), hum (small peaks at 60, 120, 180 Hz), and harmonic distortion (small peaks at 3, 5, 7 kHz).

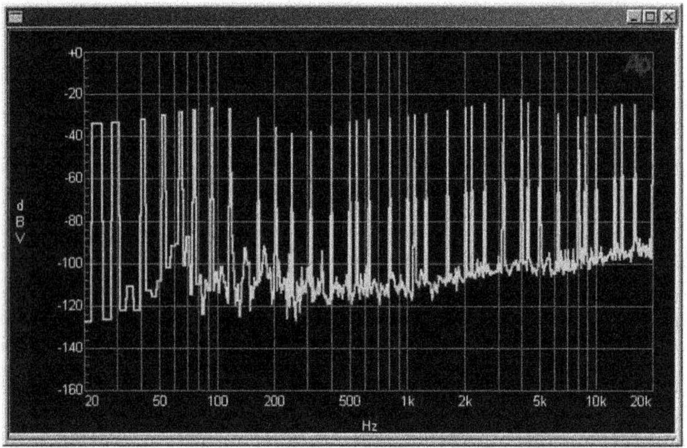

FIGURE 10.1-33 Power spectrum of a DUT output using a 36-tone multitone stimulus. The level variation across the tones shown here reflects the DUT frequency response. Phase, distortion, noise, and interchannel phase results can also be calculated from this one brief acquisition.

Photo courtesy of Audio Precision

Multitone testing is becoming important for testing the quality of digitally coded audio systems. This is because the encoding and decoding of a sinusoid may not be challenging to the channel, but the simultaneous coding of multi-sine signals may produce distortion products at significantly higher levels. One version of this method uses 1/3 octave spaced sinusoids from $20f_L$ to $fs/2$ (f_L is the lower frequency limit of the system in Hz and fs is the system bandwidth in Hz). This produces a high crest factor that is demanding of system linearity. However, care should be exercised in arranging the phases to produce a known and consistent crest factor. ITU-T Rec. O.81 provides a procedure for phases;

- *Continuous exponential sweep*: A method for measuring the linear and harmonic nonlinear impulse responses of a system. A sine tone whose frequency increases exponentially in time is applied to the DUT, while the DUT output is acquired (see Figure 10.1-34). A mathematical deconvolution operation recovers the linear impulse response of the DUT and its harmonic responses independently. The linear impulse response can be transformed into the frequency domain to recover the frequency and phase responses. The harmonic responses can also be transformed to recover the harmonic frequency responses, or power summed to recover the THD response. If desired, the impulse response can be truncated before frequency transformation to reject room reflections in acoustic measurements.

Continuous sweep techniques are very fast; a commercially available analyzer system has demonstrated the ability to make 14 different measurements using data acquired in a 1-second sweep. The sweep has a chirping sound, with a typical duration of 1000 ms or less;

- *Close-tone testing with two-sine signal, [ITU-R (formerly CCIF) method]* Unlike the SMPTE intermodulation test discussed earlier, this method uses high-frequency sinusoids of equal amplitude that are separated by a frequency difference of less than 10%, such as 13 kHz and 14 kHz. Nonlinear intermodulation distortion is characterized by the appearance at the output of frequencies that are linear combinations of the fundamental frequencies in the input signals. The primary intermodulation component in this case is 1 kHz, the frequency difference between the tones (the 27-kHz sum component is beyond audibility and is usually not considered). Harmonic components are also present but are not measured for intermodulation distortion. This distortion test method is easy to generate and measure, in comparison with the multitone method, but is still a demanding test of channel performance;
- *Time delay spectrometry (TDS)*: Like RTA, TDS can be implemented with analog circuitry, but it is more commonly done with DSP. A sine tone whose frequency increases linearly in time is applied to the DUT. The output of the DUT is applied to a bandpass filter whose center frequency follows the generator frequency. An RMS detector measures the output level of the bandpass filter. A plot of the RMS level against generator frequency reveals the frequency response of the DUT. TDS is popular in acoustic measurement, where the tracking bandpass filter is set to lag the generator by a fixed time; this allows the direct radiation from the acoustic source to be measured while rejecting room reflections, which arrive later and are rejected by the filter. The TDS signal has a chirping sound and typically lasts for several seconds;
- *Maximum length sequence (MLS)*: A method for measuring the impulse response of a system (and therefore its frequency and phase response). A pseudo-random signal is applied to the DUT, while the DUT output is acquired. A mathematical deconvolution operation is performed between the acquired and generated signals. This extracts the impulse response of the DUT. The impulse response can be transformed into the frequency domain to recover the frequency and phase responses. MLS is popular in acoustic measurement, since the recovered impulse response can be truncated before it is frequency transformed, thus preventing room reflections from corrupting the measurement. The MLS sounds noise-like and typically lasts for around 500 ms.

Converter-Specific Measurements

ADCs and DACs, being hybrid analog/digital devices, exhibit distortions and anomalies not seen in purely analog equipment. To a large extent, their performance can be measured using the traditional and nontraditional measurements described previously. However, further measurements may be needed to characterize them fully. For ADCs, these measurements include:

- *Anti-alias filter stopband attenuation*: Tones at frequencies above half the sampling rate must be

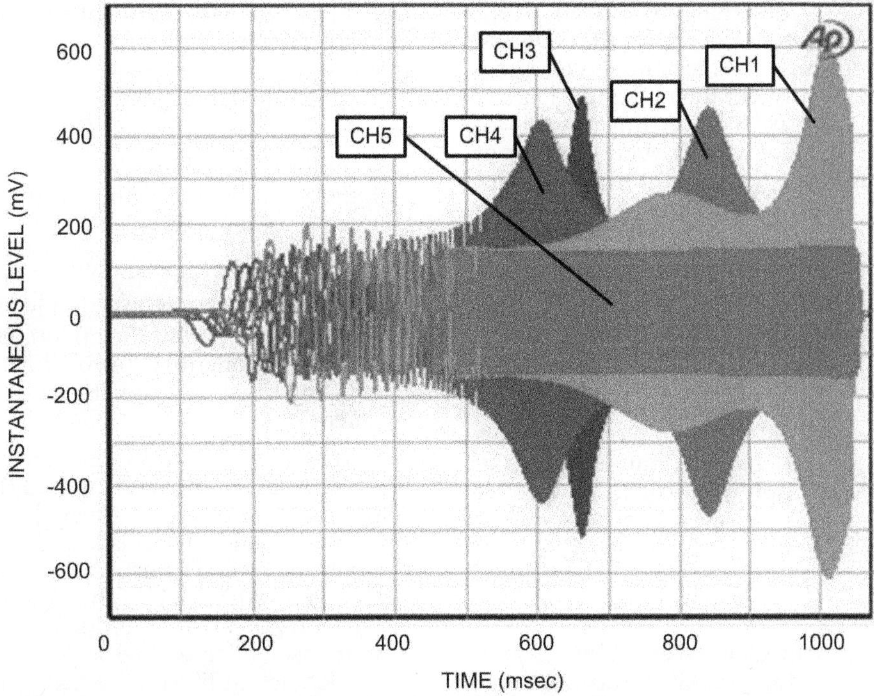

FIGURE 10.1-34 A multichannel continuous sweep acquisition, unprocessed. Each channel was subjected to different equalization, resulting in the modulation envelopes shown. Further processing of this 1-second acquisition can provide impulse response, level, frequency response, interchannel phase, distortion, group delay, and other results.

filtered out before sampling; otherwise, aliasing will occur. In this test, a full-scale analog sine wave is presented to the ADC, and the digital output level is measured as the sine wave frequency is varied. Tones in the stopband should exhibit very high attenuation;

- *Idle channel noise spectrum*: Some ADCs produce audible tones at low level when the input signal is small or nonexistent. A high-resolution FFT of the output spectrum of the ADC with the input grounded can show these tones;
- *Jitter modulation*: Jitter is variation in timing of clock edges from their nominal positions. If there is jitter on the sample clock, distortion is introduced in the sampling process, and this distortion is measurable in the digital output. Sample clock jitter is most likely to occur if the ADC is synchronized to an external clock. The external clock itself may have jitter, or the clock synchronization circuitry inside the ADC may introduce jitter of its own. An FFT of the output spectrum of the ADC when the input is driven with a sine wave can show jitter sidebands around the fundamental tone.

Measurements specific to DACs include

- *Reconstruction filter ripple and image rejection*: The images of the original signal are typically removed with a digital filter. This filter has a nonideal response that exhibits ripples in the passband and noninfinite attenuation in the stopband. To measure passband ripple, a full-scale digital sine wave is presented to the DAC, and the analog output level is measured as the sine wave frequency is varied. To measure stopband attenuation, the analog output level is measured at the known image frequencies. The passband ripple should be very low, and the stopband attenuation should be very high.

Digital Interface Measurements

The measurements described so far are concerned with the quality of the audio signal itself. It is also possible to measure characteristics of the embedded data (bit exactness) and of the digital bit stream or interface carrier signal. Errors in the data or degradation of the carrier can both cause distortion or failure in recovery of the embedded audio.

In particular, the carrier is subject to influences (primarily noise and high-frequency line loss) that can result in jitter at the receiver. A moderately jittered carrier can cause distortion in the decoded audio; in the worst situations, it may not be possible to convert the digital signal back to analog form.

Bit Exactness

A digital storage or transmission medium must be able to store or convey embedded audio without introducing errors. Digital storage and broadcast systems make use of error correction to detect and correct the inevitable errors that arise from dropouts, transmission interference, and the like. Digital transport mechanisms such as AES3, however, use a simple parity scheme to protect the embedded audio signal. The scheme used in AES3 is not capable of correcting errors, or of detecting an even number of errors in a single subframe. It is therefore important to verify that a digital link of this type is performing properly.

In a test for bit exactness, a prescribed sequence of digital samples is transported over the medium to be tested. At the receiving end, the recovered sequence is compared to the known transmitted sequence, and any errors are reported. The sequence is often pseudorandom to allow it to be generated easily and to ensure that all of the bits in the audio word are exercised. The transmitter and receiver must share the sequence generation function but do not need to be synchronized.

Interface Jitter

A digital audio receiver must recover a clock signal from the bit stream it receives. In AES3, the clock signal is recovered from the transitions in the bit stream; typically, the zero crossing of a transition is designated as the transition time. Although the transitions have a short rise time at the transmitter, the rise time at the receiver is longer because of high-frequency roll-off in the transmission line. This nonzero rise time introduces the possibility that the time at which the digital receiver detects the transition differs from the true transition time (see Figure 10.1-35). This can occur in several ways:

- Noise on the transmission line causes the transitions to cross through zero earlier or later than the true transition time. This is known as noise-induced jitter and is illustrated in Figure 10.1-36. In this figure, the bottommost trace represents the amalgamation of the bit sequences above, where each of these sequences has been "smeared" by the effect of the rise time of the transmission line; that is, the transition from 1 to 0 and 0 to 1 is no longer sharp but is now an exponential curve. The highlighted zero crossing illustrates the jitter which results from this smearing;
- In transmission lines where the rise time is comparable to the unit interval (UI), the position of the zero crossing is affected by the previously transmitted bits. This is known as intersymbol interference and affects even noise-free lines (see Figure 10.1-37). In the figure, a jittery clock signal has been captured using a signal monitor that has available a stable, jitter-free reference clock. Each individual waveform in the figure represents a successive trace of the (jittery) signal under test—the fact that the zero crossings do not line up is evidence of jitter. Note that the zero-crossing extremes of this group of waveforms are indicated by the "a" and "b" markers; jitter is often characterized by the differences in the zero-crossing extremes and, for this case, is about 300 nanoseconds.

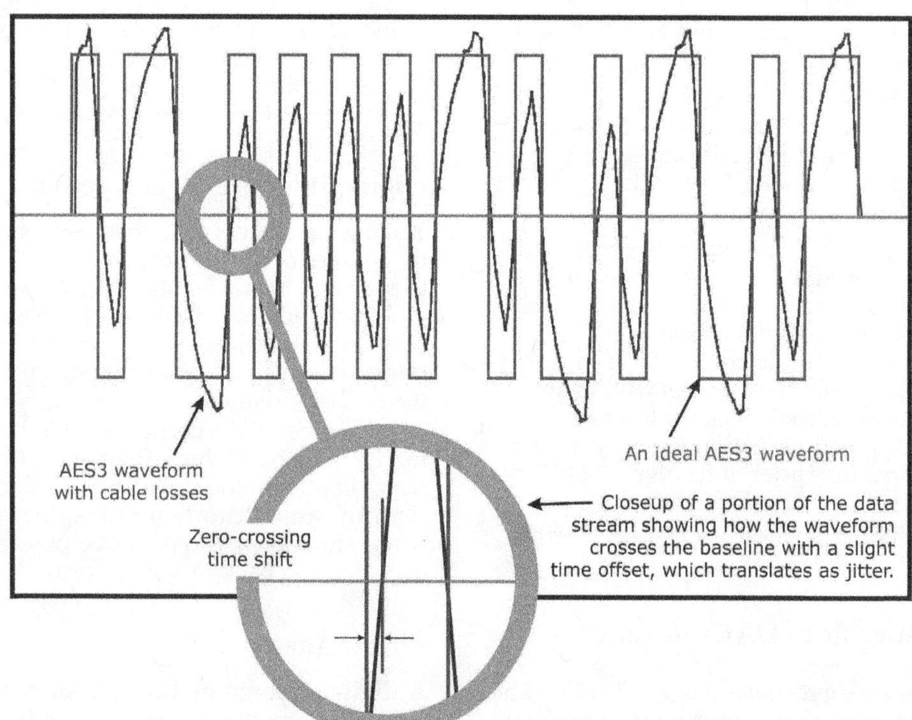

FIGURE 10.1-35 The transitions of the ideal and received waveforms differ because of cable losses, resulting in jitter.

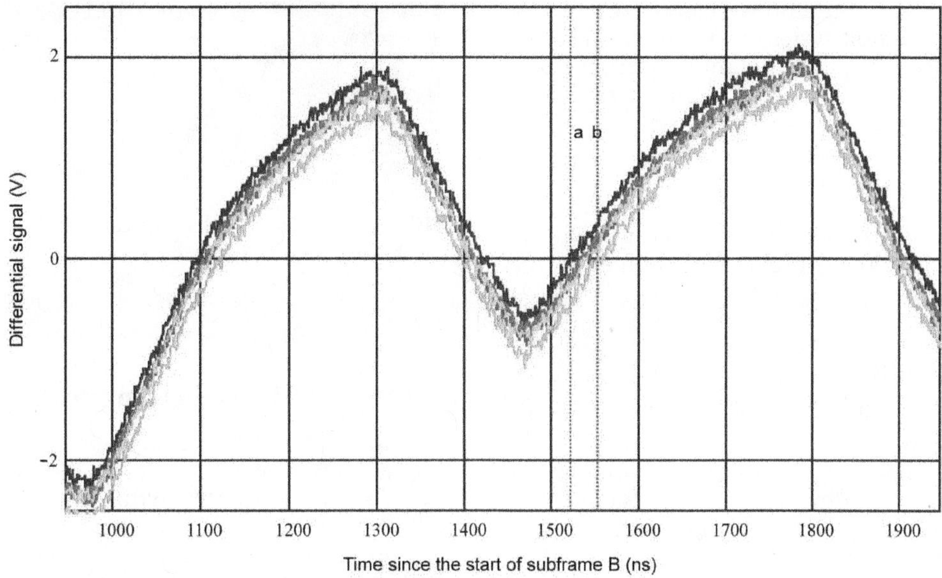

FIGURE 10.1-36 Noise-induced jitter.

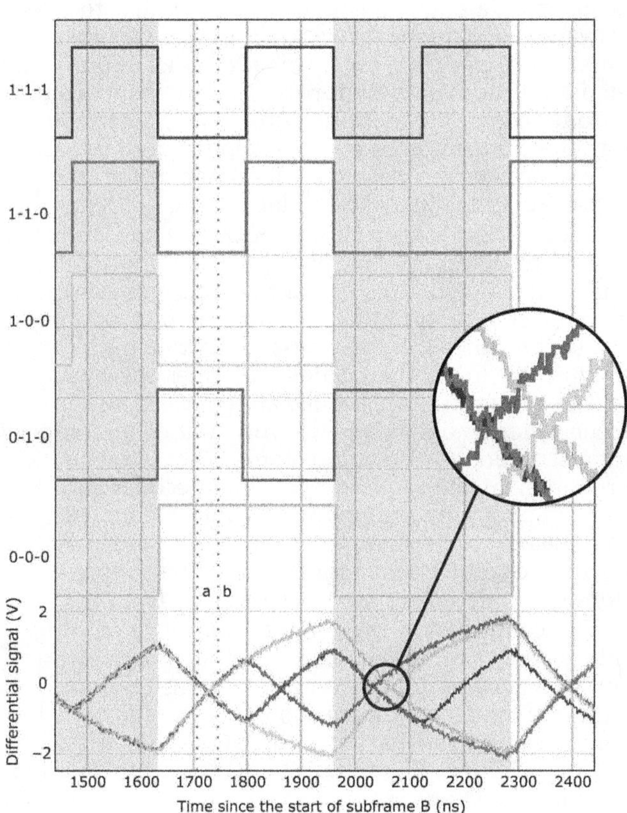

FIGURE 10.1-37 Intersymbol interference and the resultant jitter.

Therefore, because of impairments in the transmission line, the clock recovered at the receiver will contain some jitter. Digital receivers are designed to pass jitter below a certain frequency and attenuate jitter above that frequency. This characteristic allows the receiver to follow slow variations in sample rate that occur intentionally at the transmitter while rejecting high-frequency variations characteristic of line-induced jitter. The AES17 standard specifies the amount of jitter that a digital receiver must tolerate.

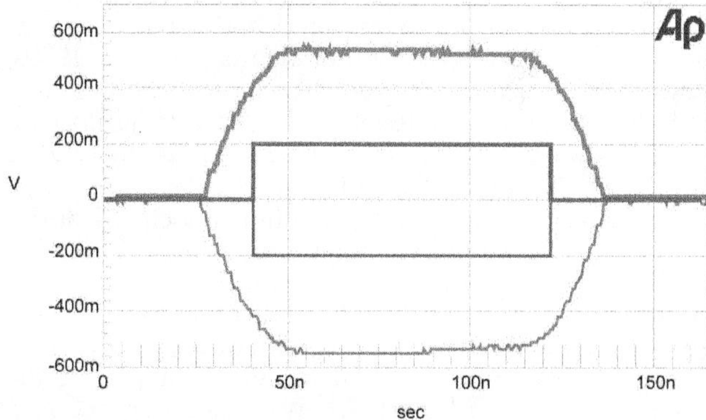

FIGURE 10.1-38 Eye pattern. The small box in the center is the AES17 minimum eye specification; the trace surrounding it is the eye pattern created from many acquisitions of a measured interface signal.

Eye Pattern

If noise or roll-off in a transmission line is severe, the clock (and therefore the embedded audio) may be partially or completely unrecoverable, leading to signal errors and dropouts. The eye pattern test allows this problem to be diagnosed quickly (see Figure 10.1-38). The signal at the receiving end is acquired for many UI (typically tens of thousands) and plotted on a 1 UI-wide graph, superimposing the many UI traces to show the worst case for pulse width, timing, and voltage swing.

Jitter causes the eye to close horizontally, whereas noise causes the eye to close vertically. The AES17 standard specifies the minimum size the eye must have for the transmission system to be deemed acceptable. If the eye closes beyond this minimum size, it is likely that data errors will occur.

ADDITIONAL TOPICS

Some additional information relevant to digital audio signal analysis not covered previously is given in the following subsections.

Multichannel Audio

Throughout the first half-century of sound technology, recorded and transmitted audio was almost exclusively monaural (single channel). Audio generators, meters, and analyzers were also monaural. Two-channel stereo arrived in recording and film sound in the 1950s, spread to FM radio in the 1960s, and was added to television in the 1980s. Audio analyzers took longer to adopt stereo; in many cases, a single-channel tester switched between stereo channels was adequate. However, some important measurements, such as crosstalk and interchannel phase, required both channels to be stimulated or analyzed simultaneously. By the 1980s, cutting-edge analyzers were stereo, and the best test and measurement systems could perform almost all the tests needed on two channels simultaneously. This brought the added benefit of reducing test time.

Alternative languages and other associated audio services are driving both television and radio broadcasters to support the transport of multichannel audio within their facilities. In addition, multichannel surround sound is established in film sound and DVDs and has been adopted by digital television standards. These so-called 5.1 and 7.1 systems offer six and eight channels of audio, respectively, with the ".1" referring to a low bandwidth channel intended for low-frequency effects in film soundtracks. The large channel counts of these systems present a monitoring and testing challenge.

When these systems are tested with stereo analyzers, it is necessary to be able to choose which pair of channels to test at any one time. It is also desirable to automatically test all the channels without moving cables. An output switcher is a demultiplexing device that allows the generator signal to be routed to any of the DUT channels. An input switcher is a multiplexing device which allows any of the DUT channels to be routed to the analyzer. Under software control, these devices allow automated testing of multichannel systems on conventional analyzers.

However, measurement of crosstalk requires stimulation or analysis of more than two channels simultaneously. In addition, the time required to test eight channels two at a time may be prohibitive. Multichannel audio analyzers are now available with the capability to generate and analyze eight or more audio signal channels simultaneously (see Figure 10.1-39). Such analyzers do not require input or output switchers.

FIGURE 10.1-39 Multichannel (eight input, eight output) audio analyzer.

Photo courtesy Audio Precision

Psychoacoustic Data Compression

In audio delivery channels such as DVD media and digital television and radio broadcast, it is desirable to maintain audio quality while minimizing bandwidth, to allow more content or more channels. An AES bit stream with a 48- kHz sampling rate has a bit rate of 6.144 Mbps and a special content that extends beyond 20 MHz [16], which is too great to be practical for direct delivery over most wireline and radio links. Psychoacoustic data compression can be exploited to avoid transmitting data that is inaudible or of low perceptual importance (see Chapter 4.8 for additional information on digital audio data compression). A perceptual audio encoder converts the input signal into a data-reduced form for transmission, and a perceptual audio decoder at the receiver reconstructs a facsimile of the original signal from the data-reduced bit stream.

The encoder uses a model of the human auditory system to determine the perceptually important content of the signal. This content is then represented in a compact way, typically making use of a frequency transform. Since the transmitted data rate (or bit rate) is typically fixed, the encoder allocates the available bits to minimize the audible difference between the reconstructed signal and the original. The compact representation of the audio signal is then transmitted or stored. At the receiving end, the decoder converts the compact representation back into a conventional (or linear) audio stream. The encoder and decoder must agree on the meaning of the compact representation so that decoding can proceed correctly.

The compression ratio is the ratio of the input bit rate to the compressor to the output bit rate. Compression systems for delivery to the consumer, such as Dolby® Digital, maximize compression ratio, whereas compression systems for transport within a facility, such as Dolby® E, maximize robustness to multiple encode/decode cycles.

A compressed digital audio signal may be carried on an AES3 connection. An extra layer of protocol, defined by the IEC61937 standard, is used to identify the signal to the receiver as compressed rather than linear.

Most audio test equipment generates and analyzes linear (uncompressed) signals. It may therefore be necessary to convert between uncompressed and compressed formats for testing. When testing a transport designed to carry Dolby E audio, for instance, the stimulus signal must be encoded with a Dolby E encoder before applying it to the transport. At the receiving end, the signal must be decoded with a Dolby E decoder before it can be accepted by the audio analyzer.

Dolby E

Digital audio can be encoded for transport using Dolby E compression, which uses perceptual coding techniques to allow one AES3 connection to carry up to eight channels of audio information in the IEC61937 layer. Dolby E sacrifices compression ratio to allow multiple encode/decode cycles without audible degradation.

Dolby Digital

Digital audio can be encoded for distribution to end users using Dolby Digital (AC-3) compression. Dolby Digital allows one AES3 interconnection to carry 5.1 or 7.1 channels of surround sound in the IEC61937 layer. The ".1" refers to a low bandwidth channel, typically used for low-frequency effects in movie soundtracks.

Embedding

An AES3 transport carries two channels of linear PCM; multiple AES3 connections must be used to carry more than two channels. Alternatively, multichannel digital audio may be transported over a single connection using embedding techniques. In embedding, multiple linear digital audio signals are placed in a high-bandwidth bit stream to be conveyed over a single connection. The connection may also carry other signals simultaneously, such as digital video and metadata.

Most audio test equipment makes use of standard audio interfaces such as S/PDIF. To interface to an embedded system, it is necessary to use an embedder at the generator end and a de-embedder at the analyzer end. If the embedded system is carrying audio in a compressed format, it may be necessary to daisy-chain equipment. For example, when the audio quality of an SDI transport carrying Dolby E compressed audio is tested, it is necessary to first compress the digital output of the audio analyzer using an outboard Dolby E encoder and then use an SDI embedder to embed the Dolby E data into the SDI bit stream. At the receiving end, an SDI de-embedder extracts the Dolby E data, and a Dolby E decoder converts it to linear digital. This digital bit stream can then be accepted by the audio analyzer.

SDI

In television studios, digital audio is often embedded with video for transport within a serial digital interface (SDI) signal. For standard definition (SD) video, the SDI format is specified in SMPTE 259M; SMPTE 292M specifies HD-SDI for high-definition (HD) video. The serial interfaces provide for eight AES3compatible 48-kHz 20- or 24-bit audio data streams (each with two channels of audio), for a total of 16 linear digital channels of audio information.

SUMMARY

Radio and TV broadcast engineers alike need to be aware of the principles and techniques involved in the analysis and characterization of audio signals. This chapter has provided an introduction into these topics for both analog and digital audio systems. DSP-based audio analyzers are a powerful tool for characterizing both analog and digital audio, accommodating:

- Traditional measurements, such as level, frequency, and total harmonic distortion plus noise (THD+N) ratio;
- Nontraditional measurements, such as power spectrum, continuous exponential sweep, and maximum length sequence (MLS);
- Converter-specific measurements, such as stopband attenuation and jitter sensitivity;
- Digital interface measurements, such as bit exactness, jitter, and eye patterns.

Digital audio does not degrade gradually, as is the case with analog, but suddenly and annoyingly. Having and regularly using proper test equipment will identify performance that is marginal before it fails and becomes difficult to find among the many components in a digital audio facility.

References

[1] Graeme, J., *Applications of Operational Amplifiers*, New York: McGraw-Hill, 1973, pp. 132–139, 202–203.

[2] McKnight, J. G. (Jay),"Some Questions and Answers on the Standard Volume Indicator ('VU meter')", 2006, www.aes.org/aeshc/pdf/mcknight_qa-on-the-svi-6.pdf.

[3] Harry, D., "Audio Program Analysis," as presented to the National Association of Broadcasters, April 1985.

[4] IEEE Standard: Recommended Practice for Audio Program Level Measurement, DOC. G-2.1.2/13, 1988.

[5] Recommendation of International Telecommunications Union Radiocommunication Sector, "Algorithms to Measure Audio Programme Loudness and True-Peak Audio Level", ITU-R BS.1770–4, 2015, www.itu.int/dms_pubrec/itu-r/rec/bs/R-REC-BS.1770-4-201510-I!!PDF-E.pdf.

[6] "Audio Levels and Loudness", Report ITU-R BS.2054–1, 2008, www.itu.int/dms_pub/itu-r/opb/rep/R-REP-BS.2054–1—2008-PDF-E.pdf.

[7] Skritek, P., and H. Pichler, "Extended Application of T.I.M. Test Procedures," preprint no. 1557, 64th AES Convention, November 1979.

[8] Takahashi, S., and S. Tanaka, "A New Method of Measuring Transient Intermodulation Distortion: A Comparison with the Conventional Method," preprint no. 1539, 64th AES Convention, November 1979.

[9] National Semiconductor Corporation, Santa Clara, CA: National Semiconductor Corporation, *Audio/Radio Handbook*, 1980, pp. 2.3–2.10.

[10] See reference 11.

[11] Rane Corporation, "Pro Audio Reference", on the web at www.rane.com/par-w.html, "weighting filters" entry.

[12] National Association of Broadcasters, Magnetic Tape Recording and Reproducing (NAB-2/65), 1965, Figure 6, p. 23.

[13] Dolby, R., D. Robinson, and K. Gundry, "CCIR/ARM: A Practical Noise-Measurement Method," *Journal of the Audio Engineering Society*, March 1979, vol. 27, pp. 149–157.

[14] Mendenhall, G., "The Composite Signal-Key to Quality FM Broadcasting," Broadcast Electronics.

[15] Graeme, J., *Designing with Operational Amplifiers*, New York: McGraw-Hill, 1977, pp. 76–77, 251–253.

[16] Robin, M., "The AES/EBU Digital Audio Signal Distribution Standard," Technical Notes—1, Miranda Technologies.

Bibliography

AES3–1992, "Recommended Practice for Digital Audio Engineering—Serial Transmission Format for Two-Channel Linearly Represented Digital Audio Data," *Journal of the Audio Engineering Society*, vol. 40, no. 3, pp. 147–165 (June 1992) www.aes.org

AES-3id–1995, "AES Information Document for Digital Audio Engineering—Transmission of AES3 Formatted Data by Unbalanced Coaxial Cable," *Journal of the Audio Engineering Society*, vol. 43, no. 10, pp. 827–844 (October 1995)

AES17, "AES Standard Method for Digital Audio Engineering—Measurement of Digital Audio Equipment," *Journal of the Audio Engineering Society*, vol. 46, no. 5, pp. 428–447 (May 1998)

Dunn, Julian, *Measurement Techniques for Digital Audio*, Audio Precision, Beaverton, Oregon (2002)

EBU Recommendations (all available at https://tech.ebu.ch/loudness):

"Loudness normalisation and permitted maximum level of audio signals", R 128, August 2011;

"Loudness parameters for short-form content", R 128s1, January 2016;

"Loudness metering specification", Tech 3341, January 2016;

"'EBU Mode' metering to supplement EBU R 128 loudness normalisation", Tech 3342, January 2016;

"Production guidelines", Tech 3343, January 2016; and

"Distribution & reproduction guidelines", Tech 3344.

IEC-60958–1:1999, "Digital Audio Interface—Part 1: General," International Electrotechnical Commission, Geneva (1999)

IEC-60958–3:1999, "Digital Audio Interface—Part 3: Consumer Applications," International Electrotechnical Commission, Geneva (1999)

IEC-60958–4:1999, "Digital Audio Interface—Part 4: Professional Applications," International Electrotechnical Commission, Geneva (1999)

IEC-61937, "Digital Audio—Interface for Non-linear PCM Encoded Audio Bitstreams Applying IEC 60958," First Edition, International Electrotechnical Commission, Geneva (2000)

Metzler, Robert, *Audio Measurement Handbook*, Audio Precision, Beaverton, Oregon (1993) www.ap.com

SMPTE 276M-1995: for Television, *Transmission of AES/EBU Digital Audio Signals Over Coaxial Cable*, Society of Motion Picture and Television Engineers, White Plains, NY, USA (1995) www.smpte.org

CHAPTER

10.2

AM Radio Performance Measurements

DAVID P. MAXSON

Isotrope, LLC
Medfield, MA

INTRODUCTION

AM radio stations are required to manage the spectrum occupied by their AM broadcast signals.[1] The FCC requires at a minimum that AM stations perform annual equipment performance measurements (EPMs) of their spectral occupancy. The rules for documenting spectral occupancy are more complex than many realize, resulting in an industry-wide tendency to make compliance measurements that are not entirely consistent with the rules. In particular, while it may already be challenging for some AM stations to identify measurement locations that are free of unwanted noise and interference, directional AM stations have an added burden of seeking out measurement sites that conform to the FCC reference levels for such measurements. In this chapter, the underlying rules are explained, and methods of measuring compliance are reviewed.

Under FCC regulations in 47 CFR §73.1590 and §73.44, AM broadcasters must maintain their emissions within limits described by a "spectral mask." Emissions of the station under test must not exceed the level set by the mask across the frequencies surrounding the station's center frequency. Farther away from the center frequency, spurious emissions, including but not limited to harmonic emissions, must also comply with the mask.

This chapter is organized with a definition of key spectrum-occupancy terms appearing first, to set the stage for a discussion of the regulations and the mask. The FCC regulations are summarized and explained.[2] Particular attention is paid to selecting a geographic location for EPMs, which can be a rather complex process that many directional stations tend to overlook. After sussing out how to select a measurement location, the methods of taking measurements are explored. To take measurements with a spectrum analyzer, it is valuable to understand how to specify a spectrum analyzer that is capable of making EPMs. Typically, stations rely on both a spectrum analyzer and a field intensity meter to complete the measurements. Causes of noncompliant EPMs are discussed.

DEFINITION OF TERMS

Measurement of the spectral occupancy of radio frequency (RF) transmissions is based on some common terminology adopted worldwide. FCC regulations in part §2.1 include the definitions here, adopted by the International Telecommunications Union (ITU). First are two ways of characterizing the bandwidth of an emission:

- *Occupied bandwidth.* The width of a frequency band such that, below the lower and above the upper frequency limits, the mean powers emitted are each equal to a specified percentage $\beta/2$ [half-bandwidth] of the total mean power of a given emission. Note: Unless otherwise specified in an ITU-R Recommendation for the appropriate class of emission, the value of $\beta/2$ should be taken as 0.5%.
- *Necessary bandwidth.* For a given class of emission, the width of the frequency band that is just sufficient to ensure the transmission of information at the rate and with the quality required under specified conditions.

Two classes of unwanted/unnecessary emissions are defined.

- *Out-of-band emission*. Emission on a frequency or frequencies immediately outside the necessary bandwidth that results from the modulation process, but excluding spurious emissions.
- *Spurious emission*. Emission on a frequency or frequencies that are outside the necessary bandwidth and the level of which may be reduced without affecting the corresponding transmission of information. Spurious emissions include harmonic emissions, parasitic emissions, intermodulation products and frequency conversion products but exclude out-of-band emissions.

There are two "domains" that refer to the portions of the spectrum where out-of-band and spurious emissions are most likely to be found, respectively. This chapter refers to these domains.

- *Out-of-band domain (of an emission)*. The frequency range, immediately outside the necessary bandwidth but excluding the spurious domain, in which out-of-band emissions generally predominate. Out-of-band emissions, defined based on their source, occur in the out-of-band domain and, to a lesser extent, in the spurious domain. Spurious emissions likewise may occur in the out-of-band domain as well as in the spurious domain.
- *Spurious domain (of an emission)*: The frequency range beyond the out-of-band domain in which spurious emissions generally predominate.

The term "occupied bandwidth measurements" is often used to describe, informally, one aspect of an EPM. However, by definition, an occupied bandwidth measurement would simply determine the upper and lower frequency bounds within which 99% of the transmitted signal power falls. A particular radio signal's occupied bandwidth under normal modulation might be described as, for example, ±9.4 kHz, while the assigned channel of the signal might be ±10 kHz. The occupied bandwidth measurement shows whether there is too much undesired energy outside the assigned channel of the radio signal. In the environment of AM frequency assignments, AM channels overlap considerably, prompting the need to look more closely at the spectral density of the transmitted signal.

Because §73.44 defines a spectral "mask" (also called a limit line), it is more sophisticated than a traditional occupied bandwidth measurement. Rather than encapsulating the 99% power spectrum in a simple bandwidth number, the mask defines the "power spectral density" (or "PSD") the station is allowed to emit across the necessary bandwidth, through the out-of-band domain and in the spurious domain.

Power spectral density is a description of the amount of received signal power per unit bandwidth across the spectrum of interest. The fundamental SI units of PSD are W/Hz. Typically, measurements are taken with a defined resolution bandwidth (300 Hz in the AM mask case) and power is typically displayed in decibels with respect to one milliwatt (dBm). The resulting PSD units employed in this chapter are dBm/300 Hz. The spectrum analyzer display will simply show dBm and a corresponding resolution bandwidth (RBW). A graph of PSD is produced across the frequency span of interest.

The PSD measurements are then compared to a reference level. The difference between the reference level (the AM "carrier" level) and each point on the PSD graph is reported in decibels with respect to carrier power (dBc). While dBm indicates absolute power levels based on the units of milliwatts, dBc indicates the difference (as a ratio) between two power levels and is unitless. Spectral occupancy is limited by the §73.44 mask and reported in units of dBc/300 Hz.

Reverting to the FCC/ITU terminology, the necessary bandwidth of the AM signal is that from center frequency to the highest modulating frequency. An AM signal is simple that way. The highest modulating frequency that is intended to be heard by the listener (10 kHz, typically) sets the bounds of the necessary bandwidth. Nominally, an AM broadcast signal's necessary bandwidth would encompass 20 kHz, from center frequency (f_c) ±10 kHz.

In a September 1986 paper,[3] Harrison Klein, PE, explored audio clipping, and to a lesser degree AM carrier pinch-off, as sources of "splatter" in the AM broadcast band. "Splatter" is the presence of significant unwanted emissions in the necessary bandwidth and in the out-of-band domain resulting from flaws in the modulation process. An occupied bandwidth measurement of a station with splatter might or might not reveal there is excessive energy outside the necessary bandwidth. However, a mask measurement would be more sensitive to splatter than an occupied bandwidth measurement because it makes the splatter "visible" in the frequency domain.

A comparison of the PSD of an AM signal with unclipped and clipped audio is provided in Figure 10.2-1.

These spectral density plots show how the excessive modulation caused by audio clipping appears. Excessive energy even within the necessary bandwidth of the signal can contribute to polluting the overlapping adjacent channels with unnecessary noise. For example, comparing the spectral densities for unclipped and clipped audio modulation in the 5- to 10-kHz offsets, the splatter components are some 10–20 dB higher than with the unclipped spectrum. This work by Klein et al contributed to the creation of the AM mask.

Finally, returning to the definitions, note the distinction between out-of-band emissions and spurious emissions. In a general sense, out-of-band emissions fall near the assigned channel of the emitter and are generally construed to be products of modulation of the carrier. Out-of-band emissions typically can be generated by "overmodulation," which in an AM broadcast plant can be produced by allowing audio frequencies higher than 10 kHz to modulate the carrier or by audio clipping, RF pinch-off and other nonlinear functions in generating the signal.

Spurious emissions, on the other hand, are generally considered to be farther from the center frequency and may tend to appear as stand-alone artifacts in the RF spectrum. Spurious emissions in an AM broadcast are not modulation-dependent and can result from such

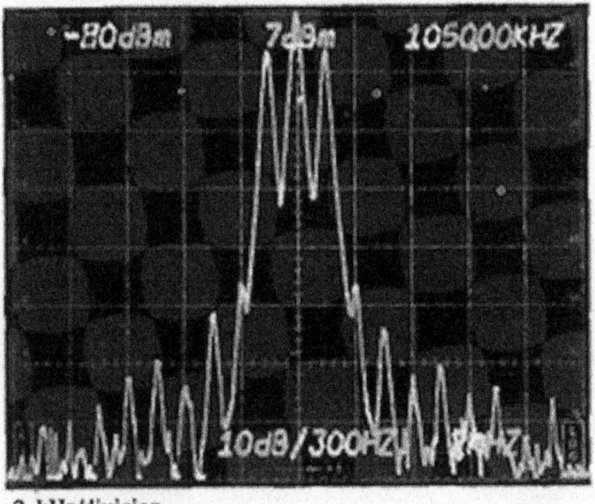

2 kHz/division

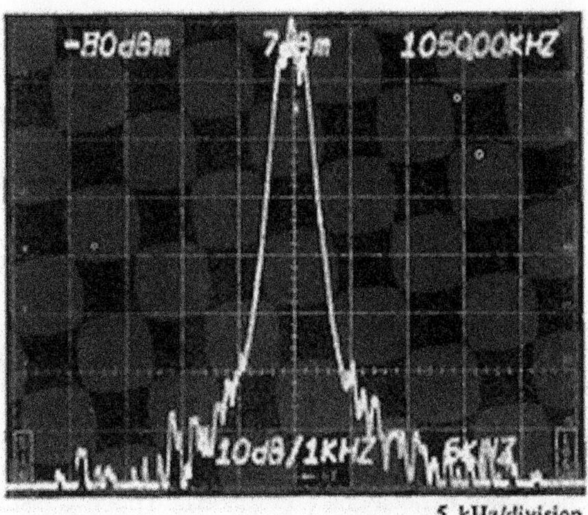

5 kHz/division

c. 100% modulation, 1 kHz tone

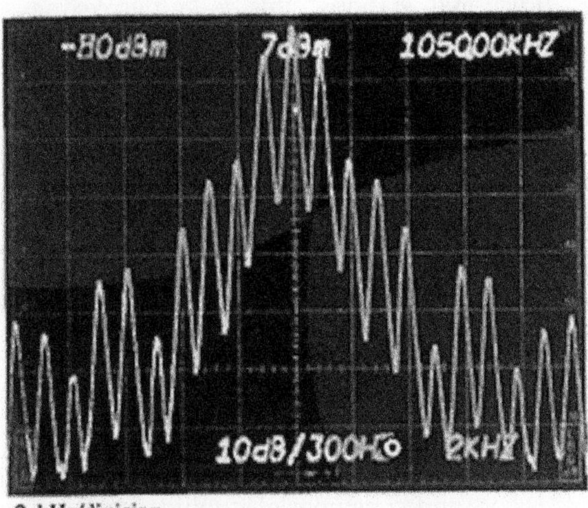

2 kHz/division

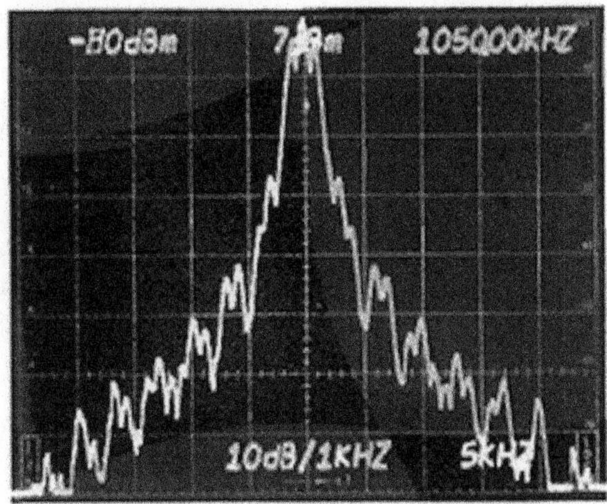

5 kHz/division

e. 100% + 3 dB, 1 kHz tone, –95% audio clip

FIGURE 10.2-1 AM Splatter Images From 1986 NAB AM Improvement Committee

mechanisms as excessive harmonic components, intermodulation of the signal with another station's signal, "birdies" produced by parasitic oscillations, arc detection and internal intermodulation components such as with pulse width modulated ("PWM") signals.

In the transmission plant and the AM antenna system, faulty components, connections and bonding can introduce arc noise, impedance mismatches, intermodulation and reradiation that affect the radiated signal. In the field, those same kinds of conditions can confound EPMs by generating localized products near the point of measurement. The best way to distinguish between a valid unwanted emission from the transmission plant and a confounding localized unwanted re-emission is by direction-finding. Selection of a measurement site requires paying attention to avoiding locations with locally generated spurious emissions.

Pulse width modulators run at a variety of frequencies ranging from 30 kHz to 175 kHz (and higher!). Unwanted PWM products are perhaps best classified as spurious emissions rather than out-of-band emissions because a) they tend to be well separated from the center frequency, b) their frequency offsets tend to be multiples of the PWM frequencies and c) they are not dependent on the modulating audio.

With this grasp of the terminology for sources of unwanted emissions, the discussion turns to the FCC regulations. Next, relevant regulations are presented in this order: §73.1590 Equipment performance measurements; §73.44 AM transmission system emission limitations; Hybrid AM IBOC mask.

REGULATION

§73.1590 Equipment Performance Measurements

The general *Equipment performance measurements* section of the FCC broadcast rules is found in 47 CFR

§73.1590. AM, FM and TV stations must conduct equipment performance measurements (EPMs) when new transmission or stereo generation equipment are installed or other modifications of the transmission system are made. AM stations are also required to conduct an annual EPM, with no more than 14 months between EPMs.

There has been some discussion in the industry about whether the AM EPM rules should be relaxed. FM and TV stations are no longer required to perform annual EPMs. In a 2014 commentary by respected broadcast engineer Mark Persons,[4] he floated a suggestion that stations with solid-state AM transmitters be excused from making annual EPMs. Persons suggested that tube aging in AM transmitters can affect a station's spectral occupancy, while solid-state transmitters do not experience any material spectral issues with component aging.

In the online comments to the Persons commentary, there was mixed support for the idea of eliminating annual EPMs for solid-state AM transmitters. Some discussion revolved around failing capacitors and leakage of pulse width modulation (PWM) products on the air and whether that kind of condition was any more likely than from a similar component failure in an FM system not required to be checked annually.

While the annual EPM requirement remains in force for all AM stations, the discussion on the Persons commentary highlights some key things to keep in mind when maintaining compliant spectral occupancy of an AM facility:

- Tubes should be monitored, maintained and replaced as needed.
- PWM transmission techniques require effective filtering to prevent spurious emissions on PWM switching frequencies.
- Capacitor failure can promote spurious emissions.

Perhaps Person's proposal does not go far enough. As this chapter discloses, the FCC method of evaluating AM broadcast spectral emissions is rather arbitrary and cannot be generalized to apply to every radial from a directional array. The directional station engineer must follow a complicated rule to determine where to take the measurement.

Annual AM EPMs as specified may seem like unproductive make-work to many broadcast engineers whose plants are stable and well maintained. Even if the annual requirement were to be lifted, it is prudent to keep an eye on the quality of the transmissions to catch events typically caused by component aging or failure before they turn into an interference complaint. Because annual EPMs are still the rule, this chapter explores how to comply.

Finally, §73.1590(d) requires EPM results to be documented:

> (d) The data required by paragraphs (b) and (c) of this section, together with a description of the equipment and procedure used in making the measurements, signed and dated by the qualified person(s) making the measurements, must be kept on file at the transmitter or remote control point for a period of 2 years, and on request must be made available during that time to duly authorized representatives of the FCC.

§73.44 AM Transmission System Emission Limitations

AM transmission system emission limitations—found in 47 CFR §73.44 of the FCC's rules—is based on the 1988 "NRSC Mask."[5] In the measurement section of this chapter, the full wording of §73.44 is presented. In summary, §73.44(a) describes how to take an emissions mask measurement with a spectrum analyzer (peak hold, no video filter, 10-minute duration, 300-Hz RBW); §73.44(b) describes the shape of the emissions mask, providing emissions attenuation levels required at frequencies offset from the carrier frequency (e.g., –25 dBc between 10.2- and 25-kHz offsets from carrier frequency); §73.44(c) enables the FCC to decide in special cases to reduce interference by requiring greater attenuation than allowed under the mask.

FCC regulation §73.44(d) establishes the reference level for AM emissions mask measurements. For directional stations, this is a perplexing and often-overlooked part of the required measurement technique. In a later section of this chapter, the setting of a reference level for measurements in compliance with (d) is addressed.

Finally, §73.44(e) contains transitional language allowing stations that installed NRSC-1 compliant equipment before July 1990 to avoid compliance measurements until 1994. NRSC-1 is the AM broadcast audio standard ("AM Preemphasis/Deemphasis and Broadcast Transmission Bandwidth Specifications"). The current version is NRSC-1-B (2012). The difference between NRSC-1 and NRSC-2 is the difference between the input to the transmitter and the output. Where NRSC-2 and §73.44 specify radio frequency emissions limitations, NRSC-1 addresses the audio going into the AM transmitter. NRSC-1 provided broadcasters and consumer electronics manufacturers with a standard way to pre/de-emphasize AM audio and limit the spectral density of the audio signals as they are put into the AM transmitters.

As NRSC-2-B explains, the purpose of NRSC-1 "is to reduce second-adjacent channel interference by band limiting AM stations to a nominal 20 kHz occupied radio frequency (RF) bandwidth. . . . Implementation of the NRSC-1-B audio standard largely achieves this purpose."

In short, the first step in establishing a spectrally clean signal that is §73.44-compliant is to ensure that the audio delivered to the transmitter is NRSC-1-B compliant.

Hybrid AM IBOC Mask

Hybrid AM IBOC signals are addressed under FCC guidance independent of §73.44.

Looking for regulatory guidance on AM hybrid IBOC operations, one starts with the FCC definition of a hybrid DAB system:

> **§73.402 Definitions**
>
> *(c) Hybrid DAB System. A system which transmits both the digital and analog signals within the spectral emission mask of a single AM or FM channel.*

The FCC provides no other emissions mask in Part 73 for hybrid AM broadcasting. One might assume that this means hybrid AM IBOC transmissions should be tested against the mask described in §73.44(b). However, digital signals are not well-characterized by capturing peak measurements. Instead, the requirements for hybrid AM IBOC operation are found in a series of Report and Order decisions of the FCC.

In particular, in the *First Report and Order In the Matter of Digital Audio Broadcasting Systems and Their Impact on the Terrestrial Radio Broadcast Service* (FCC 02–286, October 10, 2002) the FCC laid the foundation of FCC regulation of IBOC technology. In the First Report and Order, the FCC identified IBOC technology as the digital audio broadcasting technology for radio broadcasting in the United States. In doing so, it adopted an Appendix C that contained limits on hybrid AM IBOC emissions. Since then, the developer of the hybrid AM IBOC technology, iBiquity Digital Corporation (now subsumed by DTS, Inc.), modified the recommended hybrid AM IBOC emissions mask based on field experience. Strict laboratory-grade emissions limitations on the order of −100 dBc were relaxed to levels consistent with technological reality in high-power transmission systems and consistent with FCC expectations for analog emissions. In practice, the hybrid AM IBOC emissions mask provided by iBiquity and adopted by the NRSC in its NRSC-5-C standard is the mask that applies.

The making of hybrid AM IBOC EPMs is addressed in a separate section following a full discussion of §73.44 EPMs.

AM EMISSION MEASUREMENT TECHNIQUE

In the following sections various technical aspects of measuring AM broadcast emissions are discussed. First, the selection of a location for the measurements is covered. The test equipment and measurement methods for §73.44 compliance are then discussed.

Location for Emissions Measurement

In the AM broadcast spectrum, selection of a location for EPMs is constrained by several competing factors.

- Maximizing signal level by minimizing distance from station under test.
- Obtaining accurate representation of far-field signal, away from near-field distortions.
- Representative azimuth with respect to directional pattern.
- Avoiding unwanted high-power in-band signals.
- Avoiding sources of reradiation, noise and interference.

The FCC calls for the EPM to be taken "at ground level approximately 1 kilometer from the center of the antenna system." The reason for employing this distance is to harmonize the measurement with the 1-km non-directional field strength in the station's FCC records. For non-directional stations, the distance is not critical, because the reference level for the measurement is the received AM signal power at the point of measurement. Moreover, there is a general assumption that because the signal is omnidirectionally radiated, any out-of-band and spurious emissions will also be omnidirectional. Taking a measurement at one location is assumed to be sufficient to characterize the spectral occupancy of the signal in all directions.

It is more challenging to evaluate the emissions of directional AM stations. With directional stations, some calculations are necessary to arrive at the ideal spots to perform EPMs. These are addressed in a directional measurements section later in the chapter.

How Close is too Close?

Regardless whether a station is directional or not, if the station operates at lower power levels, or if there is too much noise or interference at or near the 1-km distance, it may be prudent to move the measurement location closer to the antenna site. To have the flexibility to move closer to the antenna system, it helps to understand how close to the antenna one can be and still obtain a reliable measurement.

It is best to stay out of the reactive near field of the antenna system. In the reactive near field, the propagating wave has not been fully formed, and the received signal may not be indicative of what is radiating across the service area. There are numerous rules of thumb for estimating the reactive near-field radius for antennas. A simple rule for a "small" 1/4-wavelength antenna on a ground plane is that the reactive near-field radius is a small fraction of a wavelength, where wavelength is denoted by λ.

$$\lambda/_{2\pi} = 0.16\pi$$

The longest wavelength is at the bottom of the AM band. At 540 kHz, the wavelength is 0.56 km (1822 ft). Using this rule of thumb, the reactive near field of an omnidirectional antenna ends approximately 89 meters (292 ft) from the antenna at the lowest frequency, and less at higher frequencies. The radiating near field is generally said to extend to 1 wavelength. Here it is best to avoid making measurements of pattern field strengths because the relationship between electric and magnetic radiating fields is not yet a free space relationship.

Chris Scott, experienced in EPM work and the technology required to perform EPMs, explains in a 1994 whitepaper on his website (scott-inc.com) that he has seen success with being "as close as one or two wavelengths" to an omnidirectional station. His close-in measurements "showed the same results [as at 1 km] but with fewer noise problems."

Well-known AM antenna engineer Ron Rackley, PE, reports that another way to look at minimum distance to an antenna or antenna array is to avoid the induction field in proximity to each individual tower in the array. He suggests maintaining a five-to-ten-times-tower-height separation to avoid measurement bias from inductive pickup on the measuring antenna. At 540 kHz, a 1/4-λ tower would be about 140 meters (460 ft) tall. A prudent minimum distance under this method would be five times the height, or 700 meters (2300 ft). At the other end of the band, 1710 kHz, the wavelength is reduced by a factor of 3.2, and the minimum distance might be as little as 221 meters (725 ft).

In section 4.3.2 of NRSC-2-B, the NRSC recommends selecting a location 10 wavelengths from an AM antenna system to perform measurements. In the case of a station at 540 kHz, that would be 5.6 km (3.5 mi).

As the rules of thumb mentioned suggest, it is more art than science in determining how close to the antenna system one can get before risking biased EPMs. If a clean measurement at the approximately 1-km distance can be obtained, this is the best way to proceed.

Taking an EPM of a non-directional station is relatively straightforward. No adjustments are necessary for antenna gain or for shifting off the 1-km target radius. The measured field strength of the unmodulated carrier at the measurement point is the reference for the spectral emissions measurements. Directional stations require a more complicated method.

Directional Antenna EPM Location

Measurement of directional antenna emissions under §73.44 requires some analysis to select a measurement spot. Rather than relying on the received power level of the station at the measurement location, §73.44 requires using a non-directional or a pattern RMS signal level reference. This injects a curious twist to the measurement method. The desired reference level has to be found in the field, at 1 km, because there is no tractable way after the measurement is made to adjust the mask to the required reference level, unless the requirement is interpreted selectively. This is discussed further later, following a discussion of EPMs and directional patterns in general.

Directional Antenna EPM Pattern Basics

The directional AM signal, by definition, has locations where the signal strength at 1 km is greater than the pattern's average signal strength at 1 km and other locations where the actual signal strength is below the average. The question arises, if just one measurement is to be made to evaluate the unwanted emissions of the directional station, in which part of the pattern should it be made?

If a measurement were taken in a pattern null, the received signal level at 1 km would be quite low. At this azimuth, the gain of the antenna array at the frequencies of unwanted emissions is unknown. If unwanted emissions were found to exceed the mask based on this arbitrarily low reference level found in the null, it does not necessarily indicate the unwanted emissions have the potential to cause interference. On the other hand, a measurement of unwanted emissions on the main beam of the transmitted signal would employ this, the highest radiated power, as the reference level. The thinking goes, unwanted emissions that might satisfy the mask in the main lobe might be too high in other directions.

In an effort to split the difference between a low-reference signal level and a high-reference signal level, the FCC rule chooses one in the middle: the non-directional, or alternatively the pattern RMS, signal level of the directional station.

Bear in mind that the two kinds of unwanted emissions may be affected by the antenna pattern bandwidth in different ways. Out-of-band emissions straddle the assigned channel of the station and can be skewed by variations in the pattern bandwidth. Pattern bandwidth effects on EPMs are discussed later in this chapter. To best represent out-of-band emissions performance (and performance within the necessary bandwidth as well), measurements would be made in the main lobe. Spurious emissions may fall substantially off the center frequency, where the antenna pattern is simply unknown. Because §73.44 calls for measurements at a single location, it is prudent to look carefully for spurious emissions even if below the mask, because they could be higher in other directions. Even if compliant with the mask, spurious emissions may be an indication of something wrong or about to go wrong with the transmission plant.

To illustrate how the antenna pattern can affect the occupied spectrum of the received signal, consider Figure 10.2-2.

It shows an AM station's directional pattern on its center frequency (bold line). Overlaid on the center frequency pattern are the antenna patterns at plus and minus 10 kHz. To the southwest (lower left) the patterns across the station bandwidth are quite similar. There is not much difference in the antenna gain across the channel bandwidth in this half of the coverage footprint. In contrast, the plus and minus 10 kHz patterns

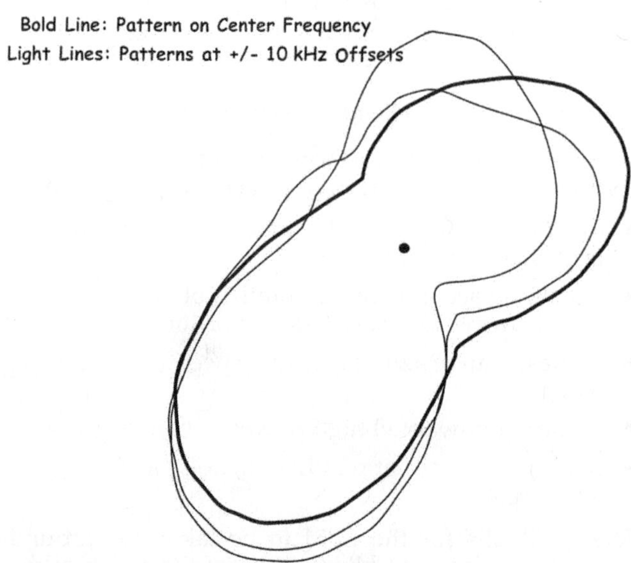

FIGURE 10.2-2 Simplified AM Pattern Bandwidth Plot

to the northeast deviate significantly (upper right) in comparison to the bold center frequency pattern. There are locations along this side of the pattern where the gain at a 10-kHz offset is substantially higher than at center frequency. At such locations one can expect to see the PSD skew higher at the offset frequency than at center frequency. In other words, the antenna array acts as a filter, shifting the "frequency response" of the transmitted signal in a particular direction. Performing a mask measurement at such a location could produce results that appear out of compliance with the mask, simply due to the pattern bandwidth response at that location.

Directional Antenna EPM Reference Level

The FCC hierarchy of preferred reference levels for the directional station EPM is stated in §73.44(d). During construction, directional stations typically have a non-directional pattern measurement taken from the reference tower at a reference transmitter power. This establishes the behavior of the ground conductivity and related factors independent of the array. From this non-directional pattern, a reference non-directional field strength at 1 km is developed. This is the preferred carrier reference for the EPM, as identified in §73.44(d)(1).

If the station's files lack non-directional information or the station otherwise has a good reason to use a different method, the next best thing, according to §73.44(d)(2), is to obtain the "RMS field strength determined from the measured directional pattern." Finally, if there is a defensible reason to not rely on this information, the last resort is §73.44(d)(3), "calculated expected field strength that would be radiated by a non-directional antenna at the station-authorized power." Bear in mind that all three choices are based on the field strength at 1 km from the center of the array.

For those unfamiliar with the RMS field strength, consider Figure 10.2-3.

It shows the night pattern of WRNI obtained from the FCC AM Query search tool. A pattern, of course, shows the directions where the antenna has more gain and where there is less gain. To figure out what the "average" gain of the antenna is, the individual points on the pattern must be averaged in a particular way. When averaging voltages, it is the general practice to square each field value of the antenna pattern, add the squared values and average them. The square root of the average is then taken. This is the essence of root-mean-square ("RMS") calculations.

$$X_{1\ldots n}\ \mathrm{V\ RMS} = \sqrt{\frac{X_1^{\,2} + X_2^{\,2} \ldots X_n^{\,2}}{n}}$$

The pattern is shown graphically on a polar plot, and the pattern point field strength values at five-degree intervals of azimuth are shown on a table. The RMS values of two of the three types of pattern (theoretical and augmented, but not standard in this example) are calculated and shown to the lower right of the polar plot shown in Figure 10.2-3.

In RF calculations (for a given impedance), the field strength squared is proportional to the power density delivered to the point. Performing an RMS on the antenna field values ensures that the resulting average is a close approximation of the received field strength if the antenna were non-directional with the same total power input.

Siting Examples for a Directional Antenna EPM

There are two approaches frequently taken by broadcast engineers for selecting EPM locations. One approach is to perform the EPM in the main lobe where there is antenna gain over the non-directional or RMS values and make an adjustment to the reference level to compensate for antenna gain. The other approach is to identify a location on the slope of the pattern where the directional antenna gain matches the non-directional or RMS gain.

Main Lobe Adjusted Measurements

When selecting a location in the main lobe, the operator is selecting a location that is likely to have the flattest pattern bandwidth response. In other words, there will be the least amount of increase or decrease in frequency response between ±10 kHz on the station's signal. The signal at this location is most representative of the designed mask response of the directional antenna system.

Because the received signal at this spot in the main lobe is higher than the reference level, the antenna gain on this bearing must be deducted from the reference level on the spectrum analyzer. This essentially shifts the mask down by an amount equal to the gain of the antenna on the radial of the measurement. One can readily visualize that shifting the mask down by, for example, 4 dB of antenna gain could drive the mask into the modulated signal's necessary bandwidth. It is assumed, when using this technique, that the mask within the necessary bandwidth is left alone. Generally, it is the mask beyond ±75 kHz that gets the deduction with this method. For example, if a station has 4 dB of pattern gain at the measurement location and a mask requirement to maintain spurious emissions below −80 dBc, the part of the mask outside ±75 kHz is lowered to a new compliance threshold of −84 dBc. This is the selective interpretation of the FCC reference level rule that was mentioned earlier. The reference level is only changed for evaluating spurious emissions (including harmonics). More interpretation might be necessary if there are out-of-band emissions that skirt below the mask transition at ±75 kHz. If the mask were adjusted for the reference level at this transition point, some stations may be driven into noncompliance. Newer AM broadcast plants have more margin transitioning past ±75 kHz, so a mask adjustment is less likely to be detrimental to such very clean facilities.

With these limitations in mind, this method of performing main lobe measurements on a directional antenna system is regularly used by some engineers.

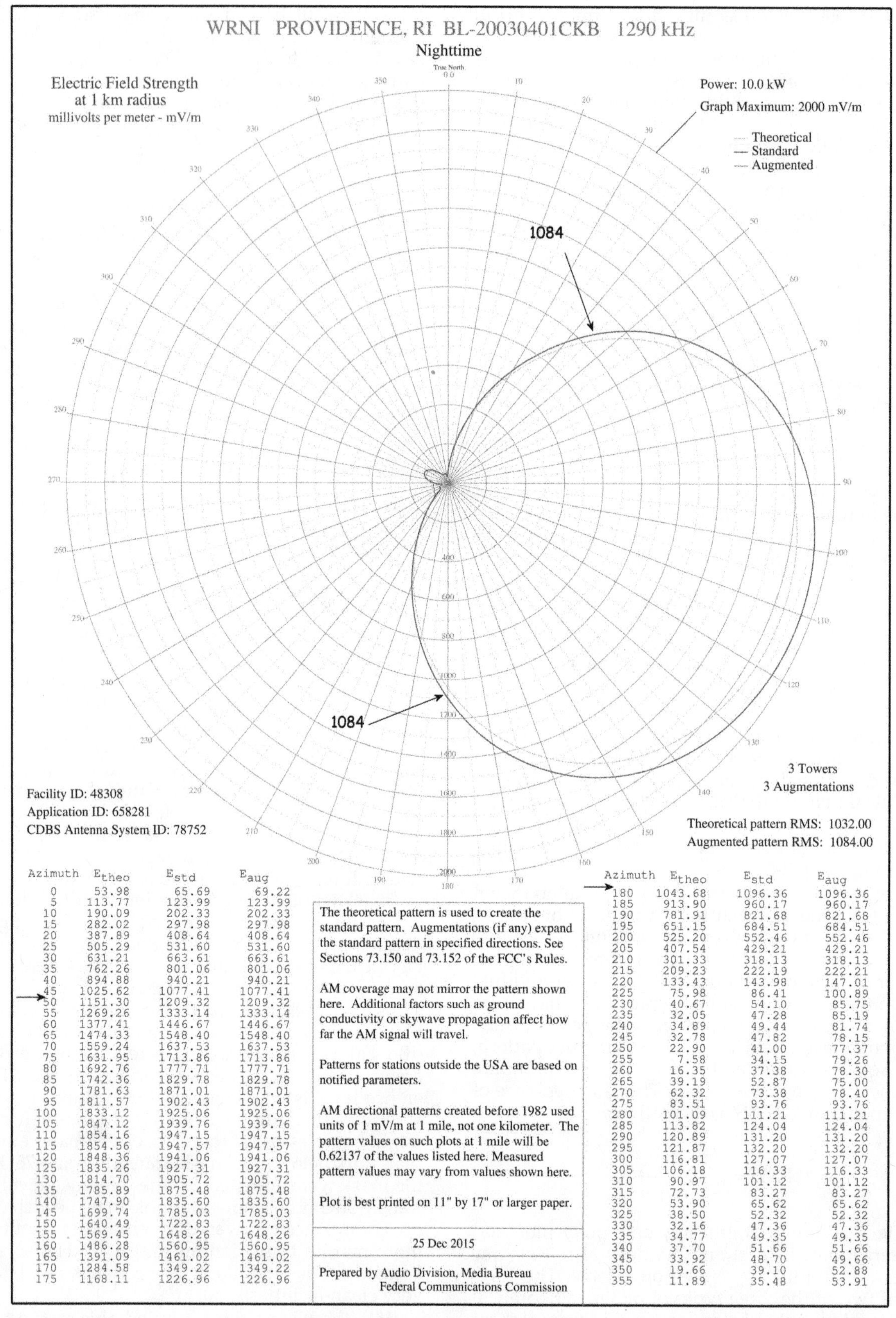

Azimuth	E_{theo}	E_{std}	E_{aug}
0	53.98	65.69	69.22
5	113.77	123.99	123.99
10	190.09	202.33	202.33
15	282.02	297.98	297.98
20	387.89	408.64	408.64
25	505.29	531.60	531.60
30	631.21	663.61	663.61
35	762.26	801.06	801.06
40	894.88	940.21	940.21
45	1025.62	1077.41	1077.41
50	1151.30	1209.32	1209.32
55	1269.26	1333.14	1333.14
60	1377.41	1446.67	1446.67
65	1474.33	1548.40	1548.40
70	1559.24	1637.53	1637.53
75	1631.95	1713.86	1713.86
80	1692.76	1777.71	1777.71
85	1742.36	1829.78	1829.78
90	1781.63	1871.01	1871.01
95	1811.57	1902.43	1902.43
100	1833.12	1925.06	1925.06
105	1847.12	1939.76	1939.76
110	1854.16	1947.15	1947.15
115	1854.56	1947.57	1947.57
120	1848.36	1941.06	1941.06
125	1835.26	1927.31	1927.31
130	1814.70	1905.72	1905.72
135	1785.89	1875.48	1875.48
140	1747.90	1835.60	1835.60
145	1699.74	1785.03	1785.03
150	1640.49	1722.83	1722.83
155	1569.45	1648.26	1648.26
160	1486.28	1560.95	1560.95
165	1391.09	1461.02	1461.02
170	1284.58	1349.22	1349.22
175	1168.11	1226.96	1226.96

Azimuth	E_{theo}	E_{std}	E_{aug}
180	1043.68	1096.36	1096.36
185	913.90	960.17	960.17
190	781.91	821.68	821.68
195	651.15	684.51	684.51
200	525.20	552.46	552.46
205	407.54	429.21	429.21
210	301.33	318.13	318.13
215	209.23	222.19	222.21
220	133.43	143.98	147.01
225	75.98	86.41	100.89
230	40.67	54.10	85.75
235	32.05	47.28	85.19
240	34.89	49.44	81.74
245	32.78	47.82	78.15
250	22.90	41.00	77.37
255	7.58	34.15	79.26
260	16.35	37.38	78.30
265	39.19	52.87	75.00
270	62.32	73.38	78.40
275	83.51	93.76	93.76
280	101.09	111.21	111.21
285	113.82	124.04	124.04
290	120.89	131.20	131.20
295	121.87	132.20	132.20
300	116.81	127.07	127.07
305	106.18	116.33	116.33
310	90.97	101.12	101.12
315	72.73	83.27	83.27
320	53.90	65.62	65.62
325	38.50	52.32	52.32
330	32.16	47.36	47.36
335	34.77	49.35	49.35
340	37.70	51.66	51.66
345	33.92	48.70	49.66
350	19.66	39.10	52.88
355	11.89	35.48	53.91

FIGURE 10.2-3 WRNI Night Pattern

FCC Public Document

Pattern Crossover Point Measurements

Figure 10.2-3 has information that could be used to establish a measurement location for a §73.44 EPM on a directional station. This figure provides calculated values of directional pattern RMS based on the licensed pattern. Note that these are the licensed values and are likely not the same as the measured values that §73.44 call for. Assume for the sake of illustration that the values on the Figure 10.2-3 are equivalent to the "calculated expected field strength of the non-directional pattern" stated in the third FCC option. To properly satisfy the rule, find the information in station records that best fits 73.44(d).

At this point, the AM station engineer has selected a reference level in the form of a non-directional field strength value (or equivalent) at 1 km. The next step is to select a measurement location that will provide that reference level. At the same time, the location selected for measurement on the pattern should be one with as broad and flat a pattern bandwidth response as possible, as explained earlier.

The next step in identifying a location to perform an EPM is to check the station antenna pattern for the locations where the actual signal strength at 1 km is equal to the selected reference level. Typically there will be two points, one on each side of the main lobe, where the actual pattern will have a signal level equal to the reference level. We will call these the "crossover points." In the example (Figure 10.2-3), the augmented pattern RMS (1084 V/m) is used as the reference level for the sake of this illustration.

The pattern is compared to the reference level. In this example, the arrows on the pattern plot indicate the locations where the pattern matches the reference level of 1084 V/m. To the nearest five degrees, the 1084 V/m crossover points are at 45 and 180 degrees. If there is a desire to be more precise, the antenna pattern table provides data point values at every five degrees, between which one can interpolate to obtain an estimate to the nearest degree. Bear in mind that at the 1 km radius, a five-degree swath is about 87 meters (286 ft) of the arc. Considering the rough nature of estimating the reference level and the actual pattern in the first place, a resolution of five degrees is likely to be about as reliable as a resolution of one degree.

Once the azimuths of the two crossover points are identified, use a map to run a radial to 1 km at each azimuth. Online mapping applications with drawing tools (length and azimuth) provide a convenient and reliable way to pinpoint the crossover point locations. Examine the two locations to estimate which is more likely to be successful. Factors such as the density of development, accessibility, absence of industrial and utility noise sources and distance from potentially challenging nearby AM stations should be considered for selecting the more likely site.

If a clean, reliable measurement location can be found at the about 1-km distance on one of the two radials, measurements would be made there. Lacking formal clarification from the FCC, here the interpretation of the FCC rule is further left to the judgment of the station. If the received level at the crossover point is not close to the reference level, determine whether it is necessary to find an alternate location that is closer to the reference level, or to accept this location because it was identified using figures of record. Once a location has been selected for one antenna pattern, stations with more than one pattern will have to identify suitable locations for performing EPMs on any other patterns.

This subsection has covered the specifics of identifying a site for performing an EPM on a directional station. In the following sections, the discussion includes both non-directional and directional EPMs.

Measurement Interference

It is assumed here that AM station engineers with field intensity meter experience are familiar with working near structures and conductive surfaces that may intermodulate, detect, or re-radiate energy. A wide array of conductive materials including such things as utility lines, guy wires, metal structures on and in buildings, towers and fences act as antennas that can confound field measurements of AM signals.

In addition, active sources of electrical and electronic noise in the AM band are to be avoided. In short, it is the field engineer's responsibility to find sites that are relatively clear of conditions that will affect the quality of the measurements. If such sources cannot be completely avoided, their impacts can be limited by ensuring the offending sources are off-axis when the measuring antenna is peaked on the station's signal.

If it is not possible to find a clean location at 1 km, or if the received signal strength is not a good match to the spectrum analyzer's dynamic range, consider moving closer to the station. Discussion of spectrum analyzer performance is continued further in later sections. For non-directional EPMs, the reference level at the new spot remains the measured unmodulated carrier level. For directional EPMs, as long as the measurement is on the same radial (azimuth) as the targeted 1-km crossover point, the measured unmodulated carrier should still represent the reference level.

Use battery-operated equipment if possible. This minimizes the possibility of picking up either electrical noise or RF interference from an external power source. Portable inverters for vehicles are notorious for generating electrical noise. If one is required, choose wisely. Consider a well-shielded true-sine-wave inverter or a battery-powered true-sine-wave UPS.

The Mask

§73.44 is reproduced here for reference.

§ 73.44 AM TRANSMISSION SYSTEM EMISSION LIMITATIONS

(a) The emissions of stations in the AM service shall be attenuated in accordance with the requirements specified in paragraph (b) of this section. Emissions shall be measured using a properly operated and suitable swept-frequency RF spectrum analyzer using a peak hold duration of 10 minutes, no video filtering, and a 300 Hz resolution bandwidth,

except that a wider resolution bandwidth may be employed above 11.5 kHz to detect transient emissions. Alternatively, other specialized receivers or monitors with appropriate characteristics may be used to determine compliance with the provisions of this section, provided that any disputes over measurement accuracy are resolved in favor of measurements obtained by using a calibrated spectrum analyzer adjusted as set forth above.

(b) *Emissions 10.2 kHz to 20 kHz removed from the carrier must be attenuated at least 25 dB below the unmodulated carrier level, emissions 20 kHz to 30 kHz removed from the carrier must be attenuated at least 35 dB below the unmodulated carrier level, emissions 30 kHz to 60 kHz removed from the carrier must be attenuated at least [5 + 1 dB/kHz] below the unmodulated carrier level, and emissions between 60 kHz and 75 kHz of the carrier frequency must be attenuated at least 65 dB below the unmodulated carrier level. Emissions removed by more than 75 kHz must be attenuated at least 43 10 Log (Power in watts) or 80 dB below the unmodulated carrier level, whichever is the lesser attenuation, except for transmitters having power less than 158 watts, where the attenuation must be at least 65 dB below carrier level.*

(c) *Should harmful interference be caused to the reception of other broadcast or non-broadcast stations by out of band emissions, the licensee may be directed to achieve a greater degree of attenuation than specified in paragraphs (a) and (b) of this section.*

(d) *Measurements to determine compliance with this section for transmitter type acceptance are to be made using signals sampled at the output terminals of the transmitter when operating into an artificial antenna of substantially zero reactance. Measurements made of the emissions of an operating station are to be made at ground level approximately 1 kilometer from the center of the antenna system. When a directional antenna is used, the carrier frequency reference field strength to be used in order of preference shall be:*

  (1) *The measure non-directional field strength.*
  (2) *The RMS field strength determined from the measured directional radiation pattern.*
  (3) *The calculated expected field strength that would be radiated by a non-directional antenna at the station authorized power.*

(e) *Licensees of stations complying with the ANSI/EIA-549–1988, NRSC-1 AM Preemphasis/Deemphasis and Broadcast Transmission Bandwidth Specifications (NRSC-1), prior to June 30, 1990 or from the original commencement of operation will, until June 30, 1994, be considered to comply with paragraphs (a) and (b) of this section, absent any reason for the Commission to believe otherwise. Such stations are waived from having to make the periodic measurements required in § 73.1590(a)(6) until June 30, 1994. However, licensees must make measurements to determine compliance with paragraphs (a) and (b) of this section upon receipt of an Official Notice of Violation or a Notice of Apparent Liability alleging noncompliance with those provisions, or upon specific request by the Commission. [47 FR 8588, Mar. 1, 1982, as amended at 49 FR 3999, Feb. 1, 1984]*

The emissions mask is described in §73.44(b). Bear in mind that the emissions mask is based on §73.44(a), which requires measuring PSD with a 300-Hz resolution bandwidth (RBW). Measuring technique is described in detail later.

The NRSC-2-B emissions mask[6] is the same as the §73.44 mask and is reproduced here as Figure 10.2-4.

The Spectrum Analyzer

The §73.44 mask measurements rely on capturing the maximum (peak) excursions of the incoming signal. Often this technique is called "max hold" or "peak hold" on spectrum analyzers. This is a feature that has been available since spectrum analyzers were able to process multiple sweeps. The key characteristics of a spectrum analyzer for this application are the ability to employ a peak detector and accumulate peaks with max hold over a ten-minute period, sufficient sensitivity to receive the station's signal off the selected antenna at 1 km from the station with a ≤ −90 dBc/300 Hz noise floor, and sufficient dynamic range to capture the incoming signal with the required noise floor without overloading the instrument or generating confounding internal spurious products.

The measurement protocol calls for having no video bandwidth filter engaged. Video bandwidth filtering tends to smooth out noisy traces. This was a key feature in the days of analog spectrum analyzers. Now, the video filter is typically set to be much wider than the RBW as a way of disabling its smoothing behavior. For a 300-Hz RBW, if the video filter cannot be disengaged, set it to ten times the RBW (3000 Hz) or more.

Analyzer Examples

As a baseline for comparison, we refer to a revered portable spectrum analyzer from the 1990s featured in many EPMs because of its combination of size, performance and affordability (the "legacy analyzer"). The legacy analyzer was barely able to perform reliable EPMs. The legacy analyzer specifications provide a good example of specifications that were barely viable for many AM EPMs at the time. Select a contemporary analyzer with better specifications if possible. Below, three recent models of spectrum analyzer ("Models 1, 2 & 3") are compared with the legacy analyzer.

Display Range

The legacy analyzer was designed with a display range of 80 dB, requiring the use of a narrow notch filter to pull down the carrier. This would enable the user to turn up the input level to obtain a virtual increase in the dynamic range of the instrument. Contemporary analyzers typically display up to 100 dB reliably. However, although an instrument may have a 100-dB vertical display, other factors must be considered to determine if the full 100-dB view will be sufficiently clean to perform an EPM.

Displayed Average Noise Level

The displayed average noise level ("DANL") specification of the legacy analyzer was -127 dBm/300 Hz, at 100 MHz. This means that with a terminated input, the analyzer would show a noise line at −127 dBm with a 300-Hz RBW while tuned to 100 MHz. The legacy analyzer's DANL is 12 dB lower (−139 dBm/300 Hz) when the internal preamplifier is activated (more about preamplification later). In Figure 10.2-5, the DANL of a contemporary handheld spectrum analyzer is shown at an AM broadcast frequency.

Note how a DANL must be accompanied by the assumptions used to produce the DANL figure. The

NRSC-2-B

Table 1: RF Mask for Analog AM Broadcast Station Spectrum Occupancy

Frequency band relative to carrier (± kHz)	Attenuation relative to carrier (dB)
0 to 10.2	0
10.2 to 20	at least 25[3]
20 to 30	at least 35
30 to 60	–at least (5 + 1 dB/kHz) from carrier[4]
60 to 75	at least 65[4]
Above 75	at least 80[4]

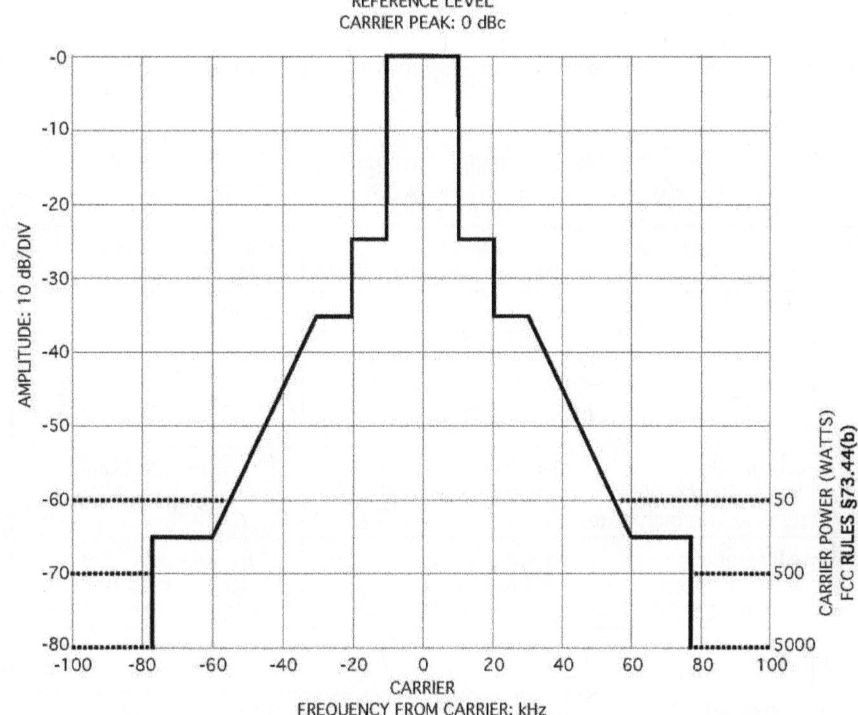

Figure 1. Analog AM Broadcast RF Emission Limits

[3] The slope of occupied bandwidth in the transition region between 10 and 11 kHz is defined by the NRSC-1-B audio standard. Accordingly, attenuation levels in the region would be 6 dB greater than described in the audio standard to adjust for carrier level reference. As of NRSC-2-A, the transition region above 10 kHz has been modified to conform with the FCC specification of 10.2 kHz.

[4] For carrier power levels between 158 and 5000 watts, the minimum attenuation from carrier reference level for emissions more than 75 kHz from carrier frequency shall be (43 + $10\log P_W$) dB (where P_W is the carrier power in watts) or as indicated in Table 1, whichever is lesser attenuation. For carrier power levels less than 158 watts, a 65 dB minimum attenuation shall apply.

FIGURE 10.2-4 NRSC-2-B AM Emission Mask

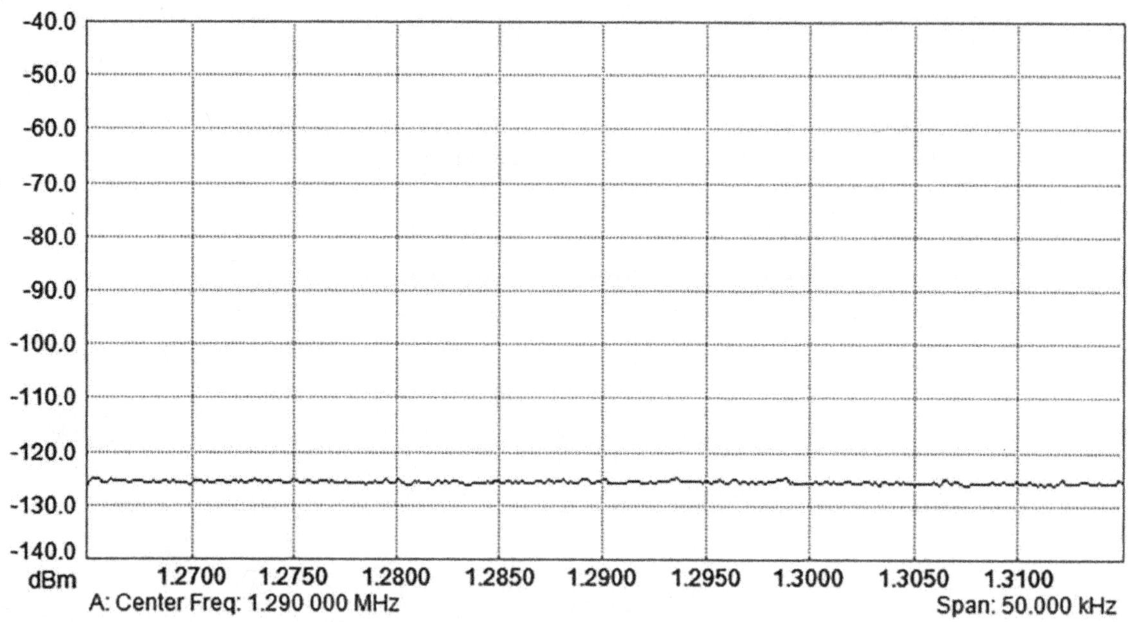

FIGURE 10.2-5 Displayed Average Noise Level of a Spectrum Analyzer at 300-Hz RBW on AM Frequencies

TABLE 10.2-1
Comparison of Spectrum Analyzer Specifications

Spectrum Analyzer Model	Legacy Analyzer	Model 1—Mid-quality, handheld	Model 2—Good quality, handheld used by author	Model 3—High quality, benchtop
Display range	*80 dB*	100 dB	100 dB	100 dB
DANL @ 100 MHz no preamp (compare with @ 1 MHz values)	–127 dBm/300 Hz	–125 dBm/300 Hz	[> 3 MHz] –112 (spec) –125 (typ) dBm/300 Hz	–129 dBm/300 Hz
DANL @ 1 MHz no preamp	About –115 dBm/300 Hz (based on tech note)	*–66(spec)/ –93(typ) dBm/300 Hz*	–125 dBm/300 Hz (from bench test)	–125 dBm/300 Hz
Phase noise from center frequency	All frequencies @ 9 kHz offset: –70 dBc/300 Hz	*500 MHz @ 100 kHz offset: –74 dBc/300 Hz*	All frequencies @ ≥ 10 kHz offset: –75 dBc/300 Hz	100 MHz @ 10 kHz offset: –117 dBc/300 Hz
Distortion—3rd order –30 dBm ref.	*–60 (spec)/ –70 (typ) dBc*	–110 dBc	–106 dBc	–146 dBc
Distortion—2nd harmonic –30 dBm ref.	*–60 (spec)/ –70 (typ) dBc*	*–60 dBc*	*–56 dBc*	–110 dBc derived from spec
1 dB compression	–15 dBm	0 dBm	–2 dBm	+15 dBm

Notes:
Table is based on published specifications unless otherwise indicated.
DANL @ 100 MHz is shown because it is published for all four instruments.
DANL @ 1 MHz is not specified for two instruments and is derived.
Italicized specifications may not be consistent with the AM mask.

DANL is actually a PSD figure of the noise floor of the instrument 1) at a specified frequency and 2) with a given RBW (and, for splitting hairs, at a specified temperature). Sometimes DANL is reported in dBm/Hz. In other cases it is reported for one or more of the RBW settings available on the instrument. For AM EPMs, the relevant RBW is 300 Hz; 300 Hz of noise bandwidth is 25 dB greater than the 1 Hz normalized value.

Contemporary analyzers have DANLs similar to the legacy analyzer. In Table 10.2-1 are the specifications for the legacy analyzer and for three more recent model analyzers. Model 1 is a medium-quality handheld unit. Model 2 is a higher-quality handheld unit. Model 3 is a high-quality benchtop unit. Each model is of a different brand. The difference in price between the handhelds and the benchtop model is about 7 dB.

As the table reveals, Model 3 is higher quality in some, but not all, respects. The high-quality benchtop unit is also equipped with many features and options that a handheld unit cannot support, so some of the cost difference is in the intended applications of the analyzers.

Note how the mid-quality handheld model (Model 1) has a higher noise floor (DANL) at 1 MHz than the other two current models. Because of this, Model 1 loses a substantial amount of dynamic range at AM broadcast frequencies. Also note how the specification of Model 1 has a respectable DANL at 100 MHz, even when it is quite poor at 1 MHz. This phenomenon of elevated noise levels at 1 MHz is not uncommon among mid- and lower-priced analyzers.

Do not rely on a VHF DANL when selecting an analyzer for AM EPMs.

This is one characteristic to look for when selecting an analyzer for AM EPMs—the 1-MHz DANL at 300-Hz RBW must be low enough to provide at least -90 dBc below the expected reference level of the received signal under test.

Third-Order Distortion

The legacy analyzer had less-than-optimal internal intermodulation specifications for EPMs. Third-order intermodulation distortion[7] of signals received by the legacy analyzer was guaranteed to be more than 60 dB below the input signal levels, with typical levels about 10 dB lower. This was often manageable if the instrument was used by a person familiar with its quirks, because one could know where to look for internally generated products based on the frequencies of the strongest input signal levels. With a strong AM signal under test and no other strong signals entering the instrument, the −60/−70 dBc third-order distortion figure was frequently not an obstacle to obtaining clean enough EPMs. Without a competing strong signal in the analyzer, the only third-order product of any magnitude would have been the third harmonic. The field strength meter was available to measure harmonic emissions when the legacy analyzer was not up to the task.

The third-order distortion products of the recent models in the table are well attenuated. Starting with a reference level of −30 dBm, the third-order products will be more than 100 dB down—well below the −80-dBc lowest limit line of the mask. It is not likely, with an input level around −30 dBm, that there will be analyzer-induced products that will taint the measurements.

Above −20 dBm (at the input to the active components of the analyzer), there may be cause for concern even on current analyzer models, because as the input power levels go up by 1 dB, the third-order products rise by 3 dB. Bear in mind that third-order (and second-order) intermodulation products in a spectrum analyzer do not vanish when operating below the compression level; these spurious products simply diminish as the input level diminishes. The goal is to find an input level that yields potential distortion products below the DANL of the instrument while still maintaining the necessary dynamic range between the signal under test and the DANL.

1 dB Compression Level

Other limitations can get in the way of an unsuspecting user. With the display reference level (top line of the display) set to −30 dBm, the 1-dB compression level of the legacy analyzer was −15 dBm, which sometimes required additional input attenuation (even with the notch filter) or use of a smaller antenna when measuring more powerful stations. Bear in mind that it is the peaks of the sampled signal that should avoid entering the compression range. As a rule of thumb, ensure the max hold peak of the AM carrier is at least 15 dB below the 1-dB compression level of the instrument.[8] The legacy analyzer specifications illustrate this practice by using the −30 dBm display reference level, with the 1-dB compression point 15 dB above that reference level. Contemporary spectrum analyzers are available with 1-dB compression levels, typically between −5 and +20 dBm. This provides more headroom for receiving stronger signals without adding attenuation.

Phase Noise

Phase noise is the noise of the analyzer local oscillator(s) that creates the "skirts" that are seen around a signal under test. If the phase noise is well below the expected signal levels under the mask, it will not materially affect measurements under the mask close to center frequency.

The legacy analyzer phase noise sidebands at ≥ ±9-kHz offset from carrier were guaranteed to be more than 70 dBc/300 Hz below the signal under test. The AM mask at this frequency is −35 dBc/300 Hz. Phase noise this low is more than sufficient for making mask measurements in the out-of-band domain.

As the frequency offset from carrier increases, the phase noise continues to drop until it merges with the DANL of the instrument. At the last mask transition point (±75 kHz), look for a phase noise that falls to less than −90 dBc/300 Hz. This ensures the phase noise is at least 10 dB below the mask and will not materially affect the readings. Based on the legacy analyzer phase noise specification, the −90-dBc/300-Hz point of the legacy analyzer phase noise is at ±90 kHz offset, not at ±75 kHz. This is a close call but was found in practice to have good enough margin at the ±75-kHz mask transition points to provide credible measurements.

Bear in mind that the phase noise creates a skirt that straddles not only the carrier at center frequency but also straddles each sideband component at its frequency. With AM EPMs, the sideband components are substantially lower than the center carrier, so the phase noise around the carrier is likely to dominate the nearby noise floor of the scan because it is singly so much higher than the power in the sidebands. Phase noise limitations can be a bigger issue when evaluating digital waveforms that maintain relatively high power levels to the edges of their channels.

The phase noise of the mid-quality Model 1 analyzer is not impressive. First, it is reported at a center frequency of 500 MHz, with no indication of how that might vary at lower frequencies. Second, it is −74 dBc at a large offset frequency of 100 kHz. This model of contemporary spectrum analyzer places the phase noise skirt uselessly above the AM mask from ±60 kHz to well outside ±100 kHz. Even if Model 1 had a potentially usable DANL, the operator might be lulled into thinking a noise skirt that is not mask-compliant is real, rather than an artifact of the instrument's phase noise. Instruments with high phase noise can artificially boost the incoming measured emissions-plus-environmental-noise above the mask.

Bear in mind that transmitting equipment also exhibits phase noise, so even if there were a perfect spectrum analyzer with no phase noise, sloped phase noise skirts will be observed on the measured spectrum. The goal is to ensure that the internal phase noise is low enough that it will not compromise the measurement.

Model 2 and Model 3 have excellent phase noise characteristics for AM EPMs. Figure 10.2-6 shows a max hold trace from the Model 2 analyzer.

Note the sloped noise floor immediately adjacent to the sidebands of the signal under test. These slopes have been highlighted with two curves inserted to show the approximate contribution of phase noise to the trace. Bear in mind that this phase noise slope is the combination of the phase noise of the signal under test and that of the measuring instrument. The phase noise of this instrument is low and is compatible with AM mask measurements.

The ideal AM EPM instrument satisfies these criteria:

1. The desired input level is suitably below the 1-dB compression point,
2. The internal intermodulation products are under −90 dBc with respect to this input level,
3. The 1-MHz displayed average noise level at 300-Hz RBW is also under −90 dBc and
4. The instrument phase noise at 300-Hz RBW is always at least 10 dB below the mask.

Antenna and Filters

Consider the size and efficiency of the antenna that will be used for the measurement. How much power will that deliver to the spectrum analyzer at the 1-km location selected for the measurement? (See the separate discussion in the "Antenna Factor" section for converting the received power level to field strength.) To avoid adding attenuation, seek an analyzer that has a 1-dB compression point high enough to avoid needing attenuation. If attenuation is necessary, make certain that the input level to the mixer, after attenuation, remains at least 15 dB or more below the 1-dB compression level (when viewing the AM carrier with a peak detector). Meanwhile, as explained in the previous section, the DANL and distortion must not be forced too close to the mask by the insertion of attenuation.

The shielded loop antenna is the preferred type because it has directional characteristics for isolating local interference and noise sources and a current loop

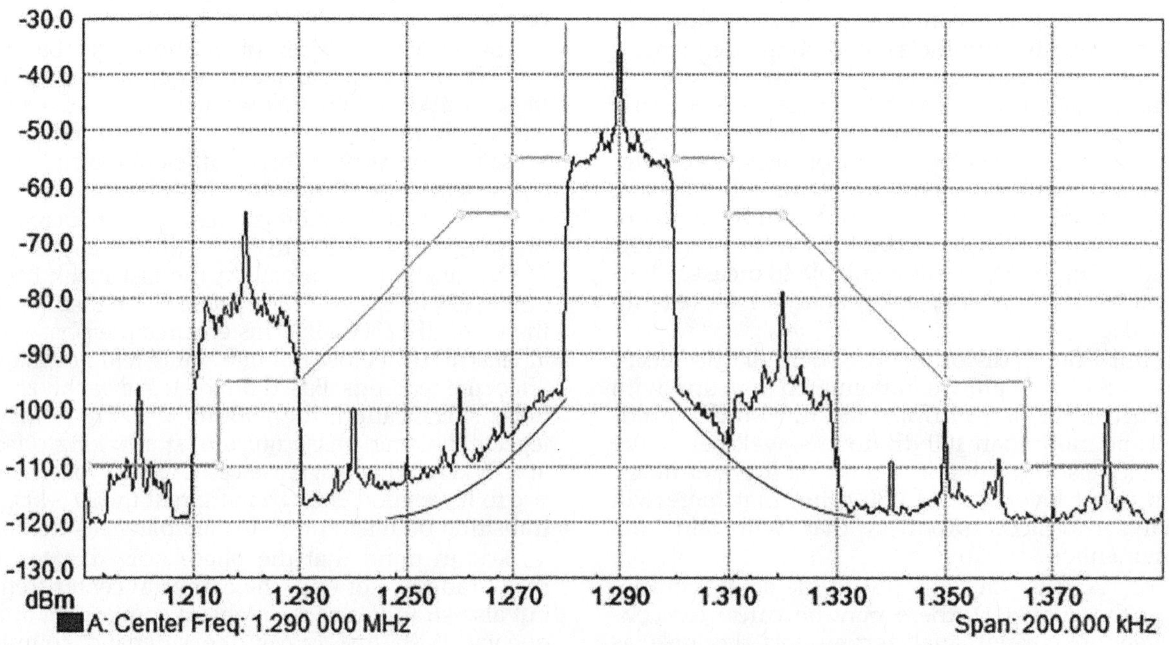

FIGURE 10.2-6 Max Hold Trace at 300-Hz RBW on the Model 2 Spectrum Analyzer

for picking up the magnetic field while being shielded to reduce electrical field noise from the environment.

If the circumstances of a particular measurement location demand, low-pass filters are available to minimize the reception of any frequencies above the AM band. This protects the instrument from inadvertent overload from unseen signals outside the AM band. And in special cases where there are multiple AM stations nearby (or a multiplex AM facility is being evaluated), consider getting notch filters for the unwanted signals.

Discussed earlier, a notch filter for the desired signal can help compensate for a limited dynamic range on a spectrum analyzer by cutting down the incoming carrier. This reduces the production of distortion products and purchases some additional virtual dynamic range. Be sure to determine where the notch filter skirts no longer affect the signal's PSD under the mask. Close-in mask measurements should be made before inserting the notch filter. With the notch filter in place, the reference level for dBc measurements remains what it was before the notch was inserted, except that the insertion loss of the pass frequencies of the notch filter must be accounted for.

Preamplifier

The signal under test can be boosted further above the instrument noise floor by activating the internal low-noise preamplifier. The small increase in noise contributed by the preamp is dwarfed by the large effective decrease in the instrument noise floor resulting from the preamp gain. There is one caveat with the use of preamplifiers: input overload. The preamp specification is not always found in the instrument specifications. Generally, instruments do not enable the preamp feature unless the reference level on the display is at or below some midlevel setting such as −40 or −50 dBm. It might require a combination of input attenuation and preamplification gain to set the input level at just the right value.

Input Level

From the perspective of minimizing instrument overload, the input power level of concern to the analyzer is the *total power into the analyzer*. The total power reaching the first active components of the analyzer (first RF stage, or the preamp if it is activated) determines whether those active components are driven into compression. With AM signals modulated to 100%, the total power is nominally 150% of the unmodulated carrier power (+1.8 dB). However, the peak voltage of a 100% modulated carrier is twice the voltage of the unmodulated carrier. The peak instantaneous power in the modulated envelope would therefore be four times the unmodulated carrier power, or +6 dB with respect to the carrier. With aggressive positive peak modulation, it might be necessary to increase this headroom to 10 dB or more.

Also bear in mind whether there are other stations nearby. If the other station(s) are received at similar or higher levels than the station under test, those unwanted signals, whether they appear on the display or not, could be silently contributing to overload of the preamp or the RF front end of the analyzer. Overload produces artifacts that might be misread as flaws in the emissions of the station under test.

Setting the Reference Level

With an understanding of the constraints on the input level as discussed earlier, the signal under test must be set to the optimal level for the instrument. The antenna should be set in a fixed position, such as on a tripod, and aimed while watching the signal level vary on the spectrum analyzer (leave max hold off and do not activate any averaging). Or before connecting the spectrum analyzer, use the external input jack on a field intensity meter to peak the shielded loop antenna on the desired signal and minimize interference. Ensure that a good peak appears and that the peak is toward the station and not toward any reradiation sources. Set the antenna and leave it in place. If there is reradiation that is making it difficult to isolate the radiated signal from interfering reradiation and noise, seek another location.

The reference level, according to §73.44, is the "unmodulated carrier" as measured at the 1-km radius. The most reliable way to establish this as the reference level is to cut the modulation and sweep with the 300-Hz RBW. With the unmodulated carrier, for all practical purposes, the same value will be obtained with either a peak detector or an RMS detector (some manufacturers call the RMS detector an average detector, although other manufacturers define average detection in a different way.). Since this is a reference level based on a continuous wave signal, it would be most appropriate to use the RMS detector. But there is more to this.

To be precise, §73.44 calls for "peak hold" measurements of the modulated signal. As mentioned earlier, peak hold and max hold are equivalent terms that describe what the instrument does when repeatedly sweeping. The peak/max hold function for accumulating sweeps is independent of the detector that is utilized. No detector is specified in §73.44. It is generally assumed that the intent to capture the peak values among repeated sweeps also means that a peak detector should be activated.

The most conservative approach, from a compliance perspective, would be to capture the reference level of the unmodulated signal with an RMS detector and capture the PSD measurement of the modulated signal with the peak detector.[9]

As an alternative, the operator may experiment with measuring the center frequency carrier level with a narrow RBW and with modulation present. A reference can be established by comparing the result with a measurement made with the unmodulated carrier. On the average, the power of the AM carrier remains fixed at the unmodulated carrier power. Experience has shown that the 300-Hz RBW on center frequency (±150 Hz) of a modulated AM signal can capture a reasonably stable carrier level reference with the right combination of analyzer settings. For future reference, document the comparison between an unmodulated carrier measurement and a narrow-band modulated carrier measurement. This will save time in the future by avoiding coordinating dead air for unmodulated carrier measurements.

Stations using modulation-dependent carrier level (MDCL) control to reduce power consumption must ensure the MDCL function does not affect the measurement of the reference signal level. Check with manufacturer instructions on how to obtain an unmodulated carrier for measuring the reference level.

Once the reference level is set, measurements can begin.

Examine the Spectrum

Many spectrum analyzers have the capability to display limit lines. If available, obtain or create a limit line according to the mask and call it up on the analyzer. Before implementing the max hold function, examine the spectrum in the necessary bandwidth and in the out-of-band domain for any obvious anomalies while the trace is repeatedly refreshing. To average out noise and random modulation components, consider using a short-term average. There could be unwanted emissions that are the result of the modulation and transmission process that are telltales of problems in the plant. For example, know the PWM frequency and look for artifacts, small or large, at multiples of that frequency, offset from center frequency. Also calculate any likely intermodulation frequencies (2A ± B) that might be present from the interaction with other nearby AM stations and examine those. Any of these emissions might be compliant with the mask but indicative of potential future problems, so a quick look before starting the mask measurement is prudent.

Select a span (the start and end frequencies on the display) to capture the desired characteristics. 20 kHz/division will capture the mask out past the ±75-kHz break points. If needed, narrower spans can help show detail where the emissions may be close to the mask. Be careful not to expand the span beyond what the 300-Hz RBW bins can cover (e.g., with a 1000-point horizontal sweep, 1000 × 300 = 300-kHz maximum span). Spectrum analyzers usually indicate when the RBW and the span are not compatible.

A larger RBW is allowed for by §73.44(a), but it is not helpful: ". . . a wider resolution bandwidth may be employed above 11.5 kHz to detect transient emissions." First, the DANL goes up with an increased RBW. In addition, the total PSD of a noise-like waveform or environmental noise will appear higher with a wider RBW. If an out-of-band emission is noisy and is not a single-frequency component that is constrained to less than 300-Hz bandwidth, widening the RBW will make that emission appear higher on the spectrum analyzer. Likewise, the goal of looking for "transient emissions" is best addressed with the prescribed practice: capture repeated sweeps over the ten-minute period with the max hold feature. In short, it is disadvantageous to use a wider RBW outside the necessary bandwidth because it will only elevate the resulting trace with respect to the mask.

The "Is-the-Distortion-Internal?" Test

A well-known technique helps determine whether a spurious signal on a spectrum analyzer is generated by the instrument or is really picked up over the air. Because internally generated spurious product levels are nonlinear with respect to the levels of the fundamental frequencies under test, changing the input attenuation will aid in the diagnosis. For example, a 5-dB reduction in the input level using the internal attenuator will leave the desired signal at the reference level on the display because the instrument adjusts the display to compensate for the attenuation setting. The signal level reaching the first active stage of the analyzer is reduced by 5 dB. The spectrum analyzer noise floor will come up 5 dB. Meanwhile, any internally generated spurious signals will go down. Third-order products will drop by 15 dB, while second-order products will drop by 10 dB.

If it turns out the spur is internally generated, revisit the gain structure of the measurement setup to better fit the signal under test within the bounds of compression, DANL and distortion. Excluding unwanted signal energy with filtering and antenna positioning may help clear things up.

Harmonic Distortion

The legacy analyzer had a second harmonic distortion specification of −66 dBc or lower. This would not have been sufficient for performing harmonic measurements with the legacy analyzer. Fortunately, the tendency was and is to use a field intensity meter with calibrated antenna to measure harmonics, leaving the spectrum analyzer to measure the mask and the nearby out-of-band emissions.

However, with an antenna properly calibrated at AM fundamental and harmonic frequencies, a clean spectrum analyzer would be up to the task of performing harmonic measurements.

In addition, analyzers with less than stellar harmonic performance can be brought in to line by notching or bandstopping the desired signal. This will reduce the analyzer's tendency to generate internal harmonic distortion. Distortion is nonlinear, so a 15-dB notch or high-pass filter on the fundamental frequency could reduce internal second harmonic distortion automatically by 30 dB. This would be enough to bring the legacy analyzer, Model 1 and Model 2 into line for harmonic measurements.

Measuring Harmonic Emissions with a Spectrum Analyzer

To make a harmonic measurement, the antenna and notch filter must be calibrated for the harmonic frequencies. Establish the reference level of the desired signal without the notch in place. Insert the notch.

Measure the harmonics. Account for any insertion loss induced by the filter at the harmonic frequencies. Use the antenna factors at the fundamental and harmonic frequencies (explained in a later subsection) to normalize all measurements to field strength. Compare the resulting fundamental measurement with each harmonic.

Finally, it is often overlooked that the mask criterion of performing PSD measurements at 300-Hz bandwidth applies to all spurious emissions, including harmonics. When using a field intensity meter to measure the harmonics, the station accepts a mild disadvantage. The full channel power of the harmonic product is measured with the field intensity meter. If one were to use a spectrum analyzer according to §73.44, that harmonic product would be swept with the 300-Hz RBW filter, resulting in lower readings. Recall that §73.44 says the gold standard for measuring spurious emissions (which includes harmonics) is the spectrum analyzer EPM method.

Measuring Harmonic Emissions with a Field Intensity Meter

Field intensity meters are capable of looking at the second or higher harmonics (depending on the fundamental frequency and the range of the instrument). It is important to note that because the harmonic frequencies are by definition far removed from the fundamental frequencies, an antenna that is calibrated at the fundamental and continuing through the harmonics must be employed. A commercial off-the-shelf AM antenna is typically not calibrated. Field intensity meters are equipped with and calibrated to antennas that cover the lower harmonics of the AM band.

Follow the instrument instructions for calibrating the instrument at the fundamental frequency. Then take a measurement of the station, making sure that the measurement is not polluted by environmental noise. Obtaining a good peak on the azimuth to the station and a good null at right angles to the station helps confirm this. This is the reference level for the harmonic measurements.

Then tune to the second harmonic and recalibrate. Verify the measurement is not polluted by local noise and record the result. Repeat for any additional harmonics sought.

If the recorded level of the fundamental and harmonics are in dB (dBV/m, dBmV/m, etc.), then subtract the fundamental measurement from the harmonic measurement(s) to obtain the value in dBc. This produces a negative number, which indicates the harmonic is less than the fundamental by so many dB. Ensure that the measurement units of the fundamental are the same as those of the harmonic measurements.

If the measurements are not recorded in dB but in V/m (mV/m, μV/m, etc.), then the fundamental and the harmonic values must be calculated this way:

$$20\log_{10}\frac{E_H}{E_f} = dBc$$

Where E_H is the voltage (field intensity) of the harmonic and E_f is the voltage (field intensity) of the fundamental. To comply with FCC Rule Section §73.44 spurious emissions, including harmonics, are required to be attenuated more than 80 below the unmodulated carrier level, which is to say, algebraically, that harmonics each should be less than −80 dBc. This applies to stations operating at and above 5 kW.

For stations operating between 158 and 5000 watts, the −80-dBc/300-Hz requirement relaxes based on the station power. Use the equation on 73.44 or read the graph on NRSC-2-B to establish the beyond-75 kHz limit level for stations operating in these power ranges.

Antenna Factor

If measuring harmonic or other spurious emissions with a spectrum analyzer, it is necessary to have a table of antenna factors for the antenna. The antenna factor (or "*k*" factor) establishes the relationship between the received signal strength indicated on the analyzer display and the actual field strength of the signals under test. While the antenna factor accounts for the gain of the antenna at each frequency of interest, it also accounts for the wavelength at the frequency of interest. Together, the gain and the wavelength combine in an equation to form the "effective area" or "effective length" of the antenna at each frequency of interest.

Adding the antenna factor (in dB/m) to the received signal voltage on the analyzer (in dBV) yields the field strength (in dBV/m). In this equation, the value "*E*" is field strength while "*V*" is the voltage read on the instrument.

$$AF_{dBm^{-1}} = E_{dBV/m} - V_{dBV} = E_{dB\mu V/m} - V_{dB\mu V}$$

The derived field strength of the carrier can be directly compared with the derived field strength of the harmonic or other spur to determine whether it is less than −80 dBc. Alternatively, antenna factors can be put into spectrum analyzers, and the analyzer can be set to read directly in dBV/m.

HYBRID AM HD RADIO SIGNALS

The procedure for measuring hybrid AM HD Radio[10] signals is documented in the NRSC G-201-A Guideline, available on www.nrscstandards.org. It provides a recommendation for taking three sets of spectrum analyzer measurements. NRSC G-201-A emphasizes that this is the NRSC-recommended practice, but it is not officially adopted by the FCC.

NRSC G-201-A recommends that the digital carriers be deactivated and a regular §73.44 EPM be performed on the modulated AM signal in the normal fashion. Next, with the digital carriers still off, perform a hybrid AM HD Radio scan, using the 100-sweep, 30-second average with RMS detector. The goal of this scan is to verify the analog modulation is confined to the 5- (or optionally 8-) kHz bandwidth required for hybrid operation. Finally, activate the digital carriers and remeasure with respect to the hybrid AM mask.

SUMMARY OF CONCLUSIONS

To summarize, AM EPMs require a spectrum analyzer that has the necessary specifications to make mask measurements. Attend to the following to determine whether an analyzer is right for the job.

- DANL
- Phase Noise
- Distortion
- 1-dB Compression Level
- Expected Received Signal Level.

Select a location at 1 km from a non-directional station or a location on the crossover point between the non-directional signal level at 1 km and the directional pattern with the same signal level at 1 km. Alternatively, select a point within the main lobe and employ spurious emissions limit adjustment equal to the AM antenna gain at that point. Verify the location is free of significant sources of noise or interference.

Establish a reference level using an unmodulated carrier. Then set the instrument to peak detection and peak (max) hold. Activate a limit line if available. Verify 300-Hz RBW setting. Set the span as desired. Verify that the video bandwidth is either turned off or is set to a value of 3 kHz or more. Run the instrument in peak hold mode for 10 minutes with normal modulation. Save the trace. Log the usual information one would expect to find in a lab report. After capturing the necessary bandwidth and the out-of-band domain beyond the ±75 kHz breakpoint of the mask, look for spurious emissions at any likely suspected frequencies, including harmonics. If using the spectrum analyzer to do so, be sure to have a calibrated antenna with antenna factors at all frequencies of interest. Otherwise, switch to the field intensity meter; take a reference level reading and measure emissions in the spurious domain, including harmonics, making certain to recalibrate the meter for each frequency under test. Apply the antenna factors to compensate for changes in antenna sensitivity with frequency.

For multipattern directional stations, repeat the process for each antenna pattern, which may require changing locations.

The generous consultations provided by Messrs. Ronald Rackley, P.E., Grady Moates and Robert Shotwell are gratefully acknowledged.

Notes

[1] This chapter provides insight based on the accumulated experience of numerous station and consulting engineers and one Alternative Broadcast Inspection Program professional. It does not constitute legal advice, which stations should obtain from their FCC counsel.

[2] AM broadcasters have additional obligations to monitor and maintain their transmission plants under FCC regulations. This chapter addresses only the equipment performance measurements requirements.

[3] Harrison J. Klein, P.E. Hammet & Edison, Inc. "On Behalf of AM Improvement Committee National Association of Broadcasters, Sept. 11, 1986—Modulation, Overmodulation, and Occupied Bandwidth: Recommendations for the AM Broadcast Industry—An AM Improvement Report from the National Association of Broadcasters." www.nrscstandards.org/sg%20ref%20docs/am_mod_overmod_1986.pdf

[4] Mark Persons, Commentary: "Adjust Rules for NRSC Measurements," *Radio World*, Feb. 3, 2014.

[5] The 1988 NRSC-2 RF emissions specification was slightly modified by the FCC when codified in §73.44. In 2007, the NRSC-2-A revised standard harmonized its RF mask with §73.44.

[6] National Radio Systems Committee, " NRSC-2-B Emission Limitation for Analog AM Broadcast Transmission," Sept. 2012.

[7] Familiar third-order products are of the form 2A ± B, where A and B are two fundamental frequencies intermodulating. The third harmonic product is also a third-order product, mathematically: 3A.

[8] For the more adventurous, the 1-dB compression point is reported at the input to the active electronics in the analyzer. If the internal input attenuation is set to 0 dB, a −15 dBm input will drive the instrument into 1 dB of compression. If the internal attenuation is set to, say, 10 dB, it requires a −5-dBm input to drive the electronics postattenuator to −15 dBm. The display on the spectrum analyzer will account for the input attenuation and show the input level as the −5 dBm that it is. Input attenuation can buy the user some headroom, at the expense of raising the noise floor of the instrument with respect to the signal under test.

[9] A cogent argument can be made for performing EPMs with the RMS detector and max hold, but this will be saved for another forum.

[10] "HD Radio" is a trademark of iBiquity Digital Corporation (owned by DTS, Inc.)

CHAPTER

10.3

AM Radio Antenna System and Field Strength Measurements

JACK S. SELLMEYER, P.E.

Sellmeyer Engineering
McKinney, Texas

INTRODUCTION

In order to demonstrate a directional antenna system is in compliance with the terms of its FCC station license, a proof-of-performance is required. Certain arrays may employ Method of Moments analysis to demonstrate compliance. This is covered in detail in Chapter 7.13 "Antenna System Moment Method Modeling." This chapter will detail use of the traditional field strength measurement techniques for arrays that do not qualify for such analysis or where a preference for such techniques is appropriate.

PURPOSES OF FIELD STRENGTH MEASUREMENTS

Ground wave field strength measurements are used for the following purposes:

1. In support of directional antenna proof of performance applications for station license.
2. In support of allocation studies for new or improved medium wave AM broadcasting facilities in areas where the FCC Conductivity Map M-3 is known or suspected of being inaccurate.
3. Preparation of AM broadcast station coverage maps.

FCC REQUIREMENTS FOR FIELD STRENGTH MEASUREMENTS

Preparation for Field Strength Measurements

The applicable sections of the FCC rules (47CFR73) are as follows:

Sections §73.151 and §73.152 of the FCC rules define field strength measurements made to establish the performance of directional antenna systems. Section §73.182 defines the Engineering Standards of Allocation for the various classes of AM broadcast stations. The effective measured field strength (RMS) at 1 km is used to establish the efficiency of the antenna system. Radial measurements are required along radials used to define the measured radiation pattern for stations employing directional antenna systems and along each of the specified radials, and on bearings centered on major and minor lobes as well as the minima of the pattern. In certain cases where array symmetry produces identical null pairs, only one of the bearings needs to be measured.

Section §73.183 defines, in detail, the methods of determining ground wave interference using measured data which crosses various conductivity zones determined from radial measurements. Section §73.184 contains the FCC ground wave field strength graphs, which show the conductivity curves versus distance from the transmitter site. These curves are used to determine the distances to the various conductivity segments along the radial, and the unattentuated radiation at 1 km from the transmitter site. There are a total of 20 graphs for frequency segments from 540 kHz through 1700 kHz, each of which is divided into two sets of curves. The top set extends to 50 km and the bottom extends to 5000 km.

Notes

The ground wave field strength charts are no longer published but are available from the FCC website in PDF, JPEG and Postscript files from www.fcc.gov/

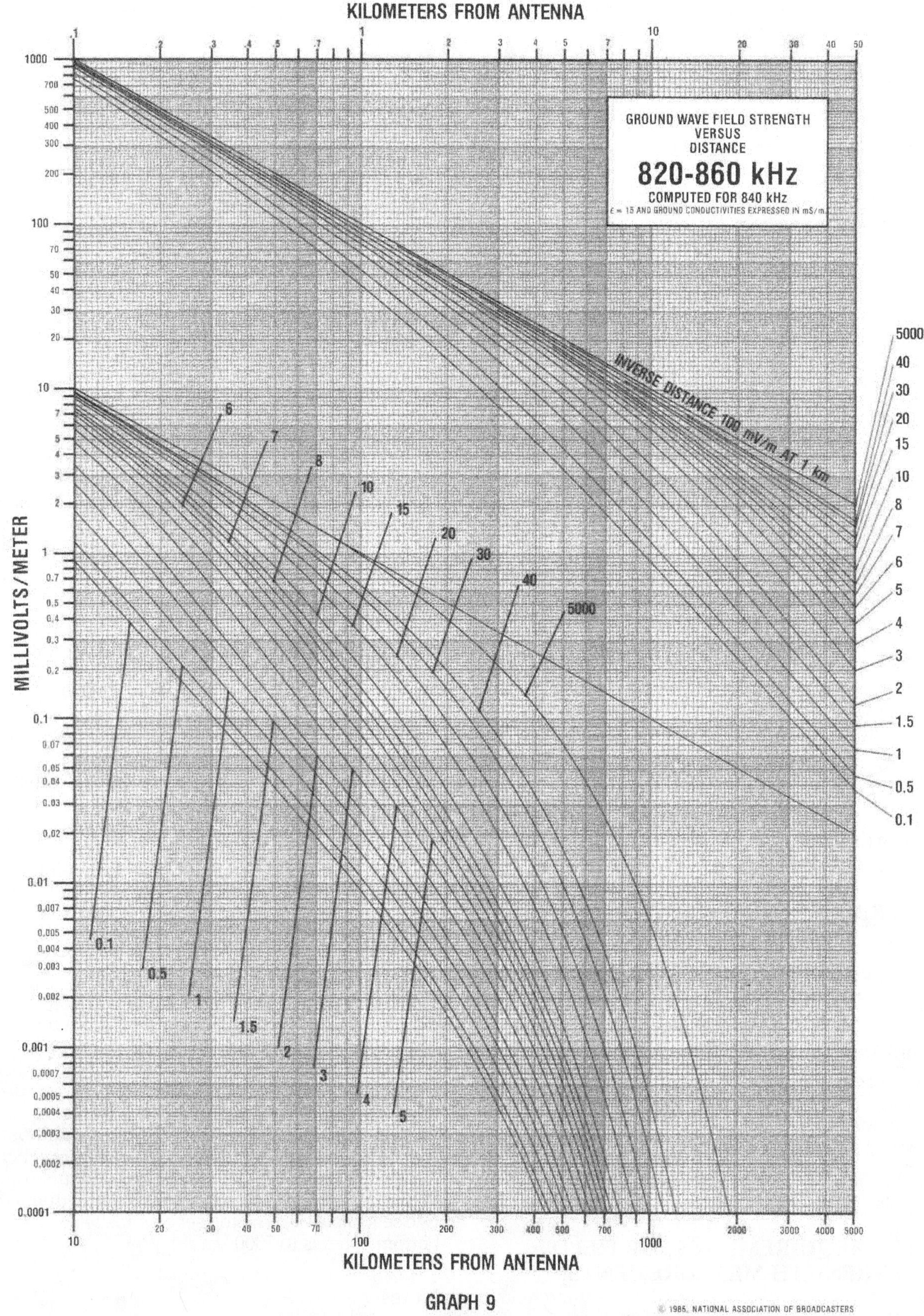

FIGURE 10.3-1 FCC Graph-9A 820 to 860 kHz

From FCC Rules—courtesy NAB

media/radio/am-groundwave-field-strength-graphs. The PDF files may be resized and printed on larger paper for ease of plotting, as described later in this chapter. Typically, the curves for the lower half of the band cover frequency spans from 20 to 40 kHz while those covering the upper half of the band cover spans between 40 and 90 kHz. The upper curves are used for analysis of measured data for directional antenna proofs-of performance. The lower curves are predominately used for daytime allocation studies where measurements over long distances are required to demonstrate that the measured conductivities are significantly lower than those shown in the FCC's Figure M-3 ground conductivity map.

Other sources for the originally published curves are post-1980 FCC rules published by the Government Printing Office or copies of Pike & Fisher's Broadcast Rules Service, which can sometimes be found at older AM transmitter sites. Each of these services contained full size 8-1/2 by 11-inch or 11 by 17-inch sets of curves that are suitable for use on a light stand for tracing the conductivity curves to suitable graph paper. An example appears in Figure 10.3-1.

FIELD STRENGTH MEASUREMENTS FOR DIRECTIONAL ANTENNA PROOF OF PERFORMANCE

Field strength measurements must be made with great care and attention to detail. Particular attention must be given to accuracy of each measurement point location. Accurate descriptions of each location as well as the distance from the center of the antenna array should be recorded on the log form together with the measured value, the time and date of the measurement and any notes peculiar to the location. The measurements taken during a full proof of performance will impact the maintenance of the array far into the future. The measurement locations will be used in the future to establish proper adjustment of the array following relocation of monitor points, repair work on the antenna system and any work above the base insulator on a tower. A little extra time spent documenting locations and making certain of the accuracy of all measurements will pay big dividends in the future.

Initial measurements in the non-directional mode are made to establish the inverse field of the antenna system in mV/m at 1 km from the array and to determine ground conductivity along each radial. This information will provide the basis for analysis of the directional data to determine the shape and size of the directional pattern in compliance with the terms of the underlying Construction Permit. It will facilitate location of the various field strength contours used for determination of coverage. Subsequent directional measurements are made at measured points beyond three kilometers and referenced to the non-directional measurements to determine the inverse distance field at 1 km of the directional pattern to establish compliance with the terms of the Construction Permit for each specified radial bearing.

Locating Suitable Measurement Points

A means of accurately locating suitable measuring points in conformance with FCC requirements is required for proper measurements. In the post-World War II period of AM radio broadcasting, USGS 7.5-minute topographical maps were used to locate the center of the directional array or the non-directional tower as required. A great deal of effort and drafting time was required to lay out each of the measurement radials. In current times this is an expensive option, but modern computerized mapping software, coupled with inexpensive GPS receivers, laptop computers and cell phones can significantly lower the cost of gathering the required measurement data. Computer analysis and plotting of the data significantly reduces the overall cost of the DA proof of performance while minimizing transcription errors.

Low-cost wide-area augmentation system (WAAS)-enabled handheld GPS receivers are readily available from Garmin, Trimble, Magellan and other manufacturers. Garmin also makes a small "puck" style WAAS-enabled GPS receiver that can be attached to the automobile windshield and connected to a laptop computer via a USB cable. Such a receiver can be used in a vehicle in conjunction with a laptop computer for navigation to locate suitable measuring points and to keep track of the measured data and point location descriptions. Garmin and others have applications available for Apple and Android cell phones that can transfer mapping data to and from Garmin Topo USA or similar mapping programs directly to the cell phone. This will reduce errors and save a large amount of time for preparation of a DA proof of performance. The cell phone, with the GPS application and the initial route mapping data transferred from the GPS receiver, can be carried to the exact measurement point, and the point location may be stored in the cell phone for direct transfer to the computer. This will facilitate integrated storage of the navigation and waypoint data along with the measured field strength measurements and notes.

The GPS receiver, the associated laptop computer and cell phone application should be set up for NAD-27 and True North reference, with distances in kilometers. Note that the array center coordinates in the FCC construction permits and licenses are specified in NAD-27 Datum. These coordinates should be used for the site reference. All tower registrations (antenna structure registrations, or ASRs) are specified in NAD-83 or WGS-84. It is important to use the proper datum reference for the array center coordinates. "The licensed transmitter site is specified in NAD-27 coordinates, therefore NAD-27 coordinates should be used in the mapping software." The exact location of each measurement is stored in the GPS receiver as an individual waypoint, along with the distance from the reference point and coordinates of the point.

Enter a route into the GPS receiver for each radial direction from the transmitter plant array center. Follow the receiver manufacturer's instructions for doing so. In general, this will involve entering the array center coordinates as the origin and a distance and bearing from this point to a point 1 or 2 km beyond the end of the last point on the radial. The GPS receiver will calculate the endpoint coordinates upon entry of a distance and a bearing. The route should be designated according to the radial bearing, i.e.: N-1.5E, meaning "North, 1.5 Degrees, East." This will generate a line with the origin at the center of the array extending to the endpoint of the radial. Individual measurement points may then be keyed in during the field measurement run along with the point number and stored as individually numbered waypoints. This procedure will store the radials and measurement point locations in the computer. Upon completion of the radial, the data is downloaded to the computer which will enable high accuracy point location data to be available for the initial measurements as well as for future measurements, in the event significant repairs to the array are required.

The computer screen will facilitate navigation to the locations that need to be measured, and the cell phone will allow off road-locations to be accessed and logged at the exact locations. Upon completion of the full radial measurements, the data from the cell phone is downloaded into the navigation/logging computer. The distance from the origin, the coordinates of the measurement points and the waypoint number are then automatically available in the computer files. This allows transfer of the data into an office computer, where it may be sent to an analysis program to automatically plot the measured data and analyze the Inverse Distance Field and the individual conductivity segments as well as produce a correct plot of the graph with all of the relevant data.

This effectively reduces the office time to plot and analyze a measured radial from a day or more to a matter of 1 to 2 hours.

Selection of Measurement Points

Measurement points should be selected with care to avoid, where possible, overhead power lines, nearby towers, large metal buildings and other objects likely to impair the accuracy of the measurement. In general, a separation of at least 100 feet from any large metallic object or power line should be maintained. The measurement points *must be accurately located on the map* and be on, or as close to, the actual radial as possible. Repeatability of the measurement location is the most important criterion in the selection of a measurement location. For this reason, a description of the actual location should be entered into the measurement log. It is common practice to locate a measurement point along a road at an easily identifiable location but to make the actual measurement in an adjacent field to avoid overhead power lines or other potential sources of reradiation. In this case the actual location should be marked by a paint stripe along the side of the road where the measurement is taken and a note recorded in the remarks column such as "25 paces west northwest of mark on road," or a similar statement that accurately portrays the actual location. The exact location should be entered as a numbered waypoint in the GPS receiver or cell phone.

The FCC Construction Permit will specify certain bearings on which a maximum radiated field limit will be specified in mV/m at 1 km. The non-directional and directional proofs of performance will demonstrate compliance with the terms of Construction Permit through a series of two sets of field strength measurements, one, using the non-directional antenna for reference and a second set of measurements using the same measurement points with the antenna system operating in the directional mode. The directional measurements are mathematically referenced to the non-directional measurements to demonstrate compliance with the terms of the Construction Permit.

Close-In Measurement Points

The "close-in" points typically begin at approximately five times the height of the towers from the center of the array. This equates to approximately 556 meters at 540 kHz (0.6 km) to 176 meters at 1700 kHz (0.2 km) for the non-directional measurements required to determine the inverse distance field for the reference field measurements. Points should be spaced according to the following list:

Distance	Interval
< 3 km	0.2 kilometers
3 km to 5 km	1.0 km
5 km to 15 km	2.0 km
>15 km	3.0 to 5.0 km

Initial GPS Settings

Set up the initial waypoints for each radial at 0.2-km intervals beginning at the appropriate frequency-related distance from the array reference point rounded to the nearest tenth of a kilometer. For instance: at 540 kHz, 556 meters rounds conveniently to 0.6 km. This will facilitate walking from the array carrying the GPS receiver or cell phone for guidance to each point. If using a cell phone, save each measurement point with a reference to the distance from the array center. A log sheet or notebook should be carried to enter the measured field at each point along with the waypoint number. It is not necessary to log a point description for each point on the close-in measurements, but permanent features at or near a particular point will facilitate future measurements. A location such as a cable terminal for an underground telephone, a power line or telephone pole number, cable junction box or a gas meter are examples of features that are not likely to change over a reasonable period of time and which will serve as good confidence builders as to the accuracy of the location many years later.

Set up additional points spaced 0.2 km out to 3 km from the array reference, for a total of 12 points in this case. Log the point number, distance, field meter reading, waypoint number, time and date for each point on the paper form and save the waypoint in the GPS receiver memory. This will allow repeatable measurements to be made at the original measurement points where required by FCC rules. Section §73.186 of the FCC rules requires a minimum of seven measurements within 3 km of the array and at least 15 measurements within 15 km of the array. It is desirable to measure more than the minimum number of points so that if one or more points are found to be abnormally high or low in the analysis process, they may be deleted from the list and still have sufficient points for an accurate analysis of the radial. A minimum of ten to 12 close in points are suggested for this reason. Similarly, a minimum of ten to 15 points beyond 3 km is suggested as well.

Close-In Measurements

The close-in non-directional measurements, commonly referred to as "walk-in or walk-out points," which originate near the array and terminate at 3 km from the center of the array, are used to determine the effective radiation efficiency of the antenna system. This is the "unattenuated field." It is specified in mV/m at 1 km from the array. The additional measurements on the radial beyond 3 km are used as the reference field for each measured radial. The ratios of the measured directional fields at each measurement point to those of the non-directional antenna are mathematically averaged to determine the directional antenna radiation on each measured radial. The directional antenna radial average is then multiplied by the unattenuated reference field determined for the non-directional antenna to determine the radiated directional antenna field. The "instrument of authorization," the Construction Permit, specifies the maximum permissible radiated fields in each direction. The measured directional radiation must be maintained at or below that value on each of the measured radials.

Drive-Out Measurements

The measurements beyond the initial 3-km "walk-out" measurements will be repeated on the directional pattern(s). These are often named "drive-out measurements" due to the distances involved. The starting distance for a directional antenna is typically ten times the spacing of the towers, about 3 km in most cases. A minimum of eight measurements are required on each of the measured radials beyond 3 km on each pattern. It is suggested that a minimum of 10 to 15 such points be measured on each radial in case analysis indicates that one or more points are not representative of the radial field as a whole. Since the majority of the drive-out measurements are on or near roads, it is suggested that the points be marked with yellow or orange paint in the form of a line on the side of the road nearest the measurement location in the likely case it is necessary to revisit the location.

Time may be saved if the transmitter pattern/power control system is accessible by remote control from a cell phone. Non-directional and directional measurements may be made at each location a minute or less apart with a single trip. Special care should be taken to provide positive identification of the measured pattern, and the common point/ND currents should be monitored remotely to be certain the power is correct for the selected operating mode, be it non-directional, directional daytime or directional nighttime, as appropriate. Often, the radiated fields will be significantly different for non-directional and directional operation on some radials but not others.

It is useful to enter manual waypoints in the GPS receiver at 1-km intervals beyond the 3-km point for reference purposes to serve as "placekeepers" for the drive-out measurements. These should carry waypoint designations such as 3.0, 4.0. . . 10.0, etc., to the end of the radial. As the exact measurement locations are entered along the route, the placekeeper entries can be modified and the actual measurement locations inserted. One or more measurements should be taken in the vicinity of each of the markers with the exact location stored as a waypoint. Point descriptions should be entered for each of these locations in the measurement log. The chosen locations should be easily accessible locations such as road crossings of the radial, and all such available locations should be measured in case one or more points plot significantly off the curve and, thus, may distort the true radiation value of the radial. Such points should be discarded from the final log. The GPS waypoint files should be saved in the project files for future use to locate the originally measured radials.

The complete computer file including point descriptions and notes should be saved and kept in a safe place for future reference. It is suggested that the map files, the measurement point GPS derived data and measurement log sheet be burned to one or more CDs and stored in different locations for future reference.

Measurement of the Field Strength

Measurements should be taken only after proper operation and calibration of the field strength meter has been verified.

First, determine that the meter has a current calibration date that will not expire while the measurement program is in process. The current FCC policy is within 2 years from the last calibration date.

Follow the meter manufacturer's instructions for calibration *exactly*. When calibrating the instrument, carefully tune the instrument to the carrier frequency of the station you are measuring. While adjusting the gain, be certain to rotate the meter 90 degrees from the source and adjust the gain slowly for the deepest null, which should be greater than 20 dB in depth.

When taking a measurement, hold the meter in the vertical plane directly in front of your eyes so that you may read the meter "head on." This will reduce the chance of error due to parallax effects. A tripod is useful for making accurate measurements but is not required. Carefully aim the FIM for maximum signal strength and check the bearing relative to the GPS indication. It should be in line toward the transmitter site. Rotate the meter 90 degrees off axis and note the drop in received signal strength. It should be at least 10 dB for a good location. It is not unusual to observe a depth of 20 dB or greater reduction in a rural area free of reradiation. Log the measured field together with the waypoint in your written log and store the waypoint in the GPS receiver. Move to the next location. Continue measurements until the end of the radial.

Spacing of Measurements

Measurements at distances greater than 3 km from the array center must be made at the same location for both non-directional and directional modes of operation. The points should be selected as noted earlier and should be spaced approximately as shown in the following table.

Distance	Spacing
3 to 5 km	1 km
5 to 15 km	2 km
> 15 km	3 to 5 km

Where the antenna system is rurally located and unobstructed measurements may be made, at least 15 measurements are required on each radial, of which at least seven measurements should be within 3 km of the antenna.

Miscellaneous Notes

Record the antenna monitor, unmodulated common point current and transmitter readings before beginning a measurement run and immediately following the measurement run. For non-directional runs, record the unmodulated antenna current (unmodulated) reading and transmitter readings before and after each run.

When making measurements for a directional antenna proof of performance, do not begin measurement runs until 2 hours after **local sunrise** and cease measurement runs by 2 hours prior to **sunset**.

On each log sheet, record the following data:

- Station Call Letters
- Frequency
- Station Location
- Operating Modes
- Power
- Meter Type
- Meter Serial Number
- Point Number
- Distance From Array
- Times of Measurements
- Date of Measurement
- Point Description/ Remarks

Remember to keep all data and notes in a neat and legible manner. It must be read by others during the analysis process and must be legible to avoid the necessity of remeasurement. Failure to follow these instructions to the letter may result in the need for remeasurement of the data. A GPS receiver and a notebook or laptop computer will significantly reduce the overall time required for the measurements program and improve the accuracy of the measurements.

A suggested log sheet format appears as Figure 10.3-2.

Plotting and Analysis of Measured Data

In the early years of the broadcast industry, the field strength measurement data gathered for directional antenna proofs of performance and allocation studies was hand plotted in ink on 8-1/2 by 11-inch log-log paper by an experienced draftsman. This was a very labor-intensive operation which is no longer a practical option due to the high costs of preparation, and our schools no longer teach this art.

Measurements for Directional Antenna Proof of Performance

The first step in the analysis process is the manual plotting of the measured field data for the non-directional and each directional pattern used by the station. This requires one sheet of 8-1/2 by 11-inch log-log graph paper for each pattern on each measured radial, as many as three sheets for measured radial for a two-pattern directive array. The field strength measured at each point is transferred to the graph in the form of a small inked circle at the appropriate location.

The FCC reference graph for the proper frequency range is taped to the light table, and the measured data sheet containing the plotted data is carefully aligned with the left and right edges of the reference sheet. As the distance increases from the transmitter site, the

CALL:		DATE:			CITY				INST:		S/N:	SHEET: OF:
FREQ:		TEMP:			STATE				COORDINATES			WEATHER:
RADIAL:	PT. NO:	DIST-KM	ND	TIME	DA-D	TIME	DA-N	TIME	N:DDD/ MM/SEC	W:DDD/ MM/SEC		DESCRIPTION

FIGURE 10.3-2 Suggested Field Measurement Log

Sellmeyer Engineering

measured data tends to fall below the inverse distance curve by varying amounts, depending on soil conductivity in the region. Groups of measured data tend to "bunch up" and track along one of the conductivity lines on the reference graph for varying distances, then show a distinct break and follow a different conductivity line for a certain distance. The "groups" with the same apparent slope are the individual conductivity segments with a specific conductivity value expressed in "millisiemens" (mS). A former expression was "millimhos" (mho). The data sheet is carefully moved up and down while maintaining alignment with the vertical edge lines until the best match of the various conductivity breaks is obtained. At this point the data sheet is taped to the table, and each conductivity group is traced in ink to the data sheet. The individual conductivity segments are labeled with the conductivity value of the segment. The inverse distance curve is then traced, and the value at 1 km is determined. This represents the radiated field at 1 km in mv/m.

Figures 10.3-3 and 10.3-4 illustrate the plotting and analysis process:

To illustrate the amount of work and detail required, a portion of the 1951 license application for the former Dallas/Ft. Worth, Texas, "share-time" operation on 570 kHz was extracted from the engineering portion of the application.

Figures 10.3-5 through 10.3-7 show a typical set of filing data in 1951 for a single radial N41.5°E of a two-pattern directional antenna system from that era. The pages show the plotted results of the non-directional, daytime directional and nighttime directional patterns along with the tabulation and plots of the measured data for this single radial. A total of 11 measured radials were required for this license application, with all three patterns measured on each radial. This required several crews and more than 6 weeks of measurement time to complete the field measurements.

All of the drafting work was done by hand including manual preparation of the data sheets and plotting of each of the measured radial maps along with plotting of each measured point location. The finished maps were then photographically reduced to 8-1/2 by 11-inch sheets for the filing. A typical proof of performance report from this era was usually bound in a "Redi Cover" hard-cover binder, which allowed for easy removal of one or more sheets for close examination. For a project of this magnitude, the overall thickness was on the order of an inch.

Substantially all of this type of work is now automated in software, with the measured radials automatically drawn using the individual waypoints taken from the GPS receiver files plotted on the appropriate segments of the USGS topographic maps, which have been digitized.

The individual radial measurements may be downloaded from the appropriate Excel spreadsheet columns from the measurement log, directly into the radial analysis program to determine the DA/ND ratios as numerical and log ratio values for inverse distance field calculations.

All of the measurement acquisition and analytical work required for the following section was done using computerized mapping, storage and analysis methods. This shortened the preparation time of the application by several hundred man hours.

Measurements for Allocation Purposes

A set of data taken, analyzed and used in a successful FCC application to relocate a pre-World War II Class III AM station from Wichita Falls, Texas, to Plano, Texas, a distance of approximately 120 miles southeast of the original site, is shown for illustration of the process of analyzing field strength measurements in an AM allocation context. The allocation work involved careful optimization of a two-pattern directional antenna system to provide optimal coverage over the entire Dallas/Ft. Worth Metro Area during daytime and nighttime hours. Fortunately, the station operated on 620 kHz, had a relatively low nighttime interference limit and was located in an area of relatively high conductivity. The Dallas/Fort Worth Metro Area is located in an area of similar conductivity with typical conductivities of 15 mS west of a north/south line running through the center of the Dallas/Ft. Worth Airport and conductivities in vicinity of 30 mS east of the line.

The field work required extensive measurements on three radials of one existing co-channel station and two radials on one adjacent channel station located east and southeast of the market to allow the daytime pattern to be optimized for best possible coverage. The resulting analysis of the data taken on the chosen stations toward the Dallas/Ft. Worth Metro Area revealed significantly lower conductivities than those shown on FCC Map M-3, a reduced size version of which is known as Map R-3, included as Figure 10.3-10 at the end of this chapter . These measurements allowed significantly higher radiated fields in areas critical to the best full market coverage.

Three radials were measured on Station WJDS, 620 kHz, licensed to Jackson, Mississippi, and two radials on KARV, 610 kHz, Russellville, Arkansas. An additional radial was run from the then-existing daytime site of KSLR, 630 kHz, licensed to San Antonio, Texas, but the conductivity toward the Dallas/Ft. Worth area shown on M-3 was in close agreement with the measured data, so no relief was possible toward the southeast in that case. This was probably due to density of stations in the DFW and San Antonio areas in the pre-World War II era. Apparently there was more measured data from the prewar era on file with FRC and FCC than was available from the more sparsely populated area toward Arkansas.

The graph of Figure 10.3-8 and the data tabulation labeled Figure 10.3-9 is the 276 degree radial from Station WJDS, 620 kHz, Jackson, Mississippi, toward the proposed 620-kHz transmitter site near New Hope, Texas.

FCC Graph 3 of the "Groundwave Field Intensity Versus Distance" files for the range from 600 kHz to 620 kHz was used for analysis of field strength measurements that resulted from traveling along three

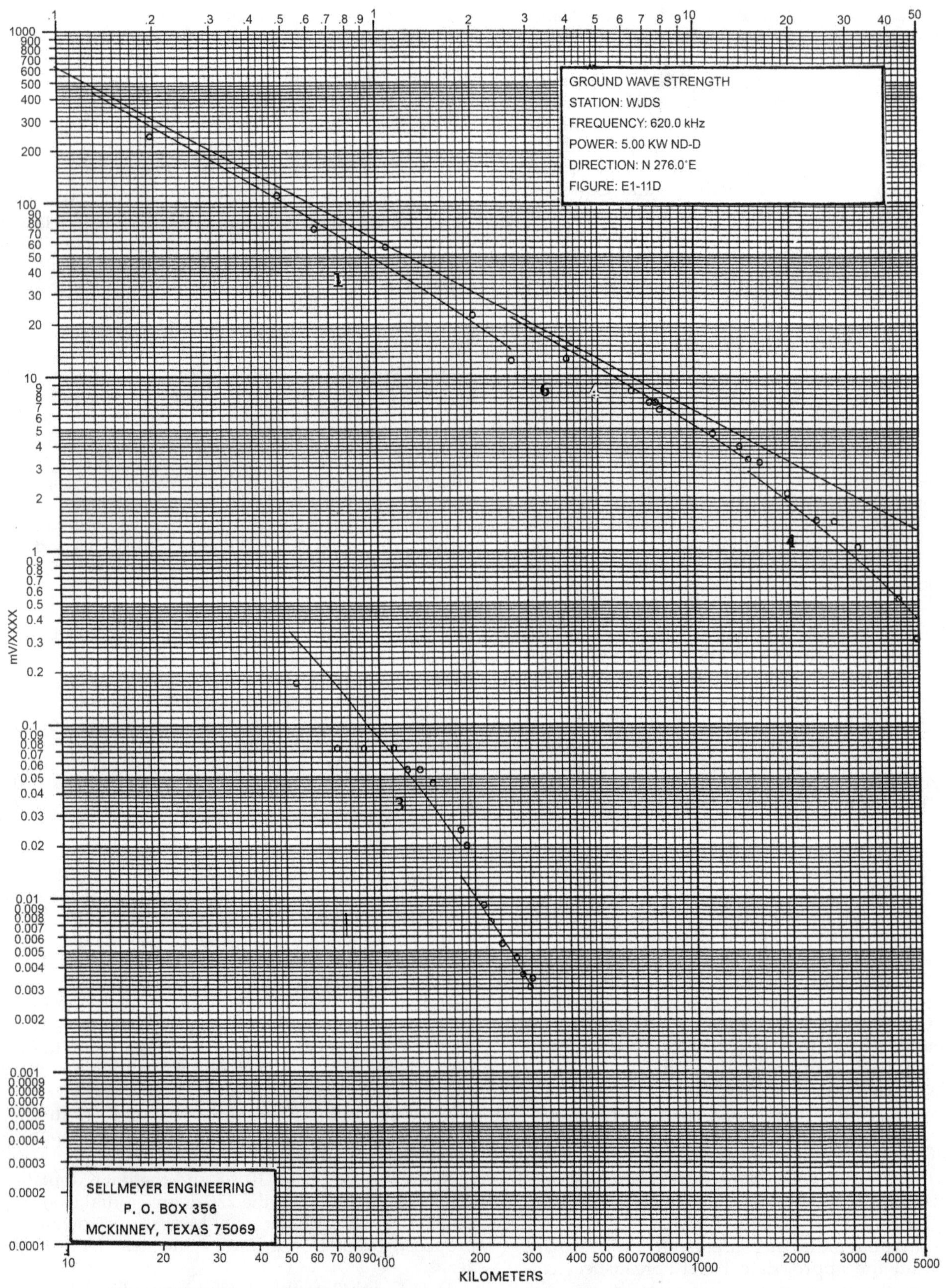

FIGURE 10.3-3 Plotted Data Sheet Ready for Analysis

Cavell & Mertz

SELLMEYER ENGINEERING

BROADCAST AND COMMUNICATIONS CONSULTING ENGINEERS
P. O. Box 356 McKinney, Texas 75069
MEMBER AFCCE
(214) 542-2056

134

EXHIBIT E1-11C

FIELD STRENGTH MEASUREMENT DATA

Azimuth: N276.0 E

Call: WJDS
Frequency: 620.0 kHz
Power: 5.000 kW
Pattern: ND-D
Inverse Distance Field: 610 mV/M

Point Number	Date	Time	Distance (km)	Field Strength (mV/m)
1	931230	0700	.20	2700.00
2	931230	0703	.50	1220.00
3	931230	0705	.65	780.00
4	931230	0716	1.10	610.00
5	931230	0809	2.05	247.00
6	931230	0087	2.70	135.00
7	931230	0802	4.00	138.00
8	931230	0820	6.40	90.00
9	931230	0844	7.28	77.00
10	931230	0839	7.61	77.00
11	931230	0841	7.82	70.00
12	931230	0851	11.38	51.00
13	931230	0919	13.80	43.00
14	931230	0924	14.70	36.00
15	931230	1001	16.00	34.50
16	931230	1007	19.50	22.80
17	931230	1020	24.00	16.00
18	931230	1051	27.20	15.70
19	931230	1107	32.10	11.20
20	931230	1222	43.00	5.60
21	931230	1232	48.90	3.30
22	931230	1345	55.10	1.90
23	931230	1539	74.00	.80
24	931229	1616	90.50	.80
25	931229	1552	112.50	.80
26	931228	1658	124.00	.60
27	931228	1647	135.50	.60
28	931228	1629	148.50	.50
29	931228	1552	181.50	.37
30	931228	1530	189.50	.22
31	931228	1500	214.00	.10
32	931228	1446	225.00	.08
33	931228	1428	243.00	.06
34	931228	1345	270.00	.05
35	931228	1330	282.50	.04
36	931228	1318	298.00	.034
37	931228	1304	303.00	.038

FIGURE 10.3-4 Completed Data Sheet Ready for Publication

Cavell & Mertz

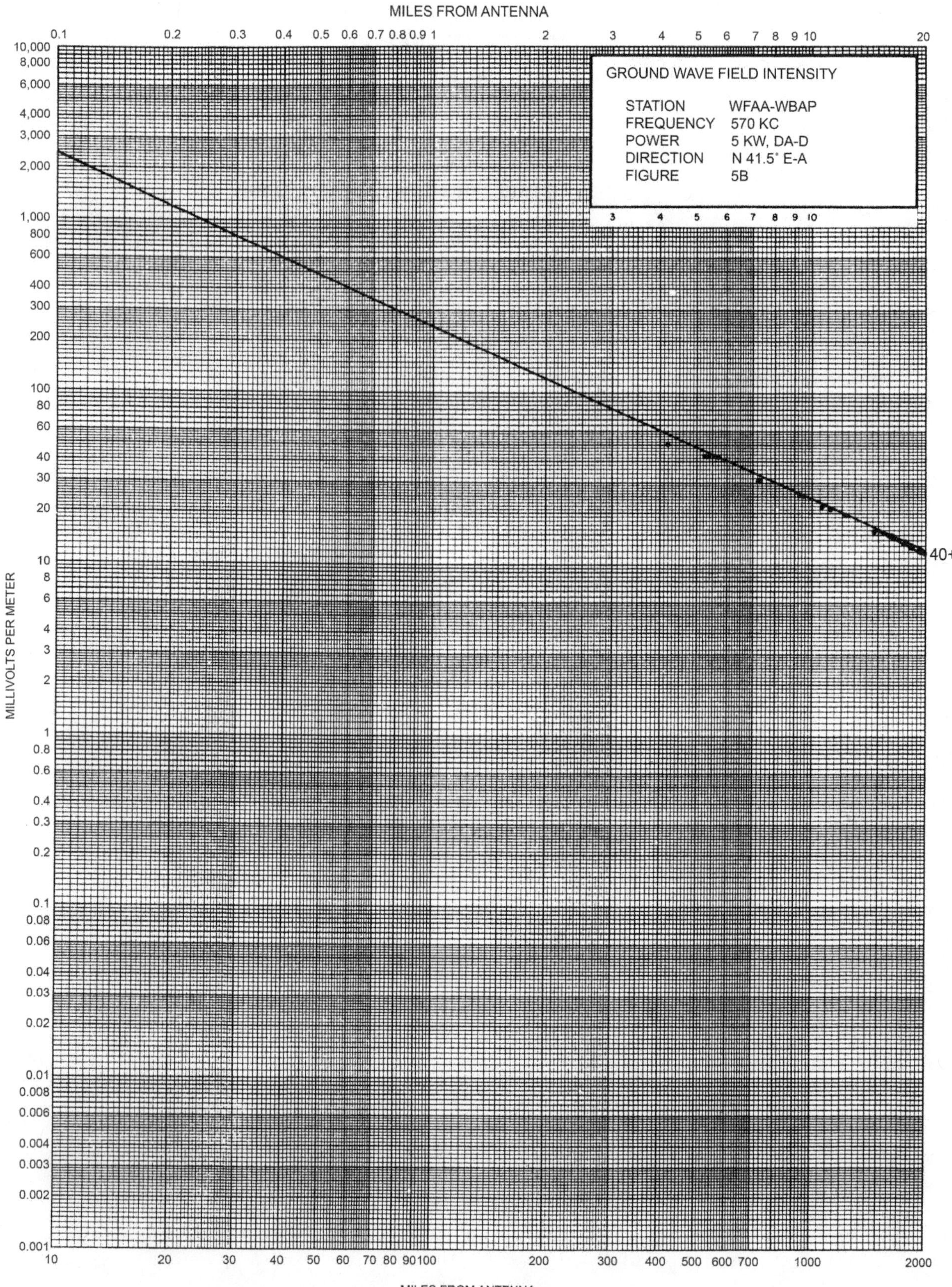

FIGURE 10.3-5 WFAA-WBAP 570 Radial N-41.5 E ND Fig.-5A

FCC Archives

RADIO STATIONS WFAA-WBAP
DALLAS-FORT WORTH, TEXAS
570 KC, 5 KW, DA-2

"A"

FIELD INTENSITY MEASUREMENTS RADIAL NORTH 41.5° EAST

		NONDIRECTIONAL		DAYTIME DIRECTIONAL		NIGHTTIME DIRECTIONAL	
POINT	MILES	MV/M	DATE	MV/M	DATE	MV/M	DATE
A-6	0.60	735.	A	-	-	-	-
A-7	0.70	630.	A	-	-	-	-
A-8	0.80	575.	A	-	-	-	-
A-9	0.90	495.	A	-	-	-	-
A-10	1.00	440.	A	-	-	-	-
A-11	1.10	400.	A	-	-	-	-
A-12	1.20	358.	A	-	-	-	-
A-13	1.30	335.	A	-	-	-	-
A-14	1.40	320.	A	-	-	-	-
A-15	1.50	290.	A	-	-	-	-
A-16	1.60	270.	A	-	-	-	-
A-17	1.70	255.	A	-	-	-	-
A-18	1.80	-	A	-	-	-	-
A-19	1.90	227.	A	-	-	-	-
A-20	2.00	217.	A	-	-	-	-
A-21	2.70	170.	A	-	-	-	-
A-22	3.32	135.	A	-	-	-	-
A-23	4.20	101.	A	50.0	B	13.3	B
A-24	5.23	85.0	A	43.2	B	12.0	B
A-25	5.37	85.0	A	43.2	B	13.0	B
A-26	5.58	83.5	A	43.2	B	13.1	B
A-27	5.70	84.0	A	42.5	B	13.4	B
A-28	7.20	62.5	A	31.3	B	8.50	B
A-29	7.30	62.5	A	31.3	B	8.50	B
A-30	9.33	49.3	A	25.3	B	7.00	B
A 31	9.48	49.3	A	25.3	B	6.95	B
A-32	10.6	40.7	A	22.0	B	6.15	B
A-33	11.1	39.8	A	21.1	B	6.25	B
A-34	12.3	37.5	A	19.7	B	6.30	B
A-35	14.6	29.2	A	15.7	B	4.95	B
A-36	16.0	29.2	A	15.0	B	4.62	B
A-37	16.2	29.2	A	15.0	B	4.65	B
A-38	16.3	28.2	A	14.6	B	4.62	B
A-39	17.3	26.2	A	13.5	B	4.35	B
A-40	17.7	24.9	A	13.7	B	4.40	B
A-41	18.1	24.9	A	12.9	B	4.00	B
A-42	18.5	24.2	A	12.9	B	4.20	B
A-43	19.2	23.0	A	12.1	B	4.13	B
A-44	19.4	22.8	A	12.0	B	4.10	B
A-45	20.0	21.3	A	11.8	B	3.75	B

NONDIRECTIONAL: 5,000 WATTS INTO CENTER TOWER
DIRECTIONAL: 5,410 WATTS INTO COMMON POINT

A: OCTOBER 1, 1951
B: NOVEMBER 14, 1951

511102

FIGURE 5D

FIGURE 10.3-6 WFAA/WBAP, Radial N-41.5 E Tabulation

FCC Archives

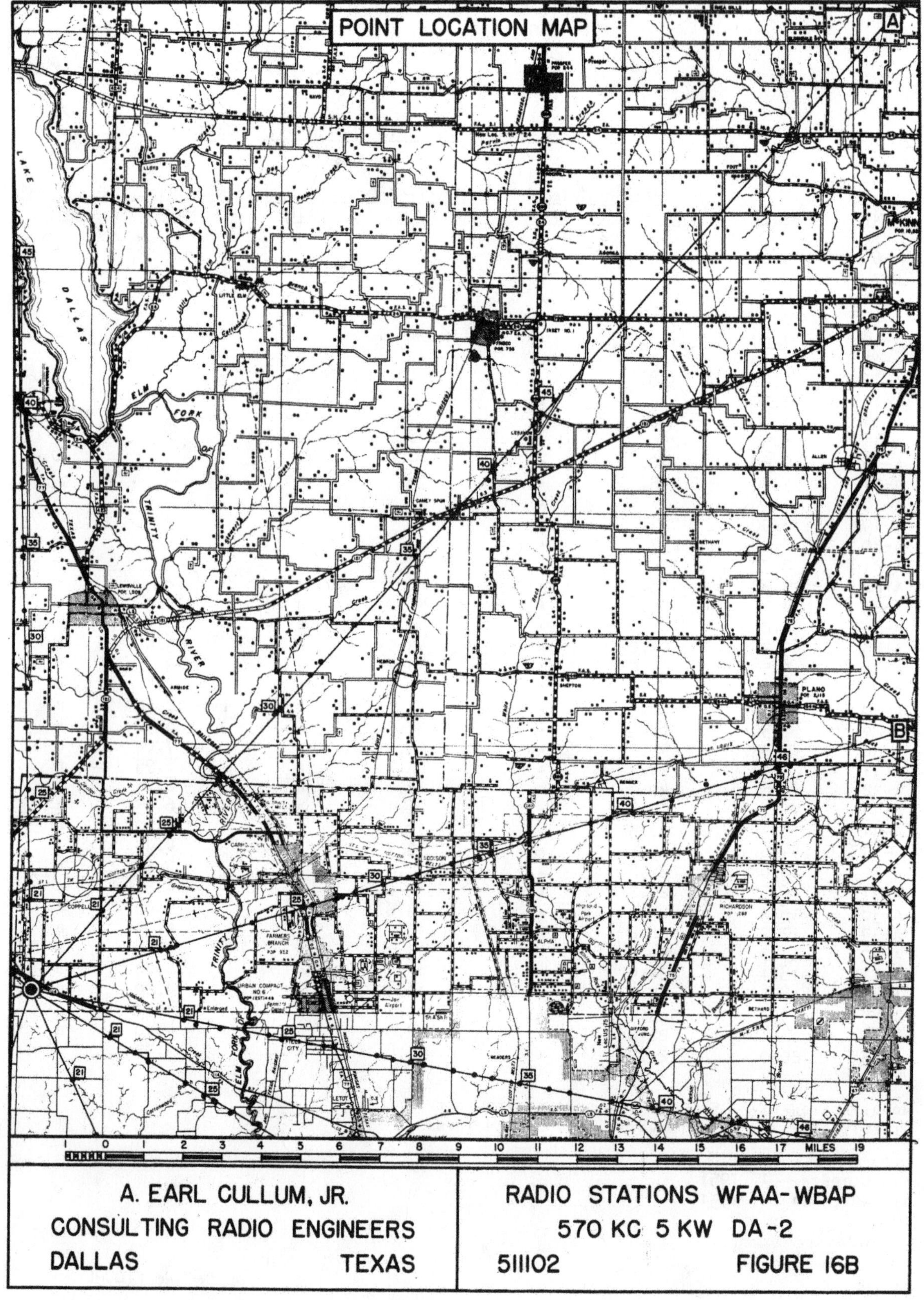

FIGURE 10.3-7 WFAA-WBAP 570 Point Location Map Fig. 16B

FCC Archives

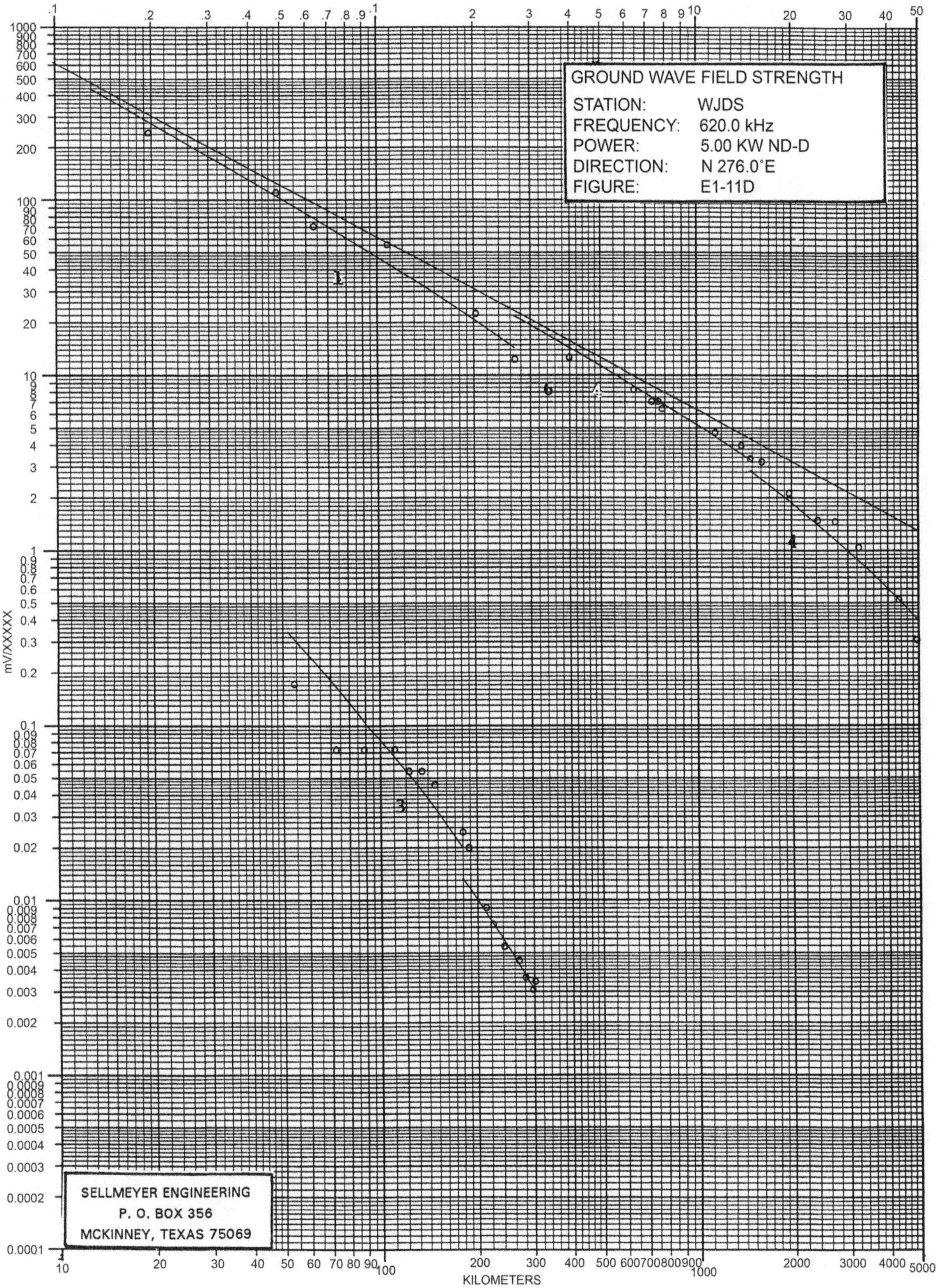

GROUND WAVE FIELD STRENGTH

STATION: WJDS
FREQUENCY: 620.0 kHz
POWER: 5.00 KW ND-D
DIRECTION: N 276.0°E
FIGURE: E1-11D

mV/XXXXX

KILOMETERS

SELLMEYER ENGINEERING
P. O. BOX 356
MCKINNEY, TEXAS 75069

FIGURE 10.3-8 WJDS-620 Radial N-276 ND Data Tab

Sellmeyer Engineering

SELLMEYER ENGINEERING
BROADCAST AND COMMUNICATIONS CONSULTING ENGINEERS
P. O. Box 356 McKinney, Texas 75069
MEMBER AFCCE
(214) 542-2056

134

EXHIBIT E1-11C

FIELD STRENGTH MEASUREMENT DATA

Azimuth: N276.0 E

Call: WJDS
Frequency: 620.0 kHz
Power: 5.000 kW
Pattern: ND-D
Inverse Distance Field: 610 mV/M

Point Number	Date	Time	Distance (km)	Field Strength (mV/m)
1	931230	0700	.20	2700.00
2	931230	0703	.50	1220.00
3	931230	0705	.65	780.00
4	931230	0716	1.10	610.00
5	931230	0809	2.05	247.00
6	931230	0087	2.70	135.00
7	931230	0802	4.00	138.00
8	931230	0820	6.40	90.00
9	931230	0844	7.28	77.00
10	931230	0839	7.61	77.00
11	931230	0841	7.82	70.00
12	931230	0851	11.38	51.00
13	931230	0919	13.80	43.00
14	931230	0924	14.70	36.00
15	931230	1001	16.00	34.50
16	931230	1007	19.50	22.80
17	931230	1020	24.00	16.00
18	931230	1051	27.20	15.70
19	931230	1107	32.10	11.20
20	931230	1222	43.00	5.60
21	931230	1232	48.90	3.30
22	931230	1345	55.10	1.90
23	931230	1539	74.00	.80
24	931229	1616	90.50	.80
25	931229	1552	112.50	.80
26	931228	1658	124.00	.60
27	931228	1647	135.50	.60
28	931228	1629	148.50	.50
29	931228	1552	181.50	.37
30	931228	1530	189.50	.22
31	931228	1500	214.00	.10
32	931228	1446	225.00	.08
33	931228	1428	243.00	.06
34	931228	1345	270.00	.05
35	931228	1330	282.50	.04
36	931228	1318	298.00	.034
37	931228	1304	303.00	.038

FIGURE 10.3-9 WJDS-620 Radial N-276 ND Data Plot

Sellmeyer Engineering

routes in excess of 300 km in length. Along this particular route, the FCC Map M-3 shows conductivities varying from 8 mS to a high of 30 mS near the proposed transmitter site. In this particular case, the measured conductivities along the five measured radials of the Arkansas stations ranged from 1 mS near the transmitter sites to a maximum of 6 mS along the entire routes. Two additional radials were measured either side of the direct site-to-site radial, with one area exhibiting a maximum conductivity of 7 mS. This made the difference between an excellent signal along the east side of the market and a poor one.

A Brief History of the Evolution of Maps M-3 and R-3 and the Development of Proofs-of Performance

Prior to April 5, 1954, the date that map M-3 became effective, a map existed within the FCC that used a legend of cross-hatching to define its conductivities rather than numbers appearing in its various zones. That map was based on sketchy field strength information supplied to the FRC and, later, to the FCC, and what the staff knew, or thought they knew, about propagation based on various soil types, landforms and terrain. The FCC did not require non-directional analysis of DA Proof radials until the early 1950s and, consequently, had little measured information in its files from which to verify the accuracy of the earlier conductivity map. As the years progressed, sufficient data were gathered to enable a collection of ground conductivity plots to be assembled and annotated. (See the National Bureau of Standards Circular 546, Effective Radio Ground Conductivity Measurements in the United States, prepared by R.S. Kirby, J.C. Harman, F.M. Capps, and R.N. Jones, USGPO, Washington, DC, February 1954.) From this collection of plots, FCC Figure M-3 and its smaller counterpart FCC Figure R-3 was developed. See Figure 1.2-4 in Chapter 1.2, Fundamentals of Broadcast Coverage.

Many of the larger prewar stations made "Field Studies" of their coverage in order to determine their practical service areas and to provide the sales departments with promotional materials to justify higher rates. A significant amount of the underlying measured data was supplied to the FRC and the FCC at the government's request to enable the design of prediction tools for estimation of coverage contours. These, in turn, were used to minimize mutual interference between AM broadcast stations. With the end of World War II, a surge in new radio stations took place, and with this surge came directional antenna systems to avoid interference to and from other stations during daytime and nighttime hours. Measurements were required on the directional antenna systems to verify the shape and size of the daytime and nighttime patterns. It quickly came to light that a means needed to be developed to accurately determine the size of the radiated pattern, or its efficiency.

The means of determining the size and efficiency of a directional antenna system was to require a non-directional proof of performance with accurately determined input power to the non-directional antenna system. The characteristics and efficiency of a single vertical radiator were well understood in this era from both a theoretical and practical standpoint. The radiated non-directional field would become the "calibration reference" for the DA pattern by use of a simple ratio of DA to ND radial measurements taken on several bearings of the directional antenna to define the shape and the size of the pattern. The radials were drawn from the center of the array, typically the non-directional antenna, to a distance of 10 to 20 miles from the array. The bearings of the radials were oriented along the minima and maxima of the DA pattern and toward the protected stations, with a minimum six to eight radials arranged as symmetrically as possible around the pattern.

The non-directional pattern efficiency was known by the electrical height of the antenna. A series of measurements were taken within 2 miles of the antenna, with spacings of approximately 2/10 of a mile along each of the measured bearings for the directional antenna to determine the RMS efficiency at 1 mile, later at 1 km, expressed in mv/m at the reference distance of 1 mile, later 1 km, when the metric standards were implemented. For determination of the radiation efficiency and pattern compliance of the directional array, additional non-directional measurements were taken along each of the radials at the same locations as the directional measurements. The DA to non-directional ratio was determined at each of the measured points. The numerical or log ratio average of the radial measurement is then multiplied by the analyzed non-directional field to determine the inverse distance field for the radial. This permits determination of compliance with the terms of the underlying Construction Permit.

Computer Analysis

Analysis of collected data can also be performed by modern software. One analytical model is described in "Systematic Bivariate Analysis of AM Field Strength Measurement Data" by Harry R. Anderson, P.E. While computer analysis can be efficient in terms of time, caution must be exercised in its use. In some cases, close-in data can result in different calculated inverse distance fields (IDFs) for slightly different measured values at short distances from the tower.

CHAPTER

10.4

FM Antenna Performance Measurements

MARTYN GREGORY
Shively Labs
Bridgton, ME

PETER LONG
Shively Labs
Bridgton, ME

INTRODUCTION

The transmitting antenna is the final link in the long chain from the studio microphone to the listener over which the broadcaster has any control, and yet it is often the least regarded portion of that chain. "Out of sight, out of mind" often best describes the engineer's approach to the antenna, and while a well-designed, well-implemented antenna installation should be able to be forgotten for the most part, the quality of its assembly and installation will determine for just how long it is likely to stay "out of mind."

FM radio antennas vary enormously both in size and complexity, although it can be generally said the two are directly proportional to one another. At one extreme we have simple low-power dipole antennas mounted singly or in arrays on a pole to serve a small community, while at the other extreme we have high-power arrays up to 160′ from top to bottom and some may be mounted well in excess of 1,000′ above ground level and serving many millions of people. Whatever the complexity or size of the antenna, however, all systems have the same two fundamental requirements: to radiate the necessary energy in the necessary directions and to do it efficiently and without incident.

This first requirement can be summed up by the term "directivity," which encompasses both the azimuthal plane radiation pattern and the elevation plane radiation pattern (VRP).

The second is more involved, being determined by the design of the system, which should both minimize losses within the system that ultimately result in energy being dissipated as heat and also minimize electrical discontinuities that result in energy being reflected. To the broadcaster, efficiency relates not only to wasted power but also to the overall reliability of the antenna system. With today's competitive market demands for minimal off-air time and energy conservation, the mechanical design of the system is as important as the electrical aspects, and of equal importance is the need to be able to demonstrate that the entire system is performing as designed. The many measurements involved will be discussed in detail throughout this chapter.

COVERAGE AND EFFECTIVE RADIATED POWER

The broadcaster's remit is to provide a service within an area as efficiently as possible, which effectively means using as little transmitter power as possible while achieving full coverage and building penetration. Wasted power is both an expensive luxury and a potential source of interference to other listeners when it falls beyond the bounds of the coverage area.

In an ideal world, and FM propagation being primarily "line of sight," the transmitting location would have a clear, unobstructed view of the service area and be located centrally within that area. In practice, licensing rules and geographical and topographical constraints quite often dictate otherwise, and it is for the radio planner to come up with a compromise between less-than-ideal site locations and an antenna "template" that will nevertheless achieve the coverage objectives. The antenna designer will turn the planning engineer's coverage template into a practical antenna comprised of one or more elements. The shape, topography and size of the service area will determine the azimuthal and elevation responses required, and these in turn will determine the necessary transmitter power to achieve adequate field strength (i.e. "coverage") throughout the area while minimizing the potential for interference beyond the service area. This latter point is equally important and is the basis for the FCC FM allocations nationwide.

The effective radiated power (ERP) is the product of the power reaching the antenna and the gain of the antenna in any particular direction.

Antenna gain has two components, due to directivity in the horizontal/azimuthal and vertical planes, and it is the product of the two maximum components of the gain, which determines peak gain and hence peak ERP for directional antennas. To complicate things somewhat further, the components of gain are considered separately, providing both horizontally and vertically polarized components (e.g., circularly or elliptically polarized antennas). Consideration of the gain in any particular azimuth and elevation will allow computation of the ERP in that direction and therefore the field strength resulting in free space at a distance from the antenna. ERP takes into account all losses between the transmitter and the input to the antenna and the horizontal and vertical radiation characteristics (or directivity) of the antenna. It is the final ERP in the required directions and the height of the antenna that ultimately determine the station's contour, or coverage area.

FM antennas vary from single-bay systems to 16-level arrays, and the ERP permutations vary equally widely.

Accurate determination of the gain is one of the most important measurement techniques that we shall be discussing in this chapter.

TESTING AND MEASUREMENT

Antenna performance testing is a complex and fairly lengthy process that begins with an assessment of the radiation patterns achievable with a given physical configuration of panels or elements. Often, this will be dictated by the dimensions of the support structure, and for side-mounted arrays, it is fair to say that the structure plays at least as big a part in the resulting pattern as does the antenna itself. These days, while antenna test ranges represent the ultimate pattern measurement approach (and in many situations, the only approach acceptable to the FCC), it is also possible to use sophisticated electromagnetic modeling software, factoring in the system geometry, structure, other appurtenances and computations of feed amplitudes and phases to produce theoretical horizontal and vertical radiation patterns on which coverage predictions can be made. The most sophisticated software, in the hands of an experienced user, can also give a very close representation of VSWR; antenna currents and voltages and most manufacturers are now making extensive use of such software to speed up the design process, which previously relied upon many mechanically complex prototypes. Whether computed azimuthal patterns (as opposed to measured patterns) are acceptable really depends upon the authority with jurisdiction over broadcast licensing and the application and class of license being sought.

Performance tests conducted during the manufacture of an antenna system and prior to its delivery and installation therefore encompass three basic elements to ensure that the requirements of the specification have been met

- Coverage—Radiation patterns
- Electrical performance—Impedance, voltage and power ratings
- Quality and integrity tests—Component assembly, grounding, voltage ratings, pressurization, etc.

MEASUREMENT OF RADIATION PATTERNS—AZIMUTHAL AND ELEVATION PLANES

Radiation patterns for FM antenna systems can be considered in two distinct categories. Antenna systems, which will be categorized as a non-directional antenna (NDA) or a directional antenna (DA), and in all cases the term "omnidirectional" refers to the azimuthal response of an NDA, whether it is horizontally or vertically polarized. A perfect NDA will have a completely circular azimuthal pattern, although to achieve this in practice with anything other than vertically polarized monopoles top-mounted in free space is difficult if not practically impossible.

Elevation patterns vary considerably from single-bay antennas to 16-bay systems. Antennas can be supplied with or without beam tilting and with or without null filling, depending upon the requirements of the broadcaster.

ELEVATION PLANE MEASUREMENTS, OR VERTICAL RADIATION PATTERNS

The elevation plane response, or VRP for a pure omnidirectional antenna, determines the effective gain of the antenna system and is directly proportional to the size, or *aperture*, of the antenna. *Physical aperture* is defined as the distance between the top of the uppermost element (or panel) and the bottom of the lowest element (or panel). Converting this distance to wavelengths at a given operating frequency results in a figure known as *electrical aperture*, from which the intrinsic gain of the antenna system can be derived mathematically.

The larger the aperture, the higher the gain of the antenna, and in principle it would appear possible to increase the aperture as required rather than use higher-power transmitters, but practical constraints limit the maximum usable gain as well as the coverage implications because the beam narrows as elevation gain increases. A series of examples shown in Figure 10.4-1 clearly illustrates this as well as the increasing need to be able to very accurately measure or, at

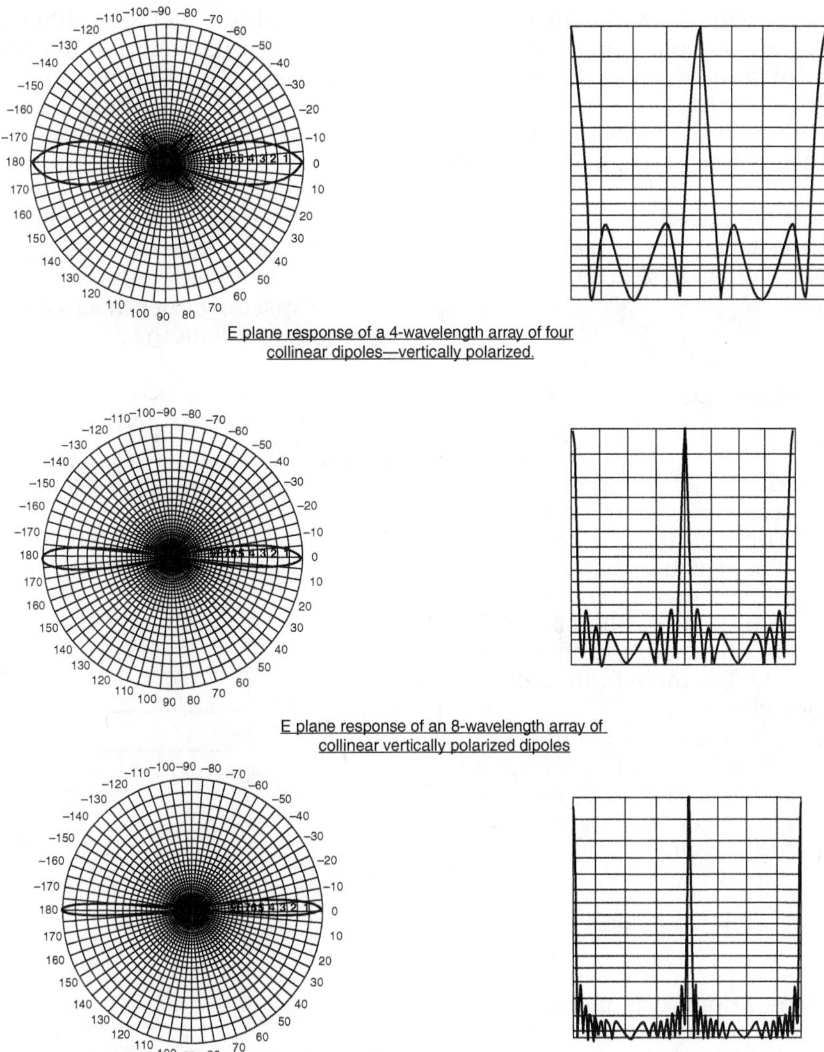

FIGURE 10.4-1 Note the requirement for increasingly accurate beamwidth measurements with increasing aperture/gain.

least, verify the response of a high gain antenna in the elevation plane.

Not only is it important to be able to verify the beamwidth and position of the various nulls, it is equally important to be able to verify any beam tilt angle and/or null filling that has been applied to the antenna.

Assuming a perfectly omnidirectional azimuthal (H plane) pattern, a long, uniformly illuminated aperture (A) of *isotropic* radiators will theoretically exhibit a gain of $2A$, and this expression is valid for apertures greater than $8l$ where the inter-bay spacing does not exceed l. With respect to a half-wave dipole that has a gain of 1.64 times that of an isotrope, the theoretical aperture gain becomes

$2A/1.64$

Maximum theoretical gain, $G = 1.22A$

(Only applicable to 8-bay, full-wave spaced arrays and higher)

While a useful rule of thumb, the gain achievable in practice is typically a little lower because of inefficiencies in illuminating the antenna aperture evenly, mutual impedances and other frustrations that generally occur in the real world.

Doubling the aperture of the antenna results in a nominal gain increase of two times. Structural limitations quickly determine the maximum antenna aperture that can be accommodated, and the largest FM antennas are generally limited to 16 wavelengths (16λ, which is around 160′ in linear distance), and apertures that large are relatively rare.

Two major problems begin to have an increasing effect on coverage. As the elevation (aperture) gain increases, the width of the main beam correspondingly decreases. Of more concern, though, is the increase in the number of side-lobes and the nulls either side of the side-lobes. In the service area, wherever deep nulls hit the ground, field strengths will be unacceptably low and a loss of coverage will occur. The main beam itself is aligned at 0° elevation angle, firing parallel to the ground. The surface of the earth is curved, whereas the beam follows a path tangential to the surface, causing a substantial proportion of the radiated energy to be lost over the horizon as wasted energy, which never reaches the service area.

Two mechanisms are employed to overcome the problems of nulls and wasted energy, namely null filling and beam tilting (see Figures 10.4-2a and 10.4-2b.

Beam Tilt

Beam tilting involves slewing the main beam downwards by a few degrees to minimize the energy falling outside the service area. It is accomplished in one of two ways—either by mechanically tilting the antenna (only feasible in limited scenarios and bringing with it another set of challenges) or by introducing a progressive phase delay down the aperture that electrically slews the beam. The example shown in Figures 10.4-2 and 10.4-2b has an exaggerated beam tilt of around 6° to better illustrate the effect. In practice, beam tilts of a degree or so are likely. Note: The main beam will slew in the direction of phase delay.

The Cartesian plot of the same antenna clearly shows the change in main beam angle and corresponding movement of the side-lobes. With this amount of beam tilt, the proportion of power "lost over the horizon" has been halved (–3dB), leaving only the problem of nulls falling within the service area.

Null Filling

The basic elevation plane response is modified to a cosecant approximation, and this results in a much smoother response either side of the main lobe, with reduction on the maximum to minimum excursions of the side-lobes (see Figure 10.4-3). This is achieved by varying a combination of the phases and/amplitudes of the feeds to each element of the aperture.

In most cases null filling is achieved by introducing a controlled perturbation of the feed phases only, which simplifies design of power dividers and antenna feed systems. It is, however, possible to adjust the amplitudes of the feeds to each element and apply a binomial distribution where extreme suppression of side-lobes and/or downward radiation is a requirement.

The spacing between successive minima or nulls in the radiation pattern is:

$$\sin^{-1}(1/A)$$

$$\text{Aperture, } A = n \times d$$

where n = number of elements, and d = the spacing between elements in wavelengths.

To remove the sharp nulls, and resulting service area field strength minima, the elevation response is

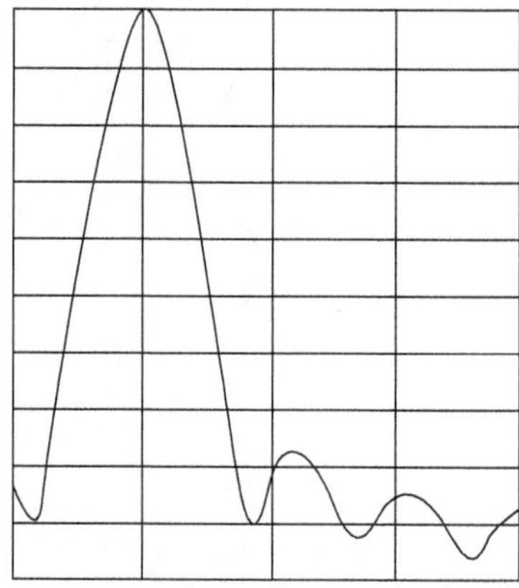

FIGURE 10.4-3 Elevation plane response of an antenna with null fill.

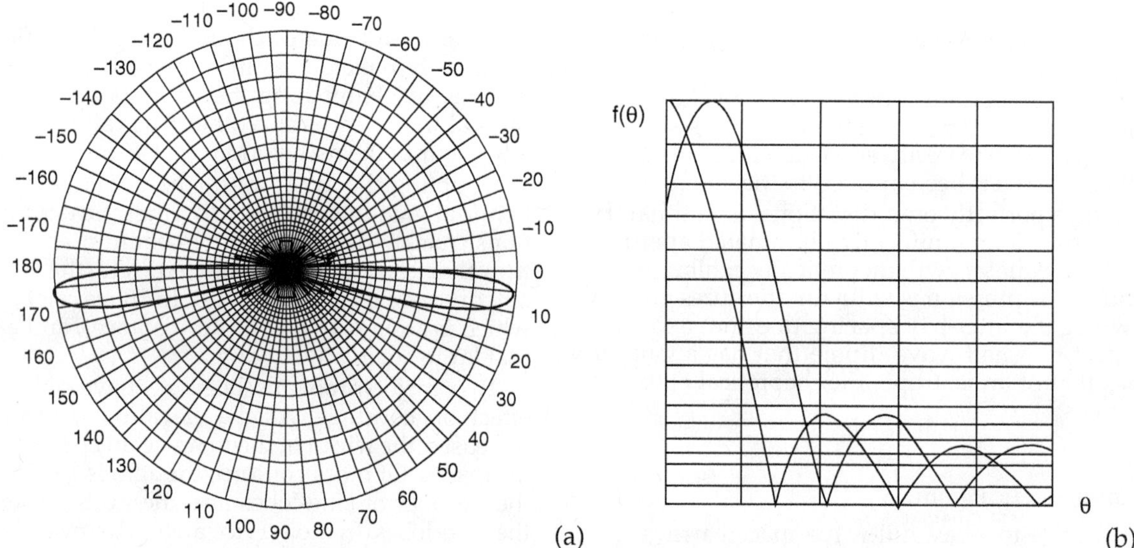

FIGURE 10.4-2A AND 2B Elevation plane response of a beam tilted array.

modified to approximate to a cosecant envelope of the form.

$$m \times \text{cosec}\,(\theta - q_r) \;/\, p\, n\, d$$

where q_r is the beam tilt angle
and m is a constant, normally greater than 0.5

Where nulls are filled in the VRP, the "extra power" is not free and in fact comes from the main beam of the response, thereby decreasing the overall available gain of the aperture. Antenna gain tables will frequently list "null filling loss" if applicable.

In summary, then, with so many options available to modify the elevation plane response, accurate measurement of the elevation plane response is as important as the measurement of the azimuthal plane response. It is achieved generally either by measurement of the feed amplitudes and phases to individual elements or by probing the radiated field from each element. In either case, the elevation response is computed based upon the measured parameters and the physical geometry of the array considering inter-bay spacing, frequency and the patterns of the individual elements. The elevation plane response is formed from the product of the element pattern and the array factor.

Radiation Pattern Measurement

Why make measurements at all? As previously discussed, with the availability of modern electromagnetic modeling software, it could be argued that antenna responses can be modeled on structures in a virtual environment equally as well as they can be determined using a pattern test range. The answer to this is that in some cases, it may be possible, and with an investment of tens of thousands of dollars in high-end simulation software, it almost certainly is possible. Whether it is a permissible approach, however, depends primarily on the regulatory authority. In the case of the Federal Communications Commission (FCC), measured patterns are still mandatory for directional FM applications, and this is certainly one way the FCC can ensure that the data is reliable, consistent and not the result of what may be a flawed or incomplete computer model of an antenna and its supporting structure. Garbage in–garbage out applies always!

Measurements of the antenna response or radiation pattern can be carried out directly in two ways, using either a purpose-designed full-size test range at the manufacturer's premises or by using a reduced size, or scale model, test range. An outdoor scale model test range is shown in Figure 10.4-4. There are advantages, of course, to both of these approaches, and as you'd expect, there are also disadvantages. Whether the antenna under test transmits or receives the signals is not important; reciprocity applies, and practices will vary from manufacturer to manufacturer.

To add further complexity, test ranges adopt different approaches to overcoming sources of error, and it is useful to have at least a working understanding of how and why these occur.

Whether an antenna test range is operating at the fundamental operating frequency of the antenna being designed, or is based upon very accurate scale models of the antennas and tower structures being designed, the fundamental sources of error are common to both and are as follows:

- Amplitude Error
- Phase Error

It is also necessary to consider the effects of any secondary reflections, whether these originate from structures "outside" the test range, including the ground, or from impedance mismatches within the system under test. Over the years, considerable time has been spent eliminating the ground reflection (shown in Figure 10.4-5 as the "Indirect Path") from elevated ranges, and from the '90s onwards, especially in Europe, with space constraints at some facilities, the use of time domain gating techniques became popular to eliminate all reflections arriving after the main contribution from the "Direct Path" (see Figure 10.4-5). The rush to adopt what was then new technology resulted in manufacturers using shortened but relatively elevated ranges so that the total length of the indirect path was greater than the direct path length, which facilitated gating it out from

FIGURE 10.4-4 An outdoor scale model test range for FM. The antenna under test is to the left, with the source antenna to the right, and the equipment building far right. Test range length as shown is 50′, and the range operates at a scale of 1:4.5. Shively Labs, Bridgton, ME.

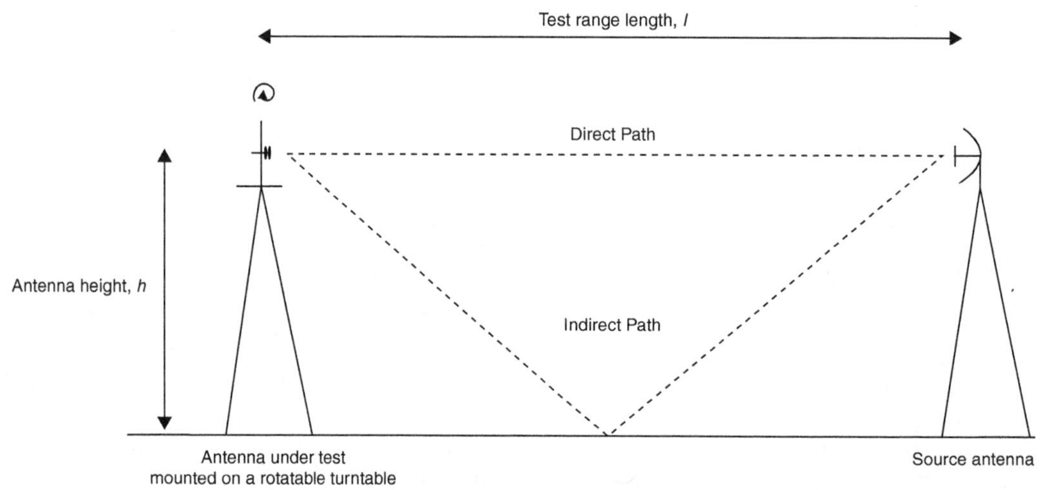

FIGURE 10.4-5 Test range direct and reflected paths.

the response of the direct path. The problems which became apparent subsequently were that reflections integral to the antenna system itself, in the case of broadband panel arrays, were also being eliminated, with the result that the patterns measured were not always indicative of the actual antenna pattern. Additionally, reducing the range length to facilitate gating out the indirect path contributions resulted in quite severe amplitude and phase errors being introduced, too, which far exceeded the relatively insignificant problems resulting from ground reflections.

Various other approaches have been tried over time, including the use of vegetation between the towers to scatter or absorb the indirect path contribution and diffraction "fences" to scatter it. One of the most important considerations, though, is to choose a source antenna that provides uniform illumination of the antenna under test but does not illuminate the ground (Figure 10.4-6 is an example of a source antenna). The beamwidth of the source antenna is therefore of paramount importance. A ground reflection, which is –30dB relative to the direct path contribution, can cause gain variations of +/– 0.25dB (arguably not too damaging, perhaps, at first sight) but it can also noticeably impact the accuracy of the side-lobe structure of the pattern measured.

So, when designing an antenna pattern measurement facility, it is first necessary to decide how to mount the antennas under test, how high above the ground they should be to reduce the effect of ground reflections and how far apart they need to be.

Traditionally, test ranges have been designed to make sure that all measurements are conducted in the far field of the antenna under test, and by definition this requires that there be a distance of at least:

$$Distance = \frac{2.D^2}{\lambda}$$

Where D = the largest dimension of the antenna under test,
and λ = wavelength at the test frequency

FIGURE 10.4-6 Source antennas—variable height and polarization. Note the different beamwidths adopted for different frequency bands.

Courtesy of SIRA Srl, Italy

A secondary restriction arises from the requirement to limit the phase error at the extremities of the aperture under test, too, and it is quite possible to satisfy the far-field distance requirement while still allowing excessive phase variation at the edges of the antenna under test.

TIP: If time domain gating is being used, ask the manufacturer to measure the antenna response with and without gating. Look for any substantial changes in the shape of the pattern to assess whether or not gating is hiding anything.

The most common errors occur when the range is either too short or the antenna under test itself is too large for the length of range available. The illustrations below illustrate how these errors can arise.

Considering the illustration in Figure 10.4-7, when the aperture, d, of the antenna under test becomes

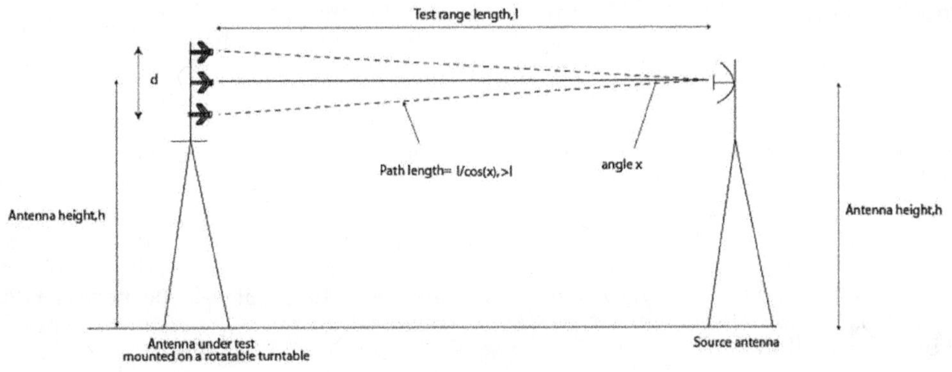

FIGURE 10.4-7 Illustration of phase error and amplitude taper.

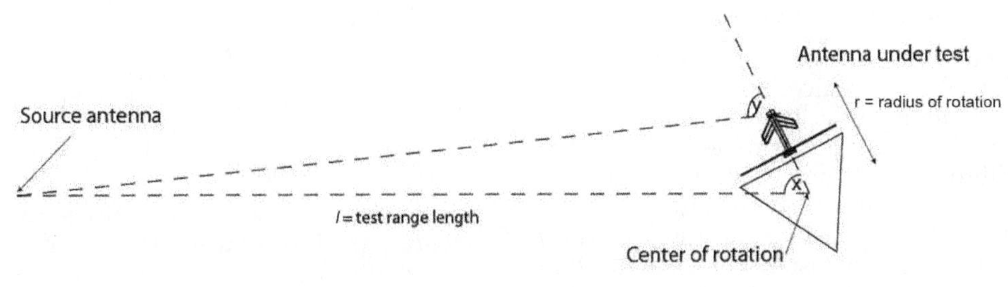

FIGURE 10.4-8 Illustration of test range amplitude error.

significant compared to the length of the test range, l; the time for the wave front to reach the center of the antenna under test will be substantially shorter than the time the wave front takes to reach the extremities of the antenna under test. As a result there will be a phase shift across the aperture of the antenna under test (known as "phase taper") that can result in significant errors if allowed to exceed values, in practice, of $\varpi/8$ radians (22.5°). To maintain the phase error below this level, the range must be long enough that $l >> d$ and the path difference between both the source and the center of the antenna under test and the source and the edges of the aperture under test is $< \lambda/16$.

Phase errors typically manifest as incorrectly measured nulls in the azimuthal pattern, and the indicated nulls are displayed as less deep than they probably are in reality.

Amplitude Error

Let's assume that the antenna under test is a single broadband panel mounted on the face of a tower and oriented radially to the tower face, as depicted in Figure 10.4-8.

The azimuthal response of the panel is a cosine pattern, $\text{Cos}^n(\theta)$, where n, in this instance, = 1, and θ becomes x.

Considering the diagram in Figure 10.4-8, it is evident that angle y, the angle between the source antenna and the antenna under test, is not the same as the angle of rotation, x. As a result, the amplitude of the measured pattern is no longer solely a function of angle x, but of angle y. As a result, the actual azimuthal pattern of the antenna under test, Cos(x), will be modified to Cos(α), where α can be shown to equal:

$$\alpha = 180 - (Tan)^{\uparrow}(-1)(([r - l\cos(x]) / \sqrt{(r^{\uparrow}2 - l^{\uparrow}2 - 2r\ l\cos(x)))}$$
$$180 - (Tan)^{\uparrow}(-1)(([r - l\cos(x]) / \sqrt{(r^{\uparrow}2 - l^{\uparrow}2 - 2r\ l\cos(x)))}$$

Make sure that the range length l is long enough to minimize amplitude errors reducing them to less than 0.5 dB in practice. The radius of rotation must be insignificant compared to the range length, l.

Test Ranges—Hands on Recommendations, by Peter Long, Shively Labs

Whether considering an authorized directional FCC pattern study or omnidirectional pattern analysis to optimize and improve coverage for an FM station, there are a number of considerations to take note of, because just mounting the antenna to the tower can cause serious problems in the resulting antenna pattern. Mounting structures come in numerous different sizes, shapes and styles. Poles can be from 3" to 36" diameter and larger and can also have climbing steps or ladders along with coax runs passing through the aperture of your antenna. Cell towers have sector frames for mounting cell antennas. Towers can verify from 12" to as much as 24′ across the face. The tower bracing (X, Z, K or zig-zag, for example) and the height of the bracing (also known as "panel height") will all have different effects on the pattern. In an attempt to control the impact of the tower design on the antenna pattern, it is also possible to have custom-designed tower sections fabricated to suit the frequency of operation of a side-mounted antenna. For a new tower build this may be a worthwhile aspect of the design, though adding to an existing structure is considerably more involved. One such option is the LAMBDA™ section, available from Electronics Research, Inc. (ERI), and is described by ERI as follows:

> The LAMBDA™ system's design considers the resonant and resulting non-reactive fractional lengths of your frequency. Each LAMBDA™ section's overall length is equal to the wavelength of the matching antenna. Every LAMBDA™ section is electrically symmetric to the antenna element's mounting position. The result is an identical, controlled energy distribution regardless of the element's position in the array.[1]

It is imperative also that any coax, waveguide, conduits, ladders and other metallic objects in or nearby the aperture of the antenna need to be considered and included in the model when doing pattern work. All of these items will to some degree or other act as aperiodic parasitic elements resulting in scattering and reradiation effects that will change the fundamental response of the antenna under test. Exact leg azimuths (true north) are needed to determine how the antenna will be mounted on the tower, to achieve the pattern requirements. Guy wire material and location need to be addressed. The location will obstruct the antenna from being mounted straight off the leg, and metallic guy wires will distort the pattern. After this information is reviewed, the manufacturer will construct a duplicate of the tower structure to run patterns. This can be done full scale or to a model scale. Using a model scale setup allows the manufacturer to run several patterns at a lower cost and more quickly than a full-size facility might permit.

In most cases it will require running several patterns to get an acceptable pattern. This will be done by changing mounting positions (leg or face), trying different antenna distances from the tower and rotating the mounts or bays. Manufacturers may also add vertical parasitic elements to optimize the pattern. If the antenna is directional, most times horizontal parasitic elements will be added.

Directional antennas should have space above and below for at least one wavelength, and this space should be kept clear of other appurtenances. Should this change in the future, due to the tower owner adding more users, it will be necessary to get a written certification from the antenna manufacturer that the additions will not significantly change the approved directional response of the antenna.

Some manufactures, as an added service when the pattern work is complete and they believe they have an acceptable pattern, will generate a field strength map for the guidance of the customer. They will do this with software from an accredited vendor. However, it is important to realize that any map is a prediction only and that many external factors can influence propagation and coverage once the signal has left the antenna.

Evaluating the Type of Test Range

There are several considerations to evaluate when deciding the type of test range: the directionality of the antennas to be tested, the testing results you require and

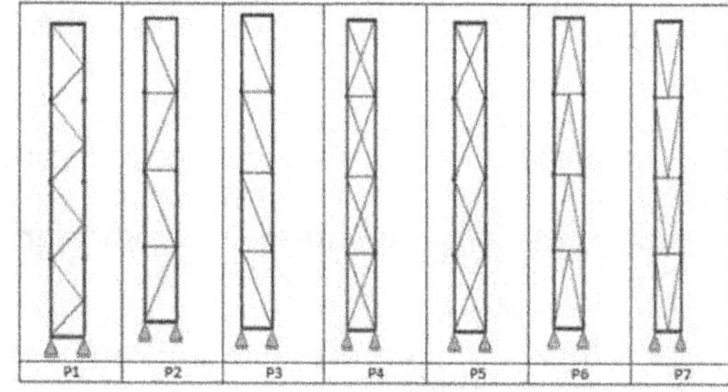

FIGURE 10.4-9 Various forms of LAMBDA™ tower sections.

FIGURE 10.4-10 Picture of an antenna at the test range, mounted on an accurate scale model of the planned tower structure. All range models should also include internal lines, coax sections, ladders and conduits as appropriate.

the range of frequencies for the antennas to be tested. Also, size, weight and other mechanical features of the antennas under test should be considered. For outdoor test ranges, considering the location is also important. The location needs to be free of any interfering RF and other signals created by reflections that could distort the accuracy and metrics of the antennas being tested.

Selecting Range Equipment and Instrumentation

Although there are several different types of test ranges, the equipment required is basically the same:

- A **signal source** to generate a signal to the antenna being tested
- A **receiving source** to acquire the signal
- A **control and positioning** system to accurately determine the azimuth angle of the antenna under test and therefore the azimuth angle of the data being transmitted
- A **processing and recording** system that will acquire the data being generated. A computer-based system with interfaces to the signal source, receiving source, and the control and positioning system will be required. The information would then be formatted and saved on the hard drive. The computer program being used to operate the range would then format the information into polar plots (**radiation patterns**), measured data, including **gain** figures, and contour plots.

Radiation Pattern

A radiation pattern is a graphical plot of the relative field strength transmitted or received by the antenna being tested. The pattern is generated by a number of points and then plotted on a polar chart. Manufactures may use different standards and formats. Some formats may have advantages in certain applications. The radiation patterns will show both the horizontal and vertical polarization. The polar plot will also show the azimuth **gain** and RMS (root mean square) of both the horizontal and vertical patterns.

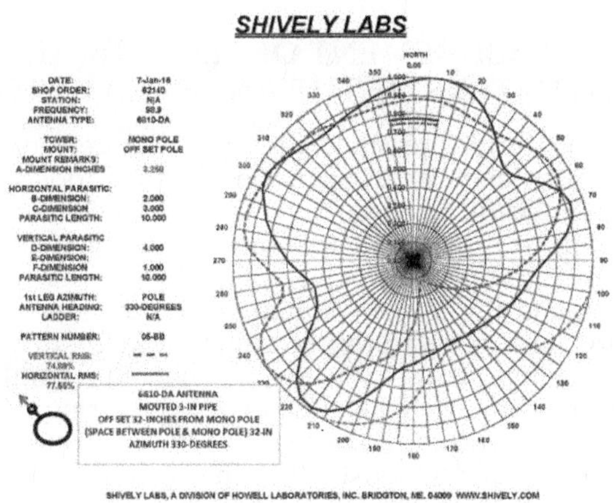

FIGURE 10.4-11 Horizontal and vertical polar plot of a directional FM antenna.

Gain

The azimuth gain of an antenna is determined by its directivity, or the pattern it generates. A high-gain antenna has a narrow, more focused pattern, whereas a low-gain antenna has a broader pattern. Patterns can be changed by adding parasitic elements or adding a structure behind the antenna like a panel. Power is not added to the antenna as gain increases, it is just focused in a narrower beam in a specific direction—analogous to comparing a 60 watt light bulb on its own to a 60 watt light bulb installed in a car headlamp. The power consumed is the same, but the pattern of light cast is completely different.

The gain figure will be required in certain regulatory authority applications such as directional antennas. The two components of antenna gain, the azimuth gain in the azimuthal radiation pattern and the elevation gain (number of antenna bays) are multiplied together to determine the antenna peak power gain.

$$\text{Azimuth Gain} = \frac{1}{RMS\ value^2} \frac{1}{RMS\ value^2}$$

In the example in Figure 10.4-11, the RMS value of the horizontally polarized component is 77.55%; the azimuth gain is therefore:

$$\text{Azimuth Gain} = \frac{1}{0.7755^2} \frac{1}{0.7755^2} = 1.663,$$

$$\text{Or, } 10\log_{10} 1.663 \text{ dB} = 2.21\text{dB}$$

Peak Gain, or Power Gain = Azimuth Gain × Elevation Gain

Types of Antenna Test Ranges

Far-Field Scale Model Range Used at Shively Labs

The scale of the test range is determined by the manufacturer and becomes a function of ease of assembly mechanically without requiring the ultimate in precision as wavelengths become tiny. A scale of 4.5 to 1 reduces all items in the model to just under a quarter of the full-scale size and is fairly typical for FM operations. The frequency at which the test range operates is then multiplied by 4.5 (ex: 98.1 MHz × 4.5 = 441.45 MHz).

The range would consist of two signal towers, a source and receiver, approximately 25′ in height above the ground and a minimum distance of 20 wavelengths (at the scaled frequency) apart. The lowest FM frequency of 88 MHz scaled to 396 MHz is used for this calculation, and it is assumed that the antenna under test is used as the transmitting antenna.

The source tower incorporates a rotating pedestal. The pedestal has a digital positioning encoder from 0 to 359 degrees, which should be accurate to 0.5 degrees and have zero backlash. A pair of crossed dipoles offset from a parabolic reflector is located on the receiver tower to determine the amount of energy in both the horizontally and the vertically polarized planes, radiated from the antenna under test.

Low-loss, very low-leakage cables are used to connect the network analyzer to the source and receiver towers, and an in-line amplifier is incorporated to partially compensate for the down-range free space path loss of almost 50 dB. This ensures that the received signals are robust and permits noise-free measurement of antenna responses down to −40dB with relative ease. A network analyzer operating in transmission mode is interfaced with a computer and the azimuthal position indicator to log the signal amplitudes every degree of rotation or less. The software used to collect the amplitude and phase measurements from the network analyzer system for the antenna under test is usually written by the manufacturer, but essentially they all operate to the same basic principles.

The software will typically generate a polar chart showing the azimuth pattern of both the horizontal and vertical polarizations. Amplitude axes can be scaled to linear or logarithmic (dB) values as required.

Full-Scale Far-Field Ranges

A full-scale far-field test range requires a much larger area to accommodate the equipment and heavy lifting equipment to handle the structures needed to perform the testing, but it ultimately operates under the same principles as the scale model range, except that the range is running at the fundamental frequency of the actual antennas and not that of the scale models.

Heavy-duty equipment is generally needed to move full-size tower sections and antennas into position onto the support structures, and quite often personnel will be required to climb the structure to make small adjustments. Due to the time involved in iteratively achieving a solution solely on the test range, operators of full-scale ranges will generally carry out much of the preliminary pattern development using very accurate computer models and then transfer the predicted antenna geometry to the test range for final validation and measurement. The basic equipment required for testing would be the same, transmitting, receiving source, control, positioning, and processing and recording.

FIGURE 10.4-12 Anechoic chamber for scale modeling FM antennas.

Full-scale ranges also include various subcategories; examples are ground-reflection test ranges and near-field test ranges, although from the perspective of FM antenna manufacturers, these will be very few and far between.

The only remaining ground-reflection range was decommissioned at the former Harris Corporation's antenna facility in Missouri in the early 2000s, shortly after being bought by a competitor.

Near-field test ranges rely on intense computational techniques but do allow measurement of relatively large antennas in a very small space by conventional standards that are used where large antennas are to be tested indoors in a relatively small space. In an indoor environment, the antenna will be placed in an anechoic chamber lined with RF absorbent materials (Figure 10.4-12) to eliminate any reflections.

A small RF sensing probe is mechanically scanned over a uniform surface[2] surrounding the antenna under test. The surface over which the probe tracks will typically be anywhere from 4 to 10 wavelengths from the antenna, and over this entire surface a dataset/matrix of amplitude and phase measurements is obtained as the probe scans. The data collected is, of course, in the near field, and E and H fields will vary tremendously. It is not possible to measure power as such because the Poynting vector ($E \times H$) is not established or stable. Transformation of this data and correction to far field is achieved by the use of Fourier transforms, requiring fairly intense processing capability. The transformed far-field data can then be displayed and analyzed in the same way as conventional far-field antenna measurements.

MEASUREMENT OF ELECTRICAL PARAMETERS OF THE ANTENNA SYSTEM

It is not normal practice for manufacturers to subject any antenna or transmission line components to any significant RF power during the testing phase, although in many instances new component designs will go through a power testing phase to prove the design, and the majority of antenna and filter manufacturers have transmitting equipment for this purpose. There are also well-documented techniques using standing wave resonators or coaxial ring resonators, which, with a modest transmitter power of, say, 20 kW, allow us to boost either the current through or the voltage across the component under test to be the equivalent of an 80-kW or even 100-kW transmitter.[3] To set up such a test, however, takes enough time and is specialized enough that it is limited to development and prototype, testing a new product rather than being a part of the regular production test and QC process.

Where coaxial components are concerned, though, a high-voltage breakdown test is typically standard for every component, and it is conducted as an integral part of the QC process using a high-voltage tester.

During this process, coaxial lines, components, hybrids, power dividers, etc., are all subjected to either a DC potential or an AC potential at low frequency, which is adjusted to the specified breakdown voltage for the size of component under test. The test potentials used range from 5.8 kW DC for 7/8 components to 52 kV for 9-3/16″ components all the way to almost 70 kV for 12″ coaxial components (which are limited to high power HF/MF facilities and scientific research). Using commercially available high-voltage test generators, the process is safe because current is limited to 1mA or much less, but the test area should still be tightly controlled, with no access for anyone not directly doing the testing. Regardless of the current, high voltages should always be treated with the utmost respect, and all safety procedures should be followed exactly.

High-voltage testing is an excellent way to show up any sharp edges, stray metal particles not cleaned out, and any insulator contamination that may have slipped past a visual inspection because the current is limited to a level which will not cause damage but nevertheless show a sudden rise in leakage should there be a problem.

Moving on from quality checking the components to commencing performance testing, it is probably fair to say that almost no other term strikes dread into broadcast engineers more than "high VSWR." It may also not be too unfair to state that no parameter is quite so universally misunderstood as VSWR. Voltage standing wave ratio is simply the resultant of the forward wave leaving the transmitter and heading towards the antenna, and the waveform resulting from any energy reflected back from the antenna towards the transmitter rather than being radiated into space. As such, VSWR is an indication of how well two different systems are impedance matched to each other, and while unity VSWR represents a perfect impedance match, such a condition rarely exists in practice. So given that perfection is unlikely to be achieved in a real-world scenario, what value of VSWR is good enough, and what range of values should we be pursuing in the assessment of an antenna's performance?

The answer to this is fairly complex (no pun intended) and depends on a number of factors, not least of which is the mode of transmission for which the system is being designed. Broadcast FM is a relatively wideband mode requiring a transmission channel bandwidth of over 200 kHz, ideally. Implicitly, it is also assumed that the antenna/ RF system provides a reasonable impedance match over this entire bandwidth, and ideally this means a low VSWR, without any sudden changes or asymmetry. The effects of high VSWR to FM transmissions are various and include:

- Degraded stereo quality, including poor L/R separation, audio/SCA crosstalk and multiplex distortions. Synchronous AM distortion is another transmission degradation that can occur.
- Increased transmission line loss.
- Excessive voltages or currents present on RF components, including transmitter filters and output stage components.
- Transmitter power reduction or even shutdown, especially in areas where when icing is prevalent during winter months.

Typical VSWR Levels for FM Antennas

Most modern transmitters will operate satisfactorily until a VSWR of around 1.4:1 is encountered. Older legacy tube-based transmitters with tunable output stages could accommodate levels up to 1.7:1 without starting to fold back their output power, but this does not mean that they were operating either as efficiently as they could or that the transmitted signal quality would be unaffected. At around 1.4:1, then, most modern transmitters will trigger a "high VSWR warning," and as VSWR increases above this threshold, they will begin to reduce power, or "foldback" their power output levels. The antenna and RF system generally need to provide plenty of headroom before this threshold and ideally present a VSWR to the transmitter of less than 1.2:1. In some situations lower than 1.1:1 is not unreasonable and is quite achievable.

Typically, for a side-mounted antenna that is designed to operate over a broad bandwidth such as 12–20 MHz, 1.2:1 is not unreasonable, while a panel antenna array may quite easily be better than 1.15:1 or even 1.1:1 over the full FM band. Narrow-band side-mounted antennas can be tuned for optimum performance over a very narrow range of channels, and as a result it is not unreasonable to expect to see VSWR below 1.1:1 and as low as 1.01:1 at carrier frequency of a single-channel antenna.

The graphs in Figure 10.4-13 show typical VSWR curves for various antennas with VSWR plotted against frequency. The two narrow-band antennas are both plotted with a 1-MHz span.

Measurement Methods

There are various methods by which the VSWR of an RF system can be determined, and the approach taken on site often depends on the system itself, availability of suitable test equipment and availability of experienced personnel. It can range from iteratively (transmitter off to adjust, then, with the tower worker clear, on to test) adjusting the antenna matching and checking the transmitter reflected power indication, to making a return loss measurement via a directional coupler, to using a manufacturer's representative or consulting engineer with a network analyzer. Where impedance matching by the use of transmission line "slugs" is required, a vector network analyzer (VNA) is without parallel in capability and efficiency. This is not to say that there are no other methods, some of which the more seasoned among us will remember from first-hand and sometimes bitter experience, but it is fair to say they take a lot longer and are unlikely to be encountered these days (think paper Smith Charts!) Arguably though, we are more fortunate today with the range of testing equipment available than we have ever been previously.

As far as factory measurements are concerned, almost universally, a VNA will be used, and this permits a very high degree of flexibility in the format of the measurement and the accuracy as well as a high degree of diagnostic information in the event the system requires further adjustment. The challenge for factory measurements is the location of the antenna, which will typically be at ground level or raised a wavelength or so above ground and therefore subject to ground effects rather than many feet above the ground in "free space" as it will be in its final location on site. For manufacturers to generate representative measurements demonstrating compliance with a specification, it is essential they provide an environment that effectively insulates the antenna from any significant interaction with the ground or nearby conductors. Nearby conductors can include such things as metal roofs, metal-clad buildings and overhead power lines, and if any of these things are close enough to an FM antenna, they will distort the accuracy of VSWR measurements. In practice, any metallic structures and, for that matter, partially conducting objects such as ourselves that are not forming part of the antenna must be further away by a distance greater than at least a wavelength. In Figure 10.4-14, for example, the operator, any visitors and the test equipment are in the test shack to the right-hand side of the photograph.

The actual configuration of the test facility depends upon the type of FM antenna under test, and typically, two different approaches are used.

Side-mounted arrays will generally be built on an elevated test rail or rack that is elevated at least 10′ above the ground and assembled using wood or another non-conducting material capable of the structural integrity required. See Fig 10.4-14 for one example.

Construction techniques must be substantial, given that the weight of a 16-bay side-mounted medium power array can easily top 1,000 pounds and occupy a linear space of 160′. Such test racks or rails are arranged with a metallic beam or "rail" running the length of the rack to which the antenna is mounted using the appropriate mounting brackets for the array. Under this arrangement, the antenna is essentially firing up into the sky, and the limited F/B ratio of side-mounted arrays reduces the energy hitting the ground and being reflected back to the array. If the structure to which the antenna is mounted while on the test rack does not fairly closely replicate the final tower, some change in

FIGURE 10.4-14 Side-mounted antenna array prepared for testing.

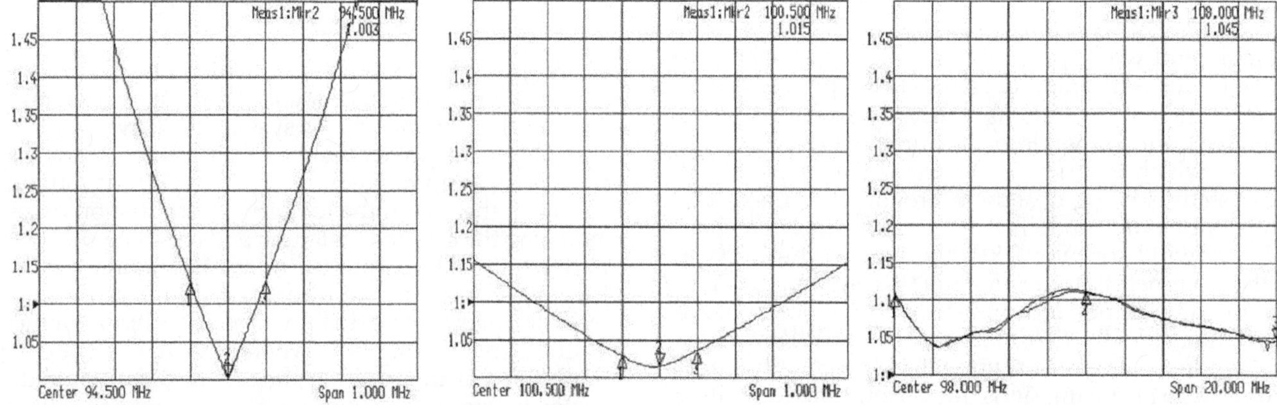

FIGURE 10.4-13 VSWR for various FM antenna types.

FIGURE 10.4-15 Broadband panel array prepared for testing.

antenna VSWR can be expected at installation, and fine matching units or transformers may be helpful to trim the final VSWR. These may or may not be included with the antenna—read the quotation carefully.

TIP: It is possible to get a level of comfort that the manufacturer's test rail meets the criteria for isolation from the ground by measuring the complex impedance of a single bay when mounted in its normal position, firing skywards. Compare this to the same antenna bay, turned through 90 degrees and firing parallel to the ground and normal to the rail. The difference in measured impedance between the two positions should be minimal, and no more than a few percent change in the reflection coefficient.

Broadband panel arrays are typically tested by laying out the panels parallel to the ground either on wooden trestles or on a custom-designed rack so that the elements point upwards—Figure 10.4-15.

Power dividers and feed system cables are located underneath the radiating panels and connected exactly as they will be in the final array. The F/B of a typical panel is better than 20 dB, and so the small amount of energy radiated to the ground starts at least at −20dB and at reflection is attenuated further by the same F/B. For systems with relatively small reflector screens, which would allow more significant diffraction around the panel edges, absorptive foam can also be positioned below the screens to provide additional attenuation of the ground reflection.

Techniques for Measurement at Site

As mentioned previously, if you are conducting tests at a manufacturer's facility, it is almost certain that they will demonstrate the antenna VSWR using a modern vector network analyzer (VNA); however, there are other methods that you may encounter on site, or even need to adopt in an emergency.

VSWR relates to the relationship between the forward wave and the reflected wave, and with this in mind, perhaps the most likely secondary method of determining VSWR is by calculation after measuring return loss. This is simply the ratio of the forward power to the reflected power, expressed in decibels.

$$\text{Return} = \text{Loss(dB)} = 10\log_{10}\left(\frac{\textit{Reflected Power}}{\textit{Forward Power}}\right)$$
$$= 10\log_{10}\left(\frac{\textit{Reflected Power}}{\textit{Forward Power}}\right)$$

It is a "loss" because (hopefully) only a small proportion of the forward power is reflected from the antenna, or the filter, or whatever load you happen to be measuring. It is therefore a negative dB value. Typical values to be encountered with FM antennas range from −16.5dB or so to under −34dB, and these correspond to a VSWR range from 1.35:1 down to 1.04:1.

Network Analyzers

Network analyzers come in two basic forms, and both are generally able to measure the transmission and the reflection characteristics of an RF network. Where FM antennas are concerned, we are generally only concerned with the reflection characteristics of the antenna, but if analyzing or troubleshooting a filter, for example, we would want to be able to see both the reflection and the transmission characteristics of the device.

In many situations, an instrument capable of displaying the amplitude characteristics of a network is all that is required, and generically these instruments are known as scalar network analyzers. However, if the phase characteristics of a network are required, then a vector network analyzer is essential. With the ability to determine phase comes the capability to measure group delay, timing and, of course, the ability to carry out complex impedance matching using transmission line slugs and impedance transformers.

For FM antenna measurements, the network analyzer can be used in one-port "reflection" mode, having first been put through a standard calibration routine using calibrated open circuit, short-circuit and load components. At this point it will be measuring S11, or reflection characteristics of the network to which it is connected, and displaying a Smith chart complex impedance representation, or a VSWR or "log mag" return loss plot—both the latter will be in Cartesian format.

While a network analyzer is the ultimate tool for directly assessing VSWR, return loss can be measured by other techniques, and by calculation, the VSWR may also be determined. There are various methods of achieving this, and two common approaches are outlined next, but both require either a directional coupler or a VSWR/return loss bridge.

Spectrum Analyzer or Power Meter and a Directional Coupler

Many stations will have a directional coupler mounted in line with the main feedline to the antenna, or in the case of combined stations, in the output of the channel combiner. Most directional couplers come with two ports, one reading forward power and one reading reflected power. By measuring the power at each port, the ratio of reflected power to forward power can be established and the return loss determined using the previously mentioned formula. (Make sure to take into account any differences in coupling factor between the forward and reflected power monitoring ports) Admittedly, this measurement applies only to the carrier frequency, but as this is where the majority of the energy is concentrated, it is probably good enough in a pinch. It will not tell you the degree of VSWR slope either side of carrier, although by carefully comparing the shape of the upper sidebands to the lower sidebands at the reflected power port, it can be seen whether substantial asymmetry exists.

The great benefit of this method is that it is completely nonintrusive and the transmitter remains on the air throughout, and as a result it can provide periodic spot check capability.

This method does require that the transmitter be isolated (locked out and tagged out) and disconnected from the feedline, in accordance with all safety procedures, but in return it is a wideband, swept measurement and will give far more information. Before connecting the analyzers (or any test instrument, for that matter), verify that there are not high levels of RF being induced by other nearby antennas, using an absorptive power meter or similar. Set the spectrum analyzer "center frequency" and "span" appropriately for the system being assessed and add any frequency markers you may want.

Using an existing directional coupler, or one you have available, the output of the spectrum analyzer tracking generator is hooked up to the through line of the directional coupler, and the input to the spectrum analyzer is connected initially to the directional coupler forward port, as shown by the dotted line. Adjust the input attenuation and reference level settings on the spectrum analyzer to establish a convenient baseline for your return loss measurement. This will act as the "0 dB" reference.

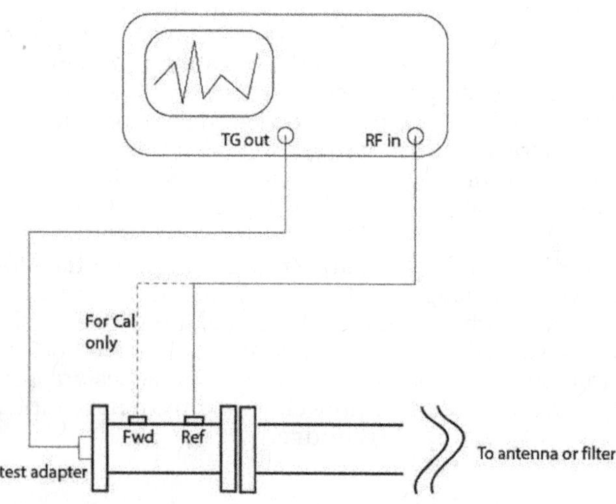

FIGURE 10.4-16 Diagram of testing setup—spectrum analyzer with tracking generator and directional coupler.

Now transfer the RF in cable from the Fwd port of the directional coupler to the Ref/ Rev port as shown by the solid line, and note that the display will be below the "baseline" established previously. This same technique can be used to equal effect with a return loss or VSWR bridge.

The difference in dB between this display and the baseline display represents the return loss measured in dB.

SITE MEASUREMENTS AND FACTORY MEASUREMENTS

Compared to the measurements of VSWR conducted at the factory, site measurements differ because they generally include tens to hundreds of feet of coaxial line—assuming measurements are being made in the transmitter building and not at the antenna. This changes the results to some degree and also produces a display that may not immediately resemble the factory data for a variety of reasons.

- First, the transmission line adds attenuation to the displayed VSWR because the response displayed on the network or spectrum analyzer screen is the result of a round trip up and back down the feedline to show the reflected component.
- Second, there are many connections between the point of measurement and the antenna, especially if a rigid transmission line is used. Each connection, adapter, flange, etc., adds a small impedance discontinuity, and what is displayed on the screen is actually the net result of everything adding and canceling. Typically the effect of all these secondary effects or secondary reflections is a ripple that appears to be superimposed on the actual VSWR plot or even a series of ripples for multiple impedance bumps. For a broadband antenna, the amplitude of the ripple will be consistent over the bandwidth of the antenna, but for a narrow-band antenna, the amplitude of the ripple will vary from quite large outside the bandwidth of the antenna to almost nothing at the center of the antenna operating bandwidth. The frequency of the ripple is a function of the distance between the test equipment and the discontinuity causing the ripple. In years gone by, this technique was actually used to assess the reflection bandwidth of a device under test and can still provide useful, though fairly approximate, distance to fault information.

An example of this is shown in Figure 10.4-17.

This measurement was taken at a radio station using a single-frequency antenna, fed via several hundred feet of rigid line. The two VSWR traces simply correspond to the antenna before and after retuning.

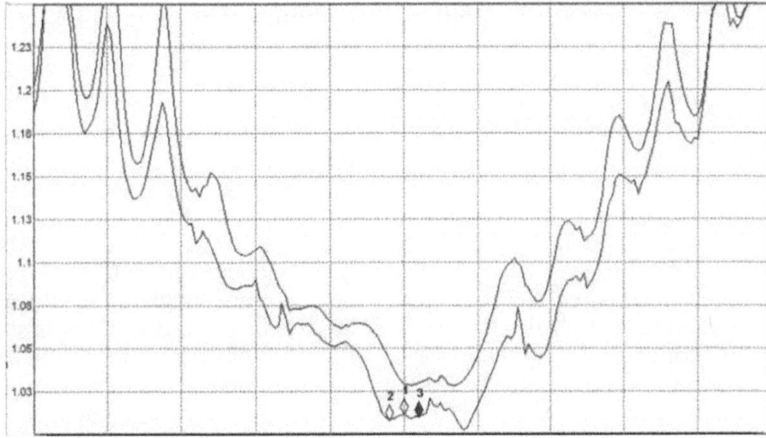

FIGURE 10.4-17 VSWR traces showing before-and-after antenna retuning.

As well as the characteristic "V" response of the VSWR, there is a clear visible ripple superimposed. The period of the ripple is consistent and repeats approximately every 400 kHz. So, what can we learn from this? Well, we know from transmission line theory that reflection coefficient repeats every half-wavelength, and it is inverted every quarter-wavelength, so given that what we are seeing on the screen is a graphical display of reflection coefficient, we can assume that the distance between adjacent maxima or minima of the ripple is a half-wavelength. At 400 kHz a half-wavelength is around 1,245′ including a very slight velocity factor correction velocity factor correction, and so whatever mismatch is causing the ripple is approximately 1,245′ away from the test equipment. A subsequent measurement using a time domain reflectometer confirmed the line length to be 1,276′ so the ripple evident on this line is due to the impedance discontinuity, as the 50-ohms transmission line transitions to the small mismatch at the antenna. As the frequency moves away from the antenna design frequency and the VSWR mismatch increases, so, too, does the amplitude of the ripple component. It's not exact, certainly, but the presence of a high level of ripple is a good indicator that there is a mismatch in the system, and the period of that ripple can give you a clue as to where.

The effect of the line loss is to attenuate the reflected signal by twice the line loss. This is because the signal from the analyzer is attenuated on its way up to the antenna, and the reflected component is attenuated by the same amount on its way back to the analyzer. For this reason the return loss measured by the analyzer should be better than the return loss at the antenna by a factor of two times the line loss.

Feedline Measurements

The de facto standard for transmission line assessments is the distance to fault (DTF) or time domain reflectometer (TDR), and the basic principle of operation has been around a very long time indeed. As outlined previously, it is possible to get a reasonable guide to within 5% or better where a significant fault might lie by adopting the approach mentioned earlier, but if the line consists of almost 2,000′ of 20′ rigid line sections, there are an awful lot of bullets and connections to consider. From experience one learns very quickly that pointing the tower crew to the wrong line section more than a couple of times while trying to isolate the faulty bullet is not the best way to gain their nomination for favorite engineer!

In its original form, the TDR generated a pulse with an extremely fast rise time and applied this to the transmission line under test. By using a cathode ray oscilloscope display, it was possible to view the initial pulse and the various reflections coming back over time to the instrument from any impedance mismatches at different points along the line. By comparing the time delay associated with the reflections compared to the main pulse, it is straightforward to convert delay to physical distance and therefore distance to fault. In addition to displaying distance to fault, the early instruments would also indicate the magnitude of the reflection and whether it represented low or high impedance compared to the transmission line Z_0.

One of the best known examples of instruments from this early era is shown in Figure 10.4-18 and was manufactured by Tektronix. While long since discontinued by the manufacturer, if operational and calibrated, it is still very useful today and is capable of providing distance to fault indication for lines up to 2,000′ long with resolutions of a foot or less. The pulse generator is fragile and cannot tolerate any RF induction on the line under test.

Analog instruments relying on a very fast (and in some cases, very fragile) pulse generator were superseded in the ′90s by instruments incorporating considerable processing power, with the result that fast Fourier transforms could be conducted on digitized frequency domain data, converting it into the time domain and emulating the analog-style of graphical display but with additional information. As years have passed and processing power has increased exponentially, all instruments with "distance to fault" capability now rely on a very fast transform and offer numerous options and capabilities compared to the early designs. The display shown in Figure 10.4-19 is the time domain measurement obtained by the TDR. It clearly shows the end of the feedline at 1,276′, with distance plotted on the x-axis and the return loss values on the y-axis.

The illustration in Figure 10.4-20 shows the same system but with the instrument set to consider only the first 200′ of line and ignore everything else. This measurement was made from a test adapter on the coaxial (Main Tx/Aux Tx) transfer switch, and while none of the peaks are worrisome or of concern, it clearly shows the various discontinuities introduced by some flanges and an elbow complex along the ground run as the line passes through the building on its way to the tower.

Using a TDR/DTF instrument, a damaged bullet or a pinched insulator can be isolated relatively quickly, hopefully before it becomes a bigger problem that could potentially take the station off the air and cause thousands of dollars of damage to a transmission line system.

While this type of measurement is necessarily intrusive in that it can only be conducted with the antenna de-energized and isolated, it provides a high level of confidence in the integrity of the system, and, if done routinely on an annual basis, any changes can be addressed and investigated before they deteriorate too far. Like any instrument, though, it can give misleading readings, and aliasing is something to look out for. If in doubt about the validity of a display indication, try retaking the measurement from a different location, and make sure the most appropriate transform option is selected in the menu. On more than one occasion, for reasons unknown, what looked like a damaged bullet has checked out completely fine when assessed from a different point on the same transmission line. Aliasing is the most likely reason.

The other key thing to remember when analyzing TDR test data is that if there is a DC short at any point in the system prior to the actual radiators themselves, the TDR will typically show that as a short-circuit or 0 dB return loss (assuming zero line loss, of course). Typical examples of this are some fine matching transformers that are DC grounded, although obviously not grounded at RF.

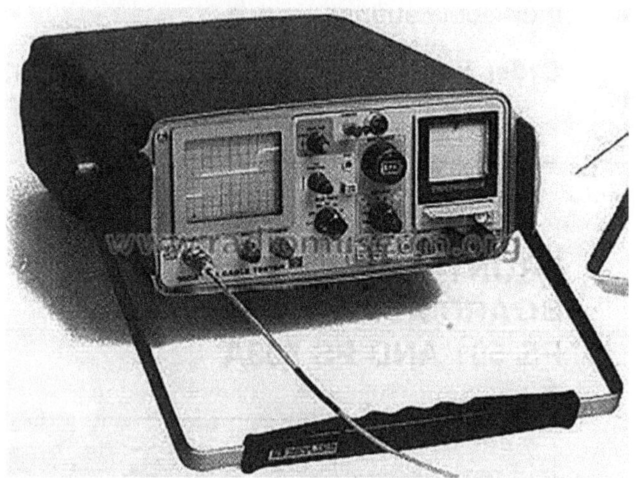

FIGURE 10.4-18 Time domain reflectometer (TDR).

MECHANICAL AND PHYSICAL EVALUATION

Of equal importance for the reliability and longevity of any antenna system is a full inspection of the mechanical aspects, including its basic design and the fabrication techniques used by the manufacturer. While not strictly falling under the category of "measurements,"

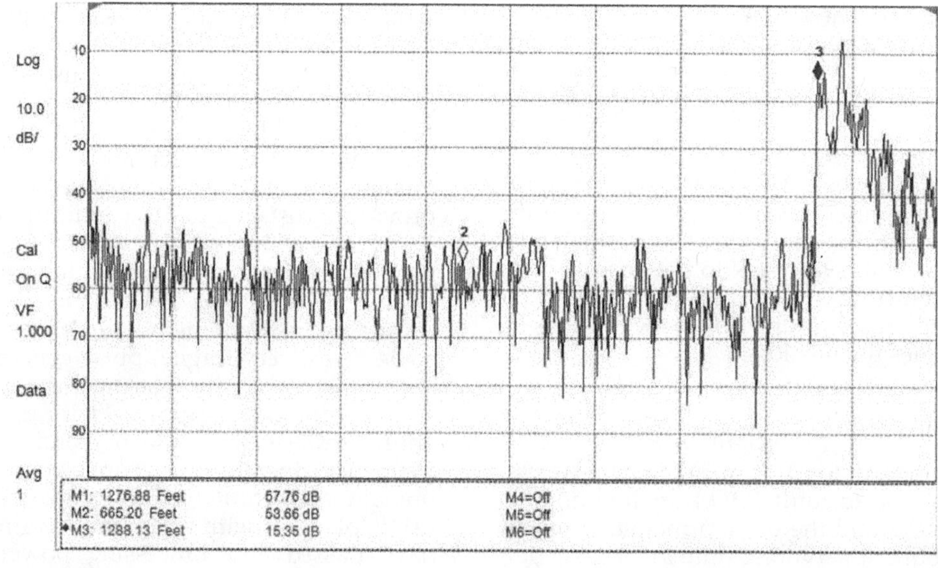

FIGURE 10.4-19 Distance vs. return loss time domain measurement of entire line.

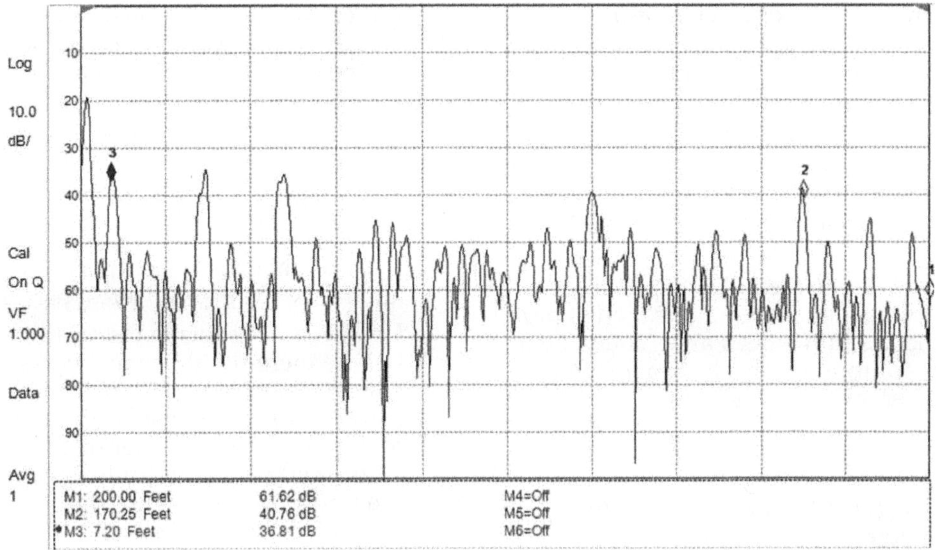

FIGURE 10.4-20 First 200 feet of transmission line time domain measurement.

mechanical inspections and checks are nevertheless an important part of any factory visit or system acceptance, just as they are an integral part of tower and antenna inspections.

Things to consider are:

- Dissimilar metals and the potential for galvanic corrosion. Manufacturers are more than aware of the corrosion that can result from the electrical activity resulting from contact between two dissimilar metals when moisture is present. Depending on your location and the climate, some combinations may be borderline acceptable in a desert location that would be unacceptable in a tropical or ocean location. The tables of galvanic potentials are readily available, so will not be reproduced here. Also, the subject is considered in detail elsewhere in this book, but some of the more notable prohibited combinations are:
    - Aluminum and copper or brass
    - Aluminum and a high proportion of stainless steel (stainless hardware is common, of course)
    - Copper and galvanized steel tower or antenna grounding straps that have not been silver soldered
- Pressurized coaxial components are typical for higher-power systems, and the long-term benefits outweigh the additional cost associated with the compressor and dehydrator equipment. Not only will a dry, climate-controlled environment prolong the life of the coaxial components, internally at least, it also provides an early warning of failing seals, damaged O-rings and punctured lines.

  In severe cases of leakage, the compressor/dehydrator will either struggle to keep up with the leaking air or will be unable to maintain pressure, but as very minor leaks develop, the air system duty cycle will simply increase slightly. Comparing compressor run time month to month and checking for increased compressor running hours is a very good indicator that something has started to spring a leak and will need attention.

  During factory tests/production testing, manufacturers will generally pressurize a system being tested, cap it off and monitor for pressure loss over a few to several hours at reasonably constant temperature. This provides the manufacturer with confirmation that no seals or O-rings have been forgotten during assembly and also gives the customer peace of mind that any leaks occurring after installation are the result of lost or incorrectly installed parts. The swept volume of the antenna system is relatively small compared to the transmission line and antenna system at site, so very small leaks during the factory testing will result in pressure drops fairly quickly.
- Cables and components should all be checked for proof of high-voltage testing, pressure testing and phase matching, if appropriate, not forgetting VSWR or return loss.
    - Make sure that cables associated with a branch-fed antenna array are all well-labeled to avoid any confusion or errors during installation and that none of the cables display overly tight bends (check cable manufacturer mechanical specifications for the minimum bending radius).
    - Are cable lengths fully documented to permit future replacement without disturbing phasing?
    - Make sure that no cables begin a bend immediately as the cable leaves the connector. Always allow, typically, at least the length of the connector as a distance from the connector before the flexible cable starts a bend.

- Verify that there are no kinks or dents in the cable outers, that there is no damage to the jacket and that the correct size of cable has been used.
- Hardware: If stainless, does it include coated nuts or some other method, such as thread lubrication or the use of different grades of stainless steel, for preventing, or at least alleviating, galling during assembly?
- Check the availability of the following: Full sets of drawings for installation and identification of part numbers should the future need arise, component calculations as appropriate, mechanical loading data for the tower owner/manager.

SUGGESTIONS FOR PERIODIC SYSTEM TESTING

Key to keeping any antenna system operating reliably and for many years is to take care of the little things and doing so before they have chance to become bigger problems. Some considerations are:

- Monitor dry air or nitrogen usage month to month, and if possible keep an informal tally of compressor run hours per month for your dehydrator system. Most shunt fed/side-mounted antenna systems will include some form of purge valve to relieve an overpressure situation, and with the exception of slight venting when the system is at its hottest midsummer compared to cooler nighttime temperatures, the hours run per month should not only follow a fairly consistent pattern, they should also not vary significantly, either. A sudden drop in run hours could indicate a dehydrator failure, perhaps (hopefully accompanied by a warning light on the unit itself), whereas a sudden increase in running hours immediately suggests that something has sprung a leak. In an ideal world, leaks would develop gradually, and you would be able to see a gradual increase in dehydrator/compressor hours alerting you gently to the fact that something is beginning to leak. Of course, in many cases the first and only warning will be when the transmitter trips during a severe lightning storm and the system loses pressure where part of the antenna system has been melted, causing the dehydrator to run continuously. Forewarned is forearmed, though, and keeping a regular eye on run times is a very useful diagnostic tool. Adding a simple flow meter to your system might help you spot changes in air movement, too.
- Looking for small but consistent change applies equally to the feed system and the antenna itself. Coax will run at a steady temperature once equilibrium has been achieved between the power flowing through it and the building ambient. Get used to how the various lines and connections feel under normal operational conditions. A change in line temperature, especially if the additional heat can be localized, can indicate a deteriorating bullet connector internally. *Note that it is always good and safe practice to use a RF hazard meter initially to make absolutely sure that there is no RF leakage before placing your hands upon a flanged connection* (it is worth checking every year that the system is still RF-tight, too, in this respect). If a few bolts have been left loose or have loosened up, you will burn yourself and it will hurt—I can attest personally to that!

  Localized hot spots in the middle of a line run (where there should be no bullets) may well suggest that a higher than normal standing wave has developed because heat will be concentrated at the current maxima (antinodes), and this would certainly warrant further checks, even if the transmitter VSWR and/or reflected power indications show no change. Better still, if you can have access to a thermal or IR camera or scanner, this allows you to check everything for temperature, from circuit breaker panels to feedline running up your tower, and record it for future reference/comparison. Diurnal cycles will still apply, so a few attempts may be needed to establish a baseline.
- Transmitter parameters, especially when comprehensive metering capability is available, can also be excellent indicators of possible change in the antenna system. Tube-based tetrode transmitters are extremely sensitive to load impedance in terms of screen grid (g2) current. A slight change in cavity loading resulting from a VSWR shift will be readily visible as an increase or decrease in screen current and has probably already been noted resulting from antennas icing over winter in more northerly latitudes. Triode amplifiers are less sensitive, but grid current feeds (and to a lesser degree, plate feeds) drifting can be noted and might be an indicator of changing load conditions again.

  Solid state amplifiers typically incorporate a myriad of metering ability as well as generally well-calibrated reflected power metering. Drain or collector current changes combined with reflected power readings changing might be indicative of further investigation, although the extent to which the transistor current might change with a small shift in load impedance also depends on the internal transformation ratios within the amplifier.
- VSWR monitoring and, ideally, forward and reverse power monitoring is invaluable. Fluctuation of a VSWR or reflected power meter with modulation can indicate reduced system bandwidth resulting from a detuned filter or a detuned/damaged antenna. The fluctuation results from the FM sidebands "riding up and down" a steep VSWR slope and is a direct cause of synchronous, or incidental, AM. Under these circumstances, fluctuation may precede an overall

increase in VSWR if the bandwidth rather than the VSWR at carrier frequency has been affected.

- It goes without saying that any increase in VSWR over time is probably indicative of a trend toward a failure or at least degradation of your antenna system. Get to know where your power metering rests most of the time and be alert for any steady changes over time. The metering available at the transmitter is not capable of telling you where in the system the degradation might be occurring, and if you spot a worrying trend, it is time to bring in some of the test equipment methods described earlier in this chapter before a loud bang and a burn mark on the feedline does that for you.

- If the station budget will stretch to annual testing, or if the system is critical enough, it is well worth obtaining a full set of system sweeps on an annual basis taken at a similar time every year, either using rented equipment or the services of a consulting engineer/manufacturer's engineer. While this is invasive in that the station must go off the air (or at least to an auxiliary system), the information obtained over a period of years is invaluable, and a slight change that might go unnoticed from 1 year to the next becomes clearly apparent when compared with data from 2 or 3 years ago. Once change is noted, the problematic portion of the antenna system can be identified and repairs completed all in the same night's work.

Never underestimate the simplicity of an old-school resistance meter, especially an analog VOM. Most antennas (unfortunately not all, though, so check with your manufacturer's data) are designed to be short-circuit at DC, and a quick resistance test can tell you whether the feedline is open at some point if you can't see the DC short at the antenna. If you do see a short, or at least some low value of resistance, remember that the feedline manufacturer publishes values of resistance per foot that can give you a reasonable level of confidence that everything is as it should be measuring through the short at the antenna and back down again.

And finally. . . one neat trick for low power users: If you are seeing a very high VSWR after the system warms up and you know that there is a short in place at the antenna but do not have immediate access to test equipment, find a high current, low voltage (AC or DC, doesn't matter) supply capable of sourcing tens of amps (filament transformer and a Variac, for example). By passing a fairly heavy current through the antenna system—through the short at the antenna—intermittent connections may become evident as the current changes due to heating effects at the intermittent connector or connection. In one instance I am aware of, a low-power FM station in Arizona was able to use this technique to not only confirm an intermittent connection but also to repair (which presumably means they managed to weld back together) the intermittent connection. Not to be recommended as a permanent solution by any means, it did nevertheless allow one LPFM to get back on the air in the interim, although for how long after this approach is not known!

Notes

[1] "LAMBDA™ Optimized Mounting System for FM Applications"—ERI (Electronics Research, Inc.), www.eriinc.com/Catalog/Towers/Specialty-Structures/LAMBDA-for-FM.aspx

[2] Typically the measurement probe will scan a planar or cylindrical surface, but spherical near-field scanning is also possible with a more complex positioning system.

[3] "High Level RF Testing Methods"—Peter S Hayes & Robert A Surette, 1988 NAB Proceedings.

CHAPTER

10.5

FM Radio Field Strength Measurements

CLARENCE M. BEVERAGE
Communications Technologies, Inc.

WILLIAM P. WEEKS
Hungry Wolf Electronics

INTRODUCTION

FM radio field strength measurements provide data that can be of significant value to management, sales, programmers and engineers at the station level and to brokers and FCC attorneys, as well. However, the true value is highly dependent on the quality of the measurement data, and that is set by the knowledge and experience of the engineer obtaining the data, the quality of the equipment used and the way that the data is organized and presented to the end user. An ability to explain the limitations and appropriate use of the measurement data will go a long way toward the users being able to accept the data, properly apply it to their needs and come to rely on it in the future.

A client came to us saying that a new contract engineer had been hired for the station and that the engineer had taken some spot field strength readings on the client's NCE FM station and also a nearby commercial band FM station. The report said that a Potomac field strength meter with companion dipole had been used and that the NCE FM station field strength seemed exceptionally low when compared to the commercial station. The NCE horizontally polarized ERP was only 14% of the vertically polarized due to channel protection requirements, while the commercial band station was circularly polarized. Thus, when using a horizontally polarized receive antenna, it would be expected that close-in field strength readings on the NCE station would appear to be far weaker than expected.

This brief story is offered to illustrate the importance of doing full research on the facilities you will be measuring before getting in the field. It hopefully also reminds us that the engineer taking measurements must have a good understanding of FM antenna types, polarization and radio propagation as well as sufficient practical transmitter experience to know things such as an FM exciter can sometimes generate spurs that are low in level but sufficiently strong to cause interference.

REASONS FOR MAKING MEASUREMENTS

FM field strength measurements were originally used solely for scientific purposes. As an example, field strength measurement data was used as the basis for the FM field strength curves found in section 73.333 of the FCC rules, which we continue to use today for determining distance to coverage and interfering contours. Going back to a 1980 edition of the FCC Rules, 73.314, "Field Strength Measurements," gives us a very specific procedure to follow for making and reporting field strength measurement data. The rule is summarized here:

1. Draw eight or more radials on topographical maps oriented to traverse representative types of terrain. Start measurements at 10 miles and repeat at 2-mile intervals to the required ending distance. Measure at accessible roads closest to mile markers.
2. Measurements to be taken at a receive antenna height of 30′ with a horizontally polarized antenna. Calibrate the equipment. Rotate receive antenna to determine if the strongest signal is coming from the transmitter site. Orient the max gain point

of the antenna at the transmitter site and take measurement for a 100′ run, with the measured field continuously recorded over the length of the run. Record median field strength along with max and minimum field. If strongest signal is found to come from a direction other than the transmitter site after the mobile run is completed, a cluster of five points should be used to gather additional FIM data, with the antenna aimed at the transmitter site and also oriented at the maximum field. A list of calibrated equipment, model number, serial number, rated accuracy and most recent calibration are required. If field calibration is employed, a detailed description of calibration equipment and procedures are to be used.

3. Grid measurements for determining city of license coverage are also described and would normally consist of 20% mobile runs and the remainder of the measurements at grid points, taking extreme care to determine if the max signal is coming from the transmitter site and providing a detailed description of the local environment because it might impact the signal level. Here also the measurements would be at a 30′ elevation with a horizontally polarized antenna.

The FCC Media Bureau no longer accepts measurement data, but that does not lessen the importance of using good engineering practice for purposes of obtaining quality data that can be relied upon when presented to the end user. Today, FM field strength measurements can be used for a number of purposes:

- To compare the signal strengths of competing stations in a market. This can be particularly important when evaluating a station for purchase or for sales and marketing.
- To evaluate the performance of a site or an antenna system at a particular site.
- To compare computed signal level to measured signal level for coverage map purposes.
- To determine the presence of interfering signals.

ADVANTAGES AND CONSIDERATION IN TAKING MOBILE FIELD STRENGTH MEASUREMENTS

Taking measurements from a vehicle with a horizontally polarized directional antenna 30′ in the air is not time efficient, in part due to the need to repeatedly lower the antenna for safety reasons. Also, the requirement to do only short mobile runs and the need to point the antenna at a particular transmitter site takes significant time.

Properly mounting an omnidirectional antenna on the roof of a vehicle frees the engineer to measure at highway speeds and with the correct software and receiver to measure multiple stations on one run. This process can be relatively inexpensive and provide meaningful data for analysis, processing and presentation.

When taking mobile measurements it is important to consider a number of factors:

- Signals levels will be lower than at 30′, and correcting for the lesser height is not an exact science.
- Lower to the ground generally means stronger reflections from surrounding objects, causing the received signal to be less representative of the true radiated signal. The reality, however, is that this is the signal level that vehicles on the highway around you are receiving. It is nevertheless important to be aware of the measuring environment and to choose your route with care. If a tractor trailer is right next to you and is between you and the transmitter site or acting as a big reflector of the signal back at you, those readings are not going to be accurate. How will you handle this—redo that portion of the run, maybe note the situation for later evaluation? Would it help to have a local driver so that you can make good notes of conditions like this while measurements are being taken?
- Measurements taken at single locations in a vehicle are generally not recommended for purposes of determining field strength. A much better approach would be to get into a parking lot or a street without a lot of overhead wires and do a 100′ or 200′ run at constant speed.
- Your software should continuously log field strength and then be capable of providing geographic coordinates and a mathematical median of the measured values. When measuring a large number of stations at highway speed, a question arises . . . at what interval are you able to gather data for each station, and is this enough to get a good representation of the signal? This is a function of receiver and software design and integration.
- It is wise to look at the locations of other high-power stations along your route before driving the route. We have observed unexpected areas of higher-than-expected signal level for the station being measured. It turned out not to be the intended signal but multiple signals from a master site mixing in the receiver. In reality, the spot was a low-signal-level area.
- Another cause of inaccurate measurements is spurious or intermod products coming from another station in the market. Listening to the station(s) that you are measuring may help confirm that the signal being measured is actually the signal of interest and not some incoming interference. When evaluating a low-power station such as an FM translator, interference from high-power stations can become a real factor, and at times turning off the desired station to evaluate other incoming signals is necessary.
- Consider measuring at least one other station while you are measuring the station of concern. This provides a bit of control and may help you to determine if a low signal level is related to a receiver malfunction or an antenna fault. Carrying a spare antenna to see if the signal level changes is a great control. It may be almost impossible to identify an

equipment malfunction in the field, so always try to look at the data before packing up to leave. Using Longley-Rice predicted signal as a comparison is another way to determine that your measurement data looks reasonable before leaving the site area.

- Switching-type DC power inverters can generate noise in the FM band frequencies, possibly unexpectedly the frequency of interest. Mobile equipment should of course be checked for poor or loose connections or damaged cables before commencing measurements. Possibly most frustrating would be completing a long mobile measurement run and discovering that the equipment was not properly recording the measurements for some reason.
- An unfamiliar rental measurement vehicle should be checked for measurement hazards, including the ability to properly support the antenna and secure the test equipment, a shape that might unusually distort the antenna pattern, unexpected noisy electronics or poor electrical connections.

The message here is hopefully clear. Yes, mobile measurements offer a lot of benefit but also expose the equipment to uncontrolled conditions that can affect the accuracy of the measurement data.

MEASUREMENTS AT FIXED LOCATIONS

Individual, single-location measurements of FM signal strength are typically unreliable. The measured signal strength may vary as much as 10 dB over the space of only a few feet. The measured signal strength may be locally attenuated or reflected by terrain, by vegetation, by structures and even by passing vehicles. Analysis of FM signal strength measurements must then be by statistical methods, which can prove to be quite reliable and repeatable. If fixed location measurements are necessary, a tight cluster of five measurements in an open space may be the best solution.

EQUIPMENT FOR FIELD STRENGTH READINGS

A few decades ago, if you were going to do FM field strength readings, your "go to" gear was a Potomac Instruments FIM-71 VHF field strength meter with companion dipole and coaxial cable, all calibrated as a system. In 1995, Z Technology, Inc. entered the scene with a moderately priced, NIST traceable, portable, programmable field strength meter. The R-507 is the company's most recent receiver product, which covers the range from 5–1000 MHz and is available with calibrated dipoles and cables. Today there are a number of FM band radios designed specifically for taking FM field strength readings, which include:

Audemat FMC-4 and FM MC5

Deva Radio Explorer II, Radio Explorer and Band Scanner GPs

Microgen TS9065, TS9085 and FMX100

A measurement system is only as good as its antenna, so let's look at antennas in some detail.

Calibrated dipole antennas in the VHF range are commercially available. These are normally used horizontally polarized, and they are directional. They are suitable for measurements at static locations and for cluster or straight-line measurement runs. In particular instances, a more directional antenna, such as a Yagi or log periodic, may be appropriate. Commercially made antennas should have a published gain at the frequency of interest. Two examples of dipole antennas are the Potomac Instruments and Z Technology calibrated antennas. Calibrated biconical and log periodic antennas are available from a number of manufactures in the EMI industry including A.H. Systems and AR, Inc.

One example of an omni horizontally polarized FM antenna that can be mounted on a vehicle with some work is the Stellar Labs unit shown in Figure 10.5-1.

Another example is the h.pol. implementation designed by Bill Weeks, taken from the amateur radio VHF "squalo" style antenna, as shown in Figure 10.5-2.

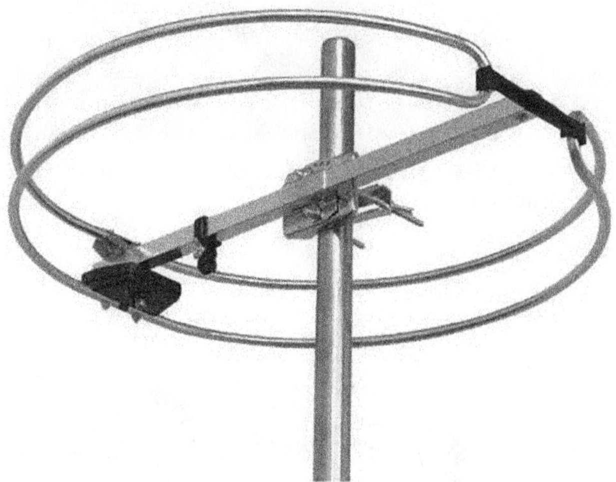

FIGURE 10.5-1 Omnidirectional horizontally polarized vehicle-mounted antenna.

FIGURE 10.5-2 Squalo style horizontally polarized vehicle-mounted antenna.

Finally we have a custom vertically polarized antenna designed by NPR Labs and constructed by Kintronic Labs, as shown in Figure 10.5-3.

This antenna was manufactured by Kintronic Labs and calibrated at the Institute for Telecommunications Science Table Mountain Test Range near Boulder, Colorado.

Mobile measurements require an antenna with an essentially non-directional horizontal radiation pattern. The most common antenna for this purpose is a quarter-wave, vertically polarized whip antenna mounted to the roof of the measuring vehicle. A vertical whip antenna mounted near the center of the metallic roof of a typical passenger vehicle is likely to vary a few dB, depending on the vehicle. A more circular pattern was achieved in the design by NPR labs and constructed by Kintronic Labs, shown in Figure 10.5-3, having a circular grid structure of approximately half-wave diameter to use as a ground plane, which are then mounted to roof racks above the vehicle roof. Figure 10.5-4 is a plot of the NPR Labs antenna pattern from the test range at four different frequencies.

FIGURE 10.5-3 NPR Labs designed vertically polarized antenna.

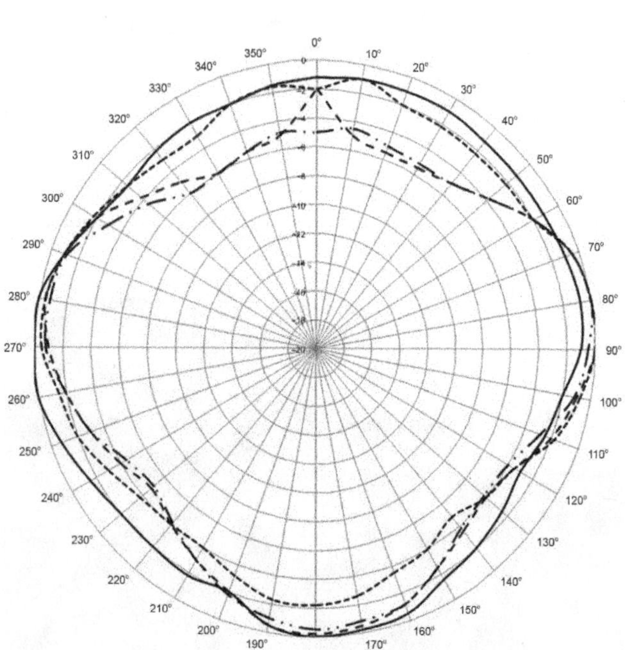

FIGURE 10.5-4 Four-frequency plot of NPR Labs antenna on test range.

Calibrated vertical whip antennas are not commonly available. A close estimate of the antenna factor can be calculated using the formula:

$$K = 20 \times \log_{10}(f_{MHz}) - 10 \times \log_{10}(G) - 29.7707$$

where:

K = Antenna factor in a 50-ohm system
f = Frequency in MHz
G = Gain of a 1/2-wave dipole = 1.64
so the antenna factor at 100 MHz = 40 – 2.148 – 29.7707 = 8.08.

The length of a quarter-wave dipole is:

L_{inches} = 1/4 of the speed of light in inches per microsecond / $Frequency_{MHz}$ = 2946.24 / F_{MHz} or 29.46 inches at 100 MHz.

If the receiver input is calibrated in dBuV and the desired results are in dBuV/m, the results will be:

dBuV + *K* + the loss of the feedline.

Typically, RG58 is used for the antenna connection in mobile measurements, and its loss is nominally 4.9 dB per 100 feet at 100 MHz.

Figure 10.5-5 depicts a standard magnet-mounted whip antenna mounted on the roof of a vehicle. Figure 10.5-6 depicts a vertically polarized whip antenna above a ground screen similar to the design developed by NPR labs.

FIGURE 10.5-5 Vertically polarized vehicle mounted whip antenna.

FIGURE 10.5-6 Replica of NPR Labs-designed antenna with whip antenna for vehicle mounting.

CALIBRATED RECEIVER SYSTEMS

Mentioned earlier are receivers designed specifically for mobile field strength measurements that include a GPS and programmable software to change frequency and store field strength and coordinates for later presentation and analysis. Figure 10.5-7 depicts a Microgen receiver and associated laptop securely mounted in the front of a truck for one-man measurement use.

Back at the time that the original Audemat FM-FMC series was introduced, Bill Weeks was using an Icom R-8500 for FM field strength measurements. This receiver was chosen because of its wide dynamic range and ability to be computer controlled. Discussing use of that receiver in multiple field strength measurement campaigns over many years and covering over 100 different stations across the country offers a good breakdown of what a custom system can consist of.

Calibrated receiver—we use the term "calibrated" first of all recognizing that the receiver AGC must be able to be turned off. Second, we use the term "calibrated" because the receiver field strength reading must be calibrated against a signal generator whose absolute signal level accuracy is known and a table of correction factors that can be stored in the receiver or the PC controlling the computer. A necessary attribute of a calibrated receiver is its ability to work under computer control to output field strength to a file at selected intervals and frequencies.

GPS—Having a receiver, interconnecting cable and antenna that are quantified doesn't get us out on the road without an accurate GPS unit. The GPS unit must have an accuracy sufficient to correlate measurement results to actual location. It must be sensitive enough to work in the environment to be measured, considering vegetation or urban canyons. It must have an update rate adequate to handle the speed of the measurement vehicle. It must be able to communicate with the computer in use. It is best if it can be mounted on the vehicle roof. One example of a device with such specifications is the GlobalSat BU-353S4.

FIGURE 10.5-7 Vehicle-mounted Microgen receiver, laptop computer and measurement system.

COMPUTER CONTROL AND LOGGING SOFTWARE

The controller of the mobile measurement system must tune the receiver to the desired frequency, retrieve and save the measured signal strength and record the corresponding GPS location at the desired intervals in time or distance. The purpose-built commercial systems such as those provided by Audemat or Deva Broadcast either perform these tasks internally or provide a custom computer program that must be run on a connected laptop. Firms such as Signal Intelligence, Inc., provide software to perform these functions with a variety of receivers. NPR Labs and others have written custom software for special purposes, such as evaluating HD radio performance.

INDICATIONS ON A MAP

The measurement locations or route will normally be predetermined based on the purposes of the campaign. If the engineer is in unfamiliar territory, some means of navigation to stay on the route may be necessary. This may need to be something like a separate GPS mapping system. All of this equipment can be very distracting, and due attention must be paid in both planning and execution to traffic safety.

ANALYSIS AND PRESENTATION OF MOBILE MEASUREMENTS

Figure 10.5-8 is a black and white signal level comparison of field strength measurements taken on a station at the original site and at the new site for purposes of confirming that the new site did provide the predicted improvement.

Figure 10.5-9 is a comparison of signal strength obtained by mobile measurements to the signal level computed using a Longley-Rice program. In both cases, the plot shows dB deviation between before and after or between measured and predicted signal levels.

The FM measurement receivers that are sold with integrated software and GPS generally allow the measurement route and data to be output in a file format that easily imports into Google Earth. This is a particularly nice presentation method, as Google Earth Professional is now a freeware product widely used and it allows the data to be seen in its entirety or with greater specificity by zooming in on the drive route. Figure 10.5-10 is an example of the plotted output from the Microgen product.

FIGURE 10.5-8 Comparison of two drive studies from old and new transmitter sites.

FIGURE 10.5-9 Comparison of drive study and Longley-Rice signal levels.

COMPARISON WITH COMPUTER PREDICTIONS (LONGLEY-RICE)

Prediction of received signal strength in the VHF range is based on the power delivered to the transmit antenna, the characteristics of the transmit antenna, the location and height above ground of the transmit antenna and the signal path to the receive location. Typically, the power is measured at the transmitter output, and losses between there and the antenna are calculated. The gain and directional characteristics of the transmit antenna may be assumed based on the design and mounting arrangements or may have been measured by the manufacturer on a test range. The antenna and associated feed system, including such things as power dividers, may or may not have been installed as designed.

The effects of the signal path to the receive location have been calculated with increasing precision over the years. The progression has moved through drawing a relatively few lines on topographic maps and picking elevations along the lines (and averaging for the directions between the lines on the map) to calculations using terrain mapped in various resolutions by radar from space shuttles. This progression has obviously been aided by the increasing computing power available to radio engineers.

Even the best terrain data currently available is imperfect at the very local level because of varying vegetation and various man-made structures. At many locations of interest the receiver is virtually below ground with respect to the incoming VHF signal, that is, below the trees or below rooftop level of surrounding structures. In these conditions the received signal may have arrived by way of multiple reflections.

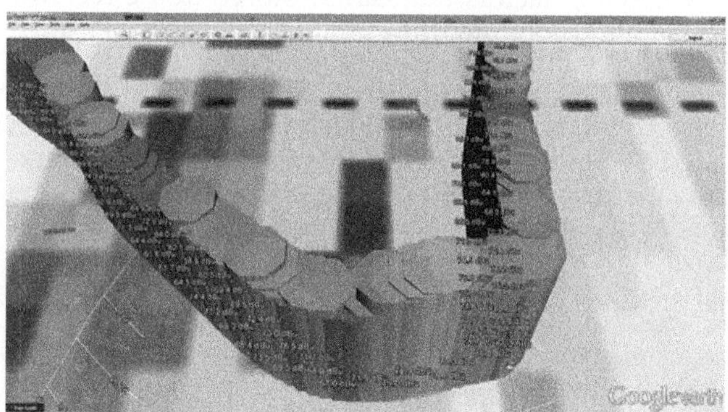

FIGURE 10.5-10 Microgen display of signal levels on Google Earth with overlay of Longley-Rice signal levels.

No currently available prediction methodologies can fully account for these local conditions, and thus caution must be used when comparing measured and Longley-Rice predicted signal levels. Despite this caveat, in open areas, Longley-Rice and measured signal levels may be found to show high levels of agreement.

IN-BUILDING FIELD STRENGTH MEASUREMENTS

In-building measurements present special difficulties. The signal strength may vary widely in different locations within the same room. This presents a challenge for both measurement and reporting. The resultant signal within a building may be influenced by varying attenuation, varying reflections and even the body of the measuring engineer.

The measuring receiver will need to be portable and battery operated and the antenna methodology carefully chosen if the results are to be consistent and repeatable. One choice of antenna is a vertically polarized helically wound (and therefore fairly small) antenna mounted on a small ground plane, such as a round metal pan. The antenna may be attached to a non-conductive pole for a consistent height above ground. The cable between the antenna and receiver should be fairly short and dressed so as not to affect the measurements inconsistently and treated with ferrite beads to prevent current flow on the cable. Figure 10.5-11 is an example, and this antenna would be a good companion with a small handheld spectrum analyzer.

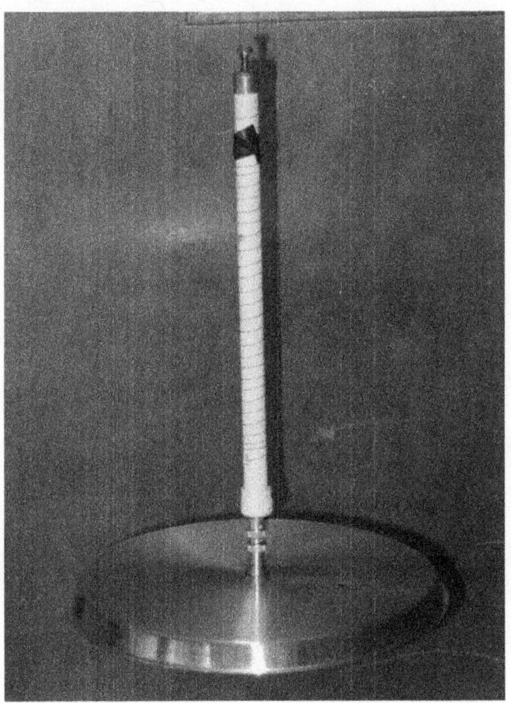

FIGURE 10.5-11 Short vertical whip antenna for indoor measurements.

EMERGING METHODS & EQUIPMENT FOR FIELD STRENGTH MEASUREMENTS

Technological changes and innovations occur at a rapid rate today and promise more and better tools. Here are just a few observations.

Drones—in its December 11, 2015, issue, *RCR Wireless* reported that an Australian company, Innovative Drone Solutions (IDS), is using Signal Hound's compact SA44B signal analyzer with an airborne drone to characterize the radiation pattern of antennas on a tower, with up to 10 stations being measured in about 25 minutes. Subsequent communications with Jason Schreiber, the RF engineer at IDS, revealed some considerations when using this method that are still developing in the United States:

- Potential need to be an unmanned aerial vehicle (UAV) certified operator.
- Sufficient lead to have the flight plan approved by aviation authorities.
- Plan contingency days due to weather.
- Need to fly in the far field to prevent EMI and receiver overload.

IMPROVED RECEIVER CHARACTERISTICS

It is now possible to build digitally tuned bandpass filters using digitally tuned capacitors such as the Peregrine PE64102. SDR receivers are now coming on the market with incredible specifications compared to today's receivers, such as 35-dBm IP3 front ends, 3.5-dB noise figure from 40–1000 MHz and 0.5 ppm frequency accuracy. As writing this chapter is wrapping up, SDR receivers are being fully controlled by a Raspberry Pi 2, and software such as SDR-J is a suite of programs that include the ability to decode DAB, FM and several shortwave modes such as AM, USB, LSB, PSK, RTTY, WeatherFax, SSTV, BPSK, QPSK, CW, NavTex (Amtor-B), MFSK, Domino, Olivia, Hell, Throb and now DRM. It can directly connect to RTL-SDR receivers as well as other hardware such as the Airspy and SDRplay.

CHAPTER

10.6

Digital Television Proof of Performance Measurements

MARK H. BRICKER AND STEVEN RAMER

President, SJ Ramer Associates

INTRODUCTION

A television transmitter proof of performance is a series of measurements performed to verify that the broadcast transmission meets the emission/quality requirements set forth by the FCC as well as industry accepted standards of operation.

For the purposes of this chapter, the measurements that verify the suitability of the transmitted signal for off-air reception by the viewing public will be referred to as in-band measurements. The measurements that evaluate the presence of signals with the potential to cause interference to other stations or services operating outside of the allocated channel will be referred to as out-of-band measurements.

The FCC requires certain performance measurements of the digital TV transmitter output signal at initial installation and at each time an equipment modification is made that changes the level of performance/"significant" change in equipment configuration is made as set forth under section §73.1590(a)(1), "Equipment Performance Measurements," of the Code of Federal Regulations.

The FCC has accepted out-of-band performance limits recommended by the ATSC in its Recommended Standard A/64 as its standard of operation/performance. The FCC has not set standards relating to the in-band operation of the transmitter output signal as long as the modulation format conforms to the A/59 standard. Good practice accepts the recommendations set forth within the A/64 standard as the performance criteria.

The proof of performance measurement data will typically be compiled into a document to comply with rules section §73.1590(d) for the retention of the performance information. This document serves a dual purpose—it demonstrates compliance to the FCC and other regulatory authorities, and it captures a baseline of performance for future reference as the standard of operation for that particular transmitter.

The format of the proof of performance document will vary depending on the engineer making the measurements; however, the pertinent data included should be common to all. Currently available 8VSB test equipment is able to provide an output that is a summary of the data and is suitable for printing and record retention. This provides an automated solution to the acquisition and collection for many of the required measurements.

TESTS

While the 8VSB digital transmission signal is more complex than an analog signal, the number of basic measurements necessary to prove its quality is reduced. Further, many of the tests are largely automated and require only the proper RF level to the input of the test instrumentation. Careful attention should be made to the sampling of the transmitter output signal. The sampled signal must be taken after any transmitter-associated RF system channel bandpass or mask filters to provide an accurate representation of the transmitted signal.

The sampled signal will typically be obtained via a directional coupler port located in a transmission line or waveguide section. Any markings indicating the power level at or coupling value of the coupled port should be considered suspect unless absolutely trusted.

It would be recommended to add substantial amounts of external padding by way of fixed attenuators before connecting an unknown sample level to a piece of test equipment. Once the level is determined with certainty, only then should any added attenuation be removed to provide adequate signal level, as required by the test instrument. Most current 8VSB and RF measurement test equipment can accept a +10-dBm/10-mW input signal without damage. Most equipment has an indicator for the input level to aid in proper setup that be used to ensure a suitable signal level is present.

The measurements that should be included for all proof of performance testing are summarized in the lists in this chapter. Following the list are descriptions of each referenced test and what is being tested, the test equipment requirements for the test and a general procedure for the test.

The test instrumentation described for use in this document are a Rohde & Schwarz ETL TV Analyzer, a Rohde & Schwarz NRP-Z51 RF Power Sensor, a Telonic-Berkeley TTR-750 bandstop filter and a Mini-Circuits NHP-1000 high-pass filter.

Out-of-Band and FCC Required Measurements

Shoulder amplitude

Out-of-band emission

Harmonic and spurious

Pilot frequency

Output power

In-Band or Signal Quality Measurements

Linear frequency response and group delay

Nonlinear amplitude and phase

Constellation and eye diagram

MER/SNR

Test Equipment Configuration for Signal Measurements

Figure 10.6-1 depicts the configuration of the required test equipment in the RF output chain of the transmitter under test. Attenuators that may be required for signal level adjustment are not shown in this graphic.

Shoulder Amplitude

The transmitter emission that lies within the first 500 kHz either side of the allocated 6-MHz channel is referred to as the shoulder. The signal in this region must be attenuated 47 dB below the transmitter reference power output for full service mask, or 36.4dB below the flat portion, or head, of the 8VSB signal. The reference power level is for the entire 8VSB signal, which includes the pilot. The difference between

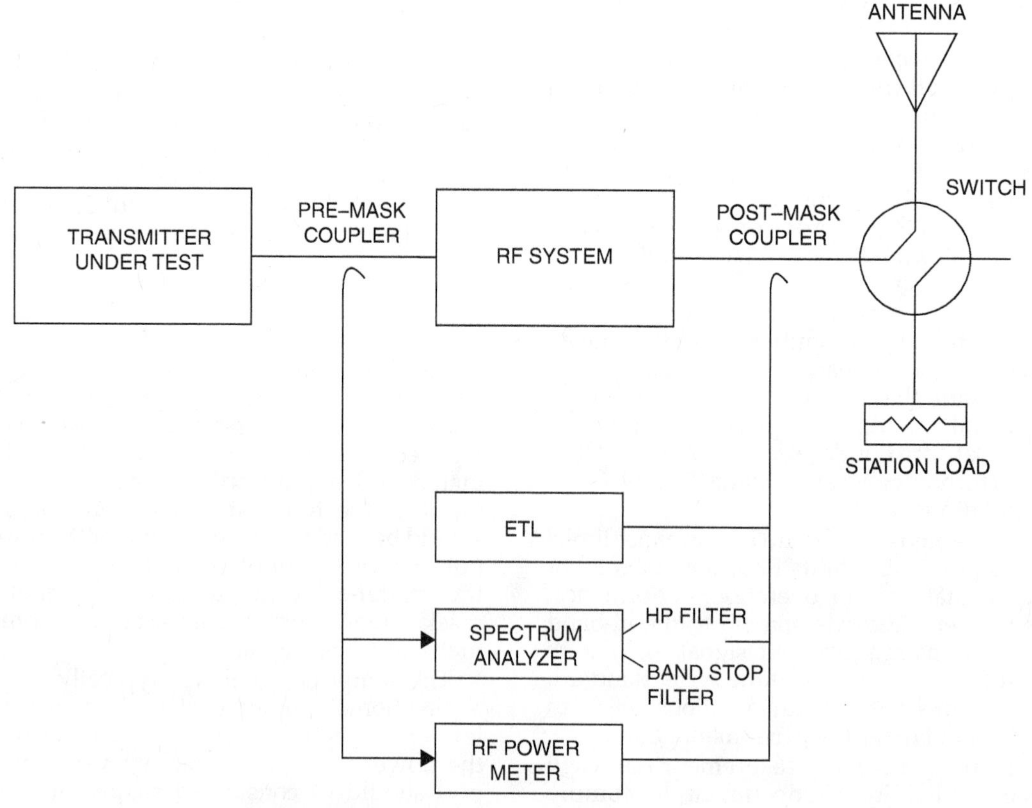

FIGURE 10.6-1 Test equipment configuration.

the reference power and head values is the difference between the signal power in the entire 6-MHz channel versus signal power in a 500-kHz measurement band, effectively a correction factor of 10.6 dB. This can be calculated as the ratio of the 500kHz bandwidth to the 8VSB bandwidth, or 10log(0.5/5.38) dB equal to 10.3 dB. The pilot adds 0.3 dB, totaling 10.6 dB. This correction factor can then be applied to the out-of-band signal as viewed on a spectrum analyzer at any resolution bandwidth, assuming both the in-band signal and out-of-band signal exhibit noise-like characteristics. This is a good assumption, as most if not all of the energy being measured out-of-band is the result of intermodulation distortion. The 10.6-dB factor means that, instead of looking for 47 dB on a spectrum analyzer, an engineer would look for the energy in the first 500 kHz from the channel edge to be 36.4 dB below the flat portion of the signal.

Test Equipment

The shoulder attenuation measurement is made using the Rohde & Schwarz ETL in either its spectrum analyzer mode or in its TV Analyzer/Receiver mode, depending on the method used, as described next.

Measurement Procedure for Spectrum Analyzer Method

For shoulders, it is convenient to use a digital spectrum analyzer with power band markers to measure each 500 kHz segment and compare each to the in-band signal power in 6 MHz. On the spectrum analyzer, a marker is placed at the center frequency of the channel being measured. A second marker is then offset from the center of channel marker by +3.25 MHz. Finally, a third marker is offset from the center of channel marker by −3.25 MHz.

1. Calculate the correct amount of external attenuation necessary to provide protection to the test equipment input circuits. This is based on transmitter output power and sample port coupling value.
2. Connect the ETL to the test port to be measured (premask filter) or (postmask filter)
3. Set the ETL to Spectrum Analyzer Mode
4. Select—FREQUENCY
5. Set the center frequency at mid-band on ATSC channel to be measured
6. Set Span to 15 MHz
7. Set Detector to RMS
8. Set the resolution bandwidth (RBW) to 10 kHz
9. Set the video bandwidth (VBW) to 100 kHz
10. Set Sweep Time to 2s
11. Set Marker 1 at center frequency
12. Set Marker 2 at +3.25MHz
13. Set Marker 3 at −3.25MHz
14. Read the marker delta values
15. Print test screen (Figure 10.6-2)

Measurement Procedure for ETL TV Analyzer/Receiver Mode

1. Calculate the correct amount of external attenuation necessary to provide protection to the test equipment input circuits. This is based on the transmitter output power and sample port coupling value.
2. Connect the ETL to the test port to be measured.
3. Set the ETL to TV Analyzer/Receiver Mode.
4. Select MENU—ADJUST ATTENUATION
5. Select MEAS—SPECTRUM
6. Print test screen (Figure 10.6-3)

Out of Band Measurements

For full-service broadcast, the amplitude of the transmitter emissions at frequencies between 0.5 and 6 MHz from the channel band edge, the lower and upper adjacent channel areas, must decrease from the 47 dB shoulder attenuation value to a level of 110 dB. The rule for out-of-band emissions is given in the FCC rules, Section 73.622(h)(1). At any frequency between 0.5 and 6 MHz from the channel edge, emissions must be attenuated no less than the value determined by the following formula:

$$\text{Attenuation in dB} \sim -11.5(\Delta f + 3.6);$$

where Δf ~ frequency difference in MHz from the edge of the channel.

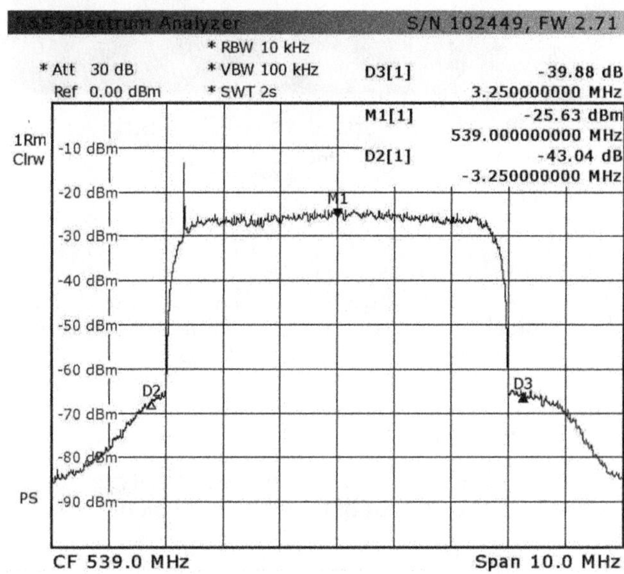

FIGURE 10.6-2 Shoulder measurement using the cursor method in Spectrum Analyzer Mode.

Rohde & Schwarz GmbH

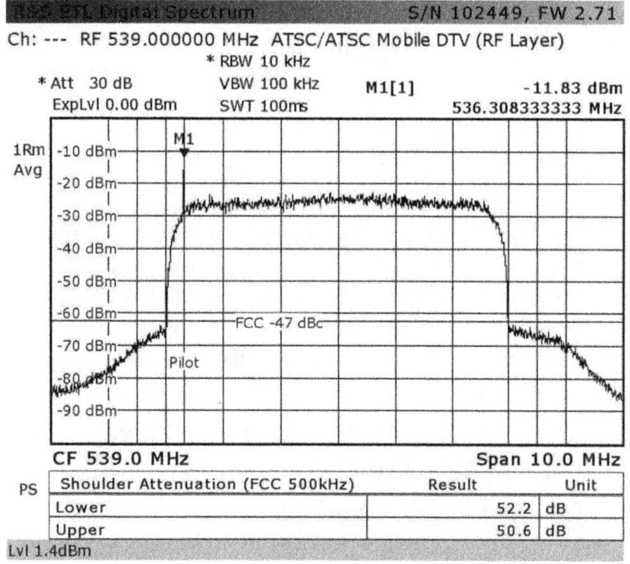

FIGURE 10.6-3 Measuring shoulder attenuation using the tangent method in ETL TV Analyzer/Receiver mode.

Rohde & Schwarz GmbH

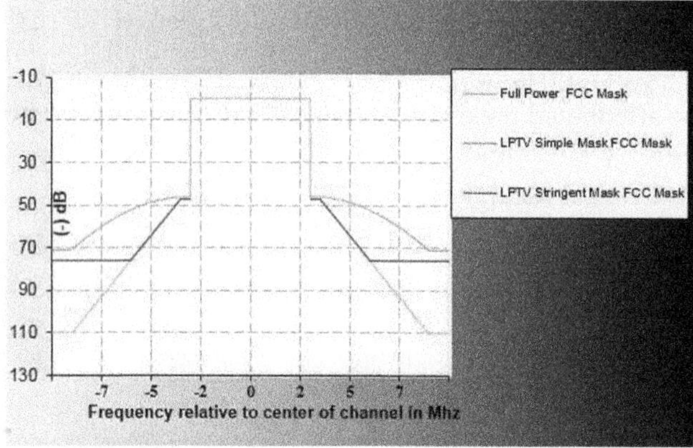

FIGURE 10.6-4 Full-service vs. dLPTV Stringent vs dLPTV simple emission masks.

Dielectric LLC

This establishes a mask for levels of out-of-band emissions, as shown in Figure 10.6-4. The simple and stringent mask characteristics are applicable to Part 74 low power and Class A operation.

Test Equipment

The Rohde & Schwarz ETL in spectrum analyzer mode is used to make the out of band measurement.

Test Procedure

Most spectrum analyzers available at transmitter sites do not have the dynamic range necessary to measure the out-of-band performance directly. An alternative method to a direct measurement involves taking measurements in two parts. The first part is to measure the transmitter emissions at a point before the final output mask filter. The second part is to measure the response of the mask. These two sets of data are then numerically superimposed to obtain the final output response of the transmitter. Taking the measurement in two parts effectively doubles the dynamic range of the measurement and allows standard test equipment to be able to demonstrate compliance to the FCC mask.

Harmonic and Spurious

All emissions from the transmitter that are further than 6 MHz from the allocated channel, both below and above, must be a minimum of 110 dB below the reference amplitude for Part 73 full service operation, 76 dB for Part 74 low power/Class A stringent mask operation or 71 dB for Part 74 low-power/Class A simple mask operation.

Harmonic signals will appear at integer multiples of the transmitter frequency, while spurious signals may appear at any frequency. Harmonic signals are typically measured for the second and third harmonic. Spurious signals should be investigated across the entire spectrum, from a somewhat arbitrary 1 MHz to the third harmonic frequency.

The high attenuation value required for harmonic and spurious signals requires that extra care be taken during testing. It is important to apply as high a signal level as possible to the spectrum analyzer for this measurement so that the maximum dynamic range is gained in the spectrum analyzer without causing overload of the analyzer input. Analyzer overload will generate intermodulation within the analyzer and provide a false indication of high harmonic level.

Lower-power transmitters may be able to use a high-power attenuator with a flat frequency response as measured across the spectrum of interest up to the third harmonic. A flat frequency response means that there are no correction factors that must be applied to the value measured.

In most cases a directional coupler is used to provide the signal sample for the measurement. Directional couplers in waveguide or coaxial line will, in most situations, exhibit a frequency-dependent coupling value, making the measurement of third—and sometimes second—harmonic and spurious frequencies in those areas subject to inaccuracies.

The optimal method of harmonic and spurious measurement would require that the frequency response of the coupler port used be characterized across the spectrum to be measured.

In the absence of this measurement, you must assume that the coupler will present a 6 dB per-octave increase to the signal so that second harmonic samples will be 6 dB lower than measured and third harmonic signals will be 9 dB lower than measured.

The tunable bandstop filter will probably have a limit to its usable highest frequency, typically about twice the frequency of its prescribed operation. For example, for a tunable bandstop filter with a 750-MHz upper stopband frequency, the maximum upper frequency to use for harmonic and spurious observation will be 1,500MHz. For evaluation of higher frequencies, a high-pass filter is used. The cutoff of this filter should be about 1,000 MHz.

Test Equipment

The Rohde & Schwarz ETL in spectrum analyzer mode, Telonic-Berkeley TTR-750 bandstop filter and a Mini-Circuits NRP-1000 high-pass filter are used for these measurements.

TEST PROCEDURE

Initial spectrum analyzer setup:

Frequency: Center frequency for the channel under test

10dB per division vertical

RBW: 10 kHz

VBW: 10 kHz

Span: 20 MHz

Sweep: 2 seconds

Amplitude Grid: REL (relative)

Amplitude RF Attenuation: 20dB

1. Connect a 30-dB attenuator to the input of the analyzer then apply the transmitter output sample.
2. Adjust the input attenuator of the spectrum analyzer to set the display of the on-channel signal to the −10-dB graticule line. This establishes the 0 dB reference amplitude at the −10 dB graticule line (See Figure 10.6-5).
3. Remove the input cable from the spectrum analyzer and insert the bandstop filter into the sample feeding the spectrum analyzer. Adjust the filter to reject the maximum amount of channel signal possible as displayed on the analyzer.
4. Remove the external 30-dB attenuator installed in step (1) and reconnect the sample to the spectrum analyzer with the bandstop filter in the circuit (see Figure 10.6-6). The 0 dB reference level (originally set at the −10-dB graticule line) is now 30 db *higher* on the display, and the −10-dB graticule line is now 30 dB lower, or −30 dB relative to the 0-dB reference. This then places the −110dB level for mask compliance at the graticule line labeled −80 dB. This graticule line now represents the maximum level for any observed harmonics or spurious signals.
5. Adjust the spectrum analyzer internal RF attenuation to 0 dB.
6. Set the spectrum analyzer center frequency (cf) step size to 10 MHz.
7. Adjust the frequency of the spectrum analyzer down to 10 MHz and then use up/down arrow keys to incrementally step the frequency upward while reviewing the display for any signal higher than the −110dB relative graticule line established in step (4).
8. Review the RF sample spectrum to twice the upper tunable frequency of the bandstop filter. Remove the bandstop filter and replace it with the high-pass filter, and continue the spectrum investigation up to the third harmonic of the transmitter channel.

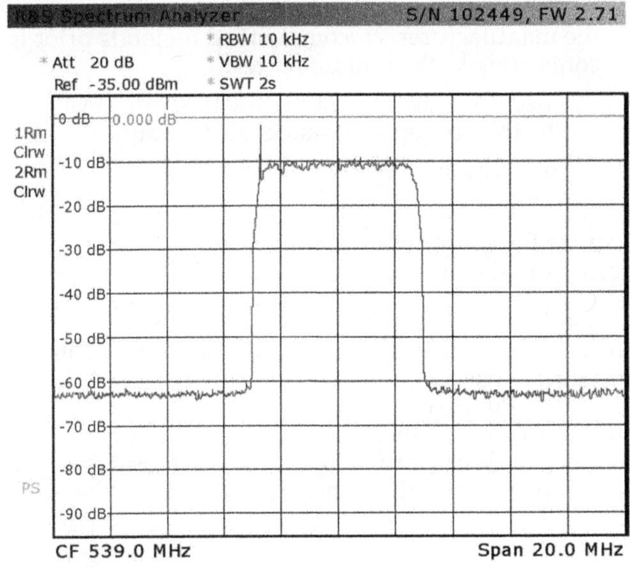

FIGURE 10.6-5 Harmonic and spurious emission measurement (reference).

Rohde & Schwarz GmbH

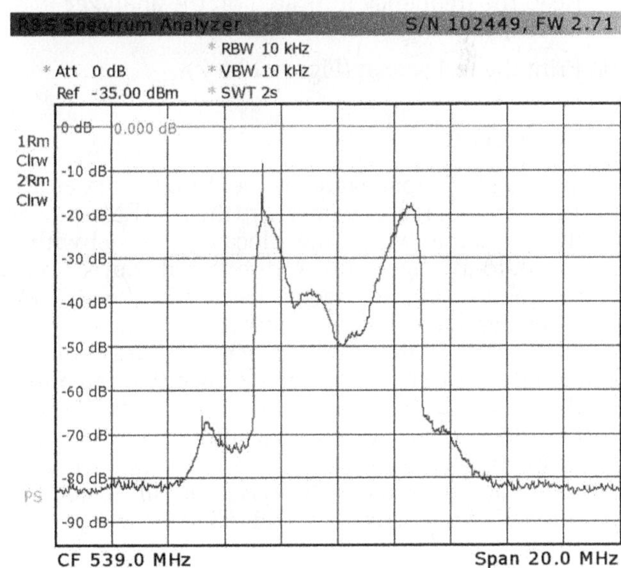

FIGURE 10.6-6 Harmonic and spurious emission measurement with bandstop filter (reference).

Rohde & Schwarz GmbH

Pilot Frequency

The pilot of a DTV signal should be 309.4405kHz from the channel lower band edge, within a tolerance of ±1 kHz as set by FCC Rules Section 73.1545(C).

Test Equipment

The Rohde & Schwarz ETL in spectrum analyzer mode and a precision 10-MHz reference, whether from an internal high-precision reference or from a global-positioning system (GPS) receiver, are used to make an accurate measurement of the pilot signal.

TEST PROCEDURE

1. Calculate the correct amount of external attenuation necessary to provide protection to the test equipment input circuits. This is based on output power and sample port coupling value.
2. Connect the R&S ETL to the test port.

Equipment settings for Pilot measurement using the Rohde & Schwarz ETL:

3. Resolution bandwidth (RBW): 10 Hz.
4. Video bandwidth (VBW): 10 Hz.
5. Set center frequency to pilot frequency for channel under test.
6. Span: 200 Hz (manual tuning of the analyzer may be required to center the pilot on the center graticule of the display).
7. Sweep time (SWT): 2 seconds.
8. Turn on equipment marker (if available) and position on peak of pilot display, as in Figure 10.6-6.
9. Read the frequency indicated on the analyzer screen.
10. Print the test screen (Figure 10.6-7).

Output Power

The power of the total signal is measured using an RF power meter that is able to measure the true RMS power of the DTV signal within the allocated channel with a high peak-to-average ratio. A directional coupler port with a known coupling factor, located in the RF system after the mask (bandpass) filter, is required. Since any uncertainty in the coupling factor is a direct uncertainty in the output power of the transmitter, care must be taken to verify the accuracy of this value. FCC Rules Section 73.15670(c)(1) establishes an output power tolerance of 80 to 110%; however, due to the rapid fall off of decode ability of the DTV signal, or cliff effect, the recommended tolerance in A/64 is 95 to 105%.

Test Equipment

A Rohde & Schwarz NRP-Z51 RF power sensor is used for transmitter RF output power measurement.

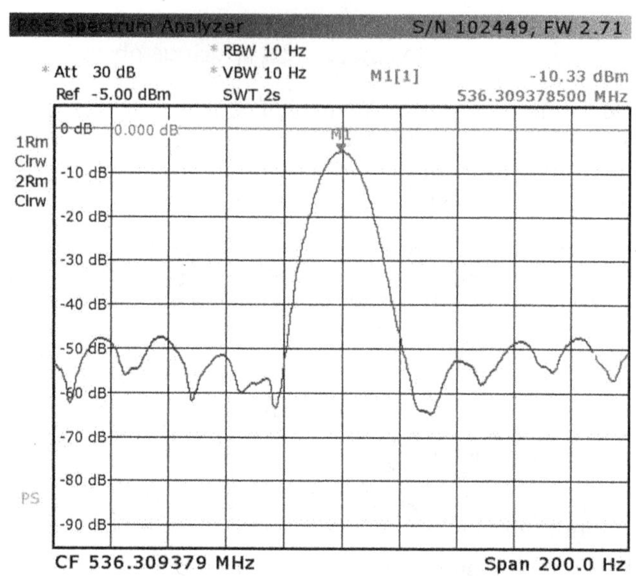

FIGURE 10.6-7 Example of pilot frequency measurement.

Rohde & Schwarz GmbH

TEST PROCEDURE

1. The measurement port forward coupling value must be known to obtain an accurate reading.
2. If this value is not known to be accurate, calibrating the port with a calibrated network analyzer should be completed prior to taking any measurements.
3. The "Signal Frequency" is the center frequency of the channel under test.
4. The "Level Offset" refers to the coupling value from the test port.
5. Always "zero" or calibrate the power sensor as per the manufacturer's recommended methods prior to connecting to the unit under test.
6. Connect the power sensor to the (forward) test port on the transmitter post mask filter output.
7. Record reading (see Figure 10.6-8).

Linear Frequency Response and Group Delay Errors

Linear frequency response errors are the deviation of the transmitted in-band signal from an ideal flat amplitude response as measured across the 8VSB waveform. Linear group delay errors are the deviation of the transmitted in-band signal from a constant delay, as measured across the same 8VSB waveform. The primary source of both of these errors is the RF system mask filter, due to its necessarily narrow-band tuning.

The remainder of the RF system presents no linear errors, because they are all wideband with respect to the transmitter channel of operation. These measurements are referred to as Amplitude and Group Delay in the R&S ETL.

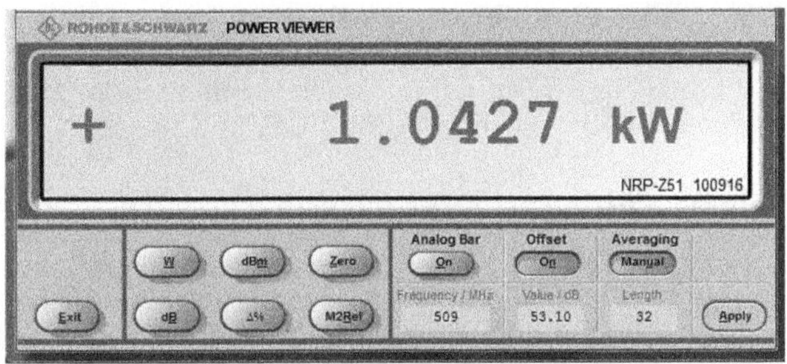

FIGURE 10.6-8 Example of output power measurement.

Rohde & Schwarz GmbH

Test Equipment

Amplitude and Group Delay are measured using the Rohde & Schwarz ETL in its TV Analyzer/Receiver mode.

TEST PROCEDURE

1. Calculate the correct amount of external attenuation necessary to provide protection to the test equipment input circuits. This is based on output power and sample port coupling value.
2. Connect the R&S ETL to the test port.
3. Switch transmitter on.
4. Select MEAS: CHANNEL ANALYSIS.
5. Select AMPLITUDE and GROUP DELAY from the menu.
6. Select MENU: ADJUST ATTENUATION.
7. Press AUTORANGE.
8. Print the test screen (Figures 10.6-9 and 10.6-10).

Figure 10.6-9 exhibits a poorly equalized linear correction defined by high amplitude and group delay error.

Figure 10.6-10 exhibits good equalization (flat response) of the linear amplitude and group delay.

Nonlinear Amplitude and Phase

Nonlinear amplitude errors are nonlinearities in the amplifier due to amplifier gain variations as a function of the signal amplitude. This would most typically be observed as compression or clipping of the output signal. This would also be referred to as AM/AM distortion.

Nonlinear phase errors are nonlinearities in the amplifier due to signal phase change as a function of the signal amplitude. This would also be referred to as an AM/PM distortion.

Test Equipment

Nonlinear amplitude and phase are measured using the Rohde & Schwarz ETL in its TV Analyzer/Receiver

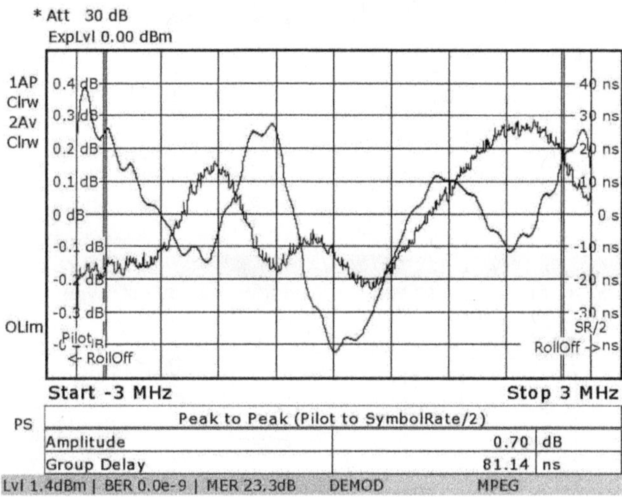

FIGURE 10.6-9 Example of high linear frequency response and group delay errors.

Rohde & Schwarz GmbH

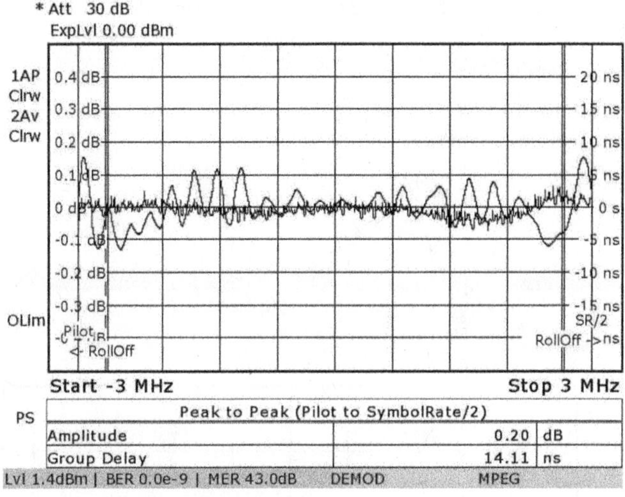

FIGURE 10.6-10 Example of low linear frequency response and group delay errors.

Rohde & Schwarz GmbH

mode. These measurements are referred to Amplitude and Phase in this instrument.

TEST PROCEDURE

1. Calculate the correct amount of external attenuation necessary to provide protection to the test equipment input circuits. This is based on output power and sample port coupling value.

2. Connect the R&S ETL to the test port.
3. Switch transmitter on.
4. Select MEAS: CHANNEL ANALYSIS.
5. Select AMPLITUDE and PHASE from the menu.
6. Select MENU: ADJUST ATTENUATION.
7. Press AUTORANGE.
8. Print the test screen (Figures 10.6-11 and 10.6-12).

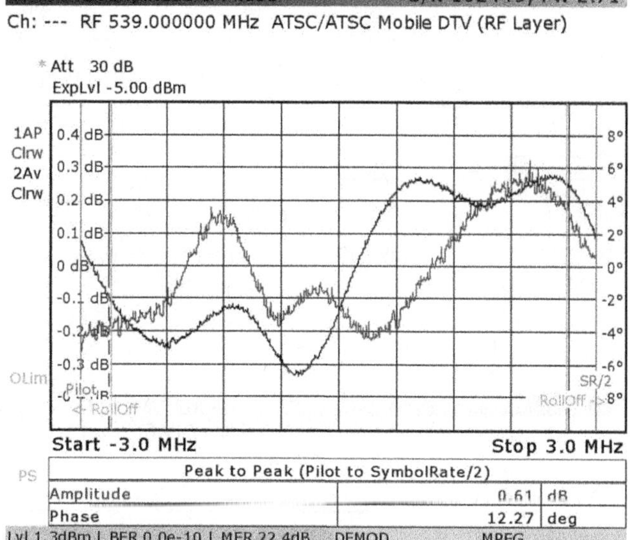

FIGURE 10.6-11 Example of high nonlinear amplitude and phase errors.

Rohde & Schwarz GmbH

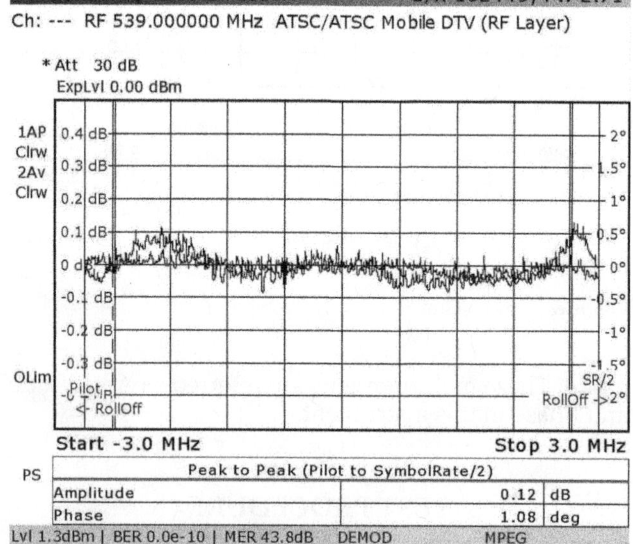

FIGURE 10.6-12 Example of low nonlinear amplitude and phase errors.

Rohde & Schwarz GmbH

Figure 10.6-11 exhibits a poorly equalized nonlinear correction defined by high amplitude and phase error.

Figure 10.6-12 exhibits good equalization (flat response) of the nonlinear amplitude and phase.

Constellation and Eye Diagram

The 8VSB eye diagram turns up in the pages of many articles on 8VSB and on the screens of many pieces of 8VSB test and monitoring equipment. The diagram is the overlay of many traces of the received RF signal amplitude at the instant of sampling.

At each sampling time, the demodulated RF amplitude assumes one of eight possible levels. The resulting display creates eight vertical "eyes." If the 8VSB signal is corrupted during transmission (Figure 10.6-13), these "eyes" will close up and disappear because the RF signal will no longer have the correct amplitude at the right instant.

The 8VSB signal constellation signal is common to many pieces of test equipment. This is a two-dimensional graphical representation of the 8VSB RF carrier amplitude and phase.

When the 8VSB RF signal is corrupted, the eight vertical lines become blurred and errors are received. Generally speaking, the thinner the lines, the lower the distortion and the higher the signal to noise (SNR).

Test Equipment

The constellation and eye diagrams are measured using the Rohde & Schwarz ETL in its TV Analyzer/Receiver mode.

TEST PROCEDURE—EYE

1. Calculate the correct amount of external attenuation necessary to provide protection to the test equipment input circuits. This is based on output power and sample port coupling value.
2. Connect the R&S ETL to the test port.
3. Switch transmitter on.
4. Select MEAS: MODULATION ANALYSIS.
5. Select EYE DIAGRAM from the menu.
6. Use MENU: ADJUST ATTENUATION.
7. Press AUTORANGE.
8. Print the test screen (see examples).

In Figure 10.6-13, the eye diagram shows significant signal impairment and poor SNR.

In Figure 10.6-14, the eye diagram shows minor signal impairment and good SNR.

TEST PROCEDURE—CONSTELLATION

1. Calculate the correct amount of external attenuation necessary to provide protection to the test equipment input circuits. This is based on output power and sample port coupling value.

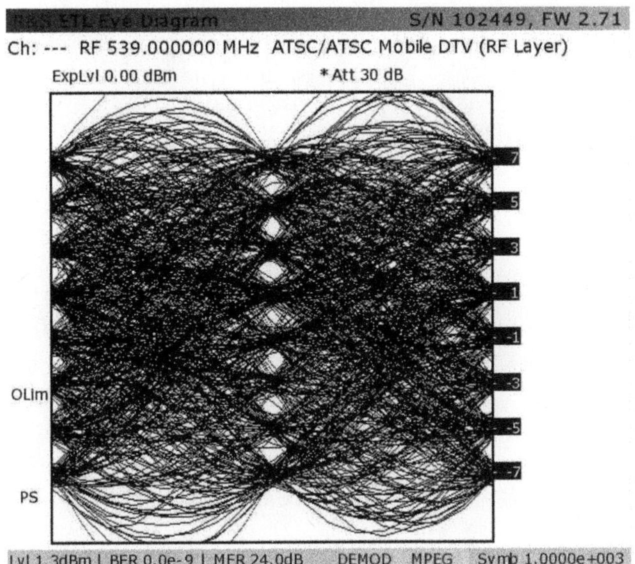

FIGURE 10.6-13 Example of eye diagram with high signal impairment.

Rohde & Schwarz GmbH

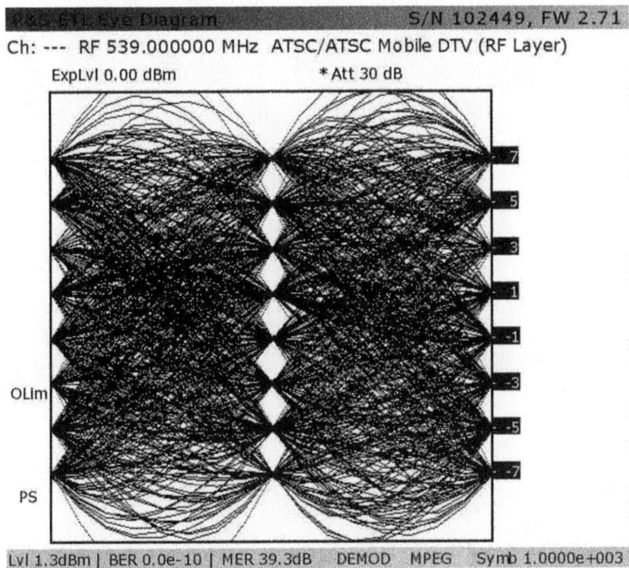

FIGURE 10.6-14 Example of eye diagram with minimal signal impairment.

Rohde & Schwarz GmbH

2. Connect the R&S ETL to the test port.
3. Switch transmitter on.
4. Select MEAS: MODULATION ANALYSIS.
5. Select CONST DIAGRAM from the menu.
6. Use MENU: ADJUST ATTENUATION.
7. Press AUTORANGE.
8. Print the test screen (see examples).

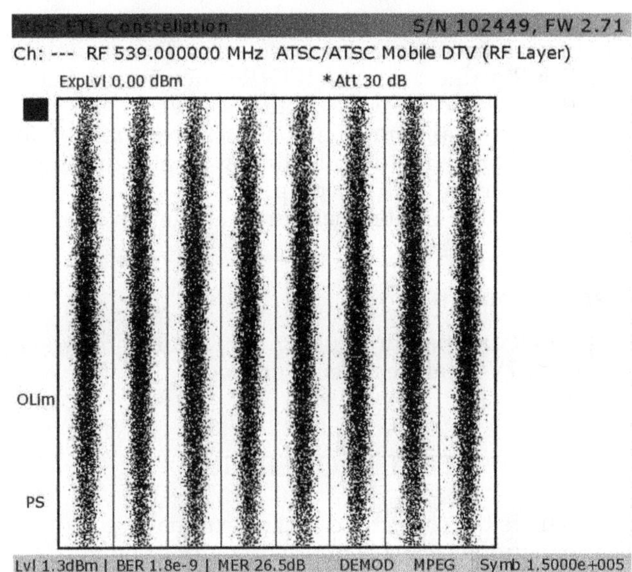

FIGURE 10.6-15 Example of constellation with high signal impairment.

Rohde & Schwarz GmbH

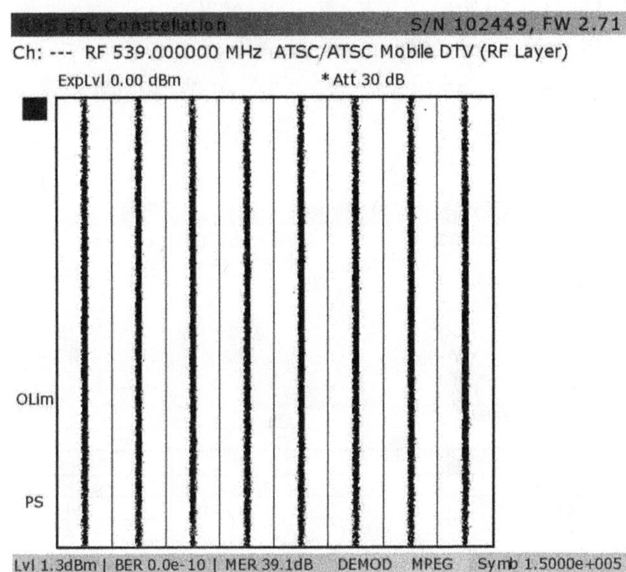

FIGURE 10.6-16 Example of constellation with minimal signal impairment.

Rohde & Schwarz GmbH

In Figure 10.6-15, the constellation shows significant signal impairment and poor SNR.

In Figure 10.6-16, the constellation shows minor signal impairment and good SNR.

MER/SNR

MER or SNR is an overall quality measure of the 8VSB signal that includes multiple sources of impairment including noise, intermodulation and amplitude or group delay variations across frequency. The minimum MER/SNR value set by ATSC Recommended Standard A/64 is 27dB. Higher values should be the goal of a transmitter alignment and exciter precorrection to provide better signal decodability at greater distances and allow for transmitter equipment operational variations over time and temperature, i.e., margin.

Test Equipment

MER/SNR are measured using the Rohde & Schwarz ETL in its TV Analyzer/Receiver mode.

TEST PROCEDURE

1. Calculate the correct amount of external attenuation necessary to provide protection to the test equipment input circuits. This is based on output power and sample port coupling value.
2. Connect the R&S ETL to the test port.
3. Switch transmitter on.
4. Select MEAS: OVERVIEW.
5. Use MENU: ADJUST ATTENUATION.
6. Print the test screen (see examples).

In Figure 10.6-17, the MER/SNR reflects minor signal impairment.

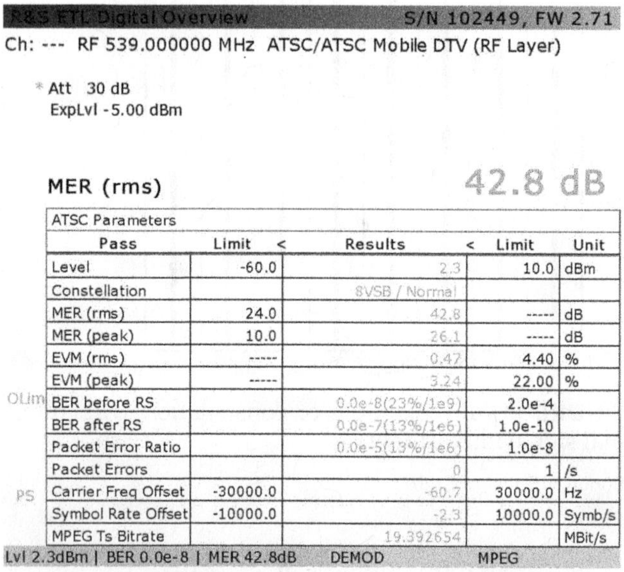

R&S ETL Digital Overview S/N 102449, FW 2.71
Ch: --- RF 539.000000 MHz ATSC/ATSC Mobile DTV (RF Layer)

* Att 30 dB
ExpLvl -5.00 dBm

MER (rms) 42.8 dB

ATSC Parameters					
Pass	Limit <	Results	< Limit	Unit
Level	-60.0	2.3	10.0	dBm
Constellation		8VSB / Normal		
MER (rms)	24.0	42.8	-----	dB
MER (peak)	10.0	26.1	-----	dB
EVM (rms)	-----	0.47	4.40	%
EVM (peak)	-----	3.24	22.00	%
BER before RS		0.0e-8(23%/1e9)	2.0e-4	
BER after RS		0.0e-7(13%/1e6)	1.0e-10	
Packet Error Ratio		0.0e-5(13%/1e6)	1.0e-8	
Packet Errors		0	1	/s
Carrier Freq Offset	-30000.0	-60.7	30000.0	Hz
Symbol Rate Offset	-10000.0	-2.3	10000.0	Symb/s
MPEG Ts Bitrate		19.392654		MBit/s

Lvl 2.3dBm | BER 0.0e-8 | MER 42.8dB DEMOD MPEG

FIGURE 10.6-17 Example of ETL TV Analyzer/Receiver mode overview measurements.

Rohde & Schwarz GmbH

CHAPTER

10.7

TV Transmission Line and Antenna System Measurements

TODD R. LONEY

Electron Dynamics, Inc.
Galiano Island, BC

STEPHEN N. HEAZLEWOOD

Broadcast Australia—BAI Communications
Melbourne

INTRODUCTION

Lord Kelvin succinctly stated, "To measure is to know" . . . of course, he also posited that "Radio has no future." Fortunately for our industry, the latter proved incorrect.

The purpose of a broadcast transmission system is to deliver radio frequency (RF) energy as efficiently as possible from the transmitter to the antenna with minimal spectral distortion of the transmission envelope. Proper testing helps to ensure good system integration and optimal performance. Rigorous testing methods allow for the comparison of results between systems or to track a system's performance over time. Testing can also identify potential problems that would ultimately result in faults. The early identification of a fault could minimize the extent of the damage and also minimize interruption of the transmission.

Transmission line and antenna testing is also colloquially referred to as "sweeping." A low-power RF signal is sent out from the test equipment via the transmission line toward the antenna, and the amount of reflected energy is measured. Any reflected energy means not all of the power is getting out of the antenna. This is analogous to light passing or reflecting off a clear or mirrored piece of glass. Any difference in impedance between circuit elements causes reflections and a loss of power transfer. In broadcast RF applications, transmission lines generally are coaxial and have a characteristic impedance that is directly dependent on the geometry of their conductors (i.e., their physical dimensions). The characteristic impedance of the transmission lines and the input and output impedances of components are typically designed to be 50 or 75 Ω. Optimal power transfer takes place when impedances are matched. Since a wave passing from one characteristic impedance to another causes a reflection, if the impedances are the same, there is no reflection. When there is a reflected wave due to an impedance mismatch, there will be waves traveling in both directions on the transmission line: the original incident wave and the new reflected wave. These waves can add both constructively and destructively, creating a standing wave pattern of nodes and antinodes.

All RF transmission systems will have some minimal reflections or mismatch due to some imperfections along the line or from the antenna. It is important that these reflections be quantified and characterized to determine if they will cause impairment to the signal or lead to failure.

RF MEASUREMENT METRICS

For transmission line and antenna testing, we are mainly interested in measuring the S_{11} parameter of a one-port network as shown in Figure 10.7-1. The quality of the S_{11} match can be represented by a variety of metrics such as reflection coefficient, return loss or voltage standing wave ratio.

- *Reflection coefficient* (Γ) is given as:

$$\Gamma = \frac{Z_L - Z_O}{Z_L + Z_O}$$

Where Z_L is the load impedance and Z_O is the characteristic line impedance. The reflection coefficient is a complex value with a range from −1 (short) to 1 (open) with magnitude ρ and angle θ. Reflection coefficient represents the relative voltage amplitude of the reflected wave with respect to the original incident wave. The locus of reflection coefficient values across the frequency sweep is plotted on a polar or Smith chart graph as shown in Figures 10.7-2A and 10.7-2B.

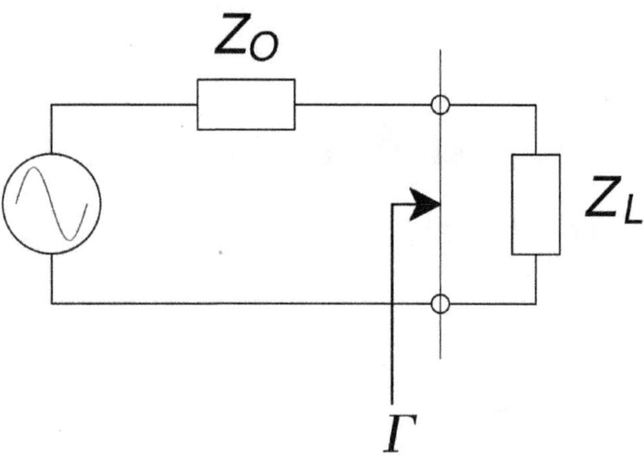

FIGURE 10.7-1 Simplified one-port circuit.

Artwork Courtesy Tom Smuts

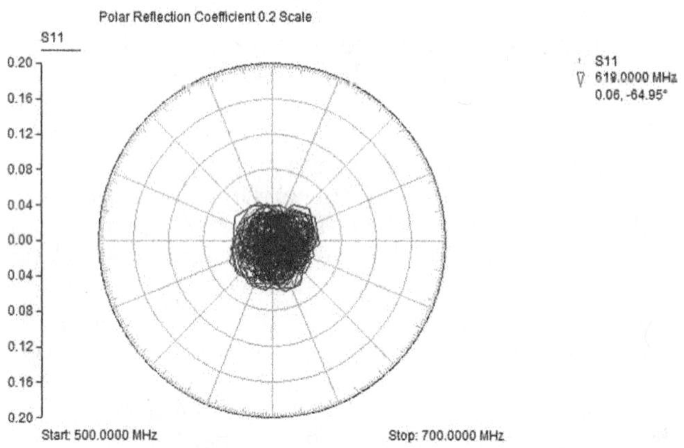

FIGURE 10.7-2A Polar impedance plot.

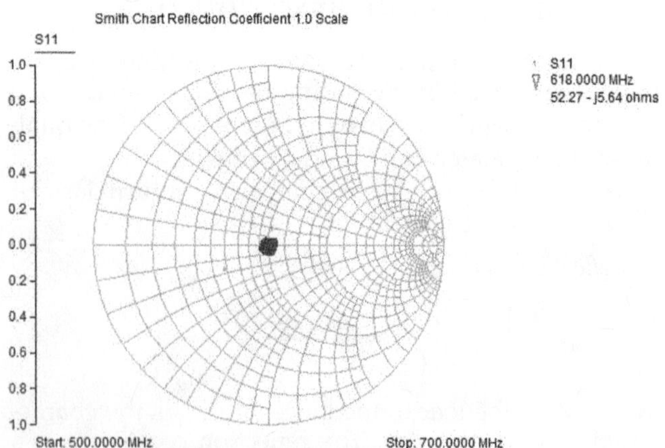

FIGURE 10.7-2B Smith chart impedance plot.

A scale of 200 mU (Γ of 0.2) or lower is a common scale for broadcast RF systems.

- *Return loss* given as:

$$RL = -20\log|\Gamma|$$

Is a measurement of the attenuated reflected signal compared to the incident signal in decibels (dB). Return loss is magnitude only ($|\Gamma| = \rho$) and is plotted on a logarithmic Cartesian graph, as shown in Figure 10.7-3.

- *Voltage standing wave ratio (VSWR)* given as:

$$VSWR = \frac{1+|\Gamma|}{1-|\Gamma|}$$

Is a measurement ratio of the highest voltage to the lowest voltage along the transmission line. Although this metric developed from obsolete slotted line measurement techniques, the value is still commonly used to describe system performance. VSWR is also magnitude only ($|\Gamma| = \rho$) and is plotted on a linear Cartesian graph as shown in Figure 10.7-4.

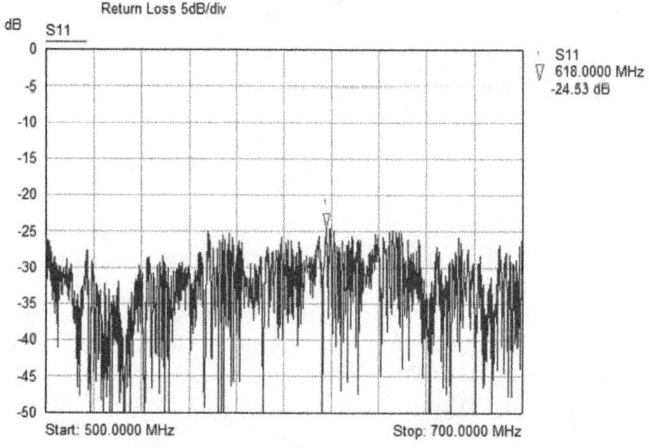

FIGURE 10.7-3 Return loss plot.

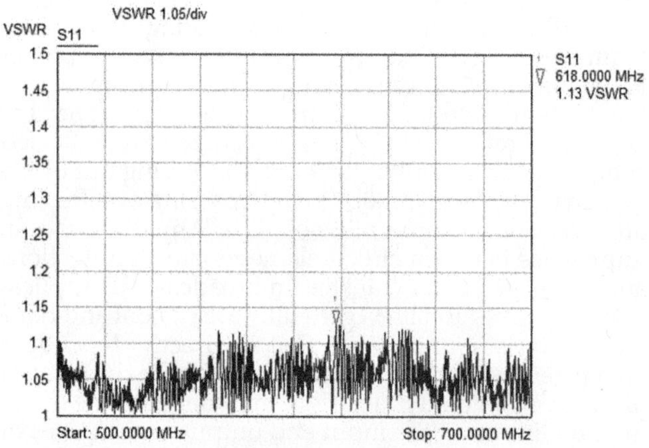

FIGURE 10.7-4 VSWR plot.

TABLE 10.7-1

VSWR	Return Loss	Reflection Coefficient	Reflected Power
1.00		0.00	0.00%
1.01	46.1	0.00	0.00%
1.05	32.3	0.02	0.10%
1.10	26.4	0.05	0.20%
1.20	20.8	0.09	0.80%
1.50	14.0	0.20	4.00%
2.00	9.5	0.33	11.10%
3.00	6.0	0.50	25.00%
∞	0.0	1.00	100.00%

Figures 10.7-2–4 all demonstrate the same measurement data in different formats. These metrics are correlative, as demonstrated in Table 10.7-1.

FIGURE 10.7-5 All-in-one calibration tool.

Photo Courtesy Spinner GmbH

EQUIPMENT

Instrumentation

The most ubiquitous piece of test equipment utilized for passive RF component and system testing is the vector network analyzer (VNA). Crucially, the VNA measures the magnitude and phase of the reflected signal response across a range of swept frequencies. The VNA contains two or more tuned receivers and a synthesized frequency source. Prior to the introduction of computerized VNA in the mid '80s, the RF engineer or technician utilized an assortment of test equipment: an RF sweep generator as a source feeding an impedance bridge or a slotted line and two diode detectors to a vector voltmeter. The results of these measurements were hand plotted on either a scalar or polar graph. Life slightly improved when the analog vector network analyzer was introduced, essentially packaging the sweep generator and tuned vector receivers into one instrument. An external test set, which included a splitter to feed a reference receiver and a high-directivity directional coupler to feed the measurement receiver, was also required. Measurement uncertainty was reduced, and one could sweep the entire frequency span of interest and measure both the return loss and reflection coefficient directly on the screen.

With the introduction of the automated or computerized VNA, the user was, for the first time, able to perform a calibration or vector error correction of the entire test setup. For one-port S_{11} measurements, the VNA calibration engine prompts the user to attach a known short/open/load (Figure 10.7-5) to the test point, and error correction of both magnitude and phase responses are characterized and stored in the instrument. This vector correction is applied to the measured data points. Today, there are many types of VNA in use, which range from high-end laboratory grade to economical, portable and even USB-connected instruments, as shown in Figures 10.7-6A-D.

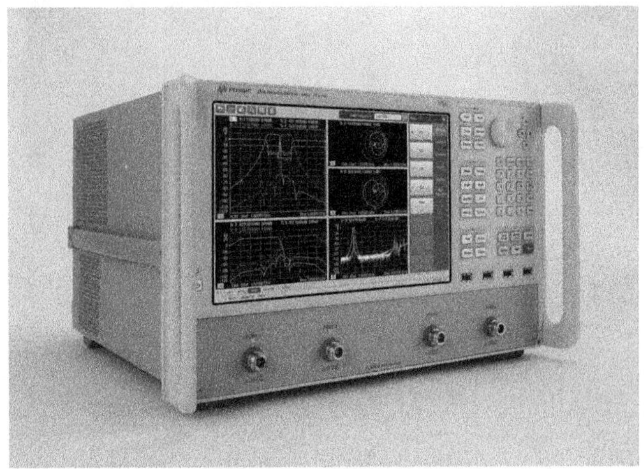

FIGURE 10.7-6A Laboratory four-port VNA.

Photo Courtesy Keysight Technologies, Inc.

FIGURE 10.7-6B Portable two-port VNA.

Photo Courtesy Keysight Technologies, Inc.

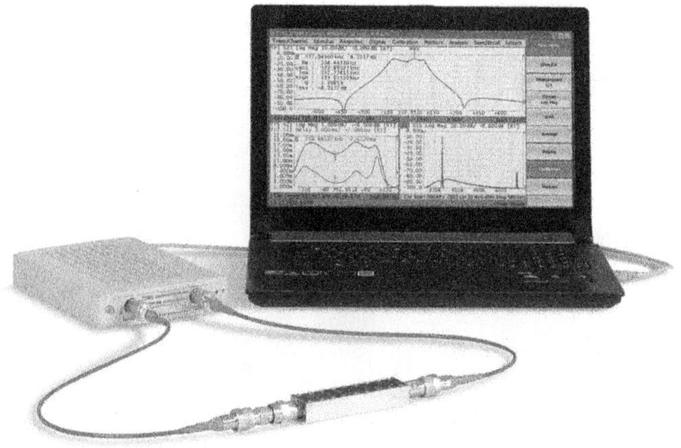

FIGURE 10.7-6C USB two-port VNA.

Photo Courtesy Copper Mountain Technologies

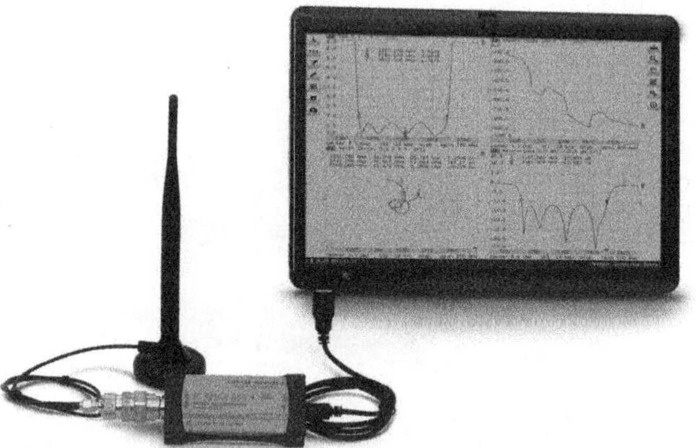

FIGURE 10.7-6-D USB one-port vector reflectometer.

Photo Courtesy Copper Mountain Technologies

In addition to frequency domain S_{11}, another technique known as time domain reflectometry provides a "distance to fault measurement" (DTF) for cable and was traditionally performed with a time domain reflectometer (TDR). A traditional TDR determines the characteristics of transmission lines by observing reflected waveforms. It sends a pulse of energy into the line and shows the energy reflected as echoes as a function of time. This is analogous to radar detecting echoes from aircraft but inside the transmission line instead of free space. By analyzing the magnitude and shape of the reflected waveform, the nature of the impedance variation in the line can be determined. The major disadvantage of this measurement is that it is performed with a DC pulse or DC step pulse and thus does not indicate any frequency-related issues.

The modern VNA has the ability to simulate the TDR by transforming the frequency domain data into time domain through a generalization of the inverse discrete Fourier transform (IDFT) called either inverse chirp Z transform (ICZT) or inverse fast Fourier transform (IFFT). With broadcast systems, antenna bandwidths are typically limited, and thus a bandpass transform

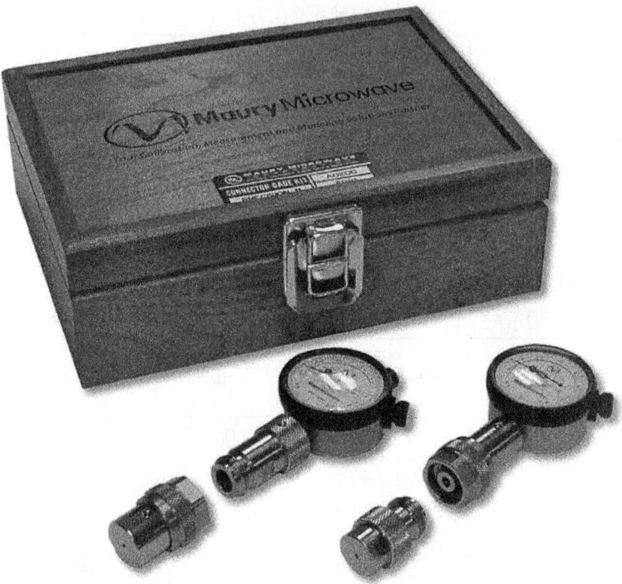

FIGURE 10.7-7 Type-N connector gage kit.

Photo Courtesy Maury Microwave Corp

is used. A bandpass transform allows the user to select the center frequency and span, whereas a low-pass step or low-pass impulse requires performance down to DC and provides data similar to a conventional time domain reflectometer.

The TDR acronym, in the context of this document, refers to the VNA Fourier transform method and is used interchangeably with frequency domain reflectometry (FDR) or distance to fault (DTF).

Test Cables

A test cable is used to connect the VNA to the line and antenna being measured. This cable must be included in the user calibration of the VNA, and no adapters (elbows, gender adapters, etc.) should be attached to it following the calibration, as they will distort the measurement accuracy. A high-quality phase-stable cable should be used so that measurements remain stable and accurate. It is quite common that a long test cable is required to reach the input to the line being swept and, in this case, an armor-jacketed test cable would be recommended. Typical high-quality cables will have a hex-shaped type-N connector and no gasket at the base. If a test cable has a gasket, it should be removed to ensure that the center pin fully seats in the test ports. A high-quality test cable will have the proper pin depth and will not damage the VNA test port. Connector gauges are common (see Figure 10.7-7) and should be used on any test cable configuration before being connected to either the VNA or calibration standards, as they will alert the user to potential issues with damage to test equipment or standards, as well as ensuring the calibration is not corrupted by a mismatch between the standard and the cable. Although the calibration will still be performed, the user can unwittingly create a new potential issue with the measurements by building in an inherent mismatch that will be applied to all measurements made on the components or system.

The performance of test cables can and should be measured regularly. To do this, a calibration is conducted on the test port of the VNA, with the test cable connected to the port and a precision load connected to the far end of the cable. The cable should have a measured return loss of 30 dB or better over the measurement band. Poor performing cables can cause significant measurement errors by making small reflections in the system, either through reflection or attenuation of the desired measurement.

Test Adapters

To ensure accurate measurements, a precision test adapter is required at the input of the line being tested and, if only the line is being swept, another precision adapter and load will be required atop the run of line. A test adapter is used to reduce the line size being swept to a 50 Ω female N-type or 7–16 DIN connector (see Figure 10.7-8).

Some companies manufacture a clever coaxial test section or direct-access unit that allows testing of the line or antenna separately. Normally, a plug connects the line to the antenna, but for testing, the plug is removed and the test section is inserted (see Figures 10.79A and B). This unit is typically installed at the antenna input. With the test section inserted, a load can be placed on the lower port to verify the transmission line only. The antenna can also be tested separately by attaching a short test cable to the lower port and performing a calibration. The cable can then be connected to the upper port to measure the antenna.

Since most VNAs are 50-Ω systems for 75-Ω transmission line, a test adapter is normally required, which includes a broadband transformer with multiple impedance steps to reduce to a female 50-Ω N-type or 7–16 DIN connector (see Figure 10.7-10).

These are generally frequency-limited devices, so care must be taken to ensure that the transformer

FIGURE 10.7-8 EIA-size rigid coax step adapters to 7–16 DIN.

Photo Courtesy Myat Inc.

FIGURE 10.7-9A 6″ rigid coax test section.

Photo Courtesy Spinner GmbH

FIGURE 10.7-9B 6″ rigid coax test section test adapter.

Photo Courtesy Spinner GmbH

FIGURE 10.7-10 Broadband 7-3/16″ 75-Ω transformer/test adapter to N-type.

Photo Courtesy Dielectric LLC

FIGURE 10.7-11 Tunable 3-1/8″ EIA test adapter to type-N.

Photo Courtesy Hitachi Kokusai Electric Comark LLC

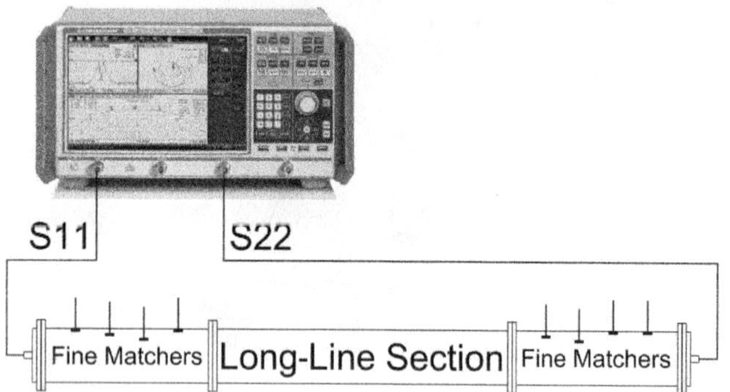

FIGURE 10.7-12 Long-line adapter tuning test configuration.

Artwork Courtesy Tom Smuts (VNA Image Courtesy of Rohde & Schwarz GmbH)

covers the operating band with enough extra bandwidth to ensure that the measurements are valid and repeatable. As a rule, this should be at least one channel above and below the frequency span of interest. There are 75-Ω calibration kits and minimum loss pads to adapt 50-Ω N-type to 75-Ω N-type connectors, but 50-Ω is typically the normalized impedance.

Traditionally, precision test adapters have a short section of line attached, with multiple nuts welded on to put tuning pins into the adapter (see Figure 10.7-11). This allows the adapter to be optimized for the frequency span of interest.

The adapters are tuned by the long-line method, which requires two adapters spaced several wavelengths apart (see Figure 10.7-12). While observing the S_{11} locus on the polar display, the far-end adapter is optimized by locating the correct tuning pin locations. The ends are reversed, and the other adapter is tuned for the best match. This is an iterative process. Once both adapters are better than 40-dB return loss (1.02 VSWR),

FIGURE 10.7-13 WR1500 waveguide sliding load test setup.

Photo Courtesy Hitachi Kokusai Electric Comark LLC

the adapters are considered tuned and will provide an accurate assessment of the system being swept.

Today, with high-frequency computer-aided engineering (CAE) solvers, test adapters are routinely designed to be broadband across the usable bandwidth of the line being tested, with better than 40 dB of return loss.

Although not as common today, waveguide is still used at UHF and higher frequencies. Tuning a waveguide to N-type test adapter requires a section of waveguide and a sliding load (see Figure 10.7-13). The adapter is considered properly tuned when the load can be slid though $\lambda/2$ and the polar locus rotates symmetrically around the normalized impedance at the center of the polar display.

VNA INSTRUMENT SETTINGS

To yield quality measurements, the VNA must have the appropriate instrument settings for the type of measurements to be performed. Instrument settings vary between manufacturers, but general settings are the same.

- S_{11} should be selected as the measurement mode for any one-port reflected measurements. Most portable units do not have full two-port S-parameters, and the RF output port is configured for reflected measurements.
- Frequency is centered on the channel being tested. For ATSC television, the span is 6 MHz. For DTF or TDR measurements, the span should be at least 200 MHz to obtain reasonable resolution.

- The number of points should be set to at least 1601 to ensure adequate TDR range.
- Output power should be set to the maximum capability of the instrument without receiving an error message.
- Measurement format can be set to the user preference, but TDR/DTF measurements will be in either the VSWR or LOG MAG formats.

Time Domain (TDR/FDR/DTF) Settings

Measurement range and resolution are two factors affecting TDR measurements. It is critical to be able to measure the complete transmission line/antenna system and ensure that the transform settings are correct for the system being measured. Many newer VNAs can show the x-axis in meters or feet, but most lab-grade instruments will give the x-axis in time.

Some useful information to remember when working with time domain, where c = speed of light:

c = 299,792,458 m/s = 299.792458 m/µs = 0.299792458 m/ns

(easiest to round to 300 m/µs or 30 cm/ns)

c = 983,571,056 ft/s = 983.571056 ft/µs = 0.983571056 ft/ns

(easiest to round to 984 ft/µs or 1 ft/ns)

For reflected measurements, these values are halved to relate to physical distance.

Range is the length of time that a measurement can be made without an alias (repetition) of the response. Alias-free range is proportional to the number of data points and inversely proportional to the frequency span. Range can be increased by selecting more data points or by reducing the frequency span.

Alias-free range (in meters) can be calculated by:

$$range = \frac{(1/\Delta f/(pnts-1))\times c\times Vf}{2}$$

Where: Δf is frequency span in Hz, $pnts$ is number of data points, c is the speed of light and Vf is velocity factor of the transmission line.

Resolution is defined as the ability to distinguish between two closely spaced responses.

Resolution (in meters) is calculated by:

$$resolution = \frac{(1.6/\Delta f)\times c\times Vf}{2}$$

Where: the constant of 1.6 represents the relative impulse width of the bandpass transform by selecting a normal (Hann) window, which offers good compromise between pulse width and side-lobe suppression, Δf is the frequency span in Hz, c is the speed of light and Vf is velocity factor of the transmission line.

Typically, frequency spans of 200 or 300 MHz are used to provide an approximate resolution of 1.2 and 0.8 meters, respectively, for rigid (air) line. Selecting 1601 data points provides approximately an 800-meter measurement range at a 300-MHz span for rigid (air) line. When an unexpected impulse is shown, to determine if it is an alias response, momentarily change the frequency span. If the suspected impulse moves in time, it is an alias response.

Many of the portable field units have the TDR or DTF settings preset to provide the user with quick access to the measurement and will calculate the alias-free range. Some have built-in transmission line tables for various manufacturers and sizes of line but are commonly targeted to the cellular network cable types.

CALIBRATION

The modern VNA has very powerful calibration engines that remove measurement uncertainty. Measurements are only as good as the calibration standards used. Either a lab-grade calibration kit or high-quality calibration tool is essential for accurate measurements.

The VNA allows for interpolation of the calibration if the measurement stimulus changes, but shows a CAL?, CALΔ or C* annunciator on the screen indicating the calibration is somehow inaccurate. Typically, the VNA calibration is performed with the widest span and highest number of points that will be used. Changing frequency, span or power level inside that calibration envelope will have little to no consequence on the accuracy of the measurement; indeed, even the normal flexure of a high-performance test cable will have a greater influence on measurement accuracy, but even this residual error is typically better than 45 dB. If the instrument drops out of CAL, the VNA will give an error and require calibration with the new stimulus.

To perform an S_{11} (reflection) calibration:

- Ensure that the VNA has been powered up long enough to meet the manufacturer's specifications.
- Choose the CAL menu, and it will provide the applicable recommended calibration steps for the instrument in use, which will typically be OSL (open/short/load), SOL (short/open/load) or SOM (short/open/match).
- Ensure that the calibration kit model in use is selected.
- Ensure that the cable is firmly connected to the VNA, and only turn the knurled or hex connector, while taking care that the standard or cable do not turn.
- Follow the prompts to apply the correct standard at the end of the high-quality test cable.
- Save the calibration and instrument stimulus state.
- With the load still attached, slightly flex the test cable and confirm that the residual error is still > 40dB. If the trace moves more than this, the cable should be replaced.
- The calibration should be checked periodically by attaching the load to confirm that the calibration is still valid.

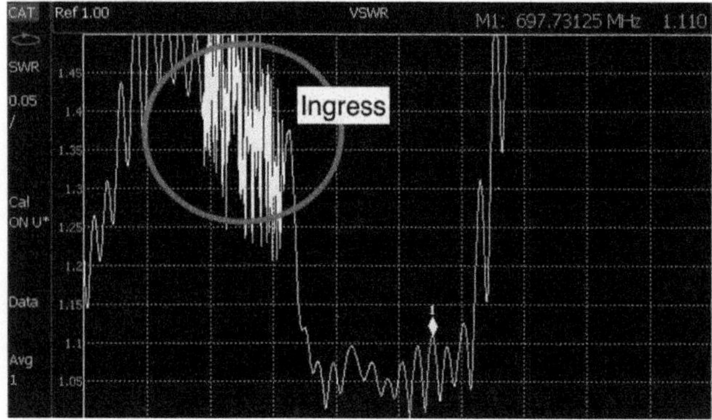

FIGURE 10.7-14A Sweep showing ingress from nearby station.

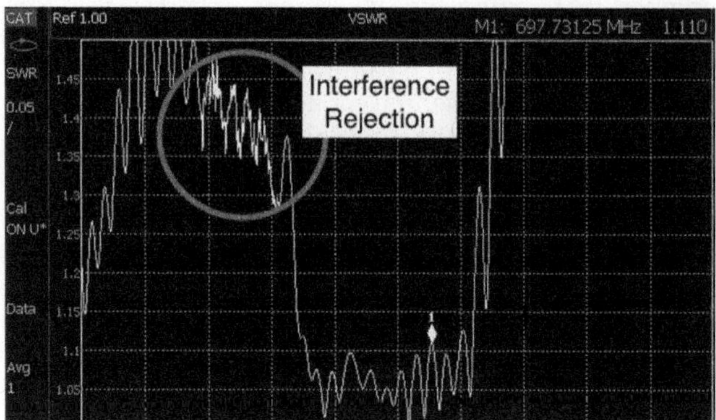

FIGURE 10.7-14B Sweep exhibiting much lower ingress after enabling interference rejection on VNA.

- Calibration standards are typically per device, and care should be taken to ensure the standards used are valid for the device. The reference plane for the open and short of the calibration standard can be significantly different between kits from the same or different manufacturers.

Note

To prevent instrument damage, prior to attaching the VNA to an unknown transmission line/antenna system, it is recommended to determine the power coming in from the line to ensure it is below the damage level of the instrument. A well-attenuated connection (10–20 dB) to a spectrum analyzer or power sensor will make for a quick and safe determination of the level.

When there is ingress from nearby transmitting stations, the trace will have significant "noise," and trace averaging can be used to reduce this effect, but it will impact the sweep speed. Trace smoothing averages out the data between points and should not be used.

Some portable units have a selectable interference rejection mode that does an effective job of reducing ingress (see Figures 10.7-14A and B). The drawback is a significant reduction in sweep speed.

At many multi-use sites and congested antenna farms, there is inadequate source power from the VNA

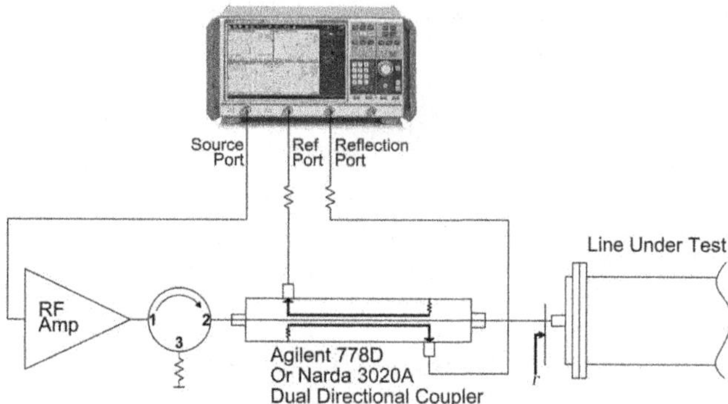

FIGURE 10.7-15 High-power VNA setup configuration.

Artwork Courtesy Tom Smuts (VNA Image Courtesy of Rohde & Schwarz GmbH)

to overcome the ingress. The VNA receiver is also vulnerable to damage. It is not uncommon to have > +30 dBm (1 watt) coming back down the line. In these cases, an external amplifier and high directivity dual directional coupler are used. The VNA will need to have access to the receivers or at least three configurable ports. A typical recommended setup is shown in Figure 10.7-15.

Great care must be taken to ensure measurement accuracy and that the configuration can be calibrated. The VNA receiver linearity must also be considered. It is recommended that the VNA manufacturer be consulted to confirm the measurement configuration is viable.

ANTENNA TESTING

This section discusses the various testing and adjustment procedures during the construction of a new antenna and transmission line system.

One of the uses of the VNA and associated test equipment is the ground testing of new antenna systems prior to their installation atop the tower.

When an antenna arrives on site, where practical, ground tests or pre-lift verification tests are requested to confirm that the antenna is performing as it was in the factory before transportation to site. The preferred testing method is to lift the antenna completely off the ground several wavelengths or more and perform return loss/VSWR and phasing checks with the antenna suspended. Due to the proximity of the ground and any other structures during this phase of testing, the return loss performance of the antenna will likely be in the region of 3–6 dB worse than the expected performance when installed. Any significant deviation from this should be investigated, as this may highlight an issue caused during transportation or preassembly on site.

Corporate-fed antenna systems are sometimes partially assembled on site, and it is imperative that feed system testing be completed to ensure that the correct cables are connected to the corresponding power dividers and elements.

Feed System Testing

Corporate-fed antenna systems can be designed to provide highly complex antenna radiation patterns in both the azimuth and elevation patterns. Often, components of the systems must be removed for transportation. These components must be reinstalled in their correct location before the antenna is hoisted into place and commissioning tests performed. Each manufacturer will have a guide as to how to field test its system prior to installation, but, in general terms, this process is the same for all manufacturers.

A brief description of the process is detailed here, but the manufacturer or specialist should be contacted for a detailed process document.

For antennas where the system has only been broken into two parts (e.g., upper and lower) for transportation, a basic feed system check should be performed.

All elements of the system or subsystem under test must be either terminated as designed or into test loads. Port 2 (S_{21}) of the VNA (if calibrated) is considered an acceptable termination.

As part of the manufacturer's handbook or manual, there will be a detailed phasing and power distribution diagram to show what the designed phase and power of each element are expected to be; there will also be a design phase frequency that has been used to validate the phase reference. This must be used as the reference frequency for all of the feed system tests. This is likely to be a frequency other than the antenna operational frequency for multichannel antennas.

Using a broadband sweep of the system covering all of the operational channels for multichannel antennas, or a minimum of 10 MHz for single-channel antennas, the VNA should be configured to view S_{11} in a scalar format such as Log/Mag and also measure S_{21} as a separate channel or trace.

Calibrate the VNA for S_{11}, then connect port 1 of the VNA to the input of the feed system and connect port 2 to a panel sample or end of cable feeding panel that will be used as reference to measure the phase differences between elements. Calibrate the VNA transmission loss S_{21} only as a through measurement, as the measurements are relative rather than absolute. The S_{21} plot is normally viewed as a polar chart so that the magnitude and phase information are readily viewed.

Locate the point used on the power and phasing diagram and note the design power and phase; this will be used as the delta reference. Move the port 2 test cable to the next panel or point under test and connect the designed load to the point used as a calibration. The difference in the power and phase from the two points should correlate with the power and phasing diagram variations when the calibration factor or delta is taken into account. Each endpoint or test point should be checked and the data validated.

Monitor the system impedance for any changes throughout the process, as removal and reconnection of cables or elements could introduce mismatches. A validation should be performed with each reconnection of the cable throughout the process. Split bullets and poor terminations will show a variation in the impedance between the time the element is disconnected and then reconnected again.

TRANSMISSION SYSTEM TESTING

In addition to the antenna, the transmission line must also be tested, preferably without the antenna connected.

A short-circuit should be installed at the end of the transmission line and a wideband sweep performed. This measurement performs two key functions. The first is to establish the electrical length of the transmission line, while the second is to establish the performance of the transmission line in isolation from the antenna. This can be used to determine the attenuation of the signal through the line (line loss) and also can be used later, during post-processing, to ensure the properties of the line can be separated from the performance of the antenna system.

The frequency span of this measurement should be as wide as possible, and for coaxial transmission lines, the cutoff frequency of each transmission line should be used as a guide to set the highest frequency in the sweep.

In many cases, the line will already be connected to the antenna and the system will be swept together. Ideally, there will be a switch or patch panel at the output of the transmitter system onto which a test adapter can readily connect (see Figure 10.7-16).

On larger UHF transmitter systems that employ waveguide, it is easiest to remove the waveguide-to-coax adapter and attach the test adapter to the first piece of coax in the transmission line.

Assuming that the antenna is connected, and before any measurements are taken, the antenna should be clear of rigging or personnel, as this would affect the overall match. It is also important that the tuners are fully withdrawn on any field matching sections. Some manufacturers have a factory-tuned matching section on the antenna input, and it should initially be

FIGURE 10.7-16 Testing transmission system through 6-1/8″ patch panel.

undisturbed. Some transmission line manufacturers also include matching adjustments in their elbows and, initially, these components should also be undisturbed.

If not already performed, the transmission line and antenna should be pressure tested and purged to the manufacturer's specifications.

Baseline data should be taken prior to any tuning adjustments. Measurement data should be saved in Touchstone (s1p) format. For single-channel ATSC television, these measurements should include:

- 6-MHz polar impedance plot
- 6-MHz return loss and/or VSWR plot
- 12- or 20-MHz return loss and/or VSWR plot
- 6-MHz time domain plot (Figure 10.7-17)
- An intermediate bandwidth span of 12- (Figure 10.7-18), 20- or 50-MHz time domain plot
- Either a 200- or 300-MHz return loss and/or VSWR plot
- A wideband 200-, 300- or 350-MHz (Figure 10.7-19) time domain plot (depending on the upper frequency limit of the coax size in use)

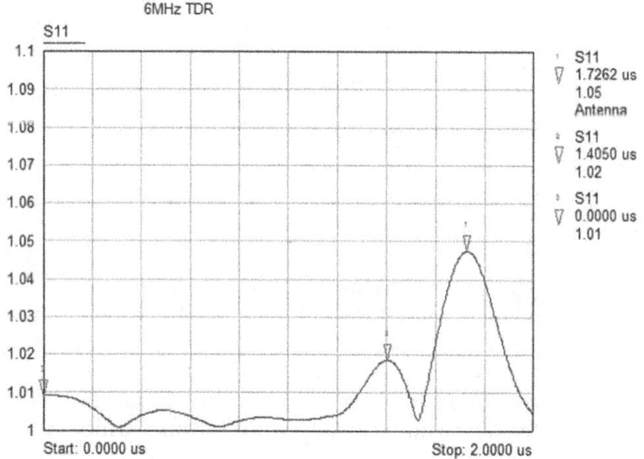

FIGURE 10.7-17 6-MHz TDR plot.

As can be seen in the previous figures, the narrow-band and intermediate bandwidth (6- and 12-MHz) TDR plots do not provide the required resolution to accurately identify the location of the mismatch but do give a general location of mismatches at the channel being tested. The wider 350-MHz bandwidth TDR plots provide much greater resolution, thus allowing each mismatch to be shown in greater detail.

It should be determined that the line is correctly assembled and all the line connections look basically the same on the wideband TDR plot. Anomalous impulses should be investigated.

Time domain gating is a useful feature in the VNA that allows the user to isolate or remove certain portions of the systems in the time domain and then transform back to frequency domain to see the match at that point in the system. This is commonly performed to look at the antenna without seeing the transmission line. Figures 10.7-20A and B show a TDR plot with the gate "flag" markers selecting the antenna, and the frequency domain plot shows the VSWR with and without the gate.

Some commercial post-processing software applications can also perform gating from s-parameter data files.

Some antenna systems have dual inputs, and a transmission line is run to feed each of the two inputs. It is crucial that the lines be phased properly to ensure the antenna radiation pattern is correct. Manufactures typically specify that the phase be within ±5° (approximately ±0.25" at 700 MHz). For complex systems like this, it is important in the planning stages to design the runs to be mechanically identical and that the line locations on the tower be oriented to prevent differential solar loading. Although the lines may mechanically be the same length, the electrical length will undoubtedly be different. This is especially true at UHF frequencies.

To accurately measure line phase, both lines must be shorted at the far end of the system. To obtain the most accurate results, these measurements should be completed at night so as to eliminate any solar loading differential on the lines. In the transmitter building, the line should be broken at the co-phased output of the

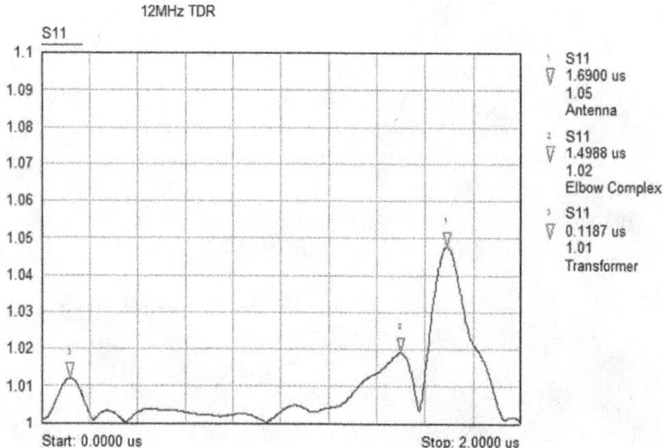

FIGURE 10.7-18 12-MHz TDR plot.

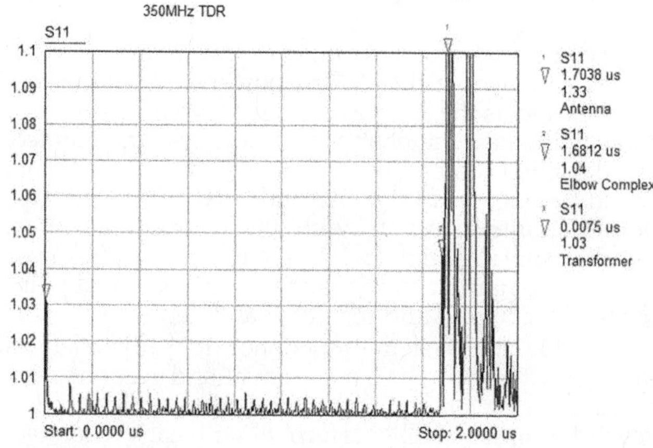

FIGURE 10.7-19 350-MHz TDR plot.

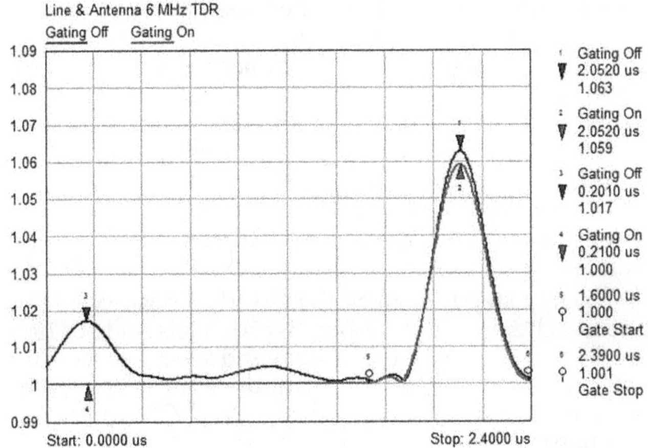

FIGURE 10.7-20A 6-MHz TDR with and without transmission line gated.

Data Courtesy Peter Eckmann Broadcast Consultants, Inc.

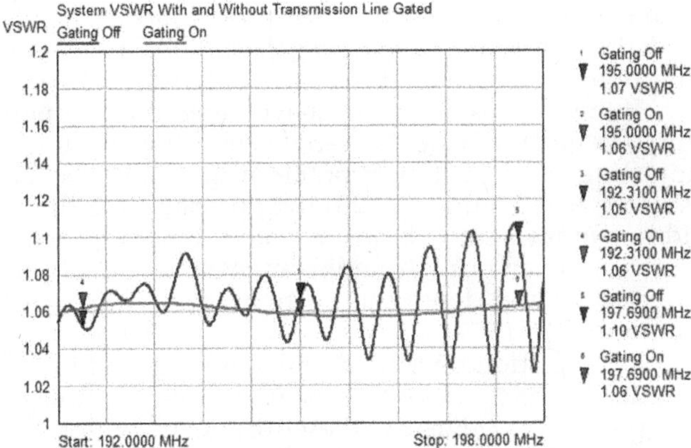

FIGURE 10.7-20B 6-MHz VSWR with and without transmission line gated.

Data Courtesy Peter Eckmann Broadcast Consultants, Inc.

power splitter. At the antenna end, the line should be broken at a co-phased location closest to the antenna. Two adapters are required to measure, and if more adapters are used, they must be phased-match sets. A short-circuit is placed at the antenna end of one of the lines and a wideband TDR sweep taken and stored in the VNA trace memory. The TDR span should be as wide as possible to get an accurate length. A frequency domain sweep of the entire antenna bandwidth should also be taken, looking at a phase trace format. This plot should also be stored in VNA trace memory and exported as S-parameters to allow post-processing and detailed analysis. The adapters should then be placed on the other line and the measurement performed in the same way. On the VNA, select the DATA & MEMORY function to see both the current and stored traces simultaneously for a direct comparison of the two. Ideally, the TDR lengths will be within a few inches of each other, or at least within $\lambda/2$. On the frequency domain phase trace, select DATAMEM and dial in a variable electrical delay to determine the length required to bring lines into phase. Once the physical phasing length is added to the system, the same measurements should be performed to confirm a successful result.

In the initial stages of line measurement, the line length variation should be confirmed with a low-frequency sweep, e.g., < 50 MHz, to confirm the variation before any decision on adjustments is made.

Another measurement technique involves the use of a vector voltmeter (VVM) to measure the phase at the end of the line. Many of the portable field units have this mode available. To accomplish this measurement, the following steps are taken:

- Phase-matched test adapters are required at the top of the line runs.
- A potential issue with the VVM method is that large variances in the transmission line length cannot be effectively resolved. The system may appear phase matched, but one transmission line is multiples of λ longer or shorter. Accordingly, each line should be previously swept with a wideband TDR into a short to confirm that the lines are within $\lambda/2$ (8.4" at 700 MHz in rigid line).
- A CW source is injected at the input of the power splitter in the transmitter building.
- The VVM is connected to the two phase-matched adapters with two phase-matched test cables, and phase is measured at the low, center and upper frequency of the antenna bandwidth.

SYSTEM OPTIMIZATION

Narrow-band TDR and Polar Method

Once the construction and installation of the new transmission system is complete and baseline data is taken, the system likely requires some optimization to improve the overall match. Ideally, there will be a fine matching section installed at the input to the antenna (see Figure 10.7-21).

Typically, the fine matcher will have four tuning probes spaced $\lambda/8$ apart, and a pair of probes will be used to improve the match by introducing an error vector to cancel the undesired reflection. For 6-MHz channels, it is easiest to look at a 6-MHz polar display and a 6-MHz TDR simultaneously and have a tower technician slightly insert each probe to see which two have the desired effect. If the tuners are spaced $\lambda/8$ apart, it will be adjacent pairs or the end two adjusters (1 and 4). Once the required adjusters are located, they should be manipulated to find the best overall far-end match.

The position and shape of the sweep locus on the polar display are very helpful in determining the mismatch location as well as being a useful aid when tuning. The N-S-E-W position of the locus relative to the origin (perfect match point at center) indicates the quality of the near-end match. A spread out and nonconcentric locus is indicative of an intermediate

FIGURE 10.7-21 Antenna input fine matcher (with pressure caps) and coaxial test section.

Photo Courtesy of Broadcast Australia & Radio Frequency Systems (RFS)

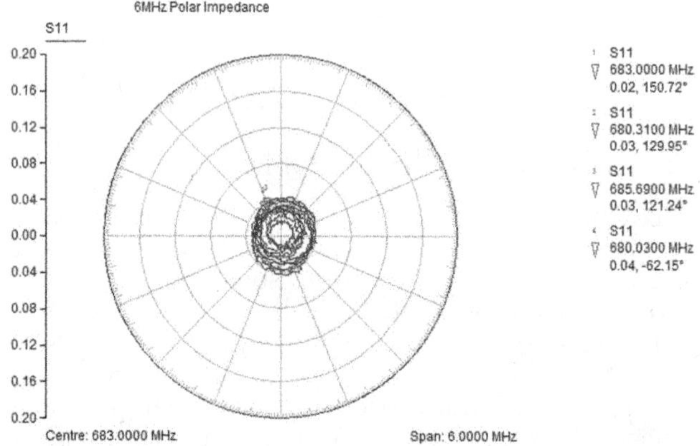

FIGURE 10.7-22 Reflection coefficient of a well-matched transmission line and antenna system.

mismatch (e.g., an elbow at the end of the horizontal run). The radius of the locus indicates the quality of the far-end match (i.e., at the antenna). The locus will ideally be small, concentric and centered, as shown in Figure 10.7-22.

Polar and Electrical Delay Method

A significant feature of the modern VNA is the ability to add or subtract artificial electrical delay to the display locus to analyze the performance of the system at a particular point in the transmission line or system. This is particularly useful when using fine tuners, as it allows the user to accurately determine if the individual tuner is optimized. The technique for finding the correct point at which to set the electrical delay depends slightly on what brand of analyzer is used, but the main principles are applicable to all popular brands of analyzer available today.

The key to this process is looking at the impedance display locus at the electrical position of the transmission line fine tuner.

The first step to finding the electrical position of the fine matcher is to ensure electrical delay is set to 0° or 0 sec. Make a sweep of the system and save it to the analyzer memory so that the memory trace and current trace are visible simultaneously (via the DATA&MEM function).

The next step is to adjust both of the center two tuning probes to their fully-in position. This will create a significant impedance change on the display. Select the polar chart format on the VNA and add electrical delay until the memory and measured trace loci are effectively the same shape and the memory trace is directly above the measured trace on the polar chart.

Adjust all tuning probes to the fully-out position and note the position of the locus relative to the origin. If the locus is not perfectly centered for the operating frequency band or band of interest, begin by adjusting the first tuning probe in until a deflection is noted (see Figure 10.7-23). At this point the direction of movement should be noted. Fine adjust the electrical delay until this adjustment moves the locus directly up and down the polar imaginary axis (see Figure 10.7-24).

With the added electrical delay now dialed to the location of the tuner on the line, when the fine matching probes are adjusted, the entire locus will move uniformly around the display in a N-S-E-W fashion but without any significant change in shape to the locus, and the best possible match for the measured frequency span is when the locus is centered on the origin. Where one probe will move the locus in one plane, the next will move the locus perpendicular to the first, e.g., the first will move down along the imaginary axis and the second will move it right to left along the real axis. The third probe will move the locus in the opposite direction of the first, and the fourth will move it in the opposite direction of the second. It is intuitive to see that only two of the adjusters should be used.

If additional optimization is required at another location, an additional factory tuner can be ordered, or field-installable tuners can be soldered onto the outer conductor of rigid coaxial lines. Capacitive rings (slugs) can be located and soldered on the inner conductor to tune out mismatches, but for the inexperienced, more time and money will likely be spent using this method than to add a factory tuner or install field tuners. Misuse of these capacitive rings can significantly degrade the voltage rating of the transmission line.

Any optimization technique must be located as close as possible to the mismatch to maintain impedance bandwidth.

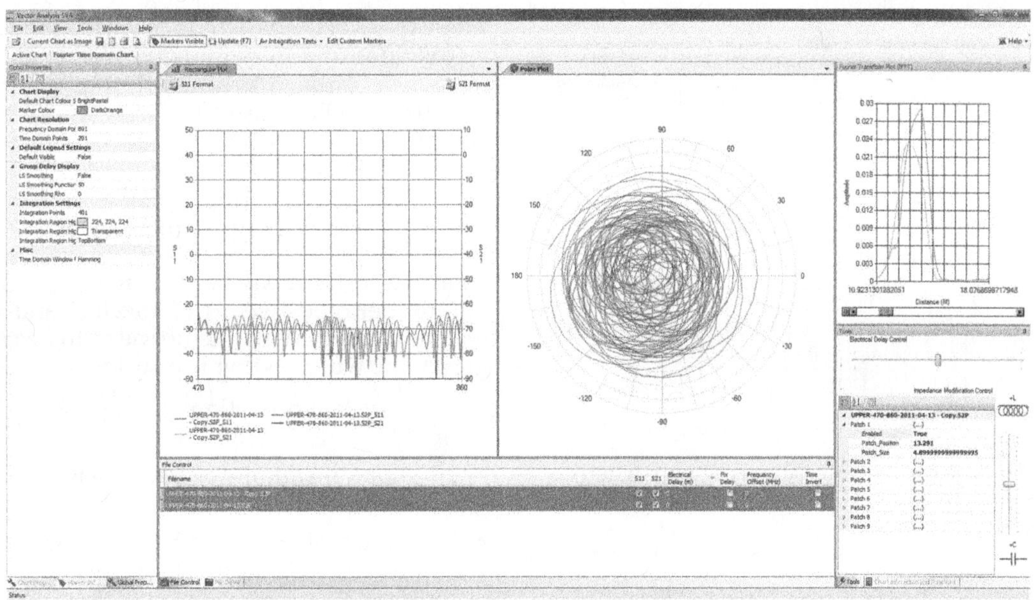

FIGURE 10.7-23 Locus with mismatch induced.

Image Courtesy of Vector Analysis

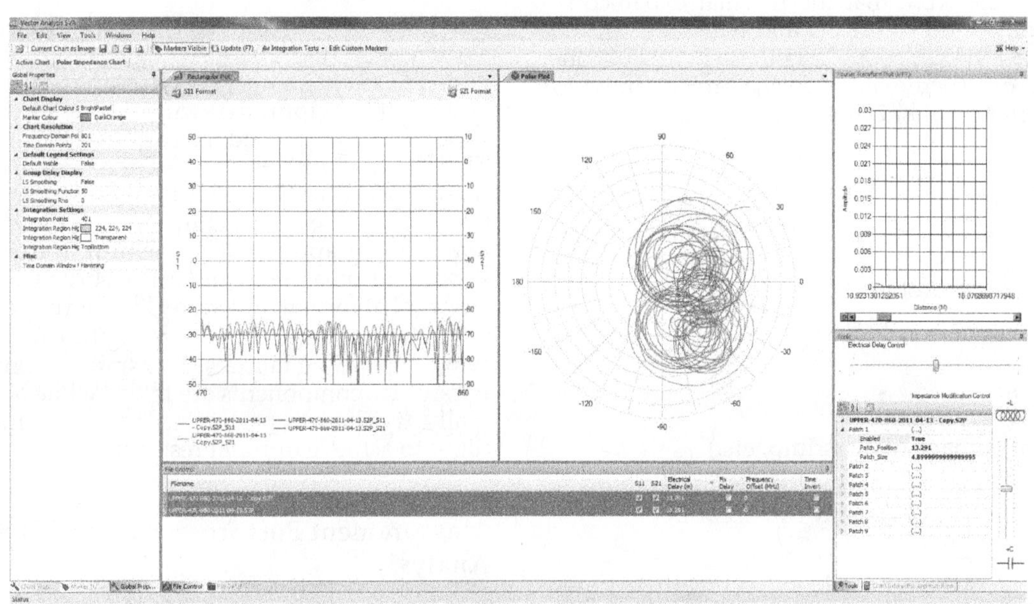

FIGURE 10.7-24 Locus with the correct electrical delay added.

Image Courtesy of Vector Analysis

A common misconception is that the system VSWR must be < 1.1:1. While a well- matched system with low VSWR is desirable, achieving such a value is not as critical as ensuring there are no deleterious impulse reflections that will cause signal impairment or damage to the transmitter, transmission line or antenna. If the individual reflection coefficients are added up, it can be clearly demonstrated that the overall peak VSWR could exceed the 1.1 VSWR value. This is especially true of broadband rigid transmission line and antenna systems.

As an example, some commonly quoted specifications of broadband system components are:

- Transmission Line VSWR 1.08 RL 28.3 ρ 0.04
- Antenna VSWR 1.1 RL 26.4 ρ 0.05

When constructive interference occurs at certain frequencies, (reflection coefficients (ρ) add) the following results:

System Summary VSWR 1.2 RL 20.92 ρ 0.09

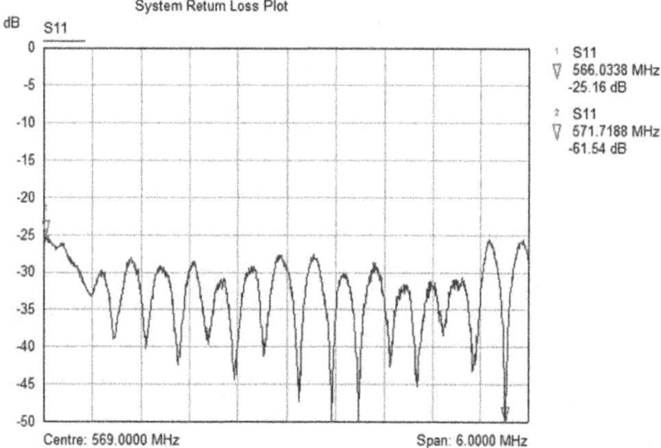

FIGURE 10.7-25 Return loss plot showing constructive and destructive interference on a 400-meter line and antenna system.

Some frequencies will have destructive interference (i.e., the reflections cancel). Typically, the system VSWR will be something of an average based on magnitude and phase of the reflections, as shown in Figure 10.7-25. Note that the thermal expansion/contraction of the transmission line will cause phase changes to the reflected wave, and these points of constructive and destructive interference can certainly go through a variation of $\lambda/2$ or more.

FAULT ANALYSIS

There are issues that occur during installation that should be caught during the commissioning of the system. These types of problems include:

- Split inner connector (bullet)
- Pinched O-ring
- Loose flange
- Mechanical damage (dent, crimp, etc.)
- Conductor sag in rigid line (insufficient supports for inner conductor, typically can occur in unflanged rigid line)

Wideband TDR analysis will aid in locating these types of issues in the transmission line. Issues in the antenna system will be more difficult to diagnose, especially in large multi-user corporate-fed antenna systems. For antenna systems that are harnessed in the field, it is recommended to separate secondary and tertiary power dividers and terminate into a good load to look at each portion of the antenna. In this way, it is easier to determine where among the many cables to investigate any installation issues.

Faults can also occur during operation for many reasons, including the following:

- Defective inner expansion connector, typically a galled watchband spring contact or a connection that has loosened.
- Pressurization loss due to a break in the outer conductor and moisture ingress has occurred.

These types of faults, left unchecked, will result in a catastrophic failure commonly referred to as a "burnout."

As with faults diagnosed during commissioning, a wideband TDR is used to locate and identify failing components.

MAINTENANCE

To avoid costly repairs and to minimize the chance of failure, periodic sweeps of the line and antenna are recommended. Many factors need to be considered to determine when testing is justified, such as:

- Is the system operating at or near the power-rating limit?
- Has the transmitter reflected meter gradually or suddenly changed?
- Does the system have a history of failures?
- Have there been strong windstorms or lightning damage that caused reflected power trips on the transmitter?
- Has there been a pressurization loss, or has the line been unpressurized?

Baseline data should have been collected when the system was commissioned. This data should be maintained to aid in identifying changes in the transmission system. It is helpful to document the instrument measurement settings used for various measurements (e.g., TDR bandwidth) as well as the measurement location(s). Make notes about ingress or any other issues that will aid the next technician who measures the system. Photographs are also helpful in documentation. Ensure all measurements are recorded in the Touchstone (s1p) format and stored for future reference.

A thermal imaging camera is often useful in identifying impending failures. Transmission lines and other passive RF components are tested while operating normally. It is best to perform outdoor testing during the night to achieve optimal resolution.

Measurement Postprocessing and Data Analysis

In addition to the basic S_{11} frequency sweep and TDR tests described in preceding sections, all modern VNAs have the ability to export the complex pair data in an industry standard format called Touchstone or SnP, so that advanced data postprocessing may be performed offline with sophisticated data analysis tools. In certain cases, such analyses may provide a better a visualization of the data and permit the detection of a previously unforeseen problem.

For single-port S_{11} measurements, the file format extension is. S1P. This file contains all the data points being measured. S1P data will contain the frequency of the data point in the first column and the following two columns will contain the real and imaginary data (complex pairs). This data can be imported to a spreadsheet program or commercial measurement software. An advantage to this is the ability to take one capture of the data and then later display it in polar, return loss or VSWR formats. Markers can be added, scales changed, etc. Perhaps the biggest advantage is the ability to overlay historical plots with current data to clearly see any changes to the transmission system.

If the VNA being used does not have the time domain option installed, the software can perform the chirp Z or Fourier transform and provide the same results as the VNA and still allow the system to be fully documented.

With advanced software tools, direct comparisons can be made of transmission lines and antennas even though they may have been measured from different points in the system with time or electrical delay offsets.

There are a few software solutions for the analysis of S-parameter data, and they are loosely categorized into two key areas: data presentation or data analysis.

Most solutions will target a specific category but will have some functionalities into the other; both of these systems can prove to be invaluable in their target application.

SoftPlot (Aphena) is an example of a presentation implementation that has some analysis functionalities; it is most often used in document presentation and data acquisition from test equipment. The software contains many trace stores, which allow for easy comparison with historical measurements. S-parameter measurements can be embedded into documents as OLE (Object Linking and Embedding) items that render in the selected format.

SVA (vector analysis) is targeted as a data analysis tool that has some presentation functionalities. With this package, the user has access to time domain analysis of measured S-parameters as well as the ability to compare directly measurements that have been taken using different parameters, such as number of points and different measurement position. A key functionality that is provided is that of a time adjustment. In the time domain, the two measurements can be overlaid such that the reflections from common points may be used as calibration points visually, and the measurement may be analyzed to show what has changed in the common area of the two plots. Another tool available to the skilled user is the ability to remove impedance mismatches selectively to assess the impact of each particular discontinuity or simulate the effect of the inclusion of an additional tuning section.

One manufacturer, CMT (Copper Mountain Technologies), provides a soft VNA interface that can perform most functions of a VNA on measured S-parameters, allowing measurements to be postprocessed with time domain functionality and reviewed in different format.

Bibliography

Dunsmore, Joel P. *Handbook of Microwave Component Measurements: with Advanced VNA Techniques*. Chichester, UK: John Wiley & Sons, Ltd, 2012

Fanton, Myron D. *Analysis and Measurement of RF System Reflections and DTV Transmission*. Electronic Research Inc, 2006

Myers, Dana W. *A Brush Up for Broadcast Engineers*. NAB / PBS / SBE Ennes Workshop, 2006

Murlatt, Ed. *Preventative Maintenance of Passive RF Systems*. SBE Chapter 22 Broadcast & Technology Expo, 2014

Pongratz, Stephan. *Understanding Cable & Antenna Analysis*. Anritsu Company, 2011

Ruck, Jeremy D. *Trade Secrets of a Guy with a Network Analyzer*. Wisconsin Broadcasters Association / SBE Conference, 2004

Stenberg, James T. and Walter Pries. *Advanced RF System Measurement Techniques*. NAB Broadcast Engineering Conference, 1989

Understanding the Fundamental Principles of Vector Network Analysis. Application Note 5965-7707E. Agilent Technologies, Inc. (Keysight), 2012

WEBSITES:

VNA Manufacturers

www.anritsu.com
www.coppermountaintech.com
www.keysight.com
www.rohde-schwarz.com

Test Cables

www.megaphase.com
www.minicircuits.com

Calibration Kits

www.maurymw.com
www.spinner-group.com
VNA manufacturers listed above also produce excellent calibration kits

Test Adapters

www.dielectric.com
www.eriinc.com
www.jampro.com
www.myat.com
www.rfsworld.com
www.spinner-group.com

CHAPTER

10.8

VHF and UHF Television Antenna Test Range Measurements

JOHN L. SCHADLER

Vice President Engineering, Dielectric LLC
Raymond, ME

Antenna measurements are extensively conducted throughout the television industry with the purpose of measuring the free space far zone performance. Pattern testing is essential to ensuring specified coverage requirements are met and signal is efficiently directed to the audience and not wasted. The purpose of this chapter is to provide a guide on the understanding, operation, design, performance and interpretation of results associated with VHF and UHF antenna test range measurements. It will explore the basic concepts and techniques used for antenna pattern measurement and evaluate the benefits and drawbacks of the various methods. Both near-field and far-field approaches as well as outdoor and indoor anechoic chamber systems will be discussed. Theory and equations governing the antenna properties and parameters are included along with complete descriptions of range designs and evaluation.

BASICS

The antenna pattern can be described as the three-dimensional distribution of radiated energy in space about the antenna. A multitude of antenna pattern measurement techniques have been developed. The first technique developed was the far-field range, where the antenna under test is placed in the far zone of a source antenna. Because of the relatively large distances that are often required for far-field testing between the source antenna and the antenna under test, it is not practical to measure the pattern of the test antenna by rotating the source antenna around the test antenna. Instead, the line of sight between the antennas is held fixed in space while the antenna under test is changed in orientation to simulate the movement around the antenna (see Figure 10.8-1) [1].

FIGURE 10.8-1 Far-field range RCA/Dielectric, Voorhees, NJ. The separation between the source antenna and the antenna under test was 5 miles.

FIGURE 10.8-2 Outdoor near-field range, Dielectric, Raymond, ME.

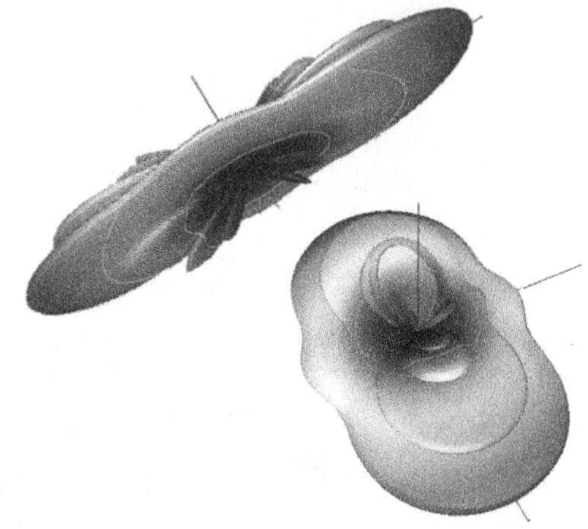

FIGURE 10.8-3 Three-dimensional radiation pattern from a typical "peanut pattern" UHF television broadcast slotted coaxial pylon style antenna.

Due to the area required to create a far-field range for large antennas, near-field techniques were developed, which allow the measurement of the field on a surface close to the antenna (see Figure 10.8-2). The data is then mathematically transformed into the far field to predict the radiation characteristics at infinity.

In addition to the basic contour of coverage, there are various pieces of information that can be determined from radiation patterns, specifically gain, beamwidth and polarization parameters.

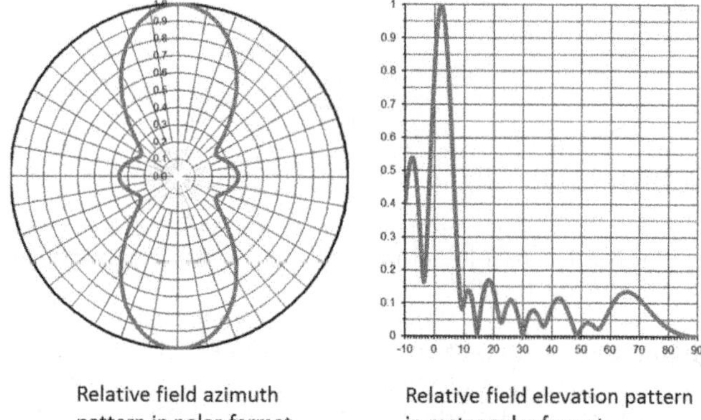

FIGURE 10.8-4 Azimuth and elevation cuts of the three-dimensional radiation pattern given in Figure 10.8-3.

RECIPROCITY

There are several forms reciprocity theorems take in electromagnetic problems. In range measurements, a consequence of the reciprocity theorem is that the radiation pattern of an antenna is the same whether it is used as a transmitting antenna or receiving antenna. Reciprocity allows the reciprocal measurement of an antenna pattern in either transmit or receive mode. In other words, reciprocity states that the source antenna and the antenna under test can be interchanged without changing the system response [3]. The mathematical proof of the theorem can be found in many references, such as [5].

RANGE MEASUREMENT PARAMETERS

Radiation Pattern

Antenna radiation patterns are graphical representations of the distribution of radiated energy as a function of direction about the antenna and are an important tool for both the antenna designer and end user. Different antenna suppliers use different standards and formats, sometimes making their interpretation confusing. They can be plotted in terms of field strength or decibels and can be absolute or relative to some reference level, with the peak of the beam often chosen as the reference. Radiation patterns can be displayed in three dimensions or at any cut in space in a rectangular or polar format [4]. Each format has its pluses and minuses, and each highlights the information differently. An example of a three-dimensional radiation pattern from a typical peanut-shaped UHF television slotted coaxial, broadcast pylon style antenna is depicted in Figure 10.8-3.

Three-dimensional representations of antenna patterns are often complex and hard to translate into coverage, especially for high-gain antennas. To simplify them, a Cartesian coordinate system (a two-dimensional system that refers to a point in space) is typically used. A two-dimensional pattern is referred to as a pattern cut. The primary planes of interest are usually the vertical axis plane, normally referred to as elevation pattern, and the plane perpendicular to the antenna axis, normally referred to as the azimuth pattern. The azimuth and elevation cuts of the three-dimensional radiation pattern shown in Figure 10.8-3 are presented in Figure 10.8-4.

The patterns in Figure 10.8-4 are plotted in relative field, with the azimuth using polar coordinates while the elevation pattern is plotted in rectangular coordinates. This is common throughout the North American television broadcast industry, but it should be noted that they can just as easily be viewed in other formats such as decibels. Breaking the radiation pattern down into azimuth and elevation cuts provides a quick reference to the response of the antenna in any direction.

Gain–Directivity

If the antenna is a point source with spherical radiation having equal energy radiating in all directions, it is said to be isotropic. If the three-dimensional radiation intensity varies from a perfect sphere, the antenna is said to be directive. In its simplest form, the directivity D can be defined as:

$$D = \frac{\text{Maximum radition intensity}}{\text{Average radiation intensity from a reference antenna}} = \frac{U_m}{U_{avg}} \quad (1)$$

The definition of directivity is based entirely on the shape of the radiation pattern and does not include the input power or antenna efficiency. The gain G quantity, which does involve the antenna's efficiency, is given by:

$$G = \eta D \quad (2)$$

Where η is the antenna's efficiency factor and is usually attributed to feed-system losses.

The directivity of an isotropic source is 1 or 0 dBi. In the television broadcast industry, directivity and gain are typically referenced to a half-wave dipole and not an isotropic source. The directivity of a half-wave dipole is 1.64 or 2.15 dBd relative to an isotropic source.

$$D_{1/2\,wave\,dipole} = \frac{D_{isotropic}}{1.64} \quad (3)$$

$$D_{1/2\,wave\,dipole}(dBd) = D_{isotropic}(dBi) - 2.15 \quad (4)$$

It is important to be able to calculate the gain and directivity from measured radiation patterns. As mentioned earlier, it is often convenient to break the three-dimensional radiation patterns into the two primary planes referred to as the elevation and azimuth patterns (see Figures 10.8-5 and 10.8-6). The directivity of each can then be calculated from the measured data. The peak directivity then simply becomes the multiplication of the two individual directivities.

$$D_p = D_A \times DE \quad (5)$$

Where D_A and D_E are the directivity of the azimuth pattern and the elevation pattern, respectively.

Since the directivity is inversely proportional to the normalized area within the radiation pattern, integration can be used to determine the directivity of measured patterns.

$$D = 4\pi \frac{1}{\iint |F(\theta,\phi)|^2 \sin\theta d\theta d\phi} \quad (6)$$

In the broadcast industry, both the azimuth and elevation patterns are typically measured in relative field, and the directivity is listed relative to a half-wave dipole.

The most convenient method to extract the directivity from measured patterns is to numerically integrate each separately and compare the calculated areas relative to a half-wave dipole. The assumption of separating the radiation patterns is that the overall peak directivity is the product of the separate integrals for the radiated power. A variety of numerical integration techniques may be employed. The most straightforward of these techniques is the rectangular rule, but others such as Simpson's rule can be used. Through numerical integration, the azimuth and elevation directivity can be calculated from the normalized measured azimuth and elevation field data using Equations (7) and (8) using the rectangular rule.

$$D_A = \frac{N}{\sum_{\varnothing=0}^{360} |F(\varnothing)|^2} \quad (7)$$

$$D_E = \frac{N}{\frac{\pi}{2} \bullet 1.64 \bullet \sum_{\theta=0}^{180} |F(\theta)|^2 \sin\theta} \quad (8)$$

Where:
N = number of equally spaced sample points
$F(\phi)$ = field strength at azimuth angle ϕ
$F(\theta)$ = field strength at elevation angle θ

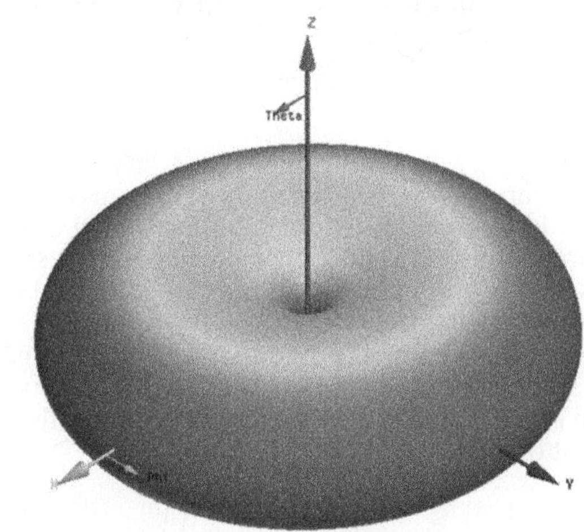

FIGURE 10.8-5 Three-dimensional radiation pattern of a half-wave dipole.

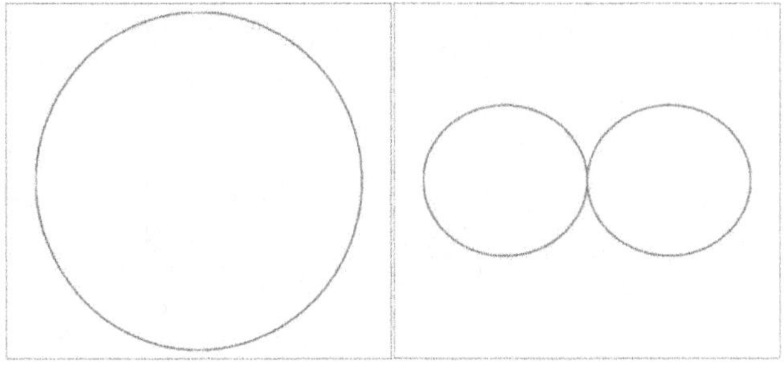

FIGURE 10.8-6 Azimuth and elevation cuts of a half-wave dipole.

Referring to Figure 10.8-6, it should be noted that the directivity of a half-wave dipole is 1 in the azimuth plane, and therefore the quantity (D_A) calculated in Equation (7) is already relative to a half-wave dipole. The elevation pattern of a half-wave dipole, shown in Figure 10.8-6, is represented by Equation (9) and has a directivity of 1.64.

$$U(\theta)=\frac{\cos\left(\frac{\pi}{2}\cos\theta\right)}{\sin\theta} \tag{9}$$

Dividing by 1.64 as shown in Equation (8) normalizes the elevation directivity to be relative to a half-wave dipole. The peak directional directivity (D_p) relative to a half-wave dipole can then be calculated using Equation (5) and specified in dBd by Equation (10).

$$Dp(dBd) = 10\log(Dp) \tag{10}$$

Beamwidth

The beamwidth is usually described as the "half-power beamwidth," which is defined as the angular separation of the points where the main beam in a pattern plane is one-half the maximum value of the peak main beam (see Figure 10.8-7).

The maximum directivity, based on an estimate of the beam area from the half-power beamwidths of the patterns in both the horizontal and vertical planes, is given by Equation 11. [6]

$$D=\frac{41,253}{\varnothing_{HP}\theta_{HP}} \tag{11}$$

Where θ and ϕ are the half-power beamwidths expressed in degrees. It can be stated that the antenna gain and beamwidth in the vertical plane are simply inversely proportional. Doubling the vertical beamwidth reduces the gain to half.

Figure 10.8-8 confirms that as null fill is applied by variations in the illumination, the resulting reduction in gain equally affects the half-power beamwidth independent of the amount of null fill.

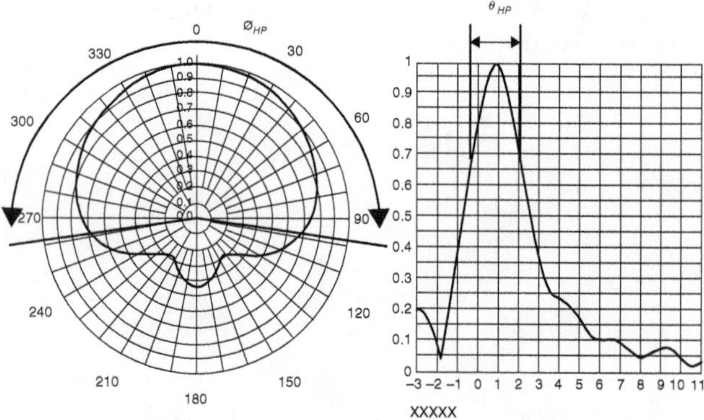

FIGURE 10.8-7 Azimuth and elevation pattern half-power beamwidths.

Polarization

The polarization state of the wave is described by the shape and orientation of an ellipse formed by tracing the extremity of the electromagnetic field vector versus time. Polarization may be classified as linear, circular or elliptical according to the shape of the ellipse (see Figure 10.8-9). Linear and circular are special cases of elliptical polarization where the ellipse becomes a straight line or a circle, respectively.

The ratio of the horizontal axis to the vertical axis represents the polarization ratio.

$$PR = 20log\left(\frac{E_x}{E_y}\right) \tag{12}$$

The ratio of the major axis to the minor axis of the polarization ellipse defines the magnitude of the axial ratio.

$$AR = 20log\left(\frac{OA}{OB}\right) = tan\left[\frac{sin^{-1}\left(sin\left(2tan^{-1}(PR)sin(\varphi)\right)\right)}{2}\right] \tag{13}$$

In Equation 13, φ represents the phase angle between the horizontal and vertical polarization ($\varphi_x - \varphi_y$). The

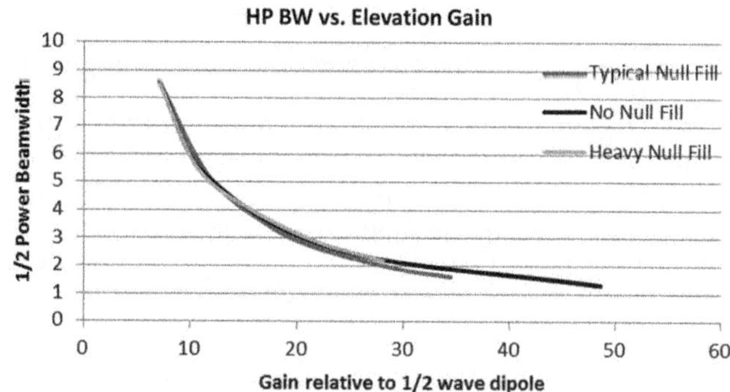

FIGURE 10.8-8 Elevation pattern half-power beamwidth vs. elevation gain.

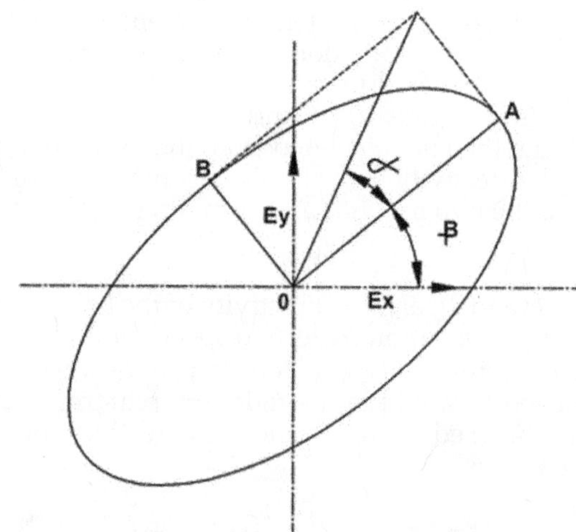

FIGURE 10.8-9 Polarization ellipse.

sense of polarization is determined by observing the direction of rotation of the electric field vector from a point behind the source. Right-hand and left-hand polarization corresponds to clockwise and counter-clockwise rotation, respectively. It should be noted that FCC regulation restricts television broadcast to right-hand circular polarization, while FM broadcast can use either right hand or left hand. Circular polarization is produced by combining two equal orthogonal linearly polarized waves with a 90-degree phase difference. If the magnitudes of the linearly polarized waves are unequal, then elliptical polarization is produced. The axial ratio can be described as an evaluation of the quality of polarization. If 90 degrees exist between the orthogonal polarizations, the axial ratio is equivalent to the polarization ratio. As the phase difference between the polarizations veer away from 90 degrees, there exists reception orientations that drop below and go above the primary horizontal and vertical polarizations, thus restricting the orientational immunity of the receiver and diminishing the "quality" of polarization.

The testing of circularly or elliptically polarized antennas can be conducted in one of two ways. Since both the polarization ratio and axial ratio vary with direction, both the antenna under test and a linearly polarized source antenna can be simultaneously rotated, thus sampling the far field at all locations and orientations. For this method to be valid, the rotation of the source antenna must be increased compared to the motion of the AUT positioner so the probe will essentially turn through at least one complete rotation at each far-field measurement point. The goal is to vary the signal source in such a way that the full polarization ellipse is measured at each far-field angle. Refer to Figures 10.8-10 and 10.8-11.

This procedure has two advantages. First, it yields a direct real-time measurement of the axial ratio and second it can be measured using a scalar network analyzer or separate transmitter and receiver since phase information is not needed for record.

The second method is to derive both the polarization ratio and axial ratio performance mathematically from the measurements of both the amplitude and phase of the orthogonal linear polarizations. This is accomplished by using Equations 12 and 13.

CP and EP Gain Calculations

The following set of equations can be used to determine horizontal and vertical gains, ERPs and required transmitter power given the individual gains of both the azimuth and elevation patterns, polarization ratio and transmission line efficiency of a circularly polarized or elliptically polarized antenna.

Definitions:

G_{hv} = Hpol elevation pattern gain

$G(vv)$ = Vpol elevation pattern gain

G_{hh} = Hpol azimuth pattern gain

G_{vh} = *Vpol* azimuth pattern gain

$$\frac{E_v}{E_h} \quad \frac{Peak\,Vpol}{Peak\,Hpol} \quad (Polarization\ ratio)$$

H_{erp} = Hpol ERP

ε = Transmission line efficiency

P_t = Transmitter power

Power available at antenna input, $P_{avi} = P_t\,\varepsilon$ (14)

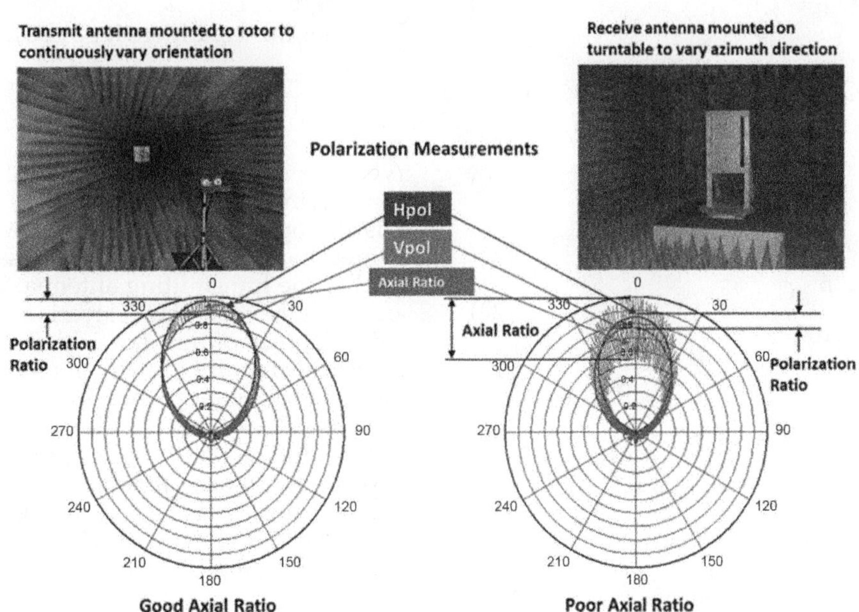

FIGURE 10.8-10 Tapered anechoic chamber Dielectric, Raymond, ME. Far-field azimuth pattern measurements of a circularly polarized and elliptically polarized UHF panel antenna.

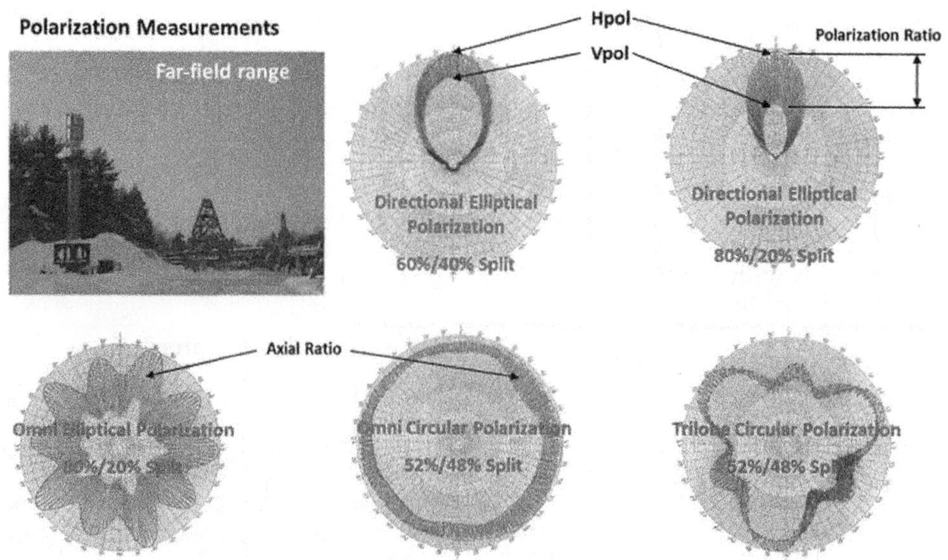

FIGURE 10.8-11 Outdoor far-field range Dielectric, Raymond, ME. Far-field azimuth pattern measurements of different types of circularly and elliptically polarized UHF antennas. In all cases, the axial ratio equals the polarization ratio, depicting the fact that 90 degrees phase exists between the horizontal and vertical polarizations.

$$\text{Hpol Gain} = G_h = G_{hv}\,G_{hh} \tag{15}$$

$$\text{Vpol Gain} = G_v = G_{vv}\,G_{vh} \tag{16}$$

Power needed at antenna input

$$P_{in} = H_{erp}\left[\frac{\left(\frac{E_v}{E_h}\right)^2}{G_v} + \frac{1}{G_h}\right] \tag{17}$$

$$\textit{Peak gain of Hpol} \quad G_{hp} = \frac{H_{erp}}{P_{in}} \tag{18}$$

$$\textit{Peak gain of Vpol} \quad G_{vp} = G_{hp}\left(\frac{E_v}{E_h}\right)^2 \tag{19}$$

$$\textit{\% Power into Hpol} \quad P_{hpol} = \frac{G_{hp}}{G_h} \tag{20}$$

$$\textit{\% Power into Vpol} \quad P_{vpol} = 1 - P_{hpol} \tag{21}$$

$$\textit{Vpol ERP} \quad V_{erp} = \frac{P_{in}P_{vpol}G_v}{100} \tag{22}$$

$$\textit{Transmitter power required}\, P_{tr} = \frac{P_{in}}{\varepsilon} \tag{23}$$

$$\textit{Transmitter power leftover} \quad P_l = P_t - P_{tr} \tag{24}$$

NEAR FIELD/FAR FIELD

The ideal far-field test environment for determining antenna performance would provide for a plane wave of uniform phase to illuminate the antenna's aperture under test. At this point the field distribution is essentially independent of the distance from the antenna. At large distances the radius of curvature of the phase front emanating from an antenna is, for all practical purposes, considered planar. As the antennas are brought closer together, a condition is reached in which the antenna under test is illuminated with a radius of curvature in the phase front, causing the edges of the antenna to receive a delayed signal. A commonly employed criterion for determining the minimum allowable separation between the source antenna and the antenna under test is to restrict the phase variation to be less than a maximum of ϖ/8, or 22.5 degrees [1].

Under this condition, it can be shown that:

$$R \geq \frac{2D^2}{\lambda} \tag{25}$$

Where D is the aperture length of the antenna in the measurement cut of interest, R is the separation and λ is the operating wavelength. When this condition is met, the receiving antenna is said to be in the far field of the transmitting antenna. Typically broadcast antennas have higher elevation gain than azimuth gain. Because of this, far-field elevation pattern measurements require a much greater separation distance then azimuth pattern measurements, as shown in Figure 10.8-12.

Even though the phase front is considered planar at very short distances for small aperture diameters, caution must be taken to ensure that inductive or radiation coupling as well as scattering and reradiation is not present between the source antenna and the antenna under test. For this reason, it is recommended that a few extra wavelengths are added to the separation. It must also be noted that this distance must be greater than the larger of the two limits for both the receive

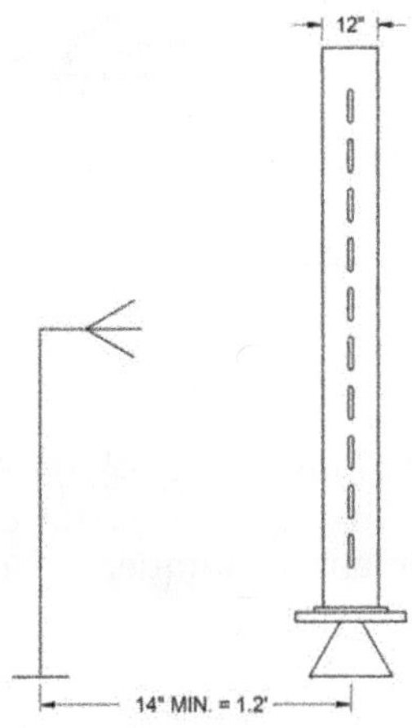

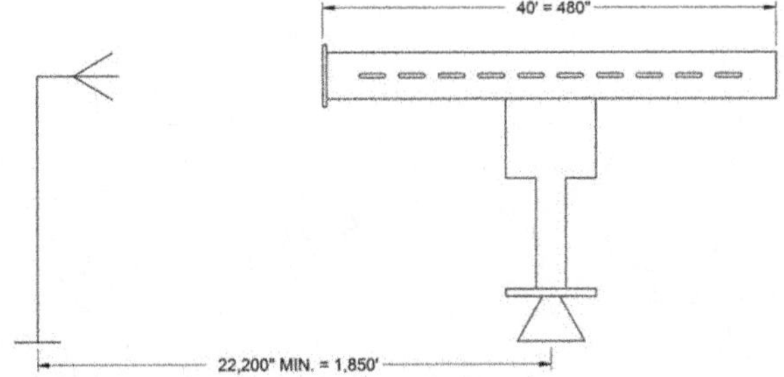

FIGURE 10.8-12 Example of minimum separation needed for a typical channel 30-UHF pylon antenna far-field measurement in both azimuth and elevation.

antenna and transmit antenna, that is, the two antennas must be in each other's far field. This is typically not a problem when measuring elevation patterns of a large array broadcast antenna aperture but may come into play when measuring the azimuth pattern, as shown in Figure 10.8-12.

ANTENNA RANGES

Antenna measurement ranges can be categorized as outdoor ranges and indoor ranges. Each can then be subcategorized as near-field or far-field ranges. Each type of range has its advantages and disadvantages as well as challenges. There are four basic subsystems need for antenna range instrumentation. They are transmitting, receiving, positioning and recording. The choice of equipment depends on the type of measurements that are required, the accuracy required and the capital and operating budgets allotted [7]. The specific measurement to be performed dictates the frequency of operation, the polarization and the pattern cuts required. There are many types of range designs that fall under the broad indoor, outdoor, far-field and near-field categories. Some include elevated, ground reflection and slant ranges as well as rectangular, tapered and compact chambers. In depth, detailed information on various range designs can be found in references [1], [2] and [7]. For the purpose of this chapter, focus will be placed on those designs that are most commonly used in the broadcast industry.

FAR-FIELD OUTDOOR RANGE DESIGN

In many instances, broadcast antenna measurement requirements only dictate the primary principle-plane amplitude patterns and the peak gain of the antenna in order to predict and verify coverage. For this, the use of a far-field outdoor range is a logical choice.

In an outdoor far-field range configuration, the test antenna is placed on a test positioner located on a support structure outside the instrumentation room. The source antenna is normally located at the opposite end of the range on a tower section. If the system is designed as an open loop configuration, the signal source is located near the source antenna and the receiver near the antenna under test to minimize signal loss. For this case, free space loss is the major contributor to the overall system loss and can be calculated using Equation (26).

$$L_{fs} = 32.45 + 20 * \log(R) + 20 * \log(f) \tag{26}$$

Where R = the range length in meters and f = test frequency in GHz.

In a closed-loop system, both the transmit and receive functions are integrated through the use of a network analyzer (Figure 10.8-13 shows both open-loop and closed-loop designs). The use of a closed-loop far-field system may not be practical if the distance between the transmit antenna and the antenna under test is significantly large. This is largely due to the cable losses over a long distance, which must be added to the free space loss in order to calculate the entire system loss.

The source antenna can be either linearly polarized or dual polarized. Polarization measurements can be made simultaneously using a dual-polarized source antenna or made by rotating a linearly polarized source antenna when attached to a positioner.

The goal in all ranges is to suppress or remove the effects of all surroundings that can cause large reflections. This is accomplished by the following factors:

1. Use of a directive source antenna with low side-lobes. It is essential that the transmitting antenna be accurately directed so that the peak of its main beam is centered on the aperture of the antenna under test. Misalignment results in an asymmetrical aperture illumination and error in the measurement.
2. Clear line of sight between the source antenna and the antenna under test.
3. Redirection or absorption of energy outside of the line of sight.
4. Adequate test height to avoid direct ground reflections

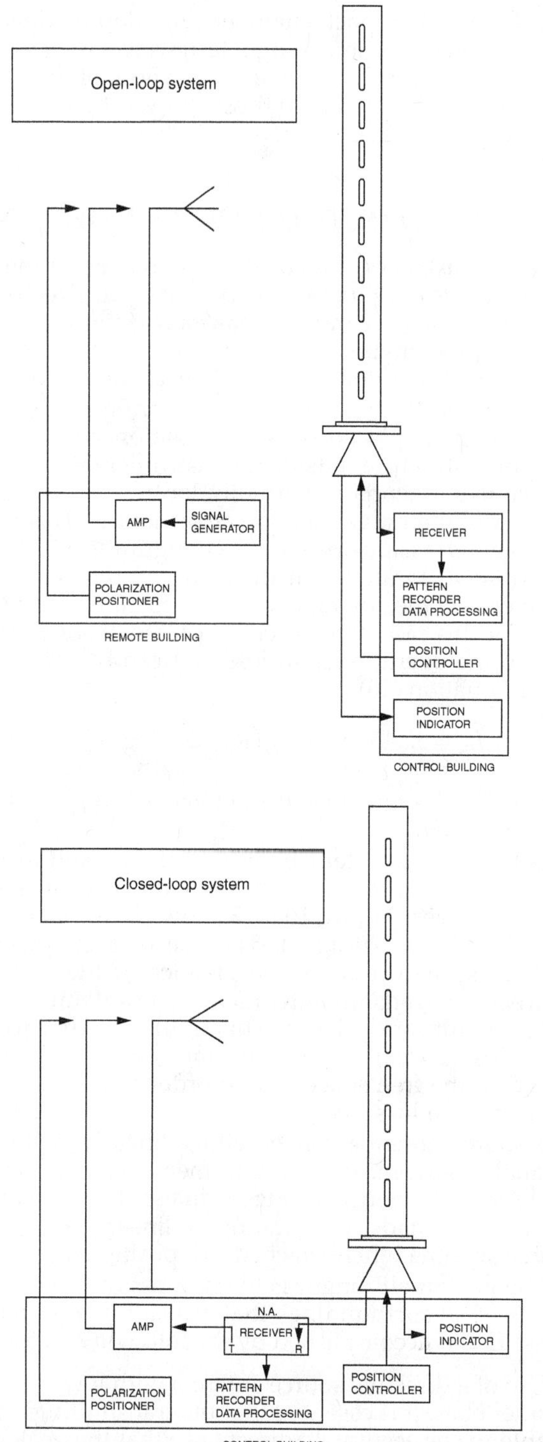

FIGURE 10.8-13 Far-field outdoor range setup. Both open-loop and closed-loop designs.

FAR-FIELD RANGE EVALUATION

An important part of antenna pattern measurements is the alignment and reflectivity of the range. Once the antenna under test is mounted on the positioner, it should be carefully aligned with the source antenna

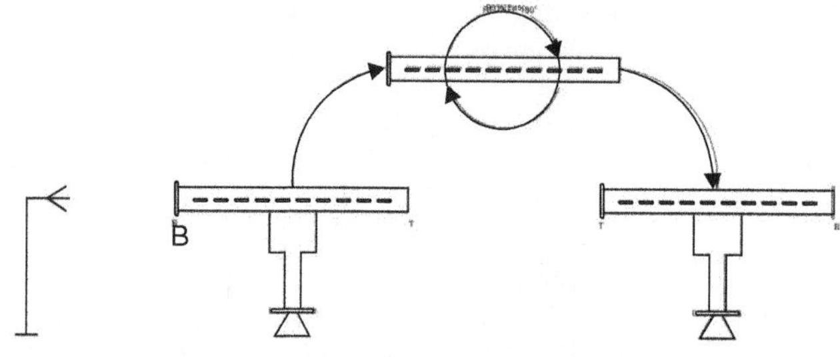

FIGURE 10.8-14 Bore sighting far-field elevation pattern measurement using main beam crossover elevation pattern alignment technique.

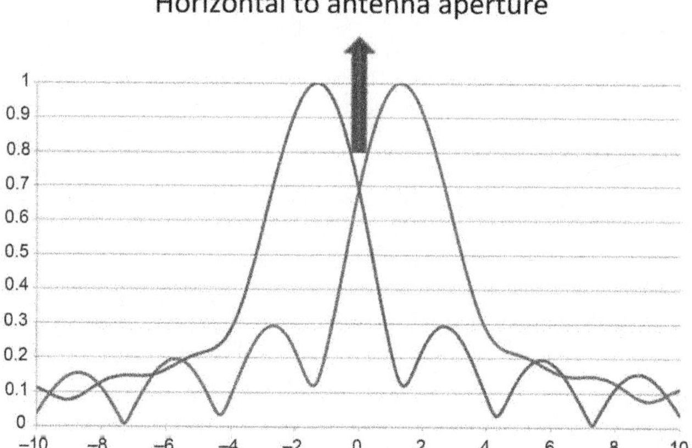

FIGURE 10.8-15 Crossover elevation patterns to determine zero degree horizontal reference direction.

and an exploratory measurement made. In general, the expected characteristics of the antenna's pattern are known, and at this point obvious problems should be revealed. Azimuth pattern measurement alignment for a system set up as shown in the top view of Figure 10.8-12 is typically achieved by a laser site and zero referenced to a particular azimuth point on the aperture. For elevation pattern alignment for a system shown in the bottom view of Figure 10.8-12, the "main beam crossover" technique can be used. This is accomplished by first measuring the elevation pattern from the top of the aperture to the bottom, then rotating the test antenna's positioner 180 degrees and measuring the same plane from bottom to top. By overlaying the two patterns, the zero or horizontal will be located at the crossover points of the main beams. Refer to Figures 10.10-14 and 10.10-15. This technique ensures the accuracy in determining the antenna's beam tilt.

Once alignment has been accomplished, the state of the incident field over the test region must be determined. The principle reason for the incident field to deviate from that which is expected from an idealized range are reflections from the range surface, surrounding objects, the positioner and the cables used to feed the antenna. Sometimes signals from external sources also pose a problem [7]. The field at a point in the aperture under test is the phasor sum of the test signal and the extraneous signals. The relative amplitudes and phases of the desired and extraneous signals will vary with position along the test aperture, causing constructive and destructive additions, thus producing a measured pattern that will depart from the free space expected pattern. The amount of departure can be calculated using Equation 27, where E_D and E_X are the field voltages of the desired and extraneous (reflections) fields for a known level of reflection. The possible error in pattern measurement due to extraneous reflections is plotted in Figure 10.8-16. For example, suppose the main beam of the test antenna is directed towards the source antenna. At the same time, a reflected signal whose level is 35 dB down from the direct signal is received by the side lobe of the test antenna, which is 25 dB down from the main lobe. For this case, $E_X/E_D = -60$ dB. From Figure 10.8-16, the error bounds in the pattern measurement is approximately ± .01 dB. On the other hand, when the main beam is moved to the direction of the reflection, the E_X/E_D ratio is only −10 dB. The error bounds on the pattern measurement at this location is −3.3 dB to +2.4 dB [7].

$$\Delta F = 20\log\left(\frac{E_D \mp E_x}{E_D}\right)(dB) \tag{27}$$

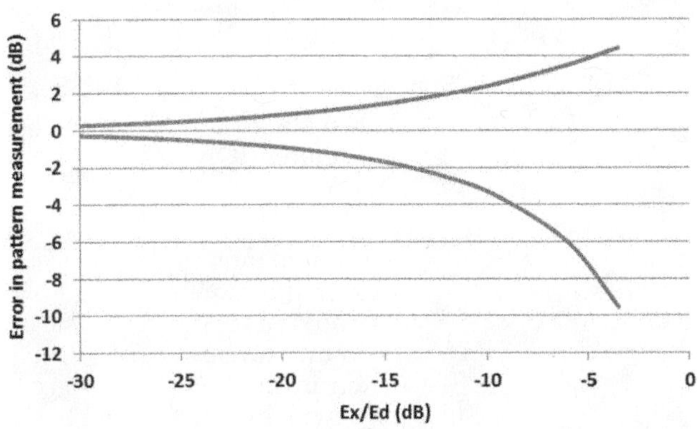

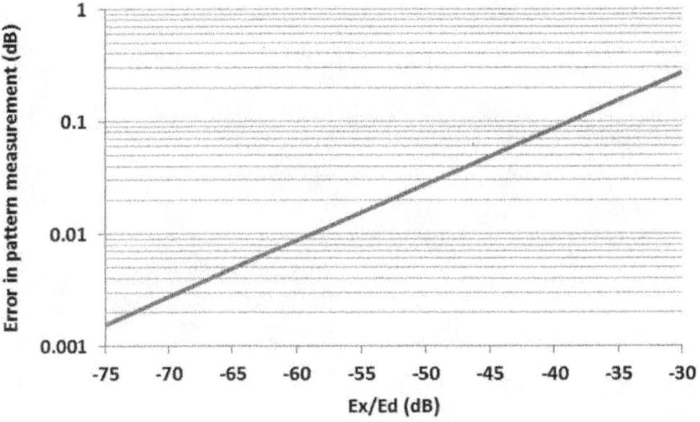

FIGURE 10.8-16 Possible error in pattern measurement due to extraneous reflections. The bottom chart uses a log scale for E_X/E_D levels below 30 dB. It should also be noted that the +/− error is approximately equivalent for E_X/E_D levels below 30 dB.

FAR-FIELD ANECHOIC CHAMBER DESIGN

The advantage of using an indoor far-field range is its all-weather capability and shielding from external noise. The obvious disadvantages of an indoor far-field anechoic chamber are test aperture size limitations and chamber construction expense.

There are two basic types of far-field anechoic chambers, rectangular and tapered. The advantage of using a tapered design is based on the principles of geometrical optics, where the goal is to minimize the amplitude and phase ripple in the test zone (see Figure 10.8-17).

Tapered anechoic chambers were created in the early 1960s after it become obvious that rectangular chambers were not suited for VHF and UHF measurements. Emerson [8] determined that tapering one end of the chamber would cause the chamber to act like an indoor ground reflection range. The wall reflections, instead of being suppressed, were used to form a uniform illumination across the test region. This is accomplished by forcing the wall images close together at the source end of the chamber. The free space range Equation (25) applies to the tapered

FIGURE 10.8-17 60′ tapered anechoic chamber, Dielectric, Raymond, ME.

chamber, and therefore the size of the test aperture sets the range length [2]. External interference and noise is eliminated by enclosing the exterior of the chamber in a metal sheath. When doing so, the expected reflectivity inside the chamber is a function of the performance of the pyramidal absorber.

Far-field anechoic chambers are instrumented essentially the same way as outdoor ranges, with range lengths being the primary difference [4]. Due to the shortened distances, closed-loop systems are generally employed. The control room is typically centrally located near the chamber.

NEAR-FIELD RANGE AND CHAMBER DESIGN

Near-field measurements are used when large antennas must be tested in a relatively small space. The advantages to near-field measurements are:

- Complete pattern and polarization information
- Any cut can be extracted from the three-dimensional pattern.
- Back projection
- Detection of anomalies
- Field can be calculated at any distance from the antenna.

In some cases, perhaps, where near-field RFR requirements may need to be predicted or a full view of the effects of mechanical beam tilt or asymmetrical panel antennas are used, pattern measurements over the entire radiation sphere of the antenna are desirable.

Near-field ranges are usually configured for automatic control. The large number of measurements required and the need to transform the near-field data to the far field requires the use of a computer system both for data acquisition and for data reduction and display. The configuration of a near-field range is similar to a very short indoor or outdoor range [4].

There are three basic types of near-field ranges: planar, cylindrical and spherical, with cylindrical near field being best suited for broadcast antennas. For this method, the source probe is scanned in one linear dimension using a single-axis positioner. The test antenna is stepped in angle-oriented parallel to the linear axis. The resulting scan maps out an imaginary cylindrical surface surrounding the test antenna's aperture (see Figures 10.8-18, 10.8-19 and 10.8-20). The data is then transformed to the far field using Fourier techniques. The resulting data can then be displayed in the same formats as conventional far-field measurements.

The near-field technique provides information that is not available from a far-field measurement. For example, since desired far-field patterns are calculated from aperture illuminations and antennas are designed accordingly, if they do not meet the specifications, it is possible to use the near-field data to help troubleshoot the problem. This can be accomplished by using a reverse transform and collapsing the data unto the antenna's aperture in order to pinpoint the issue.

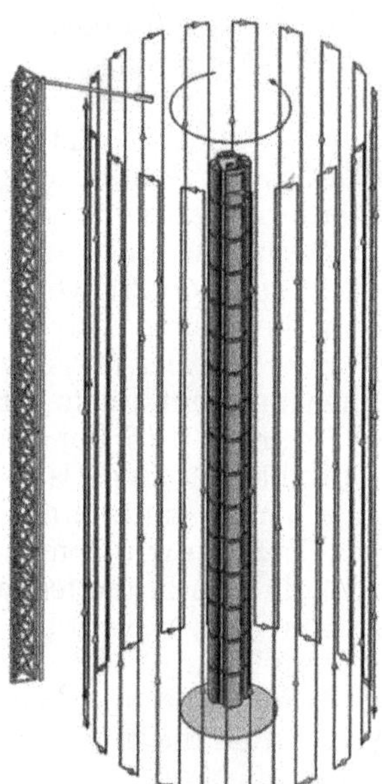

FIGURE 10.8-18 Cylindrical near-field scan map.

FIGURE 10.8-19 Outdoor cylindrical near-field range—Dielectric, Raymond, ME.

FIGURE 10.8-20 Indoor anechoic chamber near-field range—Dielectric, Raymond, ME.

References

[1] Hollis, J.S., T.J. Lyon, L. Clayton, "Microwave Antenna Measurements", Scientific Atlanta, Inc., Atlanta, Georgia, 1985.

[2] Hemming, L.H., "Electromagnetic Anechoic Chambers", IEEE Press, Wiley & Sons, Hoboken, NJ, 2002.

[3] Stutzman, W.L., Gary A. Thiele, "Antenna Theory and Design", John Wiley & Sons, Hoboken, NJ, 1981.

[4] Fordham, J., "An Introduction to Antenna Test Ranges, Measurement and Instrumentation", Microwave Instrumentation Technologies, LLC, Suwanee, GA, 2016.

[5] Balanis, C.A., "Antenna Theory Analysis and Design", Harper & Row, Publishers, New York, NY, 1982.

[6] Kraus, J.D., "Antennas", McGraw-Hill Electrical and Electronic Engineering Series, New York, NY, 1988.

[7] Lo, Y.T., S.W. Lee, "Antenna Handbook, Theory, Applications, and Design", Van Nostrand Reinhold Company, New York, NY, 1988.

[8] Emerson, W.H., U.S. Patent No. 3,308,463, March 1967.

CHAPTER

10.9

VHF and UHF Television Field Strength Measurements

DANIEL M. BARTON

President/CEO
Alive Telecom

INTRODUCTION

Field strength measurements are a useful tool for determining real-world television station coverage areas and possible incoming interference. This chapter will review a technical and practical approach to making and reviewing measurements. It is important to understand the environment of the measured signals—not only the geographical environment but also the RF and mechanical environments prior to making any measurements. Knowledgeable personnel performing measurements will have a good understanding of RF propagation, spectrum measurements and general RF safety practices.

SUGGESTED EQUIPMENT LIST

- Antenna calibrated dipole set to a half-wave of the frequency under test
- Spectrum analyzer capable of measuring to a minimum −110 dBm noise floor
- Amplifier (can be included in spectrum analyzer or external unit)
- Tested 100% shielded coax test cables
- GPS
- Laptop PC
- Test receiver
- Broadband power divider(s)
- Adjustable and fixed attenuator(s)
- Suitable vehicle with mast

TEST PLAN

Prior to any measurements, a good plan must be in place to ensure a quality return on the time invested. Start formulating the plan by reviewing the details of the transmitting antenna system under test. Such items that are important include expected azimuth and elevation pattern information. If the station owner does not have the details, you can review the filing with the FCC and confirm the original antenna specifications with the antenna manufacturer. Make sure that the licensed ERP of the station is actually what the broadcaster is broadcasting. Choose a schedule with enough time and least amount of traffic to complete the measurements (watch for potential weather issues) and confirm there is no work being done at the tower during the testing process. FCC Rule Section 73.686 can be used to obtain excellent guidance for the measurement procedure and processes, as can ATSC's Doc. A/75—"ATSC Recommended Practice: Developing DTV Field Test Plans."

The azimuth and elevation pattern data will help populate the test location map. A minimum of 36 radials should be taken from the transmit tower. Additional radials should be added for areas that include dense ground clutter or specific areas of a unique directional azimuth pattern and/or client interest. Additionally, confirm the proper peak of beam mounting orientation. Locations along the radials can be chosen to form a grid-like format, but sometimes ideal locations can be hampered by terrain or building issues. Do not choose test locations that are not safe or are on private property. Care should be taken to avoid overhead power lines and overpasses. Review a map to choose the locations closest to the target that are on open roads, parking lots or other public areas. Consider coordinating with local law enforcement agencies so that they will be aware of your activities and not unnecessarily respond to "suspicious activity" calls occasioned by your measurement work. They may also be helpful in providing information on potential problem areas, road closures, and other possibly conflicting activities. In some locations, they may be able to provide escort to help keep you and the equipment safe. In all cases, carry proper identification, including a document that identifies you as a professional and a description of the work you will be doing and the areas you will be conducting your tests. Having station personnel aware of your activity and able to verify who you are—that you are working for the station—will also be helpful. Above all, personal safety must be the first priority.

The locations along the radials should be chosen to include the peak of beam and first and second nulls. These distances are determined by the antenna mount height above sea level and elevation pattern data. Data should also be measured out past the expected area of coverage. Once the test location matrix is populated, the test equipment must be set up and "qualified."

Use a spreadsheet to document the description, serial numbers and last manufacture calibration date for the items used. Generate a flow chart for the RF test equipment system; this will help later for any troubleshooting needs. Test each individual piece used in the test system; measure and document return loss and insertion loss for the frequencies under test. The return-loss reading will qualify the equipment, and document that it is in good working order. The insertion loss numbers will be used to calculate true "off air" received signal measured values.

The vehicle should be in good working safe condition. Antenna mounts should be secure and mast checked prior to any tests. Exercise the mast with the test cable attached to safeguard against any catch or pinch points that could possibly damage the cable or mast. The activity of multiple measurements will take some time, and it is strongly suggested to confirm the cable and antenna are in good working order periodically throughout the data-collection process.

DATA COLLECTION

At each test location, choose multiple locations within the immediate area to cluster the data, a minimum of five tests per location. Antenna height is suggested by the FCC to be 9.1 meters (30 feet) above ground level. Occasionally testing is conducted at a lower height, such as 6.1 meters (20 feet). Good documentation is necessary for evaluation of the data capture. For consistency, pick one height and conduct the tests all at that same height. Orient the antenna toward the tower to receive the greatest signal strength. The FCC denotes coverage in the horizontal plane, so the dipole should be horizontally polarized. Vertical data can be collected for evaluation, but this test should be done at the same time but separate from the horizontal measurements. If measuring two signals from separate towers, reorient the test antenna before each measurement. Document the test location: street address, height above sea level, latitude and longitude, and denote what coordinate datum is used. NAD 27 or NAD 83 is typical. Include additional notes on the environment, ground clutter, time and date and the results of subjective observations on received signal quality. Adaptations of the old "TASO" style observation criteria should be considered.

Multiple measurement setups can be used to collect a full range of data. A wideband measurement can be helpful to see relative signal strengths of other emitters. Picking a secondary channel to test while testing the prime frequency can be useful for comparisons across the coverage area. When picking a second channel it is useful to pick one with similar ERP and HAAT; if the two stations happen to be on the same tower, that is even better. True field measurements have to be completed with the spectrum analyzer bandwidth set as wide as the channel. The results for most spectrum analyzers will be in dBm or dBμV, and conversions can be based on antenna efficiency and impedance. Cluster measurements[1] are performed by selecting at least five evenly distributed measuring locations to capture data over approximately nine square wavelengths (see Figures 10.9-1 through 10.9-4).

DATA REVIEW

Once all the data is collected, calculate the median value of the five tests per location and use that value in the reporting process. Include all components used in the RF test system to determine the final RF value; cable, splitter loss, antenna gain and any other values. Use exact values per channel if multiple channels tested on the same setup. If any of the data points look suspect when compared to the surrounding areas and with consideration of the terrain effects, retest the site location if possible or omit the data.

KNOW THE ENVIRONMENT

It is important to have a good understanding of the transmit antenna system being tested and the surrounding systems while testing. The broadcast industry will be moving to new technologies with multiple RF paths, which can potentially cause confusion when making field measurements. There will be a larger movement to wide or broadband antenna systems, and this generates a new opportunity for antenna reradiation in multi-tenant candelabra towers. Reradiation can occur in or out of phase with the antenna being evaluated, causing amended signal coverage in a unique fashion. The more information that is collected prior to testing, the easier the postcollection report will be.

Measurement Results

102-008.set

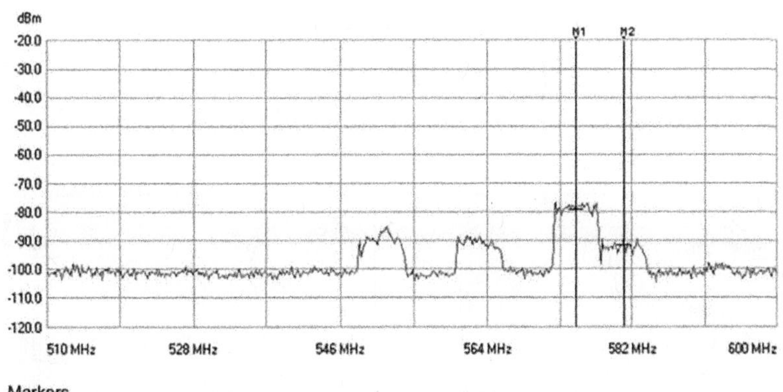

Markers

M1 575 MHz	-79.24 dBm	M2 581 MHz	-91.56 dBm

Measurement Settings

Mode:	Spectrum	Mode:	Spectrum
Center:	555 MHz	Span:	90 MHz
RBW:	100 kHz	VBW:	100 kHz
Sweep Time:	45 ms	Attenuator:	0 dB
Reference:	-20.0 dBm	Range:	100 dB
Int DC Bias:	Off		

Trace Settings

Trace Mode:	Average	Detector:	Sample

FIGURE 10.9-1 Wideband test result.

Measurement Results

102-009.set

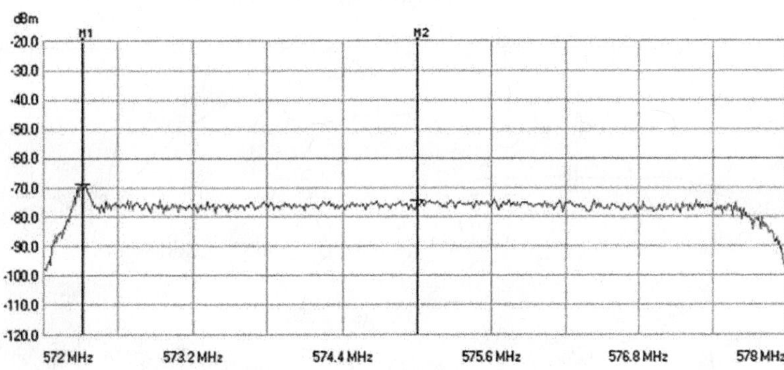

Markers

M1 572.3142857 MHz	-68.87 dBm	M2 575 MHz	-74.18 dBm

Measurement Settings

Mode:	Spectrum	Mode:	Spectrum
Center:	575 MHz	Span:	6 MHz
RBW:	100 kHz	VBW:	100 kHz
Sweep Time:	20 ms	Attenuator:	0 dB
Reference:	-20.0 dBm	Range:	100 dB
Int DC Bias:	Off		

Trace Settings

Trace Mode:	Average	Detector:	Sample

FIGURE 10.9-2 Single channel with pilot.

Measurement Results

102-111.set

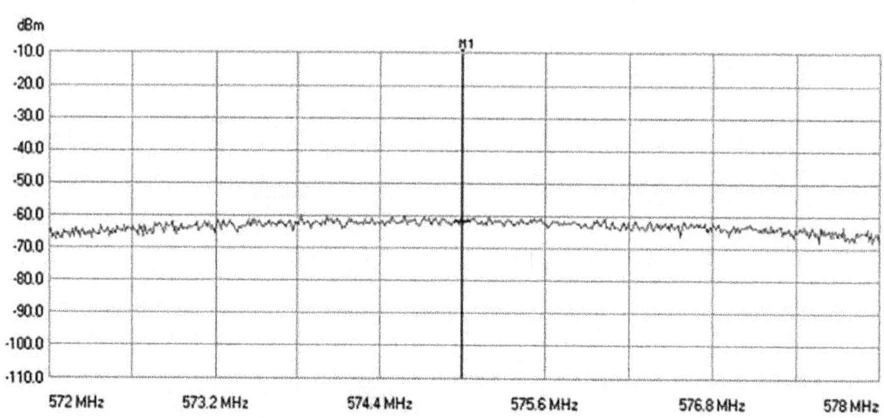

Markers

M1 575 MHz -61.55 dBm

Measurement Settings

Mode:	Spectrum	Mode:	Spectrum
Center:	575 MHz	Span:	6 MHz
RBW:	6 MHz	VBW:	6 MHz
Sweep Time:	20 ms	Attenuator:	0 dB
Reference:	-10.0 dBm	Range:	100 dB
Int DC Bias:	Off		

Trace Settings

Trace Mode:	Average	Detector:	Sample

FIGURE 10.9-3 Single occupied channel.

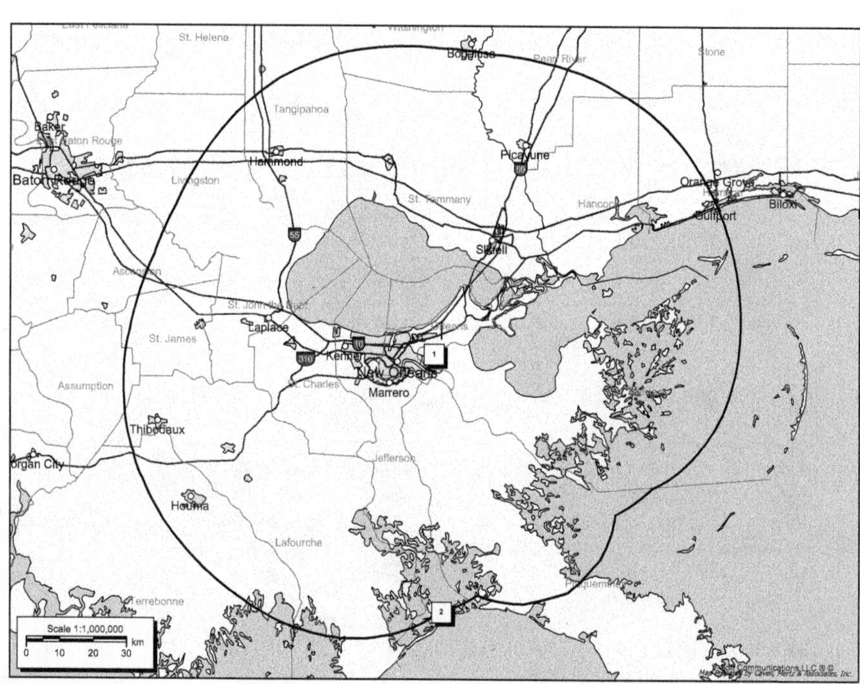

FIGURE 10.9-4 Typical-41 dBμ map of a directional UHF antenna. Some areas require more planning to fulfill testing requirements.

Map courtesy of Cavell, Mertz & Associates, Inc. and V-Soft Communications, LLC.

Ch 31 575 MHz 700 kw ERP Side Mount Directional 1120' HASL 1.25° Beam Tilt 30' Test Height							
Location	Address	Lat/Long	Radial #	Distance (ml)	Raw Value (ml)	Result (µv/m)	Notes
101-14	1890 South 2nd st	37,9683,–88,9049	15	22	–83	32	Church Parking Lot Clear
101-15	3200 South 2nd st	37,9302,–88,8852	15	25	–92	23	Public Park 15' Trees 40' away
101-16	1870 W HWY 15	37,9662,–88,5704	16	26	–91	24	Quick Mart Parking Lot Clear

FIGURE 10.9-5 Sample data collection spreadsheet.

Note

[1] "ATSC Recommended Practice: Developing DTV Field Test Plans" Doc. A/75 21 July 2001 Section 6.1.4.1

References

"ATSC Recommended Practice: Developing DTV Field Test Plans" Doc. A/75 21 July 2001

"A Study of ATSC (8-VSB) DTV Coverage in Washington, DC, and Generational Changes in DTV Receiver Performance" Project TRB-00-1 Interim Report April 9, 2001, Technical Research Branch, Laboratory Division, Office of Engineering and Technology, Federal Communications Commission

SECTION

11

ANCILLARY BROADCAST SYSTEMS

CHAPTER

11.1

Electrical Power Systems, Specification, Conditioning, Fault Protection

JOHN N. WEST

Power & Systems Innovations of Tampa, Inc.
Hernando Beach, FL

INTRODUCTION

Utility companies' generation facilities initially distribute well-regulated power to their customers; however, it does not always arrive at the customer's facility as well-regulated power.

Disturbances on the AC line are very often beyond the utility company's control. Surge events such as lightning or accidents produce system faults that are beyond the utility company's reasonable control; however, these events are damaging just the same. In addition, large load changes by customers of the utility can produce power-quality events that are disruptive and damaging in some cases. Power factor (PF) correction switching and distribution system faults all combine to produce an environment in which tight control over AC power quality is difficult to maintain. As a result of the uncontrolled disturbances and the induced surge events, the responsibility for ensuring AC power quality must rest with the users of sensitive equipment. Therefore, for the broadcasting industry, "*Caveat Emptor*" is a prudent and wise policy.

If the unpredictability of the incoming electrical power quality were not enough, conditions or issues in the electrical system of the broadcaster's facility further complicate power quality. Stuart Feldstein of SMR Research Corporation published: "At the end of 2015, the average U.S. commercial building was almost 50 years old (49.6 years). This finding is in line with prior years. When we looked at building age in 2011, and in 2014, the average was exactly 50 years old." In addition to that, Feldstein stated: "At the other end of the spectrum, mixed-use buildings had an average age of 74 years. This means the average mixed-use building was built all the way back in 1941."

Given the age of a building, how well do you think the electrical distribution system in the "average" building will support current electronic technology unless it has undergone a significant upgrade? In addition, current code requirements and proven reliable installed systems would require a total redesign and upgrade of the facility with all new electrical panel boards and wiring; in short, a new electrical system would be needed.

The sophistication of technology is growing at a geometric rate. Nobody can accurately predict what will be available in 10 years. One area of the broadcast facility's infrastructure is easy to predict: The demands placed upon the electrical system by technology in a broadcast facility are growing at an alarming rate that is increasing dramatically every year. At one time the equipment in a broadcast facility was reasonably robust. Today's modern broadcast facility has more of the look of a data center and will continue to change in that direction as IP technology's role increases in broadcasting.

UNDERSTANDING POWER QUALITY ISSUES AND EFFECTS

The first step is to understand the cause and effect of power-quality disturbances or issues. To begin the understanding process, these events need to be outlined. This will aid the reader of this article when mitigation of these issues and events are discussed.

STEP #1—THE FACILITY ELECTRICAL DISTRIBUTION SYSTEM EVALUATION & REMEDIATION

Power conditioning starts with a properly designed, flexible, robust, well-maintained electrical system, which must be sustainable for the intended use. The installation of additional equipment to compensate for an electrical system that is unsustainable is a recipe for disaster. This practice is flawed and will often come back to punish those doing so. Job one is to be 100% sure your electrical system is sustainable. If you are not qualified to determine that, hire a professional power-quality expert with a documented record as well as experience in the broadcast industry. The proper information he or she provides can be used as the basis to update your electrical distribution system.

When the necessary update is completed, the entire electrical system must be sustainable. To generate a list of devices and technologies to be evaluated for installation, site factors should be considered. You may wish to add some additional devices to the list in this chapter, which may be needed for your unique facility.

Geographic location exposure factors will include lightning exposure and topographic location (top of a mountain versus a lower exposure). Other location exposure factors will include multiple-tenant electrical distribution systems such as electrical circuits that extend outside the facility and outside circuits shared with any other function or building-specific exposure factors such as a building with a rooftop chiller that is bonded to the building steel support structure. If you are in any area of the world that has lightning, that chiller could directly impact the electrical system.

Location of the facility in relation to the utility source and grid is important; consideration needs to be given if the building is at the end of the power line. Being on a utility transformer that also serves others that may utilize equipment that produces power-quality anomalies is an important consideration; those anomalies may be imposed on your equipment if not mitigated.

Dual-feed utility source locations exist where the utility has the capability to provide a secondary utility source from a different local distribution grid. If available, this alone could significantly reduce the cost of power-quality equipment.

Now we are ready to consider and evaluate the use of power-quality devices for electrical systems and equipment that can benefit from their installation. To properly apply power-quality devices, you must understand their function and the benefit they can provide.

It is extremely important to note that surge protection devices (SPDs) are "diversion devices." They divert an over-voltage typically to neutral (utility ground) and/or to the electrical system ground. Electrical system ground is defined by the NEC in *Article 250-Grounding and Bonding*. A superior reference for grounding and bonding is the IEEE *Green Book*. A key element in the application of any SPD is to be certain the "diversion path" is both low resistance and low impedance.

This is critical for the function of Type 1 and Type 2 SPDs. Until it is known if these paths are correct, the SPD should NOT be installed. The application of SPDs lacking proper installation conditions will at best be ineffective and, at worst, cause equipment damage rather than prevent it.

STEP #2—THE SELECTION AND APPLICATION OF SPDS

Today, SPDs are not optional in a broadcast facility; they are the most basic protection after the grounding and bonding installation is completed to a high standard. They should be installed in a "layered" approach to be most effective.

Today's SPDs were formerly known as two distinct categories of devices. Lightning arrestors (LA) and transient voltage surge suppressors (TVSS) are now grouped into the SPD classification. Type 1 SPDs are what used to be the LA, and Type 2 SPDs are TVSS, which were permanently installed after the primary means of electric disconnect (the main breaker for disconnect).

UL 1449 R4.0 Type 1—Permanently connected SPDs intended for installation between the secondary of the service transformer and the line side of the service equipment overcurrent device, as well as the load side, including watt-hour meter socket enclosures and molded case SPDs intended to be installed without an external overcurrent protective device. Type 1 SPDs for use in PV systems can be connected between the PV array and the main service disconnect.

UL 1449 R4.0 Type 2—Permanently connected SPDs intended for installation on the load side of the service equipment overcurrent device, including SPDs located at the branch panel and molded case SPDs.

UL 1449 R4.0 Type 3—Point-of-utilization SPDs, installed at a minimum conductor length of 10 meters (30 feet) from the electrical service panel to the point of utilization, for example, cord connected, direct plug-in, receptacle type and SPDs installed at the utilization equipment being protected. The distance (10 meters) is exclusive of conductors provided with or used to attach SPDs.

The selection of SPD protection for a facility is as much an economic question as it is a technical one, and it is recommended that a power-quality professional assist in that process.

STEP #3—TYPES OF BACKUP SYSTEMS TO BE APPLIED TO UTILITY OUTAGES

The installation of backup power systems can be as basic as simple battery backup systems and equipment or as complete as backup generation devices. Broadcast facilities should have a high level of utility outage protection. They should be capable of continuing operation during extended power outages that could be many days or more than a week, which requires backup generation capability and may require a significant fuel source. All this depends on proper planning, design and implementation of the installation. Additionally, the organization must have a plan in place for all contingent situations that could arise during a long-term "off-grid" time period. The public depends on the broadcast community for accurate, up-to-date information during catastrophic times. Imagine going through a major hurricane without the broadcast community providing information both in a news format as well as public service announcements.

Backup Generators—Fossil Fuel (Diesel, Natural Gas, LP Gas)

The most popular source of power during utility outages is what is called a standby backup generator (see Figure 11.1-1). It will power the facility during a utility outage and all elements of the electrical system. This generator will also power battery UPS system(s). These fossil fuel generators are the most likely choice when combined with generator compatible battery UPS systems. The most common installation will not switch the neutral at the automatic transfer switch. *Please note*: The transfer switch should be installed with a robust SPD in a Type 1 configuration to protect the transfer switch from damage. The power-sensing circuit and the auto start components in unprotected generator automatic transfer switches are often damaged by power surges resulting from utility events or lightning. Without these control elements, the "automatic" transfer switch will at best become a manual transfer switch, and the generator may have to be manually started.

FIGURE 11.1-1 Standby Motor-Generator

Backup generators alone are not a complete solution to a power outage. Modern broadcast technology can be taken down (offline) and may need to be manually rebooted during even a brief power failure. However, when generators are installed and sensitive loads are powered by a battery UPS or rotary UPS, you have a dependable combination.

An important element of design and installation of standby UPS is load control. The use of control devices to schedule loads to be added to the standby motor-generator is important.

Load control will reduce the inrush current the standby motor-generator must accommodate and can therefore reduce the need for support of significant inrush current. When numerous inductive loads are powered by the facility inrush current is an important consideration.

Motor-Generator UPS

Motor-generator sets have been used successfully in power-quality applications for decades. As the name implies, it is a motor-generator (MG) set consisting of a generator that is an over-excited AC motor powered by the AC utility supply that is mechanically connected to a generator by an induction coupling that connects to the motor. This provides complete electrical isolation and filtration from any transients on the utility feed. One of the more popular models using this technology is shown in Figure 11.1-2.

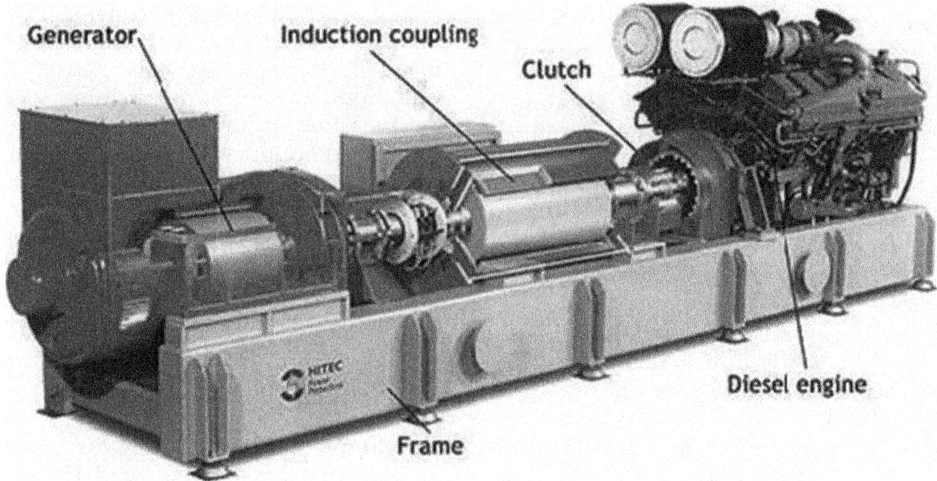

FIGURE 11.1-2 Uninterruptible Motor-Generator Set Using Synchronous Motor

The diesel engine is kept warm as well as under lubrication so it can instantly start and provide full-load power in seconds. The filtration and isolation provided by this type of motor-generator UPS is a combination of a reactor (choke) and the synchronous generator. It combines to become a filter to clean the incoming utility supply. Motor-generator UPS units are found in highly critical sites all over the world. Units such as the one shown in Figure 11.1-2 power much of the internet today.

DC Flywheel UPS

There have been significant advances in the use of flywheels to provide stored energy in the form of DC to complement or supplant batteries in a battery UPS system. These systems are typically used when the facility is less likely to be without utility power for a long period of time and when the facility operation requires instant power supply when the utility company power supply fails or when a dangerous low-voltage or phase-failure event occurs. Motor-generators will often offer operational advantages and be more cost effective than backup generation combined with DC flywheel UPS and/or battery UPS. The main drawback in applying flywheels in the past has been the fact that frequency decay occurred quickly and provided limited ride-through. By utilizing this technology on the DC side, longer outage protection can be achieved, which provides enough time to transfer over to a generator in the case of a sustained utility outage. DC flywheel UPS are often combined with battery UPS.

DC flywheels can also be used to increase the life and reliability of a battery system in a UPS, but more commonly they replace battery UPS in the power-backup plan. The DC flywheel can be adjusted to take the shorter power "hits," leaving the longer outages to be handled by the battery systems in the battery type UPS, as shown in Figure 11.1-3. This prevents the batteries from aging prematurely in remote applications that may have frequent outages. It can also prevent the loss of load in case a cell opens during battery operation if the UPS battery power is from a single string of batteries. Single strings are the weakest link in a UPS today. It is rare for a large UPS system to have a single string of batteries because most critical installations use redundant battery strings. This battery-hardening technique provides another alternative in improving reliability. DC flywheel technology, when combined with standby motor-generators, provides both power conditioning and backup power, often at a cost that is less than traditional battery UPS. Overall cost of ownership is a critical consideration, as are the space required, air conditioning issues and the associated cost of a battery UPS.

The DC flywheel UPS continues to evolve, but it has become less popular due to the introduction of modular on-line double conversion UPS. The redundancy available in these modular systems has great appeal to the buying market. They are available in sizes from 3 kVA and larger. Most of the UPS units today have full modularity. The modules include the rectifier as well as the control elements. The common term employed in these systems is N + 1 (2, 3, etc.). The N is for "normal" and the "+ number" is the levels of redundancy employed in the UPS system.

FIGURE 11.1-3 DC Flywheel UPS

Prior to the modular UPS with N + "*x*" redundancy, it was common for two or more fully load-rated UPS to be installed with a static transfer capability. Either could power the entire load if the other failed. This configuration resulted in a much greater equipment capital cost, higher maintenance, greater operation cost, more space and added heat load. The return on investment of the modular UPS is much higher, and the reliability is also greater. That is the "win-win" all need to seek in business. The large UPS systems are now often larger than 1 MW, and their footprint is significant. The modular UPS systems also have modular battery systems. The modular configurations are a big plus because in the past that redundancy required multiple strings of fully rated batteries, and the system employed resulted in a very costly way to avoid a battery failure; however, it caused UPS failure. The modularity, when combined with the intelligence and communications capabilities, makes these systems a very logical and cost-effective choice when combined with a motor-powered backup generator system with a transfer capability.

Thermal and Compressed-Air Storage UPS

Development in energy storage using compressed air to provide run times in excess of 15 minutes at 80 KW is now possible. During a power outage, compressed air is sent through a thermal storage unit and greatly expands. This heated air spins a turbine that generates the power to support the DC bus on a double conversion UPS. An example of such a unit is shown in Figure 11.1-4.

This arrangement bridges the gap until the flywheel effect of the turbine is at full RPM. This system has not yet found popularity in the United States. The one-line diagram in Figure 11.1-5 shows the major components and their function in the design. A useful byproduct of this technology is the creation of cold air, which can help in keeping the space cool during an extended outage; however, the heat of today's integrated technology is significant. Initial high hardware costs and the lack of building codes for high-pressure piping could limit the initial application of this technology.

Uninterruptible Power Supply (UPS)

The term UPS, "uninterruptible power source," can be one of many technologies used. The technologies vary and are most often determined by the needs and requirements of the load that is being powered. The term "uninterruptible" is a function of the load. Some loads can function with brief power interruptions or anomalies. Others cannot tolerate any anomalies or power loss. The most common use of the term UPS is generic. The correct term for a UPS that is battery based is "battery UPS," not just "UPS."

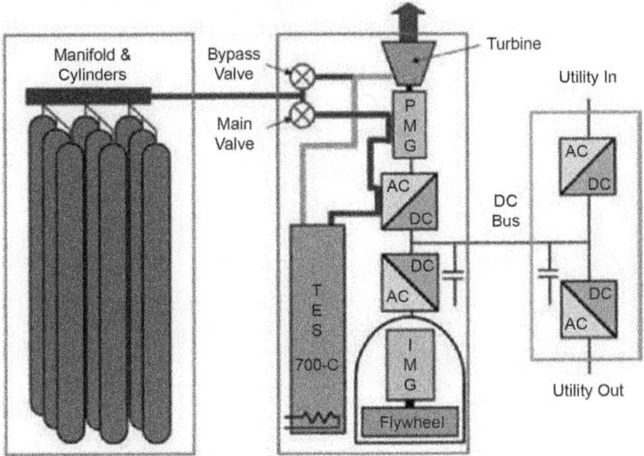

FIGURE 11.1-4 Thermal and Compressed-Air Storage UPS

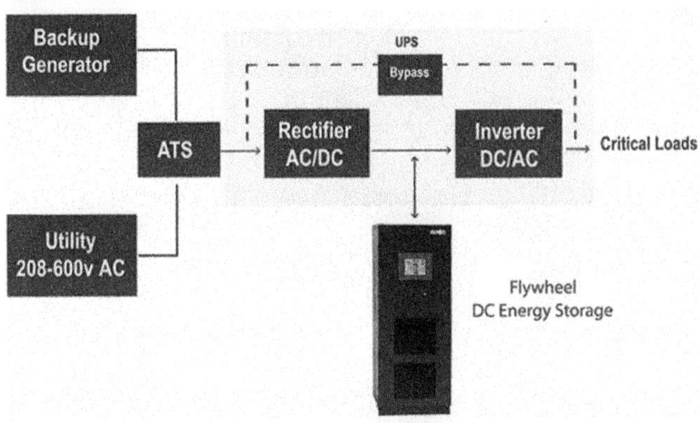

FIGURE 11.1-5 Standby Battery UPS Block Diagram

Power Supply (Battery UPS)

Battery UPS systems are used to power loads that are sensitive to power anomalies, and their processes can be upset or data lost as a result of power anomalies. Battery UPS systems that will be outlined here vary in technology; therefore, the features that are offered and the level of load isolation and power regulation varies as well.

Battery UPS systems, such as the one depicted in Figure 11.1-6, have become one of the options in powering "sensitive" loads. The unit shown is available in models that exceed 1 MW. These new generations of battery UPS systems are "smart," redundant at all levels and have a mean time between failures (MTBF) that was only possible at a much higher cost in the past. Today they are available to power small loads as well. It is not uncommon for these to be installed when the load is under 1 amp.

Locations with lower power requirements very often involve equipment critical to the overall function of the broadcast process. The smaller systems and equipment used in broadcasting can be the source of off-air time. This equipment will often be critical, and the continuity of service is essential to broadcasting. Broadcast technology and data communications systems are no more reliable than the power from which they operate. Later is a discussion of the available technologies that will aid you in selecting the battery UPS that will best meet your systems and equipment needs.

Solid-State (Static) UPS Technologies

Today, the designs in static UPS systems employ isolated gate bipolar transistors (IGBT) to create the pulse width-modulated (PWM) conversion not only from DC to AC in the inverter but also to chop the AC down to DC in the rectifier section. The main reason for the change to this technology is the speed at which the transistors can be switched. This improves the output performance of the UPS and its ability to handle overloads, step loads and harmonics generated by these loads.

UPS design employing IGBT technology changes the design configuration and both improves reliability and reduces cost. The reduction in footprint is a further benefit. That also translates to reduced heat output by the battery UPS, therefore greater efficiency. The high-speed conversion in the rectifier removes the requirement for an input filter to reduce reflected harmonics back to the service entrance. Therefore there is greater efficiency, lower component count and higher reliability of the system, especially as it relates to engine generators and the possible interaction with the input filter on the UPS.

Different Types of UPS Systems

A variety of design approaches are used to implement UPS systems, each with distinct performance characteristics. The most common design approaches are as follows:

- Standby
- Line interactive
- Standby on-line hybrid
- Standby-Ferro

- Double conversion on-line
- Delta conversion on-line

The Standby Battery UPS

The standby UPS is the least expensive, most available and common type of battery UPS. Standby UPS topology is "single conversion." In the block diagram illustrated in Figure 11.1-6 the transfer switch is set to choose the filtered AC input as the primary power source (solid line path) and switches to the battery/inverter as the backup source should the primary source fail. When that happens, the transfer switch must operate to switch the load over to the battery/inverter backup power source (dashed path). The inverter only starts when the power fails, hence the name "standby." A significant consideration of "standby" is that the equipment you are trying to protect is plugged into the AC power source with very minimal surge protection and filtration. Standby battery UPS is the low-cost and not necessarily the best option in battery UPS.

A more significant shortcoming of the topology of a standby battery UPS is it is often *not* compatible, or problematic at best, when used with many standby generators that are undersized. The author's recommendation is to consider a different topology if you will be using a generator. Some models of standby battery UPS can be "desensitized" through software or switches on the UPS. But the issue of their topology and design remains; powering nonlinear loads with a generator can result in the output waveform distortion. This distortion will most likely result in the standby battery UPS product frequently transferring back and forth from on-line to on-battery.

The Standby On-Line Hybrid Battery UPS

Standby on-line hybrid battery UPS is the most misunderstood topology (shown in Figure 11.1-7). The common belief is that the primary power path is always "on-line." The fact is the dual power paths mean it is only 50% on line (the load is powered by and running

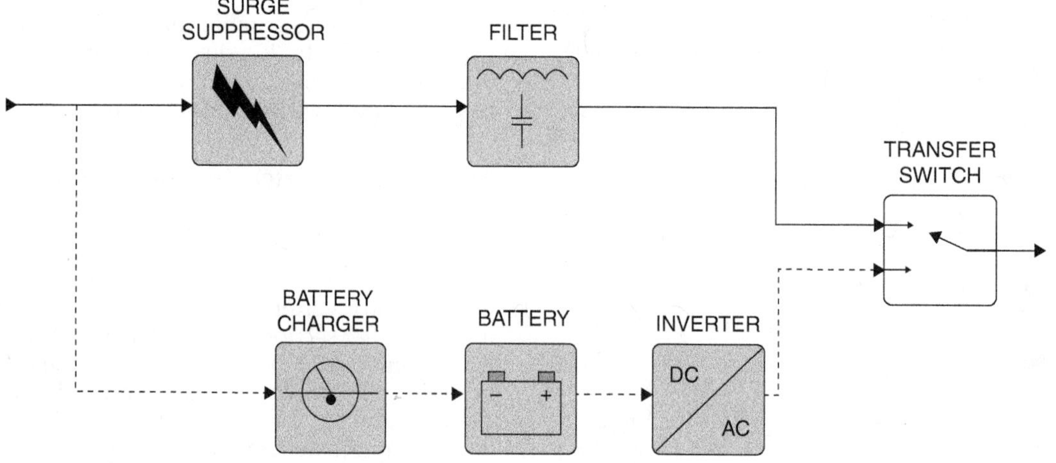

FIGURE 11.1-6 Standby Battery UPS

Schneider Electric Data Center Science Center

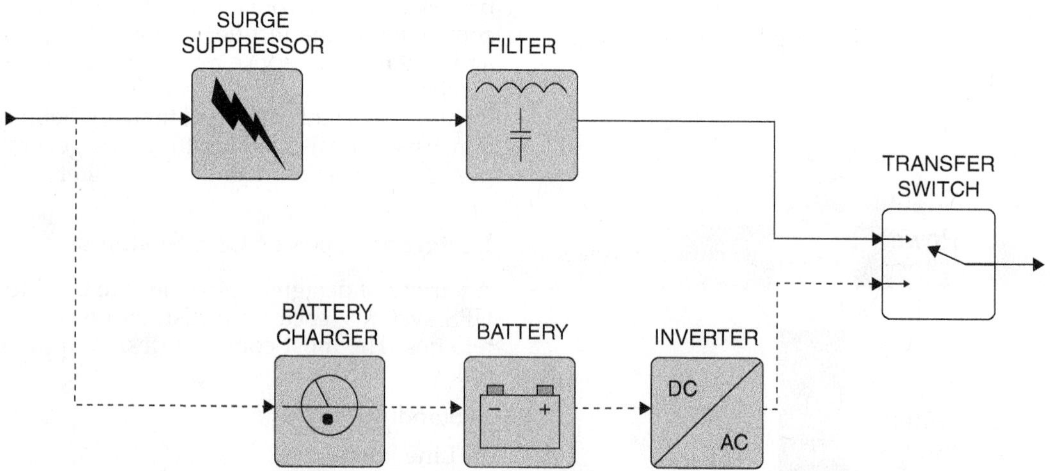

FIGURE 11.1-7 Standby On-Line Hybrid Battery UPS

Schneider Electric Data Center Science Center

on the inverter). The balance is subject to the same issues as the standby battery UPS. This topology is not recommended for the broadcast industry, as it does not offer the features or load-power reliability needed by the broadcast industry.

The Line Interactive UPS

The line interactive battery UPS, illustrated in Figure 11.1-8, is a common design used for small business because it offers some power filtration and the battery UPS is smart. The meaning of "smart" is that it can communicate the condition of the power, its status, etc. It is a very basic "smart" standby battery UPS.

The more superior line interactive battery UPS design is "smart," and it also offers power conditioning using a buck/boost circuit to stabilize the voltage of the output power. This circuit reduces "nuisance" transfers due to minor voltage fluctuations, which results in a reduction of the number of transfers from utility power to the battery and inverter circuit and will extend the life of the battery. It also creates a more stable electrical environment than a standby battery UPS, as shown in Figure 11.1-9.

The Standby-Ferro Battery UPS

The standby-Ferro UPS was once one of the dominant forms of UPS in the 3–15 kVA range. Its design depended on a special saturating transformer with three windings (power connections). The primary power path is from AC input, through a transfer switch, through the transformer and to the output. In the case of a power failure, the transfer switch is opened and the inverter picks up the output load. The field of energy of the transformer acted as a tank circuit, and the primary vendor making units of this design called it "no-break" power, meaning the load was never without conditioned power and the tank circuit held up the load during the transfer from line to inverter. This topology is still in use today.

In the standby-Ferro battery UPS design, the inverter is normally in the standby mode. It is energized when the input power deviates from the "allowable" AC input power range. This is adjustable to a degree in some models when the power gets out of the allowable input power range or fails and the transfer switch is opened. The transformer is Ferro resonant, which provides limited voltage regulation and output waveform "shaping." The transformer's tank circuit action added to the stability of the power supplied to the load. The isolation from AC power transients provided by the Ferro transformer is as good as or better than any filter available. However, it should be noted the Ferro transformer itself can create severe output voltage distortion and transients depending on the load being powered. The principal reason that standby-Ferro UPS systems are not popular today is that they can be fundamentally unstable when operating a modern computer power supply load. Loads today use "power factor corrected" power supplies that present a negative input resistance over some frequency range; when coupled

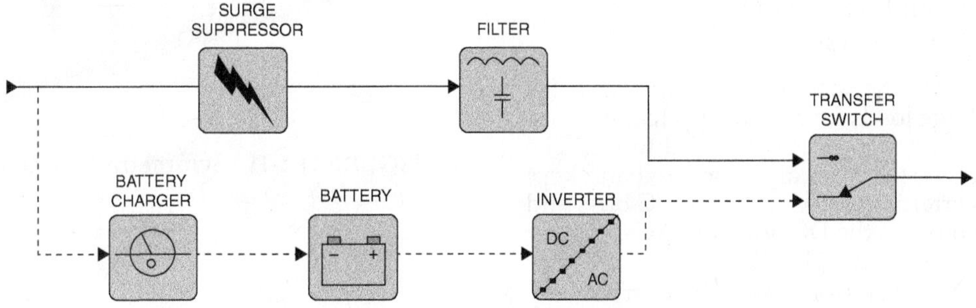

FIGURE 11.1-8 Line Interactive Battery UPS

Schneider Electric Data Center Science Center

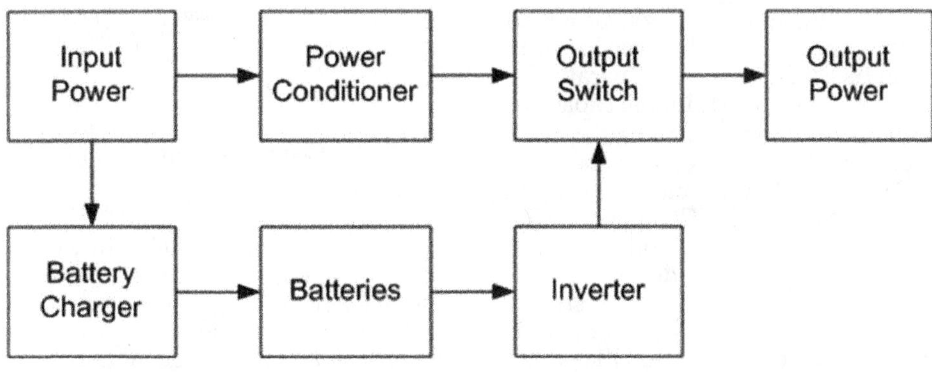

FIGURE 11.1-9 Smart Line Interactive Battery UPS Block Diagram

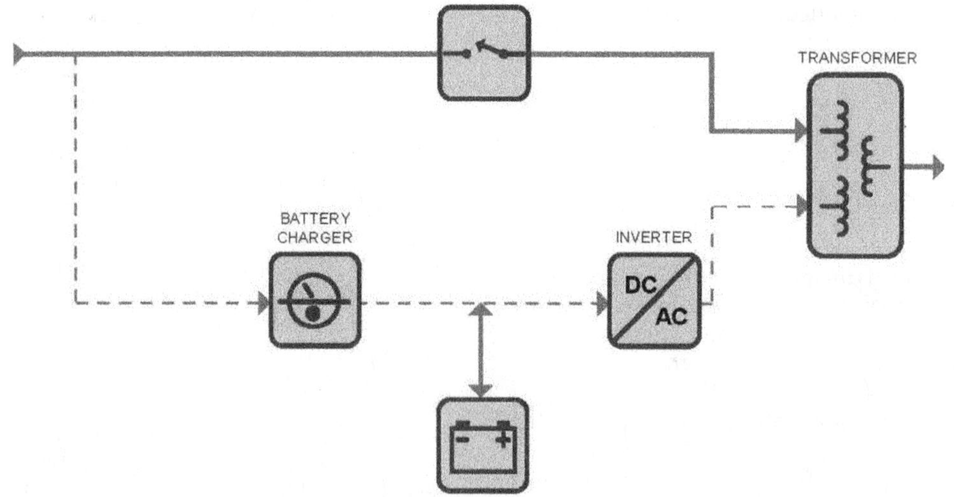

FIGURE 11.1-10 Standby-Ferro Battery UPS

Schneider Electric Data Center Science Center

with the relatively high and resonant impedance of the Ferro transformer, this can give rise to spontaneous and damaging oscillations.

The standby-Ferro generates a great deal of heat because the Ferro resonant transformer is inherently inefficient. These transformers are also large relative to regular isolation transformers, so standby-Ferro UPS systems are generally quite large and heavy. This topology is shown in Figure 11.1-10.

Double Conversion UPS (Also Known As On-Line Double Conversion Battery UPS)

This design is by far the most popular approach for small as well as large loads for critical applications (see Figure 11.1-11). Utility power is converted to DC. The DC buss is where the batteries are connected and kept charged. The inverter is connected to the DC buss and subsequently reconverts the DC power to AC to power the load, as shown in Figure 11.1-12.

The two conversions isolate the load from any and all power-quality problems on the primary side. During normal operation the load is fed from the inverter. In the event of a failure or overload condition, the static bypass would activate to keep the load running. This approach provides the highest reliability in a single-module UPS. Higher reliability is provided by the modular redundant models (such as shown in Figure 11.1-12) or can be achieved through paralleling of battery UPS for redundancy or additional capacity.

In the past, double conversion on-line battery UPS systems were the preferred technology and the most common type of UPS above 10 kVA. That has changed with today's highly sensitive loads. This battery UPS topology is available in less than 1 kVA from a number of suppliers. The block diagram of the double conversion on-line UPS is the same as the standby, illustrated in Figure 11.1-5, except that the primary power path is the inverter instead of the AC main. Wear on the power

FIGURE 11.1-11 Symmetra MW 1600-kW Frame

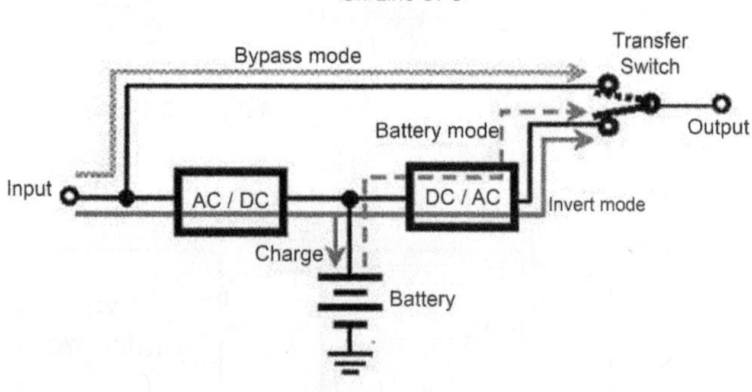

FIGURE 11.1-12 On-Line Double Conversion Battery UPS

components reduces reliability over other designs, and the energy consumed by the electrical power inefficiency is a significant part of the life-cycle cost of the UPS. Also, the input power drawn by the large battery charger is often nonlinear and can interfere with building power wiring or cause problems with standby generators.

In the double conversion on-line design, failure of the input AC does not cause activation of the transfer switch because the input AC is *not* the primary source but is rather the backup source. Therefore, during an input AC power failure, on-line operation results in no transfer time.

The on-line mode of operation exhibits a transfer time when the power from the primary battery charger/battery/inverter power path fails. This can occur when any of the blocks in this power path fail. If the inverter is subjected to sudden load changes or internal control problems, the inverter power can drop out briefly, causing a transfer.

Double conversion on-line UPS systems do exhibit a transfer time but under different conditions than a standby or line interactive UPS. While a standby and line interactive UPS will exhibit a transfer time when a blackout occurs, a double conversion on-line UPS will exhibit a transfer time when there is a large load step or inrush current. This transfer time is the result of transferring the load from the UPS inverter to the bypass line. Generally, this bypass line is built with dual silicon controlled rectifiers (SCR). These solid-state switches are very fast, so, similar to the standby and line interactive UPS, the transfer time is very brief, usually 4–6 ms.

Both the battery charger and the inverter convert the entire load power flow in this design, which causes reduced efficiency and increased heat generation. Figure 11.1-13 shows an example of an on-line double conversion battery UPS with AVR, automatic voltage regulation, when in bypass.

The Delta Conversion On-Line UPS

This UPS design, illustrated in Figure 11.1-14, is a new technology introduced to eliminate the drawbacks of the double conversion on-line design and is available in the range of 5 kVA to 1 MW. Similar to the double conversion on-line design, the delta conversion on-line UPS always has the inverter supplying the load voltage. However, the additional delta converter also contributes power to the inverter output. Under conditions of AC failure or disturbances, this design exhibits behavior identical to the double conversion on-line.

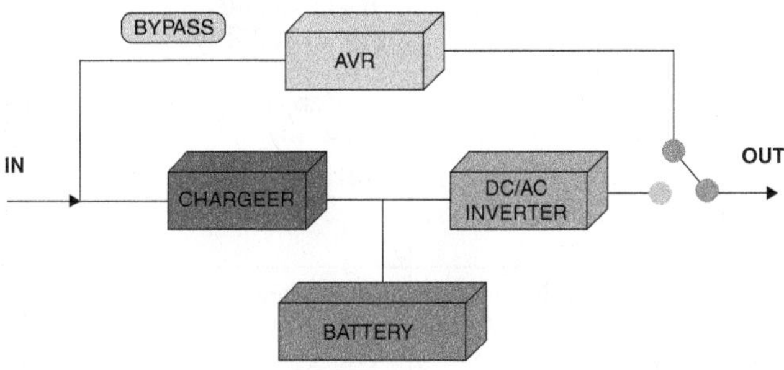

FIGURE 11.1-13 Power Flow of an On-Line UPS With AVR

A simple way to understand the energy efficiency of the delta conversion topology is to consider the energy required to deliver a package from the fourth floor to the fifth floor of a building, as shown in Figure 12.2-15. Delta conversion technology saves energy by carrying the package only the difference (delta) between the starting and ending points. The double conversion on-line UPS converts the power to the battery and back again, whereas the delta converter moves components of the power from input to the output.

In the delta conversion on-line design, the delta converter acts with dual purposes. The first is to control the input power characteristics. This active front end draws power in a sinusoidal manner, minimizing harmonics reflected onto the utility. This ensures optimal conditions for utility lines and generator systems and reduces heating and system wear in the power distribution system. The second function of the delta converter is to charge the battery of the UPS by drawing power and converting it to the appropriate DC charging voltage.

The delta conversion on-line UPS provides the same output characteristics as the double conversion on-line design. However, the input characteristics are extremely different. With full power factor correction, the delta conversion on-line design provides both input power control and output power control. The most important benefit is a significant reduction in energy losses. The input power control also makes the UPS compatible with all generator sets and reduces the need for wiring and generator oversizing. Delta conversion on-line technology is the only core UPS technology today protected by patents and is therefore not likely to be available from a broad range of UPS suppliers.

Summary of UPS Types

The following table shows some of the characteristics of the various UPS types. Some attributes of a UPS, like efficiency, are dictated by the choice of UPS type. Since implementation and manufactured quality more strongly impact characteristics such as reliability, these factors must be evaluated in addition to these design attributes.

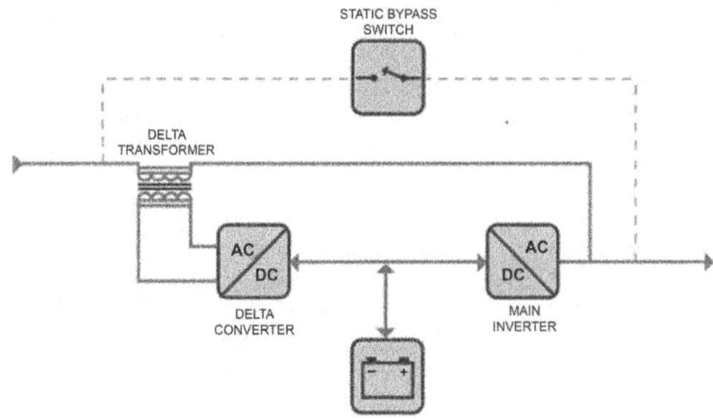

FIGURE 11.1-14 Delta Conversion On-Line UPS

Schneider Electric Data Center Science Center

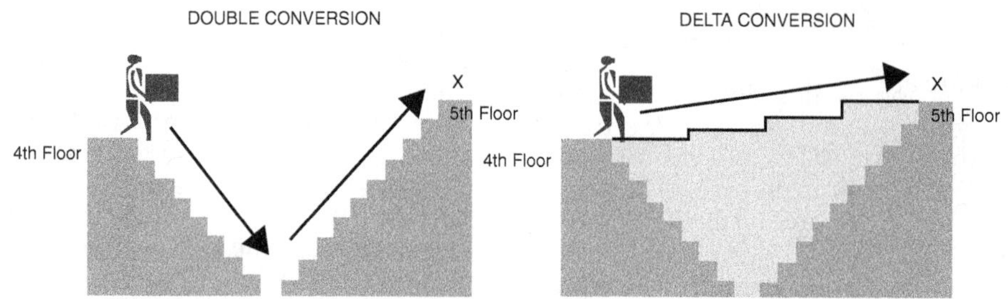

FIGURE 11.1-15 Analogy of Double Conversion vs. Delta Conversion

Schneider Electric Data Center Science Center

	Practical Power Range (kVA)	Voltage Conditioning	Cost per VA	Efficiency	Inverter always operating
Standby	0–0.5	Low	Low	Very High	No
Line Interactive	0.5–3	Design Dependent	Medium	Very High	Design Dependent
Standby On-Line Hybrid	0.5–5	High	High	Low	Partially
Standby-Ferro	3–15	High	High	Low	No
Double Conversion On-Line	5–5000	High	Medium	Low	Yes
Delta Conversion On-Line	5–5000	High	Medium	High	Yes

Schneider Electric Data Center Science Center

Use of UPS Types in the Industry

The current UPS industry product offering has evolved over time to include many of these designs. The different UPS types have attributes that make them more or less suitable for different applications, and the APC product line reflects this diversity, as shown in this table:

	Use in APC products	Benefits	Limitations	APC's Findings
Standby	Back-UPS	Low cost, high efficiency, compact	Uses battery during brownouts, Impractical over 2 kVA	Best value for personal workstations
Line Interactive	Smart-UPS, Back-UPS Pro, and Matrix	High reliability, high efficiency, good voltage conditioning	Impractical over 5 kVA	Most popular UPS type in existence due to high reliability, ideal for rack or distributed servers and/or harsh power environments
Standby On-Line Hybrid	Not used by APC	Excellent voltage conditioning	Impractical over 5 kVA, low efficiency, low reliability, high cost	Line interactive provides better reliability and similar conditioning at a better value
Standby-Ferro	Not used by APC	Excellent voltage conditioning, high reliability	Low efficiency, unstable in combination with some loads and generators	Limited application because low efficiency and instability issues are a problem, and $N + 1$ on-line design offers even better reliability
Double Conversion On-Line	Symmetra	Excellent voltage conditioning, ease of paralleling	Low efficiency, Expensive under 5 kVA	Well suited for $N + 1$ designs
Delta Conversion On-Line	Silicon, Symmetra MW series	Excellent voltage conditioning, high efficiency	Impractical under 5 kVA	High efficiency reduces the substantial life-cycle cost of energy in large installations

Schneider Electric Data Center Science Center

Conclusions

Different UPS types are appropriate for different applications, and there is no single UPS type that is ideal for all applications. With the variety of UPS topologies on the market today, these guidelines will help clear confusion about how each topology operates and the advantages and disadvantages of each.

There are significant differences in UPS design between available products on the market, with theoretical and practical advantages for different approaches. Nevertheless, the basic quality of design implementation and manufactured quality are often dominant in determining the ultimate performance achieved in the customer application.

Battery Supply

UPS systems typically are supplied with sufficient battery capacity to carry a critical load for periods ranging from 5 minutes to 1 hour or more. Longer backup time periods are usually handled by a standby diesel generator. Batteries require special precautions. For large installations, they almost always are placed in a room dedicated to that purpose. Proper temperature control is important for long life and maximum discharge capacity.

Four battery types are found in UPS systems, the first two being more common:

Sealed Lead-Calcium

A gel-type electrolyte is used that does not require the addition of water. There is no outgassing or corrosion. This type of battery is used when the devices are integral to small UPS units, or when the batteries must be placed in occupied areas. The lifespan of a sealed lead-calcium battery, under ideal conditions, is about 5 years.

Conventional Lead-Calcium (Wet Cell)

The most common battery type for large UPS installations, these units require watering and terminal cleaning about every 3–6 months. Expected lifetime ranges up to 20 years. Conventional lead-calcium batteries outgas hydrogen under charge conditions and must be located in a secure, ventilated area.

Lead-Antimony

Traditional lead-acid batteries, these devices are equivalent in performance to lead-calcium batteries. Maintenance is required every 3 months. Expected lifetime is about 10 years. To retain their capacity, lead-antimony batteries require a monthly equalizing charge.

Nickel-Cadmium

Advantages of the nickel-cadmium battery include small size and low weight for a given capacity. These devices offer excellent high- and low-temperature properties. Life expectancy is nearly that of a conventional lead-calcium battery. Nickel-cadmium batteries require a monthly equalizing charge, as well as periodic discharge cycles, to retain their capacity. Nickel-cadmium batteries are the most expensive of the devices typically used for UPS applications.

Batteries used in battery UPS systems are electro-chemical devices whose ability to store and deliver power is reduced as they age. No matter how well you maintain, store and use your batteries, they will still require replacement; it is just a matter of when. How often is a factor of the way they are applied, the conditions in which they are installed and care. If your battery UPS is installed in conditions where the temperature of the UPS battery exceeds the optimal temperature of the batteries (77°F), the battery life will be reduced.

The general service life of a standard valve regulated lead-acid (VRLA) battery is 3 to 5 years. A number of environmental, chemical and user-related factors can affect a battery's life. These are the primary mistakes to avoid and stretch the most life out of your UPS's battery.

FACTORS THAT CAUSE PREMATURE BATTERY FAILURE

Poor storage of unused batteries: A battery's lifetime begins to decrease when it is stored. That's because lead-acid batteries discharge small amounts of energy while being stored. To prolong a battery's life in storage, it should be charged every 3 to 4 months of storage. In a battery UPS that is connected to a power source, that is not an issue. But if that device or the battery for the device is stored for an extended time and you don't charge the battery, it will lose capacity in as little as 6 months. You can also prolong your unused battery's storage life by storing it at a temperature of 50°F (10°C) or less.

High ambient temperature: The rated capacity of every battery is based on an ambient temperature of 77°F (25°C). Variations in the form of increased or decreased temperature will impact the performance and lifespan of the battery. To put this in perspective, a general rule: For every 15°F above the recommended ambient temperature, the expected life of the battery is reduced by 50%.

Over-cycling: After a UPS operates on battery power during a power failure, the battery recharges for future use, an event called the discharge cycle. When a battery is installed, it is at 100% of its rated capacity. However, each discharge and subsequent recharge slightly reduces the capacity of the battery.

Improper float voltage: Every battery manufacturer will specify the charging voltage ranges for its own cell design. If a battery is consistently charged outside of these parameters, it can cause appreciable damage. This will reduce the life of the battery significantly.

Undercharging or continuous under-voltage charging (low battery voltage) will most often cause sulfate crystals to form on the battery plates. These crystals will eventually harden and reduce the available capacity of the battery over time.

The most dangerous event is overcharging. A typical UPS battery with a float voltage that is too high can cause excessive hydrogen and oxygen gases and can lead to internal drying out that, once accelerated, can cause thermal runaway—resulting in failure or even fire and explosion.

Incorrect battery application: UPS batteries are made specifically for UPS systems, just as other batteries are made specifically for their respective appliances. UPS batteries are built to deliver extremely high rates of energy for a short time. The installation of batteries in battery UPS systems that are not rated for that use is also dangerous.

Conversely, other batteries, such as telecom and switchgear batteries, are designed to run for longer periods of time, typically between 4 and 8 hours. If a user runs a telecom application with a UPS battery, it will force the battery to run for much longer than its intended purpose. This could cause the battery's plates to overheat and fail.

The most common failure mode of a VRLA battery is an open circuit, most often caused by one or more cells drying out. UPS systems typically have a *series-connected* battery system to provide a high current to the UPS system DC bus.

In "legacy" battery UPS systems (nonmodular and redundant battery UPS systems) if one cell in a string opens, it will break the current in the entire string. ***In short, just one failed battery cell can bring down your entire infrastructure unless the battery system is redundant and has the proper overload and safety systems.***

In a "nonlegacy" battery UPS. this susceptibility to UPS system failure and the implied downtime has been addressed. Modular and redundant battery UPS systems give you the option to build in multiple levels of redundancy. The term is "N + some number" that signifies the number of redundant levels within the system. The cost of this is also reduced with the modern systems available. Today, you have multiple 3 KVA on-line double conversion UPSs systems to choose from, and all will have N + 1 redundancy for a few thousand dollars. If you had searched the manufacturers in the mid-1980s, none would have a product that could meet that specification.

Regular Battery UPS Maintenance Checks Prevent Battery Failures and Downtime

Preventative battery maintenance includes visual inspection, battery voltage testing, thermal image scans and testing the torque on the battery terminal connections. If this process shows any irregularities, then it can be determined how critical the situation is and if battery replacement is required. Preventative battery maintenance should be considered battery life extension. It is very cost effective and should be considered mandatory in the broadcast industry.

Regular battery service and maintenance is critical in ensuring the reliability of a battery UPS. As with any device or system, regular maintenance is the key to protecting the investment and will foster long-term use without failure of the battery UPS. It will allow identification of an issue before it results in system failure.

The installation of battery UPS systems is intended to prevent downtime; however, lack of service and maintenance of battery UPS systems can be the cause of downtime. The benefit of the battery UPS is clear; the cost of that benefit includes proper maintenance. Lacking that, the system will just be an added point of AC power failure. The proper storage, care and maintenance of battery systems will help increase its longevity and will catch a failing battery before it fails.

Power Conditioners—Also Known as Line or Power Line Conditioners

These devices are installed to improve the quality of the power that is delivered to the equipment. These devices function to remove anomalies in the AC power that could be detrimental to the function of the connected devices. They deliver voltage at the proper level and improve the characteristic of the power delivered to the equipment they power. All offer some voltage regulations, power factor correction, limited impulse (transient) protection and noise suppression. Some models also have a voltage regulator (buck and boost feature).

The term "power conditioner" is often used but not well defined. Some manufacturers offer "plug & play" models that are designed for point-of-use applications. One instance could be electronic equipment that is powered by a circuit from an electrical panel that also supports inductive or noisy loads. The mixed loads powered by such a panel can be disruptive and cause distortion of the AC power supplied to a circuit on the panel that supports sensitive loads. The installation of a power conditioner may resolve the anomalies of the circuit and allow the connected equipment to operate without any issues.

Power conditioners vary in function to the point that the term can be misapplied. Some will improve or protect against multiple power anomalies. The purchase of these devices should include a detailed examination of the individual products being considered. Match your system and equipment needs with the features and benefits of the product you purchase.

The goal of any power delivery system is to meet the ideal specifications of AC sine wave power with an impedance of zero ohms at all frequencies. The ideal is never achieved in a power delivery system because there are factors beyond control. The list here will help you understand the issues that electronic equipment might face and assist you in making power conditioning choices. There is no one technology that is both effective and reasonable in cost. Application must be made case by case or site by site.

Peak or RMS Voltage Variations

Voltage swells: Events when the RMS voltage exceeds the nominal voltage by 10 to 80% for 0.5 cycle to 1 minute.

Voltage sag: This is the opposite of a voltage swell. A voltage sag (also called dip) is when the RMS voltage is below the nominal voltage by 10 to 90% for 0.5 cycle to 1 minute.

Voltage Flicker: This is random and or repetitive variation in the RMS voltage between 90 and 110% of nominal voltage levels. Flicker in lights is a visual indication of what the connected equipment is seeing. There is disagreement as to a detailed definition of flicker. Regardless of how it is defined, flicker is a combination of power-quality anomalies. Imagine trying to power broadcast equipment under flicker conditions.

Surges, spikes, impulses: These are the most common name for the abrupt, very brief increases in voltage. While most associate them with lightning events, the majority are caused by inductive loads being turned off, utility grid switching and accidents as well as by lightning.

Brownout: Occurs when the nominal voltage drops below 90% for more than 1 minute. The term "brownout" is used most commonly to describe the reduction in system voltage by the utility. It is also common when the utility or system operator creates a brownout in order to decrease demand or to increase system operating margins.

Overvoltage: Occurs when the nominal voltage rises above 110% for more than 1 minute. This can have catastrophic effect on connected equipment.

Frequency variations: The sinusoidal characteristics of AC power are the basis of design of all modern power supplies. Frequency variations can wreak havoc with the connected equipment and upset any and all networked devices. Variations in the frequency will dramatically impact induction motors. Generators that are not running at the proper speed will not provide power with a proper and stable frequency. Their ability to provide power with a stable frequency is critical to their value as a backup power source. The stability of the power system depends on the frequency.

Variations in the wave shape: Harmonics are the cause of most variations in the wave shape of AC power. At one time it was a major factor in electrical design. Today's tighter standards have all but eliminated the impact of harmonics.

Nonzero low-frequency impedance: This occurs when a load draws more power than the supply circuit can deliver. Most of these are short duration events because, after the inrush current requirements are met, the demand for current drops to within the normal current-carrying ability of the circuit. A good example of this anomaly is to see the lights dim when the air conditioner turns on.

Nonzero high-frequency impedance: This occurs when a load demands a large amount of current then stops demanding it suddenly. What was the "sag" in voltage when the air conditioner turned on becomes a surge, spike or impulse when it goes off. This is due to the inductances in the power supply line and the amount of demand placed upon the power supply line (circuit).

Summary

The susceptibility of electronic equipment to failure because of disturbances on the AC power line has been studied by many organizations. Most standards groups, industry groups and code organizations are composed of a multitude of voices. Many of the participants may have some agenda of their own, or their employer and their efforts are not fixed on the common good. Add to that the many opinions of the participants, and often their efforts are watered down, skirt the issues and are a compromise at best.

It is the responsibility of those involved in their facility to insure their facility's electrical infrastructure is sustainable and capable of supporting all loads properly. Backup systems must be proper and must be maintained.

The advances in technology have far outpaced the electrical systems of many facilities. The meeting to approve the purchase of new technology rarely includes a discussion of what changes, improvements or modifications of the facility's infrastructure, including the electrical system, will be required to properly support it.

There is a wide range of power backup and power conditioning equipment available. The need for or the selection of this equipment can be made only with accurate, documented data on the types of disturbances typically found on the AC power service to the facility in question. The protection equipment and systems must be chosen to match the problems. The installation and use of very basic power conditioning equipment may not be much better than operating equipment directly from the AC line. Sophisticated systems and equipment designed to shield the broadcast facility from every conceivable power disturbance may not be economically justifiable.

In addition to the equipment capital costs, it is necessary to consider the expenses associated with site preparation and installation as well as ongoing costs incurred as a result of the installation of power conditioning devices including proper maintenance. There will most often be changes to and cost associated with the air conditioning system(s) modifications. Therefore, consideration of any new technology will have broad implications.

For example, it is wise to prepare a complete life-cycle cost analysis of the protection methods proposed. The study may reveal that the long-term operating expense of one system outweighs the lower purchase price of another.

The amount of money a facility manager and management are willing to spend on protection from utility company disturbances and the other issues outlined in this article generally depends on a multitude of factors that relate to the location, geographic exposure considerations, stability of the utility and the cost of downtime. The cost of downtime should not be only the raw cost factor but should also include those costs associated with customer goodwill and the reputation of the station.

LIGHTNING HAZARD

The exposure to lightning damage in broadcast facilities cannot be overstated. Other than the cellular industry, no other industry other than utilities erects large metal objects hundreds of feet into the sky in lightning-prone environments. The towers in the broadcast industry have "hit me" written all over them. That does not imply that the towers and facilities cannot be protected from lightning damage. The protection level possible is not absolute; some damage is always possible. However, almost all damage can be mitigated. To do this, a properly designed and installed lightning protection system must be installed. This will include a proper path to low-resistance, low-impedance earth grounding (earthing). It will also require proper ground bonding of all networked or associated devices and the installation of diversion and protection devices.

Antennas and other tower-mounted devices are damaged on occasion as a result of a direct equipment lightning strike. It should be noted that equipment mounted at the top of towers is exposed to damage. Don't let anyone convince you that lightning only strikes the tallest object—it does not. Lightning's path can be a random path that you may not expect. The key to avoiding most of the damage lightning can cause is to give it a path to earth that avoids your equipment and systems. To accomplish this, a robust properly engineered lightning protection system(s) will be required. Chapter 9.2 "Lightning Protection for Towers" addresses the proper techniques for minimizing such risks, and Chapter 11.2 "Facility Grounding Practice and Lightning Protection" provides additional advice on proper facility grounding.

It is important to understand the standards and codes such as NFPA 780, UL 96, UL96A and LPI 175 should be considered minimums to be exceeded by the broadcast industry. These "minimums" should never be the final design and installation standard for any broadcast facility.

Exceeding the codes and standards will increase facility life safety, reduce equipment damage and downtime and provide a high return on investment. Lightning protection systems should be installed to what the author will call a best industry standard. There are elements in NFPA 780 and UL 96A that can be improved significantly. As an example, both standards allow 90° conductor intersections in one section of their standards and do not allow the same intersections in another section. The 90° conductor bends are not "best industry standards." Grounding is not well defined because there is no performance testing required under the standards and codes.

The key elements of avoiding lightning damage with a lightning protection system are:

- Properly sized and positioned air terminals are installed.
- Well-insulated, low-impedance and low-resistance down conductor path.
- Deep earth ground rods that have a 5-ohm resistance.
- The down conductor path is very low resistance and impedance.
- All mechanical connections are visible or installed in inspection wells.
- Ground rods with test wells that will accommodate clamp-on ground rod test equipment.
- An interconnecting counterpoise bonding all grounding to a single reference for the entire facility, and tower counterpoise, if one is installed.
- Proper bonding, following best industry standards of all equipment & systems.
- Inspection and maintenance of all the installed components of the lightning protection system.
- Numerous steps and tests are required to insure everything is properly installed and in place.

Figure 11.1-16 shows the isokeraunic map of the United States, which estimates the number of lightning days per year across the country. On average, 30 storm days occur per year across the continental United States. This graphic does not consider population density. The data also does not fully describe the lightning threat because many individual lightning flashes occur during a single storm.

There are no maps that outline the dollar amount of damage resulting from lightning. However, there is another source of data that the author suggests may parallel the financial impact of lightning.

It is logical to believe population density is a good measure of the economic impact of lightning when it is combined with the strike density. Therefore, the economic impact of lightning may well follow the data as shown in Figure 11.1-17.

The structure of a facility, construction type and elevation of the structure as well as the elevation of the site impact lightning strike frequency. Higher structures in more elevated locations tend to be subject to more frequent lightning events. Because storm clouds tend to travel at specific heights above the earth, conductive structures in mountainous areas more readily attract lightning activity. The exposure factor is a function of the size of the facility and the isokeraunic rating of the area.

The frequency of power problems is seasonal in nature. As shown in Figure 11.1-16, most problems are noted in Southeastern and Midwest areas of the United States. These high problem rates can be traced primarily to increased thunderstorm activity during the summer months, as shown in Figure 11.1-18.

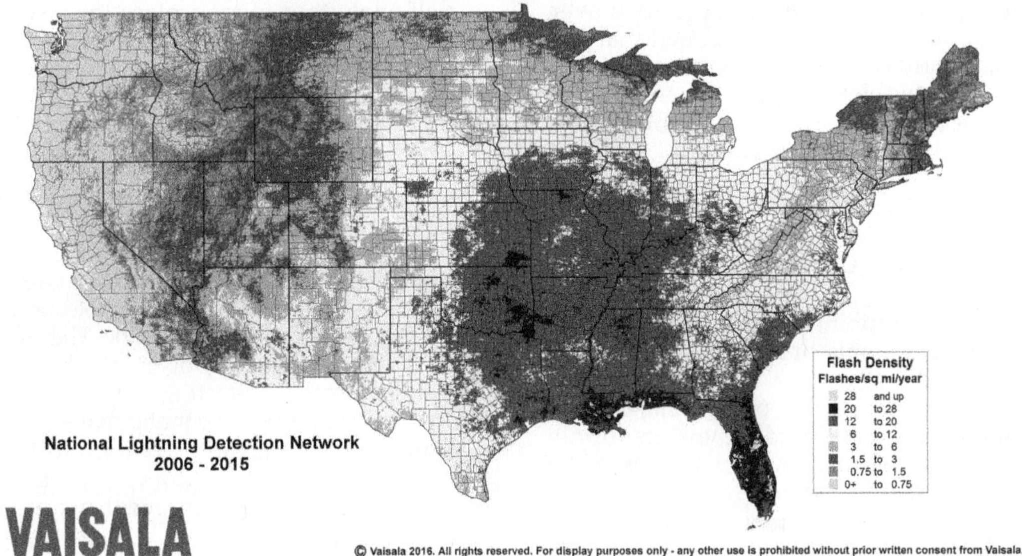

FIGURE 11.1-16 Estimated Number of Lightning Days Using Isokeraunic Map of United States

Vaisala Inc.

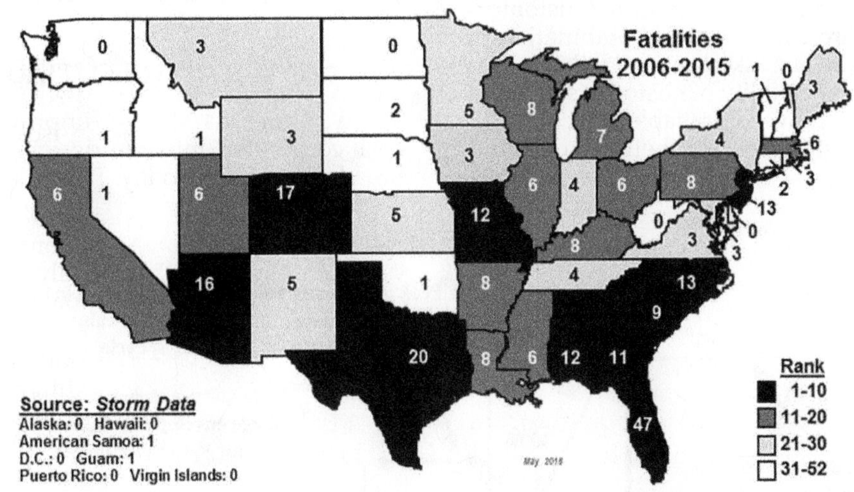

FIGURE 11.1-17 U.S. Map Depicting Lightning Related Fatalities from 2006–2015

Vaisala Inc.

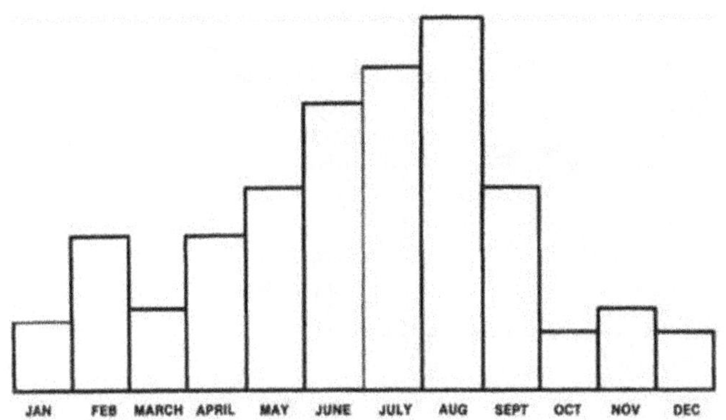

FIGURE 11.1-18 Monthly Depiction of Lightning Storm Frequency

FIPS PUBLICATION 94

In 1983, the U.S. Department of Commerce published a guideline summarizing the fundamentals of powering, grounding and protecting sensitive electronic devices. The document, known as "Federal Information Processing Standards Publication 94" (FIPS Pub. 94), was first reviewed by governmental agencies and sent to the Computer Business Equipment Manufacturers Association (CBEMA) for review. When CBEMA approved the document, the data processing industry was provided with an important guideline for power quality.

FIPS Pub. 94 was written to cover automatic data processing (ADP) equipment, which at that time constituted the principal equipment that was experiencing difficulty running on normal utility-supplied power. Since then, IEEE Standards have been issued that further outline the conditions that are required for the operation of all sensitive electronic equipment, including computer-based broadcast systems. IEEE 1100 is the benchmark when powering any and all sensitive equipment.

The CBEMA curve has been used as a guideline for its members in designing power supplies for modern electronic equipment. The CBEMA curve is shown in Figure 11.1-19.

The curve is a susceptibility profile. In the center of the chart is the acceptable operating area, and on the outside are areas that outline voltage levels that will cause equipment damage or operational failure. The CBEMA guideline states that if the voltage supply stays within the acceptable area given by the curve, the sensitive load equipment will operate as intended.

Figure 11.1-21 shows two different views regarding the cause of power-quality issues. In Figure 11.1-20, the actual events are outlined by EPRI (Electric Power Research Institute). That date does not match the customers' or utilities' perception. What is sure is that not all problems can be attributed to any one source. Some are from accidents, Mother Nature, customers and utility at the source. Others are a combination of one or more cause. The one takeaway from this is that in the end it is the responsibility of customers, the end users, to assure their power is reliable and of sufficient quality to meet the needs of their systems and equipment.

Self-analysis is a good place to start; however, it is wise to confirm the findings using the services of a professional. Often that professional is available from the utility or a supplier at little or no cost. What you don't want is to "assume," as all know where that can lead.

ADDITIONAL CONSIDERATIONS

The broadcast industry has yet to fully convert to digital technology. As such, broadcast transmitters using inductive output tubes (IOT) have a protection circuit in the event of an arc-over inside the tube. The arc-over grounds out the high-voltage DC power supply within a few milliseconds, thereby protecting the tube. This looks like a short-circuit to any source that powers this type of load.

This must be considered when sizing and selecting power-protection equipment. As a rule of thumb, oversizing the source to prevent frequent transfers to bypass can accomplish the desired result. Some systems can be set up to delay this transfer to bypass, thereby keeping the load protected during these arc-over events. Testing should be performed to simulate this event and repeated on an annual basis to assure it does not impact on-air time or damage systems.

THE FUTURE

There numerous design improvements which can be incorporated into the design and construction of broadcast facilities today. The list of broadcast facilities

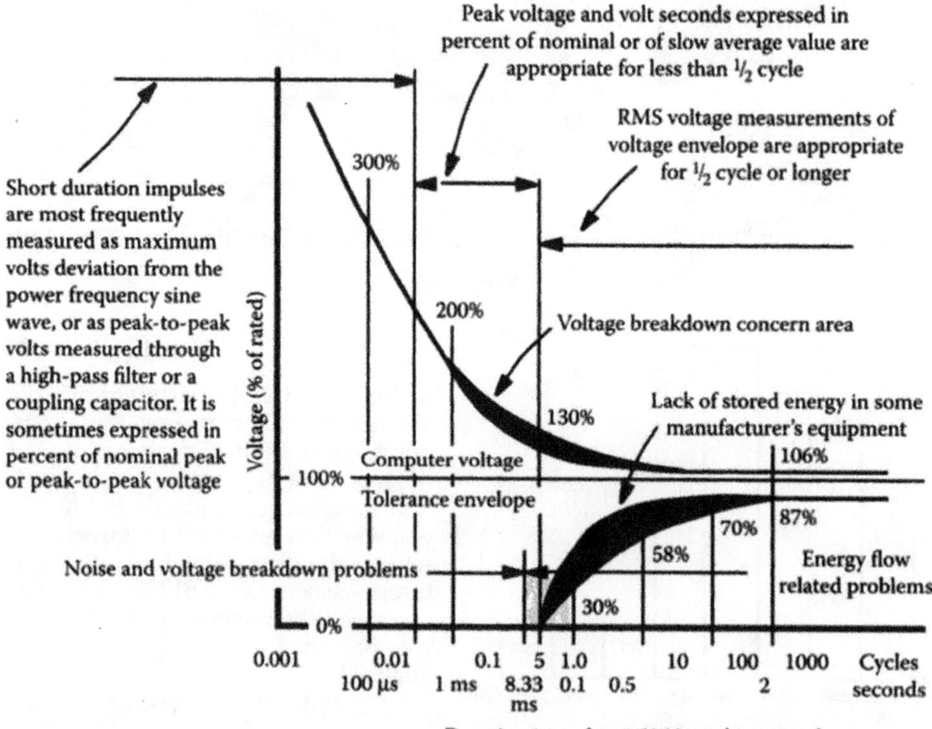

FIGURE 11.1-19 Suitability of Computer Equipment Over and Under Voltages

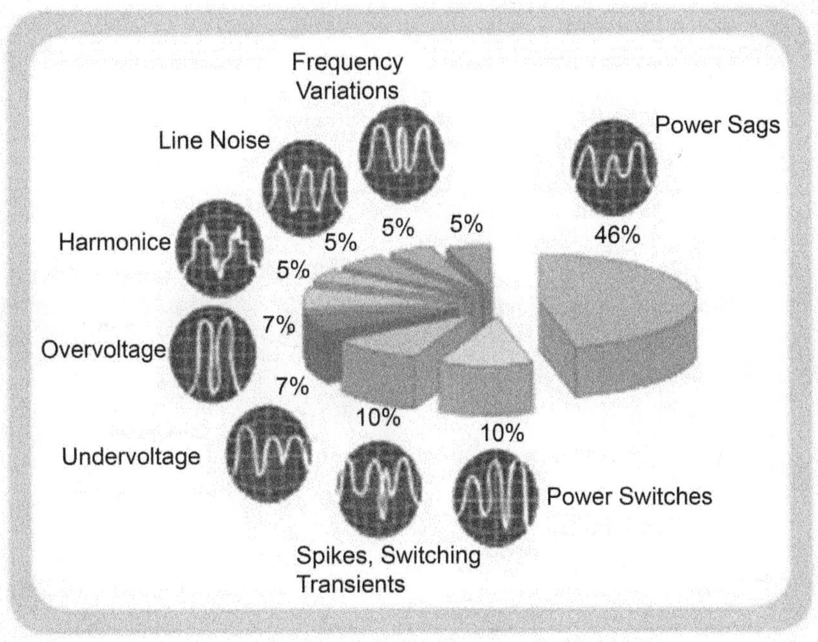

FIGURE 11.1-20 Numerical Data of Power-Quality Issues

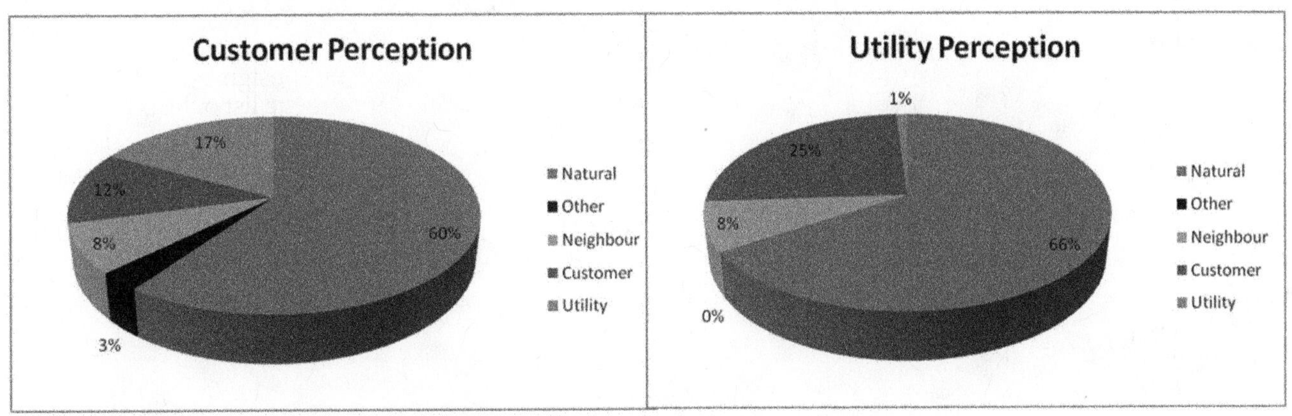

FIGURE 11.1-21 Who Is Responsible For Power Quality?

that can be improved includes studios, production facilities, transmission facilities, towers and administration facilities. The improvements have an excellent ROI, increase the facility flexibility, have no downside and improve the facilities' capability to support future technology.

Recommended Reading

AEMC Instruments. *Understanding Ground Resistance Testing*. 15th ed. AEMC Instruments, Chauvin Arnoux, 2016.

Barnett, Randy. "Power Quality Measurement and Analysis Basics." *EC&M*, March 1, 2012. http://ecmweb.com/power-quality/power-quality-measurement-and-analysis-basics

Bollen, Math H., and Irene Gu. *Signal Processing of Power Quality Disturbances*. Hoboken, NJ: Wiley-IEEE Press, 2006.

Ingale, Rajesh, and Laxman Tawade. "Detection and Comparison of Power Quality Disturbances Using Different Techniques." *International Journal of Computer Applications*, vol. 75, no. 18, August 2013.

Melhorn, Christopher J., and Mark F. McGranaghan. *Interpretation and Analysis of Power Quality Measurements*. IEEE, 1995.

Power Standards Lab. "Free Power Quality Teaching Toy." www.powerstandards.com/PQTeachingToyIndex.php.

Waggoner, Ray. "Conducting a Power Quality Site Analysis." *EC&M*, Sept. 1, 1995. http://ecmweb.com/content/conducting-power-quality-site-analysis

Credits

EPRI 2001. "Study of Typical PQ Phenomena. EPRI's Consortium for Electric Infrastructure for a Digital Society. Madison, WI, 2001.

Feldstein, Stuart A. "Research Commentary," *SMR Research*. Hackettstown, NJ. http://commbuildings.com/ResearchComm.html.

Fluke. Numerous articles on the Fluke website. http://en-us.fluke.com/training/training-library/?jobtype=15591&searchKeyWords=power+quality&page=1&filtLevel1=&selectedTab=videos

Georgia Power Co. "Results of a survey on the cause of Power Quality Problems." Chapter 1, Figure 1.1. *Electrical Power Systems Quality, Second Edition*. New York: McGraw-Hill, 2004. ProfessionalEngineer/Downloads/Electrical_Power_Systems_Quality_Second.pdf.Melhorn, Christopher J., and Mark F. McGranaghan. *Interpretation and Analysis of Power Quality Measurements*. IEEE, 1995. www.vicmyers.com/wp-content/uploads/2013/09/Power-Quality-Measurements.pdf.

PowerLogic. "Power Quality Analysis for Utilities." www.powerlogic.com/literature/3000BR1109%20Power%20Quality.pdf

Snyder Electric. "Power Quality Diagnosis Sample Report." SchneiderElectric (Hong Kong) Limited, 2005. www.schneider-electric.com.hk/documents/services-samples/PowerQualityDiagnosisSampleReport.pdf.

CHAPTER

11.2

Facility Grounding Practice and Lightning Protection

JEFF WELTON
Nautel Limited

INTRODUCTION

This chapter is intended to discuss facility grounding with an eye toward lightning protection and safety of personnel. Note that it is not intended to address electrical grounding, which is thoroughly covered by the National Electric Code (NEC) in the United States, as well as other electrical codes throughout the world, and which requires the services of licensed electricians/electrical engineers to properly interpret and implement. However, there will be a couple of references to NEC and to areas that are relevant to both disciplines.

GROUNDING VS. BONDING

Note that there is a significant distinction between these two terms. For the purposes of the NEC, grounding is the connection of something to earth, where bonding is the connection of two pieces of current carrying material together, whether through mechanical connections (conduit to equipment, or safety wires to ground lugs, for example) or through electro-mechanical connections, such as brazing, exothermic welding (cad welding) or soldering of ground conductors to each other or to ground rods.

For the purpose of this discussion, both mechanical and electromechanical types of bonding will be covered, although the preference is for an electromechanical connection wherever possible, for reasons that will be explained in detail.

National Electric Code (NEC)

It is important to note that NEC Article 250.4 (2011 edition) specifically states that earth is not an effective ground-fault current path. This is valuable information, as many inspectors and power company employees will frequently disagree with bonding of the AC entrance ground to a facility ground. However, per Article 250, these two systems most certainly should be bonded to provide a low-resistance path for any objectionable current. In the case of a radio or television transmission facility, this objectionable current could easily be created through a lightning strike, creating a hazard to personnel and posing a risk to equipment if the facility ground is not properly bonded to the AC electrical entrance ground.

Several resources exist to clarify this section of the NEC, and further discussion is beyond the scope of this chapter, but it is definitely worth remembering that, especially in a transmitter site, entrance ground and the station reference ground not only should, but must, be bonded together in order to maintain compliance with the national code. Note also that local and state codes may also apply. These circumstances are best handled by a licensed electrician well versed in all applicable codes for the specific location.

ANATOMY OF A LIGHTNING DISCHARGE

Before venturing into the physical steps to take in order to provide the best possible grounding for lightning

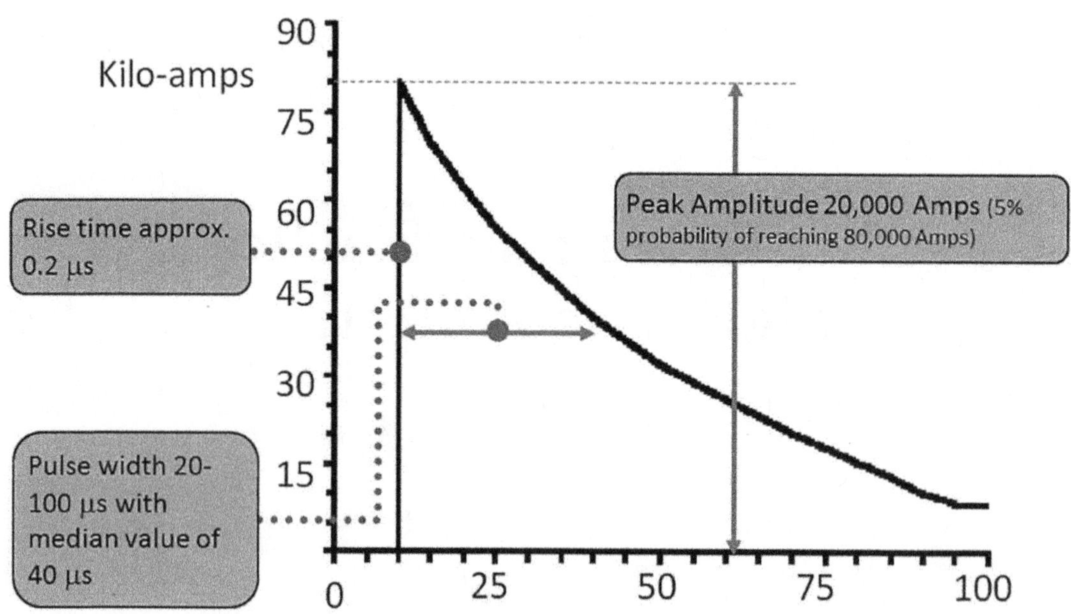

FIGURE 11.2-1 Anatomy of a Lightning Strike

Nautel Limited

protection purposes, it is useful to understand the science of a lightning discharge.

As indicated in Figure 11.2-1, a lightning strike has a very fast rise time—in the order of 0.2 to 12 microseconds, with a slower decay time having a median value of 40 microseconds. The average strike will have an amplitude of 20,000 amperes, with a 1% possibility of reaching 200,000 amperes. This rise time corresponds to a nominal frequency range of 83 kHz to 5 MHz, with the majority of strikes being in the 1 MHz area. The decay time corresponds to a nominal 25 kHz.

It is also worth mentioning that the taller a tower is, the more likely it is to get hit by lightning—and that towers, whether AM, FM or TV, contain (or are) antennas, which have a resonant frequency. It has been discovered that, during the rapid energy transfer of a lightning strike, antennas can frequently resonate briefly, placing a ringing on the strike pulse at the nominal frequency of resonance.

This means that the lightning strike contains components near the audio band in the SCA through AM bands and can contain components at the resonant frequency of the broadcast equipment on the site—in short, everything that the system is designed to pass is contained within the strike pulse. Thus, proper grounding becomes critical, especially with solid-state systems containing multiple devices, as well as switching power supplies. However, it is also important for older sites, since multiple strikes can shorten lifespan of other devices within a tube transmitter, as well as having an impact on tube life.

SINGLE POINT GROUNDING

While it is impossible to prevent lightning strikes, many things can be done to help to direct the lightning energy away from the equipment within the facility by providing a lower impedance alternate path to ground, where the energy can be safely dissipated. This is done by means of single-point grounding, also called star grounding.

Effectively, the premise of single-point grounding is that if there is only one ground path from any device in a chain to the station reference ground point and that an attractive (low-impedance) discharge path is provided, there will be no current flow through the equipment in question. It is virtually impossible to establish a pure single-point ground in a transmitter, for example, where the equipment is sitting on a somewhat conductive concrete floor, perhaps touching other racks, with conduit for electricity, control and audio connected, as well as an RF coaxial cable and the electrical safety ground. However, it is possible to ensure that the other connections are less attractive to lightning energy and the preferred ground is more attractive.

The first step is to ensure a low impedance ground path. Based on the rise time and potential ringing on a lightning strike, skin effect will come into play, so the greater the surface area of any grounding conductor, the better. This is where copper strap is a very good thing. Next is ensuring the connections made (from one strap to another, from strap to ground rod, etc.) stay reliable for as long as possible. Brazing, soldering or exothermic welding can ensure that connections stay low impedance without risk of being degraded by loose hardware, by copper compressing and creating poor connections over time, or by oxidation affecting conductivity.

At many sites, grounding is done with stranded copper cable and connected to ground rods using ground rod clamps such as the type shown in Figure 11.2-2. The problem with these is that copper is malleable and a stranded cable will compress over time, becoming loose in the clamp—sometimes to the point where it can be simply pulled out of the clamp. Thus, clamp connections need to be tightened on a regular basis, until the copper has compressed to the point where all of the air has been displaced. A preferred alternative would be either brazing or exothermic (cad) welding of the cable to the rod. This provides a more permanent low-resistance connection without the worry of it loosening over time (see Figure 11.2-3).

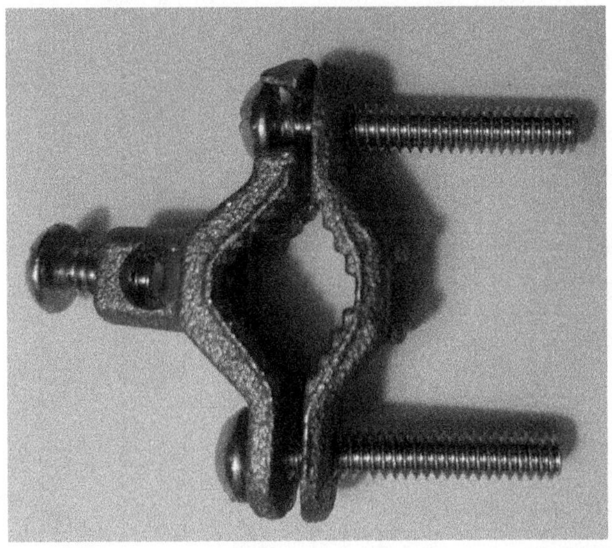

FIGURE 11.2-2 Typical Compression Type Ground Rod Clamp

J. Welton

FIGURE 11.2-3 Repairing a Loose Ground Connection Permanently

J. Welton

The next consideration is to minimize impedance of the lightning protection ground, using copper strap where possible, or the heaviest gauge wire available where strap is not feasible. Try to keep grounding runs as straight as possible (shorter length translates to less impedance; sharp turns can add inductance, thus higher impedance). Establish a station reference ground point to which all ground conductors within the building to connect, with a single conductor (again, strap is preferred) exiting the building to the ground rod(s). A bulkhead panel near the point where the coaxial cable exits the building would be an ideal solution, with the caveat that each ground run should be as short as possible (no loops) and each connection should be made to a separate point on the bulkhead panel (see Figure 11.2-4).

A point worthy of note is that there are many facilities equipped with "halo" type grounding systems, where there is a ring of copper around the inside perimeter of the building, frequently with a grounding conductor exiting the building at each corner to a ground rod. These are perfectly acceptable but must have all external ground rods connected together (preferably with strap), in order to minimize resistance between the ground rods. As well, it is critical that the single-point system described earlier be observed when connecting equipment ground points to the halo, so as to avoid any ground loops, where lightning energy could be split and control lost over the direction energy can go to dissipate.

Finally, increase the inductance of nonlightning protection ground connections (audio shields and electrical wiring, for example). This can be done by positioning equipment so that these runs are longer relative to the ground conductor. Since the ground conductor is a much bigger cable/strap, there will be gains there as well. Also, adding ferrite toroids to nonlightning protection related connections can increase

FIGURE 11.2-4 Ground Buss for Coaxial Cables

John West, Power & Systems Innovations of Tampa Inc.

their effective impedance to high-frequency common mode energy (current on either ground or feed cables but not both).

SURGE PROTECTION

Surge protectors come in many shapes and sizes, some better than others. The two most common surge protector devices encountered contain either MOVs (metal oxide varistors) or SASD (silicon avalanche suppression diodes) as the protection devices. For conventional facility protection, either is acceptable, although the faster reaction time of the SASD devices makes them better suited for areas with large quantities of highly sensitive electronics, such as server rooms. However, as switching power supplies and processor-controlled transmitters become much more commonplace, it certainly is worth considering them at the transmitter site, as well.

In addition, there are both series and shunt types of surge protectors. A shunt-type unit contains MOVs or SASDs that are connected between AC lines and ground, with each leg being fused (see Figure 11.2-5). Series-type surge protectors are connected between the AC mains and the device being protected. They will contain MOVs or SASDs from each AC line to ground at both line and load terminals and will usually have a series choke as well, for additional protection from AC-related transients.

In both cases, whether MOV or SASD, series or shunt, the protection devices have a knee, or clamping voltage, that is set to be above the level of the AC mains voltage for which they are designed to be connected. Once that knee voltage is exceeded, the device will conduct, passing excess energy through the device. Note that these devices typically have a short-circuit failure mode, so having a fuse in series with them is essential, and such fuses should never be replaced with a device that does not have the same

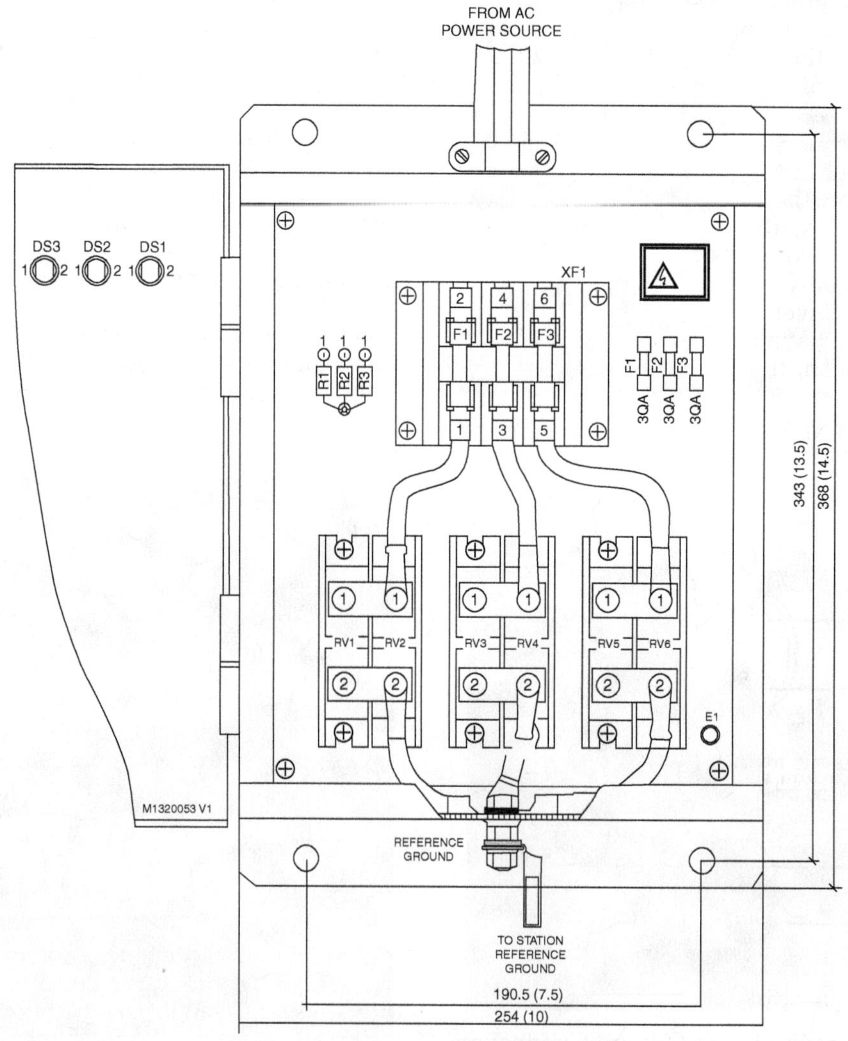

FIGURE 11.2-5 Example of a Shunt Type Surge Protector

Nautel Limited

current rating and reaction time. In addition, MOVs have a cumulative decay effect, meaning that they can pass a fixed amount of energy before failure—that level of energy will typically handle many strikes, but at some point they will fail and the fuse will blow—typically either extinguishing a status light or illuminating an alarm, depending on the design of the unit.

Less expensive surge protectors may contain MOVs but may have a small gauge wiring harness for connection to the mains entrance—remember that the smaller the gauge of wire being used, the higher the inductance, so these are less preferable for facility protection, although better than no protection at all.

As a rule, since wiring between the surge protector and the mains panel will be a higher inductance than the connection between the surge protector and the station reference ground, it is desirable to locate the surge protector as close as possible to the AC entrance panel. For shunt-type protectors, a separate breaker can be installed (rated at the next size above the protector's fuse rating). This allows the surge protector to be disconnected from the AC mains for service without the need to remove power from any other equipment in the facility and is probably the biggest advantage of shunt-type protectors over series units.

A common misconception with surge protectors is that they are a unidirectional device—protecting only from surges on the AC mains. Even in a very well-grounded facility, there will be some ground resistance from one end of the property to the other. If that resistance is only 0.1 ohm, for a median lightning strike on the tower at an amplitude of 20 kA, Ohm's law dictates that a potential of 2000 volts will exist. In this situation, the surge protection devices will clamp once the potential on the ground conductor rises to the level of their clamping voltage above the AC main's voltage. Once they begin to conduct,

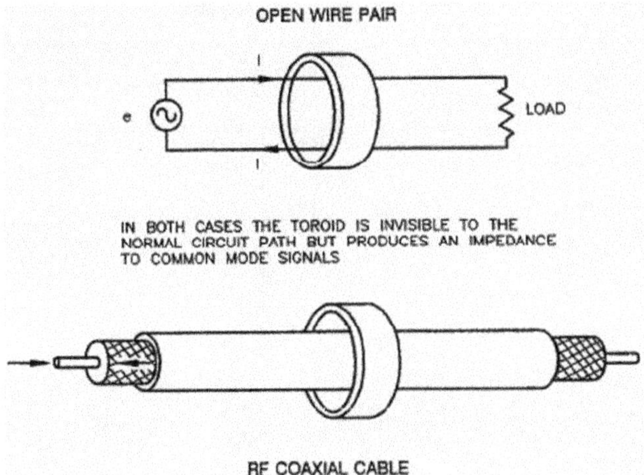

FIGURE 11.2-6 Use of Ferrite Toroids to Impede Common Mode Signals

Nautel Limited

excess energy can be passed through the protection devices to dissipate to ground via the AC lines; thus, proper surge protector selection and installation is important even at sites with well-regulated line voltage.

FERRITE TOROIDS

One aspect of lightning protection that is frequently overlooked is the ferrite toroid. A toroid is a ring of compressed carbon and ferrous material (frequently iron) that is often used as a core for winding transformers. However, when placed over a group of conductors containing equal feed and return currents, such as AC wiring or a coaxial cable, toroids are a good tool for assisting with lightning protection. An important characteristic of toroids is their permeability—effectively, the higher the permeability, the more effective the toroid will be for lightning protection, as it will saturate at lower frequencies. Since in this application the toroid is being used to protect against common mode surges (surges that create an imbalance between feed and return currents), having a saturation frequency below the broadcast frequency (on a transmitter RF output or STL antenna line, for example) is not an issue.

Essentially, when a toroid is placed over multiple conductors having equal feed and return currents (see Figure 11.2-6), it is completely inert and has no effect on the signal being transmitted, whether AC, RF or somewhere in between. However, if a common mode surge occurs on any conductor, creating a net current flow above zero, the ferrite will saturate and produce a magnetic field. This field then creates a matching potential on the remaining conductors, resulting in the feed and return conductors charging by the same amount, negating the effect of the surge. In addition, in the process of saturating, the ferrite acts as a choke, presenting a higher impedance to the surge energy. In this manner, a ferrite can be very effective at helping to prevent lightning energy from reaching the equipment being protected, both by minimizing the surge and by presenting a higher impedance so that the surge energy is more likely to pass through the surge protector, either to ground in the event of an AC transient or to the AC lines in the event of a tower strike.

Ferrite toroids should be installed on any cable connected to a device that could be damaged by lightning energy, including coaxial and power cables, as well as audio cables, remote control wiring and telephone or data lines. Assuming no ground loops or current imbalances in the site layout, ferrites have no effect on desired signals. However, in a site with less than optimal grounding, they can also provide a troubleshooting tool to indicate current imbalances due to ground loops, as any difference between feed and return current will cause them to saturate and can cause them to heat up. This is especially useful for finding ground loops in AM transmitter sites, where there may be a coaxial cable from the transmitter to the phasor, then several coaxial cables to multiple antenna tuning units or towers, with current imbalances resulting from either too many ground connections or a broken/ missing ground conductor.

DESIGNING A NEW FACILITY

When designing a new facility, the principles outlined in preceding pages can be summarized quite simply: Practice single point grounding, keep ground conductors as short as possible, bond everything that can reasonably be bonded with an electromechanical connection rather than just a compression-type mechanical connection, have a properly installed surge protector and use ferrites. Figure 11.2-7 shows a simplified schematic of a single-point grounded installation with these principles applied.

Note that, in a well-designed installation, the ground rod at the tower and the one at the building will also be bonded together, preferably with copper strap. The ferrite toroids help to isolate the transmitter from any surge energy that makes it into the building, and the surge protector allows excess energy to pass, dissipating on the AC lines. Figure 11.2-8 shows a pictorial view that depicts the path that lightning energy will tend to follow.

Note that both Figure 11.2-7 and Figure 11.2-8 depict AM transmitter sites, as shown by the spark gaps and the antenna radials; however, the concepts remain the same whether the site is AM, FM, TV or any other radio frequency transmitting facility. Figure 11.2-9 shows an overhead view of the transmitter room, including a ground ring around the building.

There are situations where ideal installations are not as easily obtainable. For example, many FM transmitting facilities are either in high rise buildings or on mountaintops, where obtaining a good connection to earth ground is more difficult. In many high rise buildings, the best ground that can be obtained is the building structure. This could require the involvement of structural engineers and the building owner to ensure that the integrity of the structure is not compromised in the process of bonding a ground conductor.

On mountain sites, where earth ground could have a very high resistance, sometimes all that can be done is to improve conductivity as much as possible by installing multiple ground rods or using chemically augmented ground rods. Chemically augmented rods are hollow rods that contain a conductive material. When buried and backfilled with more conductive material, they can help to improve ground conductivity. Something to be aware of when installing multiple ground rods is spacing—ground rods are most effective when they are at

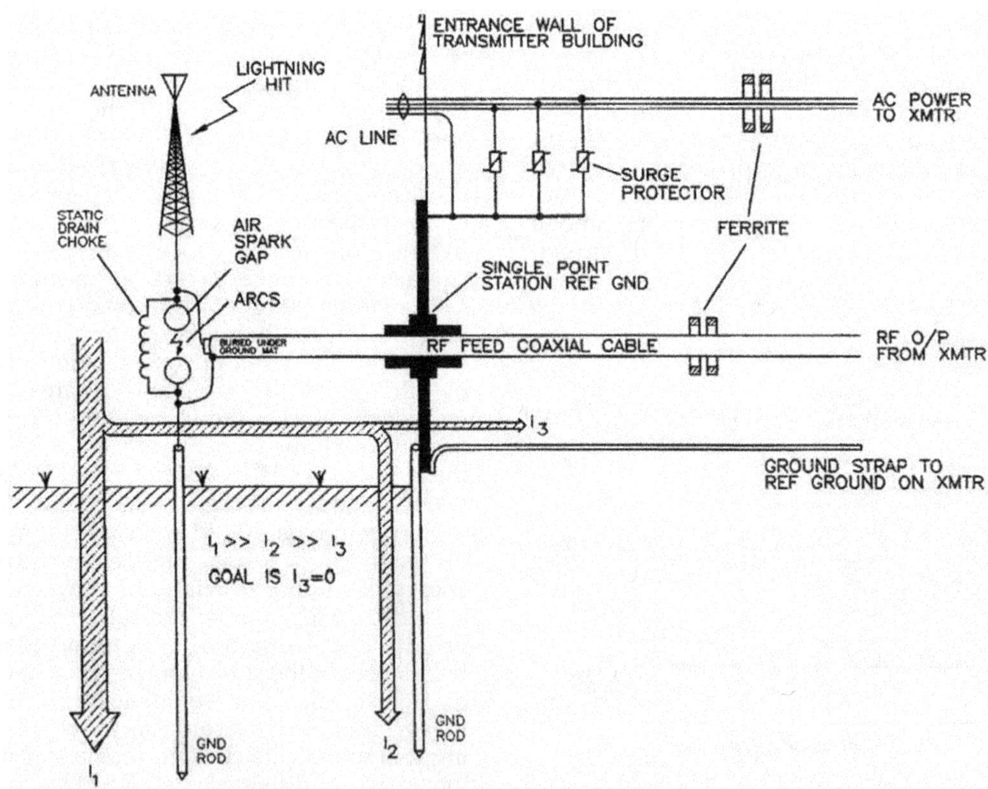

FIGURE 11.2-7 Simplified Schematic of Single-Point Ground

Nautel Limited

least double their length apart. For example, 10-foot ground rods should be spaced at least 20 feet from each other. Ground rods have a hemispherical area of influence, with the radius of the hemisphere equal to the length of the rod. Putting them too close together reduces the effectiveness of this hemisphere and can actually reduce the amount by which they improve ground conductivity. Two rods are almost always better than one—but two rods too close together are not nearly as good as two rods properly spaced.

An additional note on single-point grounding, whether for new installations or existing: In some cases, it is difficult to establish an overall single-point ground, such as a transmitter facility with multiple transmitter rooms or a studio with several production booths. In these situations, it is usually best to establish

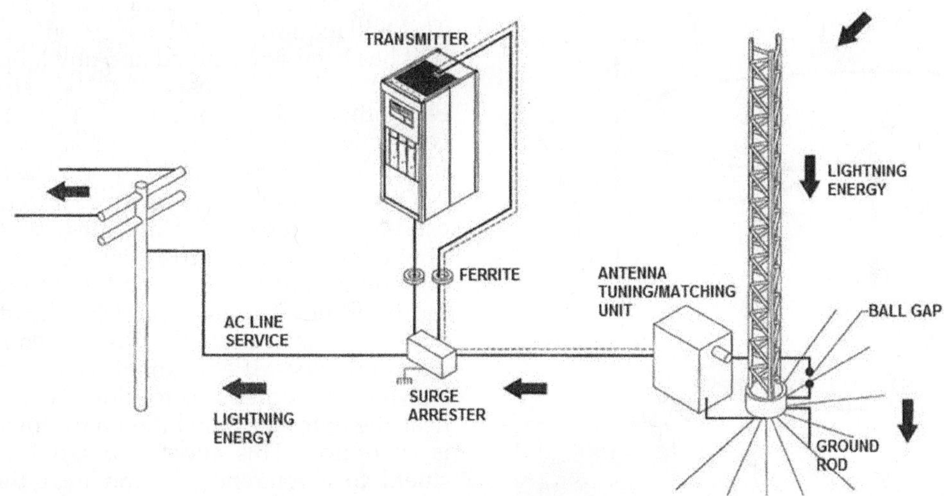

FIGURE 11.2-8 Simplified Overview of Single-Point Ground

J. Welton

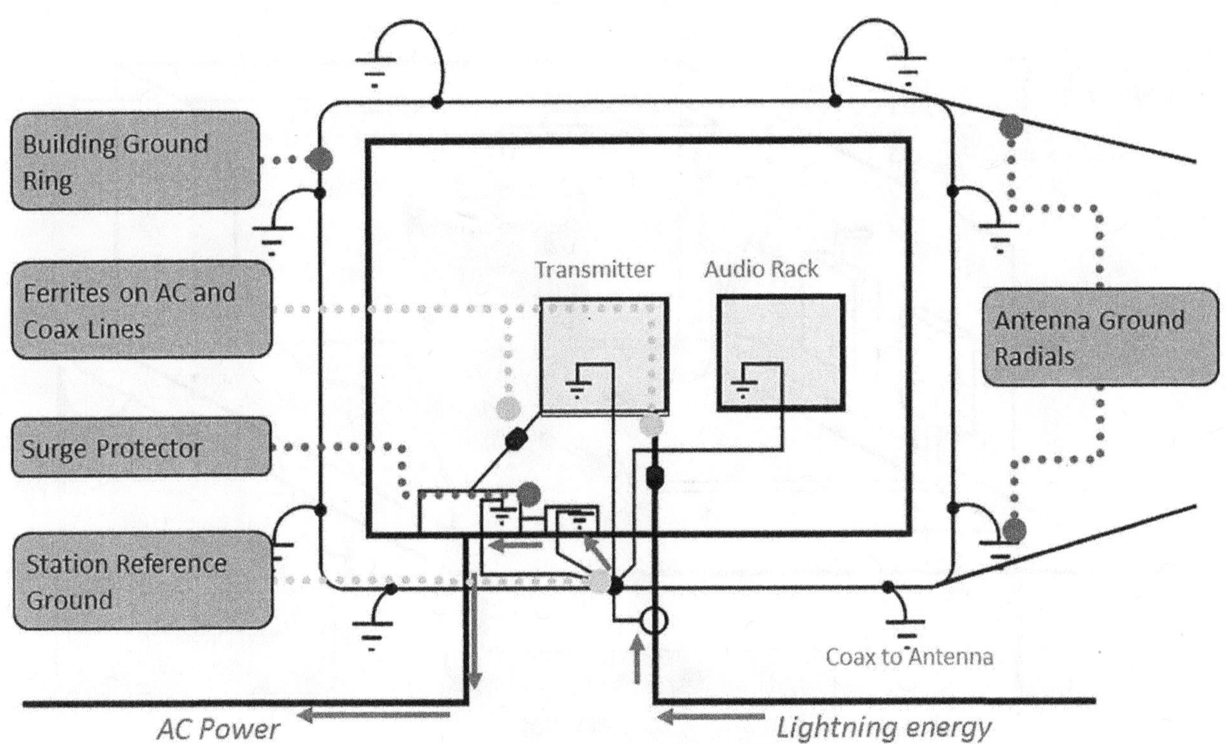

FIGURE 11.2-9 Ideal Facility Layout

J. Welton

a single-point grounding layout for each room in the facility, then to individually connect each of these rooms to a single-reference ground point for the facility as a whole. Ideally, surge protection and coaxial cable grounds will all be brought together at this point; however, sometimes compromises may need to be made and the transmitter room treated as if it were the entire facility.

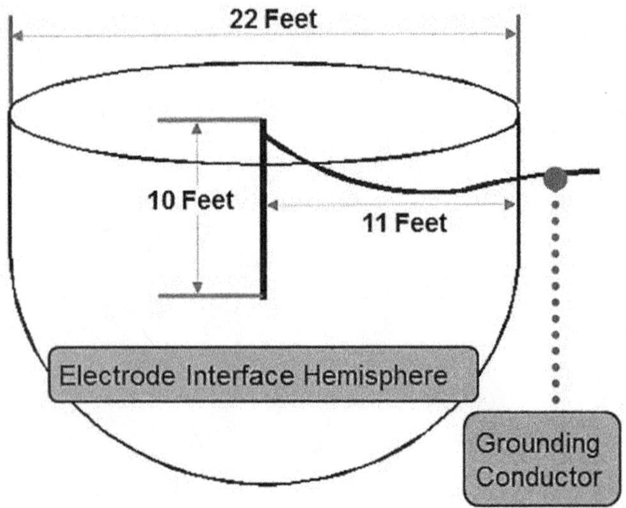

FIGURE 11.2-10 Ground Rod Interface Hemisphere
Nautel Limited

IMPROVING AN EXISTING FACILITY

In some cases, doing a full rebuild of a grounding/lightning protection system is simply not possible. In these cases, there are frequently things that can be done to improve the situation without totally rebuilding the facility. See Figure 11.2-11 for reference.

First and foremost, it is imperative to know where all ground conductors go—whether they are connected in a giant ground loop, or if they are even connected at all. This will require a visual inspection in most cases. Once that has been determined and any loops removed, any connections can be inspected—two straps folded over each other and hammered flat are not a good ground connection, even if they are screwed to the floor. Improving these would be the next step.

Once ground straps are optimized, cable routing can be considered. In many sites, AC power comes in at one end, from the road, and the coaxial cable exits the building at the other end, to the tower in back of the building. This is a far from ideal scenario, as it adds undesirable length (thus impedance) to grounding conductors. If additional coaxial cable or hardline is available, sometimes routing the transmitter output near the entrance panel then back toward the antenna is an option. This allows for grounding of the coax shield to a reference ground near the AC entrance, minimizing grounding conductor impedance.

In addition, ferrite toroids can almost always be added if not already in place. Ferrites for large diameter coaxial cable can be expensive, but if the site has

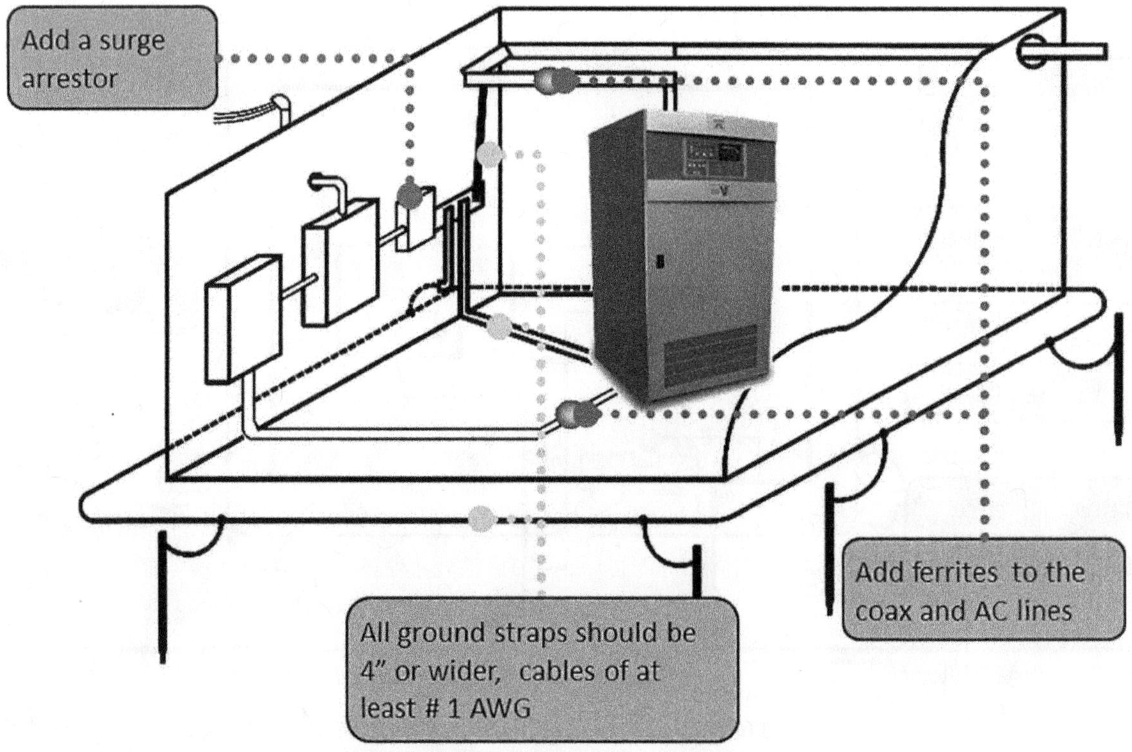

FIGURE 11.2-11 Optimizing an Existing Layout
J. Welton

hardline, ferrites to go over a 3-1/8" hardline are fairly reasonably priced. As well, if the site does not have a surge protector, one definitely should be installed.

DIFFERENCES BETWEEN STATION GROUND, ELECTRICAL GROUND AND COUNTERPOISE

One other thing that has been mentioned briefly, but not in any detail, is the difference between station ground (or reference ground), electrical ground and counterpoise (RF ground in an AM station). While these three points should ideally be bonded together, they perform significantly different tasks, although not exclusive of each other.

Station ground is quite simply the lightning protection ground, consisting of ground rods and conductors bonding the tower to the building and the contents inside. Electrical ground is for overcurrent protection and safety of personnel and equipment. It should not be interfered with in any way during the installation or improvement of lightning protection grounding. As indicated previously, the National Electrical Code has very specific requirements for electrical grounding that are well outside the scope of this chapter. Additional information may be found in Chapter 9.2 "Lightning Protection for Towers." Finally, counterpoise is the RF return path for an AM antenna system. It usually consists of several copper radials radiating out from the base of the tower. The counterpoise does virtually nothing for lightning protection but is another essential part of a properly functioning AM transmitter site and should be tied to the station reference ground wherever a radial intersects a ground conductor.

SUMMARY

Grounding for the purpose of lightning protection can be summarized quite simply. Use the largest surface area conductors feasible in short, straight runs, bond them with an electromechanical connection such as solder or brazing, establish a single-point station reference ground and avoid ground loops, which can also cause interference to the station's signal as well as providing a risk of failure due to lightning damage. Finally, have a properly installed surge protector and use ferrites to increase the impedance seen by surge energy with respect to the equipment in the facility.

Bibliography

Block, R.R., *The Grounds for Lightning and EMP Protection*, PolyPhaser Corporation, Hayden, Idaho, 1993.

Pinks, J., *Lightning Protection for Radio Stations*, Nautel Limited, Nova Scotia, CA, 1985.

C H A P T E R

11.3

Media Archives

JAMES SNYDER
Library of Congress

The audiovisual content industry stands at a crossroads at the writing of this chapter in 2016. Time is ticking on existing collections of audio and moving image content. We are at a technology crossroads spawned by fundamental changes in how content is captured, edited, distributed and archived. It is a survival crossroads for both the content that has come before and the content created from here on. The content created over the past century-plus since recording was invented lives on ephemeral carriers made of substances that don't last more than about 5 years to 5 decades. Even if something has survived the self-editing already done by how collections of content have been created, stored and, many times, discarded, the very medium on which content is recorded was and is decaying from within. With the conversion to digital the challenge is even more difficult: not only the media on which digital recordings survive over time but their survival depends on all of us doing our best to make the content self-identifying, keep the documentation and expertise to replay the content extant, and accept the reality that recordings that we want to last "forever" must be copied from time to time or they will be lost.

The challenge includes two goals: making sure born-digital content survives well into the future and converting existing physical carriers to files so they can survive as well. Just as recording technologies were an engineering challenge to create and make usable for the masses, survival of our cultural heritage is a challenge in temporal engineering: engineering content to survive for long periods of time, and keeping the systems to recover older content operating until the conversion is complete.

For nearly 120 years audio and moving images have been recorded on linear media that had to be played in real time to access the content. Audio on Edison and other types of cylinders plus various types of grooved records were the earliest recording methods. Magnetic recordings on steel wire were the first that could be edited physically. Starting with the Germans in the 1930s and later perfected by Jack Mullin and the teams at Ampex and 3M in the mid-1940s, magnetic recording on linear magnetic tape, first for audio and then in 1956 for video, revolutionized content creation, distribution and access. Early hard drives provided still frame and slow motion for sports starting in the late 1960s, but it wasn't until MPEG-2 was adopted in 1992 that video could be recorded to files that were manageable in size and could use low-cost, commodity computer and IT equipment. By the turn of the millennium in 2000 the move to file-based media began apace with XDCAM Blu-ray-based optical discs, recordable CDs, DVDs and Blu-rays and the very first direct-to-solid-state storage systems. But while most of the media was optical as the medium, the essences were still files. And files don't care what they're recorded on as long as the medium can record data.

Traditional media libraries have been organized similarly to traditional libraries of printed books and other materials: items on shelves. The big difference between traditional libraries and archives and those in many media organizations was that the media libraries were many times not as well documented and organized. While books used either the Dewey Decimal System or the Library of Congress Call Numbers (LCCN), tapes usually only have labels. If we're lucky. In some well-organized tape libraries there was a cataloging system that gave tapes unique IDs, perhaps even with machine-readable barcodes, and tied them to searchable database records. Most media organizations don't maintain their libraries of content as diligently unless there is a direct revenue stream that pays for the people and resources required. Unfortunately, this many times has meant content was lost years or decades before its cultural or revenue value was identified.

Here in the second decade of the 21st century, linear magnetic tape recording has almost entirely disappeared as a primary production format for both audio and video, and with it the traditional form of archiving tapes on shelves. Nonlinear, file-based production systems now predominate, requiring new principles to archive. We now stand at the beginning of an era where public interest and business models encourage more archiving of current materials and are increasingly mining historic audiovisual collections for commercial and noncommercial uses. With the switch to file-based workflows, we have the opportunity to fix many of the shortcomings of the previous, linear-based content storage systems. The payoffs for fixing known issues will be large, but the work required must be started from the beginning.

Linear production systems were based almost entirely on linear magnetic tape recording. The file-based systems of the future will still have the option of storage on linear magnetic tape, but it also offers magnetic recording on hard drive, magneto-optical drives, solid-state storage and a variety of optical storage systems, and the future holds the possibility of new recording methods. So the library of audio- and videotapes on shelves is changing to a library of digital content still stored on physical media but in new systems that no longer offer the simplicity of a tape on a shelf. They offer many new features but come with significant new organizational challenges as well.

Those of us with experience in managing large collections of media have, through painful experience, learned the limitations of organizing linear media. Organization started with physical objects on shelves. Descriptions were physical labels on the protective cases and the tapes themselves (if the folks who generated the content had the time or the inclination to label items in the first place during the production or distribution process) and were limited to the size of the labels. In the final decade of linear tape, features such as embedded chips were added to allow storage of additional data points about the recording, but the metadata functions inherent in files still could not be matched.

The birth of file-based audiovisual workflows in the 1990s required new ways to describe, organize and store audiovisual content. Otherwise it would have been an unmanageable mess. The ability to include descriptive data, termed "metadata" because it is data about the essence being carried, opened up a wide range of functions only dreamed of in the linear media-based era. The separation of the essence from its carrier by standardizing wrappers that could carry a multitude of different types of essences also created the ability to organize data, already in standardized wrappers, in different types of wrappers specifically designed to store data for long periods of time. Data could now be self-identifying no matter what storage medium was used, making long-term recoverability much more likely.

The challenge for those creating the new "tape library on physical shelves," which are now content libraries on different types of data storage media, is to create data sets that, when properly organized, can survive many generations into the future of not only the people managing the data sets but generations of different types of storage systems. Two key features must be enabled: structuring the content itself for long-term survival and structuring the data itself so it can survive multiple copying steps over time. At every point, standards are important, because history has shown only those items that are well documented have the greatest chance of survival long into the future. Standards such as those from SMPTE, IEEE, AES and others provide worldwide dispersion of standards documents so future generations will have a chance to know how to replay content even if the original playback devices are lost.

WORST-CASE SCENARIO

For those of us whose job it is to make sure content survives for centuries, we must consider the worst-case scenario. What is the "worst case"? For those of us who grew up with the Cold War, it was a nuclear holocaust. With climate change has come more frequent severe weather with inundating rain storms, stronger hurricanes and cyclones, more frequent tornado and other high-windspeed, damage-causing events even in areas not previously prone to wind events. But when you boil it down to the most basic challenge, weather and human-generated calamities are only part of what must be planned for.

At its most basic, the worst case is that only the recording survives, with no playback devices. Assuming we only have the media, what would it take to recover a recording? It would take documentation about how that recording was made. That means documentation on how to read the media itself, documentation on how the data is written on the media so one can figure out what is there, documentation on the content recorded and how it is structured so one can recreate playback devices, and finally documentation on the basic standards represented in those recordings so the original content can be recreated.

When might such a worst case exist? War, social upheaval, lack of funding and geological and weather disasters are all candidates. But they still happen relatively rarely. What happens more often is that recordings are orphaned or ignored for long periods of time. How many of us have boxes of stuff sitting in storage, either in our basements or in ministorage? Collecting dust. Being exposed to who knows what conditions. Much the same case exists for media collections: They're either in peoples' personal collections, or they're in commercial or company storage that may go ignored or even unidentified for long periods of time. We see these cases every day!

If you consider these cases our worst-case scenario, what are the challenges for us in designing systems to allow content to survive both natural and man-made losses? The answers are actually pretty straightforward: Choose well-documented media systems that are common at the time data is recorded. Choose data organization standards (such as SMPTE's AXF, the Archive eXchange Format) so that the data written on the media is written in a well-documented way, so that if all you have is the media and the playback device to see the bits, the data will self-identify itself. Choose content standards (audio recordings, video recordings, other types of recordings) that are well-known (whenever possible) at the time of their creation. And, finally, choose to create and maintain metadata that describes all parts of the content creation and data storage history of the content being saved.

Your challenge is to create the virtual version of a library's bookshelves. At a price that you or your organization can afford. For a period of time meaningful to you or your organization's goals. Whether you are an individual, part of a small group, commercial or noncommercial, for-profit or not-for-profit, the challenges are all the same: finding the right combination of features and costs to allow your content to survive.

KEY DESIGN FACTORS FOR ARCHIVES

Multiple Copies

One of the most important guarantors of longevity is having multiple copies of content geographically dispersed. Three copies is considered the minimum for long-term safety. Having multiple copies in different physical locations means that if there is a disaster in one location that destroys one copy, there will be at least two other copies outside the disaster zone. In the past, having multiple copies in different locations could be very expensive, especially for collections of any size. However, with "cloud" storage now prevalent, at least one copy could be local and second copy could be in the "cloud." A third copy could be with a trusted data vendor, a media storage facility or another facility owned by the same company. In each case, geographic dispersion is important so that, in an emergency such as a natural disaster or social upheaval, at least one copy will be out of the danger zone.

At least one copy should be maintained at a trusted location controlled by the content owner. This could be a collection of data tapes, hard drives or other storage devices. They could be sitting on a shelf or be under more active management in a tape robot or hard disc or solid-state array such as a SAN (storage area network) or NAS (network attached storage) system. A key factor in maintaining a local copy is how to store it: Tapes on shelves worked with audio and videotape for decades, but data tapes may not survive as well if they are not in a temperature- and humidity-controlled environment. Hard drives are sensitive electronic devices and are not manufactured to last more than about 7 years, even if they are unpowered and sitting on a shelf.

In both cases, technical obsolescence also makes long-term storage of more than 10 years difficult for data tape and hard drives (as well as other storage systems) simply because data interfaces (such as e-SATA, USB, etc.) only last from 5–20 years, depending on their adoption rate by the IT equipment industry. A key planning factor for locally held copies is whether the equipment to actually attach to a storage device and the drivers and software to read communicate with the storage device and interpret the data accurately will be available more than 10 years from the point you purchase it. After all, how many people reading this book have devices that can still read SCSI hard drives or other devices? Ten years before the publication of this book, you could still get devices to read SCSI, and 20 years ago they were common. Now, if you have a SCSI hard drive, you will need to have a specialist recover the data for you.

A second choice is with some sort of data storage facility, Locations such as salt mines where audio and videotapes have been stored successfully for decades will also work for data storage media. It could be with a service provider that holds media as its business, preferably under proper temperature and humidity controls for the media involved. For small amounts of data, it could even in be safety deposit boxes at banks or climate-controlled storage facilities. In these cases, having good records of what is being stored at each location as well as the costs and procedures for getting the content back when necessary are key decision factors. And one more thing: Actually physically moving data storage devices tends to increase data corruption over time. If physical storage is a choice for you, you must take data corruption caused by moving the media from one place to another into account. In some cases, such movement has been shown to increase bit error rates by five orders of magnitude. Not 5%. Not five times. Five ORDERS OF MAGNITUDE. As in (for example) one error every 100,000 bits, instead of one every 10,000,000,000 bits. Potential data corruption of this scale cannot be taken lightly or ignored.

A new variant in the off-site storage option is what has been dubbed "cloud' storage. "Cloud" means a commercial or noncommercial vendor that maintains its own storage systems in one or more physical locations. Content might be in one or more "cloud" locations or might be distributed across multiple locations. At the time of this chapter's writing, LOCKSS (Lots of Copies Keeps Stuff Safe: www.lockss.org) at Stanford University is an example of a noncommercial "cloud" storage system. Amazon's AWS (Amazon Web Services) and Google Cloud are commercial examples. There are others as well in both the noncommercial and commercial sectors.

In each case, content is stored under contract with the content owner for a fee. A key planning factor when using commercial data storage not under your control is that you are assuming that the vendor will exist for as long as you wish to hold the content with that vendor. You are assuming you have sufficient transmission capacity to handle the send and receive traffic your organization will require. You are also assuming that that vendor has sufficient redundancy so you will not lose any data should the hardware, software and human-based maintenance that makes up its "cloud" fail. And one key factor to ask for up front is the cost of not only storing your data but getting it back. What good is cheap data storage if it costs a fortune to get it back out of the "cloud"? All of these planning factors must be taken into consideration as the total cost of using external storage. There are plenty of good choices, but go into it with your eyes open. Caveat emptor.

Some organizations, such as group media owners or affiliated organizations, might build their own "private cloud," that is, one or more locations around the world where data could be stored for a group of program producers or owners. Similar design considerations come into play here, with the additional cost of the equipment, the software to run it and the maintenance and personnel costs now borne by the shared infrastructure organization.

SINGLE INSTANTIATION

A key design factor for longevity is single instantiation. That means that a single work or asset has a single representation of that work in long-term storage, even if there are three or more copies of that single instantiation in geographically dispersed locations. As you can imagine, given the many files, essences, versions and metadata associated with such a work or asset, the concept becomes mind-expandingly complex. However, the creation of data wrappers such as AXF make this goal attainable. Why single instantiation? When your content is stored with thousands to hundreds of millions of other items, it becomes very easy to lose track of individual files. Keeping all the content in a single "file" (or wrapper) means that data systems have less of a chance of losing the constituent parts of a work. Experience has shown that wrapping and unwrapping tends to be less processing intensive than keeping track of billions of individual files.

In the case of media, such as film scanning, where every frame is a single file, there are other benefits to single instantiation. In the SMPTE DPX standard as well as the TIFF standard, film content is scanned with one file for every frame of visual content. One 2,000-foot reel of 35mm film has 32,000 frames. One 2,000-foot reel of 16mm film has 80,000 frames. Film collections such as the Library of Congress have hundreds of millions of feet of film. That's billions of frames, meaning billions of files. One technical issue even with state-of-the-art enterprise class high-speed data processing and storage systems is that when a system contains more than 100,000,000 (one hundred million) files (the number is different depending on the system, but this is a good average number based on experience), the ability of the system to search for and find an individual file slows down considerably. Currently, many systems break their large collections into 100,000,000 file chunks to keep things moving smoothly. But that comes at a price in coordination, searchability, retrievability and making sure individual files do not get lost.

Keeping a single motion picture with its tens of thousands of frames in a single wrapper brings some very real benefits for collections that will exceed 100,000,000 files.

Also keep in mind that files generated as part of the production process may need to be saved. That means not just the camera original files, original film scans, audio files and such, but that could mean the various versions of the script, the spreadsheets for the financial transactions required to complete production, information on who owns what licensed materials in a production (music, raw footage, etc.), talent contracts, and many more types of associated data. In building a long-term archive, saving these items in a wrapper along with the content itself and its various versions is indeed attractive.

Critical associated data that MUST remain separate from content files, such as QC reports and cryptographic hash checksums, strongly imply that a wrapper must be used to keep associated content and essences together in a well-managed data archive.

STANDARDS

One of the most important aspects to succeeding in having our recordings last for hundreds of years is the ability to create and maintain the hardware and software to first create and then play back recordings in the centuries to come. The most successful formats in the analog and linear digital era were those that enjoyed wide adoption, becoming either de facto standards such as 1/4" audiotape and 2" quadruplex videotape that were documented as standards after the fact or actual standards such as DAT (digital audiotape) and SMPTE Type B & C videotape formats.

So, too, in the digital file era. File formats such as WAV and MXF have brought basic standardization to audiovisual media. AXF (the Archive eXchange Format; SMPTE 2034–1), as of the writing of this chapter, has begun the move to standardize how data are organized, identified and copied from storage medium to storage medium. Standards for describing data such as MARC, METS, SMPTE Core, Dublin Core, PBCore and the efforts of the AES and Motion Picture Academy (AMPAS) have started to standardize the vast and bewildering amount of ways of describing the thousands of hours of essences our industry creates every year.

And files are no longer just files: The wrapper concept has allowed standardized methods of creating and transporting many different types of audiovisual content and their companion descriptive data in the same package.

Likewise, if one thinks about what it takes to preserve data for the long term, one must answer a key question: If all the people of the future have to read the content of today is the ability to play the media it is recorded on (assuming the data has survived into the far future) and look at the bits recorded there with a copy of the standards documents available today, could they recover the data and know what they are looking at? The answer today is mostly "no." But very soon, given the standards already developed and this issue being worked on at the time this book was published, the answer could well be "yes." THAT is the definition of success and survivability.

One note about popular conventional wisdom. Many people and vendors at the time this book was published believed fervently that the only successful recording systems in the future will be those that mimic the past, specifically, media that allow future humans to look at recordings and actually see something other than the black and white dots representing binary data. While there is certainly an elegance and simplicity to such a belief, and early indications are that it could work, it starves us of the options for lower-cost, higher-capacity alternatives that can't be seen by the naked eye because the bits are so dense the eye cannot see them. So a future question will be: Do we sacrifice cheaper, larger capacity storage methods to ensure usefulness over hundreds of years? Or do we design high-capacity storage systems and the data sets they contain so that our future selves will be able to self-identify data even if their human eyes can't see it? Both methods have their features and drawbacks. It is not an easy question to answer. The future media will most likely need versions of both to exist.

PHYSICAL MEDIA MIGRATION

The challenge for all of us is to operate during this transition period. There are tens, if not hundreds, of millions of pieces of physical media that need to be either digitized or, if already digital, converted to file-based forms. At the very same time, content creation continues as files, and we must work through the myriad of challenges in identifying what metadata needs to be defined, how much of it can be standardized, developing methods to store metadata that can't be standardized or hasn't been defined yet—all while doing our day jobs and, in many cases, with an absolute minimum of additional funding available.

But wait . . . where are all these files to live? Physical instantiations of audio and video recordings may no longer be the norm, but the files have to live somewhere. And that somewhere is . . . physical media. The difference is the recording medium is no longer directly tied to the content recorded on it: Tape formats are no longer dedicated to a specific type of recording. File-based media can contain ANY recordings, content of any type that can be expressed as a file. Content can be on ANY type of media designed for files: magnetic tape, optical disc, various types of solid state storage—perhaps even DNA or crystals in the future.

So the next challenge beyond digitizing historic content, converting early digital content to files, and keeping up with the flood of content that must be produced every day is to organize and identify the content we are storing.

IDENTIFICATION

One of the problems with the physical media era was identifying what we were recording. If we were lucky, we had a diligent tape operator who labeled our tapes properly. If we were not, we had some cryptic handwritten label that looked like a medical doctor had written it. Or a Post-It note whose adhesive had dried up and it fell off. If there was any label at all. Sometimes we might have rundowns or producer's notes in a tape case on yellow or white pad paper. Sometimes there would be some context to the notes; sometimes we were left to guess what the person who recorded the tape actually meant. Many times we ended up with tapes with little or no labeling, and we just had to make the best of it.

In the era of file-based media there are no labels other than the file name. Some information can be included in those relatively short names, 256 characters long if we're lucky. But in the digital era, data within a file and carried alongside the file in a side-car allows us much more flexibility to identify what something is. Groups of files that are related to each other can be described and tied together using *metadata*: data that describes an essence.

While metadata is one term, it encompasses many different types of descriptive data: cataloging records, statutorily required data such as closed captioning and content ratings, labels that tell us what language or languages a production contains, production data such as camera settings or provenance data on how many processing steps content has been through. The list of potential metadata is endless. How we deal with it must be specific and standardized if it is to be valuable.

At the time of the writing of this chapter in 2016, quite a bit of work on both standardizing metadata and standardizing where it is placed within content or within the production chain has already been done. Metadata that is required by law such as closed captioning and content ratings in the United States have defined locations within the most common file standards available today. Definitions on how to define descriptive metadata exist as well: Dublin Core, PBCore, EBU Core and the evolving SMPTE Core standards all define not only what terms are standardized but what they actually mean, in varying levels of specificity. More work still needs to be done, however, since similar or identical terms currently can mean two different things between two standards. For example: "Title" can mean a number of different things if its various incarnations (title, series title, episode title, etc.) are not very clearly defined. As one might guess, the same term can mean different things depending on what content area one is working in: motion pictures, episodic television, news, audio recordings, podcasts, etc.

Technical metadata is also quite clearly defined currently because files must clearly identify what their content is or replay software or hardware cannot play the content properly. However, even here there are issues in terminology. For example, the SMPTE standard for digital video (SMPTE 125M) defined component color difference signals as Y, Cb and Cr, which of course are the digital versions of the analog Y, Pb and Pr. However, many times today YCbCr is actually described using the term for analog PAL/SECAM component color channels: YUV. This probably happened because somebody at some point needed a term for digital component color difference, was most familiar with YUV as a term for component video, and made an assumption that using that term for digital component color difference was acceptable. However, for those organizations and individuals who actually have analog PAL or SECAM content and then must digitize them, using the same term for both the analog color YUV and the digital component color difference YCbCr provides no end of confusion. For those tasked with educating the future generations of technicians and engineers, this will be equally confusing. The object lesson here is first to be precise and unambiguous when defining technical terms and second to beware of confusing terminology, such as YUV meaning YCbCr, that already exists.

Cataloging metadata has also been standardized in various ways. The library and archives communities decades ago developed their own standards for cataloging the content of every type worldwide, including audiovisual content. MARC (MAchine Readable Catalog record: www.loc.gov/marc), and METS (Metadata Editing and Transmission Standard: www.loc.gov/standards/mets) were both created by the library and archives communities and are maintained by the United States Library of Congress (www.loc.gov). Both are used widely in the library and archives communities, including some major media collections such as the Library of Congress. However, because the two standards were developed in a community completely separated from the audiovisual standards-setting, production and vendor communities, how the very standards used by media archivists diverge from the work being done in the audiovisual standards-setting, production and vendor communities is important when the records are all electronic and many times must be translated between the two different communities. This is especially important to understand when systems to automatically generate, distribute, process and populate electronic records are being designed. How things are described, by what terms they are described and how different standards use different terms to describe the same things become critical. This is a primary challenge facing our communities today.

Lastly, a key challenge is the survivability of metadata within files throughout acquisition, production, distribution, transmission and archiving processes. One key issue that has existed since the dawn of file-based workflows has been that metadata included in file headers and footers many times has been erased when a file is edited, sometimes even when a file was just opened and not even edited in any way. While the incidence of metadata loss is dropping as software coders pay attention to the importance of metadata throughout the production chain, anyone testing or validating systems (in addition to folks who are using systems for their intended purpose) should validate that metadata within files is not erased, corrupted or destroyed by any software package or pieces of equipment in a production chain. If your organization uses metadata in headers, survivability should be a part of any test of equipment.

EXTERNAL METADATA

Other types of metadata that are just starting to become familiar to those in the audiovisual community are checksums and external metadata such as QC (quality control) reports. A key aspect to the survivability of digital content over time is not only the number of separate copies that have been made of that content but knowing whether any data corruptions (usually termed "bit flips") have occurred in the content being stored. Checksums is a tool the computer hardware and software industries have used since the dawn of computing to ascertain whether the data they are processing is what it should be. In the case of audiovisual content, where data corruption may manifest itself as artifacts any ordinary person can see or hear or even the inability to play the content at all, especially robust checksums must be used. These are cryptographic hash checksums (CHCs). For decades the intelligence and data safety communities have developed and used very strong mathematical algorithms to generate unique strings of characters (hashes) that are a unique footprint of the data within a file or storage location. While their cryptographic value may have been compromised (in other words, the ability to prevent decoding by unintended personnel), many cryptographic algorithms can still be valuable as generators of unique values that call out when data has been corrupted in a file. The key feature here is that there should never be two different file sizes that generate the same hash value.

CHECKSUMS AND HASHES

The most common form of CHC used currently is MD5. Many vendors and software packages have MD5 baked into their code. An attractive feature of MD5 is that it is relatively simple to implement in computer code, which when used on systems where processing speed is a challenge can be important. However, MD5 has been shown in a number of conditions to produce the same hash value for two different file sizes. It is rare, but it can happen. That's a problem if the goal of using MD5 is to validate that a file has not changed from its original bit pattern, but more than one bit pattern can have the same MD5 value.

The next most common CHC used is the SHA (secure hash algorithm) family. SHA-1, SHA-256, SHA-512 and others generate hash values that, as of this writing, have never been shown to have the same hash value for two different file sizes. The simplest is SHA-1, used at the Library of Congress and other institutions. It is a more complex algorithm than MD5 but also produces a hash that can indicate if even a single bit has flipped in file sizes up to 2^{61}. While adoption of SHA was slow because of its processing requirements, this has disappeared as an issue as computer speeds have gotten faster.

The key to using CHCs is to create the hash value on the computer or storage system where the original file was created. Currently this tends to be on internal server storage, but increasingly in systems where shared SAN (storage attached network) or NAS (network attached storage) systems are used, the hash can be created by a different system than the one that created the content because it is acting on the original file in storage even if it wasn't the computer that created the file. Increasingly, systems are fast enough to create hashes while a file is being produced, making hash generation only seconds behind the creation of the content itself. Key factors in these features are the processing speed of the computers performing the work, the speed of the data connections between those computers and any attached storage they are using to create the content, and the ability of the storage system to both write and read data at the same time. Solid-state storage is particularly well suited to this type of processing: high write/read speeds on multiple items at the same time.

CONTENT/ESSENCE QC REPORTS

Another type of external metadata is the quality control report. Because files and their constituent essences go through so many steps on the way to a final production version and on to distribution and delivery to the end user, the ability to not only validate that a file in storage has no corruption but also make sure that the essences have the intrinsic quality expected at any stage in the production or distribution chain is very important. Quality control software allows the validation of files as they move and is especially important when creating an archive, since QC reports, along with the CHC hash values, tell the end user, perhaps decades in the future, not only that a file has or hasn't been corrupted but also what the file looked like when it went into storage. After all, if one is digitizing original analog tapes or creating digital files from digital linear media, just because the bits within a file pass a CHC check doesn't mean there wasn't a corruption when the content was originally turned into a file. Keeping a QC report right next to the content to which it applies is just as important a confidence-building assurance tool as the CHC.

There are two fundamental types of QC reports: human readable and machine readable. In 2016, the two most common forms of QC reports are PDF/A (human readable) and XML (machine readable) files.

CHCs and QC reports are a key reason why data wrappers are becoming as important as media content wrappers such as MXF: The metadata the external files contain is as important as the content itself, and making sure the files stay together through the processing, storage and long-term archiving processes is key to the usability of the content in the future.

BIT ERROR RATE: THE LESS ERRORS TO CORRECT, THE LESS MONEY SPENT CORRECTING THEM

The final but no less key factor in the survivability of data into the far future is bit error rate (BER). This is the average amount of errors that occur in any data transmission, storage or processing system. As a design factor, think of it as a direct cost of doing business: The fewer errors you have, the fewer you will have to pay to correct. Correcting bit errors is real money, minimizing the servers, transmission capacity, number of times you access a backup copy of your data, and how many backup copies of your data must be maintained (remember: minimum of three to ensure long-term viability of your data). Bit errors occur at every stage in the creation, processing, storage, retrieval, transmission and use of data. Your bit error rate is only as good as the worst link in that chain of devices and software. Know what your bit error rates are for every device and process, and plan your systems to minimize the errors in your data as much as your funding can afford. Sometimes a higher up-front cost can provide significant long-term cost benefits for the data stored.

Long-Term Data Recording

Now that we've got an idea on how to store our data sets, what about the future? Analog recordings have lasted over 120 years. Early digital recordings began over 40 years ago with the first PCM-coded audio on 3/4" videotape. These media, while by no means perfect, have lasted for decades. How do we make sure our digital recordings last? By examining the problem in its constituent parts.

For content deemed worthy of long-term survival, media must last. The ultimate question is for how long.

In this era when the capacity to record large amounts of data is still increasing, the cost of maintaining data on the same media vs. converting to media with double the capacity every few years is a major consideration. For now, copying from media to media is still less expensive than just recording everything on a static format that might last for decades or centuries. However, the cost benefit of migrating your data every few years to a higher-capacity format will disappear once the current magnetic capacity gains being made every few years no longer continue. Current guidance has this happening sometime after the year 2025, but stay tuned and pay attention. Once magnetic tape cannot continue to double its capacity every few years, if there isn't some magic new technology that can continue the trend to higher recording densities, then media where the data lasts for decades or centuries will become economically viable. Why continue to copy data sets to new media every few years if the capacity of the new media isn't increasing anymore?

Also, if one considers what it takes to record data onto an unchanging type of media, several assumptions apply: Media of that capacity will remain available into the far future (not a forgone conclusion in any media type currently available); the availability of recorders and playback units will remain the same or even grow for decades to come (certainly not an assumption indicated by the last 70 years of technological change); and the expertise and willingness of vendors to produce machines that read and recover data accurately from media intended for decades or centuries of survivability will remain, as well. The considerations are of paramount importance. Here again, bit error rate over time will be a very important planning factor and decision point for what technologies to choose.

We must accept that migration from one underlying medium to another is the most cost-effective way to preserve our content for the long term. If we accept that premise, what comes next? We must design the data recorded on the media to survive multiple copying steps. How do we do that?

Use skill sets already learned in the previous digital audio and videotape recording eras: Forward error correction and checksums for data validation. But with huge data sets, how do we KNOW that the data is good? By using advanced cryptographic hash checksums such as SHA (secure hash algorithm) to revalidate that no bits have been corrupted ("flipped"). After all, all digital recording media are analog in their nature, and the very same rules of recording and corruption apply to the digital recordings: Tape contains noise, magnetic media can be corrupted by various means, solid state can be corrupted by static electricity, and the very fact that the tiny physical locations bits are recorded on can be corrupted by the energies in the universe itself, like cosmic rays. Even particles of radiation given off by substances such as radon-emitting rock that exist in many parts of the world must be considered and understood.

THE ULTIMATE PROTECTION: MIGRATE YOUR CONTENT NOW

No discussion of archiving at this point in time would be complete without this word of warning: If you have audio or video content that needs to survive into the future for any reason that isn't already in file form, you have about ten (10!) years from now (2016 when this chapter was written) to get it copied or digitized to file. Ten years. One decade. The year 2026, from the time this chapter was written. No, I'm not kidding.

Why?

Because the ability to play back historic audio and video content, whether it be analog or digital, is quickly disappearing. Even if your organization owns playback equipment, the ability to repair it (as in spare parts and the expertise to service, align and maintain) is already gone or nearly gone for many formats. For formats where the equipment was produced within the last 10 years, you might have 10–20 years before expertise, spare parts, and used machines start to run out. The machines are no longer being manufactured. All but a couple of audio and only one videotape format playback-capable machines are being made. Only a few blank tape formats are still available. If content is going to survive for the long term, it MUST be converted to a file format sometime in the next decade. After about 2026 there may still be some limited capacity to get content converted to digital, but that capacity will be increasingly expensive unless manufacturers suddenly start building tape decks to play back the tens (if not hundreds) of millions of tapes around the world. That's not likely.

So this is your "two-minute warning," to use an American football term. You may have a decade. Start now. Or as soon as you can. But start. Identify what needs to be saved: quantities of media, locations, who owns the rights, etc. From that you can figure out what it will cost to get the media converted, either by your own people or by a vendor. If you need help, contact the folks who are already doing it: the U.S. Library of Congress and others. Go to www.digitalpreservation.gov for ideas and www.digitizationguidelines.gov for starter ideas and folks to talk to. Check out the resources at AMIA (the Association of Moving Image Archivists (www.amianet.org), IASA (the International Association of Sound and Audiovisual archives: www.iasa-web.org), the Society of American Archivists (www.archivists.org), the International Association of Film Archives (FIAF: www.fiaf.org), PrestoCentre (www.prestocentre.org), and LOCKSS (www.lockss.org). These are not the only ones, so think of these as a starting point.

But start. Start now. Don't wait. And good luck. Our future citizens, leaders, researchers and historians are depending on us to get it right.

"The views expressed are the author's and do not necessarily reflect the views of the Library of Congress".

Index

NAB ENGINEERING HANDBOOK